32nd European Photovoltaic Solar Energy Conference and Exhibition (EU PVSEC 2016)

Munich, Germany
20-24 June 2016

Volume 1 of 4

Editors:

M. Topic
P. Helm
N. Taylor

ISBN: 978-1-5108-3651-8

Printed from e-media with permission by:

Curran Associates, Inc.
57 Morehouse Lane
Red Hook, NY 12571

CURRAN ASSOCIATES INC.
proceedings
.com

Some format issues inherent in the e-media version may also appear in this print version.

Printed by Curran Associates, Inc. (2017)

For permission requests, please contact WIP – Renewable Energies
at the address below.

WIP – Renewable Energies
Sylvensteinstr. 2
81369 Munchen
Germany

Phone: +49 89 72012735
Fax: +49 89 72012791

wip@wip-munich.de

Additional copies of this publication are available from:

Curran Associates, Inc.
57 Morehouse Lane
Red Hook, NY 12571 USA
Phone: 845-758-0400
Fax: 845-758-2633
Email: curran@proceedings.com
Web: www.proceedings.com

32nd European Photovoltaic Solar Energy Conference and Exhibition

Proceedings of the International Conference held in Munich, Germany

20 June - 24 June 2016

Edited by:

M. TOPIČ
University of Ljubljana
Chairman of European Technology & Innovation Platform Photovoltaics
Ljubljana, Slovenia

N. TAYLOR
European Commission - DG Joint Research Centre
Ispra, Italy

P. HELM
WIP Renewable Energies
Munich, Germany

Coordination of the Technical Programme:
European Commission - DG Joint Research Centre
Via E. Fermi 1
21020 Ispra (VA)
Italy

Institutional Support:
European Commission
UNESCO – United Nations Educational, Scientific and Cultural Organization, Natural Sciences Sector
WCRE – World Council for Renewable Energy

Supporting Organisations:
ARE – Alliance for Rural Electrification
EREF – European Renewable Energies Federation
ESA – European Space Agency
EUFORES –The European Forum for Renewable Energy Sources
EUREC – The Association of European Renewable Energy Research Centres
European Photovoltaic Technology Platform
IEA PVPS – International Energy Agency Photovoltaic Power Systems Programme
REN21 – Renewable Energy Policy Network for the 21st Century

Institutional PV Industry Cooperation:
SolarPower Europe
SOLARUNITED

Supporting Associations:
AIE – European Association of Electrical Contractors
AREA – African Renewable Energy Alliance
CABA – Continental Automated Buildings Association
EPIC – European Photonics Industry Consortium
SASIA - Saudi Arabia Solar Industry Association

EU PVSEC realised by:
WIP
Sylvensteinstr. 2, 81369 München, Germany
Tel: +49 89 720 12 735, Fax: +49 89 720 12 791
Email: pv.conference@wip-munich.de
www.photovoltaic-conference.com
www.wip-munich.de

Proceedings produced and published by:
WIP
Sylvensteinstr. 2, 81369 München, Germany
Tel: +49 89 720 12 735, Fax: +49 89 720 12 791
Email: pv.conference@wip-munich.de
www.photovoltaic-conference.com
www.wip-munich.de

ISSN 2196-0992

32nd EUROPEAN PHOTOVOLTAIC SOLAR ENERGY CONFERENCE AND EXHIBITION
20 JUNE – 24 JUNE 2016
MUNICH, GERMANY

EU PVSEC COMMITTEES

INTERNATIONAL SCIENTIFIC ADVISORY COMMITTEE (ISAC)

Chair
G.F. De Santi, European Commission, DG JRC, Director of the Institute for Energy and Transport, Ispra, Italy

Secretary
N. Taylor, European Commission, DG JRC, Institute for Energy and Transport, Ispra, Italy

Committee Members
P. Frankl, Head of the Renewable Energy Division, International Energy Agency, France
P. Hofer-Noser, CTO, Meyer Burger Technology, Switzerland
P. Malbranche, CEA, Le Bourget du Lac, France
S. Nowak, Managing Director of NET Nowak Energy & Technology, Switzerland
W. Sinke, Program Development Manager, ECN Solar Energy, The Netherlands
P. Verlinden, Chief Scientific and Vice Chair, State Key Lab. TRINA Solar, China
E. Weber, Director, Fraunhofer ISE, Germany
P. Helm, WIP, Munich, Germany
A. Grassi, ETA-Florence Renewable Energies, Florence, Italy

CONFERENCE EXECUTIVE COMMITTEE

Conference General Chairman
M. Topič, University of Ljubljana, Chairman of European Technology & Innovation Platform Photovoltaics, Ljubljana, Slovenia

Technical Programme Chairman
N. Taylor, European Commission, DG JRC, Institute for Energy and Transport, Ispra, Italy

Committee Members
H. Ossenbrink, European Commission, DG JRC, Institute for Energy and Transport, Ispra, Italy
S. Nowak, Managing Director of NET Nowak Energy & Technology, St. Ursen, Switzerland
W. Sinke, Program Development Manager, ECN Solar Energy, The Netherlands
P. Helm, WIP, Munich, Germany
A. Grassi, ETA-Florence Renewable Energies, Florence, Italy

2016 SCIENTIFIC COMMITTEE
Topic Organisers and Paper Review Experts

Topic 1 – New Materials and Concepts for Solar Cells and Modules

G. Beaucarne, Dow Corning, Seneffe, Belgium
N.J. Ekins-Daukes, Imperial College London, United Kingdom
A. Martí Vega, UPM, Madrid, Spain
J. Poortmans, imec, Leuven, Belgium
M. Rusu, HZB, Berlin, Germany
J.A.M. Van Roosmalen, ECN, Petten, Netherlands

V. Bukauskas, Center for Physical Sciences and Technology, Vilnius, Lithuania
F. Falk, IPHT, Jena, Germany
J. Govaerts, imec, Leuven, Belgium
J.F. Guillemoles, CNRS, Chatou Cedex, France
M.A. Hernandez-Fenollosa, UPV, Valencia, Spain
J.C. Hummelen, University of Groningen, Netherlands
D. König, UNSW, Sydney, Australia
I. Konovalov, Ernst Abbe University of Applied Science Jena, Jena, Germany
C. Lévy-Clément, CNRS, Thiais, France
Y. Okada, University of Tokyo, Japan
B. Rech, HZB, Berlin, Germany
F. Roca, ENEA, Portici, Italy
V. Saly, Slovak University of Technology, Bratislava, Slovakia
H. Shirai, Saitama University, Japan
W. Warta, Fraunhofer ISE, Freiburg, Germany

Topic 2 – Wafer-Based Silicon Solar Cells and Materials Technology

O. Anspach, PV Crystalox Solar, Erfurt, Germany
D.L. Bätzner, Meyer Burger Research, Hauterive, Switzerland
R. Brendel, ISFH, Emmerthal, Germany
F. Ferrazza, Eni, San Donato Milanese, Italy
S.W. Glunz, Fraunhofer ISE, Freiburg, Germany
G. Hahn, University of Konstanz, Germany
D. Muñoz, CEA, Le Bourget du Lac, France
S. Riepe, Fraunhofer ISE, Freiburg, Germany
A.W. Weeber, ECN, Petten, Netherlands
P. Wohlfart, Singulus Technologies, Kahl am Main, Germany

A.G. Aberle, SERIS, Singapore
N. Azkona, UPV/EHU, Zamudio, Spain
J. Bagdahn, Fraunhofer CSP, Halle (Saale), Germany
R. Barinka, Fill Factory, Rožnov pod Radhoštm, Czech Republic
P.A. Basore, PV Specialist Services, Solana Beach, USA

D. Biro, Fraunhofer ISE, Freiburg, Germany
S. Bowden, Arizona State University, Tempe, USA
T.M. Bruton, TMB Consulting, Woking, United Kingdom
B. Dauksher, Arizona State University, Tempe, USA
M. Debucquoy, imec, Leuven, Belgium
T. Desrues, CEA, Le Bourget du Lac, France
D. Dimova-Malinovska, Bulgarian Academy of Sciences, Sofia, Bulgaria
S. Dubois, CEA, Le Bourget du Lac, France
T. Dullweber, ISFH, Emmerthal, Germany
P. Fath, RCT-Solutions, Konstanz, Germany
A.J. Galdikas, Modernios E-Technologijos, Vilnius, Lithuania
M. Hermle, Fraunhofer ISE, Freiburg, Germany
K. Hesse, Wacker Chemie, Burghausen, Germany
J. Horzel, CSEM, Neuchâtel, Switzerland
J. John, imec, Leuven, Belgium
S.D. Kim, SunPower, San Jose, USA
J. Libal, ISC Konstanz, Germany
A. Metz, h.a.l.m. elektronik, Frankfurt am Main, Germany
J.W. Müller, Hanwha Q CELLS, Bitterfeld-Wolfen, Germany
Y. Ohshita,TTI, Nagoya, Japan
S. Peters, Hanwha Q CELLS, Bitterfeld-Wolfen, Germany
K. Petter, Hanwha Q CELLS, Bitterfeld-Wolfen, Germany
R. Preu, Fraunhofer ISE, Freiburg, Germany
J. Rentsch, Fraunhofer ISE, Freiburg, Germany
E. Sauar, Brighterlite, Oslo, Norway
H. Schlemm, Meyer Burger, Hohenstein-Ernstthal, Germany
M.C. Schubert, Fraunhofer ISE, Freiburg, Germany
B. Terheiden, University of Konstanz, Germany
Y. Veschetti, CEA, Le Bourget du Lac, France
G.P. Willeke, Fraunhofer ISE, Freiburg, Germany
D. Wood, Hemlock Semiconductor, Hemlock, USA
A. Yamada, Tokyo Institute of Technology, Japan
D.L. Young, NREL, Golden, USA

Topic 3 – Thin Film Solar Cells and Modules

C.J. Brabec, University of Erlangen-Nuremberg, Germany
J. Cárabe, CIEMAT, Madrid, Spain
A. Meeder, Flisom, Dübendorf, Switzerland
J. Meier, PV Consultant, Berlingen, Switzerland
S. Siebentritt, University of Luxembourg, Belvaux, Luxembourg
M. Topic, University of Ljubljana, Slovenia
S.C. Veenstra, ECN, Eindhoven, Netherlands
W. Wischmann, ZSW, Stuttgart, Germany

C. Ballif, EPFL&CSEM, Neuchâtel, Switzerland
R. Birkmire, University of Delaware, Newark, USA
D. Cheyns, imec, Leuven, Belgium
S. Cros, CEA, Le Bourget du Lac, France
T. Dalibor, AVANCIS, Munich, Germany
P. Delli Veneri, ENEA, Portici, Italy
B. Dimmler, Manz CIGS Technology, Schwäbisch Hall, Germany
K. Durose, Stephenson Institute for Renewable Energy, Liverpool, United Kingdom
S. Gall, HZB, Berlin, Germany
I. Gordon, imec, Leuven, Belgium
S. Huang, UNSW, Sydney, Australia
S.J.C. Irvine, Swansea University, St. Asaph, United Kingdom
R. Klenk, HZB, Berlin, Germany
C.-F. Lin, NTU, Taipei, Taiwan
M.C. Lux-Steiner, HZB, Berlin, Germany
E. Mellikov, Tallinn University of Technology, Tallinn, Estonia
M. Meuris, imec/IMOMEC - ATFPV, Leuven, Belgium
C. Molpeceres, UPM, Madrid, Spain
J. Parisi, University of Oldenburg, Oldenburg, Germany
R. Po, Eni, Novara, Italy
M. Powalla, ZSW, Stuttgart, Germany
A. Romeo, University of Verona, Verona, Italy
V. Sittinger, Fraunhofer IST, Braunschweig, Germany
A. Slaoui, InESS, Strasbourg, France
A.N. Tiwari, EMPA, Dübendorf, Switzerland
A. Virtuani, SUPSI, Canobbio, Switzerland
M. Zeman, Delft University of Technology, Netherlands

Topic 4 – Concentrator and Space Applications

R. Campesato, CESI, Milan, Italy
G. Flamand, imec, Leuven, Belgium
F. Rubio Berenguel, Freiburg, Germany
C. Signorini, ESA-ESTEC, Noordwijk, Netherlands

I. Anton Hernandez, UPM, Madrid, Spain
A.W. Bett, Fraunhofer ISE, Freiburg, Germany
V. Khorenko, AZUR SPACE, Heilbronn, Germany
R. King, Arizona State University, Tempe, USA
G. Siefer, Fraunhofer ISE, Freiburg, Germany
T. Takamoto, SHARP, Nara, Japan
G. Timò, RSE, Piacenza, Italy

Topic 5 – Operations, Performance, Reliability and Sustainability of PV

J. Binder, ZSW, Stuttgart, Germany
M. de Wild-Scholten, SmartGreenScans, Groet, Netherlands
R. Gottschalg, Loughborough University, United Kingdom
U. Jahn, TÜV Rheinland Energy, Köln, Germany
N.M. Pearsall, Northumbria University, Newcastle upon Tyne, United Kingdom
M. Perrin, CEA, Le Bourget du Lac, France
C. Protogeropoulos, Phoenix Solar, Athens, Greece
K. Radouane, EDF EN, Paris La Defense, France
T. Sample, European Commission DG JRC, Ispra, Italy
W.G.J.H.M. van Sark, Utrecht University, Netherlands
K. Wambach, Wambach-Consulting, Aindling, Germany

C. Alonso-Tristán UBU, Burgos, Spain
M. Bächler, PerVorm, Ulm, Germany
T.R. Betts, Loughborough University, United Kingdom
P. Blanc, MINES ParisTech, Paris, France
R. Einhaus, Apollon Solar, Lyon, France
G. Friesen, SUPSI, Canobbio, Switzerland
V. Fthenakis, Columbia University, New York, USA
W.J. Gambogi, DuPont, Wilmington, USA
L. Gomes de Freitas, Federal University of Uberlândia, Brazil
G. Graditi, ENEA, Portici, Italy
M. Grottke, WIP - Renewable Energies, Munich, Germany
Y. Hishikawa, AIST, Tsukuba, Japan
T. Huld, European Commission DG JRC, Ispra, Italy
A.R. Lagunas, CENER, Sarriguren, Spain
O. Mayer, GE Global Research, Garching, Germany
M.A. Muñoz, UPM, Madrid, Spain
H. Nagel, Fraunhofer ISE, Freiburg, Germany
H. Nussbaumer, Zurich University of Applied Sciences, Winterthur, Switzerland
C. Nyman, Soleco, Porvoo, Finland
S. Oktik, Sisecam, Istanbul, Turkey
W. Palitzsch, Loser Chemie, Zwickau, Germany
A.H.M.E. Reinders, University of Twente, Enschede, Netherlands

J. Remund, Meteotest, Bern, Switzerland
J.N. Roy, IIT Kharagpur, Kharagpur, India
M. Schroedter-Homscheidt, DLR, Wessling, Germany
M. Sengupta, NREL, Golden, USA
R. Sinton, Sinton Instruments, Boulder, USA
J. Sutterlueti, Gantner Instruments, Schruns, Austria
S. Tselepis, CRES, Pikermi, Athens, Greece
M. Van Iseghem, EDF R&D, Moret-sur-Loing, France
E. Voroshazi, imec, Leuven, Belgium
A. Wade, First Solar, Berlin, Germany
J. Wohlgemuth, NREL, Golden, USA
M. Zehner, Rosenheim University of Applied Sciences, Germany

Topic 6 – PV Applications and Integration

F.P. Baumgartner, Zurich University of Applied Sciences, Winterthur, Switzerland
F. Bonemazzi, ENEL, Rome, Italy
F. Frontini, SUPSI, Canobbio, Switzerland
P. Malbranche, CEA, Le Bourget du Lac, France
A. Scognamiglio, ENEA, Portici, Italy
X. Vallvé, Trama TecnoAmbiental, Barcelona, Spain
I. Weiss, WIP - Renewable Energies, München, Germany

R. Burkhardt, Energy Depot, Kreuzlingen, Switzerland
S. Caneva, WIP - Renewable Energies, Munich, Germany
I. Da Silva, Strathmore University, Nairobi, Kenya
J.P. de Barros Duarte Fonseca, University of Cape Verde, Mindelo, Cape Verde
B. de Boer, TNO, Eindhoven, Netherlands
H. Fechner, University of Applied Sciences Technikum Wien, Breitenfurt, Austria
I.B. Hagemann, Architekturbüro Hagemann, Aachen, Germany
C. Helmke, Management Consultancy Research, Erfurt, Germany
E. Kaijuka-Okwenje, Makerere University, Kampala, Uganda
G. Mütter, Alternative Energy Solutions, Vienna, Austria
K. Peter, ISC Konstanz, Germany
T. Reijenga, BEAR-iD Sustainable Urban Planners + Architects, Gouda, Netherlands
S. Wittkopf, HSLU, Horw, Switzerland

Topic 7 – PV Economics, Markets and Policies

C. Breyer, Lappeenranta University of Technology, Finland
M. Getsiou, European Commission DG RTD, Brussels, Belgium
G. Masson, Becquerel Institute, Brussels, Belgium
T. Nordmann, TNC Consulting, Feldmeilen, Switzerland
S. Nowak, NET Nowak Energy & Technology, St. Ursen, Switzerland
E. Perezagua, Consultores de Energía Fotovoltaica, Madrid, Spain

I. Kaizuka, RTS, Tokyo, Japan
D. Mayer, MINES ParisTech,Paris, France
P. Mazzucchelli, EUREC,Brussels, Belgium
D. Stickelberger, Swissolar, Zürich, Switzerland
S. Szabó, European Commission DG JRC, Ispra, Italy

SUBJECT INDEX

Topic Code

1 New Materials and Concepts for Solar Cells and Modules
2 Wafer-Based Silicon Solar Cells and Materials Technology
3 Thin Film Solar Cells and Modules
4 Concentrator and Space Applications
5 Operations, Performance, Reliability and Sustainability of PV
6 PV Applications and Integration
7 PV Economics, Markets and Policies

Session Type

P = Plenary Session
O = Oral Session
V = Visual Session

Day Codes

A = Monday, 20 June 2016
B = Tuesday, 21 June 2016
C = Wednesday, 22 June 2016
D = Thursday, 23 June 2016
E = Friday, 24 June 2016

e.g. 3CP.1 ⇒ 3=Thin Film Solar Cells and Modules, C=Wednesday, P=Plenary Session, 1=Session 1

TABLE OF CONTENT
of EU PVSEC 2016 Proceedings Papers

New Materials and Concepts for Solar Cells and Modules

Visual PRESENTATIONS 1BV.6 New Materials and Concepts for Cells

Wafer-Based Silicon Solar Cells and Materials Technology

Visual PRESENTATIONS 2AV.2 Silicon Solar Cell Improvements and Innovation (II)

Visual PRESENTATIONS 2AV.3 Silicon Solar Cell Improvements and Innovation (III)

Visual PRESENTATIONS 2BV.7 Silicon Solar Cell Characterisation and Modelling / Manufacturing and Processing

Thin Film Solar Cells and Modules

Concentrator and Space Applications

Oral PRESENTATIONS 4CO.6 III-V-based Devices for Terrestrial and Space Applications / Concentrator and Space Systems

Visual PRESENTATIONS 5BV.1 PV Cells and Modules (I)

Visual PRESENTATIONS 5BV.3 Balance of System Components

Visual PRESENTATIONS 5CV.3 Solar Resource and Forecasting / Sustainability and Recycling

PV Applications and Integration

Plenary SESSION 6DP.2 PV as Part of Our Built Environment: Solutions for Integration into Building Envelops and Energy Systems

Oral PRESENTATIONS 6CO.12 Grid and Energy System Integration (III) - Technology Solutions

Visual PRESENTATIONS 6AV.5 PV in Buildings and the Environment

PV Economics, Markets and Policies

Visual PRESENTATIONS 7DV.4 PV Economics and Markets / PV Global Issues, Policies and Strategies

FOREWORD

The EU PVSEC 2016 in Munich concluded with the strong message that photovoltaic solar energy is here to stay. Research and innovation across the sector, together with its deployment and endless application option, are paving the way to putting PV at the center-stage of sustainable energy systems. 1,742 conference participants from 73 countries participated in this most inspiring platform. It provided a vital forum for information exchange in the field of PV solar energy research, demonstration and applications.

The EU PVSEC 2016 Proceedings give a comprehensive overview of the photovoltaic solar sector, its current status and future prospects in science, research, innovation, development and deployment on more than 5,000 pages. Selection for inclusion in the conference was made by paper review experts and topic organisers (see the listing on pages III-V), to whom we express our sincere gratitude for their comprehensive review work and overall contribution to the success of the conference. In addition to the 658 submitted papers, the proceedings includes 82 presentations (slides) shown during the plenary and oral presentations as well as 269 poster files of the visual presentations. In total this amounts to 1,009 publications!

The Proceedings constitute a powerful tool for targeted and quick information search and retrieval, enabling you to search by keyword, paper title and authors. We are sure that these features will help to simplify the use and exploit the full potential of this extensive source of information.

The Conference Proceedings are also published online. **A DOI code (Digital Object Identifier) has been assigned to each paper. This ensures unequivocal and permanent identification and full citability**. The papers can be viewed and downloaded from the EU PVSEC Proceedings website www.eupvsec-proceedings.com.

We are confident that these Proceedings will play an important role in providing a comprehensive overview of the current actors and activities in the global PV sector and that they will disseminate information on the state-of-the-art of technologies and applications. This can generate further research, add momentum to innovation and promote interest in PV worldwide.

We would like to cordially thank all authors and participants of the EU PVSEC 2016 for their contributions and look forward to welcoming you in 2017 at the next EU PVSEC.

The Editors

Joint Research Centre

the European Commission's in-house science service

Serving society
Stimulating innovation
Supporting legislation

EU PVSEC 2016 Highlights

Nigel Taylor
Technical Programme Chair

European Commission

First some stats ...

Participants: 1742

Presentations: 1053 (334 oral, 719 posters)

Parallel Events: 6

European Commission

European Commission

Silicon: strength in depth

European Commission

Q.ANTUM
Champion R&D module results

Champion module **features**

- Use current production tools
- Standard soldering technique
- Same BOM as in standard modules (high quality ARC glass, back sheet)
- Standard 60 cells module size
- No additional light-trapping schemes

- Usage of champion cell technology with module optimized ARC coating
- Half cell technology and splitted junction box
- Minimized intercell distance

- η = 19.5 % (ap)*
- V_{oc} = 41.5 V
- J_{sc} = 9.30 A
- FF = 77.4 %

- P = 301 W (ta)*

Classification	Efficiency (%)	Area (cm²)	V_{oc} (V)	I_{sc} (A)	FF (%)	Test centre (date)	Description
Si (crystalline)	23.8 ± 0.5	11562 (ap)	53.4	6.32ᵃ	81.6	AIST (1/16)	Panasonic (72 cells) [11]
Si (multicrystalline)	19.5 ± 0.4	15349 (ap)	41.53	9.299ᵃ	77.4	FhG-ISE (12/15)	Hanwha Q Cells (120 cells) [12]

M. Green et al, Prog. Photovolt: Res. Appl. 2016; 24:905–913

Current Status of High-Efficiency Q.Antum Technology with New World Record Module Efficiency of 19.5%

M. Scherff, P. Kowalzik, C. Gerber, K. Duncker, M. Junghänel, C. Fahrland, S. Kunath, S. Hörnlein, M. Schütze, L. Niebergall, B. Klöter & J.W. Müller
Hanwha Q CELLS, Bitterfeld-Wolfen, Germany

European Commission

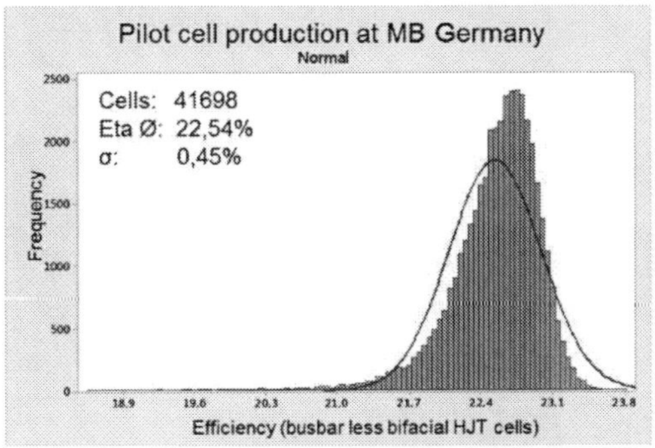

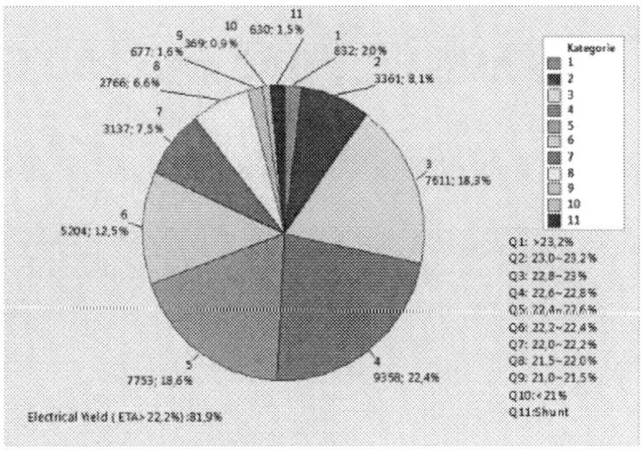

Golden Cell

Status	Cell	Area (cm²)	Eff. (%)	V_{oc} (mV)	FF (%)	J_{sc} (mA/cm²)
09/2015	Cz	239,5	23,3	742,5 ⇑	81,3 ⇑	38,64

- In this pilot production, total 41698 SWCT(busbar less) bifacial cells were processed
- Average ETA reached 22,54%
- 34617 cells (yield of 82%) are over 22,2%
- Record cell reached 23,3% cell efficiency, which is from its high V_{oc} and FF.

Silicon Heterojunction Solar Cells in Meyer Burger's Demo line: Results of Pilot Production on Mass Production Tools

J. Zhao et al

European Commission

Tipping point:

2BO.3 Heterojunction Solar Cell Concepts

Speaker gender balance for the first time!

European Commission

Vibrationally Induced Fluidization

1. CapEx, Full Cost, and Cash Cost will be lower than legacy processes

China: $/Kg on a Polysilicon Plant Basis[1]			
	Genesis™	H-FBR	HC-TCS/Siemens
CapEx	43	65	68
Full Cost	10.25	14.00	18.45
Cash Cost[2]	7.35	9.65	13.90

2. Advantages include
   - Only 1% yield loss to dust
   - Only 1 KwHr/Kg for V-FBR process unit – only 8 KwHr/Kg for entire plant
   - 10X higher reactor productivity
   - Self-seeding and in-situ particle size classification
   - On/off fluidization and "Ok" to temperature cycle
   - Continuous self-regulating product withdrawal

2BP.1.2 New Monosilane Fluid Bed Decomposition Technology for the Production of Solar Quality Silicon Feedstock, M. Dassel, SiTec, Augsburg, Germany

European Commission

Material Limits of Si from State-of-the-Art PLI Techniques

WAFER ELBA

Exact specification of loss mechanisms by ELBA

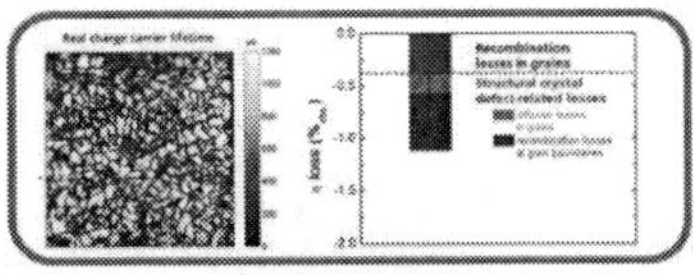

INGOT

Precise PL imaging of τ and $[Fe_i]$ on ingot level

modulum

harmonically modulated luminescence

Fraunhofer ISE

ELBA CUSTOM APP

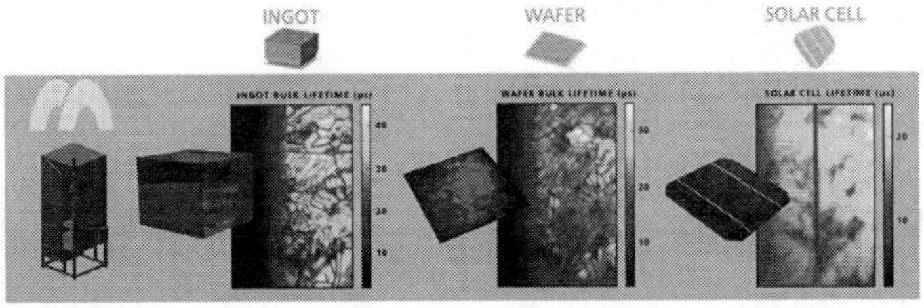

2CO.1.2 Material Limits of Silicon from State-of-the-Art Photoluminescence Imaging Techniques, F. Schindler, J. Giesecke, B. Michl, W. Warta & M.C. Schubert Fraunhofer ISE, Freiburg, Germany

European Commission

Thin film competes on efficiency

European Commission

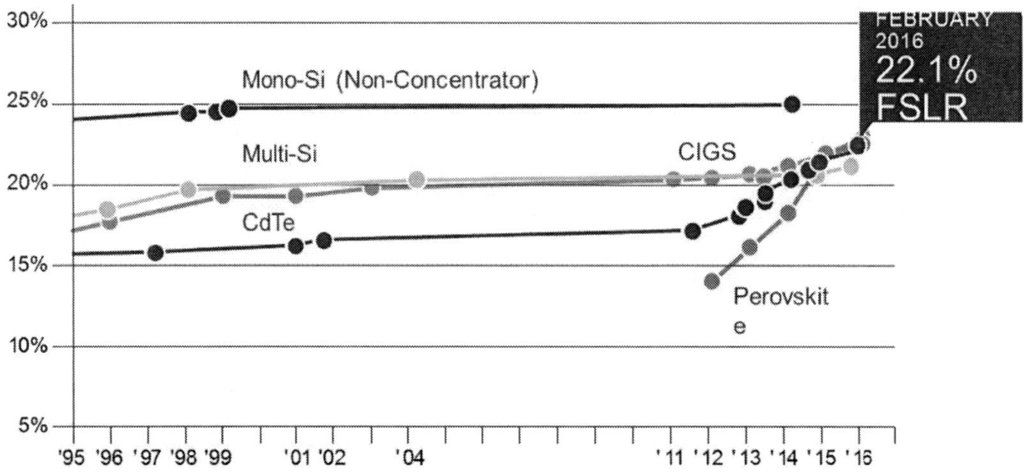

First Solar: current production module average 16.2%, top 18.2% and set to increase

European Commission

AVANCIS' Track Record of Certified Champion Efficiencies

LATEST 2016 EXTERNALLY CERTIFIED CHAMPION MODULE

STC measurement under natural sunlight according to IEC 60904-1/ IEC 60891

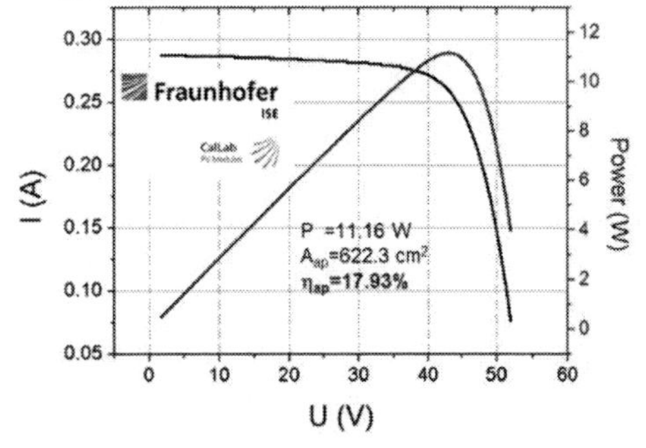

Comparison of champion modules:

	η_ap (%)	Jsc (mA/cm2)	Voc (mV)	FF (%)
2014 Champ (NREL)	16.6	38.5	621	69.4
2016 Champ (ISE)	**17.9**	39.7	627	72.2
Improvement	+7.8%	+3.1%	+1%	+4%

▶ Key improvements 2016 at a glance:

- Optimized cell width and dead area
- Improved TCO layer

▶ Major improvements on FF and Jsc

3BO.8.2 Efficiency Improvement of CIGSSe/Cd-Free Solar Module by Optimized Cell and Interconnect Design

P. Eraerds, C. Schubbert, T. Kwast, M. Grave, F. Braun, M. Algasinger, R. Lechner, T. Dalibor & J. Palm
AVANCIS, Munich, Germany

European Commission

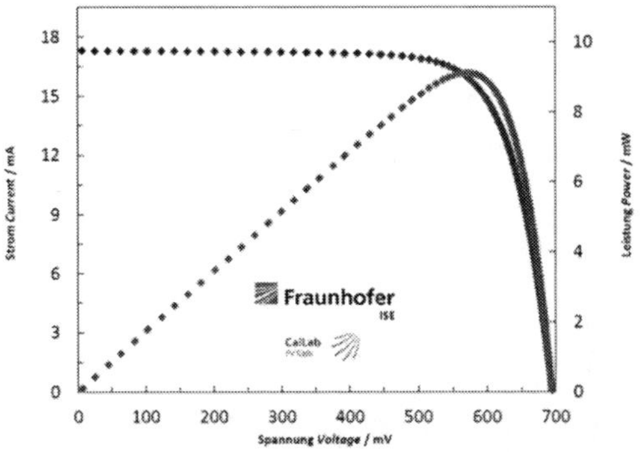

V_{OC}	= (	694.5	±	2.4	) mV
I_{SC} (Ed.2 - 2008)/3/	= (	17.28	±	0.33	) mA
J_{SC}	= (	33.58	±	0.64	) mA/cm²
I_{MPP}	=	15.94			mA
V_{MPP}	=	572.0			mV
P_{MPP}	= (	9.12	±	0.18	) mW
FF	= (	75.98	±	0.49	) %
η	= (	17.72	±	0.37	) %

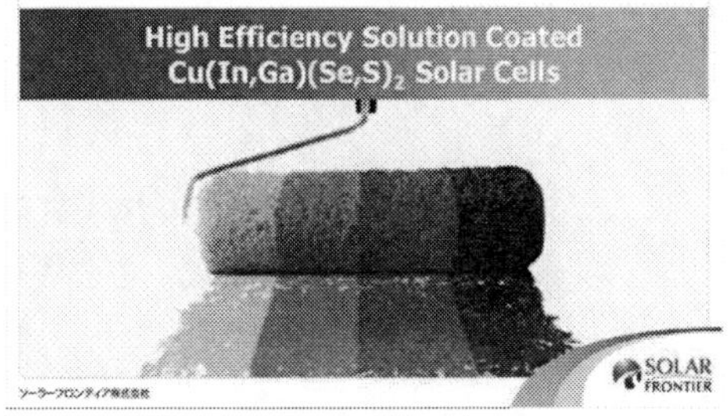

- **Eff 17.72% has been approved by Fraunhofer Institute**
- **Best result for the solution coated CIGSSe cell (Nano solar : Eff 17.1%)**

Photo conversion efficiency of 18.7% was achieved with solution coated Cu(In,Ga)(S,Se)$_2$.

3BO.8.3 High Efficiency Solution Coated Cu(In,Ga)(Se,S)2 Thin Film Solar Cells

T. Aramoto & Y. Kawaguchi
Solar Frontier, Atsugi, Japan
Y.-C. Liao, Y. Kikuchi, T. Ohashi, H. Iida & A. Nakamura
Tokyo Ohka Kogyo, Koza-Gun, Japan

European Commission

Lab scale PSC baseline development: mini-modules

- **Perovskite module development: lab scale**
  - Glass/sputtered ITO/P1 laser scribe /e-beam TiO$_2$/spin coat/Perovskite/spin coat OMeTAD/P2 mechanical scribe/evaporate Au/P3 mechanical scribe
- **Status Solliance**: **16 cm²** mini-module performance of **14,2%**

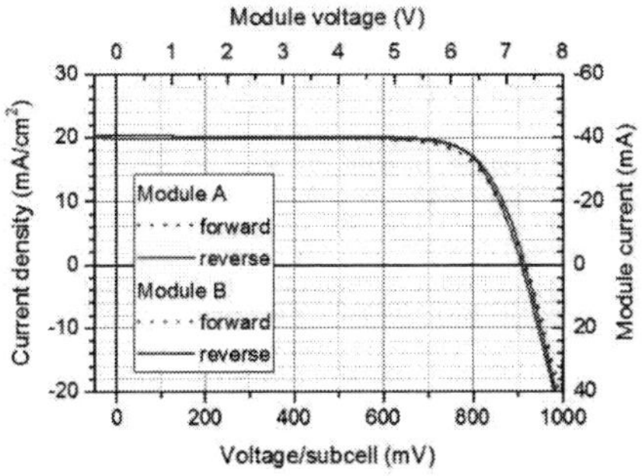

For further information:
3CO.7.5
Loss Analysis and Optimization for High Efficiency Perovskite Photovoltaic Modules
R. Gehlhaar, T. Merckx, C. Masse de la Huerta, L. Rakocevic, W. Qiu & M. Jaysankar
imec, Leuven, Belgium

Towards Roll-to-Roll Manufacturing of Perovskite Based PV Modules

R. Andriessen & P. Poodt, TNO, Eindhoven, Netherlands, T. Aernouts, imec, Leuven, Belgium. S. Veenstra, ECN, Eindhoven, Netherlands, R. Janssen & A. Createore, TU/e, Eindhoven, Netherlands, D. Vanderzande, University of Hasselt, Diepenbeek, Belgium T. Kirchartz Forschungszentrum Jülich, Germany

European Commission

Computational Screening of A_2MX_6 Compounds

Lead-free Perovskite Materials for Photovoltaic Applications

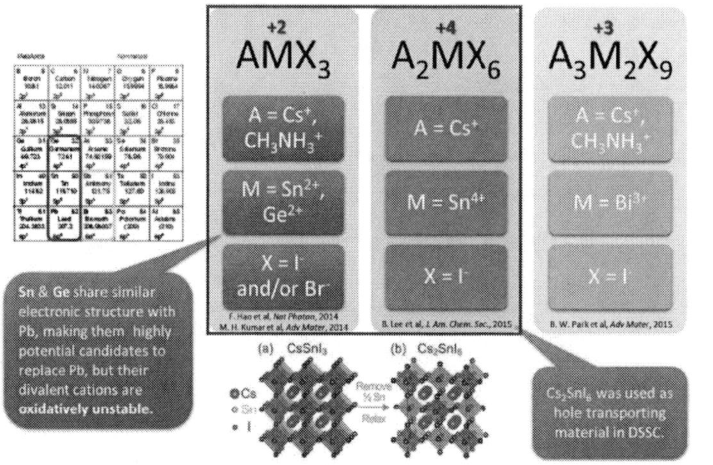

Tin(IV)-based Iodide Perovskites A2SnI6 (A = Rb+, Cs+, MA+, FA+)

Y. Chen, T. Krishnamoorthy, C. Yao, T. Baikie, S. Dharani, J.H. Fu, S.T. Chien, M. Asta, N. Mathews, L.H. Wong & S.G. Mhaisalkar
Nanyang Technological University, Singapore, Singapore & University of California, Berkeley, USA

European Commission

Concentrators:

new paradigms

European Commission

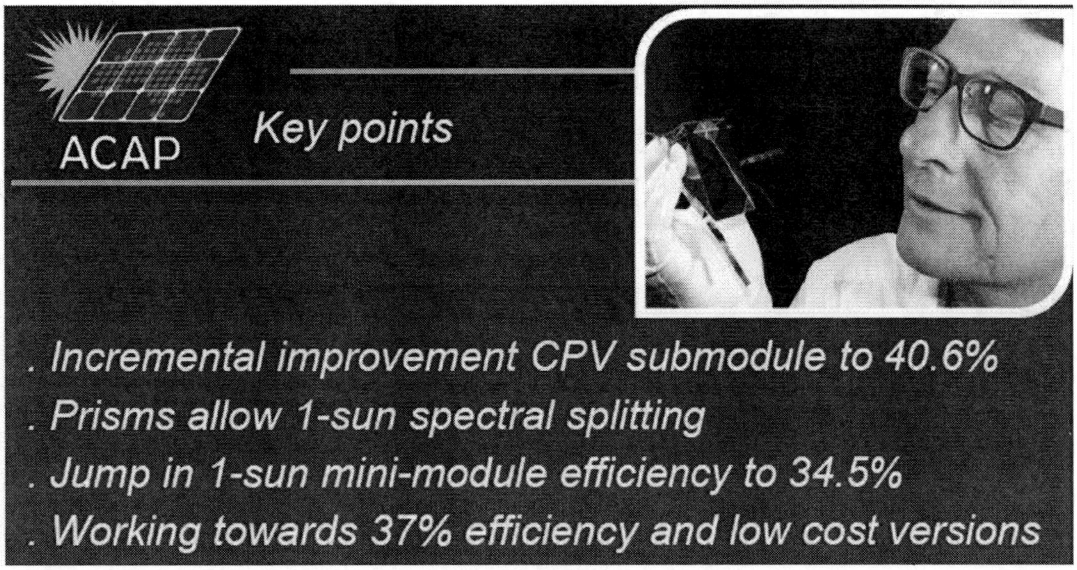

1AP.1.1 37% Efficient One-Sun Minimodule and over 40% Efficient Concentrator Submodules, M.A. Green, M.J. Keevers, B. Concha-Ramon & J. Jiang, UNSW, Sydney, Australia, P.J. Verlinden, Y. Yang & X. Zhang, Trina Solar, Changzhou, China

European Commission

4CO.6.4 Luminescent Solar Noise Barrier – Large Scale Testing and Modeling L.H. Slooff ECN, Petten, Netherlands S. Verkuilen Heijmans Wegen, Rosmalen, Netherlands M.M. de Jong & M.N. van den Donker SEAC, Eindhoven, Netherlands M. Kanellis & M.G. Debije Eindhoven University of Technology, Netherlands

European Commission

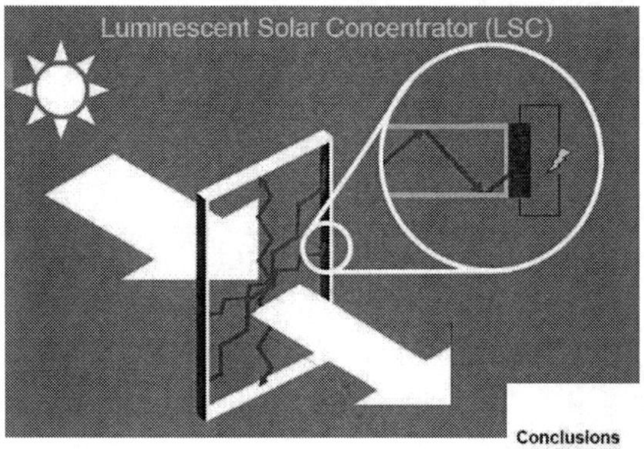

Conclusions

- Tm^{2+} doped halides discovered
  - Colorless
  - No self-absorption
  - >60% sunlight absorption
  - Expected ~15% power efficiency
- Made as both powders and thin-films
- Scattering still a challenge for BIPV

Outlook

- Not a lot known/understood (only ~4 major studies)*
  - Fundamental luminescence properties
  - Optimizing thin film growth
- Realize a lab-scale Tm^{2+} LSC prototype

1AO.2.4 Efficient Luminescent Solar Concentrators Based on Self-Absorption Free, Tm2+ Doped Halides

O.M. ten Kate, M. ten Kate, M. De Jong, O.M. ten Kate & E. van der Kolk
Delft University of Technology, Netherlands
K.W. Krämer
University of Berne, Switzerland

European Commission

Operation

European Commission

Peer-to-Peer (P2P) Forecasting:

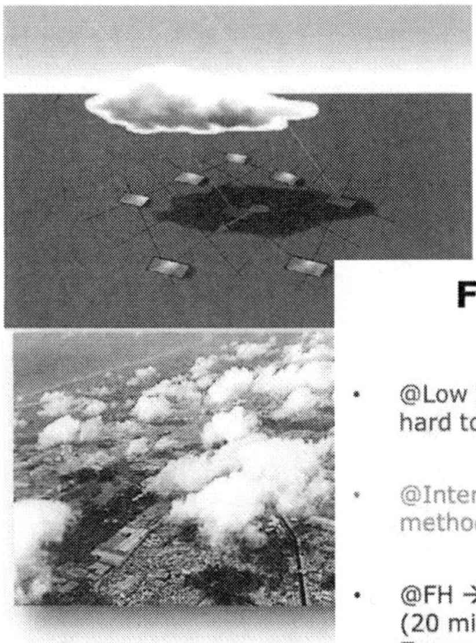

Forecast Skill vs. Forecast Horizon

- @Low FH: **persistence** hard to beat
- @Intermediate FH: P2P method shows its **value**
- @FH → T_{Obs} (20 min. in this example.): **Fewer** to **no** available periph. $k_j^*(t-\tau_{ij})$ timeseries for P2P FC

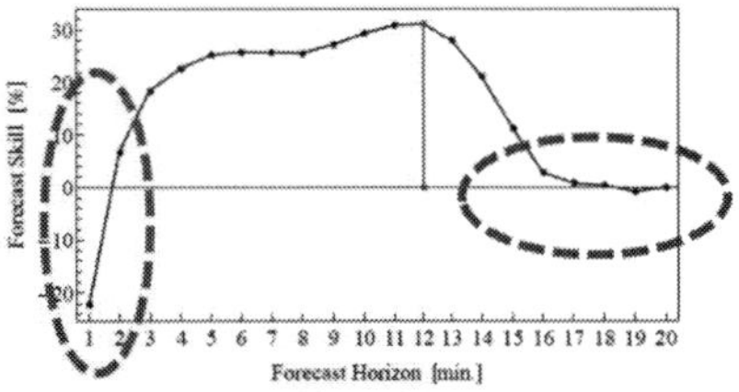

Dependence of Peer-to-Peer Solar Forecast Skill on Irradiance Variability

B. Elsinga & W.G.J.H.M. van Sark
Utrecht University, Netherlands

European Commission

Analysis of Failure Occurrence

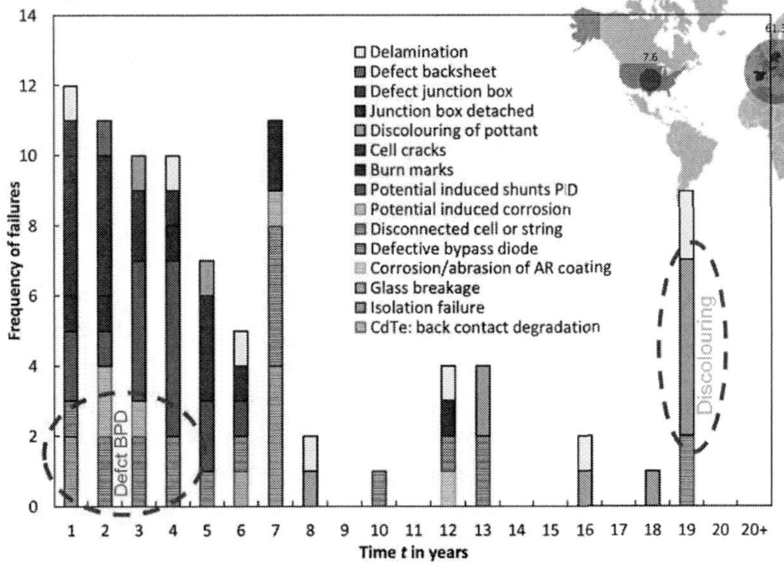

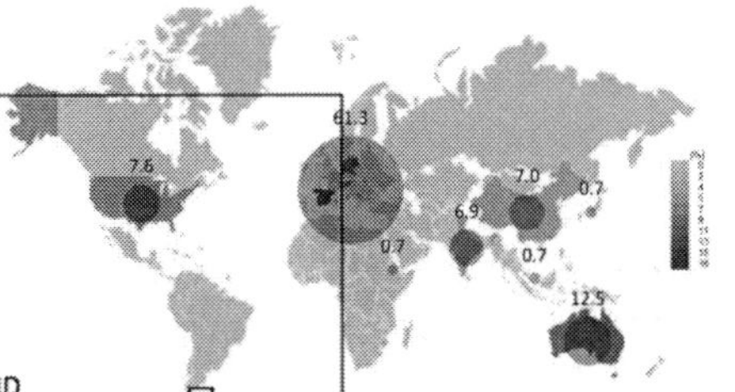

5DP.1.2 Mean Degradation Rates in PV Systems for Various Kinds of PV Module Failures

M. Köntges & S. Altmann
ISFH, Emmerthal, Germany
U. Jahn
TÜV Rheinland, Cologne, Germany

European Commission

Implementation and Monitoring

Detailed measurement

2015/16

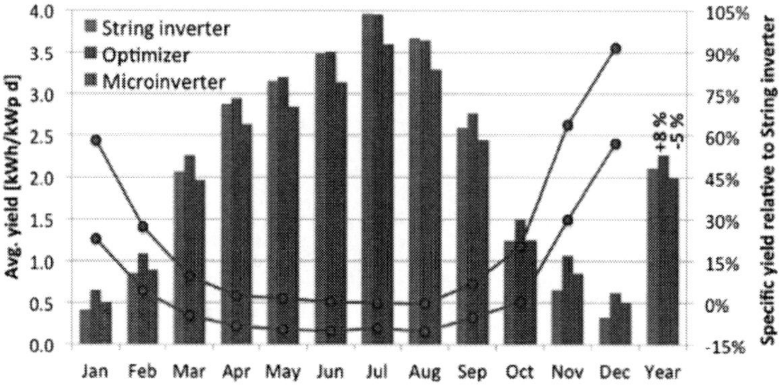

TNC

(4.6.2015 – 3.6.2016)

12

5AO.9.4 Performance of Recent Inverter Systems under Partial Shading Conditions

R. Lingel, T. Nordmann & T. Vontobel
TNC Consulting, Feldmeilen, Switzerland

European Commission

Energy Solutions

European Commission

IS "BIPV" ALREADY "VINTAGE"...?

SO... BIPV "MAINSTREAM POTENTIAL"

.TECHNOLOGICAL INTEGRATION | PV AS A PART OF A WHOLE ENERGY SOLUTION SET
.POLICY DRIVERS ARE CRUCIAL

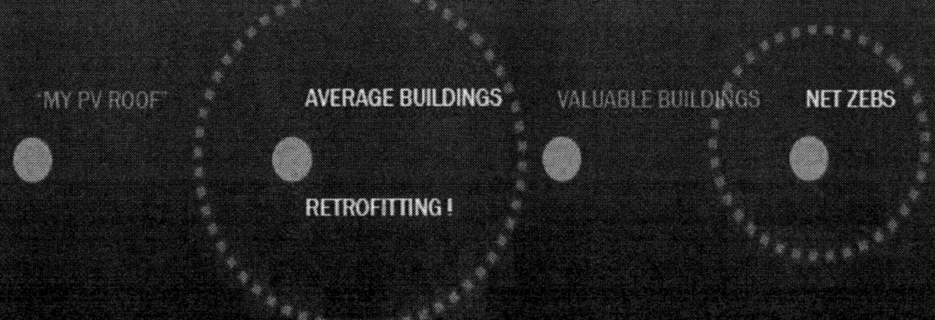

Kinetic envelope for heating cooling and light comfort + PV generation

Case Study

- Single Zone Office
- Zurich Weather Data
- 50 Panels in a Single Cluster
- Heating COP: 4
- Cooling COP: 3
- Lighting Load: 11.74 W/m^2

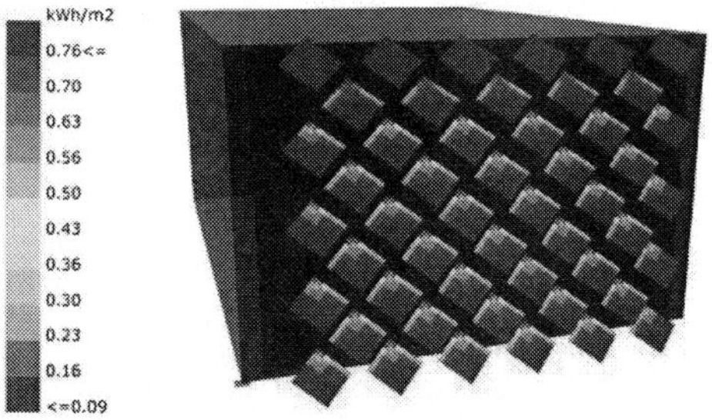

6DO.8.5 Energy Performance of PV Modules as Adaptive Building Shading Systems

J. Jayathissa, J. Schmidli, J. Hofer & A. Schlueter
ETH Zurich, Switzerland

European Commission

New metrics for PV + storage performance!

Definition of the Storage Performance Index (SPI)

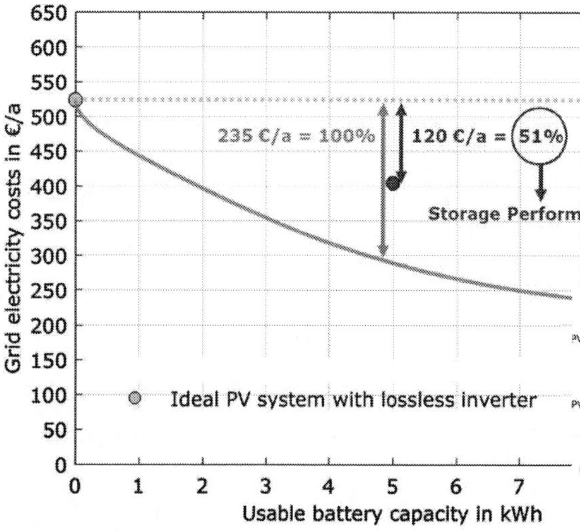

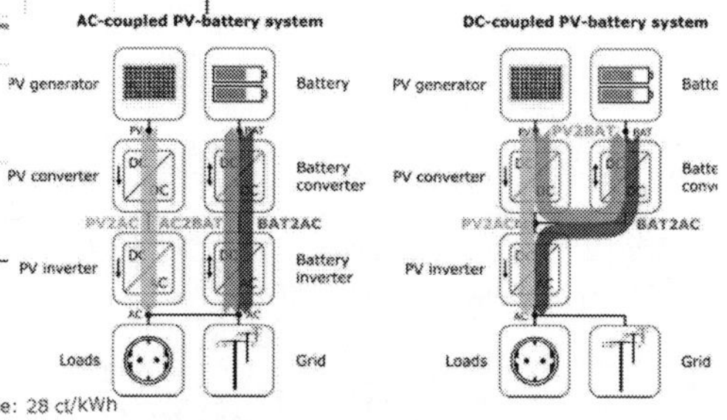

PV-generator: 5 kWp, load demand: 5 MWh/a, feed-in tariff: 12 ct/kWh, retail electricity price: 28 ct/kWh

Emerging Performance Issues of Photovoltaic Battery Systems, Johannes Weniger, Tjarko Tjaden, Joseph Bergner, Volker Quaschning, HTW Berlin - University of Applied Sciences, Germany

European Commission

Thanks!

European Commission

Start working on excellent abstracts for EU PVSEC 2017!

European Commission

We thank the EU PVSEC 2016 Sponsors

This is a pre-peer review version of the paper submitted to the journal Progress in Photovoltaics.
It has been selected by the Executive Committee of the EU PVSEC 2016 for submission to Progress in Photovoltaics.
32nd European Photovoltaic Solar Energy Conference and Exhibition

INNOVATIVE APPROACHES TO INTERCONNECT BACK-CONTACT CELLS

Jonathan Govaerts[1], Tom Borgers[1], Eszter Voroshazi[1], Shruti Jambaldinni[1], Barry O'Sullivan[1], Sukhvinder Singh[1], Maarten Debucquoy[1], Jozef Szlufcik[1] and Jef Poortmans[1,2]
[1]Imec, Kapeldreef 75, 3001 Leuven, Belgium
[2]Also part-time professor at KULeuven and UHasselt

ABSTRACT: This paper reports on a novel and highly innovative technology to interconnect back-contact cells. First a background section explains the currently developed interconnection technologies. Then our novel concept is introduced: interconnection of the cells is achieved with a woven fabric, combining both conductive wires, needed for the interconnection, and insulating wires, crucial to avoid shunting between both polarities at the backside of the cell. Weaving also allows out-of-plane stress relief features. In the third section, we explain that our technology combines the advantages of multi-wire interconnection such as low-stress and reduced resistive losses with known and low-cost materials of traditional module assembly.
We demonstrate that the technology can be tuned broadly in terms of costs and resistive losses using an optimized interconnection pattern. These loss simulations are also validated with the measured additional resistance in IBC test samples. Finally, first functional proof-of-concept with MWT cells as well as IBC cells have been fabricated.
Keywords: back-contact module, interconnection

1 BACKGROUND

As an introduction, Figure 1 illustrates the different interconnection approaches currently applied or under development for back-contact cells (independent of IBC or MWT, but related to the design of the cell metallization). The edge stringing approach [1], decouples the cell interconnect from the cell contact metallization. While an elegant approach in this respect, it also implies the cell metallization has to carry the current of the full cell. This leads to a trade-off in terms of cell metallization thickness: a low resistance requires a high thickness, however stress considerations (warping due to CTE mismatches) ask for a thin metallization. To overcome this trade-off, a busbar stringing approach [2] can help, where the interconnect metallization on top of the busbars can reduce the need for a thick cell metallization. In this approach, however, some electrical performance is lost as some area is sacrificed for the busbars, causing electrical shading, unless in the case of floating busbars [3]. To reduce electrical shading from the busbars, a point contact approach [4] is an interesting solution, but this means classical tabbing, where the conductive tab is directly placed over the cell, is compromised as it will shunt the cell. To avoid this shunting an isolator is needed. Similar to printed circuit board assembly technology, this function could be realised by a soldermask. However, the cost of such an additional layer is typically considered too high for the mainstream PV industry. One way of significantly reducing this cost is put forward by the solder-through approach, where the insulation is guaranteed by a porous insulator, e.g. a woven glass fibre sheet [5], through which a solder paste reflows and provides contact between cell contact and ribbon. Alternatively, the interconnection can be integrated in the backsheet in the so-called conductive backsheet approach [6]. Here the insulation is achieved by the encapsulant. Drawbacks here are that standard soldering is not allowed, asking for conductive adhesives, and cost and reliability are important points-of-attention.

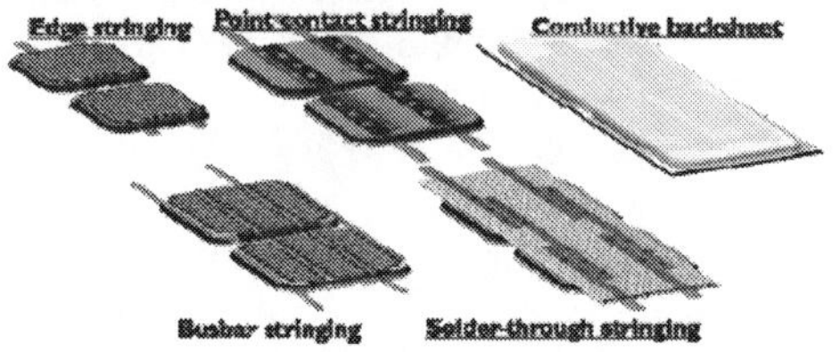

Figure 1: State-of-the-art interconnection approaches for back-contact cells

2 WOVEN WIRING

With the above background in mind, a novel concept is under development at imec [7]. In this concept, mostly evolving from the solder-through approach, a woven fabric further integrates the conductive wires, as indicated in Figure 2. Figure 3 shows that it also brings along benefits of the conductive backsheet approach, and even of the multi-wire approaches that are now slowly gaining ground in the module concepts for 2-side contacted cells [8,9].

The flexibility in cell contacting is due to allowing many distributed contacts and having the interconnect metallization on a different (2nd) level than the cell metallization. Avoiding the necessity of a separate stringing step, and the potential of solder paste for self-alignment allows a simplified module assembly, while the interconnection itself inherently has a low topography thanks to the uniformly distributed wires that are integrated in the fabric. As the interconnection can be separated from the encapsulation process, the technology also allows freedom in encapsulation system. Apart from obviously allowing glass-glass lamination, the reliability potential is high, considering the solder joints, distributed stresses (with distributed wiring), the built-in out-of-plane stress relief that can be implemented, and an effective "armouring" of the encapsulation due to the mechanical fibre (glass and metal) reinforcement of the encapsulation. An interesting aspect is also the low-cost potential by employing glass fibre material and the limited topography that allows a minimized amount of encapsulant material.

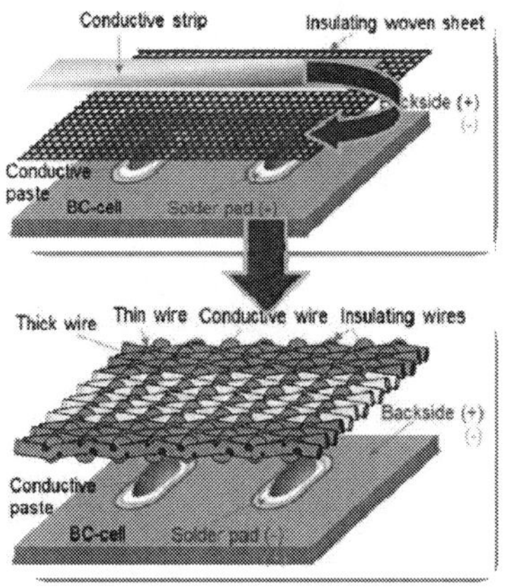

Figure 2: Interconnect tabs from the solder-through approach (top) are integrated inside the woven fabric in this concept (bottom)

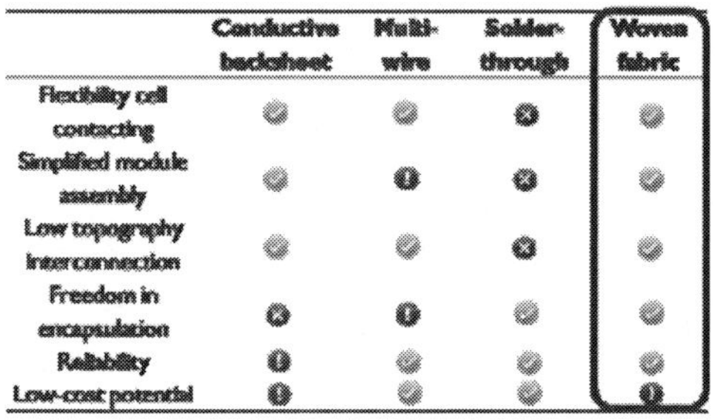

Figure 3: Qualitative comparison between interconnection approaches

3 TECHNOLOGY DESIGN AND IMPLEMENTATION

Keeping in mind the concept, the interconnection fabric needs to fulfill a few requirements. It has to incorporate both conductive and insulating wires, making sure the conductive wires do not make contact to the underlying backside of the cell, even after lamination. Also, it has to firmly fix the wires, and from an industrial point-of-view a high throughput potential is required. That is why we have considered weaving technology for the manufacturing of this fabric. In particular, so-called Leno weaving is of interest, as it allows to twist (and thus fix) the insulating wires around the conductive wires. This fabric, based on a half Leno weave to be exact, and its use on a MWT cell backside with a contact of either polarity is illustrated in Figure 4.

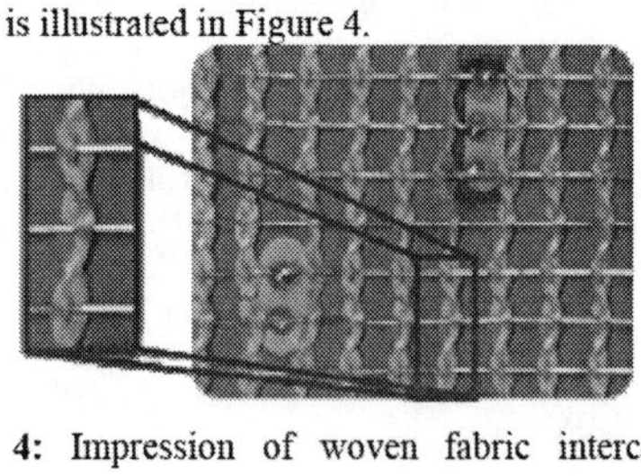

Figure 4: Impression of woven fabric interconnect technology

In Figure 4, the conductive wires are horizontally "floating" above the cell, fixed by the insulation wires that act as spacer between the cell and the conductive wires. The conductive wires consist of Sn-coated Cu and measure 200 µm in diameter, the multifilament glass fibre ranges 100-200 µm in diameter. In this example the two twisted glass fibres are identical, but using 2 different diameters would allow very local out-of-plane stress relief in the module. The solder paste dots (standard Sn62Pb36Ag2) are then applied only where a contact between cell and conductive wire is needed. Its self-alignment properties inherently reduce the risk on shunting issues. Finally, the open weave structure allows an easy penetration of encapsulant during lamination and straightforward alignment during assembly. The pitch for both conductive and insulating wire (couple) in this first fabric example is 2.5 mm.

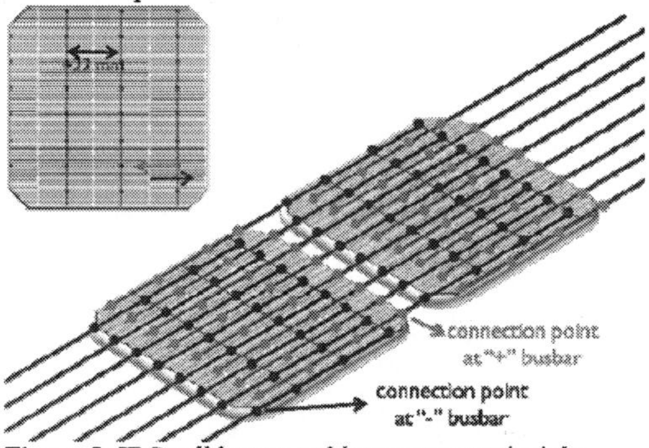

Figure 5: IBC cell layout and interconnect principle

Together with the fabric, the cell metallization (in particular contact) layout has to be (co-)designed and optimized to benefit maximally from the technology. The optimized (IBC) cell layout and interconnection principle is drawn schematically in Figure 5.

To minimize interconnect losses, the busbar layout is designed asymmetrically with the outer busbars of either polarity at the edge of the cell. The interconnect wires, with distributed contacting, can take over part of the current carrying function of the busbars, and as a result these busbars can be minimized in width, reducing dead area from electrical shading. Finally, due to the self-alignment properties of the solder paste, large contact pads are not required (compared to conductive adhesive), further increasing the cell's active area. In the final design, the 8 alternating busbars are ~0.7 mm wide and spaced ~22mm apart. Implementation of this design has recently resulted in cells with 22.5% efficiency [10].

4 INTERCONNECT LOSS SIMULATION

Further on, the interconnect wiring scheme can be optimized. To reduce electrical losses, the interconnect cross-section (both diameter and pitch or number of wires) should be as high as possible. But obviously this will be a cost trade-off: Cu wire and Sn-coating, glass fibre, encapsulant (thickness), solder paste and processing (weaving, dispensing, soldering,...). Figure 6 shows a simulation for the electrical loss and Cu material cost per cell. For the first cell and fabric with dimensions, pitch and busbar pattern as mentioned above, and assuming a uniformly generated cell current density of 40 mA/cm2 with an optimized interconnect pattern, the power loss per cell due to the interconnect is calculated to be 131 mW.

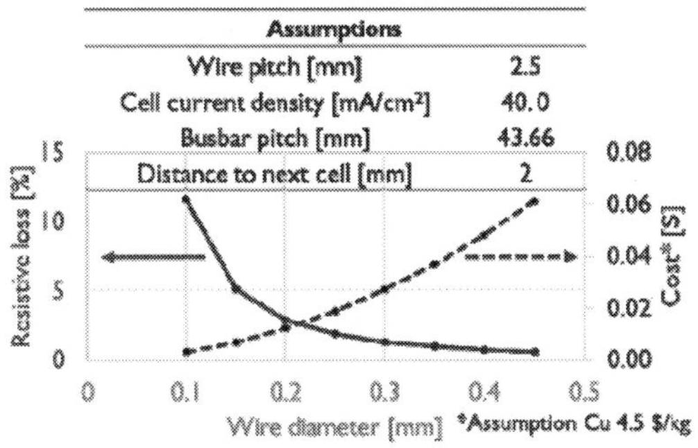

Figure 6: Illustration trade-off electrical loss and Cu material cost

As a side note it is useful to point out that the above mentioned parameters may not only impact performance and cost, but also reliability aspects (stress, humidity ingress, mechanical integrity).

5 PROCESS FLOW AND DEVELOPMENT

Figure 7 shows the process flow for the woven fabric technology, both schematically and illustrated with detailed pictures. Worth noting is how the wiring is floating above the cell metallization, and the clear self-alignment of the solder paste on the cell metallization, sometimes observed even early on after drying already.

The dispensing was carried out with a home-assembled XYZ-stage and a state-of-the-art dispenser, and proved sufficiently accurate (less than 100 um deviation for repeatability) and fast (less than 5 minutes for 120 dots) for our purpose. In the next step the fabric is placed over the cell. To keep alignment over large-area, it is helpful to put the fabric under light tension during soldering. Here we realized this by using an external removable frame.

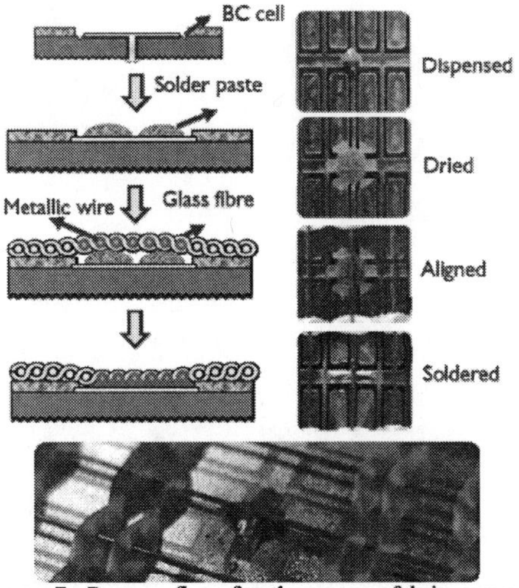

Figure 7: Process flow for the woven fabric concept: schematically (left) and illustrated with close-up pictures (right and bottom)

We could demonstrate proof-of-principle with several soldering options, contact, hot-air and proximity (ironing) soldering. The best results were achieved with hot air with an underlying hotplate at 160°C. Finally the stack can be laminated.

6 PROVING THE CONCEPT

For technology development, and checking the feasibility of the concept, we prepared test samples: isolated wafers, with a metallization stack and an interdigitated pattern identical to functional IBC cells. Part of these samples were intentionally shunted to allow series resistance measurements to deduce the resistance added by the interconnection and compare it with the simulated resistance. To this end, a 2x2-cell laminate was prepared with those test samples, shown in Figure 8, where resistance could be measured for each cell separately. The added resistance per cell was deduced to be 1.2 mΩ, which corresponds relatively well to the value of 1.4 mΩ that can be calculated from the power loss simulated above.

Figure 8: 2x2-cell module based on IBC test samples for proof-of-concept interconnection: back- (left) and front side (right) view

To also functionally demonstrate the concept, we fabricated a 6-cell module, based on diced MWT cells. Though the performance is somewhat limited due to low-quality multicrystalline Si MWT cells, manual processing (with higher risk on microcracks and non-uniform contacting) and a non-matching cell contact and interconnect fabric design, the functionality is clearly proven. All this is indicated by the IV and EL results in Figure 9.

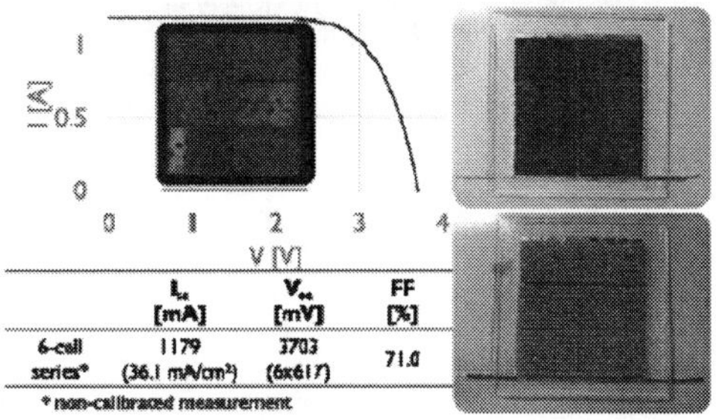

	I_{sc} [mA]	V_{oc} [mV]	FF [%]
6-cell series*	1179 (36.1 mA/cm²)	3703 (6x617)	71.0

* non-calibrated measurement

Figure 9: 6-cell laminate based on MWT cells with the new woven back-contact interconnection technology indicating functionality

7 CONCLUSION AND OUTLOOK

In this paper we introduce a novel back-contact module interconnection technology with its potential benefits and indicate how the cell and fabric design can be optimized. Further on we elaborate the process flow and technology development and we validate simulations on the added resistance due to the interconnection. Finally we demonstrate proofs-of-concept, both with IBC cells and MWT cells.

As an outlook, a more detailed cost analysis to determine the viability of the technology is necessary, next to further technology development for an industrially-compatible manufacturing process, optimizations, e.g. integrating the bussing ribbons inside the fabric, and upscaling to larger modules.

8 ACKNOWLEDGMENT

The authors gratefully acknowledge imec's industrial affiliation programme and its partners for valuable input, the European Union for EFRO funding within the SolsTHORe project, the flemish government for funding within the LENO-project and VdSWeaving for discussions and manufacturing of woven interconnect material.

REFERENCES

[1] D. Rose, S. Daroczi, Development and Manufacture of Reliable PV Modules with > 17% Efficiency. Proceedings of the 20th European Photovoltaic Solar Energy Conference. (2005).

[2] A. Halm, V.D. Mihailetchi, G. Galbiati, L.J. Koduvelikulathu,R. Roescu, C. Comparotto, R. Kopecek, K. Peter, J. Libal J, The ZEBRA cell concept – large-area n-type interdigitated back contact solar cells and one-cell modules fabricated using standard industrial processing equipment. Proceedings of the 27th European Photovoltaics Science and Engineering Conference Vol. I (2012) 567.

[3] G. Galbiati, V. D. Mihailetchi, R. Roescu, A. Halm, L. J. Koduvelikulathu, R. Kopecek, K. Peter, J. Libal Large-Area Back-Contact Back-Junction Solar Cell With Efficiency Exceeding 21%., IEEE Journal of Photovoltaics 3 (2012) 560 – 565.

[4] U. Eitner,I. Hädrich, M. Tranitz, W. Wirth Progress in HIP-MWT module technology. Presentation at MWT workshop (2013).

[5] S. Dewallef, New interconnection method for MWT cells based on tabber-stringer technology. Presentation at MWT workshop (2013).

[6] I. J. Bennett, E.E.Bende, M.J.A.A.Goris, W. Eerenstein W. An overview of developments in foil-based back-contact modules. Proceedings of the 29th European Photovoltaics Science and Engineering Conference Vol.I (2014) 5-8.

[7] T. Borgers, patent n# EP2660878

[8] S. Braun, G.Hahn, R. Nissler, C. Pönisch, D. Habermann, Multi-busbar solar cells and modules: high efficiencies and low silver consumption. Energy Procedia 38 (2013) 334 – 339.

[9] T. Söderström, P. Papet, J. Ufhei, Smart-wire connection technology. Proceedings of the 28th European Photovoltaics Science and Engineering Conference Vol I (2013) 495-499.

[10] B. O'Sullivan, S. Jambaldinni, S. Singh, M. Kyuzo, M. Debucquoy, J. Szlufcik, Loss Analysis and Design Optimization of Large-Area High-Efficiency Back-Contacted Silicon Solar Cells. IEEE Journal of Photovoltaics (2016) (accepted for publication)

Proceedings 17th European Photovoltaic Solar Energy Conference, Vol. I (2002) 903.

TRANSIENT I-V MEASUREMENT SET-UP FOR PHOTOVOLTAIC LASER POWER CONVERTERS UNDER MONOCHROMATIC IRRADIANCE

S.K. Reichmuth[1], H. Helmers[1], C.E. Garza[1,2], D. Vahle[1,3], M. de Boer[1], L. Stevens[1], M. Mundus[1], A.W. Bett[1], G. Siefer[1]
1: Fraunhofer Institute for Solar Energy Systems ISE, Heidenhofstraße 2, 79110 Freiburg,Germany
phone +49 761 4588 5388, kasimir.reichmuth@ise.fraunhofer.de,
2: Now with Nanoscribe GmbH, Hermann-von-Helmholtz-Platz 1, 76344 Eggenstein-Leopoldshafen, Germany
3: Now with Ruhr-Universität Bochum, Universitätsstraße 150, 44801 Bochum, Germany

ABSTRACT: A new I-V measurement set-up called LaserSim has been developed for photovoltaic laser power converters – photovoltaic cells optimized for monochromatic artificial light - providing irradiances of up to 89 W/cm². The evaluation of the set-up and first I-V curve and efficiency measurements are presented. Similar to flash simulators used for I-V characterization of solar cells, the LaserSim is using transient measurement routines to avoid an influence of heating of the samples during measurement. An excellent light uniformity in the designated test area was realized. A beam sampler allows for in-situ irradiance determination while recording I-V data. This is mandatory for characterization of devices showing a non-linear behavior of short circuit current with irradiance. Efficiency values under monochromatic light at 809 nm have been measured for a single junction GaAs cell (55.6 %) and a dual-junction GaAs device (53.5 %).

Keywords: photovoltaics, III-V semiconductors, laser power converter, power-by-light, characterization.

1 INTRODUCTION

Laser power converters are photovoltaic (PV) cells optimized for monochromatic artificial light sources. Laser or LED light sources are typically used for "Power-by-Light" applications. These are optical power transmission systems, where instead of using copper cabling, power and data signals are guided through optical fibers, free space laser transmission is also possible. This technology inherently allows for galvanic isolation, electromagnetic compatibility, wireless solutions and other [1]. Application examples for this technology can be found in various domains and are as diverse as structural health monitoring in wind turbines [2, 3], cortical neural sensing in brain implants [4], wireless powering for implantable sensing platforms [5], monitoring of high voltage power lines [6], optically powered video surveillance [7], submarine hydrophone and seismometer networks [8], passive optical networks [9].

Photovoltaic laser power converters reach very high opto-electrical conversion efficiencies, as the semiconductor bandgap of the absorber material can be designed to match the laser wavelength of the system [10-12] and thermalization losses occurring in solar applications can thereby be minimized. A typical semiconductor compound chosen for "power-by-light" systems is GaAs with a bandgap of 1.42 eV, which is well suited for laser wavelengths in the range 800 to 850 nm [13]. Laser power converters are intended to supply electric energy to sensors and actuators, with a power consumption ranging from the milliwatt to the watt level which can well be realized with single cell devices. However, the voltage required in standard electronic applications is often higher than what is supplied by a single PV cell with one pn-junction. A single-junction GaAs cell provides ~1 V depending on irradiance and temperature, while electronic circuitry often requires voltages between 2 and 13 V. The usage of DC-to-DC converters is possible, as well as the solution of connecting several PV cells in series using several separate optical links. A more elegant solution is the monolithic series connection of cells on chip level. Both, lateral and vertical series connection schemes exist and have been realized for PV laser power converters. In the lateral realization multi-segment cells, also known as monolithic interconnected modules or MIM are realized [14-16]. There a single-junction cell is separated into several individual segments of same size by post growth technological processing. The individual segments are then connected in series by galvanic metallization [17]. Consequently, processing becomes more complex and time consuming as the number of technological steps is increased. When integrating such multi-segment devices into power-by-light systems, fiber optical coupling to the cell is less tolerant against misalignment than in the case of a single segment device [16]. This is due to the fact that for maximum current generation, uniform illumination of the segments of the series connection is required.

The approach to realize the vertical series connection is done by epitaxial growth, as in multi-junction devices for space or concentrating photovoltaic applications. In contrast to multi-junction devices designed for the solar spectrum, a laser power converter designed to absorb monochromatic light is realized with several subcells of the same absorber material [10, 11, 15, 18]. Maximum current is reached, when all subcells generate the same current. Consequently, for the design of such series connected multi-junction cells, the thickness of the subcells has to increase from top to bottom and is calculated following the Beer-Lambert law of exponential absorption [15, 16]. The temperature dependence of the bandgap energy of the used semiconductor material introduces a current mismatch as layer thicknesses can only be adapted for one temperature [19]. However, in contrast to the MIM concept, these devices are as tolerant in respect to alignment as a standard single-segment cell. Additionally, post growth technological processing is similar as for a single-junction cell, and therefore less complicated compared to the case of the MIM concept.

However, in the emerging field of photovoltaic laser power converters, the measurement of these new devices poses a challenge. Therefore, the development of adapted and new characterization tools is required. Recent publications of 5-junction GaAs devices [18] proof the relevance of developing a standard approach for comparability of characterization and discuss repeatability and measurement accuracy [20]. PV cells are usually characterized using sun simulators with a broadband spectrum tailored to be close to the solar spectrum, where the simulator spectrum will generate the same current as under the reference spectrum. In the case of multi-junction devices made of one absorber material (i.e. only one bandgap), measurement of laser power converters under such spectrally distributed light will, however, result in significantly different performance than under monochromatic illumination. The Xenon spectrum of a typical flash simulator leads to a significant current mismatch of the respective subcells connected in series, and is therefore inadequate as it will strongly influence the I-V parameters [21]. In addition, it has been demonstrated that the behavior of these multi-junction laser power converters shows a significant influence of luminescence coupling [18, 19, 22, 23], resulting in a non-linear behavior of short circuit current with irradiance. This means that for such devices it will not be possible to determine the irradiance level from the short circuit current of the measured cell itself as it is commonly done for single-junction laser power converters (and concentrator solar cells). Therefore, an alternative way for the determination of the irradiance level in the test plane during measurement is a requirement for such a set-up for the characterization of laser power converters.

2 DESCRIPTION AND EVALUATION OF LASERSIM MEASUREMENT SET-UP

2.1 Description

The laser beam profile across the PV receiver in power-by-light applications varies with the specific system specifications (light source, fiber type and length, optical elements, fiber coupling). In order to enable general comparisons of different cells, cell concepts and cell architectures (varying in size and geometry, as well as multi-junction or multi-segment cells) [15], similar design goals as applied for solar simulators have been chosen for the new measurement set-up. As illumination cannot be reproduced for all different forms of power-by-light systems, a uniform so-called flat-top laser profile (i.e. homogenous light intensity) in the test plane for typical cell sizes from 1 to 3 mm edge length was set as design criterion.

The new I-V measurement set-up is designed to enable measurements at variable irradiances. Opposed to solar simulators a **monochromatic** light source provides illumination in the test plane: For high irradiances in the 4 mm x 4 mm test plane a 50 W laser diode system emitting light at 809 nm (FWHM: 3nm), which leads to an intensity of 89 W/cm² in the measurement plane. The laser can be current modulated to emit single pulses of e.g. 4 ms duration.

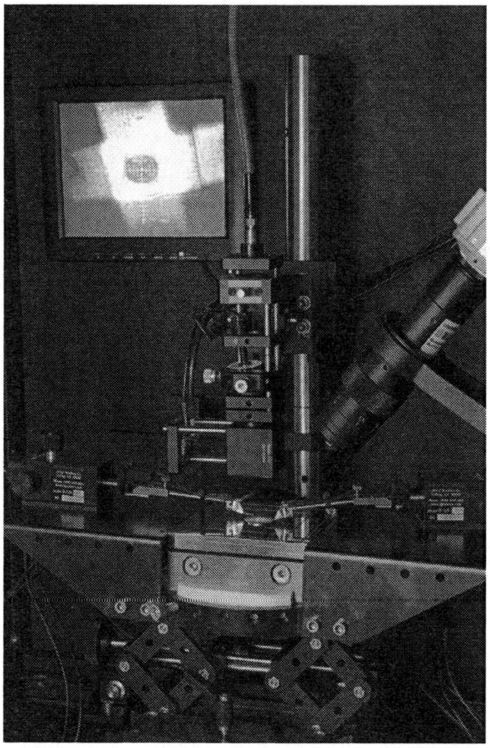

Figure 1: Photo of the LaserSim measurement set-up. The PV-sample is fixed on a temperature-controlled chuck that can be positioned by a x-y-z-stage. The sample is contacted by Kelvin probes. Above the cell the homogenizing optics are held in a cage system (see schematic in Figure 2). A digital microscope allows for accurate positioning of the cell in the light spot. A visible pilot laser (650 nm) coupled into the same fiber as the infrared main laser, is used for positioning the sample into the beam.

The homogenizing optic creating the flat-top profile is shown schematically in Figure 2. The laser beam is guided through a multi-mode fiber with 400 µm core diameter. The radiation emitted from the fiber end facet is collimated by a lens and is oriented onto a diffusor disc for initial homogenization. The main homogenization is realized in form of an "imaging multi-aperture beam integrator" as described by Dickey [24]. The specific design was realized by "Bayerisches Laserzentrum" and uses two micro-lens-arrays (MLA) from "SUSS MicroOptics". In principle, the individual lens apertures of the first MLA, that are illuminated by the collimated laser beam, are imaged onto the measurement plane by the second MLA and the Fourier lens. Overlapping these multiple images onto the measurement plane significantly reduces inherent non-uniformities present in the collimated laser beam.

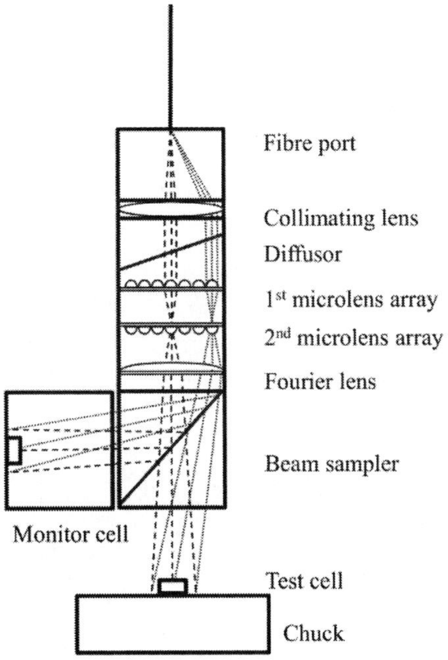

Figure 2: Schematic of homogenizing optics mounted in a cage system. Radiation emitted from the fiber is imaged multiple times onto the measurement plane by microlens arrays and a Fourier lens, thereby enabling homogeneous illumination of the PV sample. As an example single light beams are sketched into the drawing.

A beam sampler directs a small fraction of the incident radiation onto a monitor cell. The purpose of the monitor cell is twofold: On the one hand it is used for a relative correction of short term variations in intensity during the I-V measurement. On the other hand, as the beam splitting ratio or more precisely the ratio between the light intensities in test and monitor cell plane is independent of the absolute intensity, the monitor cell signal can also be calibrated for in-situ determination of the absolute intensity level in the test plane.

Similar to the flash simulator described by Siefer [25], I-V measurement data is logged using a transient recorder. After manually starting the laser pulse of 4 ms duration, the measurement is triggered after a delay of ~1.9 ms which assures stabilization of the laser. Next, the I-V measurement is performed during the following millisecond. The comparatively short illumination and measurement time is used to avoid heating of the test sample during measurement also at high laser intensities. A bipolar power supply is used as variable load in order to vary the voltage at the test sample and consequently switching the I-V curve. In order to detect capacitive or temperature effects the voltage ramp of the load is switched from I_{sc} to V_{oc} during a first measurement, followed by a second reverse measurement from V_{oc} to I_{sc} at same laser intensity settings. The two resulting I-V curves are then compared and examined for possible differences. A difference in fill factor of above 1.5 % is used as indicator for capacitive effects. In that case the measurement can be repeated with increased measurement time. However, this effect is rarely found for III-V based solar cells, but is known for Silicon based solar cells [26, 27]. Differences between V_{oc} determined from the two I-V curves – in particular a higher V_{oc} found for the I-V curve switched from V_{oc} to I_{sc} – indicate an influence from heating of the sample during measurement. This is due to the fact that for the reverse I-V curve switched from V_{oc} to I_{sc}, the V_{oc} is measured at the beginning of the measurement window (i.e. after 1.9 ms laser illumination), whereas for the forward curve (I_{sc} to V_{oc}) V_{oc} is determined at the end of the measurement window (after 2.9 ms of laser illumination). Samples are fixed with vacuum on a temperature controlled chuck, and Kelvin probes and the gold plated chuck itself are used to contact the cells electrically.

2.2 Evaluation

For the initial evaluation of the set-up, the respective standard IEC 60904-9 applicable for sun simulators was chosen as a reference [28]. The standard defines minimum requirements for equipment and allows for a solar simulator classification. Except for the monochromatic light source and thus the simulator spectrum, the set-up is based on a very similar design approach as flash based simulators in use for the characterization of photovoltaic cells or modules. In the following the results regarding temporal stability, including suitability of the intensity correction using the monitor cell signal as well as in respect to spatial uniformity are shown.

Temporal instability of irradiance in the test plane during the recording of the complete I-V curve (long term) as well as instability during the recording of one data sample of current, voltage and monitor signal (short term) show a constant behavior.

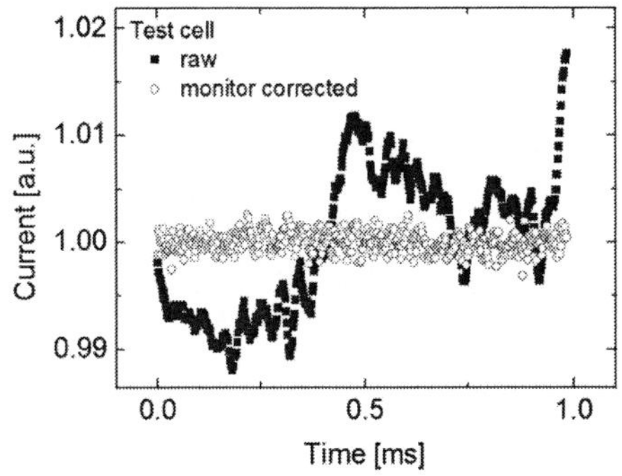

Figure 3: Temporal instability of the laser set-up, determined through the measurement of the short circuit current of a single-junction GaAs test cell (solid squares) Applying a monitor cell intensity correction the fluctuations of the raw signal are canceled out and variations are reduced to ±0.16 % (open circles).

Figure 3 shows the relative course of the short circuit current of a GaAs cell measured in the test plane without monitor cell correction (filled squares). A variation in signal and thus laser intensity for the whole measurement window well below ±2 % is found which would correspond to the requirement for a class A simulator in IEC 60904-9 in respect to long term instability. As described above the long term variations in intensity can be corrected for with the monitor cell's signal. By applying this procedure these variations almost completely vanish and reduce to ±0.16 % (open circles in Figure 3), which is even below the required 0.5 % for class A in short term instability.

Using a cell with known (calibrated) behavior of short circuit current with irradiance which is placed in the test plane, the signal of the monitor cell was calibrated. In this way the current signal measured on the monitor cell can be used for the determination of the irradiance level in the test plane. This is of special interest in the case of the measurement of non-linearly behaving cells. In particular this can be expected for multi-junction devices with strong luminescence coupling between the subcells [18, 19].

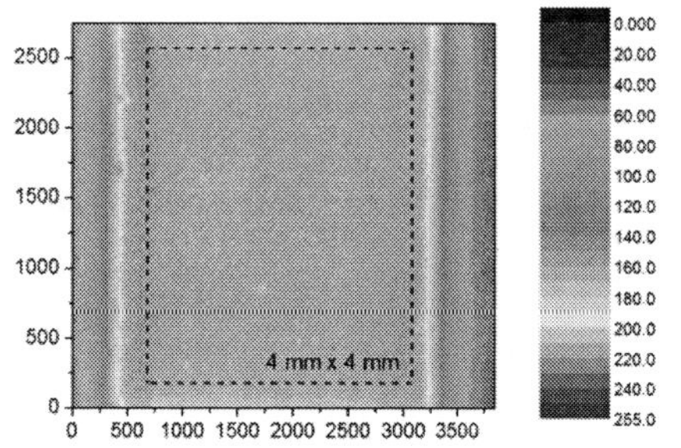

Figure 4: Contour plot of uniformity of the flat-top homogenized laser beam profile in the test plane. A 4 mm × 4 mm designated test area is indicated with dashed lines. Axis show pixel coordinates for the contour plot, whereas a color coding is used to show pixel intensity values.

The **spatial light intensity uniformity** in the test plane has a direct influence on the accuracy of irradiance determination using the monitor cell signal. When using the monitor cell for determining the intensity in the test plane, high uniformity is required for tolerant positioning of the test cell in x-y- direction. In addition, different sized test samples would need different ratios of transmission to reflection (T/R) in the beam sampler as the monitor cell is of fixed size at a fixed position.

The uniformity of the flat-top beam profile in the test plane has been assessed with a beam profiling camera sensor with a resolution of 3840 × 2748 pixels – see Figure 4. IEC 60904-9 requires dividing the test plane into at least 64 equally sized areas for determination of non-uniformity. When binning the pixels of the camera to the required minimum of 64 sub areas, a non-uniformity of 1.6 % is found for the 4 mm × 4 mm sized test plane (dashed square in Figure 4). Reducing the test plane area to 3.5 mm × 3.5 mm the non-uniformity improves to 1.2 %. Both values of uniformity fulfill the 2 % requirement for a class A simulator.

3 I-V MEASUREMENTS

For initial evaluation of the I-V measurements using the LaserSim set-up, two GaAs devices of similar size and front side grid design were measured both with a flash simulator and the new LaserSim at different intensity levels. One of the samples used is a common single-junction GaAs cell whereas the other sample is a monolithic dual-junction GaAs/GaAs device with a tunnel diode for series connection, similar to the ones presented by Schubert [15]. Both samples were chip mounted on a copper heat sink. Designated cell areas are 0.84 mm² for the single-junction and 0.87 mm² in case of the dual-junction device.

Figure 5 shows the open circuit voltage as a function of short circuit current density for the single-junction test cell. A good agreement between both set-ups is found for low current densities. However, for higher current densities >10 A/cm² (i.e. high irradiances) the open circuit voltage reveals differences between both set-ups. In particular, a lower voltage is measured with the flash simulator. This difference could be attributed to temperature effects at the flash set-up by comparing the V_{oc} values of the I-V curves switched from I_{sc} to V_{oc} and vice versa. For the flash set-up significant differences of up to 0.4 % were found at high intensities whereas for the LaserSim this difference was below 0.1 % even at highest intensity.

An explanation for the lower V_{oc} visible for the measurement performed with the flash set-up is an increased heating in the sample during the measurement. The increased heating is a result of the differences in the used irradiance sources. Obviously, the generated current in the test sample has to be the same in both set-ups. However, the irradiance of the flash simulator that is necessary to generate the same current as in the LaserSim is a factor 2.8 higher. This is due to the fact that the flash-simulator uses a Xenon flash bulb and thus a broad-band spectrum impinges on the cell. A high portion of that additional incident irradiance generates thermal energy. As an example, the high energy photons available in the broad spectrum reveal thermalization losses thus cause heating of the solar cell. The heating leads to a lower V_{oc} of the sample.

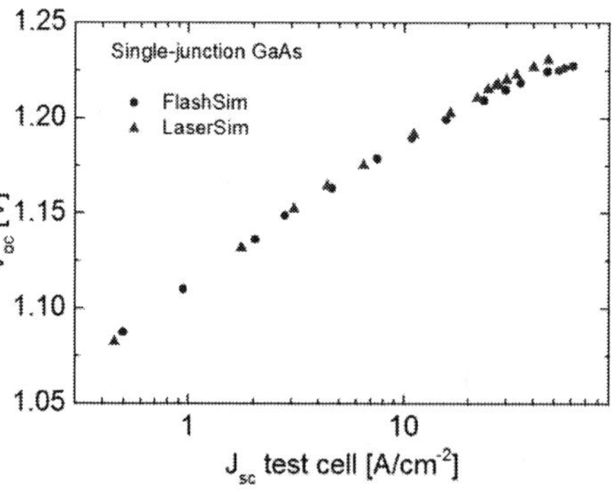

Figure 5: V_{oc} as function of J_{sc} of a single-junction GaAs cell. Measurements at the flash simulator show a decrease in values at high irradiances in comparison to measurements at the LaserSim.

Figure 6 shows the fill-factor as function of short circuit current density for both test cells measured at both set-ups. As expected for the single-junction device only small deviations below 0.5 % in fill factor between the two set-ups are found which is in the range of the measurement uncertainty and repeatability. However, for the dual-junction device higher deviations up to 1.5 % are detected for low current densities. These differences cannot be explained by measurement uncertainty or repeatability. For the 809 nm monochromatic light, the GaAs/GaAs dual-junction cell sample would be current matched at 246 K. Present measurements however were performed at room temperature (298 K). Due to the change in absorption coefficient with temperature this leads to an excess current of 17 % in the top cell. For the spectrum of the flash bulb used at the flash set-up this excess current is as high as 140 %. The significantly higher current mismatch between the two subcells leads to a shifting in operating voltage of the top cell and consequently an increase in fill factor [21]. Due to the heating of the cells at the flash set-up at high intensities and the differences found in fill factor for the GaAs/GaAs dual-junction cell, only the LaserSim measurements will be further discussed in respect to cell efficiency.

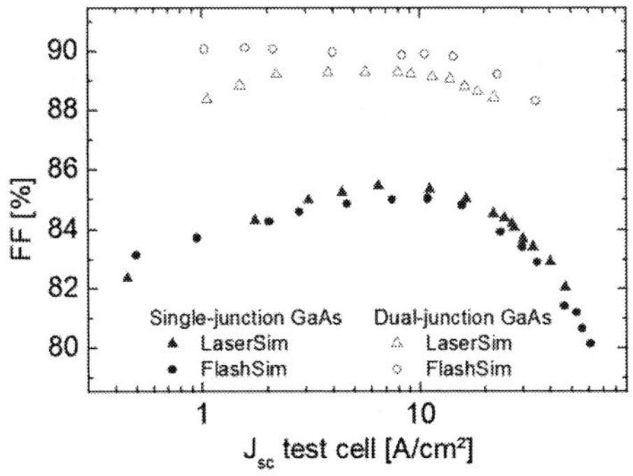

Figure 6: Fill factor as function of the short circuit current density (J_{sc}) for both test cells. Both test cells are measured at the LaserSim and the flash simulator.

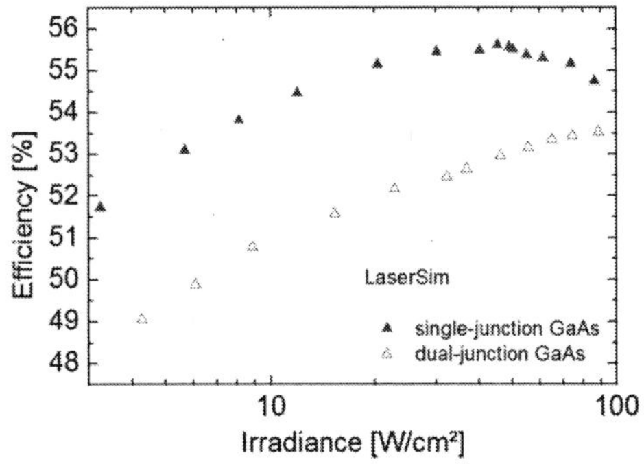

Figure 7: Efficiency as function of irradiance measured with the LaserSim set-up (809 nm). The single-junction test cell is plotted as filled symbols, whereas the dual-junction device data corresponds to the open symbols.

Figure 7 shows the efficiency as a function of irradiance for the two devices measured at the LaserSim. The relation between irradiance in the test plane and the monitor cell signal has been calibrated using the measurement of the single-junction cell. In particular, the intensity has been determined from the spectral response of the single-junction cell at 809 nm and its measured short circuit current density. This means that in the case of the single-junction cell the cell itself was used as reference for intensity determination. The linearity of the short circuit current of the cell with intensity has been verified on a similar cell structure. For the dual-junction device the irradiance in the test plane was determined using the monitor cell reading. The efficiency for the single-junction cell shows a maximum of 55.6 % at 45.3 W/cm². For the dual-junction device the course of efficiency indicates that the maximum efficiency is found for intensities above the ones that were investigated with the LaserSim set-up. The maximum found at highest intensity of 88.6 W/cm² is 53.5 %. Note that due to the splitting of current onto two subcells in case of the dual-junction cell, the current at same intensity will roughly be half the value of the single-junction device. Consequently, for the same grid design the maximum in efficiency for the dual-junction cell is expected to be found at roughly twice the intensity compared to the single-junction cell.

4 SUMMARY AND CONCLUSIONS

A new set-up for measuring PV laser power converters under laser illumination at 809 nm at varying irradiances up to 89 W/cm² has been presented. Similar to flash simulators the sample is only illuminated for a short period of time in order to avoid heating of the cell during measurement. The cell is illuminated for 4 ms in total and the I-V curve is measured in a measurement window of 1 ms. The evaluation of the set-up followed (where applicable) the procedures described in the corresponding standard for solar simulators used in photovoltaics (IEC 60904-9). Temporal stability as well as uniformity meet the requirements of a class A simulator.

The laser intensity in the test plane is calibrated using a cell with known linear behavior of short circuit current with intensity. In particular, the reading of a monitor cell, on which part of the laser beam is orientated with a beam sampler, is calibrated to enable in-situ determination of the irradiance in the test cell plane. In this way also non-linearly behaving cells can be tested. This is of special interest for cells where an influence of strong luminescence coupling is expected. Also it enables efficiency determination without the need to determine the spectral response of each test cell.

I-V curve data taken on a single-junction and a dual-junction cell, using the new LaserSim set-up and a flash set-up, have been compared. Interestingly the open circuit voltage measured using the flash simulator indicated an influence of heating at high intensities despite the short time of illumination. Such effects were not found for the LaserSim set-up. A comparison of fill factor values between the two set-ups revealed a good agreement for the single-junction cell, whereas deviations up to 1.5 % were found for the dual-junction cell. These differences could be explained by the significantly higher current mismatch between the two subcells present at the broad band flash illumination. This clearly demonstrates that such multi-junction devices require testing under appropriate laser illumination.

Finally, first efficiency measurements of two above mentioned cells were shown. For the GaAs single-junction cell a maximum efficiency under 809 nm monochromatic light of 55.6 % at 45.3 W/cm² has been determined. For the dual-junction cell a maximum efficiency of 53.5 % has been found at the highest intensity of 88.6 W/m².

5 ACKNOWLEDGEMENTS

We acknowledge A. Brand, M. Schachtner for useful discussion, O. John for his support with the beam profiling, W. Rosenfeld for writing an open-source tool for flat-top laser beam profiling (Sourceforge), M. Zimmermann for designing the optics regarding the high uniformity specifications (Bayerisches Laserzentrum) and the company DILAS for the support in adapting an industrial laser system to laboratory requirements.

6 REFERENCES

[1] J.G. Werthen, S. Widjaja, T.C. Wu, and J. Liu, in *Optical Technologies for Arming, Safing, Fuzing, and Firing*. (2005), 58710C-1-6.

[2] K. Worms, C. Klamouris, F. Wegh, L. Meder, D. Volkmer, S.P. Philipps, S.K. Reichmuth, H. Helmers, A.W. Bett, J. Vourvoulakis, C. Koos, W. Freude, J. Leuthold, and W. Stork, in *17th ITG/GMA-Symposium: Sensors and Measuring Systems*. (2014). 1.

[3] K. Worms, C. Klamouris, F. Wegh, L. Meder, D. Volkmer, S.P. Philipps, S.K. Reichmuth, H. Helmers, A. Kunadt, I. Vourvoulakis, A.W. Bett, C. Koos, W. Freude, J. Leuthold, and W. Stork, Wind Energy. in press (2016).

[4] S. Park, D.A. Borton, M. Kang, A.V. Nurmikko, and Y.-K. Song, Sensors (Basel, Switzerland). 13 (2013) 6014.

[5] M. Mujeeb-U-Rahman, D. Adalian, C.-F. Chang, and A. Scherer, Journal of Biomedical Optics. 20 (2015) 095012.

[6] F.V.B.d. Nazare and M.M. Werneck, IEEE Sensors Journal. 12 (2012) 1193.

[7] G. Boettger, M. Dreschmann, C. Klamouris, M. Hubner, M. Röger, A.W. Bett, T. Kueng, J. Becker, W. Freude, and J. Leuthold, IEEE Photonics Technology Letters. 20 (2008) 39.

[8] F. Audo, M. Guegan, V. Quintard, A. Perennou, J. Le Bihan, and Y. Auffret, Optical Engineering. 50 (2011) 045001.

[9] M. Roeger, G. Boettger, M. Dreschmann, C. Klamouris, M. Huebner, A. Bett, J. Becker, W. Freude, and J. Leuthold, Optics Express. 16 (2008) 21821.

[10] A.W. Bett, F. Dimroth, R. Löckenhoff, E. Oliva, and J. Schubert, in *Proceedings of the 33rd IEEE Photovoltaic Specialists Conference*. (2008). 1.

[11] S. Fafard, M.C.A. York, F. Proulx, C.E. Valdivia, M.M. Wilkins, R. Arès, V. Aimez, K. Hinzer and D. P. Masson, Applied Physics Letters. 108 (2016) 071101.

[12] E. Oliva, F. Dimroth, and A.W. Bett, Progress in Photovoltaics: Research and Applications. 4 (2008) 289.

[13] O. Höhn, A.W. Walker, A.W. Bett, H. Helmers, Applied Physics Letters. 24 (2016) 108.

[14] R. Peña and C. Algora, in *IEEE Transactions on Electron Devices*. (2001). 196.

[15] J. Schubert, E. Oliva, F. Dimroth, W. Guter, R. Löckenhoff, and A.W. Bett, IEEE Transactions on Electron Devices. 56 (2009) 170.

[16] H. Helmers, L. Wagner, C.E. Garza, S.K. Reichmuth, E. Oliva, S. Philipps, D. Lackner, and A.W. Bett, in *Proceedings of the SENSOR 2015 - 17th International Conference on Sensors and Measurement Technology*. (2015). 519.

[17] H. Helmers, E. Oliva, W. Bronner, F. Dimroth, and A.W. Bett, IEEE Transactions on Electron Devices. 57 (2010) 3355.

[18] M. Wilkins, C.E. Valdivia, A.M. Gabr, D. Masson, S. Fafard, and K. Hinzer, Journal of Applied Physics. 118 (2015) 143102.

[19] S.K. Reichmuth, H. Helmers, S.P. Philipps, M. Schachtner, G. Siefer, and A.W. Bett, (submitted for publication).

[20] K. Heidler, in *Proceedings of the 23rd IEEE Photovoltaics Specialists Conference*. (1993). 1107

[21] G. Siefer, C. Baur, M. Meusel, F. Dimroth, A.W. Bett, and W. Warta, in *29th IEEE Photovoltaic Specialists Conference*. (2002). 836.

[22] A.W. Walker, O. Höhn, D.N. Micha, L. Wagner, H. Helmers, A.W. Bett, and F. Dimroth, Journal of Photonics for Energy. 5 (2015) 053087.

[23] A.W. Walker, O. Höhn, D.N. Micha, L. Wagner, H. Helmers, F. Dimroth, and A.W. Bett, in *SPIE Photonics West*. (2014).

[24] F.M. Dickey, *Laser Beam Shaping*. second ed. (2014), Crc Pr Inc. 567.

[25] G. Siefer, P. Abbott, C. Baur, T. Schlegl, and A.W. Bett, in *20th European Photovoltaic Solar Energy Conference*. (2005). 495.

[26] J. Metzdorf, A. Meier, S. Winter, and T. Wittchen, in *Proceedings of the 12th European Photovoltaic Solar Energy Conference*. (1994). 496.

[27] F. Lipps, A. Zastrow, and K. Bücher, in *Proceedings of the 13th European Photovoltaic Solar Energy Conference*. (1995). 2274.

[28] IEC, *Photovoltaic devices - 60904 Part 9: Solar simulator requirements*, IEC, (2007). 34.

ABSTRACT FOR EUPVSEC 2016
TOPIC: 1.1 (FUNDAMENTAL STUDIES)

SIMULATION-BASED OPTIMIZATION FOR SOLAR CELLS AND MODULES WITH NOVEL SILVER NANOWIRE TRANSPARENT ELECTRODES

*S. Altazin[1], R. Hiestand[1], M. Fontenlos[1], B. Ruhstaller[1,2]**

[1]Fluxim AG, Technoparkstrasse 2, 8406 Winterthur, SWITZERLAND,
[2]Zurich University of Applied Sciences, Institute of Computational Physics, Technikumstrasse 9, CH-8401 Winterthur, SWITZERLAND,

1) Summary

In this contribution, we present a modelling tool allowing to efficiently optimize the design of solar cells and modules in order to maximize the power output for different application scenarios and reduce the lateral electrical losses in the transparent conducting electrodes (TCE). We investigate the use of combined metal grids and conductive polymers as well as silver nanowire materials for enhancing the lateral conductivity of the TCEs. These promising new TCE materials provide the option to tune the conductivity and transparency, therefore a smart electrode layout choice has to be made for a specific application. We present the optimization and parameter sweep calculations to illustrate the design and customization freedom.

2) Introduction and motivation

Thin film solar cell technologies offer an attractive alternative to 'standard' crystalline silicon solar cell technology. For instance, these new technologies allow for flexible and light-weight modules. Moreover, since the (semi-) transparency can be tuned, building-integrated photovoltaics (BIPV) applications can be particularly interesting. Apart from optical design aspects for applications like BIPV, thin, large-area solar cell technologies often offer the freedom to design suitable geometrical cell and interconnection layouts.

Inherently, the large-area cells and modules suffer from lateral electrical potential losses due to the limited electrical conductivity of the (semi-) transparent conducting electrode (TCE) therefore limiting the power efficiency. Moreover, the increase of conductivity of the electrode materials generally comes at the cost of reduced transparency. For example, adding some metal grids as conductivity enhancement method usually diminishes the illuminated area of the module and using a thicker TCE, for increasing the lateral conductivity, also increases parasitic absorption inside the TCE layer leading to additional losses in the module. Thirdly, for a given TCE candidate material (and layer thickness) a change in material composition always affects both the transparency and conductivity.

From the above statements we can conclude that a compromise between optical properties and conductivity of the electrodes needs to be found in order to maximize the module efficiency.

In a first part of this contribution we will briefly introduce the modelling approach and in a second part we will show how this innovative tool was used in order to optimize a PV module design.

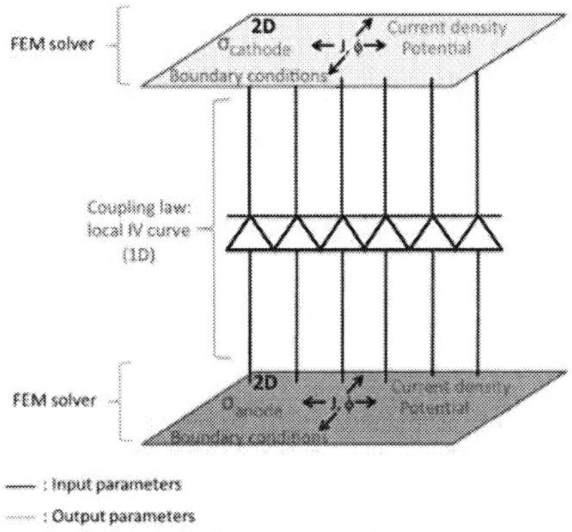

Figure 1. Sketch of FEM model for 2 large-area electrode domains as well as an electric coupling law (i.e. local IV curve) between top and bottom electrode domains.

3) Approach

From an electrical point of view, a PV module can be regarded as a stack of thin (few tens of nanometers) semiconducting layers sandwiched between an anode and cathode. One of these electrodes requires being transparent so that the light can be absorbed in the active layer of the device, however transparent electrodes are fairly bad conductors leading to an inhomogeneous potential distribution in the large area electrodes and thus to electrical losses. The electrical potential decreases significantly far from the metallic electrode boundaries. Due to the high aspect ratio (large area electrode versus the through-plane thickness) of a solar cell, the electrical model

* corresponding author email: beat.ruhstaller@fluxim.com

can be split up into two 2D FEM models, to solve the current distribution in the large area electrodes, and a 1D microscopic model that represents the current-voltage (IV) curve flowing vertically across the semiconducting layers of the solar cells and couples the two electrode domains. This alternative approach to a full 3D FEM simulation exploits the given aspect ratio to achieve fast computation times. We presented a first case study using this kind of model a couple of years ago for investigating the influence of shunts [1].
Meanwhile we have developed and implemented this approach in a commercial software tool called LAOSS [2] that allows importing a CAD drawing file for the geometry definition and performs meshing before calculating the potential distribution by the finite element method (FEM).
The used 1D IV curve for the local coupling of top and bottom TCE can either come from an IV measurement [3] of a small device (thus neglecting the potential losses in the electrodes) or from a simulation using an opto-electronic model solving both the semiconductor equations and the optical absorption in the thin film layer stack of the cell [4].

4) Results

In this contribution we mainly focus on thin film solar applications, though other wafer based solar cell technologies may equally profit from model based layout optimization.

4.1 Influence of non-ideal electrodes

In the following we illustrate how the electrode conductivity influences the current-voltage curve of solar cells (Fig.2). In the "ideal case" we just plot the IV law of the vertical coupling law between the 2D electrode domains, which presents an upper limit to the achievable current density at larger cell sizes. Result shown in Fig. 3 also demonstrate that the fill factor can be improved significantly, converging to the fill factor of the vertical coupling law for high electrode conductivity.

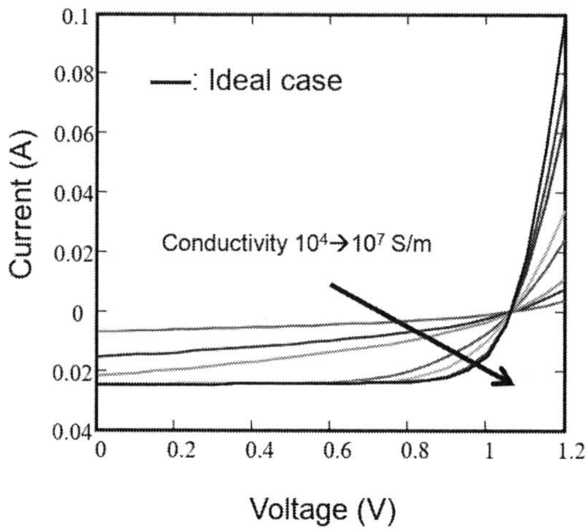

*Figure 2.Current-voltage curves for different conductivities of the TCE material of the constant and square reference solar cell (1cm*1cm).*

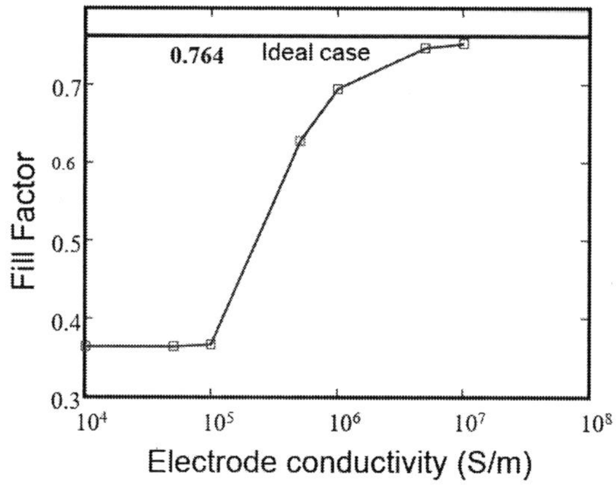

Figure 3. Plot of the fill factor (FF) of the simulated IV curves for the square 1 cm^2 solar cells from Fig. 4.

4.2 Metal finger width optimization

As demonstrated in the previous section, it is of primary importance to improve the overall conductivity of the electrodes in order to maximize the power efficiency of the solar cell. We start with looking at a popular TCE approach based on a combination of conducting polymer (mostly PEDOT:PSS) and a metal grid. An array of metal fingers (metal grid) can be added to the transparent electrode of a PV cell in order to improve its conductivity, and as a consequence to improve the device efficiency. This approach is suitable for printed solar cells on flexible substrates.
The solar cells are serially and monolithically connected to form a module. The parameter choice to

* corresponding author email: beat.ruhstaller@fluxim.com

be made concerns the number of parallel metal fingers, their finger width as well as the width of the serially connected cells.
In Fig. 4 we present a simulation result of a sweep of the width of the metal finger. This optimization finds the best compromise between two counter-acting effects: shadowing due to the metal finger and low conductivity due to the absence of a metal finger. We find an optimum finger width around 0.15 mm and reach just below 165W/m^2 compared to the "ideal case" of 200W/m^2 from the small-scale cell that has no lateral electrical losses. The considered IV current law was taken from a perovskite solar cell with high efficiency (approx. 20% PCE) measured under AM1.5 illumination. Clearly, the result would be different if a less efficient solar cell would be considered. A high-efficiency solar cell is carrying more photocurrent, thus the lateral electrical losses in the TCE are more critical. The power maximum in Fig. 2 would move to a thinner metal finger width in order to increase the illuminated area.

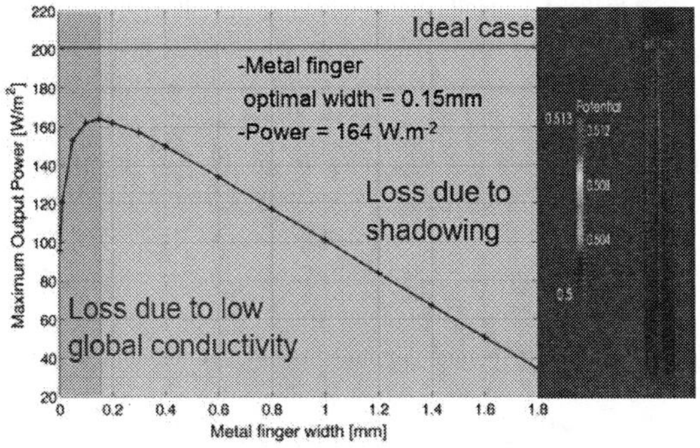

*Figure 4. Simulated solar cell power for varying width of the metal finger used to increase the conductivity of the TCE in a 2mm*18mm device. Example simulation of a single-metal-finger unit, showing the electrical potential distribution.*

4.3 Variation of silver nanowire TCE material

An attractive alternative to the above TCE approach of polymer plus printed metal grid is the use of silver nanowire (AgNw) TCEs, with higher ease of manufacturing and no planarization challenges. We present a simulation study based on properties of silver nanowire material provided by Cambrios Advanced Materials Corporation. Two sample silver nanowire materials are considered here:

Sample 1: Transparency = 90%, $R_{\square} = 15\ \Omega_{\square}$
Sample 2: Transparency = 85%, $R_{\square} = 8.8\ \Omega_{\square}$

The material with higher transparency has reduced conductivity compared to the material with reduced transparency. It has to be noted that contrary to the metal-finger approach, the transmission losses due to the AgNws are mainly the consequence of an increase of the reflection which can be tuned in the final module with the packaging layer, however this effect has not been taken into account for the calculations in this contribution. For a perovskite PV cell structure (same as in Fig. 2), we consider now the scaling of the width of a solar cell for monolithic interconnection and module formation. In Fig. 6 the calculated device power density is plotted versus the cell width. While for both nanowire material samples the power decreases with increasing cell width, sample 2 provides a better choice at large cell widths and for narrow cell widths below 1.44 cm material sample 1 allows for higher power density. The achieved power densities for the 1.44cm*1.44cm device is just below 165 mW/m^2 which is similar to the values obtained for the polymer plus metal grid approach, see Fig. 2. This shows that silver nanowire TCEs are an attractive TCE material for photovoltaic applications.

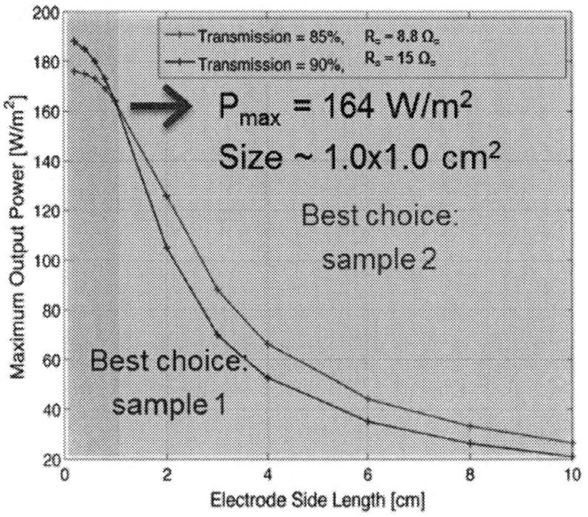

Figure 6.Simulation result of the maximum solar cell power output vs the side length of the (quadratic) solar cell for two "samples" of transparent conductive electrode material, sample material 1 with better conductivity but lower transparence than sample material 2.

4.4 Full PV module optimization

In a last step we have optimized the number of interconnected cells a module (with a length of 1m) should contain in order to maximize the power efficiency.
We have assumed that the length of the module was subdivided into many cells connected in series with a 0.5mm 'dead area width' at each connection between two adjacent cells. This optimization was performed for the two different types of electrodes presented in the previous section in order to find out which

* corresponding author email: beat.ruhstaller@fluxim.com

module architecture with which TCE would give the highest power efficiency.
In figure 7 we show how the IV curve of the full module changes with the number of cells it contains for a given area of 1m*0.4m (changing the number of cells modify the size of each individual sub-cell).

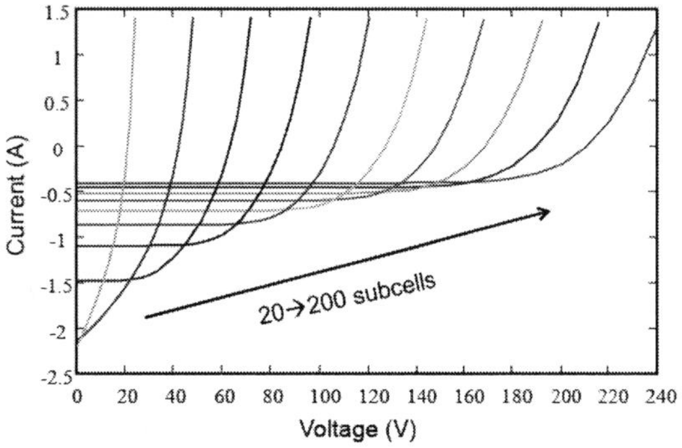

Figure 7. IV curve of the full module using sample 1

In figure 8 we show the output power as a function of the number of cells for the two different samples. We find that the maximum of power was obtained for a module with 160 cells connected in series achieving a power efficiency of 166 $W.m^{-2}$ for the high transparency TCE (sample 1).

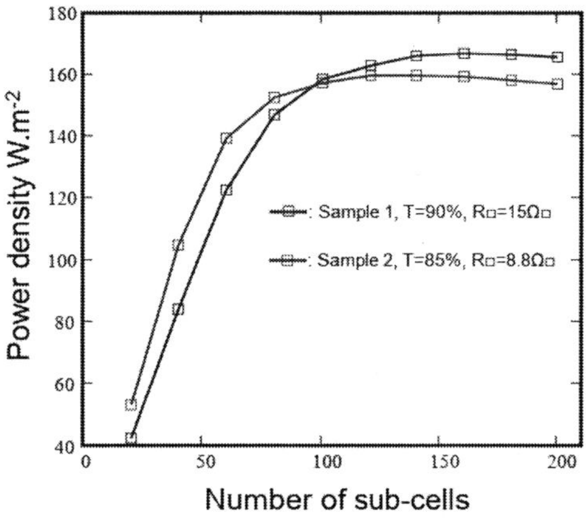

Figure 8. Power density generated by the full module as a function of the number of interconnected cells.

5) Conclusions

We presented the electrical simulation of solar cells and modules for maximizing the power conversion efficiency. Given the freedom of design in electrode layouts and TCE material properties, this optimization will have to be repeated for the specific application, e.g. whether it is BIPV or indoor PV. Moreover, depending on the solar cell technology's level of photocurrent (and cell efficiency), the lateral current transport losses are different. A design tool provides guidance in this respect and can quantify these electrical losses for different TCE properties. We have shown that the silver nanowire-based TCEs are suitable candidates for PV applications. Since their optical and electrical properties can be tuned, they can be chosen for the desired application and performance requirements. Finally we have shown how the design of a module can be optimized in order to maximize the power efficiency.

Acknowledgments
Financial support by the Swiss Commission for Technology and Innovation (CTI) is as well as through FP7 project SUNFLOWER is gratefully acknowledged.

References
[1] T. Lanz, M. Bonmarin, M. Stuckelberger, C. Schlumpf, C. Ballif, and B. Ruhstaller, "Electrothermal Finite-Element Modeling for Defect Characterization in Thin-Film Silicon Solar Modules," *IEEE J. Sel. Top. Quantum Electron.*, vol. 19, no. 5, pp. 1–8, Sep. 2013
[2] LAOSS 1.0 software by *Fluxim AG*, Switzerland
[3] SETFOS 4.3 software with Absorption and Drift-diffusion modules, *Fluxim AG*, Switzerland
[4] PAIOS 3.1 measurement platform by *Fluxim AG*, Switzerland

* corresponding author email: beat.ruhstaller@fluxim.com

OPTICAL EVALUATION OF MULTI-WIRE MODULES

Keith R. McIntosh,[1] Malcolm D. Abbott,[1] Matt B. Edwards,[2] Rhett Evans[3] and Yu Yao[4]

[1]PV Lighthouse, Coledale, NSW 2515, Australia
[2]University of New South Wales, Sydney, NSW 2052, Australia
[3]Solinno, Bulli, NSW 2516, Australia
[4]Meyer Burger AG, Thun, 3645, Switzerland

ABSTRACT: Multi-wire interconnects are currently being commercialized by several large PV companies to replace the traditional flat-tabbed busbars. Advantages of multi-wires include reduced silver usage, lower grid resistance, higher tolerance to cracks, and improved optics. This paper evaluates the optical losses of a multi-wire module and compares them to those of busbar modules. It includes the first application of ray tracing to multi-wire modules and finds that the results agree reasonably well with some published experimental mini-modules with similar wires and cells. The simulations also indicate that the optical advantage of multi-wire modules should increase slightly as the angle of incidence changes from 0 to 80°, whether for cells with random pyramids or isotexturing. We also find that the effective optical width of the wires increases as the number of wires increases, and that it is approximately 60% of the wire diameter for typical multi-wire applications; this effective width is significantly greater than what is calculated from a geometrical assessment of a singe wire.
Keywords: Ray tracing, PV module, optical losses, encapsulation, ribbons, simulation.

1 INTRODUCTION

The majority of c-Si solar cells employ busbars on their front surface to extract current from the cell. There are typically 2–5 busbars per cell and each has a width of about 1–2 mm. These busbars introduce a significant reduction in module power for two main reasons: (i) the busbars shade 2–3% of the cell's front surface and (ii) the current generated within the cell must travel laterally through the cell before reaching a busbar, as far as 2.5 cm for 3 busbars or 3.7 cm for 2 busbars; the associated lateral transport losses are reduced by a grid of thick metal fingers but these fingers also shade the front surface.

An attractive alternative to traditional busbars is the "multi-wire" approach conceived by Day4™ [1] and now being commercialized by Meyer Burger (SmartWire™), Schmid (MultiBusbar™) and GTAT (Merlin™). Multi-wires are comprised of a large number of thin circular wires, such as 15–40 copper wires of 200 μm diameter. The main advantages of multi-wires are (i) their circular cross section reduces optical shading by reflecting some light onto the cells, much via total-internal reflection at the glass–air interface (see Figure 1), and [3] (ii) the larger number of wires reduces the requirement for lateral transport, thereby reducing series resistance and/or the requirement for thick fingers [2]. Other advantages include (iii) an improved resistance to cracks and (iv) reduced silver usage [3].

The performance benefits of multi-wires are simple to understand, but a quantitative comparison to traditional busbars is not trivial. This task is greatly aided by simulation due to the inordinate number of possible configurations: these include variations in the number of fingers and wires, in the cross-sectional profiles of the fingers and wires, in the encapsulation materials, in the surface texture of the cell and glass, and in the thin-film coatings. Consequently, experiments to determine the optimal configuration are expensive, making an accurate prediction of the optical and electrical behavior of multi-wire modules highly desirable.

Although there have been many comparative studies of multi-wire and traditional busbars, the optical simulation in those studies has lacked accuracy and validation. This is understandable given that (i) a PV module is a complicated optical structure with variously textured surfaces, optical materials, thin films, metal geometries, and a scattering backsheet between cells, and (ii) until now, there has not been a publicly available ray tracer capable of rapidly simulating a PV module with multi-wires. Consequently, previous studies have been restricted to normally incident light and planar glass, and they made the rudimentary assumption that the effective shading fraction of a wire is independent of the number of wires and the type of texture. Some use 35.7% [3], others 75% [4] and 80% [2]; the appendix gives a derivation of a simple effective width.

We now present the first ray tracing study into multi-wire PV modules and compare the results to published experimental data.

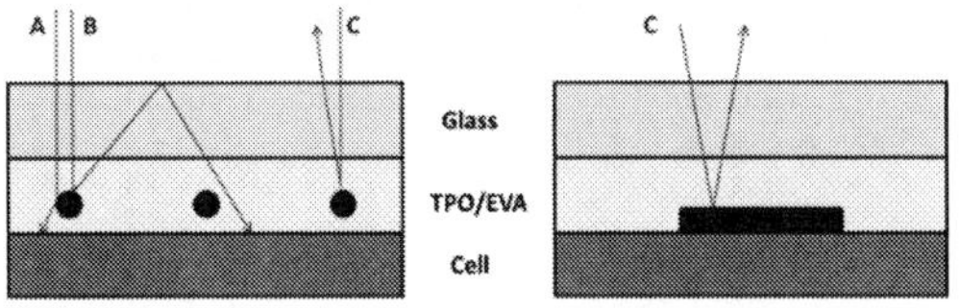

Figure 1: Cross sections of multi-wire and traditional-busbar modules. Arrows show example rays that (A) reflect from metal directly onto cell, (B) reflect from metal onto the glass-air interface and internally reflected back onto the cell, and (C) reflect from metal out of module. Although the wires appear suspended in this image, they in fact rest upon the fingers that run perpendicular to the wires.

2. RAY TRACING: RESULTS VS NUMBER OF WIRES

Ray tracing was employed to evaluate how the number of wires or busbars affects the current generation in a PV module. The simulation inputs were set at values that represent a modern module.

The inputs were as follows: The module was comprised of 2 mm of low-iron glass, 450 µm of EVA, 200 µm of Si solar cell coated with 75 nm of SiN_x on its front surface. The wires had a circular cross-section with a diameter of 200 µm and were raised 25 µm above the surface of the cell; the busbars had a rectangular cross-section with a height of 200 µm and a width of 1.5 mm and were located on the surface of the solar cell. The values for $n(\lambda)$ and $k(\lambda)$ for glass, EVA, Si and SiN_x were taken from [5], [6], [7] and the SiN_x with n = 1.96 at 632 nm from [8]. Internal reflection at the rear was simulated as Lambertian with a reflectance of 65%, approximating a scattering Al BSF. The wires and busbars were simulated as specular with a reflectance of 70% (see Section 3). Three cases were considered for the cell texture: (i) planar, (ii) upright random pyramids (rantex) with a base angle of 53°, and (iii) inverted spherical caps with a characteristic angle of 60°, which approximates industrial isotexture [9, 10]; both textures were assigned a height of 5 µm. Each simulation was performed with ~300,000 rays using the module ray tracer on the PV Lighthouse website [11]. (Although fingers and cell spacing are neglected in this study, both can be included in the module ray tracer; it can also be used to simulate wires as copper coated with InSn films, as incorporated into some SmartWires, but we did not find $n(\lambda)$ and $k(\lambda)$ data for InSn.) The ray tracing monitored the polarization of each ray as it propagated through the module, and the integrated generation currents were calculated using the AM1.5g spectrum.

Figure 2 presents the results. It plots the simulated generation current J_{gen} against the number of wires (or busbars). The trends in Figure 2 are as expected: (i) J_{gen} decreases as the number of wires (or busbars) increase due to shading; (ii) J_{gen} is greater for rantex than for isotexture, which is itself still superior to no texturing; and (iii) J_{gen} decreases much faster for busbars than for wires as the number of wires/busbars increases, primarily because their width is 1.5 mm rather than 0.2 mm. Note that these values for J_{gen} are larger than a typical module's J_{SC} because (i) they do not include fingers or cell spacing, and (ii) they do not consider recombination.

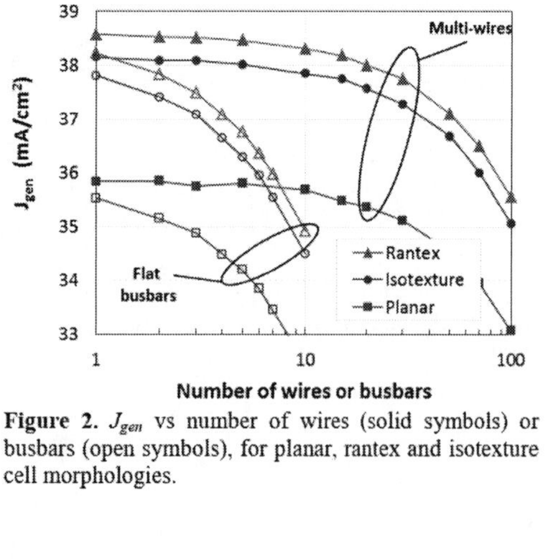

Figure 2. J_{gen} vs number of wires (solid symbols) or busbars (open symbols), for planar, rantex and isotexture cell morphologies.

With Figure 3, we come to the more novel findings of this section. It plots the effective wire width w_{eff}, which is the thickness that equivalent rectangular wires would have if they were to yield the same J_{gen}. The closed symbols show the case where the wires have a reflectance of R = 70% (as for Figure 2), and the open symbols show the case for R = 100%. The latter is consistent with the afore-published calculations of the geometrical value [3]. Also note that for all relevant λ, $n(\lambda)$ of the encapsulants is close to n ~ 1.5, the value assumed in [3] and the appendix. The figure is limited to 5 or more wires, below which the uncertainty is inhibitive.

By applying ray tracing to determine w_{eff}, we have omitted several of the major assumptions contained in the 'simple geometrical' derivation (see appendix). Specifically, the ray tracing does not neglect light reflected from wire to wire, or from the textured cell to the wires and vice versa; it also does not neglect absorption in the encapsulants.

Here are some observations from Figure 3:

1. w_{eff} increases as the reflectance of the wires decreases (due to less light reflecting from the wires onto the cell).
2. w_{eff} increases as the number of wires increases (due to secondary reflections from wires).
3. w_{eff} differs slightly from one cell texture to another, tending to be higher for planar wafers (due to differences in reflections from the cell onto the wires and vice-versa).
4. w_{eff} is within about ±10% absolute of the simple geometrical value (derived by [3] and in the appendix). Depending on the situation, the geometrical value may or may not be sufficiently accurate.
5. w_{eff} is only near the oft-used value of 70-80% [2, 4] if the reflectance of the wires is low; e.g. R ~ 0.5.
6. w_{eff} is ~60% for R ~ 0.7 and 40 wires, which is our best estimate of the current generation of a SmartWire mini-module. Naturally, that could change in the near future as the product evolves.

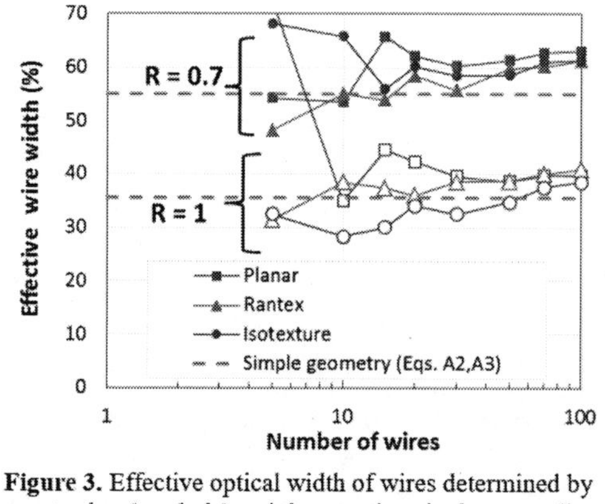

Figure 3. Effective optical width of wires determined by ray tracing (symbols) and the equations in the appendix (dashed lines) for a wire reflectance of R = 0.7 and 1.

3 RAY TRACING VS EXPERIMENT

3.1 Comparison to SmartWire mini-modules

In Figure 4, we compare the results of ray tracing to experimental mini-modules fabricated with the SmartWire technology. The mini-modules were comprised of a single solar cell tabbed with SmartWires and encapsulated in glass and TPO.

Figure 4 contains a lot of information. We describe it methodically, starting with an explanation of the y-axis, followed by the experimental results, and finally the ray tracing results.

The y-axis in Figure 4 plots the normalized J_{SC} of the mini-modules, where J_{SC} was determined using an aperture to prevent illumination falling on areas beyond the extent of the cells. It plots the normalized data to make it easier to compare the different cell types and easier to compare experiment to ray tracing. The normalized plot somewhat removes the contributions of effects unrelated to the number of wires, such as differences in cell thickness and the TPO and finger properties. The data is normalized to the 18-wire mini-module because this is the number of wires for which we have both experimental data sets.

Next examine the closed symbols, which represent the experimental data. The red closed symbols plot mini-modules with PERC cells and either 18 or 38 wires, and the blue closed symbols plot mini-modules with HJT cells and either 6, 18 or 36 wires. For each data point, there were three PERC mini-modules and ten HJT mini-modules; the symbols plot the average of those results and the error bars plot the maximum and minimum (available for PERC only).

We see from this limited set of experimental data that the J_{SC} decreases more rapidly with the number of wires for the HJT mini-modules than it does for the PERC mini-modules. The reason for this is not known. As will become evident from the ray tracing, this observation might be due to the HJT mini-modules having wires with a lower reflectance; it might also relate to differences in the IQE of the solar cells or experimental variance.

Finally, the results of ray tracing are plotted as open symbols and least-square-fit parabolic lines. There are many unknown aspects to the experimental samples that prevent an accurate simulation, making this an explorative and preliminary examination—but educational nonetheless. Specifically, we have not characterized the cell texture, the films on the front or rear of the cells, the cell thickness, or the illumination source, and we have simply applied the same inputs used in the previous section with the exception of the TPO, for which early measurements suggest its $n(\lambda)$ is about 0.05 higher than $n(\lambda)$ of EVA at all relevant λ and that its $k(\lambda)$ is similar to that in [6]. Each simulation contained 100,000 rays.

The unknowns mentioned so far introduce some error but are not expected to have a large effect on the results in Figure 4 due to the normalization of the y-axis. A more important unknown, however, is the properties of the SnIn film that coats the copper wires (i.e., the thickness, $n(\lambda)$ and $k(\lambda)$ of that film). We therefore simulate the wires simplistically as having a constant reflectance R at all λ, and we vary R to see how it affects the results. A second major unknown is the IQE of the cells, which will be quite different for the PERC and HJT cells, and which could combine with the λ-dependent wire reflectance to introduce additional uncertainty in the comparison. In any case, we simply plot the normalized J_{gen}, assuming it to be equivalent to the normalized J_{SC}.

We see from Figure 4 that the experimental data lies well within the simulated bounds of R = 0.1 and R = 1; this offers a degree of confidence to the simulation. We then find that the best-fit R is ~0.7 for the PERC and ~0.4 for the HJT mini-modules. Given the afore-mentioned unknowns, we cannot be sure how realistic this is, but an examination of pure copper coated with pure Sn suggests that R might lie between 0.65 and 0.75. This can be construed from the results of Figure 5, which were determined with OPAL 2 [12] for normally incident unpolarised light incident to a planar EVA/Sn/Cu structure, where $n(\lambda)$ and $k(\lambda)$ of Sn and Cu are taken from [13] and [14]. How the In atoms affect the properties of the InSn alloy has not been investigated.

Thus, our best estimate for the effective reflectance of the wires is R ~ 0.7, which agrees with the ray tracing that gave the best-fit to the PERC mini-modules but is higher than that for the HJT mini-modules. We emphasize again that this is a preliminary study with many unknowns in relation to the experimental samples—particularly the HJT cells.

Finally, we note that both the simulated and experimental J_{SC} has a sub-linear dependence on the number of wires; that is, the data sets in Figure 4 have a downward curve. This is expected because the increasing the number of wires leads to additional losses from secondary reflections from the wires. Those secondary losses are not accounted for in simpler investigations that employ a constant value for the effective-shading.

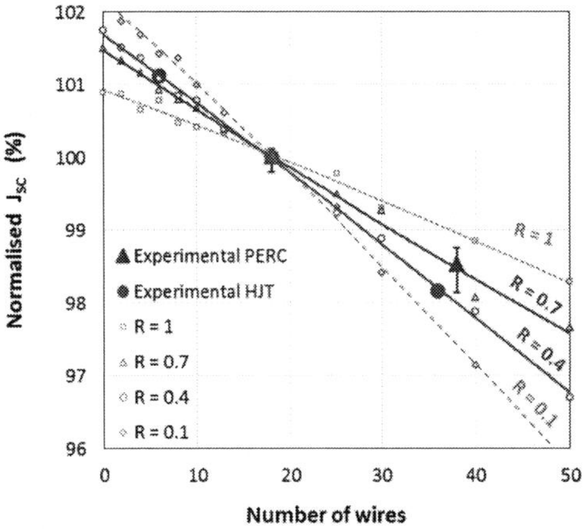

Figure 4. J_{SC} vs number of wires, where J_{SC} is normalized to the 18-wire module. Closed symbols are experimental, open symbols are ray tracing, lines are least-square parabolic fits to the open symbols.

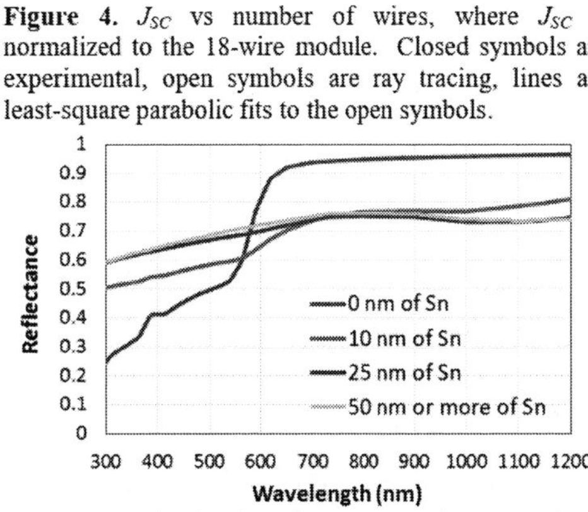

Figure 5. Simulated reflectance vs λ at a planar EVA/Sn/Cu interface for normally incident light.

Table I. Comparison of multi-wire vs 3 × 1.5 mm busbar encapsulated samples.

Data source	Cell Technology	Multi-Wire # wires	Multi-Wire diameter (μm)	Multi-wire minus Standard ΔFF (%abs)	Multi-wire minus Standard ΔJ_{SC} (mA/cm²)	Multi-wire minus Standard $\Delta\eta$ (%abs)	Reference
Simulated	c-Si standard, screen printed	14	250	+0.8	+0.3	+0.34	[3]
	c-Si standard, plating	14	250	+1.1	+0.54	+0.55	[3]
Measured	c-Si standard, screen printed	15	300	+1.2	+0.14	+0.33	[16]
	mc-Si standard, screen print	38	200	+1.9	+0.20	+0.50	[2]
	c-Si hetero-junction, printed	38	200	+2.0	+0.02	+0.44	[17]
	c-Si (standard)	38	200	+2.9	–0.44	+0.62	[4]
	c-Si (SHJ)	38	200	+3.9	+0.04	+1.09	[4]

3.2 Comparison to other published experiments

Figure 6 compares the results of ray tracing (lines) to previously published works (symbols). It plots the change in J_{SC} for SmartWire modules relative to a traditional three-busbar module (1.5 mm wide busbars). The results for simulated (open symbols) and experimental (closed symbols) are also summarized in Table I. The ray tracing was performed with the same inputs used above where the reflectance of the wires was constant and set to $R \sim 0.7$ to be consistent with the expected reflectance of the wires (see above), and where the wire diameter is either 200 and 300 μm.

Although there are many differences between the studies that prevent a precise comparison (such as differences in wire diameter, materials, and cell types), on the whole, we find that the trend between ΔJ_{SC} and the number of wires is consistent, and that the ray tracing results are similar to the experiments using 200 μm wire.

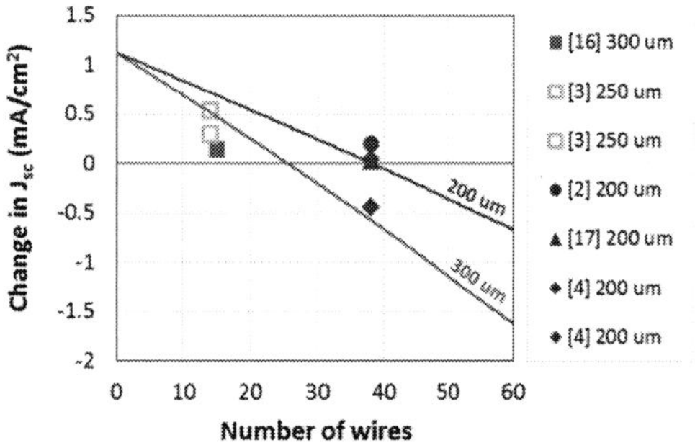

Figure 6. ΔJ_{SC} when using multi-wires rather than flat busbars. Lines are best-fit curves to ray tracing ($R = 0.7$, dia. 200 or 300 μm), symbols are results published elsewhere (see Table I). Ray tracing results for ΔJ_{SC} are almost identical for isotexture and rantex.

4 RAY TRACING: ANGLE OF INCIDENCE

Figure 7 plots the simulated optical behavior of multi-wire modules for varying angles of incidence. Specifically, it plots (a) generation and (b) the incident angle modifier (IAM) against the zenith angle θ, where IAM = J_{gen} / ($J_{gen0}\cdot\cos\theta$) and where J_{gen0} is J_{gen} at $\theta = 0$. It therefore plots the relative change in J_{gen} due to factors other than the 1/cosθ dependence of the incident light intensity. Most notably, IAM accounts for the increase in external reflection from the glass, but in Figure 7, we also see that when θ is greater than ~30°, IAM is superior for random pyramids than for isotexture (as also concluded in [15]), and slightly superior for multi-wires compared to traditional busbars. To make this clearer, Figure 7(c) compares the different tabbing approaches for only cells textured with random pyramids.

We conclude that the optical advantage that multi-wires has over traditional busbars increases slightly as the incident angle increases. At $\theta = 60°$, the IAM of multi-wires is ~0.5% absolute higher than the IAM of traditional busbar modules, whether for rantex or isotextured solar cells.

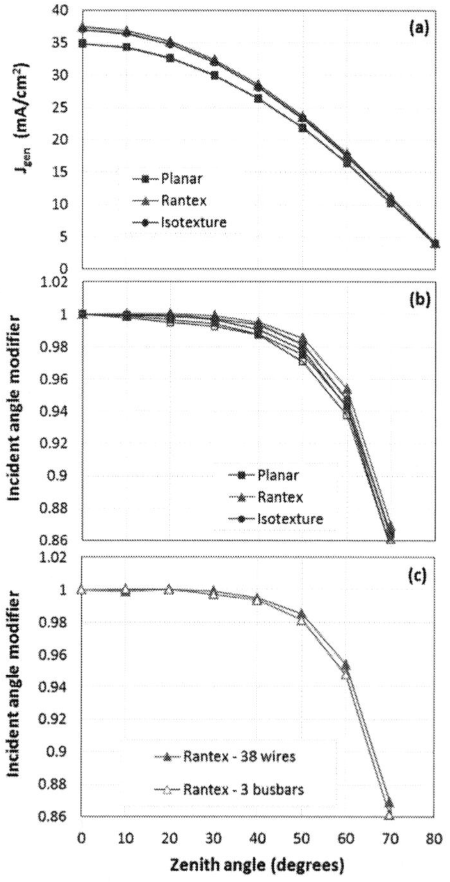

Figure 7. **(a)** Generation current density and (b) incident angle modifier (IAM) as a function of zenith angle for planar, rantex and isotexture, and (c) IAM for rantex multi-wire and busbar modules.

5 CONCLUSION

This paper investigated the optical behavior of modules with the multi-wire tabbing technology. We found with ray tracing that the effective width of wires depends on the number of wires, on the cell texturing, and on the reflectance of the wires. The effective width is similar to that determined by simple geometrical considerations (using Eqs A2 and A3) but is more accurately determined with ray tracing. We found reasonable agreement between most experiments and simulation when the reflectance of the wires was set to $R \sim 70\%$.

6 REFERENCES

[1] A. Schneider, L. Rubin and G. Rubin, "Solar cell improvement by new metallization techniques — The DAY4™ electrode concept," *4th WCPEC*, Waikoloa, pp. 1095–1098, 2006.

[2] M. Edwards, R. Evans, T. Söderström, A. Sugianto, L. Koschier and S. Wenham, "High efficiency with almost no metallization: Multiple busbar wire interconnection of plated solar cells," 28th *EU PVSEC*, Paris, 2013.

[3] S. Braun, G. Micard and G. Hahn, "Solar cell improvement by using a multi busbar design as front electrode", *Energy Procedia* **27**, pp. 227–233, 2012.

[4] T. Söderström, P. Papet and J. Ufheil, "Smart wire connection technology," *38th EU PVSEC* , pp. 495–499, 2013.

[5] K.R. McIntosh, G. Lau, J. Cotsell, K. Hanton, D. Batzner, F. Bettiol and B. Richards, "Increase in external quantum efficiency of encapsulated silicon solar cells from a luminescent down-shifting layer," *Progress in Photovoltaics* **17**, pp. 191–197, 2009.

[6] K.R. McIntosh, J. Cotsell, J. Cumpston, A. Norris, N. Powell and B. Ketola, "An optical comparison of silicone and EVA encapsulants for conventional silicon PV modules: A ray-tracing study," *34th IEEE PVSC*, Philadelphia, pp. 544–549, 2009.

[7] M.A. Green, "Self-consistent optical parameters of intrinsic silicon at 300 K including temperature coefficients," *Solar Energy Materials and Solar Cells* **92**, pp. 1305–1310, 2008.

[8] S. Duttagupta, F. Ma, B. Hoex, T. Mueller and A. Aberle, "Optimised antireflection coatings using silicon nitride on textured silicon surfaces based on measurements and multidimensional modeling," *Energy Procedia* **15**, pp. 78–83, 2012.

[9] K.R. McIntosh, T.G. Allen, S.C. Baker-Finch and M.D. Abbott, "Light trapping in isotextured silicon wafers," *Submitted to IEEE Journal of Photovoltiacs*, 2016.

[10] S. Wasmer, J. Greulich, H. Höffler, J. Haunschild, M. Demant and S. Rein, "Investigating the impact of parameter and process variations on multicrystalline silicon PERC cell efficiency," *31st EU PVSEC*, Hamburg, 2015.

[11] https://www.pvlighthouse.com.au. Last accessed 23-Jun-2016.

[12] K.R. McIntosh and S.C. Baker-Finch, "OPAL 2: Rapid optical simulation of silicon solar cells," *38th IEEE PVSC*, Austin, pp. 265–271, 2012.

[13] A. I. Golovashkin and G. P. Motulevich, "Optical and electrical properties of tin," *Sov. Phys. JETP* **19**, pp. 310–317 (1964).

[14] E. Palik, *Handbook of Optical Constants of Solids* Vol I, Academic Press, Orlando, pp. 283–286, 1985.

[15] K.R. McIntosh, M.D. Abbott and B.A. Sudbury, "Ray tracing isotextured solar cells," *Energy Procedia*, Si PV conference, Chambery, 2016.

[16] S. Braun, G. Hahn, R. Nissler, C. Pönisch and D. Habermann, "Multi-busbar solar cells and modules: high efficiencies and low silver consumption", *Energy Procedia* **38**, pp. 334 – 339, 2013.

[17] K. Hashimoto, M. Ouchi, N. Nakamura, E. Kobayashi and Y. Watabe, "Low Cost Module with Heterojunction Solar Cells Applied Gravure Offset Printing and Multi-Wire Technologies" *38th EU PVSEC* , pp. 1073–1076, 2013.

APPENDIX: EFFECTIVE WIDTH OF A SINGLE WIRE AS DERIVED BY SIMPLE GEOMETRY

The fraction of light that follows each of the paths labelled in Figure 1 was stated by Braun *et al.* [3] for a wire reflectance of $R = 1$ and a refractive index of $n = 1.5$. Here we provide a more generic equation for those fractions and the effective width of a wire.

The fraction of rays that follow Path A in Figure 1 can be shown by simple geometry to be

$$f = 1 - \frac{1}{\sqrt{2}} = 0.293, \qquad \text{(A1)}$$

where we assume an infinitely large cell, no scattering, and a single wire.

The fraction of rays that follow either Path A or Path B constitutes the rays that are reflected from the wire such that their incident angle at the air–glass interface is not within the critical angle:

$$f = 1 - \sqrt{\frac{1-\sqrt{1-\frac{1}{n^2}}}{2}}, \qquad \text{(A2)}$$

where n is the refractive index of the encapsulant that surrounds the wire (not the glass, due to Snell's law). This can be derived by noting that $\sin(\theta_c) = 1/n$ and $\sin(\theta_c/2) = 1 - f$. Notice that when $n = 1$, the above equation simplifies to equation for Path A; i.e., there is no contribution from Path B. For the commonly simulated case of $n = 1.5$, $f = 64.3\%$ [3].

The effective width of a wire (as a proportion of its actual width) is given by

$$w_{eff} = 1 - R_m f. \qquad \text{(A3)}$$

This assumes that none of the rays that are reflected by the wire within the escape cone of the glass–air interface (Path C) are returned to the cell, and that no light within Path A or B is absorbed by the encapsulation.

Thus, if one only considers Path A, $w_{eff} = 70.7\%$ for an ideal reflector ($R_m = 1$) or $w_{eff} = 73.6\%$ for $R_m = 0.9$. The references that use the higher figures of w_{eff} have either considered just Path A or taken experimentally determined values for which non-ideal effects arise [2, 4].

There is no need to neglect Path B. When both Paths A and B are considered, $w_{eff} = 35.7\%$ for $R_m = 1$ (as stated by [3] or $w_{eff} = 42.1\%$ for $R_m = 0.9$. Thus, the results of the ray tracing in Section 2 indicate that some of the aforementioned assumptions lead to significant error.

INFLUENCE OF A VERY EFFICIENT BACK REFLECTOR ON THE QUANTUM EFFICIENCY OF SOLAR CELLS

D. N. Micha[1,2], A. W. Walker[3], G. Siefer[3], F. Dimroth[3], A.W. Bett[3]

[1]CEFET/RJ campus Petrópolis, Rua do Imperador 971, Petrópolis, 25620-003, Brazil.
[2]INCT de Nanodispositivos Semicondutores – INCT/DISSE/CNPq – PUC-Rio – RJ, Brazil
[3]Fraunhofer Institute for Solar Energy Systems ISE, Division Materials–Solar Cells and Technologies, F-517, Heidenhofstraße 2, Freiburg 79110, Germany.

Very reflective back side reflectors directly below the active region of a solar cell alter the photon absorption and carrier collection dynamics in a manner that can be exploited for enhanced efficiency. The enlarged optical path increases the short circuit current, and the confinement and (re)absorption of the photoluminescence leads to a higher open circuit voltage through the photon recycling effect, which is critical in radiative limited solar cells. Furthermore, interesting features arise in the experimental external and internal quantum efficiencies (EQE and IQE) for solar cells with such reflective back side reflectors. This includes an increase in EQE magnitude compared to cells on absorptive substrates, and an extension of the IQE edge towards longer wavelengths. The former characteristic is due to an increased photon path length in the active region, which is a consequence of the high reflection occurring in the back side. The latter, however, is not a real feature of the interaction between the light and the solar cell materials, but a consequence of the way the IQE is defined and calculated. The resulting separation of the IQE and EQE near the bandedge of the solar cell thus questions the concurrent definitions for the IQE commonly referred to in the literature, which is clarified in this work. The influence of the definition of quantum efficiency in the material diffusion length extraction is also discussed.
Keywords: Photovoltaics, gallium arsenide, photon recycling, thin film III-V semiconductors, quantum efficiency.

1 INTRODUCTION

Radiative limited GaAs solar cells have shown great potential to exploit the photon recycling (PR) effect via the use of a back side reflector (BSR), which has been shown to enhance the overall power conversion efficiency compared to cells on their native substrates [1-5]. The open circuit voltage (V_{OC}) is the most influenced figure of merit due to the suppression of the radiative component of the dark current, and is expected to increase exponentially for increasing reflectivity through the effective use of PR [6].

Exploiting the PR effect requires the careful management of the light paths within the high quality crystalline device by confining the photoluminescence (PL) within the absorbing *p-n* junction for re-absorption. As the front interface to air has a small critical solid angle, the omnidirectional PL can be mostly confined through the formation of an optical cavity using a highly reflective BSR directly after the active region [6]. Furthermore, the incident sunlight path length through the cell nearly doubles as a result of the BSR, which allows for a reduction in material without significantly deteriorating the photocurrent, thus further enhancing the V_{OC} [7].

The growth of two identical solar cells, one on the native substrate and the other with a BSR directly below the active region after substrate removal, leads to measurable changes which can be observed primarily as 1) an increase in V_{OC}, due to exploitation of PR, and 2) an increase in short circuit current (J_{SC}), due to the enhanced sunlight optical path. These cells can then be compared to determine the influence of the BSR on the voltage and on the photocurrent, which can be deeply understood via the external quantum efficiency (EQE). The study on voltage has been performed previously [7-8], but a more thorough study is necessary to better understand the influence of the BSR on the external and internal quantum efficiencies. The EQE is defined as the probability of extracting a photoexcited electron-hole pair per incident photon. The internal quantum efficiency (IQE), on the other hand, has two concurrent definitions in the literature. The first is the probability of extracting photoexcited electron-hole pairs per absorbed photon, and gauges the internal carrier dynamics, and hence, the material's quality [9-11]. The second definition involves all the photons entering the cell rather than the light absorbed in the cell [12-14]. The primary difference between these two definitions is where the light is absorbed, since the second definition assumes all light entering the cell is absorbed. Since the quantum efficiency curves play a key role in the extraction of important active material parameters of the solar cells via modeling, such as the minority carrier diffusion lengths and recombination velocities at the interfaces to the active region, it is important to determine which definition is correct for this kind of analysis.

The paper is outlined as follows. Section 2 summarizes the experimental methods used in preparing traditional GaAs solar cells, i.e., on their native substrate, and those removed from said substrates in order to define the BSR directly below the active region of the solar cell; the experimental measurement setups are also concisely described. Section 3 presents a detailed analysis of the influence of a BSR on the measured EQE and IQE of solar cells grown at Fraunhofer ISE. The differences in IQE results for these two types of solar cells can help clarify the definition of this quantity in literature. Finally, the role of the quantum efficiency curves in extracting material parameters such as diffusion lengths is discussed. Conclusions are given in Section 4.

2 EXPERIMENTAL METHODS

GaAs solar cell structures were grown by low (a) pressure Metal-Organic Vapor Phase Epitaxy (MOVPE) in a standard upright fashion on 4 inch (100) p-doped

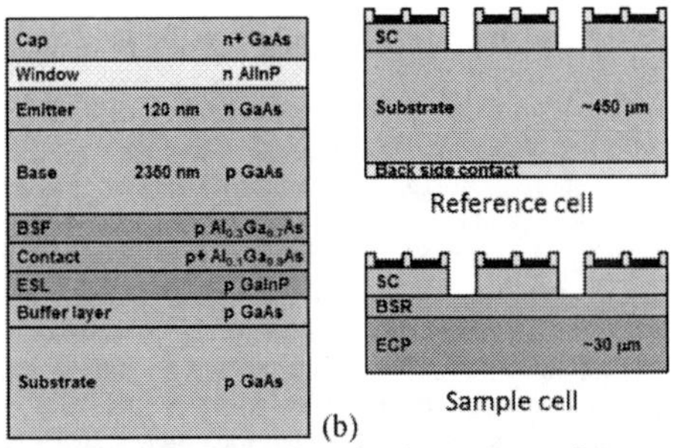

Figure 1: Summary of production process of GaAs solar cell samples. (a) MOVPE grown layers with materials and labels, and (b) design of devices on the original substrate (top) – reference cell – and as thin film with back reflector (bottom) after applied back surface technology – sample cell.

GaAs wafers with a 2° cutoff off [100] orientation. The structures consist nominally of an n+ doped GaAs cap layer, an n-doped AlInP window layer, a 120 nm n-doped GaAs emitter, a 2500 nm p-doped GaAs base, a p-doped $Al_{0.3}Ga_{0.7}As$ back surface field (BSF), a p+doped $Al_{0.1}Ga_{0.9}As$ contact layer, and a p-doped GaInP etch stop layer (ESL), as depicted in Figure 1a.

The solar cells under test were grown with the same structure and in the same epitaxial run. The reference cell, i.e., the cell kept on the original substrate, was produced under standard top-down photolithography technology, with Ohmic contacts and an MgF/TaO double layer anti-reflective coating (ARC), as shown in the top of Figure 1b. In order to produce the required light confinement for the exploitation of photon recycling, a metallic BSR with a normal incidence reflectivity of ~95% over a wide range of wavelengths was defined just below the active region in the sample cell. The metal deposition was done after the removal of the original substrate by a chemical etch step with $NH_4OH:H_2O_2$ = 5:1. The stabilization of the thin film was done using a thick (~30 µm) electroplated metal (ECP). The final shape and design of the cell is presented in the bottom of Figure 1b.

The EQE of the solar cells is measured under short circuit conditions using the differential spectral responsivity method [15]. The EQE is normalized to absolute values with the short circuit current from the light I-V characteristics obtained under AM1.5g at 25°C including a spectral mismatch correction. The front side total reflection was measured in a CARY 500i by collecting the monochromatic beam reflected by the sample using an integration sphere.

3 RESULTS AND DISCUSSIONS

3.1 Influence of BSR on the Figures of Merit of Solar Cells

Figure 2 presents the measured reflectivity – R(λ) – and external quantum efficiency – EQE(λ) – results for the solar cells processed on the original substrate and under the form of thin films (TFSC) with a BSR, referred hereafter as reference and sample cells, respectively. The measured short circuit current densities and open circuit voltages for these cells are summarized in Table I under standard measurement conditions of the AM1.5g spectrum (IEC 60904-3 with 1 kW/m^2).

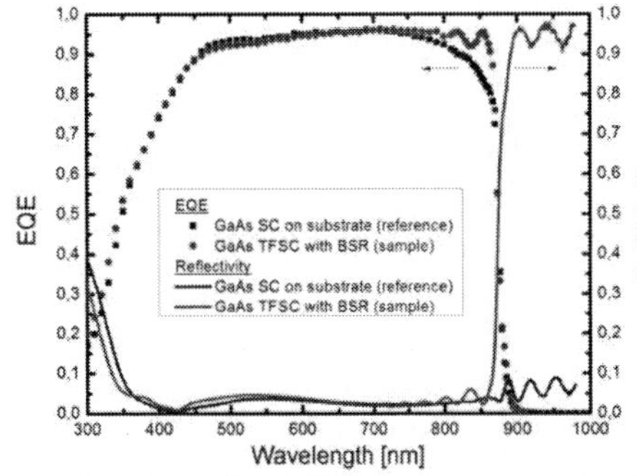

Figure 2: Spectral optical and optoelectronic characterization data for the solar cells treated in this work: EQE (left) and Reflectivity (right).

First, one can notice the effectiveness of the BSR applied to the sample cell in the reflection for wavelengths longer than the GaAs bandgap (870 nm), as demonstrated by the high reflectivity shown as red line in Fig. 2. On the other hand, the reference cell only reflects back 5 to 10% of the incoming light, therefore illustrating that the near-infrared wavelengths are absorbed by the substrate. Note that these wavelengths do not contribute to current (i.e. EQE does not increase) but instead lead to parasitical heating of the device.

TABLE I: FIGURES OF MERIT OF THE SOLAR CELLS

Sample	J_{sc} [mA/cm^2]	V_{oc} [V]
Cell on substrate (reference)	29.05	1.055
Thin film cell with BSR (sample)	29.38	1.074

The EQE of the sample cell (red circles) shows a corresponding increase for long wavelengths (760 – 870 nm) compared to the reference cell in Fig. 2. Overall, this results in an increased J_{SC} (see Table 1) of 1.1%. As the only difference between these cells is the BSR, we can conclude that this is due to the extended optical path of the incident light in the cells, resulting in more absorption and therefore carrier collection. The small differences observed in EQE for shorter wavelengths are due to differences in the reflectivity introduced by fluctuations in the ARC layer properties. These artifacts disappear when the IQE is calculated (see Fig. 3). Another fact worth commenting on is the increase of 1.8% in V_{OC}. This can be attributed to the photon recycling effect due to the formation of an optical cavity by the BSR, as detailed discussed elsewhere [7, 8].

In Fig. 3, the IQE of the solar cells is illustrated as black squares for the reference cell and red circles for the sample cell, where the IQE is computed using Equation 1

$$IQE(\lambda) = \frac{EQE(\lambda)}{1 - R(\lambda)} \quad (1)$$

One can notice two main differences: 1) an extension of the IQE towards longer wavelengths for the sample cell with BSR, and 2) an increase in magnitude close to the bandgap. At first glance, one may interpret this as the material quality of the solar cells being quite different, with the material of the reference cell on substrate being the worst due to the lower IQE. However, this is not a reasonable argument, since these cells were grown in the

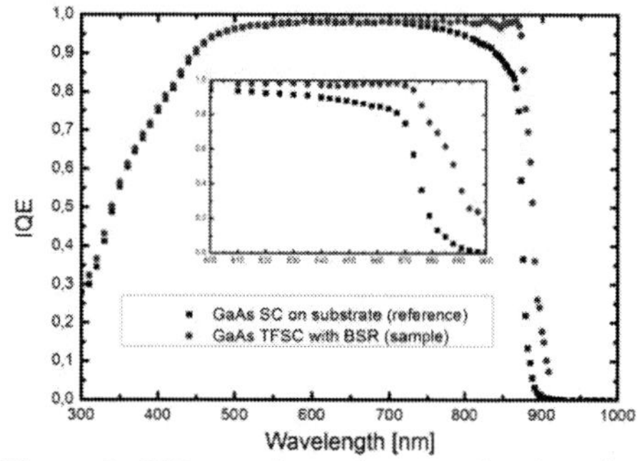

Figure 3: IQE experimental curves for the reference (black squares) and sample (red circles) cells as from Equation 1.

same epitaxial run. The materials should have the same properties. It is therefore worth investigating the definition of the IQE as given by equation (1).

3.2 Internal Quantum Efficiency for Solar Cells

The two concurrent definitions of the IQE involve the reflectivity $R(\lambda)$ and the absorptivity $A(\lambda)$. When IQE is calculated using equation 2a [9-11], one is probing the number of electron-hole pairs generated and collected in the form of electrical current for each photon absorbed inside the active region of a solar cell, and thus making use of the absorptivity. On the other hand, Equation 2b [12-14], same as Equation 1, which is the most common method, adopts the total light entering the cell, rather than the absorptivity.

$$IQE_{abs}(\lambda)=\frac{EQE(\lambda)}{A(\lambda)}=\frac{EQE(\lambda)}{1-T(\lambda)-R(\lambda)} \quad (2a)$$

$$IQE_{trans}(\lambda)=\frac{EQE(\lambda)}{T_{in}(\lambda)}=\frac{EQE(\lambda)}{1-R(\lambda)} \quad (2b)$$

For a solar cell on a thick absorptive substrate, the transmitted light $T(\lambda)$ through the whole structure of the device is practically negligible ($T\approx0$). This results in $A(\lambda)$ being equivalent as $T_{in}(\lambda)$, and thus both definitions (equations 2a and 2b) lead to identical results. However, the concepts are not generally identical because the latter definition assumes that all light entering the cell is absorbed, but it does not matter where this light is absorbed (active region vs. substrate). For a cell on substrate, complete absorption occurs exclusively in the active region of the device if and only if the substrate is part of the active region, which is the case in silicon devices for example. However, in highly absorptive materials such as direct bandgap semiconductors, absorption in the substrate is typically a source of loss, and therefore does not contribute to the photocurrent. In such a case, there could be a significant difference in the IQE computed using both definitions for the most transparent photons (i.e. with energy close to the bandgap) if the active region transmits a non-negligible quantity to the substrate. In such a case, the fraction of photons parasitically absorbed in the substrate would result in an artificially reduced IQE according to equation 2b, since the transmission into the cell, $T_{in}(\lambda)$, would be larger than the absorption solely in the active region $A(\lambda)$. The definition according to equation 2b would thus not reflect the material quality of solely the active region but instead the combined quality of the active region and the substrate. If one could extract the absorptivity solely in the active region, which is extremely difficult in practice for a GaAs solar cell on substrate, then equation 2a would more accurately reveal the behavior of the active region without losses to the substrate. The absorptivity in the active region can be extracted via optical modeling such as with the transfer matrix method (TMM) [16, 17]. Thus, one could extract more realistic material parameters for the active region by fitting the IQE to a theoretical model (discussed in section 3.3).

Another way of accessing the two different definitions expressed in equations 2 is via the experimental reflectivities. For a solar cell with an ideal BSR, the measured reflectivity (R_R) gives insight into the absorptivity of the active region. Equation 3a expresses this mathematically. For cells on an ideal absorptive substrate, the quantity ($1-R_A$), where R_A is the reflection of such cells, yields the portion T_{in} of the light entering the front side (Equation 3b).

$$A(\lambda)\approx 1-R_R(\lambda) \quad (3a)$$

$$T_{in}(\lambda)=1-R_A(\lambda) \quad (3b)$$

In other words, it is possible to obtain the IQE described by equation 2a for devices with the same active region design, whether on absorptive or reflective substrates, using the experimental reflectivities or using modeling.

Thus, one way of consolidating these definitions of IQE is by modeling the optical properties of the solar cell and accounting for them in equations 2. Figure 4a shows the IQE using equation 1 with the experimental reflectivity for the reference cell on substrate and the sample cell with the BSR. By extracting solely the active region absorption using the TMM, the IQE can be computed for the reference cell using equation 2a. The black solid line demonstrates that the IQE for both reference and sample cells has increased in the wavelength range close to the bandedge. This improved agreement outlines that transmission through the active region is not negligible, which is quite clear from the thickness of the active region (2.47 µm). Taking into account this transmission by accounting for the relative absorption results in the same IQE, which is expected based on similar material quality. Note that the base thickness had to be enlarged to 2.8 µm to the curve fit better the experimental points. To avoid regions of negligible absorption where Equation 2a surpasses unity, the EQE is forced to go to zero for values less than 1×10^{-3}. The agreement between such theoretical manipulation to obtain IQE_{abs} for the reference cell (black solid line in Figure 4a) and the experimental points for the sample cell (red circles) corroborates our point of view. Once more, we emphasize that the increase in intensity and the extension of the IQE curve towards long wavelengths is not a real physical fact but a matter of definition.

To further demonstrate the consolidation of IQE definitions, the IQE_{abs} and IQE_{trans} is computed for both cells using equations 3(a) and 3(b), i.e. the absorptivity of the active region and the transmission into the cell respectively. The results are shown in Figure 4b, where the experimental reflectivities are used to go from one IQE definition to the other for each cell. By using the reflectivity of the sample cell to calculate IQE for the reference cell (black dashed line), the curve shows the expected extension in the long wavelength. The opposite

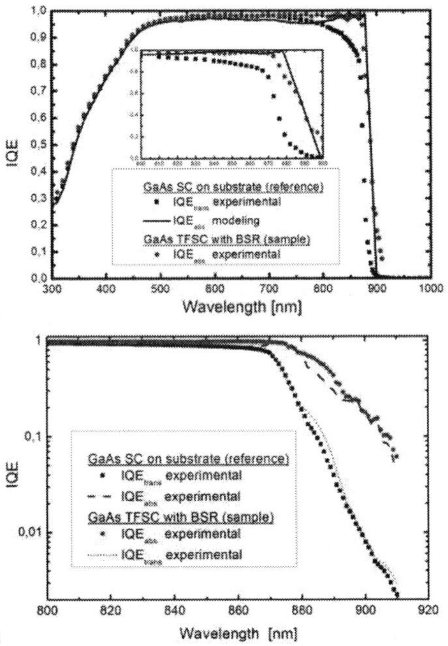

(a)

(b)

Figure 4: Illustration of ways to access the two definitions of IQE. The points are a representation of IQE obtained experimentally by the standard equation 1 (black squares represents IQE_{trans} for the reference cell and red circles represents IQE_{abs} for the sample cell). In (a), we used simulated optical properties to obtain IQE_{abs} for the reference cell (black solid line). In (b), we used the reflectivity of the solar cells to obtain IQE_{abs} for the reference cell (black dashed line) and IQE_{trans} for the sample cell (red dotted line).

method involves using the reflectivity of the reference cell to calculate the IQE for the sample cell (red dotted line), which shows the long wavelength region of the curve at the same level of the original IQE of the reference cell.

3.3 Modeling of the Quantum Efficiency to Extract of Material Diffusion Lengths

Theoretically, the QE of a solar cell depends on the active region's response to the incident light of wavelength λ, and can be separated into the contributions from the emitter, depletion region and base according to Equation 5:

$$QE(\lambda) = QE_{emitter}(\lambda) + QE_{depl.}(\lambda) + QE_{base}(\lambda) \quad (5)$$

The individual contributions of each component of Equation 5 can be described by the Hovel model [12]. Each contribution is a function of important material and structural parameters, such as the minority carrier diffusion lengths (L), the active layers thicknesses (x), the minority carrier mobility (μ), and surface recombination velocities at interfaces to the active region (S). In a recent paper, Lumb *et al.* extended the model for solar cells subject to different boundary conditions at the back side such as coherent and incoherent reflections arising from a BSR [18].

Figure 5 illustrates the models defined in Ref. 18 fitted to the experimental EQE of the reference and sample solar cells from this work, assuming the same hole and electron diffusion lengths. In the long wavelength range, a very good agreement is obtained between experimental data and the model for both cells. Due to the high agreement, the material properties of the two cells are nearly the same, as could be expected from cells originating in the same epitaxial run. This shows that the EQE can be used to properly extract the base diffusion length, since the electron diffusion length is extracted to be the same for the two different solar cells. Furthermore, this demonstrates that the IQE of the two cells should also be similar, if one interprets the IQE as gauging the material quality of the active region. If the IQE were used to extract material properties by the use of equation 1 in comparison to the modeling, then equation 2b should also be used rather than equation 2a. Otherwise, significant differences in diffusion lengths could be extracted from the modeling.

The short wavelength range of the simulations has a larger discrepancy between experiment and model due to the unknown contribution from the window layer. The models take into account the proper boundary conditions for each cell. For the reference cell, we applied the equations of the model with no reflection at the back side. For the sample cell, a weighted average between the results from the models with coherent (20%) and incoherent (80%) reflection at the back side led to the best fit. The remaining set of material parameters is shown in Table II, with mobilities from IOFFE database. Once more, a larger base thickness of 2800 nm had to be used. The reason for the overestimation of the base thickness can be the underestimation of the absorption coefficient in the modeling.

TABLE II: SET OF MATERIAL PARAMETERS USED IN FITTING EQE

Parameter	Value
Hole diffusion length (μm)	0.9
Electron diffusion length (μm)	5
Front side surface rec. velocity (cm/s)	1000
Back side surface rec. velocity (cm/s)	1
Hole mobility ($cm^2V^{-1}s^{-1}$)	270
Electron mobility ($cm^2V^{-1}s^{-1}$)	2500

4 CONCLUSIONS

With very reflective BSR applied directly below the active region of high quality crystalline thin film GaAs solar cells, a higher J_{SC} and V_{OC} was demonstrated when compared to a reference cell on its native substrate. The increased photocurrent is achieved by the enhanced path length within the solar cell, observed experimentally by an increase of up to 14.6% in EQE near the bandgap of GaAs The reported increased V_{OC} is due to the exploitation of the photon recycling effect.

The IQE of the sample cell with BSR demonstrates an increase in magnitude and an extension to long wavelengths compared to the reference cell on substrate. This is not expected for solar cells with identical material quality and structures originating from the same epitaxial run. By modeling the devices optically, these differences are shown to be matter of the definition used to calculate IQE rather than physical artifacts. In order for the IQE to gauge the material quality of the active region without losses from the substrate, the definition of the IQE requires the absorptivity from the active region, which can be accessed via optical modeling. The extraction of

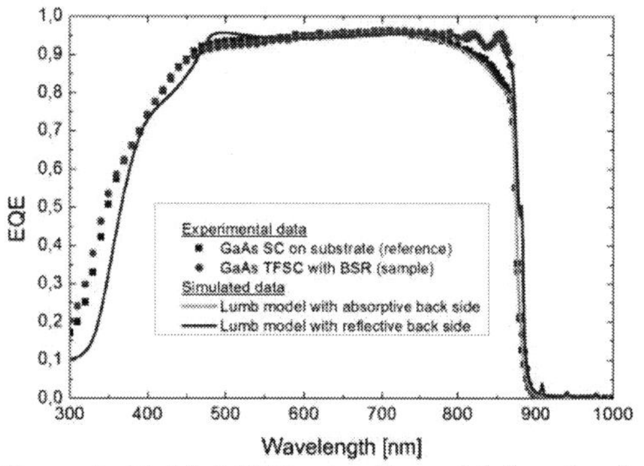

Figure 5: Modeled EQE using the model from Lumb showing a good agreement with the experimental data of the solar cells studied in this work. The same diffusion length was adopted in modeling both cells.

minority carrier diffusion lengths could be influenced by the definition of the IQE. To avoid this problem, the EQE is recommended to be used according to the proper boundary condition of the solar cell, as in the models developed by Lumb *et al.*

5 ACKNOWLEDGEMENTS

This research was supported by ESA through the project "Optimisation of Photon Density in Ultra-Thin GaAs Solar Cells" (ESTEC/Contract no 4000104852). The authors would like to thank the entire Fraunhofer ISE III-V group. Daniel Micha was partially supported by INCT-DISSE/CNPq/Brazil under the program "Ciência sem Fronteiras".

6 REFERENCES

[1] G. J. Bauhuis, "26.1 % thin film GaAs solar cell using epitaxial lift-off," Solar Energy Materials and Solar Cells, vol. 93, pp. 1488-1491, 2009.

[2] X. Wang, "Design of GaAs Solar Cells Operating Close to the Shockley–Queisser Limit," IEEE Journal of Photovoltaics, vol. 3, no. 2, 2013.

[3] B. M. Kayes, H. Nie, R. Twist, S.-. G. Spruytte, F. Reinhardt, I. C. Kizilyalli and G. S. Higashi, ""27.6% conversion efficiency, a new record for single-junction solar cells under 1 sun illumination," Proceedings of 37th IEEE Photovoltaic Specialist Conference, 2011.

[4] M. A. Steiner, "Optical Enhancement of the Open Circuit Voltage in High Quality GaAs Solar Cells," Journal of Applied Physics, vol. 113, p. 123109, 2013.

[5] M.-. C. Tseng, "Fabrication and Characterization of GaAs Solar Cells on Copper Substrates," IEEE Electron Device Letters, vol. 30, pp. 940-942, 2009.

[6] O. D. Miller, E. Yablonovitch and S. Kurtz, "Strong Internal and External Luminescenec as Solar Cells Approach the Shockley-Queisser Limit," IEEE Journal of Photovoltaics, vol. 2, no. 3, pp. 303-311, 2012.

[7] A. W. Walker, O. Höhn, D. N. Micha, L. Wagner, H. Helmers, A. W. Bett, F. Dimroth. "Impact of photon recycling and luminescence coupling on III-V single and dual junction photovoltaic devices," Journal of Photonics for Energy 5(1), p. 053087-1 – 053087-11, 2015.

[8] A. W. Walker, O. Hoehn, D. N. Micha, B. Blaesi, A. W. Bett and F. Dimroth, "Impact of Photon Recycling on GaAs Solar Cell Designs," IEEE Journal of Photovoltaics, vol. 5, no. 6, pp. 1636-1645, 2015.

[9] U. Rau, D. Abou-Ras and T. Kirchartz, Advances characterization techniques for thin film solar cells, John Wiley & Sons, 2011.

[10] B. Liu, R. Q. Png, L. H. Zhao, L. L. Chua and R. H. Friend, "High internal quantum efficiency in fullerene solar cells based on crosslinked polymer donor networks," Nature Communications, vol. 3, p. 1321, 2012.

[11] W. J. Yang, "Internal quantum efficiency for solar cells," Solar Energy, vol. 82, pp. 106-110, 2008.

[12] H. J. Hovel, Semiconductors and Semimetals, Vol. 11: Solar Cells, New York: Academic Press, 1975.

[13] M. Law, M. C. Beard and S. Choi, "Determining the Internal Quantum Efficiency of PbSe Nanocrystal Solar Cells with the Aid of an Optical Model," Nano Letters, vol. 8, no. 11, pp. 3904-3910, 2008.

[14] R. Brendel, M. Hirsch, R. Plieninger and J. H. Werner, "Ouantum Efficiencv Analvsis of Thin-Layer Silicon Solar Cells with Back Surface Fields and Optical Confinement," IEEE Transactions on Electron Devices, vol. 43, no. 7, pp. 1104-1113, 1996.

[15] J. Metzdorf, "Calibration of solar cells. 1: The Differential Spectral Responsivity Method," Applied Optics, vol. 26, no. 9, pp. 1701-1708, 1987.

[16] Emanuele Centurioni. Generalized matrix method for calculation of internal light energy flux in mixed coherent and incoherent multilayers. Applied Optics, 44(35), 7532-7539, 2005.

[17] Charalambos C. Katsidis and Dimitrios I. Siapkas. General transfer-matrix method for optical multilayer systems with coherent, partially coherent, and incoherent interference. Applied Optics, 41(19), 3978-3987, 2002.

[18] M. P. Lumb, C. G. Bailey, J. G. J. Adams, G. Hillier, F. Tuminello, V. C. Elarde and R. J. Walters, "Extending the 1-D Hovel Model for Coherent and Incoherent Back Reflections in Homojunction Solar Cells," IEEE Journal of Quantum Electronics, vol. 49, no. 5, pp. 462-470, 2013.

A THREE-DIMENSIONAL PHANTOM NODE METHOD TO STUDY COMPLEX CRACK PATTERNS IN PHOTOVOLTAIC SOLAR CELLS

1. P.R. Budarapu1, 2. J. Reinoso, 1. M. Paggi
1. IMT School of Advanced Studies Lucca, 55100 Lucca, Italy
2. Elasticity and Strength of Materials Group, School of Engineering, University of Seville, Camino de los Descubrimientos s/n, 41092 Seville, Spain

ABSTRACT: In this paper, we introduced a novel three-dimensional shell based framework, coupling the Phantom node method to study complex fracture patterns in PV Silicon cells. The developed framework has been applied to study complex fracture patterns observed in a fatigue test carried out in [15]. Based on the estimated crack opening displacement from the computational model, we are able to correlate the reasons for the development of inactive areas leading to power loss. Therefore, we report that the critical crack opening displacements results in large resistances along the crack lines, finally leading to electrically inactive areas depending on the fracture pattern..
Keywords: three-dimensional shell, phantom node method, PV Silicon cells

1 INTRODUCTION

The renewable, non-polluting and clean nature of the solar energy makes it one of the fast growing technology in the World [1]. The International Energy Agency (IEA) predicted more than 500 GW of solar power generation in the World by 2020 [2], which is about 150% increase within the next 5 years. Photovoltaic (PV) solar cells made of Silicon (Si), see Fig. 1, play a key role in converting solar energy into electricity based on the PV effect of Silicon semiconductor. Standard PV modules are laminates composed of: (i) a glass superstrate, (ii) an encapsulating polymer layer like ethylene vinyl acetate (EVA), (iii) a layer of Si solar cells, (iv) a second layer of EVA, and finally (v) a thin multi-layered back sheet [3], see Fig. 1. Two main semiconductors, called bus bars, connect the cells together.

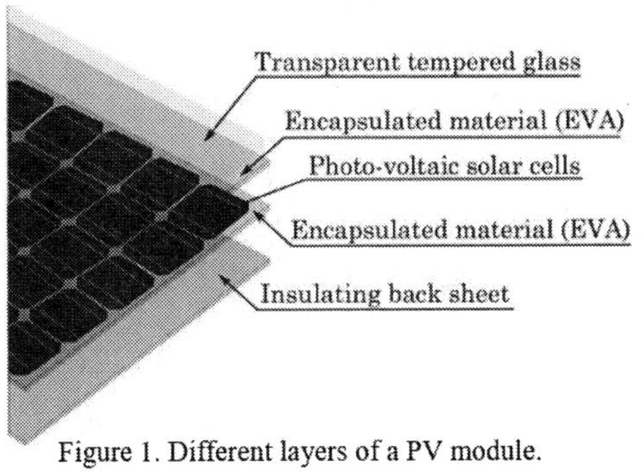

Figure 1. Different layers of a PV module.

The present theoretical and applied research has mostly focused on increasing the solar energy conversion efficiency of the cells [4, 5]. Although efficiencies up to 40% have been reached in the laboratory using single junction Gallium Arsenide (GaAs) and multi-junction concentrators [6], the technology based on mono- and poly-crystalline Si is still the most competitive on the market, due to the low price of Si semiconductor and the widely established material processing developed in the field of electronics [7]. Therefore, researchers were focused on developing new technologies [8] based on thinner and thinner silicon wafers [9–12] and thin films to save material, although these solutions will be even more prone to cracking than standard solar cells. New kinds of cell structures and coatings are also explored [13, 14] to achieve higher solar energy conversion efficiencies. In this context, research on the durability of PV modules is an area not much investigated yet. The study of durability of PV modules requires the characterization of the effect of micro-cracking induced by mechanical loads and thermal variations on the electric response of the solar panel. Therefore, in order to efficiently tackle this issue, interdisciplinary research including disciplines from structural mechanics to solid state physics is required.

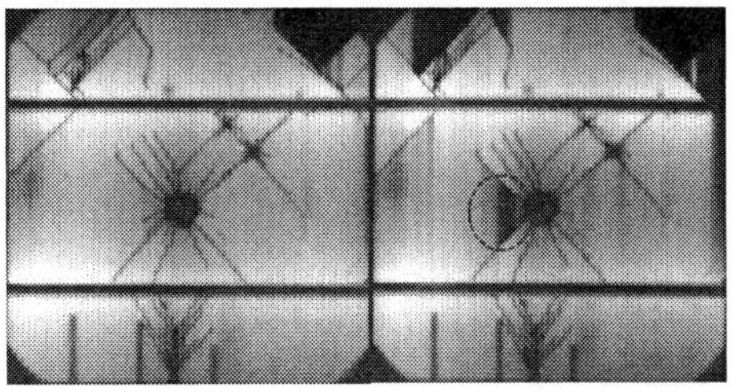

(a) (b)

Figure 2: Electric response and crack pattern of a solar cell: (a) flat; (b) bent configuration (adapted from [15]).

Due to small thickness and brittle nature of Silicon, cracks are prone to develop during production, transportation, installation and operation. Cracking hinders the progress towards the use of thinner Silicon wafers. The durability of PV devices is mainly affected by cracking in Silicon due to thermomechanical loading [16]. Cracks can interrupt the grid line deposited on solar cells, creating large portions of electrically insulated areas [17, 18]. Fig.2 shows the electrical response and crack pattern during a bending test of solar cells [15]. Dark areas of the loaded configuration in Fig.2(b) indicate the electrically inactive regions, which are not seen in the unloaded configuration in Fig.2(a). In spite of the new and efficient manufacturing techniques to reduce the number of cells/modules rejected by quality control [19, 20], it is impossible to avoid the occurrence of micro-cracking due to the brittleness of Silicon. Qualification standards IEC 61215 require passing of severe laboratory tests in an artificial climate chamber. However, micro-cracking is not used as a quantitative indicator for the quality assessment of PV modules. Kajari-Schröder et al. [21] have analyzed micro-cracking resulting from snow tests and artificial aging in the laboratory using the electroluminescence (EL) technique. Micro-cracking can lead to large electrically disconnected cell areas (see e.g. Fig. 2), with up to 21% of power-loss [17, 22–25]. Furthermore, orientation of cracks with respect to the bus bars plays a key role in determining their criticality [22]. A micro-cracked cell shows also a local increase of operating temperature near the crack, as highlighted from molecular dynamics simulations [26]. Ther-mographic studies show that cracks are acting primarily as recombination centers with a local increase in temperature around them [27]. Deformations are also induced by the night and day variation in tem-perature. Moreover, temperature affects the electric performance of the PV module, since the electrical output of the semiconductor exponentially varies with temperature [28, 29].

Paggi et al. [15] demonstrated the complex interaction between crack propagation in monocrystalline Silicon cells embedded in PV modules and the electric output, based on an experimental study using the EL technique. They proved that, due to the action of the encapsulating polymer and residual thermo-elastic stresses, cracked regions can partially recover the electric conductivity during mechanical unloading due to crack closure, apart from reporting a fatigue degradation during cyclic bending. Berardone et al. [24] have developed a one-dimensional model for simulating the electric current distribution in solar cells accounting for a distributed series resistance in the presence of partially conductive cracks. This can be included in a multi-physics and multi-scale computational approach to study the evolution of microcracking in polycrystalline Silicon (Si) solar cells in PV modules, as proposed in [29]. In this paper, we propose a three-dimensional finite element method coupled with the phantom node technique, to identify the electrically insulated areas based on a complex fracture pattern observed in the experiments.

2 MODELING AND SIMULATION

A computational efficient 8-node solid shell element [30, 31] formulation is considered in this work to discretize the solar cell layer. This finite element formulation has the advantage over solid brick element, to correctly capture bending deformation patterns of thin solar cells with just one element over the thickness, see also instability analyses in [31]. The novelty of the present work regards the extension of the shell element to study complex crack patterns in solar cells through the adoption of the Phantom node method. Any type of crack tip enrichment is not assumed in this paper, since this would drastically increase the complexity of the formulation mainly due to integration and blending. Therefore, the crack needs to cross an entire element.

Detailed mathematical formulation of the solid shell element, which can also work under finite defor-mation (Fig.3), is based on the Assumed Natural Strain (ANS) and the Enhanced Assumed Strain (EAS) methods to avoid locking pathologies [31]. The Phantom node method [32] has been employed here to model the through the thickness disconti-nuities in the solar cell. In the phantom node method, when an element is completely cut by a crack, the displacement field is continuous on each part of the cracked element, but discontinuous across the crack surface. Hence, the crack kinematics can be obtained based on the overlapping elements [33–35,37-44].

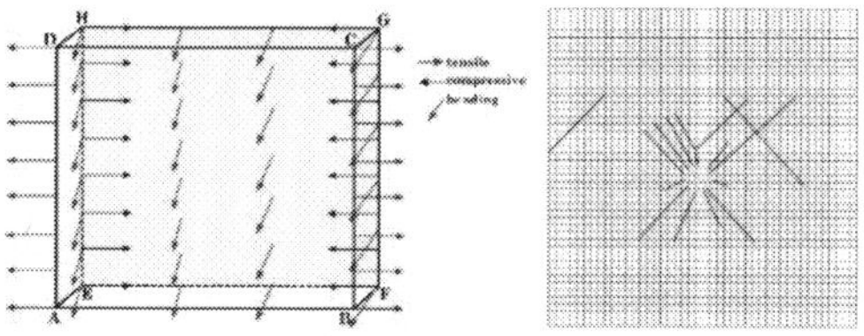

Figure 4: (a) Schematic of the Silicon cell showing the boundary conditions. (b) Projected view of the finite element discretization of the cell along with the cracks.

Due to its versatility, the proposed formulation can be suitably integrated within the global/local approach proposed in [36], where the bending displacements at the solar cell boundaries can be passed to the local model of the cracked solar cell (see Fig.4). Figure 4(a) shows a schematic of the solar cell along with the boundary conditions and the finite element discretization along with the cracks planned to analyze in this paper are shown in Fig.4(b). Note that the crack locations and orientations are ex-tracted based on Fig.2. With the present computational framework at hand, the relative crack opening displacements of the various cracks in the local model can now be accurately computed. Based on the relation between crack opening and electric resistance identified in [36] using the experimental data, it is then possible to predict voltage and current at any point along the fingers. Finally, this prediction can be compared with the EL intensity levels obtained from the EL tests (Fig. 2) using the methodology proposed in [24] for further validation.

REFERENCES

[1] F. Meillaud, M. Boccard, G. Bugnon, M. Despeisse, S. Hänni, F.-J. Haug, J. Persoz, J.-W. Schüttauf, M. Stuck-elberger, and C. Ballif *Materials Today*, p. http://dx.doi.org/10.1016/j.mattod.2015.03.002, 2015.

[2] S. Nowak tech. rep., International Energy Agency, http://iea-pvps.org/fileadmin/dam/Nowak.pdf, 2014.

[3] M. Paggi, S. Kajari-Schröder, and U. Eitner *Journal of Strain Analysis for Engineering Design*, vol. 46, no. 8, 772–782, 2011.

[4] M. A. Green, K. Emery, Y. Hishikawa, W. Warta, and E. D. Dunlop *Progress in Photovolts: Research Appli-cations*, vol. 22, pp. 701–710, 2014.

[5] D. J. You and et. al. *Progress in Photovoltaics: Research and Applications*, vol. 23, pp. 973–988, 2015.

[6] NREL tech. rep., National Renewable Energy Laboratory, U.S. Department of Energy, http://www.nrel.gov/news/press/2008/625.html, 2008.

[7] A. Goetzberger, C. Hebling, and H.-W. Schock

Material Science and Engineering: R, vol. 40, pp. 1–46, 2003.
[8] M. Ye, X. Wen, M. Wang, J. Iocozzia, N. Zhang, C. Lin, and Z. Lin *Materials Today*, vol. 18, no. 3, pp. 155–162, 2015.
[9] Y. Aya, W. Shinohara, M. Matsumoto, K. Murata, T. Kunii, M. Nakagawa, A. Terakawa, and M. Tanaka
Progress in Photovoltaics: Research and Applications, vol. 20, pp. 166–172, 2012.
[10] H. Simon, B. Grégory, P. Gaetano, B. Mathieu, E. Jordi, D. Matthieu, M. Fanny, and B. Christophe *Progress in Photovoltaics: Research and Applications*, vol. 21, pp. 821–826, 2013.
[11] M. Foti, C. Tringali, A. Battaglia, N. Sparta, S. Lombardo, and C. Gerardi *Solar Energy Materials & Solar Cells*, vol. 130, pp. 490–494, 2014.
[12] J. Alok, S. Richa, P. Hardik, P. Nilesh, and R. P. Sharma *Journal of Physics D: Applied Physics*, vol. 48, 275101, 2015.
[13] D. Das and A. Banerjee *Applied Surface Science*, vol. 345, pp. 204–215, 2015.
[14] H. Babar, E. Abasifreke, and I. Ferguson *Solar Energy Materials & Solar Cells*, vol. 139, pp. 95–100, 2015.
[15] M. Paggi, I. Berardone, A. Infuso, and M. Corrado *Scientific Reports*, vol. 4, p. 4506, 2014.
[16] N. Park, J. Jeong, and C. Han *Microelectronics Reliability*, vol. 54, pp. 1562–1566, 2014.
[17] M. Köntges, I. Kunze, S. Kajari-Schröder, X. Breitenmoser, and B. Bjørneklett *Solar Energy Materials and Solar Cells*, vol. 95, pp. 1131–1137, 2011.
[18] T. M. Pletzer, J. I. v. Mölken, S. Rißland, O. Breitenstein, and K. Joachim *Progress in Photovoltaics: Research and Applications*, vol. 23, pp. 428–436, 2015.
[19] S. Rein, K. Bothe, and B. Sattler *FVS, BSW-Solar*, pp. 120–127, 2007.
[20] A. R. Barron *Materials Today*, vol. 18, pp. 2–3, 2015.
[21] S. Kajari-Schröder, I. Kunze, U. Eitner, and M. Köntges *Solar Energy Materials and Solar Cells*, vol. 95,
pp. 3054–3059, 2011.
[22] S. K. Schröder, I. Kunze, and M. Köntges *Energy Procedia*, vol. 27, pp. 658–663, 2012.
[23] J. Käsewieter, F. Haase, M. H. Larrodé, and M. Köntges *Energy Procedia*, vol. 27, pp. 469–477, 2014.
[24] I. Berardone, M. Corrado, and M. Paggi *Energy Procedia*, vol. 55, pp. 22–29, 2014.
[25] H. Yang, H. Wang, D. Cao, D. Sun, and X. Ju *International Journal of Photoenergy*, vol. 2015, p. 251615, 2015.
[26] [Budarapu, P R and B, Javvaji, B]^{equal}, V. K. Sutrakar, D. R. Mahapatra, G. Zi, and T. Rabczuk *Journal of Applied Physics*, vol. 118, pp. 382–395, 2015.
[27] A. Sapora and M. Paggi *Comptational Mechanics*, vol. 53, pp. 845–857, 2013.
[28] J. A. Gow and C. D. Manning *IEEE Proceedins Electric Power Applications*, vol. 146, no. 2, pp. 193–200, 1999.
[29] M. Paggi, M. Corrado, and M. A. Rodriguez *Composite Structures*, vol. 95, pp. 630–638, 2013.
[30] M. Bischoff and E. Ramm *International Journal for Numerical Methods in Engineering*, vol. 40, no. 23, pp. 4427–4449, 1997.
[31] J. Reinoso, M. Paggi, and R. Rolfes *Computational Materials Science*, 2015. doi: 10.1016/j.commatsci.2015.08.031.
[32] A. Hansbo and P. Hansbo, "A finite element method for the simulation of strong and weak discontinuities in solid mechanics," *Computer Methods in Applied Mechanics and Engineering*, vol. 193, pp. 3523–3540, 2004.
[33] J. H. Song, P. M. A. Areias, and T. Belytschko *International Journal for Numerical Methods in Engineering*, vol. 67, pp. 868–893, 2006.
[34] T. Rabczuk, G. Zi, A. Gerstenberger, and W. Wall. *International Journal for Numerical Methods in Engineering*, vol. 75, no. 5, pp. 577–599, 2008.
[35] P. R. Budarapu, R. Gracie, S. P. A. Bordas, and T. Rabczuk *Computational Mechanics*, vol. 53, no. 6, pp. 1129– 1148, 2014.
[36] M. Paggi, M. Corrado, and I. Berardone *Engineering Fracture Mechanics*, 2016. doi: 10.1016/j.engfracmech.2016.01.018.
[37] PR. Budarapu, B. Javvaji, VK. Sutrakar, D. Roy Mahapatra, G. Zi, T. Rabczuk. Journal of Applied Physics, 118, 064307, 2015.
[38] YB. Sudhir Sastry, PR. Budarapu, N. Madhavi and Y. Krishna. Computational Materials Science, 96B, 459-471, 2015.
[39] PR. Budarapu, TSS. Narayana, B. Rammohan and T. Rabczuk. Applied Acoustics, 89, 128-140, 2015.
[40] SW. Yang, PR. Budarapu, DR. Mahapatra, SPA. Bordas, G. Zi and T. Rabczuk. Computational Materials Science, 96B, 382-395, 2015.
[41] YB. Sudhir Sastry, PR. Budarapu, Y. Krishna and S. Devaraj. Theoretical and Applied Fracture Mechanics, 72, 2-12, 2014.
[42] PR. Budarapu, YB. Sudhir Sastry, B. Javvaji and DR. Mahapatra. Frontiers of Structural and Civil Engineering, vol 8(2), 151-159, 2014.
[43] PR. Budarapu, R. Gracie, SW. Yang, X. Zhaung and T. Rabczuk. Theoretical and Applied Fracture Mechanics, vol 69, 126-143, 2014.
[44] PR. Budarapu, B. Javvaji, VK. Sutrakar, D. Roy Mahapatra, M. Paggi, G. Zi, T. Rabczuk. International Journal of Fracture, DOI: 10.1007/s10704-016-0115-9.

FABRICATION OF STRAIN-COMPENSATED HETEROJUNCTION GE/$SI_{1-X}C_X$ QUANTUM DOT SOLAR CELLS

Kazuhiro Gotoh[1], Ryuji Oshima[2], Takeshi Tayagaki[2], Takeyoshi Sugaya[2], Koji Matsubara[2], and Michio Kondo[1,3]

[1] Department of Innovative and Engineering Materials, Interdisciplinary Graduate School of Science and Engineering, Tokyo Institute of Technology, 4259 Nagatsuta-cho, Midori-ku, Yokohama, Kanagawa 226-8503, Japan
[2] Research Center for Photovoltaics, National Institute of Advanced Industrial Science and Technology (AIST), 1-1-1 Umezono, Tsukuba, Ibaraki 305-8568, Japan
[3] Fukushima Renewable Energy Institute, AIST (FREA), 2-2-9 Machiikedai, Koriyama, Fukushima 963-0298, Japan

ABSTRACT: Heterojunction 50-layer-stacked Ge quantum dot solar cells (QDSCs) were fabricated via molecular beam epitaxy using a strain-compensation technique. $Si_{1-x}C_x$ materials consisting of a Si matrix with a small amount of carbon (~0.05%) were used as strain-compensating layers (SCLs). In the 50-layer-stacked Ge QDs, large coalesced islands with a density of 4.0×10^8 cm^{-2} were observed for QDs stacked with Si spacer layers, whereas the formation of coalesced islands was effectively suppressed for QDs stacked with $Si_{1-x}C_x$ SCLs. The external applied bias voltage dependence on the short circuit current was calculated from external quantum efficiency measurements. This study revealed that the carrier collection efficiency Ge/$Si_{1-x}C_x$ QDSCs was superior to Ge/Si QDSCs, possibly due to the reduction of non-radiative recombination centers associated with coalesced islands. As a result, the open circuit voltage increased from 0.38 V for Ge/Si QDSCs to 0.42 V for Ge/$Si_{1-x}C_x$ QDSCs.
Keywords: Silicon, Germanium, Carbon, Quantum Dots and Solar Cell

1 INTRODUCTION

Recently, self-assembled quantum dots (QDs), which were formed via the Stranski-Krastanov growth mode in lattice-mismatched hetero-epitaxy, were employed in intermediate band solar cells (IBSCs). In IBSCs, an additional photocurrent is generated by utilizing two-step optical transitions via IB states (transitions from the valence band (VB) to the IB and from the IB to the conduction band (CB)). This current adds to the conventional direct optical transition from the VB to the CB (Fig. 1). IBSCs may ideally achieve higher conversion efficiency than the single-junction Shockley-Queisser limit of ~30% [1,2].

In the Ge QDs in the Si matrix, the photo-generated electron-hole pairs can be efficiently separated in real and k spaces, because they have type-II heterointerfaces and indirect band structures [3,4]. These characteristics may lead to an improvement of the collection efficiency of carriers. In order to realize practical QD-IBSCs, QDs should be dense, homogeneous in size, and periodically distributed to form the mini-bands utilized as IB states. Increasing the number of stacked, self-assembled QDs is a promising way to increase the total QD density, although some difficulties may arise from the accumulation of compressive strains induced by QD layers. If the host material used for burying the QDs was the same as the substrate, the QDs in a stacked configuration would be generally accompanied by an increase in size [5] and, in case their thickness exceeded a critical value, they would generate misfit dislocations. To address this issue, a strain-compensation growth technique was recently developed for GaAs-based [6-8] and InP-based QD [9] systems, in which the compressive strain induced by the QDs was compensated by the tensile strained spacer layers. In the case of Ge/Si QDs, we have developed strain-compensated Ge QDs buried in $Si_{1-x}C_x$ strain compensating layers (SCLs) with a small amount of carbon (x = 0.05%) [10]. Fig. 2 shows the schematic concept of strain compensation in multi-stacked Ge QDs on Si substrates with Si spacer layers or $Si_{1-x}C_x$ SCLs. This work characterizes the carrier collection efficiency of strain-compensated Ge/$Si_{1-x}C_x$ QDSCs.

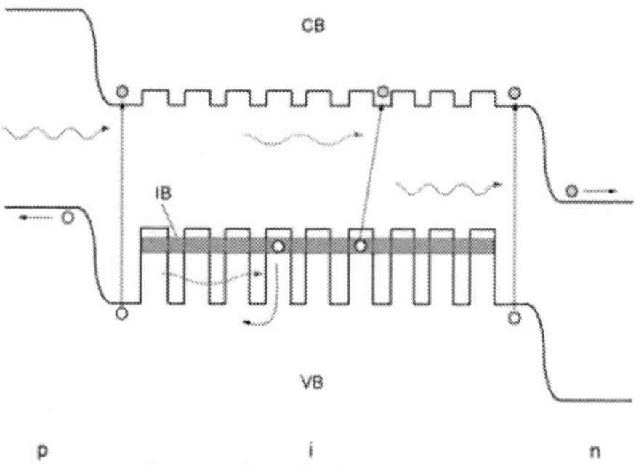

Figure 1: Schematic band diagram of Ge/Si QDSCs.

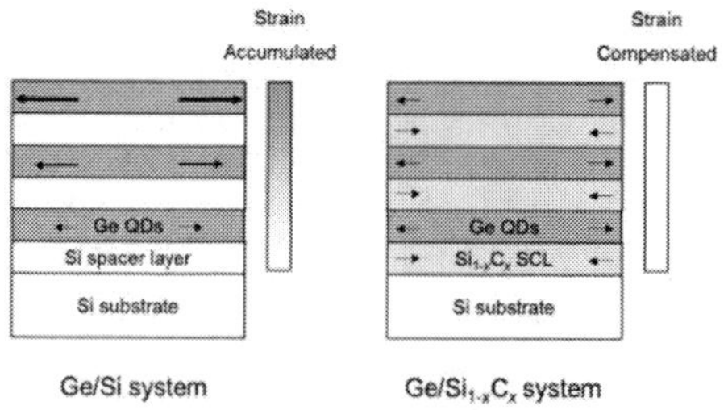

Figure 2: Schematic concept of strain compensation in multi-stacked Ge QDs on Si substrate.

2 EXPERIMENTAL

All QD samples were grown by solid-source molecular beam epitaxy (SS-MBE) on 350 μm-thick p^+-Si (001) substrates (resistivity: ~10^{-3} Ω·cm). The typical base pressure of the growth chamber was 3×10^{-10} Torr.

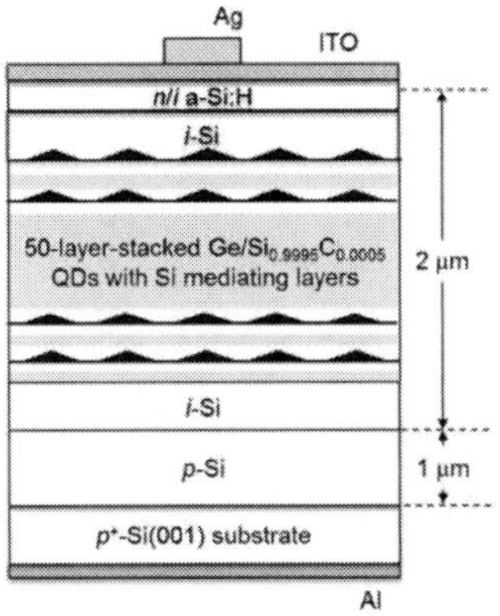

Figure 3: Schematic structure of 50-layer-stacked $Ge/Si_{1-x}C_x$ QDSCs.

Both Si and Ge sources were heated by an electron beam gun. A Ga dopant cell and a C filament cell were installed. The Si substrate was chemically cleaned by RCA cleaning methods [11] and then was set to the growth chamber. After the deposition of a 1-μm-thick *p*-type (1×10^{16} cm^{-3}) Si base layer at 700 °C, 50-layer-stacked 4.6 monolayer (ML) Ge QDs were grown at 500 °C. Composite structures of a 6-nm-thick $Si_{1-x}C_x$ SCL were sandwiched by 2-nm-thick Si mediating-layers [10], which resulted in a total spacer layer thickness of 10 nm. The growth rates of Ge QDs and $Si_{1-x}C_x$ SCLs were 2.6 Å/s and 2.0 Å/s, respectively. The C filament cell was heated at 1400 °C for the growth of $Si_{1-x}C_x$ SCLs. The C content in the $Si_{1-x}C_x$ SCLs, x, was determined to be 0.05% by X-ray diffraction measurements. A 50-layer-stacked Ge/Si QD sample was fabricated using an identical structure, except for 10-nm-thick Si spacer layers instead of $Si_{1-x}C_x$. For solar cells, multi-stacked Ge QD structures were buried by a 750-nm-thick intrinsic Si layer. The surface morphology was studied by atomic force microscopy (AFM).

Heterojunction $Ge/Si_{1-x}C_x$ QDSCs using hydrogenated amorphous silicon (a-Si:H) emitter layers were fabricated, as shown in Fig. 3. The use of a-Si:H as an emitter layer is advantageous for the suppression of interdiffusion of Ge atoms during deposition [12]. In order to form p-n junction, the deposition of a-Si:H is typically employed by plasma-enhanced chemical vapor deposition (PECVD) at a temperature of as low as 200 °C, in comparison with the conventional phosphorous dopant-diffusion process processed at ~900 °C in a furnace [13]. MBE-grown epi-layers were finished with a combination of 10-nm-thick intrinsic a-Si:H deposited at 150 °C and 20-nm-thick *n*-type a-Si:H deposited at 180 °C by PECVD. An indium tin oxide (ITO) layer was used as transparent conductive oxide (TCO); silver and aluminum were deposited for the front and rear contact electrode, respectively. The cell size was 0.25 cm^2. For the solar cell characterization, the current-voltage (*I-V*) characteristic was measured under air mass 1.5 global (AM1.5G), 100 mW/cm^2 at 25 °C. External quantum efficiency (EQE) measurements were performed at room temperature under a constant monochromatic photon irradiation of 10^{14} cm^{-2} using a standard lock-in configuration.

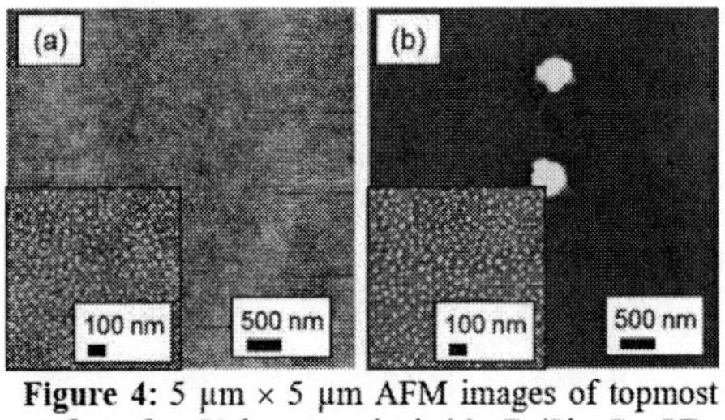

Figure 4: 5 μm × 5 μm AFM images of topmost surface for 50-layer-stacked (a) $Ge/Si_{1-x}C_x$ QDs and (b) Ge/Si QDs. The insets show 1 μm × 1 μm enlarged views for each sample.

3 RESULTS AND DISCUSSION

Fig. 4 shows the AFM images of the topmost surface measured for 50-layer-stacked$Ge/Si_{1-x}C_x$ QDs and Ge/Si QDs. Average height, average lateral size, fluctuation in lateral size and areal density were 1.5 nm, 28.1 nm, 13.6% and 6.5×10^{10} cm^{-2} for $Ge/Si_{1-x}C_x$ and, 1.8 nm, 29.3 nm, 14.5% and 4.4×10^{10} cm^{-2} for Ge/Si, respectively. Almost identical Ge QDs were observed for both samples, except for the occurrence of large coalesced islands with a lateral size of 600 nm and a density of 4.0×10^8 cm^{-2} in the Si/Ge samples. The generation of coalesced islands was probably caused by exceeding the local critical thickness, due to the accumulation of compressive strain [14,15]. These coalesced islands contained misfit dislocations, resulting in generation of non-radiative recombination centers. Although $x = 1.7\%$ would be necessary to exactly compensate the compressive strain induced by Ge QDs in the $Ge/Si_{1-x}C_x$ QD configuration [16], the partial strain-compensation with $x = 0.05\%$ could result in the suppression of the formation of coalesced islands.

Fig. 5 shows EQE spectra of (a) $Ge/Si_{1-x}C_x$ QDSC, (b) Ge/Si QDSC and (c) p^+-Si cell. For the fabrication of a p^+-Si cell, the bare p^+-Si substrate without any epilayers was processed by identical heterojunction procedures with QDSCs, and was used as a reference. Lower EQE in shorter wavelength region from 300 to 700 nm for both QDSCs, which are mainly contributed from the absorption of the Si layers, was observed compared with the p^+-Si cell. Moreover, the EQE for $Ge/Si_{1-x}C_x$ cells was slightly higher than that for Ge/Si cells in the entire wavelength range. These results suggest that carriers generated in Si layers are trapped in both QD states and non-radiative recombination centers generated in the stacked QD structure [17,18]. Thus, the

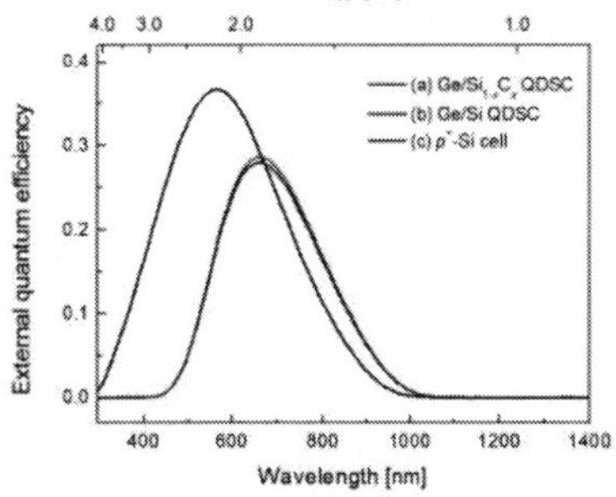

Figure 5: EQE spectra of 50-layer-stacked (a) $Ge/Si_{1-x}C_x$ QDSCs, and (b) Ge/Si QDSCs and (c) p^+-Si cell as a reference.

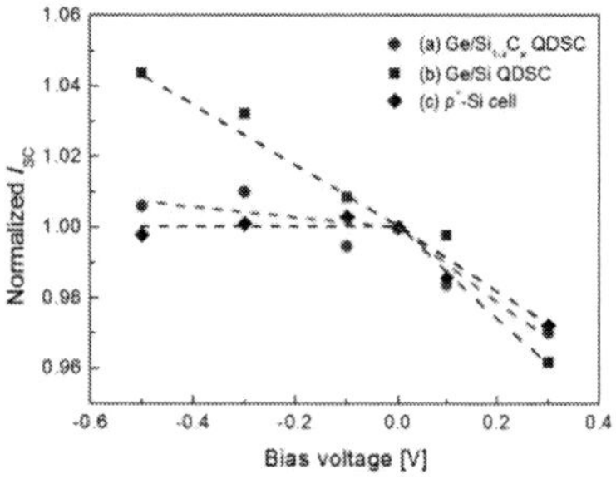

Figure 6: Normalized I_{SC} of (a) Ge/$Si_{1-x}C_x$ QDSC and (b) Ge/Si QDSC and (c) p^+-Si cell as function of bias voltage. Dot lines are guides for each sample.

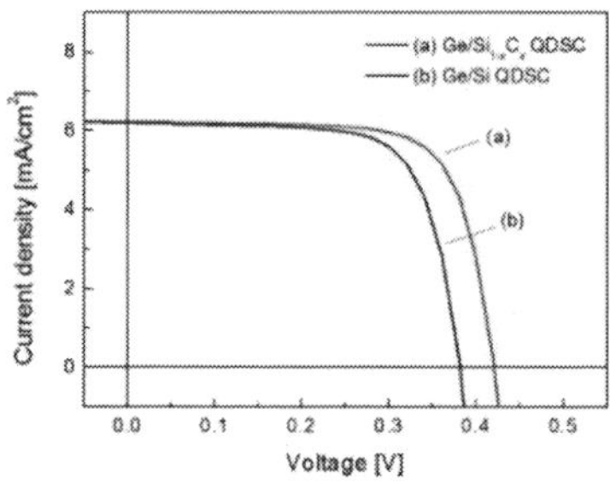

Figure 7: *I-V* curves of 50-layer-stacked Ge QDSCs grwon with (a) $Si_{1-x}C_x$ SCLs and (b) Si spacer layers.

slight increase in EQE for Ge/$Si_{1-x}C_x$ cells was probably due to the suppression of coalesced islands by strain compensation. Conversely, an increased EQE in the longer wavelength region up to 1100 nm for both QDSCs was attributed to additional absorption by the insertion of stacked Ge QD structures.

Fig. 6 shows the normalized short-circuit current (I_{SC}) calculated from the EQE measurements for the (a) Ge/$Si_{1-x}C_x$ QDSC, (b) Ge/Si QDSC and (c) p^+-Si cell as a function of the external bias voltage applied. The I_{SC} was calculated from the EQE spectrum measured at each applied voltage by using following equation:

$$I_{SC} = \frac{1}{1000}\int \frac{\lambda \cdot \mathrm{EQE}(\lambda) \cdot b_s(\lambda)}{1240} d\lambda \qquad (1)$$

where λ and $b_s(\lambda)$ are the wavelength and spectral irradiance, respectively. The I_{SC} was normalized with the 0 V value for each cell. In the forward applied bias region, a significant decrease in I_{SC} with increasing applied voltage was observed for all cells due to the weakened electric field in the intrinsic region, resulting in the enhancement of recombination in the junction. Conversely, in the reverse applied bias region, the degree of I_{SC} increase significantly depends on the kind of cell measured. Especially, the p^+-Si cell showed almost no change in I_{SC} in the reverse bias region, indicating that p^+-Si SCs contained few recombination centers in the junction. The fact that the degree of increase for the Ge/$Si_{1-x}C_x$ and the Ge/Si cells were greater than that for the p^+-Si cell indicates poor carrier collection efficiency for both QDSCs due to the increased recombination in the Ge QD structures. Since both the radiative recombination via quantum states in Ge QDs and non-radiative recombination associated with the misfit dislocations seem to be dominant in QDSCs, the smaller degree of I_{SC} increase for Ge/$Si_{1-x}C_x$ compared to Ge/Si is probably due to the reduction of non-radiative recombination centers.

Fig. 7 compares the light *I-V* curves measured for 50-layer-stacked Ge QDSCs grown with (a) $Si_{1-x}C_x$ SCLs and (b) Si spacer layers. The short-circuit current density (J_{SC}), open-circuit voltage (V_{OC}), fill factor (*FF*) and conversion efficiency (η) were 6.23 mA/cm^{-2}, 0.42 V, 0.73 and 1.91% for the Ge/$Si_{1-x}C_x$ cell and 6.20 mA/cm^{-2}, 0.38 V, 0.70 and 1.62% for the Ge/Si cell. Note that the V_{OC} was increased from 0.38 V for the Ge/Si cell to 0.42 V for the Ge/$Si_{1-x}C_x$ cell. From the diode equation, the V_{OC} is given by

$$V_{OC} = \frac{kT}{q}\ln\left(1 + \frac{I_{SC}}{I_0}\right) \qquad (1)$$

where k, T, q, and I_0 are the Boltzmann constant, temperature, elementary charge, and reverse saturation current density, respectively [19]. The I_0 for Ge/$Si_{1-x}C_x$ QDSCs was estimated from dark *I-V* measurements to be 5.48 × 10^{-7} mA/cm^2, which was smaller than that of 2.56 × 10^{-6} mA/cm^2 for Ge/Si QDSCs. Thus, the enhancement of V_{OC} is a result of the reduction of I_0, indicating suppression of recombination losses due to strain compensation. However, the difference in I_{SC} between Ge/$Si_{1-x}C_x$ and Ge/Si QDSCs was negligibly small, since the radiative recombination via quantum states in Ge QDs may play a dominant role in carrier recombination at short circuit conditions.

4 SUMMARY

We fabricated 50-layer-stacked heterojunction Ge/$Si_{1-x}C_x$ QDSCs via SS-MBE. The formation of coalesced islands was effectively suppressed via strain-compensation techniques. The EQE response extended up to 1100 nm due to the contribution of Ge QD layers compared with that of the p^+-Si cell. From the applied external bias voltage dependence on the I_{SC} calculated from the EQE, a smaller increase in I_{SC} was observed for Ge/$Si_{1-x}C_x$ QDSCs in the reverse applied bias region than that for Ge/Si QDSCs. Moreover, a reduced I_0 of 5.48 x 10^{-7} mA/cm^2 was observed for Ge/$Si_{1-x}C_x$ QDSCs. These results suggest that the strain compensation effectively reduced the non-radiative recombination centers, although the radiative recombination via quantum states in Ge QDs may play a dominant role in carrier recombination at short circuit conditions. Thus, the V_{OC} was improved from 0.38 V for the Ge/Si QDSC to 0.42 V for the Ge/$Si_{1-x}C_x$ QDSC. This initial demonstration of Ge/$Si_{1-x}C_x$ QDSCs is an important step toward the development of highly efficient, cost-effective Si-based QDSCs.

ACKNOWLEDGEMENTS

This work was supported by the New Energy and Industrial Technology Development Organization (NEDO) under the Ministry of Economy, Trade, and

Industry (METI). One of the authors (K. G) was supported by Research Fellowships of Japanese Society for Promotion of Science (JSPS) for Young Scientists (Grant-in-Aid for Fellows: grant number 26010268).

References

[1] A. Luque and A. Martí, Adv. Mater. **22** (2010) 160-174.
[2] K. Yoshida, Y. Okada and N. Sano, J. Appl. Phys. **112** (2012) 084510.
[3] A. I. Yakimov, N. P. Stepina, A. V. Dvurechenskii, A. I. Nikiforov and A. V. Nenashev, Semicond. Sci. Technol. **15** (2000) 1125-1130.
[4] S. Fukatsu, H. Sunamura, Y. Shiraki and S. Komiyama, Appl. Phys. Lett. **71** (1997) 258-260.
[5] Vinnh Le Thanh, V. Yam and P. Boucaud, F. Fortuna, C. Ulysse, D. Bouchier, L. Vervoort and J.-M. Lourtioz, Phys. Rev. B 60 (1999) 5851-5857.
[6] D. Alonso-Álvarez, J. M. Ripalda, B. Alén, J. M. Llorens, A. Rivera and F. Briones, Adv. Mater. **23** (2011) 5256-5261.
[7] R. Oshima, A. Takata and Y. Okada, Appl. Phys. Lett. **93** (2008) 083111.
[8] S. M. Hubbard, C. D. Cress, C. G. Bailey, R. P. Raffaelle, S. G. Bailey and D. M. Wilt, Appl. Phys. Lett. **92** (2008) 123512.
[9] K. Akahane, N. Yamamoto and T. Kawanishi, Phys. Status. Solidi. A **2** (2011) 425-428.
[10] K. Gotoh, R. Oshima, T. Sugaya, I. Sakata, K. Matsubara, M. Kondo, J. Cryst. Growth **425** (2015) 167.
[11] W. Kern, J. Electrochem. Soc. **137** (1990) 1887-1892.
[12] T. Tasyagaki, N. Usami and Y. Kanemitsu, J. Jpn. Appl. Phys. **51** (2012) 10NE24.
[13] W. G. J. H. M. van Sark, L. Korte and F. Roca, "*Physics and Technology of Amorphous-Crystalline Heterostructure Silicon Solar Cells*", (Springer, 2012), Chapter 1.
[14] J. Tersoff, C. Teichert and M. G. Lagally, Phys. Rev. Lett. **76** (1996) 1675-1678.
[15] R. Oshima, T. Hashimoto, H. Shigekawa and Y. Okada, J. Appl. Phys. **100** (2006) 083110.
[16] K. Akahane, N. Yamamoto, N. Ohtani, Y. Okada, M. Kawabe, J. Cryst. Growth **256** (2003) 7-11.
[17] T. Tayagaki, N. Usami, W. Pan, Y. Hoshi. K. Ooi and Y. Kanemitsu, Appl. Phys. Lett. **101** (2012) 133905.
[18] A. Alguno, N. Usami, K. Ohdaira, W. Pan, M. Tayanagi and K. Nakajima, Thin Solid Films **508** (2006) 402-405.
[19] J. Nelson, "*The Physics of Solar Cells*" (Imperial College Press, 2002), Chapter 1.

THIN GAASSB CAPPING LAYERS FOR IMPROVED PERFORMANCE OF INAS/GAAS QUANTUM DOT SOLAR CELLS

Antonio D. Utrilla,[1] Daniel F. Reyes,[2] José M. Llorens,[3] Alicia Gonzalo,[1] Irene Artacho,[4] Teresa Ben,[2] David González,[2] Žarko Gačević,[1] Álvaro Guzman,[1] Adrián Hierro,[1] and José María Ulloa,[1]
[1)]*Institute for Systems based on Optoelectronics and Microtechnology (ISOM), Universidad Politecnica de Madrid, Ciudad Universitaria s/n, 28040 Madrid, Spain*
[2)]*Departamento de Ciencia de los Materiales e IM y QI, Universidad de Cádiz, 11510 Puerto Real (Cádiz), Spain*
[3)]*IMM-Instituto de Microelectrónica de Madrid (CNM-CSIC), Isaac Newton 8, PTM, E-28760 Tres Cantos (Madrid), Spain*
[4)]*Instituto de Energía Solar, Universidad Politecnica de Madrid, Ciudad Universitaria s/n, 28040 Madrid, Spain*

ABSTRACT: The use of thin GaAsSb capping layers (CLs) is demonstrated to provide InAs/GaAs quantum dot (QD) solar cells (SCs) with improved efficiencies up to 20%. The application of such CLs allows the tunability of the QD ground state, switching the QD-CL band alignment to type II for high Sb contents and extending the photoresponse up to 1.5 µm. Moreover, the CL provides a significant contribution to the photocurrent. An improved carrier collection efficiency is also found to be a main reason behind the increase of the short-circuit current density, unveiling a negative impact of the wetting layer (WL) on carrier transport in standard InAs/GaAs QD SCs. Calculations from an 8×8 k·p method suggest the attribution of such an improvement to longer carrier lifetimes in the WL-CL structure due to the transition to a type-II band alignment in Sb-containing structures. A faster open-circuit voltage recovery under light concentration is demonstrated for high Sb content type-II QD-CL structures, leading, in turn, to faster efficiency improvements with light power.
Keywords: Solar cell, Quantum dots, GaAsSb, Capping layers, Type-II

1 INTRODUCTION

The use of thin GaAsSb capping layers (CLs) has been shown to provide improved characteristics in different optoelectronic devices based on InAs/GaAs quantum dots (QDs) such as QD laser diodes [1-3] and QD infrared photodetectors [4-5]. However, this approach has barely been applied to the field of solar cells (SCs) [6]. The application of alloyed CLs can be exploited as to its ability to modify and thereby extend the QD ground state to longer wavelengths, so that a broadened absorption spectrum is attainable in a QDSC [7]. In particular, Sb affects mainly the valence band of GaAs [8], so the GaAsSb CL reduces the hole confinement energy in the QDs [9]. Moreover, the GaAsSb CL could assist the escape of holes through a two-step mechanism [9], facilitating carrier extraction from the nanostructures. The use of higher amounts of Sb in the CL should lead to a further extended absorption edge, while switching to a type-II QD-CL band alignment for Sb contents above 17%, with the holes localized outside the QD in the CL. This is predicted to provide additional benefits [10]. Indeed, the subsequent separation of both carriers in real space and the resulting longer radiative carrier lifetime is expected to be beneficial for a more efficient extraction. Moreover, recent findings make type-II structures also promising for concentration photovoltaics. Sunlight concentration has been found to suppress the recombination activity in type-II GaSb/GaAs QDs [11], as well as to provide a fast recovery of the V_{oc} in type-II quantum rings [12].

The purpose of the present work is focused on the attainment of improved SC characteristics through the introduction of thin GaAsSb CLs in InAs QDSCs using both type-I and type-II QD-CL structures. Through this work, we also show the relevance of the modified WL-CL structure, which presents a type-II band alignment from low Sb contents. This is suggested to mitigate the negative impact of the WL, inherent to the Stranski-Krastanov growth mode, on carrier transport in standard InAs/GaAs QDSCs.

2 METHODS

The set of samples under study consists of three p-i-n junction structures grown by MBE on n+ GaAs (001) substrates under As_4-stabilized conditions. All the samples contain 10 QD layers grown at 450 °C by depositing 2.8 MLs of InAs at low growth rate. The QD layers were covered by a 20 ML-thick $GaAs_{1-x}Sb_x$ CL grown at 470 °C, with nominal Sb contents of 0, 10, and 20% (samples Sb_0, Sb_L and Sb_H, respectively). The QD-CL structures were separated by a spacer consisting of a 2.5 nm-thick GaAs layer grown immediately after the CL deposition at the same temperature, in order to avoid Sb desorption, followed by a 30 nm-thick GaAs layer grown at higher temperature (580 °C), to avoid the formation of extended defects [13]. The 10 QD layers were embedded in the middle of a 400 nm-thick intrinsic GaAs layer. The whole active region was grown after the deposition of a 500 nm-thick n-GaAs buffer and followed by an emitter layer consisting of a 350 nm-thick p-GaAs layer followed by a 150 nm-thick p^+-GaAs layer.

The structural properties were analyzed by conventional transmission electron microscopy (TEM) techniques, using a JEOL JEM-2100 LaB6 microscope operating at 200 kV, equipped with an Oxford INCA Energy 2000 electron dispersive X-ray (EDX) spectrometer.

200 µm-diameter solar cell test devices were processed by mesa etching and standard fabrication techniques, using AuGe/Au and Au/AuZn/Au for the n- and p-type contacts, respectively. AM 1.5D measurements were conducted using a solar simulator equipped with a 150W Xenon Lamp and an AM 1.5D filter combination (Newport), calibrated to a total light intensity of 90 mW/cm². A Keithley 2400 spectrometer was used as sourcemeter for photocurrent (PC) density-

voltage (J-V) measurements. PC measurements were also carried out under monochromatic illumination from a quartz halogen lamp dispersed through a 0.34 m-monochromator, using a Keithley 230 voltage source and a Keithley 617 electrometer.

3 RESULTS

Figure 1 (a) and (b) show dark field 002 TEM images of representative views containing the ten stacked QD layers in both Sb_I and Sb_{II}, respectively. By this chemically sensitive imaging mode, the addition of Sb into GaAs results in a brighter contrast, so that the CL can be well distinguished from the surrounding material. This reveals a flat growth front in Sb_I all through the structure, despite the introduction of the CL. However, the further accumulation of strain in Sb_{II} leads to a progressively uneven growth front, giving rise to the formation of extended defects. Indeed, no extended defects were found in Sb_I using 220 bright-field TEM imaging mode (Figure 1 (c)), while dislocations were found to be present in some extent in Sb_{II} (Figure 1 (d)). A CL thickness close to 5.5 nm was measured in both Sb_I and Sb_{II}, near the nominal thickness. The Sb contents, determined by EDX, were found to be 10 and 24% in Sb_I and Sb_{II}, respectively.

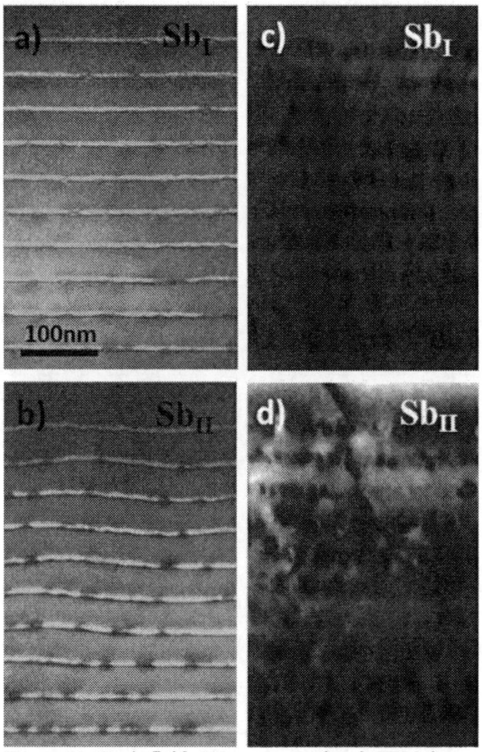

Figure 1: Dark-field 002 cross-sectional TEM images containing representative global views of the ten stacked QD layers in samples Sb_I (a) and Sb_{II} (b). GaAsSb CLs show a brighter contrast with respect to the surrounding material. (c) and (d) are 220 bright-field TEM images from similar regions in Sb_I and Sb_{II}, respectively, lighting up the presence of extended defects in Sb_{II}.

Room-temperature PC from these samples is shown in Figure 2 (top). The reference sample Sb_0, where no GaAsSb CL was employed, clearly reveals the contribution from the standard InAs QDs and the WL to the PC. The introduction of the GaAsSb CL in Sb_I extends the absorption edge to longer wavelengths as a consequence of the modified QD ground state. A strong QW-like contribution from the CL is also observed, the structure acting as a hybrid QD-QW SC. Regarding Sb_{II}, the higher Sb content further extends the photoresponse, up to 1.5 μm, while providing an enhanced PC from the CL. J-V curves under standard AM 1.5D conditions are shown in Figure 2 (bottom). The use of the GaAsSb CL in sample Sb_I induces a small reduction of 0.033 mV in the V_{oc} with respect to Sb_0 (V_{oc} = 0.585 V). Nevertheless, the J_{sc} is increased in almost a 15%. In Sb_{II}, the V_{oc} is further decreased in 0.059 V, while J_{sc} remains virtually unaltered, as compared with Sb_I. Hence, Sb_I and Sb_{II} exceed the efficiency of Sb_0 in a 20 and a 10%, respectively. Remarkably this improvement takes place despite the significant introduction of additional accumulated strain in the Sb-containing SCs.

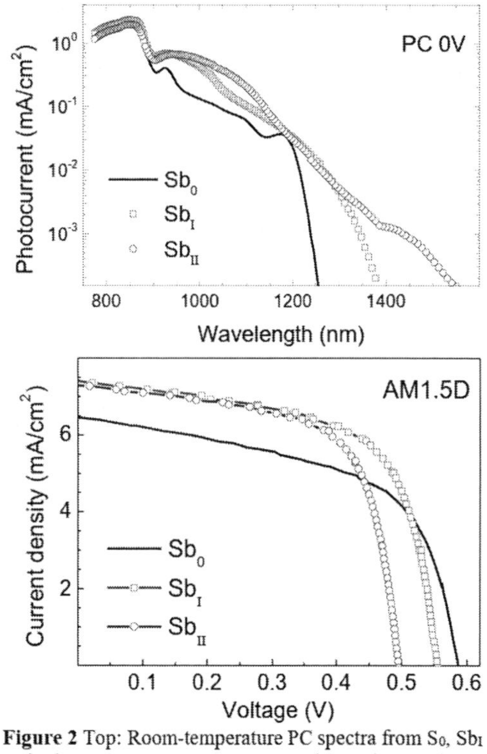

Figure 2 Top: Room-temperature PC spectra from S_0, Sb_I and Sb_{II} at 0V; Bottom: Current-voltage characteristics under 900 W/m^2 AM 1.5D conditions of Sb_0, Sb_I and Sb_{II}.

The apparent improvement of the PC in Sb_{II} with respect to Sb_I contrasts with the similar J_{sc} observed in Figure 2 (bottom). In fact, the current photogenerated from GaAs has such a significant weight in the overall PC that relatively small variations have an appreciable impact on the final device performance. This is observed in Figure 3 (left), showing a zoom of the GaAs-related PC under photovoltaic conditions in all the samples. A slight increase of the PC in Sb_I can be now well distinguished, while it decreases in Sb_{II} below that of Sb_0. The higher GaAs-related PC in Sb_I indicates that the presence of InAs QDs yields some mechanism hindering carrier collection, which is partially inhibited by the introduction of the GaAsSb CL. This suggests that the sole presence of a thin WL might be affecting carrier transport through the structure, as it has already been proposed in other devices [5], and that the modified band structure exhibited by the WL-CL structure leads to a more efficient collection. PC under reverse bias was also measured for the same range of wavelengths (Figure 3 (right)). These measurements were carried out at -3 V, where I-V measurements under monochromatic illumination with a light energy above the GaAs bandgap reveal a saturation current in all the samples and, therefore, a complete carrier collection [14]. The increase of PC in Sb_0 under reverse bias supports our hypothesis that the sole introduction of InAs QDs is enough to affect carrier transport (no PC enhancement is observed in a reference GaAs SC). The lower increase of the PC in Sb_I (13% in Sb_I vs. 21% in Sb_0) confirms a more efficient carrier collection by the introduction of the GaAsSb CL. Also PC from Sb_{II} increases, closely approaching that of Sb_I.

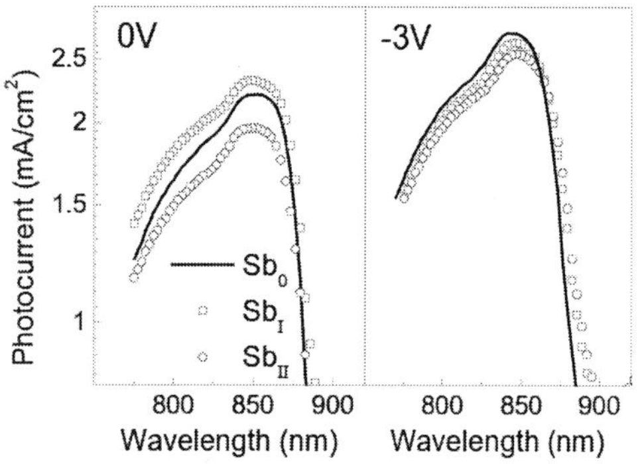

Figure 3: Zoom of the GaAs-related PC from Sb_0, Sb_I and Sb_{II} at 0V (left) and -3V (right).

To confirm the impact of the WL on the SC performance, the WL-CL structure has been theoretically modelled for all the samples considering the data obtained by TEM measurements. The Nextnano++ package was used to solve the Schrödinger equation and determine the electronic structure, estimating a ground state for the WL-CL in good agreement with the relative absorption edges observed in the PC in all the samples (Figure 4). In addition, the WL-CL structure shows a type-II band alignment from low Sb contents leading to increasing radiative lifetimes with the Sb content. This would reduce carrier recombination in the WL and therefore improve collection efficiency, as observed above in Sb_I. The lower carrier collection in Sb_{II}, could be ascribed to the deeper wells introduced for holes or, more likely, to the presence of extended defects.

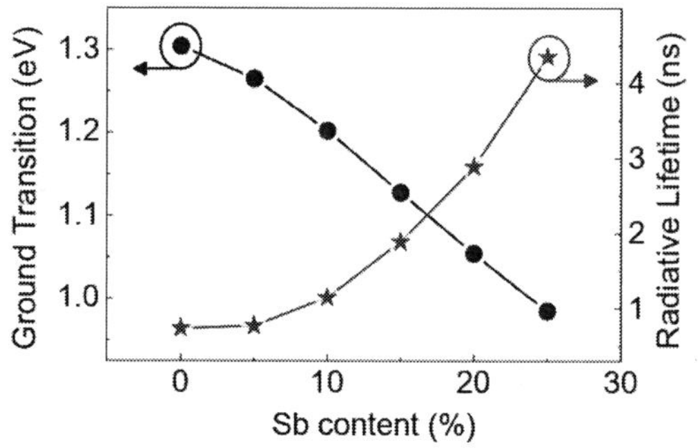

Figure 4: Dependence of the ground state transition and the radiative lifetime in the WL-CL structure with the Sb content.

The impact of light concentration on V_{oc} has also been analyzed using a He-Ne laser. Figure 5 (top) shows the evolution of the monochromatic V_{oc} as a function of the light power density. Continuous lines are linear fits, revealing slopes of constant ideality factors whereby V_{oc} recovers, while dashed lines are the expected V_{oc} evolution according to the dark ideality factors (n_{dark}) deduced from the dark current-voltage curves. Close ideality factors of 1.38 and 1.36 are found in Sb_0 and Sb_I, respectively, from the linear fits to the power-V_{oc} evolution, while it increases up to 1.84 in Sb_{II}. This value in Sb_{II} significantly differs from n_{dark}, so that the V_{oc} follows a recovery tendency faster than expected with light power in this case. For the highest light power, V_{oc} of Sb_{II} (0.716 V) is virtually reaching that of Sb_I (0.722 V), tending to exceed it for light intensities higher than those reachable by the used setup. Indeed, the highest light power density is below 10 W/cm^2, still leaving room for additional improvements under higher concentrations. The almost negligible difference of V_{oc} in Sb_{II} and Sb_I means that light concentration is avoiding the negative impact that the introduction of lower effective bandgaps within the host material may exert on it. Effects related to a strong localization of holes in type-II quantum rings has also been recently found to lead to a fast V_{oc} recovery [12]. Although this might be also happening in Sb_{II}, this only factor could not explain the achieved V_{oc}, very close to that of Sb_I. A faster Quasi-Fermi-level splitting might be behind this, induced by the saturation of the QD ground state with light power, due to the much longer radiative lifetimes in the type-II structures. Figure 4 (bottom) shows the evolution of the SC efficiency as a function of the incident light power, normalized to the value for the lower intensity. A faster efficiency enhancement is found in Sb_{II}, following the V_{oc} evolution with light power. This makes these type-II QD structures promising candidates for concentration photovoltaics, demonstrating the possibility to achieve efficiencies up to at least 16% higher as compared to the reference QD sample Sb_0.

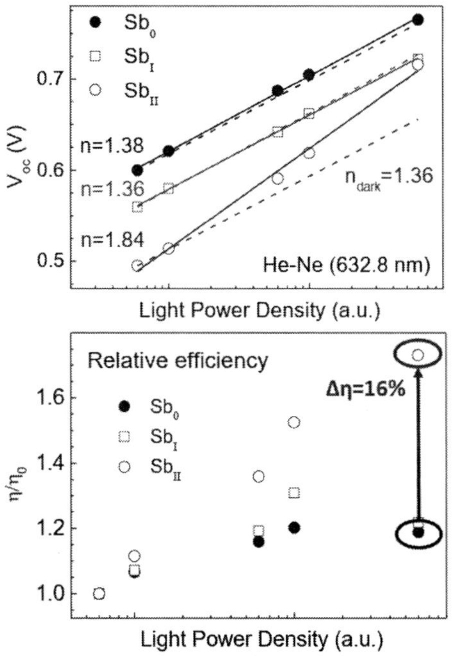

Figure 4 Top: The symbols show the V_{oc} evolution, in Sb_0, Sb_I and Sb_{II}, under He-Ne laser illumination as a function of the light power density. Solid lines are linear fits to the measured data. Dashed lines show the expected evolution for V_{oc} according to the ideality factors deduced from the dark IV curves; Bottom: relative efficiency increase with light power density.

4 CONCLUSIONS

In conclusion, we demonstrate the ability to improve QDSC performance by the application of thin GaAsSb CLs. Such CLs allows the tunability of the InAs QD ground state extending the absorption edge up to 1.5 µm, while providing an additional contribution to the PC. Although high Sb contents yield a reduction of the V_{oc}, the pronounced type-II structure formed in the WL-CL significantly increases carrier radiative lifetimes, reducing the trapping of drifting carriers in the WL and leading to a more efficient carrier collection. A faster V_{oc} recovery under light concentration is also found for high Sb content type-II QD-CL structures, reaching at least a similar value as that for the sample with a lower Sb content. The faster V_{oc} recovery leads, in its turn, to a faster efficiency enhancement. These results suggest the possibility to further improve the SC performance by the combination of GaAsSb CLs with strain-balancing techniques and points to the possibility to exploit this approach for light-concentration photovoltaics.

5. ACKNOWLEDGMENTS

We acknowledge the Spanish MICINN-MINECO for funding through project MAT2013-47102-C2-2-R. José M. Llorens and Irene Artacho acknowledge the financial support of the Comunidad Autónoma de Madrid (Grant S2013/MAE-2780) and MINECO (Grant AIC-B_2011-0806), respectively. We would also like to thank Alicia Fraile (ISOM-UPM) for processing the samples.

6 REFERENCES

[1] H.Y. Liu, M.J. Steer, T.J. Badcock, D.J. Mowbray, M.S. Skolnick, P. Navaretti, K.M. Groom, M. Hopkinson, R.A. Hogg, Appl. Phys. Lett. 86 (2005) 143108.

[2] K. Akahane, N. Yamamoto, S.-I. Gozu, N. Ohtani, Physica E 26 (2005) 395.

[3] A.D. Utrilla, J.M. Ulloa, A. Guzman, A. Hierro, Appl. Phys. Lett. 103 (2013) 111114.

[4] C.-T. Huang, Y.-C. Chen, S.-C. Lee, Appl. Phys. Lett. 100 (2012) 043512.

[5] A. Guzmán, K. Yamamoto, J.M. Ulloa, J.M. Llorens, A. Hierro, Appl. Phys. Lett. 107 (2015) 011101.

[6] W.S. Liu, H.M. Wu, F.H. Tsao, T.L. Hsu, J-I. Chyi, Sol. Energy Mater. Sol. Cells 105 (2012) 237.

[7] A.D. Utrilla, J.M. Ulloa, Ž. Gačević, D.F. Reyes, I. Artacho, T. Ben, D. González, A. Hierro, A. Guzman, Sol. Energy Mater. Sol. Cells 144 (2016) 128.

[8] R. Teissier, D. Sicault, J. Harmand, G. Ungaro, G. Le Roux, L. Largeau, J. Appl. Phys. 89 (2001) 5473.

[9] J. M. Ulloa, J. M. Llorens, M. del Moral, M. Bozkurt, P. M. Koenraad, A. Hierro, J. Appl. Phys. 112 (2012) 074311.

[10] A. Luque, P. G. Linares, A. Mellor, V. Andreev, A. Marti, Appl. Phys. Lett. 103 (2013) 123901.

[11] A. M. Kechiantz, L. M. Kocharyan, H. M. Kechiyants, Nanotechnology 18 (2007) 405401.

[12] H. Fujita, P. J. Carrington, M. C. Wagener, J. R. Botha, A. R. J. Marshall, J. James, A. Krier, K.-H. Lee, N. J. Ekins-Daukes, Prog. Photovolt. 12 (2015) 1896.

[13] S.M. Willis, J.A.R. Dimmock, F. Tutu, H.Y. Liu, M.G. Peinado, H.E. Assender, A.A.R. Watt, I.R. Sellers, Sol. Energy Mater. Sol. Cells 102 (2012) 142.

[14] H. Fujii, Y. Wang, K. Watanabe, M. Sugiyama, Y. Nakano, IEEE J. Photovolt. 3 (2013) 859.

INFLUENCE OF THE QUANTUM DOT CAPPING PROCEDURE ON THE DENSITY OF DEFECTS IN INAS/GAAS QUANTUM DOT INTERMEDIATE BAND SOLAR CELLS

D. N. Micha[1,2], E. C. Weiner[2,3], R. Jakomin[2,4], L. D. Pinto[2,3], M. P. Pires[2,5] and P. L. Souza[2,3]

[1]Coordenação de Licenciatura em Física, CEFET/RJ, Petrópolis - RJ, 25620-003, Brazil
[2]DISSE-Instituto Nacional de Ciência e Tecnologia de Nanodispositivos Semicondutores, PUC-Rio, 22451-900, Brazil
[3]LabSem/CETUC, Pontifícia Universidade Católica, Rio de Janeiro, 22451-900, Brazil
[4]Campus de Xerém, Universidade Federal do Rio de Janeiro, Duque de Caxias, 25245-390, Brazil
[5]Instituto de Física, Universidade Federal do Rio de Janeiro, Rio de Janeiro, 21945-970, Brazil

In this work, the influence of different growth conditions of the cap layer of quantum dots on the figures of merit of Intermediate Band Solar Cells is investigated. The optimized quantum dot capping procedure led to improvements in all figures of merit, namely by 7.7% in short circuit current density, 41% in open circuit voltage, 29% in fill factor and 85% in efficiency, when the temperature of the In flush process increases from 630 to 700 °C. We attribute the improvement in the devices' performance to a lower density of Shockley-Read-Hall recombination centers, as the analysis of the dark current points out. Furthermore, the short circuit current density raises by 6.9% when compared to the control solar cell (without quantum dots) due to subbandgap absorption.
Keywords: Photovoltaics, Intermediate Band Solar Cells, Quantum Dots, Capping Layer.

1 INTRODUCTION

Intermediate Band Solar Cells (IBSC) [1] are under the scrutiny of the scientific community. To date, the devices fabricated under this concept have not yet fully proved to be advantageous, as theoretically predicted.

Nonetheless, several attempts using the electronic quantum dots' (QD) confined states as intermediate levels to promote optical transitions of subbandgap energies have been tested. In some cases, figures of merit at least equal to those of equivalent solar cells without the intermediate levels were reached [2, 3].

The three key points missing so that the QD based IBSC can reach the predicted theoretical figures of merit are: 1) the increase in volume of QDs available for subbandgap absorption, which would guarantee the use of more low energy photons to increase the short circuit current density (J_{SC}), 2) the right choice of materials to match the most suitable subbandgap transitions, which would help in several critical points, such as the isolation between the intermediate band and the conduction band, as well as the reduction of loss by thermalization, 3) the better control of the density of defects produced during the QDs' formation, which would assure a higher open circuit voltage (V_{OC}). [4]

In this work, we concentrate on the third point mentioned in the last paragraph, bringing up some relevant and novel issues. We investigate the influence of the growth conditions of the capping layer of the QDs on the electrical parameters and figures of merit of the IBSC. In order to do so, several different QD cap layer deposition conditions were employed concerning the source of arsenic and its flow and the final temperature of an In flush process. Finally, we show significant improvements in all the solar cells' figures of merit and attribute that to a better control of the density of defects formed during the stacking of QD layers.

2 EXPERIMENTAL DETAILS

InAs/GaAs QD-IBSCs were produced by Metal Organic Vapor Phase Epitaxy according to the structure shown in Figure 1. The following layers were grown lattice matched to the GaAs n-type doped substrate at 630 °C and 100 mbar: 100 nm n-type $Al_{0.3}Ga_{0.7}As$ back surface field (BSF), 300 nm n-type GaAs base, 100 nm p-type GaAs emitter, 30 nm p-type $Ga_{0.51}In_{0.49}P$ window

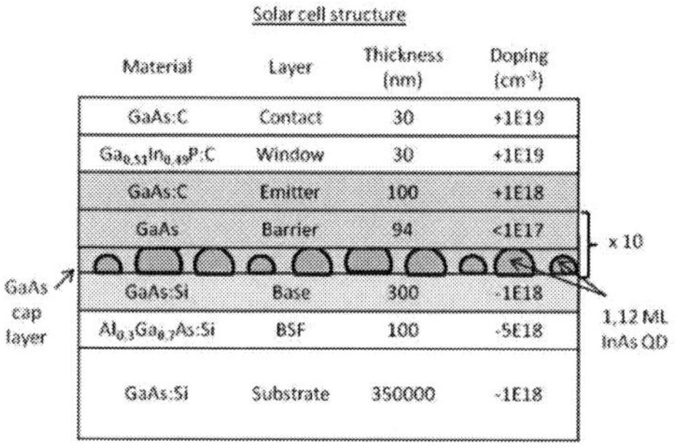

Figure 1: Scheme of the investigated QD-IBSC structure.

and 30 nm p-type GaAs contact layers using TMAl, TMGa, TMIn, PH_3 and AsH_3 as precursors, CBr_4 and DEZn as *p*-type dopants and SiH_4 as *n*-type dopant. Ten layers of 1.12 ML *n*-type doped QDs with a density around 1×10^{10} cm^{-2} were grown at 490 °C using AsH_3 and TMIn as precursors and annealed for 12 s at the same temperature. The optimal growth conditions for the deposition of QDs were published elsewhere [5, 6]. To cap the QDs, the In flush technique [7] was used, where a GaAs layer of 3 nm was deposited followed by an annealing treatment at two different temperatures. This cap layer was grown under a temperature ramp of 0,9 °C/s up to 500 °C with different flows for AsH_3 (F_A). After the growth of the capping layer, the temperature was raised up to a certain value (T_F) to flush out the top of the QDs which were higher than the cap layer thickness of 3 nm. A 94 nm thick intrinsic GaAs barrier (spacer) layer was then deposited, maintaining the same temperature. The different growth conditions (F_A and T_F) were as follows: sample 630-20 – F_A = 20 sccm and T_F = 630 °C, sample 700-20 – F_A = 20 sccm and T_F = 700 °C, and sample 700-200 – F_A = 200 sccm and T_F = 700 °C. For the reference sample, in place of the QDs, the equivalent thickness of GaAs was grown at 630 °C.

The devices were processed with a conventional photolithography technique in solar cells of 10,25 mm^2. The obtained solar cells were characterized by standard techniques, such as J x V curves under AM1.5g light spectrum, dark J x V curves and spectral response under bias light.

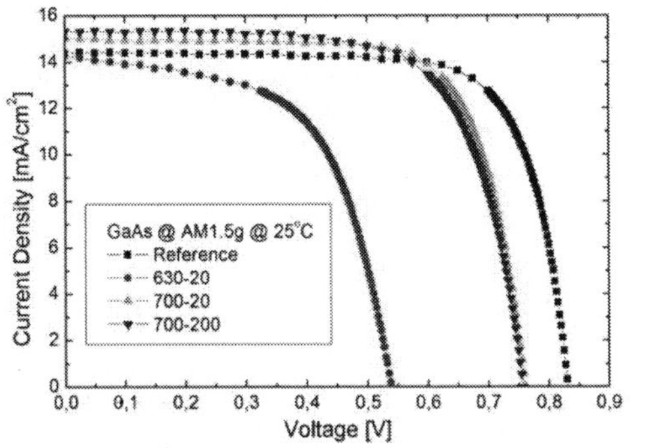

Figure 2: Light IV curves obtained at 25°C under AM1.5g insolation.

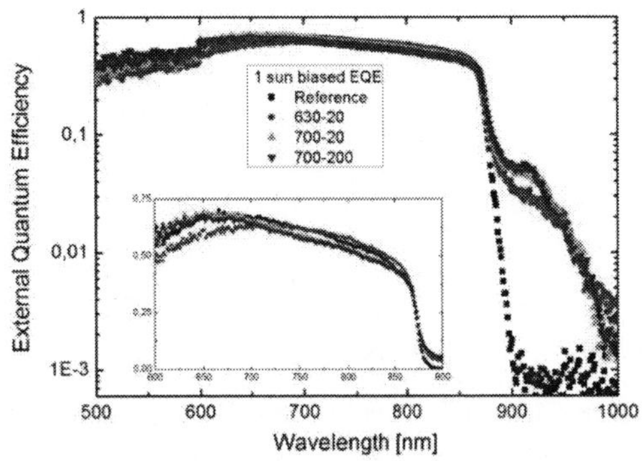

Figure 3: External Quantum Efficiency spectra obtained under 1-sun bias light and short circuit conditions. The inset is the same plot but in linear scale.

3 RESULTS AND DISCUSSION

Figure 2 shows the light IV curves for the three IBSC samples and the reference obtained with a solar simulator under AM1.5g spectrum with the device at 25°C. Table I summarizes the figures of merit of such curves. The first result to be highlighted is the huge improvement in the figures of merit of the solar cells when increasing the In-flush temperature from 630 °C to 700°C. Both samples grown at that temperature, 700-20 and 700-200, show an increment of about 2 mA/cm² in J_{SC} and around 200 mV in V_{OC} when compared to sample 630-20. Comparing the same samples to the reference, we see an increase of about 1 mA/cm² in J_{SC} but a decrement of 100 mV in V_{OC}.

TABLE I: FIGURES OF MERIT OF THE SOLAR CELLS

Sample	J_{SC} [mA/cm²]	V_{OC} [mV]	FF [%]	η [%]
Reference	14.4	831	75.1	9.0
630-20	14.3	538	59.8	4.6
700-20	15.0	760	73.3	8.4
700-200	15,3	758	69.6	8.1

The improvement in J_{SC} can be explained by the absorption of subbandgap photons promoting optical transitions between the valence band and the intermediate band, E_H, and from the intermediate band to the conduction band, E_L. The subbandgap absorption is demonstrated by the extension of the external quantum efficiency (EQE) curves towards wavelengths longer than 870 nm, corresponding to the GaAs bandgap, as shown in Figure 3. This is a clear sign of the optical activity of the QD. On the other side, the lowering of EQE for sample 630-20 in the short wavelength range indicates a poorer material quality in the emitter layer. We attribute this fact to the creation of a dislocation field introduced by the stacking of QD layers with low quality due to the low temperature of the In-flush process. One possible explanation is the incomplete evaporation of the quantum dots taller than the cap layer. This would sustain the presence of a group of incoherent nanostructures, responsible for the generation of the dislocation field.

The reduction in V_{OC} is directly related to the creation of recombination centers by the introduction of QDs. Figure 4 presents the dark current curves for the IBSC samples and the reference from which one can infer the material quality of the solar cells. To discern the major contributions for the dark current, we have used the two diode model with parasitic resistances, expressed in Equation 1, to fit the curves (dashed lines in figure 4) and extract the magnitude of each source of recombination. The values for the electrical parameters are expressed in Table II.

$$I(V) = I_{RAD}\left[e^{\frac{q(V+R_S I)}{kT}} - 1\right] + I_{SRH}\left[e^{\frac{q(V+R_S I)}{2kT}} - 1\right] + \left(\frac{V + R_S I}{R_P}\right) \quad (1)$$

I_{RAD} and I_{SRH} are the radiative and Shockley-Read-Hall (SRH) contributions for the dark current, and R_S and R_P are the series and shunt resistance, respectively.

TABLE II: ELECTRICAL PARAMETERS EXTRACTED FROM THE DARK IxV CURVES

Sample	J_{Rad} [10^{-10} mA/cm²]	J_{SRH} [10^{-3} mA/cm²]	r_S [10^{-1} Ωcm²]	r_P [10^{2} Ωcm²]
Reference	0.09	0.4	4.6	9.0
630-20	< 5	200	6.9	0.3
700-20	2	1	4.9	2.1
700-200	0.6	2	5.4	5.7

Figure 4 shows that the performance of all the investigated solar cells are limited by non-radiative recombination process. As can be seen in Table II, sample 630-20 has the highest SRH contribution for the dark current followed by samples 700-200 and 700-20 while the reference cell has the lowest value for this parameter. This result corroborates with the trends of V_{OC}, indicating once more that a careful and optimized growth of the QD layers and their surroundings is a key factor to reduce the detrimental impact that the deposition of QDs can have on the device's performance due to the unavoidable introduction of structural defects.

Worth of noting is the increase in fill factor for samples 700-20 and 700-200 when compared to sample 630-20, which is a clear consequence of the improvement in the shunt resistance of the devices (visible in the increase of the slope of the light IV curve close to J_{SC}). As discussed in Ref. 8, the energy levels introduced by defects in the middle of the bandgap can serve as vertical paths with lower resistance for electrical charges to contribute to the dark current. The improvement in FF can also be attributed to the suppression of the energy levels responsible for leakage of current, indicating a better material quality.

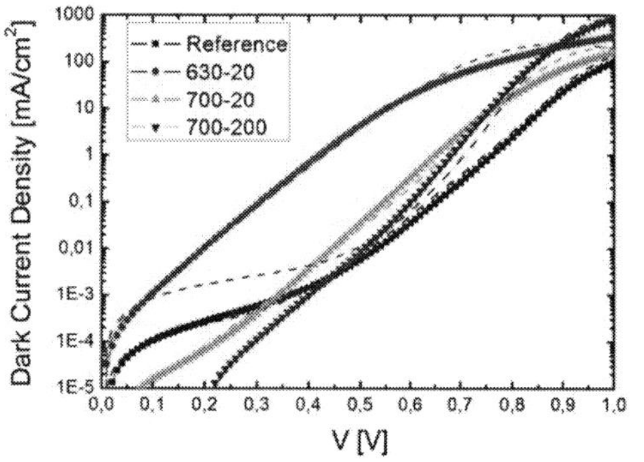

Figure 4: Dark current density *versus* voltage. The dashed lines are fittings using the two diode model including parasitic resistances.

5 CONCLUSION

In conclusion, we showed results of a study on the growth parameters of the capping process of quantum dots in QD-IBSC. As revealed by results obtained with standard characterization techniques, the increase of the In flush temperature of the QD capping layer from 630°C to 700°C has a direct influence on the quality of the subsequent layers in the stack. Therefore, better figures of merit for the final device were obtained, especially for the V_{OC} and the FF. V_{OC} clearly increases due to the reduction in the density of non-radiative recombination centers in the active regions. The enhancement in FF can also be attributed to the same factor, as a lower density of defects in the stack of QD layers leads to a higher shunt resistance. Finally, the effect of increasing the flow of AsH_3 during the growth of the capping layer was not fully understood, further probing being required.

6 ACKNOWLEDGEMENTS

The authors acknowledge the help of Godoy, A. von Zuben and N. Frateschi with the photolithographic process of the samples and of G. Torelly with the electrical measurements. This work was partially supported by FAPERJ, CNPq, FINEP and CAPES.

7 REFERENCES

[1] A. Luque and A. Martí, Physical Review Letters, vol. 78, pp. 5014-5017, 1997.

[2] C. Bailey, D. Forbes, S. Polly, Z. Bittner, Y. Dai, C. Mackos, R. Raffaelle and S. Hubbard, IEEE journal of Photovoltaics, vol. 2, no. 3, pp. 269-275, 2012.

[3] D. Guimard, R. Morihara, D. Bordel, K. Tanabe, Y. Wakayama, M. Nishioka and Y. Arakawa, "Fabrication of InAs/GaAs quantum dot solar cells with enhanced photocurrent and without degradation of open circuit voltage," Applied Physics Letters, vol. 95, pp. 203507-(1-3), 2010.

[4] A. Martí and et al., "Six not-so-easy pieces in intermediate band solar cell research," Journal of Photonics for Energy, vol. 3, pp. 31299 (1-11), 2013.

[5] D. N. Micha, R. Jakomin, M. P. Pires, R. Mourão, R. Kawabata and P. L. Souza, "A proposal for Intermediate Band Solar Cells with optimized transition energy - InAs QD / AlGaAs," in Microelectronics Technology and Devices (SBMicro), Curitiba, PR, Brazil, 2013.

[6] R. Jakomin, R. Kawabata, R. Mourão, D. Micha, M. Pires, H. Xie, A. Fischer, F. Ponce and P. Souza, "InAs quantum dot growth on AlxGa12xAs by metalorganic vapor phase epitaxy for intermediate band solar cells," Journal of Applied Physics, vol. 116, pp. 093511 (1-7), 2014.

[7] Z. R. Wasilewski, S. Fafard and J. P. McCaffrey, "Size and shape engineering of vertically stacked self-assembled quantum dots," Journal of Crystal Growth, Vols. 201-202, pp. 1131-1135, 1999.

[8] P. Lam, J. Wu, M. Tang, D. Kim, S. Hatch, I. Ramiro, V. Dorogan, M. Benamara, Y. Mazur, G. Salamo, J. Wilson, R. Allison and H. Liu, Solar Energy Material and Solar Cells, vol. 144, pp. 96-101, 2016.

OPTIMIZING ABSORPTION IN RESONANT-CAVITY-ENHANCED GERMANIUM NANOABSORBER SOLAR CELLS FOR TANDEM APPLICATION

Volker Steenhoff, Martin Vehse, Carsten Agert
NEXT ENERGY · EWE Research Centre for Energy Technology at the University of Oldenburg
Carl von Ossietzky-Str. 15, 26129 Oldenburg, Germany

ABSTRACT: Strong broadband absorption can be achieved in solar cells with only 25 nm thick amorphous germanium nanoabsorbers by taking advantage of the optical resonances generated between the electrodes of the devices. Two different configurations of such resonant-cavity-enhanced a-Ge:H solar cells have been investigated: When exploiting lowest-order resonances, broadband absorption enhancement is obtained, enabling a short circuit current density of more than 20 mA cm^{-2} in single cells and 12 mA cm^{-2} in tandem devices. However, this requires the use of an extremely thin resonator and hence thin doped layers. This complicates electrical optimization, as too thin doped layers lead to poor charge carrier extraction. The approach presented here aims to solve this problem by specifically designing devices for application in a-Si:H/a-Ge:H tandem cells. In this case, only near-infrared light needs to be absorbed. This allows utilization of longer cavities producing second-order resonances inside the relevant spectral range. As a result, thicker doped layers can be used, which improves the electrical device characteristics. In this contribution, we identify basic design rules to optimize the infrared absorption of such devices.
Keywords: PV Materials, Multijunction Solar Cell, a-Si:H/μc-Si:H, Light Trapping

1 INTRODUCTION

In silicon-based thin-film multijunction solar cells, a few hundred nanometer thick amorphous silicon (a-Si:H) absorber is generally combined with a microcrystalline silicon (μc-Si:H) bottom cell [1–4]. However, micrometer-thick bottom absorber layers are generally required to achieve current-matching, as the absorption coefficient of the latter material is significantly lower compared to that of a-Si:H.

To enable a thickness reduction and hence throughput increase, we investigate the application of amorphous germanium (a-Ge:H) as alternative bottom cell absorber material. The use of this material for application in electro-optics, and more specifically in solar cells, has been studied 20 years ago already [5–8]. However, high defect densities and the poor charge carrier transport properties resulting from this up to now hindered the fabrication of highly efficient devices.

We recently demonstrated that this problem can be partly circumvented by combining the concept of resonant-cavities with extremely thin a-Ge:H nanoabsorbers. We showed that a broadband absorption enhancement can be obtained in a-Ge:H "nanocavity" solar cells [9]. These are devices that support lowest-order resonance for wavelengths λ inside the spectral range relevant for photovoltaic operation using cavity thicknesses significantly below $\lambda/4$. Using an only about 10-20 nm thick a-Ge:H absorber layer, a short circuit current density (j_{SC}) of 20 mA cm^{-2} or more could be reached. Furthermore, we demonstrated the integration of this concept into a-Si:H/a-Ge:H tandem solar cells, showing that a bottom cell quantum efficiency comparable to μc-Si:H devices is indeed possible [10].

One drawback of this concept is that such nanocavity solar cells require very thin doped layers to enable an optimal resonance position. This complicates electrical optimization of the devices, as too thin doped layers lead to poor charge carrier extraction [11].

However, broadband absorption inside the entire solar spectrum is only required in single junction solar cells. When designing a-Si:H/a-Ge:H tandem devices, the a-Si:H top cell already absorbs the short wavelength part of the spectral range. Hence, it is sufficient if the a-Ge:H bottom solar cell absorbs near-infrared light only. This allows utilization of longer cavities producing second-order resonances inside the relevant spectral range. This means that the doped layers can be optimized more easily.

Using this approach, we were able to improve the fill factor to 64%, leading to an increase in single cell efficiency to 4% [12]. This is in spite of a significantly lower j_{SC} of 17 mA cm^{-2}, which is a result of the narrowband nature of the resonances. However, the actual potential of the devices for application in tandem solar cells does not depend on the single cell j_{SC} but only on the photocurrent generated from light in the near-infrared spectral range. In this contribution, we perform an optimization of this quantity and identify basic design rules for the optimization process.

2 METHODS

2.1 Simulation

1d optical simulations of the optical field profile depicted in Figure 1b were performed using the commercially available software package Scout/Code. Light inside the ITO front electrode has been simulated in incoherent mode. Refractive index data were either taken from literature (silver [13]), obtained by ellipsometric spectroscopy and subsequent modeling (a-Si:H), or were taken from the database provided with Scout/Code (ZnO and ITO).

2.2 Cell Fabrication

Flexible aluminum sheets with a pre-deposited lacquer and silver layer (provided by Alanod GmbH, Germany) [14] were utilized as substrates and coated with an 80 nm thick DC magnetron sputtered aluminum zinc oxide (AZO) film. The a-Ge:H solar cells were fabricated in n-i-p substrate configuration with intrinsic buffer layers separating the doped silicon layers from the a-Ge:H absorber (i.e. a-Si:H(n) / a-Si:H(i) / a-Ge:H(i) / μc-Si:H(i) / μc-Si:H(p)). A constant a-Ge:H absorber thickness of 25 nm was used for all devices. The p-buffer had a thickness of about 10 nm. All other layers were adjusted to yield the cavity length and absorber position desired for the experiments, while keeping the thicknesses of the n-layer and of the n-side buffer roughly

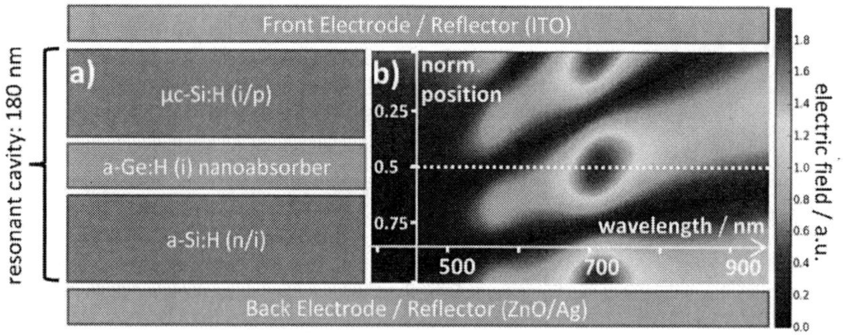

Figure 1: RCE a-Ge:H nanoabsorber solar cell: a) Sketch of the cell stack. b) Simulated optical field profile generated inside the cavity if no absorber is present (180 nm a-Si:H layer between the reflecting electrodes); the dotted line represents the geometric center of the cavity.

equal. Subsequently, an 80 nm thick DC magnetron sputtered ITO front electrode was deposited. All layers were fabricated using a cluster tool type CS400PS manufactured by Von Ardenne GmbH Dresden. Additional details on the fabrication processes can be found elsewhere [9,10].

Individual solar cells were separated using laser-scribing. Quantum efficiency (QE) spectra were determined from differential spectral response measurements at short circuit conditions.

3 RESULTS AND DISCUSSION

A simplified sketch of the cell stack under investigation is depicted in Figure 1a. While germanium and silicon both exhibit a refractive index around 4, a high refractive index jump typically occurs at the silicon/electrode interfaces. This means that a significant fraction of light is reflected at these interfaces, so that the electrodes can be considered as the (partly) reflecting mirrors of a Fabry-Pérot type cavity. By increasing the thickness of the silicon layers, the cavity length can be tuned to produce a second-order resonance for wavelengths corresponding to photon energies above the bandgap of a-Ge:H. As mentioned before, by utilizing second-order resonances instead of lowest-order ones, thicker doped silicon layers and optimized silicon buffer layers can be used, which enables significant improvement of the electrical device characteristics.

In this case, the cavity length is much higher than the absorber thickness, meaning that in addition to its resonance frequency the relative position of the absorber inside the cavity needs to be optimized. To support this optimization process, the right part of Figure 1 shows the simulated optical field profile building up inside a 180nm thick cavity under conditions where no a-Ge:H absorber is present.

It can be clearly seen that standing waves with an electric field maximum directly adjacent to the back reflector are formed inside the relevant wavelength range. The reason for this is that due to the lower refractive index of ZnO compared to silicon, light undergoes a phase shift near zero when reflected at the ZnO surface. Of course, the major contribution to the reflectance comes from the silver back electrode, which itself introduces a nonzero phase shift. However, the thickness of the ZnO layer is tuned to yield high reflectivity in the relevant spectral range, meaning that partial waves reflected at the silver surface are in phase with those from the silicon/ZnO interface. This translates to an effective overall phase jump near zero.

For a given wavelength λ, the distance between two field maxima is $\lambda/(2n)$, where n is the refractive index inside the cavity. Hence, the above implies that all field maxima are situated at positions which are multiples of $\lambda/(2n)$ away from the back electrode. At the same time, second-order resonance occurs for a cavity length of λ/n. This means that the field maximum corresponding to the resonant wavelength is always situated approximately inside the cavity center. Hence, it can be expected that optimum utilization of the field profile is always reached by placing the a-Ge:H nanoabsorber inside the cavity center.

To validate this, we fabricated RCE a-Ge:H nanoabsorber solar cells with a constant cavity thickness of 180 nm and varying absorber position. A shift of the absorber position from the front side to the back has been realized by increasing the thickness of the p-doped layer and decreasing the thickness of the n-side by the same amount. A constant absorber thickness of 25 nm was used for all samples.

Figure 2 depicts quantum efficiency (QE) results for these devices. All samples show an interference pattern with a QE maximum located at a wavelength of about 780 nm. A second local maximum, corresponding to the third-order resonance, is situated around 550 nm. As expected, the value of the QE at a specific wavelength is highly correlated with the electric field strength calculated for that wavelength at the absorber position.

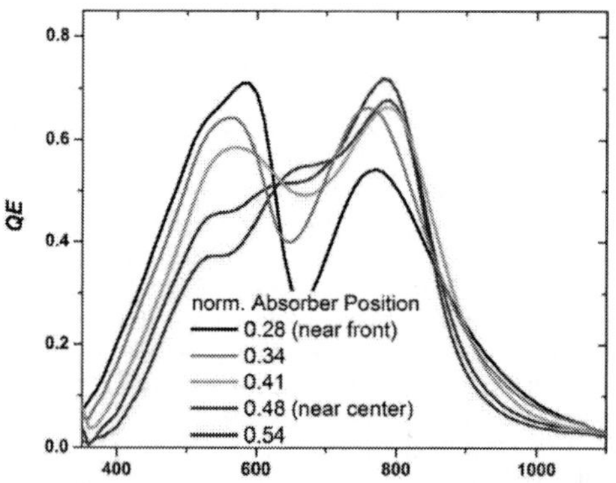

Figure 2: Influence of the absorber position on the QE for a fixed cavity length of 180 nm. Absorber position is given as distance from the front contact and normalized to the cavity length (compare Figure 1b).

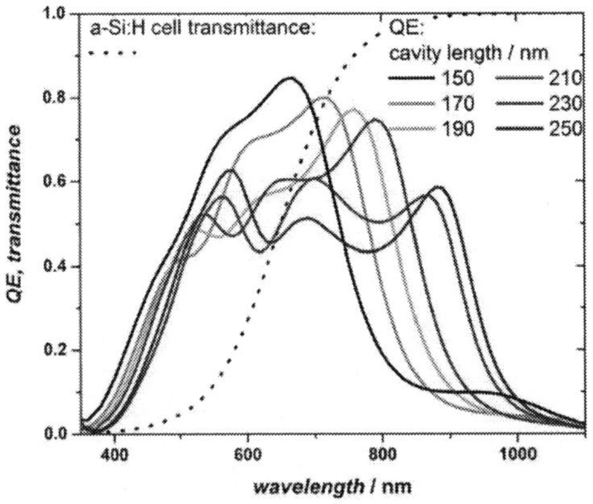

Figure 3: Influence of cavity length on the QE of RCE a-Ge:H nanoabsorber solar cells using a 25 nm thick a-Ge:H absorber layer placed in the cavity center. The approximate transmittance of an a-Si:H top cell is shown for reference (see [15]).

Hence, when shifting the absorber from the front electrode towards the center of the cavity, the peak value of the second-order maximum around 780 nm increases, while the third-order maximum is suppressed. When further increasing the distance of the absorber from the front contact, the second-order maximum decreases again, as the overlap with the electric field maximum in the cavity center is reduced again in this case. The above results confirm the assumption that a centered absorber position optimizes the absorption-enhancement induced by the second-order resonance.

This means that to optimize the absorption for tandem cell applications, it suffices to vary the length of the resonant cavity while keeping the absorber layer in its center. Figure 3 shows QE spectra of a corresponding sample set along with a curve that roughly mimics the light transmission of an a-Si:H top cell yielding a j_{SC} of about 12.5 mA cm^{-2} [15]. As expected, the spectral position of the resonance maximum can be easily tuned by changing the resonator length. Shifting this position to higher wavelengths where an a-Si:H top cell shows high transmittance increases the amount of light that can theoretically be utilized when incorporating the concept in a-Si:H/a-Ge:H tandem devices.

By weighting the QE spectra with the transmittance and convoluting the resulting spectra with the AM1.5g solar spectrum, we calculate the j_{SC} that such a-Ge:H bottom cells would yield. Corresponding results are presented in Figure 4 together with values obtained for the respective single junction cells. The results show that for the single junction case, j_{SC} stays nearly constant for all samples. However, when utilizing the devices as bottom cells, the current density is strongly improved when increasing the resonator length up to a value of 230 nm, as a higher amount of infrared light is absorbed. When shifting the resonance maximum further into the infrared range, the calculated bottom cell j_{SC} starts decreasing again due to a drop of the absorption coefficient of a-Ge:H close to its bandgap energy.

Although a single cell j_{SC} of only about 16 mA cm^{-2} was obtained in our experiments, a maximum estimated bottom cell j_{SC} of about 10 mA cm^{-2} has been reached. Keeping in mind that the transmittance function mimics the behavior of a top cell yielding 12.5 mA cm^{-2}, the above result indicates that the absorption of the devices presented here would suffice to replace a μc-Si:H bottom cell yielding a single cell j_{SC} of at least 22 mA cm^{-2}.

4 CONCLUSION

Our results indicate that resonant-cavity-enhanced a-Ge:H nanoabsorber solar cells with 25 nm thick absorber layers taking advantage of second-order Fabry-Pérot resonances can yield current densities comparable to micron-thick μc-Si:H solar cells, if integrated as bottom devices in tandem cells. In the presented configuration, which uses an ITO front contact and an AZO/Ag back reflector, the nanoabsorber should be placed inside the center of the cavity for optimum current generation, as a maximum of the electric field is produced at this position for light having the resonant wavelength.

ACKNOWLEDGMENTS

We gratefully acknowledge Alanod GmbH for providing the aluminum/lacquer/silver substrates.

REFERENCES

[1] J. Meier, P. Torres, R. Platz, S. Dubail, U. Kroll, J.A.A. Selvan, N.P. Vaucher, C. Hof, D. Fischer, H. Keppner, A. Shah, K.-D. Ufert, P. Giannoulès, J. Koehler, MRS Proc. 420 (1996) 3.
[2] T. Söderström, F.-J. Haug, V. Terrazzoni-Daudrix, C. Ballif, J. Appl. Phys. 107 (2010) 014507.
[3] H. Sai, T. Matsui, T. Koida, K. Matsubara, M. Kondo, S. Sugiyama, Y. Takeuchi, I. Yoshida, Appl. Phys. Lett. 106 (2015) 213902.
[4] S. Kim, J.-W. Chung, H. Lee, J. Park, Y. Heo, H.-M. Lee, Sol. Energy Mater. Sol. Cells. 119 (2013) 26–35.
[5] T.P. Drüsedau, B. Schröder, Solid State Phenom. 44-46 (1995) 425–462.
[6] W. Kusian, E. Günzel, R.D. Plättner, Sol. Energy Mater. 23 (1991) 303–311.
[7] E. V. Johnson, P. Roca i Cabarrocas, Sol. Energy Mater. Sol. Cells. 91 (2007) 877–881.
[8] J. Zhu, V.L. Dalal, M.A. Ring, J.J. Gutierrez, J.D. Cohen, J. Non. Cryst. Solids. 338-340 (2004) 651–654.
[9] V. Steenhoff, M. Theuring, M. Vehse, K. von Maydell, C. Agert, Adv. Opt. Mater. 3 (2015) 182–186.
[10] V. Steenhoff, A. Neumüller, O. Sergeev, M. Vehse, C. Agert, Sol. Energy Mater. Sol. Cells. 145 (2016) 148–153.
[11] D. Redfield, J. Appl. Phys. 54 (1983) 2860–2862.
[12] V. Steenhoff, M. Juilfs, M. Vehse, C. Agert, Submitt. to Nano Energy. (2016).
[13] E.D. Palik, Handbook of Optical Constants of Solids, Academic Press Inc, Orlando, FL, 1985.
[14] S. Geißendörfer, M. Theuring, T. Titz, S. Mogck, C. Pflaum, B.T. Abebe, F. Schütze, D. Wynands, U. Kirstein, A. Schweitzer, V. Steenhoff, A. Neumüller, K. Borzutzki, R.-E. Nowak, A. Philipp, P. Klement, O. Sergeev, M. Vehse, K. von Maydell, in: Proc. 29th Eur. Photovolt. Sol. Energy Conf. Exhib., 2014: pp. 1667–1670.
[15] V. Steenhoff, M. Juilfs, R.-E. Ravekes, M. Ahrlich, M. Kellermann, O. Siepmann, M. Vehse, C. Agert, Submitt. to IEEE J. Photovoltaics. (2016).

WHITE BIFACIAL MODULES – IMPROVED STC PERFORMANCE COMBINED WITH BIFACIAL ENERGY YIELD

Bas B. Van Aken*, Lars A.G. Okel, Ji Liu, Stefan L. Luxembourg and John A.M. Van Roosmalen
ECN – Solar Energy
P.O. Box 1, 1755 ZG Petten, the Netherlands
* Corresponding author, e-mail: vanaken@ecn.nl

ABSTRACT: We present a novel module design to increase the power output of bifacial modules without compromising the increased energy yield due to the bifaciality. This innovation includes a highly reflective interlayer between and around the bifacial solar cells and laminated between the front and rear encapsulant. Although bifacial solar cells in modules with transparent front and rear sides can absorb light incident at either side, the transmission of (infrared) light through the rear side reduces the Pmax. The lower Wp rating is seen as a drawback for bifacial modules. The innovative white bifacial modules show almost the same front side Isc and Pmax as a monofacial module and a 2.2% higher Pmax, due to increased Isc, compared to transparent bifacial modules. The same increase is found for the rear side Pmax. The bifacial energy yield for white bifacial modules, due to the higher Pmax for both sides, is much higher than the slightly higher ground reflected "albedo" light of regular bifacial modules.
Keywords: Bifacial, PV module, energy performance, optical properties

1 INTRODUCTION

Bifacial solar cells can absorb irradiance from both front and rear sides if transparent materials are applied on both sides of the PV module. Double glass modules (and modules with a transparent back sheet) will not benefit from the back sheet scattering, but are able to harvest the light that is incident on the rear. The annual energy yield of bifacial modules and PV systems is widely reported to be significantly higher.

Standard crystalline Si PV *monofacial modules* have a rear side of white back sheet. A part of the light that hits the white back sheet is reflected back towards the (rear of the) solar cells via one or more internal reflections. The use of a transparent rear side reduces the maximum power output compared to using a white back sheet, when the modules are measured under the standard test conditions (STC) for monofacial modules, because this scattering mechanism is replaced by transmission out of the module. This apparent lower Wp rating is seen as a drawback for *transparent bifacial modules*.

In contrast, increases of the annual energy yield for transparent bifacial modules between 5% and 30% are reported in the literature [1,2], compared to monofacial modules. The exact values depend on device parameters like cell technology and rear side metal fraction and on environmental parameters such as the fraction of diffuse light and the scattering properties of the underground (albedo). Shading of the underground by the PV system reduces the generation of ground-reflected light. The small amount of transmitted light through regular bifacial modules will only have a small, albeit positive influence on the amount of ground-reflected light.

In this work, a new, innovative module stack is introduced. To increase both the maximum power output under standard test conditions compared to transparent bifacial modules as well as increase the annual energy yield compared to both monofacial and to transparent bifacial modules, a highly reflective interlayer is applied between the bifacial solar cells and on the area between the cells and the edge of the modules. We will refer to the bifacial module with the highly reflective, white, interlayer as *white bifacial module*.

2 EXPERIMENTAL SET-UP

2.1 Bifacial solar cells

Bifacial solar cells, n-type, three bus bar, were processed on semi-square 156 mm wafers in a single batch [3]. The processing was optimised for front-side efficiency and reproducibility. The best cells were selected for module manufacturing. These cells averaged 9.16±0.07 A (average ± standard deviation) for the short-circuit current and 19.69±0.16% conversion efficiency, measured on a conductive, brass measurement chuck. A selection of these cells were also measured on the rear, averaging 7.62±0.22 A and 16.4±0.6%. These cells, based on processing optimised for monofacial application, show a bifaciality factor of 83%.

2.2 Monofacial and bifacial modules

The three full-size modules have been manufactured from hand-soldered 10-cell strings. Front and rear side glass panels are identical. Three full-size modules were made from the same batch of cells: a monofacial module with standard white back sheet; a bifacial module with front and rear 2 mm solar glass; and a white bifacial module with a highly reflective interlayer between the front and the rear encapsulant covering all non-cell area. All other manufacturing parameters, e.g. encapsulant, tabs and layout, were identical.

3 RESULTS & DISCUSSION

3.1 IV-measurements

In Figure 1 photographs of the front and rear side of the white bifacial module are shown. The rear view photo shows clearly the cross-connectors and the four terminals for the junction box connection. Note that the rear sides of these non-optimised bifacial solar cells are slightly less black than the front side.

All modules were measured under STC in a room with black walls with a Pasan module flash-tester, class AAA. For the bifacial modules, both front and rear were measured. The short-circuit currents of the modules and the module powers are summarised in Table 1. For the

Figure 1: Close up photograph of the rear (left) and front (right) view of the white bifacial, double glass module. The white area between the cells and around the aperture area are well visible.

monofacial module typical cell-to-module (CtM) changes are observed, i.e. 1% lower current and about 4% power loss due to interconnection resistive losses and the small loss in current. The transparent bifacial module also shows CtM results as expected: the 2.7% reduction in current is due to the transmission through the module (lack of scattering on the rear side panel); the same resistive losses as for the monofacial module are observed, leading to -6% Pmax CtM

Table 1: Overview of Isc and Pmax for the three modules, front and rear side measured, and cell-to-module (CtM) changes. Note: the rear side results are influenced by some outliers that do not influence the front side performance.

	Isc [A]	Isc CtM	Pmax [W]	Pmax CtM
Monofacial module	9.07	-1.0%	271	-3.9%
Bifacial module white interlayer	9.10	-0.6%	271	-3.9%
rear side	7.69	+0.9%	230	-2.0%
Bifacial module transparent	8.92	-2.7%	265	-6.0%
rear side	7.46	-2.0%	226	-3.7%

In contrast to the transparent bifacial module, the white bifacial module shows similar values for the front side Isc and Pmax as the monofacial module. Compared to the transparent bifacial module, a 2% higher Isc is observed for the white bifacial module. This results in a correspondingly 2.2% higher Pmax for the white bifacial module. The same increase is found for the rear side Pmax of the white bifacial module compared to the transparent bifacial module. All other IV-characteristics are not changed by the choice of rear side material or presence of interlayer.

Our novel bifacial module architecture combines the double glass module design including high-efficiency n-Pasha bifacial cells with the introduction of a highly reflecting white interlayer in the open area between and around the solar cells. Front and rear side incident light striking the area between the cells is scattered in all directions, whereby a large fraction will strike the front or rear surface of the bifacial solar cells, depending on the location of the scattering event. Note that the scattering takes place at about the same "depth" in the module as the front or rear surfaces of the solar cell, compared to 400 to 600 micron below the rear surface for scattering at the white back sheet in monofacial modules.

As a result, the interlayer will improve the front side Isc and Pmax, e.g. as measured under standard test conditions for monofacial modules, to values very close to those obtained for monofacial, white back sheet modules. The same argument applies for the rear side measurement of the white bifacial module.

3.2 Ray tracing

To get a better understanding of how the reflection from the back sheet or glass area behind the cell and from the intercell area contribute to the short-circuit current a simple ray tracing model was set-up. Light was incident perpendicular to the module stack either on the cell area (dark blue) or on the area between the cells (white) as shown in Figure 2. The analysis for light incident on the cell area was performed using a realistic optical model of the cell (including texture and ARC). For the light impinging on the area between the cells the solar cells was treated as a black absorber. Both cases were calculated a) with a white back sheet, simulating a monofacial module, and b) with a rear side glass panel, simulating the bifacial modules. For the second case, also a white reflective foil was placed at the intercell area between the two EVA layers. Optical parameters for the white back sheet and the white reflective foil were calculated from reflection and transmission measurements.

Figure 2: schematic view of the simulated solar cell and intercell area for the ray tracing simulation. The intercell area is defined as the white area between the red dashed box and the dark blue solar cell area.

From the ray tracing simulations, the fraction of incident light absorbed in each layer or transmitted out of the simulation stack, front and rear, was determined. From these spectrally resolved (330-1100 nm) absorption graphs, the contribution or loss to the short circuit current is calculated. In this analysis it is assumed that all light absorbed in the silicon is contributing to Isc. In the monofacial case, for light incident on the cell area only, the contribution to Jsc is 38.65 mA/cm^2; the double glass case gives 37.91 mA/cm^2. This difference of 2% is due to the reflection of mostly near infrared light back into the solar cell by the white back sheet. In contrast, for a bifacial double glass module, most of that near infrared light is transmitted through the rear glass panel and thus lost.

In

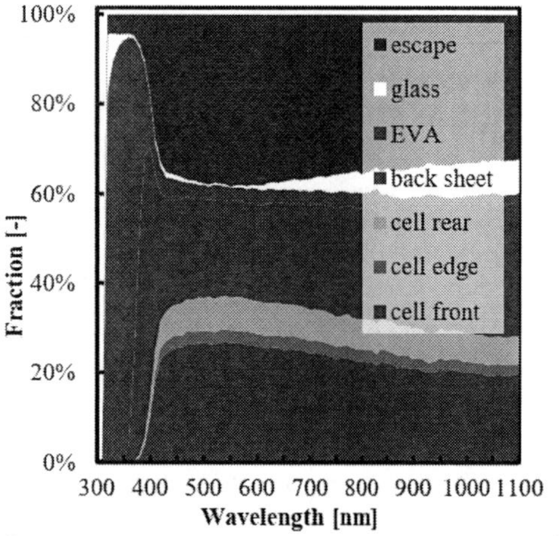

Figure 3 and

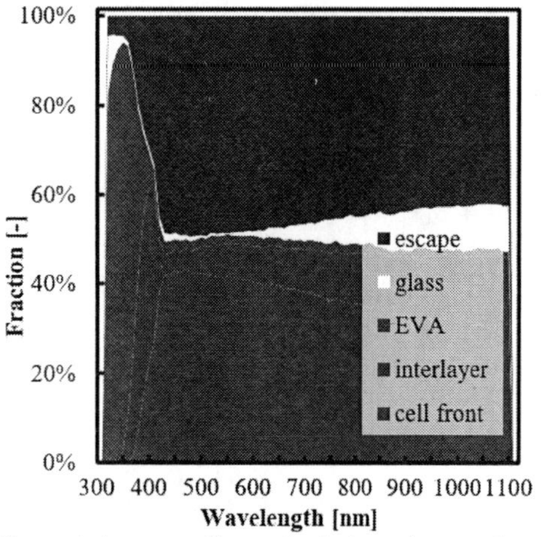

Figure 4 the spectrally resolved absorption graphs are shown for the case when the light is incident on the intercell area for (

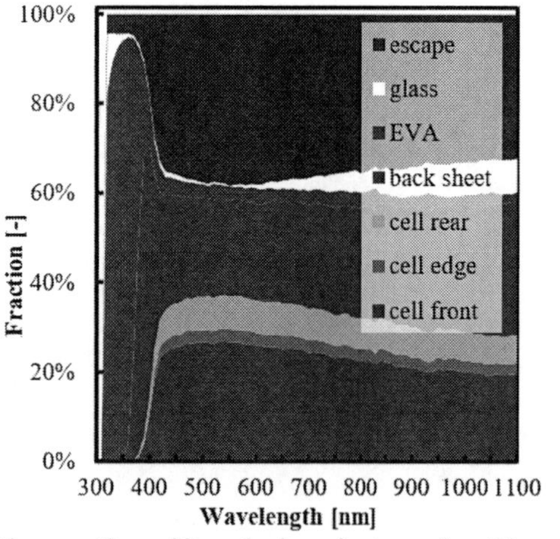

Figure 3) white back sheet and (Figure

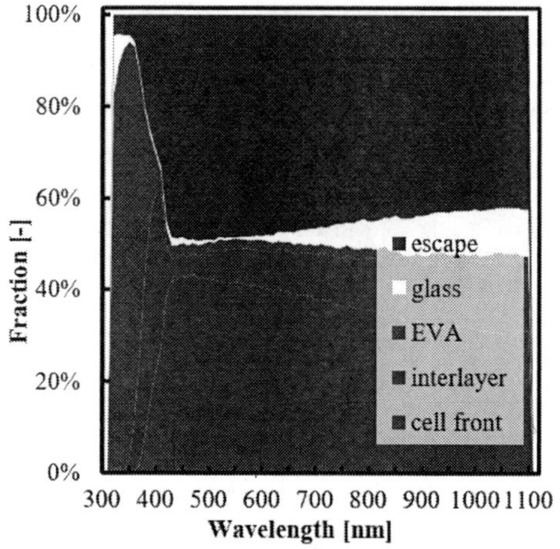

Figure 4) white interlayer. The general picture is the same, but there are still some major differences. First, the fraction of light that falls on the solar cell on any surface, green area(s), is somewhat lower, namely 32% against 36%, for the monofacial module than for the white interlayer. Secondly, the absorption in the scattering layer, red area, is much higher, namely 25% against 11%, for the monofacial module.

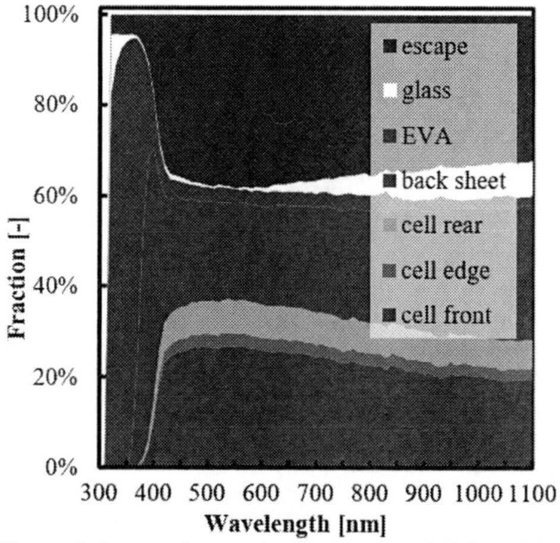

Figure 3: Ray tracing results for the monofacial module for light incident on the intercell area.

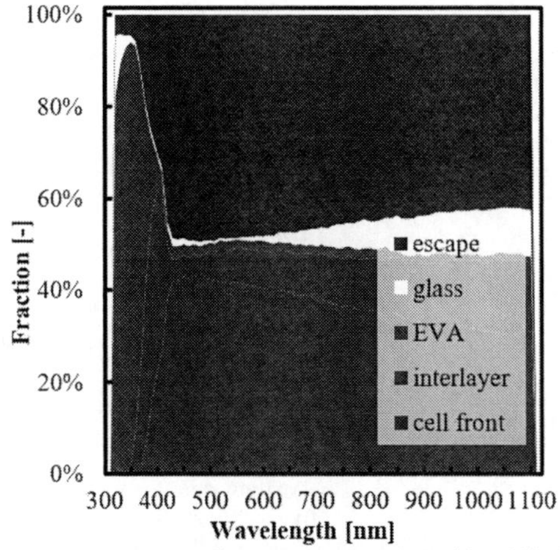

Figure 4: Ray tracing results for the white bifacial module for light incident on the intercell area.

The fractions absorbed in the layers of EVA and the glass are small and very similar, for both modules totalling 9% of the maximum photo current. Consequently, the fraction of light, incident on the intercell area, that is directly reflected or indirectly scattered out of the front glass is much higher for the interlayer module, namely 45% against 35% for the monofacial module.

The differences between the graphs follow from the two major differences in the optical stacks considered. First, the scattering takes place either 450 micron below the rear side of the cell (white back sheet) or at the same depth as the front side of the cell (white interlayer). Secondly, the reflection, which is >90% diffuse for both materials, is significantly lower for the white back sheet than for the white interlayer as was determined experimentally.

The increased absorption in the back sheet versus the interlayer is a logical consequence of the latter point. Therefore, we can say that from the light reflected of either the back sheet or interlayer a similar fraction ends up on the cell. Thus, the reduced absorption in the silicon for the monofacial case is also caused by the absorption in the scattering layer.

Since for the interlayer module the reflection takes place at a depth similar to the front side of the solar cell, all light absorbed by the solar cell is incident on the front side through internal reflection of the glass. In contrast, for the white back sheet module, about 30% of the light absorbed by the solar cell is incident on the rear side.

From the ray tracing we can also deduce whether the light absorbed by the front (rear) of the solar cell, is absorbed close to or further away from the wafer edge. Figure 5 shows the edge fraction on the front side of the solar cell for the reflector case and, for the back sheet case, the edge fractions of both the front and rear sides of the solar cell. On the front side, independent on the module layout, only 20% is absorbed on an edge with width 1.5 mm. In sharp contrast, the rear side absorbs 50% of the light that is scattered on the rear side in an edge region of only 300 microns wide from the edge and 90% for an edge that is 1.5 mm wide.

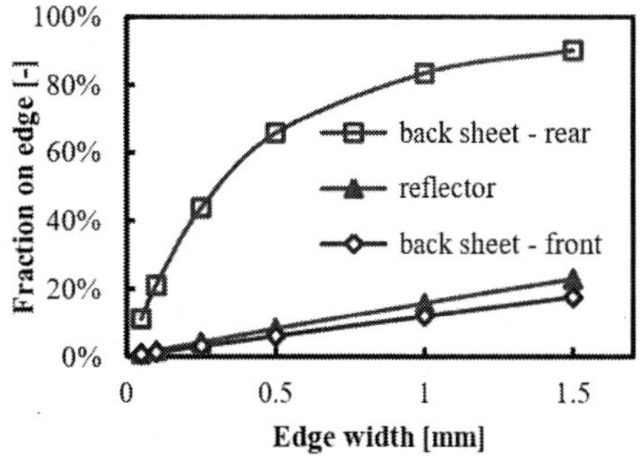

Figure 5: Fraction of the light incident on the edge of the front/rear side of the solar cell, as function of the edge width.

Knowing the contributions for light incident on the cell area and on the intercell area, the total Isc can be calculated for the three cases: a) the monofacial module, b) the bifacial double glass module and c) the bifacial module with white interlayer. The cell area contributes 9.24 A for the white back sheet and 9.06 A for the bifacial modules. The intercell area contributes nothing for the double glass module, 0.28 A for the white back sheet module and 0.36 A for the white interlayer module.

For simplicity we have assumed that all light falling onto the cell after reflection from either the back sheet or interlayer is absorbed and contributes to Isc. The intercell area contribution for the white back sheet takes into account a bifaciality factor of 85% for rear illumination. It does not take into account the fact that 50% of the rear illumination is absorbed by an edge region of 300 micron width, that may have a reduced efficiency compared to that of the bulk.

Also the scattering properties of AR-coatings and texture on either side of both glass panels is not taken into account. This will affect the contribution of the intercell area of the double glass module the most, as it is now, by default, zero. We think that the, partially quantified, arguments mentioned above, support the observed similarity in the experimentally derived Wp values for the monofacial and white bifacial modules.

3.3 Effect on annual energy yield

First, the difference between the partially transparent bifacial module and the fully non-transparent white bifacial module will be discussed. We have done annual irradiance and energy yield modelling for a typical system in the Netherlands. In general, the shadow of the modules lowers the ground-reflected albedo light by about 20%, assuming 100% blocking of the light by the modules. The transparency of a standard bifacial module, about 5%, will increase the ground-reflected albedo light, but the effect on the total annual irradiance is only a non-significant 0.1%. However, the white bifacial module shows a 2.2% better conversion efficiency, this increases the energy yield with the same fraction, on top of the increase in energy yield due to the bifaciality.

The AEY model requires an optical, a thermal and an electronic model, all adapted for bifacial irradiance. Details have been reported before [4]. The optical model is derived according to [5], where the irradiance on both tilted module surfaces is calculated in terms of contributions by the beam, by diffuse light and by ground-reflection caused by beam and diffuse light. Specific for bifacial modules is that the ground reflection has to be reduced by self-shading. The annual energy yield AEY is the summation of the calculated power output, using time resolved meteorological data. For yield predictions meteorological data at a certain location have been synthetically generated from monthly average values obtained from data bases.

Using this AEY model for a single module, the annual energy yield for the three types of modules discussed here are modelled for five locations, each simulated with albedo of 0.2 and 0.5. Results are plotted in Figure 6. Note the minimal effect of albedo for the monofacial module AEY at all locations.

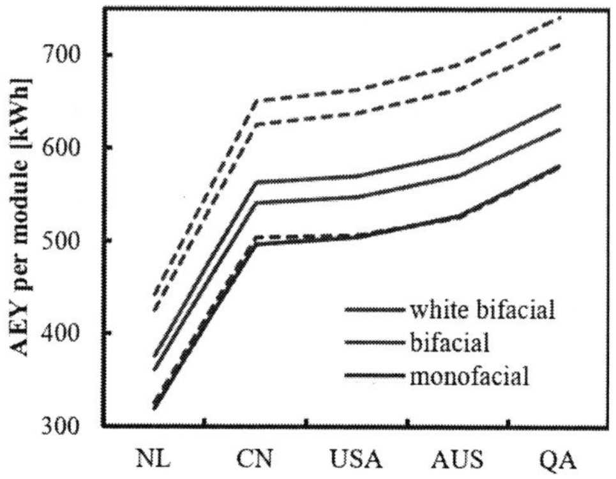

Figure 6: Modelled annual energy yield for the three different module types at five locations. Solid (dashed lines) are simulated with an albedo of 0.2 (0.5). Same colour represents same module type. Details of the locations are given in Table 2.

Table 2: overview of the five locations used for the AEY modelling. Horizontal irradiances are given in kWh/m^2.

Label	City	Country	Latitude	Longitude	Mean tempera-ture [°C]	Global horizontal irradiance	Direct horizontal irradiance	Diffuse horizontal irradiance
NL	Amsterdam	the Netherlands	N 52° 24′	E 4° 54′	10.7	1067	501	566
CN	Baoding	China	N 38° 54′	E 115° 30′	13.3	1579	1027	552
USA	Sacramento	United States	N 38° 31′	W 121° 30′	15.7	1803	1229	574
AUS	Rockhampton	Australia	S 23° 23′	E 150° 29′	22.4	2013	1279	734
QA	Doha	Qatar	N 25° 16′	E 51° 30′	28.1	2274	1500	774

For each combination of module type, location and albedo the optimal tilt angle is determined. These are plotted in Figure 7. As expected, the closer the location is to the equator, the lower the optimal tilt angle is. The combination monofacial module with low albedo yields the lowest tilt angle for each location. The high albedo simulations always yield an optimal tilt angle larger than for the low albedo value. There are no differences between the optimal tilt angle for the regular bifacial module and the white bifacial module.

Three modules, see Table 1, have been placed on ECN's roof top for outdoor characterisation. Figure 8 shows a typical power against time curve for these three modules on a sunny day. Clearly the bifacial modules give a higher power output during the whole day. The higher gain in the afternoon, is due to the slightly off-South orientation of the modules.

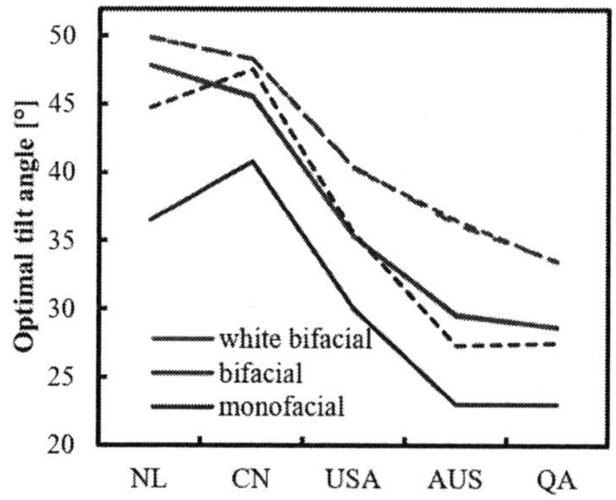

Figure 7: Optimal tilt angle for each combination of location, module type and albedo. Colours and dashes same as in Figure 6.

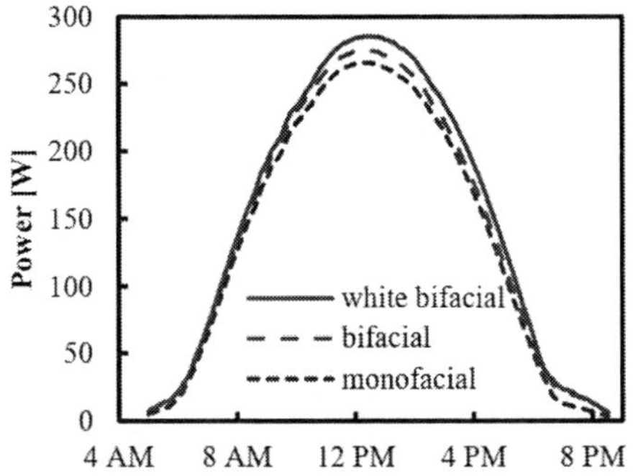

Figure 8: Example of the daily transient of the power output of the three modules on a sunny day.

Maximum power for the data shown in Figure 8 is reached around solar noon as shown in Table 3. Whereas the monofacial module was 5 W short of its rated power, the bifacial module produced 10 W more. The white bifacial showed the highest output with 14 W more than the rated power. Note that these modules were fabricated from the same batch of solar cells.

Table 3: comparison of rated power (indoor) and observed maximum power for the data of Figure 8.

	Indoor STC power [W]	observed maximum power [W]
monofacial	271	266
bifacial	265	275
white bifacial	271	285

Based on the outdoor IV data, we can estimated the daily energy yield in kWh. In Figure 9: **Daily energy yield per module for the first five days of June.**Figure 9, the daily energy yield for the three modules is plotted for the first five days of June. The energy gain for the bifacial module is >5% for sunny days, up to 8% when the total irradiance is low. Note that the albedo of this location is not high. The white bifacial module performs even better, even though the same solar cells were used for module manufacturing. The energy gain is >10%, even on sunny days, when the irradiance is dominated by direct sunlight.

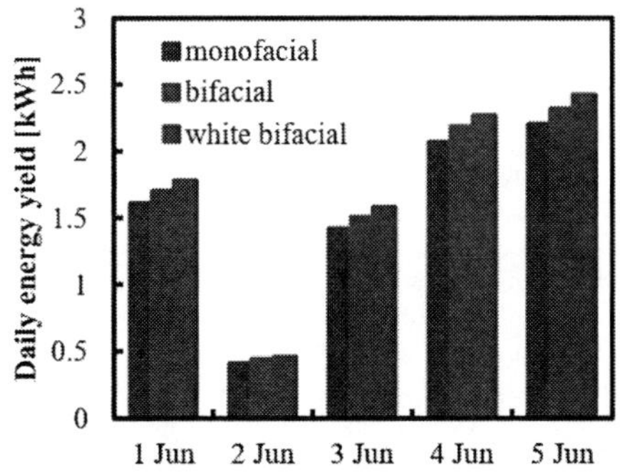

Figure 9: Daily energy yield per module for the first five days of June.

4 CONCLUSIONS

We have manufactured a bifacial double-glass module with an innovative white, highly reflecting interlayer that is positioned between and around n-Pasha solar cells. The Wp rating of this module was very similar to an identically processed monofacial white back sheet module solar cells from the same production batch. The 83% bifaciality of these front side-optimised n-Pasha solar cells was maintained in the white bifacial module. Because the white bifacial module is 0% transparent, there will be a slightly darker shadow behind the modules, compared to the 5% transparent bifacial module. This will reduce the amount of direct-light generated ground reflection by tenths of a percent of the total irradiance. In contrast, and 10× more significant, the annual energy yield of the white bifacial module will be improved by the higher front and rear Pmax.

ACKNOWLEDGEMENTS

We thank Yingli Green Energy and Tempress for their support. The modelling work has benefitted by funding through the Dutch TKI Solar Energy programme, projects BING and NChanted.

REFERENCES

[1] T. Mishima, M. Taguchi, H. Sakata and E. Maruyama, "Development status of high-efficiency HIT solar cells", *Sol En. Mat. Sol. Cells*, **95** (2011) 18-21.

[2] L. Kreinin, et al., "Experimental analysis of the increase in energy generation of bifacial over monofacial PV Modules", Proc. *26th EU PVSEC* 4CO.7.1, (2011).

[3] G.J.M. Janssen, et al., "Front side improvements for n-Pasha solar cells", *Proc. 29th EU PVSEC*, Amsterdam (2014).

[4] G.J.M. Janssen, B.B. Van Aken, A.J. Carr and A.A. Mewe, "Outdoor performance of bifacial modules by measurements and modelling", *Energy Procedia* **72** (2015) 364-373.

[5] J.A. Duffie and W.A. Beckman, Solar engineering of thermal processes, 2nd ed., John Wiley & Sons, (1991).

ENABLING SOLDERABILITY OF PVD AL REAR CONTACTS ON HIGH-EFFICIENCY CRYSTALLINE SILICON SOLAR CELLS BY WET CHEMICAL TREATMENT

H. Nagel, M. Kamp, D. Eberlein, A. Kraft, J. Bartsch, M. Glatthaar and S. W. Glunz

Fraunhofer Institute for Solar Energy Systems (ISE), Heidenhofstraße 2, 79110 Freiburg, Germany
Email: henning.nagel@ise.fraunhofer.de

ABSTRACT: Results of an electroless displacement process for the deposition of inexpensive solderable metal on physical vapour deposited (PVD) aluminum are presented. The dip – rinse – dry processing sequence for electroless metal deposition on Al is simple and easy to use because it does not harm the SiN antireflection coating and the screen-printed Ag metallization on the cells' front side. Hence, finished cells can entirely be immersed into the electrolyte. The deposited metal layer is highly porous, but forms a massive alloy with common Sn/Pb/Ag solder at low temperature of 245 °C in the presence of no-clean flux. Because the molten alloy wets Al and copper well, we obtained strong solder joints for Sn/Pb/Ag coated copper ribbons on Al. The measured average 90° peel forces of ribbons soft soldered to thermally evaporated Al on shiny etched silicon and on the rear of alkaline textured mono-crystalline Si PERC solar cells were up to 3.1 and 2.2 N per mm ribbon width, respectively. These values exceed 1 N/mm that is specified by the standard DIN EN 50461. Climate tests on single-cell glass-glass modules yielded 4.1 and 7.7 % power degradation after 200 thermal cycles from -40 °C to 85 °C and after 1000 h under damp-heat conditions at 85 °C and 85 % relative humidity, respectively. We attribute the too high power degradation after damp-heat test to residuals of the acidic flux which was used for soldering the cells in this first, older version of the developed technology.

Keywords: Evaporated Aluminum, Electroless Deposition, Cell Interconnection

1 INTRODUCTION

At present, most industrial crystalline silicon solar cells feature screen-printed rear contacts that consist of large-area Al and local Ag solder pads. During sintering of the printed pastes the Al alloys with the Si and forms a highly doped *p*-type back surface field (BSF) which is well suited for common *p*-type c-Si cells. However, printed and sintered Al rear contacts suffer from several shortcomings:

i) Thin dielectric passivation layers (like e. g. Al_2O_3) on the back of passivated emitter and rear cells (PERC) have to be protected against the printed Al by means of a capping layer (usually SiN) in order to avoid spiking of Al through the passivating film.

ii) Possible degradation of the rear surface passivation even with capping layer present due to the high sintering temperature.

iii) Increased effective surface recombination velocity in comparison to boron BSF made by other methods.

iv) Damage of *n*-doped BSF on *n*-type cells.

Hence, high-efficiency c-Si solar cells usually rely on physical vapor deposited (PVD), i. e. evaporated or sputtered, Al. It makes good ohmic contact to *p*- and *n*-type Si, provides excellent back reflection of weakly absorbed infrared light passing through the cells and features very high electrical conductivity. Good examples for c-Si solar cells featuring PVD Al rear contacts are *n*-Q.ANTUM cells presented by Hanwha Q-Cells which yielded efficiencies above 21 % [1] and the 'PassDop' cells introduced by Fraunhofer ISE which reached efficiencies up to 23.5 % on small area [2] and above 22 % on large area [3]. However, a technical challenge that is associated with PVD Al rear contacts is the series connection of the solar cells to strings. The interconnect ribbons cannot be directly soldered to the base metal aluminum because in ambient atmosphere it is covered by a very quickly formed dense Al_2O_3 film that is about 2.5 nm thick. Solutions which were already proposed for this problem have distinct pros and cons in terms of feasibility and cost:

In [4,5,6], solderable metal stacks were vacuum deposited on Al. The stacks TiN/Ti/Ag, NiV/Ag und NiSi/Ag on Al yielded high peel forces of soldered interconnect ribbons. The layers can be vacuum deposited in the same deposition system in a single pass without venting, but the costs are relatively high because several vacuum chambers and different installations for thermal evaporation and sputtering are needed in the deposition system.

A well known modification of this technique is the metallization scheme developed by Sunpower Corp. for rear-contacted c-Si solar cells [7]. Here, PVD Al is sputter coated with a TiN or TiW adhesion layer which is subsequently electroplated with copper and tin. The adhesion layer simultaneously acts as a barrier against diffusion of copper through the aluminum into the silicon substrate. This technique greatly reduces the amount of vacuum deposited metal for cost reasons but relies on electroplating equipment.

A plating-based approach was also investigated in [8]. In a first step, the aluminum is immersed in an alkaline zincate solution. Then, the deposited zinc layer is etched off in order to obtain a roughened surface and to increase the surface area. Subsequently, zinc is re-deposited in a second zincate step followed by electroplating of a Ni/Cu/Ag stack. Here, the thin plated nickel layer acts as adhesion promoter and diffusion barrier against copper. Interconnect ribbons soldered on this metal stack yielded high peel forces exceeding 1 N/mm on an average.

In [9,10] it was proposed to apply a tin coating on the aluminum rear contact by ultrasonic soldering. High-throughput systems for production of these so-called tin pads on large-area cells are available [11]. Onto tin pads interconnect ribbons can be soft soldered in standard tabber stringers. No flux is needed and no expensive metals are consumed. The technique was developed for screen-printed aluminum. Probably it is suited for PVD Al as well. However, to the best of our knowledge there is no publication about the application to PVD Al, yet.

There are more joining technologies for aluminum, for example welding and hard soldering. We expect that they are not well suited for solar cell interconnection because of the involved high substrate temperatures above 450 °C and the associated threat of electrical degradation of the dielectric rear side passivation on PERC cells. A noteworthy exception is laser-welding as demonstrated in [12] for aluminum ribbons on PVD Al. During the laser welding process heat is applied only locally at small spots for a very short time. This way the aluminum foil and the PVD Al can selectively be melted without increasing the temperature of the underlying cell above the degradation threshold. A disadvantage of this process is that special tabber stringers are necessary and that laser welding on PVD Al was only shown for aluminum interconnect ribbons and not for conventional copper ribbons.

In this work we investigated an innovative low-cost approach to make PVD Al solderable. We developed a single wet chemical treatment that deposits inexpensive solderable metal on PVD aluminum.

2 EXPERIMENTAL

Aluminum layers featuring various thicknesses were deposited by high-rate thermal evaporation in an inline vacuum system supplied by Applied Materials Inc. Different substrates were used: shiny etched and alkaline textured silicon wafers and two types of PERC cells. Subsequently, the entire Al coated wafers and cells were dipped in a wet chemical solution for 90 to 180 s at room temperature. At first the used electrolyte etches the native Al_2O_3 layer and then an about 1 µm thick solderable metal (Me) is electroless plated on the Al from the same electrolyte by means of a displacement reaction

$$3\,Me^{n+} + n\,Al \rightarrow 3\,Me + n\,Al^{3+}$$

where n is the chemical valence of the deposited metal. No external electrical contacts were attached to the cells and no external voltage was applied. The driving force for the chemical reaction is the difference in standard electrode potential between Al and the more noble Me. After rinsing and drying, conventional Sn/Ag/Pb-coated Cu ribbons were soldered to the metallization stack by contact soldering. The thickness of the ribbons' copper core and solder coat was 130 µm and about 15 µm, respectively. The ribbons were 2 mm wide and fluxed with widespread Kester 952 s in most of the experiments. The soldering temperature was varied in the range between 245 °C and 400 °C.

From 5 inch 'PassDop' cells mentioned in the Introduction we manufactured single-cell mini-modules featuring soda-lime glass on the front and on the rear. It must be noted that in this experiment we applied the electrolyte on the rear only and that we used a more reactive acidic flux on the ribbons. After soldering at a temperature of 245 °C the flux had to be washed off. The laminated modules were subjected to damp-heat and temperature cycling tests according to the standard IEC 61215.

We studied the microstructure of the films and solder joints by secondary electron microscopy (SEM) and their composition by energy dispersive X-ray spectroscopy (EDX). In order to measure the adhesion of the ribbon soldered on the wafers and cells, an automated 90° peel-force-tester was used. The *IV* curves of cells and modules were measured by means of a flasher system.

3 RESULTS

3.1 Microstructure of the deposited metal layer

The top of Figs. 1 and 2 show plan view SEM images of 3 µm Al thermally evaporated on a shiny etched Si wafer and on an alkaline / random pyramid textured rear of a CZ Si PERC cell, respectively. In the latter case a common surface passivating Al_2O_3/SiN stack was placed between the Si and the Al but in additional experiments we did not observe an effect of these dielectric layers on the microstructure of the Al. It can be seen from the figures, that on shiny etched Si the PVD Al layer consists of 1 to 3 µm large grains and that it uniformly covers the wafer surface. On alkaline textured silicon the PVD Al layer basically reproduces the shape of the random pyramids as well, but the thickness of the Al layer is very non-uniform on a microscopic scale. In the valleys between the pyramids the Al is thin or even missing at small spots, whereas on the tips of the pyramids thick Al crystals are formed, noticeable from their rectangular shape.

After immersion in the electrolyte and rinsing, on both substrates metal layers which essentially consist of separate small particles and a few very large crystallites were found, see bottom of Figs. 1 and 2. The size of the large grains lies in the range between 5 to 13 µm. Without further quantification it is already obvious from the pictures that the porosity of the Me layer on the pyramid textured sample exceed the one on the flat sample.

It can be expected that a highly porous Me layer causes undesired voids in solder joints. In order to shed light on this we electroless deposited Me layers featuring an effective thickness of about 1 µm on each side of an Al foil and soldered Sn/Pb/Ag-coated Cu ribbon at 245 °C on one side. A massive and uniform Al foil was chosen as a substrate instead of a PVD Al coated Si wafer in order exclude the effect of the microstructure of PVD Al in this

Fig. 1: Plan view SEM picture taken from 3 µm Al thermally evaporated on shiny etched Si before (top) and after (bottom) electroless metal deposition.

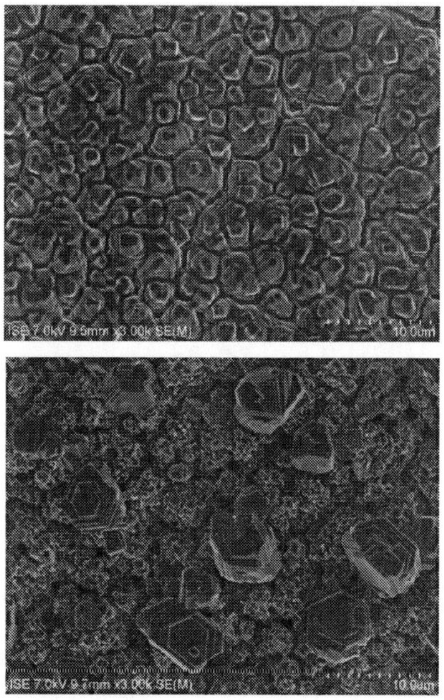

Fig. 2: Plan view SEM picture taken from 3 µm Al thermally evaporated on the Al_2O_3/SiN coated alkaline textured rear side (random pyramids) of a PERC Si cell before (top) and after (bottom) electroless metal deposition.

fundamental experiment. Fig. 3 shows the obtained cross-sectional SEM image of the solder joint between Al foil and copper ribbon together with the cross-faded result of the EDX analysis. On top of the Al foil the original porous Me layer can be seen (blue color). Obviously, the area coverage of Me on Al is quite low. Nevertheless, the solder joint region in the center of the image reveals that the Me layer completely alloyed with Sn/Pb/Ag solder. Me is distributed all over the solder, even to the Cu surface. The alloy wets aluminum and copper well which is a prerequisite for the formation of strong solder joints.

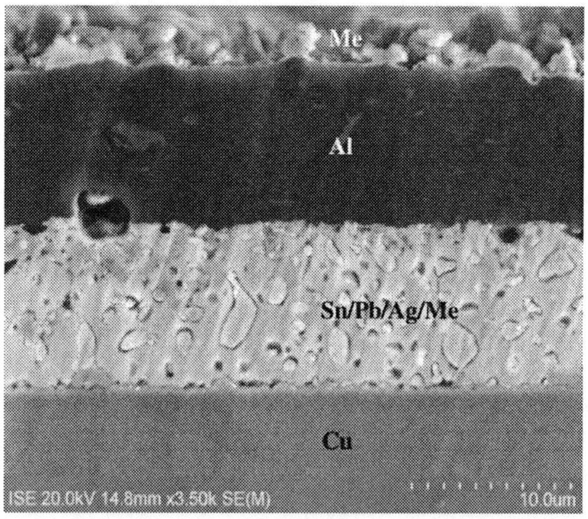

Fig. 3: Cross-sectional SEM picture taken from Al foil electroless coated on both sides with Me and soldered on the bottom side to Sn/Pb/Ag coated Cu ribbon. Cross-faded is an EDX image indicating the concentration of the elements Sn (red) and Me (blue).

3.2 Peel force of soldered interconnect ribbons

1 to 5 µm Al was thermally evaporated on shiny etched and alkaline textured 6 inch CZ Si wafers, respectively. After dipping the wafers for 1 min 45 s into the electrolyte, we soldered interconnect ribbons on the deposited Me layer at a relatively high temperature of 400 °C. The high temperature was chosen because we found that it yields more reproducible results on textured samples. The measured 90 ° peel forces normalised to the ribbon width are shown in Fig. 4 as a function of Al thickness. It can be seen from the figure, that the peel force increases with Al thickness and that there is a trend of the peel force to saturate for thicknesses exceeding 3 µm on alkaline textured Si. Below an Al thickness of 2 µm the peel force significantly drops and approaches zero at 1 µm. The reasons are that about 1 µm Al is dissolved in the electrolyte and the Me layer doesn't adhere to Si but only to Al. In general, the peel force on shiny etched Si is higher compared to alkaline textured Si because the porosity of the Me layer is lower. Independent of surface texture average peel forces of above 1 N/mm were regularly obtained for Al thicknesses exceeding 2 µm which complies with the standard DIN EN 50461.

By optimising the electroless deposition and especially the rinsing process after deposition in order to avoid residual electrolyte on the samples we could further increase the peel forces. Fig. 5 shows the force vs. distance curves for 3 µm Al thermally evaporated on shiny etched Si and on the alkaline textured rear side of a PERC solar cell covered by a common Al_2O_3/SiN passivation stack, respectively. At every location of the wafer and cell the peel force exceeded the desired 1 N/mm. Again, the peel force obtained on the shiny etched wafer was higher compared to the alkaline textured cell. The average peel forces were 2.2 and 3.1 N/mm, respectively. These results imply good adhesion of

a) electroless deposited Me on evaporated Al if the Al thickness is larger than 2 µm

b) evaporated Al on Si

c) evaporated Al on SiN

d) SiN on Al_2O_3 and

e) Al_2O_3 on Si.

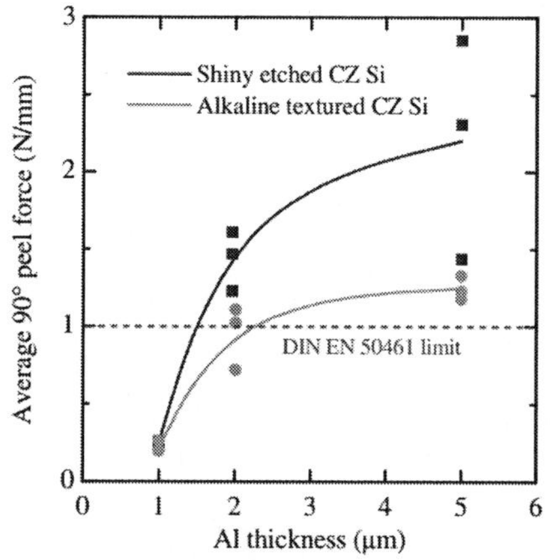

Fig. 4: Average values of the measured interconnect ribbons' 90 ° peel forces of as a function of PVD Al thickness on shiny etched and alkaline textured silicon, respectively. The Al was electroless plated with about 1 µm Me. Lines are guides to the eye.

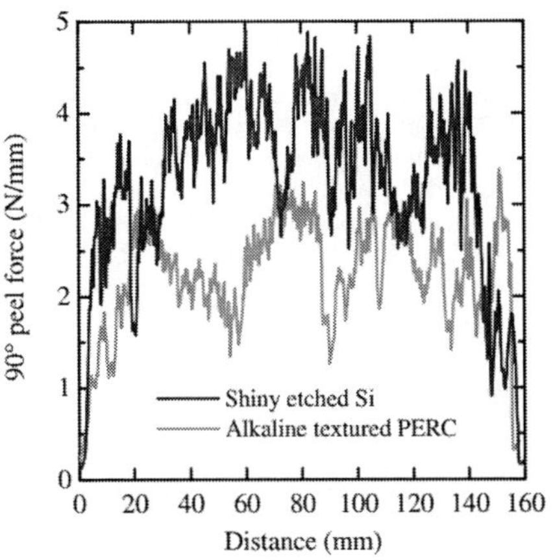

Fig. 5: Measured 90 ° peel force vs. distance curves of Cu ribbons soldered on 3 μm thick PVD Al located on the rear of a PERC c-Si solar cell and on a shiny etched c-Si wafer, respectively.

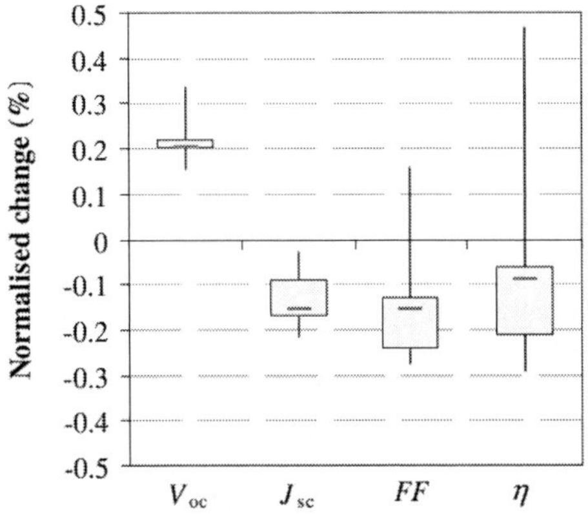

Fig. 6: Measured normalised change of electrical performance parameters caused by immersing entire cells in the electrolyte.

3.3 Impact of electrolyte on solar cell efficiency

If desired, electroless metal deposition can solely be applied on the cells' rear side by means of state-of-the-art high-throughput inline systems which are available for electroplating or for rear side etching. In this production-type equipment the cells move horizontally above the electrolyte, e. g. on rollers that are arranged in a row. The rollers are partially immersed in the wet chemical solution so that they transport the electrolyte to the rear of the cells during rotation. However, in some other cases batch processing is desired. It offers high throughput with small equipment footprint because a large number of cells are processed simultaneously. In the lab it might be the handling of choice because no inline wet bench is available.

Batch processing is associated with placing a large number of cells vertically in a carrier and immersing the entire carrier at once in the electrolyte. Inevitably both, PVD Al rear contact and front metal grid are exposed to the electrolyte if the latter was deposited before. This is generally the case for common screen-printed and fired front metallization because of the high thermal budget which would counteract the advantages of the PVD Al as described in the Introduction. Hence, we investigated the effect of the electrolyte on 6 inch *p*-type CZ Si PERC cells featuring screen-printed Ag front side metallization and SiN antireflection coating. Both sides were alkaline textured and the rear was coated by an Al_2O_3/SiN surface passivating layer stack. 3 μm Al was thermally evaporated on the rear and local back-surface fields were prepared by laser contact firing of the Al after deposition. The entire cells were dipped for 1 min 45 s in the electrolyte, rinsed and dried. *IV* curves were measured before and after dipping. Fig. 6 shows the measured change of electrical performance parameters normalised to the values before electrolyte exposure. It can be seen from the figure, that the median normalised efficiency (η) reduction was 0.1 % which transfers to less than 0.02 % absolute for these cells that feature a maximum efficiency of 20.1 %. Considering the small number of only 5 cells used in this experiment and the measurement uncertainty we conclude that the obtained small change is within the experimental error and hence neither SiN antireflection coating nor screen-printed front metal grid are harmed by the electrolyte. The latter is further supported by the fact that there was no change of measured finger resistance on the front side. However, as expected, we observed an increase of the rear contacts' sheet resistance from 11 to 13 mΩ. It corresponds to a reduction of Al thickness from 3 to 2.4 μm assuming that the conductivity of the deposited porous metal layer can be neglected. From this, the analytical calculation of the resistive power loss taking into account the used 3 busbar design yielded an efficiency reduction of 0.04 % relative.

3.4 Results of first climate tests

We manufactured two single-cell glass-glass modules from 5 inch *n*-type CZ Si 'PassDop' PERC cells featuring alkaline textured front side and shiny etched rear side, Ni/Cu/Ag plated front contact, SiN antireflection coating, phosphorous doped passivating SiN on the rear underneath about 2.5 μm thermally evaporated Al. In this case we locally deposited solderable Me only on the areas where the interconnect ribbons were soldered on the rear Al. Furthermore, we applied a strong acidic flux to the interconnect ribbons before soldering at 245 °C in order to guarantee good wetting of the Me. After soldering, the ribbons were cleaned, i. e. the flux was washed off. Table 1 shows the measured change of electrical performance parameters after damp-heat (DH) test at 85 °C, 85 % relative humidity for 1000 h and after 200 thermal cycles (TC) at temperatures from -40 °C to 85 °C. These climate tests were performed according to IEC 61215 standard. Whereas the TC test was passed (efficiency degradation was lower than 5 % relative), normalized efficiency reduction after DH test was 7.7 % mainly due to a decrease of *FF*. We attribute it to residuals of the acidic flux which presumably were not completely washed off. Hence, no-clean flux should be used as we did in the study for optimization of the peel force.

Table 1: Measured change of electrical performance parameters of single-cell glass-glass modules containing 5 inch 'PassDop' cells after temperature cycling (TC) and damp-heat (DH) test.

Test	$\Delta V_{oc}/V_{oc}$ (%)	$\Delta J_{sc}/J_{sc}$ (%)	$\Delta FF/FF$ (%)	$\Delta\eta/\eta$ (%)
TC	-0.6	-0.9	-2.7	-4.1
DH	-0.6	-0.8	-6.4	-7.7

4 SUMMARY

We developed an electroless displacement process for the deposition of solderable metal on PVD Al. By a simple dip – rinse – dry processing sequence about 1 µm of Al is dissolved in the electrolyte and at the same time a Me layer featuring an average thickness of about 1 µm is deposited. Although the microstructure of the Me layer is highly porous, the adherence to PVD Al is good. Already at a temperature of 245 °C the metal uniformly alloys with standard Sn/Pb/Ag solder using common no-clean flux. Because the alloy wets Al and copper well, we obtained strong solder joints on thermally evaporated Al. By an optimised process we achieved average 90° peel forces of 2.2 N/mm on the rear of PERC cells and of 3.1 N/mm on shiny etched Si. These values are well above 1 N/mm which is specified by the standard DIN EN 50461. In combination with SiN antireflection coating and screen-printed Ag metallization on the front side the electroless displacement process is suited for batch handling because the electrolyte does not harm the solar cells' front side. First climate tests on single-cell glass-glass modules containing 'PassDop' cells yielded 4.1 and 7.7 % power degradation after 200 thermal cycles and 1000 h damp-heat test. We attribute the too high power degradation after damp-heat test to residuals of the non-standard acidic flux which was used for soldering in this first, older version of the developed technology.

ACKNOWLEDGEMENT

The authors thank all colleagues at Fraunhofer ISE who contributed to this work and the German Federal Ministry for Economic Affairs and Energy for funding within the projects 'IdeAl' under contract number 0325889B and 'rEvolution' under contract number 0325586B. The excellent co-operation of the project partners is gratefully acknowledged.

REFERENCES

[1] V. Mertens, T. Ballmann, J. Cieslak, M. Kauert, A. Mohr, F. Stenzel, M. Junghänel, K. Suva, C. Klenke, G. Zimmermann, J. Müller, *Proceedings of the 28th European Photovoltaic Solar Energy Conference* (2013) 714.

[2] B. Steinhauser, M. Kamp, J. Bartsch, A. Brand, J. Benick and M. Hermle, *Proceedings of the 6th World Conference on Photovoltaic Energy Conversion* (2014).

[3] B. Steinhauser, M. Kamp, A. Brand, U. Jäger, J. Bartsch, J. Benick and M. Hermle, IEEE J. of Photovoltaics (2016) 419.

[4] J. Kumm, H. Samadi, P. Hartmann, S. Nold, A. Wolf and W. Wolke, *Proceedings of the 28th European Photovoltaic Solar Energy Conference* (2013) 1287.

[5] V. Jung and M. Köntges, Progress in Photovoltaics (2012) 876.

[6] V. Jung, F. Heinemeyer, M. Köntges and R. Brendel, Energy Procedia (2013) 362.

[7] W. P. Mulligan, M. J. Cudzinovic, T. Pass, D. Smith, and R. Swanson, US patent 7,388,147 B2.

[8] M. Kamp, J. Bartsch, G. Cimiotti, R. Keding, A. Zogaj, C. Reichel, A. Kalio, M. Glatthaar and S. Glunz, Solar Energy Materials & Solar Cells 120 (2014) 332.

[9] H. v. Campe, S. Huber, S. Meyer, S. Reiff and J. Vietor, *Proceedings of the 27th European Photovoltaic Solar Energy Conference* (2012) 1150.

[10] P. Schmitt, D. Eberlein, C. Ebert, M. Tranitz, U. Eitner and H. Wirth, Energy Procedia (2013) 380.

[11] www.schmid-group.com

[12] H. Schulte-Huxel, S. Blankemeyer, A. Merkle, V. Steckenreiter, S. Kajari-Schröder and R. Brendel, Progress in Photovoltaics 23 (2015) 1057.

RESULTS ON MODULE INTEGRATION OF IBC SOLAR CELLS BASED ON THE CONDUCTIVE BACKSHEET APPROACH

Andreas Halm[1], Bart de Gier[2], Andreas Schneider[1], Valentin D. Mihailetchi[1], Lejo J. Koduvelikulathu[1], Giuseppe Galbiati[1], Haifeng Chu[1], Razvan Roescu[1], Joris Libal[1], Nico van Ommen[2], Radovan Kopecek[1]

1: International Solar Energy Research Center - ISC Konstanz, Rudolf-Diesel Str.15 , 78467 Konstanz, Germany; e-mail: Andreas.Halm@isc-konstanz.de
2: Eurotron BV, Van Beukelaarweg 45, 2971 VL Bleskensgraaf, Netherlands

ABSTRACT: We present here results on 60-cell modules assembled with 6 inch n-type interdigitated back contact (IBC) solar cells which were accomplished in close collaboration between ISC Konstanz, who developed an industrial process to fabricate IBC cells which we call Zebra cells [1] and Eurotron BV, who supplies equipment for back contact module assembly and processed Zebra cells into 60-cell modules. Interconnection is realized by electrically conductive adhesive (ECA) gluing of cells to a conductive copper backsheet. The module sandwich is laminated with commercially available encapsulation material in a standard laminator. We assembled three 60-cell modules so far reaching 294 W, 298 W and 303 W peak power at 1 sun respectively. The cell to module (CTM) power losses ranges between 1.1 to 1.8% relative which is an excellent result taking into account the absence of ISC gain due to finger reflection which eases the CTM loss for modules assembled with front contacted solar cells.
Keywords: back contact module, IBC solar cell, conductive backsheet

1 INTRODUCTION

Back contact solar cells gain more and more industrial relevance due to their high efficiency potential and appealing appearance. Many research institutes and companies are investigating IBC cell technologies, e.g. [2,3,4], but until today, only Sunpower Corp. manages to profitably fabricate IBC solar cells and modules. One issue for industrial production is the module integration of such cells. E.g. due to the absence of a full screen printed Al layer on the rear side providing mechanical strength, IBC cells behave differently while assembled into a module. The use of standard stringers is currently not applicable. In this work we evaluate module assembly based on the conductive backsheet approach which was developed and introduced by Eurotron and ECN in 2008 for MWT solar cells [5]. Our work aims to prove that this concept can be successfully adapted for the industrial production of IBC modules.

In the following we show our progress on process development and device optimization which finally resulted in the assembly of a 60-cell Zebra module with 303 W power output using the conductive backsheet approach. To our knowledge this is the most powerful 60-cell module containing 6 inch IBC cells based on conductive backsheet worldwide.

2 MODULE PROCESS OPTIMIZATION

In the course of the module process optimization several small size modules containing one or 4 cells were fabricated. Initially, the process was adjusted to eliminate module failures; afterwards test modules and dedicated samples for contact resistance measurements were fabricated. Simulations were performed to study the dependence of the number of contact points on each busbar and the metal cross section provided for current transport on the expected module fill factor.

2.1 Contact resistance measurements

In order to qualify the specific contact resistance between IBC cell and conductive copper backsheet, dedicated samples were prepared. They allow to measure an overall contact resistance as depicted in figure 1. A sample consist of six parallel, 1 mm wide and 5 cm long copper tracks milled out of a conductive backsheet with 32 µm copper layer thickness. The tracks have a distance of 5 mm; a piece of silicon wafer with a screen printed busbar is placed facing towards the Cu backsheet. Prior to the placement, ECA is dispensed onto the Cu tracks and subsequently cured to interconnect busbar and copper as it is the case for a module during lamination. The contact height h of the ECA joint is controlled by wafer pieces as spacer placed in parallel between the copper tracks.

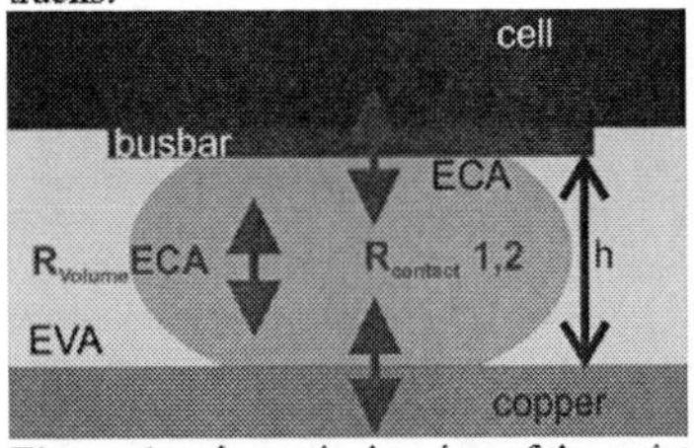

Figure 1: schematic drawing of the resistances accessible with specially prepared samples; the measured quantity is $R_{CONTACT}$ system = $R_{CONTACT\,1} + R_{CONTACT\,2} + R_{Volume}$ (h)

After sample preparation, the specific contact resistance of the system can be retrieved from 4 point probe measurements at different distances between probe and contact area in order to eliminate the series resistance introduced by the copper tracks. The sample layout chosen in this experiment allows for measuring a resistance consisting of the sum of the contact resistance between busbar and ECA, the volume resistance of the ECA and the contact resistance between ECA and copper as depicted in figure 1. Individual contributions cannot be separated but a measure for the interconnection quality for one combination of ECA, busbar paste and copper backsheet can be obtained. Figure 2 shows the measured values for different combinations of ECA and busbar paste and for two layer thicknesses h of the ECA. In one configuration the ECA contact between busbar and copper sheet is 150 µm high. For the other configuration the two components are pressed together resulting in a nearly direct contact between busbar and copper backsheet connected by a thin layer of ECA. The results show that for the investigated combinations the choice of busbar paste is more influential on the quality of the contact than the choice of the ECA. An influence of the volume resistance of the ECA cannot be observed. Comparing results with same configurations for different contact height, the samples with contact height 150 µm show even lower contact resistances than the ones with directly pressed contact.

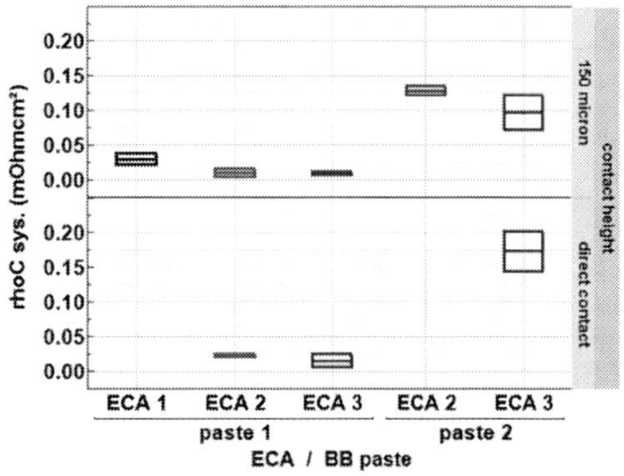

Figure 2: contact resistance measurements on the whole system busbar-ECA-copper for different material combinations and ECA layer thicknesses; the depicted dimension ρC_{system} is a combined value retrieved from the contact resistance ECA-copper, ECA-BB paste and the volume resistance of the ECA and physical dimensions of the sample.

2.2 Optimization of contact geometry

In order to minimize the fill factor losses from cell to module a simulation study was performed. The amount of contact points for each busbar on the cell's rear side was varied. Figure 3 shows the relative FF drop from cell to interconnected cell for cells with eight busbars and six, nine or twelve contact points on each busbar in dependence of the metal cross section available for current transport. The calculations are computed for an ideal case without any contact resistance and one with relatively high contact resistance between cell and the interconnection material. The results are depicted in figure 3 and show that the most influential factor to reach low FF losses is the amount of metal provided for current transport. Additionally, more contact points lead to lower losses, especially for high contact resistance. We chose nine contact points per polarity to realize FF losses of 1% relative or lower from cell to module given a low contact resistance and a high metal cross section for current transport.

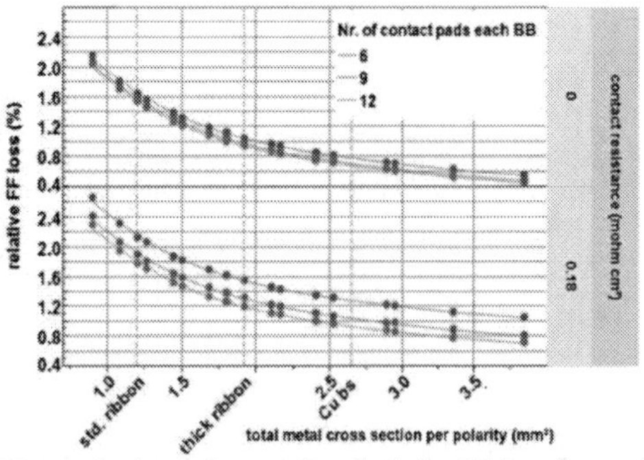

Figure 3: simulation results of relative FF loss from cell to interconnected cell versus the total metal cross section available for current transport and the number of contact points per busbar for two different contact resistances.

2.3 Results on 4-cell modules

After integrating all process optimizations, four cell test modules have been manufactured to prepare for scaling up to full size modules. IV results of the best module are shown in table 1.

	V_{OC} (V)	I_{SC} (A)	FF (%)	η (%)	P_{MPP} (W)
Cells (Σ or Ø)	2.63	9.9	79.1	21.6	20.7
Module	2.6	9.9	77.6	20.5*	20.3
CTM (%)	-0.1	-0.1	1.9	5.3	1.7

Table 1: IV results and CTM losses for a record 4-cell module assembled with conductive backsheet

3 RESULTS ON 60 CELL MODULES

Figure 4: 60-cell Zebra module assembled with conductive backsheet at Eurotron's competence center

A batch of 200 Zebra cells was processed at ISC Konstanz on M0 wafer with an area of 238.95 cm². The Zebra cells were sorted according to I_{MPP} to reach low CTM losses. IV curves were recorded with a HALM cell flasher system and 120 ms long light pulse. Table 1 shows the average IV results including standard deviation for the 180 top cells divided in three groups of 60 cells each.

group	I_{MPP} (A)		P_{MPP} (W)		$\Sigma\ P_{MPP}$ (W)
	mean	st. dev.	mean	st. dev.	
A	9.18	0.016	5.10	0.026	306
B	9.13	0.016	5.06	0.03	304
C	9.07	0.019	4.99	0.037	299

Table 2: cell binning for module assembly based on IV measurements at one sun; average I_{MPP} and P_{MPP} values plus standard deviation for group A,B,C containing 60 cells each and added power

In a joint experiment at Eurotron's competence center group A, B and C cells were processed into module A,B, C with Eurotron's back contact module assembly line and laminated on site. A commercially available 200 µm thin front encapsulation layer was used in combination with anti reflection coated module glass to minimize absorption and reflection losses. A white rear encapsulation layer promotes back reflection of transmitted light and light impinging the inactive module areas. The cell to cell gap is 1.25 mm. All three 60-cell modules were successfully fabricated. Module B was sent to Fraunhofer Callab for certification and then used as calibration reference for in house measurements of module A and C. In house measurements are performed at one sun using a HALM flash tester setup running a hysteresis measurement routine with 60 ms light pulses for each voltage sweep. Table 3 shows the IV results and relative CTM losses. The power output of module B was measured at Fraunhofer Callab with 298.3 W at one sun. Module C reaches 294 W and module A performs best with 303 W. It has a total area of 1.622 m² including its frame - this corresponds to 18.7% total area efficiency or 19.4% total aperture efficiency. This is to our knowledge the world record for an IBC cell module assembled with conductive backsheet. The CTM power losses range between 1.1% (module A) and 1.8% (module C).

		Isc (A)	Voc (V)	FF (%)	Pmpp (W)
module	A	9.97	39.3	77.1	303
CTM (%)		-0.6	0.00	1.8	1.1
module	B	9.94	39.2	76.5	298
CTM (%)		-0.7	-0.05	2.4	1.7
module	C	9.84	39.1	76.3	294
CTM (%)		-0.6	0.04	2.3	1.8

Table 3: IV results and CTM losses for three 60 cell Zebra modules assembled with conductive backsheet; results of module B are confirmed by Fraunhofer Callab

Main contribution of the CTM losses of all modules is the FF drop which is still higher than predicted by our simulations. It is partly compensated by a small gain in short circuit current. This gain results from minimized optical losses which are overweight by the bifacial gain on module level since here cells are placed against a white back reflector. On cell level though, the IV curve is recorded against a black background.

4 SUMMARY AND OUTLOOK

We prove the industrial relevance of the conductive backsheet approach for module integration of IBC solar cells by successfully assembling three 60-cell IBC modules. The champion module reaches 303 W power output at STC. All three prototypes were assembled successfully; they feature high power output and low CTM power loss which are indicators for the repeatability and reliability of the module process. The CTM analysis shows that there is still room for improvement; the gain in short circuit current can increase further by optimization of the cell's front passivation stack and the front encapsulation material in respect to reflection and absorption losses, the FF loss can be further reduced by optimizing contact formation and geometry.

With some R&D efforts the potential of our technology today is projected to 322 W for a 60-cell module by introducing minor changes:

- Cell process on M2 wafer instead of M0 (2.25% gain)
- Module assembly using only 22% efficient cells (2.8% gain)
- Increase cell spacing in the module to 4 mm (1% gain)

The resulting module is with 1.68m² slightly larger but still within market standards.

In a next step, even CBS modules can be designed capture light from both sides. This measure could drastically increase the specific energy (kWh/kW_{peak}) produced by such a module due to the bifacial nature of the Zebra cell. A first prototype which is shown in figure 5 was assembled at ISC Konstanz using half cells to minimize R_{series} losses. Its front power output is 18.1 W with a bifaciality factor (P_{front}/P_{rear}) of 0.7. Outdoor performance tests on that prototype module reveal 12% higher specific energy compared to the average of 4 standard CBS modules with white back reflector for a 6 month period (12-2015 to 05-2016) on the rooftop of ISC Konstanz on white background.

Figure 5: prototype of a Zebra half-cell bifacial module with CBS

ACKNOWLEDGEMENTS

The author wants to acknowledge the project HERCULES which has received funding from the European Union's Seventh Programme for research, technological development and demonstration under grant agreement No 608498, the project PfZ (contract no. 0325840) financially supported by the German Federal Ministry for Economic Affairs and Energy and Coveme for providing the bifacial conductive backsheet.

REFERENCES

[1] Mihailetchi V.D. et al. "High efficiency IBC solar cells fabricated on large area n-type silicon using industrial available techniques", technical digest, 21st International Photovoltaic Science and Engineering Conference 28.11.2011-2.12.2011, Fukuoka, Japan

[2] Galbiati G. et al., "Large Area Back-Contact Back-Junction Solar Cell with Efficiency Exceeding 21%", IEEE Journal of Photovoltaics, Jan 2013.

[3] Mo, C., et al. "High efficiency back contact solar cell via ion implantation." Proceedings of the 27th EU-PVSEC (2012)

[4] Zhang, X. et al. 2014, "Development of high efficiency interdigitated back contact silicon solar cells and modules with industrial processing technologies", Proc. 6th WCPEC, Kyoto, Japan

[6] M. Späth et al. "A novel module assembly line using back contact solar cells", proceedings of the 33rd IEEE PVSC 2008, San Diego, USA

SMALL UNIT COMPOUND MODULES: A NEW APPROACH FOR LIGHT WEIGHT PV MODULES

Hartmut Nussbaumer*, Markus Klenk, Nico Keller
Zurich University of Applied Science, SoE, Institute of Energy Systems and Fluid Engineering
Technikumstrasse 9, 8401 Winterthur, Switzerland
Phone*: +41 58 934 4799, Fax: +41 58 935 47 51, e-mail: hartmut.nussbaumer@zhaw.ch

ABSTRACT: A new concept for light weight solar modules is presented in this paper. The main difference to the common module type is the replacement of the frame at the laminates fringe by a lattice-like structure at the rear or at the rear and front side. Due to the smaller distances between the mechanical supporting elements the stiffness of the laminate itself can be reduced to a minimum, enabling the use of thin glass or alternative materials like for example polymer foils.

Keywords: PV module, new concept, compound, new material, construction

1 INTRODUCTION

For some PV applications the weight of PV modules is an obstacle. Today's commercial buildings are designed with little to no spare structural capacity due to cost constraints. In fact, most industrial properties built before the 1997 Uniform Building Code (UBC) change are not suitable for conventional rooftop solar installation without costly structural improvement. Therefore light weight modules may address this market segment by achieving weights, which still allow an installation on such roofs. Also for some innovative PV system solutions, e.g. PV elements for parking roofs [1], light weight PV modules may be a prerequisite. In general, a lower module weight may also result in more lightweight and thus cheaper mounting solutions increasing the safety and time-efficiency of installations. Due the tremendous price reduction of PV modules achieved in last couple of years an increasing share of the price for a PV system is related to the mounting of modules, which makes cost reductions in this field more relevant [2]. Innovative concepts based on lightweight modules and mounting structures are thus a suitable means to further reduce the total cost of PV systems.

The segment of light weight modules can be subdivided in thin film and crystalline silicon modules. Light weight thin film modules are mainly flexible structures intended for a flat installation on roof tops. However, a horizontal positioning on the ground is unfavourable with regard to energy harvest, soiling and residual water, which can lead to faster degradation of the modules. There are approaches for light weight c-Si PV modules on the basis of glass/backsheet and glass/glass laminates. The most light weight concepts for c-Si however use alternative materials to glass, e.g. ETFE as transparent medium at the front side, supported by a rigid material such as glass fibre reinforced plastic at the laminates rear.

Standard, 60 cell, crystalline silicon glass/backsheet modules typically have weight in the range of 18- 22 kg, depending on the thickness of the glass and the frame. This results in a specific weight of about 12.5 kg/m^2. The mechanical rigidness is due to the glass and a circumferential aluminium frame.

Glass/glass modules are gaining market share, due to a presumed longer lifetime and other benefits such as the potential bifaciality. Glass/glass modules may be mounted without an additional frame depending on the thicknesses of the glasses and the sub construction used. Light weight glass/glass modules have been reported from Fujipream achieving a weight of 8.2 kg using glass thickness as low as 1.1 mm and an ionomer– based encapsulant [3,4]. The size of the 215 Wp module is not given in the references, therefore a specific weight cannot be calculated.

Modules using alternative materials to glass are commercially available and achieve specific weights of as low as 2.74 kg/m^2 [5]. These semi-flexible modules use a polymeric material as front sheet and glass-fibre reinforced plastics to obtain sufficient mechanical stability.

The basic concept of standard c-Si PV modules has not changed significantly in the last couple of years and has proven its long-term reliability. However, the concept also has some inherent structural drawbacks.

Due to the laminates dimension and weight there are considerable forces which have to be considered in the module design, particularly because of the large open laminate area [6].

For glass/glass modules the glass thickness is the crucial factor which determines the modules stiffness, while for glass/backsheet modules typically also a surrounding frame provides additional stability. The basic structures of both module types cause a considerable weight and material consumption. In order to significantly reduce the weight of a standard crystalline silicon glass/backsheet module one needs to reduce the glass or the frame thickness. Reducing the thickness of the glass or the frame however drastically affects the mechanical rigidness of the module.

Currently heat strengthened solar glass is available with a thickness in the range of 2mm. According to our knowledge at present thinner solar glass is only available in form of chemically strengthened glass, with differing properties compared to the tempered standard. There are however ongoing research activities, which aim at thermally strengthened glass with 1 mm thickness [7].

In present glass/glass modules mostly glass with a thickness $\geq$ 2 mm is used to obtain sufficient stability; some of the modules with 2 mm glass thickness also use an additional frame. For the state of the art glass/glass laminate a significantly reduced glass thickness would therefore be of limited value.

Today, the typical glass thickness for standard c-Si glass/backsheet modules is 3.2 mm. Standard modules with circumferential frame have a large unsupported central laminate area. By using readily available front glass with a thickness of 2 mm, a severe dishing of the module is not avoidable with the standard setup and cannot be prevented by a surrounding frame.

Provided that cost-effective thin glass is available also innovative module concepts are needed to exploit the potential advantages.

One approach to achieve sufficient mechanical bending strength is to use a rigid supporting structure on the back of the module. This solution was repeatedly realized with ultra-lightweight honeycomb structures [8], but also repeatedly demonstrated for standard applications [9].

The current paper presents a different approach to prevent dishing for light weight c-Si laminates, by using a lattice instead of a backing plate, as shown in figure 1. This approach subdivides the module into smaller units resulting in reduced lever arms between supporting points and therefore suppressed dishing. Due to the subdivision of the modules in smaller mechanical segments the modules are named Small Unit Compounds (SUC) modules. The concept with a lattice is in-between the above presented standard c-Si module with unsupported laminate and the use of a continuous substrate plate.

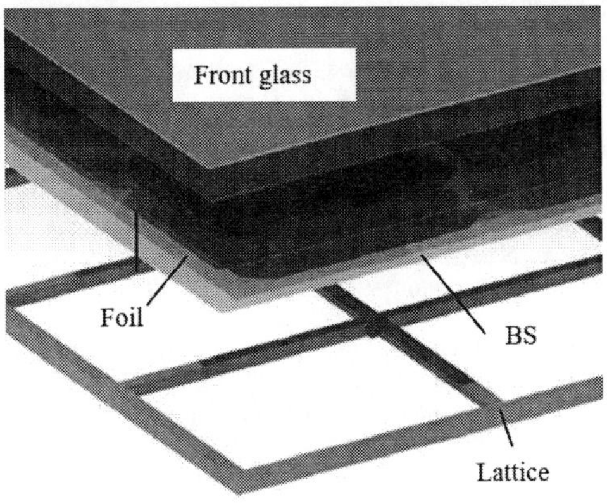

Figure 1: SUC modules with conventional glass/backsheet laminate and lattice structure at the rear.

Because of the smaller distances between the frame elements the laminates stiffness and weight can be reduced. This enables new approaches for the laminate structure, ranging from conventional layouts with reduced weight, for example glass/glass laminates using ultrathin glass, to the implementation of alternative non-glass materials. Besides the laminate also the lattice frame may be made of lightweight materials and components.

2 MECHANICAL STABILITY OF SUC MODULES

In order to get a first appraisal of the mechanical properties some trials with similar lattice-like structures were carried out. Since standard c-Si glass/backsheet modules typically use 3.2 mm glass, we decided to prepare a corresponding laminate, however with thinner 2 mm glass, resulting in lower weight and rigidness.

Several lattice-like structures were tested in combination with the described 2 mm glass/backsheet laminate. The maximum bow of the laminate was respectively measured as a function of the applied mechanical load. The mechanical load varied by using different amounts of sandbags, each weighing 12 kg (see figure 2) to a maximum of 144 kg. In the experiments the modules were supported lengthwise by tables to simulate the sub-construction. The same tests were also performed with a c-Si 60 cell standard module, with glass/backsheet lay-out, frame and 3.2 mm glass.

Figure 2: Laminate with applied mechanical load in form of sandbags on top.

The lattice was simulated by aluminum beams of quadratic shape with side lengths of 10 and 15 mm respectively and a material thickness of 1 mm. The beams were not fixed at the intersections of the perpendicular oriented beams. In figure 3 the set-up of the beams is indicated for clarification.

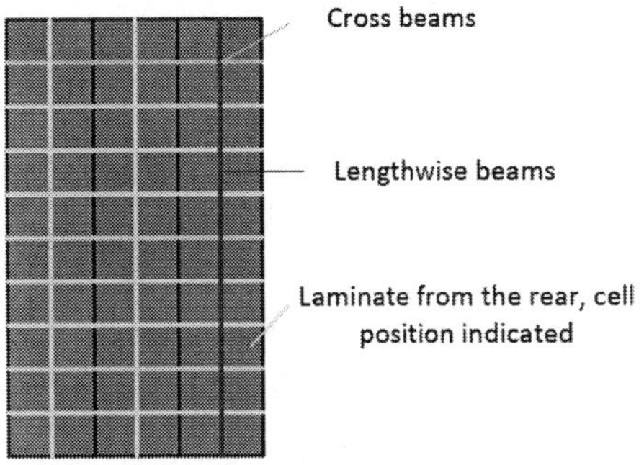

Figure 3: Schematic drawing of the aluminum beams position at the rear of the laminate.

Due to the limited strength of our test rig the maximum applied load was 882 Pa which is considerably lower than the value of 2400 Pa, as required to pass the IEC 61215. However, the results are an indication for the mechanical stiffness and can be compared to the data of a standard module.

In figure 4 the maximum bow as a function of the load for various structures is shown. The standard 60 cell glass/backsheet module with a 3.2 mm front glass and frame shows a maximum bow of about 15.5 mm at a load of 144 kg. The 2mm glass/backsheet laminate without a frame and any supporting element shows extreme bowing already for lowest load values. Therefore this thin laminate itself provides a very small contribution to the mechanical rigidness of the module. As shown in figure 3 the beams were positioned lengthwise or crosswise. Only the crosswise beams were supported by a sub-construction.

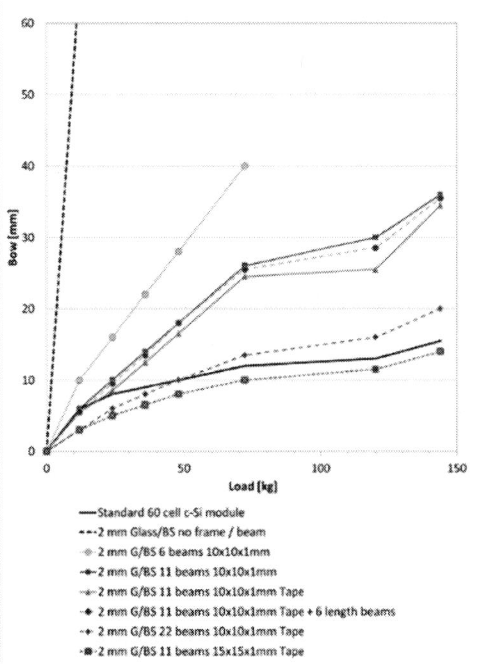

Figure 4: Maximum bow as a function of load for several lattice-like structures and a standard module. Due to the limited strength of our experimental set-up the maximum load is limited to 144 kg, which is considerably lower than the loads according to the IEC 61215.

First 6 lengthwise beams were taped to the backsheet, 11 beams were then positioned crosswise below them without any fixing. This was compared with a composition of 11 crosswise beams with and without fixation to the module. The fixation was done with a double sided adhesive tape.

As shown in figure 2 (blue lines with symbols), there is no significant difference between the three compositions. Therefore one can conclude, that only the cross beams, supported by an appropriate sub-construction have a significant influence on the mechanical rigidness of the structure as a whole. Neither the taping, nor the varied placement in direct contact with the backsheet or below the lengthwise beams showed any significant effect.

In the following experiments we have analyzed how the amount and the dimensions of cross beams affect the measurements. Reducing the amount of cross beams from 11 to 6 has a tremendous effect on the mechanical rigidness of the module as shown in figure 4.

Enhancing the cross-section from 10 x10 to 15 x 15 mm and keeping the number of cross beams constant at 11 resulted in a bow slightly lower as the bow observed for the standard module. This indicates that this structure could be a promising candidate for a light weight SUC module.

As material for the lattice aluminum or glass fibre reinforced plastics may be used, dependent on the specific requirements. First estimations, based on the elasticity module of the aluminum beams, do not indicate that a significant weight reduction can be expected by replacing the aluminum. Using plastics may however be beneficial if not only cross beams but a real lattice structure is chosen.

Based on this lattice-like structure in combination with the 2 mm glass/backsheet laminate we made an estimation concerning the obtainable weight and compared it to a standard c-Si module and the most light weight 60 cell c-Si glass/backsheet module which is currently available on the market. The results are shown in table 1.

Table 1: Weight and specific weight for different types of 60 cell c-Si modules

Type	Weight [kg]	[kg/m^2]
Standard c-Si 3.2 mm glass/bs, frame	~20 kg	12
SUC lattice at rear, 2 mm glass/bs	13.65 kg*	8.5
Light-weight c-Si Glass/bs, frame, BenQ	10.5 kg	6.5
SUC lattice at rear 0.8 mm, glass/bs	8.9 kg	5.6

*12 kg laminate + 1.65 kg (11 cross beams 15 x 15 x 1 mm) = 13.65 kg

The lastly described structure with lattice and rear and 11 cross beams (15x15x1 mm) would result in a module with a weight of about 13.7 kg. This is considerably lower than a standard module with a weight from 18 to 22 kg, but still heavier than the best in class type with only 10.5 kg [10]. Alone the laminate which was used by us has a weight of 12 kg. This means that the best in class module type also uses considerably thinner glass than the laminate prepared by us. Since the use of very thin and flexible substrates should be a major advantage of the proposed lattice concept, we estimate a potential module weight with thinner glass. According to our knowledge the thinnest available solar glass presently is chemically strengthened 0.8 mm glass from AGC. Based on this assumption a module weight of 8.9 kg may be obtained with a lattice at rear.

Even lower weights might result if the concept of a back lattice is combined with ultra-light weight non-glass laminates. In such laminates the glass would be replaced by transparent layers at the front side. There are several options for suchlike frontsheets based on different materials (ETFE, glass fibre reinforced plastic, polyester,…) which were repeatedly tested and are also used in currently available products. The combination of both approaches would result in light weight mechanically rigid PV modules, which may be used in the same way as standard modules. In addition also larger light weight modules could be an interesting option. Because the cross beams could be also seen as a part of the sub construction, it may be an option to mount them on-site to the appropriate sub construction, e.g. by silicon adhesive.

For very thin and flexible laminates it might be an option to use a lattice at the rear and front of the laminate. This structure would have some advantages for laminates with a symmetrical structure, for example foil/cell/foil as shown in figure 5 or glass/glass laminated with very thin glass.

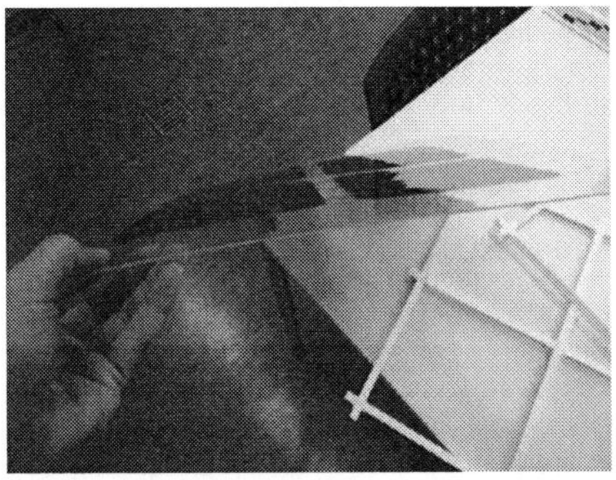

Figure 5: Foil/cell/foil laminate. Both foils are transparent which would enable the use of bifacial solar cells.

If the laminate has a symmetrical structure, a cell in the middle of the symmetric structure does not experience compression- tension forces (neutral fibre) as shown in figure 6.

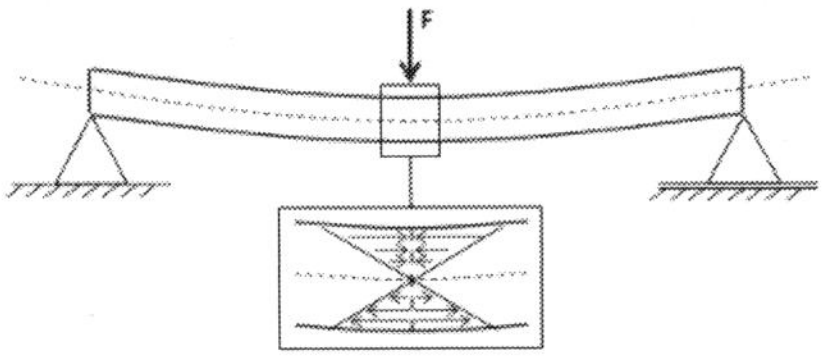

Figure 6: Schematic drawing of a solar cell (dotted red line) in the center of a symmetrical structure resulting in lowest compression- tension forces.

Pursuing the symmetrical approach with regard to the lattice one has to apply a lattice at front and rear. We have designed a demonstrator lattice out of plastic fabricated by the use of a 3D- printer.

Figure 7: Lattice at front and rear with a foil solar cell/ foil laminate using bifacial solar cells. The cross-section of the upper and lower part of the lattice has a triangular shape to reduce the shading and to enhance the light scattering from the reflective lattice material.

First results concerning the mechanical rigidness of a lattice front and rear structure are included in figure 3. The structure that was tested consisted of 11 front and 11 rear aluminum bars with cross sections of 10 x 10 mm. The maximum bow as a function of the load for this structure is comparable to the bow of a standard module.

Obviously, the front and rear lattice approach has some disadvantages. Besides the shading for low sun positions soiling may be an issue. For very thin and fragile laminates the resistivity to mechanical damage, as caused by hail, is another problem that needs to be addressed. Possibly, the front side grid structure may enable an option to address these issues. A transparent sheet may be used to protect the laminate as indicated in figure 8. At a first glance the use of such a foil seems too fragile and prone to damage. However, there are comparable solutions in architectural and BIPV applications with ETFE foil. This material has a very good transparency and an extremely high resistivity against crack-propagation. Pressure stabilized ETFE structures were also repeatedly used to include PV modules [11]. The feasibility of the proposed concept needs to be verified in experiments.

Figure 8: Lattice at front and rear with a foil/ solar cell/ foil laminate covered on both sides with an ETFE foil resulting in an ultra-light weight SUC module. The ETFE foil acts as soiling protection and may also act as a protector against hail impact (to be verified). The triangular shaped frame is the bearing for the ETFE foil.

The lattice structure may possibly also be used to generate small unit compound elements in a way that the lattice also includes electrical elements such as cables or bypass diodes. These small elements would be mounted then on an appropriate sub-construction with mechanical and electrical connectors as indicated in figure 9.

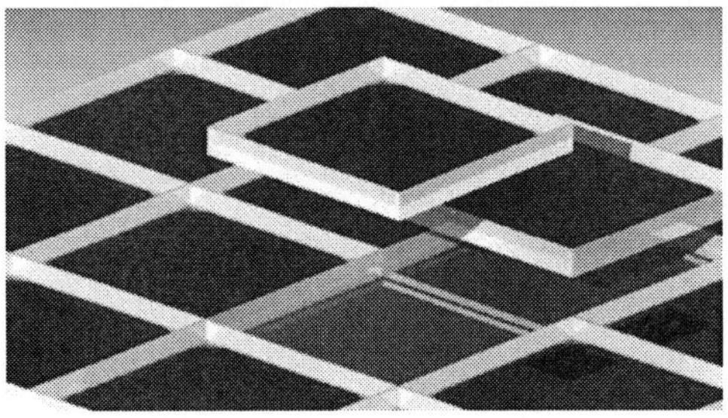

Figure 9: Small units, here a single solar cell, may be placed on an appropriate designed sub-construction hosting mechanical and electrical joints.

4 DISCUSSION, CONCLUSION AND OUTLOOK

The replacement of the frame by a lattice like structure opens the path towards the use of glass with thickness below 2 mm or the use of very thin and flexible non-glass laminates. The simplest approach is to use cross bars with sufficient stiffness, as indicated in figure 10. The cross beams can either be a part of the module or of the sub-construction. The cross beams have to be fixed on a supporting element, e.g. a perpendicular oriented bar.

This concept is an evolutionary approach towards more light weight structures. Conventional glass/backsheet and glass/glass laminate lay-outs can be used with lower glass thicknesses than the currently used ones. For standard glass/backsheet laminates the glass thickness could be lowered to readily available 2 mm as a first step. Chemically strengthened thin solar glass with 0.8 mm thickness is already available. Heat strengthened solar glass with 1 mm thickness is subject to recent research. The concept should however also be suitable for even thinner and more flexible non-glass laminates. With the lattice at rear concept specific weights of <6.5 kg/m^2 and sufficient mechanical stiffness should be feasible.

Applying the lattice-like structure not to the laminate but to the sub-construction would further lower the laminates transport weight and the transport volume. Also considerably larger modules may possibly be realized with this approach.

Figure 10: Type 2 SUC module with the subjacent supporting structure in form of cross bars.

A lattice at the rear and the front might be a suitable approach for laminates with symmetrical lay-out. Keeping the symmetrical structure of the laminate is beneficial in order to suppress tension-compression stress. Obviously, the front and rear lattice approach has some disadvantages. Besides the shading for low sun positions soiling may be an issue. For very thin and fragile laminates the resistivity to mechanical damage, as caused by hail, is another problem that needs to be addressed. Possibly, the front side grid structure may enable an option to address these issues. A transparent ETFE sheet may be used to protect the laminate, but the feasibility of the proposed concept has to be verified in experiments first.

5 ACKNOWLEDGEMENT

The development is done in the European project U-Light: Ultra lightweight PV modules, in the solar-era.net program. The development is financially supported by the Swiss Federal Office of Energy under the contract SI/501280-01.

6 REFERENCES

[1] F. Baumgartner et. al., "Solar ski lift pv carport and other solar wings cable based solutions",
Proceedings of the 27th European Photovoltaic Solar Energy Conference, Frankfurt, pp. 4343 – 4347, 2012

[2] F. Baumgartner et. al., ""Urban Plant"" Light-Weight Solar System for Parking and other Urban Double Use Applications , 4CO.12.2, 28th EUPVSEC, Paris, pp. 2897 – 2901, 2013

[3] pv magazine, News 1st of March 2012, "PV Expo Japan: solutions for a unique market"

[4] PhotonicsOnline, News October 13th, 2010, New DuPont Encapsulant and NSG Thin Glass Enable Fujipream To Build World's Thinnest Crystalline Silicon Solar Panel

[5] DAS Energy GmbH, Data sheet: Module DAS-Serie 240 – 250P

[6] A. J. Beinert et al. "Influence of photovolataic module mounting systems on the thermos-mechanical stresses in solar cells by FEM Modeling", EUPVSEC, Munich, 5BV.1.14, 2016

[7] LIMES, Light Innovative Materials for Enhanced Solar Efficiency, http://www.limes.nu/

[8] Julien Rion
Ultra-Light Photovoltaic Composite Sandwich Structures
EPFL, THÈSE NO 4138, 2008

[9] Evonik, PLEXIGLAS® for Solar Applications,
http://www.plexiglas.net/product/plexiglas/Documents/Marktsegmente/511-4-Solar-Applications-en.pdf

[10] BenQ Solar claims title for world's lightest crystalline silicon PV module
by Nitin Nampalli on March 13, 2014 in Solar Choice News, http://www.solarchoice.net.au/blog/news/benq-solar-claims-title-worlds-lightest-crystalline-silicon-pv-modules-130314g

[11] Taiyo-europe,
http://www.taiyo-europe.com/en/tensile-etfe-architecture/projectportfolio/abfallwirtschaftsbetrieb-muenchen/detail/

RECONFIGURABLE TOPOLOGIES FOR SMARTER PV MODULES: SIMULATION, EVALUATION AND IMPLEMENTATION

Pieter Bauwens[*,a], Jonathan Govaerts[b], Maro Baka[c], Francky Catthoor[b, d], Kris Baert[d], Giel Vandenbroeck[d], Hans Goverde[b, d], Dimitris Anagnostos[b, c], Jan Doutreloigne[a], Jef Poortmans[b, d, e]
[a]CMST-ELIS, imec and Ghent University, Technologiepark 15, B-9052, Ghent, Belgium
[b]imec, Kapeldreef 75, B-3001, Leuven, Belgium
[c]National Technical University of Athens, Heroon Polytechneiou 9, Zographou Campus, 157 80 Athens, Greece
[d]ESAT/ELECTA, KU Leuven, Kasteelpark Arenberg 10, B-3001, Heverlee, Belgium
[e]University Hasselt, Martelarenlaan 42, B-3500, Belgium

ABSTRACT: In a reconfigurable PV panel, the connections between groups of cells (cell-strings) can be rearranged in order to increase the overall power output under non-uniform conditions. In a traditional PV panel, current mismatch between the series-connected cells (due to irradiation differences of differential aging) limits the generated power output by either reducing the overall current or by bypassing the substring containing the limiting cells. A reconfigurable panel can respond to such a situation by changing the connections to a more favorable configuration, allowing the majority of the cells to operate at their MPP. This paper elaborates on such a topology and its components (high current switches, wiring). To get a clearer picture of the added value of such a reconfigurable panel, a detailed energy yield analysis was performed which includes the performance of those added components. The simulation model used for this analysis is calibrated with extensive measurement results. A comparison of that model and corresponding measurements will be discussed. The analysis shows that a reconfigurable PV panel can outperform the traditional panels when suboptimal conditions are predominant, as encountered for instance in BIPV. As a demonstration of the principle, a small-scale prototype was created and will be presented in this paper.
Keywords: Energy Performance, PV System, Shading, Configurable Topology

1 INTRODUCTION

In a conventional PV panel, the (e.g. 60) individual cells are place in series for practical reasons. A high-voltage low-current power is easier to manage than a low-voltage high-current power. But this approach comes with the disadvantage that when the panel is partially shaded, the current in all serially connected cells is limited to that of the poorest performing cell. The output power drops disproportionally to the affected area. A small shade on a single cell may disrupt the power output of the entire panel. Furthermore, hot-spots can arise that may permanently damage the panel. This problem is well-known and solved (or circumvented) by placing a bypass diode over a group (substring) of cells, typically 15 to 20 cells.

It is clear that this is still not an ideal solution. Those bypass diodes dissipate a significant amount of power when active. A conventional Schottky diode easily dissipates a couple of Watts when an 8-10A current is surging through it. There are alternatives available that can reduce that lost power [1, 2, 3, 4, 5] by limiting the voltage drop over the bypass element. But a more important contributor to the power loss is the fact that the bypassed substring might still generate power that is completely lost when bypassed. That substring might contain some fully illuminated cells, but even the cells that are in the shade will still produce power. A reconfigurable topology, as presented in [6, 7, 8, 9] and in this paper, can reclaim a lot of that lost power.

Such a topology introduces a lot of new components that are not lossless. It is important to have a clear picture of the performance of this topology and how it compares to the conventional panels. In this paper, we will elaborate on a reconfigurable topology, its components and resulting performance. To ensure optimal accuracy of the results, the used simulation framework will be matched against measurement results.

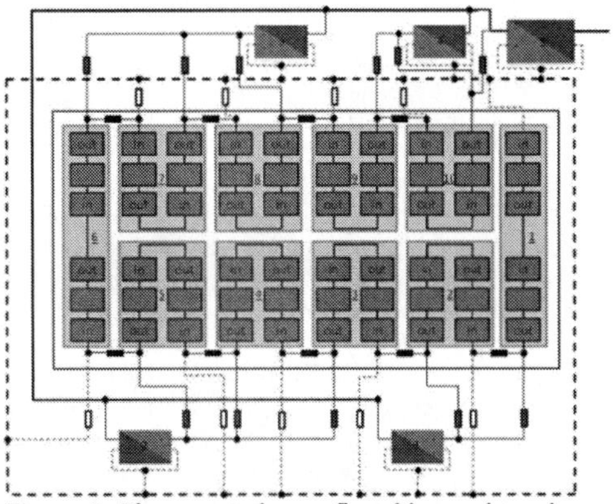

Figure 1: The proposed reconfigurable panel topology. There are 10 6-cell substrings, 4 string level DC/DC converters and one panel level DC/DC converter.

2 COMPONENTS

2.1 Topology

The reconfigurable topology that will be analyzed is depicted in Figure 1. A reconfigurable panel is divided in cell-strings. The cells within a cell-string have a fixed series connection and any mismatch occurring within such a cell-string cannot be handled. A smaller granularity seems beneficial, but also requires more components and complex wiring. The discussed simulation framework is essential to weigh these different aspects against each other as it is impossible to create all different topologies. The resulting compromise is the 6-cell cell-string topology from Figure 1. Under uniform illumination, all substrings will be connected in series. The snake-like structure of the topology ensures minimal additional resistive losses due to the added wiring, as compared to the standard wiring scheme. To ensure an efficient series connection, all neighboring cell-strings are connected through low-ohmic switches (black switches in Figure 1). The panel-level DC/DC converter maximizes the power output.

Non-uniform operating conditions are determined by a deviation between the instantaneous I-V curves of the cell-strings. In order to support operation of all cell-strings as close as possible to their MPP, parallel connection of cell-strings is recommended as the divergence of the voltage at the MPP under different operating conditions is limited. The main issue for parallel connection is the accumulations of the current which leads to increased resistive losses. In order to limit this current accumulation, and thus the resistive losses, the parallel connection of cell-strings is done in two steps. The cell-strings are divided in groups and each group is connected to a single converter. For each group, this small local DC/DC converter boosts the voltage, in order to limit the current. The local converters are then connected in parallel and accumulate the current through a low resistivity line which connects to the panel-level DC/DC converter of the module.

This topology has two configurations. The resistive losses in the parallel connection of the cell-strings are higher than in the series connection. The switch between the two configurations depends on the level of non-uniformity. The recovered power due to different operation conditions should compensate for the additional resistive and conversion losses. An in depth discussion of the topology can be found in [9]. The switches that are required to make this reconfigurable panel topology possible are discussed in the next section.

2.2 High-current switches

The most crucial feature of these switches is their low voltage drop. There will always be switches present in the current path, even under uniform illumination. It is paramount for the overall performance that the voltage drop over the switches (and thus the dissipated power) is as low as possible. We aimed at a maximum on resistance of a couple of mΩs per switch, to limit their voltage drop to a couple of tens of millivolts under a full load of 10A.

The switches are controlled by a central unit that determines the optimal configuration for the current shading scenario. Every switch has a dedicated address and can be communicated with over a common data line (2 common wires, not shown in Figure 1) for all switches. Power for all switches is also provided by the central unit (1 common wire, not shown in Figure 1). An extra challenge comes from the fact that most of the switches have a different voltage potential. The reference voltage of a switch is dependent on the location in the panel. Moreover, this reference voltage can and will change if the configuration of the panel changes. Communication and voltage supply need to be ensured irrespective of the reference voltage of the switch. A full elaboration on the method of operation of such a switch can be found in [10].

The switches are designed in the I3T50 technology of On Semiconductor. This smart-power IC technology is a 50V extension of a standard 0.35μm CMOS process. We created both a low-voltage (LV, 3.5V) and a high-voltage (HV, 50V) version. HV switches are inherently larger and thus more expensive than their LV counterpart. Limiting their amount in the panel reduces the cost. The LV version was designed for a maximum load current of 10A with an on resistance of 2mΩ (SOMBRA-LV10) and the HV version was designed for a maximum load current of 5A with an on resistance of 8mΩ (SOMBRA-HV05). Figure 2 gives an idea of the layout; the total die area here is 13mm². When two switches are placed in parallel and given the same address, they will act as one switch with half the on resistance and double the current load capacity.

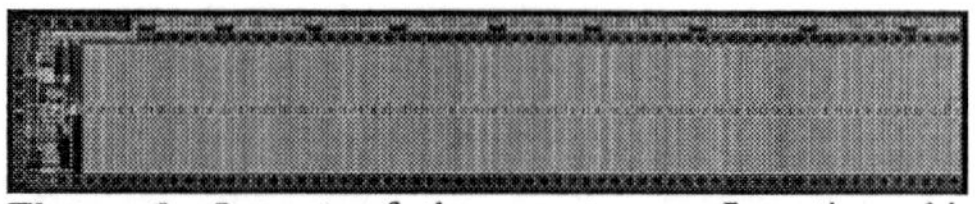

Figure 2: Layout of the smart reconfiguration chip (SOMBRA-HV05)

3 ENERGY YIELD ANALYSIS

Figure 3 shows the measured on resistance of the SOMBRA-HV05 and the SOMVBRA-LV10 chips. They both perform somewhat better than designed. For the HV version, we measured an average on resistance of 7.3mΩ at the maximum current of 5A. The LV version measured an average on resistance of 1.3mΩ at its maximum load current of 10A. These results will be used in the following energy yield simulation.

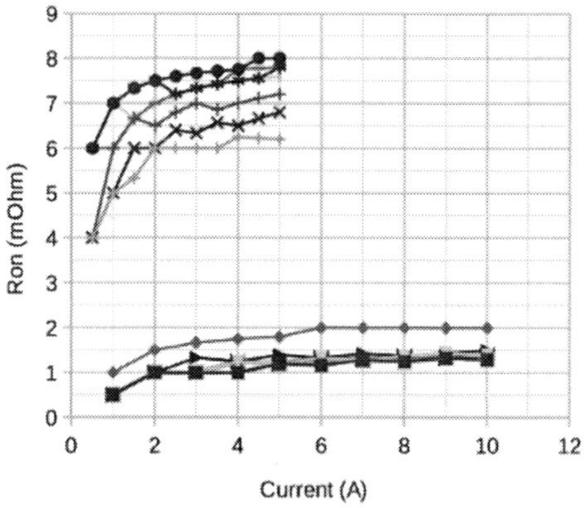

Figure 3: Measured on resistance for the SOMBRA-HV05 (7.3mΩ) and SOMBRA-LV10 (1.3mΩ) switches

The reference model in these simulations is a conventional 6-by-10-cell module, with 3 20-cell, bypassed substrings. Uniform operating conditions are expected to be frequent. The best configuration of the reconfigurable module in that case is an all series connection of the cell-string. The aim here is to have negligible losses. We minimized the number of switches on the current path and the wire length, but the total resistive losses will still be larger than in the conventional module. It needs to be mentioned that these losses are limited to about 0.4%. They depend on the current produced which in turn depends on the irradiation level and the cell's technology. The operation of the reconfigurable module can be improved by increasing the cross-section of the wires on the edges of the module and by minimizing the on resistance of the switches, as explained above. This is a trade-off between manufacturing cost and performance, but with a proper selection of the wires, the performance of the conventional module can be reached.

Table I: Comparison of conventional module and reconfigurable module with a single shaded cell (normalized power)

Cell shade density	Normalized power	
	Conventional module	Reconfigurable module
10%	0.991	0.987 (S)
20%	0.957	0.954 (S)
30%	0.890	0.911 (P)
40%	0.800	0.902 (P)
50%	0.693	0.892 (P)
60%	0.650	0.881 (P)
70%	0.650	0.871 (P)
80%	0.650	0.860 (P)
90%	0.650	0.849 (P)

Under non-uniform conditions, not all cells in the conventional module can operate at their MPP, leading to energy loss. The presence of the diode allows bypassing of "poorly" performing cells (the entire bypass section) when their operation significantly degrades the power output of the entire module. This depends on the level of non-uniformity of the module – i.e. shading pattern, shading density etc. In the conventional module, a single shaded cell affects the bypass section it belongs to and multiple peaks appear in the IV curve of the module. In the reconfigurable topology, a single cell affects the operation of one cell-string (panel in parallel configuration, P), or the operation of all the cells (panel in series configuration, S). The shading density determines whether the recovered power in a parallel configuration can compensate for the additional resistive losses. Table I compares the normalized power output for a conventional module with that of the reconfigurable module, both with a single shaded cell under different shading densities. It indicates that a shade density of about 30% justifies the switch to the parallel configuration. Under higher shading densities, more significant gains are present despite the limited size of non-uniformity. Even more significant gains arise when more bypass sections of the conventional module are affected. Under high density shades, gains up to 200% were observed. It is not trivial to provide an 'overall' expected energy gain, as this would require stochastic shade distribution.

Scenarios can occur where the reconfigurable topology underperforms. The worst case scenario is a shade that covers the middle substring of a conventional module. The granularity level of the reconfigurable topology does not allow recovering any lost power in that case, while the bypass diode of the conventional module allows isolation of the shaded cells. Here we can expect about 35% less power from the reconfigurable module. This is an extremely specific scenario and is not expected to occur frequently in reality. Overall gain from other non-uniform operating scenarios should compensate for these extremely rare cases.

In order to guarantee accuracy, the simulations are based on a module which is fitted to real data (See Figure 4). The parameters for a single cell are extracted based on a physics based model [11, 12]; the efficiencies of the local and panel-level converters are 95% and 98% respectively. Measurements of the IV curves of cell-strings are performed in a controlled environment. When comparing the measured IV curves with the simulated IV curve we notice a marginal error, which is expected due to measurement errors.

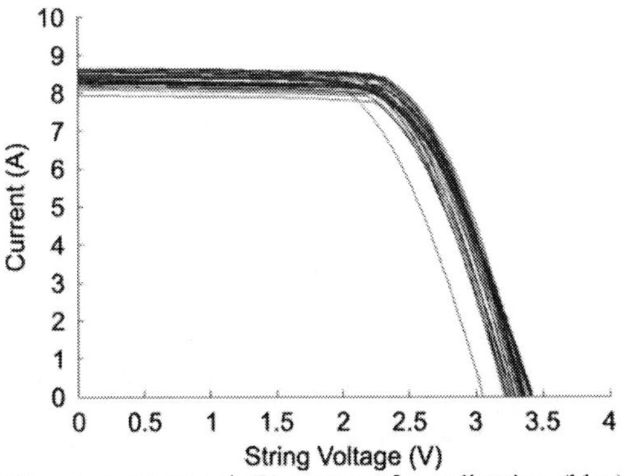

Figure 4: Measured IV curves of a cell-string (blue) compared to the simulated results of the fitted model (red)

4 DEMONSTRATOR

For demonstrating the principle of a first "smart" reconfigurable PV module, we designed mini-modules, using cells with a reduced (quarter) size. The cells are connected as 2 6-cell U-shaped strings, as indicated in Figure 5. For each string we provide a bypass diode (active), and 3 switches to allow either a series of parallel configuration of those strings. For the highest reliability potential, we want to work with glass-glass modules, a clear trend emerging in fact for all PV modules [13]. But of course, the same design could be applied to a glass-backsheet approach, and then be more tolerant to component topography.

For the first prototype we used commercial components 2-side contacted standard full Al BSF cells from TSEC (TSS62TN), active bypasses from STMicroelectronics (SPV1001N40 [5]) and FET switches from International Rectifier (IRF7769L1TRPbF). The choice for the bypass and switching components was mainly driven by thickness (minimized topography for glass-glass lamination, See Figure 6), contact pad size (allowing manual soldering) and resistive losses. These switches do not provide the necessary functionality to be employed in a full reconfigurable panel. These will be implemented in the next stage (see below).

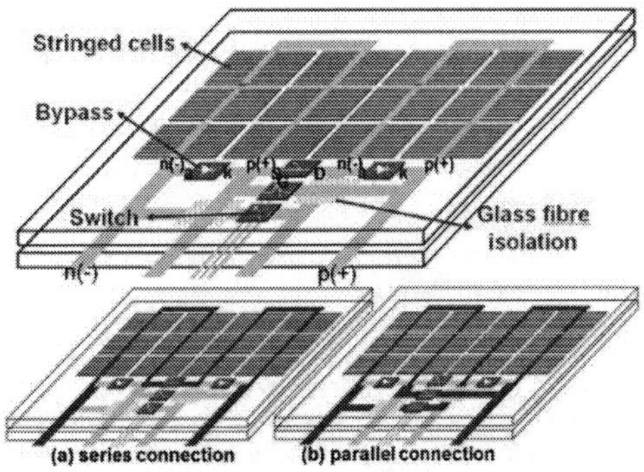

Figure 5: Design and layout for the prototype mini-modules, indicating electric paths for (a) series and (b) parallel configuration

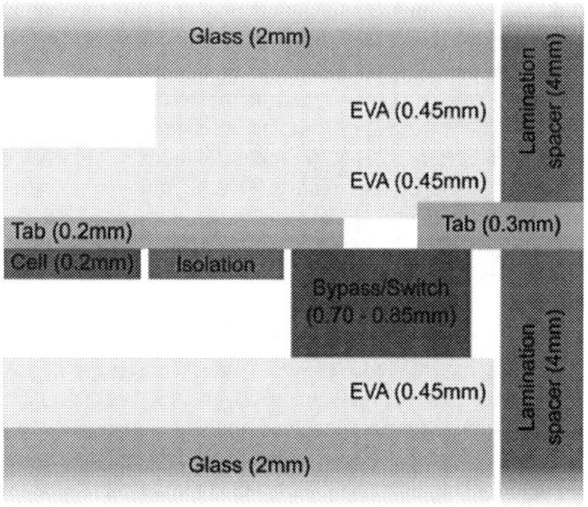

Figure 6: Schematic representation of the cross-section of the reconfigurable panel with integrated components

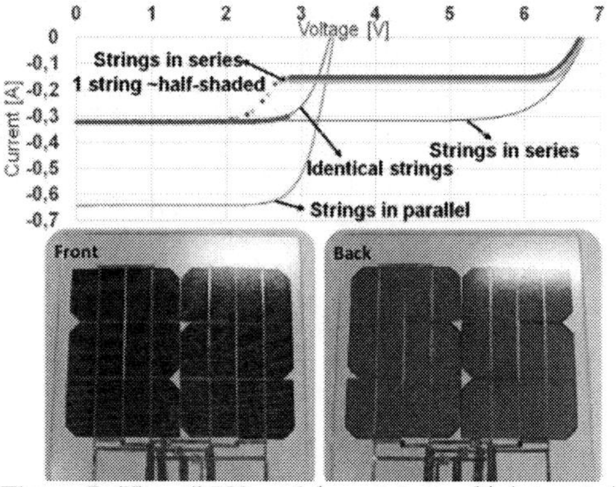

Figure 7: "Smart" PV module prototype with integrated bypasses and switches: electrical functionality (top) and pictures (bottom)

The cells are first soldered into strings, aligning the ribbons to the busbars and using a standard tabbing-stringing process (hotplate preheating @170°C and 2s contact soldering @ 238°C) and materials (flux and Sn62Pb36Ag2-coated Cu ribbons). Then the bussing ribbons and other components are manually soldered according to the designed layout.

For lamination, we used standard EVA encapsulant (EVASKY from Bridgestone) and 2-mm-thick textured glass (non-tempered) from Ducatt. A woven (black) glass fiber tissue is used for insulation where ribbons need crossover. A standard EVA lamination cycle is applied (totaling 15 min at a temperature of 150°C).

This first prototype has been tested for functionality, using an in-house developed illumination setup [14]. Partial shading has been applied on either of the strings to verify the functioning of the bypasses, and the configuration has been switched between series and parallel. The resulting proof-of-concept module and graphs are shown in Figure 7.

In the last stage, the commercial switching devices were exchanged with the custom-made SOMBRA-HV05 switches. These were not yet integrated into the panel itself, but rather connected externally. Figure 8 shows the measurement results using the same illumination setup as above. A microcontroller controls the three SOMBRA switches through the I2C protocol and is able to change the configuration from a series connection to a parallel connection and back.

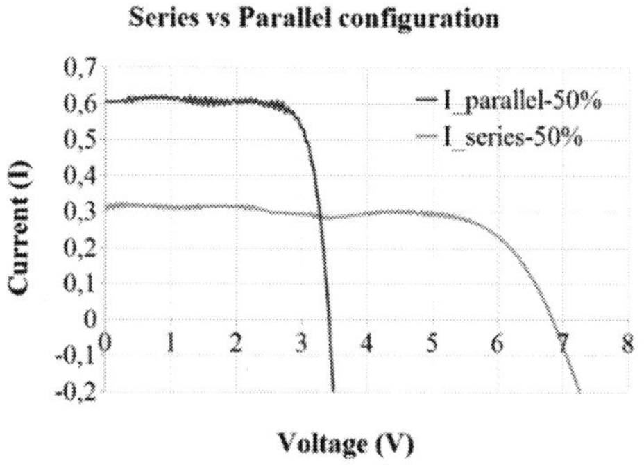

Figure 8: Measurement results of the reconfigurable mini-module equipped with the custom-made SOMBRA switches. The strings can be configured to be in parallel (blue) or in series (orange)

5 CONCLUSIONS

In this paper, we presented the reconfigurable module topology together with its components. The switches necessary to realize such a module have a measured on resistance of 7.3mΩ (HV) and 1.3 mΩ (LV). Using this information, simulation showed that under non-uniform conditions with a shade density of >30% the reconfigurable topology can outperform the conventional module. It is expected that on average, the reconfigurable module will provide a significant power gain in shade-sensitive environment such as BIPV. As a proof-of-concept, a small-scale reconfigurable module was created, with integrated commercial devices. Another version employs the custom-made SOMBRA devices. In a later stage, these devices can also be integrated.

6 ACKNOWLEDGMENTS

The authors gratefully acknowledge the financial support of the Flemish government through the funded project SBO SmartPV, as well as imec's industrial affiliation program and its partners.

7 REFERENCES

[1] P. Bauwens and J. Doutreloigne, *"Reducing Partial Shading Power Loss with an Integrated Smart Bypass"*, Solar Energy, vol. 103, pp 134-142, 2014.

[2] P. Bauwens and J. Doutreloigne, *"Low voltage drop unidirectional smart bypass elements"*, Patent US9082915, 2013.

[3] S. Anthony, *"Low Voltage Drop Closed Loop Unidirectional Electronic Valve"*, Patent US8169252 B2, 2009.

[4] Texas Instruments, *"SM74611, Smart Bypass Diode"*, [Online]. Available: http://www.ti.com/product/SM74611. [Accessed 25 January 2016].

[5] STMicroelectronics, *"SPV1001N Datasheet,"* [Online]. Available: http://www.st.com/web/en/resource/technical/document/datasheet/DM00031860.pdf. [Accessed 25

January 2016].

[6] G. Velasco-Quesada, R. Piqué-López and A. Conesa-Roca, *"Electrical PV Array Reconfiguration Strategy for Energy Extraction Improvement in Grid-Connected PV Systems,"* IEEE Transactions on Industrial Electronics, vol. 56, no. 11, pp. 4319-4331, 2009.

[7] F. Catthoor and M.-I. Baka, *"Reconfigurable PV Configuration".* Patent EP2587334A1, 2013.

[8] M.-I. Baka, A. Papanikolaou, F. Catthoor and D. Soudris, *"Configurable Module Topology to Recover Power Lost due to Current Mismatch,"* in 29th EUPVSEC, Amsterdam, 2014.

[9] M. Baka, F. Catthoor and D. Soudris, *"Smart PV Module Topology with a Snake-Like Configuration,"* in 31th EUPVSEC, Hamburg, 2015.

[10] P. Bauwens and J. Doutreloigne, "Integrated Switch for Substring Reconfiguration to Optimize Module Power under Partial Shading," in 31th EUPVSEC, Hamburg, 2015.

[11] D. Anagnostos, H. Goverde, F. Catthoor, D. Soudris, J. Driessen and J. Poortmans, *"Demonstration and validation of an energy yield prediction model suitable for non-steady-state non-uniform conditions,"* in 6th World Conference on Photovoltaic Energy Conversion, Kyoto, 2014.

[12] H. Goverde, B. Herteleer, D. Anagnostos, G. Kose, D. Goossens, B. Aldaladi, J. Govaerts, K. Baert, F. Catthoor, J. Driesen and J. Poortmans, *"Energy yield prediction model for PV modules including spatial and temporal effects,"* in 29th EUPVSEC, Amsterdam, 2014.

[13] ITRPV, *"International Technology Roadmap for Photovoltaic,"* 2015.

[14] N. Ruas, J. Govaerts, H. Goverde and F. Catthoor, *"Development and Evaluation of a Low-Cost Dynamic Illumination Setup for Advanced Photovoltaic Modules Performance Testing,"* in 31th EUPVSEC, Hamburg, 2015.

PHOTOVOLTAIC-PERFORMANCE-ENHANCING PATCH WITH COMBINED LIGHT TRAPPING AND SPECTRAL UPCONVERTING EFFECT

D. Desta[*], E. Eriksen, M. Bellettato['], R. Rizzoli['], C. Summonte['], B. R. Jeppesen, P. B. Jensen, S. P. Madsen[**], A. Nylandsted Larsen, P. Balling, Sanjay K. Ram[§]

Department of Physics and Astronomy-iNANO, Aarhus University, Gustav Wieds vej 14, 8000 Aarhus C, Denmark
[*]Department of Physics and I3N, University of Aveiro, Campus Universitário de Santiago, 3810-193 Aveiro, Portugal
[**]Department of Engineering-Nanoengineering, Aarhus University, Inge Lehmanns Gade 10, 8000 Aarhus C, Denmark
[']IMM – Consiglio Nazionale delle Ricerche, via Gobetti 101, 40129 Bologna, Italy

[§]Email of the corresponding author: sanjayk.ram@inano.au.dk and sanjayk.ram@gmail.com

ABSTRACT: In this article, we demonstrate a patch, which contains upconverting crystals of $Gd_2O_2S:Yb^{3+}$, Er^{3+} mixed with a UV curable gel, integrated with a textured back-reflector that can be externally affixed to a bifacial thin-film silicon solar cell. The patch serves to provide the benefit of light trapping as well as spectrum upconversion to the solar cell fabricated on a flat substrate. Upconverting patches with different textures were studied for their effectiveness in enhancing the short-circuit current density of the solar cell and for the relationship between the current enhancement and the intensity of the incident light. The upconverting patch is shown to enhance the solar cell performance and offers a new stick-on technique to apply upconversion-based strategies to existing solar cells. The photogenerated short-circuit current densities of the solar cell show a power-law dependence on the illuminated laser light intensity when an upconverter patch is attached to the solar cell (exponent 1.5-1.6) while a linear dependence is observed for the solar cell without the upconverter patch.

Keywords: Bifacial solar cell; Silicon thin-film solar cell; Upconversion; Light-trapping;

1 INTRODUCTION

Improving the utilization of the incident light by a solar cell is central to increasing the energy conversion efficiencies and cost effectiveness of the photovoltaic technology. Spectral upconversion (UC) is one of the new nanotechnology-based strategies that can address the issue of unabsorbed sub-bandgap light that normally passes through the solar cell without being absorbed [1]. A spectral-upconversion layer can be conveniently placed at the back of a solar cell without any change in the existing solar cell architecture. Upconversion quantum efficiencies are, however, not high enough to significantly enhance the photovoltaic performance, which is a hindrance for the commercial application of upconversion in solar cells [2]. One of the effective ways to enhance the upconversion luminescence yield is by using plasmonic nanostructures [3]. A limitation to the plasmonic route is that the small wavelength range of plasmon induced enhancement precludes the use of larger sized and typically more efficient upconverter crystals [3]. Besides, plasmonic nanostructures also involve expensive fabrication techniques and are difficult to apply effectively in the overall solar cell structure.

Light trapping using nanotextures is an established method to enhance carrier generation by increasing the absorption of above-bandgap light in solar cells [4]. Textures that are large enough to scatter and increase multiple reflection of the IR light can be useful to enhance the upconversion yield, and may in fact be more beneficial than the near-field effects of plasmonic nanostructures for larger sized upconverter crystals. In this article, we demonstrate a patch, which contains upconverting crystals of $Gd_2O_2S:Yb^{3+}$, Er^{3+} mixed with a UV curable gel, integrated with a textured back-reflector that can be externally affixed to a bifacial thin-film silicon solar cell. The absorption line of these upconverters is at 980 nm while the emission lines are around 540 nm and 660 nm [5]. These absorption and emission wavelengths are very suitable for use with wide-bandgap solar cells, such as single-junction a-Si:H, as the absorption edge of a-Si:H is between the wavelengths for absorption and emission. Upconverting patches with different textured back-reflector (BR) were studied for their effectiveness in enhancing the short-circuit current density of the solar cell and the relationship between the current enhancement and the intensity of the incident light.

2 METHOD

The upconverting material used in this study is Er^{3+} and Yb^{3+} codoped Gd_2O_2S crystals [5]. The mean crystal size is around 800 nm and their dominant upconversion mechanism is energy-transfer upconversion. These crystals were dissolved in ethanol and dispersed using a magnetic stirrer before mixing with the UV curable sol-gel lacquer.

The various nano/micron sized textures used in this study include nanowrinkles, random-pyramidal and random-inverted-pyramidal shaped features, which were engraved on one of the sides of the free-standing upconv-

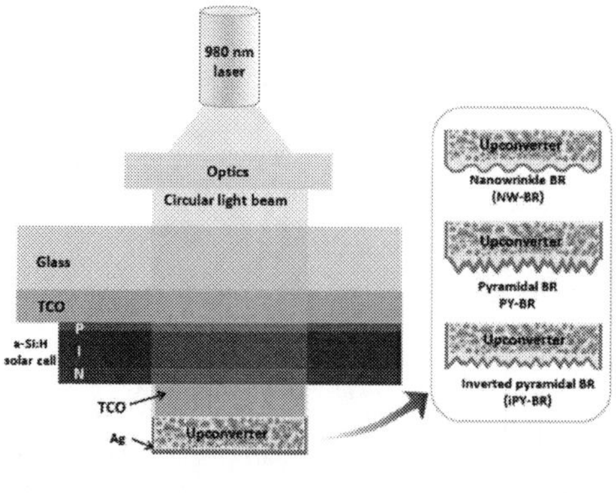

Figure 1: Schematic illustration of solar cell attached with upconverter patches with various textured back-reflector.

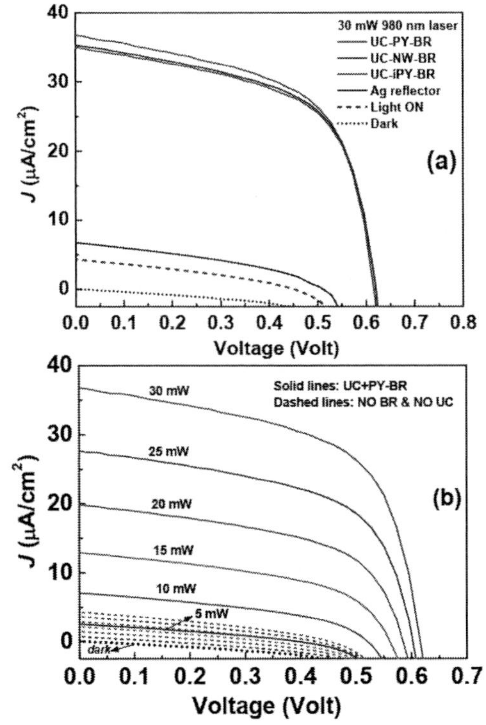

Figure 2: (a) The J-V characteristics of the bifacial a-Si:H solar cell (i-layer thickness 220 nm) when illuminated with 980 nm laser light with 30 mW (0.45 W/cm^2) with and without upconverter patches having various textured back-reflector; (b) The J-V characteristics of the bifacial a-Si:H solar cell (i-layer thickness 220 nm) when illuminated with 980 nm laser light with varying laser power from 5 mW (0.071 W/cm^2) to 30 mW (0.45 W/cm^2) with and without upconverter patches having pyramidal shaped textured back-reflector.

-erting patches. The textured side and the side walls of the patch were coated with Ag.

The p-i-n type amorphous silicon solar cells were fabricated using a cluster type plasma-enhanced chemical vapor deposition system on rf-magnetron deposited indium tungsten oxide (IWO) coated flat glass substrate. The details of the silicon thin film solar cell deposition are described in our previously reported work [4]. The 220 nm thickness of the *i*-layer was chosen after simulating the optical properties of the whole solar cell device with different absorber-layer thicknesses to find the lowest possible absorption in the cell in the wavelength range from 940 to 1000 nm due to interference. This is done to ensure the maximum possible infrared light reaching the upconverter instead of getting absorbed by localized tail states of the a-Si:H absorber layer. The flat surface configuration for the solar cell is also chosen to minimize the absorption of the 980 nm light by the silicon layers.

The solar cell *J-V* characteristics were measured with and without 980 nm laser diode illumination and with different upconverting patches.

3 RESULTS AND DISCUSSION

Figure 2 (a) shows the *J-V* characteristics of the flat superstrate PIN a-Si:H solar cell under the illumination of 980 nm laser diode with and without upconverting patches having various textured back reflectors. The current flow in the device amounts to J_{sc} = 4 μA/cm^2 due to 0.425 W/cm^2 intensity of 980 nm light without any upconverter. The origin of this photocurrent is due to the absorption in the localized tail states in the band gap of the a-Si:H layers. After attaching a flat Ag back-reflector the J_{sc} increases to ~7 μA/cm^2 with the same light intensity. However, attaching the upconverter increases the current to 37 μA/cm^2, 35.5 μA/cm^2 and 34.5 μA/cm^2 for the upconverter with pyramidal, nanowrinkle and inverted pyramidal shaped back reflector, respectively. Thus, the corresponding gains ($\Delta J_{sc}/J_{sc}$) in the current densities due to upconverter with a textured back-reflector are almost 4 times the current density with only a flat back-reflector.

Figure 2 (b) shows the dependence of the *J-V* characteristics on laser intensity for the same solar cell with and without upconverter. For clarity, results for only one type of pyramidal shaped back-reflector upconverter are shown.

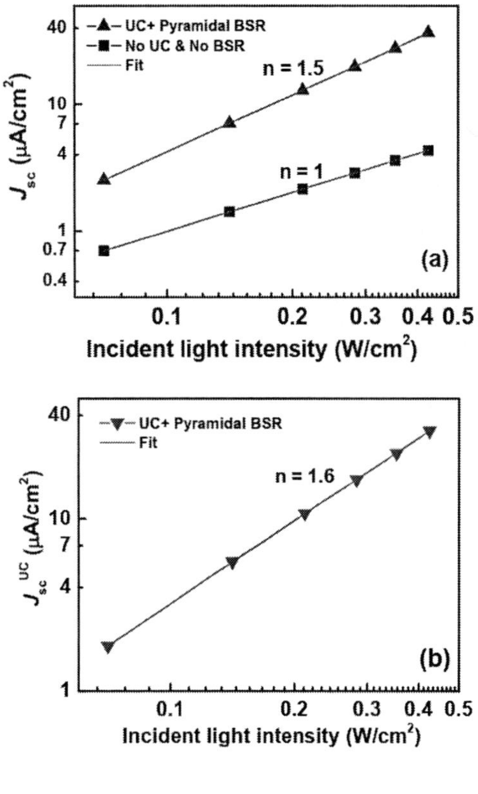

Figure 3: (a) The total photocurrent of the solar cell (J_{sc}) as a function of incident laser light intensity is shown on a log-log scale for two configurations with and without upconverter patch. The part (b) of this figure shows the dependence of the photocurrent due to the upconversion process in the solar cell (J_{SC}^{UC}) with illuminated laser light intensity.

Since the intensities of the emitted visible light (I^{UC}) from the upconverter at wavelengths 540 nm and 650 nm have a non-linear dependence with incident light intensity (I^n_{in}),

$$I^{UC} \propto I^n_{in} \quad (1)$$

we can expect a similar non-linear dependence of the photogenerated current in the presence of an upconverter at the back of a solar cell with the following behavior:

$$J_{sc}^{UC} \propto I^n_{in} \quad (2)$$

where J_{sc}^{UC} is the short-circuit current density increase due to upconversion in the solar cell. As a result, for the increase in photocurrent due to upconversion, a quadratic power dependence on the concentration factor is expected.

In figure 3 (a), the dependence of total photocurrent of the solar cell (J_{sc}) on incident laser light intensity (power density) is shown for two configurations with and without upconverter patch. The slope in the log-log plot obtained without upconverter patch is around 1, while a slope of 1.5 is obtained for the solar cell with upconverter. Panel (b) of this figure shows that the power-law dependence of the photocurrent generated solely by the upconversion mechanism in the solar cell (J_{sc}^{UC}) has an exponent of 1.6. The slope "$n < 2$" that signifies the photocurrent is not increasing quadratically with increasing incident light intensity. It is possible that the emitted light of 540 nm is getting absorbed within or near the *n*-layer before reaching the *i*-layer. This issue is currently being investigated in our ongoing work.

4 CONCLUSION

We demonstrate the feasibility of a patch, which contains upconverting crystals of $Gd_2O_2S{:}Yb^{3+}$, Er^{3+} mixed with UV curable gel, integrated with a textured back-reflector that can be externally affixed to a bifacial thin film a-Si:H solar cell. The patch serves to provide the benefit of light-trapping as well as spectrum upconversion to the solar cell fabricated on a flat substrate. Upconverting patches with different textures were found to enhance the short-circuit current density of the solar cell by almost 4 times the value when no such patches are placed at the back of the bifacial a-Si:H solar cell. The photogenerated short-circuit current densities of the solar cell showed a power-law dependence on the illuminated laser light intensity when an upconverter patch is attached to the solar cell and a linear dependence for the solar cell without upconverter. The upconverting patch is shown to enhance the solar cell performance and offers a new stick-on technique to apply upconversion based strategies to existing solar cells. A larger utilization of the solar spectra without changing the core architecture of the solar cell is thus possible with the proposed design.

5 ACKNOWLEDGEMENTS

This project is supported by the Danish Strategic Research Council under the projects "THINC" and "SUNTUNE".

6 REFERENCES

[1] T. Trupke, M. A. Green, P. Würfe, J. Appl. Phys. 92, (2002) 4117.

[2] J. de Wild, A. Meijerink, J. K. Rath, W. G. J. H. M. van Sark and R. E. I. Schropp, Energy Environ. Sci. 4 (2011) 4835.

[3] M. Saboktakin, X. Ye, S.J. Oh, S.H. Hong, A.T. Fafarman, U.K. Chettiar, N. Engheta, C.B. Murray, C.R. Kagan, ACS Nano 6 (2012) 8758.

[4] S.K. Ram, R. Rizzoli, D. Desta, B.R. Jeppesen, M. Bellettato, I. Samatov, Y.C. Tsao, S.R. Johannsen, P.T. Neuvonen,T.G. Pedersen, R.N. Pereira, K. Pedersen, P. Balling, A.N. Larsen, J. Phys. D: Applied Physics, 48 (2015) 365101.

[5] *www.phosphor-technology.com*

Electrical and Optical Performances of Silicon Solar Cells Modulated by Plasmonics Scattering of Silver and Indium Nanoparticles

Su-Han Weng[1], Wen-Jeng Ho[1]*, Yi-Yu Lee[1], Chia-Hua Hu[1], Chia-Hua Hu[1], Wei-Lien Wang[1], Yu-Jie Deng[1], and Hung-Pin Shiao[2]
[1] Department of Electro-Optical Engineering, National Taipei University of Technology
[1] No. 1, Sec. 3, Zhongxiao E. Rd., Taipei 10608, Taiwan, R.O.C.
[2]Win Semiconductor Corp, Taoyuan, Taiwan, R.O.C.
*e-mail: wjho@ntut.edu.tw

ABSTRACT: The electrical and optical performances of plasmonic silicon (Si) solar cells with a matrix comprising a pattern of silver nanoparticles (Ag-NPs) surrounded by indium nanoparticles (In-NPs) to modulate plasmonics light scattering as well as enhanced photovoltaic performance are demonstrased. The plasmonic light scattering induced by Ag-NP/In-NP matrices with various degrees of coverage were examined by electrical and optical measurements. The surface plasmon Raman scattering of Ag-NPs and In-NPs was characterized before investigating the plasmonic light scattering produced by a matrix of Ag-NPs and In-NPs in a Si solar cell. We also measured the optical reflectance and external quantum efficiency (EQE) response of various plasmonic solar cells. We achieved impressive results with regard to plasmonic light scattering through the implementation of a matrix that included Ag-NP coverage of 20% and In-NP coverage of 80% over an area of 16 mm^2. This resulted in a 9.93% increase in short-circuit current density (J_{sc}) (from 31.91 mA/cm^2 to 35.08 mA/cm^2) and a 10.12% increase in conversion efficiency (η) (from 13.04% to 14.36%), compared to a reference cells with uniformly distributed Ag-NPs.

Keywords: Plasmonic, plasmonics light scattering, indium nanoparticles (In-NPs), silver nanoparticles (Ag-NPs), silicon (Si) solar cell

1 INTRODUCTION

Low reflectance and broadband performance nanoscale-surface structures have been developed for light trapping on silicon (Si)-based photovoltaic devices [1-4]. Plasmonic effects have been applied to enhance light trapping in silicon and GaAs-based solar cells. Metallic nanoparticles (NPs) exhibiting localized surface plasmon resonance (LSPR) have been used to enhance conversion efficiency in solar cells [5-8]. Matheu et al. reported that metal and dielectric nanoparticle scattering for improved optical absorption in photovoltaic devices [9]. Li et al. presented that optical absorption enhancement in solar cells with cylindrical Ag nanoparticles by localized surface plasmons [10]. Spinelli et al. reported that arrays of silver (Ag) NPs placed on a substrate with a high refractive index could be used to enhance the coupling of light into the substrate over a broad spectral range, due to the effects of optical impedance matching [11]. An array of Ag and gold (Au) NPs on a crystalline Si solar cell was shown to improve the photovoltaic current-voltage (I-V) characteristics at wavelengths of 800 to 1050 nm via surface plasmon resonance [7,12]. However, little research has been conducted using plasmonic light scattering induced by a matrix comprising of Ag-NPs and indium (In)-NPs on silicon solar cell to enhance photovoltaic performance.

In this study, we developed novel plasmonic Si solar cells with a matrix comprising a pattern of Ag-NPs surrounded by In-NPs. This is the first study to experimentally demonstrate plasmonic light scattered-modulation in Si solar cells through the application of an Ag-NP/In-NP matrix with various degrees of various coverage. We examined the surface plasmon Raman scattering induced by Ag-NPs and In-NPs. We also compared various matrix configurations with regard to optical reflectance and the external quantum efficiency (EQE) response in solar cells. Finally, we compared the photovoltaic performance of cells with uniformly distributed Ag-NPs, matrix distributed Ag-NPs, and Ag-NP/In-NP matrices with various degrees of coverage.

2 EXPERIMENT

To characterize the Raman scattering produced by Ag-NPs, In-NPs, respectively, 4 nm and 3.8 nm thick films of Ag and In were deposited over a 250-nm thick layer of TiO_2 on glass substrates using E-beam evaporation. All deposited samples were then annealed at 200 °C for 30 min under ambient H_2 to form metal nanoparticles. To enable a comparison of plasmon-enhanced Raman scattering, samples with a configuration of TiO_2/Glass-substrate without the deposition of metallic NPs were also prepared as a reference. The spectra of Raman scattering were collected using a micro-Raman system (Ramboss 500i Micro – Raman/PL Spectroscopy, DONGWOO). A semiconductor laser operating at 473 nm with output power of 0.09 mW was used as the excitation source.

For fabricating and characterizating of cells with patterned matrix of Ag-NPs/In-NPs, a TiO_2 spacer layer (20-nm thick) was first deposited on the top surface of the bare solar cells using e-beam evaporation. To investigate the plasmonic light-scattering modulation produced by uniformly deposited Ag-NPs, we created a matrix of Ag-NPs or an Ag-NP/In-NP matrix with various degrees of coverage over bare Si solar cells, as shown in Fig. 1. In this study, we deposited Ag film to a thickness of 4-nm as well as In film to a thickness of 3.8 nm and the coverage of the Ag-NPs was either 20% or 40%. All of the metallic films were deposited on the bare solar cells through patterned photolithographic photoresists via e-beam evaporation. Following lift-off processing and annealing, Ag-NPs formed on the surface of the bare solar cells. To create cells with an Ag-NP/In-NP matrix, we first created a matrix of Ag-NPs using photolithography in conjunction with Ag film evaporation and lift-off followed by thermal annealing. Surrounding the matrix-pattern of Ag-NPs, we deposited a film of In to a thickness of 3.8 nm using e-beam evaporation followed by lift-off. Annealing at 200 °C in an RTA chamber under ambient H_2 for 20 mins resulted in the formation of Ag-NPs and In-NPs. The morphology and dimensions of the metallic NPs were characterized using electron scanning microscopy (SEM) and the size distribution of the particles was calculated from SEM images using J-image software. We then characterized the plasmonic light-scattering modulated performance of solar cells fabricated using NPs of various degrees of surface coverage by measuring the optical reflectance, external quantum efficiency (EQE) at wavelengths between 300 nm and 1000 nm, and photovoltaic current density-voltage (J-V) under AM 1.5G (1000 mW/cm^2 at 25 °C) solar simulation.

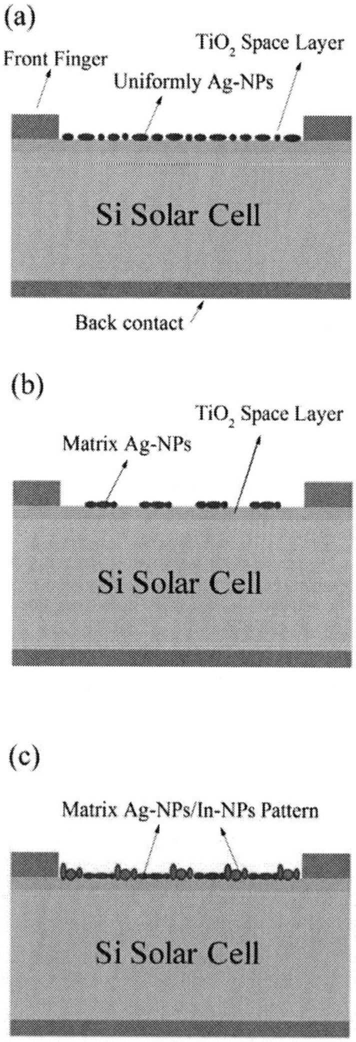

Fig. 1: Schematic diagram showing Si solar cells fabricated using (a) uniformly deposited Ag-NPs, (b) a matrix of Ag-NPs and (c) a matrix of Ag-NPs/In-NPs on a TiO_2 space layer.

3 RESULTS AND DISCUSSION

Fig. 2 presents the Raman spectra of samples with the following configurations: TiO_2/Glass, Ag-NPs/TiO_2/Glass, and In-NPs/TiO_2/Glass. Because the confinement of surface acoustic vibrational modes in nanoscale metallic particles can be experimentally observed using Raman spectroscopy. The intensity of the vibration mode was enhanced by surface plasmon resonance absorption. Three Raman signal intensity peaks at 778, 1392, and 1628 cm^{-1} in the sample of Ag-NPs/TiO_2/Glasse and two peaks at 562 and 1120 cm^{-1} in the sample of In-NPs/SiO_2/Glass were obtained which exceeding the intensity of the TiO_2/Glass sample. This may be explained by the strong local field of surface plasmon-resonance (LSPR) produced by Ag-NPs and In-NPs, respectively, with excitation wavelength of 473 nm and power of 0.09 mW.

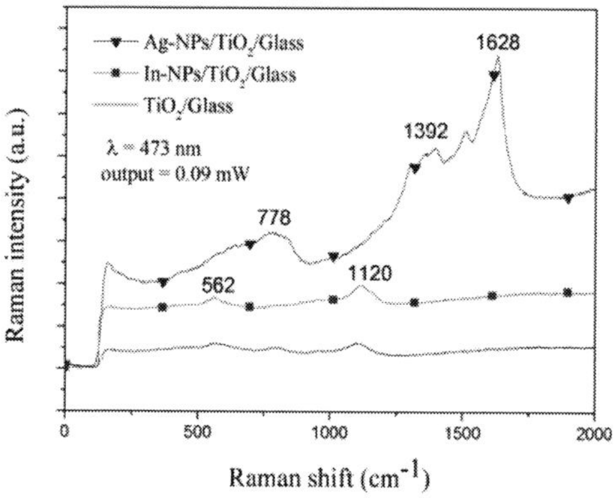

Fig. 2: Raman spectra of samples with the following configurations: TiO_2/Glass, Ag-NPs/TiO_2/Glass and In-NPs/TiO_2/Glass.

Fig. 3(a) presents the optical reflectance of the following: a bare solar cell, a cell with a TiO_2 layer, and a cell with uniformly deposited Ag-NPs on a TiO_2 layer. The reflectance of the cell with a layer of TiO_2 was lower than that of the bare cell, due to the antireflective properties of the TiO_2 on the surface of the Si. The reflectance of the plasmonic cell with Ag-NPs on the TiO_2 layer exceeded that of the cell with a TiO_2 layer at wavelengths of 350-525 nm, due to optical reflectivity of the Ag-NPs. It should be noted that at wavelengths of 600-1100 nm, the reflectance of the cell with uniformly deposited Ag-NPs on the TiO_2 layer decreased to a level below that of the cell with only a TiO_2 layer, due to the plasmonic forward scattering of photons by the Ag-NPs. Fig. 3(b) presents the optical reflectance of a cell with a uniformly deposited Ag-NPs on a TiO_2 layer, a cell with a matrix-pattern (20×20 μm^2) of Ag-NPs on a TiO_2 layer (Ag-NPs coverage of 20% or 40% over an area of 4×4 mm^2), and a cell with Ag-NPs patterned as squares covering 20% or 40% of the surface over an area of 4×4 mm^2 surrounded by In-NPs. Compared to the cell with uniform layer of Ag-NPs, the optical reflectance of the cell with a matrix-pattern of Ag-NPs was lower when the coverage of the matrix-pattern of Ag-NPs decreased. The optical reflectance of the cell with a matrix-pattern of Ag-NPs/In-NPs was far lower than that of the cell with matrix of Ag-NPs. In the cell with matrix of Ag-NPs/In-NPs, we also observed a decrease in reflectance following a decrease in Ag-NP coverage, particular at wavelengths of 350-550 nm. This demonstrates that the optical reflectance of cells can be modulated by altering the coverage of Ag-NPs and In-NPs on the surface of the cells.

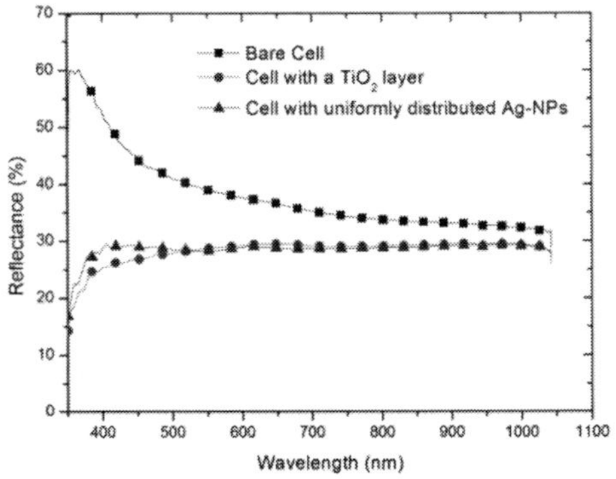

(a)

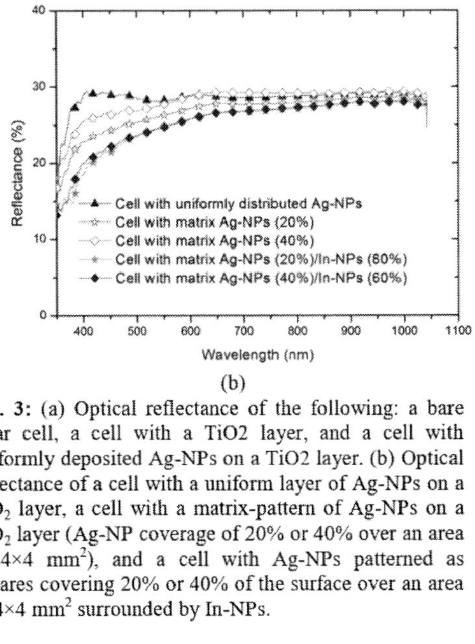

(b)

Fig. 3: (a) Optical reflectance of the following: a bare solar cell, a cell with a TiO2 layer, and a cell with uniformly deposited Ag-NPs on a TiO2 layer. (b) Optical reflectance of a cell with a uniform layer of Ag-NPs on a TiO_2 layer, a cell with a matrix-pattern of Ag-NPs on a TiO_2 layer (Ag-NP coverage of 20% or 40% over an area of 4×4 mm^2), and a cell with Ag-NPs patterned as squares covering 20% or 40% of the surface over an area of 4×4 mm^2 surrounded by In-NPs.

Fig. 4(a) presents the EQE response of a bare solar cell, a cell with a TiO_2 layer, and a cell Ag-NPs uniformly distributed on a TiO_2 layer. The cell with a TiO_2 layer presented higher EQE values than did the bare solar cell at wavelengths from 350 to 1100 nm. This finding is in agreement with the results of optical reflectance. At wavelengths below 650 nm, the cells with uniformly distributed Ag-NPs displayed EQE values lower than that of the cell with only a TiO_2 layer, which is in agreement with the results of optical reflectance obtained at wavelengths of 350-525 nm due to the reflectivity of the Ag-NPs. The EQE values increased when the wavelength of the incident photons exceeded 650 nm. This is also in agreement with the results of optical reflectance, in which it was indicated that the Ag-NPs induced plasmonic forward scattering. Fig. 4(b) illustrates the EQE response of the cell with uniformly distributed Ag-NPs on a TiO_2 layer, cells with a matrix of Ag-NPs on a TiO_2 layer, and cells with a matrix of Ag-NPs/In-NPs on a TiO_2 layer. At wavelengths from 350 to 700 nm, the Ag-NPs matrix increased the EQE values beyond those of the cell with uniformly distributed Ag-NPs. This may be due to a lower optical reflectance in the Ag-NP matrix, which is inversely proportional to the coverage of patterned Ag-NPs. We observed a broadband increase in EQE values in cells with an Ag-NPs/In-NPs matrix, exceeding that of cells with uniformly distributed Ag-NPs and cells with a Ag-NP matrix. This can be attributed to the combined effects of plasmonic light scattering, induced by In-NPs at wavelengths of 350-650 nm and by Ag-NPs at wavelengths beyond 650 nm. The EQE values of cells with 20% Ag-NPs/In-NPs exceeded those of 40% Ag-NPs/In-NPs at wavelengths of 350-700 nm. These findings demonstrate the importance of EQE response in the generation of photocurrent in solar cell devices.

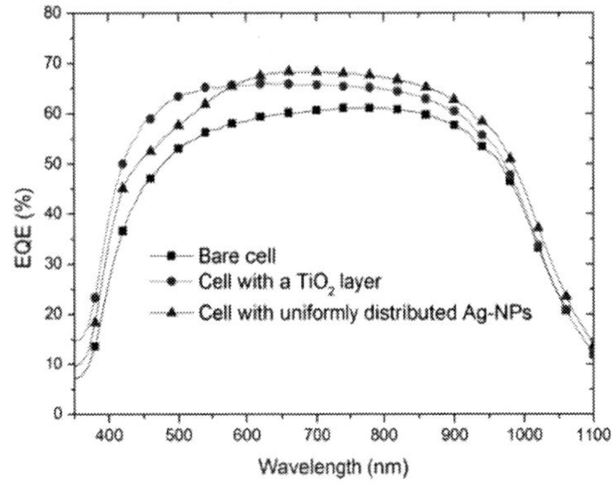

(a)

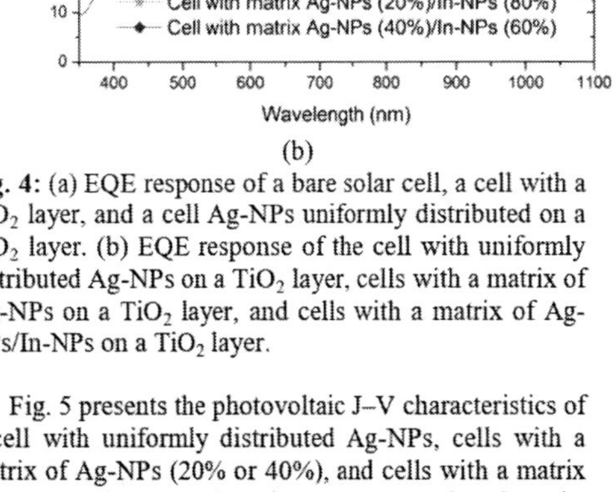

(b)

Fig. 4: (a) EQE response of a bare solar cell, a cell with a TiO_2 layer, and a cell Ag-NPs uniformly distributed on a TiO_2 layer. (b) EQE response of the cell with uniformly distributed Ag-NPs on a TiO_2 layer, cells with a matrix of Ag-NPs on a TiO_2 layer, and cells with a matrix of Ag-NPs/In-NPs on a TiO_2 layer.

Fig. 5 presents the photovoltaic J–V characteristics of a cell with uniformly distributed Ag-NPs, cells with a matrix of Ag-NPs (20% or 40%), and cells with a matrix of Ag-NPs/In-NPs (20%/80% or 40%/60%). The photovoltaic performance of cells is summarized in Table 1. The J_{sc} and η values of cells with a Ag-NP matrix or a Ag-NP/In-NP matrix are higher than in cells with uniformly distributed Ag-NPs. The J_{sc} and η values of the cell an Ag-NP/In-NP matrix exceeded those of the cell with an Ag-NP matrix. The J_{sc} and η values of the cell with a 20%/80% Ag-NP/In-NP matrix were superior to those of cells with a 40%/60% Ag-NP/In-NP matrix. This is a clear demonstration that plamonic modulation can be achieved by adjusting the coverage of Ag-NPs and In NPs in the configuration of the matrix. Photovoltaic performance can be significantly improved through the implementation of plamonic-modulated light scattering induced by a matrix of Ag-NPs and In NPs.

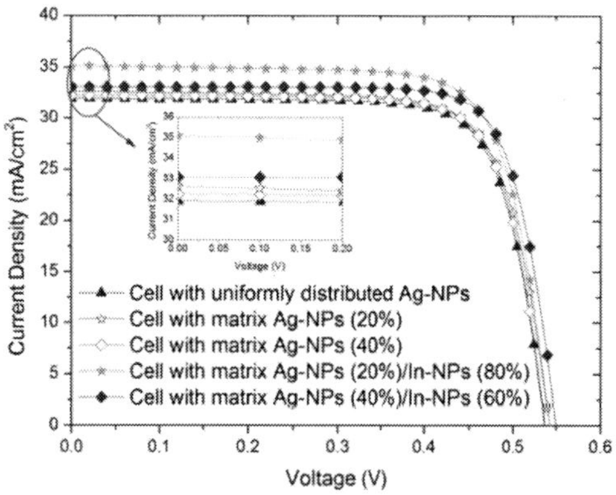

Fig. 5: Photovoltaic J–V characteristics of a cell with uniformly distributed Ag-NPs, cells with a matrix of Ag-NPs (20% or 40%), and cells with a matrix of Ag-NPs/In-NPs (20%/80% or 40%/60%).

Table 1: Photovoltaic performance of all evaluated cells summarization.

	J_{sc} (mA/cm^2)	V_{oc} (mV)	η (%)
Bare cell	26.54	529.3	10.93
Cell with a TiO_2 layer	28.63	537.2	11.95
Cell with uniformly distributed Ag-NPs	31.91	537.2	13.04
Cell with matrix Ag-NPs (40%)	32.24	537.9	13.22
Cell with matrix Ag-NPs (40%)/In-NPs (60%)	33.08	550.1	14.13
Cell with matrix Ag-NPs (20%)	32.65	542.4	13.27
Cell with matrix Ag-NPs (20%)/In-NPs (80%)	35.08	542.6	14.36

4 CONCLUSION

The plasmon Raman scattering induced by Ag-NPs and In-NPs was characterized. Plasmonic light scattering modulation in solar cells was examined by optical reflectance and EQE response. Plasmonics scattering modulation can be adjusted by varying the configuration of matrices of Ag-NPs and In-NPs. We also verified improvements in photovoltaic performance resulting from plasmonic light scattering via the implementation of an appropriate matrix of Ag-NPs/In-NPs. The application of a 20%/80% Ag-NP/In-NP matrix increased the EQE values across a broad range of wavelengths. The high J_{sc} in cells with a Ag-NP/In-NP matrix can be attributed to the combined effects of plasmonic light scattering induced by In-NPs at short wavelengths of 350-650 nm as well as by Ag-NPs at wavelengths beyond 650 nm.

5 ACKNOWLEGEMENT

The authors would like to thank the Ministry of Science and Technology of the Republic of China for financial support under Grant MOST 103-2221-E-027-049-MY3.

6 REFERENCE

[1]. S. Ji, J. Park and H. Lim, Nanoscale 4, 4603 (2012).
[2]. Y. Ou, X. Zhu, V. Jokubavicius, R. Yakimova, N. A. Mortensen, Scientific Reports 4 : 4662, 1-4 (2014).
[3]. J. H. Selj, T. T. Mongstad, R. Søndenå, E. S. Marstein, Sol. Energ. Mat. Sol. Cell 95, 2576 (2011).
[4]. R. Yu, K. L. Ching, Q. Lin, S. F. Leung, D. Arcrossito, Z. Fan, ACS Nano 5, 9291 (2011).
[5]. X. Sheng, J. Hu, J. Miche, and L. C. Kimerling, Optics Express 20(S4), A496 (2012).
[6]. D. Derkacs, W. V. Chen, P. M. Matheu, S. H. Lim, P. K. L. Yu, E. T. Yu, Appl. Phys. Lett. 93, 091107-1 (2008).
[7]. S. Pillai, K. Catchpole, T. Trupke, M. Green, J. Appl. Phys. 101, 093105-1 (2007).
[8]. H. A. Atwater and A. Polman, Nat. Mater. 9, 205 (2010).
[9]. P. Matheu, S. H. Lim, D. Derkacs, C. McPheeters, and E. T. Yu, Appl. Phys. Lett. 93, 113108 (2008).
[10]. B. Z. Li, X. X. Su, J. Q. Lin, X. Gao, J. G. Lei, H. Y. Tao, J. Modern Optics 58(9), 796 (2011).
[11]. P. Spinelli, M. Hebbink, R. De Waele, L. Black, F. Lenzmann, A. Polman, Nano Lett., 11 1760 (2011).
[12]. D. Schaadt, B. Feng, E. Yu, Appl. Phys. Lett. 86, 063106-1 (2005).

AL-BACK SURFACE FIELD-TYPE CRYSTALLINE SI-BASED SMART STACK TRIPLE-JUNCTION (INGAP/GAAS/SI) CELLS

Hidenori Mizuno,[1] Kikuo Makita,[2] Takeshi Tayagaki,[2] Toshimitsu Mochizuki,[1]
Yasuhiro Kida,[1] Takeyoshi Sugaya[2], and Hidetaka Takato[1]
[1]National Institute of Advanced Industrial Science and Technology (AIST), Fukushima Renewable Energy Institute
2-2-9 Machiike-dai, Koriyama, Fukushima, 963-0298 Japan
[2]National Institute of Advanced Industrial Science and Technology (AIST), Tsukuba Central
1-1-1 Umezono, Tsukuba, Ibaraki, 305-8568 Japan

ABSTRACT: Smart stack is a handy method to construct multi-junction architectures through the interconnection of different types of solar cells using a Pd nanoparticle array as a bonding mediator. In this study, a triple-junction cell was fabricated from a InGaP/GaAs-based double-junction and a crystalline Si-based single-junction subcell using the smart stack approach. The simplest, matured, and widely-used Al-back surface field-type structure was particularly focused on for the Si bottom cell, and despite that, a promising efficiency of 20.9% was achieved as a smart stack InGaP/GaAs/Si device [short-circuit current density (J_{sc}), open-circuit voltage, and fill factor, were 9.65 mA/cm^2, 2.75 V, and 0.790, respectively]. The EQE analysis of the smart stack cell was also carried out, which suggested that the device performance was limited by the J_{sc} generated at the GaAs subcell. Therefore, higher efficiencies are feasible through a regorous design and adjustment of each subcell, especially in terms of current matching.
Keywords: III-V Semiconductors, c-Si, Multijunction Solar Cell, Nanoparticles

1 INTRODUCTION

Fabrication of high efficiency solar cells has always been a topic of high interest in the photovoltaic (PV) research community. While the best efficiencies are actively pursued using a variety of semiconductor materials, theoretical limits (~30%) are known to exist for any single-junction cells associated with their band gaps (E_g) [1]. Multi-junction (MJ) cells can surpass those efficiency limits by utilizing the broad range of the solar spectrum more effectively than each component single-junction cell (subcell). For example, GaAs/Ge-based triple-junction cells routinely show over 30/40% efficiencies (under 1 sun/concentration) and have been utilized in space and concentrator PV applications [2]. It is known, however, that the fabrication cost of conventional MJ cells is very high because of the necessity of elaborative crystal growth processes and the usage of expensive substrates. Simple, flexible, and low-cost strategies are required for the widespread deployment of high-efficiency MJ cells in the foreseeable future.

In this respect, mechanical stack has been regarded as a convenient approach to build MJ architectures with flexible material/subcell combination and low-cost. One remarkable trend is the use of wafer bonding techniques to produce GaAs/InP-based MJ cells with current record efficiencies [3,4], and their GaAs/Si offshoots have also emerged recently [5,6]. We have also proposed a unique bonding technique, termed smart stack, as shown in Figure 1. Here, a metal (Pd) nanoparticle (NP) array was employed as a bonding mediator between top and bottom cells to allow the interface both electrically conductive and optically transparent (i.e., low resistance and minimal transmission/absorption loss) [7,8]. We have so far demonstrated a quadruple-junction cell by the smart stack of InGaP/GaAs and InGaAsP/InGaAs cells, which showed the efficiency of 31.6% [9]. Another exotic combination is a triple-junction cell made of InGaP/GaAs and CuInGaSe cells, which showed the efficiency of 24.2% [10].

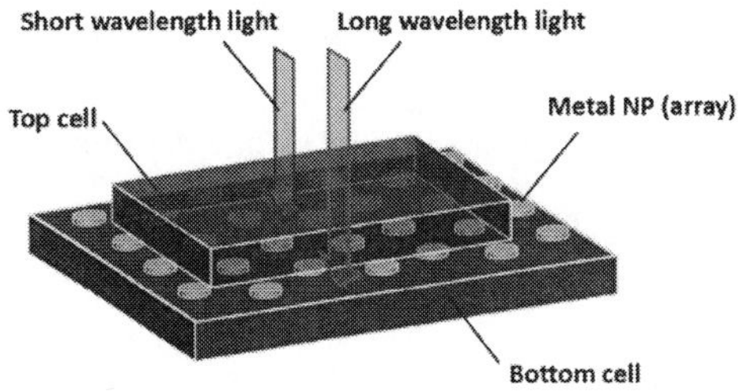

Figure 1: Schematic drawing of a smart stack cell. Reprinted with permission of ref. [7].

As an extension of our smart stack concept, we herein describe a triple-junction cell made of InGaP/GaAs and crystalline Si subcells. Becuase of the simplicity and availability (i.e., easy-fabrication and widespread use), Al-back surface field (BSF) Si cells were particularly employed in this study, although other adavanced strucutres, such as passivated emitter, rear locally-diffused (PERL) cells, are also available.[11] The External Quantum Efficiency (EQE) analysis of the fabricated smart stack cell as well as future direction towards higher efficiencies are also addressed.

2 EXPERIMENTAL SECTION

2.1 Fabrication of Bottom Cells

A single-side mirror polished, CZ-grown p-type (B-doped) Si(100) wafer with the thickness of 625 μm and resistivity of 0.5-4 Ω-cm^2 was used to prepare a bottom cell. The mirror polished side was thermally diffused with $POCl_3$ to form a n$^+$ front layer. The other side was screen-printed with Al paste and fired to simultaneously form a p$^+$ BSF layer and an electrode.

On the n$^+$ front side of this Si bottom cell, a Pd NP array was formed using a self-assembled polystyrene-block-poly-2-vinylpyridine (PS-b-P2VP) template following the similar procedure described elsewhere [8]. Figure 2 shows a typical scanning electron microscopy

(SEM) image of such a Pd NP array on the Si bottom cell. Pseudohexagonally patterned Pd NPs were observed, whose size, inter-particle distance, density, and surface coverage, were ~40 nm, ~100 nm, ~1×10^{10} cm^{-2}, and ~12%, respectively.

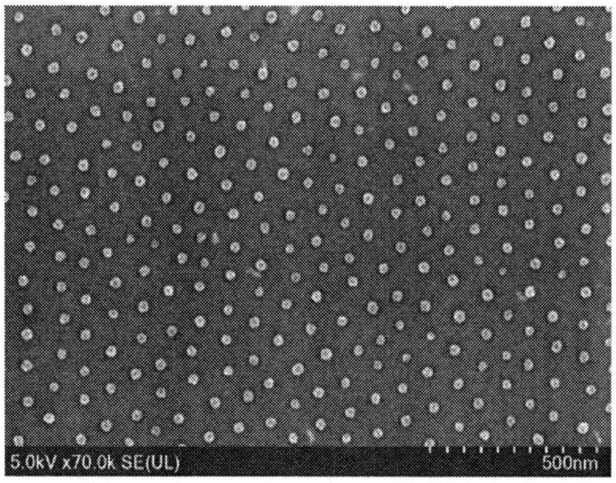

Figure 2: SEM image of the Pd NP array formed on the front side of the Si bottom cell.

2.2 Fabrication of Top Cells

A tandem InGaP/GaAs cell was grown on a GaAs growth substrate by molecular beam epitaxy. The thicknesses of InGaP and GaAs absorption layers were 0.35 and 0.5 μm, respectively. In order to allow the release of the device part from the growth substrate afterwards, an AlAs sacrificial layer was also grown between the substrate and cell. After an AuGeNi top electrode was deposited by electron beam evaporation, the entire sample was immersed into aqueous HF. Due to the selective etching of the AlAs layer by HF, the cell was separated from the substrate as a free-standing thin-film [i.e., epitaxial lift-off (ELO)].

2.3 Fabrication of Smart Stack Cells

The ELO InGaP/GaAs top cell was placed on the Pd NP-decorated Si bottom cell which was wet with H_2O. The resulting pre-stack was pressed lightly using a manual hand press equipment at room temperature for 3 h. A stable contact was formed between the bottommost layer of the ELO cell and the Pd NPs during the evaporation of the H_2O. As a result, a target smart stack InGaP/GaAs/Si (E_g = 1.89/1.42/1.12 eV) structure was obtained. Finally, a SiO_2/TiO_2 (100/70 nm) anti-reflection coating (ARC) was deposited by sputtering. The final structure (cross-section) of the smart stack cell is shown in Figure 3.

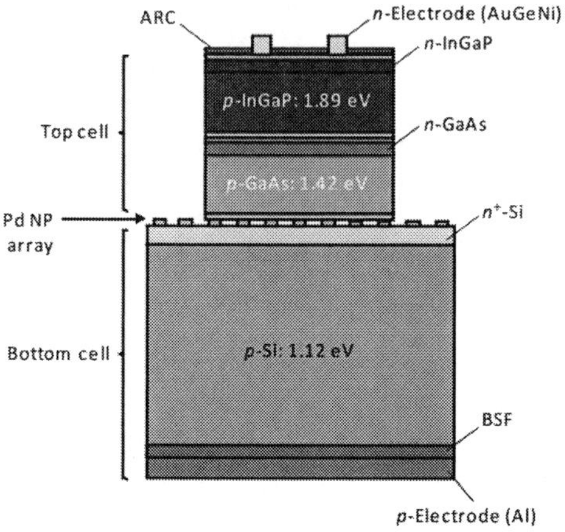

Figure 3: Schematic cross-section of the InGaP/GaAs/Si smart stack cell.

3 RESULTS AND DISCUSSION

3.1 EQE

The EQE spectra of the smart stack cell as well as each subcell were summarized in Figure 4. Over 60-80% EQEs were observed across the 400-1000 nm wavelength range as a total device. As for the subcells, it was revealed that the EQE of the Si bottom cell already starts to increase at the wavelength of around 600 nm because of the thinness of the GaAs layer (0.5 μm). This design was intentional to aim the current matching among the InGaP top, GaAs middle, and Si bottom cells, although there is still room for improvement. Indeed, the photogenerated J_{sc} of the InGaP, GaAs, and Si subcells, calculated from the corresponding spectra, were 11.1, 9.1, and 12.7 mA/cm^2, respectively, suggesting that the overall performance of the smart stack cell is mainly limited by the GaAs middle cell.

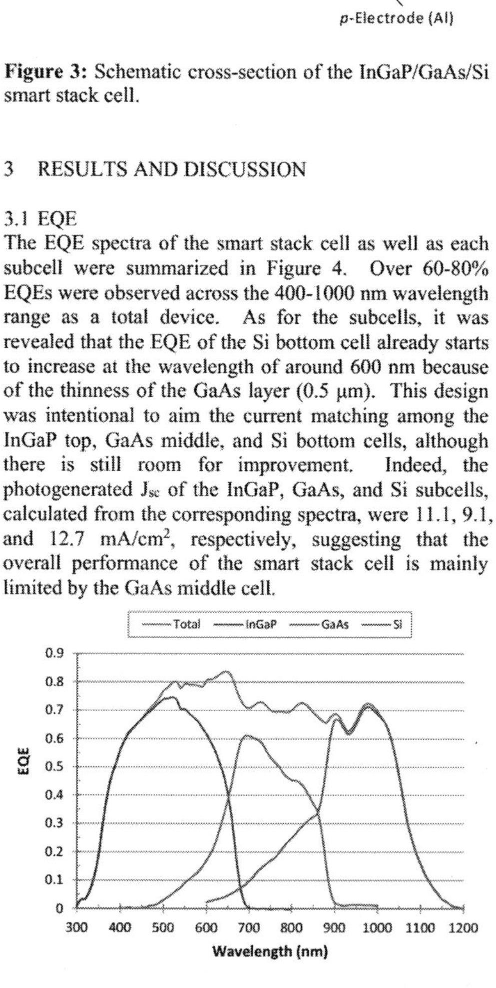

Figure 4: EQE spectra of the InGaP/GaAs/Si smart stack cell and subcells.

3.2 Current-Voltage (J-V) Characteristics

The photovoltaic performance was investigated under AM1.5G solar spectrum illumination. Since the EQE measurement showed that the short-circuit current density (J_{sc}) of the smart stack cell is limited by the GaAs subcell, no light-shielding mask was applied to the cell to restrict the light illumination (although the size of the Si cell is larger than that of the InGaP/GaAs cell, as shown in Figure 3). In this case, the designated area of the cell was 0.209 cm^2. Figure 5 shows the resulting J-V characteristics. The J_{sc}, open-circuit voltage (V_{oc}), and fill factor (FF), were 9.65 mA/cm^2, 2.75 V, and 0.790, respectively, which resulted in the efficiency of 20.9%.

It should be pointed out, however, that there is a little discrepancy between the J_{sc} of the smart stack cell (9.65 mA/cm^2, Figure 5) and the one calculated from the EQE spectra (9.1 mA/cm^2, Figure 4). Although the exact reason is unclear at this moment, one possibility could be luminenscent coupling as reported previously [12]. Further analyses and optimization of smart stack cells are intensively ongoing in our group.

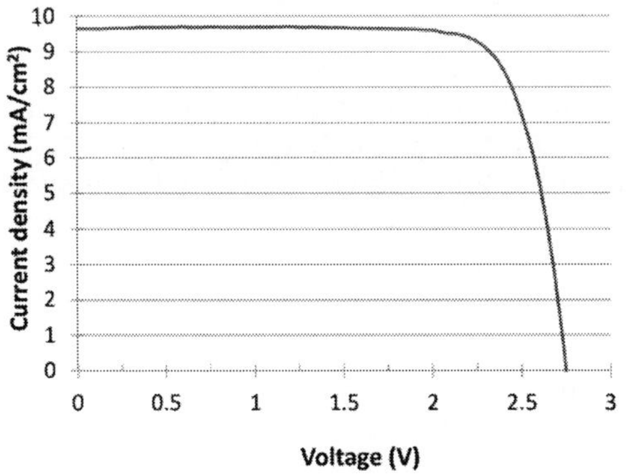

Figure 5: J-V characteristics of the InGaP/GaAs/Si smart stack cell.

4 CONCLUSION

We fabricated a triple-junction cell consisting of InGaP/GaAs and crystalline Si subcells based on the smart stack approach. An efficiency of 20.9% was reported as a tentative result despite the use of (less-advanced) Al-BSF-type Si cells. The EQE measurement of the smart stack cell suggested that the overall performance seems to be limited by the GaAs middle cell (the lowest photogenerated J_{sc}). Therefore, higher efficiencies are feasible through the further optimization of the subcell design, especially in terms of current matching [13].

5 ACKNOWLEDGEMENT

This work was supported by New Energy and Industrial Technology Development Organization (NEDO) under Ministry of Economy, Trade, and Industry (METI), Japan.

6 REFERENCES

[1] W. Shockley, H. J. Queisser, "Detailed balance limit of efficiency of p-n junction solar cells," *J. Appl. Phys.*, vol. 32, pp. 510, 1961.

[2] A. W. Bett, S. P. Philipps, S. Essig, S. Heckelmann, R. Kellenbenz, V. Klinger, M. Niemeyer, D. Lackner, F. Dimroth, "Overview about technology perspectives for high efficiency solar cells for space terrestrial applications," *Proc. 28th EUPVSEC*, pp. 1, 2013.

[3] T. N. D. Tibbits, P. Beutel, C. Grave, C. Karcher, E. Olivia, G. Siefer, A. Wekkeli, M. Schachtner, F. Dimroth, A. W. Bett, R. Krause, M. Piccin, N. Blanc, Munoz-Rico, C. Arena, E. uiot, C. Charles-Alfred, C. Drazek, F. Janin, L. Farrugia, B. Hoarau, J. Wasselin, A. Tauzin, T. Signamarcheix, T. Hannappel, K. Schwarzburg, A. Dobrich, "New efficiency frontiers with wafer-bonded multi-junction solar cells," *Proc. 29th EUPVSEC*, pp. 1975, 2014.

[4] P. T. Chiu, D. C. Law, R. L. Woo, S. B. Singer, D. Bhusari, W. D. Hong, A. Zakaria, J. Boisvert, S. Mesropian, R. R. King, N. H. Karam, "Direct semiconductor bonded 5J cell for space and terrestrial applications" *IEEE J. Photovoltaics*, vol. 4, , pp. 620, 2014.

[5] N. Shigekawa, J. Liang, R. Onitsuka, T. Agui, H. Juso, and T. Takamoto, "Current-voltage and spectral-response characteristics of surface-activated-bonding-based InGaP/GaAs/Si hybrid triple-junction cells," *Jpn. J. Appl. Phys.*, vol. 54, pp. 08KE03, 2015.

[6] S. Essig, J. Benick, M. Shachtner, A. Wekkeli, M. Hermle, and F. Dimorth, "Wafer-bonded GaInP/GaAs//Si solar cells with 30% efficiency under concentrated sunlight" *IEEE J. Photovoltaics*, vol. 5, , pp. 977, 2015.

[7] H. Mizuno, K. Makita, K. Matsubara, "Electrical and optical interconnection for mechanically stacked multi-junction solar cells mediated by metal nanoparticle arrays," *Appl. Phys. Lett.*, vol. 101, pp. 191111, 2012.

[8] H. Mizuno, K. Makita, T. Sugaya, Y. Hozumi, H. Takato, and K. Matsubara, "Palladium nanoparticle array-mediated semiconductor bonding that enables high-efficiency multi-junction solar cells "*Jpn. J. Appl. Phys.*, vol. 55, pp. 025001, 2016.

[9] K. Makita, H. Mizuno, R. Oshima, H. Sai, S. Nonaka, A. Furukawa, H. Takato, T. Sugaya, "Development of smart stacking technology for multi-junction solar cells" in *PVSEC-25*, 2015, GAS-O-13.

[10] K. Makita, H. Komaki, H. Mizuno, H. Sai, T. Sugaya, R. Oshima, H. Shibata, K. Matsubara, and S. Niki, "A 24.2%-efficiency III-V/CuInGaSe mechanical stacking multi-junction solar cell using semiconductor bonding method," *Proc. 29th EUPVSEC*, pp. 1427, 2014.

[11] H. Mizuno, K. Makita, T. Takagaki, T. Mochizuki, H. Takato, T. Sugaya, H. Mehrvarz, M. Green, and A. Ho-Baillie, "A "Smart Stack" Triple-Junction Cell Consisting of InGaP/GaAs and Crystalline Si," presented in *43rd IEEE PVSC* (poster #K45).

[12] D. J. Friedman, J. F. Geisz, and M. A. Steiner, "Effect of Luminescent Coupling on the Optimal Design of Multijunction Solar Cells," *IEEE J. Photovoltaics*, vol. 4, , pp. 986, 2014.

[13] Indeed, we have successfully achieved a 24.7% InGaP/GaAs/Si cell through the improvement of the GaAs middle cell (J_{sc}: 10.60 mA/cm^2, V_{oc}: 2.92 V, FF: 0.799).

Photoresponse properties of $BaSi_2$ film grown on Si (100) by vacuum evaporation

Cham Thi Trinh[1], Yoshihiko Nakagawa[1], Kosuke O. Hara[2], Ryota Takabe[3], Takashi Suemasu[3], and Noritaka Usami[1]

1Graduate school of Engineering, Nagoya University, Nagoya 464-8603, Japan

2Center for Crystal Science and Technology, University of Yamanashi, Yamanashi 400-8511, Japan

3Institute of Applied Physics, University of Tsukuba, Tsukuba, Ibaraki 305-8573, Japan

Email: msscmr@gmail.com

Abstract: We have succeeded in observation of photoresponsivity of $BaSi_2$ film grown on Si by vacuum evaporation method, raising its promising application in high efficiency thin film solar cell. Photocurrent was observed at photon energies larger than 1.28 eV, which corresponds to band gap of evaporated $BaSi_2$ film. The effect of deposited substrate temperature on photoresponsivity was also investigated. The obtained results indicate that $BaSi_2$ film grown at substrate temperature of 500 °C, which is single phase and free from crack, shows the best photoresponsivity. Maximum value of photocurrent was obtained at photon energy of 1.9 eV, corresponding to an external quantum efficiency of 22% under reverse applied voltage of 2 V.

Keywords: Barium silicide, evaporation, photoresponse, carrier lifetime.

1. Introduction

On development of innovative and sustainable material for thin film solar cell, we have concentrated on semiconducting $BaSi_2$ which owns many great advantages such as abundant elements, ideal band gap and high absorption coefficient [1-3]. High quality (100)-oriented $BaSi_2$ film has been epitaxially grown on Si (100) and (111) surfaces by molecular beam epitaxy (MBE) [4, 5]. The potential of the film in application to thin film solar cell has been explored [6-9]. An open circuit voltage of as high as 0.9 V and a conversion efficiency of ~ 25% are estimated for 2 μm thick $BaSi_2$ homojunction diode under AM 1.5 illumination [8]. The application of $BaSi_2$ into solar cell fabrication has gained many interesting results, particularly; an efficiency of 9.0 % has been achieved for p-type $BaSi_2$ on n-type crystalline Si (c-Si) heterojunction cell [9]. From the view point of practical application, we are focusing on growing $BaSi_2$ film by vacuum evaporation (VE) method that has high deposition rate and large-scale substrate area. In fact, single-phase polycrystalline $BaSi_2$ was successfully grown on various substrates such as Si, glass and CaF_2 by using VE [10-12]. However, there is no report of photoresponsivity for $BaSi_2$ film grown by VE method. In this study, we attempt to investigate photoresponsivity properties of $BaSi_2$ film grown on Si substrate by VE, demonstrating the possibility of the film for high-efficiency thin film solar cell.

2. Experimental procedure

We attempted to grow $BaSi_2$ thin film with 0.26 g $BaSi_2$ granule source (size <1 mm). The source was melted on heated tungsten boat at distance of 19 cm from the substrate holder in a vacuum chamber. After $BaSi_2$ was completely melt, heating current was rapidly raised to 140 A to increase deposition rate. $BaSi_2$ films were grown on substrates at temperature (T_s) of 400, 500 and 600 °C. Three types of substrate were used for investigation. The first one is p-type c-Si (100) with high resistivity (ρ>1000 Ω·cm) for sheet carrier density by Hall measurement using the van der Pauw method. The second one is intrinsic c-Si (100) with high resistivity (ρ>1000 Ω·cm) for carrier lifetime evaluation. The last one is p-type c-Si (100) with low resistivity (ρ<0.02 Ω·cm), used for photoresponsivity and other measurements. Surface morphology of the film was obtained by scanning electron microscope (SEM). Raman spectroscopy and X-ray diffraction (XRD) were used to characterize crystallinity. Carrier lifetime measurement was performed by microwave-detected photoconductivity decay (μ-PCD) method, using laser excitation with a wavelength of 349 nm and differential microwave antenna detection with frequency of 26 GHz. In order to investigate vertical photocurrent, 100-nm-thick ITO and 200-nm-thick Al were sputtered on $BaSi_2$ film and back-side Si, respectively. Sample was then annealed at 100 °C for 15 min to improve quality of ITO film. Photoresponsivity was measured by lock-in technique using a 450 W xenon lamp with a monochromator to produce monochromatic light with a constant power of 50 μW/cm^2.

3. Results and discussion

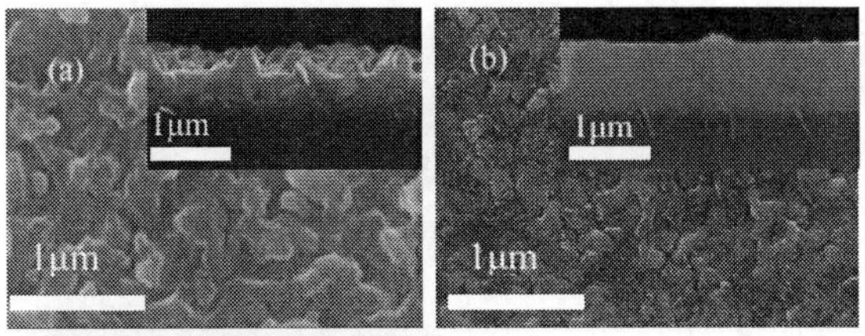

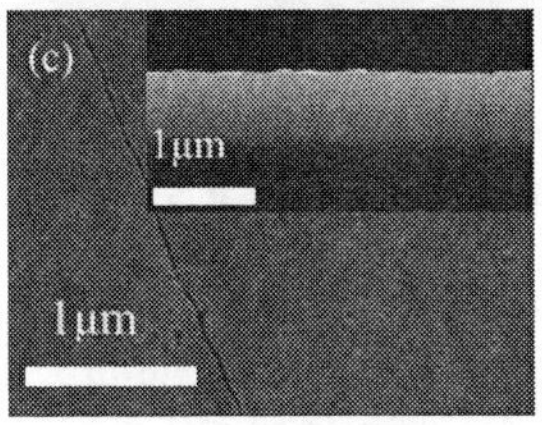

Fig. 1. Top-view and cross-sectional view SEM images of $BaSi_2$ films grown on Si substrate at T_s of (a) 400, (b) 500 and (c) 600 °C.

Top-view and cross-sectional view SEM images of $BaSi_2$ film grown at various T_s are displayed in Fig. 1. One can see that the film grown at low T_s of 400 °C has porous structure and higher roughness. Film surface becomes smoother at higher T_s. However, cracks appear on the film grown at T_s of 600 °C. The origin of crack might be related to in-plane tensile stress in the film prepared at high T_s. Thickness of $BaSi_2$ film grown at T_s of 400, 500 and 600 °C are 0.92, 0.86 and 0. 74 μm, respectively.

Fig. 2 shows Raman spectra of $BaSi_2$ film deposited at various T_s. The peaks observed from the spectra indicate the vibration of $[Si_4]^{4-}$ anions in $BaSi_2$ [13], confirming that all films contain orthorhombic $BaSi_2$. Fig. 3 shows XRD patterns of $BaSi_2$ film deposited at various T_s. Films grown at $T_s \geq 500$ °C are single-phase polycrystalline films. On the other hand, film grown at T_s of 400 °C is a mixture of $BaSi_2$ and BaSi phases, in which BaSi phase is dominant. It has been reported that the

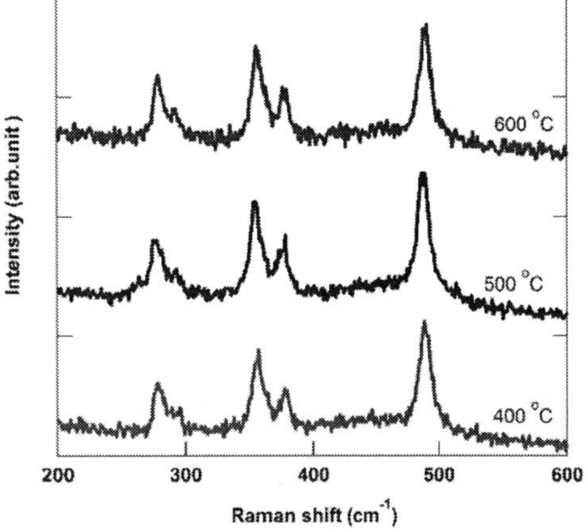

Fig. 2. Raman spectra of $BaSi_2$ film deposited at various T_s.

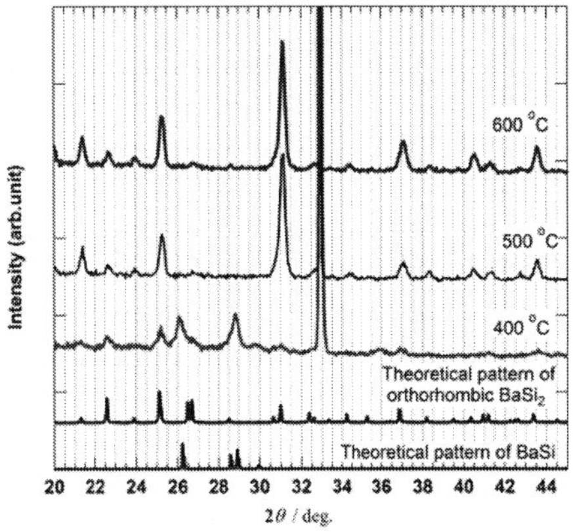

Fig. 3. XRD patterns of $BaSi_2$ film deposited at various T_s. Black line displays theoretical pattern.

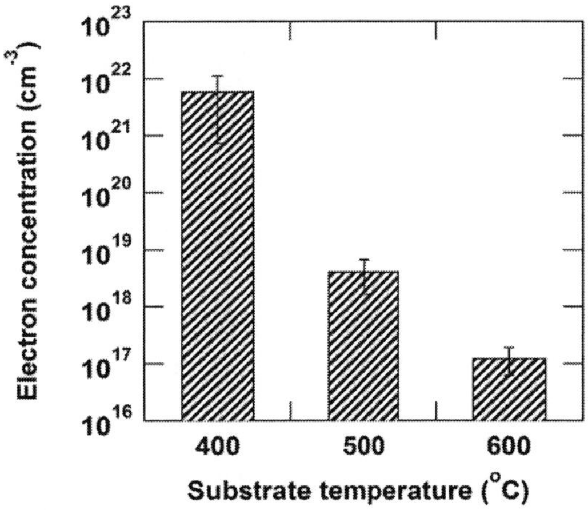

Fig. 4. Carrier concentrations of $BaSi_2$ film as functions of T_s.

initial vapor phase is Ba-rich in VE method [9]. Hara *et al.* have found that longer time is required for Ba diffusion into Si substrate to form single-phase film grown at lower T_s. As we mentioned above, we used high heating current of 140 A, therefore, the condition might not meet requirement to obtain single-phase film at 400 °C.

Carrier concentration (n) of the films grown at various T_s are shown in Fig. 4. n decreases with increasing T_s. High n of ~ 10^{21} cm^{-3} was found for $BaSi_2$ film grown at T_s of 400 °C. From XRD results, the film contains metallic phase, resulting in such a high n. Lower n at higher T_s might be due to suppression of metallic phase. However, n of evaporated $BaSi_2$ film is higher than that of MBE film. One possible reason is the number of crystal defects in evaporated film is higher than MBE film.

Fig 5 shows photoconductance decay of $BaSi_2$ films at various T_s measured at photon density of 1.1×10^{14} cm^{-2}. Like in case of MBE $BaSi_2$ film, photoconductance decays are composed of three decays: initial rapid decay, approximately constant decay and slow decay corresponding to Auger recombination, Shockey-Read Hall recombination and carrier trapping effect, respectively [14]. We estimated carrier lifetime (τ_{eff}) through the second decay. τ_{eff} are 4.8 and 2.7 μs for films grown at 500 and 400 °C, respectively. The lower value is obtained for film grown at 400 °C. It is consistent with general case when film has high defect density, thus, SRH recombination is dominant, τ_{eff} increases with increase in injection level since defect density is finite. For sample grown at 600 °C, signal is very low and there is much noise, implying low carrier lifetime.

Vertical photocurrents were measured for ITO/$BaSi_2$/p-type c-Si/Al structures. Photoresponse was obtained for films grown at T_s of 400 °C and 500 °C. For the film grown at 600 °C, photoresponsivity could not be observed. Fig 6 shows photoresponse properties of ITO/$BaSi_2$/ p-Si/Al with $BaSi_2$ grown at T_s of 400 and 500 °C under various reverse bias voltages. One can see that for all samples, the onset of photocurrent is at photon energy of 1.28 eV, which corresponds to the band gap of evaporated $BaSi_2$ film [9]. This fact indicates that photoresponsivity originates from evaporated $BaSi_2$ film. For $BaSi_2$ film grown at T_s of 400 °C, it is hard to obtain signal at higher applied voltage and there was much noise. It might

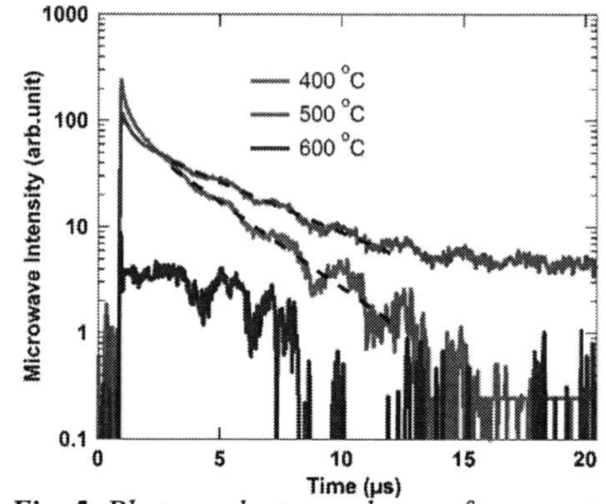

Fig. 5. Photoconductance decay of evaporated $BaSi_2$ film grown at various T_s.

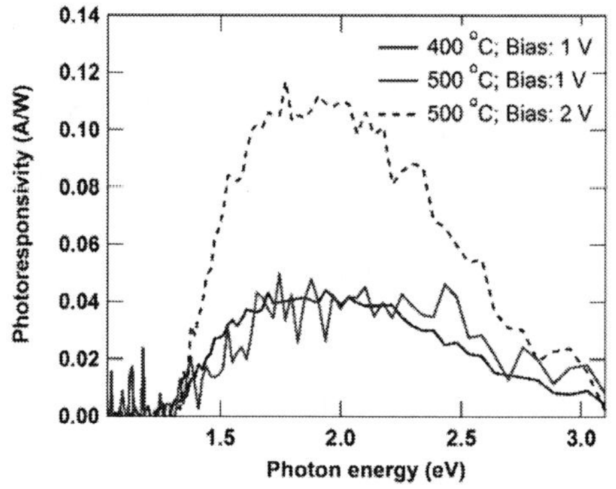

Fig. 6. Photoresponse properties of ITO/$BaSi_2$/ p-Si/Al with $BaSi_2$ grown at T_s of 400 and 500 °C under various reverse bias voltages.

be due to scattering due to voids, surface roughness or/and poor interface property of $BaSi_2$/p-Si. On the other hand, high photoresponsivity can be clearly observed for $BaSi_2$ film grown at T_s of 500 °C, as shown in Fig. 6. The maximum value of vertical photocurrent was obtained at photon energy of 1.9 eV, corresponding to an external quantum efficiency (EQE) of 22% under reverse bias of 2 V. This important result demonstrates potential application of evaporated $BaSi_2$ film for thin-film solar cell. However, EQE of evaporated film is lower than that of MBE $BaSi_2$ film. For example, maximum EQE of 17.8 % was found for 360-nm-thick MBE $BaSi_2$/p-type c-Si with tunnel junction under reverse voltage of 4 V [15]. Especially, EQE as high as 75% can be obtained for ITO/MoO_x/600-nm-thick MBE $BaSi_2$/ p-type c-Si (ρ<0.01 Ω·cm) structure under voltage of 1 V [16].

The outstanding results were achieved for MBE $BaSi_2$ film due to its high crystal quality as evidenced by long minority carrier diffusion length (10 μm) [7], long $\tau_{eff} \leq$ 11 μs [6], and potential variations around grain boundaries [17]. In order to make evaporated film become more reliable for application, further investigation to improve quality of evaporated $BaSi_2$ film should be focused. In summary, photoresponse property is successfully observed for evaporated $BaSi_2$ film, demonstrating that evaporated $BaSi_2$ film is promising absorber material for high-efficiency thin film solar cell.

4. Conclusion

We attempted to grow $BaSi_2$ thin films on Si (100) substrate at various T_s by using VE method. At $T_s \geq$ 500 °C, films are single-phase polycrystalline $BaSi_2$ film while the film grown at T_s of 400 °C is a mixture of various phases including BaSi. Additionally, high τ_{eff} of 4.8 and 2.7 μs can be obtained for the films grown at 500 and 400 °C, respectively. Moreover, photoresponsivity of evaporated $BaSi_2$ is clearly observed at energy level larger than 1.28 eV, which corresponds to the band gap of evaporated $BaSi_2$ film, indicating photoresponse signal originate from $BaSi_2$ film. The film grown at T_s of 500 °C, which is crack-free and single phase, shows the best photoresponsivity. Maximum value of vertical photocurrent was obtained at photon energy of 1.9 eV, corresponding to an external quantum efficiency of 22% at reverse applied voltage of 2 V. This study strongly confirms the potential of evaporated $BaSi_2$ film in high-efficiency thin film solar cell.

Acknowledgements

This work was financially supported by the Core Research for Evolutionary Science and Technology of the Japan Science and Technology Agency (CREST, JST), and New Energy and Industry Development Organization (NEDO).

References

[1] M. Kumar, N. Umezawa and M. Imai, *J. Appl. Phys.* **115** (203178) 2014

[2] V. E. Borisenko, *Semiconductor Silicides* (Springer, Berlin, 2000).

[3] T. Nakamura, T. Suemasu , K. Takakura , F. Hasegawa , A. Wakahara , and M. Imai , *Appl. Phys. Lett.* **81**, 1032 (2002)

[4] Y. Inomata, T. Nakamura, T. Suemasu, and F. Hasegawa, *Jpn. J. Appl. Phys.* **43** (4A) L478 (2014)

[5] K. Toh, K. O. Hara, N. Usami, N. Saito, N. Yoshizawa, K. Toko, and T. Suemasu, *J. Cryst. Growth* **345**, 16 (2012).

[6] K. O. Hara, N. Usami, K. Nakamura, R. Takabe, M. Baba, K. Toko, and T. Suemasu, *Appl. Phys. Express* **6**, 112302 (2013)

[7] M. Baba, K. Watanabe, K. O. Hara, K. Toko, T. Sekiguchi, N. Usami, and T. Suemasu, *Jpn. J. Appl. Phys.* **53**, 078004 (2014).

[8] T. Suemasu, *Jpn. J. Appl. Phys.***54**,07JA01 (2015).

[9] D. Tsukahara, H, Takeuchi, W. Du, R. Takabe, K. Toko, N. Usami and T. Suemasu, *Appl. Phys. Lett.* **108** 152101 (2016).

[10] K. O. Hara, Y. Nakagawa, T. Suemasu and N. Usami, *Jpn. J. Appl. Phys.* **54**, 07JE02 (2015).

[11] Y. Nakagawa, K. O. Hara, T. Suemasu, and N. Usami, *Jpn. J. Appl. Phys.***54**, 08KC03 (2015).

[12] K. O. Hara, J. Yamanaka, K. Arimoto, K. Nakagawa, T. Suemasu, and N. Usami, *Thin solid films* **595A**, 68 (2015).

[13] M. Somer, *Z. Anorg. Allg. Chem.* **2000 (626)**, 2478 (2000).

[14] K. O. Hara, N. Usami, K. Toh, M. Baba, K. Toko, N. Saito, and T. Suemasu, *J. Appl. Phys.***112**, 083108 (2012).

[15] Y. Matsumoto, D. Tsukada, R. Sasaki, M. Takeishi and T. Suemasu, *App. Phys. Express.* **2**, 021101, (2009).

[16] W. Du, R. Takabe, M. Baba, H. Takeuchi, K. O. Hara, K. Toko, N. Usami and T. Suemasu, *Appl. Phys. Lett.* **106**, 122104 (2015).

[17] M. Baba, S. Tsurekawa, K. Watanabe, W. Du, K. Toko, K. O. Hara, N. Usami,T. Sekiguchi and T. Suemasu, *Jpn. Appl. Phys. Lett.* **103** 142113 (2013).

SOLAR GRADE III-V SUBSTRATES FOR COST EFFECTIVE HIGH EFFICIENCY PHOTOVOLTAICS

Yan-Ting Sun[1, 3], Giriprasanth Omanakuttan[1], Carl Reuterskiöld Hedlund[2], Mattias Hammar[2], Sebastian Lourdudoss[1]
1. Laboratory of Semiconductor Materials, Department of Materials and Nano Physics, School of Information and Communication Technology, Royal Institute of Technology-KTH, Electrum 229, 164 40 Kista, Sweden
2. Department of Integrated Devices and Circuits, School of Information and Communication Technology, Royal Institute of Technology-KTH, Electrum 229, 164 40 Kista, Sweden
3. Tandem Sun AB, Lillhersbyvägen 10A, 191 45 Sollentuna, Sweden

ABSTRACT: Solar grade III-V substrate is a virtual III-V/Si substrate consisting of high crystalline quality III-V semiconductor with direct III-V/Si heterojunction on Si substrate, which can be the key to realize high efficiency Si based tandem solar cell at a competitive cost. Recently, corrugated epitaxial lateral overgrowth (CELOG) was used to fabricate high quality InP on Si with atomically abrupt interface. In this work CELOG based fabrication technology of solar grade InP substrate comprising n-InP/p-Si type II heterojunction or n^+-InP/n-Si ohmic heterojunction was developed in hydride vapor phase epitaxy (HVPE) reactor. The structural, optical and electrical properties of the InP/Si heterojunctions were systematically studied by scanning electron microscopy (SEM), photoluminescence (PL), transmission electron microscopy (TEM) and current-voltage (I-V) techniques. Photodiode devices based on solar grade InP/Si substrate were demonstrated. Exploitation of the solar grade III-V substrate technology will pave the way to achieve the Si PV cell with efficiency beyond the level of 30% at a competitive cost.
Keywords: Solar Grade III-V Semiconductor Substrates, III-V/Si Solar Cells, Cost Reduction, High Efficiency

1 INTRODUCTION

The conversion efficiency of single p-n junction solar cell such as Si solar cell is limited to ~30% according to the detailed balance theory [1]. In order to enhance the efficiency of Si solar cell beyond 30%, a Si tandem solar cell comprising wide bandgap III-V semiconductor top cell and Si bottom cell is a promising solution [2]. The key element for the success of III-V/Si tandem solar cell is the high crystalline quality virtual III-V semiconductor substrates consisting of Si active PV cell and coherent III-V/Si heterojunctions with low electrical and optical losses. However, the lattice mismatch between III-V semiconductors and Si prevents the monolithic integration of III-V top cells on Si bottom cells by introducing high density defects at the heteroepitaxial III-V/Si interface. CELOG is a unique epitaxial technology to avoid the formation of crystalline defects at III-V/Si interface despite large lattice mismatch, which is attributed to the absence of epitaxy characteristics in the III-V semiconductor growth on Si surface in a manner similar to epitaxial fusion [3], [4], [5]. The electrical properties of the III-V/Si heterojunction fabricated by CELOG can be tailored by controlling the doping profile during growth.

In addition, several III-V semiconductors and Si can form type II heterojunctions with staggered band discontinuity, which are desired by photovoltaic application for their ultrafast photogenerated charge separation [6] and increased carrier lifetime [7] as demonstrated in the photovoltaic devices based on core/shell nanowire arrays [8]. Such photovoltaic effect has not been investigated in the III-V/Si type-II heterojunction due to large lattice mismatch and associated high density defects formed in heteroepitaxy.

In this work solar grade InP/Si substrates with both p-n and ohmic heterojunctions will be investigated and utilized for photovoltaic applications. It will be shown that the crystalline quality of III-V/Si heterostructure obtained in wafer bonding approach and the flexibility in choosing favorable III-V semiconductor band gap given by heteroepitaxy can be achieved simultaneously by CELOG technology, which is desired for realizing high efficiency Si tandem solar cell at a competitive cost.

2 EXPERIMENTAL

In this work, the virtual solar grade InP substrate on Si with the coherent InP/Si heterojunction was fabricated by the self-aligned corrugated epitaxial lateral overgrowth (CELOG) of InP on Si in low pressure hydride vapor phase epitaxy (LP-HVPE) reactor. The process flow of CELOG is depicted in Figure 1 (a-d). Both n-type and p-type (001) Si substrates with 4° off cut toward [111] direction were used to fabricate solar grade InP substrates. A 50 nm thin GaAs buffer layer and a 2 μm thick seed InP layer were grown by metalorganic vapor phase epitaxy (MOVPE) on the Si substrate, which were processed to mesa structures composed of multiple circular openings with diameter of 30 μm by using plasma enhanced chemical vapor deposition (PECVD) (to deposit SiO_2), photolithography, and inductively coupled plasma (ICP) etching to expose the surface of underlying Si substrate (Figure 1 (a)). Then the sample was etched in 1 H_2SO_4:1 H_2O_2:20 H_2O (volume ratio) for 1 min to remove the residual of GaAs buffer layer in the openings on Si. A 200 nm Si_3N_4 spacer layer was deposited by PECVD to cover the surface of InP/Si template including the sidewalls of InP mesas (Figure 1(b)). CHF_3 based reactive ion etching (RIE) was applied to expose the surface of InP seed mesa and that of the circular Si openings in a self-aligned manner while keeping the Si_3N_4 spacer layers on the sidewalls of the InP seed mesa intact during etching by taking advantage of the anisotropic etching characteristics of RIE (Figure 1(c)).

The CELOG InP was grown by HVPE on the prepared InP/Si template (Figure 1(d)). Before the sample was loaded to the reactor, it was cleaned in buffered HF (BHF) diluted by iso-propanol in 1:10 volume ratio for 15 seconds. CELOG was carried out in a LP-HVPE reactor with flows of 12 sccm InCl and 120 sccm PH_3, which consists of a 5 min growth of semi-insulating

InP:Fe (SI-InP) with resistivity of $>2\times10^8$ Ω·cm and 25 min growth of unintentionally doped (UID) InP or 35 min growth of S-doped n^+-InP. The CELOG UID InP was grown on p-Si to form n-InP/p-Si heterojunction since the residual dopant is of n-type with a concentration of $\sim1.5\times10^{15}$ cm^{-3}. Sulfur doped n^+-InP with dopant concentration $>2\times10^{18}$ cm^{-3} was grown on n-Si to create a heterojunction with ohmic conduction followed by 10 min growth of unintentionally doped InP (UID-InP) [3]. The growth temperature was 590 °C, and the reactor pressure was 20 mbar. As shown in Figure 1(d), the CELOG initiated on the surface of InP mesa and extend laterally and downwards to coalesce with the exposed Si substrate surface to form a direct heterojunction. The 5 min SI-InP growth will only cover the surface of InP mesa to create electric isolation between the CELOG InP and the seed layer.

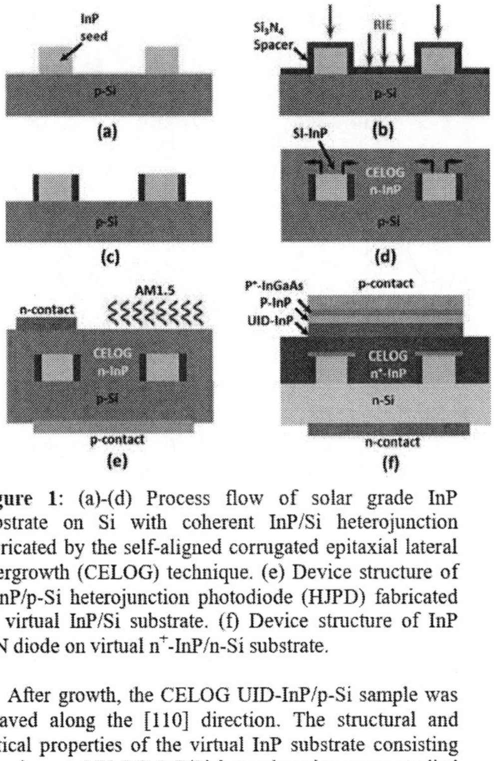

Figure 1: (a)-(d) Process flow of solar grade InP substrate on Si with coherent InP/Si heterojunction fabricated by the self-aligned corrugated epitaxial lateral overgrowth (CELOG) technique. (e) Device structure of n-InP/p-Si heterojunction photodiode (HJPD) fabricated on virtual InP/Si substrate. (f) Device structure of InP PIN diode on virtual n^+-InP/n-Si substrate.

After growth, the CELOG UID-InP/p-Si sample was cleaved along the [110] direction. The structural and optical properties of the virtual InP substrate consisting of coherent CELOG InP/Si heterojunction were studied by scanning electron microscope (SEM), μ-photoluminescence (μ-PL) equipped with a 514.5 nm Ar^+ laser, and high resolution transmission electron microscope (HRTEM).

The grown UID-InP/p-Si heterojunction was processed to heterojunction photodiode (HJPD) by depositing ohmic n- and p- contacts on InP and Si, respectively (Figure 1(e)). E-beam evaporation and lift-off techniques were used to deposit n-contact consisting of 90 nm AuGe/50 nm Ni/150 nm Au on UID-InP and 600 nm Al p-contact was deposited on Si by sputtering technique. InP PIN diode structure consisting of 500 nm p^+-InGaAs:Zn (1E19cm^{-3}) contact layer lattice matched to a 300 nm p-InP:Zn (2E18cm^{-3}) was grown by MOVPE on the virtual n^+-InP/n-Si substrate with n^+-InP/n-Si ohmic heterojunction (Figure 1(f)). The MOVPE growth was conducted in a low-pressure (100 mbar) Aixtron 200/4 reactor with TMGa, TMIn, AsH_3 and PH_3 as precursors at a growth temperature of 650°C, V/III ratio around 150 and a growth rate of 0.21 nm/s for both layers. 50nm Ti/50nm Pt/150nm Au circular p-contact with diameter of 300 μm on p^+-InGaAs and 300 nm Ag on Si were deposited by e-beam evaporation. The p-contact metal was used as etching mask in ICP etching to form the device mesa and define the junction area by removing the p^+-InGaAs, p-InP and UID-InP layers above CELOG n^+-InP. The fabricated InP/Si HJPD and InP PIN diode on Si were annealed at 380 °C for 5 min. The dark and photo current-voltage (I-V) characteristics of the devices were measured by an Agilent B1500A Semiconductor Device Analyzer equipped with a LOT LS0106 full spectrum AM1.5 solar simulator.

3 RESULTS AND DISCUSSION

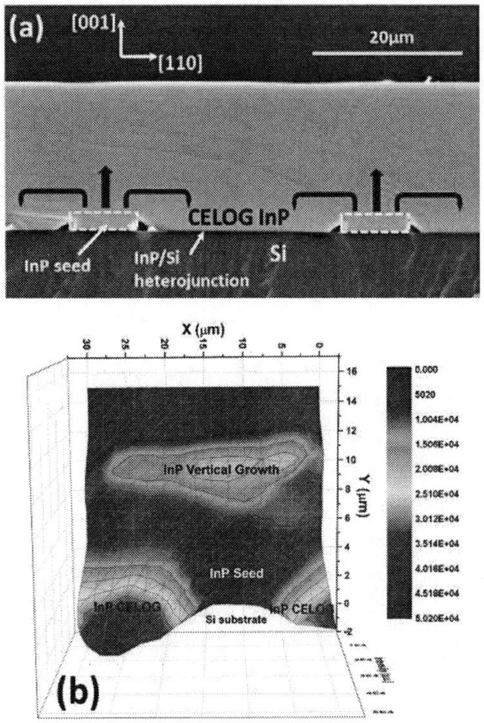

Figure 2: (a) Cross-sectional SEM image of solar grade InP substrate on Si with direct InP/Si heterojunction fabricated by CELOG technique (b) Cross-sectional PL mapping of CELOG InP/Si heterojunction. The color scale bar of PL intensity (in arbitrary units) measured at the central wavelength of 920 nm corresponding to InP bandgap is shown at the right side of plot.

The growth morphology and optical properties of the virtual InP substrate on Si with direct UID-InP/Si heterojunction were examined by cross-sectional SEM (Figure 2(a)) and μ-PL (Figure 2(b)) in the cleaved [$\bar{1}$10] view. As shown in Figure 2(a), the selective growth of InP on the surface of InP seed was achieved by protecting the sidewalls of InP mesa with Si_3N_4 spacer layer. The selective growth of InP proceeds vertically and laterally over the edge of seed mesa, which continues downwards and coalesces with the surface of Si exposed in the openings. Such CELOG InP grown on engineered InP/Si template with InP mesa can form direct InP/Si heterojunction with high crystalline quality and synergetic properties desired for high efficiency photovoltaics and photonic integrated circuits.

The crystalline quality of the initial InP seed has a high dislocation density (~10^9cm^{-2}). During CELOG process, the dislocations in the InP seed will propagate to the surface of growth and probably escape to the lateral overgrowth region. However, the dislocations cannot bend and propagate to the Si surface. Therefore high crystalline quality InP can be expected in the CELOG region close to the InP/Si interface, which is revealed by cross-sectional room temperature µ-PL mapping at center wavelength of 920 nm corresponding to the band gap energy of InP. The scanning area is 30 µm×15 µm and the step sizes are 2 µm along the X-axis and 1 µm along the Y-axis. The PL intensity contour is shown in Figure 2(b). The seed layer grown by MOVPE has low PL intensity due to high-density dislocations. When the growth thickness increases, the PL intensity increases along the vertical growth direction, which can be attributed to the dependence of threading dislocation density on the thickness of heteroepitaxial III-V layers on Si [9], [10]. The highest PL intensity is observed in the InP directly above InP/Si interface in the CELOG section adjacent to the InP seed. This is an evidence of the effective threading dislocation filtering mechanism by CELOG. The threading dislocation can escape to the lateral growth region but cannot bend to the surface of Si substrate and therefore high crystalline quality InP with direct InP/Si heterojunction can be obtained.

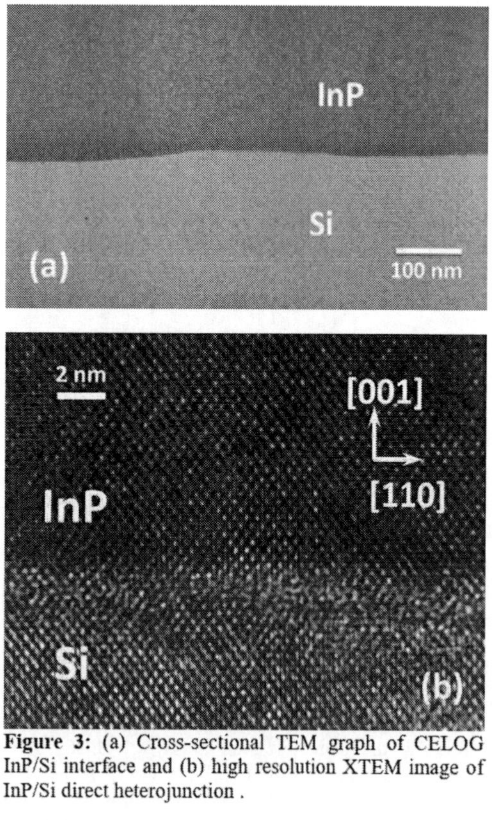

Figure 3: (a) Cross-sectional TEM graph of CELOG InP/Si interface and (b) high resolution XTEM image of InP/Si direct heterojunction .

The crystalline quality and the structural properties CELOG InP on Si were further investigated by cross-sectional TEM and high resolution TEM. The lamella consisting of CELOG InP/Si interface region was prepared perpendicular to the [$\bar{1}$10] direction by focused ion beam (FIB) etching. As shown in Figure 3(a) no threading dislocation or stacking faults defects are observed in the CELOG InP on Si in the TEM image despite 8% lattice mismatch. In the HRTEM image, a coherent InP/Si interface is revealed. No amorphous or intermediate layers can be observed at the interface of InP/Si heterojunction. The CELOG InP nucleates on the lateral growth section originated from the InP seed without nucleation on Si surface. The grown InP crystal is fused with the Si surface in a manner similar to wafer bonding at high temperature, which will facilitate the formation of atomic bonds between InP and Si crystals. Therefore CELOG combines the advantages of monolithic integration by heteroepitaxy and the high crystalline quality given by wafer bonding.

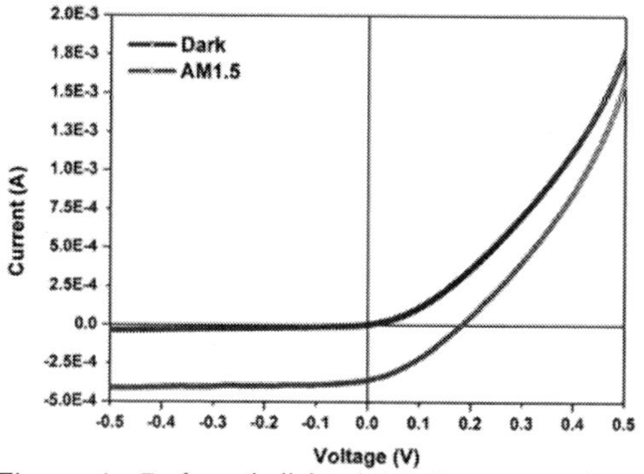

Figure 4: Dark and light (AM1.5) current-voltage characteristics of n-InP/p-Si heterojunction photodiode. The red curve is the light current measured under AM1.5 condition with an open circuit voltage of 180mV and a short circuit current density of 1.89 mA/cm^2.

The virtual InP substrate with UID-InP/p-Si heterojunction was used to fabricate photodiode, which was characterized by current-voltage (I-V) technique under AM1.5 illumination and in dark (Figure 4). The dark current density (DCD) of photodiode is affected by the crystalline quality. The InP/Si HJPD device has a dark current of 39 µA under a reverse voltage of -0.5V, which corresponds to a DCD of 0.21mA/cm^2. This value is much lower than the reported value of 23mA/cm^2 for Ge/Si HJPD fabricated by wafer bonding technique based on single crystalline Ge substrate [11], which indicates a high crystalline quality InP/Si heterojunction with low carrier generation in the depletion region under reverse bias. The photoresponse arising from the type II n-InP/p-Si heterojunction was characterized under AM1.5 illumination. The type-II InP/Si heterojunction has an effective spatial indirect band gap of 0.74 eV between the minimum of InP conduction band and the maximum of Si valence band calculated from the electron affinity model and a built in electric field is formed in the depletion region under equilibrium condition. The photo generated electron-hole pairs were separated by the built in electric field. Electrons will accumulate on InP side and holes will accumulate on Si side, which gives rise to an open circuit voltage of 180mV under AM1.5 condition as shown in Figure 4. The short circuit current is 0.35mA corresponding to a short circuit current density (J_{SC}) of 1.89mA/cm^2. A high J_{SC} is critical for achieving high conversion efficiency in PV devices. In InP/Si HJPD the band gaps of InP and Si are close, which are 1.35 eV and 1.12 eV, respectively. Considering the direct band gap characteristics of InP, most of high energy photons will be absorbed in the thick InP and few photons can reach the depletion region and Si base. In addition, no anti-reflection (AR) coating was applied on InP/Si HJPD in this work. Therefore further increase of J_{SC} and conversion efficiency can be expected by reducing the InP thickness and introducing appropriate AR coating on InP.

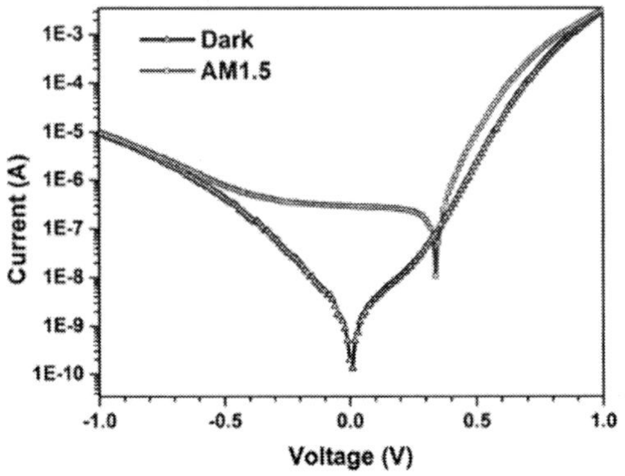

Figure 5: Dark and light (AM1.5) current-voltage (I-V) characteristics of InP PIN diode fabricated on solar grade n^+-InP substrate with ohmic n^+-InP/Si heterojunction. The red curve is the light current measured under AM1.5 condition with an open circuit voltage of 340 mV and a short circuit current density of 1.3 mA/cm^2.

Figure 5 shows the I-V characteristics of InP PIN diode fabricated on the virtual n^+-InP substrate with ohmic n^+-InP/n-Si heterojunction [3]. The PIN structure was chosen to mitigate the reduced carrier life time observed in heteroepitaxial III-V semiconductors on Si due to high density threading dislocation [12]. The photo generated carriers can be formed in the intrinsic, i.e. undoped InP and transported for long distance to the contacts. A dark current density of 43 mA/cm^2 is observed at reverse bias of -1 V. The increasing dark current with reverse bias indicates a carrier generation in the depletion region, which could arise from the dislocations. Under AM1.5 illumination open circuit voltage (V_{oc}) of 340 mV and short circuit current density of 1.3 mA/cm^{-2} are obtained (Figure 5) although only the mesa edge of InP PIN diode is exposed to the incident photons. V_{oc} of 876 mV has been reported for InP solar cell grown on p-InP substrate by MOVPE with optimized emitter structure [13]. The V_{oc} of solar cell can be affected by the density of threading dislocations in the device structure [14], which can act as generation center in the depletion region. Thus, both the low V_{oc} and the increasing reverse dark current observed in this InP PIN diode could be due to the residual threading dislocations escaped to the surface of CELOG InP. In order for III-V/Si based solar cells built on solar grade III-V substrates to achieve efficiency comparable to that fabricated on conventional III-V substrates, the dislocation density in the III-V semiconductors grown on Si should be lower than 10^5cm^{-2} [14]. As shown in Figure 2(b), the CELOG InP close to the InP/Si interface has the highest crystalline quality due to the dislocation filtration effect of CELOG. InP p-n junction solar cell with high performance can be expected by depositing the device structure on CELOG InP region after applying chemical mechanical polishing (CMP) to remove the vertical growth section.

4 CONCLUSION

In this work virtual solar grade InP substrates on Si comprising high crystalline quality InP with coherent n-InP/p-Si and n^+-InP/n-Si heterojunctions have been fabricated by self-aligned CELOG in LP-HVPE reactor, which were processed to HJPD and used for InP PIN diode fabrication. CELOG can remove the lattice matching constraint in the heteroepitaxy of III-V semiconductors on Si and avoid the formation of threading dislocations by creating the III-V/Si heterojunction in an epitaxially fused manner. The structural and optical properties of the virtual InP/Si substrate were studied extensively and the high crystalline quality InP on Si was observed by HRTEM and cross-sectional μ-PL mapping. The I-V characteristics of InP HJPD and InP PIN diode fabricated on the virtual InP/Si substrates were characterized for PV effect. With further improvement in the CELOG process high V_{oc} and conversion efficiency can be expected for the PV devices grown on the solar grade InP substrate. The technology of solar grade InP substrate on Si with direct InP/Si heterojunction demonstrated in this work is generic and can be extended to other III-V semiconductors, which are critical to achieve Si tandem solar cell with efficiency beyond 30% at a competitive cost. In the near future, solar grade III-V substrates with free choice of lattice constant can be fabricated on Si by CELOG, which will facilitate the realization of cost effective Si based multi-junction solar cells with optimum bandgap combinations for high efficiency performance.

ACKNOWLEDGEMENTS

We gratefully acknowledge the financial support from Swedish Energy Agency, Swedish Research Council (VR), EIT-KIC InnoEnergy and the umbrella program of SOLAR-ERA.NET.

REFERENCES

[1] W. Shockley and H. J. Queisser, J. Appl. Phys. 32 (1961) 510.
[2] N. Jain and M. K. Hudait, Energy Harvesting and Systems 1 (2014) 121.
[3] Y.T. Sun, H. Kataria, W. Metaferia and S. Lourdudoss, CrystEngComm 16 (2014) 7889.
[4] W. Metaferia, H. Kataria, Y.T. Sun and S. Lourdudoss, J. Phys. D: Appl. Phys. 48 (2015) 045102.
[5] Y. Sun, S. Lourdudoss, US9293625 (2016).
[6] C.J. Dooley, S.D. Dimitrov, T. Fiebig, J. Phys. Chem. C Lett. 112 (2008) 12074.
[7] C. Chuang, S.S Lo, G.D. Scholes, C. Burda, J. Phys. Chem. Lett. 1 (2010) 2530.

[8] K. Wang, J.J. Chen, Z.M. Zeng, J. Tarr, W.L. Zhou, Y. Zhang, Y.F. Yan, C.S. Jiang, J. Pern, A. Mascarenhas, Appl. Phys. Lett. 96 (2010) 123105.
[9] J. E. Ayers, L. J. Schowalter, S. K. Ghandhi, J. Cryst. Growth 125 (1992) 329.
[10] C. Junesand, M.-H. Gau, Y.T. Sun, S. Lourdudoss, I. Lo, J. Jimenez, P. A. Postigo, F. M. Morales, J. Hernandez, S. Molina, A. Abdessamad, G. Pozina, L. Hultman, P. Pirouz, Mater. Express 4 (2014) 41.
[11] F. Gity, A. Daly, B. Snyder, F. H. Peters, J. Hayes, C. Colinge, A. P. Morrison, B. Corbett, Opt. Express 21 (2013) 17309.
[12] M. Yamaguchi, C. Amano, Y. Itoh, J. Appl. Phys. 66 (1989) 915.
[13] C. J. Keavney, V. E. Haven, S. M. Vernon, Conference Record of the Twenty First IEEE Photovoltaic Specialists Conference, vol.1 (1990) 141.
[14] C. L. Andre, A. Khan , M. Gonzalez, M. K. Hudait, E. A. Fitzgerald, J. A. Carlin, M. T. Currie, C. W. Leitz, T. A. Langdo, E. B. Clark,D. M. Wilt, S. A. Ringel, Conference Record of the Twenty-Ninth IEEE, Photovoltaic Specialists Conference, (2002) 1043.

RECENT ADVANCES IN POLYMER/SILICON HETEROJUNCTION SOLAR CELLS

J. Schmidt[1,2], D. Zielke[1], R. Gogolin[1], R. Sauer[3] and W. Lövenich[3]
[1]Institute for Solar Energy Research Hamelin (ISFH), Am Ohrberg 1, 31860 Emmerthal, Germany
[2]Department of Solar Energy, Institute of Solid-State Physics,
Leibniz Universität Hannover, Appelstr. 2, 30167 Hannover, Germany
[3]Heraeus Deutschland, Chempark Leverkusen, 51368 Leverkusen, Germany

ABSTRACT: In this contribution, we first give a brief historical overview of the recent developments on polymer/silicon heterojunction cells. We then focus on our most recent results concerning: (i) the PEDOT:PSS/c-Si interface design, where the thickness of the native oxide in-between the c-Si surface and the polymer is shown to play a crucial role, (ii) a new type of heterojunction cell, where the hole-selective layer is PEDOT:PSS and the electron-selective layer is well-passivating phosphorus-doped amorphous silicon (*n*-a-Si:H), (iii) the stability of the polymer/silicon cells in ambient environment. We demonstrate for the first time that the efficiencies of BackPEDOT cells are long-term stable if a metal foil is used as rear metallization scheme, as the metal foil does not transmit any humidity in contrast to evaporated metals. (iv) We present first results of our attempt to implement a PEDOT:PSS/c-Si junction into an industrial-type large-area (15.6×15.6 cm^2) screen-printed c-Si solar cell and reach an efficiency of 19.5% in our first batch of cells.
Keywords: c-Si; Polymer Film; Heterojunction; Hybrid; Passivation; Stability; Screen-Printing

1 BRIEF HISTORICAL OVERVIEW

State-of-the-art junctions in silicon solar cells are realized by technologically advanced methods, such as high-temperature dopant diffusion or plasma-enhanced chemical vapour deposition of doped amorphous silicon layers. A technologically much less demanding approach is the deposition of carrier-selective polymers from a liquid precursor onto the silicon surface. A lot of effort has been devoted in recent years on polymer/silicon heterojunction cells based on the hole-selective polymer poly(3,4-ethylene-dioxythiophene):polystyrenesulfonate (PEDOT:PSS).

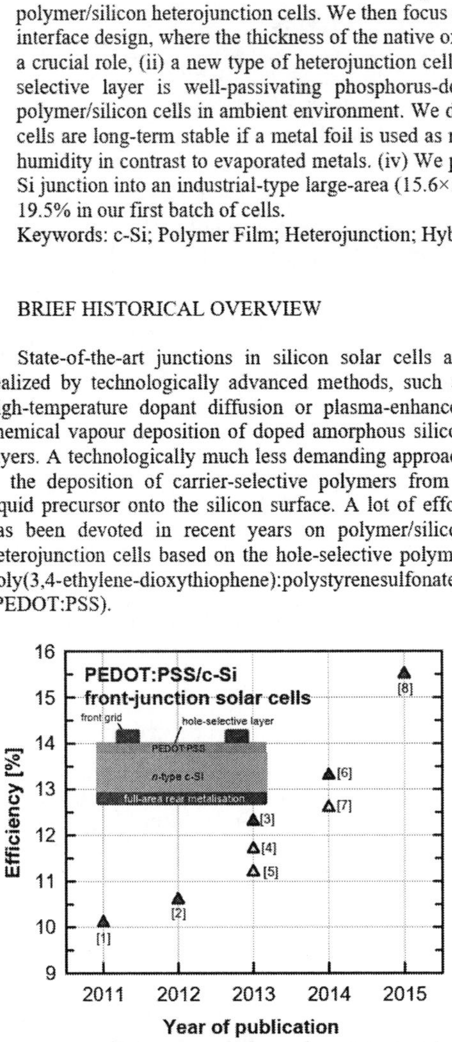

Figure 1: Efficiency evolution of PEDOT:PSS/c-Si front-junction cells. The highest efficiencies reached in the respective year are marked red.

Figure 1 shows the evolution of efficiencies since the first silicon solar cell with PEDOT:PSS on the front was presented by Avasthi et al. [1] in 2011. Although the efficiencies of the front-PEDOT:PSS/c-Si junction devices shown in Fig. 1 increased quite dramatically in recent years [1-8], open-circuit voltages were typically limited in these cells to values $V_{oc} < 600$ mV, which was assumed to be due to recombination at the PEDOT:PSS/c-Si interface. However, as we were able to show in 2013 by means of contactless lifetime measurements [3], PEDOT:PSS leads to an excellent interface passivation on crystalline silicon. We recently demonstrated [9,10] that V_{oc} values > 700 mV are achievable with the PEDOT:PSS/c-Si junction. It turned out that most of the cells shown in Fig. 1 were in fact limited by recombination at the full-area metallized rear. An additional limitation was the parasitic absorption within the PEDOT:PSS at the cell front. We hence proposed to deposit the PEDOT:PSS onto the back of the cell and based on this so-called "BackPEDOT" cell concept [10] we were able to demonstrate a record-high efficiency of 20.6% [9] and V_{oc} values up to 663 mV on device level, clearly proving that BackPEDOT cells have the potential to reach very high efficiencies.

2 PEDOT:PSS/c-Si INTERFACE DESIGN

One crucial element in the optimization of the PEDOT:PSS/c-Si interface turned out to be the silicon oxide layer in-between the polymer and the silicon substrate [10]. The preparation method and the oxide thickness have a critical impact on the interface recombination properties as well as the hole transport properties. We achieve the best passivation properties using a native oxide grown in ambient environment. As the holes have to tunnel through the native SiO_x layer, the thickness of the oxide is critical concerning the contact resistance of the polymer/silicon junction. We have optimized the interfacial oxide on test structures and then verified our results on actual BackPEDOT solar cells [11].

Figure 2 shows the *J-V* curves of two BackPEDOT solar cells [11] with conventionally processed, that is, phosphorus-diffused, front surface and evaporated contacts featuring two different pre-treatments of the silicon surface prior to the PEDOT:PSS coating. After finishing the conventionally processed front side of the solar cells all samples were dipped in a 1% HF solution for 1 minute. After this HF dip, some samples were deposited with PEDOT:PSS after 10 minutes storage in air ('thin' native SiO_x) while other samples were kept in air for 48 hours ('thick' native SiO_x). Subsequently, the

PEDOT:PSS layer was deposited by spin-coating on the entire rear at 500 revolutions per minute (rpm) for 10 seconds and subsequently 1000 rpm for 30 seconds. The samples were then annealed on a hotplate in air at 130°C for 15 minutes for drying and to remove residual solvents.

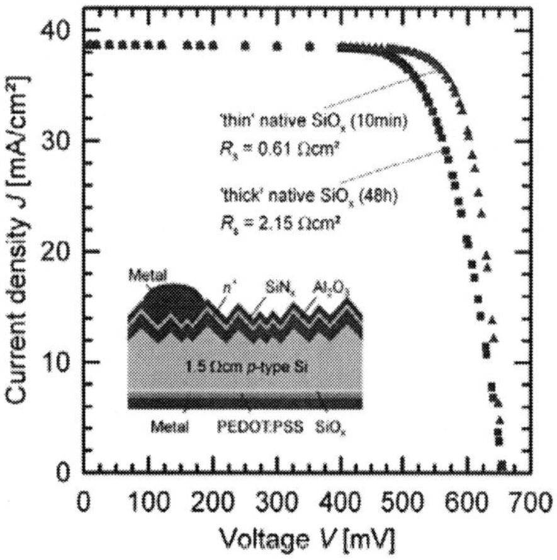

Figure 2: *J-V* characteristics of two BackPEDOT solar cells with two different SiO_x tunneling layer thicknesses. 'Thin' (< 1 nm) tunneling oxide is required to achieve a low contact resistance.

Table I: Measured one-sun solar cell parameters of the two BackPEDOT cells shown in Fig. 2.

native SiO_x	J_{sc} [mA/cm²]	V_{oc} [mV]	FF [%]	η [%]
'thin'(10 min)	38.9	657	80.6	20.6
'thick' (48 h)	38.7	657	73.1	18.6

Finally, the entire rear surface was metallized with silver by e-beam evaporation. Table I and Fig. 2 show the solar cell performances of the two best cells with a 'thin' native SiO_x (red triangles) and a 'thick' native SiO_x (blue squares) tunneling layer, respectively [11]. The J_{sc} and V_{oc} values of the two solar cells in Fig. 2 are almost identical. However, we observe a large deviation in the fill factors *FF*. The solar cell with 'thick' SiO_x shows a drastically increased series resistance R_s of 2.15 Ωcm^2, leading to a low *FF* of only 73.1% compared to 80.6% and an R_s of 0.61 Ωcm^2 for the cell featuring the 'thin' (< 1 nm) SiO_x tunneling layer. We hence conclude that a too thick SiO_x interface layer hampers the transport of holes from the *c*-Si wafer into the hole-conducting PEDOT:PSS layer and in order to achieve high efficiencies, a sufficiently thin SiO_x interface layer has to be chosen to obtain low contact resistances between PEDOT:PSS and *c*-Si.

3 TWO-SIDES HETEROJUNCTION CELLS

We have recently introduced [11] an attractive alternative cell design with a much higher V_{oc} potential compared to our standard BackPEDOT cell design as shown in Fig. 2, based on combining the PEDOT:PSS hole-selective layer with an *a*-Si:H(*n*) electron-selective layer on the other surface of the silicon wafer. A schematic cross section of such a solar cell is shown in the inset of Figure 3.

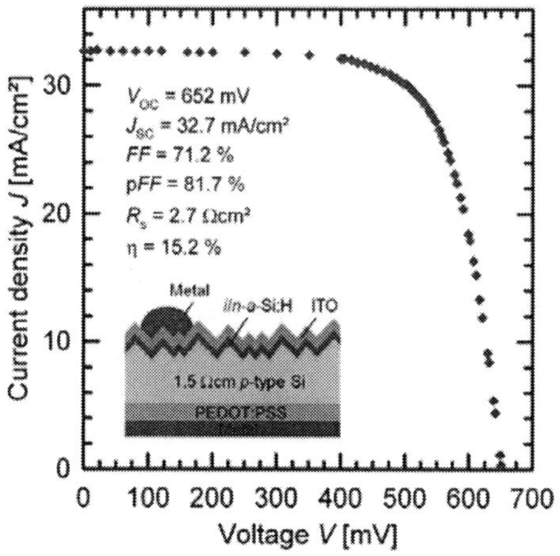

Figure 3: *J-V* characteristic of a *two-sides heterojunction* BackPEDOT solar cell.

In the cell structure shown in Fig. 3, the electron-collector is realized at the cell front using *n*-type *a*-Si:H with an interfacial intrinsic *a*-Si:H layer to achieve the best possible interface passivation (V_{oc} potential > 730 mV has been demonstrated on reference cells). The front is then coated with a sputtered ITO layer and contacted by a screen-printed low-temperature silver paste. Note that it is also possible to achieve very high V_{oc} values (~700 mV) with an *a*-Si:H(*n*) layer without interfacial *a*-Si:H(*i*) layer, which is not the case for *a*-Si:H(*p*). Hence, it might be an attractive solution to combine *a*-Si:H(*n*) as electron-selective layer with PEDOT:PSS as hole-selective layer, which would only require one single PECVD process step. The performance of a representative solar cell of our first batch of PEDOT:PSS/*c*-Si/*a*-Si:H(*n*) cells is show in Fig. 3. Although the reached cell efficiency of 15.2% is far below optimal, this is a clear proof-of-principle. However, the full potential of this cell type could not be exploited in our first batch due to a non-optimal front side preparation (too thick *a*-Si:H and ITO layers limit J_{sc}, front screen-printed contact limits R_s, non-optimal *a*-Si passivation limits V_{oc}). Further optimized cell runs are currently under way in our lab and the results will be presented in the near future.

4 LONG-TERM STABILITY

One major challenge of solar cells with polymer/silicon heterojunctions are their long-term stability. As PEDOT:PSS is known to be very hygroscopic, humidity has been found to be a main cause of efficiency instabilities during storage in ambient atmosphere [3]. We have now performed a comprehensive study of the fundamental degradation mechanisms which will be published soon [12]. The main result is that due to a humidity-related drifting of the work function of PEDOT:PSS, the barrier height for holes transported from the silicon into the PEDOT:PSS increases with time and hence the contact resistance increases, decreasing fill factor and efficiency.

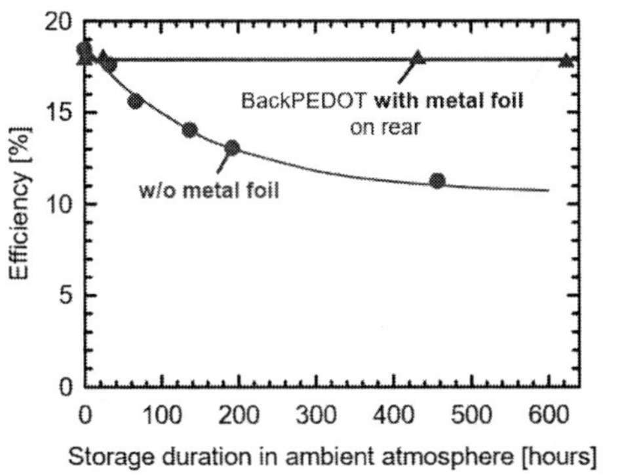

Figure 4: Stable BackPEDOT solar cell with metal foil on the rear.

To keep any humidity away from the PEDOT:PSS layer, we have developed a novel rear-side metallization scheme based on placing a metal-foil onto the entire cell rear, which also has the potential of significantly simplifying the cell processing (see also Section 5). The metal foil itself is – in contrast to the evaporated metal – blocking any humidity. Figure 4 shows a direct comparison of the efficiency evolution of a standard BackPEDOT solar cell without metal foil (green circles) and a BackPEDOT cell with metal foil on the rear (red triangles) during the first 600 hours after processing. Evaporated silver (as applied in our standard BackPEDOT cells) shows a very high permeability against humidity, whereas the metal foil effectively blocks any humidity leading to stable cell efficiencies. The demonstration of stable cell efficiencies can be considered an important milestone in the development of polymer/silicon heterojunction cells.

5 INDUSTRIAL-TYPE BACKPEDOT CELLS

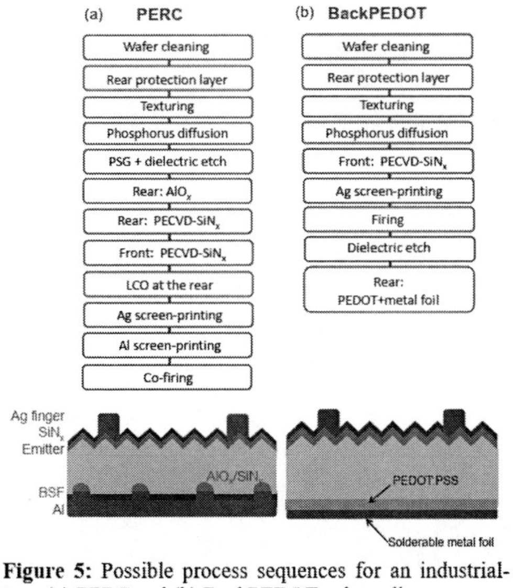

Figure 5: Possible process sequences for an industrial-type (a) PERC and (b) BackPEDOT solar cell.

One attractive option to implement the PEDOT:PSS/c-Si junction into industrial solar cells is sketched in Fig. 5. It is widely believed that the next-generation industrial silicon solar cell generation is based on the passivated emitter and rear cell (PERC) concept. A typical process sequence for an industrial-type PERC solar cell with screen-printed metal contacts is shown in Fig. 5 (a). The rear surface is passivated with an Al_2O_3/SiN_x stack which is locally opened by laser ablation. Finally, the rear is covered with a screen-printed Aluminum metal paste and fired. The process steps of Al_2O_3 deposition, SiN_x deposition, laser contact opening and the Al screen-printing could be replaced in an industrial-type BackPEDOT cell by just depositing the PEDOT:PSS onto the entire rear and attaching a solderable metal foil to it. This process sequence would reduce the number of process steps and hence has the potential of significantly reducing the process costs compared to a PERC process. An example of such a process sequence is shown in Fig. 5 (b). In our lab, we have fabricated the first batch of solar cells with screen-printed front and PEDOT:PSS/c-Si at the rear.

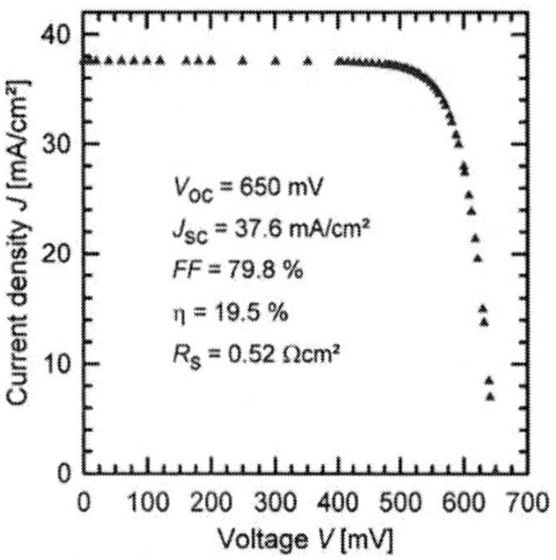

Figure 6: *J-V* curve and solar cell parameters of a large-area BackPEDOT solar cell fabricated on boron-doped Czochralski-grown silicon (15.6×15.6 cm^2). The cell front is conventionally processed (phosphorus diffusion, silicon nitride deposition, Ag screen printing and firing).

Figure 6 shows the *J-V* curve of the best cell of our first batch, fabricated on industrial boron-doped *p*-type Czochralski-grown silicon (cell size 15.6×15.6 cm^2). The reached efficiency of 19.5% is mainly limited by the non-optimal phosphorus-diffused front emitter and the non-optimized firing process. As our BackPEDOT concept works on *p*- as well as on *n*-type silicon wafers, we have also processed solar cells of the same size on phosphorus-doped *n*-type Czochralski-silicon wafers, resulting in maximum efficiencies of 18.4% due to shunting problems. Based on realistic estimates, we expect efficiencies >21% in the near future on *p*- as well as *n*-type silicon wafers, making the BackPEDOT cell concept potentially attractive for industrial solar cell production.

6 SUMMARY

- The thickness of the native SiO_x between the c-Si surface and the PEDOT:PSS plays a crucial role.
- Contact resistances which are compatible with high efficiencies have only been obtained for ultrathin native SiO_x (< 1 nm).
- Two-sides heterojunction cells with *n*-a-Si:H as electron-collector at the front and PEDOT:PSS as hole-collector at the rear show very high V_{oc} and η potential (> 22%). First experimental cells were, however, limited by process issues to V_{oc}=652 mV and η = 15.2%.
- The efficiency of BackPEDOT cells is stable during storage in darkness at ambient atmosphere if a metal foil is put on the rear, as metal foils do not transmit any humidity in contrast to evaporated metals.
- We have presented first results of our attempt to implement a PEDOT:PSS/c-Si junction into an industrial-type large-area (15.6×15.6 cm^2) screen-printed c-Si solar cell. The front surface of this cell was conventionally processed by phosphorus-diffusion, silicon nitride deposition, Ag screen-printing and firing. The highest efficiency of our first batch of this industrial-type BackPEDOT cell was 19.5%, mainly limited by the conventionally processed front.
- We estimate a realistic efficiency potential >21% for the industrial version of the BackPEDOT cell on *p*- as well as *n*-type silicon wafers.

ACKNOWLEDGEMENT

This work was funded by the German State of Lower Saxony and the German Federal Ministry of Economics and Energy (BMWi) within the research project "SiPoly" (contract no. 0325884A). The content is the responsibility of the authors.

REFERENCES

[1] S. Avasthi et al., *Adv. Mater.* **23** (2011), 5762.
[2] L. He et al., *Appl. Phys. Lett.* **100** (2012), 073503.
[3] J. Schmidt et al., *Appl. Phys. Lett.* **103** (2013), 183901.
[4] K. Nagamatsu et al., *IEEE JPV* **4** (2013), 260.
[5] Q. Liu et al., *Appl. Phys. Lett.* **102** (2013), 243902.
[6] J. Thomas et al., *Adv. Funct. Mater.* **24** (2014), 4978.
[7] M. Pietsch et al., *Appl. Phys. A* **115** (2014), 1109.
[8] Q. Liu et al., *Adv. En. Mat.* **5** (2015), 10.1002/aenm.201500744.
[9] J. Schmidt et al., *Tech. Dig. 6th World Conf. Photovolt. Solar Energy Conv.*, Kyoto, Japan (2014), p. 869.
[10] D. Zielke et al., *Sol. En. Mat. Sol. Cells* **131** (2014), 110.
[11] R. Gogolin et al., *Energy Procedia*, in press (2016).
[12] D. Zielke et al., in preparation (2016).

BACK IN THE PV GALAXY:

THE RETURN OF THE SILICONE MODULE

Guy Beaucarne[1], ShenCun Wang[2], Xiang Sun[2], Yujian Wu[3], Yanxia Huang[3], Nick Shephard[4]

[1] Dow Corning Europe SA, Rue Jules Bordet Parc Industriel Zone C, B-7180 Seneffe, Belgium
[2] BYD, Shenzhen, Guangdong, China
[3]Dow Corning(China) Holding Co.Ltd., Shanghai, China
[4]Dow Corning Corporation, Midland, USA

ABSTRACT: Early PV modules mostly used silicone as encapsulant, which resulted in very durable and reliable modules. However, silicone encapsulation was replaced at the end of the eighties by EVA for process and cost reasons. Thanks to the simultaneous emergence of a market for very robust and durable modules (long lifetime utility modules, sunbelt region modules, tropical region modules) and the development of specific silicones and encapsulation processes meeting the requirements of modern PV manufacturing, silicone encapsulated panels have become again a feasible and attractive technology option. Double glass modules with silicone encapsulation have recently been launched and are now mass produced and globally commercialized. The combination of the glass-glass structure and silicone is shown to lead to exceptional durability in accelerated aging tests.

Keywords: MWT, module technology

1 INTRODUCTION

In the early days of photovoltaics silicone was mostly used as encapsulant in PV modules. This resulted in very durable and reliable modules, as evidenced by several reports of old systems showing outstanding performance after more than 25 years of operation. Studies of module degradation over 20 to 30 years in test fields have shown that the degradation rates are much lower than typical degradation rates for EVA-encapsulated modules, in the order of 0.3 %/year against an average value of 0.8 %/year for EVA modules [1-3]. At the time, the industry's primary concern was to reduce cost. The silicone materials that were used were however products that had been developed for electronics. Moreover, the manufacturing processes were not adapted to PV needs. Therefore, the cost of silicone encapsulation was high. Once the industry established that EVA encapsulation was working adequately for the targeted products (at the time only a 20 years lifetime was envisaged, and virtually all modules were installed in temperate climate locations), it replaced silicone encapsulant for EVA at the end of the eighties.

Almost 30 years later, the PV industry is unrecognizable. PV has grown into a major industry and has started having a strong impact on global energy supply. It is anticipated to play a major role in the energy transition that is going to happen worldwide. In this context, requirements for PV module have also changed. A very large part of the worldwide module production will be installed in the Sun Belt region. Those locations are characterized by a much higher insolation (that includes large doses of UV light), sometimes hot and humid conditions (tropical regions) and sometimes hot dry conditions (desert regions). These are much harsher conditions than those in temperate regions. Modules need to be designed to perform well in those conditions and maintain that level of performance for at least 20 years. Although there are presently too few reported module degradation studies in those regions, there are indications that encapsulant degradation in traditional module structures is a problem for high insolation regions (see e.g. [4]).

Moreover, the required global PV deployment demands a sustained decrease of the PV Levelized Cost Of Electricity (LCOE). Great improvements in cell and module technology, balance of systems components, as well as installation design and procedure have resulted in drastically lower cost/Wp of PV installations. This has had a direct impact on LCOE, with prices that decreased from 11.1 to 8.7 $cents/kWh from 2013 to 2015 for moderate insolation, and even down to 4.4 $cents/kWh for high insolation countries, as calculated by the team behind the ITRPV roadmap [5]. As shown in Figure 1, further reductions in $/Wp are expected to take place, resulting in continuing the downward trend. However, there is an opportunity to substantially further reduce LCOE, in a way that can be implemented in the short term. This solution is to install PV modules that have a longer lifetime. In Figure 1, we illustrate this, comparing projected LCOE prices for 25 years lifetime modules with those for 40 years lifetime modules. This is for a system with energy yield of 1000 kWh/kWp. The assumed system costs in $/Wp are those predicted by the 2016 ITRPV roadmap [6], and the LCOE values were calculated using the simple NREL LCOE calculator [7] assuming a discount rate of 3 %.

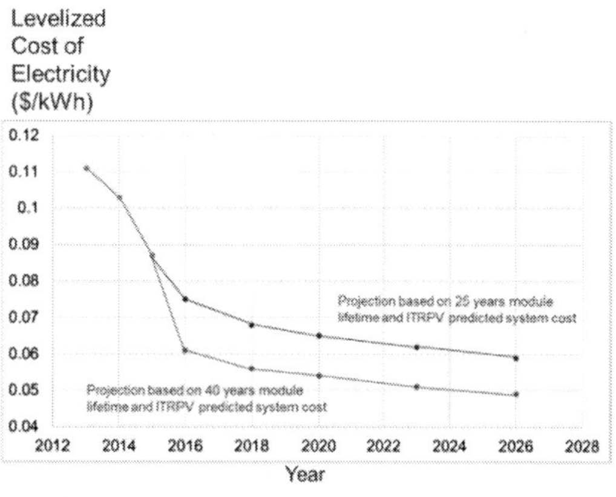

Figure 1: LCOE projections for 25 and 40 years modules, for PV systems with energy yield of 1000 kWh/kWp

To respond to the need for more durable and longer lifetime modules, the industry has started bringing double glass modules to the market. Indeed, degradation caused by water ingress is expected to be much slower in such modules than for a conventional glass-backsheet structure. Moreover a glass-glass laminate is mechanically much more robust than a glass-backsheet laminate. Finally it is easier to meet electrical requirement for 1500 V system operation with such a structure. Even though the use of EVA encapsulant is challenging for a glass-glass structure (see for instance [8]), most commercial double glass c-Si modules today use EVA. These modules typically are presented as having a service lifetime of 30 years.

In order to extend module lifetime even further and provide enhanced reliability, we propose to combine a glass-glass structure with liquid silicone encapsulation. We show that this indeed results in outstanding behavior in accelerated aging tests, while solving some typical problems encountered when making double glass modules.

2 SILICONE ENCAPSULATION

2.1 Silicone encapsulant material

The silicone encapsulant material used in this work is *Dow Corning®* PV-6212 Cell Encapsulant. In contrast to most commercial PV encapsulants, this is supplied as a liquid, not as a foil. It is dispensed in the liquid form and is cured into a solid by cross-linking during lamination. PV-6212 is a polydimethylsiloxane (PDMS : molecules with Si-O-Si-O... backbone and two CH3 groups on each Si atom, see Figure 2) that uses an addition cure system. As a result, there is no byproduct of the cross-linking reaction. PV-6212 is provided in two parts which need to be mixed just before application. Curing results in a firm but fairly soft rubber (shore A hardness 11.5), which gives protection but still allows some movement of the encapsulated pieces.

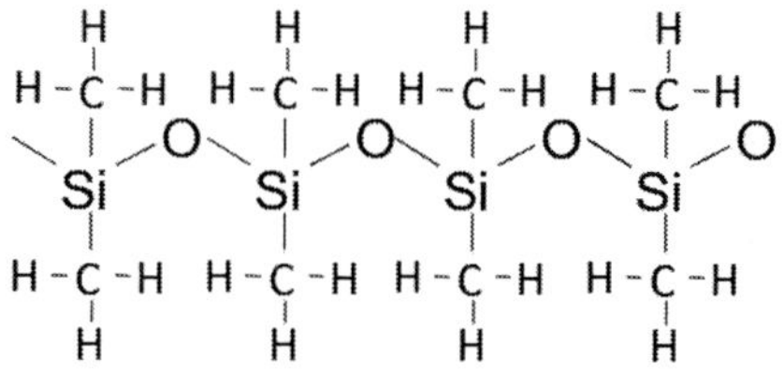

Figure 2: Chemical structure of polydimethylsiloxane silicone

The Si-O bond in the silicone backbone is much stronger than the C-C bond in the backbones of organic polymers such as EVA, with a bond energy of 452 KJ/mol vs. 346 KJ/mol. This makes the material intrinsically more stable. Moreover, cured PV-6212 has a high transparency, including in the UV range where conventional encapsulants have reduced transmittance (Figure 3). This is particularly important for new solar cell technologies which show improved response at short wavelengths. Transparency in the UV range also makes the silicone inherently more resistant to hardness and color changes caused by UV degradation.

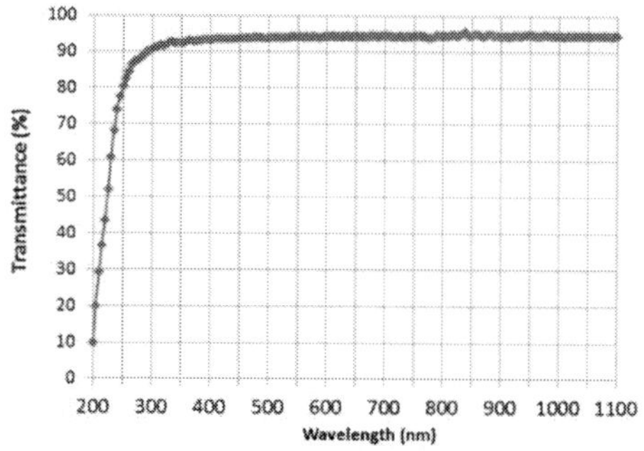

Figure 3: Transmittance of PV-6212 in function of wavelength

2.2 Encapsulation process

The encapsulation process is schematically depicted in Figure 4. It starts with dispensing PV-6212 on the first glass panel. The solar cell matrix is then placed onto the silicone. Another glass panel is coated with the encapsulant and is then turned upside down and placed onto the first panel with cell matrix. The whole sandwich is then introduced in a laminator, where vacuum is drawn and the laminate is heated up, starting silicone curing. The typical curing temperature profile is 1 min at 100 °C. Even though some time is needed for ramping up, this low temperature compared to other encapsulants does speed up the encapsulation and lamination process and can lead to lower production cycles when optimized.

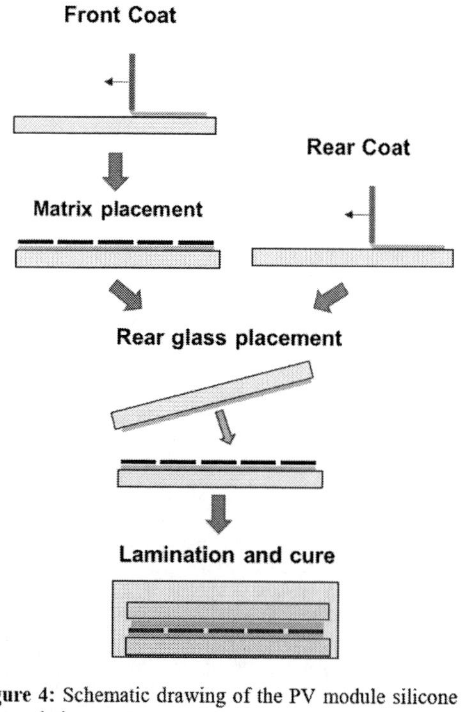

Figure 4: Schematic drawing of the PV module silicone encapsulation process

The fact that the encapsulant is applied as a liquid rather than a foil implies that some different pieces of equipment are needed than for traditional encapsulants. However, there are also some significant process advantages to a liquid encapsulant solution. First, there is no need to trim the laminate as with EVA-backsheet modules. Secondly and more importantly, a liquid silicone gives more process freedom. In all glass-glass structures, whatever encapsulant is used, a main challenge is to ensure that no air or gas bubbles remains in the laminate. To fully evacuate any air or gas, the encapsulant should be liquid. There is a race between evacuation and cure. If cure is too fast, gaseous substances will be trapped in the encapsulant (solid or too viscous). Now, an encapsulant that is supplied as a thermoplastic (such as EVA) needs to be heated up to become liquid. However heating kicks off the peroxide cure process, starting the cross-linking reactions and also generating some gaseous by-products. As a result, the risk of bubble formation is high and process control is delicate [8]. In contrast, with addition cure liquid silicone, no gaseous by-product is generated, and evacuation and curing processes can be controlled more independently.

3 MODULE DESIGN AND MANUFACTURING

A few years ago, in anticipation of the market opening for durable, long lifetime modules, BYD started a development program for a double glass module. They assessed different encapsulation options including Dow Corning's silicone solution. After careful assessment and multiple trials, BYD selected silicone encapsulation and embarked on a project to bring this PV module product type to the market.

3.1 Module concept

The structure of the module is shown in Figure 5.

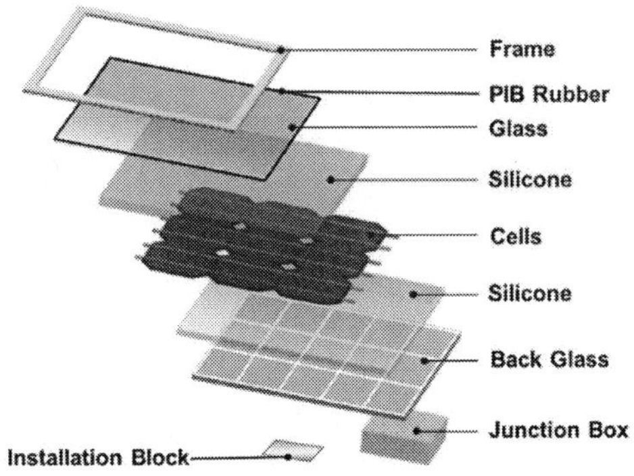

Figure 5: Schematic structure of BYD's double glass module

Solar cells for this module type are standard multicrystalline silicon solar cells. In order to capture the maximum possible amount of light, the glass is locally coated with a white reflective layer, applied in a grid pattern and aligning with the area between the cells in the final module.

In order to exclude water vapor ingress through the edges of the laminate, a PIB rubber seal is applied along the perimeter. In combination with the rear glass sheet and the silicone encapsulation, it leads to exceptional environmental and electrical module protection.

The double glass structure makes the module very rigid compared to standard laminates. As a result, a conventional Al frame is not needed. Instead, a light frame made of PPO polymer, a high temperature thermoplastic, is used to protect the glass edges from impact during transport and installation. Because the frame is non-conductive, it does not need to (and cannot) be grounded. The whole front surface of the module is electrically floating. For the junction box, a solution was chosen consisting of three separate edge junction boxes containing each one bypass diode, instead of one central junction box containing three bypass diodes.

To mount the module onto its support, a strong metal fixture is attached to the laminate. These are rectangle-shaped 'Installation blocks' made of stainless steel. They are glued directly on the rear glass sheet using structural silicone adhesive *Dow Corning®* PV-8303 Ultra Fast Cure Sealant. The technology to bond glass to structural elements creating structures that can withstand large wind loads is commonly practiced in the construction industry and benefits from decades of experience. The installation blocks are designed and placed on the laminate such that they can be bolted onto two rails that are part of the mounting structure This module concept enables easy installation, with low risk of module breakage, compared to clamping systems typically used for frameless modules.

3.2 Module manufacturing

Based on the module design, a preliminary silicone encapsulation process and an initial selection of other key components such as frame, junction box and PIB rubber, a complete manufacturing process was designed. It was first implemented, improved and fine-tuned in a pilot line. Subsequently the process was transferred to a manufacturing line and large volume production began in the BYD Shenzen manufacturing plant, China.

Pictures of one of these modules is shown in Figure 6:

a) b)

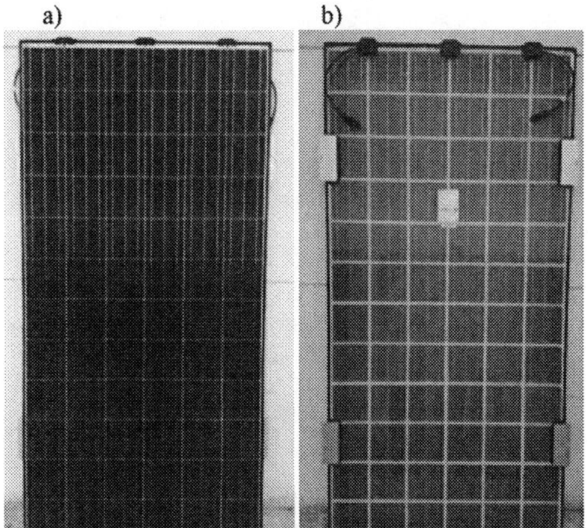

Figure 6: Pictures of a BYD double glass module : a) front, b) rear

4 MODULE PERFORMANCE, ACCELERATED AGING TESTS AND CERTIFICATION

The performance of these double glass modules is as expected for multicrystalline cells, with power bins ranging from 245 W to 265W for sixty cells modules and from 295 W to 315W for 72 cells modules.

These modules passed numerous accelerated aging tests. Table 1 shows climate chamber test data for various tests.

Table 1: Results of critical tests according to IEC61215

Test type	Test duration	Relative power change (%)
Thermal Cycling	TC200	-0.1
	TC400	-0.8
	TC600	-1.3
	TC800	-0.6
Damp Heat	DH1000	-0.5
	DH2000	-0.9
	DH3000	-2.5
Humidity Freeze	HF10	-1.2
	HF20	-1.0
	HF30	-1.4
UV exposure	UV90	-0.7
Potential Induced degradation	PID600(+)	-0.6
	PID600(-)	-1.4
Combined mechanical load+TC50+HF10	DML1000/TC50/ HF10	-0.4

After twice the duration of standard tests prescribed by IEC-61215, only a small degradation can be observed, with just about 1 % degradation, whereas the standard sets the maximum degradation after 1×IEC to 5 %. After three times IEC, all tests resulted in a degradation well below 3 %. For thermal cycling, the test was pushed to 800 cycles (4 times IEC) and resulted in 0.6 % degradation which is exceptionally low, even when comparing with other double glass module technologies (e.g. [9]). The reason may be the better protection that silicone encapsulation provides at strongly negative temperatures compared to EVA as silicone remains soft at those temperature, in contrast with EVA, which goes below its glass transition temperature and becomes very stiff [10].

The last row in Table 1 shows the result of a test combining dynamic mechanical testing (1000 cycles, 1440 Pa), 50 thermal cycles and 10 humidity freeze cycles. Thanks to the module's exceptional robustness, the mechanical load has no negative impact on performance. The subsequent TC and HF tests reveal no significant damage to the strings, with a final degradation of less than 0.4 % relative to the initial module power.

All modules still comfortably passed wet leakage tests after aging.

Potential Induced Degradation (PID) was also tested. The very stringent test involved keeping modules with metallic foil on the front surface in 85 % relative humidity /85°C conditions for 600 h under 1000V. Both positive and negative bias were tested. The degradation was only - 0.6 % for + 1000 V and -1.4 % for - 1000 V, which is outstanding for such a demanding test. A test combining 96 h at 1000 V and an additional 96 h at 1500V, resulted in a power loss of only 0.7%. Therefore, it can be concluded that this type of module is extremely PID resistant.

Hail impact was also investigated. The test consisted in projecting hail stones with 5 cm diameter at a velocity of 126 km/h onto the module and observing damage. The double glass module remained intact, whereas the same test on a conventional glass with EVA and back sheet results in substantial glass damage and module power reduction. The laminate rigidity thanks to the double glass-structure combined with some dampening effect of the silicone appears very effective in avoiding impact damage.

A set of modules were submitted for certification at TÜV Rheinland. All the tests according to IEC 61215, IEC 61730-1 and IEC 61730-2 were carried out and passed. Moreover, TÜV Rheinland did the electrical testing necessary for the new PV plant voltage standard of 1500 V, and based on positive results issued, for the first time for modules manufactured in China, a certificate confirming that these modules could safely be used in PV plants at operating voltages up to 1500 V (to be compared to present standard of 1000 V). The possibility of building systems operating at higher voltages (more modules per string) enables significant PV system cost reduction because fewer inverters are needed and the total cable length is lowered.

As a result of the outstanding reliability and durability data collected, the BYD double glass module was nicknamed the "430" module. The first digit "4" indicates that the modules are estimated to work for 40 years. The middle digit "3" means that the average annual power attenuation is as low as 0.3%. The last digit "0" indicates that the module is PID free and shows no snail tracks.

5 PV SYSTEMS

These double glass silicone modules are being installed in large PV systems in various parts of the world. Figure 7 for instance shows the picture of a 1 MW system installed in 2015 in Altenau, Germany. This type of modules can be considered by system operators as a safe technology thanks to the convincing accelerated aging test data that has been obtained, but also thanks to the decade-long track record of old PV systems equipped with silicone-encapsulated modules.

Figure 7: Picture of the PV system installed with BYD double glass silicone modules in Altenau, Germany

6 NEXT GENERATION OF GLASS-GLASS SILICONE MODULES

There are many options to build even better technologies while capitalizing on the unique advantages of silicone glass-glass modules. An option is to start using higher efficiency solar cells, such as PERC or heterojunction solar cells. Another one is to make use of the potential for bifaciality that the use of a glass rear cover offers. Bifacial cells can be introduced in the module process with minimal changes to the module assembly process.

BYD has introduced several improvements resulting in the 'PV-module 2.0'. The module is now totally frameless. The interconnection ribbons have been replaced by narrow wires with round cross-section. The cells are 'multi busbar' cells, featuring many very small solder pads instead of the conventional 3 wide busbars, onto which the wires are soldered. This structure reduces silver consumption while increasing short-circuit current and efficiency. With this structure, a power of 275 W has been obtained for a multicrystalline 60 cells module.

7 CONCLUSION

Thanks to the simultaneous emergence of a market for very robust and durable modules and the development of specific silicones and encapsulation processes meeting the requirements of modern PV manufacturing, silicone encapsulated panels have become again a feasible and attractive technology option. Glass-glass modules with silicone encapsulation have recently been launched by BYD and are now mass produced and globally commercialized. The combination of the glass-glass structure and silicone is shown to lead to exceptional behavior in accelerated aging tests, including extremely stringent PID tests. The concept enables safe module operation at 1500 V system voltage, and innovative, low cost module mounting through pad bonding.

6 REFERENCES

[1] Artur Skoczek, Tony Sample, and Ewan D. Dunlop, The results of performance measurements of field-aged crystalline silicon photovoltaic modules, Progress in Photovoltaics, 17 (2009) 227-240.

[2] Barry Ketola, Keith Mc Intosh, Ann Norris, Mary Kay Tomalia, Silicone for photovoltaic encapsulation, in Proceedings 23rd European Photovoltaic Solar Energy Conference, Valencia, 2008.

[3] Juan Lopez, A.Pozza, Tony Sample, Analysis of crystalline silicon PV modules after 30 years of outdoor exposure, in Proceedings 31st European Photovoltaic Solar Energy Conference, Hamburg, Germany, 2015.

[4] D.C. Jordan, J.H.Wohlgemuth, S.R. Kurtz, Technology and climate trends in PV module degradation in Proceedings 27th European Photovoltaic Solar Energy Conference, Frankfurt, 2012.

[5] ITRPV roadmap, Roadmap editions 2014, 2015 and 2016, www.ITRPV.net

[6] ITRPV, International Technology Roadmap for Photovoltaic (ITRPV) 2015 Results March 2016, www.ITRPV.net

[7] Webpage 'NREL Levelized Cost Of Electricity Calculator', http://www.nrel.gov/analysis/tech_lcoe.html

[8] L.-E. Perret-Aebi, H.-Y.Li, R. Théron, G. Roeder, Y. Luo, T. Turlings, R. F. M. Lange, C. Ballif, Insights on EVA lamination process : where do the bubbles come from ? in Proceedings 25th European Photovoltaic Solar Energy Conference, Valencia, Spain, 2010.

[9] Pierre J. Verlinden, Yingbin Zhang, Jianmei Xu, Yunhua Shu, Yang Yang, Yifeng Chen, Peng Quan, Shu Zhang, Guanchao Xu, Jing Mao, Feilin Ping and Zhiqiang Feng, Optimized PV modules for tropical, mountain and desert climates, in Proceedings 29th European Photovoltaic Solar Energy Conference, Amsterdam, 2014.

[10] Rafal Mickiewicz, Bialo Li, Dan Doble, Theresa Christian, John Lloyd, Adam Stokes, Carola Voelker, Matthias Winter,, A.N. Barry Ketola, Nick Shephard, Effect of Encapsulation Modulus on the Response of PV Modules to Mechanical Stress, in Proceedings 26th European Photovoltaic Solar Energy Conference, Hamburg, Germany, 2011.

INVESTIGATION OF THERMOMECHANICAL STRESS IN SOLAR CELLS WITH MULTI BUSBAR INTERCONNECTION BY FINITE ELEMENT MODELING

Li Carlos Rendler[1], Achim Kraft[1], Christian Ebert[2], Steffen Wiese[3], and Ulrich Eitner[1]
[1]Fraunhofer Institute for Solar Energy Systems ISE, Heidenhofstraße 2, 79110 Freiburg, Germany
[2]SCHMID Group | Gebr. SCHMID GmbH, Robert-Bosch-Str. 32-36, 72250 Freudenstadt, Germany
[3]Saarland University, Microintegration and Reliability, Campus A5 1, 66123 Saarbruecken, Germany

ABSTRACT: The interconnection of silicon solar cells by soldering causes thermomechanical stress due to different coefficients of thermal expansion of the materials involved, especially copper and silicon. In contrast to the current standard interconnection technology, using three to five flat ribbons to be soldered on continuous or segmented busbars on the solar cells front side and on large contact pads on the solar cells rear side, the Multi Busbar concept uses twelve to fifteen round solder coated copper wires to be soldered on a large number of small contact pads for each polarity.

In this paper we present two finite element models, one for a solar cell with a common busbar-based interconnection and a second one for a solar cell using the Multi Busbar interconnection approach to analyze and compare the distribution of the thermomechanical stress induced by the cell interconnection. The results show only a slight vertical deformation of a contacted solar cell of 0.4 mm for the three busbar design and 2.2 mm for the Multi Busbar interconnection technology due to the non-symmetric pad layout and the aluminum metallization on the rear side. However, most parts of the copper interconnector (ribbons and wires) are plastically deformed because the tensile stress exceeds the yield strength of 100 MPa. In addition, the results indicate that there is only slight tensile stress in most parts of the silicon solar cell after the soldering process. For both interconnection technologies we detect local stress peaks and, on the rear side of the solar cells near the outermost contacts, we determine areas where the tensile stress reaches levels above 200 MPa that potentially cause defects.

Keywords: FEM, Modeling, Multi Busbar, PV Materials, Thermomechanical Stress, Wire Interconnection

1 INTRODUCTION

Standard silicon solar cells are interconnected by soldering three to five solder coated copper ribbons onto screen-printed contacts on both sides of the solar cell. The most common concept uses continuous or segmented screen printed busbars on the front side combined with around six large contact pads on the rear side. However, there are alternative concepts on the market that show several advantages. Two innovative examples are Schmid's Multi Busbar (MBB) concept, working with twelve to fifteen round wires, soldered on both sides on small contact pads of a solar cell by infrared light [1-5] as well as the Smart Wire Connection Technology of Meyer Burger, using around thirty wires on both sides that are embedded in a polymer foil and interconnected during the lamination process directly with the contact fingers [6]. Both concepts use round wires instead of flat ribbons as well as adapted metallization layouts on the solar cell surfaces. The round shape of the wire interconnectors promotes a redirection of incoming light on the solar cells front surface. Using contact pads instead of continuous busbars and lowering the finger thickness results in a significant reduction of the silver consumption. By using a higher number of interconnectors on the solar cells surfaces the concept shows particular high reliabilities because there is a redundancy in the current collection paths and the series resistance increases less in case of defects like cell cracks, failure of solder joints or even broken interconnectors. Additionally, a reduction of the series resistance caused by a more homogenous current collection has been shown [1-4].

All these concepts have in common, that after the solder process the silicon solar cell and the copper interconnectors are rigidly fixed by the solder. The different coefficients of thermal expansion (CTEs) cause thermomechanical stress in the interconnected cell. In the interconnection process, where the cell is heated up to temperatures above the solder melting point, as well as during its lifetime, a solar module undergoes several temperature changes that might cause defects resulting from thermomechanical stress [7]. In addition to the CTE, the mechanical properties of the interconnector, for example the Young's modulus and the yield strength, have a significant influence on the thermomechanical stress [8]. Defects of solder joints like cracks or delamination of layers result in an increase of a module's series resistance and therefore reduce the module efficiency.

The distribution of the thermomechanical stress caused by the interconnection process is determined by means of simulation models using the finite elements method (FEM). In previous work we investigated the relevant material data that serve as input parameters for FEM simulations. Additionally, by simulating the single-side soldering of a solar cell section and showing the good agreement of experimental and model results, we demonstrated that a FEM model based on the used parameters deliver reasonable results [7]. We use two FEM models to analyze the stress distribution in solder joints after the interconnection process of a common solar cell with three busbars (3BB) as well as in a solar cell with the wire based MBB interconnection. By analyzing the respective stress level caused by the solder process it is possible to analyze if an interconnection concept is prone to defects like cracks or delamination. Additionally, we determine where such defects most likely occur, which is crucial for further development of the interconnection technology.

2 MODELING

The model geometry for both analyzed interconnection approaches (3BB and MBB) consists of a section of a silicon solar cell (156 x 10.2 x 0.175 mm³).

To reduce computational memory consumption and calculation time, only one fourth of the whole cell section is taken into account and two symmetry conditions, longitudinal and transversal, are applied to the cut planes as shown in figure 1 and 3. Both model geometries include screen-printed silver-based contact areas on both sides. Contact and redundancy fingers are not modeled since their influence on thermomechanical stress is expected to be negligible. On the rear side, we simulate a screen-printed aluminum-based metallization layer with a thickness of 8 µm (3BB) or respectively 30 µm (MBB) that includes square shaped cut-outs in the area of the contact pads as shown in figure 2 and 4.

2.1 Three busbar geometry

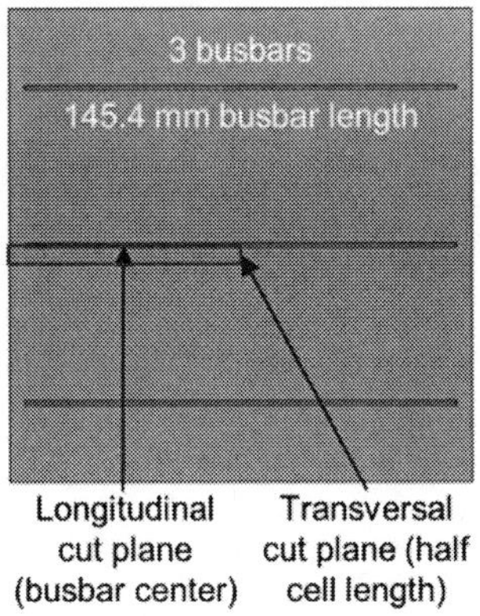

Figure 1: Schematic drawing of the front surface of a common silicon solar cell with three continuous busbars on the cell front side. The red box indicates the area used in the FEM model. Additionally, cut planes to apply symmetry conditions that reduce the model complexity are shown.

The 3BB cell section includes one quarter of a busbar (0.675 x 72.7 x 0.014 mm³) on the front side as well as half of three large contact pads (9.0 x 1.6 x 0.008 mm³) on the rear side as pictured in figure 2.

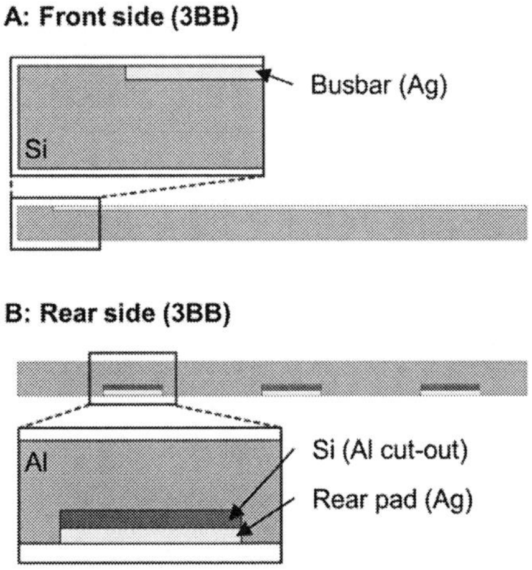

Figure 2: Metallization layout on the front (A) and the rear side (B) of the 3BB model geometry.

2.1 Multi Busbar geometry

In the MBB model we simulate a metallization geometry including 16 contact pads on the front and 20 pads on the rear side of the cell section. The physical dimensions for the MBB model are determined by numerous microsections and confocal microscope imaging shown in previous work [7].

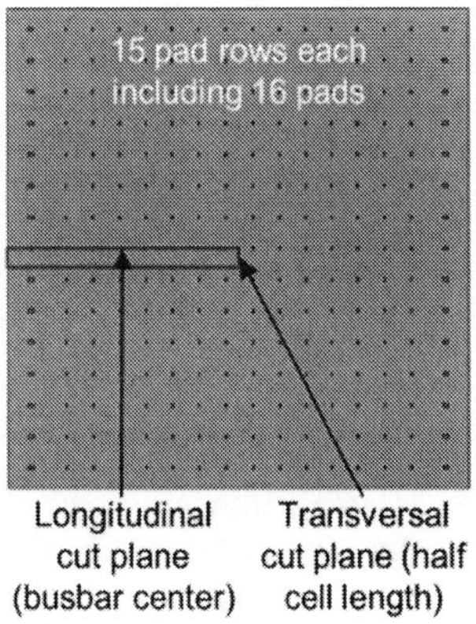

Figure 3: Schematic drawing on the front surface of a MBB solar cell with 15 pad rows, each including 16 pads. The red box shows the area that is used as geometry in the FEM model.

On the front side, the outermost pads have a size of 2 x 1 mm². The inner pads measure 0.45 x 1 mm² in size. On the rear side of the model geometry, which is displayed in figure 4, there are cut-outs in the aluminum layer and the pads of each pad group with a size of 2 x 1.6 mm² (outer pads) or respectively 0.5 x 1.6 mm² (inner pads). The height of each pad on the front side is 20 µm and 10 µm on the rear side. Round copper wires with a diameter of 300 µm are connected to the contact pads on both cell sides by a specified volume of solder (Sn62-Pb36-Ag2).

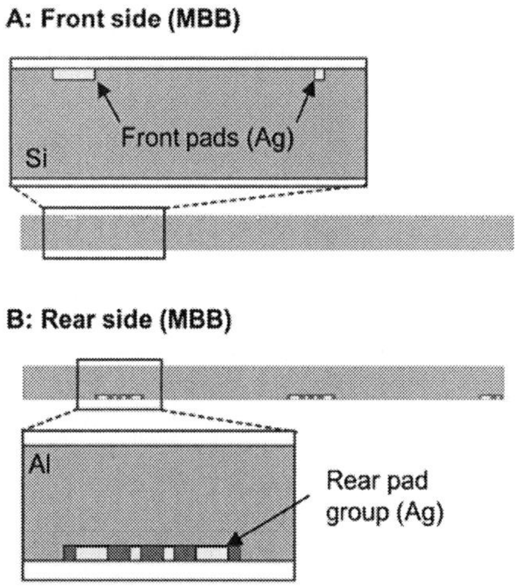

Figure 4: Metallization layout on the front (A) and rear side (B) of the geometry for the MBB model.

2.2 Simulation mesh

The used mesh for the stress simulation consists of tetrahedrons and prisms and its elements are smaller around the solder joint edges to increase the model precision in this area. Figure 5 shows the mesh that is used for the 3BB model. The front side in the area where the end of the busbar is located is displayed.

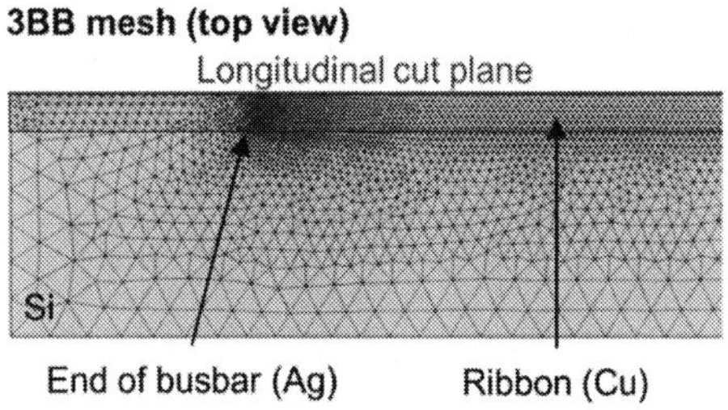

Figure 5: Details of the mesh that is used in the 3BB model. Finer mesh elements near the end of the busbar are shown. Additionally, the red line shows the top edge of the longitudinal cut plane (see figure 1) in the model geometry.

Figure 6 shows the mesh of the MBB model in the area of one pad on the front side.

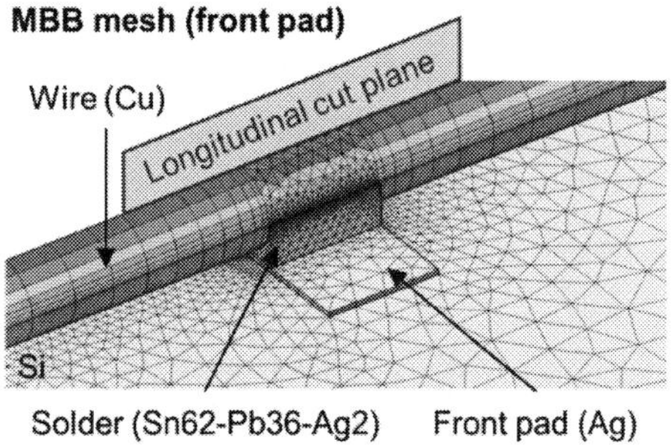

Figure 6: Details of the mesh used in the model for a solar cell with MBB interconnection. The mesh in the area of one small pad on the front side is displayed. Additionally, the red box shows how the longitudinal cut plane (see figure 3) is located in the model geometry.

2.3 Material models

The material data used in the model are measured or respectively taken from literature. A linear elastic material model is used for the silicon, the silver paste and the solder. Additionally, the silicon model includes an anisotropic Young's modulus and a temperature dependent CTE. The bilinear material model for the copper (ribbons and wires) and the aluminum paste includes plastic deformation with hardening when exceeding their yield strength. A FEM model using the material data shown in table I delivers reasonable results, which we previously verified by experimental results [7].

Table I: Material parameters used for the FEM simulation

Material	Young's mod. [GPa]	Yield strength [GPa]	Harden. gradient [GPa]	CTE $[10^{-6} \cdot K^{-1}]$
Bilinear material model				
Al paste [9,10]	6	28.3	0.06[1]	15.9
Copper	70	100	30	17.0 [10]
Linear material model				
Sn62-Pb36-Ag2 [11]	16	-	-	23.9
Ag paste [9,10]	7	-	-	10.0
Silicon [12,13]	Stiffness matrix[2]	-	-	Value table[3]

[1]Assumption of a moderate growth of the stress when increasing the strain level above the yield point
[2]The material model for silicon includes an anisotropic Young's modulus
[3]The Material model for silicon includes temperature dependency of the CTE

2.4 Temperature sweep

The simulation starts at the assumed stress-free state of 179 °C, which is the solidus temperature of the solder, and the whole solder volume is supposed to be solid. Subsequently, the model temperature decreases in steps of 1 K until the target temperature of 20 °C is reached.

3 RESULTS

Silicon is much more prone to defects caused by tensile stress than by compressive stress [14]. Our results indicate that the maximum levels of the third principal stress (compressive stress) in the copper ribbons (53 MPa) and wires (57 MPa) does not exceed their yield strength of 100 MPa. For these reasons we focus on the first principal stress (tensile stress) in the copper interconnector and the silicon caused by their CTE mismatch. Hence the first principal stress in the longitudinal cut plane of both models (see fig. 1 and 3) is shown in the following figures.

3.1 Three busbar solar cell

As shown in figure 7 the tensile stress in the ribbon is induced at the outermost contact on the cell surface – the end of the busbar on the front side and the edge of the first pad on the rear side of the solar cell section. Both interconnectors, on the front and on the rear side, undergo plastic deformation since the maximum tensile stress is about 145 MPa in the front ribbon and 143 MPa in the rear ribbon.

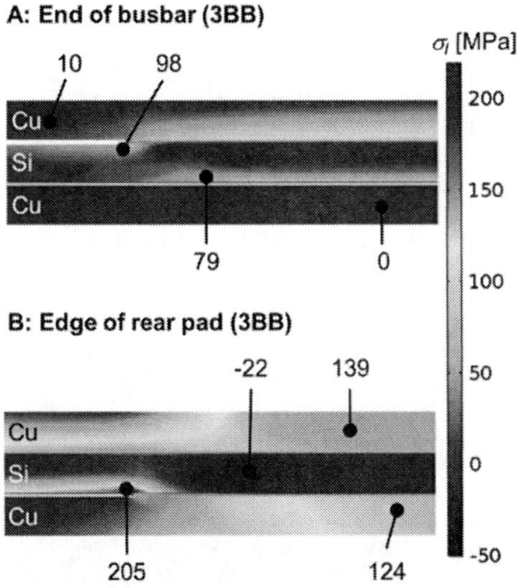

Figure 7: Distribution of the first principle stress in the longitudinal cut plane (see figure 5) of a solar cell with 3BB interconnection. The end of a busbar (A) and the edge of the first rear pad (B) are shown.

In most areas in the silicon there is only small tensile stress. However, we find a local stress peak in the range of 100 MPa near the end of the busbar. In addition, there is a stress peak in the silicon that exceeds 200 MPa adjacent to the first rear pad that can possibly provoke a crack in the silicon.

3.3 Multi Busbar solar cell

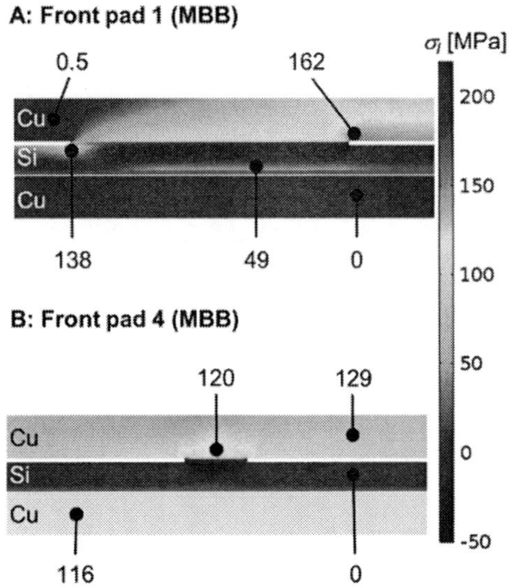

Figure 8: Distribution of the first principle stress in the cut plane (see figure 6) through a solar cell with MBB interconnection. The first pad (A) and the fourth pad (B) of a pad row on the front side are displayed.

Figure 8 and 9 show the distribution of the first principal stress in a MBB solar cell. The tensile stress in the wires is induced at the outermost contact pads on both sides of the cell section. The maximum stress reaches around 160 MPa in the front wire and about 150 MPa in the rear wire. We find nearly homogeneous stress levels of approximately 129 MPa in the front wire and 116 MPa in the rear wire between the contact pads. Additionally, we determine lower stress in the wires adjacent to contact pads on the front (120 MPa) and on the rear side (113 MPa).

The MBB model shows small tensile stress in most parts of the silicon. However, as for the 3BB model, we find local stress levels in the silicon around 140 MPa adjacent to the first front pad and above 200 MPa adjacent to the first rear pad. That may cause a crack in the silicon after the interconnection process.

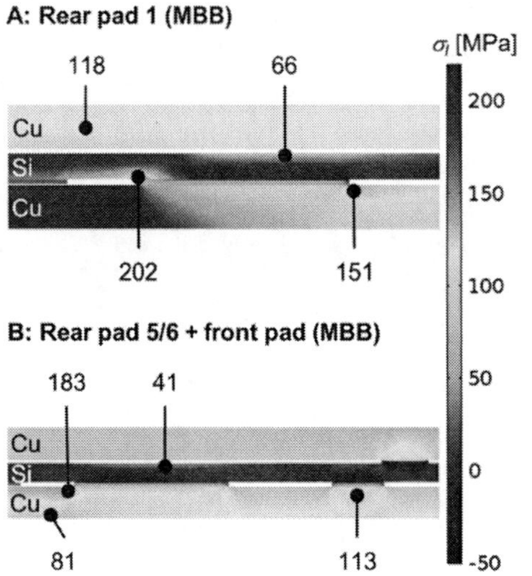

Figure 9: Distribution of the first principle stress in the cut plane (see figure 6) through a solar cell with MBB interconnection. The first pad (A) and two pads of the second pad group (B) on the rear side are displayed.

4 CONCLUSION

This work presents two FEM models to determine the distribution of the thermomechanical stress in solar cells. The models are based on previously published work that deals with the simulation of the single-side soldering of a cell section, including one MBB pad row [7]. In the first model we simulate a solar cell with a common busbar-based interconnection. With the second model the stress in a solar cell interconnected by means of a pad and wire based approach is analyzed. Both models allow the determination of maximum stress areas in a solar cell after the interconnection process.

The model results show a tensile stress in the ribbons up to 145 MPa and in the wires up to 183 MPa, which exceeds their yield strength of 100 MPa and indicates that the copper material undergoes plastic deformation after the soldering process, regardless of the interconnection technology. Adjacent to the contact pads, the stress is slightly lower than in between them. Hence, we expect cold-hardening in the copper material due to the interconnection process that certainly influences the reliability of the solder joints [8].

Due to the interconnection with two identical interconnectors on both sides, the results show only a moderate bending deformation of the cell section of about 0.4 mm with 3BB and respectively 2.2 mm with

MBB interconnection. In both cases that deformation is caused by the non-symmetrical metallization layout and we expect these bending levels to be noncritical for the lamination process.

In most parts of the silicon the tensile stress is comparatively low. However, there are stress maxima in the silicon near the outermost contacts with the copper interconnectors. On the front side we determine a maximum stress of nearly 100 MPa for 3BB interconnection and about 140 MPa for cells with MBB approach. With regards to fracture stresses in silicon published by Kaule et al. [14] these stress levels are not expected to be critical. On the rear side we determine significantly larger tensile stress due to the interconnection. We find maximum stresses in the range of 200 MPa in the 3BB model as well as in the MBB model. These critical areas are prone to defects caused by the interconnection of 3BB and MBB solar cells. Potential defects after the solder process or during its lifetime result in an increase of the series resistance of a solar module and therefore reduce the module efficiency. In case of solder joint defects, cell fracture or broken interconnectors the series resistance of solar modules with MBB concept increases less than of 3BB modules due to the large number of redundant interconnecting wires [4]. Since we find no significant difference of the mechanical stress maxima we expect the solar modules with MBB interconnection technology to show advantages in terms of long-term stability.

In further work it is possible to use both models to determine the influence of the material parameters on the thermomechanical stress by parameter studies.

5 REFERENCES

[1] J. Walter, M. Tranitz, M. Volk, C. Ebert, and U. Eitner, "Multi-wire Interconnection of Busbar-free Solar Cells," Energy Procedia 55, 2014, pp. 380-388.

[2] S. Braun, G. Micard, and G. Hahn, "Solar cell improvement by using a multi busbar design as front electrode," Energy Procedia 27, 2012, pp. 227-233.

[3] S. Braun, G. Hahn, and R. Nissler, "Multi-Busbar solar cells and modules: high efficiencies and low silver consumption," Energy Procedia 38, 2013, pp. 334-339.

[4] A. F. Dethlefsen, M. Volk, K. Ramspeck, and C. Buchner, "Reliability Testing of Multi-Busbar Modules," EUPVSEC, Amsterdam, Nederlands, 2014.

[5] S. Schindler, J. Schneider, C. Poenisch, R. Nissler, and D. Habermann, "Soldering process and material characterization of miniaturized contact structures of a newly developed multi busbar cell metallization concept," EUPVSEC, Paris, France, 2013.

[6] T. Söderström, P. Papet, and J. Ufheil. "Smart wire connection technology," EUPVSEC, Paris, France, 2013.

[7] L. C. Rendler, J. Walter, T. Geipel, M. Volk, C. Ebert, and U. Eitner, "Modelling and Verification of Mechanical Stress Induced by Soldering of Wires for Multi Busbar Interconnection", EUPVSEC, Hamburg, Germany, 2015.

[8] R. Meier, M. Pander, R. Klengel, S. Dietrich, S. Klengel, M. Ebert, and J. Bagdahn, "Reduction of soldering induced stresses in solar cells by microstructural optimization of copper-ribbons," in SPIE Solar Energy+ Technology, 2011.

[9] C. Kohn, T. Faber, R. Kübler, J. Beinert, G. Kleer, F. Clement, D. Erath, I. Reis, F. Martin, and A. Müller, "Analyses of warpage effects induced by passivation and electrode coatings in silicon solar cells," EUPVSEC,Milan, Italy, 2007.

[10] S. Wiese, F. Kraemer, E. Peter, and J. Seib, "Mechanical problems of novel back contact solar modules," EuroSimE, Cascais, Portugal, 2012.

[11] Q. Zhang, "Isothermal Mechanical and Thermomechanical Durability Characterization of Selected Pb-free Solders," Doctor of Philosophy, Department of Mechanical Engineering, University of Maryland, 2004.

[12] U. Eitner, S. Kajari-Schroeder, M. Koentges, and H. Altenbach, "Thermal stress and strain of solar cells in photovoltaic modules," in Shell-like Structures: Non-classical Theories and Applications. vol. 15, Springer, 2011, pp. 453-468.

[13] K. G. Lyon, G. L. Salinger, C. A. Swenson, and G. K. White, "Linear thermal expansion measurements on silicon from 6 to 340 K," Journal of Applied Physics, vol. 48, 1977, pp. 865-868.

[14] F. Kaule, W. Wang, and S. Schoenfelder, "Modeling and testing the mechanical strength of solar cells," Solar Energy Materials and Solar Cells, vol. 120, 2014, pp. 441–447.

PRODUCTION OF LOW COST BACK CONTACT BASED PV MODULES

M.J.A.A. Goris, B.W.J. Kikkert, J.M. Kroon, K. Rozema[1], I.J.Bennett[2], J. Verlaak[3]
ECN Solar Energy, P.O. Box 1, 1755 ZG Petten, The Netherlands.
Tel: +31 88 515 4505; Fax: +31 88 515 8214; E-mail: goris@ecn.nl
[1]Dycomet Europe, It Vegelinskampke 9, 8491 PD Akkrum, The Netherlands
[2]DSM Innovation Center, DSM Advanced Surfaces, Poststraat 1, 6135 KR Sittard, The Netherlands
[3]DSM Coating Resins, DSM Resins B.V., Ceintuurbaan 5, 8022 AW Zwolle, The Netherlands

ABSTRACT: In this paper, two developments targeted at reducing the production cost of back-contact modules are presented. The first cost reduction potential is realised by replacing the conductive copper layer with aluminium with a very thin and locally applied copper coating using cold spray technology. This results in a 2% PV module cost saving while also meeting industrial requirements for production speed and reduction of contamination. Prototype mini-modules manufactured using this method have shown very good performances and passed selected IEC reliability tests (Damp Heat and Thermal Cycling).The second cost reduction comes from reduction of encapsulant usage. By using powder coating technology, encapsulant thicknesses below 100 microns can be achieved. This not only results in the reduction of encapsulant usage as compared to standard modules, but it also increases production flexibility. Application of powder coated encapsulant results in a cost reduction of 1%. An additional cost advantage of the implementation of thinner encapsulants is a reduction of conductive adhesive of at least 60%, which can result in an additional 2% cost reduction.
Keywords: Back Contact, Cost reduction, Encapsulation, Cold Spray, Powder Coating

1 INTRODUCTION

The foil-based back-contact c-Si PV module technology offers possibilities to increase efficiencies and reduces the cell-to-module fill factor loss compared to standard tabbed front-to-back contact technology. It also offers improved design flexibility, and is compatible with any back-contact cell type [1-6]. This module technology is ready for large-scale implementation, with equipment manufacturers and material suppliers offering qualified and diverse commercial solutions [2]. In the past few years, a significant reduction of the costs per Wp for foil-based back-contact PV modules has been achieved by reducing the material and process costs [3]. Despite these cost reductions industrial implementation is still slow due to the relatively high BoM costs, the need for investment in new equipment and limited long-term field performance data. Improvements in the cost-performance ratio of high-efficiency back-contact cells such as n-type metal wrap through (n-MWT) and interdigitated back contact (IBC) [4,5], and implementation of novel module processing solutions and materials combined with extensive reliability studies on module level should lower the threshold for market introduction.

The Dutch R&D project Los Bacos, a consortium consisting of ECN, DSM and Dycomet Europe, is developing technologies to reduce the material costs of back-contact PV modules. This project focuses on reducing the amount of copper, silver and encapsulant by replacing the copper foil with aluminium foil and by applying thinner encapsulation. Aluminium foils combined with copper cold sprayed layers have been investigated in the past, resulting in reliable PV modules [7,8]. Using powder coating technology, thin thermoplastic encapsulant layers can be applied, thus reducing the amount of conductive adhesive needed to overcome the height of the encapsulant. The application of powder coating significantly improves and simplifies the application of the encapsulant since there is no need for punching or lasering holes and often troublesome handling of thin encapsulant films.

This paper reports the results of implementing two solutions, i.e. aluminium conductive foil in combination with copper cold spray and powder coated encapsulant to achieve thin encapsulant layers. The implementation of these material cost savings could have a major impact on the reduction of the overall back-contact PV module costs.

2 EXPERIMENTAL

2.1 Optimization copper cold spray process

Low pressure cold spray is a coating deposition method [8] used for several ductile metals such as copper. Copper powders (1 to 50 um in diameter) are accelerated in a supersonic gas jet to velocities up to 500–1000 m/s. During impact with the substrate, particles undergo plastic deformation and adhere to the surface. A huge benefit is that the temperature of the particle during impact on the surface is below 100 degrees C. The used equipment for cold spray is shown in Figure 1.

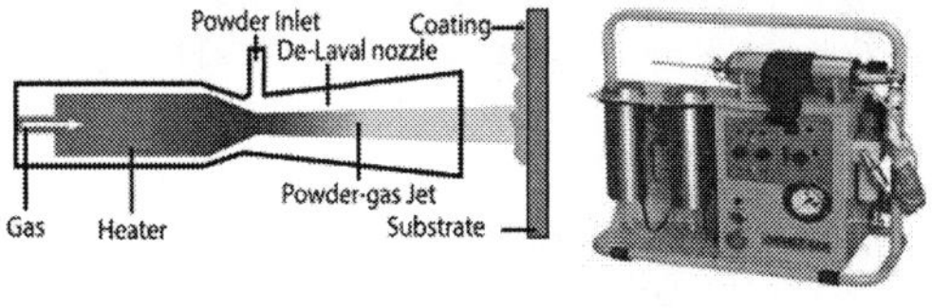

Figure 1: Schematic view of cold spray technology (left) and the used equipment, Dymet 423 (right)

In the past several mini-modules were produced using this process and were subjected to Damp Heat (DH) and Thermal Cycling (TC) tests, resulting in reliable modules with less than 5% power loss after 3 times DH and TC testing according to the IEC 61215 standard [7]. The main focus of this paper is on realizing further cost reductions while maintaining the level of reliability. There are important parameters that can be improved to reduce the cost of the copper cold spray process. One important parameter is the price of the copper powder, which is strongly related to the morphology, particle size and the powder yield during the processing of the powder. Powders from several powder suppliers were tested using standard spray conditions and deposited on aluminium based back-contact foil. Based on visual inspection of the layer and contact resistance measurements, potential powder candidates were selected; see Figure 2

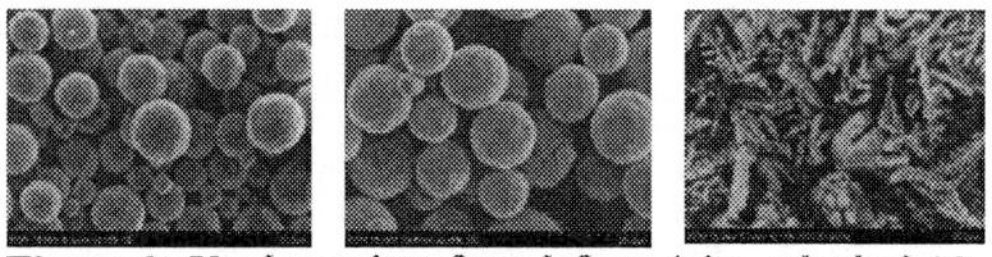

Figure 2: Used powders from left to right: spherical 10 micron (A), spherical 20 micron (B) and dendrite <25 micron (C)

Several -4-cell back-contact (MWT) mini- modules were produced using the different selected powder morphologies. The modules were measured using the PASAN IIIb flash tester and tested in DH and TC according to IEC 61215.

One powder was selected, based on lowest price and production capacity as claimed by the supplier. With this powder several parameters influencing the spray efficiency were investigated and optimized, such as powder flow, temperature, gas pressure, gas composition (compressed air or nitrogen), robot speed required to cover the dimensions of full size foil in less than 1 minute and spray nozzle geometry . The produced copper layers on aluminium based back-contact foil were visually inspected and some were selected for contact resistance measurements as shown in Figure 3. The contact areas were deposited with copper. Both foils were connected to each other using conductive adhesive. An electrically isolating encapsulant foil with punched holes was placed between the two metal foils. The contact resistance measurement was executed using a four point probe measuring method with a Keithley 2430 pulse source meter.

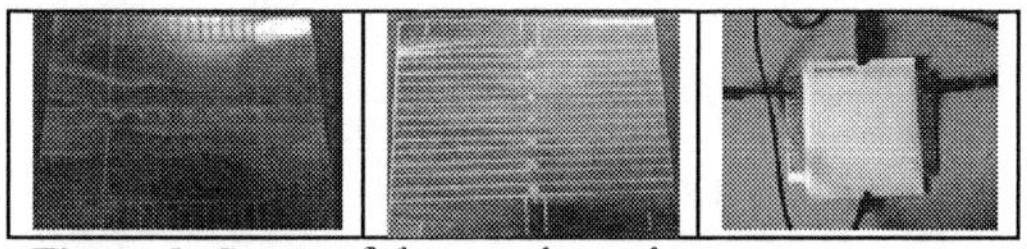

Figure 3: Set-up of the sample used to measure contact resistances. Base foil (left), tab foil (middle) and 4 point probe measurement using Keithley 2430 pulse source meter (right)

The powder particles that do not adhere to the foil were removed using a vacuum cleaner and captured in a cyclone (see Figure 4). Further research is needed to test whether this material can be re-used for secondary spraying, as starting material for other deposition techniques or as recycled material for the production of virgin copper powder.

For IBC module design more contact pads are available and a higher accuracy is required because of the small connection area. Several IBC mini-modules were produced to test the robustness of the cold spray process for IBC PV modules. For mini-modules in the MWT concept with 4x4 vias, 14 lines are sprayed, compared to 18 lines for the IBC module (see Figure 5), resulting in an increased copper material usage of around 30 %. The produced IBC modules are measured using IV, EL and DLIT.

Figure 4: Vacuum system to collect the powder that does not adhere to the foil (left) and the cyclone to separate the powder from the air flow

Figure 5: 18 lines of copper cold spray corresponding with the position of the front and back contacts

2.2 Powder coated encapsulant

The production of EVA encapsulant foils with a thickness less than < 200 micron is a technical challenge. The bottleneck for thin encapsulant layers can be solved by using the powder coating technology. Powder coating is a type of deposition that is applied as a free-flowing, dry powder. The coating is typically applied electrostatically and is then cured under heat to allow it to flow, forming a "skin". The powder may be a thermoplastic or a thermoset polymer. The used powder coating equipment is shown in Figure 6.

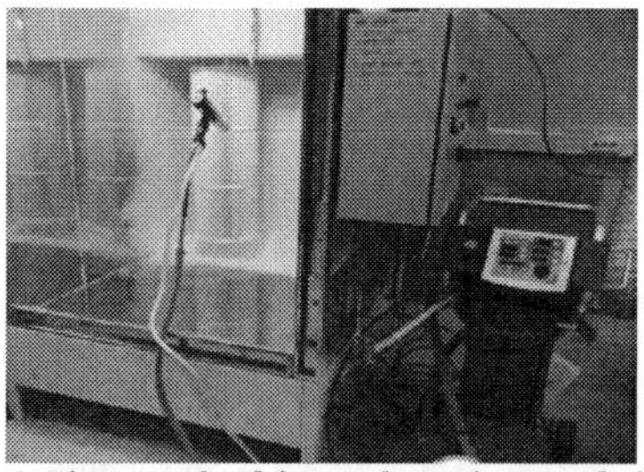

Figure 6: Photograph of the used powder coat facility (Norton Encore spray controller and spray booth)

An industrial thermoplastic powder was used for process optimization. It was known that this powder could not lead to reliable PV modules under DH conditions, because of the melting temperature of approximately 80 degrees C, but it was suitable to demonstrate the process.
With the set-up illustrated in Figure 3, powders were applied on the back side of back-contact cells. The attached powder on the contact pads was removed by vacuum using a contact point cleaner (CPC-unit), see Figure 7.
The powder is pre-heated in an IR heated belt furnace under conditions that just make the thermoplastic

material melt, and in case of thermoset powders the curing is minimal.

Figure 7: Photograph of the CPC-unit

The spraying and pre-heating conditions were optimized to achieve homogeneous layers ranging from 80 to 150 microns.

Secondly, the glass was coated with layers of 150 to 200 microns thick using the same powder. PV modules were produced using optimized conditions for the applied powders. The modules were characterized on their initial performance using the PASAN IIIb flash tester.

2.3 Reduction Conductive adhesive

One of the production methods used in contact foil technology is printing of conductive adhesive on the patterned metal foil. The height of the adhesive dot is in line with the thickness of the punched encapsulant. For stencil printing a certain ratio between hole diameter and stencil thickness is required to fulfil the release properties of the adhesive from the walls in the stencil. The height of the conductive adhesive dot is fixed by the height of the encapsulant. For a 200 micron thick encapsulant a 400 micron thick stencil is needed and for release of the adhesive, a hole diameter of 1.7 mm is needed. For an encapsulant thickness of 100 microns a stencil thickness of 200 microns is needed and a hole diameter of 1 mm is possible. The applied amount of conductive adhesive using the stencils mentioned above will be investigated.

3 RESULTS AND DISCUSSION

3.1 Optimization copper cold spray process

First optimization was done to find a relation between copper usage (extrapolated to full-size foil), observed intensity of the copper line and the contact resistance between two foils and conductive adhesive. This resulted in a copper usage of 50 g per full-size foil and the contact resistance was below 1 mΩ. Further optimisation by increasing the air pressure and temperature of pre-heating the carrier gas resulted in a copper usage of less than 20 g without influencing the contact resistance. Under these deposition conditions the spray efficiency was approximately 5-10%. The spray efficiency can be increased by activating the aluminium surface by bombarding the surface with an inert 'grit' powder. This surface roughness has a positive effect on the adhesion properties of the sprayed copper powder. The copper usage can be reduced to less than 10 g and the spray efficiency increased to 40%. At industrial processing speeds, coating full-size foils in less than 1 minute, the spray efficiency decreased to 20%. Further optimization of spray parameters, such as nozzle design and distance of the spray nozzle to the substrate, is necessary to further reduce the copper powder consumption and increase the spray efficiency. Contact resistance measurements were performed on foils with and without copper deposited with a semi-optimised cold spray process; see table I. These results show that the contact resistance between foil and conductive adhesive is significantly reduced when using copper cold spray applied on aluminium foil. The reason is that the native oxide on the aluminium is removed. An additional layer of copper cold spray on copper foil also reduces the contact resistance, probably caused by removing the surface protection layer (OSP) and increasing the contact area by roughening the surface.

Table I: Contact resistances on foils with and without cold spray copper (CS Cu)

	Average	Standard deviation
Copper foil	2.3	1.2
Copper foil + CS Cu	0.19	0.06
Aluminium foil	150	230
Aluminium foil + CS Cu	0.19	0.12

The back-contact copper foil used in back-contact PV modules is protected against corrosion by an OSP layer when stored under atmospheric conditions. For the foils applied with copper using the cold spray technology, such an OSP layer is not needed. Contact resistance measurements were performed on aged aluminium foils with copper cold spray contacts. It turned out that after one year of storage the contact resistance was still below 0.3 mΩ.

The shape and diameter of the applied powder for cold spray technology is an important parameter for cost reduction. Using different types of copper powder (A, B or C), -4-cell back-contact (MWT) mini-modules were produced with aluminium foils, characterised using the PASAN IIIb flash tester and tested under DH and TC test conditions according to IEC 61215. No differences were observed in cell-to-module (CtM) fill-factor losses when a different powder was applied, compared to modules fabricated using a copper conductive back-contact foil. For mini-modules a standard CtM loss of 2.5% is acceptable. The power and fill-factor decay as function of the testing time of the different powders is shown in Figure 8. No difference in reliability behaviour has been observed. During TC, the fill factor decay was less than 1% after 200 cycles. The power loss during TC was higher than expected, especially for the first 100 cycles, i.e. approximately 3%. Reliability data of modules in which standard copper based back-contact foils were applied and identical solar cells were used, showed the same decay during the first 100 cycles in a TC test. Investigation made clear that this decay was caused by current induced degradation (CID) that is comparable to light induced degradation (LID), which is often observed for P-type silicon solar cells [9].

Using the vacuum and separation set-up to collect the unattached copper from the aluminium surface during the cold spray process showed that it is possible to have a clean coating process without excessive contamination of the foil and surroundings of the spray unit. Wiping the foil with a piece of paper after spraying showed no copper on the paper. The unattached powder was during spraying immediately removed from the surface through suction and separated from the gas stream by using a cyclone. The collected powder is shown in Figure 9.

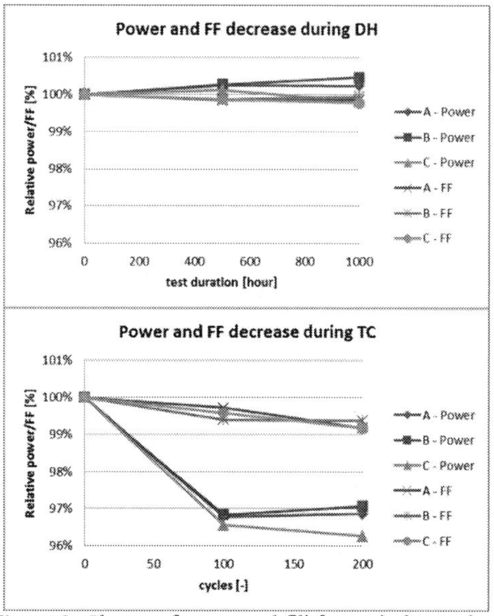

Figure 8: Change of power and fill factor during testing modules under DH (left) and TC (right) conditions

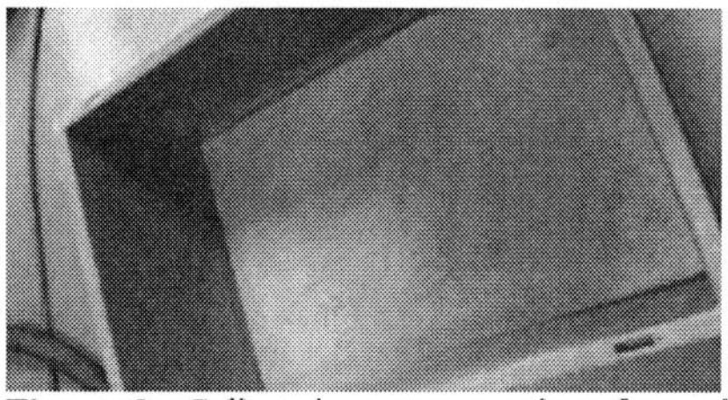

Figure 9: Collected copper powder after cold spray process

Four IBC modules (-4-cell mini modules) were produced to investigate the possibility to perform the cold spray technology also on back-contact cell concepts that require a higher accuracy in module manufacturing. Figure 10 shows a picture of a semi-fabricate of an IBC module.

The IV characteristics of The IBC modules were measured using a PASAN IIIb flash tester. The average value of the IV data is listed in table II. The module characteristics were compared with IBC modules produced with standard copper conductive back sheets (the used solar cells were slightly higher in performance).

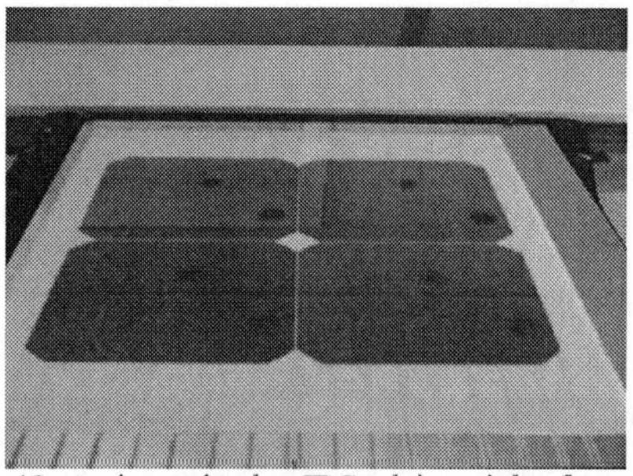

Figure 10: An in production IBC mini-module after pre-tacking, before lamination process

Table II: Average IV data of four IBC modules built with aluminium foil and copper cold spray (IBC Al+CS) compared with average IV data of 4 IBC modules built with copper foil (IBC Cu)

	Isc [A]	Voc [V]	Fill factor [%]	Effici-ency [%]	Power [W]
IBC Cu	9.622	2.603	74.6	19.55	18.69
IBC Al+CS	9.465	2.602	74.9	19.33	18.44

3.2 Powder coated encapsulant

To develop a module manufacturing process using powder coated encapsulant, a commercially available powder based on thermoplastic material was used. With this powder layers were applied on glass and solar cells. Optimisation was required to find the right deposition conditions of the powder, the uniformity in the applied layer and the conditions to pre-heat the powder to enable handling of the coated component.

After optimization, thicknesses of 150-200 µm were applied on glass and thicknesses of approximately 130 +/- 20 µm on solar cells. Further optimization is required to reduce the thickness of the powder coated encapsulant to below 100 µm with a variation of less than +/- 10 µm between the solar cells. The glass and solar cells with applied powder coated encapsulant were used to produce MWT mini-modules; see Figure 11.

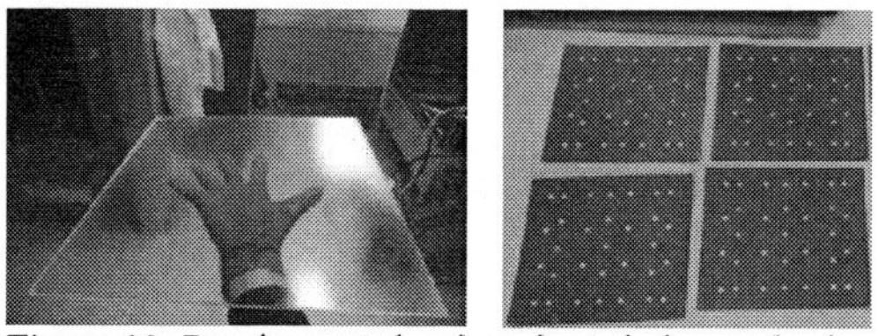

Figure 11: Powder coated and pre-heated glass and solar cells (contact areas are cleaned from encapsulant before pre-heating using the contact point cleaner)

Reference modules were produced using fast cure EVA from STR. The modules were measured using the Pasan. The average performance values of the modules are mentioned in table III.

Table III: Average IV characteristics for modules with fast cure EVA (Reference EVA) and powder coated encapsulant (thermoplastic A)

	Isc [A]	Voc [V]	FF [%]	Power [W]
EVA	8.75	2.52	75.9	16.66
Thermoplastic A	8.81	2.52	75.9	16.82

The results shows that the Isc for modules using thermoplastic A is slightly higher than for the reference modules with EVA. The current gain could be explained by the higher transmission of the applied thermoplastic or be the result of the expected lower absorption due to the thinner encapsulant thickness.

Applying encapsulant thicknesses of 100 µm or less for module manufacturing offers the possibility to print adhesive dots with a lower stencil thickness and maintain the same aspect ratio. In table III the amounts of conductive adhesive are depicted when conductive adhesive is printed using a 400 µm thick stencil and hole diameter of 1.7 mm (stencil A) compared with 200 µm thick stencils and a hole diameter of 1.0 mm (stencil B).

The adhesive dots were printed on bare solar cells. For comparison conductive adhesive has been printed on solar cells that were powder coated with a thermoplastic encapsulant with a thickness of approximately 130 μm.

Table III: Printed grams of conductive adhesive on solar cells with and without powder coated encapsulant

	Bare solar cell	Cell + encapsulant
Stencil A	0.085	0.155
Stencil B	0.018	0.036
Gain B-A	79%	78%

From these results it is clear that printing on solar cells with encapsulant roughly doubles the amount of adhesive required compared to printing on solar cells without encapsulant. If thinner stencils are used, the saving in conductive adhesive for printing on solar cells with or without applied encapsulant is approximately 78% . This indicates that the presence of encapsulant has no influence on the savings of adhesive. Comparing the printed grams on solar cells using the 400 μm thick stencil and 1.7 mm holes, approximately 60% less adhesive is consumed using the 200 μm thick stencil and 1 mm holes and printed on solar cell with powder coating.

Powder coating of encapsulant on a solar cell has technical advantages over punching holes in encapsulant foil, because of the elimination of a punch-unit. The application of encapsulant is done in the processing sequence before the printing step of the conductive adhesive. The amount of applied conductive adhesive is a factor 2 higher for cells with the encapsulant than for cells without encapsulant. The applied conductive adhesive using stencil A (required for 200 μm encapsulant thickness) on solar cells without powder coated encapsulant is compared to the combination stencil B and solar cell with encapsulant. It has been observed that with encapsulant thicknesses of 130 μm applied on solar cells there is already a saving of conductive adhesive of approximately 60%. Further savings in conductive adhesives are expected for printing on solar cells where an encapsulant with a thickness of 100 micron or lower is applied.

3.3 Calculation cost reduction PV modules after implementation copper cold spray and powder coated encapsulant

First cost calculation studies have been done using several assumptions. They reveal that the module components are contributing to approximately 1/3 of the total module cost price (see figure 15 upper figure). From this part, the conductive back-sheet, conductive adhesive and encapsulant contribute approximately 45% (see figure 15 lower figure) which amounts to 15% of the ~total module costs.

In order to reveal the cost reduction potential of the replacement of copper by aluminium, assuming that all other factors remain the same, the following calculations were made. As a starting point we assumed copper and aluminium price levels to be €3,-/m^2 and €0,75/m^2 respectively The cold spray process is optimized such that the powder consumption is approximately 3 g/m^2 (as mentioned earlier in this paper 5 g/1.6 m^2).

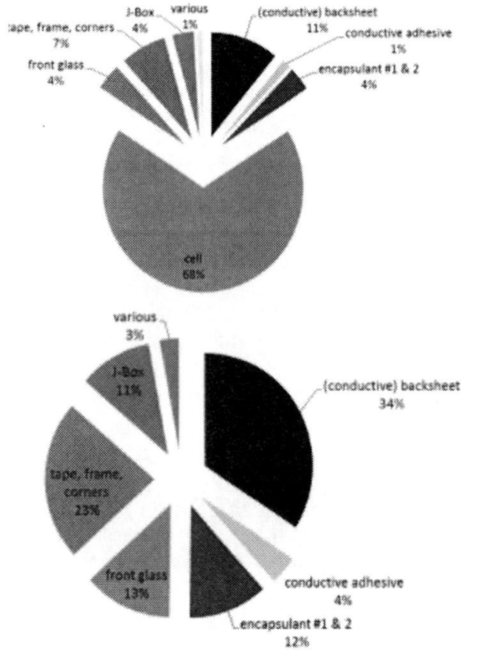

Figure 15: Build-up module cost (upper figure) and the contribution of the several module material costs (lower figure)

By using the lowest price of the copper powder mentioned in this paper, a price of €0,13/m^2 can be calculated for the copper powder, which does not include the potential revenue of recycled powder. There are potential copper powder suppliers in the market who aim at a further reduction of the copper price to a level of around €20,-/kg. Before implementing use of this powder additional research needs to be done to test deposition conditions and to find out if this material also leads to reliable PV modules.

As shown in this study with optimized powder coated encapsulant on solar cells at a thickness of 130 μm, so far a reduction of 60% in conductive adhesive consumption has been achieved. This already leads to a cost reduction of more than 2% per module. Further reduction of the adhesive is expected to be feasible after further reduction of the encapsulant thickness and optimising the stencil aspect ratio to minimize the adhesive consumption without increasing the contact resistance between solar cell contacts and conductive back-sheet and maintaining the reliability.

4 CONCLUSIONS

The work presented in this paper summarises two approaches to reducing the cost of back-contact modules made using MWT or IBC cells. The first approach concerns the use of an aluminium conductive back-sheet with a very thin and locally applied copper coating, realized by using cold spray technology. Similar performance and reliability are observed when compared to the current back-contact module build.

The second approach concerns the use of powder coating technology for the deposition of encapsulants with a thicknesses lower than 100 microns, resulting in a significantly reduced conductive adhesive consumption. The potential of this technology, in terms of processing, has been demonstrated by very good initial performances of back-contact modules using thermoplastic powder encapsulants compared to reference modules with standard EVA encapsulation. Further investigations are ongoing to test novel powder encapsulants with the aim to show improved reliability.

An overall cost saving approaching 5% should be possible by combining the two approaches, which would significantly increase the competitiveness of back-contact modules with MWT or IBC cells when compared with conventional modules.

5 ACKNOWLEDGEMENTS

This work was carried out in the TKI tender 2 framework of the Dutch Ministry of Economic Affairs, Agriculture and Innovation. Many thanks go to Eurotron for designing and building the suction/vacuuming equipment with cyclone to capture the re-bounced copper powder.

6 REFERENCES

[1] P.C. de Jong, Photovoltaic International (2010) 138.
[2] J. Govaerts, K. Baert and J. Poortmans, Photovoltaic International (2012) 116.
[3] I. Bennett, W. Eerenstein, V. Rosca, Energy Procedia 38 (2013) 329.
[4] I. Cesar, N. Guillevin, A.R. Burgers, A.A. Mewe, M. Koppes, J. Anker, L.J. Geerligs, A.W. Weeber, Energy Procedia 55 (2014) 633.
[5] I. Cesar, N. Guillevin, A.R. Burgers, A.A. Mewe, E.E. Bende, V. Rosca, B. van Aken, M. Koppes, J. Anker, L.J. Geerligs, A.W. Weeber, Proceedings 29th European Photovoltaic Solar Energy Conference, Amsterdam, 2014.
[6] Y. Wu, B.B. van Aken, G. Janssen, J. Loffler, F. Li, Y. Shen, W. Yang, J. Shi, G. Li, Z. Hu, J. Xiong, G. Coletti, Proceedings 29th European Photovoltaic Solar Energy Conference, Amsterdam, 2014.
[7] M.J.A.A. Goris, , I.J. Bennett, W. Eerenstein, Energy Procedia 55 (2014) 342
[8] V. Champagne, The cold spray material deposition process; fundamentals and applications, Woodhead Publishing ISBN 978-1-84569-181-3
[9] K.M. Broek, I.J. Bennett, M.J. Jansen, N..C.M. van der Borg and W. Eerenstein, Proceedings 27th European Photovoltaic Solar Energy Conference and Exhibition, 24-28 September 2012, Frankfurt, Germany.

NOVEL CONDUCTIVE ADHESIVE CONCEPT FOR SOLAR MODULE MANUFACTURING

Helland, Susanne[1a], Kristiansen, Helge [2], Jonsson, Per-Erik[3], Helland, Tore[1], Kalland, Erik[1], Redford, Keith[2]

[1]Mosaic Solutions AS, Dragonveien 54, N-2013 Skjetten, Norway
[2]Conpart AS, Dragonveien 54, N-2013 Skjetten, Norway
[3]SweModule AB, 670 20 Glava, Sweden
Tel: +47 64 00 40 30, Fax: +47 63 84 09 06
[a]sh@mosaicsolutions.no

ABSTRACT: Mosaic Solutions AS is developing a novel concept for conductive adhesives for use in electrical interconnection of solar modules. The goal is to provide an alternative to the current soldering process in module manufacturing, enabling lower process temperatures and reducing the amount of toxic elements, e.g. lead. The conductive adhesive is based on replacing solid silver flakes by polymer particles coated with a thin layer of silver. The main advantage of this concept, compared to conventional conductive adhesives, is a typical ten-fold reduction in silver consumption. We have investigated the electrical properties of conductive adhesives based on silver-coated polymer particles, and demonstrated the use of the conductive adhesive concept in module manufacturing. The results are promising, showing that the electrical conductivity is more than sufficient for use in solar modules, and that our conductive adhesive concept represents a promising and realistic alternative to soldering.
Keywords: Module Manufacturing, PV Module, Conductive Adhesive, Contacts

1 INTRODUCTION

The soldering process used in today's solar module manufacturing has two large drawbacks: Firstly, the solders used in PV manufacturing contain high levels of toxic lead, resulting in environmental issues related to recycling and waste management. Secondly, the soldering process requires a process temperature above 210℃, and introducing lead-free solder would further increase the temperature. The high temperature causes cell breakage, introduces micro cracking and imposes limitations on wafer thickness and cell design. As new solar cell designs emerge and the wafers are getting thinner, the demand for a reduction in the process temperature is evident.

An obvious alternative to solder is conductive adhesives, which allow significantly lower bonding temperature. Most commercially available conductive adhesives are based on silver flakes, and these conductive adhesives contain roughly 70 – 80 wt% or 25 – 30 Vol% of silver, causing this alternative to be less cost efficient.

The purpose of this work is to evaluate the possibility for replacing the costly silver flakes with silver-coated polymer particles. When using these particles as filler in a conductive adhesive, the total silver content of the adhesive is typically in the range 1 – 3 Vol%, as most of each particle is comprised of the polymer core. Still, the electrical conductivity is in the same range as conventional flake-based conductive adhesives [1]. Cross sections comparing conventional silver flake conductive adhesive and conductive adhesive based on Mosaic Solutions' technology are shown in Figure 1.

In this work, we have tested experimentally how particle properties, like particle size, silver thickness and particle concentration, influence the electrical conductivity of the adhesive. Additionally, conductive adhesives based on silver-coated polymer particles have been tested in solar module manufacturing, fabricating test modules of crystalline silicon solar cells interconnected by conductive adhesive instead of conventional soldering. The I-V response of the test modules was measured, and EL imaging was used to discover any micro cracking of the cells. The test modules were exposed to the standard reliability tests for solar modules: Temperature cycling (TC200) and damp-heat test (DH1000).

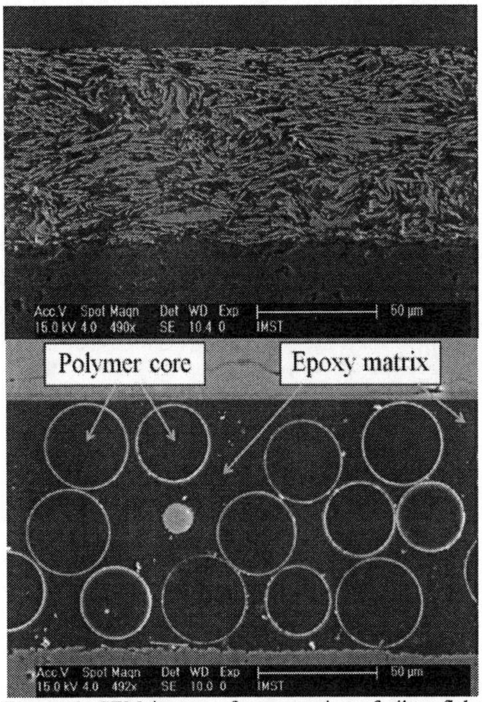

Figure 1: SEM images of cross-section of silver-flake based conductive adhesive (upper) and conductive adhesive based on silver-coated polymer particles (lower) [2].

The silver-coated polymer particles are produced with a very narrow size distribution and very homogenous properties [3]. For conductive adhesives intended for use in solar module manufacturing, we have found that particle sizes in the range 5 – 30 µm, with a 50 – 250 nm silver coating, is advantageous. A SEM image of 10 µm polymer particles with 140 nm silver is shown in Figure 2.

Figure 2: SEM image of 10 µm polymer particles with a 140 nm silver coating

Compared to silver flakes, silver-coated polymer particles have some technological advantages and fundamental differences: Firstly, reducing the use of costly silver significantly makes the use of conductive adhesives in solar module manufacturing more attractive from a cost perspective. The spherical shape of the particles gives new opportunities with respect to adhesive formulation. Sedimentation issues associated with heavy silver flakes become less prominent due to the lower density of the particles. In addition, the mono-sized spheres cause a smaller viscosity increase in the adhesive compared to the silver flakes. The result is a conductive adhesive that is easier to apply, giving new opportunities with respect to dispensing and jet printing [4]. Furthermore, the polymer core provides flexibility and reduced thermal mismatch in the adhesive itself, preventing the brittleness that is a concern for the silver flake-based adhesives.

Using this conductive adhesive technology in solar module manufacturing affects a large part of the value chain, from wafer level to power plant. A reduced process temperature enables thinner wafers for crystalline silicon-based solar cells. With respect to cell design, the lower process temperature gives room for the use of new materials that are temperature sensitive. Additionally, the application of conductive adhesives enables new module designs, e.g. back contacted solar cells. On the module side, the lower process temperature will give less cell breakage, higher quality products and better manufacturing yield. Finally, the solar energy plant will be even more environmental friendly, avoiding use of toxic lead in the modules.

2 EXPERIMENTAL DETAILS

2.1 Materials

Several types of silver-coated polymer particles with varying particle diameter, silver coating thickness and coating morphology were used in this study, and all particles were manufactured by Conpart AS. The particles have a CV (coefficient of size variation) of < 3 %, and the coating thickness is estimated from the amount of silver used during chemical plating. Slightly different plating procedures were used for the different particles, processes A and B denotes the present state-of-the art process for 10 and 30 µm particles, respectively, whereas process C is an older process. The particle specifications are summarized in Table I.

Table I: Overview of particles used in this work

Ref. name	Core size	Silver thickness	Plating process	Test
10A035	10 µm	35 nm	A	Bulk & contact resistivity
10A050	10 µm	50 nm	A	Bulk & contact resistivity
10A070	10 µm	70 nm	A	Bulk & contact resistivity
10A100	10 µm	100 nm	A	Bulk & contact resistivity
10A140	10 µm	140 nm	A	Bulk & contact resistivity
10A200	10 µm	200 nm	A	Bulk & contact resistivity
30B100	30 µm	100 nm	B	Bulk resistivity
30C230	30 µm	230 nm	C	Bulk resistivity & Test modules

Two types of adhesive matrix were used, both commercially available two component epoxy systems. They are denoted as Adhesive A and Adhesive B in this study, and the main difference between them were the degree of cross-linking. Adhesive A had a glass transition temperature (T_g) of ~90℃ and Adhesive B ~65℃, measured by DSC. Samples for resistivity measurements were cured at 150°C for 30 minutes.

2.2 Resistivity measurements

Samples were prepared by mixing particles into Adhesive B and stencil printing the adhesive onto specially designed test circuits. These test circuits were PCBs manufactured using FR4 laminate material with a standard Ni/Au surface finish on standard copper foil conductors. Bulk resistivity (or the resistivity of the conductive adhesive it-self) was obtained by printing adhesive lines with a cross section of 2 x 0.2 mm. The resistance was measured by a four wire method, and the resistivity of the material calculated based on the geometry. The sample preparation method and bulk resistivity measurement procedure is described in more detail in [5]. Contact resistance between the conductive adhesive and gold pads was measured separately using dedicated PCB substrates and corresponding PCB components, where defined adhesive contacts were made between orthogonal conductors on the substrate and components, respectively. The contact resistance was measured by a kind of the transmission line method [6].

2.3 Test module manufacturing

Test modules were fabricated using 6 multi-crystalline silicon solar cells in conventional modules based on glass, EVA based encapsulant and PET based backsheet. A custom-made lead-free ribbon was provided by Luvata, consisting of the conventional copper ribbon covered with SnAg solder. The test modules were made manually using both Adhesive A and B. Particle 30C230 was used as filler in both adhesives, with a particle content of 52 Vol%0 The resulting resistivity of both adhesives was 1.1 mΩ. The adhesive was applied by stencil printing and cured during module lamination. Sufficient handling strength before lamination was obtained by securing ribbon/cell using small dots of Araldite 90 Seconds rapid curing epoxy. Machine soldered reference modules were made for comparison.

2.4 Equipment Used

A Flack Tek Speed Mixer DAC 150 FVZ-K from Hauschild, Germany was used to mix all of the adhesives. This mixer works on the dual asymmetric centrifuge principle i.e. by spinning a high speedmixing arm at speeds up to 3500 rpm in one direction and a basket containing the sample in the opposite direction. A balance with a resolution of 1 mg was used to weigh out the different components of the adhesives.

Stripes of the formulated adhesives were printed manually onto the PCB boards or solar cells using a 0.2 mm thick laser cut brass stencils and a razor blade. The samples for electrical testing were cured in a Termaks, TS 8136 oven, which was preheated to the curing temperature and then the samples were placed in it. A Keithley 3706 System Switch/Multimeter was used to measure the resistances.

3 RESULTS AND DISCUSSION

3.1 Electrical properties of conductive adhesive

The electrical properties of the conductive adhesive based on silver-coated polymer particles were investigated, including the bulk resistivity and contact resistance.

The effect of particle size and particle concentration on bulk resistivity was measured by comparing conductive adhesives based on particles 10A140, 30B100 and 30C230, with varying volume fraction of particles. The results are shown in Figure 3. Experiments showed that for the 10 µm particle 10A140, contact was established at 35 Vol%, indicating a percolation threshold of < 35 Vol%. For the 30 µm particle 30B100, an infinite resistance was measured at 35 Vol%, indicating a percolation threshold in the interval 35 – 40 Vol%. Additionally, it was found that a higher volume fraction could be obtained for the 30 µm particles compared to the 10 µm particle, before the adhesive mixture became too dry for printing. As expected, resistivity decreases with increasing particle concentration for both particle sizes. Higher resistivity was measured for the conductive adhesive based on 30 µm particles with smallest silver thickness, and this can be explained by a lower total silver content. The conductive adhesives based on 30B100 contain around 1 Vol%, the 30C230 around 2 Vol%, and the 10A140 roughly 3 Vol% silver.

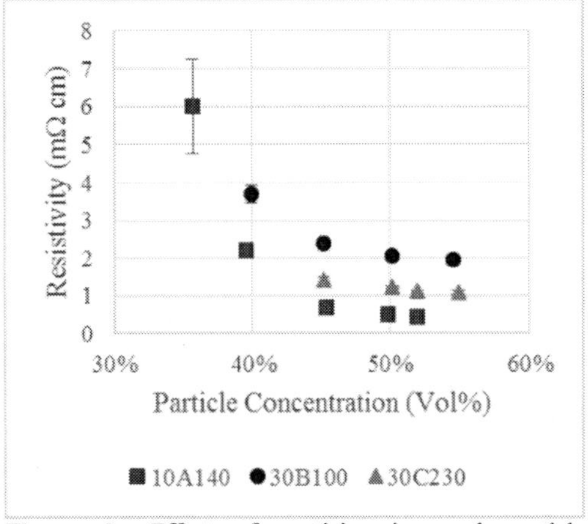

Figure 3: Effect of particle size and particle concentration on bulk resistivity of the conductive adhesive

The effect of silver thickness on bulk resistivity and contact resistance was investigated by making conductive adhesives filled with 10 µm particles with varying silver thickness. Particles 10A035, 10A050, 10A070, 10A100, 10A140 and 10A200 all were made with identical particle core material and the same plating process, and with silver thicknesses in the range 35 – 200 nm. Particle concentration was kept constant at 56 Vol% in all conductive adhesive samples. The resulting measured bulk resistivity and contact resistance are given in Figure 4. As expected, increasing the silver thickness reduces both bulk resistivity and contact resistance. However, the absolute value of the reduction is lower for silver thickness above 140 nm, and hence a silver thickness of 140 nm seems to be a good compromise between low resistivity and silver consumption. The measured contact resistance consists of the actual contact resistance between pad and conductive adhesive (times two), as well as the contribution from the bulk conductive adhesive. The bulk contribution was calculated using the measured bulk resistivity data in the upper graph of Figure 4, assuming a contact height of one stencil thickness (200 µm), which represents the maximum value possible. For all data contact resistance data points, the measured value for contact resistance is slightly higher than the estimated bulk contribution, indicating a significant contribution from contact resistance between pad and conductive adhesive. For the samples based on 10A035, with the thinnest silver coating of 35 nm, a high value and large variation in contact resistance was observed. This indicates that the coating is not homogeneous and not able to form stable contacts to the pads. 50 nm can be considered as the minimum thickness that should be made using the current plating process.

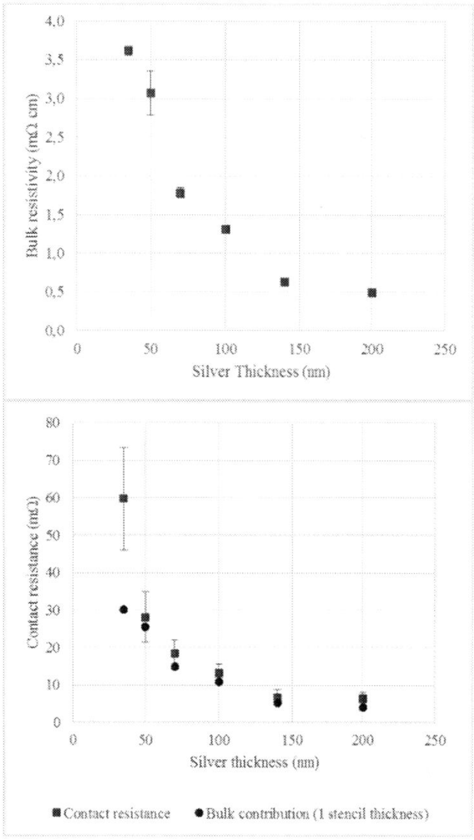

Figure 4: Effect of silver thickness on bulk resistivity (upper) and contact resistance (lower) for 10 µm particles. The bulk contribution to the contact resistance were estimated using the measured bulk resistivity data (upper), assuming a contact height of one stencil thickness.

3.2 Test module evaluation

I-V measurements were performed in a Endeas QuickSun simulator, which was calibrated for a full scale 60 cell module, giving slightly higher values than expected for a 6 cell test module. However, the obtained values can be used to compare the different test modules. Results from I–V measurements of the modules after manufacturing are shown in Table II. The results show that the I–V performance of the modules made with conductive adhesive was similar to that of the soldered reference modules, indicating a sufficient electrical conductivity of the conductive adhesive.

Table II: Results from I–V measurements on test modules

	# Panels	Max Power (W)
Adhesive A	7	25.9
Adhesive B	4	25.8
Solder	2	25.8

Results from I–V measurements of the modules after temperature cycling and damp heat testing are shown in Table III. Examples of EL images captured before and after reliability testing are shown in Figure 5 (module with Adhesive A before and after damp heat test) and Figure 6 (soldered module before and after damp heat test). None of the test modules showed any sign of crack propagation after reliability testing. The cells used for making the test modules had a large number of defects, but these were not been affected by the reliability testing. In both temperature cycling and damp heat tests, the modules with conductive adhesive had degraded more than the soldered reference module, but all within the requirement of <5 % degradation. However, modules that are soldered with a well-calibrated process, without introducing micro cracking, are known to not degrade substantially under temperature cycling and damp heat testing. Modules with adhesive B (lower T_g) degraded more than modules with Adhesive A, indicating that the T_g should be kept above 85 ℃, which is the maximum cycling temperature and the temperature of the damp heat test. El imaging analysis showed no sign of crack propagation for the modules with conductive adhesive, indicating that the mechanism for degradation is in the adhesive joint, either degradation of the adhesive itself or reduced adhesion during testing.

Table III: Results from reliability testing of test modules: Temperature cycling TC200 and Damp heat Test DH1000

	TC200		**DH1000**	
	# Panels	Loss in Max Power	# Panels	Loss in Max Power
Adhesive A	4	1.7 %	1	2.3 %
Adhesive B	2	3.0 %	1	4.2 %
Solder	1	0.5 %	1	0.1 %

Figure 5: EL images of test module made with Adhesive A before (upper) and after (lower) damp heat testing

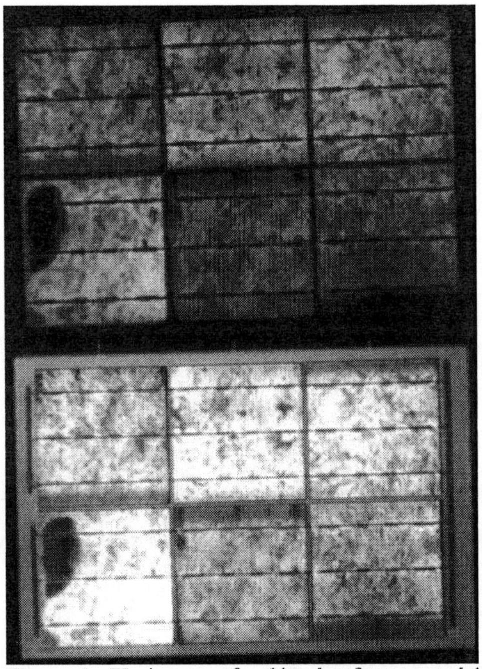

Figure 6: EL images of soldered reference module before (upper) and after (lower) damp heat testing

4 CONCLUSIONS

In conclusion, we have demonstrated that a conductive adhesive based on silver-coated polymer particles can be used in solar module manufacturing. The electrical properties of such an adhesive can be tailored by adjusting particle concentration, particle size and silver thickness. The obtained electrical resistivity is low enough to ensure no measureable resistive losses in the solar modules. Test modules have successfully been made and their reliability evaluated, showing that the production process based on conductive adhesives is gentle with respect to cell breakage and micro cracking. Further work is needed for optimising the properties of the adhesive matrix to improve reliability.

5. ACKNOWLEDGEMENTS

The authors gratefully thank the Research Council of Norway (RCN) for funding through project number 225962/E20 –"Novel conductive adhesive technology for solar industry". Partial funding has also been obtained from the European Union Seventh Framework Programme FP7- NMP-2013-LARGE-7 under grant agreement n°604668 ("Quantiheat") and by funding from the European Union Seventh Framework Programme (FP7/2007-2013) under grant agreement n° FP7- NMP-310420 ("HyperConnect"). We would also like to thank Luvata, Finland for customizing lead-free ribbon for this work.

6. REFERENCES

[1] Gakkestad, J., et al. Thermo-mechanical properties of Isotropic Conductive Adhesive filled with Metallized Polymer Spheres, Electronics Packaging Technology Conference (EPTC 2013), 2013 IEEE 15th. 2013.

[2] Nguyen, H.-V., et al., Die Shear Testing of a Novel Isotropic Conductive Adhesive—Epoxy Filled With Metal-Coated Polymer Spheres, Components, Packaging and Manufacturing Technology, IEEE Transactions on , Volume: 3, Issue: 7, Page(s): 1084 - 1093

[3] Kristiansen, H., et al., Development and Characterisation of Micrometer Sized Polymer Particles with Extremely Narrow Size Distribution, IEEE Advanced Packaging Materials 2007, ISSN 1089-8190, pp 130-134

[4] Nguyen, H.-V., et al., Spherical polymer particles in isotropic conductive adhesives – A study on rheology and mechanical aspects, Electronic System-Integration Technology Conference (ESTC), 2010 3rd

[5] Jain, S., et al. Factors affecting conduction in novel isotropic conductive adhesives filled with silver coated polymer spheres, Microelectronics Packaging Conference (EMPC), 2013 European. 2013.

[6] D.B. Scott, R. A. Chapman, C. Wei, S. S. Mahant-Shetti, R.A. Haken and T.C. Holloway, IEEE Trans. on Electron Devices, pp. 562-574, 1987

NOVEL HIGH PERFORMANCE, HIGHLY DURABLE, ANTI-REFLECTIVE COATING FOR PHOTOVOLTAIC GLASS

Brenor Brophy*[1], Sina Maghsoodi[1], Peter Gonsalves[1], Yufeng Hu[1], Mason Terry[2], Jeffrey Dee[2], Christopher Alcantara[2], Yongfeng Wang[3], Ji Qi[3], Dong Hu[3]

1. Enki Technology Inc., 2192 Bering Dr. San Jose, CA 95131, USA
*bb@enkitech.com, Phone 1 408 383 9034
2. DuPont Silicon Valley Technology Center, 965 E. Argues Ave., Sunnyvale, CA 94085, USA
3. LERRI Solar Technology Co., Ltd., No. 388 Middle Aerospace Rd., Chang'an Dist., Xi'an, China

ABSTRACT: Anti-reflective (AR) coatings address reflection losses at the PV glass surface of approximately 4% of normally incident light. These coatings reduce this loss by 1.5% to 3% and include sputtered multi-layer interference coatings and many variations of sol-gel derived, single-layer, porous silica. Early AR coatings suffered from poor field durability including delamination, corrosion and problems with contamination and soiling. In recent years, durability has improved and modern coatings are capable of long-term performance under moderate environment conditions. However, recent research has shown that these coatings remain susceptible to abrasion damage from activities such as regular cleaning, and to degradation from deployment in wet or humid environments. The structure of AR coatings create an inherent tradeoff between durability and optical performance, so more durable coatings tend to be at the lower end of the optical performance scale. In addition, the composition of the coating matrix itself can impact its long-term stability under humid or moist conditions. Now, in this work, we report on a novel method of creating structured porosity in highly durable, functionalized silica films that enable high optical performance with improved abrasion and moisture resistance. The resulting films are characterized optically and structurally and we report on their performance to standard and extended testing according to IEC specifications and to other applicable abrasion tests.
Keywords: Antireflection Coating, Durability, PV Materials

1 INTRODUCTION

Reflection losses for uncoated glass are about 4% of normally incident light. Today, most photovoltaic modules from major manufacturers have an anti-reflective (AR) coating on the front cover glass. These AR coatings can mitigate 1.5% to 3% of those losses (on an absolute basis), which increases peak power. AR coatings also improve light acceptance at higher angles of incidence and hence improves energy yield slightly more than the increase in peak power.

The development of AR coatings for photovoltaic and thermal solar collectors has a long history and several different technologies have been developed and tested. Multi-layer interference coatings, consisting of even numbers of high and low refractive index (RI) layers can achieve high optical performance and durability. However, since they are typically applied by magnetron sputtering or chemical vapor deposition processes, their costs are high. Such coatings are now only found in applications such as aerospace, where performance is paramount. Many variations of porous-silica low RI single-layer coatings based on sol-gel technology have also been tried [1]. Early coatings generated porosity by the use of colloidal silica such as Stöber particles [2, 3] or the inclusion of template materials [4]. These coatings suffered from poor field durability often from degradation due to soiling [5]. In recent years, significant development of sol-gel AR coating technology has occurred and many of these issues have been addressed [6, 7]. However, recent research has shown that many of these coatings are still susceptible to abrasion damage from common activities such as cleaning [8] and field exposure in humid locations [6].

The optimal single layer sol-gel AR coatings for glass has a refractive index given by the geometric mean of the refractive indices for glass and air.

$$n_{AR} = \sqrt{n_{GLASS} n_{AIR}} \quad \dots (1)$$

Typical low-iron patterned glass used in the solar industry has an RI of 1.52, yielding an optimal RI for an AR coating of 1.23. The AR performance of a single layer coating can be further enhanced for a specific wavelength of light by controlling the film thickness to close to one quarter of the wavelength.

$$\frac{\lambda}{4} = \frac{\lambda_0}{4 n_{AR}} \quad \dots (2)$$

Where λ_0 is the vacuum wavelength. Taking into account weighting for the solar spectrum and typical quantum efficiencies of solar cells, optimal thickness for a PV AR coating is between 100 nm to 120 nm. An AR coating with these properties would have excellent optical performance.

The RI of pure silica is approximately 1.45. However a lower effective RI may be achieved by reducing the density of the material through the introduction of porosity. A number of factors determine the ideal pore size; large pores can scatter light causing a hazy or milky appearance; small pores may adsorb water which leads to unstable RI as the water content of the film increases or decreases [9]. Reduction in silica density can also be achieved by modifying the molecular structure of the material by for example incorporating additional elements or chemical moieties; by introducing a templating material that is later removed leaving a void behind or by structural means that creates large voids. Effective Medium Theory provides several models that relate the effective RI of a film to its porosity [10] for various types of films. In the case of an AR film that is composed of a matrix material with RI n_M and pores filled with air ($RI = 1$), and a pore fraction P, then the Lorentz-Lorentz model can be used to calculate effective RI [11].

$$n_{AR} = \sqrt{\left[1 + 2P\left(\frac{n_M-1}{n_M+2}\right)\right] / \left[1 - P\left(\frac{n_M-1}{n_M+2}\right)\right]} \quad \ldots (3)$$

Achieving an effective RI of 1.23 with a matrix RI of 1.45 requires a pore fraction of approximately 45%. Coatings with this degree of porosity have very poor mechanical strength and are not suitable for solar applications. This is the inherent tradeoff between mechanical strength, porosity and hence RI. High performance solar AR coatings must optimize porosity within a tough matrix material that provides sufficient mechanical strength to withstand the abrasion stresses encountered with blowing sand or typical module operations such as washing.

Some commercial AR coatings achieve a structured porosity by coating a silica shell over a spherical bead of polymer material; then embedding the silica shell within a matrix and finally burning off the polymer bead template during the glass tempering process. This approach can produce a high degree of porosity but does not meet abrasion requirements for all environments [6], and high-temperature (> 600 °C) curing precludes the use of coating materials that are intrinsically moisture repellant.

In this paper we disclose a novel method of creating porosity within a particularly durable and intrinsically moisture repellant matrix material that is well suited to solar glass applications. Importantly, the new material is capable of processing at relatively lower (< 300 °C) temperatures, and therefore is able to achieve functional properties such as moisture and oil repellency that are unattainable with traditional solar AR coatings. This type of functionalization has been shown to impart anti-soiling functionality [12] and durability [7].

2 EXPERIMENTAL

In previous work we reported on the high abrasion resistance and durability of Enki Technology's CleanARC® 208 material [7]. The coating reported here, called CleanARC 228 uses a similar high durability matrix within which large, structured pores are now formed while retaining intrinsic moisture repellant functionalization of the matrix. In this work several sets of samples were prepared for the various test sequences outlined below. Samples are prepared using standard low-iron patterned glass as a substrate and roll-coating equipment used by the PV glass coating industry. The samples were cured by exposure to infrared radiation with a maximum substrate temperature that did not exceed 300 °C.

2.1 Physical and optical characterization

Samples were submitted for physical analysis by SEM followed by image analysis to determine film structure and porosity. Optical transmission was measured on coated Starphire® float glass using a Perkin Elmer Lambda-35 spectrophotometer.

2.2 Extended durability testing

A set of glass samples were assembled into laminates consisting of coated glass laminated to black Tedlar® back-sheet. This technique replicates the materials environment of a full-size PV module and also protects the inner-side of the glass during testing. Performance of the AR coating is determined by measuring optical reflection. The use of black back-sheet minimizes back reflection noise. Six laminates were produced and three each were submitted to 480 hours of Neutral Salt Fog (NSF) according to specification ASTM B-117 and to 40 cycles of Humidity Freeze (HF) according to specification IEC 61215 respectively. Reflection was measured with a Perkin Elmer Lambda 950 spectrophotometer at five points on each laminate prior to testing and thereafter at every 96 hours of NSF and after 10 cycles of HF. Abrasion testing according to EN 1096 using medium density felt, 4N of force and 1000 strokes was performed on a selection of the HF samples at the 20, 30 and 40 cycle read-points. All tests and measurements of this set of samples were completed at the DuPont Silicon Valley Technology Center.

2.3 Comparative accelerated testing

Another set of laminates was prepared in a similar manner to the prior set, but including glass substrates coated with two market-share leading temper-cured anti-reflective coatings in addition to samples coated with CleanARC 228. In total, 10 laminates of each type of coated glass were prepared. Two of each were retained as unstressed controls and the remaining eight were split into groups of two and submitted to the four tests outlined below. This test series was completed at a leading Chinese module company and consisted of 10 cycles of HF and 500 hours of 85 °C / 85% RH damp heat (DH) according to IEC 61215, 48 hours of pressure cooker test (PCT) according to specification JC/T 2170-2013 (2 atm. 120 °C, 99%-100% RH) and 96 hours of NSF test according to specification IEC 61701. 1000 hours of DH testing could not be completed in the time available. Optical loss was measured with a Konica-Minolta 2600d spectrophotometer.

One unstressed control sample of each coating type was subjected to abrasion testing with a Taber CS-10F Weardisc® (Part Number 131433). This is a 0.75 inch (19 mm) diameter, 0.25 inch (6 mm) thick disc of resilient rubber with embedded abrasive ceramic particles. This test is performed using the same linear abrader as the EN1096 test, but without rotation of the abrasive disc. A force of 4N is used. This test is significantly more aggressive than the EN1096 test and the expectation is that severe degradation of the film will result. However it is a useful test to determine how an anti-reflective coating will perform when tested to destruction. Five separate tests of 50, 100, 150, 200 and 250 abrasive cycles were performed on each sample. Reflection was measured with a Konica-Minolta 2600d spectrophotometer at the center-point of the abraded area before and after each test.

One unstressed control sample of each coating type was subjected to chemical resistance testing with boiling salt water, strong acids and bases according to the conditions shown in Table I below.

Table I: Chemical resistance test conditions

Chemical	Concentration	Temperature	Duration
NaCl	5 wt. %	100 °C	30 Min
H_2SO_4	1 M	25 °C	30 Min
NaOH	0.67 wt. %	25 °C	50 Min

These test conditions are considered good fast-screening tests for AR coatings in the PV glass industry. In particular the boiling salt water test acts like a highly accelerated damp heat test and is used as a process monitor when glass is coated in a production facility. Optical measurements before and after chemical testing were made with a Konica Minolta 2600d spectrophotometer.

3 RESULTS AND DISCUSSION

3.1 Physical and optical characterization

Figure 1 shows an isometric cross-section SEM micrograph of the CleanARC 228 coating. The coating is approximately 100 nm thick and contains many oblate shaped pores 20 nm to 50 nm in size. It is a "closed-cell" structured material with some pores open at the surface and some buried within the coating. Distribution of pores both vertically and horizontally is homogeneous. The matrix material between the pores is dense, without any evidence of cracking and the surface of the matrix material is smooth and continuous between pores.

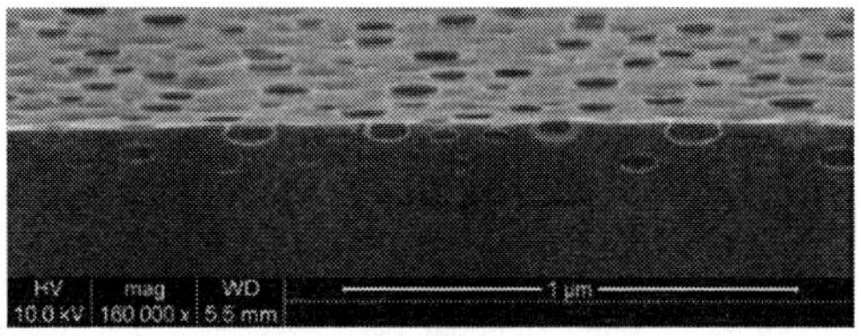

Figure 1: Scanning electron microscope cross-section micrograph showing the structure of the CleanARC 228 Coating.

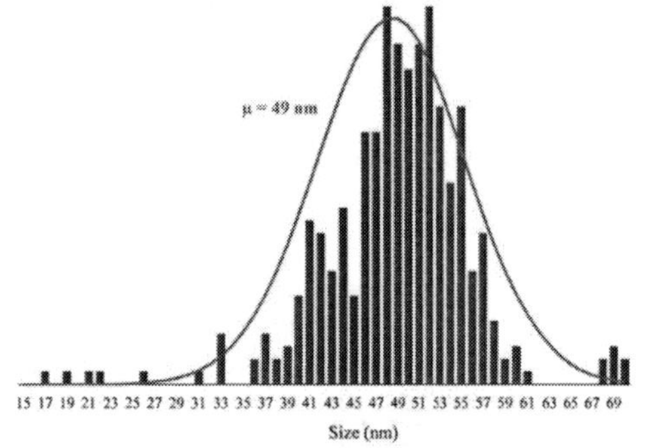

Figure 2: Histogram of pore size within the coating showing a mean of 49 nm with a standard deviation of 7 nm.

Figure 2 shows the results of an image analysis using ImageJ of a top-down SEM micrograph of the CleanARC 228 coating. It shows, a distribution of pore diameter with a mean of approximately 50 nm and standard deviation of 7 nm.

Optical performance of an AR coating is a function of its effective RI and the coating thickness. Referring back to equation 3, the effective RI depends on the porosity of the coating and the RI of the matrix material, assuming the pores are filled with air. RI may be decreased by either increasing the porosity or reducing the RI of the matrix material. Unfortunately increasing porosity reduces mechanical strength and durability. However this new coating uses a matrix material that is similar to the high-durability matrix of the CleanARC 208 material reported previously [7] and consequently has a relatively low RI when compared to the matrix materials used by other commercially available high-temperature cured AR coatings. This means in practice that equivalent optical performance can be attained at lower porosity and so with much greater mechanical strength and durability.

Figure 3 shows the transmission spectra of the CleanARC 228 coating relative to an uncoated piece of the same type of glass measured on a Perkin Elmer Lambda-35 spectrophotometer. In this case the glass used was PPG Starphire low-iron float glass. The solar weighted absolute increase in transmission over the range 380 nm to 1100 nm (T_{AM}) for this sample is approximately 2.75%, placing this coating in the top tier of commercially attainable solar PV AR coating performance.

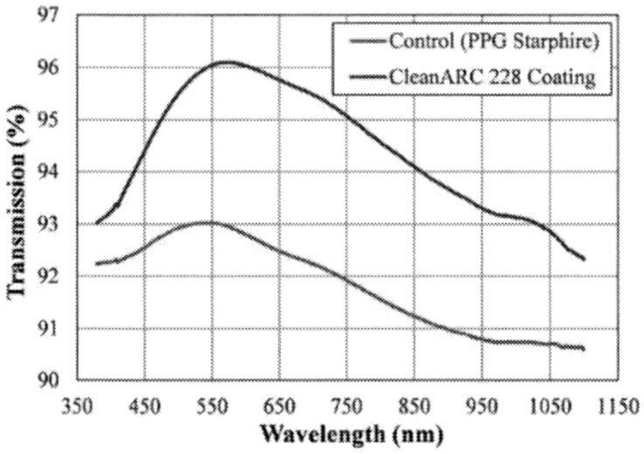

Figure 3: Transmission spectra of CleanARC 228 coating versus an uncoated control.

3.2 Extended accelerated testing

Figure 4 shows the results after 40 cycles of humidity freeze according to specification IEC 61215. This is 4 times the typical test duration. These tests were performed on laminates. Reflection was measured at 5 points on each sample at 0, 10, 20 30 and 40 cycles. The chart shows the change in solar weighted reflection measured over 380 nm to 1100 nm from its initial value. An absolute increase in reflection of 0.34% after 40 cycles is a very good result. Contrasted to other commercially available coatings tested to 20 cycles, which show a loss in absolute solar weighted transmission of 0.53% to 1.44% when subjected to the same test protocol [6].

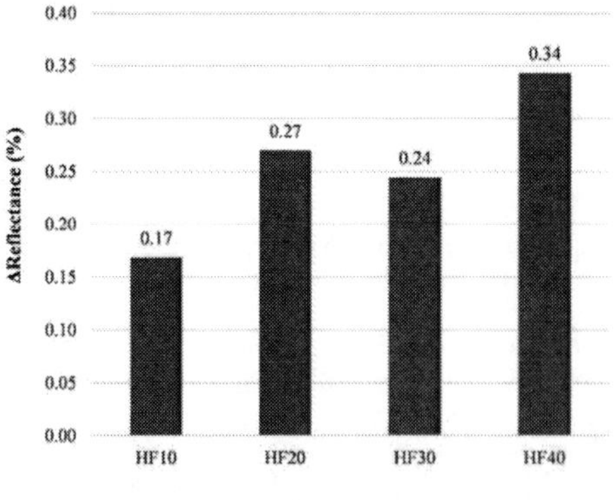

Figure 4: Extended humidity freeze test results.

Figure 5 shows the results of 480 hours of neutral salt fog testing according to specification ASTM B-117. Reflection was measured at 5 points on each sample after 0, 96, 192, 288, 384 and 480 hours of exposure. This is 5 times the typical test duration. The chart shows the change in solar weighed reflection over 380 nm to 1100 nm from its initial value. This coating also exhibits very good salt fog resistance and again compares very favorably to the performance of other commercially available coatings [6].

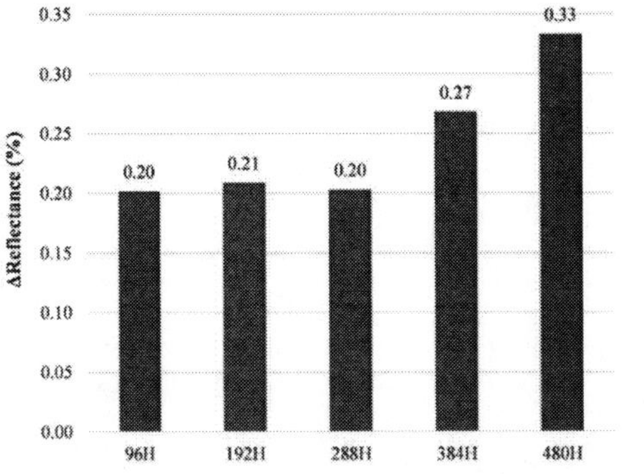

Figure 5: Extended neutral salt fog test results.

Figure 6 shows the results of abrasion testing of 1000 strokes to specification EN1096. This test was performed on one set of laminates before starting the humidity freeze test and then repeated on a fresh location of the sample at the 20, 30 and 40 cycle read-points. The chart shows the change in reflection at the center of the abraded area after 1000 abrasion strokes with a medium density felt pad and 4N of force. The results show that abrasion properties are excellent and essentially unchanged even after four times the standard humidity freeze stress.

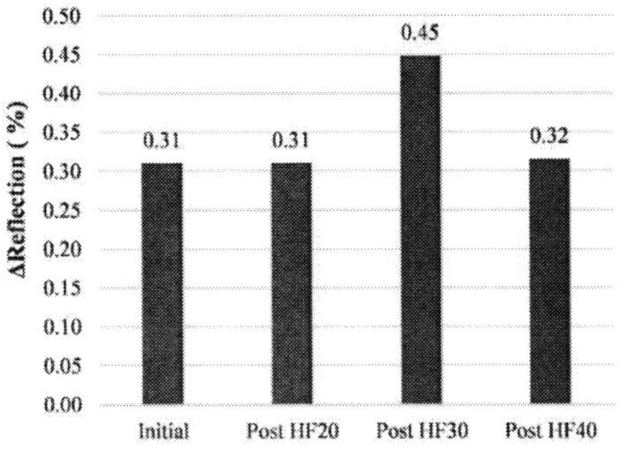

Figure 6: Abrasion test results performed to specification EN1096 before and after 20, 30 and 40 cycles of extended humidity freeze testing.

3.3 Comparative accelerated testing

Figure 7 shows the results of testing samples of CleanARC 228 against the two other commercially available anti-reflection coatings. The post-stress reflection of the coupons was measured with a Konica-Minolta 2600d spectrophotometer and the results compared to reflection measurements of equivalent un-stressed laminates. Errors bars shown represent one standard deviation.

The CleanARC 228 coatings performed very well on all four tests with only small changes that are on the order of the measurement uncertainty. Coating A performed well on HF10 and NSF96, but showed significant losses in PCT48 and DH500. Coating B performed even more poorly with significant losses on all tests except NSF96.

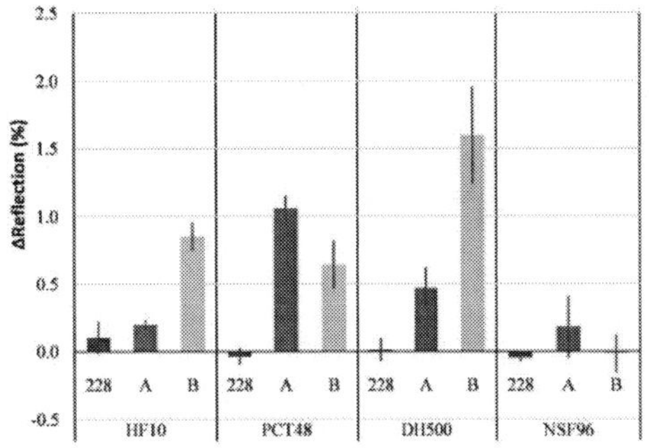

Figure 7: Results of comparative testing of CleanARC 228 and two market-share leading anti-reflective coatings labeled A and B.

Figure 8 shows the results abrasion testing with a CS-10F Weardisc. The chart shows the increase in solar weighted reflection of the coating after the indicated number of abrasive cycles in intervals of 50 up to 250 cycles.

The CleanARC 228 coating reflection increased by about 0.6% after 250 cycles of abrasion which was the lowest of the 3 samples tested. In addition, its degradation was reasonably linear. Both coatings A and B show characteristic catastrophic loss of performance wherein they lost most of their performance between two read-points. Coating A failed between 0 and 50 strokes and coating B fared slightly better failing between 150 and 200 strokes.

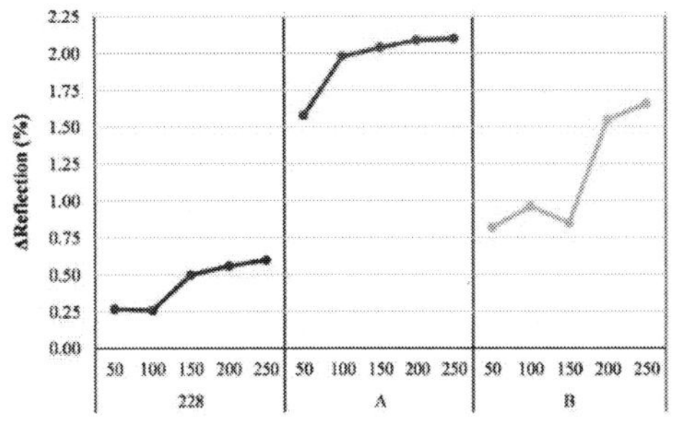

Figure 8: Results of comparative linear abrasion testing using Taber CS-10F Weardisc with 4N of force with 50 to 250 abrasion cycles.

Table II shows the starting water contact angle and the absolute change in solar weighted optical performance of the coating after chemical resistance testing for the three coatings. Water contact angle is a measure of hydrophobicity and moisture resistance, the larger the angle, the greater hydrophobicity. The table shows a clear difference in boiling salt water testing between the CleanARC 228 coating and coatings A and B which increased by 0.24% and 0.16% respectively. In the context of this test any change in performance is bad as an increase in performance typically indicates the coatings is being etched, increasing it porosity and consequently damaging its durability. All the coatings respond relatively well to a strong acid as would be expected from silica base materials. Coating A underperforms the others in resistance to a strong base.

Table II: Chemical resistance test results

Test	228	A	B
Water Contact Angle	86°	38°	66°
Boiling Salt Water	-0.06%	+0.24%	+0.16%
Acid (H_2SO_4)	-0.02%	+0.07%	+0.01%
Base (NaOH)	+0.05%	+0.17%	+0.01%

4 CONCLUSIONS

The coating described in this paper uses a new and novel method of pore formation to further reduce the refractive index within a previously proven highly durable sol-gel coating material. Unlike traditional porous solar AR coatings, this new material is functionalized for intrinsic moisture repellency. Testing indicates that this coating achieves excellent optical gain while maintaining superior chemical resistance, moisture and abrasion resistance, making it particularly well suited to solar glass applications.

5 REFERENCES

[1] Chen, Dinguo. "Anti-reflection (AR) coatings made by sol–gel processes: a review." solar energy materials and solar cells 68.3 (2001): 313-336.

[2] Suratwala, T. I., et al. "Surface chemistry and trimethylsilyl functionalization of Stöber silica sols." Journal of non-crystalline solids 316.2 (2003): 349-363.

[3] Stöber, Werner, Arthur Fink, and Ernst Bohn. "Controlled growth of monodisperse silica spheres in the micron size range." Journal of colloid and interface science 26.1 (1968): 62-69.

[4] San Vicente, G., et al. "Long-term durability of sol–gel porous coatings for solar glass covers." Thin Solid Films 517.10 (2009): 3157-3160.

[5] Gombert A., et al. "Results of Durability Testing of Antireflective Glazing." Glass Processing Days, August 2005.

[6] Pan X., et al. "Performance of Different Anti-Reflective Coated Glass for PV Modules." 31st EU PVSEC, Hamburg, 2015.

[7] Sergiu C. Pop et al. "A highly Abrasive-Resistant, Long-Lasting Anti-Reflective Coating for PV Module Glass." 40th IEEE PVSC, Denver, Colorado, 2014.

[8] Ferretti N., et al. "Reflectance Measurements for Testing the Effect of Module Cleaning on the Glass Surface." 31st EU PVSEC, Hamburg, 2015.

[9] Nielsen, Karsten H., et al. "Optical breathing of nano-porous antireflective coatings through adsorption and desorption of water." Scientific reports 4 (2014).

[10] Han, Katherine, and Chih-Hung Chang. "Numerical modeling of sub-wavelength anti-reflective structures for solar module applications." Nanomaterials 4.1 (2014): 87-128.

[11] Dash, J. G. "Between two and three dimensions." Physics Today 38 (1985): 26-35.

[12] Cuddihy, E. F., and Po B. Willis. "Antisoiling technology: theories of surface soiling and performance of antisoiling surface coatings." (1984).

MODELS FOR LAMBERTIAN OPTICS IN SI

Luigi Abenante and Massimo Izzi
ENEA, Italian National Agency for New Technologies, Energy and Sustainable Economic Development
Via Anguillarese 301, SP 064, 00123, ROMA (Italy)

ABSTRACT: We review the available models for Lambertian optics in Si solar cells with surface texture. We check these models in devices having same structure and assignments to transport and optical parameters and different depth. Based on the check, we infer that, in Si solar cells with surface texture, the standard model for Lambertian optics, where the light-propagation angle is equal to 60° at all passes of light across the considered device, is more reliable than the remaining models, in the case of thin devices.

Keywords: Optical Properties, Light Trapping, Modelling, Simulation

1 INTRODUCTION

Nearly all kinds of today's Si solar cells incorporate light-trapping (LT) structures aiming at improving optical absorption. One of the most common LT-structures is surface texture. Surface texture boosts the optical absorption in a device by increasing the propagation angle of internal light rays with respect to the device normal, θ (see Fig. 1). In the PV R&D, reference «ideal» surface-textured devices are used in simulations [1-3], where Lambertian optics is imposed because it offers the maximum light absorption effectiveness. Different models for Lambertian optics are used. In this paper, the available models for Lambertian LT are studied with reference to Si solar cells with surface texture. We first review the models.

We begin with the standard model, where θ is constant and equal to 60°. θ is constant when it has the same value at all passes of light across the considered device. In Fig. 1, for example, as it is equal to θ_1 at the first pass and to θ_2 at the second pass, θ is not constant. This model, in [1], was found to be in accordance with ray-tracing techniques.

We denote the second model as Green's model because it was presented by M. A. Green in [4]. In this approach, θ is constant and depends on both device-thickness, W, and, through the absorption coefficient, wavelength, λ. This model was used in the simulations in [2-3].

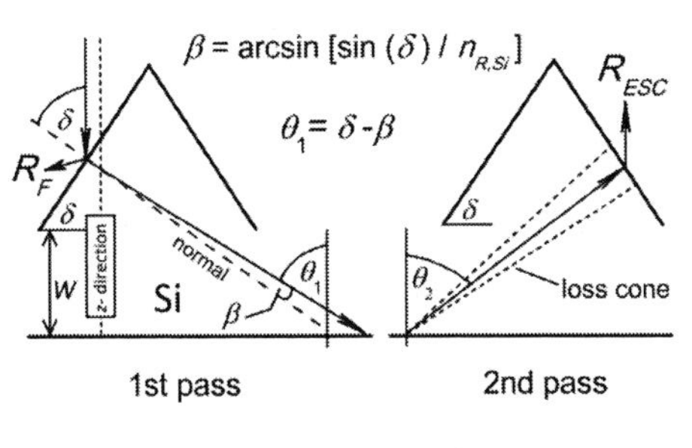

Figure 1: The first two passes of a light ray across a Si sheet with pyramidal front-surface texture. R_{ESC} and R_F are the escape and front-surface reflectance, respectively [6]. θ_1 and θ_2 are propagation angles of internal light. Pyramid dimensions are exaggerated. Relationships among internal angles are indicated, where $n_{R,Si}$ is the refractive index of Si.

The last model for Lambertian LT studied in the present work is called PC1D-model because it exploits an optical diffusion model, which is embedded in the well-known numerical simulation program PC1D [5].

The optical models are compared to each other by simulating with PC1D the light-generated current density, J_{sc}, in a set of Si solar cells of various W having same structure at same assignments to transport and optical parameters. The PC1D-model is exploited as a reference. Since this model underestimates J_{sc} in thin devices, such a comparison approach allows checking the validity of the remaining two models.

We find that the Green's model underestimates J_{sc} in thin devices even more than the PC1D-model does.

We propose a way to correct the Green's model. However, although the corrected Green's model for Lambertian optics gains in physical consistency with respect to the original Green's model, we conclude that the best option in terms of reliability is still the standard model.

2 MODELS FOR LAMBERTIAN OPTICS

2.1 Standard model

This model simply assumes $\theta = \theta_m = 60°$. Such an assignment is obtained by solving

$$e^{-\alpha_{Si} W/\cos\theta_m} = 2\int_0^{\pi/2} \sin\theta\cos\theta\, e^{-\alpha_{Si} W/\cos\theta} d\theta \qquad (1)$$

where α_{Si} is the Si absorption coefficient, for $\alpha_{Si}W \rightarrow 0$. This model can be implemented in PC1D either at textured front surface by assigning δ =75.705° with Si refractive index, $n_{R,Si}$, given by the PC1D default value (see Fig. 1) or at flat surfaces by assigning a LT absorption coefficient [1]

$$\alpha_{lt}(\lambda, W) = \alpha_{Si}(\lambda)/\cos\theta_{op} \qquad (2)$$

with θ_{op}=60°.

2.2 Green's model [4]

In this model θ_{op} to be used in (2) is a function of λ and W given by [4]

$$\theta_{op}(\lambda, W) = \cos^{-1}\left[\frac{1 + a(\alpha_{Si} W)^b}{2 + a(\alpha_{Si} W)^b}\right] \qquad (3)$$

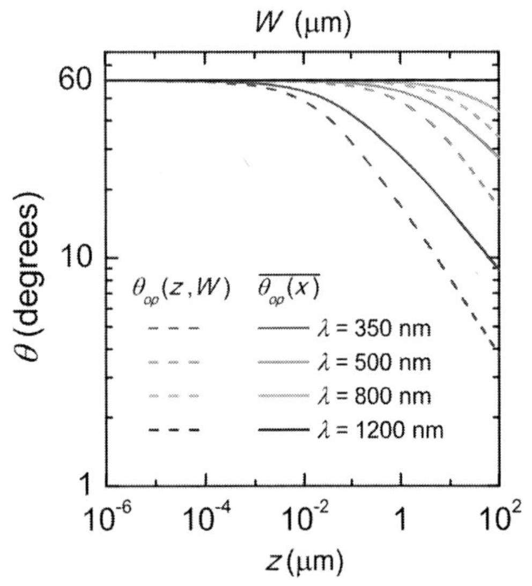

Figure 2: Please note that the graph has two abscissae. If abscissa W is considered, then the figure shows θ_{op} as calculated with both Eq. (3) at the rear surface of cells of thickness W (dashed-line curves) and Eq. (7) (full-lines curves). If abscissa z is considered, the figure shows θ_{op} at each position, inside a Si region of W=100 μm, as calculated with Eq. (5). All quantities are calculated at four λ-values.

where $a = 0.935$ and $b = 0.67$. $\theta_{op}(\lambda,W)$ is shown in Fig. 2 at four λ-values (dashed lines).

Eq. (3) is an approximation of a solution to Eq. (1) obtained under conditions for which the standard Lambertian model can be derived as well. It is worth mentioning that the standard model results from developing the exponentials in (1) in series.

2.3 PC1D-model

Triggering the diffusion model embedded in PC1D by checking "diffuse" in the "rear surface" section of the "Reflectance" dialog box assigns θ_2=60°. Implementing simultaneously the standard model assures that θ_1=60° as well. Concerning the third and subsequent internal-light passes, the PC1D diffusion model is such that, "*because only weakly absorbed photons survive the first round trip of the device thickness, subsequent photogeneration resulting from these trapped photons is assumed to occur uniformly over the device thickness*" [5]. In thin devices, however, also non-weakly absorbed photons may survive the first round trip. In the case of thin devices, consequently, assuming uniform absorption after the first round trip of the device thickness leads to underestimate light-generation.

3 CHECK

By implementing the optical models described in the previous Section according to the given directions, J_{sc}-values were calculated with PC1D at various W-values in Si solar cells having same structure and values for optical and transport parameters as the cells modelled in [2-3].

Results are shown in Fig. 3, where the discrepancies, Δ, with respect to the PC1D-model are also reported.

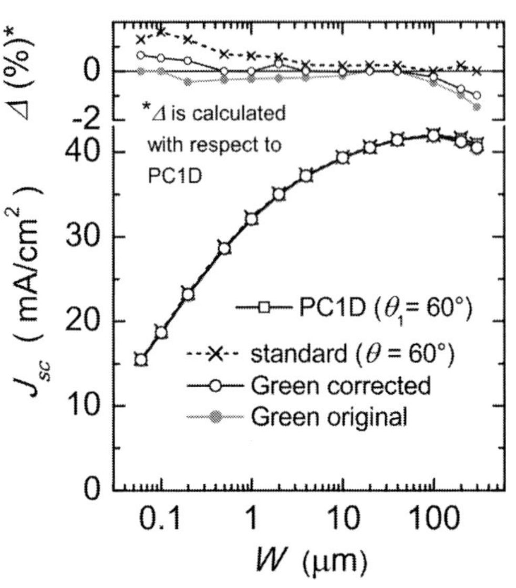

Figure 3: Comparison among currents calculated for a set of Si solar cells, which only differ in the W-value, by implementing in PC1D the modelling approaches studied in this work. The discrepancies Δ of the standard, original Green's, and corrected Green's models with respect to the PC1D-model are also reported, which were calculated by dividing the difference of the considered curves by their mean value.

As it can be seen in Fig. 3, implementing the original Green's model (lines with solid circles) results in J_{sc}-values that are equal or lower than those calculated by using the PC1D-model (lines with open squares). This means that the original Green's model at his best is physically as consistent as the PC1D-model.

In Fig. 3, it can be also observed that the agreement between the standard (crosses) and PC1D models increases with W. This is expected, because with increasing W the percentage of carriers that are generated in the first two passes of light across a device increases and, in the first two passes, θ=60° in both models.

4 DISCUSSION

In a region of thickness W, the following equation

$$e^{-\alpha_{Si}z/\cos\theta_p} = 2\int_0^{\pi/2} \sin\theta\cos\theta\, e^{-\alpha_{Si}z/\cos\theta} d\theta \qquad (4)$$

where z is defined in $(0,W]$, describes the Green's optical approach as it is presented in [4].

At $z \to 0$, Eq. (4) is solved at θ_{op}=60° [1] representing the random direction of light rays at the front surface of a textured cell (see Fig. 3 in [4]). At $z=W$, Eq. (4) reduces to Eq.(1), whose solution, according to [4], is well approximated by Eq. (3).

The following equation

$$\theta_{op}(\lambda,z) = \cos^{-1}\left[\frac{1+a(\alpha_{Si}z)^b}{2+a(\alpha_{Si}z)^b}\right] \qquad (5)$$

represents a position-dependent effective propagation angle that gives the solutions of Eq. (4) as approximated according to [4] at all z in $(0,W]$. In Fig. 1, $\theta_{op}(\lambda,z)$ is represented in a region of W=100 μm at four λ-values (dashed lines).

In the standard model for Lambertian optics, at all the z- and λ-values, $\theta_{op}(z,\lambda)=\theta_{op}(0)=60°$ is assigned. In the Green's model, instead, $\theta_{op}(z,\lambda)=\theta_{op}(W,\lambda)$ is assigned.

Fig. 2 shows that the latter assignment may only be true at the highest λ-values. At lower λ-values, consequently, this assumption underestimates light absorption, and hence the current in the considered region. The Green's model for Lambertian optics, as reported in [4], is therefore such as to underestimate currents. This explains the results in Fig. 3.

Underestimation can be corrected in a way analytically viable, if $\theta_{op}(z,\lambda)$, instead of $\theta_{op}(W,\lambda)$ as calculated with (3), is assigned a mean value taking into account absorption at all z-values in the given region. We propose here

$$\overline{\theta_{op}(x)} = \cos^{-1}\left[\frac{\int_0^x \frac{1+z}{2+z}dz}{\int_0^x dz}\right] \tag{6}$$

where, as in [4], $x=a(\alpha_{Si}W)^b$ with the values of a and b given above. Eq. (6) calculates $\overline{\theta_{op}(\lambda,x)}$ at each λ-value as

$$\overline{\theta_{op}(x)} = \cos^{-1}\left[1-\frac{\ln(2+x)-\ln(2)}{x}\right] \tag{7}$$

The angle $\overline{\theta_{op}(\lambda,x)}$ is shown by the full-line curves in Fig. 1 at W-abscissa. Obviously, it is a function of W and not a function of z.

In the same aforementioned surface-textured devices, we have re-calculated $J_{sc}(W)$ for the Green's model by assigning in (2) $\theta_{op}=\overline{\theta_{op}(\lambda,x)}$ in place of $\theta_{op}=\theta_{op}(\lambda,W)$ as given by (3). Results are reported in Fig. 3 together with the relevant curve for discrepancy Δ (lines with open circles). As it can be seen the physical consistency of the Green's model is improved at the lowest W-values.

5 CONCLUSIONS

We have compared the available analytical models for Lambertian optics in surface-textured Si solar cells. The PC1D- and Green's models appear to underestimate currents at low W-values. At low W-values, therefore, the standard model appears to be the most reliable. The Green's model, however, is liable to improvement.

REFERENCES

[1] R. Brendel, D. Scholten, Appl. Phys. A 69, (1999) 201–213/ DOI: 10.1007/s003399900069
[2] L. Abenante, J. Appl. Phys. 117 (2015) 026101 / DOI: 10.1063/1.4905182
[3] A. Bozzola, P. Kowalczewski, L. C. Andreani, J. Appl. Phys. 117 (2015) 026102 / DOI: 10.1063 /1.4905183
[4] M.A. Green, Prog. Photovoltaics Res. Appl., 10, 4 (2002) 235-241
[5] P.A. Basore, IEEE Trans. Electron Devices 37, 2 (1990) 337-343
[6] P.A. Basore, Conf. Rec. Twenty Third IEEE Photovolt. Spec. Conf. 1993 (Cat. No.93CH3283-9), (1993) 0-5.

AN INEXPENSIVE SPECTRAL SENSOR FOR MMPT IN PARTIAL SHADE

Miguel A. López-Álvarez[1,2], Steve Collins[3], and Javier Hernández-Andrés[2]
(1) Intracon, Color Science Division, SPAIN (miguel.lopez@intracon.com)
(2) The Color Imaging Laboratory, Department of Optics, Granada university, Spain (javierha@ugr.es)
(3) The Department of Engineering Science, Oxford University UK (steve.collins@eng.ox.ac.uk)

ABSTRACT: Metrics such as the Useful Fraction (UF), the Weighted Useful Fraction (WUF) and the Average Photon Energy (APE), have been proposed and successfully used to demonstrate the relationship between the spectrum of the incident radiation and solar cell performance. In this work we propose the use of a new inexpensive optical sensing technology to accurately determine the spectrum of incident light and hence the UF, WUF, APE and Photon Absorption rate in real time. These sensors could be integrated within a photovoltaic system to gather data for a maximum power point tracking (MPPT) system that can deal with challenging partial shading conditions.

Keywords: Amorphous Silicon, c-Si, Calibration, Devices, Solar Radiation, Shading,

1 INTRODUCTION

There is an output voltage at which the maximum power can be extracted from a photovoltaic (PV) system. Unfortunately, this maximum power point depends upon environmental factors, particularly irradiance and operating temperature, that are time dependent. Consequently, to ensure efficient operation of the system maximum power point tracking has to be employed. If all the individual PV cells in a module are operating under the same conditions, the necessary tracking of the resulting unique maximum power point can be achieved using one of a number of existing maximum power point tracking (MPPT) techniques [1]. However, this is not the case when some of the cells in a module are shaded, either by an object or by a passing cloud. Under these conditions several local maxima in output power can arise at different output voltages. Any MPPT strategy that fails to find the global maximum power point can lead to a loss of power of 70% at some instances and between 1% and 2% over a whole day [2].

A possible approach to avoiding these losses is based upon a comprehensive study of the characteristics of partial shaded PV systems [3]. This study showed that the maxima in output power always occur at a voltage that is approximately an integer multiple of 80% of the open circuit voltage of a module. This means that the global maximum can be found by using an existing MPPT technique to determine the maximum power available at voltages close to these known voltages [3]. Since the conditions that determine the optimum output voltage can change rapidly, the search for a new global maximum has to be undertaken regularly, possibly every 25 seconds, or when a site supervisor initiates a search [3]. The need for these regular searches or intervention by a supervisor could be avoided using sensors to determine the pattern of shading. Furthermore, since the location of the global maximum is related to the pattern of shading these sensors could provide the information needed to determine the probable location of the global maximum when conditions have changed.

Inexpensive sensors placed close to modules could determine the irradiation falling on each module. However, the performance of PV cells depends upon both the intensity and the spectrum of irradiation. Sensors are therefore required that determine a parameter that is related to the spectrum of the irradiation. One parameter that could be calculated by the sensor system is the Useful Fraction (UF) [4]. One major assumption in the definition of UF is that the spectral response (SR) of the PV cells is 100% for wavelengths that can be absorbed by the PV cell. However, not all of the incident energy that is within the cell's spectral range will be converted to useful energy. To compensate for this the Weighted Useful Fraction (WUF) was introduced [5]. For a PV cell design with a spectral response SR(λ), where λ represents the wavelength, irradiated by an illuminant with a spectrum E(λ) the weighted useful fraction (WUF) is defined as

$$WUF = \frac{\int_{cell-min}^{cell-max} E(\lambda)SR(\lambda)\,d\lambda}{\int_{observed\_radiation-min}^{observed\_radiation-max} E(\lambda)d\lambda} \quad (1)$$

Since UF is calculated assuming SR(λ) = 1 in the range of the cell absorption in Equation 1, UF is always bigger than WUF. Alternatively, Jardine and co-workers have shown a strong correlation between the average photon energy (APE) the short circuit current in single junction devices [6].

Previously, a simple multispectral imaging system that can obtain accurate estimates of the spectra of skylight from a combination of a small set of optimized Gaussian sensors and the Imai-Berns spectral estimation method has been developed [7,8]. In a similar work, the Imai-Berns method was used to estimate the spectrum of skylight based upon the responses of a 3-sensor commercial scientific CCD camera [9]. Recently, the ability to integrate Fabry-Perot filters with a full width at half maximum (FWHM) of between 10 nm and 25 nm on top of the pixels of an otherwise standard CMOS imager has been demonstrated [10]. This technology could be used to create inexpensive sensors that could be integrated as part of a MPPT system that is robust to shading of some modules in a system. The aim of the work reported in this paper is to investigate the number of sensors that would be required and the accuracy with which key parameters can be estimated.

2 DESCRIPTION OF THE EXPERIMENTAL DATA

To obtain statistical significant results, a set of 2600 spectral daylight irradiance measurements taken in Granada (Spain, 37.18ºN 3.60ºW) over two years has been used. This data was acquired with a cosine receptor attached to the spectro-radiometer and it represents the irradiance from the whole sky dome [11]. These data cover the wavelength range from 300nm to 1100nm in 5nm steps, and hence allow us to estimate the total number of photons that could be absorbed by any material that absorbed these wavelengths, including both a-Si and c-Si. However, the work that is reported assumes an absorption spectrum that represents an a-Si PV cell.

An investigation of this data showed that there were strong correlations between the UF, the WUF and the APE metrics. In contrast, the results in Figure 1 show that there is no correlation between the UF and the correlated color temperature (CCT) which can be calculated from spectral data). This suggests that CCT will not be a useful measure of the output of a PV cells irradiated with the same light as the sensor.

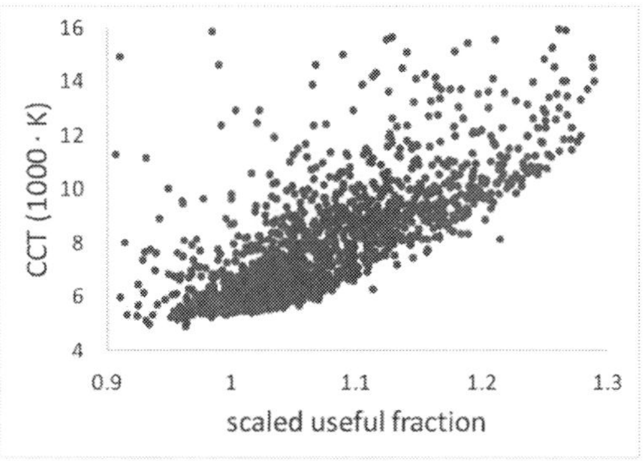

Figure 1. Color temperature in K as a function of scaled UF for our dataset.

Spectral estimation methods, like the Imai-Berns method, provide unrealistically good results when the training set of spectra (required for these methods to be used) and the test set are the same. In our case, when we use the Imai-Berns method for estimating spectra from sensors responses, and then calculate the UF, WUF or APE from those estimated spectra, we use a subset of 100 spectra extracted from the whole dataset as a training set, by maximizing a distance metric among the selected spectra. The accuracy of the spectral estimation method was then determined using a separate representative set of spectra. The selection of this subset for training was done following the method described in [9], which proved to be very effective at obtaining a training set which is representative of the results that would be obtained from the whole data set [9].

3 DESCRIPTION OF THE SYSTEM

Although the Imai-Berns method has been required previously to estimate CCT and other metrics that require spectral or colorimetric information, it may be possible to estimate UF, WUF and APE by simply combining the responses of a set of Gaussian sensors with spectral responses that peak at wavelengths that are evenly distributed across the wavelength range of interest, with a spectral separation among peaks not much higher than their spectral width. For example, Figure 2 shows the results obtained when using a weighted sum of the spectral responses of Gaussian sensors whose spectral responses are separated by 20nm and that are 20nm wide, to build an approximation of the spectral sensitivity curve of a typical a-Si cell [12]. For the case of the UF, the spectral profile to simulate is constant and equal to one, as we show in Figure 3, and can also be approximated by combining the responses of a set of sharp, equally spaced Gaussian sensors. Hence, we only need to adequately sum these sensors responses in order to obtain the total number of photons absorbed by the PV cell (numerator in Equation 1) for WUF, or the total number of photons available at the spectral range of interest for UF. We assume that the total irradiance impinging on the cell (denominator in Equation 1) can be measured by using a pyranometer or a similar device.

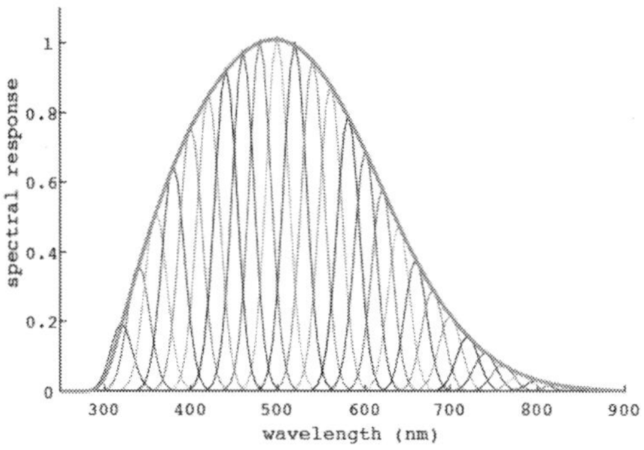

Figure 2. The solid blue curve is the sum of the weighted responses of 20-nm width Gaussian sensors, the peaks of which are separated 20nm.

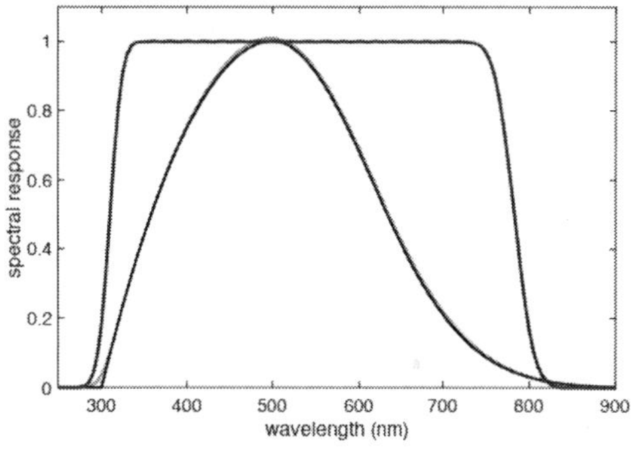

Figure 2. The solid red curve is an approximation to a constant spectral response from 300nm to 780nm (the assumed useful range for an a-Si PV cells), used for calculating UF. Solid green curve is the spectral response of a typical a-Si PV cell, while the blue curve is the estimated spectral response obtained using a weighted sum of the responses of 30 evenly spaced 20nm width Gaussian sensors, used for calculating WUF.

The results in Figures 2 and 3 suggest that summing the responses of 30 sensors (20-nm width Gaussian sensors, from 300nm to 900nm every 20nm) can give a very good estimate of the spectral response of an a-Si PV cell. The accuracy of the summed response can be obtained by calculating quantities such as the UF, WUF and APE using both the spectral response of the a-Si cell and the approximation obtained by summing the sensor responses. However, this leaves the question of the target accuracy.

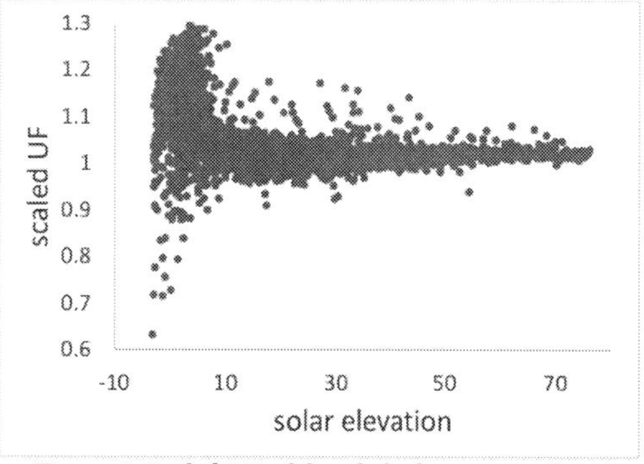

Figure 4. Scaled UF of the whole dataset as a function of solar elevation.

Analysis of our data set, Figure 4, shows that for the spectra that it contains, the UF varies by up to 20%. This suggests that a target accuracy of better than 1% would detect the larger changes in UF. In addition, the efficiency of an a-Si PV cell changes by 0.1%/K, and since the temperature of the PV cells won't be controlled an accuracy of better than 0.1% might be superfluous. A target estimation error of between 0.1% and 1% would therefore seem reasonable.

The final factor that has been taken into account in this work is the impact of noise. In particular, a noise level of 1%, equivalent to a SNR of 40dB, has been added to the data in the simulations. However, in an attempt to determine the accurate of the proposed method regardless of noise results have also been obtained using noise-free data.

4 RESULTS

As we showed in Figures 2 and 3, the responses of a set of sharp (20nm width) equally spaced Gaussian sensors can be added together to estimate the spectral responsivity of a typical a-Si cell, or a bandpass filter in the range of interest of that cell (necessary to estimate the total integrated irradiance for UF calculations). This means that it is possible estimate the UF, WUF, or the APE from the responses of the sensors in our system, without using spectral estimations as an intermediate step. In the case of UF and WUF, sensors up to 900nm approximately are necessary because their tails have some sensitivity in the a-Si range (300nm to 780nm). However, in the case of the APE, we are interested in the average energy taking into account the whole range between 300nm and 1100nm. Hence, a bigger number of sensors is necessary for estimation of this metric (shown in Table 2).

Table 1 contains the average and standard deviation error in the estimation of the UF and WUF (Table 2 shows the results for APE) for the case of a typical a-Si cell, expressed in percentage, for a SNR of infinity and a more realistic one assuming an SNR of 40dB. Unsurprisingly, the mean error in these tables represents the error in the integrated area under the spectral curves of the a-Si cell (in the case of WUF) or the band-pass filter (in the case of UF) made by approximating these curves with our weighted sum of sensor responses. For this reason, the mean error shown in the tables is independent of noise, and only depends on the number of sensors used to approximate the curve of interest by a weighted sum of Gaussian sharp sensors. This constant offset could then be easily corrected. Hence, the standard deviation should be taken as a representative metric for the error of our direct method in estimating the metrics of interest.

These results show that the responses from sets of Gaussian sensors with a FWHM of 20nm can be used to estimate the UF, WUF and APE to an accuracy of better than 1% if enough number of sensors is used. However, for this potentially cost sensitive application it is important to use the fewest possible sensors. Unfortunately, as the number of sensors is reduced the errors increase rapidly and consequently when there are only 8 sensors the errors have increased approximately by a factor of 10 compared to the case with 30 sensors.

Filters sampling the a-Si range	SNR	%error UF (mean & StDev)	%error WUF (mean & StDev)
30	Infinity	-0.45 & 0.06	-0.27 & 0.04
	40dB	-0.45 & 0.30	-0.27 & 0.32
24	Infinity	-1.9 & 0.10	-1.7 & 0.08
	40dB	-1.9 & 0.43	-1.7 & 0.63
15	Infinity	-10 & 0.31	-10 & 0.31
	40dB	-10 & 0.92	-10 & 2.00
8	Infinity	-17 & 0.94	-17 & 1.60
	40dB	-17 & 2.10	-17 & 3.30

Table 1. Accuracy of the direct method estimating UF and WUF for various numbers of 20nm-width sensors and two levels of signal-to-noise ratio.

Filters sampling the 300-1100nm range	SNR	% error APE (mean & StDev)
40	Infinity	0.0 & 0.02
	40dB	0.0 & 0.12
30	Infinity	1.7 & 0.04
	40dB	1.7 & 0.42
20	Infinity	2.0 & 0.90
	40dB	2.0 & 0.58
10	Infinity	2.8 & 0.46
	40dB	2.8 & 1.10

Table 2. Accuracy of the direct method estimating APE for various numbers of 20nm-width sensors and two levels of signal-to-noise ratio.

Previously the Imai-Berns method has been successfully used to estimate the spectrum of daylight from the responses of a limited number of sensors. This method has therefore been used to estimate the spectrum of the light falling on the sensors which can then be used to calculate the UF, WUF and APE. In addition, having spectral information of the radiation impinging on the cells permits to estimate the photon absorption rate (PAR), calculated as shown in Equation 2 (where h is Planck's constant and c is the speed of light).

$$PAR = \int_{300nm}^{1100nm} E(\lambda)SR(\lambda)\,\frac{\lambda}{hc}\,d\lambda \qquad (2)$$

which is proportional to the short circuit current that would be obtained from a PV cell with a spectral response SR(λ).

Figures 5 to 8 show the standard deviation of the percentage errors in these four parameters when the Imai-Berns spectral estimation method had been developed using 100 spectra chosen to represent the entire data set as a training set. In all cases shown, the number of used eigenvectors equals the number of sensors, since this approach have proven to provide optimum results [7]. As expected, the errors in the estimates obtained using this method increase as the number of sensors is reduced. However, using the limited data available from the sensors to first estimate the spectrum of the light irradiating the sensor, significantly improves the results that are obtained when the number of used sensors is smaller than 15. For example, when using data from 15 sensors using spectral estimation before calculating the parameters, reduces the errors in the estimates by half, approximately.

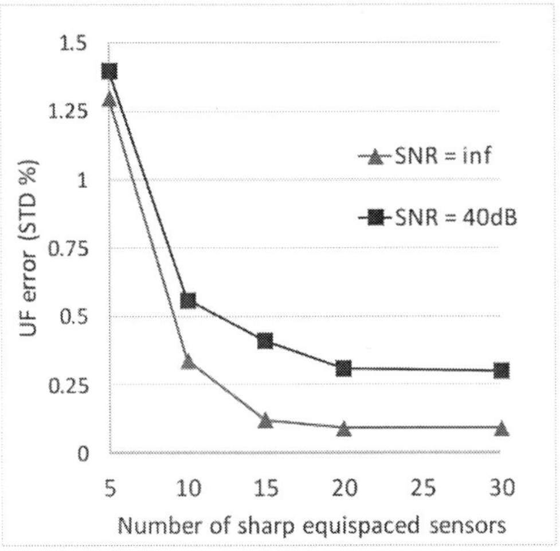

Figure 5. The UF from the Imai-Berns method for varying numbers of sensors.

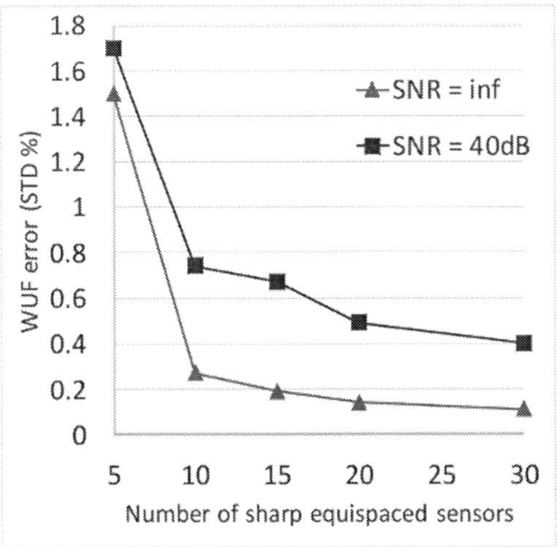

Figure 6. The WUF from the Imai-Berns method for varying numbers of sensors.

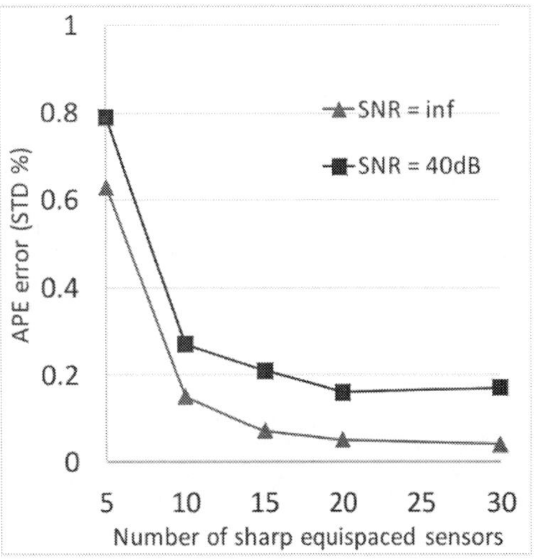

Figure 7. The APE from the Imai-Berns method for varying numbers of sensors

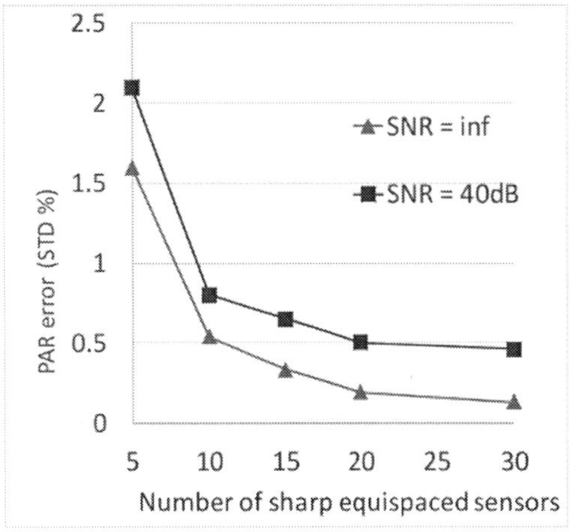

Figure 8. The PAR from the Imai-Berns method for varying numbers of sensors

5 CONCLUSIONS

Having real-time knowledge of the spectrum of the radiation impinging on a solar cell could be a useful part of an efficient maximum power point tracking system that maximized the performance of a PV system under partial shading conditions. We have shown that the spectral distribution of the available daylight in Granada is variable enough to have a significant impact on the performance of a PV cell. In particular, data gathered over two years, showed that the useful fraction, previously used to explain changes in PV cell efficiency, can vary up to 20%. Using these data, we have shown how a potentially inexpensive system, based on a small number of sensors made using a new process that allows the integration different Fabry-Perot filters onto individual photodetectors on the same silicon substrate, can provide accurate estimates of UF, WUF, APE or PAR metrics. Two different approaches have been compared: direct estimation of the PV cell responsivity from the combination of equally spaced sensors with different spectral responses, and calculating these parameters after first estimating the spectrum of the light falling on the sensor. The first approach is good enough to determine the UF and WUF parameters to an accuracy of less than 0.32% in realistically noisy situations. By using the Imai-Berns method to estimate the spectrum of the illuminating light before calculating the parameters of interest, this accuracy can be achieved with a reduced number of sensors (around 20 instead of 30). We can additionally calculate the rate at which photons will be absorbed, which is related to the short circuit current of the PV cell whose spectral response is being estimated.

The results that have been obtained suggest that a recently commercialized sensor technology could be used to sense the intensity and spectrum of light falling on each module in a PV system. This information could then be used to determine which of the possible maximum power points will be the global power point at a particular instant. Further work is required to determine how these sensors might be distributed within a set of PV modules, the number of sensors, and hence cost, required to obtain the information required by the MPPT system, and the potential benefits of using sensors to both reduce the time needed to find a global maximum power point and eliminate the need to frequently check if the current operating conditions are the optimum conditions.

6 REFERENCES

[1] T. Esram, P. L. Chapman, "Comparison of photovoltaic array maximum power point tracking techniques", in IEEE Trans. Energy Conversion, vol. 22 .2 (2007), pp. 439-449.

[2] R. Bruendlinger, B. Bletterie, M. Milde, and H. Oldenkamp, "Maximum power point tracking performance under partially shaded PV array conditions," in Proc. 21st EUPVSEC, Dresden, Germany, Sept. 2006, pp. 2157–2160.

[3] Patel, Hiren, and Vivek Agarwal. "Maximum power point tracking scheme for PV systems operating under partially shaded conditions." Industrial Electronics, IEEE Transactions on 55.4 (2008): 1689-1698.

[4] Gottschalg R, Betts TR, Infield DG, Kearney MJ. Experimental study of variations of the solar spectrum of relevance to thin film solar cells. Solar Energy Materials and Solar Cells; 79(2003): 527–537.

[5] Simon, Michael, and Edson L. Meyer. "The effects of spectral evaluation of c-Si modules." Progress in Photovoltaics: Research and Applications 19.1 (2011): 1-10.

[6] C. N. Jardine, T. R. Betts, R. Gottschalg, D. G. Infield, K. Lane, "Influence of spectral effects on the performance of multi junction amorphous silicon cells", in Europe Conference and Exhibition, Rome, 2002. http://web.etaflorence.it/102.0.html

[7] M. A. Lopez-Alvarez, J. Hernandez-Andres, E. M. Valero, and J. Romero "Selecting algorithms, sensors and linear basis for optimum spectral recovery of skylight" JOSA A, 24 (2007): 942-956.

[8] M.A. Lopez-Alvarez, J. Hernandez-Andres, and J. Romero "Developing an optimum computer-designed multispectral system comprising a monochrome CCD camera and a liquid crystal tunable filter." App. Opt. 47 (2008): 4381-4390.

[9] M.A. Lopez-Alvarez, J. Hernandez-Andres, J. Romero, F.J Olmo, A. Cazorla, and L. Alados-Arboledas "Using a trichromatic CCD camera for spectral skylight estimation", App. Opt. 47 (2008): H31-H38.

[10] Bert Geelen, Carolina Blanch , Pilar Gonzalez , Nicolaas Tack and Andy Lambrechts " A tiny VIS-NIR snapshot multispectral camera ", Proc. SPIE 9374, Advanced Fabrication Technologies for Micro/Nano Optics and Photonics VIII, 937414 (March 13, 2015); doi:10.1117/12.2077583.

[11] J. Hernández-Andrés, J. Romero, J. L. Nieves, R. L. Lee Jr., "Color and spectral analysis of daylight in southern Europe", in *JOSA A*, vol. 18 (2001):1325-1335.

[12] M. Cheggar, P. Mialhe, "Effect of atmospheric parameters on the silicon solar cells performance", in *Journal of Electron* Devices, vol. 6 (2008): 173–176.

7 ACKNOWLEDGEMENTS

This work was supported in part by the U.S. Department of Commerce under Grant BS123456, and the Color Science Division of Intracon Spain (intracon-es.com).

CONTACT-FREE RAMAN SPECTROSCOPIC MEASUREMENT OF RESIDUAL STRESS IN SILICON SOLAR CELLS CAUSED BY STRINGING

L. Neumaier[1], W. Mühleisen[1], T. Fischer[2], J. Scheurer[3], W. Pranger[4], C. Hirschl[1]
[1] CTR Carinthian Tech Research AG, Europastraße 12, 9524 Villach, Austria
[2] Teamtechnik GmbH, Planckstraße 40, 71691 Freiberg am Neckar, Germany
[3] Polytec PT GmbH, Polytec-Platz 1-7, 76337 Waldbronn, Germany
[4] Ulbrich of Austria GmbH, Industriestr. 1, 7052 Müllendorf, Austria
Corresponding author: E: lukas.neumaier@ctr.at, T: +43 4242 56300-260; F: +43 4242 56300-400
Co-Authors: E: christina.hirschl@ctr.at, wolfgang.muehleisen@ctr.at, wpranger@ulbrich.com, Thomas.Fischer@teamtechnik.com, j.scheurer@polytec-pt.de

ABSTRACT: With the current trend in the photovoltaic industry towards thinner solar cells and four (or more) busbar modules, the cell-interconnection/stringing process is becoming more challenging. Besides the possibility of damaging solar cells during the soldering process itself, thermal and mechanical stresses within this process step are also assumed to induce cracks, which are occurring later during the operation of the PV module under certain environmental conditions, such as heavy snow or wind loads and excessive temperature fluctuations. In order to produce reliable high-quality PV modules at sound costs, it is necessary to minimize the stress on the solar cells during the manufacturing process. Within this study Raman spectroscopy is utilized to analyze the stress of solar cells induced by the interconnection process. Besides investigations on manually soldered solar cells, also industrially interconnected solar cells were evaluated. Here, an industrial soldering as well as a glueing/bonding interconnection technique was analyzed. All investigated solar cells are based on three busbars technology.

Keywords: Silicon solar cell, residual stress, Raman spectroscopy, soldering, glueing/bonding

1 INTRODUCTION

The current trend within the photovoltaic industry towards four or more busbar concepts raises new challenges in terms of handling the cell-interconnection or stringing process. Thermal as well as mechanical stresses, occurring during the stringing process are also assumed to induce micro-cracks, which are occurring later during the operation of the PV module under certain harsh environmental conditions in various climate zones, such as heavy snow or wind loads and excessive temperature fluctuations. Due to the current market situation, it is essential that PV module manufacturers produce reliable high-quality PV modules at sound costs. Hence, the induced stress onto the solar cell, as one of the significant components regarding efficiency and reliability, needs to be considered during the manufacturing process. Utilizing Raman spectroscopy, the residual stress of solar cells was determined for different cell types as well as interconnection procedures.

2 APPROACH

Within this work Raman spectroscopy was utilized to determine residual stress in silicon solar cells, which is caused by thermal and/or mechanical stress during the stringing process. Here, usually the busbar (BB) ribbons are soldered onto the solar cell in a fully automated process at temperatures of around 200 - 250°C. The stress occurs because of differences in the thermal expansion coefficients of the interacting materials as well as the handling of the cells during this process (lifting, placing, positioning and pressing the solar cell) [1]. First, manual soldering-tests were conducted, where the residual stress of three solar cells was measured before and after soldering (i) one, (ii) three and (iii) six BB ribbons onto the cell. Additionally, the first cell, attached with one BB ribbon, was enhanced in a step-by-step soldering process to monitor the stress variation of additionally added ribbons on one cell. In the next step, the initial residual stress of an additional set of solar cells (mono- and multi-crystalline cells) was determined. Afterwards half of the cells were soldered at an industrial stringer and measured again at the same position. For the comparison of an alternative interconnection process to the standard soldering procedure, the other half of the cell-set was measured after the busbar ribbons were applied in a manual procedure using an electrical conductive adhesive (Polytec PT). To validate the measured data, FEM simulations for the initial stress condition of solar cells (i.e.: before the stringing process), were carried out previously.

3 RESULTS

3.1 FEM Simulations

For the validation of the Raman measurements, FEM simulations were carried out in advance. For conducting the simulations a standard multi-crystalline silicon solar cell was modeled using ANSYS® 15.0 (see Figure 1).

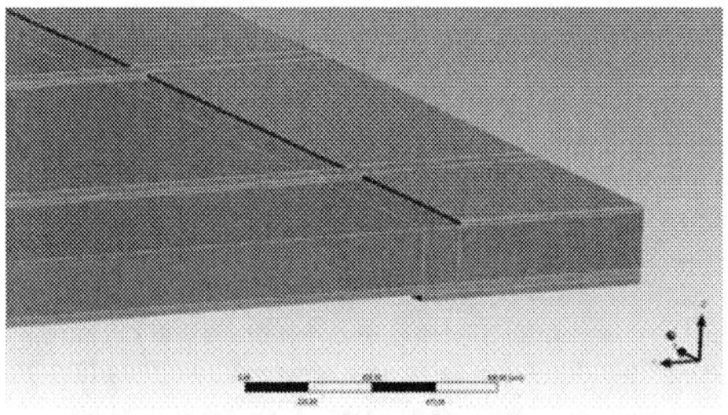

Figure 1: Section of the solar cell geometry

Simulations of the stress distribution on the solar cell surface indicated a peak of the equivalent (von Mises)

stress of around 30 MPa at the end of the front, middle busbar pad (see Figure 2).

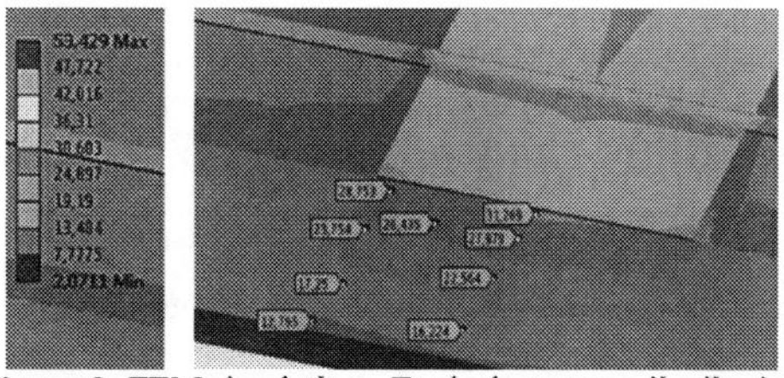

Figure 2: FEM simulation - Equivalent stress distribution in MPa

3.2 Reference Measurements

For determining the initial residual stress of the solar cell samples, a Renishaw inVia Raman Microscope RE04, equipped with a 633 nm / 18 mW excitation laser and a 20-fold microscope objective was used. All subsequent measurements, regarding stress condition monitoring, were performed using the same system and settings. Spectral acquisition was performed in "map image acquisition" mode with 1 s signal integration time and 60 co-added spectra (60 s total measurement time per point). The dataset for each sample was created by applying a 15 times 4, i.e. 60-point mapping with a grid dimension of 100 µm in x and y direction (see Figure 3, left).

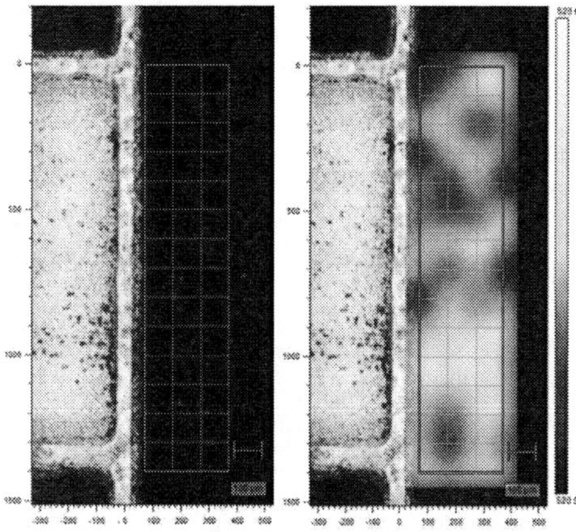

Figure 3: Raman spectroscopy – (left) Mapping at the end of the front middle BB; (right) results of the silicon peak position determination

Figure 3 (right) exemplarily displays a visualization of the measured peak position. For comparing the residual stress of the solar cells, the mean value of the respective peak position mapping (end of the front middle BB) was calculated utilizing the relationship of stress and peak shift (measurement accuracy: ± 20 MPa) [2,3]. The following plots display an overview of the average initial residual stress of a set of multi- (Mu) as well as mono-crystalline (Mo) solar cells equipped with continuous (-c) and discontinuous (-d) BB pads (see Figure 5). Regarding the solar cells based on multi-crystalline base material, cells with discontinuous BB pads show on average twice as much residual stress compared to the ones with continuous BB pads (plot: green and red line). For solar cells based on mono-crystalline base material this increase cannot be confirmed. Here both pad configurations resulted in similar mean values. Nevertheless the overall divergence in the measured stress values is larger when compared to the multi-crystalline solar cells. Additionally it should be mentioned that almost all cells show positive stress values, indicating tensile strain. Negative values of the mean stress values would indicate compressive stress.

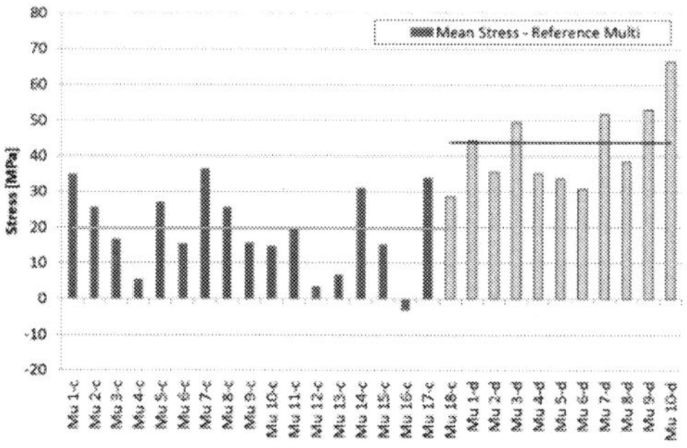

Figure 4: Initial residual stress – multi-crystalline solar cells

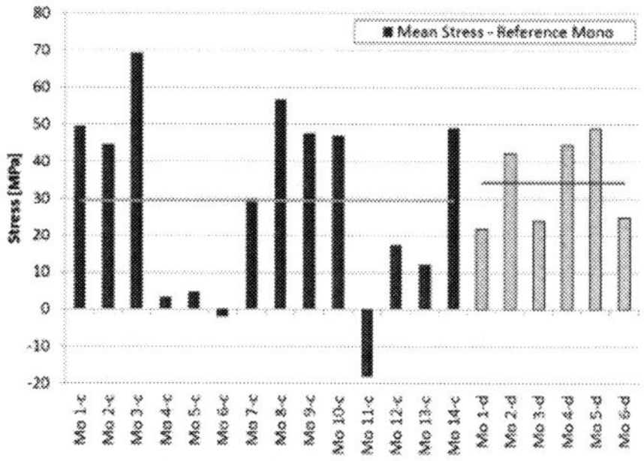

Figure 5: Initial residual stress – mono-crystalline solar cells

3.2 Measurements after soldering procedure

After the initial stress condition of the solar cells was determined, the cells were interconnected using a standard photovoltaic stringing and soldering system (teamtechnik). For this, mono- and multi-crystalline cells, equipped with continuous as well as discontinuous BB pads, were interconnected. Figure 6 displays a comparison of the stress condition before and after the soldering procedure for the multi-crystalline cells.

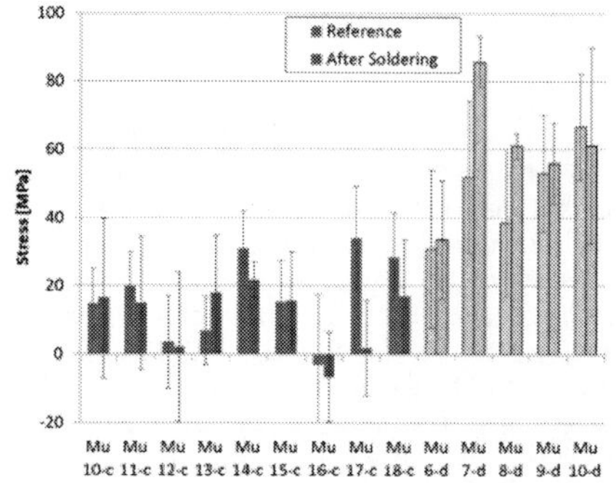

Figure 6: Comparison – Before/after soldering multi-crystalline cells

Additionally, Figure 6 shows a comparison of the stress condition before and after the soldering procedure for the mono-crystalline cells.

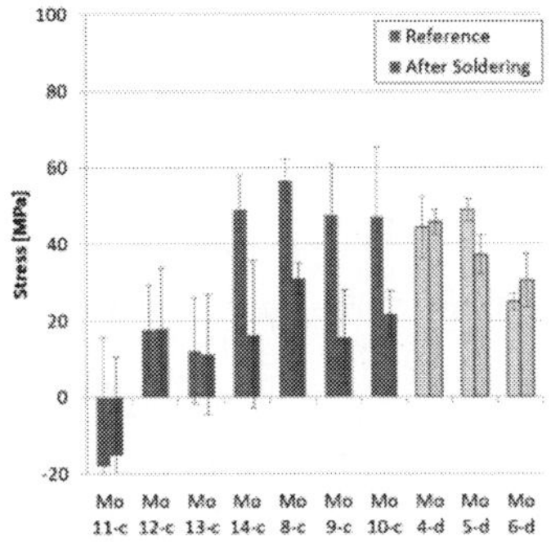

Figure 7: Comparison – Before/after soldering mono-crystalline cells

3.3 Measurements after glueing procedure

In addition to the stress determination of the solar cells after the standard soldering process, the effect on the cell-stress of an alternative procedure has been investigated. Hence, the cells were interconnected using an electrically conductive adhesive (ECA) based on silver (Polytec PT). In opposition to the fully automated soldering process, the glueing or bonding procedure was conducted manually. The BB ribbons are glued simultaneously onto the flattened cell, followed by a curing of the ECA on a heating plate at 190°C. Due to residues of the ECA in the measurement area (see Figure 8), it was not possible to get meaningful results from some of the samples (marked with a red X in the plots).

Figure 8: Residues of the ECA at the end of the middle BB pad after the glueing procedure; 20-fold objective

Figure 9 shows a comparison of the stress condition before and after the glueing procedure for the multi-crystalline cells, Figure 10 displays the results for the mono-crystalline cells.

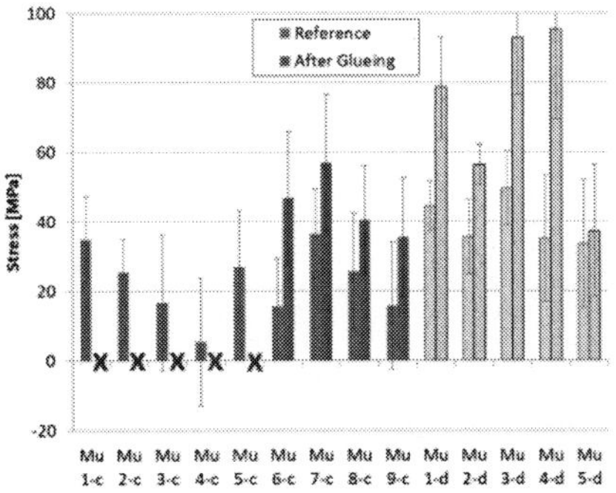

Figure 9: Comparison – Before/after glueing multi-crystalline cells

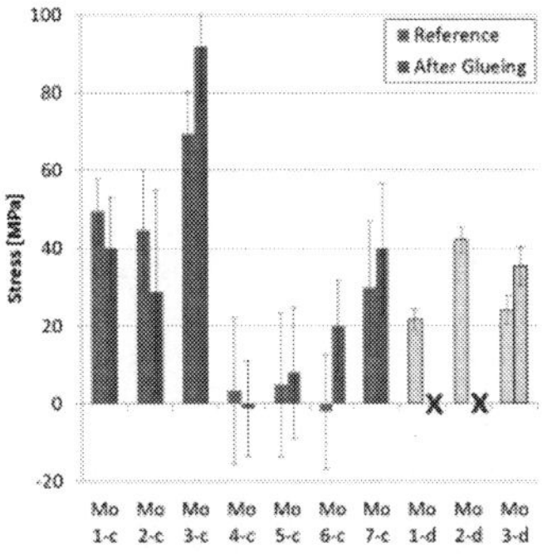

Figure 10: Comparison – Before/after glueing mono-crystalline cells

3.4 Manual soldering-tests

For determining the effect of soldering front and back BB ribbons separately onto the solar cell and furthermore compare a manual soldering procedure to the industrial one, the following tests were performed. For this, the residual stress of three mono-crystalline solar cells was determined. Afterwards (i) one, (ii) three and (iii) six BB ribbons were soldered manually onto the cells and measured again. In addition the first cell, which is only attached with one BB ribbon, was enhanced in a step-by-step soldering process, where first two additional BB ribbons were soldered on the front and second three more on the backside of the cell (see Figure 11, left). Between every step, the stress was measured to determine the effect of additionally added ribbons on the stress level of the cell (see Figure 11, right).

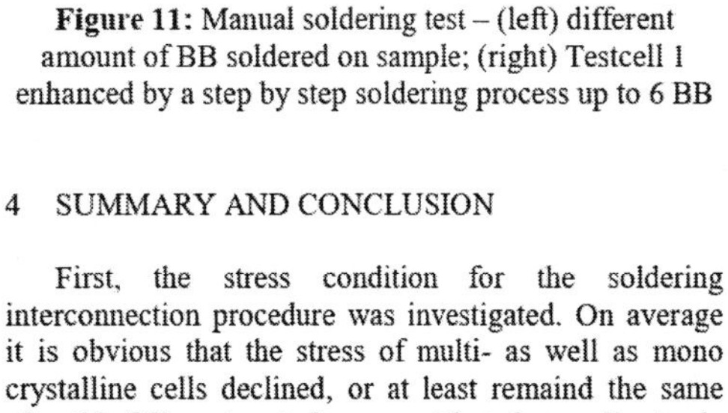

Figure 11: Manual soldering test – (left) different amount of BB soldered on sample; (right) Testcell 1 enhanced by a step by step soldering process up to 6 BB

4 SUMMARY AND CONCLUSION

First, the stress condition for the soldering interconnection procedure was investigated. On average it is obvious that the stress of multi- as well as mono crystalline cells declined, or at least remaind the same after this fully automated process. That also applies to six out of eight cells, equipped with discontinuous busbar pads. Just two of them showed a significant increase of the tensile strain, which could be resulting from differences in the random composition of the silicon crystal structure (multi-crystalline base material). In terms of the process flow, the cell is being pre-heated, pressed flat, and afterwards the solder on the BB ribbons is melted almost simultaneously. It is expected, that this process balances the internal stress condition of the cells.

Considering the glueing procedure, some of the cells were not able to be measured after the interconnection process, because of disruptive ECA residues. Hence, half of the multi-crystalline samples as well as two mono-crystalline cells could not be measured correctly according to these residues. Nevertheless it seems obvious that the stress increased slightly after the glueing process. For cells, which are equipped with a discontinuous BB pad, the stress is tending to increase even more, whereas the initial stress was already higher compared to the cells with continuous pads. For the mono-crystalline cells the results are similar, indicating a slight increase of the tensile strain. The cells "Mono 1-c" and "Mono 2-c" also showed a small amount of residues within the measurement area, which could lead to a minor falsification of the results. At this point it must be mentioned that this manually conducted glueing procedure cannot be compared directly to the fully automated soldering process. Here, the major difference lies in the fact, that the cell is put "cold" onto the heating plate, whereas in the stringing process pre-heating is applied. It is assumed, that the high temperature difference could cause a certain additional thermo-mechanical stress for the cell.

The additional manual soldering-test, where busbars were soldered onto a testcell in a step-by-step process, resulted in an overall increase of the stress. In contrast, the fully automated soldering procedure resulted in a stress reduction. This could be explained by the fact, that the front and back BB were not soldered simultaneously onto the solar cell. Hence, by first soldering the front BB onto the cell, the wafer is being bended and therefore stressed, induced by the shrinking of the BB ribbon after the cooling process. Since there is no BB soldered on the backside, there is no opposing force, which would counteract the bending of the wafer.

Nevertheless, neither cells interconnected by soldering, nor by glueing showed critical residual stress values. Here it should be mentioned, that the critical fracture stress for silicon wafer material is varying strongly. Critical stress values for silicon solar cells starting around 160 MPa and are going up to 380 MPa, strongly depending the material thickness and test method for characterizing this values [4].

5 ACKNOWLEDGEMENT

This work was conducted as part of the Austrian "Energy Research Program" project INFINITY, using results from the predecessor project InnoModu. Funded by the Austrian Climate and Energy Fund and the Austrian Research Promotion Agency (FFG), both are gratefully acknowledged.

6 REFERENCES

[1] A.M. Gabor, M. Ralli, S. Montminy, L. Alegria, C. Bordonaro, J. Woods, L. Felton; "Soldering induced damage to thin Si solar cells and detection of cracked cells in modules"; 21st EU PVSEC, 2006

[2] J. Chen and I. De Wolf; "Theoretical and Experimental Raman Spectroscopy Study of Mechanical Stress Induced by Electronic Packaging"; IEEE Transactions on Components and, 28:484–492, 2005

[3] I. De Wolf; "Micro-Raman spectroscopy to study local mechanical stress in silicon integrated circuits"; Semiconductor Science and Technology, 11:139–154, 1996

[4] S. Schoenfelder, A. Bohne, J. Bagdahn; „Comparison of test methods for strength characterization of thin solar wafer"; 37th IEEE Photovoltaic Specialists Conference IEEE, Seattle, USA, 1641 – 1646, 2011

AIR COOLING OF PHOTOVOLTAIC PANELS: A NUMERICAL APPROACH

Laura Martin-Carron*(1), Roland Becker(2), Didier Graebling(2), Robert Luce(2), Daniel Ugarte(1), André Macq(1) and Nicolás Cristi(1).
(1) SUNiBRAIN S.A.S.,
52 Blv. Gabriel Koenigs, F-31300 Toulouse, France
(2) Laboratoire de Mathématiques Appliquées
UMR CNRS n°5142, Université de Pau, France
(*) corresponding author e-mail : laura.martin@sunibrain.com

ABSTRACT: It is well known that high temperatures affect in a negative way the performance of PV Silicon modules. There are several ways of decreasing the modules temperature by external cooling, therefore increasing its performance: water and air cooling. Although we are interested in studying the way that the PV module transfers heat to the external cool water while it approaches its optimum working temperature, this work is a first step towards developing the digital tools to study water cooling of PV roof installations. The model, which is a fine modeling, studies the cooling of a PV panel by air in two different cases of convection: natural cooling in a cool environment and forced cooling with wind. The numerical simulations show the efficient heat transfer between the PV panel (modeled by a glass plate) and ambient air. We use the Navier-Stokes equations coupled with the heat equation, using the Boussinesq approximation to take into account the thermal variation of density. This allows us to take into account the varying parameters as temperature, wind speed and inclination of panel (gravity). The simulations show that the most efficient regime for free convection is when the module is not perpendicular to gravity direction. The efficiency increases with the wind speed. The most efficient case is obtained when the wind is parallel to the panel.
Keywords: Modelling, Heat transfer, Cooling, Silicon modules, Photovoltaic Panels

1 INTRODUCTION

The performance of a PV system depends on several parameters including temperature. It is well known, that the part of absorbed solar radiation that is not converted into the electricity converts into thermal energy and causes a decrease in electrical efficiency. This efficiency can be improved if panel temperature is reduced by cooling (i.e. Si-based PV modules can increase their power output by ~0.45%/K) [1-3]. This problem has been deeply studied in past years developing different cooling approaches to reduce the modules temperature to an optimal performance value. We can find, as examples, several previous studies where air or water was used as the cooling media [4-8].

This work is in the framework of a bigger project that aims to model and study heat transfer between water and Si-based PV modules, for the further optimizations in systems that use water cooling to increase performance of PV roof installations [9,10]. The work is a first step to developing the numerical tools to study the way that the PV module transfers heat to the external cool water while it approaches its optimum working temperature.

2 MODEL

The work involves two different axes. The first one is a conventional thermal study to provide orders of magnitude of various phenomena and to show the importance of the convection transfer coefficient in panel cooling. We explain the limits of this approach, making the numerical simulation essential for these physical systems. That is why the second axis bases the simulation on partial differential equations to describe the heat exchange between glass and air.

The global thermal analysis shows that the cooling of the panel is governed by convection. We perform numerical simulations based on the Navier-Stokes equations for velocities and pressure in the air. The thermal conduction in the panel and the thermal convection in the air are described by coupled heat equations. On the interface, temperature and heat flux are continuous.

The convection is due to the dilatation of the fluid associated with gravitation. The variation of density with temperature is taken into account by the Boussinesq approximation.

The overall coupled system is discretized by a stabilized finite element method.

The panel has been modelled by a glass and the interaction with air was modeled by partial differential equations justified by its physics (Fig.1). Several parameters as temperature, wind speed and inclination of panel (gravity) have been varied to study two different cases of convection: natural cooling in a cool environment (free convection) and forced cooling with wind (forced convection) [11].

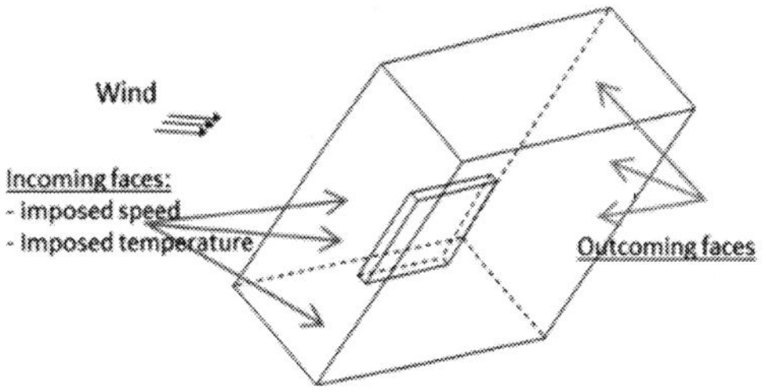

Figure 1: Computational domain

3 RESULTS

The results obtained for natural cooling (free convection) gives us the following values for dimensionless parameters, for a reference wind speed of $v_{\mathcal{L}}$=1 m.s^{-1}

$$\mathcal{R}_e = 6\text{x}10^3,\ Pr = 6\text{x}10^4 \text{ and } Gr = 5\text{x}10^9$$

The value of the Reynolds number shows that this flow is low turbulent. As the Grashof number is very large, the force of gravity is predominant over the viscous forces, which leads to a strong convection, indicating a thermal turbulence. Moreover, the Peclet number confirms that heat exchange occurs essentially by convection.

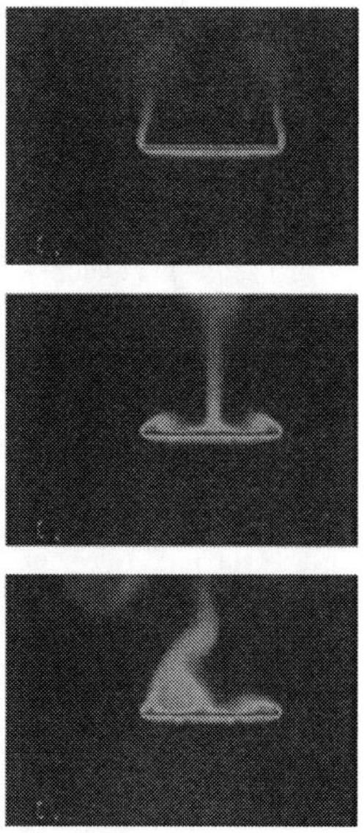

Figure 2: Free Convection. Heat transfer after 10, 50 and 100 s (panel horizontal).

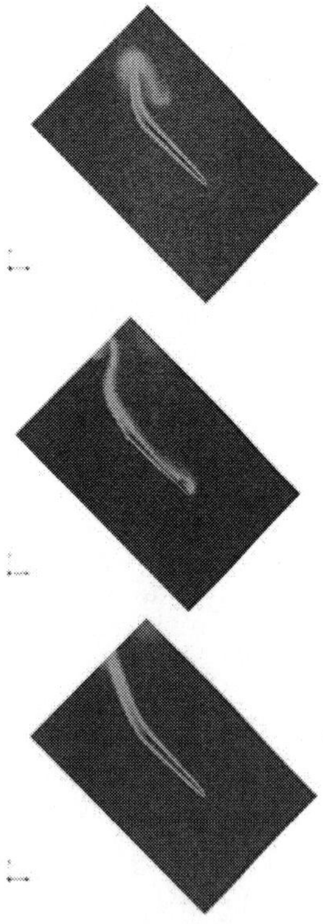

Figure 3: Free Convection. Heat transfer after 10, 50 and 100 s (panel oblique at 45°).

Figures 2 and 3 show the temperature of the free convection system for different orientations of the panel: horizontal and oblique at 45°. The heat transfer of the panel is shown after 10, 50 and 100 s. Figure 4 shows the vertical orientation, for free convection. In this case, the heat transfer of panel is shown after 5, 10 and 50s. When the module is horizontal, the formation of heat plumes is chaotic and the low speed allows air ripple effect. When the panel is no longer perpendicular to the direction of gravity, it appears a convection current that evacuates the heat through the stack effect. In these conditions, heat exchanges are more effective.

For wind cooling (forced convection), the following results are obtained for the dimensionless numbers:

- v_∞=1 m.s^{-1}: $\mathfrak{R}_e = 6\text{x}10^3$, $Pr = 6\text{x}10^4$

- v_∞=5 m.s^{-1}: $\mathfrak{R}_e = 30\text{x}10^3$, $Pr = 30\text{x}10^4$

When the speed of wind is 1 m.s^{-1} the thermal exchanges are similar to those of free convection. In the case of a wind speed of 5 m.s^{-1} the Reynolds and Peclet numbers are increased indicating a better efficiency in the convective heat transfer.

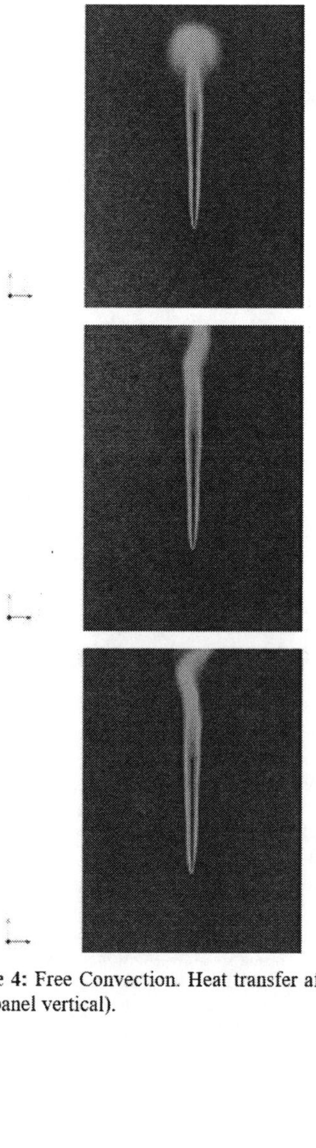

Figure 4: Free Convection. Heat transfer after 5, 10 and 50 s (panel vertical).

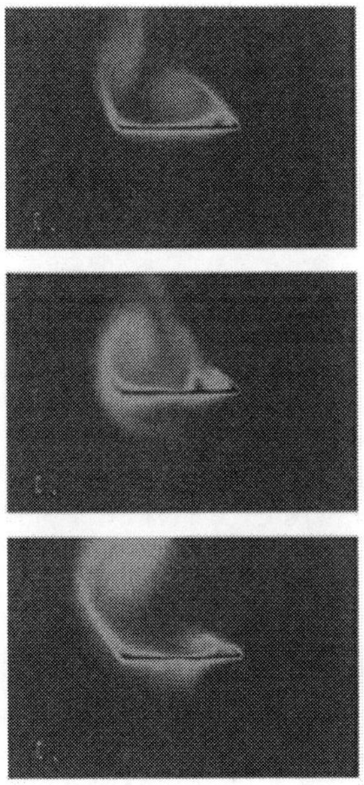

Figure 5: Forced Convection. Heat transfer after 5, 10 and 80 s for horizontal wind speed 1ms-1 (top), and 5ms-1 (bottom)

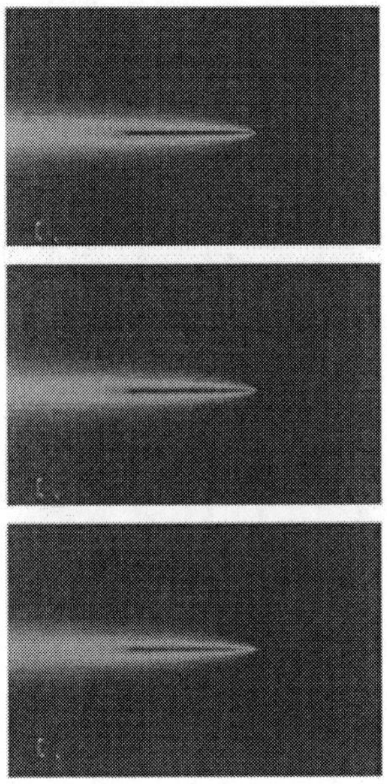

Figure 6: Forced Convection. Heat transfer after 5, 10 and 80 s for horizontal wind speed 1ms-1 (top), and 5ms-1 (bottom)

Figure 5 shows the glass panel placed horizontally with the wind direction parallel to it (from right to left) and a speed of 1 m.s^{-1}. The calculated speed v_L associated with free convection is the same magnitude as that of the wind v∞ imposed. The thermal regime is called joint, that is to say intermediate between free convection and forced convection. Simulation results perfectly match with that behavior. We find that the heat plumes are partially blown by the wind.

On the contrary, with a wind speed of 5 m.s-1 (Fig. 6), we observe no formation of heat plumes, which corresponds to a forced convection regime.

In Figure 7 the wind direction is at 45o compared to panel's one, with a speed of 5 m.s-1. For a sufficiently long time, we find a close scenario to that of the forced convection configuration. It should be noted that heat is dissipated only from the upper face under the wind. On the underside face, the hot air cannot easily drain; heat accumulates which severely limits the heat exchange between the panel and the air. For a shorter time, there is an intermediate regime where we can see a few heat plumes appear.

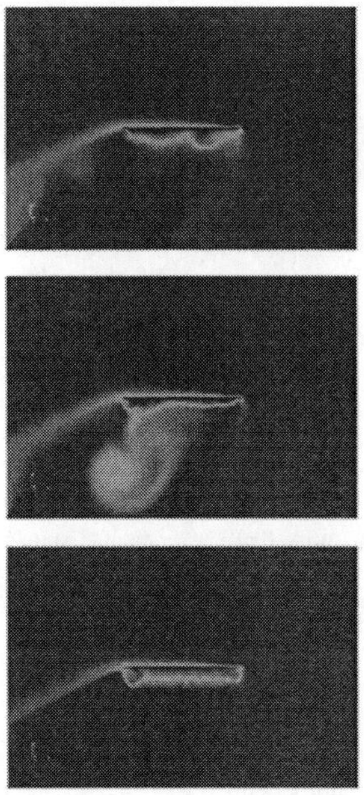

Figure 7: Forced Convection with wind speed of 5ms-1 at a direction of 45o. Heat transfer after 5, 10 and 100 s.

4 CONCLUSIONS

We have studied the cooling (heat transfer) of a PV panel by air. The panel has been modelled by a glass and the interaction with air was modeled by partial differential equations justified by its physics. Several parameters as temperature, wind speed and inclination of panel (gravity) have been varied to study two different cases of convection: free convection and forced convection.

The numerical simulations show that the most efficient regime for free convection is when the module is not perpendicular to gravity direction. The efficiency increases with the wind speed. The most efficient case is obtained when the wind is parallel to the panel.

This first numerical approach yields to results consistent with the physical approach. Our next steps in 2016 will be applying the model for water cooling and linking the model to sunlight irradiation.

5 REFERENCES

[1] Skoplaki, E. and Palyvos, J. A., "On the temperature dependence of photovoltaic module electrical performance: A review of efficiency/power correlations ", Solar Energy, Vol.83, pp. 614-626, 2009

[2] Radziemska, E. and Klugmann, E. "Thermally affected parameters of the current-voltage characteristics of silicon photocell", Energy Conversion and Management, Vol.43, p.p. 1889-1900, 2002

[3] Njomo, Donatien and Daguenet, Michel, "Sensitivity analysis of thermal performances of flat plate solar air heaters", Heat Mass Transfer, Vol. 42, pp. 1065-1081, 2006

[4] Odeh, S., Behnia, M.," Improving photovoltaic module efficiency using water cooling", Heat Transfer Eng., Vol.30 (6), pp. 499–505, 2009

[5] Krauter, S. "Increased electrical yield via water flow over the front of photovoltaic panels." Solar Energy Materials & Solar Cells, Vol. 82, pp. 131–137, 2004

[6] Kristijan Brecl, Marko Topič "Performance of PV Modules with regard to the mounting situation", 29th European Photovoltaic Solar Energy Conference and Exhibition

[7] Valeh-E-Sheyda P, Rahimi M, Parsamodghadam A, Masahi MM. "Using a wind driven ventilator to enhance a photovoltaic cell power generation", Energy Build, Vol.73, pp.115–9. 2014

[8] Y. Tripanagnostopoulos1, P. Themelis1 and J.K. Tonui2 "Natural flow building integrated Air-cooled Photovoltaics", 23rd European Photovoltaic Solar Energy Conference and Exhibition

[9] Nicolas Cristi, André Macq and Laura Martin-Carron, Evaporating pure rainwater to increase the yield of commercial-size pv arrays, Proceedings 32nd European Photovoltaic Solar Energy Conference and Exhibition (to be published), 2016

[10] André Macq, Loïc Mercier des Rochettes, Laura Martin-Carron, Nicolas Cristi, Marie-Pierre Gleizes, Carole Bernon and Pierre Glize, "A use of artificial intelligence for improving pv array performance (empirical approach)", Proceedings 32nd European Photovoltaic Solar Energy Conference and Exhibition (to be published), 2016

[11] Do, Kyu Hyung and Kim, Tae Hoon and Han, Yong-Shik and Choi, Byung-Il and Kim, Myung-Bae, "General correlation of a natural convective heat sink with plate-fins for high concentrating photovoltaic module cooling", Solar Energy, Vol.86, pp. 2725-2734, 2012

THE EFFECT OF PHOSPHORUS GETTERING ON FINE-GRAINED MULTICRYSTALLINE SILICON

K. E. Ekstrøm[1*], R. Søndenå[2], A. Autruffe[1], G. Stokkan[3], L. Arnberg[1], M. Di Sabatino[1]

[1]*Department of Materials Science and Engineering, NTNU, N-7491 Trondheim, Norway*

[2]*Institute for Energy Technology, N-2007 Kjeller, Norway*

[3]*SINTEF Materials and Chemistry, N-7465 Trondheim, Norway*

* Corresponding author. Tel.: +47 73 55 12 19. E-mail address: kai.e.ekstrom@ntnu.no

An analysis of the minority carrier lifetime of individual non-dislocated grains before and after phosphorus diffusion gettering suggests that smaller grains respond less to the gettering process than larger grains. The response is highest in the top and bottom positions of the ingot, since these positions are heavily polluted by the ingot red-zone. However, independently of position, there is a clear trend in the response decreasing approximately linearly with the smallest dimension of the grains. This can be related to increased influence from recombination active grain boundaries, which appear to affect the intragranular response up to 2mm from the boundary.

Keywords: Multicrystalline Silicon, Gettering, Lifetime, Defects.

1. INTRODUCTION

Fine-grained multicrystalline silicon, achieved through seeding and/or new temperature regimes, are generally found to be less prone to dislocation generation than conventional multicrystalline silicon [1]. The potential efficiency gain and relatively inexpensive modifications to existing infrastructure have therefore resulted in this material becoming a new standard for production of multicrystalline silicon [2]. As previously shown by Stokkan et al [3], dislocation clusters tend to terminate at random angle grain boundaries. A high density of these boundaries therefore effectively limits dislocation clusters from dominating the material to the same extent as in conventional multicrystalline silicon. A recent study by Yang et al [4] also stated that small grains and large fractions of non-coherent grain boundaries may be beneficial for stress relaxation.

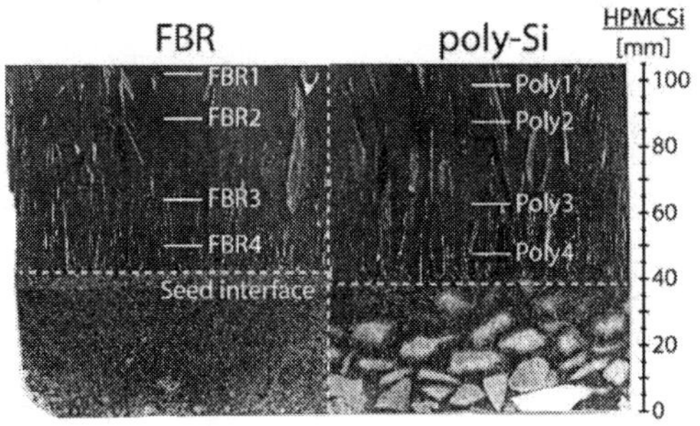

Figure 1. A vertical cut showing the two compartments of the ingot and the positions used for characterization. The left compartment was seeded by fluidized bed reactor (FBR) granules and the right compartment was seeded by polysilicon chips (poly-Si).

As-grown multicrystalline silicon contains various amounts of metallic impurities that can be very detrimental to final solar cell efficiency [5]. In their dissolved state, impurities can be removed from the material by phosphorus diffusion gettering (PDG). However, defects, such as dislocations and grain boundaries, tend to internally getter these impurities and trigger precipitation [6]. Precipitates respond poorly to phosphorus gettering [7] and will remain in the material, and increase the overall recombination activity [8, 9]. However, in materials with low dislocation densities, the detrimental effect of grain boundaries becomes more pronounced [10]. The high density of detrimental grain boundaries will therefore still reduce the potential for further increase in minority carrier lifetimes, especially for fine-grained materials, and further enhancements will therefore depend on better impurity control; e.g. by mitigating impurity sources and/or developing new and improved gettering techniques.

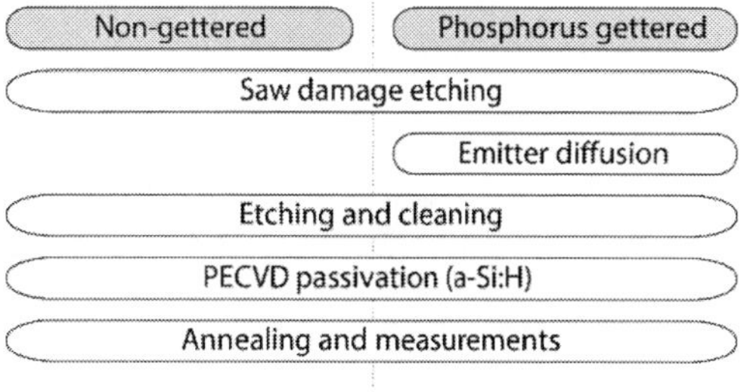

Figure 2. Wafer preparation prior to photoluminescence imaging. In order to obtain comparable surface properties, both non-gettered and PDG wafers were etched together.

The present work investigates the effect of grain size on the PDG efficacy. Photoluminescence imaging has been used to record the lifetime images from multiple positions in a seeded pilot-scale HPMC Si ingot, and to assess the materials response to PDG.

2. EXPERIMENTAL DETAILS

A seeded 12 kg pilot-scale ingot (G1, Ø=250mm) has previously been solidified in a Crystalox DS 250 directional solidification furnace [1]. The ingot was seeded by fluidized bed reactor granules and polysilicon chips in separate compartments, as clearly visible from the vertical cut in Figure 1. The ingot was wafered into 50x50mm wafers, and wafers from four different heights were selected for minority carrier lifetime analysis, i.e. positions FBR1-4 and Poly1-4 in Figure 1. Two neighbouring sister-wafers were selected from each position, and prepared according to the process scheme in Figure 2. In order to study the effect of phosphorus gettering, one of each sister-wafers went through an additional PDG step before doing any minority carrier lifetime measurements. Minority carrier lifetime maps were recorded for all wafers by photoluminescence (PL), and calibrated by Quasi Steady State Photoconductivity (QSSPC) at an injection level corresponding to 1 sun.

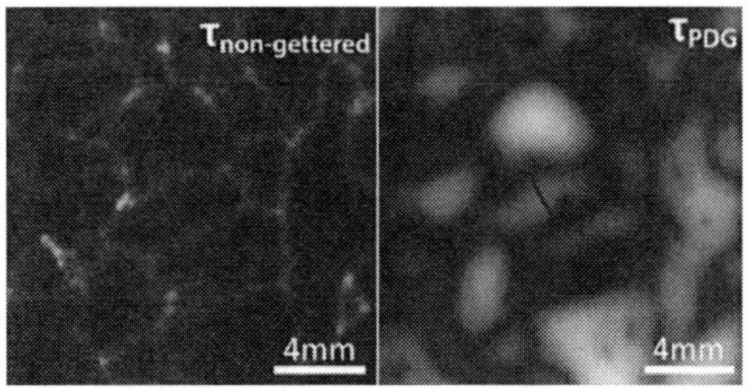

Figure 3. Identical minority carrier lifetime linescans in the smallest grain dimension, of both non-gettered and PDG state, serve as the basis for calculating averaged RPG values for individual grains.

In order to study the effect of grain size on PDG, a response to phosphorus gettering (RPG) parameter was defined as τ_{PDG} divided by $\tau_{non-gettered}$:

$$RPG = \frac{\tau_{PDG}}{\tau_{non-gettered}} \quad \text{(Eq. 1)}$$

The response is measured as a function of the smallest grain dimension, which can be visualized from Figure 3. Linescans of the grain dimension and the associated minority carrier lifetimes, before and after PDG, have all been measured from minority carrier lifetime maps by the ImageJ image processing software. Average minority carrier lifetimes have then calculated, from the linescans, as $\tau_{non-gettered}$ and τ_{PDG}, for non-gettered and phosphorus gettered wafers, respectively.

3. RESULTS AND DISCUSSION

The minority carrier lifetime images for non-gettered and PDG wafers are shown in Figure 4 (a) and (b), respectively. The average lifetimes are generally below 30µs for non-gettered wafers, but increases significantly with PDG. The microstructural differences between the compartments have been discussed in a previous paper [1].

The RPG has been calculated and plotted, as a function of grain area, for a large number of grains in Figure 5. The average RPG value is found to be one-to-two orders of magnitude larger for the bottom Poly 4 and top Poly 1 positions than for the two middle positions. This is most likely an effect of impurity segregation and the wafers being heavily polluted by the ingot red-zone, resulting in high gettering responses. The response for the two middle positions is, in contrast, relatively low. The response for Poly 1 and Poly 4 are plotted in Figure 5 (a), while grains from the two middle positions, i.e. Poly 2 and Poly 3, are plotted in Figure 5 (b). The measurements fit to linear trend lines, all showing a slight decreasing trend for smaller grain dimension. The same trend can also be seen in Figure 5 (c), where all measurements have been normalized together in terms of the total response at the respective wafers. When not containing any dislocation clusters, larger grains tend to respond well to phosphorus gettering, attaining relatively high intragranular lifetimes.

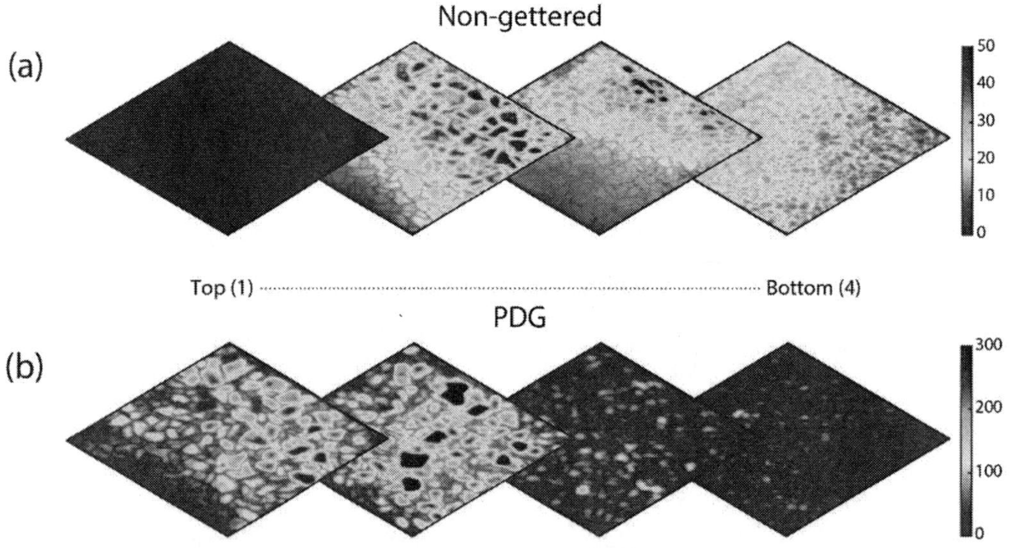

Figure 4. Minority carrier lifetime images, as recorded by photoluminescence, of neighbouring *(a)* non-gettered and *(b)* phosphorus gettered sister-wafers.

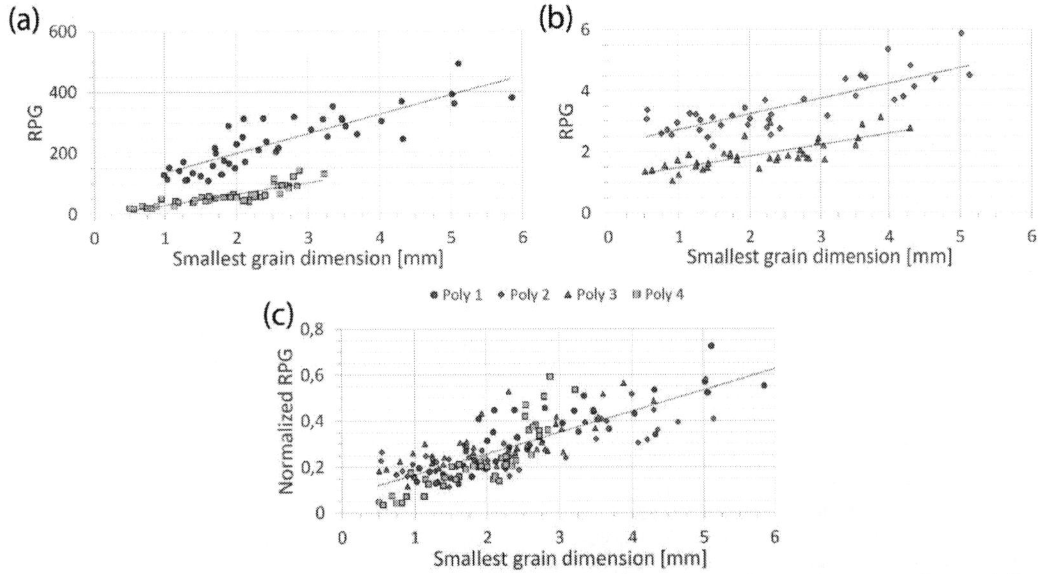

Figure 5. RPG values, as a function of the smallest grain dimension, for (a) Poly 1 in blue and Poly 4 in yellow, and (b) Poly 2 in orange and Poly 3 in green. (c) The same RPG values, but normalized into a common figure.

The intragranular minority carrier lifetime will in this case be largely determined by the concentration of intragranular point-defects, as even sparsely distributed dislocations may limit the maximum intragranular minority carrier lifetimes achievable [6, 10]. However, due to increased influence from surrounding grain boundaries, the response to phosphorus gettering weakens as the grains become smaller. A larger fraction of the generated minority carriers will in this case be generated close to recombination active grain boundaries, which increases the probability of carrier recombination. The minority carrier lifetime will in this case not be able to reach the value entitled by the concentration of intragranular point-defects.

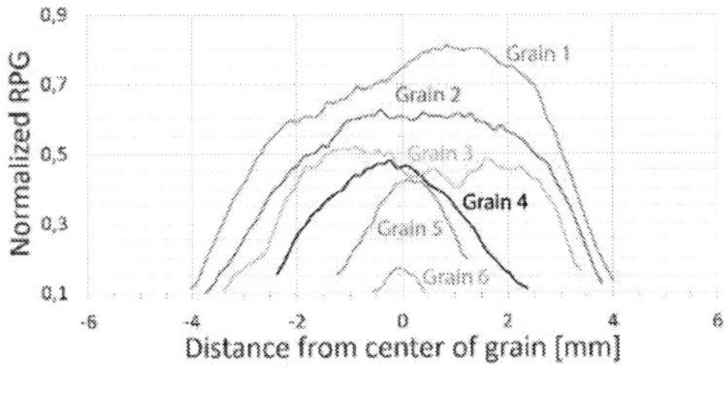

Figure 6. Linescan profiles of the normalized response in the smallest dimension of differently sized grains.

Linescans measured over the whole diameter of differently sized grains are shown in Figure 6**Error! Reference source not found.**; Grain 1 is the largest grain and Grain 6 is the smallest one. The central regions of large grains are not influenced by grain boundaries, and the central lifetime should reach a plateau value given by the concentration of intragranular point-defects. Grain boundaries will in this case only affect the periphery of the grains, and we typically see a sharp drop in lifetime close to grain boundaries. Indications on a plateau level can be found for Grain 1–3. The sharp drop in RPG close to the edges of these grains suggests that grain boundaries affect the intragranular response up to 2mm from the boundary. For the smaller Grain 4-6 this is manifested as bell-shaped linescan profiles instead, with no flat plateau level, suggesting that grains with dimensions below 4mm are severely affected by grain boundary recombination.

SUMMARY

This work suggests that grain boundaries still play a major role in limiting the minority carrier lifetime of smaller non-dislocation grains. As the grains become smaller, the influence from grain boundaries on the intragranular properties increases and the response to phosphorus gettering reduces. Grain boundaries appear to affect the intragranular response up to 2mm from the boundary, suggesting that grains with dimensions below 4mm are severely affected by grain boundary recombination.

The dependence upon small grains and a high density of recombination active grain boundaries will challenge the future potential of fine-grained multicrystalline material. Reducing the recombination activity at these grain boundaries will therefore be essential for realizing further lifetime improvements, either through advances in the gettering process or through better impurity control before and during the solidification process.

ACKNOWLEDGEMENT

This work has been funded through the Centre for Environment-Friendly Energy Research and the "Impurity Control in High Performance Multicrystalline Silicon" project, both financed by The Research Council of Norway and industrial partners in Norway.

REFERENCES

[1] Ekstrøm KE, Stokkan G, Autruffe A, Søndenå R, Dalaker H, Arnberg L, et al. Microstructure of multicrystalline silicon seeded by polysilicon chips and fluidized bed reactor granules. J Cryst Growth 441, 95 (2016).

[2] Lan CW, Lan WC, Lee TF, Yu A, Yang YM, Hsu WC, et al. Grain control in directional solidification of photovoltaic silicon. J Cryst Growth 360, 68 (2012).

[3] Stokkan G, Hu Y, Mjos O, Juel M. Study of evolution of dislocation clusters in high performance multicrystalline silicon. Sol Energ Mat Sol C 130, 679 (2014).

[4] Yang YM, Yu A, Hsu B, Hsu WC, Yang A, Lan CW. Development of high-performance multicrystalline silicon for photovoltaic industry. Prog Photovoltaics 23, 340 (2015).

[5] Green MA. Solar cells : operating principles, technology, and system applications. Englewood Cliffs, NJ: Prentice-Hall; 1982. xiv p.

[6] Autruffe A, M'hamdi M, Ekstrom KE, Schindler F, Schubert MC, Di Sabatino M, et al. Grain structure and iron precipitation in multicrystalline silicon. (To be published), (2016).

[7] Buonassisi T, Istratov AA, Peters S, Ballif C, Isenberg J, Riepe S, et al. Impact of metal silicide precipitate dissolution during rapid thermal processing of multicrystalline silicon solar cells. Appl Phys Lett 87, (2005).

[8] Kittler M, Seifert W, Knobloch K. Influence of contamination on the electrical activity of crystal defects in silicon. Microelectron Eng 66, 281 (2003).

[9] Fell TS, Wilshaw PR, Decoteau MD. Ebic Investigations of Dislocations and Their Interactions with Impurities in Silicon. Physica Status Solidi a-Applied Research 138, 695 (1993).

[10] Castellanos S, Ekstrom KE, Autruffe A, Jensen MA, Morishige A, Hofstetter J, et al. High-Performance and Traditional Multicrystalline Silicon: Comparing Gettering Responses and Lifetime-Limiting Defects. IEEE Journal of Photovoltaics (accepted), (2016).

ELECTRICAL CONDUCTION PROCESS BY IMPEDANCE SPECTROSCOPY OF THE NANOSTRUCTURED PHOTOABSORBERS Cu_2SnS_3

S. Lahlali[a], M. Belaqziz[a], L. Essaleh[b]*, H. Chehouani[a], K. Djessas[c] and J.L. Gauffier[d]
[a] Laboratoire Procédés, Métrologie, Matériaux pour l'Energie et Environnement (LP2M2E), University of Cadi Ayyad, Faculty of Sciences and Techniques, Department of Applied Physics, Marrakech, Morocco.
[b]*Laboratoire de la Matière Condensée et Nanostructure(LMCN), University of Cadi Ayyad. Faculty of Sciences and Techniques, Department of Applied Physics, Marrakech, Morocco.
[c] Laboratoire Procédés, Matériaux et Énergie Solaire (PROMES-CNRS), Université de Perpignan, Rambla de la thermodynamique, Tecnosud, 66100 Perpignan Cedex, France.
[d] Département de Génie Physique, INSA de Toulouse, 135 Avenue de Rangueil, 31077 Toulouse cedex 4, France.
sanaa.lahlali@gmail.com; mohamed.belaqziz@edu.uca.ma; lessaleh@gmail.com; l.essaleh@uca.ma
chehouani@fstg-marrakech.ac.ma; djessas@univ-perp.fr; E-mail: jean-luc.gauffier@insa-toulouse.fr

ABSTRACT: Conduction mechanism was investigated from DC and AC conductivity measurements in the bulk ternary semiconductor compound Cu_2SnS_3 that is prepared by the hydrothermal technique. The complex impedance spectroscopy in the frequency range 100Hz–1MHz and temperature range 300K–475K is used to study the electrical properties of this material. The activation energy for conduction and relaxation mechanisms is estimated. The temperature dependence of relaxation times for grain and grain boundaries contributions are analyzed. From the equivalent circuit that takes into account the distribution of relaxation times, electrical resistance for the grain and grain-boundary are determined and their temperature dependence is analyzed.
Keywords: Electrical conductivity; Impedance spectroscopy; Copper ternary compounds; hydrotermal synthesis.

1 INTRODUCTION

Many semiconducting Cu-Sn-S phases have been reported, such as Cu_4SnS_4, Cu_2SnS_3, $Cu_4Sn_7S_{16}$, $CuSn_{3.75}S_8$ and Cu_3SnS_4 [1]. Among these ternary sulfides, Cu_2SnS_3 and Cu_3SnS_4 are the most promising compounds, and these can be used in solar cells owing to their high absorption coefficient and their optimal direct band gap for solar energy conversion [2]. The ternary compound Cu_2SnS_3 has a structure similar to Cu_2ZnSnS_4. The band gap of the Cu_2SnS_3 has been reported to be between 0.93 and 1.35 eV [3], which lies in an optimal region for photovoltaic application. In addition to the great industrial application of the copper tin sulfides, these compounds are also very useful in understanding and extending the basic physical concepts of the electrical conduction mechanisms involved in elemental and binary compound semiconductors. Moreover, continuing our earlier works on bulk [4,5] and also on thin films [6,7] ternary compounds, we report on the electrical properties of the Cu_2SnS_3 (CTS for the next) compound by means of impedance spectroscopy using an alternating signal with voltage amplitude of 20 mV. We discuss the conduction mechanisms in this material. The data are analyzed using the existing theoretical models.

2 EXPERIMENTAL DETAILS

The CTS nanoparticles were prepared using the hydrothermal synthesis with high purity water as the solvent. Precursors ($CuCl_2.2H_2O$: Sigma Aldrich 99%), ($SnCl_2$: sigma Aldrich, 98%) and Na_2S ($Na_2S.9H_2O$: Sigma Aldrich 98%) were used as source materials. The formation of the single phase compound was confirmed by X-ray powder diffraction (XRD) technique. Impedance measurements were carried using the HP 4284A spectrometer in the frequency range of $10–10^6$Hz. A source of 20mV was applied to the electroded pellet samples. The temperature variation was performed using a programmable Thermolyne heater with a temperature stability of ±0.1°C. Silver electrodes were deposited on the two circular faces of the sample to get the capacitor shaped samples. Measurements of impedance spectroscopy are a useful method to analyze the contributions of grains, grain boundaries and electrodes in the material. Thus one can establish the corresponding equivalent circuit. The temperature dependence of electrical conductivity and relaxation times for each contribution can be studied and permits to determine the activation energies for the relaxation mechanisms.

3 RESULTS AND DISCUSSIONS

At relatively higher temperatures, the temperature dependence of the conductivity is dominated by the band-to-band electronic transitions. The charge carriers acquire enough thermal energy to make an inter-band transition.

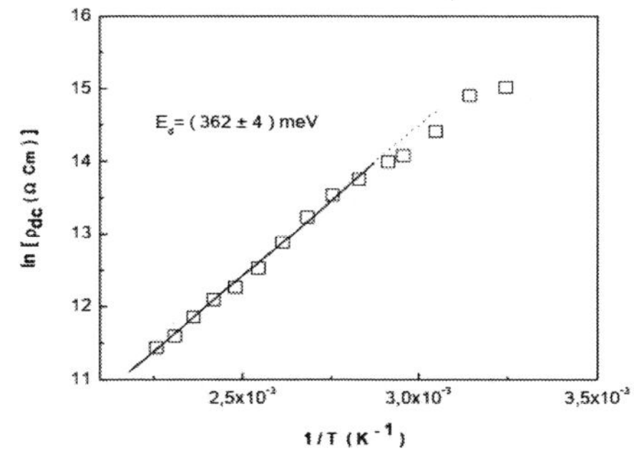

Figure 1: Variation of DC electrical resistivity as a function of 1/T for a representative sample of CTS.

The activation energy corresponding to the excitation of electrons from impurity to conduction band can be obtained in Figure 1 which represents the variation of DC electrical resistivity as a function of 1/T for a representative sample of CTS. The activation energy $E_d = (362 \pm 4)meV$ is estimated from the slope of $Ln\,(\rho)$ versus 1/T curve in agreement with values cited in literature but only in thin films [8,9].

As a function of frequency, gradual decrease in the activation energy from 360 to 213meV was observed with increase in frequency (Fig.2) from 100Hz to 1MHz, respectively.

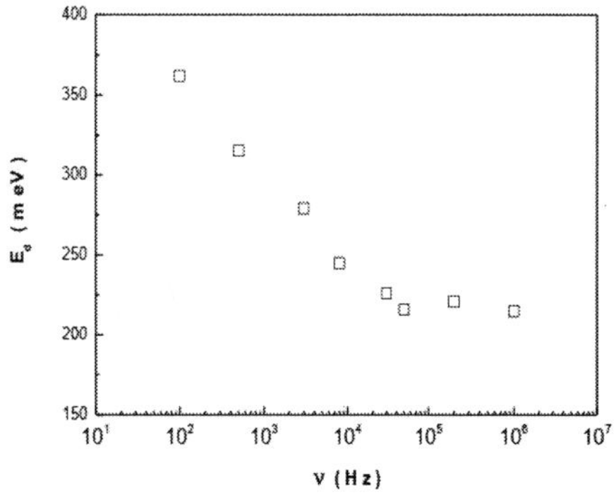

Figure 2: Frequency dependence of the activation energy for the conduction process.

Such a decrease can be attributed to the fact that the increase of the applied field enhances the charge carrier's jumps between the localized states. Accordingly, the activation energy decreases with increasing frequency [10].

In the high frequency domain, the angular frequency dependence of the AC electrical conductivity is of the form $\sigma_{ac}(\omega) = A\omega^s$ [11] where A and s are temperature-dependent. The exponent s lies usually between 0 and 1. As observed from Fig. 3, in the high frequency range, the slopes of the straight lines were found to decrease slightly when increasing temperature.

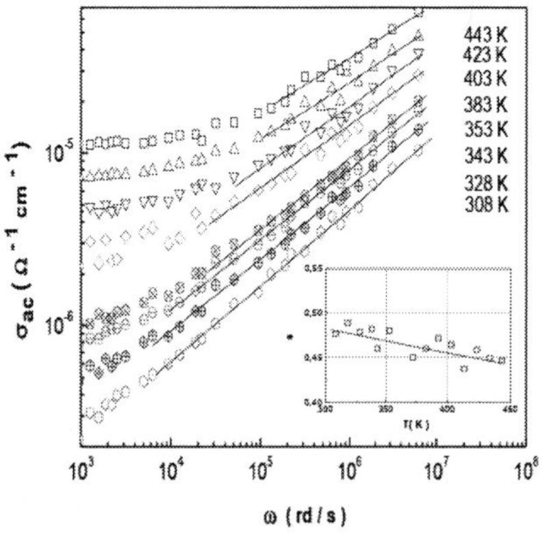

Figure 3: Angular frequency dependence of AC electrical conductivity, the straight continuous lines represents the best linear fits. The insert: The exponent s versus T.

The frequency exponent, s is obtained by least squares straight-line fit of the experimental data of Fig.3 and is plotted as a function of temperature also in Fig. 3 (the insert). It is found that s decreases from 0.492 to 0.427 with increasing temperature from 308K to 423K. This behavior was associated with a correlated barrier hopping (CBH) model for AC conductivity [12].

In order to analyze the contributions of bulk (or grain) and grain-boundary to the total electrical conductivity, we represents in fig.4 the evolution of imaginary part Z" as a function of real part Z' of the complex impedance Z* at some representative temperatures from 55°C to 170°C. An increase of temperature is accompanied by a decrease in resistance. For this purpose, equivalent circuit that takes into account the distribution of relaxation times for both two contributions is proposed. This permits to determine electrical resistances R_g and R_{gb} for these contributions and analyze their temperature dependence. This is shown in Fig. 5 in addition to the curve related to DC electrical resistivity.

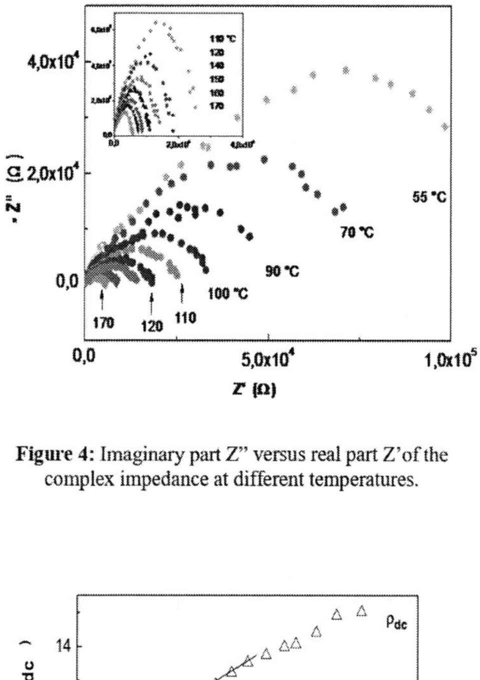

Figure 4: Imaginary part Z" versus real part Z'of the complex impedance at different temperatures.

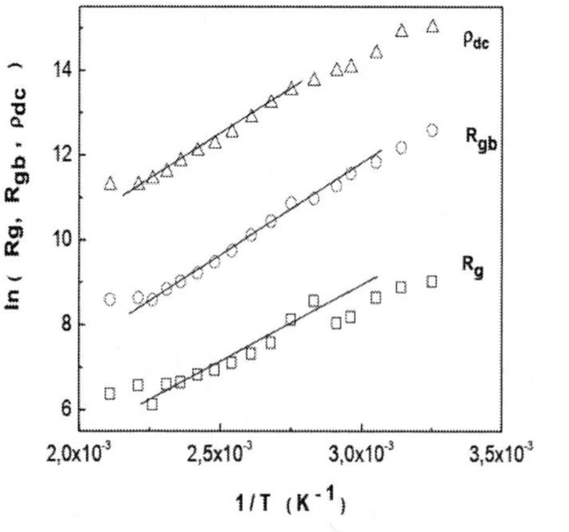

Figure 5: Variations of R_g, R_{gb} and DC electrical resistivity with temperature.

The straight lines in Fig. 5 indicate the best linear fits of the data with the Arrhenius equation $R = R_o \exp\left(\frac{E_a}{k_B T}\right)$. These lines are practically parallel indicating the same activation energy E_a of the order of 362 meV.

The relaxation times, represented in Fig.6, of different contributions (grain (τ_g) and grain boundaries (τ_{gb})) are obtained from the analysis the temperature and frequency dependence of measured complex impedance for this material. For a thermally activated relaxation process, the relaxation time (τ) generally follows the Arrhenius law $\tau_{relax} = \tau_o \exp\left(\frac{E_{relax}}{k_B T}\right)$ where τo is the relaxation time at infinite temperature, T is the absolute temperature, E_{relax} is the activation energy for relaxation, and k_B is the Boltzmann constant. The values of activation energy for relaxation for bulk and grain boundary are estimated for this material and are indicated in Fig 6.

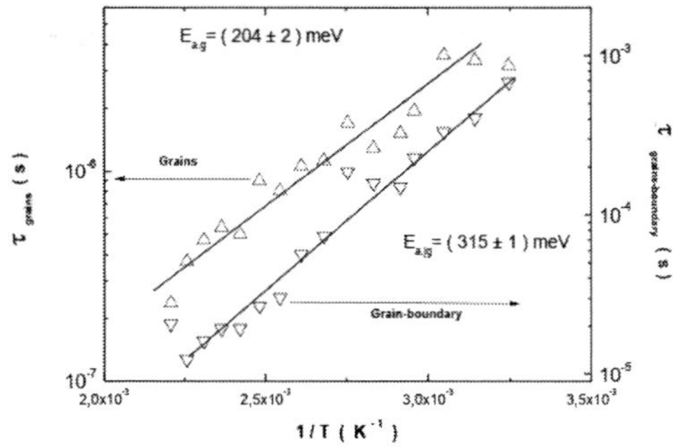

Figure 6: Variation of a relaxation time for a bulk and grain boundary with inverse of temperature, the straight continuous lines represents the agreement with the Arrhenius law

4 CONCLUSION

In conclusion to this work, the DC conductivity measurements of the bulk Cu_2SnS_3 compound outline a linearity behavior in semi-logarithmic scale depending on 1/T and indicating activation type of the charge carrier transport mechanism. Activation energy for relaxation mechanisms are estimated from the analysis of temperature and frequency dependence of complex impedance. The temperature dependence of electrical resistance for the grain and grain-boundary is analyzed. The corresponding activation energy is estimated to be of the order of 362 meV in agreement with that obtained from DC electrical resistivity.

5 REFERENCE

[1] Fernandes PA, Salome PMP, da Cunha AF, Journal of Physics D: Applied Physics, 43(2010) 215403.
[2] Bouaziz M, Amlouk M, Belgacem S, Thin Solid Films, 517(2009)2527.
[3] Dominik M. Berg, Rabie Djemour, Levent Gütay, Guillaume Zoppi, Susanne Siebentritt, Phillip J. Dale, Thin Solid Films, 520(2012)6291.
[4] Marín G, Wasim SM, Rincón C, Essaleh L, Materials Letters, 157(2015)70.
[5] Essaleh L, Wasim SM, Materials Letters, 61(2007)2491.
[6]Abounachit O, Chehouani H, Djessas K, Thin Solid Films, 540(2013) 58.
[7] Abounachit O, Chehouani H, Djessas K, Thin Solid Films, 520(2012) 4841.
[8] Bouaziz M, Ouerfelli J, Srivastava S, Bernède J, M. Amlouk, Vacuum 85 (2011)783.
[9] Kahraman S, Çetinkaya S, Çakmak HM, Çetinkara HA, Güder HS, International Journal of Materials Research, 104(2013)1020.
[10] Bekheet AE, Physica B, 403(2008)4342.
[11 Jonscher K. London: Chelsea Dielectric Press;1983.
[12] Elliott SR, Adv Phys., 36(1987)135.

INTRINSIC TRANSPORT IN NON-UNIFORMLY DOPED SI REGIONS

Luigi Abenante and Massimo Izzi

ENEA, Italian National Agency for New Technologies, Energy and Sustainable Economic Development
Via Anguillarese 301, SP 064, 00123, ROMA (Italy)

ABSTRACT: We present the exact analytical solution to minority-carrier transport in a non-uniformly doped Si region and the expression for emitter saturation current density, J_o, for the case, where diffusivity is unaltered by doping and lifetime is only determined by Auger recombination. We call this case "intrinsic transport". Numerical simulations show that, at negligible SRH recombination, assuming intrinsic transport also at doping levels lower than 5×10^{18} cm^{-3} in both *n*- and *p*-type emitters is reasonable. We apply the previous issues to fit reported data of J_o in both *n*- and *p*-type emitters as a function of sheet resistance, R_{SHEET}. To this aim, an analytical expression for $J_o(R_{SHEET})$ is derived.

Keywords: Recombination, Silicon, Modelling, Simulation.

1 INTRODUCTION

In doped Si under low-injection conditions, at the doping levels, N, where it is inversely proportional to N^2 (N>5×10^{18} cm^{-3} after [1]), minority-carrier lifetime, τ, is usually assigned the so-called Auger lifetime, τ_A, which is modeled as [1][2]

$$\tau_A = \frac{1}{C_A N^2} \tag{1}$$

where C_A is an Auger coefficient, whose value depends on the considered doping species [2]. It is worth noting that, at the aforementioned and even lower doping levels (N>1.48×10^{18} cm^{-3} in *n*-type Si after [3]), minority-carrier diffusivity, D, can be taken as N-insensitive (data in [4]). We call this N-insensitive diffusivity Auger diffusivity, D_A.

In the present work, we show that, under the condition that D=D_A and τ=τ_A, that is at intrinsic transport, the minority-carrier transport in the dark can be solved exactly in non-uniformly doped regions and an expression for J_o can be derived, which reduces to the reported solution for uniform doping [5]. The new solution can be set in a form such that J_o is a function of R_{SHEET}.

We show moreover that, at negligible SRH-recombination, minority-carrier transport can be modeled as intrinsic transport in both *n*- and *p*-type emitters with a reasonable accuracy. To this aim we have exploited the numerical calculation program PC1D6, which allows performing simulations with many available models for D and τ_A [6].

We finally check the new expression for $J_o(R_{SHEET})$ in the emitters studied in [7-8].

2 ANALYTICAL SOLUTION TO INTRINSIC TRANSPORT

If D is independent of N and lifetime is given by Eq. (1), then, in a non-uniformly doped Si region in the dark, minority-carrier transport becomes such that

$$y'' - \frac{N'}{N} y' - \frac{C_A N^2}{D} y = 0 \tag{2}$$

where $y = N\Delta m$, with Δm being the excess minority-carrier density. Equation (2) is solved by

$$y = A\sinh\left(\sqrt{\frac{C_A}{D}}\int_0^x N da\right) + B\cosh\left(\sqrt{\frac{C_A}{D}}\int_0^x N da\right) \tag{3}$$

which, at boundary conditions for an emitter in the dark, gives the emitter saturation current density, J_o, as

$$J_o = \pm q D n_{ie}^2 \sqrt{\frac{C_A}{D}} \times \frac{\sinh\left(\sqrt{\frac{C_A}{D}}\int_0^{W_{EM}} N dx\right) + \frac{S}{DN(0)}\cosh\left(\sqrt{\frac{C_A}{D}}\int_0^{W_{EM}} N dx\right)}{\cosh\left(\sqrt{\frac{C_A}{D}}\int_0^{W_{EM}} N dx\right) + \frac{S}{DN(0)}\sinh\left(\sqrt{\frac{C_A}{D}}\int_0^{W_{EM}} N dx\right)} \tag{4}$$

where the sign depends on the doping type, q is the electron charge, n_{ie} is the effective intrinsic-carrier density, S is the surface recombination velocity, x=W_{EM} is the abscissa at the interface with space-charge region, and x=0 is the surface position. At uniform doping profile and S=0 cm/s, Eq. (4) reduces to the expression for J_o included in the expression for open-circuit voltage under low-injection conditions reported in [5].

3 TRANSPORT AT NEGLIGIBLE SRH-RECOMBINATION

In Fig. 1, curves for minority-carrier diffusion length, L, as a function of N are shown. The L_{PC1D6}-curves drawn with symbols were calculated at 300 °K with the numerical simulation program PC1D6 [6] by adopting the model for mobility, μ, of Klaassen [4] and the τ_A-models of both Richter et al. [1] (circles) and Dziewior and Schmid [2] (squares) at SRH-lifetime, τ_{SRH}, equal to 10^{20} µs, i.e. at negligible SRH-recombination. The L_{PC1D6}-curves in Fig. 1 are fitted by full-line L_{FIT}-curves calculated with

$$L_{FIT} = \frac{1}{N}\sqrt{\frac{D_A}{C_A}} \tag{5}$$

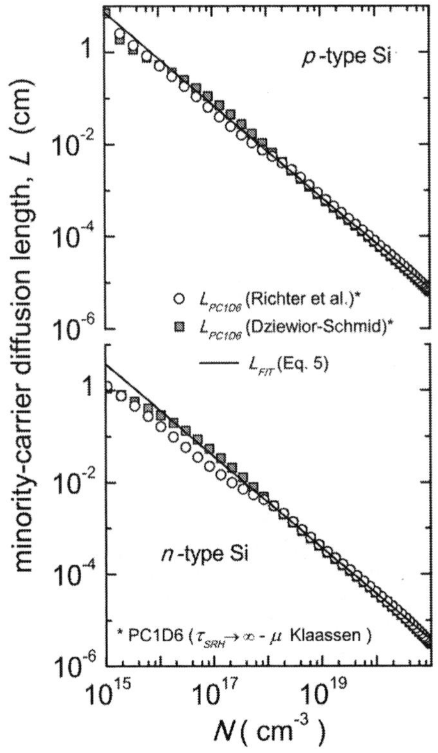

Figure 1: Theoretical curves for $L(N)$ obtained with PC1D6 [6] (symbols) at negligible SRH-recombination are compared to the relevant fit-curves calculated with Eq. (5) (full lines).

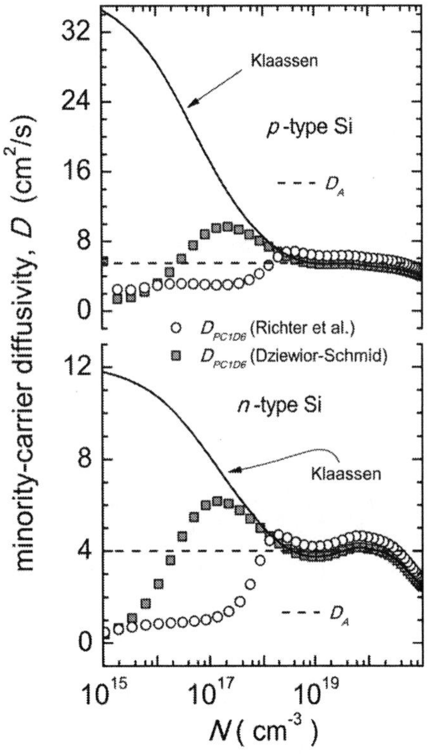

Figure 2: Theoretical curves for $D(N)$ obtained with the theoretical curves for $L(N)$ in Fig. 1 (symbols) are compared to the D_A-values relevant to the fit-curves in Fig. 1 (dashed lines), and $D(N)$-curves obtained with the μ-model of Klaassen [4] (full lines).

where C_A is assigned the value of Dziewior and Schmid [2] relevant to the considered N-species and D_A is an adjustable Auger diffusivity independent of N. The fit-values of D_A are shown in Fig. 2 (dashed-lines), together with both full-line $D(N)$-curves, which were calculated with the Klaassen μ-model [4], and $D_{PC1D6}(N)$-curves (symbols), which were calculated with

$$D_{PC1D6} = C_A \left(N L_{PC1D6} \right)^2 \qquad (6)$$

by using the L_{PC1D6}-curves in Fig. 1 and the C_A-value of Dziewior and Schmid [2] relevant to the considered doping type. The aforementioned theoretical "Dziewior-Schmid"-curves that are drawn with squares in Figs. 1 and 2 represent $L(N)$ and $D(N)$, respectively, for the case, where, at all N-values, recombination is only due to unavoidable purely collisional Auger processes [1]. Consequently, if the μ-model of Klaassen [4] is correct, the "Dziewior-Schmid" $L_{PC1D6}(N)$-curves in Fig. 1 set upper limits on $L(N)$. The $L_{FIT}(N)$-curves in Fig. 1 were calculated at D_A-values such that both the "Dziewior-Schmid" $L_{PC1D6}(N)$-curves in Fig. 1 and "Dziewior-Schmid" $D_{PC1D6}(N)$-curves in Fig. 2 are fitted at N-values, for which assuming an N-insensitive diffusivity is experimentally supported [3]. In Fig. 1, however, an N-range, which extends from N~10^{16} cm^{-3} to N~10^{18} cm^{-3} circa, can be detected, where the aforementioned "Dziewior-Schmid"-curves that are drawn with squares in Figs. 1 and 2 exceed the $L_{FIT}(N)$-curves. In that N-range, the "Dziewior-Schmid" $D_{PC1D6}(N)$-curves are consequently higher than D_A. The discrepancy in the $L(N)$-curves, though, can be shown to be only due to the fact that, in the aforementioned N-range, the Klaassen model [4] overestimates D. One can so conclude that, in the case of unavoidable purely collisional Auger processes, assuming $D=D_A$ already at $N>10^{16}$ cm^{-3} circa is reasonable.

Since, at high doping values, they exceed the "Dziewior-Schmid"-curves, the "Richter et al."-curves drawn with circles in Figs. 1 and 2 are relevant to C_A-values lower than those in [2]. Consequently, at high N-values, the Richter et al. τ_A-model [1] disagrees with the μ-model of Klaassen [4]. The behavior of the "Richter et al." $L_{PC1D6}(N)$-curves at lower N-values is instead due to the fact that, in order to explain lifetime data at low N-values, the τ_A-model of Richter et al. takes into account additional Auger processes [1]. As it can be seen, this causes a remarkable drop in D already at $N=10^{17}$ cm^{-3}.

4 APPLICATION

In [7] and [8], Yan and Cuevas report measured values of $J_o(R_{SHEET})$ (circles in Fig. 3) for both n- and p-type metal-coated emitters diffused on very-low resistivity substrates, together with values of junction depth, x_j, and $N(0)$. In those works, band gap narrowing N-functions, $\Delta E_g(N)$, were determined at both Boltzmann and Fermi-Dirac statistics by calculating J_o in the emitters with an approach that includes non-uniform doping profiles.

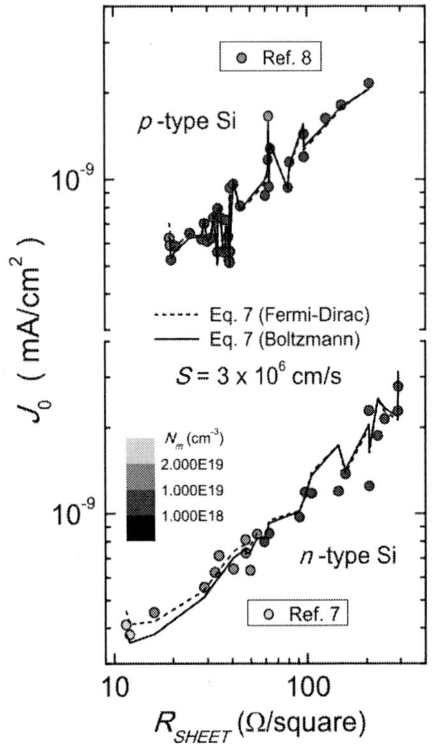

Figure 3: The curves for $J_o(R_{SHEET})$ measured in [7] and [8] (circles) are fitted with curves for $J_o(R_{SHEET})$ calculated with Eq. (7) (lines).

In the calculations in [7-8], $S=3\times10^6$ cm/s was assigned, the τ_A-model of Kerr and Cuevas [9] was adopted, and SRH-recombination in the bulk was neglected.

Since emitter doping profiles are not available, Eq. (4) cannot be used to calculate J_o in the emitters studied in [7-8].

In the case of intrinsic transport, however, from the definition of R_{SHEET} it follows that Eq. (4) can be written as

$$J_o = \pm qD_A n_{ie}^2 \sqrt{\frac{C_A}{D_A}} \times \frac{\sinh\left(\frac{V_{th}}{qD_A R_{SHEET}}\sqrt{\frac{C_A}{D_A}}\right) + \frac{S}{D_A N(0)}\cosh\left(\frac{V_{th}}{qD_A R_{SHEET}}\sqrt{\frac{C_A}{D_A}}\right)}{\cosh\left(\frac{V_{th}}{qD_A R_{SHEET}}\sqrt{\frac{C_A}{D_A}}\right) + \frac{S}{D_A N(0)}\sinh\left(\frac{V_{th}}{qD_A R_{SHEET}}\sqrt{\frac{C_A}{D_A}}\right)} \quad (7)$$

where V_{th} is the thermal voltage.

The fit-curves that are drawn with lines in Fig. 3 were calculated with Eq. (7) at the values for C_A and D_A given in the previous Section and the values for S, $N(0)$, and R_{SHEET} reported in [7][8]. To calculate n_{ie}, we exploited the value for intrinsic carrier density and analytical form of the $\Delta E_g(N)$-function, which are adopted in [7][8]. At both Boltzmann and Fermi-Dirac statistics, the analytical form of the $\Delta E_g(N)$-function is

$$\Delta E_g(N) = A \times \left[\ln\left(\frac{N}{N_{ref}}\right)\right]^b \quad (8)$$

In Eq. (8), the coefficients A, b, and N_{ref} assume values depending on the considered statistics and doping species. We calculated ΔE_g with Eq. (8) in each emitter at values for b and N_{ref} that are the same as assigned in [7][8], $N=N_m$, with N_m being a mean emitter doping value given by

$$N_m = \frac{V_{th}}{qx_j D_A R_{SHEET}} \quad (9)$$

and values for A that are around one-half of the A-values assigned in [7][8].

5 DISCUSSION AND CONCLUSION

The necessity of adjusting A is likely to be connected to the using in the J_o-calculations in [7][8] of N-dependent μ-functions despite the fact that, in those works, negligible SRH-recombination was assumed. Moreover, the τ_A-model adopted in [7][8] has been found to be liable to errors [1]. At any rate, Fig. 3 shows that Eq. (7) can fit measured data fine. Its use would avoid the need of doping profiles and N-dependent parameters in the modeling and simulation of diffused emitters. For a correct implementation of Eq. (7), however, experimentally determined Δ

$E_g(R_{SHEET})$-functions would be necessary. Such functions are unfortunately still missing.

References

[1] A. Richter, S. W. Glunz, F. Werner, J. Schmidt, and A. Cuevas, *Phys. Rev. B*, vol. 86, no. 16, p. 165202, 2012.

[2] J. Dziewior and W. Schmid, *Appl. Phys. Lett.*, vol. 31, no. 5, p. 346, 1977.

[3] A. Wang, C.H. and Misiakos, K. and Neugroschel, *Electron Devices, IEEE Trans.*, vol. 37, no. 5, pp. 1314–1322, 1990.

[4] D. B. M. Klaassen, *Solid. State. Electron.*, vol. 35, no. 7, pp. 953–959, 1992.

[5] M. A. Campbell, P. and Green, *Electron Devices, IEEE Trans.*, vol. 33, no. 2, pp. 234–239, 1986.

[6] H. Haug, A. Kimmerle, J. Greulich, A. Wolf, and E. Stensrud Marstein, *Sol. Energy Mater. Sol. Cells*, vol. 131, pp. 30–36, Dec. 2014.

[7] D. Yan and A. Cuevas, *J. Appl. Phys.*, vol. 114, no. 4, p. 044508, 2013.

[8] D. Yan and A. Cuevas, *J. Appl. Phys.*, vol. 116, no. 19, p. 194505, 2014.

[9] M. J. Kerr and A. Cuevas, *J. Appl. Phys.*, vol. 91, no. 3, pp. 2473–2480, 2002.

THE INFLUENCE OF THE EXCITON NONRADIATIVE RECOMBINATION IN SILICON ON THE PHOTOCONVERSION EFFICIENCY

A.V. Sachenko[a], V.P. Kostylyov[a], V.M. Vlasyuk[a], I.O. Sokolovskyi[a], and M. Evstigneev[b,*]
[a]V. Lashkaryov Institute of Semiconductor Physics,
NAS of Ukraine, 41 prospect Nauky, 03028 Kyiv, Ukraine
[b]Department of Physics and Physical Oceanography,
Memorial University of Newfoundland, St. John's, NL, A1B 3X7 Canada
*Corresponding author: mevstigneev@mun.ca

ABSTRACT: By comparison of the experimental dependence of bulk lifetime in silicon on the doping and excitation levels with theoretical calculations, it is shown that a new recombination channel becomes operative when Shockley-Read-Hall lifetime is below 20 ms and the density of doping impurities or the excess electron-hole pair density is of the order of 10^{16} cm^{-3}. This recombination mechanism is related to the non-radiative exciton Auger recombination assisted by the deep impurities in the bulk. The influence of non-radiative exciton recombination on the photoconversion efficiency in solar cells is analyzed. It is shown that the shorter the Shockley-Read-Hall lifetime, τ_{SRH}, the stronger its effect. In particular, for τ_{SRH} = 100 μs, this recombination channel leads to the reduction of photoconversion efficiency by 5.5 %, whereas for $\tau_{SRH} \approx 1$ μs, the photoconversion efficiency is about twice as small as the value obtained without taking non-radiative exciton recombination into account.
Keywords: Silicon Solar Cell, Recombination, Exciton Recombination

1 INTRODUCTION

In the last few decades, the photoconversion efficiency of silicon structures, such as p-n junction- and heterojunction-based solar cells (SCs), has been boosted to the value of 25 % [1-3]. It was achieved by using high-quality silicon with the nonequilibrium electron-hole pair lifetimes of the order of a few milliseconds at low doping levels, as well as by employing various passivation methods to obtain very low surface recombinations velocities [4, 5]. To further increase the photoconversion efficiency, it is necessary to optimize several parameters of the base region, in particular, doping level and thickness. As pointed out in [4], at such long bulk lifetimes, even at not very high doping and excitation levels, n_0 and Δn, respectively, radiative and band-to-band Auger recombination processes become operative, which limit the effective bulk lifetime, and the output power of the SC.

The questions related to the influence of radiative and Auger recombination in silicon on the effective lifetime have been discussed extensively in the literature, see [5] and references therein. In spite of the fact that a good agreement between theoretical and experimental findings has been achieved for doping and excitation levels exceeding 10^{16} cm^{-3}, the region below 10^{16} cm^{-3} seems to be understood to a lesser degree (here). In particular, Fossum [6], based on the experimental data from [7, 8], has proposed the following empirical expression for the reduction of the effective lifetime in silicon with doping level:

$$\tau_{eff}(n) = \frac{\tau_{max}}{1+(n/n_x)}, \qquad (1)$$

where the electron-hole pair density, $n = n_0 + \Delta n$, is the sum of the equilibrium, n_0, and the excess, Δn, contributions, and $n_x = 7.1 \cdot 10^{15}$ cm^{-3} both for p- and n-type semiconductors.

As seen from (1), the decrease of τ_{eff} begins at the doping levels of the order of 10^{16} cm^{-3}, i.e. well before the band-to-band Auger recombination starts to take place.

Finally, the PC1D code uses an empirical expression $\tau_{eff} \propto (n/n_x)^{-a}$, which takes into account, according to the code developers, the influence of crystal growth on the effective lifetime in silicon. When the first version of this code was released a few decades ago, there apparently were reasons to relate the decrease of τ_{eff} with n to the technological factors at low doping, although the fact that n_x is practically the same for n- and p-type semiconductors remained unexplained. However, as far as more modern works are concerned, such as [5], this explanation cannot be applied in view of the absence of such uncontrolled technological factors, which was possible to achieve in recent years.

As will be shown below, the region described by the empiricial formula (1) is realized in the experimental $\tau_{eff}(n)$ curves from [4, 7-9], although it is absent in the analogous data from [5]. Its presence or absence is determined by the Shockley-Read-Hall lifetime: if $\tau_{SRH} \geq 20$ ms, radiative recombinations dominates over the quadratic non-radiative one, and the decrease of $\tau_{eff}(n)$ according to $\tau_{SRH}(n)n_x/n$ is absent. In the opposite case, quadratic non-radiative recombination takes over, leading to the onset of a region described by (1). We relate this region to the non-radiative impurity-assisted Auger recombination of the excitons generated in silicon by light. In this work, the influence of this recombination channel on the photoconversion efficiency of heterojunction SCs is analyzed, and the SC design is optimized with respect to the base doping level. It is shown that the lower τ_{SRH}, the stronger the efficiency reduction due to the non-radiative exciton recombination.

2 ANALYSIS OF THE LITERATURE DATA ON τ_{eff}

In our analysis of the effective lifetime reported in the literature, especially in [5], we employ the following expression for the main quantity of interest, the effective lifetime:

$$\tau_{eff}(n) = \left(\frac{1}{\tau_{SRH}(n)} + \frac{1}{\tau_r(n)} + \frac{1}{\tau_{nr}(n)} + \frac{1}{\tau_{Auger}(n)} \right)^{-1}, \qquad (2)$$

where τ_r, τ_{nr}, and τ_{Auger} are the radiative, non-radiative, and Auger lifetimes, respectively, with

$$\tau_{nr} = \tau_{SRH} \frac{n_x}{n}, \quad (3)$$

and the Shockley-Read-Hall lifetime in n-type silicon is given by

$$\tau_{SRH}^{(n)} \cong \frac{\tau_{p0}(n_0 + n_1 + \Delta n) + \tau_{n0}(p_1 + \Delta n)}{n_0 + \Delta n}, \quad (4)$$

where $\tau_{p0} = (C_p N_t)^{-1}$, $\tau_{n0} = (C_n N_t)^{-1}$, $C_{p,n}$ are the coefficients of hole and electron capture by recombination centers (traps) of density N_t, and n_1 and p_1 are the electron and hole densities in the case when the recombination level coincides with Fermi energy.

Radiative lifetime is given by the expression

$$\tau_r^{-1} = A(n_0 + \Delta n), \quad (5)$$

where the radiative recombination parameter, obtained in [10] taking into account both band-to-band and exciton radiative recombination, varies at room temperature from $A = 6.3 \cdot 10^{-15}$ cm^3/s at $n = 10^{15}$ cm^{-3} to $4.63 \cdot 10^{-15}$ cm^3/s at $n = 10^{16}$ cm^{-3}. For numerical estimates, we will be using the value $A = 6.3 \cdot 10^{-15}$ cm^3/s, because at $n = 10^{16}$ cm^{-3} the band-to-band Auger recombination dominates in silicon, so that the inaccuracy in A will practically not affect the calculation results.

As for the band-to-band lifetime τ_{Auger} (n), we used the empirical expression given by Eq. (18) in [5].

Before analyzing $\tau_{eff}(n)$, we note that, depending on the energy of the recombination center and electron and hole capture cross section, the Shockley-Read-Hall lifetime as a function of the electron-hole pair density varies between two values and may increase, decrease, or remain practically the same in a certain range of doping and excitation levels. In the following, we will mainly consider the case, where τ_{SRH} is practically constant in the range of interest, where the doping level changes from 10^{14} to 10^{17} cm^{-3}, and the excitation level from 10^{14} to 10^{16} cm^{-3}.

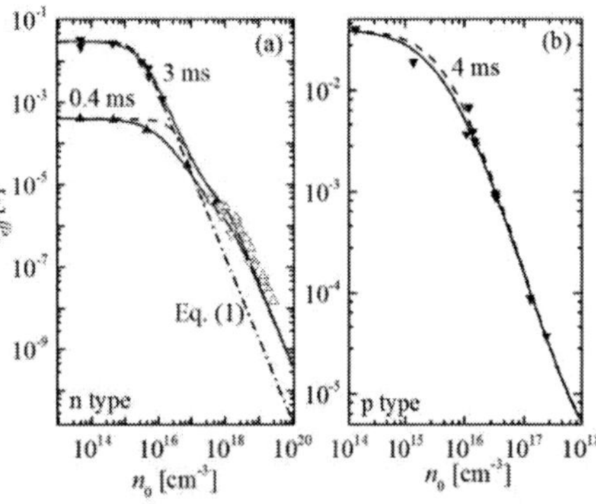

Figure 1: Effective lifetime in (a) n- and (b) p-type silicon vs. doping level. Solid lines are obtained with and dashed lines without taking the exciton recombination e ect into account. Symbols: experimental results from [5] (down filled triangles), [6] and [7] (up filled triangles), [12] (up open triangles), [11] (down open triangles). The theoretical curves were built with $n_x = 8.2 \cdot 10^{15}$ cm^{-3} for (a) τ_{SRH} = 3 ms (red) and τ_{SRH} = 0.4 ms (blue), and (b) τ_{SRH} = 4 ms. Note that for τ_{SRH} = 3 ms, the curves obtained with and without exciton recombination effect practically coincide. The dotted curve was obtained from Eq. (1).

Presented in Fig. 1 (a) and (b) are the experimental $\tau_{eff}(n)$ curves for n- and p-type semiconductors, respectively. The experimental data points are taken from [5, 7, 11, 12], and the theoretical curves are built using (2)-(5), as well as the expressions (18) and (19) for τ_{Auger} from Ref[5]. There are two fit parameters in the theory: the density of the deep levels, which determines τ_{SRH}, and n_x. In principle, they can be found separately by measuring $\tau_{eff}(n)$ at the doping levels below 10^{14} cm^{-3}. As seen in Fig. 1(a), the theoretical curves agree well with the experimental results from [7] at $n_x = 8.2 \cdot 10^{15}$ cm^{-3}, found using the least squares method. As for p-type semiconductor, Fig. 1(b), the theoretical curves built with and without taking the additional quadratic recombination channel into account do not differ much from each other.

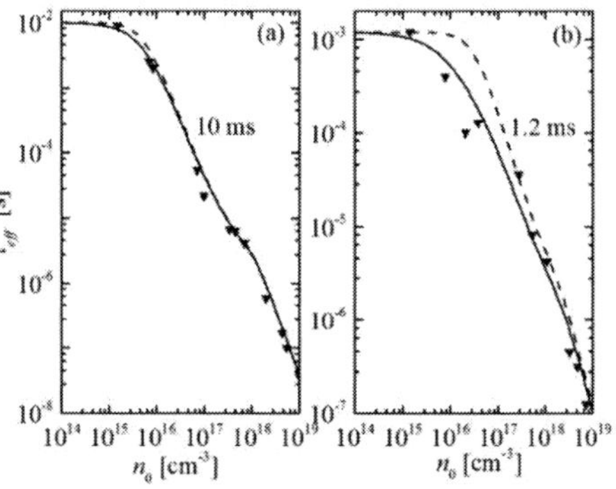

Figure 2: The reduced effective bulk lifetime τ^*_{eff} vs. doping level in (a) n- and (b) p-type silicon. The theoretical curves were obtained for (a) τ_{SRH} = 10 ms and (b) τ_{SRH} = 1.2 ms with (solid) and without (dashed) taking exciton recombination into account. Down triangles: experimental data from [9].

Fig. 2 shows the experimental results for $\tau_{Auger}(n)$ in n- and p-type silicon from [9]. The theoretical curves $\tau_{eff}(n)$ were obtained from the reduced version of (2), where the term τ_r was omitted, and Eqs. (18) and (19) from [9]. As seen in Fig. 2(a), including the additional recombination channel has practically effect on the theoretical curves for n-type silicon. At the same time, for p-type semiconductor, the theoretical curves agree with the experimental ones only when it is taken into account with $n_x = 8.2 \cdot 10^{15}$ cm^{-3}.

Shown in Fig. 3 is the effective lifetime vs. excitation level, where the experimental data points are taken from [4, 5], and the theoretical ones were obtained from (2)-(5) and the expressions (18) and (19) from [5]. In the case of n-type semiconductor (Fig. 3(a)), the theory with $n_x = 8.2 \cdot 10^{15}$ cm^{-3} agrees well with the experiment, although the theoretical curves $\tau_{eff}(\Delta n)$ for the sample with the orientation <111> are practically the same with and without exciton non-radiative recombination channel. On the other hand, in the p-type case, Fig. 3(b), the theory cannot fit the experimental data, unless the dependence $\tau_{SRH}(n)$ is taken into account. To achieve agreement, we assumed that for p-type silicon, the levels are operative, whose energy is 0.38 eV, and that $\tau_{p0} = 3.6 \cdot 10^{-2}$ s, $\tau_{n0} = 2.4 \cdot 10^{-1}$ s.

In order to find out the conditions, under which the additional quadratic recombination channel has a noticeable effect on τ_{eff}, we need to compare it to the radiative recombination channel's contribution. When the two recombination channels are of equal importance, we have the relation $A = (\tau^*_{SRH}\, n_x)^{-1}$, from which we can determine that value of Schockley-Read-Hall lifetime, τ^*_{SRH}, at which the radiative recombination starts to dominate. For $n_x = 8.2 \cdot 10^{15}$ cm^{-3}, this gives $\tau^*_{SRH} \approx 20$ ms. For $\tau_{SRH} > 20$ ms, radiative recombination dominates, whereas for $\tau_{SRH} < 20$ ms, non-radiative recombination is more important. As seen from Figs. 1-3, this criterion is applicable to all cases considered. Because the Shockley-Read-Hall lifetime in today's high-efficiency p-n- or heterojunction-based SCs is of the order of a few milliseconds, quadratic non-radiative recombination channel must be taken into account during the analysis of photoconversion efficiency.

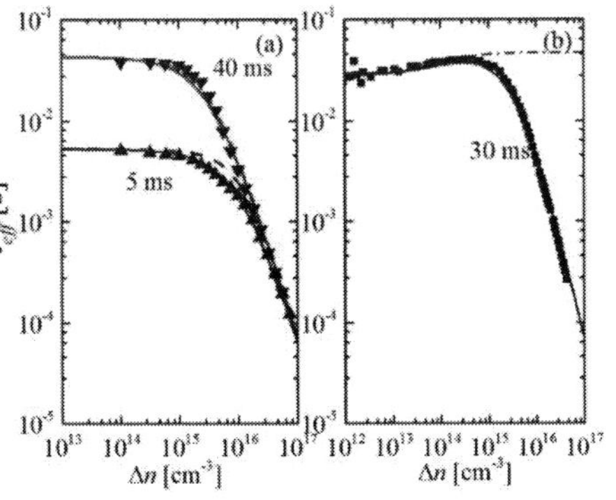

Figure 3: Effective bulk lifetime on the excess electron-hole pair density for (a) n- and (b) p-type silicon. The theoretical curves were obtained with (solid) and without (dashed) taking exciton recombination into account. The dash-dotted curve describes $\tau_{SRH}(n)$ for cobalt level. $n_x = 8.2 \cdot 10^{15}$ cm^{-3}. Experimental points in panel (a) are taken from Ref. [4] for <111> (down triangles) and <100> (up triangles) orientation, and in (b) from Ref. [5] (squares).

Also, Figs. 1-3 indicate that, at τ_{SRH} below 1 ms, the role of quadratic non-radiative recombination becomes essential. This is seen, in particular, in Fig. 1(a) for $\tau_{SRH} = 0.4$ ms, which exhibits quite a broad region that cannot be described by (1) with radiative and band-to-band Auger recombination channels only. Therefore, we need to find the physical explanation of the experimental $\tau_{eff}(n)$ curves described by (1). This explanation was proposed in [13], which is briefly reviewed here.

3 NON-RADIATIVE EXCITON RECOMBINATION IN SILICON

Sufficiently strong irradiation of silicon generates not only electron-hole pairs of density Δn, but also excitons of density n_{exc}. Although $n_{exc} \ll \Delta n$, the exciton non-radiative lifetime τ_{nr}, related to their Auger recombination assisted by deep impurity levels can in many cases be much shorter than Shockley-Read-Hall lifetime τ_{SRH}. This leads to binding of excess electron-hole pairs into excitons, which then recombine [13, 14]. This process was first considered by Hangleiter [14], who showed that both Auger and Shockley-Read-Hall recombination can be mediated by hte same deep level.

In the work [13], a phenomelogical relation between τ_{nr} and τ_{SRH} was proposed based on the following two premises. First, it was assumed that the non-equilibrium electron-hole pair density, n, and exciton density, n_{exc}, are related by

$$n_{exc} = \frac{(n_0 + \Delta n)\Delta n}{n^*}, \quad n^* = \frac{N_c N_v}{N_x} e^{-E_x/kT}, \qquad (6)$$

where N_c, N_v, and N_x are the effective densities of states in the condaction, valence, and exciton bands, and $E_x = 14.7$ meV is the binding energy of the exciton ground state. For silicon parameters at T = 300 K far away from Mott transition, $n^* = 8 \cdot 10^{17}$ cm^{-3}.

The second premise was related to the spacial correlation of an electron and a hole in an exciton [14], leading to an essential increase of their Auger recombination probability, in particular, on the impurities. The degree of correlation can be characterized by the effective density of electron-hole pairs in an exciton, $n_L = \left((4/3)\pi a_B^3\right)^{-1}$, where a_B is exhiton Bohr radius. In silicon, $a_B \approx 4.2$ nm, and $n_L \approx 3.3 \cdot 10^{18}$ cm^{-3}. This is a sufficiently high value for the non-radiative Auger recombination to be operative. The inverse lifetime of this process is taken to be

$$\tau_{nr}^{-1} = n_L G N_t, \qquad (7)$$

where G is the Auger recombination parameter, and N_t is the density of deep levels.

Let us compare the recombination fluxes due to Shockley-Read-Hall and exciton non-radiative mechanisms, $\Delta n/\tau_{SRH}$ and n_{exc}/τ_{nr}, respectively. Approximating $\tau_{SRH} \approx (C_t\, N_t)^{-1}$ and adding the two recombination fluxes using (6) and (7), we obtain:

$$\frac{\Delta n}{\tau_{SRH}} + \frac{n_{exc}}{\tau_{nr}} \approx \frac{\Delta n}{\tau_{SRH}}\left(1 + \frac{n_0 + \Delta n}{n_x}\right), \qquad (8)$$

where $n_x = \frac{n^* C_t}{n_L G}$. (8) has the same form as (1). As for the difference in the expressions for n_x between the present work and [6], it stems from the fact that the latter work neglects the influence of τ_{Auger} on $\tau_{eff}(n)$, while here we take it into account.

Estimates performed in [13] for the case when the deep levels are due to Au impurities give n_x not very different from $8.2 \cdot 10^{15}$ cm^{-3}.

Thus, this brief overview of the main results of [13] shows that the experimentally observed dependence $\tau_{eff}(n)$, see (1), can be explained by the non-radiative exciton Auger recombination on the deep impurity levels.

The estimate of the raditative recombination parameter in silicon at T = 300 K performed in [10], namely, $A = 6.3 \cdot 10^{-15}$ cm^3/s, also took exciton recombination into account. In that work, based on the experimental results from [15] on the temperature dependence of the radiative recombination parameter in silicon, the contributions of the radative band-to-band and exciton recombination effects could be evaluated separately. It was shown that at room temperature, they were comparable [10].

4 PHOTOCONVERSION EFFICIENCY OF HIGH-EFFICIENCY SCs

Next, we analyze photoconversion efficiency of a SC with an n-type base as a function of the base doping level. In this analysis, we take into account all the recombination mechanisms considered above. The theoretical results for the efficiency η, open-circuit voltage, V_{OC}, IV-curve fill factor, FF, and some other characteristics are compared to the experimental data. Our analysis employs the semiempirical model from [16], which allows one to evaluate the performance of a SC based on high-quality crystalline silicon. The feature of this model is that one of the main characteristics, the short-circuit current density, J_{SC}, is taken from the experiment as an input parameter of the theory, from which the remaining characteristics are found theoretically. This considerably simplifies the calculations, which can be used to optimize SC's performance with respect to the base doping level, N_d at given Shockley-Read-Hall lifetime τ_{SRH}, surface recombination velocity on the front and the rear surfaces, S_0 and S_d, and series resistance R_S.

Using the approach from [16], let us write the photoconversion efficiency of the cell characterized by large diffusion lengths, $L_{eff} = \sqrt{D\tau_{eff}} >> d$, and high excess electron-hole pair density under the open-circuit conditions, $\Delta n_{OC} \geq n_0$. The expression for the open-circuit voltage has the form [17]:

$$V_{OC} \cong \frac{kT}{q}\ln\frac{\Delta n_{OC}}{p_0} + \frac{kT}{q}\ln\left(1+\frac{\Delta n_{OC}}{n_0}\right), \qquad (9)$$

where kT is the thermal energy, q the elementary charge, and $p_0 = n_i^2(T)/n_0$ the equilibrium hole density in the n-type base.

At $\Delta n_{OC} \geq n_0$, the open-circuit voltage is higher than in the standard case $\Delta n_{OC} < n_0$.

(9) is a quadratic equation for Δn_{OC}, whose solution is

$$\Delta n_{OC} = -\frac{n_0}{2} + \sqrt{\frac{n_0^2}{4} + n_i^2\left(e^{qV_{OC}/kT} - 1\right)}. \qquad (10)$$

In order to determine photoconversion efficiency, we need an expression for the current-voltage curve, which is obtained by replacing V_{OC} in (10) by the applied voltage, V, and including the voltage drop on the series resistance, R_S:

$$\Delta n_{OC} = -\frac{n_0}{2} + \sqrt{\frac{n_0^2}{4} + n_i^2\left(e^{q(V - JA_{SC}R_S)/kT} - 1\right)}, \qquad (11)$$

where J is the current density, and A_{SC} the SC area.

The short-circuit current density, J_{SC}, is related to the excess carrier density on the open-circuit regime, Δn_{OC}, by the generation-recombination balance equation. For $L_{eff} >> d$, it has the form

$$J_{SC} = q\left(\frac{d}{\tau_{eff}} + S\right)\Delta n_{OC}, \qquad (12)$$

where $S = S_0 + S_d$ is the net surface recombination velocity on the front and rear surfaces.

Next, we generalize (12) to the case $V < V_{OC}$, when the current through the cell is non-zero:

$$J(V) = J_{SC} - J_{rec}(V), \qquad (13)$$

where the recombination current density is

$$J_{rec}(V) = q\left(\frac{d}{\tau_{eff}} + S\right)\Delta n(V), \qquad (14)$$

and τ_{eff} is to be found from (2)-(5) with Δn replaced by $\Delta n(V)$.

From the maximum-power condition, $d(VJ(V))/dV = 0$, we can determine V_m, whose substitution into (13) allows one to find J_m. As a result, we obtain the maximum photoconversion efficiency and the fill factor:

$$\eta = \frac{J_m V_m}{P_S}, \quad FF = \frac{J_m V_m}{J_{SC} V_{OC}}, \qquad (15)$$

where P_S is the incident radiation power density.

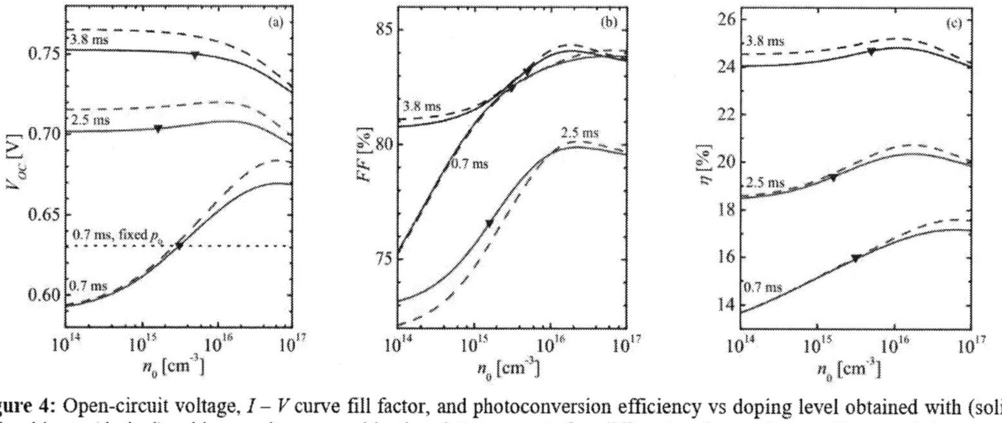

Figure 4: Open-circuit voltage, $I - V$ curve fill factor, and photoconversion efficiency vs doping level obtained with (solid) and without (dashed) taking exciton recombination into account for different values of τ_{SRH}. The symbols are the experimental points, as indicated in Table I.

Table I: Parameters from Refs. [2, 18, 19] and obtained using Eqs. (9)-(15).

Ref.	N_d, cm^{-3}	d, μm	τ_{max}, ms	J, mA/cm^2	V_{OC}, mV	FF, %	η, %	S, cm/s	A_{SC}, cm^2	R_S, Ω
[2]	$4.9\cdot10^{15}$	98	3.8	39.5	750	83.2	24.7	1.5	100	0.0027
[18]	$1.6\cdot10^{15}$	300	2.5	36	704	76.6	19.4	12	4	0.25
[19]	$3.1\cdot10^{15}$	380	0.7	30.6	631	82.5	16.0	220	2	0.03

5 ANALYSIS OF THE EXCITON NON-RADIATIVE RECOMBINATION EFFECT ON THE SC PERFORMANCE

Shown in Fig. 4 are the theoretical curves for the open-circuit voltage V_{OC}, fill factor FF, and photoconversion efficiency η as a function of the n-type base doping level. The curves were built using the formulas (12), (10), and (15), with the parameters obtained in the works [2, 18, 19]. The main parameters of these SCs are given in Table I. It was assumed that the surface recombination velocity is independent of the doping level.

In order for non-radiative exciton recombination to have an effect on the photoconversion efficiency of silicon SCs, two conditions must be fulfilled. First, in order for non-radiative recombination to be of comparable rate as the radiative one, we must have $\tau_{SRH} < 20$ ms, as discussed above. This condition is fulfilled for all of the existing p-n junction- and heterojunction-based SCs. The second condition is that the bulk recombination should dominate over the surface recombination. When discussing Fig. 4, we will pay special attention to the latter condition.

As seen in Fig. 4(a), the characteristic feature of the $V_{OC}(n_0)$ plot is that, in the low-doping region at large τ_{max} (see (1)), open-circuit voltage decreases with the dopant density. However, as τ_{max} is reduced, these curves begin to grow. At the same time, FF and η increase with the doping level even at the largest efficiency value of about 25 %. Non-radiative exciton recombination leads to a decrease of V_{OC}, FF, and η, which depends on the doping level, τ_{SRH}, and S. As seen in Fig. 4(c), at the minimal surface recombination velocities and $\tau_{SRH} \approx 4$ ms, the difference between the theoretical efficiency with and without the exciton recombination taken into account for the upper two curves is slightly reduced with increasing n_0 and is only 1.5 %. For $S = 12$ cm/s and $\tau_{SRH} \approx 2.5$ ms (see the two middle curves), the difference increases from less than 1 % at $N_d \approx 10^{15}$ cm^{-3} to 2 % at the doping level corresponding to the maximal efficiency, i.e. $n_0 \approx 2 \cdot 10^{16}$ cm^{-3}. Finally, at $S = 220$ cm/s and $\tau_{SRH} \approx 0.4$ ms (the lowest two curves), the difference between the theoretical efficiencies obtained with and without the exciton recombination increases from less than 1 % at $N_d \approx 10^{15}$ cm^{-3} to 2 % at $N_d \approx 6 \cdot 10^{16}$ cm^{-3} corresponding to the highest efficiency.

According to the parameters from Table I, for the two upper curves, the bulk recombination velocity $V_b = d/\tau_{eff}$ is 2.6 cm/s at low doping levels, and the surface recombination velocity is 1.5 cm/s, i.e. the second condition for the non-radiative exciton recombination to manifest itself is fulfilled. This is not the case for the other two pairs of curves, for which $V_b = S = 12$ cm/s (middle curve) and $V_b = 54$ cm/s, $S = 220$ cm/s (lower curve), for which surface recombination dominates.

Thus, for the parameters from Table I, the largest effect of exciton recombination on the main SC characteristics is about 2 %.

Fig. 5 shows the photoconversion efficiency on the Shockley-Read-Hall lifetime dependence, built for the parameters from the third raw of Table I for $n_0 = 2 \cdot 10^{16}$ cm^{-3}. In this case, the efficiency is maximal or close to its maximum. For the three pairs of curves, we took $S = 1.5$ cm/s (two upper curves), $S = 12$ cm/s (middle curves), and $S = 220$ cm/s (lower curves). As seen from the figure, the smaller τ_{SRH}, the larger the efficiency decrease due to the non-radiative exciton recombination. The largest difference between the curves obtained with and without its effect taken into account is observed at the smallest Shockley-Read-Hall lifetime $\tau_{SRH} = 100$ μs and is about 5.5 %.

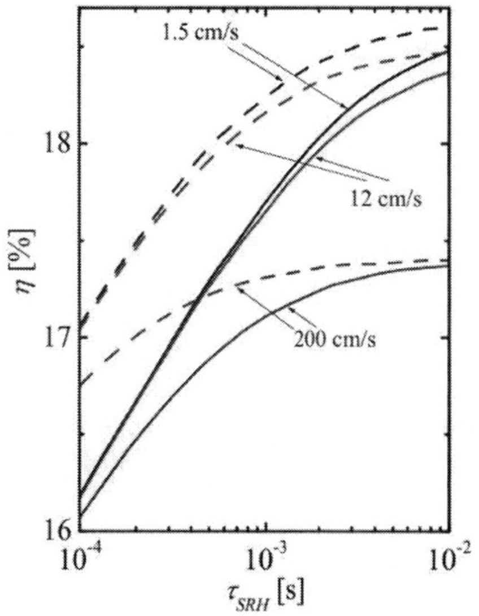

Figure 5: Photoconversion efficiency vs. Shockley-Read-Hall lifetime for SC parameters from Ref. [19] for different values of surface recombination velocity. Solid lines are obtained taking into account exciton recombination, and dashed lines are obtained without accounting for this effect.

We note that upon further decrease of Shockley-Read-Hall lifetime from 100 μs to 1 μs, exciton recombination leads to the reduction of photoconversion efficiency, which is the stronger the smaller τ_{SRH}. For instance, at $\tau_{SRH} \approx 1$ μs, this reduction reaches ≈ 25 % for n-type base and ≈ 45 % for p-type base.

In the work [20], the influence of various excitonic effects on the photoconversion in silicon structures with p-n junctions, including the non-radiative recombination, was analyzed. However, in the determination of n_x from comparison of theoretical and experimental $\tau_{eff}^{-1}(n_0)$ curves, the radiative recombination rate was underestimated, and band-to-band Auger recombination was not taken into account. This resulted in an overestimation of n_x, which was found to be $3.7 \cdot 10^{15}$ cm^{-3}.

In the work [19], τ_{Auger} was evaluated using an empirical formula from [13] and [21] and experimental dependence of Auger recombination lifetime on the doping level n_0, which were different from the ones in [5]. Namely, in the range $10^{15} < n_0 < 10^{16}$ cm^{-3}, the results for $\tau_{Auger}(n_0)$ in [5] and [16] are quite close, but for $n_0 > 10^{16}$ cm^{-3}, a discrepancy exists between the two sets of data, with $\tau_{Auger}(n_0)$ decreasing with n_0 more slowly in [16] than in [5]. As a result, calculations of the efficiency η vs. the doping level n_0 based on $\tau_{Auger}(n_0)$ curve from [16] yields a maximum efficiency at a larger value of n_0, with a somewhat higher peak efficiency. However, the conclusion about the necessity of a heavier base doping than in [2, 18, 19] to increase the SC efficiency remains valid. Moreover, as calculations show, the photoconversion efficiency at the experimentally used doping levels, obtained for different expressions of $\tau_{Auger}(n_0)$, differ by more than 2 %.

As shown in [20], calculation of the open-circuit voltage, V_{OC}, and photogenerated voltage in the maximum-power regime, V_m, should be done taking into account the thermodynamic effect of the excitons, discussed in [22-24]. This effect has to do with the reduction of the excess electron-hole pair density in the maximal-power regime, Δn_m, due to the binding of some electrons and holes into excitons, leading to the reduction of V_{OC} and V_m. For the case considered here, $\Delta n_m \geq n_0$, the following relation holds:

$$V_m \approx V_m^* - \frac{2kT}{q}\ln\left(1+\frac{n_0+\Delta n_m}{n^*}\right)+\frac{kT}{1}\ln\left(1+\frac{n_0}{n^*}\right), \quad (16)$$

where V_m* is the photogenerated voltage in the maximum-power regime without taking the thermodynamic effect of excitons into account.

Estimates based on the parameter values at which V_{OC} and V_m are maximized show that the ratio of the difference between the last two terms in (16), which take the exciton effect into account, to V_m does not exceed 0.5 % at n_0 = 10^{17} cm^{-3}, and is about 0.03 % at $n_0 = 5 \cdot 10^{15}$ cm^{-3}. This allows us to neglect the thermodynamic exciton effect.

We note that in [24], a recombination contribution of the form (1) was used in the expression for the effective lifetime in silicon, but was not connected to the nonraditative exciton recombination.

In high-efficiency heterojunction silicon SCs, exciton radiative recombination is negligible. According to [9], radiative exciton recombination parameter in silicon is $2.6 \cdot 10^{-15}$ cm^3/s, whereas the ratio of the radiative to non-radiative exciton recombination rates at $n_0 = 5 \cdot 10^{15}$ cm^{-3} and $\ \tau_{SRH} \approx 4$ ms is $4 \cdot 10^{-4}$. Thus, out of three exciton effects – the thermodynamic effect, and radiative and non-radiative recombination – it is the non-radiative Auger recombination on deep impurity levels which is the strongest.

Finally, let us sum up the criteria that allow one to describe the influence of non-radiative exciton recombination on photoconversion efficiency in silicon SCs.

Under the condition $\tau_r < \tau_{nr}$, non-radiative exciton recombination has practically no effect on photoconversion efficiency, because radiative recombination dominates. With the expressions (5) and (3) for τ_r and τ_{nr}, this condition can be formulated as $\tau_{SRH} > 20$ ms.

For $\tau_r > \tau_{nr}$ and $\tau_{nr} > \tau_{Auger}$, non-radiative exciton recombination will have a weak effect on the photoconversion efficiency. Numerically, this gives an efficiency $\eta(n_0)$ maximum as a function of surface recombination velocity in the range $n_0 \approx 10^{16}...10^{17}$ cm^{-3}. At the same time, non-radiative exciton recombination will have a weak effect on the photoconversion efficiency, if τ_{SRH} is in the range from 0.1 ms to 20 ms, in which case the maxima on the theoretical $\eta(n_0)$ curves are the same with and without taking exciton non-radiative recombination into account.

Finally, for $\tau_{nr} < \tau_{Auger}$, i.e. $\tau_{SRH} < 0.1$ ms, non-radiative exciton recombination has a strong effect on the photoconversion efficiency.

6 CONCLUSIONS

As the analysis of the dependence of experimental bulk lifetime in silicon on the doping and excitation level has shown, at $\tau_{SRH} < 20$ ms, an additional recombination channel due to non-radiative exciton recombination starts to manifest itself. The smaller τ_{SRH} the stronger its contribution. The nature of the bulk lifeteme reduction in silicon at the doping or excitation level of about 10^{16} cm^{-3} was shown to be unrelated to technological factors, but is due to the non-radiative exciton recombination on the deep levels.

Taking this channel into account did not change the conclusions of [16] regarding the optimization of the doping level. As seen in Figs. 4(b) and (c), the optimal doping level, related to the comparable contributions from Schockley-Read-Hall recombination and band-to-band Auger recombination, is about $2 \cdot 10^{16}$ cm^{-3} at the minimized surface recombination velocity S and increases to about 10^{17} cm^{-3} with S.

At the same time, at $\tau_{nr} < \tau_{Auger}$ and $\tau_{SRH} < 0.1$ ms, exciton non-radiative recombination sets a limit on the optimal base doping level in silicon SCs, at which photoconversion efficiency is the highers, to the values $n_0 < 10^{16}$ cm^{-3}.

ACKNOWLEDGMENTS

M.E. is grateful to the Natural Sciences and Engineering Research Council of Canada (NSERC) and to the Research and Development Corporation of Newfoundland and Labrador (RDC) for financial support.

REFERENCES

[1] M.A. Green, Prog. Photovolt.: Res. Appl. 17, 183 (2009).

[2] K. Masuko et al., IEEE Journ. of Photovoltaics 4, 1433 (2014).

[3] A. Jano, S. Tohoda, K. Matsuyama, T. Nishiwaki, K. Fujita, M. Taguchi, and E. Maruyama, 28th European Solar Energy Conference and Exhibition, pp. 1846-1848 (Paris, 2013).

[4] E. Yablonovitch and T. Gmiter, Appl. Phys. Lett. 49, 587 (1986).

[5] A. Richter et al., Phys. Rev. B 86, 165202 (2012).

[6] J.G. Fossum, Solid State Electron. 19, 269 (1976).

[7] D. Kendal, Conf. Phys. Appl. of Li-based diffused silicon, NASA-Goddard Space Flight Center (1969).

[8] K. Gra, H. Pieper, and G. Goldbach, Semiconductor silicon (1973).

[9] R. Häcker and A. Hangleiter, J. Appl. Phys. 75 7570 (1994).

[10] A.V. Sachenko et al., Semiconductors 40, 213 (2006).

[11] J.D. Beck and R. Conradt, Solid State Electron. 13, 93 (1973).

[12] L. Passari and N. Susi, J. Appl. Phys. 54, 3935 (1983).

[13] A.V. Sachenko, A.P. Gorban, V.P. Kostylyov, Semicond. Phys. Quantum Electronics and Optoelectronics 3, 5 (2000).

[14] A. Hangleiter, Phys. Rev. B 35, 9149 (1987); Phys. Rev. B 37, 2594 (1988).

[15] T. Trupke et al, J. Appl. Phys. 94, 4930 (2003).
[16] A.V. Sachenko, A.I. Shkrebtii, R.M. Korkishko, V.P. Kostylyov, N.R. Kulish, and I.O. Sokolovskyi, Semiconductors 49, 264 (2015).
[17] A.P. Gorban', A.V. Sachenko, V.P. Kostylyov, and N.A. Prima, Semicond. Phys. Quantum Electronics Optoel. 3, 322 (2000). A.V. Sachenko, A.P. Gorban', and V.P. Kostylyov, Ukr. J. Phys. 46, 226 (2001).
[18] R. Gogolin, R. Ferre, M. Turcu, and N.-P. Harder, Sol. Energy Mat. Sol. Cells 106, 47 (2012).
[19] A.V. Sachenko, V.P. Kostylyov, R.M. Korkishko, M.R. Kulish, V.M. Vlasiuk, and D.V. Khomenko, Semicond. Phys. Quant. Electron. Optoelectron. 18, 259 (2015).
[20] A.V. Sachenko, N.A. Prima, A.P. Gorban, and A.A. Serba, Porc. 17th European Photovoltaic Solar Energy Conference (Munich, Germany) 1 230 (2001).
[21] A.V. Sachenko, A.P. Gorban, V.P. Kostylev, and I.O. Sokolovskii, Semiconductors 41, 281 (2007).
[22] D.E. Kane and R.M. Swanson, J. Appl. Phys. 73, 1193 (1993).
[23] R. Corkish, D. S.-P. Chan, and M.A. Green, J. Appl. Phys. 79, 195 (1993).
[24] M.A. Green, 2nd World Conf. and Exhibition on Photovoltaic Solar Energy Conversion (Vienna, 1998), pp. 74-76.

UNVEILING THE EFFECT OF LEAD HALIDE ON THE THERMAL STABILITY OF PEROVSKITE SOLAR CELLS

Yangyang Du, Hongkun Cai*, Yuxiang Wu , Hongbin Wen, Zhenglong Li, Jian Xu, Like Huang, Jian Ni, Juan Li, Jianjun
Zhang

College of Electronic Information and Optical Engineering, Nankai University, Key Laboratory of Opto-electronic Information Science and Technology for Ministry of Education, Tianjin, 300071, China

ABSTRACT:

Sequential deposition methods for preparing $CH_3NH_3PbI_3$ film are becoming increasingly popular when considering processing perovskite solar cells (PSCs) under relative relaxed conditions. Correspondingly, the problem concerning PbI_2 residue receives special attention due to its deterioration to device performance reproducibility and long-term stability. In this work, we demonstrate an effective sequential thermal evaporation strategy in low vacuum (STELV) to fabricate PbI_2-free $CH_3NH_3PbI_3$ film with full coverage and good crystallization. Through a comparative study, it is observed that STELV has a stronger ability to control morphology and thickness of $CH_3NH_3PbI_3$ film, alone with good uniformity. Moreover, we demonstrate that remaining PbI_2 film should be responsible for the poor PSCs thermal stability.

1 INTRODUCTION

Organic-inorganic halide perovskites (e.g. $CH_3NH_3PbI_3$ or $CH_3NH_3PbI_{3-x}Cl_x$) have increasingly excited research community around the world as one of the most promising materials for photovoltaic application during the past few years due to their excellent characteristics such as high absorption coefficient (>10^4 cm^{-1}) [1], long carrier transport diffusion length (100-1000 nm)[2], tunable optical band gaps[3] as well as simple and low-cost spin coating technology[4-8]. Generally, high efficiency PSCs are composed of either a planar or planar/mesoporous hybrid perovskite architecture, both exhibiting comparable efficiency above 18% [4,9]. To date, the highest certified PCE value of PSCs has exceeded above 20.1%[4] and the size of monomeric PSCs has broken through 1 cm^2 as well[6]. In addition, long-term stability tests for PSCs stored in air at room temperature without encapsulation were carried out for over 1000 hrs under full sun light[6, 10], indicating enormous progresses in the development of PSCs. However, an inevitable challenge that the fabrication of efficient PSCs is usually restricted in the glove boxes filled with nitrogen as a consequence of the poor stability against water and oxygen degradation for perovskite is still unsolved [4,7,8].

Presently, a plethora of approaches on fabricating perovskite layer have been proposed in most literature reports including one-step deposition[4,7], sequential deposition [11-16] and evaporation deposition in high vacuum[17-19] in order to obtain high quality perovskite film which plays a crucial role for the PSCs performance. And it is well admitted that sequential deposition methods can offer facile perovskite morphological control and achieve better perovskite film in ambient air compared with one-step deposition [12]. High PCE over 15% has been obtained via modified sequential deposition method under relative relaxed conditions[12,16], in which PbI_2 is first deposited on a mesoporous TiO_2 substrate by spin-coating, followed by exposing it to an anhydrous isopropanol (IPA) solution of CH_3NH_3I or vapor of CH_3NH_3I. However, the problem of PbI_2 residue in these films is still a universal phenomenon, which has a negative effect on the long-term stability of device except deteriorating the reproducibility of device performance [14,16]. Furthermore, planar PSCs fabricated with conventional sequential deposition were rarely reported, which results from the fact that PbI_2 films from dimethyl formamide (DMF) solution tend to form layered and dense crystallization on a flat substrate [13,14,16], leading to the incomplete conversion from PbI_2 to $CH_3NH_3PbI_3$ when exposed to the CH_3NH_3I ingredient as a consequence of a short diffusion length between PbI_2 and CH_3NH_3I [11]. These drawbacks usually make the residual PbI_2 inevitable in the process of conventional sequential deposition methods.

In order to overcome this problem, several strategies have been reported. For instance, Choy et al. developed a modified approach via incorporating additive (4-tert-butylpyridine tBP) to form a porous PbI_2 nanostructure film contributing to the immersion of CH_3NH_3I. And they finally achieved dramatic performance improvement with 16.21% and showed a better long-term stability [16]. Likewise, zhao et al. reported a three-step sequential deposition method for preparing PbI_2-free $CH_3NH_3PbI_3$ films by thermally unstable stoichiometric PbI_2/CH_3NH_3Cl precursor films replacing the pure PbI_2 [13]. Meanwhile, relative report also demonstrated a facile morphology controllable sequential deposition for $CH_3NH_3PbI_3$ films via using a novel volume expansion adjustable PbI_2 and x CH3NH3I precusor films to replace pure compact PbI_2 [14]. With the further understanding the mechanism of sequential deposition method, the fact that the morphology of the PbI_2 layer is crucial in determining the final morphology of $CH_3NH_3PbI_3$ and photovoltaic performance was unveiled [12]. All those results indicated that successful conversion of PbI_2 to $CH_3NH_3PbI_3$ is essential to high efficiency and stable PSCs.

Here we report a facile and morphology controllable sequential thermal evaporation strategy in low vacuum (STELV) for fabricating planar $CH_3NH_3PbI_3$ film without any PbI_2 residue, in which PbI_2 and CH_3NH_3I vapour were subsequently deposited on substrate in low vacuum, followed by solid-state rapid insertion reaction during the process of annealing by means of high concentration gradient of CH_3NH_3I (shown in figure 1.a). In the first step, PbI_2 films via STELV lead to a snowflake porous construction, beneficial to the sufficient diffusion and insertion of the following CH_3NH_3I component, and hence generating tight $CH_3NH_3PbI_3$ films. In addition, the morphologies of final $CH_3NH_3PbI_3$ films can be adjusted by tuning the amount of PbI_2 powder during the first step and the evaporation rate during the deposition process. Moreover, it is experimentally demonstrated that the PSCs based on STELV lead to superior stability, apart from improved device performance.

2 EXPERIMENTAL

The neat FTO glasses substrate were washed with ethanol and acetone (with 1 : 1 volume ratio) via sonication for 15 min, and then sequentially washed with isopropanol for 15 min. Before the deposition of c-TiO_2, FTO substrates were treated in a UV-ozone clearer for 12 min. About 40 nm thick c-TiO_2 was formed on the cleaned FTO glass by directly annealing processing at 500 ℃ to the formed Ti film for 40 min, which was obtained by radio frequency (RF) magnetron sputtering from Ti target (99.9% purity) [20]. The RF power and sputtering pressure were 60 W and 0.3 Pa, respectively. In contrast, we prepared $CH_3NH_3PbI_3$ films by the methods of conventional sequential deposition methods and STELV respectively. The $CH_3NH_3PbI_3$ films based on conventional sequential deposition methods were prepared by spin-coating the solution of PbI_2 (40 wt%) on c-TiO_2 substrate at 5000 rpm for 40s in ambient air, followed by exposing it to an anhydrous isopropanol (IPA) solution of CH_3NH_3I (10 mg/ml) or CH_3NH_3I vapor and annealing at 100 ℃ for 30 min. With regard to the $CH_3NH_3PbI_3$ film fabricated by STELV, PbI_2 power was firstly evaporated on substrate via thermal evaporation in low vacuum (~ 1 Pa) with 65 A evaporation current, and subsequently superfluous CH_3NH_3I powder was evaporated on the formed PbI_2 film to make sure the rich CH_3NH_3I environment. During the process of annealing, final $CH_3NH_3PbI_3$ film was assembled with the interaction and interdiffusion of solid-state PbI_2 and CH_3NH_3I components. In order to control the morphology of $CH_3NH_3PbI_3$ film by STELV, the amount of PbI_2 powder and the evaporation current can be easily tuned. A Spiro-OMeTAD solution was prepared by dissolving 80 mg Spiro-OMeTAD in 1 ml chlorobenzene, to which 18 ul of lithium bis (trifluoromethanesulfony) imide (LiTFSI) solution (280 mg/ml acetonitrile) and 10 ul of 4-tert-butylpyridine were added. The hole-transport layer was prepared by spin-coating Spiro-OMeTAD solution at 3500 rpm for 40s. Finally, a 100 nm thick gold counter electrode was deposited by thermal evaporation and the dot area of PSCs is 0.06 cm^2. Figure 1.b represents the device structure comprising FTO/c-TiO_2/$CH_3NH_3PbI_3$/Spiro-OMeTAD/Au. The representative SEM cross-sectional image of planar PSCs is shown in figure 1.c. In addition, the real image of $CH_3NH_3PbI_3$ prepared via STELV is also provided in figure 1.d.

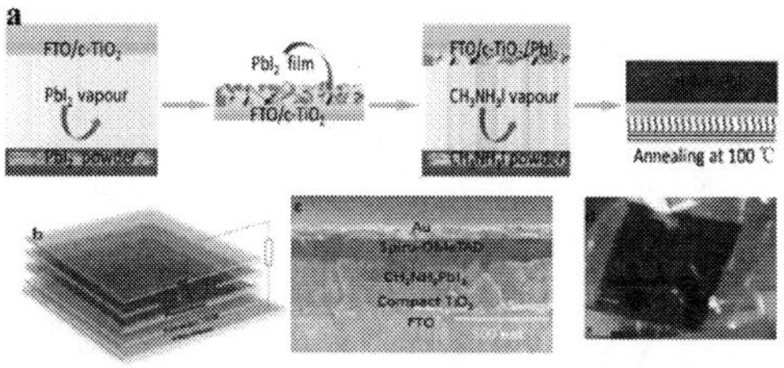

Figure 1. (a) schematic representation of fabricating $CH_3NH_3PbI_3$ film via STELV; (b) Overall device architecture of the FTO/compact TiO_2(c-TiO_2)/$CH_3NH_3PbI_3$/Spiro-OMeTAD/Au; (c) the corresponding cross-section SEM image of planar PSCs construction; (d) the real image of perovskite film via STELV.

3 RESULTS AND DISCUSSION

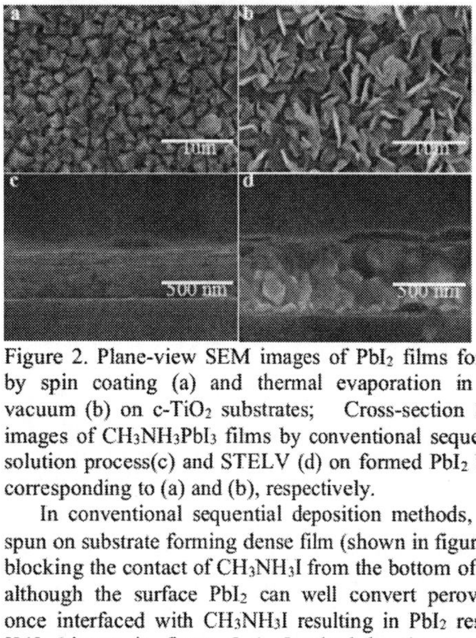

Figure 2. Plane-view SEM images of PbI_2 films formed by spin coating (a) and thermal evaporation in low vacuum (b) on c-TiO_2 substrates; Cross-section SEM images of $CH_3NH_3PbI_3$ films by conventional sequential solution process(c) and STELV (d) on formed PbI_2 films corresponding to (a) and (b), respectively.

In conventional sequential deposition methods, PbI_2 spun on substrate forming dense film (shown in figure 2.a) blocking the contact of CH_3NH_3I from the bottom of PbI_2, although the surface PbI_2 can well convert perovskite once interfaced with CH_3NH_3I resulting in PbI_2 residue [21] (shown in figure 2.c). Inspired by the previous studies [12-14], we firstly evaporated PbI_2 powder in low vacuum and obtained a snowflake porous PbI_2 film (shown in figure 2.b), and then the tight $CH_3NH_3PbI_3$ film formed with the infiltration of CH_3NH_3I vapor by high concentration gradient from surface to bottom (shown in figure 1.a). Remarkably, sufficient CH_3NH_3I vapor is necessary in the second step evaporation deposition. As figure 2.d present, high crystallization and closely packed $CH_3NH_3PbI_3$ cuboids can be observed.

As the further information to demonstrate PbI_2-free via STELV, X-Ray Diffraction (XRD) measurements of $CH_3NH_3PbI_3$ films via different methods were compared. Figure 3 shows the different XRD patterns of $CH_3NH_3PbI_3$ films corresponding to different methods. Obviously, there is an incomplete conversion as evidenced by the existence of PbI_2 (near 12.6 。) diffraction peaks in conventional sequential deposition routes no matter CH_3NH_3I IPA solution or CH3NH3I vapor. Moreover, $CH_3NH_3PbI_3$ (near 14 。) diffraction peaks also exhibited poorly peak intensities compared with STELV, which indicated inferior oriented phase not facilitating exciton transport across the perovskite above

300 nm thickness apart from vast pinholes due to $CH_3NH_3PbI_3$ dissolution.

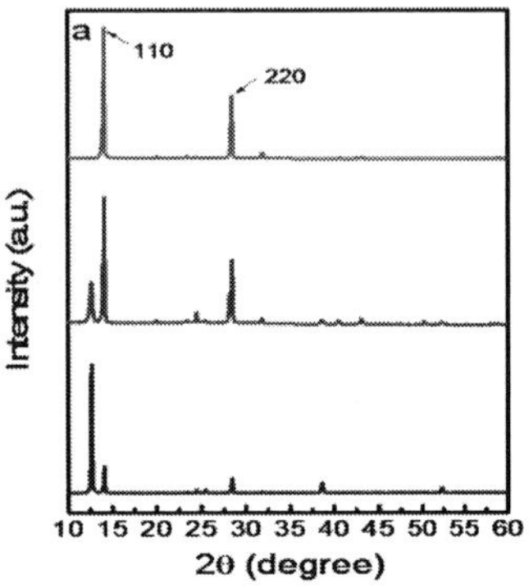

Figure 3. the XRD image of $CH_3NH_3PbI_3$ films via different methods on glass substrate, where the blue line depicts to $CH_3NH_3PbI_3$ film by spinning PbI_2 with CH_3NH_3I vapor treatment; the red line depicts to $CH_3NH_3PbI_3$ film by spinning PbI_2 with 4 min CH_3NH_3I IPA solution dipping.

Table Supplementary 1 Photovoltaic parameters derived from J-V measurements corresponding to figure 4

Device	Jsc (mA/cm2)	Voc(V)	FF (%)	PCE (%)
Spining PbI_2 +dipping CH_3NH_3I IPA	9.14	0.79	50	3.61
Spining PbI_2 + CH_3NH_3I treatment	16.02	0.94	54	8.13
STELV	19.76	0.98	64	12.47

The performance of PSCs based on $CH_3NH_3PbI_3$ films prepared by mentioned approaches without encapsulation are shown in figure 4. As can be observed from figure 4.a, the champion PCE of the device based on STELV yield a PCE of 12.47% with a V_{oc} of 0.98 v, J_{sc} of 19.76 mA cm^{-2} and FF of 64%. Compared to its counterparts based on conventional sequential deposition (shown in Table 1), the STELV-based perovskite films produces a superior PL intensity, suggesting the fewer traps state in formed perovskite film as shown in figure 4.b and thus corresponding to the higher J_{sc} value. With regard to the enhancement of V_{oc}, a more favorable recombination resistance blocking the decrease of leakage current due to compact $CH_3NH_3PbI_3$ based on STELV can illustrate this, which was demonstrated in following content. In addition, carefully contrasting the difference between figure 2.c and d, it is can be observed that $CH_3NH_3PbI_3$ film possesses better vertical growth grains by STELV compared with conventional layer growth, beneficial to the vertical carriers transport inside of the cells, thus decreasing the series resistance and increasing PSCs performance as well. Notably, all the devices were fabricated with identical other preparation conditions. Moreover, figure 4.c presents the long term stability test of the devices performance. In our case, the difference between three specimens was only the presence of fabrication of $CH_3NH_3PbI_3$ and hence the rapid efficiency decline should be attributed to the influence of PbI_2 residue. Finally, we calculated devices performance distribution by histogram (shown in figure 4.d), indicating good reproducibility with STELV. To further investigate the reason for the rapid efficiency decrease, the color change of perovskite films was observed. As we can note from figure 5, when perovskite film fabricated by conventional sequential deposition, the film color rapidly turned from dark to yellow, suggesting the degeneration of perovskite films. However, the film color of perovskite fabricated by STELV could bear long term stability even located in high temperature. It is worth noting that the remaining PbI_2 via sequential deposition method is so different from the situation of residual PbI_2 existing in grain boundary, which can directly impede the effective carrier transport and injection.

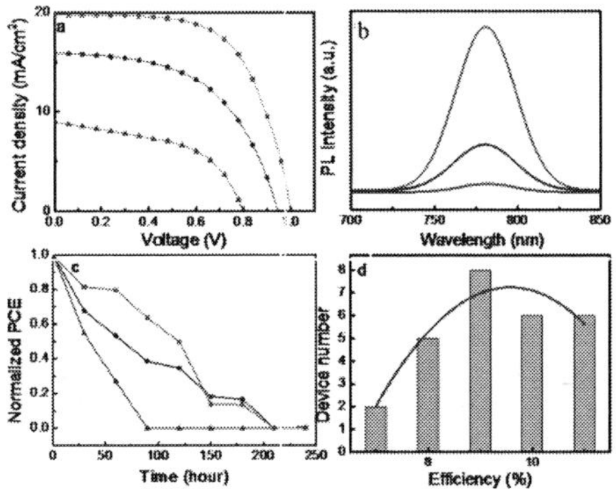

Figure 4. Efficiency measurements. J-V curves for PSCs without encapsulation (a) and corresponding EQE spectrum (b) and normalized stability test of the devices at ambient environment (c), where the purple line depicts to $CH_3NH_3PbI_3$ film by STELV; the blue line depicts to $CH_3NH_3PbI_3$ film by spinning PbI_2 with CH_3NH_3I vapor treatment; the red line depicts to $CH_3NH_3PbI_3$ film by spinning PbI_2 with 4 min CH_3NH_3I IPA solution dipping; (d) Histogram of efficiency for 27 devices based on STELV.

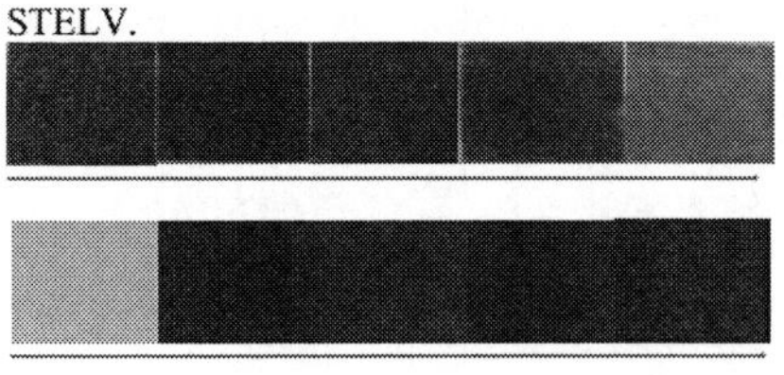

Figure 5. The color change of perovskite prepared by different methods.

4 CONCLUSION

In summary, we have successfully demonstrated an effective and facile strategy for fabrication $CH_3NH_3PbI_3$ films via sequential thermal evaporation in low vacuum. By carefully control the amount of PbI_2 and the evaporation rate, a snowflake porous PbI_2 was firstly obtain, ensuring the effective infiltration and diffusion of CH_3NH_3I vapor, and thereby assembling full coverage $CH_3NH_3PbI_3$ film without any PbI_2 residue. Compared with conventional sequential deposition, the overall performance of PSCs was improved, and so was the corresponding devices thermal stability. This study represents an important step forward in developing low-cost and stable perovskite-based photovoltaic devices.

ACKNOWLEDGE

This work was supported by National Natural Science Foundation of China (Grant No. 61377031), National Natural Science Foundation of China (No. 61504068), National Natural Science Foundation of China (Project No. 61076006).

References

[1] De Wolf S, Holovsky J, Moon S J, et al. Organometallic halide perovskites: sharp optical absorption edge and its relation to photovoltaic performance[J]. The Journal of Physical Chemistry Letters, 2014, 5(6): 1035-1039.
[2] Stranks S D, Eperon G E, Grancini G, et al. Electron-hole diffusion lengths exceeding 1 micrometer in an organometal trihalide perovskite absorber[J]. Science, 2013, 342(6156): 341-344.
[3] Filip M R, Eperon G E, Snaith H J, et al. Steric engineering of metal-halide perovskites with tunable optical band gaps[J]. Nature communications, 2014, 5.
[4] Yang W S, Noh J H, Jeon N J, et al. High-performance photovoltaic perovskite layers fabricated through intramolecular exchange[J]. Science, 2015, 348(6240): 1234-1237.
[5] Du Y, Cai H, Ni J, et al. Air-processed, efficient CH3NH3PbI3− xClx perovskite solar cells with organic polymer PTB7 as a hole-transport layer[J]. RSC Advances, 2015, 5(82): 66981-66987.
[6] Chen W, Wu Y, Yue Y, et al. Efficient and stable large-area perovskite solar cells with inorganic charge extraction layers[J]. Science, 2015, 350(6263): 944-948.
[7] Zhou H, Chen Q, Li G, et al. Interface engineering of highly efficient perovskite solar cells[J]. Science, 2014, 345(6196): 542-546.
[8] Jeon N J, Noh J H, Kim Y C, et al. Solvent engineering for high-performance inorganic–organic hybrid perovskite solar cells[J]. Nature materials, 2014, 13(9): 897-903.
[9] Rakstys K, Abate A, Dar M I, et al. Triazatruxene-based Hole Transporting Materials for Highly Efficient Perovskite Solar Cells[J]. Journal of the American Chemical Society, 2015, 137(51): 16172-16178.
[10] Berhe T A, Su W N, Chen C H, et al. Organometal halide perovskite solar cells: degradation and stability[J]. Energy & Environmental Science, 2016.
[11] Burschka J, Pellet N, Moon S J, et al. Sequential deposition as a route to high-performance perovskite-sensitized solar cells[J]. Nature, 2013, 499(7458): 316-319.
[12] Ko H S, Lee J W, Park N G. 15.76% efficiency perovskite solar cells prepared under high relative humidity: importance of PbI2 morphology in two-step deposition of CH3NH3PbI3[J]. Journal of Materials Chemistry A, 2015, 3(16): 8808-8815.
[13] Zhao Y, Zhu K. Three-step sequential solution deposition of PbI2-free CH3NH3PbI3 perovskite[J]. Journal of Materials Chemistry A, 2015, 3(17): 9086-9091.
[14] Zhang T, Yang M, Zhao Y, et al. Controllable Sequential Deposition of Planar CH3NH3PbI3 Perovskite Films via Adjustable Volume Expansion[J]. Nano letters, 2015, 15(6): 3959-3963.
[15] Chen Q, Zhou H, Hong Z, et al. Planar heterojunction perovskite solar cells via vapor-assisted solution process[J]. Journal of the American Chemical Society, 2013, 136(2): 622-625.
[16] Zhang H, Mao J, He H, et al. A Smooth CH3NH3PbI3 Film via a New Approach for Forming the PbI2 Nanostructure Together with Strategically High CH3NH3I Concentration for High Efficient Planar - Heterojunction Solar Cells[J]. Advanced Energy Materials, 2015, 5(23).
[17] Liu M, Johnston M B, Snaith H J. Efficient planar heterojunction perovskite solar cells by vapour deposition[J]. Nature, 2013, 501(7467): 395-398.
[18] Chen Y, Chen T, Dai L. Layer-by-Layer Growth of CH3NH3PbI3-xClx for Highly Efficient Planar Heterojunction Perovskite Solar Cells[J]. Advanced Materials, 2015, 27(6): 1053-1059.
[19] Chen C W, Kang H W, Hsiao S Y, et al. Efficient and uniform planar - type perovskite solar cells by simple sequential vacuum deposition[J]. Advanced Materials, 2014, 26(38): 6647-6652.
[20] Ke W, Fang G, Wang J, et al. Perovskite solar cell with an efficient TiO2 compact film[J]. ACS applied materials & interfaces, 2014, 6(18): 15959-15965.
[21] Lee Y H, Luo J, Humphry - Baker R, et al. Unraveling the reasons for efficiency loss in perovskite solar cells[J]. Advanced Functional Materials, 2015, 25(25): 3925-3933.
[22] Chen Q, Zhou H, Hong Z, et al. Planar heterojunction perovskite solar cells via vapor-assisted solution process[J]. Journal of the American Chemical Society, 2013, 136(2): 622-625.

COLLOIDAL SYNTHESIS, STRUCTURAL AND OPTICAL PROPERTIES OF $CuIn_3Se_5$ NANOCRYSTALS FOR PHOTOVOLTAICS

M. A. Ghali [1], E. M. ElNimr[1,2], Gh. F. Ali[1,2], A. M. Eissa[1]
[1]Laboratory of Nanophotonics, Physics Department, Faculty of Science, Kafrelsheikh University
33516 Kafrelsheikh, Egypt
[2]Physics Department, Faculty of Science, Tanta University
Tanta, Egypt

ABSTRACT: We report on the synthesis of *colloidal* $CuIn_3Se_5$ nanocrystals, for the first time, using simple solution-based method. The cubic phase $CuIn_3Se_5$ nanocrystals grow, in the shape of nanoparticles, with an average diameter of 35 nm. The band gap of the grown $CuIn_3Se_5$ nanocrystals is determined to be 1.33 eV and thus can be suitable for ink-coated thin film solar cells.
Keywords: Colloidal, Nanocrystals, $CuIn_3Se_5$

1 INTRODUCTION

Thin film solar cells based on effective light absorbing I-III-VI_2 semiconductors, like CuInSe (CIS) and CuInGaSe (CIGS), are well known to be stable under long term light excitation with nearly 20% solar energy conversion efficiencies [1]. Traditionally, both CIS and CIGS materials in state-of-the art devices are deposited using multi-stage co-evaporation and the two-step process of sputtering and selenization. Such type of fabrication technology is very costly and requires sophisticated and expensive vacum equipments, besides the difficulty to achieve controlled stoichiometry of CIS and CIGS materials over large devices area. The surface stoichiometry of CIS thin films as a function of composition is important for photovoltaic device performance, where "In" rich films like $CuIn_2Se_{3.5}$, $CuIn_3Se_5$ and $CuIn_5Se_8$ phases are formed due to Cu deficiency [2,3]. In many works in literature, an In-rich n-type surface layer of $CuIn_3Se_5$ phase was found at the surface of p-CuInSe layers in high-efficiency vacum based thin film solar cells. The dominant $CuIn_3Se_5$ phase is identified as ordered vacancy compounds (OVC) [4,5]. This ordering refers to the regular arrangement of preferred crystallographic sites on which defects are found. The $CuIn_3Se_5$ material shows suitable electronic energy band gap which matches the natural Sun spectrum with forbidden energy band gap of about 1.3 eV (which is close to the optimal value for solar cell applications) [6,7]. Also, $CuIn_3Se_5$ has relatively higher photoconductivity over a broad wavelength range. It was reported earlier that an improvement in the efficiency of p-CIS based solar cell with a thin 100 nm $CuIn_3Se_5$ layer can be achieved, where n-$CuIn_3Se_5$ and p-$CuInSe_2$ formed a good heterojunction [8]. The $CuIn_3Se_5$ compound is usually grown using vacum-based techniques like Co-evaporation [1], Molecular Beam epitaxy [9], Sputtering [10], etc.

Furthermore, it is widely accepted that most Cu-In-Se based solar cells which are fabricated by vacum technology may consist of two heterojunctions in series, between CIS and $CuIn_3Se_5$, and $CuIn_3Se_5$ and CdS [11]. Consequently, the ordered-defect phase $CuIn_3Se_5$ is expected to play an important role in the operation and optimization of CIS/CdS solar cell devices. Moreover, achieving high quality thin films of n-$CuIn_3Se_5$ with potential photoresposnive properties will be crucial for new organic–inorganic hybrid photovoltaic structures, where an inorganic photoabsorber layer of n-type forms an n–p junction with an organic layer of p-type (e.g., conductive polymers) [12]. A detailed study of the physical properties of $CuIn_3Se_5$ is still necessary for better understanding of device operation and further improvement of solar cells performance. Compared with vacum-based techniques, solution-based methods offer low-cost and scalability advantages.

Here, we report on the synthesis of *colloidal* $CuIn_3Se_5$ nanocrystals using simple solution-based method. The cubic phase $CuIn_3Se_5$ nanocrystals grow, in the shape of nanoparticles, with an average diameter of 35 nm. The band gap of the grown $CuIn_3Se_5$ nanocrystals is determined to be 1.33 eV. A stable ink solution from these nanocrystals was achieved.

2 EXPERIMENT

2.1 Synthesis process

$CuIn_3Se_5$ nanocrystals were synthesized by reacting CuCl, $InCl_3$ and Se in Oleylamine (OLA) at an elevated temperature of 260 °C in N_2 gas atmosphere. All manipulations were performed using standard air-free techniques utilizing nitrogen-gas filled glove box (mBRAUN, Germany). A chemical reaction between the reactants was initiated in a 25 ml three-neck round bottom flask, which was purged three times by repeated cycles of vacuum and back-filling with N_2 gas (see Fig.1). Namely, 0.04 g of CuCl, 0.11 g of $InCl_3$, 1 ml of 1 molar Se powder and 5 ml of OLA; were added to the flask under N_2 atmosphere. A 7.5 ml of the miscible anti-solvent ethanol was added to precipitate the particles before collection by centrifuge. Dark precipitate of $CuIn_3Se_5$ was observed in the solution before being re-dispersed in a non-polar solvent to form a stable dispersion.

2.2 Characterization

The size and morphology of the as-synthesized $CuIn_3Se_5$ nanocrystals were characterized using Transmission Electron Microscope (TEM, Joel 200). For this purpose, the nanocrystals were dispersed in hexane and deposited on a TEM copper grid. The crystalline structure of the $CuIn_3Se_5$ nanocrystals was determined using X-ray diffraction XRD (Shimadzu 6000-XRD), and the optical properties were measured using a UV-Vis-NIR spectrometer (Shimadzu 2600/2700). The PL

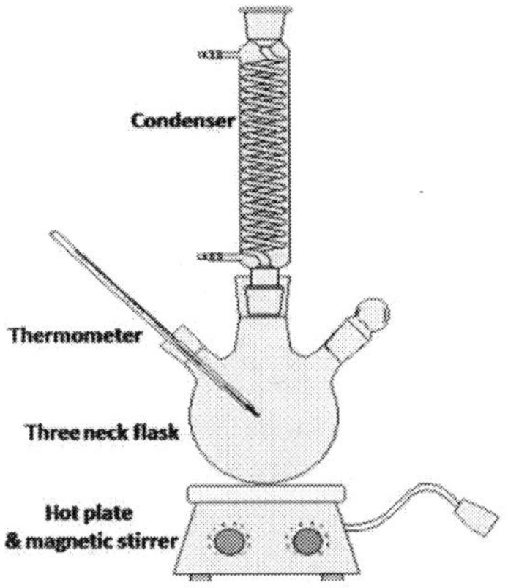

Figure 1: Three-neck flask apparatus used for the growth of colloidal $CuIn_3Se_5$ nanocrystals.

measurements were performed using our micro-PL set up (attocube DRY700) installed in the laboratory of Nanophotonics, Kafrelsheikh University using InGaN blue laser (403 nm), 55 cm grating monochromator and CCD camera.

3 EXPERIMENTAL RESULTS

3.1 Structural analysis

The crystal structures of the as-synthesized $CuIn_3Se_5$ nanocrystals have been examined using X-ray diffraction (XRD). As shown in Fig.2, the experimental XRD data match well with the standard XRD patterns for $CuIn_3Se_5$ material (ICDD 04-016-4112), indicating that $CuIn_3Se_5$ nanocrystals crystallize in the cubic structure (space group: $F\bar{4}3m$, a=5.798 Å) [13]. The peaks at 26.58°, 44.15°, 52.31°, 64.29°, and 70.84° can be indexed to the (111), (220), (311), (400) and (331) planes of Cubic $CuIn_3Se_5$, respectively. From Scherer's equation [14], $D = \frac{0.94\lambda}{\beta cos\theta}$ we estimate the average particle size to be D = 36 nm where λ is X-ray wavelength, β is the full width at half maximum (FWHM) in radians, and θ is the Bragg angle.

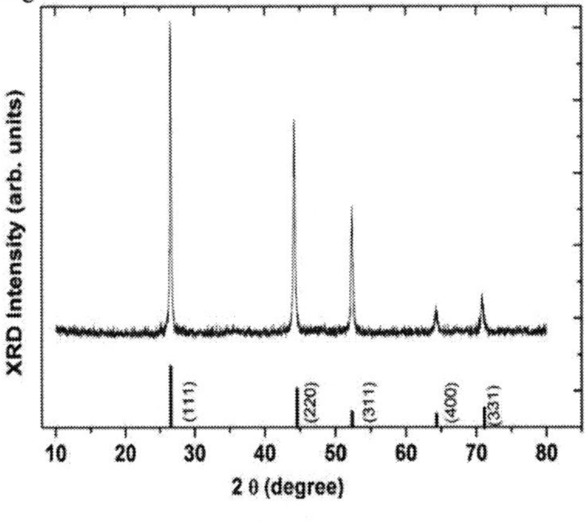

Figure 2: X-Ray Diffraction (XRD) pattern of as-synthesized $CuIn_3Se_5$ nanocrystals powder. Reference data for bulk $CuIn_3Se_5$ are shown as vertical lines.

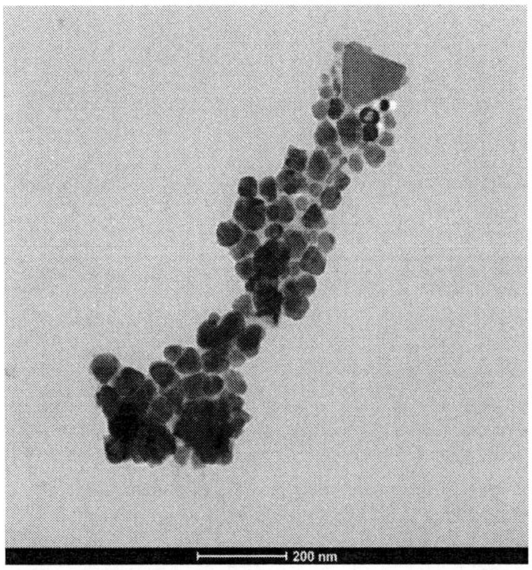

Figure 3: Transmision Electron Microscope (TEM) images of $CuIn_3Se_5$ nanocrystals with average size 35 nm. The inset shows ink-solution of the dispersed $CuIn_3Se_5$ nanocrystals.

We investigated the morphology of our grown $CuIn_3Se_5$ nanocrystals using Transmission Electron Microscopy (TEM) where a cubic-phase character of $CuIn_3Se_5$ nanocrystals can be seen in Fig.3. The as-synthesized product forms individually separated nanoparticles with diameter of 35 nm, in agreement with that calculated from the XRD data.

From independent Electron diffraction measurment (not shown), it was confirmed that $CuIn_3Se_5$ nanocrystals have cubic crystal structure. There are no observable peaks for other binary or ternary impurities such as $CuSe_2$, $InSe_2$ or other secondary phases of CuInSe system.

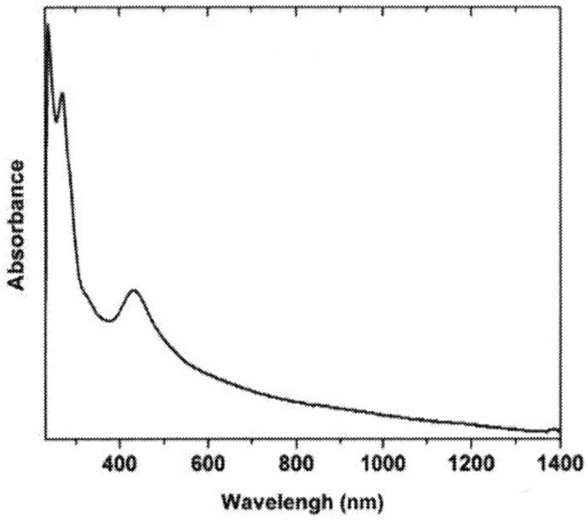

Figure 4: Absorption spectrum of $CuIn_3Se_5$ colloidal nanocrystals.

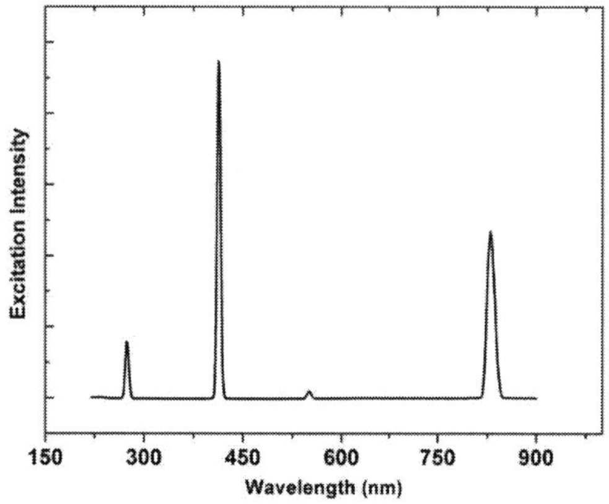

Figure 5: Room temperature photoluminscence excitation (PLE) spectrum of $CuIn_3Se_5$ colloidal nanocrystals.

3.2 Optical properties

The properties of the $CuIn_3Se_5$ nanocrystals have been studied also using optical absorption measurements. From the shown absorption spectrum in Fig. 4, we can estimate the $CuIn_3Se_5$ nanocrystals band gap by plotting $(Ah\nu)^2$ versus hν, where "*A*" is the absorbance, h Planck's constant, ν is the frequency, and extrapolating the linear portion of the spectrum in the band gap region. The band gap is determined to be 1.33 ± 0.03 eV (corresponding to λ= 930 nm, as shown in Fig.4), which is close to that well known for bulk $CuIn_3Se_5$ (E_g=1.2-1.31 eV[15]). It is worth noting in Fig.4, the sharp lines observed at 430 nm, 335 nm and 275 nm. These lines correpsopnd to optical absorptions from higher energy excited states in $CuIn_3Se_5$. The latter reflects the high optical quality of our synthesised $CuIn_3Se_5$ naocrystals.

Also, Fig. 5 shows photoluminescence excitation (PLE) spectrum from the synthesized $CuIn_3Se_5$ nanocrystals dispersed in hexane. The relatively narrow peaks shown in Fig.5, which match well with the higher energy excited states transitions observed in the absorption spectrum, reflect again a high optical quality and narrow size distribution of our synthesized $CuIn_3Se_5$ nanocrystals.

4. CONCLUSIONS

Nanocrystals of cubic phase $CuIn_3Se_5$ have been synthesized, for the first time, via a simple low-cost wet chemistry method. The as-synthesized $CuIn_3Se_5$ nanocrystals in the shape of nanoparticles were well crystallized with a nearly uniform morphology. X-ray diffraction and high-resolution TEM measurements confirm that our as-grown nanocrystals are polycrystalline cubic $CuIn_3Se_5$.

The band gap of the cubic $CuIn_3Se_5$ nanocrystals is determined to be 1.3 eV. These nanocrystals could be deposited in thin film form; and thus being suitable for potential low-cost *all*-solution processed solar cells devices.

4 REFERENCES

[1] Ingrid Repins, Miguel A. Contreras, Brian Egaas, Clay DeHart, John Scharf, Craig L. Perkins, Bobby To, and Rommel Noufi, Prog. Photovolt: Res. Appl. 16 (2008) 235.
[2] M. M. Islam, T. Sakurai, A. Yamada, S. Otagiri, S. Ishizuka, K. Matsubara, S. Niki, K. Akimoto, Sol. Energ. Mat. Sol. C. 95 (2011) 231.
[3] T. V. Kuznetsova, V. I. Grebennikov, H. Zhao, C. Derks, C. Taubitz, Appl. Phys. Lett. 101 (2012) 111607.
[4] G. Aranguren, E. Hernandez, A. Lopez Pescador, C.A. Durante Rincon, Maximo Leon, Materials Letters, 58 (2004) 573.
[5] M. León, R. Serna, S. Levcenko, A. Nateprov, A. Nicorici, J. M. Merino, and E.Arushanov, Phys. Stat. Sol. (a) 203 (2006) 2913.
[6] R .Sharma, A. Ghule, V. Taur, R. Joshi, R. Mane, J.C. Vyas, G. Cai, T. Ganesh,Sun-Ki Min, W. Lee, Sung-Hwan Han, Appl. Surf. Scien., 255 (2009) 8158.
[7] K. Laes, S. Bereznev, R. Land, A. Tverjanovich, O.Volobujeva, R. Traksmaa, T. Raadik, A. Öpik, Energ. Proced. , 2 (2010) 119.
[8] Se Hankwon, S. Park, B. Ahn, K. Yoon and J. Song, Solar Energy 64 (1998) 55.
[9] A. J. Nelson, G. S. Horner, K. Sinha, M.H. Bode, Appl. Phys. Lett. 64 (1994) 3600.
[10] Tooru Tanaka, Yasutaka Demizu, Toshiyuki Yamaguchi, Akira Yoshida, Jpn. J. Appl. Phys. 35 (1996) 2779.
[11] H. Z. Xiao, L. Chung Yang, A. Rockett, J. Appl. Phys. 76 (1994) 1503.
[12] A. Tverjanovich, S. Bereznev, E. N. Borisov, D. Kim, J. Kois, K. Laes,O.Volobujeva, A. Öpik, E. Mellikov, and Y. S. Tveryanovich, Proceedings of the Estonian Academy of Sciences, 58 (2009) 24.
[13] J. M. Merino, M. Di Michiel, M. Leo'n, Journal of Physics and Chemistry of Solids 64 (2003) 1649.
[14] B. D. Cullity and S. R. Stock, Elements of X-ray Diffraction (Prentice-Hall,Saddle River, NJ, 2001) 3rd ed., p. 264.
[15] P. Malar and S. Kasiviswanathan, Sol. Energ. Mat. Sol. C., 85 (2005) 521.

5 ACKNOWLEDGMENTS

We acknowledge financial support from Science and Technology Development Fund (STDF) in Egypt, under Contract No.4333, Young Research Grant.

FAST PROCESSING OF SOL-GEL TCO

, J. van Deelen[1], M. Rem[1], N. Arfsten[1], PJ.P. Buskens[1,2]
[1]TNO, PO Box 6235, 5600 HE Eindhoven, The Netherlands,
[2]DWI Leibniz-Institut für Interaktive Materialien, Forckenbackstrasse 50, 52074 Aachen, Germany
corresponding author: joop.vandeelen@tno.nl tel. +31 6106 25 313

ABSTRACT: TCOs are usually deposited using sputtering or chemical vapor deposition, which have a yield of typically 50-75%. The sol gel method does not need low pressure and can be done with a high precursor yield in the range of 90 – 100%. Sol gel enables also the TCO function as a planarization or texturisation layer. Major drawback and cost driver is the annealing step. The annealing process parameter requirements are highly related to the ingredients and their preparation. These factors can be tuned to get both a layer of high quality and a fast and cost effective process. We present here an annealing study with in-situ conductivity measurement to monitor the effect of temperature ramping, process time, process gas, etc. For ITO, we obtained a sheet resistance of 20 Ohm/sq for layers below 100 nm in thickness as a best result. Furthermore, the annealing process can be tuned to be within a couple of minutes.
Keywords: TCO, transparent conductors, sol-gel

1 INTRODUCTION

Thin film solar cells require transparent conductors (TCOs). Also for heterojunction cells TCOs are becoming an important part. Mainstream TCO's such as doped ZnO and ITO are usually applied by sputtering. CdTe solar cells use SnO_2F doped glass as a substrate. This is applied by in-line atmospheric pressure CVD, which gives a native texturing to the TCO. Also low pressure CVD of ZnO has shown to have such a texturing, which can be changed by the deposition parameters. For sputtered TCOs etching has been reported as a texturing method. All these ways of texturing have limits imposed by the processing.

Imprinting does not have these limitations and offers a free choice of size and shape. It is therefore interesting to investigate this method. In the past a substrate was textured and a TCO was deposited on top. This might change the TCO characteristic to sub-par quality by introduction of additional defects. If a sol gel TCO [1-4] is made and subsequently textured, it would not only save a deposition and curing step, but would also enable more control of the final texture and possibly prevent formation of defects from the texture.

On the other side of the spectrum, there is a need for planarization layers. Sol gel TCOs could be used to planarize a surface. For instance, if the top surface of a CIGS cell is very rough, the most effective way to make use of this is to get a smooth surface between the TCO and the EVA encapsulant as shown in Figs. 1 and 2.

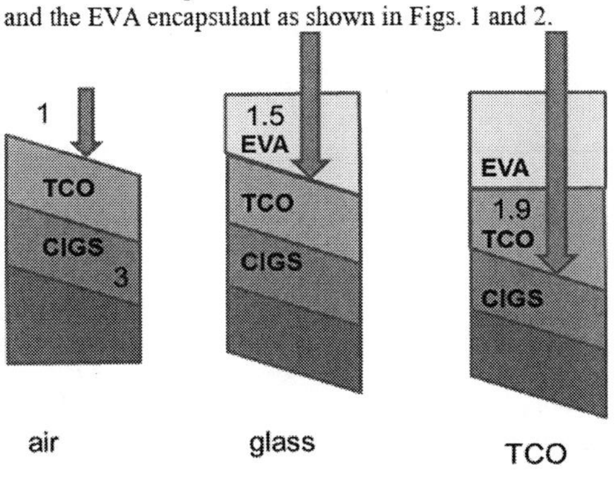

Figure 1: schematic representation of different configurations of cells with a textured CIGS.

The arrows depict up to where the light penetrates in the solar cell without changing the angle. Figure 2a shows the effect of these textures on the effective angle in the CIGS. Figure 2b shows how this angle translates into a pathlength elongation.

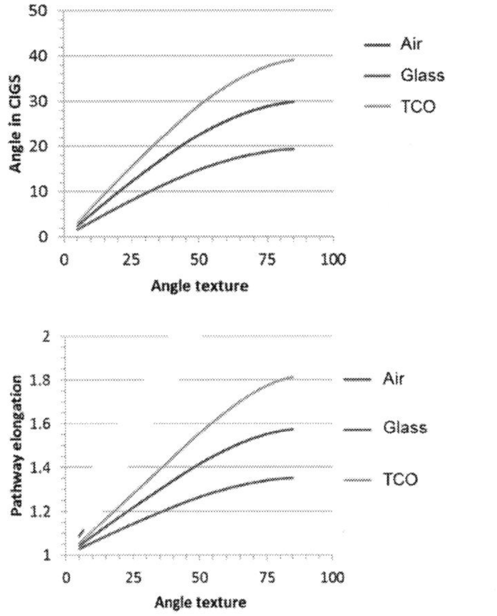

Figure 2: (a) Angle of the light path in CIGS as function of the angle of the texture. The light enters perpendicular to the overall substrate. (b) the curves of figure (a) translated to pathway elongation factor.

It is clearly seen that in an encapsulated panel, a smooth TCO EVA interface is beneficial for the path length elongation. Note that for a TCO, a back texture results in only limited path length elongation. The suggests that for accurate estimation of the back texturing, it is necessary to encapsulate the cell. It is also seen that a planarization TCO could be beneficial from an optical point of view.

This work shows the results on sol gel TCO and its annealing process that could enable these light management capabilities of imprinting and planarization.

2 EXPERIMENTAL

The TCO discussed here is ITO as a test case for other materials, such as SnO_2:F or more complex multimetal oxides. We used indium nitrate and tin chloride as metal precursors and a variety of additives such as pyridine to tune the process and optimize for the conductivity of the layer. The annealing was done using forming gas in an oven equipped with in situ resistance measurement and temperature monitoring of the sample itself, which provides accuracy at high ramping rates.

3 RESULTS

3.1 Tin concentration

The best composition was investigated by changing the tin to indium ratio [5,6]. Fig. 3 shows how there is a window between 6 and 11 at%, which gives excellent resistances.

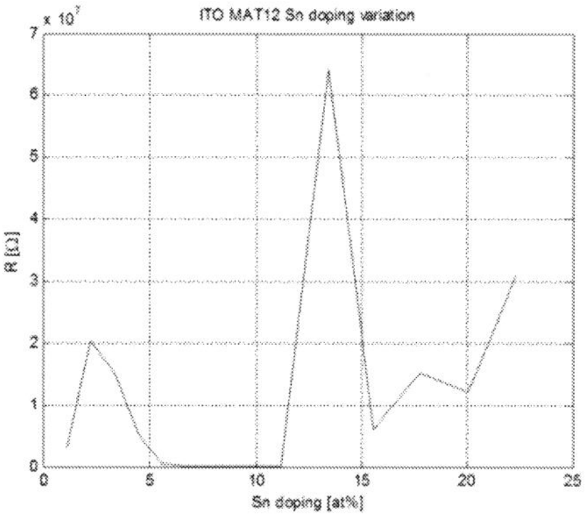

Figure 3: Series resistance as function of the Sn doping.

To analyse this window further, more detailed investigation was performed as shown in Fig. 4. For this series a resistivity of a few mΩ cm were obtained over the entire range between 5 and 10 atom % Sn over indium.

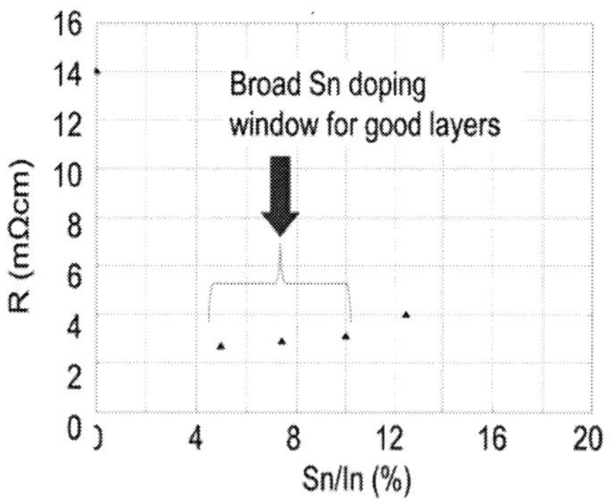

Figure 4.: Resistivity of the layer as function of the Sn/In ratio.

Within this window all the experiments were performed, leading to a lowest resistivity of 8 10-4 Ω cm.

3.2 Pyridine additive

Stability and performance of the sol are also important. For this it is necessary to put in additives. Pyridine can stabilize the sol and improve the performance of the final coating. Stability is usually seen by the time that the sol remains transparent. Addition of pyridine makes the sol stable over longer periods. For higher amounts of pyridine the sol is not as stable as it is for medium amounts of pyridine. Here we show only the electrical measurements of reasonably stable sols. If is evident from figure 5 that the best results were obtained with a pyridine volume percentage of 0.5 to 1 %. Interestingly, it was observed that 'aging' the sol for more than a day makes the final performance less sensitive to the pyridine concentration.

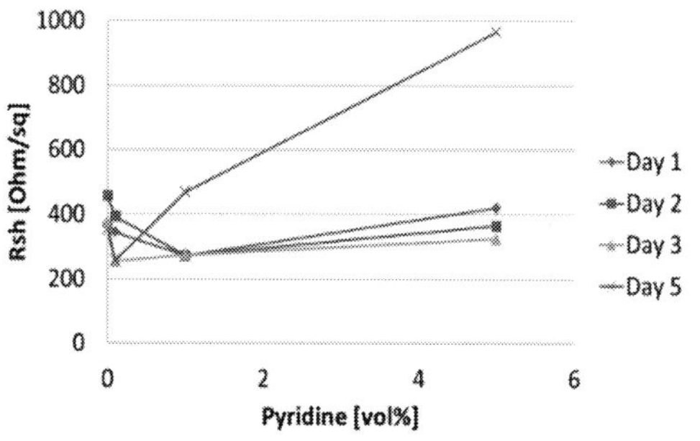

Figure 5: Sheet resistance as function of the pyridine concentration in the sol.

3.3 Annealing study

The annealing process is an important aspect of the manufacturing of the coating [7-9]. In addition to the recipe of the sol and the deposition, the coating needs to be dried and then annealed. This can take a long time and because a thermal infrastructure is needed, prolonged annealing times contribute greatly to the manufacturing cost. In order to investigate the annealing behavior in-situ probing was used. Fig. 6 show and example of the temperature profile and the obtained resistance.

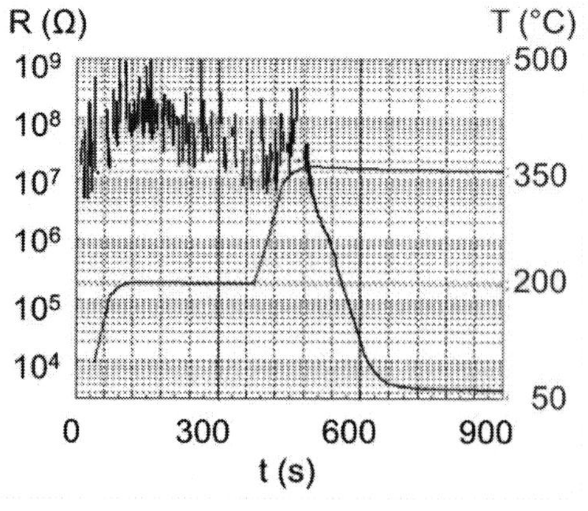

Figure 6: Series resistance (descending blue line, left axis) and the temperature (ascending green line, right axis) as function of the substrate temperature.

The temperature profile was first a heating step up to

200°C. The coating is dried, but does not show much change in electrical conductivity. However, at increase of the temperature to a modest 350°C, the coating is annealed and starts to conduct within minutes. Note that the start and end of the conductivity curve indicate a beginning and an ending of the process, while in between the temperature lowers over orders of magnitude with time at a constant annealing temperature. This trend suggests that the process could be speeded up if a higher temperature would be used. At a processing time of just a few minutes, industrial viability is improved dramatically. It shows that for thin layers of approximately 100 nm thick, the process can be performed very quickly.

For specific texturisation and planarization purposes, it is anticipated that the coating needs to be thicker and research efforts are underway in optimizing the process for various materials, recipes and applications.

4 CONCLUSIONS

Sol gel coating processes were developed giving a Ito coating of 20 Ω/sq sheet resistance for a layer thickness below 100 nm. Furthermore, a fast annealing process was developed and demonstrated via an in-situ measurement setup.

5 REFERENCES

[1] C. Thiele, e.a., *Transparent Conductive Films (TCF) 2012-2022: Forecasts, Technologies, Players*, 2012. p. 159.
[2] Hartnagel, H.L., *Semiconducting Transparent Thin Films* 1995. 358.
[3] C. J. Brinker, et al., Thin Solid Films 201 (1991) 12.
[4] R. Jahn, et al., J. Sol-Gel Sci. Technol 66 (2013) 6.
[5] C. H. Jung, et al., Thin Solid Films 520 (2012) 4.
[6] C. C. Ting, et al., Thin Solid Films 519 (2011) 7.
[7] Sato, T.,. Thermal Analysis and Calorimetry 82(3) (2005) 7.
[8] T. C. Lin, et al., Materials Science and Engineering B 129 (2006) 4.
[9] M. J. Alam, e.a., Thin Solid Films 420-421 (2002) 7.

BETAVOLTAICS. ANALYSIS OF THE ATTAINABLE EFFICIENCY FOR DIRECT-BANDGAP SEMICONDUCTORS

A.V. Sachenko[a], M.A. Evstigneev[b]*, A.I. Shkrebtii[c], R.M. Korkishko[a], V.P. Kostylyov[a], M.R. Kulish[a], I.O. Sokolovskyi[a]
[a]V. Lashkaryov Institute of Semiconductor Physics, National Academy of Sciences of Ukraine, 03028 Kiev, Ukraine
[b]Department of Physics and Physical Oceanography, Memorial University of Newfoundland, St. John's, NL, A1B 3X7 Canada
[c]University of Ontario Institute of Technology, 2000 Simcoe Street North, Oshawa, ON, L1H 7K4 Canada
*E-mail: mevstigneev@mun.ca

ABSTRACT: Conversion of energy of beta-particles into electric energy in a p-n junction based on direct-bandgap semiconductors, such as GaAs, is analyzed considering realistic semiconductor system parameters. An expression for the collection coefficient, Q, of the electron-hole pairs generated by beta-electrons is derived taking into account the existence of the dead layer. We show that the collection coefficient of beta-electrons emitted by a ^{3}H-source to a GaAs p-n junction is close to 1 in a broad range of electron lifetimes in the junction, ranging from 10^{-9} to 10^{-7} s. For the combination ^{147}Pm/GaAs, Q is relatively large (≥ 0.4) only for quite long lifetimes (about 10^{-7} s) and large thicknesses (about 100 μm) of GaAs p-n junctions. For realistic lifetimes of minority carriers and their diffusion coefficients, the open-circuit voltage realized due to the irradiation of a GaAs p-n junction by beta-particles is obtained. The attainable beta-conversion efficiency η in the case of a ^{3}H/GaAs combination is found to exceed that of the ^{147}Pm /GaAs combination.
Keywords: Betavoltaics, Direct-Gap, Efficiency

1 INTRODUCTION

Betavoltaic effect refers to the electric power production by a p-n junction bombarded by beta-particles that ionize the semiconductor material. Among the advantages of beta-batteries are their long service duration, amounting to years or even decades, and the possibility to use in the hard-to-reach areas. Betavoltaics and photovoltaics are related disciplines. In both cases, electric power results from the separation of electron-hole pairs produced by beta-electrons or photons by a p-n junction in the presence of a load in the circuit. In comparison to photovoltaics, publications in the field of the basic principles and applications of betavoltaic elements have been less numerous initially (see, e.g., [1-7]), but started to attract the attention of the researchers in the recent years [8-16].

The main task in betavoltaic design is the choice of a beta-source/semiconductor combination, which should meet certain requirements. In particular, the beta-particles produced by the source must be absorbed efficiently by the semiconductor. Within the semiconductor, the diffusion length of the electron-hole pairs generated by the beta-flux should be large enough to allow them to reach the p-n junction with as little losses as possible. Because only the relatively low-energy beta-electrons are utilized effectively (with energies varying between 5 and 70 keV) for the realistic semiconductor thicknesses, three main beta-sources are presently employed in betavoltaic applications: Tritium ^{3}H, Nickel ^{63}Ni, and Promethium ^{147}Pm [8]. The respective mean energies of the electrons produced by these sources are 5.7, 18, and 62 keV.

The efficiency, η, of a betavoltaic converter is proportional to the collection coefficient, Q, of the electron-hole pairs generated by the beta-flux. In [2, 3], Q was calculated under the assumption that the generation function of electron-hole pairs by a beta-flux is $g(x) \propto \exp(-\alpha x)$ \$g(x), where α is the effective absorption coefficient. This Ansatz, analogous to the Beer-Lambert law for the absorption of light, is an approximation, see [15, 16] for its discussion. In reality, the generation function initially increases and exhibits a maximum at some distance x_m from the surface [13-15, 17] (see the next section for more details). This implies that this exponential approximation is adequate starting from some x-value greater than x_m. The emergence of the maximum in the $g(x)$ curve is due to the fact that, initially, the primary electrons pass through the semiconductor with only weak scattering. The dead layer thickness x_m increases with the energy of the incident beta-electrons [13-15]. For GaAs, x_m is in the range 0.1 – 1 μm [17].

Although the works [2, 3] do report analytical expressions for Q (obtained under the assumption of the absence of the dead layer), the values of Q = 1 and 0.7 were used in the calculations of beta-conversion efficiency [3, 8]. While the value Q = 1 corresponds to the limiting conversion efficiency that is maximal in principle, the choice Q = 0.7 was not explained in [3, 8].

In this work, we derive an expression for Q taking the dead layer into account, and also using the realistic values of the nonradiative Shockley-Reed-Hall (SRH) recombination lifetime, τ_{SR}, for direct-bandgap semiconductors. In such materials, the values of τ_{SR} are usually short, and are in the range of 10^{-9}-10^{-7} s. We use the so obtained collection coefficient to derive the expression for the realistically attainable beta-conversion efficiency η of various combinations of beta-sources and direct-bandgap semiconductors. When calculating the efficiency, we focus on GaAs as a typical example. We show that decreasing τ_{SR} and increasing the dead layer thickness leads to a strong reduction of Q below 1, and to the corresponding reduction of the beta-conversion efficiency.

2 ANALYSIS OF THE COLLECTION COEFFICIENT

The physical picture adopted in this work is as follows. The incident beta-particles generate excess electron-hole pairs by ionizing the semiconductor. The excess electron and hole densities are practically the same, and are much smaller than the intrinsic charge carrier density. Due to the Coulomb attraction between the charge carriers of opposite signs, the excess charge carriers diffuse towards the p-n junction together in the process known as ambipolar diffusion, as explained, e.g., in chapter 10.2 of [18]. The corresponding diffusion coefficient, which is common for the excess charge carriers of both signs, is a weighted sum of the respective diffusion coefficients for electrons and holes. In the limit where the excess charge carrier density is much smaller than the majority charge carrier density, the ambipolar diffusion coefficient becomes equal the minority carriers' diffusion coefficient. For brevity, we omit the word ''ambipolar" in the rest of this paper, but keep in mind the ambipolar character of the diffusion process.

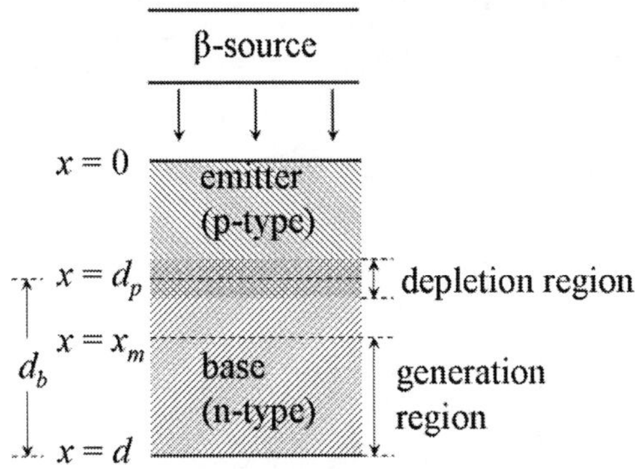

Figure 1: Schematic illustration of a p-n junction of thickness $d = d_p + d_b$, where d_p is emitter depth and d_b is base thickness. The dead layer of thickness x_m extends into the base region. Electron-hole pairs are generated in the generation region $x > x_m$, from which they diffuse into the depletion region via ambipolar diffusion mechanism. They are separated by the built-in electric field at the junction depth, $x = d_p$.

As for the generation function of the excess electron-hole pairs, it is a well-known problem in the design of semiconductor diode particle detectors that there exists a relatively thick dead layer right under the front surface, where the device is insensitive to the incident radiation, see [19] and section 10.5.1 of [20]. This dead layer, first suggested by Klein [21], owes its existence to the fact that primary beta-electrons initially travel straight into the target down to a certain distance without significant scattering. This results in a pronounced peak of the generation function, as observed, e.g., in Monte-Carlo simulations of a betavoltaic element [13]. In the open-circuit regime, typical for cathodoluminescence experiments [17] and semicondoctor detector design \[19, 20], the dead layer also manifests itself in the increased surface recombination rate as compared to the bulk semiconductor, resulting in poor collection of the electron-hole pairs produced in the dead layer.

We assume that the electron-hole pairs are generated only weakly within the dead layer, $x < x_m$, while for $x > x_m$, the generation function has the form $g(x) = I_0 \exp(-\alpha (x-x_m))$, where I_0 is the electron-hole pair generation rate in the x_m-plane, and α^{-1} is the characteristic decay length. Furthermore, we assume that $d_p < x_m$ and $S_d << D/L$, d_p being the junction depth, S_d the recombination rate on the back surface of the base, and L and D the diffusion length and coefficient of the excess electron-hole pairs generated in the base region. The sketch of our structure is summarized in figure 1. Note that we adopt the terminology from the photovoltaics literature [22] and refer to the differrent regions of the diode as "emitter" and "base".

Apart from the SRH mechanism with the lifetime τ_{SR}, the electron-hole pairs in GaAs also recombine radiatively; the characteristic time of this process is $\tau_r = (BN_d)^{-1}$, where B is the radiative recombination coefficient, and N_d is the base doping concentration. Therefore, the diffusion length in the quasineutral base region can be written as

$$L = (D\tau_b)^{1/2}, \tag{1}$$

with $\tau_b = \left(\tau_{SR}^{-1} + \tau_r^{-1}\right)^{-1}$ being the effective lifetime in the neutral base region.

Continuity equation for the excess concentration of the electron-hole pairs, Δp_1, within the dead layer (i.e., for $x < x_m$, region 1), where generation is negligible, has the form

$$\frac{d^2\Delta p_1}{dx^2} - \frac{\Delta p_1}{L^2} = 0. \tag{2}$$

In the rest of the semiconductor ($x > x_m$, region 2), the continuity equation for the excess electron-hole pair density, Δp_2, is

$$\frac{d^2\Delta p_2}{dx^2} - \frac{\Delta p_2}{L^2} = -\frac{\alpha I_0 e^{-\alpha(x-x_m)}}{D}. \tag{3}$$

The equations (2) and (3) are supplemented by the boundary conditions

$$\Delta p_1(x = d_p) = 0,\ \frac{d\Delta p_2}{dx}(x = d) = 0,$$

$$\Delta p_1(x = x_m) = \Delta p_2(x = x_m),$$

$$\frac{d\Delta p_1}{dx}(x = x_m) = \frac{d\Delta p_2}{dx}(x = x_m). \tag{4}$$

The first condition reflects the fact that the electron-hole pairs are separated at the junction depth. The second one indicates the absence of surface recombination at the back of the base. The remaining two expressions are the usual continuity conditions for $\Delta p(x)$ and $d\Delta p(x)/dx$ at $x = x_m$. The collection coefficient is then defined as the ratio of the current at the junction depth, d_p, to the pair generation rate in the plane of highest generation at $x = x_m$:

$$Q = \frac{D}{I_0}\frac{d\Delta p_1}{dx}\left(x = d_p\right). \tag{5}$$

The solution of (2) and (3) that satisfies the first two conditions (4) can be written as

$$\Delta p_1(x) = C \sinh\frac{x - d_p}{L},$$

$$\Delta p_2(x) = C' \sinh\frac{x - d}{L} + A\left(e^{-\alpha(x-x_m)} - \beta e^{-x/L}\right),$$

$$A = \frac{\alpha I_0 L^2}{D\left(1 - \alpha^2 L^2\right)},\ \beta = \alpha L \exp\left[\left(\frac{1}{L} - \alpha\right)d\right] \tag{6}$$

with constants C, C' to be determined from the remaining two conditions (4). This procedure yields:

$$Q = \alpha L \frac{\alpha L\left(\cosh\frac{d-x_m}{L} - e^{-\alpha(d-x_m)} - \sinh\frac{d-x_m}{L}\right)}{\left[(\alpha L)^2 - 1\right]\cosh\frac{d-d_p}{L}} . \quad (7)$$

If $d - x_m >> L$ and $\alpha(d - x_m) >> 1$, this expression simplifies to

$$Q = \frac{\alpha L}{1+\alpha L} e^{(d_p - x_m)/L} . \quad (8)$$

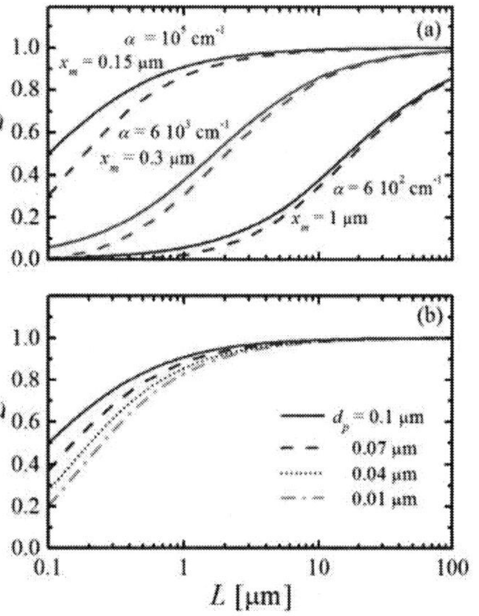

Figure 2: (a) Collection coefficient, Q, as a function of the diffusion length, L, for different absorption coefficients, α, in the limit $\alpha(d - x_m) >> 1$, $d - x_m >> L$, see (8). The values used, α = 10^5, $6\ 10^3$, and $6\ 10^2$ cm^{-1}, approximately correspond to the respective mean beta-energies of 5.7 keV (^3H), 20 keV (^{64}Ni-source), and 60 keV (^{147}Pm-source) for GaAs-based p-n junction [23]. The dashed curves are calculated for different dead layer thicknesses, x_m, and d_p = 10^{-5} cm. The solid curves are from the standard relation $Q = \alpha L/(1 + \alpha L)$, valid in the absence of the dead layer. (b) Collection coefficient (8) for different junction depth values for x_m = 10^{-5} cm and $\alpha = 10^5$ cm^{-1}, corresponding to the beta-particle energy of about 5.7 eV in the ^3H/GaAs combination.

Figure 2 shows the dependence of the collection coefficient Q on the diffusion length from (8). As seen in this figure, the strongest reduction of Q due to the presence of the dead layer is for the case of the ^3H beta-source. The smallest discrepancy in the Q-values obtained with and without taking into account the dead layer is found for the curves corresponding to α = $6\ 10^2$ cm^{-1}, realized in the case of the ^{147}Pm-source. In this case, to obtain $Q > 1/2$, one would need the diffusion length $L > 35$ μm. The values $Q \approx 1$ can be achieved only in Si p-n junctions with long minority carrier lifetimes [24].

In figure 2(b), the junction depth was varied at a fixed electron energy (and thus constant α) and dead layer thickness. As seen in this figure, the collection coefficient increases not only upon increasing L, but also upon approaching the junction depth to the x_m-value. This effect is especially important for small diffusion length L.

A further conclusion from figure 2 is that collection of the electron-hole pairs generated by the electron flux will be quite efficient in the case when the diffusion length exceeds the dead layer thickness, $L > x_m$. An alternative way to increase Q is to use deeper junctions with $d_p \approx x_m$.

Let us find the relation between the diffusion length and SHR lifetime τ_{SR} for the case of GaAs. The radiative recombination coefficient A in GaAs is an effective parameter defined by the relation $B = B_0(1 - \gamma_r)$ [25], where $B_0 \approx 6\ 10^{-10}$ cm^3/s [26], and γ_r is the photon re-absorption coefficient. In our calculations, we assumed the value B = $2\ 10^{-10}$ cm^3/s, as can be derived for poorly reflecting GaAs-based plane-parallel p-n structures without multiple reflection using the approach from [25]. In the work [26], it was shown that for realistic lifetimes τ_{SR}, the open-circuit voltage V_{OC} of GaAs-based p-n junctions increases with the base doping level, N_d, and, taking into account the interband Auger recombination, it has a maximum at $N_d \approx 10^{17}$ cm^{-3}. It is this optimal value that we use in this work.

Let us first assume that the GaAs p-n junction base is of p-type, and the diffusion coefficient of electron-hole pairs is 50 cm^2/s. Then, for $B \approx 2\ 10^{-10}$ cm^3/s, N_d = 10^{17} cm^{-3}, and lifetimes τ_{SR} = 10^{-9}, 10^{-8}, and 10^{-7} s, diffusion length L has the respective values of 2.2, 6.45, and 12.9 μm.

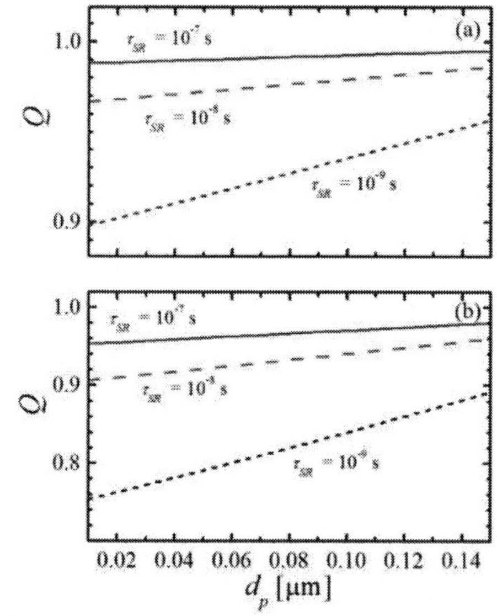

Figure 3: Collection coefficient Q of a ^3H/GaAs betavoltaic pair as a function of the junction depth for (a) p-type base and (b) n-type base.

The space-charge (or depletion) region influences the collection coefficient. However, these figures exceed the width of the space-charge region by at least an order of magnitude. Indeed, the space-charge region width can be estimated [18, 28] as $w \approx \sqrt{2\kappa k_B T \ln\left(N_d^2/n_i^2\right)/\left(q^2 N_d\right)} \approx$ 0.1 μm, where κ is the dielectric constant of GaAs, and n_i $\approx 1.8\ 10^6$ cm^{-3} the intrinsic carrier density. For this reason, we can neglect the influence of the space-charge region on the transport properties of the excess charge carriers.

Figure 3(a) shows the dependence of the collection coefficient, Q, of a pair ^{3}H/GaAs as a function of the junction depth, d_p, for these three values of τ_{SR} at x_m = 0.15 μm [17] and junction thickness d = 10 μm. As seen in this figure, Q is close to 1 for $\Delta x = x_m - d_p < 0.1$ μm. For $\Delta x > 0.1$ μm, the Q-value decreases with Δx, but remains rather large.

Presented in figure 3(b) is the collection coefficient vs. d_p for the case when the base region of the p-n junction is of the n-type. In this case, for $\tau_{SR} = 10^{-9}$, 10^{-8}, and 10^{-7} s, and $B \approx 2\ 10^{-10}$ cm^3/s and $N_d = 10^{17}$ cm^{-3}, and taking into account that $D = 7$ cm^2/s, the diffusion length L = 0.83, 2.41, and 4.83 μm, respectively. As seen in the figure, in this case Q is also quite large. For $\tau_{SR} = 10^{-7}$ and 10^{-8} s, Q is still close to 1, while for $\tau_{SR} = 10^{-9}$ s, Q exceeds 0.75 even for small Δx.

It should be noted that, because of rather strong absorption of the electrons emitted by the ^{3}H-source by the auxiliary layers of a betavoltaic element (such as protection coating or contact layers), additional reduction of the beta-generated current can take place, leading to the efficiency reduction.

Let us now analyze the collection coefficient for the ^{147}Pm/GaAs pair. In this case, according to [23], $\alpha \approx$ 600 cm^{-1}, i.e., excess electron-hole density decays much more slowly than in the ^{3}H/GaAs case. For this the inequality $\alpha L >> 1$ is always satisfied even for the shortest lifetime of 10^{-9} s. In contrast, for ^{147}Pm source, $\alpha L = 1.5$ for $L = 25$ μm, while $\alpha L = 0.06$ for $L = 1$ μm, so that Q is always notably smaller than 1.

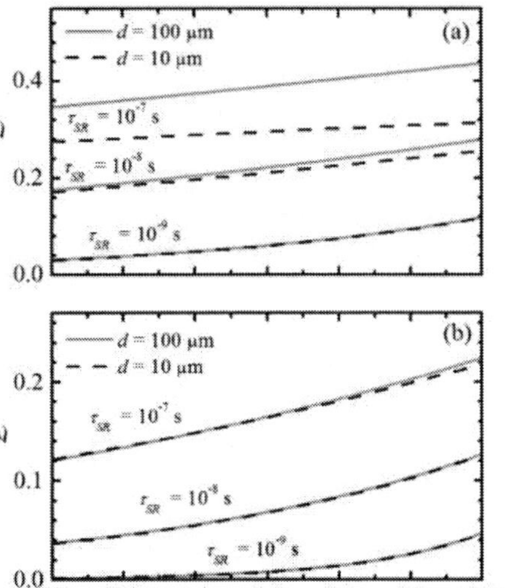

Figure 4: Collection coefficient of a ^{147}Pm/GaAs betavoltaic element as a function of junction depth for different Shockley-Reed lifetimes and element thicknesses for the case of (a) p-type base and (b) n-type base.

But this is not the only reason for the reduction of Q in realistic ^{147}Pm/GaAs structures. When manufacturing solar cells based on the direct-bandgap semiconductors, such as GaAs, full thicknesses of p-n junctions are chosen rather small (of the order of a few μm). Such structures were used in [9]. In contrast, for the ^{147}Pm/GaAs pair used in betavoltaics, the situation might be very different, especially for large values of L. In this case, the product αd will be small, so that for full absorption of beta-flux much thicker p-n junctions are required compared to those typically used in photovoltaics.

Shown in figure 4 is the collection coefficient as a function of d_p for a ^{147}Pm/GaAs pair calculated for different lifetimes τ_{SR} and junction thicknesses d of 10 and 100 μm. In this case, according to [17], x_m = 3 10^{-4} cm^3/s. Panels (a) and (b) correspond to the cases of p- and n-base conduction types, respectively. As seen in the figure, rather high values of $Q \geq 0.4$ for the ^{147}Pm/GaAs pair can be achieved only for the junction thickness $d \approx 100$ μm. Also, collection coefficient decreases dramatically as τ_{SR} decreases.

It should be noted that similar results for the attainable Q are expected for other direct-bandgap A_3B_5 semiconductors, in particular, the ones based on the three-component compounds.

3 OPEN-CIRCUIT VOLTAGE ANALYSIS

When estimating the limiting efficiency value [3, 11], we used the Shockley-Queisser approach [27], in which not only the current density, but also the open-circuit voltage, V_{OC}, is assumed to be maximal. Therefore, our next task is to calculate the open-circuit voltage, V_{OC}, with realistic values of τ_{SR}. It is given by the standard expression

$$V_{OC} = \frac{k_B T}{q} \ln \frac{N_d \Delta p^*}{n_i^2}, \qquad (9)$$

where $\Delta p^* = \Delta p\ (x = d_p + w)$ is the excess minority carrier density in the base at the boundary between the space-charge region of thickness w and the quasineutral region, N_d is the equilibrium density of the majority carriers in the quasineutral base region, and n_i is the intrinsic charge carrier density. It is related to the effective densities of states in the conduction and valence bands, N_c and N_v, as

$$n_i = \sqrt{N_c N_v} \exp\left(-\frac{E_g}{2k_B T}\right). \qquad (10)$$

We assume that both d_p and w are much smaller than the diffusion length L. This allows us to approximate

$$\Delta p(x=0) = \Delta p^*. \qquad (11)$$

Such an approximation introduces a negligible error into V_{OC} from (9) in view of its logarithmic dependence on Δp^*.

We will assume that recombination dominates in the quasineutral base region and in the space-charge region. Then, V_{OC} can be obtained using the approach from [26]. Taking into account the generation-recombination processes, we first write the continuity equation for the excess carrier density supplemented by the boundary conditions:

$$\frac{d^2\Delta p}{dx^2} - \frac{\Delta p}{L^2} - r(x)\Delta p(x) + g(x) = 0,$$

$$\frac{d\Delta p}{dx}(x = d) = 0,$$

$$D\frac{d\Delta p}{dx}(x=0)=S_0\Delta p^*, \qquad (12)$$

where the third term describes recombination processes in the space-charge region of the abrupt junction, and the last one corresponds to the beta-induced generation. The first boundary condition is consistent with our assumption $S_d << D/L$ from the beginning of the previous section, and the second one is responsible for recombination effects in the $x = d_p + w$ plane.

Integration of the continuity equation results in the balance equation for the generation-recombination currents, according to which the current density for electronic excitation is proportional to the integral of the generation term,

$$J_\beta = \int_0^d dx\frac{\Delta p(x)}{\tau_b} + q(S_0 + R_{SC})\Delta p^*, \qquad (13)$$

where q is the elementary charge. The right-hand side in (13) is responsible for the recombination in the bulk and on the front side of the emitter and within the space-charge region. The space-charge region recombination rate is given by [28]

$$R_{SC}(\Delta p^*) = \frac{L_D}{\sqrt{2}\tau_{SR}}\int_{y_{pn}}^{-1} dyN_d\left(1-y+e^y\right)^{-1/2}\times$$

$$\left[N_d e^y + n_i e^{E_r/k_BT} + b\left(\frac{n_i^2}{N_d}+\Delta p^*\right)e^{-y} + bn_i e^{-E_r/k_BT}\right]^{-1},$$

where $b = \sigma_p/\sigma_n$ is the ratio of the capture cross-sections of holes and electrons by a recombination level, E_r is the recombination level energy measured from the middle of the bandgap, y_{pn} is the dimensionless potential at the p-n boundary, L_D is the Debye length.

To evaluate the first integral in (13), we have employed the following approximative procedure. First, we write the solution of the continuity equation (12) as a sum of homogeneous and inhomogeneous parts,

$$\Delta p(x) = \frac{e^{-x/K}+e^{(x-2d)/L}}{1+e^{-2d/L}}\Delta p^* = \Delta p_i(x), \qquad (14)$$

where the homogeneous term satisfies the first boundary condition in (12) and gives the value $\Delta p(x = 0) = \Delta p^*$. The inhomogeneous contribution $\Delta p_i(x)$, with $\Delta p_i(x = 0) = 0$, is notably different from zero only within a relatively thin layer below the front surface of the emitter, where the generation-recombination processes take place. Therefore, the contribution to the integral of the second term can be neglected in comparison to the integral of the homogeneous term, allowing us to write

$$\int_0^d dx\Delta p(x) \approx \Delta p^* L\tanh(d/L). \qquad (15)$$

This approximation should produce a negligible error in V_{OC} in view of its logarithmic dependence on Δp^*. Substitution of this result into (9) taking into account that $L^2 = D\tau_b$ yields

$$J_S = q\Delta p^*\left[\frac{D}{L}\tanh\left(\frac{d}{L}\right) + S_0 + R_{SC}(\Delta p^*)\right]. \qquad (16)$$

The current density J_β is inversely proportional to the energy required to create one electron-hole pair, ς, which is approximately related to the bandgap E_g as [29]

$$\varsigma = 2.8E_g + 0.5\,\text{eV}. \qquad (17)$$

Denoting the current density in the case of Si (E_g = 1.12 eV) by J_0, the current density in the case of arbitrary bandgap can be approximated as

$$J_\beta = J_0 Q\cdot 3.64\,\text{eV}/\varsigma. \qquad (18)$$

We note that, usually, J_0 is in the 1-10^2 μA/cm^2 range [3].

The value of Δp^* found from (16) should be substituted into (9) to obtain the open-circuit voltage V_{OC}.

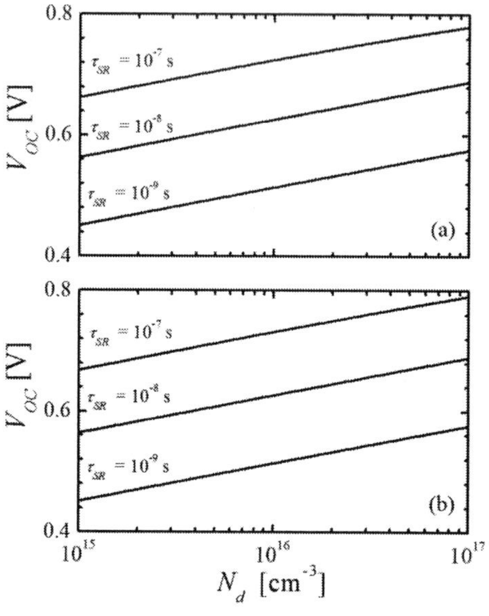

Figure 5: Open-circuit voltage as a function of the base doping level for different Shockley-Reed lifetimes in the case of (a) p-type base and (b) n-type base for E_g = 1.43 eV, T = 300 K, and J_0 = 10 μA/cm^2.

Figure 5 shows the dependence of V_{OC} of a GaAs-based p-n junction on the base doping level, N_d, neglecting the surface recombination, that is, $S_0 \approx 0$. As seen in figure 5, V_{OC} increases with N_d. On the one hand, the values of V_{OC} for the pair ^{3}H/GaAs is notably smaller than in the solar cells [26], because the beta-produced current densities are at least two order of magnitude smaller than the short-circuit current densities in photovoltaic cells. On the other hand, the open-circuit voltages in figure 5 exceed the values obtained experimentally in [9]. The reason is that, in [9], the current density J_0 was of the order of 1 μA/cm^2, whereas in our calculations, we have taken J_0 = 10 μA/cm^2. If the values J_0 = 1 μA/cm^2, N_d = 5 10^{16} cm^{-3}, and τ_{SR} = 10^{-9} s are used, we obtain V_{OC} = 0.44 V, which practically coincides with the value given in [9].

4 REFINED CALCULATION OF THE LIMITING BETACONVERSION EFFICIENCY

According to Olsen [3], the efficiency of a betavoltaic element, η, is

$$\eta = \eta_\beta\eta_C\eta_S, \qquad (19)$$

where

$$\eta_\beta = N_\beta/N_0 \qquad (20)$$

is the fraction of beta-flux that reaches the semiconductor,

$$\eta_C = (1-r)Q \qquad (21)$$

is the coupling efficiency, given by the product of absorption probability of a beta-particle (r is the electron reflection coefficient from the semiconductor surface) and collection efficiency Q of electron-hole pairs, and, finally, the semiconductor efficiency

$$\eta_S = qV_{OC}FF/\varsigma\ , \qquad (22)$$

where q is the elementary charge, V_{OC} is the open-circuit voltage, FF is the fill factor, ς is the energy necessary to generate one electron-hole pair from (17).

Let us obtain V_{OC} within the Shockley-Queisser approximation, where $\tau_{SR} \to \infty$, S_0 and $R_{SC} \to 0$, and the only recombination mechanism present is radiative recombination, characterised by the coefficient B. In this case, $V_{OC}^{\lim}$ can be found analytically from (13) and (16):

$$V_{OC}^{\lim} = \frac{k_B T}{q}\ln\frac{J_\beta}{qBdn_i^2}\ . \qquad (23)$$

The fill factor can be found using the expression from [3]

$$FF = [v_{OC} - \ln(v_{OC} + 0.72)]/(v_{OC} + 1), \qquad (24)$$

where $v_{OC}=V_{OC}/k_BT$.

To calculate the limiting beta-conversion efficiency, we take $Q = 1$, $r = 0$, $\eta_\beta = 1$, corresponding to the bidirecional source in the terminology of [3]. In this case

$$\eta_{\lim} = \frac{qV_{OC}^{\lim}FF_{\lim}}{2.8E_g + 0.5}\ . \qquad (25)$$

where $V_{OC}^{\lim}$ is given by (23).

When calculating η_{lim}, several issues may arise. First, the parameters N_c, N_v, and B are material-specific in every semiconductor. Second, when evaluating $V_{OC}^{\lim}$ and FF_{lim}, Olsen had used, for each source, concrete current density J_0 of the order of 10^2 μA/cm^2 for ^{147}Pm and 1 μA/cm^2 for ^{3}H. Finally, $V_{OC}^{\lim}$ depends on the p-n junction thickness d. Therefore, all parameters in (25) must be specified. Since such key parameters as B, N_c, and N_v, are known only for concrete semiconductors and concrete bandgap values E_g, in the best-case scenario, the dependence $\eta_{lim}(E_g)$ can be found as a set of support points for the known semiconductors with different E_g. Fitting this with a smooth curve might not be accurate enough.

In this work, we calculated η_{lim} only for the case of GaAs using (25). For $B = 2\ 10^{-10}$ cm^3/s and $d = 10$ μm gives for $J_0 = 10^2$ μA/cm^2 the value $\eta_{lim} \approx 17$ %, and for $J_0 = 1$ μA/cm^2, $\eta_{lim} \approx 14$ %. Note that the values of η_{lim} obtained here notably exceed the ones obtained by Olsen in [3, 8]. In the rest of this work, we will use the values obtained for the ^{147}Pm/GaAs and ^{3}H/GaAs combinations, respectively.

5 CALCULATION OF THE ATTAINABLE BETACONVERSION EFFICIENCY

Figure 6 shows the attainable efficiency as a function of d_p for the ^{3}H/GaAs combination, obtained from

$$\eta = \eta_{\lim} Q \frac{V_{OC}}{V_{OC}^{\lim}}\ , \qquad (26)$$

where $\eta_{lim} \approx 14$ %, Q is given by (7), V_{OC} is found from (9), and $V_{OC}^{\lim}$ from (23).

When plotting figure 6, we varied the lifetime at a constant d = 10 μm. Panels (a) and (b) correspond to the base of the p- and n-type, respectively. As seen in figure 6, the attainable efficiency values are rather high and are in the range of (6.4 - 12.5) %.

It should be noted that our results for ^{3}H/GaAs pair agree well with those given in the review [8] citing [9-11], namely, $\eta = (4 - 7)$ %. In these works, a ^{3}H-source was used with the A_3B_5-based semiconductors. But, as evident from the figures shown, the possibilities of increasing the efficiency of ^{3}H/ A_3B_5 betaconversion are far from being exhausted.

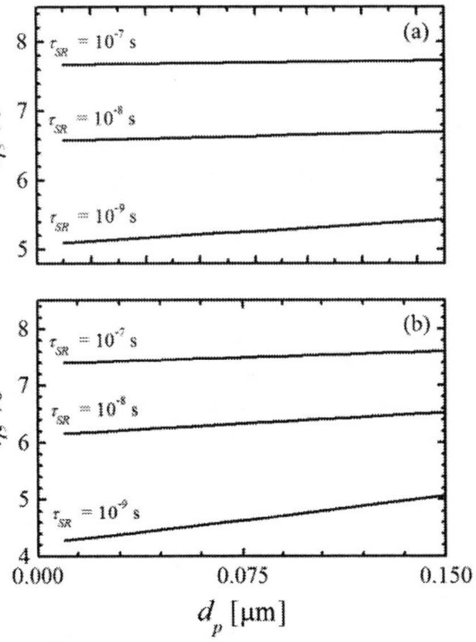

Figure 6: Beta-conversion efficiency of a ^{3}H/GaAs pair as a function of junction depth for different Schokley-Reed lifetimes for the case of (a) p-type base and (b) n-type base.

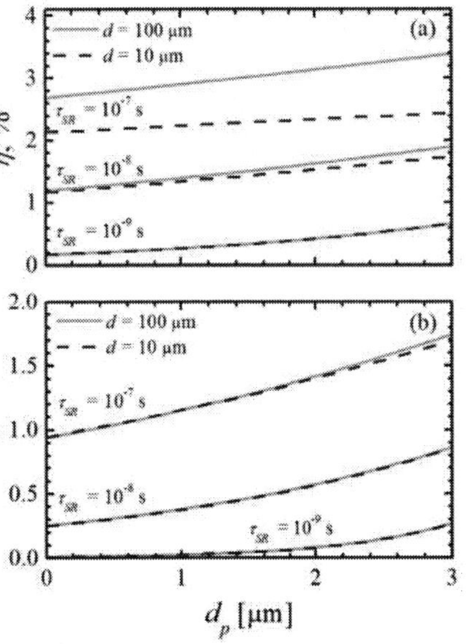

Figure 7: Beta-conversion efficiency of a \Pm/GaAs element vs. junction depth for different Schokley-Reed lifetimes and element thcknesses for (a) p-type base and (b) n-type base.

Shown in figure 7 is the attainable beta-efficiency

(26) as a function of d_p for ^{147}Pm/GaAs pair with η_{lim} = 17 %. The τ_{SR} values used were 10^{-9}, 10^{-8}, and 10^{-7} s, and GaAs thicknesses were 10 and 100 μm. Figure 7(a) and (b) correspond to the p- and n-types of the base conductivity. As seen in this figure, η reduces rather strongly as τ_{SR} is decreased. For the highest $\tau_{SR} = 10^{-7}$ s, η decreases with decreasing d. The highest efficiency attainable, η = 7.25 %, is achieved for $\tau_{SR} = 10^{-7}$ s and d = 100 μm, and the lowest value of 0.51 % is realized for $\tau_{SR} = 10^{-9}$ s and d = 10 μm.

Thus, we conclude that a ^{147}Pm/GaAs-based betaconverter is not as efficient as a ^{3}H/GaAs-based one. Perhaps, the very small efficiency of the ^{147}Pm/GaAs battery obtained in [5] is due to the small thickness of GaAs and small lifiteme τ_{SR}. The same applies also to the cases when, instead of GaAs, other direct-bandgap semiconductors are used.

6 CONCLUSIONS

Our analysis, focusing on the attainable collection coefficient Q and open-circuit voltage values V_{OC}, has revealed the following features of current collection of the GaAs-based beta-elements.

Efficient collection of the electron-hole pairs generated by a beta-flux can be achieved when the diffusion length exceeds the dead layer thickness, $L > x_m$. An alternative way to increase collection coefficient is to use deep junctions, for which $dp \cong x_m$.

Additional mechanisms responsible for the reduction of current generated by beta-electrons are possible, leading to smaller betaconversion efficiency. They may be due, for instance, to the strong absorption of the beta-electrons by auxiliary layers of a betavoltaic element.

Using the Shockley-Queisser approximation, we have derived the limiting betaconversion efficiency, $\eta_{lim}(E_g)$. Our analysis has shown that, because the main parameters affecting the efficiency are very different for different semiconductors, the $\eta_{lim}(E_g)$ curve can be build as a set of support points for semiconductors with different bandgaps, and not as a smooth curve.

^{147}Pm beta-source performs more poorly than ^{3}H-source, because the electron-hole pair generation depth in the case of ^{147}Pm-source is large, whereas the diffusion length of GaAs is small. Therefore, the majority of electron-hole pairs generated in the base recombine before reaching the p-n junction.

In the case of ^{3}H-source, the picture is different. The collection coefficient is rather high, because of the small generation depth of electron-hole pairs. Therefore, the realistic betaconversion efficiency for the ^{3}H/GaAs pair will be rather high for relevant parameters (lifitemes and diffusion coefficients) of the semiconductor.

Similar results are expected also in the case, when other direct-bandgap semiconductors are used instead of GaAs.

REFERENCES

[1] Rappaport R 1954 Phys. Rev. 93 246

[2] Pfan WG and van Roosbroeck W 1954 J. Appl. Phys. 25 1422

[3] Olsen LC 1992 Proc. XII Space Photovoltaic Research and Technology Conference p. 256; see NASA technical report 19940006935, http://hdl.handle.net/2060/19940006936

[4] Rappaport P, Loferski JJ and Linder EG 1956 RCA Rev. 17 100

[5] Flicker H, Loferski JJ and Elleman TS 1964 IEEE Transactions on Electron Devices, vol. ED-11, no. 1, 2

[6] Olsen LS 1973 Energy Conversion 12 117

[7] Olsen LC 1974 Proc. 9th Intersociety Energy Conversion Engineering Conference, p. 754

[8] Olsen LC, Cabauy P and Elkind BJ 2012 Phys. Today 65 35

[9] Andreev VM et al 2000 Conference Record of the TwentyEighth IEEE Photovoltaic Specialists Conference IEEE, Piscataway, NJ, 1253

[10] Bower KE, Barbanel' YA, Shreter YG and Bohnert GW(eds) 2002 Polymers, Phosphors, and Voltaics for Radioisotope Microbatteries (CRC Press, Boca Raton)

[11] Adams T et al 2012 Betavoltaic Power Sources for Nano and Micro Power Application, paper presented at 45th Power Sources Conference, 11-14 June 2012

[12] Landis GA, Bailey SG, Clark EB, Myers MG, Piszczor MF, and Murbach MS 2012 Proceedings of the 38th IEEE Photovoltaics Specialists Conference pp 002819-002824

[13] Oh K, Prelas MA, Rothenberger JB, Lukosi ED, Jeong J, Montenegro DE, Schott RJ, Weaver CL, and Wisniewski DA 2012 Nuclear Technology 179 234

[14] Murashev VN, Mordkovich VN, Legotin SA, Rabinovich OI, and Krasnov AA 2014 J. Nano-Electron. Phys. 6 04012

[15] Zuo G, Zhou J, and Ke G 2013 Appl. Radiat. Isot. 82 119

[16] Sachenko AV, Shkrebtii AI, Korkishko RM, Kostylyov VP, Kulish MR, and Sokolovskyi IO 2015 Solid-State Electron. 111 147

[17] Dmitruk NL, Litovchenko VG and Talat GH 1978 Surf. Sci. 72 321

[18] McKelvey JP 1984 Solid State and Semiconductor Physics (Robert E. Krieger Publishing Company, Malabar, Florida)

[19] Wall BL et al 2014 Nucl. Instr. Meth. Phys. Res. A 744 73

[20] Leo WR 1994 Techniques for Nuclear and Particle Physics Experiments (Springer-Verlag, Berlin, Heidelberg, New York)

[21] Klein CA 1966 Appl. Opt. 5 1922

[22] Chopra KL and Das S.R. 1983 Thin Film Solar Cells (Springer Science+Business Media, New York)

[23] Trischuk JM 1967 Nuclear Beta Spectroscopy Using Solid State Detectros, Ph.D. Thesis, California Institute of Technology, Pasadena, California

[24] Gorban AP, Sachenko AV, Kostylyov VP, and Prima NA 2000 Semicond. Phys. Quantum Electron. Optoelectron. 3 322

[25] Ding D, Johnson SR, Yu S-Q, Wu S-N, and Zhang Y-H 2011 J. Appl. Phys. 110 123104

[26] Sachenko AV, Kostylyov VP, Kulish MR, Sokolovskyi IO, and Shkrebtii AI 2014 Semicond. Phys. Quantum Electron. Optoelectron. 17 134

[27] Shockley W and Queisser HJ 1961 J. Appl. Phys. 32 510

[28] Sze SM and Ng KK 2007, Physics of Semiconductor Devices, 3rd Edition (Wiley)

[29] Klein CA 1968 J. Appl. Phys. 39 2029

BLACK, INFRARED REFLECTING BACKSHEETS FOR PV: WHERE AESTHETICS MEETS PERFORMANCE

S.L. Luxembourg[1], M. Kloos[1], F.J.K. Danzl[1], J. Theewis[2], J.A.M. v. Roosmalen[1]
[1]ECN Solar Energy, P.O. Box 1, NL-1755 ZG Petten, The Netherlands
[2]Eurolacke B.V. , De Riemsdijk 4, NL-4004 LC Tiel, The Netherlands

ABSTRACT: To reduce cell-to-module (CtM) losses for aesthetically appreciated all-black modules we applied a black powder coating with low infrared absorption to standard back foils used in PV modules. Upon the application of such powder coating the IR reflective properties of the underlying substrate are largely preserved. For Cu foil substrates approximately 85% reflection in near-infrared was recorded, while for a white polyamide substrate reflection was 60%. These black, IR reflecting back foils were incorporated in single-cell laminates with n-Pasha bifacial solar cells. From these structures the potential I_{sc} gain from the application of such coatings in PV modules as compared to conventional all-black modules is estimated to be 1.4 – 1.8%.
Keywords: PV module, Backsheet, Bifacial, Light Trapping, Ray Tracing

1 INTRODUCTION

The large-scale deployment of photovoltaics, in particular in the built environment, raises the importance of the aesthetics of PV modules. In this respect, the PV product portfolio is being diversified with products with different appearances and functionality. Examples of this are modules with colored cells, PV roof tiles and façade elements which should take over the functionality of conventional building elements facilitating building-integrated PV (BIPV). Currently, all-black modules are widely available and highly appreciated both for building-integrated and building-applied applications. To ensure a black appearance these modules are manufactured with a black backsheet.

In contrast, standard crystalline silicon modules incorporate a white reflective backsheet. The backsheet in the PV module is both a barrier to moist and ultraviolet (UV) radiation and acts as an electrical insulator. At the same time the use of a reflective foil as backsheet improves the output of a PV module through (diffuse) reflection of the light that falls on it. Part of the light that impinges on the exposed backsheet area in between the cells and along the outer edge of the module is re-directed towards the cells after reflection and trapping at the glass-air interface. The contribution of the light reflected from the backsheet to the output power of a PV module depends on its reflectivity and the spacing of the cells but can easily be in the range of 2 – 3% [1]. In addition, when bifacial cells are used the infrared (IR) light transmitted by the cell will be efficiently reflected from the area behind the cell, improving the IR response of the module. In the case of all-black modules the use of a black, absorbing back foil adds to the aesthetics of the module, however this comes at the price of a lower power output.

In the built environment innovative coatings are being applied to reduce the temperature of objects, such as roofs and façade elements. The application of such coatings lowers the need for air-conditioning. These coatings are available in multiple colors and black and can be applied in an electrostatic powder-coating process. To realize the cooling effect the coatings should have very limited absorption in the IR. To our opinion such coatings are of great interest for application in all-black PV modules. The combination of black, low IR absorption powder coatings with highly reflective back foils would reduce the inherent losses in all-black modules associated with the use of absorbing backsheets. Figure 1 illustrates the differences in module light management for an absorbing and reflecting (black) backsheet.

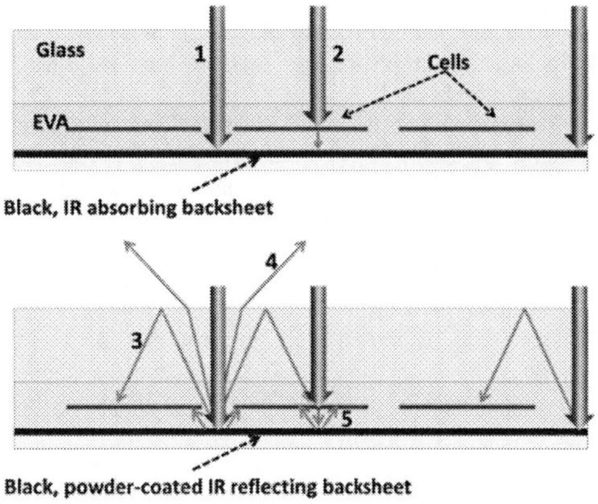

Figure 1: Illustration of the improved module light management in case of a black, IR reflecting backsheet. 1) Incident radiation absorbed by backsheet, 2) Transmitted (IR) radiation absorbed by backsheet, 3) IR part of incident radiation reflected and trapped at front, 4) As 3) but within escape cone, 5) Transmitted radiation reflected and re-entering cell through rear.

In addition, the reduced IR absorption (and IR transparency) of the black, powder coated (BPC) foils compared to conventional, absorbing black foils is thought to result in lower operating temperatures for all-black modules, which should further enhance their performance under operation conditions.

Here, we investigate the use of a black, powder coating (Alesta® Cool by Axalta) with low absorption in the IR spectral range to generate visually black but IR reflecting backsheets. For this purpose we have applied the coating to standard reflecting PV back foils, namely white polyamide and copper foil. In this study, we have focused on its effects on module light management, which we study both experimentally and in ray-tracing simulations.

2 METHODOLOGY

2.1 Experimental

Alesta® Cool, a polyester based, black powder

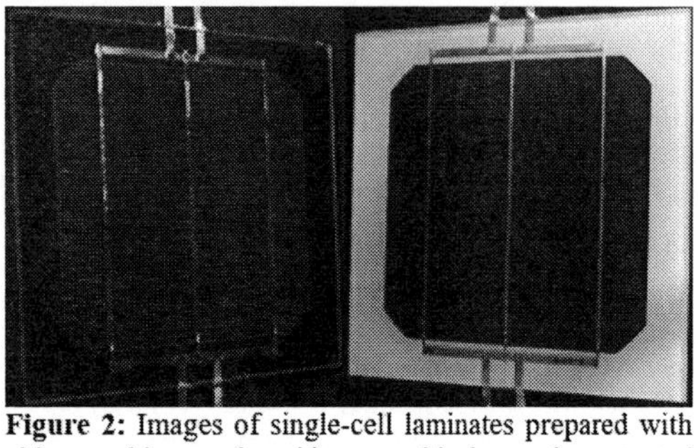

Figure 2: Images of single-cell laminates prepared with either white polyamide or black-powder coated polyamide.

coating from Axalta, was applied to standard white polyamide and Cu foils in an electrostatic spray-coating process. The powder coating was applied by a specialized company, Eurolacke B.V. Reflection measurements of the different backsheet materials were performed on a homebuilt spectrophotometer set-up which includes an integrating sphere.

Three busbar, bifacial n-Pasha cells [2] from previous baseline experiments were selected for this study. IV-characteristics of these cells were measured on a Wacom class AAA solar simulator with a brass reflective chuck. The cells were measured from front and rear to determine the bifaciality factor. From these measurements cells with very similar performances were selected for fabrication of single-cell laminates. The front-measured parameters amounted to an I_{sc} of 9.18 ± 0.02A (average ± standard deviation) and a conversion efficiency of 19.67 ± 0.09%. The rear I_{sc} and efficiency were 8.04 ± 0.07A and 17.2 ± 0.2% respectively. The bifaciality factor of this set of cells was 87%. The cells were distributed randomly over 4 groups, which showed no significant difference in performance against each other.

The selected cells were encapsulated in a 20 cm × 20 cm glass – EVA – backsheet stack. The thickness of the glass used was 3 mm, while front and rear EVA were both 450-µm thick. The black reference laminates incorporated a black absorbing backsheet. In Figure 1 images of laminates fabricated with white polyamide and black powder coated polyamide are included.

All laminates were measured under STC in a room with black walls with a Pasan module flash-tester, class AAA Different masks were applied to the laminates in these measurements to vary the exposed area of the backsheet surrounding the cells. With the masks used aperture areas of 0, 3, 5, 10 and 22 mm around and including the cell were realized.

2.2 Ray tracing simulations

With our homebuilt ray-tracing software [3] a simple model structure of the single-cell laminates was defined to simulate the trend in the I_{sc} measurements as function of the aperture size. The model includes the module optical stack but no detailed optical model of the solar cell was included. Instead, it was assumed that each ray interacting with the solar cell is absorbed. Simulation runs with varying aperture widths were performed for the different backsheets used in the experiments in the 330 – 1100 nm wavelength range. The reflection parameters for the different backsheet materials were derived from experiments. No change in refractive index was assumed between EVA and the respective backsheets. The

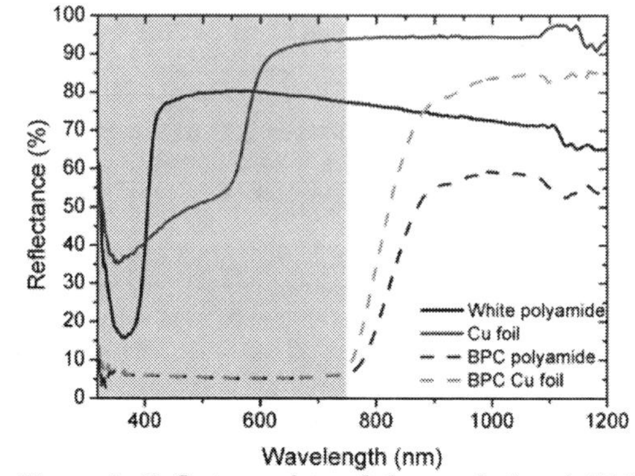

Figure 3: Reflectance data of the standard and BPC polyamide and Cu foils used in this study.

backsheet reflection was considered to be fully Lambertian. The simulation results were normalized on the measured I_{sc} values for 0 mm aperture.

3 RESULTS

The reflectance of samples of white polyamide and Cu foil were measured together with similar samples which had received the black, powder coating. The results are included in Figure 3. It is apparent that the application of the coating results in full absorption of all light in the UV and visible (VIS) range. It appears therefore, truly black to the human eye. Above 750 nm the absorption rapidly decreases. The coating becomes more and more transparent in this spectral range which is observed as an increasing reflection in Figure 3 for the coating-substrate systems considered here. When applied to a glass substrate extensive transmission is seen in this spectral range (data not shown). From the reflection and transmission data for the polyamide foils (polyamide is slightly transparent in the VIS to near-infrared (NIR) range) we find that the powder coating absorbs on the order of 10 - 15% of the incident light in the NIR range in the configurations considered here. For the BPC polyamide this amounts to a reflection of around 60% in the NIR spectral range while for BPC Cu foil reflection is as high as 85%. In addition, to the total reflectance also diffuse reflectance only spectra were recorded. It was found that the addition of the BPC did not significantly affect the ratio diffuse-to-total reflection. This ratio was at least 93% for all materials.

We performed an analysis of how the different backsheet materials affected the I_{sc} of the single-cell laminates. In Figures 4A and B the results are summarized for two cases: A) the laminates were masked to illuminate the cell area only and B) the full laminates including their 22 mm edge areas were illuminated. In addition to the copper and polyamide based backsheets also conventional black laminates were included. This latter laminates included n-Pasha cells from a different baseline run with a slightly higher efficiency and higher bifaciality factor In order, to compare them to the laminates discussed here a scaling factor was applied. This scaling factor was derived from a comparison of single-cell laminates of white polyamide and Cu foil which were fabricated for both sets of cells. Case A corresponds to the situation in which the backsheets will interact with light transmitted by the solar cell only. Since, we have used bifacial cells in this study part of the

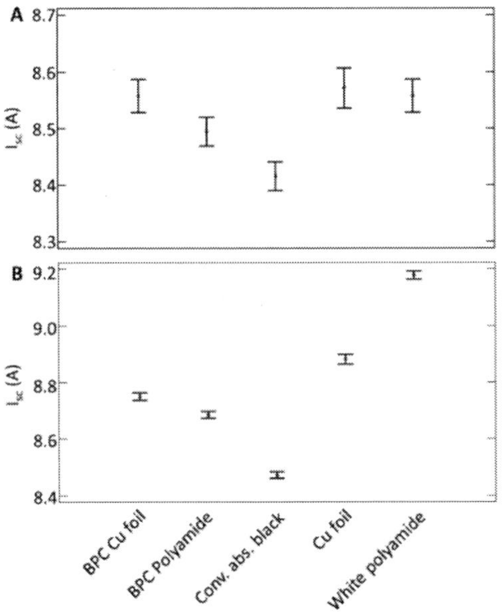

Figure 4: I_{sc} values recorded for the single cell laminates in case A) only the cell area and B) the full laminate is illuminated. In the latter case the illuminated area includes a 22 mm wide area around the cell where the backsheet is exposed to the incident light (see also Figure 2)

IR light will exit the rear of the cell and fall onto the backsheet which by reflecting the light back to the cell will aid the light management of the module. Case A corresponds to #5 in Figure 1, while case B combines #3, #4 and #5. The light that falls on the exposed backsheet area between the cells will include the VIS range as well. UV light will be filtered out by the glass and EVA layers above it. As indicated in Figure 1 only the light outside the escape cone will be trapped at the front glass-air interface.

For case A, the Cu foil, BPC Cu foil and White polyamide laminates display the highest I_{sc}. Although their IR reflection differs, the I_{sc} values found do not differ significantly among the groups and are around 8.57A. The BPC polyamide has a slightly lower I_{sc} value of 8.49A. The lower I_{sc} for the BPC polyamide as compared to the BPC Cu foil is in line with the lower IR reflection reported in Figure 3. All of the laminate configurations above clearly perform better than the conventional black laminate with an I_{sc} of 8.42A. In case B the differences in I_{sc} for the different types of laminates increase. As a consequence of the better reflection in the VIS range, the laminates incorporating the uncoated white polyamide and Cu foil outperform the all-black laminates. Although the reflection measured for the Cu foil is higher in a large part of the spectrum, the white polyamide back foil results in a higher I_{sc} when the full laminate area is illuminated. Presumably, what plays a role here is the broader range of high reflectance of the polyamide foil and a possible higher degree of Lambertianity. As was already the case when only the reflections from behind the cell were taken into account, the BPC foil based laminates clearly outperform the conventional black laminate. The BPC Cu foil yields a slightly higher I_{sc} than the BPC polyamide foil, 8.75 ± 0.01A vs 8.69 ± 0.02A, while for the conventional black

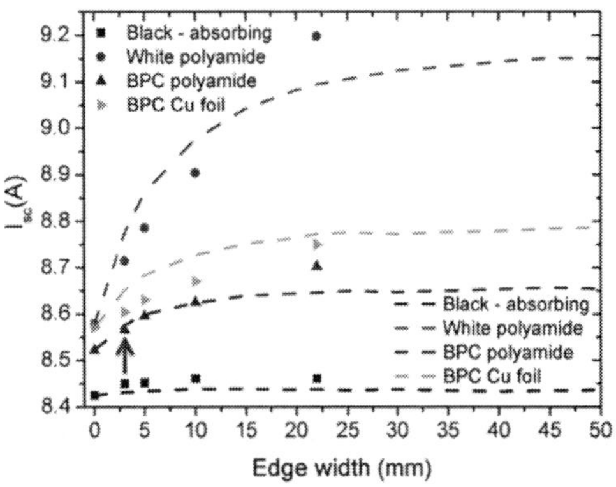

Figure 5: I_{sc} as function of the exposed edge width around the cell for different backsheets studied. The dashed lines are the result of a ray-tracing analysis using a simple model (for details see text).

laminates a (scaled) I_{sc} of 8.47 ± 0.01A is found. For each group of laminates one was selected for which a series of different sized masks was applied. In addition, ray-tracing simulations were performed to model the fraction of rays falling onto the cells after reflection from the backsheet for the different aperture areas. The measured and simulated data are included in Figure 5. In the simple ray-tracing model used here, the solar cell is modelled as a black absorber (in the 330 – 1100 nm spectral range, with an IQE 100%. This will result in an overestimation of the contribution of longer wavelength light to the I_{sc}. Therefore, it is expected that the model will overestimate the fractional increase in I_{sc} from the exposed backsheet area (in particular for the BPC foils). Furthermore, the model assumes Lambertian reflection by the backsheets and a flat interface between air-glass, any deviation from this will affect the fraction of light falling within the escape cone. Nevertheless, the simulations succeeded in reproducing the trends in the experimental data.
From Figure 5 one can observe that although the BPC Cu foil displayed significantly higher IR reflection than the BPC polyamide foil, the slope of I_{sc} vs. edge width is very similar. Compared to the I_{sc} at 0 mm edge width the I_{sc} at full laminate illumination is a factor 1.02 higher for both configurations. In addition, to the absolute reflection also the angular distribution of the reflected light is of importance for the foil's performance with respect to concentration. Therefore, we hypothesize that the BPC polyamide foil's reflection angular distribution is closer to Lambertian than for the BPC Cu foil.

Finally, we use the laminate data to make an estimate of the performance yield, which can be obtained from the application of the BPC foils. Since the typical-inter cell distances in a module are around 3 mm, we obtain from Figure 5 (indicated by the red arrow) an increase in I_{sc} of 1.4% and 1.8% for BPC polyamide and BPC Cu foil respectively with respect to the conventional black layout.

4 CONCLUSIONS

We have shown that the application of black, low IR absorption, powder-coatings to standard foils used in PV industry can be used to significantly reduce CtM losses for all-black modules. Upon the application of such foils the IR reflective properties of the underlying substrate are largely preserved. In comparison to the conventional, absorbing black foils an improvement in I_{sc} of 1.4 – 1.8% is expected on module level.

ACKNOWLEDGEMENTS
The authors acknowledge Ingrid Romijn and ECN's n-Pasha team for making available the solar cells used in this study.

REFERENCES

[1] K.R. McIntosh, R.M. Swanson, J.E. Cotter, Prog. Photovolt: Res. Appl. 14, (2006) 167.
[2] I.G. Romijn, G. Janssen, M. Koppes, J. Liu, Y. Komatsu, J. Anker, A. Gutjahr, E. Kossen, A. Mewe, K.. Tool, O. Siarheyeva, M. Ernst, Proceedings of SNEC 8th International Photovoltaic Power Generation Conference & Exhibition (2014).
[3] A.R. Burgers, L.H. Slooff, R. Kindeman, J.A.M. van Roosmalen, Proceedings of 20th EC Photovol. Solar Ener. Conf. (2005) 394.

TRIANGULAR RIBBONS FOR IMPROVED MODULE EFFICIENCY

Max Mittag, Andreas J. Beinert, Li C. Rendler, Matthieu Ebert, Ulrich Eitner
Fraunhofer Institute for Solar Energy Systems ISE
Heidenhofstr. 2, 79110 Freiburg

ABSTRACT: Different approaches like round wire interconnectors, shingled or back-contact cells have been presented in the past to improve the cell-to-module efficiency ratio by reducing the shading losses of interconnector ribbons.
We present a new cell interconnector design based on a triangular cross section to further improve modules based on interconnector ribbons. We analyze the optical behavior of the concept and compare it with rectangular ribbons and round wire interconnectors. An evaluation of the new concept using optical ray tracing is performed. Results show an advantage in optical performance of 2.35% of the new concept compared to standard interconnector ribbons under perpendicular irradiation as well as 1.94% compared to round wires. An analysis using irradiation data (DNI) shows a superior optical performance of the TriCon-Concept. We find that in an elevation tracked module using TriCon 2.32% more light reaches the cell surface over the year compared to rectangular interconnectors (5BB) and 2.02% compared to round wires.
Keywords: Module, Module Manufacturing, Optical Losses, Performance, PV Materials, Ray Tracing, Ribbons

1 INTRODUCTION

The industrial standard for the electrical interconnection of solar cells are rectangular interconnection ribbons. They are featuring the inherent disadvantage of a flat topside that reflects incident light out of the module which is then lost for power generation. Other concepts have been presented in the past to improve cell interconnection, for instance round wires or light harvesting connectors that have been proven to reflect more light onto the cell and show higher optical efficiencies [1-8].

Solar cells currently shift from three to four or even five busbars [9]. Increasing the number of busbars results in improvements of the electrical efficiency and a cost reduction of solar cells [10, 11]. New approaches like round wires are to be expected to increase the number of busbars and interconnectors even further to up to 15 per cell [5].

In order to improve the concept of round wires we present and evaluate an interconnector design that is based on a triangular shape ("TriCon") and that can be used on solar cells suitable for round wires like the Multi-Busbar Connectors (MBC) by SCHMID or the Smart-Wire Interconnection by Meyer-Burger [12]. The idea of the TriCon-Concept is to optimize the light harvesting.

We perform a ray tracing simulation to compare the optical properties of rectangular interconnectors, round wires and the triangular connector ("TriCon"). Additionally a combination of the results of the ray tracing-analysis with irradiation data of Freiburg, Germany is performed to estimate gains for a solar module using the different interconnector designs.

2 CONCEPT PRESENTATION

The interconnector as displayed in figure 1 bases on a triangular shape and features a solder coating. It has an edge length of around 400 μm. The coating thickness is identical to standard interconnectors. The corners of the triangle are rounded due to processing limitations and have a radius of around 40 μm.

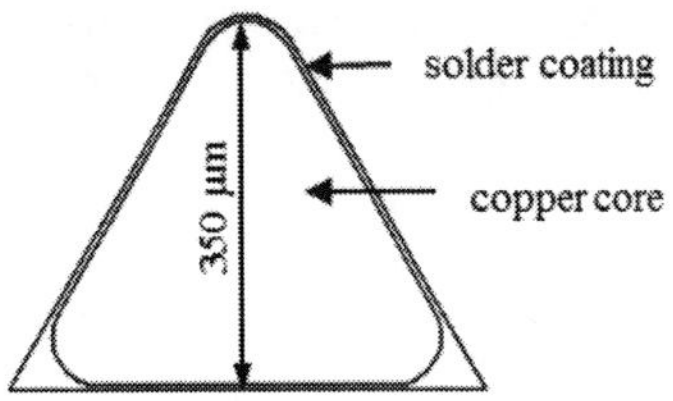

Figure 1: cross section of a soldered triangular interconnector

The dimensions of the triangle are chosen to result in an equivalent total cross section per cell. This leads to similar electrical losses compared to a round wire approach. The triangular wire can be coated with standard coating materials like SnPbAg or SnAg and is therefore compatible with industrial solar cells. The width of TriCon is chosen to be compatible to solar cells suitable for round wires.

Table I: comparison of cross sections of different interconnector technologies

	3BB	5BB	MBC15	TriCon
Number of ribbons per cell	3	5	15	15
Height of ribbon [mm]	0.20	0.20	0.35	0.35
Width of ribbon [mm]	1.50	1.00	0.35	0.40
Cross section per ribbon [mm²]	0.300	0.200	0.071	0.069
Cross section per cell [mm²]	0.900	1.000	1.060	1.039

The triangle is isosceles with internal angles of 60°. Irradiant light is reflected and reduces the cell to module losses by improving the optical performance of the module.

3 SIMULATION

2.1 Optical Analysis of Interconnector Concepts

As shown in Figures 2, 3 and 4 five different cases of reflection can be identified among the different interconnector designs:

1: Light irradiated perpendicular to the cell that is reflected out of the module.
2: Light irradiated under a non-perpendicular angle that is reflected out of the module.
3: Light irradiated under a non-perpendicular angle at the side of the interconnector / the solder that is reflected onto the cell.
4: Light irradiated perpendicular to the cell that is reflected onto the cell.
5: Light irradiated under a non-perpendicular angle that is reflected onto the cell.

Additionally a sixth important case is observed where light is reflected on the front cover after reflection on the interconnectors. All described cases are only valid for angles that are perpendicular regarding the axis of the interconnector ribbons.

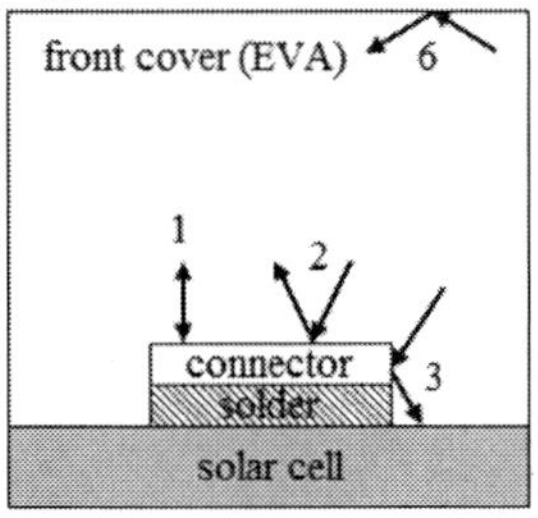

Figure 2: incident and reflected light for rectangular interconnectors

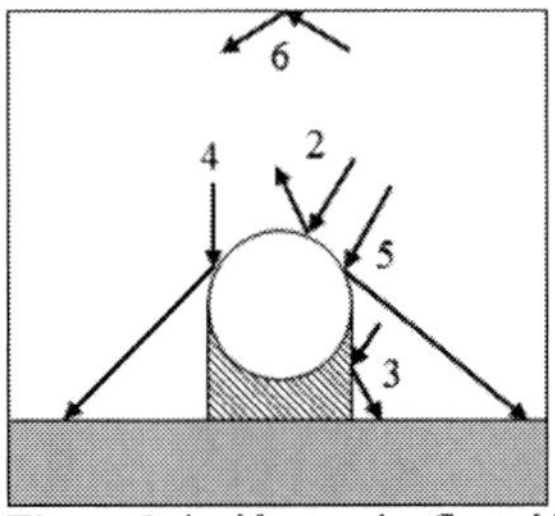

Figure 3: incident and reflected light for round wires

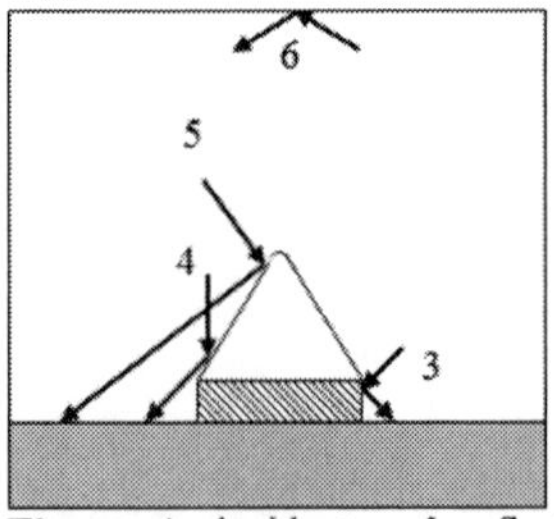

Figure 4: incident and reflected light for triangular interconnectors

The rectangular interconnector reflects a significant amount of light away from the solar cell (see Figure 2, No 1 and 2) which is lost for power generation. This loss may be reduced by additional reflections that can occur on the front interface of the module (Figure 2, No 6). Due to reflections on the side of the interconnector (Figure 2, No 3) additional gains may occur. Since perfect reflection on the side does not occur, the additional shading of the non-illuminated side of the interconnector compensates this gain. Diffuse reflections reduce the optical width of the interconnector and improve its optical performance [2, 13, 14].

For the round wire interconnector and TriCon reflections of type 1 are virtually non-existent while reflections to the solar cell or the front interface of the module increase.

Based on the geometry of TriCon reflections within the module can be divided into two groups for different angles which are illustrated in Figure 5.

The first type (a) describes light irradiated between ±30° and reflections on both sides of the triangle onto the cell. The second case occurs for light irradiated with angles above 30°. Here light is shading part of the cell (b) or is reflected back to the front surface (c) to be then reflected again (d).

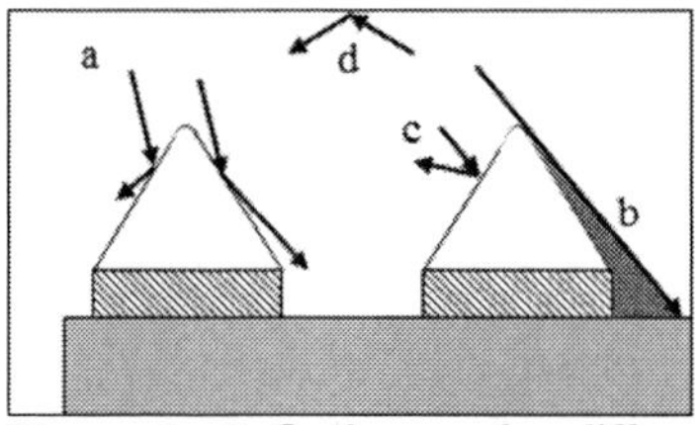

Figure 5: Reflections under different angles at a triangular connector

While reflections of type (e) are virtually non-existent for round wires a shading (Figure 6, b) of direct light occurs for every non-perpendicular incidence angle (Figure 6, d). Reflections onto the cell (b) and towards the front cover (c) are unchanged. Additional reflections on the front cover (d) might occur.

Since angles larger than approx. 42° are not possible within a module due to Snell's law of refraction, shading (b) and reflections of type (c) are not significant for TriCon.

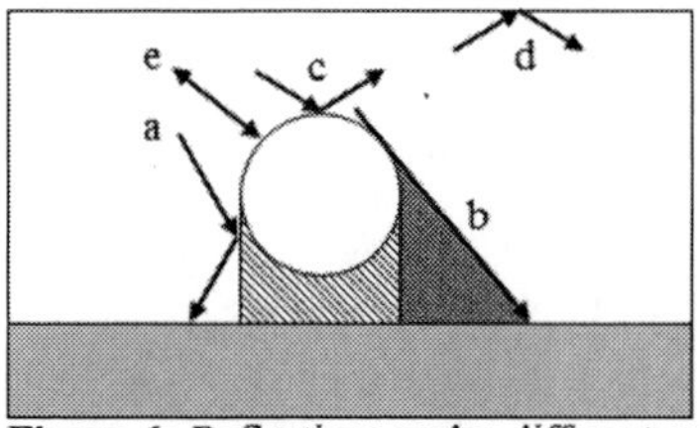

Figure 6: Reflections under different angles at a round wire connector

2.2 Ray Tracing Setup

We perform a ray tracing analysis using the finite elements method (FEM) to prove the optical advantages of the triangular interconnector. A solar cell (156 x 156 mm²) is irradiated under different angles and the reflection on interconnector ribbons is calculated.

A whole cell with the correspondent number of interconnectors is simulated. The model is a 2D model using 100.000 rays. A sweep of the incident angle from 0° to 80° is performed. The rays have a wavelength of 1000 nm. The cell is embedded in EVA that also provides the front interface of the simulation setup (Figure 2). The solar cell is modelled as a perfect absorber to independently evaluate effects caused by the interconnection. A simulation of the cell without any interconnectors is used as a reference value.

We set the reflectivity of the interconnector to be 0.80 and the reflections to be 80 % specular and 20 % diffuse. Multiple reflections on the interconnector and the front cover are computed. The thickness of the solder is set to 20 µm. All remaining geometries are shown in Table 1.

2.2 Ray Tracing Results

Figure 7 shows the results for angles between 0° (perpendicular irradiance on the solar cell) and 80°.

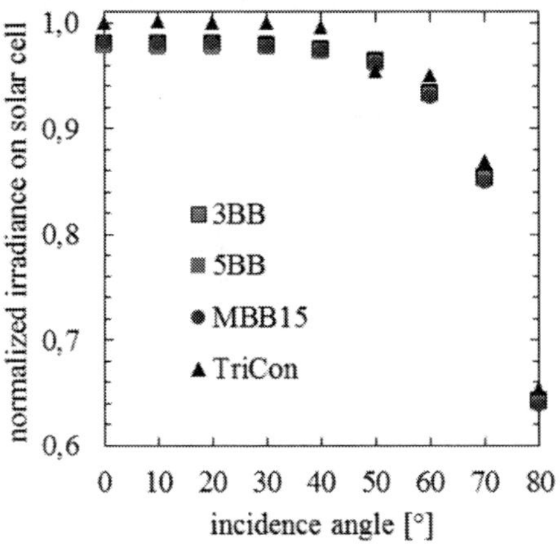

Figure 7: Optical performance of different interconnector technologies at different incidence angles, normalized irradiance on a solar cell

Simulation shows an advantage of the triangular ribbons for most angles of irradiance. Gains above 30° result from secondary reflections on the front glass. Round wires also gain from multiple reflections and increased reflections towards the cell (see Figures 3 and 4). The dip at 50° results from an optical interaction between TriCon interconnectors and the front cover interface. Nonetheless the optical properties of the TriCon-concept additionally increase the performance of the module.

The TriCon concepts shows approximately 2.35% higher efficiency than the rectangular interconnectors (5BB) perpendicular irradiation and 1.94% compared to round wires. The differences between three and five busbars result from the larger footprint of the five busbars.

2.2 Yield Simulation

Since irradiation on a solar module is incident under varying angles and radiance is not evenly distributed within these angles a calculation using irradiation data is performed [6]. We use data of Meteonorm 7 for a module mounted in Freiburg, Germany. The module is facing south and is mounted as shown in Figure 8. We calculated the incident angles for a fixed module (45°), an elevation (around horizontal axis) tracked and an azimuth (around vertical axis) tracked module. The interconnector ribbons run vertically within the module.

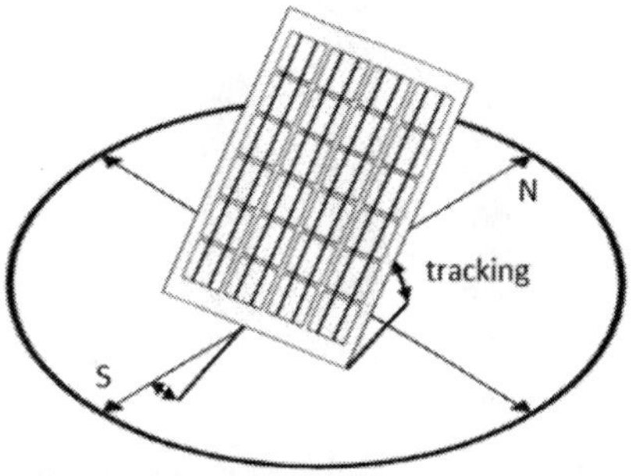

Figure 8: assumed mounting of the module with different types of tracking

Figure 9 shows the yearly direct irradiation incident within a certain range of angles. It is obvious that tracking results in lower incidence angles and therefore in lower optical losses.

To eliminate the influence of the elevation of the sun for the calculation, the module is assumed to be tracked vertically. Only changes of the azimuth are used for the yield estimation. This results in a sun path perpendicular to the axis of the interconnector ribbons and therefore in an optical behavior as shown in figures 1 to 6.

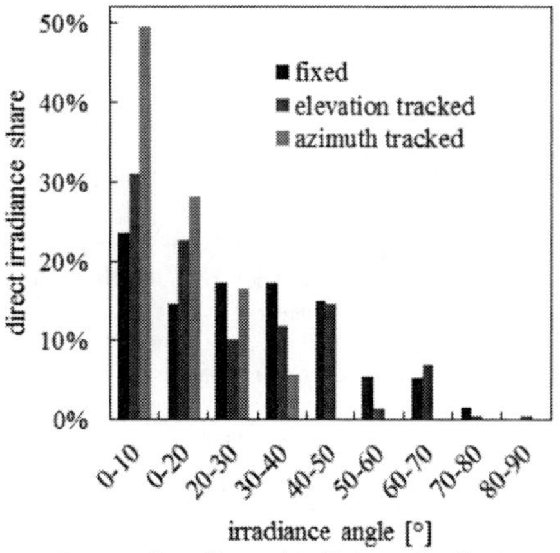

Figure 9: angular direct irradiation distribution on module plane for fixed and tracked modules in Freiburg, Germany

By combining data of Figures 7 and 9 a comparison of the different interconnector technologies is possible. It clearly indicates an advantage of the triangular interconnector design compared to other designs.

We only consider direct normal irradiance (DNI). This simplification might reduce the advantage of TriCon although Figure 7 shows an advantage of TriCon for most angles.

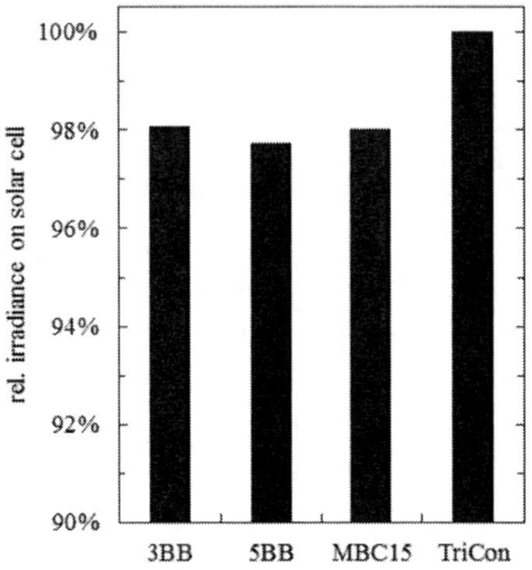

Figure 10: Optical performance of different interconnector technologies for a vertically tracked module mounted in Freiburg, facing south

Figure 10 shows a 2.02 $\%_{rel}$ higher optical efficiency of the TriCon-concept compared to the round wire approach after considering annual environmental irradiation and angular power distribution. Compared to five rectangular connectors the triangular connector gains 2.32$\%_{rel}$. Again the difference between three and five busbars only results from the increased footprint of the five busbars. Electrical advantages of the higher number of busbars are not considered for all concepts.

The optical efficiency directly relates to the power generation of the solar cell and therefore a gain can be achieved by using TriCon interconnectors.

3 SUMMARY

A new interconnector design is presented based on a triangular cross section. An analysis of the optical behavior of different types of cell interconnectors is conducted and a more complex behavior of round wires and TriCon can be observed.

A ray tracing simulation is performed to analyze the angular-dependent reflections of three interconnector concepts. Results show a superior optical performance of the TriCon ribbons. For irradiation perpendicular to the module TriCon shows a 2.35% higher optical performance than five busbars and 1.94% compared to round wires.

We conduct a yield estimation for a module using different interconnector designs using irradiation data. The yield analysis uses direct normal irradiance for Freiburg, Germany and an elevation tracked module. Results show that 2.32% more light is reaching the solar cell (optical performance) in a module using the triangular connector compared to a module with five busbar, rectangular ribbon interconnection and 2.02% more compared to round wires.

4 REFERENCES

[1] Zarmai, T., et al, "A review of interconnection technologies for improved crystalline silicon solar cell photovoltaic module assembly, Applied Energy, Volume 154, pp. 173-182

[2] Wirth, H., et al, Advances in Photovoltaics, Vol. 2, pp. 160-193, Burlington: Academic Press, 2013

[3] Eitner, U., et al, "Progress in HIP-MWT module technology", MWT-Workshop 2013, Freiburg (Germany), 2013

[4] Eitner, U., et al, „Interconnector-based module technology for thin MWT cells" 27th PVSEC, Frankfurt, 2012

[5] Walter, J., et al, "Multi-wire interconnection of busbar-free solar cells", SiliconPV 2014, 's-Hertogenbosch, 2014

[6] Braun, S., et al, "Solar cell improvement by using a multi busbar design as front electrode", SiliconPV 2012, Leuven, 2012

[7] Ebert, M., et al, "The Impact of Angular Dependant Loss Measurement on PV Module Energy Yield Prediction", 29th PVSEC, Amsterdam, 2014

[8] Schneider, J., et al, "Combined effect of light harvesting strings, anti-reflective coating, thin glass, and high ultraviolet transmission encapsulant to reduce optical losses in solar modules", Progress in Photovoltaics, Vol. 22, pp. 830-837, 2014

[9] International Technology Roadmap for Photovoltaic, ITRPV Seventh Edition 2016, VDMA

[10] Braun, S., et al, "The multi-busbar design:an overview", 4th Workshop on Metallization for Crystalline Silicon Solar Cells, Erschienen in: Energy Procedia ; 43 (2013) S. 86-92, 2013

[11] Hendrichs, M., et al, "Advanced Metallization Concepts for p-type Silicon Metal-Wrap-Through (MWT) Solar Cells", Energy Technology, Vol. 2, DOI: 10.1002/ente.201300094, 2014

[12] Faes, A., et al, "SmartWire solar cell interconnection technology." 29th PVSEC, Amsterdam, 2014

[13] Woehl, R., et al, "Analysis of the Optical Properties of Screen-Printed and Aerosol-Printed and Plated Fingers of Silicon Solar Cells", Advances in OptoElectronics, Volume 2008 (2008), Article ID 759340, 2008

[14] Stuckings, M.F., et al, "A study of shading and resistive loss from the fingers of encapsulated solar cells", Solar Energy Materials and Solar Cells, Volume 59, Issue 3, October 1999, Pages 233–242, 1999

THIN-FILM BARRIERS FOR DURABLE THIN-FILM PV MODULES

Jürgen Hüpkes[1*], A. Wrigley[1], W. Reetz[1], Nicolas Wyrsch[2], Gianluca Cattaneo[3], Fanny Sculati-Meilaud[2], A.Gerber[1]
[1]IEK5-Photovoltaik, Forschungszentrum Jülich, Leo-Brandt-Strasse, D-52425 Jülich, Germany
[2]Institute of Microengineering, Ecole Polytechnique de Lausanne, Rue de la Maladière 71B, 2002 Neuchatel, Switzerland
[3]CSEM, PV-center, Jaquet-Droz 1, 2000 Neuchâtel, Switzerland
[*]j.huepkes@fz-juelich.de, phone: +40 2461 61 2594, fax: +49 2461 61 3735

ABSTRACT: This contribution describes thin-film barrier coatings to enhance the stability of silicon thin-film solar modules. Thin-film barriers are applied to protect the modules from degradation in humid environment. Demonstrator modules were exposed to indoor and outdoor tests to evaluate failure modes by electrochemical corrosion during operation. Test results reveal some major failure modes, but we demonstrate long lifetimes under operational conditions for certain modules without conventional encapsulation. These modules without conventional encapsulation showed less than 1% relative degradation during 1000 hours of damp-heat tests with light exposure at open circuit voltage. These promising results will guide the way to new encapsulation concepts for longer lifetimes and/or new low-cost concepts. This approach might be applied to any thin-film technology and will be an important aspect for consumer electronics with less strict requirements for very long-term operation.
Keywords: stability, solar modules, thin-film barriers

1 INTRODUCTION

Standard encapsulation schemes for thin-film modules such as glass/glass have two drawbacks: (1) EVA often used as an encapsulant/adhesive releases acetic acid, which may react with the cell or its contacts, if water ingress is not sufficiently suppressed. (2) Standard encapsulation is optimized for wafer-based technology and devices with lower efficiency suffer from relatively high area cost of the encapsulation material. Though several delamination failures of thin-film PV modules have been reported [1], more recent solar modules do not suffer from such degradation anymore due to different solutions [2]. Lab tests have demonstrated non-protected thin-film Si cells/modules that survived damp-heat exposure without major degradation [3].
The PV market is dominated by silicon wafer-based technology. Thus, standards regarding measurements, regulations and standard encapsulations are defined for this technology but consequently are imposed on thin-film technologies. Unlike wafer-based technology, thin-film PV modules consist of uniform coatings on large area substrates. Thus, thin-film barrier coatings might serve as protection against moisture ingress. This is especially interesting to enhance the lifetime of PV devices or being an alternative e.g. for consumer electronics or for silicon thin-film PV modules that have demonstrated inherent stability against damp-heat environment [3].
This contribution presents functional modules on rigid glass superstrate tested under operational conditions using indoor and outdoor degradation tests. Similar approaches have been followed by other groups using barrier coatings in OPV technology [4].

2 EXPERIMENTAL

2.1 Module preparation and encapsulation

Small area modules (10x10 cm²) were prepared using our base line process at Jülich. The silicon thin-film modules consist of sputtered aluminum doped zinc oxide (ZnO:Al) or Asahi VU type fluorine doped tin oxide (SnO_2:F) front TCO, amorphous-microcrystalline silicon tandem junction solar cells and a ZnO:Al/silver back reflector. All modules were series connected via laser scribing. The P3 scribe line (back contact) reveals parts of the front contact by simultaneous ablation of the absorber and back contact. The front contact material is known to degrade under exposure to moisture. Thus, parts of the modules were protected by additional barrier coatings, namely sputtered ZnO:Al or silicon oxynitride (SiO_xN_y), plasma enhanced chemical vapor deposited amorphous silicon (a-Si) or silicon oxide (a-SiO_x), or they were left blank.

Figure 1: Photographs of solar modules on thermal insulating building materials (left) and the modules under consideration on the outdoor measurement setup (right).

In order to check the concept and get insight into the effect of the atmosphere on the various parts of the (non-encapsulated) module (such as cell itself, interconnection area and bus bar area) medium area (30x30 cm²) solar modules with a special "quasi"-encapsulation were designed. An air-space of about 7 mm between the front glass with the active layers on the back surface and the back glass enabled the air to penetrate the active layers during operation. The top and side edges of the module were sealed by a water tight glue tape. As a reference standard glass/glass encapsulation was applied. All modules were separated by laser scribing into three separate sections (A, B, C) and can be individually measured. For more details the reader is referred to the images on the poster (available for download). These modules were exposed to our outdoor test setup (see Figure 1). The center row represents the quasi-

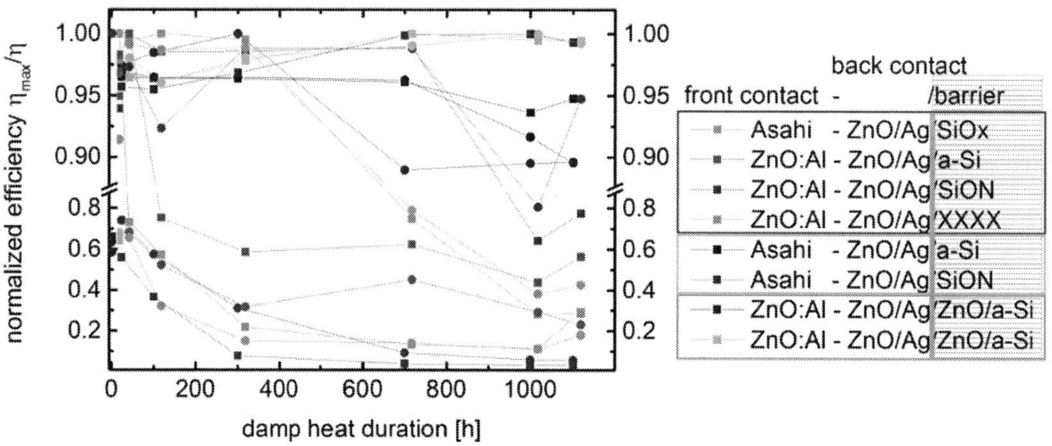

Figure 2: Normalized efficiency of solar modules with different protective barrier coatings as function of damp heat exposure time.

encapsulated modules (with air-gap and without barrier coating) and the bottom right one with glass/glass encapsulation serves as reference. The other modules, visible in the photograph, are not encapsulated at all and some of them reveal their back surface fully open to the environment, including rain runoff.

3 DAMP HEAT DEGRADATION TESTS

On small area solar modules (10x10 cm²) thin-film coatings were deposited to protect the solar modules from environmental influences. In previous studies such modules had been exposed to damp heat in the dark without electrical bias, so the modules have not been subject to electro-corrosion. To evaluate this further, new damp-heat tests were performed under AM1.5 type irradiation with ~60° inclined irradiation angle. Figure 2 shows the efficiency values of the modules as function of damp heat treatment duration measured indoors with a class A sun simulator.

The remarkable result is that some of the modules survived this harsh test for 1000 hours. Unfortunately, several modules degraded (marked in red) or nearly died after about 300 hours of irradiated damp-heat test (yellow). This was mainly the case for protective coatings that had been deposited directly onto the silver back contact. Adhesion on silver is known to be a critical issue, thus a ZnO layer between silver and the silicon protection layer led to very stable modules. Interestingly the type of front contact (Asahi U type SnO2:F or sputtered ZnO:Al) did not influence the stability to a large extent. This suggests that the main degradation occurred at the back reflector. As a result of handling issues some of the modules might have degraded due to mechanical damage rather than by corrosion during the test. This was obvious when analyzing the results of individual cells (not shown).

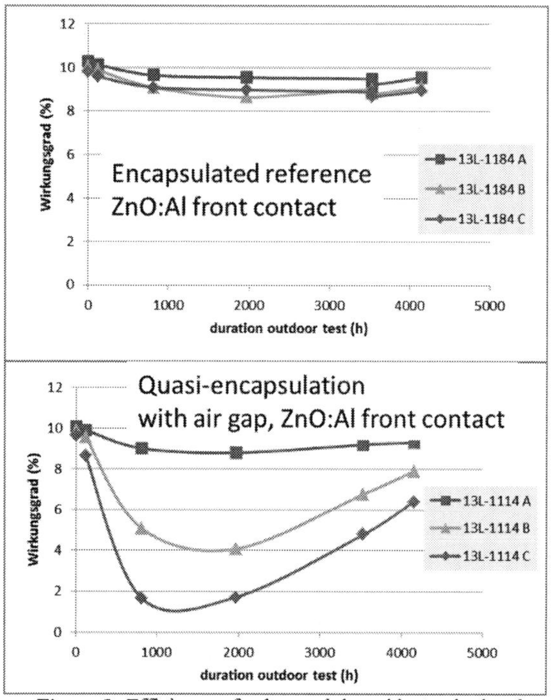

Figure 3: Efficiency of solar modules with standard and quasi-encapsulation as function of outdoor exposure time.

4 OUTDOOR DEGRADATION TESTS

Further evaluation of outdoor tests of 30x30 cm² modules has been performed. **Figure 3** shows absolute module efficiency measured indoors using a flash light sun simulator during consecutive outdoor tests that started in Dec. 2015 at the Jülich site. The encapsulated modules exhibit about 10% relative degradation during the first 1000 hours as assigned to the Staebler Wronsky effect. For the quasi-encapsulated modules the different sections can be distinguished. The section A, at the longest distande to the opening degrades to a similar extent as the encapsulated reference module. However, the module sections B and C, which are much closer to the opening, show strong degradation during the first 1000 hours of the test. These might be related to corrosion of the active layers or contacts, though their efficiency started to increase after about 2000 h. This interesting result was not investigated further due to time constrains, but the effect might be related to the maximum temperature at each day, which exceeded 10°C only for a few days during the first 2000 h and was on average 20°C for the time between May and June 2016 (see **Figure 4**).

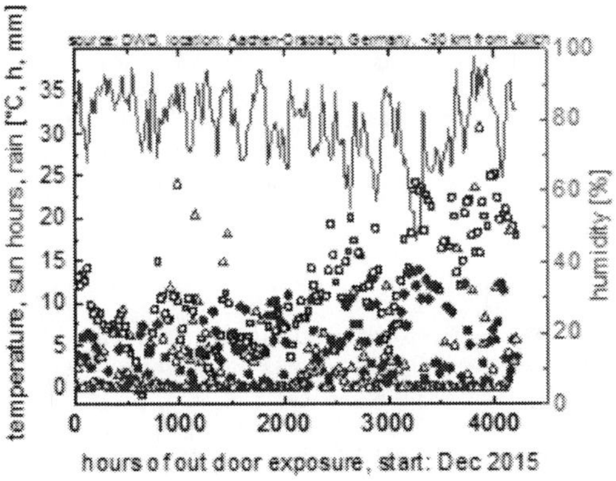

Figure 4: Weather data of the location Aachen-Orsback, at about 30 km distance to the Jülich site maximum day temperature, sun hours for each day, amount of rain and relative humidity.

5 CONCLUSIONS

We have demonstrated amorphous silicon based silicon solar modules using thin-film barrier coatings for environmental protection. Apart from some module failures, several barrier coatings provided sufficient protection against damp-heat degradation and electro-corrosion. The modules showed less than 1 % degradation after 1000 hours of damp-heat treatment. These are very promising results to enhance module lifetime and might stimulate further research on new encapsulation concepts for thin-film PV technologies including OPV.

REFERENCES

[1] G.R. Mon, R.G. Ross, 18th Photovoltaic Specialists Conference, New York, Institute of Electrical and Electronics Engineers, Inc., Las Vegas, NV, 1985, p. 1142.

[2] J.H. Wohlgemuth, D.W. Cunningham, P. Monus, J. Miller, A. Nguyen, Conference Record of the 2006 IEEE 4th World Conference on Photovoltaic Energy Conversion, 2006, p. 2050.

[3] W. Beyer, J. Hupkes, H. Stiebig, Thin Solid Films 516/2-4 (2007) 147.

[4] M.D. Clark, M.L. Jespersen, R.J. Patel, B.J. Leever, Organic Electronics 15/1 (2014) 1.

Proposed evaluation framework for exploration of smart PV module topologies

M. Baka[1], F. Catthoor[2 3], D. Soudris[1]

[1]National Technical University of Athens (NTUA),
Heroon Polytechneiou 9, Zographou Campus 157 80 Athens - Greece
[2]imec, Kapeldreef 75, 3001 Heverlee , Belgium
[3]Katholieke Universiteit Leuven (ESAT), Kasteelpark Arenberg 10, 3001 Heverlee, Belgium

ABSTRACT: The increment of the PV market has led to multiple PV installations which operate under non-ideal conditions. Heavy shading is expected in rooftop installations, Building Integrated Photovoltaics and solar powered vehicles. Demand of an increased energy-yield, despite of the non-uniform conditions render reconfigurable topologies a promising solution for such installations. Evaluation of the performance of a PV module can be performed either on the field or in a detailed simulation environment. The problem of the existing solutions is a large cost investment for constructing all potential topologies and the large time requirements for detailed simulation of the entire module. A simulation framework is proposed here which can significantly speed up the evaluation process of different reconfigurable PV module instantiations and conventional modules for different operating scenarios with a high accuracy level.
Keywords: Modelling, PV Module, Performance, Shading, Configurable Topology

I. Introduction

Solar energy is an attractive solution for future clean energy production on site. This has lead to an increase of Photovoltaic installations in urban areas, where the operating conditions of the PV array are not always optimal. In partial shading conditions, which appear frequently in rooftop installations, Building Integrated Photovoltaics (BIPV) and in solar-powered vehicles, the conventional design of a PV module cannot reach the demand for an increased energy yield. This has lead to the study of dynamic configurations which have beneficial results under non-uniform dynamic conditions [1].

Multiple options exist concerning the design of a reconfigurable module, which leads to a large number of potential topologies. In order to explore and determine the most promising configurations, a fast evaluation process is needed. Fabrication of all configurations is too cost demanding, while extensive simulations require a lot of time. In this paper we propose an evaluation process which allows significantly faster estimation of the energy-yield of reconfigurable topologies by reusing and combining simulation results.

II. Motivation

Reconfigurable topologies can produce more energy than a conventional design of a module under non-uniform operating conditions. The increased fabrication cost of reconfigurable modules however has to be taken into account and the increased energy-yield has to compensate for the additional cost. This work focuses on the evaluation of energy-yield gain where the losses of all additional components are included. Overall performance can be evaluated either in the field with extensive measurements or in a simulation environment. The multiple design options for a reconfigurable module render fabrication an unrealistic solution to compare the performance of each topology.

In [2], a type of a reconfigurable topology of a PV module is proposed. An instantiation of the parameterized template of the module is illustrated in Figure 1.

The performance of this module is evaluated in a simulation environment. The core of the simulation is the cell. The model of the cell is derived based on the physical layer. The parameters of the cells and wires have been calibrated through extensive measurements. This has lead to a detailed model where all the operating conditions (irradiation, temperature, wind) are taken into account [3]. In order to simulate reconfigurable topologies, extra elements (wires and switches) are added to this model .The final model involves both an electrical interconnection and a thermal interconnection of the cells.

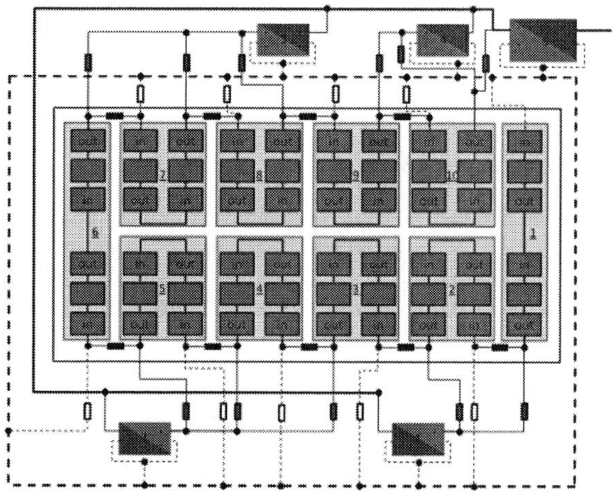

Figure 1: Instantiation of a reconfigurable module

Due to the complexity of the model -required for accuracy - simulation time has increased considerably though. Considering all potential operating scenarios and all potential instantiations of a reconfigurable module, this is a large overhead. Each simulation of a module for a shading pattern requires up toa minute to achieve the desired level of simulation accuracy. The number of shading patterns for a module with 60 cells is quite high. For any given irradiation level, even if a cell can only be completely shaded or not, restricting partial shading of cells, 2^{60} possible shading patterns are present. Of course not all of the shading patterns are realistic or relevant, but still thousands or even millions of simulations could be required to cover all possible operating conditions.

The module of Figure 1 is simply an instantiation of the general concept of a reconfigurable module. The main elements of a reconfigurable module can be parameterized which leads to a large number of potential instantiations. The components of reconfigurable topologies are:

1. The cell-strings, which consist of permanently series connected cells and are the minimum power producing element within the module
2. The supporting network:
   a. Switches and wires for interconnection of the cells
   b. Local converters for individual operation of groups of cells
   c. Module converter

The size of the cell-strings, the complexity of the supporting network, the number of converters, are all design choices that lead to different instantiations of the PV module. The framework proposed in this paper aims to speed the evaluation process by reusing simulation results.

In [4] it has been demonstrated that the strongest thermal dependences are situated between cells which are electrically interconnected within the module, thus within a cell-string. The interconnections of the cell-strings, which are outside of the module, do not result in strongly coupled cells (Figure 2).

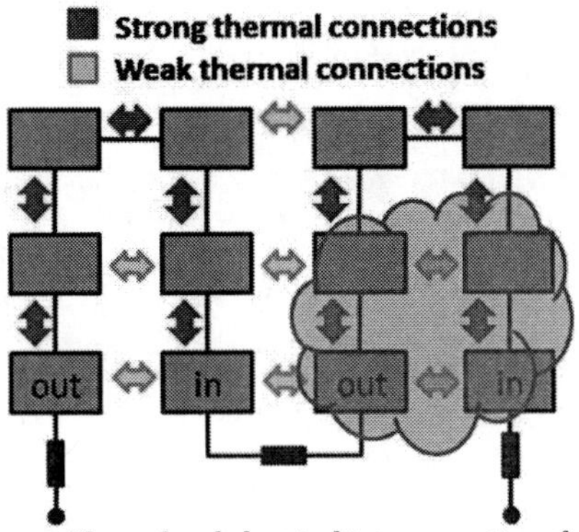

Figure 2:Thermal and electrical interconnection of cells

A comparison has been made between the results of simulating the entire module and performing MPP tracking within the same simulation environment, and combining simulated IV curves of the cell-strings. The differences of the two results are completely negligible (lower than 0.5%), and the two methods of simulating can be considered equivalent. This by itself improves simulation time, as the smaller, simpler models of the cell-strings run faster and can be executed in parallel. The combination of the results utilizes only an electrical network which is fast as well.

Further motivating this approach is the result of the comparison of measured IV curves of cell-strings in a controlled environment and a simulated IV curve of a cell-string. In Figure 3, it is shown that a marginal error is present between measured IV curves and a simulated IV curve. Some small difference is expected due to measurement errors anyway so this supports the hypothesis made above.

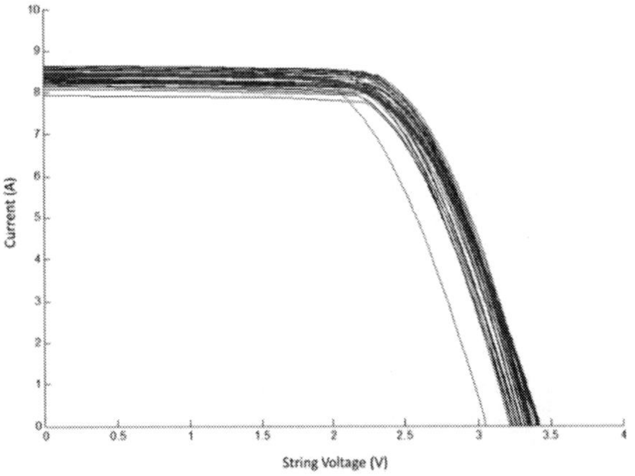

Figure 3: Measured IV curves of a cell-string (blue) compared to a simulated IV curve of a cell-string (red)

For a given selection of cell-strings, this allows to produce simulation results for various interconnection schemes of the cell-strings. Another advantage of combining results is that simulations of cell-strings which have the same geometry and the same operating conditions do not need to be performed multiple times. The simulation framework in this paper is meant to speed up the simulation time of a given cell layout for different shading patterns and different interconnection schemes. However, it should be applicable for all potential instantiations of reconfigurable topologies. In order to ensure that, the division of the module in to cell-strings has to lead to cell-strings, which are not strongly-coupled and that a limited type of cell-strings is present in each module. This is validated by describing the main principles which are considered in the design of the cell-strings.

III. CONCEPTS OF DESIGNING A SMART PV MODULE – CELL-STRING SELECTION

The aim of reconfigurable topologies is to allow the majority of cells to operate at their MPP. One main design option of the proposed configurations is the granularity level within the module. In order to reduce the initial cost of the module, the flexibility of the module is restrained to the level of a cell-string. Any mismatch within a cell-string cannot be avoided.

The selection of the size and shape of the cell-strings are crucial to the operation of the module. While determining the geometry of the cell-strings four criteria are taken into account:

1. Uniform operating conditions is the most frequent scenario. Under uniform conditions, series connection of the cells is optimal, which is the basis of the design of conventional modules. Our goal is to have negligible losses compared to a conventional module, thus a potential path with low resistivity is required to allow an efficient series connection of all cells.
2. Degree of flexibility: Both the size and the shape of the cell-strings have an important effect on the flexibility of the module. The impact of the size of the cell-string is more evident, as the operation of an entire cell-string is limited to the performance of a single cell. The importance of the shape of the cell-strings becomes clearer when patterns of casted shade are considered. Shades can come from different angles and the module should be able to be divided in both main directions.

3. Restriction of all inputs and outputs of cell-strings to the edge of the module to have limited fabrication cost
4. The size of all the cell-strings should be the same to enable potential parallel connections.

Application of these principles in the design process, results intoa limited number of types of cell-strings within a reconfigurable module. In the module of Figure 1, two typesof cell-strings are present. U-shaped cell-strings allow a better division of the module for different shading angles (second criterion). The I-shaped on the edges are designed to allow a minimal resistive path during series connection of the cell-strings (first criterion). As the interconnection of cell-strings always takes place outside of the encapsulation (third criterion), the cell-strings are never strongly thermally coupled. In conclusion, the requirements to utilize the suggested framework for all potential reconfigurable topologies are met.

IV. Proposed simulation framework

There are a large number of different operating conditions of a single PV module. Each shading pattern creates different operating conditions, leading to a different operating scenario.Simulation ofall possible scenarios requires a lot of time, while clustering shading patterns (equivalent operating scenarios) and limiting the simulations in the module level can be hard. Our proposal is to break down the simulations to the cell-strings. The potential operating scenarios of each cell-string level are reduced significantly due to the smaller number of cells. Combination of all potential operating scenarios of the cell-strings, will allow the construction of all potential scenarios of the module.

The first step is to determine the potential distinctive operating scenarios of a cell-string. In order to combine the simulation results of the cell-string to derive the power output of the entire module, the entire IV curve of each active cell-string in the configuration has to be known.

Each cell-string has a semi-infinite number of different operating conditions. However, the distinct IV curves that can be produced are finite. If different operating conditions of a cell-string lead to similar IV curves, these operating conditions can be clustered. The size of each cluster depends on the level of required accuracy. This can be determined by comparing the use of potential cluster options and detailed simulations of the entire module. This is an initial time investment, but does not need to be performed for all scenarios. It is solely performed to investigate the required distance of IV curves for a desired level of accuracy. Individual simulations have to be done for all types of cell-strings within the module. The IV curve of each cluster for each type of cell-string can be reused in the combination process to simulate the entire module and all potential operating conditions can be recreated.

This allows an exploration of different topologies for PV modules with the same arrangement of cell-strings. Fast simulation results can be obtained for different interconnection schemes of the cell-strings (series or parallel), different number of local converters, different types of wires and switches.

As the evaluation of topologies is established by comparing with a conventional module design, a fast simulation process for a conventional module is needed as well. The same framework can be applied for conventional modules, where the simulations are limited to a single bypass section.

V. Results

The main purpose of this framework is to allow fast simulations of different connection schemes of modules which share the same cell-string layout. Depending on the connection between the cell-strings, the resistive elements of the interconnection vary. As reusability is a main factor which enables the speedup of the simulation process, inclusion of resistive elements required for the interconnection of cell-strings should not be included in the simulation environment, but integrated in the combination process of the IV curves. It has to be noted here, that simulated IV curves are not continuous, but discrete. Adding resistive elements to a simulated IV curve scales the "step" of the IV curve. When the range of potential resistive values is known, the required voltage window and step can be adjusted to fit the worst case. Introducing resistive elements in the "already" simulated IV curve of a cell-string leads to a loss in accuracy level less than 0.05%.

The main available design options of electrically combining two cell-strings for a reconfigurable module are in series and in parallel. Also hybrids of these primitive options are enabled then of course. The losses of combining individual cell-strings with different irradiation levels compared to a full model IV curve are as stated in section II, under 0.5% for both series and parallel connection. This also stands for extreme irradiation differences, thus for completely different IV curves. Extending the combination to three cell-strings does not increase the loss in accuracy.

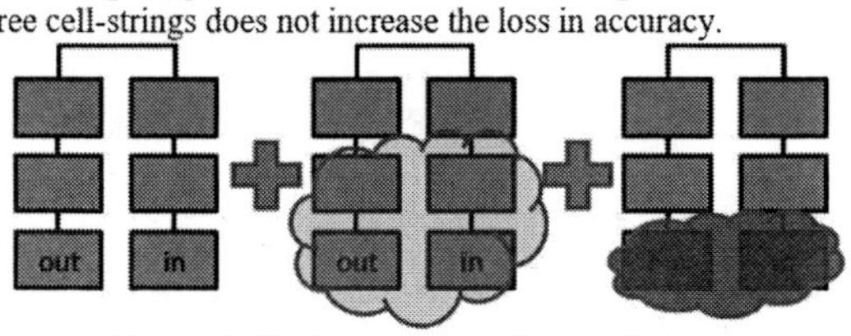

Figure 4: Shading pattern of three cell-strings

For an irradiation pattern as shown in Figure 4, the average loss in the IV curve of the series case is 0.4%, with the error at the MPP being 0.3%. In the parallel case, the average error of the IV curve is 0.1% and at the MPP 0.08%. The IV curves of the full model simulation and the combination of the individual IV curves are shown in Figure 5.

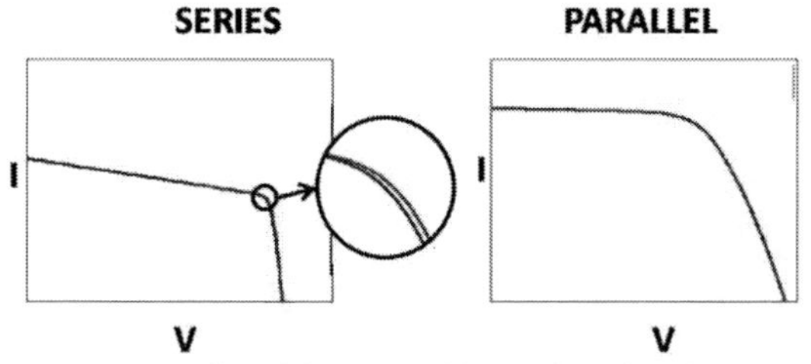

Figure 5: Full model IV curve (blue) and combined IV curve (red) of the shading pattern of Figure 4

Clearly, the IV curves in the parallel case are almost identical and no significant difference can be illustrated. In the series case however, the current is really low due to the presence of a heavily shaded cell-string. The relative error of the IV curve is 0.4%, but the absolute error is still negligible. So this supports the main assumption underlying our approach.

The simulation of individual IV curves provides information about the interconnection of the cell-strings in multiple ways (series, parallel and hybrid) without needing to repeat the simulations. The same applies for different operating conditions due to other shading scenarios. This allows significant speed up of the evaluation process and saves time from simulating the full model for all potential connections and shading patterns.

VI. Conclusions

Many PV systems, especially those installed in industrial areas, function under non-uniform operating conditions. There is a demand for an increased energy yield despite the non-ideal operating conditions. Through detailed simulation, it has been shown that reconfigurable topologies can recover power under non-uniform conditions [2]. However, multiple layouts of reconfigurable modules and multiple operating scenarios render necessary a faster evaluation process of the performance of a PV module.

The framework proposed in this paper allows to speed up the simulation process of both reconfigurable topologies and conventional ones with negligible losses in accuracy. The model has been extensively calibrated and furthermore validated with experimental results.Applying the proposed simulation framework will enable a fast identification of promising module topologies for a variety of expected shading conditions.

References

[1] G. Spagnuolo, G. Petrone, B. Lehman, C.A. Ramos Paja, Ye Zhao, and M.L. Orozco Gutierrez, "Control of Photovoltaic Arrays: Dynamical Reconfiguration for Fighting Mismatched Conditionsand Meeting Load Requests". Industrial Electronics Magazine, IEEE 9, 1 (March 2015), 62–76.

[2] M.Baka, F.Catthoor, D.Soudris, "Smart PV Module Topology with a Snake-Like Configuration", 31st European Photovoltaic Solar Energy Conference (PVSEC), Hamburg 2015.

[3] D.Anagnostos, H.Goverde, B.Herteleer, F.Catthoor, D.Soudris, J.Driesen, J.Poortmans, *"Demonstration and validation of an energy yield prediction model suitable for non-steady state non-uniform conditions"*, 6th World Conference on Photovoltaic Energy Conversion, Kyoto, 2014.

[4] H. Goverde, J. Govaerts, J. Baert, F. Catthoor, J. Driesen, J. Poortmans, *"Optical-Thermal-Electrical Model for a Single Cell PV Module in Non-Steady-State and Non-Uniform Conditions Build in SPICE"*, 28th European Photovoltaic Solar Energy Conference (PVSEC), Paris, 2013.

THE ANTI-GLARING MODULE SIMULATION, PROTO-TYPE DESIGN AND MODULE PERFORMANCE

Yi-Chia Chen, Chao-Wei Yang, Tony Lai, Ming-Yu Chen, Chien-Li Cheng
AU Optronics Corporation, No. 1, JhongKe Rd., Taichung 40763, Taiwan, R.O.C.
E-mail: Wallace.Chen@auo.com , Phone: +886 4 2460-8800 #1767, Fax: +886 4 2460-8077

ABSTRACT:
We report the 273W anti-glaring module with curved-frame structure and 0.85mm thickness glass. The optical simulation shows the reflectivity of curved module received by inspector is reduced 96.79% than flat panel by using Tracepro software. And the outdoor anti-glaring performance of this anti-glaring module has reached larger than 60% improvement than flat panel. However, the outdoor energy yield of anti-glaring module is lower (-0.19%) than flat panel. But the anti-glaring module still has a lot of benefits, like more efficient land usage, drainage and light weight (10.5 kg) which can be moved and installed quickly.

Keywords: curved solar module, anti-glaring, light weight

1 Introduction

The PV module installation requires a lot of open space and unsheltered land. Airports show a obvious opportunity for using solar technologies due to large amounts of open land. But PV modules need to be confirmed do not cause safety problems for aviation or interfere with aeronautical and airport activities. Specifically, the FAA (Federal Aviation Administration) wants to ensure PV modules do not create glint or glare [1] .

As the figure 1 showed, the glare viewed from the air traffic control tower at Manchester/Boston Regional airport is very obvious and serious [2].

In this work the performance of PV module based on two different radius of curvature technologies has been compared.

Figure 1. Glare viewed from the air traffic control tower at Manchester/Boston Regional Airport on April 27, 2012 at 8:15 a.m. (Photo from Sandia National Laboratories website) [2].

2 Model Simulation

The light reflection from PV module model of optical simulation was built upon the commercial software, TracePro [3], The benefit of such model was the capability to conduct simulation of a real size PV module.

Simulation models done based on three different radius of curvature. The system matrices are 1x1 and 20x11. The PV module dimension of system matrices is 1631x976mm (standard module) and 1623*976mm (curved module).

The optical simulation result is shown in table 1 which the information including, radius of curvature, system matrix, PV module dimensions, reflection received by inspector, reflection reducing rate.

As the Figure 2 and table 1 showed, the data shows light reflection reducing rate is 98.765% of 4000R curved module. Further simulating of 20x11 system matrix shows reflection reducing rate is 96.793%. This result shows the system matrix composed of the curved modules can effectively reduce glare caused by PV module reflection.

In order to confirm the actual situation, so we choose the 4000R (4m radius) of curvature to demo curved module.

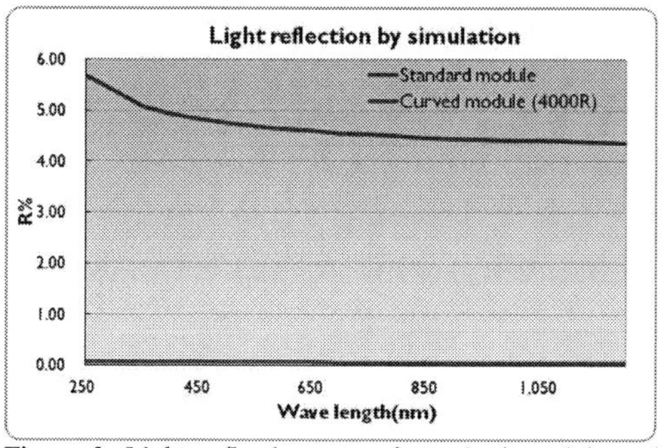

Figure 2. Light reflection rate of standard module and curved module (4000R)

Item	Standard panel	Curved panel (4000R)	
System matrix			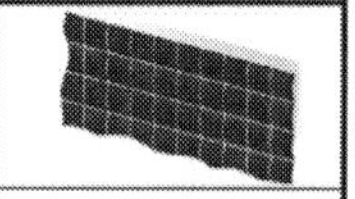
	1x1 (1631*976*0.85)	1x1 (1623*976*0.85)	20x11
Reflection received by inspector	4.680%	0.0578%	0.150%
Reflection reducing %	Baseline	98.765%	96.793%
Schematic simulation	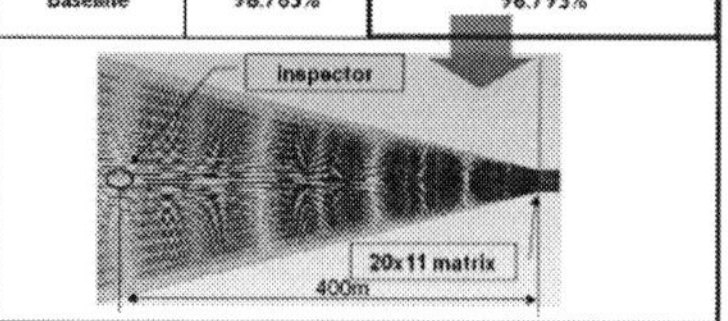		

Table 1. R% analysis and simulation for curved solar panel (matrix) with 4000R received by inspector.

3 Experiment Procedure

As the Figure 3 showed, through use flexible chemically strengthened glass (thickness 0.85mm) with EVA encapsulant, mono crystal solar cell and the PV backsheet to laminate.

In addition, we manufacture curved frame by a special mechanism, and successfully produced solar module frame curvature of 4000R.

To framing the curved frame (as Fig. 3a➔3c) and the laminated materials which include flexible chemically strengthened glass, EVA encapsulant, mono crystal solar cell and the PV backsheet, by using a special curved carrier fixture.

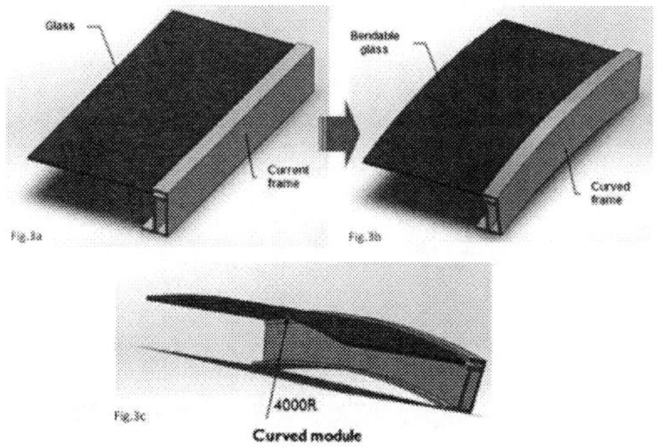

Figure 3. Fig. 3a, Fig. 3b and Fig. 3c show the experimental procedure for the curved module.

4 Result and discussion

Finally curved module could be completed successfully by above framing process. As the Figure 4 showed, the curved module with frame was produced and it's elegant appearance.

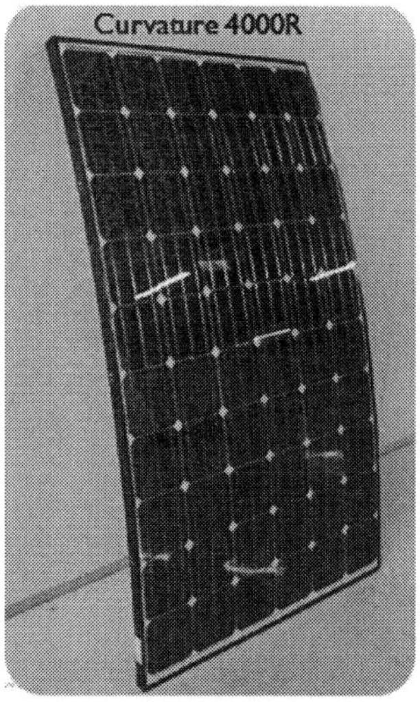

Figure 4. The curved module with frame was produced.

In order to examine the power output and solar module quality difference of standard module and curved module, power and EL inspection table was showed as table 2.

For different type module, standard-type module and curved-type module, there were almost the same power output and no difference of module quality (e.g. no cell crack). It is worth mentioning that the curved module CTM loss is slight higher than standard module about 0.5%, this result is caused by curved surface will reflect some of the direct light from the simulator.

Item	STD PV Panel		Curved(4000R) PV Panel	
ID	Sample 1	Sample 2	Sample 1	Sample 2
EL (before framing)	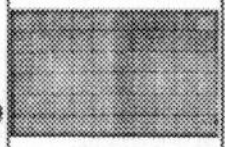	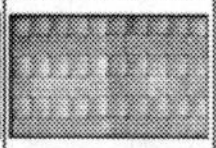	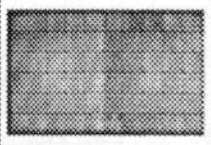	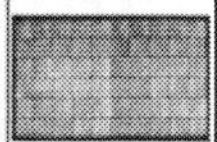
EL (After framing)				
Power (W)	274.53	274.87	273.30	273.37
CTM loss	6.13%	6.02%	6.55%	6.53%

Table 2. The measured output power by using BERGER solar simulation and electroluminescence (EL) of the standard (STD) module and curved module .

Anti-glare performance of curved module is key purpose of this work. So comparison glare situation of STD (standard) module and curved module in outdoor is important. As the Figure 5 showed, the serious glaring effect of STD module was observed when this module was placed on the rooftop with tilt angle 60 degree. But as the Figure 6 showed, the slight glaring effect of curved module was observed when this module was placed on the rooftop with tilt angle 60 degree. The curved solar module has showed obvious larger than 60% anti-glaring performance improvement.

Figure 5. The serious glaring effect of STD (standard) module.

Figure 6. The slight glaring effect of curved module.

The energy monitoring in table 3, outdoor energy yield performance for the STD and curved modules with portrait installation. The energy yield of the curved module is slight lower (-0.19%) than STD's. But this is acceptable for this work.

Type	STD	Curved(4000R)
Date	3/5~3/9, 3/11~3/15 (10 days)	
Tilt angle(°)	17	
Long-side direction	S→N	S→N
Average enery increased rate	Baseline	-0.19 %
Set up		

Table 3. Outdoor energy yield performance for the STD and curved modules with portrait installation.

Finally, in order to confirm the reliability performance of curved module. The curved solar module also carried out several reliability tests. The mechanical loading can pass 5400Pa test (ΔPower= -1.51%) as the Figure 7 showed. Other tests like thermal cycle (ΔPower= +0.59%) and damp heat (ΔPower=+2.31%) test can all pass IEC test specification.

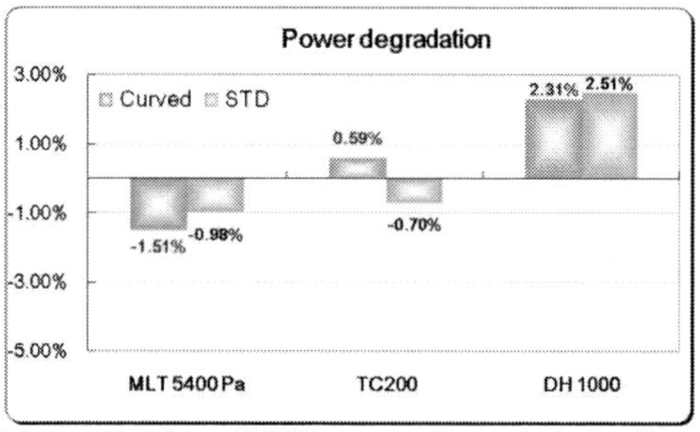

Figure 7. The reliability test (Mechanical Loading, Thermal Cycle and Damp Heat) of STD module and curved module.

4. Conclusion

The comparison glare situation of STD module and curved module in outdoor, the results show significantly reduce the impact of glare effect on curved module. And the elegant appearance of curved PV panel can more be used for airport building. An anti-glare curevature solar module has been prove it's benifit in much less reflection, usage of land, light weight (10.5kg) and self-cleaning capability (by it's curved surface).

5. References

1. Siting Solar Photovoltaics at Airports
Alicen Kandt and Rachel Romero
National Renewable Energy Laboratory
To be presented at Solar 2014
2. http://www.cnsnews.com/news/article/glare-airport-solar-panels-interfering-air-traffic-controllers
3. TracePro : Lambda Research Corporation.
http://www.lambdares.com/tracepro

LAMINATION CYCLE TIME OPTIMIZATION USING NEW POE ENCAPSULANTS

*Ignacio Fidalgo, Román Merino, Benjamín Pérez

Specialized Technology Resources España S.A., Parque Tecnológico de Asturias, Parcela 36
33428 – Llanera, Asturias, Spain

*Corresponding author: Phone: +34 985 73 23 33, Fax: +34 985 73 23 32, Email: ignacio.fidalgo@stresp.com

ABSTRACT: STR has recently developed new non-EVA encapsulants for PV modules based on polyolefin elastomers (POE's). These materials have many advantages compared to traditional EVA based encapsulants like improved thermal and chemical stability, minimized corrosive potential, lower moisture vapor transmission rate (MVTR), improved electrical properties and PID resitance [1]. However, some similarities remain, like the need for a curing approach where the encapsulant is transformed from a thermoplastic into an elastomer during lamination. In this regard, lamination cycle optimization is required in order to obtain a good compromise between sufficient curing and optimal performance and reliability of the correspondent PV module. Special focus will be place on measuring the main properties that could be affected by poor cure levels like i) creep resistance, ii) adhesion to glass, iii) adhesion retention after DH85/85 and iv) light transmission. Lamination cycle times similar to standard EVA encapsulants are desirable.
Keywords: Encapsulation, PV Materials, Polymer Film, Manufacturing and Processing, Performance.

1 INTRODUCTION

The new non-EVA POE encapsulants recently developed by STR share some similarities with traditional EVA legacy products like the necessity of a curing process to obtain optimal results, and the ability to be processed using standard lamination equipment.

However, the different properties of the base resin in both types of materials have an impact on the curing kinetics and in the crosslink density that is achieved after lamination in the final product.

For instance, a standard lamination cycle for an Ultra Fast Cure EVA encapsulant, with four minutes of evacuation time and eight minutes cure time at 150°C typically result in gel content values in the range between 75 – 80%, whereas the same cycle results in gel contents between 65 and 70% for the new POE encapsulants developed by STR.

The purpose of this study is to determine the optimal cure level for the new POE based encapsulants needed to obtain excellent performance and reliability in the final module.

Considering the higher melting point of the base resin in the POE encapsulant, and its mechanical properties, the initial hypothesis establishes that same cure levels or conversions might not be required in this new type of materials in order to obtain similar or improved performance compared to traditional EVA encapsulants.

The main goal of this study is to determine the optimal gel content range for the lamination of POE encapsulant, and as a consequence the most suitable lamination cycle needed for reaching those values.

2 DESIGN OF EXPERIMENTS

For the purpose of this study we have focused on a POE encapsulant manufactured and commercialized by Specialized Technology Resources, Inc. (STR) with the reference Photocap® 35521P HLT™. The main properties of this material are summarized in Table I.

Two main parameters have been considered during this study for optimizing the lamination cycle of the new POE encapsulants developed by STR: cure time and lamination temperature. The temperatures used in this matrix test ranged between 145°C and 155°C, whereas the cure times ranged between 6 to 12 minutes. Evacuation time was fixed to 4 minutes without pins for all the cycles described in this article.

Table I: Material properties of the POE encapsulant used in this article: Photocap® 35521P HLT™.

Properties	Units	POE
Thickness	mm	0.46
Free shrinkage	%	<10
Melting range (DSC)	°C	80-90
%T at 500 nm	%	91
$\%T_{SE}$	%	92
UV cut-off	nm	305
Adhesion to glass	N/cm	120-140
Adhesion to glass after 1000hrs DH	N/cm	80-90
10% Secant Modulus	MPa	26
Tensile Strength	MPa	15
Tensile Elongation	%	700
Volume Resistivity	Ω cm	$>1x10^{16}$
MVTR 25°C, 100%Rh	g/m^2d	5.4

Using these test conditions, a complete cure level map was generated by measuring the gel content for each cycle. The results of this study are summarized in section 3.1.

In the next sections, correlations between cure levels and key properties of the final product were established.

Special focus was placed on those properties which are good indicators of performance and reliability in the PV module, and are directly impacted by the cure level of the encapsulant.

Section 3.2 summarizes an extensive study of creep resistance using cured samples with gel contents ranging between 55 and 70%. Creep resistance of the encapsulant is crucial in order to maintain the structural integrity of the module when exposed at high temperatures, and is one of the main advantages of curable encapsulants compared to pure thermoplastics.

Adhesion to glass is another key parameter required in order to guaranty the performance and reliability of the module. Poor adhesion of the encapsulant to glass would result in delaminations and consequently in malfunctioning of the module.

In order to obtain good adhesion values, not only right after lamination, but also after accelerated climatic aging, sufficient curing is required. In this regard adhesion measurements after 1000 hours in damp heat test are a good indicator of performance in the final module.

Sections 3.3 and 3.4 summarize the results of an adhesion study performed with samples laminated using different lamination cycles, and with different gel content values.

Finally, another property that depends on the cure level of the encapsulant is the light transmission. Ethylene based polyolefins like EVA or the POE used by STR in their new products have some inherent degree of crystallinity. These microcrystalline domains have an impact on light transmission that can be minimized by converting the material into an elastomer, which is more amorphous than the original thermoplastic. As a consequence, the curing process maximizes the light transmission of the encapsulant. Section 3.5 summarizes a study of light transmission versus cure level using samples with different gel content values.

3 EXPERIMENTAL RESULTS

As previously discussed, the focus of the experiments described in this article was placed on determining the impact of different cure levels in critical properties that could directly affect the performance of the correspondent photovoltaic module. First of all, an extensive study was performed to characterize the cure levels obtained at two different positions in the PV module using different lamination cycles. This study resulted in a complete cure level map that will help to determine the optimal lamination cycle for this product. A similar approach was used for determining the values of adhesion to glass at different temperatures and cure times.

For the study of properties like creep resistance, adhesion to glass after climatic aging and light transmission, samples with different gel content values were used to determine the optimal cure level range for each property.

3.1 Cure level

For this study, samples of the new POE encapsulant Photocap® 35521P HLT™ have been cured using different temperatures and cure times.

The temperatures used in this matrix test ranged between 145°C and 155°C at 5°C intervals, whereas the cure times ranged between 6 to 12 minutes at 1 minute intervals. Evacuation time was fixed to 4 minutes without pins for all the cycles described in this article.

Samples were laminated in a 3S laminator model S1815E using solar glass with pyramidal texture and the following dimensions:

- Length: 1665 mm
- Width: 991mm
- Thickness: 3.2mm

For every cycle, samples from the center and the corner of the glass were extracted, and Teflon foils have been used to easily remove the cured POE samples after lamination.

Cure levels have been determined by measuring the insoluble fraction (gel content) after soaking the samples in toluene at 60°C for at least 20 hours, and drying the residue at 110°C for at least 24 hours.

Figure 1 shows a complete map of the results obtained in this study, both at central and corner positions.

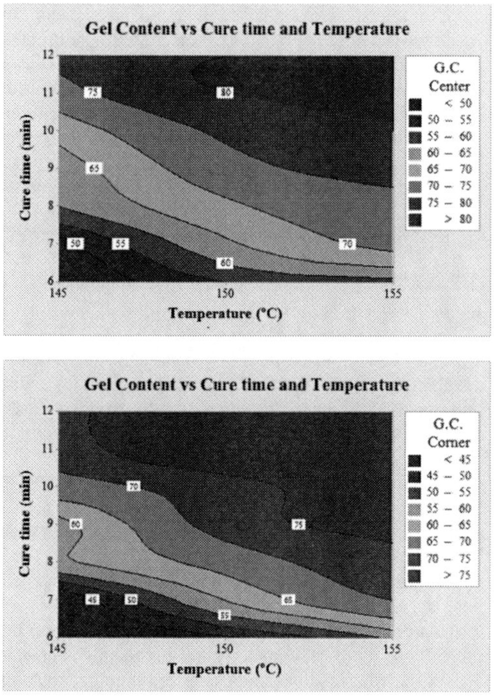

Figure 1: Cure level maps correspondent to Photocap® 35521P HLT™ at different temperatures, cure times and positions in the module.

3.2 Creep resistance

The creep resistance of an elastomer is an indicator of the tendency to flow when exposed at high temperatures. Properly cured encapsulant will not flow under those conditions, maintaining the structural integrity of the module during its life time.

For this study we have measured the creep resistance of samples with cure levels ranging between 55 and 70%. The test is performed by measuring the dimensional change (in millimeters) in a cured stripe of cured encapsulant suspended vertically in an oven at 95°C with a hanging weight equivalent to 4 times the mass of the stripe. The dimensions of the stripe are 100 mm long by 10 mm wide, and two marks are made in the stripe with a distance of 50mm as a reference to measure the elongation after the test.

Figure 2 shows the results of this study. The horizontal red line in the graph shows the creep resistance of a reference EVA encapsulant laminated according to the recommendations of the manufacturer to obtain cure levels in the range between 75 and 80%. This red line constitutes the benchmark for our POE encapsulant.

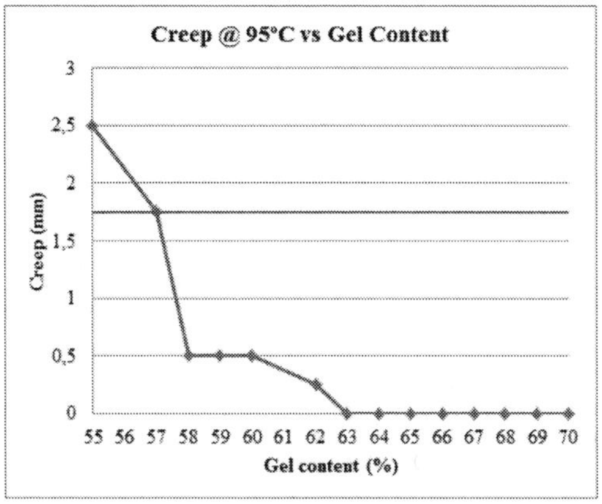

Figure 2: Dimensional change in millimeters for POE samples cured with different gel contents.

According to these results, excellent creep resistance is obtained with our POE encapsulant for gel content values higher than 60%. This is the minimum cure level we would recommend in order to laminate this POE formulation.

3.3 Adhesion to glass

Same as in section 3.1, samples of the new POE encapsulant Photocap® 35521P HLT™ have been laminated using different temperatures and cure times.

In this case, coupons made of 3mm float glass, encapsulant and PPE type backsheet have been prepared in a 3S laminator model S1815E.

A small piece of Teflon was placed between the encapsulant and the glass in order to measure the adhesion strength in this interphase.

The glass coupons had a length of 10cm and a width of 5cm, whereas the piece of backsheet had double length compared to the glass in order to be able to perform a peel test at 180°.

The temperatures used in this matrix test ranged between 145°C and 155°C at 5°C intervals, whereas the cure times ranged between 6 to 10 minutes. Evacuation time was fixed to 4 minutes without pins for all the cycles described in this study.

Figure 3 shows a map including the results of this study.

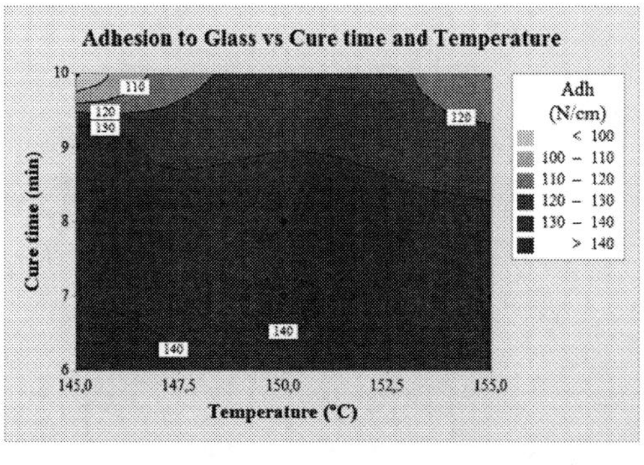

Figure 3: Adhesion map for glass/POE laminates cured using different temperatures and cure times.

According to the results obtained in this study, the values of initial adhesion to glass are excellent in a broad range of lamination cycles.

Even samples with very low cure levels resulted in very good adhesion values.

Surprisingly, the longest lamination cycles resulted in lower adhesion values (though more than acceptable) compared to the shorter cycles. This could be attributed to some short of embrittlement in the PET layers of the backsheet caused by very long lamination cycles and or high temperatures.

3.4 Adhesion to glass after damp heat aging

The adhesion retention after damp heat aging was measured in coupons laminated with cure levels ranging between 65 and 75%. These were laminated using a lamination cycle consisting in 4 minutes evacuation without pins, 8 minutes cure at 150°C in a 3S laminator model S1815E.

The adhesion coupons were prepared in the same manner as described in section 3.3. The values of adhesion retention obtained with these samples after 1000 hours in damp heat at 85°C and 85%RH were around 65%, as can be seen in Table II.

Samples with cure levels around 50% gel content also demonstrated to retain good adhesion values after 1000 hours in damp heat. In this case, the adhesion was only reduced by a half after the accelerated tests, which is a remarkable value, but mechanical properties like creep resistance is compromised at this cure level as already discussed in section 3.2, so cure levels lower than 60% cannot be recommended even though excellent adhesion values after lamination and climatic tests have been reported.

Table II: Cure level vs adhesion values before and after accelerated climatic aging and creep resistance.

Gel Cont. (%)	Initial adhesion (N·cm)	Adhesion after DH85/85 (N·cm)	Creep resistance
50 – 55	120 – 140	60 – 70	Defective
65 – 75	120 – 140	80 – 90	Excellent

3.5 Light transmission

Coupons for light transmission measurements have been prepared laminating samples of POE with solar glass using different lamination cycles. Gel content values of these samples varied between 48 and 74%.

Table III shows the measurements correspondent to two different parameters: light transmission at 500nm, and the weighted light transmission between 380nm and 1100nm.

Table III: Light transmission vs Gel content

Gel Cont. (%)	%T (@500nm)	%T (380 – 1100 nm)
48	89.5	90.4
56	90.8	91.8
59	90.2	91.2
67	90.9	91.5
69	90.6	91.6
71	90.5	91.3
73	90.6	91.5
74	90.6	91.6

These measurements have been performed with a spectrophotometer model Perkin Elmer Lambda35.

According to these results, light transmission values remain constant in the gel content range between 55 and 75%, and they can be compared with the results correspondent to traditional EVA encapsulants.

4 SUMMARY

- The curing kinetics of the new POE encapsulants developed by STR differs from those correspondent to traditional EVA based products, but equivalence in performance can be achieved for gel content values around 70%.
- Excellent creep resistance is obtained for cure levels higher than 60%.
- Measurements of key properties like adhesion to glass, adhesion retention after accelerated climatic aging and light transmission resulted in excellent performance in a broader gel content range (from 50 to 75% gel content).
- As a consequence similar processes and lamination cycle times can be used for both types of materials (POE and EVA), though the gel content targets are different: 65 to 70% for POE, and 75 to 80% for EVA.
- The standard cycle recommendation for STR POE encapsulants in order to reach the gel content target would be as follows:

Evacuation time:	**4 – 5 minutes**
Cure time:	**7 – 8 minutes**
Temperature:	**150 – 155°C**

5 REFERENCES

[1] A. Schneider, et al, paper 1BV.6.36, 31st European PV Solar Energy Conference, September 2015

A BYPASS DIODE FOR INTEGRATED SMART SOLAR CELL MODULE

H. W. Du[a], M. Gao[a], F. Xu[a], L. Zhao[a], L. F. Lu[b] and Z. Q. Ma[a]

[a)] SolarE Joint Lab., Dept. of Physics, Shanghai University 200444, PRC

[b)] SERD Center, Shanghai Advanced Research Institute, Chinese Academy of Science, Shanghai 201210, PRC

[a)] Tel: +86-21-66132402, Fax: +86-21-66134208, e-mail: zqma@shu.edu.cn

ABSTRACT: A mini-module on one silicon wafer has been fabricated through assembling a small size of SIS (semiconductor-insulator-semiconductor) device taking on a bypass diode feature in a reverse saturation current, with HIT (heterojunction with intrinsic thin film) solar cell. The architecture is that the solar cell is on the front side while the inbuilt bypass diode is on the rear side. The data showed that the output character of the integrated devices possessed obviously hermaphroditical behavior at forward and reverse biases, which was available to protect the mini-system from deterioration upon shadowing. The integrated system as an independent output power device can be further optimized against the temperature influence.

Keywords: Inbuilt bypass diode, integration system, mini-module, HIT cell, SIS device

1 INTRODUCTION

Crystalline silicon heterojunction (c-SHJ) solar cells with an amorphous Si or oxidized Si passivation layer on c-Si wafers have been attracting much more attention due to their high efficiency, low negative temperature coefficient and available production cost, especially, such as HIT (heterojunction with intrinsic thin layer) and PERC (passivated emitter and rear contact) structures, etc. These advanced technologies have brought about the conversion efficiency of 24.7 % by sole HIT and 25.6% by combination of HIT and IBC (interdigitated back contact) techniques, respectively [1, 2]. Even though the conversion efficiency more than 25% for those large area cells (> 150 cm^2) had been achieved and summarized in reference [3]. The manufacturers wish to reach a module more than 300 W through utilizing these high efficiency solar cells in the near future, and to extend the life time of the module to more than 30 years with a reliable bypass diode incorporated into the module. However, so far there is no any new idea to change the traditional method to put a bypass diode inside a solar cell like-module for the convenient operation and the intelligentized long-distance control for the next generation.

A prospect in the future, n-type crystal silicon will become significant materials for the higher efficiency a solar cell, in that the decay variation such as LID (light-induced degradation) or PID (potential-induced degradation) for p-type silicon materials could be reduced. Concerning about n-type silicon based diode with an architecture of AZO/SiOx/n-Si recently developed in our laboratory [4], we have assembled it into a silicon wafer together with HIT solar cell, because the device shows a reverse saturation current at about -2.5 V. In this way, the integrated system on a wafer could be used for abstaining the shadowing and hot spot effect of the module outside in the field. Fig.1 shows the integrated frame and the I-V properties of the HIT solar cell and the bypass diode, respectively.

2 INTEGRATED A BYPASS DIODE INTO A WAFER WITH HIT CELL

The fabrication of the integrated system was based on a normal technique for the HIT solar cell, in which a complete structure was formed like Al/Ag/ITO/(p^+)a-Si:H/(i)a-Si:H/n-Si/(i)a-Si:H/(n^+)a-Si:H/ITO/Ag/Al, while the bypass diode with the structure of Al/n-Si/SiOx/AZO/Ni/Al was subsequently coupled upon the rear side of HIT device as shown in Fig.1(a). The detail state of the art processing is as follows:

1) The core of HIT architecture with passivated and doped hydrogenated amorphous silicon films $((p^+)$a-Si:H/(i)a-Si:H/n-Si/(i)a-Si:H/(n^+)a-Si:H) on both sides of n-Si wafer has been assembled by PECVD (plasma enhancement chemical vapor deposition) technique and the thicknesses of the individual layers are exactly controlled in the nanometer scale.

2) Partial area of the a-Si:H layers on both sides was etched by RIE (reactive ion etching) method for SIS device coupled into the wafer. Where the reactive gas source used for etching was SF_6:O_2 = 36:6 (sccm) ; the radio frequency (13.56 MHz) power was 100 W; the pressure in the reactive chamber was controlled to be 20 Pa and the etching time was about 15 min.

3) Subsequently, the ITO (indium tin oxide) films with about 80 nm were simultaneously deposited on the surface of both p^+ a-Si:H and n^+ a-Si:H materials through dc magnetron sputtering deposition technique, and the contact resistance was reduced to several tens of mΩ cm^2.

4) For the bypass diode formation on the same wafer, an ultra-thin silicon dioxide layer was made by wet oxidation of the bare area into HNO_3 liquid at an appropriate temperature, and the other part of the wafer with HIT structure was closely covered by nitric acid anti-foils.

5) Immediately, aluminum doped zinc oxide (AZO) layer was deposited on rear side of the wafer with SiO_2 by dc magnetron sputtering. The deposition parameters were: 250 °C for substrate temperature; 100 W for sputtering power; 30 sccm for Ar gas flux; 1.5 Pa for the chamber pressure.

6) Then, the silicon dioxide layer on the front side of the wafer was clearly removed by 5% HF liquid.

7) At last, the electrodes of Ag/Al, Ni/Al and Al metals were thermally evaporated on ITO, AZO and the front bare silicon part, respectively. So that the SIS device was integrated with HIT cell on the same n-type silicon wafer.

The total output current for the integrated system may be expressed as the sum of the two devices as follows:

$$I = I_{sc} - I_{bp} - I_L = I_{0,sc}[\exp(qV/nkT)-1] - I_{0,bp}[\exp(-qV/nkT)-1] - I_L$$

Where the first item is the I-V relationship of HIT cell, and the second item is the I-V relationship of SIS diode and the third item is the photon generated current. Under forward bias, the second item approaches zero, while the first item tends to be zero when a negative bias is applied. At the reverse bias, the total current is the sum of the diode and photon generated currents.

I-V behavior of the integrated system has been shown in Fig.1 (d). The photovoltaic effect is obvious for the system at the first survey, which means that the integration of the two functional devices is successful for the smart mini-module on one wafer. However, when we tried to do more than two times the bypass diode was failed to bear a protective role for its too thin intermediate oxide layer. Actually, we can increase the thickness of the oxide layer to more than 2 nm during the wet oxidation process within HNO_3 liquid.

3 INDEPENDENT CHARACTERISTICS OF HIT CELL

In order to well understand the individual character of HIT cell independently, a simple output property of HIT cell with double-face structure is shown in Fig.2. The variation of current density (J) correlated to applied voltage (V) and the optoelectronic conversion parameters under 1.5 AM illumination were illustrated with the output power. Owing to the purpose of the integration with two kinds of devices for the bypass diode functional system, we did not take into account the surface texture prior to depositions of different materials. Thus, the short-circuit current density became as lower as 36.77 mA/cm^2, meanwhile, the lack of the optimization of multiplex amorphous silicon layers stacked onto n-Si substrate, as an efficient emitter, caused a lower open-circuit voltage (0.67 V) and fill factor (67%), as well as the conversion efficiency (16.4%) of the photovoltaic device presently. Whereas, the moderated values obtained from a simple processing line is enough for our integrated system.

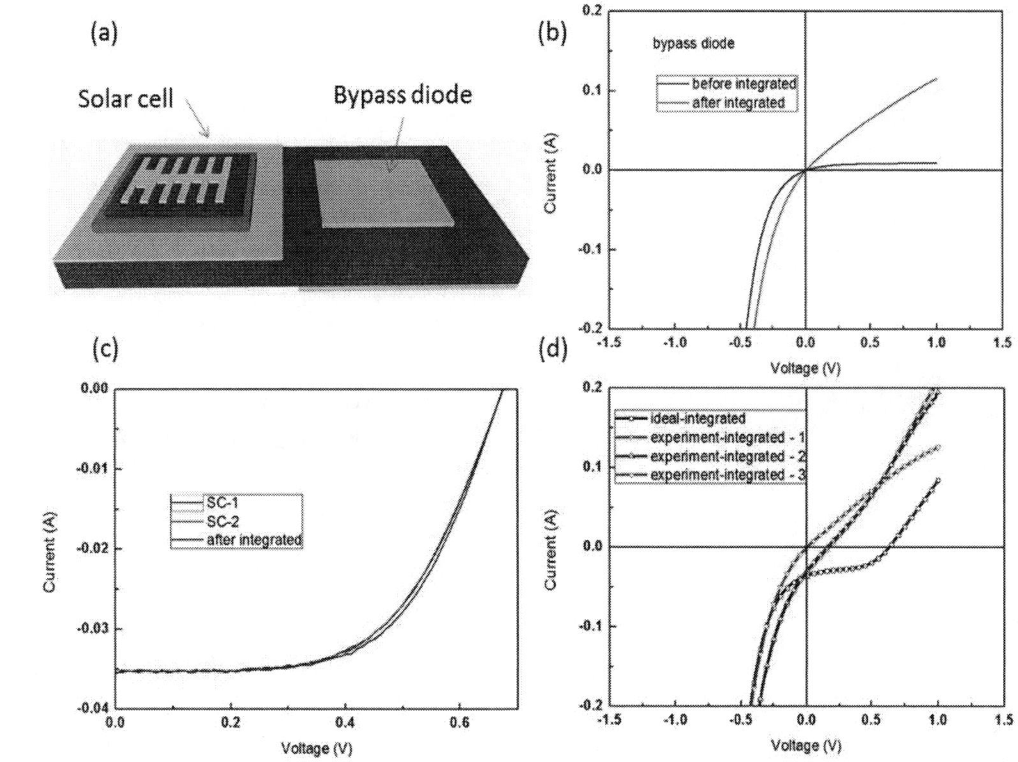

Fig.1 (a) drawing of the integrated system; (b) I-V curve of AZO/SiO_2/n-Si device in the integrated system; (c) I-V curve of HIT device in the integrated system; (d) I-V curve of the integrated system.

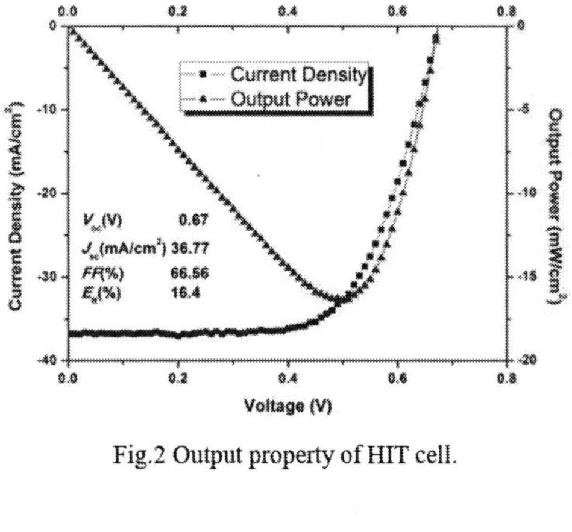

Fig.2 Output property of HIT cell.

4 THE CHARACTER OF AZO/SIOX/N-SI DEVICE

An independent investigation for the AZO/SiOx/n-Si device has been shown in Fig.3. The silicon oxide on n-Si wafer surface was formed in dry O_2 ambient. Fig. 3 showed the current density-voltage characteristics of metal/AZO/SiO_x/n-Si/metal devices with the different thicknesses of SiO_x, under dark and illumination, respectively. The different thicknesses of SiO_x layers were determined by the oxidation time (OT).

At dark condition, the I-V characteristic of the device with 1.0 min oxidation time (OT) for SiO_x layer was presented as squared hollow dots line in Fig.3, which exhibits good rectifying character with rectification ratio of 1230 at ±2.0 V. While the device was placed in illumination of halogen lamp with a power density of 30 mW/cm^2, the I-V characteristic was transformed to be an uncommon curve rather than an expected photovoltaic character, shown as the red circle dots line in Fig.3. At forward bias, the I-V characteristic still obeyed the rule of Shockley ideal diode equation with an ideal factor of 3.88, i.e., the current had an exponential increase, and the current density was in the range of several tens of $mA \cdot cm^{-2}$. However, at reverse bias, the current did not keep saturation until the bias approaching a higher value of -2.5 V, and there was a threshold voltage at -1.0 V. The same phenomenon appeared in the other two devices with the same solid structure while the different silicon oxidation times adding to 2.0 and 3.0 min were performed, respectively. The results were showed as blue and green triangle dotted line in Fig.3.

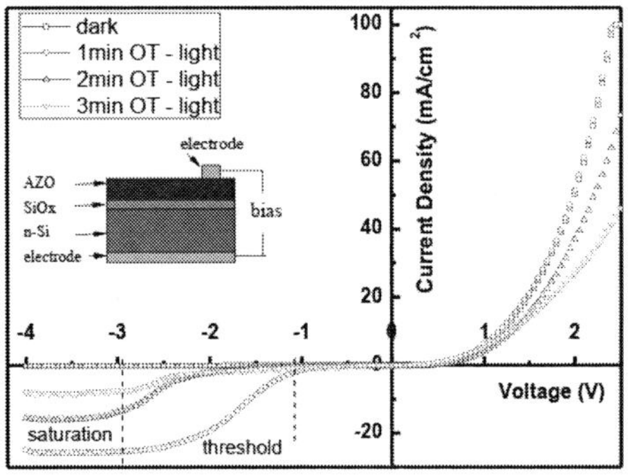

Fig. 3 J-V curves of AZO/ SiOx/n-Si device at dark and illumination.

5 *CONCLUSION*

To sum up, by using magnetron sputtering method, we have successfully assembled the AZO/SiOx/n-Si structure with bypass diode feature into high efficiency solar cells of HIT structure on one wafer. I-V test results showed that the output character of the integrated devices was obviously hermaphroditical behavior at forward and reverse biases, and the photoelectric conversion efficiency is relatively kept constant, but there was still a larger space for modification. Although the result is not ideal, however, the weak photovoltaic effect shows that the integrated devices can achieve the intelligent control of the output power for the mini-module with the high efficiency crystalline silicon cells. The design and investigated results might be applied to other similar back field and back contact beyond HIT structure solar cell, such as TOPCon (tunnel oxide passivated contact) architecture [5], not only on the MIS (metal-insulator-semiconductor) inner embedded structure and the bypass diode function in a normal p-Si solar cells with screen-printing and high temperature firing process [6].

Acknowledgements: This work was partly supported by the Natural Science Foundation (Nos. 61274067 and 60876045) of China, Shanghai Leading Basic Research Project (No. 09JC1405900), and R&D Foundation of SHU-SOENs PV Joint Lab (No. SS-E0700601).

REFERENCES:

[1] M. Taguchi, A. Yano, S. Tohoda, K. Matsuyama,Y. Nakamura, T. Nishiwaki, K. Fujita and E. Maruyama, 24.7% record efficiency HIT solar cell on thin silicon wafer, IEEE J. Photovoltaics 4, (2014) 96-99.

[2] K. Masuko, M. Shigematsu, T. Hashiguchi, T. Fujishima, M. D. Kai, N. Yoshimura, T. Yamaguchi, Y. Ichihashi, T. Yamanishi, T. Takahama, M. Taguchi, E. Maruyama, S. Okamoto, Achievement of more than 25% conversion efficiency with crystalline silicon heterojunction solar cell, IEEE J. Photovoltaics 2014, 4 (2014) 1433-1435.

[3] M. A. Green, K. Emery, Y. Hishikawa, W. Warta and E.D. Dunlop, Solar cell efficiency tables (version 47), Prog. Photovolt: Res. Appl., 24 (2016) 3-11.

[4] H. W. Du, F. Xu, L. Zhao and Z.Q.Ma, Tunneling effect of photon-assisted AZO/SiOx/n-Si heterojunction device at reverse bias, Adv. Mat. Lett., 7(5), (2016) 100-150.

[5] F. Feldmann, M. Bivour, C. Reichel, M. Hermle and S. W. Glunz, Passivated rear contacts for high-efficiency n-type Si solar cells providing high interface passivation quality and excellent transport characteristics, Sol. Ener. Mat. Sol. Cells 120, (2014) 270-274.

[6] B. Thaidigsmann, E. Lohmuller, F. Ferian, F. Clement and A. Wolf, Characterization and modeling of screen-printed metal insulator semiconductor tunnel junctions for integrated bypass functionality in crystalline silicon solar cells, J. Appl. Phys. 113, (2013) 214502.

PHASE CHANGE MATERIALS FOR HYBRID TECHNOLOGY: A REVIEW

D. González-Peña [(1,*)], M. Díez-Mediavilla [(1)], M. C. Rodríguez-Amigo [(1)], C. Alonso-Tristán [(1)]
[(1)] Research Group SWIFT (Solar and Wind Feasibility Technologies). Universidad de Burgos.
Escuela Politécnica Superior. Avda. Cantabria s/n. 09006 Burgos. Spain.
*Corresponding author: davidgp@ubu.es Phone/fax number: +34 947 258925/+34 947 258910.

ABSTRACT: The variable nature of solar energy required for its efficient use the complement of energy storage to adapt adequately the generation and production curves. As PV/T technology produces both electricity and heat energy, two storage devices should be necessary. Batteries are the most common method for storing electricity. Nevertheless, different technologies are available to store heat energy. Thermal energy storage can be classified into sensible or latent heat systems, depending if the substance that storages the energy changes its temperature during the process or its physical phase. Phase Change Materials (PCM's) involve high amount of energy with small changes in volume and temperature so they are promising materials to keep temperature almost constant and store heat energy.
In this work a complete study of different types of PCM for its application in PV/T is presented. For this purpose, different materials are classified in terms of its chemical structure and also by its use in PV/T technology. Great number of substances susceptible to be used as PCMs in a heat storage system is showed in the work by an exhaustive bibliography review. Only those PCM's with melting temperature lower than 70 ºC have been taken in mind for this work. Upper temperatures do not provide enough PV efficient increase

Keywords: PV/T, solar hybrid technology, thermal storage, PCM.

1 INTRODUCTION

According to the International Energy Agency the global primary energy consumption will be growing at a annually rate of 1.2% until year 2035 [1].

The European Union set at 20% of consumption the energy coming from renewable sources such as binding target by year 2020 and establishes in the Directive 2009/28/EC of April , 23th 2009 [2] steps to be established by countries member to encourage the use of energy from renewable sources.

Within renewable energy, solar and wind power are the most abundant in our ecosystem. In addition, the investments planned in these technologies far exceed those of other renewable energies (Fig. 1).

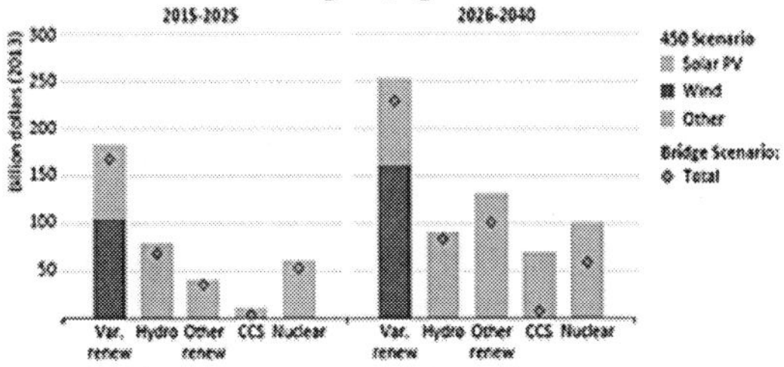

Figure 1: Renewable Energy planned investment (10^9 $) for the nest 10 years (From: IEA.) [3]

Nonetheless, these technologies have a variable character and therefore require, for an efficient use, the storing technologies that allow coupling both energy generation and consumption curves. Traditionally, storage systems for electric energy have been implemented by batteries in photovoltaic or wind installations, as shows figure 2. In the case of solar thermal energy, the most widely storage systems used are tanks where energy is stored by sensible heat, although lately, the use of Phase Change Materials (PCMs) is increasingly because of the advantages it has comparing to the tanks.

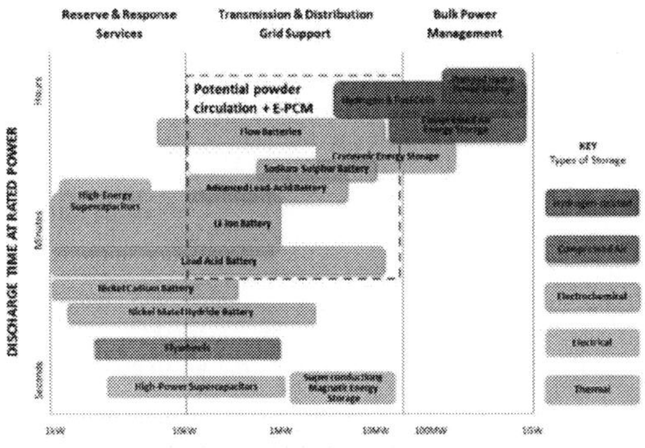

Figure 2: Energy storage systems according to form of energy stored and also to power rating and discharge period. [4]

2 SOLAR HYBRID TECHNOLOGY

The development of new technologies for the use of solar energy requires having multiple storage systems because this advances generate mora than one type of energy from the same device.

Specifically, the hybrid solar technology (PV/T) allows to transform solar energy into both electricity and thermal energy.

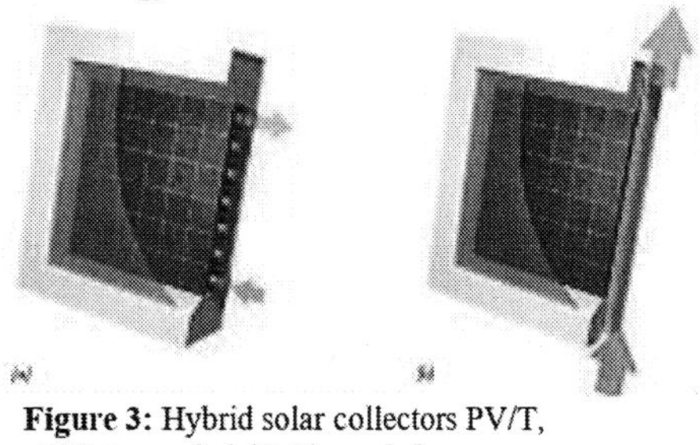

Figure 3: Hybrid solar collectors PV/T, a) Water cooled, b) Air cooled

This technology came up from the need to cool the

photovoltaic solar panels because these reduce their efficiency when operating temperature increases [5, 6]. Different PV/T designs have been developed which mainly differ in the type of fluid used to keep cool the PV cell and also their heat collector geometry [7, 8]. Figure 3 is showed both air and water PVT schemes.

Recently new solar concentrating PV/T systems [9] have been developed where thermal generation has increased because the fluid temperatures is considerably higher, although its electricity production efficiency is lower.

Due to the high variability of the solar resource, it is essential to have storage systems, which, in the case of the PV/T modules, it will be needed to have two storage systems, on the one hand electric power and on the other hand thermal energy. Nevertheless, the objective of this work is focused in the study of the thermal energy storage systems.

Unlike pure solar thermal energy, PV/T modules represent a low temperature thermal resource since it is intended to maintain the PV cell temperature as lower as possible in order to increase the PV generation. In practice, this temperature should be lower than 70C [10-12].

3 THERMAL STORAGE SYSTEMS

Currently there are several solutions for storing thermal energy. Most of these systems operate by confinement in tanks different substance, solid or liquid, which, by variation of its internal energy, allow heat storage.

These systems can be classified into two groups, according to occur or not phase change in the material confined. Latent heat systems primarily store heat by changing substances phase while sensible heat increase its temperature as storage.

The process of changing solid-liquid phase involves more energy with small changes in volume and temperature, which makes these systems predominant in the market.

The advantages presented by PCMs are very beneficial for most PV/T systems because, by making a correct choice of the PCM materials, the PV module can work in a constant temperature and this consequently increase the global efficiency.

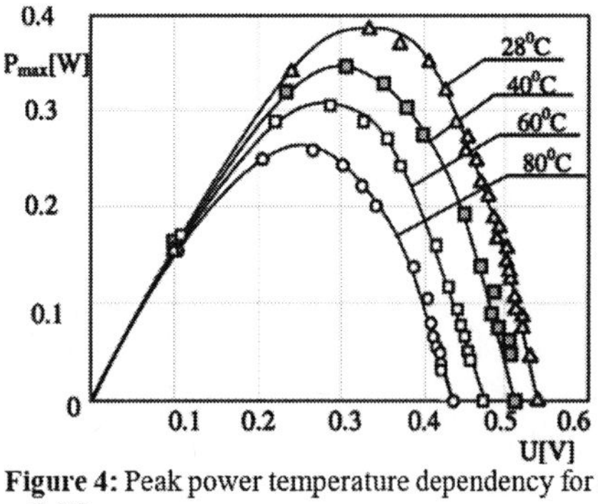

Figure 4: Peak power temperature dependency for a SI PV cell [6]

As shown in Figure 4, the PV efficiency is very dependent on the temperature at which the cell is and, for that reason, it is necessary to maintain that temperature as lower as possible and also without variations. This can be possible by the implementation of PCMs systems.

A careful choice of PCM material is essential for attaining the optimum use of energy and to keep the highest overall possible efficiency [6].

It is for this reason that an analysis of the different PCMs for their use in the solar hybrid heat storage systems is important, where the two main aspects are temperature of the phase change and also that temperature must be constant.

4 PCMS MATERIALS

Any substance of nature that is likely to undergo a phase change when its internal energy is increased can be used as PCM. Nevertheless, its physicochemical properties make some PCMs more convenient than others for each specific use.

Properties such as specific heat, phase change temperature, thermal expansion coefficient or chemical aggression are fundamental aspects when selecting the material for each application.

4.1 Classification of PCM's.

One of the most significant parameters of the PCMs is the phase change temperature. In Figure 5 a graph where different substances are classified according to their phase change temperature is shown. According to the compound used this temperature can reach values from -100 to almost 900 °C.

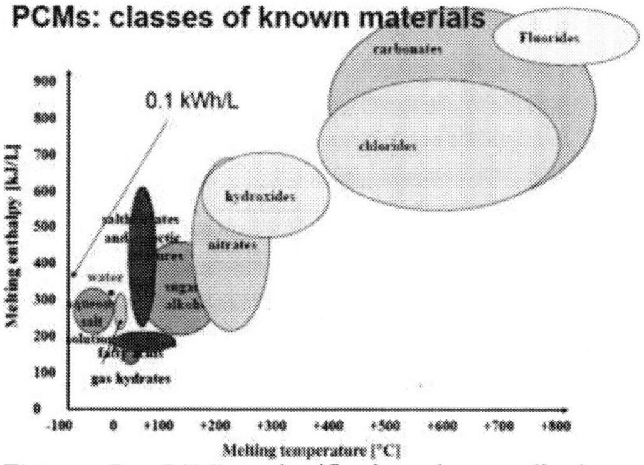

Figure 5: PCM`s classification by application temperature. [13]

It is also possible to conclude that the higher melting temperature is, the more energy per unit mass is capable of store.

Another common way of classifying PCM's is based on their chemical composition [4, 13] as shown in Figure 5. In this case there are three major groups: organic, inorganic or eutectics.

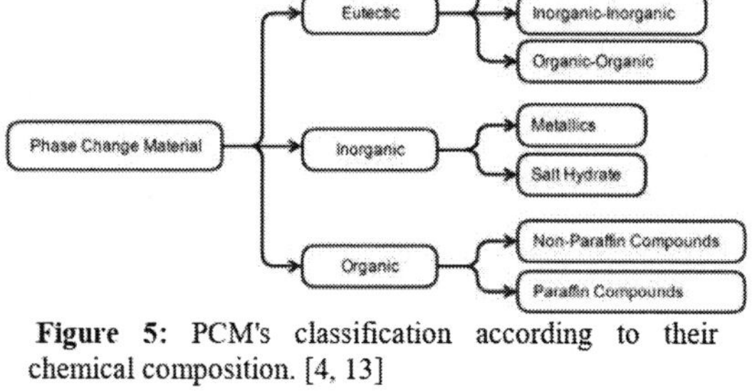

Figure 5: PCM's classification according to their chemical composition. [4, 13]

The first group corresponds to organic compounds, which consists of paraffins and fatty acids. These compounds, due to their organic origin, have not got very high phase change temperature, being generally lower than 150°C. In Table I, the most suitable compounds for PVT are ordered, all fall within the temperature range between 20 and 70 °C.

Table I: Organic PCMs substances for PVT. [14, 15]

Compound	*Melting temperature (°C)*	*Heat of fusion (kJ/kg)*
Paraffin C_{16}–C_{18}	20–22	152
Paraffin C_{13}–C_{24}	22–24	189
RT 25 HC	22–26	230
n-Octadecane	27.7	243.5
RT 28 HC	27–29	245
Paraffin C_{18}	28	244
n-Octadecane	28	245
RT 28 HC	28	245
Paraffin C_{19}	32	222
Paraffin wax	32	251
RT 35 HC	35	240
Paraffin C_{20}	36.7	246
Heneicosane	40	155.5–213
Heptadecanone	41	201
Paraffin C_{16}–C_{28}	42–44	189
RT 44 HC	44	255
Docosane	44	196.5–252
Paraffin (70 wt%) + Polypropylene (30wt%)	44.77	136.16
P116-Wax	46.7	209
Paraffin ($C_{22.2}$ $H_{44.1}$) (TG)	47.1	166
3-Heptadecanone	48	218
Paraffin C_{20}–C_{33}	48–50	189
9-Heptadecanone	51	213
Paraffin wax 53 (CG)	53	184
Paraffin wax 54 (CG)	53.32	184.48
RT 55	55	172
Paraffin C_{26}	56.3	256
Paraffin ($C_{23.2}$ $H_{48.4}$) (TG)	57.1	220
Paraffin wax 60–62	57.78	129.7
Paraffin wax 58–60	58.27	129.8
Paraffin C_{22}–C_{45}	58–60	189
RT 64 HC	64	230
Paraffin wax	64	173.6
Paraffin C_{21}–C_{50}	66–68	189
RT 70 HC	69–71	260

In the second group are grouped inorganic substances. Among such substances it is possible to work with change very high temperature phase, even above 800 °C [14]. In the range of temperatures analysed, it is possible to find substances shown in Table II.

Table II: Inorganic PCMs substances for PVT. [14, 15]

Compound	*Melting temperature (°C)*	*Heat of fusion (kJ/kg)*
Na_2SO_4 1/2NaCl $10H_2O$	20	
$FeBr_3\cdot 6H_2O$	21	105
SP 21 E	21–23	160

Table II: Continued.

Compound	*Melting temperature (°C)*	*Heat of fusion (kJ/kg)*
$Mn(NO_3)\cdot 6H_2O$	25.9	
SP 26 E	25–27	200
$CaCl_2\cdot 12H_2O$	29.8	174
TH 29	29	188
Calcium chloride	29.8	190.8
hexahydrate ($CaCl_2$	23.26	125.4
$6H_2O$)	27	
	28	86
$LiNO_3\cdot 3H_2O$	30	296
$LiNO_3\cdot 2H_2O$	30	296
SP 31	31–33	220
$Na_2SO_4\cdot 3H_2O$	32	251
Glauber's salt (Na_2SO_4	32.4	238
$10H_2O$)	32	254
$Na_2CO_3\cdot 10H_2O$	32–36	246.5
$CaBr_2\cdot 6H_2O$	34	115.5
$LiBr_2\cdot 2H_2O$	34	124
$Na_2HPO_4\cdot 12H_2O$	35–44	280
$Zn(NO_3)_2\cdot 6H_2O$	36	147
$FeCl_3\cdot 6H_2O$	37	223
$Mn(NO_3)\cdot 6H_2O$	37.1	115
$CoSO_4\cdot 7H_2O$	40.7	170
$KF\cdot 2H_2O$	42	162
$MgI_2\cdot 8H_2O$	42	133
$CaI_2\cdot 6H_2O$	42	162
$K_2HPO_4\cdot 7H_2O$	45	145
$Ca(NO_3)_2\cdot 4H_2O$	47	153
$Zn(NO_3)_2\cdot 4H_2O$	45–47	110
STL 47	47	221
$Mg(NO_3)\cdot 2H_2O$	47	142
$Fe(NO_3)_2\cdot 9H_2O$	47	155
$Na_2SiO_3\cdot 4H_2O$	48	168
$K_2HPO_4\cdot 3H_2O$	48	99
$Na_2S_2O_3\cdot 5H_2O$	48–49	201–210
$MgSO_4\cdot 7H_2O$	48.5	202
$Ca(NO_3)_2\cdot 3H_2O$	51	104
$FeCl_3\cdot 2H_2O$	56	90
$Ni(NO_3)_2\cdot 6H_2O$	57	169
SP 58	56–59	250
$MnCl_2\cdot 4H_2O$	58	151
$MgCl_2\cdot 4H_2O$	58	178
$Na(CH_3COO)\cdot 3H_2O$	58	264–267
Sodium acetate	58	230
trihydrate	58	252
($NaCH_3COO.3H_2O$)		
$Fe(NO_3)_2\cdot 6H_2O$	60	126
NaOH	64.3	227.6
$Na_3PO_4\cdot 12H_2O$	65–69	190
ClimSel C70	70–71	194
$Na_2P_2O_7\cdot 10H_2O$	70	184
$LiCH_3COO\cdot 2H_2O$	70	150

Finally we can also use mixtures of different substances. The main feature presenting these substances is to obtain a change point lower than each of the single substances phase. Table III shows the classification of the main eutectic PCMs found in the bibliography. There are bibliography from both mixtures of organic and inorganic substances.

Table III: Eutectics PCM,s.[14, 15]

Compound	*Melting temperature (°C)*	*Heat of fusion (kJ/kg)*
Inorganic		
66.6% $CaCl_2 \cdot 6H_2O$+33.3% $MgCl_2 \cdot 6H_2O$	25	127
48% $CaCl_2$+4.3% NaCl+0.4% KCl+47.3% H_2O	26.8	188
47% $Ca(NO_3)_2 \cdot 4H_2O$+33% $Mg(NO_3)_2 \cdot 6H_2O$	30	136
60% $Na(CH_3COO)$ ·	31.5	226
$3H_2O$+40% $CO(NH_2)_2$	30	200.5
61.5% $Mg(NO_3)_2 \cdot 6H_2O$+38.5% NH_4NO_3	52	125.5
58.7% $Mg(NO_3)$ ·	59	132.2
$6H_2O$+41.3% $MgCl_2 \cdot 6H_2O$	58	132
53% $Mg(NO_3)_2 \cdot 6H_2O$+47% $Al(NO_3)_2 \cdot 9H_2O$	61	148
14% $LiNO_3$+86% $Mg(NO_3)2 \cdot 6H_2O$	72	>180
Organic		
37.5% Urea+63.5% acetamide	53	n.a.
67.1% Naphthalene+32.9% benzoic acid	67	123.4

4.2 PCM's selection criteria.

As embodied in the previous section, there are plenty of alternatives for materials PCMs. The selection of a particular PCM for a specific application requires knowledge of their thermodynamic properties, kinetic, chemical and economic criteria. These aspects are shown classified in Table IV.

Table IV: PCM's selection criteria [13]

Thermodynamic properties
Use temperature rate High latent heat High Thermal Conductivity High density and specific heat Small volume changes in the phase change Low vapor pressure Eutectic
Kinetic properties
Fast nucleation rate to avoid supercooling Fast crystallization
Chemical properties
Reversibility of the melting/solidification cycle Chemical stability Non-toxic, non-flammable, non-explosive, non-corrosive
Criterios Económicos
Wide-scale availability Price

Finally, it is important to note the differences that can be found between different types of compounds. A comparison with the advantages and disadvantages of each different substances PCMs are shown in Table V.

Table V: Characteristics of the different PCM, s [13]

Advantage	**Disadvantage**
PCM organics	
Wide range of application temperature High latent heat Non-subcooling Chemical stability Good compatibility with other materials	Large volumetric changes Low conductivity Flammable
PCM inorganics	
High latent heat of fusion High thermal conductivity Small volume changes Availability Low cost	High subcooling degree Corrosives
Eutectics	
Melting point well defined High density thermal storage	Few thermophysical data

6 CONCLUSIONS

PCMs are a good alternative for storing the waste thermal energy from PV/T technology. Also, it allows to adapt the generation and production curves for solar energy making more useful for domestic use.

Main PCMs advantage is the possibility of using sensible and latent energy both with low temperature and volume variations. As storing of high amount of energy at constant temperature is possible during the melting process, higher PV efficiency is expected due to lower cell working temperature. The melting point depends on the chemical structure, so there is a great number of PCMs available to be used in an optimized PV/T installation design.

Different PCMs with melting temperature below 70 °C for applications in future designs of PV/T have been classify by physical properties, and also their thermochemical structure and availability have been studied.

Because of the wide number of different options PCMs, it is necessary to be very carefully in the right choice. Some advised have been showed in table IV and V.

Acknowledgements

This work was supported by Spanish Government grant ENE2014-54601-R. One of the authors, David González Peña, thanks to Junta de Castilla-León for economic support (PIRTU Program, ORDEN EDU/301/2015)
Also

Junta de Castilla y León

Europa impulsa nuestro crecimiento

7 REFERENCES

[1] *World Energy Outlook Special Report*, in *Energy and Climate Change*, R. Priddle, Editor. 2015, International Energy Agency: Paris, France.

[2] COM(2009), *DIRECTIVA 2009/30/CE DEL PARLAMENTO EUROPEO Y DEL CONSEJO de 23 de abril de 2009.*

[3] Aadmi, M., M. Karkri, and M. El Hammouti, *Heat transfer characteristics of thermal energy storage for PCM*

(phasechange material) melting in horizontal tube: Numerical andexperimental investigations. Energy, 2015. **85**: p. 339-352.

[4] Zhang, H., et al., *Thermal energy storage: Recent developments and practical aspects.* Progress in Energy and Combustion Science, 2016. **53**: p. 1-40.

[5] Zhang, X., et al., *Review of R&D progress and practical application of the solar photovoltaic/thermal (PV/T) technologies.* Renewable and Sustainable Energy Reviews, 2012. **16**(1): p. 599-617.

[6] Radziemska, E., *Thermal performance of Si and GaAs based solar cells and modules: A review.* Progress in Energy and Combustion Science, 2003. **29**(5): p. 407-424.

[7] Moradi, K., M. Ali Ebadian, and C.X. Lin, *A review of PV/T technologies: Effects of control parameters.* International Journal of Heat and Mass Transfer, 2013. **64**: p. 483-500.

[8] Tyagi, V.V., S.C. Kaushik, and S.K. Tyagi, *Advancement in solar photovoltaic/thermal (PV/T) hybrid collector technology.* Renewable and Sustainable Energy Reviews, 2012. **16**(3): p. 1383-1398.

[9] Xu, X., et al., *Performance and reliability analysis of hybrid concentrating photovoltaic/thermal collectors with tree-shaped channel nets' cooling system.* IEEE Transactions on Components, Packaging and Manufacturing Technology, 2013. **3**(6): p. 967-977.

[10] Chow, T.T., et al., *Performance evaluation of photovoltaic-thermosyphon system for subtropical climate application.* Solar Energy, 2007. **81**(1): p. 123-130.

[11] Bahaidarah, H., et al., *Performance evaluation of a PV (photovoltaic) module by back surface water cooling for hot climatic conditions.* Energy, 2013. **59**: p. 445-453.

[12] Li, G., et al., *Experimental study on a novel hybrid photovoltaic/thermal solar system.* Zhongguo Dianji Gongcheng Xuebao/Proceedings of the Chinese Society of Electrical Engineering, 2013. **33**(17): p. 86-89.

[13] Zhou, D., C.Y. Zhao, and Y. Tian, *Review on thermal energy storage with phase change materials (PCMs) in building applications.* Applied Energy, 2012. **92**: p. 593-605.

[14] Zalba, B., et al., *Review on thermal energy storage with phase change: Materials, heat transfer analysis and applications.* Applied Thermal Engineering, 2003. **23**(3): p. 251-283.

[15] Khan, Z., Z. Khan, and A. Ghafoor, *A review of performance enhancement of PCM based latent heat storage system within the context of materials, thermal stability and compatibility.* Energy Conversion and Management, 2016. **115**: p. 132-158.

AISOVOL PROJECT: A PHOTOVOLTAIC GENERATION SOLUTION AS AN ALTERNATIVE CONSTRUCTION MATERIAL

A.B. Cueli[1], J. Bengoechea[1], A. Murillo[(1,4)], M.J. Rodríguez[1] , A.R. Lagunas[1], C. Montes[2], A. Linares[3], E. Llarena[2], O. González[2], D. Molina[2], A. Pío[2], L. Ocaña[2], C. Quinto[2],M. Friend[2] & M. Cendagorta[2].
[1] NATIONAL RENEWABLE ENERGY CENTRE (CENER)
C/ Ciudad de la Innovación 7, 31621 Sarriguren (Navarra), Spain
Telephone: +34 948 25 28 00; Fax: +34 948 27 07 74
[2]Instituto Tecnológico y de Energías Renovables, S. A. (ITER)
[3]Agencia Insular de Energía de Tenerife (AIET)
Pol. Industrial de Granadilla, s/n
E 38600 Granadilla de Abona, Spain
[4]Universidad Pública de Navarra
C/Sadar, s/n, 31006 Pamplona, Navarra
Email: abcueli@cener.com

ABSTRACT: AiSoVol project consists on the development, manufacturing and testing in a controlled environment of a new concept of photovoltaic modules, conceived for facilitating their use as architectural elements. Thus, the modules will be fabricated by encapsulating its constituent electrical elements (cells and interconnections) with lamination techniques at low curing temperatures and using transparent thermoplastic instead of tempered glass, as well as binding materials structurally strengthened by a grid made of high-tenacity yarns, such as the ones used in sail technologies and thus avoiding the need for aluminum frames. This type of solution will provide, due to the nature of the constituent materials, lighter PV modules which can adapt easier to different surfaces. In this paper the methodology followed to define a suitable lamination process compatible to obtain reliable AiSoVol photovoltaic modules is described. Results achieved from materials and processes validation will pave the way for manufacturing the prototypes in next project phase.

Keywords: BIPV, encapsulation, polymer film, polycarbonate.

1. INTRODUCTION

At this initial phase of the AiSoVol project the focus has been put on the materials validation and its compatibility with processes involved in PV module manufacturing. Study of thermal, optical, physical, electrical and mechanical properties of several transparent thermoplastics raised the polycarbonate (PC) as best option to be used as superstrate for the AiSoVol BIPV modules. The rest of the package constituent materials are the encapsulant, PV cells, reinforcement grid made of high-tenacity yarns and the backsheet. The sublayer structure of this new BIPV product differs notably from that of conventional PV modules. What is investigated in this work is the degree of compatibility among combinations of these materials when laminated together employing the same techniques and equipment available at PV module manufacturing lines.

2. EXPERIMENTAL

Two different lines are used to study the feasibility of manufacturing reliable BIPV crystalline silicon modules based on alternative packaging materials. First approach is focused on defining the lamination process required to produce samples fulfilling minimum requirements regarding cure degree of the encapsulant, adhesion force between layers and visual aspect.

The second aspect studied in this work is the susceptibility of polymeric materials to suffer UV degradation when used for outdoor applications. This effect has been quantified by determining the evolution of the optical properties on artificially aged AiSoVol samples.

2.1. Adjustment of lamination process

The lamination parameters must be set taking into account the limitations posed by the use of polycarbonate as front cover. The polymeric films used as encapsulant and backsheet are products widely used in conventional modules: the encapsulant used in this study is an ultra-fast cure commercial EVA (ethylene vinyl acetate) and the backsheet is a black colored PPE (polyester-polyester-EVA). The reinforcement grid is made of Polyester and HMPE (High Modulus Polyethylene) fibers embedded between two layers of polyester taffeta. A picture of this material is shown in figure 1.

Figure 1. Reinforcement grid

Regarding EVA, the curing process guidelines provided by manufacturers are generally specified for typical PV modules with glass as front cover and no information is available for other concept designs, so recommended curing times and temperatures must be reconsidered. Different sets of samples have been produced varying temperature and time in the lamination process. Polycarbonate sheets of two different thicknesses, namely 2 mm and 4 mm, have been used and lamination parameters have been specifically fixed for each material.

Three diagnosis tests are conducted to evaluate the quality of the samples obtained by changing lamination parameters (temperature and time):

a) Visual appearance: A detailed visual inspection is carried out to detect the presence of visual defects detectable by naked eye as bubbles, wrinkles, delaminations, cracks and any other conspicuous condition.
b) Peel-off test: The procedure followed to measure the adhesive interlayer force is based on standard ISO 813 which determines the adhesive strength between polymeric materials bonded on a rigid substrate.
Sample preparation includes pre-cutting a strip of 10 mm width and at least 100 mm length with a release sheet inserted into the bonded interface to be analyzed. The moving grip of the peel-off machine is fixed to the release sheet and the pull force is applied at a moving rate of (50 ± 5) mm/min making an angle of 90° with the surface.
c) Degree of cure of EVA: Differential Scanning Calorimetry (DSC) technique is used to characterize the degree of crosslinking within EVA encapsulant. The residual enthalpy method, described in future standard as IEC 62788-1-6 Ed.1 currently in draft version, has been used as guidance. According to this method the exothermic peak (around 130 °C to 160 °C in our fast cure EVA) is due to the heat release from crosslinking reaction in EVA and the area below the curve relates to the percentage of un-crosslinked EVA. Therefore, the gel fraction of crosslinked EVA can be estimated using the following equation:

$$Ge = \frac{hu - ht}{hu}100$$

Where, *Ge* represents the degree of cure for the enthalpy method (%), *hu* the measured specific enthalpy of EVA of an uncured reference specimen ($J \cdot kg^{-1}$) and *ht* the measured specific enthalpy of EVA of the test specimen ($J \cdot kg^{-1}$). The recommended low and high limits of integration stated in IEC document are 100 °C and 200 °C respectively, however the DSC analysis for the uncured EVA employed in this study shows that a (110-190) °C range matches better with its specific thermal properties.

Although there are not standardized criteria regarding minimum EVA crosslinking degree and peel-off force, it is well accepted by PV industry that EVA gel content above 70% (measured using the primary method) and EVA-glass adhesion strength higher than 60 N/cm are good enough to obtain PV modules suitable for long term operation. The DSC technique used in this work is a secondary method that can be directly applied for quality control when the same EVA formulations are used. In case that DSC results of different encapsulants are to be compared the use of the primary method is recommended. DSC correlation to the gel content primary method can be obtained for encapsulants with curing degrees above 60% so this would be considered as minimum threshold for crosslinking analysis.

2.2. Optical characterization

Polycarbonate samples laminated with EVA have been optically characterized, including the measurement of the spectral transmittance (from 385 nm to 1600 nm) and the yellowness index (YI), which was obtained from it. This optical characterization allowed quantifying the influence of ultraviolet weathering on polycarbonate samples laminated with EVA.

3 RESULTS AND DISCUSSION

3.1 Lamination process

The lamination process has been evaluated in two different steps depending on the number of layers present in each sample. Initially the process is validated for type A laminates which include the PC superstrate, two EVA foils and the backsheet (PPE). Once the laminate fulfills the quality requirements defined through visual, mechanical (peel test) and curing degree (DSC), the rest of the materials (solar cell, extra EVA sheet and reinforcement grid) are included in the package system to obtain the adequate process parameters for the type B multilayer configuration.

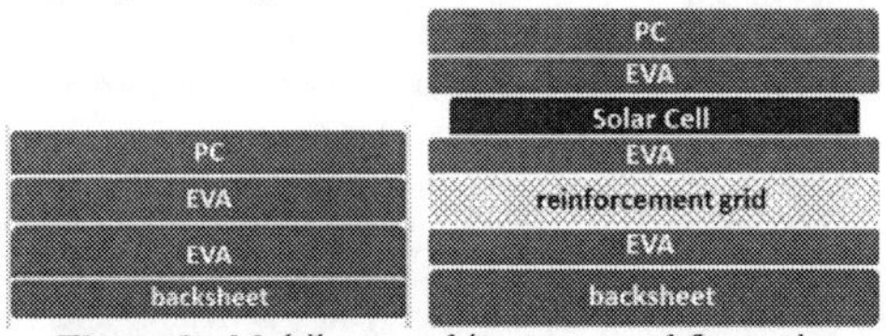

Figure 2: Multilayer architectures used for testing: laminate type A (left side) and type B (right side).

For the purpose of this work, all samples have been manufactured in a (16.5 x16.5) cm size format. The PC sheets studied are 2 mm and 4 mm thick consequently, different processes times and temperatures are used. They are shown in the text table.

Sample Code	Process	Package
16	115 °C-20.5'	Type A (2mm PC)
15	125 °C-12.5'	Type A (4mm PC)
17	125 °C-30'	Type A (4mm PC)

Table I: Process parameters and package description for the first set of laminates

Id.	Inspection	DSC (%)	Peel test (N/cm)	
Code No.	Visual aspect	Cure degree	PC-EVA	PPE-EVA
16	OK	37	84	60
15	OK	44	50	-
17	OK	48	100	65

Table II: Results on first set of laminates evaluated

Analysis of the diagnostic tests conducted over samples 15 to 17 shows that there is not enough crosslinking degree of the encapsulant for the selected process temperatures. Even increasing cycle time, the curing degree of EVA does not increase further. The peel test performed over two interfaces shows that minimum adhesion force of 50 N/cm has been achieved in all cases.

A second set of samples is laminated using higher temperatures for the curing cycle: 135 °C for samples with 2 mm thick PC sheet and 140 °C for the 4 mm thickness.

Sample Code	Process	Package
23	135 °C-12.5'	Type A (2mm PC)
24	135 °C-17.5'	Type A (2mm PC)
25	135 °C-22.5'	Type A (2mm PC)
26	140 °C-12.5'	Type A (4mm PC)

Table III: Process parameters and package description for the second set of laminates

Id.	**Inspection**	**DSC (%)**	**Peel test (N/cm)**	
Code No.	Visual aspect	Cure degree	PC-EVA	PPE-EVA
23	OK	60	99	-
24	OK	72	138	-
25	Bubbles	80	112	-
26	OK	69	120	117

Table IV: Results on second set of laminates evaluated

Characterization results from this second round (samples 23 to 26) show that the more suitable lamination processes for each package system are 135 °C and 17.5 minutes for the sample with 2 mm thick PC sheet and 140°C during 12.5 minutes for the one with the 4mm thick PC sheet. For these two recipes the DSC results meet the acceptance criteria stablished and the adhesion force PC-EVA is above the minimum required for glass-EVA interfaces. A maximum curing degree of 80% is reached for sample 25 but visual defects appear in the encapsulant. The adhesion force between encapsulant and backsheet has been measured only for one sample and the value is similar to the one obtained for EVA-PC.

Later on, these processes are reproduced for type B laminates including the rest of material layers and components in each package: solar cells, extra EVA sheet and reinforcement grid. The characterization of the samples is restricted, in this case, to visual inspection and DSC analysis of the EVA foil extracted between the reinforcement grid and the backsheet.

Sample Code	**Process**	**Package**
37	135 °C-17.5'	Type B (2mm PC)
38	140 °C-12.5'	Type B (4mm PC)

Table V: Process parameters and package description for the third set of laminates

Id.	**Inspection**	**DSC (%)**
Code No.	Visual aspect	Cure degree
37	Bubbles	68
38	OK	60

Table VI: Results on third set of laminates evaluated

The sample laminated with the 2 mm thick PC sheet does not meet the visual criteria. The presence of bubbles indicates that temperature of the EVA layer bonded to the PC frontsheet is higher than before even though the same lamination process parameter has been used. Value of crosslinking degree attained for the EVA analyzed is however lower than the one measured for the type A sample laminated using the same parameters what evidences the presence of a temperature gradient across the stack during lamination process.

In order to determine the temperature differences among front (hotter) and back (colder) side of the encapsulant, irreversible temperature labels are placed at each interface during lamination. These indicators are valid to establish temperature ratings in 6 °C steps, the results show that differences among front and back EVA sheets could range from 1 °C to 5 °C.

A final tuning recipe is proposed to define proper lamination parameters for each complete package. The curing time is increased to 15 minutes for the sample with 4 mm thick PC to obtain higher crosslinking degree in the cold side of the sample. In the case of the package with 2 mm thick frontsheet, the curing temperature is lowered to 131 °C and both configurations (type A and type B) are constructed to perform all diagnosis tests over them.

Sample Code	**Process**	**Package**
44	140 °C-15'	Type B (4mm PC)
45	131 °C-17.5'	Type A (2mm PC)
46	131 °C-17.5'	Type B (2mm PC)

Table VII: Process parameters and package description for the fourth set of laminates

Id.	**Inspection**	**DSC (%)**	**Peel test (N/cm)**	
Code No.	Visual aspect	Cure degree	PC-EVA	PPE-EVA
44	Bubbles	-	-	-
45	OK	58	114	97
46	OK	61	-	-

Table VIII: Results on fourth round of laminates tested

Visual defects appear between PC and solar cell (hotter EVA) in sample 44 so the previous process, at 140°C and 12.5 minutes, is chosen as the best one. No DSC analysis is carried out over this sample.

The results of peel-off test conducted on sample 45 fulfill minimum adhesion strength requirements for PC-EVA (114 N/cm) and PPE-EVA (97 N/cm) interlayers.

The laminate 46 obtained does not present bubbles and the curing degree measured by DSC is 61%.

The picture below shows small bubbles detected around buses of sample 44.

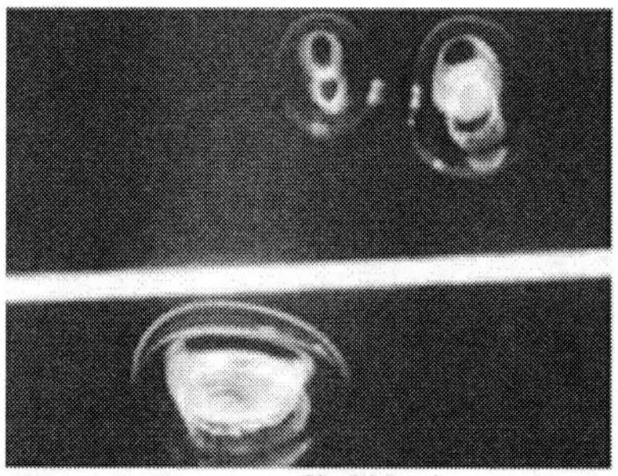

Figure 3: Presence of bubbles in sample 44

Evaluation of type B packages for both PC thicknesses show that maximum degree of cure measured for the EVA bonded to the backsheet (colder side) is around 60%. The attempts to increase this value either using longer lamination times or higher temperatures, cause the bubbles to appear in the hotter side of EVA. In previous laminates experiments the presence of bubbles has been detected for a curing degree around 80%.

In order to estimate the curing degree of the encapsulant adhered to the frontsheet in type B packages where EVA extraction is not possible further analysis are performed.

In the following graph it is presented the curing degree attained for each lamination process used for type A laminates with 2mm thick PC sheet as front cover.

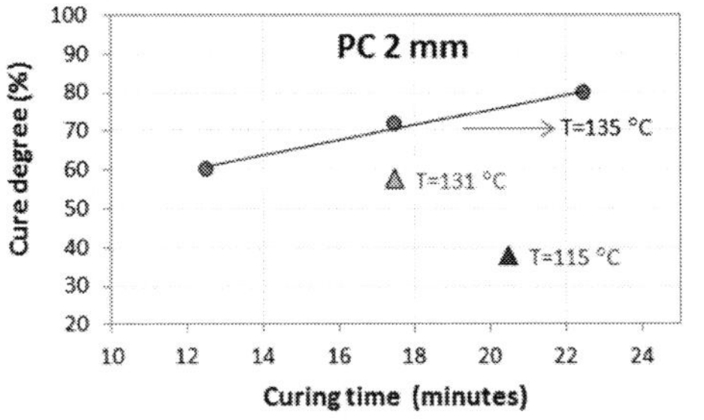

Figure 4: DSC results for different lamination process

Results presented in figure 4 show a linear behavior of curing degree versus lamination time for the same target temperature (135 °C). For the specific case studied, every 5 minutes step in curing time represents a 10% increment in crosslinking degree determined by DSC. On the other hand, a temperature decrement of 4 °C for the same lamination time (17.5 minutes) results in a 14% reduction of the curing degree what represents a 3.5% variation per °C. Taking into account that temperature differences between both EVA sides in type B packages could range from 1°C up to 5 °C it is expected curing degrees to vary accordingly.

3.2 Effect of ultraviolet weathering

Polymers often show high sensitivity to ultraviolet light, and as a result they suffer significant degradations when exposed to sunlight. With the aim of investigating durability aspects of outdoor exposure, 3 polycarbonate samples (2mm thick) laminated with EVA using different lamination conditions have received a total ultraviolet irradiation of 28 kWh/m^2 (94% corresponding to UVA and 6% to UVB). Polycarbonate side was placed facing the incident ultraviolet irradiation.

In the following table the lamination conditions for each sample together with the results of the optical characterization before and after the ultraviolet weathering test are shown. The transmittance (T) corresponds to the averaged values between 385nm and 1600nm.

Id.	Lamination		Before		After UV	
			T(%)	YI	T(%)	YI
07	125 °C	12.5'	86.2	2.50	85.8	3.18
08	120 °C	20.5'	86.0	2.59	85.9	3.05
09	115 °C	20.5'	85.8	2.51	85.9	3.09

Table IX: Lamination conditions and optical characterization of polycarbonate samples laminated with EVA before and after UV irradiation

As can be appreciated from the previous table, the initial transmittance shows minor differences among 3 different lamination conditions examined and slight decrease of this parameter is observed after the ultraviolet test. On the other hand, the yellowness index was clearly influenced by the ultraviolet weathering test, increasing its value for the 3 samples. The different lamination conditions did not show relevant influence in the optical characteristics of the 3 samples or in the ultraviolet test effect.

4 CONCLUSIONS

The BIPV module lamination processes for 2mm and 4mm thick polycarbonate (PC) frontsheets proposed in the AiSoVol project have been adjusted and validated.
A fine control of the lamination parameters has been achieved for each specific package so that time and temperature can be individually set in order to obtain specific curing degrees of the encapsulants.
Selected processes for each package system are 131 °C and 17.5 minutes for the sample with 2 mm thick PC sheet and 140°C during 12.5 minutes for the one with the 4mm thick PC sheet. The results of the peel test for type A samples laminated with these parameters are 114 N/cm in the PC(2mm)-EVA interlayer and 97 N/cm and 117 N/cm in the EVA-backsheet with 2mm PC as frontsheet. The peel forces measured in the type A sample with 4mm thick PC are 120 N/cm and 117 N/cm for the same interfaces. The degree of cure for the type B laminates is around 60% for both PC thicknesses analysed.

Temperature gradients through the complete package during lamination processes have been measured. Correlation between lamination parameters (time and temperature) and the curing degree for type A laminates with 2mm thick PC sheet has been quantified. An increase of one °C in lamination temperature implies 3.5% improvement in the curing degree and 5 minutes longer cycles at a fixed temperature represents 10% higher curing rates.

With the aim of investigating durability aspects of outdoor exposure, polycarbonate samples laminated with EVA have undergone an ultraviolet weathering test. In this case, 3 polycarbonate samples (2mm thick) laminated with EVA using different lamination conditions have received a total ultraviolet irradiation of 28 kWh/m^2, (94% corresponding to UVA and 6% to UVB). As a result of this test the transmittance showed a slight decrease after the ultraviolet test. On the other hand, the yellowness index was clearly influenced by the ultraviolet weathering test, increasing its average value for the 3 samples. The different lamination conditions applied in this case did not show relevant influence in the optical characteristics of the 3 samples or in the ultraviolet test effect.

5 ACKNOWLEDGEMENT

The AiSoVol project is funded by the Spanish Ministry of Economy and Competitiveness, having been selected from R & D + i State Program oriented to societal challenges. The authors would like to acknowledge Marcos del Rio and Asier Turumbay for their support with chemical and mechanical measurements respectively.

5 REFERENCES

[1] C. Hirschl *et al.*, Solar Energy Materials and Solar Cells, Vol. 116, pp. 203 – 218, (2013).
[2] Wu, Dan. Doctoral Thesis: Investigation of the reliability of the encapsulation system of photovoltaic modules. https://dspace.lboro.ac.uk/2134/18051 (2016)
[3] IEC FDIS 62788-1-6© IEC 2016. Materials used in PV modules-Measurement procedures. Part 1-6: Encapsulants-Test methods for determining the degree of cure in Ethylene-Vinyl Acetate

AN EXPERIMENTAL INVESTIGATION INTO PASSIVE TEMPERATURE REGULATION OF A NOVEL WICPV SYSTEM WITH PHASE CHANGE MATERIAL

Shivangi Sharma[1, a)], Nazmi Sellami[2], Asif Tahir[1] and Tapas K Mallick[1, b)]

[1]*Environmental and Sustainability Institute, University of Exeter, Penryn (Cornwall), TR109FE, United Kingdom*
[2]*Heriot-Watt University, PO Box 294345, Dubai International Academic City, Dubai, United Arab Emirates*
[a)] Corresponding author: ss664@exeter.ac.uk
[b)] T.K.Mallick@exeter.ac.uk

ABSTRACT: A novel Window-Integrated Concentrated Photovoltaic (WICPV) system proposed suitable for window integration, has been fabricated and characterised at 1000 Wm^{-2} (higher irradiance)and 500 Wm^{-2} (lower irradiance). This paper investigates the use of a commercially available organic phase change material (PCM) RT®50 for limiting the temperature rise in the system within indoor testing conditions. The results demonstrate an average of 7.4°C reduction in the system temperature at 1000 Wm^{-2} and 4.8°C at 500 Wm^{-2}. It also resulted in 17% increase in electrical power efficiency at 1000 Wm^{-2} and 6.5 % at 500 Wm^{-2} within the PCM incorporated system as compared to a naturally ventilated system. Further, differential scanning calorimetric technique has been applied to assess the thermophysical properties of the PCM as any erroneous data may affect the prudent selection of PCM for particular applications. The outcome of the DSC analysis performed on the PCM indicates 10.6 % underestimation in the reported melting peak temperature value of the PCM.

Keywords: BICPV, PCM, Passive cooling, Thermal control

1 INTRODUCTION

A Windows-Integrated Concentrated Photovoltaic (WICPV) system is a special case of Building-Integrated Concentrated Photovoltaic (BICPV) systems and is based on the concentration of sunlight employing refractive concentrator. This increases the magnitude of effective solar radiation available for electricity generation and reduces the capital cost of the system by reducing the amount of costlier (compared to the concentrator material) silicon cell required. The WICPV system under investigation is based on Square Elliptical Hyperboloid (SEH) design of concentrators focussing sunlight on smaller areas of $1cm^2$ silicon solar cells. This distinctive optical concentrator (with a geometrical concentration ratio, x6) design aids high efficiency of the system and delivers hassle-free power from window[1]. These can be easily integrated into glazing façades and have advantages in; (i) maintaining a level of transparency of the windows allowing the daylight penetration (ii) producing low-cost electricity increasing the energy efficiency of buildings and (iii) extremely low solar heat gain[2]. The concentrating elements can contribute towards saving more than 60% of the solar cells used in conventional PV[1]. Originally developed by Sellami et al. using clear plastic back-plate, the WICPV system in this research has been transformed using intermittent Aluminium strips within an acrylic back plate recessed configuration with thin copper tubes at the back side, to enhance the heat transfer rate.

The factors affecting the WICPV system electrical power output are the material, type and quality of the solar cells used, the concentration level achieved by the concentrator, the incident solar radiation and the operating temperature[3]. For a given WICPV system working under constant irradiance, the temperature increase limits the efficiency of the system. During the process of concentration, as the concentrating element focusses sunlight onto a smaller aperture, the temperature of the system rises even further.

The two central approaches for controlling WICPV temperature are active and passive thermal control mechanisms. While the former uses external power source dubbed as parasitic losses, and are composed of moving mechanical parts, the latter does not, thereby making it more cost effective and lucrative alternative to active cooling. To limit the operating temperature of the WICPV system using passive cooling mechanism, one can either opt for free or natural air convection cooling on the front and back side of the panel, or modify the system's back-plate architecture by using back plate fins or by absorbing the excess heat in PCM. The latter solution includes the use of PCMs placed in a reservoir attached on the back-side of the WICPV system.

PCM with good thermo-physical and heat transport properties, such as high latent heat and high thermal conductivity offer promising technique to passively regulate silicon solar cell temperature used within WICPV systems, thereby limiting electrical power losses associated with temperature rise. However, accurate assessment of these PCM properties play vital role in assessing their suitability for the given application as well as justify the quantity of the material required for effective heat transfer and absorption from the system.

Differential scanning calorimetry (DSC) is the standard measurement method used for the determination of thermo-physical properties of the PCM and thermal characterisation of the PCM[4]. It is used to identify chemical or physical changes resulting in either emission or absorption of thermal energy associated with phase changes, namely, via exothermic and endothermic processes.

In this work, an experimental investigation has been carried on the effectiveness of using PCM for WICPV temperature rise control and possibility of electrical efficiency increment. DSC thermograms are also obtained for the PCM sample at a constant heating rate. These are analysed for the characterisation of melting and solidification process and the value and range of data so obtained are compared with the manufacturer provided data and presented within this paper.

2 EXPERIMENTAL METHOD

There are eight silicon solar cells soldered in series and six of these soldered strips are then connected in parallel (shown by the blue lines in Fig. 1) and attached to the SEH concentrator (shown by elliptical circles in schematic; Fig. 1) on top via a thin layer of encapsulant for optical coupling and cementing force. The SEH concentrator is manufactured using urethane and the details around manufacturing process can be referred to from the literature [5, 6]. They are then assembled on thin (1.2 cm wide) Aluminium metal strips housed inside the intermittently spaced acrylic grooves within the back plate assembly. This design is selected to study the effect of heat removal from the solar cells through conduction and convection in the metal back plate and copper tubes filled with the PCM. At the same time it allows the passage of light (theoretically at least 50%) through the system because of the transparency provided by acrylic material.

Temperature regulation of this in-house fabricated WICPV system with solid-liquid PCM is experimentally evaluated using paraffin wax based PCM Rubitherm RT®50. This organic PCM exhibits low super cooling, low vapour pressure, good thermal and chemical stability, and self-nucleating behaviour. Indoor characterisation without and with PCM took place under *Wacom® Super Solar Simulator* providing highly collimated illumination; WICPV system being characterised at 0° from horizontal.

NETZSCH DSC 214 *Polyma* by has been used for thermal characterisation of PCM RT®50 and produces relevant thermograms. PCM samples of 4.0 ± 0.015 mg mass have been used with a dynamic heating program with heating rate of 10°C /min between 10°C and 65°C and cooling rate of 5°C/min between 65°C and 10°C range from the manufacturer supplied PCM melting range data. This heating rate has been selected as experimentally to simulate the similar heating rate observed within the WICPV system under the solar simulator using 1000 Wm^{-2}.

The DSC instrument has a good sensitivity and high accuracy for measuring thermo-physical properties of PCM with rapid response. The time constant for the empty system is 0.8 s +/- 0.1 s, while the time constant for pan + sample is < 5.4 s. The temperature was calibrated with Indium before the DSC measurement. The samples for the measurement were arbitrarily selected from the mixture, and at least three samples were measured for every type. The samples were sealed in a standard aluminium DSC sample crucible. The reference crucible which was an empty crucible and sample crucible with the PCM. The difference of the heat flows between the two crucibles was recorded under nitrogen gas flow of 40.0 ml/min.

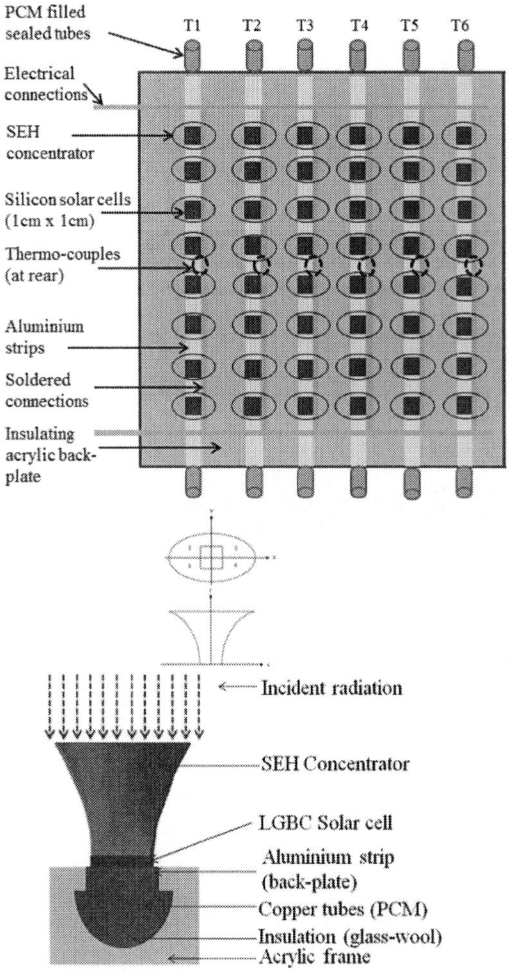

Figure 1: Schematic of the WICPV system with PCM filled in the copper tubes (Passive cooling) and the cross-sectional view

3 RESULTS AND CONCLUSIONS

3.1 Electrical characterisation

The results from photoelectric short circuit current, open circuit voltage and maximum electrical power curves showed an increase in relative electrical efficiency by 17 % at 1000 Wm^{-2} (higher irradiance)and and 6.5 % at 500 Wm^{-2} (lower irradiance) using PCM as compared to the non PCM system. The average temperature reduction achieved within the thermocouple channels (T1, T2, T3, T4, T5, and T6) were (1.4°C, 0.9°C, 2.5°C, 4.9°C, 6.2°C, 7.4°C) as shown in Fig 2 (a). The highest temperature generated within the WICPV system without PCM was 69.9°C while it was 62.7°C with PCM, thereby a difference of 7.2°C. Experimentally obtained data is tabulated in Table I. Higher average cooling within the thermocouple channels (T1, T2, T3, T4, T5, and T6) was observed for lower irradiance level with PCM (3.4°C, 4.3°C, 5.6°C, 6.6°C, 6.9°C, 8.6°C), the photoelectrical current increase at 500 Wm^{-2} wasn't in proportion for the equal experimental duration. This could be attributed to the melting range of the used PCM (45-51°C, main peak: 49°C) [7]. The maximum temperature attained by one of the centrally located channels (T3) at lower irradiance was 51.9°C towards the end of the experiment time of 2.5 hours, means the entire melting of the PCM may not have completed; increasing the experimental time for lower irradiance levels may increase electrical power gain. The average temperature within this channel remained at 39.5°C with PCM while 45.0°C without PCM. At 1000 Wm^{-2}, the highest temperature generated was 69.9°C (51.9°C for at 500 Wm^{-2}) within the system without PCM and 62.7°C (46.5°C for at 500 Wm^{-2}) with PCM. Experimentally obtained data is tabulated in Table I.

Table I: Comparison of experimentally obtained data for the photoelectrical and thermal parameters without and with PCM RT®50

Irradiance (W/m^2)	Parameter (average value)	Without PCM	With PCM
1000	I_{sc} (A)	0.49	0.57
	V_{oc} (V)	4.38	4.48
	P_m (W)	1.60	1.94
	Temp Max(°C)	59.7	55.7
500	I_{sc} (A)	0.28	0.28
	V_{oc} (V)	4.45	4.55
	P_m (W)	0.94	1.01
	Temp Max(°C)	46.1	39.5

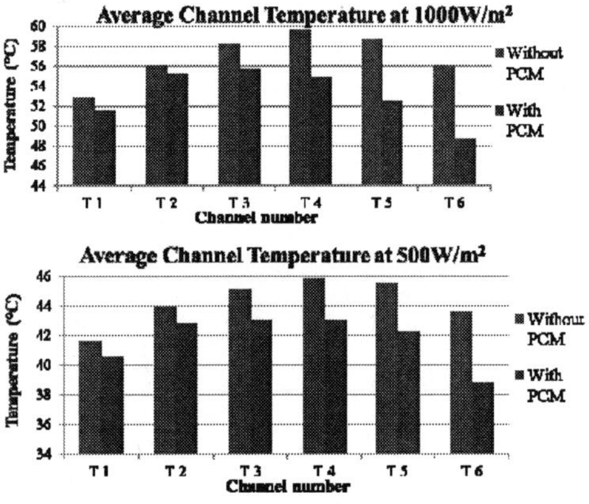

(a)

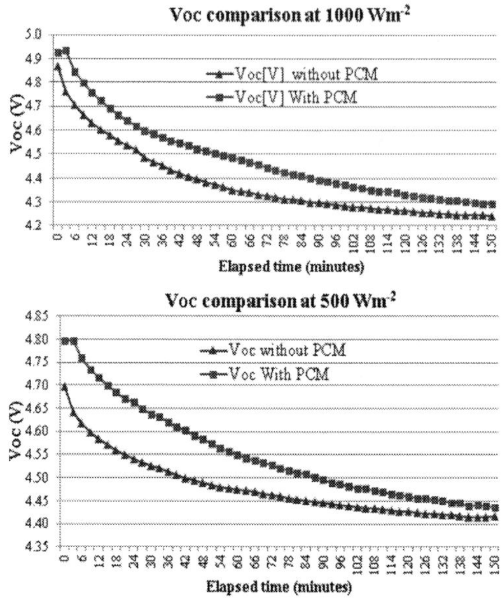

(b)

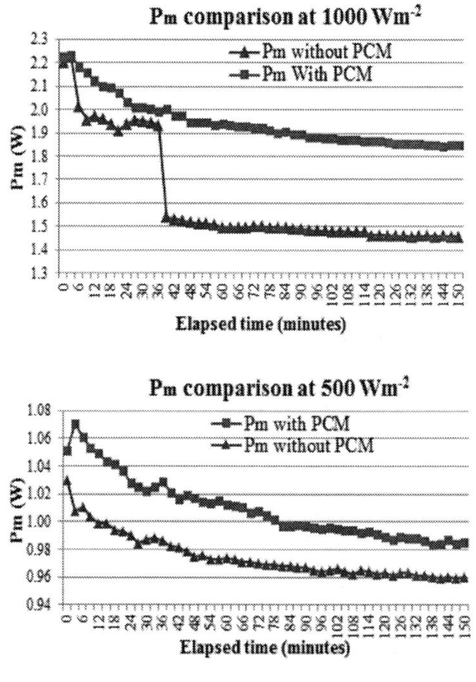

(c)

Figure 2: Comparison of the (a) average channel temperature, T_i (where i is the channel number), (b) open circuit voltage, (V_{OC}) and (c) maximum power (P_m) without and with PCM RT®50 at higher and lower irradiance

3.2 Thermal characterisation

The DSC thermograms plotted against temperature and time are plotted in Fig. 3. While the melting part of the curve shows the melting DSC signal (in red), being endothermic in nature, the solidification curve represents the exothermic cooling DSC signal (in blue), hence plotted as negative of signal intake value. The values of the manufacturer supplied data can be found in the reference[7]. However the actual values, as calculated from the DSC measurements were (48.4–56.6°C, main peak: 54.2°C) thereby revealing a remarkable 10.6 % underestimation in the reported melting peak temperature value of the PCM. It is to be noted, however, that the manufacturer data is based on a 3-layer-calorimetery method which allows bigger samples (typically of the order of 100 g). This is done in order to calculate the bulk PCM thermal properties for higher mass of PCM usage in possibly building applications. However, for these PCM to be applied in BICPV/WICPV applications, it may be highly rewarding to re-measure the values using DSC systems which normally take much lower amounts (typically of the order of 10 mg). The melting and congealing temperature range and peak data for PCM is given in Table II.

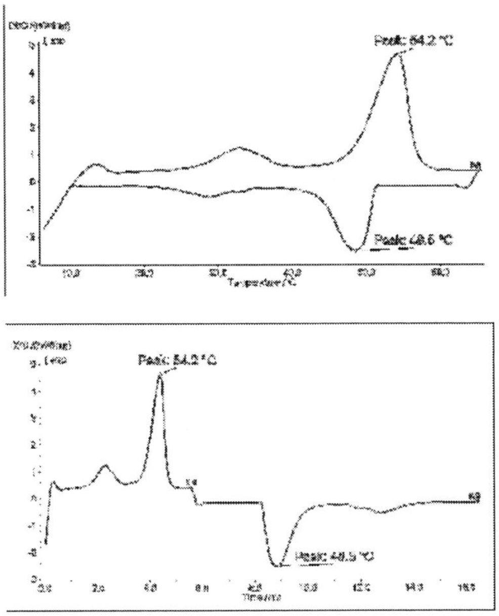

Figure 3: DSC thermograms for 4 mg specimen of PCM RT®50 heated at 10°C /min (endothermic process in red) between 10°C and 65°C and cooled to 10°C (exothermic process in blue) w.r.t. Temperature and Time

Table II: Comparison of supplier given data and DSC analysis obtained data for melting temperature data for PCM RT®50

Property	Supplier data	DSC data	Difference (%)
Melting area (°C)	45.0 - 51.0	48.4 - 56.6	7.8 -10.9
Main melting peak (°C)	49.0	54.2	10.6
Congealing area (°C)	51.0 - 46.0	51.1 - 44.2	0.19 -(-4.07)
Main congealing peak (°C)	50.0	48.5	3.1

The use of PCM also increases the temperature distribution uniformity throughout the module. The temperature distribution throughout the module has been studied and it's found that the temperature difference between the centre of the module is higher by 15.8% (without PCM) while 6.7% (with PCM) as compared to the temperature towards the edges of the module at higher irradiance. The respective data for lower irradiance indicates 13.1% and 7.1% variation, implying that PCM cooling almost halved the temperature variance within the module as compared to the natural ventilation case.

This experimental data could be used for designing heat sinks or PCM containment with variable dimensions within the same module, such as thicker tubes containing more material could be employed only for the middle sections of the panels thereby saving material costs required to maintain the uniform cooling throughout. The given copper tube configuration which could practically fit within the system could only contribute towards limited cooling material and hence time. Designs with bigger tube diameter may produce better results in terms of cooling and electrical efficiency of the WICPV systems. Also, the DSC analysis could be performed with different sample sizes as well as heating rates to determine more precise values with statistical analysis using mean, standard deviation and uncertainty of PCM thermo-physical properties.

4 ACKNOWLEDGMENTS

Financial support from EPSRC (EP/K03619X/1) is gratefully acknowledged.

5 REFERENCES

1. Sellami, N. and T.K. Mallick, *Optical characterisation and optimisation of a static Window Integrated Concentrating Photovoltaic system.* Solar Energy, 2013. **91**(0): p. 273-282.
2. Sellami, N., *PhD Thesis - Design and characterisation of a novel translucent solar concentrator*, in *Institute of Mechanical, Process and Energy Engineering*. 2013, Heriot-Watt University: Edinburgh. p. 307.
3. Ma, T., et al., *Using phase change materials in photovoltaic systems for thermal regulation and electrical efficiency improvement: A review and outlook.* Renewable and Sustainable Energy Reviews, 2015. **43**: p. 1273-1284.
4. Dumas, J.-P., et al., *Interpretation of calorimetry experiments to characterise phase change materials.* International Journal of Thermal Sciences, 2014. **78**: p. 48-55.
5. Sharma, S., et al., *Performance enhancement of a Building-Integrated Concentrating Photovoltaic system using phase change material.* Solar Energy Materials and Solar Cells, 2016. **149**: p. 29-39.
6. Sarmah, N., B.S. Richards, and T.K. Mallick, *Design, development and indoor performance analysis of a low concentrating dielectric photovoltaic module.* Solar Energy, 2014. **103**(0): p. 390-401.

7. GmbH, R.T., *Technical Data sheet RT50*, http://www.rubitherm.eu/media/products/datasheets/Techdata_-RT50_EN.PDF, Editor. 2013: Rubitherm Technologies GmbH - Sperenberger Str.5a - 12277 Berlin. p. 1.

EFFECTS OF LUMINESCENT COUPLING IN PEROVSKITE/C-SI MULTIJUNCTION SOLAR CELLS WITH NANOSTRUCTURED INTERLAYER

Takeshi Tayagaki,[1] Yasuyoshi Kurokawa,[2] and Noritaka Usami[2]
[1]National Institute of Advanced Industrial Science and Technology (AIST)
1-1-1 Umezono, Tsukuba 305-8568, Japan
[2]Graduate School of Engineering, Nagoya University, Nagoya 464-8603, Japan
e-mail: tayagaki-t@aist.go.jp

ABSTRACT: Light management in tandem solar cells has gained much attention for achieving high conversion efficiency. Here, we investigate the effects of nanostructured interfaces on the optical properties such as optical coupling of luminescent tandem solar cells by using electromagnetic wave simulation. We calculated the angle-dependent Fresnel coefficients at the nanostructured interface in lead-halide perovskite/c-Si tandem solar cells. The simulation results show that the luminescent coupling between the perovskite and c-Si sub-cells can be improved using a short-period nanostructured interface. A luminescent coupling efficiency greater than 95% is obtained for the nanostructured interface with height less than 1 μm.
Keywords: Luminescent Coupling, Tandem Solar Cells, Light Management, Optical Coupling, Photon Recycling

1 INTRODUCTION

Light management for designing the optical properties of solar cells has gained much attention for reducing optical loss and achieving high conversion efficiency [1,2]. To reduce the surface reflection loss and increase light trapping efficiency, surface textures formed using alkali etching have been widely used for crystalline silicon (c-Si) solar cells [3]. With the recent trends of thinning of c-Si wafers, more efficient light trapping using novel concepts and optical designs have been theoretically and experimentally explored [4–6]. The assessment of light management also results in a significant increase in conversion efficiency of thin-wafer GaAs solar cells through photon recycling [7,8]. In addition to single-junction solar cells, the design of optical coupling between sub-cells, such as luminescent coupling, has received increased attention in multijunction solar cells to boost the performance further [1,9]. In luminescent coupling, the luminescence of the top sub-cell is transmitted into the bottom cell and is used for photocurrent generation in the bottom sub-cell [Fig. 1(a)]. In contrast, for photon recycling, the top-cell luminescence that is reflected at the interface can be re-absorbed by the top cell. In series-connected tandem solar cells, the current density is determined by the minimum current density among the sub-cells. Therefore, the lowest sub-cell current density can be increased through luminescent coupling, through which the current density in tandem solar cells increases [Fig. 1(b)]. The luminescent coupling can adjust the balance between the generated photocurrent in each sub-cell, which may also contribute to the recovery of performance of the worst-performing cell under spectral mismatch conditions when tandem solar cells are used for terrestrial applications [10].

Silicon-based tandem solar cells have attracted much attention for boosting the performance of widely used crystalline silicon (c-Si) solar cells because the conversion efficiency of c-Si solar cells has increased to a level close to their theoretical limit. As a promising type of low-cost tandem solar cells, lead-halide perovskite/c-silicon multijunction solar cells have attracted much attention. Perovskite solar cells have demonstrated unprecedented progress in efficiency, and multiple groups have reported efficiencies of approximately 20% after only several years of active research [11]. Even though the radiative efficiency of perovskite solar cells is low compared to the radiative efficiency of III-V semiconductors (>0.9), efficient luminescence has been reported in lead-halide perovskite [12]. Thus far, a 2-terminal perovskite/silicon multijunction solar cell fabricated using a silicon-based tunnel junction has been demonstrated [4]. The bandgap energies of a typical lead-halide perovskite ($CH_3NH_3PbI_3$) and c-Si are 1.6 eV and 1.1 eV, respectively, which determine the photocurrent generation in the perovskite and c-Si sub-cells of 25 and 19 mA/cm^2, respectively [5]. In practice, lower current generation in the c-Si sub-cell due to the weak optical absorption of c-Si in the near-infrared region may cause a larger current mismatch of ~5-10 mA/cm^2, which may be lost with increasing dark current through recombination current generation in the perovskite sub-cell because forward bias voltage is applied to the perovskite sub-cells under the operating condition.

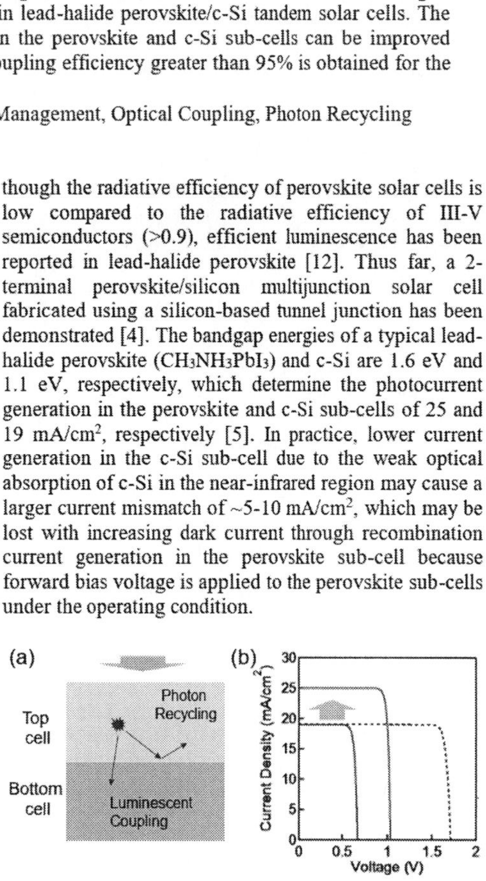

Figure 1: (a) Schematic illustration of luminescent coupling and photon recycling in multijunction solar cells. (b) Schematic of current-voltage characteristics in the perovskite (red) and c-Si (blue) sub-cells and the perovskite/c-Si multijunction solar cells (broken curve). Arrows indicate the operating point in the perovskite cell at the current matching condition. Luminescent coupling increases the current density of the bottom c-Si cells.

To align the generated photocurrent among the sub-cells in series-connected tandem solar cells, several approaches have been proposed thus far, including the tuning bandgap energy of the sub-cells, thinning of the sub-cell thickness, four-terminal device configuration [6], and bifacial application [7]. Here, we focus on the luminescent coupling in the tandem solar cells because it also provides a critical advantage for the terrestrial use of tandem solar cells. Efficient luminescent coupling can adjust the current generation in the sub-cells of the tandem solar cells, through which the additional photocurrent in the sub-cell is transmitted through the junction via luminescence and reabsorbed to generate a photocurrent in the neighboring cells [9]. If this luminescence is re-absorbed by the bottom Si solar cell, we can expect that the current density in c-Si cells increases, which results in an enhanced total current density in the perovskite/c-Si series-connected tandem solar cells. However, compared to conventional top-cell absorbers, such as InGaP and GaAs, the refractive index of lead-halide perovskite is so low that the refractive-index mismatch at the interface causes the optical reflection of luminescence. Antireflection for the luminescence at the interface may be necessary to use the luminescent coupling more efficiently. To achieve efficient luminescent coupling, understanding the optical properties including incident-angle dependence will be critical for optical design for efficient luminescent coupling.

In this study, we investigated the effects of nanostructured interfaces on the optical coupling in luminescent tandem solar cells by using electromagnetic wave simulation. To understand the luminescent coupling efficiency, we calculated the incident-angle dependence of the interface transmission and reflection, $T(\lambda, \theta)$ and $R(\lambda, \theta)$, respectively, which determines the efficiency of luminescent coupling from the perovskite sub-cell to the c-Si sub-cell. Nanostructured interfaces were introduced in the perovskite/c-Si tandem solar cells to achieve efficient luminescent coupling. Our findings indicate that the luminescent coupling between the perovskite and c-Si sub-cells can be significantly improved using a short-period nanostructured interface.

2 SIMULATION OF OPTICAL PROOERTIES

We investigated the optical properties of the perovskite/c-Si tandem solar cells by using the rigorous coupled wave analysis (RCWA) [13,14]. A schematic of the nanostructured interface used in the simulation is illustrated in Fig. 2(a). In the model, nanostructured textures are formed at the interface of perovskite and c-Si. In Fig. 2, we used half spheres, and the depth and period of surface dips are fixed at 100 and 200 nm, respectively. For the optical parameters, we used the refractive indices of c-Si and lead-halide perovskite [15,16]. Here, we neglect TiO_2 nanocrystals because the refractive index of TiO_2 is almost the same as that of lead halide perovskite ($n \sim 2.6$) at ~800 nm. For simplicity, we neglected the absorption at the interface.

Figure 2(b) shows the typical interface reflectance for normal incident light on the perovskite/c-Si layer with the nanostructured and flat interface; the reflectance values are the averages of the transverse-electric (TE) and transverse-magnetic (TM) modes. Due to the small index difference between lead-halide perovskite and c-Si, a part of the incident light is reflected ($R \sim 0.05$) at the flat interface. In the nanostructured interface, reflectance for normal incident light decreases in a wide wavelength range. For luminescent coupling, the reflectance for light with a large incident angle is more important because the direction of luminescence exhibits random distribution.

Figures 2(c) and 2(d) show the incident-angle dependence of the reflectance and transmission for light with 800-nm wavelength. Here, we use incident light with 800-nm wavelength because the lead-halide perovskite $CH_3NH_3PbI_3$ shows luminescence at ~1.5 eV [12]. For simplicity, we neglect the optical absorption here. The reflectance of the flat interface increases from 0.05 for normal incidence up to ~0.3 with increasing incident angle, which means that the light with a large incident angle is reflected at the interface and is not transmitted to the c-Si sub-cell. Because of the enhanced reflectance, the transmission decreases with increasing incident angle, resulting in the reduced transmission of luminescence from the top perovskite layer. Note that the GaAs/Si interface exhibits a small index difference, resulting in small reflectance even for a large incident angle. In contrast, the nanostructured interface exhibits an enhanced transmittance even for light with a large incident angle, which implies that the nanostructured interface can improve the luminescent coupling. Consequently, the enhanced coupling efficiency may increase current generation in the bottom Si sub-cell through the luminescent coupling from the top perovskite layer.

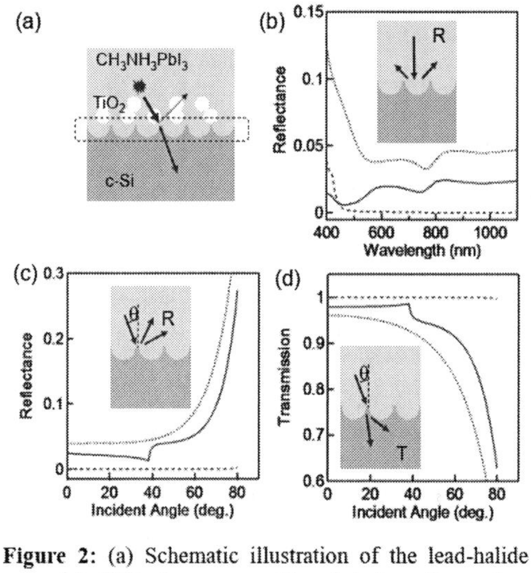

Figure 2: (a) Schematic illustration of the lead-halide perovskite/TiO_2 layer on c-Si substrate with surface nanostructures. (b) Reflection spectra for normally incident light in at the perovskite/c-Si interface with (solid curve) and without a nanostructured interlayer (broken curve). (c) Calculated effective angle-dependent Fresnel coefficients for the reflection $R_i(\theta)$ and (d) transmission $T_i(\theta)$ at the perovskite/c-Si interface with and without a nanostructured interface for light with the 800- nm wavelength light. Broken curves show these results for a GaAs/c-Si interface.

3 SIMULATION OF LUMINESCNET COUPLING

To investigate systematically the optical properties of the nanostructured interface, we calculated the effective Fresnel coefficients for reflection $R = R(\lambda,\theta)$ and transmission $T = T(\lambda,\theta)$ with varying shape of the nanostructured interface, where λ and θ are the light wavelength and angle of incident light, respectively. Note that in single-junction solar cells, the probability densities of escape from the front, P_{esc}, and of reabsorption, P_{abs}, are determined by the effective-energy- and angle-dependent Fresnel coefficients for specular reflection and transmission at the front and back interfaces [7].

Figure 3(a) shows the angle-dependent transmission for the perovskite/c-Si interface with nanostructured interfaces of varying period (P). Here, we used triangle textures instead of half-circle textures to vary the texture height and period systematically. The height was set to 1 μm. As shown in the figure, transmission at a larger incident angle increases with decreasing period. This indicates that efficient luminescent coupling is achieved for shorter-period nanostructures. Figure 3(b) shows the transmission with varying height for a period of 100 nm. The transmission for a wide incident-angle range increases with increasing nanotexture height, which implies that the insertion of a nanostructured interlayer between the sub-cell with index mismatch can improve the optical coupling between the sub-cells and that a large height is required for efficient luminescent coupling.

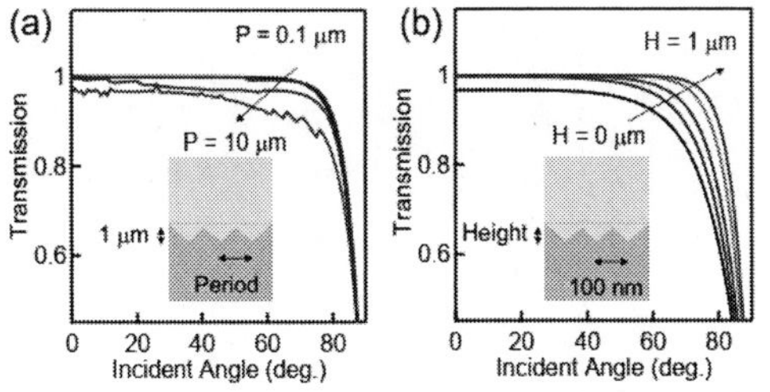

Figure 3: Angle-dependent transmission for the perovskite/c-Si interface with and without nanostructured interfaces having (a) a height of 1 μm and different periods and (b) period of 100 nm and different heights.

Based on the angle-dependent Fresnel spectra, we discuss the luminescent coupling effect from the perovskite sub-cell to the c-Si sub-cell. Here, we estimated the luminescent coupling efficiency in the perovskite/c-Si solar cell with the nanostructured interface. To evaluate the transmission efficiency of the luminescence at the top layer into the bottom layer, we calculated the transmission efficiency for incident light with random distribution [Fig. 4(a)], T, as

$$T = \int_0^{\pi/2} T(\theta)\sin\theta d\theta \tag{1}$$

where θ is the angle of incidence. From Eq. (1), we obtained a transmission efficiency of 0.81 for the flat perovskite/c-Si interface without a nanostructured interface. This result indicates that a part of the luminescence from the perovskite layer is not transmitted into the c-Si layer.

Figure 4(b) summarizes the calculated transmission efficiency as a function of period P for different nanostructured interfaces of height H (= 0, 0.1, 0.2, 0.5, 1, 2, 5, and 10 μm). The black circles and broken line indicate the results for the nanostructured interface with the same height as period, i.e., $H = P$. For an elongated nanotextured interface with height greater than the period ($H > P$), the transmission efficiency for luminescence increases with nanostructure height. In the case of shallow nanostructures ($H < P$), transmission efficiency decreases with increasing period. Thus, the interface structure can be designed to increase the optical coupling between the perovskite layer and c-Si cell.

To understand the mechanism of enhanced transmission efficiency of luminescence, the transmission efficiencies are compared as a function of the surface angle and height. Figure 4(c) shows the transmission efficiency as a function of the tilt angle $\Theta = \tan^{-1}(H/P)$ for different nanostructure periods. The transmission efficiency for larger tilt angles depends on the nanostructure height, which means that the transmission efficiency is not simply governed by ray optics. In ray optics regime, transmission efficiency is determined only by the surface tilt angle and is independent of the nanostructure height.

Figure 4(d) shows the transmission efficiency as a function of the nanostructure height. For all the nanostructure periods ranging from 0.1 to 50 μm, the transmission efficiency increases with nanostructure height. This tendency can be understood based on the effective medium approximation, in which a nanostructure of larger height shows a more gradual change of refractive index from low- to high-index materials. This result shows that the optical properties are primarily governed by a wave optics mechanism, rather than by ray optics. Furthermore, a coupling efficiency greater than 90% is obtained for the nanostructured interface with height less than 1 μm, which indicates that the nanostructured interface will be useful to increase the luminescent coupling efficiency in the perovskite/c-Si tandem solar cells.

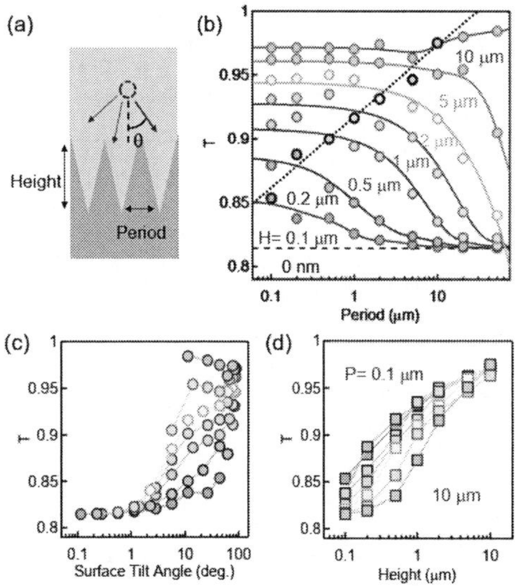

Figure 4: (a) Schematic illustration of transmission efficiency of luminescence. (b) Calculated transmission efficiency as a function of period for interface nanostructures of different heights (H = 0, 100, 200, 500 nm, and 1, 2, 5, 10 μm). (c) Tilt-angle dependence of the transmission efficiency for different nanostructure periods. (d) Height dependence of the transmission efficiency.

In the lead-halide perovskite/c-Si tandem solar cells, the interfaces are not required to be flat. Textured interfaces have not been designed for monolithic multijunction solar cells, because conventional monolithic tandem solar cells require a flat interface. However, recent progress in mechanical stacking may provide the opportunity to use non-flat interfaces with a small textured interface for realizing highly efficient solar cells through luminescent coupling. This approach of using a nanostructured interface to enhance the luminescent coupling will be applied to various types of luminescent tandem solar cells in the future.

4 CONCLUSION

We reported the impact of nanostructured interlayers on optical coupling in perovskite/c-Si tandem solar cells. Short-period, large-height nanostructures at the interface are so effective in reducing the interface reflection for enhancing luminescence, that the transmission efficiency reaches a level greater than 95%. The modification of interlayer structures is a critical light management approach to improve the performance of luminescent tandem solar cells.

Acknowledgment

This work was supported by the Advanced Low Carbon Technology Research and Development Program (ALCA) of the Japan Science and Technology Agency (JST).

References

[1] A. Polman and H.A. Atwater, Nat. Mater. 11 (2012) 174.
[2] E.R. Martins, J. Li, Y. Liu, V. Depauw, Z. Chen, J. Zhou, and T.F. Krauss, Nat. Commun. 4 (2013) 2665.
[3] J. Zhao, A. Wang, P.P. Altermatt, S.R. Wenham, and M.A. Green, Sol. Energy Mater. Sol. Cells 42 (1996) 87.
[4] N. Usami, W. Pan, T. Tayagaki, S.T. Chu, J. Li, T. Feng, Y. Hoshi, and T. Kiguchi, Nanotechnology 23 (2012) 185401.
[5] T. Tayagaki, Y. Kishimoto, Y. Hoshi, and N. Usami, Appl. Phys. Express 7 (2014) 122301.
[6] T. Tayagaki, Y. Hoshi, Y. Kishimoto, and N. Usami, Opt. Express 22 (2014) A225.
[7] M.A. Steiner, J.F. Geisz, I. Garcia, D.J. Friedman, A. Duda, and S.R. Kurtz, J. Appl. Phys. 113 (2013) 123109.
[8] X. Sheng, M.H. Yun, C. Zhang, A.M. Al-Okaily, M. Masouraki, L. Shen, S. Wang, W.L. Wilson, J.Y. Kim, P. Ferreira, X. Li, E. Yablonovitch, and J. A. Rogers, Adv. Energy Mater. 5 (2015) 1400919.
[9] C. N. Eisler, Z. R. Abrams, M. T. Sheldon, X. Zhangbcd, and H. A. Atwater, Energy Environ. Sci. 7 (2014) 3600.
[10] A.S. Brown and M.A. Green, Conf. Rec. Twenty-Ninth IEEE Photovolt. Spec. Conf. 2002. 868 (2002).
[11] Y. Zhao and K. Zhu, Chem. Soc. Rev. 45 (2015) 655.
[12] S.D. Stranks and H.J. Snaith, Nat. Nanotechnol. 10 (2015) 391.
[13] L. Li, J. Opt. Soc. Am. A 14 (1997) 2758.
[14] DiffractMOD http://optics.synopsys.com/rsoft/
[15] Handbook of Optical Constants of Solids, Ed. E. D. Palik (Academic Press, New York, 1985). (1985).
[16] P. Loeper, M. Stuckelberger, B. Niesen, J. Werner, M. Filipic, S.-J. Moon, J.-H. Yum, M. Topic, S. De Wolf, and C. Ballif, J. Phys. Chem. Lett. 6 (2015) 66.

COPPER IODIDE – HOLE SELECTIVE CONTACT FOR THE HOT CARRIER SOLAR CELL

Simon Chung, Robert Patterson, Santosh Shrestha, Gavin Conibeer
School of Photovoltaic and Renewable Energy Engineering, Australian Center for Advanced Photovoltaics,
UNSW Australia, Sydney, New South Wales 2052, Australia

ABSTRACT: In a hot carrier solar cell the collection of hot carriers requires electrical contacts with a narrow energy pass bandwidth, known as energy selective contacts to minimise energy lost as heat via the increase in entropy at the contacts. However due to the large effective mass of holes compared to electrons majority of the energy from absorbed photons are taken up by electrons, 94.3% in the case for an InAs absorber. Therefore a carrier selective contact can be used rather than an energy selective contact for the collection of holes in a hot carrier solar cell with minimal energy loss from the hot hole population cooling. Copper iodide (CuI) has been identified as a material which can potentially be used to collect holes. Preparation of CuI films is done by solution processes, dissolving CuI powder in a solvent and spin coating on the substrate. The material properties such as optical properties, crystal structure and film morphology are studied in the context of its suitability for use as an electron blocking and hole collecting layer for the hot carrier solar cell.
Keywords: CuI, hole collector, hot carriers

1 INTRODUCTION

Hot carrier solar cells (HCSC) [1] aim to circumvent the Shockley-Queisser limit [2] by significantly reducing the energy loss due to carrier thermalisation. With the reduction of carrier thermalisation losses, materials with narrow and at the limit with no bandgap can be used as absorbers in a photovoltaic device. The HCSC operates by collecting hot carriers before majority of its energy lost through thermalisation processes. The collection of hot carriers requires electrical contacts with a narrow energy pass bandwidth to minimise energy lost as heat via the increase in entropy at the contacts [3]. The fabrication of these energy selective contacts (ESC) can be challenging for example energy selectivity can be achieved via the use of resonant tunneling through nanometer thin layers of materials creating a double barrier quantum well structure [4]. Due to the large effective mass of holes compared to electrons majority of the energy from absorbed photons are taken up by electrons, 94.3% in the case for an InAs absorber [5]. The complexity of a HCSC is reduced by only having one ESC and utilising a hole collecting contact rather than an ESC for the extraction of holes forgoing the thermal energy in the hot hole population.

The HCSC operates in principle by thermal diffusion of the hot carriers out of the absorber through the contacts. A hole collecting contact for the hot carrier solar cell must be able to collect holes while also blocking electrons from diffusing into the contact. Materials being investigated for use as p-type contacts in various organic and perovskite based solar cell technologies rely on materials with a high electron affinity. These materials such as MoO_x and spiro-OMeTAD are materials with high electron affinity allowing holes to be collected by a low laying electron band. However for the HCSC electrons will also be able to be collected by these materials. A p-type hole collecting material with a low electron affinity and also with a wide bandgap would be able to function as such a material. The wide bandgap block electrons from being collected by the material while holes are collected by within the valence band of the material.

Copper iodide has been identified as a material which can potentially be used as a hole contact for HCSCs. CuI is a p-type semiconducting material with a bandgap reported to be 2.9 - 3.1 eV depending on synthesis methods. It also has a low electron affinity, with reports between 1.9 – 2.1 eV [6,7]. The combination of the wide bandgap and low electron affinity allow holes to be collected via the valence band while blocking electrons due to the large bandgap. For some applications the bandgap of around 3 eV may be sufficient to be used as a p-type transparent window layer for solar cells. Figure 1, shows an illustration of the band alignment of a HCSC utilising CuI for a hole contact.

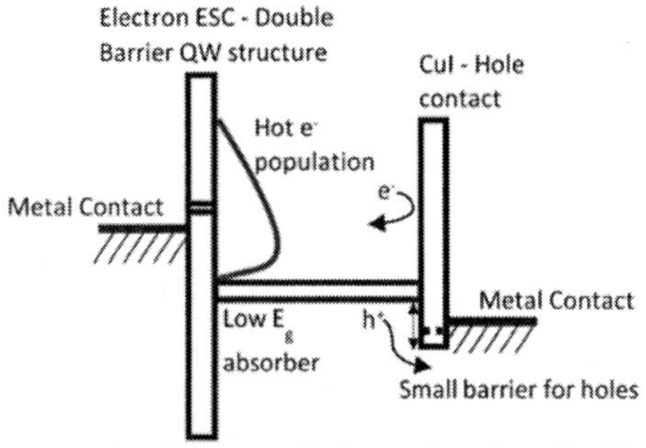

Figure 1: Band diagram of a hot carrier solar cell with an energy selective contact for hot electron collection and CuI as a hole contact.

In such a device as illustrated in Figure 1, the hot electrons generated from absorption of photons are collected by an ESC such as from double barrier resonant tunneling type contact [4]. The hot holes which contain less thermal energy due to its relatively high effective mass would have a distribution closer to the band edge of the absorber. The hot holes will be collected by thermionic emission over a hetereojunction with the hole collector. The barrier height between the valence bands in the hetereojunction is a crucial aspect for its ability to efficiently collect holes. This type of hole contact will allow hot holes of energies above the barrier height to be collected unlike with the ESC carriers with energies above the collection energy are reflected back into the absorber to minimise entropy generation.

Work reported here is the initial preliminary work in optimising the spin coating process for good uniform CuI films to be used in a device. Subsequently CuI heterojunctions with varying doped Si substrates will be used to investigate the hole collecting and electron blocking ability of CuI and its feasibility to be used as a hole contact for the HCSC.

2 EXPERIMENTAL

The deposition of CuI was performed with solution processing. A CuI solution of was prepared by dissolving from 16 - 20 mg/mL of CuI powder (>99.5%, Sigma-Aldrich) in acetonitrile and stirred for 3 hours at room temperature. CuI films were spin coated onto quartz substrates with various spinning speeds from 500 rpm to 4000 rpm and annealed for 10 mins on a hot plate at 120 °C to allow the solvent to dry. Optical spectroscopy was used to study the optical bandgap and the transmissivity of the films deposited on quartz substrates. X-ray diffraction was used to study the crystallography of the CuI film. CuI films were characterised with scanning electron microscopy for studying the film morphology. From X-ray reflectivity measurements the CuI film thickness is estimated to be around 13 nm when spun at 1500 rpm and faster.

3 RESULTS

3.1 Optical spectroscopy

The transmittance and reflectance of the CuI thin films deposited on quartz were measured using a 150 mm integrating sphere with a PerkinElmer LAMBDA1050 UV/Vis/NIR spectrophotometer. The absorption was calculated from 1-*R*-*T*, where *R* is the measured reflectance and *T*, transmittance.

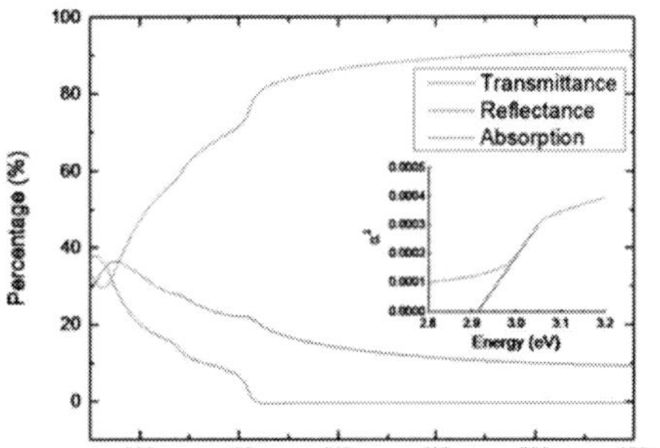

Figure 2: Measured transmittance, reflectance and absorption of a CuI film on a quartz substrate with optical spectroscopy. Inset, Tauc plot for determining the optical bandgap of the CuI film.

Figure 2 shows the measured transmittance, reflectance and absorption of a typical CuI film. From optical measurements the film is found to be transparent at wavelengths from around 430 nm and longer. The high tranmisivity in the visual spectrum from 430 nm and longer leads it to be potentially used as a transparent p-type conducting window layer. The bandgap for a direct semiconductor can be estimated by the Tauc plot where the relation $\alpha^2 = C(h\nu - E_g)$ is plotted. E_g is the material's bandgap, $h\nu$ is the photon energy, C is a constant and α is the absorption coefficient was calculated by $\alpha = -\ln T/d$, where d is the film thickness and T the transmittance as measured from the spectrometer. By extrapolating α^2 to the x-axis E_g, can be determined. Utilising a Tauc plot as shown in the inset of Figure 2, an optical bandgap of 2.9 eV was calculated which is lower than the previously quoted bandgap value of 3.1 eV [7,8].

3.2 X-ray Diffraction

The crystallinity of the deposited films was studied using x-ray diffraction (XRD) with a PANalystical Empyrean X-ray diffractometer. Figure 3 shows a ω-2θ scan of the CuI film. Sharp CuI(111) and Cu(222) peaks are visible located at 25.5° and 51.5° 2θ. The broad peak around 22.3° is associated with the substrate. The strong peak of the CuI crystallities in the [111] directions show the film primarily crystallizes in the [111] direction. Glancing incidence XRD was also performed. It showed no features, providing further evidence majority of the crystallites in the CuI films are orientated in the [111] direction. A lattice constant of 6.04 Å was calculated from Bragg's law utlising the CuI(111) peak location. The lattice constant agrees closely with that from the literature value of 6.088 Å [9]. Depositions made with solutions with a higher CuI concentration such as 20 mg/mL and with slow spin speeds of less than 1000 rpm addition cystal orientations such as CuI(220) and CuI(311) are observed in the XRD measurements. Films deposited from a CuI solution with a lower concentration such as around 16 mg/mL produced films with better crystallinity.

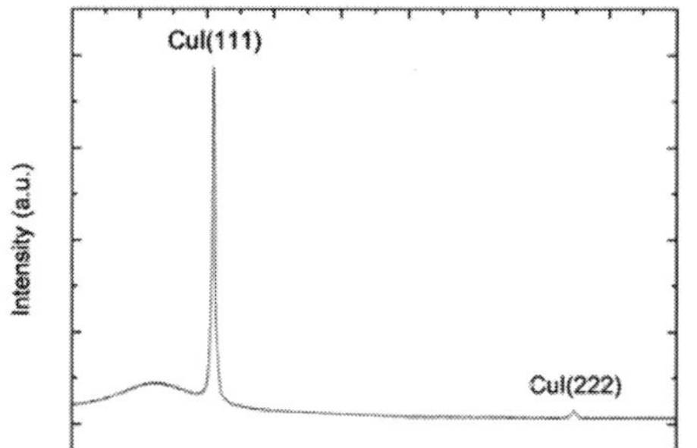

Figure 3: XRD ω-2θ scan of CuI on quartz substrates showing sharp CuI(111) and CuI(222) peaks located at 25.5° and 51.5° 2θ.

3.3 Surface Morphology

Scanning electron microscopy was used to investigate the surface morphology of the CuI films. A FEI Nova NanoSEM450 was used to perform the electron microscopy. Figure 4 shows a SEM micrograph of a CuI film with deposited with two different spin speeds, 1500 rpm and 4000 rpm. From the micrographs it shows there are lateral features on the order of several hundred nanometers with porous structures between these features. The faster spin speed of 4000 rpm had smaller structures and also wider porous structures. Within the features individual crystallites can be observed. The sizes of these crystallites are several hundreds of nanometers and greater. In these crystallites there are pores around 40 nm and more pronounced with faster spin speeds. The film morphology between samples with different spin speeds shows that with faster spin speeds the CuI crystallites are spread out further apart and with a larger number of pores in the film. The faster spin speeds does not allow sufficient amount of CuI to remain on the substrate leading to poor surface coverage. While films deposited with a slow spin speed such as at 500 rpm leads to films with better surface coverage, however the film uniformity is not as good, also CuI crystallites of different orientations starts to form, which may have detrimental effects on its electrical properties. The step of annealing the sample to dry the solution is crucial, by allowing the film to dry at room temperature leads to large porous structures of several hundred nanometers.

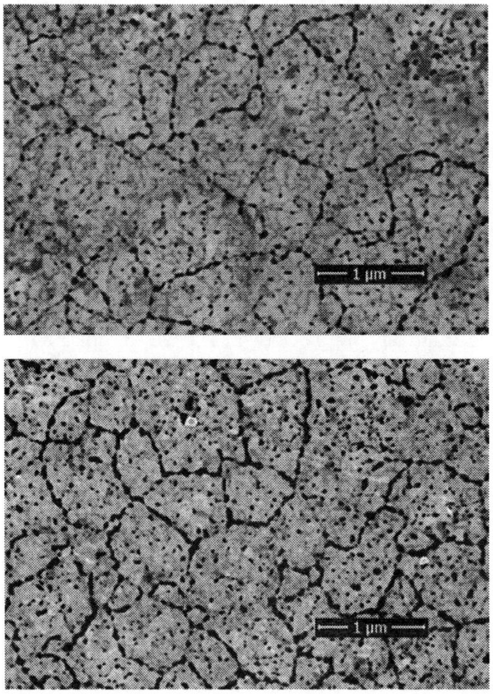

Figure 4: SEM micrographs of CuI film deposited with a concetration ration of 16 mg/mL at spin speeds of 1500 rpm (top) and 4000 rpm (bottom).

4 DISCUSSION

Practical realisation of the HCSC faces many challenges. One of the technical challenges is the requirement of ESCs to maintain a narrow energy selectivity band. For example the fabrication of thin layers required for resonant tunneling requires precise control. The idea of using a hole selective contact and only utilising an ESC for electrons is attractive approach to simplify the HCSC device. CuI which has been investigated as a hole contact for organic and perovskite based photovoltaics may also be utilised as a hole contact for the HCSC. CuI as a hole contact is applicable to for HCSCs because of the low electron affinity and large bandgap which is able to block the transport of the hot electrons out the absorber.

The simplicity depositing CuI thin films via solution processing methods leads it to be a promising material to be used as a hole collector. Preliminary experiments have been performed to study the quality of the films produced by the spin coating method. The peaks observed from XRD of the CuI films shows the film crystallizes easily. However the films are quite thin at around 13 nm and from the SEM micrographs there exist some porous structures in the film. If the CuI films are to be used in devices further optimisation of the deposition process is required for good surface coverage and uniformity to prevent the shunting of the CuI layer.

5 CONCLUSION AND FUTURE WORK

The HCSC device can be simplified at the cost of forgoing the efficient collection of hot holes. The electrical properties and easy of depositing CuI via solution processing methods leads it to be a good material to collect holes from a HCSC. Ongoing work is being performed to CuI material properties such as its electrical properties and its ability to collect holes. Initial devices fabricated to investigate the hole collecting properties of CuI will be made with a CuI/Si hetereojunction. The electrical properties of silicon is well known, CuI/Si devices will be used as a test case for its hole collecting ability.

ACKNOWLEDGEMENTS

'This program has been supported by the Australian Government through the Australian Renewable Energy Agency (ARENA). Responsibility for the views, information, or advice expressed herein is not accepted by the Australian Government. The authors acknowledge the facilities, and the scientific and technical assistance of the Electron Microscope Unit and the Solid State and Elemental Analysis Unit of the Mark Wainwright Analytical Centre, The University of New South Wales.

REFERENCES

[1] G. Conibeer, S. Shrestha, S. Huang, R. Patterson, H. Xia, Y. Feng, et al., Sol. Energy Mater. Sol. Cells. 135 (2014) 124–129.
[2] W. Shockley, H.J. Queisser, J. Appl. Phys. 32 (1961) 510.
[3] Y. Feng, P. Aliberti, B.P. Veettil, R. Patterson, S. Shrestha, M.A. Green, et al., Appl. Phys. Lett. 100 (2012) 053502.
[4] S.K. Shrestha, P. Aliberti, G.J. Conibeer, Sol. Energy Mater. Sol. Cells. 94 (2010) 1546–1550.
[5] D. König, Y. Takeda, B. Puthen-Veettil, Appl. Phys. Lett. 101 (2012) 153901.
[6] S.A. Mohamed, J. Gasiorowski, K. Hingerl, D.R.T. Zahn, M.C. Scharber, S.S.A. Obayya, et al., Sol. Energy Mater. Sol. Cells. 143 (2015) 369–374.
[7] K. Zhao, G.O. Ngongang Ndjawa, L.K. Jagadamma, A. El Labban, H. Hu, Q. Wang, et al., Nano Energy. 16 (2015) 458–469.
[8] S. Inudo, M. Miyake, T. Hirato, Phys. Status Solidi. 210 (2013) 2395–2398.
[9] D. Huang, Y.-J. Zhao, S. Li, C.-S. Li, J.-J. Nie, X.-H. Cai, et al., J. Phys. D. Appl. Phys. 45 (2012) 145102.

ZNTE:O INTERMEDIATE BAND MATERIAL PREPARATION BY MOLECULAR BEAM EPITAXY AND ION IMPLANTATION

Cai Liu*†1, Nan Tang†2, Aobo Ren†3, Wei Li†4, Lili Wu†5, Jingquan Zhang†6, Bing Li†7, Lianghuan Feng†8
*corresponding author
†Institute of Solar Energy Materials and Devices, College of Materials Science and Engineering, Sichuan University, No.24 South Section 1, Yihuan Road, Chengdu, China, 610064
*Tel: +86 (0) 28 8541 2542
Email: 1 liucaicn@foxmail.com, 2 15884550342@163.com, 3 182478747@qq.com, 4 waylee2000@sohu.com, 5 wulily77@163.com, 6 zh_jq2000@263.net, 7 libing70@126.com, 8 lh.feng@263.net

ABSTRACT: Highly mismatched alloy of ZnTe:O for intermediate band solar cell application was conventionally prepared by in-situ molecular beam epitaxy alloying, or ion implantation into ZnTe single crystal wafer. In this study, an unique method to realize ZnTe:O intermediate band material was employed, combing ZnTe molecular beam epitaxy and subsequent oxygen ion implantation and final laser pulse ablation. Oxygen band related emissions had been observed from laser pulse annealed ZnTe:O films. Poly-crystalline ZnTe also showed potential for intermediate band solar cell application.
Keywords: ZnTe:O, poly-crystalline, intermediate band, molecular beam epitaxy, ion implantation

1 INTRODUCTION

The concept of intermediate band (IB) solar cells is very attractive and promising since it was originally proposed in 1960s[1] and developed in recent decades[2]–[6]. As its ideal limit efficiency could be as high as 63% under concentrated sunlight[6], it has been intensively studied in recently years, both theoretically and experimentally[7]. There are many feasible approaches to realizing intermediate band and ZnTe:O is a representative material system of the approach of highly-mismatched alloy (HMA)[7], [8]. The photoluminescence through intermediate band in ZnTe:O was observed and the photovoltaic device based on it was demonstrated firstly by Wang et al [9], [10]. After that there were very few experimental reports about ZnTe:O based IB solar cells[11]–[15]. Generally, two methods had been applied for material preparation in existing studies on ZnTe:O. One of them was co-alloying through molecular beam epitaxy (MBE) to form ZnTe:O HMA [16], which is the method for demonstrating solar cells. The other method was employing ion implantation (IIP) to single crystal ZnTe wafers, for the observation of intermediate band absorption phenomena as well as device demonstration. The former is not easy to achieve high crystal quality while the latter is not suitable for large area application. Both of them are high-cost, which is of no problem for research but not for scaled production in the future.

In this study we prepared ZnTe:O IB materials by utilizing the advantages of the two methods above, by firstly growing crystalline film via MBE and then doping it via IIP and finally annealing the oxygen implanted film using laser pulse ablation. There are several merits of this new approach. The first is that the oxygen was incorporated into ZnTe by implantation under non-equilibrium condition and would be free from the limit of solid solubility and oxygen concentration can be tuned flexibly. Moreover, the laser pulse ablation afterward is also of great freedom to activate the implanted oxygen, modulate the properties of IB and recover the partial destroyed crystal lattice as well.

2 EXPERIMENTAL

Intermediate band material of ZnTe:O was prepared by combing MBE and IIP. Poly-crystalline ZnTe matrix material with deliberate low crystal quality was grown under non-optimal MBE growth conditions on epi-ready GaAs(100) substrate. The growth procedures and conditions are summarized in Table 1. Prior to loading to growth chamber, GaAs substrate was baked at 400 °C for 4 hours for surface desorption and 650 °C for 30 min to eliminate surface oxidation. The base pressure and growth pressure were 7×10^{-8} Pa and 2×10^{-7} Pa respectively. Zn source was pre-flowed for 10 s to avoid possible reaction between Te and GaAs. Growth time for all samples was

Table 1: Sample preparation procedures and conditions.

Sample	Step 1 Substrate treatment	Step 2 Growth		Step 3 Oxygen ion implantation	Step 4 Pulse-laser annealing
		Common conditions	Te source temperature		
A	Desorption at 400 °C for 4 h; Oxide removal at 650 °C for 30 min.	2×10^{-7} Pa; $T_{substrate}$=330 °C ; T_{Zn}=260 °C ; Zn pre-flow 10 s.	290 °C	See Table 2.	KrF laser at 248 nm; pulse width 20 ns; energy of 90 mJ/pulse; energy density 0.4 J/cm^2。
B			300 °C		
C			310 °C		

Table 2: Oxygen ion implantation procedures and conditions.

Implantation sequence	Energy (keV)	Depth of peak concentration (nm)	Dose (cm^{-2})		
			S1	S2	S3
Step 1	10	10	3.44×10^{14}	1.72×10^{15}	3.44×10^{15}
Step 2	25	50	4.99×10^{14}	2.50×10^{15}	4.99×10^{15}
Step 3	69	125	1.11×10^{15}	5.54×10^{15}	1.11×10^{16}
O atom concentration (cm^{-3})			2.50×10^{20}	1.25×10^{21}	2.50×10^{21}

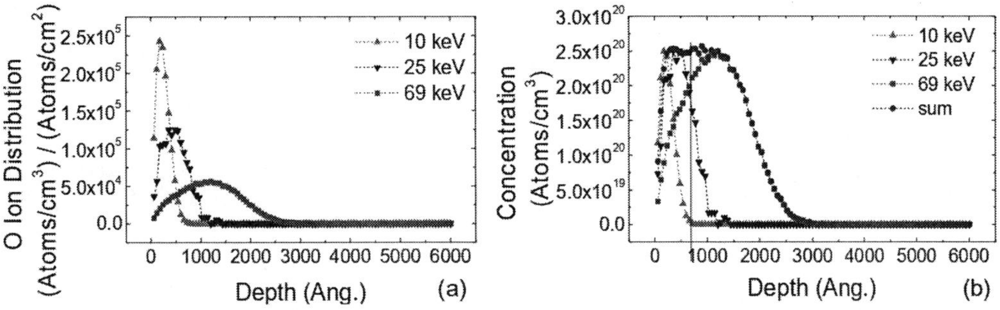

Figure 1: SRIM simulation of implanted oxygen concentration profile in ZnTe.

fixed to be 90 min. As shown in Table 1, the cell temperature of Te source was varied to tune the quality of ZnTe. After growth, oxygen ion implantation was applied to form ZnTe:O which was followed by pulse-laser annealing (PLA) to improve the crystal quality and enhance intermediate band formation. X-ray diffraction (XRD) and scanning electron microscopy (SEM) were employed to evaluate the crystal quality of ZnTe films. PL was conducted to observe the possible emission which intermediate band might be involved.

In order to achieve uniform oxygen concentration distribution, varied ion energy with corresponding does was implanted following the sequence from low to high energy, as shown in Table 2 and Figure 1. Oxygen ion implantation and its distribution in ZnTe was simulated using the software of The Stopping and Range of Ions in Matter (SRIM)[17].

3 RESULTS

3.1 ZnTe crystal information

The crystal quality of as-grown ZnTe films was characterized by XRD, as indicated in Figure 2. Since there was only the peak of ZnTe (200) and (400) in ω-2θ scanning from 10 to 70 degrees of 2θ, sample A had the relatively best quality compared with sample B and C. There were some peaks of other planes like ZnTe (111) emerged in ω-2θ scanning of sample B and C, which was

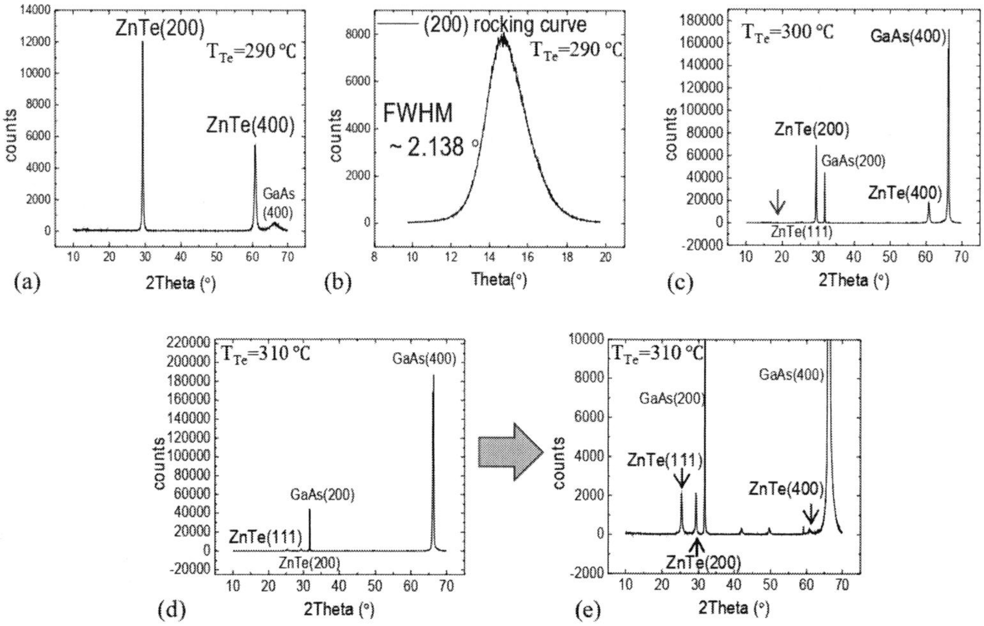

Figure 2: (a) ω-2θ scanning and (b) rocking curve of sample A, (c) ω-2θ scanning of sample B, (d) and (e) ω-2θ scanning of sample C.

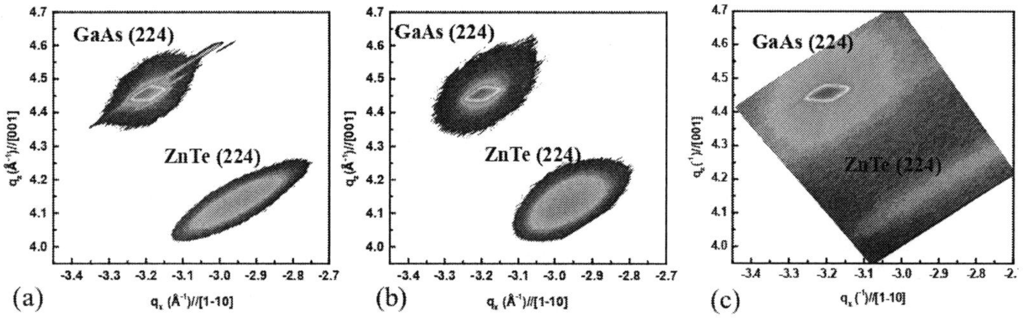

Figure 3: RSM of (a) sample A before oxygen implantation, (b) sample A after oxygen implantation (c) sample B after oxygen implantation.

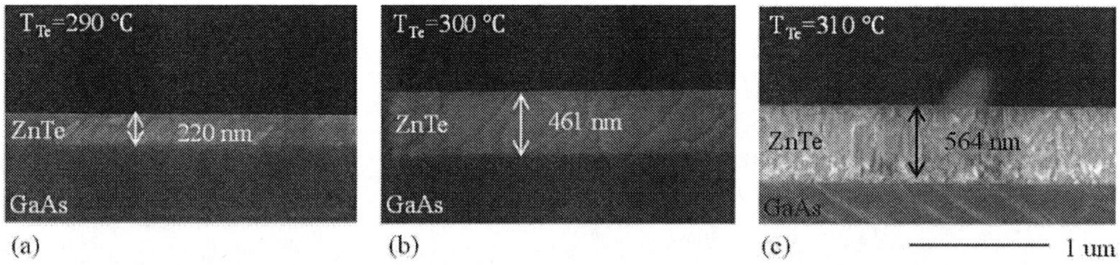

Figure 4: SEM of (a) sample A, (b) sample B and (c) sample C without oxygen implantation.

an indication of two major facet orientation to (200) and (111). Despite tending to be single crystal, the quality of sample A was very low with full-width half of half maximum (FWHM) of 2.138°.

Reciprocal space mapping (RSM) of (224) plane of both ZnTe film and GaAs substrate was performed for sample A before and after oxygen implantation and for sample B after implantation, to check the epitaxy of ZnTe on GaAs as well as the influence of implantation on crystal quality, as illustrated in Figure 3. Firstly the lattice constant of ZnTe (100) films of all measured samples was calculated from the RSM results, the in-plane and out-of-plane values were around 6.000 Å and 6.085 Å respectively, both of which were very close to its bulk value of 6.101 Å. This proved that ZnTe films grown GaAs(100) substrate were almost fully relaxed and ZnTe films more likely tended to be poly-crystal but not single-crystal. From Figure 3a and Figure 3b, the RSM spot of ZnTe (224) in sample A became bigger after oxygen IIP, which is a piece of evidence that IIP worsens the crystal quality of ZnTe matrix. Comparison between Figure 3b and Figure 3c showed that the crystal quality of implanted sample B was much worse than implanted sample A, which is consistent with the conclusion of XRD scanning. SEM images of cross section in Figure 4 give more intuitive impression of the crystallinity of all the samples without implantation. The growth rate varied drastically as the Te source temperature was elevated from 290 °C to 310 °C. In the same 90 min, thickness of sample A to C was 220 nm, 461 nm and 564 nm respectively, which means the growth rate of sample B and C was 2 and 2.5 times of that of sample A. More importantly, the morphology of cross section of sample A is well single-crystalline and that of sample C shows likely poly-crystallinity.

3.2 Photoluminescence of ZnTe:O

Photoluminescence characterization of pulse-laser annealed ZnTe:O was performed at room temperature to detect the possible luminescence phenomena related to intermediate band in ZnTe:O. The excitation wavelength of laser is 371 nm. The typical oxygen emission band in ZnTe:O in the range from 1.7 eV to 2.0 eV[18], [19] was observed while the near band edge emissions disappeared in all samples, which evidenced that oxygen has been incorporated into ZnTe matrix effectively, despite of varied ZnTe crystal quality. Since there is very few published data on PL of ZnTe:O prepared by combing MBE-ZnTe and oxygen IIP, thus here the result is compared with the study on MBE co-alloyed ZnTe:O. Reported oxygen content in ZnTe via MBE co-alloying typical ranged from 0.04% to 1.34%[20]. If oxygen concentration is as high as approximately 1%, the oxygen emission band would be negligible and ZnTe band edge emission is strong[14]. The oxygen band emission position here is nearly the same as the MBE co-alloyed ZnTe:O sample with oxygen content of 0.46% which is corresponding to higher oxygen partial pressure during growth, with not only O-Te substitution but also increased interstitial oxygen compared with samples with higher oxygen content [14]. This mechanism is more likely reasonable in oxygen ion implanted ZnTe. Oxygen atoms are taking interstitial positions with much more possibility during ion implantation than MBE co-alloying. The band edge emission was also disappeared in as-grown samples while weak peaks at P_1 also occurred (not shown here). This may be due to high-density defects or impurities, like the yellow emission in GaN. After oxygen implantation, another emission P_2 emerged which relates to oxygen incorporation.

Another noticeable feature of Figure 5 is, emission intensity of both P1 and P2 was proportional to ZnTe crystal quality, while the thickness was almost doubled and tripled from sample A to B and C respectively. In spite of tending to be poly-crystalline, samples C also yielded similar emissions to that of single-crystalline ones. This is very attractive. The published researches on ZnTe:O intermediate band materials and devices are exclusively based on high-quality MBE-ZnTe epitaxial films or single-crystal ZnTe wafers. The effect of ZnTe matrix crystal

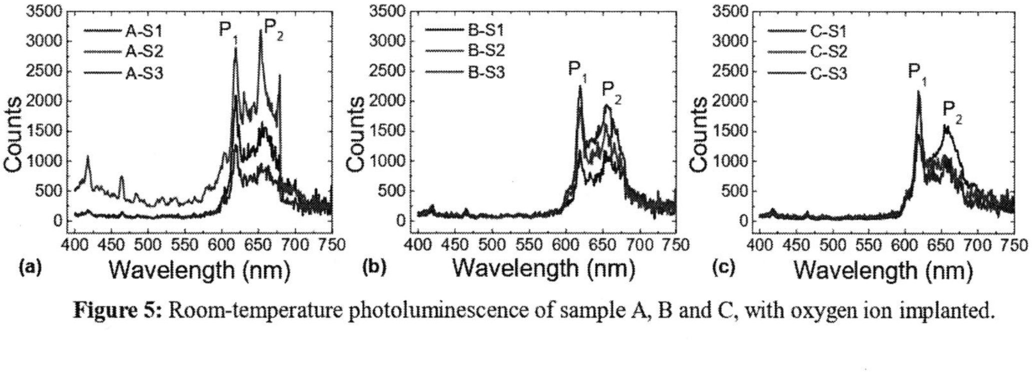

Figure 5: Room-temperature photoluminescence of sample A, B and C, with oxygen ion implanted.

quality on intermediate band formation has never been considered. In our group, great effort of co-alloying oxygen in poly-crystalline ZnTe, for example using radio frequency magnetron sputtering, turned out to be hardly achievable. This is the first report of oxygen band emission from oxygen implanted poly-crystalline ZnTe, which showed that poly-crystalline ZnTe is also suitable for intermediate band material fabrication. This is encouraging that the combination of independent ZnTe deposition and following oxygen ion implantation emerges to be an effective approach to intermediate band material, which is beneficial to future cost reduction.

4 CONCLUSION

A new strategy of using MBE to grown ZnTe films with following oxygen ion implantation and pulse-laser annealing, which is of great intermediate band modulation freedom, was adopted to prepare ZnTe:O intermediate material in this study. Effects of crystal quality of ZnTe matrix on intermediate band formation in ZnTe:O have been considered in this study. Oxygen band emissions had been observed from all of the samples, even from the sample which was tending to be poly-crystalline. This suggested that high-quality poly-crystalline ZnTe may be also suitable for intermediate band material preparation, and the combination of independent ZnTe film deposition and ion implantation later on is a promising approach to achieve it.

5 REFERENCES

[1] M. Wolf, "Limitations and Possibilities for Improvement of Photovoltaic Solar Energy Converters: Part I: Considerations for Earth's Surface Operation," *Proc. IRE*, vol. 48, no. 7, pp. 1246–1263, Jul. 1960.

[2] K. W. J. Barnham and G. Duggan, "A new approach to high‐efficiency multi‐band‐gap solar cells," *J. Appl. Phys.*, vol. 67, no. 7, pp. 3490–3493, Apr. 1990.

[3] M. J. Keevers and M. A. Green, "Efficiency improvements of silicon solar cells by the impurity photovoltaic effect," *J. Appl. Phys.*, vol. 75, no. 8, pp. 4022–4031, Apr. 1994.

[4] A. Luque and A. Martí, "Increasing the Efficiency of Ideal Solar Cells by Photon Induced Transitions at Intermediate Levels," *Phys. Rev. Lett.*, vol. 78, no. 26, pp. 5014–5017, Jun. 1997.

[5] A. Luque and A. Martí, "A metallic intermediate band high efficiency solar cell," *Prog. Photovolt. Res. Appl.*, vol. 9, no. 2, pp. 73–86, Mar. 2001.

[6] A. Luque and A. Marti, "PHOTOVOLTAICS: Towards the intermediate band," *Nat. Photonics*, vol. 5, no. 3, pp. 137–138, Mar. 2011.

[7] Y. Okada, N. J. Ekins-Daukes, T. Kita, R. Tamaki, M. Yoshida, A. Pusch, O. Hess, C. C. Phillips, D. J. Farrell, K. Yoshida, N. Ahsan, Y. Shoji, T. Sogabe, and J.-F. Guillemoles, "Intermediate band solar cells: Recent progress and future directions," *Appl. Phys. Rev.*, vol. 2, no. 2, p. 021302, Jun. 2015.

[8] W. Walukiewicz, W. Shan, K. M. Yu, J. W. Ager, E. E. Haller, I. Miotkowski, M. J. Seong, H. Alawadhi, and A. K. Ramdas, "Interaction of Localized Electronic States with the Conduction Band: Band Anticrossing in II-VI Semiconductor Ternaries," *Phys. Rev. Lett.*, vol. 85, no. 7, pp. 1552–1555, Aug. 2000.

[9] W. Wang, A. S. Lin, and J. D. Phillips, "Intermediate-band photovoltaic solar cell based on ZnTe:O," *Appl. Phys. Lett.*, vol. 95, no. 1, p. 011103, Jul. 2009.

[10] W. Wang, A. S. Lin, J. D. Phillips, and W. K. Metzger, "Generation and recombination rates at ZnTe:O intermediate band states," *Appl. Phys. Lett.*, vol. 95, no. 26, p. 261107, Dec. 2009.

[11] T. Tanaka, K. M. Yu, P. Stone, J. W. Beeman, O. Dubon, L. A. Reichertz, V. M. Kao, M. Nishio, and W. Walukiewicz, "DEVELOPMENT OF ZNTE(1-X)O(X) INTERMEDIATE BAND SOLAR CELLS," in *35th Ieee Photovoltaic Specialists Conference*, 2010, pp. 2974–2977.

[12] T. Tanaka, T. Mochinaga, K. Saito, Q. Guo, M. Nishio, K. M. Yu, and W. Walukiewicz, "Synthesis and optical properties of ZnTe1-xOx highly mismatched alloys for intermediate band solar cells," in *2012 38th IEEE Photovoltaic Specialists Conference (PVSC)*, 2012, pp. 000109–000112.

[13] Z. Kang, G. Ran, Y. Jian-Dong, G. Shu-Lin, R. Fang-Fang, Z. Shun-Ming, H. Shi-Min, T. Kun, T. Dong-Ming, Y. Yi, Z. Rong, and Z. You-Dou, "Effect of

oxygen implantation on microstructural and optical properties of ZnTe:O intermediate-band photovoltaic materials," *Acta Phys. Sin.*, vol. 63, no. 23, p. 237103, Dec. 2014.

[14] L. Q. Zhou, C. Chen, H. Jia, C. Ling, D. Banerjee, J. D. Phillips, and Y. Wang, "Oxygen Incorporation in ZnTeO Alloys via Molecular Beam Epitaxy," *J. Electron. Mater.*, vol. 43, no. 4, pp. 889–893, Jan. 2014.

[15] J. Kim, W.-J. Yin, J. Kang, Y. Yan, S.-H. Wei, and M. M. Al-Jassim, "Creating intermediate bands in ZnTe via co-alloying approach," *Appl. Phys. Express*, vol. 7, no. 12, p. 121201, Nov. 2014.

[16] J. Phillips, A. Teran, C. Chen, E. Antolin, I. Ramiro, E. Lopez, E. Hernandez, I. Artacho, C. Tablero, A. Marti, and A. Luque, "Intermediate band solar energy conversion in ZnTeO," in *Photovoltaic Specialists Conference (PVSC), 2013 IEEE 39th*, 2013, pp. 1640–1643.

[17] "James Ziegler - SRIM & TRIM." [Online]. Available: http://www.srim.org/. [Accessed: 05-Feb-2016].

[18] C. Chen, J. Zheng, K. Nguy, F. Naab, and J. D. Phillips, "Distinguishing Optical Behavior of Oxygen States and Native Deep Level Emission in ZnTe," *J. Electron. Mater.*, vol. 43, no. 4, pp. 879–883, Jan. 2014.

[19] Y. Nabetani, T. Okuno, K. Aoki, T. Kato, T. Matsumoto, and T. Hirai, "Photoluminescence properties of ZnTeO and ZnSeO alloys with dilute O concentrations," *Phys. Status Solidi C*, vol. 3, no. 4, pp. 1078–1081, Mar. 2006.

[20] T. Tanaka, S. Kusaba, T. Mochinaga, K. Saito, Q. Guo, M. Nishio, K. M. Yu, and W. Walukiewicz, "Molecular beam epitaxial growth and optical properties of highly mismatched ZnTe1−xOx alloys," *Appl. Phys. Lett.*, vol. 100, no. 1, p. 011905, Jan. 2012.

INTEGRATED POWER AND DATA TRANSCEIVER DEVICES FOR POWER-BY-LIGHT SYSTEMS – A CONCEPT STUDY

Henning Helmers#, David Lackner, Gerald Siefer, Eduard Oliva, Frank Dimroth, and Andreas W. Bett

Fraunhofer Institute for Solar Energy Systems ISE, Heidenhofstraße 2, 79110 Freiburg, Germany
#: corresponding author: henning.helmers@ise.fraunhofer.de; phone: +49 761 4588-5094

ABSTRACT: In optical power transmission systems photovoltaic cells are used as power converters for monochromatic light. In case of powering remote sensors, supplementary to the optical power link also bidirectional data transmission between the two sides of the optical link is required. It is appealing to integrate the functionalities of the optical power link as well as the data up- and downlink into one. In this work, different concepts on how this integration and combination of functionalities can be realized are explored and possible system and device architectures for integrated transceiver devices are discussed, including different examples of practical realizations based on advanced semiconductor fabrication techniques.

Keywords: III-V semiconductors, laser power converter, optical power transmission, optical communication

1 INTRODUCTION

Optical power transmission is an elegant way to power sensor electronics in demanding and challenging environments. Application examples for this technology can be found in various domains and are as diverse as structural health monitoring in wind turbines [1, 2], cortical neural sensing in brain implants [3], wireless powering for implantable sensing platforms [4], monitoring of high voltage power lines [5], optically powered video surveillance [6], submarine hydrophone and seismometer networks [7], passive optical networks [8]. Advantages of power-by-light technology, compared with conventional copper wiring, include inherent galvanic isolation, electromagnetic compatibility, magnetic resonance compatibility, lightning protection, explosion protection, and low weight. Central components of the system are a laser as light source and a photovoltaic laser power converter [9]. Typically the optical power is transmitted via an optical fiber. However, even wireless power transmission is feasible.

Since optical power transmission is often used to supply sensor electronics, in addition to the power link bidirectional data transmission between base station and remote sensor is desirable [1-8, 10-16]. To realize a purely optical power and data link, multiplexing[1] of three functionalities at both respective sides (base and remote) are required, namely power link transmitter and receiver (P-Tx, P-Rx), data downlink transmitter and receiver (Dd-Tx, Dd-Rx), and data uplink receiver and transmitter (Du-Rx, Du-Tx), located at base station and remote appliance, respectively (compare Figure 1). P-Rx is typically a photovoltaic (PV) laser power converter. Depending on the required power level and system architecture various laser types can be used as P-Tx, laser diodes are very common; furthermore light emitting diodes (LED) are possible. For Dd-Rx and Du-Rx photo diodes are used (PIN diodes, avalanche diodes). For Dd-Tx and Du-Tx both LEDs and vertical-cavity surface-emitting lasers (VCSEL) are commonly used.

[1] It is remarked that the term ‘multiplexing’ in this work is referring to the integration of power and up- and downstream data signals, and thus is used somewhat differently than in telecommunication.

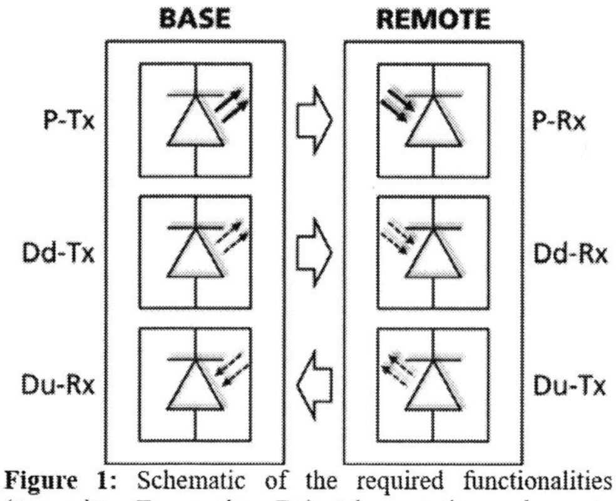

Figure 1: Schematic of the required functionalities (transmitter Tx, receiver Rx) at base station and remote appliance for power link (P-), data downlink (Dd-), and data uplink (Du-). Thick and dashed arrows indicate the power beam and data stream, respectively.

This work explores different possibilities to realize these functionalities on system and device level, and practical ways to implement them are discussed. For implementation combinations of advanced semiconductor architectures and processing tools are considered, such as monolithic multi-junction stacking, bi-facial epitaxial growth, multi-segment lateral separation, mechanical stacking/device bonding, wafer bonding, or the use of quantum wells for bandgap engineering.

2 ARCHITECTURES FOR POWER+DATA

2.1 Space division multiplexing

A straightforward solution for the integration of functionalities is space division multiplexing (SDM). This means that one power link and two data links are realized in three independent channels, namely one optical link connects laser (P-Tx) and PV cell (P-Rx), and one is used respectively for data up- and downlink (Du-Tx/Dd-Tx: VCSEL/LED, Du-Rx/Dd-Rx: photo diode).

2.2 Wavelength division multiplexing

Especially for long transmission distances, it is

beneficial to couple all channels into the same fiber. That way installation cost and weight related to the fiber itself can be reduced. This can be realized by applying common wavelength division multiplexing (WDM) techniques. That means that for the respective multiplexed transmitter devices different wavelength devices are used (with appropriate receivers), e.g. 830 nm for P-Tx with a GaAs based photovoltaic laser power converter as P-Rx, 1310 nm for data downlink (Dd-Tx, Dd-Rx), and 1550 nm for data uplink (Du-Tx, Du-Rx). Specifically for the bidirectional data link existing optical transceiver solutions can be applied.

2.3 Power beam as carrier for data uplink

To reduce power consumption on the remote side, the data uplink can be realized by taking part of the power beam and reflecting it in a controlled way for the data uplink. In other words, part of the power (down-)link is used as carrier for the data uplink. This can be implemented by the use of optical micro-electro-mechanical systems (MEMS) or micro-opto-electro-mechanical systems (MOEMS), such as optical switches or micro-mirror devices. Thereby, only a fraction of the power that would be required for a VCSEL or LED transmitter is needed to power the micro device. Using a mirror with only small reflectivity the data uplink can be realized continuously without compromising the power link. If a fully reflecting mirror is applied, the multiplexing between power downlink and data uplink can also be realized in the time domain (TDM), e.g. 95% of the time the power beam is directed down to the PV receiver and converted to electricity, and during 5% of the time the power beam is used for the data uplink by modulated back reflection. The back reflection can be coupled into a separate fiber (SDM) to be guided to a photo diode (Du-Rx) at the base station. It can also be coupled back into the fiber of the power downlink and e.g. an optical circulator can be used to direct the data uplink to the Du-Rx.

2.4 Amplitude modulation

An elegant way to integrate power and data downlink is using amplitude modulation, or more precisely adding a modulated signal (data downlink) to the DC power beam. That way, the data stream can be read off directly from the PV cell output by appropriate electronic circuitry, thus it combines P-Rx and Dd-Rx functionalities. This approach can further be combined with WDM. That way a fully optical transceiver with all three desired functionalities at each side can be realized.

3 INTEGRATED TRANSCEIVER DEVICES

3.1 PV cells as data receiver

High bandwidths photo diodes are typically thin p-n junction devices optimized for minimal RC time constants, i.e. maximal cutoff frequencies. On the contrary, PV laser power converters are comparatively thick devices optimized for full absorption of the impinging light; thus compromising response time. A device with the purpose of combining both functionalities, power and data reception (P-Rx, Dd-Rx), needs to be optimized for data rate and bandwidth as well as absorption and light-to-electricity conversion efficiency.

3.2 PV-LED transceivers

Since PV cells and LEDs are in principle the same devices, namely a p-n junction, PV cells used as P-Rx can also be used for data transmission Du-Tx by operation in forward bias, i.e. in electroluminescence mode. On the contrary, an LED as Du-Tx could in principle also be used as photovoltaic receiver Pd-Rx. The emission wavelength of a PV-LED device is determined by the bandgap of the emitting semiconductor material, in the case of a PV cell the absorber material. It is noted that the optimum operating point with respect to wavelength is typically some few tens of nm below the emission wavelength determined by the bandgap [17]. Thus, if this combination is applied e.g. at the remote side, the wavelength difference can be used by WDM techniques to direct the light on the base side with respect to P-Tx and Du-Rx.

3.3 PV-LED transceivers with quantum wells

An interesting idea building on the previous thought is to make use of a quantum well (QW) structure [18, 19]. Here the absorbance properties from the bulk material and performance of the PV under monochromatic light (e.g. in terms of spectral response and output voltage) are only marginally influenced; though, by tuning the QW properties the luminescence spectrum, i.e. the emission peak wavelength, can be tuned to a reasonable degree of freedom to photon energies well below the absorption edge of the bulk material. Consequently, P-Rx, Dd-Rx, and Du-Tx can be realized with a single device: By implementing amplitude modulation on the power link (see Sec. 2.4) power and data downlink are both realized using the bulk PV material as receiver (e.g. 808 nm to 850 nm range laser light for the downlink to a GaAs based device). The data uplink transmitter is realized by modulated forward biasing of the same device to generate the data signal from electroluminescence from the QW (e.g. ≥880 nm for the uplink emitted from GaInAs QW). At the base station wavelength division de-multiplexing can be used to direct the respective optical stream to Du-Rx.

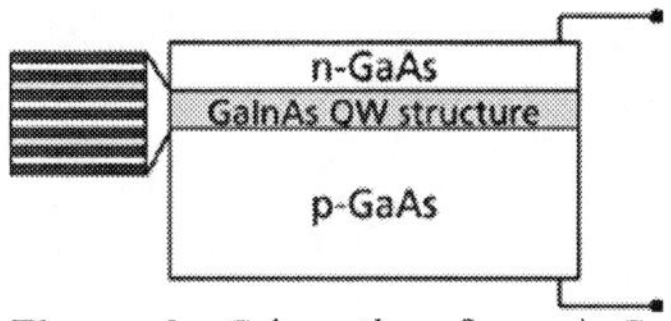

Figure 2: Schematic of a n/p-GaAs PV cell for absorption of e.g. 830 nm laser light (P-Rx+Dd-Rx) with integrated GaInAs quantum well structure for emission of the data upstream on a higher wavelength (e.g. 880 nm) as Du-Tx.

3.3 Vertical integration

By application of advanced semiconductor fabrication techniques for vertical integration, individual functionalities (i.e. LED, VCSEL, PIN photo diode, photovoltaic cell) can also be integrated into a combined device. This can be done monolithically by epitaxial growth of multi-junction devices, eventually added by appropriate etch stop layers, tunnel diodes, absorber and lateral conduction layers, and Bragg reflectors (similar to e.g. monolithically integrated VCSEL-PIN transceivers for optical communication [20]). In the analogous fashion, wafer-bonding of separately grown structures can be applied for vertical integration, eventually combined with multi-junction growth. Moreover, bi-facial epitaxial growth techniques allow for realization of two devices with different functions on the two sides of a single wafer. That way e.g. a higher wavelength Du-Tx can be realized on the rear side of the P-Rx(+Dd-Rx) device. Finally, vertical integration can also be realized on device level by mechanical stacking of individual devices, or by direct device bonding.

Example device architectures are depicted in Figures 3 to 5 and detailed in the respective captions.

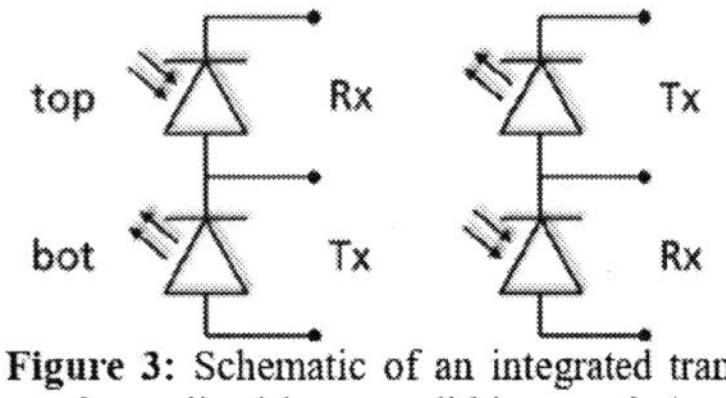

Figure 3: Schematic of an integrated transceiver which can be realized by monolithic growth (vertical stacking) and three terminal contact scheme, e.g. realized by partial selective etching of the top device. Using amplitude modulation Rx can be an integrated power and data downlink receiver. Light is assumed to come in and go out on the top of the figure. Thus, the top device layers must be transparent for the bottom device wavelength (lower bandgap material): $E_{g,bot} < E_{g,top}$. Alternatively, this scheme can also be realized by mechanical stacking of individual devices. Finally, in the same fashion the third functionality can be added to the stack.

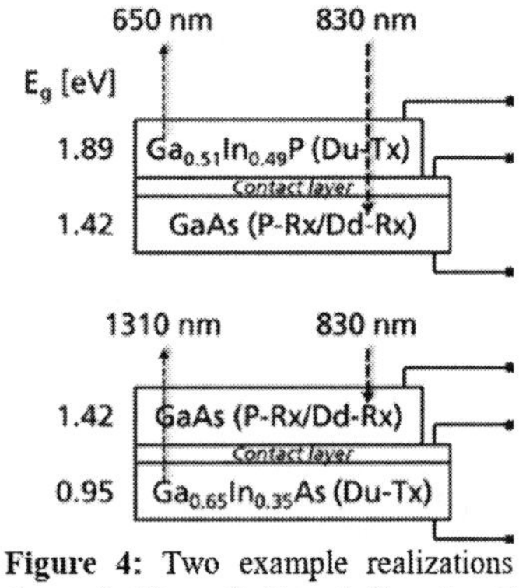

Figure 4: Two example realizations of the approach shown in Figure 2. *Top*: A $Ga_{0.51}In_{0.49}P$ based transmitter Du-Tx (650 nm) can be grown lattice-matched on top of a GaAs based receiver (P-Rx and Dd-Rx with amplitude modulation). The upper Du-Tx as well as the intermediate contact layer are transparent for the 830 nm range downlink beam. The transparent and conductive contact layer is implemented as common terminal for the upper junction's back and lower junction's top contact. It is accessed from the front side of the device. In addition, the bottom layer of the GaInP p-n junction can be of additional use to support lateral conduction and current distribution in the top layer of the GaAs PV cell p-n junction. *Bottom*: A $Ga_{0.65}In_{0.35}As$ based Du-Tx emits at 1310 nm (0.95 eV). Above the combined receiver P-Rx + Dd-Rx (amplitude modulation) is realized based on GaAs for an 830 nm range power and data downlink, which is transparent for the 1310 nm emission. The device can be realized by inverted growth techniques on a GaAs substrate, i.e. first the GaAs receiver is grown upside down, and afterwards the ternary GaInAs transmitter is grown based on a metamorphic approach with a step-graded buffer which can also additionally be used as intermediate lateral conduction and contact layer. The lower bandgap of the Du-Tx compared to the previous example has the advantage that it requires a lower driving voltage for the data uplink.

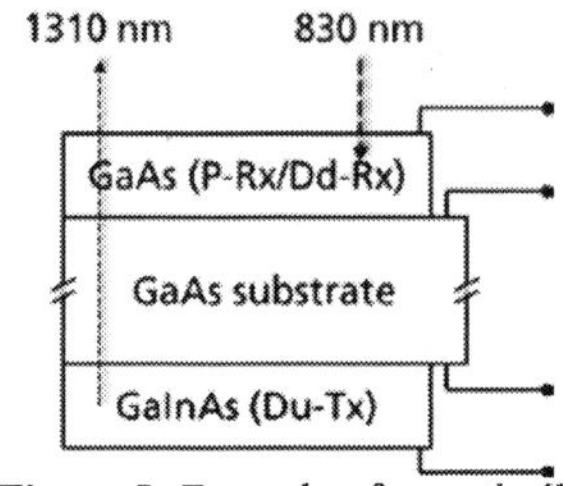

Figure 5: Example of a vertically integrated transceiver based on bi-facial growth. The P-Rx+Dd-Rx is realized as a GaAs based PV cell on a GaAs substrate, well matching a 830 nm modulated power and data downlink. The data uplink is transmitted by a GaInAs based Du-Tx on the rear side of the substrate, which emits at 1310 nm and consequently the substrate and the PV cell are transparent for the light of the data uplink. The transmitter can be realized e.g. as a VCSEL.

3.4 Lateral integration

Multi-segment device architectures (also known as monolithic interconnected modules, MIMs [21-23]) are well known for PV laser power converters [24-28]. The benefit of multi-segment PV converters is that due to the series connection higher output voltages can be delivered at device level – as strongly desirable for efficient power management for reliable power supply of sensor electronics at e.g. a constant voltage of 5 V DC. The multi-segment concept is based on epitaxial growth of the semiconductor structure on a semi-insulating substrate – as an alternative solution an appropriate blocking diode that suppresses current flow from the active layers to a conductive substrate could be implemented inside the structure. The active layers are laterally split into electrically separated segments by isolation trenches. Interconnection of the segments and/or contacting of individual segments (both +/- terminals) are realized on the front side by a combination of a structured dielectric and front side metallization.

Based on the multi-segment approach, all desired functionalities can be integrated into a single device, namely by using a single p-n junction structure as (multi-segment) PV laser power converter (P-Rx), in reverse bias as photo detector (Dd-Rx), and in forward bias as LED (Du-Tx). It is noted that the combination of PV cell (P-Rx) and LED (Du-Tx) has been proposed previously [29]. Figure 6 shows three examples as schematic drawings of how this approach could be realized.

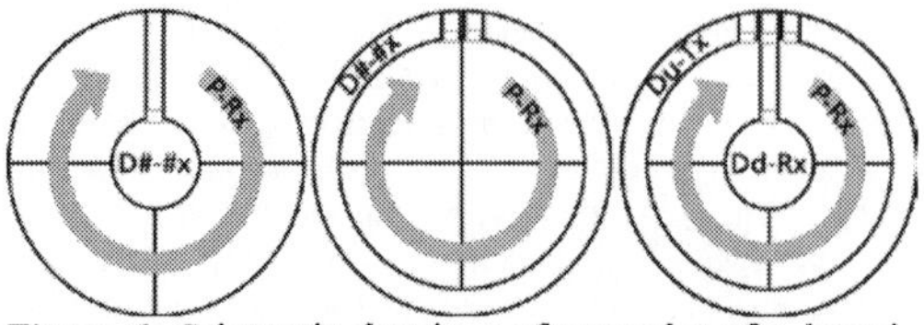

Figure 6: Schematic drawings of examples of a lateral integration of a data functionality into a multi-segment PV laser power converter (P-Rx); the gray arrow indicates the series connection of the PV segments to

increase the output voltage. The term "D#-#x" is used to indicate that the respective segments could be used for data transmission (Du-Tx) as well as reception (Dd-Rx), depending on usage (forward biasing LED mode or reverse biased photo detection), or even both multiplexed in the time domain (TDM). Electrical contacting (not shown here) can be realized outside the circular active area, respective conductive paths are sketched by white connections to the outside of the circle. *Left*: A central segment is used for the data link, whereas as an example four outer series connected segments are used for conversion of the power beam. *Middle*: The data segment is moved to the outer part of the active area. That way the usually high power in the center of the beam can be absorbed by the power receiver for electricity generation. Hence, this design is considered to be beneficial for power transmission, but might deteriorate signal-to-noise ratio of the data link. *Right*: Combination of the previous two with data transmitter in the outer region and data receiver in the center of the active area.

3.5 Areal splitting

Splitting of the device area can be beneficial, especially for multi-junction architectures. Here, e.g. upper junctions can be partly removed (by selective etching) to enable direct light exposure of lower layers. That way, the design rule of full transparency of the upper layers can be circumvented to gain more freedom in device design. Figure 7 shows an example of a 3-junction structure with partly removed layers.

In addition, devices of different can be integrated to still guarantee operational functionality of their individual purpose. E.g. a small sized VCSEL, LED or photo diode can be bonded onto a comparatively larger photovoltaic receiver. That way, the small sized device is fully functional, whereas the PV cells absorber area is only marginally diminished (compare Figure 8).

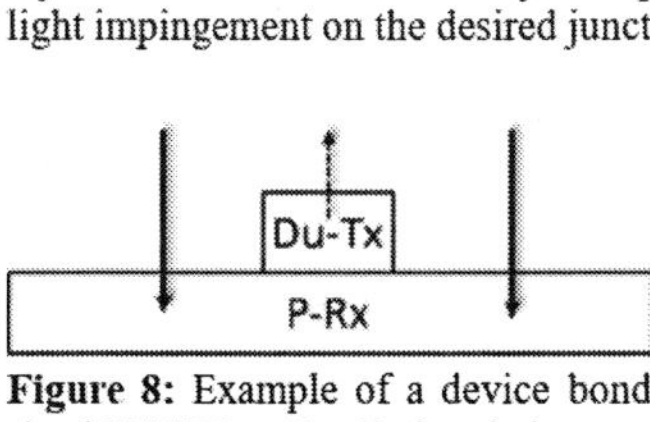

Figure 7: Example of a 3-junction device with partial removal of the upper junctions. That way, the upper layers do not need to be fully transparent to guarantee light impingement on the desired junction.

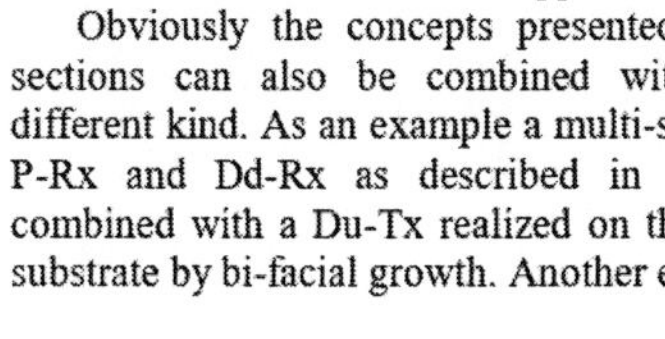

Figure 8: Example of a device bonding of e.g. a small sized VCSEL as Du-Tx bonded to a comparatively larger PV cell as P-Rx.

3.6. Combinations of different approaches

Obviously the concepts presented in the previous sections can also be combined with each other in different kind. As an example a multi-segment device for P-Rx and Dd-Rx as described in Sec. 3.4 can be combined with a Du-Tx realized on the rear side of the substrate by bi-facial growth. Another example is a multi-junction structure as shown in the bottom of Fig. 4, where the upper P-Rx is composed of a multi-segment device including one segment for Dd-Rx. The lateral conduction layer of the P-Rx can then also be used for contacting the Du-Tx beneath. Many more combinations are possible.

4 CONCLUSION

Different possibilities to realize purely optical power and bidirectional data transmission have been presented and discussed. It was shown that a diversity of alternative solution exists to combine the desired functionalities into a multi-purpose integrated transceiver device.

REFERENCES

[1] K. Worms, C. Klamouris, F. Wegh, L. Meder, D. Volkmer, S.P. Philipps, S.K. Reichmuth, H. Helmers, A.W. Bett, J. Vourvoulakis, C. Koos, W. Freude, J. Leuthold, and W. Stork, in Proceedings 17th ITG/GMA-Symposium: Sensors and Measuring Systems (2014) Nürnberg, Germany.

[2] K. Worms, C. Klamouris, F. Wegh, L. Meder, D. Volkmer, S.P. Philipps, S.K. Reichmuth, H. Helmers, A. Kunadt, I. Vourvoulakis, A.W. Bett, C. Koos, W. Freude, J. Leuthold, and W. Stork, Wind Energy (2016).

[3] S. Park, D.A. Borton, M. Kang, A.V. Nurmikko, and Y.-K. Song, Sensors (Basel, Switzerland) 13(5) (2013) 6014. DOI: 10.3390/s130506014

[4] M. Mujeeb-U-Rahman, D. Adalian, C.-F. Chang, and A. Scherer, Journal of Biomedical Optics 20(9) (2015) 095012. DOI: 10.1117/1.jbo.20.9.095012

[5] F.V.B.d. Nazare and M.M. Werneck, IEEE Sensors Journal 12(5) (2012) 1193. DOI: 10.1109/jsen.2011.2163709

[6] G. Boettger, M. Dreschmann, C. Klamouris, M. Hubner, M. Röger, A.W. Bett, T. Kueng, J. Becker, W. Freude, and J. Leuthold, IEEE Photonics Technology Letters 20(1) (2008) 39. DOI: 10.1109/LPT.2007.912695

[7] F. Audo, M. Guegan, V. Quintard, A. Perennou, J. Le Bihan, and Y. Auffret, Optical Engineering 50(4) (2011) 045001. DOI: 10.1117/1.3560542

[8] M. Roeger, G. Boettger, M. Dreschmann, C. Klamouris, M. Huebner, A. Bett, J. Becker, W. Freude, and J. Leuthold, Optics Express 16(26) (2008) 21821. DOI: 10.1364/oe.16.021821

[9] R. Pena and C. Algora, in Proceedings 20th European Photovoltaic Solar Energy Conference (2005) Barcelona, Spain.

[10] H. Kirkham and A.R. Johnston, IEEE Transactions on Power Delivery 4(4) (1989) 1997. DOI: 10.1109/61.35623

[11] W.B. Spillman Jr., D.H. Crowne, and D.W. Woodward, Optics and Lasers in Engineering 16(2–3) (1992) 105. DOI: 10.1016/0143-8166(92)90003-P

[12] S.J. Pember, C.M. France, and B.E. Jones, Sensors and Actuators A: Physical 47(1–3) (1995) 474. DOI: 10.1016/0924-4247(94)00944-D

[13] P. Karioja, R. Jurva, K. Keränen, K. Tukkiniemi, J. Lammasniemi, A. Tervonen, and M. Englund, *Optical-To-Electrical Power Conversion and Data Transmission Module*, in *Applications of Photonic Technology 2*, Lampropoulos and Lessard, eds. (1997). 325. DOI: 10.1007/978-1-4757-9250-8_53

[14] C. Klamouris, G. Bottger, M. Hubner, M. Dreschmann, K. Paulsson, A.W. Bett, T. Kueng, J. Becker, W. Freude, and J. Leuthold, in Proceedings European Conference on Optical Communications (2006) Cannes, France. DOI: 10.1109/ecoc.2006.4801363
[15] F. Audo, A. Pérennou, M. Guégan, V. Quintard, Y. Querrou, J.L. Bihan, and Y. Auffret, in Proceedings Ocean & Coastal Observation: Sensors and observing systems, numerical models & information Systems (2010) Brest, France.
[16] I. Haydaroglu and S. Mutlu, Journal of Microelectromechanical Systems 24(1) (2015) 155. DOI: 10.1109/jmems.2014.2323202
[17] O. Höhn, A.W. Walker, A.W. Bett, and H. Helmers, Applied Physics Letters 108(24) (2016). DOI: 10.1063/1.4954014
[18] C. Rohr, P. Abbott, I.M. Ballard, D.B. Bushnell, J.P. Connolly, N.J. Ekins-Daukes, and K.W.J. Barnham, *Quantum wells in photovoltaic cells*, in *Next Generation Photovoltaics: High Efficiency Through Full Spectrum Utilization*. (2004). 91.
[19] K.W.J. Barnham, I.M. Ballard, B.C. Browne, D.B. Bushnell, J.P. Connolly, N.J. Ekins-Daukes, M. Fuhrer, R. Ginige, G. Hill, A. Ioannides, D.C. Johnson, M.C. Lynch, M. Mazzer, J.S. Roberts, C. Rohr, and T.N.D. Tibbits, *Recent Progress in Quantum Well Solar Cells*. Nanotechnology for Photovoltaics, ed. Tsakalakos. (2010). 187. DOI: 10.1201/9781420076752-c5
[20] A. Kern, S. Paul, W. Schwarz, D. Wahl, R. Blood, and R. Michalzik, IEEE Photonics Technology Letters 23(15) (2011) 1058. DOI: 10.1109/lpt.2011.2153843
[21] P.G. Borden, Applied Physics Letters 35(7) (1979) 553. DOI: 10.1063/1.91176
[22] H. Helmers, E. Oliva, W. Bronner, F. Dimroth, and A.W. Bett, IEEE Transactions on Electron Devices 57(12) (2010) 3355. DOI: 10.1109/TED.2010.2076190
[23] R. Löckenhoff, F. Dimroth, E. Oliva, A. Ohm, J. Wilde, D. Faiman, S. Biryukov, V. Melnichak, S. Kabalo, D. Bokobza, and A.W. Bett, Progress in Photovoltaics: Research and Applications 16(2) (2008) 101. DOI: 10.1002/pip.778
[24] S. Wojtczuk and T. Parodos, in Proceedings 13th Space Photovoltaic Research and Technology Conference (1994) Bedford, UK.
[25] R. Peña and C. Algora, IEEE Transactions on Electron Devices 48(2) (2001) 196. DOI: 10.1109/16.902716
[26] D. Krut, R. Sudharsanan, W. Nishikawa, T. Issiki, J. Ermer, and N.H. Karam, in Proceedings 29th IEEE Photovoltaics Specialists Conference (2002) New Orleans.
[27] J. Schubert, E. Oliva, F. Dimroth, W. Guter, R. Löckenhoff, and A.W. Bett, IEEE Transactions on Electron Devices 56(2) (2009) 170. DOI: 10.1109/TED.2008.2010603
[28] H. Helmers, L. Wagner, C.E. Garza, S.K. Reichmuth, E. Oliva, S.P. Philipps, D. Lackner, and A.W. Bett, in Proceedings SENSOR 2015 - 17th International Conference on Sensors and Measurement Technology (2015) Nürnberg, Germany. DOI: 10.5162/sensor2015/D1.4
[29] R. Peña and C. Algora, in Proceedings 2nd World Conference on Photovoltaic Energy Conversion (1998) Vienna, Austria.

FREE THE BANDGAP!
SERIES-PARALLEL CONNECTION OF TANDEM CELLS

Matthew Stocks[1]*, Yao Xiong Loo[1] and Niraj N. Lal[1,2]
[1]PV Group, Research School of Engineering, Building 32, The Australian National University
[2]Department of Materials Science and Engineering, Monash University, Clayton, VIC
*matthew.stocks@anu.edu.au (phone) +61 2 61259876

ABSTRACT: Series-parallel interconnection of tandem cells is an alternate approach to standard monolithic interconnection design. By matching string voltage instead of cell current, a broad range of top cell bandgaps can be effectively combined to produce near-ideal efficiency. Annual energy yield using spectrum and weather data collected by NREL was simulated for tandem module designs. The series-parallel approach is extremely tolerant of spectrum and temperature variation, achieving a simulated yield of >99% of the ideal four-terminal tandem for a given bandgap. Importantly, a wide range of bandgaps (1.4eV to 2.0 eV) also achieved energy yields greater than 95% of that achieved for the ideal bandap (~1.65eV). The series-parallel approach provides broad design flexibility, particularly for the top cell, to enable relatively free choice of the best material for efficiency and manufacturability rather than being constrained solely by current matching.

Keywords: Tandem, Silicon, Energy Performance, Perovskite

1 TANDEM CELLS

Efficiency improvement is a major contributor to the reduction of PV costs by providing savings across all module components as well as some balance of systems costs. While silicon dominates the PV market, the record efficiencies for silicon devices have improved by only 1% in the last 15 years [1]. As a result, the gap between record and commercial devices has closed and further opportunities for increasing performance are becoming more challenging to find.

Tandem cells provide the opportunity to increase efficiency beyond the limits of single junction devices, and have been the source of increased interest in recent years. There is a strong push towards examining tandem solar cells with silicon as the "bottom" cell that absorbs the lower energy photons [2, 3, 4, 5]. Adding a suitable "top" cell of a higher bandgap has potential to push efficiencies over 30% under standard test conditions (STC) [6].

Ideal efficiency for two different cells can be achieved in the four-terminal (4T) arrangement where the top and bottom cell are separately wired to operate independently of each other. This arrangement requires power control such as maximum power-point tracking for two circuits, thereby increasing implementation costs. Practical tandem modules are much more likely to have a two terminal arrangement in line with conventional single junction modules.

The most common approach for connecting the two or more cells is in series (ST). This approach has the advantage that the top cell can be grown or deposited directly on the bottom cell without needing additional wiring for interconnection of the cells.

However, this approach has a major limitation: the need to match current of the top and bottom cells. The bandgap of the top cell (as well as optical design) needs to be chosen such that the current in the two devices at any instance are similar. If this is not the case, the device with the lower current limits the power in the other cell.

There is a limited collection of semiconductors that have a bandgap around 1.7eV to suitable match of cell current to the silicon bottom cell at STC. The efficiency of two potential candidates, amorphous silicon [7] and antimony sulphide [8] have not been sufficient to enable higher tandem performance than a good silicon device due to a combination of low voltage and poor carrier transport. The bandgap of lead-based perovskites can be modified to 1.7eV through alterations of the I/Br halide ratios, but comes at the cost of increased complexity and lower efficiency [9]. The GaAs family of blended III-V semiconductors can produce sufficient efficiency but at relatively high costs.

This paper considers an alternate approach for the interconnection of the devices which provides greater flexibility in the choice of bandgap of the top cell, while producing energy yields approaching that of the four-terminal arrangement.

2 SERIES-PARALLEL CONNECTION

The interconnection approach considered in this work is shown schematically in Figure 1. Cells of similar type are first connected in series, and then the strings are connected in parallel.

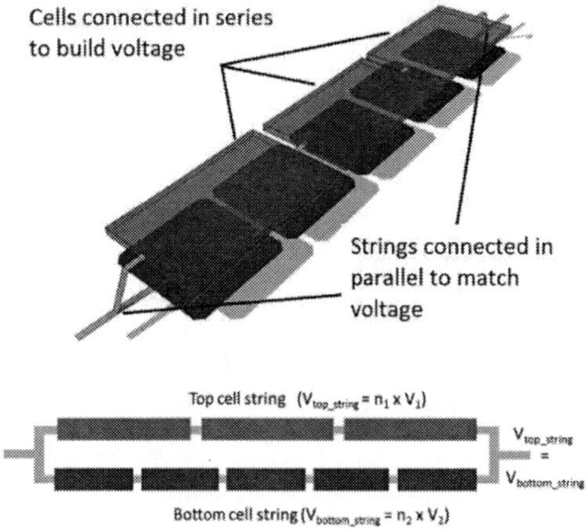

Figure 1. Series-parallel arrangement of tandem cells

This approach relies on *voltage* matching between the strings rather than *current* matching of the two (or more) types of cells. The number of cells connected in series depends on the voltage of the top and bottom cells in inverse proportion to their voltage. For example, if the top cell has a maximum power voltage of 1.0V while the bottom string has a voltage of 0.6V, then 3 top cells should be connected in series for every five bottom cells that are connected in series for a matched string voltage of 5.0V. Since this ratio can be easily adjusted, a wide variety of materials have the potential to work as an effective top material with silicon independent of their bandgap.

The idea of interconnecting tandem cells comprising a parallel element is not new. However, previous examples comprise connecting a single top cell in parallel with one or more different bottom cells. Guo [10] and Sista [11] connected a single bottom in parallel with a single top cell which requires similar top and bottom cell voltages. Fraas [12] connected three GaSb bottom cells in series then connected a single GaAs top cell in parallel. Trupke and Würfel [13] connected the two lower bandgap cells in a three cell tandem in series and then connected this string in parallel to the top cell. All of these approaches are constrained in the choice of bottom cells to match the voltage of the top cell. In contrast, this approach provides considerably more freedom in choice of the bandgap of the top cell by connecting multiple top cells in series. This provides significant yield benefits as demonstrated in the following section.

3 ANNUAL ENERGY YIELD MODELLING

Annual energy yield modelling was undertaken to determine the effectiveness of the series-parallel tandem (SPT) arrangement in comparison to the ideal four-terminal and the usual series-connected arrangement under "real world" conditions.

The model was based on the approach taken by White [5], and extended to include

- Incorporation of real-world spectral data instead of standard AM1.5G spectrum
- Module temperature adjustments based on weather conditions
- Full cell current-voltage curves to enable simulation of cell interconnection

Absorption for each bandgap is simulated with the analytical expression:

$\alpha(\lambda) = \alpha_0 \left(\frac{E_\phi(\lambda) - E_g}{kT}\right)^{1/2}$, where $\alpha_0 = 10^5$ (cm^{-1}) is the characteristic absorption coefficient for the material, $E\phi(\lambda) = hc/\lambda$ is the incident photon energy at wavelength λ, k is the Boltzmann constant, and T = 298 K is the cell temperature.

The cell is modelled with p-i-n structure, with voltage based on the luminescent efficiency. Further details can be found in White [6] and Lal [14].

3.1 Spectral data

One year's worth of spectral data at NREL's site at Golden, Colorado, was obtained from the NREL website [15]. This data is collected by a spectroradiometer at a resolution of 3.215nm over the range 350 to 1050 nm at one minute intervals.

The recorded spectral range is less than the full light spectrum. Whilst having minimum impact on power of the simulated device, the reduced spectral range does affect the reported incident light power. Efficiency was therefore not considered in this work, only energy yield. Irradiance of the standard AM1.5G spectrum in this spectral range is only 760W/m^2 instead of 1000W/m^2. The NREL-measured spectrum was therefore weighted by a factor of 1.32 when simulating module temperature (see Section 3.2 below).

3.2 Module temperature

Temperature changes affect cell voltage more significantly than current, so module temperature was simulated to more accurately evaluate the relative series-parallel tandem cell performance. The mathematical model of TamizhMani [16] was used to determine module temperature from the ambient temperature T_a, incident irradiance G_{inc}, and wind speed v_w as shown in Equation (1). This formulation is applicable for the study since it is independent of both the solar cell technology and the location.

$$T_m = 0.943T_a + 0.028G_{inc} - 1.528v_w + 4.3 \quad (1)$$

The weather data was also obtained from the same NREL test site so that the weather and spectrum data are well correlated.

3.3 Interconnection

The relative number of cells that were required to be connected in series was determined from the maximum power point of the tandem cells for a nominal initial condition. This was chosen as one of STC (default), NOCT or the "average" conditions at the NREL site (see Section 4.2).

Thickness of the top device was then optimised to produce maximum efficiency for the four-terminal arrangement under the chosen operation conditions. The maximum power voltage for each of the strings was then determined.

The number of top cells and bottom cells was chosen to obtain the minimum difference between the maximum power voltage of the top cell string and the bottom cell string with a constraint of < 40V (typical Si module V_{mp}). For example, for a bottom cell V_{mp} of 0.63V and a top cell V_{mp} of 1.08V, 60 bottom cells and 35 top cells would result in a matched string V_{mp} of 37.8V.

Since the number of cells in the top string differs from that in the bottom string, the area of the top cells therefore needs to be scaled to match area of the bottom cells in inverse proportion to the relative number of cells in the string. As the energy yield results are reported per square metre, the area of a bottom cell in the above example is assumed to be 166cm^2 while the area of a top cell would be 285cm^2.

Smaller cells could be connected in multiple parallel strings to achieve the same relative area. For example, if the top cell area was restricted to ~60cm^2 for yield or performance reasons, then 4 or 5 strings of smaller top cells could be connected in parallel to the bottom string to achieve the same outcome. Modules with many parallel connections provide additional advantages including increased tolerance of partial shading and long term reliability benefits as demonstrated with SLIVER module technology [17].

4 RESULTS

4.1 Series-parallel modelling

Figure 2 is a plot of the simulated energy yield for the

four-terminal (4T), series-connected (ST) tandem and series-parallel (SPT) cells over a year. The series-parallel arrangement was found to produce high annual energy yield, >99% of the four-terminal arrangement for the wide range of bandgaps simulated (1.4eV to 2.29eV). Maximum energy yield is achieved with a top cell having 1.65eV bandgap.

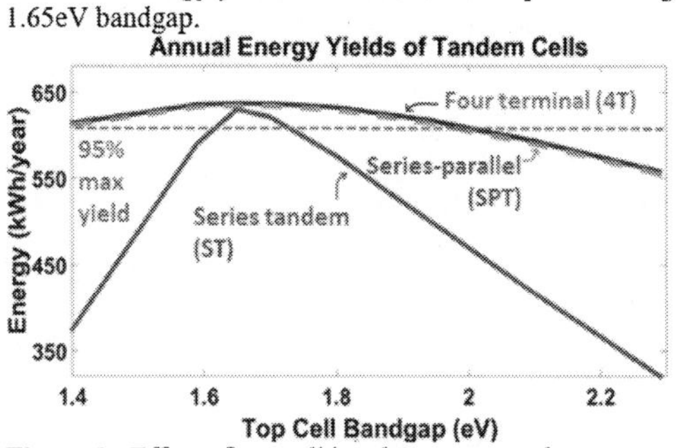

Figure 2. Effect of top cell bandgap on annual energy for four terminal (4T), series (ST) and series-parallel (SPT) tandem cells. The grey line represents 95% of maximum yield.

The extent of this increased flexibility for the series-parallel tandem design can be considered in the context of the range where high performance can be expected. Cells from all bandgaps between 1.4eV and 2.0eV produce energy yields within 95% (grey line in Figure 2) of the best four-terminal efficiency. In contrast, this 95% threshold is only achieved for series-connected tandems for a much narrower bandgap range of between 1.61eV and 1.74eV.

4.2 Impact of temperature

Variations in temperature have a greater impact on cell voltage than current, so the effect of such on power generated by the series-parallel interconnection is more significant than the series-connected tandem. Figure 3 is a plot of the difference between power generated by the series-parallel and four-terminal cells as temperature is varied for a fixed AM1.5G spectrum.

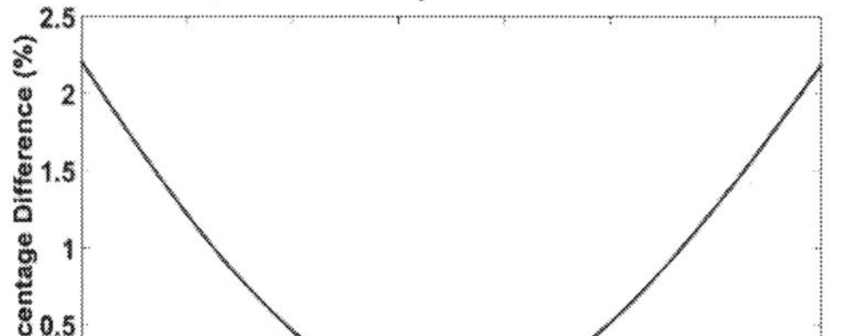

Figure 3. Effect of temperature on SPT power relative to 4T cell. Cell strings of SPT cell were matched at STC (25°C)

As expected, the performance is matched at the design condition of 25 °C. As the module temperature deviates from this temperature, the yield difference increases. The power loss relative to four-terminal efficiency is ~2% at 35°C above or below the reference temperature.

Since modules usually operate at temperatures well above STC, two additional design constraints were considered. One was the average daylight conditions for the NREL site while the other was NOCT. The bandgap of the top cell was modelled at the top of the simulated range (2.29eV) where the relative difference due to temperature change was expected to be greatest.

Table 1. Annual energy yield difference variation between 4T and SPT cells with alternate design conditions.

String matching design conditions	**STC (Default)**	**NREL Average**	**NOCT**
Irradiance (W/m^2)	1000	670	800
Air temperature(°C)	N/A	13.6	20
Wind velocity (m/s)	N/A	2.4	1
Calculated module temperature (°C)	25	32	44
Annual energy yield difference	**0.49%**	**0.20%**	**0.21%**

As shown in the bottom row of Table 1, both the new series-parallel design conditions further reduced the already low difference to the ideal, four-terminal tandem efficiency by more than 50%.

4.3 Impact of illumination changes

The series-parallel design is very tolerant of variations in illumination conditions. Spectrum and intensity variation causes significantly less impact on series-parallel performance since the voltage of the strings varies proportionally to the log of the cell string current. This can be seen clearly in Figure 4 with temperature effects included or removed from the model.

Figure 4a shows the variation in performance for a range of different top cell bandgaps over a clear winter day. Figure 4b shows the results over the same day with module temperature fixed at the design condition (25°C). This clearly shows that temperature is the dominant cause of mismatch between the two strings with losses reduced by approximately one order of magnitude. Similar results were found for other illumination conditions including cloudy days with temperature effects dominating the, albeit small, losses.

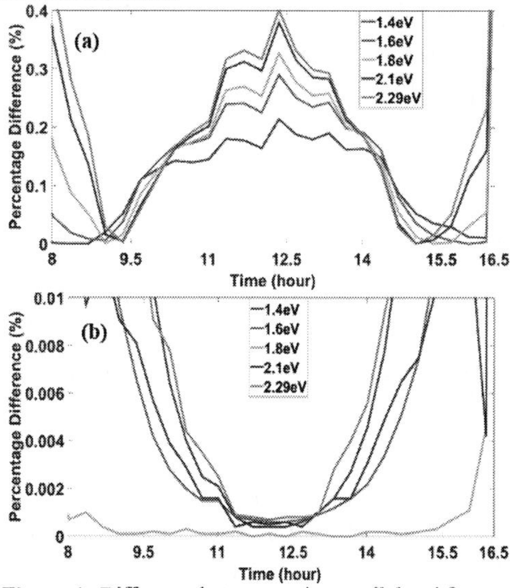

Figure 4. Difference between series-parallel and four-terminal tandem performance for a clear winter day for a range of top cell bandgaps. (a) Full model. (b) Model with fixed module temperature (25°C). Note the different vertical scales.

5 DISCUSSION

There are a number of features of the series-parallel interconnection approach that are desirable for future tandem module designs.

5.1 Increased freedom with bandgap choice

The most significant finding of this study is the broad range of top cell bandgaps which can deliver high energy yield when configured with series-parallel interconnection.

An example of the potential benefit of this approach is the perovskite/silicon tandem. There has been considerable interest in combining perovskite cells with silicon to achieve efficiencies >30% since the bandgap of the perovskite can be tuned by modifying the halide ratio [18]. The best perovskite cells have been based on lead iodide, with efficiencies >20% demonstrated with a bandgap of 1.55eV which is too low for effective series connection. Attempts to achieve the ideal bandgap of around 1.7eV for matching with silicon cells has led to the development of complicated blends of five materials to achieve stability [9].

In contrast, the series-parallel approach enables the use of the most efficient devices relatively independent of the top cell bandgap. Use of the lead iodide perovskite would lead to minimal (~1%) power loss compared to an equivalent quality device with ideal bandgap.

Liu [3] also modelled annual energy yield for series connected GaAs/Si tandems using NREL and NUS spectral and weather data. The bandgap of GaAs is 1.42eV which is towards the bottom end of our modelled range. This work showed that the annual yield of the series connected arrangement was ~17% below the four terminal yield. In contrast, this work has shown a series-parallel connection scheme would have achieved yields >99% of four terminal.

5.2 Manufacturing flexibility

A potential advantage of the series-parallel approach is the ability to separate the manufacturing of the two cell types, which allows for the top cell to be manufactured independently from the silicon cell. This decouples the yield of the manufacturing of the two cell types unlike monolithic interconnection where a failure in either cell leads to rejection of both cells.

Furthermore, the size of the top and bottom cells can be designed independently. While silicon cells are relatively large (~150 to 250cm^2) with a square or pseudosquare geometry, other devices tend to be smaller or have a high length to width aspect ratio to manage lateral transport losses. The only constraint is the number of cells required to be connected in series, which may be significant and requires extensive connections.

The series parallel approach would be particularly attractive for top cells which could be processed directly on glass. All of the processing of the top cell and the cell interconnections could be completed on the glass superstrate which would subsequently laminated to the underlying silicon cells in a similar fashion to standard modules.

6 CONCLUSION

Series-parallel interconnection of tandem cells is a promising alternative to series-connection. We simulated the annual energy yield of four terminal, series connected and series-parallel connected device under real world illumination and weather conditions. This demonstrated that energy yields >95% of ideal can be achieved for four terminal and series-parallel arrangements for bandgaps between 1.4 and 2.0eV. In contrast, monolithic series-connected tandems are constrained by current matching which significantly limits the range of acceptable bandgaps to within 0.05eV of the optimum bandgap for similar performance. The series-parallel approach literally 'frees the bandgap' to provide increased design flexibility for the rapid development of high-efficiency tandem solar cell modules.

Acknowledgement

This project has been supported by the Australian Government through the Australian Renewable Energy Agency (ARENA). The Australian Government through ARENA is supporting Australian research and development in solar photovoltaic and solar thermal technologies as part of its commitment to making renewable energy technologies more affordable and increasing the amount of renewable energy used in Australia. The views expressed herein are not necessarily the views of the Australian Government, and the Australian Government does not accept responsibility for any information or advice contained herein.

References

[1] NREL, "Best Research-Cell Efficiencies," [Online].http://www.nrel.gov/ncpv/images/efficiency_chart.jpg. [Accessed 9th June 2016].

[2] Hong Kong Polytechnic University, "PolyU Develops Perovskite-Silicon Tandem Solar Cells with the World's Highest Power Conversion Efficiency," 12th April 2016. [Online]. http://www.polyu.edu.hk/web/en/media/media_releases/index_id_6208.html. [Accessed 9th June 2016].

[3] Lui, et al., "The realistic energy yield potential of GaAs-on-Si tandem solar cells: a theoretical case study," *Optics Express*, vol. 23, pp. 382-390, 2015.

[4] McGehee, et al. "Modeling low cost hybrid tandem photovoltaics with the potential for efficiencies exceeding 20%," *Energy Environ. Sci.*, vol. 5, p. 9173, 2012.

[5] Green "Commercial progress and challenges for photovoltaics," *Nature Energy*, vol. 1, pp. 1-4, 2016.

[6] White et al., "Tandem Solar Cells Based on High-Efficiency c-Si Bottom Cells: Top Cell Requirements for >30% Efficiency," *IEEE Journal of Photovoltaics*, vol. 4, pp. 208-214, 2014.

[7] Matsui et al., "Development of highly stable and efficient amorphous silicon based solar cells," 28th European PV Conference, Paris, 2013.

[8] Sang Hyuk Im, et al., "Toward Interaction of Sensitizer and Functional Moieties in Hole-Transporting Materials for Efficient Semiconductor-Sensitized Solar Cells," *Nano Letters*, vol. 11, no. 11, pp. 4789-4793, 2011.

[9] McMeekin, et al. "A mixed-cation lead mixed-halide perovskite absorber for tandem solar cells," *Science*, vol. 351, no. 6260, p. 151, 2016.

[10] Guo, et al "Solution-Processed Parallel Tandem Polymer Solar Cells Using Silver Nanowires as

Intermediate Electrode," *ACS Nano*, vol. 8, no. 12, pp. 12632-12640, 2014.

[11] Sista, et al. "High efficiency polymner tandem solar cells with three-terminal structure," *Advanced Materials*, vol. 22, no. 8, pp. 77-80, 2010.

[12] Fraas, et al. "Over 35-percent efficient GaAs/GaSb tandem solar cell," *IEEE Transactions Electron Devices*, vol. 37, no. 2, pp. 443-449, 1990.

[13] Trupke and Würfel, "Improved spectral robustness of triple tandem solar cells by combined series/parallel interconnection," *Journal Applied Physics*, vol. 96, no. 4, pp. 2347-2351, 2004.

[14] Lal et al., "Optics and Light Trapping for Tandem Solar Cells on Silicon," *IEEE Journal of Photovoltaics*, vol. 4, no. 6, pp. 1380-1386, 2014.

[15] NREL, "NREL 2015, Baseline Measurement System (BMS)," [Online]. Available: https://www.nrel.gov/midc/srrl_bms/#DOI. [Accessed 25th March 2015].

[16] TamizhMani, et al. "'Photovoltaic Module Thermal/Wind Performance: Long-Term Monitoring and Model Development for Energy Rating, *NCPV and Solar Program Review Meeting*," 2003.

[17] Stocks, et al "The Technical and Performance Advantages of Production SLIVER module," in *4th World Conference on Photovoltaic Energy Conversion*, Waikoloa, Hawaii, 2006.

[18] Noh, et al., "Chemical Management for Colorful, Efficient, and Stable Inorganic - Organic Hybrid Nanostructured Solar Cells," *Nano Letters*, vol. 13, no. 4, pp. 1764-1769, 2013.

INFLUENCE OF GaAsSb CLADDING LAYER ON THE OPTICAL PROPERTIES OF InAs/GaAsSb QUANTUM DOTS

Z. Zhang[1], Y.Liao[1], N. N. Faleev[2], P.Reece,[3] S. P. Bremner[1]

[1]School of Photovoltaic and Renewable Energy Engineering, University of New South Wales, Sydney 2052, Australia
[2]Solar Power Lab, School of Electrical, Computer and Energy Engineering, Arizona State University, Tempe, AZ, 85287, USA
[3]School of Physics, University of New South Wales, Sydney 2052, Australia

ABSTRACT: The optical properties of InAs quantum dots with GaAsSb cladding layers of different alloy composition have been studied by photoluminescence techniques. Power-dependent photoluminescence data suggest a transition from type-I to type-II could be achieved by GaAsSb cladding layer at Sb composition around 11% for these samples. At low Sb compositions, good crystal quality was evident from temperature-dependent photoluminescence, with a small energy barrier inferred, while high-level dislocation and defects exist under higher antimony contents. The results show the interplay between the tendency of Sb in the buffer layer to blue-shift the emisison, whilst Sb in the capping layer red-shifts emission. With the additional presence of strain relaxation through formation of defects at higher Sb content, it is clear that designing a structure to achieve a zero valence band offset must take into account the strain distribution throughout the structure.

Keywords: III-V Semiconductors, Quantum Dots, Epitaxy, Photoluminescence

1 INTRODUCTION

Self-assembled (SA) InAs/GaAs quantum dots (QDs) have been drawing great attention due to the potential applications in novel optoelectronic devices [1, 2], like photodetectors, 1.5-um wavelength emission laser diodes and photovoltaic devices. Compared with conventional designs, novel devices like the Intermediate Band Solar Cell (IBSC) [3] and hot carrier solar cell [4], are predicted to provide pathways to ultra-high efficiency by offering an ability to tailor their structural and optical properties. In particular, IBSCs have limiting efficiencies exceeding 60% under maximum concentration of sunlight, and 50% under AM1.5 [5], providing a possible realization with high conversion efficiencies to overcome the Shockley-Queisser (SQ) limitation[6]. Self-assembled quantum dots with delta-function-like density of states provide a promising way to form the intermediate band (IB) lying between the valence band (VB) and the conduction band (CB) through the ability of delocalizing enery levels[7]. However, the valence band offset (VBO) existing between InAs QDs and bulk GaAs could give rise to the open-circuit voltage loss [8], meaning designs that ensure a VBO of zero should be used [9]. The large (7%) lattice mismatch between InAs and GaAs leads to strain accumulation during the growth complicating optical propoerties and can lead to degradation in the performance of an IBSC due to defect formation.

The use of GaAsSb barriers, or cladding layers, has been suggested as a way to minimize the VBO for an IB solar cell, with the InAs/GaAsSb system being shown to form a small valence band offset [10], with the type I to type II heterostructure transition observed in the 12% to 14% Sb composition range [11]. Additionally, the presence of a GaAsSb buffer in the QD system leads to a higher areal density of QDs with greater size uniformity, while previous reports for samples capped by thin GaAsSb showed improvement in photoluminescence (PL) intensity and a peak red-shift indicating the capping layer provides strain relief to the QDs [12]. However, too high an antimony content has been shown to lead to strain relaxation via extended crystalline defects like threading dislocations that can lead to the suppression of QD formation [13].

2 EXPERIMENTAL

All the QD structures studied were grown on GaAs (100) substrates in a GEN930 MBE system. The Sb_2 and As_2 fluxes were supplied by valved cracker sources and the substrate temperature was determined through calibrating an Ircon pyrometer to the de-oxidation temperature of the GaAs substrate observed by Reflective High Energy Electron Diffraction (RHEED). Approximately 17 nm GaAsSb buffer and 17 nm GaAsSb capping layer were grown at a growth rate of 0.4 monolayers per second (ML/s) under various Sb_2/As_2 ratios. Substrate temperature was set to 490 °C to fabricate QDs by deposition of approximate 2 monolayers InAs at a growth rate of ~ 0.035 monolayers per second. All samples were finished with a 100 nm GaAs cap with 10 nm grown at 490 °C and the remainder grown at 600 °C.

After removal from the MBE system the thicknesses and alloy compositions were checked by ω-2θ rocking curves (RCs) using a Philips X'Pert High Resolution X-ray Diffractometer with CuKa1 radiation and a 4 bounce Ge(220) monochromator in the double crystal axis alignment.

Samples were then mounted in a closed-cycle He cryostat with a built-in temperature controller, with excitation provided by an argon ion laser with wavelength of 488 nm. PL emission from the sample was relayed to a SPEX 270 M scanning spectrometer captured by a liquid nitrogen cooled InGaAs detector array by EOSystems. The spectral resolution of the system is approximately 2 nm.

The structural characteristics of InAs/GaAs QDs with GaAsSb cladding layers under variable antimony composition were investigated by cross-sectional high-resolution transmission electron microscopy (HRTEM)

using a JEOL-JEM-3000F microscope operated at 300 kV.

3 RESULT AND DISCUSSION

X-ray data was collected in the vicinity of the GaAs (004) substrate peak for GaAsSb cladding layers of various Sb composition displayed in Figure 1. Fig. 1 (a) shows the ω-2θ measurement for the 3% Sb GaAsSb sample. Good vertical coherence of epitaxial layers was indicated by the clearly visible interference fringes in the vicinity of GaAs peak in the range over one degree. Such behaviour was also detected from the GaAsSb cladding layer (both buffer and capping) samples under 11% (displayed as Fig. 1(b)) and 13% antimony composition, suggesting good epitaxial crystal quality under these Sb contents. By contrast, the results for the 20% Sb content sample displayed as Fig. 1(c), indicate a significant deterioration of the vertical coherence compared to the samples with lower antimony composition. No extended interference fringes can be discerned, suggesting a diminution of crystal quality as Sb content iincreases in the quantum dots system, with a high-density of defects and dislocations expected within the structure[13].

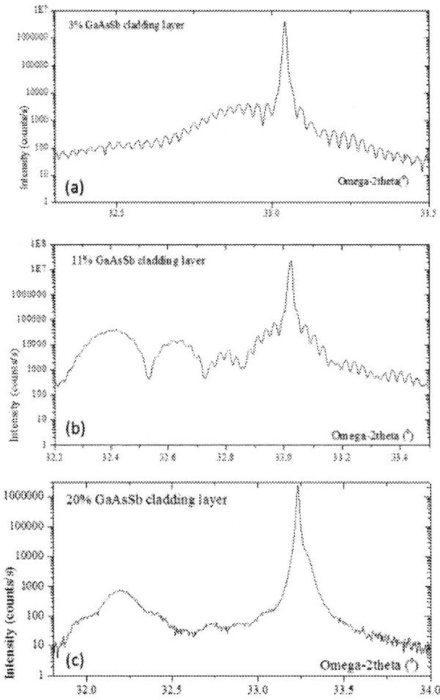

Figure 1: Experimental double crystal, ω-2θ curves in the vicinity of GaAs (004) reflection, with antimony contents of 3% (a), 11% (b), and 20% (c), respectively.

Figure 2 shows PL spectra for GaAsSb cladding layer (both buffer and cap) samples taken at 20 K with Sb compositions 0%, 3%, 11%, 13%, and 20%. For GaAsSb cladding layers with 3%, 11% and 13% Sb content there is a pronounced enlargement of PL intensity compared with the InAs/GaAs reference sample (0% Sb). Such behaviour is similar to the PL results of InAs QDs buffered with GaAsSb layer [15]. The increase seen in PL emission is consistent with previous observations that the presence of the Sb in the GaAsSb buffer acts as a growth surface surfactant, reducing surface energy during growth of QDs[14] with a higher areal density of QDs. This explanation is further supported by the observation of the blue shifted in PL for samples of 3% and 11% compared with the 0% Sb content. This result is consistent with a reduction in the QDs size due to the surfactant function of GaAsSb layer.

With increasing Sb composition from 3% to 20%, a continuous red-shift is observed in GaAsSb cladding layer samples, from approximate 1010 nm to 1100 nm, which was observed in the InAs QDs with GaAsSb capping layer [16]. Such behaviour is attributed to the GaAsSb capping layer acting as a strain relieving relieving layer for the InAs QDs, and thus red-shifting the peak position. Furthermore, the continuous red-shift and PL intensity reduction as the Sb content increases from 11% to 20% is suggested to be, at least partially, explained by the heterostructure transition from type-I to type-II in the interface of InAs quantum dots and GaAsSb layers under higher Sb composition [17], where the electrons in the InAs QDs are likely to recombine with holes in the GaAsSb layers instead of those spatially located in the InAs.

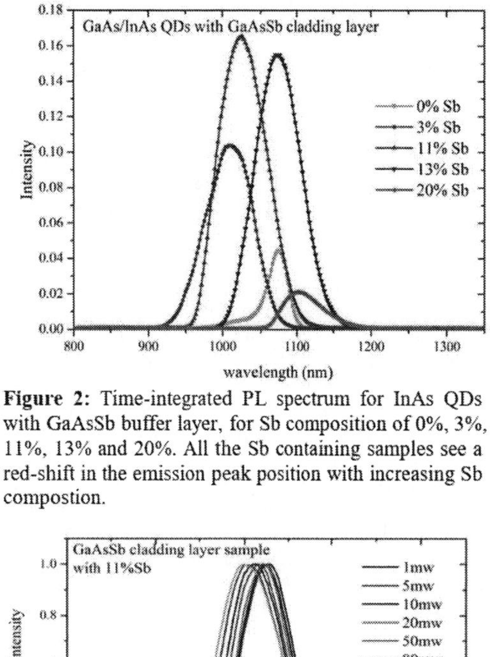

Figure 2: Time-integrated PL spectrum for InAs QDs with GaAsSb buffer layer, for Sb composition of 0%, 3%, 11%, 13% and 20%. All the Sb containing samples see a red-shift in the emission peak position with increasing Sb compostion.

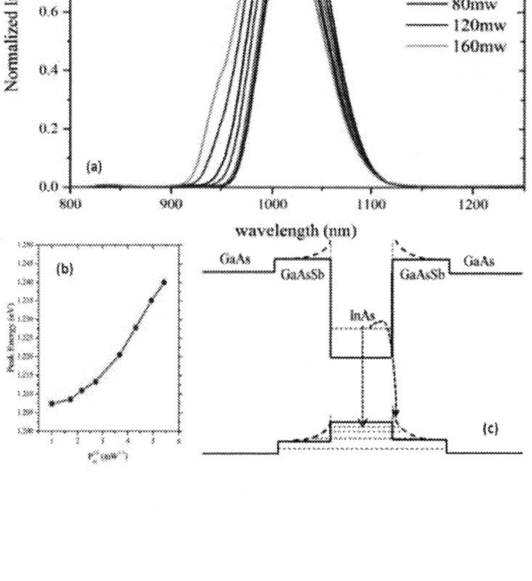

Figure 3: (a) Normalized Power-dependent PL results for cladding layer samples with 11% Sb, (b) Relationship between peak energy and $P^{1/3}_{ex}$, and (c) Deduced band structure of InAs quantum dots cladded by GaAsSb.

The power-dependent photoluminescence (PDPL) for all samples were measured under different incident laser power settings from 1 mW to 160 mW. The normalized spectra for InAs QDs with cladding layer of Sb composition 11% are plotted in Figure 3(a). Interestingly, the PDPL peak position for the QDs in the 11% Sb cladding layer sample experiences a blue-shift from 1030 nm to 1000 nm when increasing laser power emission from 1 mW to 160 mW. The peak energy was found to increase linearly as a function of $P^{1/3}_{ex}$, where P is the laser power level, as shown in figure 3(b). This power dependence suggests a typical type-II band structure between InAs QDs and GaAsSb layer [11]. At low temperatures, the Type-II transition between electrons in the QDs and holes in the GaAsSb leads to a band bending and thus a blue-shift emission due to the spatially separated confined carriers [18]. These results are consistent with the report of Tang et al. [19], where a significant enhancement of InAs multilayer QDs carrier lifetimes by increasing Sb composition (from 9.8% to 10.2%) in GaAsSb cladding layer was observed through testing time-resolved photoluminescence decay curves. Those results indicated the type-I to type-II transition could happen at a Sb composition around 10% when the GaAsSb-buffered InAs QDs is covered with GaAsSb cap instead of typical approximate 14% in GaAsSb capping layer [12].

This lowering of the transition Sb content can be explained by the Sb segregation onto the InAs QDs surface after the growth of GaAsSb buffer layer. After depositing InAs, the surface-resident Sb is likely to participate in the GaAsSb capping layer growth, in which way the Sb composition experiences a significant sudden increase at the interface of InAs QDs and GaAsSb capping layer. The sudden enlarged Sb composition, which is larger than the measured value, therefore forms a valence band valley at the interface between InAs and GaAsSb displayed in the fig. 3(c), leading to a possible type-II transition pathway in terms of the 11% Sb cladding layer. Such type-II band structure system is formed at higher Sb composition situations, leading to a peak position red-shift at lower recombination energy with decreased PL efficiency which is also observed in figure 2.

Further investigations by temperature dependent PL (TDPL) are also under way with the results for the 11% GaAsSb cladding layer sample presented in Fig. 4. The following expression for the emission intensity as a function of temperature with a band edge spike was used to analyse the QD TDPL results [20],

$$I = \frac{G_M}{1+\frac{K}{1+\exp(-E_b/kT)}\left[1+\sum_i B_i \exp(-E_i/kT)\right]}, \quad (1)$$

Where k is Boltzmann's constant, T is the temperature, E_i are the activation energies for non-radiative processes, and E_b is the energy barrier for carrier to be captured into the QDs, which is apparent during low temperature. For the sample with 11% Sb composition, a value of 6.88 ± 0.36 meV for E_b was found, while the data was well fitted with two activation energies for carrier escape from the QD bound states. The hump in intensity seen around 100 K from TPDL curve fitting suggests a potential barrier is present at the interface between InAs quantum dots and the GaAsSb layers, thus preventing carriers to be captured by dots at the low temperature range as is displayed in Fig. 3(c). As the temperature increases, carriers have greater thermal energy and so can overcome the energy barrier, which means they can then contribute to an enhancement of the PL signal.

Meanwhile, with the energy barrier present in the interface between InAs QDs and GaAsSb capping layer, it is likely that the spike in the conduction band structure leads to the rising of the valence band at its corresponding position. Therefore, a valence band valley is formed, giving a pathway for electrons to recombine, via a type-II band structure.

The E_1 value of 70 meV of InAs QD in 11% Sb cladding layer corresponds well to the energy difference between ground state and first excitation state of QDs in valance band, while the E_2 with a larger value around 290 meV is the thermal activation energy of the InAs QDs, which, when added to the ground state energy approximately equals the GaAs band gap value. Such results suggest the sample has good crystal quality in the immediate vicinity of the QDs and fewer parallel pathways exist for carriers to escape before recombining, which was also observed in other samples with lower Sb composition [21]. As Sb composition increased to approximately 20%, no potential barrier could be seen, though high dislocation and defect densities are assumed to be inside the structure, with far lower activation energies than the values of expected. Such results indicate that the electrons are likely to non-radiatively recombine through the defects in the vicinity of InAs QDs instead of radiatively recombining and emitting a PL signal.

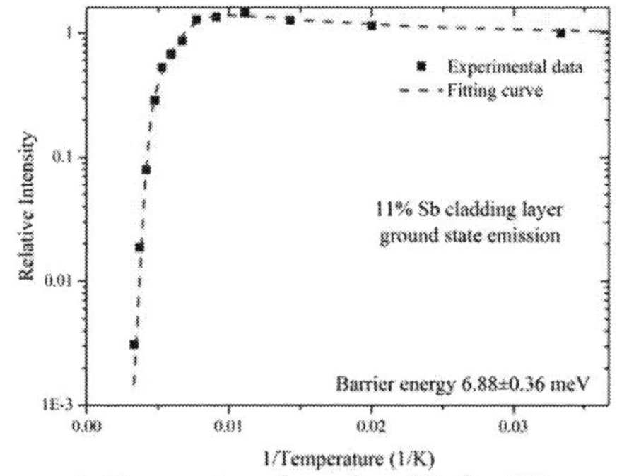

Figure 4: Temperature dependent PL for QDs ground state emission with 11% Sb GaAsSb cladding layer, barrier energy of 6.88±0.36 meV.

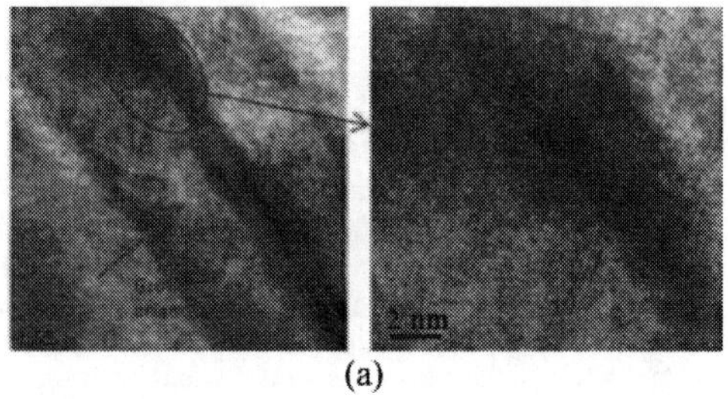

(a)

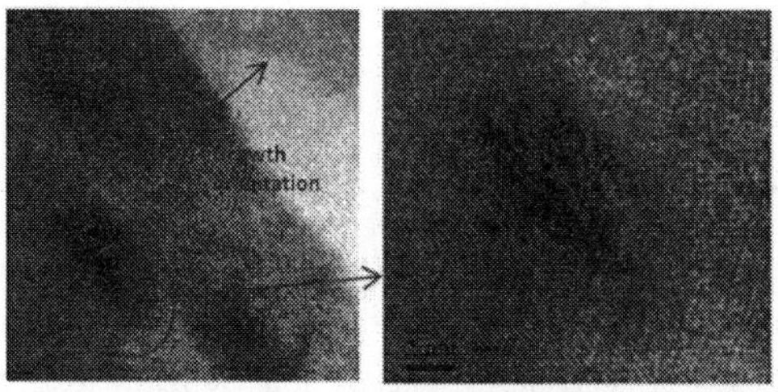

(b)

Figure 5: Cross-sectional TEM results of InAs QDs with GaAsSb cladding layers with Sb composition of (a) 11% and (b) 20%, a magnified section showing a quantum dot is shown for each sample.

Cross-sectional TEM images were taken to analyse the InAs quantum dots structures cladded by GaAsSb with bright-field results for the 11% and 20% Sb content samples shown in Figure 5. As expected these results show better crystalline properties for the structure with 11% Sb GaAsSb cladding layer (Fig. 5(a)) compared with 20% Sb (Fig. 5(b)). With the dark contrast indicating quantum dots trapped in the middle of GaAsSb buffer and cap, a clearly ordered atom distribution along the growth orientation was observed, even across the quantum dots area, in the enlarged graph. Such results indicates the InAs QDs of high crystal quality with dense and uniform distribution, which is consist with the PL results, seeing large PL integrated intensity at lower wavelength range, despite a relatively large FWHM indicating a relatively large variation in quantum dot size. Figure 5(b) shows a typical [1-10] HRTEM image of InAs quantum dot with 20% Sb GaAsSb cladding layer along with with an image zoomed in on the QDs. Under high resolution view, a large density of dislocations and defects were observed in this sample. In terms of the QDs displayed, the disorder of atom distribution above QDs suggests a high density of crystalline defects in the vicinity of the QDs for 20% Sb content. The high defect density is consistent with the PL results of 20% Sb sample, which showed a significant intensity reduction compared with samples with lower antimony content. Furthermore, such conclusions are also supported by the TDPL results, which inferred a significant density of non-radiative escape paths for carriers, consistent with a large density of energy levels within the cladding layer band gap due to crystalline defects.

4 CONCLUSION

We have studied the optical properties of InAs QDs with GaAsSb cladding layers with various Sb composition from 3% to 20% by PL. The GaAsSb cladding layer sees the emission red shifted, with increasing antimony content. Power-dependent photoluminescence results suggests a transition from type-I to type-II could be achieved by GaAsSb cladding layer at Sb composition around 11%. TDPL results indicates good crystal quality and energy barrier at low Sb conposition, while high-level dislocation and defects under high Sb incorporation, which could also be observed from HRTEM images. Such results show that determining the exact Sb composition to realize zero VBO between InAs and GaAsSb depends on many design factors. The Sb present in the buffer layer blue-shifts emission, while the Sb present in the capping layer red-shifts emission, with defect formation due to strain relaxation further complicating considerations. Therefore the strain condition throughout the structure is central when designing quantum dots solar cell systems.

5 REFERENCES

[1] Y. Arakawa, H. Sakaki, Applied Physics Letters, 40 (1982) 939-941.

[2] A.J. Shields, M.P. O'Sullivan, I. Farrer, D.A. Ritchie, R.A. Hogg, M.L. Leadbeater, C.E. Norman, M. Pepper, Applied Physics Letters, 76 (2000) 3673-3675.

[3] A. Martí, N. López, E. Antolín, E. Cánovas, C. Stanley, C. Farmer, L. Cuadra, A. Luque, Thin Solid Films, 511–512 (2006) 638-644.

[4] M.A. Green, Progress in Photovoltaics: Research and Applications, 9 (2001) 123-135.

[5] S.P. Bremner, M.Y. Levy, C.B. Honsberg, Applied Physics Letters, 92 (2008) 171110.

[6] W. Shockley, H.J. Queisser, Journal of Applied Physics, 32 (1961) 510-519.

[7] A. Marti, E. Antolin, P.G. Linares, A. Luque, Journal of Materials Chemistry, 22 (2012) 22832-22839.

[8] N.J. Ekins-Daukes, C.B. Honsberg, M. Yamaguchi, Conference Record of the Thirty-first IEEE Photovoltaic Specialists Conference, 2005., 2005, pp. 49-54.

[9] S.N. Dahal, S.P. Bremner, C.B. Honsberg, Progress in Photovoltaics: Research and Applications, 18 (2010) 233-239.

[10] H.Y. Liu, M.J. Steer, T.J. Badcock, D.J. Mowbray, M.S. Skolnick, F. Suarez, J.S. Ng, M. Hopkinson, J.P.R. David, Journal of Applied Physics, 99 (2006) 046104.

[11] K.-Y. Ban, D. Kuciauskas, S.P. Bremner, C.B. Honsberg, Journal of Applied Physics, 111 (2012) 104302.

[12] H.Y. Liu, M.J. Steer, T.J. Badcock, D.J. Mowbray, M.S. Skolnick, P. Navaretti, K.M. Groom, M. Hopkinson, R.A. Hogg, Applied Physics Letters, 86 (2005) 143108.

[13] S.P. Bremner, K.-Y. Ban, N.N. Faleev, C.B. Honsberg, D.J. Smith, Journal of Applied Physics, 114 (2013) 103511.

[14] H. Shimizu, S. Saravanan, Applied Physics Letters, 88 (2006) 041119.

[15] K. Yamaguchi, T. Kanto, Journal of Crystal Growth, 275 (2005) e2269-e2273.

[16] J.M. Ripalda, D. Alonso-Álvarez, B. Alén, A.G. Taboada, J.M. García, Y. González, L. González, Applied Physics Letters, 91 (2007) 012111.

[17] C.Y. Jin, H.Y. Liu, S.Y. Zhang, Q. Jiang, S.L. Liew, M. Hopkinson, T.J. Badcock, E. Nabavi, D.J. Mowbray, Applied Physics Letters, 91 (2007) 021102.

[18] N.N. Ledentsov, J. Böhrer, M. Beer, F. Heinrichsdorff, M. Grundmann, D. Bimberg, S.V. Ivanov, B.Y. Meltser, S.V. Shaposhnikov, I.N. Yassievich, N.N. Faleev, P.S. Kop'ev, Z.I. Alferov, Physical Review B, 52 (1995) 14058-14066.

[19] D. Tang, Y. Kim, N. Faleev, C.B. Honsberg, D.J. Smith, Journal of Applied Physics, 118 (2015) 094303.

[20] D.P. Popescu, P.G. Eliseev, A. Stintz, K.J. Malloy, Semiconductor Science and Technology, 19 (2004) 33-38.

[21] T. Bray, Y. Zhao, P. Reece, S.P. Bremner, Journal of Applied Physics, 113 (2013) 093102.

IMPACT OF SEEDING INDIUM TIN OXIDE ON ZNO NANOROD ANTIREFLECTIVE LAYER IN SILICON HETEROJUNCTION SOLAR CELLS

Maike Ahrlich, Oleg Sergeev, Maren Juilfs, Alex Neumüller, Martin Vehse, Carsten Agert
NEXT ENERGY · EWE Research Centre for Energy Technology at the University of Oldenburg,
Carl-von-Ossietzky-Str. 15, 26129 Oldenburg, Germany

ABSTRACT: In this work we investigated the influence of fabrication procedure of indium tin oxide (ITO) seed layers, acting as front contacts in silicon heterojunction solar cells, on the growth of ZnO antireflective layers. Furthermore, ZnO nanorod growth out of $ZnCl_2$ and $Zn(NO_3)_2$ solution was compared. It was observed that for both $ZnCl_2$ and $Zn(NO_3)_2$ nanorod layers, the ITO sheet resistance affects the density and size of resulting nanorods. The most effective antireflective nanorod layers grew on an ITO layer with a sheet resistance of approximately 50 $\Omega_\square$. In this case, the nanorod layers had a suitable density, which was neither too low nor too high. The application of such a nanorod antireflective layer in a SHJ solar cell led to a relative gain in the short circuit current density by 0.6 %.
Keywords: Silicon Heterojunction, Nanostructures, Electrodeposition

1 INTRODUCTION

Amorphous/crystalline silicon heterojunction (SHJ) solar cells are a powerful candidate for future photovoltaic applications, because they combine high efficiency (25,6 % [1]) with reduced fabrication cost, if compared to conventional homojunction silicon solar cells [2]. As is the case for homojunction silicon solar cells, there are also optical losses such as reflection losses in SHJ solar cells.

In order to reduce these reflection losses, most SHJ solar cells are built on wafers with pyramid-like textured surfaces. However, wafer surface texturing is a rather complicated procedure with several drawbacks, such as increased defect densities at the surface and hence increased surface recombination losses.

Recently, ZnO nanostructure layers were implemented as broadband antireflection coating on top of SHJ solar cells with front contacts based on indium tin oxide (ITO) [3] and aluminium doped zinc oxide [4]. The use of such nanostructured layers could possibly substitute wafer texturing and provide the use of flat wafers with other related advantages. Especially for very thin wafers (< 90 µm), for which surface texturing would be hard to realize, efficient light trapping techniques are highly important, since light absorption in low energy spectral regions is less than in thick wafers.

In this work, we report on the formation of ZnO nanorod layers on differently fabricated ITO front contacts of SHJ test structures and solar cells in order to investigate the impact of ITO properties on the morphology and antireflective behavior of ZnO nanorods and to further improve the reduction of reflection. For the formation of ZnO nanorod layers electrochemical deposition was used, as a low-cost and industrially applicable method.

2 EXPERIMENTAL

The SHJ test structures and solar cells investigated in this work are depicted in Fig. 1. They were fabricated on n-type non-textured and polished FZ c-Si wafers. Prior to deposition, the substrates were immersed to HF (1 %) for 3 min to remove the native oxide. The intrinsic and doped a-Si:H and µc-Si:H layers were grown by plasma enhanced chemical vapour deposition from SiH_4:H_2 mixture, adding B_2H_6 or PH_3 for p- and n-doping. ITO and Ag films were deposited by DC magnetron sputtering and electron beam evaporation, respectively. For the SHJ test structures, the deposition temperature of the sputtered ITO front contact was varied. Additionally, half of the test structures have been annealed at a temperature of 190 °C for 90 min after nanorod deposition. Tab. I shows the used ITO fabrication parameters.

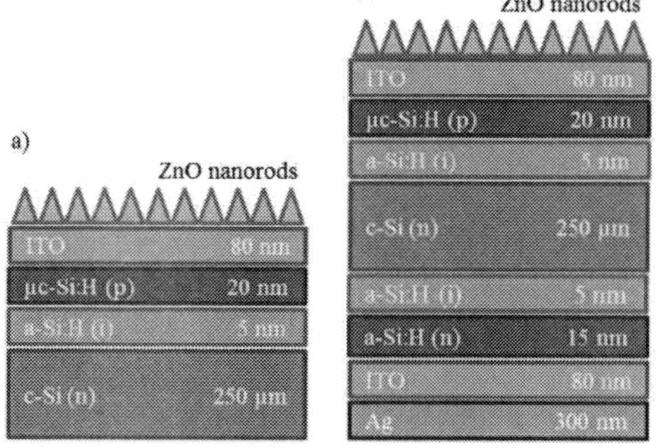

Figure 1: Schematic illustration of used SHJ a) test structures and b) solar cells.

Table I: ITO fabrication parameters for SHJ test structures

Sample name	Deposition temperature [°C]	annealed
ITO_{NT}	25	no
ITO_{NT}^T	25	yes
ITO_{HT}	200	no
ITO_{HT}^T	200	ves

Electrochemical deposition of ZnO nanorods was performed in a three-electrode setup with the SHJ test structure or solar cell as working electrode, an Ag/AgCl reference electrode and a Pt counter electrode, using a VERSASTAT 4 potentiostat system. The ZnO nanorod layers were grown on each kind of SHJ test structure twice. First, according to the method published by Peulon and Lincot [5], an aqueous solution of 0.5 mM $ZnCl_2$ and 0.1 M KCl, saturated with molecular oxygen, was used as electrolyte. Potential, temperature and time during deposition was -0.82 V, 80 °C and 1800 s, respectively.

Second, deposition was carried out using an aqueous solution of 0.02 M $Zn(NO_3)_2$ as electrolyte, as proposed firstly by Izaki and Omi [6]. Here, a potential of -0.915 V, a temperature of 70 °C and a deposition time of 480 s was applied.

3 RESULTS AND DISCUSSION

3.1 Nanorod morphology

Fig. 2 and Fig. 3 show scanning electron microscopy (SEM) images of the resulting ZnO nanorod layers grown out of $ZnCl_2$ and $Zn(NO_3)_2$ solution on the differently fabricated ITO films. As the ZnO nanorod films were deposited directly on an ITO film, the nanorods did not grow epitaxial and up-right, but rather tilted. The nanorods exhibit a hexagonal shape, as can be seen in the enlarged depiction of a single nanorod on the ITO_{NT} film in Fig. 2. This indicates a good crystallinity of the nanorod films.

Tab. II contains the diameter and density of the $ZnCl_2$ nanorods grown on different ITO films, which were deduced from the SEM images. Here, the value for the nanorod diameter was averaged over 80 nanorods, whereas the density was determined as the amount of nanorods per 25 μm^2. The ITO sheet resistance, measured prior to nanorod deposition, is also listed in Tab. II.

Table II: Sheet resistance $R_\square$ of the differently pre-treated ITO films and diameter and density of the nanorods grown on the ITO films out of $ZnCl_2$ solution.

Sample name	$R_\square$ [$\Omega_\square$]	nanorod diameter [nm]	nanorod density [nanorods/μm^2]
ITO_{NT}	61	368	0.3
ITO_{NT}^T	53	181	2.3
ITO_{HT}	50	144	5.1
ITO_{HT}^T	47	139	12.4

It can be observed that with decreasing ITO sheet resistance, which is caused by increasing heat treatment of the ITO films, the nanorod density increases. Simultaneously, the nanorod diameter as well as the overall crystal size decreases. The increasing nanorod density was expected, due to a higher conductivity of the ITO film, which led to a higher electrochemical current during ZnO nanorod deposition and hence a higher density of nucleation sites. The decreasing nanorod size indicates that the amount of deposited zinc oxide is similar for all ITO films. If there is a lack of nucleation sites, the few existing nanorods will grow comparably big. In contrast, the nanorods will remain comparably small, if the same amount of zinc oxide has to be used for formation of many nanorods.

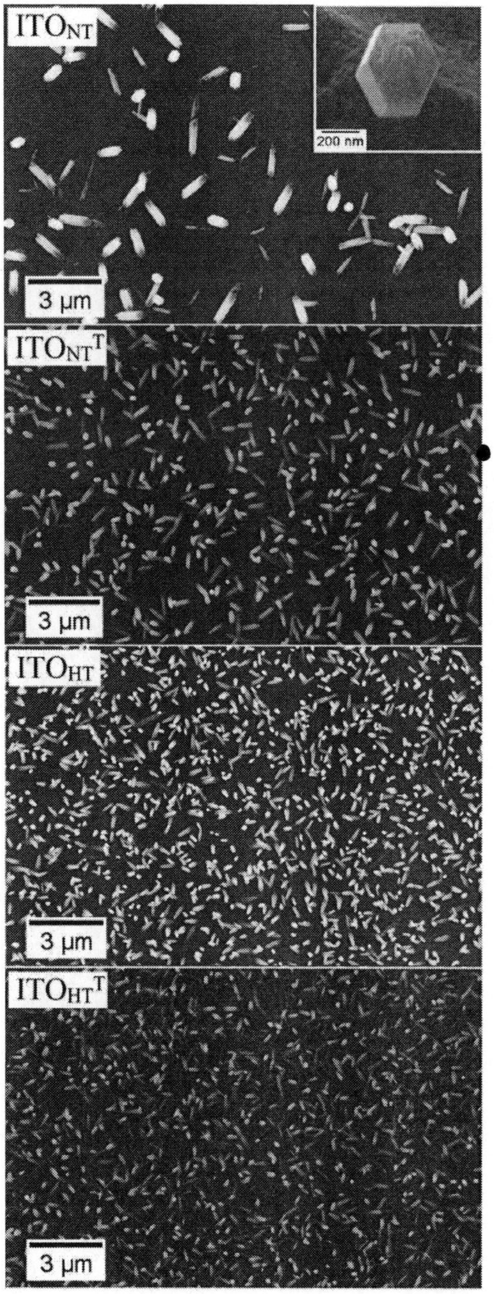

Figure 2: SEM images of ZnO nanorod arrays grown out of $ZnCl_2$ solution on differently fabricated ITO layers.

The diameters and densities of the $Zn(NO_3)_2$ nanorods depicted in Fig. 3 were determined the same way as for the $ZnCl_2$ nanorods and listed in Tab. III, together with the corresponding ITO sheet resistances.

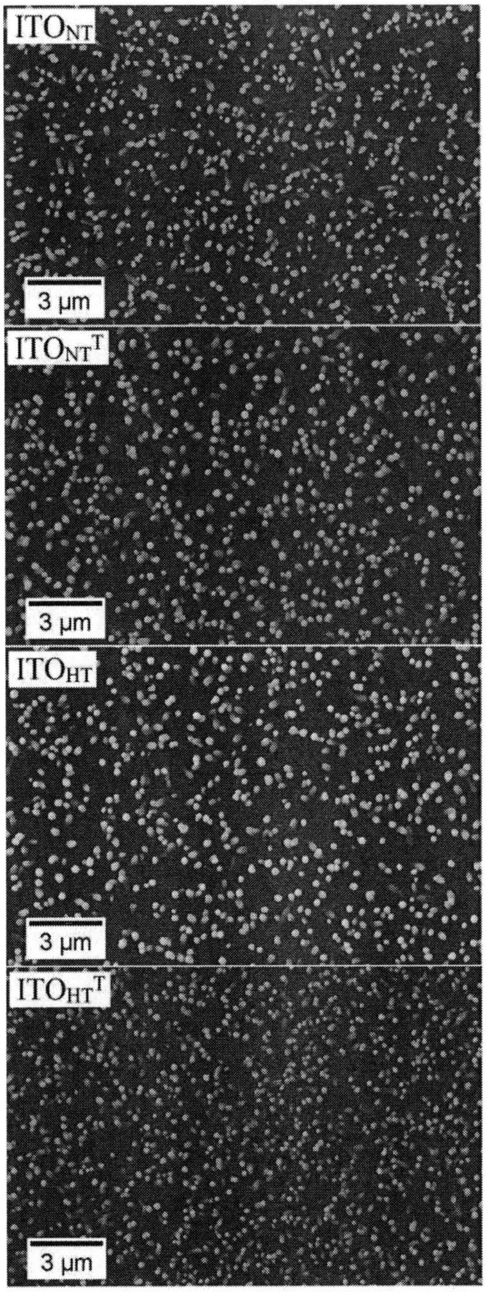

Figure 3: SEM images of ZnO nanorod arrays grown out of $Zn(NO_3)_2$ solution on differently fabricated ITO layers.

The impact of ITO heat treatment or sheet resistance in the case of the $Zn(NO_3)_2$ nanorods is not so pronounced as for the $ZnCl_2$ nanorod case. Nevertheless, with decreasing ITO sheet resistance, an increase in the nanorod density and a slight tendency of a decrease in the nanorod size was also observed in this case.

Table III: Sheet resistance $R_\square$ of the differently fabricated ITO films and diameter and density of the nanorods grown on the ITO films out of $Zn(NO_3)_2$ solution.

Sample name	$R_\square$ [$\Omega_\square$]	nanorod diameter [nm]	nanorod density [nanorods/µm²]
ITO_{NT}	63	212	3.0
ITO_{NT}^{T}	48	214	3.7
ITO_{HT}	50	220	3.3
ITO_{HT}^{T}	43	182	4.2

3.2 Antireflective properties of ZnO nanorods

In Fig. 4, reflection curves of the $ZnCl_2$ and $Zn(NO_3)_2$ nanorod layers, deposited on SHJ test structures with different ITO front contacts, can be seen. The reflection curves of the corresponding test structures without nanorods are also presented.

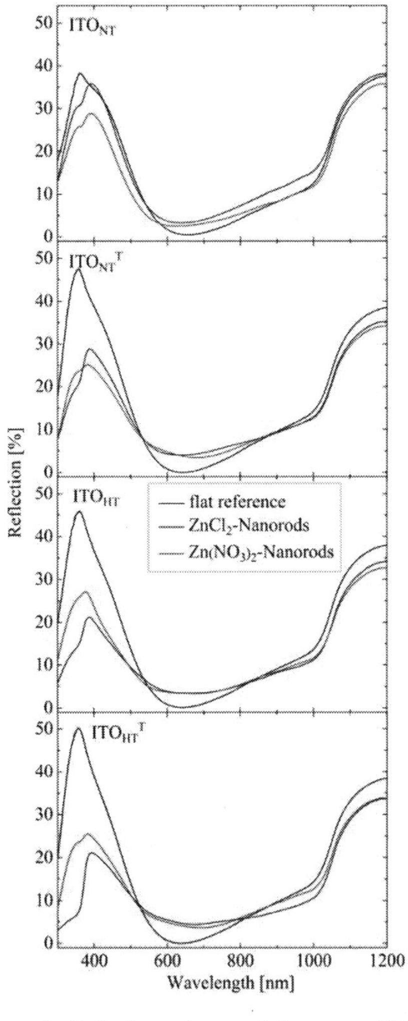

Figure 4: Reflection of nanorod layers on SHJ test structures with differently fabricated ITO, compared to flat reference.

In general it can be seen, that the application of the ZnO nanorods causes a reduction of reflection. Only in the wavelength range around 650 nm, where antireflection is most pronounced for the flat SHJ test structure due to the optimized ITO layer thickness of 80 nm, the antireflection condition is weakened by the use of the nanorod layer and hence the reflection is increased.

For evaluating the effectiveness of this ZnO nanorod antireflection coating, it is important to take the AM 1.5G solar spectrum into account. Therefore, the average reflectivities R^w of the nanorod and flat structures, weighted with the AM1.5G spectrum, were calculated according to the following formula and listed in Tab. IV.

$$R^W = \frac{\int_{380}^{1200} S(\lambda) \cdot R(\lambda) d\lambda}{\int_{380}^{1200} S(\lambda) d\lambda}$$

Here, $S(\lambda)$ and $R(\lambda)$ denote the AM1.5G spectrum and the measured reflection as function of incident wavelength λ, respectively.

Table IV: Average weighted reflectivities R^w (380-1200 nm) of flat and nanorod SHJ solar cells, grown out of $ZnCl_2$ and $Zn(NO_3)_2$ solution, respectively.

Sample name	R^w_{flat} [%]	R^w_{Cl} [%]	R^w_{NO} [%]
ITO_{NT}	11.58	13.08	10.37
$ITO_{NT}{}^T$	11.96	11.59	10.79
ITO_{HT}	11.59	9.46	9.60
$ITO_{HT}{}^T$	11.75	10.26	10.77

According to Tab. IV, the use of a nanorod layer led to a reduction in the weighted reflectivity for almost all SHJ test structures. This is due to the formation of an effective medium at the air/ITO interface by the ZnO nanorod layer, which acts as a layer with effective refractive index. This weakens the abrupt transition in refractive indices at the interface [7], which is responsible for reflection according to the Fresnel equations. The reflection is increased only for the SHJ test structure with ITO_{NT} front contact and $ZnCl_2$ nanorods. This is because the antireflective effect of the ITO layer is affected in a comparably wide wavelength range, as can be seen in Fig. 4.

Furthermore it can be seen in Tab. IV that the weighted reflectivity of the nanorod test structures initially decreases with increasing ITO heat treatment, but afterwards rises again. The reason for this is that the reflection curves exhibit two opposing tendencies with increasing ITO heat treatment, as can be seen in Fig. 5 for both $ZnCl_2$ and $Zn(NO_3)_2$ nanorods. On the one hand, the overall reflectivity decreases with increasing ITO heat treatment. Only at wavelengths around 600 nm, where antireflection is most pronounced due to the antireflective effect of ITO, reflection increases with increasing ITO heat treatment. Since it was shown before that an increase in the ITO heat treatment goes along with an increasing nanorod density, it can be concluded, that the nanorod density should be neither too high nor to low, to ensure an optimized reflection behavior.

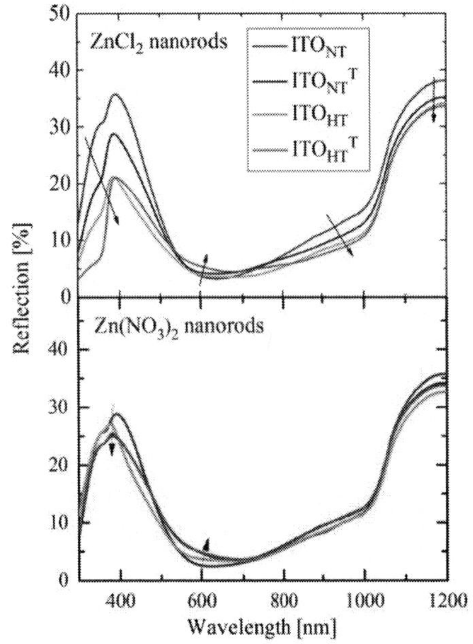

Figure 5: Reflection curves of test structures with different ITO layers and $ZnCl_2$ (top) and $Zn(NO_3)_2$ nanorods (bottom). The arrows indicate increasing ITO heat treatment.

In our investigations, the lowest weighted reflection of 9.46 % was observed for the SHJ test structure with ITO_{HT} front contact and $ZnCl_2$ nanorods, as can be seen in Tab. IV. Therefore, this configuration was chosen for the implementation in a solar cell, which is described in the following chapter.

3.3 Solar cell performance

Fig. 6 displays the external quantum efficiency (EQE) and total absorption (1- Reflection R) of the SHJ solar cell with and without ZnO nanorods. While the overall reflectivity decreases by the use of ZnO nanorods, the reflectivity at around 600 nm is enhanced, as was also the case for the SHJ test structures.

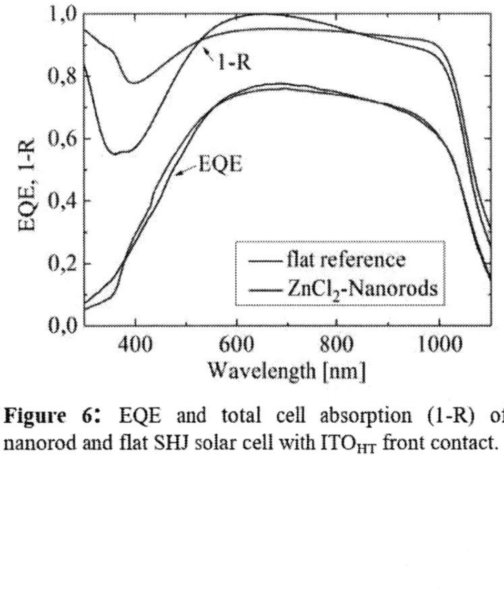

Figure 6: EQE and total cell absorption (1-R) of nanorod and flat SHJ solar cell with ITO_{HT} front contact.

In general, a direct correlation between the reflection and EQE curves is recognizable. The EQE is enhanced at those wavelengths, where the reflection is decreased by the ZnO nanorods and vice versa. Losses in the EQE for wavelengths <380 nm are related to parasitic absorption of the electrochemically deposited ZnO. All in all, a relative increase in the short circuit current density by 0.6 % could be achieved by the use of the ZnO nanorods.

4 CONCLUSION

In this study, we investigated the impact of ITO sheet resistance, which was influenced by deposition temperature and annealing, on the morphology and antireflective properties of the resulting nanorod antireflective layers.

It was observed that the nanorod arrays become denser with increasing ITO conductivity, whereas the crystal size decreases. These tendencies were recognizable for both $ZnCl_2$ and $Zn(NO_3)_2$ nanorods. Regarding the antireflective behavior, two opposing trends revealed: In the wavelength range around 650 nm, where antireflection for the flat structures was most pronounced due to the optimized ITO thickness, reflection was increased with increasing ITO conductivity or nanorod density, whereas for the other wavelengths, the reflection decreased. Nevertheless, almost all test structures showed a reduction in weighted reflectivity by use of the ZnO nanorod layer.

The nanorod layers deposited on an unannealed ITO layer fabricated at 200 °C (sheet resistance: ~ 50 $\Omega_{\square}$) showed the best antireflective properties. These nanorod layers had a suitable density, which was neither too low nor too high. The application of such a nanorod antireflective layer in a SHJ solar cell led to a relative gain in the short circuit current density by 0.6 %.

Since electrochemical deposition as fabrication technique is cost-effective and can be easily transferred to industrial scale, the concept is promising as a process step in industry to improve light absorption in silicon heterojunction solar cells in the future.

ACKNOWLEDGEMENTS

The authors thank Martin Kellermann for support with PECVD depositions. Financial support from the German Estonian Ministry of Higher Education and Research (Project IUT19-28) as well as German Federal Ministry of Economics and Technology (BMWi, reference number 0325825H, HERA), are gratefully acknowledged.

REFERENCES

[1] K. Masuko, M. Shigematsu, T. Hashiguchi, D. Fujishima, M. Kai, N. Yoshimura, T. Yamaguchi, Y. Ichihashi, T. Yamanishi, T. Takahama, M. Taguchi, E. Maruyama, S. Okamoto, IEEE Journal of Photovoltaics 4 (2014) 1433.

[2] S. de Wolf, A. Descoeudres, Z.C. Holman, C. Ballif, Green 2 (2012) 7.

[3] M. Petrov, K. Lovchinov, M. Mews, C. Leendertz, D. Dimova-Malinovska, Physica Status Solidi A 212 (2015) 166.

[4] M. Ahrlich, O. Sergeev, M. Juilfs, M. Kellermann, A. Neumüller, M. Vehse, Proceedings 31st European Photovoltaic Solar Energy Conference (2015).

[5] S. Peulon, D. Lincot, Advanced Materials 8 (1996) 166.

[6] M. Izaki, T. Omi, Applied Physics Letters 68 (1996) 2439.

[7] L. Rayleigh, Proceedings of the London Mathematical Society s1-11 (1879) 51.

ELECTRIC PROPERTIES OF NANOCRYSTALLINE DIAMOND THIN FILM DEPOSITED ON ACTIVE SUBSTRATE SOLAR CELL STRUCTURE

Martin Kusko*, Milan Perný**, Vladimír Šály**, Michal Váry**, Juraj Packa**
* Fill Factory, Televizní 2618, 756 61 Rožnov pod Radhoštěm, Czech Republic
**Slovak University of Technology
Faculty of Electrical Engineering and Information Technology
Ilkovičova 3, SK-81219 Bratislava, Slovak Republic

ABSTRACT: Photochemical splitting of water for hydrogen production by photocatalysis is very promising technology for the future covering of energy demands. Physical and chemical properties and resistance to corrosion predestine boron doped diamond to be widely used in electrochemical applications. In this study we present the results of electrical characterization of layered structure of boron doped thin film diamond deposited on the surface of photovoltaic monocrystalline silicon pn junction. As a results of AC measurements, the equivalent circuit was proposed and its elements were calculated and discussed. Two relaxation processes were identified and also activation energy of carrier transport determined.
Keywords: photovoltaics, water splitting, complex impedance

1 INTRODUCTION

Photovoltaic energy is often required to be stored, either using the common batteries or other mechanisms. Convenient way can bethe generation of hydrogen and its storing as a gas. On the other hand, the hydrogen is the medium which has started to be one of the fuel substitutions of the oil powered engines. So it is very appropriate to pay more attention to its generation, e.g. via photovoltaic electricity.

The hydrogen is the most easily generated by water electrolysis (water splitting). Water splitting triggered by the sun light became interesting topic since an application of titanium dioxide (TiO_2) as a photo-catalytic electrode material for oxygen evolution which absorbs UV part of spectrum and together with counter-electrode activates the splitting process powered by sun radiation [1]. Photochemical splitting of water for hydrogen production by photocatalysis is very common and often studied technique [2]. It is due to the potential of this process while economic and environmental benefits are expected. As a photocatalysis material many materials have been developed during the long period from the very first experiments. Except other material properties, it is already known, that the photocatalytic activity for water splitting is dependent on the crystallinity and the particle size of the used material which are dependent on the preparation conditions.

The dissociation process of water molecule follows the equation

$$H_2O \rightarrow H_2 + 1/2O_2 \qquad (1)$$

where water molecule H_2O is liquid phase and produced hydrogen and oxygen are gases. The process is accompanied with an energy demand 1.23 eV. Theoretically, the minimum voltage needed to force splitting reaction is 1.23 V [1]. In consideration of the ohmic losses and processes at the electrode surface that voltage has to be higher [3] up to 1.8 V. If the solar cell based on crystalline silicon is applied to generate that voltage, single cells generating not more than 0.7 V have to be connected in series to increase the output voltage.

Diamond can be applied for very wide areas and for very special applications, while it provides advantageous mechanical properties – thanks to extreme hardness and the resistance against wearing [4], optical properties – stand-alone diamond substrates (in defined shape) can be used as the transparent windows from VIS to far IR radiation [5], electrical properties – the most common electrical structures are made by MWPE CVD techniques also for easy doping, mainly boron doped diamond (BDD). By proper doping well semiconductor and conductivity properties can be achieved [6] and even biochemical applications are within interest while carbon is the base material of all live species on Earth. This predetermines the highly pure synthetic (doped) diamond for biological sensing of molecules or specific cells [7]. Boron doped diamond is material with suitable electrode properties and is attracting considerable interest [8, 9]. Above mentioned physical and chemical properties and high chemical stability and resistance to corrosion based option for boron doped diamond to be widely used in electrochemical applications or sensor device technology. Advantage of using a boron-doped diamond (BDD) electrode for hydrogen production and wastewater treatment in a single electrochemical cell was also try out and presented eg in [10]. The BDD electrode exhibited high anodic potential, generating high oxidation state radicals that facilitated oxidation of toxic waste organic compounds. Moreover, because of widening of potential windows, the rate of hydrogen evolution at the cathode was significantly increased.

Synthetic diamond seems to be promising material with convenient properties and fulfilling essential requirement for photo-electrode material with potential to avoid unwanted photo-corrosion as secondary phenomenon which occurs during water splitting redox processes on most of semiconductor and metal materials. Synthetic diamond, which has been intensively studied during the last decades also as electrode material, especially the nanocrystalline diamond (NCD), has been mostly prepared by the CVD technique and its modified alternatives like Hot Filament CVD, Microwave Plasma Enhanced CVD or others [11]. Regarding the NCD thin film deposition, the pre-treatment of substrate surface is necessary for heteroepitaxial growth. This step is fundamental, simply because the nucleation sites represent seeds for consequent growth of diamond crystals which gradually grow, at some point joint together and create homogenous layer [12]. Nucleation also determines other properties like grain size, orientation, adhesion or reflection (transparency). Generally, for the NCD growth process, it is necessary to ensure presence of gas mixture consisting of hydrocarbon-based precursors (like CH_4) in hydrogen atmosphere where low concentration of these precursors, usually between 1 – 4 %, ensures proper process of sp^3 hybridization. The main and important role of hydrogen presence is etching of graphitic phase for ensuring the formation of diamond structure. Positive impact onto the layer quality and the growth rate has the presence of oxygen. The conductivity of the diamond layer depends, of course, on the concentration of the dopants (doping level).

In this work we present the results of electrical characterization of nanocrystalline diamond layer doped with boron deposited on special solar cell structure (active substrate) by CVD technique. Nanocrystalline diamond is typically denoted to structure with grain size about 10 to 100 nm. The Si substrate with the pn+ junction played a role of the "active substrate" (SC structure) as the potential voltage generator in the Si/B-NCD heterostructure. The diamond in combination with crystalline silicon solar cell substrate has the potential in application for water splitting application. The substrate is prepared in the term of future water splitting tests as an electrically active substrate able to transfer the created potential through the boron-doped NCD layer after illumination and finally splits the water.

The electrical measurement, DC measurement and AC measurement were performed, impedance spectroscopy has been used to study electrical and physical properties and obtained resultshas been discussed. Impedance spectroscopy technique is successfully used in electrochemistry and frequent also as a tool to characterize transport properties in semiconducting and dielectric structures. The complex impedance in the form of Nyquist plot was used also to study e.g. the interface of substrate (titanium in this case) and BDD [9].

2 EXPERIMENTAL

The 2x2 cm^2 cuts from standard single crystalline p-type Cz wafers of resistivity 3 Ω.cm and thickness of 190±15 μm (depends on etching process) have been used for NCD deposition. After the preparation of pn junction, structures were estimated to have shown photovoltaic behaviour. The pre-treatment has been optimized for successful deposition of NCD. The chemicals used for NCD synthesis consisted of fundamental mixture, so as:

- Hydrogen (H_2),
- Methane (CH_4),
- Carbon Dioxide (CO_2),
- and the dopant source Trimethylboron (TMB).

Various structural configurations and technology steps were estimated to get the optimal substrate for deposition of NCD. The optimized substrate structure selected for our study consists of phosphorous diffused n+ emitter. Silicon nitride (a-SiN_x:H) with thickness 75 nm was deposited by PECVD (450 kHz) on the surface of the structure as antireflection coating(ARC). This layer functions alsoas the barrier against potential contamination from Mo base. For the lifetime (LT) investigation the Semilab model WT-2000 PV using principle of Microwave Photoconductance Decay (MW PCD) was used, which uses IR laser source with wavelength 904 nm (penetration depth is approximately 30 μm). Fig. 1 shows the map of lifetime distribution in pn+ junction over the representative sample. The average LT value of all used wafers was 12.55 μs. The average lifetime 12.14 μs was obtained on the sample shown in Fig. 2. In closer view one can recognize small dispersion of LT's from minimal to maximalvalues. The red areas in lateral direction can be related to the marks after wafer slicing. At very top of the wafer one can also recognize the label which means that picture shows first wafer of the batch. The lifetime was measured again after the deposition of NCD. The influence of substrate temperature during the deposition of NCD on LT is shown in Fig. 2.

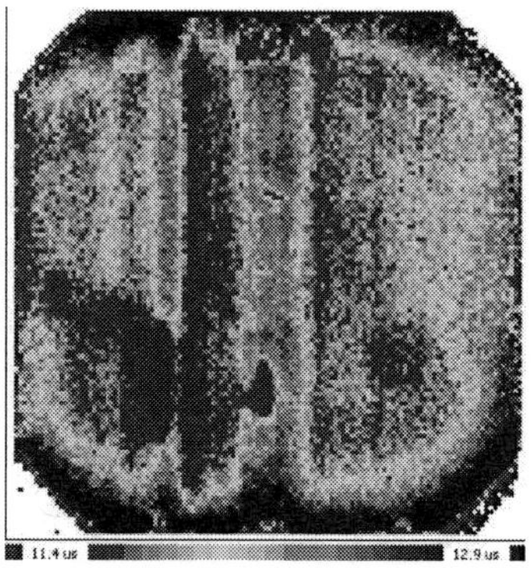

Figure 1: Example of Minority Carrier Lifetime (MCLT) distribution map of wafer chosen from batch of p-n+ /SiN_x structures (before NCD deposition)

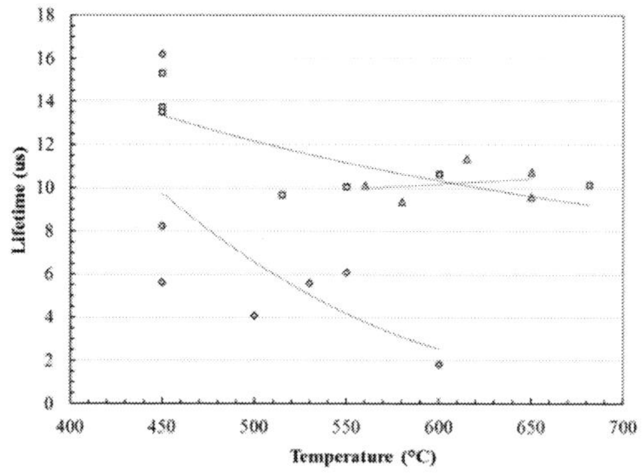

Figure 2: The minority carrier lifetime dependences as a function of processing temperature – deposition of NCD

With regard to aforegoing results we decided to choose the sample pn+ with the surface protected by SiN_x layer for electrical characterization. More about preparation and structural characterizationof prepared structures one can find in [13].

DC measurements were performed using Keithley 2440. AC impedance spectroscopy characterization was performed by SOLARTRON Analytical module. The measurements have been performed on biased samples at various temperatures and at AC frequencies from the interval 10 Hz to 1 MHz. The equivalent circuit which results from complex impedance plot fit offers the view on electric transport properties of studied structure.

3 RESULTS AND DISCUSSION

Barrier behavior of analyzed samples was confirmed from depicted DC characteristics measured at temperatures from interval starting at room temperature 298 K to 400 K (Fig. 3). The reverse and forward diode behavior is evident. The shape under the forward bias is apparently influenced by high series resistance.

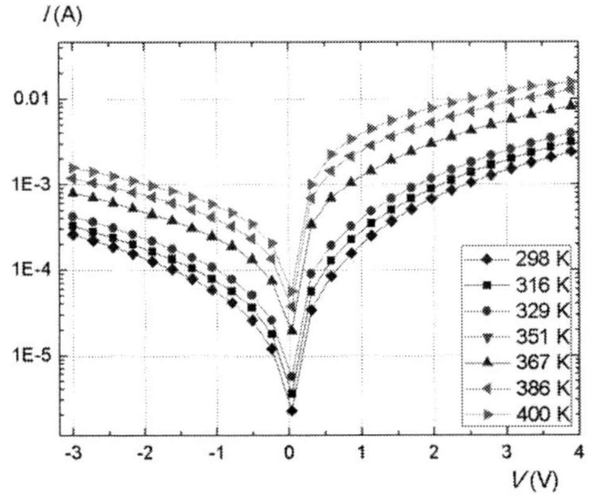

Figure 3: Temperature dependence of dark biased I-V characteristics

The increased bias voltage in reverse direction enlarges the space charge region thickness while the positive superposed bias voltage influences the injection of minority carriers. Process, in both cases, results in change of equivalent circuit parameters and so the shape of the complex impedance plot as one can see in Fig. 4.

Measured impedance response of Si/B-NCD heterostructures illustrated in a series of impedance diagrams in Fig. 4 cannot be fitted by a simple representation of connected parallel RC elements, how it is common for the behavior of pn junction, in complex plain depicted as single arc. The different approach should be applied. Radius of the arc decreases with an increase of bias voltage and temperature in Fig. 4.The presence of the second arc related to another carrier transport process in the impedance spectrum is visible particular at higher temperature 386 and 400 K – the impedance diagram at +1 V in Fig. 4. For the understanding of transport mechanisms and processes we have searched for the appropriate AC equivalent circuit of the structure. The AC equivalent circuit in Fig. 5 of the structure was proposed using the numerical methods in order to determine the conductive processes. The combination of passive components resistor R in parallel with constant phase element CPE, here denoted as P optimally reflects (fits) the AC response [14].

Fig. 6. shows dependence of obtained parallel resistances on the DC bias voltage. The voltage dependence of resistance indicates two different types of resistors. Voltage-dependent resistor R_3 and voltage independent resistor R_2 were obtained. Voltage-independent dependence of R_2 can be assigned to standard ohmic behavior while the voltage-dependent R_3 is related to the injection of the electronic charge. The temperature dependence of the parallel equivalent resistance in Fig. 7 completes the comprehensive information about the electrical properties of the structure. The activation energy E_a was calculated using the well-defined linear dependence of conductance $1/R_3$ on temperature – $\ln(1/R_3) = f(1/T)$ and Arrhenius equation. E_a is dependent on the bias voltage in this case.

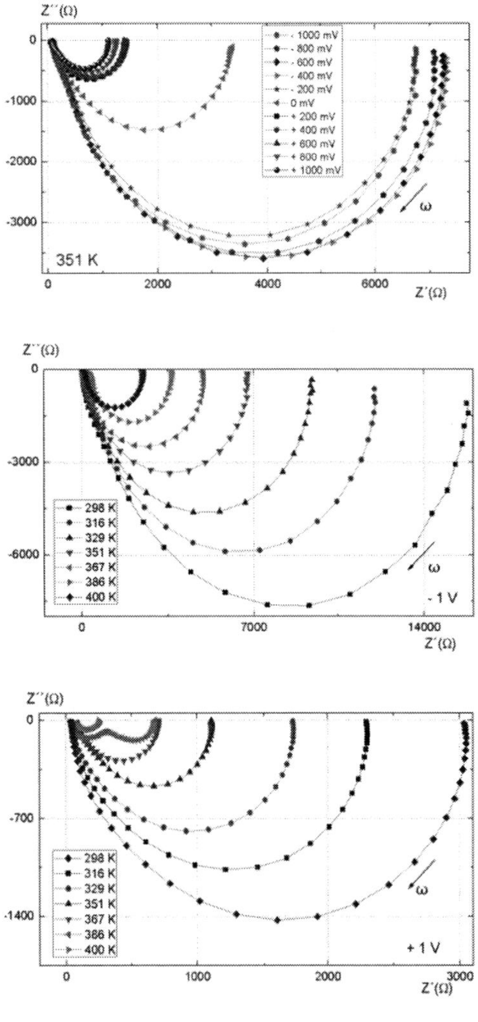

Figure 4: Complex impedance diagrams at various bias voltages and temperatures

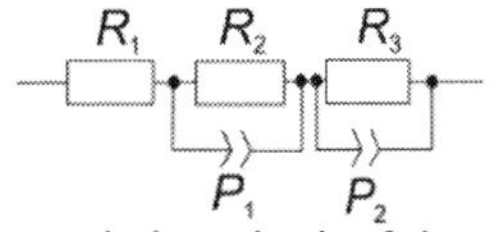

Figure 5: AC equivalent circuit of the structure Al/a-Sic/-Si(n)/Al

Arrhenius dependence shows the presence of two linear regions I. and II. (an example in Fig. 8) which are related to two different activation processes. The voltage dependence of calculated E_a is shown in Fig. 9 for both temperature intervals. E_{a1} is related to region I., while E_{a2} to region II. In our case the dependence at forward bias is not linear as it might to be for barrier junction, e.g. presented in [15].

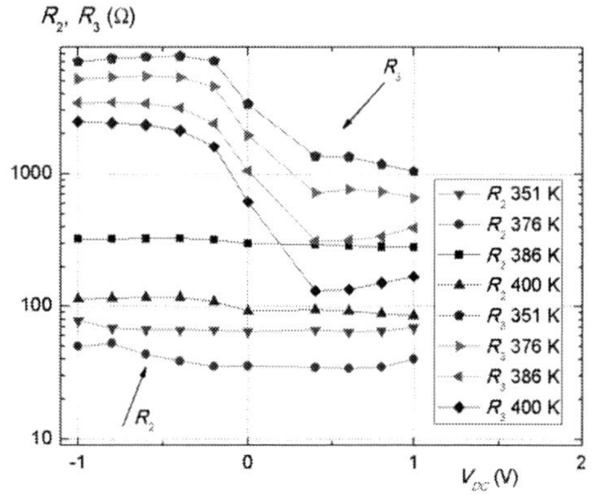

Figure 6: AC equivalent circuit resistances as a function of applied DC bias

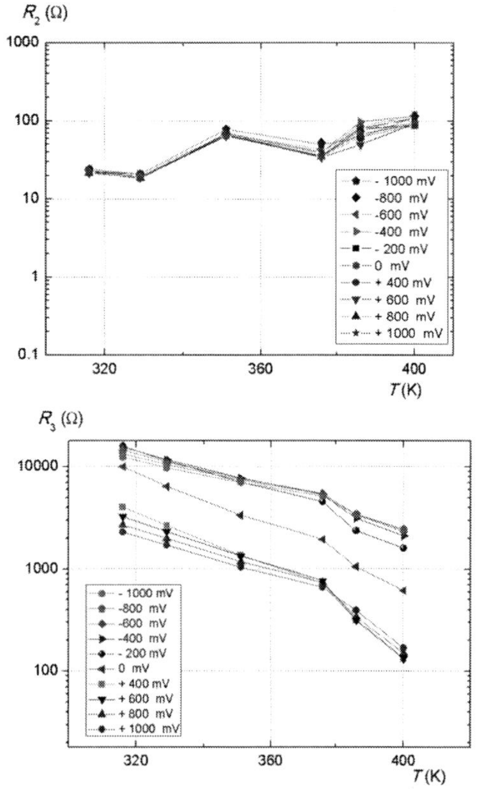

Figure 7: AC equivalent circuit resistance as a function of temperature for different applied DC bias

4 CONCLUSIONS

Obtained results represent alternative approach to the silicon pn junction application, here in combination with diamond boron-doped nano-layer. The Si substrate with the pn+ junction plays a role of the "active substrate" (SC structure) as the potential generator in the Si/B-NCD heterostructures. The substrate in the form of silicon photovoltaic pn junction was optimized to get maximal lifetime of minority carriers. The LT investigation brought solutions for MCLT degradation issue. At first highly doped p+ layer was prepared which prevents recombination of minority carriers close to the surface and secondly, the addition of passivating and ARC SiN_x

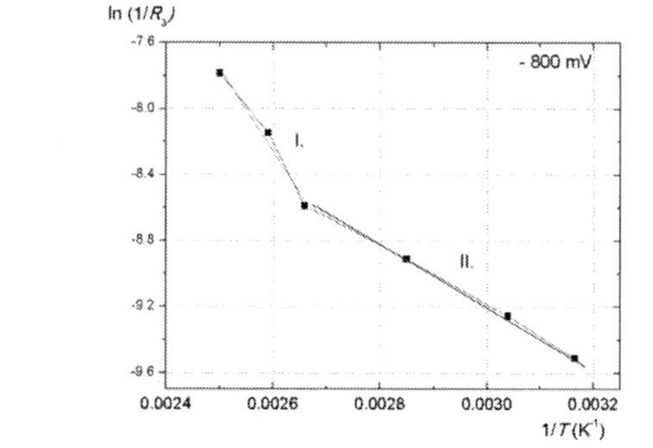

Figure 8: Arrhenius dependence plotted at reverse bias 0.8 V

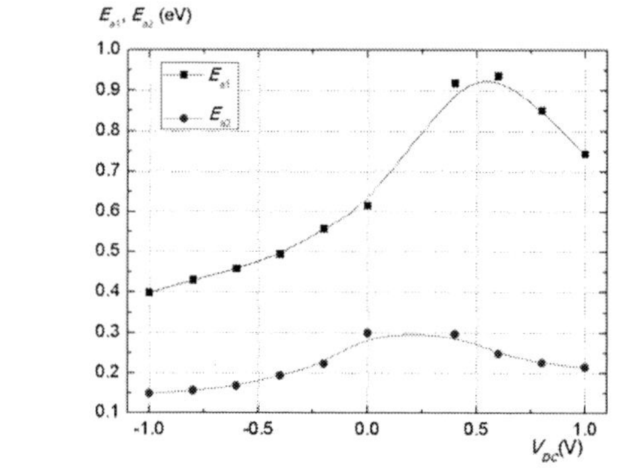

Figure 9: Voltage dependence of activation energies for forward and reverse bias

layer as barrier to contamination from the base of the CVD chamber was deposited. NCD electrode was deposited to form B-NCD/Si SC and electrically characterized in this paper. Two relaxation processes were identified using the AC complex impedance response, numerically analyzed and equivalent circuit proposed. Ohmic and barrier mechanisms were proposed to reflect the transport properties. Ohmic behaviour results in voltage-independent R_2 with the value lower than 100 Ω while the resistance R_3 depends on the bias. The conductance $1/R_3$ is thermally activated and activation energy was calculated. Two different activation energies E_{a1} and E_{a2} were calculated for two regions I. and II. resulting from Arrhenius plot.

Acknovledgement: Partially this research has been supported by the Slovak Research and Development Agency (APVV) under the Contract No APVV-0443-12.

References

[1] A. Fujishima and K. Honda, Electrochemical photolysis of water at a semiconductor electrode. Nature (London), 238 (1972), 37.

[2] A. A. Ismail, D. W. Bahnemann, Photochemical splitting of water for hydrogen production by photocatalysis: A review. Solar Energy Mat.&SolarCells. 128 (2014), 85.

[3] K. Maeda, K. Domen, Photocatalytic Water Splitting: Recent Progress and Future Challenges. J. Phys. Chem.

Lett. 1 (2010), 2655.
[4] J&M: Diamond cutting tools, Available from: http://www.diamondtool.com/dctcatalog.pdf.
[5] Diamond materials: The CVD diamond booklet, Fraunhofer IAF, 2008, available at: www.diamond-materials/downloads/cvd_diamond_booklet.pdf.
[6] eVince Technology, Avialable from: http://www.evincetechnology.com/devices.html
[7] B. Rezek, M. Krátka, E. Ukraintsev, O. Babchenko, A. Kromka, A. Brož and M. Kalbacova: Diamond as a functional material for bioelectronics and biotechnology, New Perspectives in Biosensors Technology and Applications, Prof. P. A. Serra (Ed.), ISBN: 978-953-307-448-1, In Tech.
[8] M. C. Granger et. al., Standard electrochemical behavior of high-quality, boron-doped polycrystalline diamond thin-film electrodes. Anal. Chem. 72, (2000), 3793.
[9] Y. He, W. Huang, R. Chen, W. Zhang, H. Lin, Improved electrochemical performance of boron-doped diamond electrode depending on the structure of titanium substrate. 758, (2015), 170.
[10] J. Jiang, M. Chang, P. Pan, Simultaneous hydrogen production and electrochemical oxidation of organics using boron-doped diamond electrodes. Environ. Sci. Technol. 42, (2008), 3059.
[11] K. Okada, Plasma-enhanced chemical vapor deposition of nanocrystalline diamond. Sci. Technol. Adv. Mat. 8, (2007), 624.
[12] F. Fendrychetal., Growth and characterization of nanodiamond layers prepared using the plasma enhanced linear antennas microwave CVD system, J. Phys. D: Appl. Phys. 43 (2010), 374018.
[13] P.Ashcheulov, et al.,Nanocrystallinediamond on Si solar cells for direct photoelectrochemical water splitting. Phys. St. Sol. A-Applications and material science, Vol. 211,(2014), 2347.
[14] A.R.West, D.C. Sinclair, N. Hirose, "Characterization of Electrical Materials, Especially Ferroelectrics, by Impedance Spectroscopy",Journal of Electroceramics 1, 1997, pp. 65-71.
[15] Mei Wang, Dengjing Wang, Junjie Ma, Ruwu Wang, Yunbao Li, Measure the barrier height of manganite p–n heterojunction by activation energy measurement, Materials Science in Semiconductor Processing, 29, (2015), 213.

DIODE PROPERTY OF METAL AND/OR Si NANOPARTICLE EMBEDED LIQUID SOURCE SiO_2 ON Si

Hiroshi Nagayoshi[1], Hirotomo Demura[1], and Alexander Ulyashin[2]

1 Tokyo National College of Technology
1120-2, Kunugida-machi, Hachoji, Tokyo, Japan
2 SINTEF Materials and Chemistry, Metallurgy Department, Forskningsveien 1,
Blindem, NO-0314 Oslo, Norway

This paper describes a junction property of nanoparticle(NP) embedded SiO_2/Si. P^+Si NPs and/or combination with Ag particles were tested. A liquid source SiO_2 layer was obtained using Perhydropolysilazane (PHPS), which shows good passivation property on Si. After spin coating deposition, NP embedded SiO_2 layers were obtained using a hot steam annealing at 600°C. The p^+SiNPs embedded SiO_2/Si(n) hetero structure showed rectifying property. A forward bias current under illumination was lower than the dark, indicating a memory effect, which can be explained by hole trap of floating Si NPs in the SiO_2 film. Using this property, selective contact was tested using the combination of p^+ Si NP and Ag particle. An Ag particle embedded SiO_2 was an ohmic contact. On the other hand, the SiO_2 film including both p^+Si NPs and Ag particle showed rectifying property with no memory effect. In addition forward bias current increased by illumination.

Keywords: Silicon nanoparticle, Passivation, PHPS, SiO_2, c-Si, Selective electrode

1. INTRODUCTION

Materials with nano particles (NPs) form a new class of artificial materials with tuneable electronic properties controlled at the nanoscale. Their properties discover a new route for novel electronic, optical, and optoelectronic applications. The combination of nanoparticles with liquid source SiO_2 is an interesting option. Nanoparticle embedded SiO_2 using Perhydropolysilazane (PHPS) has a potential to create new approaches in photovoltaic application such as a new type selective electrode. We have been investigating the passivation effect of liquid source SiO_2 prepared by using the PHPS, showing good interface property with silicon surface, which is suitable for solar cell device applications. In addition, thick pure SiO_2 layer of more than 10 μm could be easily obtained by low temperature process, that means the SiO_2 mede from PHPS has a variety of potential applications like MEMS devices, composite materials with the combinations of nano particles, etc. [1-3] The purpose of this work is to create a new type thin contact structure for Si solar cells with low process temperature as well as low cost. A liquid source SiO_2 matrix using PHPS can be prepared less than 450°C, showing good passivation property. Compare to the ordinally organic binder materials, quite stable against ultra violet light and heat. This paper describes a junction property of nanoparticle(NP) embedded SiO_2/Si. P^+Si NPs and/or combination with Ag particles were tested.

2. EXPRIMENTAL

The SiO_2 was prepared by using the PHPS. The PHPS is an organosilicate polymer consisting of cyclic $(SiH_2\text{-}NH)n$ units, which changes into pure SiO_2 by heat treatment lower than 450°C and does not contain carbon in the reaction. A non catalyst type PHPS was used as a binder material. The PHPS was coated by spin coater. The film thickness was controlled by rotation speed, time and concentration of the PHPS. A spray coat is also applicable. To obtain thin layer, the density of PHPS was controlled by adding xylene solvent. After the spin coat, the samples were dried at room temperature. A SiO_2 formation process using the PHPS is shown as follows.

$$(\text{-}SiH_2NH\text{-}) + 2\,H_2O \rightarrow (\text{-}SiO_2\text{-}) + NH_3 + 2H_2 \qquad (1)$$

This reaction does not contain organic component like sol-gel source materials, releasesing H_2 and NH_3 in the process, which might contribute to improve the SiO_2/Si interface peoperty. A pure PHPS needs around 450°C of reaction temperature to complete the reaction. By using catalyst such as amine or vanadium, SiO_2 films can be obtained even at room tempreature. A SiO_2 passivation layer thickness was controlled by the concentration of PHPS in xylene solvent. After dried at room temperature, the hot steam annealing was carried out at 600 °C, 1 hour. This is a optimized condition to obtain good passivation effect.[1-2] SiNPs and/or metal particles are added into the PHPS. The used P+ silicon nanoparticle has large size distribution from 10nm to 1μm. Raman spectra suggested that most are crystalline. Fifure 1 shows the schematic of the sample image. Semi transparent Au electrodes, 2mm in diameter were prepared by evaporation. A CZ N-type <100> 1 Ωm wafer was used as a substrate. As shown in Fig.1, large size particles are making pn junction. Small size particles are electrically floating in the film.

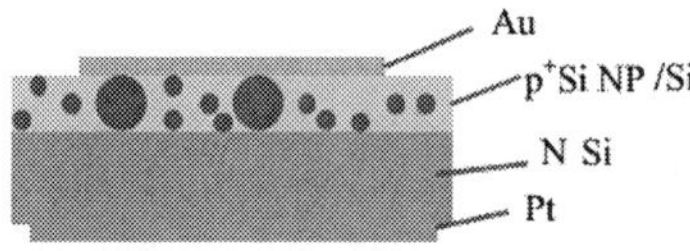

Fig.1 Schematic of the sample SiP+NPs/SiO_2 on N-type silicon.

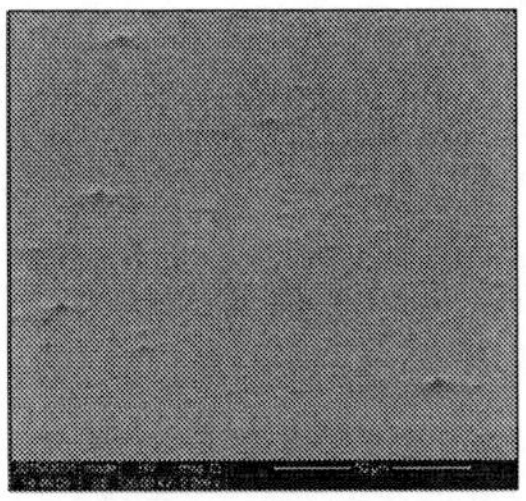

Fig.2 Surface view of nanoparticle(NP) embedded SiO_2 surface.

Figure 2 shows the SEM image of the sample sarface. We can observe small projections on the surface. Total junction area is essentially quite small. Figure 3 shows the IV characteristics as a function of p+ silicon nanoparticle concentration. We confirmed clear rectifying property as shown in the figure, however the dependency of silicon nanoparticle was not clear yet.

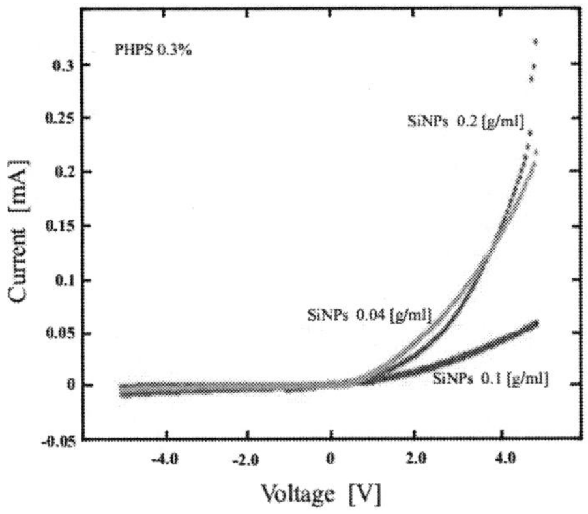

Fig.3 I-V characteristics of as a function of silicon nanoparticle concentration.

Figure 3 shows the photo response of the sample. There was no photovoltaic effect caused by defects at the point contact between N-Si and the P^+ SiNP. The forward bias current decreased by illumination. After the first photo response measurement, the photo response disapperared. The most part of the surface under the Au electrode has MOS structure. The property of point contact PN junction strongly affected by a surface field of surrounding MOS area. At the reverse bias condition, the surface of silicon including point contact area become depression mode, showing very low reverse current even when the point contact has defects. On the other hand, under the forward bias condition, the surface become accumulation mode except pn junction. The band bending of the surface is opposite direction to the electric field of PN junction. The P^+ silicon NPs have large size distribution. Hence lots of small size p^+ silicon nanoparticles are electrically flowating as shown in Fig.5. The floating particles near the surface works as a hole trap. The decrease of forward bias current by illumination can be explained by hole injection into the floating silicon NPs under illumination. The trapped hole causes a built in field in the film, which canceles the effect of external bias. The charge remaines after the phto IV measurement.

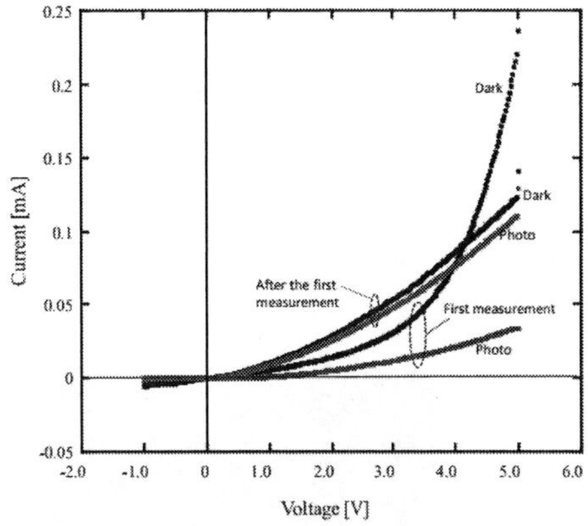

Fig.3 Photo response of nanoparticle embedded SiO_2 surface.

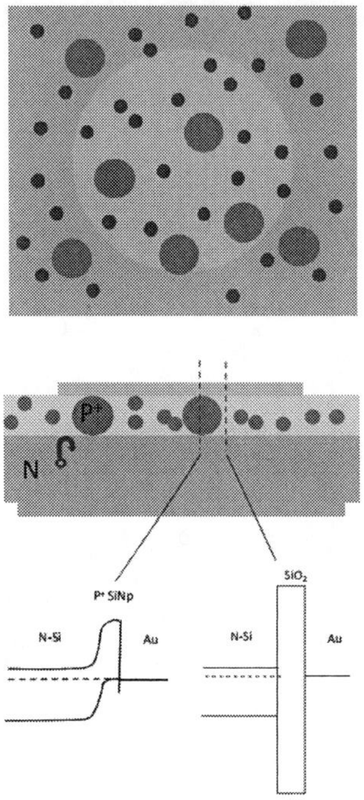

Fig.4 Schematic of the contact structure.

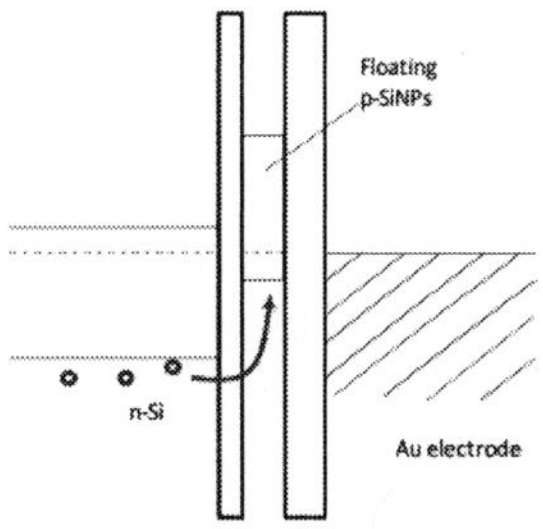

Figure 5 Image of hole trapping by floating p+ silicon nanoparticle in SiO_2.

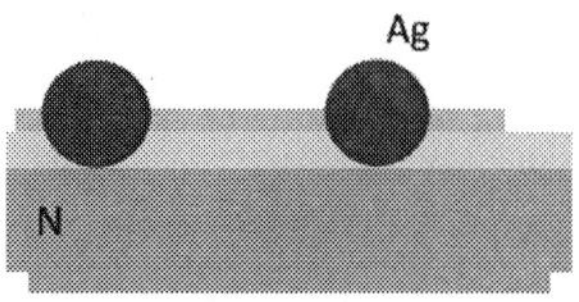

Figure 6 Schematic of Ag/SiO_2contact structure.

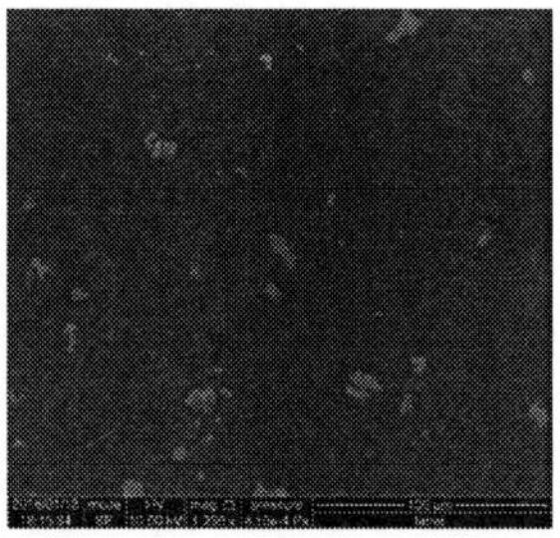

Figure 7 SEM image of Ag/SiO_2 on silicon surface.

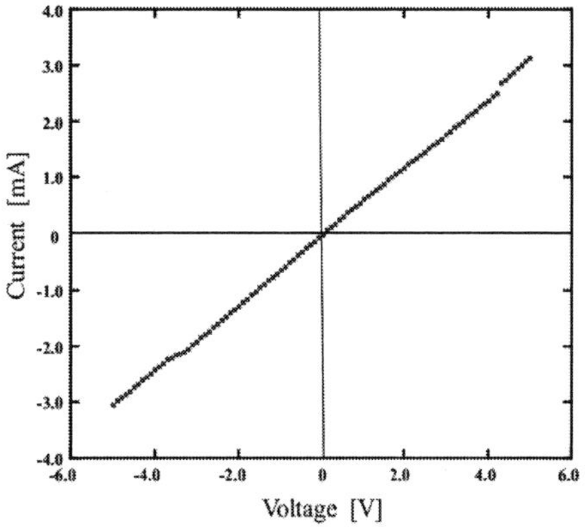

Figure 8 IV characteristics of the sample shown in Fig.6.

By adding metal particle into the silicon nanoparticle embeded layer, we can create nanoscale MIS-IL structure. The Ag particles was extracted from a conductive paste, having size distribution from 1 to 20 μm. Figure 6 and 7 shows the schematic structure and IV characteristics of Ag particle embeded SiO_2 film, respectively. The contact showed an ohmic property as shown in Fig.8. Figure 9

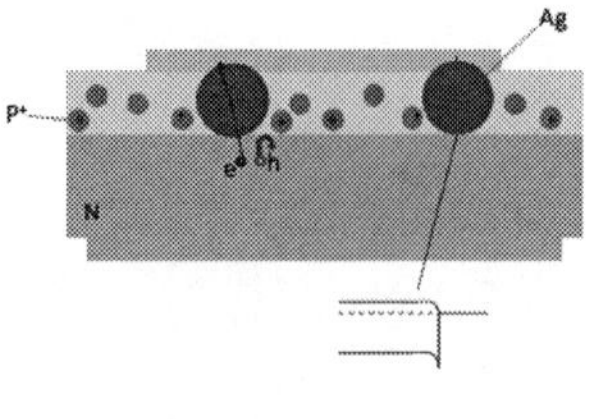

Fig.9 Schematic of Ag, p+SiNPs/SiO_2 composite structure.

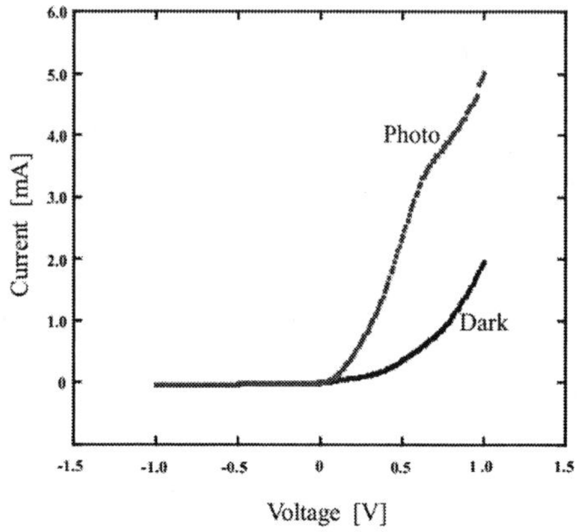

Fig.10 I-V of Ag,p+SiNPs embeded SiO_2 /N-Si diode.

shows the schematic of Ag and p^+SiNPs embeded SiO_2 structure. The film thickness was 100nm. The Ag particles are enoughly larger than the film thickness. On the other hand, most P^+ silicon nanoparticle are smaller than the film thickness. In this case, IV characteristics become rectifying property as shown in Fig.10. photo responce was also measured. When the negative bias (minus voltage was applied to the Au electrode) condition, the diode current was zero. In this condition, whole surface area including a contact point between N-Si and Ag become depression mode, showing insulationg charactristics. On the other hand, in the forward bias condition, Ag particle works as a drain of accumulated surface electrons. The increase of current under the illumination is caused by the increase of photocarrier and conductibity modulation.

SUMMARY

We have investigated the junction property of nanoparticle embedded SiO_2/Si. The p^+SiNPs embedded $SiO_2/Si(n)$ structure showed rectifying property. The forward bias current decreased by illumination caused by charge trap in SiO_2 layer. The SiO_2 film including both p^+Si NPs and Ag particle showed rectifying property with no memory effect. The forward bias current increased by illumination. Those results suggested that the point contact property was strongly affected by the condition of surrounding MOS structure.

REFERENCES

[1] H. Nagayoshi and C. Hagiwara, Proceedings on 30th European Photovoltaic Solar Energy Conference and Exhibition, (2014)1000-1002.
[2] Hiroshi Nagayoshi, and Chihiro Hagiwara , Energy Procedia , Volume 77, August 2015, Pages 822-826.
[3] Hiroshi Nagayoshi, and Takuya Murooka, Energy Procedia, Volume 77, August 2015, Pages 242-247

ENHANCED LIGHT SCATTERING AND HYDROPHOBICITY OF GLASS WITH UPRIGHT NANOPYRAMID STRUCTURE FOR SOLAR CELLS USING UV NANOIMPRINT LITHOGRAPHY

Amalraj Peter Amalathas, Maan M Alkaisi
The MacDiarmid Institute for Advanced Materials and Nanotechnology, Department of Electrical and Computer Engineering, University of Canterbury, Private Bag 4800, Christchurch 8140, New Zealand.
E- mail: amalraj.peteramalathas@pg.canterbury.ac.nz; maan.alkaisi@canterbury.ac.nz

ABSTRACT: The optical path length and the intensity of the incident light reaching the solar cell could be enhanced by incorporating antireflective and light scattering nanostructures. In this work, the upright periodic nanopyramid structures were fabricated on the surface of a glass substrate using UV nanoimprint lithography, which was employed in this case as a protective layer for solar cells, in addition to its antireflection coating function. The haze value was significantly increased for upright nanopyramid patterned glass, especially, in the wavelength range 300-600 nm, which implies that strong light scattering by upright nanopyramid structure. By replacing a bare protective glass with upright nanopyramid patterned glass, the power conversion efficiency of the monocrystalline Si and commercially made polycrystalline Si solar cells were substantially enhanced about 10.89% and 8.22% due to the scattering and prolong the optical path length by the upright nanopyramid structures. In addition, the surface of the patterned glass exhibited hydrophobic properties due to the surface nanostructures and the self-assembled monolayer coating. The high hydrophobicity provided the solar cell with a self-cleaning functionality.
Keywords: encapsulation, photovoltaic, performance, nanopyramid, self-cleaning

1 INTRODUCTION

Generally, solar cell modules are installed in an outdoor environment for the vast majority of applications. Therefore, protective glass is commonly used for preventing damage from dust, moisture and external shock [1, 2]. However, the device performance can be reduced both through optical reflection by the protective glass and through scattering or absorption by contaminants [3, 4]. The amount of the incident light reaching the solar cell could be enhanced by incorporating antireflective and light scattering nanostructures. The light scattering effect improves the light harvesting of the solar cells as a result of prolonged optical path length within the solar cells and thus, increasing the conversion efficiency.

In order to enhance the efficiency and self-cleaning effect of the solar cells, the fabrication of various nanoscale structures with anti-reflective and scattering effects have been extensively studied by numerous research groups [5-8].There are several methods for obtaining nanoscale structures including electron beam lithography, focused ion beam lithography (FIB), laser interference lithography (LIL), nanoimprint lithography (NIL), nanosphere lithography (NSL) and Block copolymer lithography (BCPL). However, the UV nanoimprint lithography (UV- NIL) is an emerging technique for fabricating nanoscale structures on large surfaces with simple, high throughput, low cost and high resolution manufacturing technology [9].

In this work, the upright periodic nanopyramid structures were fabricated on the surface of the protective glass substrate using UV nanoimprint lithography. The transmittance and contact angle of the bare glass were measured along with current-voltage and external quantum efficiency measurements of monocrystalline Si and polycrystalline Si solar cells. This was contrasted with cells with a protective layer of glass patterned with upright nanopyramid to verify the effectiveness of this layer as anti-reflective, light scattering and self-cleaning functions.

2 EXPERIMENTAL METHODS

2.1 Preparation

The upright periodic nanopyramid structures were formed on UV curable resist coated glass substrate by means of UV nanoimprint lithography. The OrmoStamp resist (micro resist technology GmbH) was used as a UV-curable resist. The periodic inverted nanopyramid master Si mold was fabricated by laser interference lithography and subsequent pattern transfer by combined reactive ion etching and KOH wet etching [10]. A 1H,1H, 2H, 2H-Perfluorooctyl trichlorosilane (also known as F13-TCS solution from Sigma-Aldrich) based hydrophobic self-assembled monolayer(SAM) was deposited on Si mold in order to improve the anti-sticking property.

2.2 UV nanoimprint process

Prior to coating, the glass substrate was normally cleaned with acetone, methanol, and isopropyl alcohol (IPA) solvents in an ultrasonic bath and then rinsed with deionized water and finally dried with Nitrogen gas. Next, the substrate surface was treated using oxygen plasma to enhance optimum adhesion at the interface between the glass and the OrmoPrime08. The OrmoPrime08 from micro resist technology was used as an adhesion promoter solution based on organofunctional silanes. After that, the substrate was baked using an oven at 200°C for 30 minutes and cooled down to room temperature immediately before coating. OrmoPrime08 was deposited onto the glass substrate by spin coating at a spin speed of 4000 rpm for 60 sec. The spin-coated film was then baked on a hot plate at 150°C for 5 minutes and cooled down to room temperature. Finally, the OrmoStamp was spun coated onto the OrmoPrime08 layer coated substrate at a spin speed of 6000 rpm for 60 sec and prebaked on a hot plate at 80°C for 2 minutes.

To perform the imprint experiments, a vacuum operated manual imprint tool was attached to the UV illumination source of the Karl Suss Mask Aligner (MA-6) system. This created a vacuum environment in order to reduce the air bubbles trapped in-between the mold and resist during the imprint process. The Si master mold/UV curable resist/glass plate was loaded into the imprint tool.

A vacuum pressure of 4 mbar and the Mask Aligner (MA-6) system were then activated and the resist was cured under a UV exposure for 4 minutes using 4.4 mW/cm^2 illumination intensity with 365 nm UV source at room temperature. Finally, a manual de-molding process was utilized by applying gradual force using a scalpel at one corner of the mold in order to delaminate between the two material surfaces. The upright nanopyramid pattern was successfully replicated from the Si mold to the glass plate with high fidelity. Fig. 1 illustrates the schematic diagram of overall imprint process steps. After the UV nanoimprint process, F_{13}-TCS based SAM was coated onto the upright nanopyramid patterned glass plate in order to form a more hydrophobic surface.

2.3 Characterization techniques

The morphology of the pattern samples was analyzed by scanning electron microscope (SEM) (JEOL 7000F FE-SEM) and atomic force microscope (AFM, DI3000). The total and diffuse transmittance of patterned glass and the bare glass were measured using a UV-visible spectrophotometer at room temperature, with an integrating sphere over the wavelength range of 300-1200 nm. The contact angle of water droplets on the patterned glass and the bare glass were measured using a contact angle goniometer (with Edmund Scientific Camera). In order to measure the effect of the periodic upright nanopyramid patterns, the patterned glass substrate, and the bare glass substrate were used as a protective layer on a monocrystalline Si and commercially made polycrystalline Si solar cells. The current density-voltage characteristics were measured using a Keithley 2400 source measure unit and a solar simulator (ABET Sun3000) with AM 1.5G filter under illumination of 1 sun.

3 RESULTS AND DISCUSSION

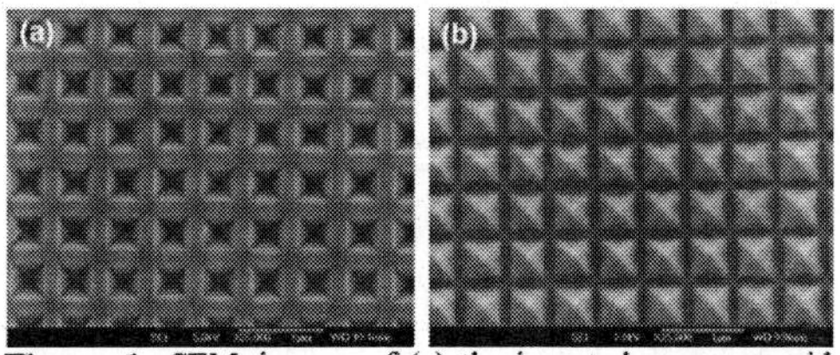

Figure 1: SEM images of (a) the inverted nanopyramid Si master stamp with dimension of 450 nm wide and 310 nm height and (b) the upright nanopyramid patterned glass

The SEM images of the periodic inverted nanopyramid Si master stamp and the upright nanopyramid patterned glass substrate are shown in the Fig. 1. It can be clearly seen that the inverted nanopyramid structure was faithfully transferred onto UV curable resist coated glass substrate with high fidelity using UV nanoimprint process. Fig. 2(a) illustrates the comparison of the total and diffuse transmittance of the glass substrates without and with upright nanopyramid patterns, which were measured using an integrating sphere with the incoming light entered from the patterned glass substrate. According to Fig.2 (a), the total transmittance of the patterned substrates was slightly lower than that of the bare glass in the wavelength range of 450-800 nm, which may be caused by the diffraction losses. In the Fig. 2(a), the diffuse transmittance of the patterned glass substrate was increased up to 24% in the visible wavelength region due to the nanostructured substrate while the bare glass substrate shows almost no diffuse transmittance over a wide wavelength range. In addition, the haze value (H) is defined by the ratio of the diffuse transmittance to the total transmittance, which indicates the scattering properties of the samples. When the incident light passed through the bare glass, the H value (Fig. 2(b)) was close to zero. In contrast, the H value was significantly increased for patterned glass, especially, in the wavelength range 300-600 nm, which implies that strong light scattering by the upright nanopyramid structure. This scattering changed the propagation direction of light from normal to the oblique incidence in the solar cell. As a result, the optical path length of the incident light was elongated and hence the light absorption in the active layer was also improved by the patterned glass. Indeed, the high haze optical property due to the light scattering effect would positively enhance the power conversion efficiency of the solar cells with the upright nanopyramid structures on glass compared to the bare glass.

(a)

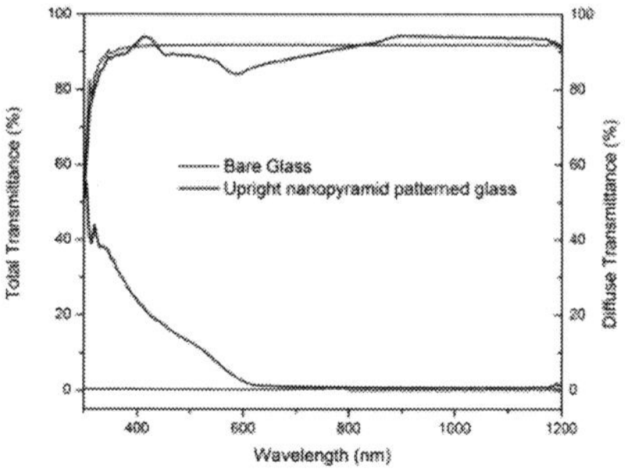

(b)

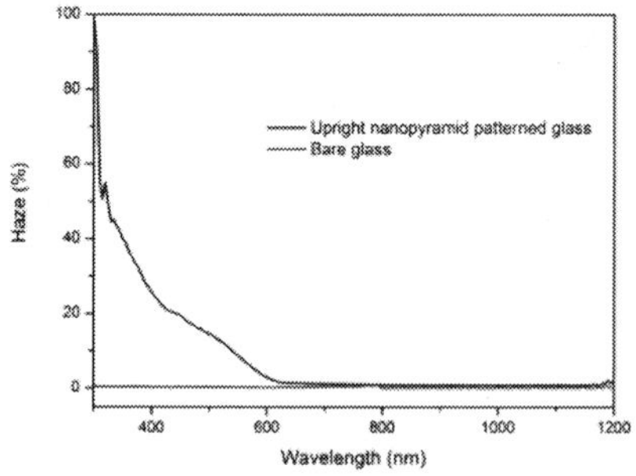

Figure 2: (a) Total and diffuse transmittance spectra and (b) optical haze spectra of the bare glass and the upright nanopyramid patterned glass as a function of wavelength

In order to verify the effect of the periodic upright nanopyramid patterns, the patterned glass substrate, and the bare glass substrate were used as a protective layer on a monocrystalline Si and commercially made polycrystalline Si solar cells. Fig. 3 shows the I-V characteristics of the monocrystalline Si and polycrystalline Si solar cells with and without the inverted nanopyramid patterned protection glass. There was no significant change in the open circuit voltage (V_{OC}), but a significant enhancement in the short-circuit current density (J_{SC}) was observed as expected. The J_{SC}

values of the monocrystalline and polycrystalline Si solar cell without protection glass were 34.38 and 32.51 $mAcm^{-2}$, respectively. The J_{SC} values of monocrystalline and polycrystalline Si solar cell were decreased to 31.60 and 29.56 $mAcm^{-2}$ with a bare protection glass, respectively. This decrease in J_{SC} indicates that the protective glass reduces the amount of photon incident on the solar cell. However, by replacing the bare glass with the upright nanopyramid patterned glass, J_{SC} values were increased to 32.39 and 30.21 $mAcm^{-2}$ for monocrystalline and polycrystalline Si solar cell, respectively. As a result, the use of the patterned glass as a protective glass, the conversion efficiency of the monocrystalline Si and polycrystalline Si solar cell were increased about 10.89% and 8.22% compared to the bare glass as a protective layer, respectively.

(a)

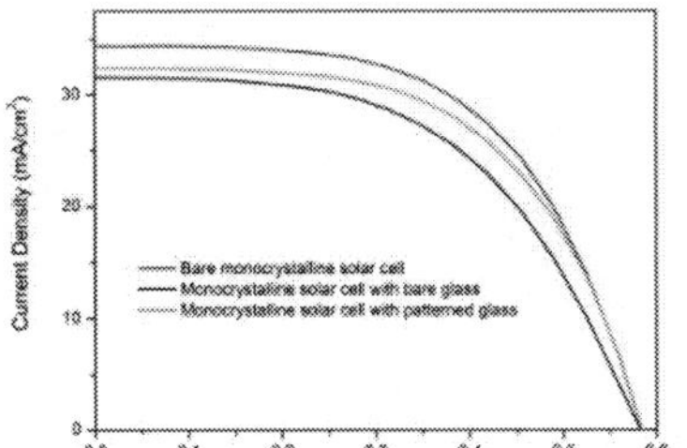

(b)

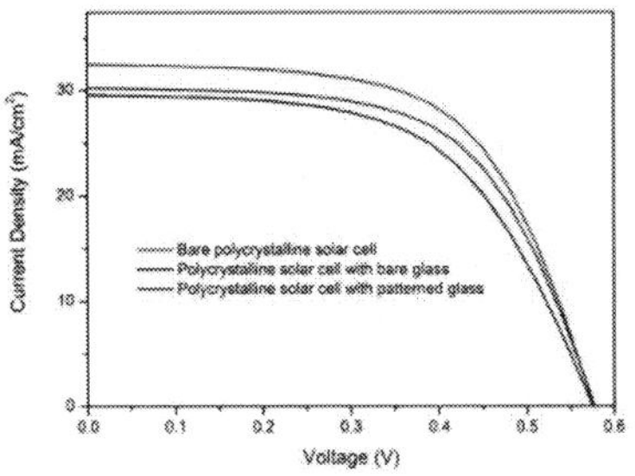

Figure 3: (a) Current density – voltage characteristics of (a) monocrystalline Si and (b) polycrystalline Si solar cells with and without the inverted nanopyramid patterned on protective glass under illumination

From these results, the periodic upright nanopyramid structure offers a better-graded index medium to the incident light, compare to the bare glass, which can reduce the Fresnel reflectance and scatter more incident light into the solar cells and prolong the optical path length, therefore improving the light trapping and increasing the overall conversion efficiency.

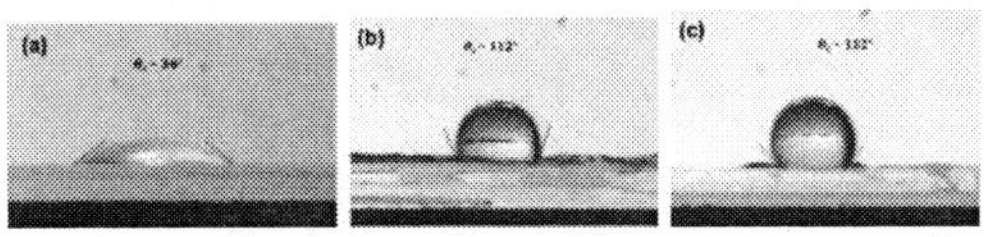

Figure 3: Contact angles of water droplets on (a) bare glass, (b) patterned glass and (c) SAM-coated patterned glass

In outdoor environments, the productive glass layer of the solar cell can be easily contaminated by dust particles, which interfere with incident light into the cell absorption layer and thus reducing the device performance. Therefore, a self-cleaning function at the top surface of the protective glass layer is very essential to maintain the cell performance. Fig. 4 shows the contact angles values of water droplets on bare glass, nanopyramid patterned glass and SAM-coated nanopyramid patterned glass for the tested solar cells. In Fig. 4, the contact angle of the glass substrate was increased from 36 to 112° after the formation of upright nanopyramid patterns, which implies high hydrophobicity. Moreover, the surface became more hydrophobic after the formation of upright nanopyramid patterns and SAM coating. In this case, the contact angle of the SAM-coated patterned glass substrate was increased to 132°. As a result, solar cells with this nanostructured glass substrate can utilize the self-cleaning functionality induced by the high hydrophobic surface properties in addition to their antireflection properties

4 CONCLUSIONS

The periodic upright nanopyramid structures with antireflective and self-cleaning properties were fabricated on glass substrates by UV nanoimprint lithography. UV-NIL is simple nanolithography process with low cost that can be used at industrial scale production for solar cells. By replacing a bare protective glass with upright nanopyramid patterned glass, the power conversion efficiency and short circuit current density of the monocrystalline Si and polycrystalline Si solar cells were substantially enhanced due to light scattering and prolong the optical path length by the upright nanopyramid structures. Moreover, the surface of the patterned glass exhibited hydrophobic properties due to the surface nanostructures and the self-assembled monolayer coating, and this hydrophobicity provided the solar cell with a self-cleaning functionality. The results demonstrate that the periodic upright nanopyramid structure may be applied to various types of solar cells and optical systems that utilize a protective layer and transparent substrates such as glasses, quartzes, and plastics.

ACKNOWLEDGEMENTS

The authors would like to thank Annette Koo, Callaghan innovation for transmittance measurements, Jonathan Halpert and Elijah Peach, Victoria University of Wellington, New Zealand for EQE measurements, Hari Murthy for SEM measurements, Gary Turner and Helen Devereux from the Nanofabrication Laboratory, University of Canterbury, New Zealand for providing technical assistance. Amalraj PA acknowledges UC Doctoral Scholarship, University of Canterbury, New Zealand.

REFERENCES

[1] G. Jorgensen, K. Terwilliger, J. DelCueto, S. Glick, M. Kempe, J. Pankow, F. Pern, T. McMahon, Moisture transport, adhesion, and corrosion protection of PV module packaging materials, Solar Energy Materials and Solar Cells, 90 (2006) 2739-2775.

[2] J.H. Shin, K.S. Han, H. Lee, Anti-reflection and hydrophobic characteristics of M-PDMS based moth-eye nano-patterns on protection glass of photovoltaic systems, Progress in Photovoltaics: Research and Applications, 19 (2011) 339-344.
[3] L.K. Verma, M. Sakhuja, J. Son, A. Danner, H. Yang, H. Zeng, C. Bhatia, Self-cleaning and antireflective packaging glass for solar modules, Renewable Energy, 36 (2011) 2489-2493.
[4] J. Son, S. Kundu, L.K. Verma, M. Sakhuja, A.J. Danner, C.S. Bhatia, H. Yang, A practical superhydrophilic self cleaning and antireflective surface for outdoor photovoltaic applications, Solar Energy Materials and Solar Cells, 98 (2012) 46-51.
[5] H.J. Gwon, Y. Park, C.W. Moon, S. Nahm, S.-J. Yoon, S.Y. Kim, H.W. Jang, Superhydrophobic and antireflective nanograss-coated glass for high performance solar cells, Nano Research, 7 (2014) 670-678.
[6] K.-S. Han, H. Lee, D. Kim, H. Lee, Fabrication of anti-reflection structure on protective layer of solar cells by hot-embossing method, Solar Energy Materials and Solar Cells, 93 (2009) 1214-1217.
[7] J. Zhu, C.-M. Hsu, Z. Yu, S. Fan, Y. Cui, Nanodome solar cells with efficient light management and self-cleaning, Nano letters, 10 (2009) 1979-1984.
[8] G. Kang, K. Bae, M. Nam, D.-H. Ko, K. Kim, W.J. Padilla, Broadband and ultrahigh optical haze thin films with self-aggregated alumina nanowire bundles for photovoltaic applications, Energy & Environmental Science, 8 (2015) 2650-2656.
[9] L.J. Guo, Nanoimprint lithography: methods and material requirements, Advanced Materials, 19 (2007) 495-513.
[10] S. Sivasubramaniam, M.M. Alkaisi, Inverted nanopyramid texturing for silicon solar cells using interference lithography, Microelectronic Engineering, 119 (2014) 146-150.

Improvement of short circuit current of single junction amorphous silicon solar cells by incorporating nanoparticle as back reflector

Sourav Mandal*, Sukanta Dhar, A. K. Barua
Centre of Excellence for Green Energy and Sensor Systems, Indian Institute of Engineering Science & Technology, Shibpur, Howrah 711103, India
*Address for correspondence: E-Mail: 4souravmandal@gmail.com

ABSTRACT: Thin-film solar cells employ a back reflector (BR) for a more efficient use of the long wavelength light. This paper deals with the prospect of using indium tin oxide (ITO) nanoparticle (NP) as back scatterer/reflector in single junction amorphous silicon solar cells. As the ITO NP's have comparatively low dissipative loss and tuneable optical properties. NP's of ~200nm size is developed on the back side of the complete a-Si solar cell by deposition of ~20-30nm thick ITO by DC magnetron sputtering followed by hydrogen plasma treatment in PECVD. Various characterizations like FESEM, TEM and AFM are performed to analyse the morphology, chemical composition, optical characteristics of the NP's. The p-i-n structured solar cell with 8.17% of efficiency is fabricated in a conventionally used cluster tool PECVD system. The short circuit current (I_{sc}), open circuit voltage (V_{oc}) and fill factor (FF) of the solar cell are 13.17mA, 0.885V and 70.10% respectively. ITO nanoparticles at the back surface of the solar cell significantly enhances the short circuit current by ~9.5% to 14.42mA, (V_{oc}-0.884V, FF-70.82%) and due to that increase the overall efficiency of the device reaches to 9.02%.

The results reported show a novel method of enhancing short circuit current of a-Si based solar cells which will be explored for further enhancement of the effect by varying nanoparticle size.

Keywords: nanoparticle, solar cell, back reflector

1. INTRODUCTION

One of the great challenges of modern photovoltaics is the effective capture of light. Although from thousands of years many techniques are being used for better capture of light, modern photovoltaics is now offered a rich landscape of tools and methodology of how this field can be amplified and expanded. High efficiency amorphous silicon solar cells requires use of good quality back reflector layer (BRL). Different types of BRL's and their properties have been reported [1,2]. Back reflectors in thin film solar cells, are of great value and importance. Especially when considering multiple passes of light through the solar cell structure, as it is the goal of light trapping. However, the science of back surface reflectance is dictated by a number of important aspects, in addition to the reflective properties. These aspects include the back contact technology, material conductivity, passivation and the material adherence properties [3,4]. The combination of these features gives rise to a complex interplay, between the optical and electrical functionality. These requirements, often lead to the use of metals as back reflectors, because of their conductive and reflective material qualities. In practice, the quality of conductance and reflectance varies between metal and semiconducting materials, often the reflectance properties are substandard.

Here we explore a new method of light trapping by using metal oxide nanoparticles in single junction a-Si solar cells. The exposure of the Indium Tin Oxide (ITO) coated films in the hydrogen plasma results in formation of nanostructured particles [5,6]. These nanoparticles were used in the back of a-

Si solar cells for back reflection of the higher wavelength of the spectrum.

2. CONCEPT OF METAL OXIDE NANOPARTICLE AS BRL IN a-Si SOLAR CELL

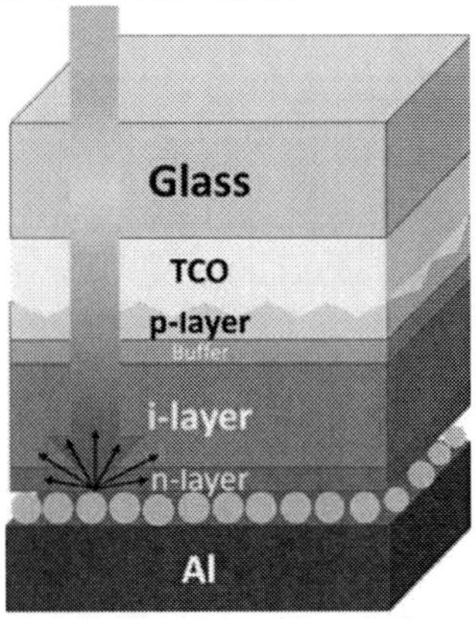

Figure 1: Schematic illustration of the layer structure of a-Si single junction solar cell containing ITO nanoparticles embedded in the Al layer at the back of p-i-n solar cell.

Figure 1 illustrates the structure of the proposed solar cell to have ITO nanostructures at the back surface of a p-i-n a-Si solar cell. The fundamental structure is TCO coated glass (Asahi U)/p-a-SiO:H/a-SiO:H buffer layer/i-a-Si:H/n-a-Si:H/Al. The transmittance of the ITO nanoparticles was less than the thin ITO coated glass substrates but the reflection is more. We used the superior property of reflection of the ITO nanoparticles in solar cells to use it as back reflector.

3. FABRICATION OF ITO NANOPARTICLES

ITO films can be deposited by using a variety of methods including direct current and radiofrequency sputtering [7,8], reactive evaporation [9], pulsed laser ablation [10], electron-beam evaporation [11], spray pyrolysis [12] and sol-gel technique [13]. Studies on ITO films, grown by the above mentioned techniques, revealed that the electro-optical properties of such films are highly sensitive towards the method of preparation, composition, and partial pressure of oxygen, substrate temperature, substrate type and annealing processes. The microstructural changes could affect the optical and electron transport properties. Amongst the different techniques available, we preferred dc sputtering. Sputtering shows the best performance in terms of surface morphology, high deposition rates, low growth temperature, scaling-up ability, and excellent electro-optical performances [14].

A thin ITO layer with 20-30nm of thickness was deposited and the hydrogen plasma treatment (in PECVD system) on that layer forms the nanoparticles. The sputtering chamber temperature was kept fixed at 120°C. The size of the particle depends on the thickness of the ITO layer. The hydrogen plasma process temperature was at 120°C.

4. APPLICATION TO SOLAR CELLS

Amorphous silicon single junction solar cells (area 1cm^2) were fabricated with and without nanoparticles. Commercially available Asahi U type substrates were used for the fabrication of the p-i-n solar cells. After proper cleaning of the substrates the layer were deposited by using cluster tool PECVD (Supplied by HHV, Bangalore) system. After completion of the p-i-n structured solar cells, the cells are loaded in the sputtering chamber for the deposition of the thin ITO layer. Then the solar cells with the thin ITO coating were again put into the PECVD system for the synthesis of nanoparticles by hydrogen plasma treatment. Then the back contact of the solar cells were made by Aluminium. I-V characteristics of the solar cells were measured by solar simulator under AM 1.5 i.e. 100 mW/cm^2 light intensity. The spectral response was measured by using Bentham PVE300.

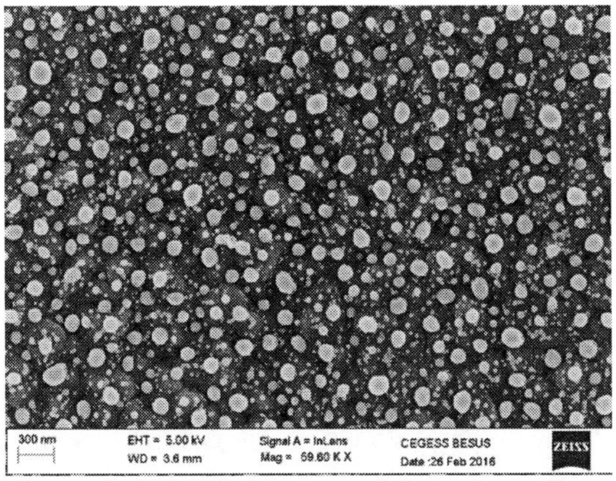

Figure 2: FE-SEM image of the solar cell with ITO nanoparticle.

The FE-SEM image of the back surface of a-Si solar cell with thin ITO nanoparticles is shown in Figure 2. The surface coverage of the nanoparticles are around ~56% area of the solar cell. We can vary the surface coverage of the nanoparticle by varying the particle size which is varied with plasma treatment time.

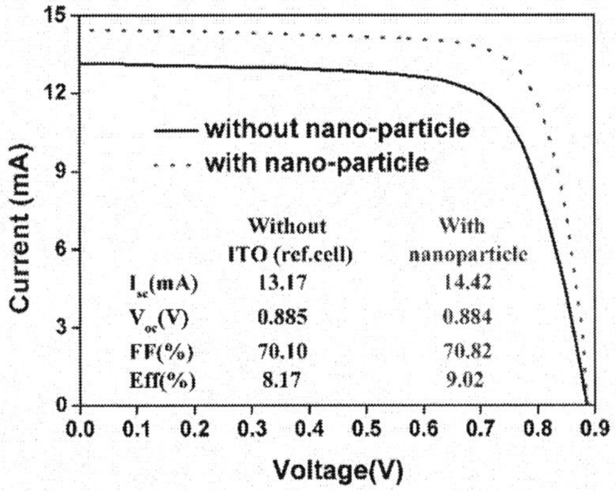

Figure 3: I-V characteristics curves of the solar cells without ITO layer and with ITO nanoparticles

I-V characteristics curve of the solar cells with and without ITO nanoparticles are shown in figure 3. Due to the back reflection from the ITO nanoparticles the Short circuit current (I_{sc}) was increased by ~9.5%. The Fill factor (FF) and open circuit voltage (V_{oc}) remains almost same.

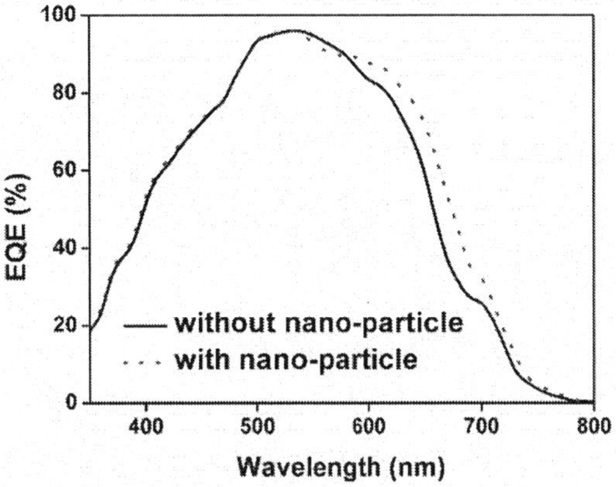

Figure 4: The EQE curve of the a-Si solar cells with and without ITO nanoparticles.

For better understanding of the response of the solar cells in different wavelength regions the EQE curve is shown in Figure 4. In the EQE curve, it is observed that the red response of the solar cell with the ITO nanoparticles is better that the cell without that. The back reflection from the ITO nanoparticle is responsible for the effect. Therefore that part of the spectrum is absorbed more in the active region of the solar cell and improves I_{sc} without effecting other parameters.

5. SUMMARY

Single junction a-Si solar cell with ITO nanoparticles in the back, display better performance than reference solar cell with standard structure. In addition, the optical responses of these cells confirms the increased light trapping in the device. The short circuit current of the devices were increased above the values of the reference cell indicating that parasitic absorption processes related to the ITO nanoparticles. The nanoparticles have a lesser strength of transmission than backscattering or reflection processes. Open circuit voltage and fill factor are comparable to the reference cell. The negligible impact of ITO nanoparticles on the other parameters of the cells, is an encouraging feature. In conjunction with the trends observed on single junction solar cells with variable ITO thicknesses, the results presented in this work suggest that careful tailoring of the ITO nanoparticle size and shape could be an effective approach for further improvement.

REFERENCES

[1] S. Mandal, G. Das, S. Dhar, R.M. Tomy, S. Mukhopadhyay, C. Banerjee, et al., Mater. Chem. Phys. 157 (2015) 130–137. doi:10.1016/j.matchemphys.2015.03.027.

[2] F. Leblanc, J. Perrin, J. Schmitt, J. Appl. Phys. 75 (1994) 1074–1087. doi:10.1063/1.356489.

[3] J. Muller, B. Rech, J. Springer, M. Vanecek, Sol. Energy. 77 (2004) 917–930. doi:10.1016/j.solener.2004.03.015.

[4] S.S. Hegedus, R. Kaplan, Prog. Photovoltaics Res. Appl. 10 (2002) 257–269. doi:10.1002/pip.408.

[5] A.G. Ulyashin, K. Maknys, S. Diplas, A. Olsen, S. Jorgensen, B.G. Svensson, Conf. Rec. 2006 IEEE 4th World Conf. Photovolt. Energy Conversion, WCPEC-4, IEEE, 2007: pp. 134–137. doi:10.1109/WCPEC.2006.279382.

[6] R. Banerjee, S. Ray, N. Basu, A.K. Batabyal, A.K. Barua, J. Appl. Phys. 62 (1987) 912–916. doi:10.1063/1.339699.

[7] G.H. Guai, Q.L. Song, Z.S. Lu, C.M. Li, Surf. Coatings Technol. 205 (2011) 2852–2856. doi:10.1016/j.surfcoat.2010.10.062.

[8] L. Kerkache, A. Layadi, A. Mosser, J. Alloys Compd. 485 (2009) 46–50. doi:10.1016/j.jallcom.2009.06.103.

[9] A. Amaral, P. Brogueira, C. Nunes de Carvalho, G. Lavareda, Opt. Mater. (Amst). 17 (2001) 291–294. doi:10.1016/S0925-3467(01)00051-9.

[10] J.H. Kim, B. Du Ahn, C.H. Lee, K.A. Jeon, H.S. Kang, G.H. Kim, et al., Thin Solid Films. 515 (2007) 3580–3583. doi:10.1016/j.tsf.2006.11.006.

[11] D. Raoufi, Appl. Surf. Sci. 255 (2009) 3682–3686. doi:10.1016/j.apsusc.2008.10.020.

[12] A. V. Moholkar, S.M. Pawar, K.Y. Rajpure, V. Ganesan, C.H. Bhosale, J. Alloys Compd. 464 (2008) 387–392. doi:10.1016/j.jallcom.2007.09.138.

[13] H.Y. Valencia, L.C. Moreno, A.M. Ardila, Microelectronics J. 39 (2008) 1356–1357. doi:10.1016/j.mejo.2008.01.036.

[14] K. Ellmer, J. Phys. D. Appl. Phys. 33 (2000) R17–R32. doi:10.1088/0022-3727/33/4/201.

I-V DOUBLE EXPONENTIAL MODELING IN PC1D6

Luigi Abenante and Massimo Izzi
ENEA, Italian National Agency for New Technologies, Energy and Sustainable Economic Development
Via Anguillarese 301, SP 064, 00123, ROMA (Italy)

ABSTRACT: PC1D6, as well as PC1D, allows fitting I-V curves with the double-exponential approach. In addition, it allows implementing up-to-date physical models for band-gap narrowing (BGN) and other parameters. We use PC1D6 together with an analytical iterative I-V fitting approach to check the influence of physical models on the assignments to parameters in fitting data of external-quantum efficiency (EQE) and I-V measurements. Data relevant to an actual PERC solar cell are exploited. We find that the chosen physical model only determines the SRH-lifetime value to be assigned in the I-V fits performed with PC1D6.

Keywords: Evaluation, Recombination, Modelling, Simulation

1 INTRODUCTION

PC1D6 [1] is a modified version of the widely used numerical simulator PC1D that preserves all the features of PC1D but, in addition, allows implementing Fermi-Dirac statistics and several up-to-date models for crystalline Si including band gap narrowing (BGN), carrier mobility and intrinsic recombination by means of configuration text files. The double diode model

$$I = I_1\left[\exp\left(\frac{V-R_sI}{n_1}\right)-1\right]+I_2\left[\exp\left(\frac{V-R_sI}{n_2}\right)-1\right]+\frac{V-R_sI}{R_{shunt}} \quad (1)$$

where I_1 and I_2 are the saturation currents in the quasi-neutral regions and space-charge regions, respectively, R_s is the series resistance, R_{shunt} is the shunt resistance, and $n_1 \approx 1$ and $n_2 \approx 2$ are constants, is generally adopted to fit measured I-V curves so as to extract values for the parameters that are very informative in order to assess the studied device.

In the present work we show how to implement Eq. (1) in PC1D6 and, of course, in the original version of PC1D. We then use PC1D6 together with a reported I-V fitting approach [2] to detect the differences in the assignments to the fit-parameters in (1) implied by changing physical models in PC1D6. This is performed by fitting the EQE-curve and I-V curves measured both in the dark and under illumination of a PERC solar cell. We finally draw conclusions about the influence of physical models on assignments.

2 PROCEDURES.

Fig. 1 shows how to implement an R_{shunt} and an SCR in PC1D. It only remains to assign the R_s-value after clicking "base contact" in the parameter view and Eq. (1) is fully implemented. I_1 is automatically calculated by the program. The procedure to check physical models is:

a) For given device and configuration file, we first fit the measured EQE-data with PC1D6 to obtain starting values for SRH recombination at surfaces and in the bulk.
b) Then, we exploit the iterative approach described in [2] to fit I-V data with Eq. (1).
c) The obtained values for I_2, R_s, R_{shunt}, and n_2 are inserted into PC1D6 and the resulting PC1D6 I-V curve is compared to the I-V data. If necessary, changes in the adjustable parameter-values are made till the I-V data are fitted again with PC1D6.

For a given cell, steps a) and c) are repeated for each physical model. Since, in PC1D6, I_1 cannot be assigned, the procedure above allows checking the consistency be-

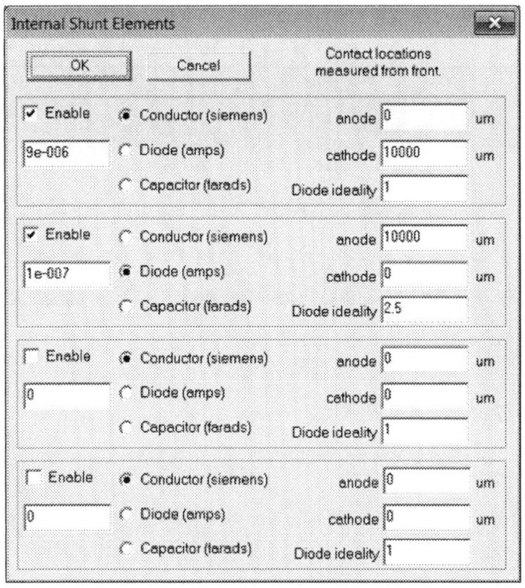

Figure 1: Example of implementation in PC1D6 of a space-charge region (Diode) and a shunt resistance (Conductor).

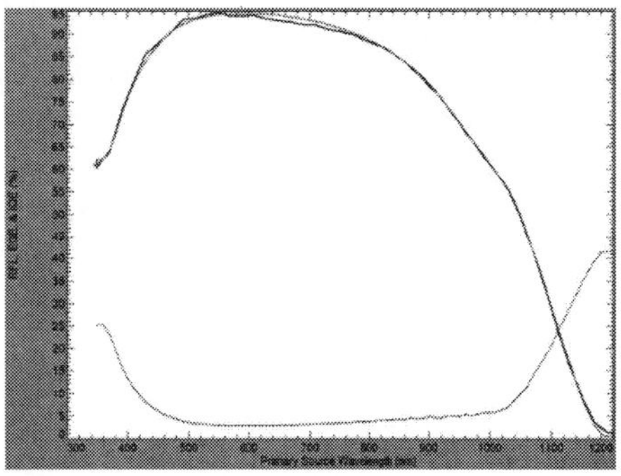

Figure 2: PC1D6-fit to the EQE-data of the Solsonica PERC-cell exploited in this work.

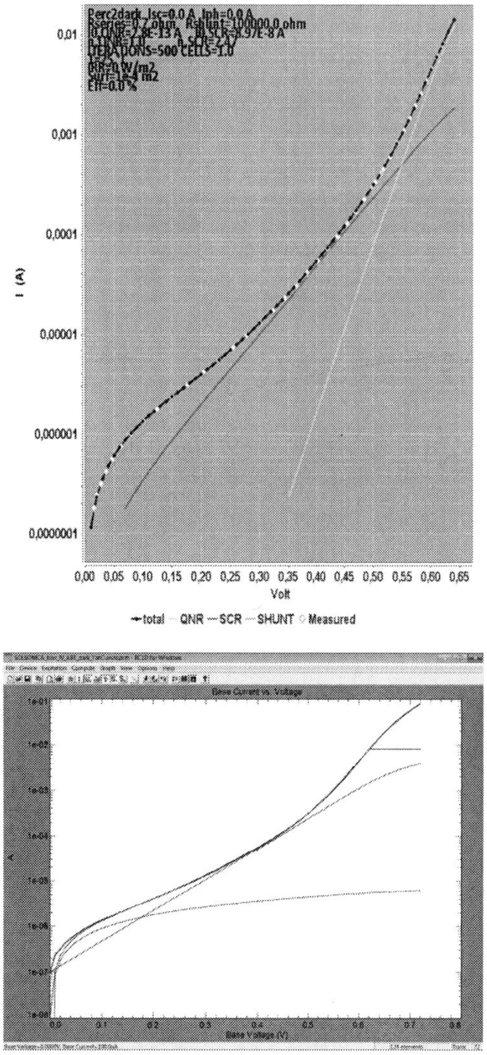

Figure 3: Fit to the measured dark I-V data exploited in this work. The fits were performed with the iterative approach in [2] (upper plot) and PC1D6 (lower plot). Data regard a PERC-cell produced by Solsonica.

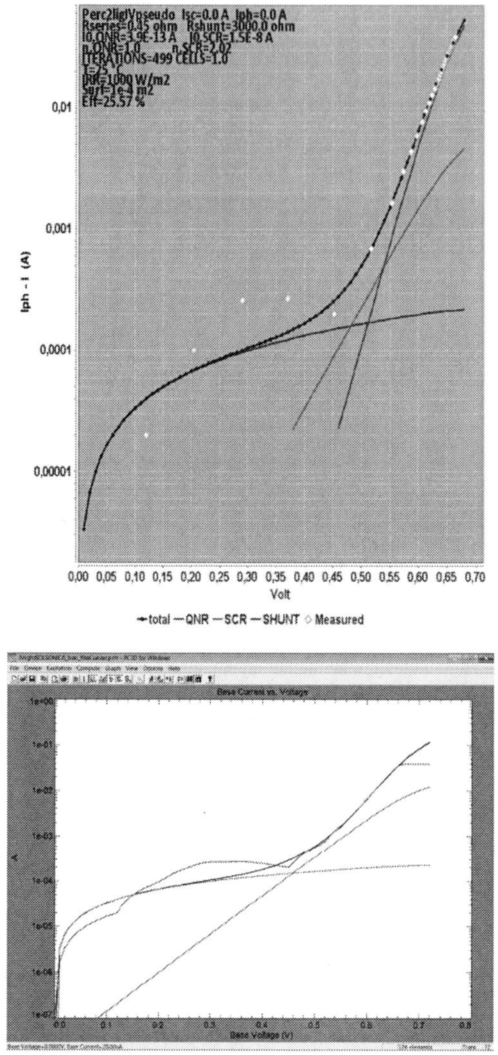

Figure 4: Fit to the measured pseudo-dark I-V data exploited in this work. The fits were performed with the iterative approach in [2] (upper plot) and PC1D6 (lower plot). Data regard a PERC-cell produced by Solsonica.

tween the assignments to the remaining parameters, which are made in the fitting approach in [2] and PC1D6.

3 APPLICATION

We have applied the approach described in the previous Sections to compare two configuration files by exploiting measured data relevant to a PERC-cell produced by Solsonica. The first configuration file includes the original models used in PC1D and will be denoted as "PC1D". The other includes the BGN model of Yan and Cuevas [3], the model for intrinsic recombination of Kerr and Cuevas [4], and the mobility model of Klaassen [5] together with the remaining models used in PC1D6 by default. The second configuration file will be denoted as "Yan-Cuevas-Kerr". In Fig. 2, the fit to the EQE-data is shown as performed by PC1D6. Such fit does not depend on the used physical model. In Figs. 3 and 4, the fits to the dark and pseudo-dark I-V data are shown as performed with the iterative approach in [2] and PC1D6.

The PC1D6-curves shown in Figs. 3 and 4 were obtained by using the "Yan-Cuevas-Kerr" configuration file. Fits of comparable accuracy result from PC1D6 by using the "PC1D" configuration file. All the assignments to the adjustable parameters that have been necessary to obtain such fits are reported in Table I.

4 DISCUSSION

As it can be seen in Table I, all the PC1D6-fits to the I-V data are relevant to assignments to the parameters in (1) that are the same as those made in the iterative approach in [2]. It can be observed moreover that the assignments to SRH-lifetime made to fit I-V data differ from

Table I: Summary of the assignments in the fits performed in this work

		FIT						
		EQE with PC1D6	I-V with approach in [2]		I-V with PC1D6			
			Dark	Pseudo-dark	Dark		Pseudo-dark	
Configuration file		Either			PC1D	Yan-Cuevas-Kerr	PC1D	Yan-Cuevas-Kerr
PARAMETER	Hole lifetime (μs)	800			150	450	60	100
	Electron lifetime (μs)	3			150	450	60	100
	S front (cm/s)	20000			20000	20000	20000	20000
	S back (cm/s)	0			0	0	0	0
	I_1 (A)		2.8×10^{-13}	3.9×10^{-13}				
	I_2 (A)		8.97×10^{-8}	1.5×10^{-8}	8.97×10^{-8}	8.97×10^{-8}	1.5×10^{-8}	1.5×10^{-8}
	n_2		2.47	2.02	2.47	2.47	2.02	2.02
	R_s (Ω)		0.7	0.45	0.7	0.7	0.45	0.45
	R_{shunt} (Ω)		10^5	3000	10^5	10^5	3000	3000

those necessary to fit the measured EQE-data. However, since, as aforementioned, EQE-fits are independent of the adopted models, the analysis of this result exceeds the scope of the present work. The same can be said of the difference between the SRH lifetime in the dark and under illumination, which is detected by using both configuration files in PC1D6.

What is most relevant to the present work is detecting that the only parameter, which is affected by the adopted set of physical models is the SRH-lifetime in the I-V fits performed with PC1D6. In particular, both in the dark and under illumination the "Yan-Cuevas-Kerr" configuration forecasts a lower SRH recombination than the "PC1D" configuration.

5 CONCLUSION

We have presented an approach to fit I-V curves with PC1D6 that uses the double exponential I-V model. This approach, which avails itself also of a reported iterative fitting procedure [2], has been useful to check the weight of physical models on the assignments to fit-parameters in simulating the solar-cell performance.

We have considered two sets of physical models for Si, one of which is the original set included in PC1D.

We have applied the approach to measured data of a PERC solar cell fabricated by Solsonica.

Based on our results we conclude that:

1. the parameters in equation (1) are independent of physical models;
2. the EQE-fit is independent of physical models;
3. assignments to SRH-recombination are different in fitting EQE-data and fitting I-V data;
4. the only parameter, which is determined by the chosen set of physical models, is the SRH-lifetime in fitting I-V data with PC1D6.

REFERENCES

[1] H. Haug, A. Kimmerle, J. Greulich, A. Wolf, and E. Stensrud Marstein, "Implementation of Fermi–Dirac statistics and advanced models in PC1D for precise simulations of silicon solar cells," *Sol. Energy Mater. Sol. Cells*, vol. 131, pp. 30–36, Dec. 2014.

[2] L. Abenante, M. Izzi, and L. Serenelli, "Modified Double-Exponential Model for Dark Current in Sili con Solar Cells," in *22nd European Photovoltaic So lar Energy Conference, 3-7 September 2007, Milan, Italy*, 2007, pp. 280–282.

[3] D. Yan and A. Cuevas, "Empirical determination of the energy band gap narrowing in highly doped n+ silicon," *J. Appl. Phys.*, vol. 114, no. 4, p. 044508, 2013.
D. Yan and A. Cuevas, "Empirical determination of the energy band gap narrowing in p+ silicon heavily doped with boron," *J. Appl. Phys.*, vol. 116, no. 19, p. 194505, 2014.

[4] M. J. Kerr and A. Cuevas, "General parameterization of Auger recombination in crystalline silicon," *J. Appl. Phys.*, vol. 91, no. 3, pp. 2473–2480, 2002.

[5] D. B. M. Klaassen, "A unified mobility model for device simulation—I. Model equations and concen tration dependence," *Solid. State. Electron.*, vol. 35, no. 7, pp. 953–959, 1992.

BRAGG ENHANCED PLASMONIC LUMINESCENT DOWNSHIFTING FOR PV DEVICES

J. Walshe*[a], H. Ahmed[b], S.J. McCormack[b], and J. Doran[a],

[A] School of Physics and Dublin Energy Lab, Dublin Institute of Technology, Dublin, Ireland.
[b] School of Engineering, Trinity College Dublin, Dublin, Ireland
* Corresponding author. Email: james.walshe@mydit.ie Tel:+353866620283

ABSTRACT: Selectively reflecting Bragg structures comprised of simple quarter-wave coatings offer a mechanism to further improve upon the enhancements achieved with plasmonic enhanced luminescent down shifting (pLDS) layers. Theoretical modelling using a transfer matrix approach demonstrates the ability of simple one- or two-layer reflecting structures to offer enhancements in the performance of a PV device consisting of a 9μm-90μm pLDS layer deposited on top of a silicon cell. Enhancements of 10.8% - 21% are reported, with the enhancement depending upon a number of design parameters (number of layers, spectral region of maximum performance and the optical properties of the constructing materials). Design of thin film enhanced pLDS layers are presented as well as establishing initial design parameters for these enhanced structures.

Keywords: Spectral losses; Photovoltaics; Plasmonic luminescent down-shifting layers; Simulation

1 Introduction

The efficiency with which electromagnetic radiation can be converted into useful work in a solar cell is limited by a number of factors [1,2]. The limited spectral response (SR) of photovoltaic (PV) cells to the wideband solar spectrum, creates fundamental spectral losses which represent the largest percentage of losses in every developed PV technology [2]. The spectral performance of a PV cell is very much dependent on the technology upon which the cell is based. However, a common trend across all technologies is a reduction in efficiency with shorter wavelengths [2].

In recent years PV research has resulted in an increase in the efficiency achievable across all generations of the technology. Using luminescent materials to capture low energy photons, previously poorly utilized by cells, and re-emit it a longer wavelength where cells are more efficient, led to the development of luminescent down-shifting (LDS) layers as a potential solution to increase the efficiencies achievable [2]. This eliminates the need to alter the active material of a PV device, offering a passive method of improving the short wavelength response across a wide range of technologies currently available [2-7].

The enhancement offered by the LDS approach is limited by the poor optical properties (low luminescent quantum yield, re-absorption losses, florescent quenching etc.) of the luminescent and encapsulating materials investigated thus far [2,6]. A novel approach of exploiting the plasmonic interaction between a specific combination of metal nanoparticles (e.g. Ag) and lumophore, has opened up the possibility of including materials which were previously deemed non-viable for the LDS concept due to their limited optical properties, with the inclusion of metal nanoparticles greatly enhancing the luminescent properties of materials [6-10].

The shift between the absorbance and emission peaks (Stokes shift) of the lumphore provides an opportunity to recycle some of the photons, which did not undergo luminescence in the LDS layer. A spectrally selective Bragg reflector structure that transmits photons within a lumophore's emission range while reflecting those within its absorption range, could further impvoe upon the enhancements achieved with the LDS and PLDS concepts.

2 Method and Materials

2.1 Transfer Matrix Model

In this investigation, we implement the transfer matrix method (TMM) to investigate the effect of selectively reflecting Bragg layers on the performance of a silicon cell fitted with a 9μm-90μm LDS layer in two different configurations (figure 1). Such structures could potentially allow for further enhancements when combined with plasmonic enhanced luminescent down shifting (pLDS) layers. A central objective during this initial investigation was to establish initial design parameters for such structures.

The TMM provides an efficient procedure for determining the reflectance and transmittance of a multilayer dielectric structure, with each layer specified by a thickness and refractive index. The model can incorporate both absorption and dispersion within a layer through the use of complex refractive index.

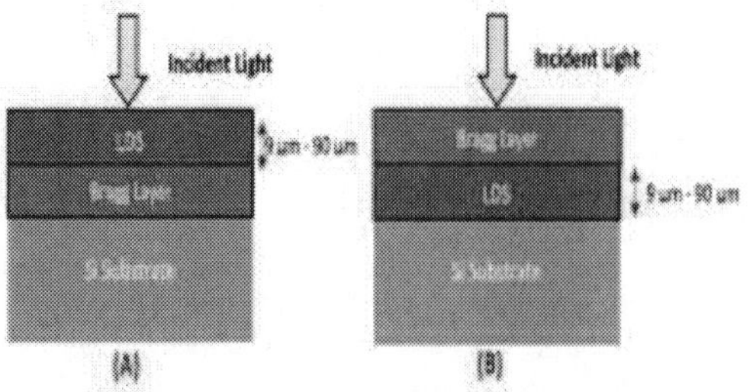

Figure 1: Schematic diagram of two possible configurations for Bragg enhanced LDS structures (A) the reflecting structure is designed to have high reflectivity across the lumophore within the LDS layers absorption range, reflecting light back into the LDS layer and allowing light that did not undergo fluorescence another opportunity to do so. While in (B) the reflecting structure is designed to selectively transmit only a certain portion of the

incident spectrum into the LDS layer where it can undergo fluorescence. The incident media was taken to be air $n = 1$

Selectively reflecting Bragg layers were designed using both single and double layer anti-reflection coatings (ARC's) to produce maximum reflection suppression at a series of wavelengths across the spectral range investigated (300nm – 1100nm), allowing the optimal materials and corresponding thicknesses to be determined for each structure investigated. Anti-reflection coatings can consist of a single layer or multilayer structure, with multilayer coatings offering the possibility of a greatly reduced broadband reflectance at one or more reflectance minima [11-12]. Optimum refractive indices for the layers comprising the reflecting structure were determined. Based upon the values determined, a range of suitable materials, whose refractive indices values lay both above and below the ideal values, were considered. The materials utilized in the study are listed in table I.

Table I: List of the materials used to design selectively reflecting structures [13-15]

Single layer (n)	**Double layer (n-n)**
MgF_2 (1.38)	MgF_2-SiO (1.38, 1.85)
SiO (1.85)	MgF_2-CeO_2 (1.38, 2.2)
CeO_2 (2.2)	MgF_2-ZnS (1.38, 2.3)
ZnS (2.3)	SiO-Si_3N_4 (1.46, 2.02)
Ideal (1.97)	MgF_2-Ta_2O_5 (1.38, 2.13)
	MgF_2-TiO_2 (1.38, 2.61)
	Ideal (1.57, 2.47)

To allow a realistic representation of the effect of the addition of such coatings onto an LDS-cell structure the incident solar spectrum was taken as the AM1.5G standard[16] and the spectral response of the silicon cell [6] factored into the model. The performance of the underlying silicon was determined using equation 1, which takes into account the spectral response (*SR*) of the cell, the incident irradiance (*I*) and the transmittance coefficient (*T*).

$$\int_{300nm}^{1100nm} SR(\lambda)\, I(\lambda)\, T(\lambda)\, d\lambda \qquad (1)$$

As a first approximation all layers and substrates were modelled as non-absorbing non-dispersive media, and all structures were modelled under normal incidences conditions. However, to get a more realistic picture of what is occurring within these structures and to allow a more accurate determination of the benefit they offer to existing PV technologies, these effects will need to be accommodated into the model.

3 Results and Discussion

Figure 2 shows the wide range of reflection suppression capability offered through the use of simple single-layer or double-layer Bragg structures. A Bragg structure can be designed to produce a single reflectance minimum at a particular wavelength of interest or a series of minima at different wavelengths, with the structure design tailored to a specific lumophore's spectral properties to maximize the further enhancement offered by this additional structure.

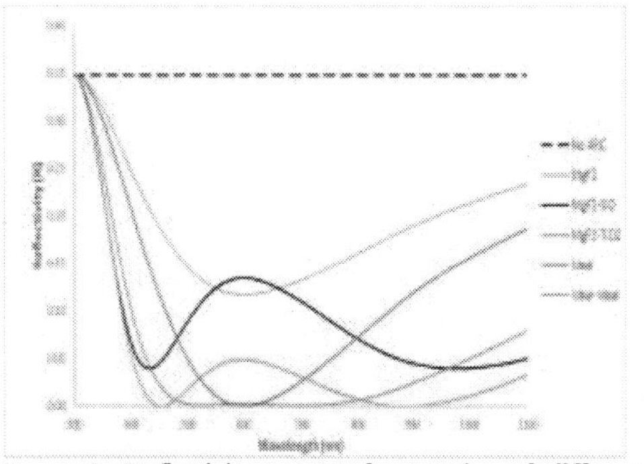

Figure 2: Reflectivity spectra for a series of different Bragg structures designed to produce maximum reflection suppression at 600nm. Depending upon the number of layers within the structure and the refractive indices of each layer a single reflection minimum or two minima can be produced, each of varying bandwidth. The most suppression offered is by the MgF_2-TiO_2 combination producing complete suppression over a bandwidth of 500 nm -- 750 nm.

The performance of the underlying silicon substrate for each structure composed of a series of different materials was evaluated using equation 1. Using the TMM method to determine the overall transmittance (T) spectra for a given structure, the product of the T spectra, incident spectra and the spectral response of the cell was integrated to provide a quick method of determining the performance of the cell over the entire spectral range investigated (300 nm – 1100 nm). An example of the ease at which the optimal design parameters of such structures can be determined using this method is provided in figure 3, whereby it is clearly visible which material combination and wavelength of maximum suppression yield the best spectral performance for the silicon substrate.

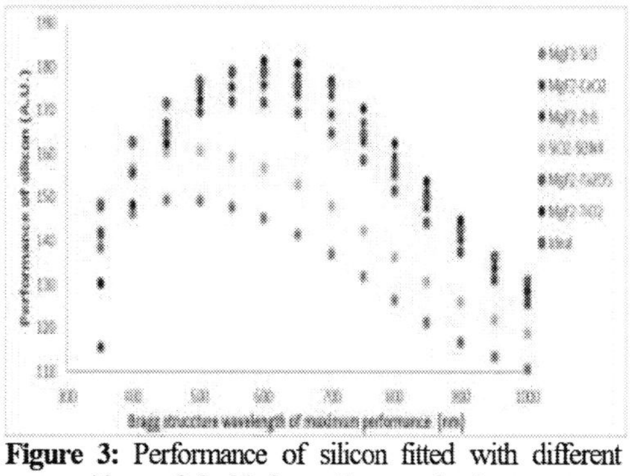

Figure 3: Performance of silicon fitted with different compositions of double-layer Bragg reflecting structures designed to produce maximum refection suppression at a series of wavelength across the spectral range investigated (300nm – 1100nm). All performance values presented take into account the base performance of silicon and such all structures presented in the figure show improvements over bare silicon of 30% - 50%.

One of the most common materials used to encapsulate both the luminescent material and/or metal nanoparticles in a rigid layer capable of being directly fitted to a photovoltaic cell while offering some level of protection from the environment is poly-methyl-methacrylate (PMMA). The effect of implementing a Bragg reflecting layer in two different configurations on the performance of a pLDS-silicon device was investigated, assuming the effects of both the luminescent material and metal nanoparticles had little effect on the layer's optical properties. The pLDS layer was represented as a single PMMA layer (n=1.496), with two thickness values investigated (9μm and 90μm). LDS/pLDS layers are typicaly fabricated to a thickness of 90μm[6], while the effects of thinner layers on the performance of PV cells have not to the authors knowledge been invesitgated.

Figure 4 shows the results of the calculations of the enhancement achieved with single layer Bragg structures when combined with a pLDS layer. Enhancements with the pLDS fitted on-top of the Bragg structure produced enhancements of 11% - 18%, while structures with the pLDS layer fitted below the Bragg layer produced a decrease from the performance of a bare silicon device. This is due to the ordering of the refractive indices within the structure, and the associated phase shifts being introduced into the reflection components responsible for the reflection suppression capabilities of the Bragg structure.

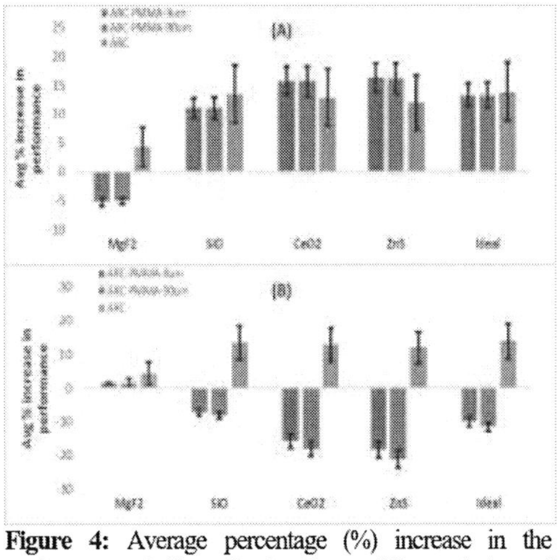

Figure 4: Average percentage (%) increase in the performance of silicon fitted with two different designs of Bragg enhanced pLDS layer, with the pLDS layer represented as a uniform layer of PMMA: (A) pLDS layer fitted above single layer Bragg structures (B) pLDS layer fitted below single layer Bragg structures (directly above silicon substrate). The percentage increases presented are based upon a comparison with the overall performance of the same thickness pLDS layer fitted directly onto a silicon substrate. The uncertainty in the improvements reported is due to the variation in the performance observed when designing a particular structure to minimize reflections at a specific part of the spectrum investigated (300nm – 1100nm).

Similar levels of enhancement (12.7% - 21%) were observed with double layer Bragg reflecting structures, as shown in figure 5. Again the configuration in which the PLDS layer is placed directly onto the silicon substrate and hence below the Bragg reflecting layer results in a large decrease in performance across the range of materials investigated. Based on Figures 4 & 5 the configuration in which a Bragg reflecting structure is fitted directly on-top of the pLDS-Silicon device can be discarded in favor of a design incorporating the Bragg reflecting structure below the pLDS layer.

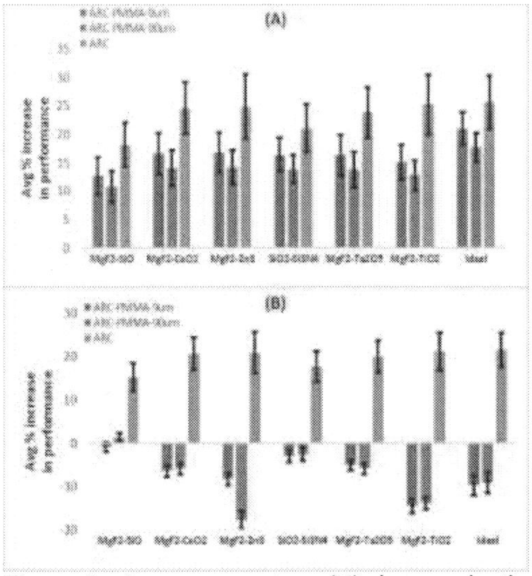

Figure 5: Average percentage (%) increase in the performance of silicon fitted with two different designs of Bragg enhanced pLDS layer, with the pLDS layer represented as a uniform layer of PMMA: (A) pLDS layer fitted above double layer Bragg structures (B) pLDS layer fitted below double layer Bragg structures (directly above silicon substrate). The percentage increases presented are based upon a comparison with the overall performance of the same thickness pLDS layer fitted directly onto a silicon substrate. The uncertainty in the improvements reported is due to the variation in the performance observed when designing a particular structure to minimize reflections at a specific part of the spectrum investigated (300nm – 1100nm).

4 Conclusions

Initial investigation into the further enhancements possible with Bragg enhanced plasmonic down shifting has demonstrated the ability of such structure to increase the spectral performance of silicon within the 300nm – 1100nm region of the spectrum. Initial design parameters for such structures have been established as well as the further enhancement possible with such structures (10.8% – 21%), with the level of enhancement dependent upon the structure design and materials used.

5 Acknowledgements

The authors would like to acknowledge the financial support of the Dublin Institute of Technology through the Fiosraigh Dean of Graduate Studies programme.

6 References

[1]. Gp, E. (2009). Fundametnal Losses in Solar

Cells. Progress in Photovoltaics: Research and Applications, 19(3), 21–25.

[2]. Klampaftis, E., Ross, D., McIntosh, K. R., & Richards, B. S. (2009). Enhancing the performance of solar cells via luminescent down-shifting of the incident spectrum: A review. *Solar Energy Materials and Solar Cells, 93*(8), 1182–1194. http://doi.org/10.1016/j.solmat.2009.02.020

[3]. Kettle, J., Bristow, N., Gethin, D. T., Tehrani, Z., Moudam, O., Li, B., ... Krebs, F. C. (2016). Printable luminescent down shifter for enhancing efficiency and stability of organic photovoltaics. Solar Energy Materials and Solar Cells, 144(November), 481–487.

[4]. Ahmed, H., Mccormack, S. J., & Doran, J. (2016). External Quantum Efficiency Improvement with Luminescent Downshifting Layers :Experimental and Modelling. International Journal of Spectroscopy, 2016, 1–8.

[5]. Ross, D., Klampaftis, E., Fritsche, J., Bauer, M., & Richards, B. S. (2012). Increased short-circuit current density of production line CdTe mini-module through luminescent down-shifting. Solar Energy Materials and Solar Cells, 103, 11–16.

[6]. Ahmed, H. (2014). Materials Characterization and Plasmonic Interaction in Enhanced Luminescent Down-Shifting Layers for Photovoltaic Devices. Dublin Institute of Technology.

[7]. Ahmed, H. a., Walshe, J., Kennedy, M., Confrey, T., Doran, J., & McCormack, S. J. (2013). Enhancement in solar cell efficiency by luminescent down-shifting layers. *Advances in Energy Research, 1*(2), 117–126.

[8]. Ahmed, H., Doran, J., & McCormack, S. J. (2016). Increased short-circuit current density and external quantum efficiency of silicon and dye sensitised solar cells through plasmonic luminescent down-shifting layers. Solar Energy, 126, 146–155.

[9]. Chandra, S., Doran, J., McCormack, S. J., Kennedy, M., & Chatten, A. J. (2012). Enhanced quantum dot emission for luminescent solar concentrators using plasmonic interaction. Solar Energy Materials and Solar Cells, 98, 385–390.

[10].Ahmed, H., Doran, J., & McCormack, S. J. (2016). Increased short-circuit current density and external quantum efficiency of silicon and dye sensitised solar cells through plasmonic luminescent down-shifting layers. *Solar Energy, 126*, 146–155.

[11].Lee, C. C. (2015). *Thin film optical coatings. Topics in Applied Physics* (Vol. 129).

[12].Raut, H. K., Ganesh, V. A., Nair, A. S., & Ramakrishna, S. (2011). Anti-reflective coatings: A critical, in-depth review. *Energy & Environmental Science, 4*(10), 3779.

[13].M. N. Polyanskiy. "Refractive index database," Retrieved July 1, 2016, from http://refractiveindex.info

[14]. SOPRA. (n.d.). database. Retrieved May 28, 2016, from http://www.sspectra.com/sopra.html

[15].Green, M.A. and Keevers, M., Optical properties of intrinsic silicon at 300 K, *Progress in Photovoltaic Journal*, 3(3).189-192. 1995

[16]. Emery, K. (2010). Reference Solar Spectral Irradiance: Air Mass 1.5. Retrieved June 1, 2016, from http://rredc.nrel.gov/solar/spectra/am1.5/#refs

THE INFLUENCE OF NEUTRONS AND Xe-IONS FLUX ON c-Si – a-SiC PHOTOVOLTAIC DEVICE

Milan Perný*, Michal Váry*, Vladimír Šály*, Miroslav Mikolášek*, Jozef Huran**
*Slovak University of Technology
Faculty of Electrical Engineering and Information Technology
Ilkovičova 3, SK-81219 Bratislava, Slovak Republic
Phone: (421 2) 6029 1746, Fax: (421 2) 6542 0415, vladimir.saly@stuba.sk
**Slovak Academy of Sciences, Institute of Electrical Engineering
Dúbravská cesta 9, 841 04 Bratislava, Slovakia

ABSTRACT: Amorphous silicon carbide in various forms, thin films or bulk structures, is a longer period used in microelectronics, power electronics, sensors and mechanical engineering. Thin films of n-doped amorphous silicon carbide (a-SiC:H) in combination with crystalline silicon representing heterostructure for the solar cells were object of this study. AC measurements and subsequent analyzes were used to estimate the impact of Xe and neutron irradiation on electrical properties of studied heterostructure. Impact of Xe and neutron irradiation has been quantified by using AC equivalent circuits and the corresponding voltage dependence of AC equivalent circuit elements. The influence of Xe-ions and neutron bombardment of a-SiC:H/c-Si heterostructure which resulted in the changes of elements of equivalent circuit – parallel resistance and capacitance – have been identified and explained.
Keywords: Silicon carbide, PECVD, impedance, heterojunction

1 INTRODUCTION

Free neutrons are anywhere in the nature coming either from natural source like cosmic rays or from artificial source emitted at nuclear fission or nuclear fusion or from any number of different nuclear reactions. Neutron radiation is a kind of ionizing radiation which consists of free neutrons. Neutron is a particle without electric charge but with quite large mass. The reaction of neutron flux with solid matter is through elastic collision with atomic nuclei, often resulting in the damage of crystalline structure. Electronic devices can be at specific application exposed to neutron radiation of various intensities [1]. Inert gas, like xenon, ion flux has been used as technological tool in ion bombardment process [2] or implanted into material to influence its properties [3]. Complex impedance measurement and analysis of measured data also called impedance spectroscopy (IS) is technique often being applied mainly in electrochemistry and is attractive and meaningful also in other fields of research. The application of IS to study the behavior of electronic devices with pn junction, like photovoltaic solar cells, is very common. The analysis of measured and fitted data based on proposed equivalent circuit allows the study of electric properties and electronic transport within investigated structure.

The pn junction structure constitutes capacitive impedance $|Z|e^{i\varphi}$ with a phase shift φ between voltage and current of applied ac signal. Complex impedance $Z = Z' + iZ''$, where i is imaginary unit. The parallel connection of resistance R (dynamic resistance) and capacitance C (depletion region capacitance) represents simple approximation and equivalent circuit model. Both elements R and C are modulated by bias voltage and models the behavior of pn junction. Complex real structures need deeper description and next/more electronic elements connected in various way to formulate either transport mechanisms.

2 EXPERIMENTAL

We have studied the influence of irradiation of photovoltaic device hetero-structures based on crystalline silicon and amorphous silicon carbide heterojunction.

Xe-ion flux irradiation has been applied in our experiment. The flux of Xe ions was changed from 5.10^8 to 5.10^{11} cm^{-2}. The results obtained for the chosen sample irradiated with Xe-ion flux 5.10^{11} cm^{-2} are presented here. The fluence of neutrons was varied from $\approx 10^{11}$ to $\approx 10^{15}$ cm^{-2}. The results obtained for the chosen sample irradiated with neutron fluence 10^{13} cm^{-2} are presented also here. Thin a-SiC:H layer was prepared by plasma CVD deposition. A crystalline silicon wafer of (100) orientation was used as the substrate for the growth. The deposition was from the gas mixture of SiH_4, CH_4, H_2 and Ar, common for all prepared samples. More details one can find in [4]. The area of prepared solar cell structures was 1 x 1 cm.The electrical behaviour of samples and the influence of irradiation was studied using an impedance spectroscopy within wide range of ac frequencies from 10 Hz to 1 MHz. The impedance spectroscopic measurement can help to explore the interface characteristics to some extent. The impedance spectroscopic analysis was performed by following Cole-Cole principle [5] in which the real and imaginary components of the complex impedance (Z*) are plotted for sinusoidal input signals. Impedance spectroscopy has been particularly suitable for studying the properties of junctions, interfaces, and contacts.

3 RESULTS AND DISCUSSION

The complex impedance was measured using small 25 mV AC signal before irradiation and after neutron and xenon irradiation. The obtained results are presented using Nyquist plot and shown in Fig.1. The influence of neutron irradiation is easy to quantify. Measured impedance spectra in the form of Nyquist plot in our case are represented by two semicircles each for a different frequency interval. The presence of two semicircles, respectively two different relaxation processes, was confirmed by numerical simulations.

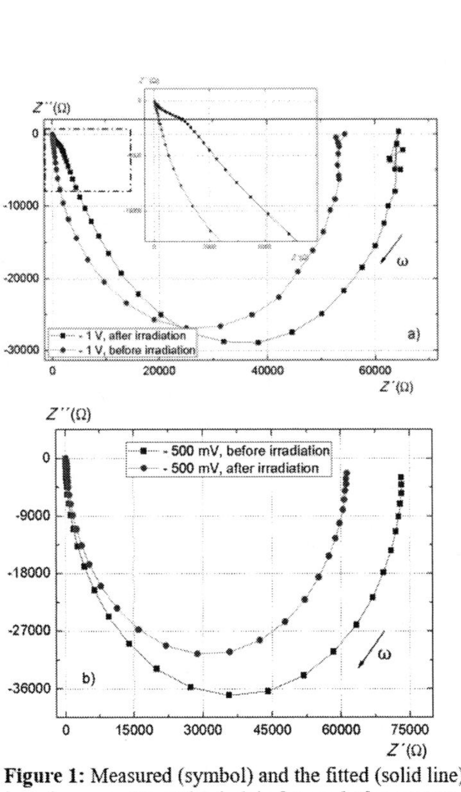

Figure 1: Measured (symbol) and the fitted (solid line) impedance spectra at the dark before and after neutron a) and Xe b) irradiation

Numerical simulation for more detailed assessment of the processes in the structure was necessary to carry out in order to determine the individual parameters of the AC equivalent circuit. Elements of the equivalent circuit have a direct relation with electrical transport processes in the structure. Series resistance R_1 is related to resistance of contacts and electrode system (grid on top and whole area bottom). The effect of neutron irradiation was quantified by comparison of the value of the parallel equivalent resistances R_2, R_3, the constant phase elements P_1, P_2 resulting from the AC equivalent circuit shown in Fig.2 and 3. More details about the equivalent AC circuit of a-SiC/c-Si heterostructure and its correlation to electric transport process can be found in [6, 7].

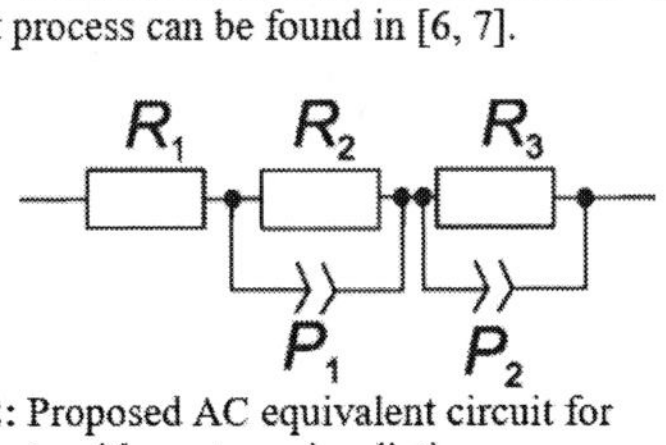

Figure 2: Proposed AC equivalent circuit for experiments with neutrons irradiation

Another type of equivalent circuit had to be chosen for the experiments with irradiation by Xe. Inductance L in the AC equivalent circuit represents inductance of leads and top finger grid system. Series resistance R_1 is related to resistance of contacts and electrode system (grid on top and whole area bottom). Parallel combination of resistance R_2 (transport resistance) and capacitance C_2 is linked to space charge region of the pn hetero-junction situated within the structure. Second combination of resistance R_3 (recombination resistance) and constant phase element (CPE) P_3 should be associated with traps and defects in the structure [8, 9].

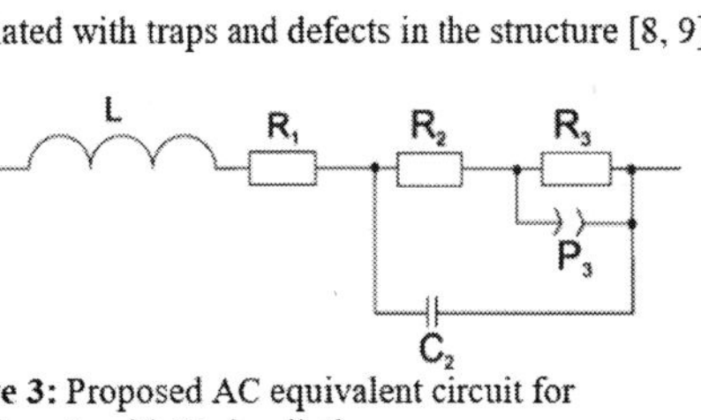

Figure 3: Proposed AC equivalent circuit for experiments with Xe irradiation

For further clarification impedance response of constant phase elements at applied bias between –1 V and 0.5 V is shown in Fig.4. In case of CPE capacitance a slight increase of capacity P_2 which is supposed to be associated with the creation of new recombination centres was observed as a consequence of neutron bombardment.

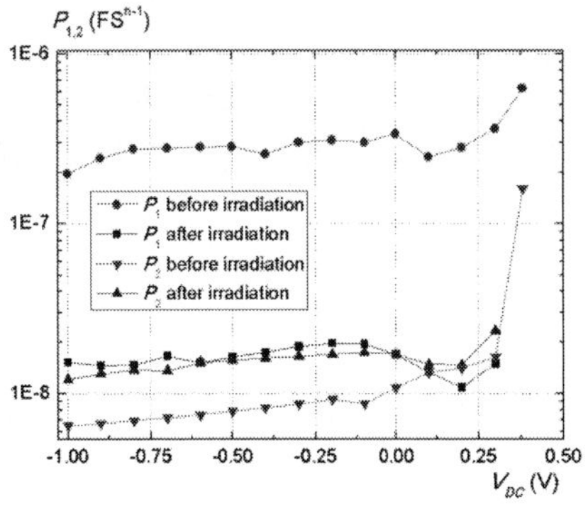

Figure 4: Impedance results for CPE capacitances P_1, P_2 as a function of applied bias before and after neutron irradiation

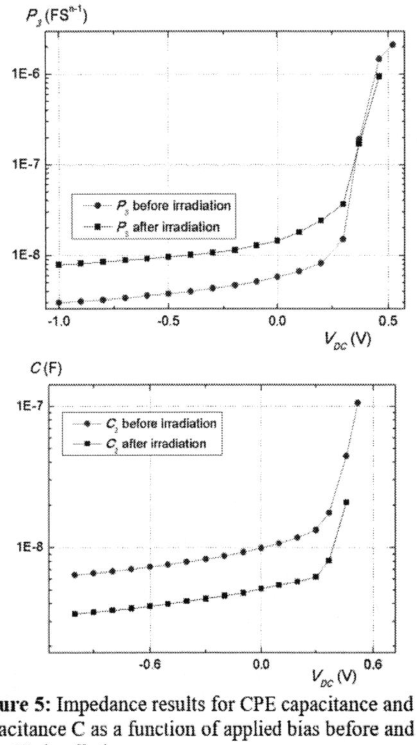

Figure 5: Impedance results for CPE capacitance and capacitance C as a function of applied bias before and after Xe irradiation

For better clarification of impact of Xe irradiation, impedance response of constant phase element capacitance and C_2 capacitance at applied bias between –1 V and 0.5 V is shown in Fig. 5

Fig. 6 shows the dependence of obtained parallel resistances R_2 and R_3 on the DC bias voltage before and after Xe irradiation.

As it is shown in Fig. 7, both parallel resistances R_2 and R_3 of AC equivalent circuit are at higher forward bias voltage strongly dependent on applied DC voltage. They decrease toward negligible values and behaviour of the structure is controlled by series resistance. The effect of the Xe irradiation regarding R_2 and R_3 was minimal.

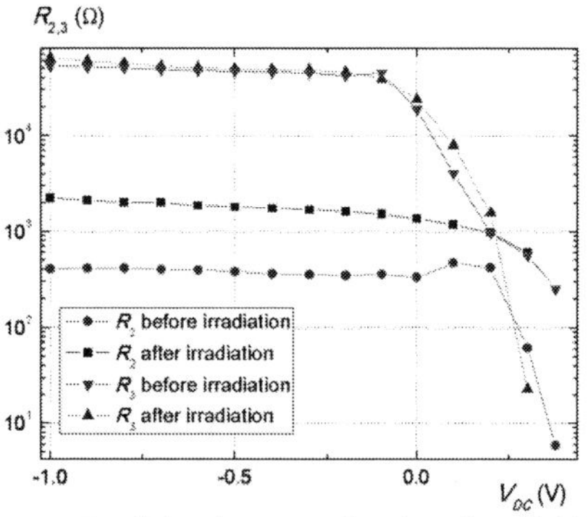

Figure 6: Parallel resistances a function of applied bias before and after neutron irradiation

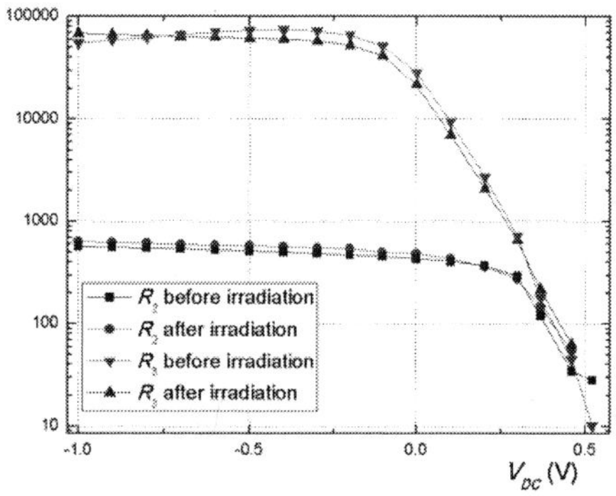

Figure 7: Parallel resistances a function of applied bias before and after Xe irradiation

4 CONCLUSIONS

The influence of the radiation of xenon ions on the behavior of diode heterostructures SiC-Si intended for photovoltaic application was studied. DC characterization and impedance spectroscopy on devices a-SiC/c-Si over a range of applied biases was performed. An equivalent circuit, consisting of a series resistance, a transport resistance, a device capacitance, a recombination resistance and a CPE was proposed and used to identify and quantify the underlying physical mechanisms. The structural defects induced by Xe particles and neutrons affect mainly trapping mechanism illustrated by the influence of radiation on CPE included into equivalent circuit.

Acknowledgements

We thank M.V. Bulavin, Join Institute for Nuclear Research Dubna, Russia, for neutron and xenon irradiation of samples. This work was supported by the Scientific Grant Agency of the Ministry of Education of the Slovak Republic and of the Slovak Academy of Sciences (Projects VEGA 1/0651/16) and Slovak Research and Development Agency under the contracts APVV-0443-12).

References

[1] B. Simic, D. Nikolić, K. Stanković, L. Timotijević and S. Stanković. Damage induced by neutron radiation on output characteristic of solar cells, photodiodes and phototransistors. Int. J. Photoenergy 2013, Article ID 582819.

[2] K W. Shindo, T. Ohmi. Ion energy, ion flux, and ion mass effects on low-temperature silicon epitaxy using low-energy ion bombardment process. J. Appl. Phys. 79, (5) 1996, p. 2347.

[3] J. R. Kaschny, L. Amaral, M. Behar, D. Fink. Ion radiation induced diffusion of Xe implanted into a polymer film. J. Appl. Phys. 72, (11) 1992, p. 5139.

[4] M. Perný et al. Electrical Properties of Solar Heterojunction a-SiC/c-Si Irradiated by Neutrons. In Power Engineering 2016. Renewable Energy Sources 2016: 6th International Scientific Conference OZE 2016. Tatranské Matliare, Slovakia, May 31-June 2, 2016. Bratislava : Slovak University of Technology in Bratislava, 2016, ISBN 978-80-89402-81-6

[5] K. S. Cole and R. H. Cole, Dispersion and absorption in dielectrics I. Alternating current characteristics, Journal of Chemical Physics, vol. 9, pp. 341–351, 1941.

[6] J. Packa, et al. Accelerated ageing of a-SiC:H/c-Si photovoltaic heterostructures. In EPE 2016 : 17th International scientific conference on electric power engineering. Prague, Czech Republic. May 16-18, 2016. Prague: Czech Technical University in Prague, pp. 215-219, 2016.

[7] M. Perný, et al. AC impedance spectroscopy of Al/a-SiC/c-Si(p)/Al heteostructure under illumination. In Journal of Electrical Engineering. Vol. 65, No. 3, pp. 174-178, 2014.

[8] B. J. Leever, C. A. Bailey, T. J. Marks, M. C. Hersam, and M. F. Durstock In situ characterization of lifetime and morphology in operating bulk heterojunction organic photovoltaic devices by impedance spectroscopy, Adv. Energy Mater. 2, pp. 120-128, 2012.

[9] Bisquert and G. Garcia-Belmonte, On Voltage, Photovoltage, and Photocurrent in Bulk Heterojunction Organic Solar Cells J. Phys. Chem. Lett. 2, pp. 1950-1964, 2011.

EFFECTS OF TEMPERATURE AND POST DEPOSITION ANNEALING ON SNS POLYCRYSTALLINE THIN FILM GROWTH

Simone Di Mare°, Andrei Salavei°, Daniele Menossi°, Fabio Piccinelli*, Elisa Artegiani°, Arun Kumar†, Gino Mariotto†, Paolo Bernardi^ and Alessandro Romeo°.

* Department of Biotechnology, University of Verona, Strada Le Grazie 15, 37134 Verona, Italy.

†Raman Laboratory, Department of Computer Science, University of Verona, Strada Le Grazie 15, 37134 Verona, Italy.

^Department of Neurosciences, Biomedicine and Movement Sciences, University of Verona, Strada Le Grazie 15, 37134 Verona, Italy.

°LAPS-Laboratory for Applied Physics, Department of Computer Science, University of Verona, Ca' Vignal 1, Strada Le Grazie 15, 37134 Verona, Italy, tel.:+39-045-8027974, e-mail address: alessandro.romeo@univr.it

ABSTRACT: The limited amount in nature of Te and In and the use of toxic elements like Cd and Se in CdTe and $CuInGaSe_2$ solar devices (despite they have no effect on the environment) could be overcome with alternative thin films. SnS is one of the most promising materials due to its electrical and physical properties, but it is already demonstrated that high quality SnS thin films are required to enhance solar cells efficiency.
In the present work, we have found out a threshold temperature, above which the absorber grows highly crystallized and compact avoiding the porous structure observed previously. The properties of this differently grown film is presented and compared with low substrate temperature layers. Moreover the study of a new post deposition treatment, which merges our $CdCl_2$ methodologies with new non-toxic salts (KI or $SnCl_2$), more suitable for SnS, is presented.

Keywords: Thin Film, Annealing, Crystallization.

1 INTRODUCTION

Despite the big improvement in efficiency for $CuInGaSe_2$ and CdTe based solar cells [1][2], demonstrating the high feasibility of these technologies, there is still ongoing research for new absorbers: low cost thin films made of abundant and non-toxic materials.

Among different compounds available, SnS is a promising compound due to its simple stoichiometry, its low temperature growth and the abundance of its elements. However record efficiencies are still very low and stoichiometric growth, free of secondary phases, is not easy to achieve [3][4].

It has been already reported [5] that SnS structure and morphology strictly depends on the substrate temperature. In our case, we observe a threshold above which the material grows highly crystallized, compact and smooth. Below this temperature the layer is more porous and shows the presence of porous material on the surface.

Another possibility to increase crystal quality is to apply a post deposition treatment. Previously we have discussed the effect of $CdCl_2$ and Freon gas on the SnS structure [6], whereas in this present work the recrystallization of SnS crystals is achieved by annealing the absorber layer at high temperature under a flux of molten KI or $SnCl_2$ salts.

2 EXPERIMENTAL

2.1 SnS preparation

Tin sulfide solar cells in superstrate configuration were made by vacuum evaporation (VE). SnS pieces, 99.99 % pure, from Huizhou Tian Yi Rare Material Co. Ltd., are used for the deposition of the absorber layer on a stack of glass/ITO/ZnO/CdS. On a previously cleaned soda lime glass substrate, ITO (400 nm) and ZnO (100 nm) front contact layers are made by subsequent RF sputtering in Ar/O_x atmosphere heated at 400 °C. The window layer (namely 50 nm of CdS) is deposited at a substrate temperature of 100 °C and then annealed in vacuum for 30 minutes at 450 °C. Prior to CdS deposition, transparent conductive oxide (TCO) is also annealed at 450 °C for 30 minutes in vacuum.

SnS is evaporated at pressures below 10^{-5} mbar in a dedicated chamber with special Knudsen sources. The evaporation is controlled by monitoring the source temperature with thermocouples and by a water-cooled IL150 quartz crystal thickness monitor from Intellemetrics Ltd. The crucible is usually filled with ~ 2 g of SnS, which is enough for processing two different runs; In each run we are able to produce 2 samples of 3 cm x 3 cm size. After this, the crucible is emptied and refilled again. Since it has been observed a difference in performance after each run, more will be explained in the next chapter.

SnS thin films, 2 μm thick, were prepared at different substrate temperatures in the range between 210 °C and 390 °C. The back contact is made by depositing 50 nm of Au by VE at 25 °C substrate temperature in a dedicated chamber.

The SnS thin film morphology was studied with a NT-

MDT Smena-A Atomic Force Microscopy (AFM) in semi-contact mode. The crystalline structure was analyzed with a Thermo ARL X'TRA powder diffractometer (in Bragg-Brentano geometry, equipped with a Cu-anode X-ray source (Kα, λ=1.5418 Å) and a Peltier Si (Li) cooled solid state detector). Raman spectra were collected with a Horiba Jobin-Yvon LabRam HR800 microprobe setup (in backscattering geometry) by using an exciting radiation at 632.8 nm (He-Ne laser). The current–voltage characteristics were measured at room temperature with a Keithley 2420 SourceMeter, using a calibrated halogen lamp with an irradiation of 1000 W/m^2. Energy Dispersive X-ray analysis (EDX) was applied by a Philips FEI XL30 Scanning Electron Microscope (SEM).

2.2 Post deposition treatment

The $SnCl_2$ and KI salts were diluted into different solvents: $SnCl_2$ salts were dissolved in isopropyl alcohol (0.6 M) and in ethanol (2.2 M), while KI salts were dissolved in methanol (0.95 M) and in ethanol (0.15 M). A known amount of solution (typically 100 μl) is then deposited on the sample surface, and the solvent is evaporated in order to leave a thin layer of salt. The sample is then annealed into a furnace in air with a temperature ramp of 20 minutes followed by 30 minutes at the treatment temperature (typically 350 °C for $SnCl_2$ and 450 °C for KI). After the treatment, the samples are rinsed with de-ionized water in order to remove the salt residues.

3 RESULTS AND DISCUSSIONS

3.1 *As deposited SnS films*

The morphologies of SnS films deposited with the substrate heated to 270 °C and to 400 °C are shown in Fig. 1a and Fig. 1b respectively.

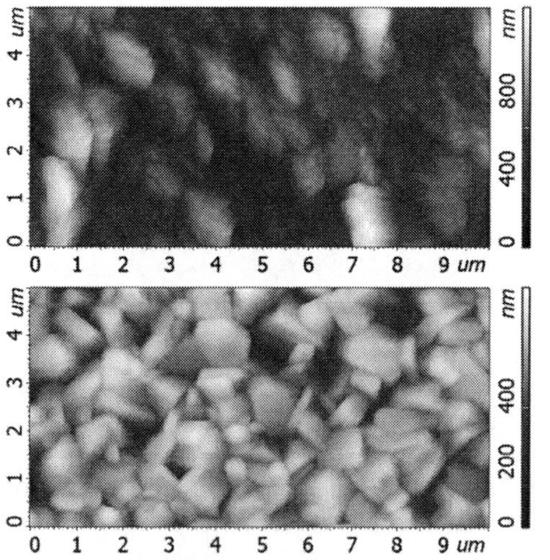

Figure 1: AFM pictures of SnS thin films deposited at 270 °C (a) top and at 400 °C (b) bottom.

Samples grown at low temperature exhibit small-stacked grains (the layer has reduced adhesion) forming larger clusters. On the other hand, samples grown at high temperature show a very compact structure with rectangular grains.

Increasing T_{sub} results in very different morphologies. At 210 °C (not shown) and 270 °C samples have a similar inhomogeneity in grain sizes: where very big grains (over 1 μm) and smaller grains (hundreds of nm) are present at the same time; on the other hand if the shape of bigger grains do not change at the two T_{sub} values, the shape of smaller grains changes from small circular grains to elongated grains. At 330 °C (not shown) an improved homogeneity of size distribution, and small circular grains (< 500 nm) are found. Finally, at 400 °C, a very different morphology is found: big rectangular grains of over 1 μm are observed, but smaller grains also appear.

In order to correlate these different morphologies with the crystalline structure of samples we collected the XRD patterns of samples deposited at 400 and 200 °C, see Fig. 2(a). Both samples are compatible with the orthorhombic phase of tin sulfide (JCPDS card 39-0354). In the sample grown at 400 °C the (111) peak (2Θ = 31.6 °) predominates, indicating a stronger preferential orientation of crystallites. The sample grown at 270 °C exhibits several peaks with comparable intensity, which suggests a structure with several orientation of grains.

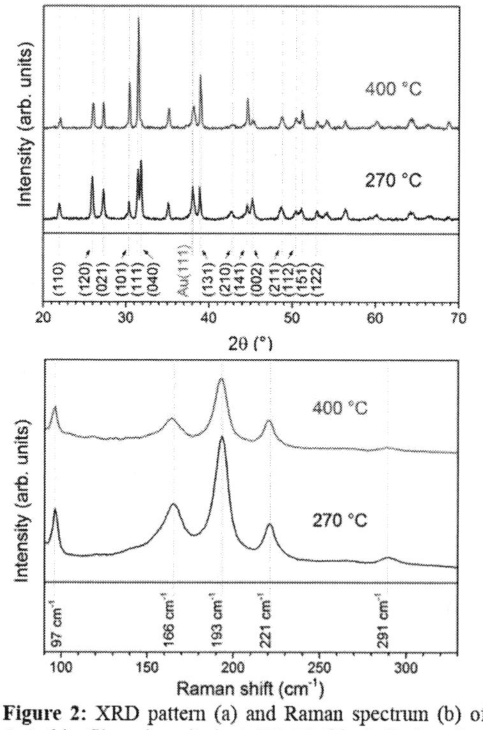

Figure 2: XRD pattern (a) and Raman spectrum (b) of SnS thin films deposited at 270 °C (black line) and at 400 °C (red line).

However the peaks reveal a very stoichiometric structure irrespective of the substrate temperature.

Comparing XRD spectra of 1st run and 2nd run samples after refilling of the source (not shown here), no significant differences could be found. Moreover the surface of the as-deposited samples was analyzed by micro-Raman spectroscopy, in order to analyze the presence of SnS_2 and Sn_2S_3, and also to verify the existence of any connection between the different morphologies and their structures.

The Raman spectra are shown in Fig. 2(b): the peak positions are in good agreement with those for a SnS single crystal [7], and the absence of any signal in the region immediately above 300 cm^{-1} indicates that the layers have no secondary phases. In fact no signals around 310 cm^{-1} is detected, showing that the SnS thin films do not have secondary phases such as Sn_2S_3 (main mode at 307 cm^{-1}) and SnS_2 (312 cm^{-1}) [7].
The effects of the varying T_{sub} are only slightly visible in the Raman measurements. The main change between the two spectra is a different relative intensity of the vibrational modes at 164 cm^{-1} and at 221 cm^{-1}. The most intense peak is the 193 cm^{-1} mode, and it also shows a small shift as substrate temperature is varied: this mode belongs to a symmetric stretching of the Sn-S bond [7]. At 270 °C the 166 cm^{-1} mode has higher intensity than the one at 221 cm^{-1}, while at 400 °C the opposite occurs; these modes belong to two different lattice vibrations, the interaction along the inter-layer direction and the Sn-S bonding symmetric stretching respectively. A clear trend is visible in the region between 97 cm^{-1} and 166 cm^{-1} modes, where increasing T_{sub} a plateau appears.

3.2 *Post Deposition Treatment*

After annealing a strongly recrystallized surface is formed on the layer (see Fig. 3). EDX analysis shows that this surface is strongly Sn-rich. In order to remove this surface different etchings have been applied: water and ammonia solution are the most successful in our case.

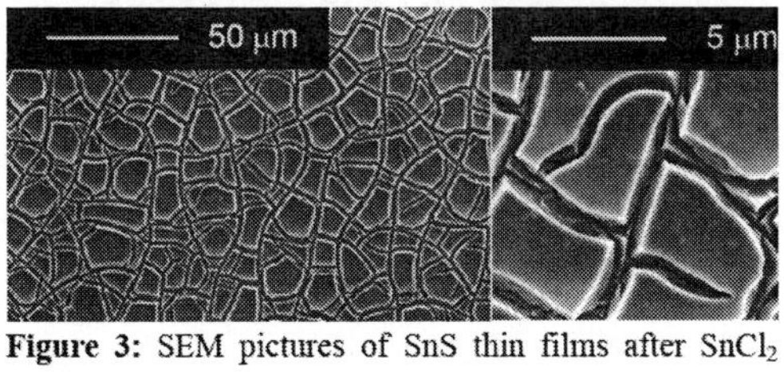

Figure 3: SEM pictures of SnS thin films after $SnCl_2$ treatment and before etching.

In Fig. 4 the morphology of SnS thin films treated with $SnCl_2$ and etched is depicted. The sample exhibits small regular grains with size of 0.5 microns.

The $SnCl_2$ post deposition treatments have been performed with a fixed volume of 100 µl of $SnCl_2$ solution. The effect is of general rearrangement of the grains without a particular enlargement, on the other hand the recrystallization is proven by the disappearance of the smaller grains observed in figure 1b (as deposited case).

No AFM picture of $SnCl_2$ treated layer of SnS deposited at low temperature is shown since, in this case, the change in the structure is very intensive and there is a major loss of material: the layer is yellow and an analysis with EDX shows an excess of sulphur. Most probably the low substrate temperature, during deposition, delivers SnS grains with lower stability, which are easily transformed by the post deposition treatment.

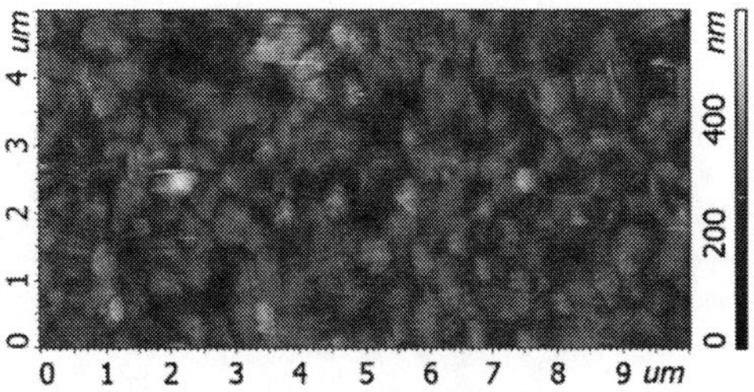

Figure 4: AFM pictures of SnS thin films deposited at 400 °C and treated with $SnCl_2$-EtOH.

In Fig. 5 the morphology of the thin film treated at 450 °C with 300 µl of KI-EtOH and etched is shown; again the morphology, compared to the as deposited case, does not change drastically in terms of grain size, but there is a clear rearrangement; as demonstrated by the different shape. Again the morphology becomes more regular where the smallest grains are strongly reduced in number and most of them are in the range of about 1 micron for the 400 °C-deposited samples and 0.5 micron for the 210 °C-deposited samples.

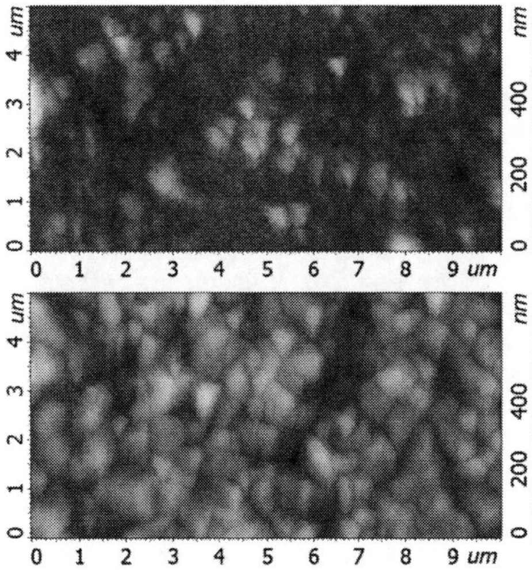

Figure 5: AFM pictures of SnS thin films deposited at 210 °C and at 400 °C and treated with KI-EtOH.

The XRD patterns of the treated samples are shown in Fig. 6: all samples exhibit the main diffraction peaks of orthorhombic SnS, but with a clear change in their structure. The pattern related to KI-EtOH treated layer deposited at 210 °C show that the main (111) and (101) peaks disappear. On the other hand the KI-EtOH treated layers, previously deposited at 270°C, show still a slight (111) main orientation; moreover, few new peaks appears. The pattern of $SnCl_2$-EtOH treated layers deposited at 400 °C is more randomly oriented compared to the previous cases.

We analyzed the SnS batch material purchased from Huizhou Tian Yi Rare Material Co. Ltd before and after deposition. XRD spectra show pure orthorhombic SnS in both cases; Raman and SEM-EDX analysis show instead sulfur excess in the batch material before its use, which vanishes after the use. As a matter of fact, when the SnS crucible is heated up to its deposition temperature for the first time after refilling, a large pressure increase (from below 10^{-7} mbar to over 10^{-5} mbar) with a very weak deposition rate is observed; after a few minutes the evaporation rate reduces to zero and the chamber pressure returns to its starting value.

We decided to deposit two different sets of samples in order investigate the case of a non-stable starting material and its effects on SnS thin films. Although only minor variations were highlighted by AFM, XRD and Raman spectroscopy techniques comparing the two sets of samples, the electrical behavior changes significantly. The best working devices are those deposited at 270 °C for both sets, despite the large inhomogeneity of grain sizes. We elsewhere reported a device with 1.5 % power conversion efficiency at a similar substrate temperature [6]. The performance of the 2nd run sample improved the results of the 1st run for a factor of 2.5 and this trend has been seen at every T_{sub} value.

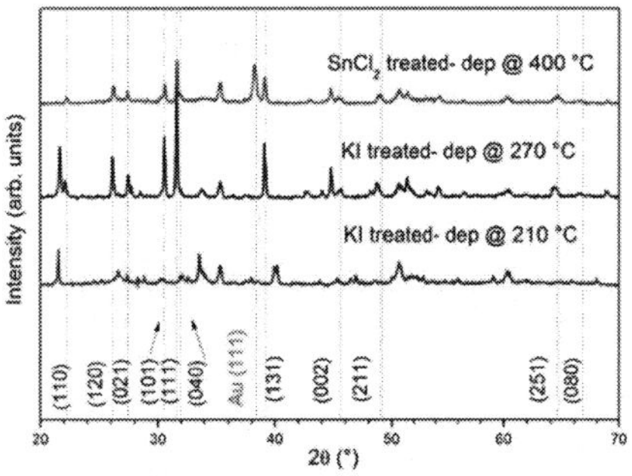

Figure 6: XRD pattern of SnS thin films: as deposited (black line) and treated with $SnCl_2$-EtOH (green line) and KI-EtOH (blue line).

4. CONCLUSIONS

SnS on superstrate configuration have been grown at different temperatures.

The morphology, shape and size of the grains, is strongly dependent on the substrate temperature, as well as a slight difference in orientation is registered. Also from Raman spectroscopy is confirmed that the structure is very similar for both temperatures, even if the prevalence of a different vibrational mode confirms a slightly different orientation of the crystals. However most evident is the low adhesion at low temperatures.

SnS is deposited with no secondary phases, but a difference in (anyway low) efficiency is observed between the first and second run after refilling.

A different recrystallization is pursued by applying KI and $SnCl_2$ in solution and annealed in air. The process reveals a recrystallization of the grains similar to $CdCl_2$ treatment for CdTe, in terms of increase in random orientation of the crystal structure. A tin-rich surface is produced with very large plates that can be removed by water or ammonia etching: a large sulphur loss is observed on the surface after recrystallization treatment so the surface becomes almost Sn metallic. After etching of the surface a recrystallized structure with a different morphology is revealed with no enlargement of the grain size but with different crystal orientation. KI-treatment is more aggressive and for this reason the SnS layers have to be deposited at high substrate temperature to avoid large material consumption and re-evaporation.

References

[1] First Solar, First Solar Achieves yet Another Cell Conversion Efficiency World Record., (2016). http://investor.firstsolar.com/releasedetail.cfm?ReleaseID=956479.

[2] Kai Marau; Akane Yamiya, Solar Frontier Achieves World Record Thin-Film Solar Cell Efficiency: 22.3%, (2015). http://www.solar-frontier.com/eng/news/2015/C051171.html.

[3] P. Sinsermsuksakul, K. Hartman, S. Bok Kim, J. Heo, L. Sun, H. Hejin Park, et al., Enhancing the efficiency of SnS solar cells via band-offset engineering with a zinc oxysulfide buffer layer, Appl. Phys. Lett. 102 (2013) 053901. doi:10.1063/1.4789855.

[4] P. Sinsermsuksakul, L. Sun, S.W. Lee, H.H. Park, S.B. Kim, C. Yang, et al., Overcoming efficiency limitations of SnS-based solar cells, Adv. Energy Mater. 4 (2014) 1400496–02. doi:10.1002/aenm.201400496.

[5] A. Schneikart, H. Schimper, A. Klein, W. Jaegermann, U. Darmstadt, F. Materialwissenschaft, et al., Fabrication and Characterization of Vacuum Evaporated SnS x Solar Cells, (2012) 2012.

[6] S. Di Mare, A. Salavei, F. Piccinelli, A. Romeo, Analysis of SnS Growth and Post Deposition Treatment by Congruent Physical Vapor Deposition, Proc. EU PVSEC 2015. (2015) 1349–1352. doi:10.4229/EUPVSEC20152015-3DV.3.22.

[7] H.R. Chandrasekhar, R.G. Humphreys, U. Zwick, M. Cardona, Infrared and Raman spectra of the IV-VI compounds SnS and SnSe, Phys. Rev. B. 15 (1977) 2177–2183. doi:10.1103/PhysRevB.15.2177.

MINORITY CARRIER LIFETIME ENHANCEMENT OF C-Si/TiO_2 HETEROJUNCTION BY POST DEPOSITION ANNEALING

Swasti Bhatia[1], Shraddha Kothari[1], Neha Raorane[2], Saurabh Lodha[1], Pradeep R. Nair[1], Aldrin Antony[3]*

[1]Department of Electrical Engineering, Indian Institute of Technology Bombay
[2]Centre of Excellence in Nanoelectronics (CEN), Indian Institute of Technology Bombay
[3]Department of Energy Science and Engineering, Indian Institute of Technology Bombay
Mumbai - 400076, India
*phone: +91 2225764400, Fax: +91 2225723480, email: aldrinantony@iitb.ac.in

Abstract

Titanium Oxide has recently been demonstrated as an electron selective layer because of its asymmetric band offsets with conduction and valence band of silicon [1]. In order to be used as an electron extracting layer in non P-N junction solar cells TiO_2 thin films also need to passivate the silicon dangling bonds at the interface. In this work we report reduction of surface recombination velocity by TiO_2 films deposited by Thermal Atomic Layer Deposition (T-ALD) and Plasma Atomic Layer Deposition (P-ALD). The as deposited sample with TiO_2 deposited by thermal ALD showed an implied V_{OC} of 568 mV and a high surface recombination velocity (SRV) of more than 5000 cm/s. The post annealing of this samples yielded a high implied V_{OC} of 658 mV and surface recombination velocity of less than 120 cm/s.
Keywords: Passivation, ALD, TiO_2

1 INTRODUCTION

Titanium oxide has emerged as a promising material that could enable diffused junction free solar cells, due to its electron selective nature as well as passivating capabilities. Recently, the electron selective nature of TiO_2 films has been explored by fabrication of Si/TiO_2 heterojunction [1][2]. In order to function as an electron selective contact and yield a high efficiency cell, TiO_2 films must also passivate the silicon surface. Hence, fabrication of a passivating TiO_2 film and the phenomena which make the same possible are of significant interest to the community. While low surface recombination velocities have been achieved by Liao et al. [3], Yu et al [4] and Gad et al. [5], the passivation mechanism is not entirely understood. While Thomson et al. [6] suggest a field effect passivation due to negative charge in TiO_2, Sahasrabudhe et al. [7] have suggested chemical passivation due to formation of Si-O-Ti species at the interface. This study indicates a contribution of oxygen vacancies in TiO_2 films and the better surface coverage towards the improvement of passivation.

Apart from the recent interest in TiO_2 as an electron selective contact, it has been a material of investigation for its use in resistive random access memory. Films prepared with and without plasma have been compared from a memory perspective, with an emphasis on parameters like dielectric constant and oxygen migration and breakdown electric field [6][7]. However, these parameters are not of primary importance from a solar cell perspective. In this report we present a comparative study of TiO_2 films prepared via ALD in the presence and absence of plasma, from the perspective of solar cells. The highlight of this study is the improvement of open circuit voltage by more than 90 mVolts and an SRV of less than 120 cm/sec.

2 EXPERIMENT

P type silicon wafer of 1-4 Ohm-cm resistivity and (100) orientation was used for the experiment. After standard RCA cleaning in NH_4OH + H_2O_2 solution and HCl + H_2O_2 solution in DI water, a thin layer of TiO_2 was deposited via Thermal-Atomic Layer Deposition (T-ALD) and Plasma-Atomic Layer Deposition (T-ALD), each on different wafers. The substrate temperature was 200°C throughout the process for both the processes. The deposition was started only after the wafer achieved thermal equilibrium with the substrate. The precursors used in T-ALD process were TDMAT for titanium and H_2O for oxygen. The deposition was carried out by pulsing water and TDMAT alternatively with a wait period of 10 seconds after each pulse of TDMAT and 15 seconds after each pulse of H_2O, till 8nm thick film of TiO_2 was obtained. The P-ALD process used TDMAT as the titanium precursor and O_2 as a source of oxygen. Remote argon plasma was present in the chamber during the oxygen flow for each cycle. TiO_2 film of 8nm was deposited via P-ALD.

Post deposition, one set of samples, prepared by P-ALD as well as T-ALD underwent an annealing for 20 minutes at 350°C in forming gas (5% H_2 + 95% N_2) ambient.

3 CHARACTERIZATION

The samples were tested for minority carrier lifetime by using Quasi Steady State Photoconductance Decay (QSSPC) [8], before and after annealing. An X-Ray Diffraction scan over a range of 10-90 degrees was carried out to ascertain the phase of the TiO_2 films and the effect of annealing on the same. It was found that both the films were amorphous before and after annealing. In order to further probe into the passivation mechanism of TiO_2 films, surface images were taken using Atomic Force Microscopy and X-Ray Photoelectron Spectroscopy was done on the samples. Electrical characterization of films was also done by depositing metal contacts on TiO_2 and silicon. The thickness of the films was verified by XRR measurement.

4 RESULTS

4.1 Implied open circuit voltage and Surface Recombination Velocity (SRV)

Silicon wafers with 8nm of TiO_2 deposited by T-ALD and P-ALD on both sides of the wafer were tested for lifetime of minority carriers. Surface recombination velocity was extracted from effective lifetime measured. The lifetime at the interface was probed by doing the QSSPC measurement with UV light to find the exact value of surface recombination velocity. A consistent surface recombination velocity (SRV) of less than 120 cm/sec is achieved by annealed T-ALD films. An enhancement of implied open circuit voltage is observed in both the T-ALD as well the P-ALD samples by, post annealing. However, the enhancement in T-ALD samples is more compared to P-ALD samples.

Table 1: Post anneal values are at a carrier concentration of 10^{15} carriers per cm^3 and the values for as deposited samples are at carrier concentration of 10^{14} carriers per cm^3.

Sample	As deposited iVoc (mV)	As deposited SRV (cm/sec)	After annealing iVoc (mV)	After annealing SRV (cm/sec)
P-ALD	537	>5000	612	<300
T-ALD	564	>5000	658	106

4.2 X-Ray Diffraction and Raman Spectroscopy

The surface recombination velocity of T-ALD reduces from more than 5000 cm/sec to 106 cm/sec by post annealing. Since TiO_2 is expected to crystallize with exposure to temperature near 400°C, an XRD scan was required to know if there is any phase transformation. Raman scattering and XRD scan rule out any crystalline nature in the as deposited and annealed films. This could be explained by the fact that an annealing in the presence of hydrogen would only reduce the oxygen content of the films, making it non-stoichiometric and hence preventing crystallization.

4.3 Atomic Force Microscopy

The AFM images indicate more uniformity in the T-ALD films as compared to the P-ALD films. It is also observed that the films become more uniform post annealing. Hence, better uniformity of the film could be one reason behind the lifetime enhancement post-anneal. Uniformity could be an important factor since the films are only 8 nm thick.

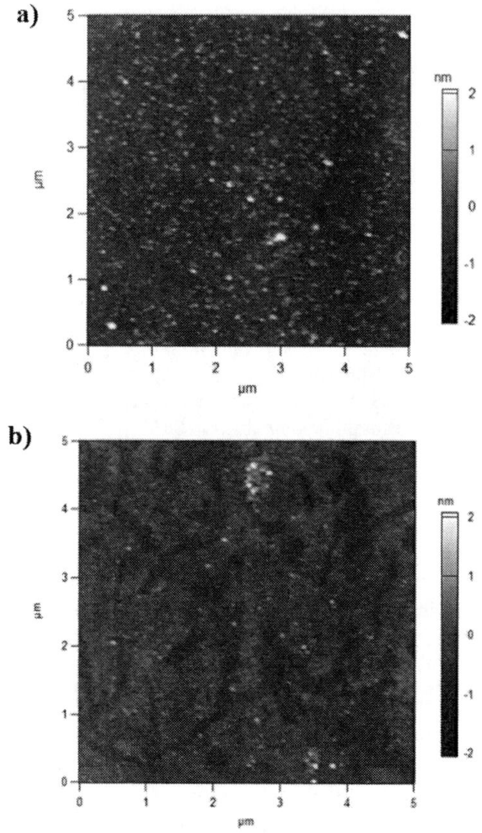

Figure 1: a) AFM image of P-ALD sample after annealing b) AFM image of T-ALD sample after annealing.

4.4 X-Ray Photoelectron Spectroscopy

X-ray Photoelectron spectrum had two titanium peaks at 458.6 for Ti 2p3/2 and at 464.5 for Ti 2p1/2. The oxygen 1s peak is shown in figure 2.

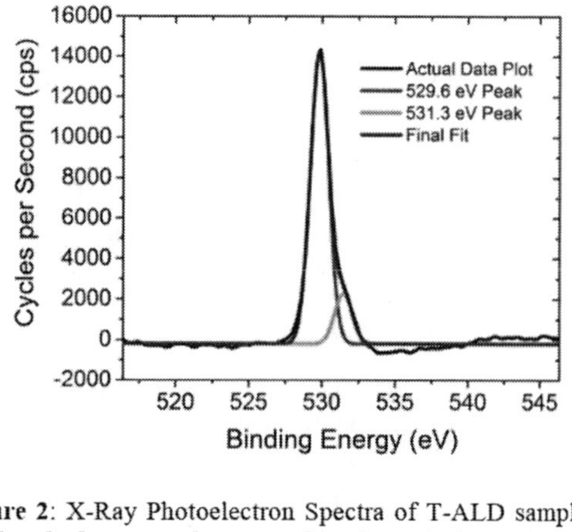

Figure 2: X-Ray Photoelectron Spectra of T-ALD sample that has the lowest surface recombination velocity.

Figure 2 shows the spectra of oxygen O1s level. The oxygen peak on de-convolution results in two peaks at 529.6 eV and 531.3 eV. The second peak 531.3 eV indicates the presence of oxygen vacancies. The percentage of area under the peak at 531.3 eV increases after annealing which implies an increase in oxygen vacancies, which would be expected after annealing in hydrogen. Table 2 shows the effect of annealing on the percentage of oxygen vacancies. It is clear that the P-ALD samples have fewer oxygen vacancies as compared to the T-ALD samples. It is observed that the oxygen vacancies increase on annealing in both the samples which would be expected considering a reduction of films in hydrogen at elevated temperatures.

Table 2: Percentage of area under the peak at 531.3 eV. The peak corresponds to oxygen vacancies.

Sample	Oxygen vacancies (as deposited)	Oxygen vacancies (annealed)
P-ALD	6.42 %	7.15%
T-ALD	8.45%	9.70%

Diodes were fabricated with TiO_2 on P type, CZ, Silicon wafers of 4-7 Ω-cm resistivity and (100) orientation. The current voltage characteristic of the same is shown in Figure 3. Aluminum film of 350 nm thickness, deposited by thermal evaporation was used to make contact with the TiO_2 layer. The best devices presented a reverse saturation current of 0.373 $\mu A/cm^2$.

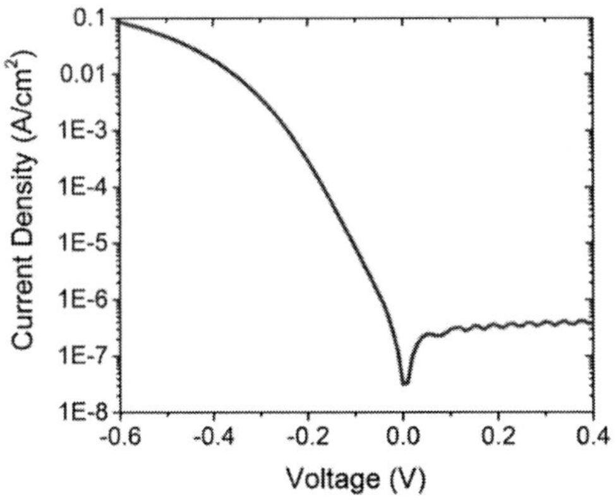

Figure 3: J-V characteristic of TiO_2 diode on P type silicon. Voltage is applied to TiO_2 contact.

4 CONCLUSION

The XPS and minority carrier lifetime results suggest a relation between passivation quality and the number of oxygen vacancies. A larger number of vacancies are associated with better lifetime. The result is interesting since a less stoichiometric film would be associated with lower lifetime values. This suggests that annealing in non-oxygen ambient leads to increased oxygen vacancies in the films, which might be contribute to reduction of surface recombination velocity. This effect along with uniform coverage as achieved by thermal ALD is responsible for good passivation with the lowest SRV of 106 cm/seconds and an implied V_{OC} of 658 mVolts.

ACKNOLEDGEMENT

This work was supported by Indo-US Solar Energy Research Institute (SERIIUS), National Centre for Photovoltaic Research and Education (NCPRE), Indian Ministry of New and Renewable Energy (MNRE) through the project AMANSI. Pradeep R. Nair acknowledges seed grant funding from IIT Bombay.

REFERENCES

[1] Avasthi, Sushobhan and McClain, William E. and Man, Gabriel and Kahn, Antoine and Schwartz, Jeffrey and Sturm, James C., Applied Physics Letters, 102, 203901 (2013).

[2] Nagamatsu, Ken A. and Avasthi, Sushobhan and Sahasrabudhe, Girija and Man, Gabriel and Jhaveri, Janam and Berg, Alexander H. and Schwartz, Jeffrey and Kahn, Antoine and Wagner, Sigurd and Sturm, James C., Applied Physics Letters, 106, 123906 (2015).

[3] Liao, Baochen and Hoex, Bram and Aberle, Armin G. and Chi, Dongzhi and Bhatia, Charanjit S., Applied Physics Letters, 104, 253903 (2014).

[4] I. Yu, Y. Wang, H. Cheng, Z. Yang, and C. Lin, Int. J. Photoenergy, vol. 2013, pp. 1–7, 2013.

[5] K. M. Gad et al., IEEE Journal of Photovoltaics, vol. 6, no. 3, pp. 649-653, May 2016.

[6] Thomson, A. F. and McIntosh, K. R. (2012), Prog. Photovolt: Res. Appl., 20: 343–349.

[7] Girija Sahasrabudhe, Sara M. Rupich, Janam Jhaveri, Alexander H. Berg, Ken A. Nagamatsu, Gabriel Man, Yves J. Chabal, Antoine Kahn, Sigurd Wagner, James C. Sturm, and Jeffrey Schwartz, Journal of the American Chemical Society 2015 137 (47), 14842-14845.

[8] Sinton, R. A.; Cuevas, A. Appl. Phys. Lett. 1996, 69, 2510.

IMPACT OF MINORITY CARRIER LIFETIME AND TEMPERATURE ON SiC BASED REAR CONTACT SiGe SOLAR CELL FOR CONCENTRATOR PHOTOVOLTAIC (CPV) APPLICATIONS

Rahul Pandey, Apurva Jain, Ajay Kumar and Rishu Chaujar*

Microelectronics Research Laboratory, Department of Engineering Physics
Delhi Technological University, India
rishu.phy@dce.edu*

In this work, the impact of material quality i.e. impact of minority carrier lifetime on $SiC/Si_3N_4/SiO_2$ passivated 10μm thick rear contact SiGe solar cell have been obtained. Result shows, the photovoltaic parameters are less affected by minority carrier lifetimes (τ). Also, as τ changes from 10^{-2} s to 10^{-5} s, change in J_{SC}, $V_{OC,}$ FF and PCE are 0.4%, 0.5%, 2.2% and 3.2%, respectively. Further, the application of rear contact SiGe solar cell for concentrator photovoltaic (CPV) has also been discussed. At 300 K operating temperature, we have obtained the open circuit voltages 627mV, 646mV and power conversion efficiencies of, 15.7%, 16% at 5-sun (0.5 $W.cm^{-2}$) and 10-sun (1$W.cm^{-2}$) conditions, respectively for $SiC/Si_3N_4/SiO_2$ coated SiGe cell. This, results in 6.8 % and 8.8% higher power conversion efficiency (PCE) at the concentration levels of 5, 10 suns compared to 1-sun condition (14.7%, PCE). Also, under 10-sun illumination, 9the PCE of 16.4% and 16.1% at 20°C and 25°C temperature, respectively have been obtained. All the simulation have been done using Silvaco atlas device simulator.

Keywords: Atlas, Rear contact, Solar cell, SiGe, Photovoltaic, Silicon carbide (SiC)

1 INTRODUCTION

Previously, novel 14.8% efficient 10μm thick rear contact SiGe solar cell device with $SiC/Si_3N_4/SiO_2$ based highly absorbing passivation layer has been proposed. The device shows optical coupling in near UV/ Visible part of the wavelength spectrum due to the presence of SiC. Discussed device was 10-15 times thinner than conventional rear contact silicon solar cell [1-2]. Also, we have shown that SiGe solar cell significantly saves the material cost as compared to silicon solar cell [3]. The incorporation of germanium into silicon leads to a reduction in band gap [4-5] and enhancement in the absorption co- efficient at longer wavelengths [6-7]. Here we have obtained the impact of minority carrier lifetime on optical and electrical characteristics of the rear contact SiGe solar cell. Also, the application of $SiC/Si_3N_4/SiO_2$ coated SiGe cell, and SiC/SiO_2 coated SiGe cell for concentrator's photovoltaics (CPV) cell application has been discussed i.e. cell operating at more than 1 sun (0.1$W.cm^{-2}$). Since the device has rear contact architecture, it is easier to interconnect and can be placed closer together in the modules, no need of space between cells.

In this work, the impact of material quality i.e. impact of minority carrier lifetime on $SiC/Si_3N_4/SiO_2$ based 10μm thick rear contact SiGe solar cell has been discussed. Result shows, the photovoltaic parameters are less affected by minority carrier lifetimes (τ). Also, as τ changes from 10^{-2} s to 10^{-5} s, change in short circuit current density (J_{SC}), open circuit voltage (V_{OC}), fill factor (FF) and power conversion efficiency (PCE) are 0.4%, 0.5%, 2.2% and 3.2%, respectively. Whereas, for 200μm thick devices, as τ changes from 10^{-2} s to 10^{-5} s, change in power conversion efficiency is approximately 70%. At 300 K operating temperature, the open circuit voltages 627mV, 646mV and power conversion efficiencies of, 15.7%, 16% is achieved at 5-sun (0.5 $W.cm^{-2}$) and 10-sun (1$W.cm^{-2}$) conditions, respectively for $SiC/Si_3N_4/SiO_2$ coated SiGe cell. This, results in 6.8 % and 8.8% higher power conversion efficiency (PCE) at the concentration levels of 5, 10 suns compared to 1-sun condition (14.7%, PCE). Also, under 10-sun (1$W.cm^{-2}$) illumination, we have obtained the PCE of 16.4% and 16.1% at 20°C and 25°C temperature, respectively. Desired irradiance and temperature can be selected depending upon the application. In future *N*-type and *P*-type, SiGe wafers based rear contact architecture can be discussed instead of i-SiGe. This paper concludes $SiC/Si_3N_4/SiO_2$ based rear contact SiGe solar cell can be a good candidate for a cost-effective solution for energy efficient applications. All the simulation have been done using Silvaco atlas device simulator. This paper shows proposed device can be a good candidate for a cost-effective solution for energy efficient concentrator applications.

2 DEVICE STRUCTURE AND MODEL DESCRIPTION

The simulated device structure is shown in Fig. 1. It contains an intrinsic SiGe wafer having a length in y-direction equal to 10 μm and in x-direction equals to 100 μm, without optical absorber coating on top. The $Si_{0.9}Ge_{0.1}$ was used during the simulation. The width of *N+/P+* regions are 30 and 45 μm, respectively whereas the depth is 100nm and 300nm, respectively. The device consists of ohmic contacts, and the doping concentration of *N+ /P+* regions are $4x10^{20}$ cm^{-3}. The concentration dependent mobility, parallel field mobility, band gap narrowing, Shockley-Read-Hall recombination and Auger recombination, models, are used during simulation. Previously, superior absorption is achieved in $SiC/Si_3N_4/SiO_2$ coated SiGe wafer compared to other

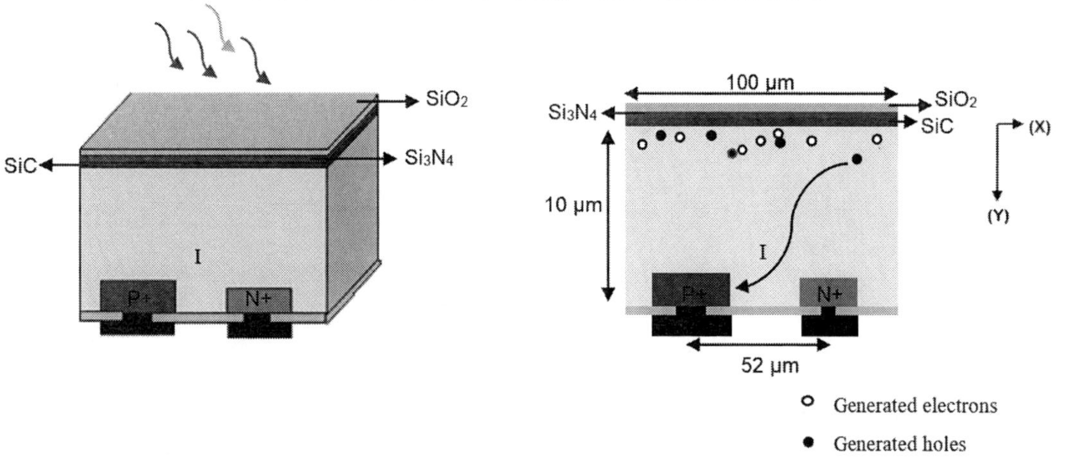

Figure 1. Schematics of rear contact SiGe solar cell with SiC/ Si_3N_4 / SiO_2 based highly absorbing passivation layer on top. I represents the intrinsic SiGe region

passivation layers. Furthermore, results also indicate that SiC/SiO_2 coated SiGe wafer shows approximately same response as $SiC/Si_3N_4/SiO_2$ coated SiGe. This indicates SiC/SiO_2 coated SiGe can be used instead of $SiC/Si_3N_4/SiO_2$ coated SiGe in order to avoid complexity shown in reference [1-2].

3 RESULTS

3.1 Impact of minority carrier lifetime on $SiC/Si_3N_4/SiO_2$ coated SiGe cell, and SiC/SiO_2 coated SiGe cell

Proposed rear contact SiGe solar cell can be fabricated with inexpensive, less-pure material i.e. low minority carrier lifetime materials. Whereas, minority carrier lifetime need to be very large (> 1 ms), in conventional si solar cells with rear contact design. Because their thickness is normal >150 µm. This results, in the need of high-quality material and increased module cost. Discussed device is 15 times thinner than the convention rear contact. Thus, lower diffusion length is required in order to collect the generated carriers. Results presented in this section confirmed the benefits of thin (10µm) rear contact SiGe solar cell device. Further, Fig. 2 shows the impact of minority carrier lifetime on EQE of the devices. We have obtained the EQE of the devices with two different minority carrier lifetimes (10^{-5} s and 10^{-2} s), and the same has been presented in Fig. 2. Normally, the blue response of solar cell is reduced due to front surface recombination, whereas the midway response is reduced due to lower diffusion length and the red response is reduced due to the transmitted photons and the back surface recombination. The result clearly indicates that lower wavelength response is independent of minority carrier lifetime. Also, with 10^{-5} s minority carrier lifetime, slightly lower EQE is obtained in wavelength >550 nm compared to 10^{-2} s minority carrier lifetime devices. Fig. 3 shows the impact of τ on the current density - voltage (J-V) characteristics of the devices the J-V curve is also not affected by τ variation. Further, we have also obtained the impact of minority carrier lifetime on device photovoltaic parameters: short circuit current density (J_{SC}), open circuit voltage (V_{OC}), fill factor (FF) and power conversion efficiency (PCE). The results are presented in Fig. 4(a-d). Result shows, the photovoltaic parameters are less impacted by minority carrier lifetimes (τ). Also, as τ changes from 10^{-2} s to 10^{-5} s, change in J_{SC}, V_{OC}, FF and PCE are 0.4%, 0.5%, 2.2% and 3.2%, respectively. Whereas, for 200µm thick devices, as τ changes from 10^{-2} s to 10^{-5} s change in power conversion efficiency is approximately 70%. All the results of this section conclude that performance of 10µm thick rear contact SiGe is independent of material quality. The performance of 10^{-5} s, τ device is approximately same compared to 10^{-2} s, τ device. This facilitates the use of less pure material, hence reducing the module cost.

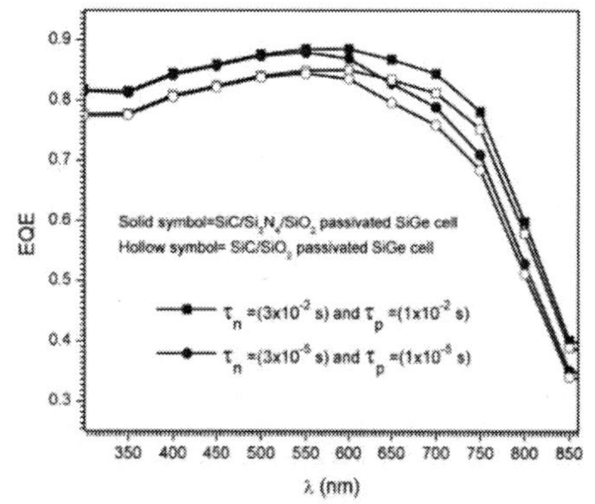

Figure 2. Spectrum response i.e. EQE of rear contact SiGe solar cell with SiC/ Si3N4 / SiO2 based highly absorbing passivation layer on top with two different minority carrier lifetimes (10^{-5} s and 10^{-2} s)

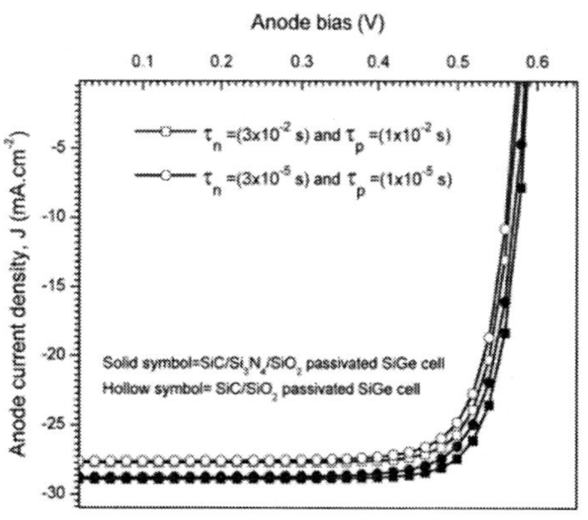

Figure 3. Electrical characteristics i.e. J-V curve under illumination at two different minority carrier lifetimes (10^{-5} s and 10^{-2} s)

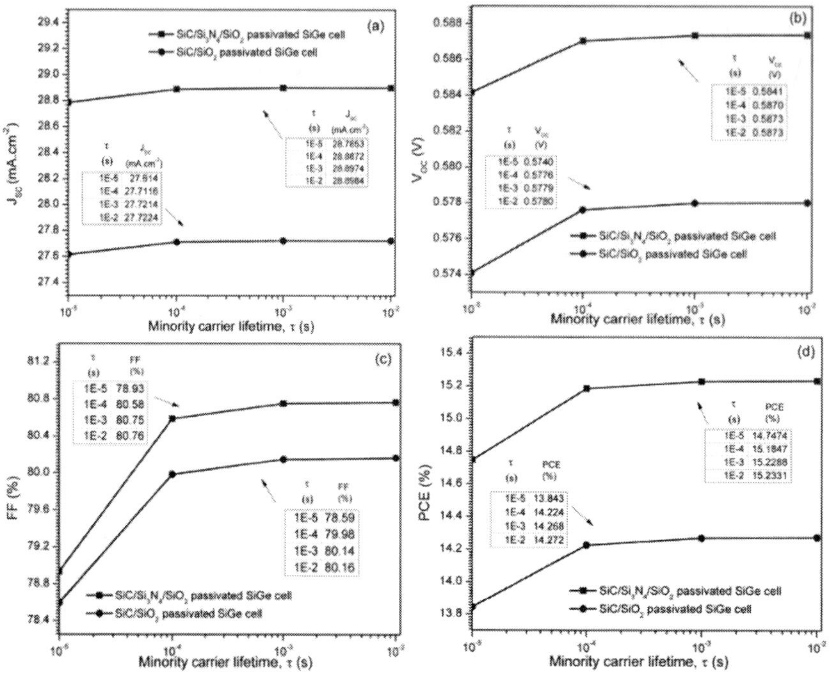

Figure 4. Impact of minority carrier lifetime on device photovoltaic parameters: (a) short circuit current density (J_{SC}), (b) open circuit voltage (V_{OC}), (c) fill factor (FF) and (d) power conversion efficiency (PCE)

3.2 The impact of solar irradiance and temperature on rear contact SiGe solar cell.

In this section, we have discussed the application of rear contact SiGe solar cell for concentrator's photovoltaics (CPV) cell application. Altering the irradiance on a solar cell varies the photovoltaic parameters, including J_{SC}, V_{OC}, FF, PCE, series and shunt resistances. The Jsc of a solar cell depends linearly on irradiance and thus indicates that device operating under 10 suns would have 10 times the J_{SC} as the same device under one sun operation. However, this will not result in increased PCE, since the incident power also increases linearly. The efficiency benefits arise from the open-circuit voltage, as V_{OC} increases logarithmically with light intensity [8]. Here, we have obtained the impact of temperature and irradiance on proposed device, and the same has been presented in Fig. 5(a-c). We have simulated the $SiC/Si_3N_4/SiO_2$ coated SiGe cell, and SiC/SiO_2 coated SiGe solar cell devices up to 10-suns (1 $W.cm^{-2}$) condition at room temperature (300K). Further, the cell operating at higher concentration levels requires complex cooling system, and overall cost of the PV module will be increased. Previously, a silicon-based concentrator solar cell with the efficiencies range of (15-27.5) % has been reported under the concentration level of (20-150) suns [9-13]. In silicon-based concentrator cells, thickness >100μm is required in order to achieve high efficiency. This results in higher $/Watt. Furthermore, we have obtained the open circuit voltages 627mV, 646mV, and power conversion efficiencies of, 15.7%, 16% at 5-sun and 10-sun conditions respectively, for $SiC/Si_3N_4/SiO_2$ coated SiGe cell shown in Fig. 5a. This, results in 6.8 % and 8.8% higher PCE at the concentration levels of 5, 10 suns compared to 1-sun condition (14.7%). Further, the desired concentration level can be selected from the Fig. 5a, depending upon the applications. Moreover, the impact of temperature on device parameters: V_{OC} and PCE for different conditions (1-sun and 10-suns) have been analyzed. Fig. 5 (b-c), shows that device performance degrades as the temperature increases since intrinsic carrier concentration increases at higher temperatures.

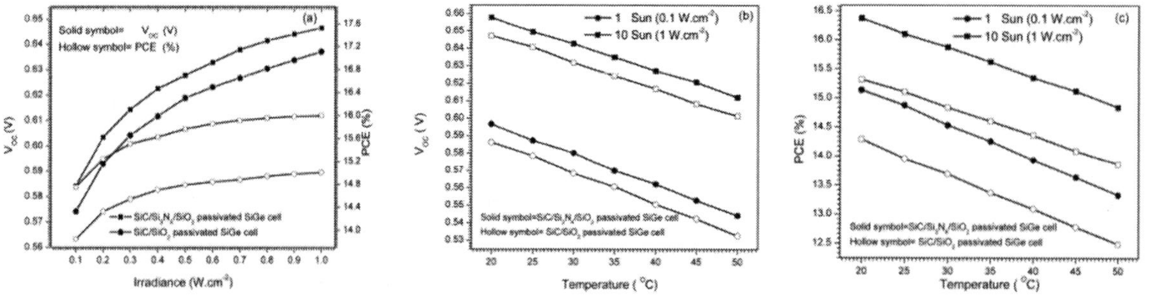

Figure 5. Impact of irradiance and temperature: (a) effect of irradiance on V_{OC} and PCE (b) effect of temperature on V_{OC} for 1 sun and 10 suns, respectively (c) effect of temperature on PCE for 1 sun and 10 suns, respectively.

Therefore, bandgap decreases, which further results in decrease V_{OC} , as shown in Fig. 5b. The V_{OC} values 596 mV, 658 mV and PCE values of 15.1%, 16.4% is obtained at 20^{0}C for 1-sun, and 10-sun conditions respectively, whereas V_{OC} values 544 mV, 612 mV and PCE values of 13.3 %, 14.8 % is obtained at 50^{0}C for 1-sun, and 10-sun conditions respectively. Results show, as the temperature increases from 20-45^{0}C, 9.9% decrease in PCE is obtained for flat plate cell, whereas 7.3% decrease in PCE is obtained for the same cell operating at 10-suns, as shown in Fig. 5c. Further, all the results of this section conclude that PCE of 16.4% is obtained at 10-suns in a sub 10μm thick device with the cost effective material. This suggests that the proposed device can be a cost-effective candidate for flat plate as well as low concentrator applications

IV CONCLUSIONS

This paper analyzed the impact of material quality i.e. impact of minority carrier lifetime on photovoltaics device parameters. Result shows, the device parameters are less affected by minority carrier lifetime. This concludes proposed device structure can be fabricated with less pure low-cost materials. Also, the device has rear contact architecture which is suitable for concentrator photovoltaic applications. Therefore, we have analyzed the $SiC/Si_3N_4/SiO_2$ based rear contact SiGe device for more than 1-sun irradiance condition, along with the impact of operating temperature on device parameters. The V_{OC} values 596 mV, 658 mV and PCE values of 15.1%, 16.4% is obtained at 20^{0}C for 1-sun, and 10-sun conditions respectively, whereas V_{OC} values 544 mV, 612 mV and PCE values of 13.3 %, 14.8 % is obtained at 50^{0}C for 1-sun, and 10-sun conditions respectively. Desired irradiance and temperature can be selected depending upon the application. In future *N*-type and *P*-type, SiGe wafers based rear contact architecture can be discussed instead of i-SiGe. This paper concludes $SiC/Si_3N_4/SiO_2$ based rear contact SiGe solar cell can be a good candidate for a cost-effective solution for energy efficient applications.

ACKNOWLEDGMENT

The authors would like to thank Microelectronics Research Lab, Department of Engineering Physics, Delhi Technological University to carry out this work.

REFERENCES

1. R. Pandey and R. Chaujar. (2015). Novel Back-Contact Back-Junction SiGe (BC-BJ SiGe) Solar Cell for Improved Power Conversion Efficiency" Microsystem technologies, DOI: 10.1007/s00542-015-2552-1. (In press)
2. R. Pandey and R. Chaujar. (2014). TCAD Analysis of Novel Back Contact-Back Junction SiGe (BC-BJ SiGe) Solar Cell for Improved Optical and Electrical Characteristics. 1st International Conference on "Microelectronics, Circuits and Systems, pp-127-129, Kolkata, India
3. R. Pandey and R. Chaujar. (2015) .TCAD Analysis of Silicon-Germanium (SiGe) Based Back-Contact Back-Junction (BC-BJ) Solar Cell-an Alternative for Silicon-Based Cells" TechConnect World Innovation, June 14-15 (2015). Available: http://www.techconnectworld.com/World2015/a.html?i=481
4. D. V. Lang, R. People, J. C. Bean and A. M. Sergent. (1985) .Measurement of the Band Gap of GexSi1–x/Si Strained- Layer Heterostructures. Applied Physics Letters, 47(12), pp. 1333-1335. doi:10.1063/1.96271
5. N. Usami, T. Ichitsubo, T. Ujihara, T. Takahashi, K. Fuji- wara, G. Sazaki, and K. Nakajima. (2003). Influence of the Elas- tic Strain on the Band Structure of Ellipsoidal SiGe Coherently Embedded in the Si Matrix. Journal of Applied Physics, 94(2), pp. 916-920. doi:10.1063/1.1580194
6. R. Braunstein, A. R. Moore and F. Herman. (1958). Intrinsic Optical Absorption in Germanium-Silicon Alloys," Phy- sical Review, 109(3), pp. 695-710. doi:10.1103/PhysRev.109.695
7. K. H. Jun, J. K. Rath and R. E. I. Schropp. (2002) .Enhanced Light-Absorption and Photo-Sensitivity in Amorphous Silicon Germanium/Amorphous Silicon Multilayer. Solar Energy Materials & Solar Cells, 74(1-4), pp. 357-363. doi:10.1016/S0927-0248(02)00095-8
8. P. Würfel. (2008, Jul.). Physics of Solar Cells: From Principles to New Concepts [Online]. Available: http://as.wiley.com/WileyCDA/WileyTitle/productCd-3527618554.html
9. R.A. Sinton, Y. Kwark, J.Y. Gan and R.M. Swanson. (1986). 27.5-percent silicon concentrator solar cells. Electron Device Letters, IEEE. [Online]. 7(10). pp. 567-569. Available: http://ieeexplore.ieee.org/xpls/abs_all.jsp?arnumber=1486299.
10. A. Cuevas, R. A. Sinton, A. Ronald, N.E. Midkiff and R.M. Swanson. (1990, Jan). 26-percent efficient point-junction concentrator solar cells with a front metal grid. Electron Device Letters, IEEE. [Online]. 11(1). pp. 6-8. Available: http://ieeexplore.ieee.org/xpls/abs_all.jsp?arnumber=46913.
11. M.D. Lammert and R. J. Schwartz. (1977). The interdigitated back contact solar cell: A silicon solar cell for use in concentrated sunlight. Electron Devices, IEEE Transactions.[Online]. 24(4). pp. 337-342. Available: http://ieeexplore.ieee.org/xpls/abs_all.jsp?arnumber=1478930.
12. Z. Fuzu, S. Wenham and M.A. Green. (1995). Large area, concentrator buried contact solar cells. *Electron Devices, IEEE Transactions*. [Online]. *42(1)* pp. 144-149. Available: http://ieeexplore.ieee.org/xpl/articleDetails.jsp?arnumber=370024 .
13. J. Zhao, A. Wang and M.A. Green. (1990). 24% efficient PERL structure silicon solar cells. Photovoltaic Specialists Conference.

COMPARATIVE STUDY OF THE EFFECTS OF RARE EARTH IONS DOPED $BISRFEO_3$ NANOMULTIFERROIC

Maha Ayman[1], M.M.Abdelaal[2]
[1]Physics Department, German University in Cairo (GUC), Cairo, Egypt
Maha_ayman_2010@yahoo.com
[2] Physics Department, Faculty of Girls, Ain Shams University, Cairo, Egypt.
Mervatabdelaal2025@yahoo.com

ABSTRACT: We report a comparative study of rare-earth ions substituted $BiSrFeO_3$ (BSFO) crystal for; BSFO, $Bi_{0.57}Sr_{0.4}Y_{0.03}FeO_3$, $Bi_{0.57}Sr_{0.4}Nd_{0.03}FeO_3$, $Bi_{0.57}Sr_{0.4}Ce_{0.03}FeO_3$, and $Bi_{0.57}Sr_{0.4}La_{0.03}FeO_3$, respectively are successfully synthesized by using flash autocombustion reaction. X-ray diffraction analysis reveals a crystal transformation from rhombohedral to hexagonal upon substitution rare earth ions and a clear hexagonal grain growth habit is seen by SEM microstructure. The doping rare earth ions to BSFO improved the magnetic properties specially Y^{3+} and Nd^{3+} ions. We argue that the addition of RE is likely to suppress the spiral spin modulation and at the same time increase the canting angle which favors the enhanced ferromagnetism. Excess Bi is expected to act as point defects and occupy interstitial positions which in turn surrounded by oxygen vacancies and is likely to promote defect driven ferromagnetism with enhanced spin and electric polarization.
Keywords: Nanoparticles, Rare-earths, Multiferroic, Bismuth ferrite.

1 Introduction

Multiferroic materials are single-component materials or composites exhibiting two or more ferroic features such as ferromagnetism, ferroelectricity, or ferroelasticity effects. Recent resurgence of research interest for multiferroic [1–4] materials revived the possibility of strong coupling between the magnetic and ferroelectric order parameters of some promising materials and are likely to be used around room temperature (RT) for possible device design and applications specially in frontier field of spintronic [5] devices, sensor, multiple state memory elements [6], electric field controlled ferromagnetic resonance devices, and transducers with magnetically modulated piezoelectricity [7]. $BiFeO_3$ (BFO) is one of the very few known magnetoelectric system that exhibits antiferromagnetic ordering ($T_N \approx 380^{\circ}C$) and ferroelectric behavior with high ferroelectric Curie temperature ($T_C \approx 830°C$) [8]. It is well documented [9] that due to canting of Fe sub-lattice moment; BFO shows the presence of weak magnetism. It is the primary goal of the researchers over the globe to improve the desired magnetic properties, without altering the ferroelectric properties of the original BFO materials system. To address the issue of enhancing desired magnetic property specially aiming towards ferromagnetic property, a suitable doping in BFO might be the right approach [10–12] to tailor this material in which the linear magneto-electric coupling would exist [13]. It has also been suggested that the spiral spin structure can be changed with the application of a higher magnetic field [14], epitaxial constraint [15] and doping [10]. As far as doping in BFO is concerned, one way of doping is by doping of BFO with rare earth elements [11, 12] which are promising to enhance multiferroics properties.

Several authors tried to improve the magnetic properties of BFO by suitably doping with rare earth elements like La^{3+},Nd^{3+},Gd^{3+},Sm^{3+},Pr^{4+} [16–18] where the dopants produce magnetization by suppressing the spiral spin structure of BFO. Li et al. [19] found that LaFeO3 materials exhibited higher photocatalytic activity. Following this, BiFeO3 materials were shown to exhibit visible light photocatalytic properties [20].

The present work aimed to perform a comparative study of ionic radius of different rare earth; Y, Nd, Ce, and La on the physical properties of $Bi_{0.57}Sr_{0.4}RE_{0.03}FeO_3$ nanomultiferroics which opens a new era of applications as combustion catalysts and sensors, cathode material for solid oxide fuel cells operating at temperatures below 750°C, and electrode material based ceramic electrolytes.

2 Experimental techniques

2.1 . Samples preparation

$Bi_{1-x-y}Sr_xR_yFeO_3$; x=0.4 and y=0.03 multiferroic powders were synthesized by using flash autocombustion reaction [21]. The main advantage of combustion synthesis among many wet methods is that it offers a unique route via a highly exothermic redox reaction between metal nitrates and an organic fuel to produce metal oxides with stoichiometric ratio very near to that chemically mentioned. This method produces a homogenous product in a short time without the use of expensive high temperature furnaces. Analytical grade Bismuth Nitrate;[$Bi(NO_3)_3$. $5H_2O$], Strontium Nitrate; $Sr(NO_3)_2$, Iron Nitrate; $Fe(NO_3)_3.9H_2O$, different rare earth metal ions (R) as Lanthanum nitrate; $La(NO_3)_3$, Yttrium nitrate ;$Y(NO_3)_3$, Cerium nitrate; $Ce(NO_3)_3$, Neodynium nitrate; $Nd(NO_3)_3$, and Urea; CO $(NH_2)_2$ that acts as the reducing agent [22] all were used as raw materials for the synthesis of these ferrites. The nitrates and urea were mixed in desired stoichiometric proportions and dissolved separately in minimum amount of triply distilled water and adding gradually with the mixed solution. The mixture was slowly heated and stirred using a hot plate magnetic stirrer till a self-propagating combustion manner to obtain fluffy powder. The obtained nanopowders were heated at 500°C for 4 hrs using a heating/cooling rate of 4°C/min. Cold pressing for this composite was carried out using a uniaxial press of value 8×10^5 N/m^2 into a pellet form.

2.2 Characterization

A rich variety of information can be extracted from X-ray Diffraction (XRD) measurements using diffractometer model Proker D_8 with CuK_α radiation (λ = 1.5418°A) in a wide range of Bragg's angle (20–80°) at room temperature. From the position and shape of the lines, one can obtain the unit cell parameters and micro-structural parameters (crystallite size, micro-strain, etc.) respectively. The relation between the average crystallite size (CS-in Å) and the full width at half maximum (FWHM-in rad) is given by the Scherrer's formula:

$$L = \frac{k\lambda}{\beta \cos\theta}$$

Where, λ is X-ray wavelength = 1.54056A, β is the full width of half maximum (FWHM), and k is the shape factor constant = 0.89. X-ray diffraction of these samples was carried out using Philips Pu 1390 channel control Co-kα target and filter Fe of wavelength (λ=1.791Å) for all samples to assure the formation of the samples. Also from the figure it can be seen that, from Miller indices (hkl) values of the diffraction peaks were referred to hexagonal axes rather than rhombohedral axes where the lattice parameters a and c are calculated using the equation of hexagonal.

Scanning electron microscope (SEM) FEI company-Quanta FEG 250- Holland was used to reveal the surface morphology and microstructure of the prepared samples. The dc magnetic susceptibility measurements were carried out using Faraday's method from room temperature up to 700K at three different magnetic field intensities (1340, 1660, 1990 Oe). In addition, magnetization measurement of the nanoparticles were performed using a vibrating sample magnetometer (VSM; 9600 – 1LDJ; USA) at room temperature. The dielectric constant and ac electrical conductivity was carried out in a wide range of temperature 300K ≤ T ≤ 900K temperature in the frequency range of 100 Hz up to 100 kHz using RLC Bridge Model PM6306 Fluka.

3 Results and discussion

3.1 Structural properties

Figure (1) illustrates the XRD pattern for the samples $Bi_{1-x-y}Sr_xRE_yFeO_3$; (BRSFO) at y = 0.03 and x = 0.4 nanomultiferroic system. The XRD patterns are in excellent accord with the powder data $BiFeO_3$ (ICDD card number 86-1518). It is reveal that, the effect of Bi substitution by Y^{3+}, Nd^{3+}, Ce^{3+}, and La^{3+} ions on the structure of BSRFO indicated that Generally, for all samples, the second phase peaks of the undoped sample attributed to Fe or Bi rich phases, $Bi_2Fe_4O_9$ and $Bi_{25}FeO_{40}$ (asterisk in Figure 1) were routinely observed as shown in previous results [23]. However, the remain peaks in the XRD traces are related to perovskite-structured phases, while the second phase peaks apparently disappear in the XRD data of the doped samples. Besides, RE-doped $BiFeO_3$, all peaks are indexed according to the R3c cell of $BiFeO_3$. The lattice parameters deduced for pure $BiFeO_3$ hexagonal unit cell were found to have values a = 5.5795 Å and c = 13.8359 Å.

The presence of rare earth in bismuth strontium ferrites indicates that, the incorporation of RE^{3+} ions into BSFO lattice crystal structure cause the merging of the two intense diffraction peaks at 31.75° and 32.06° which corresponds to the lattice plane (104) and (110) merges and thus the peak shift appears at 32°. This peak shift occurs due to the structural distortion when doping of smaller ionic radius of RE^{3+} (Y^{3+}=1.04, Nd^{3+}=1.12, Ce^{3+} =1.15, and La^{3+} = 1.21 nm) instead of higher ionic radius Bi^{3+} (1.23 nm) ions which were also reported by several authors [24, 25]. Rare earth substitution did not affect the crystal orientation of the BSFO ferrites although the peaks have shifted towards higher angle. This suggests that the addition of rare-earth can either eliminate oxygen vacancies or avoid compositional fluctuations of Fe^{3+}–Fe^{2+} oxidation state influencing the appearance of secondary phases. These results agree well with Bellakki et al [26-28]. Moreover, the crystallite sizes were calculated using Scherrer's formula (L = 0.9λ/βcosθ) for the given XRD patterns and average crystallite sizes were found to be about 22.43, 13.27, 14.12, and 21.66nm for samples $Bi_{0.57}Sr_{0.4}Y_{0.03}FeO_3$,$Bi_{0.57}Sr_{0.4}Nd_{0.03}FeO_3$,$Bi_{0.57}Sr_{0.4}Ce_{0.03}FeO_3$, and $Bi_{0.57}Sr_{0.4}La_{0.03}FeO_3$, respectively.

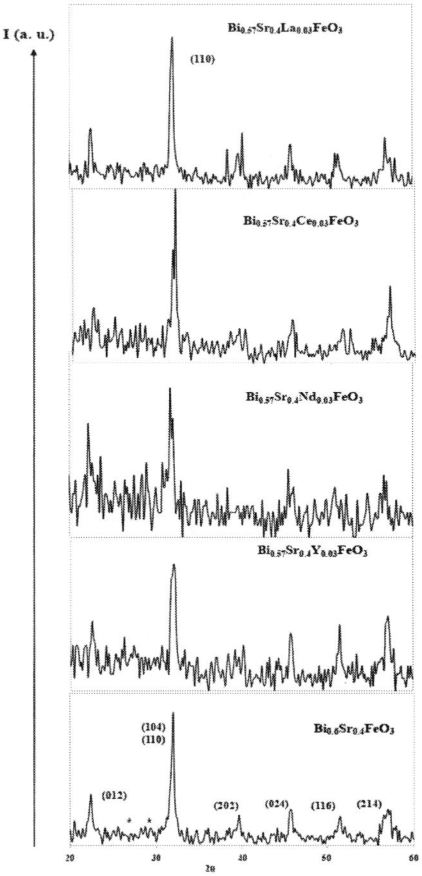

Figure (1): XRD patterns of the $Bi_{0.57}Sr_{0.4}RE_{0.03}FeO_3$ nanomultiferroics at different rare earth ions RE=None, Y^{3+}, Nd^{3+}, Ce^{3+}, and La^{3+} ions.

Table (1): XRD structure parameters of $Bi_{0.57}Sr_{0.4}RE_{0.03}FeO_3$; RE= None, Y, Nd, Ce, and La nanomultiferroics system.

Sample	a=b (nm)	c(nm)	c/a(nm)	V(nm)	L(nm)
$Bi_{0.6}Sr_{0.4}FeO_3$	5.5795	13.8359	2.4798	373	25.95
$Bi_{0.57}Sr_{0.4}Y_{0.03}FeO_3$	5.5552	13.8675	2.4963	370.6	22.43
$Bi_{0.57}Sr_{0.4}Nd_{0.03}FeO_3$	5.6343	13.8014	2.4495	379.4	13.27
$Bi_{0.57}Sr_{0.4}Ce_{0.03}FeO_3$	5.5782	14.1598	2.5384	381.5	14.12
$Bi_{0.57}Sr_{0.4}La_{0.03}FeO_3$	5.5836	13.7915	2.47	372.3	21.66

The lattice parameters were computed on the basis of hexagonal unit cell and reported in Table (1). It is clear that, the change in the lattice parameters at different ionic radii of the rare earth elements. From the Table, it is indicated that Y^{3+} ions has the smallest volume. Besides, the stability of perovskite compound was determined by calculating the tolerance factor (t), given as:

$$t = r_A + r_O / \sqrt{2}\,(r_B + r_O)$$

Where, r_A was the average ionic radii of Bi^{3+} and RE^{3+} ions, r_B was the average ionic radii of Fe^{3+} and Fe^{4+} ions, r_O was the ionic radius of O^{2-} ion. They have reported a continuous decrease in value of t indicating a structural transition [29]. Lattice parameters are also slightly modified on doping as a result of difference in ionic radii of host and dopant ions.

3.2 Microstructural analysis

The gain understanding of microstructural features of $BiSFO_3$ ferrite and $BiSFO_3$ doped rare earth samples are shown in Figure (2). SEM was performed show typical the SEM micrographs of $Bi_{0.57}Sr_{0.4}Y_{0.03}FeO_3$, $Bi_{0.57}Sr_{0.4}Nd_{0.03}FeO_3$, and $Bi_{0.57}Sr_{0.4}Ce_{0.03}FeO_3$ samples, recorded at room temperature. The microstructures were overall dense with small grains, well-interlinked and randomly arranged though out the sample surfaces besides some of the particles are agglomerated with a high interconnection of grains. In other words, SEM microstructure exhibits clear hexagonal grain growth habit. Addition of excess, doping of RE causes structural distortion as well as suppresses spiral spin structure in BSFO; as a result both ferroelectric and ferromagnetic properties get enhanced.

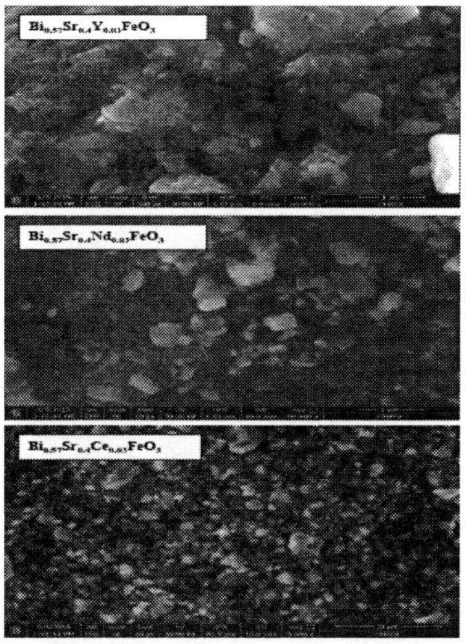

Figure (2): Typical SEM images of $Bi_{0.57}Sr_{0.4}RE_{0.03}FeO_3$ nanomultiferroic system; RE = Y^{3+}, Nd^{3+}, and Ce^{3+}

3.3 Magnetic Properties

The data shows that all investigated samples were G-type antiferromagnetic with a slight canting resulting in a weak ferromagnetic moment. Besides, the magnetic susceptibility decreases with increasing magnetic field for all samples. This is because the magnetic susceptibility is inversely proportional to the magnetic field according to $\chi_M = M/H$. The contribution to the Néel temperature (T_N) was due to the superexchange interaction between antiferromagnetically coupled Fe^{3+} ions (Fe^{3+}–O^{2-}–Fe^{3+}) through the oxygen anions. Different values of χ_M and T_N were obtained for the different samples and reported in Table (2). For all samples, χ_M decreases slowly with increasing temperature until reaching T_N. Moreover, from the Table it is clear that, the values of χ_M different with the ionic radius of the doping ion at room temperature where $Bi_{0.57}Sr_{0.4}Y_{0.03}FeO_3$ has the highest magnetization (0.26 emu/g) while $Bi_{0.57}Sr_{0.4}La_{0.03}FeO_3$ has the smallest magnetization (0.08 emu/g). As the ionic radii change, the values of μ_{eff}, T_N and χ_M, are changed and the optimum rare earth has the highest physical properties is $Bi_{0.57}Sr_{0.4}Y_{0.03}FeO_3$. Finally we can conclude that the, yttrium doped BSFO possesses enhanced photocatalytic activity along with multiferroic properties which makes it suitable for device applications.

Table (2): Effect of ionic radius of RE^{3+} ions on the magnetic parameters of $Bi_{0.57}Sr_{0.4}RE_{0.03}FeO_3$ nanomultiferroic system; RE= None, Y^{3+}, Nd^{3+}, Ce^{3+}, and La^{3+} ions.

Sample	T_N(K)	M (emu/g)	μeffe 1340Oe	1660Oe	1990Oe
$Bi_{0.6}Sr_{0.4}FeO_3$	604	0.162	11.6	9.51	8.69
$Bi_{0.57}Sr_{0.4}Y_{0.03}FeO_3$	710	0.26	24.09	20.8	17.8
$Bi_{0.57}Sr_{0.4}Nd_{0.03}FeO_3$	610	0.167	22.37	19.64	16.3
$Bi_{0.57}Sr_{0.4}Ce_{0.03}FeO_3$	530	0.135	22.88	20.37	18.72
$Bi_{0.57}Sr_{0.4}La_{0.03}FeO_3$	620	0.08	11.73	11.11	10.09

The antiferromagnetic transition temperature (T_N) was identified and shown and listed in Table (2). It is clear that, the highest value of (T_N) at $Bi_{0.57}Sr_{0.4}Y_{0.03}FeO_3$ (710K) and the smallest T_N at $Bi_{0.57}Sr_{0.4}Ce_{0.03}FeO_3$ (530 K). This variation is due to the different ionic radius for the doped elements. The change in the Néel temperature with ionic radii of the doping transition metals is directly correlated with the change of the tolerance factor as reported in Table (2) and the change of Fe-O-Fe angle (θ). This angle controls the superexchange interaction between antiferromagnetically coupled Fe cations through oxygen anions. These results indicated that, the enhancement effect of the ionic radii of the doped ions on the structure and the physical properties especially Y^{3+} doped ions of the investigated multiferroics.

The magnetization versus applied magnetic field (M–H) loops of all the samples measured at 300 K for pure and doped $Bi_{0.57}Sr_{0.4}RE_{0.03}FeO_3$ nanomultiferroics; RE= Y^{3+}, Nd^{3+}, Ce^{3+}, and La^{3+} ions are shown in Figure (3). It is reveals that, all the loops confirm weak ferromagnetic behavior, which is greatly enhanced for doped samples (Table 3). Saturation level of loops is also improved. $BiFeO_3$ having G-type antiferromagnetic spin ordering, shows weak magnetism due to spin canting from perfect antiparallel direction by symmetry. The improvement in magnetic behavior can be explained as follows: the spin cycloid structure of BSFO may be disturbed due to change in Fe–O–Fe angle on incorporation RE^{3+} ion in BSFO lattice and/ or RE–O–Fe like spin coupling between rare earth-transition metal ions [30]. Further, increment in magnetization on transition metal ion RE^{3+} doping may be attributed to ferrimagnetic like super-exchange interaction between RE^{3+} and Fe^{3+} ions due to the difference in their magnetic moments [31, 32].

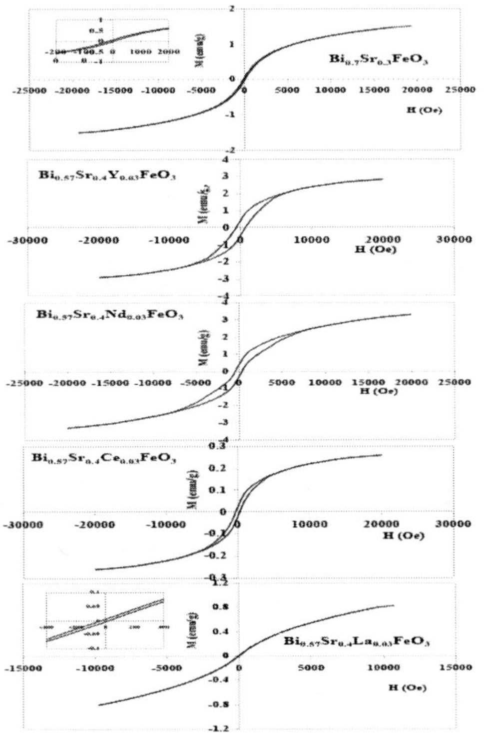

Figure (3): Hystersis loop for $Bi_{0.57}Sr_{0.4}RE_{0.03}FeO_3$ nanomultiferroics at different ionic radius of RE= None, Y^{3+}, Nd^{3+}, Ce^{3+}, and La^{3+} ions.

In other words, it is observed that, the hysteresis curves seen at all the samples unsaturated M-H behavior of the solid solutions suggest their antiferromagnetic nature with weak ferromagnetism. The values of saturation magnetization (M_s), the remanent magnetization (M_r), and magnetic coercive field (H_c) of the investigated samples are summarized in Table (3) indicated that the values of these parameters change with different doped ions where, $Bi_{0.57}Sr_{0.4}Nd_{0.03}FeO_3$ has the highest saturation magnetization (3.31 emu/g) higher than the undoped sample (0.24 emu/g) and highest magnetic moment η_B (0.163) while undoped sample has η_B (0.012). this indicated that, Bismuth ferrite is antiferromagnetic, but although each Fe^{3+} ion is surrounded by six other Fe^{3+} ions with opposite angular magnetic spin directions, their spins do not annul completely, so there are some resulting magnetic moments. These moments have cycloidal ordering with period of around 64 nm, which means that particles smaller than 64 nm are weekly ferromagnetic [33, 34]. Besides, the hysteresis loop shows a small but nonzero remnant magnetization (M_r). The spontaneous magnetization should result from the spiral magnetic ordering partially destroyed in the BSFO ferrite [35]. The magnetization curve of $Bi_{0.6}Sr_{0.4}FeO_3$ (RE = 0.0) multiferroic showing antiferromagnetic behavior has been reported earlier. Whereas $Bi_{0.57}Sr_{0.4}RE_{0.03}FeO_3$ (RE= Y^{3+}, Nd^{3+}, Ce^{3+}, and La^{3+} ions) exhibit a weak ferromagnetism, with remnant magnetization (M_r) values of 0.43, 0.39, 0.38, and 0.046 emu/g respectively. These values are higher than that of undoped BSFO (0.037 emu/g) sample [36]. The values of M_r increase with doped Y^{3+} ions because RE^{3+} ions substitution can suppress or destroy the space, modulated cycloidal spin structure of RE : BSFO ferrite. The other factor, which could be responsible for increase in M_r with substitution, is size effect [37]. The considerable increase in the remnant magnetization could result from improved crystallinity and a lack of secondary phases in these multiferroics. Several reasons contributing .o the weak ferromagnetism in RE : BSFO multiferroics have been reported: (1) The enhanced magnetization with doping RE^{3+} is mainly attributed to the gradual collapse of the space modulated spin structure [38] together with the increased degree of canting and magnetic interactions between magnetic ions giving rise to long range canted antiferromagnetic order. (2) The higher remnant magnetization in the Y: BSFO may also be attributed to the spin interaction between Y^{3+} and Fe^{3+}or Fe^{2+} ions [38]. (3) The particle size (or) crystallite size also plays a role in enhancing magnetization. The reduced particle size modified the long range spiral order [38]. Furthermore from the table it is clear that, all the samples have coercivity (Hc) higher than undoped sample except $Bi_{0.57}Sr_{0.4}La_{0.03}FeO_3$ while $Bi_{0.57}Sr_{0.4}Y_{0.03}FeO_3$ multiferroic has the highest coercive field (H_c) of 635.9 Oe which higher five times on the undoped sample (163.44 Oe). According to the above results, the high coercivity is related to the radii of rare earth ions are larger than that Fe^{3+}ions. Hence, the symmetry of crystal will be decreased after doping by rare earth ions. The low symmetry of crystal will lead to strong magnetocrystalline anisotropy, which is one reason for dopes samples to have high coercivity relative to the undoped sample.

Table (3): Magnetization parameters for the nanomultiferroics $Bi_{0.57}Sr_{0.4}$ $RE_{0.03}$ FeO_3; at different ionic radius of RE= None, Y^{3+}, Nd^{3+}, Ce^{3+}, and La^{3+} ions.

Sample	M_s (emu/g)	M_r (emu/g)	H_c(Oe)	η_B(BM)
$Bi_{0.6}$ $Sr_{0.4}FeO_3$	0.24	0.037	163.4	0.012
$Bi_{0.57}$ $Sr_{0.4}Y_{0.03}FeO_3$	2.87	0.43	635.9	0.140
$Bi_{0.57}$ $Sr_{0.4}Nd_{0.03}FeO_3$	3.31	0.35	403.06	0.163
$Bi_{0.57}$ $Sr_{0.4}Ce_{0.03}FeO_3$	0.56	0.022	282.35	0.028
$Bi_{0.57}$ $Sr_{0.4}La_{0.03}FeO_3$	0.11	0.046	25.95	0.005

4 Conclusion

Substitution of small amount of La^{3+}, Ce^{3+}, Nd^{3+}, and Y^{3+} ions for $BiSrFeO_3$ nanomultiferroics is seen to improve both the magnetization and the dielectric constant as following :

Secondary phases observed in pure BFO are significantly suppressed in the substituted samples and no structural transition is observed with co-substitution and the rare earth doping improves the particle morphology. Magnetic measurements were carried out at room temperature and that the hysteresis loops shows a weak ferromagnetism and the highest value at Y^{3+} ion doping which consider the optimum rare earth for wide applications.The enhancement of saturation magnetization for the Nd^{3+} ion doping may be attributed to its great amount of Fe^{2+} and suppressed periodic self-spinning structure. These changes in the magnetic properties could be more useful in terms of the use of RE doped BSFO nanomultiferroics for magneto optic applications and in addition these results suggest that the improved magnetic property in the BSFO systems could push the rare earths dopant toward multiferroic applications. The structural distortion causes increase in canting angle and variation of Fe^{2+} ion can also lead to increased magnetization. Meaning that the, yttrium doped BSFO possesses enhanced photocatalytic activity along with multiferroic properties which makes it suitable for device applications.

5. References

[1] Q.H. Jiang, F.T. Liu, C.-W. Nan, Y.-H. Lin, M.J. Reece, H.X. Yan, H.P. Ning, Z.J. Shen, Appl. Phys. Lett. **95** (2009) 012909.

[2] S. Hunpratub, P. Thongbai, T. Yamwong, R. Yimnirun, S. Maensiri, Appl. Phys. Lett. **94** (2009) 062904.

[3] X.-Y. Zhang, Q. Song, F. Xu, C.K. Ong, Appl. Phys. Lett. 94 (2009) 022907.

[4] A. Lahmar, S. Habouti, M. Dietze, C.-H. Solterbeck, M. Es-Souni, Appl. Phys. Lett. **94** (2009) 012903.

[5] J. Wang, J.B. Neaton, H. Zheng, V. Nagarajan, S.B. Ogale, B. Liu, D. Viehland, V. Vaithyanathan, D.G. Schlom, U.V. Waghmare, N.A. Spaldin, K.M. Rabe, M. Wuttig, R. Ramesh, Science **299** (2003) 1719.

[6] Y. Wang, C.W. Nan, Appl. Phys. Lett. **89** (2006) 052903.

[7] N.A. Hill, J. Phys. Chem. B **104** (2000) 6694.

[8] W. Eerenstein, N.D. Mathur, J.F. Scott, Nature (London) **442** (2006) 759.

[9] G.A. Smolenskii, L. Chupis, Sov. Phys. USP **25** (1982) 475.

[10] F. Bai, J. Wang, M. Wutting, J.F. Li, N. Wang, A. Pyatakov, A.K. Zvezdin, L.E. Cross, D. Viehland, Appl. Phys. Lett. **86** (2005) 032511.

[11] C. Ederer, N.A. Spaldin, Phys. Rev. B **71** (2005) 060401.

[12] D. Lee, M.G. Kim, S. Ryu, H.M. Jang, S.G. Lee, Appl. Phys. Lett. **86** (2005) 222903.

[13] A. Singh, V. Pandey, R.K. Kotnala, D. Pandey, Phys. Rev. Lett. **101** (2008) 247602.

[14] I. Sosnowska, T. Peterlin-Neumaier, E. Steichele, J. Phys. C **115** (1982) 4835.

[15] S.W. Cheong, M. Mostovoy, Nat. Mater. **6** (2007) 13.

[16] Gruverman A, Kholkin A. Nanoscale ferroelectrics: Processing, characterization and future trends. Rep Prog Phys; **69** (2006) 2443- 2474.

[17] Ahn CH, Rabe KM, Triscone JM. Ferroelectricity at the nanoscale: local polarization in oxide thin films and heterostructures. Science; **303** (2004) 488-491.

[18] T. Slimani Tlemcani, T. El Bahraoui, M. Taibi, A. Belayachi, G. Schmerber, A. Dinia, M. Abd- Lefdil, J. Sol- Jel Sci. Technol., **37** (2015) 673.

[19] Li S, Jing L & Fu W, Yang L, Xin B & Fu L, Mater Res Bull, **42** (2007) 203.

[20] Gao F, Chen X, Yin K, Dong S, Ren Z, Yuan F, Yu T, Zou Z & Liu JM, Adv Mater, **19** (2007) 2889.

[21]M.A.Ahmed, N. Okasha, S.I. El-Dek, Nanotechnology **19** (2008) 065603.

[22] M.A. Ahmed, J. Magn. Magn. Mater. **322** (2010) 763.

[23] M. Kumar and K. Yadav. J. Phys.: Condens. Matter, **18** L (2006) 503.

[24] J. Yang, X. Li, J. Zhou, Y. Tang, Y. Zhang, Y. Li. Factors controlling pure-phase magnetic $BiFeO_3$ powders synthesized by solution combustion synthesis. J Alloys Compd., **509** (2011) 7.

[25] Rojas-George, G, Silva, J, Castañeda, R, Lardizábal, D,Graeve, O.A, Fuentes L,etal. Modifications in the rhombohedral degree of distortion and magnetic proper- ties of Ba-doped $BiFeO_3$ as a function of synthesis methodology. Mater. Chem. Phys., **146** (2014) 73.

[26] M. B. Bellakki, & V. Manivannan, Citrate-gel synthesis and characterization of. yttrium-doped multiferroic $BiFeO_3$. Journal of Sol-Gel Science and Technology, **53** (2010) 184.

[27] A. Mukherjee, S. M. Hossain, M. Pal, & S. Basu, Effect of Y-doping on optical properties of multiferroics $BiFeO_3$ nanoparticles. Applied Nanoscience. DOI 10.1007/s13204 (2012) 0114-8.

[28] A. Gautam, P. Uniyal, K. L. Yadav, & V. S. Rangra, Dielectric and magnetic properties of $Bi_{1-x}Y_xFeO_3$ ceramics. Journal of Physics and Chemistry of Solids, **73** (2012) 188.

[29] R. Palai, R. S. Katiyar, H. Schmid, P. Tissot, S. J. Clark, J. Robertson, S. A. T. Redfern, G. Catalan, and J. F. Scott, "B Phase and c-b Metal- Insulator Transition in Multiferroic $BiFeO_3$," Phys. Rev. B, **77** [1] 014110–11 (2008).

[30] Li J-B, Rao GH, Liang JK, Liu YH, Luo J, Chen JR., Magnetic properties of $BiFe_{1-x}Cr_xO_3$ synthesized by a combustion method. Appl. Phys. Lett., **90** (2007) 162513.

[31] Nguyen Van Minh, Dao Viet Thang, Journal of Alloys and Compounds **505** (2010) 619.

[32] R. Safi, H. Shokrollahi, Physics, chemistry and synthesis methods of nanostructured bismuth ferrite ($BiFeO_3$) as a Ferro electromagnetic material, Prog. Solid State Chem.,40 (2012) 6.

[33] G. Catalan, J. F. Scott, Physics and applications of bismuth ferrite, Adv. Mater. 21 (2009) 2463.

[34] D. Lebeugle, D. Colson, A. Forget, M. Viret, P. Bonville, J.F. Marucco, S. Fusil,Room-temperature coexistence of large electric polarization and magnetic order in $BiFeO_3$ single crystals. Phys. Rev. B **76** (2007) 024116.

[35] C. Anthonyraj, M. Muneeswaran, P. Jegatheesan, N.V. Giridharan, V. Sivakumar, G. Senguttuvan, Effect of annealing time in the low-temperature growth of BFO thin films spin coated on glass substrates. J. Mater. Sci.: Mater. Electron. **24**, (2013) 4148.

[36] D. Haiyang, C. Zhenping, L. Tao, L. Yong, Microstructure and properties of Sm-substituted $BiFeO_3$ ceramics. J. Rare Earth. **30** (2012) 1123.

[37] Y.H. Lee, J.M. Wu, C.H. Lai, Influence of La doping in multiferroic properties of $BiFeO_3$ thin films. Appl. Phys. Lett. **88** (2006) 042903.

[38] T. J. Park, G. C. Papaefthymiou, A. J. Viescas, A. R. Moodenbaugh, S. S. Wong, Size-dependent magnetic properties of single-crystalline multiferroic $BiFeO_3$ nanoparticles.J Mater Sci: Mater Electron, **7** (2007) 766.

[39] X. Xu, T. Guoqiang, R. Huijun, X. Ao, Structural, electric and multiferroic properties of Sm-doped $BiFeO_3$ thin films prepared by the sol–gel process. Ceram. Int. **39** (2013) 6223.

NEW MONOSILANE FLUID BED DECOMPOSITION TECHNOLOGY FOR THE PRODUCTION OF SOLAR QUALITY SILICON FEEDSTOCK

Mark W. Dassel
SiTec GmbH / Seattle Center for Applied Research and Product Development
26272 Twelve Trees Lane NW, Suite A, Poulsbo, Washington 98370 USA

ABSTRACT: SiTec is developing a novel vibrationally fluidized bed reactor (V-FBR) that can significantly lower overall costs compared to current practice, using undiluted monosilane to produce polysilicon granules for solar use. This FBR is compact, self-seeding, operates at high pressure, and demonstrates very low dust levels at high conversion efficiencies. A commercially prototypic pilot reactor is in the early stages of scale-up testing, with good results.
Keywords: monosilane decomposition, vibrationally fluidized bed reactor, FBR, Genesis, monosilane fluidized bed reaction, lined reactor, granular product, fluidization, heterogenous decomposition, poly-powder, scavenging

1 SUMMARY

SiTec technology will reduce monosilane fluid-bed reactor (FBR) capital and operating costs by approximately 45% each, compared to legacy FBR decomposition processes. Called "Genesis™," SiTec's technology fluidizes the granular reaction bed by mechanically induced vibration as opposed to gas induced (a.k.a., "hydraulic") fluidization used by legacy processes. SiTec calls its development "vibrationally fluidized bed reaction" or V-FBR. Legacy FBR processes fluidize reaction beds by blowing a mixture of monosilane gas (diluted to 5% to 20% concentration in hydrogen recycle) up through a bed of small silicon particles. The feed gases lift and fluidize the bed. Decomposition and particle growth occur when monosilane gas contacts hot granules in the bed. Resultant epitaxial growth increases particle size until sufficiently large to be harvested. Attempts to increase hydraulic fluid bed reactor productivity by operating at high pressure and high monosilane feed concentration result in high rates of dust formation and low yield. In marked contrast, SiTec's Genesis™ process operates with 100% monosilane feed – no hydrogen dilution, and at high pressure, with less than 2% dust formation. SiTec recently started up a commercially prototypic pilot reactor at its Seattle Center for Applied Research and Product Development. The pilot plant is being used to perfect its mechanically fluidized beds, prove commercial design and product quality, and verify the fast reaction rates, in situ self-seeding, and high yields demonstrated in earlier lab test work.

2 BACKGROUND

2.1 A Historical Overview of Monosilane Fluidized Bed Reaction

There are currently two main process routes to polysilicon (a.k.a., "legacy" processes):

1. The conversion of metallurgic silicon to refined trichlorosilane, followed by decomposition to rod form polysilicon in a Siemens type reactor, and
2. The conversion of metallurgic silicon to refined monosilane, followed by decomposition to granular form polysilicon in a fluidized bed reactor, where fluidization is due to gases rising through the bed – referred to herein as hydraulically fluidized beds or H-FBR.

In addition to these main routes, monosilane is converted to very high quality rod form in Siemens reactors (commercial), and TCS is decomposed to granular form factor in a fluidized bed reactor (believed to be pre-commercial).

Work on the granular process began in earnest in the 1970's by Union Carbide, at its Washougal, Washington State, U.S.A. laboratory and semi-works facility, and by Albemarle Corporation at its plant site in Pasadena, Texas. Albemarle Corporation subsequently sold its granular process technology and the Pasadena facility to MEMC, and this plant was successfully operated until recent times. The monosilane source for the Pasadena plant was unique in that it was based on a silicon tetrafluoride process. The Pasadena plant used lined reactor technology to mitigate the migration of metal ions from the reactor walls into the granular product.

Union Carbide stopped its granular process development in favor of monosilane decomposition in Siemens reactors. This approach was successfully practiced at both its Washington State and Montana U.S.A. plant sites. Union Carbide subsequently sold its technology and plant facilities to Komatsu, and Komatsu to REC Inc. (now operated as REC Silicon). Between 2000 and 2015, REC Silicon resurrected Union Carbide's early efforts to convert monosilane to granular polysilicon in an H-FBR, based on perceived economic advantages. REC Silicon subsequently installed significant monosilane H-FBR capacity at its Washington State, U.S.A. plant site. Its plants use unlined reactor technology. Instead a barrier of polysilicon is formed on interior reactor surfaces. This in situ barrier is at least partially successful in preventing the migration of metal ions into the granular product. REC's monosilane process – as was Union Carbide's original – is based on redistribution of crude TCS. REC Silicon recently developed a lined H-FBR process and, in conjunction with Shaanxi Non-Ferrous Tian Hong New Energy Company, is building a plant based on this technology in China.

2.2 Key Issues with Legacy Monosilane H-FBR Decomposition Processes

Issues include:

- Excessive monosilane yield loss to dust,
- Requirement to make seed ex situ (i.e., outside the H-FBR),
- Thermal heat load is increased as much as 10X due to hydrogen diluent,
- Heat transfer limits reactor size and efficiency,

- Relatively high FBR system capital cost, and
- Problematic product quality.

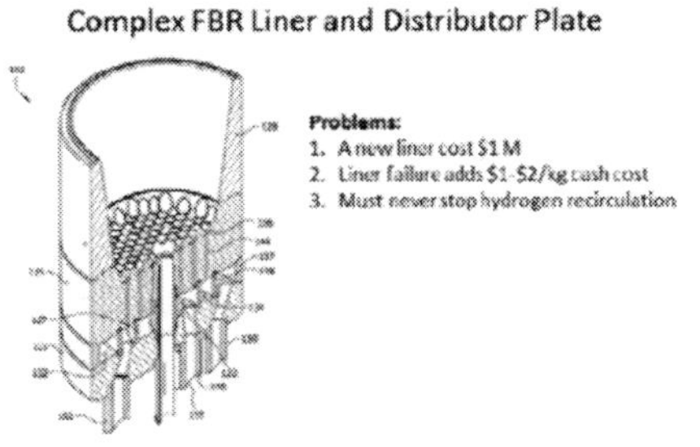

Figure 1: A conceptual drawing of a lined reactor. Liners typically comprise graphite and silicon carbide. [1]

The common factor in these issues is the use of reactant and diluent feed gases to fluidize the reaction bed.

Fluidization is essential to uniformly distribute monosilane to hot granular surfaces where heterogeneous decomposition occurs, and to prevent the formation of agglomerates which can plug the bed.

Legacy H-FBR processes fluidize the granule bed with feed gases consisting of monosilane and hydrogen diluent. Hydrogen diluent is added to:

1. Help fluidize the bed, and
2. Minimize the conversion of monosilane feed gas to dust (also known as "poly-powder").

A 1:5 to 1:20 monosilane to hydrogen feed gas ratio is used. Thus the dominant fluidization gas is hydrogen. Pure monosilane feed is avoided in H-FBRs because it results in very high dust formation rates (e.g., > 25% of monosilane to dust) and because, without hydrogen diluent, there is insufficient hydraulic flow to lift and fluidize the bed.

Even with hydrogen dilution, 15 ±5 % of the monosilane feed is converted to dust. The gas rate required for fluidization blows this dust out of the reactor. The gas also blows seed out of the reactor – to the extent it is made in situ. Typically all material made in the FBR smaller than approximately 85 microns in size is lost as dust in the off-gas filters. As a result, on-purpose seed is made ex situ. Seed is required to replace harvested granular product. Ex situ manufacture requires product classification, size reduction, and seed recycle. These ex situ steps are capital and cost intensive, and introduce impurities into the H-FBR.

Hydrogen gas has an inherently high heat capacity, thus hydrogen dilution significantly increases H-FBR heat load. The addition of heat to an H-FBR is difficult because the heat is added to the non-process side of the liner system on the circumference of the reactor. Heat exchangers in the fluidized bed itself would be more efficient but would increase meal contamination, so this is not practiced. Scale-up to larger diameter reactors is difficult because the surface area for heat transfer increases only by the diameter whereas the internal reaction volume – and concomitant heat load – goes up by the square of the diameter. As a result, legacy H-FBR reactors are usually ≤ 1 meter in diameter and limited to around 500 to 1,000 metric annual tons (MTA).

H-FBRs are typically tall vessels with a large height to diameter ratio (e.g., 10:1). Extra reactor volume, in the form of additional height, is added to increase heat transfer from the heated wall to the granules in contact with the wall, and for granule de-entrainment. Consequently, much of the reactor volume – as much as 2/3rds – is relatively non-reactive.

Legacy H-FBRs run near atmospheric pressure. Theory, however, suggests reactor productivity (as production per unit volume per unit time) increases with increasing pressure. Attempts to increase hydraulic fluid bed reactor productivity by operating at higher pressure have reportedly resulted in high rates of dust formation and low yield. Notwithstanding, efforts to increase reaction pressure by an order of magnitude continue, and one major producer has announced plans to build and startup a high pressure reactor in hopes of increasing reactor productivity.

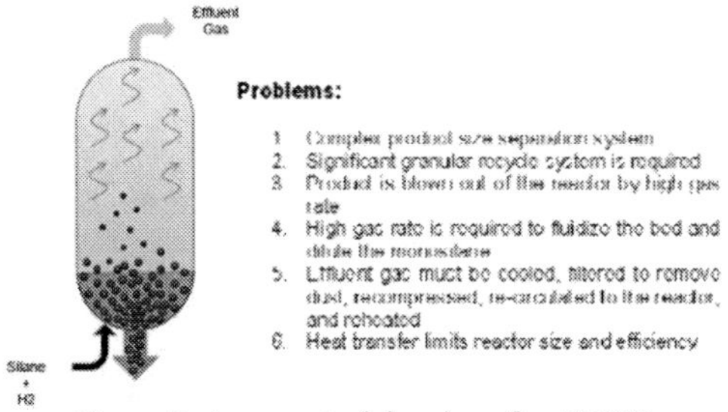

Figure 2: A conceptual drawing of an H-FBR.

2.3 Factors Affecting Dust Formation in an H-FBR System

Monosilane decomposition chemistry proceeds in two regimes:

1. Heterogeneous, and
2. Homogeneous.

Heterogeneous decomposition is desired and produces granules. Seed-stock is required for this to occur. The decomposition reaction occurs at temperatures above 500 C, and typically in the 600 C to 700 C temperature range.

Homogeneous decomposition produces dust, also known as "poly powder." Poor contact with hot granular surfaces is a necessary condition for homogeneous decomposition. Further, the following conditions promote homogeneous decomposition:

- High temperature (e.g., >600 C), and/ or
- High monosilane molar density.

Large gas bubbles formed in H-FBRs contribute to dust formation. Their formation is inherent in the physical process as is well documented in the literature.

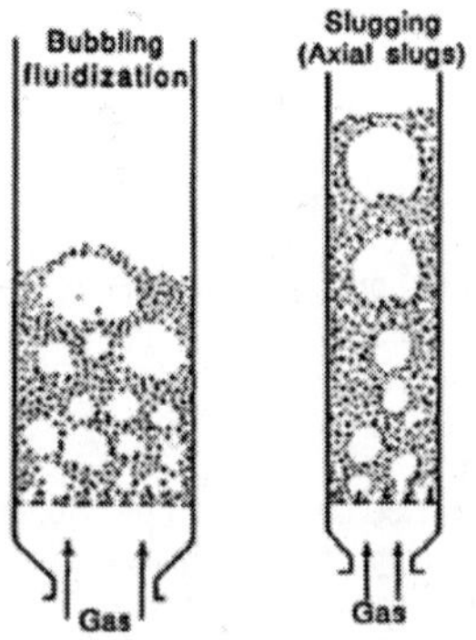

Figure 3: Bubble formation in legacy style H-FBRs. [2]

Bubble size approaches the inner diameter of the fluidized bed. Poor resultant contact with hot granular surfaces combined with high bubble temperature and unreacted monosilane gas produce excessive amounts of dust.

2.4 Factors Affecting Product Quality in an H-FBR System

Considerable effort is devoted to improving granular product quality. Unlined systems produce granules with relatively high metals content. Current generation lined systems are better, but fall short of high electronic grade quality. Factors impacting quality include:

1. Granule contact with metal surfaces,
2. Granule contact with impure gases, and
3. Impurities introduced as a result of ex situ seed generation.

Impurity content due to metal contact is proportional to surface area, and a function of temperature. Legacy systems operate at 600 C to 700 C. At these temperatures metal ions migrate from metal to granules by surface-to-surface contact. The addition of extra reactor height to promote heat transfer increases metal transport.

Liners are used to manage metal contamination. In one system, silicon carbide is applied to the process side (i.e., inside) of a graphite liner. This mitigates metals transport but introduces carbon – also an impurity – into the reaction zone. As a result, carbon content in granular product is orders of magnitude greater than that in rod form polysilicon produced from monosilane.

The carbon impurity may be sourced by carbon migration through the graphite/ silicon carbide composite liner and/or by hydrogen gas introduced into the pressurized plenum surrounding the liner system. Hydrogen purge gas introduced into the plenum may transport carbon from the non-process side (i.e., the outside) of the liner into the process side of the reactor.

However, the greatest potential source of metal impurity in granular product is the granule product classification and ex situ seed generation system. Product removed from the H-FBR is classified to separate particles smaller than 400 microns and larger than 3,000 microns. The large particles are size reduced to create reactor seed-stock, combined with the small particles, and this is recycled back to the reaction zone. Elimination of these requirements would significantly improve granule product quality; however, the inherent physics controlling hydraulic fluidization (i.e., high required gas rates) have so far precluded the achievement of this goal.

3 SITEC'S NEW MONOSILANE DECOMPOSITION TECHNOLOGY

3.1 Vibrationally Induced Fluidization

SiTec is developing a novel silane decomposition process that will produce solar and electronic-grade polysilicon. The new process is called Genesis™. Genesis™ fluidizes a granular reaction bed by mechanically induced vibration as opposed to the gas induced (a.k.a., "hydraulic") fluidization used in legacy processes.

Production cost of the new technology is forecast to be lower than that of legacy H-FBR technology and that of refined TCS/Siemens CVD.

Advantages include high reactor productivity, self-seeding, in-situ particle size classification, <2 % yield loss to dust formation, and lower CapEx.

Genesis™ is being developed at lab and pilot-reactor scale at SiTec's Seattle Center for Applied Research and Product Development (Seattle Center). Genesis™ reactors run at 10 to 20 BARG operational pressure and with 100% silane feed gas. Hydrogen diluent is not used. Ninety-nine percent conversion-per-pass is achieved in pilot-plant demonstrations. Micro-photographs show scavenging contributes to improved silane yield and reactor productivity, and that seeds – formed in-situ – are tight clusters of micron-sized polysilicon spheres. Liquid-like motion is achieved in the operating reactors – completely decoupled from gas velocity in and through the bed.

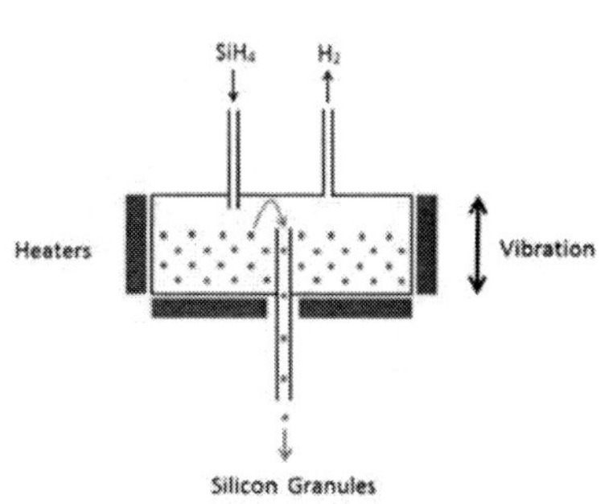

Figure 4: Simplified sketch of the Genesis™ reactor.

Vibrational rates and amplitudes (mid-point to peak) comprise 1,800 to 2,200 cycles per minute, and 0.75 to 1.25 mm.

3.2 Purpose of Development, Approach, Innovation, and Scientific Relevance

The purpose is to develop a new production means for lower cost, high quality solar-grade polysilicon granules. Genesis™ reactor technology is positioned to dramatically reduce the cost of converting silane gas to polysilicon granules, compared to existing processes. SiTec forecasts that polysilicon plant operating costs will be greatly reduced and achieve savings that will strengthen the entire solar energy value chain.

SiTec's technological approach is to fluidize the granular reaction bed by mechanically induced vibration. SiTec calls its development "vibrationally fluidized bed reaction" or V-FBR. This novel approach to fluidization has been demonstrated in bench tests and at pilot scale to greatly improve reactor throughput and yield, while simplifying reactor operation.

Legacy H-FBR processes fluidize reaction beds by blowing a mixture of hydrogen and monosilane gas up through a bed of small silicon particles ranging in size from 85 to 3,000 microns. The feed gases lift and fluidize the bed. Decomposition and particle growth occur when monosilane gas contacts hot granules in the bed. Resultant epitaxial growth increases particle size until sufficiently large to be harvested. Large particles sink to the bottom of the H-FBR reactor where they are removed batch-wise. In contrast, in a V-FBR the large particles move to the top of the bed where they are removed in a continuous, self-regulating, overflow process.

Figure 5: Commercially prototypic pilot reactor operated at the Seattle Center for Applied Research and Product Development

Legacy H-FBR processes typically run close to atmospheric pressure with 5:1 to 20:1 hydrogen dilution. Yield loss to dust – a waste product – typically runs 10% to 20%. This dust is called "poly-powder" in the industry and:

- Is micron and sub-micron in size,
- Sticks to surfaces,
- Has low density (in the 0.1 to 0.2 grams per cc range), and
- Bridges instead of pouring.

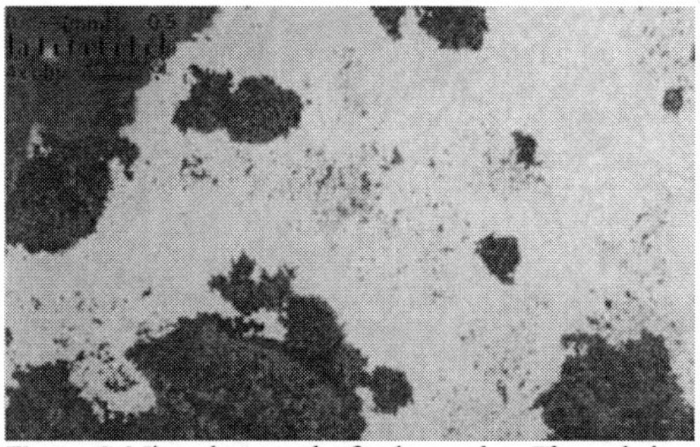

Figure 6: Microphotograph of poly-powder. The scale is in the upper left hand corner. The large clumps are filter cake removed from the reactor off-gas filters. The filter cake is comprised of individual, separable fine particles measuring under a few microns in size.

In marked contrast, the V-FBR process operates with 100% monosilane feed, no hydrogen dilution, and at high pressure (e.g., 10 BARG) with low dust formation. Seeds formed *in situ* range from 5 microns to 50 microns in size, are dense, and have a compact, spherical morphology. They are free-flowing and do not stick to surfaces or bridge.

Test work on the new fluidization technology is being performed in a commercially prototypic pilot reactor at SiTec's Seattle Center. The pilot reactor is being used to perfect the vibrationally fluidized bed, prove commercial design and product quality, and to further study the fast reaction rates, self-seeding, and high yields demonstrated in test work at the Seattle Center.

SiTec's V-FBR reactor is designed to run at temperatures up to 700 C and high pressure (up to 20 BARG) with 100% monosilane in reactor feed gas. These conditions make for high conversion rate and economical, efficient, small-sized reactors. SiTec believes these same conditions, if attempted in a legacy H-FBR process, would convert around one-half of silane feedstock to dust.

Dust formation in SiTec's reactor is <2% and is minimized by keeping the reactant gas in close contact with hot granules at all times. The film thickness between monosilane reactant molecules and hot granules is less than 400 microns because, with vibrational fluidization, there are no gas bubbles. By way of contrast, as a first approximation, the separation distance in a conventional hydraulic bed can be as much as 250X to 1,000X as great due to the formation of large-sized gas bubbles.

Seeds formed in the V-FBR reactor tend to stay in the bed because exit gas velocity is relatively low – only $1/50^{th}$ to $1/100^{th}$ as high as in a legacy H-FBR. Exit gas velocity is low because the V-FBR is operated at high pressure and because silane feedstock is not diluted with hydrogen. In comparison, the high exit gas velocity in a legacy H-FBR blows a substantial fraction of seed particles smaller than 85 microns out of the bed, necessitating the installation of costly *ex situ* seed generation and recycle facilities.

3.3 Product Morphology and Size Range

Figures 7 and 8 show material produced in the pilot reactor. Figure 8 compares Genesis™ product morphology with commercial polysilicon granules sold by legacy producers.

Figure 7: Material produced in the Genesis™ pilot reactor.

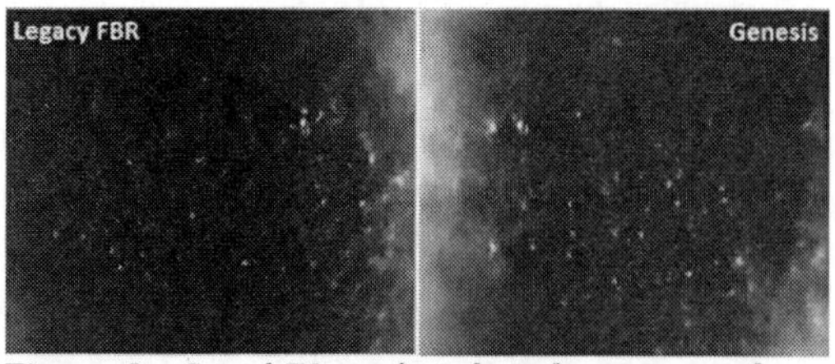

Figure 8: Genesis™ product has the same surface morphology as legacy granules. This photograph is of a single granule (about 2 mm in diameter). The surface texture is believed to be a result of concurrent epitaxial deposition and scavenging of small micron-sized particles produced *in situ*.

3.4 In Situ Seed Formation

Genesis™ is unique among all fluid bed reaction processes in that it makes and retains seed, in the reaction mass, sufficient for on-going operations (i.e., does not require supplemental *ex situ* seed manufacture). Figures 9 and 10 are photographs of seed made in the Genesis™ pilot-reactor.

The seeds measure 10 to 30 microns in diameter, are spherical in shape, free-flowing, and non-sticky. Each seed is comprised of an aggregate of smaller particles measuring a few microns in size.

Figure 9: Microphotograph of seed made in the Genesis™ pilot-reactor.

Figure 10: Photograph of the same seeds as in Figure 10 but at higher magnification. The 1 to 2 micron size particles making up the individual seeds are clearly visible.

3.5 Product Withdrawal System

The Genesis™ reaction system employs a unique product withdrawal system. In legacy FBRs, product is withdrawn from the bottom as underflow. In Genesis™, product is withdrawn from the top surface of the reaction bed in an overflow standpipe. Overflow works well in a V-FBR because the top surface of the bed, while highly fluidized, is relatively flat and not prone to the surging flow seen in H-FBRs. Further, and critically, large particles move to the surface in a vibrationally fluidized bed. The Genesis™ product withdrawal system is designed in a way that particles sized smaller than 400 microns stay in the reaction bed, while large particles are preferentially removed.

3.6 Reaction Kinetics

A reaction rate model, see Figure 11, has been fitted to Genesis™ laboratory data collected at temperatures ranging from 500 C to 650 C, and pressures ranging from 2 BARG to 12 BARG. While the activation energy for the reaction rate constant is similar to that reported by others (200 KJ/mole), the overall kinetics are faster for a given temperature. This is attributed to greater space time from operation at higher pressure. Further, the yield to product is higher because the amount of dust (a.k.a., poly-powder) made is much lower.

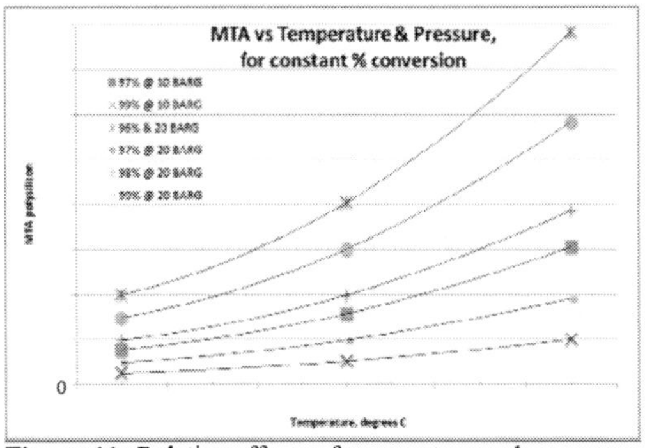

Figure 11: Relative effects of temperature and pressure on reactor productivity. For example, in the operating ranges studied, doubling the pressure roughly doubles productivity for a given size reaction mass.

Photographic evidence indicates that Genesis™ runs with a high degree of scavenging – i.e., dust is incorporated into product granules.

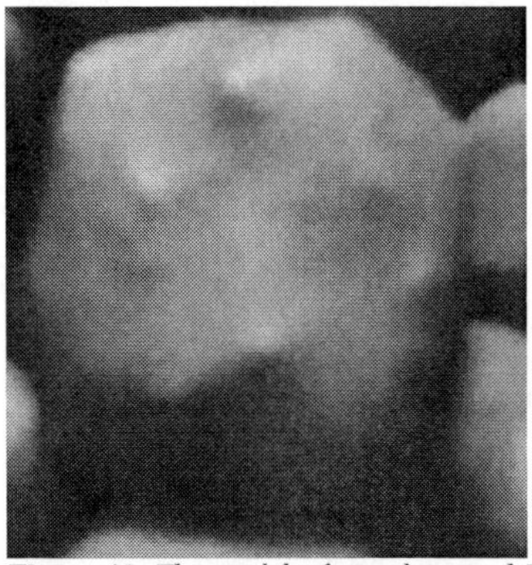

Figure 12: The particle shown is around 2.5 mm in size. Scavenged particles appear as bumps on its surface.

3.7 Conclusions and Path Forward

V-FBR technology brings the following advantages compared to conventional hydraulic fluidization (H-FBR):

1. Fluidization is decoupled from gas flow meaning hydrogen dilution is not required
2. Reactor heat load is reduced because hydrogen dilution is not required
3. Lower reactor heat load enables an 80% savings in decompositional electrical demand
4. Capital cost is lower because reactors are more efficient and small in size, expensive hydrogen recycle equipment is eliminated, and off-gas cooling and filtration loads are reduced in size by an order of magnitude
5. Yield is improved by up to 15% (e.g., from 85% to 99%) because <2 % of monosilane feed is converted to poly-powder
6. The monosilane plant is 15% smaller for a given polysilicon production rate due to improved yield
7. The technology is self-seeding; *ex situ* seed generation facilities are not required

8. Superior quality
9. Easy "on-off" operation

SiTec's innovative approach to fluid-bed silane decomposition should significantly lower silicon feedstock cost and, at the same time, break the quality barrier long associated with legacy H-FBR product.

Total polysilicon plant investment and energy consumption, including the monosilane feedstock plant, will be reduced by approximately 35% each compared to legacy H-FBR processes. Cash cost in China will be reduced to only $7.25/kg polysilicon*, compared to $9.50/kg cash cost for legacy monosilane/FBR polysilicon plants, and compared to $14/kg for legacy refined TCS/Siemens plants.

**Based on 8.5 cents per KwHr electricity price, $10/million BTU natural gas price, $2.25 per kg metallurgic silicon price, $43/kg total CapEx for a 10,000 MTA plant (including monosilane mfg.), and 2% maintenance cost factor.*

Extra steps to eliminate metal contamination are economical because SiTec's reactor is small in size and highly productive. SiTec's goal is quality solar-grade feedstock containing low metals content <10 ppbw. Quality will be inherently superior compared to legacy H-FBR because *ex situ* seed generation and recycle is not necessary.

Because V-FBR fluidizes by simple mechanical means, fluidization is "on-off". The complicated startup and shutdown protocols of legacy FBR processes are eliminated. Time to steady-state reaction from a cold start is a few hours, compared to legacy FBR processes which may require days. Heating is simple electrical resistance applied with direct conduction or radiation. Scale-up is easier with SiTec's V-FBR reactor because heat load is low. Electrical requirements are forecast to be 1 kwhr/kg polysilicon in commercial reactors.
Vibrational silane decomposition is an important new technology. When commercialized it will be a substantial improvement over legacy H-FBR processes.

4 REFERENCES

[1] Erk, Henry F., *Methods for reducing the deposition of silicon on reactor walls using peripheral silicon tetrachloride,* U.S. Patent No. 2011/0158888 A1, June 20, 2011.

[2] Kunii, D., Levenspiel, O., *Fluidization Engineering, Second Edition,* Butterworth-Heinemann, 1991, p.2

[3] Daleffe, R.V., Ferreira, M.C., and Freire, J.T., *Effects of binary particle size distribution on the fluid dynamic behavior of fluidized, vibrated and vibrofluidized beds*, Brazilian Journal of Chemical Engineering, **25**, No. 1 (January – March 2008) pp.83-94

[4] Laroche, C., Douady, S., and Fauve, S., *Convective flow of granular masses under vertical vibrations*, Le Journal de Physique, **50** (April 1989) pp.699-706

[5] Ford, Kenneth J., Gilchrist, James F., and Caram, Hugo S., *Transitions to vibro-fluidization in a deep granular bed*, Powder Technology, **192** (2009) pp.33-39

[6] Wassgren, C.R., Brennen, C.E., and Hunt, M.L., *Vertical vibration of a deep bed of granular material in a container,* Journal of Applied Mechanics, **63** (September 1996) pp.713-719

[7] Clement, E., Labous, L., and Vanel, L., *Granular Packing Under Vibration*, Europhysics News (May/June 1998) pp.107-111

CRYSTALMAX SILICON FOR HIGH EFFICIENCY/ LOW-COST SOLAR CELLS

Raphaël Cabal*[1], Nicolas Enjalbert[1], Etienne Pihan[1], Nelly Plassat[1], Gautier Fortin[2], Lotfi Bounaas[2], Sébastien Dubois[1]
[1] CEA, LITEN, INES, 50 av du Lac Léman, BP332, Technolac, 73370, Le Bourget du Lac, France
[2] ECM Green Tech, 109 rue Hilaire du Chardonnet, Technisud, 38100, Grenoble, France

ABSTRACT: This paper aims at presenting the CRYSTALMAX technology developed from the collaboration of ECM Green Tech crystallization furnace manufacturer with CEA-INES research institute. CRYSTALMAX technology provides fully <100> oriented 650 kg silicon ingots, obtained by seed-assisted cast solidification. Resulting 6^{inch} full square wafers feature low production costs that could be further reduced by reduced kerf loss offered by diamond wire sawing, the recycling of Cz-seeds, or the use of SoG_M feedstock. P-type CRYSTALMAX wafers were processed as conventional Al-BSF solar cells. The variation of the cells performances along the ingot's height are detailed. Correlations between wafer impurity (O_i; C_s) content, cell performances, effective carrier diffusion length are presented. Device stability towards prolonged light exposure is also evaluated.

Keywords: silicon, monocrystalline, seed-assisted, cast, diamond wire

1 INTRODUCTION

One way to reduce solar €/W is to maintain cell efficiency using cheaper silicon with a quality satisfying PV conversion demands. Among the different crystallization methods and related materials, quasi-mono silicon grown from seed-assisted cast solidification represents a good trade-off between material cost & quality, yet still showing limitations such as the presence of extended defect-rich domains. In the literature, side bricks from industrial quasi-mono/monolike ingots contain a significant multicrystalline area [1]. The presence of such multi domains generates difficulties in terms of solar cell processing technology (e.g. surface texturing), and affects the charge carrier lifetime. R&D programs between CEA-INES and ECM Green Tech led to the development of high quality, fully <100>-oriented silicon ingots, obtained by seed-assisted cast solidification. This so-called CRYSTALMAX technology has been developed thanks to ECM Green Tech specific furnace designs enabling the control of thermal gradients and solid-liquid interface shape. As a result, Gen6 ingots are today available, exhibiting 100% monocrystalline structure without parasitic multicrystalline grains nucleation (Fig.1).

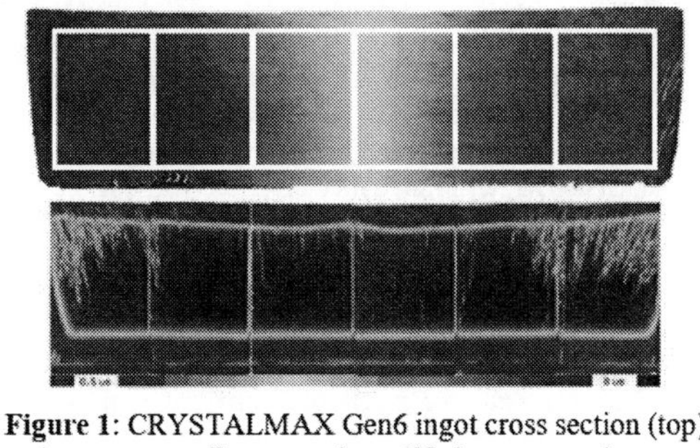

Figure 1: CRYSTALMAX Gen6 ingot cross section (top) and corresponding carrier lifetime mapping on unpassivated bricks (bottom)

The originality of CRYSTALMAX solidification process concerns the convex solid/liquid interface control allowing the growth of mono ingots without significant multicrystalline areas [2]. Resulting silicon differs from mono-Cz material by: (1) the presence of dislocated domains in ingot upper part, accentuated in side bricks (as observed in Fig. 1); (2) A lower oxygen concentration ($[O_i]< 6.10^{17}cm^{-3}$ against $>7.10^{17}at/cm^3$ in conventional Cz); (3) A higher carbon concentration ($[C_s]< 4.10^{17}cm^{-3}$)

2 ECONOMICAL CONCERN

2.1 Cost analysis

CRYSTALMAX technology already features a production cost (<0.6$/wafer) on both p-type and n-type. Besides, further wafer cost reduction is still possible as evidenced by the wafer cost analysis results plotted in Fig.2. Among the different parameters considered in the present analysis one can emphasize the strong impact of kerf loss and wafer thickness reductions. Both could be allowed by using Diamond Wire sawing. Upsizing CRYSTAMAX ingot would also be beneficial. As silicon Cz seeds represent a significant cost of the CRYSTALMAX crystallization, the recycling process of these seeds is considered. From previous calculation costs, it was found that the seed recycling was relevant if applied two times, leading thus to a cost reduction of around $0.7c\$/W_p$ at cell level. Further recycling has only marginal effect. Nevertheless, the resulting silicon quality has to be maintained: indeed, the gain offered by the 1st recycling occurrence is counterbalanced by a cell efficiency reduction of 0.25% abs.

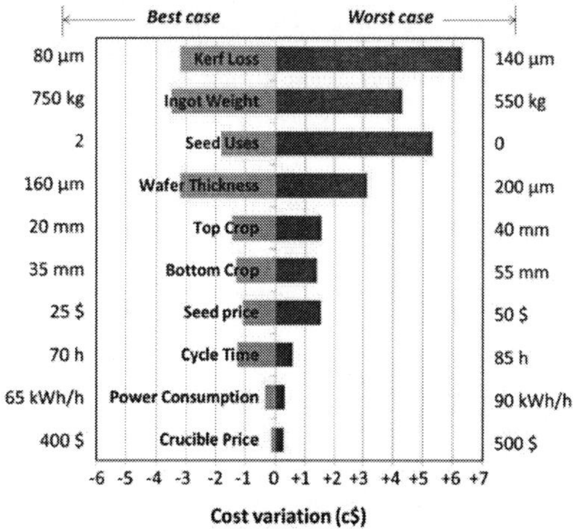

Figure 2: CRYSTALMAX wafer cost tornado analysis

Eventually the use of SoG_M (for Solar Grade Si purified by metallurgical route) feedstock during crystallization, instead of classical Electronic Grade (EG), may also be of interest - in a longer term - for wafer cost reduction.

2.2 CRYSTALMAX Diamond Wire sawing

Contrarily to HP multi material, the compatibility of CRYSTALMAX silicon with Diamond Wire (DW) sawing with decent yield performances has been validated on significant volume. A set of 8 bricks sampled in different positions within two Gen6 ingots was cut using this process. The target thickness for this cut was 190 μm. Wafers were optically inspected in order to extract process yield statistics described by geometrical and mechanical parameters. Table I summarizes the results obtained out of the cuts and emphasizes the remarkable achievements on CRYSTALMAX with a general yield >97%. These results underlines the compatibility of CRYSTALMAX with a sawing process that gives rise to a significant wafer cost reduction.

Table I: Average DW sawed wafer thickness and TTV inspected parameters; general yield after DW sawing.

Thickness (μm)	TTV (μm)	General yield (%)
195.2	12.6	97.1

3 EXPERIMENTAL

3.1 Ingot properties

For the present study, 650kg Gen6 p-type CRYSTALMAX ingots are considered. For the ingots growth, the chosen Cz seeds were 25 mm thick full square 156.7x156.7 mm² and <100> oriented. The first (reference) ingot was grown using EG feedstock, and is referred to as "EG" in the rest of the paper. The second ingot was grown from an EG feedstock, using twice recycled Cz-seeds and is then referred to as "R2". Boron was used as p-type dopant for a resistivity adjustment in the 1-2 Ω.cm range in EG and R2 ingots. The third ingot considered here was grown using exclusively Solar grade purified by metallurgical route feedstock (B=0.3ppmw; P=1.5ppmw) logically referred to as "SoG_M" in what follows. FTIR analyses of interstitial oxygen (O_i) and substitutional carbon (C_s) concentrations were conducted on side bricks of the three ingots. Corresponding [O_i] and [C_s] concentration variations along the ingots are plotted, respectively in Fig. 2 and Fig. 3.

Figure 3: [C_s] variation along the ingot's height for EG, R2 and SoG_M CRYSTALMAX ingots

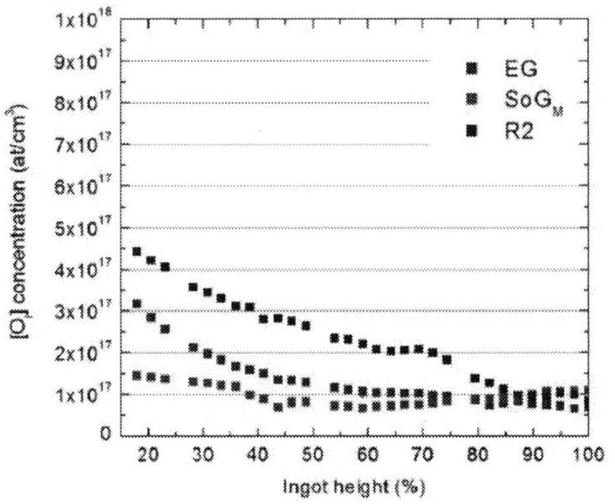

Figure 3: [O_i] variation along the ingot's height for EG, R2 and SoG_M CRYSTALMAX ingots

As a CRYSTALMAX reference, EG material features both low oxygen and carbon contents. The effect of using twice recycling Cz-seeds is marginal on carbon, but leads to an increased oxygen content, as can be observed on R2 FTIR spectra. Evidence is given here that the recycling seed procedure does not lead to any enhanced carbon contamination of the ingot. The slightly higher oxygen content of R2 is due to EG feedstock variation, not to the seed recycling. Eventually, the SoG_M ingot exhibits the lower oxygen concentration, while higher carbon content is probably coming from the degraded feedstock used in this configuration. In any case, [O_i] remains below 4.10^{17}at/cm³. After proper cropping of the bottom and top surfaces of the ingot, the resulting material yield is of 64%. The bricks were then cut into 200μm thick 243cm² full-square wafers using Diamond Wire sawing technique (as detailed in previous section).

3.2 Solar cell processing

Al-BSF solar cells were then fabricated with CRYSTALMAX p-type 243cm² wafers. After alkaline texturing of the silicon wafers, a PECVD SiO_x diffusion barrier is deposited on the rear side, prior to $POCl_3$ diffusion in conventional Lydop tube furnace, leading to an emitter sheet resistance of 70 Ω/sq. After the removal of $POCl_3$-induced phosphorus glass (on the front) and diffusion barrier (on the rear) in an HF-dip, the silicon nitride (SiN) antireflective coating (75nm thick) is deposited on the front by direct low-frequency PECVD. Eventually, rear blanket Al and front silver (Ag) grid contacts are screen-printed then co-fired. The solar cells I-V characteristics were measured under STC (25°C; AM1.5G; 1000W/m²) in a class AAA solar simulator. P-type Cz (1-3 Ω.cm, 243cm²) wafers were included as an extra reference material. The variation of the PV conversion efficiency along the ingots height is plotted in Fig.4. Considering a Cz reference providing an average efficiency of 18.6%, one should first notice that CRYSTALMAX reference ingot (EG) shows competitive performances (18.3%). In addition, evidence is given here that seed recycling does not alter CRYSTALMAX performance since cells made from R2 ingot exhibit an average efficiency (18.2%) comparable to cells made from initial EG ingot. Such an efficiency loss (0.1%abs.) makes the recycling of Cz-seeds a valid approach for cost reduction. In the case of SoG_M, cell performance is limited at first order by a lower J_{sc} that is expected to be due to a lower bulk quality (higher impurities content) of SoG_M

wafers, resulting in a lower minority carrier effective diffusion length (L_{eff}). In order to verify this assumption, L_{eff} mappings of our cells have been achieved by means of LBIC (combined with reflectance measurements) using a Semilab WT2000 tool. Two wafer positions (30% and 90% of total height) have been considered in order to determine the influence of dislocations propagating from bottom to the top of the ingot.

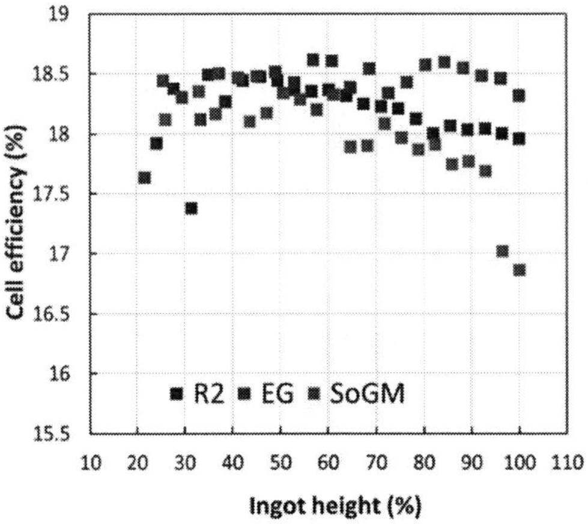

Figure 4: Al-BSF cells PV conversion efficiency along CRYSTALMAX Gen6 ingots height.

Table II: 243cm² cells average I-V parameters under STC (25°C; AM1.5G; 1000W/m²) for center bricks

Material	V_{oc} (mV)	J_{sc} (mA/cm²)	FF (%)	η (%)
Cz(1-3 Ω.cm)	635.9	37.2	78.8	18.6
C_{max}-EG	634.7	36.6	78.6	18.3
C_{max}-R2	630.0	36.9	78.1	18.2
C_{max}-SoG_M	628.0	35.8	79.2	17.8

L_{eff}>2W (W being the wafer thickness, here 180µm) is a well-known bulk quality requirement for Al-BSF devices. The L_{eff} distribution on cells made from EG and SoG_M CRYSTALMAX ingots is plotted in Fig. 5. In each case, one cell was taken at 30% of total ingot height (and another at 90%). The cells are then referred to as "bottom" and "top" respectively. For such positions, bottom and top cells feature quite similar O_i and C_s content, but the sub-grain domains size and density are higher in top cells. The cells fraction area (F.A.) meeting the demand for L_{eff}>2W was calculated from L_{eff} distribution. This fraction area, as well as the average L_{eff} values, are listed in Table II.

Bulk quality of CRYSTALMAX EG ingot is comparable to Cz in terms of average L_{eff} (~600µm) and F.A.(>77%) The use of SoG_M feedstock causes a reduction of cell average L_{eff} which drops down to 377µm, against ~549µm for EG at same wafer position. Nevertheless the average L_{eff} remain higher than 2W (=360µm) in any case. The propagation of dislocations from the bottom to the top of the CRYSTALMAX ingot has only little effect on average L_{eff} value. In terms of success cell fraction area (F.A.) top wafers exhibit a shift down of 3% (EG) to 10% (SoG_M) L_{eff} distribution of cells made from R2 ingot (not plotted here) is nearly the same as EG cell distribution.

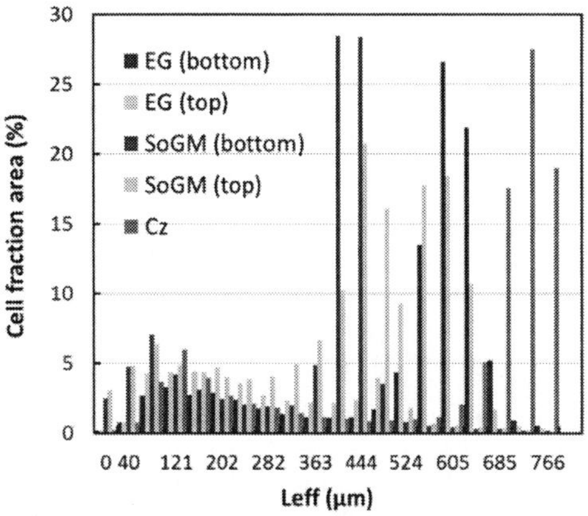

Figure 5: Minority carrier effective diffusion length distribution for Al-BSF cells made from EG (blue) and SoG_M (red) CRYSTALMAX ingots, and reference Cz (green).

Table III: Cells average L_{eff} and cell fraction area (F.A) for which L_{eff} > 2W

	Cz	EG_{bottom}	EG_{top}	$SoG_{Mbottom}$	SoG_{Mtop}
L_{eff}(µm)	612	549	513	377	367
F.A.(%)	78.5	77.3	73.0	70.3	60.5

3.3 LID study

Reference material such as Cz and HP mc-Si are both known to exhibit Light Induced Degradation (LID). The LID effects in Cz are related to the carrier-induced generation of B-O complexes, which curb wafer bulk carrier lifetime, and degrade corresponding cell V_{oc} [3, 4]. The LID effects in HP-multi are rather attributed to the presence of metallic impurities such as copper [5]. CRYSTALMAX silicon behavior towards prolonged light exposure was investigated. The above-mentioned Al-BSF solar cells were used for the LID study. Again, bottom (30%) and top (90%) positions are both considered. The cells submitted first a prolonged (here 10 days) light exposure (0.2sun) at 50°C. The cell V_{oc} was monitored during 10 days of continuous exposition to light. Cell IV measurement was done after degradation (2). Both V_{oc} and efficiency losses observed on CRYSTALMAX are lower than what is monitored on our Cz reference.

Table IV: Relative losses in efficiency (Δη%) after LID

Material	Bottom (30%)	Top (90%)
C_{max}-EG	-0.8%	-2.0%
C_{max}-R2	None	-1.7%
C_{max}-SoG_M	-0.5%	-0.9%
Cz (1-3 Ω.cm)	-2.6%	

This result is in agreement with the significantly lower O_i content exhibited by CRYSTALMAX, and also with literature [6]. The LID effects is systematically higher on wafers coming from the upper part of the ingot despite the O_i content is lower in this region. One could think other LID mechanisms involving metallic impurities come into play here. No accentuated LID could be observed on R2 ingot compared to referent EG. SoG_M exhibit slightly lower relative losses compared to EG. The relative efficiency losses remain ≤ 2% (the current industrial specification on Al-BSF architecture) on any CRYSTALMAX material studied, for both bottom and top located wafers.

3.4 Highly-efficient p-type and n-type devices

One of the key drivers for the deployment of CRYSTALMAX is to show its compatibility with advanced solar cells technologies. The potential of CRYSTALMAX material to reach high efficiency was therefore demonstrated by processing p-PERC and n-PERT solar cells. In this way one of the differentiating advantages of CRYSTALMAX technology to produce high quality boron-doped and phosphorus-doped ingots was highlighted. In each case, 200 wafers were regularly picked along the soldified silicon height in order to test a representative sample of respectively a center and an edge brick. Solar cells were fabricated in two different production lines through industrial processes running either monocrystalline Cz or multicrystalline silicon material. Average electrical performances reached with those devices are listed in Table V. Both architectures show remarkable efficiency >20% emphasizing the potential of CRYSTALMAX to be processed on future leading technologies.

Table V: 243cm² cells average I-V parameters under STC (25°C; AM1.5G; 1000W/m²) for p-PERC and n-PERT solar cells.

Architecture	V_{oc} (mV)	J_{cc} (mA/cm²)	FF (%)	η (%)
p-PERC	660.1	39.3	78.8	20.4
n-PERT	645.6	40.4	77.8	20.3

CONCLUSION

In this paper was introduced a promising / DW sawing compatible silicon material, obtained from seed-assisted cast solidification, and providing fully monocrystalline full square wafers. The main technical characteristics of CRYSTALMAX silicon ingots were presented, as well as the related paths of cost reduction. Three p-type CRYSTALMAX ingots have been studied: a reference grown from EG feedstock; a child ingot grown using same feedstock but twice re-used Cz-seeds; a third ingot using SoG_M feedstock. For each ingot, the carbon and oxygen contents/profiles were presented. Solar PV devices (Al-BSF architecture) were fabricated using 6 inch wafers from different positions along these ingots, providing average efficiency up to 18.3% (against 18.6% on Cz(p) reference). It is worth noticing the cell efficiency was not altered by the use of twice recycled Cz-seeds. Material bulk quality (after the cell fabrication) was estimated in terms of minority carrier effective diffusion length. C_{MAX}-SoG_M exhibits L_{eff} > 360μm, compatible with Al-BSF structure. C_{MAX}-EG features L_{eff} > 500μm which opens the door to higher efficiency cell architecture such as PERC. Again, the L_{eff} is not altered by the use of recycled seeds. Eventually, CRYSTALMAX stability towards prolonged illumination was evaluated by means of solar cell ageing tests. In any case, the devices exhibit reduced low light-induced efficiency losses. It should be interesting to carry out a similar study on CRYSTALMAX solar devices involving PERC architecture, which is known to feature accentuated light-induced degradation.

REFERENCES

[1] X.Zhang et al., "Characteristics and value enhancement of cast silicon ingots", Solar Energy Materials & Solar Cells 139 (2015) 27 -33.

[2] E.Pihan et al., 6th International Workshop on Crystalline Silicon for Solar Cells, Aix les bains, 10 October 2012.

[3] J. Schmidt, A. G. Aberle, and R. Hezel, in Proceedings 26th IEEE PVSC, Anaheim, CA, USA, September 30-October 3 (1997), pp. 13–18.

[4] S. W. Glunz, S. Rein, W. Warta, J. Knobloch, and W. Wettling, Sol. Energy Mater. Sol. Cells 65, 219–229 (2001).

[5] J. Lindroos, H. Savin, Sol. Energy Mater. Sol. Cells 147, 115–126 (2016).

[6] T. Dullweber, et al. "Silicon Wafer Material Options for Highly Efficient p-Type PERC Solar Cells", 2013 IEEE 39th Photovoltaic Specialists Conference (PVSC)

OPTIMIZED GRAIN SIZE OF SEED PLATES FOR HIGH PERFORMANCE MULTICRYSTALLINE SILICON

Patricia Krenckel, Stephan Riepe, Florian Schindler, Theresa Strauch
Fraunhofer Institute for Solar Energy Systems ISE
Heidenhofstrasse 2, D-79110 Freiburg, Germany

ABSTRACT: Two silicon ingots with different seed plate configurations were crystallized by directional solidification. Seed plates with high performance multicrystalline structure varying in their mean grain area were used as well as seed plates with comparable mean grain area varying in their grain structure (standard and high performance multicrystalline). All grown materials show an increase of grain size and saturation at different levels with ingot height. Additionally, a dependence of the growth behavior on the mean grain area was observed implying a faster coarsening for smaller grains. The material grown on the standard multicrystalline seed plate shows a larger area fraction limited by dislocations than the material with high performance multicrystalline seed plates, showing less dislocation structures on samples with smaller grains. The minority carrier lifetimes measured on passivated samples after a boron diffusion, as used in high efficiency solar cell processes for n-type silicon, showed a benefit for the material grown on a high performance multicrystalline seed plate with smaller grains compared to the other materials. The harmonic mean lifetime values are even higher than corresponding ones for high performance multicrystalline material nucleated on granular beads. The same trend was seen in ELBA, a combination of injection-dependent lifetime measurements on processed samples with PC1D simulations, which estimates material-related efficiency losses of only 0.4 % compared with the solar cell limit of an n-type solar cell structure on material nucleated on the fine grained seed plate with high performance multicrystalline structure.

Keywords: Crystallization, Seeded growth, Defects, Multicrystalline Silicon

1 INTRODUCTION

Currently, the solar cell market is dominated by the so-called high performance multicrystalline silicon (HPM), which holds a market share of about 40 % in 2015 and will replace the share of standard multicrystalline silicon completely [1]. The homogeneous and randomly oriented grain structure of HPM silicon reduces the evolution of dislocation clusters due to the overgrowth of dislocated grains [2]. One method for the production of HPM silicon is the use of seed materials, especially granular silicon beads [3], another one is the nucleation at the crucible coating. Both lead to the typically uniform grain size distribution at the ingot bottom.

HPM silicon as established in the industry reaches average solar cell efficiencies in the range of 19 %. The low dislocation densities result from the homogeneous, fine grain structure [4]. Due to the fact that dislocations are no longer the efficiency limiting factor in most parts of these ingots, the fraction of grain boundaries should be reduced to increase the material quality for higher solar cell efficiencies. However, an increase of the mean grain area even with sustained homogeneity of the grain structure could lead to higher dislocation densities and clusters, which would again limit the efficiency. Thus, the understanding of the trade-off between the amount of grain boundaries and dislocations is essential to find an optimum grain structure for high efficiency solar cells.

Previous experiments have shown that the same low dislocation density can be obtained for the use of small seed crystals as well as for the use of seed plates with the typical HPM grain structure, which allow a grain size variation [5]. For HPM silicon ingots produced with seeded growth on small sized silicon material, the properties of the seed material determine the initial grain structure and sizes of the crystallites at the bottom of the ingot. A variation of the mean grain area or their distributionin the initial grain structure of the ingot is possible by the use of seed plates with a pre-defined grain structure. The comparison of two different seed materials in one crystallization experiment generates wafers of comparable impurity concentrations, only varying in their grain structure. This procedure allows an evaluation of the optimum seed configuration for highly efficient solar cells.

This work aims for the development of HPM silicon ingots with continuously low dislocation densities and optimized grain structure based on the use of seed plates. Two ingots with different kinds of seed plates, varying in structure and mean grain area, were crystallized. Calculating solar cell efficiency potentials via ELBA analyses [6; 7], an optimized seed plate configuration is estimated.

2 EXPERIMENTAL PROCEDURE

2.1 Seed configuration

The seed materials for the ingots described in the following section were different silicon plates with a thickness of 30 mm. The seed plates were cut from previously crystallized ingots at Fraunhofer ISE. For each of the two crystallized ingots, four different seed configurations were used but only two of them are discussed in this work.

The first ingot was crystallized with two seed plates with comparable mean grain area varying in their grain distribution and homogeneity. One seed plate was taken from a previously crystallized HPM ingot. The second seed plate features a standard multicrystalline structure with the same mean grain area. These parts are labeled "HPM medium" and "mc medium" in the following.

For the second ingot, two seed plates from different brick heights from the HPM grown ingot used for the first ingot in this work were used. Due to their different height positions in the previously crystallized ingot, they are comparable in structure but vary in grain size. One finely grained seed plate was cut from the bottom part, and one larger grained from the top part. The crystallized parts from these seeds are labeled "HPM small" and "HPM large" in the following.

2.2 Directional solidification

Two n-type multicrystalline silicon ingots of G2 size, equivalent to 75 kg silicon and an ingot height of 220 mm, have been crystallized by directional solidification. For both ingots, pre-coated industrial quartz crucibles were used. The seed arrangement explained in the previous section was placed at the crucible bottom and covered with high purity silicon feedstock. Both ingots were charged with highly phosphorous doped monocrystalline silicon chunks to achieve a resistivity of approximately 1 Ωcm. For both crystallizations, the same thermal process including the heating up, melting from sides and top, crystallization from bottom to top and cooling down was used. For the main growth process after the nucleation phase, a growth velocity of 8 mm/h was achieved.

2.3. Material analysis

A centered brick of the size of 156 x 156 mm^2 was cut from each ingot. After polishing of the brick sides and a bottom and top cut off, the bricks were wafered with a wafer thickness of 210 µm. Automated photoluminescence measurements were performed on all "as cut" wafers. The grain structure was depicted by optical analyses of the wafer surface. Image analysis provides different structural parameters as mean grain area, weighted percentiles of grain areas, cumulative length of all grain boundaries, and the aspect ratio of the grain shape [8].

Selected wafers from four different heights of the second ingot were reduced to 125 x 125 mm^2 for further processing. From each height, one wafer was boron diffused as used for emitter formation in an n-type high efficiency solar cell process [9]. For the preparation of lifetime samples, the diffused layers were etched back and all selected wafers received a silicon nitride passivation layer on both sides. These wafers were used for spatially resolved lifetime measurements with a calibration via quasi-steady-state photoluminescence [10]. The minority carrier lifetime was harmonically averaged on specific areas. Further analysis of the efficiency limiting bulk recombination, so called ELBA, was applied [6; 7]. The combination of injection-dependent lifetime measurements on these processed samples with PC1D simulations of the TOPCon cell concept gives indications of the efficiency potentials of solar cells on this material [11]. The parameters used in the PC1D simulations are taken from [9] (cell limit without bulk recombination and with isotextured boron-diffused front side: 20.9 %).

All shown material analyses were evaluated for the wafer parts corresponding to the different seed configurations separately.

3 RESULTS

3.1 Grain structure

As a first and most common parameter, the mean grain area was calculated for the different seed configurations over ingot height, see Figure 1. All examined configurations show a growing mean grain area over ingot height, so called coarsening of grains. The seed plate with the smallest grain size, HPM small, starts at smaller grain sizes, both other HPM seeds at a comparable grain size. At an ingot height between 130 mm and 150 mm, saturation of the mean grain size value is visible for all materials.

As an alternative parameter, the weighted median indicating a grain size at which all smaller grains account for one half of the wafer area and all larger grains for the other half is less affected by small grains [8]. For the HPM seeded material, both parameters show the same trend, which illustrates the homogeneity of the grain structure. Only the multicrystalline seed leads to a larger variation in grain size distribution.

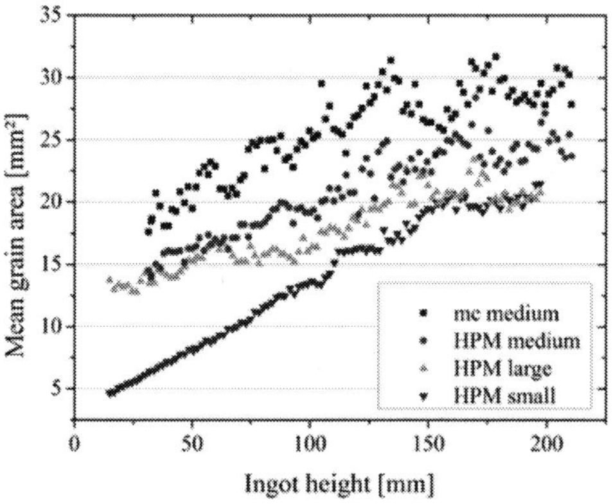

Figure 1: Mean grain area of all different seed configurations over ingot height.

The consideration of the height profiles of the percentiles 30 and 70, indicating the grain size at which all smaller grains account for 30 % respectively 70 % of the wafer area, depicts the coarsening of grains up to an ingot height of approx. 150 mm, see Figure 2. Starting from that ingot height, the grain structure of both materials seems to be very similar for the upper part. The other percentiles show the same behavior over ingot height. Additionally, a dependence of the coarsening on the mean grain size is visible. The smallest HPM seed plate induces a stronger coarsening at the beginning than all other seed plates, which show a comparable coarsening.

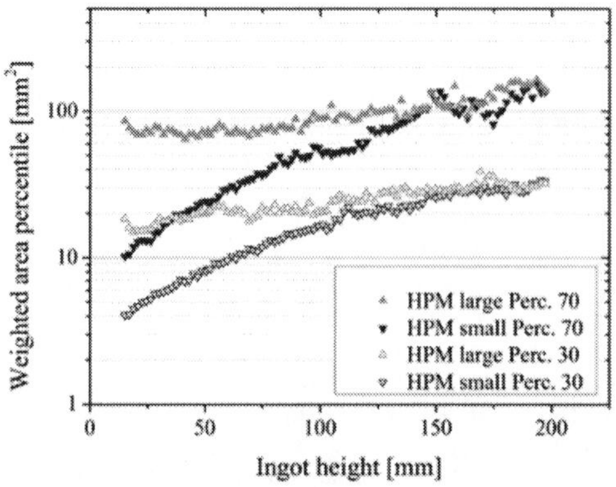

Figure 2: Weighted area percentiles 30 and 70 for the seed configurations HPM small and HPM large.

The cumulated grain boundary length normed on the considered wafer area decreases with ingot height correlating with increasing grain sizes. The smallest seed configuration leads to the fastest coarsening and thus the strongest reduction of grain boundaries. The HPM seeds show a comparable amount of grain boundaries at the ingot top, whereas the mc seed has a constantly lower grain boundary length due to some very large grains.

Besides the size of grains, the shape is an important parameter due to its correlation with grain boundaries. The aspect ratio is calculated by the ratio of minor and major axis of an ellipse fitted on each grain, giving a value between zero and one. The mean aspect ratio for the HPM seeds decreases with ingot height whereas it increases for the mc seed, see Figure 3. HPM material starts with homogeneous, round grains enlarging and narrowing to ingot top, whereas grains in mc material are inhomogeneous and irregular. For grains with an aspect ratio smaller than 0.25 a good correlation with twin grains is given [8]. With this optical identification, a higher share of twinned grains for the mc seed (more than 30 % of the relevant wafer area) than HPM seeds (between 10 % and 20 % of the relevant wafer area), is found. The increase of the mean aspect ratio of the mc seed for the second half of the ingot is correlated to a decrease in the area share of twinned grains.

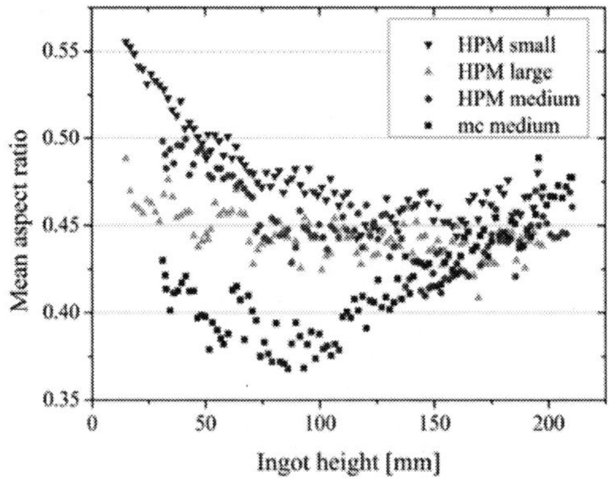

Figure 3: Height profile of the mean aspect ratio for all seed plates.

3.2 PL images and dislocations

The spatial resolution of recombination active structures such as grain boundaries and dislocations can be seen in photoluminescence (PL) images indicated by darker regions. The main percentage of the recombination in the ingot middle can be ascribed to grain boundaries for all different seed configurations, see Figure 4. Of course, the larger the grains are the smaller is the area share of the grain boundaries and thus the recombination induced by them. The area fraction of dislocated material is the largest for the mc medium seed. It is comparably low for all investigated HPM seeds, even though it seems a bit smaller for the investigated wafers from larger HPM seeds.

Considering the evolution over the whole ingot height, the area affected by dislocations is higher for the ingot bottom and top. Dislocations outlast longer and multiply very fast in large grains in the mc seed region. In comparison, the area affected by dislocations is smaller for bottom and top for HPM seed plates. In the bottom and top region, more dislocations are visible in both HPM samples with larger grains, HPM large and HPM medium.

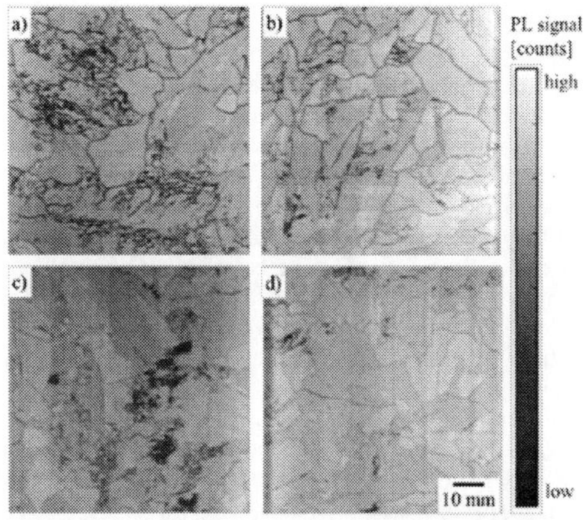

Figure 4: PL images of relevant wafer areas with a) mc medium, b) HPM medium, c) HPM large, and d) HPM small at 70 % ingot height.

3.2 Lifetime and efficiency potential

To estimate the influence of the seed configurations on the material quality, minority charge carrier lifetime images at passivated wafer areas from the material with HPM large and HPM small seed plates were measured, see Figure 5a) and b). A higher square root harmonic mean lifetime of 330 µs is observable in HPM small compared to 210 µs in HPM large. This is due to the visible dislocation structures in the material with larger grains,

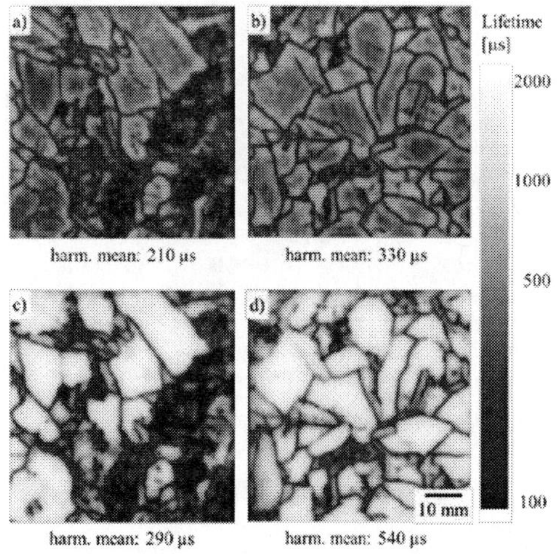

Figure 5: Minority charge carrier lifetime mappings at an illumination of 0.05 suns of relevant wafer areas at 70 % ingot height and corresponding harmonic mean values. Initial lifetimes at passivated samples of a) HPM large and b) HPM small and lifetimes at samples after boron diffusion of c) HPM large and d) HPM small. Note the logarithmic scaling.

As one process step towards the final solar cell, a boron diffusion has been applied. This process enables a much higher lifetime in the grains compared with the as initial, of up to 2 ms for some larger grains, see Figure 5c) and d). Recombination active areas, grain boundaries and dislocations, are still visible. The harmonic mean lifetime was increased for both samples. The HPM large material reaches a mean lifetime of 290 µs, the HPM small material performs even better with 540 µs.

Based on injection-dependent lifetime images in combination with PC1D cell simulations, the efficiency potential for these materials combined with a high efficiency cell concept for n-type silicon was calculated via ELBA analyses. The used solar cell model features an efficiency limit without bulk recombination of 20.9 % for samples with an isotextured front surface [9]. The material nucleated on the finely grained HPM seed plates features higher efficiency potentials than the material nucleated on the larger grained HPM seed plate for all considered ingot heights, see Figure 6. The efficiency potential increases even for the highest ingot position, whereas the efficiency potential of the sample on HPM large stays constant.

In detail the material nucleated on the larger grained HPM seed features a global efficiency potential of 20.2 %, whereas the material nucleated on the finely grained HPM seed performs a global efficiency potential of 20.5 % on the relevant wafer area at 70 % ingot height. This means that only 0.4 % of the absolute efficiency is lost due to material-related recombination for the applied solar cell model with a cell limit of 20.9 %. The HPM small material is even slightly better than material nucleated on granular beads in a completely comparable process, providing an efficiency potential of 20.4 % at 70 % ingot height.

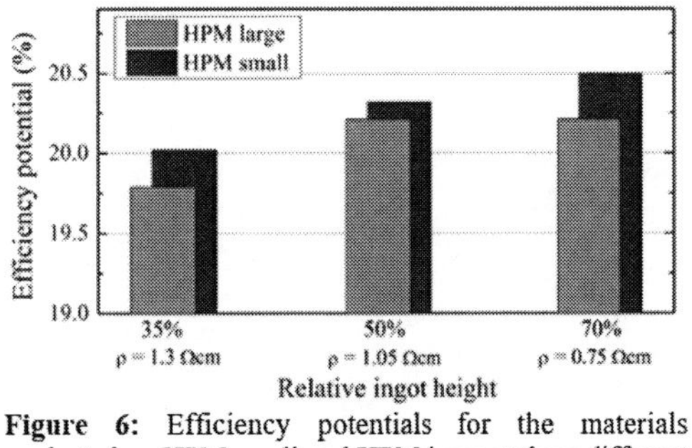

Figure 6: Efficiency potentials for the materials nucleated on HPM small and HPM large at three different ingot heights. The resistivities ρ at the considered ingot heights are noted below.

4 DISCUSSION

The saturation of the mean grain area as well as the described percentiles at the same ingot height for all HPM seeds as shown in Figures 1 and 2 suggest a dependence of the coarsening behavior on crystallization conditions, which were the same for both ingots. The growth velocity can be excluded as a possible reason because previous experiments with the same thermal conditions have shown a fairly constant growth velocity for the whole crystallization excepting bottom and top region. Furthermore, spatially resolved resistivity measurements at the brick side, not shown here, exhibit a flat and homogeneous form of the phase boundary at this height, which is relatively stable over a large part of the ingot height. This cannot explain the saturation of the grain coarsening. One possibility is that this effect is owed to the relative ingot height; i.e. the height of melt in front of the phase boundary influencing thermal conditions. Yet none of our previous experiments on G1 sized ingots have shown such saturation. This should be investigated in further experiments. A more probable factor is the crystallized ingot height, which would account for the different behavior for G1 sized ingots. The existence of an energetic minimum of the grain structure could lead to saturation in grain coarsening. Further experiments with higher ingots could cast more light on this relation.

The area share of twins increases for the HPM parts with ingot height and reaches a constant level. This trend can be explained by a growth model which proposes a grain elongation with ongoing growth [12]. Samples grown on the finer grained seed material, HPM small, start with a lower share of twins and end up at the same area share like the other samples, even the multicrystalline seed material, which is similar to results in the literature [8]. The grain structure of all different seed configurations develops to an area share of twins of between 15 % and 20 % at ingot top, which matches with reported experiments as well [3; 8].

The trend that smaller grains reduce dislocation structures faster and more efficient in the growth process, as illustrated by PL images, could be seen in the minority charge carrier lifetime images, as well. Even though the investigated material with larger grains is less limited by grain boundaries, the higher amount of dislocation structures reduces the harmonic mean lifetime in the initial state. The boron diffusion used for a high efficiency solar cell process intensifies this tendency. The gettering effect enhances the bulk lifetime inside the grains. Impurities concentrated at structural defects, grain boundaries and dislocations, are released into the bulk material due to the high temperature step and only partly gettered during the boron diffusion and in the subsequent cooling. Hence, regions with high initial lifetimes improve and regions with lower lifetimes deteriorate.

The material with smaller grains outperforms the larger grained material at all considered ingot heights. The improvement of the efficiency potentials from 35 % ingot height to 50 % ingot height is visible for both materials, HPM large and HPM small, and can be explained by the increase of the mean grain size, which reduces the area fraction of grain boundaries and thus recombination. Another factor for the increase of the efficiency potential is the decreasing resistivity as a lower resistivity has a positive effect on the efficiency potential for this cell concept in the range of the investigated lifetimes. The difference in the efficiency potential between the material nucleated on HPM small and HPM large can be entirely attributed to the crystal structure characteristics. This effect is most dominant for the highest ingot position, where the mean grain area of both seed configurations is comparable but the area fraction of dislocation structures is different. Hence, the benefit of the material nucleated on the HPM small seed is due to a lower dislocation density based on a different grain development showing that the dislocation density does not depend on the mean grain area only but on the evolution of the grain structure during crystallization.

At 70 % ingot height, he efficiency losses due to the material quality for the finely grained HPM seed plate are very low with 0.4 $\%_{abs}$. Therefore, the seed plate with smaller grains seems to be the appropriate seed material for high efficiency solar cells.

Compared to previously crystallized HPM silicon with the same thermal conditions nucleated on granular silicon beads, the HPM small shows higher lifetimes as well [5]. Especially for the essential high temperature step, the wafer from the granular seed process features lifetimes of 350 µs harmonically averaged on the relevant wafer area at illumination of 0.05 suns compared to the wafer from the HPM small process with 540 µs. The use of seed plates with small HPM grain structure seems to be beneficial for higher efficiencies featuring a global efficiency potential of 20.5 %. There are two possible reasons for this effect. On the one side, the different geometry of the seed material, i.e. the difference between a completely solid seed plate and a loose granular structure, might have an influence. On the other side, the mean grain area of the investigated HPM small seed plate might be the optimum trade-off between larger grains to reduce recombination due to grain boundaries and grains small enough to guarantee a constantly low dislocation density to reduce recombination due to dislocations. This correlation was not further examined in this work and would require further experiments with seed plates.

5 CONCLUSION

A dependence of the coarsening of grains on the ingot height, probably determined by the already crystallized ingot height, was observed with saturation at about 150 mm ingot height. The mean grain area seems to affect the grain coarsening as well, showing a faster coarsening for smaller grain structures. The development of recombination active dislocation structures depends on this grain size and its distribution. The material nucleated on seed plates with HPM structure shows smaller areas affected by dislocations than that on the mc seed plate, as reported as benefit of HPM materials.

The HPM seed plate with the smallest initial grain size features the highest lifetimes in the initial and boron diffused state as well as the highest efficiency potential of 20.5 %. Lifetimes and efficiency potential are slightly higher compared to similarly crystallized HPM material with granular seeds.

6 ACKNOWLEDGEMENTS

The authors would like to thank Wacker Polysilicon for the silicon materials and express their gratitude to the crystallization, cell processing and characterization teams at ISE for their support and input in many valuable discussions. This work was funded by the German Federal Ministry for Economic Affairs and Energy within the research projects "THESSO" (grant number 0325491) and "multiTOP" (grant number 0324034).

7 REFERENCES

[1] A. Metz, et al., "International Technology Roadmap for Photovoltaic (ITRPV)", July 2015.

[2] G. Stokkan, Y. Hu, Ø. Mjøs, and M. Juel, "Study of evolution of dislocation clusters in high performance multicrystalline silicon", *Solar Energy Materials and Solar Cells* **130** (2014) 679.

[3] Y. M. Yang, A. Yu, B. Hsu, W. C. Hsu, A. Yang, and C. W. Lan, "Development of high-performance multicrystalline silicon for photovoltaic industry", *Prog. Photovolt: Res. Appl.* **23** (2015) 340.

[4] D. Zhu, L. Ming, M. Huang, Z. Zhang, and X. Huang, "Seed-assisted growth of high-quality multi-crystalline silicon in directional solidification", *Journal of Crystal Growth* **386** (2014) 52.

[5] S. Riepe, P. Krenckel, F. Schindler, and C. Schmid, "Development of multicrystalline silicon for 20 % efficient n-type solar cells", *31st European Photovoltaic Solar Energy Conference and Exhibition (*2015).

[6] B. Michl, M. Kasemann, W. Warta, and M. C. Schubert, "Wafer thickness optimization for silicon solar cells of heterogeneous material quality", *Phys. Status Solidi RRL* **7** (2013) 955.

[7] B. Michl, M. Rüdiger, J. A. Giesecke, M. Hermle, W. Warta, and M. C. Schubert, "Efficiency limiting bulk recombination in multicrystalline silicon solar cells", *Solar Energy Materials and Solar Cells* **98** (2012) 441.

[8] T. Strauch, M. Demant, P. Krenckel, S. Riepe, and S. Rein, "Key Parameters for Brick Characterization of Grain Structure via Image Processing based on Optical Measurements of as-cut Si Wafers", *43rd IEEE Photovoltaic Specialist Conference (*2016).

[9] F. Schindler, et al., "High-Efficiency Multicrystalline Silicon Solar Cells: Potential of n-Type Doping", *IEEE J. Photovoltaics* **5** (2015) 1571.

[10] J. A. Giesecke, M. C. Schubert, B. Michl, F. Schindler, and W. Warta, "Minority carrier lifetime imaging of silicon wafers calibrated by quasi-steady-state photoluminescence", *Solar Energy Materials and Solar Cells* **95** (2011) 1011.

[11] F. Feldmann, M. Bivour, C. Reichel, H. Steinkemper, M. Hermle, and S. W. Glunz, "Tunnel oxide passivated contacts as an alternative to partial rear contacts", *Solar Energy Materials and Solar Cells* **131** (2014) 46.

[12] R. R. Prakash, T. Sekiguchi, K. Jiptner, Y. Miyamura, J. Chen, H. Harada, and K. Kakimoto, "Grain growth of cast-multicrystalline silicon grown from small randomly oriented seed crystal", *Journal of Crystal Growth* **401** (2014)

THERMAL DECOMPOSITION OF MONOSILANE OBSERVED IN A FREE SPACE REACTOR

Guro Marie Wyller, Thomas Preston, Trygve T. Mongstad, Hallgeir Klette, Ørnulf Nordseth, Dag Lindholm, Werner O. Filtvedt, Erik S. Marstein
Institute for Energy Technology (IFE)
P.O. Box 40, NO-2027 Kjeller

ABSTRACT: Thermal decomposition of monosilane (SiH_4) is an important process step in the production of poly-Si in fluidized bed reactors (FBRs) and centrifugal chemical vapor deposition reactors (CCVDRs). These two reactor technologies both offer energy-saving alternatives to the Siemens reactor, which is the most commonly used reactor technology in today's market. We have studied thermal decomposition of SiH_4 diluted in hydrogen (H_2) in a free-space reactor (FSR) with varied inlet concentrations of SiH_4 and varied reactor temperature. The reactor exhaust was studied with an advanced combination of Gas-Chromatography and Mass-Spectrometry (GC-MS) allowing for detailed chemical analysis. For an inlet concentration of 10 vol. % SiH_4 in H_2, we observe a peak in the outlet concentration of higher order silanes as function of reactor temperature between 500 and 550 °C. When the inlet concentration of SiH_4 is lowered to 1 vol. %, the peak is found between 600 and 650 °C. Our results, which are in line with earlier reported observations, are relevant for the further optimization of industrial poly-Si production reactors utilizing SiH_4.

Keywords: Silane, Silicon purification, GC-MS

1 INTRODUCTION

Today's most frequently used production path for polysilicon and solar grade silicon (SG-Si) relies on thermal decomposition of trichlorosilane ($SiHCl_3$, "TCS") in a Siemens-type reactor. This reactor, which is operated batch-wise, requires heating of silicon (Si) rods on which decomposition of TCS is designed to occur. The reactor wall, on the other hand, must be cooled in order to hinder TCS from decomposing on the wall surface. During reactor operation, only about one fourth of the TCS gas introduced to the reactor forms solid silicon [1]; the rest undergoes a competing reaction forming silicon tetrachloride ($SiCl_4$). This byproduct is recycled to form new TCS that is subsequently re-introduced to the reactor. Nevertheless, the combination of the low chemical yield with batch-wise operation and high demands for heating and cooling makes the Siemens process relatively expensive, both in terms of cash and energy.

Fluidized bed reactors (FBRs) (see Figure 1) are one of the most promising alternatives to the Siemens reactor [2], [3]. In these reactors monosilane (SiH_4) is commonly used as the source of Si atoms. In addition to SiH_4, hydrogen or nitrogen gas is injected at high velocity from the reactor bottom. This high velocity gas creates a turbulent zone and fluidizes the Si particles (seed particles) that are injected from the reactor top. A particle is said to be fluidized when the drag force created by the gas flow is at same order of magnitude as the weight of the particle [3], meaning that the particle will "float" in the reactor.

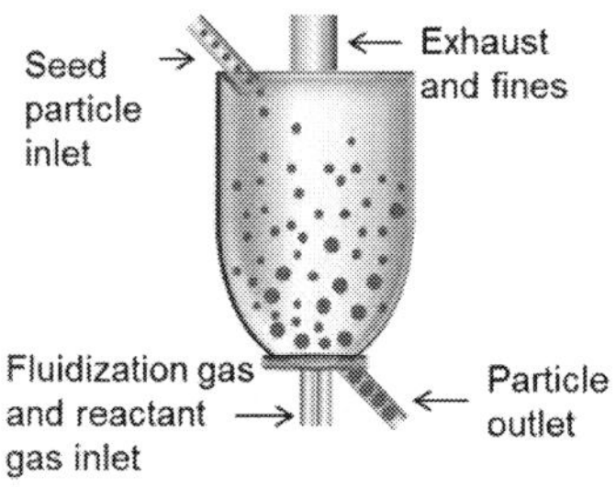

Figure 1: Fluidized bed reactor.

When the seed particles are heated to a temperature above the decomposition temperature of SiH_4, the SiH_4 gas will decompose. The decomposition is desired to take place heterogeneously at the gas-solid interface, meaning that Si will deposit on the seed particles, causing them to grow. When the weight of the particles exceeds the drag force from the fluidizing gas, i.e. when they have reached a suitable size, they will sink to the reactor bottom from where they can be harvested.

Homogeneous decomposition of gaseous SiH_4 is a competing process to the desired heterogeneous decomposition. When homogeneous decomposition takes place, fine Si dust (fines) is created. The production of this fine dust can be a major challenge for FBR operation. The fines are transported out with the off-gas from the process and cannot be sold as high quality polysilicon. In addition, the fines can lead clogging of piping valves and filters.

Recently, the centrifugal chemical vapor deposition reactor (CCVDR, see Figure 2) was introduced as a second SiH_4-based alternative to the Siemens process. This method utilizes centrifugal acceleration to separate Si from H_2 upon thermal decomposition of SiH_4, and has proven the ability to produce polysilicon at low cost and with very low dust production [4]. There is, however, still room for improvement of the process by improving our fundamental understanding of the thermal decomposition process of SiH_4.

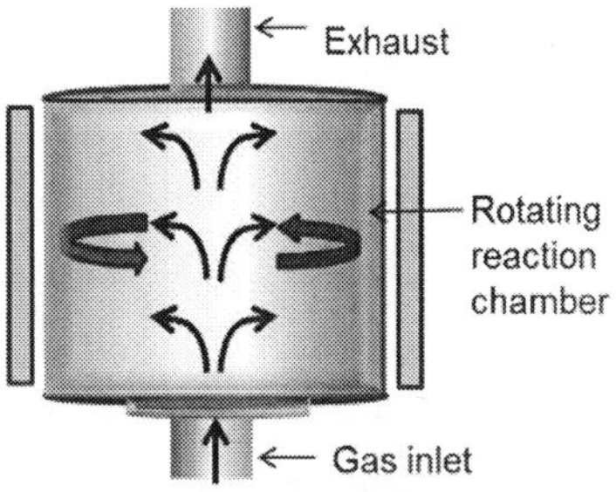

Figure 2: Centrifugal chemical vapor deposition reactor.

Highly simplified, the thermal decomposition of SiH_4 can be expressed as the SiH_4 molecule splitting off two H_2 molecules to produce solid Si according to the following overall formula:

$$SiH_4(g) \rightarrow 2H_2(g) + Si\,(s). \qquad [1]$$

In reality, however, the reaction proceeds through a complicated kinetic reaction mechanism, that is yet to be understood in detail. There seems to be broad agreement ([5]–[9]) that the first step within the mechanism of homogeneous decomposition of SiH_4 is

$$M + SiH_4 \leftrightarrows M' + H_2 + SiH_2. \qquad [2]$$

M here represents a third body that may transfer energy into and out of the system. M and M' differ, in other words, only in energy content, not identity.

Further silane chain growth requires a balance between hydrogen elimination from silanes to form silyl radicals, and new Si-Si bond formation by reaction of silyl radicals with silane (e.g. [8], [10], [11]). As soon as the number n of Si atoms in a molecule exceeds 3, branching and ring formation give rise to a large number of isomeric structures. Moreover, the elimination of H_2 to form silenes (species containing double bonding) and silylenes (divalent radical species) further increases the number of reactions and chemical species that must be considered in a reaction network [11].

An extensive amount of work has been done in order to survey the decomposition process, both by experiments (*e.g.* [7], [12]–[14]) and by modelling (*e.g.* [9], [11], [15], [16]). Nevertheless, there seem to be discrepancies among the suggested models as well as poor agreement between models and experimental work.

2 EXPERIMENTAL APPROACH

At IFE (Institute for Energy Technology), a suite of characterization tools for reactor operation, including optical exhaust monitoring for particle detection and an advanced GC-MS system (gas chromatography combined with mass spectrometry) is available [17], [18]. The laboratory infrastructure also includes a CCVDR, two free-space reactors (FSRs) and a laboratory scale FBR.

2.1 Reactor

This experiment was conducted in the FSR outlined in Figure 3, recently constructed at IFE. The horizontal reactor of stainless steel has an inner diameter of 50 mm and is 3.1 m long. The length of the heating, reaction, and cooling zone are 1.0 m, 0.1 m, and 1.0 m, respectively.

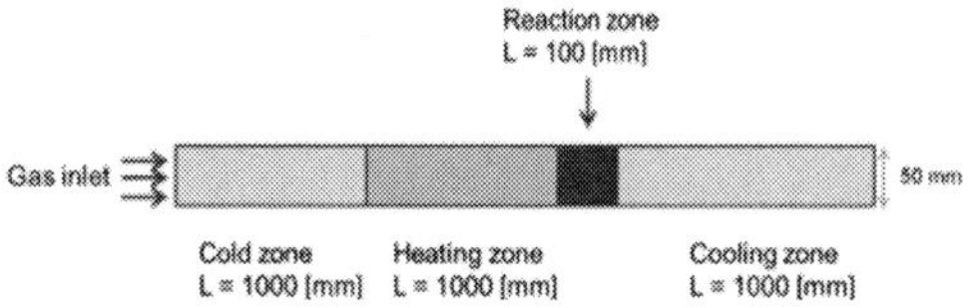

Figure 3: Simplified sketch of the free-space reactor in which the experiments were conducted.

The reactor used for this experiment differs significantly in size and shape from the FSR that has been utilized for SiH_4 decomposition experiments at IFE in the past [17]. This new, smaller reactor allows for better control of process parameters, including gas temperature and gas flow. The gas temperature is monitored by four sensors outside the reactor tube as well as four sensors placed in the center of the reactor. The flow and composition of the reactant gas is controlled and monitored by mass flow controllers.

In this experiment the preheating zone temperature was kept constant at 300 °C, whereas the reaction zone temperature was varied from 350 °C to 700 °C. Concentrations of 10 vol. % and 1 vol. % SiH_4 in H_2 were used as reagent gas in the experiment. The total gas flow was kept constant at 2 SLM (standard liter per minute), giving theoretical residence times of about 1.6 to 2.6 seconds, depending on reactor temperature. The reactor was kept at atmospheric pressure.

2.2 GC-MS (Gas Chromatography – Mass Spectrometry)

The reactor exhaust was sampled and analyzed with a tailor-made GC-MS system, outlined in Figure 4.

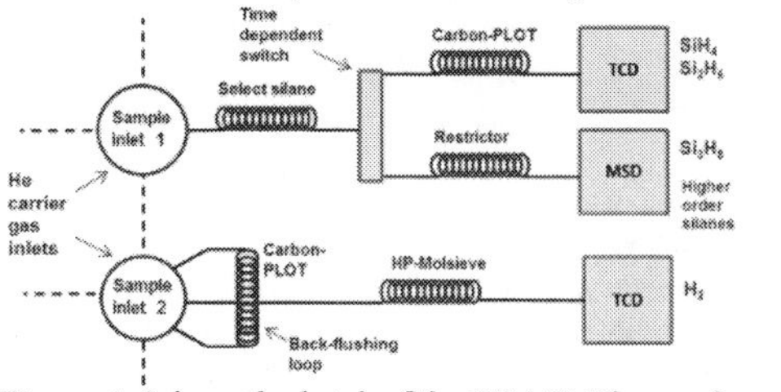

Figure 4: Schematic sketch of the GC-MS. The species that are detected in each of the detectors are indicated in red.

In the GC-MS, the separation of the analyte gas is achieved by an Agilent 7890B GC system, with injection of exhaust gas into two parallel gas streams, both using helium as carrier gas. The first stream, which is dedicated to the detection of silanes uses a Select Silanes GC column (Agilent) to separate silanes of order $n \geq 3$ according to their molecular mass and shape. SiH_4 and disilane (Si_2H_6), on the other hand, are not well separated by this column. Therefore, a time-dependent switch ensures that the compounds that elute from the column first (SiH_4 and Si_2H_6) are sent to a Carbon-PLOT (porous layer open tubular) column to be further separated before they are detected in a TCD (thermal conductivity detector). Using a TCD protects the MS from the high concentrations of mono- and disilane in our experiments.

The TCD is a simpler and more robust detector than the mass spectrometer and can more easily be replaced when degraded by silane decompositions. Compounds with three or more silicon atoms (trisilane (Si_3H_8), tetrasilane (Si_4H_{10}), etc.) are well-separated by the Select Silanes column and are sent through a restrictor column before they are ionized and detected in the MSD (Agilent 5977A Series quadrupole mass spectrum detector with 70 eV electron ionization).

The second gas stream is dedicated to separation and quantification of H_2 gas, obtained by a Carbon-PLOT column with a back flushing system, combined with a HP-Molsieve (Agilent). After being separated from the gas mixture, H_2 gas is lead to second TCD for quantification.

3 RESULTS AND DISCUSSION

3.1 Interpretation of GCMS results

Figure 5 shows an example of the data obtained from one GC-MS measurement. Each higher order silane elutes from the GC system at one specific time, thus creating a well-defined peak in the total ion chromatogram shown in blue along the left axis. The mass spectra corresponding to each peak in the total ion chromatogram are shown in green along the right axis.

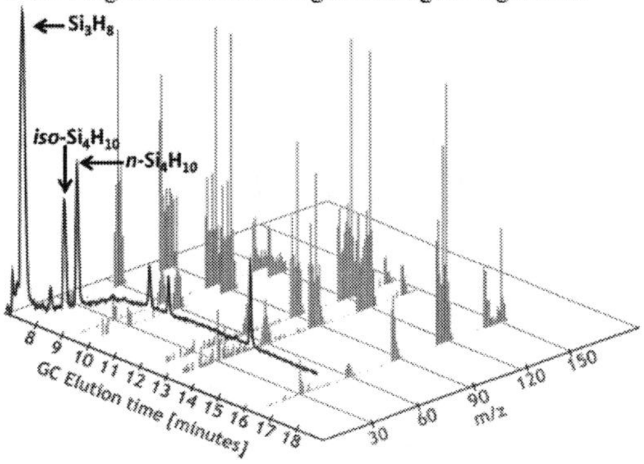

Figure 5: Visualization of the 2 dimensions (GC elution time and m/z values) of the data obtained by the GC-MS. The data along the blue axis is the total ion count, and the data along the green axis is the abundance of the fragments. Both intensities are plotted in linear arbitrary units.

To the best of our knowledge, the 70 eV electron ionization mass spectra of only the smallest silanes (Si_2H_6 and Si_3H_8) and have been published so far [19], [20]. This work therefore represents a new level of detail in mass spectrometric analysis of inorganic compounds and there are no existing entries in a database predicting the mass spectra of the higher order silanes. Because of the hard ionization (70 eV electron ionization), extensive fragmentation of the silanes occurs. The molecular ion is, in most cases, overshadowed by the fragment ions in the mass spectra. A silane with n Si atoms appears as fragments with m/z ("mass-to-charge-ratio") values corresponding to $k \leq n$ Si atoms with a number of attached H atoms ranging from 0 to $2k+2$. The latter corresponds to a fully saturated higher order silane. The observed fragmentation patterns have some similarities to that of hydrocarbons. In the case of silanes however, the fragmentation appears to be even more extensive. Matching mass spectral and chromatographic data to silane species, and especially to isomers, is therefore challenging. So far, trisilane (Si_3H_8), *normal*-tetrasilane and *iso*-tetrasilane (Si_4H_{10}) are identified (see Figure 4). Work is in progress to identify the numerous silane isomers with $n > 4$.

3.2 Concentration of higher order silanes as function of reactor temperature

Figure 6 shows normalized concentrations of SiH_4, Si_2H_6, Si_3H_8 and the two tetrasilane isomers (n-Si_4H_{10} and iso-Si_4H_{10}) for an inlet SiH_4 concentration of 10 vol. %. The normalization is done for each species individually. Their relative concentrations should therefore not be compared.

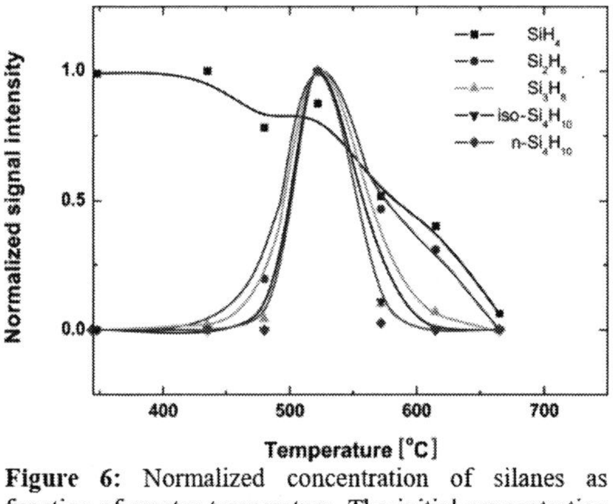

Figure 6: Normalized concentration of silanes as function of reactor temperature. The initial concentration of SiH_4 in H_2 was set to 10 vol. %. The lines are guides to the eye.

The black line clearly indicates that the concentration of SiH_4 decreases as the reactor temperature is increased. This decrease is reasonable, because increasing reactor temperature causes an increasing amount of SiH_4 to decompose and form higher order silanes or even silicon particles. Further, Figure 6 indicates that the concentrations of all species, except from SiH_4, attain a maximum value between 500 and 550 °C. The concentrations seem to quickly decrease both at higher and lower temperatures. This decrease suggests that di-, tri-, and tetrasilanes at temperatures higher than about 550 °C contribute to the development of even larger silanes or silicon particles. Therefore, the reactor is depleted for these species. These findings are in line with observations that we [17], Slootman and Parent [7], and Onischuk et al. [6] have reported earlier; all based on studies conducted in different free space reactors.

Figure 7 compares the concentrations of SiH_4 and Si_3H_8 in the reactor exhaust as function of reactor temperature, for inlet SiH_4 concentrations of 10 vol. % and 1 vol. %. Interestingly, for 1 vol. % SiH_4 inlet concentration, the peak in the Si_3H_8 concentration in the exhaust has shifted approximately 100 °C to the high temperature side, as compared to the 10 % case. The highest concentration appears between 600 and 650 °C. As in Figure 6 all higher order silanes seem to have their peak concentrations at approximately the same temperature at equal inlet concentrations; for clarity only Si_3H_8 is included in the figure.

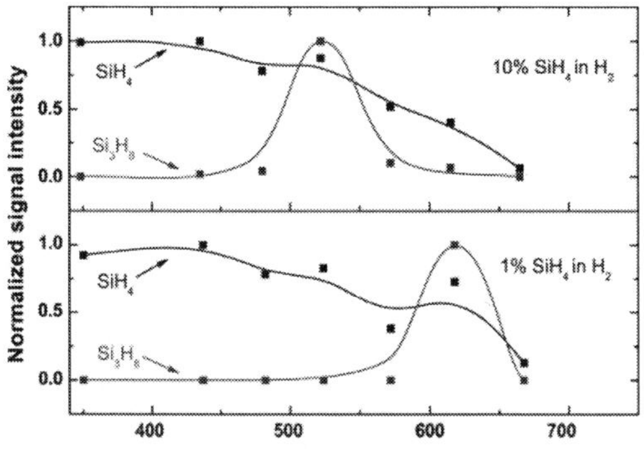

Figure 7: Comparison of normalized concentrations of SiH_4 and Si_3H_8 in reactor exhaust as function of temperature, for inlet concentrations of 10 vol. % and 1 vol. % SiH_4 in H_2. The lines are guides to the eye.

The shift of the elevated concentration to higher temperatures at lower SiH_4 inlet concentrations can be explained by the displacement of the equilibrium reaction [2], which is endothermic when proceeding from left to right, as written in Equation 2. When SiH_4 is removed from the system, heat must be added (*i.e.* the temperature must be increased) in order to initiate the first step of the thermal decomposition.

Slootman and Parent [7] and Onischuk et al. [6] both report a clear peak in the concentration of Si_2H_6 as function of reactor temperature during thermal decomposition of SiH_4. For inlet SiH_4 concentrations at the order of 1 %, they observe maximum Si_2H_6 exhaust concentrations for temperatures about 550 °C, which is less than the temperature by which our highest concentration is observed. This discrepancy leads us to hypothesize that the combined heat transfer by convection and conduction in our reactor is insufficient to heat the reagent gas to the desired temperature quickly enough to initiate the reaction within the residence time. Temperature gradients within the reactor will be further discussed in the following subsection.

It should also be noticed that the peak that we observed between 600 and 650 °C for 1 vol. % SiH_4 inlet concertation is in the concentration of Si_3H_8 (and higher order silanes), whereas the observations of Onischuk et al. and of Slootman and Parent are based on the Si_2H_6 concentration. Because the TCD with which our Si_2H_6 signal is measured is less sensitive than our MSD, we were unable to quantify the Si_2H_6 concentration in the exhaust for the case of 1 vol. % SiH_4 inlet concentration. Based on Figure 6, one would expect to find the maximum concentration of all higher order silanes for one given SiH_4 inlet concentration at a similar temperature. We have, however, no evidence that the temperatures at which the concentration of each different higher order silane attains its maximum value cannot change relative to each other as the SiH_4 inlet concentration is varied. Our temperature steps (50 °C) might be too large to detect such shifts. Future experiments will be conducted with higher temperature resolution, allowing for a more detailed understanding of the concentrations of individual higher order silanes as function of temperature.

3.3 Temperature distribution within the reaction zone and its influence on residence time

In order to obtain a more in-depth understanding of the temperature distribution within the reactor, a computational fluid dynamics model using the software (SiSim) [21] has been applied. Figure 8 shows the modelled temperature distribution in a cross-section of the reactor for a heating zone set temperature of 300 °C and a reaction zone set temperature of 650 °C. The total gas flow is 2 SLM, and the SiH_4 inlet concentration is 1 vol. %. These reactor conditions correspond to the conditions for the Si_3H_8 peak concentration in the lower part of Figure 7.

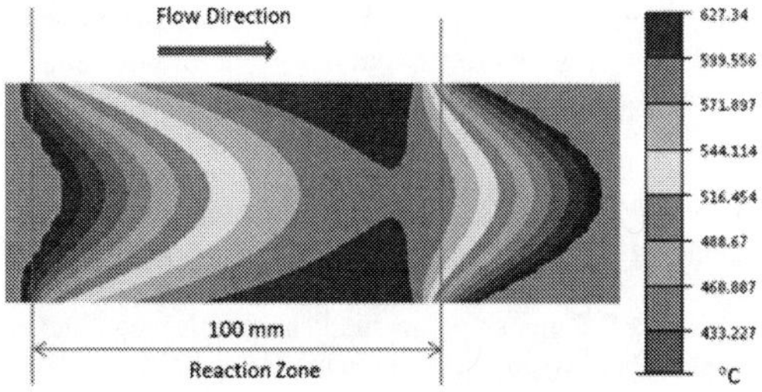

Figure 8: Modelled temperature map in reactor cross section.

The figure reveals that the gas flow through the reactor combined with limited heat transfer from the reactor wall to the reactor center causes a shift of about 50 °C from the set temperature to the maximum reactor center temperature. Our temperature measurements from the reactor core compared to those from the reactor wall show a similar trend. Figure 8 further illustrates how limited heat transfer reduces the effective length of the reaction zone and therefore the effective residence time in the high temperature region. Because of the low Reynolds number, which is equal to 5 at the inlet, the fluid flow is laminar and the heat transfer coefficient between the reactor wall and the gas low. For the modeled reactor conditions, the axial region in which the temperature in the reactor center lies within a 20 °C range from its maximum value is no more than 2.5 cm, i.e. one fourth of the reaction zone length. The effective residence time in this temperature range is therefore only one fourth of the theoretical residence time.

Based on our model, we have reason to believe that our measured reactor center temperature is in the upper range of the temperatures that the gas actually attains on its way through the reactor. There might therefore be a discrepancy between the measured temperatures (as shown in Figures 6 and 7) and the temperatures at which the observed thermal decomposition actually takes place. This discrepancy can partly explain the apparent difference between our results and the results of Slootman and Parent [7] and Onischuk et al. [6] who find the highest concentrations of higher order silanes at about 550 °C for an inlet SiH_4 concentration at the order of 1 %.

3.4 Influence of residence time on the nucleation process

Residence time is, in addition to SiH_4 inlet concentration and reactor temperature, an important parameter for the thermal decomposition of SiH_4. Ideally, in our experiment, we would like to keep our reactor operating at a constant residence time. However, as the reactor temperature is increased, thermal gas expansion will increase the volumetric flow through the reactor and thus reduce the effective residence time. It is reasonable to assume that a decreased residence time requires an increased temperature for decomposition to take place. Therefore, the inherent decrease in residence time that will take place when reactor temperature is increased will influence measurements of decomposition as function of temperature. The influence of residence time on the thermal decomposition process will be further addressed in a future publication.

3.5 Implications for reactor operation

Our findings on the exhaust concentrations of higher order silanes as functions of reactor temperature and SiH_4 inlet concentration are of relevance for the operation of silicon production reactors. In the operation of FBRs, heterogeneous decomposition of SiH_4 is desired, whereas homogeneous decomposition is unwanted because it causes dust production and reactor clogging. The decrease in the exhaust concentration of higher order silanes at high temperatures is probably a signature of higher order silane molecules bonding to each other to produce nanoparticles/dust. Operating a reactor under these conditions is therefore unfavorable, if fines formation is to be avoided. The relatively narrow temperature region for which a high concentration of higher order silanes is measured (between 500 and 550 °C for 10 vol. % SiH_4 inlet concentration and between 600 and 650 °C for 1 vol. % SiH_4 inlet concentration) is, on the other hand, likely to be a favorable temperature for reactor operation. The higher order silanes that are produced at these temperatures are "sticky" (attach easily to surfaces) and are therefore likely to contribute positively to heterogeneous decomposition and thus growth of the seed particles, as well as scavenging of generated fines. A similar result was published by Filtvedt et al. [3] reporting that too high reactor temperatures applied during FBR operation seem to promote fines production over heterogeneous growth.

4 CONCLUSION

The outlet concentration of higher order silanes in the exhaust of a free-space reactor, in which thermal decomposition of SiH_4 takes place, shows a clear increase in a narrow reactor temperature range. The reactor temperature that produces maximum higher-order silane concentrations increases with decreasing inlet SiH_4 concentration. For an inlet concentration of 10 vol. % SiH_4 in H_2, we observe a significant increase in the concentration of higher order silanes in the exhaust between 550 and 550 °C. For 1 vol. % inlet concentration, the increased concentration is found at temperatures about between 600 and 650 °C. The elevated concentrations of higher order silanes that are observed for certain reactor temperatures probably indicate that at these temperatures, higher order silanes are produced at a higher rate than they are rebuilt to form silicon particles.

ACKNOWLEDGEMENT

This work has been funded by the Research Council of Norway through "The Norwegian Research Centre for Solar Cell Technology", NFR Project No. 192839.

REFERENCES

[1] A. Luque and S. Hegedus, *Handbook of Photovoltaic Science and Engineering*. Chihcester: Wiley, 2003.

[2] W. O. Filtvedt, M. Javidi, A. Holt, M. C. Melaaen, E. Marstein, H. Tathgar, and P. a. Ramachandran, "Development of fluidized bed reactors for silicon production," *Sol. Energy Mater. Sol. Cells*, vol. 94, no. 12, pp. 1980–1995, 2010.

[3] W. O. Filtvedt, T. Mongstad, A. Holt, M. Melaaen, and H. Klette, "Production of silicon from SiH_4 in a fluidized bed, operation and results," *Int. J. Chem. React. Eng.*, vol. 11, no. 1, pp. 1–12, 2013.

[4] W. O. Filtvedt, H. Klette, S. Sørensen, and J. Filtvedt, "Low cost/high quality silicon production by centrifuge CVD reactor upscaled and hot harvest," in *EU PVSEC XXXI 2BO.1.2*, 2015.

[5] E. L. Petersen and M. W. Crofton, "Measurements of High-Temperature Silane Pyrolysis Using SiH_4 IR Emission and SiH_2 Laser Absorption," *J. Phys. Chem. A*, vol. 107, no. 50, pp. 10988–10995, 2003.

[6] A. A. Onischuk, V. P. Strunin, M. A. Ushakova, and V. N. Panfilov, "On the pathways of aerosol formation by thermal decomposition of silane," *J. Aerosol Sci.*, vol. 28, no. 2, pp. 207–222, Mar. 1997.

[7] F. Slootman and J.-C. Parent, "Homogeneous gas-phase nucleation in silane pyrolysis," *J. Aerosol Sci.*, vol. 25, no. 1, pp. 15–21, 1994.

[8] A. A. Onischuk and V. N. Panfilov, "Mechanism of thermal decomposition of silanes," *Russ. Chem. Rev.*, vol. 70, no. 4, pp. 321–332, 2001.

[9] A. A. Onischuk, A. I. Levykin, V. P. Strunin, M. A. Ushakova, R. I. Samoilova, K. K. Sabelfeld, and V. N. Panfilov, "Aerosol formation under heterogeneous/homogeneous thermal decomposition of silane: experiment and numerical modeling," *J. Aerosol Sci.*, vol. 31, no. 8, pp. 879–906, Aug. 2000.

[10] M. R. Mahajan, "Numerical Modeling of Particle Nucleation , Growth , and Transport during Atmospheric Pressure Chemical Vapor Deposition of Silicon from Silane," 1999.

[11] M. T. Swihart and S. L. Girshick, "Thermochemistry and Kinetics of Silicon Hydride Cluster Formation during Thermal Decomposition of Silane," *J. Phys. Chem. B*, vol. 103, no. 1, pp. 64–76, 1999.

[12] Z. M. Qian, H. Michiel, A. Van Ammel, J. Nijs, R. Mertens, and Interuniversitair, "Homogeneous Gas Phase Nucleation of Silane in Low Pressure Chemical Vapor Deposition (LPCVD)," *J. Electrochem. Soc. SOLID-STATE Sci. Technol.*, vol. 135, no. 1, pp. 2378–2379, 1988.

[13] F. C. Eversteijn, "Gas-phase Decomposition of siilane in a horizontal epitaxial reactor," *Philips Res. Repts*, vol. 26, pp. 134–144, 1971.

[14] J. O. Odden, P. K. Egeberg, and A. Kjekshus, "From monosilane to crystalline silicon, Part I:

Decomposition of monosilane at 690–830K and initial pressures 0.1–6.6MPa in a free-space reactor," *Sol. Energy Mater. Sol. Cells*, vol. 86, no. 2, pp. 165–176, 2005.

[15] H. Y. Dang and M. T. Swihart, "Computational Modeling of Silicon Nanoparticle Synthesis: I. A General Two-Dimensional Model," *Aerosol Sci. Technol.*, vol. 43, no. 3, pp. 250–263, 2009.

[16] S. Nijhawan, P. H. McMurry, M. T. Swihart, S.-M. Suh, S. L. Girshick, S. A. Campbell, and J. E. Brockmann, "An experimental and numerical study of particle nucleation and growth during low-pressure thermal decomposition of silane," *J. Aerosol Sci.*, vol. 34, no. 6, pp. 691–711, 2003.

[17] G. M. Wyller, T. J. Preston, H. Klette, Ø. Nordseth, T. Mongstad, W. O. Filtvedt, and E. S. Marstein, "Critical Nucleation Concentration for Monosilane as Function of Temperature Observed in a Free Space Reactor," *Energy Procedia (submitted)*, vol. 00, no. 1876, 2016.

[18] T. Mongstad, H. Klette, Ø. Nordseth, T. J. Preston, G. M. Wyller, and W. O. Filtvedt, "New technologies for silicon production from monosilane; Pilot testing and on-line process monitoring aiding the development," *Silicon Chem. Sol. Ind. XIII*, pp. 1–10, 2016.

[19] J. Simon, R. Feurer, A. Reynes, and R. Morancho, "Thermal dissociation of disilane : quadrupole spectrometry investigation," vol. 24, pp. 51–59, 1992.

[20] N. Fujiki, Y. Nakatani, K. Inoue, M. Okuyama, and Y. Hamakawa, "Highly oriented polycrystalline Si film on quartz grown from Si3H8 by Thermal and Photo-CVD," *Jpn. J. Appl. Phys.*, vol. 28, no. 5R, pp. 829–835, 1989.

[21] D. Mortensen, D. Lindholm, K. Friestad, B. R. Henriksen, H. G. Fjær, M. Rudshaug, and E. A. Sørheim, "Crystallization furnace modeling including coupled heat and fluid flow, stresses and deformations," *Energy Procedia*, vol. 38, pp. 597–603, 2013.

ON THE PARAMETERS THAT IMPACT THE PERFORMANCE OF DIAMOND WIRE IN THE PRODUCTION OF SILICON WAFERS

K. Sunder* and O. Anspach
PV Crystalox Solar Group, Gustav-Tauschek-Str. 2, D-99099 Erfurt, Germany
*Phone: +49 361 60085306, Fax: +49 361 60085109, E-mail: kirsten.sunder@pvcrystalox.com

ABSTRACT: During the production of silicon wafers with diamond wire the wire is moved regularly back and forth with a small fresh wire supply. As a consequence diamond wire is used very intensively and passes several times the wire web, contrary to the single pass during the slurry-based process with unidirectional wire movement. This results in a nonuniform wear and subsequently performance reduction of the diamond wire along the wire web that becomes visible in a complex wire bow behavior. Particularly designed diamond wire cuts and measurements of wire bow reveal that besides the uniformly increasing wire abrasion due to sawing of silicon, the wire is additionally damaged during the first contact with silicon, the winding on and off the wire spools and the sawing of beam material. It is shown that different diamond wires are more or less sensitive to damage during the first contact with silicon and that different beam materials result in more or less wire wear. At last, a cut without winding the wire on and off the wire spools suggests that a stable wire tension force has a major impact on diamond wire performance.
Keywords: diamond wire, crystalline silicon wafer, wire performance, multi-wire sawing

1 INTRODUCTION

For the production of silicon wafers via multi-wire sawing with fixed abrasives a thin steel wire plated with diamond particles is wound several hundred times around wire guiding rollers. The resulting wire loops form a wire web with a width of up to 1000 mm. This wire web is moved with high speed while pressing silicon blocks into it. At the position of each wire loop material is removed and the silicon wafers remain. To use the high-performing, but also expensive diamond wire economically the wire web has to be moved regularly back and forth with a small fresh wire supply per cycle (pilgrim mode) and every consecutive cut has to start with an already worn wire web.

The aim of the process is the production of high quality wafers with as low diamond wire consumption as possible. As a consequence the performance of a diamond wire should be measured in wafer quality as a function of wire consumption. When the wire consumption decreases, every wire loop has to cut a larger silicon length. This increasing silicon contact length is the major reason for increasing diamond wire wear and results in an increase of wire bow [1]. The increase of wire bow goes along with a linear decrease of wafer quality in terms of increasing total thickness variation (TTV) [2]. Thus the wire bow is a qualitative measure for both wire consumption and wafer quality and therewith ideal for rating the impact of different parameters on diamond wire performance.

Further parameters that increase wire wear and therewith reduce diamond wire performance are the first contact with silicon at the beginning of a cut [1], the winding of wire on and off the wire spools [3] and the sawing of beam material [4]. Figure 1 illustrates exemplarily the dependency of these parameters on table and wire web position for a cut with constant sawing parameters a) before and b) after forward movement of the wire web by w_p. Wire loops that have cut beam material during a previous cut and/or suffered first contact with the silicon block are colorized in yellow. Wire loops that were wound on and off the wire spools are marked in dashed green. And wire loops that suffered the maximal number of rewindings on the fresh wire spool are given in violet. The figure reveals that the

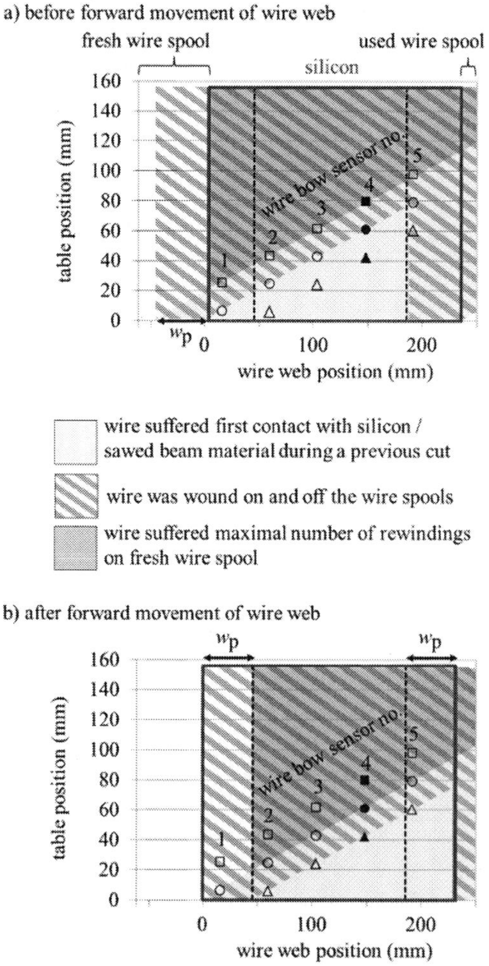

Figure 1: Parameters that increase wire wear as a function of table and wire web position for a cut with one silicon block a) before and b) after movement of the wire web by w_p. The symbols indicate for the positions of five wire bow sensors at which table positions a change of wire bow can be expected. Sensor 4 is discussed in detail in the text.

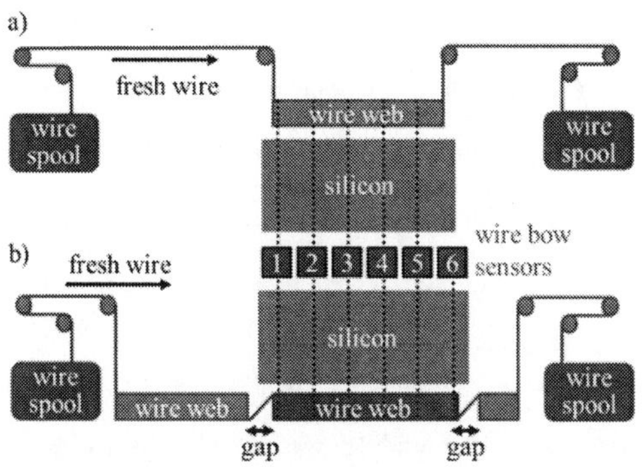

Figure 2: Setup for two comparable cuts with diamond wire that a) is frequently wound on and off the wire spools and b) stays at the wire web during the whole cut. The setups serve to investigate the impact of rewinding on diamond wire performance. Both cuts were performed with wire F [2].

parameters change with table and wire web positions. Hence the impact of different parameters on wire performance can be followed and compared with each other via wire bow measurements at eligible wire web positions as indicated by the symbols in Figure 1.

The purpose of this work is to summarize and prove the parameters that reduce diamond wire performance and to evaluate their relevance. Special designed diamond wire cuts were carried out to investigate the impact of winding the wire on and off the wire spools and to follow the impact of different beam materials on diamond wire performance. Furthermore previous experiments [2] were taken into account to judge the sensitivity of different diamond wires on the mentioned parameters. At last it is reasoned that the wire tension force has a superior impact on diamond wire performance.

2 EXPERIMENTAL

In this work all cuts were performed on a Meyer Burger DS 271 wire saw with PEG 200 as coolant and multicrystalline silicon blocks with a cross section of 156x156 mm². The wire bow was measured during all cuts via inductive sensors at five to six equidistant positions along the wire web. Every cut started with a fresh wire web and worked with proportional feed and wire speed. The maximal feed speed was 800 µm/min, the maximal wire speed 18 m/s and the wire acceleration 5 m/s² for all cuts.

During each back-and-forth movement of the wire web (pilgrim cycle) a certain length of wire is sawing at each position of the wire web. This wire length comprises a certain number of wire loops that form a wire web with a certain width. This wire web width w_p depends on the pilgrim length l_p, the wire loop length l_{wl} and the wire guiding roller pitch p_{wgr} as follows:

$$w_p = \frac{l_p}{l_{wl}} \cdot p_{wgr} \quad (1)$$

In [1] it was shown that the wire bow at a certain wire web position depends on the silicon contact length of the wire loops that are sawing at this position. In this work not only the silicon contact length is considered as a performance-reducing parameter, but also the first contact with the silicon block and the winding of wire on and off the wire spools are taken into account. Significant values for these impacts on sawing performance are determined as a function of table position for the wire loops that are sawing below each wire bow sensor as follows:

i. The fraction of wire loops that suffered first contact with the silicon block is calculated analytically based on the sawing parameters.
ii. The fraction of wire loops that were wound on and off the fresh wire spool is calculated analytically based on the sawing parameters.
iii. The mean number of rewindings is calculated by:

$$\overline{n_{rw}} = \sum_{i=1}^{n_{wl}} n_{rw}^i / n_{wl} \quad (2)$$

where n_{rw}^i is the number of rewindings for wire loop i and n_{wl} is the number of wire loops within w_p. The maximal number of rewindings on the fresh wire spool is a function of the pilgrim lengths:

$$n_{rw}^{max} = l_p^{forth} / \left(l_p^{forth} - l_p^{back} \right) \quad (3)$$

iv. The mean silicon contact length is calculated from the sawing parameters by iteration [1].

2.1 Impact of rewinding

To investigate the impact of winding the diamond wire on and off the wire spools on the sawing performance two cuts each with one silicon block were conducted that differed in the way of wire supply. As illustrated in Figure 2a) for one cut the wire was supplied directly from the wire spools to the wire web below the silicon block and was therefore repeatedly wound on and off the wire spools. Thanks to automatic regulation the wire entered the wire web with always the same wire tension force of 28 N. For the other cut (Figure 2b) the wire was supplied from additional parts of the wire web. Here the wire tension force depended on wire bow and wire elongation [5].

Both cuts were conducted with pilgrim lengths of 300 m / 293 m. The resulting width w_p that is covered by a pilgrim length is shown in Figure 1 relative to the wire web width for the cut with wire supply directly from the wire spools. The figure also illustrates which wire loops suffered first contact with the silicon block (yellow areas), were wound on and off the wire spools (green dashed areas) or suffered the maximal number of rewindings on the fresh wire spool (violet areas) as a function of table position. For a better understanding the wire web condition a) before and b) after forward movement of the wire web is given.

The symbols in Figure 1 mark on the one hand the wire web positions of the wire bow sensors. On the other hand they give the table positions at which the fraction of wire loops that suffered first contact with the silicon block, the fraction of rewound wire loops and the number of rewindings change. At these table positions a change of wire performance and therewith of wire bow can be expected.

As illustrated in Figure 2 the cuts a) with and b) without rewinding differed with 230 mm and 246 mm, respectively, slightly in the width of the wire web below the silicon block. As a consequence only the bow measurements at sensor 1 to 4 are comparable for the two cuts.

2.2 Impact of diamond wire type

In order to increase the data base several cuts that were already discussed in [2] are also taken into account. All cuts worked with identic sawing profiles with pilgrim lengths of 500 m/ 493 m. In all cases the wire was supplied directly from the wire spools to the wire web below the silicon block that had a width of 200 mm. Ten diamond wires (A-J) from six suppliers were used that partly differed in core diameter and diamond size distribution. Depending on the supplier's suggestions wire tension forces between 20 and 28 N were applied (see [2]).

As the wire tension force has a strong impact on wire bow [6] a qualitative comparison of the performance of these diamond wires requires the consideration of wafer quality as done in [2]. In this work the focus is on wire bow development to judge the sensitivity of different diamond wire types on various performance reducing parameters. The cuts described in section 2.1 and section 2.3 were run with wire F and wire A, respectively.

2.3 Impact of beam material

Four cuts with beam/silicon/beam/silicon sandwiches and different beam materials (1 - 4) and one cut with sole silicon were performed. The sawing profiles, wire supply and wire web width were identical for all cuts with those described in section 2.2. The thicknesses of the silicon and beam sandwich slices were (40 ± 2) mm and (12.0 ± 0.5) mm to (16.0 ± 0.5) mm, respectively.

3 RESULTS

3.1 Impact of rewinding

Figure 3a) illustrates exemplarily the measured wire bow development at sensor 4 as a function of table position for the cut illustrated in Figure 1 and Figure 2a). The wire bow measurement at sensor 4 has, similar to the not-shown measurements at sensor 2 and 3, an s-shape.

Figure 3b) gives the calculated development of the parameters that reduce the diamond wire performance and therewith increase the wire bow. The dashed lines mark the table positions at which the fraction of wire loops that suffered first contact with the silicon block, the fraction of rewound wire loops and the mean number of rewindings start to change or reach constant values. These table positions are highlighted in Figure 1 by filled symbols. The development of all parameters is described in the following:

i. At the beginning of the cut all wires loops within the wire web and within one pilgrim length on the fresh wire spool suffer first contact with the silicon block. As a consequence the fraction of wire loops that suffered first contact (red line in Figure 3b) starts at 100 %. At a table position of 60.0 mm (filled circle in Figure 1b) the first wire loop that was not involved in the beginning of the cut reaches wire bow sensor 4. Henceforward the percentage of wire loops that suffered first contact with the silicon block decreases linearly until it becomes zero at a table position of 79.9 mm (filled squares in Figure 1).
ii. At the wire web position of wire bow sensor 4 the sawing wire loops are not wound on and of the wire spools. For this the fraction of rewound wire loops is zero (green line in Figure 3b) until the first wire loop from the fresh wire spool arrives at a table position of 42.2 mm (filled triangle in Figure 1b). Henceforward three more rewound wire loops per cycle reach the sensor. Thus the fraction of rewound wire loops increases linearly until it reaches 100 % at a table position of 60.0 mm (filled circle in Figure 1).
iii. The first rewound wire loop that reaches wire bow sensor 4 at a table position of 42.2 mm (filled triangle in Figure 1b) was wound only once on and off the fresh wire spool, the next three rewound wire loops were wound two times on and off the fresh wire spool and so on. This combination of fraction of rewound wire loops and number of rewindings per wire loop results in an s-shaped increase of the mean number of rewindings (violet line in Figure 3b) until it reaches a maximum at table position 79.9 mm (filled squares in Figure 1).
iv. Due to the applied homogeneous sawing profile the mean silicon contact length (orange line in Figure 3b) increases linearly with table position until it reaches a constant value at a table position of 79.9 mm.

Figure 4 shows the comparison of wire bow measurements at a) sensor 1, b) sensor 2 and c) sensor 4 for the two cuts that were performed with (green) and without (red) winding the diamond wire on and off the

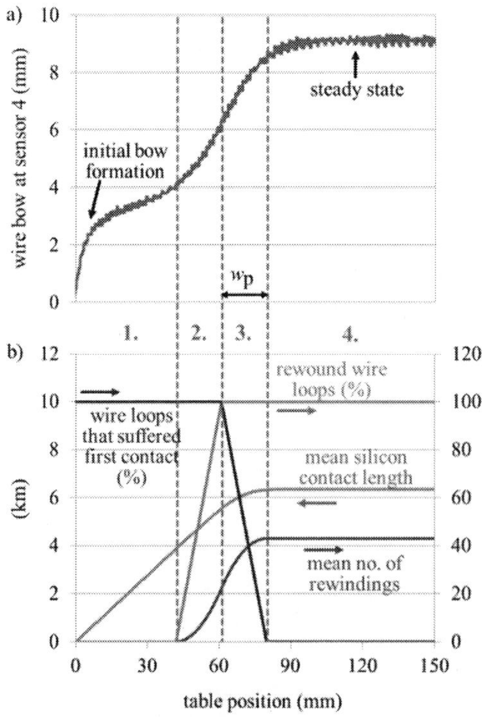

Figure 3: a) Wire bow at sensor 4 as a function of table position together with b) development of parameters that reduce the diamond wire performance and therewith increase the wire bow for a cut with winding the wire on and off the wire spools. The cut was performed with wire F [2].

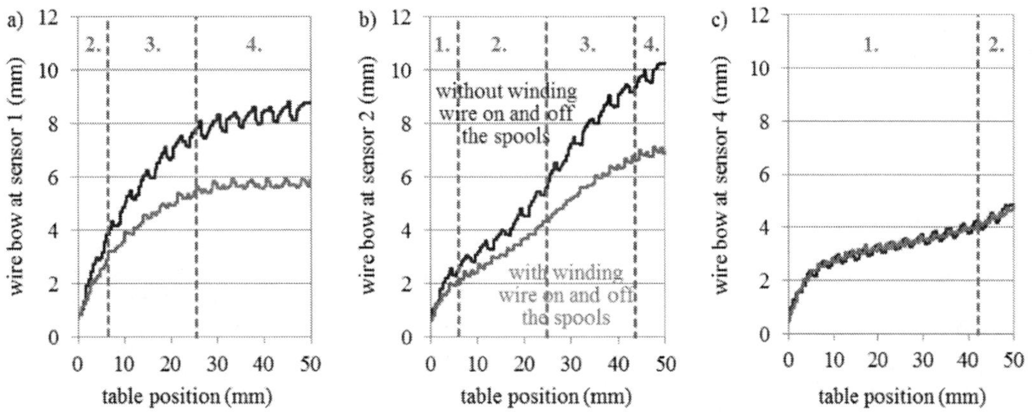

Figure 4: Development of wire bow with table position at a) senor 1, b) sensor 2 and c) sensor 4 for a cut with (green) and without (red) winding the wire on and off the wire spools. As long as no rewound wire loops are sawing below the sensors (1st section) the wire bow values correspond for both cuts. With increasing number of rewound wire loops and increasing number of rewindings (2nd and 3rd section) the wire bow measurements spread. All measurements reach a constant value as soon as the number of rewindings becomes constant (4th section).

wire spools (see Figure 2). Due to identic sawing profiles and wire bow sensor positions the development of the fraction of wire loops that suffered first contact with the silicon block and of the mean silicon contact length are identical for both cuts (see Figure 3b). The dashed lines in Figure 4 mark analogous to Figure 3 the table positions at which the parameters that reduce diamond wire performance start to change or reach constant values. For clarity the sections formed by these lines are labeled with numbers. Within the 1st section, where the wire loops were not wound on and off the wire spools, the wire bows are identical for both cuts. By contrast within the 2nd to 4th section the wire loops that were not wound on and off the fresh wire spool (red) result in higher wire bows than those that were rewound (green). Nevertheless all wire bow measurements reach constant values within the 4th section.

3.2 Impact of diamond wire type

Figure 5a) shows exemplarily the wire bow development at sensor 3 as a function of table position for three cuts that started with fresh wire webs and worked with identic sawing profiles, but different diamond wires together with b) the calculated parameters that reduce the diamond wire performance. The dashed lines in Figure 5 mark the table positions at which the fraction of wire loops that suffered first contact with the silicon block, the fraction of rewound wire loops and the mean number of rewindings start to change or reach constant values. The sections formed by these lines are labeled with numbers analogous to Figure 3.

Figure 5a) reveals that different diamond wires not only result in different steady wire bows below sensor 3 as discussed in [2], but also show different wire bow developments. Wire F, which was also used during the comparable cuts with and without winding the wire on and off the wire spools, shows the s-shape known from Figure 3a). By contrast wire A and C reach their steady states with a square root growth and an overshoot, respectively.

The development of parameters that reduce the diamond wire performance in Figure 5b) is comparable to those given in Figure 3b). Differences lay in the widening of the 2nd and 3rd section and the increase of the maximal number of rewindings both by a factor of 1.7. This results from the increase of pilgrim lengths from 300 m / 293 m to 500 m / 493 m as can be followed via Equation (1) and (3), respectively.

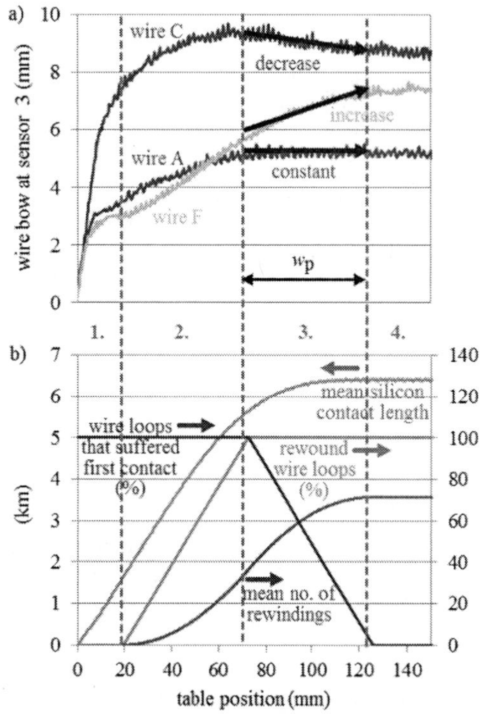

Figure 5: a) Wire bow at sensor 3 as a function of table position together with b) development of parameters that reduce the diamond wire performance and therewith increase the wire bow for three cuts with identical sawing profiles, but different diamond wires. The labeling of the wires originates from [2].

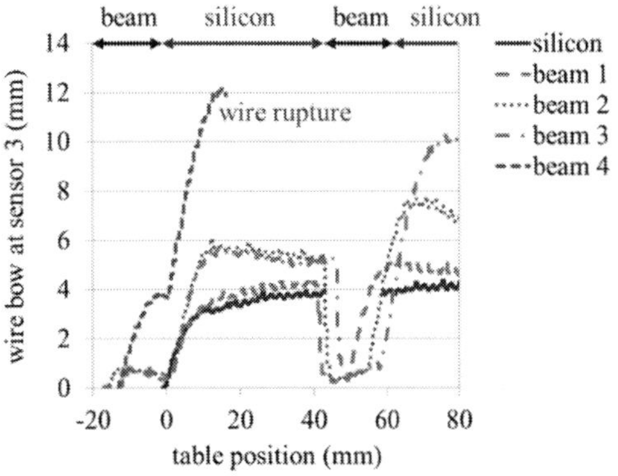

Figure 6: Development of wire bow at sensor 3 during cuts with sole silicon (red line) and beam/silicon/beam/silicon sandwiches (dashed lines) with four different beam materials. The smaller the wire bow the smaller the diamond wire wear caused by beam material. All cuts were performed with wire A [2].

3.3 Impact of beam material

Figure 6 shows exemplarily the wire bow measurement at wire bow sensor 3 for four cuts with beam/silicon/beam/silicon sandwiches (dashed lines) and sole silicon (red line), respectively. To facilitate the comparison the beginning of the first silicon slice is shifted to table position zero for all cuts. Furthermore the wire bow measurement for sole silicon is split in two parts that accord with the two silicon slices within the sandwiches.

The figure reveals that the wire bows are much smaller during cutting of beam 1 to 3 than during cutting of silicon or beam 4. In detail beam 1 to 3 result in equal wire bows within the beam materials, but different wire bows within the silicon slices. The differences get more pronounced when more beam material and silicon is cut. At the end of the cut the wire bow for beam 1 is with around 5 mm only slightly increased compared to the cut with sole silicon that results in a wire bow of 4 mm. By comparison the cuts with beam 2 and 3 result in clearly enhanced wire bows of around 7 mm and 10 mm, respectively.

Beam 4 results in very high wire bows for both the cutting of beam material and of silicon. As a consequence the diamond wire broke before finishing the first silicon slice.

4 DISCUSSION

4.1 Impact of rewinding

The s-shaped wire bow measurement and the development of parameters that reduce diamond wire performance in Figure 3 are sectioned by dashed lines in four parts. The development of wire bow within each section can be explained as follows:

1. All wire loops that cut below the bow sensor have suffered first contact with the silicon block and were not wound on and off the wire spools. After the formation of an initial wire bow that is required for sawing the wire bow increases linearly with increasing silicon contact length.
2. Wire loops that were wound on and off the wire spools reach the sensor. The fraction of rewound wires and the mean silicon contact length increase linearly due to the homogeneous sawing profile. By contrast the mean number of rewindings shows a quadratic increase with table position as the rewound wires have suffered different numbers of rewindings. Since all performance-reducing parameters increase the wire bow increases continuously until the curve reaches its reflection point.
3. All wires loops that are sawing below the bow sensor were rewound, but still not all of them have suffered the maximal number of rewindings. Consequently the mean number of rewindings still increases as does the silicon contact length. By contrast the number of wires that suffered first contact with the silicon block decreases linearly due to the homogeneous sawing profile. In total the wire bow still increases, but less strong as within the 2nd section.
4. All wire loops that are sawing below the bow sensor were maximally rewound, suffered no first contact with the silicon block and have cut the same silicon length. As a consequence a constant wire bow value (steady state) is reached after a short response time.

In summary the s-shape of the wire bow measurement in Figure 3a) can be described by means of the parameters given in Figure 3b). This implicates the relevance of these parameters for diamond wire sawing.

To evaluate the relevance of rewinding the wire bow measurements at a) sensor 1, b) sensor 2 and c) sensor 4 for the two cuts with (green) and without (red) winding the diamond wire on and off the wire spools are illustrated in Figure 4. The identic wire bows within the 1st sections were expected as the respective wire loops were not wound on and off the wire spools for both setups and indicate a proper experimental procedure. By contrast the expected increase of wire bow for the rewound wire loops (green) within the 2nd to 4th section was not found. The contrary development in Figure 4 indicates that another parameter must have a major impact on diamond wire performance. This parameter is proposed to be the wire tension force.

It is known from multi wire sawing with loose abrasives that the steel wire experiences a plastic elongation when it stretches during first bow formation [5]. A reduction of wire tension force is the consequence. When the elongated wire loops are wound on and off the wire spools the wire tension force is set back to the original 28 N by automatic regulation. By contrast elongated wire loops that are not wound on and off the wire spools keep the reduced wire tension force. Thus the unexpected higher wire bows for the wire loops that were not rewound (red lines in Figure 4) can be explained by reduced wire tension forces.

As the increase of wire bow with decreasing wire tension force results in a linear decrease of wafer quality in terms of TTV [2] the wire tension force is another parameter that impacts diamond wire performance. The fact that the impact of wire tension force on wire bow overwhelms the impact of rewinding indicates the relevance of wire tension force for diamond wire cutting. This finding is also supported by the improvement of wire bow and TTV at the used wire side that is generally observed for diamond wire cuts [1]. Hence automatic regulation within the wire web should lead to improved diamond wire performance.

To clarify the relevance of rewinding on diamond wire performance the impact of wire tension force has to be excluded. For this cuts without winding the diamond wire on and off the wire spools and simultaneous regulation of the wire tension force would serve.

In any case a reduction of the maximal mean number of rewindings would lead to an improvement of diamond wire performance. In practice this can be realized on the one hand by increasing the fresh wire supply per cycle that is given by the difference of forward pilgrim length l_p^{forth} and backward pilgrim length l_p^{back} or by decreasing the pilgrim level, i.e. decreasing both l_p^{forth} and l_p^{back} (see Equation (3)). To keep at the same time the wire consumption in terms of meter per wafer unchanged these measures would have to go along with an increase of wafers per cut. This can be realized by a decrease of wafer thickness or an increase of silicon load.

4.2 Impact of diamond wire type

Figure 5 reveals that different diamond wires result in different shapes of the wire bow development even if the parameters that reduce diamond wire performance are identical. This indicates that different diamond wires are more or less sensitive to different performance-reducing parameters.

Within the 1st and 2nd section in Figure 5 all wire bows as well as several parameters that increase diamond wire performance increase and no conclusions can be drawn. By contrast within the 3rd section the wire bow for wire F increases, while it is already constant for wire A and decreases for wire C. At the same time the number of wire loops that suffered first contact with the silicon block decreases while all other performance-reducing parameters increase. Hence the constant or even decreasing bow values within the 3rd section indicate a medium and high performance reduction caused by the first contact with silicon for wire A and wire C, respectively.

Table 1 summarizes the development of wire bow within the 3rd section together with the concluded weight of wire performance reduction caused by first contact with the silicon block for all wires investigated in [2]. For 60 % of these wires the first contact with the silicon block is of minor relevance.

4.3 Impact of beam material

A beam material for diamond wire sawing should on the one hand be easy to cut to facilitate the finish of the cut. On the other hand it should not damage the diamond wire and therewith reduce its sawing performance, because the diamond wire that finishes one cut is used at the beginning of the previous cut. The very small wire bows during cutting of beam 1 to 3 illustrated in Figure 6 indicate that the sawing of these materials is much easier than the sawing of silicon or beam 4. The even more important impact of the beam materials on diamond wire wear and therewith on wire bow and sawing performance can be followed by comparing the wire bow developments within the silicon slices.

Within the first silicon slice at a table position of 0 to 40 mm the wire bow development after sawing of beam 1 equals that of sole silicon and therefore indicates that beam 1 hardly reduces the diamond wire performance. By contrast beam 2 and 3 resulted in increased wire bow values within the first silicon slice, indicating a reduction of diamond wire performance. At last the sawing of beam 4 resulted in a very high wire bow within the first silicon slice that even caused a wire rupture. As a consequence beam 4 is not suitable for diamond wire sawing.

Table 1: Weighting of the performance reduction of the diamond wires evaluated in [2] caused by first contact with the silicon block as deduced from the wire bow development within the 3rd section.

wire	bow development in 3rd section	performance reduction caused by first contact with the silicon block
A	constant	medium
B	constant	medium
C	decrease	high
D	increase	low
E	increase	low
F	increase	low
G	increase	low
H	increase	low
I	increase	low
J	constant	medium

Looking at the second silicon slice at a table position of 60 to 80 mm another difference for the remaining beam materials becomes apparent. While beam 1 still results in a small increase of wire bow compared to the wire bow for sole silicon, the increase of wire bow for beam 3 becomes more pronounced than for beam 2. This leads to the conclusion that beam 1 is the best candidate for diamond wire sawing followed by beam 2 and beam 3.

5 CONCLUSIONS

To optimize the production process of silicon wafers via the multi wire sawing process with fixed abrasives it is important to understand which parameters have an impact on the sawing performance of the applied diamond wire. For an economical production process the diamond wire is moved back and forth with a small fresh wire supply per cycle and every consecutive cut starts with an already worn wire web. This results in a variety of parameters that reduce the diamond wire performance and change with wire web and table position.

In this work special designed diamond wire cuts were performed to enable the differentiation and evaluation of these parameters by means of wire bow measurements. It was confirmed that the first contact with silicon at the beginning of a cut [1], the winding of wire on and off the wire spools [3] and the sawing of beam material [4] reduce the diamond wire performance.

The comparison of wire bow measurements from former diamond wire cuts [2] revealed that different diamond wires are more or less sensitive to performance reduction caused by the first contact with the silicon block (see Table 1).

Furthermore it was shown that the performance reduction due to winding the wire on and off the wire spools can be more than equalized by keeping the wire tension force at the desired value. For this the refreshment of wire tension force within the wire web is recommended. In addition a reduction of the maximal number of rewindings can economically be achieved by increasing the fresh wire supply or decreasing the pilgrim level as long as the number of wafers per cut is simultaneously increased. The latter affords a decrease of wafer thickness or an increase of silicon load.

At last it was shown that commercial available beam materials are more or less suitable for diamond wire cutting. Within the test one beam material caused a wire rupture while another hardly reduced the diamond wire performance.

In summary the diamond wire process can practically be improved by applying diamond wire that is not sensitive to the first contact with the silicon block, by decreasing the wafer thickness, be increasing the silicon load, by refreshing the wire tension force within the wire web and by using an optimal beam material.

6 ACKNOLEDGEMENT

The authors thank Wilfried Büchner and Matthias Rothe for technical support during the diamond wire cuts.

The authors gratefully acknowledge the financial support of the German Federal Ministry for Economic Affairs and Energy (BMWi) within the program "Innovationsallianz Photovoltaik" in the project DiaSiP.

7 REFERENCES

[1] K. Sunder, H. Uhle, S. Knöppel and O. Anspach, Proceedings of 28th European Photovoltaic Solar Energy Conference (2013) 1491.
[2] K. Sunder and O. Anspach, Proceedings of 29th European Photovoltaic Solar Energy Conference (2014) 588.
[3] J. Peguiron, C. Heiniger and S. Habegger, Proceedings of 29th European Photovoltaic Solar Energy Conference (2014) 386.
[4] T. Rist, M. Krappitz, S. Mee, H. Lagger, R. Kübler, G. Kleer, H. Kaden and O. Naumann, Proceedings of 29th European Photovoltaic Solar Energy Conference (2014) poster.
[5] D. Meißner, B. Hurka, J. Zeh, R. Köpge, T. Wagner, H.-J. Möller, S. Schönfelder and O. Anspach, Solar Energy Materials and Solar Cells (2014) 346.
[6] B. Hurka, J. Zeh and O. Anspach, Proceedings of 25th European Photovoltaic Solar Energy Conference (2010) 1591.

THE INFLUENCE OF SURFACE QUALITY ON DIAMOND WIRE SAWN MULTI-CRYSTALLINE SILICON WAFER

Teng-Yu Wang[1*], Jui-Chung Hsiao[1,2], Cheng-Yao Cheng[3], Jyun-Yi Liu[3], Wen-Hsin Lin[3], and Pei-Chen Yu[2]
[1]Green Energy and Environment Research Labs, Industrial Technology Research Institute, Hsinchu, Taiwan, R.O.C.
[2]Department of Photonics and Institute of Electro-Optical Engineering, National Chiao Tung University, Hsinchu, Taiwan, R.O.C.
[3]Green Energy Technology Inc., Taoyuan, Taiwan, R.O.C.

ABSTRACT: Diamond wire-saw is widely using in single-crystalline silicon wafer slicing process. The vantages of diamond wire technology are low cost and high production speed. However, the diamond-wire saw is not applied on multi-crystalline wafer slicing process. The reason is the saw mark on the surface of multi-crystalline wafer caused by slicing will affect the texturing process. Therefore, a surface treatment for the diamond wire sawn multi-crystalline wafer is necessary. In this study, the grinding and sandblasting process was used for surface treatment of diamond wire sawn multi-crystalline wafer. The wafer quality and the solar cell efficiency were used to evaluate the effect of surface treatment process. The experimental results reveal significantly improved the energy conversion efficiency. With surface treatment process, the saw mark on the wafer surface could be removed. The reforming surface can have better texturing structure and enhance the solar cell characteristics. The energy conversion efficiency is 18.4% with surface treatment, which is comparable with the slurry sawn solar cell (18.3%). The cost of the surface treatment is lower than the traditional slurry wire technology.
Keywords: Multicrystalline Silicon; Diamond Wire Saw; Etching.

1 INTRODUCTION

The multi-crystalline silicon wafer is sliced by SiC slurry, and the following textured step used the acid solution for the commercial process. This is because each grain on the multi-crystalline wafer have different atomic packing density, each grain will have different etching speed and become anisotropic etching if the alkali solution is used as the texturing solution. The acid solution mixed with nitric acid and hydrofluoric acid have higher etching rate and suitable for multi-crystalline silicon wafer isotropic texturing. As the diamond wire saw was used for wafer slicing, the fixed abrasive (diamond powder) will leave saw marks on the wafer surface [1]. The saw marks will affect the texturing and the light absorption loss will reduced the solar cell efficiency [2]. Cao et al. shows that the metal assistance etching can apply on the diamond wire sawn multi-crystalline wafer [3]. The energy conversion efficiency could be comparable to the solar cell fabricated by slurry sawn wafer. Lippold et al. used sulfuric acid, nitric acid, and hydrofluoric acid mixtures as the texturing solution to adjust the etching speed [4]. Chen et al. used vapor blast etching method to remove the saw mark on the wafer surface [5]. It can remove the round-run fringes, and at the same time create a well textured surface, which reduces the light reflection. In this study, sand blasting method was used as the surface treatment method to remove the saw mark on the wafer surface.

2 EXPERIMENTAL

540 kg boron-doped p-type multi-crystalline silicon ingot was growth by casting method in this study. The resistivity of silicon ingot was controlled in the range of 1-3 Ω-cm. After crystal growth, the ingot was cut to bricks and sliced to wafers by diamond wire-saw. The obtained silicon wafer is 6-inch in edge and 180 μm in thickness. After wafer slicing, the saw mark on the wafer surface was removed by grinding and sandblasting.

After all the saw marks were removed, the wafers were dipping in HNO_3/HF solution at 10°C for texturization. The emitter was formed on the front surface by diffusion using $POCl_3$ at 850°C. A SiNx film was deposited on the front surface of the wafer by plasma-enhanced chemical vapor deposition (PECVD). The front and rear electrode were printed using a silver paste and aluminum paste, which was co-fired in an IR furnace at 965°C. The energy conversion efficiency of the solar cells was measured by a solar simulator system (QuickSun 120CA, Endeas Co.). Fig. 1 is the process flow diagram for solar cell fabrication.

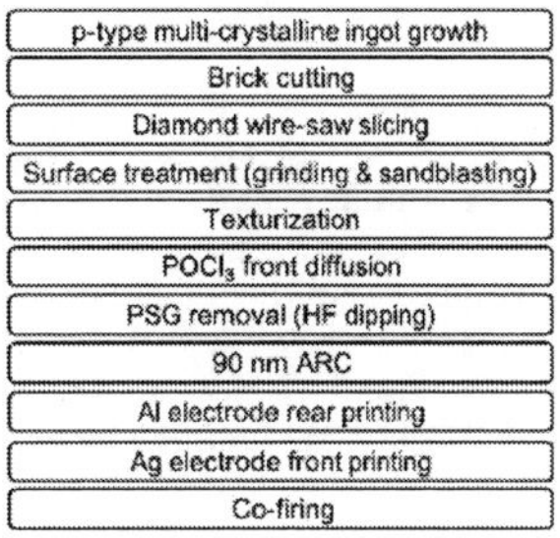

Figure 1: Process flow diagram for solar cell fabrication.

3 DISCUSSIONS

Texturization is an important step in solar cell fabrication. The texturing step can reduced the light reflection on the wafer surface, it can also removes the surface damage caused by wafer slicing. For the slurry sawn wafer, the saw damage is easy to be removed in texturing step. However, the saw damage on the diamond sire sawn wafer is deeper and usually residual on the wafer surface after texturing. For the single crystalline silicon wafer, the adjustment of texturing solution can totally remove the saw damage caused by diamond wire. The problem on the diamond wire sawn multi-crystalline silicon wafer is still unsolved. Each grain on the multi-crystalline wafer has different orientation and atomic packing density, therefore the cutting resistance is always change during the slicing process. For the slurry cut process, the SiC abrasive particle is not fixed on the wire and could be rolling between the wire and the cutting surface. The diamond wire saw combines the diamond cutting particle on the steel wire by resin or electroplating. The diamond particle is fixed and cannot rolling during cutting process, therefore, a deep scratch will form by the abrasive. These scratches will cause problem in solar cell fabrication.

In this study, different surface treatment methods were used to remove the surface damage on diamond wire sawn wafer. In the beginning, the as-cut wafer surface was retreated by grinding. Fig. 2 is the solar cell characteristics varied with grinding speed. The energy conversion efficiency (Eff) and short circuit current density (Jsc) have best value with 70 cm/sec grinding speed. Considering the process complexity, a more convenient method, the sand blasting, was selected.

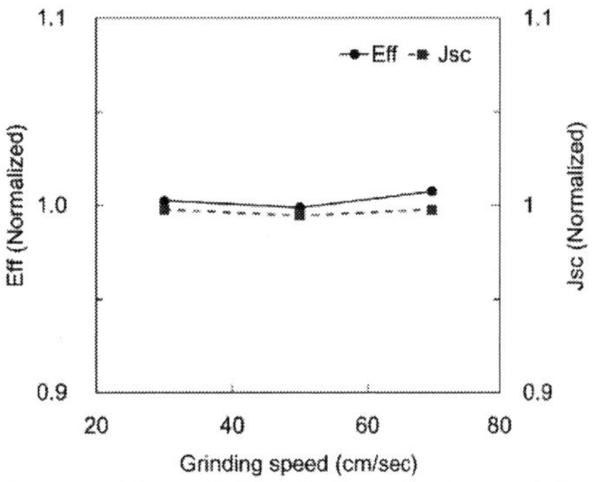

Figure 2: The variation of solar cell characteristics with different grinding speed

Fig. 3 is the solar cell characteristics varied with sandblasting pressure. The Eff and Jsc were decreased while the pressure was increased. It was assumed that the higher pressure will cause deeper damage, and the quality of the wafer will decrease. Fig. 4 shows the influence of different sandblasting times. With sandblasting, the Eff and Jsc could be improved. However, the solar cell will become worse if 3 times sandblasting treatment was used. It is assumed that excessive surface treatment will cause new damage and reduced the performance of solar cell, on the other hand, the cost will also increase.

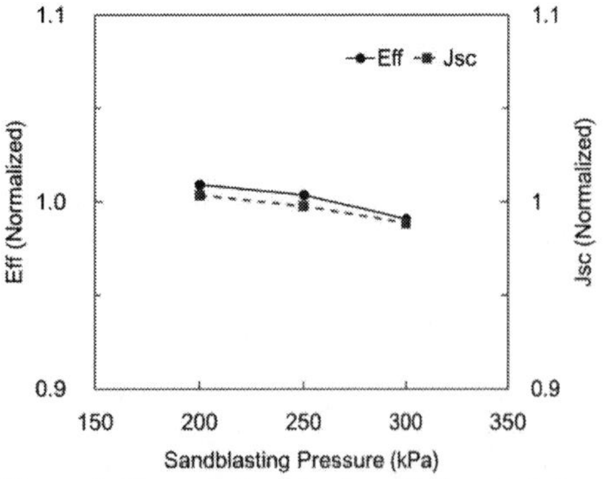

Figure 3: The variation of solar cell characteristics with different sandblasting pressure.

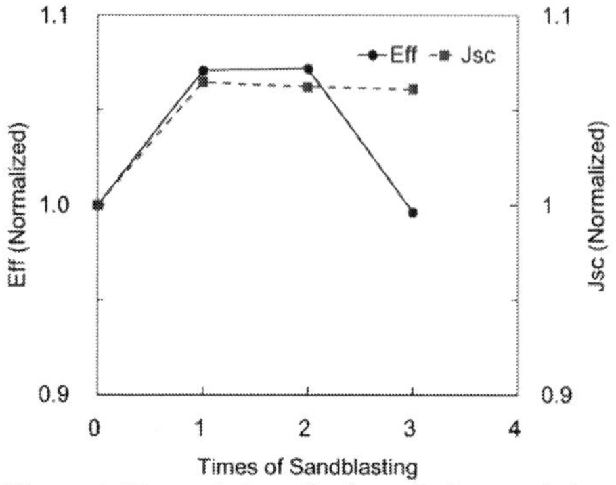

Figure 4: The variation of solar cell characteristics with different sandblasting times.

The reflectance of diamond wire sawn wafer is higher than slurry sawn wafer, which mean the surface of diamond wire sawn wafer is smoother. However, the energy conversion efficiency is worse if the diamond wire sawn wafer was directly used for solar cell fabrication. Fig. 5 shows the surface morphology of as-cut wafer. It was found that even the diamond wire sawn wafer have smoother surface, a lot of holes were scattered on the surface. These holes were caused by the fixed abrasive, diamond particle, and cannot be removed by regular acid etching. Sometime the hole even became bigger and deeper after texturization. With sand blasting process, the surface of diamond wire sawn wafer became similar to slurry sawn wafer.

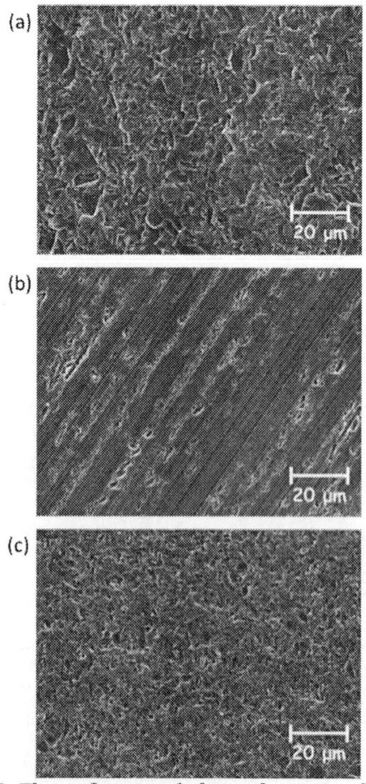

Figure 5: The surface morphology of as-cut wafer from (a) slurry sawn; (b) diamond wire sawn; (c) diamond wire sawn with surface treatment.

Fig. 6 shows the surface morphology of silicon wafer after acid texturization. The surface of diamond wire sawn wafer is too smooth and cannot have good texturing structure. Some of the surface damages became pin holes after texturization. If the surface was reforming by the surface treatment process after wafer slicing, the diamond wire sawn wafer can have the similar surface condition with slurry sawn wafer. Therefore, a good texturing structure could be obtained with the regular acid solution. Fig. 6(a) and 6(c) shows the textured surface of diamond wire sawn wafer with surface treatment and slurry sawn wafer. The surface morphology is almost the same.

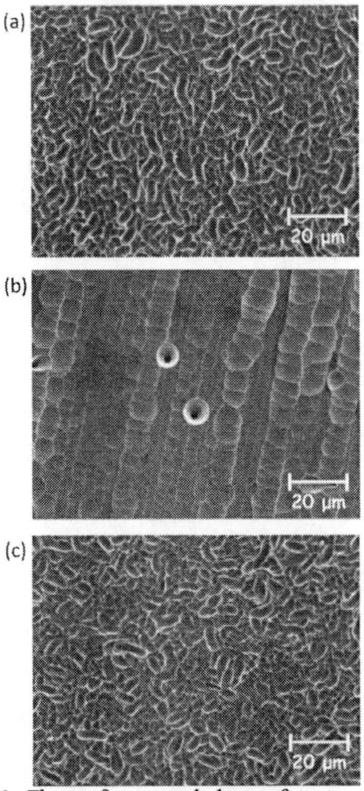

Figure 6: The surface morphology of textured wafer from (a) slurry sawn; (b) diamond wire sawn; (c) diamond wire sawn with surface treatment.

Fig. 7 shows the I-V curves of solar cell fabricated with different sawn wafer. Compared with the efficiency of slurry sawn solar cell (18.3%), the diamond wire sawn solar cell is only 17.7%. This is because the reflective surface due to bad texturing will reduce the ability of light capturing. If the surface treatment process was applied to the diamond wire sawn wafer, the solar cell characteristic could be recovered. The characteristics of surface treatment solar cell and slurry sawn solar cell are very close. Although the surface treatment will increase the additional costs, the final cost of wafer is reduced. In our estimate, replace the SiC abrasive and the steel wire with diamond wire can reduce 5 cent per wafer production. The cost increase by additional surface treatment process is 1 cent per wafer. Therefore, the total cost reduction is 4 cent per wafer.

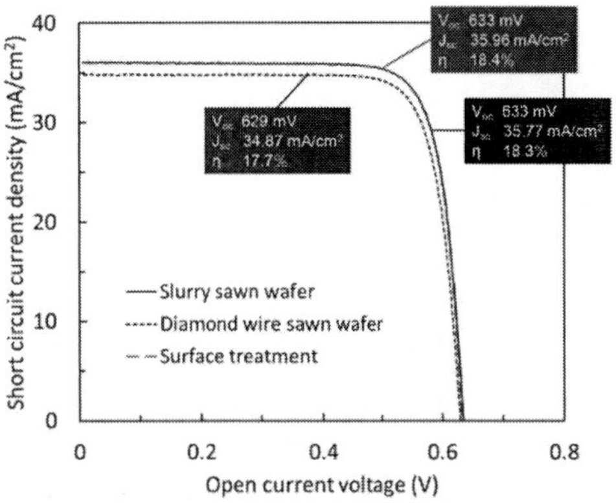

Figure 7: The I-V curves of solar cell fabricated with different sawn wafer.

4 CONCLUSIONS

The effect of surface treatment on multi-crystalline silicon sawn wafer was compared in this study. With grinding and sandblasting, the saw mark on the wafer surface could be removed. The reforming of wafer surface can resolve the texturing problem for solar cell fabrication. The characteristics of solar cell could be enhanced with optimized surface treatment condition. The energy conversion efficiency of diamond wire sawn solar cell is 18.4%, which is comparable with the SiC-slurry sawn solar cell.

ACKNOWLEDGEMENT

The financial support provided by Bureau of Energy is gratefully acknowledged.

REFERENCE

[1] X. Wang, P. Wang, X. Li, and D. Yang, Solar Energy Materials & Solar Cells 98 (2012) 337.

[2] Y. F. Zhuang, S. H. Zhong, Z. G. Huang, W. Z. Shen, Solar Energy Materials & Solar Cells 153 (2016) 18.

[3] F. Cao, K. Chen, J. Zhang, X. Ye, J. Li, S. Zou, and X. Su, Solar Energy Materials & Solar Cells 141 (2015) 132.

[4] M. Lippold, F. Buchholz, C. Gondek, F. Honeit, E. Wefringhaus, and E. Kroke, Solar Energy Materials & Solar Cells 127 (2014) 104.

[5] W. Chen, X. Liu, M. Li, C. Yin, and L. Zhou, Materials Science in Semiconductor Processing 27 (2014) 220.

3 DIMENSIONAL DIRECT WAFER® PRODUCT WITH LOCALLY-CONTROLLED THICKNESS

Adam Lorenz, Jasmin Hofstetter, Harry Malkasian, Laureen Sanderson, Frank van Mierlo
1366 Technologies
6 Preston Court, Bedford, Massachusetts, USA 01730

ABSTRACT: To reduce the amount of silicon used and increase the efficiency achievable by photovoltaic wafers, manufacturers have pursued methods of reducing wafer thickness. While wire sawing can be used to produce wafers thinner than the standard 180-200 micron thickness, these thin wafers have reduced mechanical integrity and break during cell fabrication, electrical interconnection and encapsulation in modules. As such, standard industry wafer thickness has remained between 180-200 microns. 1366 Technologies' Direct Wafer® process can meet the industry's anticipated need for thinner wafers without compromising wafer strength. The process, which forms a wafer directly from molten silicon, has certain advantages over wire-sawn wafers, including the ability to locally-control wafer thickness. This local control can create a thick border where the perimeter of each wafer is of greater thickness than the remainder of the wafer, resulting in a strong, thin wafer able to withstand typical manufacturing stresses. This elegant solution can be delivered at a cost below <\$0.40/wafer and silicon usage below 1.5 g/W.
Keywords: kerfless, thin, Direct Wafer, multicrystalline silicon, manufacturing and process

1 PURPOSE OF THE WORK AND APPROACH

The photovoltaic industry operates in a wafer thickness domain limited by wafer breakage during standard manufacturing processes, particularly during value-added processes in cell fabrication and during interconnection. While wafers have reduced thickness over time to lower silicon cost and optimize efficiency, breakage and poor back surface properties of the dominant BSF cell design has resulted in the industry settling on a thickness between 180-200 microns. Thick wafers use more silicon, a key factor in wafer manufacturing costs. Reducing the amount of silicon wasted, as well as the amount of silicon used, will dramatically reduce the cost of wafers and, in turn, the overall cost of photovoltaics. While manufacturers have expressed a desire to move to thinner wafers not only for their cost advantages but for their potential role in realizing new cell architectures or module advancements, they have been unable to do so using their existing manufacturing lines.

Industry is trending towards PERC-type solar cells and thinner module glass, and both are synergistic with thin wafers and their significant advantages, specifically higher voltage levels and flexibility. Though, again, their lack of mechanical integrity makes it difficult for manufacturers to deliver against future product road-maps.

The increased breakage of thinner wafers has several origins, including

1. Edge stress. During cell fabrication, wafers often break by propagation of a defect from an edge of the wafer. Cracks starting from an edge are also a problem during cell making and module fabrication.

2. Stringing stress. The bus wires or interconnect ribbons must be bent down from the top surface of the cell so as to wrap under the adjacent cell. This bent wire adds to local stress near the edge of the cell during lamination.

Because 1366's Direct Wafer process allows for local control of wafer thickness, it enables entirely new wafer configurations that reduce the volume of silicon required for each wafer without sacrificing strength, durability or performance and increase the strength of wafers without unduly increasing their cost, weight, size, rigidity or other properties. This capability makes it possible for manufacturers to realize advancements in cell architecture and module fabrication.

To demonstrate feasible methods by which Direct Wafer technology can enable reduced silicon usage while simultaneously maintaining or even improving cell efficiency, we conducted preliminary modeling and simulations and continue to run numerous trials with Direct Wafer products with locally-controlled thickness manufactured in our demonstration facility, Fig. 1. We explored scenarios associated with wafer breakage – specifically edge and stringing stresses. Within each load case scenario, modeling was conducted to establish relative comparisons among three types of Direct Wafer products: 1) Direct Wafer, standard thickness; 180 µm 2) Direct Wafer, thin; 80 µm and 3) Direct Wafer with local thickness control; 80 µm / 180 µm. We also conducted 2D PERC and IBC cell simulations with PC-2D [1]: The default PERC and IBC models provided on the PC-2D web site [2] were used but base thickness and bulk minority carrier lifetime were varied.

Figure 1: Thin Direct Wafer product with thick border achieved by locally controlling thickness during growth

2 SCIENTIFIC INNOVATION AND RELEVANCE

In this paper, we introduce the concept of a thin wafer with a thick border, demonstrate its mechanical integrity and propose a geometry that allows for maximum benefit without significant increase in breakage rates during manufacturing processes. A section view of a wafer with thick border is provided in Figure 2.

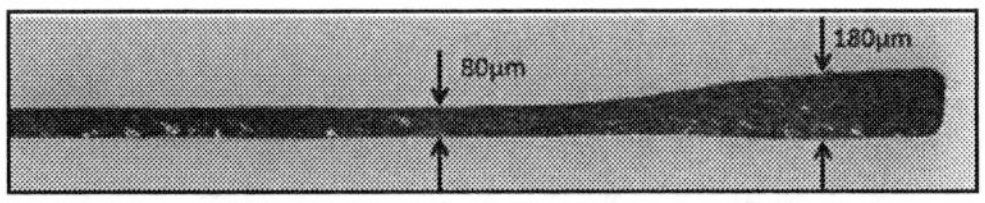

Figure 2: Example of locally-controlled wafer thickness.

The invention of a wafer that is thinner than a standard 180-200 micron thick wafer in certain controlled regions, but is also strong and robust enough to be used in conventional, or nearly conventional photovoltaic applications has significant implications for the industry. It provides manufacturers with a solution to reduce silicon usage without compromising existing standards or quality and makes it possible to realize industry advancements in cell architecture or module features. Because the Direct Wafer process avoids the waste associated with sawing, the silicon usage can be less than 1.5 g/W. This enables the dominant crystalline silicon PV supply chain to leverage all of its existing infrastructure and continue down the cost roadmap via a wafer cost below $0.40/each.

3 SIMULATION RESULTS

3.1 Solar cell performance

Fig. 3 shows how PC-2D simulations predict the effect of decreased wafer thickness on efficiency of a PERC-type solar cell. The bulk electron lifetimes are indicated at low injection levels close to the maximum power point, which are lower than lifetimes typically measured by μPCD but are better representative of operating performance. For a bulk lifetime of 50 μs, an efficiency maximum is obtained at 120 μm wafer thickness. For higher electron lifetimes ≥100 μs, efficiencies can be maintained at high levels for decreased material cost with wafers of 160 μm thickness. Larger efficiency benefits of decreased wafer thickness could be achieved through improved surface passivation and light management, e.g., through improved internal reflectance.

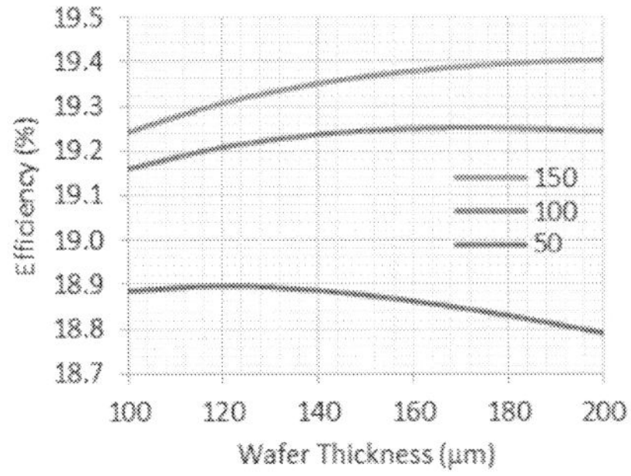

Figure 3: PC-2D simulation of efficiency as a function of wafer thickness for different bulk electron lifetimes (in μs) in a p-type PERC solar cell with 1.3 Ω-cm bulk resistivity

In addition to improved absolute efficiencies for thinner PERC cells with moderate-lifetime wafers, the efficiency distribution is expected to tighten by 0.1 % absolute in a wafer batch of variable bulk lifetime when reducing wafer thickness. A tighter efficiency distribution helps to reduce cell-to-module losses and therefore, leads to increased module power.

Even larger efficiency benefits of thinner wafers are expected for several advanced cell concepts like for example heterojunction cell concepts with passivated contacts or the Interdigitated Back Contact (IBC) cell structure, as demonstrated by PC-2D simulations in Fig. 4.

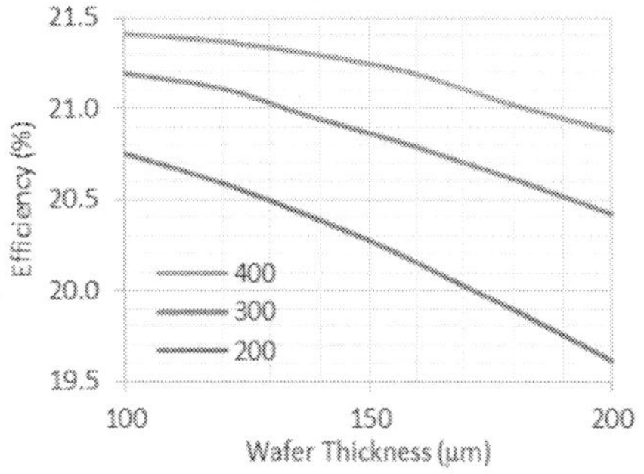

Figure 4: PC-2D simulation of efficiency as a function of wafer thickness for different bulk hole lifetimes (in μs) in an n-type IBC solar cell with 2.3 Ω-cm bulk resistivity

IBC cells are currently being produced industrially on n-type, mono-crystalline silicon wafers and are selected as just one example of possible future cell architectures. In principle, it is straight forward to grow thin, n-type silicon wafers of any desired thickness with Direct Wafer technology. N-type silicon also is much less sensitive to common metal impurities and therefore, several times higher minority carrier lifetimes are usually obtained in n- over p-type silicon wafers of equal material quality. Fig. 4 shows a clear trend of increasing efficiency with decreasing wafer thickness for all simulated bulk hole lifetimes between 200 and 400 μs.

3.2 Edge stress

Using the concept of locally controlled thickness, and the results from the uniformly thick wafer simulation, a model of a wafer which is 80 μm thick in the interior area, but increases to 180 μm thickness at the perimeter was constructed. This model was compared to a thin wafer without the border feature as shown in Fig 5.

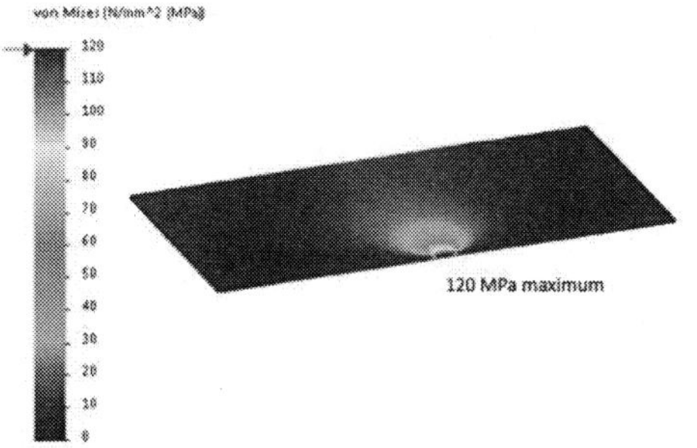

Figure 5: Edge load on 80 μm wafer

Fig. 6 shows the frame of 180 μm thickness from the wafer perimeter inward 1.5 mm. This configuration allows the stress field to dissipate before reaching the thinning region of the wafer.

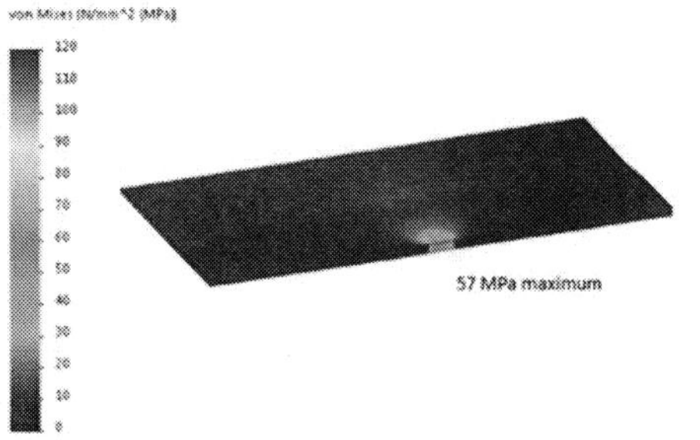

Figure 6: Edge load on 80 µm thin wafer with locally controlled 180 µm thick frame

Here we have a wafer which uses only 58% of the silicon used in the 180 µm wafer and whose thickness is 80 µm for approximately 95% of its normal area. Typical fracture stresses for multi-crystalline silicon wafers is between 120-150 MPa, so it is important to stay well below 100MPa to avoid cracking.

3.3 Stringing stress

Fig. 7 displays modeling results for the case of a stress imposed by the copper interconnect ribbon, indicating that when wafer thickness is uniformly reduced to only 80 µm, stress values increase significantly and exceed the critical strength limit. Using the same locally controlled thick frame configuration as was used in the edge-load scenario, breakage problems can be alleviated.

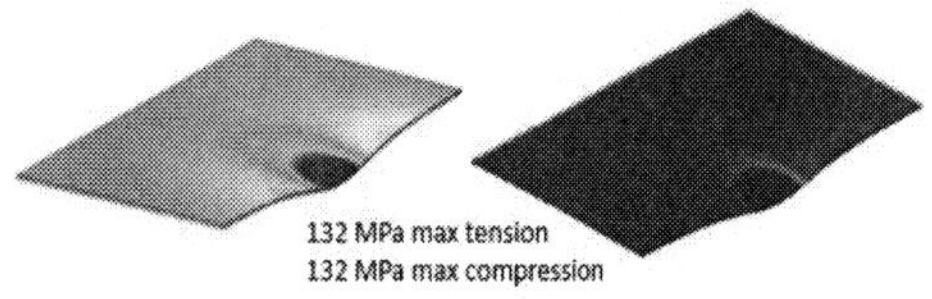

Figure 7: Stringing stresses on 80 µm wafer

Figure 8 shows that for the wafer of 80 µm thickness with 180 µm thick frame and 1.5 mm wide frame, the imposed stress is reduced by more than a factor of two.

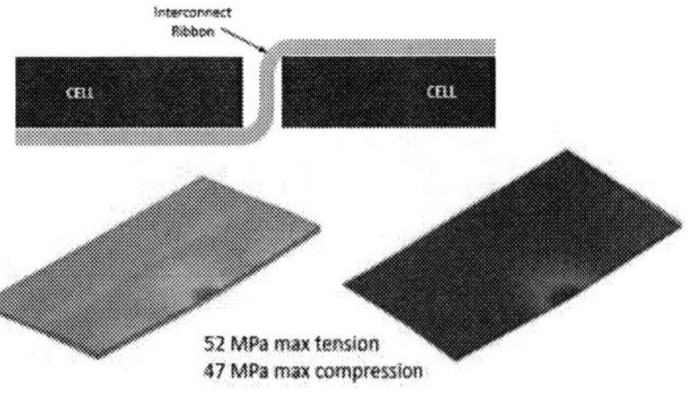

Figure 8: Stringing stress on 80 µm thick wafer with locally-controlled thickness including 180 µm thick frame.

4 CONCLUSIONS

In PERC-type solar cells, a positive effect of decreased wafer thickness on efficiency is expected in particular for moderate minority carrier lifetimes observed in multicrystalline silicon wafers. At the same time, no considerable efficiency losses are observed for higher-lifetime wafers when decreasing the wafer thickness down to ~160 µm, making Direct Wafer products with locally-controlled thickness an enabler for a thin PERC product. Future improvements to PERC design are expected to improve efficiency by further reduction of surface recombination and improved light trapping. Both of these trends will by synergistic with thinner wafers. Direct Wafer products with locally-controlled thickness also have the potential to enable advanced cell architectures like n-PERT, hetero-junction or IBC cells that bear the potential of even larger efficiency gains when using thinner wafers.

Comparative stress analysis through finite element modeling has demonstrated that silicon photovoltaic wafers of 80 µm thickness would likely fail if subjected to typical stress events found in the wafer, cell, and module manufacturing process. Furthermore, it has been shown that employing locally controlled thickness features such as a thick border or frame significantly reduces peak tensile stresses to levels which are below recognized critical levels. Other factors including soldering stresses on thin wafers could also be alleviated by locally increasing the wafer thickness in the region directly underneath the solder contact point.

A test condition not yet analyzed is the stress imposed on the wafer during module mechanical loading. Recent industry trends moving to larger area (72 cell) and thinner glass (<3.2mm for cost savings) will result in larger deflection for the common 6000 Pa wind and snow test loads. The increased flexibility of a thinner wafer will result in lower stress and avoid cell cracking during these conditions with larger and thinner glass.

The potential benefits of 3D Direct Wafer product are being tested at cell and module manufacturers to determine what features provide the optimal benefit to performance and cost. The unique capabilities of the Direct Wafer process is an enabler for the crystalline silicon industry to continue its dominant market share position and deliver progressively lower cost PV solutions for years to come.

5 REFERENCES

[1] Paul A. Basore and Kirsten Cabanas-Holmen, *IEEE Journal of Photovoltaics*, **1**, 1, July 2011, pp. 72-77, 10.1109/JPHOTOV.2011.2166376.

[2] www.pc2d.info

This work was partially supported by the U.S. Department of Energy under Award Number DE-EE0006808.

MULTIPLE REUSE OF THE SILICON SUBSTRATE IN A POROUS SILICON BASED LAYER TRANSFER PROCESS

A. Hajijafarassar[1,4], K. Van Nieuwenhuysen[1], I. Sharlandziev[1], V. Depauw[1], H. Sivaramakrishnan Radhakrishnan[1], T. Bearda[1], M. Debucquoy[1], I. Gordon[1], J. Szlufcik[1], Y. Abdulraheem[5], J. Poortmans[1,2,3], L. Magagnin[4]

[1]*IMEC, Kapeldreef 75, B-3001 Belgium*
[2]*KUL, Departement Elektrotechniek – ESAT, Kasteelpark Arenberg 10, B-3001 Leuven, Belgium*
[3]*UHasselt, Martelarenlaan 42, 3500 Hasselt, Belgium*
[4]*Politecnico di Milano, Dip. Chimica, Materiali e Ing. Chimica Giulio Natta, 20133 Milan, Italy*
[5]*Electrical Engineering Department, College of Engineering & Petroleum, Kuwait University, PO Box 5969, 13060 Safat, Kuwait*

ABSTRACT: Layer transfer process based on a porous silicon template is one of the promising replacements for the conventional wafer sawing technologies. For the process to be cost-effective, one challenge is the multiple recycling of the Si substrate used as a template for epitaxial growth. The Si substrate after detachment has different types of defects. Therefore, reusing the substrate without surface reconditioning induces low detachment yields and generates low quality epitaxial Si-foils. The current work presents the first steps taken into the realization of successful substrate re-use. Firstly, possible defects and undesirable features on the surface after the detachment process are identified. Secondly, a wet-chemical reconditioning process is developed. Finally, the effectiveness of this reconditioning process is evaluated by monitoring the evolution of the foil quality through different generations. Until now, 5 generations of epitaxial foils have been demonstrated at imec.
Keywords: Photovoltaic, Silicon, Porous Silicon, Epitaxy, Recycling, Etching, Defect Density

1 INTRODUCTION

Lifting off a thin layer of high quality epitaxial silicon is one of the promising replacements for wafer sawing to create Si substrates in order to lower the material cost per m^2 [1], [2] and [3].

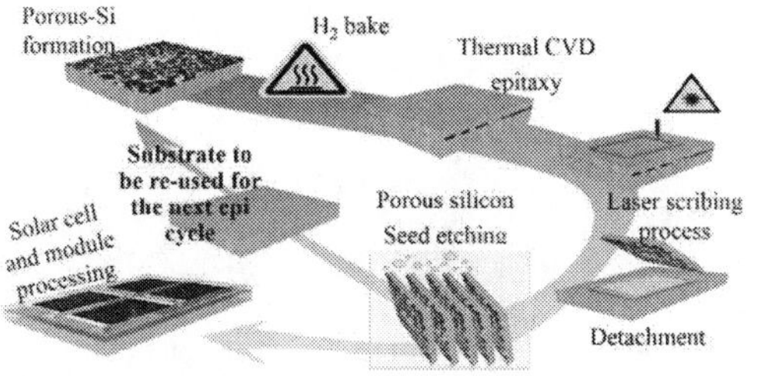

Figure 1: The process flow proposed to process epitaxial foils into devices and recycling the parent substrate.

At imec, the foil is created by a porous silicon (PSi) based layer transfer process. In this process, a PSi stack consisting of a low- and high-porosity layer is etched on a Si wafer. During a high temperature bake before epitaxial growth, this stack is reorganized such that the low porosity layer has small voids and the high porosity layer becomes a long lateral void, supported by Si pillars [4], [5]. On top of this, the absorber of the solar cell is grown epitaxially by atmospheric pressure chemical vapor deposition (APCVD). Afterwards, the foil is detached from this parent substrate at the detachment layer formed in the PSi and then bonded to a glass carrier or processed in a free-standing way into cells. The main steps of this process are shown schematically in Figure 1.

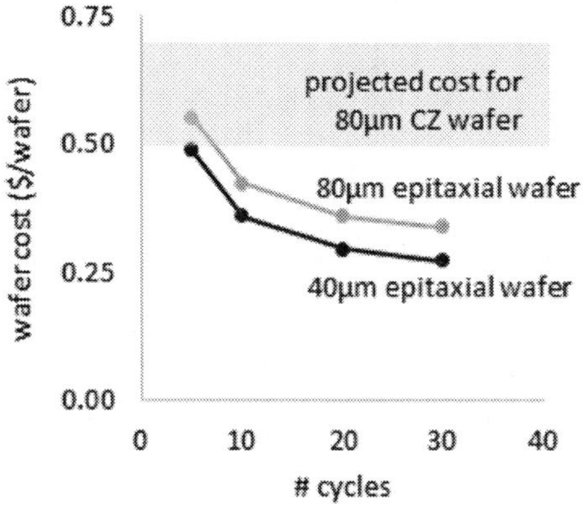

Figure 2: Estimated variation of the wafer cost as a function of the number of re-use cycles of the parent substrate [6].

To make this process cost effective, the silicon parent substrate used as a template for epitaxial growth should be reused several times. A first cost calculation performed at imec [6] shows the estimated variation of the wafer/foil cost as a function of the number of re-use cycles compared to the cost of a 80µm Cz wafer (Figure 2). It's clear that recycling of the parents between 10 and 30 times should be reached to lower the substrate cost substantially. Currently, there are a few studies in literature reporting on multiple substrate reuse of the parent substrate [3], [7] [8]. However, details of the reconditioning process have not been described in detail.

Before using the parent in the next lift-off cycle, the substrate surface must be conditioned to achieve high detachment yield and foil quality, as will be explained in this paper. Hao et al. [3] reported up to 50 reuses. Horbelt et al [7] and Steckenreiter et al [8] use KOH as the etchant to recondition the parent surface. In this work, we present a two-step reconditioning process to achieve: i) high detachment yield, ii) high foil quality, and iii) minimal silicon removal. Scanning electron microscopy (SEM), optical roughness measurements, defect etching and lifetime measurements are used to evaluate the effectiveness of our method.

2 APPROACH AND RESULTS

2.1 Experimental

We use highly boron doped p^+ (10^{19} cm^{-3}) Cz-grown Si wafers as the substrate. The details of the PSi etching process and epitaxy can be found in [3] and [4]. After detachment, the substrates undergo full RCA cleaning followed by reconditioning of the surface using wet etchants (see Section 2.3). The reconditioned substrates are then used in the next cycle of PSi and epitaxial growth.

The surface morphology is studied using SEM, and optical roughness measurements. The quality of foils from various generations is characterized by defect density and minority-carrier lifetime measurements. Defect etching is performed using Secco etchant. For lifetime measurements, surfaces are passivated with a-Si:H. Effective lifetime is measured using Quasi Steady State Photo Conductance (QSSPC) measurements.

2.2 Identifying the main challenges in reuse

Reusing the parent substrate without reconditioning leads to highly defected epitaxial foils with low detachment yield. Two possible reasons were identified. Firstly, the surface of the parent substrate after detachment showed a decrease in boron concentration, as measured by Spreading Resistance Probe (SRP). This could be caused by segregation of B into the PSi stack, during the reorganization step (Figure 3). Since the anodization process is highly dependent on the doping concentration of Si [9], this decrease in B concentration would lead to an un-optimized PSi stack, resulting in poor detachment.

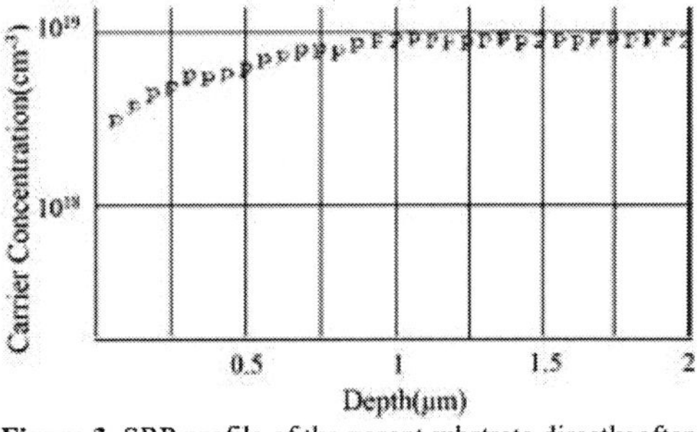

Figure 3: SRP profile of the parent substrate directly after detachment of the epitaxial foil.

Secondly, different kinds of irregularities are present on the surface after detachment. Some result from the breakage of pillars, e.g. pits or 'mushrooms', typically in the order of 1μm (Figure 4) [10]. These defects are inherent to the layer transfer process, and since those features have small sizes, they are relatively easy to remove. However, when the crack propagation during the detachment process does not occur in the separation layer, but is deviated into the epitaxial layer, a hillock is formed after detachment [10]. This type of defect is typically much larger than pillar remnants and are the most detrimental defects in the re-cycle process.

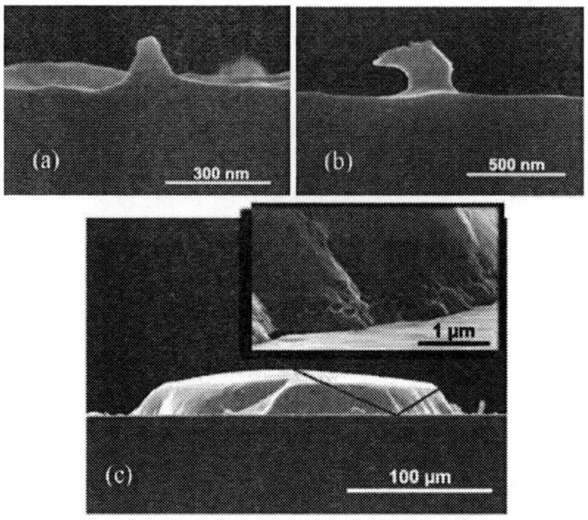

Figure 4: XSEM images of different types defects after detachment: (a) broken pillar (b) mushroom (c) hillock.

In addition, the entire front surface of Si substrate is not anodized owing to the design of the wafer holder at imec. This would leave a 0.5 cm wide non-porosified area at the borders which leads to the presence of non-detachable Si foil on the borders i.e. the flap (Figure 5). The flap area would increase over each recycling step which can have adverse effects on epitaxial growth of higher generations e.g. formation of blisters. However, this hurdle could be overcome by adapting the design of the porosification set-up by sealing the wafers from the rear-side. We will not discuss further the influence of the edges and flaps in this paper.

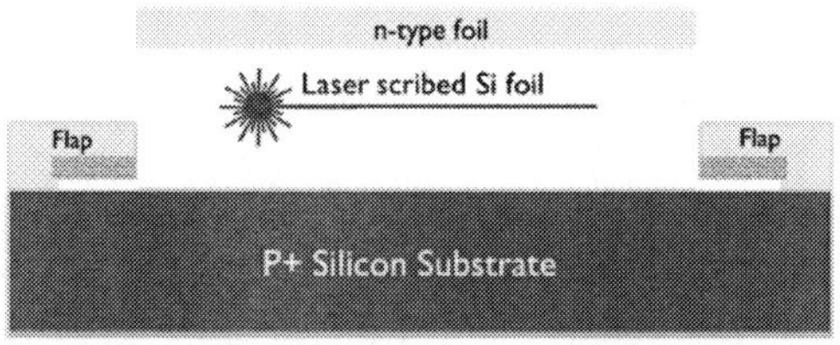

Figure 5: Schematic representation of non-detached n-type Si foil at the borders of the Si substrate (flaps).

2.3 Multiple reuse of silicon substrate

2.3.1 Acid-based Chemical Polishing (CP)

Due to the abovementioned reasons, the Si substrate after each detachment must be reconditioned.

In order to avoid the region with a decreased B concentration, at least 1 μm of the surface is etched off. Acid-based chemical polishing (CP) can be used for this purpose and for removing minor irregularities (see Section 2.2). Sweeping between different concentrations in the ternary system of $HF:HNO_3:H_2O$, i.e. CP (Chemical Polishing) triangle [11] established that only a narrow range of compositions can be used for polishing i.e. those CP solutions where etching is dominated by mass transfer effects rather than reaction kinetics. In this case, a uniform oxide forms on the surface, and etching is controlled by the diffusion of HF from the bulk solution through the stagnant layer present on top of the surface. This leads to higher etching rate of the peaks compared to valleys due to their smaller diffusion length. Thus, only the region of highly concentrated HNO_3 is of interest for our reconditioning process, which smoothen surface features, as shown in Figure 6 [12].

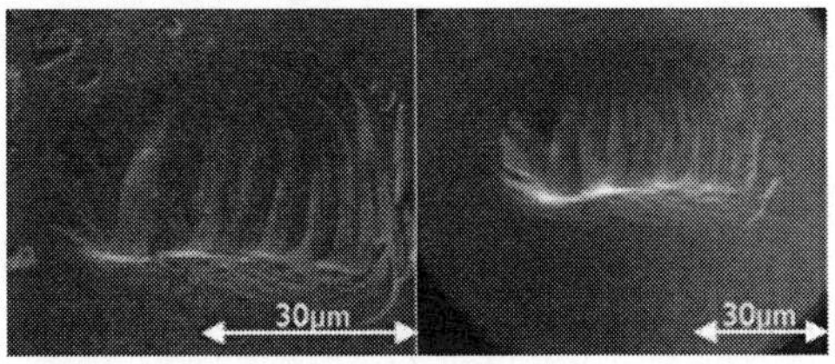

Figure 6: XSEM images of a pit before (left) and after (right) chemical polishing.

Besides the region with the decreased B concentration also hillocks need to be removed from the surface during the reconditioning process. Hillocks are difficult to remove by CP solutions because of their size (10-500 µm) and the isotropic nature of the etchant. Furthermore, if a hillock is not completely removed, subsequent porosification of this area will fail because of the presence of n-type epitaxial silicon at the surface (Figure 7). As a consequence, the reconditioning process must be selective towards n-type Si to remove these large defects without losing large amounts of Si from the parent substrate at the same time.

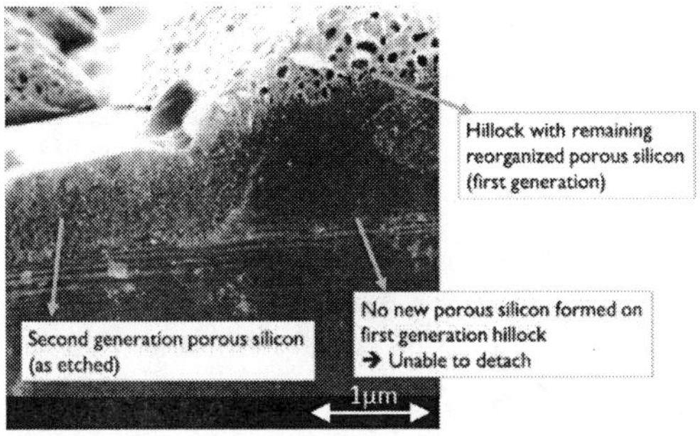

Figure 7: Second generation porosification on a reconditioned surface with a remaining hillock.

2.3.2 Acid and base chemical polishing

To allow selective removal of the hillocks, the parent surface reconditioning was also performed by combining two etchants in separate steps: (1) anisotropic tetramethyl ammonium hydroxide (TMAH) which is selective toward n-type Si[13], and a CP solution which etches isotropically. Water was used as a diluent in both cases. This two-step process satisfies both (i) selective removal of n-type silicon remnants on hillocks and (ii) isotropic smoothening of the surface. As such, the removal of detrimental hillocks is easier than when only 1 type of etchant is used. As a result of this combination, it has been observed that the samples reconditioned with 2 etchants resulted in higher detachment yields in higher generations of re-use compared to using only CP solutions. However, it is clear that the size of the hillocks should remain sufficiently small to be removed during wet-etching reconditioning of the surface since the TMAH etching time should be also limited in time. This will further be investigated statistically by recycling the parents in larger batches.

For the results presented in this paper, several CP solutions are tried out on different sets of wafers. Initially, CP 1:8:1 and CP 1:8:4, roughly having etching rate of 3-4µm/min per side, are used for their good polishing effect. RMS values around 3 nm are measured after 10um Si removal (Figure 8). However, those CP solutions have an exothermal reaction and with several wafers in the bath, the temperature of the solution is rising fast resulting in an unstable etching rate. The amount of Si etched for these highly HF concentrated solutions is too high to be industrially feasible.

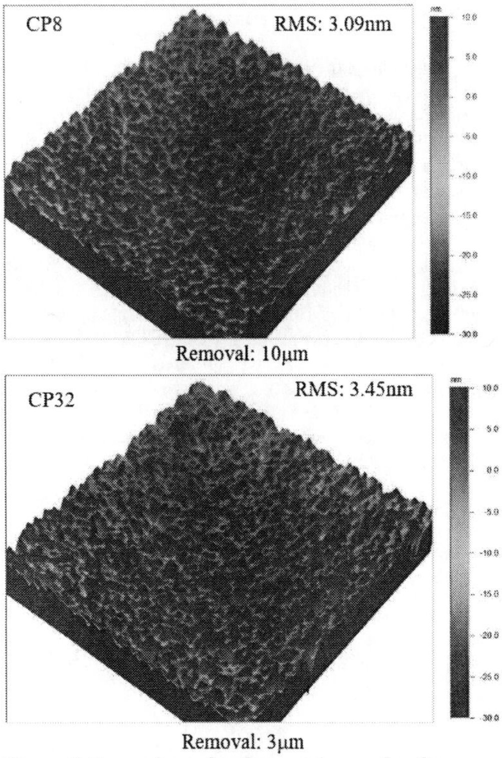

Figure 8: Comparison of surface roughness of wafers etched by CP8 and CP32. Similar surface roughness can be obtained for both solutions. Minimal silicon removal of 3µm per wafer side is obtained by CP32.

Furthermore, due to rather vigorous reaction, the emitted bubbles as the reaction products, result in a wavy surface specially at the borders of the wafer(Figure 9). These traces are disadvantageous for both detachment yield and the generated foil quality.

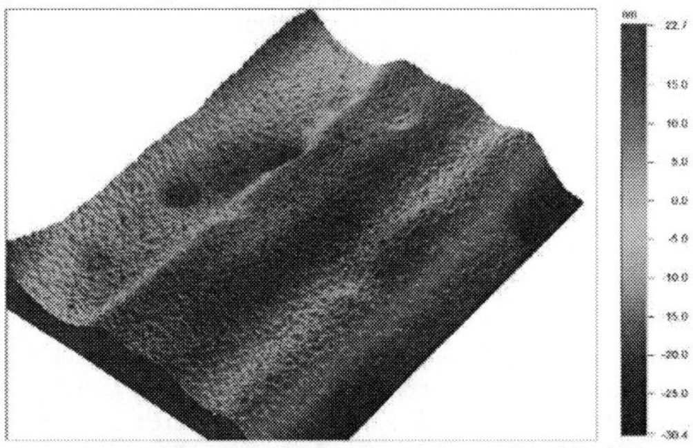

Figure 9: Surface waviness caused by emitted bubbles after 10µm removal by CP8.

As an alternative, CP32 having more controlled

etching rate of 0.5µm/min per side, is evaluated resulting in similar smoothness of the surface by only removing 3µm on each side of the wafer. However, it is clear that a two-step etching solution combining CP32 and TMAH with will be more sensitive to the presence of hillocks which should be almost completely removed by the TMAH solution alone. This combined etching still need to be tested on large batches.

2.4 Quality evolution of foils with different generations

To evaluate the crystalline quality, defect etching was performed on epitaxial foils produced from the re-use process involving CP solutions only. A clear increase in density of stacking faults could be observed until the third generation of re-use, as shown in Figure 10, but for higher generations, the stacking fault density stabilized around $2x10^4$ defects/cm^2. Effective lifetimes of different foil generations are measured to be in the range of 100 – 150 µs at the injection level of 10^{15} cm^{-3}. The relative lifetimes of higher generation foils with respect to the 1st generation are plotted in Figure 10b. Until now, parents treated with the two-step process manifest more stable lifetimes than those treated only with CP solutions. The absence of correlation between lifetime and defect density may be a sign that the foil crystalline quality is currently not the dominating recombination factor. Further tests are needed to reproduce these results.

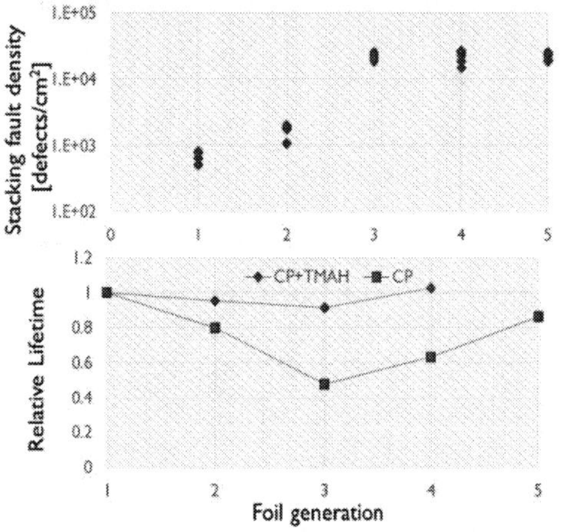

Figure 10: a) Stacking fault density evolution through different foil generations recycled CP. b) Relative lifetime values of epitaxial foils with respect to the 1st generation having different treatments.

3 SUMMARY

First steps are taken into the realization of reusing the parent substrate for the fabrication of silicon foils of ~40-50 µm thickness. A two-step reconditioning process is developed which led to effective defect removal, stabilization of detachment yield and minority-carrier lifetime. The reconditioning process will further be optimized for minimal silicon removal.

4 ACKNOWLEDGEMENTS

This work was funded partially by Kuwait Foundation for the Advancement of Sciences under project code: P115-15EE-01. Part of this work is also funded by the H2020 EU projects CABRISS and CHEETAH. Finally, the authors also gratefully acknowledge the financial support of imec's industrial affiliation program for Si-PV.

5 REFERENCES

[1] R. Brendel, "Review of Layer Transfer Processes for Crystalline Thin-Film Silicon Solar Cells," *Japanese Journal of Applied Physics,* p. 4431–4439, 2001.

[2] H. Sivaramakrishnan Radhakrishnan and V. Depauw, Thin Film Solar Cells Fabrication, Characterization and Applications, J. Poortmans, Ed., John Wiley & Sons, 2016.

[3] R. Hao and T. S. Ravi, "High Efficiency Solar Cells on Direct Kerfless 156 mm Mono Crystalline Si Wafers by High Throughput Epitaxial Growth," in *Photovoltaic Specialist Conference (PVSC), 2014 IEEE 40th,* 2014.

[4] H. Sivaramakrishnan Radhakrishnan and R. Martini, "Improving the Quality of Epitaxial Foils Produced Using a Porous Silicon-based Layer Transfer Process for High-Efficiency Thin-Film Crystalline Silicon Solar Cells," *IEEE Journal of Photovoltaics,* vol. 4, pp. 70-77, 2014.

[5] K. Van Nieuwenhuysen and V. Depauw, "High quality epitaxial foils, obtained by a porous silicon based layer transfer process, for integration in back contacted solar cells," in *EUPVSEC,* 2012.

[6] T. Bearda and I. Gordon, "Thin Epitaxial Silicon Foils Using Porous-Silicon-Based Lift-Off for Photovoltaic Application," in *MRS,* Phoenix, 2016.

[7] R. Horbelt and T. Barbara, "MANIFOLD USE OF GROWTH SUBSTRATE IN THE POROUS SILICON - LAYER TRANSFER - PROCESS," in *Conference Record of the Thirty-first IEEE Photovoltaic Specialists Conference,* 2005.

[8] V. Steckenreiter, "Reconditioning of silicon substrates for manifold re-use in the layer transfer process with porous silicon," in *22nd International Photovoltaic Science and Engineering Conference,* 2012.

[9] V. Lehmann and R. Stengl, "On the morphology and the electrochemical formation mechanism of mesoporous silicon," *Materials Science & Engineering B,* vol. 69, pp. 11-22, 2000.

[10] H. Sivaramakrishnan Radhakrishnan, "Kerfless layer-transfer of thin epitaxial silicon foils using novel multiple layer porous silicon stacks with near 100% detachment yield and large minority carrier diffusion lengths," *Solar Energy Materials and Solar Cells,* vol. 135, pp. 113-123, 2015.

[11] B. Schwartz and H. Robbins, "Chemical Etching of Silicon," 1976.

[12] M. S. Kulkarni, "Acid-Based Etching of Silicon Wafers: Mass-Transfer and Kinetic Effects," *The Electrochemical Society,* 2000.

[13] J. . T. L. Thong and W. . K. Choi, "TMAH etching of silicon and the interaction of etching," *Sensors & Actuators: A. Physical,* 1997.

IMPACT OF HIGH-TEMPERATURE PROCESSES ON CARRIER LIFETIME OF N-TYPE CZ SILICON

S. Werner[1], A. Wolf[1], S. Mack[1], E. Lohmüller[1], R.C.G. Naber[2]
[1]Fraunhofer Institute for Solar Energy Systems ISE, Heidenhofstraße 2, 79110 Freiburg, Germany
[2]Tempress Systems, Radeweg 31, 8171 MD Vaassen, The Netherlands
Phone: +49 761 4588 5049, e-mail: sabrina.werner@ise.fraunhofer.de

ABSTRACT: We investigate the impact of high-temperature processes on phosphorus-doped n-type Czochralski-grown silicon (Cz-Si) wafers. Wafers from five ingots are subjected to high-temperature process sequences with different temperatures, times, and gas atmospheres. The process sequences are specifically chosen to represent different routes for the fabrication of high-efficiency n-type Cz-Si solar cells. As $POCl_3$ diffusion is known for its effective gettering of metal impurities, it is preferably used as the last high-temperature step. However, our results show that other process sequences, e.g. with BBr_3 diffusion as last high-temperature step, enable high charge carrier lifetimes in the range of a few milliseconds as well. For some materials and process sequences, ring-shaped defect structures are observed while for others increased charge carrier lifetimes by a factor up to 2.6 are found compared to their initial values prior to high temperature processing.
Keywords: n-type, silicon, high-temperature processes, lifetime, precipitates, oxygen, $POCl_3$ diffusion, BBr_3 diffusion, nitrogen

1 INTRODUCTION

Phosphorus-doped n-type Czochralski-grown silicon (Cz-Si) wafers are often used in the fabrication of high-efficiency silicon solar cells due to the absence of light-induced degradation [1]. For processing n-type Cz-Si solar cells, typically high-temperature process steps form both the boron-doped emitter and the highly phosphorus-doped surfaces. These doped layers can be realized by, e.g., high-temperature tube furnace diffusion processes using BBr_3 and $POCl_3$ as liquid dopant precursors. Another option is the use of high-temperature tube furnace processes in O_2 or N_2 gas atmosphere, either to anneal crystal damage after ion-implantation or to drive in dopants from previously deposited or printed sources. However, these high-temperature processes may lead to a decrease in charge carrier lifetime in some n-type Cz-Si materials due to the formation of oxygen precipitates [2] or other defects. To overcome this issue, one may select either n-type Cz-Si material that does not degrade during high-temperature processing or process sequences that do not lead to a degradation of the bulk lifetime of n-type Cz-Si wafers.

This work presents a comprehensive study of the key components: n-type Cz-Si wafer materials and processing sequences. Therefore, we investigate n-type Cz-Si wafer sets from different ingots and suppliers with respect to their lifetimes before and after different high-temperature tube furnace process sequences.

2 APPROACH

2.1 High-temperature processes

For the fabrication of high-efficiency n-type silicon solar cells, different combinations of high-temperature processes are imaginable to form the phosphorus- and boron-doped regions. $POCl_3$ diffusion and ion-implantation of boron with subsequent high-temperature annealing or, on the other hand, BBr_3 diffusion and subsequent $POCl_3$ diffusion are two exemplary options.

For the material evaluation of n-type Cz-Si wafers, various combinations and sequences of BBr_3, $POCl_3$, O_2, and N_2 tube furnace processes at different peak temperatures T_{Peak} and peak times t_{Peak} are used; see Table I. Both the $POCl_3$ diffusion process and the BBr_3 diffusion process feature an *in-situ* oxidation [3–5] and are frequently applied in the PV-TEC laboratory at Fraunhofer ISE. The processes "O_2 low" and "N_2 low" have a similar thermal budget as the $POCl_3$ diffusion process, but for the process atmosphere we look at two extreme cases with pure O_2 or pure N_2 gas atmosphere. The processes "O_2 high" and "N_2 high" feature a higher thermal budget as the diffusion processes. Such processes might be applied to drive in the dopants from a previously deposited layer or to anneal/redistribute implanted dopants.

2.2 Sample preparation

Various n-type Cz-Si wafers with an edge length of 156 mm and different base resistivities ρ_{Base} are investigated with respect to their lifetimes before and after different high-temperature tube furnace process sequences; see Fig. 1. Therefore, two experiments are carried out. The first experiment focuses on the investigation of a wide range of different materials in high-temperature process sequences while the second experiment focuses on a broader investigation of high-temperature process sequences. Thus, in the first experiment wafers from five different ingots from four different suppliers (Mat1 – Mat5) are subjected to five high-temperature process sequences. In the second experiment, which includes eight additional high-temperature process sequences, only two materials (Mat1 and Mat2) are used. More details about the applied process sequences will follow with the discussion of the results. Each group contains four wafers per material.

First, ρ_{Base} of the wafers are measured by means of inductive coupling using an inline tool [6]. After random

Table I: Peak temperature T_{Peak}, peak time t_{Peak}, and the gas atmosphere of the used high-temperature processes.

Process name	T_{Peak} (°C)	t_{Peak} (min)	Gas atmosphere
$POCl_3$	860	12	$POCl_3$, O_2, N_2
BBr_3	955	20	BBr_3, O_2, N_2
O_2 low	860	12	pure O_2
O_2 high	1000	60	pure O_2
N_2 low	860	12	pure N_2
N_2 high	1000	60	pure N_2

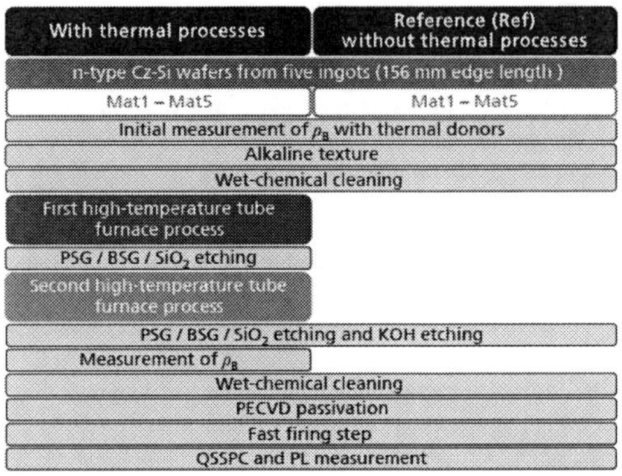

Fig. 1. Schematic process sequence used for the material evaluation of n-type Cz-Si wafers from five different ingots and four different suppliers (Mat1 – Mat5) in different high-temperature tube furnace process sequences (PSG: phosphosilicate glass, BSG: borosilicate glass, ρ_{Base}: base resistivity, PECVD: plasma-enhanced chemical vapor deposition, QSSPC: quasi-steady-state photoconductance, PL: photoluminescence). In a second experiment, which addresses additional high-temperature processes, only two materials (Mat1 and Mat2) are used.

pyramid formation in alkaline solution and wet-chemical cleaning, the wafers are subjected to one or two selected high-temperature processes such as BBr_3, $POCl_3$, O_2, or N_2 tube furnace processes at different T_{Peak} and t_{Peak} (see Table I). In between those two processes, the possibly formed phosphosilicate glasses (PSG), borosilicate glasses (BSG) or SiO_2 layers (depending on the used processes) are removed in HF solution. The two subsequently carried out high-temperature processes are referred to as "process1 + process2" (e.g. "BBr_3 + O_2 low"). The double-sided highly-doped regions after the initial $POCl_3$/BBr_3 diffusion are neither removed nor coated by diffusion barrier layers for the second high-temperature process. One reference group per material is not subjected to any high-temperature process.

After the high-temperature sequences, all surface oxide layers are removed in HF solution and the diffused and textured surfaces are subsequently etched back and smoothened in KOH solution (silicon removal of about 10 µm from each side resulting in a final wafer thickness $W \approx 150$ µm, also for the reference group). The used high-temperature processes dissolve any thermal donors which might have existed before. To determine information about the content of thermal donors per material, another measurement of ρ_{Base} is performed at this stage of the experiment.

Following wet-chemical cleaning, the alkaline saw-damage etched surfaces are passivated with a layer stack consisting of silicon-rich oxynitride (SiRiON) [7] and silicon nitride (SiN_x) deposited by PECVD. After activating the passivation layers in a fast firing furnace at a set peak temperature of 700°C, the effective charge carrier lifetime τ_{eff} is measured by QSSPC at five positions over each wafer and spatially resolved PL images are recorded. As all wafers are identically wet-chemically etched and passivated, we assume the surface recombination velocity to be very similar for all wafers. Therefore, we conclude that all correlations found for the measured effective lifetimes can be related to the silicon bulk.

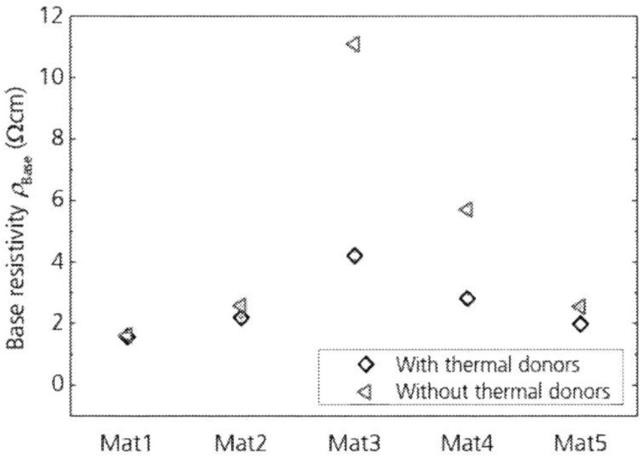

Fig. 2. Base resistivity ρ_{Base} of the wafers from five different n-type Cz-Si ingots investigated in this work with and without thermal donors, measured by means of inductive coupling [6].

3 RESULTS

3.1 Base resistivity with and without thermal donors

The base resistivities ρ_{Base} of the five n-type Cz-Si materials investigated in this work are shown in Fig. 2 with and without thermal donors. Thermal donors can be correlated to the oxygen content of the Cz-Si [8]. For Mat1, ρ_{Base} is determined to be $\rho_{Base} \approx 1.6$ Ωcm with and without thermal donors. Thus, the content of thermal donors within this material is marginal. In contrast, ρ_{Base} for Mat3 increases from $\rho_{Base} \approx 4.2$ Ωcm with thermal donors to $\rho_{Base} \approx 11.1$ Ωcm without thermal donors, which comes along with a high content of thermal donors. The different materials have been chosen such that the investigation includes materials with different amounts of thermal donors to examine if they affect the material properties after exposure to different high-temperature processes.

3.2 Experiment 1: material variation

Fig. 3a) shows the effective charge carrier lifetimes τ_{eff} measured by QSSPC at an injection level $\Delta n = 10^{15}$ cm^{-3} for the five materials and five different high-temperature process sequences after firing. The reference group "Ref" processed without any high-temperature steps is also stated. It is evident that the materials Mat3 and Mat4 with higher ρ_{Base} show higher τ_{eff} throughout the experiment due to less recombination active centres.

For easier interpretation, we calculate the lifetime factor β by dividing τ_{eff} per material and process sequence by the effective charge carrier lifetime $\tau_{eff,Ref}$ measured for the reference group per respective material:

$$\beta = \frac{\tau_{eff}}{\tau_{eff,Ref}}. \qquad (1)$$

Fig. 3b) shows β for all groups.

Compared with $\tau_{eff,Ref}$, the measured τ_{eff} increases for all materials after a single $POCl_3$ diffusion by $1.6 \leq \beta \leq 2.4$ or after sequential BBr_3 + $POCl_3$ diffusions by $1.2 \leq \beta \leq 1.9$. Mat1 shows τ_{eff} up to $\tau_{eff} \approx 3$ ms after single $POCl_3$ diffusion and Mat3 and Mat4 even up to $\tau_{eff} \approx 6$ ms.

With BBr_3 diffusion being the second process (sequences: $POCl_3$ + BBr_3 and O_2 low + BBr_3), τ_{eff} is on the

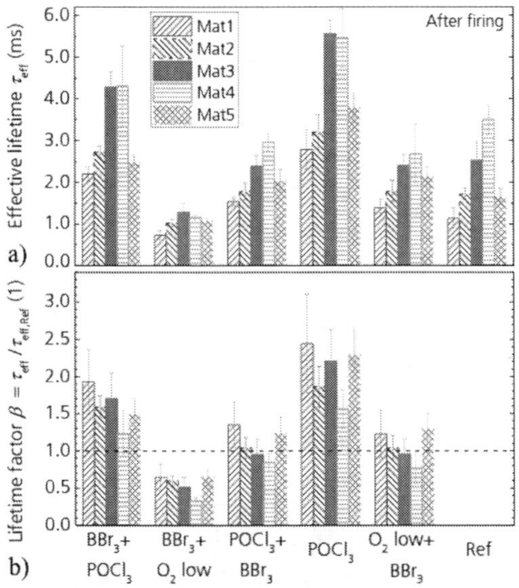

Fig. 3. a) Effective charge carrier lifetimes τ_{eff} measured by QSSPC at an injection level $\Delta n = 10^{15}$ cm^{-3} for five materials and five high-temperature process sequences as well as the reference group without any high-temperature process. The error bars represent the standard deviation of τ_{eff} measured on three to four samples at five positions each. The measurement positions are exemplarily marked in the first PL image in Fig. 4. b) Lifetime factor β as given in Eq. (1).

same level as the group Ref independent of whether there has been a $POCl_3$ diffusion or process O_2 low before BBr_3 diffusion. The factor β is found to be $\beta \approx 1$ for most of the materials or even $\beta \approx 1.3$ for Mat1 and Mat5.

Only the process sequence $BBr_3 + O_2$ low significantly reduces τ_{eff} by $0.3 \leq \beta \leq 0.7$ compared with group Ref. This degradation might be correlated with the O_2 low process as for all other combinations with BBr_3 diffusion the factor β is found to be $\beta \geq 1$ (see discussion above). A more detailed discussion will follow when discussing the PL images.

In the literature, it is reported that BBr_3 diffusion might lead to a significant degradation of τ_{eff} [9]. Here, we demonstrate that our optimized BBr_3 diffusion process is suitable for sequential diffusion processes. In combination with $POCl_3$ diffusion, τ_{eff} is at least the same as for group Ref, no matter which process order is used. When using sequential diffusion processes with $POCl_3$ first and then BBr_3, the phosphorus atoms can be driven-in even deeper during the BBr_3 diffusion. Our results indicate that both orders of the two diffusion processes are suitable for cell manufacturing, but for cell concepts that heavily rely on high bulk lifetimes, such as back contact back junction cells, $POCl_3$-last processing does appear to be advantageous.

Single $POCl_3$ diffusion as well as $POCl_3$-last processing lead to highest β values in this investigation. This might be correlated to a high gettering effect initiated by $POCl_3$ diffusion. In n-type silicon, impurities as Co, Cr, Ni and Fe might be recombination active and reduce τ_{eff} [10]. Phosphorus gettering is known to reduce these impurities in silicon wafers [11–14]. However, it is

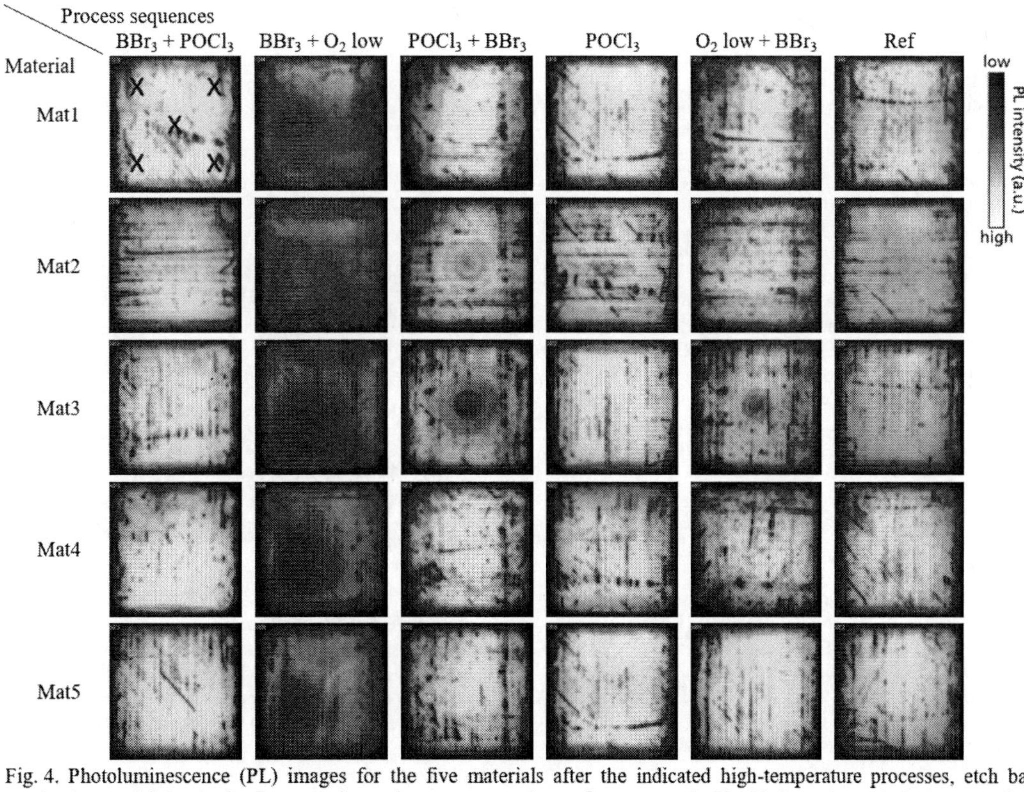

Fig. 4. Photoluminescence (PL) images for the five materials after the indicated high-temperature processes, etch back, passivation, and firing in the first experiment (one representative wafer per group). The PL intensity scale is the same for all images. The reference group Ref is also shown. At the top left PL image, the positions of the performed QSSPC measurements in Fig. 3 and Fig. 5 are exemplarily marked.

not clear which type of impurity is relevant for the n-type wafers used in our investigation.

However, all materials show the same trends for the different process combinations. Most of the materials and most of the process sequences show $\tau_{eff} > 1$ ms and $\beta \geq 1$. Mat4 shows the lowest β for all process sequences compared with the other materials. In comparison with Mat1, Mat2, and Mat5, the content of thermal donors is much higher (see Fig. 2). Thus, the thermal donors might affect τ_{eff} for Mat4. However, the high content of thermal donors for Mat3 does not seem to affect τ_{eff}.

In Fig. 4, PL images for the five materials after the high-temperature processes, as indicated in Fig. 3, are shown for one representative wafer per group after firing. Lines with lower PL intensity parallel to the wafer edges are caused by an insufficient saw-damage etching. Other scratches are attributed to wafer handling. The group Ref without any high-temperature processing is also shown in the very right column.

As already discussed, the process combination $BBr_3 + O_2$ low shows a degradation of τ_{eff} for all materials (Fig. 3). This can also be seen from the PL images. Wafers processed with $BBr_3 + O_2$ low show a dark area with lower PL intensity for all materials (Fig. 4). These dark structures might originate from an interaction of boron dopants and silicon interstitials injected by the O_2 low oxidation causing lower τ_{eff} compared with the reference group Ref. As these dark structures are visible in every material, they might be attributed to the process sequence and not to the material properties.

Mat2 and Mat3 show slight ring structures for the process sequence $POCl_3 + BBr_3$, and Mat3 also for the sequence O_2 low + BBr_3. Such ring structures are typically formed by oxygen precipitate formation [15]. These ring structures seem not to cause a significant degradation of τ_{eff} for Mat2. The QSSPC measurement performed in the wafer's center results in a similar τ_{eff} as the measurements performed at the other four positions. The wafers from Mat3 behave different. The QSSPC measurement in the wafer's center shows about 10% lower τ_{eff} compared to the measurement at the other four positions for both sequences with BBr_3-last processing.

In general, Mat3 with high content of thermal donors behaves similar as Mat1 with marginal content of thermal donors. Only for the process sequences with BBr_3-last processing small ring defects and a slight degradation of τ_{eff} occur in the wafer's centre for Mat3. But these ring defects are also visible for Mat2 which does hardly contain any thermal donors. Hence, thermal donors do not seem to significantly influence the materials' properties after exposure to specific high-temperature processes.

3.3 Experiment 2: high-temperature process sequence variation

For Mat1 and Mat2, the influence of the order of high-temperature processes with higher or lower temperature first as well as the applied gas atmosphere on τ_{eff} is investigated in more detail. The used process sequence is the same as in Fig. 1 and the measured τ_{eff} after firing are shown in Fig. 5a). Again, we calculate the lifetime factor β as given in Eq. (1); see Fig. 5b).

Again, it is obvious that the $POCl_3$ diffusion process being the second high temperature process shows an increased τ_{eff} compared with the reference group Ref. The factor β is calculated to be $\beta = 2.6$ and $\beta = 1.8$ for Mat1 and Mat2, respectively, when O_2 high is the preceding

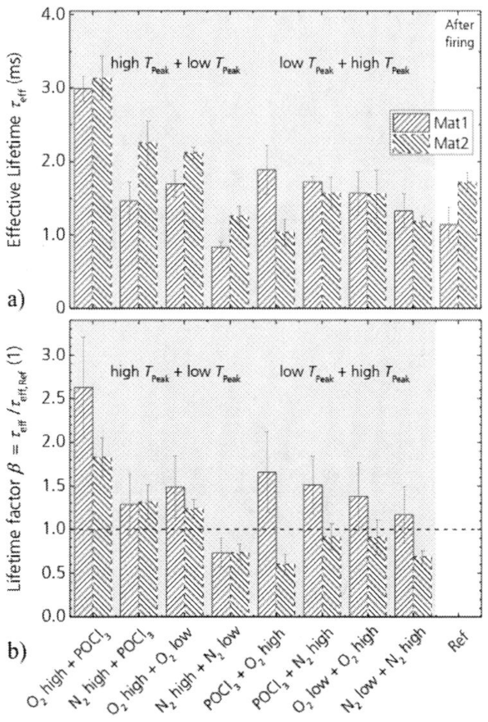

Fig. 5. a) Effective charge carrier lifetimes τ_{eff} measured by QSSPC at an injection level $\Delta n = 10^{15}$ cm^{-3} for two materials and the stated eight high-temperature process sequences, including the reference group without any high-temperature process. The error bars represent the standard deviation of τ_{eff} measured on three to four samples at five positions each. The measurement positions are exemplarily marked in the first PL image in Fig. 4. b) Lifetime factor β calculated as given in Eq. (1).

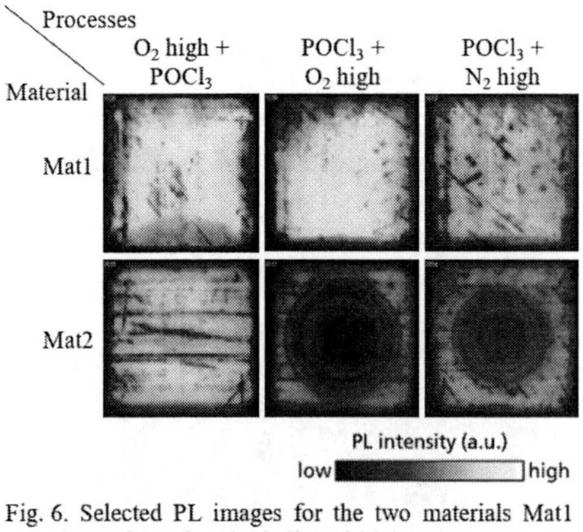

Fig. 6. Selected PL images for the two materials Mat1 and Mat2 taken after the indicated high-temperature processes, wet-chemical etch back, passivation and firing in the second experiment (one representative wafer per group). The PL intensity scale is the same as in Fig. 4.

high-temperature process. These β values are very similar as those calculated for Mat1 and Mat2 after a single $POCl_3$ diffusion (see section 3.2). But when N_2 high is the preceding high-temperature process before $POCl_3$

diffusion, β is reduced to $\beta = 1.3$ for both materials. Thus, the process O_2 high seems advantageous compared to N_2 high within the process sequence with following $POCl_3$ diffusion.

However, most of the process sequences with the high-temperature process at low temperature first and the process at high-temperature second yield $1.3 < \beta < 2.6$, only the sequence N_2 high + N_2 low shows a reduced τ_{eff} with $\beta \approx 0.7$ for both materials. The reason for this is not clear yet and further investigations are necessary.

Mat2 shows reduced τ_{eff} after process sequences with the higher temperature last (either O_2 or N_2 gas atmosphere). The factor β is found to be $0.6 < \beta < 0.9$ for Mat2 while $1.2 < \beta < 1.6$ for Mat1. The origin can be found in ring-like structures as apparent from the PL images in Fig. 6. For Mat2, O_2 high as well as N_2 high cause dark ring structures when being the second high-temperature process. The PL intensity in the ring area is even lower for O_2 high than for N_2 high. This is not only valid when $POCl_3$ is the first process but also for the groups with O_2 low + O_2 high or N_2 low + N_2 high (PL images not shown). As already mentioned, such ring structures are typically formed by oxygen precipitate formation [15]. But no ring structures are visible when O_2 high is the first process followed by either $POCl_3$ diffusion or O_2 low.

Furthermore, a lot of point-like structures appear especially for Mat1 after the process sequences featuring N_2 high as first process, as can be seen in the PL images in Fig. 7a) for representative wafers of material Mat1. The point-like structures are obvious for the sequences N_2 high + $POCl_3$ as well as N_2 high + N_2 low. These point-like structures might be attributed to oxide stacking faults [16]. But as $\beta = 2.6$ for Mat1 and the process sequence N_2 high + $POCl_3$, these point-like structures seem not to necessarily reduce τ_{eff}. When changing the process order to N_2 low + N_2 high, no point-like structures are visible.

Fig. 7b) shows the enlargement of the area marked in Fig. 7a). The PL images are taken as-cut, after process N_2 high, after $POCl_3$ diffusion and finally after passivation and firing. It is obvious that the point-like structures occur right after the N_2 high process. They stay at the same positions after $POCl_3$ diffusion as well as after passivation and firing. This indicates that the point-like structures can be related to the silicon bulk and not to the surface. The reason for these point-like structures as well as their impact on τ_{eff} need to be further investigated.

4 SUMMARY AND CONCLUSION

The present work shows the impact of high-temperature processes at different temperatures, times, and gas atmospheres on n-type Czochralski-grown silicon wafers. Symmetric samples for lifetime measurements have been processed from wafers from five different ingots with different base resistivities and different contents of thermal donors. The high-temperature processes include BBr_3, $POCl_3$, O_2, or N_2 tube furnace processes, which are typically used for fabrication of high-efficiency n-type silicon solar cells.

The five materials react very differently to the high-temperature process sequences. Some process sequences have a positive impact on the effective charge carrier lifetimes τ_{eff} of these wafers, while other wafers show reduced τ_{eff}.

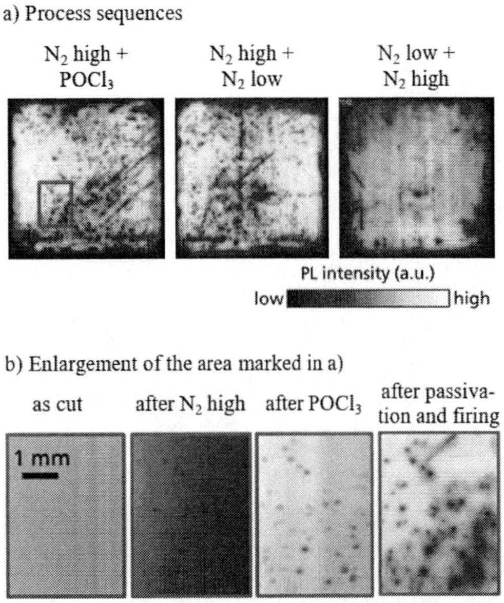

Fig. 7. a) Selected PL images for the material Mat1 taken after the indicated high-temperature process sequences, wet-chemical etch back, passivation, and firing in the second experiment (one representative wafer per group). The PL intensity scale is the same as in Fig. 4 and Fig. 6. b) Enlargement of the area marked in a). The PL images are taken as-cut, directly after the N_2 high tube furnace process, after $POCl_3$ diffusion and finally after passivation and firing. Different PL systems are used and the scaling is different.

$POCl_3$-last processing has a positive impact on τ_{eff} for all materials. This might be related to a $POCl_3$ gettering that reduces recombination active impurities in the silicon wafers. After single $POCl_3$ diffusion, this results in $3\ \text{ms} \leq \tau_{eff} \leq 6\ \text{ms}$ for all investigated materials.

Our optimized BBr_3 diffusion process is suitable for sequential diffusion processes. In combination with $POCl_3$ diffusion, τ_{eff} is at least the same as for the reference group (no high-temperature processing), no matter which process order is used.

For process sequences with a moderate temperature in the first and a high temperature in the second process, some of the investigated wafers show ring structures, which likely originate from oxygen precipitate formation. These materials do not show any ring structures when changing the order of the high-temperature processes. Besides ring structures, also point-like structures are observed for some wafers which are subjected to process sequences with pure N_2 gas atmosphere at 1000°C as first process. These point-like structures might be attributed to oxide stacking faults, but they seem not to necessarily reduce τ_{eff}.

An appropriate order of the high temperature processes enables final charge carrier lifetimes that maintain or even exceed the values initially measured without any high-temperature processing.

ACKNOWLEDGEMENT

The authors would like to thank all colleagues at the Fraunhofer ISE Photovoltaic Technology Evaluation

Center (PV-TEC), especially U. Belledin, S. Maier and J. Stoffel.

REFERENCES

[1] K. Bothe, R. Sinton, J. Schmidt, "Fundamental boron-oxygen-related carrier lifetime limit in mono- and multicrystalline silicon", *Prog. Photovolt.: Res. Appl.*, vol. 13, pp. 287–296, 2005.

[2] A. Usami, Y. Fujii, K. Monoka, "The effect of swirl defects on the minority carrier lifetime in heat-treated silicon crystals", *J. Phys. D: Appl. Phys.*, vol. 10, no. 6, pp. 899–910, 1977.

[3] S. Werner, E. Lohmüller, S. Maier et al., "Process optimization for the front side of p-type silicon solar cells", *Proc. 29th EU PVSEC*, Amsterdam, Netherlands, 2014, pp. 1342–1347.

[4] A. Wolf, A. Kimmerle, S. Werner et al., "Status and perspective of emitter formation by $POCl_3$-diffusion", *Proc. 31st EU PVSEC*, Hamburg, Germany, 2015, pp. 414–419.

[5] S. Werner, E. Lohmüller, U. Belledin et al., "Optimization of BBr_3 diffusion processes for n-type silicon solar cells", *Proc. 31st EU PVSEC*, Hamburg, Germany, 2015, pp. 637–641.

[6] M. Spitz, U. Belledin, S. Rein, "Fast inductive inline measurement of the emitter sheet resistance in industrial solar cell fabrication", *Proc. 22nd EU PVSEC*, Milan, Italy, 2007, pp. 47–50.

[7] Seiffe J., Weiss L., Hofmann M. et al., "Alternative rear surface passivation for industrial cell production", *Proc. 23rd EU PVSEC*, Valencia, Spain, 2008.

[8] R. C. Newman, "Thermal donors in silicon: oxygen clusters or self-interstitial aggregates", *J. Phys. C: Solid State Phys. (Journal of Physics C: Solid State Physics)*, vol. 18, no. 30, pp. 967–972, 1985.

[9] M. A. Kessler, T. Ohrdes, B. Wolpensinger et al., "Charge carrier lifetime degradation in Cz silicon through the formation of a boron-rich layer during BBr_3 diffusion processes", *Semiconductor Science and Technology*, vol. 25, no. 5, pp. 1–9, 2010.

[10] J. Schmidt, B. Lim, D. Walter et al., "Impurity-related limitations of next-generation industrial silicon solar cells", *IEEE J. Photovoltaics*, vol. 3, no. 1, pp. 114–118 ,

[11] S. M. Myers, M. Seibt, W. Schröter, "Mechanisms of transition-metal gettering in silicon", *J. Appl. Phys.*, vol. 88, no. 7, pp. 3795–3819, 2000.

[12] M. Seibt, A. Sattler, C. Rudolf et al., "Gettering in silicon photovoltaics: current state and future perspectives", *phys. stat. sol. (a)*, vol. 203, no. 4, pp. 696–713, 2006.

[13] M. B. Shabani, T. Yamashita, E. Morita, "Metallic impurities in mono and multi-crystalline silicon and their gettering by phosphorus diffusion", *214th ECS Meeting*, Honolulu, HI, 2008, pp. 179–193.

[14] J. Schön, H. Habenicht, W. Warta et al., "Chromium distribution in multicrystalline silicon: comparison of simulations and experiments", *Prog. Photovolt: Res. Appl.*, pp. 676–680, 2012.

[15] A. Borghesi, B. Pivac, A. Sassella et al., "Oxygen precipitation in silicon", *J. Appl. Phys.*, vol. 77, no. 9, pp. 4169–4244, 1995.

[16] J. D. Murphy, M. Al-Amin, K. Bothe et al., "The effect of oxide precipitates on minority carrier lifetime in n-type silicon", *J. Appl. Phys.*, vol. 118, no. 21, p. 215706, 2015.

IMPLEMENTATION OF N^+ AND P^+ POLO JUNCTIONS ON FRONT AND REAR SIDE OF DOUBLE-SIDE CONTACTED INDUSTRIAL SILICON SOLAR CELLS

R. Peibst[1,2], Y. Larionova[1], S. Reiter[1], M. Turcu[1], R. Brendel[1,3], D. Tetzlaff[2], J. Krügener[2], T. Wietler[2], U. Höhne[4], J.-D. Kähler[4], H. Mehlich[5], S. Frigge[5]

[1]Institute for Solar Energy Research Hamelin (ISFH), Am Ohrberg 1, 31860 Emmerthal, Germany
[2]Institute for Electronic Materials and Devices, Leibniz Universität Hannover, Schneiderberg 32, 30167 Hannover, Germany
[3]Institute for Solid State Physics, Leibniz Universität Hannover, Appelstrasse 2, 30167 Hannover, Germany
[4]Centrotherm Photovoltaics AG, Vahrenwalder Str. 269A, D-30179 Hannover, Germany
[5]Meyer Burger (Germany) AG, An der Baumschule 6-8, D-09337 Hohenstein-Ernstthal, Germany

Abstract: We present building blocks for double-side contacted cells with poly-Si on passivating interfacial oxides (POLO) junctions for both polarities, fabricated by a lean process flow. For this purpose, we evaluate p^+ and n^+ POLO junctions utilizing ~1.7 nm thin wet chemically grown (ozone diluted in di-ionized water) and ozone grown interfacial oxides on different surface morphologies. We achieve excellent passivation quality on damaged etched (100) surfaces with record low J_0 values of 0.6 fA/cm^2 (implied open circuit voltage $V_{oc,impl}$ 748 mV) for n^+ POLO junctions and of 5 fA/cm^2 ($V_{oc,impl}$ 729 mV) for p^+ POLO junctions. However, on alkaline textured surfaces, ~6 times higher J_0 values are obtained. We compare *ex-situ*-doped poly-Si (intrinsically deposited and subsequently ion implanted) with *in-situ*-doped poly-Si layers. For POLO junctions formed on textured surfaces by utilizing wet chemical oxides and low-pressure chemical vapor deposition (LP-CVD) of 20 nm *in-situ* n^+-doped poly-Si, we obtain J_0 values down to 2.4 fA/cm^2. Also with plasma-enhanced chemical vapor deposition (PE-CVD) of *in-situ*-doped amorphous Si and subsequent crystallization, we obtain comparable results. A transparent conductive oxide (TCO), preferably temperature stable, seems to be required to support the limited lateral conductivity of POLO junctions with poly-Si layer thicknesses ≤ 20 nm. We find that the conductivity of indium tin oxide (ITO) strongly decreases upon firing, while the initial conductivity can be maintained even for firing temperatures of 800°C when capping the ITO with a thin SiN_x layer. Our recent cell precursors (156 mm × 156 mm Cz *n*-type wafers with an n^+ POLO junction on an alkaline textured front-side and a p^+ POLO junction on a damage-etched rear-side) exhibit a promising $V_{oc,impl}$ value of 732 mV, and a total J_0 value of the doped surfaces of 12 fA/cm^2. In combination with the high pseudo fill factor of 85.3 % and with a short-circuit current density of 40 mA/cm^2 as calculated by ray tracing simulations, the corresponding pseudo efficiency is 25.0 %.

Keywords: Carrier selective contacts, Polycrystalline Silicon, Passivation, Silicon Solar Cell

1 INTRODUCTION

A progressive approach to address the strongest limitation for the efficiency of today's industrial solar cells – recombination in the passivated and metallized regions [1] – is to utilize the excellent passivation quality of carrier selective junctions formed by doped **pol**ycrystalline-Si layers on passivating interfacial **o**xides (POLO junctions). Saturation current densities (J_0) values down to 1 fA/cm^2 have been demonstrated with these junctions on planar test structures [2], and efficiencies up to 25.1% have been achieved on small area lab-type cells with n^+ TopCon junctions on a planar rear side [3]. GIT and ECN recently attempt to transfer this "hybrid approach" (n^+ POLO/TopCon rear, "conventional" boron emitter front) on industrial scale [4,5]. Recombination in the passivated and screen-print metallized regions of the conventional boron emitter was found to limit the cell efficiency to 20.7 % – 21.2 % so far [4,5]. In order to exploit to the full efficiency potential of POLO junctions, the front-side recombination has to be addressed as well. This could be achieved by applying the excellent passivation quality of POLO junctions on both sides of the cell.

Although this approach offers numerous inherent advantages, also as compared to amorphous-Si:H/*c*-Si heterojunctions (less parasitic absorption in the Si layer, compatibility with high-temperature screen-print metallization and thus potentially lower Ag consumption), it is also accompanied with new challenges. The passivation quality of POLO junctions for both polarities has to be optimized (at least for one polarity on textured surfaces), doping of the poly-Si layers with opposite polarities on the front and rear side has to be realized, parasitic absorption in the poly-Si on the front has to be minimized by reducing the layer thickness, and, as we show below, lateral conductivity possibly has to be provided by temperature stable transparent conductive oxides. Since we are also targeting a lean process flow, we furthermore attempt to substitute the previously-used thermal oxidation for interfacial oxide growth by more elegant processes, and investigate different approaches to realize thin n^+ and p^+ poly-Si layers on the two sides of the cell

The targeted cell structure is sketched in Fig. 1.

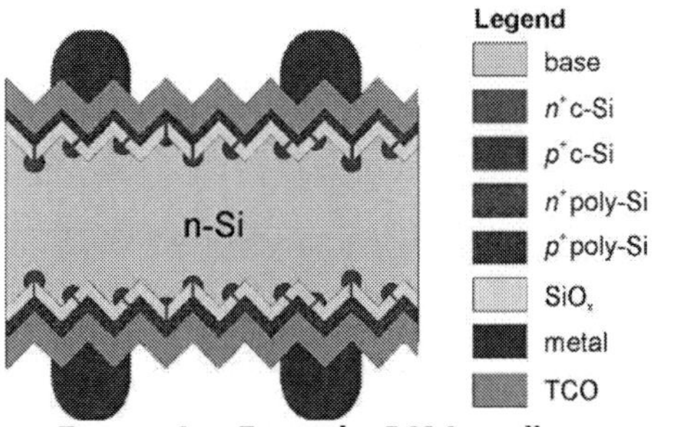

Figure 1: Targeted POLO cell structure. Modifications are described in the text.

Surface	Poly-Si doping type	H treat-ment	Chemical oxide (~1.7 nm)		Ozone oxide		Thermal oxide (~1.7 nm)	
			J_0 [fA/cm^2]	iV_{oc} @ 1 sun [mV]	J_0 [fA/cm^2]	iV_{oc} @ 1 sun [mV]	J_0 [fA/cm^2]	iV_{oc} @ 1 sun [mV]
planar	n^+	FGA	1.5	742	0.5	743	1.1	742
		SiN_x	0.6	747	0.6	748	0.8	749
textured	n^+	FGA	29	693	13	715	11	727
		SiN_x	6	735	4.4	738	3.8	741
planar	p^+	FGA	8	718	11	712	6	721
		SiN_x	5	729	5	727	3.8	734
textured	p^+	FGA	71	669	112	661	54	681
		SiN_x	40	690	48	684	27	700

Table 1: Passivation quality of ex-situ-doped poly-Si layers on different interfacial oxides grown on different Si surface morphologies. Here, intrinsic poly-Si (thickness 50 nm or 150 nm) is deposited by LP-CVD and subsequently doped by ion implantation with optimized doses. For all groups, an optimization of the annealing process is performed.

2 EXPERIMENTAL DETAILS

As substrate material for the optimization of the POLO junction formation, as well as for the cell precursors, we use 156 mm × 156 mm *n*-type Cz wafers (4 Ωcm, (100), as-cut thickness 170 µm). After saw damage removal in KOH and alkaline texturization of some wafers, a full RCA cleaning sequence (including the last HF dip) is performed. The interfacial oxide is grown by three different methods: (*i*) by a short dry thermal oxidation, yielding 1.7 nm oxide thickness as measured with ellipsometry on polished reference samples, (*ii*) wet chemically in de-ionized water with ozone, also yielding 1.7 nm oxide thickness, (*iii*) by ozone using the UV lamp based inline system *SQi* from Meyer&Burger Germany. For *ex-situ*-doped Si layers, intrinsic poly-Si is deposited by Low-Pressure Chemical Vapor Deposition (LP-CVD) in an *E2000* furnace from Centrotherm. Subsequently, the poly-Si is doped by phosphorus, boron or BF_2 ion implantation with optimized doses comparable to those given in Ref. [2]. For the textured samples or samples with lower poly-Si thickness (50 nm instead of 150 nm), the doses are adjusted accordingly in order to obtain a comparable doping concentration in the poly-Si.

We furthermore evaluate the POLO junction formation for *in-situ*-doped Si layers. For this purpose, Centrotherm developed a LP-CVD deposition process for *in-situ* n^+-doped poly-Si. We evaluate this process on symmetric, double-side-textured test structures based on 156 mm × 156 mm *p*-type Cz wafers. In this experiment, we use the ~1.7 nm thin wet chemically grown interfacial oxide. The thickness of the *in-situ* n^+-doped Si layers is 20 nm. Alternatively, 25 nm *in-situ* n^+- and p^+-doped amorphous Si layers are deposited by Plasma-Enhanced Chemical Vapor Deposition (PE-CVD) in a vonArdenne *CS 400P* cluster system.

After Si deposition, all samples are annealed for 30 min at different temperatures from 800 °C to 1050 °C in order to form POLO junctions. The optimum annealing temperature for the samples with wet-chemically- or ozone-grown oxides observed is 900 °C. One should note that the final poly-Si layer thickness is slightly reduced as compared to the as-deposited thickness, e.g. due to a partial oxidation during the annealing process. For example, the final poly-Si thickness for the PECVD-deposited samples was found to be ~17 nm according to TEM measurements. A low junction resistance is verified for all samples by the method of Römer *et al.* [6]. A forming gas anneal is performed, which slightly improves the passivation quality. Eventually, we also investigate the influence of an additional supply of hydrogen from SiN_x layers deposited on top of the poly-Si. The recombination behavior is analyzed by quasi-static photoconductance decay (QSSPC) measurements using a Sinton WCT-120 lifetime tester. Saturation current densities J_0 are determined according to the method of Kane and Swanson [7] and under Auger correction according to the model of Richter [8].

For the indium tin oxide (ITO), we deposit 75 nm-thick films by reactive DC magnetron sputtering process from an In_2O_3/SnO_2 (5 wt.%) target (vonArdenne *CS 400P* cluster system). Argon, Oxygen and Hydrogen are used as process gases. The deposition temperature is 320 °C. For electrical resistivity measurements, we deposit ITO on 6-inch single side-polished planar *n*-type Cz silicon wafers covered with an approximately 220 nm thick SiO_2 layer grown in a wet oxidation process. These thick oxide layers are electrically insulating and thereby enable four point probe measurements (4PP) to determine the sheet resistance of our ITO layers. Finally, half of the ITO samples are fully covered with a 10 nm thick SiN_x (refractive index n = 1.9) layer deposited by PE-CVD (Roth & Rau *SiNA* system). Firing is performed in a conveyor belt furnace from Centrotherm.

3 RESULTS AND DISCUSSION

3.1. SURFACE MORPHOLOGY AND INTER-FACIAL OXIDE (*EX-SITU*-DOPED LP-CVD POLY-SI)

Since we target a lean process flow, we first investigate methods to grow the interfacial oxide without inducing an additional process step. Thus, we replace thermal oxidation either by wet chemical oxide growth (cleaning is anyhow necessary), or by using the *SQi* inline UV lamp system from Meyer&Burger for ozone enhanced oxidation. Since our targeted cell structure in Fig. 1 also comprises POLO junctions on the front-side, we investigate test structures with planar and textured surfaces. For this investigation, *ex-situ*-doped poly-Si layers are used. For all interfacial oxides, we obtain very low J_0 values on planar (100) surfaces (Tab. 1). In particular, the J_0 value of 5 fA/cm^2 (implied open circuit voltage of 729 mV) for p^+ POLO junctions on a wet chemically grown oxide after hydrogenation is remarkably lower than the respective values reported so far. For n^+ POLO junctions, the excellent passivation quality on planar surfaces enables implied open circuit voltages up to 747-749 mV for the different interfacial oxides. These values are measured on ~150 µm thick substrates. For thinner wafers, $V_{oc,impl}$ should have exceeded 750 mV. A general finding is that the thermal budget of the oxide break-up step has to be adapted to the interfacial oxide. For the thermally grown oxide with 1.7 nm thickness, annealing temperatures of 1000°C yield the best result, while for both alternative oxides, the optimum annealing temperature is only 900 °C. A second general finding is that passivation quality on textured surfaces is worse than on planar surfaces. While for n^+ POLO junctions, promising J_0 values of 4.4 fA/cm^2 are obtained on an interfacial oxide grown under ozone atmosphere in an inline *SQi* system, the lowest J_0 values for p^+ POLO junctions are 27 fA/cm^2 (40 fA/cm^2) for thermally grown (wet chemically grown) interfacial oxide within the investigated range of annealing temperatures from 800 °C - 1050 °C. As shown by Larionova *et al.* [9], the poorer passivation quality of p^+ POLO junctions on textured surfaces is possibly due to the different passivation quality of the interfacial oxide on (100) and (111) surfaces (i.e., by different Si/SiO$_x$ interface state densities for the different surface orientations). Our cell concept shown in Fig. 1 is therefore modified so that the p^+ POLO junction is formed on a planar rear surface.

3.2. *IN-SITU*-DOPED POLY-SI

While subsequent doping of intrinsically deposited Si layers works well (see above) and is especially suited for an optimization of the doping concentration within the poly-Si, it would be desirable to further simplify the cell process flow by integrating the doping of the poly-Si into the Si deposition process. Both deposition techniques , LP-CVD and PE-CVD, offer the possibility for *in-situ* doping.

Centrotherm developed a LP-CVD process for the deposition of *in-situ* n^+-doped Si layers on 156 mm × 156 mm wafers. For a ~1.7 nm thin wet chemical interfacial oxide grown on textured surfaces, we obtain an excellent J_0 value of 2.4 (10) fA/cm^2 after hydrogenation utilizing SiN$_x$ (FGA). The poly-Si thickness is only 20 nm in this experiment. The optimum annealing temperature is determined to 900 °C. We also verified a homogeneous lifetime distribution across 156 mm × 156 mm *p*-type Cz wafers.

LP-CVD deposition is extremely conformal, which, on the other hand, complicates a single-side deposition of *in-situ*

Surface		H treat-ment	Thermal oxide (1.7 nm) J_0 [fA/cm^2]	Thermal oxide (1.7 nm) iV_{oc} @ 1 sun [mV]
planar	n^+	FGA	0.9	741
		SiN$_x$	0.4	749
textured	n^+	FGA	10	719
		SiN$_x$	6	727
planar	p^+	FGA	12	712
		SiN$_x$	8	723
textured	p^+	FGA	85	659
		SiN$_x$	68	667

Table 2: Passivation quality of 25 nm thin in-situ-doped Si layers, deposited by PE-CVD on thermally grown interfacial oxides (1.7 nm).

-doped layers by means of back-to-back processing of the wafers. However, for the targeted cell structure as shown in Fig. 1, different doping polarities for the poly-Si on the cell front side and rear side have to be realized. Therefore, we furthermore investigate PE-CVD, which offers the possibility for single-side deposition of *in-situ*-doped layers. Table 2 shows the passivation quality of POLO junctions obtained after PE-CVD deposition of *in-situ*-doped amorphous Si layers (25 nm thickness) and subsequent crystallization during the oxide break-up step. We obtain comparable passivation quality as with LP-CVD and subsequent ion implantation: 0.4 fA/cm^2 for n^+ and 8 fA/cm^2 for p^+ POLO junctions on planar surfaces after hydrogenation by SiN$_x$. Again, we obtain higher J_0 values on textured surfaces, in particular for p^+ POLO junctions. This is consistent to our hypothesis regarding the different interface state densities of SiO$_x$ on Si (100) and Si (111) [9].

3.3. OPTICAL TRANSPARENCY AND IMPLICATIONS ON SHEET RESISTANCE

As reported previously [10], the poly-Si layer thickness has to be restricted to values ≤ 20 nm in order to limit the photo current losses due to parasitic absorption to values below 1 mA/cm^2. While adjusting the ion implantation parameters and using BF$_2$ instead of boron as implant species, we successfully reduce the thickness of the poly-Si layers deposited by LP-CVD from 150 nm to 50 nm while maintaining comparable junction quality. For the *in-situ*-doped poly-Si layers deposited by either by LP-CVD or PE-CVD, the thickness is already in the targeted range.

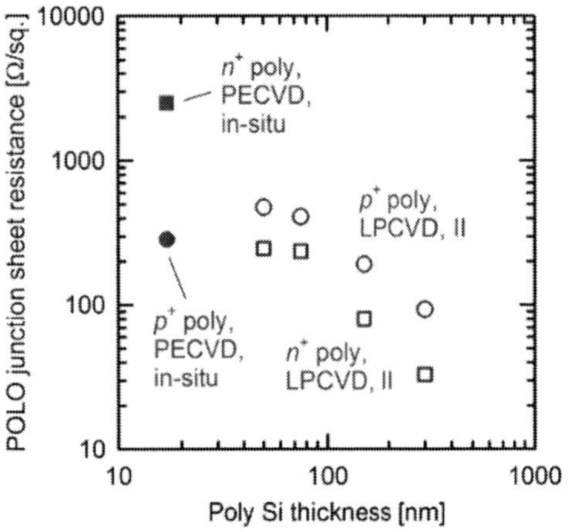

Figure 2: Sheet resistances for POLO junctions with optimized passivation and current transport properties as a function of the poly-Si layer thickness.

A drawback of the reduction of the poly-Si thickness to values of 20 nm and below is an increase of sheet resistance R_{sheet} above 1000 Ω/sq. for n^+ POLO junctions and to several hundred Ω/sq. for p^+ POLO junctions (Fig. 2). Thus, additional lateral conductivity should be provided by a transparent conductive oxide (TCO) at least for the polarity of the emitter (unless considering a rear emitter configuration in combination with a full-area rear-side metallization based on Physical vapor Deposition (PVD)).

3.4. TEMPERATURE STABILITY OF TCO

Since we are targeting screen-print metallization with-high temperatures pastes, we investigate the temperature stability of TCO layers, starting with ITO as the most common candidate. We anneal the ITO layers in a belt firing furnace with peak temperatures up to 850 °C whereas one non-annealed sample serves as reference. Subsequently, we determine the sheet resistance with four point probe measurements. Figure 3 shows that the sheet resistance of

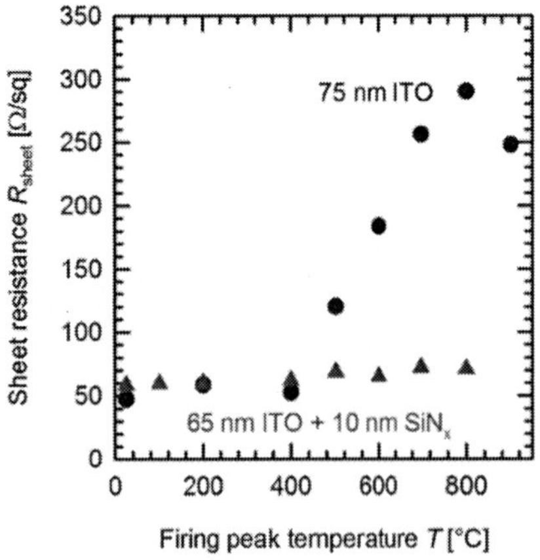

Figure 3: Influence of firing temperature T in air atmosphere to the sheet resistance R_{sheet} of ITO with and without SiN_x capping layer.

ITO films increases from 47 Ω/sq. to 274 Ω/sq. upon firing. In order to prevent such an increase, we use a 10 nm thick silicon nitride layer which functions as protective layer. Figure 3 shows that we are able to keep the sheet resistance of ITO films close to its initial value with a SiN_x capping layer. This simple method ensures a good lateral conductivity and is quite promising for solar cell applications. Moreover, the high temperature treatment improves the transparency of the TCO.

3.5. CELL PRECURSORS

Figure 4 shows the injection level dependency of the effective lifetime measured by QSSPC after junction formation on a 156 mm × 156 mm n-type Cz cell precursor with n^+ POLO junctions on the textured front side and p^+ POLO junctions on the planar rear side. According to ray-tracing simulations [10], the 50 nm-thin poly-Si layers deposited by LP-CVD enable photo current densities of 40 mA/cm^2. The $V_{oc,impl}$ value of 732 mV (total J_0 value of both doped surfaces = 12 fA/cm^2) is promising. The implied pseudo fill factor $pFF_{,impl.}$ of 85.3 % shows an excellent recombination behavior. The resulting pseudo efficiency is 25.0 %.

In order to finish these precursors to final cells, we still have to further develop our back-end process. In particular, the SiN_x capping layer on top of the TCO complicates the metallization sequence. However, we do not consider this aspect as a fundamental road blocker.

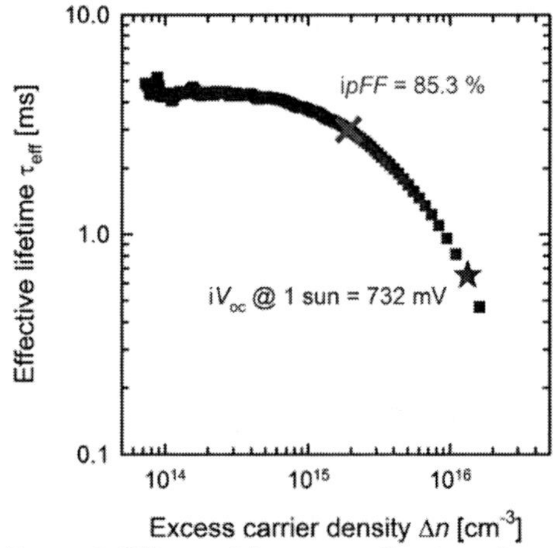

Figure 4: Effective lifetime as a function of excess carrier density, measured on a cell precursor with an n^+ POLO junction on the textured front side and a p^+ POLO junction on a planar rear side. The 50 nm thin poly-Si layers are deposited intrinsically by LP-CVD, and subsequently doped by P and BF_2 ion implantation.

4 CONCLUSIONS

We propose to integrate POLO junctions for both polarities in industrial double-side contacted solar cells. We demonstrate excellent passivation quality for *ex-situ*-doped n^+ POLO junctions on planar and textured surfaces utilizing wet chemically-grown or ozone based interfacial oxides, reaching

J_0 values down to 0.6 fA/cm^2 on planar and down to 4.4 fA/cm^2 on textured surfaces. For p^+ POLO junctions, J_0 values down to 5 fA/cm^2 are obtained on planar surfaces. However, the passivation quality on textured surfaces is found to be worse (40 fA/cm^2) for p^+ POLO junctions with wet-chemically grown interfacial oxides. For our cell concept, this issue implies the necessity to form the p^+ POLO junctions on a planar rear surface. For POLO junctions on textured surfaces utilizing wet chemical oxides grown and LP-CVD-deposition of 20 nm *in-situ* n^+-doped poly-Si, we obtain J_0 values down to 2.4 fA/cm^2. Also with PE-CVD deposition of *in-situ*-doped amorphous Si and subsequent crystallization, we obtain comparable results. A temperature stable TCO can support the limited lateral conductivity of POLO junctions with $\leq$ 20 nm poly-Si thickness while still facilitating screen-print metallization with high-temperature pastes. We find that the conductivity of indium tin oxide (ITO) strongly decreases upon firing, while the initial conductivity can be maintained even for firing temperatures of 800°C when capping the TCO with a 10 nm SiN_x layer. Our recent cell precursors (156 mm × 156 mm Cz n-type wafers with an alkaline textured front-side and a damage-etched rear-side) with 50 nm n^+ poly-Si on the front and 50 nm p^+ poly-Si on the rear exhibit a promising $V_{oc,impl}$ value of 732 mV, and a total J_0 value of the doped surfaces of 12 fA/cm^2. In combination with the high pseudo fill factor of 85.3 % and a short-circuit current density of 40 mA/cm^2 as calculated by ray tracing simulations, the corresponding pseudo efficiency is 25.0 %.

5 ACKNOWLEDGEMENTS

We thank B. Gehring, R. Winter, H. Fischer, A. Mohr, G. Glowatzki, B. Koch, R. Zieseniß and R. Gogolin for their help with sample processing, and R. Reineke-Koch and M. Rienäcker for valuable discussions. Furthermore, we thank the German Ministry for Economic Affairs and Energy for funding under grant and 0325702 (POLO).

6 REFERENCES

[1] R. Brendel, T. Dullweber, R. Peibst, C. Kranz, A. Merkle, D. Walter. Breakdown of the efficiency gap to 29% based on experimental input data and modeling. Progress in Photovoltaics: Research and Applications DOI: 10.1002/pip.2696, (2015).

[2] U. Römer, R. Peibst, T. Ohrdes, B. Lim, J. Krügener, T. Wietler, R. Brendel. Ion implantation for poly-Si passivated back-junction back-contacted solar cells. IEEE J Photovoltaics **5**, pp. 507–514 (2015).

[3] S.W. Glunz, F. Feldmann, A. Richter, M. Bivour, C. Reichel, H. Steinkemper, J. Benick, M. Hermle. The Irresistible Charm of a Simple Current Flow Pattern – 25% with a Solar Cell Featuring a Full-Area Back Contact. Proceedings of the 31st EU-PVSEC, pp. 259-263 (2015).

[4] M. K. Stodolny, M. Lenes, Y. Wu, G.J.M. Janssen, I.G. Romijn, J.R.M. Luchies, L.J. Geerligs. N-type polysilicon passivating contacts for industrial bifacial n-PERT cells. Presented at the 6th SiliconPV conference, Chambery, submitted to Energy Procedia, (2016).

[5] Y. Tao, V. Upadhyaya, C.-W. Chen, A. Payne, E. L. Chang, A. Upadhyaya, A. Rohatgi. Large area tunnel oxide passivated rear contact n-type Si solar cells with 21.2% efficiency. Progress in Photovoltaics: Research and Applications **24**, pp. 830–835, (2016).

[6] U. Römer, R. Peibst, T. Ohrdes, B. Lim, J. Krügener, E. Bugiel, T. Wietler, R. Brendel. Recombination behaviour and contact resistance of n^+ and p^+ poly-crystalline Si/mono-crystalline Si junctions. Solar Energy Materials and Solar Cells **131**, pp. 85–91, (2014).

[7] D.E. Kane, R.M. Swanson, Proceedings of the 18th IEEE Photovoltaic Specialists Conference, p. 578, (1985).

[8] A. Richter, S. W. Glunz, F. Werner, J. Schmidt, A. Cuevas. Improved quantitative description of Auger recombination in crystalline silicon. Physical Review B **86**, p. 165202, (2012).

[9] Y. Larionova; R. Peibst, M. Turcu, S. Reiter, R. Brendel, D. Tetzlaff, J. Krügener, T. Wietler, U. Höhne, J.-D. Kähler. Optimization of p^+ poy-Si / c-Si junctions in wet-chemically grown interfacial oxides and on different wafer morphologies. 2CO.4.3, this conference.

[10] S. Reiter, N. Koper, R. Reineke-Koch, Y. Larionova, M. Turcu, J. Krügener, D. Tetzlaff, T. Wietler, U. Höhne, J.-D. Kähler, R. Brendel, R. Peibst. Parasitic absorption in polycrystalline Si-layers for carrier-selective front junctions. Presented at the 6th SiliconPV conference, Chambery, submitted to Energy Procedia, (2016).

PROCESS DEVELOPMENT OF SILICON HETEROJUNCTION INTERDIGITATED BACK-CONTACTED (SHJ-IBC) SOLAR CELLS BONDED TO GLASS

Menglei Xu[1], Twan Bearda, Hariharsudan Sivaramakrishnan Radhakrishnan, Shashi Kiran Jonnak, Miha Filipič, Valerie Depauw, Kris Van Nieuwenhuysen, Yaser Abdulraheem[2], Maarten Debucquoy, Ivan Gordon, Jozef Szlufcik, Jef Poortmans[1,3]

IMEC, Leuven, Belgium
[1]Also with KU Leuven, Leuven, Belgium
[2]Kuwait University, Safat, Kuwait
[3]Also with Universiteit Hasselt, Hasselt, Belgium

ABSTRACT: In imec's i^2-module concept, silicon heterojunction interdigitated back-contacted (SHJ-IBC) solar cells are fabricated on monocrystalline foils bonded to glass. The proposed technology allows for cell processing on thin wafers mechanically supported by the glass, increasing the yield of processing such thin wafers. A process sequence for SHJ-IBC cell fabrication that can be applied to bonded thin foils is described. We investigated and optimized individual process steps on thick wafers. Then the developed steps were integrated into a process flow to fabricate solar cells on wafers with different thicknesses and bonding agents. On wafers with a thickness of 190 μm, functional cells with efficiencies of 22.6% and 21.7% were made on freestanding and silicone bonded wafers, respectively. On thin wafers of 57 μm, our best SHJ-IBC cell on an EVA bonded wafer exhibits excellent V_{oc} of 740 mV and efficiency of 20.0%, which demonstrates the high potential of the i^2-module concept.
Keywords: Amorphous Silicon, Heterojunction, Silicon Solar Cell, interdigitated back-contacted, Superstrate Processing

1 INTRODUCTION

The cost of silicon material accounts for approximately 61% of today's solar cell cost and 30% to 40% of the total c-Si solar module cost [1]. Hence, to reduce the silicon consumption, fabrication of solar cells on thin monocrystalline silicon foils is being studied by several groups [2-4].

Imec proposed the i^2-module concept, which allows for large-scale module level processing of interdigitated back-contacted (IBC) solar cells on 40 μm thin silicon foils bonded to glass [5]. The glass delivers mechanical support to the thin foil and thus significantly reduces the foil breakage during cells processing. The choice for an IBC device allows to process the junction and the contacts after bonding. Certain constraints are imposed to the cell process mainly due to the bonding and the glass. For that reason, an a-Si:H heterojunction process is chosen for junction formation as that allows to keep the processing temperature low.

In this contribution, we report the recent progress in the process development of silicon heterojunction IBC (SHJ-IBC) solar cells. An integration process is presented, which is fully compatible with thin foils bonded to glass. Functional cells were fabricated on both freestanding and bonded n-type float zone (FZ) wafers with different thicknesses. Two different bonding adhesives, i.e., silicone and Ethylene Vinyl Acetate (EVA), are evaluated at cell level. The factors limiting the cell performance are identified and the main losses of the cells are discussed in detail.

2 RESULTS AND DISCUSSION

2.1 Integration process

All cells, with an area of 3.97 cm², were processed on 190 μm and 57 μm thick n-type FZ wafers (6 inch, <100>, 3Ω-cm).

The wafers were exposed to Sulfuric acid/Ozone mixture (SOM) and following $HF:HCl:H_2O$ (2:5:93) solution for cleaning. Tetramethyl ammonium hydroxide (TMAH) based solution was used for thinning down and texturing of the wafers. As shown in Figure 1, the textured front side underwent PECVD a-Si:H passivation and SiN_x antireflection coating, followed by bonding to glass using EVA or silicone encapsulant. We have reported previously that plasma-silicone interaction during a-Si:H deposition has a detrimental impact on passivation quality [6]. Hence, silicone bonded wafers were treated with Ar plasma to form a more densely crosslinked layer by local modification of the exposed silicone surface [6]. Then the depositions of a-Si:H (i/n+) and a stack of SiO_x/a-Si:H were performed. This SiO_x layer is a sacrificial mask for subsequent lift-off process and the top a-Si:H layer is used to protect SiO_x during HF dip prior to a-Si:H (i/p+) deposition. Photolithography and wet etching were used to define the back surface field (BSF) area. The emitter was passivated by a-Si:H

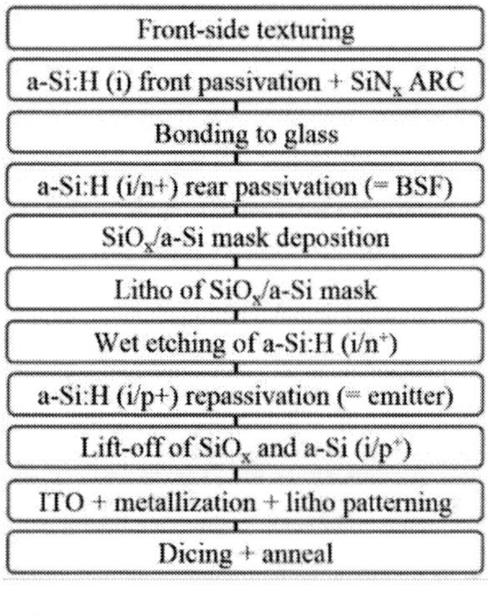

Figure 1: Integration process flow of SHJ-IBC cells on bonded substrate.

(i/p+) and patterned by self-alignment lift-off process, which was carried out in dilute HF. A stack of indium-doped tin-oxide (ITO) and Cu was deposited to form the rear side metal contact. Thermal annealing was applied to the finished cells in order to improve the contact behaviour. To investigate the influence of the presence of bongding agents and glass on the cell performance, identical cells were also fabricated on freestanding wafers without bonding to glass.

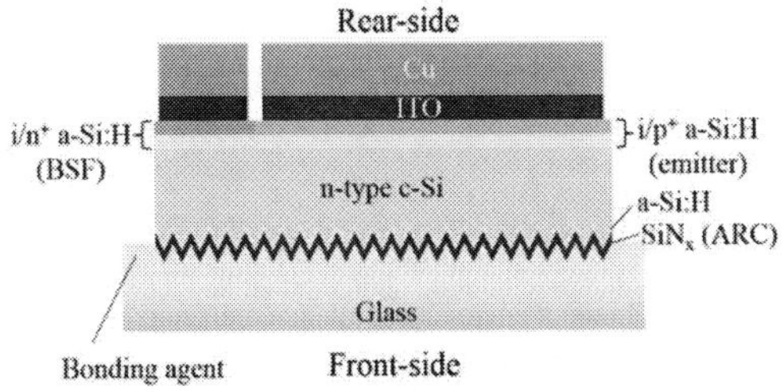

Figure 2: The cross-sections (not drawn to scale) of the bonded SHJ-IBC solar cells are depicted in the diagram.

2.2 Device results

The cell results are shown in Table I. For the best cells before anneal, we observe an improvement of $\Delta\eta$ = 0.5% on freestanding cells with respect to silicone bonded cells. This is due to an absolute increase ΔV_{oc} = 8 mV and ΔFF = 0.7%, but more significantly due to an increase ΔJ_{sc} = 1.1 mA/cm^2. Figure 3 compares the external quantum efficiency (EQE) of cells on freestanding wafers with that on silicone and EVA bonded wafers. The difference in EQE suggests that glass and silicone at the front side lead to additional reflectance and absorption losses for a wide range of wavelengths. Nevertheless, a high J_{sc} of 41.2 mA/cm^2 was still achieved on silicone bonded cells.

Table I: Best Cell results on freestanding and bonded wafers

Bonding Thickness	Anneal	J_{sc} (mA/cm^2)	V_{oc} (mV)	FF (%)	η (%)
freestanding 190 µm	before	42.3	735	69.7	21.5
	after	41.8	730	74.8	22.6
Silicone 190 µm	before	41.2	727	70.4	21.0
	after	40.8	734	73.1	21.7
EVA 190 µm	before	40.7	718	68.6	20.0
	after	39.9	724	71.6	20.7
EVA 57 µm	before	38.8	738	67.5	19.0
	after	38.6	740	71.7	20.0

For the cells on EVA bonded thick wafers, the best J_{sc} is 40.7 mA/cm^2 that is 0.5 mA/cm^2 lower than that of silicone bonded cells. As shown in Figure 3, this result can be attributed to the light absorption loss at short wavelength due to the UV blocker contained in EVA. And for the EVA bonded cells on 57 µm thin wafer, a J_{sc} of 38.8 mA/cm^2 was achieved, which is further limited by the light absorption loss at long wavelength.

The excellent V_{oc} of 734 mV and 740 mV were obtained on the silicone bonded and EVA bonded cells after anneal, respectively, similar to the V_{oc} of cells without bonding agent. It indicates that the impact of bonding agents on the surface passivation quality has been prevented. Such achievement has been made possible by several adaptations in the process:

- The new encapsulants silicone and EVA were selected to replace the previously used PV 6100 [7], because these alternative bonding agents were proven to be more resistant to chemicals and plasma treatments.
- An optimized curing/outgassing step reduces outgassing of bonding agents during a-Si:H deposition.
- The Ar plasma improves crosslinking of the silicone, and hence reduces the plasma-silicone interaction.
- The SiO_x hard mask (~680 nm) also covers the bonding agents during rear side process and thus further reduces the interaction of plasma /chemicals and bonding agents.
- Optimum a-Si:H deposition conditions were chosen for surface passivation in presence of different bonding agents.

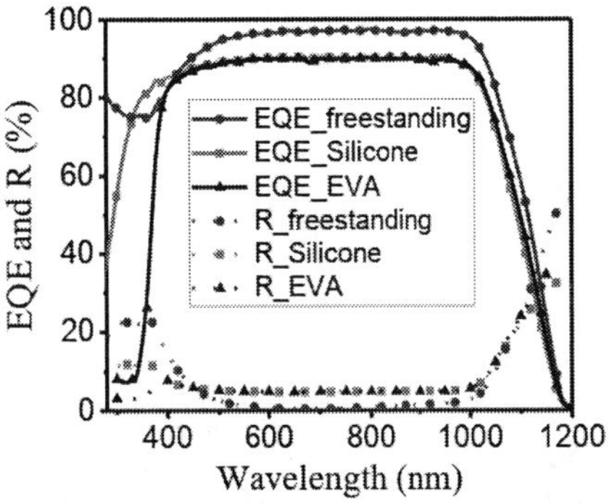

Figure 3: External quantum efficiency of a cell on freestanding wafer compared to that on bonded wafers. The thicknesses of the wafers are 190 µm.

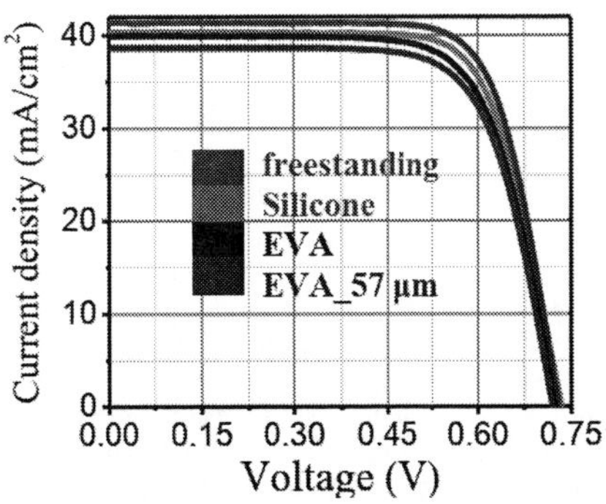

Figure 4: Light J-V curve of the best SHJ-IBC cells after anneal on freestanding and bonded wafers.

Table II: FF loss analysis of a cell on a freestanding wafer before and after anneal

Anneal	Cell parameters				FF loss analysis	
	R_s (Ω.cm^2)	R_{sh} (Ω.cm^2)	FF (%)	pFF (%)	ΔFF_{Rs} (% absolute)	ΔFF_{Rsh}
before	2.8	2.2×10^3	68.0	84.1	13.5	0.6
after	1.6	2.1×10^3	73.8	82.5	7.9	0.6

For all cells before annealing, the efficiencies are strongly limited by the relatively low FF (≤70.4%). Thermal anneals of the finished cells at temperatures below 200°C were performed under air ambient, which yielded a significant improvement in FF. For cells on freestanding wafers, the highest FF of 74.8% has been achieved after annealing at 150°C. A detailed FF loss analysis is shown in Table II [8]. The series resistance R_s was determined by the Bowden method at different light intensities [9]. The shunt resistance R_{sh} was determined by a linear fit of the dark I-V curve. The pseudo fill factor (pFF) was extracted from Suns-V_{oc} measurements. The absolute FF loss ΔFF_{Rs} was calculated to be 13.5%, while ΔFF_{Rsh} was only 0.6%. This reveals that the drop of the measured FF was primarily caused by series resistance. After annealing, R_s reduced significantly from 2.8 Ω.cm^2 to 1.6 Ω.cm^2, where corresponding ΔFF_{Rs} reduced from 13.5% to 7.9%.

3 CONCLUSION AND OUTLOOK

We presented an integration process for module-level processing of SHJ-IBC solar cells, which is compatible with thin silicon foils bonded to glass. Individual process steps have been investigated and optimized. To demonstrate the developed process, highly efficient cells on 190 μm and 57 μm thick wafers were fabricated. Issues limiting the cell performance were identified. Current development is the introduction of thin epitaxially grown wafers instead of thin FZ wafers [10]. Next steps will be directed towards improvement of fill factor and proces simplifications, including investigations and integration of lithography-free emitter patterning approach, dry etching of a-Si:H (already demonstrated in [11]) and selective depositon of a-Si:H to replace lift-off.

ACKNOWLEDGEMENTS

The authors gratefully acknowledge the financial support of imec's industrial affiliation program for Si-PV. This work was also partially funded by Kuwait Foundation for the Advancement of Sciences under project code: P115-15EE-01. The research leading to these results has received funding from the European Union's Seventh Programme for research, technological development and demonstration under grant agreements n° 609788 (Cheetah), and also funding from the European Union's Horizon 2020 research and innovation programme under the Marie Sklodowska-Curie grant agreement No 657270.

REFERENCES

[1] SEMI, International Technology Roadmap for Photovoltaic (ITRPV) (2016)
[2] J. Govaerts et al., Energy Procedia 77 (2015) 871
[3] S. Schoenfelder et al., Energy Procedia 38 (2013) 942
[4] L. Wang et al., IEEE J. Photovoltaics 4 (2014) 1397
[5] J. Govaerts et al., Sol. Energy Mater. Sol. Cells 113 (2013) 52
[6] S.N. Granata et al., Sol. Energy Mater. Sol. Cells 124 (2014) 48
[7] T. Bearda et al., 31st EUPVSEC (2015) p365
[8] A. Khanna et al., IEEE J. Photovoltaics 3 (2013) 1170
[9] S. Bowden et al., 17th EUPVSEC (2001) p1802
[10] K Van Nieuwenhuysen et al., 28th EUPVSEC (2012) p2471
[11] H. Sivaramakrishnan Radhakrishnan et al., 43rd IEEE PVSC (2016) conference proceedings

STATUS OF THE EU FP7 HERCULES PROJECT: WHAT IS THE POTENTIAL OF N-TYPE SILICON SOLAR CELLS IN EUROPE?

D. Muñoz[1], P.J. Ribeyron[1], S. Harrison[1], Ch. Allebe[2], A. Descoeudres[2], M. Despeisse[2], Ch. Reichel[3], S. W. Glunz[3], R. Peibst[4], A. Merkle[4], O. Nielsen[5], I. Martin[6], V. Mihailetchi[7], Th. Södeström[8], B. Demaurex[8] S. de Wolf[9], H. Mehlich[10], J. Zhao[10], J. Alvarez[11], J. Dupuis[12], E. Macron[13], B. de Gier[14], M. Taillan[15], F. Korsos[15] and L. Korte[16]

[1]CEA-LITEN, Institut National de l'Energie Solaire (INES), Savoie Technolac, F-73377 Le Bourget du Lac, France
[2] CSEM, PV-center, Jaquet Droz-1 , 2000 Neuchâtel, Switzerland
[3] Fraunhofer Institute for Solar Energy Systems (ISE), Heidenhofstrasse 2, 79110 Freiburg, Germany
[4] Institute for Solar Energy Research Hamelin (ISFH), 1 Am Ohrberg, 31860 Emmerthal , Germany,
[5]NorSun AS,Karenslyst Allé 9C,NO-0278 OSLO, Norway
[6] Departament d'Enginyeria Electrònica, Universitat Politècnica de Catalunya, C/Jordi Girona 1-3, Mòdul C4, 08034 Barcelona, Spain.
[7]International Solar Energy Research Center - ISC Konstanz, Rudolf-Diesel Str.15 , 78467 Konstanz, Germany
[8]Meyer Burger AG,Schorenstrasse 39, CH-3645 Gwatt (Thun) Switzerland
[9] École Polytechnique Fédérale de Lausanne (EPFL), Institute of Microengineering (IMT), Photovoltaics and Thin Film Electronics Laboratory, Rue de la Maladière 71b, 2000 Neuchâtel, Switzerland
[10] Meyer Burger (Germany) AG, Gewerbegebiet Am Sachsenring II, An der Baumschule 6-8, Germany, 09337 Hohenstein-Ernstthal
[11]CNRS UMR 8507, Université Paris Sud, Université Paris 6 et Centrale Supelec. Paris, France
[12] 1/ EDF R&D, 6 quai Watier, 78400 CHATOU Cedex, France, 2/ IRDEP, Institut de Recherche et Développement de l'Energie Photovoltaïque, 6 quai Watier, 78400 CHATOU Cedex, FRANCE
[13] AYMING société par actions simplifiée (SAS), Business Line Grant Solutions: 55 avenue René Cassin 69009 Lyon (France)
[14] Eurotron BV, Van Beukelaarweg 45, 2971 VL Bleskensgraaf, The Netherlands
[15]Semilab Co. Ltd.,Prielle Kornélia u.2.,1117 Budapest, Hungary
[16]Helmholtz-Zentrum Berlin, Institute Silicon Photovoltaics, Kekuléstr. 5, 12489 Berlin, Germany

Corresponding Author: *Delfina Muñoz (delfina.munoz@cea.fr)*

ABSTRACT: The concept proposed by the HERCULES project is to develop innovative n-type monocrystalline c-Si device structures based on both sides' contacted silicon heterojunction (SHJ) solar cells, on interdigitated back-contact (IBC) solar cells with alternative junction formation, as well as on hybrid concepts (homo-heterojunction). These concepts are the most promising technologies to reach ultra-high efficiencies with industrially relevant processes. The HERCULES strategy is to transfer the developed processes to industrial scale by considering all major cost drivers of the entire manufacturing process chain. The final objective is to obtain both high-efficiency solar cells and modules, based on adequately simple process sequences. It requires strong expertise and equipment mastering in order to provide a commercially viable technology that can be implemented by the European PV industry. To this end, the HERCULES consortium takes advantage of its leading European expertise in both research and production. In this paper, major results are presented concerning the first 30 months in material, solar cell development, module integration and pilot line activity. The strong improvements achieved in all the value chain demonstrate the strong need of collaboration within all partners to reach very ambitious objectives from the optimization of n-type solar cells & modules up to production level.
Keywords: heterojunction, IBC, Silicon, solar cells, industrialization

1 INTRODUCTION

The concept proposed by the HERCULES project is to develop innovative n-type monocrystalline c-Si device structures based on both sides' contacted silicon heterojunction (SHJ) solar cells, on interdigitated back-contact (IBC) solar cells with alternative junction formation, as well as on hybrid concepts (homo-heterojunction). These concepts are the most promising technologies to reach ultra-high efficiencies with industrially relevant processes. The HERCULES strategy is to transfer the developed processes to industrial scale by considering all major cost drivers of the entire manufacturing process chain. The final objective is to obtain both high-efficiency solar cells and modules, based on adequately simple process sequences. It requires strong expertise and equipment mastering in order to provide a commercially viable technology that can be implemented by the European PV industry. To this end, the HERCULES consortium takes advantage of its leading European expertise in both research and production.

The key objectives of the HERCULES project are to:

1) Develop ultra-high-efficiency (η > 21%) modules at the pilot line scale;
2) Reduce production and investment complexity and demonstrate costs of 0.4 €/Wp at a 500 MW/year commercial plant level;
3) Increase the lifetime of modules up to 35 years;
4) Demonstrate ultra-high-efficiency solar cells

The HERCULES project started in November 2014 and in this work we will present the most relevant results concerning R&D activities and industrial pilot line production obtained up to project month 24.

The complete value chain is taken into account in the HERCULES project. First of all, a specification for n-type c-Si material parameters and wafer thickness has been developed to obtain high efficiency devices. NorSun has been able to obtain effective lifetimes >3ms on ingots with very low oxygen content.

Furthermore, strong efforts have been made on improved p/n junction formation by alternative methods (high and low temperature approaches) combining high passivation quality (implied Voc>725mV) with very low j_0<30fA/cm^2. Several round robin experiments were performed to combine and characterize the best junction/passivation layer and/or laser treatments. Different approaches on the optimization of TCO and metallization and their interaction have been explored with successful results in terms of performance, but also of stability: low silver consumption, indium free TCOs, high mobility materials, new seed layers for Cu-plating and their interaction with layers, new interconnection

approaches (smartwire, bilevel metallization). A first workshop on module interconnection was performed to establish how to homogenize the calculations on cell to module (CTM) losses for all proposed technologies.
Record solar cells have been obtained: IBC with efficiencies over 24% at the R&D level and over 22% at the industrial level, SHJ surpassing 23% and IBC SHJ close to 23%. In-depth characterization and 2D simulation have been proposed to understand physical mechanisms limiting efficiency for the different approaches. More than 26000 cells have been fabricated in the Meyer-Burger pilot line with efficiencies well over 22%. In the ISC pilot line, 1200 IBC cells with median efficiency over 21.5% have been produced. High efficiency modules have been fabricated based on both technologies: up to 329W for SHJ and 318W for SHJ half cells module and up to 303W for IBC modules. Finally, first realistic cost calculations based on HERCULES data for IBC and SHJ show that 0.4€/Wp is a realistic goal for the project, and that a European technology is possible.

2. UNDERSTANDING THE ROLE OF MATERIAL

One of the key points to obtain high efficiency solar cells is the quality of the n-type silicon material. In the frame of the HERCULES project, Norsun has provided different set of wafers coming from different ingots for material quality evaluation. Different tests have been performed considering the different structures available in the HERCULES consortium, mainly SHJ, PERT and IBC cells. On the one hand, the impact of the thickness has been evaluated in terms of mechanical tests and solar cell impact. On the other hand, the impact of different wafers coming from the whole ingot has been also studied in terms of material specification and solar cell behavior (see figure 1).

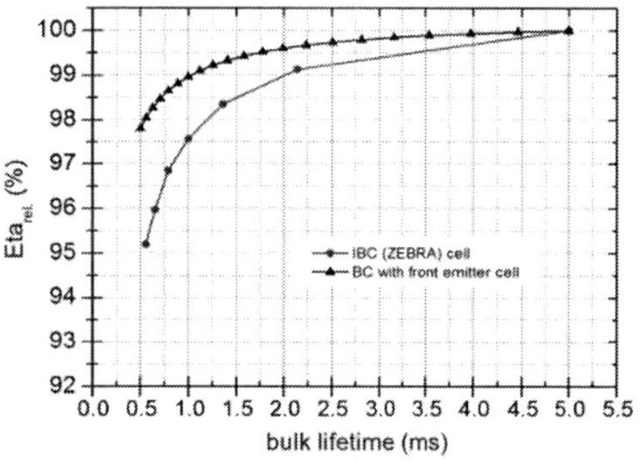

Figure 1: Efficiency dependence of SHJ and OBC technology in function of the effective lifetime of the wafers form a Norsun ingot.

Moreover, Norsun has been able to obtain complete ingots with effective lifetime surpassing 3ms, with very low oxygen concentration (Figure 2).

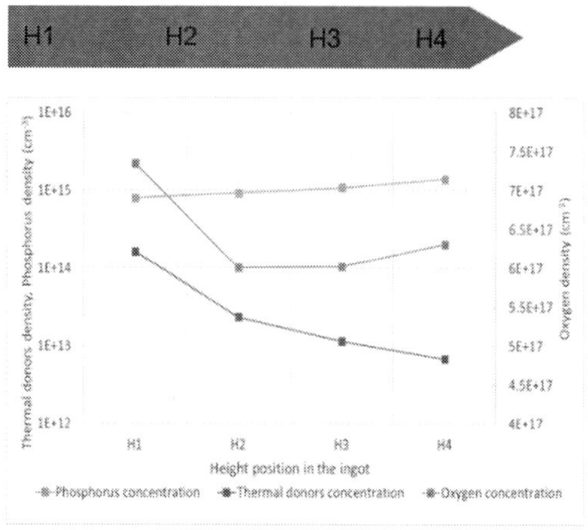

Figure 2: Measurements of a whole ingot oxygen, phosphorus and thermal donor content form a Norsun ingot.

3. SOLAR CELL RESULTS

Concerning solar cells development, lot of work has been done up to now to improve efficiency but also to simplify the process to reduce manufacturing costs. Different paths for junction formation have been explored:

1. Ion implantation (II)
2. Laser contacts (LC)
3. Heterojunction layers (SHJ)

In all cases, solutions with very promising values of J0e and low recombination velocities have been obtained as summarized in Table 1.

Table 1: j0e, iVoc and passivation values obtained by the HERCULES consortium with the different approaches for the junction formation.

		J0e fA/cm^2	Rsh Ω□	Seff cm/s	iVoc mV
II	emitter	5-10			
	BSF	<8			
LC	Al2O3/SiCx:B + laser UV	<20	<100		
	SiNxP + laser UV		<50	3500	
SHJ	a-Si:H				740
	a-SiOx /a-SiCx				730
	laser annealed				720

Moreover, innovative approaches like the MoOx and WOx contacts have been tested to overcome a-Si:H limitations in terms of parasitic absorption in the front side. Strong characterization has been performed to identify the main limitations of new materials to reach high efficiency devices. For instance, the correlation between WOx band bending and Voc and FF has been put in evidence as seen in figure 3.

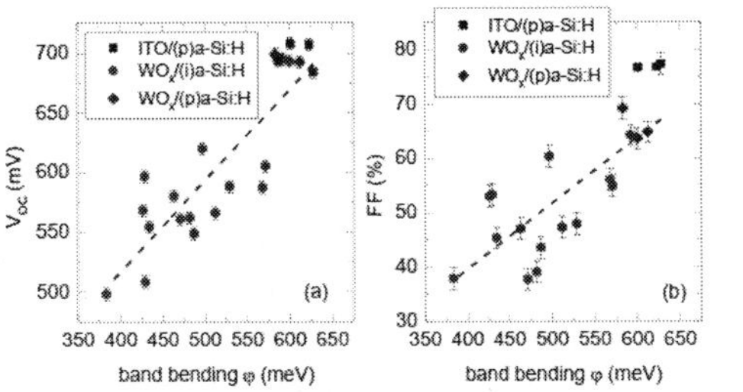

Figure 3: Correlation between band bending measurements of different WOx layers and solar cells parameters.

Metallization and contacts have been deeply studied to improve efficiency but specially to reduce fabrication costs guaranteeing reliability. Strong efforts have been done for the busbar screen printing technology to decrease Silver consumption while maintaining high efficiencies and reliability. Alternative metallization schemes have been also tested as Cu-plating and the so-called smart-wire (SW) approach by MeyerBurger (MB). Cu-plated cells have been fabricated and record efficiencies have been obtained for SHJ cells approaching 23% efficiency with a 4BB metallization. At the same time, the SW technology has allowed to fabricate cells around 23% efficiency with the BusBarless measurements and very few silver content (~50mg). New Transparent Conducting Oxides (TCO) have been also developed. Indium-free cells with an efficiency loose of <0.5% have been also demonstrated using ZnO:Al as front and back TCO. Record mobility over 120 cm2/V·s have been also obtained optimizing Hydrogenated Indium Oxide materials as shown in table 2. The integration of those materials allowed the HERCULES partners to approach 23% efficiency large area SHJ cells thanks to the best optical/electrical compromise.

Table 2: Electrical characterization of different TCO obtained by the HERCULES consortium.

	N_e [cm^{-3}]	μ [cm²/Vs]	ρ [μΩcm]
AZO (dep. @ 180 °C)	2.62E+20	32.7	728
ITO (annealed 5min@200°C in air)	2.43E+20	54.5	472
IO:H (as dep. @ 25°C)	3.86E+20	39.9	405
IO:H (annealed 1h@180°C in vacuum)	1.70E+20	126.6	290

Finally, lot of work has been also developed to find alternative ways to pattern the back side of the IBC solar cells, especially for the IBC-SHJ devices. Industrial approaches as masking, laser and chemical etching have been tested by different partners to pattern amorphous Silicon and TCOs. Even if good results have been demonstrated at the lab scale for all technologies, the solution with metallic masks seems to be the most encouraging for a simple, industrial process flow in terms of cost. Nevertheless, the up-scaling of the IBC-SHJ is still a challenge.

In figures 4, 5 and 6 we show the development performed by the HERCULES consortium form the beginning of the project to improve the efficiency of the IBC, SHJ and IBC-SHJ technologies respectively, in small area devices and in the large area devices.

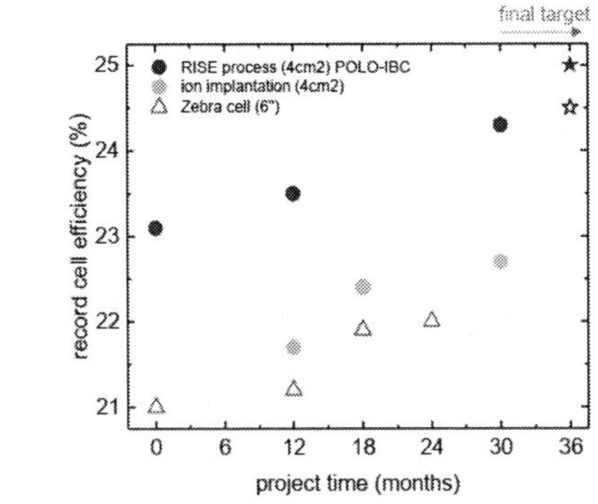

Figure 4: Evolution of efficiency for the IBC technology during the HERCULES project up to M30.

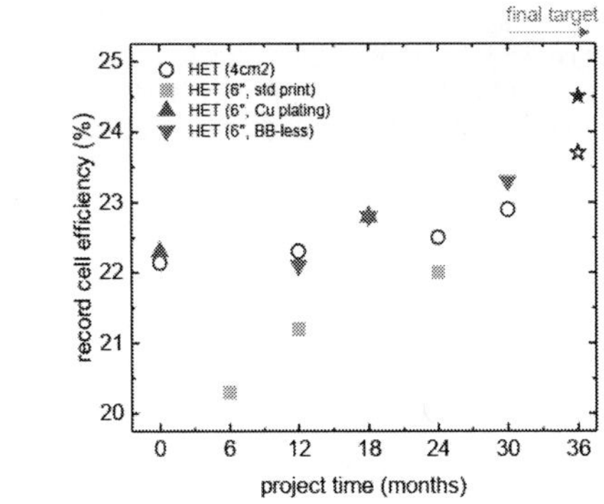

Figure 5: Evolution of efficiency for the SHJ technology during the HERCULES project up to M30.

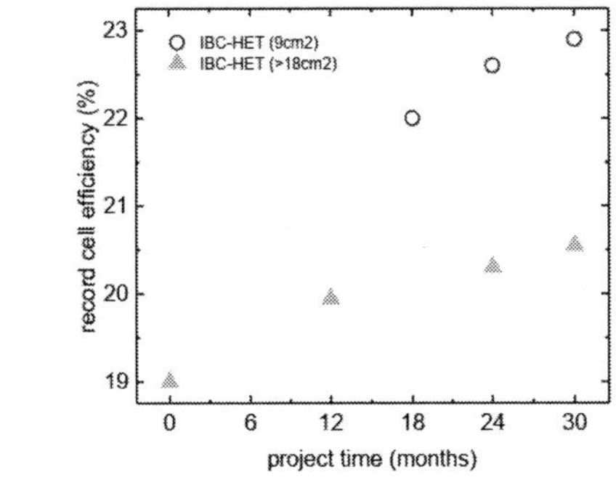

Figure 6: Evolution of efficiency for the IBC-SHJ technology during the HERCULES project up to M30.

4. MODULE RESULTS

Concerning the module activity, IBC and SHJ cells have been integrated into modules by CEA-INES, Eurotron and MB, for indoor and outdoor measurements. The best results for 60-cells modules are shown in Table 3.

Furthermore, the best prototype modules reach 20.3% efficiency for 4-cells IBC and 21% efficiency for 20-cells SHJ cells with the SW technology, opening the path for new record modules in the next months to come.

Table 3: Measurements on the best 60-cells modules within the HERCULES consortium.

cells	Partner	Description	Pmpp (W)	Isc (A)	Voc (V)	FF (%)
Zebra IBC	Eurotron + ISC	White reflective rear encapsulant with holes, conductive adhesive + back foil	303	9,97	39,3	77,1
SHJ Bifacial	CEA-INES	4BB conductive glue textured ribbon	308	9,18	43,9	76,2
SHJ Bifacial	MB	Smartwire (18 wires)	327	8,91	44	78,2

Moreover:

- Long reliability has been proven
- Low Cell-to-Module losses (<5% for all technologies) have been demonstrated.
- Best encapsulant for each technology identified
- Outdoor monitoring on-going to establish LCOE calculations and analyze real performance of the different technologies

5. PILOT LINE RESULTS

The two key solar cell technologies, Zebra IBC and SHJ have been rump up to pilot line production at the Megacell facilities and at the MB Germany site.

Concerning IBC-Cells, ISC-Konstanz achieved:

- Ramp up and pilot production of >1200 n-type 6 inch IBC cells
- ≥75% bifaciality at cell level
- best cell in pilot line 22%
- best 60 cells module power of 303 Wp

Results from production are shown in figure 7.

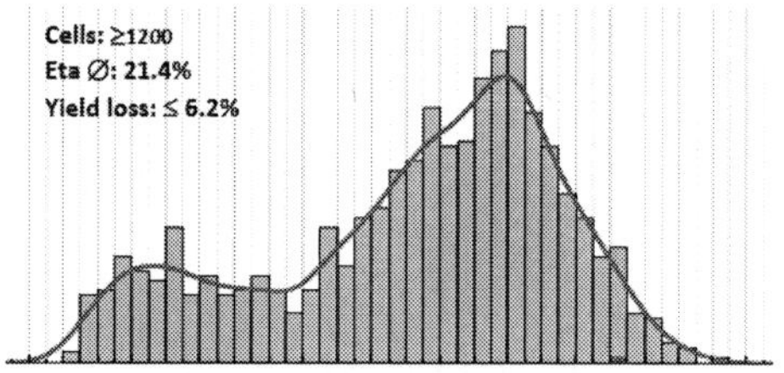

Figure 7: Production efficiency repartition for >1200 IBC Zebra cells by ISC Konstanz.

Concerning heterojunction by Meyerburger, main results are summarized below:

- Ramp up and process improvement in pilot line for busbar less, bifacial SHJ cells: more than 100 000 cells produced
- Busbarless GT measurement in cooperation with ISE Cal Lab
- Full automatic wafer handling with
- PECVD tact time: 84s
- Golden cell in production 23,3% efficiency

And an example of daily production is shown in figure 8.

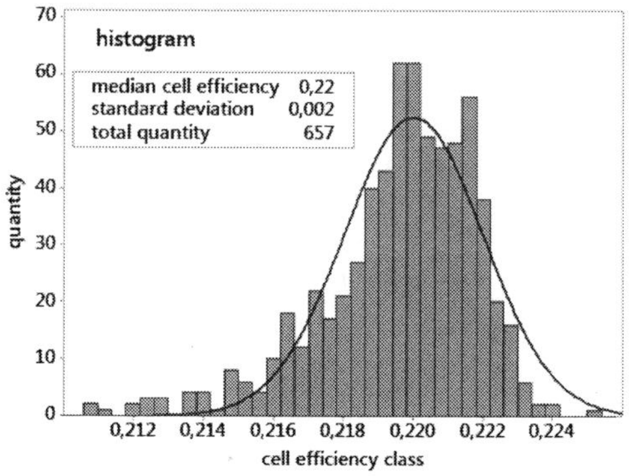

Figure 8: Production efficiency repartition for a daily fully automated production at MB pilot line.

Moreover, the module pilot-line of MB is Switzerland has fabricated SHJ modules combining SHJ with the Smart-wire technology with the following characteristics:

- Bifacial silicon heterojunction solar cells (M0 wafers)
- Interconnected with18 wires (200µm diameter)
- Glass-Glass encapsulation
- Decentralized junction box (shadow and resistive losses reduced)

Modules up to 320W efficiency have been produced with a bifaciality over 91% as shown in figure 9, demonstrating the potential of the DHJ/SW technology for high efficiency production.

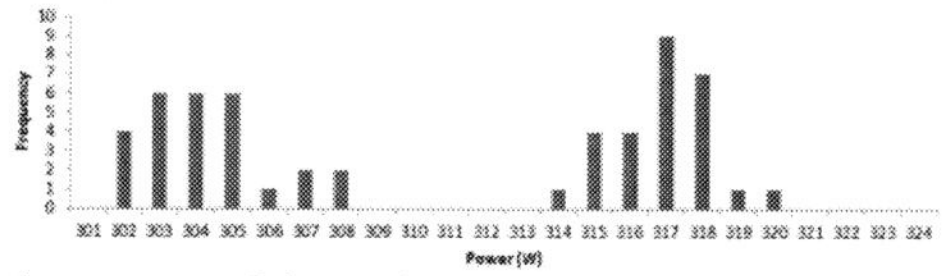

Figure 9: Module production by MB comparing front with black rear reflector (blue) and front with white rear reflector (red). Record modules up to 320W have been demonstrated.

6. CONCLUSIONS

During the first 30 months of the HERCULES project, we have shown a strong evolution concerning the understanding, optimization, fabrication and demonstration of n-type solar cells and modules. Strong characterization and simulation have been used to understand and surpass solar cell limitations improving not only efficiency but also cost. IBC and SHJ cells have been successfully rump up to production with efficiencies surpassing 21% and 22% respectively demonstrating the advantage of the n-type technologies for high efficiency devices compatible with low cost fabrication approaches. The collaboration of all partners has been crucial to reach the ambitious goals of the HERCULEs project.
In the next period, analysis on production cost, sustainability and feasibility will be done to demonstrate the high potential of n-type based modules.

ACKNOWLEDGEMENTS

This work was supported by the European Union through the HERCULES project which has received funding from the European Union's Seventh Program for research, technological development, and demonstration under Grant Agreement No. 608498.

EMITTER SATURATION CURRENTS OF 22 FA/CM² APPLIED TO INDUSTRIAL PERC CELLS APPROACHING 22% CONVERSION EFFICIENCY

T. Dullweber[1]*, H. Hannebauer[1], S. Dorn[1], S. Schimanke[1], A. Merkle[1], C. Hampe[1], R. Brendel[1,2]

[1]Institute for Solar Energy Research Hamelin (ISFH), Am Ohrberg 1, D-31860 Emmerthal, Germany
[2]Department of Solar Energy, Institute of Solid-State Physics, Leibniz University of Hanover, Appelstrasse 2, D-30167 Hanover, Germany

ABSTRACT: Passivated Emitter and Rear Cells (PERC) are currently being introduced into mass-production. The conversion efficiency of industrial *p*-type PERC cells is limited by the emitter saturation current density of around 90 fA/cm² of conventional homogeneously $POCl_3$ diffused emitters. In this paper we investigate two alternative emitter formation technologies. The first approach named in-situ oxidation inserts a short thermal oxidation in-between the phosphorus silicate glass deposition and the drive-in of a conventional homogeneous $POCl_3$ diffusion thereby reducing the phosphorus surface concentration. The second approach named Gas Phase Etch Back (GEB) selectively removes around 40 nm of the highly doped surface of the $POCl_3$ diffused emitter by the reactive gas phase of the wet chemical rear polishing bath. Whereas the conventional $POCl_3$ emitter exhibits a phosphorus surface doping concentration of 3×10^{20} cm^{-3}, the in-situ oxidation and the GEB process reduce the doping concentration to 7×10^{19} cm^{-3} and 4×10^{19} cm^{-3}, respectively. Accordingly, the emitter saturation current density is reduced to values of 22 fA/cm² (in-situ oxidation) and 28 fA/cm² (GEB). This compares with 89 fA/cm² for the reference $POCl_3$ diffusion. Whereas the reference $POCl_3$ emitter limits the PERC conversion efficiency η to 21.1% and the open circuit voltage V_{oc} to 655 mV, the in-situ oxidation improves the PERC IV parameters up to 21.5% and 669 mV. The highest efficiency of 21.6% is obtained with the selective GEB emitter. When solving series resistance issues with the most advanced GEB emitter, the measured V_{oc} and J_{sc} values allow for PERC conversion efficiencies of up to 21.9%.

Keywords: silicon solar cells, PERC, emitter, phosphorus doping, in-situ oxidation, gas phase etch back

1 INTRODUCTION

Passivated Emitter and Rear Cells (PERC) are currently being introduced into mass-production by several leading solar cell manufacturers. The International Technology Roadmap for Photovoltaics (ITRPV) predicts a market share of 35% of PERC silicon solar cells in the next few years [1]. Record efficiencies up to 22.1% have been reported for industrial-type PERC cells [2]. However, with the strongly reduced carrier recombination at the PERC cells rear side, the conversion efficiency of industrial p-type PERC cells is mainly limited by the high emitter saturation current density J_{0E} of around 90 fA/cm² of conventional homogeneously $POCl_3$ diffused emitters [3,4]. In this paper we investigate two alternative emitter formation technologies which both strongly reduce the phosphorus surface concentration thereby decreasing Shockley-Read-Hall and Auger recombination. The first approach named in-situ oxidation inserts a short thermal oxidation in-between the phosphorus silicate glass (PSG) deposition and the drive-in of a conventional $POCl_3$ diffusion. The interfacial oxide limits the phosphorus diffusion from the PSG into the silicon wafer thereby reducing the phosphorus surface concentration of the homogeneously doped emitter. Previous literature [5,6] reported industrially feasible emitter saturation current densities down to 35 fA/cm² by simulation and on experimental test structures using the in-situ oxidation approach. However, no PERC solar cells results with in-situ oxidized $POCl_3$ emitter have been published so far. In this paper we demonstrate saturation current densities down to 22 fA/cm² with in-situ oxidized emitters as well as PERC cell efficiencies up to 21.5%. The second approach applied in this paper named GEB (Gas Phase Etch Back) removes around 40 nm of the highly doped surface of the $POCl_3$-diffused emitter by the reactive gas phase of the wet chemical rear polishing bath which is a required PERC process step. We form a selective GEB emitter by depositing an organic etch barrier with an inkjet prior to the gas phase etch back. We introduced the GEB emitter technology in 2013 [7] and since then strongly improved its industrial feasibility as well as its technical performance from 20.3%-efficient GEB PERC cells in [7] to 21.6%-efficient PERC cells with GEB emitter reported in this paper.

2 EMITTER TECHNOLOGIES AND PERC SOLAR CELL PROCESSES

We use 2 to 3 Ωcm, 156x156 mm², boron-doped Czochralski-grown (Cz) silicon wafers. The process flow of the PERC cells with in-situ oxidized emitters as shown in Table I, column 1, is identical to the ISFH baseline PERC process with conventional $POCl_3$-diffused emitter which is described in detail in [8]. Here we just highlight the most important process steps. After cleaning and damage etching, the rear side of the wafer is coated with a protection layer which acts as etching and diffusion barrier in the following alkaline texturing and phosphorus diffusion. The recipe of the conventional $POCl_3$ diffusion is modified to include a short thermal oxidation step using O_2 gas in-between the PSG deposition and the drive-in as displayed in Fig. 1a). We vary the in-situ oxidized $POCl_3$ emitter properties by applying different oxidation times of 5 min and 10 min. We name the modified $POCl_3$ recipes in-situ 1 (5 min oxidation) and in-situ 2 (10 min oxidation). After texturing and phosphorus diffusion, the PSG with the interfacial oxide as well as the rear protection layer are removed by wet chemistry. Then, the rear side is passivated using a stack of ALD AlO_x and PECVD SiN_y, and the front side is passivated with PECVD SiN_y. Line-shaped laser contact openings (LCO) are formed on the rear side using a picosecond laser. The Ag front grid is printed with a dual print process, meaning that first the 5 busbars are screen-printed with a non-firing through Ag paste and then the Ag fingers are printed with a stencil using a firing-through Ag paste. Then the PERC rear side is screen-printed full-area with a commercially available Al paste and a furnace firing step completes the PERC cell process. A schematic drawing of the resulting PERC solar cell with in-situ oxidized emitter is shown in Fig. 2a).

For one split group named in-situ 1+ with 5 min in-situ oxidation, we modify the PERC process sequence above by applying 1 Ωcm p-type Cz wafers in combination with a rear Al finger grid instead of the full-area Al rear layer. The rear Al finger grid is screen-printed aligned to the LCOs, enables bifacial applications, and increases the aluminum back surface field (Al-BSF) depths thereby slightly increasing the open circuit voltage V_{oc} as has been shown in [9] for these so-called PERC+ solar cells.

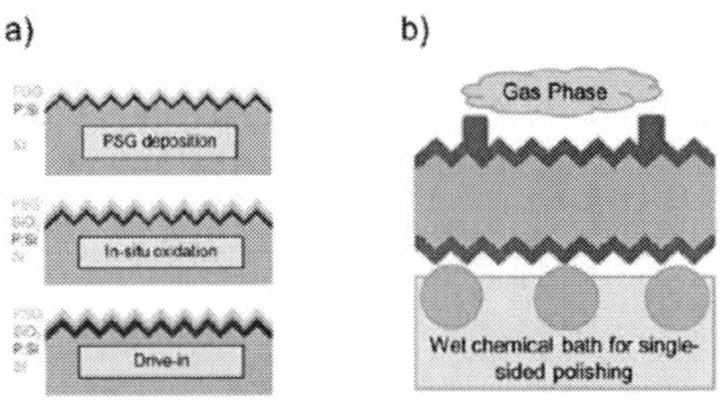

Figure 1: Process schematics of the emitter technologies investigated in this paper. a) In-situ oxidation: modified homogeneous $POCl_3$ diffusion applying an in-situ oxidation step. b) Gas Phase Etch Back (GEB): selective emitter etch back by the reactive gas phase of the acidic rear polishing bath. Both technologies aim at reducing the emitter phosphorus surface concentration thereby decreasing charge carrier recombination leading to increased PERC cell efficiencies.

The process sequence of the PERC cells with GEB emitter as summarized in Table I, column 2, starts with texturing and a conventional $POCl_3$ diffusion without in-situ oxidation. Then we deposit an organic etch barrier with 100 µm wide fingers on top of the PSG layer by inkjet printing at areas where later the Ag fingers are screen-printed. We also print the organic etch barrier material at the wafer edges thereby reducing the wet chemical wrap around during the following wet chemical polishing step. The recipe of the acidic rear polishing bath is adjusted to form a highly reactive gas phase which during the rear polishing process etches the front side emitter thereby removing the PSG layer and about 40 nm of the emitter doping as sketched in Fig. 1b). We vary the GEB emitter properties by applying different $POCl_3$ diffusions as well as different polishing times resulting in two slightly different final phosphorus doping profiles named GEB 1 and GEB 2. The etch barrier is removed in the wet clean following the polishing step. Afterwards the process flow continues as in the first process sequence with surface passivation of both wafer sides, laser contact opening, screen-printing of the Ag and Al contacts, and firing. A schematic drawing of the resulting PERC solar cell with selective GEB emitter is shown in Fig. 2b). Whereas in the first publication of the GEB process [7] we were applying a separate PSG etch step between phosphorus diffusion and organic etch barrier printing, the further developed GEB process as described above works without a separate PSG etch step thereby simplifying the GEB process sequence and increasing its industrial applicability.

We choose these two different PERC process flows because the process with rear protection layer avoids any impact of the rear polishing process on the front side emitter and hence allows a direct assessment of the as-diffused emitters. On the other hand, the GEB PERC process flow utilizes the gas phase of the polishing bath to form a lean single-step selective emitter PERC process.

Table I: Column 1: The PERC processes with homogeneous $POCl_3$ diffusion as well as with in-situ oxidized emitter apply a rear protection layer deposited prior to phosphorus diffusion. Hence, no rear polishing is required which may impact the emitter performance by the reactive gas phase and which hence allows a direct comparison of the in-situ oxidized emitter to a conventional $POCl_3$ emitter. PERC cells applying the GEB emitter as shown in column 2 receive an inkjet-printed etch barrier at the Ag finger area. The following rear polishing recipe is modified to form a highly reactive gas phase which removes around 40 nm of the emitter surface in-between the printed etch barrier fingers hence forming a selective emitter. The etch barrier is removed in the cleaning step post polishing.

PERC with in-situ oxidation	PERC with GEB
Cleaning	Texturing
Rear protection layer	$POCl_3$ diffusion
Texturing	Etch barrier, Inkjet
$POCl_3$ with in-situ oxidation	Rear polishing with GEB
PSG etch / clean	clean
Rear AlO_x/SiN_y	Rear AlO_x/SiN_y
Front SiN_y	Front SiN_y
Laser ablation (LCO)	Laser ablation (LCO)
Ag screen printing	Ag screen printing
Al screen printing	Al screen printing
Firing	Firing

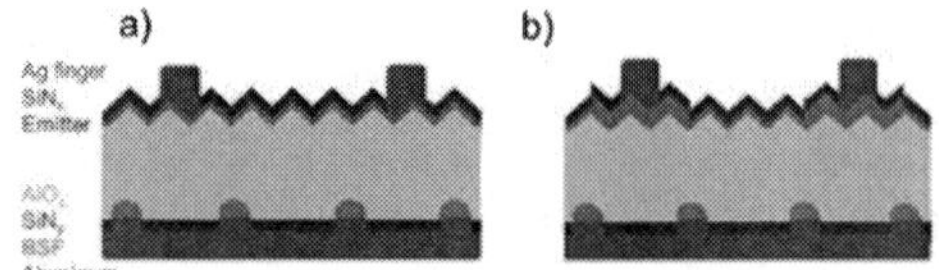

Figure 2: Schematic diagrams of the resulting industrial PERC solar cells with a) homogeneously doped $POCl_3$ emitter with in-situ oxidation (Table I, column 1) and with b) GEB selective emitter (Table I, column 2).

3 PROPERTIES OF THE IN-SITU OXIDIZED AND GEB PHOSPHORUS EMITTERS

We determine the electrically active phosphorus doping profiles by electrochemical capacitance-voltage (ECV) measurements of planar test wafers. Figure 3a) displays typical ECV profiles of three emitter technologies. Whereas the conventional $POCl_3$ emitter without in-situ oxidation exhibits a phosphorus surface doping concentration of 3×10^{20} cm^{-3}, the in-situ oxidation process with 10 min oxidation time (in-situ 2) reduces the doping concentration to 7×10^{19} cm^{-3}. The in-situ 1 diffusion with 5 min oxidation time results in a surface doping concentration of around 1×10^{20} cm^{-3}. The GEB 1 process obtains a further reduced phosphorus surface concentration of 4×10^{19} cm^{-3} whereas the GEB 2 process exhibits the lowest surface doping concentration of 2×10^{19} cm^{-3}.

a)

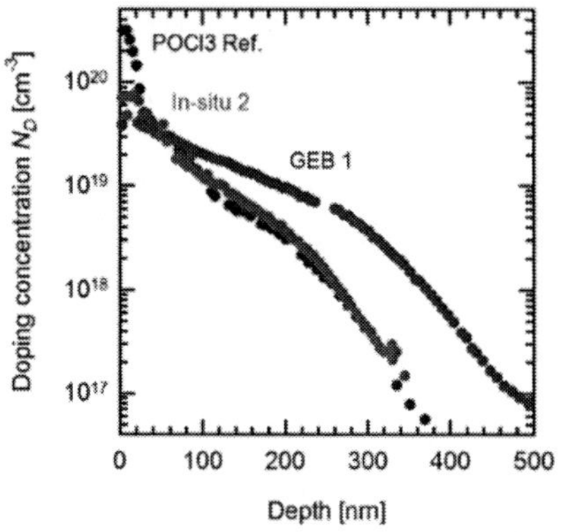

b)

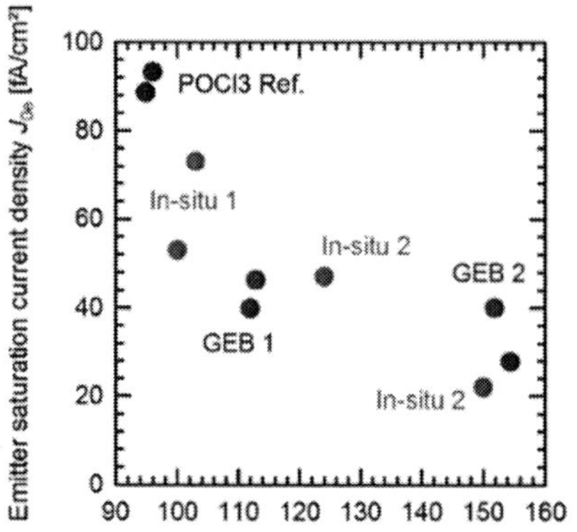

Figure 3: a) Electrochemical capacitance-voltage (ECV) measured doping profiles of the conventional reference $POCl_3$ emitter, the in-situ oxidized emitters and the Gas Phase Etch Back (GEB) emitters. Whereas the conventional $POCl_3$ emitter exhibits surface doping concentrations of 3×10^{20} cm^{-3}, the in-situ oxidation and the GEB process reduce this value to 7×10^{19} cm^{-3} and 4×10^{19} cm^{-3}, respectively. b) Accordingly, the J_{0E} is reduced down to 22 fA/cm^2 (in-situ oxidation) and 28 fA/cm^2 (GEB) compared with 89 fA/cm^2 for the reference $POCl_3$ emitter.

In order to measure the emitter saturation current density and the emitter sheet resistance, we use 200 Ωcm float-zone (Fz) wafers and process both wafer sides with texturing, $POCl_3$ diffusion (reference, in-situ 1, in-situ 2), PECVD-SiN_y-passivation and furnace-firing. In case of the GEB emitters, after double-sided texturing and $POCl_3$ diffusion the Fz wafers receive the GEB process without inkjet printed etch barrier thereby etching back the front-side emitter while simultaneously removing the rear side emitter and slightly polishing the rear pyramids. Then the front side emitter is passivated with PECVD SiN_y, whereas the polished rear side is passivated with an AlO_x/SiN_y stack. Finally, the test wafers are furnace-fired. We determine the emitter saturation current density J_{0E} as well as the emitter sheet resistance R_{sheet} by Quasi-Steady-State Photoconductance (QSSPC) measurements using an intrinsic carrier concentration $n_i = 8.6\times10^9$ cm^{-3} and an excess carrier density $\Delta n = 3\times10^{15}$ cm^{-3}. As shown in Fig. 3b), the reduced phosphorus surface concentrations of the in-situ oxidized emitters in-situ 1 and 2 as well as the GEB 1 emitters decrease J_{0E} to values in the range of 40 to 70 fA/cm^2 already for moderate R_{sheet} values around 110 Ω/sq compared with 89 fA/cm^2 for the reference $POCl_3$ emitter. When increasing R_{sheet} to 150 Ω/sq, we obtain excellent J_{0E} values down to 22 fA/cm^2 (in-situ 2) and 28 fA/cm^2 (GEB 2) which are one of the lowest J_{0E} values reported so far for industrially applicable $POCl_3$ emitters with industry typical textured and PECVD SiN_y-passivated surface.

4 PERC SOLAR CELLS WITH IN-SITU OXIDIZED EMITTER AND GEB EMITTER

The current-voltage (IV) parameters of the industrial PERC solar cells are measured as-processed under standard testing conditions (STC) and are shown in Fig. 4 in dependence of the different emitter formation technologies. Whereas the conventional reference $POCl_3$ emitter limits the PERC conversion efficiency η to 21.1% and the open circuit voltage V_{oc} to 655 mV, the in-situ 1 oxidation improves the PERC IV parameters up to 21.3% and 662 mV. The in-situ 1+ PERC+ cells with rear Al grid and 1 Ωcm Cz wafers obtain slightly higher efficiencies up to 21.5% due to higher V_{oc} values up to 669 mV. The in-situ 2 emitter increases the short circuit current density up to 40.1 mA/cm^2. However, the efficiency of 21.1% is limited by a lower fill factor FF of 79.6% resulting from a high specific contact resistance of the screen-printed Ag contacts to the in-situ 2 emitter with low phosphorus surface concentration as determined by transmission line measurements (TLM). The highest efficiency of 21.6% is obtained with the selective GEB 1 emitter with V_{oc} values up to 668 mV and FF up to 81.6% as shown in Fig. 4. The highest V_{oc} values up to 671 mV are obtained with the GEB 2 emitter. However, the FF of the GEB 2 emitter is reduced to 77.8% thereby decreasing the efficiency to 20.8%. The decline in FF is caused by a high series resistance of 1.0 Ωcm^2 for the GEB 2 cells compared with 0.4 Ωcm^2 for the GEB 1 cells. This is surprising since the specific contact resistance should be similar for both emitter types since the Ag pastes contact the non-etched-back area. Also, the Ag finger spacing should be able to tolerate the emitter sheet resistance range of 100 to 150 Ω/sq of the GEB 1 and GEB 2 emitters. Hence, further analysis is required to understand the high series resistance values of the GEB 2

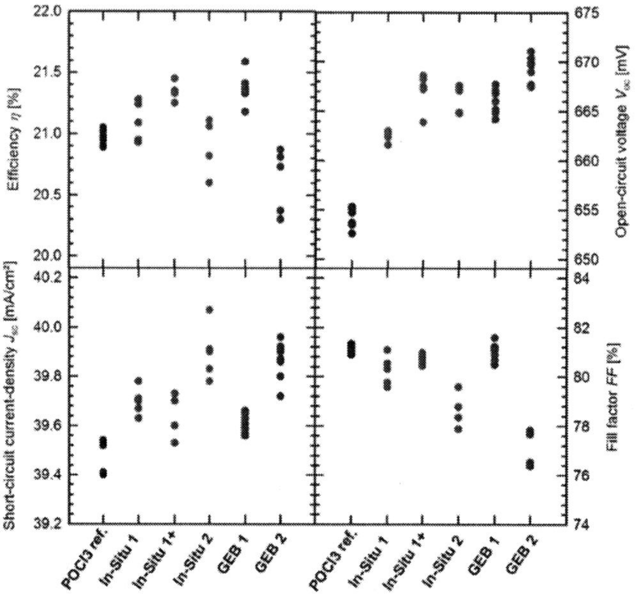

Figure 4: IV parameters of industrial PERC solar cells measured under standard testing conditions (STC) in dependence of the different emitter formation technologies. The advanced in-situ and GEB emitters increase the PERC conversion efficiency η up to 21.6% compared to the reference $POCl_3$ emitter with 21.1% primarily due to a strongly increased open circuit voltage V_{oc}.

emitter. Assuming that the series resistance issues of the GEB 2 emitter could be solved and similar FF values could be achieved as for the GEB 1 emitter, the GEB 2 PERC cells would have an efficiency potential of 21.9% due to their high measured V_{oc} = 671 mV and J_{sc} = 40.0 mA/cm² values.

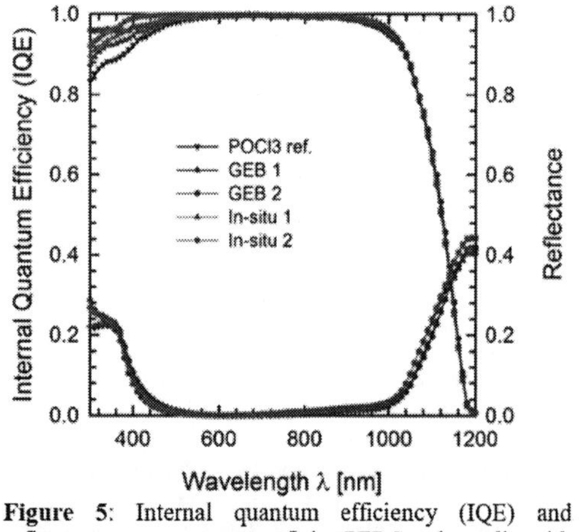

Figure 5: Internal quantum efficiency (IQE) and reflectance measurements of the PERC solar cells with different emitter technologies. The increased IQE values in the blue wavelength regime clearly demonstrate the reduced emitter recombination of the advanced in-situ oxidized and GEB emitters compared to the conventional POCl3 emitter.

We further analyze the PERC solar cells by measuring their reflectance and spectral response and calculate the internal quantum efficiency (IQE) as shown in Fig. 5. Whereas the reflectance measurements look similar, the IQE clearly shows the impact of the different emitter technologies in the blue wavelength range between 300 nm and 500 nm. Whereas the conventional $POCl_3$ diffusion exhibits an IQE value of 0.84 at 300 nm wavelength, the in-situ 1 and GEB 1 emitters improve this value to 0.89. The in-situ 2 and GEB 2 emitters achieve the highest IQE values around 0.94 at 300nm wavelength clearly demonstrating the reduced emitter recombination due to the reduced front surface phosphorus concentration.

5 CONCLUSIONS

The conventional $POCl_3$ emitter exhibits a phosphorus surface doping concentration of 3×10^{20} cm^{-3}, which results in a relatively high emitter saturation current density J_{0E} of 89 fA/cm² limiting the PERC conversion efficiency to η = 21.1% and the open circuit voltage to V_{oc} = 655 mV. The advanced emitter formation processes in-situ oxidation and GEB reduce the surface doping concentration to 7×10^{19} cm^{-3} and 4×10^{19} cm^{-3}, respectively. Accordingly, the emitter saturation current density J_{0E} is reduced down to 22 fA/cm² (in-situ oxidation) and 28 fA/cm² (GEB) obtained with emitter sheet resistances R_{sheet} around 150 Ω/sq. The in-situ oxidation improves the PERC efficiency up to 21.5% and the V_{oc} up to 669 mV. The highest efficiency of 21.6% is obtained with the selective GEB emitter. When solving series resistance issues with the most advanced GEB emitter, the measured V_{oc} = 671 mV and J_{sc} = 40.0 mA/cm² would support PERC conversion efficiencies up to 21.9%. IQE measurements confirm the reduced emitter recombination of the in-situ oxidized and GEB emitters which demonstrate IQE values up to 0.96 at 300 nm wavelength. In terms of process costs, the in-situ oxidized emitter requires no additional process step compared to a conventional $POCl_3$ diffusion but just slightly increases the $POCl_3$ process time. Implementation of the GEB emitter into a PERC production process requires one additional inkjet process step to deposit the organic etch barrier as well as an adjustment of the polishing process recipe to increase the reactivity of the gas phase.

ACKNOWLEDGEMENTS

We thank the German Federal Ministry for Economic Affairs and Energy for funding part of this work under contract no. 0325753D (HighPERC). We thank the company Sun Chemical Ltd., UK, for supplying us with the etch barrier material used for the GEB process.

REFERENCES

[1] Metz A. et al., "International Technology Roadmap for Photovoltaic" (ITRPV.net), 2014 Results, April 2015. Available at http://www.itrpv.net/Reports/Downloads/2015/

[2] Trina Solar, press release, December 16th 2015. Available at http://www.prnewswire.com/news-releases/trina-solar-announces-new-efficiency-record-of-2213-for-mono-crystalline-silicon-solar-cell-300193727.html

[3] C. Kranz, J. H. Petermann, T. Dullweber, R. Brendel, presented at the 5th SiliconPV conference, Konstanz, Germany (2015).

[4] R. Brendel, T. Dullweber, R. Peibst, C. Kranz, A. Merkle, D. Walter, Progress in Photovoltaics: Research and Applications, DOI: 10.1002/pip.2696, (2015).

[5] A. Piechulla, S. Denzer, V. D. Mihailetchi, T. Zhou, W. Jooß, S. Eisert, R. Harney, in Proc. 31st Eur. Photovoltaic Solar Energy Conf., Hamburg, Germany, 2015, pp. 420 – 424.

[6] A. Wolf, A. Kimmerle, S. Werner, S. Maier, U. Belledin, S. Meier, D. Biro, in Proc. 31st Eur. Photovoltaic Solar Energy Conf., Hamburg, Germany, 2015, pp. 414 - 419.

[7] H. Hannebauer, T. Dullweber, S. Wyczanowski, K. Weise, F. Delahaye, O. Doll, I. Köhler and R. Brendel, in Proc. 28th Eur. Photovoltaic Solar Energy Conf., Paris, France, 2013, pp. 752 – 756.

[8] H. Hannebauer, T. Dullweber, U. Baumann, T. Falcon, R. Brendel, Phys. Status Solidi RRL **8**, 675-679 (2014).

[9] T. Dullweber, C. Kranz, R. Peibst, U. Baumann, H. Hannebauer, A. Fülle, S. Steckemetz, T. Weber, M. Kutzer, M. Müller, G. Fischer, P. Palinginis, H. Neuhaus, Prog. Photovolt: Res. Appl., 2015, DOI: 10.1002/pip.2712.

Towards a 300 W_P p-type HIP-MWT-Module – Simulation, Experimental Results and Costs

A. Spribille[1], T. Savisalo[2], A. Kraft[1], D. Eberlein[1], M. Ebert[1], T. Fellmeth[1], S. Nold[1], H. Pantsar[2], F. Clement[1]
[1]Fraunhofer Institute for Solar Energy Systems ISE, Heidenhofstraße 2, 79110 Freiburg, Germany
[2]Valoe Oyj, Insinöörinkatu 8,FI-50150 Mikkeli, Finland
Phone: +49 761 4588 535, e-mail: alma.spribille@ise.fraunhofer.de

ABSTRACT: In this work we are aiming at the goal of fabricating a cost-effective HIP-MWT module exceeding 300W. In order to accomplish this goal HIP-MWT (high-performance metal wrap through) silicon solar cells [1, 2] are fabricated on industrial PERC (passivated emitter and rear cell) precursors. Simulation of the optimal metallization layout for MWT based on measured parameters show cell efficiencies up to 21.5%. The consequentially fabricated HIP-MWT solar cells reach maximum efficiencies of 21.4%. The in parallel processed H-pattern reference cells reach maximum efficiencies of 21.2%. The cell efficiencies show a reduced advantage for MWT than in similar experiments, which is due to the tapered busbars of the reference cells allowing nearly the same short circuit currents. Anyhow, combined with a module interconnection based on back contact foils a cell-to-module (CTM) loss of 2 % is demonstrated which allows module power over 300 W_P. Due to a power advantage of about 15W in comparison to H-pattern modules the cost of ownership calculation shows a cost advantage of the HIP-MWT module of 3.2 %. Simulation, experimental results and cost calculation show an advantage for HIP-MWT technology over the H-pattern reference leading to the conclusion that MWT is a more cost-effective concept.
Keywords: HIP-MWT, module, cost of ownership

1 INTRODUCTION

Reaching higher output powers per module is a major goal of solar cell and module development. At the same time the cost per wattpeak has to be kept constant or should even be reduced. Applying the MWT concept on PERC solar cells (HIP-MWT [1, 2]) enables to achieve both at once. HIP-MWT solar cells not only allow higher cell efficiencies due to less shading by metallization on the front side; they also enable module interconnection based on conductive backsheets, which can result in lower cell-to-module losses. In contrast to standard H-pattern modules the conductive cross-section is independent of shading related losses.

2 APPROACH

As the cells are fabricated on partly processed PERC (passivated emitter and rear) wafers provided by Gintech the most crucial part in the process sequence of HIP-MWT solar cells is the metallization layout, which needs to be optimized as a trade of between series resistance losses and shading. For the optimization of the metallization layout we use the tool GridMaster developed at Fraunhofer ISE [4]. The via drilling process is inspired by the optimized metallization layout and can be carried out within the same laser system as the local contact opening leading to the same amount of process steps for HIP-MWT and H-pattern solar cells (s. Fig 1).

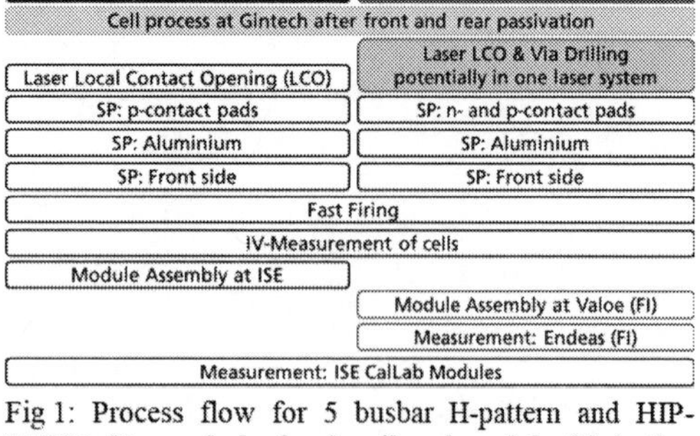

Fig 1: Process flow for 5 busbar H-pattern and HIP-MWT5 (5 pseudo-busbars) cell and module fabrication on industrial PERC wafers. The PERC wafers are processed until rear side passivation and front side anti-reflection coating at Gintech. The HIP-MWT module is build using a coated conductive backsheet to ensure the electrical isolation between the two polarities.

This optimized MWT metallization is applied on passivated wafers from Gintech. In parallel H-pattern cells are fabricated on the same wafers. Both cell types are assembled into 60 cell modules for direct comparison. To determine the cost-effectiveness of our 300W_P module approach a cost of ownership calculation of both cell and module concepts is applied employing SCost [4].

3 SIMULATION OF CELL EFFICIENCIES

The simulation of the cell efficiencies is carried out using GridMaster [3].

Fig 2: Simulation in GridMaster [3] of HIP-MWT and H-pattern cell efficiencies. The MWT5 metallization layout features 8 n-pads per contact row on the rear side. Further parameters for the simulation are: 50 µm printed finger width, pseudo-busbar width between 150 µm and 220 µm, via resistance of 2 mΩ, base resistivity of 1 Ωcm, thickness of 170 µm, sheet resistance of 100 Ωsq.

The simulation results presented in Fig 2 and table 1 show that HIP-MWT cells with efficiencies of up to 21.5 % are realistic. The j_{SC}-advantage of 0.3 mA/cm^2 for HIP-MWT cells in comparison to H-pattern cells is opposed by the fill factor loss of 0.4 % due to the MWT metallization layout. The combination of both effects results in very similar cell efficiencies for both concepts. Nevertheless this simulation shows the situation only on cell level; the module interconnection of the H-pattern cells with 0.8 mm wide tapes will results in additional shading of 1.3 % reducing the j_{SC} of the H-pattern module and in additional series resistance losses within the tabs.

Considering the cell area of 243 cm^2 the simulated cell efficiency of the HIP-MWT cells translates to a cell power of 5.2 W; 60 cells theoretically provide 312 W_P allowing a rather high cell-to-module loss of 4 % while still enabling a 300 W_P module.

Table I: Simulated cell results of H-pattern and HIP-MWT cells

Simulated parameters	Eta [%]	V_{oc} [mV]	J_{sc} [mA/cm²]	FF [%]	R_S [Ohmcm²]
HIP-MWT	21.5	663	40.4	80.2	0.53
H-pattern	21.4	662	40.1	80.6	0.45

3 EXPERIMENTAL RESULTS

3.1 Cell results

Based on the simulation described in chapter 2 H-pattern and HIP-MWT cells were fabricated on passivated wafers from Gintech. The cell layouts are shown in Fig 3. The enlargement of the HIP-MWT rear side shows one n-pad with surrounding aluminium (spacing 350 µm) and the alloyed LCO (local contact opening) lines. As the vias are largely filled with pad paste they are not visible in the image.

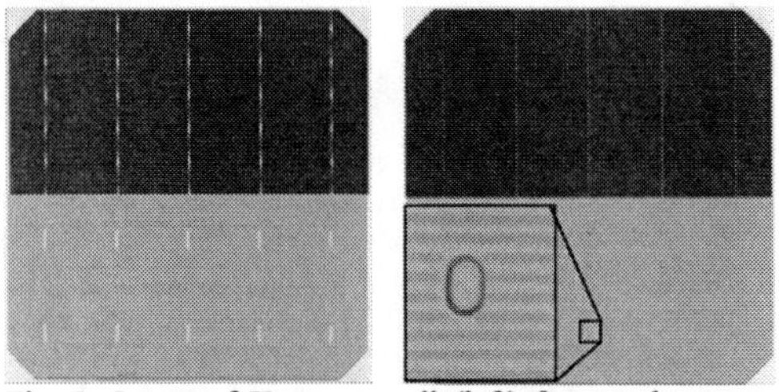

Fig 3: Image of H-pattern cell (left) front and rear as well as image of HIP-MWT cell (right) front and rear with a zoom of one n-pad. Detailed metallization parameters are listed in table II.

Table II: Input parameters GridMaster [3]

Input parameters	
Finger width	50 µm
Finger height	14 µm
Via resistance	2 mOhm
Spec. contact resistance	1.7 mOhmcm²
Base resistivity	1 Ohmcm
Emitter sheet resistance	100 Ohm/sq.
# of n-pads for HIP-MWT	5 x 8
n-pad size for HIP-MWT	1.5 x 2.5 mm
# of vias for HIP-MWT	80
Front side shading HIP-MWT	4.4 %
Front side shading H-pattern	5.7 %

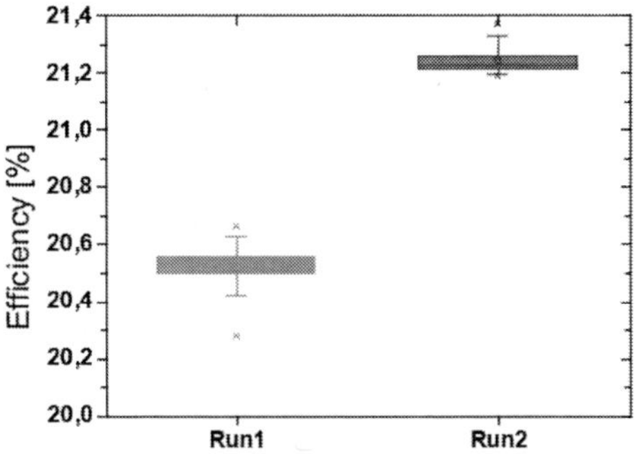

Fig 4: HIP-MWT cell efficiency results for first and second cell run. The open circuit voltage has improved 12 mV, the short circuit current 0.7 mA/cm² from first to second run due to the improved quality of the passivated wafers.

Fig 4 shows the cell efficiencies of the HIP-MWT cells conducted within two runs. For the first cell run passivated wafers with lower efficiency potential have been used. The metallization layout stayed the same in both runs, while the screen printing process was improved in run 2 leading to more homogeneous fingers. The improved passivated wafers in combination with the improved screen printing process result in a mean efficiency increase of about 0.7 %. The homogeneous fingers allowed to fabricate the second cell run with single screen print using only 75 mg of silver paste on the front side.

In parallel with the second HIP-MWT cell runs H-pattern 5 busbar cells with tapered busbars have been fabricated

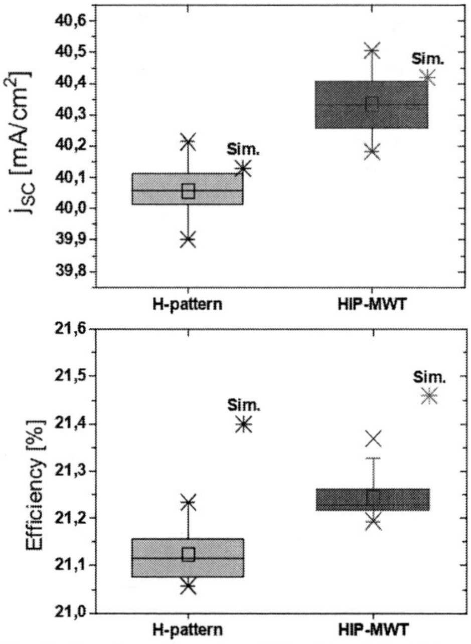

Fig 5: Experimental results of H-pattern (97 cells) and HIP-MWT (93 cells) for j_{SC} and efficiency (box plots). The stars represent the simulation results.

The experimental results of the H-pattern cells in direct comparison with the HIP-MWT cells are shown in Fig 5 also including the according simulation results (stars in graph). The experimentally achieved j_{SC}-values are in very good accordance with simulation results. The difference in the efficiency is caused by the fill factor, which is shown in Fig 6. The best solar cells achieve the simulated fill factors values. The deviation is caused by increased series resistance values, which is most likely caused by not perfectly homogeneous screen print. Overall the high fill factor values show, that a single step screen printing is sufficient. The overall deviation in efficiency is very small showing similar process stability for H-pattern and HIP-MWT cells.

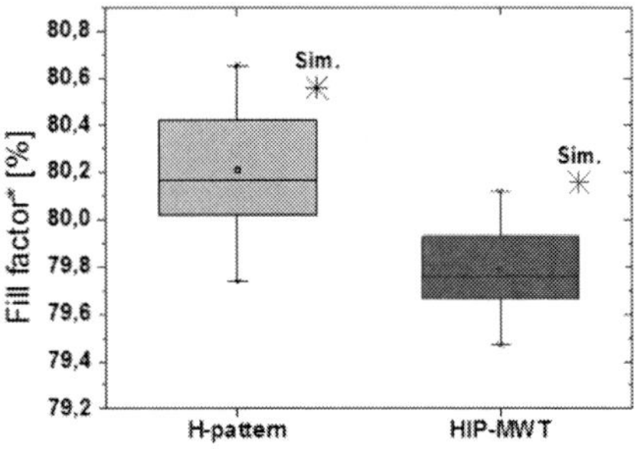

* FF of HIP-MWT cells corrected based on GridMaster simulation

Fig 6: Measured fill factor values of H-pattern and HIP-MWT cells (box plots). The stars represent the simulation results.

3.2 H-pattern module results

Table III: H-pattern cell and module results as well as cell-to-module loss.

	I_{sc} [A]	U_{oc} [V]	P_{mpp} [W]	FF [%]	Eta [%]
60 cells	9.7^{Min}	39.4^{Σ}	307.5^{Σ}	$80.1^{Ø}$	$21.1^{Ø}$
Module @ ISE*	9.6	39.2	**287.7**	76.6	18.4
Cells-to-module loss	-1.3%	-0.5%	-6.5%	-4.4%	-12.9%

*measured at Fraunhofer ISE CalLab Modules

The 60 best H-pattern cells were built into a module at Fraunhofer ISE using 0.8 mm wide interconnectors. For the H-pattern module a moderate cell-to-module loss of 6.5 % in power was achieved. The j_{SC}-loss of 1.3 % corresponds directly to the additional shading due to the interconnectors on top of the tapered busbars.

3.3 HIP-MWT module results

Out of both HIP-MWT cells runs one HIP-MWT module each was build. The first HIP-MWT show the summed power of 297 W_p. The HIP-MWT module shows a power output of 291 W_P leading to a CTM of only 2 %. Of the second cell run with a mean efficiency of 21.2 % again the 60 best cells were built into a module by Valoe (see Fig 7). The 60 cells sum up to a power of 309 W_P. Unfortunately the contacting between cells and conductive backsheet has not been ideal resulting in a measured module output power of 282 W_P.

		P_{mpp} [W]
Run 1	60 cells	297^{Σ}
	HIP-MWT module*	291
	Cells-to-module loss	*-2.0_{rel}*
Run 2	60 cells	309^{Σ}
	Cells-to-module loss Run1	*-2.0_{rel}*
	Expected module Power	***303***
	HIP-MWT module**	282

* Measured at Endeas (FI)
** Measured at Fraunhofer ISE CalLab Modules

Fig 7: HIP-MWT module results of the first and the second run as well as the cell-to-module loss conducted from the first run.

The electroluminescence image (Fig 8) shows some inhomogeneity mainly caused by increased series resistance.

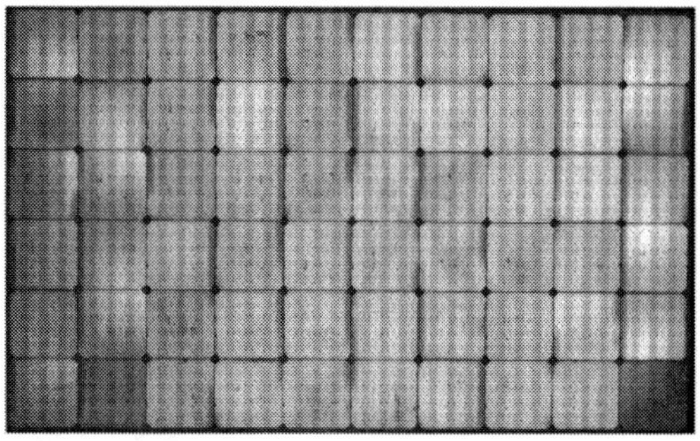

Fig 8: Electroluminescence image of the HIP-MWT module of run 2.

Nevertheless, if the same CTM as in the first run would have been realized a module exceeding the 300 W_P is achievable with the measured cell efficiencies.

4 COST OF OWNERSHIP CALCULATION

A bottom up Cost of Ownership (CoO) calculation is carried our using SCost [8] developed at Fraunhofer ISE. A combined cell and module production facility based in Europe on a green field site is taken as basis. The calculation in SCost is based on SEMI E35 CoO standard [11]. No capital costs are assumed; furthermore the calculation is carried out without inclusion of overhead costs for R&D (research and development) as well as SG&A (Selling, General and Administrative Expenses).

4.1 Cell production costs

The process sequence shown in Fig 1 is taken as basis for the CoO calculation of the cell production costs; these costs only include the production costs for the cells without the costs for the wafer. Furthermore the cell efficiencies of the experiment, shown in Table II are applied.

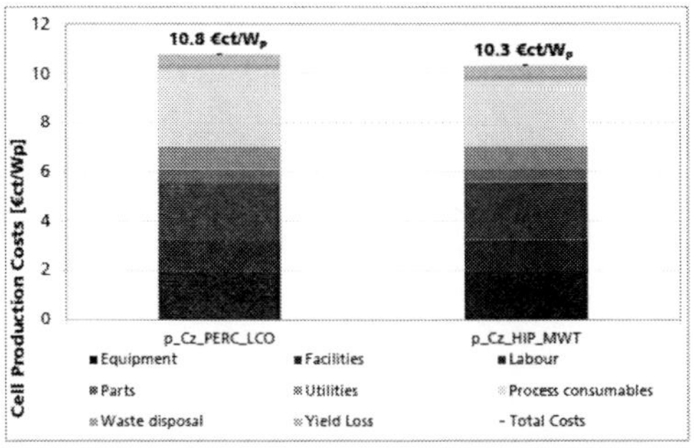

Fig 9: Cell production costs (without wafer costs) based on the Cost of ownership calculation carried out with SCost [8]. A silver price of 525 €/kg is taken from [7].

Fig 9 shows a cost advantage of 0.5 €ct/W_P for the HIP-MWT cells in comparison to the H-pattern cells. This advantage is caused by the higher cell efficiency of the HIP-MWT cells. In addition the HIP-MWT cell process uses 35 mg less silver paste on the front side (MWT: 75 mg; H-pattern: 110 mg), which is a large cost driver for solar cells.

4.2 Module costs

The cost of ownership calculation of the module production is based on the process sequence shown in Fig 10. Furthermore the achieved module output of 288 W_P of the H-pattern and the expected 303 W_P of the HIP-MWT module is applied. As over 70 % of the module production costs are material costs a detailed comparison of the module material costs is performed, see Fig 11.

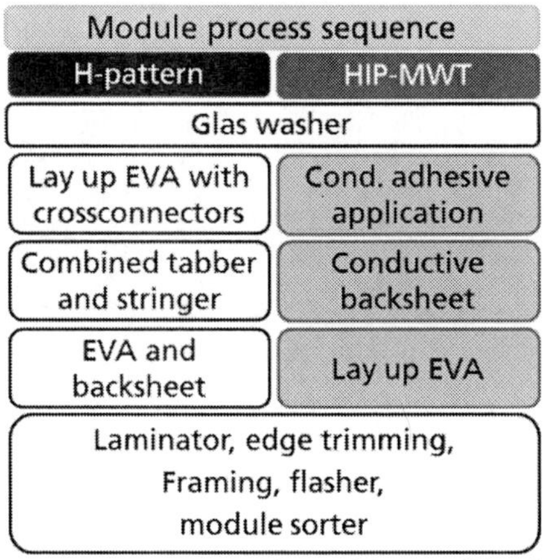

Fig 10: Process sequence for module assembly of H-pattern and HIP-MWT modules. The process sequence is taken as a base for the cost of ownership calculation.

The absolute module material costs are higher for MWT modules as the conductive backsheet with 10 € per module (6.25 €/m^2) is an additional cost driver. The lower costs for the conductive adhesive in comparison to the inter- and cross connectors of the H-pattern module partly revokes the costs of the conductive backsheet.

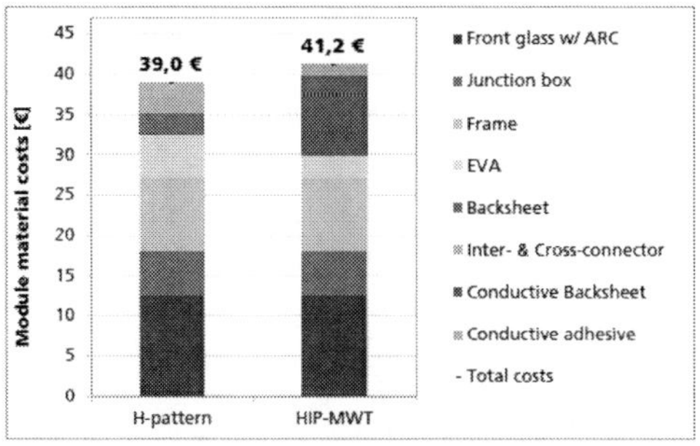

Fig 11: Module material costs of H-pattern and HIP-MWT modules. Front glass, junction box and frame are equal for both concepts. The costs for encapsulant of MWT is only 50 % of H-pattern as the rear EVA is included in the conductive backsheet. The largest cost factor for MWT is the conductive backsheet with 6.25 €/m^2. For the conductive adhesive a consumption of 2.3 g/module and a price of 600 €/kg is assumed.

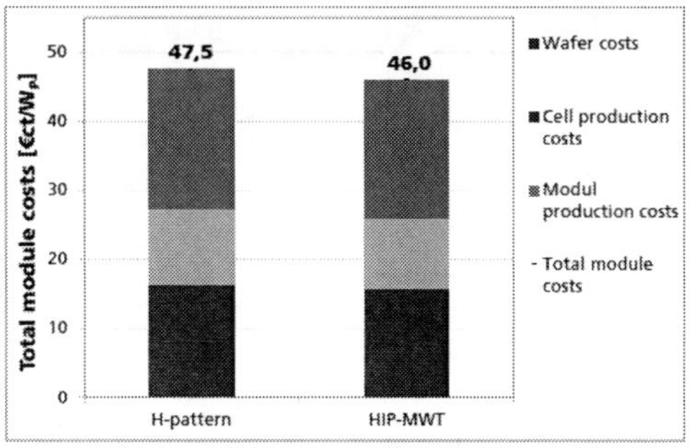

Fig 12: Total module costs subdivided into wafer, cell production and module production costs per wattpeak. Results are based on the Cost of ownership calculation carried out in SCost [4] using a wafer price of 79.09 €ct [8].

Fig 12 shows the total module costs for H-pattern and HIP-MWT subdivided into wafer, cell production and module production costs. In all categories the HIP-MWT technology shows a cost advantage over H-pattern due to the higher module output power: 0.8 €ct/W_P on wafer level; 0.5 €ct/W_P on cell level and 0.2 €ct/W_P on module level. Resulting in an overall cost advantage for HIP-MWT of 1.5 €ct/W_P translating to a relative cost advantage of 3.2 %.

5 SUMMARY AND CONCLUSION

With a cell simulation carried out in GridMaster the possibility of fabricating HIP-MWT cells with 21.5 % using passivated wafers from Gintech was displayed. In a first HIP-MWT run on passivated wafers with a lower efficiency potential a 291 W_P module was built. A cell-to-module loss of only 2 % in power is demonstrated. The second HIP-MWT cell run shows a mean efficiency of 21.2 % exceeding the first cell by 0.7 %abs.. In combination with 2 % CTM this opens up the possibility of manufacturing a p-type module with over 300 W_P. The second cell run also featured reference H-pattern cells and resulted in maximum efficiencies of 21.4 % for HIP-MWT cells and 21.2 % for H-pattern cells. The experimental results correspond well with the simulation. The best 60 cells of each concept were assembled into 60 cell modules resulting in 288 W_P for the H-pattern module. Due to challenges in the module assembly of the second HIP-MWT, a 300 W_P module is only theoretically shown.

Finally the Cost of Ownership calculation shows a cost advantage of 3.2 % for the HIP-MWT technology.

ACKNOWLEDGEMENTS

The authors would like to thank Valoe for the good cooperation.

The authors would like to thank Gintech Energy Corporation for providing the precursors.

The authors would like to thank Toyo Aluminium K.K. for providing the aluminum paste.

The authors would like to thank all colleagues at the Fraunhofer ISE Photovoltaic Technology Evaluation Center (PV-TEC).

REFERENCES

[1] Dross F, van Kerschaver E, Allebe C, Van der Heide, A., Szlufcik J, Agostinelli G, Choulat P, Dekkers, H. F. W., Beaucarne G. Impact of rear-surface passivation on MWT performances. In: *4th World Conference on Photovoltaic Energy Conversion Hawaii: 2006*; 2006, p. 1291–4.

[2] Thaidigsmann B, Spribille A, Plagwitz H, Schubert G, Fertig F, Clement F, Wolf A, Biro D, Preu R. HIP-MWT - A new cell concept for industrial processing of high-performance metal wrap through silicon solar cells. In: *26th EU PVSEC*. Proceedings; 2011, p. 817–20.

[3] Fellmeth T, Clement F, Biro D. Analytical modeling of industrial-related silicon solar cells. *IEEE Journal of Photovoltaics* 2014;4:504–13, doi:10.1109/jphotov.2013.2281105.

[4] S. Nold, N. Voigt, L. Friedrich, D. Weber, I. Hädrich, M. Mittag, H. Wirth, B. Thaidigsmann, I. Brucker, M. Hofmann, Jochen. Rentsch, R. Preu. Cost Modelling of Silicon Solar Cell Production Innovation along the PV Value Chain;2012.

[5] Haedrich I, Wiese M, Thaidigsman B, Eberlein D, Clement F, Eitner U, Preu R, Wirth H. Minimizing the optical cell-to-module losses for MWT-modules. *Energy Procedia* 2013;38:355–61, doi:10.1016/j.egypro.2013.07.289.

[6] SEMI E35-0312 - Guide to Calculate Cost of Ownership (COO) Metrics for Semiconductor Manufacturing Equipment

[7] http://www.lbma.org.uk/pricing-and-statistics; 10th June, 2016

[8] http://pvinsights.com/Price/SolarWaferPrice.php; 13th June, 2016

PLASMA PROCESS ANALYSIS OF ICP-PECVD OF ALOX LAYERS FOR C-SI SURFACE PASSIVATION

M. Hofmann[1], M. Jäcklein[1], P. Saint-Cast[1], D. Wagenmann[1], B. Cord[2], T. Dippell[2], M. Dörr[2], T. Schütte[3], M. Siemers[4]
[1]Fraunhofer Institute for Solar Energy Systems ISE, Heidenhofstrasse 2, 79110 Freiburg, Germany
[2]SINGULUS Technologies AG, Hanauer Landstrasse 103, 63796 Kahl am Main, Germany
[3]PLASUS GmbH, Lechstrasse 9, 86415 Mering, Germany
[4]Fraunhofer Institute for Surface Engineering and Thin Films IST, Bienroder Weg 54E, 38108 Braunschweig, Germany

ABSTRACT: The detailed analysis of the ICP-PECVD plasma process using spectroscopic plasma monitoring and the ex-situ analysis of the manufactured layers and interfaces to the c-Si substrates provide insight into the interdependency of passivation mechanisms and process conditions. Plasma process simulations based on the particle-in-cell Monte Carlo method additionally allowed getting insight into electron density and electron energy distributions. The combination of these measures lead to highly passivating very thin AlO_x layers of thicknesses below 5 nm allowing to lower the cost of ICP-PECVD AlO_x deposition.
Keywords: passivation, PECVD, AlO_x, plasma monitoring

1 INTRODUCTION

In 2015, more and more passivated emitter and rear cell (PERC) –type [1] solar cell production facilities were installed on industrial level. This trend is expected to go on and lead to a significant market share of PERC-type cells in the near future. One very attractive route for the implementation of the industrial rear passivation technology is the application of plasma-enhanced chemical vapor deposition (PECVD) of aluminum oxide (AlO_x) [3] [5]. Its attraction is driven by the industrially proven PECVD technology already applied for decades for the front side anti-reflection coating (ARC) using silicon nitride (SiN_x) layers. PECVD will stay in use for the ARC application; therefore it is very logic to use the same technology for the AlO_x deposition as well. On the other hand, despite a successful proof in literature for a very high surface passivation quality possible by PECVD AlO_x the competition with atomic layer deposition (ALD) of AlO_x layers is still ongoing.
PERC-type cells which are becoming increasingly important in industrial production but also almost all other kinds of advanced crystalline silicon solar cell concepts (PERL, PERT, IBC, ...) are depending on an effective surface passivation technology for reaching highest energy conversion efficiencies. Built-in negative charges in AlO_x on c-Si are creating a beneficial band bending at the p-type c-Si PERC rear surface.
This paper will focus on the analysis of PECVD-AlO_x layers deposited by an inductively coupled plasma (ICP) using in-situ and ex-situ experimental and simulative methods in order to increase the understanding of the plasma process and to enhance passivation quality, especially for very thin layers below 5 nm. The reduction of AlO_x layer thickness will allow to further reduce the process cost for AlO_x passivation layers, mainly due to an increase in wafer throughput per hour and a decreasing use of consumables per wafer.

2 EXPERIMENTAL

ICP-PECVD allows very effective gas dissociation [2] due to the very high plasma density and therefore high ionization of the gas species. In this work, a SINGULUS Singular ICP-PECVD tool was applied for the fabrication of thin AlO_x layers. Sketches of the tool can be found in Figure 1 and Figure 2. Figure 2 elucidates the way the wafers are taken through the deposition tool. A carrier with four 156x156 mm^2 wafers enters the vacuum through the load lock and then is moved clockwise to the IR station and onwards to the AlO_x and the three SiN_y stations. Here, the ICP-sources operating at 13.56 MHz are mounted and are process gas-separated by individual vacuum locks.
The deposition experiments and carrier lifetime investigations were carried out using float-zone (FZ) monocrystalline silicon wafers (p-type, boron doped) with a resistivity of 1...1.5 Ω cm and a thickness of about 250...300 μm.

Figure 1: SINGULUS Singular PECVD tool applied in the investigations here [10].

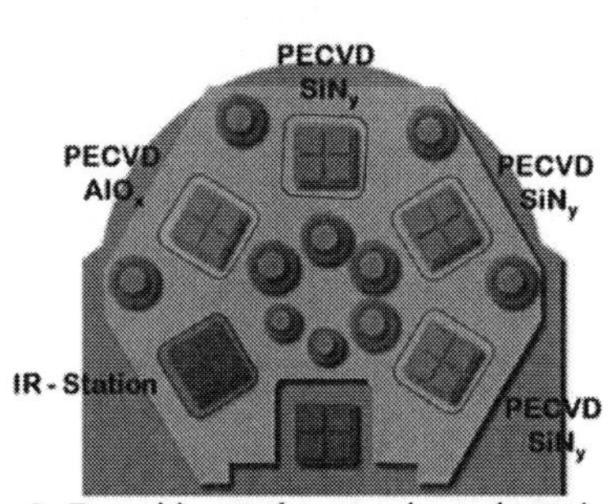

Figure 2: Deposition station top view schematic of the Singular PECVD tool showing the load lock at the bottom of the image, IR station for wafer heating and the deposition chambers for AlO_x and SiN_y. Since the SiN_y layers are typically prepared much thicker than the AlO_x layer, three SiN_y stations are used sequentially. For all process stations, the same process time is available due to the synchronized clockwise movement of the 4 wafers from one process station to the other. See also [10].

Plasma analysis applying spectroscopic plasma monitoring (SPM) technique based on optical emission spectroscopy (OES) provides deep time-resolved insights into the dynamics of the plasma species. Therefore, it can be recognized what kinds of species are present and how to control their ratios. In the investigations presented here, a PLASUS EMICON 3MC system was applied. The analysis of the emitted spectra was partly performed using the actinometry method where the spectra are normalized to a non-reactive gas which was in our case the emission from the argon gas being kept at constant partial pressure throughout the experiments.
The combination of this in-situ characterization method with ex-situ analysis of the created layers is of benefit to achieve a correlation of parameters and an optimization. Contactless capacitance-voltage (COCOS, applying a Semilab SDI CFORS tool), spectroscopic ellipsometry (SE, applying a J. A. Woollam M-2000 tool) and carrier lifetime measurements (QSSPC, WCT-120 tool by Sinton Instruments) were performed. Detailed characteristics of the plasma technology were elucidated by particle-in-cell Monte Carlo (PIC-MC) simulations of the ICP-PECVD process chamber [4] providing local electron temperature and electron density distributions in the PECVD reactor that can facilitate an efficient deposition process.
Since PERC-type and other solar cell types applying passivated surfaces are becoming increasingly important not only in academia but also industrially, a thorough understanding of the AlO_x deposition process is of high relevance and allows for stable and highly efficient processes.
One goal of the investigations presented here is the reduction of the AlO_x layer thickness from 20-30 nm to 5 nm or even less still allowing the same excellent surface passivation properties.

3 RESULTS

3.1 Plasma simulations

For the simulation of the AlO_x deposition plasma process, the setup of the process chamber was used as sketched in Figure 3 which is in accordance with the actual reactor geometry.

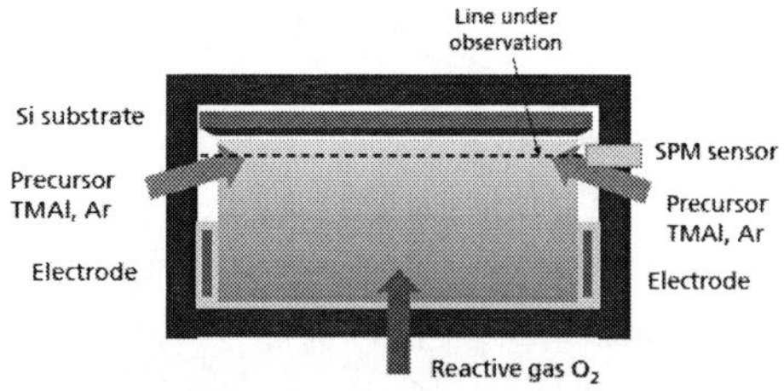

Figure 3: Schematic drawing of the setup of the AlO_x deposition chamber. The deposition takes place via the plasma which is located below the substrates. The reactive gas (oxygen-containing in the case of AlO_x deposition) is passing the inductively coupled plasma (ICP) and gets dissociated. The precursor gas (trimethylaluminum TMAl) is entering the deposition chamber near the substrates. A spectroscopic plasma monitoring (SPM) sensor is installed to observe the plasma close to the substrate.

PIC-MC plasma simulations are one fundamental tool to understand the physics occurring in the PECVD process.
Figures 4 and 5 show simulated electron density and electron energy distributions for a 2D cross section through the reaction chamber. The simulations were performed with argon as working gas. Argon is often used for benchmarking and with argon experimental plasma measurements are easily done to compare against the simulations. Additionally, the applied power was about 1/100 of the reality due to limitations in the simulation. However, it is expected that the simulation results still allow qualitatively correct interpretations.
In the conducted simulations, a variation of the process pressure was applied with p=2 Pa in Figure 4 and p=10 Pa in Figure 5. Increasing the gas pressure leads to higher plasma density and lower electron energy. This higher plasma density causes more shielding of the inner plasma volume resulting in a stronger plasma confinement in close vicinity of the electrode. Electron densities up to $3 \times 10^{15}\,m^{-3}$ or $9 \times 10^{15}\,m^{-3}$ and low maximum electron energies of 3.4 eV or 2.5 eV for process pressures of p=2 Pa or p=10 Pa, respectively, were results of the simulation. Low electron energies are typical for an ICP –plasma generating low voltages and thus avoiding high energy ions which could damage the cell surface. The electron densities are relatively low which can be attributed to the low RF power (13.56 MHz) applied in the simulation compared to the real power used in the experimental setup. In the electron density distribution, one can find the "secondary coil" shape of the ICP setup.

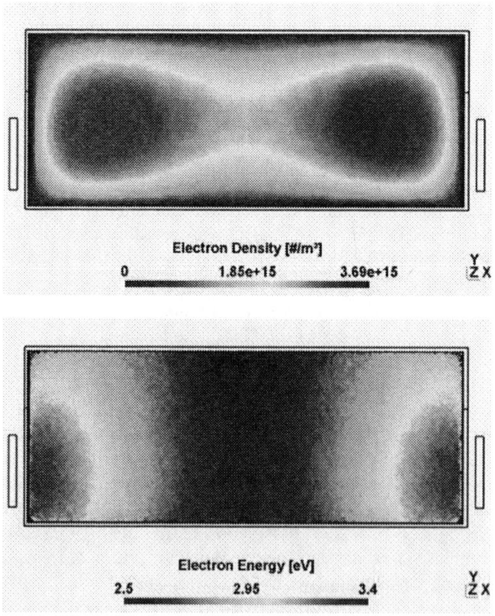

Figure 4: To investigate general plasma behaviour, intense simulations were carried out using the particle-in-cell Monte Carlo (PIC-MC) method.
The upper and lower parts show the two dimensional distributions of electron density and electron energy, respectively, in the reactor vessel (horizontal width about 500 mm) for a gas (Ar) pressure of p=2 Pa.

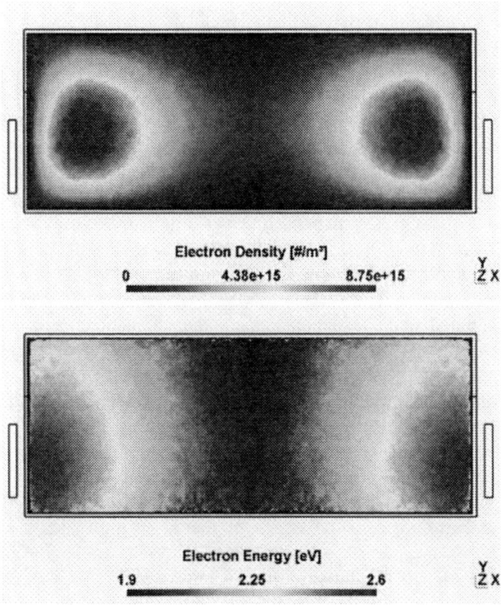

Figure 5: PIC-MC simulation results of electron density (top) and electron energy (bottom) for a process pressure of p=10 Pa. Please compare to Figure 4.

3.2 In-situ plasma monitoring investigation

Spectroscopic plasma monitoring (SPM) based on optical emission spectroscopy was used to observe the plasma created for the deposition of AlO_x layers. A gas mixture of trimethylaluminum (TMAl, $Al(CH_3)_3$), oxygen (O_2), and argon (Ar) was applied. In the photon spectrum, a lot of optical emission lines could be detected and assigned to various atoms or molecules in the plasma with the main contributions from species like Ar, H, O, Al and AlO. Since Ar is inert and an Ar emission line is separated from all others, one can use it as base for an actinometric analysis of the data. Actinometry was presented by Coburn and Chen. More information can be found in [6]. Shortly, the Ar signal can be used to normalize the other signals and to allow an easier cross-interpretation over different parameter sets including start-up and shut-down characteristics.

One example of an investigated process can be found in Figure 6 and Figure 7. Figure 6 shows a typical spectrum of the process with the spectral lines of the observed species. In Figure 7 the temporal behavior of the species over process time are presented where the top graph shows the data as acquired and the bottom graph's data underwent an actinometric calculation.

In the case presented here with each deposition cycle taking 7.5 s we find that in the first 1.8 s after switching on the plasma, start-up conditions occur for the formation of the different gas species after the plasma ignition. A peak of H is most prominent. H must be coming from cracked TMAl molecules. Since H is beneficial for interface passivation, enough visible H in the spectra can be interpreted as a first positive sign for the surface passivation properties of the layer. At the same time, the amount of detected O species is relatively slowly increasing to a saturation value. One might wonder why the oxygen signal is so small in the start-up phase. We see that during start-up AlO species are providing an increased emission behavior showing that the formation of these species is one drain for oxygen. Thus, a very effective decomposition of TMAl in the beginning of the plasma process and/or the presence of a large amount of TMAl seems to occur which provides bonding partners for oxygen in form of AlO, OH and presumably other species such as CO_x as well. This is leading to stable process conditions after about 1.8 s.

The very steep curves in the lower graph at t=0 s are due to actinometric calculations and should not be over-interpreted.

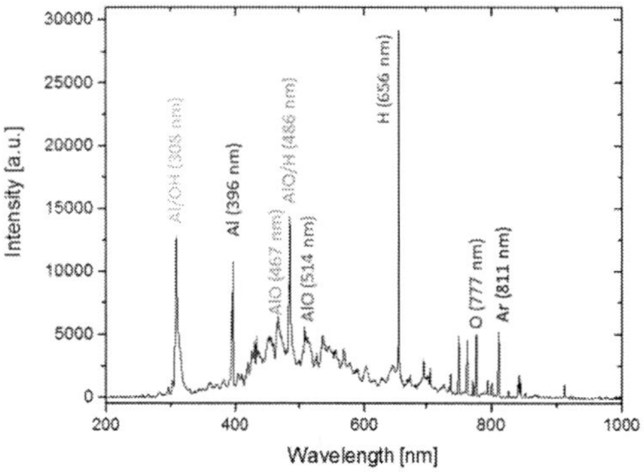

Figure 6: An example of a plasma monitoring spectrum.

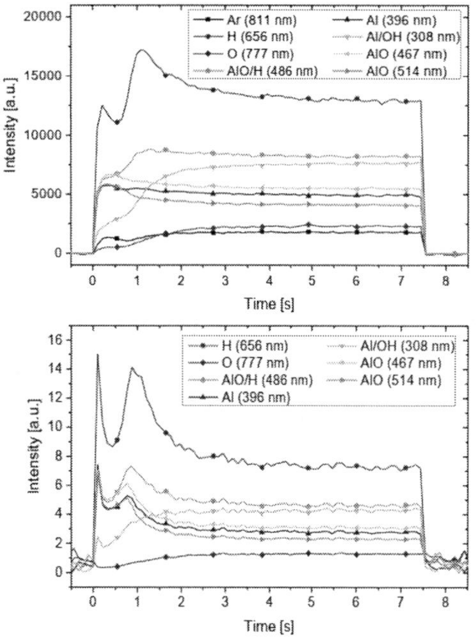

Figure 7: An example of gas dynamics within one deposition cycle is shown as acquired with spectroscopic plasma monitoring (SPM) technique. The intensity on the y-axis equals the relative counts of photons at the mentioned spectral lines. The different species can be identified and separated. Aluminium and/or oxygen-containing species create the AlO_x layer. Hydrogen plays an additional crucial role in the creation of highly passivating layers due to its ability to lower surface defect density. Since Argon is inert, one can use it as reference gas for further analysis based on actinometry.
Top: SPM data as acquired.
Bottom: SPM data after actinometry based on Ar.

3.3 Ex-situ investigation of layer and interface properties

PECVD process parameter variations were carried out with simultaneous plasma monitoring and deposition of passivating AlO_x layers on wafer samples. One of these investigations was focussing on an O_2 flux variation. Here, the O_2 flux was changed from 20% to 100% (relative units) while keeping all other process parameters constant. The process pressure was increased slightly due to the higher total gas flux. Figure 8 shows an overview of the lifetime results. A strong increase of lifetime was found for relatively low oxygen fluxes. In Figure 9, plasma monitoring spectra of the O line (right) and the H line (left) in dependency of the O_2 gas flux at constant flux of the aluminum precursor in the PECVD process, after actinometric data analysis, are displayed. The oxygen signal (λ=777 nm) is increased as expected for the process with higher O_2 gas flux. However, the H signal is decreased with increasing O_2 flux.

We assume that the decreased H signal correlates to the observed lifetime reduction, probably due to a reduced level of H incorporation in the passivating AlO_x layer. Also the AlO signal is increased with increasing O_2 flux (not displayed as graph here) indicating a more prominent presence of AlO compounds already in the gas phase. The influence of these compounds on the AlO_x layer structure, the interface defect density (D_{it}) and the negative charge density (Q_{tot}) is subject to further investigation.

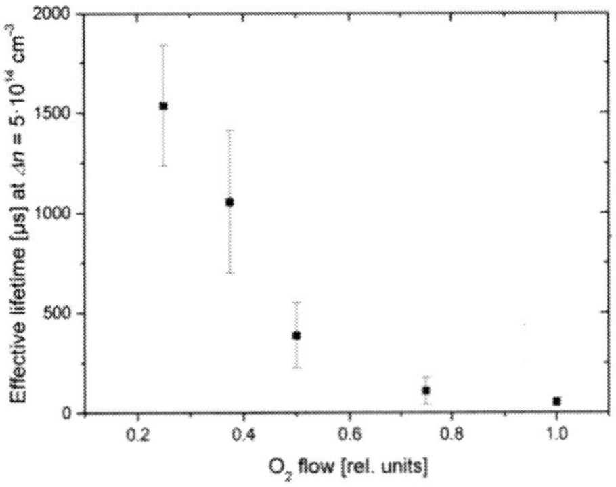

Figure 8: Effective carrier lifetime as a function of the O_2 gas flux (given in relative units).

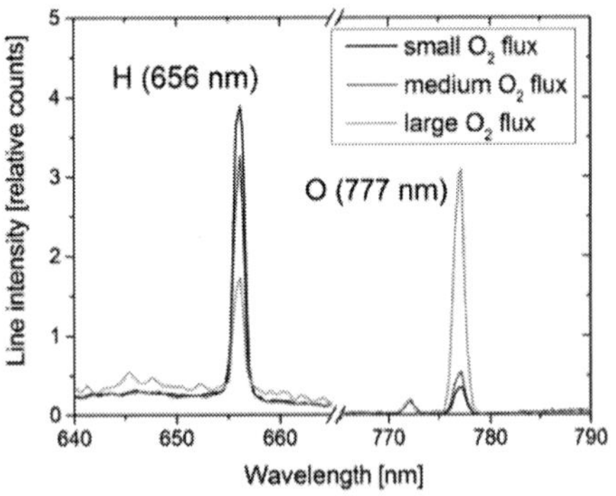

Figure 9: Plasma monitoring spectra of the O line (right) and the H line (left) in dependency of the O_2 gas flux at constant flux of the aluminum precursor in the PECVD process, after actinometric data analysis.

Just like the variation of the O_2 flux, also other process parameters can play a crucial role on the passivation results. However, it is possible to find robust high-quality processing setups.

Contactless capacitance-voltage (COCOS) measurements were performed to investigate the physical mechanisms behind the passivation results. Figure 10 provides an overview. The exact details of the process conditions are not given in this paper. However, one can see the range of results. In Figure 10, the lifetime results are additionally represented in the colors and heights of the data points. The surface preparation process variation is labelled with a small letter, the deposition process variation with a capital letter. Deposition process H allows excellent surface passivation almost independent of the preconditioning of the wafer surface (within the applied frame here). The passivation quality of layers deposited by process A is significantly depending on the surface preparation.

Thus, one can find combinations of surface preconditioning and passivation that allow high quality.

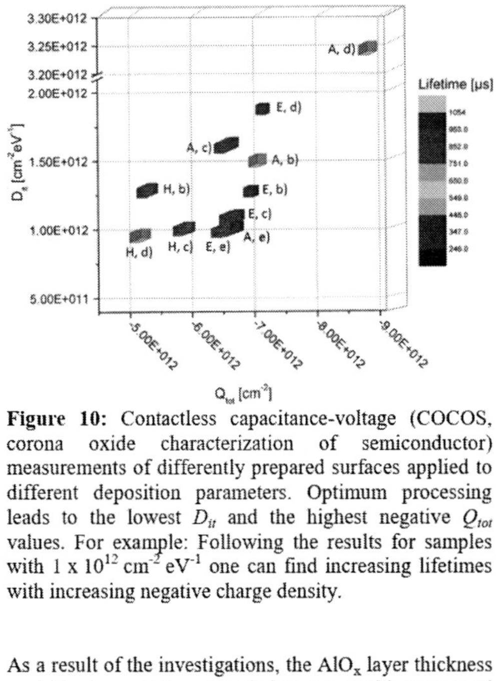

Figure 10: Contactless capacitance-voltage (COCOS, corona oxide characterization of semiconductor) measurements of differently prepared surfaces applied to different deposition parameters. Optimum processing leads to the lowest D_{it} and the highest negative Q_{tot} values. For example: Following the results for samples with 1 x 10^{12} cm^{-2} eV^{-1} one can find increasing lifetimes with increasing negative charge density.

As a result of the investigations, the AlO_x layer thickness could be lowered to ranges below 5 nm with very good passivation results as shown in Figure 11 which were comparable to thick ICP-PECVD AlO_x layers of about 18 nm thickness and to a variety of ALD passivation results from the literature [7-8] [11]. MW-PECVD AlO_x layers were reported to show significant lifetime decrease starting at ~10 nm [9].

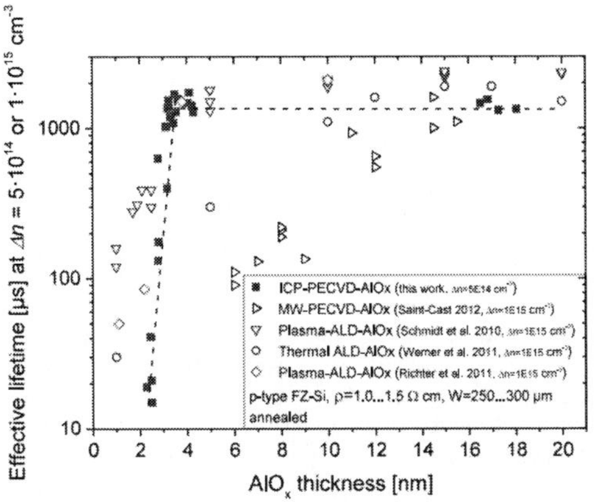

Figure 11: Carrier lifetimes of c-Si wafers with AlO_x layers of thicknesses in the range up to 20 nm on both sides after annealing (FZ-Si substrates 1.0…1.5 Ω cm, p-type, 250…300 µm). A slight variation of the O_2 flux provided the data behind the red ICP-PECVD symbols shown. It seems evident that the deposition process is quite robust in this parameter range. With changing O_2 flux, still excellent passivation properties were possible. Lifetimes of about 1.7 ms reflect a surface recombination velocity of about 5 cm/s. For comparison, also ALD-AlO_x and PECVD-AlO_x passivation layers on comparable substrate types from literature [7-9] [11] are shown.

4 CONCLUSION

Using the knowledge from the manifold analyses it was possible to deposit ICP-PECVD AlO_x layers with a thickness of only 3-4 nm that provide excellent surface passivation with a surface recombination velocity of ~5 cm/s which is comparable to the passivation quality of thicker (18 nm) ICP-PECVD-AlO_x layers and of thin ALD-AlO_x layers and would allow excellent solar cell efficiencies. A relatively robust process could be found that is capable to compensate for potential variations in gas flow during deposition. In-situ spectroscopic plasma monitoring and ex-situ analyses revealed the dynamic behavior of plasma species which can be used to understand and optimize the PECVD process. Contactless CV measurements confirmed that low interface trap densities (D_{it}) and high densities of negative charges (Q_{tot}) in the AlO_x layer are possible and lead to an excellent surface passivation. Simulations of the plasma revealed that the electron density and temperature distribution strongly depend on the process pressure and that the shape of the distribution function is based on the reactor and electrode geometry.

ACKNOWLEDGEMENT

This work was partly funded by the German Federal Ministry for the Environment, Nature Conservation and Nuclear Safety (Contract Number 0325754D).

6 REFERENCES

[1] Blakers, A. W.; Wang, A.; Milne, A. M.; Zhao, J.; Green, M. A. (1989): 22.8% efficient silicon solar cell. Applied Physics Letters 55 (13), pp. 1363–1365. DOI: 10.1063/1.101596.

[2] Lieberman, M. A.; Lichtenberg, A. J. (1994): Principles of plasma discharges and materials processing: John Wiley & Sons.

[3] Saint-Cast, P.; Kania, D.; Hofmann, M.; Benick, J.; Rentsch, J.; Preu, R. (2009): Very low surface recombination velocity on p-type c-Si by high-rate plasma-deposited aluminum oxide. Applied Physics Letters 95 (15), p. 151502. DOI: 10.1063/1.3250157.

[4] Pflug, A.; Siemers, M.; Melzig, T.; Schäfer, L.; Bräuer, G. (2014): Simulation of linear Magnetron Discharges in 2D and 3D, Surface and Coatings Technology 260 (2014) 411-416.

[5] Veith, B.; Dullweber, T.; Siebert, M.; Kranz, C.; Werner, F.; Harder, N.-P.; Schmidt, J.; Roos, B.F.P.; Dippell, T.; Brendel, R. (2012): Comparison of ICP-AlOx and ALD-Al2O3 layers for the rear surface passivation of c-Si Solar cells. In: Energy Procedia 27 (2012) 379-384.

[6] Coburn, J.W.; Chen, M., Optical emission spectroscopy of reactive plasmas: A method for correlating emission intensities to reactive particle density, Journal of Applied Physics 51 (6) (1980) 3134–3136.

[7] Schmidt, J.; Veith, B.; Werner, F.; Zielke, D.; Brendel, R. (2010): Silicon surface passivation by ultrathin Al2O3 films and Al2O3/SiNx stacks, Proceedings of the 35th IEEE Photovoltaic Specialist Conference, pp. 885-890.

[8] Werner, F.; Veith, B.; Zielke, D.; Kühnemund, L.; Tegenkamp, C.; Seibt, M.; Brendel, R.; Schmidt, J.: Electronic and chemical properties of the c-Si/Al2O3 interface, Journal of Applied Physics 109, 113701 (2011)

[9] Saint-Cast, P: Passivation of Si Surfaces By PECVD Aluminum Oxide, PhD Thesis, University of Konstanz, 2012

[10] Roos, B.F.P.; Dippell, T.; Hohn, O.; Wohlfart, P.; Beier, B.; Veit, B.; Siebert, M.; Dullweber, T.: ICP-PECVD Production Tool For Industrial AlOx and Si-based Passivation Layers, 27th EUPVSEC, pp. 1684-1687, 2012

[11] Richter, A.; Benick, J.; Hermle, M.; Glunz, Stefan W.: Excellent silicon surface passivation with 5 Å thin ALD Al2O3 layers: Influence of different thermal post-deposition treatments, Phys. Status Solidi RRL 5, No. 5–6, 202– 204 (2011)

Al_2O_3 Passivation for Copper Plated 15.6x15.6 cm^2 IBC Cells

Shruti Jambaldinni[1], Manabu Kyuzo[2], Barry O'Sullivan[1], Sukhvinder Singh[1], Emanuele Cornagliotti[1], Bartosz Zielinski[1], Maarten Debucquoy[1], Jozef Szlufcik[1], Jef Poortmans[1,3,4]

1. imec, Leuven, Belgium, 2. Kyocera corporation, Japan, 3. KU Leuven, Belgium, 4. U Hasselt, Belgium

Shruti Jambaldinni, imec, Kapeldreef 75, 3001 Leuven, Belgium, Phone: +32-16288449

ABSTRACT: Today, the silicon solar cells with the highest efficiency are interdigitated back contact solar cells. At imec, the cell concept for 15.6x15.6 cm^2 IBC cells relies on a rear design with multiple busbars and a Cu plated rear metallization. Previously, we reported efficiencies up to 21.9% by introducing Cu plated contacts. Further developments in the rear design to overcome FF and J_{sc} losses were implemented and reported as well, showing IBC cells with 22.5% efficiency. In a next optimization round, the passivation quality at the rear of these IBC cells needed to be improved. To do so, we demonstrate in this paper the application of a rear Al_2O_3 based passivation stack on both emitter and BSF. A 10 nm ALD Al_2O_3 - PECVD SiO_x stack was implemented and a thermal activation step (firing) post deposition was applied and optimized to guarantee a good passivation on both the emitter and the BSF. Best cells with this process showed an efficiency of 22.8% and V_{oc} up to 688mV.

Keywords: Interdigitated back-contact cells, IBC, passivation, copper plated, aluminum oxide

1 INTRODUCTION

Excellent surface passivation with the help of passivation stacks including Al_2O_3 have been demonstrated for p-type and to a smaller extent n-type silicon surfaces for a wide range of doping concentrations [4,5,6]. This can be achieved particularly by a thermal activation step after the deposition of the passivation stack [6]. For p$^+$ regions, the thermal activation step is typically an anneal for several minutes at 400-450°C [4]. For n$^+$ regions, better passivation can be obtained by a thermal step of a few seconds at T> 800°C, similar to a contact firing step [5]. These passivation stacks have been implemented mainly in 2-side contacted cells (for example an ALD Al_2O_3 - PECVD SiO_x stack to passivate the boron emitter [3]). Also in IBC cells Al_2O_3 passivation stacks have been applied (for example to passivate both emitter and BSF by ALD Al_2O_3 - SiN_x [7] or by ALD Al_2O_3 - PECVD SiO_x [8]). As the passivation of the emitter is one of the larger recombination losses in our IBC cells (See Figure 3.) and an ALD Al_2O_3 - PECVD SiO_x stack would provide an excellent passivation for this emitter, this paper focuses on improving the efficiency of our IBC process [1] [2] by introducing Al_2O_3 passivation at the rear.

2 EXPERIMENTAL

As IBC cells have both an emitter and a BSF on the rear, the impact of Al_2O_3 for phosphorus diffused BSF had to be studied before implementing this in the device processing scheme. In this section we compare the ALD Al_2O_3 - PECVD SiO_x stack to our standard thermal SiO_2 passivation in terms of J_0 and lifetime.
For this experiment, symmetrical n$^+$/n/n$^+$ structures were used. A 10-nm thick layer of Al_2O_3 using ALD and SiO_x using PECVD were deposited [3]. High temperature firing steps were conducted at 800°, 850°, 900° and 950°C. This could help to reduce the negative fixed charge content of the Al_2O_3 layer and accommodate for n$^+$ passivation [6]. A comparison of J_0 and lifetime values before and after this firing step was established. For reference an n$^+$/n/n$^+$ structure with thermal SiO_2 and FGA at 450°C was included as well [1].

The results are displayed in Figure 1. Firstly, these indicate an improved lifetime and low J_0 value after the firing step for all temperatures. Secondly, it can be seen that 800-850°C gives the highest lifetime and lowest J_0 value when compared with standard SiO_2 passivation and other firing temperatures. Hence, 10 nm ALD Al_2O_3 - PECVD SiO_x and 800°C firing temperature were applied for both emitter and BSF passivation.

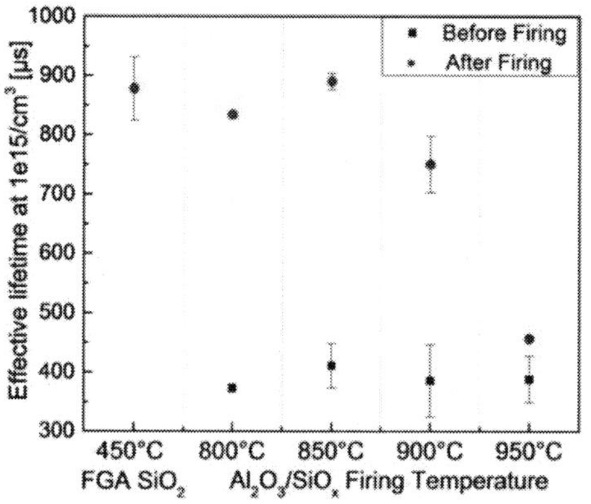

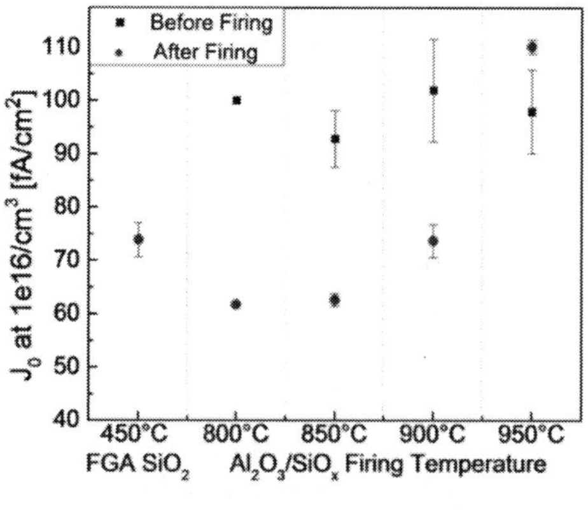

Figure 1: Comparison of minority carrier lifetime and J_0 for a BSF passivated with SiO_2 and with ALD Al_2O_3 - PECVD SiO_x fired at different temperatures

3 PROCESS FLOW

The complete baseline process scheme for IBC cells with standard SiO_2 passivation at the rear has been reported before [2]. The difference in processing with Al_2O_3 comes post-FSF formation. The rear oxide utilized for passivation and activation of BSF and emitter regions was selectively etched and subsequently ALD Al_2O_3 deposition was performed. Further a PECVD SiO_x capping layer was deposited for enhanced optical performance [3]. Firing at 800°C was implemented on selected cells.
BSF and emitter contact points were defined using a laser ablation process. Consequently, metallization scheme with Cu plating was applied as described by Singh et al [1]. Finally, for a good electrical contact between the metal and doped regions and to recover from damage induced during metallization, a forming gas anneal at 400°C was executed.

4 RESULTS AND DISCUSSION

4.1 I-V Results

In the experiment, three splits in processing were compared, namely (i) the reference process with rear SiO_2 passivation, (ii) the rear Al_2O_3 - SiO_x passivation with firing at 800°C and (iii) the rear Al_2O_3 - SiO_x passivation without firing. Table I reports the best and the average *I-V* parameters obtained in each case.
Best cell efficiency of 22.8% was achieved for an IBC cell with rear Al_2O_3 - SiO_x fired at 800°C. V_{oc} values up to 690 mV were measured (a cell with efficiency of 22.7%), a major improvement over the reference process. Lower performance was observed for cells without firing.

Table I: Experimental I-V results for 15.6x15.6 cm² IBC device wafers.

Cell type		J_{sc} (mA/cm²)	V_{oc} (mV)	FF (%)	η (%)
Rear SiO_2	Best	41.1	683	80.2	22.5
	Average of 5	40.8	682	78.8	21.9
Rear Al_2O_3-SiO_x (firing)	Best	41.2	688	80.3	22.8
	Average of 4	40.8	688	79.8	22.4
Rear Al_2O_3-SiO_x (no firing)	Best	40.7	678	79.7	22.0
	Average of 3	40.4	678	79.3	21.7

With the help of monitor wafers, being symmetrical test wafers processed together with the cells, a comparison of lifetime and J_0 values for the emitter and BSF region for the different passivations is shown in Table II. These confirm a threefold improvement in J_0 for the p^+ emitter with fired dielectric Al_2O_3 - SiO_x stack compared to SiO_2. For the BSF a similar quality of passivation is obtained.

Table II: Minority carrier lifetime and J_0 for monitor wafers processed together with the cells.

Type	Emitter		BSF	
Monitor Wafer	**J_0 (fA/cm²)**	**τ_{eff} @ 1e15 cm^{-3} (ms)**	**J_0 (fA/cm²)**	**τ_{eff} @ 1e15 cm^{-3} (ms)**
Thermal SiO_2	36.0	1.2	83.6	0.6
Al_2O_3-SiO_x (firing)	10.0	2.9	84.0	0.5
Al_2O_3-SiO_x (no firing)	12.1	2.8	194.5	0.2

The low $J_{0,emitter}$ leads to the higher V_{oc} for devices with fired Al_2O_3 passivation on the rear. The high $J_{0,BSF}$ for the devices with non-fired AlO_x passivation leads to their lower V_{oc}. The high $J_{0,BSF}$ is also the reason for their low J_{sc}, as can be noticed in the LBIC cross-sections of Figure 2 extracted at 850 nm. A higher $J_{0,BSF}$ is an indication of increased recombination over the BSF regions and as such decreases the current harvested over the BSF fingers (electrical shadowing).

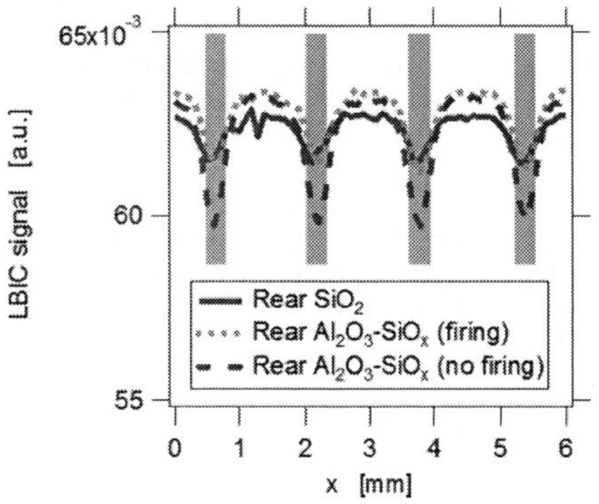

Figure 2: LBIC cross-section for the three types of cells. The LBIC signal (850 nm) is measured perpendicular to the fingers of the IBC design. Gray zones indicate the location of the BSF fingers

4.2 Simulations

The measured J_0 values from Table II are applied as simulation input to support the analysis of the measured I-V results. We use Quokka [9] to simulate a unit cell of our IBC cells and extend those simulations with analytical formulas to include the R_{series} losses in the metallization of fingers and busbars, to include the J_{sc} losses related to the wide diffused BSF busbars and to include the R_{series} losses related to the wide diffused emitter busbars [10].

In Table III, those simulated results are compared to the best measured cells. The trends and differences between the three types of devices are fitted well, confirming the effect of the measured J_0 values on the IV characteristics.

Table III. Comparison between the best measured cells and the simulated I-V results.

Cell type		J_{sc} (mA/cm^2)	V_{oc} (mV)	FF (%)	η (%)
Rear SiO_2	Best Cell	41.1	683	80.2	22.5
	Simulated	41.1	685	80.0	22.5
Rear Al_2O_3-SiO_x (firing)	Best Cell	41.2	688	80.3	22.8
	Simulated	41.1	690	79.9	22.7
Rear Al_2O_3-SiO_x (no firing)	Best Cell	40.7	678	79.7	22.0
	Simulated	40.5	682	79.5	22.0

These simulations are used to breakdown the power lost in our large area IBC cells with the use of FELA [11] power loss analysis incorporated in Quokka simulator. In Figure 3 and 4, the power loss analysis is given for the best cell with rear Al_2O_3 - SiO_x (firing) and compared with reference SiO_2 passivation. Improvement in emitter passivation with fired rear Al_2O_3 - SiO_x provides significantly lower recombination loss. Bulk recombination apart from the intrinsic bulk recombination (bulk background lifetime of 2.5 ms taken as simulation input) sets for one of the main recombination losses, although further improvement of all passivated and contacted regions would lead to gains as well. For the resistive losses, the lateral transport in the bulk towards emitter and BSF is limiting. Reduction of the pitch can be beneficial here. This pitch is currently set by the alignment accuracy of our different patterning steps. A reduction of the wafer resistivity could reduce these resistive losses as well but is typically accompanied by an increased bulk recombination.

Figure 3: Breakdown of the recombination losses for the best IBC cell with rear Al_2O_3-SiO_x (firing) compared to SiO_2 rear passivation

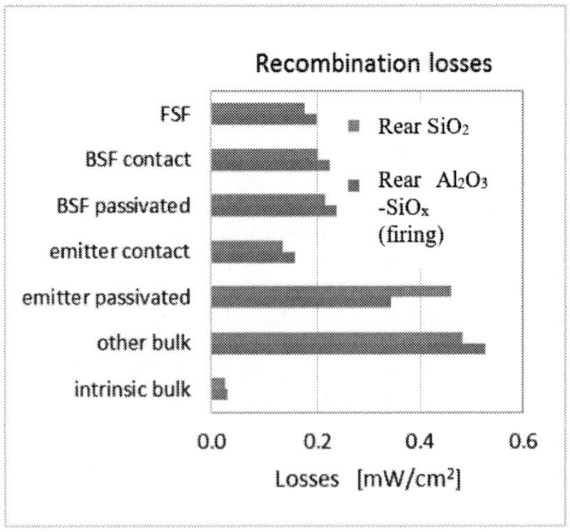

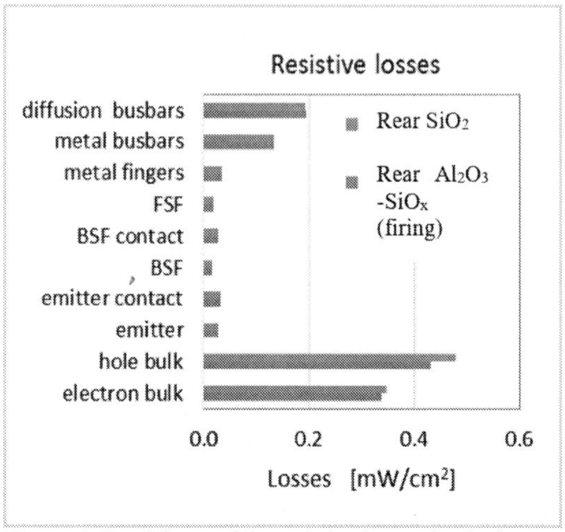

Figure 4: Breakdown of the resistive losses for the best IBC cell with rear Al_2O_3-SiO_x (firing) compared to SiO_2 rear passivation

5 CONCLUSION

An Al_2O_3 based dielectric stack was successfully implemented on large area IBC cells. This aided in upscaling the efficiency to 22.8% by improving in particular the V_{oc} of the devices. It was observed that a short thermal activation step improves the passivation quality of n^+ BSF without compromising the properties of the p^+ emitter. A lower V_{oc} and a decrease in current due to recombination losses at the BSF was observed when this thermal step was omitted.

6 ACKNOWLEDGEMENTS

The authors gratefully acknowledge the financial support of imec's industrial affiliation program and of the European Union's Seventh Program for research, technological development and demonstration under grant agreement No 308350 (Cu-PV).

7 REFERENCES

[1] Singh et al., 31st EUPVSEC (2014).

[2] O'Sullivan et al., IEEE Journal of Photovoltaics, **6**,4,(2016), p810.

[3] Cornagliotti et al., IEEE Journal of Photovoltaics, Vol.5, No.5. (2015).

[4] Hoex et al., Phys. Status Solidi RRL 6 (1),4 (2012).

[5] Duttagupta et al., IEEE Journal of Photovoltaics (2013).

[6] Richter et al., Journal of applied physics (2014).

[7] Peibst et al., 28th EUPVSEC (2013).

[8] Mueller et al., Applied Physics Letters 105, 103503 (2014).

[9] Fell et al., IEEE Transactions on Electron Devices 60,733-738 (2013).

[10] Magnone et al., IEEE Journal of Photovoltaics 5, 552-558 (2015).

[11] R. Brendel et al., Appl. Phys. Lett., vol. 93, pp. 173503-1–173503-3, 2008.

FORMATION AND EVOLUTION OF THE SIO_XF_Y MASKING LAYER CAUSED BY PLASMA TEXTURING

Maria Gaudig[1,2,*], Volker Naumann[2], Jens Hirsch[1,2], Martina Werner[2], Stephan Großer[2], Christian Hagendorf[2], Norbert Bernhard[1,2], Dominik Lausch[2]

[1] Anhalt University of Applied Sciences, Technologies of Photovoltaics Group, Bernburger Str. 55, 06366 Köthen
[2]Fraunhofer Center for Silicon Photovoltaics CSP, Otto-Eißfeldt-Str. 12, 06120 Halle (Saale)
*m.gaudig@emw.hs-anhalt.de, +49 (0) 3496 67 2362

ABSTRACT: Black silicon (b-Si) is of particular interest for the PV industry since it can significantly reduce reflection losses. A promising preparation method of b-Si surface textures is maskless plasma etching. However, the detailed mechanism of maskless plasma texturing in an SF_6/O_2 atmosphere is still unclear. In particular, the self-masking process is not well understood. For that reason, the change of the elemental composition of the Si surface caused by the plasma process is investigated by X-ray photoelectron spectroscopy (XPS) to discuss the formation of the self-masking layer. Evidence of a fluorine containing oxide layer was found on the surface of b-Si for the first time. Furthermore, it is shown that fluorine desorbs quickly out of the top layer under ambient air, which makes it challenging to characterize the self-masking layer. These results are an important contribution towards the understanding of the formation of plasma etched b-Si and the resulting oxide layer.
Keywords: Plasma texturing, Selfmasking layer, Absorption, Passivation

1 INTRODUCTION

Maskless SF_6/O_2 plasma texturing of silicon for the fabrication of black silicon (b-Si) is a widely investigated method with applications for photovoltaics [1-4]. The main benefits of this dry method against the standard wet chemical texturing are:

- a strongly enhanced absorption without additional deposition of an antireflection layer,
- the independence against surface quality and crystal orientation and
- the possibility to texture thin solar cells due to reduced material removal.

Challenges of an increased effective area on the front side of a solar cell are an increased surface recombination and the contact formation to the front fingers. Both issues could already be solved by Al_2O_3 passivation with atomic layer deposition [5] and by applying an interdigitated back-contacted (IBC) solar cell concept [1,2], respectively.

However, the detailed mechanism of maskless plasma texturing in an SF_6/O_2 atmosphere is still unclear, its understanding is yet of interest regarding the further optimization of the texturing process. In particular, the self-masking process through the SiO_xF_y layer is not well understood, which shall be responsible for the initial roughening. A general model towards the formation of b-Si by maskless plasma etching is already described by different groups [6,7]: It is supposed that F and O radicals from the plasma react with silicon to form a thin SiO_xF_y layer which prevents the etching at some points on the surface. At certain conditions, a first small roughening is boosted resulting in growing textures. The SiO_xF_y layer also acts as passivation layer of the built texture sidewalls protecting them against under etching during the plasma treatment.

So far, no experimental investigations to proof this description are provided. Therefore, this contribution deals with this topic by analyzing the chemical composition of plasma textured Si surfaces by transmission electron microscopy (TEM) and X-ray photoelectron spectroscopy (XPS). It was possible to detect F on top of the surface, when the wafer was not in contact to air, hinting on the formation of a SiO_xF_y layer during the plasma process. Moreover, we show that F desorbs quickly when b-Si is exposed to air. This observation explains the challenging F detection by different analysis method where the wafer surface has to be exposed to air.

With the presented results, a first time progression is proposed describing the evolution of the SiO_xF_y masking layer caused by plasma texturing.

2 EXPERIMENTAL

Mono crystalline FZ silicon wafers are used with a thickness of 250 µm, resistivity of 2 Ωcm (p doped) and <100> orientation. The samples are textured in a plasma etch chamber of Oxford Instruments by an inductively coupled plasma (ICP) process using SF_6 and O_2 as gases at 5 °C. This configuration without any capacitive plasma coupling allows etching without DC bias leading to a minimum ion impact on the surface. These plasma conditions generate parabolic formed pits with dimensions of about 400 nm in height and width, as illustrated in **Figure 1** [8].

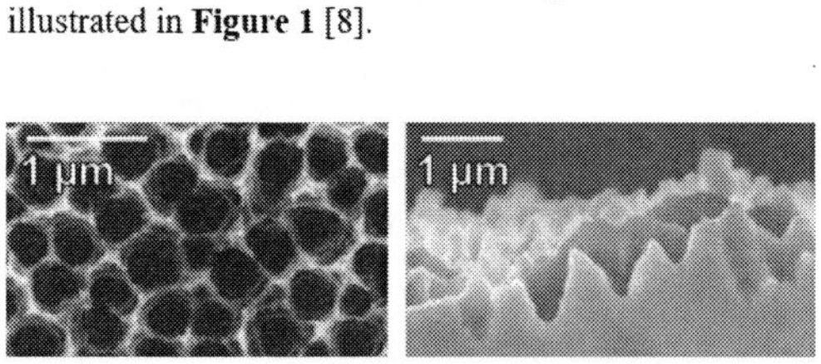

Figure 1: SEM pictures of a textured silicon surface by an ICP plasma process prepared at 5 °C.

For the investigation of the elemental composition, TEM combined with energy dispersive X-ray spectroscopy (EDX) and XPS analysis are applied. The lateral resolution during EDX measurement is about (1 - 2) nm. XPS measurements are performed with a monochromatic Al K-α X-ray source. The X-ray spot is 0.2 mm² and thus, larger than the texture dimensions. Therefore, it is not possible to resolve the tips and valleys of the texture forms by XPS. The information depth for planar silicon is 3.4 nm, thereby making the XPS investigations extremely surface-sensitive. The used binding energy interval is between 1100 eV and -5 eV. Further, high-resolution spectra of Si 2p peaks area acquired in the binding energy interval between 108 and 96 eV. Shirley background subtraction is applied for determination of the peak areas.

To observe the influence of the ambient air on the elemental composition of the surface, the textured wafers are separated after plasma texturing in samples

- which are **exposed to the ambient** atmosphere and
- which are **not in contact to the ambient** before they are investigated. Here, the transport was realized under nitrogen atmosphere. Introduction of samples into the vacuum chamber of the XPS instrument took place within 2 minutes after opening the N_2-flooded transport container.

3 RESULTS

3.1 TEM analysis

TEM measurements combined with EDX are performed with the objective to investigate the top surface layer of plasma textured silicon. In particular, the elemental composition is of interest as well as its potential distribution depending on the texture position (tip, valley or sidewall). In **Figure 2**, two TEM images of a tip cross section are illustrated.

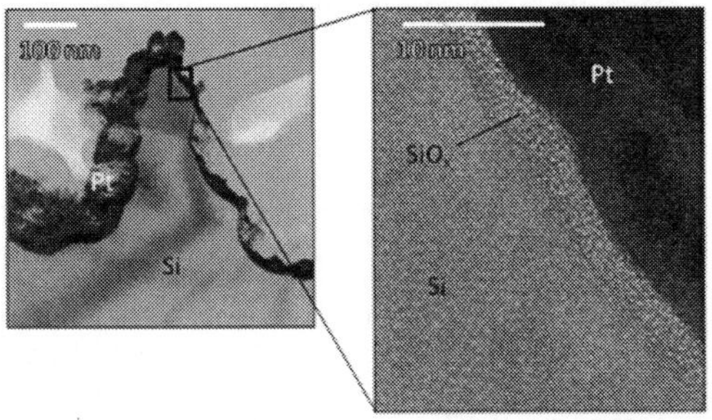

Figure 2: TEM images of an ICP textured silicon surface. The black region consists of Pt which is deposited to protect the surface during TEM preparation. The red line in the right image shows the Si-SiO_x interface.

An amorphous layer is observed on top of the surface with an average thickness of about (3.0 ± 1.2) nm. In the right TEM image, this layer is highlighted by the red line. On the left of this line, the lattice planes of the crystalline silicon can clearly be seen; whereas on the right of the red line, no long-range order exists. A thickness distribution depending on the texture position is not observed. The determination of the layer thickness was challenging due to blurred interfaces or overlaps to the Si crystal (explaining the high indicated error). Furthermore, this thickness is rather overestimated because the cross sections are not perfectly parallel to the layer stack.

Elemental analysis by EDX attests that this amorphous layer consists of silicon oxide (SiO_x). F was also detected in traces. Nevertheless, a fluorine containing oxide layer cannot be proved with these investigations. Besides, C and Pt was detected on the surface of the wafer. Both contaminations can arise during sample transport or sample preparation for TEM analysis, respectively.

3.2 Elemental composition by XPS

Through XPS spectra, quantitative elemental composition is determined and illustrated in **Figure 3** of:

- planar silicon (left),
- plasma textured silicon not in contact to air (middle) and
- plasma textured silicon exposed to air (right).

To realize no contamination by air molecules, the transport from the plasma chamber to the XPS tool was in a nitrogen atmosphere.

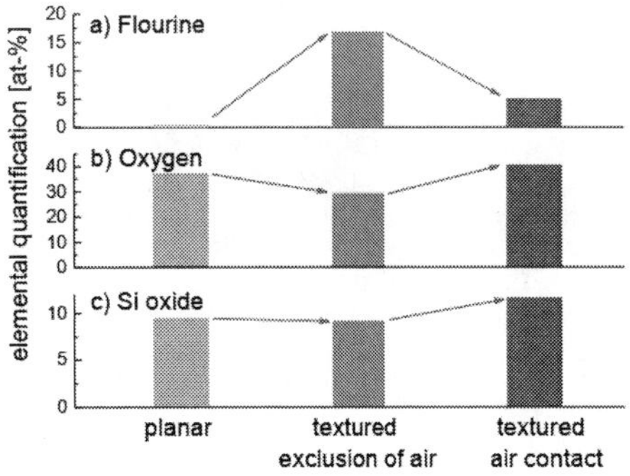

Figure 3: Elemental quantification of F, O and Si oxide (Si which is bounded in the oxide layer) for planar silicon (left), plasma textured silicon excluded to the ambient (middle) and plasma textured silicon in contact to the ambient.

The XPS spectra were acquired with the analyzer aligned parallel to the sample normal. The signal is collected from a large area compared to the dimensions of the structured surface. Therefore, influences of inclined regions are not considered separately but lead to an effectively reduced information depth for the textured samples. The main part of the observed information depth consists of Si, O (diagram b) and for the textured samples of F (diagram a). The major share of Si atoms within the information depth is arranged in the elemental Si of the substrate, bounded in a Si-Si matrix. The other part is bounded in the Si oxide layer on top of the sample (diagram c), as it is illustrated in **Figure 4** for a planar wafer.

Further elements like C, N and S are also detected in small fractions, which can be explained by contaminations from the plasma chamber, transport boxes or ambient air during sample introduction into the XPS instrument.

The elemental composition (all following percentage values referred to information depth) is changed during plasma treatment and again when it is exposed to air. The grey arrows demonstrate the trends in principle. The greatest modifications can be observed for F: Whereas the planar wafer exhibits less than 1 % of F, the amount increases to about 17 % during plasma texturing and decreases again to 5 % when the textured wafer was exposed to the ambient for approximately 5 hours. Further XPS analyses of plasma textured silicon showed that after approximately 10 hours exposure to air, the F concentration is further reduced to a constant value less than 1 %. At the same time, the O (and Si oxide) fraction is reduced from 37 % (10 %) for the planar wafer to 29 % (9 %) after plasma texturing and increased again to 41 % (12 %) after exposure to air, respectively. For evaluation of the Si oxide concentration, the effective area of the samples has to be taken into account, when assuming a complete covering of the silicon by an oxide layer.

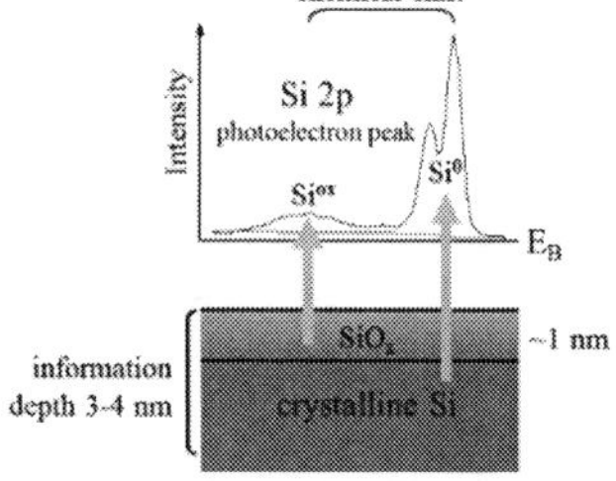

Figure 4: Information depth by XPS characterization of a planar silicon wafer with a thin silicon oxide on the surface. The Si atoms bounded in the elemental Si and those bounded in the SiO_x can be resolved in the XPS spectrum (intensity vs. binding energy E_B) through the chemical shift of the Si 2p photoelectron peak.

4 DISCUSSION

From the TEM and XPS results, the time progression is presented in this section explaining the modification of textured silicon surface during and after plasma texturing. For that purpose, in **Figure 5**, the elemental development in the diagrams in Figure 3 is extended manually to a potential elemental composition after about 10 hours air exposure (striped bars). The values are estimated from different further XPS investigations of textured silicon samples which are exposed more than 5 hours to air. In the following, this time progression for F, O and the Si oxide part is explained.

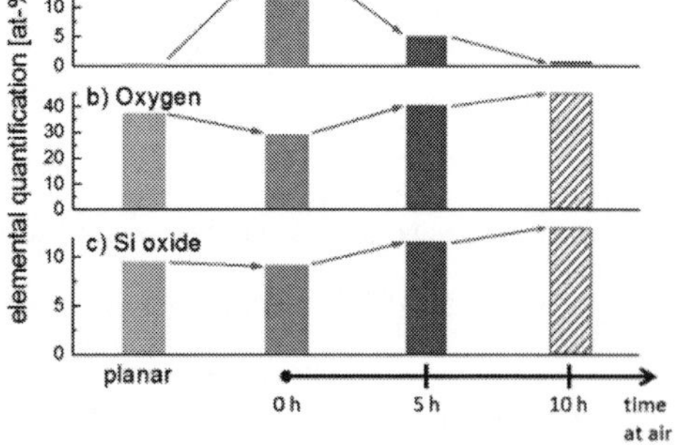

Figure 5: Progression of the elemental surface modification (F, O and Si oxide) of plasma textured silicon wafers depending on the time exposure to air. The orange elemental bars on the left side belong to a planar, untreated silicon wafer.

Planar silicon

The starting point is the planar silicon wafer (Figure 5, left), which is in contact with the ambient atmosphere before characterization by XPS. The wafer is covered with a native, thin oxide layer, as it is measured in Figure 5 c) for the Si oxide fraction. This fraction can be used to calculate the thickness of the layer to approx. (0.8 ± 0.2) nm [9], with presumption of a flat surface and homogenous layer thickness. Furthermore, a thin adsorbate layer (< 1 nm) is deposited on the wafer under ambient atmosphere, which cannot be prevented. It consists mainly of water and other inorganic or organic contamination. These elements are also detected during XPS (and TEM/EDX) analyses explaining the observation of C, N or F for the planar wafer.

Textured silicon (time at air = 0 h)

When silicon is plasma textured in an SF_6/O_2 atmosphere, F is incorporated and partly replaces O in the O-Si-O network as can be followed from the increased F content and decreased O fraction. We suggest the formation of a SiO_xF_y layer on top of the silicon because the further diffusion of F in the silicon lattice is unlikely [10]. Assuming that all F atoms are homogeneously distributed in the SiO_xF_y, the ratio of the F atoms against the Si atoms which are bounded in the oxide can be calculated to approx. 1.9 by using the values of Figure 5 a) and c) (blue bars). With that, the O to Si ratio is approx. 1.05 for an expected stoichiometric composition of the oxide layer. As result, the ratio x:y for the SiO_xF_y can be designated to approx. 1:2. That means that one divalent O atom is replaced by two monovalent F atoms in comparison to SiO_2. This configuration reduced the O-Si-O connection in the crystal resulting in increased pore sizes and a reduced density of the layer.

An oxide thickness calculation is not possible for textured surfaces by XPS. Nevertheless, it can be estimated by comparing the Si fraction which is built in the oxide to the fraction of the planar wafer. Figure 5 c) shows only a slight decrease of the Si oxide fraction which means that the thickness of the oxide layer is in a similar range as for the planar wafer (less than 1 nm). In comparison, during TEM investigation, a higher oxide thickness of approx. 3 nm was observed. However, with this characterization method, the wafer was exposed to air which can lead to a further oxidation. We estimate that the 'real' value for the Si oxide thickness is closer to 1 nm than to 3 nm [11]. This means that F influences about ten monolayers of the silicon surface during plasma texturing. It is assumed that firstly F species (as radicals or ions) are deposited on the Si surface because of the high sticking coefficient, forming a SiF_x layer[12]. Then O radicals can react with this layer to SiO_xF_y.

Nevertheless, no thickness or elemental distribution of the SiO_xF_y layer is detected by XPS and TEM depending on the position at the tips, bottoms or sidewalls of the textures. In particular, for the ICP texture formation, a distribution was expected regarding the plasma texturing mechanism. A physical component, e.g. through ions, is minimized for this inductive plasma process, which is the usual explanation for the growing texture features by maskless plasma texturing [13,14]. Therefore, a chemical process may be dominant for the ICP texturing which can be characterized by the passivation layer SiO_xF_y.

Textured silicon (time at air > 0 h)

When the wafer is exposed to humid air, the elemental content changes again. F desorbs from the surface, whereas the O content increases again. We assume that water molecules are the cause of this effect: They are adsorbed at the surface and react chemically with the SiO_xF_y by cracking the Si-F bonds. The adhesion of water molecules to the Si-F bonds is thereby enhanced due to the increased pore sizes of the layer. This hydrolysis process leads to desorption of H-F molecules and building of Si-O-Si bonds in the layer. In [15], ab-inito calculations demonstrate that this reaction

$$2\ Si–F + H_2O \rightarrow Si–O–Si + 2\ HF$$

between water and a SiO_xF_y system are possible leading to the rearrangement of SiO_xF_y to SiO_x accompanied with the formation of HF and an emission of thermal energy (0.7 eV).

Nitrogen does not influence the SiO_xF_y explaining that F is detected for the textured wafers which are transported under nitrogen atmosphere [16].

5 CONCLUSION

The morphological changes of SF_6/O_2 plasma textured silicon surfaces are investigated by TEM/EDX and XPS. TEM/EDX analyses reveal an amorphous, thin oxide layer covering the textures completely. With XPS, a higher F concentration could be determined on the surface of textured wafers in contrast to untreated planar samples; the F concentration decreases again as soon as the textured wafer is exposed to the ambient air. Following assumptions can be made with these results:

i. F is completely incorporated in the silicon oxide layer forming SiO_xF_y during plasma texturing.
ii. A thickness of about 1 nm and the ratio of approx. $x{:}y \approx 1{:}2$ are indicated for this layer, when assuming a homogeneous distribution.
iii. The F leakage after air exposure can be explained by hydrolysis reactions leading to a desorption of F.

6 ACKNOWLEDGEMENT

This work was carried out within the joint research graduate school “StrukturSolar” of Anhalt University of Applied Sciences and Martin-Luther-University Halle-Wittenberg which was funded by the German Ministry of Education under identification code 03SF0417A/B.

4 REFERENCES

[1] P. Repo, J. Benick, V. Vähänissi, J. Schön, G. von Gastrow, B. Steinhauser, M.C. Schubert, M. Hermle, and H. Savin, Energy Procedia 38 (2013) 866.

[2] A. Ingenito, O. Isabella, and M. Zeman, ACS Photonics 1 (2014) 270.

[3] J. Yoo, G. Yu, and J. Yi, Sol. Energy Mater. Sol. Cells95 (2011) 2.

[4] M. Gaudig, J. Hirsch, Th. Schneider, A.N. Sprafke, J. Ziegler, N. Bernhard, and R.B. Wehrspohn, J. Vac. Sci. Technol. A 33 (2015) 05E132.

[5] M. Otto, M. Kroll, Th. Käsebier, R. Salzer, A. Tünnermann, and R. B. Wehrspohn, Appl. Phys. Lett. 100(2012) 191603.

[6] H. Jansen, M. de Boer, R. Legtenberg, and M. Elwenspoek, J. Micromech. Microeng. 5 (2) (1995) 115.

[7] R. Dussart, T. Tillocher, P. Lefaucheux, and M. Boufniche, J. Phys. D Appl. Phys. 47 (2014) 123001.

[8] J. Hirsch, M. Gaudig, N. Bernhard, and D. Lausch, Appl. Surf. Sci. 374 (2016), 252.

[9] M. P. Seah, and S. J. Spencer, Surf. Interface Anal. 35 (2003) 515.

[10] Ch. G Van de Walle, F. R. McFeely, and S. T. Pantelides, Phys. Rev. Lett. 61 (1988) 1867.

[11] S. Richter, K. Kaufmann, V. Naumann, M. Werner, A. Graff,S. Großer, A. Moldovan, M. Zimmer, J. Rentsch, J. Bagdahn, and Ch. Hagendorf, Sol. Energ. Mat. Sol. C. 142 (2015) 128.

[12] X. Mellhaoui, R. Dussart, T. Tillocher, P. Lefaucheux, P. Ranson, M. Boufnichel, and L. J. Overzet, J. Appl. Phys. 98 (2005) 104901.

[13] H. Jansen, M. de Boer, R. Legtenberg, and M. Elwenspoek, J. Micromech. Microeng. 5 (2) (1995) 115.

[14] M. Schnell, R. Lüdemann, and S. Schaefer. 28th IEEE Anchorage, AK (2000) 367.

[15] H. Yang, and G. Lucovsky, J. of Non-Crystalline Solids 254 (1999) 128.

[16] Y.-H. Kim, H.-H. Kim, S.-K. Lee, and H. J. Kim, J. Korean Physical Society 33(1998) 179.

HOW TO DEAL WITH THIN WAFERS IN A HETEROJUNCTION SOLAR CELLS INDUSTRIAL PILOT LINE: FIRST ANALYSIS OF THE INTEGRATION OF CELLS DOWN TO 70μM THICK IN PRODUCTION MODE

S.Harrison, O.Nos, A.Danel, D.Munoz, JP.Rakotoniana, J.Gaume, C.Roux, PJ.Ribeyron
CEA-LITEN, Institut National de l'Energie Solaire (INES)
Savoie Technolac, F-73377 Le Bourget du Lac, France

ABSTRACT: In this paper we present the first results obtained at CEA-INES for the integration of thin and ultra-thin silicon wafers in the current silicon heterojunction (SHJ) LabFab Pilot Line production flow. Impact of the cell thickness reduction on final device characteristics is first detailed. Then overall compatibility of such wafers with current production constraints and actual equipment's configuration is analyzed, identifying main breakage and throughput issues. Finally, improvement paths are proposed to enhance both efficiencies and reliability on the overall cell fabrication flow demonstrating also the potential compatibility of current SHJ production line for wafers as thin as 80μm.
Key Words: heterojunction, amorphous silicon, thin wafers, SmartWire, pilot-line.

1 INTRODUCTION

Amorphous/crystalline silicon (a-Si:H/c-Si) heterojunction (SHJ) solar cells is one of the most promising technology for the future production of high efficiency solar photovoltaic modules. This is emphasized by the world record cell efficiencies obtained on back contact and both sides contacted SHJ cells [1, 2, 3] for single junction silicon based solar cells. Nevertheless, in order to make this technology even more competitive at an industrial level, further production cost reductions are mandatory. In this context, the development of SHJ cell production processes compatible with thinner c-Si wafers appears as an important asset since wafer production costs (including material and sawing) constitute actually about 30 to 35% of the total cost of an industrial PV module [4, 5, Fig. 1]. With its unique low temperature processes, symmetric active layer and metallization configuration added to its outstanding surface passivation properties, SHJ technology is indeed particularly well adapted to thin wafer constraints.

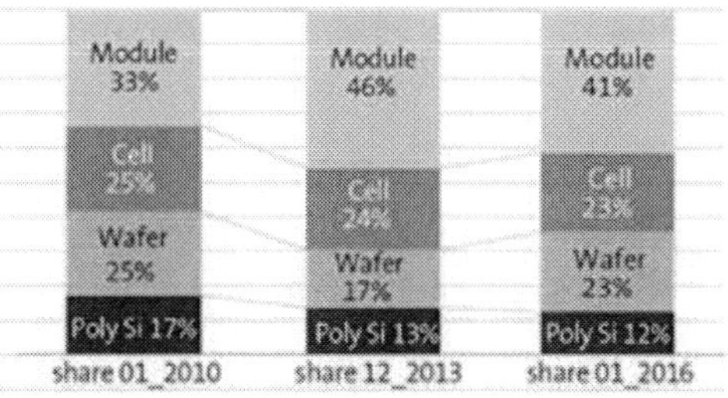

Figure 1: Extract from ITRPV 2016. Repartition of costs for a final silicon based PV module. Silicon wafer represents approximatively 35% of total module cost.

Furthermore, there is already a need for innovative modules and specific applications (BIPV, light weight modules, flexible modules…etc) are thus particularly interested in the use of such thin wafers.
In this paper we present the main production results obtained in our industrial pilot line for both sides contacted SHJ solar cells on standard (130 to 160 μm final thickness) and thin (68 to 127 μm final thickness) c-Si wafers.

2 APPROACH

The CEA-INES "LabFab" pilot line is actually capable to produce 130 to 160 μm final thickness SHJ cells at high throughput (1200 wafers/hour on 156 pseudo-square wafers/hour) with typical average efficiency of 21% for 4 BusBar metallization configuration (4BB) [6] and 22% for BusBarLess metallization configuration (BBL). This is illustrated in Figure 2 with populations of several thousands of bifacial cells (4BB and BBL cells*).

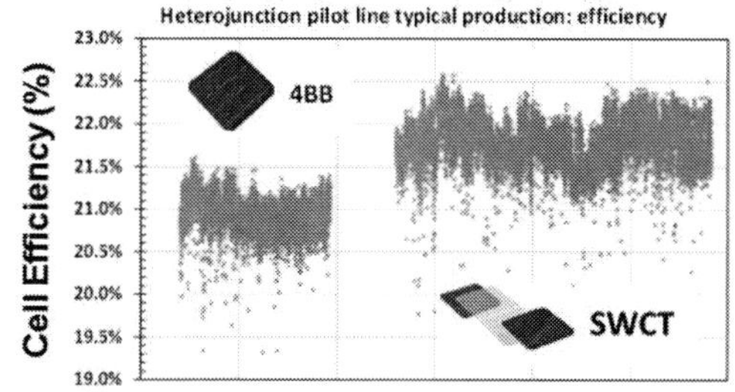

Figure 2: Example of weekly activity on commercial 156PSQ c-Si substrates: 1200wph continuous production mode, 4BB Ag print bifacial cells, double-comb IV test

Global breakage rate on the line is below 1.5% for typical standard thickness (160μm). To learn what happens when thinner cells are elaborated on the line, two kinds of thin wafers have been produced and processed firstly without any adaptation of the production line: standard 170 to 200μm as-cut wafers (high volume) and 120 μm as-cut wafers provided by EU CHEETAH project partners** (low wafer volume), both thinned down by KOH etch. Efficiencies and breakage rate were monitored step by step for these 70 to 130 μm wafers. Main issues encountered in our production process chain were then pointed out to improve the breakage rate by adjusting some key automation parameters and to state on practical limitation with conventional cassettes, belts and pickers handling design.
First module integration results are also presented as well as first device results with optimized industrial process-flow, tuned to limit the defectivity induced by the use of the thin wafers.

3 SOLAR CELL RESULTS

As mentioned previously, set of wafers with different thicknesses are processed firstly with no modification/adaptation of the current production flow. Cell results are presented in Fig. 3.

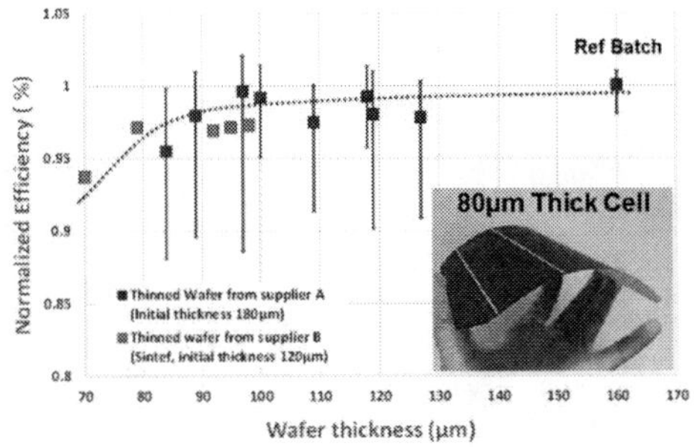

Figure 3: Impact of cell thickness reduction on final device characteristics.

For cell thicknesses ranging between 90 µm and 160 µm, the final cell efficiency remains approximately the same, this considering for all wafers the same process-flow and recipes optimized at 160 µm. This is mainly due to an improvement of the open circuit voltage for thinner cells, which compensates lower photo-generated current (due to a reduced total absorption in the infrared region of the incident light). For cells thinner than 90 µm, a drop in efficiency is obtained for 4BB cells and mainly due to a decrease of Jsc, but also Fill Factor - FF (higher series resistance, higher defectivity rate - scratches), figure 4. We can also notice a sensible higher efficiency spreading for the thinner wafers, also linked to an overall defectivity increase (mostly surface scratches appearing at PL analysis). We can also highlight that the record cells are obtained for devices in the 90-100µm range, proving the compatibility of such thin wafers with very high efficiencies.

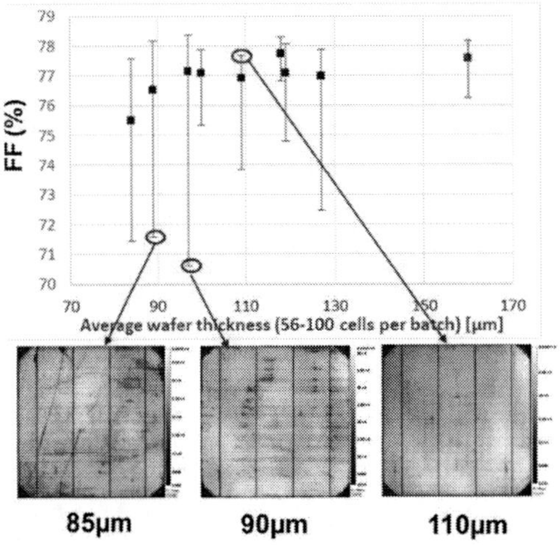

Figure 4: Cell FF is the most impacted parameter by the wafer thickness reduction. PL analysis reveal mainly an increased surface defectivity (scratches).

The thinnest wafers processed were 70µm thick SHJ solar cells with a best promising efficiency of 18.9%. It is worth to notice that manual handling was necessary for this wafer thickness because of the too high breakage rate. This will be detailed in the following section.

Temperature behavior, bifacial coefficient and low illumination properties of thin cells (80 and 90µm) were also measured and compared to similar SHJ 160µm thick cells. As shown in Fig. 5, thin and standard cells responses are similar. Thus, same final module outdoor productivity (kWh/Wp Ratio) is expected when moving towards thinner wafer thicknesses.

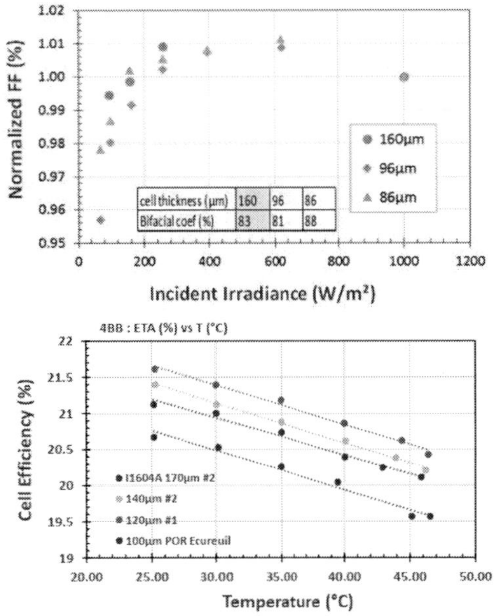

Figure 5: Complementary solar cell characterizations. No cell behavior modification expected when moving towards thinner wafer thicknesses.

Additionally, breakage rate against wafer thickness has been monitored along the whole process chain (Fig. 6). Specific previous wafer flexion tests demonstrate that thin wafers are initially not more fragile than the reference ones. First integration with standard line settings allowed us to identify the main bottlenecks for the production of thin wafers in the standard SHJ production. Strong increase of breakage rate below 100µm is thus mainly related to automation issues: wafer transfer between deposition chambers (too high picker speed) and wafer pick and place in and out of the cassettes (excessive wafer bowing). Below 80µm, the current production line is not capable at all to handle wafers.

Most impacted process areas are deposition and metallization where a lot of wafer manipulation is needed (carrier transfer, flip, wafer transfer). Breakage issues are in all cases linked to the excessive wafer bowing. For pick and place wafer transfer, the bowing associated to the vibrations allows air to enter between the wafer and the picker leading to a sudden drop of the wafer during the movement. For cassette loading and unloading, the wafer bowing leaves to little space for the end effector between two cassette slots, leading to either breakage of defectivity (scratches).

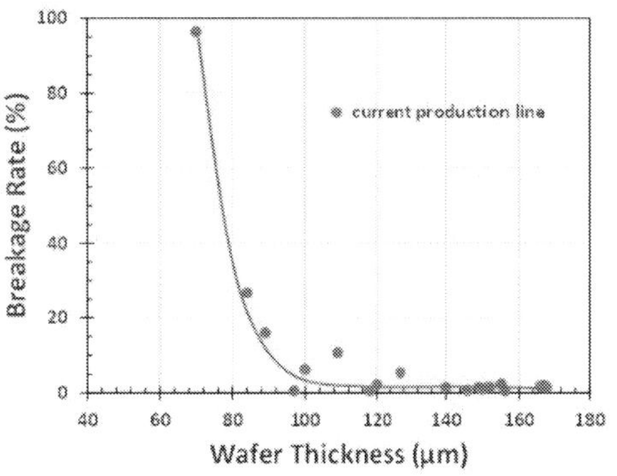

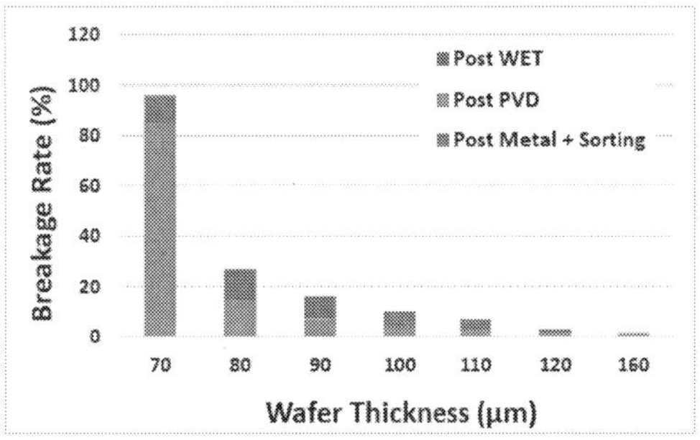

Figure 6: Impact of wafer thickness reduction of production line breakage rate (without modification of the current production flow). Below 80µm, automatic transfer is not possible at all. Wafer automation (transfer, unload/reload in cassette) are the main breakage cause, explaining why deposition and metallization areas are the most impacted.

Finally, the line overall capability to process thin wafers can be summarized as presented in table 1. Down to 100µm thickness, very limited precautions are needed to be able to process wafers with performances and defectivity comparable to our standard products. Below 100µm, specific line upgrade would be necessary, but actual set of equipment remains compatible with the thin wafer constraints. 80µm nevertheless seems to be currently the limit, to process wafers below 80µm thick, overall line and tool modification must be considered (single side WET etch and clean tool, new transfer systems, different pocket configuration on trays during deposition…etc.)

Table 1: Summary of current production capability and associated critical steps with regards to targeted device thickness. Without line major upgrade, 80µm thick wafer is currently the actual limit for current tool and processes.

Thickness Range	Line Status	Critical Step(s)
160µm → 120µm	No Problem	-
120µm → 100µm	Minor Adjustments needed	✓ Sticking to avoid after HFLast
100µm → 80µm	Major Adjustment needed	✓ Sticking to avoid after HFLast ✓ Automation+Handling
80µm → 60µm	Not compatible avec current line and equipements	✓ All processes

4 MODULE RESULTS

In addition the production trials, different set of thin wafers have been tested up the module level. Several 2x2 mini-modules (white Tedlar backsheet) have been realized and fully characterized with wafer thicknesses in the 85 to 160µm range.
EL images and electrical characteristics of the different modules are presented in figure 7.

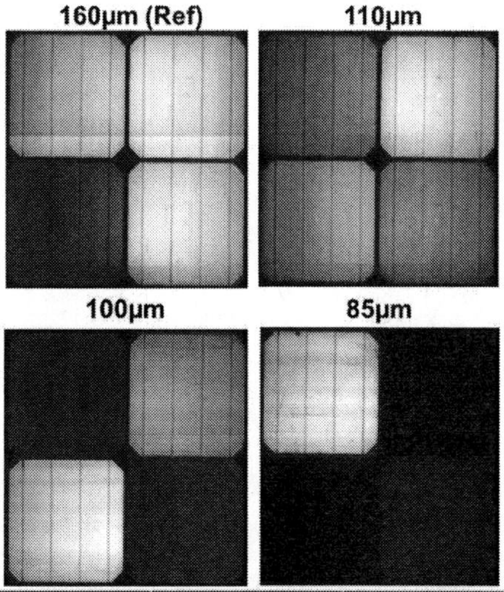

Thickness (µm)	Pmax (W)	Cell Efficiency (%)
160	18.19	19.03
110	18.42	19.27
100	18.38	19.23
85	16.8	17.58

Figure 7: EL and electrical measurements of 2x2 mini-modules realized with sets of different cell thicknesses.

Functional modules down to 85µm thick have thus been realized. We can observe as for the cells, quite stable module power and efficiencies down to 100µm thick wafers. Below this thickness, performances start to drop, and first module defect to appear (edge cracks, line cut) at the EL observation. We can also notice again a higher performance spread of the initial incoming wafers, as translated in the different EL intensities measured between several wafers within the same module.
It is worth to note that first automation tests have also been realized on the production stringer, with initial 100µm thick wafers. No problem was observed neither in breakage rate or machine throughput. So 60 cells fully automatized modules are planned to be fabricated soon.
Reliability tests of the mini-modules has also to be done in order to validate fully the compatibility of the actual module fabrication scheme with very thin silicon wafers.

5 LINE ADJUSTMENTS

Taking into account learning from the first set of experiments, several line adjustments have been performed to improve both breakage rate and global cell defectivity. It is worth to note that overall throughput is obviously impacted by these specific trials and cannot be kept at its nominal value. Different line modifications are:

- Systematic Double pitch during WET (only 50 wafers in a 100-slot carrier) especially when hydrophobic surfaces are handled. Loading remains automatic. Actual limit for 156 PSQ M0 size is around 70μm
- Decrease of 50 to 70% of the picker speed in the deposition tool automation. No adjustments on belts or wheels. .No modification for the loader cassette. Indeed, despite the wafer bending because unloading starts by the bottom of the cassette, there is currently no problem with the end effector movement between wafers.
- Unloading: with current slot pitch on J&R carriers, using a single pitch automation (only mode available on our tool), 80μm is really the limit. Standard unloading conditions kept for wafers > 85μm, mostly all manual unloading performed for wafers < 85μm.
- ITO deposition with edge exclusion validated down to 100μm thick wafers. Lower thicknesses still to be evaluated. Under 100μm, deposition is still performed on standard full trays with manual masking of the wafers (specific masks manually added on the trays).
- Metallization: again automation is the critical parameter to adjust. Especially for loading of cassettes for the final paste curing (not enough space for the end effector). Without short term possibility to modify the cassette design, loading of the cassettes is done manually for wafers <100μm. Printing parameters are also optimized for such wafers. For example, because of dense back side layout used in our cells, there is a higher risk of sticking of the very thin wafers on the printing screen.
- IV + sorting: double comb design used for BusBar Cells. Grid-Touch measurement used for SmartWire cells. No breakage observed on both automatized measurement systems, no modification done.

With these simple line adjustments, successful reduction of total line breakage rate was obtained, figure 8.

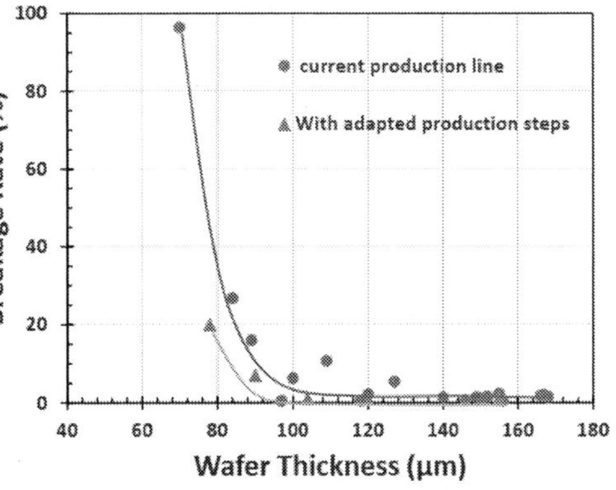

Figure 8: Improved breakage rate with specific production line adjustments.

These results prove that our current line are theoretically fully compatible with the processing of thin wafers down to 80μm given some simple equipment modification. As mentioned previously, total line throughput is impacted (slower wafer robotic, less wafers per carrier) but simple design modifications on cassettes or pickers could be imagined to keep the line at nominal value (added Teflon tit on retention bar for WET cassettes, new picker design with vacuum edge contact points, wider lateral cassette ergots for cassettes used for unloading...etc). Below 80μm, our first tests show that the current line and equipment would require major upgrades or modifications, especially in automation and handling system to still be able to process such thin wafers.

6 SMARTWIRE PRELIMINARY CELL RESULTS

Along with line adjustments, SmartWire (SWCT) technology compatibility with thin wafers has also been evaluated. All process steps are identical with BusBar Cells up to metallization. SWCT should be even more favorable for thin wafers, as two printing steps only instead of three for Busbar cells are necessary. Furthermore, lower quantity of paste is deposited on wafers. First set of thin cells have been tested with this technology. Main preliminary results of SWCT cells produced with line adjustments as presented in the section 5, are presented in Figure 9.

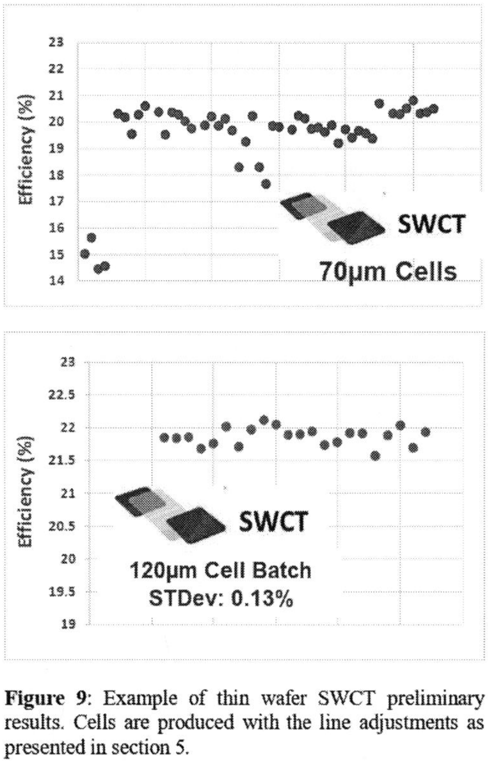

Figure 9: Example of thin wafer SWCT preliminary results. Cells are produced with the line adjustments as presented in section 5.

As we can see in figure 9, both SWCT and line adjustments are electrically validated by the realization of 120μm cell batches with highly improved variability. Efficiencies close to 22% average (Best cell at 22.1%) are obtained, with a standard deviation of only 0.13%. Line breakage rate improvements also translates in the

realization of complete 70µm thick SWCT batches with limited defectivity and breakage compared to first trials presented in section 3. Efficiencies close to 21% were obtained for the best wafers (Best cell at 20.9%). Record Voc (not shown in figure 9) were also obtained on 100µm thick wafers, with measured Voc up to 742mV.
With the validation of this new line protocol, new extensive set of wafers with different thicknesses will be processed to improve the efficiencies and optimize further the production line for the thin wafers.

7 CONCLUSIONS

Using our industrial SHJ pilot line, we evaluated the impact of processing thin and ultra-thin wafers. Firstly, without specific line modification, we identified the main breakage and defectivity sources. Automation is currently the main limiting factor, especially for wafer < 100µm thick. But cassette pitch and design are also critical for the loading and unloading of very thin wafers. Nevertheless, wafers as thin as 70µm have been produced, and overall impact of wafer thickness on final device characteristics widely analyzed.
In a second time, with specific line adjustments, we were able to strongly improve the production line breakage rate. The results obtained demonstrate that we properly identified the main line issues, and that our current production line could be compatible with very thin wafers, down to 80µm. But below this thickness limit, all our tests show that up to now, without major tool modifications, current automation and deposition system would not be suitable for mass production with high throughput.
Finally, the potential cost reduction due to the use of thinner wafer in a SHJ industrial production line (80 MW capacity, Busbar Cell configuration) have also been evaluated using an internal cost model similar to [4]. As shown in Fig. 10, considering also our current pilot line preliminary results, the optimal cell thickness range is expected to be between 90 µm and 100 µm. For thinner wafers, main issues are efficiency drop and increased breakage rate not compensated by cheaper substrate costs. Our preliminary test show that our current production line is perfectly compatible with such requirements and in line with future cost and efficiency ITRPV roadmaps.

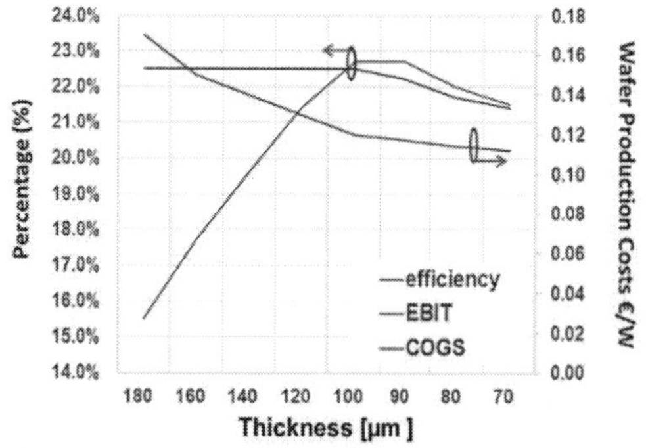

Figure 10: Evaluation of thickness impact on wafer production costs (in-house cost model). Optimum thickness is estimated in the 90-100µm range to maximize the final product EBIT (Earning before Income Taxes).

* Acknowledgments to Meyer Burger. ITO deposition and BBL print and IV test were performed in short loop on Meyer Burger Germany equipments.
** In the frame of the CHEETAH program, SINTEF provides 120µm as-cut 156PSQ wafers. Acknowledgement to FP7 EC grant 609788 and to CHEETAH partners.

[1] K. Masuko, M. Shigematsu, T. Hashiguchi, D. Fujishima, M. Kai, N. Yoshimura, T. Yamaguchi, Y. Ichihashi, T. Mishima, N. Matsubara, T. Yamanishi, T. Takahama, M. Taguchi, E. Maruyama, S. Okamoto **'Achievement of More Than 25% Conversion Efficiency With Crystalline Silicon Heterojunction Solar Cell'**, IEEE Journal of Photovoltaics, **4** (2014), pp. 1433–1435

[2] M. Taguchi, A. Yano, S. Tohoda, K. Matsuyama, Y. Nakamura, T. Nishiwaki, K. Fujita, E. Maruyama **'24.7% record efficiency HIT solar cell on thin silicon wafer'**, IEEE J. Photovolt., **99** (2013), pp. 1–4

[3] K.Yamamoto. et al, "25.1% efficiency Cu metallized heterojunction crystalline Si solar cell", 25th International Photovoltaic Science and Engineering Conference, Busan, Korea, November 2015.

[4] International Technology Roadmap for Photovoltaic, July 2015, www.itrpv.net

[5] A. Louwen, W. van Sark, R. Schropp, A. Faaij **'A cost roadmap for silicon heterojunction solar cells'**, Solar Energy Materials & Solar Cells, **147** (2016), pp. 295-314

[6] A. Danel, S. Harrison, B. Martel, C. Roux, K. Aumaille, A.S. Ozanne, M. Vandenbossche, N. Rey, A. Valla, F. Medlège, B. Novel, B. Commault, P. Lefilastre, R. Varache, P. Carroy , S. Bancel, F. Ozanne, D. Muñoz, D. Heslinga, E. Gerritsen, P.J. Ribeyron **'Recent Progress on the CEA-INES heterojunction solar cell pilot line'**, Proc. of the 31st European Photovoltaic Solar Energy Conference and Exhibition, Hamburg, Germany (2015), pp. 279-283

MASS PRODUCTION OF HIGH EFFICIENCY SILICON HETEROJUNCTION SOLAR CELLS: A LOW-COST APPROACH BY UPGRADING GEN8.5 THIN FILM SOLAR LINE

Liwei Li * [1,2], Lin Zhang[1,2], Zhan Xu[1,2], Xianfei Fang[1,2], Guanchao Zhao[1,2], Shibin Gu[1,2], Xiaorang Tian[1,2]
Baosheng Li[1,2], Rong Yang[1,2], Yuan Meng[1,2], Ted Guo[1,2]
[1]ENN Solar Energy Co., Ltd., [2]State Key Laboratory of Coal-based Low Carbon Energy
No.106 Huaxiang Road, Langfang, Hebei, P. R. China 065001
Email: liliwei@enn.cn

ABSTRACT: In this paper, we report ENN's results in R&D and mass production of high efficiency silicon heterojunction solar cells. Using large-size industrial equipment and mass production compatible processes, ENN has developed hybrid silicon technology (HST) and achieved 23.4% heterojunction solar cell efficiency. We have successfully developed a low-cost approach to mass production, i.e., converting existing Gen8.5 ($5.7m^2$) thin film solar line to high efficiency HST fab by upgrading and modification of key equipment and line automation. We have achieved average cell efficiency of 21.6% in pilot production and demonstrated 317 Wp power output using standard 60-cell (156 mm) modules. ENN's unique technology route to mass production of high efficiency heterojunction solar cells has comprehensive competition advantages such as low capital cost, high production capacity, proven equipment performance & reliability, and fast ramp-up, etc. This provides a viable path for more manufacturers to penetrate into this area, quickly commercialize low-cost heterojunction technology, and further increase its competitiveness.
Keywords: Heterojunction, silicon solar cell, PECVD

1 INTRODUCTION

Owing to higher electricity generation and less BOS cost, high efficiency solar modules are the best fit to the fast growing residential and other distributed PV markets. Silicon heterojunction technology (SHJ), combining advantages of high efficiency and relatively simple processes, is the leading candidate among various high efficiency solar cell technologies. Panasonic has pioneered development of this technology and achieved record conversion efficiency up to 25.6% [1]. Over the past a few years, SHJ has attracted extensive interests in PV industry and cell efficiencies in the range of 22% to 25% have been demonstrated by more manufacturers and research institutes [2-4]. To be truly competitive against mainstream silicon cell technologies, SHJ needs to achieve both high efficiency and low-cost mass production. However, few companies could address both and realize large-scale mass production due to lack of feasible low-cost technology approaches.

ENN Solar, once one of the leading silicon thin film solar panel manufacturers, has developed high efficiency SHJ solar cells. ENN's hybrid silicon technology (HST) employs low-cost, mass production compatible processes and uses large-area industrial equipment. For mass production, we have converted existing Gen8.5 thin film line to a high efficiency HST fab by upgrading key equipment and integrating wet-bench, screen printing, and line automation. This forms a unique, feasible low-cost technology approach, through which high efficiency has been demonstrated by pilot production. In this paper, we will report ENN Solar's results in HST R&D and pilot production.

2 DEVELOPMENT OF HIGH EFFICIENCY HST SOLAR CELLS

The key to high efficiency HST solar cell is achieving high Voc through high quality passivation of silicon wafer surface to significantly reduce surface defects and increase minority carrier lifetime (MCLT). Typical cell processing steps consist of wafer texturing, cleaning, and deposition of intrinsic and doped a-Si:H thin layers, formation of TCO as conducting and ARC layer, and Ag grid as electrodes on both sides of the wafers. To achieve high efficiency, critical factors for HST solar cell processing include excellent light trapping, high-quality passivation, and minimum carrier collection loss, etc. Below we will discuss key structures and processes affecting cell performance and how we achieved over 23.0% efficiency using mass production compatible equipment and technologies.

2.1 Wafer texturing & cleaning

Wet processing is the first step to fabricate HST solar cells, and it should offer a low-reflection, sufficiently smooth and ultra-clean wafer surface with low structural defect density. The pyramid shape and size control at texturing step is of great importance in achieving above goals. It is found that an etching rate transition point exists when isotropic etching shifts from {100} plane dominated to {111} dominated, and pyramid size has a linear correlation with etching amount at the transition point as shown in Figure 1. Through process variation, a wide range of pyramid sizes are achieved. As plotted in Figure 2, we can see strong correlation among pyramid sizes and surface reflection (R) & MCLT. Evidently, middle-sized pyramids are desirable as they result in low R and high MCLT.

Compared to traditional diffused-junction silicon solar cells which pursue low reflection rate and usually employ small-sized pyramids, HST solar cells prefer relatively larger pyramid sizes for higher MCLT. However, appropriate pyramid size alone is not sufficient to achieve high MCLT and Voc. Sharp 'peaks', narrow 'valleys', small imperfect structures embedded among pyramids, as well as micro-roughness on different facets, could degrade a-Si:H coverage and passivation quality. A proven technology addressing these issues is an isotropic etching step using chemical based polishing, which can make pyramid surface 'round' and 'smooth', and remove small imperfect structures. Figure 3 compares typical SEM images of as-textured pyramid valley and post-polishing pyramid valley, clearly showing the 'rounding' effect that makes the valley more 'open'. The open valley could be helpful for contamination removal in subsequent cleaning steps and improve a-Si:H coverage. We have further developed special cleaning instead of conventional RCA approach, featuring precise control of wafer surface potential for thorough removal of metallic ion contamination.

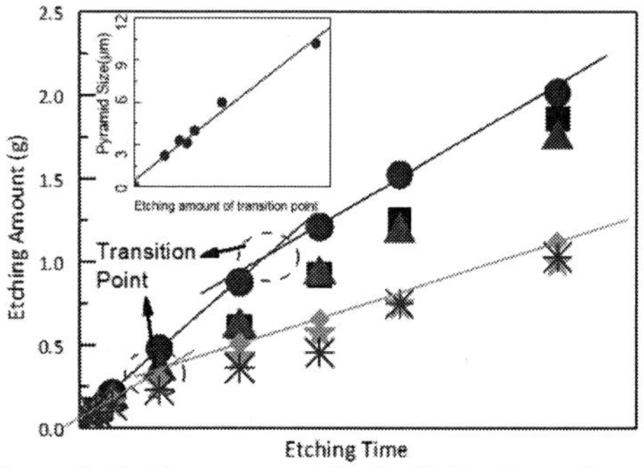

Figure 1: Etching rate transition in alkaline isotropic etching (inset: etching amount vs. pyramid size)

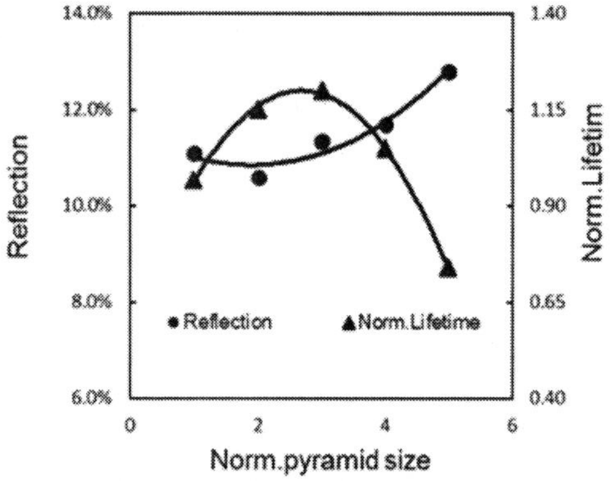

Figure 2: Light reflection and minority carrier lifetime vs. wafer pyramid sizes

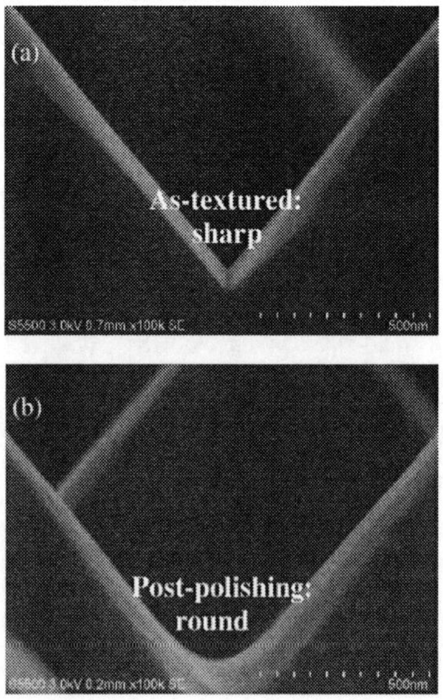

Figure 3: Typical SEM images of pyramid valley: as-textured and post-polishing

In order to evaluate the effects of chemical polishing and special cleaning processes, as well as their process compatibility, three groups of HST solar cells were processed: Group A employed normal texturing and conventional RCA cleaning, Group B added polishing between texturing and conventional RCA cleaning, and Group C combined addition of polishing and special cleaning. MCLT and Voc parameters are compared among the three groups, suggesting that polishing brings ~30% gain in MCLT and 1.0% gain in Voc, while special clean enhances MCLT by ~25% and Voc by ~0.8%, respectively. It is evident that combining chemical polishing and special cleaning leads to excellent wafer surface cleaning and passivation quality.

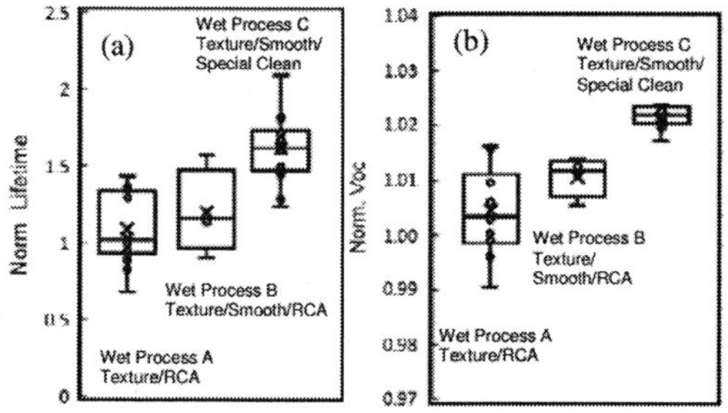

Figure 4: Comparison of (a) minority carrier lifetime and (b) open circuit voltage among different wet processes

2.2 PECVD a-Si:H deposition

As the core process, PECVD a-Si:H deposition is crucial for high-quality passivation. First of all, local epitaxial growth must be avoided during a-Si:H deposition as it significantly degrades passivation quality. Figure 5 shows TEM images of c-Si/a-Si:H interfaces with and without local epitaxial growth. In Figure 5(a), local epitaxial growth destroys sharp interface and introduces structural defects. Low deposition rate is recognized to induce epitaxial growth, however, high deposition rate could be detrimental too as it results in poor a-Si:H quality. Carefully tuning a-Si:H deposition processes, e.g., appropriate deposition rate, deposition temperature, and H/Si dilution rate, etc., a sharp, epitaxial-free a-Si:H/c-Si interface is achieved, as shown in Figure 5(b). Based on this, comprehensive experiments and optimization efforts have been conducted, and totally ~3X MCLT increase (in Figure 6(a)) and >3% Voc improvement (in Figure 6(b)) are achieved.

2.3 TCO and Ag grid electrodes

TCO and Ag grid electrodes are the major back-end processes. Their effects seem straightforward but very important, mainly related to electricity and light losses.

For TCO, bulk properties of resistivity and transmittance are essentially competing against each other. Trade-off between them occurs when we do process optimization. By pursuing high-mobility, low-carrier-density materials, such as IWO and IHO, aforementioned trade-off can be greatly reduced. Besides, work function match with doped a-Si:H is a rule of thumb to obtain ohmic-contact. Combining this with tunneling mechanism by virtue of highly doped a-Si:H, high-quality a-Si:H/TCO ohmic-contact can be obtained.

Figure 5: TEM images of c-Si/a-Si:H interface with (a) and without (b) local epitaxial growth

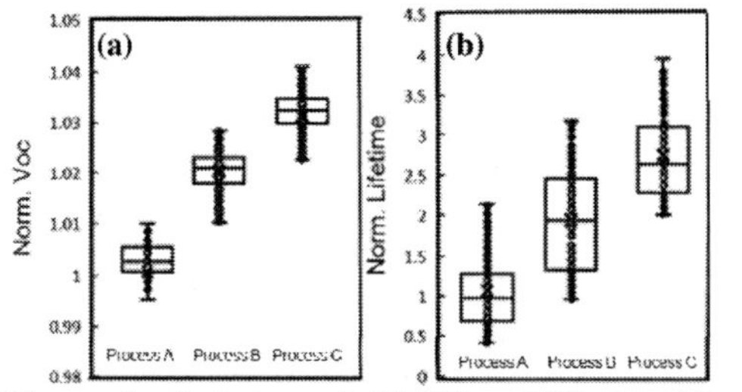

Figure 6: Evolution of PECVD a-Si:H deposition processes towards higher (a) MCLT and (b) Voc

For Ag grid electrodes, high conductivity and less light shielding are preferred to achieve both high FF and high Jsc. Double printing might be a better option due to the use of special Ag paste, which is subject to low temperature (~200℃) processing and has limited finger conductivity. It is found that double printing of low-temperature Ag paste, combined with narrow finger design, yields >0.2% cell efficiency improvement in SHJ solar cells, mainly due to line resistance reduction and FF increase (Figure 7). This efficiency improvement resulting from double printing is generally higher than that in diffused junction c-Si solar cells which use high-temperature, high-conductivity paste. Furthermore, since a-Si:H, TCO, Ag paste, and almost all interfaces are sensitive to thermal processes, post-print baking /annealing is also very important. It needs to be fully optimized to achieve high solar cell efficiency and module reliability.

2.4 HST solar cell performance

ENN has conducted HST solar R&D mainly on a large scale industrial production line converted from thin film silicon solar line and employed production compatible processes. By comprehensively optimizing solar cell structures, processes, materials and integrating all of them together, we have achieved HST solar cell efficiency up to 23.4% in Figure 8. The solar cell efficiencies were third-party verified (RETC, US and CETC, China).

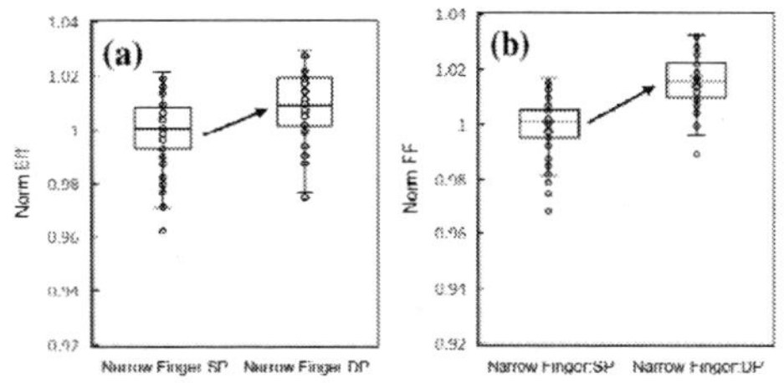

Figure 7: Double printing improvements on HST solar cell (a) efficiency and (b) fill factor

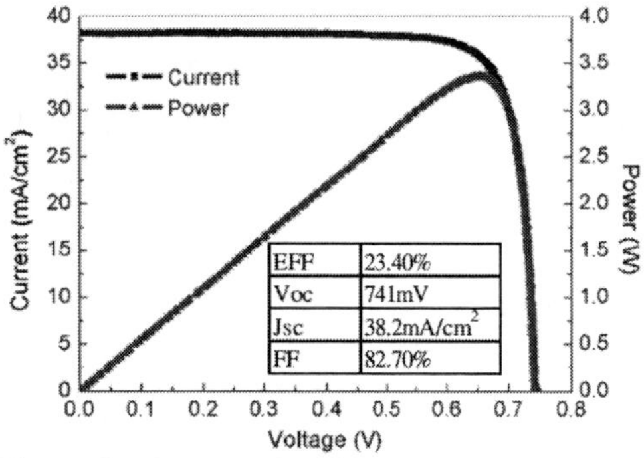

Figure 8: IV-curves of HST solar cell (Standard Test Conditions)

3 LOW-COST PRODUCTION APPROACH: THIN FILM LINE UPGRADE

Though SHJ technology has long shown great potential in achieving high cell & module efficiency, its production capacity has been low and further expansion is slow, mainly due to difficulties in developing high efficiency solar cell technology, as well as high capital cost to build SHJ production line.

Through modification & upgrading of major thin film silicon PV production tools, ENN has developed a low-cost SHJ production approach. For the first time in the world, ENN successfully converted existing TF PV line to HST production line (Figure 9 and 10) and realized volume production.

Key advantages of this approach includes: 1) utilization of existing, advanced TF deposition systems, leading to cell efficiency improvement from 10% to >21%; 2) low capital cost, around 1/3 of new line cost; 3) high capacity, greater than 4X TF capacity; and 4) field-proven core equipment performance and reliability. Major tasks to convet the line are: 1) modification of core process equipment PECVD & PVD; 2) design of wafer transfer & material handling systems; 3) integration of thin-film & c-Si production equipment; and 4) full line integration and production automation. Combining these with ourh high efficiency HST solar cell technology, we have successfully converted existing TF line (60MW, ETA~10%) to high efficency, high capacity HST line (200MW, ETA>21%).

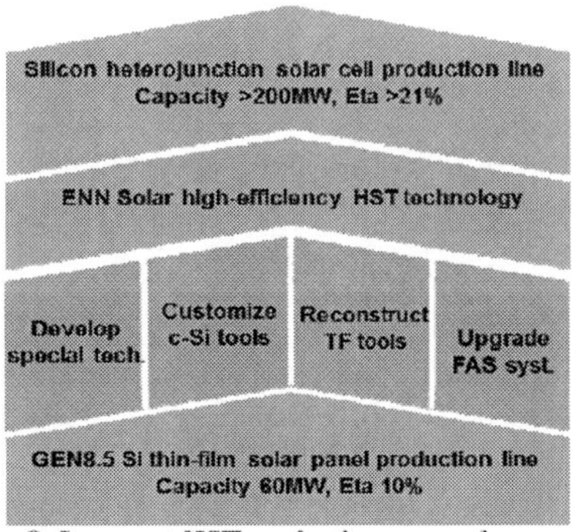

Figure 9: Low-cost HST production approach: convert TF line to HST line

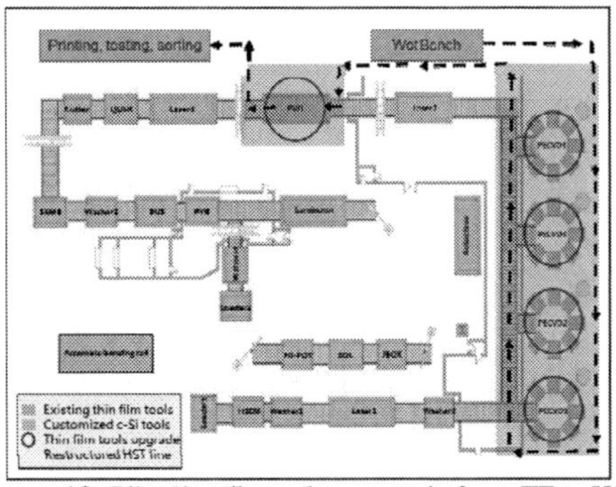

Figure 10: Pilot line floor plan: upgrade from TF to HST

4 PILOT PRODUCTION OF HST SOLAR CELLS

Figure 11 shows a typical batch in the 156 mm HST solar cell production line at ENN, with an average solar cell efficiency of 21.6%. The cell efficiency is tightly distributed. Different phases in Figure 12 shows rapid solar cell efficiency improvement in pilot production, through successful technology transfer from our R&D to production line and continuous production process flow optimization. Daily efficiency level is consistent within the same phase (A, B, or C). Wafers chosen from phase C (21.6% average efficiency) are assembled to standard 60-cell modules (156mm wafer), output power of 317 Wp is obtained as sown in Figure 13.

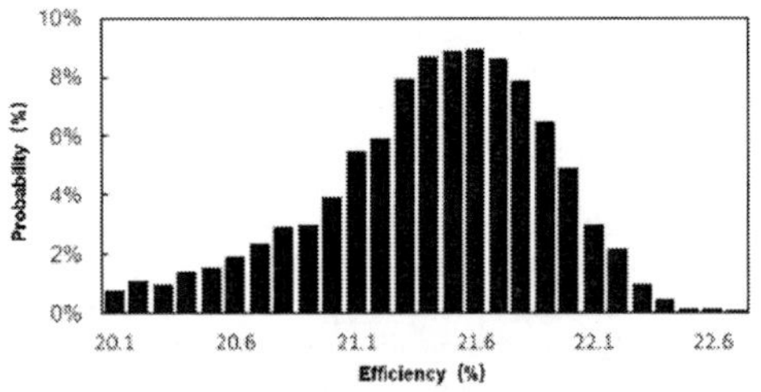

Figure 11: Efficiency distribution of HST solar cells in pilot production

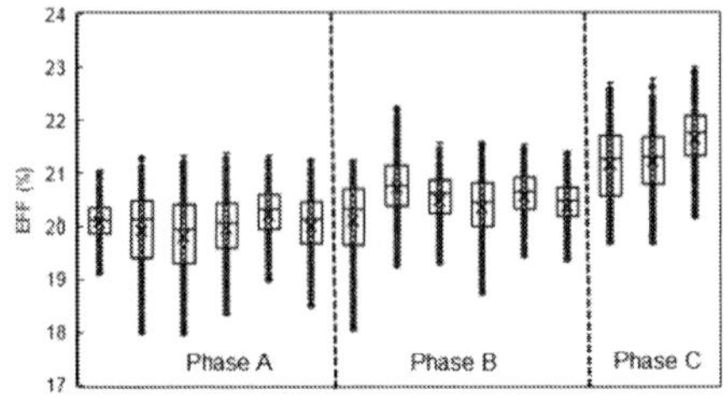

Figure 12: HST Solar cell efficiencies in different pilot production phases

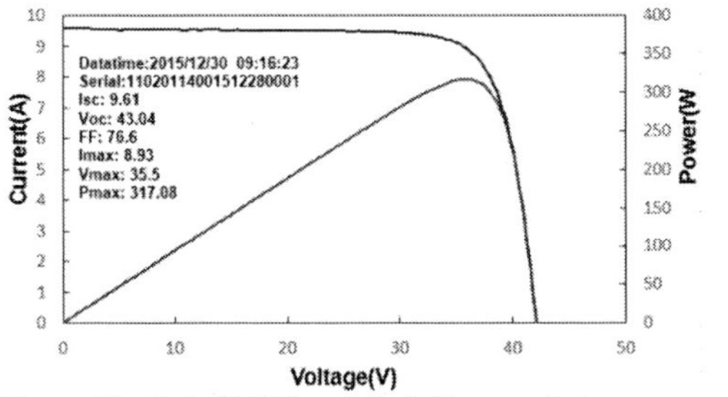

Figure 13: 60-Cell HST module (156mm wafer) output curves

5 CONCLUSIONS

ENN Solar has developed high-efficiency silicon heterojunction solar cell technology — Hybrid Silicon Technology (HST), with 23.4% solar cell efficiency achieved. We have successfully established a low-cost approach to HST mass production, i.e., converting existing Gen8.5 thin film solar line to high efficiency HST fab by upgrading and modification of key equipment and line automation. We have achieved average cell efficiency of 21.6% in pilot production and demonstrated 317 Wp power output using standard 60-cell (156 mm) modules. This provides a viable path for more PV manufacturers to penetrate into high efficiency SHJ area, commercialize low-cost heterojunction technology, and further increase its competitiveness.

6 REFERENCES

[1] K. Masuko et al, IEEE Journal of Photovoltaics, Vol.4 (2014) 1433.
[2] D. Adachi et al, Applied Physics Letters, Vol.107 (2015) 233506.
[3] E. Kobayashi et al, Applied Physics Letters, Vol.106 (2015) 223504.
[4] A. Danel et al, 31st European Photovoltaic Solar Energy Conference and Exhibition, (2015) 279.
[5] B. Hoex et al, Journal of Applied Physics, Vol.104 (2008) 113703.
[6] C. Leendertz et al, Applied Physics Letters, Vol.98 (2011) 202108.
[7] S. Olibet et al, Physical Review B, Vol.76 (2007) 035326.
[8] G. V. Elzakker et al, Thin Solid Films, Vol. 511 (2006) 252.

PERC Solar Cells and its Road to Industry

Jian Wu, Xusheng Wang*, Guoqiang Xing

Canadian Solar Inc., 199 Lushan Road, SND, Suzhou, China, 215129
**Corresponding Author: Xusheng WANG, Xusheng.wang@canadiansolar.com*

Abstract

Passivated Emitter and Rear Contact (PERC) is gaining acceptance in solar cell manufacturing due to its potential for high efficiency for p-type wafers and easy integration into existing lines. For p-type mono solar cells, the cell efficiency gain arrives 0.6~1.0%, while for multi solar cell is 0.5~0.8%, which is very attractive to the industry. We report the batch cell efficiency of 19.0% for multi PERC, and 20.70% for mono PERC solar cells in pilot run.

As for PERC mass production, the effective and reliable AlOx deposition tool is the most important issue we have to consider. Nowadays, ALD and inline PECVD AlOx tools are the only two choices in the market. Here we compare the passivation effect and cell efficiency of both tools, to find little difference, but the former seems a little bit better. High-deposition rate PECVD AlO_x shows the advantage of higher throughput and uptime, which is more acceptable by customers.

Boron-oxygen complex is the origin of the light induced degradation (LID) of the solar cells. LID is the cell efficiency bottleneck because of the bulk recombination even if the silicon surface is well passivated. In this paper, we report the LID regeneration evaluation test by thermal and light illumination. Cell level tests reveal the relative degradation is around 2.0% for mono-PERC and 1.1% for mc-PERC, by means of LID regeneration, which is acceptable.

The combination of cell efficiency, AlOx tool choice and LID regeneration gives a direction to industrialization the PERC technology.

PURPOSE OF THE WORK

Photovoltaic (PV) technologies are challenged to reduce solar cells' cost and increase their efficiency. Most of the state-of-the-art industrial solar cells are using p-type Si as the base material. Featuring an aluminum back-surface field (Al-BSF) on the rear surface, they suffer from rear surface recombination and a low internal reflectance. Additionally, the bowing induced for thin wafers limits the reduction of silicon material used on the wafer level. In order to achieve higher conversion efficiency, the quality of the rear-surface passivation and reflectance needs to be higher than what Al-BSF can provide. Therefore the application of PERC structures (see Fig.1) to industrial solar cell production would allow an increase of the efficiency while decreasing the wafer thickness. In a PERC cell, the full-area aluminum is replaced by a dielectric film covered with an Al film. The dielectric layer is locally opened to provide contact between the Al and the Si wafer.

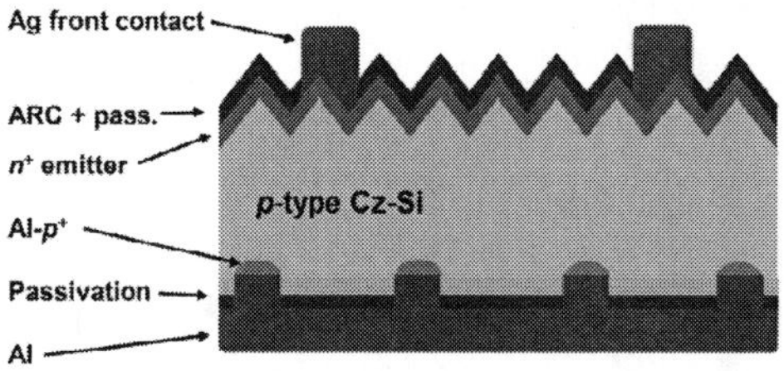

Figure 1: Schematic of PERC solar cell

A wide range of materials can be applied for rear surface dielectric [1] (SiO_2 [2], a-Si [3], a-

SiC_x [4]). Recently, lab-type plasma-enhanced chemical vapor deposited (PECVD) [5] and high-deposition rate PECVD AlO_x layers [6] have shown to provide excellent surface passivation on p-type surfaces as well, eliminating the throughput restrictions of conventional ALD systems. Alternatively, very thin ALD Al_2O_3 capped by PECVD SiN_x or PECVD SiO_x, which also show very good surface passivation quality [7], could be used for industrial production.

Boron-oxygen complex is the origin of the light induced degradation (LID) of the solar cells. The cell efficiency is limited by the bulk recombination even if the silicon surface is well passivated. In other words, the extra efficiency gain by excellent surface passivation, PERC, will be reduced by bulk material LID.

In this paper, we reveal the PERC process on mono and multi silicon material. Cell efficiency is evaluated and we compare the passivation effect and cell efficiency of ALD and inline PECVD AlOx tools, to find the difference in passivation. We also report the LID regeneration evaluation test by thermal and light illumination.

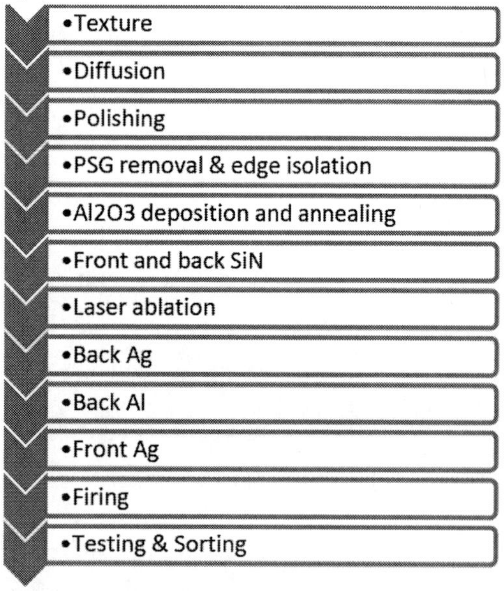

Figure 2: PERC process flow

APPROACH

Cell Efficiency for mono and mc PERC solar cells:

Thousands of mc and mono wafers were processed in the pilot-line, to fabricate PERC solar cells. The as-cut wafers went through conventional acid solution to form the textured surface. One step of single side $POCl_3$ diffusion was used to form a lightly doped homogenous emitter. After diffusion, rear surface was polished for the preparation of PERC structure. After PSG removal, a thin layer of ALD-Al_2O_3 was deposited on rear surface, and PECVD SiNx was deposited on both front and rear surfaces. A green laser (532 nm) were used for the local opening on the rear dielectric stack layers. Finally, screen printing and co-firing as the metallization steps were carried out to form emitter contact and BSF field.

Fig.2 shows the process flow sequences. Process details were reported in our previous publications [1, 6]. Cell Efficiency reaches 19.0% by PERC approach on multi silicon wafer, while the reference group reaches 18.3% by traditional process. Efficiency gain ~0.7% was achieved on cell level, see table I.

Lot	V_{oc} (mV)	I_{sc} (A)	FF(%)	R_s (mΩ)	R_{sh} (Ω)	EFF(%)	I_{rev1}(A)
Ref	634.5	8.761	80.11	1.7	123.69	18.30	0.18
PERC-1	642.4	9.131	78.9	2.34	72.19	19.02	0.21
PERC-2	642.4	9.122	78.87	2.43	80.7	18.99	0.213
PERC-3	644.3	9.129	78.78	2.49	106.07	19.04	0.165

Table I. PERC cell efficiency on multi wafer.

For mono cells, the efficiency reaches 20.7% by PERC approach, while the reference group 19.6%. Efficiency gain ~1.1% was achieved on cell level, see table II.

Lot	V_{oc} (mV)	I_{sc} (A)	FF(%)	R_s (mΩ)	R_{sh} (Ω)	EFF(%)	I_{rev1}(A)
Ref	639.8	9.26	80.47	1.59	171.9	19.63	0.10
PERC-1	659.4	9.603	78.72	2.66	141.9	20.53	0.18
PERC-2	659.7	9.594	78.61	2.74	141	20.49	0.19
PERC-3	659.9	9.610	79.31	2.33	164.2	20.71	0.12

Table II. PERC cell efficiency on mono wafer.

AlOx tool evaluation:

As for PERC mass production, the effective and reliable AlOx deposition tool is the most important issue we have to consider. Nowadays, ALD and inline PECVD AlOx tools are the only two choices in the market. The following paragraph will compare their passivation behavior and PERC efficiency

First of all, a non-contact capacitance-voltage (CV) measurement was performed in order to extract the total fixed charge density (Q_{fix}) and the interface defect density (D_{it}) in Table III. These data demonstrate a good field passivation, in both cases.

Passivation	Q_{fix}[cm^{-2}]	D_{it} [$cm^{-2}ev^{-1}$]
PECVD-AlO_x+SiN_x	-3.61×10^{12}	1.13×10^{12}
ALD-AlO_x+SiN_x	-3.95×10^{12}	9.44×10^{11}

Table III: A table presenting total fixed charge (Q_{fix}) and interface defect density (D_{it}) for single aluminum oxide layer and aluminum oxide with silicon nitride stack layers.

To examine the Al_2O_3 passivation further, effective minor carrier lifetime was characterized using the Sinton QSSPC method. With the measured lifetime data, the effective surface recombination velocity (SRV) Seff was determined by

$$S_{eff} = \frac{W}{2}\left(\frac{1}{\tau_{eff}} - \frac{1}{\tau_{bulk,intrinsic}}\right) \qquad (1)$$

The effective SPV S_{eff} is calculated by Eq.(1) with the polished wafer passivated by ALD-AlOx/SiNx and PECVD-AlOx/SiNx. After firing, the calculated S_{eff} is 12 cm/s for ALD-AlOx/SiNx and 20 cm/s for PECVD-AlOx/SiNx @$\Delta n=1\times10^{15}cm^{-3}$.

Lot	V_{oc} (mV)	I_{sc} (A)	FF(%)	R_s (mΩ)	R_{sh} (Ω)	EFF(%)	I_{rev1}(A)
ALD	659.9	9.588	79.0	2.15	150.5	20.59	0.14
PECVD	659.2	9.575	79.0	2.14	147.5	20.53	0.14

Table IV. PERC cell efficiency on mono wafer of ALD and inline PECVD AlOx tools.

To verify the passivation effect further, two splits of mono PERC cell, following the process flow sequences in Fig.2, were employed to demonstrate the cell efficiency difference, with the details in table IV.

Here we find very slight difference in passivation effect and cell efficiency of both tools. However, high-deposition rate PECVD AlOx shows the advantage of higher throughput and uptime, which is more acceptable by customers.

Light induced degradation and regeneration:

Boron-oxygen complex is the origin of the light induced degradation (LID) of the solar cells. The cell efficiency is limited by the bulk recombination even if the silicon surface is well passivated. In other words, the extra efficiency gain by excellent surface passivation, PERC, will be reduced by bulk material LID. The total light generated boron-oxygen pairs are direct proportion to the boron concentration and oxygen concentration squared. The generation rate is only related to the boron concentration. That's the reason why we suffer a lot of efficiency gain on mono silicon material, which prevents it to go to mass production.

LID can be reduced by regenerating the silicon bulk lifetime by directly passivating the generating B-O complex only if simultaneously anneal the H-rich sample at elevated temperature [8-10]. Hydrogen is believed to play a key role in the B-O deactivation process. The regeneration rate is also limited by high boron concentration or high oxygen concentration.

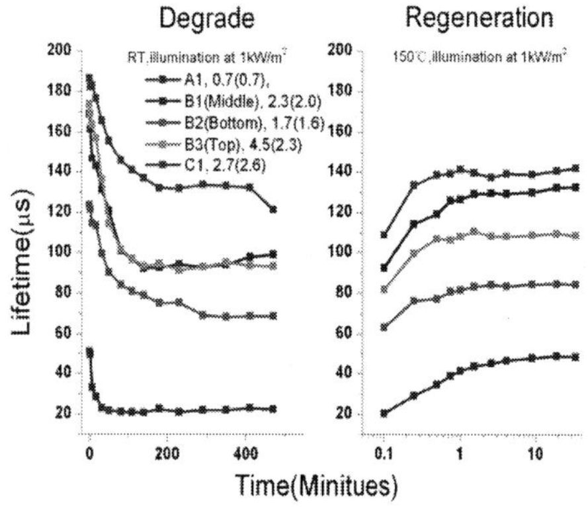

Fig. 3 LID and regeneration lifetime-time curve of various resistivities. A, B, C stands for different ingot. The data in branklet are the resistivity measured after the oxygen thermal donor dissolved.

Fig.3 reveals the lifetime regeneration data with different zone in an ingot in our previous work [11], which implies some part of lifetime recovery by light illumination at elevated temperature. To verify further, more mono and mc PERC cells were made and tested by IV tester initially. After the light regeneration and subsequent degradation tests under 1 sun illumination for 72h, we get the relative degradation data, shown in table V.

Table V shows the light induced degradation in cell could be less than relative 3% for mono PERC cells, and less than relative 2% for mc PERC. The approach could find a way to go to industry if the degradation can be controlled within a reasonable value, say 2~3% on module.

Cell	State	Efficiency	Relative Degradation
mc	Initial	18.98%	1.07%
	After LID	18.78%	
mono	Initial	20.59%	2.00%
	After LID	20.18%	

Table V. PERC cell efficiency before light illumination and after LID regeneration and light illumination.

Conclusions

In conclusion, we present PERC approach on both mono and multi silicon material. Efficiency gain of 1.10% and 0.7% were achieved on mono and multi cell level. ALD and inline PECVD AlOx show very slight difference in the passivation effect and cell efficiency; however high-deposition rate PECVD AlO_x shows the advantage of higher throughput and uptime, which is more acceptable by customers. LID regeneration process could help to reduce the light induced degradation in cell to less than relative 3% for mono PERC cells, and less than relative 2% for mc PERC. The above approaches give a direction to industrialization the PERC technology.

REFERENCE

[1] J.Wu, R.Zhu, X.Meng, L.Jiang, X.Wang, L.Zhang, "19.2% efficient H-patterned mono silicon metal wrap through solar cells", 26th European Photovoltaic Solar Energy Conference and Exhibition, Hamburg, 2011, pp. 1004-1007.

[2] E. van Kerschaver et al., "A Novel Silicon Solar Cell Structure with Both External Polarity Contacts on the Back Surface", Proceedings of the 2nd World Conference

on Photovoltaic Energy Conversion, Vienna, Austria, 1998, pp.1479-1482

[3] I. J. Bennett, et al., "World Record Module Efficiency for Large and Thin mc-Si MWT Cells", 24th EUPVSEC, Hamburg, 2009, pp. 3258-3261.

[4] Libby, P. A. and Williams, F. A., "Turbulent Reacting Flows", Academic Press, New York (1994).

[5] F.Clement et al. "Processing and comprehensive characterization of screen-printed mc-Si metal wrap through (MWT) solar cells", Proceeding of the 22nd European Photovoltaic Solar Energy Conference 2007, pp.1399-1402.

[6] R. Zhu , X. Wang, L. Zhang, "A New Solution of forming Selective Emitter with Etch Back Technology", 27th European Photovoltaic Solar Energy Conference and Exhibition, Frankfurt, 2012, pp. 1870-1872.

[7] E. V. Kerschaver, R. Einhaus, J. Szlufcik, J. Nijs, and R. Mertens, "A novel silicon solar cell structure with both external polarity contacts on the back surface," in Proc. 2nd World Conf. Photovolt. Energy Convers., Vienna, Austria, 1998, pp. 1479–1482.

[8] G. Krugel, W. Wolke, J. Geilker, S. Rein and R. Preu. Energy Procedia 8, 47, 2011.

[9] B. Lim, S. Hermann, K. Bothe, J. Schmidt, and R.Brendel, Appl. Phys. Lett. 93, 162102 (2008).

[10] B. Lim, K. Bothe, and Jan Schmidt, J. Appl. Phys.107, 123707 (2010).

[11] X.Meng, X.Wang, and L.Zhang, "Fast regeneration of the B-O LID and the positive effect of the SiNx capping layers" 28th European Photovoltaic Solar Energy Conference and Exhibition, Paris, 2013, pp. 1723-1725.

Cu-plated electrodes with green nano-laser opening metal contact on n-type silicon solar cells

*Kuang-Chieh Lai**, *Shu-Yen Liu, Yueh-Lin Lee, Ming-Shiou Lin, Yun-Kuo Tsao, Chia-Chih Chuang, Chi-Chun Li and Chien-Chun Wang*
Advance Technology Development Department,
Motech Industries, Inc, No.96, Dashun 7th Rd., Xinshi Dist., Tainan City 74145, Taiwan
* Corresponding author; E-Mail: Jay_Lai@motech.com.tw

ABSTRACT: The manufacturing of front junction n-type silicon solar cells developed at Motech is presented in this paper. The cell features a rear surface field formed by phosphorous implant and a fully diffused BBr_3 front emitter with thermal oxide passivation. Low damage laser process conditions are demonstrated to from the metal contact open by visible wavelength laser system. In our study, the green 532nm nanosecond laser (ns-laser) is used to ablate the passivation layer on the front-side of the solar cells. Our process combines laser ablation for the patterning of the dielectrics on front-side with a front metallization based on Ni/Cu forward bias plating and sintered metal paste for the rear contacting electrode. The performance results show better cell's efficiency in comparison to common double sides sintered metal paste electrodes. Large-area (238.9 cm^2) silicon solar cells manufactured on 170 μm thick, 3Ω cm, n-type Cz wafers using this technology, reach confirmed efficiencies up to 21.31%; with Voc of 654 mV, a jsc=40.8 mA/cm^2 and FF of 79.8 %.

Keywords: n-type, silicon, solar cells, nano-laser processing, forward bias, copper plating

1 INTRODUCTION

It has been well recognized that n-type Cz-silicon solar cells offer higher efficiency potential than p-type Cz-silicon solar cells, as n-type Cz-silicon has higher tolerance for common metallic contaminations such as iron [1], a potential for high diffusion lengths and don't suffer light induced degradation (LID), which significantly impacts the performance of solar cells made with boron-doped silicon wafers [2-3].

N-type silicon solar cells are often made into bifacial structures, which can produce even higher output power, with contribution from the rear side. Bifacial photovoltaic (PV) systems can be used for specific applications and areas such as white painted rooftops, vertical noise barriers along highways or railways, desert, and snow area. More importantly, bifacial n-type cells produced by industrial processes are cost competitive.

So far, at Motech we have applied our n-type know-how for the manufacturing of passivated emitter, rear totally diffused (n-PERT) cells with efficiencies up to >20.6% [4].

2 EXPERIMENTS

A schematic diagram of solar cells is shown in Figure 1. We used 6'' n-type CZ wafers with thickness of 170um and resistivity of 3~5 Ω-cm as starting materials to fabricate the solar cells. The rear side was textured with random pyramids and phosphorous implant on the rear–side surface formed a rear surface field to enhance passivation. Amorphous $SiNx/SiO_2$ thin films (~80 nm) were deposited on the rear side of cells in a PECVD reactor to passivate the doped silicon surface. The front side was textured with random pyramids and Boron thermal diffusion on the front–side surface formed a front surface field to enhance passivation. Amorphous $SiNx/SiO_2$ thin films (~80 nm) were deposited on the front side of cells in a PECVD reactor to passivate the doped silicon surface. The traditional sintered metal paste is used on rear side electrode. The sintering thermal budget could improve the passivation. The ARC is then ablated with a green ns-laser to form the front pattern. The front metallization is performed in cluster equipment from Motech, with a sequence as follows: forward bias Ni+Cu and Sn immersion plating. Finally, a rapid thermal anneal (RTA) is performed in a belt furnace from Despatch to improve the front contact resistance and form a nickel silicide.

The process flow for this front junction cell can be seen on Figure 2. The cells are manufactured on ~170 μm thick, n-type Cz-Si wafers. After texturing, the wafers are cleaned and a phosphorus implant are applied to form a n+ BSF with an Rsheet= 50 ± 5Ω/sq. The boron diffusion the BSG is removed and the dopants are driven-in through a wet-oxidation step. The resulting boron emitter features a sheet resistance of 75 ± 5 Ω/sq. A SiNx coating is deposited by PECVD on the front side as an Anti-Reflective Coating (ARC). A SiNx layer is also deposited on the rear side (on top of the thermal oxide).

3 RESULTS AND DISCUSSIONS

Figure 3 shows the front-side metal finger and laser open contacts by SEM analysis. In laser contact open processes, we expect the surface doping profile and silicon surface morphology will be changed by green 532nm laser source. Because of the dielectric films is ablated via underneath silicon to absorb laser energy and then to be ablated. So the silicon surface will form silicon melt surface. We think the profile change will not affect cell performance. If only the laser conditions can be optimized to avoid silicon defects or to release laser hot effect. However, the UV wavelength laser source can be a better choose to open the dielectric layers, even if dielectric layers on texture silicon surface can be ablated well by UV laser system [5]. In our study, figure 3(a) shows the surface morphology of SiNx/Si surface after green 532nm laser patterns on passivated dielectric films. The opening line contact is around 20μm width. The plated front-side finger on laser open area is about 46μm width and 12μm height. The SEM cross-section view shows in figure (c-d). The plated thin Sn protection layer on Cu shows in figure (e). This optimized condition is used in n-PERT solar cell processes. The performance of n-PERT solar cells is shown in table 1.

The plated best cell I-V curve shows in figure 4. The cell's efficiency for front plated electrodes show better current density (Jsc) and open-circuit voltage (Voc). Improvement of Jsc could be attributed to fine-line Cu-plated finger to increase incident light into cells. Furthermore, the finger width can be shrank about

5~10μm. However, the Voc improvement might cause that the cell has been no sintered in metallization processes and no metal paste inter-diffused affects, so that the plated metallization electrodes has better passivation performance.

4 CONCLUSION

In this study, the forward bias plating Ni/Cu with 532nm ns-laser contact opening processes was employed to the front-side of n-PERT silicon solar cells. We optimized the green ns-laser in order to locally remove dielectric layers from the front-side surface of n-PERT silicon solar cells. The effects of these opening procedures on the surface topography and cell efficiency were investigated. The contact opening on textured front-side silicon surface shows no obvious negative effect or efficiency drop by using green 532nm laser technology in comparison with sintered metal paste processes. Finally, combining the developments of Motech's n-PERT and Cu-plated technology work has resulted in double-side contacted, front junction, n-type solar cells with independently confirmed efficiencies of 21.31% on large area wafers.

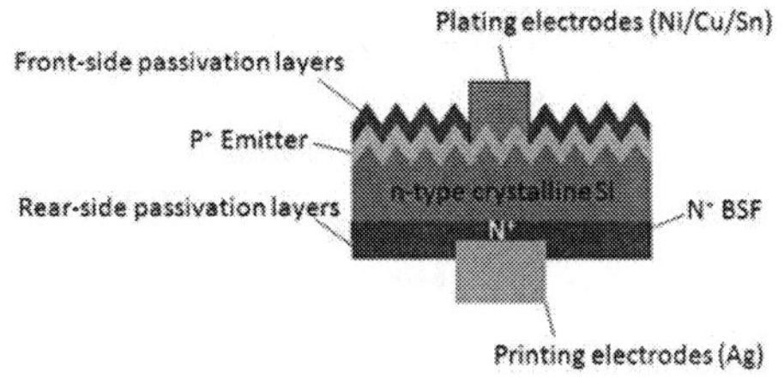

Figure 1 A schematic cross-section of n-type silicon solar cells

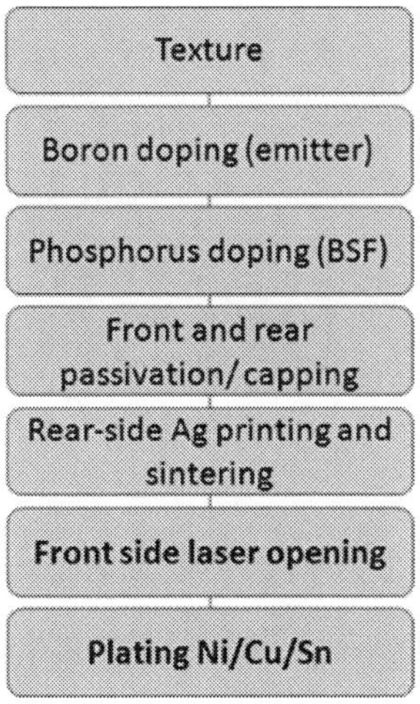

Figure 2 The flow chart with the processing steps

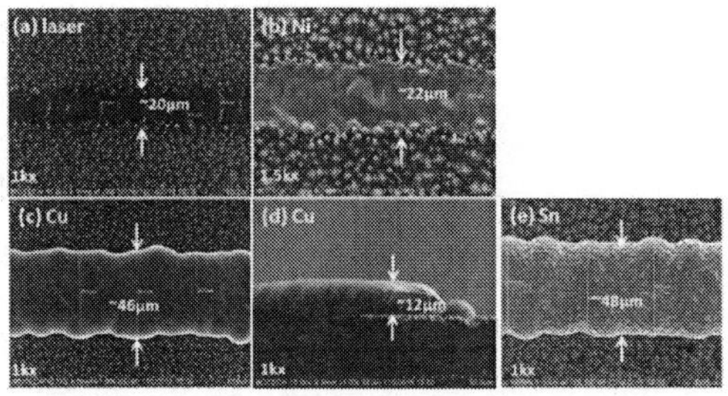

Figure 3 Front-side metal finger and laser open contacts by SEM analysis: (a) green ns-laser contact open width; (b) Ni plated finger width; (c) Cu plated finger width; (d) Cu plated finger height; (e) Sn plated finger width.

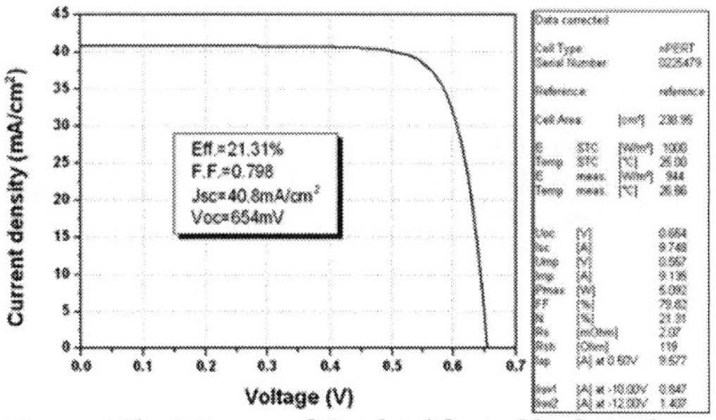

Figure 4 The *I-V* curve of Cu-plated front-side electrodes for n-PERT solar cells

Table 1 Cell efficiency with different metallization front electrodes

Conditions	Eff. (%)	FF	Voc (mV)	Jsc (mA/cm²)
Sintered Ag paste	20.63	0.813	647	39.24
Plated Ni/Cu/Sn	21.31	0.798	654	40.80

5 REFERENCES

[1] I. Romijn, *Photovoltaics International Edition 15*, p.81, 2012,.

[2] D. Song, J. Xiong, Z. Hu, G. Li, H. Wang, H. An, B. Yu, B. Grenko, K. Borden, K. Sauer, T. Roessler, J. Cui, H. Wang, J. Bultman, A. H. G. Vlooswijk and P. R. Venema, "Progress in n-type Si solar cell and module technology for high efficiency and low cost," in *38th IEEE Photovoltaic Specialists Conference*, 2012.

[3] J. Schmidt and R. Hezel, "Light-induced degradation in Cz silicon solar cells: Fundamental understanding and strategies for its avoidance," *in 12th Workshop on Crystalline Silicon Solar Cell Materials and Processes*, 2002.

[4] Po-Tsung Hsieh, Ping-Kuan Chang, Chih-Jeng Huang, Zih-Wei Peng, Wen-Tai Chung, Yuan-Jun Lin, Chun Kuo, Chi-Chun Li, "Bifacial Variety of N-type PERT Solar Cells", in *41th IEEE Photovoltaic Specialists Conference*, 2015.

[5] Sven Kluska, Jonas Bartsch, Andreas Büchler, Gisela Cimiotti, Andreas A. Brand, Sybille Hopman, Markus Glatthaar, "Electrical and mechanical properties of plated Ni/Cu contacts for Si solar cells", Energy Procedia 77 (2015) 733 – 743

40 KHZ PECVD OF ALO_X/SIN_X STACKS DEMONSTRATED IN INDUSTRIAL HIGH EFFICIENCY PERC PRODUCTION

Thomas Pernau*, Jens-Uwe Fuchs, Viet Xuan Nguyen, Andreas Nickel, Ulrich Walk, Wolfgang Jooß**
centrotherm photovoltaics AG, Johannes-Schmid-Str. 8, D-89143 Blaubeuren, Germany
Keith Hsu, Jenson Chen, Steffen Wiebicke
Centrotherm Photovoltaics Asia Pte. Ltd., 16F-2, No.8, Ziqiang S. Rd., Zhubei City, Hsinchu County, 30264 Taiwan, ROC
Hsing Hua Wu , Kuang-Hui Hung ,Kang Cheng Lin ,Walt K. W. Huang
Gintech Energy Corporation, No. 21, Kebei 1st Rd., Hsinchu Science Park, Jhunan Township, Miaoli County 350, Taiwan, R.O.C.
*corresponding author: thomas.pernau@centrotherm.de
**now with RCT Solutions GmbH, Turmstraße 20, 78467 Konstanz, Germany

ABSTRACT: Standard direct plasma (40 kHz) batch-type PECVD machines can be equipped with TMA (Trimethylaluminium) and N_2O supply to enable aluminium oxide (AlO_x) / silicon nitride (SiN_x) passivation stacks in one deposition run. The present paper is the summary of our process development for PERC structures and the implementation at an industrial production line of PERC cells at Gintech. Very good results were achieved on multicrystalline (19.3% average) as well as monocrystalline cells (21% average).
This paper starts with the description of the TMA supply system, TMA plasma chemistry and properties. Then the optimisation of deposition procedures with lifetime tests and cell runs is illustrated. The optimisation resulted in a low surface defect density D_{it}<3E10/(cm²·eV) and good effective surface charge Q_{tot}<-1.2E12/cm², corresponding to a low surface recombination velocity S_{pass} < 10 cm/s in the passivated regions. Our investigations indicate that 40 kHz plasma has a better activation efficiency of the precursors TMA and N_2O as compared to SiH_4 and NH_3. The net deposition rate for a layer stack of 18 nm AlO_x and 140 nm capping SiN_x is 3 nm/min, which is slightly higher than 2.6-2.8 nm/min for pure SiN_x.

Keywords: aluminium oxide, PECVD, passivation

1 HARDWARE

1.1 40 kHz direct plasma PECVD machine and boats

The layers were deposited in an industry-standard 40 kHz tube reactor. The only requirement is N_2O supply as an additional non-toxic process gas and a direct stainless steel pipe connection in between TMA cabinet and process gas inlet of the reactor chamber. To avoid complications, a separate inlet for TMA was installed. A showerhead is neither required nor recommended, as it might result in obstructions to the TMA flow.

For the reactor principle please see the more detailed description in reference [1]. Wafers are held by parallel graphite plates that are conductive. The uniform and quasi-DC electric field in between the opposing plates ensures that the plasma discharge occurs only in between the plates, so the layer deposition is mainly confined to the wafers and plates but not on the chamber walls or other structures. The machine can use standard graphite boats that are well proven for SiN_x deposition. As the main component of the AlO_x/SiN_x stacks is still standard SiN_x, there are no changes in boat cleaning & maintenance required.

1.2 The TMA cabinet

A heated cabinet is added at the side of the machine to provide gaseous TMA at a vapour pressure of 40 mBar, please see figure 1. Carrier gas is not required. The gas cabinet can hold two TMA source bottles sized 0.5-20 litres. Each bottle is in a separate sub-cabinet and heated to a temperature T_1 < 50°C. The individual sub-cabinets allow refill of replacement of the bottle not in use, without interruption of the TMA flow. The TMA is taken from the gas phase, i.e. no dip tube required for TMA delivery and no risk of liquid TMA escaping in case of malfunction or misuse. Bottles with dip tubes can be used to connect to an automatic liquid refill system.

The evaporated TMA goes into a third compartment heated to T_2>T_1 where the distribution valves and the vapour flow controllers are located. This is to avoid condensation in these components. After valve manifold D, the gas line is kept at much lower pressure as compared to the TMA vapour pressure so no more risk of condensation.

For safety reasons and user friendliness the TMA cabinet has its own PLC and display unit. The user is supported by automatic level control and purging.

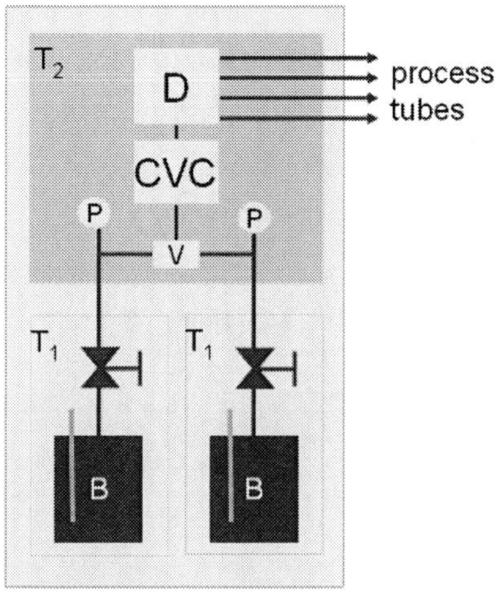

Figure 1: simplified view of the TMA cabinet. Only the TMA flow path is shown. Valve V is to select one of the bottles. Pressure in the bottle is monitored by pressure transducer P. Flow is adjusted by vapour flow controller CVC. Valve manifold D distributes the TMA flow to the process chambers.

2 PROCESS

2.1 deposition process flow

The process flow is very similar to a standard SiN_x deposition recipe, with the main difference that AlO_x is deposited at a lower temperature as compared to SiN_x. The process sequence is as follows:

1. loading
2. temperature equalisation / pre treatment
3. AlO_x deposition
4. heat up / post treatment
5. SiN_x capping deposition
6. purging
7. unload

During temperature equalisation, a pre-treatment can be done to improve the temperature and hence layer uniformity. The AlO_x deposition itself is just a few minutes. While heating up to the capping deposition temperature, a post treatment is done to improve the AlO_x layer properties. Most of the process time is spent on heating before capping deposition and the capping itself. The capping can be done faster than classical front side ARC, but not faster than 15 nm/min. In total, the process time is 45-50 min. depending on boat size and fine tuning. This corresponds to a net deposition rate of 3 nm/min, whereas front side ARC recipes can reach only 2.6 nm/min. For more details on capping SiN_x optimisation please see references [2] and [3].

2.2 layer properties and optimisation

This section refers to the process flow given in section 2.1 and discusses the individual steps in detail.

SiN_x deposition recipes typically use ammonia (NH_3) pre-plasma for heating purposes and surface conditioning. The ammonia pre-plasma has been found incompatible with any combination of AlO_x layers. The damage incurred is at least 5 mV in V_{oc}. The pre-plasma can even etch thermally grown thin oxide layers that are used for reducing the surface dopant concentration.

As an alternative, a less aggressive pre-treatment was introduced. The result of the pre-treatment can be seen in table I. The pre-treatment roughly doubles the interface defect density D_{it} so a compromise in between layer homogeneity (by improved temperature homogeneity) and cell voltage needs to be found. For multicrystalline cells, the voltage reduction from pre-treatment is neglible.

	Dit [1/(cm²·eV)]
w/o pre-treatment	4.43E+10
with pre-treatment	8.23E+10

Table I: influence of pre-treatment on surface properties.

In the 40 kHz direct plasma PECVD system, the AlO_x deposition can be done starting at the thermal self-decomposition temperature of trimethylaluminium (TMA). The reason for that can be seen in in-situ mass spectrometry measurements, see figure 2. E.g. at 275°C and with N_2O present, the TMA molecule is fully dissociated (peak at 57 amu missing). The plasma alone cannot fully dissociate the TMA molecule if no reactant is present. This is the first indicator that for the 40 kHz PECVD process, thermal activation of TMA is relevant.

The next indicator for thermal activation of TMA is the change of layer properties when varying the deposition temperature, see table II.

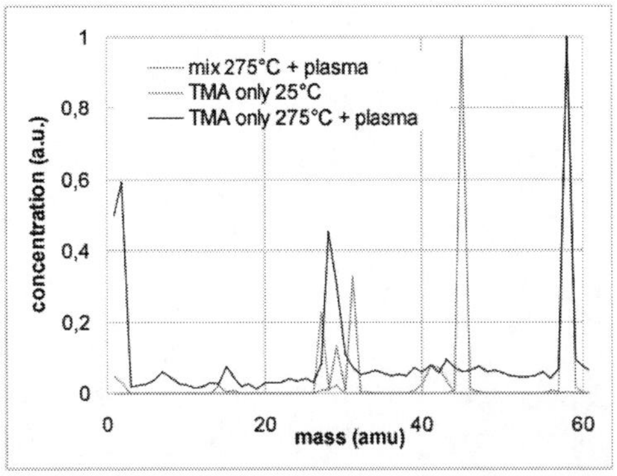

Figure 2: mass spectrometry data recorded at 275°C compared to NIST data at 25°C [4]. At 275°C and with N_2O present, the TMA molecule is fully dissociated. The sampling position is at the pump connection of the process chamber.

Except temperature all other process parameters have been fixed. Though there is only minor change in refractive index, the effect on effective surface charge and interface defect density is obvious. The temperature dependency can be compensated with the gas composition (not shown), higher temperature allows to reduce the TMA content.

T	THK	RI	Qtot	Dit
[°C]	[nm]		[cm⁻²]	[cm⁻²eV⁻¹]
norm.	20	1.64	-1.10E+12	3.00E+10
+ 50K	30	1.65	-8.50E+11	8.00E+10

Table II: AlO_x layer properties related to deposition temperature.

We used 2-4 Ohm*cm p-type Cz lifetime test wafers with KOH polished surface to monitor our layer development. An example of our best layer is shown in figure 3. The measured lifetime corresponds to a surface recombination velocity S_{pass} < 10 cm/s for the passivated regions.

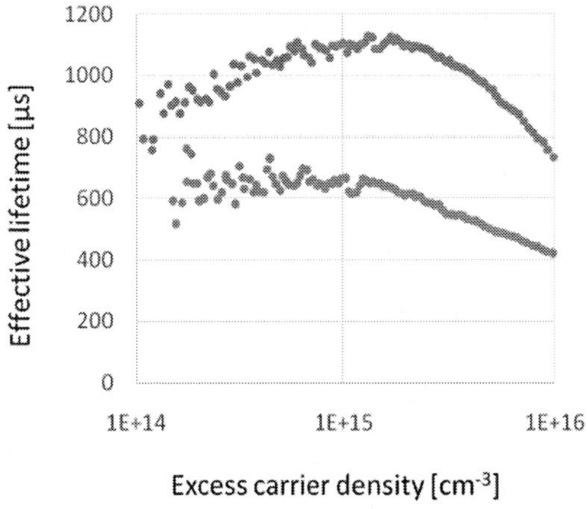

Figure 3: QSSPC measurement of our best AlO_x/SiN_x layer. Blue – good wafer quality, orange – medium wafer quality. The KOH polished p-type Cz lifetime test wafers are from one ingot, top and bottom position.

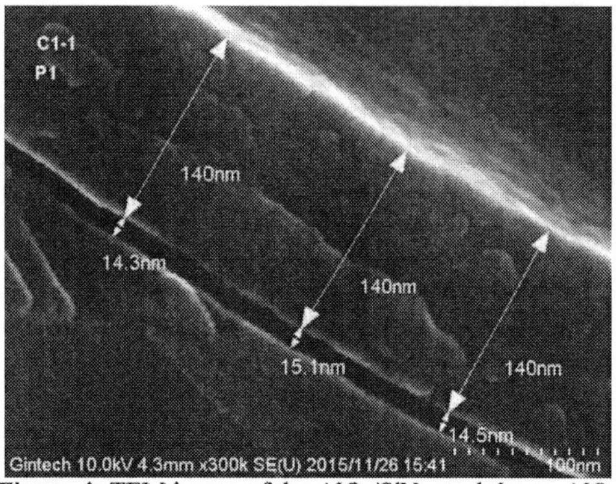

Figure 4: TEM image of the AlO_x/SiN_x stack layer. AlO_x layer can be identified as the darkest layer.

The layer optical/physical properties have been studied in detail by TEM microscopy. In Gintech's cell production line, samples from arbitrary test positions have been drawn to compare the microscopic structure. The reason for this procedure was to make sure that no passivation flaws from locally insufficient AlO_x can occur. This is a typical failure mode of inline microwave PECVD layers. One sample is shown in figure 4. The AlO_x thickness variation identified from TEM images is 14.3-17.5 nm. This variation is considered uncritical so local passivation failures are unlikely to occur with 40 kHz PECVD.

In Figure 4, there is some lighter contrast at the interface between Si and AlO_x. According to [5], it could be the presence of a SiO_x interface layer in between Si and AlO_x. To identify this layer, close-up TEM images and ToF-SIMS measurements have been performed at Fraunhofer CNT. By scanning from top to bottom, the ToF-SIMS measurement still detects unchanged high O concentration in the layer where Al concentration is already reduced, see table III. In the same region, already quite high Si content is detected. This indicates that an interface oxide in between AlO_x and Si is present. From the quantitative evaluation of the SIMS results the interface oxide can be identified as a sub-oxide SiO_x with $x \approx 1.6$

The oxygen saturation of the interface layer sub-oxide might be the important parameter for layer quality. We suggest that the oxygen concentration in the interface layer can be modified by the oxidiser supply during AlO_x layer growth, but also in a post-deposition AlO_x layer oxidation. Some hints on the oxidiser supply and resulting layer properties are also given in [6]. This is the key to grow electronically correct and well-passivating layers with thickness of less than 10 nm. Investigations in this topic are ongoing.

Depth (nm)	layer	[Al] counts	[Si] counts	[O] counts
0-20	AlO_x	8E+3	---	2E+5
20-25	SiO2	5E+3	5E+3	2E+5
30-inf	Si	5E+2	8E+3	2E+4

Table III: ToF SIMS counts (atom concentration) of the AlO_x layer shown in Figure 5.

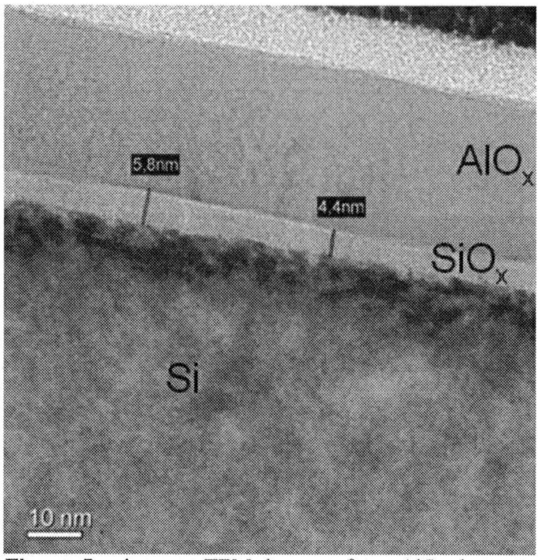

Figure 5: close-up TEM image of an AlO_x layer to identify the interface layer in between AlO_x and Si.

2.4 cell process

The 40 kHz AlO_x/SiN_x deposition has been implemented at Gintech as a replacement to microwave inline PECVD. The process flow is shown in figure 4. The process flow, esp. laser ablation and printing, didn't require any adaptions.

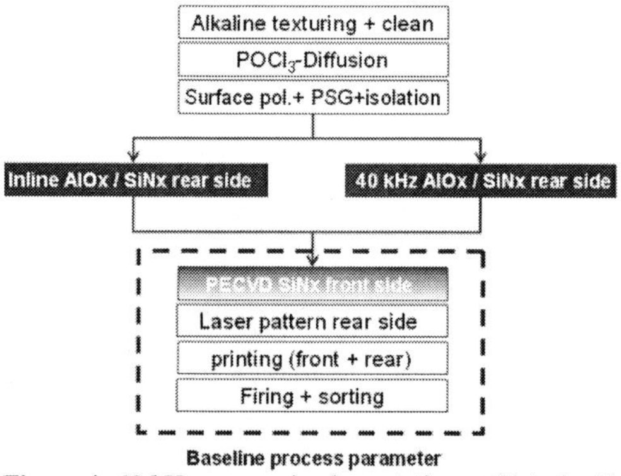

Figure 4: 40 kHz process implementation at Gintech. 40 kHz AlO_x/SiN_x deposition just replaces inline 2.4 GHz AlO_x/SiN_x deposition without the need of further process adaptions.

2.5 cell results

Solar cells were processed in 240 wafer boats, so the typical group size is 240 cells per parameter. Typically, 4-5 parameter groups per week were processed for R&D purposes. The cell results of the best groups are summarised in table IV. The best mono group was processed without pre-treatment. The deposition run took 55 min as the lack of pre-treatment increases process runtime. The expected recipe runtime with proper pre-treatment is expected to be less than 50 minutes.

The multi cells were processed with pre-treatment, but other recipe steps not yet fully optimised. Also for this recipe, the expected runtime after optimisation is less than 50 minutes.

	Eff Max [%]	Eff avg [%]	Voc [mV]	Isc [A]	FF [%]
Mono best group	21.18	20.97	656	9.79	79.8
Multi best group	19.66	19.33	643	9.19	79.4

Table IV: cell results (best groups)

In addition to these cell runs, lifetime samples have been processed in 308 wafer boats in the centrotherm lab, to make sure that the process is also compatible with larger boats.

3 SUMMARY AND OUTLOOK

3.1 Summary

Industry-standard 40 kHz PECVD reactors have been equipped with a TMA supply unit to enable AlO_x layer deposition. SiN_x/AlO_x layer stacks have been optimised to reach effective lifetimes of 1.2 ms on symmetrical lifetime samples, indicating a surface recombination velocity S_{pass} < 10 cm/s for the passivated regions. SiN_x/AlO_x layer stacks have been deposited with 40 kHz PECVD as an alternative to layer stacks from inline microwave PECVD. The AlO_x layer thickness was in the 14-18 nm range as confirmed by TEM measurements. The cell results obtained with these layers are comparable to the performance reached with 2.4 GHz PECVD.

3.1 Outlook

We found indications that the SiO_x interlayer in between AlO_x and bulk Si has some room for improvements, as it is not yet perfectly stoichiometric. Along with the optimisation of the interlayer, we are confident that the 40 kHz AlO_x deposition technology can be further developed to reduce the passivation layer thickness to less than 10 nm, as indicated in [6]. This enables better hydrogen penetration through the AlO_x layer and hence better efficiency of the regeneration process [7].

For compatibility reasons, the capping layer of stacks has been unchanged since its introduction back in 2009 [2]. By migrating the advancements in front side SiN_x to the rear side, we expect to reduce the heat-up time before capping and the capping deposition time itself. This will result in a further reduction of the overall process time.

4 ACKNOWLEDGEMENTS

The authors would like to thank all colleagues and friends that have contributed with their ideas, suggestions and hard work.

This work was partly funded by the German Federal Ministry of Economics and Energy (BMWi) within the research project "HELENE" (contract no. 0325777F).

5 REFERENCES

[1] Tong Zhou et al., "Different Dielectrics Deposited by PECVD for the Industrial Manufacturing of PERC Solar Cells", Proc. 31st European Photovoltaic Solar Energy Conference and Exhibition, 2AV.3.23

[2] K. A. Münzer et al., „Advanced Rear Side Technology for Industrial High Efficiency Solar Cells", Proc. 26th European Photovoltaic Solar Energy Conference and Exhibition, 2CV.3.66

[3] K. A. Münzer et al. "Towards 19.5 % Industrial Crystalline Silicon Solar Cells", Proc. 27th European Photovoltaic Solar Energy Conference and Exhibition, 2AO.1.5

[4] NIST Chemistry WebBook http://webbook.nist.gov/cgi/cbook.cgi?ID=C75241&Mask=200#Mass-Spec

[5] Q. Qiao, H. L, J. Ge, X. Xi, R. Chen, J. Yang, J. Zhu, Z. Shi and J. Chu, Appl. Surface Sci. 205 (2014) 439

[6] Marc Hofmann et al, "Plasma process analysis of ICP-PECVD of AlOx layers for c-Si surface passivation", Proc. 32nd European Photovoltaic Solar Energy Conference and Exhibition, 2BO.4.4

[7] J. Appl. Phys. 113, 194503 (2013); doi: 10.1063/1.4804310, see Fig. 3

MEASUREMENT OF RESIDUAL STRESSES IN LARGE SILICON SAMPLES USING THE DISSECTION METHOD

H. Behnken*, T. Bähr*, K. Dadzis+, F. Kropfgans+, L. Sylla+, T. Richter+
*: ACCESS e.V., Intzestraße 5, D-52072 Aachen, Germany
+: Solar World Innovations GmbH, Berthelsdorfer Straße 111A, D-09599 Freiberg, Germany

ABSTRACT: A method for residual stress measurements on large silicon samples is described. It combines strain gauge measurements in combination with the dissection of the gauge area. We report about tests and our experience with the technique and apply it on flat boards cut out of large ingots. Measurements on several boards from the same ingot are used to estimate the accuracy and the source of possible errors. The measurements are accompanied by thermal and thermomechanical simulation. Comparing the measurement results with the prediction of the simulation the accuracy of both measurement and simulation model can be assessed.
Keywords: Crystalline Silicon, residual stress, simulation, measurement technique

1 INTRODUCTION

The development of industrial crystallization processes for silicon ingots is usually supported by thermal simulations to get deeper insight into the temperature field within the furnace, or to study the effects of material properties, part geometries or process parameters, and finally to correlate them with the yield of the process.

The inhomogeneous temperature field within the ingot during the crystallization and the cooling process generally induces thermal stresses. Knowing the temperature history of the ingot and the mechanic behavior of the Si-material, the stress development can be simulated.

At higher temperatures, stresses partially release by plastic strain combined with dislocation multiplication. After complete cool down, the ingot remains with a distribution of residual stress which can lead to material breakage during further process steps (sawing, wafer handling, cell process). With increasing ingot size, the impact of residual stresses can be expected to become larger. Thus, one aim of process development should be the reduction of final residual stresses.

Several simulation models are available and have been applied to predict temperature fields in the furnace as well stress and strain fields and dislocation density in the ingot. To deliver realistic results such a simulation model has to reflect all essential conditions within the furnace, and the material data of all parts have to be known over the entire temperature range of the process. However, for real applications in nearly any case boundary conditions have to be simplified and some data are not known or uncertain. Therefore, to prove a model as a realistic description of a certain furnace process, it has once to be validated by comparison with experimental data.

It is a standard procedure to perform temperature measurements by thermocouples at various positions within the furnace to validate the thermal part of the model. In some furnaces the position of the crystallization front can be monitored, too.

The mechanical model has to be validated by means of experimentally measured residual stress data on the ingot and its interior. But, up to now there is no standard method to do this. The possibilities to measure residual stresses within voluminous samples like Si ingots are rare. In a previous work [1] we tested and evaluated several measurement techniques like hole drilling, birefringence or ultrasonic methods. Finally we decided to use the dissection method and applied it to flat boards cut out of G1 and G4 sized silicon ingots.

In the present project the task is the simulation of residual stresses in quasi-mono Si ingots of G5 size. The simulation should elaborate the influence of process parameters on the residual stress state of the ingots, and the model had to be validated. In the following we report about strain measurements in G5 sized flat silicon samples, the stress evaluation and the comparison with simulation results.

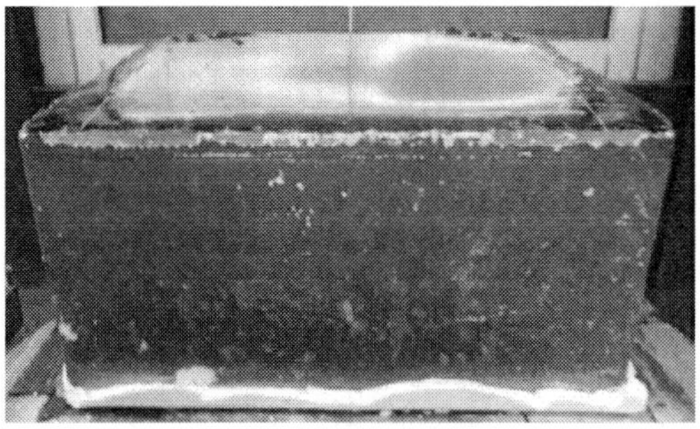

Figure 1: Quasi-mono Si-ingot of G5 size.

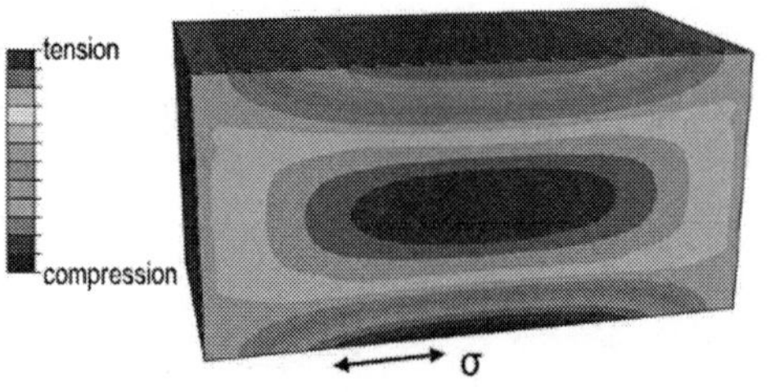

Figure 2: Typical distribution of simulated stress (half cut of the ingot) in an idealized process.

2 EXPERIMENTAL PROCEDURE

2.1 Selection of the measurement technique

Figure 1 shows an example of the G5 ingots and Figure 2 the typical stress distribution of one lateral stress component after a cooling process as expected by the simulation. Usually there is tensile stress in the center region compensated by compressive stress at top and bottom regions. The simulation assumes a homogeneous Si-material and therefore describes the macroscopic distribution of stress which varies rather smoothly over several cm. The level of residual stress is expected to be of the order of 10 MPa.

To get experimental data for validation of the simulation model the dissection method in combination with strain gauges is appropriate for following reasons:

- it averages the strain over several cm^2
- it gives directly the components of residual strain
- its sensitivity can be estimated to < 1 MPa
- recorded strains can be easily converted into stress.

Therefore, the results of such stress determinations can directly be compared with the simulation results.

2.2 Dissection method

Stress determination by cutting and dissection procedures is a common method in material testing, namely in metallic structural components [2]. The idea of the method is to fully release the stress within the gauge area and to determine and record the related strains. Strain gauges are fixed at the sample surface, and the specimen is cut around the strain gauge leading to a nearly complete stress relaxation of the cut-out region. This method can only be applied on flat specimens which can be cut completely. For the present study we prepared center boards of some mm thickness out of the G5 ingots.

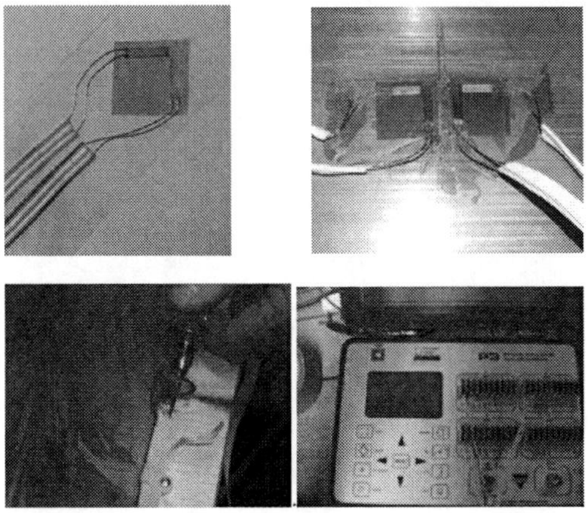

Figure 3: Experimental procedure of the dissection method: Strain gauge and its application to the sample surface, cutting out of the strain gauge region, registration of the strain relaxation during cutting.

A typical measurement setup is shown in Figure 3: The strain gauge contains two resistances positioned 90 ° to each other. Straining the gauge in one direction varies the respective resistance. The gauge is glued onto a flat board so that any additional strain of the Si material results in a change of the resistances which is detected via Wheatstone bridges in the recording box. The gauge area is cut out of the board by a hand-hold cutting disc. The recording box contains the resistance bridges, on which the electrical resistance of the sensors in the strain gauge is detected and directly transferred into strain data.

2.3 Tests and experience

The principal applicability of the method has already been examined in our previous study [1], but to establish a reliable measurement procedure some cycles of learning are needed. We performed numerous tests on different samples to enhance practicability and measurement reproducibility. Essential points are the gluing and the temperature control. Imperfect gluing leads to partly delaminated gauges and incorrect results. Slight temperature changes of the Si-material lead to strains exceeding the measurement effect. If temperature of the gluing cement exceeds about 40°-50°C the measured strain data become unreliable. Failure of the wire connection leads to wrong or even completely lost data.

In a first test procedure we performed bending tests of Si bars prepared with a strain gauge and obtained good agreement between measured bending stresses and the evaluation of strain gauge data.

Finally the following recommendations for a standardized measurement procedure for crystalline silicon can be stated:

- Polish the as-cut surface using ethanol and a fine grinding paper.
- Clean the polished surface by acetone.
- Once opened use a glue tube for not more than 2 weeks.
- Apply 2 neighbored gauges at critical positions.
- Use strain gauges with already connected wires.
- Strain gauge should have similar expansion coefficient than silicon
- Use a new cutting disc for each 2-3 measurement positions.
- Gauge temperature should kept below 40 °C during the strain recording.
- Monitor the strain data until the Si-material reaches again the previous environment temperature.

2.4 Stress relaxation by separation of center boards

The dissection method is not applicable to a voluminous ingot but needs the separation of a flat specimen. On the other hand, each sample separation out of the ingot volume changes the initial stress distribution which was intended to be determined. The elastic relaxation can be calculated by subsequent simulation of the separation process, once the residual stress distribution of the entire ingot is known. Full boards cut out of the ingot remain with zero stress normal to their surface whereas the lateral stress distribution remains similar to the original one on a reduced level of about 60%-70%. Therefore center boards are appropriate samples for stress determination.

2.5 Choice of measurement positions
Nearly all of our stress measurements were made at positions on the board where maximum or minimum stresses were expected from simulation results. But each dissection of a gauge area again changes the local stress distribution. The distance between the positions must be large enough to leave the stresses at neighboring positions almost unchanged. Figure 4 shows the effect of gauge area dissection on the local stress field. As a practical hint the distance between the position should not be less than 10 cm. Furthermore, the positions where a low stress level is expected and such cross-interaction may introduce a larger error should be measured first.

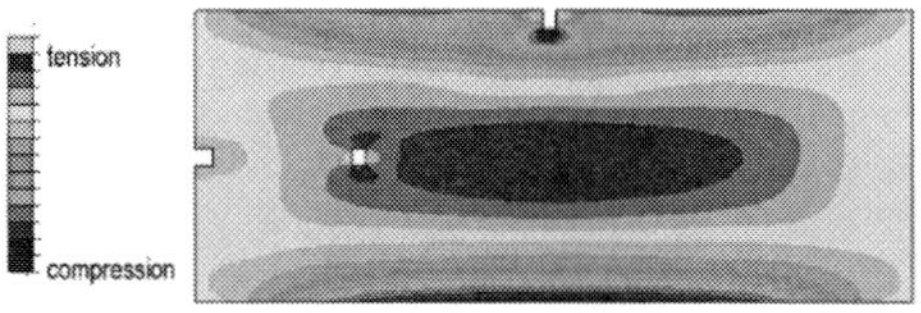

Figure 4: Effect of strain gauge area dissection on a typical stress distribution of an idealized process in a center board.

2.6 Strain measurements on center boards
Before the measurements all strain gauges are prepared at all measurement positions on the center board. For each measurement the respective strain gauge has to be connected with the bridge amplifier. Figure 5 shows the board before and after the measurements. Each measurement takes about 30 min including cutting and thermal homogenization before the final strain value is read out.

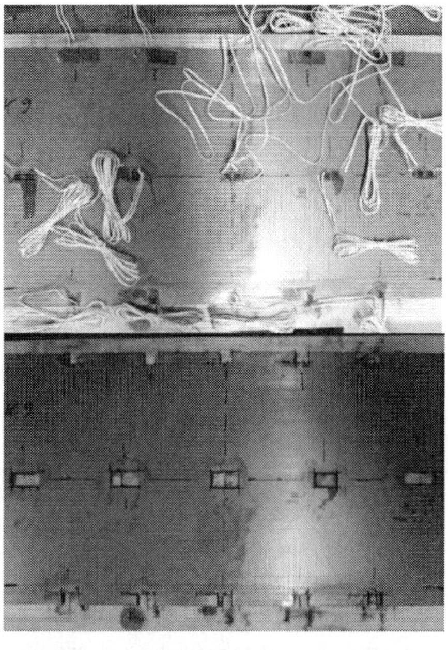

Figure 5: G5 center board (5 mm thick) before and after the strain measurements.

3 PLANE STRESS EVALUATION

Due to the small thickness of the board (5 mm) and the small gauge area compared to the board surface a plane stress state can be assumed within the measurement area which is fully released by the dissection procedure. The shear components of both, stress and strain tensor are zero.

$$\sigma = \begin{pmatrix} \sigma_x & 0 & 0 \\ 0 & 0 & 0 \\ 0 & 0 & \sigma_z \end{pmatrix} \qquad \varepsilon = \begin{pmatrix} \varepsilon_x & 0 & 0 \\ 0 & \varepsilon_y & 0 \\ 0 & 0 & \varepsilon_z \end{pmatrix} \tag{1}$$

Stress and strain tensors are connected by Hooke's law:

$$\sigma = C\,\varepsilon \tag{2}$$

with C the symmetric stiffness tensor of rank 4 which is usually displayed as a symmetric 6x6 matrix [3]. In the coordinate system of the cubic Si crystal only 3 independent components exist, and the structure of C is given as

$$C = \begin{pmatrix} c_{11} & c_{12} & c_{12} & \cdot & \cdot & \cdot \\ \cdot & c_{11} & c_{12} & \cdot & 0 & \cdot \\ \cdot & \cdot & c_{11} & \cdot & \cdot & \cdot \\ \cdot & \cdot & \cdot & c_{44} & \cdot & \cdot \\ \cdot & \cdot & \cdot & \cdot & c_{44} & \cdot \\ \cdot & \cdot & \cdot & \cdot & \cdot & c_{44} \end{pmatrix} \tag{3}$$

with c_{11}=165 GPa, c_{12}=63 GPa, and c_{44}=79 GPa. [4] Transferred from the crystal to the real specimen coordinate system all components of c_{ij} may appear depending on the crystal orientation. However, the tensor still remains symmetric.
The components ε_x and ε_z are directly determined by the measurement, the unknown component ε_y can be eliminated by the knowledge of σ_y=0. So there is a linear relation between the in-plane stress and strain components:

$$\begin{pmatrix} \sigma_x \\ \sigma_z \end{pmatrix} = \begin{pmatrix} q_{xx} & q_{xz} \\ q_{zx} & q_{zz} \end{pmatrix} \begin{pmatrix} \varepsilon_x \\ \varepsilon_z \end{pmatrix} \tag{4}$$

After inserting Eqs. (1) and (3) into Eq. (2) the matrix is given by:

$$q = \begin{pmatrix} c_{11} - \frac{c_{12}^2}{c_{22}} & c_{13} - \frac{c_{12}c_{23}}{c_{22}} \\ c_{13} - \frac{c_{23}c_{12}}{c_{22}} & c_{33} - \frac{c_{23}^2}{c_{22}} \end{pmatrix} \tag{5}$$

Note that the components of c and thus the evaluation parameters q depend on the crystal orientation. For isotropic materials the c_{ij} could be expressed by Young's modulus E and Poisson ratio ν, and the above matrix q would simplify to:

$$q = \frac{E}{1-\nu^2}\begin{pmatrix} 1 & \nu \\ \nu & 1 \end{pmatrix} \tag{6}$$

Table I: Parameters of the plane stress evaluation depending on the crystal orientation with respect to the specimen system.

Crystal Orientation		**Evaluation Parameters**
z || <hkl>	x || <mno>	q_{xx} q_{xz} q_{zx} q_{zz} [GPa]
100	010	141 39 39 141
110	001	144 51 51 187
110	-110	169 11 11 169
111	0-11	178 37 37 192
111	2-1-1	178 37 37 192
isotropic (single crystal properties averaged over all orientations)		170 37 37 170

Table I lists the evaluation parameters for several silicon crystal orientations with the nomenclature sketched in Figure 6. For each orientation the stiffness components c_{ij} are first transferred from the crystal system to the specimen system and then inserted into Eq.(5).

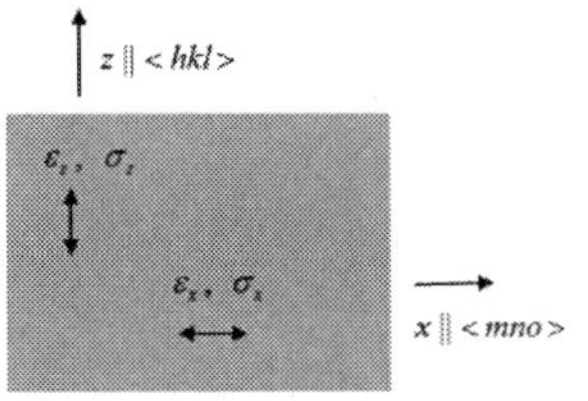

Figure 6: Description of the crystal orientation with respect to the specimen system.

4 SIMULATION MODEL

The stress measurements are accompanied by thermomechanical simulation of the whole furnace process including the crystallization of the ingot and the subsequent cooling procedure. The simulation model includes all essential components of the furnace and typically considers the applied temperature and power control [5]. The mechanical material behavior is described according to the Alexander-Haasen model [6]. The parameters of this model, especially the activation energy and the hardening parameter has been adapted in [7] according to recent results of high-temperature deformation tests on the actual quasi-mono Si-material. The anisotropy of the elastic and plastic behavior has been taken into account [8]. With this model the distribution of residual stresses in the ingot can be predicted. The remaining stresses in the finally prepared samples after cutting the ingot into center boards are derived by a subsequent simulation of the sample separation.

5 RESULTS AND DISCUSSION

Stresses were determined both by simulation as well as by measurement on several quasi-mono Si center boards. The following figures show the comparison of simulated and measured stress values. Measurements were performed at positions near the top, at mid height and near the bottom of the boards (see Fig. 5). The results are plotted along paths from the left edge to the right edge.

Three neighboring boards cut out of the same ingot (Process A) had been available and were evaluated. They should have nearly identical stress states. Figure 7a samples the results for all three boards.

It can be stated that the levels of simulation and measurement are fitting very good. In some cases the measured data even reflect the detailed profile of the simulated curves.

The vertical stress at top and bottom must be zero due to equilibrium conditions. The measured data (circles) reflect this condition well. Same holds for the lateral stress at the left and right side of the board. The respective simulation results show small deviation from zero because the paths are not taken at the very rims.

The lateral stresses at top and bottom show some more deviation between simulation and measurement. One reason might be the vertical stress gradient which is higher in these regions, but this has to be analyzed in further studies.

The measurement conditions for the results on the three neighbored boards vary slightly from each other. They include more or less matured measurements using different types of strain gauges, different cutting speeds, and different states of the cutting disc. Therefore, the scattering of the experimental data reflects the sum of several experimental errors.

The results in Figure 7b belong to a board of a second ingot which has experienced a different cooling history (Process B). The experimental data were determined using standardized identical measurement parameters as they are recommended in section 2.3. However, the number of data is too low to assess the effect on the scattering. But in summary a very good coincidence between simulation and experiment can be stated. The overall experimental accuracy of the stress measurements can be estimated as 1 MPa.

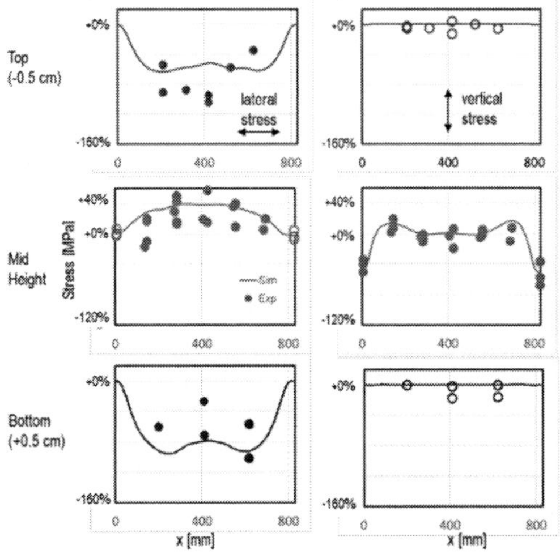

7a: Ingot of process A. The data were measured on three neighbored boards cut out of the same ingot.

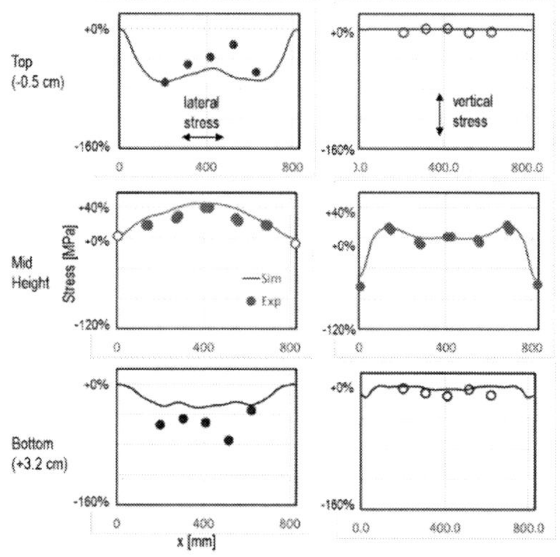

7b: Ingot of process B. The experimental data were determined on a single center board

Figure 7: Comparison of simulated (lines) and measured (circles) residual stresses. The diagrams show the lateral stress (left hand side) and vertical stress component (right hand side) along horizontal paths near the top, at mid height, and near the bottom. Stresses are given in relative units.

6 CONCLUSIONS

The dissection method has proved to be a reliable and feasible tool for the determination of macroscopic stress distributions in large flat silicon samples like G5 center boards. Combined with thermo-mechanical process simulation the stress level within the original ingot can be reconstructed. The accuracy of the dissection method allows to evaluate typical stress distributions in Si-boards and should be sufficient to detect the impact of process variations. Thus, it can serve for stress determination in ingots as well as for the validation of simulation models.

REFERENCES

[1] T. Bähr, H. Behnken, M. Lohan, M. Hollatz, D. Oriwol, 27th EU PVSEC., Frankfurt/Main, 2012, 642-646.
[2] Handbuch für experimentelle Spannungsanalyse, Chr. Rohrbach (Hrg.), VDI-Verlag, Düsseldorf. 1989.
[3] J. F. Nye, Physical Properties of Crystals, Oxford University Press, 1985.
[4] A. George, in: R. Hull (ed.), Properties of crystalline silicon, INSPEC, London, 1999, 98-103.
[5] T. Bähr, H. Behnken, J. Henker, A. Müller: 22nd EU PVSEC., Milan, 2007, 984-987.
[6] H. Alexander, P. Haasen: Solid State Physics 22 (1968), 27-158.
[7] K. Dadzis, H. Behnken, T. Bähr, D. Oriwol, L. Sylla, T. Richter, J. Crystal Growth, 2016, in press.
[8] H. Behnken:, Proc. 24th EU PVSEC, Hamburg, 2009, 1281-1285.

ACKNOWLEDGEMENTS

Dr. Bernhard Freudenberg (Coburg) is gratefully acknowledged for numerous discussions. The authors would like to thank the Federal Ministry for the Environment (BMU) for the funding received within the grant 0325805 J (project ENOWA II).

This is a pre-peer review version of the paper submitted to the journal Progress in Photovoltaics.
It has been selected by the Executive Committee of the EU PVSEC 2016 for submission to Progress in Photovoltaics.

MATERIAL LIMITS OF SILICON FROM STATE-OF-THE-ART PHOTOLUMINESCENCE IMAGING TECHNIQUES

Florian Schindler[a,b], Johannes Giesecke[a], Bernhard Michl[a], Jonas Schön[a], Patricia Krenckel[a], Stephan Riepe[a], Wilhelm Warta[a], Martin C. Schubert[a]
[a] Fraunhofer Institute for Solar Energy Systems (ISE), Heidenhofstr. 2, D-79110 Freiburg, Germany
[b] Freiburg Materials Research Center (FMF), Albert-Ludwigs-Universität Freiburg, Stefan-Meier-Str. 21, D-79104 Freiburg, Germany
Ph +49 761 4588 5321, Fax +49 761 4588 9250, Email: florian.schindler@ise.fraunhofer.de

ABSTRACT: This work reports on state-of-the-art silicon material characterization by calibrated photoluminescence imaging (PLI). PL imaging techniques allow for a characterization of a large variety of material properties, ranging from bulk and surface recombination properties of bare silicon ingots, to crystal structure, dopant concentration and bulk lifetime of silicon wafers up to material limiting impurities. In combination with solar cell simulations, injection-dependent PLI of processed wafers provides a method for determining the material's efficiency potential. In this contribution, we present two methods for imaging the concentration of interstitial iron at passivated silicon slices and at unpassivated silicon ingots. The latter is achieved by our latest progress in PL-based charge carrier bulk lifetime measurements with highest sensitivity, giving access to the bulk recombination properties of the ingot. Additionally, we present an improved analysis of specific loss mechanisms in silicon based on de-smeared injection-dependent lifetime images of processed wafers, which is demonstrated on the example of the bulk-related efficiency limitations in high-performance multicrystalline silicon. This latest progress rounds off the established techniques, enhancing the potential of PL based imaging techniques for a comprehensive assessment of silicon material quality, from limitations in bare silicon ingots to efficiency loss mechanisms in processed samples.

Keywords: Photoluminescence imaging, iron imaging, modulated PL, efficiency limits, multicrystalline silicon

1 INTRODUCTION

Photoluminescence-based characterization techniques have been successfully established in silicon material characterization labs during the past decade, both in research and industry. Its remarkably high sensitivity and non-destructive character together with its simplicity gave rise to the development of a large variety of applications, ranging from measurements of material parameters like crystal defects, doping concentration, charge carrier lifetime and concentration of specific impurities to also solar cell properties like local current and voltage, series resistance and interface recombination. An overview of the majority of these measurement techniques is given in [1].

In this contribution, we demonstrate how to identify material limits of silicon by state-of-the-art photoluminescence imaging techniques. We include our latest advances and innovations which enable a deeper insight into the material limitations directly after crystallization by iron imaging at vertical silicon slices and bare silicon ingots. The exact identification and quantification of bulk-related efficiency losses after selected solar cell processing steps is extended to separating between recombination- and diffusion-related losses.

2 DEVELOPMENT OF PL-BASED MEASUREMENT TECHNIQUES

Improvement of silicon solar cell efficiencies requires a fundamental knowledge of material limitations, many of which can be assessed by already established photoluminescence-based measurement techniques. Measurements at bare silicon ingots allow for a determination of bulk lifetime and surface recombination velocity [2], as well as for a quantification of interstitial iron concentrations [3]. Further, PL imaging of unpassivated silicon wafers allows for determination of the most recombination active wafer regions, such as dislocation clusters and edge regions in multicrystalline silicon, and, thus, enables a first quality rating of the material [4]. Apart from areas of strongly recombination active crystal defects, the charge carrier density in as-cut wafers is limited by recombination at the unpassivated surfaces. Given a sufficiently high and uniform surface recombination velocity, the PL signal is determined by the majority charge carrier concentration. Under low-level injection, this enables a spatially resolved determination of the doping concentration [5]. PL imaging at surface-passivated wafers [6] combined with different calibration techniques [7, 8] gives access to spatially resolved bulk recombination properties and, in combination with metastable defect imaging [9], allows for measuring the concentrations of interstitial iron [10] and chromium [11] in p-type silicon with a detection limit in the range of 10^{10} cm^{-3} as well as a relative concentration of the boron-oxygen defect [9, 12]. Another approach is based on the growth of thermal donors and provides images of the interstitial oxygen concentration [13]. As these methods exploit the different recombination activity of metastable defects (Fe_i/FeB, Cr_i/CrB, $BO_{degraded}$/$BO_{annealed}$) or the impact of defect redistribution on resistivity (creation of thermal donors for O_i-imaging), they are not suited for the identification of initially unknown impurities. Access to such defects can be obtained by making use of the temperature and injection dependence of charge carrier lifetime. In combination with PL imaging, this approach enables a spatially resolved identification of further material limiting impurities [14].

In order to deepen the understanding of silicon material limits, our recent research in PL-based characterization techniques focused on measuring the distribution and concentration of interstitial iron across silicon ingots after crystallization. The first approach presented in section III.A is based on lifetime measurements at passivated vertical silicon slices. The development of a method for iron imaging at unpassivated silicon ingots, making use of highly-sensitive charge carrier bulk lifetime measurements at unpassivated ingots [2, 15], is discussed in section III.B. This renders unnecessary the preparation of vertical slices. Finally, the innovations in efficiency-limiting bulk recombination analyses (ELBA) [16, 17] in combination with a de-smearing of PL images for quantification of specific loss mechanisms in the material are sketched in section III.C. We are now able to separate local recombination from diffusion effects quantitatively. Thus, we prove that grain boundaries account for the major part of efficiency losses in high-performance multicrystalline silicon, comprising losses due to recombination of charge carriers at the grain boundaries and due to diffusion of charge carriers from grains to grain boundaries, where they can subsequently recombine. This knowledge could be of importance for a further optimization of crystallization parameters for the production of high performance multicrystalline silicon in order to reduce material-related efficiency losses.

3 RESULTS

3.1 Iron imaging at silicon slices

Iron imaging at passivated silicon wafers is a well-established method for spatially resolved determination of interstitial iron concentration in silicon bulk material [9, 10, 18]. This method is well suited for tracking the interstitial iron concentration in wafers along the solar cell processing chain, as measurements can be performed at passivated lifetime samples at any stage of processing. Thus, the impact of high-temperature steps on the concentration of interstitial iron can be assessed. However, if the interstitial iron distribution along the ingot height shall be determined, horizontal wafers of several ingot positions or a vertical wafer are required. Alternatively, a thick vertical silicon slice can be cut off from the side of the ingot before wafering. Thus, a vertical silicon sample is obtained, while horizontal wafering of the ingot is still possible. The procedure for iron imaging at passivated silicon wafers can be transferred to passivated vertical slices. However, the larger thickness of vertical slices (~ 2 mm) implies some modifications, which are outlined in the following.

We compare a p-type 1 Ωcm surface-passivated silicon wafer with a thickness of d = 200 µm and a constant injection-independent bulk lifetime of τ = 100 µs with a 2 mm thick surface-passivated silicon slice featuring the same doping and bulk lifetime. The front- and back-surface recombination velocities are assumed to be $S_f = S_b$ = 1 cms^{-1} for both samples, and a typical reflection of $R(790\text{ nm}) = 0.27$ for aluminum oxide passivated samples was used in the calculations. The steady-state excess charge carrier density profile $\Delta n(z)$ in z-direction along the sample depth (perpendicular to the surfaces) can be calculated by solving the steady-state continuity equation

$$-D_e \frac{\partial^2 \Delta n(z)}{\partial z^2} + \frac{\Delta n(z)}{\tau} = G(z) \quad (1)$$

with D_e the diffusion coefficient for electrons and $G(z)$ the monoexponential generation rate for an irradiation of 1 sun with a 790 nm laser at the front surface of the samples in combination with the following boundary conditions:

$$D_e \left.\frac{\partial \Delta n}{\partial z}\right|_{z=0} = S_f \Delta n(z=0) \quad (2a)$$

$$D_e \left.\frac{\partial \Delta n}{\partial z}\right|_{z=d} = -S_b \Delta n(z=d) \quad (2b)$$

Further details on the solution of these equations can be found in [19]. Figure 1 shows the $\Delta n(z)$-profiles for both samples.

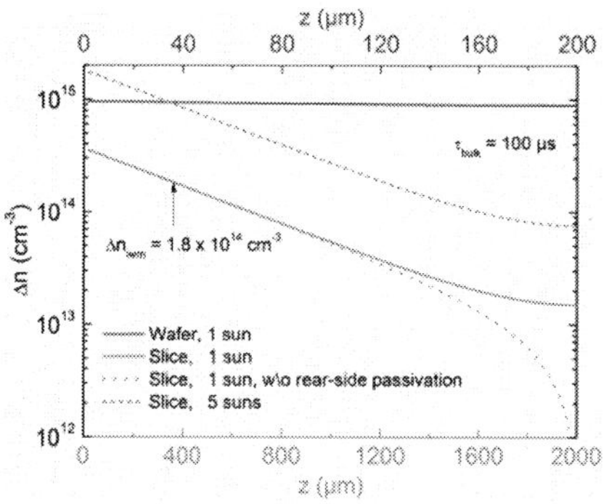

Figure 1: Comparison of excess charge carrier profile across the thickness of a surface-passivated silicon wafer with a thickness of 200 µm and a silicon slice with a thickness of 2 mm featuring double- or single-side surface-passivation (constant bulk lifetime 100 µs, illumination 1 sun or 5 suns with a 790 nm laser, R = 0.27).

While excess charge carriers distribute virtually homogeneously across the sample thickness of passivated silicon wafers (solid cyan curve), the charge carrier density decreases exponentially towards the backside of the silicon slice (solid orange curve), which is attributed to the thickness (2 mm) being significantly larger than the minority charge carrier diffusion length (~ 519 µm) and the exponentially decreasing generation rate. Thus, for passivated silicon wafers, the excess charge carrier density does not have to be regarded as a function of depth, which considerably simplifies the evaluations according to the following relation between the local PL-signal ϕ_i, measured by PL-imaging at pixel i, the local excess charge carrier density Δn_i and net doping concentration n_{net}

$$\phi_i = a \cdot \Delta n_i \cdot (\Delta n_i + n_{net}) = a \cdot G \cdot \tau_i \cdot (G \cdot \tau_i + n_{net}) \quad (3)$$

which can be directly converted into the local charge carrier lifetime τ_i for a known steady-state depth-averaged generation rate G. The calibration factor a is obtained from modulated photoluminescence measurements as described in [8, 19]. In the case of thin wafers with diffusion lengths typically not much less than wafer thickness, it is convenient to use a depth-averaged excess carrier generation rate G with incident photon flux density j_γ and front reflection $R(790\text{ nm})$ according to:

$$G = (1 - R(790\text{ nm})) \cdot j_\gamma / w \quad (4)$$

For our evaluations, we make use of a 1 sun equivalent of

$$j_{\gamma,1\text{sun}}(w = 200\ \mu\text{m}, T = 300\text{K}) = 2.5 \cdot 10^{17}\text{cm}^{-2}\text{s}^{-1} \quad (5)$$

Consequently, the calibration of PL-images directly delivers a correct pair of values for $\tau_i(\Delta n_i)$. For thick slices, where the charge carrier density becomes a function of depth, the situation is more complex. For calibrating PL-images of thick slices, we proceed as follows. First, the PL-image is calibrated to lifetime values in the same way as for wafers [8, 19]. For low-level injection conditions, which hold for all of our evaluations, this procedure delivers a correct value for τ_i. However, the obtained injection level Δn_i strongly differs from the correct injection level related to the measured lifetime, as the arithmetic depth average of excess carrier generation rate is only appropriate for uniform excess carrier density depth distributions. For evaluating the interstitial iron concentration, the exact injection level Δn_i relating to the measured τ_i is required. Therefore, the correct averaging across $\Delta n(z)$ has to be applied. As discussed in detail in [20-22], the intensity-weighted mean along the sample depth correctly accounts for the contribution of $\Delta n(z)$ to the measured PL-intensity. Thus, we obtain the correct injection-level Δn_i related to the measured lifetime τ_i by simulating $\Delta n(z)$ according to equations (1) and (2) with the measured lifetime τ_i as input and taking the intensity-weighted mean along the z-direction. Note that this approach assumes an injection-independent charge carrier lifetime for the simulation of $\Delta n(z)$. This is a good approximation, as measurements are performed in low level injection (maximum charge carrier densities are below $5 \cdot 10^{12}\text{ cm}^{-3}$ for all measurements presented in this work, where even the Fe_i-limited lifetime features a weak injection dependence only) and, more importantly, due to the intensity-weighted mean the majority of the signal stems from regions of largest charge carrier density close to the front surface with a small spread in Δn.

In addition to relating a correct $\tau_i(\Delta n_i)$ to each pixel of the PL-image, for vertical slices the net doping concentration changes with position. By applying either spatially resolved four-point-probe or eddy-current measurements, we account for this variation in net doping concentration in the evaluations. All further calculations are identical with the evaluation for interstitial iron imaging at silicon wafers.

As can be seen from Figure 1, at identical illumination conditions the charge carrier density is significantly lower for thick slices than for thin wafers. Thus, care has to be taken to ensure a complete splitting of iron-boron pairs for preparing the Fe_i-state, which is obtained by illuminating the samples with an illumination intensity of approximately 5 suns. This delivers an average excess charge carrier density comparable to that obtained on wafer level with an illumination intensity of 1 sun (*cf.* dashed orange curve in Figure 1).

While the strong decay of $\Delta n(z)$ along the depths for thick samples complicates the evaluations, it comes along with an advantage for sample preparation. For a thickness of approximately 2 mm, it renders unnecessary the rear-surface passivation. The charge carrier profile $\Delta n(z)$ for a slice without rear-surface passivation is included in Figure 1 (dotted orange curve). Differences compared with the double-side passivated slice only become significant for depths > 1 mm. However, due to the very low charge carrier density towards the rear side accompanied by a very low detector sensitivity for PL emission at such depth, these differences render negligible for the intensity-weighted mean, which is identical for both cases in our example ($\Delta n_{\text{iwm}} = 1.8 \cdot 10^{14}\text{ cm}^{-3}$).

Figure 2 shows the interstitial iron concentration measured at a vertical slice of half a G1-sized mc-silicon ingot. The two insets show 2-D simulations of the dissolved iron concentration with Sentaurus Process [23] taking the precipitation at crystal defects during the crystallization process into account. The Fe distributions are taken from our experimentally verified simulations of the Fe contamination during the crystallization process [24]. These simulations are performed independently of the measurements, as explained in detail in [24, 25].

Both the absolute values of measured and simulated $[Fe_i]$ as well as the width of the edge-region affected by solid-state diffusion of iron from the crucible into the crystal during ingot cooling are in excellent agreement. Local differences related to the crystal structure are observed at certain grain boundaries and dislocation clusters, which can be explained by the constant dislocation density of $8 \cdot 10^3\text{ cm}^{-2}$ used for the simulations [24-26]. This is a realistic dislocation density in grains of mc silicon [26]. However, in areas of dislocation clusters, the actual dislocation density is significantly larger. This effect is not taken into account in the simulations. Slight differences in the width of the edge region between measurement and simulation in the upper part can be attributed to deviations of the real temperature profile in the rectangular crucible from the radially symmetrical temperature profile used as input for the simulations. Additionally, the width of the real crucible slightly widens towards the top, whereas the simulations are based on vertical crucible walls.

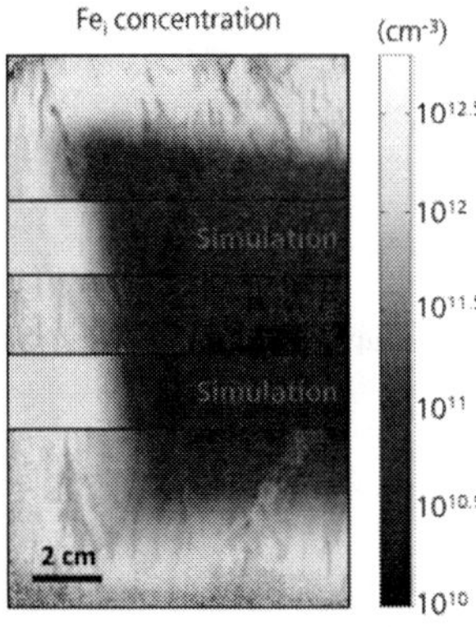

Figure 2: Interstitial iron concentration measured at a vertical slice of half a G1-sized mc-Si ingot in comparison with 2-D simulations of [Fe$_i$]. Large concentrations of interstitial iron stemming from solid-state diffusion into the crystal during ingot cooling are found in the regions close to the crucible walls (left and bottom), whereas large concentrations in the top region are attributed to back-diffusion of segregated iron from the ingot top.

3.2 Measuring bulk lifetime and interstitial iron concentration at unpassivated silicon ingots

Preparation of vertical silicon slices can be avoided completely, if bulk lifetime and interstitial iron concentration can be measured at unpassivated silicon ingots. Lacking surface passivation, polished silicon ingot side facets feature surface recombination velocities in the range of about $2 \cdot 10^5$ cm·s^{-1} [2]. Bulk lifetime contrast of excess carrier density and thus PL intensity is therefore largely disguised by surface recombination. This renders steady-state bulk carrier lifetime approaches much prone to systematic error due to e.g. lateral optical blurring [27], lateral surface morphology variations, an uncertainty of temperature-dependent band-to-band optical absorption properties, or variations and uncertainty of the minority carrier diffusion coefficient.

Dynamic decay time techniques such as harmonically modulated luminescence overcome such limitation, because they allow us to directly investigate the decay time of surviving excess carriers within the bulk rather than to implicitly infer lifetimes from cumulative excess carrier density vastly dominated by surface recombination. Equipped with a detailed understanding of harmonically modulated lifetime [28, 29], we are able to precisely and accurately determine bulk lifetime from such harmonically modulated decay time measurement design.

Figure 3 illustrates such capability via the most extreme case of a virtually undoped and unpassivated float zone silicon ingot. While measurement raw data would not even allow recognizing any harmonic modulation in the PL signal (mostly due to thermal noise originating from the photodetector), the depicted spectrally filtered signal allows determination of a harmonically modulated lifetime of 243 ± 11 µs (790 nm irradiation wavelength), which corresponds to a bulk lifetime of 612 ± 30 µs. Cumulative excess carrier density (proportional to PL intensity here) would only vary by ± 2 % within this uncertainty range, which would render steady-state approaches highly susceptible to systematic error even in view of minor perturbations.

Notably, the high bulk lifetime contrast of harmonically modulated luminescence on unpassivated ingots opens novel pathways in material investigations involving highest bulk lifetimes beyond the performance limit of state-of-the-art wafer surface passivation techniques. Rather than taking the risk of possibly insufficient surface passivation disguising bulk lifetime information, a more accurate alternative would be to investigate unpassivated bulk samples.

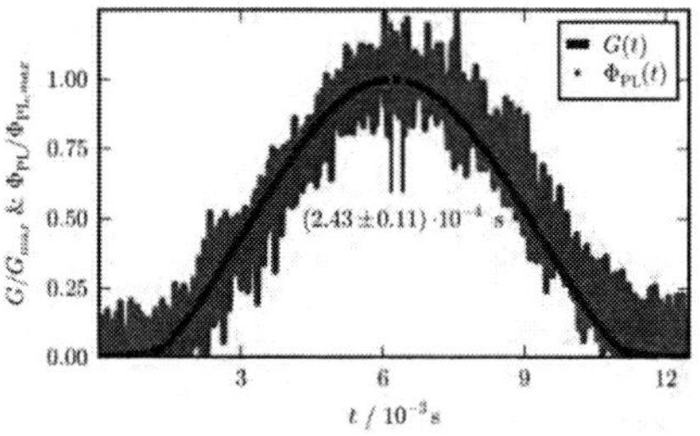

Figure 3: Extreme case of a harmonically modulated PL lifetime measurement (G: excess carrier generation rate, Φ_{PL}: PL intensity) on a virtually undoped float zone silicon ingot without surface passivation. Even here, bulk lifetime can be inferred from the harmonically modulated lifetime at an astonishing precision.

It should also be noted that the harmonically modulated approach proved eligible for accurate bulk lifetime calibration of steady-state PL images [8] – as particularly detailed for unpassivated ingots in [15], and as used in the following. Figure 4 exemplarily shows a calibrated lifetime image at low injection level of the lower left corner of an unpassivated high-performance mc-Si ingot in the Fe$_i$-state after splitting of FeB-pairs. Together with a lifetime image in the FeB-state, this is used as input for imaging the concentration of interstitial iron at unpassivated silicon ingots, as explained in the following.

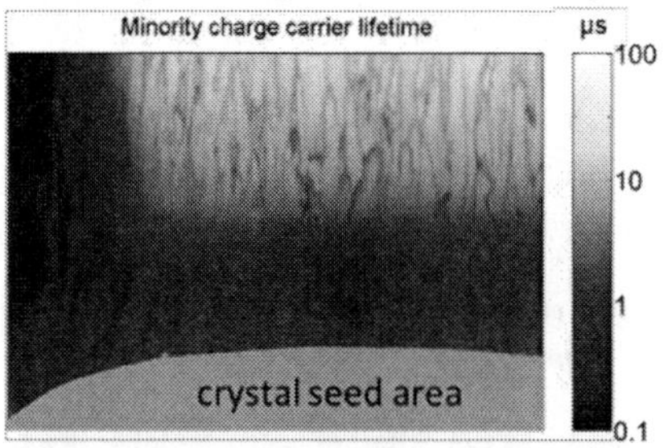

Figure 4: Calibrated bulk-lifetime image of the lower left corner of an unpassivated high-performance mc-Si ingot side facet according to a harmonically modulated calibration approach detailed in [15].

Recently, Mitchell *et al.* presented a method for measuring the interstitial iron concentration directly on the side face of silicon ingots by spectral PL intensity ratio imaging [3]. Here, we present an approach for Fe-Imaging on bare silicon ingots, also based on FeB-pairing and –splitting, in combination with the aforementioned ingot bulk lifetime calibration by modulated PL, featuring an exceptionally high sensitivity. This technique relies on a careful pairing and splitting of FeB-pairs. The largest challenge is to reach sufficiently high injection levels for FeB-splitting, which we achieve by laser-illumination with an irradiation intensity of approximately 20 suns. In order to achieve sufficiently high injection levels, the ingot is illuminated with two lasers of different wavelengths (790 nm and 940 nm). Besides acting as additional illumination source, the 940 nm laser features a larger penetration depth in silicon (~ 55 μm) compared with the 790 nm laser of the measurement setup (~ 11 μm). Due to the lacking front-surface passivation, the deeper penetration depth significantly increases the excess charge carrier density, which is shown in Figure 5: $\Delta n(z)$ is increased by a factor of ~ 5 for illumination with the 940 nm laser (red dashed line compared with orange dashed line). While at standard measurement conditions of 1 sun illumination the charge carrier density is below $10^{13}\ \mathrm{cm}^{-3}$ for a bulk lifetime of 100 μs, injection levels sufficiently high for FeB-splitting are achieved by an illumination with ~ 20 suns in combination with the deeper penetration depth of the 940 nm laser.

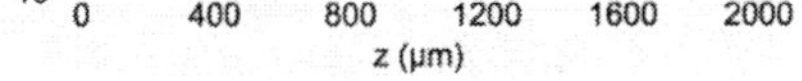

Figure 5: Comparison of excess charge carrier profiles in the first 2 mm of an unpassivated silicon ingot at measurement conditions of 1 sun and at FeB-splitting conditions at 20 suns for illumination with the 790 nm or 940 nm laser (constant bulk lifetime 100 μs).

Figure 6 shows images of the interstitial iron concentration and the fraction of recombination due to Fe_i measured on the side face of an unpassivated multicrystalline silicon ingot by this method. In order to validate the measurements, we performed independent 2-D simulations with Sentaurus Process as in the previous section. In the simulations, the Fe diffusion from the crucible walls into the solid and liquid silicon, segregation during crystallization and back-diffusion is modeled as for standard mc silicon ingots [3, 24] using the measured Fe concentrations in the crucible and coating as input parameters. We estimate that before the melting process around 30% of the seed layer volume is gas-filled voids. As soon as the melting process starts liquid silicon flows into these voids and re-crystallizes. The re-crystallization is not a continuous process and thus the Fe concentration in the re-crystallized silicon of the seed layer is not reduced by strong segregation from solid to liquid silicon. In consequence the seed layer has a quite high Fe concentration (~$3\times10^{14}\ \mathrm{cm}^{-3}$) and acts as a source for Fe diffusion into the liquid and solid silicon above the seed layer.

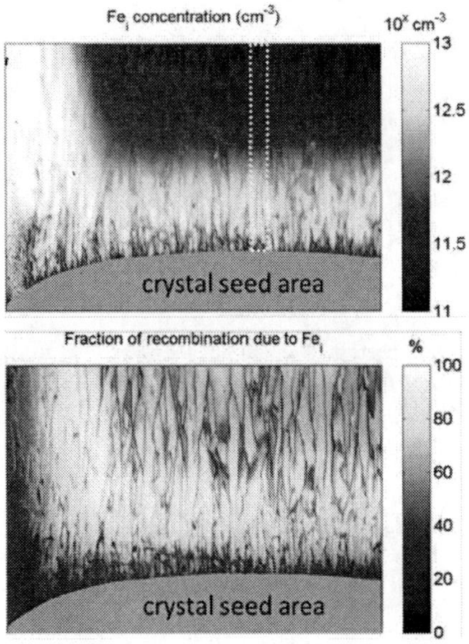

Figure 6: Image of the interstitial iron concentration and the fraction of recombination due to Fe_i, measured on the side face of an unpassivated high-performance mc-Si ingot.

Figure 7 shows an averaged line-scan of the measured interstitial iron concentration as indicated by the white rectangle in the [Fe_i]-image of Figure 6, together with two simulations. The red dotted curve is a simulation of the vertical profile of the interstitial iron concentration, showing the region of solid-state in-diffusion of iron from the crucible bottom and the crystal seed area up to an ingot height of ~ 4 cm (distance from seed). Thereafter, the interstitial iron concentration increases again due to segregation of iron dissolved in the silicon melt. In contrast, the measured iron profile (black symbols) shows a continuous decrease with increasing ingot height. This can be attributed to the distance of the side facet where the iron concentration was measured from the crucible wall being smaller (3.3-3.6 cm) than the width of the region additionally affected by solid-state in-diffusion of iron from the facing crucible wall (*cf.* Figure 8), which leads to an additional two-dimensional increase in iron concentration across the whole measured area. Accounting for this effect in the simulations results in the green dashed curve, which is in good agreement with the measured interstitial iron concentration and validates the method. The decrease in the measured values close to the seed can be attributed to an increase in grain boundary and dislocation density (see Figure 4) which is not taken into account in the simulations. Slight deviations between measurement and simulations for ingot heights larger than 2.5 cm can be related to the uncertainty of the distance of the side facet from the crucible wall, leading to an uncertainty in the simulated interstitial iron concentration.

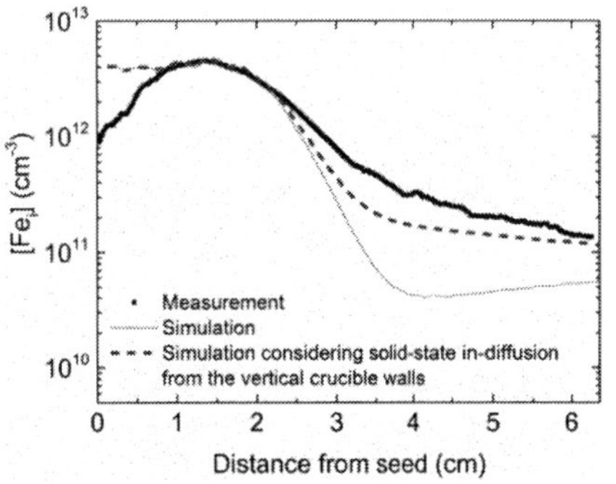

Figure 7: Measured and simulated interstitial iron profiles close to the seed area of a high-performance multicrystalline silicon ingot.

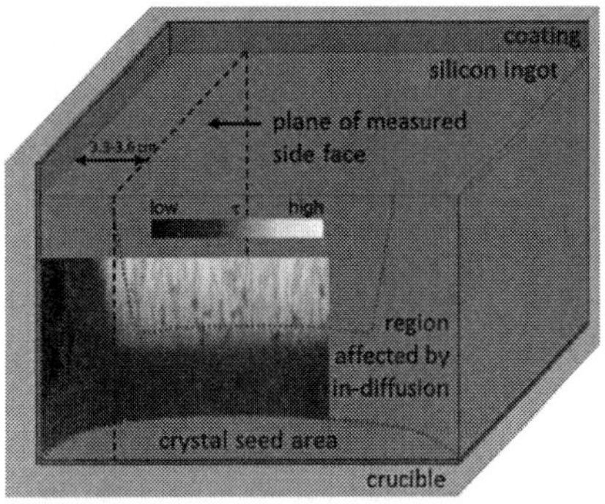

Figure 8: Illustration of the plane of the measured side face with a distance of 3.3-3.6 cm from the crucible wall. In the lower part of the ingot, this plane is significantly affected by solid-state in-diffusion of iron from the facing crucible wall as indicated by the dotted line, resulting in the shift from the red dotted to the green dashed Fe_i-profile in Figure 7

3.3 Specification of bulk-related efficiency loss mechanisms

An "Efficiency Limiting Bulk Recombination Analysis" (ELBA) [16, 17] combines injection-dependent charge carrier lifetime images obtained from calibrated PLI of wafers processed according to a certain cell process with PC1D solar cell simulations of a specific solar cell concept and, thus, allows for a prediction of the material-related efficiency potential for the cell concept under investigation. Further, it enables a quantification of the specific loss mechanisms in the bulk [30, 31], separating between losses due to homogeneously distributed recombination centers, structural crystal defects and impurities in the edge region. In this paper, we present an advanced procedure of this loss analysis which enables an in-depth understanding of the physical origins for material-related efficiency losses by separating between recombination- and diffusion-related losses.

In a homogeneously illuminated silicon wafer or solar cell, minority charge carriers diffuse from areas of low recombination to such of high recombination, which leads to a charge carrier redistribution and, thus, a smearing of PL images. On the one hand, lateral carrier diffusion is a physical effect also occurring in a solar cell under working condition and, therefore, needs to be accounted for in a correct prediction of solar cell efficiencies. On the other hand, it conceals the real recombination properties of the material and impedes an exact identification of the specific loss mechanisms. The approach presented here combines an ELBA analysis based on images of the effective charge carrier lifetime comprising lateral carrier diffusion with an ELBA analysis based on de-smeared images of the real charge carrier lifetime. The de-smeared lifetime images are calculated from the effective lifetime images using the continuity equation, as explained in [32].

In order to highlight the effect of lateral carrier diffusion, Figure 9 shows a calibrated PL-image [8] of the effective minority charge carrier lifetime (comprising lateral carrier diffusion) of an n-type high-performance multicrystalline silicon wafer after boron and phosphorus diffusion at an illumination of 0.03 suns (upper image), which suggests that the material features grains of varying material quality.

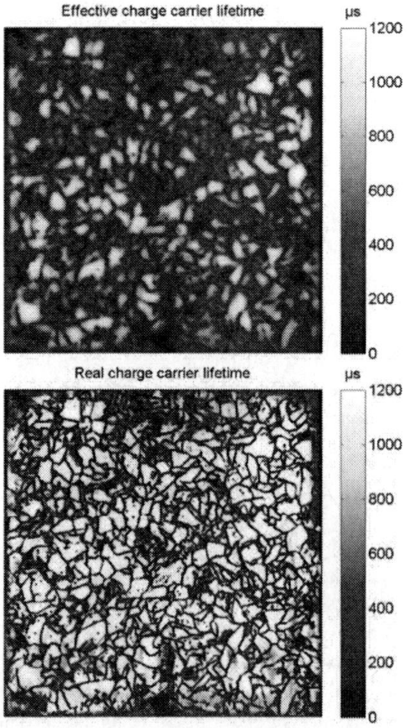

Figure 9: Minority charge carrier lifetime in a processed (B- and P-diffusion) n-type high-performance multicrystalline silicon wafer including lateral charge carrier diffusion (top) or not (bottom) at an illumination of 0.03 suns.

A calculation of the real local minority charge carrier lifetime (Figure 9, lower image) reveals that the inner-grain lifetime is virtually constant across the whole wafer. Thus, differences in the effective lifetime between different grains can be attributed to lateral charge carrier diffusion. Additionally, the de-smeared image shows that the main challenge for optimizing high-performance multicrystalline silicon is to reduce recombination active structural crystal defects, which not only lead to decreased charge carrier lifetimes at the structural defects themselves, but also significantly reduce the charge carrier lifetime in good grains due to lateral carrier diffusion.

For a quantification of loss mechanisms, effective efficiency maps (η_{eff}) are obtained from a first ELBA analysis based on injection-dependent effective lifetime images. The global efficiency potential for the entire wafer compared with the efficiency limit of the solar cell without recombination losses gives access to the total material-related losses, comprising recombination losses and diffusion losses in grains as well as recombination losses at structural crystal defects. Together with de-smeared efficiency maps ($\eta_{de\text{-}smeared}$) obtained from a second ELBA analysis based on injection-dependent de-smeared lifetime images, a separation of these loss mechanisms becomes possible: Relating the efficiency potential in the grains obtained from $\eta_{de\text{-}smeared}$ to the cell limit quantifies the recombination losses in the grains, whereas the efficiency potential in the grains obtained from η_{eff} additionally includes the losses due to lateral charge carrier diffusion to recombination active structural crystal defects. Finally, the difference between the total losses obtained from η_{eff} and the recombination and diffusion losses in the grains assesses the losses due to recombination at structural crystal defects. Figure 10 shows such a loss analysis on the example of n-type high-performance multicrystalline silicon after a solar cell process including a boron diffusion or a boron and a phosphorus diffusion for a high-efficiency solar cell concept featuring a cell limit of 23.1%. The advanced analysis based on de-smeared lifetime images reveals that approximately one third of the inner grain losses are actually due to lateral charge carrier diffusion to recombination active structural crystal defects. Note that this value depends on the size of the chosen grain as well as the chosen border between grain and surrounding grain boundaries. Nevertheless, this would only alter the ratio of structural crystal defect-related losses attributed to diffusion losses or recombination losses (shift the margin between the hatched and the double-hatched gray parts of the columns as indicated by the arrows in Figure 10), while the total structural crystal defect-related losses and the recombination losses in grains remain unaffected by the choice of the specific grain and border. Being responsible for diffusion losses in grains and direct recombination losses at grain boundaries, structural crystal defects account for the major part of efficiency losses in n-type high-performance multicrystalline silicon. This highlights the importance of reducing the recombination activity of grain boundaries in high-performance mc-Si. One possibility is the application of a phosphorus gettering diffusion, which approximately cuts in half the recombination losses at grain boundaries (c.f. Figure 10). Improving the crystallization process or applying a hydrogenation step would be further steps to close the gap between efficiencies of multi- and monocrystalline silicon solar cells.

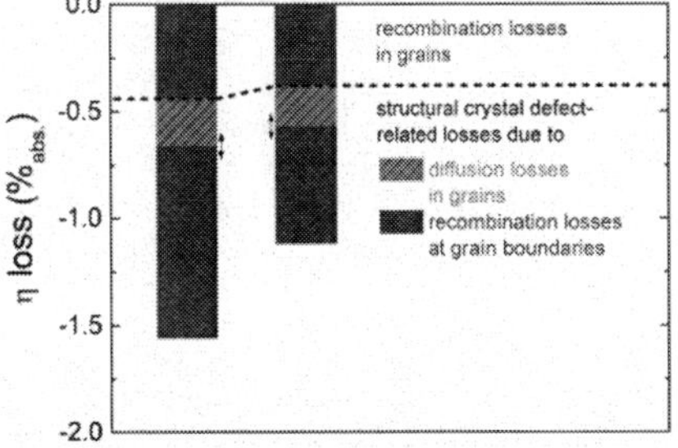

Figure 10: Advanced quantification of loss mechanisms in n-type high-performance multicrystalline silicon.

4 CONCLUSION

This contribution elaborates on progress with the investigation of silicon material limits via state-of-the-art photoluminescence techniques. The very high bulk lifetime sensitivity of harmonically modulated luminescence allows most accurate investigation of highest quality bulk samples and detailed material studies such as metastable defect imaging on the level of unpassivated ingots. Separating effects of lateral carrier diffusion in ELBA gives rise to a refined assessment of multicrystalline performance limits due to structural crystal defects.

5 ACKNOWLEDGEMENT

F. Schindler would like to thank P. Phang from the ANU for fruitful discussions about de-smearing of PL images.

Measurements of Fe_i-concentrations at silicon slices and simulations were supported by the German Federal Ministry for the Environment, Nature Conservation and Nuclear Safety and by industry partners within the research cluster "SolarWinS" under contract 0325270G.

Crystallization of high-performance multicrystalline silicon as well as measurements of Fe_i-concentrations at unpassivated silicon ingots and corresponding simulations were supported by the Deutsche Forschungsgemeinschaft (DFG) under project number BO3498/1.

Development of advanced PL imaging methods for specification of bulk-related efficiency losses in high-performance multicrystalline silicon were supported by the German Federal Ministry for Economic Affairs and Energy within the research project "multiTOP" under grant number 0324034.

The content is the responsibility of the authors.

REFERENCES

[1] B. Michl, M. Padilla, I. Geisemeyer, S. T. Haag, F. Schindler, M. C. Schubert, and W. Warta, IEEE Journal of Photovoltaics **4** (2014) 1502.

[2] J. A. Giesecke, R. A. Sinton, M. C. Schubert, S. Riepe, and W. Warta, IEEE Journal ofPhotovoltaics **3** (2013) 1311.

[3] B. Mitchell, D. Macdonald, J. Schon, J. W. Weber, H. Wagner, and T. Trupke, IEEE Journal of Photovoltaics **4** (2014) 1185.

[4] J. Haunschild, I. E. Reis, T. Chipei, M. Demant, B. Thaidigsmann, M. Linse, and S. Rein, Solar Energy Materials & Solar Cells **106** (2012) 71.

[5] S. Y. Lim, M. Forster, X. Zhang, J. Holtkamp, M. C. Schubert, A. Cuevas, and D. Macdonald, IEEE Journal of Photovoltaics **3** (2013) 649.

[6] T. Trupke, R. A. Bardos, M. C. Schubert, and W. Warta, Applied Physics Letters **89** (2006) 1.

[7] T. Trupke, R. A. Bardos, and M. D. Abbott, Applied Physics Letters **87** (2005) 184102.

[8] J. A. Giesecke, M. C. Schubert, B. Michl, F. Schindler, and W. Warta, Solar Energy Materials and Solar Cells **95** (2011) 1011.

[9] M. C. Schubert, H. Habenicht, and W. Warta, IEEE Journal of Photovoltaics **1** (2011) 168.

[10] D. Macdonald, J. Tan, and T. Trupke, Journal of Applied Physics **103** (2008) 073710.

[11] H. Habenicht, M. C. Schubert, and W. Warta, Journal of Applied Physics **108** (2010) 034909.

[12] S. Y. Lim, F. E. Rougieux, and D. Macdonald, Applied Physics Letters **103** (2013) 092105.

[13] T. Niewelt, S. Lim, J. Holtkamp, J. Schön, W. Warta, D. Macdonald, and M. C. Schubert, Solar Energy Materials and Solar Cells **131** (2014) 117.

[14] L. E. Mundt, M. C. Schubert, J. Schön, B. Michl, T. Niewelt, F. Schindler, and W. Warta, IEEE Journal of Photovoltaics **5** (2015) 1503.

[15] J. A. Giesecke, M. C. Schubert, F. Schindler, and W. Warta, IEEE Journal of Photovoltaics **5** (2015) 313.

[16] B. Michl, M. Rüdiger, J. Giesecke, M. Hermle, W. Warta, and M. C. Schubert, Solar Energy Materials & Solar Cells **98** (2012) 441.

[17] B. Michl, M. Kasemann, W. Warta, and M. C. Schubert, Physica Status Solidi RRL – Rapid Research Letters **7** (2013) 955.

[18] M. C. Schubert, H. Habenicht, M. J. Kerler, and W. Warta, Solid State Phenomena **156-8** (2010) 407.

[19] J. Giesecke, *Quantitative Recombination and Transport Properties in Silicon from Dynamic Luminescence*. (Springer, 2014).

[20] S. Bowden and R. Sinton, Journal of Applied Physics **102** (2007) 124501.

[21] S. Herlufsen, D. Hinken, M. Offer, J. Schmidt, and K. Bothe, IEEE Journal of Photovoltaics **3** (2013) 381.

[22] F. D. Heinz, J. Giesecke, L. E. Mundt, M. Kasemann, W. Warta, and M. C. Schubert, Journal of Applied Physics **118** (2015) 105706.

[23] Synopsys, (2012).

[24] M. C. Schubert, J. Schön, F. Schindler, W. Kwapil, A. Abdollahinia, B. Michl, S. Riepe, C. Schmid, M. Schumann, S. Meyer, and W. Warta, IEEE Journal of Photovoltaics **3** (2013) 1250.

[25] F. Schindler, B. Michl, J. Schön, W. Kwapil, W. Warta, and M. C. Schubert, IEEE Journal of Photovoltaics **4** (2014) 122.

[26] J. Schön, H. Habenicht, M. C. Schubert, and W. Warta, Journal of Applied Physics **109** (2011) 1.

[27] D. Walter, A. Liu, E. Franklin, D. Macdonald, B. Mitchell, and T. Trupke, in Proceedings of the 38th IEEE Photovoltaic Specialists Conference, Austin, USA (2012) 307.

[28] J. A. Giesecke and W. Warta, Applied Physics Letters **104** (2014) 082103.

[29] J. A. Giesecke, S. W. Glunz, and W. Warta, Journal of Applied Physics **113** (2013) 073706.

[30] F. Schindler, B. Michl, A. Kleiber, H. Steinkemper, J. Schon, W. Kwapil, P. Krenckel, S. Riepe, W. Warta, and M. C. Schubert, IEEE Journal of Photovoltaics **5** (2015) 499.

[31] F. Schindler, *Electrical Material Properties and Efficiency Limits of Compensated and Multicrystalline Silicon for Solar Cells*, PhD thesis, Albert-Ludwigs-Universität Freiburg im Breisgau (2015).

[32] S. P. Phang, H. C. Sio, and D. Macdonald, Applied Physics Letters **103** (2013) 192112.

LASER HYDROGENATION OF LASER DOPED AND GROOVED STRUCTURE

Sisi Wang, Ly Mai, Alison Wenham, Ziv Hameiri, Catherine Chan, Brett Hallam, Adeline Sugianto, Chee Mun Chong, Jingjia Ji, Zhengrong Shi and Stuart Wenham

School of Photovoltaics and Renewable Energy Engineering, University of New South Wales, Sydney, NSW 2052, Australia.

ABSTRACT: Both Buried Contact Solar Cells (BCSC) and Laser Doped Selective Emitter (LDSE) solar cells have shown great success in large-scale manufacturing. Recent studies show that plated contacts can be more durable than screen-printed contacts. A new concept of laser doping with grooving in a simultaneous step is introduced to form narrow grooves with heavily doped walls with the metal contact subsequently formed by plating. This process capitalizes on the benefits of both BCSC and LDSE cells. The width of the groove is only about three to five microns. The narrow width and corresponding steep walls of the groove make it possible to deposit the antireflection coating (ARC) after the laser grooving process such that the groove walls remain uncovered to allow nucleation of subsequent metal plating. Formation of laser-induced defects is minimized because thermal expansion mismatch between the ARC and silicon is avoided during the laser doping process. This structure leads to greatly enhanced adhesion of the plated contact due to it being buried underneath the silicon (Si) surface. Although the laser doping and grooving process avoids thermal mismatch between SiNx and Si, crystallographic defects are formed during the laser grooving process. Laser-activated hydrogenation is used to passivate the laser-induced defects formed during the laser–doping and grooving process. Through this. excellent results have been achieved in terms of improved carrier lifetime and implied V_{oc} values.

Keywords: Laser doping, Laser damage, Adhesion, Plating, Czochralski

1 INTRODUCTION

The University of New South Wales developed the Buried Contact Solar Cell (BCSC) in 1980s [1, 2]. Laser grooving of Si to form buried contact solar cells has previously shown a range of benefits in terms of device performance, use of copper (Cu) plated metal contacts, and suitability for a large scale manufacturing in the form of BP Solar's Saturn Technology. BCSC have shown better durability than screen printed solar cells [3]. However, the manufacturing process is complicated and with a high thermal budget with laser grooving followed by damage etching, groove diffusion and subsequent preparation for plating.

The Laser Doped Selective Emitter (LDSE) solar cell is another technology with plated contacts that has also been manufactured in large-scale [4]. In this method, a spin – on doping source is applied on to the dielectric layer, which performs as an antireflection coating and surface passivation layer, and more importantly, as a mask for the subsequent metal plating process. Laser scribing melts the silicon locally where metal contact will be formed to incorporate doping source into the Si. In this way, selective emitters are formed together with a self-aligned plating mask [5]. Despite the success of the conventional LDSE solar cells, one of their drawbacks has been shown to be laser-induced defects caused by thermal expansion mismatch between the Si and the overlying dielectric layer [6]. Another disadvantage has been shown to be relatively weak adhesion strength of the metal contacts, which may peel off during module production.

A new laser doping process that simultaneously forms grooves was developed to capitalize on the benefits of the BCSC solar cells but with the simplicity of the LDSE solar cells. Laser doping and grooving is used to form grooves and heavily dope the groove walls in a single step that minimizes the formation of laser induced defects in laser doped solar cells. The narrow width of the groove allows the dielectric layer to be deposited after the laser doping process. By applying the dielectric layer after the laser grooving step, thermal expansion mismatch between the Si and the dielectric layer is avoided to minimize laser induced defects. Since the metal contact is buried under the Si surface, the adhesion of metal contacts is also improved significantly compared to the standard LDSE solar cell. The laser grooving process also makes it possible to control the width and aspect ratio of the metal contacts.

Although the thermal expansion mismatch is avoided, there still appears to be laser damage, probably caused by the higher laser power required for the grooving process in comparison to simply melting it for laser-doping or perhaps defects formed by the rapid crystallization during cooling. Either way, the damage is likely to be crystallographic defects near the junction region. To passivate the laser damage, a laser hydrogenation tool is used to control the charge state of hydrogen atoms in the silicon to increase their mobility and reactivity [12] to passivate defects in the sample.

2 MINIMIZING LASER INDUCED DEFECTS

A Q-switch laser is used to form narrow grooves on industrial Czochralski (Cz) Si wafers while simultaneously heavily doping the groove walls to form selective emitter regions. A dielectric layer is subsequently deposited which only covers the top of the Si surface leaving the doped groove walls relatively exposed. Light induced plating (LIP) is used to nucleate and plate metal contacts in the grooves.

The conventional LDSE solar cells have localised heavily doped regions formed by simultaneously melting Si surface and doping the melted area by laser. The laser doping process is performed after the deposition of dielectric layer so that only the laser-doped Si is exposed in preparation for the self-aligned plating of the metal contacts to the heavily doped silicon. Despite the success of the LDSE solar cell in achieving numerous performance records on *p*-type Cz silicon, one of its drawbacks is the laser induced defects caused by thermal expansion mismatch between Si and the dielectric layer on top. This mismatch puts the Si surface under tension during the laser doping process, which creates defects in the Si due to the silicon being very weak under tension [6].

Laser doping with grooving makes it possible to avoid laser induced defects caused by thermal expansion mismatch. This is because the innovative process developed in this work is able to form grooves that are very narrow, therefore making it possible to deposit the dielectric layer after the laser grooving process. This is because the subsequently deposited dielectric only coats the exposed top surface of the Si while minimal dielectric deposits on the steep groove walls allowing nucleation of subsequent metal plating. By applying the dielectric layer after the laser grooving step, thermal mismatch is avoided as is also the laser induced defects.

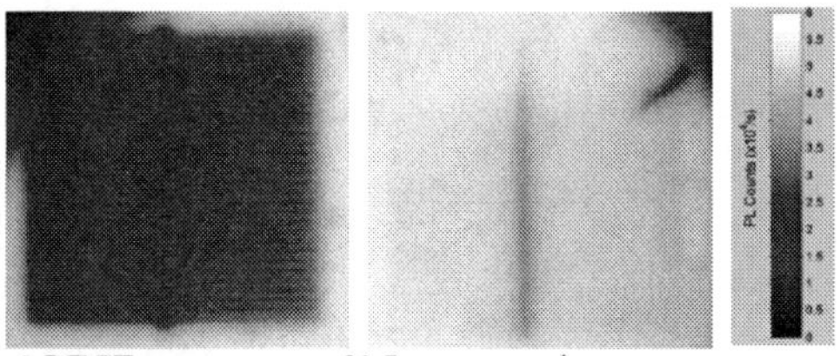

a) LDSE precursor **b)** Laser grooving

Figure 1: PL images using the same scale of a wafer where a) the laser doping process was done after SiN_x deposition and b) the laser grooving was done before the SiN_x deposition

Figure 1 presents photoluminescence (PL) images [7] of different regions of the same wafer before metal plating with screen printed aluminum (Al) on the rear, where Pattern A (**Figure 1 a**) is a standard LDSE cell precursor where the laser doping process was applied after silicon nitride (SiN_x) deposition. Pattern B (**Figure 1 b**) in comparison represents the new technology being reported in this paper whereby the laser doping with grooving is carried out before the SiN_x deposition. These PL images have been deconvolved to negate the effects of photon smearing and improve clarity by the method of Teal and Juhl [8, 9]. As can be seen in Pattern A, the laser doping process after dielectric layer deposition significantly degrades the PL response, probably due to formation of significant recombination through laser induced defects; Pattern B for the new technology shows that the absence of the dielectric during the laser doping process, reduces the laser induced damage with a minimal deterioration in the PL response and hence lifetime of the cell compared to the non-laser-processed areas. This result is perhaps surprisingly good given that the groove wall surfaces are exposed and unpassivated following the dielectric deposition, suggesting that the groove walls are sufficiently heavily doped to prevent the groove wall surfaces from contributing significantly to the device total dark saturation current. This is an important result given that these exposed surfaces will be metallised which will maintain the very high recombination velocity at this interface.

3 LASER DOPING WITH GROOVING

Figure 2 is a scanning electron microscope (SEM) image of a groove cross-section similar to that used in **Figure 1b**. The laser beam diameter used is 10 to 15 µm. It can be seen that the melted Si is pushed up to the sides and leaves only a small gap during re-solidification during the cool down process. The groove is 3 to 5 µm wide and 10 to 15 µm deep. The pyramids next to the melted region are clearly visible, which confirms that the laser induced stress and defects are kept to a minimum level.

Table 1 shows SEM and electron beam induced current (EBIC) images of four typical grooves. When the groove is very deep, although the heavy doping along the groove wall is continuous, there is not enough junction depth to electrically isolate the high recombination velocity surfaces within the grooves from the active regions of the cell. This will decrease cell voltage due to increase in dark saturation current. The second and third groove profiles are the preferred structures, as can be clearly seen that the widths of the grooves are very narrow and yet the junctions are quite deep within the groove walls as revealed by the EBIC scans. These deeper junctions have been shown to achieve significantly higher voltages by reducing the dark-saturation current. If too much silicon is lost from the molten region, the groove opening becomes too wide and again, the junction will be too shallow. The grooves with wide opening will also have additional SiNx deposited into the groove therefore necessitating some etching back of the SiNx to expose the silicon surface and allow nucleation for subsequent plating.

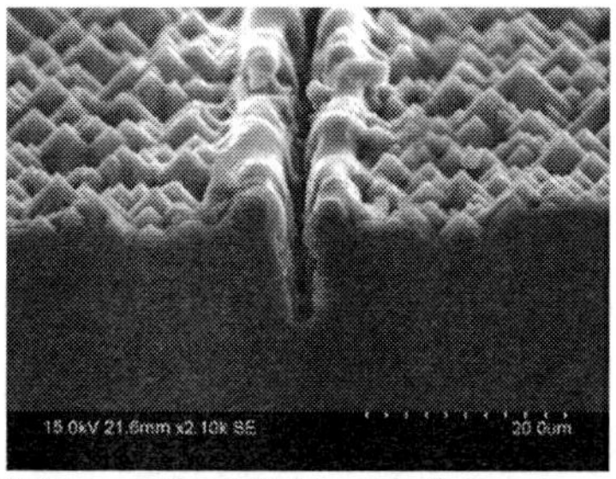

Figure 2 Cross section SEM image of a groove

Table 1 Four typical grooves with laser-doped walls formed with decreasing laser-power that produces grooves of correspondingly reduced depth.

SEM	EBIC
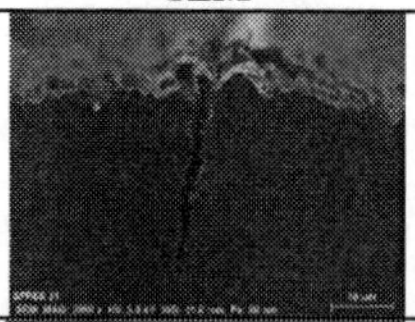	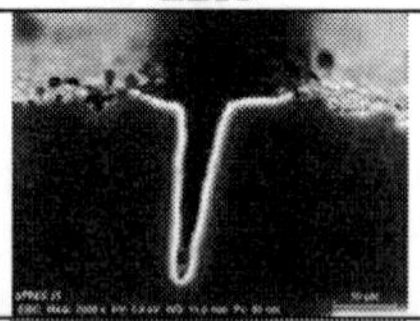
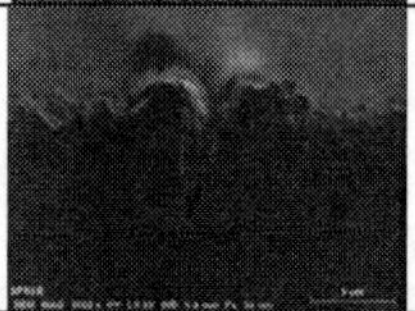	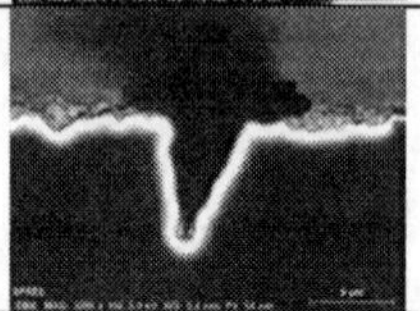
	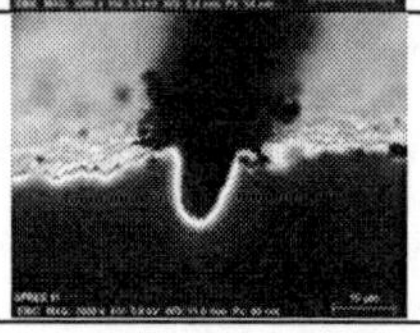

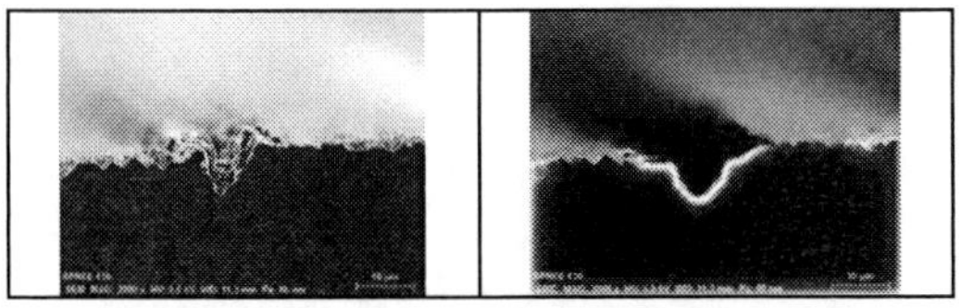

4 LIFETIME TEST STRUCTURES

Lifetime test structures were fabricated to investigate the damage caused by the laser grooving process. Four laser conditions were performed to compare with the non-laser-processed samples in the same batch. 1.6 Ωcm industrial p-type Cz wafers were used for the experiment. The recombination in the samples was characterized by measuring the effective lifetime as a function of excess minority carrier density using photoconductance-based lifetime tetser (WCT-120, Sinton Instruments) and analyzing the measured data using the generalized method [10].

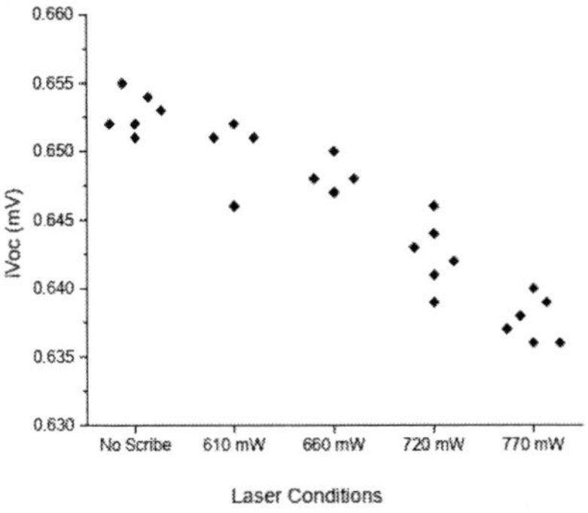

Figure 3 iV_{oc} as a function of laser grooving power with all samples using the same laser pattern

Figure 3 plots the implied open circuit voltage (iV_{oc}) at one-sun for different grooving conditions juxtaposed with samples with no laser processing to act as controls. There is only 5 to 8 mV degradation in iV_{oc} in the group using the lowest laser power for the grooving compared to the non-scribed group. This degradation in implied voltage is much lower than that caused by the LDSE process. [11] This confirms that the laser doping with grooving process has the potential to greatly reduce laser induced defects compared to standard LDSE cells. As the average laser power increases, more energy is delivered into the Si resulting in either increased damage or increased exposed unpassivated surface or both and therefore iV_{oc} reduces. Another interpretation could be that as groove depth increases with power, the increased volume of heavily laser-doped silicon in the groove walls could also be lowering iV_{oc} values.

5 LASER HYDROGENATION

Lifetime test structures were also used to test the ability to hydrogenate laser-induced defects in laser – doped and grooved cells. Four laser grooving conditions were performed to compare with the non-laser-processed samples in the same batch. Laser hydrogenation at 300 °C with 2.11×10^{19} $cm^{-2}s^{-1}$ photon flux illumination was used to passivate defects in these wafers. By using such high intensity illumination, the diffusivity and reactivity of interstitial hydrogen atoms were increased significantly because of its transition to the minority charge states [12]

. **Figure 4** shows that the laser hydrogenation improved wafer lifetime. The higher the laser grooving power, the more improvement was achieved by hydrogenation. This suggests that the low values of iV_{oc} of the high laser power group may be caused primarily by laser-induced defects, which were then able to be largely passivated by the hydrogenation process. For comparison, **Figure 5** presents PL images of a laser doped and grooved wafer with relatively deep grooves before and after laser hydrogenation. It is clear that the grooved regions were quite uniformly improved.

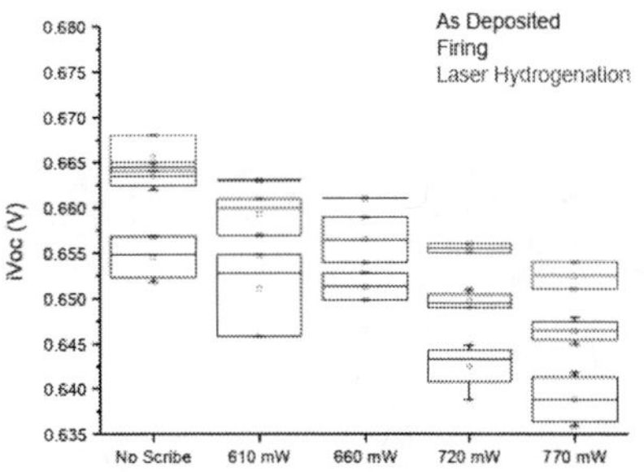

Figure 4 iV_{oc} of samples using different laser grooving conditions before and after hydrogenation

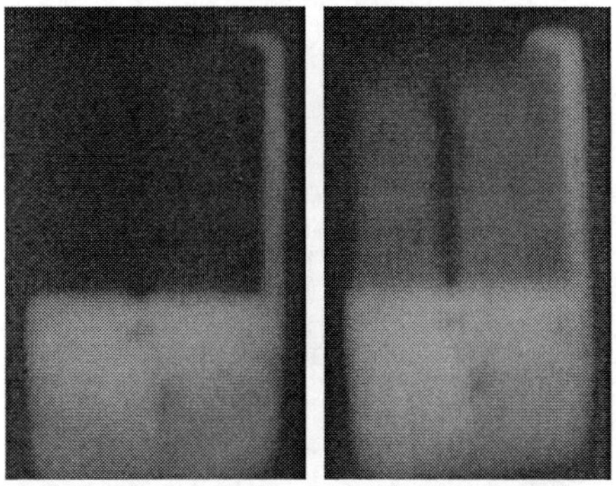

a) Before hydrogenation **b)** After hydrogenation

Figure 5: PL of a wafer before and after hydrogenation with relatively deep grooves that create increased damage

6 METAL CONTACT AND ADHESION

Light induced plating was performed to test the laser doped grooves' ability to nucleate metal plating and form the metal contacts. After laser doping with grooving, instead of having SiNx deposited on both sides of the wafer, only the front surface of the samples was passivated by SiN_x while the rear was screen printed with Al. **Figure 6** illustrates that Cu completely fills the groove as well as having a good aspect ratio. One advantage of this process is that it could be possible to control the width and aspect ratio of the metal contacts. Since only the groove walls are unpassivated after dielectric layer deposition, the width of the metal contacts could be controlled to be as narrow as the groove width (typically 3-5 microns) or much wider by changing the metal plating duration. Having narrow metal contacts would reduce shading loss significantly, but of course would increase resistive losses within the Cu unless the approach was combined with another metallisation scheme such as the smart – wire technology from Meyer Berger [13].

Another advantage of the laser doping with grooving method is improved adhesion of metal contacts. Since the metal contacts are buried under the Si surface instead of on top of it, it is much harder for the contacts to peel off from the Si. **Figure 6** shows that with this approach, the metal can be buried up to 15 µm deep beneath the Si top surface. **Figure 7** illustrates the bottom of a peeled metal finger. It is clear that chunks of Si were removed together with plated metal. This indicates that the Si is forming good and strong mechanical contact with the metal and the adhesion of the overall contact is excellent. Further testing on adhesion is being conducted and quantified for comparison with other more conventional metallization schemes such as screen-printing metallization.

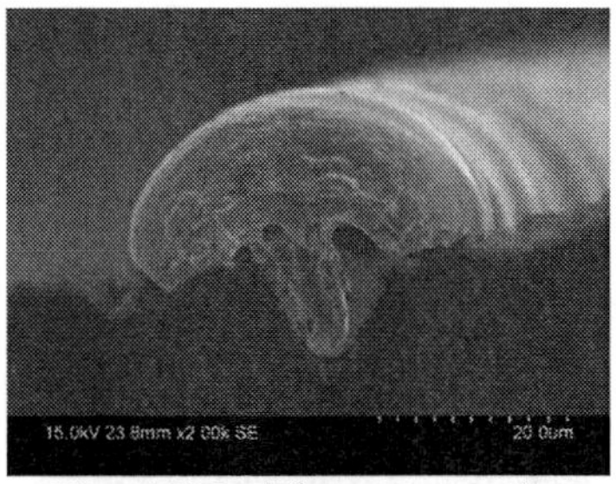

Figure 6 Metal finger cross-section

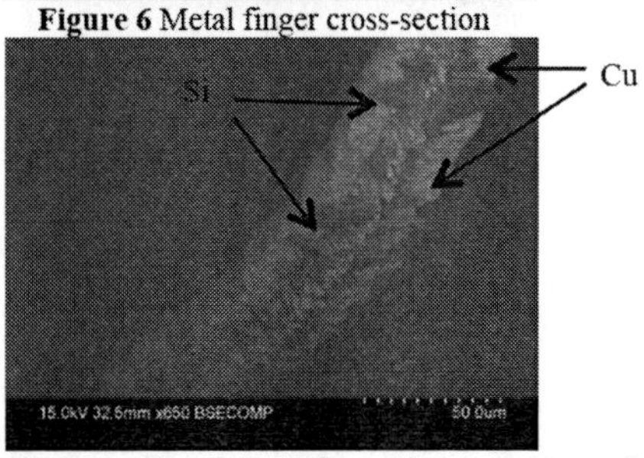

Figure 7 Bottom of peeled Cu finger showing pieces of Si removed together with metal

The adhesion strength of the metal contacts was tested using a stylus-based adhesion tester developed by Young [14]. A weighted stylus scans across the cell while the horizontal force required to cut off or dislodge the contacts is measured. The adhesion strengths of laser doped and grooved cells, commercial screen printed cells and LDSE cells are compared. The adhesion strength of each metal finger is normalized to that of the averaged screen printed contact. **Figure 8** plots the normalized data. It is observed that the adhesion strength of laser doped and grooved contacts is more than twice that of the laser doped cell and is slightly superior to the screen printed metal contact.

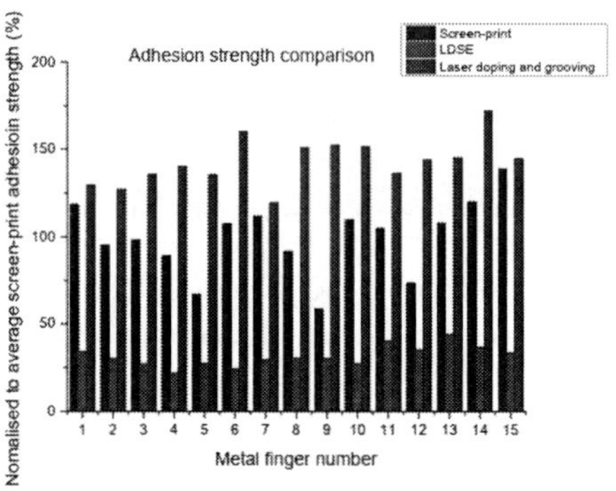

Figure 8: Normalised adhesion strengths of typical screen-printed cells, LDSE cells and laser doped and grooved cells determined using a stylus-based tester [14]

7 CONCLUSION

Laser conditions are optimized to form grooves with heavily doped walls in a simultaneous step. Early experimentation has demonstrated that the concept works in terms of reliably forming the doped grooves and plated metallization with improved adhesion, high voltages and high current densities. However further optimization is required to reduce the resistive losses to facilitate high fill factors and high efficiencies. The innovative laser doping simultaneously with grooving in a way that provides the ability to deposit the dielectric layer after the laser process not only greatly simplifies the technology, but also makes it possible to avoid laser induced defects due to thermal mismatch. Laser hydrogenation was shown to further improve the carrier lifetime in laser doped and grooved samples, by passivating crystallographic defects caused by the laser grooving process. In addition, the improved adhesion from the buried metal contacts in the grooves appears to make the metal contacts and hence the solar cells more robust and durable.

8 FURTHER WORK

The laser doping with grooving process has been recently integrated into device fabrication with impressive results in terms of device voltages and currents and end-of-processing minority carrier lifetimes. At this early stage it is perhaps not surprising that parasitic resistive losses are high, with future work and optimization expected to minimize such losses and facilitate the achievement of high performance devices.

A symmetrical PERC structure with similar p-type laser grooves on the rear has considerable appeal and also will be a focus of future work.

The laser hydrogenation process is being optimized and incorporated into the cell fabrication process.

8 ACKNOWLEDGMENT

The funding support of the Australian Research Council (ARC) and the Australian Renewable Energy Agency (ARENA) for aspects of this work are acknowledged with appreciation. The funding of laser-based work from the Institution of Electronics and Technology (IET) is also gratefully acknowledged.

9 REFERENCE

[1] S. Wenham, "Buried-contact silicon solar cells,"*Progress in Photovoltaics: Research and Applications,* vol. 1, pp. 3-10, 1993.

[2] C. M. Chong, S. R. Wenham, and M. A. Green, "High-efficiency, laser grooved, buried contact silicon solar cells," *Applied Physics Letters,* vol. 52, pp. 407-409, 1988.

[3] F. C. M. Alonso-Abella, A. Alonso and D. González, "TOLEDO PV PLANT 1 MWp – 20 YEARS OF OPERATION," presented at the 29th European Photovoltaic Solar Energy Conference and Exhibition, 2014.

[4] Z. Wang, P. Han, H. Lu, H. Qian, L. Chen, Q. Meng, *et al.*, "Advanced PERC and PERL production cells with 20.3% record efficiency for standard commercial p-type silicon wafers," *Progress in Photovoltaics: Research and Applications,* vol. 20, pp. 260-268, 2012.

[5] S. R. Wenham and M. A. Green, "Self aligning method for forming a selective emitter and metallization in a solar cell," ed: Google Patents, 2002.

[6] Z. Hameiri, L. Mai, T. Puzzer, and S. R. Wenham, "Influence of laser power on the properties of laser doped solar cells," *Solar Energy Materials and Solar Cells,* vol. 95, pp. 1085-1094, 4// 2011.

[7] T. Trupke, R. A. Bardos, M. C. Schubert, and W. Warta, "Photoluminescence imaging of silicon wafers," *Applied Physics Letters,* vol. 89, p. 044107, 2006.

[8] A. Teal and M. Juhl, "Correcting the inherent distortion in luminescence images of silicon solar cells," in *Photovoltaic Specialist Conference (PVSC), 2015 IEEE 42nd,* 2015, pp. 1-5.

[9] M. K. J. David N. R. Payne, Michael E. Pollard, Anthony Teal and Darren M. Bagnall, "Evaluating the Accuracy of Point Spread Function Deconvolutions Applied to Luminescence Images," presented at the 43rd IEEE Photovoltaic Specialists Conference, Portland OR, 2016.

[10] H. Nagel, C. Berge, and A. G. Aberle, "Generalized analysis of quasi-steady-state and quasi-transient measurements of carrier lifetimes in semiconductors," *Journal of Applied Physics,* vol. 86, pp. 6218-6221, 1999.

[11] Z. Hameiri and Z. Hameiri, "Laser-doped selective emitter and local back surface field solar cells with rear passivation," S. Wenham and A. Sproul, Eds., ed: University of New South Wales. Photovoltaics & Renewable Energy Engineering, 2010.

[12] P. Hamer, S. Wang, B. Hallam, S. Wenham, C. M. Chong, A. Wenham, *et al.*, "Laser illumination for manipulation of hydrogen charge states in silicon solar cells," *physica status solidi (RRL) – Rapid Research Letters,* vol. 9, pp. 111-114, 2015.

[13] Y. Yao, P. Papet, J. Hermans, S. T, derstr, H. Mehlich, *et al.*, "Module integration of solar cells with diverse metallization schemes enabled by SmartWire Connection Technology," in *Photovoltaic Specialist Conference (PVSC), 2015 IEEE 42nd,* 2015, pp. 1-5.

[14] T. L. Young, K. Hee, A. J. Lennon, R. J. Egan, O. Wilkie, and Y. Yao, "Design and characterization of an adhesion strength tester for evaluating metal contacts on silicon solar cells," in *Photovoltaic Specialist Conference (PVSC), 2014 IEEE 40th,* 2014, pp. 2550-2553.

HIGH EFFICIENCY, INDUSTRIALLY-RELEVANT N-CZ SI PV VIA PROCESS-TOLERANT WAFERS AND TUNNELING PASSIVATED CONTACTS

Benjamin G. Lee, San Theingi, Vincenzo LaSalvia, Jeffery Aguiar, William Nemeth, Matthew R. Page, David L. Young, Paul Stradins
National Renewable Energy Laboratory, Golden Colorado 80401 USA

ABSTRACT: High lifetime, process-tolerant wafers and low-recombination, passivated contacts for both electrons and holes, can enable high efficiency and industrially-relevant Si PV with n-type Cz material. We develop and apply an O-cluster dissolution treatment to make n-Cz wafers immune to thermal processing, maintaining high lifetime >3 ms. For low-recombination contacts, we advance tunneling passivated contacts (TPCs) based on a thin tunneling SiO_2 and doped n- or p-type poly-Si. n-type TPC that are in-situ P-doped during PECVD growth reach implied V_{oc} = 735 mV, while the performance of the corresponding p-type B-doped TPC is at 700 mV. We further develop ion-implantation doping to facilitate patterning of n- and p-type TPCs for interdigitated back *passivated* contact cells [1]; implied V_{oc} of 730 and 694 mV are achieved for ion-implanted n- and p-type TPCs, respectively. Finally, we produce an IBC cell with passivated contacts with V_{oc} = 693 mV, J_{sc} = 36.4 mA/cm^2, FF = 69.1% and efficiency of 17.4%.

Keywords: tunneling passivated contacts, tabula rasa, ion implantation, interdigitated back contact, silicon solar cell

1 INTRODUCTION

We investigate n-type Cz silicon solar cells with tunneling passivated contacts (Fig. 1), aiming for high efficiency devices that could be achieved by industry. We make advances to address 3 key challenges: maintaining high bulk lifetime in the Cz wafer through high temperature processing (e.g. B diffusion); development of n- and p-type tunneling passivated contacts (TPCs); dopant patterning to facilitate interdigitated back contact (IBC) cells using both n- and p-type interdigitated rear TPCs.

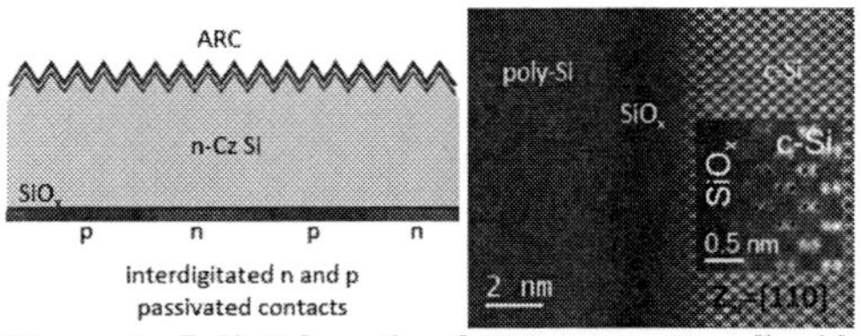

Figure 1: (left) Schematic of IBC n-type Cz cell with interdigitated n and p-type tunneling passivated contacts (TPCs). (right) HAADF STEM cross-section of a TPC, showing the poly-Si and thin tunneling SiO_x layer.

Significant **bulk lifetime degradation** is frequently observed in n-Cz wafers after lengthy high temperature processes (e.g. B drive-in). Typical n-Cz wafers have oxygen clusters that will grow during high temperature processing. Metals such as Fe and other impurities strongly attach (~0.8 eV) to O precipitates [2], making the degradation persistent and difficult to recover by gettering. These are highly active recombination centers killing the bulk lifetime; we see lifetime dropping from ~4 ms to <100 µs. To overcome this obstacle, we adapt a Tabula Rasa (TR) thermal treatment known from the IC industry [3] to dissolve O clusters prior to any diffusion processes (Fig. 2). This allows a high lifetime to be maintained in the wafers after thermal processing. The TR treatment also facilitates the gettering of interstitial impurities, since these no longer cluster at O precipitates.

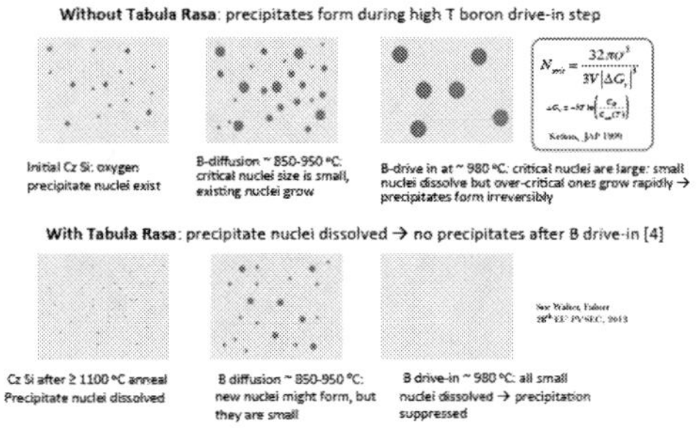

Figure 2: Explanation of how the Tabula Rasa improves n-Cz bulk wafer lifetime.

Tunneling passivated contacts (TPCs) are being pioneered by a number of groups [5-10]; our approach uses a thin 1.5 nm SiO_2 thermally grown at 700 °C and PECVD-deposited a-Si:H, which is crystallized into poly-Si at 850 °C [9,10]. We develop both n- and p-type TPCs, and investigate why good p-type TPCs are more difficult to achieve [8]. Furthermore, we test both (standard) beamline and Plasma Immersion Ion Implantation (PIII) [11] to implant various species for patterned n- and p-type passivated contacts that could be used in IBC cells.

2 HIGH LIFETIME WITH TABULA RASA TREATMENT

Tabula Rasa treatments successfully dissolved O-clusters, eliminating precipitate rings seen in untreated wafers (Fig. 3) and resulting in bright PL and high bulk lifetime >3 ms [12]. Moreover, the high bulk lifetime is maintained after cell processing steps, enabling high implied V_{oc} of 717 mV after B-emitter formation, and 724 mV for cell-like device with B-emitter and rear n-type TPC. Both n- and p-type in-situ doped (during PECVD deposition) TPCs were formed on n-Cz wafers and shown to be compatible with Tabula Rasa, achieving implied V_{oc} of 735 and 700 mV, respectively.

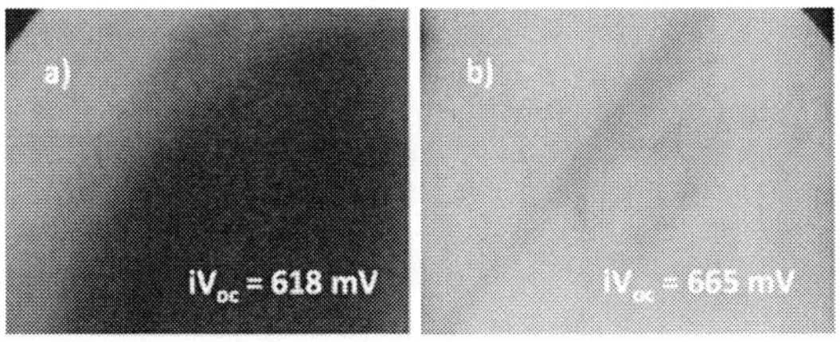

Figure 3: PL maps of n-Cz wafers after B-diffusion and deep drive in. Tabula Rasa treated sample (b) is bright and uniform, while non-treated wafer (a) is darker with O-precipitate rings. The iV_{oc} values from Sinton measurements are more than 40 mV higher with TR treatments (surface passivation from post-diffusion oxide only). Further refinements and proper surface passivation (using ALD Al_2O_3) brings the iV_{oc} to ~ 717 mV in symmetric B-emitter test structures.

3 Ion Implantation Doping of Passivated Contacts

Symmetric lifetime test structures are used to quantify the passivation quality of the TPCs. These test samples have either an n-type or a p-type TPC on both sides of the n-Cz wafer piece. Previously we have obtained implied V_{oc} of up to 735 mV and 700 mV respectively for in-situ doped n-type (P-doped) and p-type (B-doped) symmetric TPC test samples [1, 12, 13]. We were also able to achieve similarly good iV_{oc} results using ion implantation, with 730 and 694 mV for n- and p-type TPCs respectively (Fig. 4).

Both beamline implantation and Plasma Immersion Ion Implatation (PIII) are found to give good passivation quality. We investigate PIII as an alternative to beamline implanters because of several potential advantages. PIII has lower costs and CapEx, as compared with beamline. Additionally, PIII can have excellent throughput, since it can provide high doses and deposition over large areas. Finally, PIII is well-suited for implanting into thin poly-Si layers, since it can implant at low energy (several keV or less), while still maintaining high current (dose), which is a problem with beamline.

P and B were implanted with a beamline system, while PH_3, B_2H_6 and BF_3 were done with PIII. iV_{oc} obtained from ion implantation were quite similar to in-situ doped results, except for the poor performance of the BF_3 implant (Fig. 4). All samples underwent the Tabula Rasa treatment, and therefore have high bulk lifetime > 3 ms.

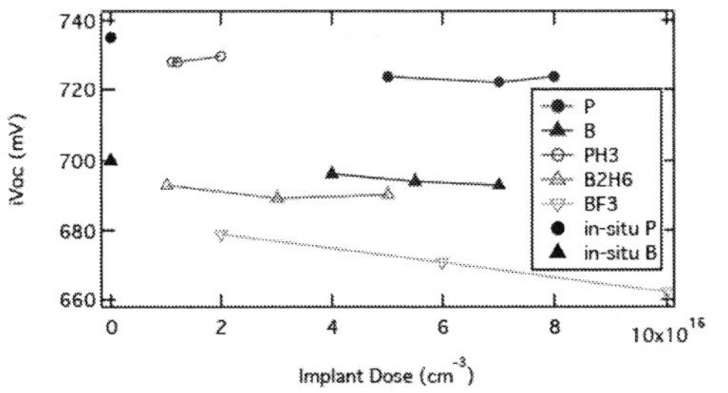

Figure 4: implied V_{oc} of symmetric n-Cz test samples, showing high V_{oc} potential of n-type and p-type tunneling passivated contacts (TPCs), doped either in-situ during PECVD deposition or ion-implanted.

We determine the contact resistivity $\rho_{contact}$ of the n-type TPC (through the tunneling oxide) is ~30 mΩ-cm^2, which is sufficiently low to not adversely affect cell fill factor. Contact resistivity was determined by the transmission line method (TLM). The contact resistivity of the p-type TPC is harder to determine since there is a p-n junction, but measurements of contact resistivity with this TPC on p-type wafers gives similar results.

4 Cell Device Fabrication

4 cm^2 area cell devices were fabricated with n-Cz wafer, employing the Tabula Rasa treatment. We fabricated 2 different types of cells. First, cells with B front emitter and rear n-type TPC. Secondly, IBC cells with interdigitated p-type and n-type TPCs on the rear. The Tabula Rasa treatment is applied to the wafers prior to the cell processing.

For the **cells with B front emitter**, the deep diffused emitter is formed in three steps: 1) B_2O_3 layer deposition at 850°C (BBr_3 + O_2); 2) Shallow in-diffusion of B at 950°C (N_2); and 3) Deep drive-in of B and oxidation at 980°C for 4.5 h (N_2 + O_2 (full or part time)). This results in a >1 µm deep doping profile with sheet resistance R_{sh} ~ 130 Ω/sq. and surface [B] ~10^{19}cm^{-3}. This diffused B-emitter has quite good performance – in symmetric n-Cz test samples after surface passivation with Al_2O_3/SiN_x we get iV_{oc} = 717 mV. 2 x 2 cm^2 cells are defined by a concentrated KOH mesa-etch of the front and simultaneous planarization of the back of the cell. A ~1.5 nm thick tunneling SiO_x is grown thermally at 700 °C. A 50 nm thick n+ a-Si:H layer is PECVD-deposited onto the tunneling SiO_x. An anneal at ~850 °C crystallizes a-Si:H into n+ poly-Si. The whole front of the wafer including the B-emitter cell mesas is passivated by a Al_2O_3 (ALD)/SiN_x (PECVD) layer stack and FGA at 400 °C. Cells are finished with photolithographically-defined front grid and full area rear metallization (with evaporated Ag).

For the **IBC cells**, the first step after the Tabula Rasa treatment is the growth of the thin tunneling SiO_x. Then, a 50 nm thick intrinsic a-Si:H layer is PECVD-deposited onto the tunneling SiO_x. Next, masked ion-implantation is performed into the a-Si:H, to define n-type and p-type regions. N-regions are 270 µm wide, p-regions are 630 µm, with undoped gap of 50 µm in between, and period of 1 mm for the rear IBC pattern. Currently the masking is done with photoresist patterned by photolithography. After implantation, the dopants are driven into the a-Si:H layer by annealing at ~850 °C, which also crystallizes the a-Si:H. Next, 10 nm Al_2O_3 is deposited by ALD on both sides of the cell, and ~75 nm SiN_x is deposited on the cell's front side as an anti-reflection coating. The surface passivation from the Al_2O_3 is activated with a 20 min anneal at 400 °C in forming gas. Finally, the rear metal contacts (Ag or Al) are deposited, patterned via photolithography.

5 Cell Results

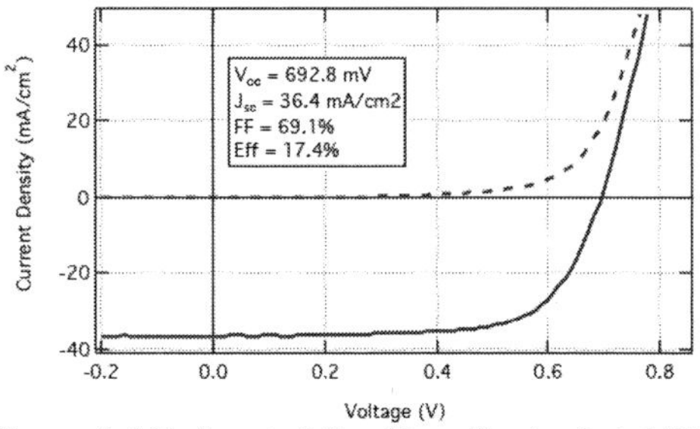

Figure 5: I-V characteristic of beamline implanted IBC cell with passivated contacts.

Front B-emitter cells were fabricated that had V_{oc} = 693 mV, J_{sc} = 39.6 mA/cm^2, FF = 78.4% and 21.5% efficiency [10]. By comparison, the IBC cells had V_{oc} = 693 mV, J_{sc} = 36.4 mA/cm^2, FF = 69.1% and 17.4% efficiency. The maintenance of high V_{oc} in the IBC cells is a good sign that the interdigitated passivated contacts can have excellent passivation. The lower J_{sc} in the case of the IBC cells is because those cells were made with planar Si wafers. This is instead of wafers with textured front surface, as in the case of the front B-emitter cells. This increases the reflectivity (and decreases light trapping), leading to lower photocurrent. The low FF in the case of the IBC cells is an issue that we are still working to understand and resolve. One possibility is that the p-type passivated contact has higher contact resistivity to the n-Cz wafer than anticipated.

6 Conclusions

A process-tolerant method for making low-cost, high-efficiency n-Cz Si solar cells was outlined and demonstrated. Tabula Rasa treatments successfully dissolved O-clusters, giving high bulk lifetime that is maintained after cell processing steps, enabling high implied V_{oc} of 717 mV after B-emitter formation. Both n- and p-type in-situ doped (during PECVD deposition) TPCs were formed on n-Cz wafers and shown to be compatible with Tabula Rasa, achieving implied V_{oc} of 735 and 700 mV, respectively. Ion implantation was also able to achieve similarly good iV_{oc} results, with 730 and 694 mV for n- and p-type TPCs respectively. Contact resistivity $\rho_{contact}$ of the n-type TPC (through the tunneling oxide) is ~30 mΩ-cm^2 for both in-situ and implantation doping, which is sufficiently low not to adversely affect cell fill factor. 4 cm^2 area cell devices were fabricated with n-Cz wafer, employing the Tabula Rasa treatment. Front B-emitter cells were fabricated that had V_{oc} = 693 mV, J_{sc} = 39.6 mA/cm^2, FF = 78.4% and 21.5% efficiency. By comparison, the IBC cells we made had V_{oc} = 693 mV, J_{sc} = 36.4 mA/cm^2, FF = 69.1% and 17.4% efficiency. Further work will be done, particularly to improve the fill factor of the IBC devices. The maintenance of high V_{oc} in the IBC cells is a good sign that the interdigitated passivated contacts can have excellent passivation.

7 Acknowledgements

Funding for this work was provided by US Department of Energy through EERE contracts SETP DE-EE00025783 and SETP DE-EE0006336 (FPACE-II) and under Contract No. DE-AC36-08GO28308. The U.S. Government retains and the publisher, by accepting the article for publication, acknowledges that the U.S. Government retains a nonexclusive, paid up, irrevocable, worldwide license to publish or reproduce the published form of this work, or allow others to do so, for U.S. Government purposes.

References

[1] Young, D.L., W. Nemeth, V. LaSalvia, R. Reedy, S. Essig, N. Bateman, and P. Stradins, "Interdigitated Back Passivated Contact (IBPC) Solar Cells formed by Ion Implantation," Journal of Photovoltaics, vol. 6, p. 41, 2016.

[2] P. Zhang, H. Vainola, A. A. Istratov, and E. R. Weber, "The thermal stability of iron precipitates in silicon after internal gettering," Physica B: Condensed Matter, 2003.

[3] D. C. Walter, B. Lim, K. Bothe, V. V. Voronkov, R. Falster, and J. Schmidt, "Effect of rapid thermal annealing on recombination centres in boron-doped Czochralski-grown silicon," APL, vol. 104, no. 4, p. 042111, 2014.

[4] J.D. Murphy, R.E. McGuire, K. Bothe, V.V. Voronkov, R.J. Falster, "Minority carrier lifetime in silicon photovoltaics: The effect of oxygen precipitation", Sol. Energy Mater. Sol. Cells, 120, pp. 402 - 411, 2014.

[5] E. Yablonovitch, T. Gmitter, R. M. Swanson, and Y. H. Kwark, Appl. Phys. Lett. vol. 47 (11), p. 1211, 1985.

[6] J.Y. Gan, R.M. Swanson, Proceedings of the 22nd IEEE Photovoltaic Specialists Conference, Kissimmee, Florida, USA, p. 245, 1990.

[7] F. Feldmann, M. Simon, M. Bivour, C. Reichel, M. Hermle, and S. W. Glunz, "Carrier-selective contacts for Si solar cells," Appl. Phys. Lett., vol. 104, no. 18, p. 181105, 2014.

[8] R. Peibst, U. Romer, Y. Larionova, H. Schulte-Huxel, T. Ohrdes, M. Haberle, B. Lim, J. Krugener, D. Stichtenoth, and T. Wutherich, "Building blocks for back-junction back-contacted cells and modules with ion-implanted poly-Si junctions," IEEE PVSC 2014, pp. 0852–0856, 2014.

[9] W. Nemeth, D. L. Young, H.-C. Yuan, V. LaSalvia, A. G. Norman, M. Page, B. G. Lee, and P. Stradins, "Low temperature Si/SiOx/pc-Si passivated contacts to n-type Si solar cells," presented at the Photovoltaic Specialist Conference (PVSC), IEEE PVSC 2014, pp. 3448–3452, 2014.

[10] P. Stradins, S. Essig, W. Nemeth, et al., "Passivated Tunneling Contacts to N-Type Wafer Silicon and Their Implementation into High Performance Solar Cells," WCPEC-6, 2014.

[11] Chu, P.K., Plasma Phys. Control. Fusion, vol. 45, p. 555, 2003.

[12] V. LaSalvia, W. Nemeth, M.R. Page, E.L. Warren, et al., "Implementation of tunneling passivated contacts into industrially relevant n-Cz Si solar cells," IEEE PVSC 2015.

[13] B. Nemeth, D.L. Young, M.R. Page, V. LaSalvia, S. Johnston, R. Reedy and P. Stradins, "Polycrystalline silicon passivated tunneling contacts for high efficiency silicon solar cells," Journal of Materials Research, vol. 31, p. 671, 2016.

A ROUTE TOWARDS HIGH EFFICIENCY N-TYPE PERT SOLAR CELLS

Weiyuan DUAN*, Shengzhao YUAN, Yu SHENG, Wenhao CAI, Yifeng CHEN, Yang YANG, Pietro P. ALTERMATT, Zhiqiang FENG and Pierre J. VERLINDEN
State Key Laboratory of Photovoltaic Science and Technology, Trina Solar, No. 2 Trina Road, Trina PV Park, New District, Changzhou, Jiangsu Province, China 213031
*Corresponding author: phone: (+86) 186 6115 3520, e-mail: weiyuan.duan@trinasolar.com

ABSTRACT: In this work, a route towards high efficiency n-type solar cells with a Passivated Emitter and Rear Totally Diffused (PERT) structure done by a combination of characterization and modeling is presented. Low cost industrial processes such as tube diffusion and plating have been used for the fabrication. Optimization result on emitter profile shows that shallower junction depth emitter with 93 ohm/sq sheet resistance brings a 0.23% absolute efficiency gain. A champion efficiency of 22.40% has been achieved on 6 inch CZ wafers after further optimization of front finger and rear pitch. The semicondutor device simulation software Sentaurus is used to reveal that injection-dependent lifetime due to shallow defects enhance the fill factor in n-type solar cells.
Keywords: n-PERT, diffusion, plating, Sentaurus.

1 INTRODUCTION

N-type silicon solar cells have been receiving growing attention due to their higher bulk lifetime and lower sensitivity toward light induced degradation compared to p-type silicon solar cells. In particular, n-type PERT (Passivated Emitter and Rear Totally Diffused) solar cells with an improved rear side passivation have produced noticeable results in recent times. In 2014, Fraunhofer ISE has reported a solar cell efficiency of 22.7% for a 2x2 cm^2 n-PERT cell fabricated using ion-implantation [1]. IMEC fabricated a 22.5% n-PERT cell featuring Ni/Cu/Ag front contacts and a rear emitter on a 6-inch wafer [2]. In this paper, we introduce a high efficiency and low-cost front emitter n-PERT cell with an optimized boron diffused profile and plated metallization.

Reports have indicated that current commercial printing pastes are difficult to apply on boron diffused emitters in mass production as there is a tendency for local spiking of aluminum [3,4], or there is a tendency for high contact resistance in the range of several mΩcm^2 [5,6] caused by a contact-inhibiting glass layer in the contact area. Also, it appears that for front emitter n-PERT solar cells, optimizing diffusion profiles is critical to achieve high efficiency. A good balance between the recombination loss in the front diffusion layer and the resistive loss of lateral currents flowing along the emitter and out to fingers through contacts is necessary.

One way to reduce the recombination loss under the contacts and to reduce the risk of aluminum spiking is to narrow the front metal fingers using plating instead of screen printing. Plating has been demonstrated as a promising metallization technology to achieve good contacts with strong adhesion even for a surface with low doping concentration [7]. When fingers become narrower, an emitter with higher sheet resistivity can be applied because the resistive losses along the emitter are reduced for a smaller finger pitch. In this paper, n-PERT cells are fabricated with Ni/Cu/Ag plated fingers on two different types of boron diffused emitters. Results show that cells with a shallower emitter have a higher J_{sc}, while V_{oc} and R_s are nearly the same. A champion efficiency of 22.40% has been achieved on a 6-inch Cz wafer.

2 EXPERIMENT

The n-type PERT cells in this work have a front junction structure, so they are not as sensitive to bulk lifetime as rear-junction cells. They were fabricated on 5-inch Cz n-type wafers in the begining and then the process was transferred to 6-inch, 180μm-thick wafers with a resistivity of 1 – 5 Ωcm. The process flow mainly includes texturing, boron diffusion, single-side polishing, phosphorous diffusion, passivation, laser process, PVD and plating. A schematic of the n-type PERT cell is shown in Fig.1. The as-fabricated cell structure has textured pyramids, a homogenous emitter, stack dielectrics and plated fingers on the front side, and a polished surface, a homogenous BSF, stack dielectrics and PVD metal films on the rear side.

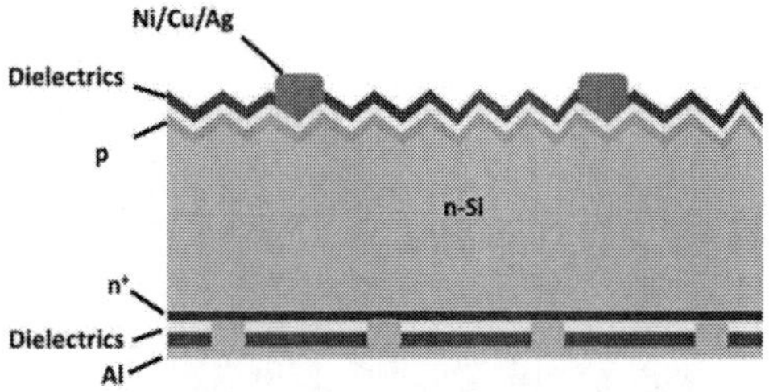

Figure 1: Cross-sectional schematics of an n-type PERT cell

In the boron diffusion process, gas flow, temperature and drive-in time are varied to form two types of emitters, A and B. Test wafers with both sides' surface textured are used to measure J_0 values of passivated and unpassivated surfaces. The dopant profile is measured using the electrochemical capacitance-voltage (ECV) technique. Surface doping concentration (N_s), sheet resistivity (R_{sheet}), junction depth and J_0 values are shown in Table I.

3 RESULTS AND ANALYSIS

3.1 n-PERT cell performance

The IV parameters of two cells were measured at Fraunhofer ISE CallLab. Table II shows the performance parameters of the two groups (the champion cell in each group). In both groups, V_{oc} is about 683 mV and R_s is smaller than 0.5 $\Omega.cm^2$ which indicates that boron doped emitters in both groups have good surface passivation and good contact resistivity. It can be inferred that the surface doping concentration as low as 10^{19} cm^{-3} is sufficient for plated fingers to obtain a low contact resistivity, which was measured to be 7×10^{-4} $\Omega.cm^2$ using the the TLM method.

Table I: Parameters of different emitters, measured by CalLab.

	A	B
N_s (cm^{-3})	1.06×10^{19}	1.08×10^{19}
Junction depth (μm)	1.1	2.0
R_{sheet} (Ω/□)	93	54
$J_{0\_passivated}$ (fA/cm^2)	28.0	32.7
$J_{0\_unpassivated}$ (fA/cm^2)	3000	2500

Table II: IV parameters of cells with Emitter A and B

	A	B
V_{oc} (mV)	683.7	683.0
J_{sc} (mA/cm^2)	40.34	39.89
FF (%)	80.77	80.92
Eff. (%)	22.28	22.05
R_s ($\Omega.cm^2$)	0.45	0.41

Fig.2 shows a SEM image of a plated contact finger. It has three layers of metal, where Ni acts as the contact material and diffusion barrier, Cu is the main electric conduction layer and Ag is used for conduction and later protection. Fine line fingers (< 35 mm) has been achieved after plating, which can reduce the shading loss significantly.

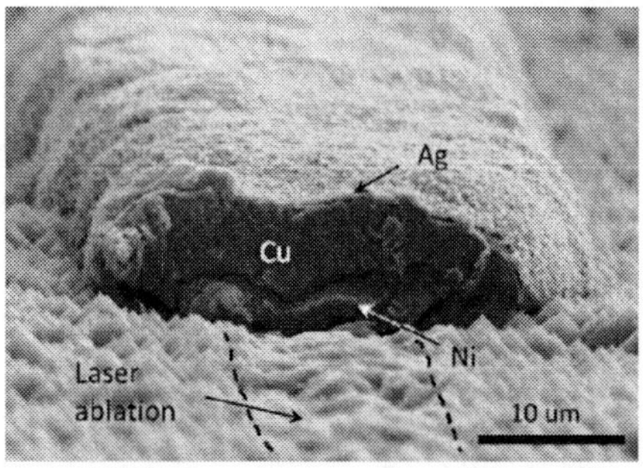

Figure 2: SEM image of cross-section of a plated finger.

Similar V_{oc} values were found for cells with emitter A and emitter B, as shown in Table II. This is mostly due to the similar total J_0, as indicated by the measurements in Table I. The ratio of unpassivated area formed by laser ablation is less than 1%. Fig. 3 is an example of a simulated total J_0, using EDNA and Gaussian profiles. The surface recombination velocity is extracted from the measured J_0 of boron diffused profiles with different N_s. It can be seen that when the ratio of unpassivated contact area to the total contact area is 0.79%, the simulated total J_0 changes very little across a wide range of junction depth from 0.8 to 2.0 um. This simulation matches well with our experimental results.

The higher efficiency for emitter A is mainly due to a higher J_{sc}. This is due to a higher collection efficiency of carriers generated in the shallower emitter, which can be seen in the IQE and reflectance curves shown in Fig. 4. The two cells show similar reflectance in the whole wavelength range, however, the cell with emitter A shows a higher IQE in the short wavelength compared to the cell having the deeper emitter B. Since the sheet resistivity of emitter A is nearly double of that of emitter B, the FF is a little higher for emitter B. Nevertheless, the current gain for emitter A dominates the efficiency gain.

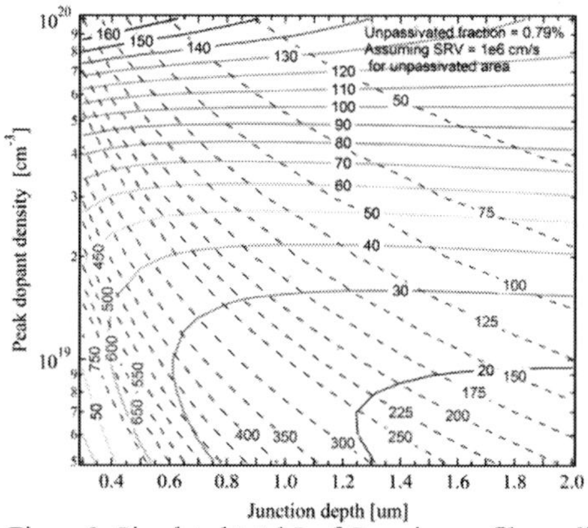

Figure 3: Simulated total J_0 of Gaussian profiles, solid lines in rainbow color are J_0 in units of fA/cm^2, and blue dashed lines are the sheet resistivity in units of Ω/sq.

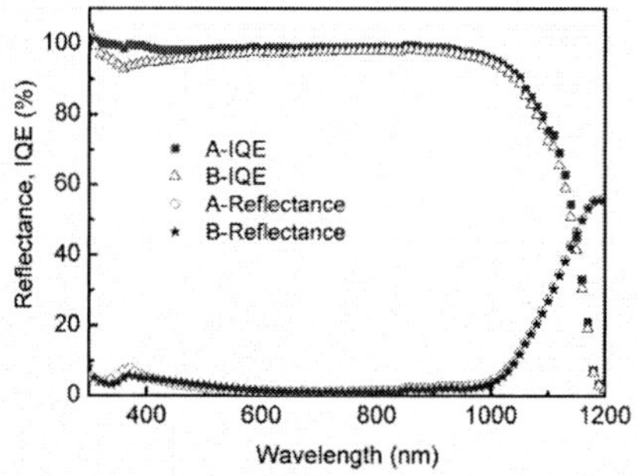

Figure 4: IQE and reflectance curves of n-PERT plated cells with a shallow (A) and deep (B) boron emitters.

3.2 Champion efficiency for 6-inch cells

Based on the above optimization of the emitter, the front finger number and rear pitch were futher optimized. Then the high efficiency process was transfered to 6-inch wafers. Fig. 5 shows the IV curve of the champion cell using emitter A with an optimized metallization structure, measured with the new Sinton FCT450 flash tester. This cell has a higher J_{sc} and V_{oc} compared to the former cells due to metallization structure optimization.

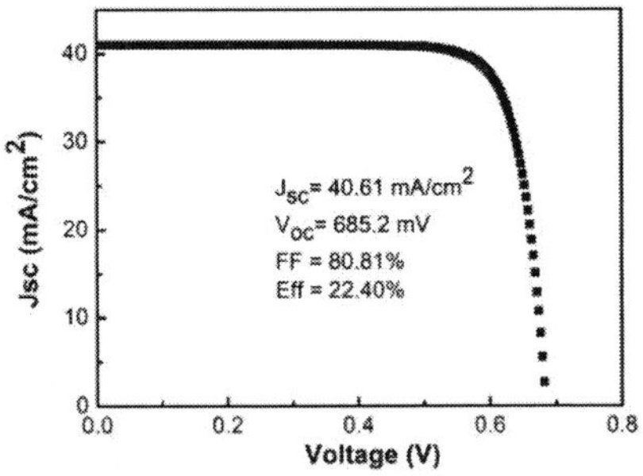

Figure 5: I-V curve of the champion cell on a 6 inch wafer, measured with the new Sinton FCT450 flash tester.

4 MECHANISM OF HIGH FF FOR N-TYPE CELLS

In the above cell results, a good FF was obtained for n-type PERT cells. It is known that FF depends on V_{oc} and R_s [8]. However, if the recombination losses do not increase in a purely exponential manner with bias, such recombination losses affect FF heavily as well. In n-type Si materials containing shallow defects (e.g. due to oxygen-related thermal donors), lifetime decreases with increasing injection density. The n-type cell's J_0 is larger near V_{oc} than near V_{mpp}. This leads to smaller V_{oc} values compared to V_{mpp} values, and hence to an increase of FF = V_{mpp} J_{mpp} / (V_{oc} J_{sc}). In the following, we assess by how much FF may increase maximally by first looking at the mechanism of SRH recombination and then by performing numerical device simulations of an exemplary PERT cell.

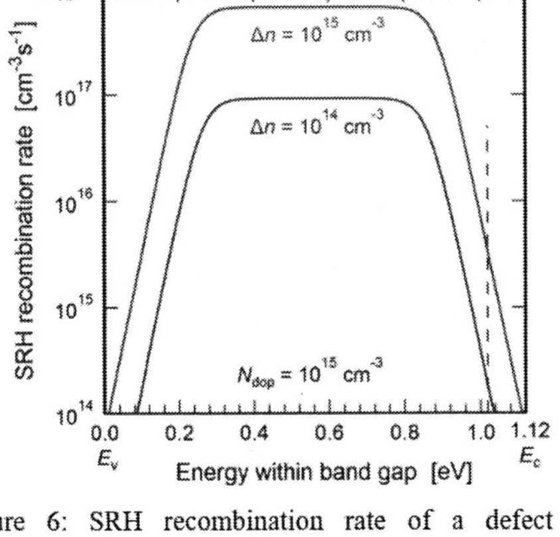

Figure 6: SRH recombination rate of a defect in dependence of its energy E_d within the band gap, for the two indicated injection densities.

In SRH theory, defect states in the band gap (with a given pair of capture cross-sections σ_n, σ_p) cause the highest recombination rate if their energy level E_d is situated deep in the band gap. If E_d is thought to move towards a band edge, the recombination rate hardly changes but then suddenly drops exponentially, see Fig. 6. This is so because the captured carrier becomes likely to be thermally emitted back into the nearby band and this competes with recombination [9]. Hence, there exists a 'plateau' of the highest recombination rate bordered by exponentially decaying 'slopes' as indicated in Fig. 6. In order to understand the behavior in n-type, it is important to note that the width of the plateau increases with increasing injection density. If a shallow defect's E_d is situated near a slope, it becomes more recombination active with higher injection density Δn, and this lowers the lifetime of the material with higher Δn, see Fig. 7. Varying all the involved parameters reveals that (i) with higher dopant density N_{dop}, the plateau extends towards the band edges, and hence the shallow defects become more recombination active; (ii) the asymmetry of σ_n, σ_p hardly affects the recombination behavior except if σ for the carrier of the nearby band is very large; and (iii) the difference in lifetime between V_{mpp} and V_{oc} is maximal at E_d < 100 meV and low N_{dop}.

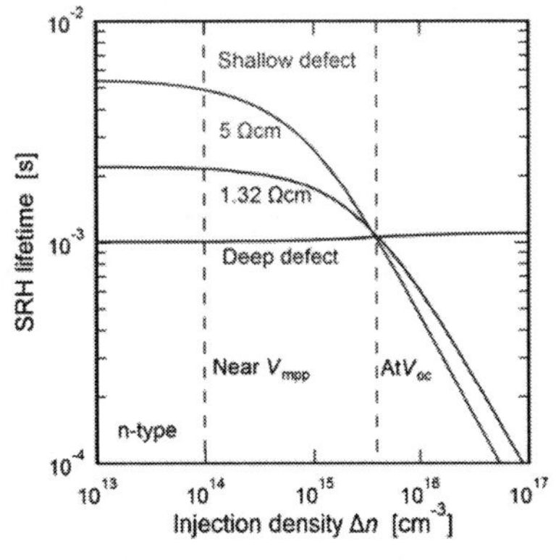

Figure 7: Injection dependent SRH lifetime in case of a deep defect or a shallow defect in n-type material having the indicated resistivity.

The semiconductor device simulation software Sentaurus was used to quantify the maximal increase in FF due to shallow defects in an exemplary PERT cell on 1.32 Ωcm material by choosing E_d, σ_n, and σ_p such that the lifetime varies strongest with injection density. The simulated FF of the entire cell (including metallization) drops from 80.59% to 79.83% if the lifetime-limiting defect is changed from a shallow defect to a deep defect (the lifetime at V_{oc} is chosen unaffected because a change in V_{oc} would affect FF additionally). This is a change in FF by nearly 0.8% absolute. In the case of this exemplary PERT cell, its efficiency decreases from 22.23% to 20.01%, i.e. by about 0.2% absolute.

In the following, this effect is investigated in our fabricated PERT cells. The Sinton FCT450 tester allows to measure both N_{dop} in the base region and the effective lifetime τ_{eff} in dependence of Δn in the finished cell, as shown in Fig. 8. The SRH lifetime in the base τ_{SRH} was extracted from τ_{eff} with a least-square fit of

$$\frac{1}{\tau_{eff}} = \frac{1}{\tau_{bulk}} + \frac{1}{\tau_{surf}} = \frac{1}{\tau_{SRH}} + \frac{1}{\tau_{intr}} + \frac{1}{\tau_{surf}}$$
$$= \frac{1}{\tau_{SRH}} + \frac{1}{\tau_{intr}} + (J_{0,em} + J_{0,BSF}) \frac{\Delta n}{qWn_{i,eff}^2}$$

where we choose Ref. [10] for τ_{intr}, the measured J_0 values from above, n_i from Ref. [11] at the temperature set by the flash tester, and $n_{i,eff}$ from Ref. [12] by including injection-dependent band gap narrowing in the base. We fit the extracted τ_{SRH} by assuming a deep defect at midgap and a shallow defect at 100 meV below the conduction band edge. With these two defects as input parameters, and all the other device parameters characterized above, we reproduce the measured PERT cell very precisely with Sentaurus.

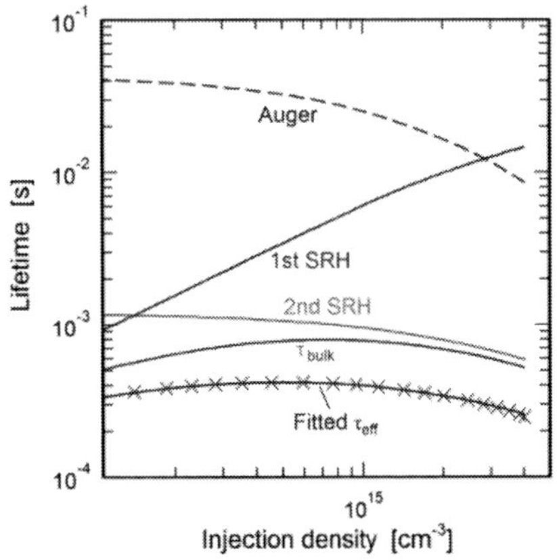

Figure 8: Effective lifetime of the PERT cell, measured with the Sinton FCT450 flash-tester, and fitted with two SRH defects,Auger lifetime, and the J_0 values of the emitter and the BSF [10].

The recombination losses in each device region are shown in Fig. 9. As expected from a well optimized cell (having efficiency of 22.28%), all device parts contribute rather equally to the total recombination losses at mpp: the emitter is slightly dominating, followed tightly by the base and the BSF. Notice that all curves are in parallel between V_{mpp} and V_{oc} (despite a lower base lifetime near V_{oc}) because the losses in the emitter and the BSF are both driven by Δn of the base, i.e. both diffusions draw a recombation current from the base according to

$$j = J_0\left(\frac{pn}{n_{i,eff}^2} - 1\right) \propto J_0 \Delta n$$

Below V_{mpp}, the base losses do not go in parallel to the emitter and BSF losses any more due to the shallow defects.

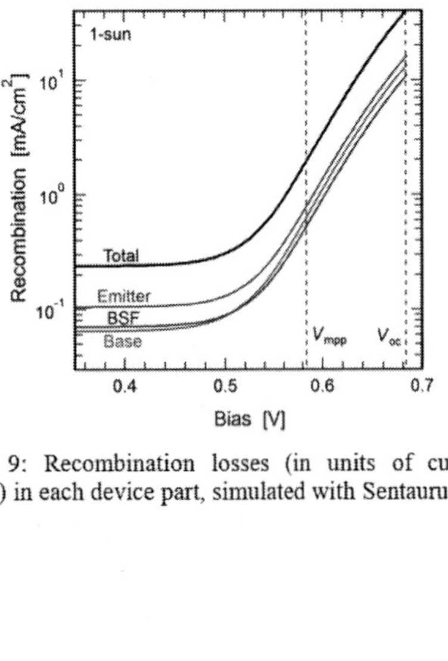

Figure 9: Recombination losses (in units of current density) in each device part, simulated with Sentaurus.

5 CONCLUSIONS

In this paper, it is demonstrated that by optimizing the boron emitter, good passivation and good contacts could be achieved for n-PERT solar cells with plated Ni/Cu/Ag front metallization. A total-area cell efficiency of 22.40% was achieved on a 6 inch wafer after further metallization structure optimization. Sentaurus simulations reveal that the injection-dependent lifetime behavior due to shallow defects in n-type Si material increases FF of n-type PERT solar cells.

ACKNOWLEDGEMENT

This work was supported by the scientific and technological achievement transformation of Changzhou City under the Project Number of CC20150005.

REFERENCES

[1] J. Benick, B. Steinhauser, R. Muller, J. Bartch, M. Kamp, A. Mondon, S. Glunz, 40th IEEE PV Specialists Conference (2014) 3637.

[2]http://www.pvtech.org/news/intersolar_2015_imec_tops_n_pert_cell_efficiency_of_22.5.

[3] A. Edler, V.D. Mihailetchi, C. Comparatoo, L.J. Koduvelikulathu, R. Kopecek, R. Harney, T. Boescke, J. Lossen, Proc. 27th EU PVSEC (2012) 1429.

[4] A. Edler, V.D. Mihailetchi, R. Kopecek, R. Harney, T. Boescke, Energy Procedia 8 (2011) 493.

[5] R. Kopecek, 3rd n-PV Workshop, Chambery, France, 2013.

[6] R. Lago, L. Perez, H. Kerp, I. Freire, I. Hoces, N. Azkona, F. Recart, J.C. Jimeno, Prog. Photovolt. 18 (2010) 20.

[7] J. Bartsch, A. Mondon, M. Kamp, A. Kraft, Proc.27th EU PVSEC (2012) 604.

[8] E. Sanchez and G.L. Araujo, Solar Cells 20 (1987) 1.

[9] J. G. Simmons and G. W. Taylor, Phys. Rev. B4 (1971) 502.

[10] A. Richter, S.W. Glunz, F. Werner, J. Schmidt, A. Cuevas, Phys, Rev. B86 (2012) 165202.

[11] R. Couderc, M. Amara, M. Lemiti, J. Appl. Phys. 115 (2014) 093705.

[12] A. Schenk, J. Appl. Phys. 84 (1998) 3684.

HIGH SPEED DISPENSING – A HIGH-THROUGHPUT METALLIZATION TECHNOLOGY FOR >21% PERC TYPE SOLAR CELLS

M. Pospischil[a*], M .Kuchler[a], M. Klawitter[a], I. Lacmago[a], S. Tepner[a], R. Efinger[a], M. Linse[a], D. Witt[a], S. Gutscher[a], A. Brand[a], M. König[b], L. Wende[c], F. Clement[a] and D. Biro[a]

a) Fraunhofer Institute for Solar Energy Systems ISE, Heidenhofstraße 2, 79110 Freiburg, Germany
b) Heraeus Deutschland GmbH & Co. KG, Heraeusstr. 12-14, 63450 Hanau, Germany
c) ASYS Automatisierungssysteme GmbH, Benzstr. 10, 89160 Dornstadt, Germany

ABSTRACT: In this study, high speed dispensing with a novel 6" dispensing print head is applied to different screen printing pastes. The novel print head features a homogeneous distribution of Ag-paste from an inlet supply with a diameter of $D_{in} = 2mm$ to an outlet slot with a width of $w_s = 160\ mm$. A first printing test followed by geometrical characterization of the dispensed contacts revealed a distribution of the contact finger width of $w_f = 35 \pm 1\ \mu m$ across the whole wafer ($156x156mm^2$) which demonstrates highest precision of the device. The print head is directly integrated in an inline feasible dispensing platform, developed by ASYS GmbH. Respective nozzle plates can be customized to any desired front side grid regarding nozzle diameter, nozzle pitch and the total number of nozzles n. In the actual version, the print head uses two strokes to print two adjacent contact fingers using 50 nozzles while moving up and down at a speed of up to $v_y = 700mm/s$. In a first solar cell test sequence with the new 6" print head, a maximum cell efficiency of $\eta = 21.2\%$ on industrially preprocessed Cz-PERC samples was reached which demonstrates a successful launching of the print head. A follow up experiment with an industrial reference paste demonstrated similar results on the same material. Here, for the first time, a nozzle opening of only $D = 35\mu m$ allowed for a reduction of the wet paste consumption down to $m_{wet,Ag} = 50mg$ for the contact grid (N = 100) which is a reduction of $\Delta m/m = -15\%$ compared to screen printed references. At the same time, more homogeneous contact fingers allow for a reduced grid resistance of $\Delta R_{Grid}/R_{Grid} = -10\%$.

Keywords: Silicon Solar Cell, Metallization, Dispensing

1 INTRODUCTION

1.1 Advantages of dispensed contact grids

The dispensing technology, as described by Specht et al. [1], offers a contactless, high-throughput single-step metallization process significantly reducing finger width and thus shading losses. Record cell efficiencies of *20.6%* on $125x125mm^2$ *FZ p-type* material using dispensing technology on MWT-PERC (Metal Wrap Through – Passivated Emitter and Rear Cell) solar cells, featuring a selective emitter structure were presented by Lohmüller et al. [2]. Screen printing pastes can directly be applied for dispensing in a first step, although the shape of dispensed contact fingers can be varied in a wide range by adapting paste rheology as described in previous studies e.g. [3].
By this approach, the relative effective width of dispensed contact fingers can be reduced by $\Delta EW = -25\%_{rel.}$ on cell level compared to screen printed contact geometries due to improved back scatter characteristics. Furthermore, a substantially improved contact homogeneity requires less paste than screen printed contacts when assuming a similar grid resistivity [4, 5]. All in all, a relative efficiency increase of $\Delta\eta \sim +2\%_{rel.}$ obtained by increases in j_{sc}, V_{oc} and FF at a reduced wet paste laydown of $\Delta m = -15\%_{rel.}$ can be reached by dispensing commercially available Ag-printing pastes [6].

1.2 Previous approaches to increase throughput rates

An increase of throughput rates is crucial in order to enable industrial production cycles comparable to commercial screen printing metallization lines, currently reaching 2000 wafer/h. A first approach of parallel dispensing for solar cells was patented by Hanoka in 1992 [7]. At former Schott Solar AG, continuously operating parallel dispensing units were in use for many years in their production lines. Here, a special approach was developed by von Campe and Schwirtlich that prevented spreading of high aspect ratio contact fingers on the wafer [8]. Chen et al. [9] developed a parallel print head applying several n-scrypt smart pumps, each feeding up to three parallel aligned single nozzles [10, 11].

The Co-Extrusion approach as patented by Fork et al. [12] and applied for solar cell metallization by Beutel et al. [13] and Richter et al. [14] at Solar World Innovations GmbH includes a parallel print head that enables ultra-fine contact lines of down to $w_f = 23.7\ \mu m$ by using a sacrificial medium surrounding and focusing the Ag-paste inside its 200-220 µm wide nozzle openings. Here, the pilot production tool employs two print heads in a dual printing lane and reaches a rated throughput of 2700 wafers/h. Although, more than 100.000 solar cells have been printed with this tool and even an intermittent printing of pseudo square wafers has been demonstrated, several aspects prevented a broad application so far: Due to the laminar sheath flow inside the nozzle, a special Ag-paste with adapted rheology and hence a reduced Ag content of only $74\%_{wt.}$ is required. Furthermore, an active distance control between print head and wafer limits printing velocity and the grid design is fixed to the configuration of the print head that contains channels and valves for two media at each nozzle.

1.3 Parallel dispensing at Fraunhofer ISE

Within the research project GECKO, parallel dispensing using a ten nozzle print head was demonstrated [15] reaching line widths of 27 µm and successfully applied for solar cell metallization [4, 6]. Here, the focus was to achieve a homogeneous distribution to a variable number of dispensing nozzles by using a single paste supply. In a second step a new 6'' wide print head with 50 nozzles was developed, enabling a potential throughput of 2000 cph. This study focuses on further cell processing with the new print head and the evaluation of different paste systems.

2 DEVELOPMENT OF PRINT HEADS

2.1 The 6" Parallel Print Head (Crocodile)

After a successful implementation of a rheological paste model including yield stress, shear thinning and wall slip phenomena [15, 16], new CFD-simulations were started aiming on the development of a central fed print head with an operating width of 160 mm, see Fig. 1. Compared to previous ten nozzle versions, this led to an enlargement of the internal paste volume by a factor of 16. During subsequent fabrication and testing of the novel prototype, several other mechanical aspects were challenging, especially sealing of movable parts that allow for automated in-situ cleaning sequences of the nozzle plates by opening up the pneumatic cylinders. To get a complete grid metallization the print head scans the cell back and forth with a scan speed of around 650 mm/s enabling industrial throughput numbers. With respect to the desired grid design, nozzle plates can be customized regarding outlet diameter, nozzle pitch as well as the number of nozzles. Due to experiences with the previous ten nozzle versions in various solar cell experiments, the current nozzle plate design was optimized using Gridmaster [17].

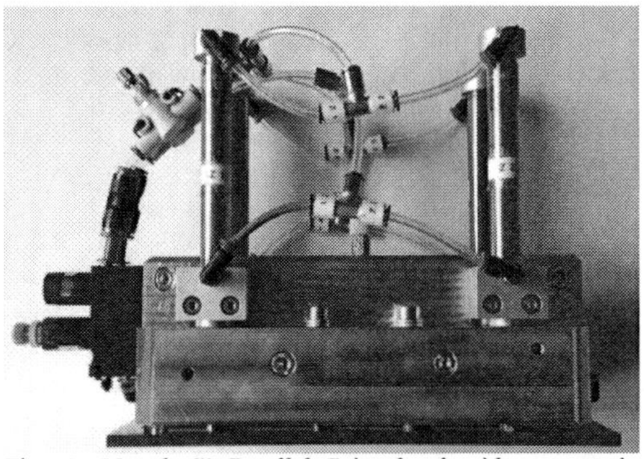

Fig. 1: Novel 6" Parallel Print head with pneumatic clambs for cleaning and maintenance.

2.2 Processing of PERC solar cells for paste evaluation

In this study, the new print head was applied for cell processing with commercially available screen printing pastes from different manufacturers. Passivated solar cells were applied to local contact opening on the back side and subsequently equipped with screen printed Al on the rear side.
In a first experiment, single screen printed reference cells were then compared with dispensed samples that were equipped with five tapered busbars using the same (contacting) paste from Heraeus. A variation of the peak firing temperature was conducted followed by IV-characterization. A schematic of the process flow is depicted in Fig. 2

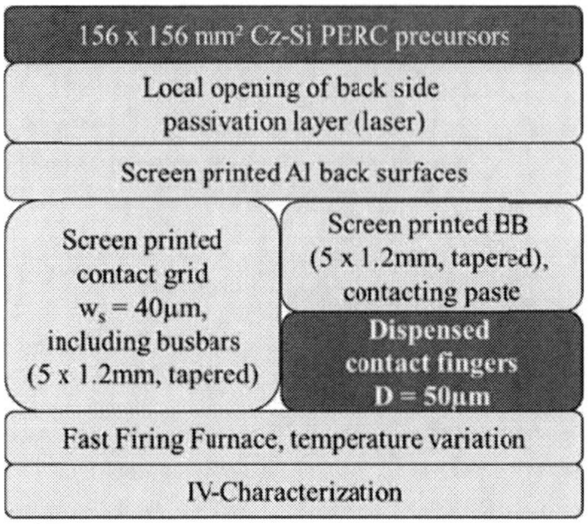

Fig. 2: Process flow of solar cell processing applying a screen printing paste from Heraeus. Dispensed solar cells with screen printed busbars (contacting, same paste) are compared with single screen printed references.

In a second experiment, a similar approach was chosen, but now with five tapered, non-contacting busbars on the front side (dual print). The contact grid was then applied by dispensing or screen printing in case of the reference group, see Fig. 3 using a reference paste from a different manufacturer. The group with dispensed contact fingers and a nozzle opening of $D = 35\mu m$ was fired after determining ideal firing conditions with the reference samples.

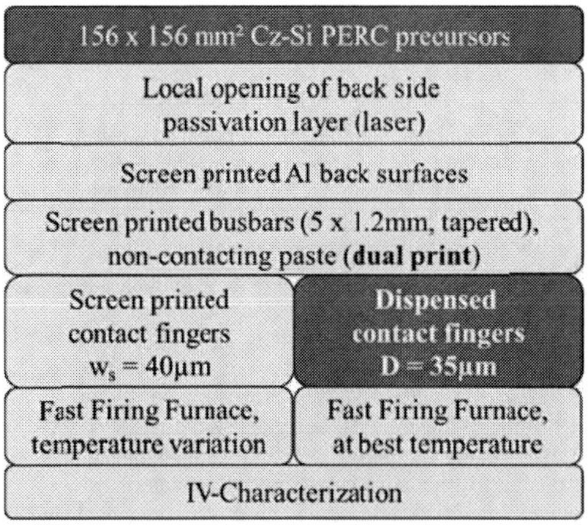

Fig. 3: Process flow of solar cell processing using an industrial reference paste. Dispensed solar cells are compared with screen printed references both using non-contacting busbars (dual print approach).

3 RESULTS

3.1 First Printing tests

First test sequences with the novel 6" print head aimed at the demonstration of a homogeneous paste flow distribution as predicted by CFD-simulations (Fig. 4). For this reason, simulated deviation of the mass flow rate $\Delta\dot{m}_i/\dot{m}$ from the average mass flow rate $\dot{m}$ was extracted for each nozzle at paste velocities of 100, 300, 500 and 700 mm/s, respectively.

Subsequently, the print head was equipped with a nozzle plate containing 50 nozzles with a diameter of D = 50µm and applied for test prints. The resulting contact finger width of every third contact finger was subsequently measured by means of confocal laser scanning microscopy (CLSM) at three positions each. Here, an average finger width of $w_f = 35 \pm 1\mu m$ was reached across the full width of the print head (Fig. 4).

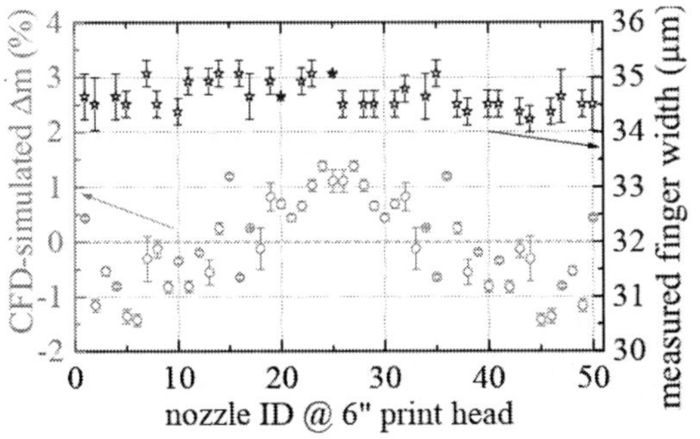

Fig. 4: Lateral deviation of simulated mass flow rate and measured finger width on 6"wafer.

Hence, a homogeneous paste distribution from an inlet channel with a diameter of D_{in} = 2 mm to an outlet slot width of w_s = 160 mm was successfully demonstrated with a commercially available screen printing paste delivered by Heraeus. This underlines the precision of rheological characterization as well as CFD modelling and last but not least, applied fabrication processes.

3.2 Printed contact geometries

As known from previous studies, the shape of dispensed grid lines strongly depends on paste rheology and nozzle configuration. Hence, some pastes allow for a contact width of $w_f = 35 \pm 1\mu m$ or even below at high aspect ratios of up to $AR_o = 0.8$ that can be reached with a nozzle opening of $D = 50\mu m$, see e.g. [4, 5]. In this study, this behavior was observed for the Heraeus paste applied for first printing tests, see section 3.1.

However, other screen printing pastes do have lower yield stresses and hence lead to lower aspect ratios. In the second cell experiment (Fig. 3), an industrial reference paste was evaluated, that could be homogeneously dispensed with $D = 35\mu m$ nozzle openings, see Fig. 5.

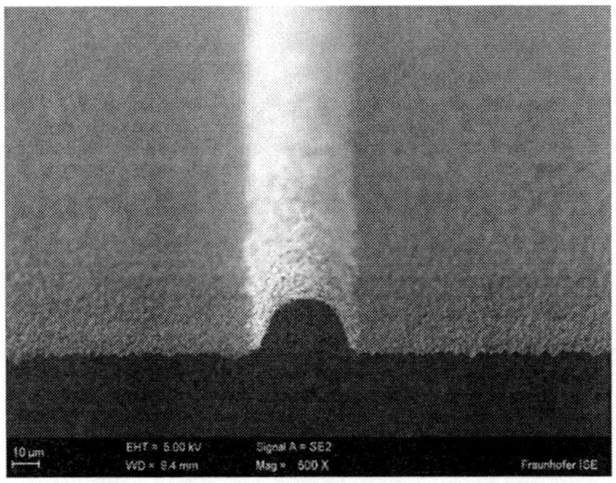

Fig. 5: SEM-image of a dispensed contact finger on a Cz-PERC solar cell. A homogeneous contact shape is crucial for an efficient usage of Ag-printing pastes.

However, due to its pronounced spreading behavior, the respective finger width turned out to be in the same region as the applied nozzle diameter with $w_f = 35 \pm 1\mu m$ as illustrated in Fig. 6

Fig. 6: SEM-image of a dispensed contact finger cross section on a Cz-PERC solar cell. An aspect ratio of $AR_o = 0.55$ ensures sufficient conductivity at limited shading of the cell.

Nevertheless, to achieve a further reduction of paste consumption in mass production, even smaller nozzle openings are needed. In this study, the wet paste laydown of a dispensed contact grid (without busbars) was reduced to a minimum of $m_{Ag,wet} = 50mg$ as later indicated in **Table II**.. From an economic perspective, this is a promising result as it demonstrates the capability of dispensing to keep pace with silver reduction, although generally applying contact geometries with higher aspect ratios than screen printing.

3.3 Solar cell results with different printing pastes

The application of commercially available screen printing pastes for dispensing enables to increase solar cell efficiencies due to an improvement of the resulting shape of the contact fingers.

However, the challenge is to keep pace with a steady reduction of line width of the reference technology – screen printing. In this study, screen printed reference groups were printed with screen openings of $w_s =$

$40\mu m$ resulting in printed line widths of $w_{f,SP} = 45 \pm 3\mu m$.

In the first experiment (Fig. 2), a respective comparison was conducted for a commercial screen printing paste from Heraeus. As mentioned before, high yield stresses lead to higher aspect ratio and hence smaller finger widths of dispensed contact fingers in comparison to the reference paste when applying similar nozzle openings. Losses in *FF* in comparison to the screen printed reference group however were subject to printing instabilities that prevented higher cell efficiencies for the dispensed group as well as the application of smaller nozzle diameters. Generally, a high level of *FF* demonstrates outstanding contacting behavior of this paste. Consequently, the screen printed group (SP40) as well as the dispensed group (D50) reached cell efficiencies of $\eta = 21.2\%$ despite the fact, that contacting busbars were used in this experiment, see **Table I**.

Table I: Best cell results with an industrial paste manufactured by Heraeus. Here, busbars were printed with the same paste as the respective contact grids.

Heraeus paste	R_{Grid} *(Ω/m)*	*FF (%)*	J_{sc} *(mA/cm²)*	V_{oc} *(mV)*	η *(%)*
SP40	66.6	80.3	40.2	660	21.2
D50	24.5	80.1	40.3	659	21.2

In the second experiment (Fig. 3), a dual print approach using an industrial reference paste for the contact fingers was chosen. When looking at the *FF*, the values are considerably lower than in the first experiment which may be due to non-ideal contacting behavior at the chosen firing conditions. An increase of V_{oc} by $3mV$ is due to the dual print approach applied in this experiment. Again, the dispensed group with nozzle openings of $D = 35\mu m$ benefitted from an improved contact shape which is visible when looking at the resulting short circuit currents j_{sc} in **Table II**.

Table II: Best cell results with an industrial reference paste in dual print approach. The wet paste laydown only accounts for the contact grid with N = 100 fingers.

Ref. paste	m_{Ag} *(mg)*	R_{Grid} *(Ω/m)*	*FF (%)*	J_{sc} *(mA/cm²)*	V_{oc} *(mV)*	η *(%)*
SP40	60	65.8	78.9	40.3	663	21.0
D35	50	61.2	79.3	40.5	662	21.2
D35*	50	59.1	79.7	40.4	663	21.3

*) Group was separately fired at ideal firing conditions.

An aspect that becomes more and more relevant at decreasing line widths is the homogeneity of the contact fingers on the wafer. Here, one can see that dispensed contacts with a reduced wet paste laydown of $\frac{\Delta m}{m} = -15\%$ additionally led to a lower grid resistivity ($\frac{\Delta R_{Grid}}{R_{Grid}} = 7 - 10\%$) than screen printed samples that suffer from mesh marks. Finally, samples with dispensed contact fingers outreached screen printed reference cells by $\Delta\eta = 0.2\% abs.$ reaching $\eta = 21.2\%$ at similar firing conditions (D35) and $\eta = 21.3\%$ (D35*) at ideal firing temperatures (see **Table II**.).

4 CONCLUSIONS AND OUTLOOK

In this study, parallel dispensing technology was enhanced towards industrial throughput rates by increasing the operational width of the print head to 16 cm reaching an average finger width of $w_f = 35\pm1$ µm across the whole 6" wafer. First solar cell processing led to a maximum cell efficiency of $\eta = 21.2\%$ on industrially preprocessed Cz-PERC samples which demonstrates a successful launching of the print head. In a second experiment, a dual print approach was chosen that increased the maximum cell efficiency to 21.3% on the same material but with a different screen printing paste. In this batch, the application of nozzle openings with a diameter of just $D = 35\mu m$ allowed for a reduction of the wet paste consumption down to $m_{wet,Ag} = 50mg$ for the contact grid which is a reduction of $\frac{\Delta m}{m} = -15\%$ compared to screen printed references at a reduced grid resistance of $\frac{\Delta R_{Grid}}{R_{Grid}} = 7 - 10\%$.

All in all, similar cell efficiencies on Cz-PERC were reached by dispensing industrial screen printing pastes of two different manufacturers.

Actual work focuses on further hardware advances, including a precise control of the paste flow and the demonstration of process stability with the new print head.

5 ACKNOWLEDGMENTS

The authors would like to thank all co-workers at the Photovoltaic Technology Evaluation Center (PV-TEC) and the mechanical workshop at Fraunhofer ISE for processing of the samples.

6 REFERENCES

[1] M. Pospischil *et al.*, IEEE JPV **4**, 498 (2014).
[2] M. Pospischil *et al.*, Energy Procedia **43**, 111 (2013).
[3] E. Lohmüller *et al.*, IEEE Electron Device Letters **32**, 1719 (2011).
[4] T. Fellmeth, F. Clement, and D. Biro, Photovoltaics, IEEE Journal of **4**, 504 (2014).
[5] M. Pospischil *et al.*, Energy Procedia **55C**, 693 (2014).
[6] M. Pospischil *et al.*, in 39th IEEE PVSC (IEEE, Tampa, 2013).
[7] M. Pospischil *et al.*, in Photovoltaic Specialists Conference (PVSC), 2013 IEEE 39th2013), pp. 2250.
[8] J. Specht *et al.*, in Proceedings of the 25th European Photovoltaic Solar Energy Conference and ExhibitionValencia, Spain, 2010), pp. 1867.
[9] D.-H. Kim *et al.*, Materials Science and Engineering: B **177**, 217 (2012).
[10] M. Pospischil *et al.*, Energy Procedia **8**, 449 (2011).
[11] M. König *et al.*, Energy Procedia **27**, 1 (2012).

FLIP-FLOP CELL INTERCONNECTION ENABLED BY AN EXTREMELY HIGH BIFACIAL FACTOR OF SCREEN-PRINTED ION IMPLANTED N-PERT SI SOLAR CELLS

Henning Schulte-Huxel[1], Fabian Kiefer[1], Susanne Blankemeyer[1], Robert Witteck[1], Malte R. Vogt[1], Marc Köntges[1], Rolf Brendel[1,2], Jan Krügener[3], Robby Peibst[1,3]
[1]Institut für Solarenergieforschung Hameln (ISFH), Am Ohrberg 1, D-31860 Emmerthal, Germany
[2]Institute of Solid-State Physics, Leibniz Universität Hannover, Appelstr. 2, 30167 Hanover, Germany
[3]Institute for Electronic Materials and Devices, Leibniz Universität Hannover, Schneiderberg 32, D-30167 Hanover, Germany

ABSTRACT: We present bifacial fully ion implanted and screen-printed *n*-PERT cells, fabricated either by applying a single co-anneal process to cure the implant damage or by applying two separate anneals after boron/BF_2 and phosphorous implant, respectively. In the first case of boron implant and co-anneal our best cells achieve an independently measured front (rear) side efficiency of 21.0 % (20.43 %) and for the boron implant and separate anneal the efficiency is 21.5 % (21.31 %). To the best of our knowledge these values are the highest efficiencies reported so far for fully ion implanted and screen-printed bifacial *n*-PERT cells. We furthermore show that light treatment of boron-implanted and co-annealed *n*-PERT cells increases the cell efficiency by 0.6 $\%_{abs}$. This diminish the efficiency gap to separately annealed cells. We measure a bifacial factor of 99.4 % that is the highest value reported so far for any high-efficiency Si cell. The high bifaciality enables an adapted module interconnection scheme called here Flip-Flop, which is based on a front-to-front and rear-to-rear interconnection of cells with alternating orientation (n^+ or p^+ side facing up). Based on the measured *IV*-characteristic of a cell with a bifacial factor of 97 % ("conservative scenario") we demonstrate that the Flip-Flop interconnection scheme has the potential for a module efficiency improvement of 0.5 $\%_{abs}$ on aperture area (as compared to the conventional "all cells emitter up" configuration) despite the 3 % current mismatch. We experimentally demonstrate a monofacial 16-cell Flip-Flop module that achieves an aperture area efficiency of 20.5 %.
Keywords: Bifacial solar cells, *n*-PERT cells, ion implantation, module integration, Flip-flop module, planar interconnection, light soaking

1 INTRODUCTION

There is growing interest in PV industry in bifacial solar cells [1], either for application in bifacial modules [2] or to raise the current generation in standard modules with a highly reflective back sheet [3,4]. These concepts are based on a conventional solar cell interconnection process with 2-5 cell interconnection ribbon (CIR) or wires, conducting the current from the front side of one cell to the rear side of the adjacent cell.

In the standard cell interconnection process, a solder coated copper CIR is soldered to the front side of one cell and lead to the next cell, where it is soldered to its rear side (Fig. 1(a)). The requirement to lead the CIR from the front to the rear side has two draw-backs: (*i*) the mechanical stress at the wafer edge, which has the potential to induce cracks into the solar cell [6], and, in consequence, (*ii*) the requirement of a gap between the cells with a width of 3-4 mm. A large cell-to-cell distance increases the module size and thus decreases the module efficiency. Since numerous contributions to the system costs scale with the module size, smaller modules offer the possibility to decrease the levelized costs of electricity (LCOE).

An alternative interconnection concept has been proposed by Kopecek et al. [7]: n-type and p-type cells (not necessarily bifacial) connected alternating in a string. This allows a planar interconnection, where the cell interconnectors (tabs or wires) conduct the current either only on the front side or only on the rear side from one cell to another (Fig. 1(b)). However, it might be challenging to process high-efficient *n*-type and *p*-type cells in the same production line. In parallel to this work Augusto *et al.* presented a similar approach using *n*-type cells with front and rear side emitter, however still employing two different cell types [8]. Alternatively, bifacial solar cells can be placed alternating with the p^+ and the n^+ side facing upwards. However, this demands cells with a similar energy conversion efficiency for both sides, which ideally equals a bifacial factor of unity. For example, the *n*-type passivated emitter, rear totally doped (PERT) solar cell, which is recently widely investigated [5, 9–13] exhibit a full area BSF which provides lateral conductivity and thus enables similar finger-to-finger distances as on the front-side. Thus, a fine line screen print can be applied on both sides of the solar cell [5]. Together with the high stable minority carrier lifetimes of *n*-type silicon, *n*-PERT cells with bifacial factors > 98 % [5] can be developed. This is the case for our ion-implanted and screen-printed *n*-PERT solar cells with efficiencies up to 21.5 % and bifacial factors exceeding 99 %. Those solar cells give the opportunity to realize the interconnection scheme mentioned above, which we call "Flip-Flop design". One should remark that any other cell concept with sufficiently high bifaciality factors, such as bifacial heterojunction cells, would also be compatible with the Flip-Flip interconnection scheme.

Similar to the concept of Kopecek et al. [7], the Flip-Flop design proposed here offers several advantages: the distance between the cells is not limited by the CIR connecting the front and the rear side of the adjacent cells, allowing a smaller cell spacing and thus higher module efficiencies. The reduced current path in the CIR decreases the resistive losses and permits thinner CIR, reducing the shading losses on the cell. Also, the mechanical stress at the edge of the solar cell by the CIR is reduced. For small cell gaps the CIR requires a crimp, which significantly reduces the module lifetime [14]. Cell interconnection degradation is one of the most dominant failure mechanisms in the field [15].

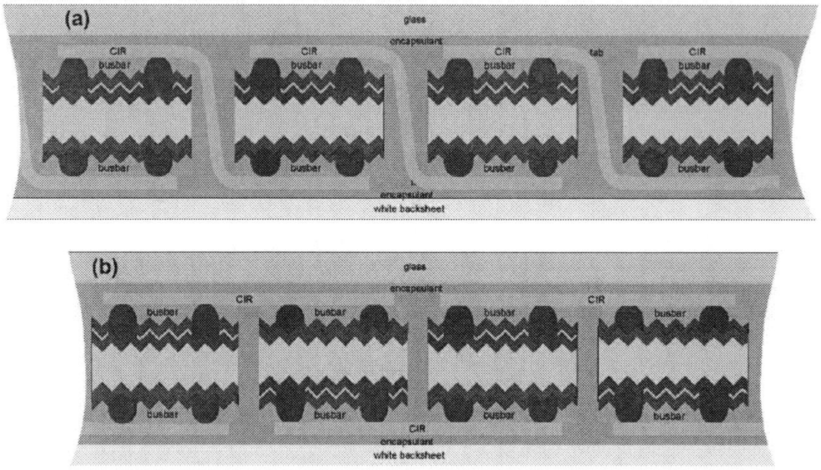

Figure 1: Schematic drawing (a) of a conventional cell interconnection scheme with CIR from front to rear side, and (b) of the Flip-flop cell interconnection scheme with CIR only from front to front and rear to rear side. (c) Photograph of our first demonstrator Flip-Flop module.

We present a 16-cell demonstrator module produced with this new interconnection process with an aperture area efficiency of 20.5 %. This promises a large potential for further improvement of the module efficiency.

2 EXPERIMENTAL DETAILS

The solar cells are processed on 156 × 156 mm² *n*-type pseudo-square Cz silicon wafers with a base resistivity of 3 Ωcm. First, saw damage removal is performed, the wafers are double-side textured and wet chemical cleaned. We use ion implantation and subsequent anneal to form the n^+ and p^+ regions. For the latter, we evaluate implant and anneal for elementary boron and BF_2 with a dose of $1.75 \cdot 10^{15}$ cm² [16]. BF_2 is more attractive for industrialization due to the simplified implanter. technology (no mass separator [17]) and due to the reduced thermal budget required for annealing (950°C instead of 1050°C) [16]. After implantation and annealing, a stack of Al_2O_3/SiN_x is deposited on the p^+ emitter side and a SiN_x layer on the n^+ side for the surface passivation. We use a double print for the metallization on both sides of the cell with a finger opening of 30 µm in the stencil and 5 busbars. The emitter is contacted with an Ag/Al paste, and the n^+ side is contacted with an Ag paste, both pastes commercially available. The cells are fired in a conveyor belt furnace. We measure the *IV* characteristics of the solar cells from both sides using a *LOANA* system from pvtools on a brass chuck. The champion cells were independently measured at F-ISE CalLab.

The module is made of cells strips with a size of 62.5×156 mm². Therefore, the cells are laser-cut from the n^+ side into rectangular solar cells without the pseudo-square edges. One should remark that the rectangular geometry is desirable for high module efficiencies, but not absolutely necessary in general for the Flip-Flop concept. Also, full square wafers would have enabled a rectangular geometry without cutting the half cells from 78 mm width down to 62.5 mm. No degradation of the cell performance is observed due to cutting with ns-laser pulses. Instead, we determine a relative increase in efficiency (illumination from the emitter side) of 0.6 $\%_{rel}$ in average comparing the initial cell performance and the rectangular half cells. This deviation is within the measurement uncertainty for the *IV*-tester of 1.0 $\%_{rel}$. After current matching, the wafers are connected into 8-half-cell strings using the Flip-Flop interconnection process with interconnector CIR with a cross-section of 0.2 × 1.3 mm, which are covered with a highly reflective structured film. The cell spacing and the string-to-string distance is 1 mm. The cells are laminated with an ethylene-vinyl acetate (EVA) lamination foil with an improved UV-transparency, a 3.2-mm-thick front glass with anti-reflection coating (F solar) and a *Coveme PYE 3000* white back sheet. The *IV* characteristic of the module is measured with an A^+ flasher from halm while using a long (25 ms) flash in order to take into account the high diffusion capacitance of the cells. Module measurements are performed in-house so far.

We determine the spectral miss match factor for this module according to the norm IEC 60904-7 and use this to correct the *IV*-characteristics according to the norm 60891 correction method 1.

3. RESULTS AND DISCUSSION

3.1. Bifacial n-PERT cells

Table I lists the *IV* parameters of our current champion *n*-PERT cells with B or BF_2 implant for emitter formation, either annealed in a single co-annealing step or by separate annealing steps for B, BF_2 and P implant. To our knowledge, these are the highest values published so far for industrial ion implanted, fully screen printed *n*-PERT cells. In particular, the bifacial factor > 99 % is the highest value reported for any high efficiency cell concept [9].

This is the essential basis for the Flip-Flop module concept. For the demonstrator module, however, we use cells from a previous run (exclusively co-annealed cells) with slightly lower efficiencies and bifacial factors, see Table I. For a sufficient number of comparable cells, we also mixed up split groups with B implanted emitter and with BF_2 implanted emitter.

TABLE I. *IV*-parameters (full-area measurement on 239 cm^2) of our current *n*-PERT champion cells with B or BF_2 implant for emitter formation, either annealed in a single co-annealing step or separate annealing steps for B/BF_2 and P implant. Additionally, the average values of the cells used for the module are given. *Independently measured by F-ISE CalLab.

	Emitter	Anneal	V_{oc} [mV]	J_{sc} [mA/cm^2]	*FF* [%]	η [%]	$f_b = \eta_{rear}/\eta_{front}$ [%]
Best cells	B	co-anneal	665*	39.8*	79.3*	21.0*	97.3
	B	separate anneal	671*	40.2*	79.7*	21.5*	99.4
	BF_2	co-anneal	658*	39.9*	78.4*	20.6*	97.7
	BF_2	separate anneal	664*	40.4*	77.6*	20.8*	98.6
Average values of cells for module	B	co-anneal	654	39.9	78.3	20.4	95.3
	BF_2	co-anneal	652	40.3	77.1	20.3	98.3

Since we wanted to use an UV-transparent EVA foil, we studied first the UV stability of our *n*-PERT cells by illuminating the Al_2O_3 passivated side of the cells with halogen lamps under approximately 1000 W/m^2 irradiation.

Here, we made an unexpected observation, which is reported in the following.

Figure 2 shows the preliminary results of our experiments. For the cells with the BF_2 implanted emitter no significant change in efficiency ($\Delta\eta < 0.1\ \%_{abs}$) after 24 hours of illumination is observed. In case of the cell featuring a boron implanted emitter we measure a significant increase in efficiency of 0.6 $\%_{abs}$. This increase in efficiency is due to an improvement of the open circuit voltage V_{oc}, as well as due to an improvement of the pseudo fill factor, resulting from lower J_{01} and J_{02} values compared to the initial state. A similar behavior for the two cell types is also observed for groups of 8 cells each (B emitter average: $\Delta\eta = 0.6\ \%_{abs}$; BF_2 emitter average: $\Delta\eta = -0.04\ \%_{abs}$). Since both cell groups feature the same passivation stack, we attribute the improvement in case of the B implanted cells to a curing of defects in the bulk and /or in the emitter. However, further research is required to understand the underlying mechanisms. Work in this direction is ongoing, and the results will be published elsewhere.

3.2. Flip-Flop module

In order to demonstrate the potential of our Flip-Flop interconnection scheme for module efficiency improvement, we perform simulations based on *IV* curve measured on a representative half cell (Tab. 2, first rows). One should note that the bifacial factor of this cell of 97.4 % is slightly lower than our average value (98.3 %), thus representing a "conservative" scenario. According to the conventional interconnection scheme, the cell side with the highest efficiency faces the sun. In the hypothetical case of a loss-free interconnection, the full efficiency potential of the cell is transferred into the module, when referring to the cell area only (see Table. II third row interconnected conventionally (emitter up), designated area). Usually, the cells feature a cell spacing of 4 mm to avoid damage at the cell edges due to mechanical stress. The cell spacing has the potential to

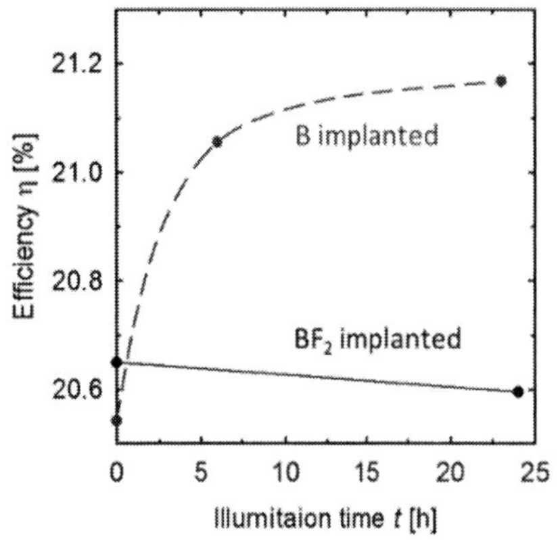

Figure 2 Efficiency of *n*-PERT solar cells with a B or BF_2 implanted emitter in dependence on illumination time.

increase the current of the module, however, it also enlarges the module size and increases resistive losses in the CIR. The first effect can be taken into account as follows: We use the probability for the collection of light reflected from the back sheet between the cells as back sheet light recovery probability *k* as given in Ref. [18] for the best white back sheet tested in that work. However, one has to remark that these values are determined for monofacial cells, which cause some uncertainty when transferring them to a module with bifacial cells. For a cell spacing of 4 mm, *k* is 0.44. Thus, a cell spacing of 4 mm for the conventional interconnection scheme results in a reduction of the efficiency to 19.4 % when referring to the aperture area (cell and back sheet area). The Flip-Flop approach, where half of the cell face the sunny side with the surface having the highest efficiency and the other halve with the opposite surface, results in a decrease in module performance due to a slight current mismatch even when assuming loss-free interconnection. By interpolating the *IV*-characteristics as measured with illumination from the emitter and from the n^+ side and adding the voltage for equal currents, we generate an *IV*-

Table. II Calculated 2-cell module efficiencies for the conventional emitter up and Flip-Flop interconnection scheme. The calculations are based on the measured *IV* data from a cell with a moderate bifacial factor of 97.4%, representing a "conservative" scenario for the Flip-Flop approach.

Module design	Measurement	η [%]	I_{sc} [A]	V_{oc} [mV]	FF [%]	P_{mpp} [W]	Cell spacing [mm]	k
Single cell	Emitter up, full area (97.5cm^2)	20.3	3.94	651	77.3	1.983	-	-
	n^+ up, full area (97.5cm^2),	19.8	3.85	650	77.3	1.930	-	-
Emitter up	designated area	20.3	3.94	1302	77.3	3.965	-	-
	aperture area	19.4	4.10	1302	77.3	4.126	4	0.44
Interconnected Flip-Flop	designated area	20.0	3.85	1300	78.0	3.908	-	-
	aperture area	19.9	3.90	1300	78.0	3.959	1	0.59

characteristics of the series connected cells. When referring to the cell area (designated area), the resulting efficiency of the Flip-Flop module is 20.0 %. For smaller cell spacing, the recovery factor k increases [18]. The Flip-Flop approach enables a cell spacing of 1 mm, which is considered in the last calculation (Table II sixth row). For 1 mm, we determined a $k = 0.59$. The resulting Flip-Flop module efficiency for 1 mm distance (aperture area) is 19.9 %, which is 0.5 $\%_{abs}$ higher compared to the case of the conventional interconnection scheme using the same cells. This is a further benefit of our Flip-Flop interconnection scheme besides the reduction of mechanical stress in the cell.

The efficiency for the emitter up configuration depends crucially on the cell spacing. Figure 3 shows the calculated aperture efficiencies for cell spacing d between 1 and 6 mm based on the cell data give in Table II. The k values are calculated for each d according Ref. [18]. The bifacial factor has no impact when the emitter of all cells is facing the sunny side.

In case of the Flip-Flop configuration we keep the cell spacing constant at 1 mm and vary the bifacial factor by shifting the current

$$I(f_b) = I_0 - I_{sc} \times (1 - f_b), \qquad (1)$$

where I_0 is the measured front side current and $I(f_b)$ is the calculated current for the bifacial factor f_b.

For a cell spacing of 3 mm in case emitter up the same efficiency can be achieved with a bifacial factor as

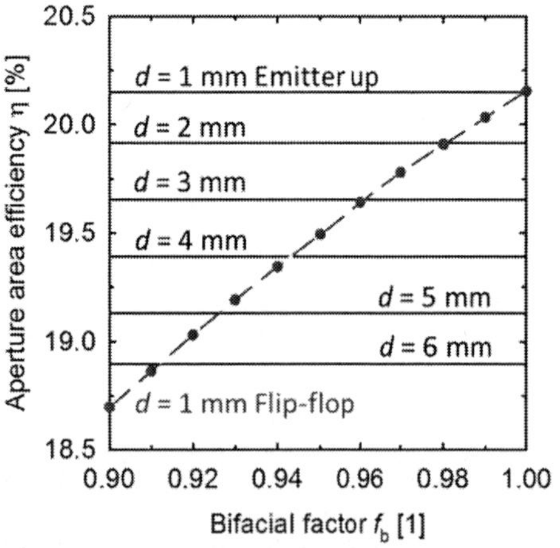

Figure 3 Aperture area (cell and back sheet area) efficiencies for emitter up configuration in dependence on the cell spacing d and for the Flip-Flop configuration with in dependence on the bifacial factor f_b.

low as 96 %. In the case of our average f_b of 98.3 % the Flip-Flop configuration with 1 mm cell spacing is superior to the emitter up configuration with 2 mm cell spacing, which is a challenging cell spacing for the standard configuration (emitter up), since the CIR must be led from the front side a cell to the rear side of an adjacent cell.

Figure 4 shows the experimental demonstrator Flip-Flop module with sixteen cells connected in series. The *IV*-parameters of the module and the expected characteristics based on the *IV* data of the individual cells are given in Table III for the case of no series resistance losses as well as including the resistive losses due the cell and string interconnect CIR. These calculations include no optical losses or gains due to changed absorption and reflection in the module. We expect a module efficiency on designated area (cell area only) of 20.2 % and a fill factor of 77.7 %.

We determine an efficiency of 20.2 % on the aperture area (cell and back sheet). This is a very promising result for a first demonstrator module made of cells with "only" 20.3 % front-side and 19.8 % rear-side efficiency.

Comparing the expected and the measured performance of the module, we observe no significant change in V_{oc} due to the interconnection. The reduction of the fill factor of 0.5 $\%_{abs}$ is moderate and can be explained by the resistive losses in the CIR (see Table III, lines 2 and 3). The short circuit current of the module is increased compared to the cells, which is caused by the light recovery form the interconnects, the fingers, and the back sheet as well as the reduced front side reflection at

Figure 4 Photograph of our demonstrator Flip-Flop module with sixteen cells connected in series.

Table. III Measured *IV*-parameters (aperture area) of a proof-of-concept Flip-Flop module consisting of 16 bifacial solar cells before and after light treatment. For comparison, the expectations based on the measured *IV* curves of all interconnected cells are also shown (no current gain due to back sheet and no losses due to the encapsulation compared to the cell *IV*-measurements assumed).

	η [%]	I_{sc} [A]	V_{oc} [V]	FF [%]	P_{mpp} [W]	A [cm²]
Calculated form cell *IV*-curves	20.3	3.87	10.45	78.2	31.62	1560
Calculated form cell *IV* with R_s	20.2	3.87	10.45	77.7	31.44	1560
Measured after lamination	20.2	3.92	10.47	77.7	31.89	1578
Measured after light soaking	20.5	3.93	10.54	78.2	32.39	1578

the cell surface, which overcompensates the losses due to absorption and reflection of the front glass and EVA. The current could be further increased by using a back sheet with a high infrared reflectivity to benefit from the light transmitted through the cell. The used back sheet has a low reflection (~75 %) in the wavelength regime between 900 nm and 1200 nm [19].

As shown above on in section 3.1, an exposure to light has the potential to enhance the cell performance. The cells received no light treatment before interconnection, since the effect was discovered after finishing the module. Therefore, we irradiate the module for 25 hours. By this the open circuit voltage and the fill factor can be increased resulting in an aperture efficiency of 20.5 %.

4. SUMMARY

We present bifacial *n*-PERT cells with an extremely high bifacial factor of close to unity (99.4 %) and a module concept benefiting from this property. The cells are fully ion implanted and screen-printed. They are fabricated either by applying a single co-anneal process to cure the implant damage or by applying two separate anneals after boron or BF_2 and phosphorous implant, respectively. In the first case, our best cells achieve an independently confirmed front (rear) side efficiency of 21.0 % (20.43 %), in the latter case of 21.5 % (21.31 %). To our knowledge, these values corresponds to the highest efficiencies reported so far for this cell type. Moreover, the bifacial factor of 99.4 % is to our knowledge the highest value reported for any Si cell concept. Laser cutting of these cells shows no negative impact on the cell performance.

We introduce a novel interconnection scheme for bifacial cells, which we call Flip-Flop. In the Flip-Flop module scheme every second cell is flipped upside-down, in order to interconnect them in a "planer manner". Thereby the sunny sides of them are interconnected by a CIR enabling the series interconnection emitter contact of one cell and the base contact of the next cell. This offers several advantages: the distance between the cells is not primarily limited by the interconnection of the individual cells, allowing a smaller cell spacing and thus higher module efficiencies. The reduced resistive losses in the interconnectors allow the usage of thinner CIR, reducing the shading losses on the cell.

We show that our cells with an average bifacial factor of 98.3 % lead to higher module efficiencies in the Flip-Flop configuration with a cell spacing of 1 mm compared to a conventional interconnection of the same cells with 4 mm cell spacing.

Our first proof of concept module reaches an aperture efficiency of 20.2 %. The efficiency increases to 20.5 % after light treatment. The underlying effects enabling this efficiency gain need to be investigated further on cell level.

The proposed Flip-Flip interconnection can be applied to any cell concept besides *n*-PERT with sufficiently high bifaciality (compare Fig. 3), such as, for example, amorphous/crystalline Si heterojunctions cells. It is compatible with both interconnection via cell interconnection ribbons and via multiple wires.

ACKNOWLEDGMENT

The authors would like to thank Sabine Kirstein, Peter Giesel and Andreas Klatt for their help with sample processing, David Hinken for the spectral mismatch correction of the module.

We gratefully acknowledge the support by the German Ministry for Economic Affairs and Energy under grants 0325480 (CHIP) and 0325641 (PERC-2-Module) as well as by the state of Lower Saxony.

References

[1] International Technology Roadmap for Photovoltaic, www.itrpv.net (2015)

[2] U. A. Yusufoglu *et al.*, Energy Procedia **55**, pp. 389–395 (2014).

[3] J. P. Singh *et al.*, IEEE J. Photovoltaics **5**(3), pp. 783–791 (2015).

[4] T. Dullweber *et al.*, Proc. 30th EU PVSEC, Hamburg, Germany, pp. 341 - 350 (2015).

[5] F. Kiefer *et al.*, Sol. Energ. Mat. Sol. Cells **157**, pp. 26–330 (2016).

[6] M. W. P. E. Lamers *et al.*, Prog. Photovolt: Res. Appl, **20**(1), pp. 62–73 (2012).

[7] R. Kopecek *et al.*, Proc. 21st EU PVSEC, Dresden, Germany, pp. 111–115 (2006).

[8] A. Augusto *et al.*, presented at 43rd IEEE PVSC, Portland, OR (2016)

[9] S. Gonsui *et al.*, Proc. 28th EU PVSEC, Frankfurt, Germany, pp. 718–721 (2012).

[10] A. Frey *et al.*, Proc. 29th EU PVSEC, Amsterdam, the Netherlands, pp. 656–660 (2014).

[11] P. Rothhardt *et al.*, IEEE J. Photovoltaics **4**(3), pp. 827–833 (2014).

[12] A. Lanterne *et al.*, Prog. Photovolt.: Res. Appl. **23**, pp. 1458–1465 (2015).

[13] V. D. Mihailetchi *et al.*, Energy Procedia **77**, pp. 534–539 (2015)

[14] M. Pander *et al.*, Proc. 28th EU PVSEC, Frankfurt, Germany, pp. 2851–2857 (2012).

[15] E. Hasselbrink *et al.*, 39th IEEE PVSC, Tampa, FL, pp. 0007–0012 (2013).
[16] J. Krügener *et al.*, Sol. Energ. Mat. Sol. Cells **142**, pp. 12–17 (2015).
[17] V. Bhosle *et al.*, Proc. of the 31st EU PVSEC, Hamburg, Germany, pp. 425–428 (2015).
[18] M. Köntges *et al.*, to be presented at 32nd EU PVSEC, Munich, Germany (2016).
[19] M. Vogt, PhD thesis, Leibniz Universität Hannover, Hannover (2015).

PROGRESS WITH ROTATIONAL PRINTING FOR THE FRONT SIDE METALLIZATION OF SILICON SOLAR CELLS

A. Lorenz[1], C. Gredy[1], M. Lehner[2], R. Greutmann[3], H. Brocker[3], J. Rohde[4], A. Senne[5], T. Strauch[1], H. Reinecke[6], F. Clement[1]
[1]Fraunhofer Institute for Solar Energy Systems (ISE), Heidenhofstr. 2, 79110 Freiburg, Germany
[2]Lehner Engineering GmbH, Ebnetstrasse 18, 9032 Engelburg, Switzerland
[3]Gallus Ferd. Rüesch AG, Harzbüchelstrasse 34, 9016 St. Gallen, Switzerland
[4]Zecher GmbH, Görlitzer Str.2, 33098 Paderborn, Germany
[5]ContiTech Elastomer-Beschichtungen GmbH, Breslauer Str. 14, 7154 Northeim, Germany
[6]Albert-Ludwigs-Universität, Institut für Mikrosystemtechnik, Georges-Köhler-Allee 101, 79110 Freiburg, Germany

ABSTRACT: The current cost pressure within the photovoltaic industry requires innovative approaches to increase productivity and reduce production costs. The use of highly productive rotational printing methods offers the potential to increase the throughput of the cost-intensive metallization process by the factor 2 to 3. Within this work, we present two innovative rotational printing technologies – flexographic printing and rotary screen printing – for the front side metallization of Silicon solar cells. We show that both technologies can realize contact fingers with a lateral finger resistance R_L well below 10 Ω/cm in one printing step which is sufficient for busbarless solar cells with multi-wire interconnection. For flexographic printing, we found a strong impact of the anilox roller properties on geometry and lateral resistance of the contact fingers. With respect to flexographic printing, narrow contact fingers with a mean lateral resistance down to R_L = 5.8 Ω/cm could be printed in a single printing step using a large dip volume of V_D = 16.5 cm³/cm². For rotary screen printing, an even lower lateral finger resistance down to R_L = 1.1 Ω/cm could be achieved. Yet, further process optimization is required to reduce paste spreading and thus the printed finger width.
Keywords: Silicon Solar Cells, Manufacturing and Processing, Metallization, Rotation Printing

1 INTRODUCTION

1.1 Rotational printing technology for solar cells

Though competition and high cost pressure in the industrial fabrication of crystalline Silicon (c-Si) solar cells demands new approaches to decrease costs and increase productivity. Considering the fabrication process of solar cells, front and rear side metallization is a particularly cost-intensive production step. To date, flatbed screen printing (FSP) is still the predominant metallization technique for c-Si solar cells and will also play an important role in the future. However, new technological approaches can help to increase the throughput of metallization lines considerably. The use of rotational printing technologies which are investigated by Fraunhofer ISE since 2011 [1] is a promising approach to meet this demand. Rotational printing methods are well established and widely used for a wide range of industrial applications. These usually roll-to-roll based printing methods can realize a very high printing speed of up to 800 m/min. on web-based materials [2]. This printing speed is obviously not possible for solar cell metallization lines due to the necessity of a precise delivery and alignment of the Si wafers. Yet, using rotational printing methods will be able to increase the throughput by the factor 2 to 3 compared to current screen-printing metallization lines. However, a rotary metallization line must be able to meet the challenging requirements of solar cell front and rear side metallization, namely ultra-fine contact fingers with sufficient lateral conductivity on the front side and a homogeneous Aluminium (Al) layer on the rear side. Within this work, we present two rotational printing methods for the front and rear side metallization of Si solar cells. We investigate the feasibility of both technologies with focus on the front side metallization of busbarless Si solar cells which are intended for multi-wire interconnection.

1.2 Flexographic printing technology

Flexographic printing is a widely used rotational printing method, mainly in package printing on roll-to-roll materials like cardboard, paper or foil. A low-cost flexible relief printing plate is used as image carrier (Figure 1). The plate is applied on the printing cylinder on top of a stack of compressible foam tape to support a homogeneous ink transfer. The low-viscous silver ink is transferred from the ink chamber onto the so-called anilox roll, a steel cylinder with chromium or ceramic surface and finely engraved micro-wells. Depending on the line screening of the wells and the so-called dip volume V_D [cm³/m²], the anilox roller transfers a specific theoretical amount of ink on the printing plate. Line screening l can vary between 100 and 700 lines/cm and (nominal) dip volume V_D between 16.5 and 1.0 cm³/m². Excessive ink is removed by a doctor blade before the anilox roller wets the elevated areas of the printing plate with a uniform layer thickness which is continuously printed on the substrate.

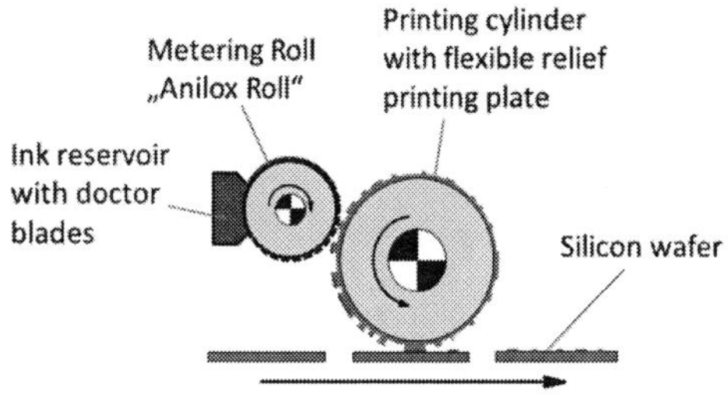

Figure 1: Schematic of a flexo printing unit for solar cell metallization

Previous studies have shown that flexographic printing is able to realize the front side metallization of busbarless solar cells for multi-wire interconnection with less than 5 mg of silver [3]. Further work [4,5] demonstrated very narrow contact fingers down to 23 µm in average on silicon solar cells. However, realizing narrow contact fingers with a lateral finger resistance R_L below 10 Ω/cm has so far proven to be a major challenge. Thus, reducing the lateral finger resistance is an important goal to prove the feasibility of this promising technology for the metallization of busbarless solar cells used for multi-wire interconnection like Meyer Burger's SmartWire Connection Technology (SWCT). [6].

1.3 Rotary screen printing technology

Rotary screen printing (RSP) is able to combine the technological advantages of thick film metallization with the potential high throughput of rotational printing. To date, this technology is primarily used on web-based materials (i.e. label or textile printing [7]) with a printing speed of up to 100 m/min. First attempts to use this technology for solar cell metallization date back to the year 1999 [8]. Yet – to the best knowledge of the authors – no results on Si solar cells have been published so far.

Similar to flatbed screen printing, a woven screen mesh covered with a partly open emulsion layer is used as printing form. However, in contrary to the flat screen used in FSP, cylinder-shaped screens are used for RSP. The meshes of such screens can consist of up to 400 threads per inch (mesh count). However, due to stability reasons of the cylinder screen, thread thickness is increased (minimum 25 to 30 µm) compared to flat screens. FSP requires two printing sequences. Firstly, the open areas of the screen are filled using a metal flood bar. Secondly, the paste is pressed through the openings of the flat screen using a flexible squeegee. Using RSP, the paste is constantly pressed through the screen openings by a fixed squeegee within the rotating screen cylinder (Figure 2). A pre-filling of the screen is thus not required. RSP requires a lower paste viscosity compared to FSP pastes to ensure a good paste transfer through the screen openings. The ability to transfer thick film metallization patterns makes RSP interesting for both – front and rear side metallization of c-Si solar cells.

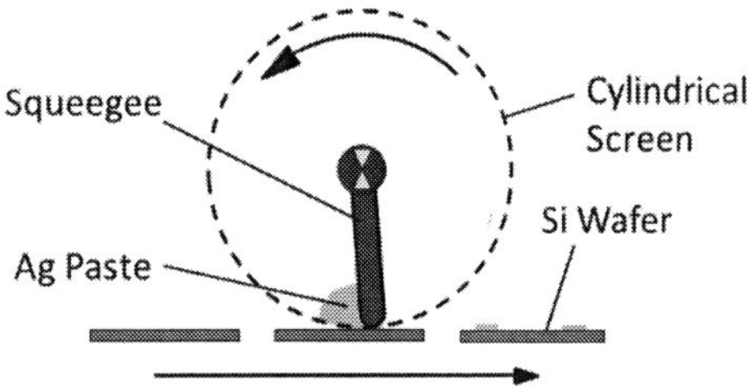

Figure 2: Schematic view of a rotary screen printing unit for solar cell metallization

1.4 Aim of the Experiment

A modern industrial FSP metallization line is able to realize a finger width of w_f = 45-55 µm and a lateral finger resistance of 0.3 to 0.5 Ω/cm [9]. While a narrow finger width is generally favorable to reduce j_{sc}-losses, lateral finger resistance R_L strongly depends on the solar cell interconnection concept. Traditional H-pattern solar cells with 3 busbars require a lateral finger resistance of 0.3 to 0.5 Ω/cm to ensure a small series resistance contribution of the grid. To date, flatbed screen printing is thus still the most suitable metallization method for such solar cells. However, modern interconnection approaches using 5-busbar solar cells or wire-interconnected busbarless solar cells allow considerably lower lateral resistances. R_L should not exceed 1.5 Ω/cm for 5-busbar cells and 10 Ω/cm for busbarless solar cells interconnected with 18 wires by SWCT [6]. The aim of this experiment is to evaluate the current opportunities of flexographic printing and rotary screen printing with respect to solar cell front side metallization. It will be clarified whether both technologies are able to meet the challenging demands regarding finger width and lateral finger resistance.

2 EXPERIMENTAL SETUP

2.1 Feasibility study for flexographic printing

The experiment is carried out using industrially pre-produced Czrochalski-grown p-type Si (Cz-Si) precursors with 156 mm edge length. The precursors have an n-type emitter with a sheet resistance of R_{sh} ≈ 85-90 Ω/sq. and a SiNx anti-reflection coating (ARC) on the front side. A roll-to-flat flexographic printing platform *Nissha Angstromer S15* is used to apply the front side metallization of the solar cells. This machine has a vacuum substrate holder to fix the wafer during the printing process. The position of the vacuum substrate holder perpendicular to the axis of the printing cylinder (z-position) is adjusted by a micrometer spindle. Each wafer is positioned manually on the vacuum table. Printing speed is kept constant at v_p ≈ 300 mm/s during the experiment. The optimum printing pressure (= minimal pressure which is required to print the whole printing image without missing areas) is determined by iteratively adjusting the vacuum substrate holder in z-direction. The identified optimum printing pressure is then kept constant during the printing process. All wafers are dried in a cabinet drier at T = 200°C for 2-5 minutes directly after printing. Subsequently, a contact firing step is carried out using a fast firing furnace with a peak set temperature of T_{FFO} = 830°C.

A ceramic band anilox roller with 3 differently engraved sections is fabricated for the experiment. Each section has a different screening and dip volume V_D (see Table 1).

Table 1: Parameters of the band anilox roller

	Well Geometry	Screening [l/cm]	Dip Volume V_D [cm³/m²]
Band 1	Hexagonal	700	1.0
Band 2	Hexagonal	340	6.3
Band 3	Hexagonal	100	16.5

A test layout with 3 identical sections (corresponding to the bands of the anilox roller) is fabricated using a *ContiTech Laserline* CSC flexographic printing plate. Each section contains identical finger elements with nominal widths w_n between 5 and 50 µm and measurement patterns for contact resistance measurement using transfer-length method (TLM) [10]. Using this setup, all sections can be printed on one wafer using the 3 anilox bands in parallel (Figure 3).

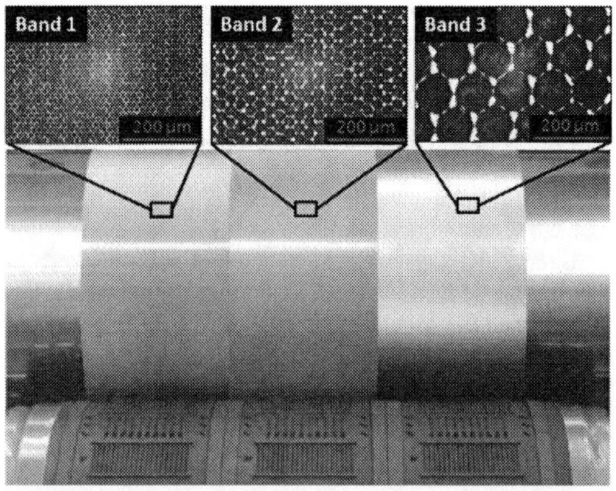

Figure 3: Experimental setup to enable a parallel printing of the test form using 3 anilox band sections in parallel (shown in the microscopic images above).

The silver ink is developed in-house and contains silver particles for metal-semiconductor contact formation, lead glass as a sintering additive, solvents, synthetic resin as binder and further additives to adjust viscosity, dispersion and electrical contact formation to the emitter. The formulation of the ink is based on a previously developed aerosol jet ink [11] and has been gradually optimized for flexographic printing in previous experiments [5,12,13].

2.2 Feasibility study for rotary screen printing

The feasibility study for rotary screen printing is carried out on a *Gallus EM 280* label printing machine (Figure 4). This machine is designed to print on continuous web materials like foil or paper web. A commercially available Ag-paste for solar cell front side metallization is diluted iteratively until the viscosity is adequate to ensure a homogeneous paste transfer. Every wafer is fixed individually on the foil web before each print run to enable the transport of the wafers through the rotary screen printing unit. Printing speed is set to $v_p \approx 170$ mm/s. A test layout comprising contact finger elements with a nominal width between 20 and 55 µm is printed on multicrystalline Si wafers (mc-Si) with SiN_x-ARC.

Figure 4: Rotational screen printing unit of the Gallus EM 280 label printing machine.

2.3 Characterization of contact finger geometry

The three-dimensional geometry (width w_f, height h_f, cross-section area A_f) of all contact fingers is measured using an *Olympus Lext* confocal microscope (amplification factor 500x).

On the flexo-printed test patterns, 10 fingers are measured within each of the 3 anilox band sections resulting in 30 measurements per wafer and approx. 210 measurements in total.

On the test patterns printed with rotary screen printing, 7 fingers per wafer are measured on four wafers, resulting in 28 measurements in total.

All microscopic images are analyzed using the image analysis algorithm *FineUp 1.4* which has been developed at *Fraunhofer ISE* [14]. This algorithm enables a precise and reproducible quantification of relevant contact finger geometry parameters like finger width w_f, finger height h_f and finger cross-section area A_f.

2.4 Electrical characterization of contact fingers

Using four-point measurement method, absolute finger resistance R_f is measured on the same contact fingers which have been previously analyzed by confocal microscopy. Subsequently, lateral finger resistance per unit length R_L [Ω/cm] is calculated with the known distance d_{pin} of the measurement pins.

3 RESULTS AND DISCUSSION

3.1 Flexo-printed contact fingers

Figure 5 shows the relation of nominal finger width w_n (in the digital test layout) and flexo-printed finger width w_f.

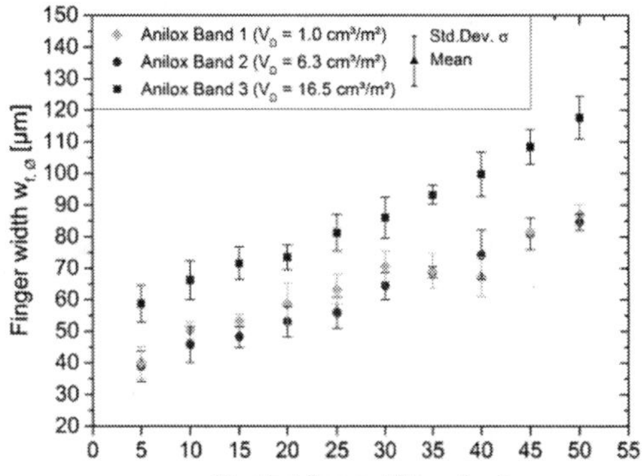

Figure 5: Printed finger width w_f depending on the anilox band section and nominal finger width w_n

It is clearly visible that w_f increases linearly with w_n which can be expected as broader finger elements usually lead to broader fingers on the substrate. The results furthermore reveal that w_f is considerably broader than the corresponding finger elements on the plate. This strong increase primarily originates from plate deformation and ink spreading effects during the printing process. Bould *et al.* [15] deeply investigated these effects which they classified as *(flank) barreling*, *(surface) extension* and *ink spreading* (Figure 6).

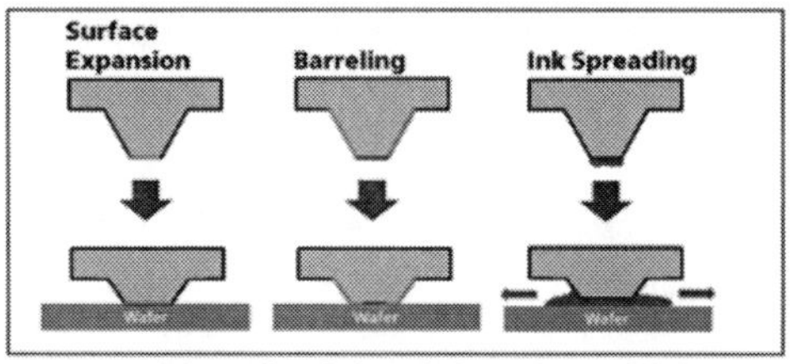

Figure 6: Schematic of plate deformation and ink spreading effects of the fine line elements on the printing plate under pressure (modified illustration based on [15]).

Particularly barreling and surface extension effects strongly depend on the mechanical stability of the finger elements on the plate. Broad finger elements obviously have a greater stability against deformation. This explains why broad finger elements (w_n = 50 µm) only show a moderate gain (percentage increase of w_f compared to w_n) of 170 to 240 % after printing. Narrow finger elements (w_n = 5 µm) are less stable and thus show a considerably stronger gain of 780 to 1180 %. Despite of this stronger gain, narrowest fingers are obtained with the smallest finger elements (w_n = 5 µm). These findings correspond to the results of earlier studies [13] and underline the necessity to minimize the width and – in parallel – increase the stability of the finger elements on the plate. This could be realized by optimizing the three-dimensional finger geometry on the plate within the *Laserline* process [16]. The results furthermore reveal a considerable influence of the anilox dip volume V_D on finger width w_f. In theory, increasing the line screening l of the anilox roller enables the printing of narrower lines. Thus, one would expect smaller fingers for band section 1 (l = 700 l/cm) compared to band section 2 (l = 340 l/cm) and 3 (l = 100 l/cm). While this effect can be observed for fingers printed with band 3 (increased finger width), band 1 and 2 reveal closely comparable results. A close investigation of the vacuum table revealed a minor axial deflection of the table in the center which decreased the effective printing pressure in section 2 slightly. It is very likely that the width of fingers printed with band 2 would thus be in between band 1 and 3 without this effect. The effect of the anilox line screening on w_f can thus be confirmed in general. Additionally, the ridge and flanks of the fine line finger elements might be covered with much more ink if the area and volume of the micro-wells increases with higher dip volume V_D. This might lead to an "over-inking-effect" in case of band 3 and thus to stronger ink spreading. A smaller dip volume and higher line screening is thus beneficial to decrease finger width w_f.

Figure 7 shows the relation between w_n and mean finger cross-section area A_f of the printed contact fingers. It is clearly visible that A_f is primarily influenced by the dip volume V_D of the anilox roller. Yet, the dip volume of band 1 and 2 seems to be too low to realize a sufficient finger height and thus cross-section area. Measurement data and a close visual inspection of the microscopic images revealed that only band section 3 is capable to realize an adequate finger height h_f in one printing step. Fingers printed with this band revealed a finger height of $h_f \approx$ 6-7 µm. In contrary, finger height is considerably lower in band 1 and 2 ($h_f \approx$ 0.5 to 3 µm). While finger cross-section area A_f in band 3 also increases linearly with w_n, this is not the case for band 1 and 2.

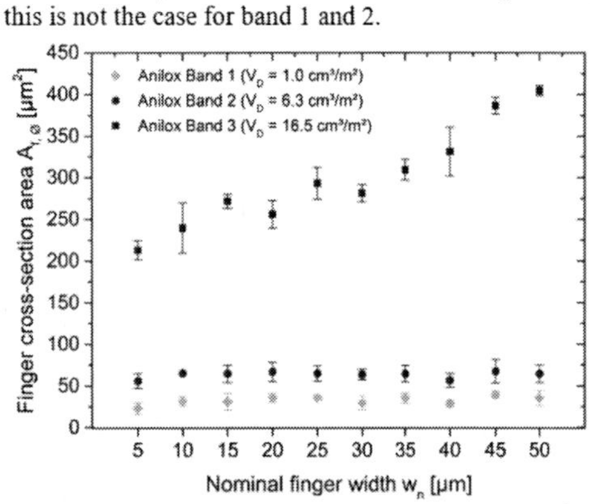

Figure 7: Mean finger cross-section area A_f depending on anilox band section and nominal finger width w_n

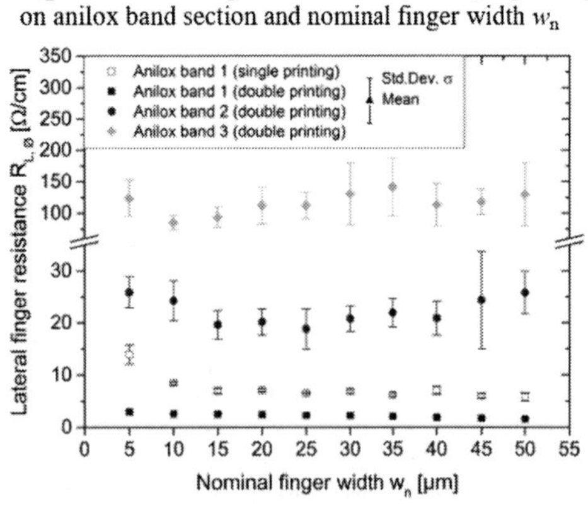

Figure 8: Lateral finger resistance R_L depending on anilox band section and nominal finger width w_n

Figure 8 illustrates the determined mean lateral finger resistance R_L depending on anilox band section and nominal finger width w_n. It is clearly visible that R_L is strongly influenced by the dip volume V_D of the anilox band section. Fingers printed with band sections 1 and 2 in one printing step did not obtain meaningful R_L-values at all. A close inspection of microscopic images revealed that the height and uniformity of fingers printed with band 1 and 2 varied considerably which led to multiple constrictions and interruptions. Thus, realizing uniform contact fingers with an adequate height requires at least a double printing step for these band sections. Figure 8 thus shows R_L-results of double printed contact fingers for band 1 and 2. However, a double printing step still let to very high R_L-values in band 1. While R_L-values of (double printed) fingers in band 2 are considerably lower, they are still by the factor 2-3 too high for the front side metallization of multi-wire solar cells.

However, fingers printed with band 3 obtained considerably lower and very promising R_L-values between 5.8 Ω/cm (w_n = 50 µm) and 13.9 Ω/cm (w_n = 5 µm) in a single printing step. The R_L-values of fingers printed with band 3 decreases moderately with increasing w_n and remains relatively stable for fingers broader than $w_n \geq$ 15 µm. By applying a double printing step with band 3, R_L can be further decreased to 2,9 Ω/cm (w_n = 5 µm) and 1.6 Ω/cm (w_n = 50 µm). In summary, only anilox band section 3 (large dip volume) proved to be capable to realize contact fingers with sufficiently low R_L-values in a single printing step. The results demonstrate for the first time that the target of flexo-printed fingers with a lateral resistance $R_L \leq$ 10 Ω/cm can be realized in one printing step using a large anilox roller dip volume.

To assess the grid-related *FF*-loss of a flexo-printed front side grid, the theoretic contribution of lateral finger resistance R_L to the series resistance ($r_{s,grid}$) and the resulting *FF*-loss (ΔFF_{grid}) of a virtual solar cell with SWCT interconnection is calculated using models based on [17–19]. The calculation assumes that all fingers of the front side grid are printed homogeneously and would achieve the same geometric and electric results. A flexo-printed busbarless front side grid with 100 contact fingers and wire-interconnection using 18 wires is assumed. The previously shown experimental results (R_L and w_f) of flexo printed contact fingers printed with w_n = 10 µm are taken as input parameters. Losses due to transition resistance between contact fingers and wires are neglected. The results are shown in Table 2.

Table 2: Calculated series resistance contribution $r_{s,grid}$ and fill factor loss ΔFF_{grid} of a virtual flexo-printed solar cell with SmartWire-interconnection (18 wires)

	R_L [Ω/cm]	$r_{s,grid}$ [Ω/cm²]	ΔFF_{grid}
Band 1[a]	85.3	7.95	-45.31
Band 2[a]	24.3	0.23	-1.29
Band 3[b]	8.5	0.08	-0.45
Band 3[a]	2.9	0.03	-0.15

[a]double printing step [a]single printing step

The results of the calculations again underline the necessity to use an anilox roll with a large dip volume V_D. Using an anilox roller with a smaller dip volume like band 1 and 2 leads to unacceptably high R_L-values and thus induces considerable *FF*-losses even when applying a double printing step.

To evaluate the shading impact of the printed front side grid, the theoretic j_{sc}-loss is calculated with known photo-generated current j_{ph} and total percentage of shading on the front side $A_{sh\%}$ (caused by the printed front side grid and the wires). The calculation is carried out for an assumed SWCT interconnection with 18 wires (wire diameter d_{wire} = 200 µm). An effective shading width of 59 % is assumed for the round wires [20]. A photo-generated current density of j_{ph} = 40.9 mA/cm² is taken as basis. Reflection and absorption effects due to module encapsulation on j_{sc} are not considered. It is furthermore assumed that the flexo-printed front side grid is printed homogeneously with the same finger width on the whole wafer. The calculated theoretical j_{sc}-loss for an optimal grid printed with band 1, 2 and 3 and w_n = 5 µm is shown in Table 3.

Table 3: Calculated total shading percentage and resulting j_{sc}-loss for a virtual wire-interconnected solar cell with flexo-printed front side grid using anilox band 1 to 3

	w_f [µm]	$A_{sh\%}$ [%]	Δj_{sc} [mA/cm²]
Band 1[a]	50	4.4	-1.8
Band 2[a]	46	4.2	-1.7
Band 3[b]	66	5.4	-2.2

[a]double printing step [b]single printing step

The broader mean finger width of anilox band section 3 leads to an increased j_{sc}-loss due to more shading. Considering the fact that suitable contact fingers with adequately low R_L can only be printed with band 3 it must be the main focus of further development to decrease the finger width. A possible path to achieve this goal is an optimization of the ink rheology to prevent excessive spreading.

3.2 Rotary screen-printed contact fingers

Figure 9 shows the printed mean finger width w_f of rotary screen-printed contact fingers in relation to the nominal screen opening of the fingers (further referred to as nominal width w_n). A visual inspection of the fingers revealed that uninterrupted contact fingers can be printed down to a nominal finger width of w_n = 30 µm. The smallest mean finger width was determined with w_f = 68 µm (w_n = 30 µm). Fingers printed with a smaller screen opening showed multiple interruptions and constrictions.

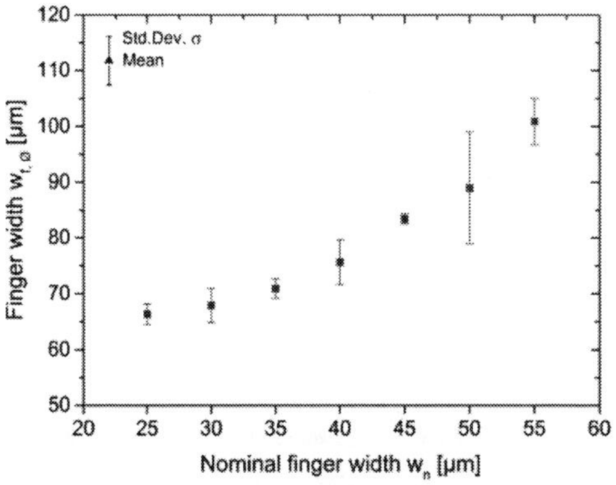

Figure 9: Relation between nominal finger width w_n in the test layout and width w_f of rotary screen printed fingers.

Compared to fingers printed with FSP using the same screen opening, w_f of rotary screen printed fingers is about 1.7 to 2 times broader. The origin of this gain might be explained by a stronger spreading of the diluted Ag-paste. A good measure for the spreading tendency of a screen printing paste is the yield stress τ_y [21]. Pospischil *et al.* have shown that τ_y has a considerable impact on the geometry of screen-printed contact fingers [22]. Yield stress τ_y of screen printing paste can be measured using a profiled parallel disk-plate geometry in order to prevent slip effects as published by Pospischil *et al.* [23]. Figure 10 shows the yield stress of a commercially available Ag-paste before and after dilution to an adequate viscosity for rotary screen printing. The results show that yield stress τ_y of the diluted paste is considerably lower which could explain an increased spreading tendency on the wafer surface directly after printing. Optimizing the rheological properties of the Ag-paste must thus be the main route to obtain narrower contact fingers in RSP.

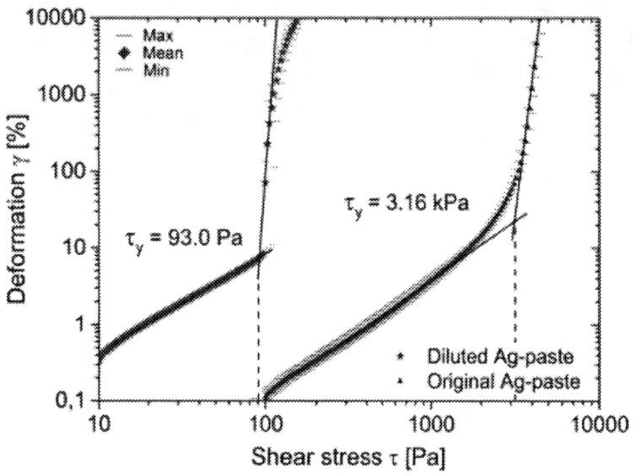

Figure 10: Yield stress τ_y of the diluted Ag-paste for rotary screen printing and the original non-diluted Ag-paste for flatbed screen printing.

Finger height h_f and cross-section area A_f increases linearly with w_n as more paste is transferred with broader screen openings. Finger resistance R_L decreases with increasing cross-section area (Figure 11). Fingers printed with a screen opening of w_n = 30 µm or more could be printed without interruptions and thus obtained relatively low mean lateral resistances between $R_{L,max}$ = 3.6 Ω/cm (w_n = 30 µm) and $R_{L,min}$ = 1.1 Ω/cm (w_n = 55 µm).

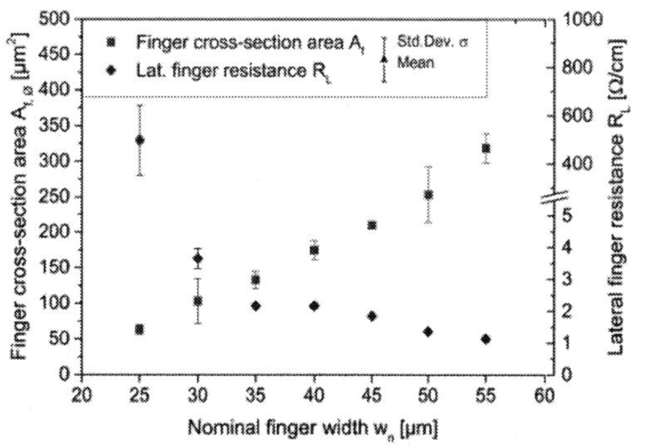

Figure 11: Finger cross-section area A_f and lateral finger resistance R_L depending on the nominal finger width respectively the nominal screen opening w_n.

Series resistance contribution $r_{s,grid}$ and fill factor loss ΔFF_{grid} of a virtual busbarless solar cell with rotary screen printed grid and SmartWire interconnection (18 wires) is calculated for 3 screen openings w_n. Again, a front side grid with 100 homogeneously printed contact fingers is assumed. The further assumptions are similar to the previously described calculation in section 3.1. The results are shown in Table 4.

Table 4: Calculated series resistance contribution $r_{s,grid}$ and fill factor loss ΔFF_{grid} of a virtual rotary screen-printed solar cell with SmartWire-interconnection (18 wires)

w_n	R_L [Ω/cm]	$r_{s,grid}$ [Ω/cm²]	ΔFF_{grid}
30	3.6	0.03	-0.19
45	1.8	0.02	-0.10
55	1.1	0.01	-0.06

The calculation shows that a front side grid printed with rotary screen printing does contribute to a minor extend to the total series resistance of a wire-interconnected solar cell.

Furtermore, total percentage of shading on the front side $A_{sh\%}$ and the resulting j_{sc}-loss Δj_{sc} of grid and wires are calculated as described beforehand. The results are shown in Table 5.

Table 5: Calculated total shading percentage $A_{sh\%}$ and resulting j_{sc}-loss for a virtual wire-interconnected solar cell with rotary screen-printed front side grid and SmartWire-interconnection (18 wires)

w_n	w_f [µm]	$A_{sh\%}$ [%]	Δj_{sc} [mA/cm²]
30	68	5.5	-2.2
45	83	6.5	-2.7
55	101	7.6	-3.1

In contrary to *FF*-losses, shading losses are much more relevant due to the relatively broad mean finger width w_f. Thus, further optimization in the field of rotary screen printing must focus on a considerable reduction of the printed finger width which could be achieved by a rheological optimization of the printing paste.

4 CONCLUSION

Within this work, we showed that both approaches – flexographic printing and rotational screen printing – are able to realize a suitable front side metallization for busbarless solar cells with wire interconnection.

With focus on flexo printing, a considerable impact of the anilox roller on the geometry and lateral resistance of the printed contact fingers has been shown. Using an anilox roller configuration with a very high dip volume of V_D = 16.5 cm³/m², contact fingers with a promising mean lateral resistance down to R_L = 5.8 Ω/cm could be printed in a single printing step. Calculations of j_{sc}-loss and *FF*-loss revealed that contact fingers printed with a smaller dip volume can reduce shading due to a smaller finger width but also induce considerably higher *FF*-losses as the lateral resistance was considerably higher even after applying a double printing step. It is thus necessary to use an anilox roller with a large dip volume if the grid should be applied with only one printing step. It was further shown that a narrow nominal width in the digital data (and respectively on the plate) is beneficial to print narrow contact fingers despite of a stronger deformation of such fine elements under pressure. The results confirm the ability of flexographic printing to apply an adequate front side grid on busbarless solar cells for multi-wire interconnection. Yet, further development is still required and must focus on realizing narrower contact fingers with comparably low R_L-values. The path to follow is a careful selection of the anilox roller specification, an optimization of the contact finger stability on the plate and a further optimization of the ink formulation and rheology.

The second experiment focused on rotary screen printing technology. It has been shown that uninterrupted contact fingers can be printed even with a narrow nominal screen opening of w_n = 30 µm. It is possible to use commercially available Ag-paste, yet the paste has to diluted to a lower viscosity. It was further shown that the dilution strongly affects the yield point τ_y of the paste and – in consequence – leads to strong paste spreading. Fingers printed with an opening of 30 µm showed in consequence a significantly broader finger width of w_f = 68 µm. Rotary screen printed fingers obtained low mean lateral resistances between $R_{L,max}$ = 3.6 and $R_{L,min}$ = 1.1 Ω/cm. While the *FF*-loss of a theoretic rotary screen-printed grid proved to be very small with respect to multi-wire interconnection, the relatively broad finger width w_f would induce a considerable j_{sc}-loss. Further optimization should thus focus on realizing narrower contact fingers by adjusting the rheological properties of the paste.

In summary, we successfully demonstrated the potential of two innovative rotational printing methods for the front side metallization of busbarless solar cells with multi-wire interconnection.

5 ACKNOWLEDGEMENTS

This work was partly supported by the German Federal Ministry of Education and Research (BMBF) within the funding program Photonics Research Germany under the contract number 13N13512 (Rock-Star). The authors further thank *Gallus Ferd. Ruesch AG*, *Zecher GmbH* and *ContiTech Elastomer-Beschichtungen GmbH* for the support of this work.

6 REFERENCES

[1] Frey M, Clement F, Dilfer S, Erath D, Biro D. Front-side Metallization By Means Of Flexographic Printing. *Energy Procedia* 8, 2011, p. 581–586.

[2] Windmöller & Hölscher. Technical Data Sheet Vistaflex C. Available at: http://www.wuh-lengerich.de/de/printing/druckmaschinen/flexodruckmaschinen/vistaflex-c/; 2015 [accessed 09.01.2015].

[3] Lorenz A, Gredy C, Senne A, Beyer S, Yao Y, Papet P, Ufheil J, Reinecke H,Clement F. Rotational Printing Technology for the Metallization of Highly Efficient Busbarless Solar Cells [Submitted for publication 2016].

[4] Thibert S, Jourdan J, Bechevet B, Mialon S, Chaussy D, Reverdy-Bruas N, Beneventi D. Study of the high throughput flexographic process for silicon solar cell metallisation. *Progress in Photovoltaics* 2015.

[5] Lorenz A, Kalio A, Hofmeister GT, Nold S, Kraft A, Bartsch J, Wolf A, Dreher M, Clement F, Biro D. Flexographic Printing – High Throughput Technology for Fine Line Seed Layer Printing on Silicon Solar Cells. *Proc. of the 28th EUPVSEC* 2013, p. 1017–1023.

[6] Söderström T, Papet P, Yao Y, Ufheil J. SmartWire Connection Technology; 2014.

[7] Gallus Ferd. Rüesch AG. Gallus Screeny – Innovativer Siebdruck. Available at: http://www.gallus-group.com/de/DesktopDefault.aspx/tabid-397/600_read-2147; 2014 [accessed 27.03.2015].

[8] Hahne P. Innovative Druck- und Metallisierungsverfahren für die Solarzellentechnologie. PhD Thesis, Fernuniversität Hagen, 2000.

[9] ITRPV. International Technology Roadmap for Photovoltaic: 2015 results. 7th ed; 2016.

[10] Kontermann S, Hörteis M, Ruf A, Feo S, Preu R. Spatially resolved contact-resistance measurements on crystalline silicon solar cells. *Phys. Status Solidi (a)* 2009, doi:10.1002/pssa.200925077.

[11] Hörteis M, Mette A, Richter P, Fidorra F, Glunz S. Further progress in metal aerosol jet printing for front side metallization of silicon solar cells. *Proc. of the 22nd EUPVSEC* 2007, p. 1039–1042.

[12] Lorenz A, Kalio A, Barnes Hofmeister T, Kraft A, Bartsch J, Clement F, Reinecke H, Biro D. Developing a high throughput printing technology for silicon solar cell front side metallisation using flexography. *J. Print Media Technol. Res.* 4, 2014, p. 227–240, doi:10.14622/JPMTR-1408.

[13] Lorenz A, Senne A, Rohde J, Kroh S, Wittenberg M, Krüger K, Clement F, Biro D. Evaluation of Flexographic Printing Technology for Multi-busbar Solar Cells. *Energy Procedia* 67, 2015, p. 126–137, doi:10.1016/j.egypro.2015.03.296.

[14] Strauch T, Demant M, Lorenz A, Haunschild J, Rein S. Two Image Processing Tools to Analyse Alkaline Texture and Finger Geometry in Microscope Images. *Proc. of the 29th EUPVSEC* 2014.

[15] Bould DC, Claypole TC, Bohan MFJ. An investigation into plate deformation in flexographic printing. *Proceedings of the Institution of Mechanical Engineers, Part B: Journal of Engineering Manufacture* 218, 2004, p.1499–1511

[16] A. Lorenz, C. Gredy, A. Senne, S. Beyer, Y. Yao, P. Papet, J. Ufheil, H. Reinecke, F. Clement. Flexo-printed busbarless solar cells for multi-wire interconnection. To be published in Energy Procedia 2016.

[17] Fellmeth T, Clement F, Biro D. Analytical Modeling of Industrial-Related Silicon Solar Cells. *IEEE J. Photovoltaics* 4, 2014, p. 504–513, doi:10.1109/JPHOTOV.2013.2281105.

[18] Fellmeth T. Silicon Solar Cells for the Application in Low Concentrator Systems-Development and Characterization. PhD Thesis, Eberhard Karls Universität, Tübingen, 2014.

[19] Mette A. New concepts for front side metallization of industrial silicon solar cells. PhD Thesis, Albert-Ludwigs-Universität, Freiburg, 2007.

[20] Witteck R, Hinken D, Schulte-Huxel H, Vogt MR, Muller J, Blankemeyer S, Kontges M, Bothe K, Brendel R. Optimized Interconnection of Passivated Emitter and Rear Cells by Experimentally Verified Modeling. IEEE J. Photovoltaics 2016;6:432–9, doi:10.1109/JPHOTOV.2016.2514706.

[21] Mezger TG. Das Rheologie-Handbuch: Für Anwender von Rotations- und Oszillations-Rheometern. Farbe und Lack Edition. 4th ed. Hannover: Vincentz Network; 2012.

[22] Pospischil M, Specht J, Konig M, Horteis M, Mohr C, Clement F, Biro D. Paste Rheology Correlating With Dispensed Finger Geometry. *IEEE J. Photovoltaics* 4, 2014, pp. 498–503, doi:10.1109/JPHOTOV.2013.2278657.

[23] Pospischil M, Klawitter M, Kuchler M, Specht J, Gentischer H, Efinger R, Kroner C, Luegmair M, König M, Hörteis M, Mohr C, Wende L, Lossen J, Weiß M, Doll O, Koehler I, Zengerle R, Clement F, Biro D. Process Development for a High-throughput Fine Line Metallization Approach Based on Dispensing Technology. Energy Procedia 43, 2013, pp. 111–116

Bifacial n-PERT cells (Bi-PERT) with plated contacts for multi-wire interconnection

E. Cornagliotti, R. Russell, L. Tous, A. Uruena, F. Duerinckx, M. Aleman, P. Choulat, A. Sharma, J. John, J. Szlufcik
imec, Kapeldreef 75, 3001 Leuven, Belgium
Corresponding author: Emanuele.Cornagliotti@imec.be, Tel: +3216287834, Fax: +3216281097

ABSTRACT: In this paper we introduce bifacial n-type PERT cells (Bi-PERT) with ultra-thin (< 3 μm) and narrow (< 20 μm) Ni/Ag plated fingers on both the n^+ and p^+ side of the cell. This cell architecture is designed for multi-wire interconnection and has the potential to deliver low Cost of Ownership at module level (< 0.40 $/$W_P$), as it can profit from bifacial gain, a low cost metallization sequence and the high efficiency potential of multi-wire interconnection.

We describe the Bi-PERT cell features and show that the metallization sequence here proposed is well suited for integration with the multi-wire technology as it features finger section well below 60 μm² and no highly conductive Cu inter-layer. We also demonstrate a self-aligned fully plated metallization sequence able to form Ni/Ag electrodes on both sides of the cell with a specific contact resistance (ρ_c) below 1 mΩ·cm² for both the n^+ and p^+ side. We benchmark this novel technique to a seed-and-plate sequence where the plating seed is obtained by physical vapor deposition of Ni (PVD-and-plate).

Finally, we present light I-V results for rear-junction and front-junction Bi-PERT cells and report efficiency up to 21.0% for cells featuring the fully plated Ni/Ag metallization sequence and 22.6% for cells fabricated with the PVD-and-plate sequence.
Keywords: bifacial cells, n-PERT, plating, multi-wire

1 INTRODUCTION

Bifacial cells are an effective solution to increase the energy yield of a photovoltaic system under a wide range of operating conditions [1] [2] [3] [4] [5]. When integrated into glass-glass modules, bifacial gains well above 10% have been reported with only 20% albedo [6]. Bifacial cells can also be integrated into traditional glass-backsheet modules. In this configuration they can profit from the scattering of long wavelength light at the backsheet which compensates for the lack of a rear metal reflector as well as from reflection of light in the gaps between the cells (which is subsequently internally reflected at the front glass-air interface, or is directly absorbed at rear side of the cell) [7]. A few companies are present since many years in the bifacial cell market (Yingli, PVGS, Panasonic etc.), but several new ones have recently entered the market, offering modules both in the glass-glass configuration (MegaCell, Solarworld, Motech, Sunpreme, etc.) and/or glass-backsheet configuration (LG, Silevo, Mission Solar, Tetrasun, etc.).

Imec has, in recent years, developed large area mono-facial n-PERT cells with efficiencies up to 22.5% (independently confirmed) using both diffused [8] or epitaxial [9] emitters. In order to further reduce the Cost of Ownership (CoO), a bifacial evolution (named Bi-PERT) of the mono-facial n-PERT has been developed. Bi-PERT cells feature simple n^+/p^+ doping decoupling, high quality front and rear passivation and self-aligned Ni/Ag plated contacts on both sides. These busbar-less cells are designed to be interconnected by multi-wire technology. Multi-wire interconnection features high efficiency potential [10] due to distributed carrier collection which allows reducing finger conductivity (hence cross-section) without compromising series resistance losses. This allows to reduce recombination losses and shading losses (also thanks to the light recycling from the round wires). Two industrial technologies are available: the Smart-Wire-Connection-Technology [11] by Meyer Burger (SWCT) and Multi-Busbar Connector from Schmid [12]. Imec is also developing a multi-wire technology where solder-coated metal wires are incorporated in a woven fabric with encapsulant ribbons [13]. Its advantage lies in the fact that these woven fabrics enable the integration of thermal stress relief features while allowing the interconnection and encapsulation to take place in one process step.

In this paper, we will first describe the details of the processing sequence of a Bi-PERT cell, focusing on the surface passivation and the laser patterning. Then, we highlight the advantages of the metallization and interconnection architecture herein proposed. Finally, we discuss the opportunity to adopt a front or rear emitter configuration and we present the I-V results of the best cells obtained.

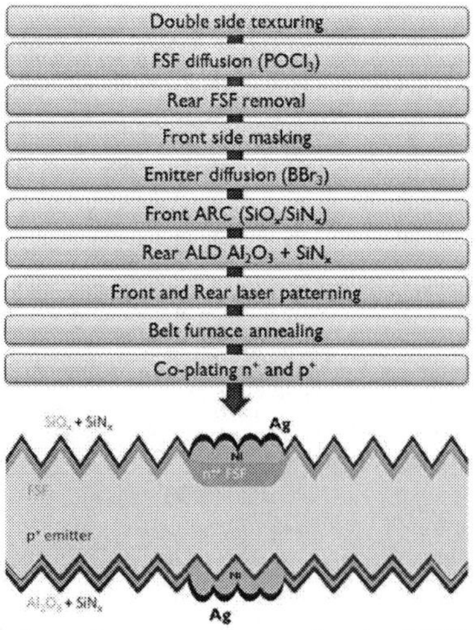

Figure 1: Processing sequence for a rear-junction (RJ) Bi-PERT cell.

2 EXPERIMENTAL

1.1 Processing sequence

The processing sequence for a rear-junction (RJ) Bi-PERT cell is described in Figure 1. After KOH saw-damage removal and texturing, wafers are subjected to a $POCl_3$ diffusion to form the front-surface-field (FSF). The diffused layer is subsequently selectively removed from the rear side by means of single side wet etching. After a front side masking step and BBr_3 diffusion, the front and rear passivation is applied. A stack of PECVD SiO_x and PECVD SiN_x is applied to the front n^+ surface, while a stack of atomic-layer-deposited (ALD) Al_2O_3 and PECVD SiN_x is employed for the rear side p^+ passivation. An alternative option consists in growing a thin layer of thermal silicon dioxide (SiO_2) and cap it by PECVD SiN_x on both the n^+ and p^+ side. Laser patterning is then carried out: at the front side, a laser doping process based on the use of phosphorous-containing spin-on dopant (SOD) is adopted [8]. At the rear side, a pico-second (ps) laser ablation process is used. Finally, a defect annealing step is performed in a belt furnace before cells are subjected to the metallization sequence.

Imec has also developed a front-junction (FJ) version of the Bi-PERT cell. The processing sequence is similar to the one reported in Figure 1, with the exception that the BBr_3 diffusion is performed the before $POCl_3$ diffusion. Passivation, patterning and metallization are the same for both FJ and RJ cells.

Table 1: Passivation of n^+ and p^+ surfaces. $J_{0,Auger}$ estimated with EDNA 2 [14] using the actual phosphorous and boron profiles. $J_{0,rad}$ not considered here.

Passivation scheme	$R_\square$ ($\Omega/\square$)	$J_{0,Auger}$ (fA/cm^2)	$J_{0,SRH}$ (fA/cm^2)	$J_{0,total}$ (fA/cm^2)
SiO_x/SiN_x (as dep.) **n^+ textured**	>250	5	130	135
SiO_x/SiN_x (annealed) **n^+ textured**	>250	5	10	15
Al_2O_3/SiN_x (annealed) **p^+ textured**	~95	22	23	45
Al_2O_3/SiN_x (annealed) **p^+ semipol.**	~95	22	15	37
Al_2O_3/SiN_x (annealed) **p^+ planar**	~95	22	< 1	22

1.2 Front and rear passivation

The front side passivation is provided by a stack consisting of a thin (< 25 nm) PECVD SiO_x layer capped by a PECVD SiN_x layer. Deposition of this SiO_x/SiN_x stack is performed sequentially in the same PECVD tube, without breaking the vacuum. The presence of a thin SiO_x interfacial layer improves the passivation compared to SiN_x passivation only. However this passivation needs to be activated via the belt furnace anneal which ensures densification [15] and release of hydrogen from the capping SiN_x layer [16] (see Table 1).

The rear passivation consists of a stack of ALD Al_2O_3 (< 10 nm) capped by PECVD SiN_x. The belt furnace anneal does not improve nor degrade the passivation quality of this stack. On the other hand, the morphology of the rear surface strongly affects the total dark saturation current ($J_{0,total}$). The SRH component of the dark saturation current ($J_{0,SRH}$) is reduced from ~ 23 fA/cm^2 in case of textured surface to below 1 fA/cm^2 in case of planar surfaces. With a low-moderate level of chemical polishing ("semipol." → ~ 4 µm etch) $J_{0,SRH}$ is reduced to 15 fA/cm^2. An improvement of the V_{oc} is therefore expected from the introduction of a polishing step on the p^+ surface. In general, for a bifacial cell it is sensible to assume that the rear surface should be textured, in order to maximize the light absorption from the rear side. However, as this can cause an increase in $J_{0,total}$ (as shown in Table 1), the opportunity to perform a moderate planarization step needs to be carefully assessed in view of the encapsulation mode and target application (bifacial or mono-facial module and intensity of the rear side illumination). It can also be concluded that the passivation is limited by the Auger component of the p^+ passivation which is due to the higher doping level of the emitter (~ 95 $\Omega/\square$) compared to the FSF (> 250 $\Omega/\square$).

1.3 Patterning

Patterning of the front and rear dielectrics is performed via ps-UV laser processing. On the n^+ side we employ a laser doping process [8] which ensures low contact recombination ($J_{0,laser\text{-}n++}$ ~ 500-700 fA/cm^2) and increases the surface phosphorous concentration to enable low contact resistance with Ni. On the p^+ side we can make use of either a laser doping process (in this case we use the thin Al_2O_3 layer as doping source [17]) or a laser ablation process. However, since the surface concentration of the p^+ emitter is, at the moment, high enough not to limit the contact resistance, we currently employ a laser ablation process. We envisage using the laser doping process on p^+ once we decrease the boron surface concentration, in an effort to reduce $J_{0,Auger}$. On both sides (n^+ and p^+), the patterning results in very narrow openings (< 15 µm).

2 METALLIZATION AND INTERCONNECTION

2.1 Metallization sequence

Two possible plating-based solutions are proposed for the simultaneous metallization of both n^+ and p^+ surfaces. In one case, after laser patterning, a Ni seed layer is deposited by PVD (sputtering) on both wafer surfaces. After a sintering step, during which Si reacts with Ni to form a Ni-silicide in the regions where the dielectrics have been ablated by the laser, the unreacted Ni on the passivated areas is selectively etched and fingers are thickened by electro-less Ni (e-Ni) and capped by a thin layer of immersion Ag (i-Ag). This process (also known as "salicidation" or "self-aligned silicidation process") is named henceforth "PVD-and-plate". We have also developed a second solution, where the seed Ni layer is deposited simultaneously on n^+ and p^+ by plating, instead of PVD. Subsequently, the plated seed is also thickened by an e-Ni/i-Ag sequence. This second process (which therefore makes exclusively use of plating) has the potential to largely reduce CoO, since it does not require the investment cost for PVD equipment. Also, no electrical contact needs to be applied to the cell at any stage of the metallization sequence and hence high throughput batch-processing can be envisaged. Using the fully plated sequence we could obtain specific contact resistance (ρ_c) below 1 mΩ·cm^2 for both the n^+/Ni and p^+/Ni interface.

2.2 Multi-wire interconnection

Multi-wire interconnection allows distributed carrier collection which intrinsically reduces series resistance (r_s) of a solar cell, as compared to a 5 busbar grid [18]. Alternatively, one can profit from multi-wire technology to reduce the finger cross-section or the conductivity of the finger material without affecting the total series resistance of the solar cell. Therefore, by using laser patterning and Ni/Ag plating we can fabricate ultra-thin (< 3 μm) and ultra-narrow (< 20 μm) fingers without the drawback of an increased r_s. Thanks to the absence of busbars and to the narrow laser-patterned fingers, both the front and rear metal contact fraction are kept to values of around 1.2-1.5 % (this value corresponds approximately also to the optical shading, because of the limited plating thickness).

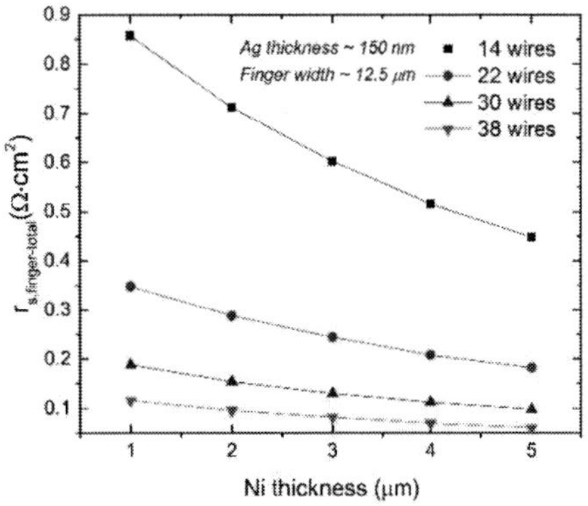

Figure 2: Total finger contribution (front and rear) to cell series resistance ($r_{s,finger-total}$) for different Ni thickness and different number of wires. Cell dimension: 15.6×15.6 cm^2. Finger width ~ 12.5 μm. Ag thickness is around ~ 150 nm.

As shown in Figure 2, for a Ag thickness of 150 nm, and for a finger width of 12.5 μm, total finger contribution (front and rear) to series resistance ($r_{s,finger-total}$) is limited to < 0.2 Ω·cm^2 even for Ni thickness below 2 μm (as long as the number of wires is higher than 30). It is therefore not necessary to include Cu as a high conductivity inter-layer. Thin electrodes also bring beneficial effects in terms of plating time and contact adhesion. Note that the thin (~150-200 nm) Ag capping has a preponderant role in determining the finger conductance due the low resistivity of Ag (ρ=1.6·10^{-8} Ω·m) compared to the one of Ni (ρ ~ 9·10^{-7} Ω·m).

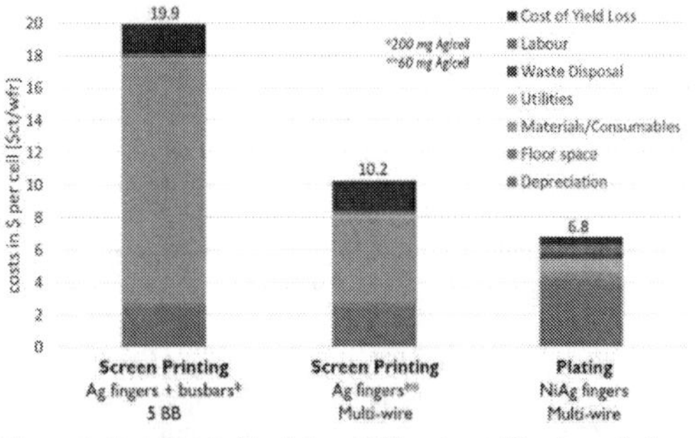

Figure 3: Cost ($ct./cell) of three different metallization solution for Bi-PERT cells: Ag screen printing and 5 busbar grid, Ag screen printing and multi-wires, Ni/Ag plating (fully plated sequence, no PVD) and multi-wires.

3 COST

The proposed architecture enables a substantial reduction in CoO. Bi-PERT cells feature a simple processing sequence: high temperature steps are minimized to one boron diffusion (BBr_3) and one phosphorous diffusion ($POCl_3$) with no long thermal oxidation/dopant drive-in steps, while preserving very low surface recombination velocity at both the n^+ and p^+ surface.

When using traditional Ag screen-printing technology, the use of multi-wire enables the reduction of finger width thus reducing the consumption of Ag and the cost associated to it (Figure 3). When Ni/Ag plating (the fully plated sequence) is used, instead of Ag screen-printing, an additional reduction in cost/cell is obtained, driven by savings on Ag consumption. This results in a total cost for the metallization sequence (including laser patterning) of 6.8 $ct./cell, compared to 19.9 $ct./cell for an Ag screen-printing based metallization of a 5 busbars grid.

When integrating Bi-PERT cells in a module using SWCT, we calculate that this architecture can potentially result in CoO at module level of 0.36 $/W$_p$ in glass-glass configuration (assuming front side cell efficiency ~ 21.5%, 90% bifaciality, 10% albedo) and 0.39 $/W$_p$ in glass-backsheet configuration (21.5% front side cell efficiency), while mainstream p-type technology (mono-PERC – 21.5% cell efficiency) lies currently around 0.40-0.42 $/W$_p$.

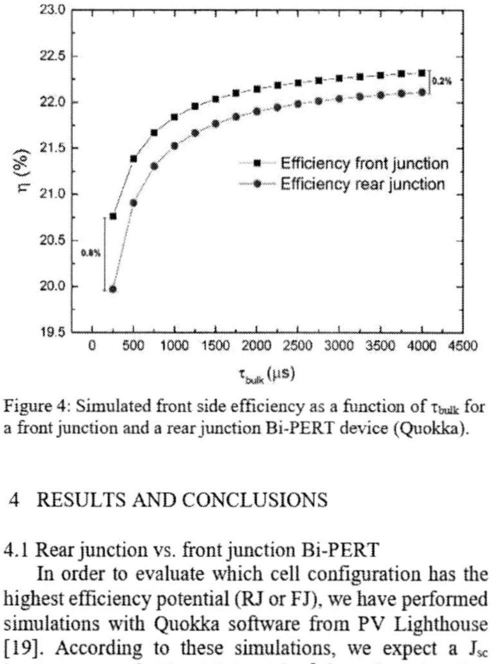

Figure 4: Simulated front side efficiency as a function of τ_{bulk} for a front junction and a rear junction Bi-PERT device (Quokka).

4 RESULTS AND CONCLUSIONS

4.1 Rear junction vs. front junction Bi-PERT

In order to evaluate which cell configuration has the highest efficiency potential (RJ or FJ), we have performed simulations with Quokka software from PV Lighthouse [19]. According to these simulations, we expect a J_{sc} improvement of at least 0.4 mA/cm^2 for a front junction device, even in the case of very high bulk lifetime (> 4 ms), which results in an increased efficiency potential for a front junction device (Figure 4). The explanation can be found in the fact that carrier collection is more effective in a front junction device. In Figure 5 we report the internal quantum efficiency (IQE) of a rear junction Bi-PERT, measured both from the front side (i.e. emitter on dark side) and from the rear side (i.e. emitter on the sunny side). Despite the asymmetric surface morphology (rear side is subjected to inline single side FSF removal), it can be noticed that, when emitter (p^+) is on the sunny side, the carrier collection is more effective across the wavelength range which accounts for most of the photo-generated current (400-900 nm).

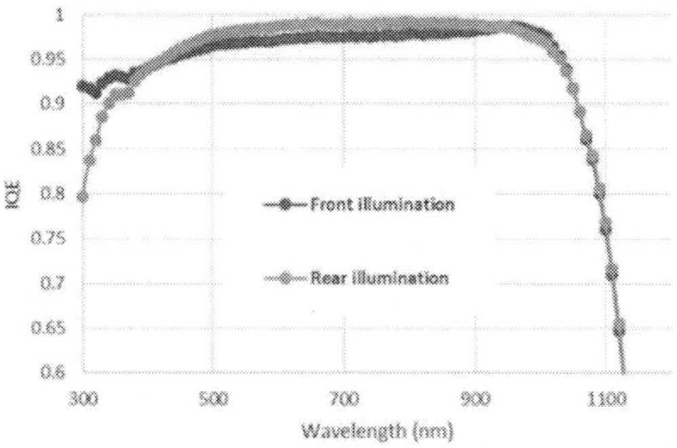

Figure 5: IQE of rear junction Bi-PERT as measured from front side (blue curve, i.e. emitter on the dark side) or rear side (green curve, i.e. emitter on the sunny side).

4.2 I-V results

Table 2 reports the best cell efficiency for a FJ device fabricated using a plated seed, compared to one fabricated using a PVD seed (PVD and plate). The cells were measured with a Grid^{TOUCH} system [20] (with 30 wires on the front and 30 on the rear for current extraction) on a low-reflection black chuck. J_{sc} is corrected to remove wire shading [21].

Table 2: Best cell efficiency (239 cm^2) for a FJ Bi-PERT cell with fully plated metallization (plated seed) compared to one with PVD-and-plate sequence. Standard Test Condition (STC) with front illumination, low-reflection chuck. Measured with Grid^{TOUCH} (30/30 wires). J_{sc} not taking into account wire shading.

Seed	FJ/ RJ	Illu mina tion	J_{sc} (mA/cm^2)	V_{oc} (mV)	FF (%)	η (%)
Plated	FJ	Front	41.3	660	76.9	21.0
		Rear	*39.8*	*659*	*77.3*	*20.3*
PVD	FJ	Front	41.1	672	80.7	22.3

Cells with a plated seed reached an efficiency of 21% (with J_{sc} bifaciality ~ 96%) while cells with PVD seed could reach 22.3%. The difference mainly lies in the V_{oc} and FF. The lower FF for cells with plated seed is attributed to a difference in pFF (~ 83.5% for the PVD seed vs. 81% for the plated seed). We attribute the reduced pFF for cells with plated seed to an increased recombination activity in the contact area which also has a negative impact on the V_{oc}. The exact underlying mechanisms as well as mitigation routes are currently under investigation.

Table 3: Best cell efficiency (239 cm^2) for a Bi-PERT cell with rear junction, planar rear and PVD-and-plate metallization sequence. STC with front illumination, low-reflection chuck. Measured with Grid^{TOUCH} (30/30 wires). J_{sc} not taking into account wire shading.

Seed	FJ/ RJ	Illu mina tion	J_{sc} (mA/cm^2)	V_{oc} (mV)	FF (%)	η (%)
PVD	RJ	Front	41.5	681	79.9	22.6

In Table 3 the best cell efficiency for a rear junction Bi-PERT with PVD seed is reported. In this case, the V_{oc} is increased to 681 mV due to the introduction of a rear side polishing step (Table 1). Bifaciality for RJ cells is close to 100% as the increased reflectance of the rear surface (due to the single side inline wet etch step) is compensated by an increased carrier collection due to the presence of the emitter. Further improvement of the V_{oc} for these cells can originate from an optimization of the boron emitter profile (Table 1), and from a reduction of the PVD sputtering damage [22] to the passivation layer (expected V_{oc} increase ~ 10 mV).

4.3 Conclusions

In this paper we have introduced a bifacial cell concept which conjugates a simple and low cost processing flow with high module efficiency potential, when associated with multi-wire interconnection technology. We demonstrated a self-aligned fully plated metallization sequence able to form Ni/Ag electrodes on both sides of the cell with a low specific contact resistance (ρ_c).

A best cell efficiency of 22.6% (measured with Grid^{TOUCH}) was reported for a RJ Bi-PERT cell with PVD seed. 21% was reported for a FJ cell with plated seed.

Future work is aimed at improvement of the performance of cells with plated seed to match the performance of the ones with PVD seed, and at integration of cells into laminates (and associated reliability testing). Moreover, together with a detailed analysis of the measurement uncertainties at cell level (e.g. accuracy of J_{sc} and FF), an accurate estimation of the cell-to-module losses/gains for this type of cell/interconnection scheme is also required.

ACKNOWLEDGEMENTS

The authors would like to thank the financial support of the partners of imec's Industrial Affiliation Program for Photovoltaics (IIAP-PV).

This work receives the support of the European Union, the European Regional Development Fund ERDF, Flanders Innovation & Entrepreneurship and the Province of Limburg

REFERENCES

[1] R. Kopecek, Y. Veschetti, E. Gerritsen, A. Schneider, C. Comparotto, V. D. Mihailetchi, J. Lossen and J. Libal, "Bifaciality: One small step for technology, one gian leap for kWh cost reduction," *Photovoltaics International,* vol. 26, pp. 32-45, December 2014.

[2] U. A. Yusufoglu, T. H. Lee, T. M. Pletzer, A. Halm, L. J. Koduvelikulathu, C. Comparotto, R. Kopecek and H. Kurz, "Simulation of Energy Production by Bifacial Modules with Revision of Ground Reflection," *Energy Procedia,* vol. 55, pp. 389-395, 2014.

[3] S. Wang, O. Wilkie, J. Lam, R. Steeman, W. Zhang, K. S. Khoo, S. C. Siong and H. Rostan, "Bifacial Photovoltaic Systems Energy Yield Modelling," *Energy Procedia,* vol. 77, pp. 428-433, 2015.

[4] F. Fertig, S. Nold, N. Wohrle, J. Greulich, I. Hadrich, K. Krauss, M. Mittag, D. Biro, S. Rein and R. Preu, "Economic feasibility of bifacial

silicon solar cells," *Progress in Photovoltaics,* 2016.

[5] I. Romijn, B. van Aken, J. Anker, A. R. Burgers, A. Gutjahr, B. Heurtault, M. Koppes, E. Kossen, M. Lamers, D. S. Saynova, C. J. Tool, L. Fang, X. Jingfeng, L. Gaofei, X. Zhuo, W. Hongfang, H. Zhiyan, P. R. Venema and A. H. Vlooswijk, "Industrial Implementation of Efficiency Improvements in n-Type Solar Cells and Modules," *27th European Photovoltaic Solar Energy Conference and Exhibition,* pp. 533-537, 2012.

[6] R. Kopecek, I. Shoukry and J. Libal, "Cost/kWh thinking and bifaciality: Two allies for low-cost PV of the future," *Photovoltaics International,* vol. 30, pp. 88-97, December 2015.

[7] J. P. Singh, S. Guo, I. M. Peters, A. G. Aberle and T. M. Walsh, "Comparison of Glass/Glass and Glass/Backsheet PV modules using bifacial silicon solar cells," *IEEE Journal of Photovoltaics,* vol. 5, no. 3, pp. 783-791, 2015.

[8] A. Uruena, M. Aleman, E. Cornagliotti, A. Sharma, J. Deckers, M. Haslinger, L. Tous, R. Russell, J. John, Y. Yao, T. Soderstrom, F. Duerinckx and J. Szlufcik, "Beyond 22% Large Area n-Type Silicon Solar Cells with Front Laser Doping and a Rear Emitter," *31st European Photovoltaic Solar Energy Conference and Exhibition,* pp. 410-413, 2015.

[9] M. Recaman, I. Kuzma Filipek, F. Duerinckx, Y. Li, E. Cornagliotti, A. Uruena, A. Sharma, L. Tous, R. Russell, A. Hajjiah, J. Szlufcik and J. Poortmans, "Advanced Doping Profiles by Selective Epitaxy in n-Type PERT Cells," *31st European Photovoltaic Solar Energy Conference and Exhibition,* pp. 433-439, 2015.

[10] R. Witteck, D. Hinked, H. Schulte-Huxel, M. R. Vogt, J. Muller, S. Blankemeyer, M. Kontges, K. Bothe and R. Brendel, "Optimized interconnection of passivated emitter and rear cells by experimentally verified modeling," *IEEE Journal of Photovoltaics,* vol. 6, no. 2, pp. 432-439, 2016.

[11] T. Soderstrom, P. Papet and J. Ufheil, "Smart Wire connection technology," *28th European Photovoltaic Solar Energy Conference and Exhibition,* pp. 495-499, 2013.

[12] S. Braun, G. Micard and G. Hahn, "Solar cell improvement by using a multi busbar design as front electrode," *Energy Procedia* , vol. 27, pp. 227-233, 2012.

[13] T. Borgers, E. Voroshazi, J. Govaerts, J. Szlufcik and J. Poortmans, "Multi-wire interconnection technologies weaving the way for back contact and bifacial PV modules," *IEEE PVSC,* 2016.

[14] "EDNA 2," *www.pvlighthouse.com.*

[15] Y. Li, M. Recaman, B. Zielinski, M. Debucquoy and J. Poortmans, "Process simplification and improvement of rear side of IBC cells by means of PECVD SiOx and epitaxy," *31st European Photovoltaic Solar Energy Conference and Exhibition,* pp. 466-469, 2015.

[16] J. Chen, E. Cornagliotti, X. Loozen, E. Simoen, J. Vanhellemont, J. Lauwert, H. Vrielink and J. Poortmans, "Impact of firing on surface passivation of p-Si by SiO2/Al and SiO2/SiNx/Al stacks," *Journal of Applied Physics,* vol. 110, no. 12, 2011.

[17] E. Cornagliotti, A. Uruena, B. Hallam, L. Tous, R. Russell, F. Duerinckx and J. Szlufcik, "Large area p-type PERL cells featuring local p+ BSF formed by laser processing of ALD Al2O3 layers," *Solar Energy Materials and Solar Cells,* vol. 138, pp. 72-79, 2015.

[18] S. Braun, G. Hahn, R. Nissler, C. Ponisch and D. Habermann, "Multi-busbar Solar Cells and Modules: High Efficiencies and Low Silver Consumption," *Energy Procedia,* vol. 38, pp. 334-339, 2013.

[19] A. Fell, "A Free and Fast Three-Dimensional/Two-Dimensional Solar Cell Simulator Featuring Conductive Boundary and Quasi-Neutrality Approximations," *IEEE Transactions on Electron Devices,* vol. 60, no. 2, pp. 733-738, 2013.

[20] N. Bassi, C. Clerc, Y. Pelet, J. Hiller, V. Fakhfouri, C. Droz, M. Despeisse, J. Levrat, A. Faes, D. Batzner and P. Papet, "GridTOUCH: Innovative Solution for Accurate IV Measurement of Busbarless Cells in Production and Laboratory Environments," *29th European Photovoltaic Solar Energy Conference and Exhibition,* pp. 1180-1185, 2014.

[21] V. Fakhfouri, C. Droz and J. Rochat, "Measurement accuracy and precision: the GRIDtouch contacting system," *Meyer Burger - White paper,* 2014.

[22] S. Singh, B. O'Sullivan, M. Kyuzo, S. Jambaldinni, L. Tous, R. Russell, M. Debucquoy, J. Szlufcik and J. Poortmans, "Application of seed and plate metallization to 15.6cm x 15.6cm IBC cells," *Photovoltaics International,* pp. 62-70, 2015.

LASER FORMED ANCHOR POINTS FOR COPPER PLATING ADHESION ON AL-BSF AND PERC CELLS

Alison Wenham[1*], Chee Mun Chong[1], Sisi Wang[1], Jingjia Ji[1], Zhengrong Shi[1], Ly Mai[1], Adeline Sugianto[1], Stuart Wenham[1], Allen Barnett[1] and Martin Green[1]
[1]School of Photovoltaics and Renewable Energy Engineering, the University of New South Wales, Sydney, NSW 2052, Australia
*Tel: +61 2 9385 6848, Fax: +61 2 9385 7762, email: a.wenham@unsw.edu.au

ABSTRACT: Batches of Al-BSF and PERC cells (including both PERD and PERL rear cells) have been produced on Suntech's Pluto production line. On the front side, laser formed anchor points of varying densities have been superimposed on laser-doped fingers for greatly improved adhesion of plated contacts. Even on a high voltage PERL structure, 10% coverage of anchor points can result in negligible efficiency loss when coupled with deep laser doped fingers, and only 0.1% absolute loss compared to highest efficiency PERC structure with no anchor points; apparently a small loss compared to the improvements in plating adhesion and reliability.
Keywords: c-Si, Copper Plated Contacts, Adhesion, Metallization

1 INTRODUCTION

Many experts predict that copper plating has the potential to replace silver contacts in PV devices if key concerns can be addressed. For example, according to the ITRPV [1], "Despite a continual reduction of silver consumption at the cell manufacturing level, silver might still be replaced on a large scale by a more cost-effective material. Copper (Cu), applied with plating technologies, is the envisioned substitute. ... it is then expected to account for around 30% of the market in 2020. Technical issues related to reliability and adhesion have to be resolved before alternative metallization techniques can be introduced. Appropriate equipment and processes also need to be made ready for mass production."

Light Induced Plating (LIP), which uses the cell's own generated current to drive the plating reaction, can be fast, requires no physical front contact, needs little chemical maintenance [2][3], and has already been demonstrated to be suited to large-scale high-throughput inline processes, with proven equipment available from both RENA and Kuttler [4-6]. Also, most concerns with copper usage have been resolved with the achievement of plated modules passing IEC 61215 thermal and damp-heat testing, and showing superior durability to SP modules over 20 years operation in the field [3,4,6-10].

Never-the-less, adhesion still seems to be a topic of concern for plated contacts. The challenge is to achieve reliable adhesion with simple commercial processes. The adhesion can be impacted by many things, not just the plating metals and surface to be plated, but even prior process steps [11]. If the process is done right however, it is possible to have contacts stronger than SP contacts, with such good adhesion that wafer breakage occurs before adhesion failure [4,9,12,13].

Forming a Nickel Silicide has been reported to provide excellent adhesion [6,9,13], however only if well formed. For example, in 2013 Mondon et al [14] found voids between their sintered Nickel (Nickel Silicide) and subsequent plated Nickel, which became an easy point of failure. In 2014, Mondon et al [9] reported excellent adhesion using a limited Nickel source for sintering (no etch back required which would oxidise the surface) before plating the Ni/Cu/Ag stack.

Others have looked at surface features or roughening prior to plating. LCP chemical etching of small anchor points to be filled with plating had no reliable benefit [15], but roughness by UV ps laser has shown good adhesion [9].

In this work, laser formed anchor points are used for improved adhesion [16], whereby holes are ablated using a q-switched laser [17], on top of laser doped fingers (Fig. 1.). The focus is to assess the impact these anchor points have on the performance of cells made on Suntech's Pluto production line [18]. Anchor points are first applied to Pluto's original technology, which incorporates a full Al-BSF rear structure since over 90% of total modules produced use this rear [19].

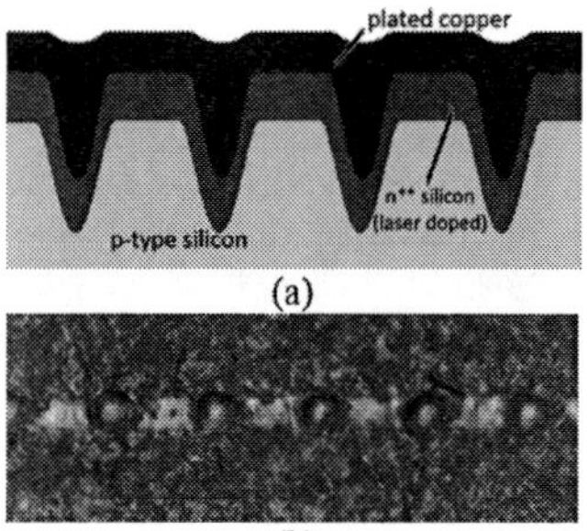

(a)
(b)

Figure 1: a) side view of an n++ laser doped finger with anchor points for the plated copper; b) a top view microscope image of a laser doped finger with anchor points that are 10 microns in diameter

In the near future, average commercial efficiencies will move above 20% as the industry moves towards the Passivated Emitter and Rear Cell (PERC) structure. PERC makes up the majority of new commercial capacity installed since 2014 and is expected to be the dominant cell technology in global production by 2020 [20]. As a comparison, and to also assess performance implications on higher efficiency structures, anchor points are also formed on 2 iterations of next generation Pluto from the PERC family [19]:

1. PERD - Passivated Emitter, Rear Direct-contact (no BSF)
2. PERL - Passivated Emitter, Rear Locally-doped.

2 EXPERIMENTAL RESULTS

2.1 Full Aluminium BSF Rear

Cells were produced on the 0.5 GW Pluto production line [1], using 155 cm2 p-type commercial grade Cz wafers, 1-3 Ω-cm and 180 μm thick after texturing. A light emitter diffusion of ~100 Ω/□ is coated with a very thin ~10nm thermal oxide, and PECVD SiNx deposition. Aluminium was screen printed to create a full aluminium rear BSF and contact. On the font, a Yuemao 532nm Nd:YAG q-switched laser was used to pattern the phosphoric acid coated dielectric, and dope the silicon with phosphorus to form a finger and bus bar patterned selective emitter. After a second coat of phosphoric acid, the same q-switched laser was used with modified speed power and frequency to introduce anchor points by high powered laser pulses ablating the silicon periodically. Cells with varied densities of anchor points along the fingers were produced in batches of 1,000 cells. Cells were plated using 30 MW capacity Kuttler LIP baths to form a stack of ~1 µm Ni, Cu bulk ~9-11 µm, and silver capping. The results can be seen in Table 1.

Table 1: Al-BSF Cells with varied anchor point density

Hole Coverage	Jsc (mA/cm^2)	Voc (mV)	FF (%)	Efficiency (%)
0% (none)	38.5	638	79.3	19.5
10%	38.5	636	78.9	19.3
50%	38.5	622	76.9	18.4
90%	38.6	608	74.2	17.4
Typical Pluto*	38.6*	636*	79.7*	19.6*

* independently confirmed by SERIS

2.2 PERD rear

Cells produced were similar to the Al-BSF cells, until rear metallization. An SiO_2, SiNx stack is deposited on the rear and line patterned using a 532nm Nd:YAG laser, with 2mm spacing. Aluminium is directly sputtered onto the rear and sintered at 400 °C to directly contact the p-type base. The front laser doping is the same as previously with batches of 20 cells with the same anchor point variations. The results are shown in Table 2.

Table 2: PERD Cells with varied anchor point density

Hole Coverage	Jsc (mA/cm^2)	Voc (mV)	FF (%)	Efficiency (%)
0% (none)*	40.9*	666*	74.4*	20.3*
0% (none)	40.4	666	75.3	20.3
10%	40.5	661	74.8	20.0
50%	40.6	642	72.9	19.0
90%	40.7	624	72.2	18.3

* independently confirmed by SERIS

2.3 PERL rear

To implement the PERL structure, cells were produced again in a similar way, until rear metallization. Similar to PERD, a rear dielectric is deposited and patterned. This time, aluminium is screen-printed on the rear and fired at 700 °C to form a local BSF in the openings. The front laser doping in this structure is done using a Coherent 532nm Nd:YAG Quasi-Steady-State laser. The benefit of this laser is that it is able to melt the silicon to much greater depths (and widths extending underneath the silicon nitride layer) to therefore form much deeper and wider heavily doped lines for the selective emitter structure. This was expected to subsequently make it easier for the alignment and containment of the anchor points within the laser-doped lines. The q-switched Yuemao laser is still used for anchor point formation with batches of 20 cells with the same variations. The results are shown in Table 3.

Table 3: PERL Cells with varied anchor point density

Hole Coverage	Jsc (mA/cm^2)	Voc (mV)	FF (%)	Efficiency (%)
0% (none)	39.4	686	74.9	20.2
10%	39.5	686	74.5	20.2
50%	39.4	684	72.5	19.5
90%	39.5	679	68.9	18.5

3 DISCUSSION

The performance parameters from Tables 1, 2 and 3 have been graphed in Fig. 2. for comparison.

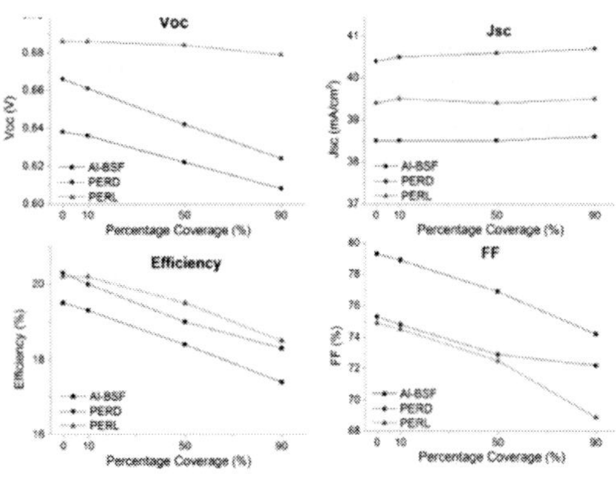

Figure 2: Charts displaying performance parameters (Voc, Jsc, Efficiency and FF) for cells with increasing density of anchor points on different rear structures (Al-BSF, PERD and PERL).

Not surprisingly, the PERL cell with a rear local BSF to shield the metal has the highest voltage at 686 mV with 10% or less anchor points, while the full Al-BSF structure is the lowest. The Jsc in the PERD cell is highest at 40.4 mA/cm^2, well ahead of the PERL structure at 39.4 mA/cm^2, and Al-BSF is again the lowest. Looking at the reflectance curves in Fig. 3., the PERD cells show a dramatic increase in 1200 nm light reflected from the cell. This results from a superior internal rear reflector created by the sputtered aluminum displaced from the silicon by the dielectric layer. Perhaps surprisingly, the internal rear reflector created by the PERL structure creates a much worse internal reflector despite having an equivalent dielectrically displaced metal rear reflector. It appears that high temperature firing damages the interface between the Al and the dielectric as the reactive Al reacts with the dielectric. In comparison, the sputtered Al in the PERD cells is only sintered at 400 °C, apparently preserving the quality of the Al interface with the dielectric and hence its reflection properties. This is an important result of relevance to the approach used by the majority of manufacturers of PERC cells where similar damaging high temperature firing of Al paste is used for rear contact formation. As expected, the Al-BSF structure has even worse internal rear reflection resulting from the direct contact between the Al paste and the silicon.

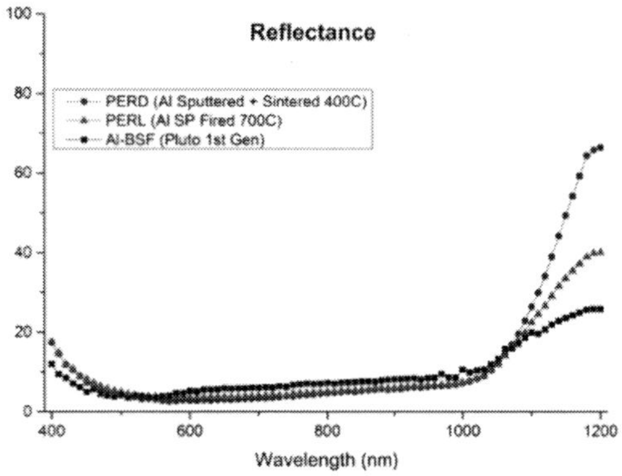

Figure 3: Reflectance curves for Pluto cells with Al-BSF, PERD and PERL rear structures; the differences at 1200nm indicate the difference in quality of rear reflector

The Al-BSF structure only wins when it comes to FF (above 79 %), while the two variations of PERC cells are similar at around 75 %, with the reduction being primarily due to the limited rear contact geometry resulting in higher series resistance. Overall, the two variations of PERC's have similar efficiencies; with no anchor points, PERD wins by 0.1% absolute. Both are clearly ahead of Al-BSF by ~1% absolute despite the rear contact design that caused high series resistance losses.

Looking at Fig. 2., the trends are generally the same for all device structures as the anchor point density increases; the Voc drops, The Jsc remains fairly stable (or increases slightly), the FF declines and the overall efficiency drops as a result. However, surprisingly the Voc of the highest voltage structure (PERL) is the least affected by increasing hole coverage. As the Coherent laser forms deeper and wider heavy doped regions (and is only used for the PERL cells), it is likely that the deeper doping prior to anchor point formation helps shield the damage formed by the anchor points, more than simply having phosphoric acid applied prior to formation of anchor points for simultaneous doping. This would help explain the much higher voltages achieved when high densities of anchor points are used. However although misalignment of the anchor points should be rarer for the PERL cells due to the wider laser-doped lines, when it does occur due to the misaligning of the two different laser systems then the laser damage can locally harm the junction, explaining the deterioration of FF for high densities of anchor points. Such damage however, as long as not too excessive, does not appear to cause significant deterioration in the open circuit voltages. Further, laser manufacturers such as Dr LASER in China have confirmed the possibility of housing and aligning both laser heads in one system so that the laser doping and anchor point processes could be done simultaneously without automatic alignment, which would also add simplicity to the processing.

Recently reported results have shown using a Meyer Berger Peel tester that ~20% coverage of anchor points triples the strength of plated contacts and actually results in measurably greater adhesion than screen printed contacts [21]. Other non-quantitative results have found even 10% spacing to be effective for improved adhesion [17]. Based on these results, the PERL structure used here seems most suited to anchor points with no loss in efficiency with 10% coverage of anchor points, and only 0.1% absolute efficiency worse than PERC with no anchor points. This loss appears tolerable for the added adhesion and durability it brings. The PERD and Al-BSF structures lose 0.3 % and 0.2 % respectively with 10 % coverage, which may still be justified for improved strength. However, introducing the deeper finger doping to the PERD and full-area BSF cells would also appear preferable to help shield the laser damage and reduce the probability of misalignment between the laser-doped lines and anchor points. This would likely reduce or eliminate the efficiency lost by these structures when anchor points are applied.

4 CONCLUSIONS

Anchor points formed by a laser prior to plating can significantly increase the adhesion of copper plated contacts, even above that of screen-printed contacts. They can be easily introduced to a laser-doped selective emitter (LDSE) cell structure, by a second laser scribe with a pulsed laser. Increasing densities of anchor points degrades the Voc and FF of cells and therefore also the efficiencies. However, using a continuous-wave laser for deep wide heavily doped fingers can shield most of this damage so that high voltage PERL cells only lose 0.1% in efficiency with 10% coverage of anchor points. Further, a laser system with two laser heads aligned provides a simple solution to facilitate automatic alignment that also reduces processing steps.

5 ACKNOWLEDGEMENTS

This Program has been supported by the Australian Government through the Australian Renewable Energy Agency (ARENA), the Australian Research Council (ARC), the UK Institution of Engineering and Technology (IET) and the Australian Centre for Advanced Photovoltaics (ACAP). The views expressed herein are not necessarily the views of the Australian Government, and the Australian Government does not accept responsibility for any information or advice contained herein.

5 REFERENCES

[1] International Technology Roadmap for Photovoltaics, *ITRPV 2015 Edition*, http://www.itrpv.net/Reports/Downloads/2015/

[2] A. Mette, C, Schetter, D. Wissen, S. Lust, S.W. Glunz and G. Willeke, Increasing the Efficiency of Screen Printed Silicon Solar Cells by Light Induced Silver Plating, *IEEE 4th World Conference*, 2006

[3] A. Lennon, Y. Yao and S, Wenham, *Progress in Photovoltaics: Research and Applications*, 2012, DOI: 10.1002/pip.2221

[4] A, Letize, D. Cullen, B. Lee, K. Crouse, N. Bay and J. Horzel "Pilot Scale Production and Reliability testing of Solar Cell Modules Based on a Low Cost Copper Electroplating process", *29th EU PVSEC*, 2014

[5] Z. Wang, P. Han, H. Lu, H. Qian, L. Chen, Q. Meng, N. Tang, F. Gao, Y. Jiang, J. Wu, W. Wu, H. Zhu, J. Ji, Z. Shi, A. Sugianto, L. Mai, B. Hallam and S. Wenham, *Progress in PV: Research and Appl.* 2012

[6] J. Horzel, Y. Shengzhao, N. Bay, M. Passig, D. Pysch, H. Kühnlein and P. Verlinden, "Industrial Si Solar Cells with Cu-Based Plated Contacts", *IEEE Journal of PV*, Vol 5. No. 6, 2015

[7] R. Russel, L. Tous, H. Philipsen, J. Horzel, E. Cornagliotti, M. Ngamo, P. Choulat, R. Labie, J. Beckers, J. Bertens, M. Fujii, J. John, J. Poortmans and R. Mertens, A Simple Copper

Metallisation Process for High Cell Efficiencies and Reliable Modules, *27th EU PVSEC,* 2012

[8] J. Horzel, Y Shengzhao, N. Bay, D. Pysch, H. Kühnlein and P. Verlinden, "Cu based solar cell contacts formed by high throughput inline plating technology", *31st EU PVSEC,* 2015

[9] A. Mondon, J. Bartsch, M. Kamp, A. Brand, B. Steinhauser, N. Bay, J. Horzel, M. Glatthaar and S.W. Glunz "Plated Nickel-Copper Contacts on C-Si: from Micro-electronic Processing to Cost Effective Silicon Solar Cell Production", *29th EU PVSEC,* 2014

[10] M. Alonso-Abella, F. Chenlo, A. Alonso and D. González "Toledo PV plant 1 MWp 20 years of operation", *EU PVSEC,* 2014

[11] J. Horzel , N. Bay, T. Knarozovski, A. Träger, M. Passig, J. Burschik, M. Sieber, A. Hoffmann, B. Latzel, D. Pysch, C. Schmitt, R. Sastrawan, H.Kühnlein, K.Vosteen, H. Nussbaumer, R. Boehme, H. Schillinger, A. Letize, B. Lee & T. Soederstroem "Different Concepts of Implementing a Directly Plated Front Side Metallization into Industrial Solar Cell Mass Production", *28th EU PVSEC,* 2013

[12] D. Kray, N. Bay, G. Cimiotti, S. Kleinschmidt, N. Kösterke, A. Löse, M. Sailer, A. Träger, H. Kühnlein, H. Nussbaumer, C. Fleischmann and F. Granek "Industrial LCP Selective Emitter Solar Cells with Plated Contacts", *35th IEEE PVSEC,* 2010

[13] A. Wenham, C.M. Chong, S. Wang, R. Chen, J. Ji, Z Shi, L. Mai, A. Sugianto, S. Wenham, A. Barnett and M. Green, "Overcoming Issues with Copper Plated Contacts for Large Scale Manufacturing" *EU PVSEC,* 2015

[14] A. Mondon, M.N. Jawaid, J. Bartsch, M.Glatthaar, S.W. Gluz "Microstructure analysis of the interface situation and adhesion of thermally formed nickel silicide for plated nickel–copper contacts on si solar cells", *Solar Energy Materials and Solar Cells,* vol.117, pp. 209-213, 2013

[15] J. Bartsch, A. Mondon, M. Kamp, A. Kraft, M. Wendling, M. Mehling, N. Wehkamp, M. N. Jawaid, A. Lorenz, F. Clement, C. Schetter, M. Glatthaar and S. W. Glunz, "Progress with Multi-Step Metllization Processes Featuring Copper as Conducting Layer at Fraunhofer ISE" *27th EU PVSEC,* 2012

[16] A. Wenham "High Efficiency Laser Doped Solar Cells" *Undergraduate Thesis, UNSW,* 2009

[17] A.M. Wenham, M.A. Green, S.R. Wenham, "Improved Metal Adhesion" *Patent: WO 2010099564 A1,* Mar 2009

[18] Z. Shi, S. Wenham, & J. Ji "Mass Production of the Innovative PLUTO Solar Cell Technology" Invited Plenary Session Paper, *34th IEEE PV Specialists Conference,* 2009

[19] M.A. Green "The Passivated Emitter and Rear Cell (PERC): From conception to mass production" *Solar Energy Materials & Solar Cells,* vol. 143, 2015 pp.190–197

[20] M.A Green "Commercial progress and Challenges for Photovoltaics" *Nature Energy,* vol 1. January 2016

[21] A. Wenham, C.M. Chong, S. Wang, R. Chen, J.J. Ji, Z. Shi, L. Mai, A. Sugianto, M. Abbott, S. Wenham, A. Barnett and M. Green "Copper Plated Contacts for Large-Scale Manufacturing" *43rd IEEE PVSC,* June 2016

N-PERT SOLAR CELLS WITH PASSIVATED CONTACT TECHNOLOGY BASED ON LPCVD POLYSILICON AND FIRE-THROUGH CONTACT METALLIZATION

R.C.G. Naber[1]*, M. Lenes[1], A.H.G. Vlooswijk[1], J.R.M. Luchies[1], J. Wang[2], F. Zheng[2], J. Lin[2], Z. Zhang[2]
1. Tempress Systems BV, Radeweg 31, 8171 MD Vaassen, The Netherlands
2. Shanghai Shenzhou New Energy Development Co. Ltd., No. 505 Jiangyue Road, 201112, Shanghai, China
* E-mail: rnaber@tempress.nl; Tel. +31 578 699 200

ABSTRACT: We present an n-PERT solar cell technology with a passivated rear side contact obtained by depositing thin LPCVD polysilicon layers and contacting these layers with standard fire-through silver pastes. The ability to use standard contact metallization processes for this application is a novel development which we hope will significantly enhance the potential for passivated contact technology based on polysilicon layers. By adding the polysilicon deposition step to an existing process flow that is in use for the large-scale production of n-PERT solar cells and by using standard Cz wafer material we obtained an efficiency of 20.8% and a V_{OC} of 667 mV for a 239 cm^2 solar cell. Compared to normal production cells without polysilicon the results show an average cell efficiency enhancement of 0.4-0.6% (absolute) and characterization of the cells indicates that the 20.8% cell efficiency will increase to more than 21% after further optimization of the polysilicon and anti-reflection coating layer thicknesses alone.

Keywords: passivation; n-type; silicon solar cell

1 INTRODUCTION

In recent publications by Fraunhofer ISE and others, passivated contact technology based on polysilicon layers has been demonstrated to bring large cell efficiency advantages in a way that is compatible with high-temperature processing, but it is still an open question how the technology could be implemented in industrial solar cell production [1-3]. Most publications about the topic envision a full-area contact on the rear side, but implementing such a technology in production may not be trivial because it will require non-standard contact metallization techniques (such as metal vapour deposition) to keep the passivation performance of the thin polysilicon layers intact. We took a different approach for the currently presented work because we wanted to make use of standard fire-through silver pastes in order to minimize the technological complexity and to enable bifacial module applications. A critical question then is whether fire-through pastes can be used to contact thin polysilicon layers without any major adverse effects on their passivation performance, caused by damage to the tunnelling layer underneath the polysilicon layer during the fire-through contacting process. We demonstrate here that for LPCVD polysilicon layer thicknesses as low as 70 nm our contact metallization technique gives high V_{OC} values, which demonstrates that the tunnelling layers remain undamaged. Compared to a thickness of 140 nm we saw a significant increase in cell efficiency, which we propose is due to a reduction in parasitic absorption losses inside the polysilicon layer.

Next to demonstrating the metallization technique the goal of our work was also to show that LPCVD polysilicon-based passivation technology can be integrated into a process flow that is already in use for solar cell production. The process flow in question normally uses a Phosphorus implantation step for the creation of a back surface field, which we use here to dope the polysilicon layer instead. The rest of the process flow included standard production processes such as BBr_3 diffusion, thermal oxidation and PECVD SiN_x:H deposition that largely remained the same as that used in production. In this way we demonstrate that the LPCVD polysilicon technology can be applied in an existing production process without any unexpected effects and yield a significantly higher average cell efficiency.

2 EXPERIMENTAL

In figure 1 we present the solar cell structure that we developed. It is identical to our standard n-PERT cell except for the fact that the back surface field region is replaced by a stack consisting of a tunnelling layer and an n-type doped polysilicon layer. During the contact firing step the silver paste on the rear side etches through the SiO_2/SiN_x stack to contact the polysilicon layer, while the front side contact is formed at the same time and in the same way as for the normal production cells.

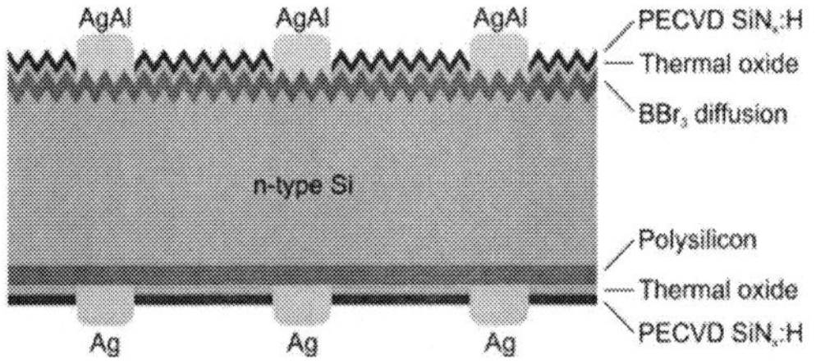

Figure 1. Structure of the developed n-PERT solar cells

The solar cell fabrication process flow that is presented in figure 2 is similar to the flow that is used in production at Shanghai Shenzhou New Energy Development (SSNED) except for the addition of an LPCVD polysilicon deposition step [4-7]. All process steps were performed on Tempress equipment, except for the wet chemical and implantation process steps. SSNED provided standard production samples (239 cm^2 n-type Cz) that were processed up until the rear side etch step and then the subsequent process steps were performed by Tempress using the facilities at ECN Solar Energy. The LPCVD process included an oxidation step to form the tunneling layer underneath the polysilicon layer and the subsequent doping step consisted of a blanket implantation without any need for masking. Dopant activation and surface oxidation was done in a tube furnace with single-slot wafer loading and a 2.4 mm pitch. PECVD SiN_x:H was deposited with a direct plasma system to a thickness of ~80 nm on the rear, while the front side deposition reached ~65 nm due to the use of an incorrect recipe. Printing/firing tests were done at ECN and at SSNED with industrial screen-printing pastes and a four-busbar screen design.

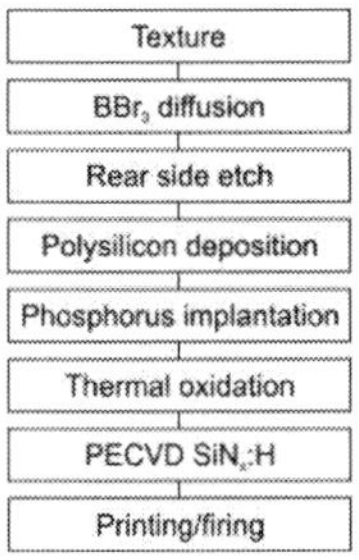

Figure 2. Overview of the process flow that was used to make the presented n-PERT solar cells.

3 RESULTS AND DISCUSSION

3.1 Polysilicon layers

In figure 3 we present a typical ECV dopant profile for LPCVD polysilicon layers with a thickness of 140 nm that were doped by ion implantation and thermal annealing. The sharp transition near the tunneling layer indicates that the layer functions as an effective diffusion barrier for Phosphorus ions. Due to the large concentration contrast a strong work function difference is created towards the bulk that helps to increase the contact selectivity towards majority carriers. The lack of onward diffusion also makes it easier to achieve a surface concentration of ~3E20 cm^{-3} that is suitably high for the combination with fire-through contact metallization.

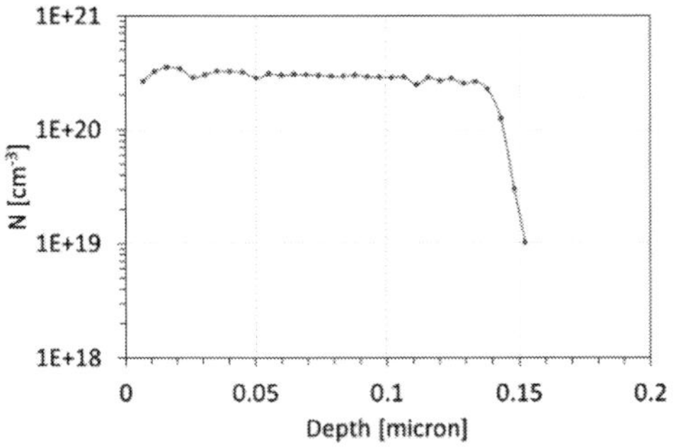

Figure 3. Dopant profile for a 140 nm polysilicon layer doped by Phosphorus implantation, as measured by ECV on a Cz solar cell half-fabricate.

Figure 4 presents QSSPC measurements on a symmetric lifetime sample with the same polysilicon thickness and dopant profile as for figure 3. The lifetime of 7.2 ms at an injection level of 1E15 cm^{-3} and the J_0 of 1.8 fA/cm^2 per side extracted near to 1E16 cm^{-3} injection level illustrate the high passivation quality that can be obtained with n-type doped LPCVD polysilicon layers. Before the application of PECVD SiN_x:H layers and simulated contact firing the lifetime was ~10% lower, so the hydrogenation of the tunnel oxide is helpful but not strictly necessary to reach a good performance. The near-linear injection level response of the measurement indicates that the passivation performance remains high under low-light conditions, which may prove to have some beneficial effect on the overall energy yield of the installed solar module. By contrast, amorphous silicon passivated surfaces tend to have a drop in lifetime at lower injection levels, according to several reports [8].

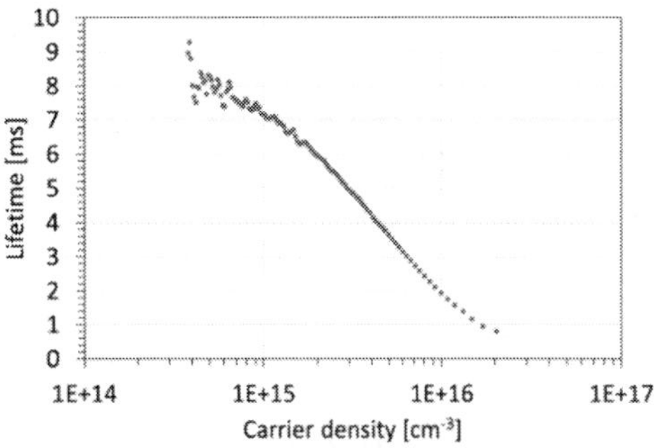

Figure 4. QSSPC measurement for a 5 Ω-cm polished Cz wafer passivated on both sides by a 140 nm n-type doped polysilicon layer and capped with PECVD SiN_x:H.

3.2 N-PERT solar cells

To check the quality of the solar cell half-fabricate samples we performed a QSSPC measurement on a sample before and after simulated contact firing. After firing the lifetime, implied V_{OC} and total J_0 became 554 μs, 689 mV and 63 fA/cm^2, respectively, but the real values could have been slightly higher or lower depending on the resistivity of the wafer, which we assume here as 2 Ω-cm. The lifetime was significantly lower before firing, which indicates that the front side passivation was improved by hydrogenation of the thermal oxide layer on the Boron emitter, since the polysilicon layer passivation was found to change by a relatively small amount after hydrogenation.

As a next step we performed a printing experiment on samples with a 140 nm polysilicon layer where the printing process was provided by ECN Solar Energy and executed in accordance with their baseline process for n-PERT solar cells. The high V_{OC} and fill factor results presented in Table I indicate that the polysilicon passivation performance is not adversely affected by the fire-through contacting process and that the doping level is suitable for the formation of a low contact resistance towards the silver paste. It is promising that the cell efficiency already reached 20.8% in an early attempt and it suggests that there is ample room for further improvement beyond 21%. The gap of ~20 mV between the real and implied V_{OC} is probably mainly due to the front side contact metallization, as the addition of aluminium to the used silver paste is known to cause a significant increase of the recombination losses near the contacts [9]. It is added to help lower the contact resistance towards the Boron emitter.

Group	#	Property	I_{SC} [A]	V_{OC} [mV]	FF [%]	η [%]
Firing setting 1	8	Average	9.28	660	79.6	20.40
		Best cell	9.29	663	80.0	20.62
Firing setting 2	6	Average	9.29	663	79.7	20.54
		Best cell	9.32	667	80.1	20.81

Table I. IV results for 239 cm^2 n-PERT solar cells with a 140 nm polysilicon layer, measured with a class A solar simulator and a reference cell that was calibrated by Fraunhofer ISE CalLab. For the 'best cell' of group 2 an optical mismatch correction was applied, which increased the as-measured efficiency by 0.049% (absolute).

We tried to make an assessment of the attainable cell efficiency with some simple process tuning alone. Using laser ellipsometry we measured that the front side nitride on the samples was ~20% thinner than the optimal value of 80 nm. According to optical simulation in OPAL2 a correct thickness should yield a 0.7% relative increase in light absorption [10]. Together with a polysilicon layer thickness reduction to 70 nm, which gives another 0.7% relative increase according to results presented below, the assessment suggests that we can enhance the obtained 20.8% efficiency to more than 21% by layer thickness tuning alone.

As a next step we performed a printing experiment in the n-PERT production line at SSNED in Shanghai. The test samples with 140 and 70 nm polysilicon layers were prepared alongside the samples used for the results presented in Table I. Table II presents current-voltage data that was measured on the industrial flash tester at SSNED, which uses a different measurement standard than the results presented in Table I. Compared to the standard production cells with a back surface field region the results represent an average cell efficiency increase of 0.4-0.6%, which is mainly due to a 10-20 mV increase of the V_{OC}. These results clearly demonstrate the large cell efficiency enhancement that is attainable in an industrial environment by the use of passivated contact technology.

Group	#	Property	I_{SC} [A]	V_{OC} [mV]	FF [%]	η [%]
70 nm poly-Si	18	Average	9.80	663	77.9	21.18
		Best cell	9.79	666	79.0	21.58
140 nm poly-Si	18	Average	9.76	662	77.7	21.04
		Best cell	9.75	662	79.1	21.39

Table II. IV results for 239 cm^2 n-PERT solar cells with two different polysilicon thicknesses, as measured by the industrial flash tester at SSNED.

The results in Table II indicate that a thinner polysilicon layer can be used to obtain a significant efficiency enhancement, which is in large part due to an increase of I_{SC}. We propose that this is mainly due to lower free carrier absorption losses, as also discussed below. The use of such thin layers could bring the risk that there are more constraints on the metallization process, but we have been able to reproduce the results in Table II several times using different rear side pastes and on samples that were processed almost fully at SSNED (except for the LPCVD polysilicon deposition step) so it appears to be a robust solution. With regards to the paste it is also to be expected that there is some potential for optimization of the formulation by paste manufacturers, to be able to contact even thinner layers for example. Since the polysilicon passivates the contact the optimal characteristics of the printed paste can be different from that for a paste that needs to contact a doped surface layer.

Apart from a cell efficiency advantage a thinner polysilicon also increases the LPCVD process throughput. The present tests were done with a partially filled wafer boat but if the used process settings were applied to fully loaded boats in a system that encompasses 4 tube furnaces then the system throughput would become more than 3000 wafers per hour, so the LPCVD process has a clear potential to reach industrially relevant throughput levels.

In Table III we present bifaciality and pseudo-fill factor measurements on one sample from each group presented in Table II. The data shows that the use of a 70 nm polysilicon layer significantly improves the bifaciality, which mainly reflects a change in J_{SC}. Using the optical modelling results in a report by Reiter *et al.* about light absorption losses in polysilicon it can be estimated that an additional thickness of 70 nm should lead to a J_{SC} loss of about 3 mA/cm^2, which is about twice as much as what we observed [11]. The absorption loss measured by Reiter *et al.* is therefore more than enough to explain the increase in bifaciality for thinner polysilicon. Considering the fact that the rear side does not have a textured surface, the obtained value of 83.9% is relatively high. For a cell with a rear side texture the bifaciality should therefore become more than 90%, which should be beneficial for certain bifacial module applications, although it probably also leads to some lowering of the front side efficiency.

Cell type	η bifaciality [%]	Pseudo-FF [%]
70 nm poly-Si	83.9	82.2
140 nm poly-Si	80.7	82.1

Table III. Solar cell efficiency bifaciality and pseudo-fill factor results for one solar cell from each group in Table II, as measured by a class A solar simulator and Sinton Suns-Voc, respectively.

3.3 Module integration and reliability

To test the applicability of the polysilicon passivated contact technology for module applications we prepared two modules using n-PERT cells with a 70 nm polysilicon layer and performed degradation tests. The cells that were used for the test were from an earlier stage of the development work and had a somewhat lower cell efficiency for that reason. The strings were prepared using standard soldering and the modules had glass on the front- and rear side to enable bifacial applications. The results presented in Table IV show that the modules passed the standard LID and PID tests, which indicates that the introduction of polysilicon technology does not lead to any major module performance stability issues. The quality norm for LID and PID we use here is a final power loss that is less than 5%, which is in accordance with IEC 61646 and 62804, except that the PID test used a higher temperature than prescribed in the PID certification procedure.

Module	15 kWh/m^2 LID test	60 kWh/m^2 LID test	96h PID test
1	-0.74%	-0.83%	-
2	-	-	-3.6%

Table IV. Potential and light induced degradation test results for two 60-cell bifacial modules that incorporate the presented N-PERT solar cells, as measured by SSNED. PID testing was with 85 °C / 85% RH / -1000 V.

4 CONCLUSION

We fabricated n-PERT solar cells in order to demonstrate the use of standard fire-through contact metallization in combination with polysilicon passivated contact technology and to show that the technology can be integrated into an existing solar cell production flow. In a first printing experiment we obtained a cell efficiency of 20.8% and a V_{OC} of 667 mV for samples with a 140 nm polysilicon layer thickness and in a second printing experiment we obtained an average V_{OC} of 662-663 mV for samples with a 70 and 140 nm polysilicon layer thickness. These results demonstrate that the passivation performance of the polysilicon stack remains intact even when ultra-thin polysilicon layers are used in combination with fire-through silver paste. By optimization of the polysilicon and anti-reflection coating layer thicknesses alone we should be able to increase the 20.8% to more than 21% cell efficiency.

The solar cells are suitable for bifacial module applications as their bifaciality can reach 84%. We assembled a first set of bifacial modules and performed PID and LID testing on them to show that the polysilicon passivated contact technology does not lead to any unexpected issues in the module assembly process and does not lead to any major module power stability issues.

5 REFERENCES

[1] S.W. Glunz *et al.*, "The irresistible charm of a simple current flow pattern – 25% with a solar cell featuring a full-area back contact", Proc. 31st EU-PVSEC, Hamburg, Germany, 2015.

[2] Y. Tao *et al.*, "Large area tunnel oxide passivated rear contact n-type Si solar cells with 21.2% efficiency", Prog. Photovolt: Res. Appl., vol. 24, pp. 830, 2016.

[3] U. Römer *et al.*, "Ion implantation for poly-Si passivated back-junction back-contacted solar cells", IEEE J. Photovolt., vol. 5, pp. 507, 2015.

[4] R.C.G. Naber *et al.*, "Industrial p- and n-type silicon solar cell processes using customized solutions for ion implantation and annealing", 27th EU-PVSEC, Frankfurt, Germany, 2012.

[5] G. Jin *et al.*, "Ion implantation technology for high efficiency solar cell manufacturing", 9th CSPV, Jiangsu, China, 2013.

[6] Z. Zhang *et al.*, 8th SNEC Conference, Shanghai, China, 2014.

[7] M. Lenes *et al.*, "LPCVD polysilicon passivated contacts for different solar cell concepts", 6th SiliconPV, Chambéry, France, 2016.

[8] S. Olibet *et al.* "Properties of interfaces in amorphous/crystalline silicon heterojunctions", Phys. Status Solidi A, pp. 1, 2010.

[9] Edler *et al.*, "Metallization-induced recombination losses of bifacial silicon solar cells", Prog. Photovolt: Res. Appl., vol. 23, pp. 620, 2014.

[10] K.R. McIntosh *et al.*, "OPAL 2: Rapid optical simulation of silicon solar cells," Proc. 38th IEEE PVSC, Austin, Texas, 2012.

[11] S. Reiter *et al.*, "Parasitic absorption in polycrystalline Si-layers for carrier-selective front junctions", 6th SiliconPV, Chambéry, France, 2016.

6 ACKNOWLEDGEMENTS

We thank Dr. Stodolny from ECN Solar Energy for his help with the measurements presented in Table I and III and Mr. Qian, formerly at SSNED, for his help with the results presented in Table II and IV.

20% N-PERT SOLAR DEVICE IN ONLY 7 STEPS: THE SOLENNA(3) CONCEPT

Raphaël Cabal*, Bernadette Grange, Rémi Monna, Yannick Veschetti, Sébastien Dubois
CEA, LITEN, INES, 50 av du Lac Léman, BP332, Technolac, 73370, Le Bourget du Lac, France

ABSTRACT: Bifacial solar cell n-PERT concept involves boron and phosphorus doped regions requiring both proper surface passivation. In order to reduce n-PERT technology related €/W, one way considered is to simplify the cell process-flow. In order to do so, we developed passivating/ anti-reflective/ doping SiO_xN_y:B and SiN_x:P layers. The co-anneal of these multifunctional layers allows for emitter & BSF simultaneous formation, subsequent doped regions passivation and provides in the same time good optical properties. We present our ultimately simplified "$SOLENNA_{(3)}$" technology, involving only 7 processing steps. Furthermore, the influence of the wafers initial properties (P and O_i concentrations, Thermal History Index) on the performances of the SOLENNA(3) cells is studied. By further optimization of our front SiO_xN_y:B, large area (239cm²) LID-free n-PERT devices could be fabricated with a PV conversion efficiency of 20.3%.

Keywords: silicon; passivated emitter rear totally diffused; solar cell; co-diffusion; boron silicate glass; passivation; doped silicon nitride

1 INTRODUCTION

The crystalline silicon photovoltaics (PV) industry has been dominated by the conventional Al-BSF cell architecture. This is easily explained by the fact Al-BSF has so far always shown the best compromise in terms of €/W_p. On the other hand, bifacial PV and related bifacial n-PERT technologies seem to be a better option for a significant reduction of levelized cost of electricity (LCOE in €/kWh). Over the last years, academics and industrials have developed several simplified 20%-efficient n-PERT concepts [1]. But an even more simplified bifacial process would be required in order to offer a competitive cost per W_p. To achieve this goal, PECV-deposited doping layers have been developed. A SiO_xN_y:B layer was developed as a boron doping source for n-PERT emitter formation, and a SiN_x:P layer was developed as a phosphorus doping source for n-PERT Back Surface Field formation. For PERT architecture, a basic $p^+/n/n^+$ structure is required. Compared to gaseous diffusion route [2], the co-anneal of SiO_xN_y:B and SiN_x:P layers makes this $p^+/n/n^+$ structure much simpler in terms of process. The integration of SiO_xN_y:B and SiN_x:P as sacrificial doping layers was detailed in a previous paper, leading to $SOLENNA_{(2)}$ devices processed in 11 steps, exhibiting 20% efficiency on Czochralski (Cz) Si and up to 19.8% on Mono-Like substrates [3].

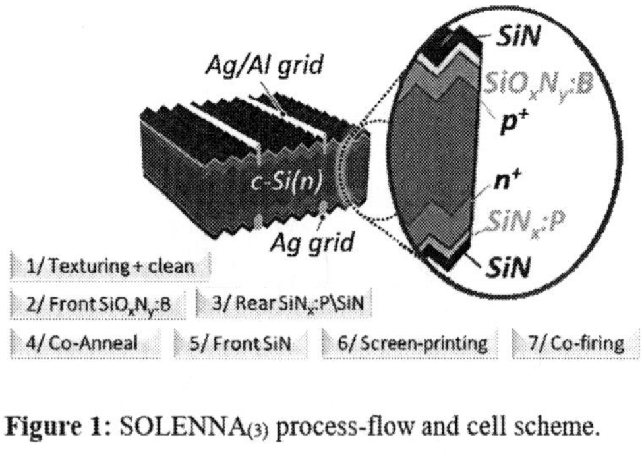

Figure 1: $SOLENNA_{(3)}$ process-flow and cell scheme.

In the present paper, we integrate SiO_xN_y:B and SiN_x:P as multifunctional layers ensuring simultaneously doping, passivation, and antireflective properties. As a consequence, the layers are not removed during the process, which is thus reduced to only 7 steps, as defined in Fig.1. A focus is made here on doping and subsequent passivating properties of SiO_xN_y:B and SiN_x:P after a co-diffusion anneal, with the aim of improving $SOLENNA_{(3)}$ cell efficiency up to 20%, i.e. to the state of the art of bifacial n-PERT. For such an efficiency, assuming a 100MW capacity, and despite a +10% extra cost of n-type over p-type wafer, $SOLENNA_{(3)}$ was already found to provide a cell process cost of 15.6€/W_p, competing with Al-BSF and PERC technologies [4].

2 SIO_xN_y:B AND SIN_x:P PROPERTIES

2.1 Doping layers

Among the different ways of simplification of n-PERT processing, dielectric doped layers are of prime interest since they permit the use of one co-diffusion step reducing both process cost and thermal budget. The most known approach used for making co-diffused PERT devices is based on the annealing of a solid boron source (BSG for boron silicate glass) during a specific $POCl_3$ diffusion cycle [5]. In the present study, the boron doped source developed is a silicon oxinitride (SiO_xN_y) in which boron is incorporated through the use of H_2-diluted diborane. The PECVD deposition process is done at low frequency, involving a mixture of SiH_4, N_2O, and B_2H_6. Instead of $POCl_3$ doping, we also developed a phosphorus doped layer SiN_x:P using a mixture of SiH_4, NH_3, and H_2 diluted PH_3. Depending on deposition time, different layer thicknesses were coated on 239cm² Cz samples. Anticipating the integration of both doping layers into the $SOLENNA_{(3)}$ process flow depicted in Fig.1, the coated samples were annealed at 940°C (<1h). The resulting boron emitter and phosphorus BSF sheet resistance (determined by 4PP) variation is plotted in Fig.2, as well as the final active doping profiles (ECV) selected for coming cell processing. Considering a satisfying process target of Rsheet < 100 Ω/sq, SiO_xN_y:B doping requires a minimal thickness around 20nm to be effective. Further thickening of the layer only improves doping non-uniformity (not plotted here) from 3% to <1.5%. SiN_x:P phosphorus doping shows more steady variation: depending on layer thickness, the R_{sheet} varies in the range of 55 Ω/sq down to 20 Ω/sq. a thickness of 10nm is sufficient for doping concern. Non-uniformity remains below 8% in any case. The selected profiles are <0.4μm deep for the emitter and <0.6µm deep for the BSF.

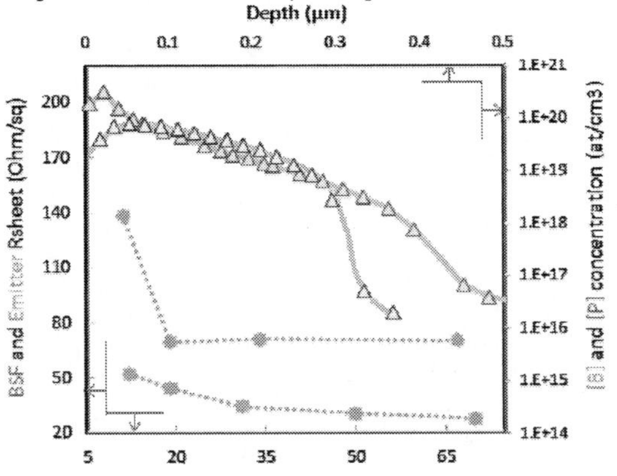

Figure 2: R_{sheet} variation against doping layer thickness; boron emitter (green) and phosphorus BSF (purple) active doping profiles

2.2 Passivating layers

In order to get rid of classical n-PERT surface passivation steps such as thermal oxidation or aluminum oxide coating, our SiO_xN_y:B and SiN_x:P layers should also provide satisfying dopant-rich surface passivation properties. Such features were monitored on textured symmetrical samples exhibiting either $p^+/n/p^+$ (emitter passivation) or $n^+/n/n^+$ (BSF passivation) structures. The $p^+/n/p^+$ samples were obtained by annealing (940°C; <1h) double-side SiO_xN_y:B coated Cz(n) wafers; the $n^+/n/n^+$ samples were obtained similarly on double-side SiN_x:P coated Cz(n) wafers. QssPC setup was used to extract J_0 (dopant rich region saturation current density) from the symmetrical samples. The n-PERT baseline (using gaseous BCl_3 and $POCl_3$ diffusions) features emitter and BSF target current densities of 80fA/cm² and 150fA/cm² respectively. SiO_xN_y:B layer initially featured poor passivation properties (J_0 > 1400fA/cm²) that were hardly reduced by the classical SiN anti-reflective capping. This can be explained by the high boron content of our layer which results in a highly defective interface with probable BRL at the interface with silicon [6].

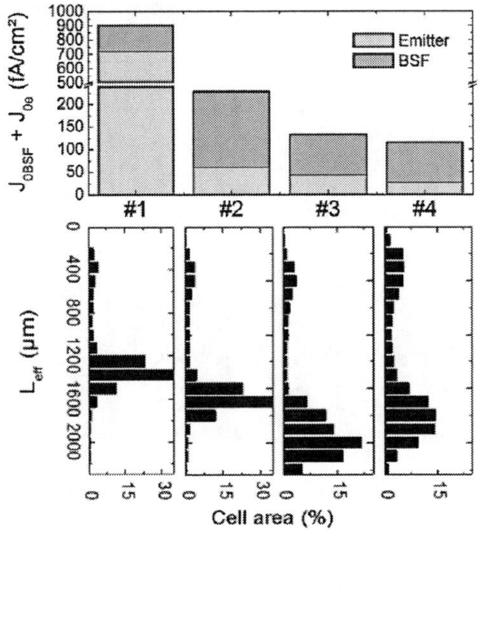

Figure 3: Evolution of J_{0e} + J_{0BSF} for different layer combinations (top); corresponding SOLENNA$_{(3)}$ solar cell L_{eff} distribution (bottom)

The improvement of SiO_xN_y:B passivating properties was made possible by reducing both boron content in the layer, and layer thickness. Such PECVD conditions tuning, combined with a modified co-annealing curb emitter J_0 down to 54 fA/cm². Passivating properties of SiN_x:P are slightly improved by the integration of an upper layer of SiN:H (337fA/cm²). Plasma and co-anneal conditions also had to be modified in order to obtain J_0 values below 180fA/cm². These raw J_0 extracted from symmetrical samples QSSPC were divided by 2 in order to consider the recombination of only one side of the device. The resulting J_{0e} and J_{0BSF} are plotted in Figure 3.

For each passivating configuration (#1 to #4), complete SOLENNA$_{(3)}$ solar cells were elaborated (see the next section for the detailed processing). The aforementioned improvements of front side (J_{0e}) and rear side (J_{0BSF}) passivation were confronted to the L_{eff} (effective hole diffusion length) distribution extracted from LBIC measurements (combined with reflectance measurements) on corresponding solar cells (see Fig.3). The first batch of cells (#1) exhibited an average L_{eff} ~1330µm. This value increased to ~1860µm, after for combination (#3). Despite the further reduction of front side J_{0e} obtained on the last batch (#4), no gain in L_{eff} was observed, this parameter being rather limited by bulk carrier lifetime and recombination on the rear side.

3 CELL PROCESSING AND LATEST RESULTS

3.1 Layer combination #3

A first batch of solar devices using SiO_xN_y:B and SiN_x:P was elaborated using 200µm thick Cz(n) 243cm² wafers with a resistivity 3-4 Ω.cm. After alkaline texturing and RCA cleaning, SiO_xN_y:B and SiN_x:P\SiN layers are deposited on wafer front and rear side. Samples are then annealed (940°C; <1h). Anrireflective SiN is deposited on sample front side, and front and rear silver based grids are screen-printed. Eventually, the samples are cofired in a belt furnace. Final device front and rear passivations are here granted by SiO_xN_y:B\SiN stack and SiN_x:P\SiN stack, respectively. Except the initial texturing and cleaning, no chemical step is involved during the process. This first batch corresponds to the layer combination referred to as "#3" in Figure 3, i.e. J_{0e} = 43fA/cm² and J_{0BSF} = 90fA/cm².

The resulting SOLENNA$_{(3)}$ solar cells I-V parameters have been measured on a class AAA solar simulator under standard test conditions (STC), and are listed in Table I. The best #3 solar cells provide an efficiency of 19.8%. Further improvement are still required to close the gap to 20%. At first order, the #3 devices are limited by the $J_{sc} \leq 38.6$mA/cm².

Table I: SOLENNA$_{(3)}$ I-V parameters under STC (25°C; AM1.5G; 1000W/m²) – layer combination #3

243cm²	V_{oc} (mV)	J_{sc} (mA/cm²)	FF(%)	η(%)
Average	644.1	38.6	79.1	19.7
best	645.0	38.5	79.6	19.8

3.1 Wafer specifications

Despite Cz n-type allows to get rid of LID effects, it

still features oxygen (~10^{18} at/cm^3). Such a high oxygen content, combined with high temperature processing steps (co-anneal at 940°C e.g.) can induce silicon oxide precipitation, which is known to be detrimental to the cell efficiency [7]. For industrialization purpose, we chose to evaluate the compatibility of SOLENNA$_{(3)}$ technology with different Cz material. Three ingots (A, B, C) from one Cz manufacturer are considered. In the case of ingot A, wafers were taken from five different batches (A1, A2, A3, A4, A5) which are assumed to be five different positions along the ingot. In the case of ingot B and C, the wafer batches were chosen for their specific ([O$_i$];[P]) detailed later. The resistivity (after thermal donor killing) is ranging from 3.5 Ω.cm to 8.5 Ω.cm. The wafers were processed as SOLENNA$_{(3)}$ devices using afore mentioned #3 combination. Table II gives an insight into resulting cell efficiencies depending on wafer resistivity windows. The efficiency variation versus wafer resistivity is not significant, so that no wafer specification can be expressed at this stage.

Table II: SOLENNA$_{(3)}$ solar cell efficiency for different wafer resistivity windows.

ρ (Ω.cm)	Ingot of concern	η range (%)
[3.5 – 5]	C	[19.3 - 19.9]
[6 – 7]	A1, A2	[19.2 – 19.8]
[7 – 8]	A3	[19.1 – 19.7]
[8 - 8.5]	A4, A5, B	[19.0 – 19.7]

The wafers had their oxygen (O$_i$), phosphorus (P) and initial thermal donor (TD) concentrations determined using the OxyMap technique, first introduced in [8] and regularly extended [9,10].]. For the [O$_i$] measurement, the OxyMap technique was calibrated with respect to Fourier Transform InfraRed (FTIR) spectroscopy using the industry-custom ASTM Test Standard F121-83 (i.e. with a calibration factor of 2.45x10^{17} cm^3.cm^{-1} or 4.90 ppma.cm).A qualitative Thermal History Index (THI expressed in minutes) can be calculated from TD and O$_i$ concentration, in order to estimate the cooling velocity after crystallization, which is a function of experimental height in the ingot [11]. Wafers with higher THI correspond to regions located closer to the ingot seed. Ingot A exhibits a decrease of [P] from batch 1 to 5, occurring concurrently with an increase of both [O$_i$] and THI (see Fig.4). As a result, batch A1 is assumed to be closer to the tail of the ingot, while A5 is closer to the seed. Every batch from ingot A feature relatively low oxygen content ([O$_i$]< 6.10^{17}at/cm^3). On the opposite, ingots B and C show both high oxygen content ([O$_i$]> 9.10^{17}at/cm^3) and similar THI (~130min). Wafers from ingots B and C correspond probably to regions close to the seed. Nevertheless, the phosphorus doping between ingot B ([P]~5.5.10^{14}at/cm^3) and C ([P]~1.3.10^{15}at/cm^3) is quite different, as suggested previously by the resistivity values (Table II). The higher doping observed on batch of ingot C may at first suggest a location closer to the tail (since phosphorus is known to segregate/accumulate in the lastly solidified region). Nonetheless, an intentional higher doping of Si feedstock in the case of ingot C is a more plausible explanation since both Oi/THI values suggest a wafer location near the seed.

For further understanding of wafer influence on cell performance, we focused on the cell open circuit voltage (V_{oc}), because the larger variations between ingots were observed on this parameter. Cell V_{oc} are confronted to initial wafer ([O$_i$], [P], THI) characteristics (see Fig.4). For "low-O" Cz (ingot A), wafers with higher phosphorus doping (A1) exhibit the lower V_{oc} (640mV): considering the relatively low THI (20min), the A1 batch is rather low in the ingot, so that high metallic impurity concentration can come into play. For "high-O" Cz ingots (B,C) whose bulk lifetime is potentially degraded by silicon oxide precipitation in the silicon matrix, the worst V_{oc} (<638mV) is observed on wafers with low phosphorus doping (ingot B) with PL images featuring consistently O-rings.

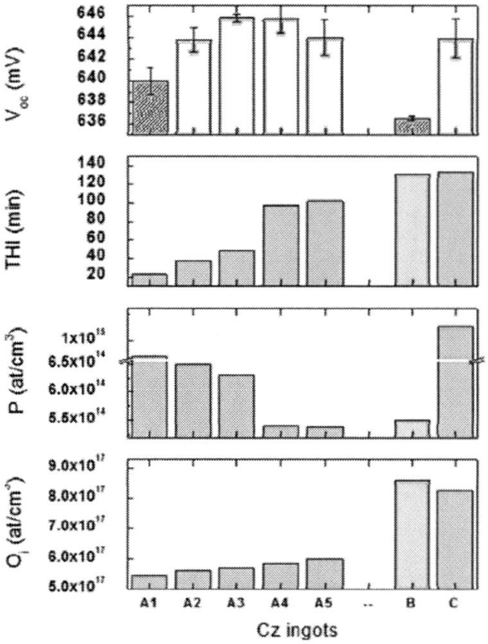

Figure 4: Wafer [O$_i$], [P], and THI for ingots A (green), B (orange) and C(blue). Resulting solar cell V_{oc}: the worst V_{oc} results are emphasized (red-filled bars).

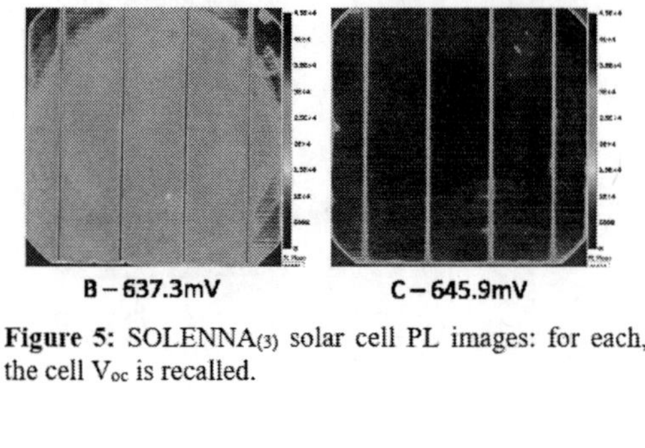

Figure 5: SOLENNA$_{(3)}$ solar cell PL images: for each, the cell V_{oc} is recalled.

On the opposite, wafers from ingot C provide good cell V_{oc} (>644mV). The reason of that remains unclear: proper Cz pulling conditions resulting in a shortened vacancy-rich region (known to catalyze oxygen precipitation) may come into play since no O-rings could be observed on batch C PL image (see Fig.5). Additionally the higher base doping of batch C induces a higher junction built-in voltage, which may also contribute to maintain a high cell V_{oc}.

Even if the crossed-influences of P, Oi and THI remain to be clarified, the present results already suggest Cz(n) wafer specifications: intermediate THI (40 to 100min in the present study, though such values may be different for other Cz ingots lengths) should be preferred, as well as [Oi] below 9.10^{17}at/cm^3. No simple specification can be given on [P] at this stage. In terms of resistivity, one should underline the necessity to perform post thermal donor killing resistivity measurement: in the range investigated, the [3.5 – 7] Ω.cm resistivity window is advantageous for $SOLENNA_{(3)}$ processing.

3.3 Optimized layer combination #4

A second batch of solar devices was fabricated on 239cm^2 wafers, with a resistivity of 3-4 Ω.cm. This time, the devices uses the combination of SiO_xN_y:B and SiN_x:P layers referred to as "#4" in Figure 3. This #4 configuration offers a reduced J_{0e} (27fA/cm^2) compared to #3. The resulting $SOLENNA_{(3)}$ solar cells I-V parameters are listed in Table II. Thanks to the improved emitter passivation, a significantly higher V_{oc} is reached (+7mV). As could be inferred from the previous L_{eff} distribution showing no improvement from combination #3 to #4, the J_{sc} gain observed with #4 devices is rather due to an improved front optics occurring concurrently with the J_{0e} reduction. Consequently to SiO_xN_y:B modifications, the solar cell effective reflection is reduced from 5.0% to 3.9% (in the 300-1200nm range) due to a lower reflectance in the short-wavelength (λ<650nm) domain, i.e. the front side of the cell. Fill factor (FF) values being not altered by these process modifications, the best #4 devices exhibit efficiencies of 20.3%.

Table III: $SOLENNA_{(3)}$ I-V parameters under STC (25°C; AM1.5G; 1000W/m^2) – layer combination #4

239cm^2	V_{oc} (mV)	J_{sc} (mA/cm^2)	FF(%)	η(%)
Average	651.0	39.0	79.2	20.1
Best	651.8	39.2	79.4	20.3

4 CONCLUSION

Features of PECV-deposited SiO_xN_y:B and SiN_x:P multifunctional layers have been detailed. Firstly, the use of such layers as doping sources was evaluated. Significant (R_{sheet}<100Ω/sq) and homogeneous (u<8%) doping have been demonstrated with relatively thin doping layers (<40nm). Preferred boron emitter and phosphorus BSF doping profiles are presented. Passivating properties of SiO_xN_y:B and SiN_x:P (alone or SiN-capped) layers was evaluated after co-anneal, on resulting $p^+/n/p^+$ and $n^+/n/n^+$. Saturation current densities obtained on such (textured) are below 50 fA/cm^2 and 100 fA/cm^2 for the emitter and BSF, respectively.

These multifunctional layers were combined (#3) following $SOLENNA_{(3)}$ process-flow, for B-emitter & Ph-BSF co-diffusion, and subsequent doped surface passivation. Such devices were processed using n-type Cz wafers with significant differences in terms of resistivity (3.5 to 8.5 Ω.cm), [P] ($5.5.10^{14}$ at/cm^3 to $1.3.10^{15}$at/cm^3) and [O_i] (5.10^{17} to 9.10^{17} at/cm^3). Even if influences of [O_i], [P] or THI is not quite clear yet, wafers cumulating high [Oi] and high THI should be preferably avoided. Up to 19.9%-efficient $SOLENNA_{(3)}$ solar cells were obtained. By choosing then proper substrates, and optimizing further SiO_xN_y:B passivating properties, it was possible to reduce J_{0e} <30 fA/cm^2. This optimized layer was integrated (#4) into $SOLENNA_{(3)}$ process-flow. By doing so, we were able to elaborate up to 20.3%-efficient screen-printed solar devices on 239cm^2 Cz with only seven steps.

REFERENCES

[1] B. Lim at al. "Simplifying the fabrication process flow of n-type PERT solar cells: recent progress at ISFH", Photovoltaic Internationals, Edition n°26, December 2014

[2] T. Blévin, et al. Development of industrial processes for the fabrication of high efficiency n-type PERT cells, Solar Energy Materials and Solar Cells 131 (2014), pp 24–29.

[3] R. Cabal, et al. "20% PERT Technology Adapted to n-Type Mono-Like Silicon: Simplified Process and Narrowed Cell Efficiency Distribution", 29th European Photovoltaic Solar Energy Conference and Exhibition, 2014, pp.648 – 652

[4] R. Cabal, et al. "Multifunctional dielectric layers for the fabrication of ultra-simplified n-PERT c-Si solar cells", Silicon-PV 2016.

[5] P. Rothhardt, et al. 2014, "19.9 % efficient bifacial n-type solar cell produced by co-diffusion-CoBiN", Proc. 29th EU PVSEC, Amsterdam, The Netherlands, pp.653–655.

[6] N. Wehmeier et al. 2015, "21.0%-efficient co-diffused screen printed n-type silicon solar cell with rear-side boron emitter" Phys. Status Solidi RRL, 1–5 (2015).

[7] J. Haunschild, et al."Cz-si wafers in solar cell production: efficiency limiting defects and material quality control", Photovoltaics International 15 (2012), pp.40-46

[8] J. Veirman, et al. "A fast and easily implemented method for interstitial oxygen concentration mapping through the activation of thermal donors in silicon", Energy Procedia 8(2011)41.

[9] M. Cascant, et al., "Investigating the spatial distribution of Oxygen-related defects in Czochralski Silicon wafers and ingots", in Proceedings of the 8th SNEC conference, Shanghai,2014.

[10] J. Veirman et al., "Mapping of oxygen-related defects in silicon for high efficiency solar cells: application to the prediction of LID losses", in Proceedings of the 28th EUPVSEC, (Munich, Germany),2013.

[11] J. Veirman, et al., "Thermal History Index as a bulk quality indicator for Czochralski solar wafers", Solar Energy Materials and Solar Cells (2016) – in press.

21.7% LARGE AREA *N*-PERT SILICON SOLAR CELLS USING SCREEN-PRINTED ALUMINIUM WITH OPEN CIRCUIT VOLTAGE OF 690mV

Jia Chen [a*], Filip Duerinckx [a], Emanuele Cornagliotti [a], Angel Uruena [a], Loic Tous [a], Monica Aleman [a], Richard Russell [a], Patrick Choulat [a], Sukhvinder Singh [a], Jinyoun Cho [a,b], Joachim John [a], Izabela Kuzma Filipek [a], Michael Haslinger [a], Ivan Gordon [a], Jef Poortmans [a,b,c] and Jozef Szlufcik [a]

[a] *Imec, Kapeldreef 75, 3001 Heverlee, Belgium*
[b] *KU Leuven, Kasteelpark Arenberg 10, 3001 Heverlee, Belgium*
[c] *University Hasselt, Martelarenlaan, Martelarenlaan 42, B-3500 Hasselt, Belgium*

ABSTRACT: Large area *n*-type rear-junction Passivated Emitter Rear Totally-diffused (PERT) solar cells with efficiency of 22.5% were recently reported by imec using rear sputtered Al. To develop a more industrially relevant process, screen-printing of Al is also currently investigated as the rear metallization. In this case, an Al p^+ emitter is locally formed by Al-Si alloying during a firing step. The quality of this p^+ emitter is crucial to the performance of the final devices. However, this is a challenging and less controlled process in terms of thickness of the doping layer and possible formation of voids. In this work, the contact geometry and the firing are optimized and their impact on the void and Al p^+ layer formation is studied. Using the optimized processing parameters, large area *n*-PERT solar cells with screen-printed Al with efficiency up to 21.7% are fabricated. These cells demonstrate high values for the open circuit voltage (V_{oc}) up to 690mV. This proves the feasibility of using screen-printed Al for high performance *n*-PERT cells.

Key Words—*n*-type, Solar Cell, Screen Printing

* Corresponding author. Tel.: +32 16 28 31 15; E-mail address: Jia.Chen@imec.be.

1. INTRODUCTION

Crystalline silicon (c-Si) solar cells are currently dominating the photovoltaic market, and this will continue to be the case for the foreseeable future. To further improve their performance, advanced *n*-type c-Si wafer solar cells featuring a dielectric passivation on the rear side, such as the *n*-type rear emitter Passivated Emitter Rear Totally-diffused (PERT) structure, have been receiving increasing attention for industrial application in recent years. Imec has recently fabricated large-area *n*-PERT solar cells with efficiencies up to 22.5% using rear sputtered Al [1]. To increase the industrial relevance of this process, we are also investigating the process of using screen-printing of Al to replace the more expensive sputtering of Al as the rear metallization.

n-PERT solar cells with screen-printed Al have been explored in the recent years. An efficiency of 21.4% was reported last year for this type of solar cells with epitaxially grown silicon wafers with built-in boron rear side emitter [2]. While using screen-printed Al, an Al p^+ emitter is locally formed by Al-Si alloying during a firing step. However, this is a challenging and difficult to control process in terms of doping profile uniformity and thickness of the doping layer. A thin and inhomogeneous Al p^+ layer and possible formation of voids (similar to the voiding issue in *p*-PERC solar cells [3, 4]) could significantly affect the performance of the device in terms of higher recombination at the rear local contact and result in an increase of the resistive losses [5]. Normally a large contact size has to be introduced to ensure a more uniform p^+ layer formation [5], which is usually at the expense of a high contact fraction. Therefore, the open circuit voltage (V_{oc}) reported for *n*-PERT solar cells with screen-printed Al was limited untill now to values below 675mV [6].

In this work, we report the progress of *n*-PERT rear junction solar cells with screen-printed Al. Compared to the processing flow for cells with sputtered Al [1], screen-printing of Al and firing was applied to replace sputtering of Al and annealing. The impact of firing on the front surface passivation is studied. The rear contact geometry and the firing conditions are optimized, and their impact on the void and Al p^+ layer formation is analyzed by photoluminescence (PL), electroluminescence (EL) imaging and scanning electron microscopy (SEM). Using the optimized processing parameters, large area *n*-PERT solar cells with screen-printed Al with efficiency up to 21.7% are fabricated. These cells show remarkable values for V_{oc} up to 690mV.

2. APPROACH AND EXPERIMENTS

Test structures and *n*-type 6 inch monocrystalline (Cz) silicon solar cells were fabricated according to the process flow of Fig. 1(a) in Imec's pilot line for Si solar cells. After a saw damage etch, a boron diffusion was applied at both the front and the rear. The resulting boron emitter had a sheet resistance of 95 ± 5 Ω/sq. A silicon oxide layer (SiO_2) was grown on both sides of the wafer by wet - oxidation. The B-dopants were also driven in during this step. Subsequently, the front oxide was removed by HF and the front surface was textured in a KOH-based solution. This was followed by a $POCl_3$ diffusion for the creation of a front surface field (FSF). The PSG was removed by an HF dip, followed by a dry oxidation. An antireflective coating (ARC) layer of silicon nitride (SiN_x) was deposited on the front side of the sample by PECVD. The rear oxide was removed by HF. The emitter is then passivated by Al_2O_3 deposited by spatial ALD and capped with PECVD SiO_x and SiN_x. The rear dielectric stack was ablated locally by a nanosecond (*ns*) laser.

The wafers were split into two groups. One group was fabricated into test samples and the other group was processed into full solar cells. The rear dielectric stack of the test samples was opened with different laser patterns (lines with different line widths or points with different radii) before screen-printing of the Al paste. While varying the contact size, the contact pitch was adapted so that the effective emitter resistance was kept unchanged. Then the test samples were fired at two peak firing temperatures,

840°C or 885°C.

For the group of solar cells, their rear dielectric stack was opened with the optimized contact fraction followed by a laser-doping step on the front SiN_x layer to create a selective FSF. The samples received a screen-printing of the Al paste on the rear, followed by a firing step. The front metallization is based on a Ni seed which is selectively plated on the laser-doped front pattern, followed by Cu plating and a short Ag immersion step. Finally a short, low temperature sintering was executed to form a Ni silicide on the front surface to decrease the contact resistance. The resulting structure for the solar cell is shown in Fig. 1(b).

The test samples were characterized by PL imaging to study the effect of the rear local p^+ formation. To better understand the contact formation, the samples were studied by means of cross-sectional scanning electron microscopy. Upon completion of the solar cells, the one-sun current-voltage characteristics were measured in house under standard testing conditions. The cells were also characterized by PL and EL, Suns-V_{OC} and spectral response (SR) measurements.

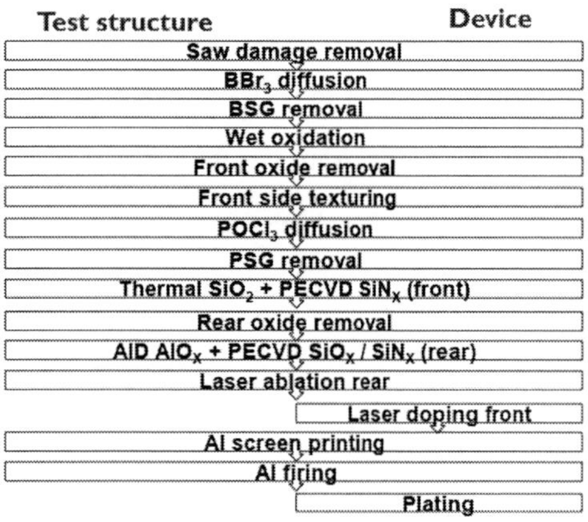

(a)

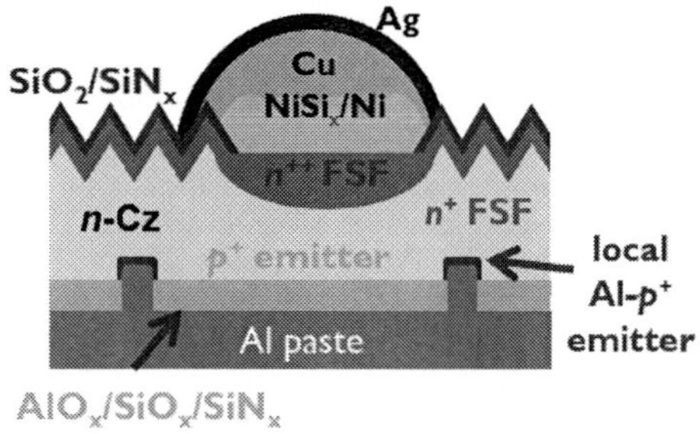

(b)

Fig.1. (a) Experimental flow of *n*-PERT test samples (top) and solar cells (bottom) with screen-printed Al. (b) A schematic of the final *n*-PERT solar cell

3. RESULT AND DISCUSSION

3.1 Impact of firing on front surface passivation

Since the annealing after sputtering of Al [1] is replaced by a firing step after screen-printing of Al in this work, the impact of firing on the front surface passivation was first investigated. Non-metallized symmetrical lifetime samples were fabricated to extract the saturation current densities of the passivated front surface, $J_{0,pass}$. These samples were textured and received a $POCl_3$ diffusion, followed by a dry oxidation and anti-reflective coating (ARC) layer of silicon nitride (SiN_x) on both sides. $J_{0,pass}$ of these samples with different thermal treatments was extracted from their effective lifetime measurements [7] and shown in Table I. There is no change in $J_{0,pass}$ before and after the annealing. However, $J_{0,pass}$ decreases after the firing step. In addition, in our previous work, it was shown that firing is also more effective at reducing the saturation current densities at front laser doped contact areas, $J_{0,laser}$ [1]. This is also shown in Table I.

Table I: Impact of firing on $J_{0,pass}$ and $J_{0,laser}$ [1]

	$J_{0,front,pass}$ [fA/cm^2]	$J_{0,laser}$ [fA/cm^2]
Before	13	~12000
Annealing	13	~7500
Firing	10	~3000

3.2 Optimization of the rear opening geometry

Test samples were fabricated in a similar way to the structure of the final device, except that there is no front metallization (see Fig.1a). Each test sample contained 16 boxes where the rear dielectric film was locally opened with different laser patterns by laser ablation as shown in Fig.2a. The observed differences in the PL intensity of different boxes can exclusively be attributed to changes at the rear of the test samples. The absolute PL intensity of the regions with laser patterns was normalized by the PL intensity of the non-ablated region of the same test sample (Ref box), to avoid possible differences in the PL intensity due to the wafer quality, bulk doping level or dielectric passivation quality.

Fig.2b shows the difference in the normalized PL intensities between test samples with different opening geometries fired at different peak temperatures. At both peak firing temperatures, lines with narrower widths have shown a higher PL intensity. The normalized PL intensities for all opening geometries increases, when the peak firing temperature increases from 840°C to 885°C. Cross-sectional SEM images of the rear contacts with the narrowest line width were examined. A thicker Al-p^+ layer was found for the samples fired at the higher temperature. However, voids were also much more frequently observed for test samples fired at the higher temperature.

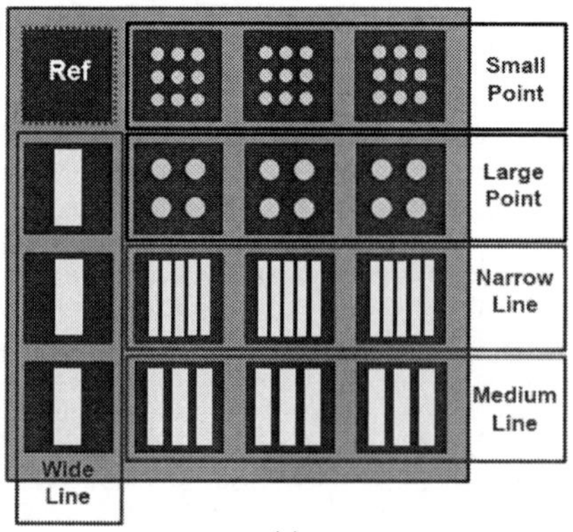

(a)

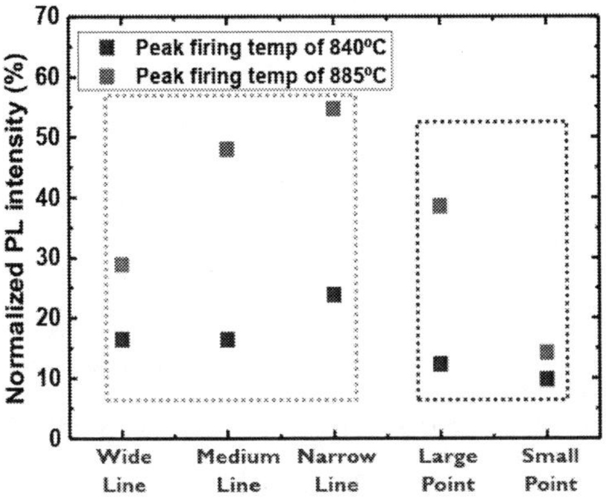

(b)

Fig.2. (a) Schematic illustration of the test structure with 16 boxes. (b) Normalized PL intensity images of the test samples with different laser patterns fired at peak temperatures of 840°C or 885°C.

3.3 Solar cells result

n-PERT solar cells with screen-printed rear Al and different peak firing temperatures were fabricated on 156 mm ×156 mm *n*-type Cz silicon wafers using the process flow shown in Fig. 1(a). In Section 3.2, it was found that the PL intensity was highest for line contacts with narrow line width. Therefore, cells with narrow lines were fabricated. Both firing temperatures of 840 and 885 °C were applied on the cells. The measured one-Sun *I-V* parameters of the final solar cell devices are summarized in Table II. In the first attempt, the highest efficiency is obtained in the group fired at 885 °C. The champion solar cell reached an efficiency of 21.3% with *FF* of 78.7%, J_{sc} of 39.6 mA/cm^2 and V_{oc} of 684 mV. However, a higher average efficiency was observed for the group fired at 840 °C. The difference in the average efficiencies of these two groups is primarily attributed to the difference in their *FF*. As shown in Table II, a higher R_S, calculated using the double-light method [8], and a higher J_{02}, fitted from Suns-V_{OC} method, are the main causes of the decrease in the *FF* of cells fired at 885 °C. As explained in Section 3.2, the voids are more likely formed at a higher peak firing temperature, and they have effectively reduced the actual contact area, so carriers have to travel an additional distance laterally at the rear surface, increasing the actual emitter resistance. The decrease in the effective metal fraction also causes the rear contact resistance to increase further. The voids can also affect the uniformity of the p^+ layer, which increases the recombination at the space charge region, and thus increases the J_{02} of the devices.

With further process optimization, we recently achieved a highest efficiency of 21.7% with V_{oc} of 690 mV in a 2nd experiment. For comparison, the parameters of one of our *n*-PERT solar cells with sputtered Al is shown in Table II, which was fabricated with similar processing flow except for the rear metallization and firing step [1]. Despite having lower electric parameters in all aspects, the *n*-PERT cells with screen-printed contact have shown a remarkably high V_{oc}. The biggest issue is the decrease in the *FF*. From a *FF* loss analysis [10], it can be attributed mainly to an increase of the J_{02} loss.

Table II: Average and best parameters of *n*-PERT solar cells with screen-printed Al fired at different peak temperatures. For comparison, the parameters of one of our *n*-PERT solar cells with sputtered Al is also shown [1].

Processing parameters		Cell parameters						
Rear Al metal.	Peak firing temp. [°C]	Area [cm^2]	J_{sc} [mA/cm^2]	V_{oc} [mV]	*FF* [%]	*Eff* [%]	R_S [Ωcm^2]	J_{02} [A/cm^2]
First attempt:								
Screen Print	840	239	39.6±0.1	681±1	78.5±0.1	21.2±0.1	0.61±0.1	8.9±2×10^{-9}
Screen Print	840	239	39.7	682	78.4	21.2	0.68	6.7×10^{-9}
Screen Print	885	239	39.6±0.1	682±2	77.2±0.8	20.8±0.3	0.72±0.1	1.6±4×10^{-8}
Screen Print	885	239	39.6	684	78.7	21.3	0.62	7.8×10^{-9}
Further development:								
Screen Print	840	239	39.8±0.1	689±1	78.6±0.4	21.5±0.1	0.49±0.0	-
Screen Print	840	239	39.8	690	78.8	21.7	0.50	9.6×10^{-9}
Best sputtered cell:								
Sputter	N.A.	239	40.0	696	80.4	22.4	0.45	2.4×10^{-10}

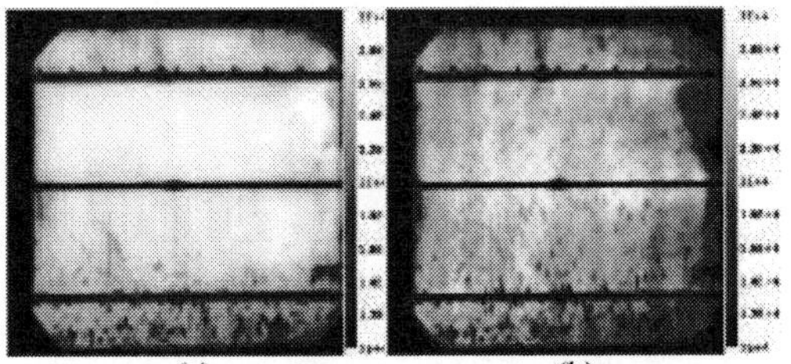

(a) (b)

Fig.3. Example of (a) PL and (b) EL images of a 'bad' cell with peak firing temperature of 885 °C

In Fig.3, the PL and EL images of a 'bad' cell fired at 885 °C are shown. This cell has a *FF* of 76.5%, J_{SC} of 39.7 mA/cm^2 and V_{oc} of 681 mV. Much more irregular discontinuous dark lines were observed in the EL image, which are likely caused by the formation of voids at the rear [4]. The non-uniform rear contact formation has smaller impact on the rear passivation from the PL image. Very small difference in the V_{oc} is observed among all the cells.

In Fig.4, Internal quantum efficiency (IQE) and reflection curves are plotted for cells with screen-printed Al and sputtered Al. A better IQE in the short wavelength range can be observed for the cell with screen-printed Al, mostly due to better FSF passivation and lower front laser damage as shown in Section 3.1. However, there is still room for improvement for the cell with screen-printed Al optically and electrically in the wavelength between

1000nm to 1200nm. The lower IQE in this range can be attributed to the higher rear contact fraction and lower quality of the Al local emitter. The lower rear internal reflectance is likely caused by lower reflectance of Al paste containing precipitated Si after firing compared with the reflectance of the sputtered Al.

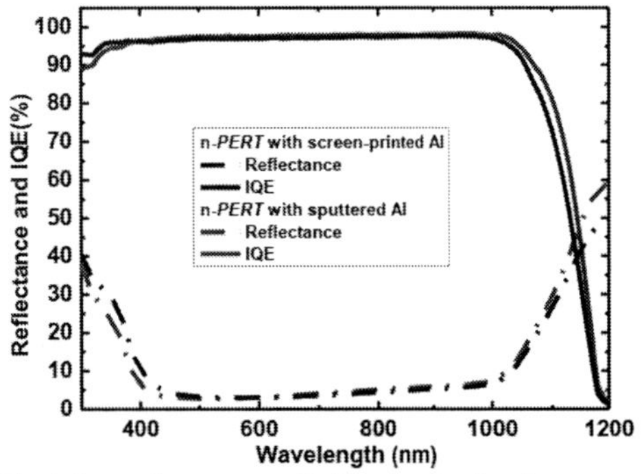

Fig.4. Internal quantum efficiency and reflection curves of cells with screen-printed Al and sputtered Al

4. CONCLUSION AND OUTLOOK

In conclusion, we report the progress of *n*-PERT rear junction solar cells with screen-printed Al. Compared with the processing flow for cells with sputtered Al [1], screen-printing of Al and firing was applied in this work to replace sputtering of Al and annealing. The firing step has a positive impact on the front surface passivation in terms of a lower $J_{0,pass}$ and $J_{0,laser}$. Using test structures, the contact geometry and the firing conditions were optimized, and their impact on the void and Al p^+ layer formation was studied. Using the optimized processing parameters, large area *n*-PERT solar cells with screen-printed Al with efficiency up to 21.7% were fabricated, with V_{oc} up to 690mV.

Future work will focus on the further optimization of rear local contact formation to reduce the contact fraction and on the characterization of the local Al doped p^+ regions to investigate the cause of the high J_{02} loss.

5. ACKNOWLEDGEMENTS

The authors gratefully acknowledge the financial support of the PV4FACADES project (Solar-Era.net) and Imec's industrial affiliation program for Photovoltaics (IIAP-PV).

6. REFERENCES

[1] A. Uruena de Castro, M. Aleman, E. Cornagliotti, A. Sharma, J. Deckers, M. Haslinger, L. Tous, R. Russell, J. John, Y. Yao, T. Söderström, F. Duerinckx and J. Szlufcik, "Beyond 22% large area n-type silicon solar cells with front laser doping and a rear emitter," *Proc. 31st EUPVSEC, Hamburg, Germany*, pp. 410-413, 2015

[2] V. Mertens, S. Bordihn, A. Mohr, K. Petter, J.W. Müllerm and D.J.W. Jeong, "Over 21.5 % efficient fully screen printed n-type solar cell on epitaxially grown silicon wafers with built in boron rear side emitter," *Proc. 31st EUPVSEC, Hamburg, Germany*, pp. 1000-1002, 2015

[3] E. Urrejola, K. Peter, H. Plagwitz, and G. Schubert, "Silicon diffusion in aluminum for rear passivated solar cells," *Applied Physics Letters*, vol. 98, p. 153508, 2011

[4] J. Chen, Z.H.J. Tey, Z.R. Du, F. Lin, B. Hoex and A.G. Aberle, "Investigation of screen-printed rear contacts for aluminum local back surface field silicon wafer solar cells," *IEEE Journal of Photovoltaics*, vol.3, no.2, pp.690-696, 2013

[5] B. Lim, T. Brendemuhl, M. Berger, A. Christ and T. Dullweber, "n-PERT back junction solar cells: An option for the next industrial technology generation?", *Proc. 29th EUPVSEC, Amsterdam, The Netherlands*, pp. 661-665, 2014

[6] B. Lim e, T. Brendemuhl, T. Dullweber and R. Brendel, "Loss analysis of n-type passivated emitter rear totally diffused back-junction silicon solar cells with efficiencies up to 21.2%," *IEEE Journal of Photovoltaics*, vol.6, no.2, pp.447-453, 2016

[7] D.E. Kane and R.M. Swanson, "Measurement of the emitter saturation current by a contactless photoconductivity decay method," *Proceedings of the18th IEEE Photovoltaic Specialists Conference*, New York, IEEE, pp. 578–583, 1985

[8] M. Wolf and H. Rauschenbach, "Series resistance effects on solar cell measurements," *Advanced Energy Conversion*, vol. 3, pp. 455-479, 1963

[9] M. Rauer, C. Schmiga, M. Glatthaar and St. W. Glunz, "Alloying from screen-printed aluminum pastes containing boron additives," *IEEE Journal of Photovoltaics*, vol.3, no.1, pp.206-211, 2013

[10] A. Khanna, T. Mueller, R. A. Stangl, B. Hoex, P. K. Basu and A. G. Aberle, "A fill factor loss analysis method for silicon wafer solar cells," *IEEE Journal of Photovoltaics*, vol. 3, no. 4, pp. 1170-1177, 2013

INDUSTRIAL N-TYPE BIFACIAL CO-DIFFUSED REAR EMITTER SOLAR CELLS WITH BORON SILICATE GLASS AS DIFFUSION SOURCE AND PASSIVATION

Nadine Wehmeier[1i], Sarah Kajari-Schröder[1], Till Brendemühl[1], Anja Nowack[1], Rolf Brendel[1,2], and Thorsten Dullweber[1]

[1] Institute for Solar Energy Research Hamelin (ISFH), Am Ohrberg 1, 31860 Emmerthal, Germany
[2] Dep. Solar Energy, Institute for Solid State Physics, Leibniz University Hanover, Appelstr. 2, 30167 Hanover, Germany

ABSTRACT: We fabricate *n*-type rear emitter cells with a boron silicate glass (BSG) layer deposited by means of plasma-enhanced chemical vapor deposition (PECVD). The multi-functional stack consisting of the BSG layer and a PECVD silicon nitride (SiN_z) capping layer acts both as a boron diffusion source and as the rear side passivation achieving saturation current densities of J_0=40 fA/cm². For solar cell fabrication, the rear emitter is diffused from the BSG/SiN_z stack while the front surface field (FSF) is diffused from $POCl_3$ gas in a co-diffusion process. We present a novel n^+ n p^+ solar cell concept and name it **BiCoRE** (**Bi**facial **Co**-diffused **R**ear **E**mitter) cell that applies a screen-printed Al grid on the rear side for bifaciality. The process flow and complexity of the BiCoRE cells are very similar to industrial Passivated Emitter and Rear Cells (PERC). The BiCoRE cells processed on the very first batch demonstrate conversion efficiencies up to η=20.6% when illuminated from the front side, up to 16.1% rear-side efficiency and a bifaciality of 78%.
Keywords: *n*-type Silicon Solar Cell, Bifacial, PECVD, Co-diffusion

1 INTRODUCTION

Solar cell concepts based on *n*-type silicon attract significant research interest due to the advantages over *p*-type silicon: *n*-type Cz-Si does not suffer from light-induced degradation (LID) [1,2,3] and is less sensitive to metal impurities [4]. For *n*-type Passivated Emitter and Rear, Rear Totally doped Back Junction (PERT BJ) cells applying separate boron and phosphorus furnace diffusions and an metallization scheme based on plating, efficiencies of up to 22.5% have been demonstrated [5] and with screen-printed contacts recently 21.8% have been reached [6]. But the *n*-PERT cells fabricated with sequential boron and phosphorus diffusions have a higher process complexity compared to industrial screen-printed Passivated Emitter and Rear Cells (PERC) based on *p*-type wafers. The latter are recently introduced into production [7] and yield R&D record efficiencies of up to 22.13% [8]. Recently, a bifacial PERC solar cell named "PERC+" has been introduced [9] by applying a screen-printed Al finger grid instead of the full-area Al layer. The resulting PERC+ cells demonstrate front side efficiencies up to 21.5% comparable to monofacial PERC cell and rear side efficiencies up to 16.7% [9,10].

For *n*-PERT cells, due to the boron diffusion, an additional high-temperature step is needed that is both cost- and time-intensive. Therefore, alternative boron diffusion sources are investigated with the aim of process simplification. High-efficiency *n*-type solar cells with a process complexity comparable to that of *p*-PERC cells shall be achieved so that an easy integration in existing production lines is possible.

We evaluate boron silicate glass (BSG) layers as solid diffusion sources that are deposited by single-sided Plasma-Enhanced Chemical Vapor Deposition (PECVD). We apply a co-diffusion process to form both the boron rear side emitter and the phosphorus Fornt Surface Field (P FSF) in a single high-temperature step. The integration of CVD-deposited diffusion sources and co-diffusion in *n*-PERT cell fabrication has been applied by different groups [11,12,13]. For a co-diffused *n*-PERT BJ cell with screen-printed contacts, we recently reached an efficiency of 21.0% [14].

For further process simplification, an additional use of the BSG layer as rear side passivation is evaluated. This would substitute the AlO_x passivation layer, and eliminate the necessity to remove the BSG wet-chemically after diffusion. Recently, Engelhardt et al. reached saturation current densities of 50 fA/cm² for R_{sheet}=60 Ω/□ on BSG passivated test structures [15]. In this work, we investigate the passivation quality of PECVD BSG layers capped with SiN_z and integrate it into the fabrication of a novel PERT cell type based on *n*-type Cz-Si which we name Bifacial Co-diffused Rear Emitter cell (BiCoRE) [16]. BiCoRE cells, as shown in Fig. 1, apply a screen-printed Al finger grid on the rear side to enable bifaciality. Further advantages of the Al grid compared to a full-area Al metallization, such as reduction of Al paste consumption and an improvement of the quality of the local Al-doped p^+ regions, have been described recently for the PERC+ cell [9] and account equally for the BiCoRE cell. A PECVD BSG/SiN_z diffusion source is applied that is not removed after co-diffusion but remains on the rear side emitter as a passivation layer and also as anti reflection coating (ARC). We demonstrate 20.6% efficient BiCoRE cells with a significantly reduced number of process steps compared to separately diffused PERT cells.

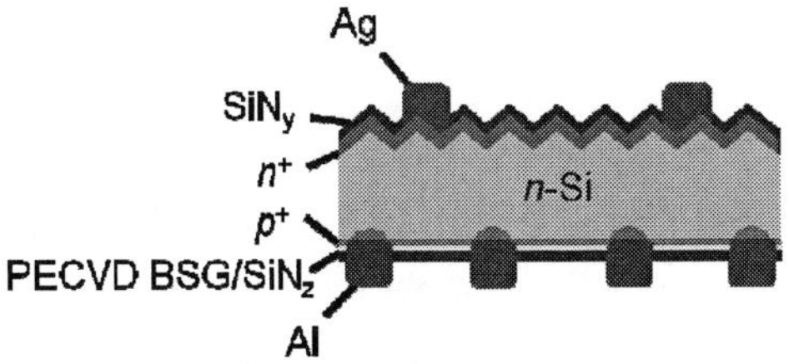

Figure 1: Schematic of the **BiCoRE** (**Bi**facial **Co**-diffused **R**ear **E**mitter) cell with PECVD BSG/SiN_z as rear-side diffusion source and passivation [16].

2 BSG DEPOSITION AND CO-DIFFUSION

The multifunctional PECVD layers are deposited in a cluster system (CS 400 P, von Ardenne) using an inductively coupled plasma (ICP) source. The precursor gases silane SiH_4, nitrous oxide N_2O and diborane B_2H_6 (3 \% diluted in 97 \% H_2) are used for the deposition of 40 nm-thin BSG layers. The BSG is capped by a 120 nm-thick PECVD silicon nitride SiN_z layer deposited in a SiNA tool (Meyer Burger). The boron drive-in from the PECVD BSG diffusion source is performed in a $POCl_3$ diffusion furnace (Tempress). Figure 2 shows the temperature profile of the two-stage co-diffusion process [17].

At first, the boron drive-in from the solid PECVD BSG diffusion source is performed at a temperature of 950 °C in N_2 ambient for 60 min. Second, at a lower temperature of 829 °C, a $POCl_3$ gas is inserted into the furnace tube and phosphorus is diffused from the gas phase on the uncapped wafer side. The PECVD layer stack of BSG and SiN_z acts both as a boron diffusion source and as a diffusion barrier against phosphorus during the co-diffusion. Applying a co-diffusion allows to introduce two dopants into the silicon wafer in only one high-temperature process.

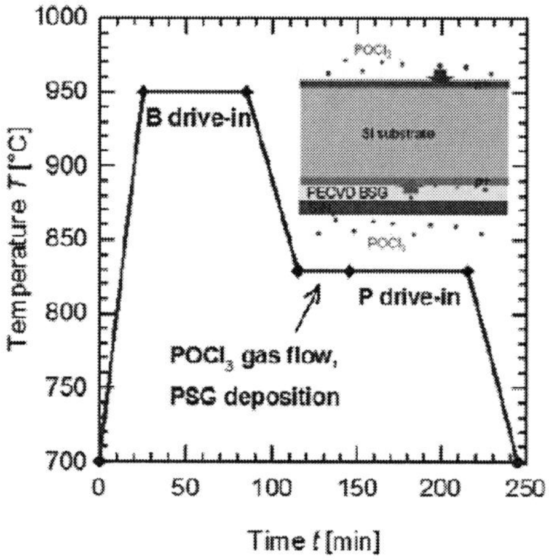

Figure 2: Temperature profile of the two-stage co-diffusion [17]. The boron drive-in from the PECVD BSG diffusion source is performed at 950 °C in N_2 ambient. In the same high temperature process, a $POCl_3$ diffusion is performed at 829 °C. The BSG layer is capped with a SiN_z layer to prevent phosphorus in-diffusion.

3 EMITTER AND PASSIVATION QUALITY

We evaluate the passivation quality of the PECVD BSG capped with either SiN_z or with PECVD silicon oxynitride (SiO_xN_y) by processing symmetrical samples with the material, deposition and diffusion parameters and the same processing steps as the BiCoRE cells. To compare different passivation stacks, we also process symmetrically boron-diffused samples that are AlO_x/SiN_y passivated after BSG-removal corresponding to the rear side passivation of the *n*-PERT BJ reference cells.

We measure the quasi steady state photoconductance (QSSPC) of these samples with a photoconductance lifetime tester (WCT-120, Sinton Instruments) using an intrinsic carrier concentration value of Si of $n_i = 8.6\times10^9$ 1/cm³. The measured data is evaluated by the Kane & Swanson method at a minority carrier density of 8×10^{15} 1/cm³ to extract the saturation current density J_0 [18]. Simultaneously, the boron sheet resistance R_{sheet} is measured inductively and the implied open circuit voltage $V_{oc,impl}$ is extracted.

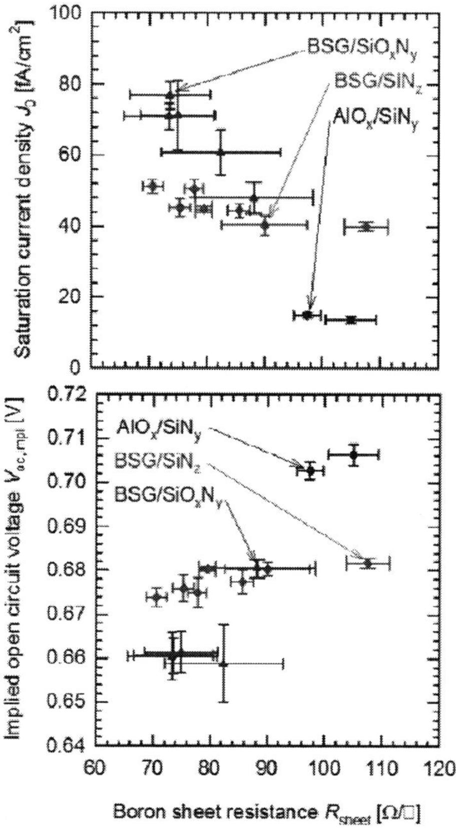

Figure 3: Saturation current density J_0 and implied open circuit voltage $V_{oc,impl}$ values for PECVD BSG and AlO_x/SiN_y passivation stacks after firing [19].

Figure 3 shows J_0 and $V_{oc,impl}$ in dependence of R_{sheet} for the different passivation stacks [19]. The errorbars correspond to the standard deviation derived from measurements at nine different positions per 156 × 156 mm² wafer. We achieve mean J_0 values of 40 to 51 fA/cm² and $V_{oc,impl}$ of 674 to 682 mV at boron sheet resistances R_{sheet} of 71 to 108 Ω/□ with BSG/SiN_z stacks as diffusion source and passivation. For PECVD BSG/SiO_xN_y layers, the standard deviation of the measured R_{sheet} is increased due to the higher thickness inhomogeneity of the 300 nm-thick PECVD SiO_xN_y layer on 156 × 156 mm² area compared to a 120 nm-thick SiN_z layer. For this stack, we also obtain higher mean J_0 values of 48 to 77 fA/cm² and corresponding lower $V_{oc,impl}$ of 659 to 681 mV with increased standard deviations compared to the BSG/SiN_z stack. All J_0 data is measured both before (not shown in Fig. 3) and after firing at 850°C. No degradation upon firing is observed. This proves the firing stability of the BSG passivation. By removing the PECVD BSG and capping layer in HF after co-diffusion, RCA cleaning and passivating with an

AlO_x/SiN_y stack, lowest J_0 results of 14 fA/cm² and highest $V_{oc,impl}$ values of 706 mV are reached. The BSG/SiN_z stack as emitter passivation does not yet reach the high passivation quality of AlO_x/SiN_y stacks [20]. However, on cell level the impact on cell performance is currently negligible (see below).

4 BICORE SOLAR CELL PROCESS

The process flow for the fabrication of BiCoRE cells and of monofacial Co-diffused Rear Emitter (CoRE) cells is shown in Fig. 4 (left column). 156 × 156 mm² *n*-type Cz silicon wafers with (100) orientation with a starting thickness of (170±20) µm and a resistivity of 5 to 6 Ω cm are damage-etched in KOH and RCA-cleaned. The PECVD BSG layer with a thickness of 40 nm is deposited on the rear side. It is capped with 120 nm PECVD SiN_z (SiNA tool). During the following texturing, the BSG/SiN_z stack acts as etching barrier. During the $POCl_3$-co-diffusion (see Fig. 2) the B emitter is diffused from the solid BSG diffusion source and the phosphorus FSF is diffused from the gas phase. After PSG etching in HF, only the front side needs to be passivated by 100 nm PECVD SiN_y, the BSG/SiN_z stack is kept as rear side passivation. We perform laser contact opening (LCO) and metalize the cell by Ag screen-printing on the front side with a 5 busbar layout. We screen print an Al grid on the rear side in order to achieve bifacial (BiCoRE) cells or full area Al for monofacial (CoRE) cells. Details about the applied LCO and screen-printed metallization are discussed in [16]. Co-firing finishes the cells.

For comparison of different rear side passivations, we also process BiCoRE cells with a stack of 5 nm-thin atomic layer deposited (ALD) AlO_x and 100 nm PECVD SiN_y. The AlO_x/SiN_y stack is deposited on the p^+ rear-side emitter after removing the BSG/SiN_z stack after co-diffusion.

Figure 3 compares the lean BiCoRE cell process with BSG-passivation to *n*-PERT BJ reference cells that are processed with separate B and P diffusions and with industry-typical *p*-PERC and bifacial *p*-PERC+ cells [9,10]. The BiCoRE cells have a very similar process flow as industrial PERC cells so that they can be implemented in existing production lines with only minor changes to the tool set. The BiCoRE process complexity is comparable to that of PERC and PERC+ cells but due to the *n*-type substrate wafer they potentially may obtain higher efficiencies.

n-BiCoRE	*n*-PERT BJ reference	*p*-PERC, *p*-PERC+
Saw damage removal & wafer cleaning	Saw damage removal & wafer cleaning	Saw damage removal & wafer cleaning
	B-diffusion	
	BSG etch	
PECVD BSG+SiN_z	Rear protection layer	
Texturing	Texturing	Texturing
Co-diffusion	P-diffusion	P-diffusion
PSG etch	PSG+dielectric etch	Polishing+PSG etch
	Rear: AlO_x+SiN_y pass.	Rear: AlO_x+SiN_y pass.
Front: SiN_y pass.	Front: SiN_y pass.	Front: SiN_y pass.
Laser contact opening	Laser contact opening	Laser contact opening
Front: Ag screen-print.	Front: Ag screen-print.	Front: Ag screen-print.
Rear: Al screen-print.	Rear: Al screen-print.	Rear: Al screen-print.
Co-firing	Co-firing	Co-firing
10 steps	**13 steps**	**10 steps**

Figure 4: The process flow for *n*-BiCoRE cells applying co-diffusion and BSG/SiN_z passivation are compared to *n*-PERT BJ reference cells using separate furnace diffusions and to *p*-PERC and bifacial *p*-PERC+ cells [9,10].

5 CELL RESULTS

Table 1 summarizes the cell results of the best *n*-type BiCoRE, monofacial CoRE and the PERT BJ reference cells measured on a brass chuck. 20.6% efficiency is demonstrated for the new BiCoRE cell concept [16] and 20.6% efficiency is reached for the monofacial CoRE cells, both with BSG rear side passivation. For the AlO_x-passivated BiCoRE cells, a slightly higher V_{oc} of 671 mV is measured due to the better passivation quality of AlO_x/SiN_y stacks on boron emitters (see Fig. 4) but the efficiency is comparable. A comparison of the BiCoRE and the monofacial CoRE cells shows that the bifacial cells reach slightly higher short circuit current densities J_{sc}=39.2 mA/cm² and about 0.4mV higher V_{oc} values. A bifaciality of up to 78% is measured for the BiCoRE cells both on a brass and a black (not shown in Table 1) measurement chuck. Compared to the bifacial *n*-PERT BJ reference cell reaching 21.1% efficiency and 78% bifaciality, the best 20.6% efficient *n*-BiCoRE cell suffers from the high series resistance R_s=1.2 Ω cm² that reduces the fill factor to 78.4%. We attribute the increased R_s to the P FSF. Obviously, the co-diffusion needs to be optimized aiming for lower and more homogeneous P sheet resistances that can effectively be contacted by Ag screen-printing.

Table I: Results of the best *n*-type BiCoRE and monofacial CoRE cells with BSG/SiN_z and AlO_x/SiN_y rear side passivation, respectively, compared to *n*-type PERT BJ reference cells with separate diffusions. All measurements are performed in-house on a brass chuck.

Cell type	Diffusion	Passivation rear side	Metallization rear side	η [%]	J_{sc} [mA/ cm²]	V_{oc} [mV]	*FF* [%]	R_s [Ω cm²]
BiCoRE	PECVD BSG co-diffusion	BSG/SiN_z	bifacial, Al grid	20.6	39.2	668	78.5	1.0
BiCoRE		AlO_x/SiN_y		20.6	39.2	671	78.4	1.2
CoRE		BSG/SiN_z	monofacial, full Al	20.6	38.8	664	80.1	0.60
CoRE		AlO_x/SiN_y		20.7	39.0	668	79.2	0.98
PERT BJ	$BBr_3/POCl_3$	AlO_x/SiN_y	bifacial, Al grid	21.1	39.5	674	79.2	0.96
PERT BJ			monofacial, full Al	21.0	39.1	670	80.2	0.71

6 SUMMARY AND CONCLUSION

We develop a novel n-type bifacial solar cell concept which we name Bifacial Co-diffused Rear Emitter cell (BiCoRE). Compared to n-PERT cells that apply separate BBr_3 and $POCl_3$ diffusions, the BiCoRE cell fabrication is significantly simplified by applying a co-diffusion process and a multifunctional PECVD stack. The PECVD BSG/SiN_z acts as a boron diffusion source resulting in boron emitters with homogeneous R_{sheet} values around 90 Ω/□ applicable in solar cell fabrication. The BSG capped with SiN_z is also used as rear side passivation saving BSG removal and additional AlO_x/SiN_y deposition steps. Lowest published saturation current densities of J_0=40 fA/cm² and highest $V_{oc,impl}$ values of 682 mV are demonstrated. The BSG/SiN_z stack as emitter passivation does not yet reach the high passivation quality of AlO_x/SiN_y stacks, but as the impact on cell performance is currently negligible and as it significantly simplifies the solar cell fabrication, it is an attractive option for the industrialization of the BiCoRE cells. BiCoRE cells processed on the very first batch reach conversion efficiencies up to η=20.6% and a bifaciality of 78%. Reference bifacial n-PERT BJ cells with separate diffusions and AlO_x/SiN_y rear side passivation are processed as well and demonstrate front side efficiencies up to 21.1% and rear side efficiencies up to 16.5% corresponding to a bifaciality of 78%. The BiCoRE cells suffer from increased series resistances up to 1.2 Ω cm² that are attributed to the P FSF. We assume that with an optimized $POCl_3$-co-diffusion, an efficiency level of 21% as for the reference cells can be achieved. The BiCoRE process flow and complexity is comparable to that of industrial PERC cells. Thus the BiCoRE cell is a potentially very attractive candidate for p-PERC solar cell manufacturers to migrate to LID-free n-type Cz-Si wafers and bifacial solar cells with only minimal changes to their existing production tools and process recipe set up.

7 ACKNOWLEDGEMENTS

We thank Miriam Berger for cell processing, Sarah Spätlich for diffusion optimization and David Sylla for laser processes. This work was funded by the German Federal Ministry for Economic Affairs and Energy under Grant 0325880A (PERC 2.0).

8 REFERENCES

[1] Y. Gassenbauer, K. Ramspeck, B. Bethmann, K. Dressler, J.D. Moschner, M. Fiedler, E. Brouwer, R. Drössler, N. Lenck, F. Heyer, M. Feldhaus, A. Seidl, M. Müller, A. Metz, IEEE J. Photovolt. 3, 125-130 (2013).
[2] J. Schmidt, A.G. Aberle, and R. Hezel, Proc. 26th IEEE Photovolt. Specialists Conf., pp. 13-18 (1997).
[3] T. Dullweber, C. Kranz, U. Baumann, R. Hesse, D. Walter, J. Schmidt, P. Altermatt, R. Brendel. Proc. 39th IEEE Photovolt. Specialist Conf., pp. 3074–3078 (2013).
[4] J. Schmidt, K. Bothe, R. Bock, C. Schmiga, R. Krain, R. Brendel, Proc. 22nd EUPVSEC, pp. 998-1001 (2007).
[5] A. Urueña, M. Aleman, E. Cornagliotti, A. Sharma, J. Deckers, M. Haslinger, L. Tous, R. Russell, J. John, Y. Yao, T. Söderström, F. Duerinckx, J. Szlufcik, Proc. 31th EUPVSEC, pp. 410-413 (2015).
[6] S. Bordihn et al., Status of Industrial Back Junction n-type Si Solar Cell Development, Proc. 6th International Conference on Silicon Photovoltaics, Chambéry, France (2016).
[7] T. Dullweber, S. Gatz, H. Hannebauer, T. Falcon, R. Hesse, J. Schmidt and R. Brendel, Prog. Photovolt: Res. Appl. 20, 630-638 (2012).
[8]http://www.prnewswire.com/news-releases/trina-solar-announces-new-efficiency-record-of-2213-for-mono-crystalline-silicon-solar-cell-300193727.html
Trina Solar, press release (16.12.2015).
[9] T. Dullweber, C. Kranz, R. Peibst, U. Baumann, H. Hannebauer, A. Fülle, S. Steckemetz, T. Weber, M. Kutzer, M. Müller, G. Fischer, P. Palinginis, and H. Neuhaus, Prog. Photovolt: Res. Appl., (2015), DOI: 10.1002/pip.2712.
[10] T. Dullweber, C. Kranz, R. Peibst, U. Baumann, H. Hannebauer, A. Fülle, S. Steckemetz, T. Weber, M. Kutzer, M. Müller, G. Fischer, P. Palinginis, H. Neuhaus, in: Proceedings 31st Europ. Solar Energy Conf., Hamburg, Germany, pp. 341-350 (2015).
[11] P. Rothhardt, S. Meier, K. Jiang, A. Wolf, D. Biro, Proc. 29th EUPVSEC, pp. 653-655 (2014).
[12] A. Frey, J. Engelhardt, S. Fritz, S. Gloger, G. Hahn, B. Terheiden, Proc. 29th EUPVSEC, pp. 656-660 (2014).
[13] R. Cabal, B. Grange, L. Bounaas, R. Monna, N. Plassat, E. Pihan, Y. Veschetti, Proc. 29th EUPVSEC, pp. 648-652 (2014).
[14] N. Wehmeier, B. Lim, A. Nowack, J. Schmidt, T. Dullweber, and R. Brendel, Phys. Status Solidi RRL, 1–5 (2015).
[15] J. Engelhardt, A. Frey, S. Gloger, G. Hahn, B. Terheiden, Applied Physics Letters 107, 042102 (2015).
[16] T. Dullweber, N. Wehmeier, A. Nowack, T. Brendemühl, S. Kajari-Schröder, R. Brendel, PSS A, submitted (2016).
[17] N. Wehmeier, B. Lim, A. Merkle, A. Tempez, S. Legendre, H. Wagner, A. Nowack, T. Dullweber, and Pietro P. Altermatt, IEEE JPV, Vol. 6, No. 1, pp. 119-125 (2016).
[18] D. E. Kane, R. M. Swanson, Proc. 18th IEEE Photovolt. Specialists Conf., pp. 578-583 (1985).
[19] N. Wehmeier, A. Nowack, B. Lim, T. Brendemühl, S. Kajari-Schröder, J. Schmidt, R. Brendel, and T. Dullweber, presented at the 6th SiliconPV Conference, Chambery, France, accepted for publication in SOLMAT (2016).
[20] J. Schmidt, B. Veith, and R. Brendel, Phys. Status Solidi RRL 3, 287-289 (2009).

[i] corresponding author, e-mail: wehmeier@isfh.de, phone: +49-5151-999-325

A QUANTITATIVE MEASURE FOR THE CARRIER SELECTIVITY OF CONTACTS TO SOLAR CELLS

R. Brendel[1,2], M. Rienaecker[1], and R. Peibst[1,3]

[1]Institute for Solar Energy Research Hameln (ISFH), Am Ohrberg 1, 31860 Emmerthal, Germany
[2]Institute for Solid State Physics, Leibniz Universität Hannover, Appelstrasse 2, 30167 Hannover, Germany
[3]Institute for Electronic Materials and Devices, Leibniz Universität Hannover, Schneiderberg 32, 30167 Hannover, Germany

Abstract: We discuss a physically motivated definition for a quantitative measure of the selectivity of electron and hole contacts. We define the selectivity $S_{10} = \log_{10}(V_{th} /(\rho_c \times J_c))$ to depend on the contact resistance ρ_c, the recombination current density J_c of the contact, and the thermal voltage V_{th}. A high selectivity relies on a highly asymmetric equilibrium carrier concentration of majority and minority carriers in the contact. The maximum efficiency η_{max} increases with the selectivity S_{10}. This increase is linear until the efficiency starts to be limited by radiative recombination. We give analytic equations for calculating the maximum efficiency $\eta_{max}(S_{10})$ of a crystalline Si cell that is ideal except for either one or two contacts. Achieving the maximum efficiency η_{max} requires optimized areal fractions $f_{e,max}$ and $f_{h,max}$ for the electron and the hole contacts, respectively . We give analytic equations for these contact fractions.

Keywords: Carrier selective contacts, selectivity in photovoltaics, loss analysis, silicon solar cell.

1 INTRODUCTION

Carrier selective contacts have to have a low recombination current density J_c in combination with a low contact resistance ρ_c for allowing high power conversion efficiencies of the solar cell. Such contacts are of major importance for highly efficient crystalline Si solar cells and have made good progress in quality and understanding [1,2,3,4]. A highly selective junction should thus carry an as low as possible minority carrier current and an as high as possible majority carrier current. The former is a recombination current; the latter is the sum of the current into the metal contact plus the recombination current. "High selectivity" of a contact has often been used as a qualitative term meaning "a contact that allows high efficiencies".

This contribution reports on a recently introduced quantitative definition of a contacts selectivity [5,6] and gives analytic equations that allow to calculate the ultimate efficiency limit of a crystalline Si solar cell with electron and/or hole selective contacts.

2 DEFINTION OF SELECTIVITY OF A CONTACT

The motivation for our definition of the selectivity S is to compare the resistances ρ_m and ρ_M that the contact imposes to minority (m) and majority (M) carriers respectively.

The selectivity

$$S = \frac{\rho_m}{\rho_M} \tag{1}$$

shall thus be the ratio of these two resistances (both measured in Ωcm^2). For calculating these resistances we have to consider the current-voltage curve for majority carriers and minority carriers flowing into the contact region.

The minority carrier recombination current density into the selective contact is

$$J_m = J_c\left(e^{\frac{V}{V_{th}}} - 1\right) \tag{2}$$

and depends on the contacts recombination current factor J_c (unit is fA/cm^2). The symbol V_{th} denotes the thermal voltage at 25 °C. The driving force for this current is the splitting of the quasi-Fermi levels V at the contact. The minority carrier resistance V/J_m is voltage-dependent and is thus not suitable for characterizing the quality of the contact. A similar situation holds when quantifying the recombination activity of a contact. That is best done by the voltage-independent factor J_c rather than by the total recombination current density J_m. We therefore use the inverse slope

$$\rho_m = \left(\left.\frac{dJ_m}{dV}\right|_{V=0}\right)^{-1} = \frac{V_{th}}{J_c} \tag{3}$$

at zero voltage for defining the resistance ρ_m. For the majority carriers we assume a good contact that has a linear current-voltage characteristic.

$$J_M = \frac{V}{\rho_c}. \tag{4}$$

The driving force is the difference of the quasi-Fermi energies V of the majority carriers on both sides of the contact. This linear relation yields the resistance

$$\rho_M = \left(\left.\frac{dJ_M}{dV}\right|_{V=0}\right)^{-1} = \rho_c \tag{5}$$

for the majority carriers, where ρ_c (unit is Ωcm^2) denotes the contact resistance. Using the motivating equation (1) we find the relation

$$S = \frac{\rho_m}{\rho_M} = \frac{V_{th}}{J_c \rho_c} \tag{6}$$

that we use for defining the selectivity. In addition we define a logarithmically scaled selectivity

$$S_{10} = \log_{10}\left(\frac{V_{th}}{J_c \rho_c}\right) \tag{7}$$

to have smaller numbers.

The contact with parameters (ρ_c, J_c) does not have to cover the full area of the solar cell. The area fraction f_c can be smaller than unity for so-called local contacts or it can be larger than unity for contacts on structured surfaces. Local contacts enhance the contact resistance and reduce the recombination factor of the cell. The effective contact parameters are thus ($\rho_c / f_c, f_c \times J_c$). According to (6) and (7) the selectivities S and S_{10} do not change when replacing contact parameters with effective device parameters. We therefore do not need to distinguish an effective selectivity in the device from the contact selectivity.

3 ASYMMETRIC CARRIER CONCENTRATIONS CAUSE HIGH SELECTIVITY

Figure 1 sketches a spatially homogeneous p$^+$-type back surface (BSF) layer of thickness W in between the metal and the p-type bulk Si. This is a simple example for a carrier selective contact that we discuss here. The recombination of this contact is quantified by an electron diffusion length L_n and the recombination velocity S_n at the metal interface. The well-known recombination current factor of this contact is

$$J_c = q\frac{n_0 D_n}{L_n}\frac{\sinh\left(\frac{W}{L_n}\right)+\frac{S_n L_n}{D_n}\cosh\left(\frac{W}{L}\right)}{\cosh\left(\frac{W}{L_n}\right)+\frac{S_n L_n}{D_n}\sinh\left(\frac{W}{L_n}\right)} \quad (8)$$

where n_0 denotes the equilibrium minority carrier density. Since the contact resistance

$$\rho_c = \frac{1}{q\, p_0\, \mu_p} W \quad (9)$$

is inversely proportional to the equilibrium majority carrier concentration p_0 the selectivity S as given in (6) is proportional to p_0/n_0. Here μ_p denotes the hole carrier concentration. Depending on the doping (or the band bending in other structures) the ratio p_0/n_0 easily reaches 18 orders of magnitude. Thus a large carrier concentration asymmetry is a key for a high selectivity.

To become more specific we first assume negligible recombination in the volume of the BSF and expand (8) in $W/L_n \ll 1$. We then distinguish the transport-limited case with $D_n \ll S_n W$ and the recombination-limited case with $D_n \gg S_n W$. The result for the recombination current factor of these two cases is

$$J_c = \begin{cases} \frac{q D_n n_0}{W} & \text{for } L_n \gg W \text{ and } D_n \ll S_n W \\ q n_0 S_n & \text{for } L_n \gg W \text{ and } D_n \gg S_n W \end{cases}. \quad (10)$$

Using (6) we calculate the non-logarithmic selectivity

$$S = \frac{p_0 \mu_p}{n_0 \mu_n} \propto \frac{p_0}{n_0} \quad (11)$$

for the transport limit and

$$S = \frac{p_0\ \mu_p}{n_0\left(\frac{S_n W}{V_{th}}\right)} \propto \frac{p_0}{n_0}. \quad (12)$$

for the recombination limit of J_c.

In the transport limit (11) the selectivity is simply given by the ratio of equilibrium conductivities. Please note that the mobilities of electrons and holes have the same order of magnitude in Si. In the recombination limit (12) the selectivity is also proportional to p_0/n_0. Here, a small n_0 value causes a large selectivity since only few electrons are available for recombination.

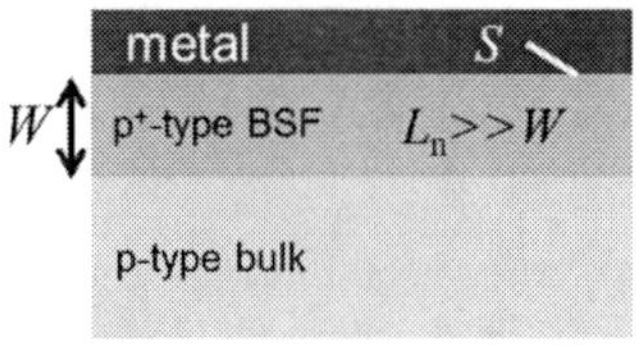

Figure 1: Schematics of a full area carrier selective contact. Here we consider a homogeneously doped p-type back surface field layer in between the metal and the p-type Si wafer.

4 SELECTIVITY AND EFFICIENCY

The areal fraction f_c of the carrier selective contacts is an important degree of freedom for optimizing the solar cell efficiency [7]. The replacement of the full area back surface field layer with local contacts of PERC cells is a current practical example.

4.1 Numerical efficiency limit for a single selective contact

The efficiency of an ideal crystalline Si solar cell that is limited by intrinsic recombination, has Lambertian light trapping, no surface or contact recombination, no resistive losses, and has the ideal thickness of W = 110 µm was calculated in [8] to be 29.4% for an intrinsic Si wafer.

We now raise the question what the efficiency of such an ideal cell is if it has a single selective contact with effective parameters ($\rho_c / f_c, f_c \times J_c$). All other assumptions are identical to those in [8] except that we assume an n-type Si wafer with a resistivity of 2 Ωcm here.

To start we consider a cell with a single contact that introduces losses while the other is assumed to be ideal and thus loss free. We calculate the one sun, AM1.5G efficiency by using the current-voltage curve

$$J = J_{sc} - q r_i W - f_c J_c\left(\exp\left(\frac{V + J\rho_i + J\rho_c / f_c}{V_{th}}\right) - 1\right). \quad (13)$$

The symbol J_{sc} denotes the photogeneration current density and is 43.6 mA/cm^2 at zero voltage. We account for the weak dependence of J_{sc} on the voltage V due to free carrier absorption. The symbols r_i and ρ_i denote the intrinsic recombination rate and the (negligible) intrinsic total resistance of the wafer at a spatially constant quasi-Fermi level splitting V. The material models for band gap narrowing, mobility, free carrier absorption, Auger recombination, and radiative recombination are identical to those chosen in [8].

From (13) we learn that the device efficiency depends on ρ_c / f_c and $f_c \times J_c$ or alternatively on S_{10} and ρ_c / f_c. For every value of S_{10} there is thus an effective contact resitaince

$$\rho_{max}(S_{10}) = \rho_c / f_c \quad (14)$$

that gives the maximum efficiency $\eta_{max}(S_{10})$. The optimum contact fraction therefore is

$$f_{max} = \rho_c / \rho_{max}(S_{10}) . \quad (15)$$

The red lines in Figure 2 show iso-lines of the efficiency η as calculated for a full area contact (f_c=1). The black lines are iso-lines of the optimum contact fraction f_{max} that a contact (ρ_c, J_c) has to have to maximize the cells efficiency to $\eta_{max} \geq \eta$. The blue lines give the selectivity S_{10} of a contact (ρ_c, J_c). The iso-η lines show the expected pattern of recombination limited efficiency for small contact resistances and resistive limited efficiency for large contact resistances. Both these regions are separated by the f_{max} = 1 line.

Figure 2 allows to find the optimum efficiency η_{max} that is feasible with contact parameters (ρ_c, J_c): (i) Find the point (ρ_c, J_c) of the contact in the plot and follow it's blue iso-S_{10} line to the intersection point with the black f_{max} = 1 line. (ii) Read off the efficiency η at this intersection point. This is the maximum efficiency η_{max} that the contact (ρ_c, J_c) and any other contact with the same selectivity S_{10} can support. (iii) This maximum efficiency η_{max} is possible at an effective contact resistance ρ_c/f_c that equals the ρ_c value of the intersection point. The ideal contact fraction f_{max} follows from (15).

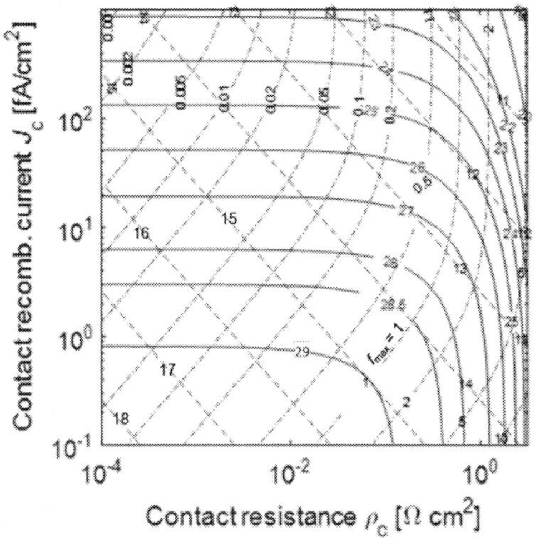

Figure 2: A contact resistance ρ_c and a recombination current factor J_c of a full area contact (f_c =1) of a cell with the current-voltage curve (13) limit the cell efficiency to η (red solid lines, numbers in %). The optimum contact fraction f_{max} (black dash dotted lines) allows in general for a higher efficiencies $\eta_{max} \geq \eta$ than at f_c = 1. The selectivity S_{10} (blue broken lines) is calculated using(7).

4.2. Analytical efficiency limit for a single selective contact

Analytic approximations of two functions allow determining η_{max} of a contact (ρ_c, J_c) without using Figure 2. On the f_{max} = 1 line in Figure 2 the maximum efficiency η_{max} is a function

$$\frac{\eta_{max}(S_{10})}{\%} = \left((2.452 \; S_{10} - 4.240)^{-19.52} + (29.21)^{-19.52} \right)^{-\frac{1}{19.52}} \quad (16)$$

of the selectivity S_{10}. Figure 3 shows a plot of this function. The efficiency η_{max} increases linearly with the selectivity S_{10} with a slope of 2.45% when being limited by the contact in the region S_{10} < 13. This linear dependence justifies introducing the logarithmically scaled selectivity (7). In the region S_{10} > 13 the device runs into radiative limitation and η_{max} saturates. The other function required is the resistance

$$\frac{\rho_{max}(S_{10})}{\Omega \text{cm}^2} = \left(0.6271^{-1.805} + \left(1.620 \times 10^6 \; e^{-\frac{S_{10}}{0.8819}} \right)^{-1.805} \right)^{-\frac{1}{1.805}} \quad (17)$$

along the black f_{max} = 1 line in Figure 2 and is plotted in Figure 4. Both approximations (16) and (17) fit the numeric results to better than 3% in the range 8 ≤ S_{10} ≤ 18. The optimum contact fraction f_{max} leading to maximum efficiency η_{max} for a contact of given selectivity S_{10} can be calculated applying (15).

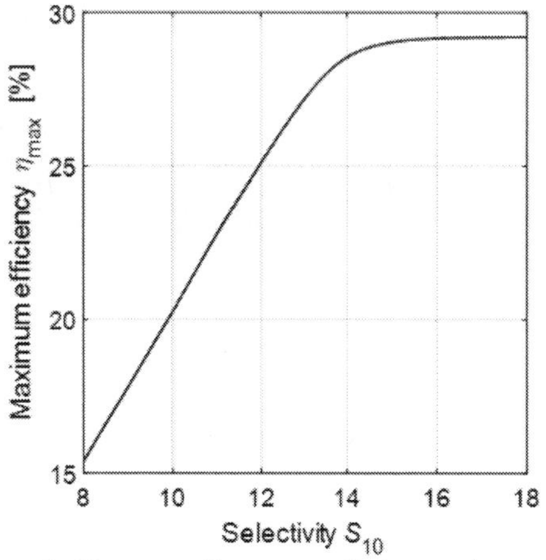

Figure 3: Maximum efficiency $\eta_{max}(S_{10})$ according to (16) assuming an ideal radiatively limited, n-type, 2 Ωcm wafer of thickness W = 110 µm with Lambertian light trapping (J_{sc} = 43.6 mA/cm^2) with a single contact of selectivity S_{10}. The other contact is ideal.

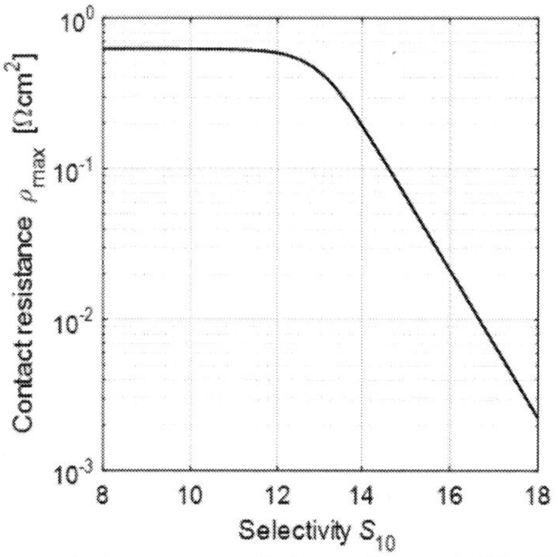

Figure 4: Resistance $\rho_{max}(S_{10})$ according to (17) leading to maximum efficiency $\eta_{max}(S_{10})$ when assuming an ideal radiatively limited, n-type, 2 Ωcm wafer of thickness W = 110 µm and Lambertian light trapping (J_{sc} = 43.6 mA/cm^2) with a single contact of selectivity S_{10}. The other contact is ideal.

4.3 Numeric calculations for two selective contacts

Solar cells have an electron and a hole selective contact with the selectivities

$$S_{10,e} = \log_{10}\left(\frac{V_{th}}{J_{c,e}\,\rho_{c,e}}\right) \tag{18}$$

and

$$S_{10,h} = \log_{10}\left(\frac{V_{th}}{J_{c,h}\,\rho_{c,h}}\right), \tag{19}$$

respectively. The combined selectivity controlling the device performance is

$$S_{10,e\&h} = \log_{10}\left(\frac{V_{th}}{\left(f_{c,e}J_{c,e} + f_{c,h}J_{c,h}\right)\times\left(\rho_{c,e}/f_{c,e} + \rho_{c,h}/f_{c,h}\right)}\right) \tag{20}$$

since resistances and recombination currents add when assuming flat quasi-Fermi levels. We learned in the previous section that the maximum efficiency η_{max} corresponds to the maximum selectivity $S_{10,e\&h}$. We therefore search for the areal fraction $f_{e,max}$ of the electron and $f_{h,max}$ of the hole contacts that maximize $S_{10,e\&h}$ and at the same time give the optimum resistance

$$\rho_{max}\left(S_{10}\right) = \rho_{c,e}/f_{c,e} + \rho_{c,h}/f_{c,h}\,. \tag{21}$$

As a result of this optimization problem we find that the maximum achievable non-logarithmic selectivity is

$$S_{e\&h,max} = \left(S_e^{-0.5} + S_h^{-0.5}\right)^{-2} \tag{22}$$

and equivalently the maximized logarithmic selectivity is

$$S_{10,e\&h,max} = \log_{10}\left(\frac{V_{th}}{J_{c,e}\rho_{c,e} + J_{c,h}\rho_{c,h} + 2\sqrt{J_{c,e}\rho_{c,e}J_{c,h}\rho_{c,h}}}\right). \tag{23}$$

This maximum combined selectivity $S_{10,e\&h,max}$ is achieved at contact fractions

$$f_{e,max} = \frac{\rho_{c,e} + \sqrt{\dfrac{\rho_{c,h}J_{c,h}\rho_{c,e}}{J_{c,e}}}}{\rho_c\left(S_{10,e\&h,max}\right)} \tag{24}$$

and

$$f_{h,max} = \frac{\rho_{c,h} + \sqrt{\dfrac{\rho_{c,e}J_{c,e}\rho_{c,h}}{J_{c,h}}}}{\rho_c\left(S_{10,e\&h,max}\right)} \tag{25}$$

and can be calculated using (17). The maximum efficiency for two contacts thus is $\eta_{max}(S_{e\&h,max})$ applying (23) and (16). For application examples of these equations we refer to [5].

4.4 Limitations of the model

Please note that the above model assumes an otherwise ideal cell and does not account for the spreading resistance in the wafer that arises from partial contacts. This is a good approximation for small scale contacts that require negligible lateral transport. This small scale assumption was chosen here to arrive at that upper efficiency limit.

Another simplification of our model is the assumption of perfect passivation of the non-contact regions. This assumption was also made to derive the upper efficiency limit η_{max}. However, a substantial reduction of the total recombination by reducing the areal fraction of the contact requires dielectric passivation layers with better passivation than that of the selective contact. Such layers are currently not available for contacts with J_c values in the order of magnitude of 1 fA/cm^2. Numerous numerical tools are available to handl e such additional loss effects and some of them could also easily be integrated in the model presented here.

5 LIMITING EFFICIENCIES FOR CARRIER SELECTIVE CONTACTS REPORTED IN THE LITERATURE

Many different material systems may be used to form carrier selective junctions on silicon. Figure 5 shows some recent results collected from the literature [9- 23]. Red symbols correspond to electron contacts and green symbols to hole contacts. The largest selectivities and thus the largest limiting efficiencies η_{max} of the data collected here are achieved for poly Si on oxide (POLO) contacts.

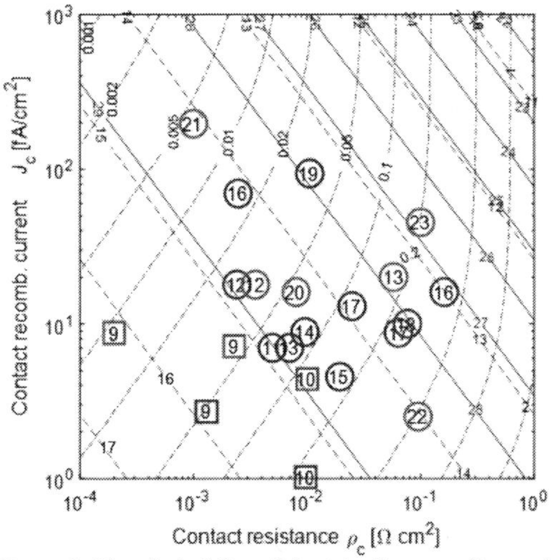

Figure 5: Blue dashed line: Selectivity S_{10} according to (7). Red solid line: Maximum efficiency η_{max} according to (16) when assuming an ideal radiatively limited wafer of thickness W= 110 μm with Lambertian light trapping (J_{sc} = 43.6 mA/cm^2). Black dash-dotted line: Contact fraction f_{max} that allows maximum efficiency η_{max}. Symbols: Data reported in the literature. Red symbols are electron contacts and green symbols are hole contacts. Square symbols are data for polycrystalline Si on silicon oxide (POLO) by ISFH. The numbers in the symbols refer to the reference number at the end of the paper. We translated some of the recombination data form recombination velocities S to recombination factors J_c.

Please note that the model discussed here only addresses the ultimate efficiency limit for any given contact pair ($\rho_{c,e}$, $J_{c,e}$) and ($\rho_{c,h}$, $J_{c,h}$). For this ultimate efficiency it is only the combined selectivity $S_{10,e\&h}$ that is important.

Experimental Si cells are mostly limited by recombination rather than by resistance. Among two contact pairs with the same combined selectivity the one with the smaller $J_{c,e\&h}$ is therefore likely to be of higher practical relevance. For double-side contacted cells, the transport losses due to the lateral current towards metal fingers also add to the "intrinsic" contact resistance ρ_c. Also, the assumed J_{sc} value of 43.6 mA/cm^2 is hard to reach for non-perfectly transparent electron or hole collecting layers if used on the front side of the cell with non-negligible metallized front-side fraction. Further important issues such as the process complexity, long term stability, material cost and environmental issues are of course also not covered by the new figure of merit S_{10} of a carrier selective contact.

6 SUMMARY

The well-known theory for calculating the upper efficiency limit of crystalline Si solar cells [8] assumes perfect contacts that have zero contact resistance ρ_c and zero recombination current density J_c. This paper extends this theory to contacts with parameters (ρ_c, J_c) that have resistive and recombination losses. We found that the maximum efficiency $\eta_{max}(S_{10})$ only depends on the selectivity S_{10} of the contact and is achieved at an optimized contact fraction of $\rho_c/\rho_{max}(S_{10})$. Analytic approximations for the functions $\eta_{max}(S_{10})$ and $\rho_{max}(S_{10})$ have been given for the case of a radiatively-limited and 110 μm-thick Si wafer with a resistivity of 2 Ωcm and Lambertian light trapping. We finally showed that the selectivity is to a large extend controlled by an asymmetric equilibrium carrier concentration in the contact that can be created by doping and/or band bending and/or the choice of the materials.

6 ACKNOWLEDGEMENTS

This work was supported by the State of Lower Saxony via the project vOx under contract No 74ZN1354.

REFERENCES

[1] A. Cuevas, T. Allen, J. Bullock, Y. Wan, D. Yan and X. Zhang, in Proc. 42nd IEEE Photovoltaic Specialists Conf. (IEEE, New York, 2015).

[2] H. Steinkemper, F. Feldmann, M. Bivour and M. Hermle, Photovoltaics, IEEE Journal of **5** (5), 1348-1356 (2015).

[3] R. Peibst, U. Römer, et al., IEEE Journal of Photovoltaics **4** (3), 841-850 (2014).

[4] U. Wuerfel, A. Cuevas and P. Wuerfel, *Charge carrier separation in solar cells*, Photovoltaics, IEEE Journal of 5, 461-469 (2015).

[5] R. Brendel and R. Peibst, Selectivity and efficiency in crystalline silicon photovoltaics, IEEE JPV, submitted March 5, 2016.

[6] M. Rienaecker, M. Bossmeyer, A. Merkle, U. Roemer, F. Haase, J. Kruegener, R. Brendel and R. Peibst, *Junction resistivity of carrier selective polysilicon on oxide junctions and its impact on the solar cell performance*, in Proc. 43rd IEEE PVSC 2016, in press. Invited Conference paper in IEEE Journal for Photovoltiacs (2016), submitted June 2016.

[7] D. L. Young, W. Nemeth, S. Grover, A. Norman, B. G. Lee and P. Stradins, in *Proc. 40th IEEE Photovoltaic Specialist Conf.* (2014), pp. 1-5.

[8] A. Richter, M. Hermle and S. W. Glunz, *Reassessment of the limiting efficiency for crystalline silicon solar cells*, IEEE Journal of Photovoltaics **3**, 1184-1191 (2013).

[9] M. Rienaecker, M. Bossmeyer, A. Merkle, U. Roemer, J. Krügener, R. Brendel, R. Peibst, *Junction resistivity of carrier selective polysilicon on oxide junctions and its impact on the solar cell performance*, IEEE-JPV, 2016 in press.

[10] U. Roemer, R. Peibst, T. Ohrdes, B. Lim, J. Kruegener, T. Wietler and R. Brendel, *Ion implantation for poly-Si passivated back-junction back-contacted solar cells*, Photovoltaics, IEEE Journal of 5, 507-514 (2015).

[11] D. Yan, A. Cuevas, Y. Wan and J. Bullock, *Silicon nitride/silicon oxide interlayers for solar cell passivating contacts based on PECVD amorphous silicon*, phys. status solidi (RRL) **9**, 617–621 (2015).

[12] J. Y. Gan and R. M. Swanson, *Polysilicon emitters for silicon concentrator solar cells*, in Conference Record of the 21st IEEE Photovoltaic Specialists Conference (1990), pp. 245.

[13] F. Feldmann, priv. communication 2016 to M. Rienäcker

[14] F. Feldmann, M. Bivour, C. Reichel, M. Hermle and S. W. Glunz, *Passivated rear contacts for high-efficiency n-type si solar cells providing high interface passivation quality and excellent transport characteristics*, Solar Energy Materials and Solar Cells **120**, 270-274 (2014).

[15] D. Yan, A. Cuevas, J. Bullock, Y. Wan and C. Samundsett, *Phosphorus-diffused polysilicon contacts for solar cells*, Solar Energy Materials and Solar Cells **142**, 75-82 (2015).

[16] J. Bullock, M. Hettick, J. Geissbühler, A. J. Ong, T. Allen, C. M. Sutter-Fella, T. Chen, H. Ota, E. W. Schaler, S. DeWolf, C. Ballif, A. Cuevas and A. Javey, *Efficient silicon solar cells with dopant-free asymmetric heterocontacts*, Nature Energy **2**, 15031 (2016).

[17] X. Yang, Q. Bi, H. Ali, K. Davis, W. V. Schoenfeld, K. Weber, *High-Performance TiO_2-Based Electron-Selective Contacts for Crystalline Silicon Solar Cells*, Adv. Mater (2016), doi:10.1002/adma.201600926.

[18] Y. Wan, C. Samundsett, J. Bullock, T. Allen, M. Hettick, D. Yan, P. Zheng, X. Zhang, J. Cui, J. McKeon, A. Javey, and A. Cuevas, *Magnesium Fluoride Electron-Selective Contacts for Crystalline Silicon Solar Cells*, ACS Applied Materials & Interfaces (2016), in press.

[19] D. L. Young, W. Nemeth, S. Grover, A. Norman, B. G. Lee and P. Stradins, *Carrier-selective, passivated contacts for high efficiency silicon solar cells based on transparent conducting oxides*, in Proc. 40th IEEE photovoltaic specialist conf. (2014), pp. 1-5.

[20] D. Yan, A. Cuevas, J. Bullock, Y. Wan and C. Samundsett, *Phosphorus-diffused polysilicon contacts for solar cells*, Solar Energy Materials and Solar Cells **142**, 75-82 (2015).

[21] J. Bullock, A. Cuevas, T. Allen and C. Battaglia, *Molybdenum oxide MoO_x: A versatile hole contact for silicon solar cells*, Applied Physics Letters **105**, 232109 (2014).

[22] A. Cuevas, T. Allen, J. Bullock, W. Yimao, Di and Z. Xinyu, *Skin care for healthy silicon solar cells, in Photovoltaic specialist conference*, 42nd IEEE PVSC 2015, pp. 1-6.

[23] R. Gogolin, D. Zielke, W. Lövenich, R. Sauer, and J. Schmidt, *Silicon Heterojunction Solar Cells Combining an a-Si:H Electron-collector with a PEDOT:PSS Hole-collector*, Energy Procedia, (2016), in press.

OPTIMIZATION OF P^+ POLY-SI / C-SI JUNCTIONS ON WET-CHEMICALLY GROWN INTERFACIAL OXIDES AND ON DIFFERENT WAFER MORPHOLOGIES

Yevgeniya Larionova[1], Robby Peibst[1,2], Mircea Turcu[1], Sina Reiter[1], Rolf Brendel[1,4], Dominic Tetzlaff[2], Jan Krügener[2], Tobias Wietler[2], Uwe Höhne[3] and Jan-Dirk Kähler[3]
[1]Institute for Solar Energy Research Hamelin, Am Ohrberg 1, D-31860 Emmerthal, Germany,
tel.: +495151999423, email: y.larionova@isfh.de
[2]Institute of Electronic Materials and Devices, Leibniz Universität Hannover, Schneiderberg 32, D-30167 Hannover, Germany
[3]centrotherm pv AG, Vahrenwalder Str. 269A, D-30179 Hannover, Germany
[4]Institute for Solid State Physics, Leibniz Universität Hannover, Schneiderberg 32, D-30167 Hannover, Germany

ABSTRACT: In this paper, we study the passivation quality of *p*-type poly-Si / *c*-Si junctions annealed at different temperatures. We evaluate wet-chemically grown (ozonized in DI-H_2O) and thermally-grown interfacial oxides and compare boron ion implantation into intrinsic poly-Si with different doses to *in situ* p^+ doping during the LPCVD Si deposition process. The experimental results show that decreasing the doping concentration of the p^+ poly-Si layer and increasing the annealing temperature improve the passivation quality of the p^+ poly-Si / *c*-Si junctions. Additionally, we explore the passivation quality of *p*-type poly-Si / *c*-Si junctions on different surface morphologies and orientations, i.e. on planar (100) and (111) as well as alkaline textured Si surfaces. Textured surfaces show a strongly increased J_{0e}, possibly due to a worse *p*-type poly-Si passivation on (111) surfaces. The emitter saturation current density J_{0e} of 8 fA/cm² achieved on planar (100) surfaces is the lowest value reported for p^+ poly-Si / *c*-Si junctions with wet-chemically grown oxides so far.
Keywords: Polycrystalline Silicon, Contact, Passivation, Silicon Solar Cell

1 INTRODUCTION

Recently, numerous *n*-type Si solar cells with an electron-selective junction formed by a thin passivating dielectric (mostly SiO_2) and a highly n^+ doped polycrystalline (at least partially) silicon layer have been presented [1-5]. As already shown by Gan and Swanson in 1990 [6], these junctions offer excellent surface passivation, enabling high open circuit voltages V_{oc}, combined with low junction resistance allowing high fill factors (*FF*).

Besides the already well-established n^+ poly-Si / *c*-Si junction, it would be desirable also to utilize p^+ poly-Si / *c*-Si junctions. These junctions could either be applied in in novel cells with carrier-selective junctions for both polarities, such as the POLO (**pol**ysilicon on **o**xide) cell structure [7], or to replace the local rear contacts of today's industrial *p*-type PERC cells with a full-area hole collecting layer [8]. However, the formation of excellent p^+ poly-Si / *c*-Si junctions is considered challenging by many groups [1, 9]. In-diffusion of boron into the interfacial oxide and into the substrate, as well as blistering in the case of Si layers deposited by Plasma Enhanced Chemical Vapor Deposition (PECVD) at low temperatures [1] are considered as possible reasons for a moderate passivation quality. As a consequence, most of the cell structures reported so far only feature n^+ poly-Si / *c*-Si junctions on a planar rear side while utilizing conventional boron front-side emitters. However, also for p^+ poly-Si / *c*-Si junctions, low $J_{0,poly}$ values down to 4.4 fA/cm^2 have already been demonstrated on planar (100) surfaces [9]. In this case intrinsic poly-Si layers were deposited by Low-Pressure Chemical Vapor Deposition (LPCVD) on thermally grown interfacial oxides and subsequently doped by boron ion implantation. The question arises whether this excellent junction quality can also be obtained by simpler processes, e.g. with wet-chemically grown interfacial oxides and *in situ* p^+ doping instead of ion implantation. Furthermore, a transfer to textured surfaces would also desirable for cell concepts such as the bifacial POLO cell [7].

In this paper, we study the passivation quality of *p*-type POLO (**pol**ysilicon on **o**xide) junctions after different annealing processes. We evaluate wet-chemically grown (ozonized in DI-H_2O) and thermally-grown interfacial oxides and compare ion implantation with different doses to *in situ* doping during the LPCVD Si deposition process. Additionally, we explore the passivation quality of *p*-type POLO junctions on different surface morphologies and orientations, i.e. on planar (100) and (111) as well as on alkaline textured Si surfaces.

2 EXPERIMENTS

Symmetrical lifetime samples are prepared on planar and alkaline textured (156 × 156) mm^2 pseudosquare *n*-type 4 Ωcm Cz silicon wafers. The samples have a thickness of 160 µm after saw damage removal. The damage-etched and RCA cleaned wafers receive either a (1.7 ± 0.2) nm thick thermal oxide in an oxidation furnace, or a (1.7 ± 0.2) nm thick wet-chemically grown oxide. The latter is prepared by using ozonized DI-H_2O with a constant ozone concentration of 290 g/m^3 pumped into a 50 l DI-H_2O-reservoir during 15 min at room temperature. In all cases, the interfacial oxide thickness was measured by ellipsometry on polished references. After oxide growth, 150 nm-thick silicon layers are deposited on both sides of the samples by LPCVD (E2000 from centrotherm) at ~550 °C in two different ways: i) *in situ* boron doped, and ii) intrinsic with a subsequent doping by B ion implantation. For planar samples, the blanket B implantation is performed with doses of 5 × 10^{14} cm^{-2}, 1 × 10^{15} cm^{-2}, or 2 × 10^{15} cm^{-2} and an energy of 10 keV. In order to achieve the same doping concentrations within the poly-Si for the textured samples, the implant doses were scaled with respect to the surface area (8.5 × 10^{14} cm^{-2}, 1.7 × 10^{15} cm^{-2} and

3.4×10^{15} cm^{-2}).

Since the electrical properties of the POLO junctions strongly depend on the post-deposition thermal treatment, the annealing is conducted in an oxidation furnace for 30 min at temperatures between 800 °C and 1000 °C. After a forming gas anneal (FGA), we measure the open circuit voltage $V_{oc,impl}$ and the emitter saturation current density J_{0e} of the POLO junctions with a WCT-120 lifetime tester (Sinton Instruments), using the method of Kane and Swanson [10]. We also perform electro-chemical voltage profiling (ECV) measurements in order to study the amount of boron diffused from the poly-Si into the *c*-Si wafer. Furthermore, a low specific junction resistance for the p^+ POLO junctions is verified on *p*-type test wafers using the method of Römer *et al.* [9].

3 RESULTS AND DISCUSSION

Figure 1 shows the influence of the boron implant dose and of the annealing temperature on the implied circuit voltage $V_{oc,impl}$ at one sun illumination measured on planar (100) samples. For the samples with thermal oxide (Fig. 1(a)), the $V_{oc,impl}$ values increase with increasing annealing temperature and reach their maximum after annealing at 1000 °C for the two lowest doping concentrations. Only for the highest dose (2×10^{15} cm^{-2}), the maximum in $V_{oc,impl}$ can be observed at 900 °C. The optimum annealing temperature shifts towards lower temperatures by increasing dopant concentration. At the same time the maximum $V_{oc,impl}$ value slightly decreases with increasing dopant concentration (for the optimum annealing temperature). This trend is even more pronounced if the *in situ* boron-doped samples are taken into account. This might be consistent with the findings for the ion-implanted samples since the boron concentration within the poly-Si is highest for the *in situ* boron–doped layers (see ECV measurements in Fig. 2). The ECV measurements performed on p^+ POLO junctions annealed at 1000°C also show an increased in-diffusion of boron from the poly-Si into the *c*-Si with increasing B concentrations in the poly-Si (Fig. 2). It is unclear so far whether this is due to the higher initial doping gradient between the higher p^+ doped poly-Si layers and the *c*-Si substrate (which might imply boron diffusion through a still continuous interfacial oxide), or whether this indicates an enhanced local breakup of the interfacial oxide for higher B concentrations in the poly-Si [11,12].

A maximum $V_{oc,impl}$ value of 721 mV and a minimum saturation current density J_{0e} of 6 fA/cm² are obtained for the planar (100) samples with thermally grown interfacial oxide, implanted with the lowest investigated boron dose of 5×10^{14} cm^{-2} and annealed at 1000 °C (Fig. 1(a)). Although for thicker interfacial oxides (~2.4 nm) an optimal ion dose of 1×10^{15} cm^{-2} has been reported [9], our present results on samples with the thinner interfacial oxide (~1.7 nm) imply a further increase in $V_{oc,impl}$ for lower doses.

Figure 1(b) shows the corresponding behaviour of $V_{oc,impl}$ measured on the planar (100) samples with wet-chemically grown oxide. Analogously to the observations with thermally-grown interfacial oxides, the maximum in $V_{oc,impl}$ seems to shift towards lower annealing temperatures and to slightly decrease with increasing boron doping concentration in the poly-Si layer. An excellent maximum $V_{oc,impl}$ value of 718 mV and a minimum J_{0e} value of 8 fA/cm² are achieved for the lowest investigated boron dose of 5×10^{14} cm^{-2} after annealing at 900 °C.

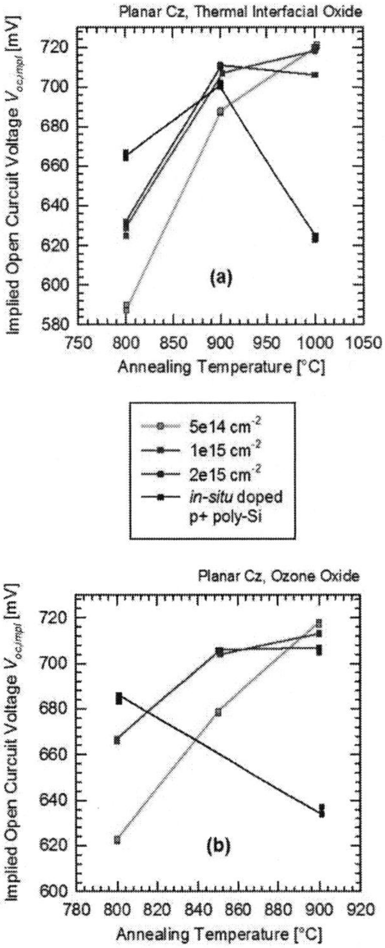

Figure 1: Temperature dependence of the implied open circuit voltage $V_{oc,impl}$ measured on planar (100) samples with *p*-type POLO junctions for thermally (a) and wet-chemically (b) grown interfacial oxides for different boron implant doses.

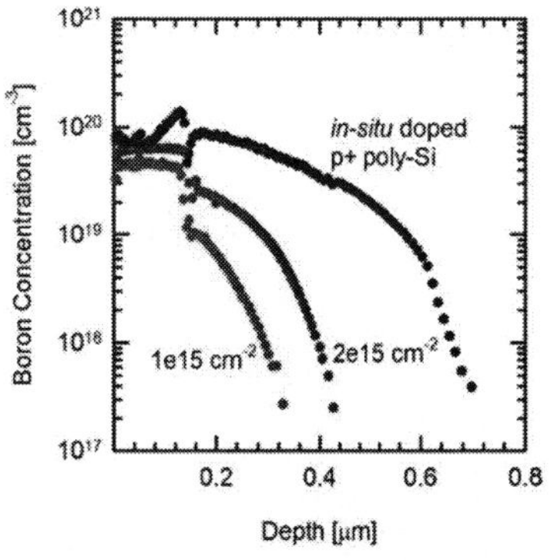

Figure 2: ECV profiles of planar (100) samples with *p*-type POLO junctions for thermally grown interfacial oxide for different boron implant doses after annealing at 1000 °C.

The passivation quality of p^+ POLO junctions on textured surfaces with thermally grown interfacial oxide and different boron doping concentrations is shown in Fig. 3(a). A qualitatively similar dependency of $V_{oc,impl}$ on annealing temperature and doping concentration as observed on the planar (100) samples is obtained. However, the absolute level of p^+ POLO junction passivation for the textured samples is dramatically decreased compared to the planar ones. The highest $V_{oc,impl}$-value is 681 mV, with a corresponding J_{0e} value of 54 fA/cm² for the lowest dose (8.5×10^{14} cm^{-2}) and for an annealing temperature of 1000 °C. On the textured samples with wet-chemical oxide (Fig. 3(b)), a maximum $V_{oc,impl}$ value of 669 mV (J_{0e} = 71 fA/cm²) is achieved for the same dose at 900 °C.

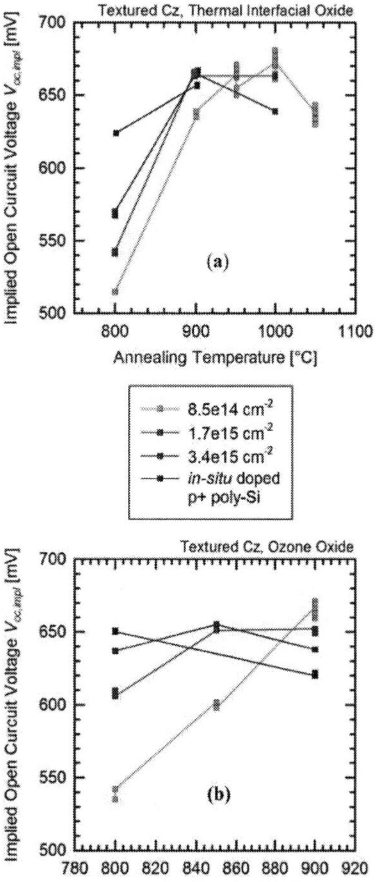

Figure 3: Temperature dependence of the implied open circuit voltage $V_{oc,impl.}$ measured on textured samples with *p*-type POLO junctions for thermally (a) and wet-chemically (b) grown interfacial oxides for different boron implant doses.

In contrast to the planar samples with (100)-oriented surfaces, the textured samples exhibit mainly (111) surfaces and edges of the texture pyramids (corresponding to <110> directions). Both, (111) planes and <110> edges, may contribute to a reduced passivation quality compared to planar (100) planes. Furthermore, the oxide thickness on both orientations will be different from that on planar (100) surfaces due to different oxidation rates.

To check the passivation quality of p^+ POLO junctions on (111) surfaces, we prepare additional lifetime test structures on planar *n*-type (111) 6 Ωcm Cz silicon wafers. On these wafers SiO_2 (of 1.7 nm thickness) is thermally grown, followed by LPCVD deposition of 150 nm intrinsic silicon. Subsequently the samples are ion-implanted with a boron dose of 1×10^{15} cm^{-2} on planar surfaces and annealed at 1000 °C. Figure 4 shows the comparison in the emitter saturation current densities measured for (111) planar surfaces, (100) planar surfaces and (100) textured surfaces. For (111) planar and (100) textured surfaces, very similar J_{0e} values are obtained, much higher than for (100) planar surfaces. This finding implies that the passivation of textured surfaces is mainly reduced due to the poor passivation quality of the interfacial oxide on (111) surfaces.

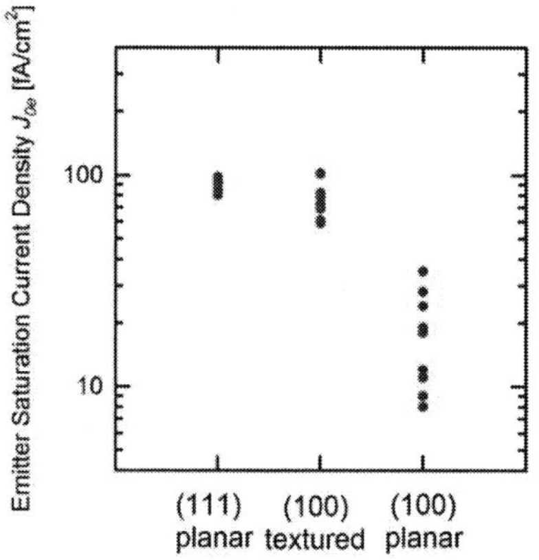

Figure 4: Emitter saturation current density J_{0e} measured on symmetric test structures after thermal growth of the interface oxide, LPCVD deposition of intrinsic Si, boron implantation and subsequent annealing at 1000 °C. Samples feature (111) planar surfaces, (100) planar surfaces and (100) textured surfaces. Boron implantation dose is 1×10^{15} cm^{-2} for planar (111) and (100) surfaces, 1.7×10^{15} cm^{-2} for textured surface.

4 CONCLUSIONS

We experimentally analyzed the passivation quality of *p*-type poly-Si with underlying interfacial oxide, thermally or wet-chemically grown on planar (100) and (111) as well as textured (100) surfaces at different annealing temperatures. Our results show that the optimum annealing temperature shifts towards lower temperatures, if the boron concentration within the poly-Si layer increases. Simultaneously, the maximum open circuit voltage $V_{oc,impl.}$ decreases with increasing doping concentration. For wet-chemically (thermally) grown oxides, emitter saturation current densities J_{0e} of 8 fA/cm² (6 fA/cm²) are achieved for the lowest used boron dose of 5×10^{14} cm^{-2} on planar (100) surfaces. To our knowledge, this is the best passivation quality of a p^+-doped poly-Si / *c*-Si (POLO) junction with wet-chemically grown oxide reported so far. Additionally, we observe a dramatically reduced passivation quality of *p*-type POLO junctions on textured (100) surfaces in comparison to planar (100) ones. A comparison of *p*-type poly-Si junctions prepared on planar and textured (100) and on planar (111) surfaces shows that *p*-type POLO junctions also exhibit a strongly reduced passivation quality on planar (111) surfaces, indicating that this issue is correlated with a difference in passivation quality of the interfacial oxide on different surface orientations.

ACKNOWLEDGEMENT

This work is financially supported by the Federal Ministry for Economic Affairs and Energy (BMWi) under contact number 0325702.

REFERENCES

[1] W. Nemeth, V. LaSalvia, M. R. Page, E. L. Warren, A. Dameron, A. G. Norman, B. G. Lee, D. L. Young, P. Stradins, Proceedings 42nd IEEE Photovoltaic Specialist conference (2015).

[2] A.D. Upadhyaya, Y.-W. Ok, E. Chang, V. Upadhyaya, K. Madani, K. Tate, E. Cho, B. Rounsaville, V. Chandrasekaran, V. Yelundur, A. Gupta, A. Rohatgi, Proceedings 42nd IEEE Photovoltaic Specialist conference (2015).

[3] D. Yan, A. Cuevas, J. Bullock, Y. Wan, C. Samundsett, Solar Energy Materials & Solar Cells 142 (2015) 75.

[4] Y. Tao, E. L. Chang, A. Upahhyaya, B. Roundaville, Y.-W. Ok, K. Madani, C.-W. Chen, K. Tate, V. Upadhyaya, F. Zimbardi, J. Keane, A. Payne, A. Rohatgi, Proceedings 42nd IEEE Photovoltaic Specialist conference (2015).

[5] M. Hermle, F. Feldmann, J. Eisenlohr, J. Benick, A. Richter, B. Lee, P. Stradins, A. Rohatgi, S. W. Glunz, Proceedings 42nd IEEE Photovoltaic Specialist conference (2015).

[6] J. Y. Gan, R. M. Swanson, Proceedings 21st IEEE Photovoltaic Specialist conference (1990) 245.

[7] R. Brendel, T. Dullweber, R. Gogolin, H. Hannebauer, N.-P. Harder, J. Hensen, S. Kajari-Schroeder, R. Peibst, J. H. Petermann, U. Römer, J. Schmidt, H. Schulte-Huxel, V. Steckenreiter, Presentation of 28th European PV Solar Energy Conference and Exhibition (2013).

[8] F. Feldmann, M. Bivour, C. Reichel, M. Hermle, S. W. Glunz, Solar Energy Materials & Solar Cells 120 (2014) 270.

[9] U. Römer, R. Peibst, T. Ohrdes B. Lim, J. Krügener, T. Wietler, R. Brendel, IEEE Journal of Photovoltaics 5 (2) (2015) 507.

[10] D. E. Kane, R. M. Swanson, Proceedings of the 18th IEEE Photovoltaic Specialists Conference (1985) 578.

[11] G. R. Wolstenholme, N. Jorgensen, P. Ashburn, G. R. Booker, Journal of Applied Physics 61 (1987) 225.

[12] N. Jorgensen, J. C. Barry, G. R. Booker, P. Ashburn, G. R. Wolstenholme, M. C. Wilson, P. C. Hunt, Inst. Phys. Conf. Ser. 76 (1985) 471.

HIGH VOLUME MANUFACTURING OF HIGH EFFICIENCY CRYSTALLINE SILICON SOLAR CELLS WITH SHIELDED METAL CONTACTS

Oliver Schultz-Wittmann, Adrian Turner, Bonne Eggleston, Denis De Ceuster, Dominik Suwito, Emmanuel van Kerschaver, Simeon Baker-Finch and Victor Prajapati
First Solar Inc.
1035 Walsh Avenue, Santa Clara, CA 95050, USA
oliver.schultz-wittmann@firstsolar.com

First Solar has launched mass production of high-efficiency crystalline silicon solar cells. The manufacturing sequence employs a new technology that eliminates recombination at the metal contacts. This "shielding" of the contacts allows a new degree of freedom for the metal grid design, i.e. the optimization is only a trade-off between shading and resistive losses. Contact recombination is so low that it doesn't need to be factored in anymore. In combination with a fine line copper metallization process sequence, median efficiencies over 21% are achieved on a 100 MW_p pilot line. 72-cell PV modules with standard soldering and interconnection technology have power bins (-0 to +5W) of 360 to 370 W_p.
Keywords: c-Si, High-Efficiency, n-type, Passivated Contacts, Bifacial

1 INTRODUCTION

Contact recombination is a major loss mechanism in high-efficiency crystalline silicon solar cells. Typical approaches to reduce contact recombination include (i) reducing contact area; and (ii) applying (usually a-Si:Si) heterojunction technology. In the former case, the positive benefit of reduced contact recombination with reduced area is counterbalanced by increased resistive losses, so a more complex process technology or sequence is often applied. A common example is a localized diffusion for selective emitter or local back surface field. In the case of heterojunction technology, contact passivation comes at the cost of (i) strict low temperature requirements for cell metallization and module manufacturing; and (ii) a new – often problematic – toolset for manufacturing (including CVD of transparent conductive oxides [1, 2].

The trade-offs listed above are largely eliminated by the application of recent advances in contact passivation, such as a solution featuring a thin oxide in combination with a thin polysilicon layer [3, 4]. In one manifestation, this approach allows for full area metallization of the rear side without significant contact recombination.

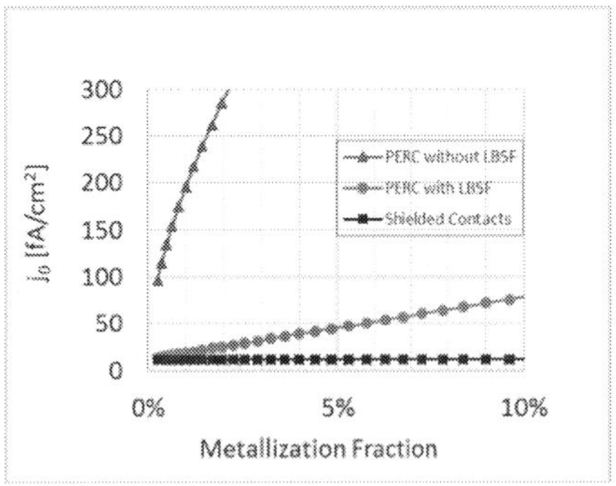

Figure 1: Surface recombination calculated for point contact rear side using the Fischer model [8] for PERC with/out Local Back Surface Field (LBSF) and constant recombination for shielded contacts.

To illustrate the nature of the passivated contact advantage, surface recombination $j_{0,back}$ was calculated for a point contact structure in a PERC cell [Figure 1] for various contact configurations. Comparing the PERC approaches, it is clear that the local BSF is of very high importance. Even then, a minimal contact fraction is required to maintain low j_0. However, the need to maintain low series resistance means that, for example, even in the world record multicrystalline silicon solar cell [6] recombination at the contacts is $j_{0,contact}$ = 500 fA/cm^2. In contrast, having shielded contacts allows for unlimited contact area. For example, a very high fill factor of 83.2% was reported for a full area rear contact while maintaining a V_{oc} value of 718 mV and achieving 25.1% efficiency [5]. The high FF can be attributed to the full area contact and the optimized current flow within the cell, the high V_{oc} value is a result of the passivation of the metal contacts.

First Solar has developed, and placed into 100 MW pilot production, a proprietary solution that has shielded contacts. The surface of our solar cells are passivated by a layer that is electrically conductive. This allows the metallization to be applied directly on the passivation layer, i.e. the metal is not in direct contact with the bulk silicon anymore. Hence, the cell design and the manufacturing processes need to be adjusted. This paper will describe the design of the TetraCell®, an optimized cell structure with shielded contacts on front and back.

2 CELL DESIGN

2.1 Cell Structure

A cross-section of a typical PERC cell design with localized base contacts is shown in Figure 2.

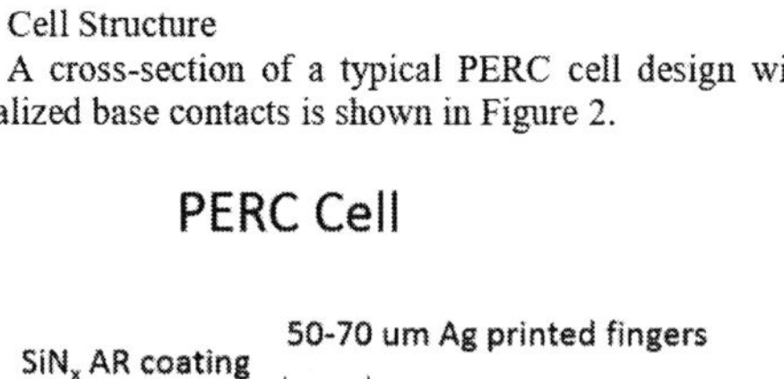

Figure 2: Cross section (not to scale) of a PERC solar cell with localized base contacts at the rear.

This structure is on the path to become a dominant technology for high-efficiency crystalline silicon solar cell manufacturing. This is due to the availability of high-volume manufacturing equipment for the additional process steps as well as the option to upgrade older manufacturing lines. However, despite low CAPEX expansion and impressive results on R&D activities resulting in champion cells of 21.2% on multicrystalline silicon and 22% for monocrystalline silicon [6, 7], this structure has still some shortcomings. Those can be found in the requirement to have special solder pads on the rear for cell interconnection as well as the need for a re-generation of the boron-oxygen complex to maintain the performance over the long term [9]. Replacing the rear side with passivated contacts eliminates contact recombination and the lack of LBSF allows for full area contacting and a low resistance current flow pattern [5]. However, if intended to be used for bifacial module application, standard H-Grid metallization patterns can be applied on both sides.

We have developed a cell structure that implements:

- electrically shielded contacts on the front and back
- double-side texture for superior light collection from the rear
- fine-line copper plated metallization that is directly solder able

on high-quality *n*-type Czochralski silicon wafers. The structure is shown in Figure 3 and will be discussed in the following sections.

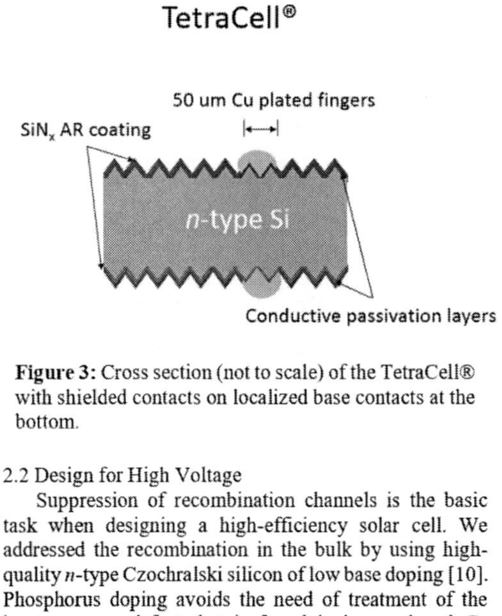

Figure 3: Cross section (not to scale) of the TetraCell® with shielded contacts on localized base contacts at the bottom.

2.2 Design for High Voltage

Suppression of recombination channels is the basic task when designing a high-efficiency solar cell. We addressed the recombination in the bulk by using high-quality *n*-type Czochralski silicon of low base doping [10]. Phosphorus doping avoids the need of treatment of the boron-oxygen defect that is found in boron doped Cz silicon and low base doping reduces susceptibility to metallic impurities, hence more than 1 ms bulk lifetimes are measured. The surfaces are passivated by our conductive passivation layer on either side. On phosphorus doped, random pyramid textured silicon the passivation is characterized by a $j_{0,front}$ of less than 10 fA/cm^2. On textured boron doped surfaces $j_{0,back}$ of less than 35 fA/cm^2 is obtained. A total value of less than 45 fA/cm^2 including all metallization on both surfaces yields a median V_{oc} of 696 mV in high volume production and up to 719 mV for full area R&D cells.

2.3 Design for High Fill Factor

Especially for high-efficiency cells several challenges with respect to the series resistance are common:

- Lowly doped diffusions are preferred because of the improved spectral response, but this increases the lateral emitter resistance and the contact resistance.
- Lowly doped base material is preferred because of its insensitivity to metallic contamination, but this increases the lateral resistance towards the base contacts.
- Small contact areas reduce the contact recombination but increase the contact resistance.

Therefore, the series resistance requires a special focus when designing a high-efficiency cell. We implemented several design features that enable a low series resistance:

- A back junction design allows for a lowly doped front side because the base doping and the high carrier injection at V_{mp} provide significant lateral conductance (compare with [10] and Figure 4).
- Narrow metal contact spacing enabled by a fine line patterning technology and shielded contacts
- Highly conductive metallization grid as a result of Cu plating

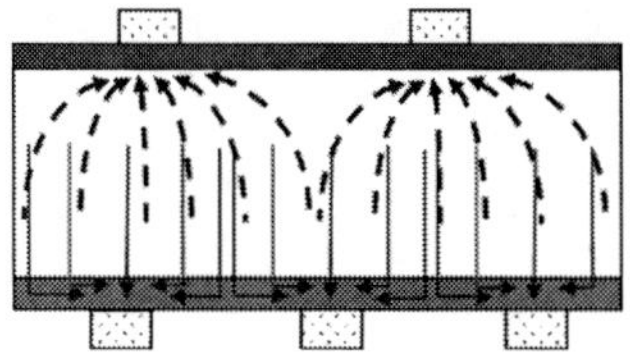

Figure 4: Cross section (not to scale) of a back junction solar cell with shielded contacts, lowly doped base and lowly doped front and back on localized base contacts at the bottom.

Consequently, median FF values of 80% are achieved in high volume manufacturing.

2.4 Design for High Current

Besides the need to keep any doping low to keep parasitic absorption at a minimum, especially reduced shading of the metal grid is of high importance. This is counterbalanced by the need for high metallization fraction for a low series resistance. We developed a highly and diffusely reflective surface finish of the plated fingers. This recovers the majority of the photons reflected by the fingers when encapsulated in a module, they have an effective shading of only 32%.

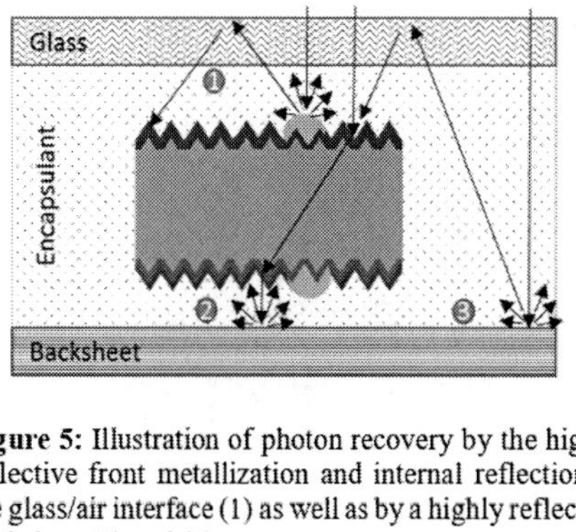

Figure 5: Illustration of photon recovery by the highly reflective front metallization and internal reflection at the glass/air interface (1) as well as by a highly reflective backsheet (2) and (3).

This means that in the module a 45 um finger width is effectively only shading like a 14 um wide finger (see (1) in Figure 5 for illustration). Even though we have moderate J_{sc} values of 37.9 mA/cm^2 on cell level, this results in a significant boost of more than a 1 mA/cm^2 after encapsulation. Likewise, the reflectance of the backsheet in the module has an impact on the current. Those optimizations are targeting optimum performance at the module level, even if cell efficiency is not as high as it would be if efficiency at cell level was the optimization target.

2.5 Pilot Production

The above detailed cell design has been implemented onto a high volume 100MW$_p$ pilot cell production line. Currently, this pilot manufacturing is achieving IV parameters as shown in Figure 6. Excellent efficiency yields of > 99% @ 20.3% cutoff are routinely achieved.

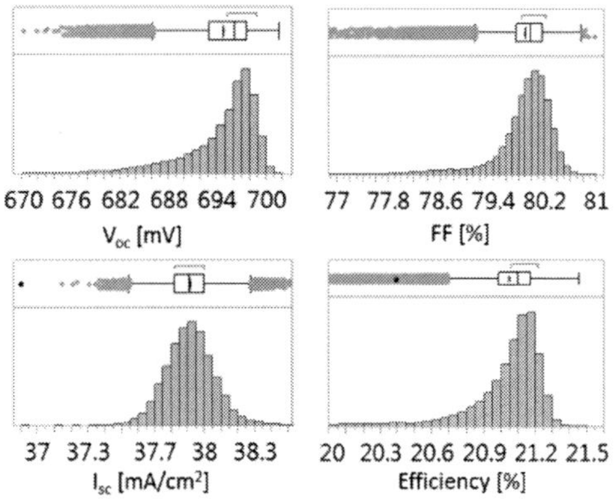

Figure 6: Production data from our pilot line in Kulim, Malaysia. The median efficiency is 21.1%.

Those cells are being soldered to strings with standard copper ribbons and assembled to modules with 72 cells. With a Cell-To-Module (CTM) factor of about CTM = 0.98 (module power divided by sum of cell power), we obtain modules of 360 to 370 W$_p$. The modules pass or exceed all accelerated reliability testing requirements and are certified for UL, IEC and JET.

3 PRODUCT ROADMAP

Following the same design rules described above, process modifications and improvements have been developed in our R&D lab in Santa Clara, California. 22% cell efficiency has been measured at Fraunhofer ISE CalLab (see Figure 7).

Individually, we have been able to significantly improve our surface passivation and increase V_{oc} to 719 mV. Improvements and optimization of the optical properties have led to j_{sc} values of more than 39 mA/cm^2. Those improvements have then been tested in our pilot line in Kulim, Malaysia on more than 60,000 cells and median values of $\eta = 21.9\%$, $V_{oc} = 701$ mV, $j_{sc} = 39.3$ mA/cm^2 and FF = 79.5% were obtained (see Figure 8).

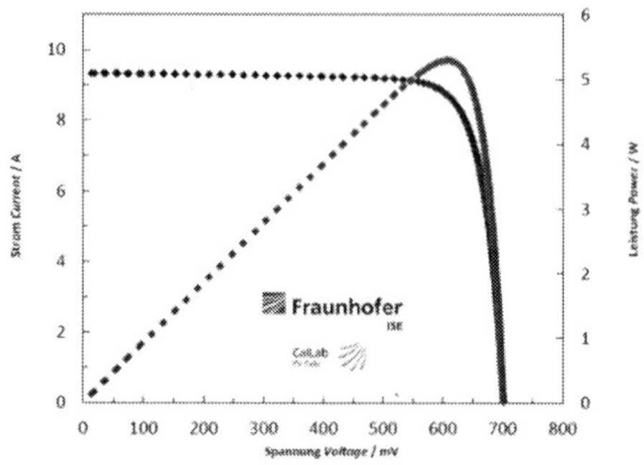

Figure 7: Measurement of improved cells from our R&D lab in Santa Clara, California. Light IV data was measured by Fraunhofer ISE CalLab at V_{oc} = 702mV, J_{sc} = 38.8 mA/cm^2 and FF = 81.2%, resulting in an efficiency of η = 22.0%.

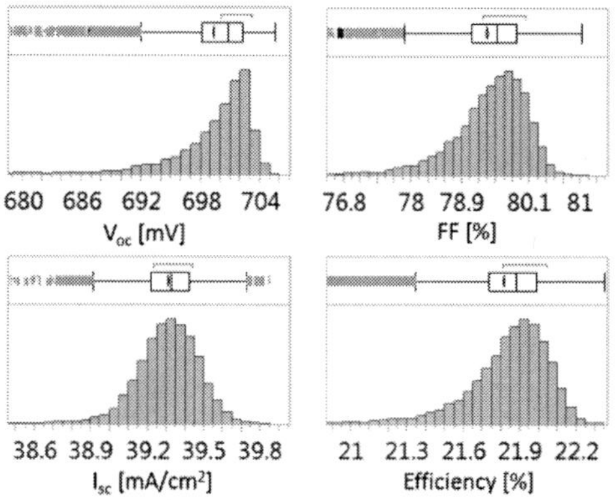

Figure 8: Efficiency distribution of 60,000 cells made in our pilot line in Kulim, Malaysia including process improvements developed in our R&D lab in Santa Clara, California. The median efficiency is 21.9%.

Also, reduction of the series resistance and reduction of optical losses in the module have been addressed in our module R&D laboratory in Berlin, Germany. Keeping the general design of our production modules with tabbing/stringing of full area cells with three busbars,

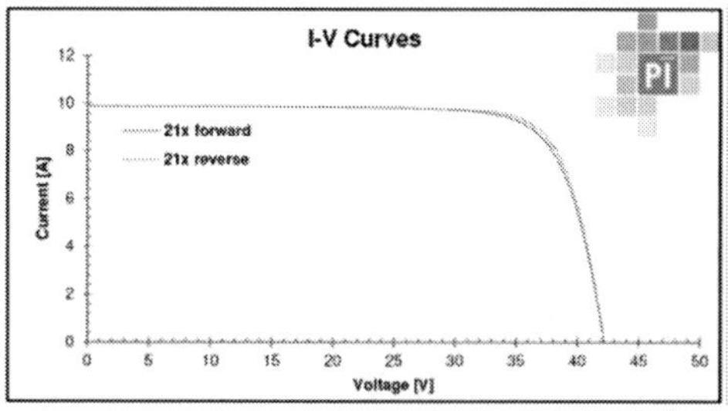

Figure 9: 330 W$_p$ measured by PI Berlin on a 60 cell module (1.64 m^2) module. Hysteresis effects of our high voltage/capacitance module were taken into account.

top values of 330 W_p/20.1% efficiency for a 60 cell module have been independently measured by PI Berlin (see Figure 9). The implementation into manufacturing is pending full qualification for our certified products.

4 BIFACIAL MODULES

Our standard module configuration has a white backsheet on the rear. Replacing this with a sheet of glass results in a bifacial module. Due to the excellent bifacial properties of the cell design, a bifaciality factor $\varphi_{bifacial}$

$$\varphi_{bifacial} = \frac{Pmax, back}{Pmax, front} \quad (1)$$

of 93% is obtained flashing either side under STC while the other side is covered by a non-reflective black cloth. First outdoor testing of those modules together with our monofacial modules at our test site in Mesa, Arizona, shows an energy yield gain of around 10% under realistic conditions with fixed tilt installation.

Besides the gains in energy yield, excellent durability results without measureable degradation after 600 thermal cycles (TC 600) and 3000 hours of Damp/Heat (DH 3000) are obtained in the glass/glass configuration. The mechanical stability is due to the low stress configuration of the cell being placed in the neutral plane between the two sheets of glass, the resistance against corrosion is explained by the impermeability of glass by water vapor. The relevance of such test can be observed in our very first installation of a bifacial module at NREL (Golden, Colorado) in 2012. It showed degradation rates of 0.08%/year after three years of continuous outdoor exposure, significantly outperforming other manufacturers' silicon modules Si 1 and Si 2 (Figure 10).

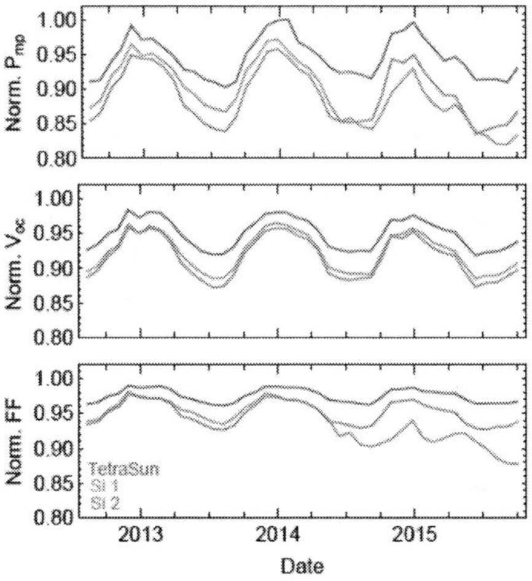

Figure 10: Long-term data collection of our bifacial glass/glass module installed at NREL in Golden, Colorado. Besides high energy yields, excellent performance without signs of degradation were measured, significantly outperforming other manufacturers' silicon modules Si 1 and Si 2.

SUMMARY

FirstSolar has successfully implemented TetraSun's shielded contact technology, the TetraCell®, into high volume manufacturing. Excellent results for pilot production prove the feasibility of this concept on an industrial level. Certified modules are being shipped and installed in selected markets.

Our R&D results developed in cooperation between our locations in the USA, Germany and Malaysia validate our path to cell efficiencies well in excess of 22% and module efficiencies higher than 20%.

ACKNOWLEDGEMENTS

The authors would like to thank the entire TetraSun team for their contributions. The help of NREL and Fraunhofer ISE for characterization and testing of our cells and modules is gratefully acknowledged.

REFERENCES

[1] M. Tanaka et al, "Development of HIT solar cells with more than 21% conversion efficiency and commercialization of highest performance hit modules," in *Proceedings of the 3rd World Conference on Photovoltaic Energy Conversion (WCPEC '03)*, vol. 1, pp. 955–958, Osaka, Japan, May 2003.

[2] B. Strahm et al, "Progress in Silicon Hetero-Junction Solar Cells Development and Scaling for Large Scale Mass Production Use", in *Proceedings of the 25th European Photovoltaic Solar Energy Conference and Exhibition* pp.1286-1289, September 2010, Valencia, Spain

[3] F. Feldmann et al, "A PASSIVATED REAR CONTACT FOR HIGH-EFFICIENCY n-TYPE SILICON SOLAR CELLS ENABLING HIGH VOCS AND FF>82%" in *Proceedings of the 28th European PV Solar Energy Conference and Exhibition*, pp.988-992, October, 2013, Paris, France

[4] Udo Roemer et al, "Recombination Behaviour and Contact Resistance of n+ and p+ polycrystalline Si/mono-crystalline Si junctions", *Solar Energy Materials & Solar Cells* **131** (2014) 85-91

[5] S.W. Glunz et al, THE IRRESISTIBLE CHARM OF A SIMPLE CURRENT FLOW PATTERN – 25% WITH A SOLAR CELL FEATURING A FULL-AREA BACK CONTACT, in *Proceedings of 31st European Photovoltaic Solar Energy Conference and Exhibition*, pp. 259-263, September 2015, Hamburg

[6] Weiwei Deng et al, "Development of high-efficiency industrial p-type multi-crystalline PERC solar cells with efficiency greater than 21%", *6th International Conference on Silicon Photovoltaics, SiliconPV 2016*, 7th -10th March 2016, Chambery, France

[7] Solar World press release January 14, 2016.

[8] B. Fischer, "Loss analysis of crystalline silicon solar cells using photoconductance and quantum efficiency measurements", *Dissertation*, Universität Konstanz (2003)

[9] B. Hallam et al, "Modelling kinetics of the boron oxygen defect system", *6th International Conference on Silicon Photovoltaics, SiliconPV 2016*, 7th -10th March 2016, Chambery, France

[10] O. Schultz-Wittmann et al, "HIGH EFFICIENCY SOLAR CELLS FROM N-TYPE CZ SILICON in *Proceedings of 28th European Photovoltaic Solar Energy Conference and Exhibition*, pp. 1004 - 7, September 2013, Paris

N-TYPE POLYSILICON PASSIVATING CONTACTS FOR INDUSTRIAL BIFACIAL N-PERT CELLS

M.K. Stodolny[1], M. Lenes[2], Y. Wu[1], G.J.M. Janssen[1], I. Romijn[1], J.R.M. Luchies[2], L.J. Geerligs[1]
[1] ECN Solar Energy, P.O. Box 1, NL-1755 ZG Petten, The Netherlands
[2] Tempress Systems, Radeweg 31, 8171 MD Vaassen, The Netherlands

ABSTRACT: We present a high-performance bifacial n-type solar cell with n^+ polysilicon (polySi) back side passivating contacts and fire-through screen-printed metallization, processed on 6" Cz wafers. The cells were manufactured with low-cost industrial process steps yielding a best efficiency of 20.7%, and an average V_{oc} of 674 mV. We analysed effects of variation of doping level, thickness, and oxide properties of the n-type polySi/SiO_x layers, as well as hydrogenation from a PECVD SiN_x:H coating, which led to recombination current densities down to ~2 fA/cm^2 and ~4 fA/cm^2 on planar and textured surface, respectively. The results are novel in four aspects: the cells are bifacial, they are full 6" size while employing LPCVD for the polySi deposition, the polySi passivating contact is metallized by fire-through screen printing paste, and hydrogenation is done by PECVD-deposited and fired SiN_x:H. Analysis shows that the wafer bulk lifetime in the cell is high and that the V_{oc} of the cell is limited by the J_o of the uniform diffused boron emitter and its contacts. Ways to improve the efficiency of the cell to > 22% are indicated.
Keywords: polysilicon, passivating contact, carrier selective contact, LPCVD, n-type solar cell, bifacial

1 INTRODUCTION

The combination of a thin oxide (SiO_x) and doped polysilicon (polySi) to obtain low recombination junctions, originating from early work on bipolar transistors [1], was demonstrated in the 1990s to be a viable candidate for creating passivating contacts to crystalline silicon (cSi) solar cells [2,3].
A key element in the passivating contact structure is the thin oxide layer which has a reasonably low interface recombination velocity, supported by a hydrogenation step (typical models estimate that a recombination velocity (S) of ~10^3-10^4 cm/s is sufficiently low [4, 5]). This thin oxide layer serves also as a tuneable (e.g., by thickness) diffusion barrier, which is indispensable for keeping most of the dopant within the polySi layer thus avoiding creation of a typical diffused junction in the wafer. The doped polySi layer, due to the induced field effect, reduces minority carrier density at the interface with the wafer while providing good conductance for the majority carriers to the contacts applied on the polySi.
The combination of the interfacial thin oxide with doped polySi provides a low transmission of minority carriers assuring therefore a minimal recombination in polySi and at the metal contact. Together with the low level penetration of dopants in the wafer, this results in excellent passivation [6, 7]. The oxide/polySi stacks can be contacted by metal, which is in principle possible without increasing the recombination of minority carriers generated in the cSi wafer, although increase of recombination can in practice occur due to various effects [8].
Recent progress on cells with such contacts, achieving conversion efficiencies above 25% on small area cells [4,9], and demonstrating in excess of 21% cell efficiency on 6" monofacial cells [10], merit the need to investigate industrialization of the concept.
The purpose of the work in the present paper is to understand practical optimization routes, using solely high volume production proven equipment, for the n-type polySi passivating contact as well as to demonstrate potential industrial cells with such a back contact. In this work we employ a polySi layer grown in a low pressure chemical vapor deposition (LPCVD) furnace which is subsequently doped by means of $POCl_3$ diffusion.
Specifically, optimization of the thin oxide properties, polySi thickness and phosphorous (P) doping level, hydrogenation from SiN_x:H, and contacting of the polySi are investigated. The evaluation of the feasibility of firing-through screen-printed metallization on polySi (without adding significant contact recombination) is included. The n-type polySi with screen printed contacts is integrated on the back of a cell with a boron emitter made by an industrial diffusion process, thus leading to a novel low-cost but high performance solar cell concept.
High efficiency solar cells made with LPCVD based polySi layers and screen printed fire-through contacts have several advantages, amongst others they can be realized by modest adaptation of conventional screen-print solar cell processes. LPCVD is a well-established high-throughput industrial process that can be easily implemented in state of the art solar cell manufacturing lines.

2 EXPERIMENTAL

2.1 LPCVD polySi layers

PolySi layers were produced in a high-throughput industrial LPCVD furnace and subsequently doped by means of $POCl_3$ diffusion [11]. An LPCVD process has the benefit of creating very conformal and pinhole-free layers. This ensures that the underlying interfacial oxide is protected against subsequent doping steps and chemical treatments. For mass production, batches of up to a few hundred wafers can be processed simultaneously with excellent process uniformity. The thin oxide was produced by thermal oxidation (called Th.Ox) or wet chemical oxidation (Nitric Acid Oxidation of Silicon, called NAOS). A reliable absolute thickness measurement is not well feasible, but the thickness was estimated from spectroscopic ellipsometry on mirror polished wafers to be 13 to 15 Å. The high repeatability and tuning of this thickness at the single Å level is possible as well as necessary and will be addressed in the near future.

The effect of variations of process parameters on the recombination parameter J_o (the dark saturation current) was characterized at lifetime level by photoconductance measurements (Sinton WCT-120 tool) and photoluminescence (PL), and related to the dopant profile measured by ECV (electrochemical capacitance-voltage). The best process parameters in terms of resulting J_o were applied at cell level.

2.2 Investigation of polySi doping and passivation

Symmetrical lifetime samples were fabricated on 5 Ω·cm n-type 6" Czochralski (Cz) wafers. The thickness of chemically polished wafers was 150 µm and of textured wafers 175 µm. After the wet-chemical oxide or thermal oxide was grown on both sides, a 70 or 200 nm thick intrinsic polySi layer was deposited by LPCVD, followed by $POCl_3$ diffusions at 3 different temperatures ranging from 830°C to 870°C, resulting in polySi layers with different dopant concentrations and sheet resistances (R_{sheet}, measured using a Sherescan tool), as well as different phosphorous leakage through the thin oxide into the wafer (measured using ECV). The polysilicon thickness determined by ECV profiling compares well with scanning electron microscope (SEM) images. For textured surfaces, the polysilicon thickness is best determined from SEM images.

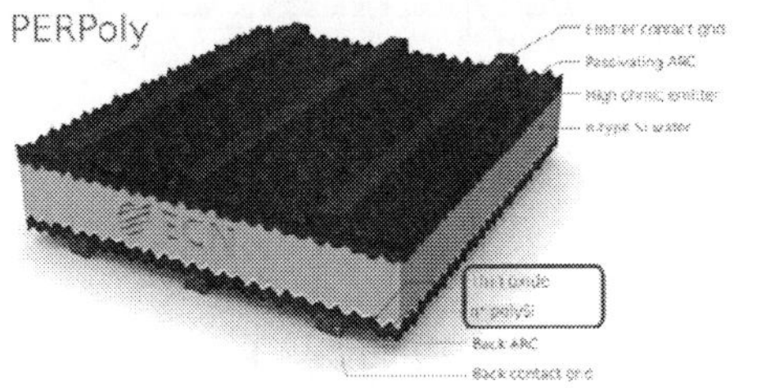

Figure 1: Schematic drawing of the bifacial n-type solar cell design of this paper, featuring n-polySi\SiO$_x$ contacts, named PERPoly (Passivated Emitter and Rear Polysilicon).

2.3 Application in industrial bifacial n-type cells

239 cm^2 n-type solar cells were produced according to standard industrial process steps with the exception of replacing a diffused Back Surface Field (BSF) with n-type doped polySi/SiO$_x$ contacts, as shown in Fig.1. Such device can be called PERPoly (Passivated Emitter and Rear Polysilicon). In the cell process tests we have so far only employed the 200 nm polySi thickness. The cell processes stay close to existing industrial n-type technology, by employing only PV process tools that are proven for low-cost high-throughput production.

n-type cells were produced on 6" Cz wafers of 5 Ohm·cm. An industrial type uniform diffused emitter with a sheet resistance of 70 Ohm/sq was employed on the front side. The emitter was passivated with Al_2O_3 deposited by ALD (Atomic Layer Deposition (Levitrack tool from Levitech)), coated with PECVD SiN$_x$:H. The back side polySi was also coated with SiN$_x$:H. The metal grid was screen-printed using fire-through pastes both to contact the B-emitter on front and polySi on back. The co-firing process was optimized and the settings are similar to the industrial standards.

3 RESULTS AND DISCUSSION

3.1 PolySi doping and passivation

Fig. 2. shows the resulting active phosphorus concentration profiles, measured with ECV on the chemically polished samples, for three different diffusion

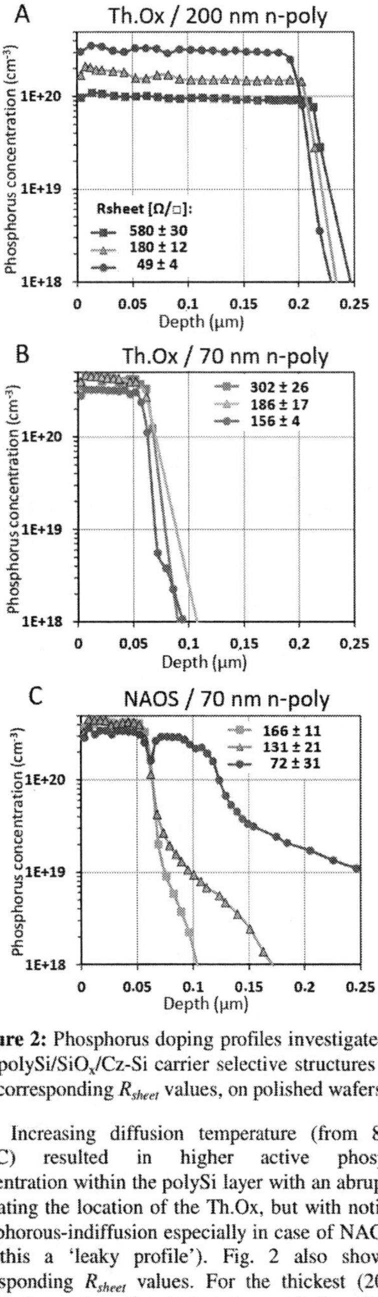

Figure 2: Phosphorus doping profiles investigated for the polySi/SiO$_x$/Cz-Si carrier selective structures with the corresponding R_{sheet} values, on polished wafers.

runs. Increasing diffusion temperature (from 830 to 870°C) resulted in higher active phosphorus concentration within the polySi layer with an abrupt drop indicating the location of the Th.Ox, but with noticeable phosphorous-indiffusion especially in case of NAOS (we call this a 'leaky profile'). Fig. 2 also shows the corresponding R_{sheet} values. For the thickest (200 nm) polySi increasing the temperature of the diffusion resulted in an increased dopant concentration. However, for the thinner (70 nm) polySi similar concentrations were found for all diffusion processes applied, despite the varying temperature. This leads us to believe the doping level in the thick polySi layers is limited due to the amount of phosphorus available, e.g. in the phosphosilicate glass. Despite similar active dopant concentrations, the thin polySi layers still show a decrease in sheet resistance (from 302 to 156 Ω/sq) with increasing temperature. This indicates that higher diffusion temperatures are resulting in higher charge carrier mobilities, which is likely due to continued crystallization of the polySi during the diffusion. We note that the ECV measurements on polysilicon may be subject to effects from the granular structure not observed in monocrystalline silicon, and the ECV measurements have not been compared with other measurements, e.g. secondary ion mass spectroscopy (SIMS).

Fig. 3 presents the evolution of the recombination parameter J_o (the dark saturation current) for the investigated n-polySi/SiO_x structures through various subsequent processing steps, again for chemically *polished* wafers. All J_o, iV_{oc} (implied open circuit voltage) and lifetime measurements presented here were obtained with the Sinton WCT-120 tester on symmetric lifetime samples.

Average J_o per side <10 fA/cm^2 was achieved already without any particular hydrogenation steps, i.e. right after doping. We attribute the higher initial J_o of the 200 nm thick high R_{sheet} layers at least partly to the lower doping level which increases sensitivity to the interfacial D_{it}, which was also suggested in other study [12]. For the 70 nm layers with Th.Ox there is little variation of doping level, and therefore little variation of initial J_o. The doping level for all those layers is in the same range as the heaviest doped 200 nm layer. As expected from the Debye length of only a few nm in the polySi, no difference in J_o is observed for 70 and 200 nm polySi layer thickness. For the 70 nm thick layers on top of the NAOS oxide, the initial J_o (before hydrogenation) is probably dominated by effects of the phosphorus leakage through the thin oxide. A significant 'leaky diffusion profile' in the case of structures with the NAOS interfacial oxide and $POCl_3$ diffusion at highest temperature (72 Ω/sq) resulted in high J_o. This is partly due to a significant Auger recombination in the wafer near to the surface (we calculated about 70 fA/cm^2 for the ECV profile shown in Fig. 2) and the effect of the damaged oxide increasing the amount of interface defect states, which could result in at most about 275 fA/cm^2. Interestingly, the moderate 'leaky diffusion profiles' of NAOS/polySi with R_{sheet} of 131 and 166 Ω/sq showed still relatively low J_o values.

Improvements in J_o were observed after SiN_x:H deposition. PECVD (plasma enhanced chemical vapour deposition) SiN_x:H resulted in improved J_o in particular for the 200 nm n-poly layers with low doping level and R_{sheet} of 580 and 180 Ω/sq. We attribute this to hydrogenation of the SiO_x/wafer interface. However, the hydrogenation from SiN_x:H did not completely mask effects from variations in D_{it} or doping level. Firing of the SiN_x:H layer did not significantly change J_o. Removal of the SiN_x layer by wet chemical etching did not change J_o, as might be expected since this should not change the hydrogen passivation of defects at the SiO_x/wafer interface, protected by the polySi top layer. This indicates that the passivation of the outer surface of the polySi is irrelevant, validating the concept of the passivating contact.

In conclusion, it should be noted that low sheet resistances could be combined with excellent J_o for some of the investigated thin polySi layers. This allows us to use a partially contacted back (open grid on the back of the cell, resulting in a bifacial cell) which is a practically useful novelty in the field of polySi.

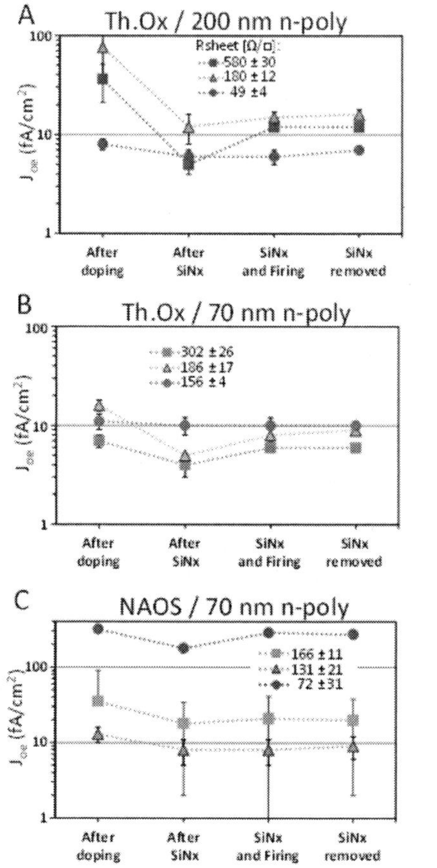

Figure 3: Evolution of surface passivation quality of studied polySi/SiO_x carrier selective structures on *polished* Cz-wafers in a sequence of process steps. All J_o values per side are average of up to 15 points on 3 wafers and error bars depict standard deviations.

Table I: Best passivation characteristics obtained on *polished* wafers. Symmetric wafer structures with n-polySi/SiO_x on both sides. "Lifetime"= effective minority carrier recombination lifetime (@$\Delta n=10^{15}$ cm^{-3}).

Cell type (polished, Th.Ox)	iV_{oc} (mV)	$J_{oe\ n\text{-}poly}$ (fA/cm^2)	*Lifetime* (ms)
After doping (200 nm, 49 Ω/sq)	~731	~3.4	~5.0
After SiN_x (70 nm, 186 Ω/sq)	~743	~2.2	~6.6
After SiN_x and Firing (200 nm, 49 Ω/sq)	~734	~5.2	~9.4
After SiN_x removed (200 nm, 49 Ω/sq)	~732	~3.0	~12.3

Table I shows a summary of the best passivation characteristics measured after these industrially relevant process steps on chemically *polished* wafers (note that table I gives best single measurement results, not best average results). After the phosphorus doping to 3×10^{20} cm^{-3} already a very low J_o of ~3 fA/cm^2 was achieved without any particular hydrogenation steps. Deposition of SiN_x:H reduced the best value of J_o further to ~2 fA/cm^2. Firing seemed to improve the effective lifetime but somewhat increased the best value of J_o to ~5 fA/cm^2.
Fig. 4 shows the evolution of average J_o per side on *textured* wafers through industrial process steps relevant for a solar cell production. The n-polySi/SiO_x capped with SiN_x:H was highly robust and showed on average J_o~10-20 fA/cm^2 per side after firing, thus proving that a direct implementation of n-poly into an industrial solar cell production is feasible. Some degradation of J_o after fire-through of the contact grid on the n-polySi is visible, which will be discussed in section 3.3.

Table II shows the best passivation characteristics measured on *textured* wafers (again, please note these are best single measurements, not averages). Similar results were obtained for Th.Ox and NAOS, with 200 nm thick polySi layers after firing of SiN_x. The achieved recombination currents of ~4 fA/cm^2 are to our knowledge the best n-polySi passivation results on textured Cz-wafers so far reported.

Table II: Best passivation characteristics obtained on *textured* wafers. Symmetric wafer structures with n-polySi/SiO_x on both sides. "Lifetime"= effective minority carrier recombination lifetime (@$\Delta n=10^{15}$ cm^{-3}).

Cell type (polished, Th.Ox)	iV_{oc} (mV)	$J_{oe\ n\text{-}poly}$ (fA/cm^2)	*Lifetime* (ms)
After SiN_x and Firing (Th.Ox/200 nm, 39 Ω/sq)	~732	~4.2	~6.2
After SiN_x and Firing (NAOS/200 nm,71 Ω/sq)	~731	~4.3	~5.6

3.1 Application of polySi in industrial bifacial n-type cells

Table III presents an overview of the best *I-V* results of screen-printed H-pattern contacts on both sides of bifacial cells obtained in two consecutive experimental runs.
The cells showed ~82% bifacial performance (with back full area 200 nm thick n-poly; 46 Ω/sq). This bifacial ratio is about 10% less than for equivalent n-PERT cells with diffused BSF [13], and this loss is in agreement with the estimated absorption of short wavelength photons in the polySi.
Analysis of the half-fabricates (cells without metallization) and cells shown in Table III indicated that performance limitations due to the diffused B-emitter and contact at the front side were dominant in both runs. In Run1 the sub-optimal bulk lifetime (~750 µs on average, possibly due to lack of phosphorus gettering) also played a role, resulting in an average iV_{oc} for half fabricated cells without contacts $iV_{oc}^{(\text{half-fab. without contacts})}$ of ~680 mV. In contrast, in Run2 the bulk lifetime was greatly improved (~3 ms on average) resulting in an average $iV_{oc}^{(\text{half-fab. without contacts})}$ of ~693 mV. Optimization of the cell process was carried out in Run2 to further exploit the excellent potential of the n-doped polySi/SiO_x back contact, yielding the best cell efficiency of 20.72% (spectral mismatch corrected, FhG ISE calibrated n-PERT reference cell, AAA Wacom system, in house) with cell V_{oc} of 675 mV. The cell efficiency distribution over the 6 cells with NAOS/n-polySi (71 Ω/sq) was 20.68 with a standard deviation of 0.05. In Run2 the Th.Ox was apparently slightly thicker than in Run1 as the FF was ~1% abs. lower than in Run1. Lower FF in Run 1 with NAOS/n-polySi may be attributed to high Rsheet (158 Ω/sq).The FF with NAOS/n-poly back contact in Run 2 (71 Ω/sq) is similar to typical *FF* for equivalent n-PERT cells with diffused BSF, and it seems there is no significant series resistance loss (more than about 0.1 Ohm·cm^2) in the polySi/NAOS/wafer junction. However, a full series resistance breakdown analysis is not yet completed for these cells.

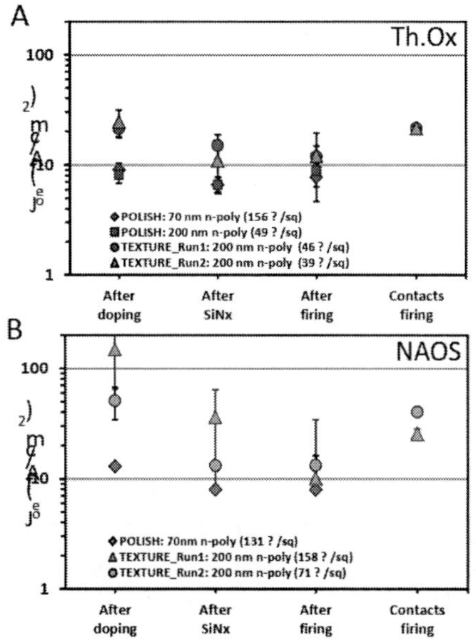

Figure 4: Evolution of average J_o per side of studied polySi/SiO_x carrier selective structures on *textured* as well as chemically polished wafers for cell production relevant steps. All J_o values are average of up to 15 points on 3 wafers and error bars depict standard deviations.

Comparing J_o or iV_{oc} of 'half-fab. without contacts' with J_o or iV_{oc} of 'cell with contacts' in Table III and with the J_o with contacts in Fig. 4, shows that the losses related to the diffused B-emitter front side and front contact were dominant. Note that the J_o and iV_{oc} values given in Table III are best values (i.e., for best location on the wafer), while the V_{oc} for the cells with contacts are an average over the wafer due to the contact grids.
Furthermore some J_{sc} loss is due to non-contributing interband and free carrier absorption (FCA) in the polySi. The FCA in the n-polySi was evaluated by ray tracing analysis to be approx. 0.9 mA/cm^2 for 200 nm thickness and 3×10^{20} cm^{-3} phosphorus doping level. We intend to reduce this FCA by decreasing the polySi thickness and doping level. Table III illustrates the opportunity, through

the experimental variation of J_{sc} as a function of the polySi doping level.

Table III: Parameters of textured bifacial solar cells with 200 nm rear polySi. Best cell results are shown as well as best J_o, iV_{oc} and best bulk lifetime obtained on half-fabricates (obtained from Quasi-steady state photoconductivity decay QSSPCD, light I-V and SunsVoc measurements).

A	J_{oe} *n-poly w/o contacts** (fA/cm^2)	J_o *F+B half-fab. w/o contacts** (fA/cm^2)	J_o *F+B cell w/ contacts**# (fA/cm^2)	iV_{oc} *half-fab. w/o contacts** (mV)	iV_{oc} *cell w/ contacts**# (mV)
Run1_Th.Ox 46 Ω/sq	9.7	64.7	~88	689	~679
Run1_NAOS 158 Ω/sq	7.9	63.9	~96	691	~672
Run2_Th.Ox 39 Ω/sq	5.1	58.9	~94	697	~675
Run2_NAOS 71 Ω/sq	7.7	56.2	~98	699	~680

B	V_{oc} (mV)	J_{sc} (mA/cm^2)	FF (%)	pFF (%)	η (%)
Run1_Th.Ox 46 Ω/sq	669	37.5	78.7	81.4	19.75
Run1_NAOS 158 Ω/sq	665	38.2	77.8	81.9	19.77
Run2_Th.Ox 39 Ω/sq	673	38.4	77.8	82.8	20.09
Run2_NAOS 71 Ω/sq	675	38.8	79.1	82.8	20.72

3.3 Outlook

Based on the obtained experimental results a qualitative description of key features of the passivating contact structures can be given.

The thin interfacial oxide layer needs to be thin enough not to limit the transmission of the majority carriers (allowing high *FF*) but should be dense/close enough to serve as a diffusion barrier, such that most of the doping is realized in the polySi layer. Slight dopant in-diffusion into the c-Si wafer (such as in the moderately leaky profile from Fig. 2c) is not harmful for the passivation properties and can be additionally beneficial for the transport of the majority carriers (perhaps avoiding transport via a tunneling mechanism only, allowing transport partially via less restrictive pinholes/percolation pathways [2]).

The polySi layer needs to be sufficiently high doped to allow adequate lateral transport for the application of bifacial grid metallization and also to provide a passivation boost by reducing the minority carrier density through the "field effect passivation". The thickness of polySi needs to be sufficient to block the penetration of the fire-through paste to the thin oxide interface. In our case 200 nm was sufficient, but we expect that thinner polySi will also be suitable. We have not yet attempted to make cells with the 70 nm polySi layers reported in this work. Also, the polySi thickness needs to be sufficient to allow the growth of phosphosilicate glass as a phosphorus dopant source, which is later on removed. On the other hand the thinner and the less doped the polySi layer, the lower parasitic absorption losses.

For the textured samples the hydrogenation of the interface defects is more critical, i.e., necessary for reaching very low J_o, compared to polished surfaces.

Fig. 5 presents a roadmap towards achieving 22%, which we have estimated assuming solely using industrial process equipment and materials. Firstly, the parasitic absorption in polySi can be reduced by making the layer thinner and less doped. Secondly, total J_o can be further reduced to 50 fA/cm^2 by implementing a higher R_{sheet} emitter and less penetrating fire-through pastes. Last but not least, the *FF* may be further improved by optimizing the thin interfacial oxide, and optimizing the front contacting grid design.

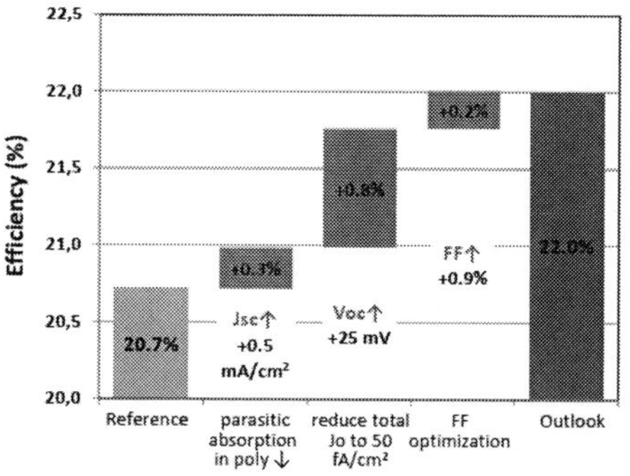

Figure 5: Roadmap towards 22% PERPoly cell.

4 CONCLUSIONS

We presented studies of n-type polySi passivating contacts and their application on the back side of a high-performance bifacial n-type solar cell with fire-through screen-printed metallization, processed on 6" Cz wafers. The cell design is tentatively called PERPoly (Passivated Emitter Rear Polysilicon contact). The polySi/SiO_x passivating carrier selective contact structures were produced with a LPCVD-based process. The cell manufacturing further comprised, in addition to the LPCVD step, only processing on a small number of industrial tools, comparable to current n-PERT process flows on the market. A best efficiency of 20.7% was achieved together with very high average cell V_{oc} of 674 mV. As an added benefit, the cells are bifacial with bifaciality factor >0.8. To our best knowledge these are the first published results on 6" cells employing LPCVD for the polySi, and also the first published results of cells employing a polySi passivating contact with fire-through screen-printed metallization.

The polySi/SiOx passivating contact layers were investigated in detail by varying interfacial thin oxide growth method, polySi thickness and doping profile. Excellent passivation has been obtained both on polished and textured surfaces with recombination current densities J_o of ~2 fA/cm^2 and ~4 fA/cm^2, respectively (after hydrogenation from SiNx). The effect of the fire-through grid on the J_o of the 200 nm thickness n-polySi contact was evaluated to be in the range of 10-30 fA/cm^2. These results show the high potential of this technology to augment current cell processes, with large performance headroom for the future. Reaching 22% seems feasible by a number of improvements, especially on the emitter side. This brings the use of polySi passivating contacts closer to becoming a reality in low-cost industrial solar cell processing.

4.2 Acknowledgements

Part of this work was performed in the projects PV4facades (TEMW140009) and NexPas (TEZ0214002), which receive funding from the Topsector Energie of the Dutch Ministry of Economic Affairs. The authors acknowledge Teun Burgers for the ray tracing analysis of FCA in the polySi.

4.3 References

[1] P. Ashburn and B. Soerowirdjo, IEEE Transactions on Electron Devices, Vol. ED-31, no. 7, pp. 853-860, July 1984.

[2] E. Yablonovitch, R. M. Swanson and Y. H. Kwark, Proceedings of the 17th IEEE Photovoltaic/Spec. Conf., pp.1146 -1148, 1984.

[3] J.Y. Gan and R.M. Swanson, 21st IEEE Photovoltaic Specialists Conference, Vol. 1, pp. 245-250, 1990.

[4] H. Steinkemper, F. Feldmann, M. Bivour and M. Hermle, "Numerical Simulation of Carrier-Selective Electron Contacts Featuring Tunnel Oxides", IEEE Journal of Photovoltaics, Vol. 5, NO.5, pp.1348-1356, 2015.

[5] R. Varache et al., "Investigation of selective junctions using a newly developed tunnel current model for solar cell applications", Solar Energy Materials & Solar Cells 141 (2015) 14–23.

[6] M. Green Silicon Solar Cells. Advanced Principles and Practice. University of New South Wales, 1995. ISBN 0 7334 0994 6. Section 9.4. Contact Recombination.

[8] W. Nemeth et al., "Implementation of Tunneling Passivated Contacts into Industrially Relevant n-Cz Si Solar Cells", proceedings of the 42nd IEEE Photovoltaic Specialist Conference (PVSC), 2015, IEEE, to be published

[9] S. W. Glunz, F. Feldmann, A. Richter, M. Bivour, C. Reichel, H. Steinkemper, J. Benick, M. Hermle, 31st European Photovoltaic Solar Energy Conference and Exhibition, September 2015, Hamburg.

[10] Y. Tao, V. Upadhyaya, C.-W. Chen, A. Payne, E. L. Chang, A. Upadhyaya, A. Rohatgi, Prog. Photovolt: Res. Appl. (2016) published online, doi: 10.1002/pip.2739.

[11] M. Lenes, M. K. Stodolny, Y. Wu, L. J. Geerligs, J. R.M. Luchies, PV Asia, 29 Oct., 2015, Singapore

[12] Yan, D. et al. 2015, "Phosphorus diffused polysilicon contacts for solar cells", Sol. Energy Mater. Sol. Cells, Vol. 142, pp. 72–82.

[13] Gaby J.M. Janssen, Bas B. Van Aken, Anna J. Carr, Agnes A. Mewe, Outdoor performance of bifacial modules by measurements and modelling, Energy Procedia 77 (2015) 364–373.

UV AND VISIBLE RAMAN SPECTROSCOPY CHARACTERIZATION OF WAFER BASED N+/P JUNCTIONS AFTER UNIQUE DIFFUSION PROCESS

C. Lien*, C. F. Hsieh, E. Y. Wang, H. S. Wu and T. C. Wu
Photovoltaic Metrology laboratory
Center for Measurement Standards, Industrial Technology Research Institute, Hsinchu, Taiwan 310, R.O.C.

ABSTRCT: UV and visible multi-wavelength Raman spectroscopy was used to monitor the Si wafer-based p-n junction properties in photovoltaic (PV) devices. The crystallization quality and dopant profiles could be extracted by comparing characteristics of Raman spectrum such as intensity and peak shift with secondary ion mass spectroscopy (SIMS). For as-implanted wafers, broad amorphous Si peak around 475 cm^{-1} was observed and the shape changed with the degree of ion implantation damage. After rapid thermal anneal (RTA), pure Si peak at 520 cm^{-1} with small offsets for different implantation dose was observed, which means Raman spectroscopy could be a very promising and non-destructive measurement tool for monitoring quality of wafer-based ion-implanted silicon.
Keywords: Multi-wavelength Raman spectroscopy; Secondary ion mass spectroscopy (SIMS); Crystallization; Annealing; Dopant concentration

1 INTRODUCTION

P-n junction is the basic structure of commercial PV devices. The operation principle of PV device is the separation of electron-hole pairs stimulated by photon energy. The excitons are drifted from one side to another driven by built-in potential formed in p-n junction [1]. Typically, the p-n junction of commercial solar cell is formed by n/p Si substrate doped with boron/phosphorus ions. The quality and depth of p-n junction possess a significant impact on the performance and reliability of PV devices [2]. As a result, the technique of detecting crystallinity and dopant depth profile in p-n junctions should be applied on product quality control.

The conventional method to obtain dopant depth profile is SIMS. Unfortunately, this method is destructive, time consuming, and vacuum environment is required during the test. Thanks to the rapid progress of laser technique, Raman spectroscopy has been proposed as a promising candidate for in-line monitoring of junction properties after diffusion or annealing process [3,4]. The properties like dopant concentration, junction depth, lattice defects and crystallinity would expect to be obtained through Raman spectroscopy [5,6,7,8,9,10]. In this work, UV and visible Raman spectroscopy was introduced to investigate the Si wafer-based junction properties before and after annealing process. This technique will be applied to solar industry or semiconductor foundries in the near future.

2 EXPERIMENTAL

Phosphorus ions (P31+) with the dose of 5.0×10^{13}, 1.0×10^{14}, 5.0×10^{14}, 1.0×10^{15}, and 5.0×10^{15} cm^{-2} were implanted into p-Si (100) wafers. The initial junction depth of as-implanted Si wafer is about 20 nm, it will extend after the annealing process. No preamorphization implantation (PAI) was done. Part of wafers were then annealed by RTA under N_2 ambient with 950 °C annealing temperature and 30 s holding time.

The wafers were characterized by SIMS and multi-wavelength Raman spectroscopy before and after RTA, respectively. The apparatus configuration is shown in Figure. 1. Diode-pumped solid-state lasers were used for Raman spectroscopy with the beam lines of 405, 532, and 671 nm. The beam divergence and diameter at the aperture are 1.2 mrad and 2.0 mm respectively. The components of Raman spectroscopy include auto-z focusing module, wide field imaging system, HORIBA Syncerity CCD deep cooled cameras and HORIBA iHR550 spectrometer.

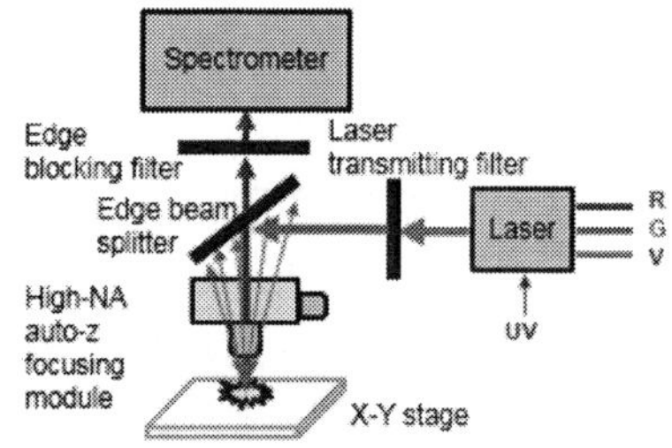

Figure 1: Schematic configuration of multi-wavelength Raman spectroscopy.

Since the specific requirement of optical path and setup, UV Raman measurement was outsourcing. For visible Raman spectroscopy, all spectrum could be measured at once in the system by switching edge filter for each particular beam.

3 RESULTS AND DISCUSSION

The UV and visible Raman spectroscopy of as-implanted samples of dose 5.0×10^{15} cm^{-2} are shown in Figure. 2(a). The Raman probing depth of the 325 nm UV line and 405, 532, 671 nm visible lines are approximately about 5, 100, 1000 and 3800 nm respectively (Figure. 2(b)). Raman peak broadening and shape change depend on degree and depth of ion implantation damage to the Si lattice.

The Raman spectrum of 325 and 405 nm beam lines show the obviously broad amorphous Si peak at 475 cm^{-1} due to severe surface lattice damage after ion-implantation. For UV line beam, the spectrum shows a relatively low signal/noise ratio due to the absorption of short wavelength light and the quality of laser system. On the other hand, since the probe depth of 532 and 671 nm beam lines are much deeper than junction depth, only pure Si signals at 520 cm^{-1} from deep bulk Si are observed and dominated. The c-Si Raman peak at 520 cm^{-1} corresponds to the triply degenerated, transverse optical (TO) phonon modes at k=0. The broad Raman peak of a-Si is a result of contributions from all TO phonon modes.

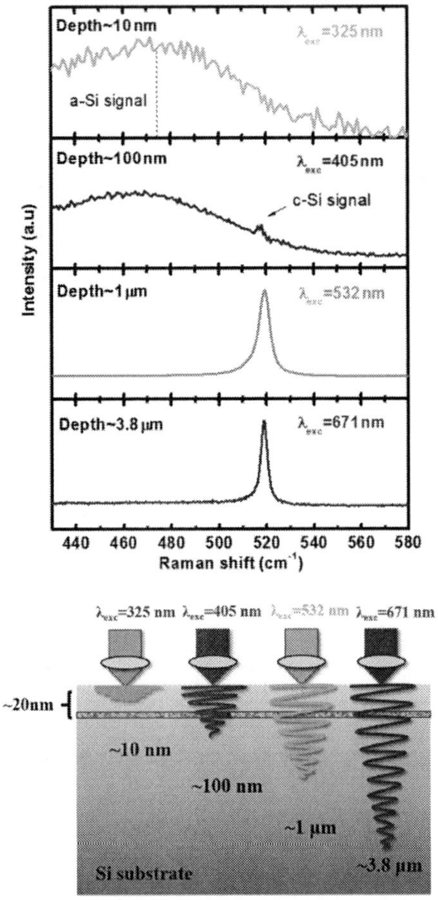

Figure 2: (a) Raman spectroscopy under different excitation laser. (b) The corresponding probe depth with different wavelength.

The dopant depth profiles analyzed by SIMS for annealing wafers with various implantation dose were shown in Figure 3. It could be observed that quite different depth profile could be distinguished for different dose level. Figure 4 is the Raman spectrum measured by 405 nm excitation laser for the annealing wafers. First, the a-Si peak made by lattice damage disappears after annealing process, which means the phase transition for the silicon crystal from amorphous phase to polycrystalline phase happened during the annealing process. Second, all wafers show nearly pure Si peak at 520 cm^{-1} associating with the small peak shift toward the higher wavenumber side. This phenomenon could be interpreted as the effect of recrystallization and the occupancy of phosphorus atoms in substitutional sites. In addition to express the degree of Raman shift, we can define a new parameter called Raman peak shift ratio Δ_{Raman} (%) as (1):

$$\Delta_{Raman}(\%) = \frac{\Delta_{Raman,f} - \Delta_{Raman,i}}{\Delta_{Raman,i}} \quad (1)$$

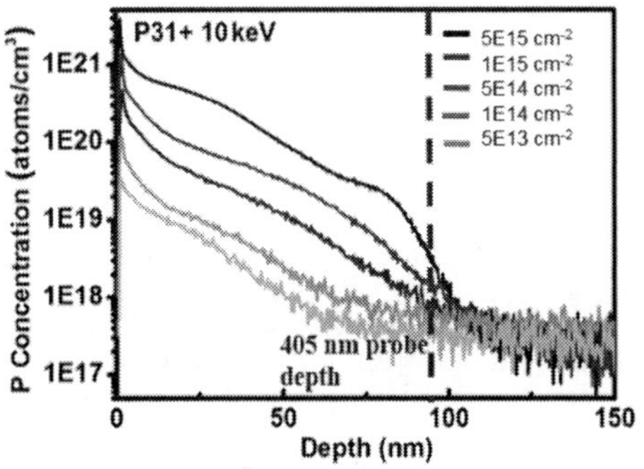

Figure 3: SIMS analysis for different dose level silicon wafer after annealing process

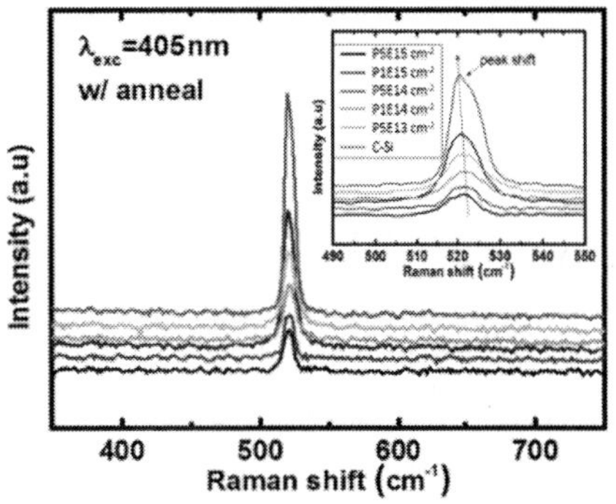

Figure 4: Raman spectroscopy of P31+ implanted Si wafer with various dose under 405 nm excitation laser.

The roughly proportional of Raman peak shift ratio and dopant concentration was observed (Figure.5). It is believed that the correlation is not general for each kind of dopant and process. Therefore, the model could be established according to the request of user for specific wafer-based condition.

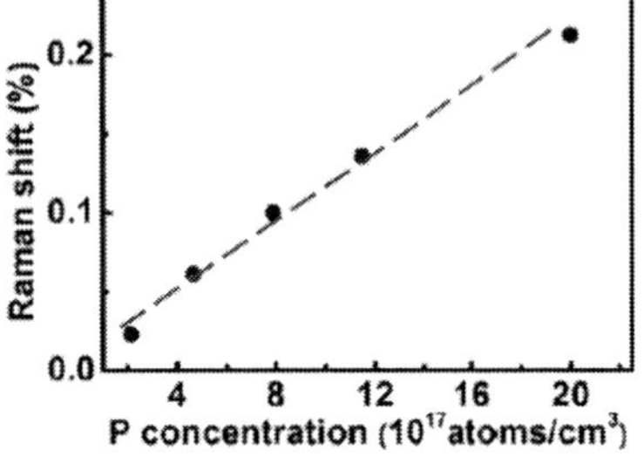

Figure 5: Correlation between Raman shift ratio and dopant concentration.

4 CONCLUSIONS

In this work, UV and visible multi-wavelength Raman spectroscopy are used to detect the crystallization quality and dopant concentration of ion-implanted Si wafer. Based on the information given by SIMS, the correlation between the Raman peak shift ratio and dopant concentration can be obtained for the specific dopant kind and annealing process. It is believe that the correlation modeling is not general for any dopant or annealing process, so further more verifications will be performed in the following experiments.

Raman spectroscopy is a very promising tool for non-contact, non-destructive and in-line monitoring of wafer-based product quality control. Then this technique could be used to analyze the performance and reliability in solar industry and semiconductor foundries.

5 ACKNOWLEDGEMENT

The authors would like to thank the division of National Applied Research Laboratories (NARLabs) - National Nano Device Laboratories (NDL) for supporting the ion-implantation apparatus and SIMS analysis.

6 REFERENCES

[1] Chenming Calvin Hu, "Modern Semiconductor Devices for Integrated Circuits", Prentice Hall, (2010)
[2] W. Wu; P. R. Li; L. Zhang; J. Li; Y. F. Wang; Z. Q. Ma, 2009 16th IEEE International Symposium on the Physical and Failure Analysis of Integrated Circuits, pp. 616-618 (2009)
[3] D. Kirillov, R. A. Powell, and D. T. Hodul, J. Appl. Phys. 58, 2174 (1985)
[4] Noboru Nakanoa, Louis Mat-Ale, Rafael Reif, J. Appl. Phys. 72, 3641 (1992)
[5] Masashi Fukumoto, Hiroaki Minami, Noriyuki Hasuike, Hiroshi Harima, Masahiro, Yoshimoto and Woo Sik Yoo, RTP '06. 14th IEEE International Conference on Advanced Thermal Processing of Semiconductors (2006)
[6] Masahiro Yoshimoto, Hiroshi Nishigaki, Hiroshi Harima, Toshiyuki Isshiki, Kitaek Kang and Woo Sik Yoo, J. Electrochem. Soc., volume 153, issue 7, G697-G702 (2006)
[7] Woo Sik Yoo, Takeshi Ueda, Toshikazu Ishigaki, Kitaek Kang, Masashi Fukumoto, Noriyuki Hasuike, Hiroshi Harima, and Masahiro Yoshimoto, J. Electrochem. Soc., volume 158, issue 1, H80-H84 (2011)
[8] Woo Sik Yoo, Takeshi Ueda, Toshikazu Ishigaki, Kitaek Kang, Kyoung Bong Rouh, Yong Seok Eun, Choon Hwan Kim and Hyo Sang Kang, AIP Conference Proceedings ; 1496(1):156-159 (2012)
[9] Woo Sik Yoo, Toshikazu Ishigaki, Takeshi Ueda, Kitaek Kang, Noriyuki Hasuike, Hiroshi Harima and Masahiro Yoshimoto, Ext. Abs. the 13th International Workshop on Junction Technology (2013)
[10] Woo Sik Yoo, Kitaek Kang, Takeshi Ueda, Toshikazu Ishigaki, 20th International Conference on Ion Implantation Technology (IIT) (2014)

ULTRAFAST LIFETIME REGENERATION IN AN INDUSTRIAL BELT-LINE FURNACE APPLYING INTENSE ILLUMINATION AT ELEVATED TEMPERATURE

Dominic C. Walter[1], Thomas Pernau[2] and Jan Schmidt[1,3]

[1]Institute for Solar Energy Research Hamelin (ISFH)
Am Ohrberg 1, D-31860 Emmerthal, Germany
[2]centrotherm photovoltaics
Johannes-Schmid-Straße 8, D-89143 Blaubeuren, Germany
[3]Department of Solar Energy, Institute of Solid-State Physics, Leibniz Universität Hannover
Appelstraße 2, D-30607 Hannover, Germany

ABSTRACT: Solar cells made of boron-doped, oxygen-rich Czochralski-grown silicon (Cz-Si) wafers suffer from a degradation in conversion efficiency upon illumination. This effect, sometimes labeled "light-induced degradation" (LID), is caused by a defect center within the silicon bulk, often referred to as BO defect, which activates its recombination properties under illumination leading to a degradation of the carrier lifetime. However, the BO defect center can be permanently deactivated, i.e. the lifetime can be permanently recovered, under illumination at *elevated* temperature. Within this contribution, we focus on transferring our lab-type BO deactivation process, using a hotplate and a halogen lamp, into an ultrafast high-throughput process, applying an industrial-type belt-furnace especially designed for this very purpose by centrotherm photovoltaics AG. Using this industrial furnace, inline processing in a solar cell production line is feasible. Our investigation clearly shows that a permanent recovery of the lifetime with an industrial belt-furnace can be achieved. In addition, we prove this result also for a shorter lab-type version of the longer industrial belt-furnace. The lifetimes reached after processing can be comparable to the lifetimes obtained on the same Cz-Si materials after applying our lab-type approach. Excellent stability of the lifetime is verified for up to 1000 hours of illumination at room temperature.
Keywords: BO defect center, lifetime, regeneration, permanent deactivation

1 INTRODUCTION

In boron-doped, oxygen-rich, Czochralski-grown silicon (Cz-Si), illumination – or more general , the injection of excess carriers – leads to a degradation in the charge carrier lifetime [1], a phemomenon sometimes labeled "light-induced degradation" (LID). This lifetime degradation can strongly affect the conversion efficiency of silicon solar cells made on this material and may lead to a severe degradation in efficiency of up to 10 % relative [2]. A multiplicity of studies have been carried out to identify the defect being responsible for the lifetime degradation [3, 4] and it has been concluded that the simultaneous presence of oxygen and boron in the silicon material is directly related to the magnitude of degradation. Hence, the corresponding defect center is often referred to as 'BO defect'.

Since almost a decade, it has been known that the BO defect can be permanently deactivated under illumination at elevated temperatures [5]. Naturally, this deactivation is accompanied by a permanent recovery of the carrier lifetime. After this recovery process the lifetime shows no or only a very weak degradation upon illumination at room temperature and also at elevated temperatures [6]. Hence, it can be expected that also solar cells show no lifetime-related degradation of the conversion efficiency after the process of permanent recovery has been applied.

Within this contribution, we transfer our lab-type deactivation process into an ultrafast high-throughput process, applying an industrial-type belt-furnace [7] especially designed for this very purpose by centrotherm photovoltaics AG. Our lab-type deactivation process, using a hotplate and a halogen lamp, is adopted within the belt-furnace, which enables inline processing in a solar cell production line. We perform lifetime investigations which clearly show that a permanent recovery of the lifetime within the industrial belt-furnace can be achieved. Additionally, this result can also be obtained for a shorter lab-type version of the longer industrial belt-furnace.

2 EXPERIMENTAL DETAILS

In this study, four different boron-doped Cz-Si materials are examined with resistivities between 1 and 5 Ωcm. Sample processing includes the removal of the surface damage with an aqueous solution of KOH. Afterwards, the surfaces are cleaned using a standard RCA cleaning sequence which is followed by a phosphorus diffusion (850°C, 1:00 h), resulting in an n^+-region on both surfaces. Before a fast-firing process is applied, the phosphosilicate glass (PSG) as well as the n^+-region are removed using hydrofluoric acid (HF) and KOH. Subsequently, the surfaces are passivated with a stack of aluminum oxide Al_2O_3 (10 nm) and silicon nitride SiN_x (70 nm, n=2.05), deposited via plasma-assisted atomic layer deposition (plasma-ALD) and plasma-enhanced chemical vapor deposition (PECVD), respectively.

The fast-firing process is performed in a commercially available belt-line firing furnace from centrotherm (DO-FF-8.600–300) at 850 °C set-peak temperature and 7.2 m/min belt speed. Please note that the actual temperature of the wafer, which has been measured with a temperature probe attached to lifetime samples of the same material processed in parallel, is with an average peak temperature of 760°C significantly lower. After the sample processing, all samples were illuminated at room temperature with a light intensity of 0.1 suns using a halogen lamp for 115 hours. Afterwards, the lifetime is measured before the permanent deactivation of the BO defect center is performed.

Lifetime measurements are performed using the WCT-120 from Sinton Instruments. If not stated otherwise, all reported lifetimes are extracted at a fixed injection density of Δn=0.1×p_0, with Δn being the excess carrier density and p_0 the hole concentration measured in darkness. In our case p_0 equals the doping concertation N_{dop} as only non-compensated boron-doped silicon is investigated.

3 RESULTS AND DISCUSSION

In the following, the results obtained on the lab-type furnace as well as the results obtained on the industrial-type belt-furnace will be discussed separately.

3.1 Lab-type regenerator

The short lab-type belt-furnace, designed for R&D, exposes the samples to high-intensity illumination at elevated temperature for approximately 15 s. Within this short process window, the lifetime samples reach a maximum temperature of ~220°C, which has been determined with a temperature probe attached to an identical sample processed in parallel.

We have measured the effective lifetime after permanent BO deactivation within the lab-type belt-furnace and checked the stability of the lifetime afterwards. In Fig. 1, the lifetime results for the group of samples permanently deactivated in the short belt-furnace are shown (orange squares). Compared to the lifetime after complete formation of the BO defect center τ_d=(144±12) µs, the lifetime increases by one order of magnitude to τ_{0p}=(1380±370) µs during the belt-furnace treatment. Only a minor degradation of this lifetime value under further illumination at room temperature can be observed. After 1000 h of illumination at a light intensity of 0.1 suns, a fully stabilized lifetime level of τ_{1000h}=(1050±170) µs is reached. As can be seen in Fig. 1, the characteristic lifetime values τ_d, τ_{0p} and τ_{1000h} are virtually identical to the lifetime values observed on a sample permanently deactivated in our lab-type regeneration process, using a halogen lamp and a hotplate (green triangles in Fig. 1). Hence, this experiment clearly demonstrates that the permanent deactivation of the BO

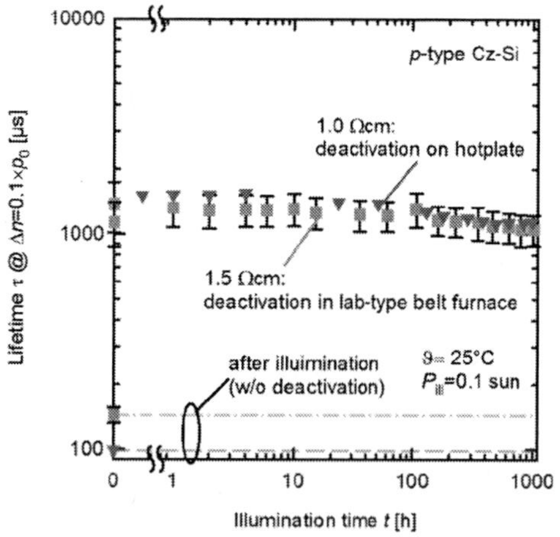

Figure 1: Lifetime τ versus illumination time *t* at room temperature (~25°C) and ~0.1 mW/cm² light intensity. Orange squares correspond to the lifetime samples regenerated in the lab-type belt furnace (average over 5 samples). Green triangles correspond to a sample regenerated on a hotplate.

center can in fact be performed within seconds in an industrially feasible way. Further experiments performed on a large number of lifetime samples using the longer industrial-type belt-line furnace emphasize the validity of this statement, as it is shown in the next paragraph.

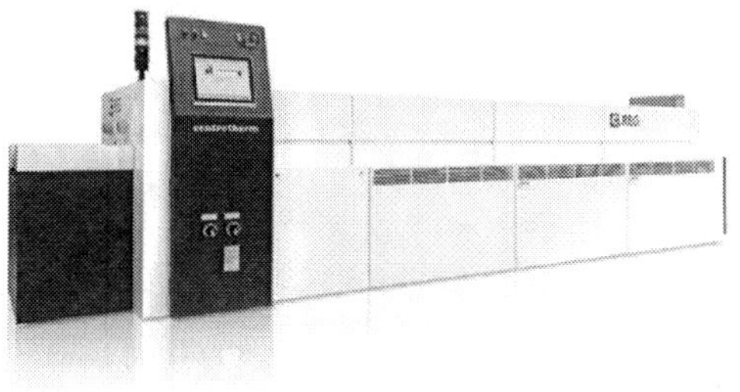

Figure 2: The inline regeneration tool C.REG 6.200 from centrotherm used within this investigation is suitable for solar cell production lines.

3.2 Industrial-type regenerator

After sample processing and subsequent illumination, the lifetimes of the various samples were measured. Afterwards, the permanent deactivation was performed in the industrial-type inline belt-furnace. This industrial-type belt-furnace is commercially available (centrotherm C.REG 6.200) [7] (see Fig. 2).

Within our study we varied the belt speed, the set temperature as well as the illumination intensity in the C.REG 6.200. The tested conditions are listed in Table I. Please note that the temperature value gives only a set value and that the actually measured temperature with a temperature probe attached to samples processed in parallel reaches for the 270°C set temperature a maximum temperatures of only 237°C and for the 300°C set temperature only 256°C. Solar cells processed with these conditions reach higher maximum temperatures because of sub-bandgap light to heat conversion. At 270°C set temperature, the cell temperature typically reaches 260°C. At 300°C set temperature, up to 285°C cell temperature can be reached.

Table I: Four different test conditions investigated in the current study.

	Illumination intensity [%] (zone 1 – 4)	**Belt speed [m/min]**	**Temperature [°C]**
P1	100, 100, 100, 100	9	270
P2	90, 90, 90, 90	8	270
P3	100, 100, 100, 30	9	300
P4	90, 90, 90, 30	8	300

The industrial-type belt furnace for the inline-application exposes the samples for approximately 32 s to the deactivation conditions (see Table I). The lifetimes measured on four different materials are shown in Fig. 3.

As can be seen in Fig. 3, before the process in the belt-furnace is applied, the lifetime after illumination depends, as expected, on the base resistivity, more precisely the doping concentration of the Cz-Si materials. After the regeneration process, the lifetime measured on the different materials is independent of the applied process and also independent of the doping concentration at a lifetime of τ_{0p}≈450 µs. Note that this finding is in contrast to our previous results [8], where we have observed an

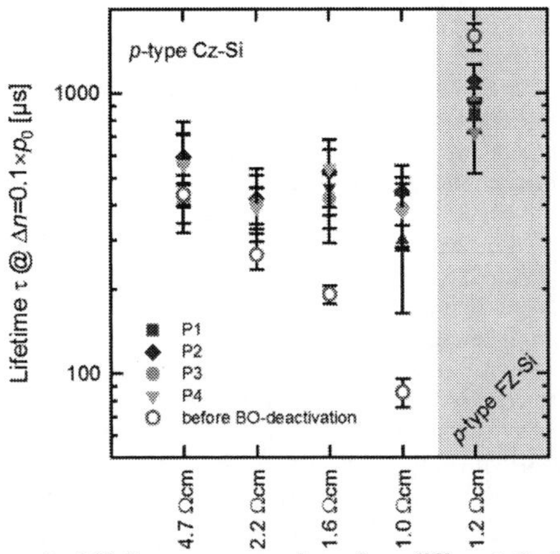

Figure 3: Lifetime τ measured on four different Cz-Si materials of this investigation. Before BO deactivation (open green circles) a clear dependence on the base resistivity can be observed. After BO deactivation (filled colored symbols) the measured lifetime is independent of the base material and independent of the applied process. On the very right, the lifetimes measured on FZ-Si references, processed in parallel, are also displayed.

increasing lifetime with decreasing doping concentration on a much higher lifetime level.

On float-zone silicon (FZ-Si) reference wafers, processed in parallel with a base resistivity of 1.2 Ωcm, we observe in average a smaller lifetime τ_{0p}=(920±160) µs after the belt-furnace regeneration process compared to the value τ=(1600±180) µs measured before. This decreasing lifetime indicates that the surface passivation quality suffered under the regeneration conditions (i.e. high illumination intensity at increased temperature) applied in the belt-furnace.

According to Sproul [9] the surface recombination velocity S can be calculated for sufficiently low surface recombination velocities via

$$S = \frac{W}{2} \times \left(\frac{1}{\tau} - \frac{1}{\tau_{int}}\right) \quad (1)$$

with W being the sample thickness, τ the measured effective lifetime and τ_{int} the intrinsic lifetime. Inserting the corresponding values for our FZ-Si reference samples, we obtain an average surface recombination velocity of S≈17 cm/s after the regeneration process. With this value, the observed independence of the lifetime on the doping concentration can be consistently explained.

Figure 4 shows the lifetime measured after process P2 plotted versus the doping concentration of the base material. For comparison the intrinsic lifetime limit according to Richter et al. [10] is also displayed. Using these values, as well as the thickness of the Cz-Si samples under investigation and the determined surface recombination velocity, an "expected" effective lifetime, assuming that the measured lifetime is dominated by the surface recombination, can be calculated (orange circles in Fig. 4). As can be seen, the calculated lifetime is not depending on the doping concentration p_0. As the actually measured lifetime is on the same level as the calculated lifetime, it is highly likely that the lifetime after the belt-furnace regeneration process is mainly limited by the surface recombination. Hence, it can be concluded that the

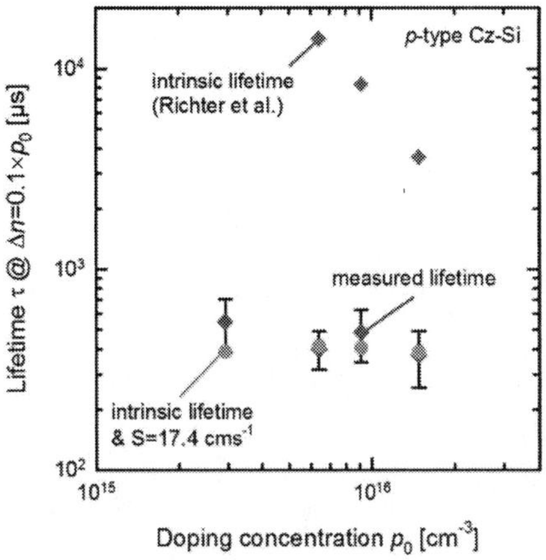

Figure 4: Measured lifetime τ of process P2 (blue diamonds) plotted versus the doping concentration p_0. Green diamonds give the intrinsic lifetime limit according to Richter et al. [10], while the orange circles show the calculated lifetime using Eqn. 1, with the intrinsic lifetime limit, the correct thickness and the S-value of 17.4 cm/s. The calculated lifetime is on the same level as the measured lifetime indicating that measured lifetime is limited by surface recombination.

lifetime within the silicon bulk is much higher than the measured effective lifetime. This suggests that the permanent deactivation of the BO defect center worked as well within the long inline belt-furnace as in the shorter belt-furnace.

One might speculate about the reasons for the reduced surface passivation quality after the belt-furnace process. A possible reason for the damage of the surface passivation quality is the intense illumination for the longer time span compared to the short regeneration furnace, where we did not observe such an increase in S. However, photoluminescence images of the samples under investigation show features which might originate form scratches introduced to the surface by either sample handling before the regeneration process or the metal belt within the furnace. If the latter is the case, the reduction of the surface passivation quality is not expected to be limiting the conversion efficiencies of solar cells treated within in the furnace if the solar cell if fully metalized on the rear side.

3.3 Lifetime stability after regeneration in the industrial-type regenerator

Although the surface recombination velocity S seems to be reduced after processing within the inline industrial-type regenerator, lifetime investigations show that the passivation quality is stable upon further illumination at room temperature (see Fig. 5(A)). Hence, we investigated the stability of the lifetime after the regeneration process by exposing the lifetime samples to a light intensity of 0.1 suns at room temperature. In Figure 5 the change of the measured lifetime is plotted versus the illumination time after regeneration for five different Cz-Si materials which have been regenerated in the inline-furnace using process P2. Only one sample (with the highest measured lifetime) from each group, consisting of five samples, is shown. However, please note that all samples from the same material show the same trend for the lifetime after regeneration.

As can be seen in Fig. 5(A), some materials, in this case with a base resistivity of 1.6 and 4.8 Ωcm, show a very stable lifetime upon illumination at room temperature. However, the lifetime of some other materials, in this case the 1.0 Ωcm sample, degrade while the lifetime of the 2.2 Ωcm material increases (see Fig. 5(B)).

We attribute these essentially different lifetime behaviors to different background defects which are expected to be present within the different Czochralski-grown materials and limit the lifetime to values usually lower than the lifetime of defect-lean float-zone silicon. Although the detailed nature of these defects is so far unknown, it is very likely that they are not related to the BO defect which causes the typical light-induced degradation in boron-doped Cz-Si, as for example no clear trend depending on the boron (or hole) concentration can be observed. Hence, further experiments are needed which study the lifetime limiting defects once the limitation due to the BO defect has been overcome.

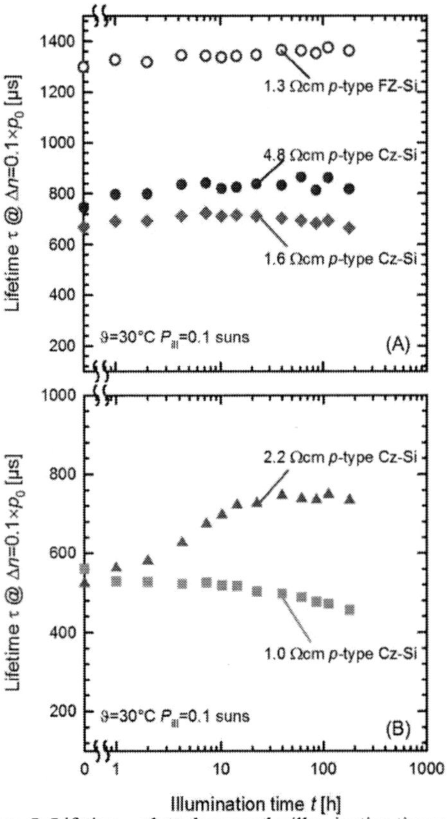

Figure 5: Lifetime τ plotted versus the illumination time t at room temperature and 0.1 suns light intensity for five Cz-Si materials treated with process P2. Above (A) two examples of Cz-Si materials featuring a stable lifetime after regeneration are shown, below (B) two examples of Cz-Si materials with changing lifetimes are shown. The FZ reference sample (A) features a stable lifetime.

4 SUMMARY

Within this contribution, our lab-type deactivation process of the boron-oxygen-related defect center, using a hotplate and a halogen lamp, has been transferred to industrial belt-line furnaces which use intense illumination at elevated temperatures.

In the first experiment, lifetime samples regenerated in a short lab-type regenerator, especially designed for research applications, which exposes the samples to high intense illumination for 15 s, have been examined. Immediately after regeneration, high lifetimes of $\tau(\Delta n=0.1\times p_0)\approx1.4$ ms on 1.0 Ωcm material have been measured which only showed a minor degradation upon further illumination at room temperature. After 1000 h of illumination an average lifetime of $\tau_{1000\,h}(\Delta n=0.1\times p_0)=(1050\pm170)$ µs was measured. This experiment clearly proves that the deactivation of the BO defect can indeed be performed in an industrially feasible way, with only 15 s process time.

In a second experiment, a longer inline version of the regenerator, which exposes the samples for approximately 32 s to the regeneration conditions, has been examined. The regeneration conditions have been varied in four different regeneration processes with regard to the belt speed, the temperature and the illumination intensity. On a large number of Cz-Si materials with different base resistivities between 1 and 5 Ωcm, a constant average lifetime of $\tau(\Delta n=0.1\times p_0)\approx450$ µs independent of the regeneration process and the base material was obtained directly after the regeneration. This constant lifetime was consistently explained by an increased surface recombination velocity, which was observed on float-zone reference samples processed in parallel. This finding indicates that the bulk lifetime within the different Cz materials is significantly higher than the measured effective lifetime. Therefore, we can again conclude that the deactivation of the BO defect can be performed within this inline belt-furnace in an industrially feasible way.

Investigations of the lifetime stability at room temperature and 0.1 suns light intensity revealed essentially different behaviors of the lifetime for different materials. While some materials featured a perfectly stable lifetime after regeneration, other materials showed an increasing lifetime, others a decreasing lifetime. We attribute these different behaviors to yet unidentified different background defects which are expected to be present within the standard Cz-Si materials. In order to identify the responsible defects, further experiments are needed which study the lifetime-limiting defects once the limitation due to the BO defect has been overcome.

ACKNOWLEDGEMENT

This work was funded by the German State of Lower Saxony and the German Federal Ministry of Economics and Energy (BMWi) within the research project "Upgrade Si-PV" (contract no. 0325877B). The content is the responsibility of the authors.

5 REFERENCES

[1] H. Fischer, W. Pschunder, Proceedings 10th IEEE Photovoltaic Specialists Conference (1974), p. 404

[2] J. Knobloch, S. W. Glunz, D. Biro, W. Warta, E. Schäffer, W. Wettling, Proceedings 25th IEEE Photovoltaic Specialists Conference (1996), p. 405

[3] J. Schmidt, A. G. Aberle, R. Hezel, Proceedings 26th IEEE Photovoltaic Specialists Conference (1997), p. 13
[4] S. W. Glunz, S. Rein, W. Warta, J. Knobloch, W. Wettling, Proceedings 2nd World Conference – Photovoltaic Solar Energy Conference (1998), p. 1343
[5] A. Herguth, G. Schuber, M. Kaes, and G. Hahn, Proceedings 21st European Photovoltaic Solar Energy Conference (2006), p. 530
[6] D. Walter, B. Lim, V. Voronkov, R. Falster, and J. Schmidt, Proceedings 29th European Photovoltaic Solar Energy Conference (2014), p. 555
[7] T. Pernau, O. Romer, W. Scheiffele, A. Reichart, and W. Jooß, Proceedings 31st European Photovoltaic Solar Energy Conference (2015), p. 918
[8] D. C. Walter, B. Lim and J. Schmidt, Progress in Photovoltaics: Research and Applications (2016), DOI: 10.1002/pip.2731
[9] A. B. Sproul, Journal of Applied Physics **76**(5), 2851-2854 (1994)
[10] A. Richter, S. W. Glunz, F. Werner, J. Schmidt and A. Cuevas, Physical Review Letters B **86**(16), 165202 (2012)

APPLICATION OF HIGH EFFICIENCY EMITTERS TO MULTI-CRYSTALLINE SILICON

Moonyong Kim, Hongzhao Li, David Payne, Stuart Wenham, Brett Hallam and Malcolm Abbott

School of Photovoltaic and Renewable Energy Engineering, University of New South Wales, Sydney, NSW 2052, AUSTRALIA

ABSTRACT: The impact of the emitter diffusion and firing process on gettering and hydrogenation of multi-crystalline silicon is investigated with the aid of photoluminescence imaging and further image processing. Whilst the use of high efficiency emitters is critical to improve surface passivation, multi-crystalline wafers using high-efficiency emitters suffered from a significantly reduced lifetime in the impurity-affected zone. For such diffusions, a pre-gettering step was required to increase the lifetime in this region. A pre-gettering step was also beneficial for such regions when using a screen print emitter. However the pre-gettering process was detrimental to the lifetime of other regions of the material for both emitter diffusions. Heavily diffused wafers showed a strong response to hydrogen passivation, particularly for non-gettered wafers. However, certain grains and the impurity-affected zone were reduced in quality after firing, suggesting a poisoning of the silicon.
Keywords: multi-crystalline silicon, recombination, hydrogen, gettering, passivation

1 INTRODUCTION

The design of high efficiency emitters is often described as a trade-off between recombination and resistive losses. However on cheaper forms of silicon, it is also important to consider the impact that emitter diffusions can have on the on gettering and hydrogenation. These additional considerations are particularly important, as cheaper forms of silicon tend to have more chemical impurities and crystallographic defects that they rely more on passivating the impurities and defects [1]. Phosphorus dopants in the emitter can cause a gettering effect for impurities [2]. This mechanism segregates some of the impurities from the bulk region to the silicon surface, hence improving the bulk minority carrier lifetime.

Recent improvements in front side silver pastes have allowed solar cells to use more lightly doped emitters as a method for improving efficiencies. By changing the Ag paste, the optimal point for the trade-off between emitter recombination and contact resistance has shifted dramatically [3],[4]. Laser doped selective emitter is also an alternative approach to resolve the contact resistance problem [5].

A major implication for these cheaper substrates is that a change of the emitter diffusion process could alter the effectiveness of gettering in the material. Furthermore, with reduced surface recombination, the bulk recombination can have a larger impact in determining the device performance.

There are number of literature in the areas of hydrogen passivation [6],[7] from firing and phosphorus gettering [8],[9] to improve the performance of the crystalline silicon wafers. Effective hydrogen passivation was well demonstrated especially along grain boundaries on mc-Si [10]. Furthermore, it has been suggested that the presence of phosphorus emitters can improve the efficacy of the hydrogen passivation during firing and dissolution [11], [12] and gettering of impurities can simultaneously occur during firing [13]. However, there has not been much of study about the influence of phosphorus emitter concentration to the efficacy of the hydrogen passivation.

This paper investigates the impact of heavily diffused and high-efficiency emitters, pre-gettering processes and firing on multi-crystalline silicon material. The multi-crystalline silicon wafers were analysed in two different regions: 1) the impurity-affected zone where high concentration of precipitated and interstitial forms of metallic impurities are present and 2) the unaffected region, which is the remaining region of the wafers. Additionally, the paper demonstrates an advanced interpretation of photoluminescence images, in which image processing is used to dramatically highlight the impact of device processing to different regions of silicon wafers.

2 EXPERIMENTAL DETAILS

Symmetrical lifetime test structures and screen-printed solar cells were fabricated on p-type 156 mm × 156 mm multi-crystalline silicon (mc-Si) wafers (source GCL). Half of the samples were gettered using a heavy $POCl_3$ diffusion (approximately 40 Ω/sq) before the surfaces of all substrates were removed using alkaline texturing to keep a similar thickness of the wafers. Samples were then $POCl_3$ diffused with either a heavily diffused emitter (approximately 65 Ω/sq) or lightly diffused, high-efficiency emitter (approximately 100 Ω/sq) recipe that also includes a thermal oxidation (see).

Table I: Recipe for diffusions

Recipe	Heavily diffused	High-efficiency
PSG growth	795 °C/30 mins 425 sccm $POCl_3$ 600 sccm O_2	770 °C/25 mins 400 sccm $POCl_3$ 600 sccm O_2 6.2 slm N_2
Phosphorus diffusion	No	850 °C/25 mins 7.5 L N_2
Drive-in	885 °C/30 mins 40 sccm O_2 7.35 slm N_2	880 °C/40 mins 5 L O_2
PSG removal	2% HF dip/5 mins	
Oxidation	No	890 °C/15 mins

The heavily diffused emitter represents a typical screen-printed solar cell emitter and the lightly diffused emitter represented a selective emitter with a low phosphorus surface concentration. Lightly doped emitters additionally had an oxidation process at 890 °C for 15 minutes to lower the peak concentration of the emitter to more effectively contrast the lightly diffused emitter to heavily diffused emitter. Silicon nitride (SiNx:H) depositions were performed on both sides of all samples to form lifetime structures using PECVD. For investigating the impact of gettering and firing, the non-gettered and gettered wafers' results were compared both before and after firing.

The samples were characterised using Sinton WCT-120 for quasi-steady-state photoconductance (QSSPL) [14] and photoluminescence imaging (PL) [15]. The PL images were deconvoluted using a point spread function [16], [17] and were calibrated to implied Voc maps[18]. Using the contrast of iVoc maps, maps of the change in iVoc, ΔiVoc, were generated to highlight the improvement and reduction in iVoc by gettering and firing impacts on heavily diffused and high-efficiency emitter wafers.

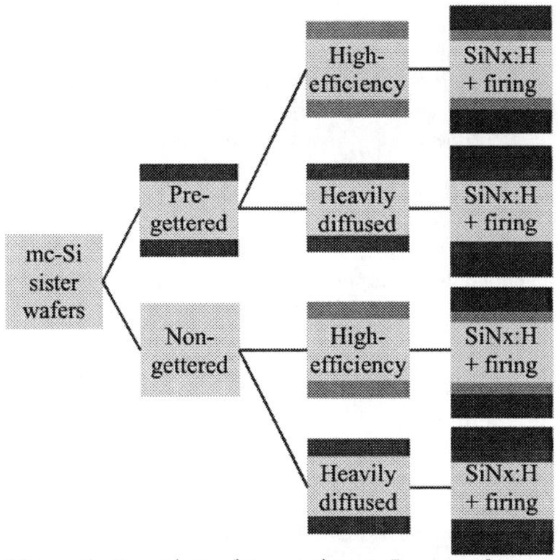

Figure 1: Procedures for experiment. Pre-gettering was done by forming emitter with less than 40 Ω/sq followed by chemical etch in HF for PSG and subsequent alkaline solution to remove the emitter. Heavily diffused emitter has 65 Ω/sq while high-efficiency emitter has 100 Ω/sq with an oxide layer. Firing with SiNx:H was done at 800°C.

3 RESULS AND DISCUSSION

3.1 Response to phosphorus pre-gettering on different emitters

The effectiveness of the gettering by the emitter was determined by giving an additional phosphorus gettering, so called pre-gettering process on the wafers before the emitter formation. The iVoc maps are shown in Figure 2. The right side of the edge, or impurity-affected zone on samples with high-efficiency emitters experienced a substantial improvement by incorporating a pre-gettering process. Using ΔiVoc map as shown in Figure 3, the localised improvements and detrimental effects with the incorporation of a pre-gettering process were analysed separately. The improvements shown in the ΔiVoc maps likely indicates the regions where gettering was not effective during emitter formation. The reduction in ΔiVoc on the pre-gettered samples indicates that the pre-gettering process can also be detrimental for the majority of the wafer within the grains.

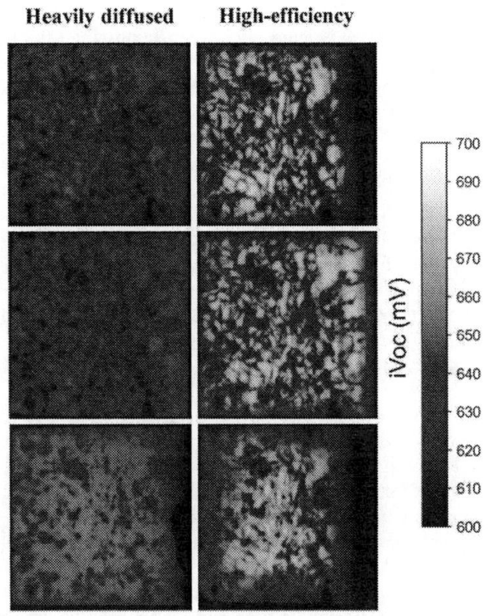

Figure 2: iVoc map of multi-crystalline silicon wafers with SiNx depositions without pre-gettering or firing **(top)**, with pre-gettering but no firing **(middle)**, and without pre-gettering but firing **(bottom)**. Note the dark spot on the right side of fired sample is due to broken region of the wafer.

The high-efficiency emitter experienced a substantial improvement in iVoc of up to 60 mV in the impurity-affected zone. In contrast, samples with heavily diffused emitters experienced much smaller increases in iVoc of approximately 10 mV. This may indicate that the standard screen print diffusion is much more effective at gettering than a high efficiency emitter.

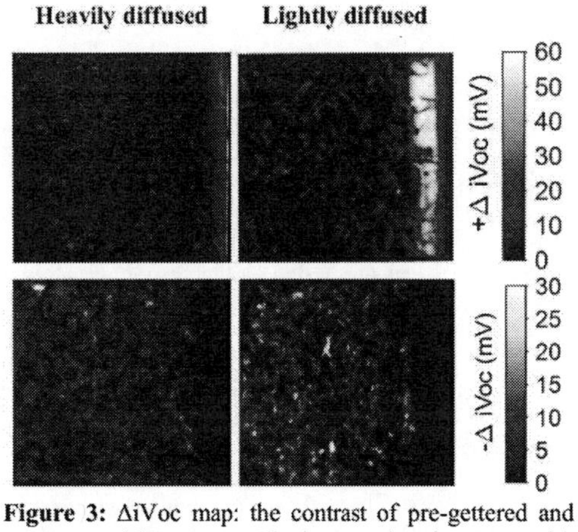

Figure 3: ΔiVoc map: the contrast of pre-gettered and non-gettered iVoc maps. The improved region **(top)** with pre-gettering shows the region that was not effectively gettered by its emitter. The harmed region **(bottom)** shows the detrimental effects by pre-gettering process.

The bulk lifetime in this region is shown in Figure 4. A substantial improvement of bulk lifetime was observed in the impurity-affected zone from approximately 50 µs to about 170 µs on samples with a high-efficiency emitter. In contrast, a much smaller increase was observed on samples with a heavily diffused emitter (140 µs to 160 µs). This gives further evidence that the heavily diffused emitter is more effective at improving the bulk lifetime through gettering than a high efficiency emitter.

To investigate the cause of the improvements in bulk lifetime, the concentration of interstitial iron [Fe_i] was measured for samples with both emitters, with and without pre-gettering processes (see Figure 4). The [Fe_i] was significantly reduced in impurity-affected zone of both types of emitters with the incorporation of a pre-gettering process, from $2x10^{10}$ /cm^3 to below detection limit and from $1x10^{11}$ /cm^3 to $2x10^{10}$ /cm^3 on samples with heavily diffused emitter and high-efficiency emitter, respectively. However this was most evident for the high efficiency emitter. Without pre-gettering, the [Fe_i] for samples with a high efficiency emitter was approximately $1x10^{11}$ /cm^3 in the impurity affected zone.

This shows pre-gettering can effectively improve the impurity-affected zone of wafers using high-efficiency emitters. In addition, heavily diffused emitters can much more effectively getter impurities than high-efficiency emitters when no pre-gettering process is used. However, it may also be possible to change the process for the formation of a high efficiency emitter to include a heavy diffusion that is effective at gettering, with a subsequent etch back of the emitter to yield a high-efficiency emitter with low surface concentration and simultaneously remove the gettered impurities [19].

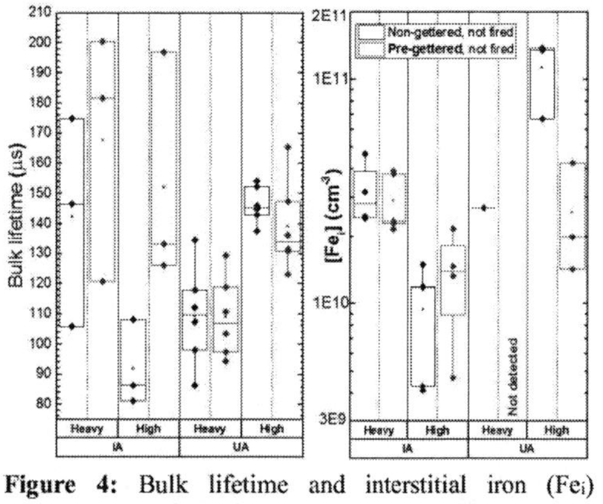

Figure 4: Bulk lifetime and interstitial iron (Fe_i) concentration of multicrystalline silicon wafers + SiNx depositions with heavily diffused emitter (**Heavy**) and high-efficiency emitter (**High**) in impurity-affected zone (**IA**) and in unaffected region (**UA**) with and without pre-gettering (**Non-gettered**).

The reduction in iVoc with pre-gettering was shown on both types of emitter (see Figure 3). The reduction was likely due to detrimental effects during the pre-gettering process such as the release of impurities from precipitated sites, despite of the initial intention of incorporating the process being to improve the bulk by removing impurities. The bulk lifetime was not affected by incorporating a pre-gettering process in most regions of the wafer, although samples with high-efficiency emitters demonstrated minor increases in [Fe_i] with pre-gettering. Possible explanation of less effective gettering in unaffected region is due to precipitated iron that can limit the gettering efficacy [20] and the precipitated iron may lead to interstitial iron form during subsequent pre-gettering process. This implies the detrimental effects could have also occurred during emitter formation as both are done in high temperature, above 800°C where precipitated impurities, such as iron can dissolve into interstitial form [21].

While the ΔiVoc maps, indicated a reduced gettering effect for samples with high-efficiency emitters, the bulk lifetime was still higher than that of samples with heavily diffused emitters before firing (with or without pre-gettering). The possible explanation is that while phosphorus gettering during emitter formation may have been more effective on heavily diffused wafer, detrimental effect such as dissolution of impurities like how pre-gettering process had shown reduction in ΔiVoc map (see Figure 3) may have simultaneously occurred during diffusion, causing over all bulk lifetimes to drop.

Another possible reason is that during the drive-in process the phase of the metallic impurities may have been changed to more precipitated form as the temperature of the driven-in process is the temperature, which can effectively precipitates iron, as most common metallic impurities. This only occurred in lightly diffused wafer, to diffuse phosphorus deeper into the junction and reduce the peak concentration.

It appears to be the detrimental effects may overwhelm the benefits of more effective gettering with heavily diffused wafers. Furthermore, other recent studies showed the effectiveness of gettering can be improved by optimising the diffusion process [22],[23]. This indicates that not only the doping profile of the emitter but how the wafer was processed to form an emitter may be more important.

3.2 Response to firing on different emitters

The ΔiVoc maps were generated to locally compare the changes before and after firing without pre-gettering on samples with different types of emitters in Figure 5.

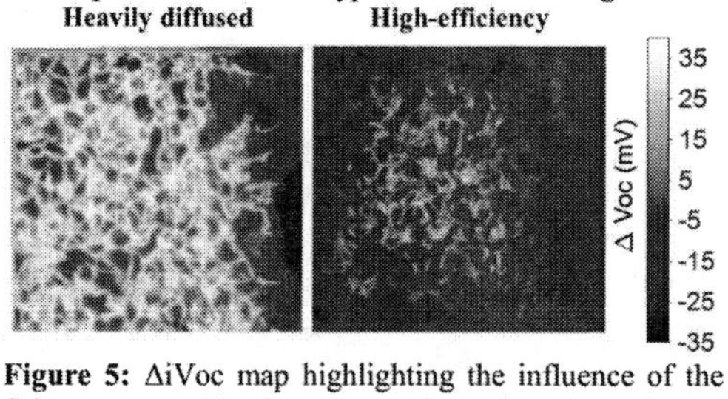

Figure 5: ΔiVoc map highlighting the influence of the firing process. The images are from the contrasting of iVoc maps of SiNx:H deposited non-gettered samples before and after firing through belt furnace above 800 °C on both types of emitters. Positive values indicate improvements in material quality with firing and negative values indicated a reduction in material quality. The dark region on the mid-right edge on non-gettered, heavily diffused wafer is due to comparison of broken region.

The firing process, which allows hydrogen passivation has shown more improvement on the heavily diffused wafer than on the lightly diffused wafer. The improvement along grain boundaries were approximately 35 mV on heavily diffused wafers while it was only about 15 mV or less on lightly diffused wafers. The impurity-affected zone on samples with both emitter types suffered from a reduction in iVoc. However a larger area with reduced iVoc values occurred on wafers with high-efficiency emitters, is likely due to dissolution of impurities, which has been reported previously [13]. Slight increases in the [Fe_i] were observed (see Figure 6) on samples with high efficiency emitters after firing, whilst reductions were observed for samples with heavily diffused emitters. The effect of impurity dissolution during firing appears to be similar for both type of emitters although samples with heavily diffused emitter experienced a hydrogen passivation along some of the grain boundaries in the impurity-affected zone, leading to improved iVoc values and bulk lifetimes.

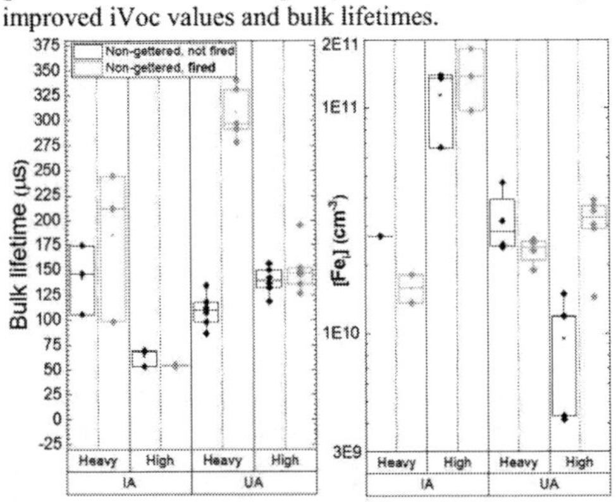

Figure 6: Bulk lifetime and interstitial iron (Fe_i) concentration of multicrystalline silicon wafers with no gettering **(Non-gettered)** + SiNx depositions with heavily diffused emitter **(Heavy)** and high-efficiency emitter **(High)** in impurity-affected zone **(IA)** and in unaffected region **(UA)** before and after firing.

The bulk lifetime of samples with high-efficiency emitters was initially higher than that of samples with heavily diffused emitters before firing. There was a significant improvement in the iVoc of samples with heavily diffused wafer after firing for both gettered and non-gettered samples. However, there was little or no change in bulk lifetime of samples with high-efficiency emitters as shown in Figure 6. Samples with high-efficiency emitters showed increases in [Fe_i] after firing, while heavily diffused wafers showed a reduction in iVoc (see Figure 6). This again indicates an improved effectiveness of gettering from the heavily diffused emitter, which can provide additional external gettering during the firing process [11]. A previous study also demonstrated that the presence of emitter could provide protection against dissolving metal precipitates into the bulk [12].

It is not fully understood why samples with high-efficiency emitters did not experience a large improvement in bulk lifetime. However, it is possible that benefits from hydrogen passivation, as clearly shown by the passivation of grain boundaries during firing, was counteracted by negative effects of increased [Fe_i].

Implications of these findings may be that the emitter concentration can influence the rate of dissolution of metal precipitates and effectiveness of gettering during firing. However, considering the type and the quality of multi-crystalline silicon wafers can also influence the response to the gettering [24] and hydrogen passivation [11], it may not be applicable to all types of multi-crystalline silicon. More studies on the relationship between the emitter diffusion properties and the efficacy of hydrogen passivation need to be conducted, as hydrogen passivation is one of the major factors to improve the performance of multi-crystalline during firing more than external gettering thermally.

4 CONCLUSION

High efficiency emitters were demonstrated to have less effective gettering properties than heavily diffused emitters. Without a pre-gettering process, wafers using high efficiency emitters suffered from substantial reductions in the bulk lifetime of impurity-affected zones compared to that of wafers using heavily diffused emitters. The incorporation of a phosphorus pre-gettering process improved the material quality in the impurity-affected zone on samples with both emitter types, but particularly for samples with lightly diffused emitters. This was supported by reductions in the [Fe_i] when using a heavily diffused emitter compared to a high efficiency emitter, and again with the incorporation of a pre-gettering process. However, the iVoc was reduced in some other regions with the incorporation of the pre-gettering process. This suggests that in some instances, the beneficial effects of gettering may be overwhelmed by detrimental effects such as the dissolution of impurities from precipitate sites. This may bring a possibility to optimise the high-efficiency emitter process to improve gettering whilst retaining the high-efficiency properties.

Both beneficial and detrimental effects were also observed during firing due to a number of possible mechanisms. This includes external gettering of impurities to the emitter, the dissolution of precipitated impurities, and hydrogen passivation. In the unaffected regions, the larger iVoc improvement after firing occurred on heavily diffused wafers. This may be due to a combination of effective gettering during firing as indicated by a reduction in [Fe_i] as well as hydrogen passivation of the bulk and surface regions. The reduced improvement of iVoc for lightly diffused samples may have been due to the detrimental effects of a dissolution of impurities from precipitated sites with less effective external gettering counteracting the benefits of hydrogen passivation. In the impurity affected regions of wafers with both emitter types, the detrimental effects dominated over the benefit of hydrogen passivation.

4.1 Acknowledgements

The authors would like to acknowledge the Australian Government through the Australian Renewable Energy Agency (ARENA) and the Australian Research Council (ARC) for the financial assistance. ARENA is supporting Australian research and development in solar photovoltaic and solar thermal technologies to help solar power become cost competitive with other energy sources. The views expressed herein are not necessarily the views of the Australian Government, and the Australian Government does not accept responsibility for any information or advice contained herein.

4.2 References

[1] F. Warchol, S. Martinuzzi, and P. Isabelle, "Hydrogen passivation of defects in multicrystalline silicon solar cells," *Sol. Energy Mater. Sol. Cells*, vol. 80, no. 3, pp. 343–353, 2003.

[2] A. Bentzen and A. Holt, "Overview of phosphorus diffusion and gettering in multicrystalline silicon," *Mater. Sci. Eng. B*, vol. 159, pp. 228–234, 2009.

[3] V. Shanmugam, J. Cunnusamy, A. Khanna, P. K. Basu, Y. Zhang, C. Chen, A. F. Stassen, M. B. Boreland, T. Mueller, B. Hoex, and A. G. Aberle, "Electrical and microstructural analysis of contact formation on lightly doped phosphorus emitters using thick-film ag screen printing pastes," *IEEE J. Photovoltaics*, vol. 4, no. 1, pp. 168–174, 2014.

[4] V. Shanmugam, A. Khanna, P. K. Basu, A. G. Aberle, T. Mueller, and J. Wong, "Impact of the phosphorus emitter doping profile on metal contact recombination of silicon wafer solar cells," *Sol. Energy Mater. Sol. Cells*, vol. 147, pp. 171–176, 2016.

[5] J. S. Kim, H. W. Lee, J. E. Shin, E. J. Lee, J. Y. Choi, J. M. Shim, D. J. Oh, K. Y. Cho, J. H. Kong, J. K. Seo, J. S. Kim, S. H. Lee, H. S. Lee, B. Hallam, and S. R. Wenham, "Improved front side metallization for laser doped solar cells," in *Photovoltaic Specialists Conference (PVSC), 2011 37th IEEE*, 2011, pp. 001091–001094.

[6] A. Liu, C. Sun, and D. Macdonald, "Hydrogen passivation of interstitial iron in boron-doped multicrystalline silicon during annealing," *J. Appl. Phys.*, vol. 116, no. 19, p. 194902, 2014.

[7] A. Azzizi, L. J. Geerligs, and D. Macdonald, "Hydrogen passivation of iron in crystalline silicon," *19th EUPVSEC*, 2004.

[8] S. A. McHugo, H. Hiesimair, and E. R. Weber, "Gettering of metallic impurities in photovoltaic silicon," *Appl. Phys. A Mater. Sci. Process.*, vol. 64, no. 2, pp. 127–137, 1997.

[9] A. Cuevas, M. Stocks, S. Armand, M. Stuckings, A. Blakers, and F. Ferrazza, "High minority carrier lifetime in phosphorus-gettered multicrystalline silicon," *Appl. Phys. Lett.*, vol. 70, no. 8, p. 1017, 1997.

[10] J. Chen, D. Yang, Z. Xi, and T. Sekiguchi, "Electron-beam-induced current study of hydrogen passivation on grain boundaries in multicrystalline silicon: Influence of GB character and impurity contamination," *Phys. B Condens. Matter*, vol. 364, no. 1, pp. 162–169, 2005.

[11] P. Karzel, M. Ackermann, L. Gröner, C. Reimann, M. Zschorsch, S. Meyer, F. Kiessling, S. Riepe, and G. Hahn, "Dependence of phosphorus gettering and hydrogen passivation efficacy on grain boundary type in multicrystalline silicon," *J. Appl. Phys.*, vol. 114, no. 24, 2013.

[12] J. Tan, D. MacDonald, N. Bennett, D. Kong, A. Cuevas, and I. Romijn, "Dissolution of metal precipitates in multicrystalline silicon during annealing and the protective effect of phosphorus emitters," *Appl. Phys. Lett.*, vol. 91, no. 4, p. 043505, 2007.

[13] J. F. Lelièvre, J. Hofstetter, A. Peral, I. Hoces, F. Recart, and C. del Cañizo, "Dissolution and gettering of iron during contact co-firing," *Energy Procedia*, vol. 8, pp. 257–262, 2011.

[14] R. A. Sinton and A. Cuevas, "Contactless determination of current–voltage characteristics and minority-carrier lifetimes in semiconductors from quasi-steady-state photoconductance data," *Appl. Phys. Lett.*, vol. 69, no. 17, pp. 2510–2512, 1996.

[15] T. Trupke, R. A. Bardos, M. C. Schubert, and W. Warta, "Photoluminescence imaging of silicon wafers," *Appl. Phys. Lett.*, vol. 89, no. 4, p. 044107, 2006.

[16] A. Teal and M. Juhl, "Correcting the inherent distortion in luminescence images of silicon solar cells," *2015 IEEE 42nd Photovolt. Spec. Conf.*, pp. 1–5, 2015.

[17] D. Payne and S. Wenham, "Evaluating the accuracy of point spread function deconvolutions applied to luminescence images," in *In Proceedings of 43rd IEEE PVSC.*

[18] B. Hallam, B. Tjahjono, T. Trupke, and S. Wenham, "Photoluminescence imaging for determining the spatially resolved implied open circuit voltage of silicon solar cells," *J. Appl. Phys.*, vol. 115, no. 4, p. 044901, 2014.

[19] A. Metz, D. Adler, S. Bagus, H. Blanke, M. Bothar, E. Brouwer, S. Dauwe, K. Dressler, R. Droessler, T. Droste, M. Fiedler, Y. Gassenbauer, T. Grahl, N. Hermert, W. Kuzminski, A. Lachowicz, T. Lauinger, N. Lenck, M. Manole, M. Martini, R. Messmer, C. Meyer, J. Moschner, K. Ramspeck, P. Roth, R. Schönfelder, B. Schum, J. Sticksel, K. Vaas, M. Volk, and K. Wangemann, "Industrial high performance crystalline silicon solar cells and modules based on rear surface passivation technology," *Sol. Energy Mater. Sol. Cells*, vol. 120, pp. 1–9, 2013.

[20] D. P. Fenning, J. Hofstetter, M. I. Bertoni, G. Coletti, B. Lai, C. del Cañizo, and T. Buonassisi, "Precipitated iron: A limit on gettering efficacy in multicrystalline silicon," *J. Appl. Phys.*, vol. 113, no. 4, p. 044521, 2013.

[21] T. Buonassisi, A. A. Istratov, S. Peters, C. Ballif, J. Isenberg, S. Riepe, W. Warta, R. Schindler, G. Willeke, Z. Cai, B. Lai, and E. R. Weber, "Impact of metal silicide precipitate dissolution during rapid thermal processing of multicrystalline silicon solar cells," *Appl. Phys. Lett.*, vol. 87, no. 12, p. 121918, 2005.

[22] S. Joonwichien, I. Takahashi, S. Matsushima, and N. Usami, "Comparison of phosphorus gettering effect in faceted dendrite and small grain of multicrystalline silicon wafers grown by floating cast method," *Jpn. J. Appl. Phys.*, vol. 54, no. 8S1, p. 08KD11, 2015.

[23] A. Zarroug, Z. Ben Hamed, L. Derbali, W. Dimassi, and H. Ezzaouia, "Enhanced phosphorus gettering of impurities in p-type Czochralski silicon through a variable temperature processing (VTP)," *J. Cryst. Growth*, vol. 422, pp. 69–74, 2015.

[24] S. Castellanos, K. E. Ekstrom, A. Autruffe, M. A. Jensen, A. E. Morishige, J. Hofstetter, P. Yen, B. Lai, G. Stokkan, C. del Canizo, and T. Buonassisi, "High-Performance and Traditional Multicrystalline Silicon: Comparing Gettering Responses and Lifetime-Limiting Defects," *IEEE J. Photovoltaics*, vol. 6, no. 3, pp. 632–640, 2016.

COMPREHENSIVE STUDY OF INTERMETALLIC COMPOUNDS IN SOLAR CELL INTERCONNECTIONS INCLUDING LEAD-FREE, LOW MELTING POINT SOLDERS

Torsten Geipel[a], Monja Moeller[b], Achim Kraft[a], Ulrich Eitner[a]
[a] Fraunhofer Insitute for Solar Energy Systems ISE, 79110 Freiburg, Germany
[b] Kiel University, Institute of Material Science, 24098 Kiel, Germany

ABSTRACT: Intermetallic compounds (IMC) in solder bonds are commonly considered critical for the reliability of interconnections. The microstructure and thermal aging characteristics of solder bonds of crystalline silicon solar cells are investigated, whereby two solders, Sn60Pb40 and a lead-free, low melting point alternative Sn41Bi57Ag2 are considered. Solder bonds of front side busbars are formed with a semi-automatic tool and cross sections are prepared, which are characterized with confocal laser microscopy, scanning electron microscopy and energy dispersive x-ray spectroscopy (EDX). The kinetic of IMC growth is modeled with a diffusion model, and the parameters are obtained with systematic thermal aging studies and the Arrhenius relationship. The model is capable of processing non-isothermal temperature profiles and is used to simulate IMC growth during PV module reliability tests like thermal cycling 600 (TC) and damp heat for 3000 h. Furthermore, the thicknesses of the IMCs are estimated for a PV module in an outdoor location. It is found that solder joints undergo strong ripening processes during thermal aging. In particular, Sn41Bi57Ag2 solder joints show faster IMC growth, grain coarsening and Sn-penetration into the busbar than Sn60Pb40 bonds. Sn-penetration is observed already in the as-soldered condition and EDX confirms the existence of significant amounts of diffused Sn around cavities in the busbar. Large portions of the busbar are consumed by the Ag_3Sn IMC after 155 h at 130 °C and further aging of the Sn41Bi57Ag2 joint leads to Bi segregation and, in some cases, complete depletion of Sn from the solder matrix. The simulation of IMC growth shows a final IMC thickness after TC 600 of 1.2 µm and 1.5 µm; furthermore, a thickness after 3000 h at 85°C of 2.6 µm and 3.7 µm respectively for Ag_3Sn in Sn60Pb40 and Ag_3Sn in Sn41Bi57Ag2 bonds. The prognosis of IMC growth in a PV module in the outdoor location estimates an Ag_3Sn thickness of 1.2 µm and 1.3 µm for Sn60Pb40 and Sn41Bi57Ag2 respectively due to the rare occurrence of temperatures above 60 °C.

Keywords: PV Module, Module Integration, Metallization, Reliability

1 MOTIVATION, GOALS AND APPROACH

Reliability of PV modules directly impacts the levelized cost of electricity (LCOE) of PV systems [1]. The series interconnection of solar cells into strings via soldering, which is part of the module assembly process, determines the reliability of the final product in such a way that inappropriate soldering conditions or material selection damages solar cells during the process and subsequent aging [2–4]. Furthermore, fatigue, coarsening as well as voiding in solder bonds are commonly observed failure modes in field-aged PV modules [5–7]. In order to more efficiently prevent such solder bond failures in PV modules this work intends to increase the understanding of thermal aging phenomena in solder bonds with a focus on intermetallic compounds (IMC).

IMCs are typically considered critical because of their brittleness and resulting inability to withstand thermomechanical fatigue [8]. The goals are to describe microstructural changes during thermal aging, to quantify the IMC growth and to simulate it under various aging conditions. Not only a standard lead-based solder alloy is considered in the investigation but also a lead-free, bismuth-based low melting point solder, that may be of relevance for the interconnection of temperature-sensitive heterojunction solar cells [9]. A direct comparison between the standard solder and the low-melting point solder is to be made.

The approach is to solder interconnector ribbons on the front side busbars of solar cells, which are then processed into metallographic cross sections. Those are characterized in terms of solder and IMC microstructure. Then, the cross section samples are isothermally aged at various temperatures and time steps. Subsequently, the measurement of the IMC growth at different temperatures allows the extraction of the kinetic parameters of an IMC growth model. Eventually, the model is used to make predictions during accelerated aging and under outdoor conditions.

2 MATERIALS AND METHODS

2.1 Solder bonds and solder materials

Soldering is the metallurgical joining of two or more substrates using an interjacent material (solder), which is able to melt and solidify at significantly lower temperatures than the substrates [10]. The process of soldering involves heating of the substrates and the solder, the removal of oxides and the activation of the substrates' surfaces with the flux and passing the melting temperature of the solder, which then is able to wet and spread on the substrates. Thin layers of the substrates are dissolved in the molten solder, and the molten alloy with the foreign atoms near the interface subsequently leads to the formation of intermetallic layers [11]. The microstructure obtained after solidification is strongly influenced by the cooling dynamics. The initial structure changes during thermal aging and even storage at room temperature. The microstructure determines the mechanical properties of the bond [12].

Thin and uniform intermetallic layers are an indication for a good metallurgical bond [11]. These layers tend to grow temperature driven and, due to their brittleness, reduce the fatigue resistance of the bonds [8,13–15]. Possible intermetallic compounds, their stoichimetry and elemental compositions are displayed in the respective alloy's phase diagram [16,17].

Table I. Ribbon material, alloy coatings and processing conditions.

	Sn60Pb40	Sn41Bi57Ag2
Ribbon dimensions [mm × mm]	1.5 × 0.2	1.5 × 0.18
Solder coating thickness [µm]	15 – 25	15 – 25
Solder melting temperature [°C]	183 – 187 °C	139 °C
Set soldering temperature [°C]	250	200
Set heating chuck temperature [°C]	175	135
Soldering time [s]	1.2	1.2
Downholding time [s]	4	4

The tin-lead alloy with the eutectic or near-eutectic compositions of 63 wt.% Sn and 37 wt.% Pb or 60 wt.% Sn and 40 wt.% Pb (Sn60Pb40) is mostly used in today's PV module manufacturing. Although Pb is a toxic material, banned by the European Parliament from many electronic products through the RoHS legislation [18], SnPb solder has many advantages in terms of its relatively low melting point of 183 °C, low cost, superior wetting behaviour and outstanding ductility, that easily absorbs thermomechanical stress [14].

Alloys based on tin-bismuth are considered as replacement to reduce thermomechanical stress during the soldering process of thin solar cells or for the interconnection of temperature-sensitive solar cells [19,20]. They are often used in the eutectic composition of 43 wt.% Sn and 57 wt.% Bi. A minor addition of Ag improves the mechanical properties of the alloy by refining its microstructure [21,22]. In this work Sn41Bi57Ag2 is used. The alloy has a melting temperature of 139 °C, the benefits of costs lower than indium-tin alloys, and it is a lead-free material. Bismuth is often deemed disadvantageous with regard to segregation, brittleness, fatigue life, the high homologous temperature at operation and negative thermal expansion during the melting/solidification phase transition [23–27]. However, not all researches find negative effects on joint strength before or after aging [28,29], which means that the situation is more complex and SnBi(Ag) has the potential to complement SnPb in specific applications.

2.2 Modeling of intermetallic phase growth

The growth of intermetallic phases during thermal aging is considered to be a diffusion-driven process. If a random walk treatment of Fick's second law is made, the mean square diffusion distance of atoms during solid state aging $\bar{x}^2$ in one dimension is expressed with equation 1 [30].

$$\bar{x}^2 = 2Dt \tag{1}$$

Hereby, D stands for the temperature dependent diffusion coefficient and t for the time of thermal aging.
The diffusion coefficient is typically described by the Arrhenius relationship:

$$D = D_0 \exp\left(-\frac{Q}{RT}\right) \tag{2}$$

D_0 is the pre-exponential factor, Q is the activation energy of the growth process, R is the gas constant and T is the absolute temperature. Combining equations

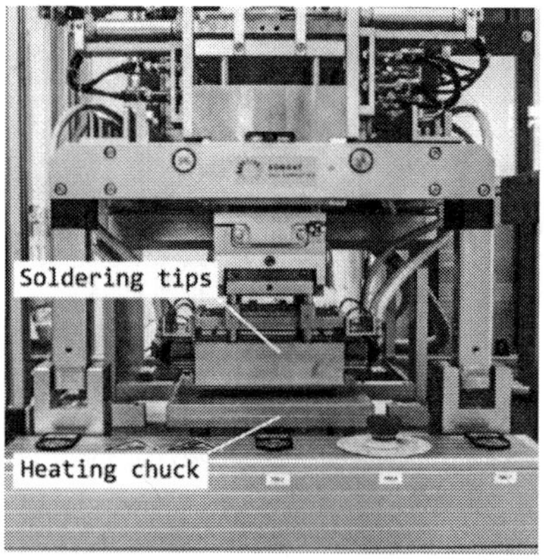

Figure 1. Semi-automatic soldering station.

(1) and (2) yields to the intermetallic layer thickness x after thermal aging:

$$x = \sqrt{2D_0 t \exp\left(-\frac{Q}{RT}\right)} + x_0 \tag{3}$$

with x_0 as the initial IMC thickness.

In order to make predictions of the intermetallic layer thickness, equation 3 requires the parameters D_0, Q and x_0 as an input, which are obtained by systematically measuring the IMC thickness at different temperatures and constant time intervals. Having those, the natural logarithm is taken of equation 3, and a rearrangement leads to a linear relationship between $\ln(x-x_0)$ and $1/T$ as shown in equation 4.

$$\ln(x - x_0) = \frac{\ln(2D_0 t)}{2} - \frac{Q}{2R}\frac{1}{T} \tag{4}$$

The slope of equation 4 gives the activation energy. Further non-linear fitting of the experimental data of thickness x to equation 3 with t as the dependent variable and Q, T, x_0 as constant parameters eventually leads to the extraction of D_0.

As equation 3 only allows predictions for a constant temperature, a numerical solving routine is necessary if non-isothermal conditions are considered.

2.3 Sample preparation and characterization

Copper ribbons with two different solder coatings, Sn60Pb40 and Sn41Bi57Ag2, are soldered on the front side busbars of industrial, mono-crystalline, screen-

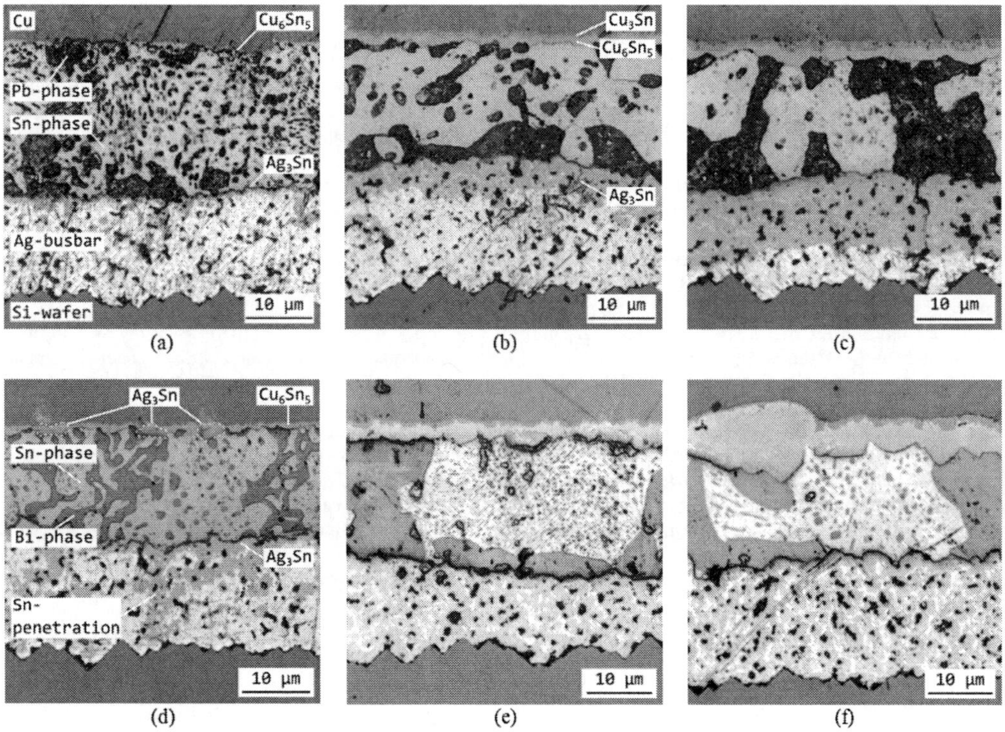

Figure 2. (a) CLM image of Sn60Pb40 solder bond in the initial state, (b) after 155 h at 130 °C, (c) after 500 h at 130 °C. (d) Sn41Bi57Ag2 solder bond in the intial state, (e) after 155 h at 130 °C, (f) after 500 h at 130 °C.

printed aluminium back-surface-field solar cells by means of a semi-automatic soldering station (Figure 1). The ribbon properties and process parameters are given in Table I.

Afterwards, metallographic cross sections are prepared [31]. These cross sections are thermally aged in a furnace under nitrogen flow at four temperatures between 85 °C and 150 °C depending on the solder melting temperature. Aging the embedded cross section samples allows to characterize nearly the same position in the interconnection before and after the thermal treatment. The time intervals to log the IMC thickness are 15 h, 85 h and 155 h. Selected samples are monitored for 500 h and 750 h. Before the microscopic inspection the samples are mechanically polished.

The samples are characterized with confocal laser microscopy (CLM), scanning electron microscopy (SEM) and electron dispersive x-ray spectroscopy (EDX). The thickness of the IMC is determined by positioning 30 measurement lines perpendicular to the layer, which are distributed across the entire cross section. An average value and a standard error are calculated.

3 RESULTS AND DISCUSSION

3.1 Microstructural changes and IMC growth

The metallographic cross section of the Sn60Pb40 solder bond in the initial condition is shown in Figure 2a. It is observed that Sn and Pb crystals with a diameter of 1 µm to 10 µm are uniformly distributed within the solder matrix. Occasionally, small sized Sn regions enclosed by Pb grains are found resulting from the secondary crystallization of Sn during the cooldown [10]. A Cu_6Sn_5 intermetallic layer with a thickness of 0.3 µm to 0.5 µm is found at the interface to the copper ribbon, which has a scallop-type morphology [32]. Between solder matrix and Ag-busbar there exists an Ag_3Sn intermetallic layer with a thickness of ≈0.7 µm. This layer is usually uniform, but excessive soldering temperature or time may lead to plate-like protrusions [14,33]. There are dark gray areas in the upper region of the busbar often located around lead-glass particles or cavities (black regions). These areas are afflicted with local Sn-penetration as will become evident in section 3.2.

After 155 h of thermal aging at 130 °C shown in Figure 2b, we find the grains of Sn and Pb to coarsen to diameters between 10 µm to 30 µm. Furthermore, in some locations fine cracks between the grains are identified. The IMC layers show notable growth after the thermal treatment. First, an intermediate IMC between Cu_6Sn_5 and Cu is formed, which is Cu_3Sn with a thickness of ≈0.3 µm. The Cu_6Sn_5 IMC measures between 1 µm to 1.5 µm. At the interface to the busbar the Ag_3Sn layer increases to 4 – 6 µm. Sn-penetration proceeds in such a way that it almost reaches the wafer surface.

The microstructure after 500 h at 130 °C shows further grain coarsening and IMC growth. The Ag-busbar is largely consumed by Ag_3Sn with a thickness of 10 – 12 µm. The total busbar thickness is 15 µm.

The initial state of the Sn41Bi57Ag2 solder bond is depicted in Figure 2d. The Bi and Sn grains are homogeneously dispersed in the solder matrix in a similar manner as in the case of Sn60Pb40. We measure that Cu_6Sn_5 and Ag_3Sn layer thicknesses are slightly higher with 0.6 µm and 0.8 – 0.9 µm respectively. The Ag-content in the solder matrix is largely found as Ag_3Sn clusters with 1 – 2 µm in diameter beaded at the interface to the Cu ribbon. Moreover, small Bi grains with a diameter of < 2 µm are distributed in the larger Sn grains. Furthermore, areas with Sn penetration extend in the busbar with up to 5 – 10 µm locally.

Thermal aging proceeds faster in the Sn41Bi57Ag2 solder joint as shown in Figure 2e. Notably, the Bi and Sn grains coarsen to form large, continuous clusters. However, the initially existent, small-sized Bi particles in the Sn grains now appear to form an even more finely grained matrix, whereby the diameters of the Bi grains are below 1 µm. The Cu_6Sn_5 layer is 2 µm to 3 µm and the Cu_3Sn ≈1 µm. We observe the Ag_3Sn to fill the entire busbar already after 155 h of thermal aging at 130 °C.

The effects of further ripening of the Sn41Bi57Ag2 bond become obvious in Figure 2f. After 500 h at 130 °C thermal aging, the IMCs possess an excessive thickness. The Cu_6Sn_5 phase reaches a thickness of 8 – 10 µm. It is observed that in some samples, especially when the solder layer is thin (5 – 10 µm), the solder matrix is practically depleted of Sn with only the Bi component remaining. This circumstance is assumed to cause a reduced fatigue life of the joint due to the brittleness of Bi, which is not able to absorb high mechanical stress without crack initiation. Judging from these microstructural observations we suspect lower reliability of Sn41Bi57Ag2 joints compared to standard Sn60Pb40 solder joints.

The development of the IMC thickness with time and temperature is shown in Figure 3. The experimentally determined thickness with CLM is plotted as symbols. Simulations according to the approach proposed in section 2.2 are shown as lines. The IMC thickness of Cu_6Sn_5 and Cu_3Sn at the Cu interface of the Sn41Bi57Ag2 solder bond is combined to a single phase growth. The reason is that Cu_3Sn can grow in expense of Cu_6Sn_5 in some instances as reported in reference [34], which leads to a stagnation of the Cu_6Sn_5 thickness. In turn, this stagnation strongly increases the error during the extraction of the activation energy, which is why we decide to simplify the IMC growth for this solder and IMC to a combined Cu_6Sn_5 + Cu_3Sn growth.

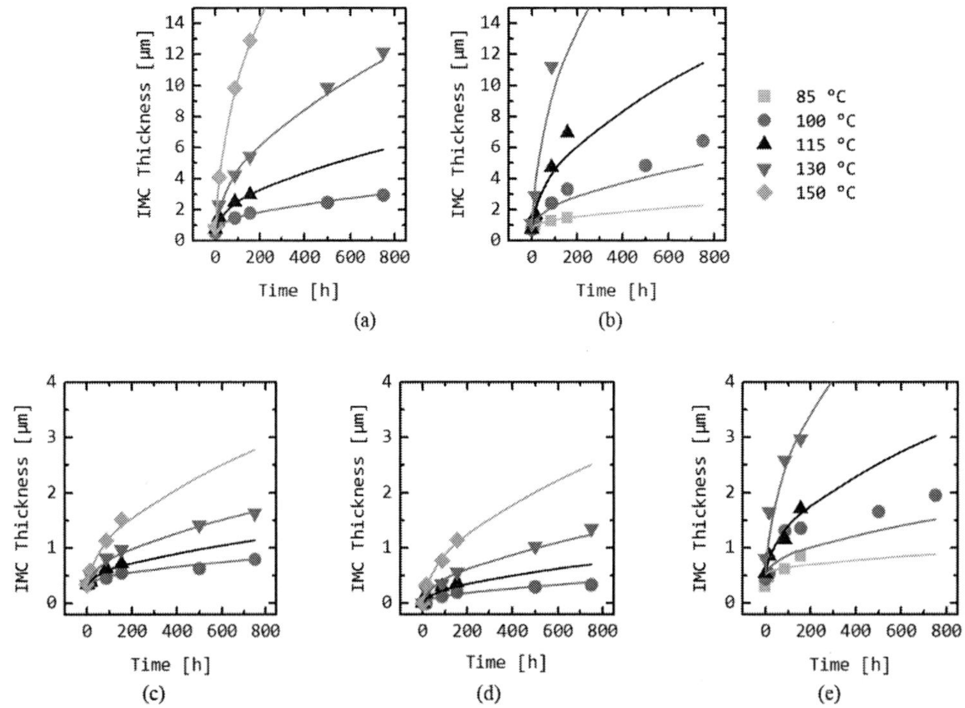

Figure 3. (a) Growth of Ag_3Sn in Sn60Pb40 solder bonds at different aging conditions, b) Ag_3Sn / Sn41Bi57Ag2, c) Cu_6Sn_5 / Sn60Pb40, d) Cu_3Sn / Sn60Pb40, e) Cu_6Sn_5 + Cu_3Sn / Sn41Bi57Ag2. Symbols represent measurements, lines represent simulated phase growth. Adapted from reference [38].

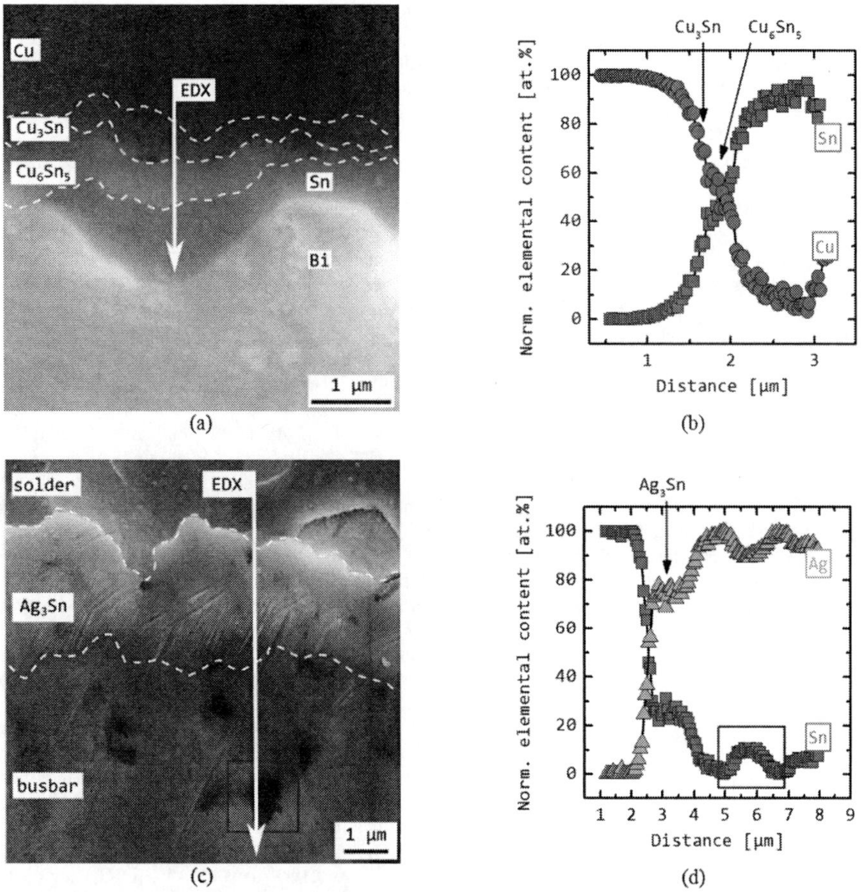

Figure 4. (a) SEM of the interface to the Cu ribbon of a Sn41Bi57Ag2 joint aged at 100 °C for 155 h, (b) EDX line scan across the interface that is shown in (a), (c) SEM of the interface from Sn41Bi57Ag2 to the busbar after 85 h at 100 °C. The red rectangle highlights a cavity in the busbar, (d) EDX line scan of the interface that is shown in (c). Adapted from references [38,39].

3.2 EDX analysis of the interfaces

The intermetallic layers at the interfaces are investigated in more detail using SEM and EDX techniques. A SEM detail of the interface at the Cu ribbon is depicted in Figure 4a. Both intermetallic phases appear as relatively uniform layers. An arrow indicates the range of a quantitative EDX line scan across this interface, which is shown in Figure 4b. The normalized elemental content in at.% of only the elements Sn and Cu is plotted versus the distance across the interface. In the region of Cu_6Sn_5 an elemental content of 40 at.% to 60 at.% Cu and respective Sn is found, which is close to the theoretically predicted content of 54.55 at.% Cu and 45.45 at.% Sn.

The Cu_3Sn phase is more difficult to verify with EDX, since no distinct plateau, where the elemental content would match the theoretically predicted 75 at.% Cu and 25 at.% Sn can be identified in the line scan. However, this is due to limitations in the resolution of the measurement.

Figure 4c shows a SEM detail of the Ag3Sn layer of a Sn41Bi57Ag2 solder bond aged 85 h at 100 °C and Figure 4d the EDX line scan across this interface accordingly. The red rectangle marks a special area of interest, which is a cavity in the busbar. The EDX line scan shows a plateau at the region of the Ag_3Sn intermetallic layer. There, an elemental content of 70 at.% to 75 at.% Ag and 25 at.% to 30 at.% Sn is determined, that is close to the theoretical stoichiometric ratio of the IMC. Unexpectedly, a significant amount of Sn is found around the cavity in the busbar highlighted by the red rectangle. Approximately 90 at.% Ag and 10 at.% Sn is measured in this region, which is assumed to be, according to the phase diagram, either (ζAg) or (Ag) with up to 10 at.% Sn in solid solution [17]. This finding corresponds to the gray areas around the cavities in the microscopic images that were mentioned in section 3.1 and interpreted as Sn- penetration. Therefore, we conclude that Sn is present in deeper regions of the busbar. It is likely that defects and cavities in the busbar promote the diffusion of Sn atoms from the solder into the busbar. Sn-penetration is often proposed as the reason for metallization adhesion loss after soldering or prolonged aging [35,36]. Therefore, this phenomenon is regarded as critical for interconnection reliability.

3.3 Simulation of IMC growth

The measured IMC thickness is converted to Arrhenius

plots according to equation 4. An example is provided for the case of Ag_3Sn using Sn41Bi57Ag2 in Figure 5. Theoretically, the three time series should result in the same slope, ergo activation energy, if a single reaction type and an independency of the activation energy on the progress of reaction is assumed. Obviously, the slopes of time series 85 h and 155 h are parallel, and the 15 h time series shows some deviation, which is similarly found in other cases. The reason for this deviation could be rooted in another kind of chemical reaction besides diffusion such as a phase-boundary or interface controlled process that is dominant in these early stages of ripening [37]. For practical reasons a weighted average of the slopes is taken in order to extract a single activation energy of the process. The activation energies and pre-exponential factors are reported in reference [38,39].

The quality of prediction of the simulation is quantified with the mean absolute error (MAE) between simulated data and experimental data [40]. In most cases a MAE < 1 µm is obtained indicating a good predictability.

The kinetic parameters are used to calculate the phase growth that we assume to occur during PV module reliability tests thermal cycling from –45 °C to 85 °C (TC) and a constant temperature of 85 °C, as present in damp heat tests [41]. The predicted IMC growth after TC

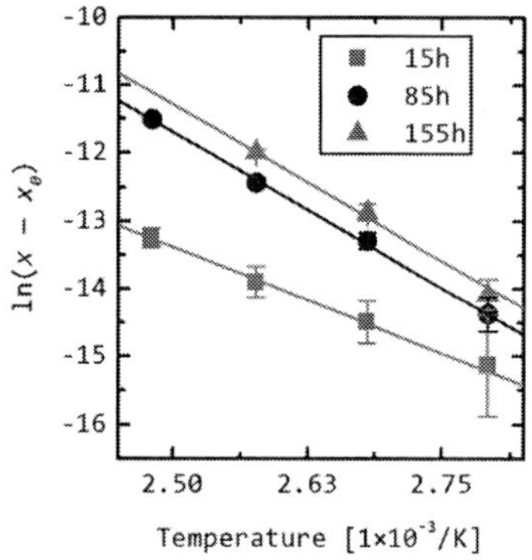

Figure 5. Arrhenius plot of the Ag3Sn growth using Sn41Bi57Ag2 solder. Adapted from reference [38].

600 (cycles) with a cycle time of 2.8 h is shown in Figure 6a. For the Ag_3Sn layer a thickness of 1.5 µm with Sn41Bi57Ag2 and 1.2 µm with Sn60Pb40 is calculated. Owing to the extended time periods at low temperatures during this test the IMC thickness remains quite limited.

The result for 3000 h at a constant temperature of 85 °C is shown in Figure 6b. Note that any possible effects of humidity are not included in the calculation. After 3000 h an Ag_3Sn thickness for Sn41Bi57Ag2 of 3.7 µm compared to 2.6 µm for Sn60Pb40 is calculated. The results indicate that a notable phase growth takes place during a constant temperature treatment of 85 °C.

Furthermore, the model is applied to make a prognosis of the phase growth within a PV module in an outdoor location. Module temperature data of a crystalline silicon PV module installed in Freiburg, Germany is chosen. The temperature curve over the course of one year recorded at one hour time intervals is shown in Figure 7a. This temperature curve is iterated 25 times in the numerical algorithm to simulate a module life of 25 years.

The progression of the Ag_3Sn phase using Sn60Pb40 during the first year is shown in Figure 7b. Since the temperature measurement starts in the summer period, the phase growth is initiated rapidly, slows down in winter and reoccurs in spring the following year. Thus, the phase growth is described as a step-wise increase each year. The prognosis for 25 years is shown in Figure 7c, where this step-wise increase becomes apparent over a longer time scale. Globally, the curve still follows a parabolic shape as consequential from the theoretical background. Surprisingly, a very limited Ag_3Sn phase growth after 25 years of only 1.3 µm and 1.2 µm for Sn41Bi57Ag2 and Sn60Pb40 is calculated respectively. This is, however, plausible if it is considered that the module temperature rarely exceeds 60 °C. Due to the relatively high activation energy of the process, phase growth becomes very slow at these temperatures compared to thermal aging at elevated temperatures. PV modules installed in desert climates can, of course, show a stronger intermetallic phase growth.

The proposed model is able to make predictions of the IMC growth even for non-uniform temperature profiles, which occur in real outdoor conditions and is therefore highly applicable in practice. Judging from the forecasts based on temperature profiles of typical

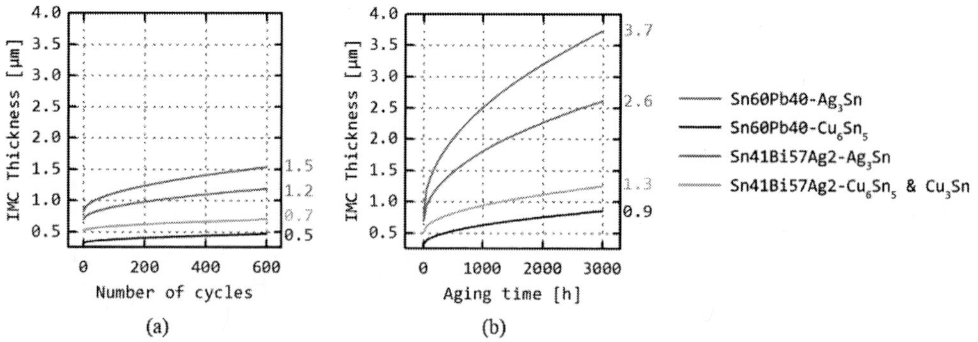

Figure 6. (a) Simulated phase growth of a PV module in a thermal cycling test, (b) simulated phase growth during constant 85 °C. Adapted from references [38,39].

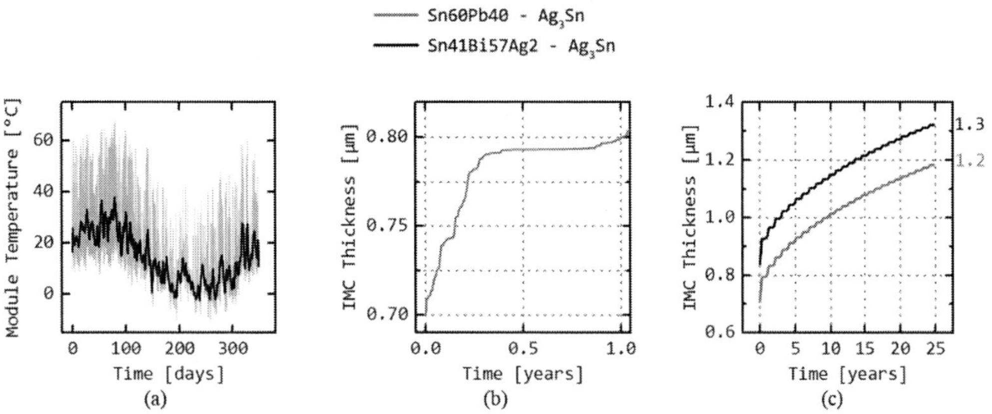

Figure 7. (a) Development of the module temperature during one year (June 01, 2012 to May 15, 2013) in Freiburg, Germany. The gray curve is temperature data recorded at a one-hour intervals. The black curve is a floating average for easier readability. (b) Predicted IMC growth of Ag_3Sn using Sn60Pb40 during the course of the first year. (c) Predicted IMC growth during the course of 25 years. Adapted from references [38,39].

accelerated aging tests and an outdoor location, it is assumed that phase growth directly at the interfaces is rather limited in many situations. Additionally, the differences between Sn41Bi57Ag2 and standard Sn60Pb40 are estimated to be not large. Probably, other observed aging mechanisms such as grain coarsening of the solder matrix or Sn-penetration may be dominant factors having an impact on reliability. Supplementary reliability tests of Sn41Bi57Ag2 soldered PV modules and outdoor testing need to be done in order to further evaluate the potential of the material.

5 CONCLUSION

The microstructure and IMC growth of Sn60Pb40 as well as Sn41Bi57Ag2 solder joints is studied. The uniform and thin IMCs formed during the soldering process at the interfaces to the Cu ribbon and the cell metallization indicate stable metallurgical bonds. They are 0.3 – 0.6 µm and 0.7 – 0.9 µm in thickness for Cu_6Sn_5 and Ag_3Sn respectively. However, it is confirmed with EDX that dissolved Sn exists around cavities in the busbar already in the as-soldered condition. Furthermore, it is observed that IMC growth and other ripening processes during thermal aging such as grain coarsening and Sn-penetration proceed faster in Sn41Bi57Ag2 solder joints owing to the high homologous temperature of the material in many situations. In particular, an entire consumption of the busbar by Ag_3Sn is found after relatively short aging times of 155 h at 130 °C, and in some cases even a depletion of Sn in the solder matrix with only Bi remaining after 500 h for the Sn41Bi57Ag2 solder. The observed ripening processes are assumed to cause commonly observed reliability problems such as metallization adhesion loss and a reduction of fatigue life.

The kinetics of IMC growth are described by a model based on diffusion. The model is parameterized by the Arrhenius approach and the necessary systematic thermal aging experiments. A good predictability of MAE < 1 µm in many cases is achieved.

The model is applicable to many practical situations including non-isothermal temperature conditions and allows the development of customized accelerated aging tests. Examples for model application to predicting the phase growth during accelerated aging tests of PV modules results in 1.2 µm for Ag_3Sn in Sn60Pb40 bonds and 1.5 µm for Ag_3Sn in Sn41Bi57Ag2 joints after TC 600. Furthermore, 2.6 µm of Ag_3Sn in Sn60Pb40 and 3.7 µm of Ag_3Sn in Sn41Bi57Ag2 after 3000 h at 85 °C are predicted. Eventually, a prognosis of phase growth during 25 years of module life in the outdoor location Freiburg, Germany leads to a final thickness of 1.2 µm for Ag_3Sn using Sn60Pb40 and 1.3 µm using Sn41Bi57Ag2.

REFERENCES

[1] M.A. Green, Nat. Energy 1 (2016), 1–4.

[2] T. Geipel, J. Moeller, Y. Zemen, S. Pingel, J. Berghold, Proceedings of the 26th European Photovoltaic Solar Energy Conference and Exhibition (2011), 3517–3522.

[3] J. Wendt, M. Träger, M. Mette, A. Pfennig, B. Jäckel, Proc. of the 24th EUPVSEC. 24th EUPVSEC (2009), 2705–2708.

[4] A.M. Gabor, Proceedings of the 21st European Photovoltaic Solar Energy Conference and Exhibition (2006), 2042–2047.

[5] D.L. King, M.A. Quintana, J.A. Kratochvil, D.E. Ellibee, B.R. Hansen, Prog Photovoltaics 8 (2000), 241–256.

[6] M.A. Quintana, D.L. King, T.J. McMahon, C.R. Osterwald, Proceedings of the 29th IEEE Photovoltaic Specialists Conference (2002), 1436–1439.

[7] G. Cuddalorepatta, A. Dasgupta, S. Sealing, J. Moyer, Tolliver, T.Loman, J., Prog Photovoltaics 18 (2010), 168–182.

[8] P.L. Tu, Y. Chan, J. Lai, IEEE Trans. Compon. Packag. Manuf. Technol. B 20 (1997), 87–93.

[9] A. Faes, M. Despeisse, J. Levrat, J. Champliaud, N. Badel, M. Kiaee, T. Söderström, Y. Yao, R. Grischke, M. Gragert, J. Ufheil, P. Papet, B. Strahm, G. Cattaneo, J. Cattin, Y. Baumgartner, A. Hessler-Wyser, J. Balliet, Proceedings of the 29th

European Photovoltaic Solar Energy Conference and Exhibition (2014), 2555–2561.
[10] Klein Wassink, R. J. Weichlöten in der Elektronik [Soldering in Electronics], 2nd ed., Leuze, Saulgau/Württ (1991).
[11] T. Laurila, V. Vuorinen, J.K. Kivilahti, Materials Science and Engineering: R: Reports 49 (2005), 1–60.
[12] P.G. Harris, K.S. Chaggar, M.A. Whitmore, Soldering & Surface Mount Tech 3 (1991), 20–33.
[13] Parent, J. O. G., Chung, D. D. L., I.M. Bernstein, J Mater Sci 23 (1988), 2564–2572.
[14] K. Zeng, K.N. Tu, Materials Science and Engineering: R: Reports 38 (2002), 55–105.
[15] D.R. Frear, P.T. Vianco, Metall and Mat Trans A 25 (1994), 1509–1523.
[16] N. Saunders, A.P. Miodownik, Bulletin of Alloy Phase Diagrams 11 (1990), 278–287.
[17] I. Karakaya, W.T. Thompson, Bulletin of Alloy Phase Diagrams 8 (1987), 340–347.
[18] Official Journal of the European Union, Directive 2011/65/EU of the European Parliament and of the Council of 8 June 2011 on the restriction of the use of certain hazardous substances in electrical and electronic equipment (recast) (2011).
[19] B. Lalaguna, P. Sánchez-Friera, I.J. Bennett, D. Sánchez, L.J. Caballero, Alonso J., Proceedings of the 22nd European Photovolatic Solar Energy Conference and Exhibition (2007), 2712–2715.
[20] B. Lalaguna, P. Sánchez-Friera, H. Mäckel, D. Sánchez, J. Alonso, Proceedings of the 23rd European Photovoltaic Solar Energy Conference and Exhibition (2008), 2705–2708.
[21] M. Mccormack, H.S. Chen, G.W. Kammlott, S. Jin, Journal of Elec Materi 26 (1997), 954–958.
[22] W.-R. Myung, M.-K. Ko, Y. Kim, S.-B. Jung, J Mater Sci: Mater Electron 26 (2015), 8707–8713.
[23] M.J. Burek, S. Jin, M.C. Leung, Z. Jahed, J. Wu, A.S. Budiman, N. Tamura, M. Kunz, T.Y. Tsui, Acta Materialia 59 (2011), 4709–4718.
[24] P.L. Liu, J. Shang, Scripta Materialia 53 (2005), 631–634.
[25] X. Hu, Y. Li, Z. Min, J Mater Sci: Mater Electron 24 (2013), 2027–2034.
[26] K. Suganuma, Current Opinion in Solid State and Materials Science 5 (2001), 55–64.
[27] Z. Mei, J. Morris, JEM 21 (1992), 599–607.
[28] R. Strauss, S. Smernos, Circuit World 10 (1984), 23–25.
[29] L.E. Felton, C.H. Raeder, D.B. Knorr, JOM 45 (1993), 28–32.
[30] R.J. Borg, G.J. Dienes An Introduction to Solid State Diffusion, Academic Press, San Diego (1988).
[31] D. Eberlein, P. Schmitt, P. Voos, Practical Metallography 48 (2011), 239–260.
[32] H.K. Kim, H.K. Liou, K.N. Tu, Appl. Phys. Lett. 66 (1995), 2337.
[33] P.T. Vianco, R.D. Wright, P.F. Hlava, J.J. Martin, Metall and Mat Trans A 37 (2006), 1551–1561.
[34] A.C.K. So, Y.C. Chan, J. Lai, IEEE Trans. Comp., Packag., Manufact. Technol. B 20 (1997), 161–166.
[35] P. Schmitt, P. Kaiser, C. Savio, M. Tranitz, U. Eitner, Enrgy Proced 27 (2012), 664–669.
[36] W.A. Crossland, L. Hailes, Solid State Technology (1971), 42–47.
[37] J. Šesták, G. Berggren, Thermochimica Acta 3 (1971), 1–12.
[38] T. Geipel, M. Moeller, A. Kraft, U. Eitner, Sol Energ Mat Sol C (submitted).
[39] T. Geipel, M. Moeller, A. Kraft, U. Eitner, Energy Procedia (submitted).
[40] D.G. Mayer, D.G. Butler, Ecological Modelling 68 (1993), 21–32.
[41] IEC 61215: 2005-04, Crystalline silicon terrestrial photovoltaic (PV) modules - Design qualification and type approval (2005).

OF APPLES AND ORANGES: WHY COMPARING BO REGENERATION RATES REQUIRES INJECTION LEVEL CORRECTION

Svenja Wilking, Sebastian Ebert, Cornelius Beckh, Axel Herguth, and Giso Hahn
Department of Physics, University of Konstanz, 78464 Konstanz, Germany
Phone (+49) 7531 883731, Fax: (+49) 7531 883895, e-mail: Svenja.Wilking@uni-konstanz.de

ABSTRACT: Different injection levels make the comparison of BO Regeneration kinetics in different Si wafers or solar cells difficult, *e.g.*, when samples made of different materials featuring different background lifetime levels are to be compared. A correction factor is presented that is able to compensate for these differences in injection level during Regeneration. This is used to investigate the influence of metallic impurities, interstitial oxygen and different dopants on BO Regeneration kinetics. It is found that a $POCl_3$ gettering step only affects Regeneration if the overall H concentration within the wafer is low. Interstitial oxygen is found to be less detrimental to Regeneration than assumed previously. The influence of different dopants is shown in a process leading to a stable regenerated lifetime, and the measured Regeneration rates can be compared because of the applied injection correction. A numerical fit optimization method is presented to separate the influence of the different dopants. It is shown that net doping concentration p_0 is not relevant for BO Regeneration whereas B as well as Ga slow down the Regeneration process. An explanation within the H-model of BO Regeneration is given.

Keywords: Czochralski, BO defects, Boron-Oxygen, degradation, regeneration, hydrogen

1 MOTIVATION

1.1 BO defects

Boron doped crystalline Si with high oxygen content, *e.g.*, Cz-Si, shows severe lifetime degradation under carrier injection due to boron-oxygen related(BO) defects [1-3]. The BO defects can be deactivated using the so-called Regeneration mechanism that is triggered by applying a combination of slightly elevated temperatures (typically 60-230°C) and carrier injection [4], applied on hydrogenated Si wafers or solar cells [5-7].

1.2 The H-model of BO-Regeneration

The explanations of BO Regeneration kinetics in this paper are based on the H-Model of BO Regeneration [7-9] assuming that the deactivation of recombination active BO defects is caused by their being passivated with hydrogen(H). This makes a high H concentration within the Si bulk desirable. H is usually bound at temperatures typically used for BO Regeneration (60-230°C) [10]. Hence, it has to be released from one of its bound states under carrier injection (*e.g.*, 1 sun illumination) to become atomic and mobile. BH pairs were suggested as a possible H source under Regeneration conditions [8] as they can be split more easily by carrier injection [11,12]. Additionally, carrier injection plays an important role concerning the charge states of hydrogen: The degraded BO defects [13] as well as atomic hydrogen in p-type Si [14] are both positively charged and would therefore suffer from coulombic repulsion. However, the minority charge states H^0 and H^- form in higher concentrations under carrier injection [15] and may be able to interact with the degraded BO defects leading to the deactivation of their recombination activity.

Instead of this direct H-passivation of BO defects, H could also be catalytically involved in the BO Regeneration process. Based on the data available until now this cannot be excluded because a catalytic process could depend on the H concentration and charge state under Regeneration conditions in the same way as if direct H-passivation of BO defects is assumed. We point out that the injection correction presented in this paper is independent of the model used to describe the physics of the BO Regeneration reaction.

1.3 Injection dependence of BO Regeneration rates

Herguth *et al.* showed already in 2008 that BO Regeneration is accelerated by increasing illumination intensity [16,17]. Very probably, this is due to different injection levels within the differently illuminated samples. Until now this fact has usually been ignored when the Regeneration kinetics of different samples has been compared: If they feature different effective minority carrier lifetimes this causes different injection levels despite constant illumination. Hence, Regeneration rates measured in different samples can only be compared directly if temperature and illumination intensity are kept constant and if anything affecting minority carrier lifetime (material, surface passivation and many processing steps) is the same for all samples. But unfortunately, varying all those parameters is key to understanding the physics of BO Regeneration as well as the possibilities of technological optimization of Regeneration processes. Thus, injection level effects have to be compensated for by adjusting the injection level within different samples during regeneration either experimentally using different illumination intensities or by correcting the effect of those differences mathematically.

Since the experimental solution is hardly feasible, the first part of this paper aims at presenting an empiric correction function that can be used to compensate for different injection levels during Regeneration treatment.

In the second part of the paper, the influence of different impurities on Regeneration kinetics is investigated. The new correction method is used to compare samples differing in the concentration of (1) metallic impurities, (2) interstitial oxygen and (3) dopants B, P and Ga. New data is presented and the results given in literature concerning this topic are discussed.

2 EXPERIMENTAL DETAILS

The material used for the different investigations is described in the respective sections of the paper. For all wafers a $POCl_3$ gettering step is applied except for denoted reference samples. The diffused layer is removed thereafter in hot NaOH. After RCA cleaning, plasma enhanced chemical vapor deposition (PECVD) SiN_x:H is deposited on both sides (75 nm each). Finally, all samples are fired in a belt furnace (~800°C peak sample temperature) in order to activate the SiN_x:H surface passivation and to release H from the layer into the Si bulk.

From every material, the so processed lifetime samples undergo the whole Regeneration procedure consisting of (1) BO annealing, (2) complete BO degradation and (3) BO Regeneration at conditions specified within the respective sections. Illumination is done by a halogen lamp. After Regeneration, lifetimes are (4) stability checked and (5) a final BO annealing is done to exclude instable surface passivation. After each step and every few minutes during Regeneration, the effective minority carrier lifetime τ_{eff} is measured and transformed into normalized defect concentrations N*(t) using

$$N^*(t) = \frac{1}{\tau(t)} - \frac{1}{\tau_{annealed}} \quad (1)$$

with τ(t) being the effective minority carrier lifetime at time t and $\tau_{annealed}$ being the lifetime in the annealed BO state where it is not affected by BO defects. N*(t) is fitted using a single exponential function

$$N^*(t) = k \cdot \exp(-R_R.t) + N_0 \quad (2)$$

with constants k and N_0. This yields the characteristic Regeneration rate R_R that is used to compare the Regeneration kinetics of different samples.

The fact that N*(t) can be fitted using a single exponential expression does not mean that the Regeneration rate is a constant during the whole Regeneration process. Even if the rate increases, *e.g.*, by factor 10 between the beginning and the ending of the Regeneration treatment, the extracted defect evolution can still be fitted by a single exponential function within measurement uncertainty. Hence, the measured Regeneration rate can either be a constant or an effective value representing a time dependent Regeneration rate. These considerations are equally valid for all samples, thus this effect does not affect the comparability of different samples regarding their (possibly effective) Regeneration kinetics.

3 RESULTS AND DISCUSSION

3.1 Correction of different injection levels

The injection level during the Regeneration treatment depends on external illumination intensity as well as recombination rate of the carriers within the Si bulk and/or at the Si surface. Hence, the injection level Δn increases with increasing generation rate G (being proportional to the illumination intensity I) and with increasing effective lifetime τ_{eff}. This means

$$\Delta n \sim G \cdot \tau_{eff} \sim I \cdot \tau_{eff} \quad (3)$$

assuming the injection level being constant during the whole Regeneration process. This is virtually impossible because lifetime changes during the process. Because of the fact that this happens likewise in all samples, their kinetics can be compared anyway without any problems. But if samples differ in the degraded (*e.g.*, when samples feature different concentrations of B or O) or the annealed lifetime (*e.g.*, when samples feature different material quality or surface passivation), Δn is systematically shifted during the whole measurement. Thus, the average Δn_{av} is shifted as well. This automatically causes higher (higher Δn_{av}) or lower (lower Δn_{av}) Regeneration rates. Both cases are represented by an average lifetime

$$\tau_{av} = \frac{1}{2}(\tau_{annealed} + \tau_{degraded}) \quad (4)$$

Altogether, Δn_{av} can then be represented by

$$\Delta n_{av} \sim I \cdot \tau_{av} \quad (5)$$

For the empiric determination of factors correcting for different injection levels, Regeneration measurements are carried out on identically processed lifetime samples (Cz-Si, B doped, 1.5 Ωcm) at different illumination intensities I (between 1.0 and 2.7 suns) and different temperatures T (100-200°C). All samples have an effective minority carrier lifetime of ~33 µs in the degraded and ~220 µs in the annealed state, hence the average lifetime is τ_{av} = 126.5 µs. All samples were first BO-annealed (200°C, 10 min, dark) and then degraded completely (40°C, 0,1 suns, ~70 h) before the start of the Regeneration measurement.

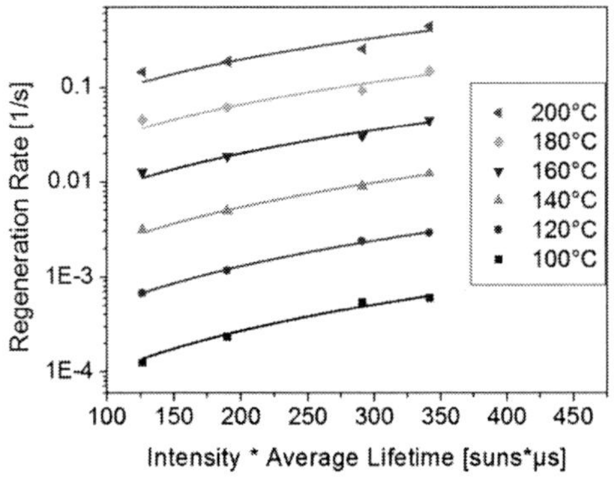

Figure 1: Regeneration rates measured at different injection levels and temperatures. The lines are parabolic fits.

Fig. 1 shows the measured Regeneration rates R_R plotted over $I^*\tau_{av}$ (in suns*µs) with parabolic fits according to

$$R_R(I, \tau_{av}, T) = C \cdot (A(T) \cdot I + \tau_{av} \cdot I^2) \quad (6)$$

The y-intercept has to be zero because we assume that no Regeneration occurs without carrier injection. Fitting parameter A is temperature dependent according to

$$A(T) = 176 \text{ suns} \cdot \text{µs} \cdot exp\left(\frac{T}{78\text{ K}}\right) \quad (7)$$

with T being the temperature in K.

Since there is only one free fitting parameter ("C"), a corrected R_{R1} can be calculated from a single measured data point R_{R0} using

$$R_{R1} = R_{R0} \cdot \frac{A(T) \cdot I_1 \cdot \tau_{av,1} + (I_1 \cdot \tau_{av,1})^2}{A(T) \cdot I_0 \cdot \tau_{av,0} + (I_0 \cdot \tau_{av,0})^2} \quad (8)$$

where subscript 0 denotes the measured values and subscript 1 the corrected ones. This may not be valid in high injection. An uncertainty of ~25% is to be added to the measurement uncertainty of the uncorrected value. More data would be needed to minimalize this additional uncertainty.

A correction like this one should always be applied when the BO Regeneration rates of samples featuring different average injection levels are to be compared. Reasons for these differences can be, *e.g.*, different illumination intensities during Regeneration, different surface recombination velocities due to different surface passivation or different bulk lifetime due to different material composition. Comparing Regeneration rates of samples illuminated by different intensities is the most obvious case where an injection level correction in required. Differences in the effective minority carrier lifetime are less obvious and will be discussed in the following sections.

3.2 Application I: Correction of different surface recombination velocities

The samples presented in this section were all made from the same material but differ in surface passivation layer (SiN_x:H vs. a stack of thermal SiO_2 and SiN_x:H) as well as in the presence of a diffused emitter layer (50Ω/□) that has been etched off in some of the samples.

Table I: Samples used to investigate the influence of different surface passivation and the presence of an emitter layer on BO Regeneration kinetics.

Sample	A	AE	B	BE
Surface Passivation	SiN_x:H	SiN_x:H	SiO_2+ SiN_x:H	SiO_2+ SiN_x:H
Emitter Layer	no	yes	no	yes

After BO-annealing and degradation, all samples undergo a Regeneration procedure at 130°C, 1.2 suns. The measured Regeneration rates are between 0.4 h^{-1} and 1.9 h^{-1} (see Figure 2 for details), thus they differ by more than factor 4. At first glance, one could reason that samples covered by a stack of silicon oxide and silicon nitride regenerate faster than samples featuring only silicon nitride passivation. In addition, the presence of an emitter layer seems to slow down the process. But are these differences in the Regeneration rate really caused by the presence of the SiO_2 interlayer or by the emitter being somehow involved in the Regeneration of BO defect? Or can everything be explained by the simple fact that the different sample structures cause different injection levels during Regeneration though illumination intensity was kept constant?

Indeed, the highest background lifetime (not affected by BO defects) as well as the highest Regeneration rate are found in the samples B (without emitter layer, passivated by a stack of SiO_2 + SiN_x:H) (τ_{av} = 190 µs, R_R = 1.9 h^{-1}). What if these samples had the same average lifetime as samples A (without emitter layer, passivated only by SiN_x:H)? Applying the correction function (8) to the measured R_R using the average lifetime values given in Figure 2 reduces R_R to the same value (within measurement uncertainty) that has been measured in samples A (τ_{av} = 120 µs, R_R = 1.1 h^{-1}). Thus, the higher R_R measured in samples B can be explained by the injection level being higher compared to samples A – despite constant illumination intensity. Further applying the correction function (8) using the average lifetime values of samples AE and BE, respectively, results in further reduced R_R. The rates that have been deduced by injection correction from samples A and B (Fig. 2, green diamonds) are identical to the measured values in samples AE and BE. There may be small systematic differences, but more data is needed to investigate if there is a small real effect of an emitter layer in Regeneration kinetics or if all differences can be explained by the obvious different background lifetime caused by recombination active defects within the emitter layer.

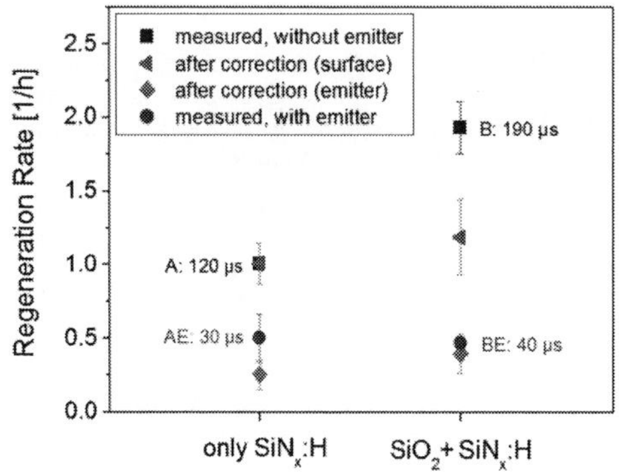

Figure 2: Regeneration rates (black squares) measured in lifetime samples passivated with SiN_x:H (samples A & AE, left) or with a stack of thermal SiO_2 and SiN_x:H (samples B & BE, right). The average lifetime values are given. The injection correction shows the reason for the different rates: If samples B had an average lifetime of 120 µs (as samples A) they would feature the same Regeneration rate as samples A (blue triangles: rates after correction to τ_{av} = 120 µs). If the average lifetime was 30 µs or 40 µs, respectively, as measured in the samples with emitter, the Regeneration rate would be even lower (green diamonds: rates after correction to τ_{av} = 30 µs (A or 40 µs (B)). The values calculated by injection correction from the values measured without emitter (measured: A & B, black squares; after correction: green diamonds) coincide with the measured values (AE & BE, red circles) within measurement uncertainty.

3.3 Application II: Correction of different bulk lifetimes

3.3.1 Interstitial oxygen

Si containing more oxygen normally has lower background lifetime (measured in the annealed BO state) than samples containing less oxygen. In addition to that, lifetime in the degraded BO state decreases by increasing $[O_i]$ [18]. Since R_R is extracted from a single exponential fit to N*(t), the lifetime values at the beginning of the Regeneration process strongly affect the calculated R_R: Lower lifetime in the degraded state automatically results in reduced R_R due to the lower injection level in samples containing high $[O_i]$. Hence, the degraded as well as the annealed lifetime differ for samples that differ in $[O_i]$ and one has to pay attention to the fact that the same illumination intensity leads to different injection levels during Regeneration.

Lim *et al.* showed in [19] that interstitial oxygen O_i decelerates BO Regeneration with $R_R \sim [O_i]^{-1.5\pm0.2}$. This is the measured result without any injection correction. Now, the injection correction (8) is applied using the lifetime values published in [19] (see Table) and the injection corrected Regeneration rates are shown in (Fig. 3). Plotting the corrected Regeneration rates over $[O_i]$ on double logarithmic scale reveals a slope of 0.8±0.1 meaning $R_R \sim [O_i]^{-0.8\pm0.1}$. Hence, interstitial oxygen is less detrimental than previously assumed.

We can confirm this by our own data (black squares in Fig. 3) also yielding a slope of -0.8±0.2. Considering the huge measurement uncertainty due to limited access to Cz silicon differing in $[O_i]$ the real influence of $[O_i]$ on BO Regeneration kinetics could be even weaker. More data is needed to quantify this influence.

Table II: Lifetime values and Regeneration rates $R_{R,meas}$ measured on samples featuring different oxygen content (taken from [20]) as well as the Regeneration rates corrected to τ_{av} = 200 μs (last column).

$[O_i]$ (10^{17} cm^{-3})	τ_{max} (μs)	τ_{min} (μs)	τ_{av} (μs)	$R_{R,meas}$ (h^{-1})	$R_{R,corr}$ (h^{-1})
2.67	690	90	390	1.18	**0.60**
3.61	575	55	315	0.71	**0.45**
4.98	405	30	218	0.40	**0.37**

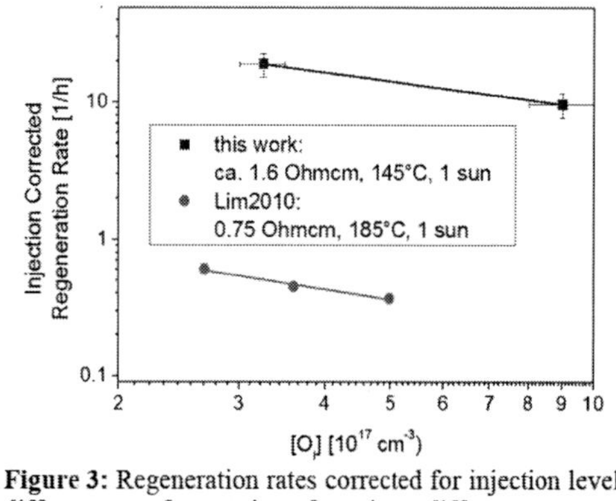

Figure 3: Regeneration rates corrected for injection level differences of samples featuring different oxygen concentrations. Both linear fits reveal a slope of -0.8.

3.3.2 Phosphorus gettering

Lim et al. presented in [21] that a $POCl_3$ gettering step accelerates BO Regeneration by a factor 5 from $R_R = 0.02\ h^{-1}$ without diffusion to $R_R = 0.11\ h^{-1}$ after $POCl_3$ diffusion. This is still valid after the correction of injection level differences during Regeneration according to (8). The measurement was done at 135°C, 0.7 suns. Our samples regenerated at 130°C, 0.6 suns (Fig. 4) seem to contradict these results. When comparing the Regeneration rates of samples that had received a $POCl_3$ gettering step ($R_R \approx 2.5\ h^{-1}$) with undiffused reference samples ($R_R \approx 2.0\ h^{-1}$), only slightly different Regeneration rates are found.

However, there is another striking difference between both investigations: The Regeneration conditions are very similar but the Regeneration rates showed by Lim et al. are a factor 100 lower than the ones presented here. This is very probably due to lower H content within the wafers because they did not receive a high temperature firing step. On the other hand, different metallic impurities that can be partly removed by a $POCl_3$ gettering step are known to trap H in Si. Hence, they can reduce the amount of H available during BO Regeneration and therefore decelerate the Regeneration process. Comparing the two data sets ([21] and this paper) suggests that H trapping by metallic impurities is especially harmful to BO Regeneration if the overall H concentration within the wafer is low.

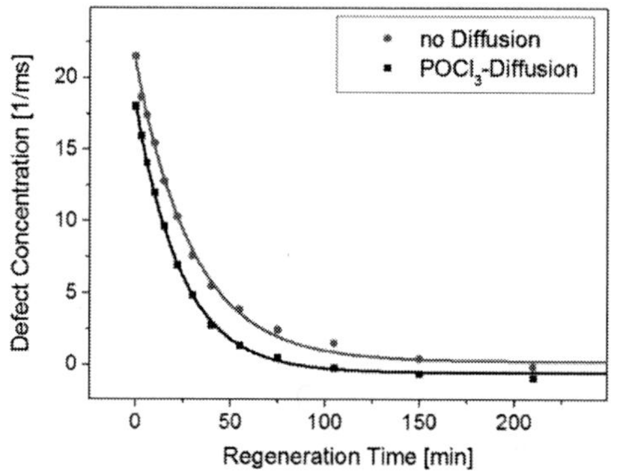

Figure 4: Evolution of the normalized BO defect concentration during Regeneration at 130°C, 0.6 suns for lifetime samples after a $POCl_3$-gettering step has been applied (black squares) or not (blue circles). They show almost the same Regeneration kinetics.

3.3.3 Dopants in compensated Si

Lim et al. showed in [22] that the BO Regeneration rate decreases with increasing B concentration. Compensated p-type silicon has then been used to determine whether this is an effect of the net doping or the absolute B concentration [22,23]. No clear answer has been found yet and it is not clear whether the measured difference in the Regeneration rates really originates from different net dopings / B concentrations or if it could also be an injection effect caused by differences in the background lifetime that automatically come with different dopant concentrations.

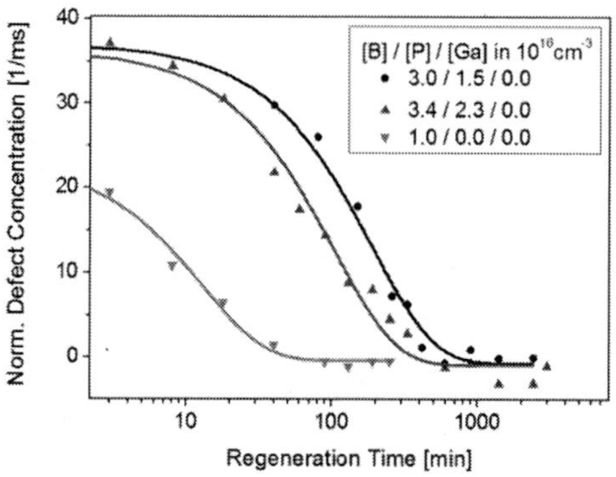

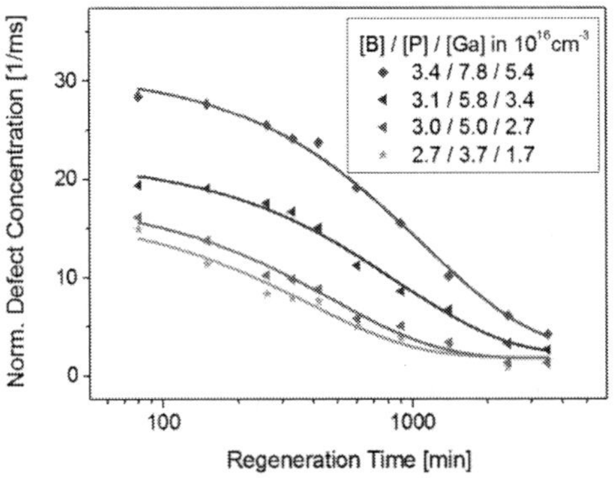

Figure 5: Evolution of the normalized BO defect concentration during Regeneration at 130°C, 1.1 suns for samples containing different amounts of B, P and Ga.

Here, we present data on the BO Regeneration kinetics of p-type Cz wafers doped with differing amounts of B, P and Ga. Details are given in Fig. 5. [B] was determined in the uncompensated samples by four point probe resistivity measurements. [B], [P] and [Ga] in the co-doped wafers were calculated from their ingot position using Scheil's equation. $[O_i]$ was determined using FTIR spectroscopy with the calibration factors by Borghesi *et al.* [24]. The concentrations are $[Oi] = (6.2 \pm 0.6) \cdot 10^{17}\ cm^{-3}$ in all compensated samples and $[O_i] = (8.0 \pm 0.2) \cdot 10^{17}\ cm^{-3}$ in the uncompensated ones. These slight differences were not mathematically corrected because they affect the measured Regeneration rates by less than 20%; thus they are not relevant in this study. Some of the results shown here has already been published in [25].

Regeneration is done at 130°C, 1.1 suns. All lifetimes are stable after Regeneration (tested at 45°C, 0.2 suns); thus no accidental BO annealing has occurred. The measured Regeneration rates are corrected for injection differences according to equation (8). The resulting rates R_R are shown in Fig. 6, plotted over the net doping concentration p_0 (left), over the boron concentration [B] (middle) and over the total acceptor concentration [B + Ga] (right).

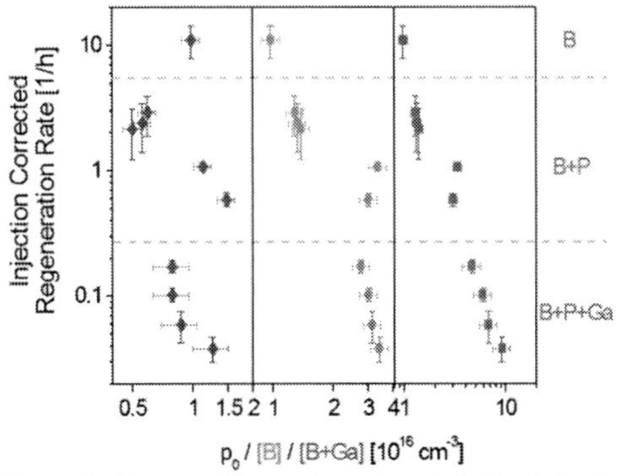

Figure 6: Regeneration rates corrected for injection differences plotted over the net doping p_0 (left), over the boron concentration [B] (middle) and over the total acceptor concentration [B + Ga] (right). Each part contains p-type samples doped with B (upper part), with B + P (middle) and with B +P +Ga (lower part).

Lim *et al.* have shown in [22] that R_R should form a straight line when plotted over the most relevant entity in double logarithmic scaling. Fig. 6 shows that this is definitely not the case if p_0 is chosen as x-axis. $R_R([B])$ rather resembles a straight line. The best match is found when R_R is plotted over the total acceptor concentration [B + Ga]. This means that additionally to B, Ga has negative impact on BO Regeneration as well. In fact, no Regeneration is found in samples containing more than $10^{17}\ cm^{-3}$ Ga.

Due to the limited data set it is not clear whether choosing other entities as x-axis would lead to similar results. *E.g.*, P could also affect R_R and B and Ga may not be harmful to the same extent. Since the samples compensated with P also contain higher [B] than the uncompensated ones and those additionally doped with Ga feature even higher [B] and [P], it is difficult to separate the effects of the respective dopants.

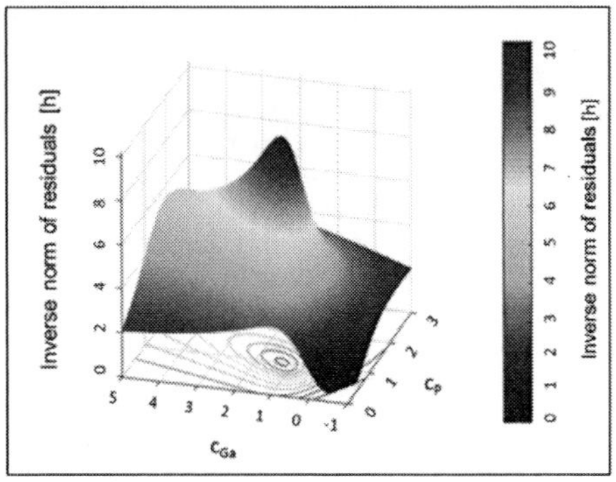

Figure 7: Inverse norm of residuals of the linear fits resulting from different combinations of weighting factors c_P and c_{Ga} for [P] and [Ga] relative to [B]. The dark red region corresponds to the highest inverse norm of residuals and thus to the lowest deviation between fit and data points.

This problem is addressed using a fit optimization approach. It is known that all Regeneration rates should form a straight line when plotted in double logarithmic scaling [5]. Hence, we try to find the best function X which is the entity that will be plotted on the x-axis. X consists of a linear combination of [B], [P] and [Ga]:

$$X = [B] + c_P \cdot [P] + c_{Ga} \cdot [Ga] \quad (8)$$

c_P and c_{Ga} are weighting factors representing how detrimental P and Ga are relative to B. Due to the limited number of data points no possible quadratic effects are considered here. If the optimal combination of c_P and c_{Ga} is found, the norm of residuals of a linear fit on $R_R(X)$ on double logarithmic scaling is minimized; the inverse norm of residuals is maximized (see Fig. 7).

We find that the best fit results from a combination of $c_P = 1.0 \pm 0.3$ and $c_{Ga} = 1.3 \pm 0.5$. The uncertainty interval (10% deviation from the optimum) hence includes the case of all three dopant being equally harmful to BO Regeneration. This is shown in Fig. 8 giving a slope of -1.98 ± 0.03. Thus, $1/R_R$ depends quadratically on the sum of all dopant concentrations. In the best case shown in Fig. 6 (R_R plotted over [B + Ga]), the fit uncertainty is a factor 10 higher. The single data point that does not fit in a straight line has been ignored during the fit optimization procedure although the reasons for its strange position are unclear.

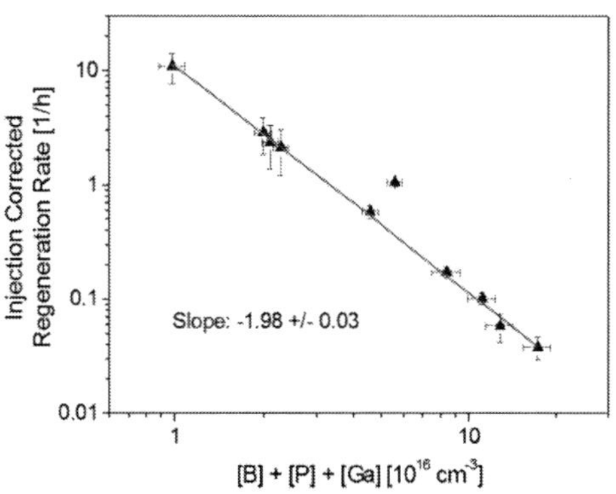

Figure 8: Fit resulting from an optimized weighting factor combination representing a similar impact of B, P and Ga.

Due to the large uncertainty concerning c_P and c_{Ga}, more data is needed to come to a definite conclusion about how harmful P and Ga are for BO Regeneration. In fact, the weighting factors within the interval $0 \leq c_P \leq 1$ and $1 \leq c_{Ga} \leq 2$ lead to less than 10% deviation from the optimal fit. Thus in the most extreme case it is even possible that P does not affect BO Regeneration at all while Ga could be clearly more harmful than B.

Considering these possibilities, different conclusions can be drawn from this finding. Based on the data presented here it is not clear whether P has negative or positive or no impact at all. Additionally, almost all information that is available on the behavior of P in Si comes from studies in n-type Si and it is not clear if these results can be used to explain experimental data from compensated p-type Si. Thus we refrain here from speculating about the possible influence of P on the Regeneration process (see [9] for a detailed discussion including the influence of P on the net dopant concentration, therefore on the quasi Fermi level under illumination and therefore on the H^- concentration during Regeneration) and focus on the negative impact of Ga additionally to B. The following hypotheses are based on the assumption that the influence of Ga is due to the same physical reasons that the influence of B. We also assume that mobile H atoms are involved in the BO Regeneration process with the concentration of mobile H during Regeneration limiting the Regeneration rate. To be more exact, the concentration of the minority charge state H^- (or possibly H^0) is assumed to determine the actual Regeneration rate [9,26].

Based on this, Ga could influence BO Regeneration kinetics in two different ways either by reducing its concentration or its mobility. First, the fact that H is typically bound at the temperatures used for Regeneration treatments [10] (here: 130°C) means that it needs to be released from a bound state due to the carrier injection applied during Regeneration. BH pairs were found to be a possible H source under Regeneration conditions because the feature a relatively low dissociation energy (1.28 ± 0.03 eV) that is even lowered under carrier injection [8,27]. The dissociation energy of GaH pairs is higher in the dark (1.40 ± 0.03 eV) [8]. No studies on how this value may change under carrier injection have been done yet but the GaH pair may stay more stable than BH. In this case the forming of GaH pairs (*e.g.*, during cool down from a high temperature step) would reduce the amount of H that can later become mobile during Regeneration.

This may explain why more Ga could be more detrimental to BO Regeneration than B but it cannot explain the negative impact of B: changes in the concentration of B could hardly affect the amount of H available during Regeneration (see [9] for details). Hence, the impact of the concentration of acceptor atoms may rather be due to their influence on the H mobility during Regeneration: they may be temporary H traps that can reduce the effective H mobility and therefore the frequency at which H atoms encounter BO defects they can interact with. Assuming this mechanism, anything being able to trap hydrogen will reduce the Regeneration rate: Acceptor atoms compete directly with the equally positively charged BO defects in the degraded state and can thus directly reduce the amount of H^- atoms available for BO passivation. Atoms binding to other H charge states (H^+ or H^0) would reduce the overall concentration of atomic H. Because of the fact that there is an equilibrium distribution of H charge states at a given temperature and injection level [15], this would reduce the H^- concentration and thus the Regeneration rate indirectly. (The H charge state relevant for BO Regeneration could also be H^0 [26] that may be able to interact with almost any impurity leading to direct negative impact of these impurities on BO Regeneration kinetics.)

Combining those aspects we conclude that impurities being able to reduce the mobility of the H species relevant for BO Regeneration (possibly H^-) under Regeneration conditions reduce the BO Regeneration rate, *e.g.*, by temporary trapping. Impurities like Ga could additionally trap H and thus reduce the amount of mobile H available for BO Regeneration.

More data is needed to decide which processes dominate during BO Regeneration. Regeneration measurements at different temperatures on differently doped material could reveal interesting information on how the activation characteristics of BO Regeneration depend on material properties. Regeneration studies on Si co-doped with B and Ga (without P) would be especially informative.

4 SUMMARY

It has been explained why the comparison of Regeneration rates measured in different samples requires experimental or mathematical correction of the injection level that can vary even if the external illumination intensity during Regeneration is kept constant. The proposed correction function includes variations that come from different background lifetimes within the silicon bulk (measured in the annealed BO state) as well as different lifetimes in the degraded BO state due to variations in the concentration of B or O_i. The injection level correction has been applied to allow the investigation how interstitial oxygen, phosphorus gettering steps and different dopants affect BO Regeneration kinetics. It has been found that the impact of O_i on BO Regeneration may be less detrimental than previously assumed. Apparently, gettering steps are only beneficial if poorly hydrogenated silicon is used for investigating BO Regeneration kinetics. The possible role of different dopants is very complex and a numerical method has been presented that can be used to discriminate the influences of different impurities being present at the same time. Additionally to B, Ga was found to slow down BO Regeneration kinetics and may be even more harmful than B. Further data is needed to quantify the influence of P reliably.

5 ACKNOWLEDGEMENTS

We would like to thank ApollonSolar for providing the compensated material and Stanislav Stoyanov for characterizing most of the lifetime samples. Parts of this work have been funded by the German Federal Ministry of Economic Affairs and Energy (0325450A, 0325581, 0325877C).

6 REFERENCES

[1] H. Fischer and W. Pschunder, "Investigation of photon and thermal induced changes in silicon solar cells," in *Photovoltaic Specialists Conference, 10 th, Palo Alto, California*, 1974, pp. 404–411.

[2] J. Knobloch, S. Glunz, D. Biro, W. Warta, E. Schäffer, and W. Wettling, "Solar cells with efficiencies above 21% processed from czochralski grown silicon," in *5th IEEE Photovoltaic Specialists Conference, 1996*. IEEE, 1996, pp. 405–408.

[3] J. Schmidt and K. Bothe, "Structure and transformation of the metastable boron-and oxygen-related defect center in crystalline silicon," *Physical review B*, vol. 69, no. 2, p. 024107, 2004.

[4] A. Herguth, G. Schubert, M. Kaes, and G. Hahn, "A new approach to prevent the negative impact of the metastable defect in boron doped Cz silicon solar cells," in *2006 IEEE 4th World Conference on Photovoltaic Energy Conference*, 2006.

[5] K. Muenzer, "Hydrogenated silicon nitride for regeneration of light induced degradation." in *24th EU PVSEC, Hamburg, Germany*, 2009.

[6] G. Krugel, W. Wolke, J. Geilker, S. Rein, and R. Preu, "Impact of hydrogen concentration on the regeneration of light induced degradation," *Energy Procedia*, vol. 8, pp. 47–51, 2011.

[7] S. Wilking, A. Herguth, and G. Hahn, "Influence of hydrogen on the regeneration of boron-oxygen related defects in crystalline silicon," *Journal of Applied Physics*, vol. 113, no. 19, p. 194503, 2013.

[8] S. Wilking, C. Beckh, S. Ebert, A. Herguth, and G. Hahn, "Influence of bound hydrogen states on bo-regeneration kinetics and consequences for high-speed regeneration processes," *Solar Energy Materials and Solar Cells*, vol. 131, pp. 2–8, 2014.

[9] S. Wilking, "Das Wasserstoff-Modell der Bor-Sauerstoff-Regeneration," Ph.D. dissertation, University of Konstanz, to be published.

[10] S. Pearton, J. Corbett, and T. Shi, "Hydrogen in crystalline semiconductors," *Applied Physics A*, vol. 43, no. 3, pp. 153–195, 1987.

[11] T. Zundel and J. Weber, "Boron reactivation kinetics in hydrogenated silicon after annealing in the dark or under illumination," *Physical Review B*, vol. 43, no. 5, p. 4361, 1991.

[12] C. Seager and R. Anderson, "Two-step debonding of hydrogen from boron acceptors in silicon," *Applied physics letters*, vol. 59, no. 5, pp. 585–587, 1991.

[13] S. Rein and S. Glunz, "Electronic properties of the metastable defect in boron-doped czochralski silicon: Unambiguous determination by advanced lifetime spectroscopy," *Applied Physics Letters*, vol. 82, no. 7, pp. 1054–1056, 2003.

[14] A. Tavendale, D. Alexiev, and A. Williams, "Field drift of the hydrogen-related, acceptor-neutralizing defect in diodes from hydrogenated silicon," *Applied Physics Letters*, vol. 47, no. 3, pp. 316–318, 1985.

[15] C. Herring, N. Johnson, and C. G. Van de Walle, "Energy levels of isolated interstitial hydrogen in silicon," *Physical Review B*, vol. 64, no. 12, p. 125209, 2001.

[16] A. Herguth, G. Schubert, M. Käs, and G. Hahn, "Investigations on the long time behavior of the metastable boron–oxygen complex in crystalline silicon," *Progress in Photovoltaics: Research and Applications*, vol. 16, no. 2, pp. 135–140, 2008.

[17] A. Herguth and G. Hahn, "Kinetics of the boron-oxygen related defect in theory and experiment," *Journal of Applied Physics*, vol. 108, no. 11, p. 114509, 2010.

[18] K. Bothe, R. Sinton, and J. Schmidt, "Fundamental boron–oxygen-related carrier lifetime limit in mono- and multicrystalline silicon," *Progress in photovoltaics: Research and Applications*, vol. 13, no. 4, pp. 287–296, 2005.

[19] B. Lim, K. Bothe, and J. Schmidt, "Impact of oxygen on the permanent deactivation of boron–oxygen-related recombination centers in crystalline silicon," *Journal of Applied Physics*, vol. 107, no. 12, p. 123707, 2010.

[20] B. Lim, F. Rougieux, D. Macdonald, K. Bothe, and J. Schmidt, "Generation and annihilation of boron-oxygen-related recombination centers in compensated p-and n-type silicon," *Journal of Applied Physics*, vol. 108, no. 10, p. 103722, 2010.

[21] B. Lim, K. Bothe, and J. Schmidt, "Deactivation of the boron–oxygen recombination center in silicon by illumination at elevated temperature," *physica status solidi (RRL)-Rapid Research Letters*, vol. 2, no. 3, pp. 93–95, 2008.

[22] B. Lim, A. Liu, D. Macdonald, K. Bothe, and J. Schmidt, "Impact of dopant compensation on the deactivation of boron-oxygen recombination centers in crystalline silicon," *Applied Physics Letters*, vol. 95, no. 23, p. 232109, 2009.

[23] B. Lim, "Boron-oxygen related recombination centers in crystalline silicon and the effects of dopant-compensation," Ph.D. dissertation, University of Hannover, 2011.

[24] A. Borghesi, B. Pivac, A. Sassella, and A. Stella, "Oxygen precipitation in silicon," *Journal of Applied Physics*, vol. 77, no. 9, pp. 4169–4244, 1995.

[25] S. Wilking, M. Forster, A. Herguth, and G. Hahn, "From simulation to experiment: Understanding BO-regeneration kinetics," *Solar Energy Materials and Solar Cells*, vol. 142, pp. 87–91, 2015.

[26] M. Gläser and D. Lausch, "Towards a quantitative model for BO-regeneration by means of charge state control of hydrogen," *Energy Procedia*, vol. 77, pp. 592–598, 2015.

[27] T. Zundel and J. Weber, "Dissociation energies of shallow-acceptor-hydrogen pairs in silicon," *Physical Review B*, vol. 39, no. 18, p. 13549, 1989.

The Development of In-Line Regeneration Tool for the Effective Suppression of Light-Induced-Degradation on P-type Silicon Solar Cells

Kuo-Yi Yen, Jun-Rui Huang, Yi-Feng Lin, Spencer Su, Sean H.T. Chen, and Li-Wei Cheng
Motech Industries INC. Taoyuan Branch
No. 1560, Sec. 1, Zhongshan Rd., Guanyin Dist., Taoyuan City 328, Taiwan (R.O.C.)

ABSTRACT: Recently, many research institutes have reported that generation treatment of illumination and heating is a useful method to transform boron–oxygen defects into a stable regenerated state for suppression of light induced degradation (LID) on p-type PERC Cells. Based on considerations of the feasibility for mass production, the in-line regeneration tool is of much competitiveness and superior properties, including less extra manpower for operation, no additional wafer handling and cost of automations. However, the idea of compact design through the compromise between regeneration conditions and belt speed in coordination with firing furnace is the major challenge for the development of in-line tool. In order to realize the concept, series of experiment conditions, in terms of illumination intensity and duration, have been investigated in this work. The average efficiency of regeneration treated cells with optimal condition shows as less as 0.1% efficiency loss relatively, compared to that of non-regeneration treated cells. Furthermore, the LID results, which are conducted at one sun illumination condition for more than 60 hours, show that regeneration treated cells have significant improvement in degradation behavior with less than 2% degradation while non-regeneration treated cells have as much as 3.6% efficiency degradation. Therefore, the in-line regeneration treatment is a promising process technique to alleviate LID on p-type PERC cells.
Keywords: Boron-oxygen, light induced degradation, regeneration, PERC

INTRODUCTION

The bulk lifetime degradation of light-induced degradation (LID) is found to have a larger detrimental effect on novel high efficiency PERC cells, which the impact of LID increases when a low back surface recombination velocity (BSRV) is achieved [1-5]. The LID issue has become problematic and an obstacle for further adoption of PERC cell concepts. It has been demonstrated that the boron-oxygen-related defects can be permanently deactivated under a light-induced regeneration treatment, which combines actions of temperatures and an external illumination source. [6-14]. The enhanced regeneration treated PERC cells through an optimized regeneration treatment is accomplished that the relative efficiency loss is less than 2% degradation [15,16]. In order to make efficiency regeneration feasible in the cell production, the idea of in-line regeneration tool has attracted attention that the function of regeneration treatment is integrated into metallization firing furnace. It can be expected that a broader adoption of the developing in-line LID prevention technology will lead to a renewed interest in the PERC optimization technologies. In addition, the inline regeneration technology eliminates the need for wafer handling and saves floor space. However, the maintained furnace throughput speeds and desired furnace length result in limiting regeneration duration of light exposure. An investigation of regeneration duration and light intensity which is in order to find fast and stable regeneration conditions will be further presented by using a prototype tool. Based on these test results, the key findings have been transferred into the in-line mass production regeneration tool. Finally, the LID distributions of the pilot run performance by using an in-line regeneration tool are demonstrated.

EXPERMENT

P-type PERC cells with a resistivity ranging from 1~3 ohm-cm and a thickness of 200 μm are used in this work. A regeneration treatment is performed following firing process by using in-line regeneration tool. The in-line regeneration tool has advantages and challenges as compared with stand-alone regeneration tool. As shown in table 1, in-line regeneration tool don't need extra manpower and added cost of automation equipment, and it don't make extra breakage rate due to without additional wafer handling steps. Therefore, the influence of introducing regeneration process to production line can be minimized by using in-line regeneration tool. However, the belt speed coherence between firing and regeneration processes (Belt speed range of firing process for PERC cells is about from 9~11 cm/s) causes limited duration variation of regeneration process. In order to integrate in-line regeneration tool into firing system, the length of in-line regeneration tool is expected to reach shorter than 3 meters. Therefore, the duration of regeneration process have to be less than 30 seconds. Initially, the prototype in-line regeneration tool was used to find out the relation between light intensity and duration of regeneration process and confirm in-line regeneration treatment is useable. The test conditions of light intensity and duration are form 1C to 4C suns (C is a constant) and from 15 to 30 seconds. The generation effect of in-line regeneration treated cells is confirmed by light soaking of one sun intensity for more than 60 hours. Finally, the mass production in-line regeneration tool was introduced to production line of PERC cell to suppress light induced efficiency degradation.

Table 1: Advantages and challenges of an in-line regeneration tool.

In-line regeneration tool	
Advantage	a. Without manpower requirement b. Saves the cost of automation equipment c. Without additional wafer handling steps
Challenge	a. Less feasibility of regeneration duration b. Only supply one screen-printing line

RESULT AND DISCUSSION

The influence of cell efficiency due to introducing in-line regeneration treatment is an important issue, which needs to be investigated by using prototype in-line regeneration tool to keep high efficiency and narrow efficiency distribution like that of non-regeneration treated PERC cells as shown in figure 1(b). Figure 1(a) shows the distributions of relative differences between average efficiency (~20.41%) of non-regeneration treated cells and efficiency of cells with different regeneration conditions. It can be observed that the light intensity and duration increase with cell efficiency increasing, except conditions of 4C suns, and the optimal condition is 2C suns for 30s. Figure 2 shows LID distributions of cells with different regeneration conditions after light soaking. It shows that the increased light intensity and duration of regeneration condition can reduce LID of cells, except a condition of 4C suns for 30s, and two regeneration conditions of 2C suns for 30s and 4C suns for 15s can make LID distribution of cells meets the target (> -2%). Based on the results of efficiency difference and LID distributions, regeneration condition of 2C suns for 30s not only make LID distribution of cells meets the target (>-2%) but also keep high efficiency of cells.

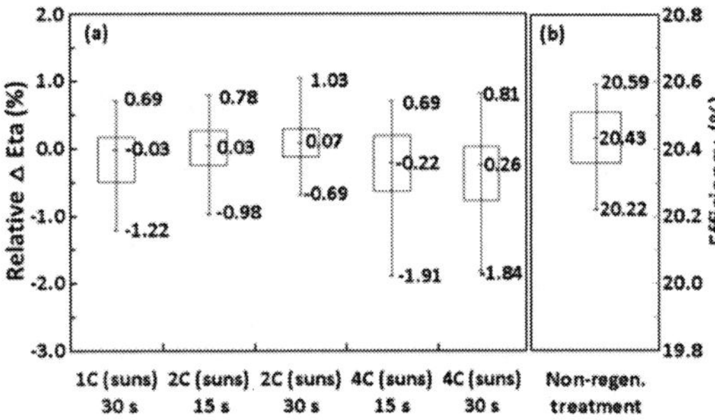

Figure 1: (a) Distributions of relative efficiency difference between efficiency of each regeneration treated cell and average efficiency (20.41%) of non-regeneration treated cells, and (b) efficiency distribution of non-regeneration treated cells.

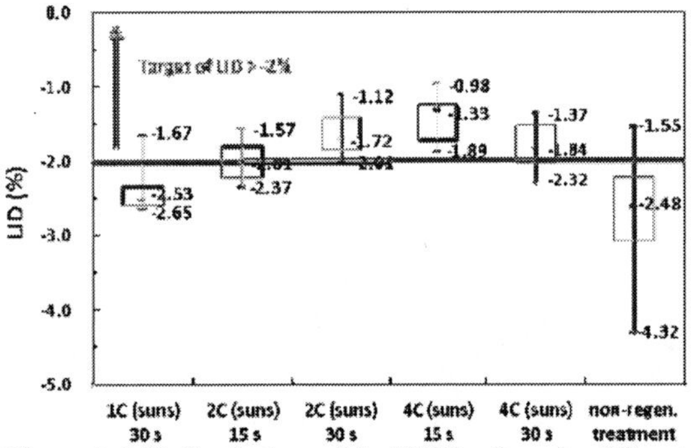

Figure 2: LID distributions of the PERC cells under different regeneration conditions by using a prototype tool.

As the results of prototype in-line regeneration test, the condition of 2C suns for 30s is very suitable to be introduced into mass production. However, the belt speed of firing furnace is increased to obtain higher efficiency of PERC cells in mass production. Consequently, the duration of 24 seconds was used in mass production in-line regeneration tool test. In mass production in-line regeneration tool test, two conditions were used, one keeps light intensity of 2C suns, the other increase light intensity to 3C suns for compensating the lost regeneration time. Table 2 shows electrical properties of non-regeneration treated cells and regeneration treated cells with two different conditions. The average efficiencies of cells with two different regeneration treatments only lose 0.6% and 0.1% relatively as compare with that of non-regeneration treated cells, and the maximum and minimum efficiencies of cells with two different regeneration treatments show narrow efficiency distributions. Figure 3 shows LID distributions of non-regeneration and regeneration treated cells after light soaking. All of regeneration treated cells have less than 2% relative efficiency degradation while non-regeneration treated cells have as much as 3.6% relative efficiency degradation.

Table 2: The conversion efficiency results of non-regeneration and regeneration treated cells by using an in-line regeneration tool.

Group		Eta (%)	Uoc (V)	Isc (A)	FF (%)
Non-regen.	Max.	20.85%	0.6666	9.54	79.57
	Avg.	20.61%	0.6642	9.50	79.30
	Min.	20.23%	0.6568	9.43	79.30
2C (suns) 24 s	Max.	20.69%	0.6640	9.50	79.34
	Avg.	20.48%	0.6635	9.47	79.16
	Min.	20.14%	0.6616	9.43	78.37
3C (suns) 24 s	Max.	20.85%	0.6670	9.55	79.43
	Avg.	20.59%	0.6638	9.49	79.36
	Min.	20.41%	0.6592	9.43	79.73

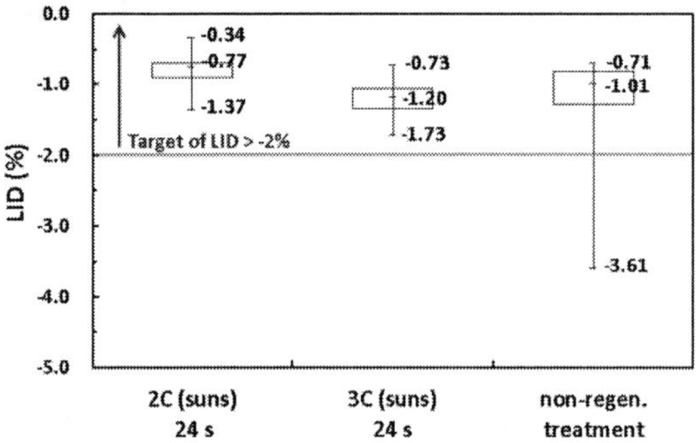

Figure 3: LID distributions of non-regeneration and regeneration treated cells.

CONCLUSION

The in-line regeneration tool has the advantages, including less extra manpower for operation, no additional wafer handling and cost of automations as compared with those of the stand-alone regeneration tool. In addition, the constraint of regeneration duration (< 30s) of in-line tool design has been tackled by adjustments of regeneration treatment conditions. In the prototype regeneration tool test, except conditions of 4C suns, the increased light intensity and duration of generation treatment is useful to sustain high efficiency and alleviate impact of LID on PERC cells. In the mass production regeneration tool test, the average efficiencies of cells with regeneration treatment only lose 0.1% relatively as compared with non-regeneration treatment, and all efficiencies of regeneration treated cells have less than 2% relative degradation after light soaking. The in-line regeneration is a promising process technique to suppress LID of p-type PERC cells and has been introduced into mass production successfully.

REFERENCE

[1] K. Both, and J. Schmidt, "Electronically activated boron-oxygen-related recombination centers in crystalline silicon", J. Appl. Phys., vol. 99, pp.013701, 2006.

[2] B. Lim, K. Both, and J. Schmidt, "Impact of oxygen on the permanent deactivation of boron–oxygen-related recombination centers in crystalline silicon", J. Appl. Phys., vol. 107, pp.123707, 2010.

[3] B. Sopori, P. Basnyat, S. Devayajanam, S. Shet, V. Mehta, J. Binns, and J. Appel, "Understanding light-induced degradation of c-si solar cells", Proceedings 38th IEEE Photovoltaic Specialists Conference, 2012.

[4] K. Petter, C. Fahrland, Y. Ludwig, and F. Kersten, "Light induced degradation of solar cells made from high performance multi wafers", Proceedings 29th European Photovoltaic Solar Energy Conference, 2014.

[5] F. Fertig, K. Krauß, and S. Rein, "Light-induced degradation of PECVD aluminium oxide passivated silicon solar cells", Phys. Status Solidi RRL 9, No. 1, pp.41-46, 2015.

[6] B. Damiani, A. Ristow, A. Ebong, and A. Rohatgi, "Design optimization for higher stabilized efficiency and reduced light-induced degradation in boron-doped czochralski silicon solar cells", Prog. Photovolt: Res. Appl. vol 10, pp185-193, 2002.

[7] A. Herguth, G. Schubert, M. Kaes, and G. Hahn, "Avoiding boron-oxygen related degradation in highly boron doped CZ silicon", Proceedings 21th European Photovoltaic Solar Energy Conference, 2006.

[8] B. Lim, K. Bothe, and J. Schmidt, "Modeling the generation and dissciation of the boron oxygen complex in B-doped CZ Si", Proceedings 33th IEEE Photovoltaic Specialists Conference, 2008.

[9] G. Krugel, W. Wolke, J. Geilker, S. Rein, and R. Preu, "Impact of hydrogen concentration on the regeneration of light induced degradation", Energy Procedia, vol 8, pp47-51, 2011.

[10] F. Fertig, J. Greulich, J. Broisch, D. Biro, and S. Rein, "Stability of the regeneration of the boron-oxygen complex in silicon solar cells during module integration", Solar Energy Materials & Solar Cells, vol 115, pp189-198, 2013.

[11] F. Wolnya, T. Webera, M. Müllera, and G. Fischera, "Light induced degradation and regeneration of high efficiency Cz PERC cells with varying base resistivity", Energy Procedia, vol. 38, pp.523-530, 2013.

[12] F. Fertig, J. Broisch, D. Biro, and S. Rein, "Stability of the regeneration of the boron–oxygen complex in silicon solar cells during module certification", Solar Energy Materials & Solar Cells, vol 121, pp157-162, 2014.

[13] M. Cascant, N. Enjalbert, R. Monna, and S. Dubois, "Influence of various p-type czochralski silicon solar cell architectures on light-induced degradation and regeneration mechanisms", Proceedings 29th European Photovoltaic Solar Energy Conference, 2014.

[14] S. Wilking, J. Engelhardt, S. Ebert, C. Beckh, A. Herguth, and G. Hahn, "High speed regeneration of BO-defects: Improving long-term solar cell performance within seconds", Proceedings 29th European Photovoltaic Solar Energy Conference, 2014.

[15] J. Huang, Y. Lin, K. Liang, S. Su, S. Chen, and L. Cheng, "Investigation of light induced regeneration phenomena on p-type CZ PERC cells", Proceedings 42th IEEE Photovoltaic Specialists Conference, 2015.

[16] K. Yen, S. Su, K. Liang, S. Su, S. Chen, and L. Cheng, "Impact of light-induced degradation on p-type multi-crystalline silicon PERC cells", Proceedings 42th IEEE Photovoltaic Specialists Conference, 2015.

This is a pre-peer review version of the paper submitted to the journal Progress in Photovoltaics.
It has been selected by the Executive Committee of the EU PVSEC 2016 for submission to Progress in Photovoltaics.

DEGRADATION AND REGENERATION IN MC-SI AFTER DIFFERENT GETTERING STEPS

Annika Zuschlag, Daniel Skorka, Giso Hahn
University of Konstanz, Department of Physics
78457 Konstanz, Germany

ABSTRACT: Light and elevated temperature induced degradation (LeTID) affects significantly the performance of multicrystalline (mc) Si PERC solar cells, and underlying mechanisms of LeTID are still unknown. In this work LeTID and following regeneration of an industrial mc-Si PERC solar cell is compared to differently processed minority charge carrier lifetime samples under illumination (1 sun) and elevated temperature (75°C).
LeTID on cell level reveals the same kinetics compared to lifetime samples. Varying the processing sequence has a significant effect on LeTID of lifetime samples. Ungettered samples with fired SiN_x:H surface passivation show a very strong LeTID and regeneration effect, with degradation kinetics being similar for all wafer areas irrespective of initial material quality. In contrast, regeneration sets in earlier in good quality areas.
Differently gettered samples are less sensitive to LeTID, while overall degradation and regeneration behaviour is strongly influenced by applied gettering sequences. Al-gettered samples show a more pronounced degradation effect than P-gettered samples, leading to the assumption that P-gettering is more effective in the reduction of LeTID sensitive defects.
A model is presented, describing LeTID in boron as well as gallium doped mc-Si being based on impurities that can be gettered and redistributed during high temperature steps.
Keywords: Degradation, Regeneration, LeTID, Multicrystalline silicon, Lifetime

1 INTRODUCTION

Light and elevated temperature induced degradation (LeTID) is a recently discovered effect on multicrystalline (mc) Si PERC (passivated emitter and rear contact) solar cells [1-3]. This effect can cause efficiency losses up to almost $10\%_{rel}$ under illumination and elevated temperature [3] and cannot be attributed to standard BO complex formation or FeB pair dissociation [1]. In mc-Si PERC solar cell concepts, a high diffusion length (or high minority charge carrier lifetime, respectively) can be used to further increase cell efficiency to values >21% [4]. In contrast, for standard full area Al back surface field cells the high bulk material quality does not lead to higher efficiencies due to the limiting recombination at the cell's rear side. Therefore, LeTID can be a showstopper when applying PERC technology to mc-Si for reaching higher efficiencies.

Also a regeneration process after degradation has been observed by Kersten et al. [3], resembling the behaviour known for BO-correlated carrier induced degradation and regeneration [5, 6, 7], but on a totally different time scale (as already mentioned in [3]). The kinetics of degradation and regeneration is influenced by illumination and temperature, while the underlying mechanisms causing the degradation and regeneration of mc-Si PERC solar cells are still unknown.

Recent investigations to clarify the parameters influencing the degradation and regeneration process are mainly based on solar cell or even on module level. Investigations on lifetime samples are still rare. Kersten et al. [3] have demonstrated the degradation of lifetime samples with different surface passivations, while Krauss et al. [8] used spatially resolved minority charge carrier lifetime maps to calculate a so-called effective defect lifetime allowing an imaging of effective defects activated upon degradation. Bredemeier et al. [9] have recently published a lifetime study on the influence of different firing temperatures on the lifetime degradation and regeneration.

The advantage of lifetime studies is that a more flexible design of experiment is possible (*e.g.*, regarding applied process steps/sequences), which can be useful to identify further parameters influencing degradation and regeneration behaviour. This work demonstrates minority charge carrier lifetime degradation and regeneration on mc-Si wafer level. The different gettering sequences result in different contamination levels and impurity distributions. Due to its flexibility, the approach chosen in this contribution based on lifetime samples allows to gain a more detailed fundamental understanding of the underlying mechanisms of LeTID.

2 EXPERIMENTAL

An industrial mc-Si PERC solar cell and several lifetime samples are investigated to compare the degradation and regeneration process on solar cell and lifetime level. The investigation of mc-Si PERC solar cells allows to link and compare the results to previously published results on LeTID, while the studies on differently processed lifetime samples are focused on identification of process steps influencing degradation and regeneration behaviour of mc-Si material.

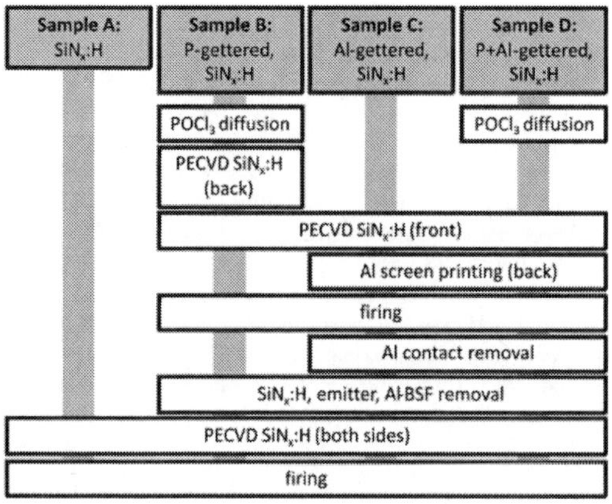

Figure 1: Process sequence of lifetime samples.

The mc-Si wafers (1.5 Ωcm boron doped) used for lifetime samples originate from the same mc-Si ingot and the same ingot height as the investigated standard industrial 156x156 mm^2 mc-Si PERC solar cell, leading to a comparable material quality of the initial material. Sister wafers (5x5 cm^2) with comparable grain and defect structure were used to compare the influence of different process steps on degradation and regeneration behaviour of the lifetime samples. A scheme of the applied process sequences is given in Fig. 1. The wafers were chemically etched to remove saw damage. Sample A was not further processed and only a surface passivation was realized by firing of a plasma-enhanced chemical vapour deposition (PECVD) SiNx:H layer. For samples B, C, and D different gettering sequences were applied. Sample B was gettered by a $POCl_3$ diffusion (55 Ω/□), while sample C was gettered by screen-printing and firing of an Al contact on the backside. P- and Al-gettering was combined in case of sample D. Additionally, a PECVD SiN_x:H layer was applied to all bare wafer surfaces to protect the wafers from contamination during the following firing step. Afterwards, the SiN_x:H layers as well as the Al back surface field and the emitter were removed. The samples were surface passivated again by firing (measured wafer temperature: 730°C) of a PECVD SiN_x:H layer (same deposition parameters as before) leading to an additional hydrogenation of the investigated samples.

For degradation, lifetime samples as well as the solar cell are held at a temperature of approx. 75°C on a hot plate under illumination with halogen lamps (0.9±0.05 suns for lifetime samples, 1±0.05 sun for the solar cell). During the degradation and regeneration process the solar cell is characterized in situ by automated open circuit voltage (V_{oc}) measurements. For lifetime samples, effective minority charge carrier lifetime (τ_{eff}) is measured repetitively by the fast and self-calibrated time resolved photoluminescence imaging (TR-PLI) method [10, 11] at room temperature, resulting in a series of spatially resolved lifetime maps for each sample over degradation time. The first τ_{eff} data point of the lifetime samples is measured directly (approx. 2 min) after the last firing step consistent with standard solar cell characterization, which takes usually place directly after firing in industrial production lines.

For a statistically relevant analysis of areas differing in material quality, the TR-PLI lifetime maps of each sample over time are aligned and an array of 2,500 areas (each 150x150 µm^2 in size) at fixed positions is distributed over the TR-PLI lifetime map of the 5x5 cm^2 mc-Si samples to get good statistics. Average τ_{eff} values within these areas are extracted over degradation time and further analyzed. This method allows tracking of changes in τ_{eff} of different sample areas under degradation conditions. The advantage of this approach is that the spatially resolved information of τ_{eff} maps over degradation time can be easily displayed. The τ_{eff} data for each 150x150 µm^2 area on the samples is plotted over time and coded by a rainbow colour bar based on the areas' lifetime values at the beginning of the degradation experiment.

3 RESULTS AND DISCUSSION

Continuously measured V_{oc} data of an industrial mc-Si PERC solar cell are shown in Fig. 2. It is obvious that this PERC solar cell is LeTID sensitive and the V_{oc} data shows a degradation of approx. 12% rel. after 200 h followed by regeneration as also observed previously, *e.g.*, in [3]. After 1500 h the solar cell recovers almost completely. As already discussed in, *e.g.*, [1, 3] the observed degradation and regeneration cannot be fully explained by BO-correlated degradation or FeB pair dissociation processes.

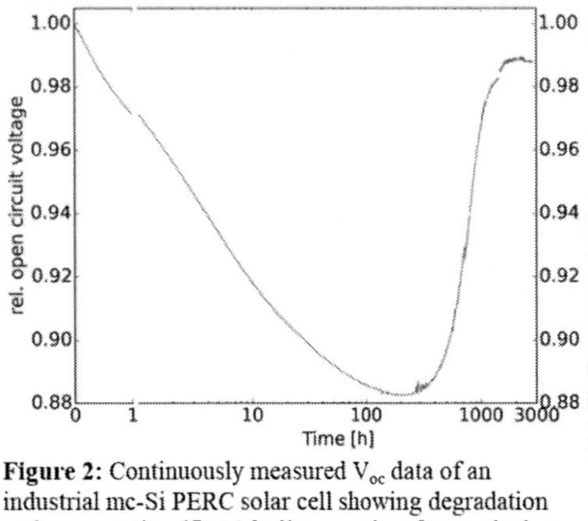

Figure 2: Continuously measured V_{oc} data of an industrial mc-Si PERC solar cell showing degradation and regeneration (first 1 h: linear scale, afterwards: log scale).

Fig. 3 shows the behaviour of harmonically averaged τ_{eff} for the four different processed lifetime sister samples A-D. All samples show fast degradation within the first minutes which could be attributed to FeB- and/or BO-based degradation effects. After longer periods, significant differences are measured for samples A, B, C, and D regarding the strength of the degradation, the range of minimum and maximum τ_{eff} in dependence of the degradation state, and the regeneration behaviour which can be clearly observed for sample A and also slightly for samples C and D. The ungettered sample A shows the strongest degradation but also the strongest relative regeneration. Applied gettering steps (samples B, C, D) reduce the degradation effect significantly. While samples with Al-gettering step (sample C and D) show a lifetime minimum due to degradation, no clear minimum could be observed in case of the only P-gettered sample B.

First investigations on a set of gallium doped mc-Si lifetime samples with and without P-gettering step result in a slower degradation at the beginning of the experiment, most probably due to missing BO- or FeB-induced degradation, strengthening the hypothesis that other impurities are responsible for the so called LeTID effect, as already observed by Ramspeck *et al.* [1] on solar cell level. Also for gallium doped lifetime samples, the degradation in ungettered samples is more pronounced compared to P-gettered samples [12].

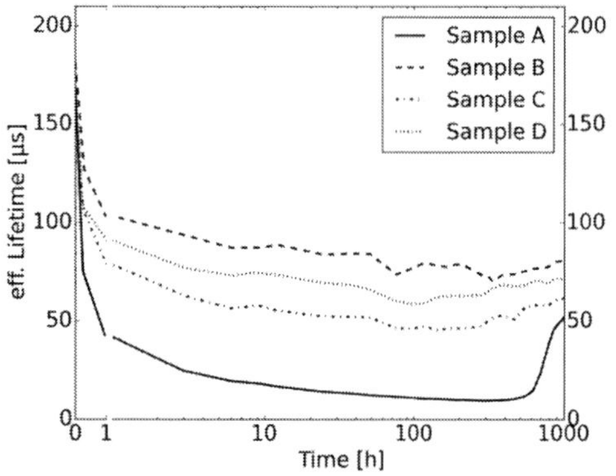

Figure 3: Harmonically averaged τ_{eff} data for differently processed mc-Si wafers. Degradation and regeneration behaviour of lifetime samples can be strongly influenced by previously applied gettering sequences (first 1 h: linear scale, afterwards: log scale).

A comparison to the degradation behaviour of the solar cell shows that the maximum degradation level for average τ_{eff} in Fig. 3 and the V_{oc} value in case of the solar cell (Fig. 2) is reached after approximately the same degradation time using (almost) the same illumination level and elevated temperature. This leads to the assumption that the underlying degradation and regeneration effect can be studied either on lifetime or on solar cell level.

The disadvantage of average τ_{eff} analyses is that information on local lifetime distribution measured by spatially resolved TR-PLI gets lost. Fig. 4 shows exemplarily τ_{eff} maps of sample A at three different states: at the beginning of the degradation, in the degraded state, and during the regeneration process. The whole series of these spatially resolved TR-PLI measurements have been aligned and analyzed as described above.

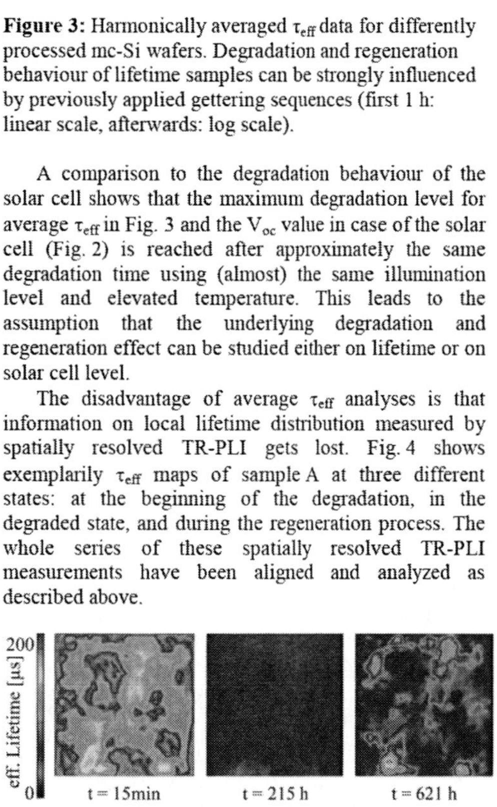

Figure 4: Spatially resolved TR-PLI measurements of sample A at the beginning of the applied degradation process, in the degraded state and during regeneration.

Further analyses of the degradation and regeneration behaviour will be discussed based on Fig. 5. Each single line (guides to the eye connecting single data points) in Fig. 5 represents one of the 2,500 small 150x150 μm^2 areas on the lifetime samples.

The applied colour code, based on the initial lifetime of each sample at the beginning of the experiment, shows that for all four samples the relative τ_{eff} distribution is maintained during the experiment, which means that independently of the applied process sequence, sample areas with the highest initial τ_{eff} show also the highest τ_{eff} at the maximum degradation level as well as during the regeneration process. This holds true even if the range of lifetime is narrowed down during degradation or is spread again during regeneration.

In case of sample A (not gettered, Fig. 5a), a very strong and fast degradation of the whole sample is observed. As already mentioned for harmonic average τ_{eff} data, FeB- and/or BO-induced degradation could be involved additionally to LeTID in the first few minutes, but the degradation behaviour cannot be fully explained based on FeB- and/or BO-induced dynamics. Especially sample areas of higher initial τ_{eff} suffer from the applied degradation conditions, and relative degradation in these areas is stronger than in sample areas with lower initial τ_{eff}.

The initial lifetime distribution of sample A (approx. 30–350 µs) narrows down to a very small range (approx. 5–30 µs) at maximum degradation level. This strong degradation even in areas of initially high τ_{eff} could be explained by an approximately homogenous formation of defects over the wafer area. The degradation kinetics is very similar in all sample areas, independently of the initial τ_{eff}. After approximately 300 h a regeneration process sets in, with regeneration in areas with higher initial τ_{eff} (red lines) starting significantly earlier than in areas of lower initial τ_{eff}, indicated by a black dashed line in Fig. 5a, and also by the inlay and the differently coloured arrows in Fig. 5a. The earlier onset of regeneration in areas of initially high τ_{eff} is the reason why in these regions the minimal visible τ_{eff} level is reached earlier compared to areas with lower initial τ_{eff}. But it has to be assumed that initially good τ_{eff} areas show very similar degradation kinetics (time constants) than initially poor τ_{eff} areas. The different starting points for observable regeneration are remarkable, especially regarding the narrow τ_{eff} distribution at around 200 h. The different regeneration behaviour leads to the assumption that the regeneration is not only a reversal of the first degradation reaction and that the underlying mechanism is more complex. It is not yet clear whether τ_{eff} recovers completely.

Up to now it is not yet clear whether the observed degradation within the first hours under illumination originates from bulk defect formation or changes in surface passivation quality. Therefore, an additional sample is prepared exactly like sample A, and the same degradation conditions are applied leading to a comparable degradation behaviour as observed for sample A. After approx. 300 h, the SiN_x:H surface passivation layer is etched back and a chemical surface passivation (iodine ethanol, [13]) is applied. The previously measured spatial distribution of low τ_{eff} values after 300 h degradation is thereby confirmed, verifying that the observed strong degradation is a bulk effect and not related to changes in SiN_x:H surface passivation quality. There are hints that SiN_x:H surface passivation quality slightly decreases after several hundred hours under illumination at 75°C. This might influence the long time regeneration behaviour (>1000 h), but cannot explain the observed degradation.

The P-gettered sample B (Fig. 5b) shows only a slight lifetime reduction within the first minutes under illumination at elevated temperature and is therefore significantly less sensitive to LeTID than sample A. In contrast to the ungettered sample A discussed above, the τ_{eff} range (approx. 40–360 µs) is only slightly narrowed down and stays nearly constant during the experiment.

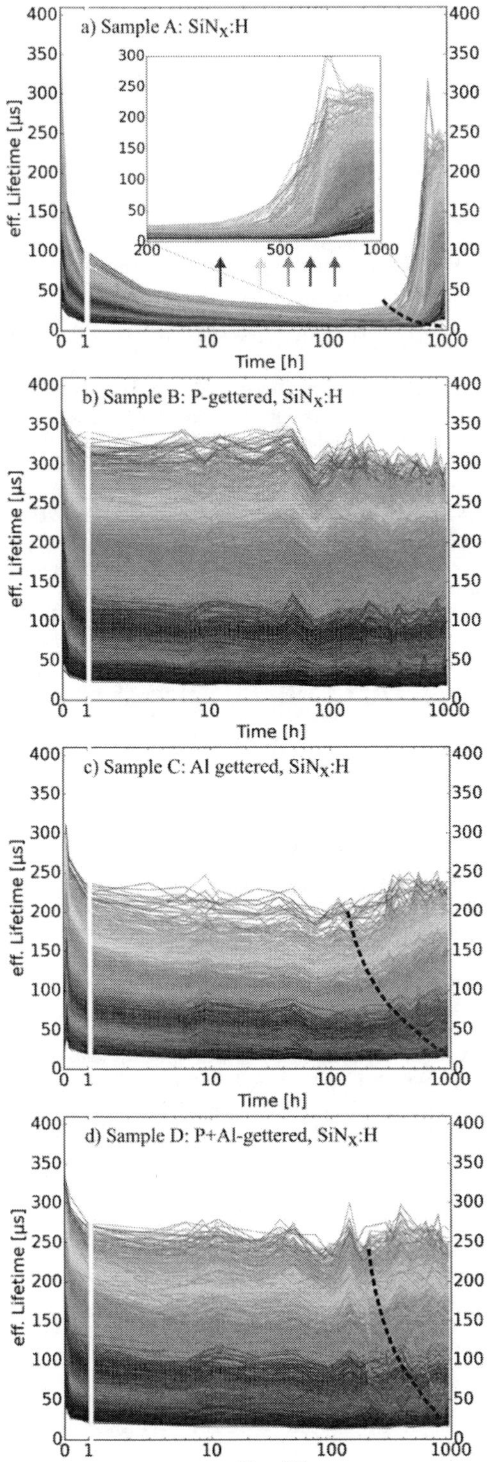

Figure 4: τ_{eff} of differently processed sister wafers (first 1 h: linear scale, afterwards: log scale). Each line in the graph represents an area of 150x150 µm² in the spatially resolved TR-PLI lifetime measurements and is colour-coded to its value at the beginning of degradation. Black dashed lines indicate beginning regeneration.

The data shows no clear maximum degradation level and no regeneration, and τ_{eff} stays almost constant in the different sample areas. Areas of lower initial lifetime after gettering show more relative degradation than areas of higher material quality.

In case of sample B, the repetition of the PECVD SiN_x:H deposition and firing as shown in Fig. 1 has been performed to get a similar temperature load compared to the Al-gettered samples C and D. If the degradation conditions are applied to a P-gettered sample after the first firing of the SiN_x:H passivation layer, comparable results to sample B are observed on an additional wafer set with comparable initial material quality. If the investigated sample area is strongly dominated by extended defect clusters of lower material quality, a more pronounced degradation could be observed also for P-gettered samples (not shown). This hints towards a complex interaction of gettering efficiency on local defect structures (as seen, *e.g.*, in [14]) and the effect on degradation behaviour.

In contrast to sample B, the Al-gettered sister sample C shows a significantly different degradation and regeneration behaviour. Overall degradation and regeneration behaviour is closer to the ungettered sample A than to the P-gettered sample B. As observed for sample A, the Al-gettered sample C shows also a maximum degradation level and a narrowing of the τ_{eff} range, but by far not as significant as sample A. The regeneration process again sets in earlier for sample areas with higher initial τ_{eff} as indicated by a black dashed line in Fig. 5c.

For sample D, both Al- and P-gettering were applied. Degradation is stronger than in case of sample B but less pronounced compared to sample C. Also a beginning regeneration can be hinted as indicated by a black dashed line in Fig. 5d. Therefore, the degradation and regeneration behaviour of sample D is a mix of the observed results of the only P-gettered sample B and the only Al-gettered sample C.

Due to the expected different contamination level or impurity distribution of the gettered samples B, C, and D compared to sample A, it is assumed that impurities strongly influence the degradation and regeneration behaviour. A P-gettering step seems to be more efficient to remove degradation sensitive defects than an Al-gettering step. After P-gettering, each following temperature step leads to a redistribution of (internally or externally) gettered degradation relevant defects. Besides firing steps, the temperature loads during surface passivation processing steps have to be taken into account. In [15], P-gettered lifetime samples with different surface passivation treatments (including stack systems like AlO_x+SiN_x:H and a thermal SiO_2 covered by SiN_x:H) are compared. It can be demonstrated that for samples with additional temperature load after the P-gettering step a more pronounced LeTID and following regeneration occurs. The effect is most pronounced for a thermal SiO_2 step with the highest temperature load, followed by samples with additional AlO_x deposition, while samples with single PECVD SiN_x:H layer show a similar behaviour as discussed above for sample B.

It has to be mentioned that the firing conditions (set parameters) are kept constant but lead to a slightly different temperature load of the bulk material during the first firing step for sample B compared to samples C and D due to the screen-printed Al back side of samples C and D. The lower wafer temperature of the P-gettered sample B compared to the additional Al gettered sample D during the first firing step may lead to a slightly different and less LeTID sensitive impurity distribution for sample B. As shown in [9], different firing temperatures generally have an influence on degradation behaviour.

The earlier onset of regeneration for LeTID in areas of higher τ_{eff} resembles the behaviour of regeneration of BO-correlated degradation. There it could be shown that a higher excess charge carrier density Δn in the Si bulk leads to a faster regeneration process [6, 7]. Interestingly, the onset of LeTID regeneration starts earlier in the best quality areas of sample C (after approx. 200 h), having higher τ_{eff} as compared to the best areas of sample A (after approx. 400 h) as indicated by the dashed line in Fig. 5, with sample D lying in between. Note that all samples are sister wafers with comparable extended defect structure and similar surface passivation quality. As higher τ_{eff} leads to higher Δn for constant illumination, this leads to the conclusion that Δn might play a decisive role for LeTID regeneration.

The next step in the analogy between regeneration for LeTID and BO regeneration would be the assumption that hydrogen plays an important role, too. As shown for BO regeneration, the presence of hydrogen in the Si bulk seems to be a prerequisite for the regeneration to occur and its concentration influences the regeneration kinetics [16]. The significantly different time scale could, *e.g.*, be explained by the different effective diffusion constant of hydrogen at relatively low temperatures in mc-Si compared to monocrystalline Si due to trapping (trap assisted diffusion). But this hypothesis still has to be confirmed in future experiments.

4 MODEL FOR LeTID AND CONSEQUENCES

Effective gettering seems to be a key component to minimize the impact of LeTID in mc-Si. Under the assumption that degradation in mc-Si at longer timescales (>1 h at 75°C) is triggered by the concentration of a specific impurity (or a mix of impurities) that is more or less homogeneously distributed over the wafer area, a first rough model can be introduced.

In areas of good material quality (low density of extended defects like grain boundaries and/or dislocations) where gettering is very effective, LeTID can be suppressed very effectively in lifetime samples. In contrast, in areas of poorer crystal quality (more extended defects), gettering is less effective as impurities might also be present in form of precipitates that are harder to getter externally, because they first have to be dissolved, and dissolved impurities might be gettered internally again at extended defects. This can be seen qualitatively for sample B (Fig. 5b), where good sample areas show less relative degradation than areas of poorer quality, and is also observed in [17] for a different set of samples and different processing sequences.

A consequence of this first model is that high temperature steps after the gettering steps might lead to a redistribution of gettered impurities from their initial gettering sites (*e.g.*, the emitter region, internal gettering sites). Indeed, it could be observed that a thermal oxidation step carried out after a P-gettering step "reactivates" LeTID and following regeneration [15].

The model presented here is most probably not yet able to explain LeTID and regeneration behaviour in all its facets, but gettering seems to play an important role and should be considered in further experiments to deepen the understanding of LeTID in mc-Si.

The regeneration observed for LeTID might be related to the presence of hydrogen in the Si bulk, being more mobile under illumination, maybe due to the change of its charge state [18, 19].

Another consequence from the proposed model is that a back-to-back P-diffusion process as normally carried out in industrial cell processing might lead to a stronger LeTID effect due to the less effective gettering.

5 SUMMARY AND OUTLOOK

The influence of different solar cell gettering steps on LeTID and regeneration has been demonstrated using spatially resolved data from TR-PLI. It could be shown that the behaviour of lifetime samples resembles the behaviour of PERC solar cells fabricated from the same mc-Si material and can therefore be regarded as relevant for PERC-type solar cell processing.

The applied data processing allows the analysis of different sample areas over degradation time in a statistically relevant way and is not limited to selected sample areas or the comparison of only two degradation or regeneration states.

For ungettered samples, it could be shown that degradation kinetics are very similar for all sample areas irrespective of their initial quality, while regeneration sets in first in good quality areas. Gettering drastically reduces LeTID while the strength of degradation and the following regeneration is influenced by the applied gettering sequence. P-gettering is more efficient than Al-gettering for removing or restructuring LeTID sensitive defects, but gettering efficacy depends also on the underlying crystal defect structure.

It could be shown that LeTID is a bulk degradation effect occurring in boron as well as gallium doped mc-Si. Therefore, it cannot be attributed to BO- or FeB-related defects alone.

The influence of the different gettering sequences on the degradation and regeneration process leads to the conclusion that the concentration and distribution of (metal) impurities play a major role for LeTID and regeneration. Therefore, the industrial PERC solar cell processing sequence should be optimized concerning effective gettering and avoidance of redistribution of gettered impurities during following temperature steps. In addition, a high concentration of hydrogen in the correct binding state might help to speed up the regeneration process (as for BO regeneration [20]), when LeTID cannot be avoided completely.

This long time experiment is still running to further analyze the regeneration process and the long time stability of the regenerated state and the surface passivation. Also, additional lifetime samples with further process sequences are under investigation.

6 ACKNOWLEDGEMENT

The authors would like to thank Lisa Mahlstaedt and David Sperber for help during sample processing and Axel Herguth for help during sample characterization and fruitful discussions. Part of this work was funded by the German Federal Ministry of Economic Affairs and

Energy and by industrial partners within the research project "SolarLIFE" (0325763B) and within 0325581. The content is the responsibility of the authors.

7 REFERENCES

[1] K. Ramspeck, S. Zimmermann, H. Nagel, A. Metz, Y. Gassenbauer, B. Birkmann, A. Seidl, Light induced degradation of rear passivated mc-Si solar cells. Proc. 27th EU PVSEC, 2012, 861-865.
[2] F. Fertig, K. Krauss, S. Rein, Light-induced degradation of PECVD aluminium oxide passivated silicon solar cells, Phys. Stat. Sol. RRL 9(1) (2014) 41-46.
[3] F. Kersten, P. Engelhart, H.C. Ploigt, A. Stekolnikov, T. Lindner, F. Stenzel, M. Bartzsch, A. Szpeth, K. Petter, J. Heitmann, J. Müller, Degradation of multicrystalline silicon solar cells and modules after illumination at elevated temperature, Sol. En. Mat. & Sol. Cells 2015 142 (2015) 83-86.
[4] M.A. Green, K. Emery, Y. Hishikawa, W. Warta, E.D. Dunlop, Solar cell efficiency tables (version 47), Progr. Photovolt.: Res. Appl. 24(1) (2016) 3-11.
[5] A. Herguth, G. Schubert, M. Kaes, G. Hahn, A new approach to prevent the negative impact of the metastable defect in boron doped Cz silicon solar cells, Proc. 4th WCPEC, 2006, 940-943.
[6] A. Herguth, G. Schubert, M. Kaes, G. Hahn, Investigations on the long time behavior of the metastable boron-oxygen complex in crystalline silicon, Progr. Photovolt.: Res. Appl. 16 (2008) 135-140.
[7] A. Herguth, G. Hahn, Kinetics of the boron-oxygen related defect in theory and experiment, J. Appl. Phys. 108 (2010) 114509.
[8] K. Krauss, F. Fertig, D. Menzel, S. Rein, Light-induced degradation of silicon solar cells with aluminium oxide passivated rear side, En. Procedia 77 (2015) 599-606.
[9] D. Bredemeier, D. Walter, S. Herlufsen, J. Schmidt, Lifetime degradation and regeneration in multicrystalline silicon under illumination at elevated temperature, AIP 6 (2016) 035119.
[10] D. Kiliani, G. Micard, B. Steuer, B. Raabe, A. Herguth, G. Hahn, Minority charge carrier lifetime mapping of crystalline silicon wafers by time-resolved photoluminescence imaging, J. Appl. Phys. 110 (2011) 054508.
[11] D. Kiliani, A. Herguth, G. Micard, J. Ebser, G. Hahn, Time-resolved photoluminescence imaging with electronic shuttering using an image intensifier unit, Sol. En. Mat. & Sol. Cells 106 (2012) 55-59.
[12] A. Zuschlag, D. Skorka, G. Hahn, Comparison of degradation and regeneration kinetics in differently doped mc-Si materials, to be published.
[13] K. Pollock, J. Junge, G. Hahn, Detailed investigation of surface passivation methods for lifetime measurements on silicon wafers, IEEE J. Photovolt. 2(1) (2012) 1-6.
[14] S. Gindner, P. Karzel, B. Herzog, G. Hahn, Efficacy of phosphorus gettering and hydrogenation in multicrystalline silicon, IEEE J. Photovolt. 4(4) (2014) 1063-1070.
[15] A. Zuschlag, D. Skorka, G. Hahn, Degradation and regeneration analysis in mc-Si, Proc. 43rd IEEE PVSC, 2016, in press.
[16] S. Wilking, A. Herguth, G. Hahn, Influence of hydrogen on the regeneration of boron-oxygen related defects in crystalline silicon, J. Appl. Phys. 113 (2013) 194503.
[17] D. Skorka, A. Zuschlag, G. Hahn, Spatially resolved degradation and regeneration kinetics in mc-Si, this conference.
[18] P. Hamer, B. Hallam, S. Wenham, M. Abbott, Manipulation of hydrogen charge states for passivation of p-type wafers in photovoltaics, IEEE J. Photovolt. 4(5) (2014) 1252-1260.
[19] C. Sun, F.E. Rougieux, D. Macdonald, A unified approach to modelling the charge state of monatomic hydrogen and other defects in crystalline silicon, J. Appl. Phys. 117 (2015) 045702.
[20] S. Wilking, A. Herguth, G. Hahn, Influence of bound hydrogen states on BO-regeneration kinetics and consequences for high-speed regeneration processes, Sol. En. Mat. & Sol. Cells 131 (2014) 2-8.

MEASURES FOR ELIMINATING LIGHT-INDUCED LIFETIME DEGRADATION IN MULTICRYSTALLINE SILICON

Dennis Bredemeier[1], Dominic C. Walter[1], Sandra Herlufsen[1], and Jan Schmidt[1,2]
[1]Institute for Solar Energy Research Hamelin (ISFH), Am Ohrberg 1, D-31860 Emmerthal, Germany
[2]Department of Solar Energy, Institute of Solid-State Physics, Leibniz University Hanover, Appelstr. 2, D-30167 Hanover, Germany

We examine the light-induced degradation and regeneration in multicrystalline silicon (mc-Si) under illumination at elevated temperature. Samples are treated with process steps typically applied in industrial solar cell production. We observe a pronounced degradation in lifetime after rapid thermal annealing (RTA) at 900°C. However, we detect only a weak lifetime instability in mc-Si wafers which are RTA-treated at 650°C. We pay particular attention to the regeneration of the lifetime which takes place under prolonged illumination at elevated temperature. It is found that this regeneration can be significantly accelerated by increasing the temperature. Additionally we have tested the stability of the regenerated state under illumination and elevated temperature. We show that the regeneration cannot be further accelerated by increasing the applied light intensity, as the regeneration rate does not depend on the illumination intensity for intensities exceeding 1 sun. We hence propose three measures for elimating the light-induced degradation: (i) the adaption of the firing profile, (ii) the growth of metal-reduced mc-Si blocks, and (iii) by regeneration at illumination at increased temperature.
Keywords: multicrystalline silicon, light-induced degradation, regeneration

1 INTRODUCTION

The degradation of solar cells fabricated on multicrystalline silicon (mc-Si) has recently attracted great interest, since Ramspeck et al. [1] showed that those cells with Al_2O_3/SiN_x-passivated rear surface lose up to 6% relative in their efficiency at an illumination intensity of 40 mW/cm^2 (0.4 suns) and at an elevated temperature of 75°C on a timescale of about 400 hours. On solar cells with a full-area aluminum back surface field (Al-BSF) the degradation was observed to a lesser extent. Since they were not able to explain this degradation by known light-induced degradation phenomena such as the boron-oxygen defect activation [2, 3, 4] or the iron-boron pair dissociation [5, 6], there is currently no fundamental understanding of the underlying defect physics. Similar results were more recently reported by Fertig et al. [7, 8], who also observed a pronounced degradation in efficiency on mc-Si passivated emitter and rear cells (PERC) on even larger timescales at 0.15 suns and 70°C. Kersten et al. [9] showed that on large timescales the degradation is followed by a full regeneration in the cell efficiency after approximately 150 hours of illumination at 0.3 suns and 95°C. Most recently, we have reported that the degradation behavior can be avoided by decreasing the peak-firing temperature during rapid-thermal annealing (RTA) [10]. We proposed a defect model based on the dissolution of metal precipitates during firing (RTA), which explains both the degradation and regeneration and is consistent with the experimental findings [10].

2 SAMPLE PREPARATION

To investigate ways of avoiding light-induced degradation in mc-Si we conduct here experiments on mc-Si lifetime samples and implement industrial-type processing steps. We implement a process flow which is designed to represent a typical industrial-type PERC rear-side process. We use block-cast boron-doped mc-Si wafers with a resistivity of 1.3 Ωcm and a size of 15.6 × 15.6 cm^2. The mc-Si material was grown in a G1 lab-size crucible [11]. The as-cut mc-Si wafers are first cleaned with a surface-active agent and subsequently etched in a potassium hydroxide solution to remove the saw damage. A phosphorus diffusion is then performed in a quartz-tube furnace at a process temperature of 853°C resulting in n^+-layers on both wafer surfaces with a sheet resistance between 50 and 60 Ω/sq. The n^+-layers are chemically removed by a solution of hydrofluoric acid and nitric acid using a chemical polishing process. The surfaces are then passivated by an Al_2O_3/SiN_x stack, where the Al_2O_3 layer is deposited by spatial atomic layer deposition (ALD) in an InPassion LAB System (SolayTec) with a thickness of 5 nm. The SiN_x layers are deposited using an industrial-type plasma-enhanced chemical vapor deposition (PECVD) process (Roth&Rau, SiNA) resulting in layers with a thickness of 100 nm and a refractive index of 2.05. The wafers are divided into two groups. The first group of wafers receives an RTA firing treatment at a set peak temperature of 900°C, while the other group is fired at a reduced set peak temperature of 650°C using an industrial conveyor belt furnace (centrotherm photovoltaics, DO-FF-8.600-300). The belt speed is set at 6.8 m/min. All wafers are stored in the dark at room temperature to prevent any degradation prior to the actual experiments.

3 EXPERIMENTAL RESULTS

Injection-dependent carrier lifetimes are measured using the photoconductance decay (PCD) and the quasi-steady-state photoconductance (QSSPC) techniques [12] within a circular area with a diameter of 18 mm located in the upper right corner of the samples, averaging the effective carrier lifetime over this area. The samples are illuminated with halogen lamps at 1 sun light intensity at 75°C. The effective carrier lifetimes are measured at defined timesteps. As can be seen in the lifetime measurements in Fig. 1, which represents the typical behavior of the processed samples, the samples fired at a set peak temperature of 900°C show a pronounced degradation in lifetime. The observed lifetime degradation can be subdivided into two stages. The first stage is finalized within less than 2 hours and is characterized by a fast and pronounced reduction in lifetime. This stage is followed by a slower exponential degradation, terminating within ~100 hours. Under prolonged illumination at elevated temperature the lifetime starts to regenerate and finally reaches lifetimes even higher than the initial state. Note that we cannot exclude that the initial lifetime measurement has been performed in a slightly degraded state. If this has been the case, the regeneration would not reach necessarily higher lifetimes than the initial ones, but just comparable lifetimes to the initial state. We find that the samples fired at the reduced set peak temperature of 650°C show only a very weak lifetime instability and show a much higher overall lifetime [10].

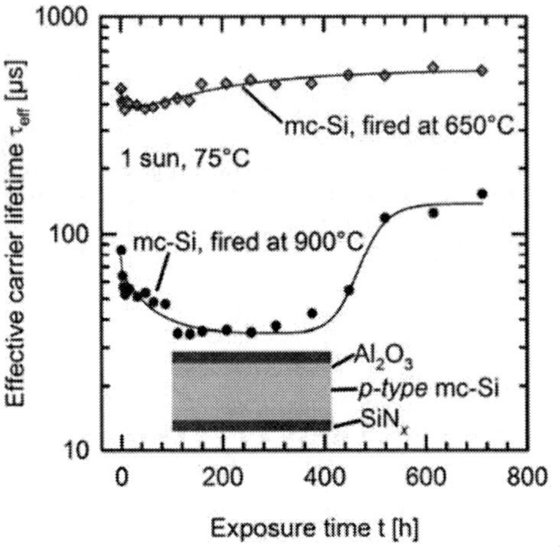

Fig. 1. Effective carrier lifetime measured at $\Delta n = 10^{15}$ cm^{-3} by the PCD (diamonds) and the QSSPC (circles) techniques of mc-Si lifetime samples plotted versus the exposure time of illumination at 1 sun at 75°C. The solid lines are guides to the eyes.

In our recent paper [10], we have proposed a defect model where we attributed the lifetime degradation and regeneration to the dissolution of metal precipitates during the firing step. From this model we conclude that either an adapted firing profile, which does not dissolve the precipitates or a reduced metal content in the as-grown mc-Si could eliminate the observed lifetime instability.

In order to speed up the regeneration of the lifetime, we also investigate the lifetime evolution at higher temperatures. We process neighboring mc-Si samples of a comparable mc-Si ingot in the same way as described above, however we use plasma-assisted atomic layer deposition (plasma-ALD) to deposit 10 nm of Al_2O_3 underneath the SiN_x-layers. As a last process step, the wafers were cut into 5 × 5 cm^2 samples. The samples are illuminated at 1 sun light intensity using a halogen lamp on a hotplate at 185°C or 300°C, respectively. We measure the effective carrier lifetime in the initial state and at defined timesteps during illumination at the respective temperature. As shown in Fig. 2, the lifetime evolution proceeds significantly faster with increased temperature. At a temperature of 185°C there is an initial degradation in lifetime, followed by a regeneration starting after approximately 30s. As already observed in Fig. 1, the lifetime finally saturates at an even higher value compared to the initial state after approximately 8 hours of exposure. For an even higher temperature of 300°C, the lifetime starts to regenerate with a much higher time constant as in the 185°C case. After approximately 10 minutes of exposure, the lifetime also saturates again at a higher value compared the initial state. However, the stabilized lifetime of 230µs is slightly lower in the 300°C case compared to 315µs in the 185°C regeneration experiment.

To investigate the stability of the regenerated state, we expose the regenerated samples to our standard degradation conditions at 1 sun light intensity and at 75°C. Fig. 3 shows that the lifetime of both samples regenerated at 185°C and 300°C remain stable and are well above the initial lifetime value before degradation and regeneration. In the case of 185°C we investigated approximately 1000 hours and in the case of 300°C approximately 100 hours.

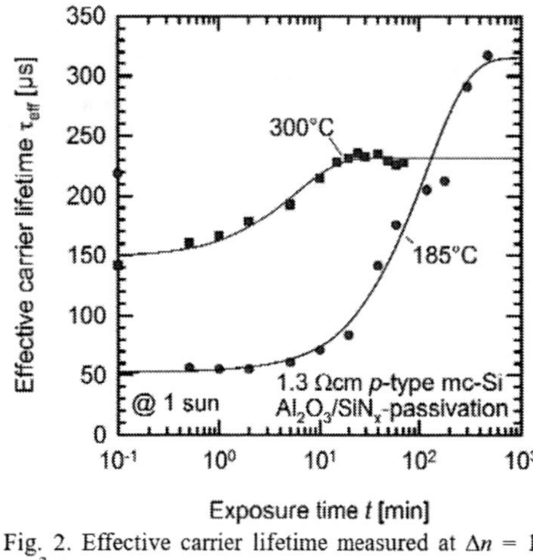

Fig. 2. Effective carrier lifetime measured at $\Delta n = 10^{15}$ cm^{-3} by the PCD technique of mc-Si lifetime samples plotted versus the exposure time of illumination at 1 sun light intensity and at elevated temperature at 185°C or 300°C on a hotplate respectively. The solid lines are exponential rise to maximum fits to the regeneration.

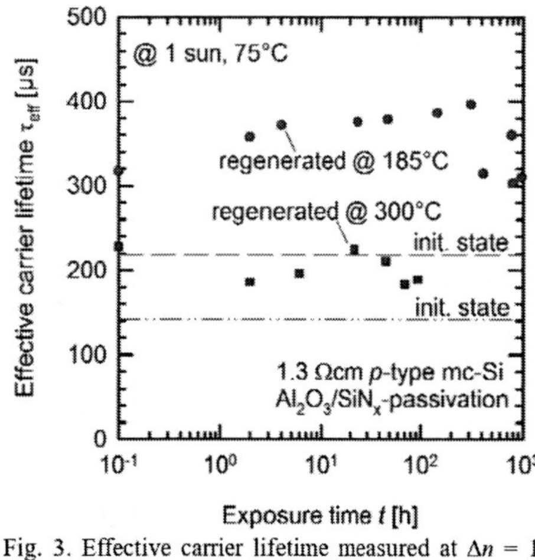

Fig. 3. Effective carrier lifetime measured at $\Delta n = 10^{15}$ cm^{-3} by the PCD technique of mc-Si lifetime samples previously regenerated at 1 sun light intensity at an elevated temperature of 185°C or 300°C on a hotplate,

respectively.

To examine the regeneration of the lifetime on our samples under increased illumination intensity, we measure the dependence of the regeneration rate on the light intensity at constant temperature. The samples are illuminated at 1 sun, 2 suns or 3 suns light intensity with a halogen lamp on a hotplate at 300°C. Note that all 5 x 5 cm^2 samples are neighboring samples. As can be seen in Fig. 4, there is an initial degradation in lifetime, followed by a very fast regeneration process in the range between 20s and 40s of exposure time. Note that this very fast regeneration process has not been observed before, since e.g. in Fig. 2, the timesteps between the measurements were chosen too large. After the fast regeneration, lifetimes very comparable to the initial state are reached. From this point onwards, there starts a regeneration process whose regeneration time constant does not depend on the applied light intensity. Within the uncertainty of the exponential rise to maximum fits displayed in Fig. 4, the regeneration time constant equals 9 h^{-1} to 9.4 h^{-1} at a temperature of 300°C. By increasing the light intensity during regeneration treatment, no change in the timespan needed to saturate the lifetime is observed.

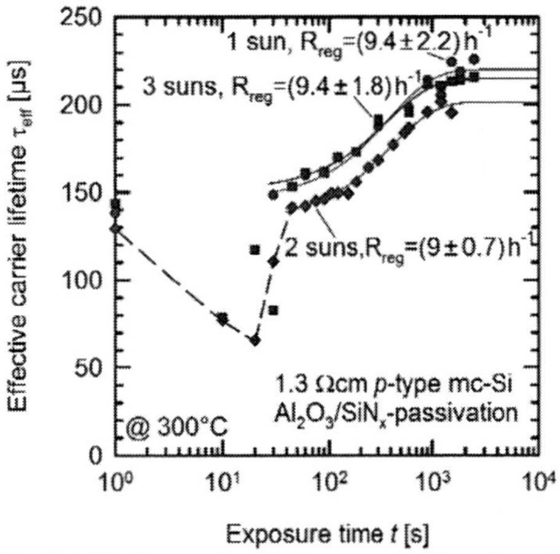

Fig. 4. Effective carrier lifetime measured at $\Delta n = 10^{15}$ cm^{-3} by the PCD technique of mc-Si lifetime samples at an elevated temperature of 300°C and at light intensities of 1 sun, 2 suns or 3 suns. The dashed line is a guide the eyes. The solid lines are exponential rise to maximum fits with the respective time constant.

4 CONCLUSIONS

In conclusion, we have shown that a high-temperature RTA treatment has a strong impact on the lifetime degradation and regeneration behavior of mc-Si wafers under illumination at elevated temperature (100 mW/cm^2, 75°C). Samples treated during a fast-firing step with a set-peak temperature of 900°C showed a pronounced light-induced lifetime degradation and regeneration, whereas the lifetime of mc-Si samples treated with a maximum peak temperature of 650°C remained stable. We have also shown that the regeneration of the lifetime can be significantly increased by illumination of the samples at a temperature of 185°C and 300°C, respectively. After complete regeneration the lifetime reaches in both cases values even higher compared to the initial state. In addition to that, we have tested the stability of the regenerated state under 1 sun light intensity and at 75°C. In both cases, the lifetimes remained stable and well above their initial values over the whole exposure time. Finally we have shown that the regeneration process cannot be accelerated by increasing the applied light intensity from 1 sun to 2 suns or 3 suns, respectively.

From a practical point of view, our experimental results suggest three easy-to-implement approaches to avoid light-induced degradation in mc-Si PERC solar cells, that is (i) by reducing the peak firing temperature below the critical temperature required to trigger the degradation, (ii) by reducing the metal concentration in the mc-Si material, or (iii) by regeneration at illumination at increased temperature.

ACKNOWLEDGEMENTS

This work was funded by the German State of Lower Saxony and the German Federal Ministry of Economics and Energy (BMWi) and by industrial partners within the research project "SolarLIFE" (contract no. 0325763C). The content is the responsibility of the authors.

REFERENCES

[1] K. Ramspeck, S. Zimmermann, H. Nagel, A. Metz, Y. Gassenbauer, B. Birkmann, and A. Seidl, *Proceedings of the 27th European Photovoltaic Solar Energy Conference* (WIP, Munich, 2012), p. 861.

[2] J. Schmidt, A. G. Aberle, and R. Hezel, *Proceedings of the 26th IEEE Photovoltaic Specialists Conference* (IEEE, New York, 1997), p. 13.

[3] S. W. Glunz, S. Rein, W. Warta, J. Knobloch, and W. Wettling, Proceedings of the 2nd World Conference on Photovoltaic Energy Conversion (WIP, Munich, 1998), p. 1343.

[4] J. Schmidt, and K. Bothe, Phys. Rev. B **69**, 024107 (2004).

[5] L. J. Geerligs, and D. Macdonald, Appl. Phys. Lett. **85**, 5227 (2004).

[6] J. Schmidt, Prog. Photovolt. **13**, 325 (2005).

[7] F. Fertig, K. Krauß, and S. Rein, Phys. Status Solidi RRL **9**, 41 (2014).

[8] K. Krauss, F. Fertig, D. Menzel, and S. Rein, Energy Procedia **77**, 599 (2015).

[9] F. Kersten, P. Engelhart, H.-C. Ploigt, A. Stekolnikov, T. Lindner, F. Stenzel, M. Bartzsch, A. Szpeth, K. Petter, J. Heitmann, and J.W. Müller, Solar Energy Materials and Solar Cells **142**, 83 (2015).

[10] D. Bredemeier, D. Walter, S. Herlufsen, and J. Schmidt, AIP Adv. **6**, 035119 (2016).

[11] W. Kwapil, A. Zuschlag, I. Reis, I. Schwirtlich, S. Meyer, R. Zierer, R. Krain, F. Kießling, M. Schumann, C. Schmid, and S. Riepe, *Proceedings of the 27th European Photovoltaic Solar Energy Conference* (WIP, Munich, 2012), p. 627.

[12] R. A. Sinton, and A. Cuevas, Appl. Phys. Lett. **69**, 2510 (1996).

IMPACT OF AL_2O_3 AND SIN_x PASSIVATION LAYERS ON LeTID

Friederike Kersten[a], Johannes Heitmann[b], Jörg W. Müller[a]
[a] Hanwha Q CELLS GmbH, Sonnenallee 17-21, 06766 Bitterfeld-Wolfen, Germany
[b] TU Bergakademie Freiberg, Institute of Applied Physics, Leipziger Straße 23, 09599 Freiberg, Germany

ABSTRACT: In this work the carrier lifetime evolution of different passivation layers under illumination and at elevated temperatures are investigated. Multicrystalline silicon lifetime samples were treated by implementing typical industrial processing steps. The degradation was found to depend strongly on surface passivation type, but is independent of the surface doping and oxide charge. The influence of the passivation layer on the silicon bulk lifetime degradation is investigated. After reaching a degradation maximum, the lifetime samples feature a regeneration phase. Long-time degradation measurements without excess carrier injection at elevated temperatures show that the carrier lifetime degrades also in darkness. Capacitance-voltage measurements show that the oxide charge is not influenced by the degradation, but a high decrease in carrier lifetime were measured. Firing experiments show that lifetime samples passivated after firing procedure are not prone to degrade.
Keywords: Multicrystalline Silicon, Degradation, Passivation.

1 INTRODUCTION

Multicrystalline silicon (mc-Si) solar cells and modules with PERC (Passivated Emitter and Rear Cell) concept can show unexpected strong light induced degradation levels at elevated temperatures. A new light and elevated temperature induced degradation effect (LeTID) has been detected, which cannot be explained by the activation of boron-oxygen (B-O) complexes or the division of iron-boron (FeB) pairs [1-3]. The degradation mechanism called LeTID, can cause a module power degradation level of above 10% [3] and might be a potential obstacle for the industrial realization of the PERC technology. Hanwha Q CELLS has shown that their Q.ANTUM technology is a unique technology differentiating from mc-Si PERC and is not effected by LeTID [3]. Currently, there is no physical understanding about the degradation effect and it is still unclear if only the rear side of the solar cell [2] or the Si bulk material [3] is affected. In addition, an interaction of surface passivation layer and bulk material could be the root cause of degradation under LeTID conditions. In this work the impact of different passivation layers and surface doping types are investigated to gain a better understanding of the specific root cause of the degradation effect.

2 EXPERIMENTAL

Neighboring LeTID-sensitive block cast boron-doped (1-2 Ω cm) mc-Si wafers (5×5 cm^2) with comparable grain and defect structure are processed to lifetime samples by implementing typical industrial processing steps (see Fig. 1). After a wet chemical saw damage removal a phosphor diffusion is performed resulting in n^+-layers on both wafer surfaces. To analyze the influence of the doping at the wafer surface two process flows were implemented. At one half of the samples the emitter was chemically removed (w/o emitter) and the other half keeps its n^+-doping region (w/ emitter). After phosphor glass removal the samples get a symmetrically passivation and receive a final firing step. Two different passivation types, deposited in two different tools, are used. Al_2O_3/SiN_x stacks were deposited in tool A only. The two kind of SiN_x passivated samples were produced in two different deposition tools (tool A & B).

To gain a better understanding of the interaction of passivation layer and bulk material additional lifetime samples were processed. The samples undergo the $POCl_3$ diffusion, chemical emitter removal, firing step and in this case the passivation layer deposition after firing procedure.

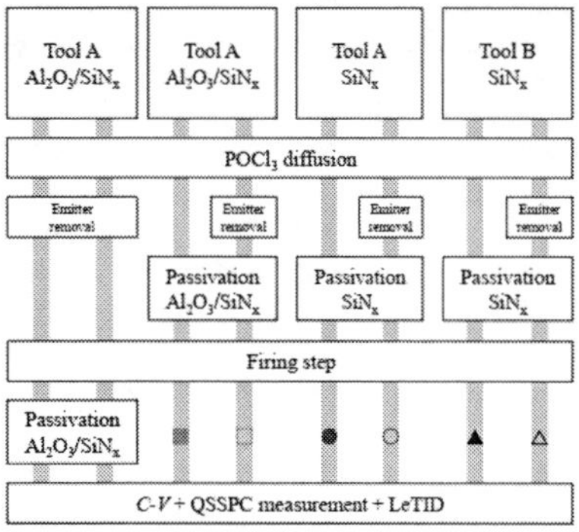

Figure 1: Process sequence of the investigated lifetime samples. The symbols in the last row are standing for the used symbols in Fig. 3-4.

The LeTID treatments are carried out at elevated temperature of 75°C and an excess carrier injection is provided by LED illumination at 300 W m^{-2}. Quasi steady state photo conductance decay (QSSPC) measurements were used to characterize the carrier lifetime τ_{eff} during time expired degradation. τ_{eff} values are carried out at $\Delta n = 10^{14}$ cm^{-3}. The long-time degradation behavior of SiN_x-passivated samples w/ emitter at elevated temperatures are also investigated without excess carrier injection.

To determine the evolution of fixed charge density Q_f before and after 24 h LeTID treatment capacitance-voltage (*C-V*) experiments were carried out using a metal-oxide semiconductor model as described in [4].

In Table 1 all passivation layers and their surface properties are collected. τ_{eff} values were measured before (t = 0 h) and after 1000 h of LeTID treatment.

3 LeTID CHARACTERISTICS AND OBSERVATION ON LIFETIME SAMPLES

The results of the lifetime measurements in Fig. 2

Table 1: Passivation layers and their surface properties. τ_{eff} was extracted at $\Delta n = 10^{14}$ cm^{-3}.

Lifetime sample	Symbol	Deposition tool	Emitter	τ_{eff} before LeTID [µs]	τ_{eff} after ~ 1000 h LeTID [µs]
Al_2O_3/SiN_x	□ & ■	A	w/o & w/	76 & 74	5 & 7
SiN_x	○ & ●	A	w/o & w/	63 & 72	7 & 7
SiN_x	Δ & ▲	B	w/o & w/	30 & 57	2 & 7
SiN_x (dark at 75°C)	▼	B	w/	56	34

show that only the samples, which are passivated before firing, degrade severe during the first 24 h degradation procedure (70 → 9 µs). Samples with a passivation layer deposited after the firing step are not prone to degrade.

In Fig. 3 the long-time τ_{eff} measurement are plotted for different surface passivation types (deposition tool A & B) w/o and w/ emitter. Similar degradation rates were determined for all samples with n^+-doped emitter as well as for samples with passivation layers deposited by using deposition tool A (~70 → 7 µs). In comparison to the n^+-doped samples and tool A samples the SiN_x-passivated samples deposited in tool B w/o emitter show a slightly higher defect generation. In this sample τ_{eff} degrades from 30 µs up to 2 µs after 1000 h. The degradation of the SiN_x sample w/ emitter in darkness is obviously slower but shows also a decrease of τ_{eff} = 34 µs after 1000 h (τ_{eff} measurement not presented here).

Fig. 4 shows the extended long-time degradation (expansion of Fig. 3) and regeneration behavior of all passivation layers w/o emitter. After reaching a degradation maximum all investigated samples feature a regeneration phase.

In Fig. 5 the Q_f results of *C-V* measurement for passivation layers present during firing procedure before and after 24 h LeTID treatment are presented. LeTID sensitive Al_2O_3/SiN_x stack w/o emitter samples exhibit the same degradation behavior as results shown in Fig. 4. There is only a slightly decrease in negative Q_f after 24 h LeTID treatment, but a high decrease in τ_{eff}. Furthermore, Q_f and τ_{eff} were determined for a LeTID controlled lifetime sample with defect engineered process sequence [3]. No decrease of Q_f could be recognized. It was impossible to measure the interface trap density D_{it} of the mc-Si lifetime samples, because of high trapping of accumulation capacitance during conductance measurement.

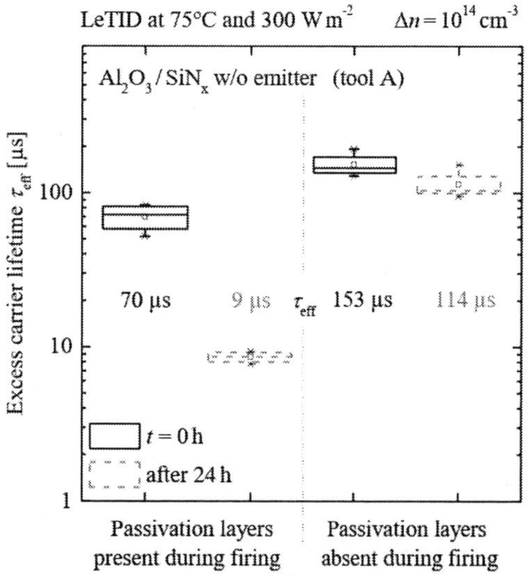

Figure 2: τ_{eff} before and after LeTID treatment of passivation layers, which are present or absent during firing.

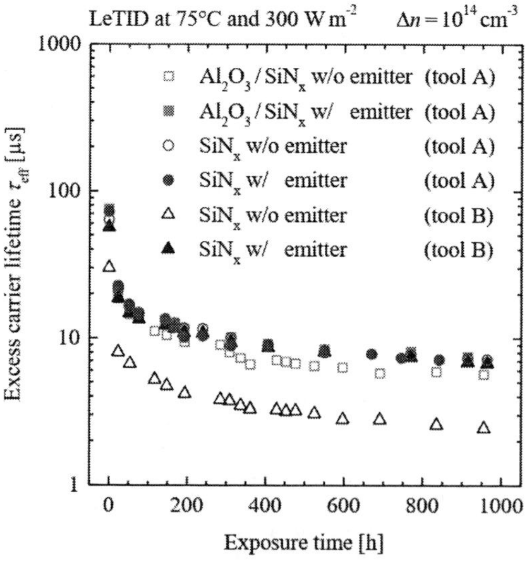

Figure 3: Long-time degradation behavior of different passivation layers with and without n^+-doped emitter.

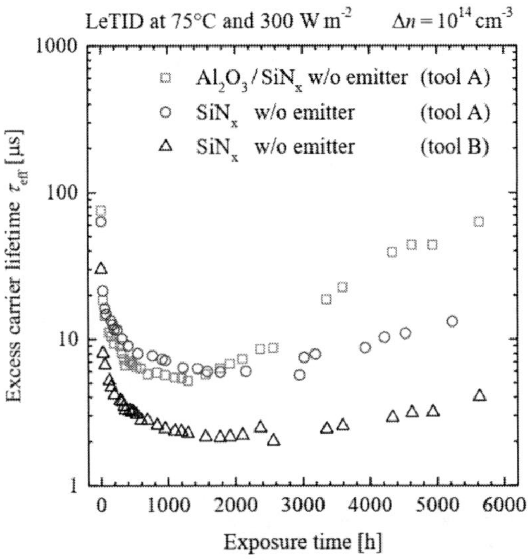

Figure 4: Extended long-time degradation and regeneration behavior of different passivation layers.

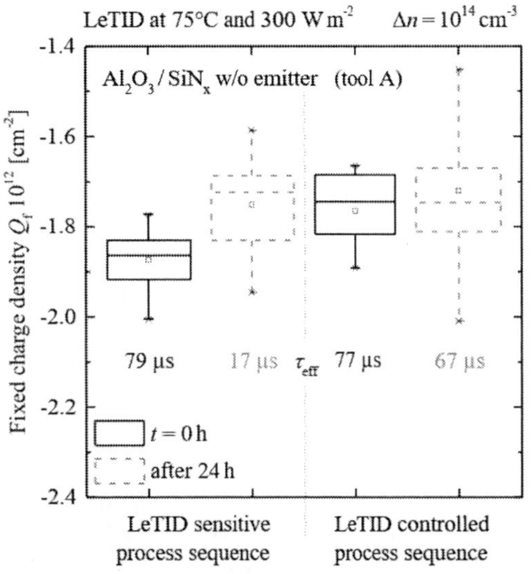

Figure 5: Q_f before and after LeTID treatment of a sensitive sample and LeTID controlled sample.

4 CONCLUSION

All investigated surface passivation layers, which are present during firing step, show a different time-dependent lifetime degradation. The observed severe lifetime degradation strongly depends on the type of surface passivation, but is independent of surface doping and oxide charge. Same degradation behavior for SiN_x (positive surface charge) or Al_2O_3/SiN_x stack (negative surface charge) were determined. After reaching a degradation maximum, the lifetime samples feature a regeneration phase. On extended time scale the Al_2O_3/SiN_x stack almost recovered to initial lifetime value. Degradation also occurs without illumination at elevated temperatures. The degradation rate is obviously slower and up to now it is unclear if it is the same degradation effect as with enhanced excess carrier densities. *C-V* measurements show no significant decrease of fixed charge density, but carrier lifetime are prone to degrade. Furthermore, a correlation with the silicon bulk material is observed. Strong degradation for samples with passivation layers present during firing step were determined. Only slight degradation of τ_{eff} with passivation layers absent during firing step and final passivation were observed. The slightly degradation of τ_{eff} is related to B-O degradation in mc-Si [5]. This represents an interaction of passivation layer with mc-Si bulk material during firing step. However, this avoidance of degradation is unfeasible for industrial production, because it is impossible to passivated solar cell rear side after metallization and firing step. Up to now there is no physical explanation for the degradation mechanism and further work is ongoing in order to create a deeper understanding of the LeTID effect on mc-Si lifetime samples and solar cells.

5 ACKNOWLEDGEMENTS

Special thanks to the staff of the Reiner Lemoine Research Center and pilot line at Hanwha Q CELLS. This work was financial supported by German Federal Ministry for Economic Affairs and Energy (BMWi) under contract no. 0325775 ("AdmMo").

6 REFERENCES

[1] K. Ramspeck, S. Zimmermann, H. Nagel, A. Metz, Y. Gassenbauer, B. Brikmann and A. Seidl, "Light induced degradation of rear passivated mc-Si solar cells", Proceedings of the 27th European Photovoltaic Solar Energy Conference, Frankfurt, Germany, 2012, pp. 861-865.

[2] F. Fertig, K. Krauß and S. Rein, "Light-induced degradation of PECVD aluminium oxide passivated silicon solar cells", Journal of Physica Status Solidi RRL, 1-6 (2014).

[3] F. Kersten, P. Engelhart, H.-C. Ploigt, A. Stekolnikov, T. Lindner, F. Stenzel, M. Bartzsch, A. Szpeth, K. Petter, J. Heitmann and J. W. Müller, "Degradation of multicrystalline silicon solar cells and modules after illumination at elevated temperatures", Journal of Solar Energy Materials and Solar Cells 142, 83-86 (2015).

[4] S. M. Sze and K.K. Ng, "Physics of Semiconductor Devices" (Wiley, Hoboken, New Jersey, 2007).

[5] K. Bothe, Ron Sinton, Jan Schmidt, "Fundamental Boron-Oxygen-related Carrier Lifetime Limit in Mono- and Multicrystalline Silicon", Progress of Photovoltaics: Research and Applications 13, 287-296 (2005).

EVALUATION OF PASSIVATED SURFACE OF SILICON WITH LASER TERAHERTZ EMISSION MICROSCOPE (LTEM)

Toshimitsu Mochizuki[1*], Akira Ito[2], Hidetoshi Nakanishi[2], Jonathon Mitchell[1], Katsuto Tanahashi[1], Iwao Kawayama[3], Masayoshi Tonouchi[3], Katsuhiko Shirasawa[1], Hidetaka Takato[1]
[1]Fukushima Renewable Energy Institute, The National Institute of Advanced Industrial Science and Technology
2-2-9 Machiikedai, Koriyama, Fukushima 963-0298, Japan
[2]SCREEN Holdings, Co., Ltd.
322, Furukawa-cho, Hazukashi, Fushimi-ku, Kyoto, Kyoto 612-8486, Japan
[3]Institute of Laser Engineering, Osaka University
2-6, Yamadaoka, Suita, Osaka 565-0871, Japan
e-mail : toshimitsu-mochiduki@aist.go.jp

ABSTRACT: Terahertz emission from silicon wafers with silicon-oxide passivation lasers were measured using a laser terahertz emission microscope (LTEM) with strong correlation between THz waveform and the surface potential being observed. The surface potential were electrically tuned by an indium-tin oxide transparent electrode disk on the top and evaluated by capacitance-voltage (C-V) characteristics. The sample were excited by Ti-Sapphire laser pulses and the waveform of THz emission were measured by the THz time-domain spectroscopy techniques using LTEM. The observed waveform changed with the external bias and inverted near the flatband voltage, and changes similar to the C-V characteristics were shown in the peak amplitude. These results indicate that the LTEM can measure and map the surface band bending of Si by analysing the waveform of THz emission. We expect that the LTEM can be applied as an optical inspection tools of passivation layers of solar cells.
Keywords: c-Si, Charactarisation, SiO_2, Passivation, Spectroscopy

1 INTRODUCTION

Laser Terahertz Emission Microscope (LTEM) measures and maps the internal electric field of semiconductors and dielectric polarisation of insulators by measuring the THz emission excited by femtosecond laser pulses[1-2]. We have applied the LTEM to evaluate silicon solar cells, of which good surface passivation layers are essential part, to achieve and maintain high efficiency[3-4].

On the silicon solar cells, passivation layers of various materials including SiO_x, SiN_x, and AlO_x are deposited and their field-effect passivation characteristics are often evaluated by the capacitance-voltage (C-V) and voltage-charge measurement techniques. The surface band bending at these silicon surfaces causes transient currents by charge separation of photoexcited carriers when excited by femtosecond laser pulse. We expect that the bending can be quickly evaluated and mapped using LTEM.

In this work, we propose the use of LTEM to evaluate the passivation surfaces of Si solar cells by measuring waveforms of THz emission from Si wafers with SiO_x layers along with C-V curves. We have compared the obtained waveforms with the electrically measured C-V curves and observed strong correlation between them. We also measured the THz emission from plane n-type Si wafers with oxidation layer with different thermal treatments to evaluate the surface passivation with LTEM.

2 EXPERIMENTS

Figure 1 shows the structure of the measured sample, which is an Si wafer coated in a wet oxidised layer. The oxidation is done at 900 °C in nitrogen gas and water vapour. The thickness was measured using an ellipsometer. The indium tin oxide (ITO) transparent electrodes at the top and the Al electrode at the bottom are evaporated, then the sample connected to a DC voltage source to tune the surface band bending of the silicon while allowing for the LTEM measurements. The electrodes were also connected to a Keysight B1500A device parameter analyser to measure C-V curve at 1 kHz. We measured both n- and p-doped sample. The other parameters of the sample were shown in Table 1.

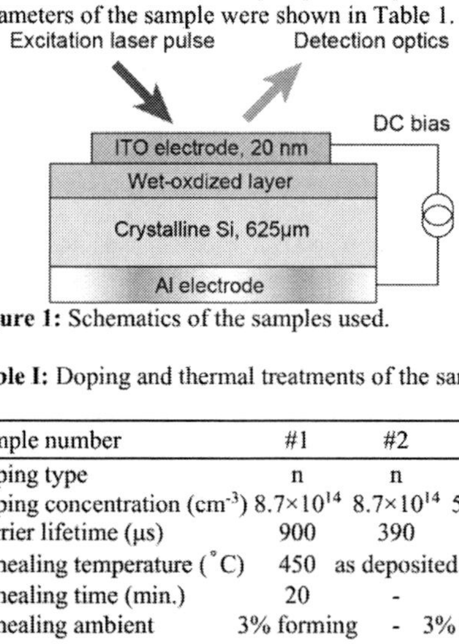

Figure 1: Schematics of the samples used.

Table I: Doping and thermal treatments of the samples.

Sample number	#1	#2	#3
Doping type	n	n	p
Doping concentration (cm^{-3})	8.7×10^{14}	8.7×10^{14}	5.4×10^{15}
Carrier lifetime (µs)	900	390	13
Annealing temperature (°C)	450	as deposited	450
Annealing time (min.)	20	-	20
Annealing ambient	3% forming	-	3% forming

Figure 2 shows the schematic of the THz time-domain spectra measurements using LTEM. Femtosecond pulses from the Ti:Sapphire laser are focused on the sample and the LT-GaAs based THz detector. The sample is photoexcited at 80 MHz by 800 nm source at an incident angle of 45°, covering a 10 mm spot size. Averaged power density for 100 fs pulse width was 127 mW/cm^2. The THz emission is focused on the detector with the pair of elliptic mirrors. The sample is placed on a 3-axis stage, which enables mapping of THz waveform. Measurement of the THz waveform is done by sweeping the delay stage and detecting the THz signal with the lock-in amplifier. In this experiment, the sample is aligned so that the excitation laser are focused on the ITO electrode.

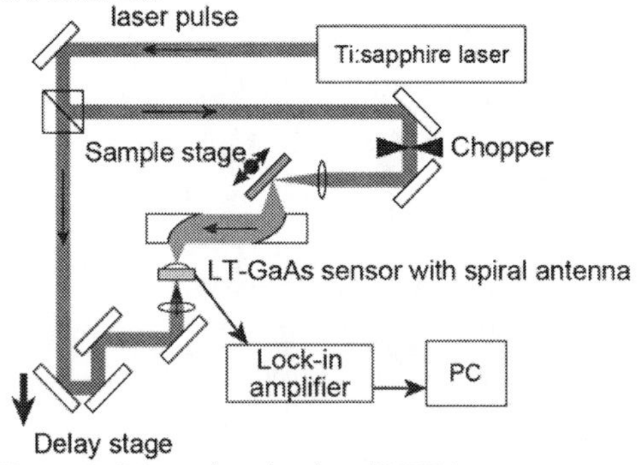

Figure 2: Schematics of optics of LTEM.

3 RESULTS AND DISCUSSION

3.1 The measured THz waveforms

Figure 3 shows waveforms of laser terahertz emission of sample No. 1, the n-type sample with forming gas anneal. In the range of -10 to -2 V and 0 to 10V, the waveforms were insensitive to the applied bias voltage. The waveforms rapidly changed with the applied bias voltage, with waveform inversion being observed.

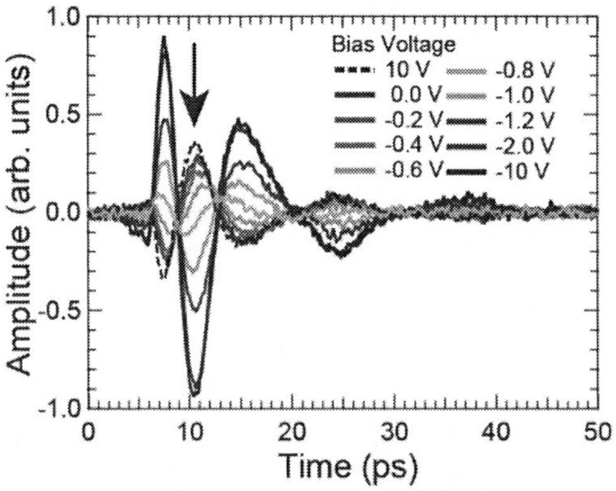

Figure 3: Waveforms of laser THz emission from #1 n-type Si. Amplitude at 10.5 ps as indicated with the black arrow is used to evaluate the amplitude of THz emission in the present work.

Figure 4 shows the amplitude at 10.5 ps and the electrically measured capacitance at various bias voltages. The THz amplitude shows strong positive correlation with the capacitance and crosses the zero near the flatband voltage -0.98 V. The other samples, not shown here, showed quite similar results including stable amplitude at large bias voltages, sharp change near the flatband voltage, and waveform inversion across the flatband voltage. The capacitance of p-type sample showed negative correlation between the THz amplitude.

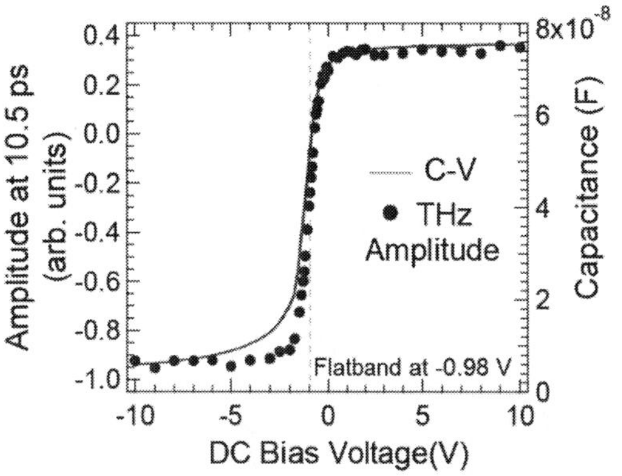

Figure 4: THz amplitude and electrically measured C-V curve of the n-type sample.

3.2 The surface potential and THz amplitude

With the MOS structure and uniform doping, we calculated the surface potential as the integrated band bending of the samples by Capacitance method.
We used the capacitance curve of the region where the fermi level at the surface is below the midgap, to evaluate the interface trapped charge density with an assumption that the density is symmetric with respect to the midgap.

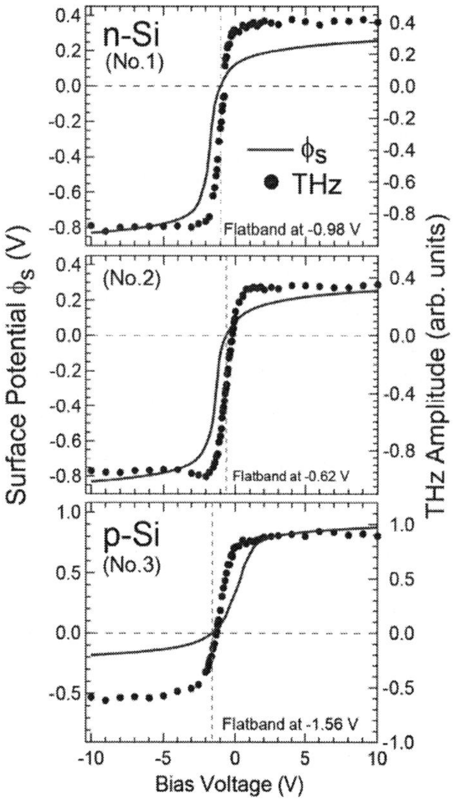

Figure 5: The surface potential calculated from the C-V curve and THz amplitude.

Figure 5 shows the surface potential with red lines

and THz amplitude with black circles. The THz amplitude showed strong positive correlation with the surface potential for all samples. The obtained relation shows that the laser THz emission from the samples is mostly determined by the surface band bending.

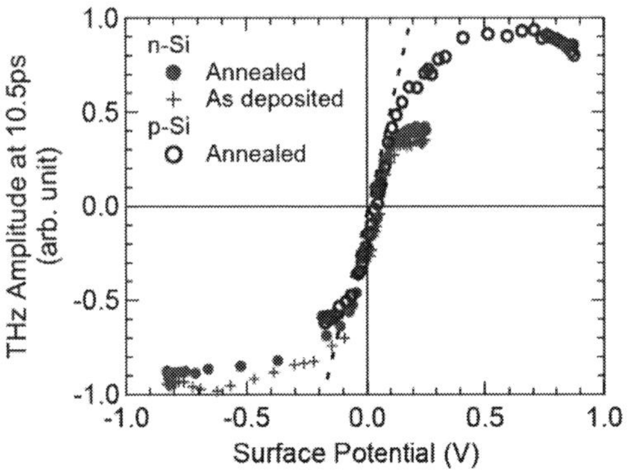

Figure 6: The relation between the surface potential and the amplitude of laser THz emission of the all samples measured.

Figure 6 summarise the relation between the surface potential and the THz amplitude for all samples and biases. The n-type samples are shown in red symbols and the p-type is shown in blue circles. In the range of -0.1 to 0.1 V, the THz amplitude of all samples is close to the dashed line, indicating the proportional relation. With the simplest estimate using Drude model, the peak amplitude of the THz radiation is proportional to the surface potential, which is consistent to the experimental results in this range. With larger positive or negative surface potential, THz amplitudes becomes insensitive to the surface potential.

3.3 Summary of the measurements on the MOS samples

We observed strong correlation between the surface potential and the amplitude including a proportionality in a limited range. With the present results, we expect that LTEM can be applied as optical mapping tool of internal field such as surface field of semiconductor devices including solar cells. Currently, we are looking into a more accurate explanation of THz amplitudes and waveforms to expand the range of quantitative evaluation of internal field using LTEM.

3.4 Evaluation of passivation of Si surfaces using LTEM

We measured and mapped the amplitude of laser THz emission from Si surfaces with oxide passivation layers. The sample structures are similar to the n-type samples described in the Sec. 2, but without any electrodes. The measurement conditions are also the same as described in the Sec. 2, measuring and mapping the THz amplitude at 10.5 ps.

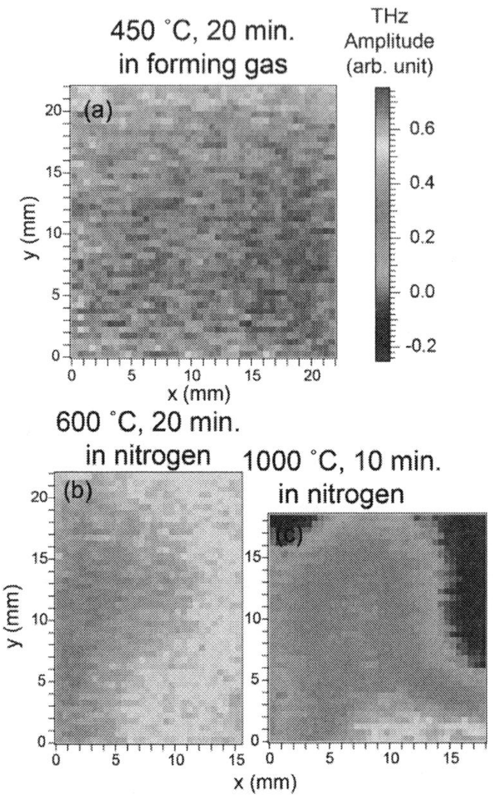

Figure 7: THz images of the Si surfaces with oxide passivation layers. The minority carrier lifetime at 3 $\times 10^{15}$ cm^{-3} minority carrier density measured by QSS-PC method using Sinton WCT-120 of the sample (a), (b), and (c) are 850 $\pm$ 50 μs, 15 $\pm$ 3 μs, and 80 $\pm$ 40 μs, respectively. The uncertainty are evaluated from the spatial inhomogeneity.

Figure 7 shows the THz images of the oxidised c-Si surfaces, with different thermal treatments after the oxidation. Fig. 7(a) shows the sample, which was annealed in 3% forming gas at 450°C to maximise the lifetime. The observed THz image showed strong signal with spatial homogeneity. Fig. 7(b) and 7(c) show the sample annealed in nitrogen ambient at high temperatures to intentionally degrade the surface passivation. The observed signal were weaker than the sample with forming gas anneal and were spatially inhomogeneous, showing weaker and inhomogeneous band bending at the surfaces thus lower quality of the field-effect surface passivation. We also observed the local negative signals in Fig. 7(c). This indicates that the negative charge density such as metallic contamination effectively exceeded the positive fixed charge density at the Si/SiO_x interface. We are currently looking into more detailed analyses by comparing the LTEM image and lifetime image measured with methods with higher spatial resolution such as μ-PCD for further analyses.

4 SUMMARY AND PERSPECTIVE

In summary, we measured Si-SiOx interface with electric biases using LTEM, and found strong correlation between surface potential and the THz amplitude including a proportionality in a limited range. We also evaluated the oxide passivation of n type c-Si with LTEM image. The present results indicate that the LTEM can be a tool to map the surface potential in silicon devices such as solar cells and useful to evaluate the surface passivation.

Currently, we are looking into a more accurate explanation of THz amplitudes and waveforms.

We expect that LTEM can be also applied to diagnose the surfaces of devices made from other semiconductors.

REFERENCES

[1] M. Tonouchi, Nat. Photonics, **1** (2007) 97. [2] H. Murakami, et al., J. Phys. D, **47** (2014) 374007. [3] H. Nakanishi, et al., Appl. Phys. Express, **5** (2012) 112301. [4] H. Nakanishi, et al., AIP Advances, **5** (2015) 117129.

INVESTIGATION OF AL_2O_3 PASSIVATION LAYERS BY PHOTOLUMINESCENCE IMAGING UNDER APPLIED VOLTAGE

Halvard Haug[1]*, Bao Yameng[2], Hele Savin[2], Erik Stensrud Marstein[1]
[1] Institute for Energy Technology, Instituttveien 18, 2007 Kjeller, Norway,
[2] Department of Micro and Nanosciences, Aalto University, Tietotie 3, Espoo 02150, Finland

*Corresponding author: E-mail: halvard.haug@ife.no Phone: +47 63 80 60 00

In this work, the passivation properties of Al_2O_3 thin films deposited by atomic layer deposition on p-type Si substrates has been studied by photoluminescence imaging under applied bias (PL-V). By measuring the carrier lifetime as a function of voltage this method enables investigations of the recombination properties of the Al_2O_3/c-Si interface under various band bending conditions in a simple and reproducible manner. The results were interpreted in the framework of the extended Shockley-Read-Hall theory using the fixed charge density Q_f and surface recombination velocity parameters S_{n0} and S_{p0} as fitting parameters. The extracted Q_f values were also found to be in good agreement with results from capacitance-voltage measurements. A high carrier lifetime above 1.5 ms was obtained with the traditionally used precursors Trimethylaluminum (TMA) and ozone (O_3), producing layers with a large negative Q_f of -4×10^{12} cm^{-2}. Samples were also deposited using a combination of TMA and the low-cost alternative precursor dimethylaluminium chloride (DMACl), also providing good passivation with carrier lifetimes over 1 ms on Cz substrates. For these samples, Q_f, S_{n0} and S_{p0} were all observed to decrease with increasing deposition temperature from 100 °C to 300 °C, with the best passivation quality obtained at 300 °C. For all the investigated samples S_{p0} was found to be larger than S_{n0} by a factor between 2 and 5.
Keywords: Aluminum oxide, ALD, Passivation, Photoluminescence, Capacitance, Characterization

1 INTRODUCTION

Al_2O_3 is a widely used material for surface passivation of crystalline silicon (c-Si). Thin films of the material can be deposited with a high, negative fixed charge density, making it well suited as a rear side passivation layer in industrial PERC (passivated emitter and rear cell) solar cell production using p-type c-Si wafers [1]. Different thin film deposition techniques can be used, but atomic layer deposition (ALD) typically produces the lowest surface recombination velocity (SRV) [2]. However, the passivation properties are strongly dependent on the processing parameters and on post-deposition annealing, and in-depth electrical characterization is therefore of high importance both for process optimization and device simulation. In this work we provide examples of such characterization, attempting to obtain a fundamental understanding of the passivation properties of ALD Al_2O_3 layers. By a combination of a novel experimental technique and simple simulations of the recombination mechanisms we extract the defining recombination parameters and show how these can be related to processing conditions.

2 EXPERIMENTAL DETAILS

2.1 Sample preparation

Double-side polished p-type Cz Si(100) wafers with an acceptor doping concentration of $N_A = 5.5 \times 10^{15}$ cm^{-3} and a thickness of 395 µm were used for the experiments. The wafers were passed through a RCA1 cleaning and a 30s dip in 5% HF solution prior to film deposition. Al_2O_3 layers were deposited on both sides using a Beneq TFS-500 thermal ALD instrument with 200 cycles. Two sets of samples were investigated: Sample A was deposited at 200 °C using the conventional Al_2O_3 precursors Trimethyl-aluminum (TMA) and ozone (O_3), whereas sample set B was deposited using a combination of TMA and dimethylaluminium chloride (DMACl) and H_2O at three temperatures, 100, 200, and 300 °C. DMACl has previously been shown to be a promising precursor for Al_2O_3 passivation layers [3], as it has a similar vapor pressure as the traditionally used TMA, but has the advantage of a lower cost. For the TMA/DMACl samples the films were deposited using 2 cycles of TMA+H_2O and 2 cycles of DMACl and H_2O, repeated 50 times, giving a total number of 200 cycles. A post-deposition anneal was performed in N_2 atmosphere at 400 °C for 30 minutes. The samples were then passed through a simulated contact firing process at 800 C for 3 seconds, using a rapid thermal annealing (RTA) tool including wafer temperature measurements. Ohmic contacts were made along one side of the samples by grinding down the Al_2O_3 layer and depositing Al by thermal evaporation on parts of the sample, which was then heated to 450 °C for 10 min. Two different types of electrodes were deposited onto the Al_2O_3 layers, again by thermal evaporation of Al through a shadow mask: 7 mm × 7 mm electrodes were fabricated for the lifetime vs. voltage (PL-V) measurements and 0.45-2.25 mm diameter circular electrodes were fabricated for the capacitance - voltage (C-V) measurements on the opposite side of the ohmic contacts.

2.1 The PL-V method

The photoluminescence (PL) imaging under applied voltage (PL-V) method has previously been described in detail in Refs. [4], [5]. Steady state carrier lifetime measurements were carried out by PL imaging [6], [7], using a LIS-R1 setup from BT imaging with an excitation wavelength of 808 nm and a constant photon flux of 1×10^{17} cm^{-2}s^{-1}. The PL intensity was calibrated to the carrier lifetime using a quasi-steady state photoconductance (QSSPC) measurement [8] of a reference region without metal electrodes. An external

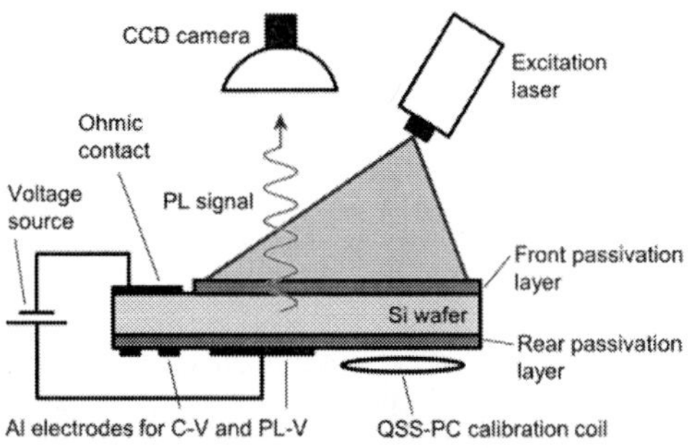

Figure 1. Schematic overview of the PL-V setup, used to measure the carrier lifetime versus voltage curves shown in Figs. 2 and 3. C-V curves can be measured on different electrodes deposited on the same sample.

voltage source was connected to the sample in order to obtain lifetime images while a DC voltage was applied over the rear side passivation layer. The effective carrier lifetime was then calculated from the PL intensity averaged over the central part of the electrode area, as described in Ref. [6]. Before calculating the effective lifetime, the PL intensity in the region over the rear side electrodes was divided by a sample-specific correction factor of 1.15 in order to correct for the enhancement of the signal caused by the higher rear side reflectivity of the metallized regions. This factor is determined prior to the measurements by measuring the signal enhancement from an electrically decoupled Al mirror placed below the sample. A sensitive ammeter was connected to the circuit in order to monitor and avoid any leakage currents disturbing the measurements. An automated measurement setup was used for the experiment, allowing for relatively fast measurements with finely spaced voltage steps. By alternating the voltage polarity and gradually increasing the absolute value it was also possible to avoid any additional charge buildup or hysteresis effects in the measured curves.

2.2 Capacitance-voltage measurements on MIS structures

The fixed charge density Q_f near the Al_2O_3 / c-Si interface was extracted from high frequency (1 MHz) C-V curves measured using a Keithley 4200-SCS semiconductor characterization system. Notice that both the PL-V and C-V measurements use a metal-insulator-semiconductor (MIS) structure, and in both cases the surface potential is determined by the charge balance throughout the system. We neglect the charge of the interface traps Q_{it} and assume $Q_{Si} + Q_g + Q_f = 0$, where Q_{Si} is the induced charge at the Si surface, Q_g is the effective charge induced by the gate electrode, and Q_f is the fixed charge density in the insulating layer. Q_g is given by [10]

$$Q_g = \frac{\varepsilon_0 \varepsilon_i}{d_i q}(V - \phi_{ms} - \psi_s), \qquad (1)$$

where ε_0 is the free space permittivity, ε_i is the dielectric constant of the Al_2O_3 layer, d_i is the layer thickness, q is the elementary charge, V is the applied voltage, ϕ_{ms} is the metal-semiconductor work function difference and ψ_s is the surface potential. Assuming a work function of 4.1 eV for Al, we calculate a value of $\psi_{ms} = -0.85$ V for all the samples. At flat band conditions, $\psi_s = 0$ and $Q_{Si} = 0$, giving $Q_g = -Q_f$. Q_f can thus be calculated using Eq. (2) by setting V equal the flat band voltage V_{fb}, which can be extracted from the C-V curve following the procedure described in [11]. ε_i can in principle be calculated directly from the capacitance measured in accumulation conditions, which should equal the insulator capacitance $C_i = \varepsilon_0\varepsilon_i/d_i q$. However, the uncertainty in this value was too large in our measurements, and we have therefore used a typical value of $\varepsilon_i = 7.5$ for all samples, both for the C-V analysis and in the extended SRH model described below. The film thickness was measured by variable-angle spectroscopic ellipsometry for each sample. For sample A $d_i = 19.7$ nm, and for the sample set B d_i = 17.2, 20.7 and 15.8 for deposition temperatures 100, 200 and 300 °C, respectively.

2.3 Contactless C-V

COCOS-measurements were performed using a PV2000 tool from Semilab, as described in Ref. [12].

3 SIMULATING THE CARRIER LIFETIME WITH VARYING SURFACE POTENTIAL

In order to obtain quantitative information about the recombination properties of the interface traps the PL-V results were compared to an extended Shockley-Read Hall (SRH) recombination model [13]. The simplified equation including one effective midgap defect level was found to sufficiently describe the observations. The rear effective SRV is then given from standard SRH theory as

$$S_{SRH} = \frac{1}{\Delta n} \frac{S_{n0} S_{p0}(n_s p_s - n_i^2)}{S_{n0}(n_s + n_i) + S_{p0}(p_s + n_i)}, \qquad (2)$$

where n_i is the intrinsic carrier concentration and S_{n0} and S_{p0} are the SRV parameters of electrons and holes. The surface concentrations of electrons and holes n_s and p_s are strongly dependent on the surface potential ψ_s The Girisch formalism [3], [16] was used to numerically find a self-consistent value for ψ_s for any given value of V, Q_f and ψ_{ms}. In addition it was necessary to add an extra contribution to the rear side SRV in order to correctly describe the flattening of the lifetime vs. voltage curves under large surface band bending:

$$S = \begin{cases} S_{SRH} + S_{min,n} \text{ for } \psi_s < 0 \\ S_{SRH} + S_{min,p} \text{ for } \psi_s > 0 \end{cases}, \qquad (3)$$

A similar added SRV contribution has also been observed in previous works by other groups [14]. The front SRV was calculated from the measured lifetime in the reference region without metal contacts and used as input for the simulations. Finally, the effective lifetime was calculated for each voltage using the full asymmetric equations described in Ref. [15]. The bulk lifetime is not negligible in the Cz wafers used for the experiments, and was described by one midgap defect with the SRH parameters $\tau_{n0} = 5 \times 10^{-4}$ s and $\tau_{0p} = 1 \times 10^{-2}$ s.

4 RESULTS AND DISCUSSION

3.1 TMA + O_3 precursors

Fig. 2 shows the results of the PL-V measurement on sample A, together with a representative C-V curve measured on the same sample. The simulation model described in section 3 above was fitted to the experimental data, with the best fit simulation parameters given in Table I. Changes in the voltage applied to the gate electrode can be considered as changes in the "added" gate charge Q_g, (Eq. 1), as indicated on the top x-axis. As more positive charge is added to compensate for the negative fixed charge Q_f present in the film the lifetime decreases. When $Q_g \approx -Q_f$ the field effect passivation is completely removed and the SRV is purely determined by chemical passivation. Close to this point the SRV reaches a maximum value as $n_s \approx p_s$, causing the lifetime to go through a minimum. As shown by the dashed line in Fig. 2 this lifetime minimum corresponds reasonably well with the flatband voltage obtained from the C-V measurements. On the left side of the lifetime minimum the surface is accumulated ($p_s \gg n_s$), and the recombination is limited by S_{n0}. On the right side the surface is inverted ($p_s \ll n_s$), and it is S_{p0} which determines the recombination behavior. The slope on each side of the minimum therefore provides independent information on S_{n0} and S_{p0}, and thus also the ratio between them, which in turn is equal to the ratio of the effective capture cross sections of the surface defects ($S_{n0}/S_{p0} = \sigma_n/\sigma_p$). We find a larger capture cross section for holes for all samples, with $\sigma_n/\sigma_p \approx 0.2$ in this case. We also observe that the minimum SRV in inversion ($S_{min,p}$) is lower than the one in accumulation ($S_{min,n}$). This might be related to the larger value for σ_p, or could be explained by additional recombination in the inversion layer, e.g. caused by subsurface defects. A higher SRV in inversion has also been observed in previous studies using corona charging [16], [17].

The model described in section 3 generally gives a good description of the measurement, but it severely overestimates the SRV (and thus underestimates the lifetime) in the depletion region close to the lifetime minimum. Parts of this discrepancy may be caused by

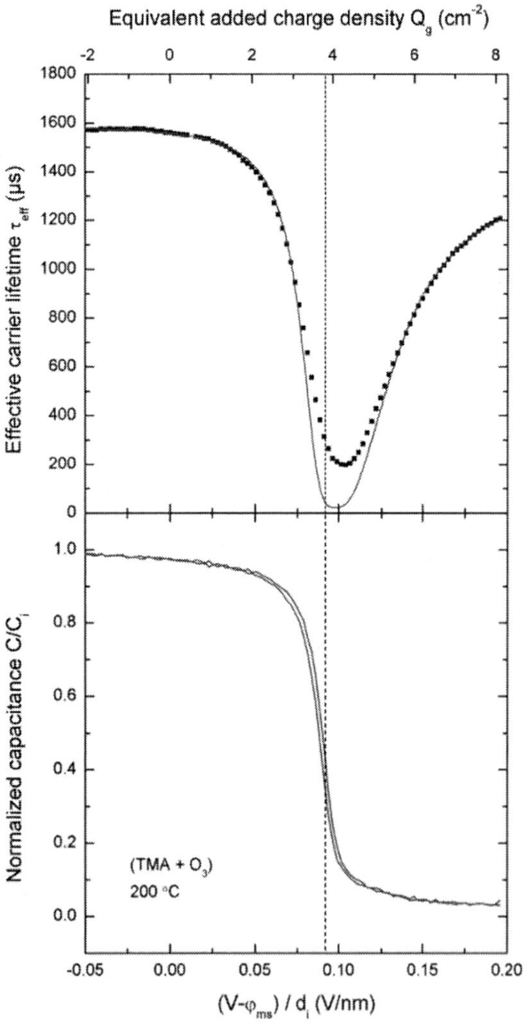

Figure 2: Top: PL-V measurement on the sample deposited with TMA + O_3 precursors. Measured (symbols) and simulated (line) effective carrier lifetime is shown as a function of voltage, corrected for the metal-semiconductor work function difference and divided with the insulator thickness. The top axis show the equivalent added charge density, calculated using Eq. 1.
Bottom: Representative high frequency C-V measurement (solid line) and the extracted flat band voltage V_{fb} (dashed vertical line). The top axis shows the equivalent added charge density Q_g, calculated using Eq. (2).

Table I: Best fit parameters used to obtain the simulated curve shown in Fig. 1.

Parameter		Unit	Value
S_{n0}	SRV parameter for electrons	cm/s	2.8×10^3
S_{p0}	SRV parameter for holes	cm/s	1.5×10^4
$S_{min,n}$	Minimum SRV, $V - \phi_{ms} < 0$	cm/s	8.2
$S_{min,p}$	Minimum SRV, $V - \phi_{ms} > 0$	cm/s	13.5
Q_f	Fixed charge density	cm^{-2}	-4.03×10^{12}

Table II: Measured Q_f values from the C-V measurements. The value and standard deviation are obtained after averaging over ~15 measurements of different pads at different parts of the sample.

Q_f	Fixed charge density	cm^{-2}	-3.87×10^{12}
	Standard deviation	cm^{-2}	0.23×10^{12}

experimental errors, since the lifetime in the area affected by the voltage is different than the surrounding region, which may cause lateral carrier diffusion or intensity from neighboring pixels being detected in the camera. As discussed in a previous paper [5] this does however not fully explain the difference, indicating that the simplified model does not correctly describe the recombination behavior when the SRV is large.

3.1 TMA/DMACl + H_2O precursors

Fig. 3 shows an example of PL-V measurements used to investigate a process variation, namely the deposition temperature used for sample set B. Note that any measured lifetimes influenced by leakage current at higher voltages has been discarded, which explains the different cut-off in the experimental data for each sample. The best fit parameters for each temperature are given in Table III, and the same values are also shown graphically in Fig. 4. Note that the voltage axis of Fig. 2 is shifted by ϕ_{ms} and divided by d_i to simplify the relation with Q_g (top axis of Fig. 3) and comparison between different film thicknesses. The PL-V curves shift towards the left with increasing deposition temperature and at the same time they become narrower. This results in a decrease in the negative Q_f, as well as a decrease in S_{n0} and S_{p0}. The field effect passivation is thus weaker at higher T, whereas the chemical passivation is stronger, with the total SRV being lowest at 300 °C. As seen in the bottom part of Fig. 3 the lifetime minimum for each curve again corresponds well with V_{fb} from the C-V measurements, which is also reflected in a good match between the Q_f values in Tables III and IV. In contrast to these clear trends $S_{\mathrm{min,n}}$ appears to be highest for the middle sample deposited at 200 °C, causing this sample to have the lowest measured lifetime when no voltage is applied. The reason for this is not understood, but could also be an error caused by differences in the bulk lifetime or in the front SRV. This issue can however hopefully be clarified by repeating the experiment using FZ substrates.

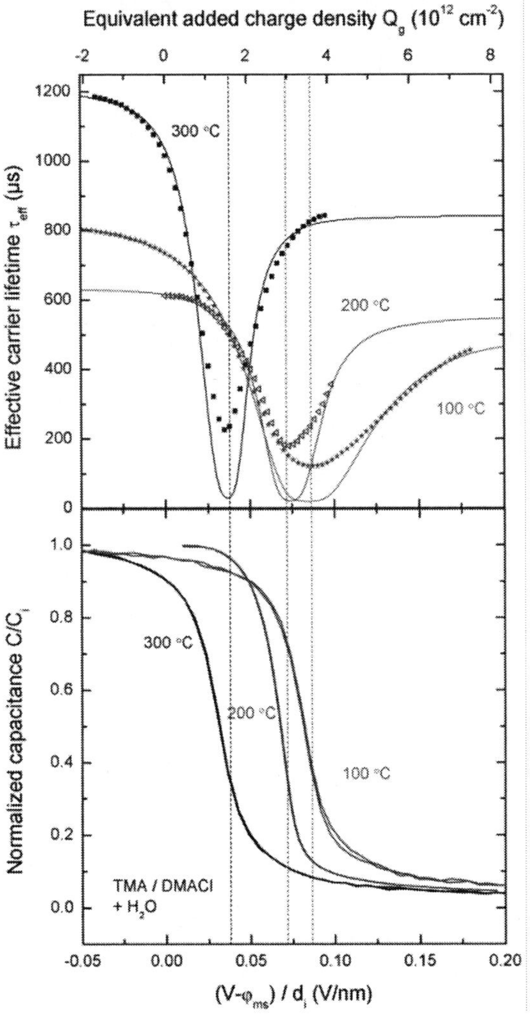

Figure 3: Top: PL-V measurements of from sample set B, deposited using TMA / DMACl + H_2O precursors. Measured (symbols) and simulated (lines) carrier lifetime is shown as a function of voltage, corrected for the metal-semiconductor work function difference and divided by the insulator thickness for easier comparison between the films of different thickness.
Bottom: Representative high frequency C-V measurements (solid line) and the extracted flat band voltage V_{fb} (dashed vertical line) for each sample.

Table III: Best fit parameters used to obtain the simulated curves shown in Fig. 2.

Parameter	Unit	100 °C	200 °C	300 °C
S_{n0}	cm/s	2.1×10^4	5.5×10^3	1.6×10^3
S_{p0}	cm/s	6.9×10^4	1.1×10^4	1.8×10^3
$S_{min,n}$	cm/s	12.4	19.1	6.9
$S_{min,p}$	cm/s	36.7	(21.5)	17.7
Q_f	cm^{-2}	-3.3×10^{12}	-2.9×10^{12}	-1.4×10^{12}

Table IV: Measured Q_f values from C-V measurements on MIS structures (the value and standard deviation are obtained after averaging over ~15 measurements of different pads at different parts of the sample) and Q_f and D_{it} values from contactless C-V.

	Unit	100 °C	200 °C	300 °C
C-V on MIS structures:				
Q_f	cm^{-2}	-3.4×10^{12}	-3.0×10^{12}	-1.6×10^{12}.
Std.dev.	cm^{-2}	0.17×10^5	0.22×10^{12}	0.32×10^{12}
Contactless C-V:				
Q_f	cm^{-2}	-2.1×10^{12}	-3.2×10^{12}	-1.4×10^{12}
D_{it}	eV^{-1}cm^{-2}	1.5×10^{11}	3.6×10^{11}	2.6×10^{11}

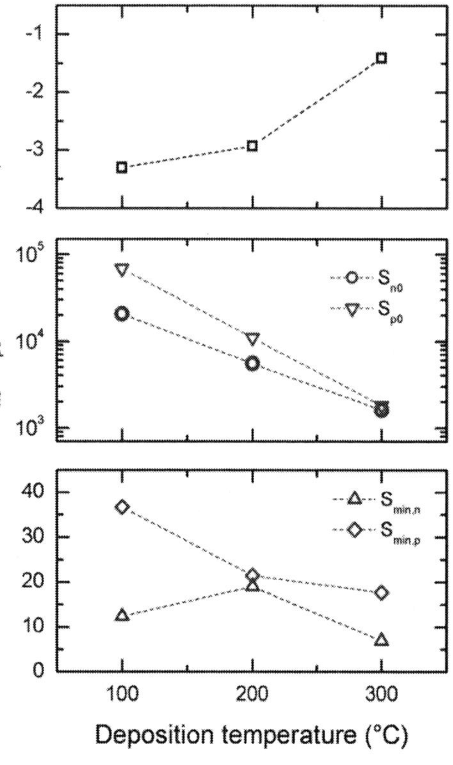

Figure 4: Extracted fit parameters from the PL-V measurements shown in Fig. 3, with values given in Table III.

We find a reasonably good agreement between the Q_f values measured on MIS structures and the ones measured using contactless C-V, except for the sample deposited at 100 °C, for which the contactless C-V gives an lower absolute value. The reason for this is not known, but could be caused by the additional storage and annealing used to produce the ohmic contacts for the MIS structures.

5 SUMMARY AND FURTHER WORK

In this paper, PL-V and C-V measurements have been used for in-depth characterization of selected Al_2O_3 layers deposited by ALD. We show this method can be used to go a step further than a simple lifetime measurement and obtain a deeper understanding of the recombination behavior, without the need for complicated sample preparation or time-consuming measurements.

In further work we wish to continue to use the automated PL-V setup to perform more measurements on a large range of ALD layers, preferably using FZ substrates and larger film thickness in order to avoid any bulk lifetime contributions and leakage currents, respectively.

ACKNOLEDGEMENTS

The authors want to thank Yacine Boulfrad for help with ALD deposition of the TMA + O_3 samples. This work has been funded through the internal project Authors from Aalto University acknowledge funding from Tekes project Nanosolar and provision of facilities at Micronova Nanofabrication Centre.

REFERENCES

[1] B. Hoex, J. J. H. Gielis, M. C. M. van de Sanden, and W. M. M. Kessels, "On the c-Si surface passivation mechanism by the negative-charge-dielectric Al2O3," J. Appl. Phys., vol. 104, no. 11, p. 113703, 2008.

[2] G. Dingemans and W. M. M. Kessels, "Status and prospects of Al2O3-based surface passivation schemes for silicon solar cells," J. Vac. Sci. Technol. A Vacuum, Surfaces, Film., vol. 30, no. 4, p. 040802, 2012.

[3] S. Li, Y. Bao, M. Laitinen, T. Sajavaara, M. Putkonen, and H. Savin, "Excellent silicon surface passivation using dimethylaluminium chloride as Al source for atomic layer deposited Al2O3," Phys. status solidi, vol. 212, no. 8, pp. 1795–1799, Aug. 2015.

[4] H. Haug, Ø. Nordseth, E. V. Monakhov, and E. S. Marstein, "Photoluminescence imaging under applied bias for characterization of Si surface passivation layers," Sol. Energy Mater. Sol. Cells, vol. 106, pp. 60–65, Jun. 2012.

[5] H. Haug, S. Olibet, O. Nordseth, and E. Stensrud Marstein, "Modulating the field-effect passivation at the SiO2/c-Si interface: Analysis and verification of the photoluminescence imaging under applied bias method," J. Appl. Phys., vol. 114, no. 17, p. 174502, Nov. 2013.

[6] T. Trupke, R. A. Bardos, M. C. Schubert, and W. Warta, "Photoluminescence imaging of silicon wafers," Appl. Phys. Lett., vol. 89, no. 4, pp. 44103–44107, 2006.

[7] S. Herlufsen, J. Schmidt, D. Hinken, K. Bothe, and R. Brendel, "Camera-based photoluminescence imaging of crystalline silicon wafers," in 24th European Photovoltaic Solar Energy Conference, 2009, pp. 913–917.

[8] R. A. Sinton, A. Cuevas, and M. Stuckings, "Quasi-Steady-State PhotoConductance, A new method for solar cell material and device characterization," in 25th IEEE Photovoltaic Specialists Conference, 1996, p. 457.

[9] H. Haug, O. Nordseth, E. Monakhov, and E. S. Marstein, "Investigation of Carrier Recombination at the SiO2/c-Si Interface by Photoluminescence Imaging Under Applied Bias," IEEE J Photovoltaics,., vol. 4, no. 1, pp. 374–379, 2014.

[10] R. B. M. Girisch, R. P. Mertens, and R. F. De Keersmaecker, "Determination of Si-SiO2 interface recombination parameters using a gate-controlled point-junction diode under illumination," IEEE Trans. Electron Devices, vol. 35, no. 2, pp. 203–222, 1988.

[11] D. K. Schroder, Semiconductor material and device characterization, 3rd ed. John Wiley & Sons, Inc., 2006.

[12] M. Wilson, J. D'Amico, A. Savtchouk, P. Edelman, A. Findlay, L. Jastrzebski, J. Lagowski, K. Kis-Szabo, F. Korsos, A. Toth, A. Pap, R. Kopecek, and K. Peter, "Multifunction metrology platform for photovoltaics," in 37th IEEE Photovoltaic Specialists Conference, 2011, pp. 1748–1753.

[13] A. G. Aberle, S. Glunz, and W. Warta, "Impact of illumination level and oxide parameters on Shockley--Read--Hall recombination at the Si-SiO2 interface," J. Appl. Phys., vol. 71, no. 9, pp. 4422–4431, 1992.

[14] S. Mack, A. Wolf, C. Brosinsky, S. Schmeisser, A. Kimmerle, P. Saint-Cast, M. Hofmann, and D. Biro, "Silicon Surface Passivation by Thin Thermal Oxide/PECVD Layer Stack Systems," IEEE Journal of Photovoltaics, vol. 1, no. 2. pp. 135–145, 2011.

[15] A. B. Sproul, "Dimensionless solution of the equation describing the effect of surface recombination on carrier decay in semiconductors," J. Appl. Phys., vol. 76, no. 5, pp. 2851–2854, 1994.

[16] F. Werner and J. Schmidt, "Manipulating the negative fixed charge density at the c-Si/Al<inf>2</inf>O<inf>3</inf> interface," Applied Physics Letters, vol. 104, no. 9. p. 91604, 2014.

[17] J. Schmidt, F. Werner, B. Veith, D. Zielke, S. Steingrube, P. P. Altermatt, S. Gatz, T. Dullweber, and R. Brendel, "Advances in the Surface Passivation of Silicon Solar Cells," Energy Procedia, vol. 15, no. 2011, pp. 30–39, Jan. 2012.

ADVANCED OPTICAL CHARACTERISATION OF INDUSTRIAL PECVD SILICON NITRIDE LAYERS

Nino Borojevic[1], Ziv Hameiri[1] and Saul Winderbaum[2]

[1]The University of New South Wales, Sydney, NSW 2052, Australia

[2]Shamash Australia Pty Ltd, Mount Barker, SA 5251, Australia

ABSTRACT: Deposition of silicon nitride layers on top of silicon wafers using industrial plasma enhanced chemical vapor deposition systems is a major component in today's manufacturing of silicon solar cells. Previous work has demonstrated non-uniformity in the optical properties within the bulk of dynamically deposited layers using such in-line systems. A gradient in the optical properties within the layer emphasizes the need for optical characterization as a function of the layer's depth. This paper presents an advanced method to characterize dynamically deposited silicon nitride layers, where the modelling of the experimental ellipsometry spectra is performed by grading of the Tauc-Lorentz optical parameters along the normal of the layer. Significant improvements are demonstrated over the standard, more commonly used Tauc-Lorentz modelling procedure; mainly by revealing of the depth profiles of the layer's optical properties.

Keywords: Characterisation, PECVD, silicon nitride

1 INTRODUCTION

Deposition of silicon nitride (SiN_x) layers on top of silicon (Si) wafers using industrial plasma enhanced chemical vapor deposition (PECVD) systems is a major component in today's manufacturing of Si solar cells. Reduction of optical losses and passivation of the Si wafer's surfaces and bulk are the two major functions of the SiN_x layer [1]. Optical losses (front surface reflection and parasitic absorption in SiN_x) are influenced by the spectra of the layer's refractive index (n) and extinction coefficient (k), which is commonly determined via analysis of spectroscopic ellipsometry measurements. Spectroscopic ellipsometry determines the relative phase change of polarized light reflected from a sample's surface [2]. The measured values are ellipsometric angles, psi (Ψ) and delta (Δ), which represent the change in amplitude and phase shift, upon reflection, respectively [2]. The ellipsometric data are fit to optical models using optical functions (dispersion formulae) for the material layers; these optical functions provide the n and k spectra.

A standard optical model to fit ellipsometric data resulting from SiN_x/Si stacks is a four-medium model; air/roughness/SiN_x/Si. The surface roughness is often modelled using a Bruggeman effective medium approximation made up of 50% voids (air) and 50% of the underlying SiN_x [3]. The SiN_x layer is commonly modelled using the Tauc-Lorentz dispersion formula, defining the imaginary part of the dielectric function $\varepsilon_2(E)$ by [4, 5]

$$\varepsilon_2(E) = \frac{AE_0C(E-E_g)^2}{(E^2-E_0{}^2)^2+C^2E^2}\frac{1}{E} \qquad E > E_g$$

$$= 0 \qquad E \leq E_g, \quad (1)$$

where E is the photon energy and the four constants the amplitude (A), the Lorentz resonant energy (E_0), the broadening factor (C) and the optical bandgap (E_g), are used as fitting parameters. The real part of the dielectric function $\varepsilon_1(E)$ is determined by Kramers-Kronig integration [4, 5]:

$$\varepsilon_1(E) = \varepsilon_1(\infty) + \left(\frac{2}{\pi}\right) P \int_{E_g}^{\infty} \frac{E'\varepsilon_2(E')}{E'^2-E^2}\,dE', \qquad (2)$$

where P is the Cauchy principal part of the integral and $\varepsilon_1(\infty)$ is an additional fitting parameter which is set to 1 during the fitting process [6]. Surface roughness and SiN_x layer thicknesses are also used as fitting parameters. Since its first application in 1996 until today, this approach has been the most frequent one to characterize optical properties of both, statically and dynamically deposited SiN_x layers [7-15]. This approach is defined in this study as the "standard" fitting method.

In 2005 a study on the optical properties of dynamically deposited SiN_x layers was conducted with a static deposition inside an industrial in-line PECVD system [16]. The system includes microwave (MW) antennas that are used for plasma excitation. The deposition characteristics were analyzed as a function of position with respect to the plasma source using wafers located at different distances from the MW antenna. The refractive index (at 630 nm) and the layer thickness were measured across all wafers; the results demonstrated that a layer with increasing n is deposited at the plasma cloud edges and a layer with n close to the target value is deposited everywhere else (central bulk of the plasma). The thickness measurements resulted in the opposite trend, showing a maximum deposition rate underneath the antenna, which is slowly decreasing towards either side and being lowest at the edges of the plasma cloud (approximately following a Gaussian shape). The study outcome was a clear indication of the wafer being affected by different local deposition kinetics as it is passing through different plasma sections, which results in non-uniform buildup of the SiN_x layer. Similar results were reported in a later study [17].

Despite these findings, dynamically deposited SiN_x layers continued to be characterized using the standard Tauc-Lorentz method, without any information about the layer's depth profile [13-15]. Fit results and the quality of the optical modelling are also rarely found in the literature, even for studies which did consider variation in the layer buildup [18]. This work demonstrates an improved fitting method to characterize optical properties of dynamically deposited SiN_x layers, including a comparison to the results achieved with the standard

Tauc-Lorentz fitting method. The advanced method demonstrates more accurate optical modelling of the experimental ellipsometry spectra, revealing more precise optical properties of the dielectric layers, particularly throughout the layer's depth.

2 EXPERIMENTAL DETAILS

An industrial in-line PECVD system (MAiA, Roth&Rau) with two parallel plasma sources was used for the deposition of SiN_x layers in this study. The layer was deposited on the front side of double-sided polished Si wafers and the deposition parameters were adjusted to result in a typical *n* used for encapsulated solar cell antireflection coatings (2.09, measured at 633 nm). Spectroscopic ellipsometry was carried out on the deposited wafers with a M2000 spectroscopic ellipsometer from J.A. Woollam. Ellipsometric spectra (Ψ, Δ) was measured at three angles of incidence (65°, 70°, and 75°) over a spectral range of 300–1000 nm. The optical modelling and fitting of the experimental data was performed with the WVASE32 software package [19], applying the standard method and an advanced method, described below.

The setup of the plasma sources in the PECVD system used in this study is identical to the systems used in Refs. [16] and [17]. Therefore similar *n* and deposition rate trends were assumed as a function of position with respect to the plasma source. This resulted in an approximation of the deposited layer structure as a stack of six sublayers [shown in Figure 1], in accordance of the wafer movement in the reactor. The approximated structure was modelled in WVASE, accompanied by the grading of three of the four Tauc-Lorentz parameters (A, E_0, E_g) along the normal of each sublayer to match the experimental data. Grading of the parameter C and the surface roughness did not have any influence on the fitting quality; hence the former was fitted as a constant for the whole bulk SiN_x layer and the latter was excluded from the advanced optical model. The graded profiles of A, E_0 and E_g and the fitted C constant were used to calculate the dispersion formulae at different intervals along the layer's depth, enabling the extraction of the *depth* profiles of the layer's optical constants at every single wavelength.

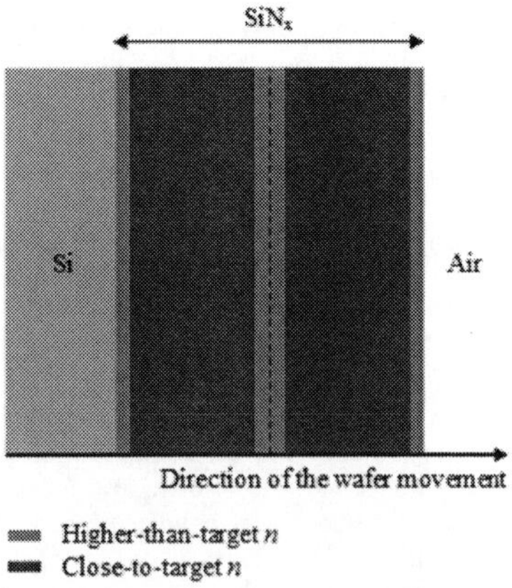

Figure 1: Approximation of the deposited SiN_x layer structure according to wafer movement in the reactor.

3 RESULTS AND DISCUSSION

Figures 2 (a) and (b) show the fits between the experimental and modelled Ψ and Δ spectra for the standard and the advanced fitting method, respectively. Grading of the Tauc-Lorentz fitting parameters matches the experimental data more closely, demonstrating the model's ability to describe the experimental data very well. A more detailed description of the advanced fitting procedure used in this study and the corresponding fitting results may be found in [20].

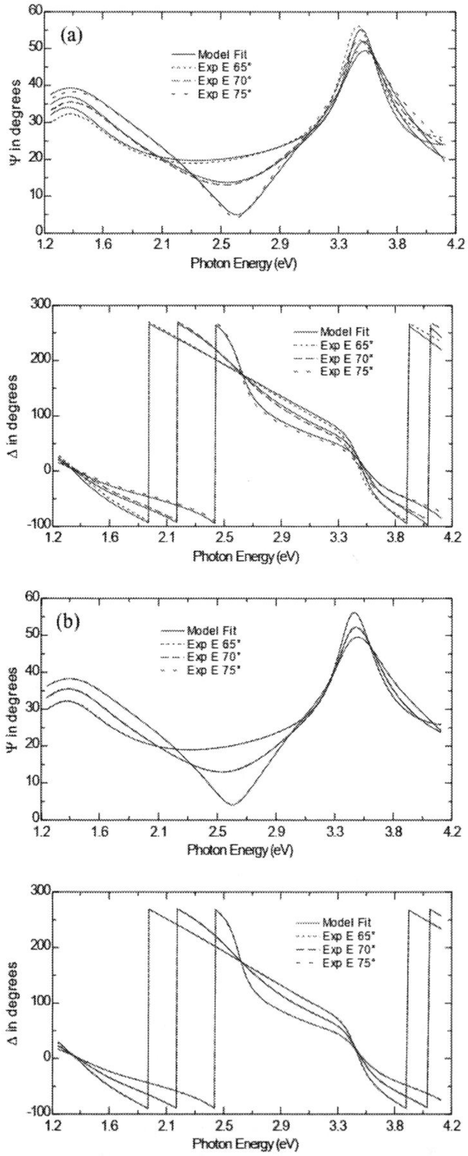

Figure 2: Fits between the experimental and modelled Ψ and Δ spectra for (a) the standard and (b) the advanced fitting method.

The depth profiles of *n* (at 633 nm) and *k* (at 400 nm) are displayed in Figure 3, showing good agreement with the data reported in Ref. [16] and demonstrating once more that thin layers with an increased Si content are deposited at the edges of the two plasma clouds, while the rest of the layer has *n* close to the target value.

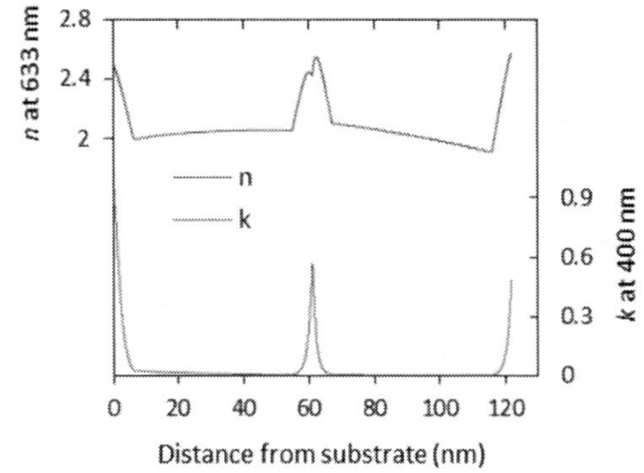

Figure 3: Depth profile of the film's *n* at 633 nm and *k* at 400 nm.

Using the depth profiles extracted via the advanced optical modelling, weighted average *n* and *k* were calculated at every single wavelength to obtain the effective dispersion of the two optical constants of the SiN_x layer. Figure 4 (a) presents a comparison of these and the optical constants obtained via the standard Tauc-Lorentz fit. An increase in *n* and *k* towards shorter wavelengths is shown for the data resulting from the advanced fit, which can be attributed to the thin Si-rich sections of the SiN_x layer that the standard fitting method was not able to reveal. These sections are contributing to an increase in the film's absorption of short-wavelength light, as shown by the *k* dispersion.

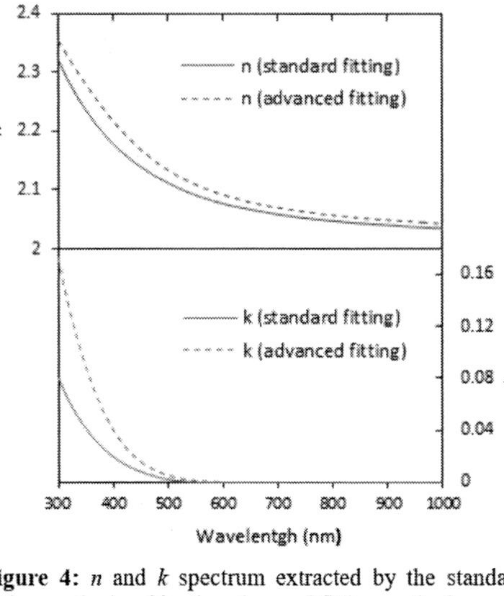

Figure 4: *n* and *k* spectrum extracted by the standard fitting method and by the advanced fitting method.

The data from Figure 4 demonstrates the significance of the advanced fitting method in determining the SiN_x layer's optical losses, particularly the parasitic absorption in the 300–500 nm spectral range. Furthermore, using the absorption data, the method can enable more accurate analysis of emitter losses in Si solar cells coated by dynamically deposited SiN_x layers (i.e. majority of today's industrial solar cells) [21, 22].

4 CONCLUSIONS

An advanced method for detailed optical characterization of SiN_x layers, deposited dynamically with industry-used PECVD in-line systems, was presented. The concept of the method is modelling of experimental ellipsometry spectra by grading of Tauc-Lorentz optical parameters along the normal of the deposited SiN_x layer. Significant improvements are demonstrated over the standard, more commonly used Tauc-Lorentz modelling procedure by more accurate fitting of the experimental data and revealing of the depth profiles of the layer's optical properties. The depth profiles obtained via the advanced fitting method demonstrated that a thin Si-rich layer is deposited at the edges of the plasma cloud surrounding the MW antennas, while the bulk of the plasma results in the deposition of a layer with *n* close to the target value. This information facilitates more accurate characterization of the layer's optical losses, and in general provides a better understanding of these SiN_x layers used so extensively in today's photovoltaic industry.

5 ACKNOWLEDGMENTS

The authors thank Jianing Sun and Tom Tiwald from J.A. Woollam for their advice regarding WVASE modelling. The authors acknowledge support from the Australian Government through the Australian Renewable Energy Agency (ARENA, Project 2014/RND097). The views expressed herein are not necessarily the views of the Australian Government, and the Australian Government does not accept responsibility for any information or advice contained herein. Ziv Hameiri acknowledges the support of the Australian Research Council (ARC) through the Discovery Early Career Researcher Award (DECRA, Project DE150100268).

7 REFERENCES

[1] A. G. Aberle, "Overview on SiN surface passivation of crystalline silicon solar cells," Solar Energy Mater. Solar Cells, 65(1-4), (2001) 239.

[2] H. Fujiwara, *Spectroscopic Ellipsometry: Principles and Applications*, John Wiley & Sons, (2007).

[3] D. A. G. Bruggeman, ""Berechnung verschiedener physikalischer Konstanten von heterogenen Substanzen,"Ann. Phys. (Leipzig) 24, (1935) 636.

[4] G. E. Jellison, Jr. and F. A. Modine, "Parameterization of the optical functions of amorphous materials in the interband region," Appl. Phys. Lett., 69, (1996) 371.

[5] G. E. Jellison, Jr. and F. A. Modine, "Erratum," Appl. Phys. Lett., 69, (1996) 2137.

[6] G. E. Jellison, Jr., F. A. Modine, P. Doshi and A. Rohatgi, "Spectroscopic ellipsometry characterization of thin-film silicon nitride," Thin Solid Films, 313-314, (1998) 193.

[7] P. Doshi, G. E. Jellison, Jr. and A. Rohatgi, "Characterization and optimization of absorbing

plasma-enhanced chemical vapor deposited antireflection coatings for silicon photovoltaics," Applied Optics, 36(30), (1997) 7826.

[8] J. Hong, W. M. M. Kessels, W. J. Soppe, A. W. Weeber, W. M. Arnoldbik and M. C. M. van de Sanden, "Influence of the high-temperature "firing" step on high-rate plasma deposited silicon nitride films used as bulk passivating antireflection coatings on silicon solar cells," J. Vac. Sci. Technol. B, 21, (2003) 2123.

[9] C. S. Cook, T. Daly, R. Liu, M. Canonico, Q. Xie, R. B. Gregory and S. Zollner, "Spectroscopic ellipsometry for in-line monitoring of silicon nitrides," Thin Solid Films, 455 –456, (2004) 794.

[10] J-F. Lelievre, A. Kaminski, J-P. Boyeaug, R. Monna and M. Lemiti, "Optical properties of PECVD and UVCVD SiN_x:H antireflection coatings for silicon solar cells," in *Proceedings of the 32nd IEEE Photovoltaic Specialists Conference* (2005) 1111.

[11] D. N. Wright, E. S. Marstein, A. Rognmo and A. Holt, "Plasma-enhanced chemical vapour-deposited silicon nitride films; The effect of annealing on optical properties and etch rates," Solar Energy Mater. Solar Cells, 92 (2008) 1091.

[12] J-F. Lelievre, E. Fourmond, A. Kaminski, O. Palais, D. Ballutaud and M. Lemiti, "Study of the composition of hydrogenated silicon nitride SiN_x:H for efficient surface and bulk passivation of silicon," Solar Energy Mater. Solar Cells, 93 (2009) 1281.

[13] S. C. Siah, B. Hoex and A. G. Aberle, "Accurate characterisation of silicon nitride films on rough silicon surfaces by ellipsometry," Energy Proceedia, 8, (2011) 122.

[14] S. Duttagupta, F. Ma, B. Hoex, T. Mueller and A. G. Aberle, "Optimised antireflection coatings using silicon nitride on textured silicon surfaces based on measurements and Multidimensional Modelling," Energy Procedia, 15, (2012) 78.

[15] S. Duttagupta, F. Lin, M. Wilson, M. B. Boreland, B. Hoex and A. G. Aberle, Prog. Photovolt.: Res. Appl. 22, (2014) 641.

[16] S. Winderbaum, A. Cuevas, J. Tan, R. Dunn, P. Pohl, J. Schmidt, J. Han and K. Roth, "Unveiling the differences between dynamic and static deposition of PECVD silicon nitride for solar cells," in *Proceedings of the 20th European Photovoltaic Solar Energy Conference* (2005) 1301.

[17] H.-P. Sperlich, D. Decker, P. Saint-Cast, E. Erben and L. Peters, "High productive solar cell passivation on Roth&Rau MAiA® MW-PECVD inline machine – a comparison of Al_2O_3 , SiO_2 and SiN_x-H process conditions and performance," in *Proceedings of the 25th European Photovoltaic Solar Energy Conference* (2010) 1352.

[18] M. Lamers, K. Butler, P. E. Vullum, J. Harding and A. Weeber, "Characterization of a-SiN_x:H layer: Bulk properties, interface with Si and solar cell efficiency," Phys. Status Solidi A 210(4), (2013) 658.

[19] J. A. Woollam, B. Johs, C. M. Herzinger, J. Hilfiker, R. Synowicki and C. L. Bungay, "Overview of variable-angle spectroscopic ellipsometry (VASE): I. Basic theory and typical applications," Proc. SPIE CR72, (1999) 3.

[20] N. Borojevic, Z. Hameiri and S. Winderbaum, "Advanced optical modelling of dynamically deposited silicon nitride layers," Appl. Phys. Lett., submitted.

[21] M. Abbott, G. Scardera, K. R. McIntosh and A. Meisel, "Simulation of emitter doping profiles formed by industrial $POCl_3$ processes," in *Proceedings of the 39th IEEE Photovoltaic Specialists Conference* (2013), 1383.

[22] B. Fischer, J. Müller and P. P. Altermatt, "A simple emitter model for quantum efficiency curves and extracting the emitter saturation current," in *Proceedings of the 28th European Photovoltaic Solar Energy Conference* (2013) 840.

ON THE STABILITY OF DIELECTRIC PASSIVATION LAYERS UNDER ILLUMINATION AND TEMPERATURE TREATMENT

D. Sperber, F. Furtwängler, A. Herguth, G. Hahn
University of Konstanz, Department of Physics
University of Konstanz, 78457 Konstanz, Germany

ABSTRACT: Different dielectric layers were deposited onto boron doped floatzone silicon wafers and the stability of the effective lifetime was investigated at elevated temperatures ranging from 75°C to 250°C with and without illumination. It was found that samples fired in a belt furnace show an instable behavior on a timescale of minutes to months. The short term behavior of a sample is strongly influenced by the peak firing temperature whereas the long term behavior appears to be less influenced by this parameter. Different chemical cleaning procedures during sample preparation show no significant influence on stability. Via corona charging, the changes at 250°C in darkness could be associated with changes in chemical passivation quality. It could be shown that a non-fired sample shows a similar, but less pronounced instability whereas an annealed sample is stable under the given treatment conditions.
Keywords: Passivation, Stability, Degradation, Silicon-Nitride, Lifetime

1 MOTIVATION

Light induced degradation (LID) mechanisms [1] such as boron-oxygen related degradation [2] or mc-PERC (passivated emitter and rear cell) degradation [3,4,5] strongly affect the minority carrier lifetime of silicon solar cells. Additionally, the effective lifetime of minority carriers τ_{eff} is strongly influenced by the quality of surface passivation. To differentiate between effects happening in the bulk and at the surface of a lifetime sample, a detailed knowledge of the behavior of passivation layers during degradation treatments is necessary. This is especially important for long term experiments, examining, *e.g.*, the long term stability of the regenerated state of boron-oxygen related defects [6]. On the other hand, a significant degradation in surface passivation quality can itself lead to a decrease in the conversion efficiency of solar cells, especially in concepts like PERC that rely heavily on the long term stability of the rear side passivation.

To investigate the stability of dielectric passivation layers, differently processed lifetime samples were investigated at elevated temperature and illumination conditions. It is important to note that the samples used in this study were processed using silicon nitride deposition recipes that are also used for typical solar cells and received a fast firing step with sample temperatures similar to those reached in solar cell processes.

2 EXPERIMENTAL

2.1 Preparation of the samples

The base material used was p-type floatzone (FZ) material with a specific resistivity of 1 Ωcm. All samples received a chemical cleaning step in a cleanroom environment followed by a dip in hydrofluoric acid (HF). Some samples also received a thin chemically grown silicon oxide (SiO_x) layer after the HF dip. All samples were then coated with hydrogenated silicon nitride (SiN_x:H). For this purpose, three different plasma enhanced chemical vapor deposition (PECVD) systems were used, one using a remote plasma (type 1) and two using a direct plasma technique (type 2 and type 3). If not stated otherwise, the samples were then fired in a fast firing belt furnace with measured peak sample temperatures reaching temperatures between 730°C and 830°C. After the firing step, the samples were stored in darkness until measurement series were carried out.

2.2 Measurement setups

The samples were treated at temperatures of 75°C, 150°C or 250°C. Some samples were simultaneously illuminated using a tungsten halogen incandescent lamp while others were kept in the dark.

For the repeated measurement of the effective lifetime during a treatment, two types of photo conductance decay (PCD) setups were available: For some samples, a Sinton lifetime tester (WCT 120) was used and the degradation treatment was shortly paused while the measurement was carried out at room temperature (RT-PCD). On other samples, PCD measurements were carried out at elevated temperature, so that only a short break in illumination occurred during measurements (ET-PCD). Because of the elevated temperature during these measurements, the absolute values of the measured τ_{eff} differ from the ones measured by RT-PCD, but the curve shape of a sample's evolution over time shows the same qualitative behavior in both measurements. All PCD measurements were evaluated at an injection level of $\Delta n = 1\cdot10^{15}$ cm^{-3}.

For corona charging experiments, a setup comparable to the one described in [7] was used.

3 RESULTS

3.1 What is already known

As has already been shown earlier, SiN_x:H layers can show strong changes in passivation quality at elevated temperature and illumination [8]. An example is shown in Fig. 1 where it can be seen that τ_{eff} of two differently processed samples is far from stable during a treatment at 75°C under illumination. It is interesting to note that the effective lifetime curves start at rather different values but show a similar qualitative behavior in the short term and converge in the long term behavior. It was already concluded that differences in the short term behavior are mainly caused by different peak firing temperatures.

Another important observation is that the process under investigation can be significantly sped up by increasing the temperature or illumination intensity as can be seen in Fig. 2. Using a chemical repassivation, it was also shown that the loss in τ_{eff} of the blue sample in

Fig. 2 after around 500 h is probably caused by a decrease in surface passivation quality, as would be expected because of the FZ material with high minority carrier bulk lifetime.

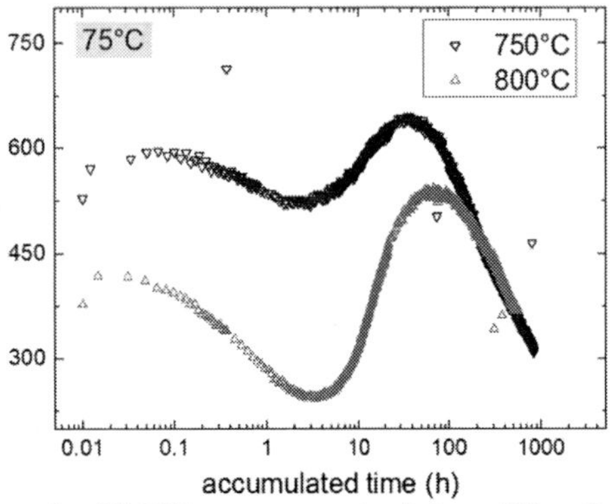

Figure 1: ET-PCD measurement of two differently processed samples coated with SiN_x:H and treated at 75°C and 0.65 suns illumination (after [8]). The peak sample temperatures during firing are given in the legend.

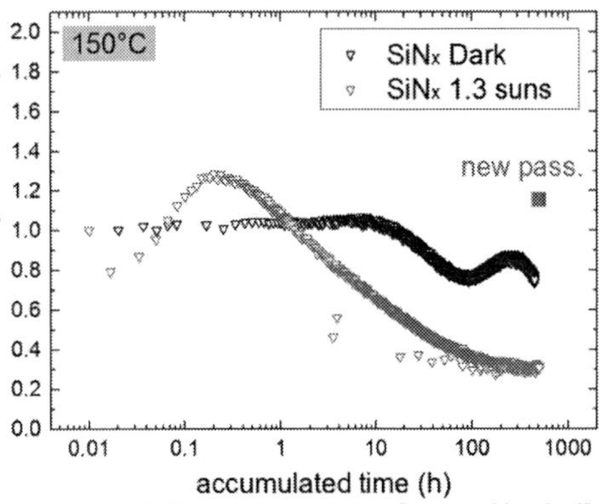

Figure 2: ET-PCD measurement of two identically processed samples coated with SiN_x:H, fired at 750°C peak sample temperature and treated at 150°C at different illumination conditions (after [8]). The blue sample was afterwards repassivated using quinhydrone (blue square).

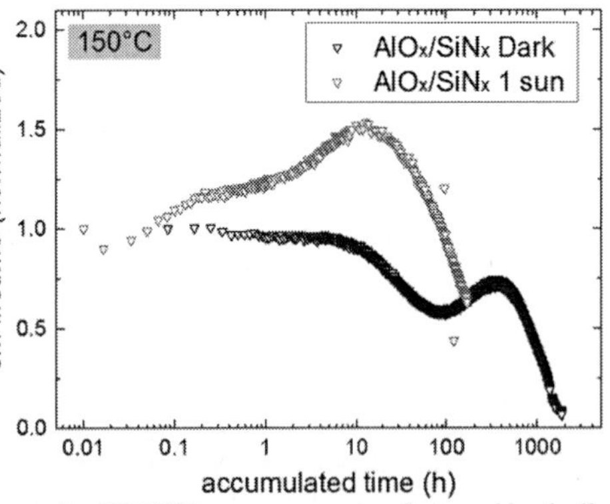

Figure 3: ET-PCD measurement of two identically processed samples coated with AlO_x:H/SiN_x:H, fired at 750°C peak sample temperature and treated at 150°C at different illumination conditions (after [8]).

Interestingly, also dielectric layer stacks consisting of a thin aluminum oxide (AlO_x:H) layer and a SiN_x:H layer show a similar characteristic curve shape as shown in Fig. 3. In darkness, the time scale of sample evolution is similar for both types of passivation. Under illumination, however, the AlO_x:H/SiN_x:H sample degrades much slower compared to the pure SiN_x:H passivated sample in Fig. 2. Still, illumination generally speeds up the process under investigation at 150°C.

This is different at 250°C where illumination only leads to a small acceleration of sample evolution compared to a treatment in darkness as can be seen in Fig. 4.

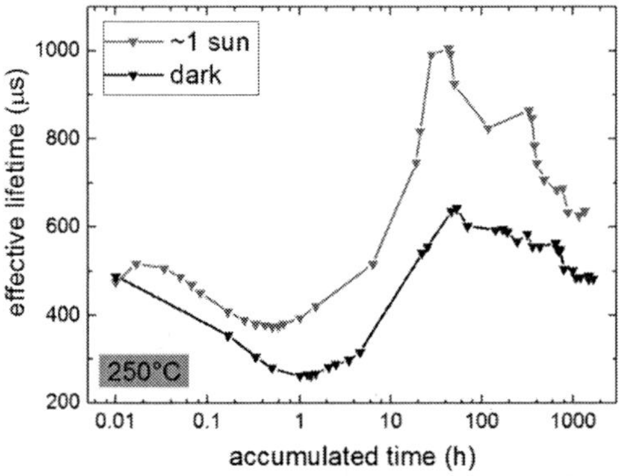

Figure 4: RT-PCD measurement of two identically processed samples coated with SiN_x:H, fired at 750°C peak sample temperature and treated at 250°C at different illumination conditions (after [8]). Lines only serve as a guide to the eye.

3.2 Combined treatment at 75°C and 150°C

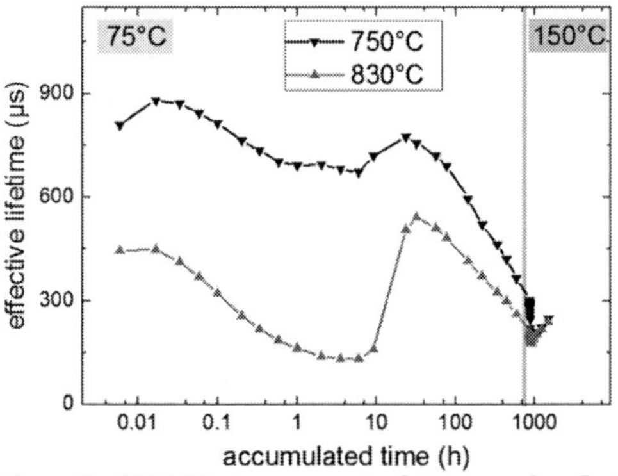

Figure 5: RT-PCD measurement of two samples first treated at 75°C and 1 sun illumination intensity (after [8] with additional new data points). Later, the temperature was increased to 150°C as indicated by the blue line. Both samples were processed identically: they first received an RCA clean and a thin chemically grown SiO_x layer before a type 2 SiN_x:H was deposited. Only the measured peak sample temperature of the firing step was varied as indicated in the legend. Lines only serve as a guide to the eye.

Using a measurement of two identically processed but differently fired samples, it was already concluded that the peak firing temperature has a strong influence on the short term behavior of a lifetime sample exposed to elevated temperature and illumination conditions [8]. But independent of the peak firing temperature, the samples show a drastic decline in measured τ_{eff} after around 30 h of treatment time that is further investigated in this study (Fig. 5).

After around 860 h of treatment time, the temperature was increased to 150°C. The further sample evolution is again shown in more detail in Fig. 6. It seems that increasing the treatment temperature leads to a faster sample evolution as the decline in τ_{eff} gets much steeper immediately after the temperature change.

Interestingly though, the samples do not reach a saturation value but again traverse a second minimum in τ_{eff}. This gives rise to the idea that the minimum observed at 250°C after 1 h (Fig. 4) might in fact be associated with the minimum after around 900 h in Fig. 6. This assignment will be investigated more closely in a future publication. After passing this second minimum, the τ_{eff} curves converge and show a very similar behavior. This again promotes the idea that especially the first part of sample evolution is strongly influenced by the firing conditions while in the long run the peak firing temperature is of minor importance.

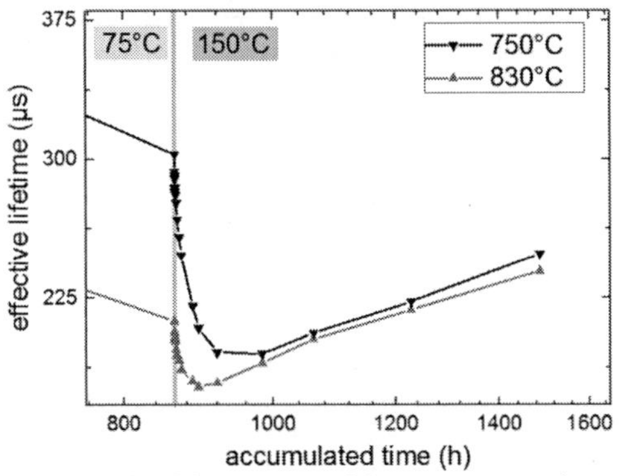

Figure 6: The right most part of the measurement shown in Fig. 2 in close up view. The blue line again indicates the switch from 75°C treatment temperature to 150°C while illumination intensity was kept constant at 1 sun. Lines only serve as a guide to the eye.

3.3 Influence of cleaning steps on stability

To investigate a possible influence of the chemical cleaning step on sample stability, three samples were processed using different cleaning steps because these might have a strong influence on the chemical passivation behavior of a sample. However, in Fig. 7 it can be seen that the qualitative curve shape at 250°C remains the same independent of the cleaning type used.

The Piranha cleaned sample (black) was also investigated with the corona charging technique and it could already be shown that also the chemical passivation quality shows a minimum after around 1 h of treatment [8]. For a further comparison of the different cleaning types, also the HCl+HF cleaned sample was investigated by corona charging and the result is shown in Fig. 8. Like the Piranha cleaned sample, also the HCl+HF cleaned sample shows a clear minimum in chemical passivation quality during the degradation treatment as indicated by τ_{eff} after corona charging of the surface. It can also be seen that the chemical passivation quality still increases after long treatment times while the overall τ_{eff} drops, suggesting a loss in charge of the silicon nitride layer.

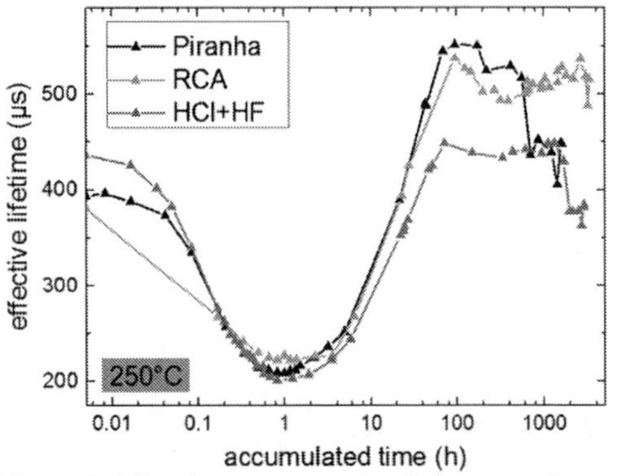

Figure 7: RT-PCD measurements of three samples at 250°C in darkness (black data after [8]). The samples received different chemical cleaning steps as given in the legend. All samples received a type 2 SiN_x:H deposition after the cleaning step and were fired at 800°C peak sample temperature. Lines only serve as a guide to the eye.

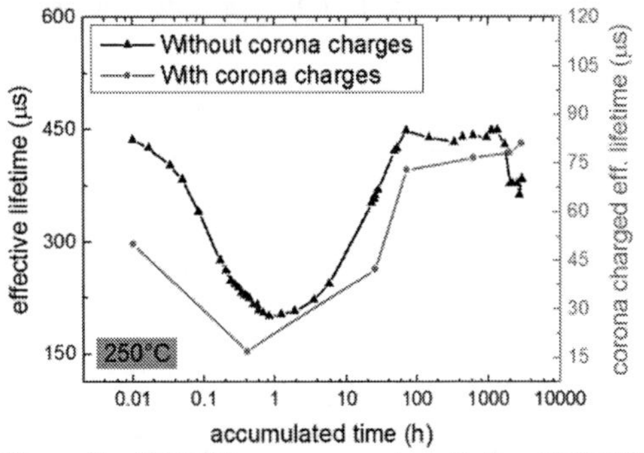

Figure 8: RT-PCD measurements of the HCl+HF cleaned sample of Fig. 4 before and after corona charging during a treatment at 250°C in darkness. For each corona charging measurement point, corona charges were deposited on the sample surface in small consecutive charging steps until a minimum in τ_{eff} was reached, which corresponds to a compensation of the field effect passivation mechanism. Therefore, the τ_{eff} values after charging (blue) correspond to the chemical passivation quality.

3.4 Influence of different high temperature steps

As it was found out that the firing step exerts a strong influence on the sample stability, some samples were identically processed and only the high temperature step after the SiN_x:H deposition was varied. Of the samples shown in Fig. 9, one did not receive a high temperature step at all, while another one was fired at 730°C peak sample temperature. A third sample was instead annealed at 420°C for 30 min in a nitrogen atmosphere.

As can be seen, the annealed sample shows a very stable behavior compared to all samples previously seen. This does not change when the temperature is increased to 150°C. The fired sample on the other hand shows a very instable behavior starting with a strong increase in effective lifetime already after 1 min of treatment time. This is probably related to the light induced curing (LIC) effect already observed in [9]. Afterwards, the sample shows only a very weak first minimum, probably because of the rather low firing temperature. The decrease in τ_{eff} after around

100 h of treatment is very pronounced and gets accelerated when the temperature is increased to 150°C.

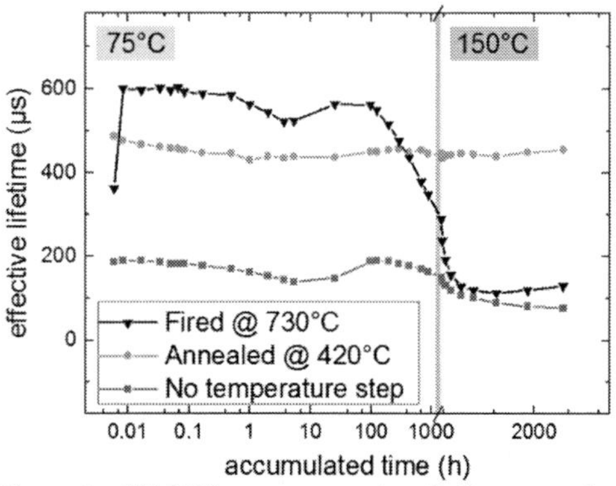

Figure 9: RT-PCD measurements of three samples treated at 75°C and 150°C with illumination between 0.6 and 1 suns. The samples differed in high temperature steps and were otherwise identically processed. After an RCA clean they received a thin chemical oxide before deposition of a type 3 SiN_x:H.

Interestingly, the sample without high temperature step (besides the PECVD step) also shows a similar qualitative curve shape as the fired sample but in a much less pronounced way. Still, it is remarkable that also a non-fired sample expresses some degree of instability under the given treatment conditions. Since it could be shown that the peak firing temperature mainly has an influence on the short term behavior but a fired sample shows a different long term behavior compared to a non-fired sample, it is assumed that other parameters of the firing process besides peak temperature influence the stability of the investigated dielectric layers.

4 SUMMARY

In summary, it could be shown that fired lifetime samples show strong instabilities in τ_{eff} when subjected to elevated temperature and illumination treatments. Because of the used FZ material and differences in the behavior of different passivation layers, it is assumed that these changes are caused by changes in dielectric passivation quality.

More evidence was found for the hypothesis that the peak firing temperature has a strong influence on the short term behavior of a sample whereas in the long run, the peak firing temperature is of less importance. Nevertheless, the firing step itself amplifies the instable long term behavior as could be seen in a comparison of a fired and a non-fired sample. It is therefore assumed that other details of the firing profile besides peak temperature may have an influence on the long term behavior. Finally, a non-fired but tempered sample shows a very stable behavior at elevated temperature and illumination.

Because the observed changes in passivation quality can be strongly pronounced, it is definitely advised to check for these effects when performing degradation experiments targeted on effects happening in the silicon bulk.

ACKNOWLEDGEMENTS

The authors would like to thank L. Mahlstaedt, B. Rettenmaier, J. Engelhardt, S. Joos and L. Kraus for technical support. Part of this work was supported by the German Federal Ministry for Economic Affairs and Energy under contract numbers 0325763B, 0325581 and 0324001.

REFERENCES

[1] J. Lindroos, H. Savin, Review of light-induced degradation in crystalline silicon solar cells. Sol. En. Mat. Sol. Cells 147 (2016) 115.
[2] K. Bothe, J. Schmidt, Electronically activated boron-oxygen-related recombination centers in crystalline silicon, J. Appl. Phys. 99(1) (2006) 013701.
[3] K. Ramspeck, S. Zimmermann, H. Nagel, A. Metz, Y. Gassenbauer, B. Birkmann, A. Seidl, Light induced degradation of rear passivated mc-Si solar cells, Proc 27th EUPVSEC (2012) 861
[4] F. Fertig, K. Krauß, S. Rein, Light-induced degradation of PECVD aluminium oxide passivated silicon solar cells, Phys. Stat. Sol. RRL 9(1) (2015) 41.
[5] F. Kersten, P. Engelhart, H.C. Ploigt, A. Stekolnikov, T. Lindner, F. Stenzel, J.W. Müller, Degradation of multicrystalline silicon solar cells and modules after illumination at elevated temperature, Sol. En. Mat. Sol. Cells 142 (2015) 83.
[6] A. Herguth, G. Schubert, M. Kaes, G. Hahn, A new approach to prevent the negative impact of the metastable defect in boron doped Cz silicon solar cells, Proc 4th WCPEC (2006) 940.
[7] S.W. Glunz, D. Biro, S. Rein, W. Warta, Field-effect passivation of the SiO_2-Si interface, J. Appl. Phys. 86 (1999) 683.
[8] D. Sperber, A. Herguth, G. Hahn, Instability of dielectric surface passivation quality at elevated temperature and illumination, En. Procedia (2016) in press
[9] S. Joos, A. Herguth, U. Hess, J. Ebser, S. Seren, B. Terheiden, G. Hahn, Light induced curing (LIC) of passivation layers deposited on native silicon oxide, En. Procedia 27 (2012) 349.

TWO-DIMENSIONAL CHARACTERIZATION OF PHOSPHORUS-IMPLANTED EMITTER AND PHOSPHORUS-DIFFUSED EMITTER OF SILICON SOLAR CELL USING SUPER-HIGHER-ORDER SCANNING NONLINEAR DIELECTRIC MICROSCOPY

Kotaro Hirose[1], Katsuto Tanahashi[2], Hidetaka Takato[2], Norimichi Chinone[1], and Yasuo Cho[1]
[1]Research Institute of Electrical Communication, Tohoku University,
Sendai, 980-8577, Japan
TEL:+81-22-217-5529,FAX:+81-22-217-5529,e-mail:yasuocho@riec.tohoku.ac.jp
[2]Fukushima Renewable Energy Institute, National Institute of Advanced Industrial Science and Technology,
Koriyama, 963-0298, Japan

ABSTRACT: Two-dimensional carrier distributions in phosphorus (P)-implanted emitter and P-diffused emitter were analyzed using super-higher-order scanning nonlinear dielectric microscopy (SHO-SNDM). The carrier distribution was clearly visualized and quantified using calibration sample. P-type, n-type and depletion layer was discriminated from local capacitance-voltage characterization obtained from SHO-SNDM measurement. The n-type region and depletion layer distribution depended on the surface texture. The position of p-n junction was estimated to be near the p-type region in depletion layer. The n-type region and depletion layer in P-implanted emitter were thicker than those in P-diffused emitter. Experimental results indicate that SHO-SNDM gives the detailed analysis of carrier distribution and is useful method for evaluation of solar cell.
Keywords: Impurities, Silicon, Scanning nonlinear dielectric microscopy, Capacitance, Surface texture

1 INTRODUCTION

Ion-implanted emitter has been proposed instead of Ion-diffused emitter [1]. Replacement of diffusion process with ion implantation gives a simplification of fabrication process and low cost solar cell.

Ion implantation process has been used for ultra-large scale integrated circuit (ULSI). Ions are implanted in the flat Si surface in ULSI. On the other hand, ions are implanted in the textured Si surface in solar cell. Thus, the estimation of carrier distribution in solar cells is difficult and hence the directly measurement of carrier distribution is needed.

Scanning probe microscopy techniques are suitable for the direct measurement of carrier distribution. Two-dimensional analysis of carrier distribution in P-implanted emitter was performed using scanning capacitance microscopy (SCM) [2], which detects the variation of tip-sample capacitance with the sensitivity of 10-19 F. The relationship between the carrier distribution in emitter and surface texture was reported. The emitter was thicker at convex position of surface texture [3].

Scanning nonlinear dielectric microscopy (SNDM) [4] can also measure the variation of tip-sample capacitance. However, SNDM has higher capacitance sensitivity of 10-22 F [5] than SCM. Super-higher-order (SHO-) SNDM [6], which is extended version of SNDM, can measure local capacitance-voltage (C-V) curve at each pixel of acquired image [7]. Local C-V curve measurement gives both the discrimination of p-type, n-type and depletion layer [7] and estimation of p-n junction positon [8]. In this paper, we applied SHO-SNDM to phosphorus (P)-implanted emitter as well as P-diffused emitter. Thus, we discriminated the type of semiconductor and estimated the position of p-n junction. Furthermore, P-diffused emitter and P-implanted emitter were compared and analyzed.

2 EXPERIMENTAL METHOD

2.1 Specimen

Both P-diffused and P-implanted emitter solar cell were measured. P-diffused emitter and P-implanted emitter were fabricated on p-type single crystalline Si wafer. Then, SiNx layer was formed on the textured surface. In addition, SiO_2 layer was deposited on the SiNx layer by sputtering in order to protect from corner sag due to polishing. These solar cells were cross sectioned. The surface of cross-section was finished by using chemical-mechanical polish. We measured the cross section of these solar cells using SNDM and SHO-SNDM. In addition, we estimated p-n junction position from SHO-SNDM data.

2.2 SHO-SNDM

Figure 1 shows SNDM set-up. The tip-sample contact force is kept constant using optical lever method of atomic force microscopy with commercial SPM controller (SPI4000, Hitach High-Technologies Co., Ltd.). The variation of tip-sample capacitance ΔC_s is measured using SNDM. The capacitance sensor, which is called SNDM probe, consists of conductive cantilever tip and GHz-range free-running active LC oscillator. The frequency deviation of SNDM probe Δf_p is proportional to ΔC_s. When the tip-sample capacitance C_s is modulated by the application of ac voltage $V(t) = V_p \cos \omega_p t$ with the frequency of several tens kHz from the metal stage, the relationship between

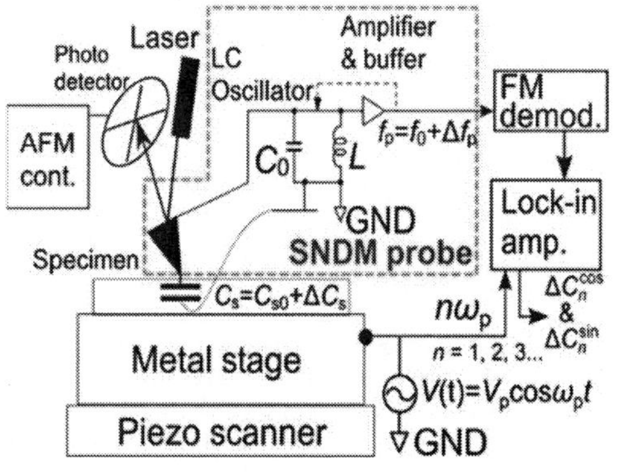

Figure 1: Schematic diagram of SNDM setup.

ΔC_s and Δf_p described as a function of time with a same fundamental frequency of applied sinusoidal voltage and can be written by

$$-\Delta f_p(t) \propto \Delta C_s(t) = \sum_{n=1}^{\infty} \left(\Delta C_n^{\cos} \cos n\omega_p t + \Delta C_n^{\sin} \sin n\omega_p t\right), \quad (1)$$

where $\Delta C_n^{\cos}$ and $\Delta C_n^{\sin}$ are the Fourier coefficients. Thus, $\Delta C_n^{\cos}$ and $\Delta C_n^{\sin}$ are obtained by extracting the n-th order harmonic component in $\Delta C_s(t)$, which is proportional to a demodulated FM signal of SNDM probe, for which we used commercial FM demodulator (MS616B, Anritsu Co., Ltd.) and a lock-in amplifier (HF2LI, Zurich Instrument Co., Ltd.). SNDM detects the first order cosine coefficient $\Delta C_1^{\cos}$ only, which is nearly equal to the first order derivative of C_s, ie $\partial C_s/\partial V$. The $\partial C_s/\partial V$ reflects the information of carrier polarity and density. The polarity of $\partial C_s/\partial V$ correlates with the carrier polarity. Therefore, SNDM can visualize carrier distribution two-dimensionally. On the other hand, SHO- SNDM detects higher order coefficients as well as fundamental component. In this paper, we detect the coefficients from 1st to 6th order. The ΔC_s is obtained as a function of voltage from a combination of $\Delta C_s(t)$, which is calculated from $\Delta C_n^{\cos}$ and $\Delta C_n^{\sin}$, and $V(t)$. Therefore, SHO-SNDM can obtain local ΔC_s-V curve at each pixel [7]. If ΔC_s-V curve is a monotone increasing function, the tip is on p-type region. In contrast, when ΔC_s-V curve is a monotone decreasing function, the tip is on n-type region. On the other hand, V-shape ΔC_s-V curve is obtained at depletion layer region. The p-type, n-type and depletion layer regions are distinguished from the shape of ΔC_s-V curve [7].

We measured the cross section of P-diffused and P-implanted emitter Si solar cell using SNDM and SHO-SNDM. An ac voltage was applied with the amplitude of 1.5 V and the frequency of 30 kHz. The cantilever tip was PtIr coated with the radius of 25 nm (EFM, Nanosensors). In addition, we quantified the carrier concentration in emitter region by comparing the obtained data and reference data. The reference sample was fabricated simultaneously under the same polishing condition for the cross-sectioned solar cell sample and was measured under the same measurement condition.

2.3 Determination of P-N Junction Position

Local ΔC_s-V curve is obtained using SHO-SNDM in the depletion layer area. The location of p-n junction position can be estimated from ΔC_s-V curve. The voltage which gives minimum $\Delta C_s(V)$ is defined as V_{min}. V_{min} is the valley position of V-shape ΔC_s-V curve. At first, for the sake of simplicity, we suppose tip has a same work function as the sample. When the concentrations of p-type and n-type dopant are same, the p-n junction position is the center of depletion layer. As shown in Fig. 2(a), V_{min} depends on the tip position. As the tip approaches the p-type, V_{min} decreases. When the tip is on the p-n junction position (i.e., the center of depletion layer), $V_{min} = 0$. Next, as shown in Fig. 2(b), we consider p+-n junction. The junction position is near the p+ region. In this case, when the tip is on the center of depletion layer, $V_{min} > 0$. As the tip approach the junction position, V_{min} gets closer to the value of 0.

Therefore, the p-n junction position can be estimated from the position where V_{min}=0. We have to note that we need to investigate whether V_{min}=0 exactly gives the junction position. Actually, the work functions of the tip and sample are different and hence there is a contact potential difference (CPD). Thus, considering the CPD, we can estimate p-n junction position from the position where the V_{min} equals to the CPD. From the obtained SHO-SNDM data, we estimated the junction position of P-diffused and P-implanted emitter Si solar cell.

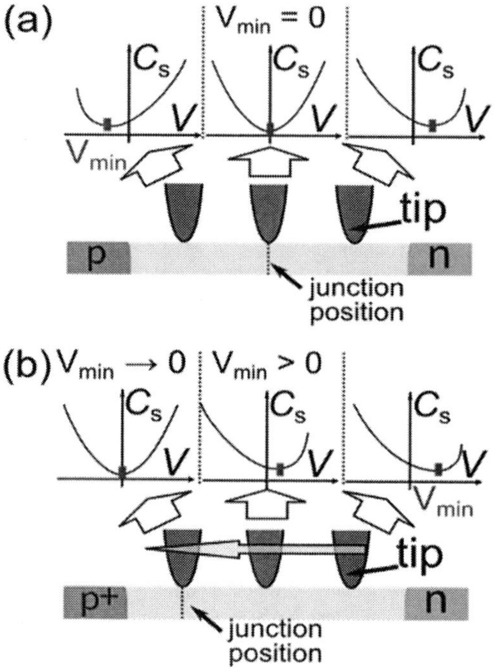

Figure 2: V_{min} value depends on the tip position. (a)P-type carrier density is same as n-type one. The p-n junction position is the center of depletion layer, where $V_{min} = 0$. (b) P-type carrier density is higher than n-type one. The p^+-n junction position is located near the p^+-type region in depletion layer. As the tip approach the position of p^+-n junction from n-type region, V_{min} gets closer to 0.

3 RESULTS AND DISCUSSION

Figure 3 shows topography of (a) a P-diffused emitter and (b) a P-implanted emitter. The dashed line indicates the edge of the emitter. The interface between emitter and SiN_x layer was visualized because polishing rate was different between emitter and SiN_x layer. Figure 4 shows SNDM image of (a) a P-diffused emitter and (b) a P-implanted emitter. The magnitude of output voltage has a

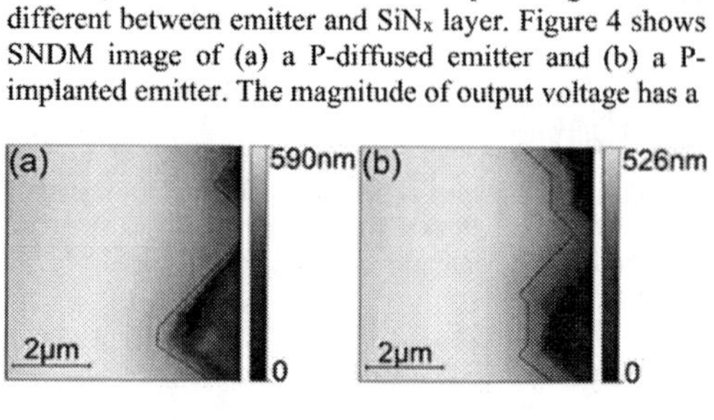

Figure 3: Topography of (a) a P-diffused emitter and (b) a P-implanted emitter.

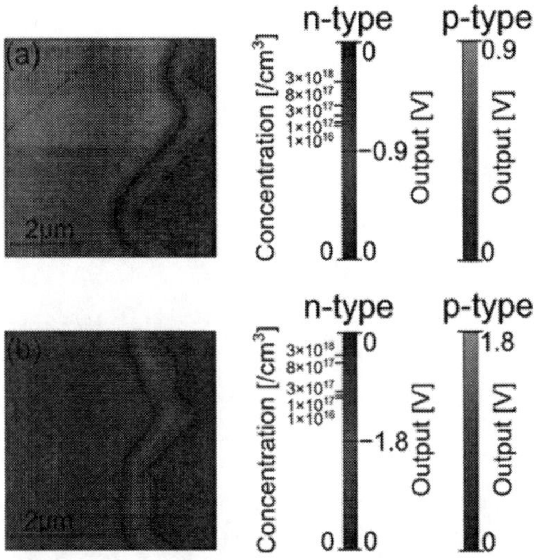

Figure 4: SNDM images of (a) a P-diffused emitter and (b) a P-implanted emitter. Electron concentration only from the emitter surface to center of emitter was quantified by the measurement of standard sample for calibration.

peak near the center of emitter (light blue region) because $\partial C_s/\partial V$ has the peak at a certain concentration. The electron concentration from $3\text{x}10^{18}$ to $10^{16}/cm^3$ (but only in the position from the surface of the emitter to the center of emitter with peak SNDM signal) was quantified from measurement of calibration sample. The emitter distributions were almost parallel to the surface texture. However, the emitter region was thicker at the convex position of surface texture than at both the concave position and side wall of surface texture. The thickness from the surface to the SNDM signal peak position in P-Implanted emitter was larger than that in P-diffused emitter. These results are similar to a previous report [3]. The local ΔC_s-V curves were reconstructed from SHO-SNDM data (but not shown in this paper). P-type, n-type and depletion layer were discriminated from SHO-SNDM data. Figure 5 shows the discrimination results in (a) a P-diffused emitter and (b) a P-implanted emitter. The depletion region was thinner at convex surface position than at other regions.

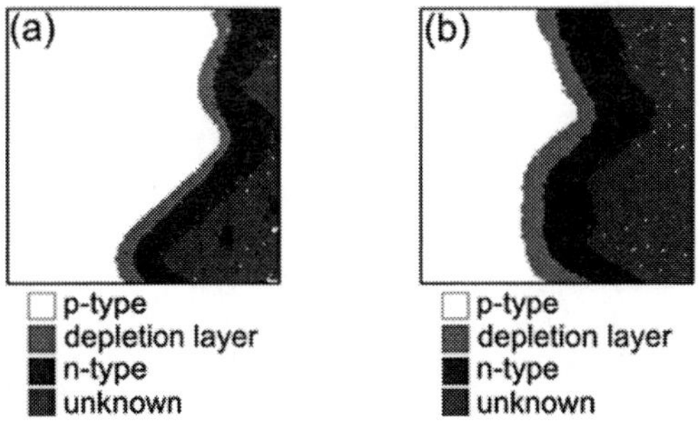

Figure 5: Discrimination of p-type, n-type and depletion layer of (a) a P-diffused emitter and (b) a P-implanted emitter.

This is because the electron density near the depletion layer at the convex position was higher than that at other regions. The thicknesses of n-type regions were ~ 630nm and ~ 730 nm at side wall of texture surface of P-diffused emitter and P-implanted emitter, respectively. Furthermore, the thicknesses of depletion layer were ~ 300 nm and ~ 450 nm at the same area of P-diffused emitter and P-implanted emitter, respectively. Both n-type region and depletion layer of P-implanted emitter were thicker than those of P-diffused emitter.

The p-n junction position was estimated from SHO-SNDM data. We assumed CPD equals to -1.0 V. Thus, junction position was given by the position where V_{min} = − 1.0 V. Figure 6 shows the V_{min} image of (a) a P-diffused emitter and (b) a P-implanted emitter. The p-n junction position was emphasized using yellow color. The line profile of line X-X' in Fig. 6(a) and line Y-Y' in Fig. 6(b) are shown in Fig 7(a) and 7(b), respectively. The p-n junction positions were near the p region in depletion layer of both P-diffused and P-implanted emitter. The p-n junction positions from p-type region were ~ 20 nm and ~ 40 nm in P-diffused emitter (X-X' line) and P-implanted emitter (Y-Y' line), respectively.

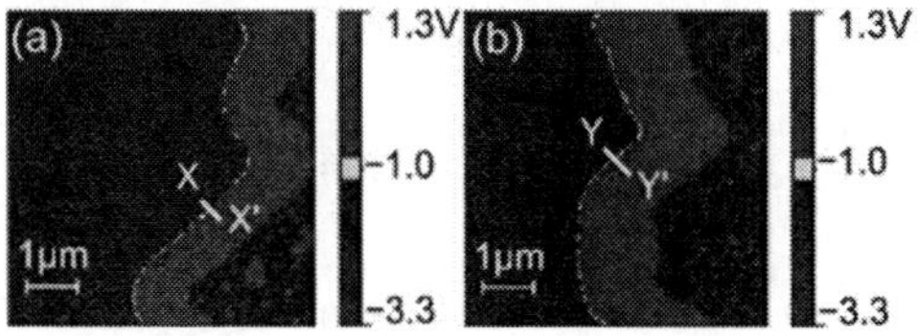

Figure 6: V_{min} images of (a) a P-diffused emitter and (b) a p-implanted emitter. The p-n junction position was emphasized by yellow color.

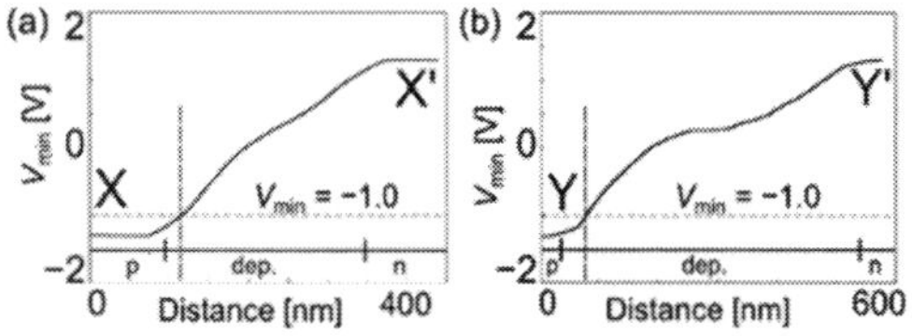

Figure 7: V_{min} line profiles of (a) line X-X' in Fig. 6(a) and (b) line Y-Y' in Fig. 6(b).

4 CONCLUSIONS

P-diffused emitter and P-implanted emitter were evaluated using SHO-SNDM. The electron densities in emitters were quantified by measurement of calibration sample. The emitter was thicker at convex surface position than side wall and concave position. Local ΔCs-V curve, which is reconstructed by SHO-SNDM, gives the discrimination of p-type, n-type and depletion layer. The n-type region and depletion layer region were thicker in P-implanted emitter than P-diffused emitter. The position of p-n junction was estimated near p-type region depletion layer. Therefore, we conclude that SHO-SNDM gives the detailed analysis of carrier distribution and is useful method for evaluation of solar cell. Further studies are needed in order to reveal the dependence of carrier distribution on the fabrication condition.

ACKNOWLEDGEMENT

This research was partially supported by a Grant-in-Aid for Scientific Research S (No. 23226008) from the Japan Society for Promotion of Science.

References

[1] Y. Tao, Y. W. Ok, F. Zimbardi, A. D. Upadhyaya, J. H. Lai, S. Ning, V. D. Upadhyaya, and A. Rohatgi, "Fully ion-implanted and screen-printed 20.2% efficient front junction silicon cells on 239 cm 2 n-type cz substrate," IEEE J. Photovoltaics, vol. 4, no. 1, pp. 58–63, 2014.

[2] J. J. Kopanski, "Scanning capacitance microscopy measurements and modeling: Progress towards dopant profiling of silicon," J. Vac. Sci. Technol. B Microelectron. Nanom. Struct., vol. 14, no. 1, p. 242, 1996.

[3] K. Tanahashi, M. Morita, Y. Kida, and N. Suzuki, "Two dimensional carrier distribution in phosphorus-implanted emitter observed by scanning capacitance microscopy," in 31st European Photovoltaic Solar Energy Conference and Exhibition, 2015, pp. 35–38.

[4] Y. Cho, A. Kirihara, and T. Saeki, "Scanning nonlinear dielectric microscope," Rev. Sci. Instrum., vol. 67, no. 6, pp. 2297–2303, 1996.

[5] Y. Cho, "Scanning nonlinear dielectric microscopy," J. Mater. Res., vol. 26, no. 16, pp. 2007–2016, 2011.

[6] N. Chinone, K. Yamasue, Y. Hiranaga, K. Honda, and Y. Cho, "Lateral resolution improvement in scanning nonlinear dielectric microscopy by measuring super-higher-order nonlinear dielectric constants," Appl. Phys. Lett., vol. 101, no. 21, p. 213112, 2012.

[7] N. Chinone, T. Nakamura, and Y. Cho, "Cross-sectional dopant profiling and depletion layer visualization of SiC power double diffused metal-oxide-semiconductor field effect transistor using super-higher-order nonlinear dielectric microscopy," J. Appl. Phys., vol. 116, no. 8, p. 084509, 2014.

[8] N. Chinone, Y. Goto, and Y. Cho, "A study on method for determination of pn-junction in SiC power transistor using super-higher-order scanning nonlinear dielectric microscopy (in Japanese)," in The 35th Annual NANO Testing Symposium, 2015, pp. 47–50.

A SIMULATION STUDY OF RESISTIVE EFFECT OF TUNNEL JUNCTION REAR CONTACTS IN BIFACIAL N-PERT SILICON SOLAR CELLS

Chih-Ming Wei*, Yong-Han Lin*, Chi-Chun Li, and Chia-Chih Chuang
Motech Industries, Inc., Science Park Branch,
No. 96, Dashun 7th Rd., Xinshi Dist., Tainan 74145, Taiwan
*E-mail: alex_wei@motech.com.tw; yonghanlin@gmail.com

ABSTRACT: The resistive effect of tunnel junction (TJ) contacts to the rear side of bifacial *n*-PERT Si solar cells has been investigated by simulation with the implementation of a quantum tunneling model. By placing an ultra-thin oxide layer between the BSF and the metal fingers, we calculated, respectively, the TJ resistivity (from a separate MOS structure) and the electrical properties of the cell as a function of oxide thickness w. As expected for a tunneling process, the TJ resistivity is nearly unchanged for $w \leq 10$ Å and increases exponentially when $w \geq 10$ Å. Because of the power dissipation introduced by the TJ resistance, the efficiency loss as w increases is mainly dictated by the loss of FF for $w \leq 17$ Å. With w being further increased, the deterioration of J_{SC} becomes more pronounced, giving an additional loss of efficiency. However, since the exact dependence of passivation quality on w is lacking, a constant SRV value at the SiO_2/Si interface is assumed for all w's. As a result, V_{OC} is practically not affected. If such a dependence for a specific passivation process is learned and implemented in the model, our method could be helpful in optimizing solar cell properties.
Keywords: tunnel junction, contact, PERT

1 INTRODUCTION

Metal-semiconductor (MS) interfaces are known to be one of the major sources that give rise to Si solar cell efficiency losses. At the interface, a large number of electronic defect states exist and are highly recombination active [1]. Minimizing such carrier recombination occurred at MS interfaces hence becomes increasingly pivotal towards high-efficiency Si solar cells. One typical method to this end is to reduce the area of the contact region, e.g., the implementation of partial rear contacts [2]. A more advanced method is to create the so-called carrier-selective contacts [3], as previously elucidated and demonstrated by Yablonovitch in 1985 [4].

A carrier-selective contact is often a layered structure with a tailored bandgap that can suppress minority carrier recombination and facilitate majority carrier transport. A very successful example is the HIT (Heterojunction with Intrinsic Thin-layer) cells invented by Sanyo Electric [5], which achieve very high efficiencies by utilizing Si heterojunctions based on hydrogenated amorphous Si (a-Si:H). A recent progress in this respect is the introduction of a scheme called TOPCon (Tunnel Oxide Passivated Contacts) [6]. This new approach employs an ultra-thin tunnel oxide and a layer of wide-bandgap doped/poly Si, which function in the same manner as a-Si:H layers in HIT cells. Since no a-Si:H layers are involved, TOPCon thus offers a compatible version that can be processed at high temperatures. Of particular interest is an alternative and similar method that applies only an ultra-thin oxide layer as a contact-passivating layer [7, 8, 9]. However, the concept of this more simplified method is different from that of the TOPCon scheme as it contains no wide-bandgap layers which play the role of selecting carriers as in the latter case.

The above-described usage of an ultra-thin oxide layer inserted into the MS interface leads to a question regarding its detrimental resistive effect on the electrical properties of the fabricated cells. In this work, we perform numerical device simulations to study this effect by placing an ultra-thin oxide layer of thickness w between the back-surface field (BSF) and the metal fingers at the rear side of a bifacial *n*-type passivated emitter and rear totally-diffused (*n*-PERT) Si solar cell. To determine the resistance contribution due to the inserted oxide layer, we calculate the tunnel junction (TJ) resistivity ρ_j (in unit of mΩcm^2) as a function of w from a separate metal-oxide-semiconductor (MOS) structure. With the implementation of a quantum tunneling model in these two structures, the variations of the electrical properties of the cell with w and ρ_j are accordingly derived.

2 SIMULATION METHOD

A commercial two-dimensional device simulator Atlas (from Silvaco, Inc.) is used in this work. The structures of the bifacial *n*-PERT Si solar cell and the separate MOS structure used in our simulations are displayed (not to scale) in Fig. 1(a) and 1(b), respectively. Figure 1(c) shows the applied depth profiles of the B-doped emitter and the P-doped BSF, as indicated. The

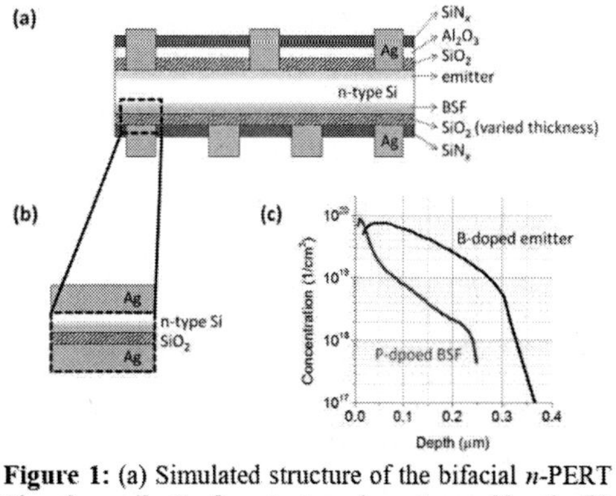

Figure 1: (a) Simulated structure of the bifacial *n*-PERT Si solar cell. Surface texture is not considered. (b) Separate MOS structure corresponding to the tunnel junction (TJ) contact at the rear side of the cell for the calculation of TJ resistivity ρ_j. (c) Doping profiles of the B-doped emitter and the P-doped BSF, as indicated.

Table I: Values of the relevant parameters used in the simulation.

Parameter	Value	Unit
Bulk lifetime	500	μs
Wafer resistivity	2	Ω cm
Wafer thickness	190	μm
Front side:		
Finger width	55	μm
Finger pitch	1730	μm
SiO_2 thickness	5	nm
Al_2O_3 thickness	5	nm
SiN_x thickness	70	nm
SRV	1000	cm/s
Rear side:		
Finger width	65	μm
Finger pitch	1150	μm
SiN_x thickness	80	nm
SRV	1000	cm/s
Ag/Si SRV	10^6	cm/2
Effective mass in SiO_2	0.4	m_0

values of the physical parameters relevant to our simulations are listed in Table I. For other parameters that are not listed, default values in Atlas are used. In Fig. 1(b), the thickness of the *n*-type Si (doped with the same BSF profile) is set to be 1 μm. Although the resistance of this layer is included in our calculated ρ_j values, its contribution is negligible (see the discussion below). To have current through the insulating oxide layer, the self-consistent versions of the quantum tunneling models for electrons and holes are enabled in the simulator. In addition, because the dependence of passivation quality on w is not known, a constant value of surface recombination velocity (SRV) at the SiO_2/Si interface is assumed for all w's for simplicity. All the Ag/Si interfaces are also taken to be ohmic and the corresponding specific contact resistivity is ignored as well.

3 RESULTS AND DISCUSSION

Figure 2 shows the variation of ρ_j with w for the simulated MOS structure. The value of ρ_j is determined from the calculated current-voltage (*I-V*) curve at a small bias (a linear regime within ±0.1 V). Clearly, for $w \lesssim 10$ Å, ρ_j changes insignificantly. However, with w being further increased, an exponential increase of ρ_j results. This behavior can be readily understood by the textbook model of direct quantum tunneling. If we neglect the band-bending effects and approximate the oxide layer by a rectangular barrier of height ϕ (with respect to the Fermi level E_F), the band diagram for our simulated MOS structure can be simplified as shown in the inset of Fig. 2. From Ref. 10, the probability P_t for an electron with energy E tunneling through a rectangular barrier of height U and width w is given by

$$P_t = \left[1 + \frac{U^2 \sinh^2(\alpha w)}{4E(U-E)}\right]^{-1}, \quad (1)$$

where $\alpha = \sqrt{2m(U-E)}/\hbar$, m is the effective mass of

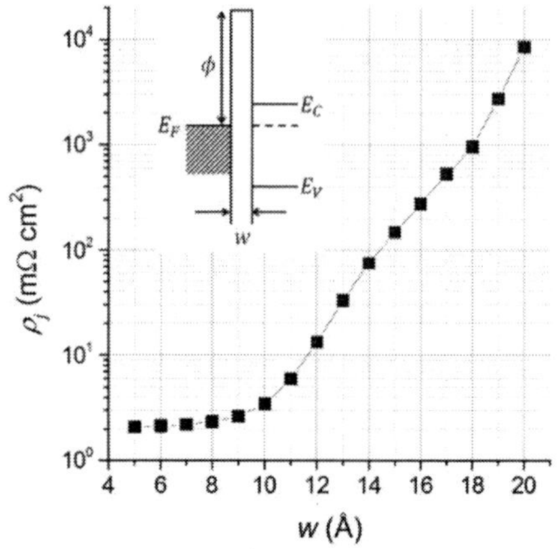

Figure 2: Tunnel junction resistivity ρ_j as a function of oxide thickness w. Inset: a simplified band diagram for the simulated MOS structure shown in Fig. 1(b). The oxide layer is approximated as a rectangular barrier of height ϕ (with respect to the Fermi level E_F) and width w. E_C and E_V are the energy levels at the conduction and valence band edges, respectively.

the electron, and $\hbar$ is the reduced Planck constant. (Note that, in our case, $E = E_F$ and $U = E_F + \phi$.)

With a known tunneling probability for the one-electron state, the tunneling current density J_t can be summed up by considering all available states and written as

$$J_t = \beta \int F_o N_o P_t (1 - F_d) N_d dE \, , \quad (2)$$

where β is a material-dependent parameter, and F_o, N_o, F_d, and N_d are the Fermi-Dirac distributions and the densities of states in the originating side and the destination side, respectively. As a simple approximation, the integrand in Eq. (2) can be replaced by the value right at E_F since only those states in the vicinity of E_F are important for electron transport. For a small applied voltage V (i.e., to the first order), Eq. (2) can then be expressed as $J_t \propto P_t V$, that is, $\rho_j \propto P_t^{-1}$. From Eq. (1), for a small value of w, $P_t \approx 1$, meaning that ρ_j remains nearly unchanged. On the other hand, for a large value of w, P_t is apparently dominated by the second term in the bracket and becomes $\approx [\exp(2\alpha w)]^{-1}$, therefore giving an exponential increase of ρ_j as displayed in Fig. 2.

As mentioned previously, the *n*-type Si layer in Fig. 1(b) has a thickness of 1 μm. For a layer with a resistivity of 2 Ωcm (see Table I) and a uniform current distribution flowing perpendicular to its plane, the resistance-area product (RA) of this layer can be calculated to be 0.2 mΩcm² (using $RA = \rho L$, where ρ is the bulk resistivity and L is the length of the conducting path, i.e., the layer thickness in this case). Obviously, this value is an order of magnitude smaller than the ρ_j values and thus can be safely neglected. (Note that it will be even smaller if the BSF profile is taken into account.)

Figure 3 shows the calculated *J-V* curves under AM1.5G illumination for the simulated bifacial *n*-PERT Si solar cell with varied w, as indicated. In Fig. 4, we plot, respectively, the derived output parameters open-circuit voltage (V_{OC}), short-circuit current density (J_{SC}), fill factor (FF), and solar cell efficiency (η) as a function of w. For reference, we show in Fig. 5 these parameters as a function of ρ_j.

It can be seen that V_{OC} is practically not affected by the insertion of an ultra-thin oxide layer. This is not surprised because in our simulations an identical passivation quality is assumed for all w's. Using a TJ contact is now equivalent to attaching an additional series resistive element R_j to the solar cell, which apparently has no impact on V_{OC} at open-circuit. However, on the contrary, this resistive element can deteriorate J_{SC} and FF considerably. As can be seen in Fig. 4, with increasing w, the decrease of FF is significant when $w \gtrsim 10$ Å while that of J_{SC} becomes dramatic for $w \gtrsim 17$ Å. As a result, the dependence of η on w is mainly dictated by the behavior of FF.

Such a feature that the deterioration of FF takes place at a smaller value of w than in the case of J_{SC} can be simply explained from the point of view of the power dissipation coming along with the TJ resistance R_j. By definition, $\mathrm{FF} = P_{MP}/V_{OC}I_{SC}$, where P_{MP} is the maximum power that can be delivered by the cell. Assuming a small change of I_{SC} (i.e., I_{SC} is nearly constant), the change of FF then takes the form $\Delta\mathrm{FF} = \Delta P_{MP}/V_{OC}I_{SC}$, where ΔP_{MP} is the change of P_{MP}, written as $\Delta P_{MP} = I_{SC}{}^2 R_j$. Putting all these together, we arrive at the expression $\Delta\mathrm{FF} = I_{SC}R_j/V_{OC}$, which clearly shows that FF already varies directly with R_j before J_{SC} being affected.

The almost constant values of J_{SC} for $w \lesssim 17$ Å perhaps can be best understood as follows. Considering the single-diode model for a solar cell, the I-V relationship is given by

$$I = I_{ph} - I_0\left[\exp\left(\frac{V + IR_s}{nk_BT/q}\right) - 1\right] - \frac{V + IR_s}{R_{sh}}, \qquad (3)$$

where I_{ph} is the photocurrent, I_0 is the diode saturation current, n is the diode ideality factor, R_s is the series resistance, R_{sh} is the shunt resistance, k_B is the Boltzmann constant, T is the temperature, and q is the magnitude of the electron charge. At short-circuit ($I = I_{SC}$ and $V = 0$), Eq. (3) becomes

$$I_{SC} = I_{ph} - I_0\left[\exp\left(\frac{I_{SC}R_s}{nk_BT/q}\right) - 1\right] - \frac{I_{SC}R_s}{R_{sh}}, \qquad (4)$$

giving the dependence of I_{SC} on R_s. Inspection of Eq. (4) reveals that both the exponential diode term and the linear parasitic term cannot be changed notably if R_s is not substantially large. This is because, in the former case, the pre-factor (the saturation current density J_0) is extremely small, typically having the value of $\approx 10^{-12}$ A/cm², while in the latter case, $R_{sh} \gg R_s$ (normally, at least 4–5 orders of magnitude larger). These two properties therefore render J_{SC} rather insensitive to the increase of w in the range of not sufficiently large values.

4 CONCLUSION

The usage of an ultra-thin oxide layer inserted into the MS interface as a contact-passivating layer raises a question regarding its detrimental resistive effect on the electrical properties of the fabricated solar cells. In this work, we studied this effect by numerical simulation. We considered a bifacial n-PERT Si solar cell with its rear contacts being replaced by MOS TJs. By employing a self-consistent quantum tunneling model to a separate MOS structure and to the cell, we calculated the TJ resistivity ρ_j and the electrical property of the cell as a function of w, respectively. We found that ρ_j is nearly unchanged for $w \lesssim 10$ Å and increases exponentially when $w \gtrsim 10$ Å, as expected for a tunneling process. Moreover, the η loss of the cell as w increases is found mainly dictated by the loss of FF for $w \lesssim 17$ Å. With w being further increased, the deterioration of J_{SC} becomes

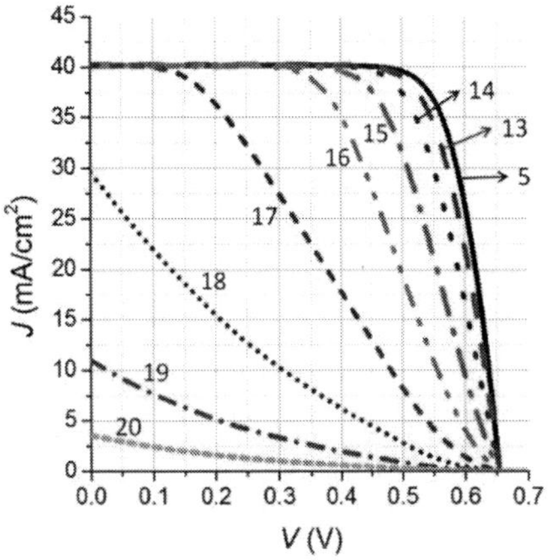

Figure 3: Calculated J-V characteristics under AM1.5G illumination for the simulated bifacial n-PERT Si solar cell with varied oxide thickness w, as indicated.

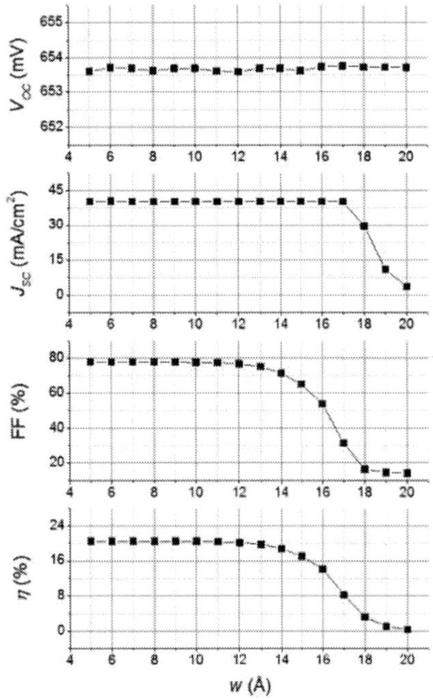

Figure 4: Output characteristics of the simulated bifacial n-PERT Si solar cell as a function of oxide thickness w.

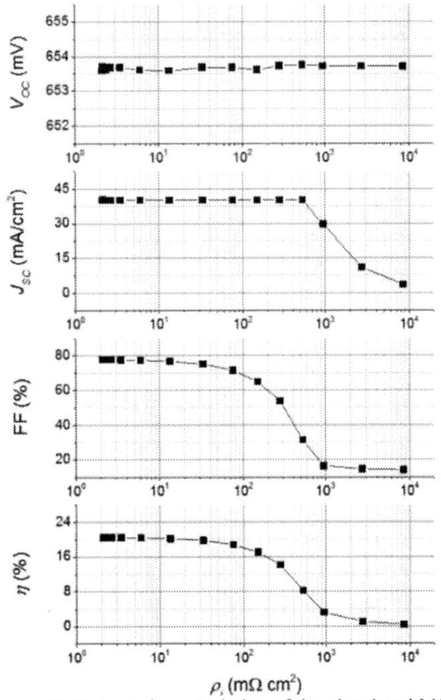

Figure 5: Output characteristics of the simulated bifacial n-PERT Si solar cell as a function of tunnel junction resistivity ρ_j.

more pronounced, giving an additional loss of η. These behaviors can be accounted for by considering the power dissipation associated with ρ_j and the application of the single-diode model. However, in this study, since the exact dependence of passivation quality on w is lacking, a constant value of SRV at the SiO_2/Si interface is assumed for all w's. As a result, the obtained V_{OC} value is practically not affected. If such a dependence for a specific passivation process is learned and implemented in the model, our method could be helpful in optimizing solar cell properties.

REFERENCES

[1] J. Benick, B. Hoex, M. C. M. van de Sanden, W. M. M. Kessels, O. Schultz, and S. W. Glunz, Applied Physics Letter 92 (2008) 253504.
[2] A. Cuevas, IEEE Journal Photovoltaics 2 (2012) 485.
[3] P. Würfel, Physics of Solar Cells - From Principles to New Concepts, WILEY-VCH Verlag GmbH & Co. KGaA, Weinheim, 2005.
[4] E. Yablonovitch, T. Gmitter, R. M. Swanson, and Y. H. Kwark, Applied Physics Letters 47 (1985) 1211.
[5] T. Kinoshita, D. Fujishima, A. Yano, A. Ogane, S. Tohoda, K. Matsuyama, Y. Nakamura, N. Tokuoka, H. Kanno, H. Sakata, M. Taguchi, and E. Maruyama, Proceeding 26th European Photovoltaic Solar Energy Conference (2011) 871.
[6] F. Feldmann, M. Bivour, C. Reichel, M. Hermle, and S. W. Glunz, Solar Energy Materials Solar Cells 120 (2014) 270.
[7] D. Zielke, J. H. Petermann, F. Werner, B. Veith, R. Brendel, and J. Schmidt, Proceeding 26th European Photovoltaic Solar Energy Conference (2011) 1115.
[8] J. Deckers, E. Cornagliotti, M. Debucquoy, I. Gordon, R. Mertens, and J. Poortmans, Energy Procedia 55 (2014) 656.
[9] J. Bullock, D. Yan, A. Cuevas, B. Demaurex, A. Hessler-Wyser, and S. De Wolf, Proceeding 40th IEEE Photovoltaic Specialist Conference (2014) 3442.
[10] S. M. Sze, and K. K. Ng, Physics of Semiconductor Physics, 3rd edition, John Wiley & Sons, Inc., Hoboken, New Jersey, 2007.

MODELING AND CHARACTERIZATION OF MULTICRYSTALLINE SILICON BLOCKS BY QUASI-STEADY-STATE PHOTOCONDUCTANCE

Mohsen Goodarzi,[1] Daniel Chung,[2] Bernhard Mitchell,[2] Thorsten Trupke,[2] R. A. Sinton,[3] and Daniel Macdonald[1]
[1]Research School of Engineering, The Australian National University, Canberra ACT 2601, Australia
[2]School of Photovoltaic and Renewable Energy Engineering, UNSW, Sydney, NSW 2052, Australia
[3]Sinton Instruments Inc., Boulder CO USA. http://www.sintoninstruments.com

ABSTRACT: A detailed knowledge of the distributions of carrier lifetimes, impurities and crystal defects in silicon ingots is key for understanding and improving wafer quality, as well as solar cell processing steps. In this work, we have applied a quasi-steady-state photoconductance tester developed for use on ingots for the measurement of lifetimes and dissolved iron concentrations along a p-type multicrystalline silicon block. The iron concentrations are determined by lifetime measurements taken before and after flashing the block to break the iron-boron pairs. The measured iron profile along the block is in qualitative agreement with those reported in previous work, and conforms to the expected trend due to impurity segregation. The impact of non-uniform carrier profiles during the block measurements on the extraction of the Fe profiles are discussed and quantified based on simulations of the quasi-steady-state measurement conditions. The lifetimes are also compared with a calibrated photoluminescence image on the same block, which shows similar qualitative trends, but lower magnitudes due to the lower injection level in the photoluminescence measurement.
Keywords: QSSPC, Carrier Lifetime, Iron concentration, Silicon block.

1. INTRODUCTION

The Quasi-Steady-State Photoconductance (QSSPC) method is well established for measurement of the minority carrier lifetimes on silicon wafers [1]. However, the application of QSSPC methods to ingots, blocks and boules is not as widespread, although it has some attractive advantages in terms of early-stage characterization. For example, knowledge about specific impurities and crystal defects at the ingot level can result in more reliable decisions in regards to the cropping of ingots and blocks for wafering purposes. It would be also possible to sort the wafers at the start of processing and tune the cell process for different quality wafers coming from the different sections of the block [2]. Moreover, accurate details about these characteristics may provide valuable information regarding the manufacturing process to produce higher efficiency or yields in final cells. There have been numerous studies performed on wafers which were cut from different parts of an ingot [3], to provide information about the electronic quality. There have however been fewer studies at the ingot level prior to wafering. Some notable examples include the development of a modified QSSPC tester for applications to ingots and blocks [4], which is also applied in this work, and the recent development of PL-based methods for ingot characterization [5, 6]. This study is focused on the further development of QSSPC methods for applications to ingots, ultimately aiming towards identifying useful correlations between data at the ingot level, such as carrier lifetime and dissolved iron concentrations, and the final cell efficiency.

2. EXPERIMENTAL METHODS AND RESULTS

2.1. Lifetime and Iron Concentration

QSSPC measurements were performed with a BCT-400 measurement system from Sinton Instruments, in QSSPC measurement mode with a 1000-nm IR-Pass Schott glass filter. The sample was an uncropped rectangular block cut from the central part of a standard industrial p-type multicrystalline silicon ingot, with a square base of 155× 155 mm and a height of 267 mm. There were 20 QSSPC measurement points starting at 10 mm from the bottom edge up to 10 mm from the top edge with a 13 mm gap between each point.

Multiple flashes with a Broncolor Picolite flash with a peak power of 1600 W were used in order to break the FeB pairs in the block. Zoth and Bergholz [7] showed that the interstitial iron concentration in silicon can be calculated from:

$$[Fe] = C\left(\frac{1}{\tau_1} - \frac{1}{\tau_0}\right) \qquad Eq.\ 1$$

Where τ_1 and τ_0 are the measured lifetimes before and after breaking FeB pairs, respectively. The Fe concentration and factor C have been calculated using the method developed in Ref [8].

Lifetime results were also compared with lifetimes measured at UNSW using the PL Intensity Ratio (PLIR) technique [6] which is based on the ratio of two photoluminescence images taken with different spectral filters in front of a CCD camera. This method does not require a separate measurement of the absolute doping density across the height of the brick, and also does not require analysis of the absolute PL signal.

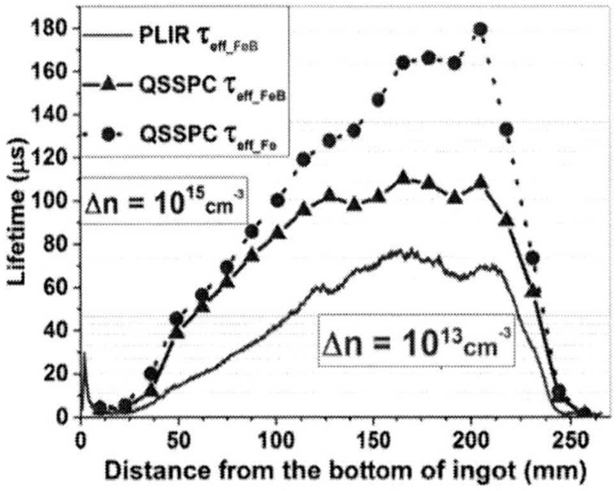

Figure 1: Lifetime vs Block position

Figure 2: PLIR lifetime image of the block (bottom of brick on the left). The scale bar has units of μs.

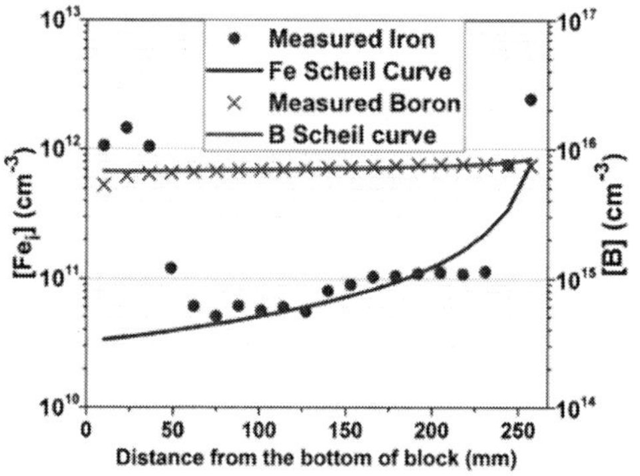

Figure 3: Iron and boron concentration vs block position

Figure 1 shows the QSSPC lifetimes measured along the block both before and after breaking the FeB pairs. The average injection level used here was $\Delta n_{avg} = 10^{15}$ cm^{-3} resulting in an increase in lifetime after breaking the FeB pairs, due to the presence of the lifetime crossover point at around 10^{14} cm^{-3} [9]. The extracted values of τ_{other} are also shown, which represent the lifetime due to all other defects, and these are expected to reflect the lifetime after removal of dissolved Fe due to gettering during cell processing [3]. Also shown are the average carrier lifetimes along the same block as measured by PLIR prior to breaking the FeB pairs. These were measured at a lower injection level, of around $\Delta n_{avg} = 10^{13}$ cm^{-3}. Nevertheless, the trend in the PLIR lifetime data is similar to the QSSPC data. The full PLIR lifetime image is shown in Figure 2 illustrating higher lifetimes towards the top of the block.

Figure 3 shows that the interstitial Fe concentration determined from the QSSPC data firstly rapidly decreases along the growth direction from the bottom of the ingot, and then experiences a steady increase in the middle of the ingot. Such a trend has earlier been reported by numerous authors, for example Sinton et al [4] and Mitchell et al [6], and can be explained by solid-state in-diffusion of Fe from the crucible and impurity segregation into the liquid phase during crystallisation, respectively. The boron concentrations determined from the dark conductance measurements are also shown in Figure 3.

To confirm that the difference in the lifetime before and after flashing the block is due to the presence of dissolved Fe, the sample was left in the dark to allow the Fe-B pairs to re-form. QSSPC lifetime measurements were performed during this relaxation period. It has been shown [8] that such lifetime data exponentially decays with an association time constant of τ_{assoc} for wafers containing interstitial Fe. The same approach can be used for ingot measurements, in principle. The results are illustrated in Figure 4 for the collected data from one set of the measurement points on the block.

Tan *et al.* [10] showed the relation between the Fe-B pair formation rate ($\frac{1}{\tau_{assoc}}$) with the dopant density and the temperature can be expressed as:

$$\tau_{assoc} = \frac{5.7\times10^5}{N_A}\, T\, exp(\frac{0.66}{kT}) \qquad \text{Eq. 2}$$

Using a dopant density of 7.2x 10^{15} cm^{-3} as measured by the dark conductance and the block temperature (308 K) results in an expected Fe-B pair formation time of approximately 26 mins. This compares well with the measured re-pairing time of 28 mins extracted from fitting the lifetime data in Figure 4 (note that the data is for a single point where lifetime was higher than the block average). This confirms that the observed changes in lifetime are indeed due to the presence of interstitial Fe.

2.2. Comparison to Scheil Equation

To confirm the extracted Fe and B profiles in our study, the results were compared with Scheil's segregation model for impurities and dopants during ingot growth. The boron profile along the block was extracted from the resistivity profile. Figure 3 presents the expected concentration of boron, with a segregation coefficient of 0.8, and iron, with a segregation coefficient of 6.2×10^{-3} [11], along the ingot as a function of the ingot height respectively. Since the Fe concentration depends on mechanisms other than segregation at the two ends of the ingots, the extracted values below 25% of the total height and above 80% of the total height of the ingot were not used in fitting the Scheil model. The boron and Fe profiles are in reasonable agreement with those expected from the Scheil equation, and with those reported in other studies [12].

3. MODELING OF QSSPC MEASUREMENTS

Cuevas [13] proposed a model to simulate Quasi-Steady State (QSS) photoconductance measurements in silicon wafers, utilizing numerical analysis in order to determine Δn profiles for typical wafer thicknesses. Here we use a similar approach to model QSSPC measurements in much thicker samples such as slabs, bricks and blocks. It should be noted that the key difference is the sample thickness, whereby minority carrier currents do not reach the rear surface, and so the carrier profiles are always highly non-uniform.

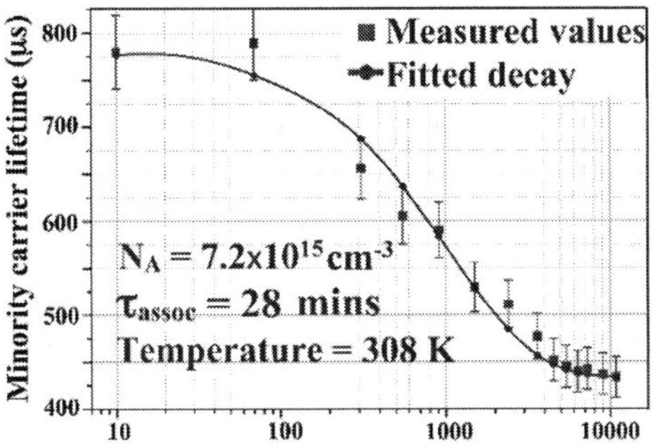

Figure 4: Reduction in lifetime during the re-pairing of FeB pairs

A first step is to simulate the generation profile within the sample depth for different light sources. Carrier recombination and diffusion are then simulated to allow the steady-state excess carrier density Δn profile to be determined. From these profiles, the average lifetime and excess carrier density extraction is performed in the same way as implemented in the QSSPC tool, producing a full simulation of the measurement process.

An example of the modeled generation profile and two excess carrier profiles are shown in Figure 5. Clearly the Δn profile in an ingot is non-uniform, and so an average value is required for the data analysis. Bowden and Sinton [2] introduced methods to estimate Δn_{avg} and W_{eff} (the effective thickness) which are utilized in both the QSSPC tool, and in the simulations used here.

The model is explained in detail in an upcoming publication, where it is shown how the Δn profile can be used to determine the value for Δn_{avg} and to ultimately calculate the effective bulk lifetime (τ_{eff}) at different average injection levels (Δn_{avg}), where τ_{eff} is:

$$\frac{1}{\tau_{eff}} = \frac{1}{\tau_{SRH_{Fe}}} + \frac{1}{\tau_{SRH_{FeB}}} + \frac{1}{\tau_{Auger}} + \frac{1}{\tau_{Other}} \quad \textit{Eq. 3}$$

Interstitial Fe and FeB pair recombination parameters can be used in the model to calculate the relevant injection dependent Shockley Read Hall lifetimes. These values can be changed to simulate lifetime measurements taken on ingots before and after breaking the Fe-B pairs, in order to estimate the interstitial iron concentration. Figure 5 shows excess carrier density profiles for a sample with [Fe] = 5×10^{11} cm^{-3} illuminated under 10 suns. The lifetime crossover point is evident at the expected value of 10^{14} cm^{-3}.

The accuracy of the Bowden and Sinton approximations are validated in our upcoming work in detail. The simulation tool allows us to demonstrate that the errors resulting from the non-uniform carrier profiles in the calculation of the interstitial iron concentration are relatively small (< 40%) in most practical cases. It is also shown the value of τ_{other}, Δn_{avg} and [Fe] can affect the accuracy in different ways. However, [Fe] itself is the most significant source of error, which can introduce up to 40% error in the typical Fe concentration range.

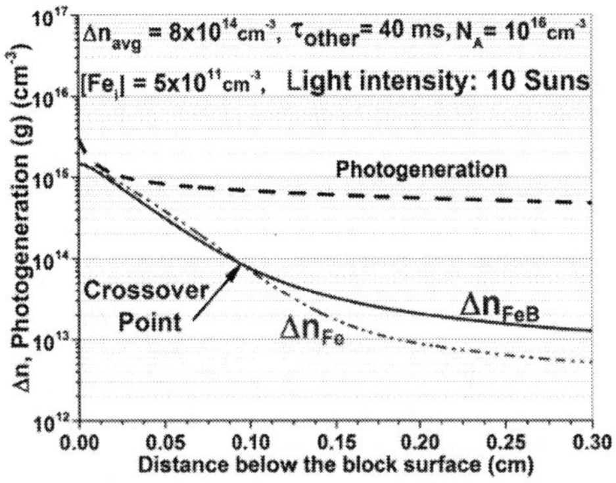

Figure 5: Excess carrier density and photogeneration vs distance below surface of block

4. CONCLUSIONS

This paper presents our preliminary results on applying the QSSPC method for silicon ingot characterization for solar cells. Comparison with PL lifetime imaging on the same ingot reveals notably higher lifetime values measured with QSSPC, due to the large difference in the injection levels at which each measurement was performed.

The main findings in terms of lifetime and dissolved Fe concentrations along the ingot are qualitatively similar to those previously reported [4]. The measured FeB re-pairing time is consistent with that expected from the dopant concentration, and confirms the presence of dissolved Fe and its effect on the minority carrier lifetime in the ingot. Extracted iron and boron concentrations were in reasonable agreement with the Scheil equation for segregation.

We have developed a simulation tool which models the generation and excess carrier profiles depth-wise, in order to calculate the effective lifetime determined by the QSSPC tool along the block. The results, which will be published in detail elsewhere, indicate that the non-uniform carrier profiles only introduce small errors in the resulting Fe concentrations (usually less than 40%), and that these can be corrected for in any case.

Future work will attempt to study the usefulness of the lifetime and dissolved Fe analyses at the ingot level for various types of ingots, with the final aim to establish correlations between the ingot level data and final cell performance.

ACKNOWLEDGEMENTS

This project has been supported by the Australian Renewable Energy Agency (ARENA) through project RND009.

5. REFERENCES

[1] R. A. Sinton and A. Cuevas, "Contactless determination of current–voltage characteristics and minority-carrier lifetimes in semiconductors from quasi-steady-state photoconductance data," *Applied Physics Letters,* vol. 69, p. 2510, 1996.

[2] S. Bowden and R. A. Sinton, "Determining lifetime in silicon blocks and wafers with accurate expressions for

carrier density," *Journal of Applied Physics,* vol. 102, p. 124501, 2007.

[3] L. Geerligs, "Impact of defect distribution and impurities on multicrystalline silicon cell efficiency," in *Photovoltaic Energy Conversion, 2003. Proceedings of 3rd World Conference on*, 2003, pp. 1044-1047.

[4] R. Sinton, T. Mankad, S. Bowden, and N. Enjalbert, "Evaluating silicon blocks and ingots with quasi-steady-state lifetime measurements," in *Proceedings of the 19th European Photovoltaic Solar Energy Conference, Paris, France*, 2004, pp. 520-523.

[5] B. Mitchell, T. Trupke, J. W. Weber, and J. Nyhus, "Bulk minority carrier lifetimes and doping of silicon bricks from photoluminescence intensity ratios," *Journal of Applied Physics,* vol. 109, p. 083111, 2011.

[6] B. Mitchell, D. Macdonald, J. Schon, J. W. Weber, H. Wagner, and T. Trupke, "Imaging As-Grown Interstitial Iron Concentration on Boron-Doped Silicon Bricks via Spectral Photoluminescence," *IEEE Journal of Photovoltaics,* vol. 4, pp. 1185-1196, 2014.

[7] G. Zoth and W. Bergholz, "A fast, preparation-free method to detect iron in silicon," *Journal of Applied Physics,* vol. 67, pp. 6764-6771, 1990.

[8] D. Macdonald, L. Geerligs, and A. Azzizi, "Iron detection in crystalline silicon by carrier lifetime measurements for arbitrary injection and doping," *Journal of Applied Physics,* vol. 95, pp. 1021-1028, 2004.

[9] D. Macdonald, T. Roth, P. N. K. Deenapanray, T. Trupke, and R. A. Bardos, "Doping dependence of the carrier lifetime crossover point upon dissociation of iron-boron pairs in crystalline silicon," *Applied Physics Letters,* vol. 89, p. 142107, 2006.

[10] J. Tan, D. Macdonald, F. Rougieux, and A. Cuevas, "Accurate measurement of the formation rate of iron–boron pairs in silicon," *Semiconductor Science and Technology,* vol. 26, p. 055019, 2011.

[11] G. Coletti, D. Macdonald, and D. Yang, "Role of Impurities in Solar Silicon," in *Advanced Silicon Materials for Photovoltaic Applications,* ed: John Wiley & Sons, Ltd, 2012, pp. 79-125.

[12] D. Macdonald, A. Cuevas, A. Kinomura, Y. Nakano, and L. J. Geerligs, "Transition-metal profiles in a multicrystalline silicon ingot," *Journal of Applied Physics,* vol. 97, p. 033523, 2005.

[13] A. Cuevas, "Modelling silicon characterisation," *Energy Procedia,* vol. 8, pp. 94-99, 2011.

GENUINE BIFACIAL SIMULATION AND OPTIMIZATION OF AN MC-SILICON PERC SOLAR CELL

Nico Wöhrle, Aina Alapont Sabater, Johannes Greulich
Fraunhofer Institute for Solar Energy Systems ISE, Heidenhofstraße 2, 79110 Freiburg, Germany
Telephone: +49 761 4588 5964, fax: +49 761 4588 7621, e-mail: nico.woehrle@ise.fraunhofer.de

ABSTRACT: Bifacial silicon solar cells gain significantly in importance as they provide a plus in power generation for various illumination scenarios without significantly complicating the manufacturing process. These scenarios can be numerous and are hard to be fully covered in experimental setups. It is suitable to perform computer simulations to gain deeper understanding of the cell characteristics under wide parameter variations and to optimize the layout towards different applications. For this purpose, a numerical simulation model has been developed, which allows a genuine bifacial simulation, i.e., independently selectable illumination windows for both sides. The model is used to simulate a broad variation of different bifacial mc-Si PERC cell layouts under different illumination scenarios. It is shown that physical effects of the front and rear side grid do not couple in good approximation, so both grids can be adjusted independently. Further the simulations show that in case of the presented mc-Si PERC cell, it is possible to find a sweet spot of 110 front fingers and 130 rear fingers where the cell provides close-to-ideal power output under all applied bifacial illumination conditions as well as a broad range of minority carrier lifetimes and base resistances.
Keywords: bifacial solar cell, multi-crystalline, PERC, Sentaurus Device simulation, PV Lighthouse

1 INTRODUCTION

The topic of bifacial passivated emitter and rear cells (biPERC, [1–3]) is supposed to be the next step in PERC [4] evolution after monofacial PERC technology successfully entered industrial mass production. With a predicted market share of 30% for bifacial silicon modules in 2026 [5] it is time to push advance-development of bifacial PERC technology. Simulation studies are no niche in PV R&D anymore but contribute substantially to the development in solar cell physics and technology. As bifacial simulation is a new field, there is a need for respective models, especially as bifacial characterization techniques are only just being evolved.

We want to provide the models for genuine bifacial simulation and show an extensive study on this topic in order to indicate the solar cell's behavior in power generation under several illumination scenarios for a large number of design variations and to give guidelines for the design of bifacial multi-crystalline (mc) PERC. The Sentaurus Device [6] models presented here are working truly bifacial, i.e., with two separately configurable illumination sources. Such simulations may generate a resilient basis for the upcoming discussion about bifacial measurement standards and rear side illumination spectra as the rear side AM1.5G spectrum may be easily replaced with optional albedo reflection in the simulation.

2 APPROACH

We equip the standard Sentaurus Device 2D models for solar cell simulation [7] with a second illumination source at the rear side which is adjustable in spectrum and intensity independent from the front illumination. The approach suits 3D models likewise. Using this model, we perform simulations with up-to-date input parameters for a bifacial p-type mc-PERC solar cell as described in section 4. The cell features a lately (at Fraunhofer ISE) developed phosphorus-doped emitter with sheet resistance $R_{sh} = 85\ \Omega/sq$ and a screen-printed aluminum-boron back surface field (AlB-BSF) [8] beneath line-shaped laser contact openings (LCO [9]). The large parameter space for this study comprises independent intervals for the front- and rear-side finger numbers, three different rear-side finger widths, two base resistances, four bulk minority carrier lifetimes, and five different illumination conditions (for cell and module optics) representing a south-facing orientation with different rear intensities and an east/west-facing orientation.

3 MODEL

3.1 Technical setup

As for most Sentaurus Device simulations, the modeling is a two-step process: Starting with a pure optical ray tracing simulation, a spectrally and depth-resolved carrier generation profile is created. This carrier profile is injected into an electrical model where the voltage is ramped through the contacts generating the illuminated current-voltage curve.

To extend this routine towards bifacial illumination a second, independent optical simulation is conducted with a rear side illumination source permitting all different kinds of rear side spectra (albedo ground reflection, diffuse ambience radiation, AM spectra with optional intensities). Several generation profiles are joint in one electrical model, each one arranged according to its designated illumination source. The detailedness of illumination regions can be chosen as far as it is necessary or suitable. In this work, we differentiate not only between front and rear side generation but also between front side generation above rear contacts, above passivated rear metallization and above rear passivation (open rear side). Fig. 2 demonstrates a step-wise implementation of generation profiles into the electrical symmetry element. We call this modelling approach "genuine" bifacial simulation as the achievable degree of precision is unique so far and helpful for optimizing the layout of bifacial solar cells.

3.2 Illumination scenarios

Using the bifacial simulation model developed here, different illumination scenarios shall be used for the optimization of a bifacial mc-PERC solar cell presented in section 4. The following two scenarios represent a simple approach of a bifacial solar cell/module in two spatial orientations, visualized by Fig. 2:

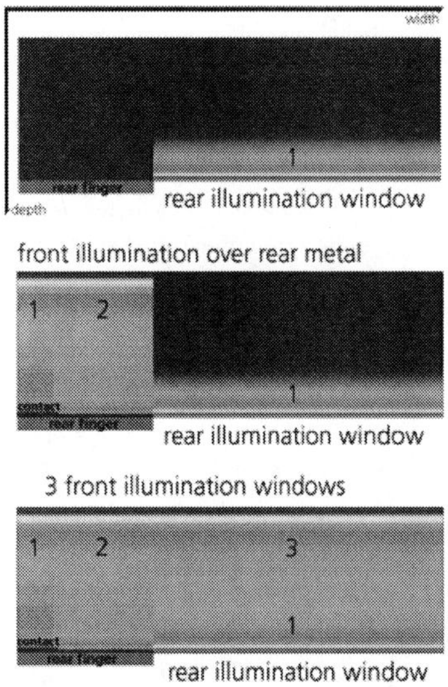

Fig. 1. a) Electrical symmetry element with implemented generation profile arising from rear illumination.
b) additional implementation of front generation over the rear contact and the metal finger (with passivation).
c) additional implementation of front generation over the rear side passivation.

1. "Standard" elevation towards the south (1000 W/m² front illumination) with different illumination intensities from the rear side: 100, 250, 500 W/m² (each with AM1.5G spectrum)
2. Vertical elevation towards east/west represented by averaging a) 750 W/m² front and 250 W/m² rear and b) 250 W/m² front and 750 W/m² rear illumination (each with AM1.5G spectrum)

For module optics a both-sided encapsulation of 0.45 mm EVA and 3 mm glass are added as specified in Table I.

3.3 Discussion

Besides the described advantage of a detailed differentiation of optical generation and adaption to numerous illumination conditions, there is currently one discussable weakness of the model: As can be seen in Fig. 1, the borders of the generation profiles in the different cell regions are vertical to the cell surface, which neglects e.g. generation beneath fingers due to angular incidence of light. We currently seek a technical option to solve this. Note that optical refraction due to texturing and multiple light bounces beneath the open rear side region are fully included in the optical simulation and therefore appropriately represented in the optical generation profile. Note that the electrical model in Fig. 1 shows carrier generation, not light paths.

4 OPTIMIZATION OF AN MC-PERC SOLAR CELL

We want to use the presented model to show an optimization of a bifacial multi-crystalline PERC with LCO contacts and a local AlB-BSF on a 156 mm large wafer. The cell's features are listed in Table I. The minority carrier lifetime is set to $\tau_{SRH} = 100$ µs if not stated otherwise.

As there are currently no norms for efficiency calculation with bifacial illumination, all results are given

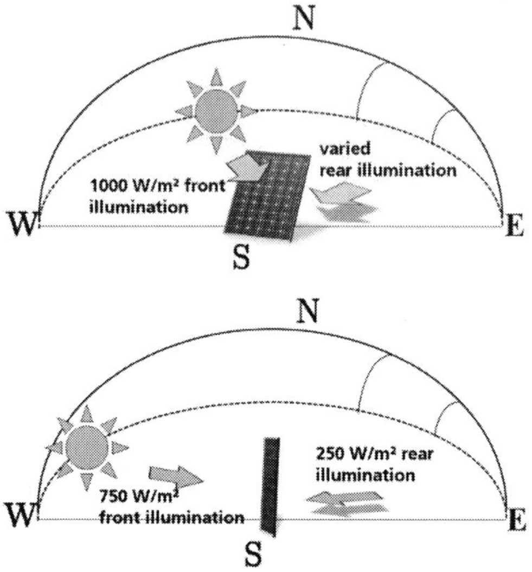

Fig. 2. a) Illumination scenario *"south"*: front illumination 1000 W/m², varied rear illumination (100, 250, 500 W/m²).
b) Illumination scenario *"east/west"*: 750 W/m² front illumination, 250 W/m² rear illumination and vice versa.

in output power density units P_{out} (mW/cm²).

4.1 Front side grid

The number of front fingers is varied between $n_{f\_FS} = 80$ and 140 while the number of rear side fingers is kept fixed at $n_{f\_RS} = 100$ in this step. Illumination scenario 1 (see section 3.2) is applied. Two different finger widths (75, 100 µm) are tested for the rear side; the base resistivity is 1.0 and 2.5 Ωcm.

A. Cell optics (no encapsulation)

Fig. 3 shows the resulting differences in output power in dependence of the finger variations for different rear

Table I: Simulated mc-PERC parameters;

characteristic angle ω, layer thickness d, depth z, specific finger resistance ρ_f, specific contact resistance ρ_c, optical constants n & k, finger width w_f, finger height h_f, emitter sheet resistance R_{sh}

front		rear	
iso-texture	$\omega = 60°$, z = 3.6 µm	iso-texture	$\omega = 60°$, z = 3.6 µm
j_{0e}	75 fA/cm²	j_{0bsf}	350 fA/cm²
$j_{0e,met}$	1200 fA/cm²	d_{bsf}	3 µm
R_{sh}	85 Ω/sq	d_{Al2O3}/d_{SiNx}	10/70 nm
d_{SiNx}	75 nm	S_{pass}	10 cm/s
front grid		rear grid	
h_{f_front}	20 µm	h_{f_rear}	25 µm
w_{f_front}	54 µm	w_{f_rear}	varied
ρ_{f_front}	5.94 µΩcm	ρ_{f_rear}	20 µΩcm
ρ_{c_front}	6 mΩcm²	ρ_{c_rear}	3 mΩcm²
encapsulation			
d_{EVA}	0.45 mm	n_{EVA}	see [10]
		k_{EVA}	see [10]
d_{glass}	3 mm	n_{glass}	1.5
(borosilicate glass)		k_{glass}	See [11]

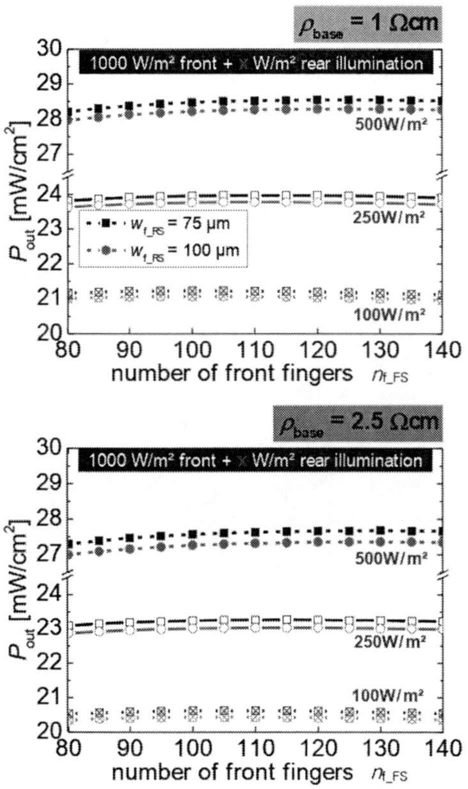

Fig. 3. Output power P_{out} as a function of the number of front fingers, calculated with cell optics for two rear finger widths w_{f_RS} = 75 µm and 100 µm. The cell is illuminated by constant front (1000 W/m²) and three different rear intensities (100, 250, 500 W/m²).

illumination intensities. The layout-induced deviations in P_{out} over n_{f_FS} are below 0.3 mW/cm² for all plotted lines. In a more reasonable interval between 90 and 120 front fingers, they do not exceed 0.15 mW/m². It can be stated that in said interval the chosen number of front fingers will not have a substantial influence on P_{out} and is to a large degree independent from base resistivity. Moreover, it is quasi-independent from rear side illumination intensity. The front and rear grid optimization is therefore uncoupled.

B. Module optics (with encapsulation)

The same counts for module optics (with EVA and glass encapsulation). The encapsulation leads to a quite constant offset in optical generation compared with cell optics depending on the additional rear illumination. The variations ΔP_{out} in the output power range on module level between -0.15 and -0.4 mW/cm² for rear illumination intensities of 100 and 500 W/m², respectively, which is not plotted for clarity reasons.

4.2 Optimization of rear side capping

The rear side optimization begins with the thickness of the passivation capping as it is independent from the grid layout. The thickness of the SiN_x passivation layer is varied between d_{SiNx} = 30 nm and 120 nm to find the optimal value for different illumination conditions. d_{Al2O3} = 10 nm is kept constant. Due to the comfortable interface this optical simulation is performed with the

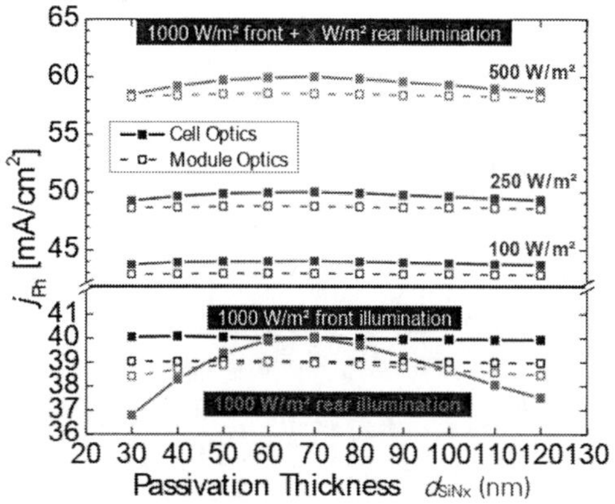

Fig. 4. Photocurrent density as a function of the thickness of the SiN_x capping layer within an Al_2O_3/SiN_x based rear side passivation stack with fixed Al_2O_3 thickness. Both sides are iso-textured. The lower third shows pure front illumination (black symbols) and pure rear illumination (green symbols). The upper two thirds show front plus rear illumination for varied rear intensities. Cell optics are plotted with solid lines, module optics with dashed lines.

PV Lighthouse "Module ray tracer" [12]. Other cell parameters remain unchanged to the previous simulations. As the ray tracer only supports front illumination, the cell structure is flipped to simulate rear illumination. Simultaneous illumination is calculated by addition of the integrated charge carrier generation rate j_{Ph} of different scenarios.

A. Cell optics

As visualized in the lower part of Fig. 4, the j_{Ph} generated by pure front illumination (black symbols) is almost independent from the SiN_x thickness of the rear side. In contrast j_{Ph} from rear illumination (green symbols) heavily depends on the SiN_x thickness with its peak at 70 nm.

The joint calculation with 1000 W/m² front and varied rear illumination (upper part of Fig. 4) therefore also shows its maxima at d_{SiNx} =70 nm, while the peak gets more pronounced the higher the portion of the rear illumination is.

B. Module optics

Adding 0.45 mm EVA and 3 mm glass does not shift the peak at d_{SiNx} = 70 nm but flattens it strongly. The maximum deviation in j_{Ph} for pure rear illumination (green symbols) in the SiN_x thickness interval between 30 nm and 120 nm decreases from 3 mA/cm² (on cell level) to 0.5 mA/cm² (on module level). For the joint calculation of front and rear illumination, a deviation of $\Delta j_{Ph} < 0.3$ mA/cm² over the regarded d_{SiNx} interval remains.

4.3 Optimization of rear side grid

As a next step, the rear side grid layout is optimized under consideration of several illumination conditions taking into account cell and module optics.

A. Cell optics

Again illumination scenario 1 (section 3.2) is applied to calculate the output power of the PERC cells as a function of the number of rear fingers, the width of the rear fingers being varied threefold with w_{f_rear} =75, 100 and 200 µm and base resistance of the substrate being

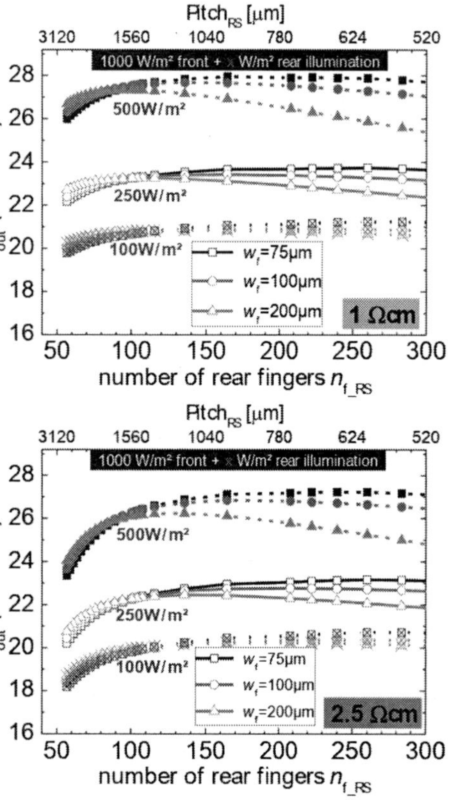

Fig. 5. Output power P_{out} on cell level as a function of the number of rear fingers for different finger widths w_{f_rear} = 75, 100, 200 µm and different base resistivities ρ_{base} = 1 and 2.5 Ωcm. The cell is illuminated by constant front (1000 W/m²) and three rear intensities (100, 250, 500 W/m²).

varied twofold with ρ_{base} = 1 and 2.5 Ωcm The results are plotted in Fig. 5. For w_{f_rear} = 75 µm (black symbols) and 100 µm (green symbols), the optima are located between 200 and 120 fingers for ρ_{base} = 1 Ωcm and between 260 and 160 fingers for ρ_{base} = 2.5 Ωcm and decrease with rear illumination intensity. While the intervals may seem large, the curves are very flat in these intervals showing a maximum deviation of only 0.3 mW/cm². For w_{f_rear} = 200 µm (orange symbols), which is the more common Al-finger thickness at the moment, the peaks shift towards lower finger numbers in the range of n_f = 150-100 due to higher shading of the rear side and thus lower j_{SC} towards increasing n_f.

The main differences between ρ_{base} = 1 and 2.5 Ωcm are a) a steeper drop in *FF* towards lower finger number due to increased spreading resistance for 2.5 Ωcm, which is visible in the larger bend of P_{out} between 50 and 100 fingers and b) a general offset in V_{OC} by about 10 mV leading to ~0.8 mW/cm² power loss for wafers with higher base resistivity caused by lower built-in voltage.

B. Module optics

As already seen in the front optimization (section 4.1), including module optics provides a general offset in j_{Ph} and thus P_{out} of -0.2 to -0.4 mW/cm² with increasing rear intensity, but does not alter the observed shapes.

4.4 Lifetime variation and east/west illumination

Illumination scenario 2 (see section 3.2) and a variation of bulk minority carrier lifetime between τ_{SRH} = 30 and 500 µs round up the simulation study which is shown in Fig. 6. This one is only performed for cell optics as module optics are not expected to yield different results than in the previous sections.

The east/west illumination scenario shows the largest dependency on the number of rear fingers and rear finger thickness, as the proportion of rear illumination is the highest in this case. Nevertheless, if the minority lifetime is higher than 250 µs, it almost yields the power output of the 1000 W/m² front illumination when choosing the optimal finger number of $n_{f\_RS} \approx$ 100-120. Lower bulk lifetimes increase the gap between front illumination and bifacial illumination (each with 1000 W/m² sum).

Another result is that the ideal rear layout for monofacial applications is a very high finger number exceeding the examined range (n_{f_RS} > 300). This means a rear layout optimized for monofacial application in this way would not be suited for bifacial application. However, a rear layout with n_{f_RS} = 110 rear fingers which is optimized for all examined bifacial scenarios, is very well suited for monofacial application as well as the deviations in P_{out} are below 0.3 mW/cm² between 100 and 300 fingers.

5 DISCUSSION

An important side note and motivation for the development of the simulation model is the non-existence of bifacial measurement norms. The relevant question of suitable measurement conditions is nevertheless beyond the scope of this publication. However, the simulations provide additional scientific insight into the underlying physical phenomena which may promote this ongoing discussion. The optimization of the mc-PERC layout is our first result which is directly applicable to bifacial cell technology.

The four steps of the simulation experiment reveal a remarkable robustness of the cell concepts towards varying material and illumination parameters if the layout is chosen correctly. The very small impact of the number of front fingers, as shown in Fig. 4, originates from the fact that the increase in *FF* (less lateral emitter resistance) compensates to a large degree for the decrease in V_{OC} (higher contact recombination) and j_{SC} (more shading) level. As these are all emitter or front surface effects neither base resistance nor rear illumination have qualitative influence on this behavior.

The rear capping layer optimization of this study focusses solely on optics. As the front illumination has negligible influence, the optimum always stays at the rear side optimum of 70 nm (together with 10 nm Al_2O_3), which is expected. Notice that this value may change if the spectral distribution of the rear illumination changes, as it is an optimized layer thickness for AM1.5G. Nevertheless, the peak is strongly flattened by module optics (Fig. 4), that this stack could be a decent choice for different albedo reflections as well. The flat peak indicates that the capping layer thickness may be adjusted for best electrical properties.

The optimization of the rear grid shows that in general the P_{out} deviation over finger number is small in a reasonable interval between 100 and 200 fingers especially for realistic rear illumination intensities

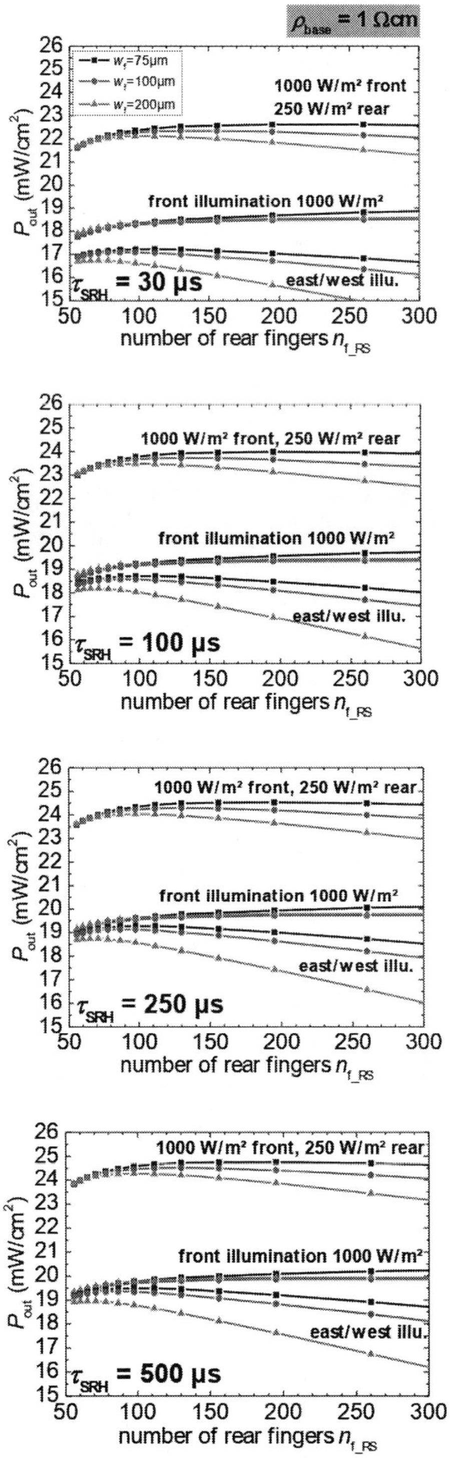

Fig. 6. Output power P_{out} on cell level as a function of the number of rear fingers. Lifetime variation τ_{SRH} = 30-500 µs for illumination scenario 1 (1000 W/m² + 250 W/m²), scenario 2 (east/west illumination) and monofacial illumination (1000 W/m²). Calculated with cell optics. Colors indicate different rear finger widths.

between 100 and 250 W/m². Thus, the exact finger number should not be the biggest concern and can be chosen according to individual preferences. Higher finger number has advantages towards process reliability, e.g. in case of finger interruptions. Less fingers may save costs. Also not much care has to be taken for base resistance concerning the layout as it mainly influences the quality of the power curve for low finger number – aside from the peak region – due to increasing spreading resistance. Two relevant aspects are observed: Firstly, for applications where the rear to front illumination ratio is large (e.g. east/west scenario, visible in Fig. 6), the rear finger number should be corrected towards smaller values. Secondly, thin Al fingers on the rear are desirable as they allow for more rear fingers, which improves the power output for higher base resistance material due to the reduced spreading resistance in the base. This can be beneficial for base resistance gradients over a group of wafers.

Finally, the variation of the SRH lifetime of the minority carriers between 30 and 500 µs and the introduction of the east-west scenario confirms the robustness of the chosen layout. There is no qualitative deviation to the optimum of 100-150 rear fingers which has been found in section 4.3. Moreover, the simulations show that the output power decreases significantly with decreasing bulk lifetime for all bifacial application, while it remains quite stable for monofacial application in the investigated lifetime range. The reason is that the investigated device exhibits a pn-junction at the front side, which leads to a reducted collection efficiency of carriers which are generated at the rear side compared to generation at the front due to higher recombination losses in the base.

6 CONCLUSION

The successful implementation of several independent illumination sources in one simulation setup was shown to provide a genuine bifacial simulation model. The introduced simulation model for bifacial mc-PERC cells has been used for a comprehensive parameter variation.

As a first result, we showed that, in an interval of 80-140 front fingers, front and rear finger numbers can be varied independently without coupling effects under 1000 W/m² (AM1.5G) front illumination in addition with 100 to 500 W/m² (AM1.5G) rear illumination. This counts for base resistances from ρ_{base} = 1 to 2.5 Ωcm as well as the applied rear side finger widths from 75 to 200 µm. This conclusion may ease the development of metal grid layouts to a great extent.

Further, we provided plots for variations of parameter sets that cover a wide range of possible cell configurations. One of the key findings was, that there is a sweet spot of rear side finger numbers at 110 fingers (ρ_{base} = 1 Ωcm) and 120 fingers (ρ_{base} = 2.5 Ωcm), which suits all investigated bulk lifetimes, finger widths and illumination scenarios. Thus, it is possible to build a bifacial mc-PERC solar cell that universally suits most irradiance scenarios with a max. loss of 0.35 mW/cm² to the single individual optimum configurations. We also showed that lifetime variations have negligible influence on the optimal cell layout. Base resistance variations have an impact but only a small one. The rear finger number should be optimized especially when dealing with high rear to front illumination ratios. Further, thin fingers of around 100 µm should be a development goal, as such a layout allows for larger finger counts on the rear side. The outcome for the optimal design of the examined mc-PERC cell with 85 Ω/sq phosphorus-doped emitter and LCO rear contacts shows to be a 1 Ωcm base with high lifetime, 110 silver front fingers with 54 µm width and 5 busbars with a width of 500 µm. The rear side layout should have 100 Al-fingers of 200 µm width (today's standard) or 120 fingers of 100 µm or even 75 µm width as a development goal. These design rules for the rear side are lifetime-invariant and apply for cell optics as well as module optics with EVA and glass encapsulation.

ACKNOWLEDGEMENTS

The authors acknowledge the funding of the German Federal Ministry for Economic Affairs and Energy (BMWi) in the frame of the project "BiZePS" (contract number 0325909) and thank the associated project partners h.a.l.m. Elektronik GmbH and SolarWorld Innovations GmbH for the excellent cooperation.

REFERENCES

[1] Y. Chevalier and I. Chambouleyron, "Getting more power out of silicon," in *Proceedings of the 1st European Commission Conference on Photovoltaic Solar Energy*, pp. 977–986.

[2] K. Krauß, F. Fertig, J. Greulich, S. Rein, and R. Preu, "biPERC silicon solar cells enabling bifacial applications for industrial solar cells with passivated rear sides," *Phys. Status Solidi A*, vol. 213, no. 1, pp. 68–71, 2016.

[3] T. Dullweber *et al*, "PERC+: industrial PERC solar cells with rear Al grid enabling bifaciality and reduced Al paste consumption," *Prog. Photovolt: Res. Appl*, pp. n/a, 2015.

[4] A. W. Blakers, A. Wang, A. M. Milne, J. Zhao, and M. A. Green, "22.8% efficient silicon solar cell," *Applied Physics Letters*, vol. 55, no. 13, p. 1363, 1989.

[5] SEMI, "International Technology Roadmap for Photovoltaic (ITRPV)," Seventh Edition, Mar. 2016.

[6] Synopsys, "Sentaurus Device User Guide," vol. H-2013.03.

[7] P. P. Altermatt, "Models for numerical device simulations of crystalline silicon solar cells—a review," *J Comput Electron*, vol. 10, no. 3, pp. 314–330, 2011.

[8] M. Rauer, C. Schmiga, M. Glatthaar, and S. W. Glunz, "Alloying From Screen-Printed Aluminum Pastes Containing Boron Additives," *IEEE J. Photovoltaics*, vol. 3, no. 1, pp. 206–211, 2013.

[9] R. Preu *et al*, "Laser ablation - a new low-cost approach for passivated rear contact formation in crystalline silicon solar cell technology," in *16th EU PVSEC*, 2000, pp. 1181–1184.

[10] K. R. McIntosh *et al*, "An optical comparison of silicone and EVA encapsulants for conventional silicon PV modules: A ray-tracing study," in *2009 34th IEEE Photovoltaic Specialists Conference (PVSC)*, pp. 544–549.

[11] *PV Lighthouse - Refractive Index Library.*

[12] PV Lighthouse, *Module ray tracer*, 2016.

LIGHT INDUCED DEGRADATION IN PERC SOLAR CELLS AND COMPARISON WITH OXYMAP

J. Arumughan[1], B. Martel[2], G. Raymond[2], S. Schmitt[1], J. Theobald[1], C. Picoulet[3], X. Brun[3], R. Kopecek[1]

[1]International Solar Energy Research Center - ISC Konstanz, Rudolf-Diesel Str.15, D-78467 Konstanz, Germany
[2] CEA/LITEN/DTS, INES, 50 avenue du Lac Léman, 73375 Le Bourget du Lac, France
[3] AET Technologies, 8 Chemin de Malacher, Inovallée, 38240 Meylan, France
e-mail: jayaprasad.arumughan@isc-konstanz.de

ABSTRACT: The PERC solar cells performances fabricated using different base resistivities and oxygen concentrations are found to be correlated. Samples with $[O_i]$ concentration of 10.69 ppma and a base resistivity of 1.75 Ω cm reaches median efficiency of 19.4%. Samples with similar $[O_i]$ concentration, but higher base resistivity have lower median efficiencies, between 19.1% and 19.3%. However samples with base resistivity 1.5 Ω cm and $[O_i]$ concentration over 12 ppma reaches lowest efficiency. The OxyMap measurements were performed on twin wafers. The FTIR $[O_i]$ measurements and OxyMap measurements are comparable. The LID prediction results are in trend with the measured LID on PERC solar cells.
Keywords: Light induced degradation, PERC solar cells, OxyMap.

1 INTRODUCTION

Photovoltaic market is a fast growing and the compound annual growth of photovoltaic installations is 44% during the period of the 2000 to 2014 [1]. The market share of the crystalline silicon technology is more than 90% and it is still dominant in the market. In the year 2013, total PV production using the crystalline technology is more than 35.2GWp [1]. Even though the standard solar cell architecture is dominant in mass production, the PV industry is thriving in different ways to achieve cost effective solar cells with higher efficiencies such as improvement in emitters, metallization techniques and advanced cell architectures. In this scenario, Passivated emitter, rear cell (PERC) is an excellent candidate for commercialization due to its similarity to the state of the art aluminum BSF solar cells. PERC solar cells can be fabricated in the existing standard solar cell production lines without much technical complexity. Therefore PERC cells can be considered as the next evolutionary step in solar cell development for efficiencies higher than 20% in mass production [2].

In this work, we study the performance and characterization of PERC solar cells prepared using two different oxygen concentrations, base resistivities and thicknesses. This paper discusses the different aspects of real time LID and prediction of LID using OxyMap tool.

2 EXPERIMENTAL

2.1 Material used in this study

The solar cells were prepared using 156 mm x 156 mm pseudo square, p-type, Czochralski grown, boron doped wafers with a resistivity in the range of 1.5 ohm-cm to 2.5 ohm-cm. The interstitial oxygen concentration was measured on thick slices by FTIR (ASTM F121-83), close to the wafers used for the study. The characteristics different groups of wafer included in this study are given below in table 1. In total, we have four different groups. Please note that the wafer group I and group II originates from the same ingot but different thicknesses. Group I has a wafer thickness 175μm and Group II has a wafer thickness of 190μm. The oxygen concentration of Group I is 10.94 ppma and for Group II it is 10,74 ppma. The resistivity of these two wafer groups is ~ 2.5 Ω-cm. The wafer group III has similar oxygen concentrations like groups 1 and group 2, but a varying resistivity, 1.75 Ω-cm. Wafer group IV has the lowest resistivity, 1.5 Ω-cm, among all other groups and contains the highest concentrations of oxygen, >12ppma. This group is used as the reference wafers for solar-cell-process control. Both Group III and Group IV have same wafer thickness, 190μm. The boron concentration of the groups I, II, III and IV are 5.33×10^{15} cm^{-3}, 5.25×10^{15} cm^{-3}, 7.70×10^{15} cm^{-3} and 7.89×10^{15} cm^{-3}.

Table I: Materials studied

Group	[B] (ppma)	$[O_i]$ (ppma)	ρ (Ω-cm)	τ (μs)
I	0.106	10.9	2.4	142
II	0.105	10.9	2.5	152
III	0.154	10.5	1.7	196
IV	0.158	12.0	1.5	129

2.2 Solar cell processing

The process sequence for processing PERC solar cells are described below. The process can be summarized in the following steps. 1), Alkaline pyramidal texture, 2) Single side polishing of rear side, 3) pre-diffusion clean and P-Diffusion, 4) Emitter removal on rear side and edge isolation, 5) phosphorous glass removal, 6) Thermal oxidation and oxide removal, 7) rear side passivation using PECVD SiN_x, 8) PECVD SiN_x deposition on front side, 9)Laser ablation and laser damage etch, 10) front side metallization: Silver and rear side metallization aluminum. The metallization is achieved by a screen printing process and 10) contact co-firing in a belt furnace. More details of PERC cells are available in literatures [3-5].

3 RESULTS AND DISCUSSIONS

3.1 Solar cell parameters

Figures 1 to 4 shows the solar cell parameters of different groups used in this study. V_{oc} of the groups I and II are identical due to the similar material properties. But the V_{oc} values are lower for groups III and the lowest for the group IV. This could be attributed to the bulk lifetime of the charge carriers. The distribution in short circuit current densities is quite tight for all groups, as seen in the figure 2. The fill factor values are found

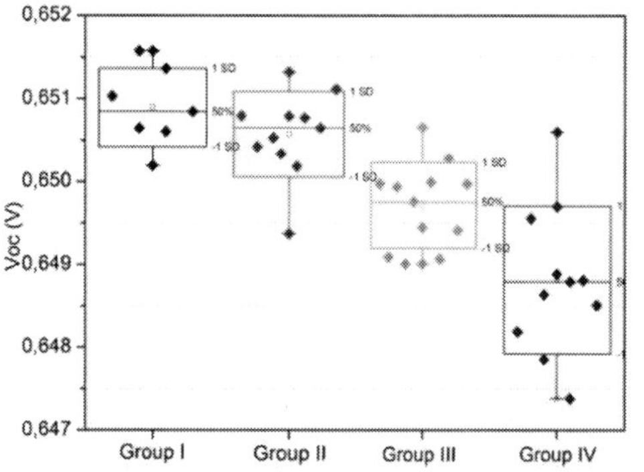

Figure 1: Open circuit voltage values of different PERC solar cells. Samples with highest oxygen concentration have lower V_{oc} values.

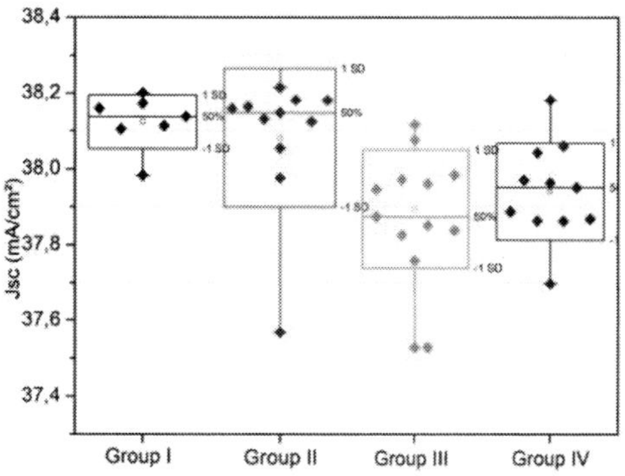

Figure 2: Short circuit current density of different PERC solar cells.

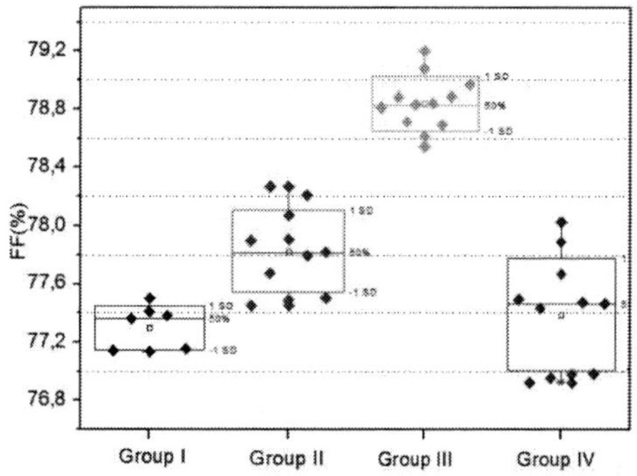

Figure 3: Fill factor vales of different PERC solar cell groups.

to be the highest for the lower base resistivity materials. However, it found to be the lowest for the fourth group with highest $[O_i]$ concentration. Due to the highest fill factor values the group III emerges as the group with highest median efficiency. From the graphs it seems that the groups with lower [Oi] concentration and lower base resistivity have added advantage PERC solar cells performance.

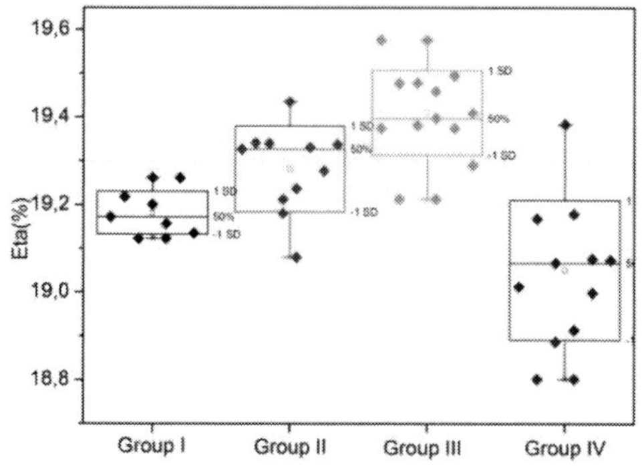

Figure 4: Efficiency of different PERC solar cells.

The spectral response curves of different groups are given in the figure 5. The curves are quite similar for all the solar cell groups except in the range of 950 nm to 1150 nm. In this range, the group IV solar cells with highest oxygen concentration deviate from other groups.

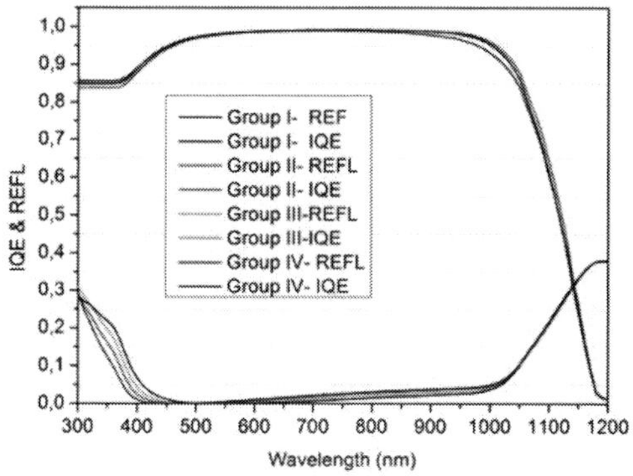

Figure 5: IQE and reflectance curves of different materials used in this study.

3.1 LID Measurements

In order to measure the LID on PERC solar cells the ''Degradtest'' equipment is used [6]. All the solar cells are kept at 25°C and illuminated under a halogen light source of intensity ~ 1 sun. The cells were kept under illumination for about 120 hours, and the LID has reached a stable level. Following graph shows the LID observed in the PERC solar cells. The LID was measured by keeping the PERC solar cells on a temperature controlled platform kept at 25°C. The intensity of light was kept at ~ 1 sun. Figure 6 shows the LID curves of all solar cells. The difference in different solar cell parameters before and after 120 hrs LID is given in Table III. Please note that the initial V_{oc} of all solar cells are less than the V_{oc} measured using I-V measurements. This is because of the homogeneity of illumination. As seen from the figure the PERC solar cells with lower resistivities degrade higher than the samples with higher base resistivities.

In addition, the group IV PERC cell with highest $[O_i]$ concentration shows the highest degradation. This could be explained with the current knowledge on the formation of B-O complex formation under illumination [9-10].

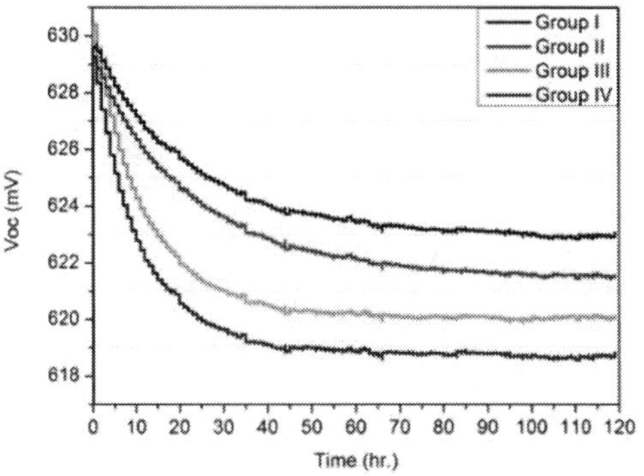

Figure 6: Light induced degradation of all solar cell groups at 25°C.

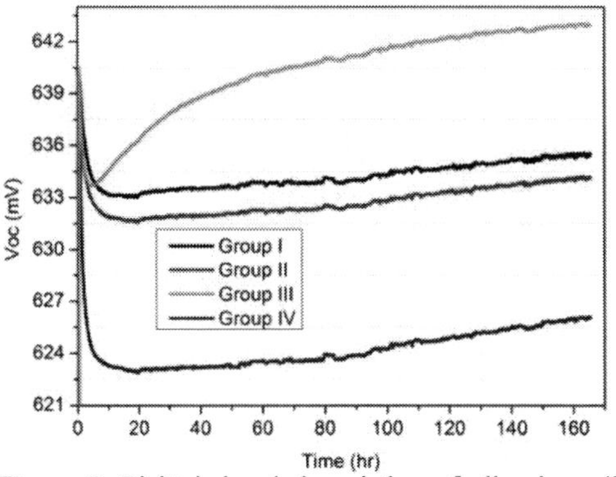

Figure 7: Light induced degradation of all solar cell groups at 70°C.

In order to 'simulate' the performance of these cells in warm climatic regions, we have done the LID at elevated temperature, 70°C. This temperature represents an approximate module temperature in warm climatic regions. From figure 7 it is clear that all groups of PERC cells show tendency of regeneration. This tendency is strongest in group 3 where the base resistivity and $[O_i]$ concentration are low. Low temperature regeneration has already been underlined by Herguth [11] on Al-BSF solar cells and our results shows that it could also be noticed in our PERC solar cells. Furthermore, we made regeneration studies at 120°C under illumination (~ 1 sun) on the cell groups.

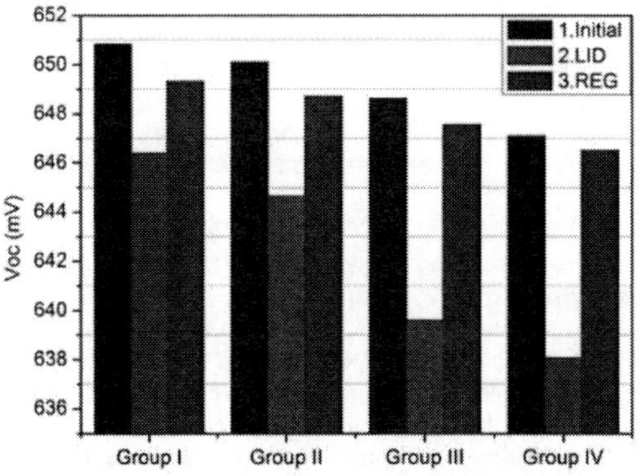

Figure 8: Absolute change in V_{oc} PERC cells initial, after degradation and regeneration (120°C, 1 sun, 4 hours).

Figure 8 shows that the initial V_{oc} values, V_{oc} value after 120 hours of LID and V_{oc} values after regeneration. From the figure it is clear that the regeneration helps to regain the efficiencies close to the initial levels. Our experiments show that regenerated PERC solar cells retains the efficiencies and they do not subject to further LID Process. The regeneration further suggests that the drop in V_{oc} values is higher for Groups III and IV rather than the Groups I and II. This could be due to the formation of higher rate of B-O oxygen complexes, due to the higher dopant concentration, in these groups. This mechanism could be explained according to the current regeneration theory of BO complex generation and regeneration [12].

3.1 OxyMap measurements

This protocol shown in the following figure 9 applied to 4 wafers (one of each groups). Measurements were carried out on 181 points across the wafer diagonal. Two annealing steps are included in the procedure. The first one aims at generating a controlled amount of thermal donors ([TD]). It is conducted at around 450°C and lasts typically for some tens of minutes up to a few hours. The second anneal leaves the wafer TD-free and is carried out above 650°C. Three resistivity (ρ) measurements are done before, between and after the annealing steps. The extraction of $[O_i]$ relies on the change in ρ before and after the first anneal. The exact dopant concentration (here the boron concentration, [B]) is obtained from the last ρ measurement when the wafer is expected to be TD free, with the help of standard mobility models. From [B] and $[O_i]$, OxyMap is proposing a semi-empirical model to predict the future cell efficiency degradation after the formation of the B-O complexes. This model is mainly based on PC1D simulations which take into account the carrier lifetime degradation law depicted in [13] which is a function [B] and $[O_i]$.

.

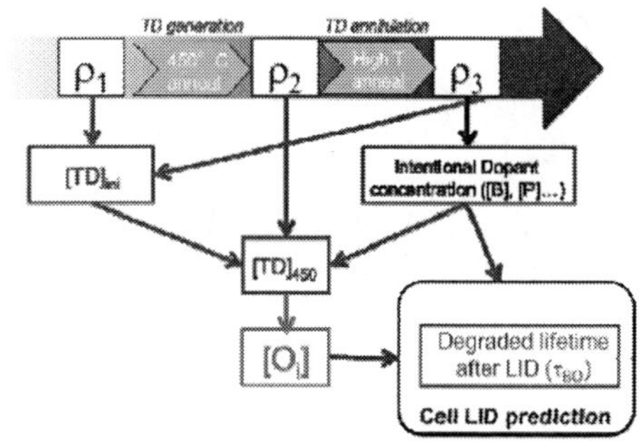

Figure 9: OxyMap measurement procedure. ρ = resistivity. $[TD]_{ini}$ = Thermal Donors concentration formed during growth. $[O_i]$ = Interstitial Oxygen content. LID = Light-Induced Degradation arising from the formation of boron-oxygen complexes in boron-doped wafers.

The interstitial oxygen concentrations were evaluated both by FTIR (Fourier Transformed Infra-Red Spectroscopy) on thick slices at the very tail part of the ingot and by OxyMap on wafers selected close to the FTIR samples. The comparison is given in Table II.

Table II: Comparison of $[O_i]$ measured using FTIR ASTM F121-83 and OxyMap calibrated on SEMI MF 1188 – 1107 and corrected to ASTM F121-83.

Group	$[O_i]$ ppma (FTIR ASTM F12183)	$[O_i]$ ppma (OxyMap)
I	10.94	10.59
II	10.75	10.17
III	10.69	11.53
IV	>12	12.87

The measurements comparisons were done at the center of the ingot (and wafer). We can notice a very good agreement between both measurement techniques. The remaining slight differences could first be explain by a difference of location between OxyMap and FTIR samples and thus be justified by common fluctuations of the $[O_i]$ through the ingot height. In addition, there is a difference between the FTIR standard used to calibrate OxyMap measurements (SEMI MF 1188-1107) and the one used for the FTIR measurements (ASTM F121-83). In order to match both standards, a correction factor of 0.78 was applied to OxyMap $[O_i]$ values. However, because the differences between both standards are not only limited to a difference on the calibrations factors (but also on signals treatments specifications) slight differences on $[O_i]$ could remain.
Tables III and IV are presenting the LID loss results, respectively for the predicted and measured values.

Table III: Prediction of light induced degradation in PERC solar cells using OxyMap technique

Group	ΔJ_{sc}(%)	ΔV_{oc}(%)	FF (%)	$\Delta \eta$(%)
I	-0.39	-0.96	-0.45	-1.80
II	-0.37	-0.91	-0.44	-1.71
III	-0.57	-1.31	-0.35	-2.22
IV	-0.70	-1.57	-0.38	-2.65

Table IV: Measured I-V parameters of solar cells after LID at 25°C.

Group	ΔJ_{sc}(%)	ΔV_{oc}(%)	FF (%)	$\Delta \eta$(%)
I	-0.90	-0.68	-1.88	-3.51
II	-1.06	-0.84	-1.83	-3.73
III	-1.27	-1.39	-1.45	-4.05
IV	-1.48	-1.39	-1.18	-4.02

First we can notice that both predicted and measured LID loss follow the same trend. Indeed, the higher the interstitial oxygen and boron concentrations, the higher the degradation. However, a rather constant difference of efficiency loss appears between the predictions and the measurements. In this study, the OxyMap model seems to underestimate the LID loss. This difference in efficiency loss is mainly coming from J_{sc} and FF loss underestimations.

The observed differences might be due to discrepancies between the semi-empirical model (based on 1-dimension numerical simulations taking into account the properties of the PERC cells fabricated at CEA-INES), used in OxyMap predictions and the real cell architecture used for these experiments. Thus, the sensitivity of the lifetime degradation associated to the formation of B-O complexes on the cells performances might be different. However, in order to overcome this difference, correction factors are implemented in OxyMap tool in order to reduce the noticed amplitude difference. Consequently the user can recalibrate the original model in order to match his own cell behaviors when the material undergoes carrier lifetime degradation due to the B-O complex formation. Figure 10 depicts the relative efficiency loss correction that could be implemented in the tool, as a function of the [B] and the $[O_i]$.

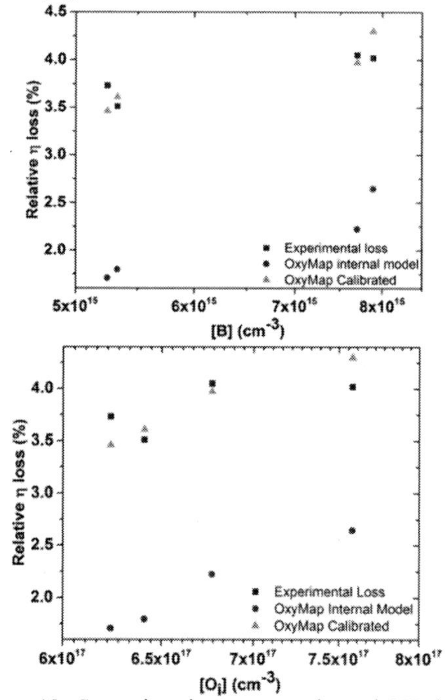

Figure 10: Comparison between experimental LID loss, OxyMap internal model predictions and OxyMap calibrated model.

After calibration, the OxyMap models suits well to the experimental loss. The slight remaining differences are coming from the fact that the model is keeping the proportionality between the LID loss and the [B] and the $[O_i]$, whereas experimental data slightly deviate from this physical trend.

Another aspect to be pointed out is the significant amplitude of the measured LID loss (close to 1% absolute efficiency loss), even if the bulk silicon feature rather low interstitial oxygen and boron concentrations. Other degradation mechanisms might occur during the LID test which would enhance the influence of the lifetime degradation of the cells electrical performances.

7 SUMMARY AND CONCLUSION

We process and characterize PERC solar cells with an emphasis on LID. As expected higher interstitial oxygen concentration and dopant concentration is leading to larger LID loss. However, the 'Regeneration process' carried out on PERC solar cells bring back the cell efficiencies close to the initial levels, and the regained cell efficiencies remain stable upon further LID process. The OxyMap Characterization technique was tested on twin wafers. The correlation between FTIR and OxyMap $[O_i]$ values in good. Even though the predicted LID by OxyMap is in trend with the measured values; there is a slight deviation between both which could be explain by deviations between the simulated and the fabricated PERC cells. However, corrections factors are proposed in OxyMap tool in order to adapt the initial model to the process specifications.

8 ACKNOWLEDGEMENTS

Authors would like to thank Dr. Thomas Pernau for fruitful discussions, Rafael Mrczak and Lejlja Hildebrand for processing of PERC cells, and Sascha Grani for characterization.

9 REFERENCE

[1] Photovoltaics report, Fraunhofer ISE, 2015 www.ise.fraunhofer.de p.2

[2] Martin A. Green. The Passivated emitter rear cell (PERC): From conception to mass production. Solar Energy Materials and Solar Cells 143 (2015) 190 -197.

[3] S. Scmitt, C. Peter, S. Olibet, J. Theobald, A. Schneider. Influence of various process variations on the fabrication of PERC solar cells. 28th European Photovoltaic Solar Energy Conference and Exhibition 2013.

[4] E. Urrejola, R. Petres, J. Glatz-Reichenbach, K. Peter, H. Plagwitz, G. Schubert, *Proceedings 26thEU PVSEC*, Hamburg (D), 2011.

[5] A. W. Blakers, A. Wang, A. M. Milne, J. Zhao, and M. A. Green, Applied Physics Letters, vol. 55, 1363 (1989).

[6] http://www.pse.de/test-equipment/photovoltaic-modules/solar-simulator-for-cells/

[7] Veirman J, Dubois S, Enjalbert N, Lemiti M. A Fast and Easily Implemented Method for Interstitial Oxygen Concentration Mapping Through the Activation of Thermal Donors in Silicon. *Energy Procedia 8* 2011; **8**, 41–46, DOI: 10.1016/j.egypro.2011.06.099.

[8] Martel B, Veirman J, Cascant M, Enjalbert N, Tomassini M, Peyronnet R, Stadler J, Fayard E, Dubois S, Raymond G, Brun X, Bonnard P. Spatial characterization of interstitial oxygen and its related defects in Czochralski silicon wafers and ingots: a way to improve the material and device quality. *IEEE Photovoltaic Specialists Conference* 2015; pp.1-3, DOI: 10.1109/PVSC.2015.7355686

[9] S. Glunz, S. Rein, W. Warta, J. Knobloch, W. Wettling On the degradation oof Cz-Silicon solar cells Proceedings 2nd WCPEC, 1998, 1346-3.

[10] F. Wolny, T. Weber, G. Fischer, A. Herguth, S. Wilking Enhanced stable regeneration of high efficiency Cz-PERC cells. Energy Proceedia 77 (2015) 546 – 550.

[11] A. Herguth, G. Schubert, M. Kaes, G. Hahn, Proceeding IEEE conference (2006)

[12] S. Wilking, C. Beckh, S. Ebert, A. Herguth, G. Hahn Solar Energy Materials and Solar cells 131, 2014, 2-8.

[13] K. Bothe, R. Sinton, J. Schmidt, "Fundamental Boron-Oxygen-related Carrier Lifetime Limit in Mono and multicrystalline Silicon" *Progress in Photovoltaic: Research and Application* 2005; 13:287–296

WHY MULTI-BUSBARS AND FUTURE EMITTERS REQUIRE FURTHER SHRINKING OF FINGER LINE WIDTH

Lejo J. Koduvelikulathu[(a)], Jan Lossen[(a)], Dominik Rudolph[(a)], Misha Matusovksy[(b)], Giora Dishon[(b)]
(a) International Solar Energy Research Center, Konstanz, Rudolf Diesel Str-15, 78467-Konstanz, Germany
(b) Utilight Ltd.,Yavne, Israel

ABSTRACT: Co-development of silver printing pastes, screens with finer meshes, and advanced printing machines, enabled for a steady shrinkage of finger widths. Advancement on glass frits allowed for contacting of lower doped emitters, improving the cell parameters from the better passivation and low Auger recombinations of these advanced emitters. We evaluate in this paper, what the optimum finger line width would be, for cells having high number of busbars and advanced emitters. A grid optimization study is performed on standard reference p-type PERC using GRIDDLER software, varying finger line widths and aspect ratio for the different number of busbars. *An optimum finger width is obtained for a given number of busbar and finger aspect ratio, beyond which further smaller finger line widths, the cell efficiency drops due to the large number of fingers required which results in increased optical shading and voltage losses.* With advance emitters (high sheet resistance) similar observations are made. In other words, it is advantageous to have fine finger line widths for cells with high number of busbars and advanced emitters. Further, we show using the patented novel Pattern transfer technology (PTP™) very small finger line widths with good aspect ratio can be achieved. PTP™ is a non-contact printing technology based on laser transfer hence has the potential to overcome the limitations of screen printing technology and increase cell performance in the future.

Keywords: Simulation, Grid optimization, Fine line printing, Busbars, GRIDDLER

1 Introduction

Screen printing is the dominant industrial technique to define fingers and busbar metallization of solar cells since more than two decades. The technology is simple, robust and cost-effective. Co-development of silver printing pastes, screens with finer meshes, and more advanced printing machines, enabled for a steady shrinkage of finger width [1]. However, for cells with 4 or 5 busbars (BB), it was calculated recently [2] that it is not advantageous to have fingers with cross sectional area below 300 µm^2. Advancements on the glass frits allowed for contacting of lower doped emitters, improving the cell parameters from the better passivation and low Auger recombinations. Though, further reductions of emitter losses come at the expense of increased lateral resistance arising from the lower doped regions, and thus require the finger pitch to be reduced further. On the other hand, interconnection schemes with a higher number of busbars (BB) became viable [3]. Many companies already adapted their production lines to four or five busbars, but also higher number of bus bars and grids of multi busbars are used [4]. This relaxes the requirements in regard to line conductivity. We investigate how this shifts the optimum finger line width anew towards much finer lines, and how Pattern-Transfer Printing (PTP™), a patented technology developed and commercialized by Utilight, particular eligible for depositing ultra fine lines, can increase cell performance in the future.

2 APPROACH

2.1 Grid model of standard PERC cell

Grid optimization study is performed using GRIDDLER simulations [5]. An in-house produced standard passivated emitter rear contact (PERC) cell is used as the reference, with cell efficiency similar to what is found in an industrial line nowadays. The GRIDDLER model is set using experimental cell parameters. The experimental cell and grid parameters of the standard PERC cell are summarized in Table 1.

Table I: Experimental parameters of modeled PERC

Experimental electrical parameters	
Jsc (mA/cm^2)	38.9
Voc (mV)	655
FF (%)	78.9
Eta (%)	20.1
Emitter sheet resistance	80 Ω/□
Emitter recombination current (J_{oe})	120 fA/cm^2
Emitter contact recombination current (J_{metal})	1200 fA/cm^2
Experimental grid parameters	
Number of Busbar (BB)	3BB, 1.1 mm wide
Line Resistivity (R_L)	0.45 (Ω/cm)
Contact resistivity (ρ)	3.4 mΩ-cm^2
Measured front finger width	60 µm
Aspect ratio (AR)	0.25

2.2 Reference PERC grid optimization

With grid model of the experimental reference PERC cell parameters using GRIDDLER-PRO software, simulations of IV parameters are performed. Figure 1 shows the optimization of the number of front fingers (#Fingers) for the experimental reference PERC cell.

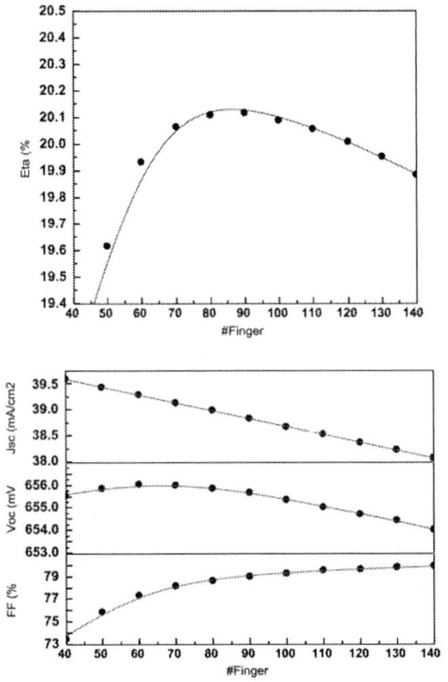

Fig.1: Grid optimization run: IV cell parameters variations for varying #Fingers. The optimum number of finger of this cell is around 85 fingers.

With increasing number of fingers, the shading increases, and therefore short current density (Jsc) decreases. Further as the fraction of metalized region increases, the resulting open circuit voltage (Voc) drops [6]. However, with increase in number of fingers, the grid resistance and the lateral resistance in the emitter reduce, improving the fill factor. Thus, optimum number of finger is a tradeoff between Jsc, Voc and FF.

Analogous to this optimization of the reference PERC cell, a series of front grid optimization simulations scenarios are performedwith varying aspect ratio, finger line width and number of busbars. Our aim is to figure out the optimum finger widths in each scenario. Additionally, grid and finger width optimization runs are done on a modelled PERC cell with an advanced emitter of higher sheet resistance and reduced recombination current, as it is expected that the trend of future industrial trend will go in this direction [7-8]. The model assumes emitter sheet resistance of 120 Ω/□ and Joe 70 fA/cm^2.

3 SIMULATION RESULTS

3.1 Effect of aspect ratio

Considering the reference PERC (3BB, 60µm wide front fingers) with better aspect ratio, cell efficiency improves from the better finger line resistivity. At the same time, paste consumption also increases. The realization of higher aspect ratio is limited by technology. Figure 2 summarizes the efficiency dependence with varying front finger aspect ratio.

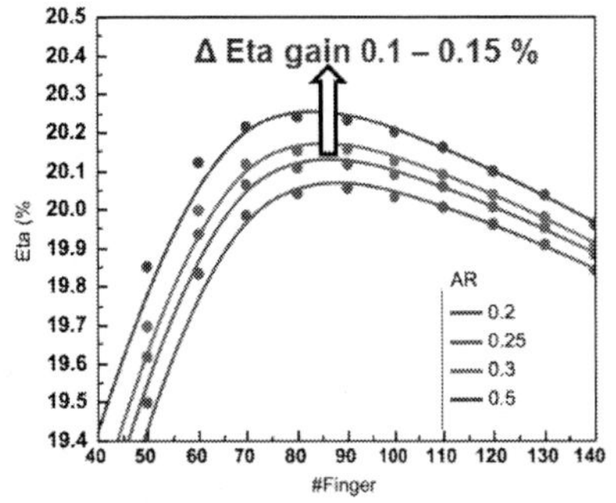

Fig. 2: Efficiency dependence for different aspect ratio

3.2 Effect of increase in number of busbars

Figure 3 summaries efficiency dependence considering 6 and 18 BBs. Additionally, for each busbar the effect of aspect ratio variation is evaluated. Similar percentage of busbar shading is assumed for 3 and 6 BBs. For the 18BB, an effective shading percentage is considered, due to the inherent nature of such busbar architecture [4], however, in terms of absolute shading, the value is still higher than for 3 BB.

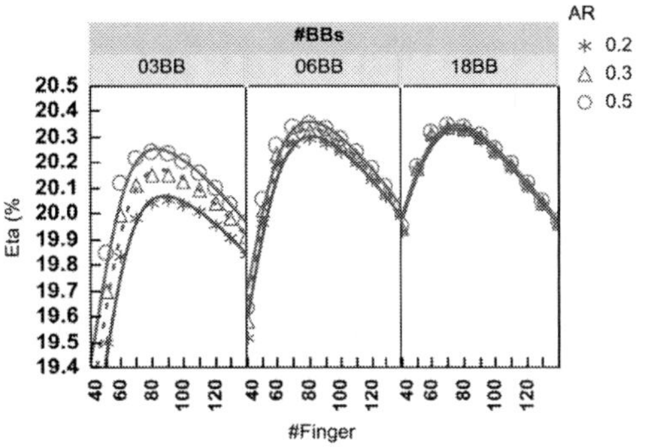

Fig.3: Efficiency dependence varying number of busbars and front finger aspect ratio. (60µm wide fingers)

With increasing number of busbars, effective finger length reduces, hence series resistance reduces [9]. With high number of busbars, due to reduced series resistance, the cell efficiency is optimum at a slightly reduced number of fingers compared to the reference cell. This is also illustrated in Figure 4, showing the optimum number of fingers required at the simulated optimum cell efficiencies, for the different number of busbars and aspect ratio. Also, with higher number of busbars, the dependence on aspect ratio is reduced drastically.

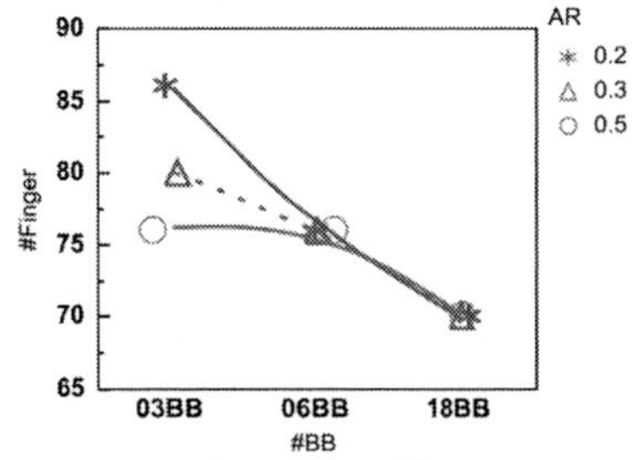

Fig. 4: Decrease in number of fingers at optimum efficiencies for high number of BBs

3.3 Effect of smaller finger line width

Next the effect of line width reduction is investigated assuming that it is possible to maintain the aspect ratio. Finger line resistivity then depends inversely to the square of its width. Therefore, for narrow finger widths, a high number of fingers is required. Past the optimum finger width, the cell efficiency drops again due to increased finger shadings and voltage losses. As shown in figure 5, for a 3BB cell, the optimum finger line width is in the range of 40 to 50 µm, for the considered aspect ratio. The optimum efficiency and finger line width is evaluated in the following section for higher number of busbars.

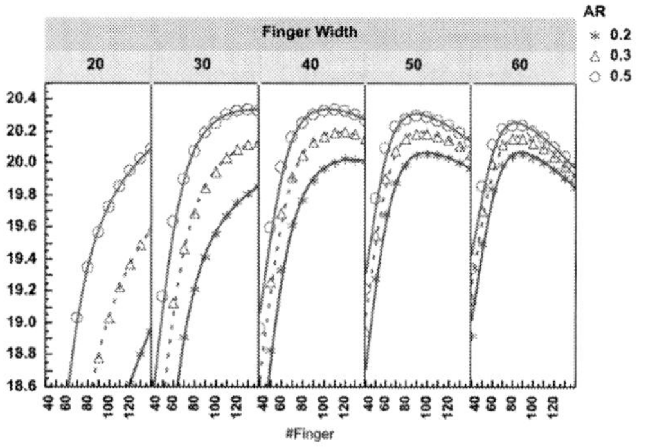

Fig. 5: Efficiency dependence on varying finger line widths

3.4 Effect of small finger line width with varying number of busbars.

Reducing finger widths requires a high number of fingers, as shown in figure 6. For a low number of busbars the increase in the number of fingers is drastically, but for more busbars the optimum number of fingers is only moderate to values around 140 fingers.

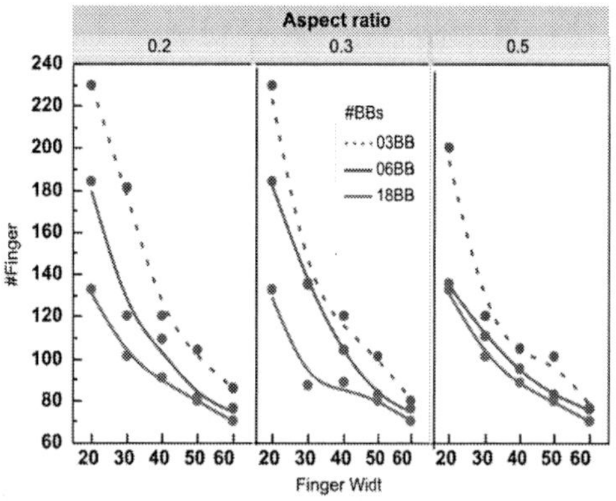

Fig 6: Simulated number of fingers with reducing finger line widths

Figure 7 shows the simulated efficiencies, at the optimum number of fingers, for different finger line widths, aspect ratio and number of busbars. With increase in number of busbars maximum cell efficiencies are reached at smaller finger line widths. For the 6BB case the optimum finger width is, depending on AR, between 30 and 20 µm, for the 18 BB case the optimum is even lower 20 µm .

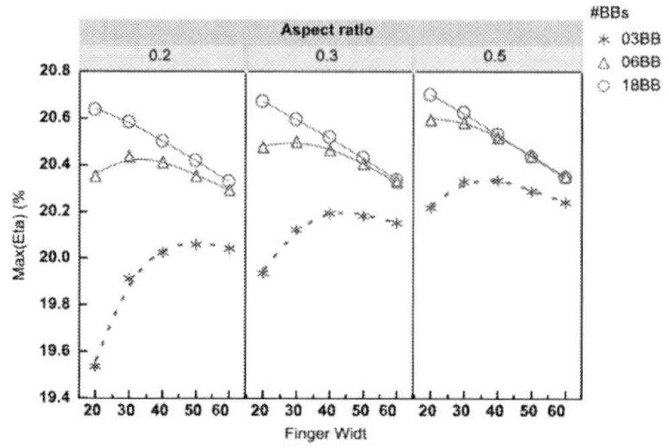

Fig. 7: Simulated efficiency: Dependence on finger line widths, aspect ratio and number of busbars.

Additionally, the dependence on aspect ratio is also reduced. Figure 8 shows the optimum finger line widths at optimum cell efficiency.

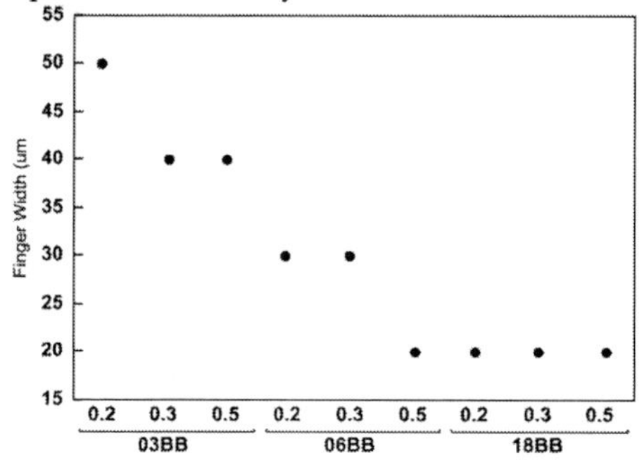

Fig. 8: Optimum finger line widths for varying number of BBs and aspect ratio, within the calculated parameter range of 20 to 60 µm

A potential increase of at least ~ 0.6 % $_{(abs)}$ in cell efficiency is predicted comparing 3BB and 18BB cells. Thus having smaller finger width is desirable and advantageous for high number of busbars.

In terms of front finger Ag paste lay-down, with the optimum finger line width, for a given number of busbar, a potential of 30 – 70% reduction in Ag consumption is calculated, comparing reference 3BB with 6BB and 18 BB respectively. Figure 9 shows the front finger Ag paste lay-down, considering an aspect ratio of 0.3.

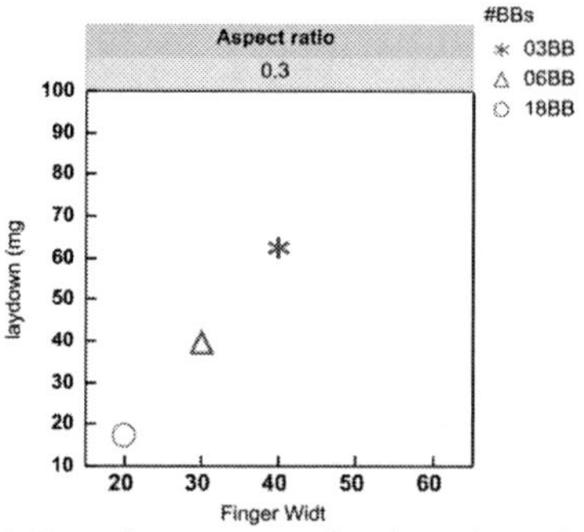

Fig. 9: Front finger Ag paste lay-down for optimum finger widths and number of fingers with varying number of busbars

3.5 PERC grid model with advance future emitters

Within the ITRPV roadmap 2015, industrial trend of having advanced emitters with high emitter sheet resistance is foreseen. With high emitter sheet resistance, better passivation and low emitter recombination current is obtained.

With this perspective, we model the reference standard PERC taking into account for an emitter sheet resistance (R_{sheet}) of 120 Ω/□. An emitter recombination current (Joe) of 70 fA/cm^2 is assumed for the simulation study [8]. Another series of grid optimization is performed varying aspect ratio, reducing finger widths and number of busbars, for the optimum finger width.

The reductions of emitter recombination losses, come at the expense of increased lateral resistance arising from the lower doped regions, and thus require the finger pitch to be reduced further, and therefore increase in the number of required optimum fingers. This is illustrated in Figure 10, comparing reference (80 Ω/□) with 120 Ω/□.

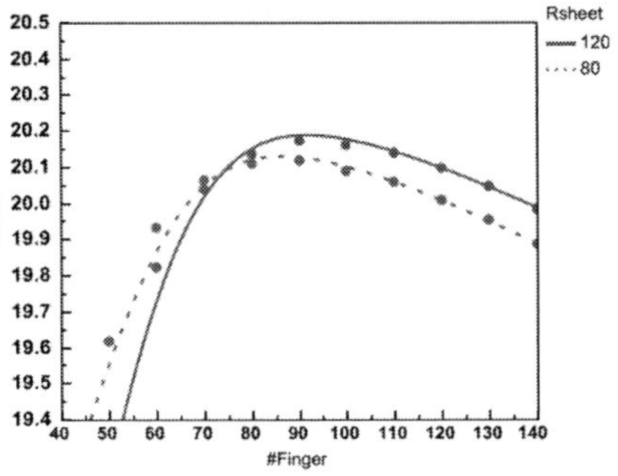

Fig. 10: Grid optimization of the reference and a cell with increased sheet resistance.

With a high efficient reference PERC, the gain in efficiency by reducing emitter recombination would probably be higher, compared to the PERC reference considered in this study, as the present reference cell is limited in the bulk and rear side passivation.
Figure 11 summarizes efficiency dependence for the two emitters with varying finger line widths, aspect ratio and number of busbars.

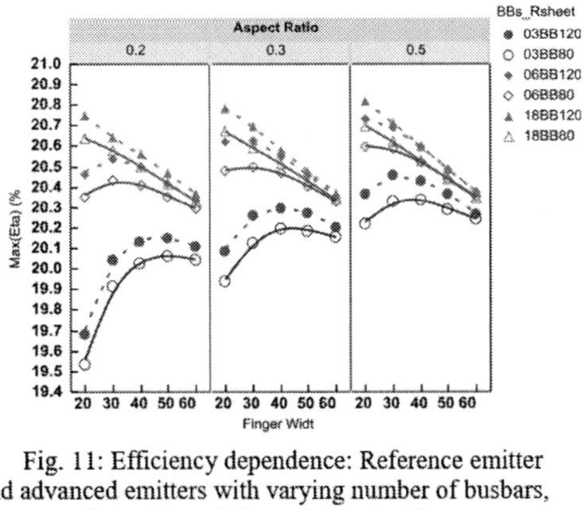

Fig. 11: Efficiency dependence: Reference emitter and advanced emitters with varying number of busbars, finger line widths and aspect ratio.

With high sheet resistance emitters, the optimum finger line width tends to shift towards smaller line width, as it can be seen in Figure 11 for the 3BB case with aspect ratio 0.5. For high emitter sheet resistance, with high number of busbars the number of optimum required fingers can be minimized and additional efficiency gains obtained from smaller finger line widths, due to reasons explained in section 3.4. A potential increase of at least ~ 0.7 % $_{(abs)}$ in cell efficiency is predicted comparing 3BB and 18BB cells. This emphasizes the need for advance high sheet resistance emitters in further reducing the finger line widths for increased number of busbar.

4 PATTERN TRANSFER PRINTING (PTP^{TM})

A potential printing technology for overcoming the limitations of screen printing technology and realizing smaller finger widths with good aspect ratio is Pattern transfer Printing (PTP^{TM}). PTP^{TM} is a non-contact advanced metallization printing technology patented, developed and commercialized by the company Utilight [10]. The technology is based on laser induced deposition from a polymer substrate, and allows for much finer, uniform finger line widths with good aspect ratio.

Figure 12 shows laser scanning microscopy images of a PTP^{TM} printed finger lines with opening width < 20μm wide. Measured average finger width is 25μm wide with an aspect ratio of 0.45. A cross-section (CS) area of ~ 200μm^2 has been measured. Thus, realization of smaller front finger line widths with high aspect ratio and low front CS area is possible.

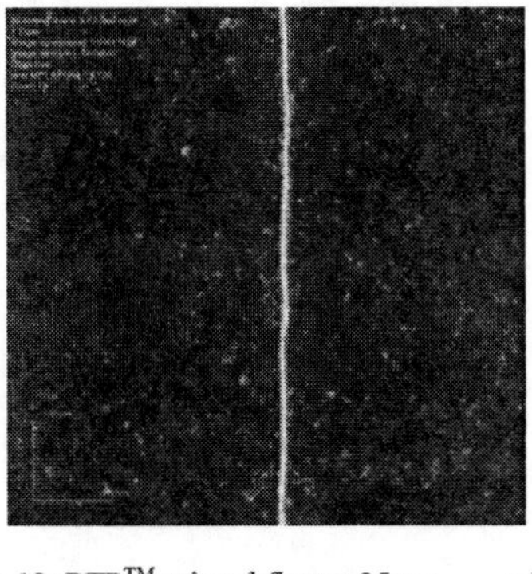

Fig.12: PTP^{TM} printed finger: 25μm $_{(avg.)}$ wide fingers, AR: 0.45, CS: ~200μm^2

4 CONCLUSION

From the grid optimization simulation studies on a reference PERC cell performed using GRIDDLER, with varying number of busbars, finger line width and aspect ratio, the optimum cell efficiencies are realized at finer finger widths with increase in the number of busbar. Additionally, the dependence of aspect ratio reduces when going to high number of busbars. With advanced emitters of high emitter sheet resistance, required number of fingers increases. However, with high number of busbars, the optimum efficiencies are obtained at smaller finger line widths. Figure 13 presents briefly a short summary of the results and conclusion from the grid optimization studies presented in this paper.

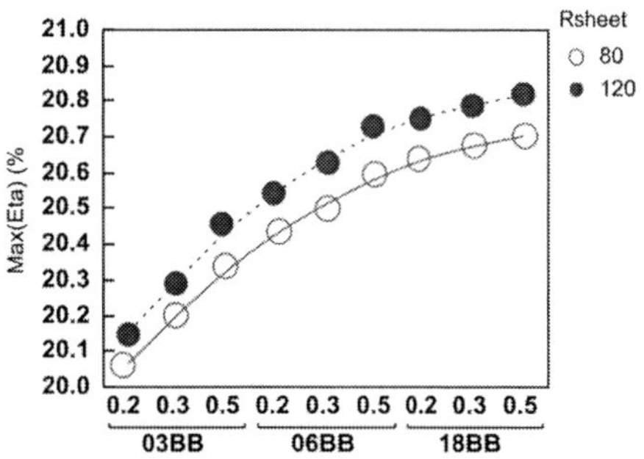

(a) Optimum simulated efficiency

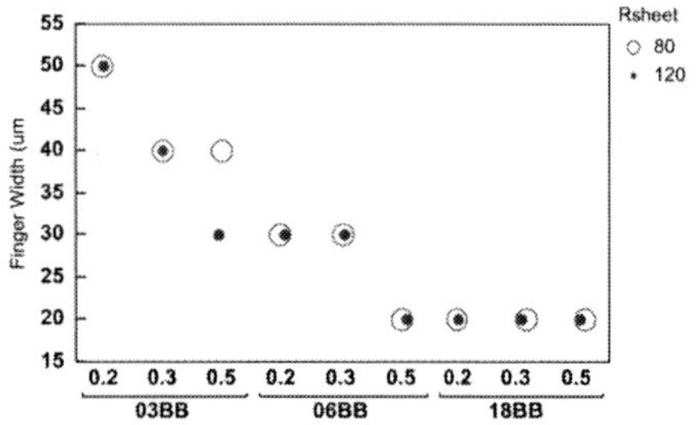

(b) Optimum finger line widths for the simulated optimum efficiencies.

Fig. 13: Summary of the presented simulation study

In other words, for multi-busbar and advanced emtters, it is advantageous to have fingers with fine fingers widths. PTP^{TM} technology demonstrates potential capabilities in overcoming the limitations of screen printing technology. Possible realization of of smaller finger line widths with high aspect ratio with low cross sectional area is demostrated.

5 ACKNOWLEDGMENT

The work is financed under Project KosmoS, Code: 0325822E. The authors acknowledge and would like to thank the German funding agency: German Federal Ministry for Economic Affairs and Energy (BMWi) and the support from Utlight Ltd.

The authors would also like to thank ISC colleagues for the fruitful discussion and suggestions on this topic

6 REFERENCES

[1] Arno Stassen & Weiming Zhang, White paper, Heraeus Precious Metals, www.heraeus.com
[2] Y. Chen et.al, 6th Metallization workshop, 2016
[3] A. Faes et.al, 29th EU-PVSEC, pp.2555-2561
[4] T. Söderström et.al, White paper, Meyer Burger
[5] J. Wong, Proc. 39th IEEE PVSEC, 2013 & J. Wong & R. Sridharan, Griddler 2.5 Pro, http://www.seris.sg/Seris/OurServices/Griddler.html
[6] A. Edler et.al. Prog. Photovolt: Res. Appl. 2015; 23
[7] B. Min et. al., 31st EU-PVSEC, 2015
[8] International technology roadmap 2015 results
[9] S.Braun et. al, ," Energy Procedia, vol. 38, 2013
[10] J. Lossen et. al , 5th Metallization workshop, Energy Procedia 67, 2015

MODELLING THE LONG-TERM BEHAVIOUR OF BORON-OXYGEN DEFECT PASSIVATION IN THE FIELD USING TYPICAL METEOROLOGICAL YEAR DATA (TMY2)

Brett Hallam, Jose Bilbao, David Payne, Catherine Chan, Moonyong Kim, Daniel Chen, Nicholas Gorman, Malcolm Abbott and Stuart Wenham
School of Photovoltaic and Renewable Energy Engineering, UNSW Australia
Tyree Energy Technologies Building, UNSW, NSW 2052, AUSTRALIA

ABSTRACT: The theoretical long-term stability of modules with prior boron-oxygen defect passivation and the in-situ passivation of boron-oxygen defects in the field are investigated using typical meteorological year (TMY2) data for locations of Hamburg, Sydney, Tucson and Wuhan. Results show that a worst-case scenario of the long-term stability for ground-mounted systems installed in desert locations, destabilisation of the passivated boron-oxygen defects is of no concern within 40 years. It is also shown that roof-top photovoltaic systems with a 100 mm air gap and building integrated photovoltaic systems with no ventilation will be more prone to a potential destabilisation of the passivated defects rather than large centralised field arrays with ground-mounted systems due to the higher module operating temperatures. The modelling presented is supported by experimental data, with no de-stabilisation occurring for extended light-soaking (1000 hours at 70 °C or 400 hours at 160 °C). A best-case scenario is also shown for the potential of defect passivation in the field, showing a strong dependence on the location, mounting structure, and date of installation. In open circuit, a rapid passivation can occur within days for locations such as Sydney if installed in summer, whilst in cool climates, passivation can take several months.
Keywords: light-induced degradation, carrier-induced degradation, boron-oxygen, hydrogen passivation, stability

1 INTRODUCTION

Crystalline silicon solar cells fabricated on boron-doped Czrochralski (Cz) grown silicon substrates are subject to a degradation of performance when exposed to carrier injection [1]. In 2006, Herguth *et al.* presented a method to passivate the boron-oxygen (B-O) defects based on illuminated annealing, hence permanently eliminating the B-O related carrier-induced degradation (CID) [2]. For conventional screen-printed solar cells with a full area aluminium back-surface-field (Al-BSF), efficiency losses can approach 1% absolute [3]. However, the percentage of degradation can also depend on the solar cell structure. High efficiency solar cell structures such as the passivated emitter with localised rear contacts (PERC) solar cell have a stronger dependence of efficiency on the bulk lifetime within the silicon. Therefore, high efficiency solar cell architectures are prone to larger potential efficiency losses due to CID, up to 10% relative [4]. With predictions of the photovoltaic (PV) industry being dominated by PERC solar cell structures in future years [5], B-O related CID could have a significant impact on global PV electricity generation.

By 2025, predictions anticipate a global installed capacity for PV of more than 1 TW, with approximately 1500 TWh of PV based electricity generation [6]. Assuming a 30% usage of silicon substrates subject to the formation of B-O defects with a 5% relative performance loss, this could equate to an additional 22.5 GWh of electricity being required from other electricity sources. If such electricity is generated using fossil fuels, this could equate to a reduction in the potential mitigation of greenhouse gas emissions of approximately 20,000 tonnes CO_2 equivalent (CO_{2e}) (based on an average of 888 tonnes CO_{2e} per GW of electricity production from coal [7]).

Furthermore, warranties issued by solar panel manufacturers require that a certain level of performance be maintained for typically 25 years. Due to the potential destabilisation of passivated B-O defects in the field, questions over long-term stability have been raised for PERC based modules. Here we perform a worst-case scenario numerical simulation of photovoltaic modules in operation. This is based on experimentally obtained values for the destabilisation reaction, and module temperatures obtained from simulations on System Advisor Model (SAM) (version 2015.06.30) from the National Renewable Energy Laboratory (NREL) [8-9] using typical meteorological year data (TMY2). We also extend the study to investigate the potential of passivating B-O defects in the field using the same principle.

2 METHODOLOGY AND EXPERIMENTAL DETAILS

2.1 Mathematical model

In this work, we adopt a 3-state model for the B-O defect system as has been used by several authors in the past [10-13]. This model uses reactions of defect formation, defect dissociation, defect passivation and destabilisation. The attempt frequency from state i to state j (ν_{ij}), where i,j = A,B or C, and corresponding activation energy (Ea_{ij}) for the reactions are shown below in Table I.

Table I: Reaction rates in the boron-oxygen defect system

Mechanism	ν (/s)	E_a (eV)	Ref.
Defect formation	4×10^3	0.475	[14]
Defect dissociation	1×10^{13}	1.32	[14]
Defect passivation	1.25×10^{10}	0.98	[15]
Destabilisation	5×10^9	1.25	[15]

State A is the 'dark annealed' recombination inactive (unstable) state, B is the recombination active 'degraded' state and C is the 'passivated' recombination inactive (stable) state. In the literature, experimental demonstrations of permanent recovery of B-O related CID has been preceded by degradation (e.g. [2]). Accordingly, the model requires defect formation before passivation, with no direct transitions between states A and C. The fraction of the defect population at any given time is denoted by ($N_i(t)$). A diagram of the B-O defect system is shown in Fig. 1. The corresponding ordinary different equations (ODE) system is shown in Equations 1-3, with the equation for the reaction rate (k_{ij}) from state i to state j shown in Equation 4.

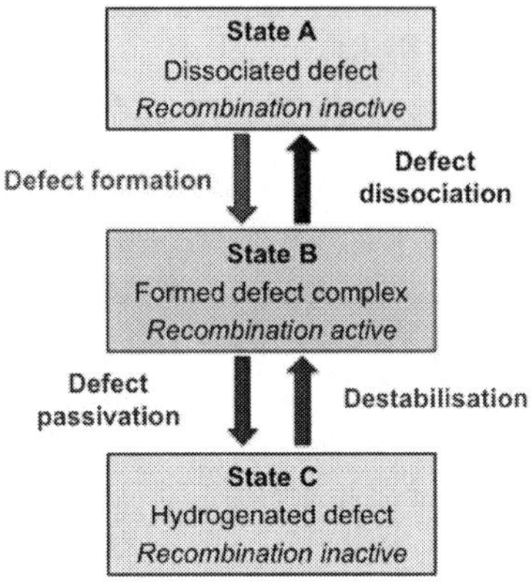

Figure 1: Diagrammatic representation of the B-O defect system.

$$\frac{\delta N_A}{\delta t} = \kappa_{BA} \cdot n_B - \kappa_{AB} \cdot n_A \quad (1)$$

$$\frac{\delta N_B}{\delta t} = \kappa_{AB} \cdot n_A + \kappa_{CB} \cdot n_C - (\kappa_{BA} + \kappa_{BC}) n_B \quad (2)$$

$$\frac{\delta N_C}{\delta t} = \kappa_{BC} \cdot n_B - \kappa_{CB} \cdot n_C \quad (3)$$

$$\kappa_{ij} = v_{ij} \cdot e^{\frac{-E_a}{k_b T}} \quad (4)$$

The reactions in the B-O defect system are extremely temperature dependent; hence, in order to assess the long-term stability of CID mitigation in modules, it is important to understand the operating temperatures of modules in the field. The three key parameters that influence the module operating temperature are the ambient temperature, solar insolation, and module ventilation (dependent on local wind conditions and the module's mounting). All of these differ substantially across the globe and throughout the year.

For the simulations performed in this work, four different locations are considered, as listed in Table II, with very different climates.

Table II: Locations studied in this work

City, Country	Latitude	Longitude
Hamburg, Germany	53.5653°N	10.0014°E
Sydney, Australia	33.8650°S	151.2094°E
Tucson, USA	32.2217°N	110.9264°W
Wuhan, China	30.5833°N	114.2833°E

Typical meteorological year (TMY2) data was used for each of the locations, in hourly intervals for an entire year. Three different mounting schemes for the system are investigated in this work. Firstly, a building integrated photovoltaic system (BIPV) that is mounted flush with the roof tilted at 15°. This system has no airflow at the back of the modules to assist with cooling, and should experience the highest operating temperatures. Secondly, rack-mounted system installed on a roof tilted at 15° (Rack-flush), with a 100 mm air gap at the back of the modules to allow some ventilation for cooling. Thirdly, a ground-mounted system (Ground-mount), tilted at the latitude towards the equator, representative of large centralized PV arrays. This system has full ventilation at the rear of the module to assist with cooling, and hence, should experience the lowest operating temperatures. Simulations for the output power, efficiency and temperature of the modules in operation were obtained using the SAM. Module efficiencies are assumed to be 17.56% with no B-O defects present and a 5% relative loss in efficiency due to CID (0.88% absolute).

Fig. 2 shows a diagrammatic representation of the model used in this work, whereby the TMY2 data and module properties are fed into the SAM. The predicted module operating temperature and reaction rates are then fed into the B-O defect model, taking account a potential dependence of the defect passivation rate on the illumination intensity (I). The output of the B-O model is then used to predict the efficiency (η) as a function of time.

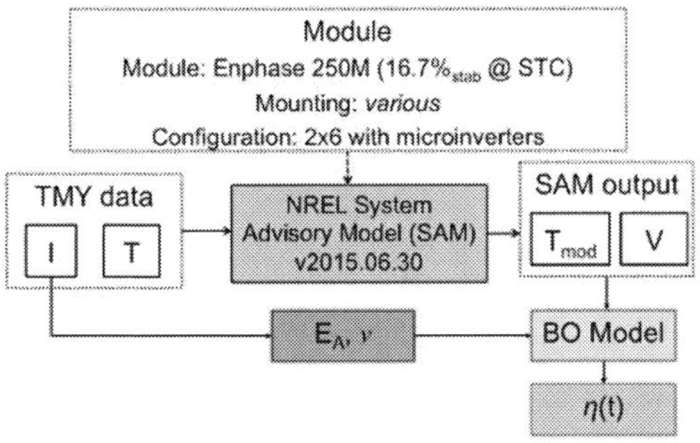

Figure 2: Diagrammatic representation of the model used in this work.

2.2 Experimental details

Test structures for experimental verification of destabilisation and passivation were prepared using commercial grade 156 mm × 156 mm pseudo-square boron-doped Cz wafers (1.6 Ω-cm, N_A = 9.1x10^{15} /cm^3). Wafers were subjected to anisotropic (alkaline) texturing followed by acidic neutralisation, with a resultant total wafer thickness of approximately 180 µm. This was followed by an RCA clean and HF dip, then a light diffusion in a $POCl_3$ tube furnace to achieve a sheet resistance of approximately 250 Ω/□ to aid in surface passivation. Subsequently, hydrogenated silicon nitride (SiN_x:H) with a refractive index of 2.08 at 633 nm was deposited onto both surfaces of the wafers using a Roth & Rau MAiA plasma-enhanced chemical vapour deposition (PECVD) system. Samples were then fired to release hydrogen from the dielectric layers to enable defect passivation. Light soaking was performed under a halogen lamp with a light intensity of approximately 60 ± 1 mW/cm^2/s at either 70 °C (total duration 1000 hours) or 160 °C (total duration 400 hours). Effective minority carrier lifetimes (τ_{eff}) were measured periodically, as a function of minority carrier density using a WCT-120 (Sinton Instruments) quasi-steady-state photoconductance (QSS-PCD) lifetime tester [16]. The measured data was analysed using the generalised technique [17]. Where relevant, lifetimes are given at a single injection level with an excess carrier concentration (Δn) of 9.12 × 10^{14} /cm^3, which is equivalent to 0.1 ×N_A to highlight the degradation due to the formation of boron-oxygen defects.

3 RESULTS AND DISCUSSION

3.1 TMY2 data and module operating temperatures

Example TMY2 data is shown for Sydney in Fig. 3(a), with ambient temperatures of up to 40 °C experienced. Fig. 3(b) shows a distribution of the hourly operating temperatures for ground-mounted systems installed in the various locations. As shown, Tucson has the highest temperature distribution, whilst Hamburg has the lowest temperature distribution. For ground-mounted systems, the operating temperature can reach in excess of 60 °C. The operating temperature of Rack-flush and BIPV systems (not shown) is substantially higher than that of Ground-mount systems, and can reach temperatures in excess of 80 °C.

(a)

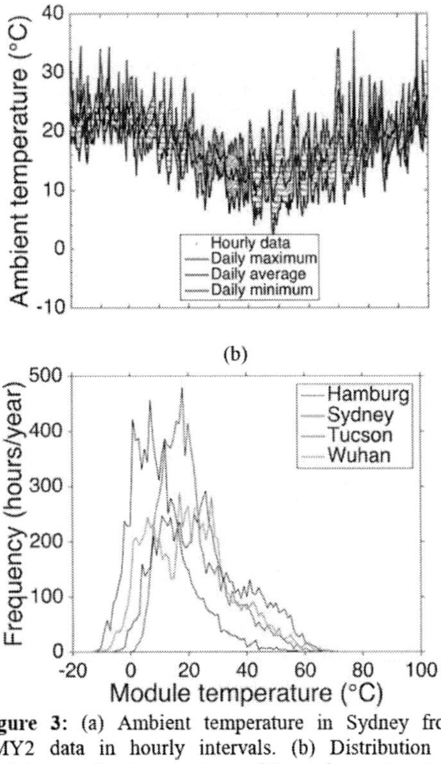

Figure 3: (a) Ambient temperature in Sydney from TMY2 data in hourly intervals. (b) Distribution of module operating temperatures of Ground-mount systems installed in a variety of locations (see legend).

3.2 Long term stability

Due to the different operating temperatures of modules in different locations, the risk of destabilisation is very location dependent. Fig. 4 shows the predicted destabilisation of the passivated B-O defects for the worst-case scenario, where the modules are in the dark. Hence, for this sample, once destabilised, the defects cannot be re-passivated. Here, the extent of destabilisation is limited to less than 0.06% absolute for all studied locations. However, in other mounting structures, such as for BIPV, destabilisation could potentially cause a loss of 0.4% absolute in Tucson (see Ref. [18]).

However, in reality, any condition that causes the modules to heat substantially will also be under illumination and enable the passivation reaction. This provides a natural level of protection against destabilisation. A recent study under typical module operating conditions by Lee et al. [19] demonstrated that on modules that had no attempt to passivate B-O defects prior to encapsulation, the modules would self-repair in the field. In that study, a full recovery was observed at 60 °C in open circuit after < 200 h. A reduced rate of recovery was observed at 50 °C under maximum power point conditions, with 40% of the V_{OC} recovered after 700 h. The rate constants demonstrated in that work are sufficient to maintain passivation of the B-O defect over the lifetime of the module.

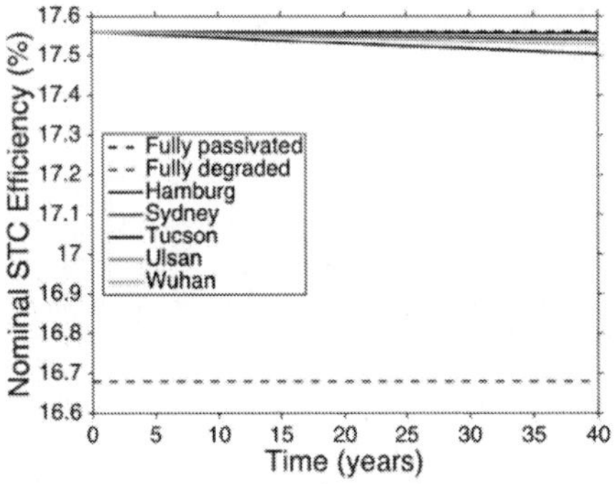

Figure 4: Distribution of module operating temperatures of Ground-mount systems installed in a variety of locations (see legend).

Experimental results for extended light soaking under 0.6 suns at elevated temperatures, give further evidence that any destabilisation of boron-oxygen defect passivation may not be a concern for modules operating in the field. Light soaking fired lifetime test samples (with no prior attempt to form- or passivate the defects) for 1000 hours at 70 °C or 400 hours at 160 °C show an initial degradation of effective minority carrier lifetime as the defects first form and a subsequent recovery of the lifetime attributed to defect passivation. After this recovery, no net destabilisation of the passivated defects occurred. At such temperatures, any net destabilisation would be observed as a reduction in lifetime. Fig. 5 shows an example evolution of minority carrier lifetime as a function of time for 70 °C light soaking for 1000 hours.

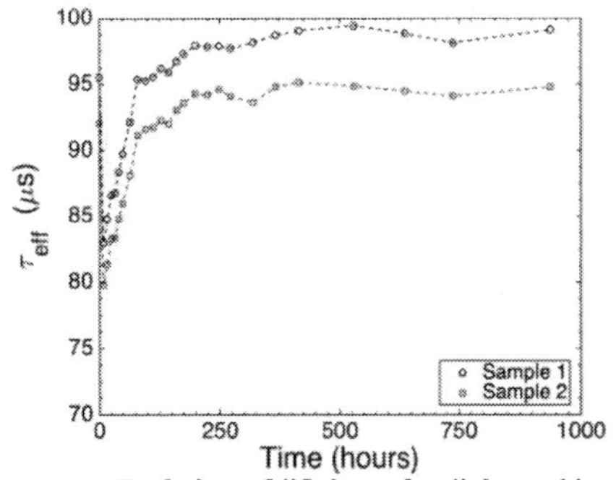

Figure 5: Evolution of lifetime after light soaking (0.6 suns) in open circuit at 70 °C.

For Ground-mount systems in Tucson, the module operates at a temperature of more than 70 °C only one hour per year. However, for building-integrated systems, the modules may operate at a temperature in excess of 70 °C for almost 400 hours per year. Hence for BIPV systems, destabilisation of B-O defects could be a concern if there is an on-going evolution of hydrogen in the silicon to reduce the availability for re-passivation. However, an application with modules operating under high temperatures is not desired in any case, as a result of the loss of efficiency of due to temperature effects (normally around 0.4%/°C relative for c-Si).

3.3 Passivation in the field

Similarly, the time to passivate defects in the field is strongly dependent on the location and type of mounting used for the system. Fig. 6 shows a theoretical simulation for potential passivation in the field in Sydney for a BIPV system assuming an attempt frequency for passivation (ν_{BC}) linearly dependent on the illumination intensity in open circuit conditions. It highlights the initial degradation of the module, which is relatively independent of the date of installation. Within 1 day, a significant degradation of the modules has occurred. In contrast, the recovery of the efficiency due to the passivation of the defects is extremely dependent on the date of installation. For systems installed in summer, for high-temperature days, the majority of defects can already be passivated, whilst for systems installed in the lead up to winter (such as in May), it can take several months to reach the same level of passivation. For systems that have much lower module operating temperatures such as a Ground-mount system in Hamburg, a much longer timescale is observed for the recovery of the efficiency. Regardless of the date of the installation, recovery of the efficiency will take more than 1 year. The differences in reaction rate for passivation purely due to the temperature are much more significant than any assumed influence of illumination intensity. It is expected that the reduced carrier concentration of modules under normal operation compared to that in open-circuit, will greatly reduce the passivation rate of defects. However, in reality, when systems are installed, they may be in open-circuit conditions for several days.

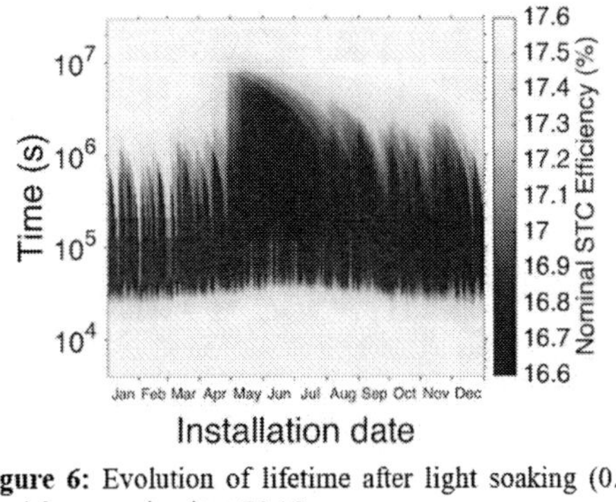

Figure 6: Evolution of lifetime after light soaking (0.6 suns) in open circuit at 70 °C.

4 CONCLUSIONS

Typical meteorological year (TMY2) data and the system advisory model (SAM) were used to investigate the long-term behaviour of boron-oxygen defect passivation in the field. The risk of destabilisation is shown to be location and mounting dependent. As a worst-case scenario (in the dark), whereby no re-passivation of destabilized defects can occur, destabilisation is no concern for ground-mounted systems (such as centralized PV) due to the substantially lower operating temperatures when compared to other mounting methods. Less than 0.06% absolute efficiency degradation would be expected over a 40-year time frame. BIPV and roof mounted systems with a small air gap will be more prone to a potential destabilisation of the passivated defects. However, in reality, an on-going re-passivation of the destabilised defects is likely due to operations in the field under illumination. The modelling presented is supported by experimental data, with no de-stabilisation occurring for extended light-soaking (1000 hours at 70 °C or 400 hours at 160 °C). A best-case scenario is also shown for the potential of defect passivation in the field, showing a strong dependence on the location, mounting structure, and date of installation. In open circuit, a rapid passivation can occur within days for locations such as Sydney if installed in summer, whilst in cool climates, passivation can take several months. Other locations such as Hamburg experience much longer timescales of recovery.

ACKNOWLEDGEMENTS

This Program has been supported by the Australian Government through the Australian Renewable Energy Agency (ARENA) and the Australian Centre for Advanced Photovoltaics (ACAP). The views expressed herein are not necessarily the views of the Australian Government, and the Australian Government does not accept responsibility for any information or advice contained herein. The authors would also like to thank the commercial partners of the ARENA 1-060 advanced hydrogenation project, and the UK Institution of Engineering and Technology (IET) for their funding support for this work through the A.F. Harvey Engineering Prize.

[1] H. Fischer and W. Pschunder, Proceedings of the 10th IEEE Photovoltaic Specialists Conference, 1973, p. 404.
[2] A. Herguth, G. Schubert, M. Käs and G. Hahn, Proc. 4th IEEE World Conference on Photovoltaic Energy Conversion (2006), pp. 940-943.
[3] B. Hallam, P. Hamer, S. Wang, L. Song, N. Nampalli, M. Abbott, C. Chan, D. Lu, A. Wenham, L. Mai, N. Borojevic, A. Li, D. Chen, M.Y. Kim, A. Azmi, & S. Wenham, *Energy Procedia*, **77**, pp. 799-809 (2015).
[4] J. Knobloch, S. Glunz, V. Henninger, W. Warta, W. Wettling, F. Schomann, W. Schmidt, A. Endrös, K. Münzer, Proc. 13th EU PVSEC, 1995, p. 9.
[5] ITRPV Working Group and others. "International technology roadmap for photovoltaics (itrpv.net):Results-2014", 2015.
[6] IEA. "Snapshot of Global PV Markets 2014: Report IEA PVPS T1-26:2015." International Energy Agency, (www.iea-pvps.org) (2014).
[7] World Nuclear Association, "Comparison of lifecycle greenhouse gas emissions of various electricity generation", 2011.
[8] W. De Soto, S.A. Klein and W.A. Beckman, Solar Energy 80(1), 78-88 (2006).
[9] T. Neises, "Development and validation of a model to predict the cell temperature of a photovoltaic cell", MS Thesis, University of Wisconsin-Madison, 2011.
[10] A. Herguth and G. Hahn, Journal of Applied Physics 108(11), 114509 (2010).

[11] B. Hallam, M. Abbott, N. Nampalli, P. Hamer, & S. Wenham, IEEE Journal of Photovoltaics, **6**, pp. 92-99 (2015).

[12] S. Wilking, M. Forster, A. Herguth and G. Hahn. Solar Energy Materials and Solar Cells. **142**, pp. 87-91 (2015).

[13] B. Hallam, M. Abbott, N. Nampalli, P. Hamer, S. Wenham. Journal of Applied Physics **119**, 065701 (2016).

[14] K. Bothe and J. Schmidt, Journal of Applied Physics 99(1), 013701 (2006).

[15] S. Wilking, C. Beckh, S. Ebert, A. Herguth and G. Hahn, Solar Energy Materials and Solar Cells **131**, pp. 2-8 (2014).

[16] Sinton RA, Cuevas A, Stuckings M. Proceedings of the 25th IEEE Photovoltaic Specialists Conference, (1996), p. 457–460.

[17] Nagel H, Berge C, Aberle AG. Journal of Applied Physics. 1999;86(11):6218–6221.

[18] B. Hallam, M. Abbott, J. Bilbao, P. Hamer, N. Gorman, M. Kim, D. Chen, K. Hammerton, D. Payne, C. Chan, N. Nampalli, S. Wenham. Energy Procedia (proceedings of the 6th International Conference on Silicon PV), *accepted for publication*, 2016.

[19] K. Lee M. Kim, J. Lim, J. Ahn, M. Hwang, E. Cho. Proc. 31st European PVSEC, 2015, p. 1837.

22.5% N-PERT SOLAR CELLS ON EPITAXIALLY GROWN SILICON WAFERS

I. Kuzma-Filipek, M. Recamán, F. Duerinckx, R. Hao[1], TS. Ravi[1], E. Cornagliotti, P. Choulat, A. Sharma, J. John, M. Aleman, L. Tous, R. Russell, A. Uruena, J. Szlufcik & J. Poortmans[2]
imec, Kapeldreef 75, B-3001 Leuven, Belgium
[1] Crystal Solar Inc., 3050 Coronado Drive., Santa Clara, CA-95054, USA
[2] imec, Kapeldreef 75, B-3001 Leuven, Belgium; KU Leuven, B-3001 Leuven, Belgium and UHasselt, B-3590 Diepenbeek, Belgium
Corresponding author: phone: +32 1628 8937, fax: +32 1628 1501, e-mail: Izabela. KuzmaFilipek@imec.be

ABSTRACT: This work reports on the fabrication of advanced nPERT (passivated Emitter, Rear Totally diffused) cells by imec on epitaxial wafers from Crystal Solar obtained by direct gas to wafer technology. Crystal Solar has developed a process to grow full-epitaxial monocrystalline wafers of ~180-200μm thick with a limited stacking fault defect density of <1 defect/cm^2. The specific growth sequence included the in-situ growth of optimized p+ epitaxial emitter and n-base. EDNA2 simulations were used to determine the total dark saturation current density for these epitaxial emitters based on depth and doping level. Based on this, the experimental doping range of the epitaxial emitter was between $1 \cdot 10^{17}$ to $5 \cdot 10^{18}$ cm^{-3}. Imec applied its nPERT process that consisted of a selective n++ Front Surface Field (FSF), Al_2O_3 emitter passivation and plating sequence suitable to contact lowly doped surfaces. The devices were evaluated using QSS-PL measurements, spectral response, light IV characterization. The potential impact of epitaxial defects on cell performance was examined.
The best performance was realized for a device with a lowly doped emitter passivated by ALD Al_2O_3/PECVD SiOx reaching a Voc of ~700 mV and an independently confirmed top efficiency of 22.5 % (238.9 cm^2).
Keywords: epitaxy, high-efficiency, n-type

1 INTRODUCTION

Following Paul Basore in his reflections on the future trends in silicon photovoltaics, it has become obvious and necessary to consider both performance and cost in order to focus efforts on developments that will lead to a net decrease in the cost of electricity produced [1]. Kerfless technologies such as gas to wafer technologies [2] could be a potential answer for achieving an improved PV performance without increase in manufacturing cost per unit area. By avoiding the classical steps of producing polysilicon chunks, ingot formation and wafer slicing (with considerable kerf losses), the wafer manufacturing cost can be decreased by approximately 50% [2].

Additionally the technology could offer cost saving on device level by skipping the number of cell processing steps needed to fabricate and isolate the doping regions of the cells. According to our internal calculations, the saving done by using a build-in emitter and other doping layers and related to them savings in cleaning, masking and etching steps would result in a reduced cell manufacturing cost of roughly 40% on cell level. Coupled with the 50% reduced wafer manufacturing cost as detailed above, this would translate in 25% cost savings on the module level.

In the past years epitaxial technology became more recognized in PV field either by its kerfless potential to grow full wafer equivalent directly from the gaseous phase [2] as explained above but also by specific application of epitaxially grown doped layers e.g. emitters, with doping profiles and thickness uniformity tailored down to the nanometer scale [3]. It was shown that ultra-thin or box profile epitaxial emitters can be achieved with J_{0e} values < 5fA/cm^2 and implied V_{OC} at the level of ~750 mV for conventional nPERT (Passivated Emitter and Rear Totally diffused) devices passivated by Al_2O_3 [3]. Recently full area nPERT devices on Cz-Si wafers were achieved with rear epitaxial emitter reaching 694 mV and 22.5% efficiency (independently confirmed by ISE CalLab), confirming the potential of this technology [3]. Similarly to these, very good results were also reported on fully screen printed n-type solar cells on epitaxially grown silicon wafers with built-in Boron rear side emitter [4] or for ultrathin kerfless epitaxial foils for industry-leading high-performance smart PV modules [5]. The latter brings additionally the potential advantages of integrated fabrication of cells and modules, which is a new revolutionary field.
Hence the epitaxial wafer growth technology offers already the advantage of in-situ growth of the emitter layer together with the absorber (base) without the need for additional steps such as glass removal or dopant activation. It goes without saying that epitaxial growth can completely replace the need for high temperature multiple diffusions or drive-in steps in the silicon solar cell fabrication flow. Additionally, since the emitter can be grown without constraints for doping and thickness, no risk of e.g. junction shunting exists due to laser damage or firing through. Moreover, the emitter can be applied on a single side of the wafer without need for masking or isolation steps. Those advantages were used in the definition of a common experiment between Crystal Solar as an epitaxial wafer supplier and Imec as a high efficiency solar cell integrator. The results of the first iteration are presented in this paper.

2 METHODOLOGY

2.1 Fabrication method of the kerfless wafer

The process to fabricate full epitaxial wafers from gaseous phase involves i) anodic etching of a silicon parent substrate to form a porous silicon release layer, ii) epitaxial growth of 180-200 μm silicon including the n-type base and in-situ grown thick p-type emitter, iii) mechanical separation of the epitaxial wafers from the parent substrate and iv) recovering of the parent substrate for its subsequent reuse, as presented in Fig. 1. The resistivity of the base is approximately 5.5 Ohm cm and two sets of emitter doping were tested in the first iteration with Imec, with respectively a "low" and "medium" doping level. Both were in the range of $1 \cdot 10^{17}$ to $5 \cdot 10^{18}$ cm^{-3}. The medium doped emitter as used in this experiment refers to the doping level of $>1 \cdot 10^{18}$ cm^{-3}, whereas the lowly doped emitter refers to the doping $<1 \cdot 10^{18}$ cm^{-3}. The epitaxial wafers were investigated by scanning electron microscopy

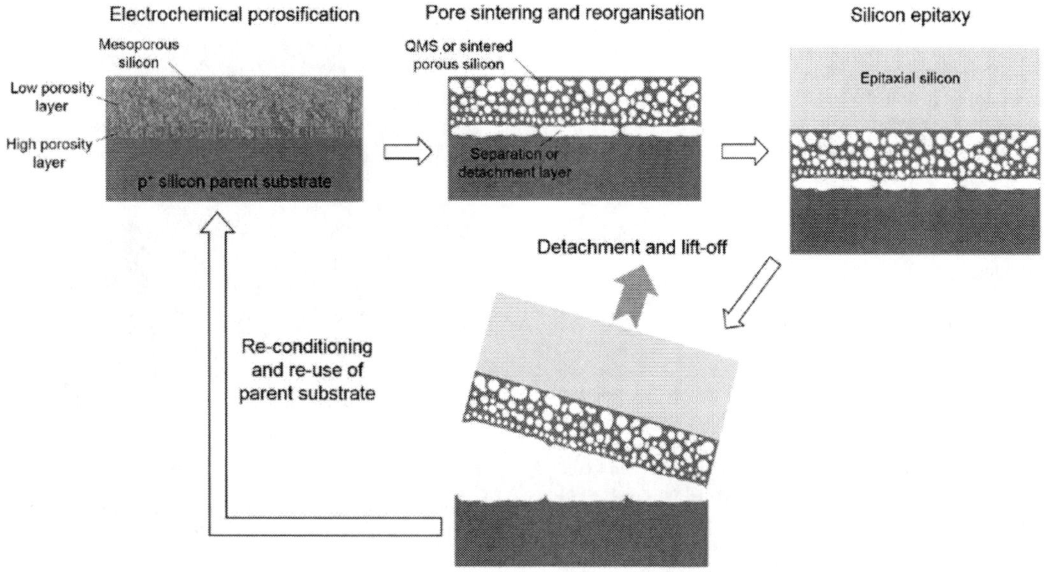

Figure 1 Schematic of the sintered mesoporous silicon-based lift-off of epitaxial silicon similar to the one used by Crystal Solar technology. Drawing not up to scale and after [6].

and transmission electron microscopy (TEM) for their crystalline silicon quality.

2.2 Fabrication method of the devices

Upon entering imec's cell processing line, the epitaxial wafer is double side textured with a single side deep p+ emitter as presented in Fig.2. The emitter side was set to be the rear side of the device. The rear surface is then polished resulting in the final thickness of the emitter of >20 µm for both emitter groups and subsequently the front surface field (FSF) n+ is formed by means of $POCl_3$ diffusion. The FSF is a shallow and lowly doped region present in the passivated area allowing reduction of Auger recombination in that region. The n+ layer is removed from the emitter side and from peripheral regions of the device by wet etching in an inline tool, followed by the dry oxidation step. The final sheet resistance of the FSF at this stage is >350 Ohm/sq. The antireflection SiN_x layer is deposited on the front with the minimum reflectivity of 0.12% at 650nm. As shown in figure 2, the next step required removal of the thin oxide from the rear and application of Al_2O_3/SiO_x as a rear passivation stack. The rear dielectrics were subsequently patterned using laser ablation with a 0.2% opening fraction and AlSi 1% was sputtered at the rear and annealed, before formation of the selective front surface field n++. This selective FSF was done by means of laser doping from an external source of phosphorous and in that step patterning and local doping

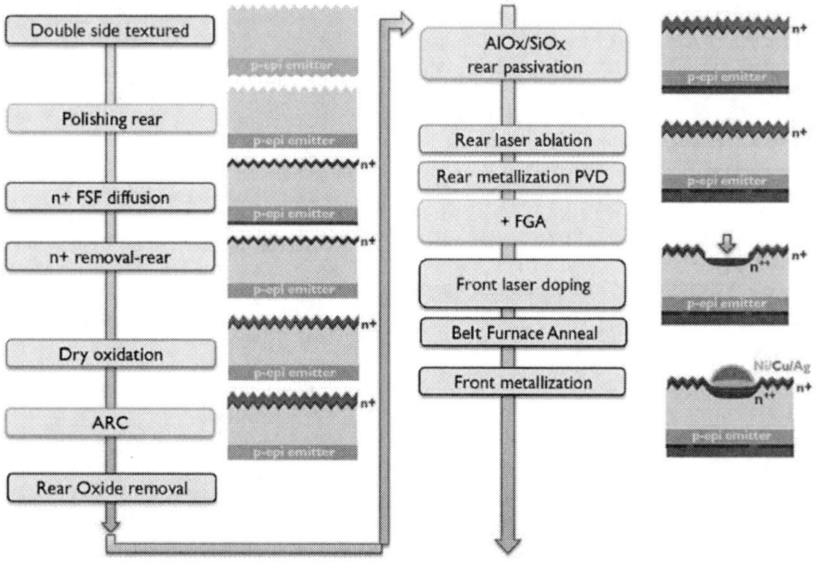

Figure 2. Solar cell integration flow of Crystal Solar epitaxial wafers using Imec's advanced nPERT technology.

to the underlying silicon is done in a single step. Underneath the metal contacts, a deeper doping region is formed, securing very good ohmic contact to the front metal. Both the doping as well as the depth of such in-situ laser doped region can be large as it does not contribute to the carrier generation. After the selective FSF is formed, a belt furnace anneal is performed and plating of the front contacts is done with a Ni/Cu/Ag metal stack. Graphical illustration of the experimental flow is presented in Fig. 2. The resulting nPERT devices were characterized by means of QSS-PL, light IV, shunt resistance and spectral response.

3 RESULTS AND DISCUSSION

3.1 Epitaxial wafer crystal quality

First observation of the epitaxial wafers has pointed out a number of defects on one side which were visible by eye. The defects were mainly present at the edges of the wafer and were randomly distributed, and not related to the type of emitter doping used. After detailed observation of wafer surface by scanning electron microscopy (SEM), it became obvious that those defects are stacking faults tetrahedral with a side wall dimension of >200 µm, see Fig 3. The fact that this dimension is roughly 1.45*epi-thickness, would indicate that those defects nucleate at the porous Si/epi interface. Although one might think that the density of these defects might be related to the number of re-uses of the parent substrate, no clear correlation between the parent substrate re-usage and the epitaxial layer quality (reflected through the V_{OC} level of the devices) was noticed (see Table 1) (for the lowly doped emitter). A V_{OC} approaching 700 mV was reached for cell fabricated from the substrate which was re-used 10 times (Table 1). This statement was also confirmed by photoluminescence (PL) observations coupled with light IV measurements. Devices with a higher number of defects (but still < 1 defect/cm^2) were measured by PL and light IV. Afterwards, the regions of defects were cut off from the device and measured again. No improvement in V_{OC}, or J_{SC} were observed after cutting off "defected" regions. In some cases only a slight improvement of FF <1% absolute value was measured indicating that stacking fault might be a source of increased series resistance due to non-conformal dielectric layer deposition and subsequent non-uniform patterning. This hypothesis must however be confirmed on large group of samples. Photoluminescence (PL) of chosen epitaxial devices with high density of stacking faults can be seen in Fig 4. In that figure, two cells fabricated from the 8th and 18th parent substrate generation were shown with no clear indication of the effect of defect densities being higher for the older generation of epitaxial layers or lower for the younger generation. Also bulk lifetime measurements reveal repeated values of ~4 ms at an injection level of $1 \cdot 10^{15}$ cm^{-3}.

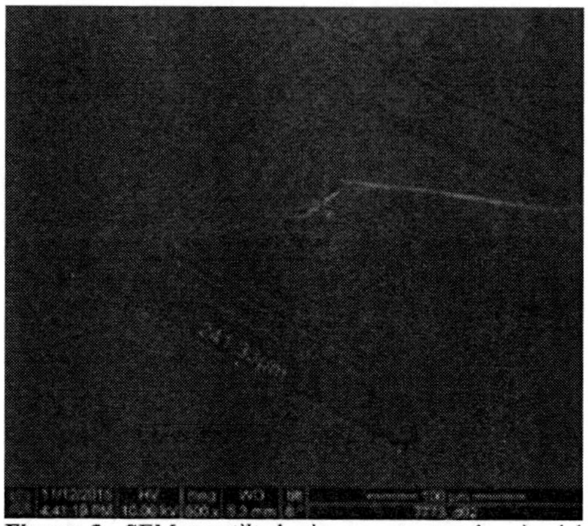

Figure 3. SEM top-tilted view on textured epitaxial surface with large epitaxial stacking fault.

TEM investigations confirmed defect free and high quality Si outside of stacking faults, allowing the fabrication of high quality devices.

Table I. Parent substrate re-usage times vs corresponding V_{OC} of the fabricated epitaxial nPERT cell from such parent substrate.

#Substrate re-use times	2	8	9	10	15	18
Voc [mV]	695	693	694	699	696	692
Record cell				x		

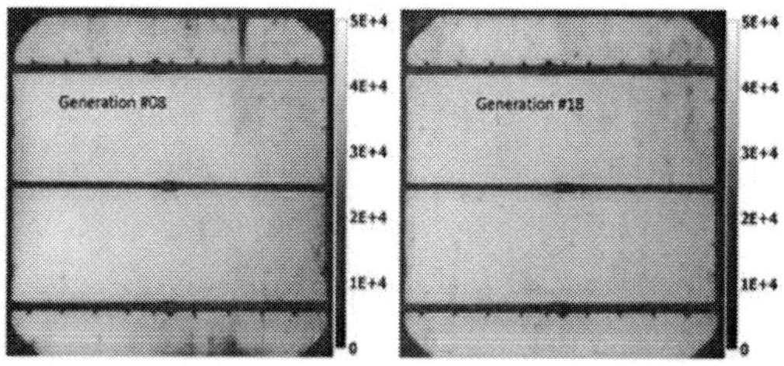

Figure 4. PL at open circuit voltage of two finished devices: left: device fabricated from substrate re-used 8 times, right: device fabricated from the substrate re-used 18 times.

3.2. Device characterization

The average and best cells parameters of Crystal Solar wafers using Imec's nPERT platform for both investigated emitters are listed in Table 2. A very good specific rear contact resistance <0.1 mΩcm^2 was reached for both types of cells with medium and lowly doped emitter. The total series resistance of the cells was also excellent with values of approximately 0.5 Ωcm^2 in average for devices with both type of emitters. In addition, a good shunt resistance > 100 Ω was measured.

Table II. IV results for the fully epitaxial device with 3 bus bar pattern fabricated using the n-type PERT process flow by Imec (238.9 cm²) as presented in Figure 2.

Emitter doping	J_{SC} (mA/cm²)	V_{OC} (mV)	FF (%)	η (%)
Medium doped	38.8±0.4	677±2	78.9±0.4	20.7±0.2
Best	39.3	678	78.8	21.0
Lowly doped	39.8±0.1	695±3	79.4±0.6	22.0±0.2
Best	39.7	699	80.4	22.3
ISE CalLab	**40.2**	**700**	**79.9**	**22.5**

As can be seen from Table 2, J_{SC} and V_{OC} of the medium doped emitter group is strongly limited due to the free carrier absorption (FCA) and Auger recombination in comparison to the group with lowly doped emitter. This is confirming the simulation results for a deep emitter with a doping level > 10^{18} cm^{-3}, presented in the next section. In case of a lowly doped emitter with a doping level <$1\cdot10^{18}$ cm^{-3}, there is little influence of the FCA and Auger recombination. This can be seen in Figure 5, where internal quantum efficiency of the two epitaxial wafers with epitaxial rear emitters in comparison with Cz-Si wafer with diffused B-doped emitter is shown. Clearly the cell with lowly doped emitter performs better at long wavelength range due to reduced FCA and its performance is similar to the reference which is Cz-Si device with diffused B-doped rear emitter. Outstanding currents and V_{OC} were reached for this set of epitaxial wafers with build-in emitter and the best cell reached V_{OC} up to 700mV. This is an outstanding open circuit voltage for the devices without any next-generation passivation scheme, like passivated contacts. The top cell of 22.5% was independently confirmed by ISE CalLab.

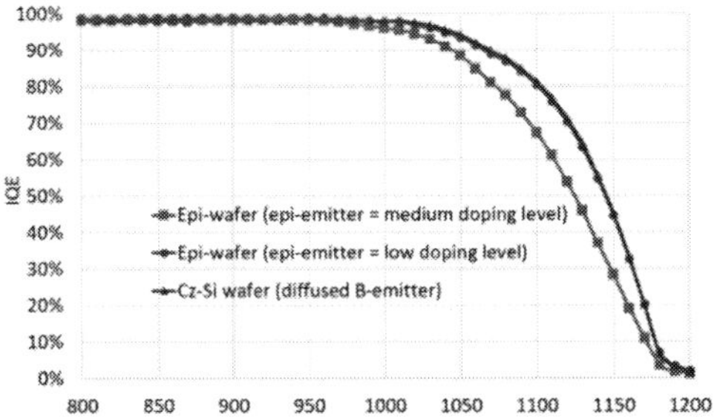

Figure 5. Spectral response of two epitaxial wafers with medium and lowly doped emitter in a comparison to the Cz-Si wafer with diffused B-doped emitter .

3.3. Build-in epitaxial emitter simulations

In house simulations using EDNA2 from PV Lighthouse [7] were used to explain the differences obtained in these two sets of devices with box profiles of lowly and medium emitter doping level and a relatively deep junction >20 µm. The emitter depth is denominated as Z_f in Figure 6. The medium doped emitter as used in this experiment refers to the doping level of >$1\cdot10^{18}$ cm^{-3}, whereas the lowly doped emitter refers to the doping <$1\cdot10^{18}$ cm^{-3}.

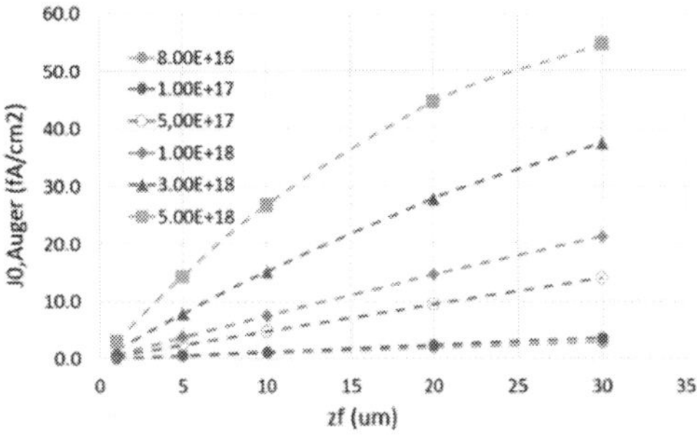

Figure 6. Auger recombination as function of emitter depth for various doping levels (EDNA2 simulations using Richter's model [8] for Auger recombination, base resistivity of 5Ωcm)

Figure 6 shows that J_{0Auger} is low when $N_{peak} < 1\cdot10e^{18}$ cm^{-3} (with N_{peak} being constant throughout the depth in case of a box profile epitaxial emitter) even for very deep emitters. Above $1\cdot10e^{18}$ cm^{-3}, Auger contribution quickly increases with emitter depth, even for shallower emitter depths. This behavior can explain the V_{OC} differences between the two sets of emitter doping as obtained in this experiment and presented in Table 2.

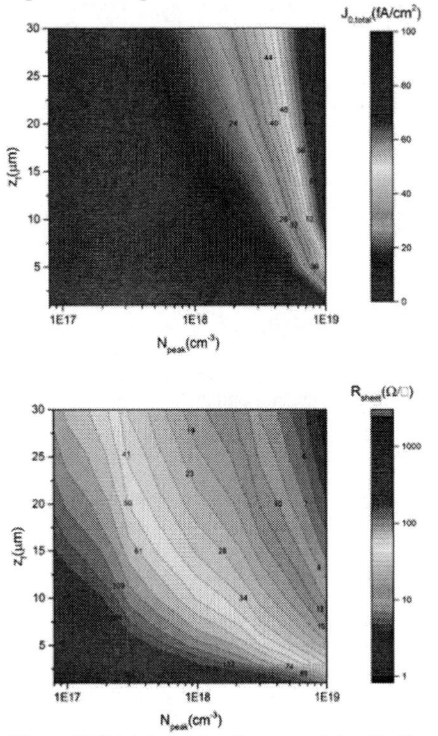

Figure 7. Total dark saturation current density ($J_{0,total}$) and sheet resistance (R_{sheet}) from EDNA simulations for epitaxial box profile emitters with depth Z_f and doping level N_{peak}.

Figure 7 shows the maps of J_{0total} and Rsheet for epitaxial box profile emitters with depth Z_f and doping level N_{peak}. The simulations indicate that low emitter Rsheet < 100Ω/sq are easily obtainable with deep emitters even at low N_{peak} <$1\cdot10e^{18}$ cm^{-3}. Such doping levels provide low J_{0total} and low Rsheet at the same time. Therefore no compromise needs to be made between emitter resistance on the one hand and recombination under the contacts on the other hand which is a condition unique to these epitaxial emitters. This leads to devices with both high FF and high Voc as shown in table 2.

4 CONCLUSIONS

In this paper we demonstrate the high efficiency potential of kerfless, epitaxial wafers grown by Direct Gas to Wafer technology developed by Crystal Solar. Epitaxial wafers with a thickness of 180-200 µm are fabricated by in situ growth of both p+ (emitter) and n- (base) regions. No additional wafer treatment is needed to activate the doping. Epitaxial wafers reveal a low density of stacking faults, around which very good crystal quality of silicon is confirmed by TEM. Integration of the epitaxial wafers into nPERT devices was straightforward and the in-situ emitter growth allowed a reduction in the number of process steps (no masking, no etching, no doping activation). There is also no negative correlation found between the number of re-use cycles of the parent substrate and the level of V_{OC} obtained from those substrates for nPERT devices. An advanced passivation scheme (Al_2O_3/SiO_x) and low surface recombination velocity, as well as effective patterning and contacting schemes at the rear and front of the device provided by Imec technology has led to very promising first integration results. Best solar cells have reached a V_{OC} of 700 mV and top efficiency of 22.5 % (independently confirmed by ISE CalLab) on full area 238.9 cm^2 devices. Further developments will concentrate on the in-situ growth of the FSF, leading to increasing cost savings. Internal cost calculations have shown that a 25% reduction in manufacturing cost is feasible at the module level by a reduction in wafer and cell process cost. Moreover, process optimizations and the use of an in-situ epitaxially grown FSF, should lead to cell efficiencies >23%.

5 ACKNOWLEDGEMENTS

The authors would like to acknowledge the financial support of Imec's industrial affiliation program for PV.

References

[1] P. A. Basore, " Understanding manufacturing cost influence on future trends in silicon photovoltaics", IEEE Journal of Photovoltaics, vol4, No 6, November 2014.
[2] R. Hao, T.S. Ravi, V. Siva, J. Vatus, D. Miller, J. Custodio, K. Moyers, Crystal Solar Inc, " High quality kerfless silicon mono-crystalline wafers and cells by high throughput epitaxial growth", Materials Research Society, Symp. Proc. Vol 1666, pp: 48-58, 2014
[3] M. Recamán Payo, I. Kuzma-Filipek, F. Duerinckx, Y. Li, E. Cornagliotti, A. Uruena, A. Sharma, L. Tous, R. Russell, A. Hajjiahl, J. Szlufcik, J. Poortmans, „ Advanced doping profiles by selective epitaxy in n-type PERT cells", EUPVSEC Hamburg 2015.
[4] V. Mertens, S. Bordihn, A. Mohr, K. Petter, J. W. Müller, D. J. W. Jeong, R. Hao, T. S. Ravi, " 21.4 % efficient fully screen printed n-type solar cell on epitaxially grown silicon wafers with built-in Boron rear side emitter", EUPVSEC 2015
[5] MM. Moslehi, P. Kapur, J. Kramer, V. Rana, S. Seutter, A. Deshpande, T. Stalcup, S. Kommera, J. Ashjaee, A. Calcaterra, D. Grupp, D. Dutton, R. Brown, "World record 20.6% efficiency 156 mm × 156 mm fullsquare solar cells using low-cost kerfless ultrathin epitaxial silicon & porous silicon lift-off technology for industry-leading high-performance smart PV modules". PV Asia Pacific Conference (APVIA/PVAP), 24 October 2012.
[6]H. S. Radhakrishnan, " Lift-off approaches for thin crystalline silicon" , in book Thin Film Si solar cells edited by J. Poortmans, 2015.
[7] K.R. McIntosh and P.P. Altermatt, "A freeware 1D emitter model for silicon solar cells," Proc. 35th Photovoltaic Specialists Conference, Honolulu, pp. 2188–2193, 2010
[8] A. Richter, S. Glunz, F. Werner, J. Schmidt, A. Cuevas, "Improved quantitative description of Auger recombination in crystalline silicon", Physical Review B, 86, 2012., pp: 165202/1-14.

CO-DIFFUSION FOR P-TYPE PERT SOLAR CELLS USING APCVD BSG LAYERS AS BORON DOPING SOURCE

S. Meier[1], S. Wiesnet[1], S. Mack[1], S. Werner[1] S. Maier[1], S. Unmüßig[1], C. Demberger[2], H. Knauss[2], D. Biro[1], A. Wolf[1]
[1]Fraunhofer Institute for Solar Energy Systems ISE, Heidenhofstraße 2, 79110 Freiburg, Germany
[2]Gebr. SCHMID GmbH, Robert-Bosch-Str. 32-36, 72250 Freudenstadt, Germany
Phone: +49 761 45 88 5059; Fax: + 49 761 45 88 9250; Email: sebastian.meier@ise.fraunhofer.de

ABSTRACT: In this work, we investigate boron-doped back surface field (BSF) formation from borosilicate glass (BSG) deposited by atmospheric pressure chemical vapor deposition (APCVD). Carrier lifetime samples with different boron concentration in the BSG are exposed to high-temperature processes with varying temperature and gas atmospheres. The development aims at producing p-type Si-based passivated emitter and rear totally diffused (PERT) cells in a $POCl_3$ co-diffusion process. Our experiments confirm the critical impact of the O_2 concentration on boron diffusion from APCVD BSG layers. For different peak temperatures Si surface oxidation formation need to be balanced carefully. Moderate diffusion temperatures of 875°C (30 min) yield a BSF doping with R_{sh} = 131 Ω/sq and a recombination current density of $J_{0,BSF}$ = 21 fA/cm² (planar surface, AlO_x/SiN_x-passivation). Finally, 156 mm sized PERT cells fabricated from p-type Czochralski-grown Si wafers using the developed co-diffusion process show peak efficiencies of 20.6%, confirming the feasibility of our co-diffusion approach.
Keywords: APCVD, boron-doping, BSG, PERT, co-diffusion

1 INTRODUCTION

Bifacial passivated emitter and rear (PERC) solar cells allow for conversion efficiencies beyond 21% while enabling bifacial module designs [1]. The bifacial operation allows for increasing the energy yield or a broadening of the energy output over a day, depending on the physical surrounding and module placement [2]. Currently, bifacial PERC cells reach rear efficiencies of 16.7%, where the high shading loss of the rear aluminum grid is a major limitation [1]. One approach to reduce these losses is to use a passivated emitter and rear totally diffused (PERT) cell structure that allows significantly higher rear finger spacing due to additional lateral conductance of the back surface field (BSF). In addition, PERT cells tolerate high resistivity wafers, which reduce light induced degradation (LID) due to boron-oxygen complex recombination [3]. In state of the art processing emitter and BSF of PERT solar cells are formed using sequential diffusion processes with $POCl_3$ and BBr_3 as liquid dopant precursor [4–6]. The additional diffusion step compared to PERC manufacturing implicates additional process complexity and cost. To avoid this additional effort, we present a fast and cost-effective approach to form the phosphorus-doped emitter and boron-doped BSF in one single co-diffusion process. Therefore we investigate co-diffusion from $POCl_3$ atmosphere and a pre deposited borosilicate glass (BSG) layer by atmospheric pressure chemical vapor deposition (APCVD). Main advantage of this approach is its industrial feasibility while keeping the flexibility to independently adjust doping profiles [7].

This paper first discusses the development of a $POCl_3$-based co-diffusion tube furnace process suitable for the fabrication of p-type PERT solar cells. Finally, we prove the co-diffusion concept by fabricating monofacial PERT solar cells from p-type Czochralski-grown silicon (Cz-Si) wafers.

2 APPROACH

2.1 Boron-doping from APCVD BSG layers

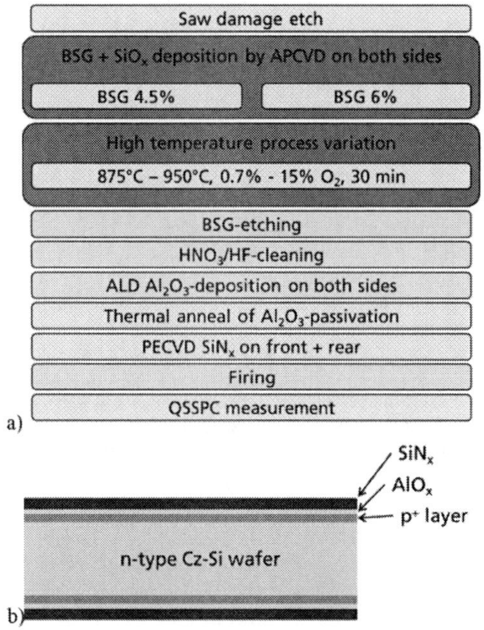

Fig. 1: a) Process sequence and b) sample design for symmetric samples used to characterize boron diffused surfaces made from APCVD BSG layers.

We first investigate boron-doping for BSF formation from APCVD BSG layers with respect to sheet resistance, doping profile and carrier recombination. Fig. 1 represents a) the process scheme and b) the sample design used to develop boron diffusion from APCVD BSG layers.

We use n-type Cz-Si wafers with base resistivity ρ_B = 2.2 Ωcm (determined after thermal processing) and an edge length of 156 mm for symmetric carrier lifetime samples. Another set of n-type Cz-Si wafers is selected for four point probe (4pp) and electrochemical capacitance voltage (ECV) measurements. All wafers receive a saw damage etch. Then, the deposition of the

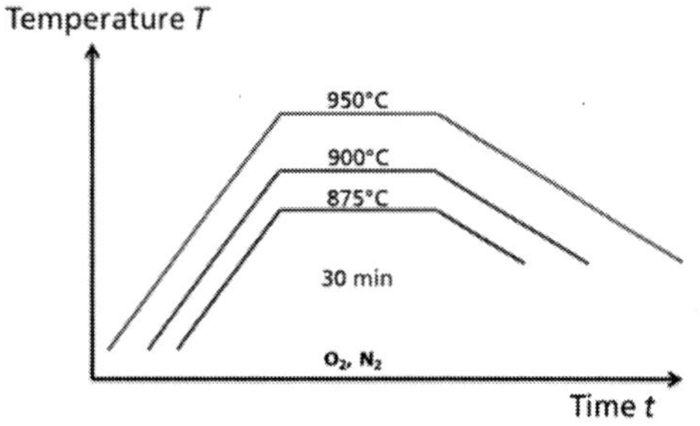

Fig. 2: Schematic of the applied high temperature processes. Peak temperatures $T_{Peak} = 875°C$, $T_{Peak} = 900°C$, and $T_{Peak} = 950°C$ are used for a plateau time of 30 minutes. During whole diffusion process the O_2 concentration is kept constant. Different O_2 concentrations from 0% to 3% are investigated.

BSG/SiO_x layer stack in one deposition is realized by APCVD on both sides of the wafer. We apply two different boron-contents of the BSG layer (nominally 4.5% and 6%). Then, diffusion temperatures $T_{Peak} = 875°C$, $T_{Peak} = 900°C$, and $T_{Peak} = 950°C$ (all with 30min plateau) at different O_2 concentrations in the atmosphere are applied. Fig. 2 presents a scheme of the high temperature processes.

Following the high temperature step, the BSG/SiO_x layers are removed in HF solution and 4pp and ECV measurements yield sheet resistance R_{sh} and doping profile, respectively. The symmetric carrier lifetime samples receive wet chemical cleaning with a HNO_3/HF sequence, both-sided surface passivation by atomic layer deposition (Fast-ALD) of an Al_2O_3 layer and subsequent out gassing as well as a plasma enhanced chemical vapor deposited (PECVD) SiN_x capping. A fast firing process activates the passivation. Finally quasi steady state photo conductance (QSSPC) measurement determine the dark saturation current $J_{0,BSF}$ using the procedure from Ref. [8].

2.2 Co-diffusion

Based on a high temperature process with $T_{Peak} = 875°C$, a $POCl_3$-based co-diffusion process is developed. Fig. 3 shows a) the process scheme and b) the sample design.

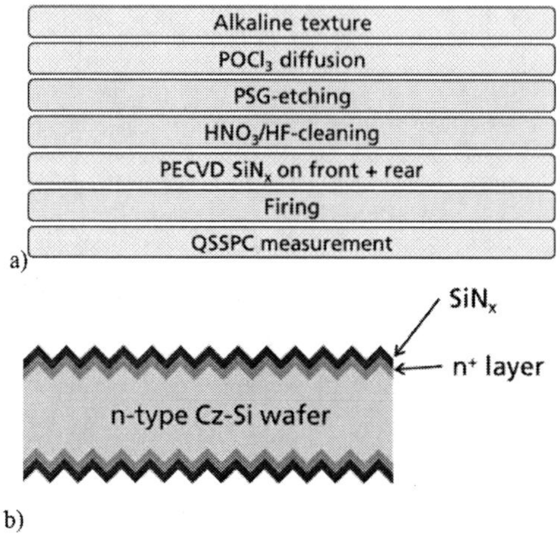

Fig. 3: a) Process sequence and b) sample design used to characterize phosphorus diffused surfaces from $POCl_3$-based co-diffusion process with respect to carrier recombination.

First, n-type Cz-Si wafers with $\rho_B = 2.2\ \Omega cm$ (determined after thermal processing) and an edge length of 156 mm are alkaline textured. Then, a $POCl_3$-based diffusion with $T_{Peak} = 875°C$ (30min plateau) is applied (see Fig. 4).

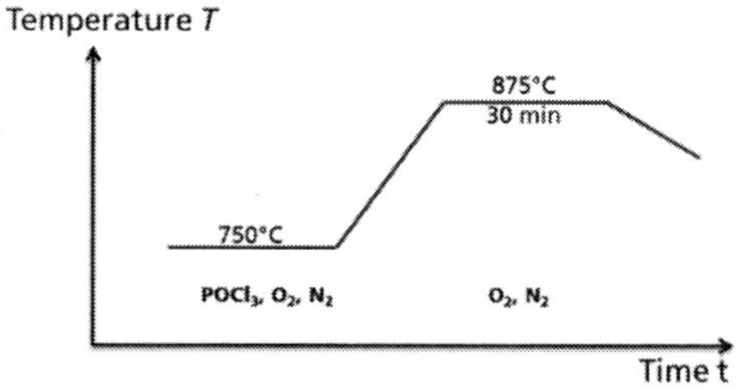

Fig. 4: Schematic of the investigated $POCl_3$-based co-diffusion process.

After phosphosilicate glass (PSG) removal, 4pp and ECV measurements are realized using saw damage etched n-type Si-reference wafers. These wafers are processed in parallel to alkaline textured Si wafers. The textured symmetric n-type Cz-Si lifetime samples receive wet chemical cleaning with a HNO_3/HF sequence, PECVD SiN_x passivation and firing. Finally, QSSPC measurements determine emitter saturation current J_{0e}; see Ref. [8].

2.3 Solar cell fabrication

Fig. 5 represents a process sequence for the fabrication of p-type PERC and co-diffused PERT solar cells.

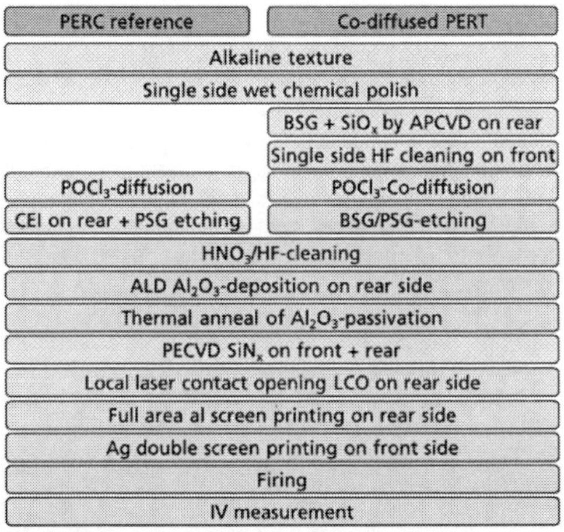

Fig. 5: Process sequence for fabricating PERC (left) and PERT (right) type solar cells.

P-type Cz-Si wafers with an edge length of 156 mm, $\rho_B = 2.9\ \Omega$cm and an initial thickness $W = 180$ µm are textured in an alkaline solution. In a second step the rear side texture is smoothed by a single side wet chemical polish.

For the PERT cells, a BSG/SiO_x layer system is deposited on the polished rear surface by APCVD in one single process, followed by a single side HF cleaning sequence for the non-deposited front side. The wafers are then subjected to a $POCl_3$-based co-diffusion tube furnace process to simultaneously form a phosphorus doped emitter on the front side and a boron doped BSF on the rear side of the PERT cell. An HF bath then removes the formed PSG and BSG layers.

In contrast, the PERC reference cells are directly subjected to a standard $POCl_3$-diffusion process [9]. The rear side of PERC cell is then subjected to wet chemical edge isolation (CEI) followed by front PSG removal.

PERC and PERT cells thus differ in diffusion sequence; however, after diffusion their fabrication is similar. Rear passivation is realized by an Al_2O_3 layer deposited by Fast-ALD followed by thermal anneal of the Al_2O_3 layer. Then a SiN_x layer is deposited on top of the Al_2O_3 layer. The phosphorus doped emitter at the front side is passivated by a SiN_x layer deposited by PECVD. Then, local contact opening (LCO) of line shaped rear contacts is realized by a laser process and a full-area aluminum paste is applied by screen-printing. Note that for both PERC and PERT cells an identical line pitch of 700 µm is chosen (optimized for the PERC structure). The emitter side is subjected to a double screen printing step using a commercially available silver paste. Final steps are contact formation by a firing step in a commercial fast firing oven (variation of peak temperature) and IV measurement using an industrial cell tester.

3 RESULTS

3.1 Influence of peak temperature on boron concentration

Fig. 6 presents boron doping profiles for BSG layers with 6% boron concentration from diffusions with $T_{Peak} = 875°C$, $T_{Peak} = 900°C$, and $T_{Peak} = 950°C$, all with an O_2 concentration of 0.7% in the diffusion atmosphere (see Fig. 2 for the schematic of the high-temperature processes). The used samples are fabricated as described in section 2.1, see Fig. 1 and Fig. 2.

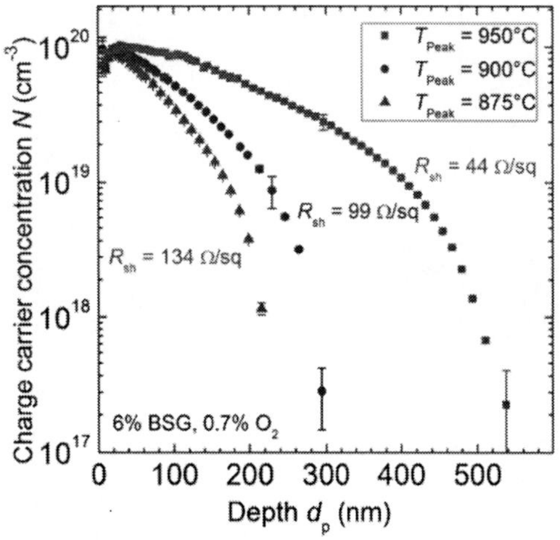

Fig. 6: Boron doping profiles measured by ECV (wafer center) on SDE surfaces for BSG layers with 6% concentration and diffusion processes with different peak temperatures T_{Peak}, all with an O_2 concentration of 0.7% in the diffusion atmosphere. The measured sheet resistances R_{sh} (center, 4pp) are given as well.

With increasing temperature, the profile depth d_P increases from $d_P \approx 240$ nm at $T_{Peak} = 875°C$ to $d_P \approx 540$ nm at $T_{Peak} = 950°C$. The maximum carrier concentration N_{max} for the diffusions with $T_{Peak} = 875°C$ and $T_{Peak} = 900°C$ is $N_{max} \approx 9\times10^{19}$ cm^{-3} and increases slightly to $N_{max} \approx 1\times10^{20}$ cm^{-3} for $T_{Peak} = 950°C$. Accordingly, the sheet resistance R_{sh} reduces from $R_{sh} = 134$ Ω/sq. at $T_{Peak} = 875°C$ to $R_{sh} = 44$ Ω/sq. at $T_{Peak} = 950°C$. All profiles show a slightly depleted surface concentration due to segregation of boron into the oxide surface layer [10,11].

3.2 Influence of O_2 concentration on boron concentration

The impact of the O_2 concentration during the thermal process on boron diffusion from APCVD BSG layers is discussed for T_{Peak} = 875°C. Doping profiles resulting after these diffusion processes are shown in Fig. 7, see Fig. 1 and Fig. 2.

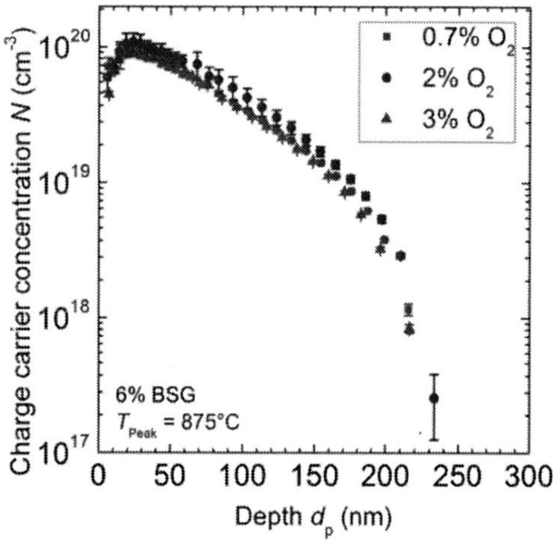

Fig. 7: Boron doping profiles measured by ECV (wafer center) on SDE surfaces for BSG layers with 4,5% and 6% concentration and high-temperature processes with T_{Peak} = 875°C and different O_2 concentrations in the process atmosphere.

All measured doping profiles show a high agreement. The boron doping is characterized by $N_{max} \approx 9\times10^{19}$ cm^{-3} and $d_P \approx 240$ nm. However, the highest oxygen concentration of 3% shows the strongest surface depletion, which is reasonable, since high oxygen concentrations enhance the oxide growth and boron segregation into the oxide [11,10].

The R_{sh} measurements by 4pp (see Table I) reveal slight differences of the processes.

Table I: Sheet resistance measurements (4pp) of the boron-doping on 156mm n-type wafers with 400 measurement points per wafer by 4pp. The given data is organized in R_{sh} on the center of the Si wafer, average R_{sh} over the wafer surface and standard deviation.

sample	R_{sh} center [Ω/sq]	R_{sh} average [Ω/sq]	R_{sh} std. dev. [Ω/sq]
4.5% BSG 0.7% O_2	128	142	41
6% BSG 0.7% O_2	134	131	11
6% BSG 2% O_2	114	116	9
6% BSG 3% O_2	135	163	71

For the process with 0.7% O_2 content, increasing the boron concentration from 4.5% to 6% reduces the average R_{sh} = (131 ± 41) Ω/sq to R_{sh} = (131 ± 11) Ω/sq and increases the homogeneity of R_{sh}. If the O_2 concentration is increased to 2% and higher, no boron diffusion can be observed for a BSG layer with 4.5%. In contrast, for a BSG layer with 6% content, R_{sh} slightly reduces from R_{sh} = (131 ± 11) Ω/sq to R_{sh} = (116 ± 9) Ω/sq, whereas a further increase from 2% to 3% O_2 concentration again increases R_{sh} = (163±71) Ω/sq. The inhomogeneous distribution of R_{sh} is observed on the border of the Si wafer.

In conclusion, high O_2 concentrations tend to increase the sheet resistance. Additionally the non-uniformity of the sheet resistance over the Si-wafer is increased if the O_2 concentration is increased. We attribute these effects to a thermally grown SiO_2 layer at the interface Si-wafer to BSG layer. This SiO_2 layer acts as diffusion barrier against boron from BSG layer and even leads to complete decoupling of the BSG layer for higher oxygen concentrations.

3.3 Dark saturation current densities for diffusion processes with peak temperature of 875°C.

Dark saturation current densities $J_{0,BSF}$ for boron BSF diffusion processes with T_{Peak} = 875°C (section 2.1 Fig. 2 and Fig. 3) are shown in Fig.8.

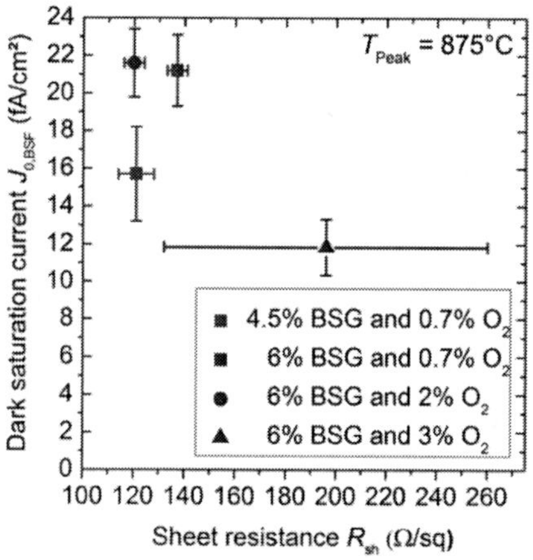

Fig. 8: Dark saturation current density $J_{0,BSF}$ (planar surface, AlO_x/SiN_x-passivation, average and standard deviation of five measurement points per wafer) over average value of sheet resistance R_{sh} (average R_{sh} measured by means of inductive coupling using an inline tool [12]) for diffusion processes at T_{Peak} = 875°C.

The observed recombination currents are very low and range from $J_{0,BSF}$ = (22 ± 2) fA/cm² at R_{sh} = (121 ± 7) Ω/sq to $J_{0,BSF}$ = (12 ± 1) fA/cm² for R_{sh} = (196 ± 64) Ω/sq. The graph shows average R_{sh} measured by means of inductive coupling using an inline tool [12].

With respect to a homogeneous distribution of the sheet resistance 4.5% boron concentration and 0.7% O_2 concentration as well as 6% boron concentration and 3% O_2 concentration are not well suited for the fabrication of PERT cells. Thus, for the fabrication of co-diffused PERT cells 6% boron content and a diffusion process with an O_2 concentration of 0.7% are selected.

3.3 Characterization of full co-diffusion process

Based on 6% boron concentration in the BSG layer and 0.7% O_2 content in the diffusion atmosphere the final co-diffusion process shown in Fig. 4 is developed. The resulting doping profiles measured by ECV are shown in Fig. 9.

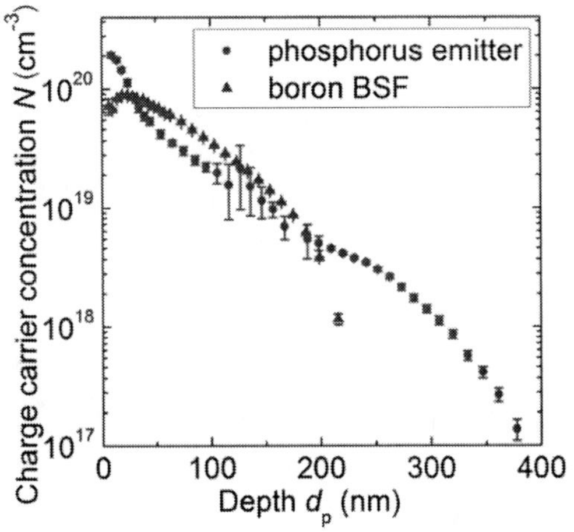

Fig. 9: ECV measurement of the co-diffusion process for simultaneous boron diffusion (red symbols) from an APCVD BSG/SiO_x-layer stack and phosphorus diffusion (green symbols) from a $POCl_3$ ambient.

The samples are processed as described in section 2.1 Fig. 3. The Phosphorus doped emitter is characterized by $N_{max} \approx 2\times10^{20}$ cm^{-3} and $d_P \approx 380$nm. This results in $R_{sh} = (90 \pm 2)$ Ω/sq. measured by means of inductive coupling using an inline tool [12] and $J_{0e} = (79 \pm 15)$ fA/cm^2 for an alkaline textured and passivated (SiN_x) surface.

3.3 Comparison PERC and co-diffused PERT solar cells

Based on the developed co-diffusion process (Fig. 4) monofacial PERT cells are fabricated and compared to monofacial PERC cells based on standard $POCl_3$ diffusion process developed at Fraunhofer ISE [9]. Fig. 10 illustrates the different cell structures which are fabricated according to the process plan in Fig. 5.

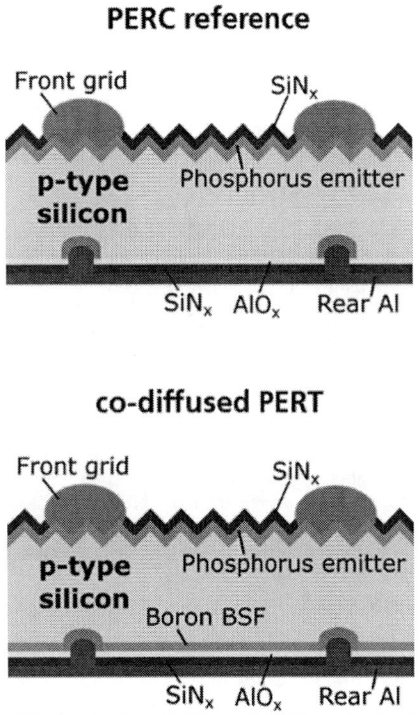

Fig. 10: Investigated PERC and PERT cell structures fabricated according to the process plan in Fig. 5. PERT cells are fabricated with a $POCl_3$-based co-diffusion process (see Fig. 4). For the PERC cells a standard $POCl_3$ based diffusion process is used [9].

The results of IV measurements using an industrial cell tester for the different cell types are presented in Table II.

Table II: IV data (as processed) for PERC and PERT solar cells with an edge length of 156 mm and $\rho_B = 2.9$ Ωcm using an industrial cell tester. Shown are the median values of best peak firing temperature groups.

Cell type	V_{oc} [mV]	J_{sc} [mA/cm^2]	FF [%]	η [%]	R_p [kΩcm^2]	R_s [Ωcm^2]
PERC reference	653	39.6	79.3	20.5	109	0.5
PERT co-diff.	649	39.4	80.3	20.5	52	0.4

Median values of best groups of PERC and PERT cells are shown. The PERC reference cells show 4 mV higher V_{oc} and 0.2 mA/cm² higher J_{sc}. This is attributed to an increased rear surface recombination of the PERT cells. Control samples of this batch showed an increased $J_{0,BSF} = (152 \pm 35)$ fA/cm^2, much higher than observed in previous runs (see Fig. 8). The lateral conductance of the BSF increases the FF of the PERT cells by 1%$_{abs}$ to $FF = 80.3\%$. Thus, both cell types show the same efficiency level of $\eta = 20.5\%$. Peak efficiencies are 20.7% for PERC and 20.6% for PERT.

4 SUMMARY AND CONCLUSION

We investigate boron doping from APCVD BSG/SiO_x layer stacks. At moderate diffusion conditions of 875°C for 30 min, average sheet resistances of R_{sh} = 116 to 163 Ω/sq are observed with low recombination parameters of $J_{0,BSF}$ = 22 to 12 fA/cm² respectively (planar surface Al_2O_3/SiN_x passivation). With these conditions, we develop a $POCl_3$-based co-diffusion process leading to a simultaneous phosphorus-diffusion of R_{sh} = (90 ± 2) Ω/sq. measured by inline-R_{sh} measurement and J_{0e} = (79 ± 15) fA/cm². Finally, we prove the co-diffusion concept by fabricating monofacial PERT solar cells with peak efficiencies of 20.6% and 20.5% (median value) on 156mm 2.9 Ωcm p-type Cz-substrates.

In conclusion, co-diffusion using APCVD BSG layers and $POCl_3$-diffusion is a promising approach for PERT cell fabrication, enabling high quality boron BSF formation with excellent recombination properties. Relevant applications are bifacial PERT cells with increased rear efficiency due to reduced shading as well as minimization of LID due to high tolerance of PERT cells to high resistivity wafers.

5 ACKNOWLEDGEMENTS

The authors thank development team of APCVD of Gebr. SCHMID GmbH and all colleagues at the Fraunhofer ISE Photovoltaic Technology Evaluation center (PV-TEC).

The authors would like to thank TOYO for the provision of aluminum pastes for the rear side of PERC and PERT solar cells.

S. Meier gratefully acknowledges the scholarship of the "Heinrich Böll Stiftung".

This work was funded by the German Federal Ministry for Economic Affairs and Energy within the research project "KoKoBi" under contract 0325875B.

6 REFERENCES

[1] T. Dullweber, C. Kranz, R. Peibst et al., "PERC+: industrial PERC solar cells with rear Al grid enabling bifaciality and reduced Al paste consumption", *Prog. Photovolt: Res. Appl.*, pp. n/a-n/a, 2015.

[2] T. S. Böscke, D. Kania, A. Helbig et al., "Bifacial n-Type Cells With >20% Front-Side Efficiency for Industrial Production", *IEEE J. Photovoltaics*, vol. 3, no. 2, pp. 674–677, 2013.

[3] K. Bothe, R. Sinton, J. Schmidt, "Fundamental boron-oxygen-related carrier lifetime limit in mono- and multicrystalline silicon", *Progress in Photovoltaics: Research and Applications*, vol. 13, pp. 287–296, 2005.

[4] A. Teppe, C. Gong, K. Zhao et al., "Progress in the industrial evaluation of the mc-Si PERCT technology based on boron diffusion, Constance, 2015.

[5] B. Lim, F. Kiefer, N. Wehmeier et al., "Simplifying the fabrication of n-PERT solar cells: Recent progress at ISFH", *Photovoltaics International*, 2015.

[6] B. Lim, F. Kiefer, T. Brendemühl et al., "Bifacial n-PERT solar cells with 20.4% efficiency applying dual print, Hertogenbosch, 2014.

[7] P. Rothhardt, S. Meier, K. Jiang et al., "19.9 % efficient bifacial n-type solar cell produced by co-diffusion-cobin, Amsterdam, 2014, pp. 653–655.

[8] A. Kimmerle, J. Greulich, A. Wolf, "Carrier-diffusion corrected J0-analysis of charge carrier lifetime measurements for increased consistency", *Solar Energy Materials and Solar Cells*, vol. 142, pp. 116–122, 2015.

[9] S. Werner, E. Lohmüller, S. Maier et al., "Process Optimization for the Front Side of p-Type Silicon Solar Cells, Amsterdam, The Netherlands, 2014.

[10] K. Taniguchi, K. Kurosawa, M. Kashiwagi, "Oxidation Enhanced Diffusion of Boron and Phosphorus in (100) Silicon", *J. Electrochem. Soc.: Solid-state Science and Technology*, vol. 127, no. 10, p. 6, 1980.

[11] A. S. Grove, O. Leistiko, C. T. Sah, "Redistribution of acceptor and donor impurities during thermal oxidation of silicon", *J. Appl. Phys.*, vol. 35, no. 9, p. 2695, 1964.

[12] M. Spitz, U. Belledin, S. Rein, "Fast inductive inline measurement of the emitter sheet resistance in industrial solar cell fabrication, Milano, 2007, pp. 47–50.

PILOT PRODUCTION OF 6" IBC SOLAR CELLS YIELDING A MEDIAN EFFICIENCY OF 23% WITH A LOW-COST INDUSTRIAL PROCESS

Zhonglan LI*, Yang YANG, Xueling ZHANG, Wei LIU, Yan CHEN, Guanchao XU, Xin SHU, Yifeng CHEN, Pietro P. ALTERMATT, Zhiqiang FENG and Pierre J. VERLINDEN
State Key Laboratory of Photovoltaic Science and Technology, Trina Solar, No. 2 Trina Road, Trina PV Park, New District, Changzhou, Jiangsu Province, China 213031
*Corresponding author: phone: (+86) 158 6183 0537, e-mail: zhonglan.li@trinasolar.com

ABSTRACT: The results of pilot production of Interdigitated Back Contact (IBC) solar cells at Trina Solar with a low-cost industrial process are presented. The 6" IBC pilot line was established in 2015 and continuous improvements were implemented over one year. In recent batches, the median cell efficiency has reached 23.0%, and the best cell efficiency was 23.2%. Advanced IBC cells with finer geometries and smaller contacts reached an efficiency of 23.5%. Further cell efficiency improvements and cost reduction are under development.
Keywords: interdigitated back contact, pilot production, 6inch, industrial process.

1 INTRODUCTION

The Interdigitated Back Contact (IBC) solar cell is one of the highest efficiency solar cell designs. The IBC concept was first introduced by Lammert and Schwartz in 1975 [1, 2]. Due to its higher cost of manufacturing, the IBC cell was originally designed for high concentration PV systems. With the improvement of low-cost manufacturing processes, IBC cells for terrestrial one-sun application became affordable. The conversion efficiency of IBC cells has reached 25.2% at one sun [3].

Sunpower is the leader of IBC cells in mass production, having the most advanced production line of 5" IBC solar cells. The highest efficiency reached by Sunpower is 25.2% [3]. Trina Solar has been developing high efficiency IBC cells for about 5 years. The Trina-ANU collaborative development project produced a small IBC cell for demonstration with an efficiency of 24.4% [4]. An industrial version of the IBC cell made on 6" wafers with only low-cost processes developed by Trina Solar reached an independently confirmed efficiency of 22.9% [5]. As far as we know, Trina Solar is one of the very few companies to set up a pilot production line for high-efficiency 6" IBC cells. The process and efficiency in the pilot line is continuously improving. IBC cells produced by Trina Solar powered the solar racing car from Osaka Sangyo University (OSU Model S) which won the "2015 FIA ALTERNATIVE ENERGIES CUP Solar Car Race" held at the Suzuka International Racing Circuit in Japan in August 2015 [6].

The median efficiency of IBC cells produced in the pilot line in recent times is 23.0% while the best IBC cell efficiency is 23.2%. 6" IBC cells with finer geometries and smaller contacts have reached an efficiency of 23.5% [7].

2 CELL PROCESSING

IBC cells are fabricated using 190 ± 10 μm thick, 156mm×156mm n-type CZ wafers with resistivity in the range of 2-10 Ω.cm. As-cut wafers are chemically polished to remove the saw damage and cleaned before the BBr_3 diffusion is applied to form an emitter. BSF regions are patterned with a low-cost screen printing process, followed by groove etching to remove the p-n junction in BSF areas, while the other regions are protected. An n^+ $POCl_3$ diffusion is used to form the BSF. The front side is textured by an alkaline etchant to form random pyramids. A shallow FSF is created on the front surface by $POCl_3$ diffusion. A front passivation and an antireflection layer of SiN_x are deposited directly on top of the FSF, while the rear side is passivated by dielectric stacks. The small emitter and BSF contacts are formed by low cost screen-printing technology. Metal fingers are screen printed and fired as in a conventional solar cell. The simplified process flow is illustrated in Fig.1.

The latest process is integrated with a $POCl_3$ gettering step on the front surface during the BSF diffusion. After boron diffusion and oxidation, the front side oxide is removed and the emitter on the front is etched during the groove etching step. So when doing the BSF diffusion, the front surface recieves as a gettering process.

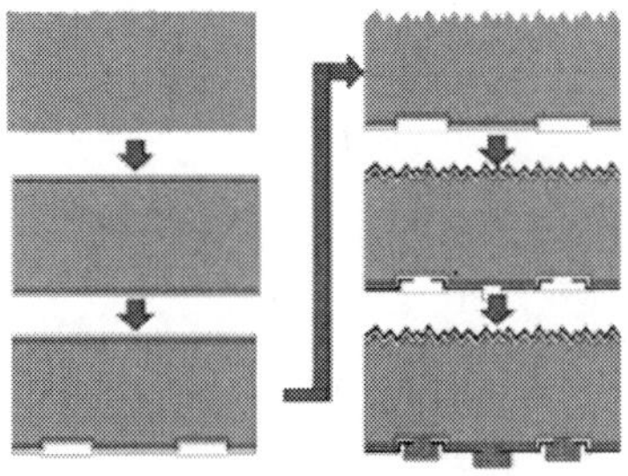

Fig.1: Process flow of Trina Solar IBC solar cells

3 RESULTS AND DISCUSSIONS

3.1 IBC cell performance

Table I shows the electrical performance gain with the gettering process relative to the baseline process - the gettering process improves the cell efficiency by 0.2% absolutely.

Table I: Gain in cell performance with gettering process

	ΔJ_{sc} (mA/cm^2)	ΔV_{oc} (mV)	ΔFF (%)	ΔEff. (%)
Gain	0.1	2.0	0.2	0.2

IBC cells in the Trina Solar's pilot line have recently reached a median efficiency of 23.0% with gettering process integrated, while the champion efficiency is 23.2%, shown in table II. In a more recent experiment, IBC cells with smaller contacts and a finer geometry reached an efficiency of 23.5% [7]. The IV performance of cells was independently measured at Japan Electrical Safety & Environment Technology Laboratories (JET). The report is shown in Fig. 2.

Table II: Performance of 6" IBC cells in the pilot line

	J_{sc} (mA/cm^2)	V_{oc} (mV)	FF (%)	Eff. (%)
Median	42.0	680.9	80.4	23.0*
Champion cell in pilot line	42.0	682.9	81.0	23.2
Best cell with small contacts	42.1	689.9	80.9	23.5*

*independently measured at Japan Electrical Safety & Environment Technology Laboratories (JET).

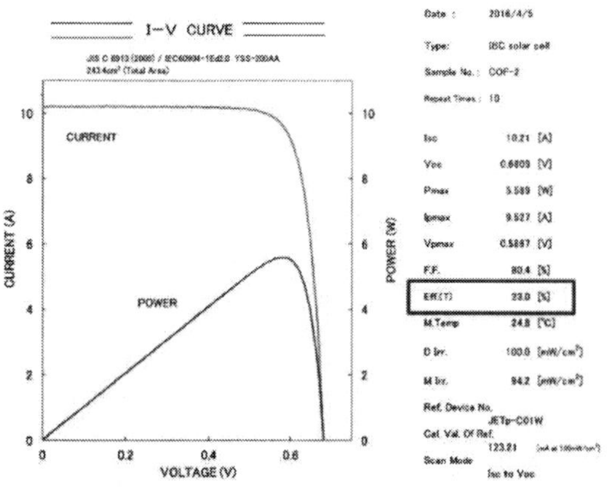

Fig.2: I-V curve of an 6" IBC cell with median efficiency, independently measured at JET.

The efficiency distribution in the pilot production is shown in Fig. 3. More than 75% of cells have an efficiency higher than 23.0%. Cells below 23.0% are mainly due to minor process fluctuations and higher oxygen content in some wafers.

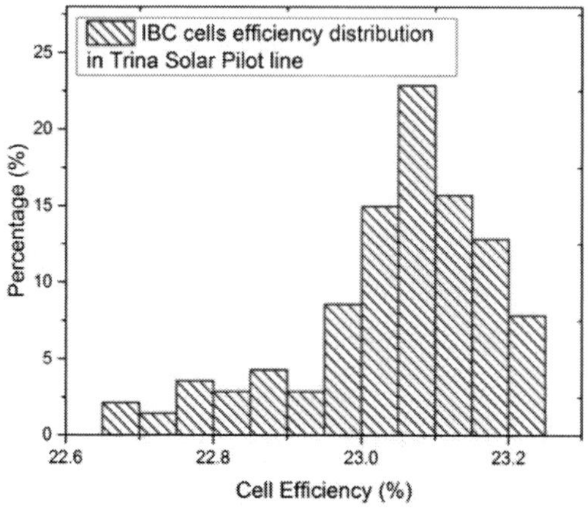

Fig.3: Efficiency distribution of 6" IBC cells in a batch in Trina Solar's pilot line.

3.2 Characterization

The photoluminescence (PL) in open-circuit condition, the electroluminescence (EL) at different bias voltages and the extracted local series resistance map were measured by a BT Imaging tool, as shown in Fig. 4. From the PL image, some scratches on the surface can be observed, which demonstrates that more careful handling procedures during processing must be applied. The cell edge is a little darker than the center, which may be caused by the non-uniformity during diffusion. The EL and local series resistance images show some non-uniform pattern, which is caused by the punctual contact of measurement probes during testing and by the series resistance of the busbars.

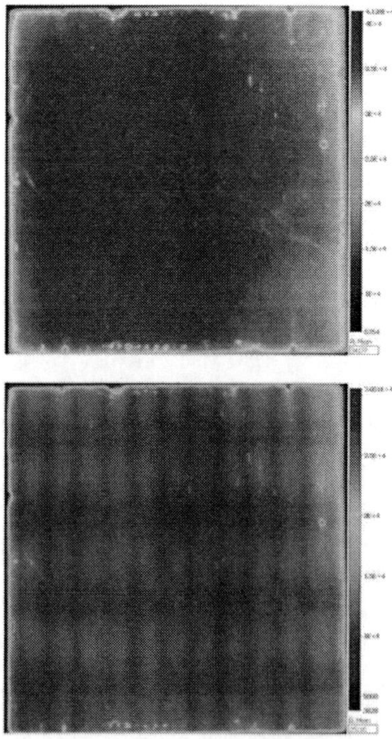

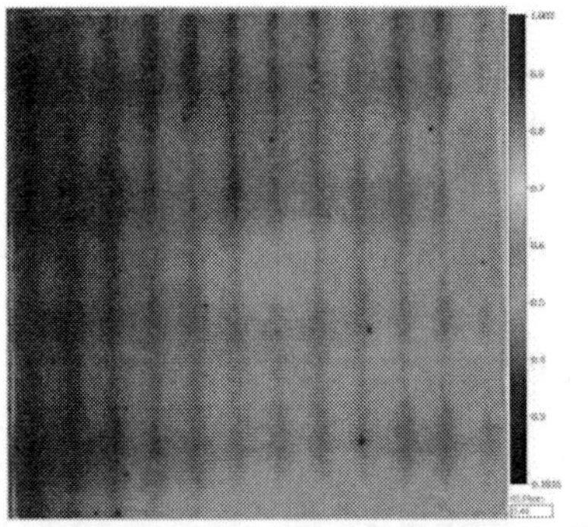

Fig. 4: Photoluminescence, electroluminescence and local series resistance mapping of a 6" IBC cell.

Internal quantum efficiency (IQE) curves of cells with and without gettering process were measured by the PV Measurement instrument and are shown in Fig. 5. Good front surface passivation leads to a very high IQE at short wavelengths. The IQE slightly exceeds unity near 400 nm due to impact ionization in the direct-gap ($\Gamma_{25'}$ – Γ_{15}) channel of the Si band structure [8]. It decreases at lower wavelengths due to surface recombination, but reaches near unity at 300 nm again due to the $X_4 - X_1$ channel. The IQE between 500nm and 1100nm is improved in the gettered cell, in accordance with the gain in J_{sc}, indicating a better bulk lifetime.

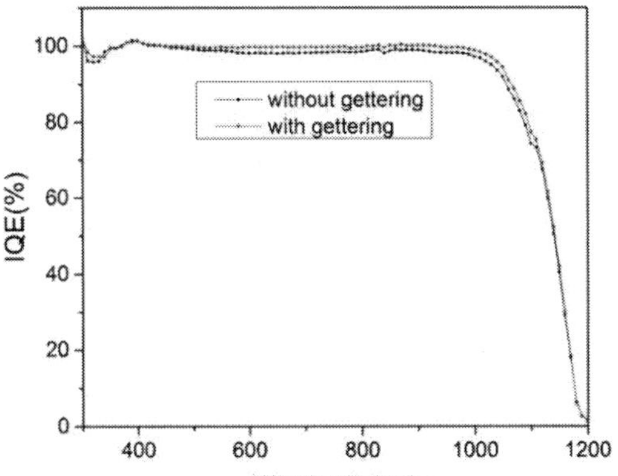

Fig.5: IQE of the cell with and without gettering process.

3.3 Simulation and power loss analysis

The simulation of an median IBC cell was performed using the simulator Quokka, because the approximations involved in the Quokka tool are sufficiently small for the IBC cell design for optimization and power-loss analysis. Quokka allows inputting J_0, bulk lifetime and contact resistivity directly into the software, and estimates the solar cell performance [9]. Table III shows simulated and measured cell performances. The simulated cell performance is in a very good agreement with measured data. Because all the input parameters like J_0, the bulk lifetime, and contact reistivities were independently measured, and the simulated IV parameters concide with the measured one without adjusting any other parameters, there is no internal compensation of losses. For this reason, the simulations can be used for a reliable loss analysis and future road map.

Table III: Simulated and measured IBC cell performances

	J_{sc} (mA/cm^2)	V_{oc} (mV)	FF (%)	Eff. (%)
simulated	42.0	681	80.6	23.1
measured	42.0	680.9	80.4	23.0

Fig. 6 shows the resistive power loss, and Fig. 7 shows the power loss due to recombination. The resistive loss is dominated by the bulk due to a relatively high bulk resistivity and a rather large pitch, which represent a present limit of screen-printing technology. The external series resistance loss indicates that the finger design and printing can be improved further. The emitter sheet resistance also has a potential for further optimization. By using finer geometries, reducing the pitch and contact size, we demonstrated that the maximum efficiency can be as high as 23.5% [7]. The recombination-related power loss is rather equally distributed over the various device regions, as is expected from a rather well optimised cell design. The analysis indicates that it is most effective to decrease the recombination in the passivated parts of the BSF, which means that the passivated J_{0_BSF} would need to be decreased. With the current J_{0_BSF} value of 240 fA/cm^2, there is much room to optimize the diffusion profile of the BSF. Additionally, J_0 of the emitter under both the passivated and the contacted areas, as well as J_0 of the FSF have room for further optimization.

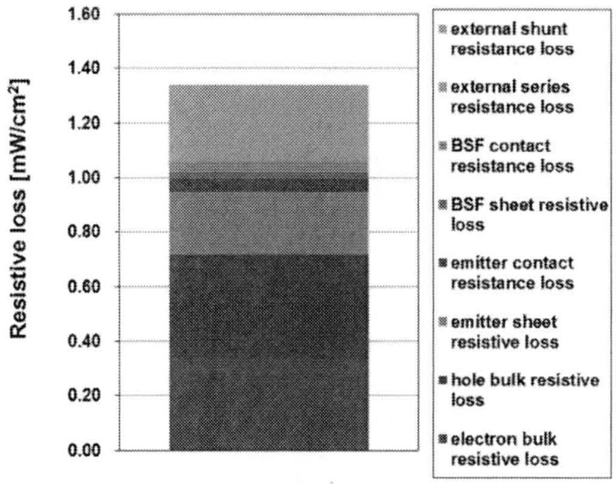

Fig. 6: Resistive power-loss analysis

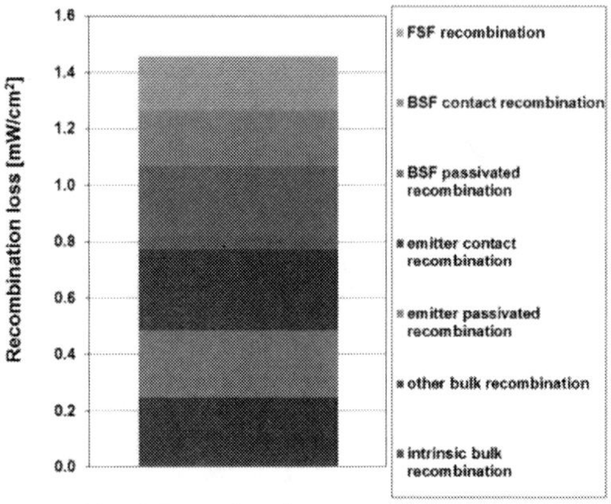

Fig. 7: Power-loss from recombination

In another publication, our team has demonstrated that decreasing the pitch and distributing a larger number of contacts with smaller size contribute to obtaining higher efficiency [7]. A cell efficiency of 23.5% was achieved using ultra smaller contacts, while all other processing steps were kept the same as in the pilot line.

4 CONCLUSIONS

A median efficiency of 23.0% was achieved in Trina Solar's IBC pilot line using a conventional low-cost technology. The power loss analysis indicates that further optimization should focus on decreasing J_0 of the emitter and the BSF in passivated areas as well as in contact areas. To decrease resistive power losses, the substrate resistivity has to be evaluated and optimized. The metallization has also room for further optimization. Future processing developments in the pilot line will focus on cost reduction, while maintaing high efficiency and excellent electrical performance.

ACKNOWLEDGEMENT

This work was supported by the scientific and technological achievement transformation of Changzhou City under the Project Number of CC20150005.

REFERENCES

[1] R.J. Schwartz and M.D. Lammert. Silicon solar cells for high concentration applications, in Technical Digest of the International Electron Devices Meeting, Washington, DC, 350-2 (1975).

[2] M.D. Lammert and R.J. Schwartz. The interdigitated back contact solar cell: a silicon solar cell for use in concentrated sunlight, IEEE Transactions on Electron Devices, ED-24 (1977) 337.

[3] M. A. Green, K. Emery, Y. Hishikawa, W. Warta and E. D. Dunlop. Solar cell efficiency tables (version 47). Prog. Photovolt: 24 (2016) 3.

[4] E. Franklin, K. Fong, K. McIntosh, A. Fell, A. Blakers, T. Kho, D. Walter, D. Wang, N. Zin, M. Stocks, E. Wang, N. Grant, Y. Wan, Y. Yang, X. Zhang, Z. Feng, P. J. Verlinden. Design, fabrication and characterisation of a 24.4% efficient interdigitated back contact solar cell, Prog. Photovolt. 24 (2014) 411-427.

[5] X. Zhang, Y. Yang, W. Liu, K. Zhang, Y. Chen, Z. Li, G. Xu, H. Jiao, S. Zhang, Z. Feng, P. J. Verlinden. Development of high efficiency interdigitated back contact silicon solar cells and modules with industrial processing technologies, 6th World Conf. on PV Energy Conversion 2014, p. 509.

[6]http://ir.trinasolar.com/phoenix.zhtml?c=206405&p=irol-newsArticle&ID=2074749

[7] G. Xu, Y. Yang, X. Zhang, W. Liu, Z. Li, Y. Chen, X. Shu, S. Chen, P. J. Verlinden, Z. Feng, P. P. Altermatt. 6 inch square IBC cell with efficiency above 23.5% fabricated with conventional technology, 43th IEEE PVSC, 2016 (to appear).

[8] S. Kolodinski, J. H. Werner, T. Wittchen, H. J. Queisser. Appl. Phys. Lett. 63 (1993) 2405.

[9] https://www.pvlighthouse.com.au/quokka2

CO-DIFFUSED BACK-CONTACT BACK-JUNCTION SILICON SOLAR CELLS WITH A NOVEL SCREEN-PRINTED BORON-DOPING PASTE

J.D. Huyeng[1], R. Efinger[1], A. Spribille[1], R. Keding[1], A. Wolf[1], O. Doll[2], F. Clement[1], D. Biro[1]

[1]Fraunhofer Institute for Solar Energy Systems (ISE), Heidenhofstrasse 2, D-79110 Freiburg, Germany
[2]Merck KGaA, Frankfurter Str. 250, D-64293 Darmstadt, Germany

ABSTRACT: Back-contact back-junction silicon solar cells feature high conversion efficiency and have since their introduction been of interest for the photovoltaic community. The interdigitated back-contact design reduces shading losses during illumination but is more complex compared to conventional cell concepts, hampering the industrial integration. This issue has been addressed in research conducted by Fraunhofer ISE and Merck KGaA with the introduction of a screen-printable boron-dopant paste. It allows for cost-reduction in established process routes and enables novel co-diffusion processes with a huge potential. The use of screen-printing avoids high investment costs and makes use of a widely adopted technology. The boron emitter formed during a $POCl_3$:O_2:N_2 co-diffusion process has been characterized by secondary ion mass spectroscopy (SIMS) and quasi-steady state photo conductance (QSSPC). Besides that, a proof-of-principle cell batch has been fabricated based on reference work, with a mean efficiency of 20.9 % over 9 cells (aperture size 4 cm²).
Keywords: Silicon Solar Cell, High-Efficiency, Screen Printing, Interdigitated Back-Contact, Co-diffusion

1 MOTIVATION AND STATE OF THE ART

A concept of interdigitated back-contact (IBC) silicon solar cells was developed by Schwartz and Lammert [1] for concentrator applications. The inherent advantage of the cell architecture is a metal-free front side, which reduces shading losses during illumination. Their prototypes reached an energy conversion efficiency of 17.0 % at 50 suns illumination intensity.

The idea was picked up by other researchers, notably the group of Swanson at Stanford, who shifted the focus to one-sun applications. Since the early 2000's IBC solar cells have been fabricated with active areas exceeding 90 cm². The concept was commercialized under the name Maxeon [2] as a product of the 1985 founded SunPower cooperation. While their early adoption provided them with a lead for many years, contributions to the progression of IBC cells have lately also been made by a number of other companies. Panasonic managed to build todays record holding silicon solar cell with an IBC architecture and the use of silicon heterojunctions. Panasonic's IBC cell features an conversion efficiency of 25.6 % [3] on 143.7 cm². SunPower's latest generation of Maxeon cells utilize passivated contacts, to achieve a mean efficiency of 25.2 % on 153.5 cm², while a number of fabrication details remain confidential [4]. Trina Solar in cooperation with the Australian National University build an IBC solar cell with 24.4 % conversion efficiency based on classical diffusion processes on 4 cm² [5] and recently announced the transformation to 243.36 cm² sized wafers with a resulting efficiency of 23.5 % [6].

The research at Fraunhofer ISE has covered basic research of the cell concept [7, 8], advanced designs [9] and alternative fabrication routes [10] in clean-room environment, as well as industrial integration within the photovoltaic technology evaluation center (PV-TEC [11]). A recent cooperation with Merck has shown the feasibility of inkjet-printed boron-inks [12].

The key issue with industrial integration has often been identified in the increased production costs, as the interdigitated rear-side design is more complex than the industry-dominating Al-BSF design. Following the vast majority of published lab scale fabrication routes, commercialization of IBC solar cells would require an expansion of existing production lines with new equipment, which – besides investment cost – often leads to higher fabrication costs and questionable scaling capabilities. In contrast, the introduction of a process sequence based on established equipment in existing production lines would be an adequate method to bypass these restrictions.

Earlier work by Woehl [13] and Kimmerle [14] explored the use of screen-printed Al-paste, as used in standard processes. Keding [15] investigated the utilization of plasma enhanced chemical vapor deposition (PECVD) for phosphorus- and boron-doped silicate glass coatings (PSG/BSG), as PECVD is already established for the deposition of passivation layers and anti-reflection coatings (ARC). This work will focus once more on the integration of screen-printing (SP) into an industrial feasible process route, by introducing a novel boron-rich paste, to innovate the fabrication of the rear side pattern.

Table I: Results of former work at Fraunhofer PV-TEC.

		Woehl [13] Al-paste	Keding [16] PECVD
A	(cm²)	16.65	4.0
V_{OC}	(mV)	647	663
J_{SC}	(mA/cm²)	39.4	40.8
FF	(%)	78.4	78.0
η	(%)	20.0	21.1

A key element, being able to further reduce the fabrication costs, is the introduction of a co-diffusion module, as described by Keding and Rothhardt [17]. The necessary prerequisites and the experimental setup to investigate the feasibility of this approach are described in the following.

2 EXPERIMENTAL SETUP

All experiments were performed under semi-industrial conditions in the PV-TEC on Czochralski (Cz) grown large-size (243 cm²) silicon wafers featuring an n-type base doping (resistivity 3 Ωcm) if not stated otherwise.

2.1 Co-diffusion

The so-called co-diffusion process reduces the

number of necessary process steps as exemplarily shown in Fig. 1. As dopants are simultaneously driven into the material from different sources, the number of cost-intensive high-temperature steps is minimized (compare left to middle column). In general, this also benefits the wafer quality as less handling and chemical treatment is involved. Additionally, co-gettering has been shown to be advantageous compared to a single diffusion species [18].

	sequential diffusion	co-diffusion (PECVD)	co-diffusion (SP)
1	texture	texture	texture
2a	SiOx – front and rear	PSG/SiOx – rear	
2b	structuring	structuring	
3	diffusion – BBr_3	BSG - rear / B-paste	screen printing – B-paste
3b	removal and cleaning		
4a	SiOx – front and rear		
4b	structuring		
4c	diffusion – $POCl_3$		
4d	removal and cleaning		
5a	SiOx – rear		
5b	diffusion – $POCl_3$	diffusion – $POCl_3$	diffusion – $POCl_3$
6	removal and cleaning	removal and cleaning	removal and cleaning
7	passivation – front and rear	passivation – front and rear	passivation – front and rear
8a	structuring	structuring	
8b	metallization - PVD	metallization - PVD	metallization - SP
8c	structuring	structuring	
9	annealing	annealing	annealing

Figure 1: Schematic process routes.
Orange (dashed) highlighted boxes refer to high-temperature steps. Green (dotted) highlighted boxes include the use of screen-printed boron-paste. The co-diffusion based on PECVD (middle column) offers a vast reduction in process complexity [15]. This paper describes an alteration in process step 3, where boron-paste substitutes BSG. A further reduction is shown in the right column if all PSG layers are substituted by a single $POCl_3$ diffusion process.

To enable this approach, the used sources must not interfere with each other. It has been shown that certain PECVD or APCVD layers can be used with each other (solid diffusion sources [15]) or in combination with a precursor atmosphere [19–21], like gaseous mixtures consisting of phosphorus oxychloride, oxygen, and nitrogen ($POCl_3$:O_2:N_2).

Ultimately, the use of a screen-printed doping source may enable further reduction of process complexity and cost (compare middle and right column), with the metallization possibly also screen-printed. SCost calculations [22] are currently under investigation to evaluate this process in detail, but have already indicated that this should be competitive to both-sides contacted passivated emitter and rear (PERC) concepts.

2.1 Boron-paste investigation

For first investigations concerning the designated novel boron-paste, full area and structured samples were prepared. The application was performed with an industrial screen-printer on large-size wafers. The samples were then treated in an industrial diffusion furnace with a $POCl_3$:O_2:N_2 atmosphere, resulting in the simultaneous formation of highly doped *n*- and *p*-type structures from thermally grown PSG and boron-paste, respectively.

The printing results were analyzed with an optical microscope. The resulting doping structures were visualized with infrared absorption imaging (Sheet-Resistance Imaging [23]), after removal of boron-paste and PSG layers.

The boron doping profile was measured with secondary ion mass spectroscopy (SIMS).

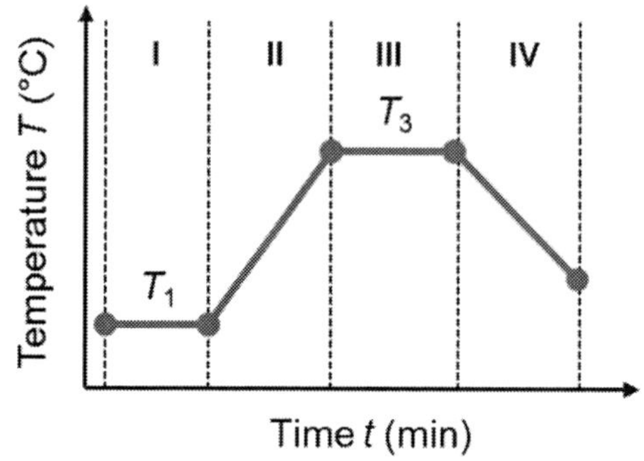

Figure 2: Schematic diffusion process.
A simplified time-temperature diagram is given to illustrate the used diffusion process. The timeframes I-IV refer to different set-temperatures and may in general feature different gas flows to *e.g.* enable the growth of PSG layers from a $POCl_3$:O_2:N_2 atmosphere.

2.2 Passivation quality

To evaluate the effectiveness of surface passivation, symmetrical lifetime samples were fabricated in the same process. Here both sides were covered with screen-printed boron-paste before diffusion. After diffusion, paste removal and cleaning, the samples were coated with atomic layer deposited AlO_x (ALD AlO_x [24]) and PECVD deposited ARC SiN_x on both sides.

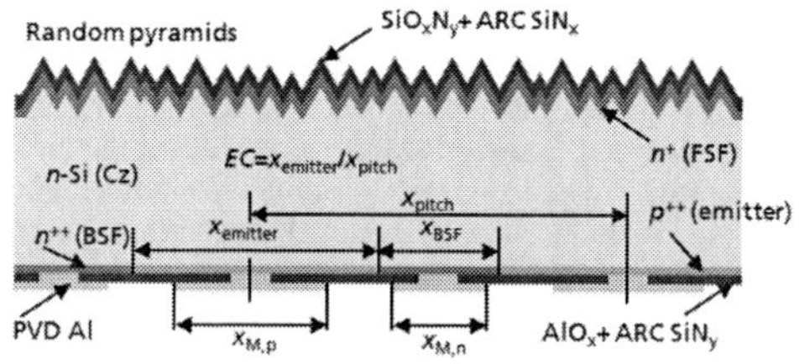

Figure 3: Schematic sample structure.
Arrows and markers give geometrical dimensions used in the text. Colors refer to differently doped or coated layers, as indicated. The front side texture is sketched by random pyramids. All dimensions are not to scale.

The samples were measured with Quasi-steady state photo conductance (QSSPC) on a Sinton WCT-120 setup [25]. The dark saturation current density J_0 was evaluated at high-injection (see details below), according to Kimmerle *et al.* [26].

2.3 Proof-of-principle cell batch

A first batch of (2 x 2) cm² solar cells was fabricated on (15.6 x 15.6) cm² wafers. The front and rear side were covered with PECVD deposited PSG, adjusted for front- and back-surface field (FSF/BSF) formation. The rear side was patterned with a mask-and-etch process involving an inkjet-patterning process module as described by Keding *et al.* [27].

Instead of a PECVD BSG layer a full-area screen-printed boron-paste layer was added on the rear side. Samples were treated in the same diffusion furnace as above.

After diffusion, dopant layer removal and cleaning, the rear side was passivated with a stack consisting of

ALD AlO_x and ARC SiN_x, while the front side was passivated with a $SiriO_xN_y$/ARC SiN_x stack [28]. This is the same setup as used in previous work [15], to emphasize the comparability of boron-paste performance to PECVD BSG.

The rear side's passivation stack and the physical vapor deposited (PVD) Al layer utilized as electrode material was patterned according to the inkjet-pattering process module, as well. Finished solar cells were then tested with an in-house solar simulator and compared with a measurement at the photovoltaic calibration laboratory at Fraunhofer ISE (CalLab) [29].

3 RESULTS

The above described experiments were conducted during fabrication of the proof-of-principle cell batch in order to minimize deviations between test samples and solar cells. If not otherwise stated, measurements were conducted in-house on regularly tested and calibrated equipment.

3.1 Sheet-Resistance Imaging

Sheet resistance imaging as introduced by Isenberg *et al.* allows for spatial analysis of doping structures [30]. The infrared absorption images revealed a distinct contrast between screen-printed regions and prior-to-diffusion bare surface of several hundred counts. The dimensions correlate well with the optically accessible features of the print. The signal was more blurred in a given region compared to a PECVD layer, but measurements were dominated by artifacts in this regard.

3.2 Secondary Ion Mass Spectroscopy

The element sensitive SIMS measurements revealed boron and phosphorus concentrations after $POCl_3$:O_2:N_2 diffusion in a boron-paste covered region after a standard cleaning procedure, as given in Fig. 4. The measurements were performed by CiS GmbH [31].

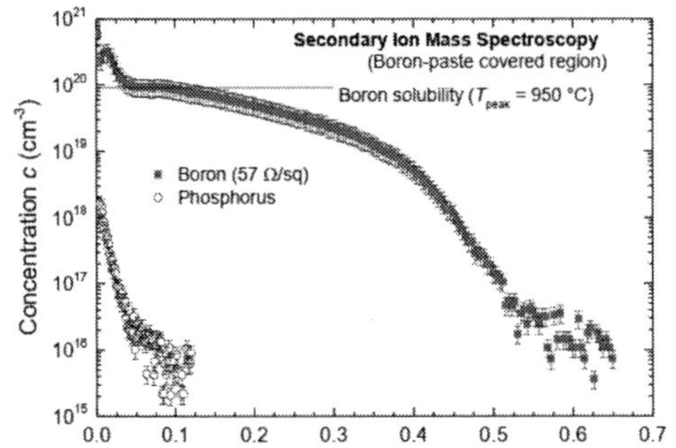

Figure 4: SIMS measurement.
The SIMS signal is calculated for boron and phosphorus species from two measurements on the same sample structure and plotted against the calculated sputter depth. Active boron solubility according to [32]. Details on interpretation are given in the text.

The measured surface concentration of boron exceeds its solubility level in silicon, hinting at remaining precipitates referred to as boron-rich layer [20]. Depth and the resulting sheet resistance of the dopings are comparable to those diffused from BSG, enabling a substitution in this aspect. Boron was never overcompensated by phosphorus, as its concentration is at least two orders of magnitude lower. The used paste can therefore be deemed as a sufficient diffusion barrier during $POCl_3$ diffusion (see Sec. 4).

To avoid the formation of a recombination active boron-rich layer at the surface, the cleaning and the diffusion process have been adjusted in the following.

3.3 Quasi-Steady State Photo conductance

The extracted J_0-values are depicted in Fig. 5. Measurements were performed at different sample positions on multiple wafers. Resulting mean, median and standard deviation values are given by the enclosing box. As a comparison, the mean values of BSG and PSG diffusion are given, as used in the reference process. The BSG and PSG (BSF) regions were passivated alike, while the PSG (FSF) was passivated with $SiriO_xN_y$ and ARC SiN_x, to reproduce the passivation scheme of the solar cell (see Sec 2.3).

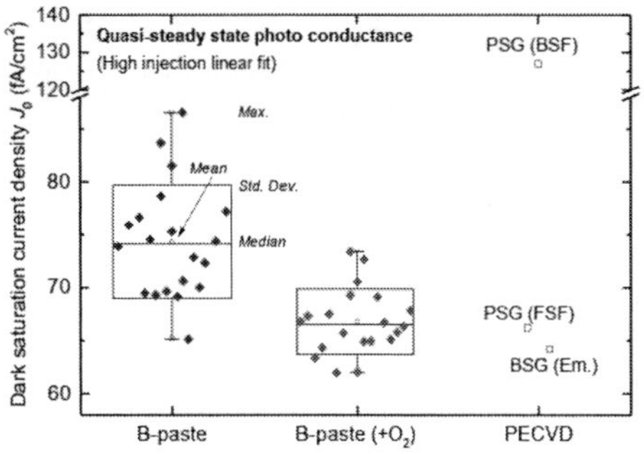

Figure 5: QSSPC measurement results.
The high-injection linear fit results from multiple wafers and different sample positions are given and statistically analyzed. Reference values are given on the right. Details on interpretation are given in the text. (Samples were measured at room temperature, $N_D = 1.3 \times 10^{15}$ cm^{-3}, $n_i = 9.2 \times 10^9$ cm^{-3}, $\Delta n = 1.8\text{-}2.0 \times 10^{16}$ cm^{-3}).

The resulting values are between 65.2 fA/cm² and 86.5 fA/cm² for the standard process. With additional oxidation during cooling (*cf.* Fig. 2, frame IV), an enhancement can be achieved in the mean as well as in the standard deviation, *i.e.* giving a more homogenous process. Its mean value is comparable with the BSG reference, fueling the interchangeability. The given PSG values have been used in numerical simulations (see Sec. 4.1).

3.4 I-V Characteristics

The fabricated cells achieve the same performance level as the reference process [16] within error margins, as could be expected from the emitter characterization. The validity of the average value has been verified by a measurement at Fraunhofer ISE CalLab, perfectly agreeing within the given statistical and systematical errors. These results demonstrate the feasibility of screen-printed boron-rich paste on solar cell level using industrial equipment.

The ideal fill factor FF_0 according to Green [33] could be calculated to 83.9 %, while the pseudo fill factor pFF as measured with a Sinton sunsVoc setup was 82.3 %, for the CalLab confirmed cell. The difference pFF-FF, related to the series resistance, is 3.8 %$_{abs}$ and is thus greater than the difference FF_0-pFF of 1.6 %$_{abs}$, related to leakage currents.

Table II: IV characteristics of proof-of-principle cell batch (one-sun condition)
Solar cells were fabricated using screen-printed boron-paste and co-diffusion (see Sec. 2.3). The cells were measured with a busbar shading aperture, resulting in an active area of 4 cm² [4].

		Average (9 cells)	FhG ISE CalLab
V_{OC}	(mV)	656 ±3	654 ±2
J_{SC}	(mA/cm²)	41.0 ±0.2	40.0 ±0.8
FF	(%)	77.6 ±0.5	78.9 ±0.5
η	(%)	20.9 ±0.3	20.7 ±0.4

Accordingly, the structure is currently rather limited by series resistance than leakage current, which resembles earlier results from the reference process. The series resistance could be decreased by *e.g.* adjusting the architecture. Leakage currents might be further limited by adjusting the pattern of the passivation stack, manipulating the expansion of the space charge region at silicon's near-surface, or both [19].

4 NEXT STEPS

As shown in the previous paragraph the screen-printed boron-rich paste can be used in the fabrication of IBC solar cells, substituting a cost-intensive PECVD deposition step. Furthermore, the paste has been shown to act as a diffusion barrier during a $POCl_3$:O_2:N_2 involving diffusion process, enabling an even narrower process route (*cf.* Fig. 1), which is the focus of future work.

A simulation of the current preliminary characteristics of the demonstrated cell vehicle shows the influence of the chosen geometry and concludes the investigations at this point.

4.1 Simulation results

Numerical simulations were performed with the freely available QUOKKA 2 [34]. Non-ideal material properties like bulk contaminants were mostly neglected, to focus on the process potential. FSF and BSF regions were modelled according to the reference work, the emitter was adjusted with the measured sheet-resistance and dark saturation current density. Recombination at contacted regions was simulated by EDNA 2 [35] and the profile shown in Fig. 4. The surface recombination velocity at the interface metal/silicon was assumed to be 10^7 cm/s.

A sweep of geometry parameters was performed, keeping the width of the BSF region constant, *i.e.* varying pitch and emitter width and therefore the emitter coverage. Additionally, the metal finger resistance was adjusted according to a geometrical relation.

A relative improvement in conversion efficiency of about 1 $\%_{rel}$ was achieved within simulation, clearly dominated by the fill factor *FF*. A similar improvement could also be achieved by replacing the front surface diffusion by a $POCl_3$:O_2:N_2 process as demonstrated by Tanvir *et al.* [36], according to the simulations.

The development of the simplified process outlined above (see Fig. 1) is expected to yield both: a cheaper fabrication sequence and increased device performance, as understanding of the specific requirements of screen-printed boron-paste grows.

4.2 Challenges for co-diffusion

The challenge in focus of future experiments will be the formation of *p*- and *n*-type doping structures with high resolution and edge sharpness, via co-diffusion. Screen-printed boron-paste will be combined with a $POCl_3$ based diffusion process, in order to fully exploit the huge cost saving potential (see Fig. 1, right column). The involved diffusion sources may not interfere with each other, while also providing sufficient emitter and BSF diffusions to be contacted with screen-printed metallization. Furthermore, the front side diffusion is, in general, different from the rear side BSF, which has in other work been addressed by etch-back experiments [37].

5 SUMMARY

A novel screen-printed boron-doping paste has been shown to enable an industrially feasible process route involving a co-diffusion module in order to fabricate highly-efficient IBC solar cells. In first experiments the fundamental prerequisites of the intended boron-rich paste were investigated and the usability was demonstrated on solar cell level, yielding devices with an average efficiency of 20.9 %. Upcoming experiments will focus on the compatibility of boron-paste and other diffusion sources.

The authors would like to thank S.W. Glunz for fruitful discussion, E. Schäffer for I-V measurements, the whole PV-TEC team at Fraunhofer ISE and the development teams at Merck.

6 REFERENCES

[1] R. J. Schwartz and M. D. Lammert, "Silicon solar cells for high concentration applications," in *1975 International Electron Devices Meeting*, pp. 350–352.

[2] W. P. Mulligan *et al*, "Manufacture of solar cells with 21% efficiency," in *19th EU PVSEC*, 2004, pp. 387–390.

[3] K. Masuko *et al*, "Achievement of more than 25% conversion efficiency with crystalline silicon heterojunction solar cell," *IEEE Journal of Photovoltaics*, vol. 4, no. 6, pp. 1433–1435, 2014.

[4] M. A. Green, K. Emery, Y. Hishikawa, W. Warta, and E. D. Dunlop, "Solar cell efficiency tables (version 47)," *Prog. Photovolt: Res. Appl*, vol. 24, no. 1, pp. 3–11, 2016.

[5] E. Franklin *et al*, "Design, fabrication and characterisation of a 24.4% efficient interdigitated back contact solar cell," *Prog. Photovolt: Res. Appl*, vol. 24, no. 4, pp. 411–427, 2014.

[6] *Trina Solar sets 23.5% IBC cell conversion efficiency record for screen printed process.* [Online] Available: http://www.pv-tech.org/news/trina-solar-sets-23.5-ibc-cell-conversion-efficiency-record-for-screen-prin. Accessed on: May 25 2016.

[7] J. Dicker, J. O. Schumacher, W. Warta, and S. W. Glunz, "Analysis of one-sun monocrystalline rear-contacted silicon solar cells with efficiencies

of 22.1%," *J. Appl. Phys,* vol. 91, no. 7, p. 4335, 2002.

[8] F. Granek, M. Hermle, C. Reichel, O. Schultz-Wittmann, and S. W. Glunz, "High- efficiency back-contact back-junction silicon solar cell research at Fraunhofer ISE," in *23rd EU PVSEC*, 2008, pp. 991–995.

[9] C. Reichel, F. Granek, M. Hermle, and S. W. Glunz, "Back-contacted back-junction n -type silicon solar cells featuring an insulating thin film for decoupling charge carrier collection and metallization geometry," *Prog Photovoltaics*, no. 5, pp. 1063–1076, 2012.

[10] R. Müller *et al,* "Analysis of n-type IBC solar cells with diffused boron emitter locally blocked by implanted phosphorus," *Solar Energy Materials and Solar Cells*, vol. 142, pp. 54–59, 2015.

[11] D. Biro *et al,* "PV-Tec: Retrospection to the three years of operation of a production oriented research platform," in *24th EU PVSEC*, 2009, pp. 1901–1905.

[12] D. Stüwe *et al,* "Inkjet-printed diffusion sources," in *29th EU PVSEC*, 2014, pp. 976–979.

[13] R. Woehl *et al,* "20% Efficient screen-printed and aluminum-alloyed back-contact back-junction cells and interconnection scheme of point-shaped metalized cells," in *37th IEEE Photovoltaic Specialists Conference Seattle: 2011*, 2011, pp. 48–52.

[14] N. Wöhrle, A. Kimmerle, J. Greulich, and A. Wolf, "All-Diffused Back-Contact Back-Junction Solar Cell With Aluminum-Alloyed Emitter—Experiment and Simulation," (English), *IEEE Journal of Photovoltaics*, vol. 6, no. 3, pp. 641–648, http://ieeexplore.ieee.org/iel7/5503869/7456349/07442767.pdf?arnumber=7442767, 2016.

[15] R. Keding *et al,* "Co-diffused back-contact back-junction silicon solar cells without gap regions," *IEEE Journal of Photovoltaics*, vol. 3, no. 4, pp. 1236–1242, 2013.

[16] R. Keding, *Co-diffused Back-Contact Back-Junction Silicon Solar Cells*. Stuttgart: Fraunhofer Verlag, 2015.

[17] R. Keding *et al,* "Silicon doping performed by different diffusion sources aiming co-diffusion," in *27th EU PVSEC*, 2012, pp. 1906–1911.

[18] J. Schön, M. C. Schubert, W. Warta, H. Savin, and A. Haarahiltunen, "Analysis of simultaneous boron and phosphorus diffusion gettering in silicon," *Phys. Status Solidi A*, vol. 207, no. 11, pp. 2589–2592, 2010.

[19] R. Keding *et al,* "$POCl_3$-based co-diffusion process for n-type back-contact back-junction solar cells," in *29th EU PVSEC*, 2014, pp. 676–680.

[20] P. Rothhardt, C. Demberger, A. Wolf, and D. Biro, "Co-diffusion from APCVD BSG and POCl3 for industrial n-type solar cells," *Energy Procedia*, no. 38, pp. 305–311, 2013.

[21] S. Meier *et al,* "Co-Diffusion for P-Type PERT Solar Cells Using APCVD BSG Layers as Boron-Doping Source," in *32nd EUPVSEC 2016*

[22] S. Nold *et al,* "Cost Modeling of Silicon Solar Cell Production Innovation along the PV Value Chain," in *27th EU PVSEC*, 2012, pp. 1084–1090.

[23] J. Isenberg, D. Biro, and W. Warta, "Fast, contactless and spatially resolved measurement of sheet resistance by an infrared method," *Prog Photovoltaics*, vol. 12, no. 7, pp. 539–552, 2004.

[24] A. Richter, J. Benick, and M. Hermle, "Boron Emitter Passivation With Al2O3 and Al2O3 SiNx Stacks Using ALD Al2O3," *IEEE J. Photovoltaics*, vol. 3, no. 1, pp. 236–245, 2013.

[25] A. L. Blum, J. S. Swirhun, R. A. Sinton, and A. Kimmerle, "An updated analysis to the WCT-120 QSSPC measurement system using advanced device physics," in *28th EU PVSEC*, 2013, pp. 1521–1523.

[26] A. Kimmerle, J. Greulich, and A. Wolf, "Carrier-diffusion corrected J0-analysis of charge carrier lifetime measurements for increased consistency," *Solar Energy Materials and Solar Cells*, vol. 142, pp. 116–122, 2015.

[27] R. Keding *et al,* "Diffusion and characterization of doped patterns in silicon from prepatterned boron- and phosphorus-doped silicate glasses," in *26th EU PVSEC*, 2011, pp. 1385–1389.

[28] J. Seiffe *et al,* "Surface passivation of crystalline silicon by plasma-enhanced chemical vapor deposition double layers of silicon-rich silicon oxynitride and silicon nitride," *J. Appl. Phys,* vol. 109, no. 3, p. 34105, 2011.

[29] *CalLab PV Cells / CalLab PV Modules — Fraunhofer ISE*. [Online] Available: https://www.ise.fraunhofer.de/en/service-units/callab-pv-cells-callab-pv-modules. Accessed on: Jun. 06 2016.

[30] J. Isenberg, M. C. Schubert, D. Biro, A. Froitzheim, and W. Warta, "Sheet resistance imaging (SRI) - A contactless and spatially resolved method for the determination of doping inhomogeneities," in *20th EU PVSEC*, 2005, pp. 674–677.

[31] CiS Forschungsinstitut für Mikrosensorik und Photovoltaik GmbH, *CiS :: Willkommen bei CiS*. [Online] Available: http://www.cismst.org/. Accessed on: May 25 2016.

[32] D. Nobili, "Solubility of B in Si," *Properties of Silicon*, p. 384, 1987.

[33] M. A. Green, "Solar cell fill factors: General graph and empirical expressions," *Solid State Electron*, vol. 24, no. 8, pp. 788–789, 1981.

[34] A. Fell, "A free and fast three-dimensional/two-dimensional solar cell simulator featuring conductive boundary and quasi-neutrality approximations," *IEEE Trans. Electron Devices*, vol. 60, no. 2, pp. 733–738, 2013.

[35] K. R. McIntosh *et al,* "An Examination of Three Common Assumptions Used to Simulate Recombination in Heavily Doped Silicon," in *28th EU PVSEC*, 2013, pp. 1672–1679.

[36] N. B. Tanvir *et al,* "Codiffusion Sources and Barriers for the Assembly of Back-Contact Back-Junction Solar Cells," *IEEE J. Photovoltaics*, vol. 5, no. 6, pp. 1813–1820, 2015.

[37] A. Kimmerle, "Diffused Surfaces For Crystalline Silicon Solar Cells: Process Development, Characterization, and Modeling," Dissertation, IMTEK, Albert-Ludwigs-University, Freiburg, 2015.

5" LASER-IBC SOLAR CELLS WITH 22.0% EFFICIENCY

E. Hoffmann[1], M. Dahlinger[1], K. Carstens[1], S. Wansleben[2], R. Zapf-Gottwick[1] and J.H. Werner[1]
[1]Institute for Photovoltaics and Research Center SCoPE, University of Stuttgart
Pfaffenwaldring 47, 70569 Stuttgart, Germany
[2]SolSol GmbH
Haydnstraße 23a, 12167 Berlin, Germany

ABSTRACT: This contribution presents our latest progress on 5x5 inch solar cells yielding a maximum efficiency $\eta = 22.0\%$. The cell process utilizes four laser irradiation steps for emitter and back surface doping as well as for the ablation of the rear passivation layer and structuring of the metallization. The improvement from the previously measured $\eta = 20.8\%$ is achieved by busbar-tapering, improved cleaning and higher area utilization. Photoluminescence proves the homogeneity and accuracy of all four laser processes. Electroluminescence illustrates the improvement achieved by the tapered of the busbars and suggests that even higher efficiencies will be possible.

Keywords: Laser, High Efficiency, Low Cost

1 INTRODUCTION

The Institute of Photovoltaics (*ipv*) recently developed a laser processed interdigitated back contact (IBC) solar cell which yields an efficiency $\eta = 23.2\%$ on an area $A = 2x2$ cm² [1]. By implementing four laser irradiation steps, the fabrication process avoids any photolithographic masking steps. The laser dopes both, emitter and back surface field (BSF), locally ablates the passivation layer for contact formation, and, finally, structures the metallization. Compared to traditional IBC processes, which require multiple photolithographic masking steps, the *ipv* process is versatile, fast and has the potential of a module production cost around 49 €cent/Wp.

In order to be industrially compatible and cost competitive, the process needs to be implemented on industrial sized 6 inch (15.6x15.6 cm²) pseudo square wafers, while yielding an efficiency $\eta > 22\%$. The upscaling onto the larger cell sizes requires homogeneous doping / passivation across the whole area and several busbars on the active cell area. Additionally, all four laser processes require an accuracy of at least ± 45 µm on the area $A = 15.6x15.6$ cm², to maintain the high cell efficiency.

This work proves the areal homogeneity and precision of the laser IBC process on an area $A = 12.5x12.5$ cm² (156 cm²), corresponding to 5 inch square solar cells. By tapering the busbars. the short circuit current rises and, along with a higher fill factor, the (maximum) efficiency increases by $\Delta\eta = 1.2\%$ within the last year from $\eta = 20.8\%$ (2015) to $\eta = 22.0\%$ (2016). The characterization of the cells promises an even higher efficiency in the future with the next optimization step, and the further upscaling onto 6 inch wafers. Measuring the current/voltage characteristics of the cells remains challenging; the currently used measurement chuck affects the fill factor, implying that even the present cells probably have an even higher efficiency.

2 LASER IBC SOLAR CELL

Figure 1a) shows the schematic cross-section of the laser IBC solar cell without busbars and figure 1b) shows the process flow. We use 6 inch n-type Czochralski silicon wafers. The first two laser irradiation steps dope the emitter and BSF areas by locally melting the silicon. In the molten silicon, the respective dopant diffuses from a precursor layer, being either a sputtered boron layer or a phosphor silicate glass. Although we already use 6 inch wafers, at present, the lasers only irradiate an area of 12.5x12.5 cm², as the boron precursor layer is not yet homogeneously thick on 6" wafers. The third laser process ablates point contacts in the passivation layer on the rear. Here, the contact to the subsequently evaporated aluminum (Al) layer forms. The fourth laser process structures an etch resistant layer on the evaporated full-area Al layer. Phosphoric acid then separates the Al layer into contact fingers, by etching the Al beneath the ablated masking layer. Finally, a laser separates the processed area of 156 cm² by laser scribing from the remaining 6 inch wafer. However, in future, this final laser cutting process will not be part of the actual production process. A detailed description of the process can be found in previous work [1].

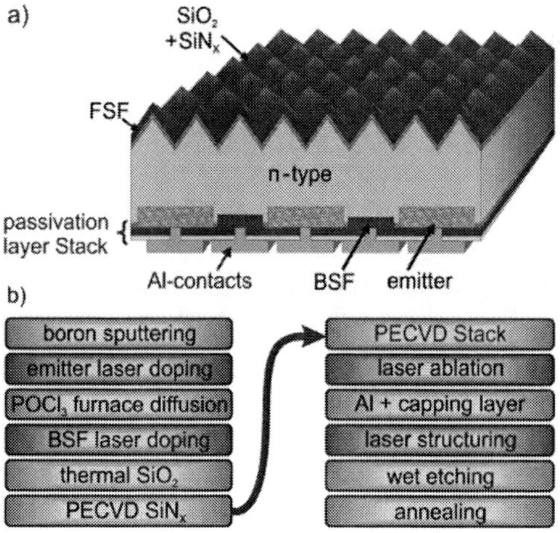

Figure 1: a) Schematic cross section of the laser interdigitated solar cell and b) the simplified process flow. Four laser processes are applied for doping of both the emitter and back surface field (BSF), laser ablation of the rear passivation stack and structuring of the metallization. The detailed process flow has been describes previously.

The metallization of the 156 cm² sized cells requires several busbars on the active cell area. For now, we use three busbars of each polarity. The busbars on the active cell area require additional doping of the silicon above each busbar according to its respective polarity. Without this doping above the busbars there is the possibility of a parasitic diode or even short circuit, if the passivation layer fails to insulate the wafer's surface [2]. However, a perfectly insulating layer is challenging to achieve and always poses a risk in the solar cell production.

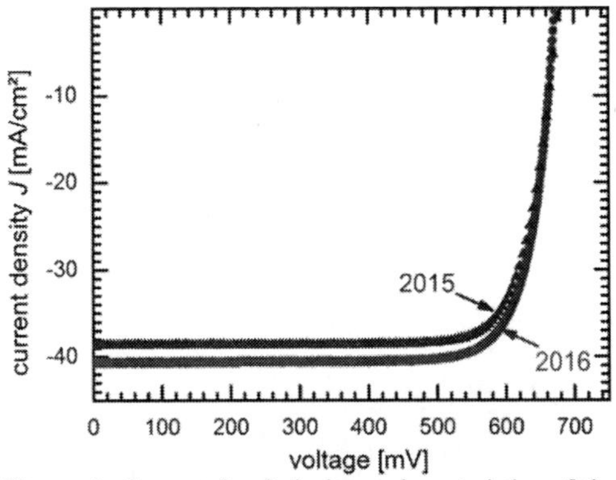

Figure 2: Current density/voltage characteristics of the illuminated 156 cm² sized solar cells from 2015 and 2016. With $\eta = 22.0\%$, the present cell (2016) gained an efficiency $\Delta\eta = 1.2\%$.

On the previously processed 5 inch solar cells (2015) the busbars width was constant across the complete 5 inch cell. The present design features tapered busbars, with a reduced busbar area fraction (positive and negative), thus lowering electric shading and series resistance [3].

3 CELL RESULTS

Figure 2 compares the illuminated current density/voltage (J/V) characteristics of the record 5 inch solar cell from 2016 with the best 5 inch cell from 2015 with $\eta = 20.8\%$. For the in-house measurement, the measurement chuck was calibrated with a certified 2x2 cm² sized cell. The measuring chuck contacts each busbar of the 5 inch solar cell with three pin pairs. A glass pane on top of the cell counters the pressure applied by the pins. The pane, howeve,r increases the reflection and also absorbs short wavelength radiation. The calibration with the 2x2 cm² sized cell took reduced illumination and different heat dissipation of the chuck into account. We compensate the reduced transmission by increasing the light intensity correspondingly. The calibration cell and the 5 inch cells exhibit the same optics and quantum efficiencies.

Table 1 presents the J/V parameters extracted from the illuminated J/V characteristics. Compared to the 156 cm² sized cell from 2015, both short circuit current $\Delta J_{SC} = 1.7$ mA/cm² and fill factor $\Delta FF = 0.5\%$ contribute to the efficiency gain of $\Delta\eta = 1.2\%$. The rise in J_{SC} stems mainly from decreased electrical shading by the tapered busbars as more generated carriers are collected.

The tapered busbars affect the FF in three ways: i) the reduced doped area under the positive busbar lowers the series resistance R_S, while ii) the lower busbar diameter could increase R_S. iii) Furthermore, reduced non-ideal recombination in the busbar region increases the pseudo fill factor, hence improves the fill factor. Overall we have a fill factor gain of $\Delta FF = 0.5\%$.

The V_{OC} decreases by $\Delta V_{OC} = 2$ mV, even though we improved wafer cleaning and handling. A possible reason lies in a lower bulk charge carrier lifetime of the record cell; the mean $V_{OC} = 673\pm 2$ mV of the batch exceeds the $V_{OC} = 672$ mV of the record cell. The lower $\Delta V_{OC} = 10$ mV compared to the 4 cm² sized cell is explained by laser cutting of the active cell out of the processed wafer. Similar to laser scribing, the laser cutting induces defects [4]. Additionally, the edges remain unpassivated. Both effects increase the saturation current density, hence lower V_{OC}. However, when manufacturing 6 inch cells in the future, this step will be obsolete.

Table 1: J/V parameters of the record 4 cm² solar cell, the first 156 cm² solar cell from 2015, the present record 156 cm² solar cell from 2015 and mean J/V parameters of the corresponding run with 14 cells.

A [cm²]	η [%]	V_{oc} [mV]	J_{sc} [mA/cm²]	FF [%]
4 (2015)	23.2	682	41.3	82.5
156 (2015)	20.8	674	38.9	79.8
156 (2016)	22.0	672	40.6	80.3
156 (2016)	21.3±0.5	673±2	40.7±0.4	78.1±1.3

4 DISCUSSION

Figures 3a),b) shows the photoluminescence (PL) image of 5 inch cells from the years a) 2015 and b) 2016. The images are captured under different conditions; thus, the intensity is not comparable. Both images prove the accuracy and homogeneity of the laser processes as none of the cells are shunted. The low current saturation densities of emitter $J_{0,E} = 46$ fA/cm² and BSF $J_{0,BSF} = 73$ fA/cm² [5] indicate a low defect density. The speckles with low PL-intensity, visible only in fig. 3a), originate from insufficient cleaning, which we improved in the 2016 process. The negative busbars appear bright in both images, as the silicon above the busbars is only lightly doped by the $POCl_3$ diffusion and features a low current saturation density $J_0 = 7$ fA/cm². The images show the changed shape of the busbars, from the constant width in the year 2015 to the tapered design in 2016.

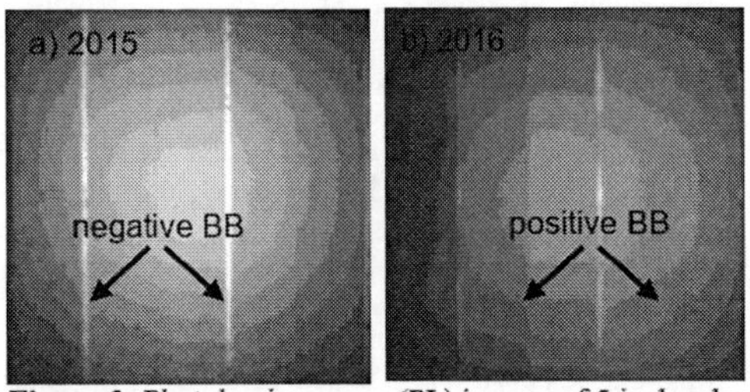

Figure 3: Photoluminescence (PL) images of 5 inch solar cells from a) 2015 and b) 2016. Both images prove sufficient accuracy of the laser processes. The 2016 image shows no dark speckles due to improved cleaning and tapered busbars.

Figures 4a), b) show the electroluminescence (EL) images of the 156 cm² cells from 2015 and 2016. Again the luminescence intensity between the pictures is not comparable. In the EL images the negative busbars show low luminescence, illustrating the effect of electrical shading. The darker positive busbars show additional series resistance, caused by the emitter doping above the busbars. The tapered busbars (2016) reduce both effects. In figure 4a) the luminescence intensity remains constant in direction of the busbars, indicating that the busbars do not affect the series resistance significantly. In figure 4b) the luminescence intensity changes along the busbars. At the widest point of the busbar, the pin pairs contact the solar cell and inject the carriers. As the busbar width decreases, the resistance rises and less carriers are injected into the silicon. This implies that the present design is not fully optimized yet. More contact pins will reduce the series resistance during the IV-measurement and yield a higher fill factor. In 2015 we still experienced interrupted fingers, as the etch resistant layer was not stable across the complete wafer surface. An improved laser process now reliably protects the Al against the etchant, thus we detect no finger interruptions on the area A = 12.5x12.5 cm^2.

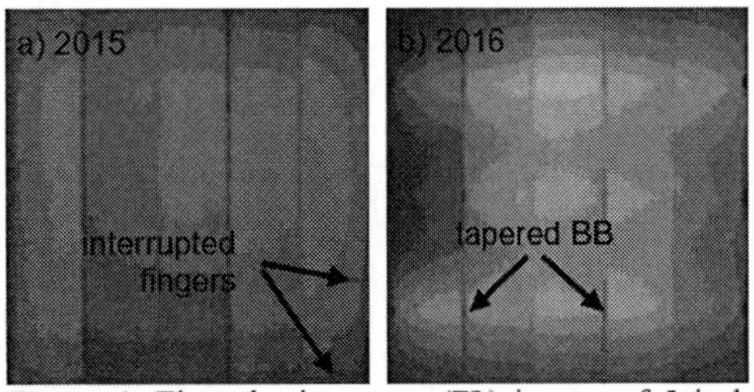

Figure 4: Electroluminescence (EL) images of 5 inch solar cells from the years a) 2015 and b) 2016. The busbars show low EL and illustrate electric shading and higher local series resistance. By tapering the busbars on the 2016 cell, the short circuit current rises and fill factor rises. Furthermore, the 2016 cell shows no interrupted fingers.

4 CONCLUSION

The four laser processes necessary for the production of laser IBC solar cells are implemented on 5 inch solar cells yielding η = 22.0%. The laser processes provide high accuracy and homogeneity on an area A =12.5x12.5 cm^2. At present, due to the geometry of the sputtering target, the boron precursor layer still exhibits a radial inhomogeneity in thickness. This varying precursor thickness layer hinders the processing of 6 inch cells. With new deposition equipment, which secures the homogeneous thickness on larger areas, 6 inch solar cells will be possible in the near future. The processing of 6 inch wafers also reduces perimeter effects, mainly because the active cell is not cut from the remaining wafer. Therefore, we will avoid laser damage and non-passivated edges.

The tapering of the busbars increased both J_{SC} and FF. However, the doped silicon above the busbars still limits the efficiency by electric shading and additional series resistance. A further optimization of the design with more contact pads, will lower the series resistance during the J/V-measurement. With continuing improvements of doping profiles, contact formation and the use of high lifetime wafer material, we will reach η > 22% also on standard 6 inch cells.

A detailed cost of ownership calculation shows that a module production cost of 49€cent/W$_P$ will be possible, even assuming a 100 MW$_P$ production line, located in Germany. The fast and versatile laser processes, along with the low-cost Al-metallization promise a competitive product in the near future.

ACKNOWLEDGMENT

The authors gratefully thank L. Beisel, H. Moldenhauer, B. Lutz, L. Bauer, and B. Winter for technological support in processing of the solar cells. This work is supported by the German Federal Ministry of Economics and Technology (BMWi) under project number 0325714SA.

REFERNCES

[1] M. Dahlinger, K. Carstens, E. Hoffmann, S.Wansleben, J. R. Köhler, R. Zapf-Gottwick and J. H. Werner, "23.2% Effciency with Laser Processed IBC Solar Cells", *in Proc. 31th Europ. Photovolt. Solar Energy Conf.*, (WIP, Munic, 2015), pp. 462-465.

[2] P. Magnone, M, Debucquoy, D Giaffreda, N. Posthuma, c. Fiegna, "Understanding the Influence of Busbars in Large-Area IBC Solar Cells by Distributed SPICE Simulations.", IEEE J. Photovolt. **5**, 552-558 (2015).

[3] C. Reichel, M. Reusch, F. Granek, M. Hermle and S. W., "Decoupling Charge Carrier Collection and Metallization Geometry of Back-Contacted Back-junction Silicon Solar Cells by using Insulating Thin Films," *in 35th IEEE Photovolt. Specialists Conf. (IEEE, New York, 2010)*, pp. 001034-001038.

[4] A. Hauser, G. Hahn, M. Spiegel, H. Feist*, O. Breitenstein, J.P. Rakotoniaina, P. Fath, E. Bucher "Comparison of Different Techniques for Edge Isolation" *in Proc. 17th Europ. Photovol. Solar Energy Conf.*, (WIP, Munic, 2001), pp. 1739-1742.

[5] Kai Carstens, Morris Dahlinger, Erik Hoffmann, Jürgen R. Köhler, Renate Zapf-Gottwick, Jürgen H. Werner, "Universal Passivation for p++ and n++ Areas on IBC Solar Cells", Energy Procedia **77**, 779-785 (2015).

NANOSTRUCTURED MGO-DOPED TIO_2 AEROGELS FOR ENHANCED MONOCRYSTALLINE SILICON SOLAR CELLS

Fanchao Meng[a,*], Aorrapum Nutasarin[a], Zahir Dehouche[a] and George R. Fern[b]
[a]College of Engineering, Design and Physical Sciences, Brunel University, Uxbridge, UB8 3PH, UK
[b]Wolfson Center for Materials Processing, Brunel University, Uxbridge, UB8 3PH, UK
Correspondence
*Fanchao Meng, College of Engineering Design and Physical Science, Brunel University, Uxbridge, UB8 3PH, UK
E-mail: fanchao.Meng@brunel.ac.ukTel: +44(0)7943630647Fax: +44 (0) 1895 269807

ABSTRACT: In this study, we investigate the impact capacity of high surface area TiO_2 and MgO-doped TiO_2 aerogel nanomaterials to enhance the energy conversion performance of commercial single-junction silicon solar cells.TiO_2 and MgO-doped TiO_2 aerogels were fabricated via a single step colloidal homogeneous precipitation sol-gel method in high pressure hydrogen reactor. TiO_2 based nanoparticles (NPs) were encapsulated in Ethylene Vinyl Acetate (EVA) binder and applied onto the surface of the cells using blade screen printing technique. This report compares the electrical and optical characteristics of mono-crystalline cells before and after coating. The specific surface area, micro-structural, composition and optical properties of the nanoaerogels were determined with BET, XRD, EDX, TEM, SEM and UV/Vis spectrometry. The MgO–doped TiO_2 aerogel with 2 mol% content of MgO (1:50 ratio) was found to be more UV-active nanomaterial for improving the light absorption and electrical characteristics of monocrystalline silicon solar cells than the undoped TiO_2.The experimental results showed an optimum relative increase of 6.12% in the mono-crystalline cell power conversion efficiency.
Keywords: Monocrystalline silicon solar cells, nanoparticles (NPs), sol-gel synthesis, high pressure hydrogen reactor, high surface area MgO-doped TiO_2 nanoaerogel.

1 INTRODUCTION

The development of high-efficiency and low-cost silicon solar cells is important for the generation of electric power to satisfy the continuous worldwide energy demand in the future. Nanostructured materials have showed the great contribution to improve the light capturing capability of silicon PV devices [1]. As one of the most widely used AR (anti-reflection) coating materials, nanocrystalline TiO_2 has potential to greatly reduce the fabrication costs of commercial solar cells and also improve the cells' electrical performance.

Thermalization losses usually limit the performance of solar cells. When the energy of illuminating photons is higher than the substrate (silicon solar cell), the excess energy of the active photons will convert into heat as a result of insufficient response of light in certain wavelength range [2]. TiO_2 is an outstanding UV-absorbing dielectric material which has relatively high refractive index and brilliant transparency in the wide wavelength range from visible to infrared light. With the help of TiO_2 AR (ant-reflection) coating, this loss can be largely reduced. So far the Screen-printing of TiO_2 has been regarded as a widespread industrially applied method because of the fast-printing technique and of the coating facility with fine controlling of the position and thickness. Moreover, this method is also economically suitable for the deposition of TiO_2 nanomaterials onto the surface of solar cells [3,4]. Ethylene Vinyl Acetate (EVA) as a polymeric elastomer usually provides a very good host environment for the encapsulated materials whist also exhibiting high transparency, good resistance, satisfied thermal stability, and strong adhesion to glass substrate. This matrix material is commonly served as an excellent encapsulation for the PV cell, trapping light through absorption [5,6].

Sol-gel synthesis method can be used to fabricate materials with various shapes such as porous structures, dense powder and thin film. This method is also an advisable way to fabricate nanoporous materials as it shows potential for scale-up and is reasonably economic. Moreover, the resultant materials from this synthesis method possess high homogeneity with enhanced physical and chemical characteristics [7].

We selected $Mg(NO_3)_2{\cdot}6H_2O$ as the precursor of dopant due to the excellent effect of MgO, on the photocatalytic activity of semiconductor which may result in a further enhancement of photon absorption [8]. Also, we synthesized nanostructured TiO_2 based aerogels via a colloidal homo-precipitation sol-gel reduction process in high pressure hydrogen reactor as this method shows the potential of obtaining TiO_2 nanomaterials with superior surface area and homogeneity of chemical and structural properties in comparison with other synthesis methods [9,10]. In this paper, we investigate the great potential of pure TiO_2 and MgO doped TiO_2 nanoaerogels to improve the power conversion efficiency of commercial single junction monocrystalline solar cell. We also demonstrate the formation of that anti-reflection layer on top of the monocrystalline solar cell by using doctor blade screen printing technique. In addition to the investigation of morphological analysis of TiO_2 based nanocomposite, the electrical and optical characteristics of coated cells are also evaluated.

2 EXPERIMENTAL

2.1 Materials

Methanol (99%, Alfa Aesar, UK), 2-Propanol (99%+, Alfa Aesar, UK), P-Xylene (99%, Alfa Aesar, UK), Titanium (IV) chloride (99.9%, Sigma Aldrich, USA) and Magnesium nitrate hexahydrate (99.999%, Sigma Aldrich, USA) were used throughout the experiments.

2.2 Synthesis of TiO_2 based aerogels

To obtain nanostructured TiO_2 based aerogels, we selected $TiCl_4$ as a precursor of TiO_2 and $Mg(NO_3)_2{\cdot}6H_2O$ as the precursor of MgO dopant. Typically 1.4mL of $TiCl_4$ was gently injected to a glass beaker containing 150mL of Methanol and 63.6mg of $Mg(NO_3)_2{\cdot}6H_2O$ salt to obtain a 1/50 molar ratio of magnesium to titanium in the synthesized nanoparticles. This molar ratio is suggested by a study which different molar ratios of MgO-doped TiO_2 nanoparticles were synthesized using the similar process and it was found that MgO-TiO_2 mixed oxides of 1:50 ratio exhibited the optimum performance in UV-active photocatalyst [11]. The mixture of that solution was vigorously stirred for 15mins at room temperature and then poured into an autoclave (Parr Model 5500) high pressure compact stirred reactor and H_2 gas was introduced into the autoclaves at a pressure of 20 bar (2Mpa). The autoclave was then heated to a setting temperature of 200°C. Once this setting point is achieved, the pressure of autoclave was promptly observed at around 68-70 bar (6.8~7Mpa). The autoclave was subsequently kept at 200 °C for 2 hours and the pressure inside that autoclave was then found to be at 80 bar (8Mpa). The auto was cooled down to room temperature and the remaining H_2 gas in that autoclave was released completely. The autoclave was then disassembled and the solution was poured back to a glass beaker. The solid aerogel obtained was then washed with methanol twice and dried in fume hood at room temperature and finally calcined in a furnace at 400°C for 10 hours under a static air environment. The synthesis process for pure TiO_2 aerogel is similar with that of MgO doped TiO_2.The only difference between those two cases is the containing of Mg dopant.

2.3 Preparation of the deposited layer

The prepared pure TiO_2 and MgO doped TiO_2 nanopowders were ready to be used in the process of layer application. Prior to that process, 15mg of each synthesized powder was measuredand then dispersed in P-Xylene using ultrasonic bath. The solution with different concentrations of dispersed powder (0.2mg/ml, 0.3mg/ml, 0.4mg/ml, 0.6mg/ml and 0.8mg/ml) was then well mixed with Ethylene Vinyl Acetate (EVA) and eventually formed a mixture which the weight concentration of EVA is fixed at 15 wt% in order to obtain the best concentration of TiO_2 based nano-powder to EVA that contribute to an optimum enhancement of power conversion efficiency in commercial solar cells.

2.4 Layer Application

We selected mono-crystalline silicon solar cell as the substrate for the layer application process. In a particular experiment, XS156B3 Mono-crystalline silicon solar cells from Motech were introduced with a surface area of 15.6×15.6cm^2. Prior to deposition, the cells were cut into three equal sizes using CTR TMX90 laser machine. The mixture of MgO-TiO_2/EVA is blade printed on top of pyramids texture of Si_3N_4 anti-reflection layer of the solar cells, with a coated area of 3×3cm^2. The coated cells were then cured in a furnace at 190°C for 5mins. After the heat treatment, the applied mixture of MgO-TiO_2 and EVA formed a layer which well deposited on the surface of solar cell.

2.5 EVA Layer

The Ethyl Vinyl Acetate (EVA) from Dupont was selected as a polymeric encapsulation for TiO_2 based nanomaterials. This EVA sheet possesses high transmittance capabilities and it has a thickness of 0.5mm. Various weight concentrations of EVA/P-Xylene solutions were obtained by dissolving EVA foils in P-Xylene. The optimum result occurs in 15wt% of EVA which shows homogeneous layer property and good optical performance. Hence, this weight concentration was chosen to mix with TiO_2 based nanomaterials.

3 MEASUREMENT

The electrical characteristics of solar cells were tested via visible light source at temperature of 25°C and light intensity of 1000W/m^2. The surface illuminated area for coated and uncoated part of the solar cell is 8.04cm^2. In a particular experiment, the Keithley 2601B source-meter with embedded TSPExpress software was employed in order to obtain the electrical characteristics of solar cells before and after coating, as well as plotting comparative current versus voltage (I-V) and power versus voltage (P-V) curves. The curing temperature of EVA is determined using Perkin Elmer single-furnace Differential Scanning Calorimetry (DSC), where an EVA sample is heated from -20°C to 220°C at 8 degrees per min for two cycles while testing the heat flow.

The optical characteristics of coated and uncoated solar cells were tested using Perkin-Elmer Lambda 650S UV-Vis spectrophotometer with spectral wavelength from 200nm to 800nm. The specific surface area of synthesized TiO_2 aerogels were measured by Nitrogen gas Physisorption at 77 K using Quantachrome Nova surface area analyzer. The morphological analysis of deposited layers were obtained using a ZEISS SUPRA 35 VP filed emission scanning electron microscope (SEM). The Energy dispersive X-Ray (EDX) analysis is used to confirm the presence of chemical elements in deposited layers. The crystalline phases of the synthesized powders were characterized by pXRD patterns with 2θ angle in the range of 5-100° using a Bruker D8 Advance X-ray powder diffractometer. Prior to the measurement, the diffractometer was calibrated using an aluminium oxide line position standard from Bruker and the LaB_6 NIST SRM 660a line profile standard. The x-ray was Cu K_α radiation with fitted source of nickel-filtered copper.

4 RESULTS AND DISCUSSION

4.1 DSC measurement

The curing temperature of EVA raw material was determined by DSC measurement, where the EVA sample is heated from -20 to 220 degrees Celsius at 8 degrees per minute for two cycles while examining the heat flow.

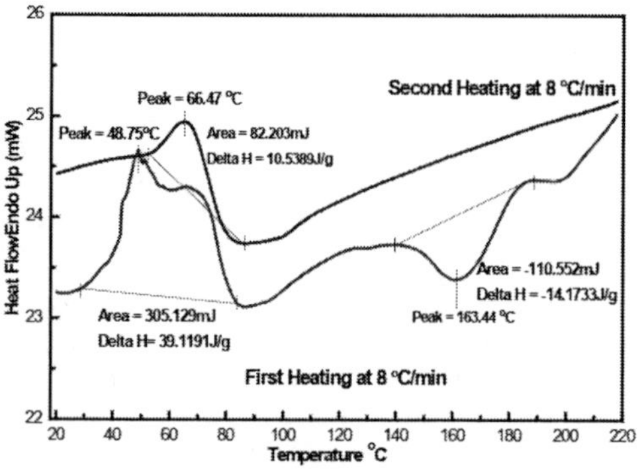

Figure 1: The first (green) and second (blue) heating curve of raw EVA material

It can be seen from Figure 1 that the first heating curve shows an endothermic peak at 48.75°C with an enthalpy value of 39.12J/g. The exothermic peak occurred at 163.44°C with the curing enthalpy of 14.17J/g. The second heating curve shows a lower melting peak of 10.54J/g in comparison with the first scan (39.12J/g). The exothermal peak was not observed in this curve, meaning that the EVA raw material has been fully cured after the first heat treatment at 220°C. The exothermic region is completed at 190 degree Celsius (186°C to be precise). Therefore, 190°C would be the optimum temperature for the cell to be fully cured.

4.2 Coated Solar Cell Characterization

Figure 2 and Figure 3 show the current versus voltage (I-V) curves and the power versus voltage (P-V) curves for the mono-crystalline silicon cell before and after coating with pure TiO_2 and MgO doped-TiO_2 nanoparticles at the maximum enhancement power conversion efficiency.

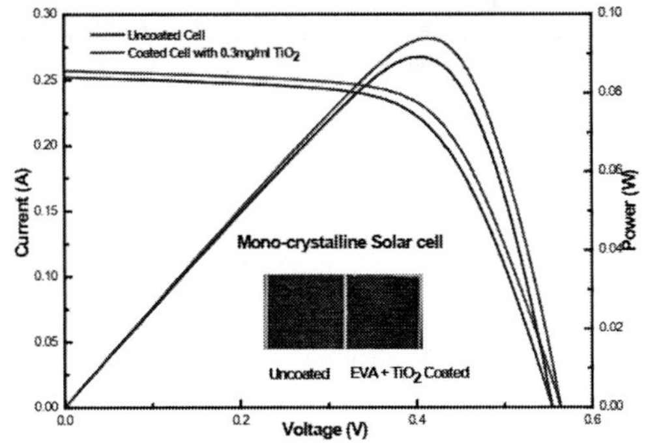

Figure 2: I-V and P-V curves of uncoated and coated cell with EVA+TiO_2 under 25°C and solar irradiance of 1000W/m^2

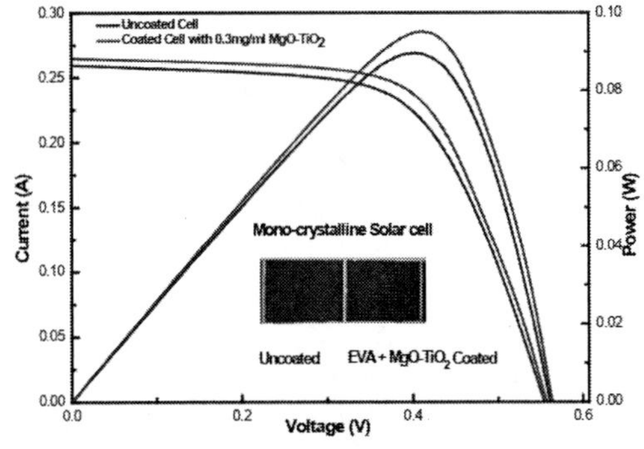

Figure 3: I-V and P-V curves of uncoated and coated cell with EVA+ MgO-doped TiO_2 under 25°C and solar irradiance of 1000W/m^2

It can be seen from figure 2 that the short circuit current (Isc) for coated cell with pure TiO_2 nanoparticles increased from 254mA to 257mA, the open circuit voltage (Voc) increased from 556mV to 564mV. Other electrical parameters such as maximum power (Pmax) and fill factor (FF) also exhibit varying degrees of enhancement. This coating provides an overall relative enhancement of 5.14% (from 11.08% to 11.65%) in power conversion efficiency. In case of the cell that coated with MgO doped-TiO_2 nanocomposite at 0.3 (mg/ml) concentration (figure 3), the Isc increased from 262mA to 265mA, the Voc slightly increased from 555mV to 559mV, a more significant increase in power conversion efficiency of 6.12% was found (from 11.11% to 11.79%).

The detailed characteristics of uncoated and coated cells with TiO_2 based nanoparticles at different concentrations are summarized in Table 1 and Table 2.

EVA+TiO_2 Coating	*Isc (A)*	*Voc (V)*	*Pmax (W)*	*FF*	*η (%)*
Uncoated	0.256	0.540	0.086	0.622	10.71
15% +0.2mg/ml	0.257	0.549	0.089	0.634	11.07
Uncoated	0.254	0.556	0.089	0.632	11.08
15% +0.3mg/ml	0.257	0.564	0.094	0.645	11.65
Uncoated	0.256	0.543	0.087	0.624	10.79
15% +0.4mg/ml	0.252	0.550	0.089	0.639	11.06
Uncoated	0.254	0.552	0.088	0.627	10.89
15% +0.6mg/ml	0.250	0.561	0.088	0.628	10.93
Uncoated	0.256	0.545	0.089	0.642	11.08
15% +0.8mg/ml	0.254	0.549	0.087	0.622	10.79

Table I: Electrical results of the uncoated and cell coated with 15wt% EVA + TiO_2 at 1000W/m^2 and 25°C

EVA+MgO-TiO_2 Coating	*Isc (A)*	*Voc (V)*	*Pmax (W)*	*FF*	*η (%)*
Uncoated	0.263	0.560	0.092	0.624	11.36
15% +0.2mg/ml	0.264	0.562	0.094	0.636	11.64
Uncoated	0.262	0.555	0.090	0.618	11.11
15% +0.3mg/ml	0.265	0.559	0.095	0.642	11.79
Uncoated	0.260	0.548	0.089	0.626	11.11
15% +0.4mg/ml	0.262	0.556	0.093	0.642	11.54
Uncoated	0.263	0.551	0.090	0.620	11.12
15% +0.6mg/ml	0.257	0.562	0.092	0.640	11.46
Uncoated	0.261	0.560	0.091	0.621	11.20
15% +0.8mg/ml	0.258	0.565	0.092	0.632	11.38

Table II: Electrical results of the uncoated and cell coated with 15wt% + MgO doped TiO_2 at 1000W/m^2 and 25°C

It can be found that the cells coated with MgO doped-TiO_2 nanoparticles contributed to the enhancement of power conversion efficiency at all tested concentration. Typically, 0.28%, 0.68%, 0.43%, 0.34% and 0.18% is obtained at 0.2mg/ml, 0.3mg/ml, 0.4mg/ml, 0.6mg/ml and 0.8mg/ml respectively. The only reduction in power conversion efficiency occurs in the case of cell which coated with pure TiO_2 nanoparticles at 0.8mg/ml which may due to the shading effects.

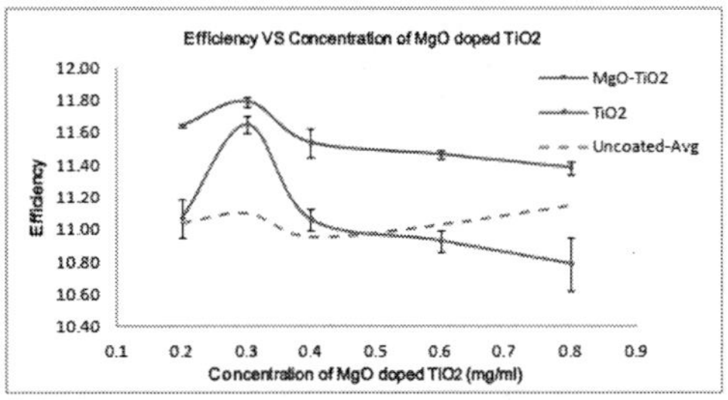

Figure 4: Electrical efficiency comparison for the raw cells and cells coated with TiO_2 based nanomaterials

Figure 4 shows the different level of enhancements in conversion efficiency of the coated cells as a function of deposited TiO_2 nanoparticles at varying concentration. It can be seen that the maximum enhancement in power conversion efficiency is achieved at 0.3mg/ml concentration of TiO_2 based nanoparticles. The results demonstrated that the concentration of dispersed TiO_2 nanoparticles encapsulated in EVA binder has a strong influence on the amount of light absorbed and thus the power conversion efficiency of the cells.

By comparing electrical performances of uncoated, EVA and TiO_2 based nanoparticles coating at different concentrations, the enhancement in power conversion efficiency is attributed to the effects of both EVA binder and TiO_2 nanocomposite. The EVA encapsulated MgO-TiO_2 with 1:50M coating provides a maximum enhancement of 0.68% absolute efficiency at concentration of 0.3mg/ml. For the pure TiO_2 coating with 0.3mg/ml, this increase is not as much as MgO doped TiO_2, which is 0.57%. Table 3 compares quantitatively the two different levels of effects on solar cells when coating with pure TiO_2 and film and MgO doped TiO_2 at optimum concentration.

Parameter / Substrate	Isc (A)	Voc (V)	Pmax (W)	FF	%η	Efficiency enhancement %	Relative Efficiency change%
Uncoated TiO_2	0.254	0.556	0.089	0.632	11.08	0.57	5.14
TiO_2	0.257	0.564	0.094	0.645	11.65		
Uncoated MgO-TiO_2	0.262	0.555	0.09	0.626	11.11	0.68	6.12
MgO-TiO_2	0.265	0.559	0.095	0.642	11.79		

Table III: Comparison of maximum enhancement on electrical characteristics between cells coated with 0.3mg/ml TiO_2 and 0.3mg/ml MgO-TiO_2

4.3 Morphological, structural and optical properties

Figure 5 shows a cross-section SEM image of the solar cell coated with TiO_2 based nanomaterial. The image indicates a good dispersion of segregated nanoparticles. The image shows also the pyramidal texture of the raw cell surface, and the coated layer of MgO doped-TiO_2 plus EVA with an average thickness of 9μm.

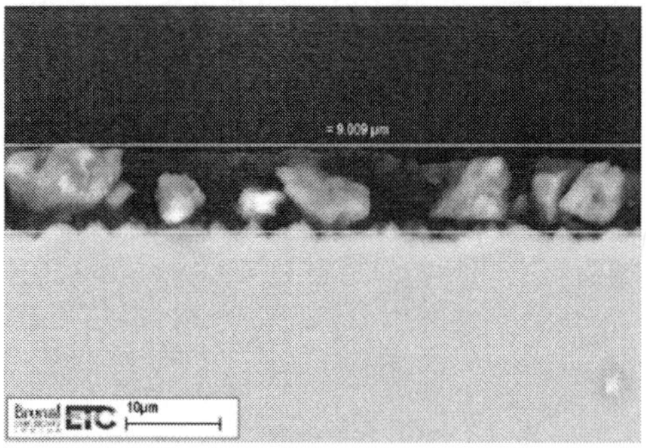

Figure 5: Cross section SEM image of the coated cell with EVA+MgO doped TiO_2

The high resolution TEM images (Figure 6) of a single nanoparticle show ordered nanoscale patterns with lamellar lattice morphology.

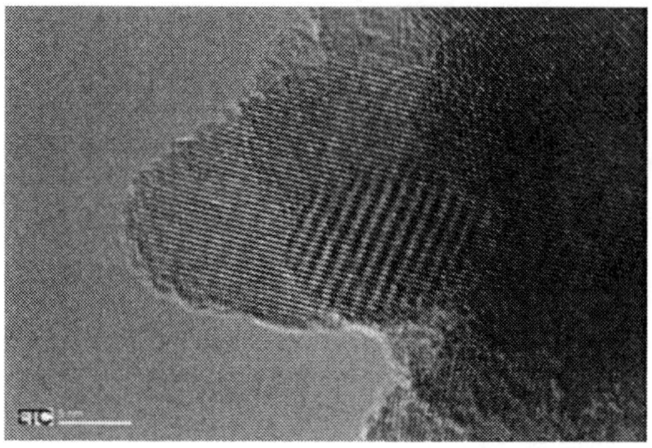

Figure 6: TEM image of nanostructured MgO-TiO_2 composite

Figure 7 shows the energy dispersive X-ray analysis spectrums from the entire SEM scanned area of coated cells with enhanced electrical characteristics. The EDX spectrums confirm the presence of expected chemical elements such as carbon, oxygen, titanium and magnesium in the deposited layers.

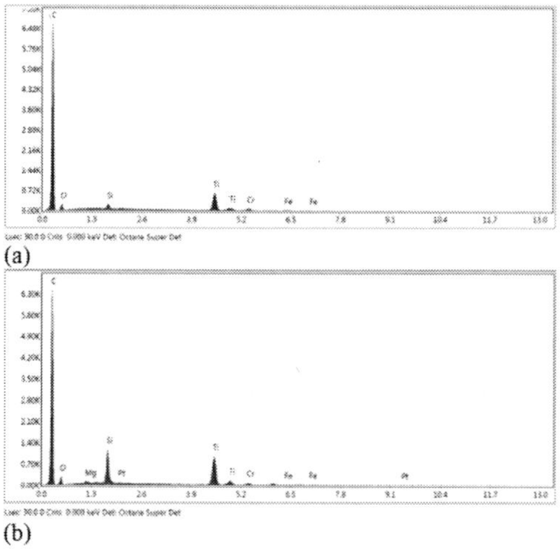

(a)

(b)

Figure 7: EDX spectrums of coated mono-crystalline solar cells; (a) EVA+TiO_2, (b) EVA+MgO-doped TiO_2

Figure 8 shows the reflectance and absorption spectra of the uncoated, EVA coated, EVA+TiO_2 coated and EVA+MgO-doped TiO_2 coated silicon solar cells. The results show that coated cell with enhanced electrical characteristics exhibit less overall reflectance in comparison with uncoated one in the spectral wavelength of middle and long ultraviolet, visible and near infrared region.

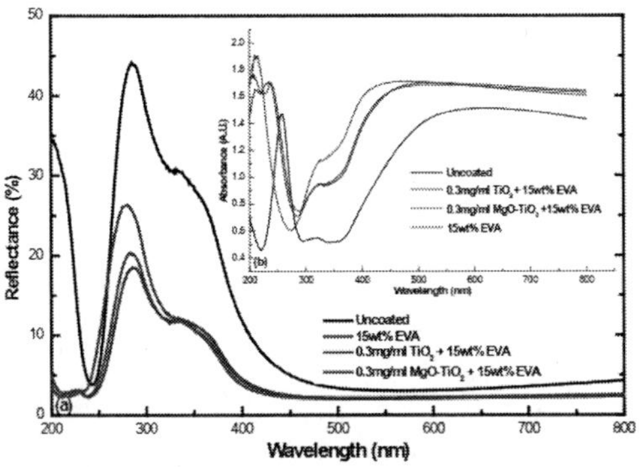

Figure 8: Comparison of (a) reflectance spectra and (b) absorption spectra for uncoated and coated cells

It can be seen from the transmission measurements (Fig9) that MgO doped TiO_2 coating at 0.3mg/ml concentration exhibits slightly higher transparency in comparison with corresponding pure TiO_2 coating and EVA coating.

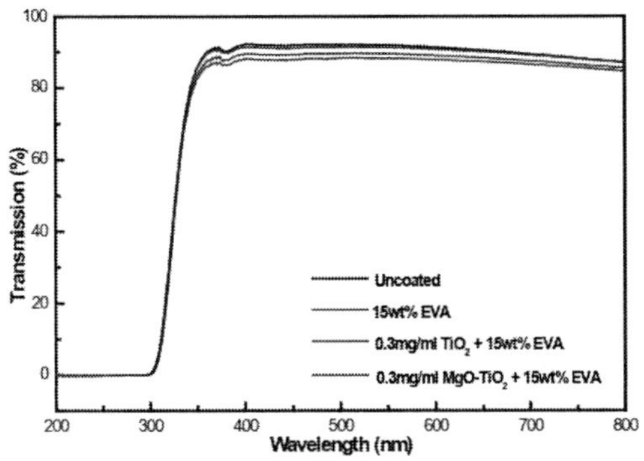

Figure 9: Transmission spectra for uncoated and coated cells

Figure 10 depicts the X-ray diffraction pattern of the MgO-TiO_2 and pure TiO_2 powders annealed at 400°C. All the peaks in the patterns are consistent with the characteristic peaks of anatase TiO_2 crystal matching PDF 03-065-5714. However, the pXRD patterns of MgO-TiO_2 (1:50M) did not show the presence of MgO as the quantity of added dopant is very low and it could be well dispersed throughout the TiO_2. However, the presence of a small amount of Mg on the surface of mixed oxides was confirmed by the previous EDX analysis (Fig.7). The peaks broadening of MgO-doped TiO_2 XRD spectrum indicates a slight structural change suggesting a reduction in the crystallite size. To determine the average size (τ_{XRD}) of those nanoparticles, we used Scherrer's equation [12]:

$$\tau = \frac{0.9\lambda}{\beta\cos\theta}$$

where λ is the wavelength of the incident X-rays (0.15406 nm), β is the full-width half-maximum (FWHM), and θ is the diffraction angle of the considered diffraction peak. The first peak (2θ = 25°C) of each diffraction pattern indicates the most stable plane of TiO_2 and MgO-TiO_2 respectively. The average crystallite size of TiO_2 nanoparticles is found to be 7.8 nm. The MgO-doped TiO_2nanoaerogels show slightly lower value but very comparable crystallite size of 5.6nm.

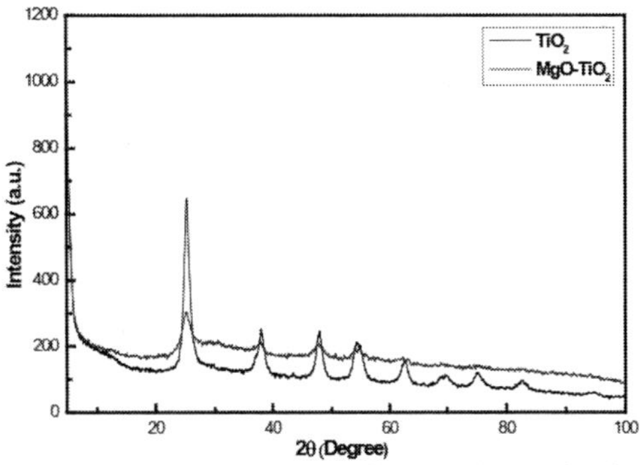

Figure 10: XRD patterns of MgO-TiO_2 and TiO_2 powders annealed at 400°C

The surface area for TiO_2 based aerogels were calculated using the Brunauer-Emmett-Teller (BET) equation. It can be seen from the table 4 that TiO_2 aerogels synthesized in this work show a specific surface area of 154.263m^2/g in comparison with 82m^2/g prepared by studied methods [7]. In case of MgO-doped TiO_2 aerogels, a larger specific surface area of 231.364 was observed.

Materials	*S_{BET} (m^2/g)*	*Crystallite size (nm)*
TiO_2 nanoaerogel (this work)	154.263	7.8
MgO-TiO_2 nanoaerogel (this work)	231.364	5.6
TiO_2 [7]	82	20

S_{BET}: the specific surface area obtained using Brunaur-Emmett-Teller equation

Table IV: Structural properties of TiO_2 based nanoaerogel

5 CONCLUSION

In this study, we successfully synthesized high surface area nanostructured TiO_2 based aerogels and investigated their capacity to improve the energy performance of photovoltaics by applying TiO_2 based nanomaterials dispersed in a polymeric matrix onto the active surface of commercial single junction mono-crystalline solar cells using blade screen printing technique.

The electrical and optical characteristics results demonstrate that doping nanostructured TiO_2 with MgO yields increased electrical efficiency and substantial increase of overall photons absorption. EVA encapsulated TiO_2 nanoparticles enhance electrical performance, but not as much as MgO doped-TiO_2. Moreover, the results demonstrated that the concentration of dispersed TiO_2 based nanoparticles has a strong influence on the electrical efficiency of the coated cells. An optimum relative increase in power conversion efficiency of 6.12% in visible light has been obtained with MgO doped-TiO_2 nanocomposite at 0.3 (mg/ml) concentration. This enhancement can be further improved by optimizing the coating procedure.

6 ACKNOWLEDGEMENTS

We are thankful to Dr. Ashley Howkins from Experimental Techniques Centre for SEM, TEM analyses and Dr. Lorna Anguilano from ETC for XRD analysis. We also express here our gratitude to Dr. Anna Howson from Quantachrome UK Ltd for the BET tests. The authors similarly acknowledge the helpful discussion with Dr Paul Harris from Wolfson centre.

7 REFERENCES

[1] Lee, Y.Y., Ho, W.J., Chen, Y.T. Performance of plasmonic silicon solar cells using indium nanoparticles deposited on a patterned TiO_2 matrix. *Thin Solid Films***2014**; **570(B)**, pp. 194-199, DOI: 10.1016/j.tsf.2014.05.022.

[2] Dhankhar, M., Singh,O.P., Singh,V.N. Physical principles of losses in thin film solar cells and efficiency enhancement methods. *Renewable and Sustainable Energy Reviews*2014; **40**:214-223, DOI: 10.1016/j.rser.2014.07.163.

[3] Richards, B.S.,*NOVEL USES OF TITANIUM DIOXIDE FOR SILICON SOLAR CELLS* 2002;Ph.D. University of New South Wales.

[4] Chen, P., Comte, P., Nazeeruddin, M.K., Liska, P., Péchy, P., Grätzel, M.*et al* .Fabrication of screen-printing pastes from TiO_2 powders for dye-sensitised solar cells.*Progress in Photovoltaics: Research and Applications***2007**;**15 (7)**, pp. 603–612, DOI: 10.1002/pip.768

[5] Yuwawech, K.,Wootthikanokkhan, J., and Tanpichai, S.Enhancement of thermal, mechanical and barrier properties of EVA solar cell encapsulating films by reinforcing with esterified cellulose nanpfibres. *Polymer Testing*2015;**48**: 12-22, DOI: 10.1016/j.polymertesting.2015.09.007

[6] Stark, W., Jaunich, M., Bohmeyer,W., Lange, K. Investigation of the crosslinking behavior of ethylene vinyl acetate (EVA) for solar cell encapsulation by rheology and ultrasound. *Polymer Testing*2012;**31(7)**,pp. 904-908, DOI: 10.1016/j.polymertesting.2012.07.004.

[7] Parayil, S.K., Psota, R.J., and Koodali, R.T.Modulating the textural properties and photocatalytic hydrogen production activity of TiO_2 by high temperature supercritical drying. *International Journal of Hydrogen Energy*2013; **38(25)**, pp. 10215-10225, DOI: 10.1016/j.ijhydene.2013.06.015.

[8] Kaviyarasu, K., Premanand, D., Kennedy, J., Manikandan, E. SYNTHESIS OF Mg DOPED TiO_2 NANOCRYSTALS PREPARED BY WET-CHEMICAL METHOD: OPTICAL AND MICROSCOPIC STUDIES. *International Journal of Nanoscience*2013;**12(5)**, pp. 1-6, DOI: 10.1142/S0219581X13500336

[9] Keshmiri M, Mohseni M, Troczynski T. Development of novelTiO_2 sol-gel-derived composite and its photocatalyticactivities for trichloroethylene oxidation. *Applied CatalysisB: Environmental* 2004;**53(4)**, pp. 209-219, DOI: 10.1016/j.apcatb.2004.05.016.

[10] Yin, H., Wada, Y., Kitamura, T., Kambe, S., Murasawa, S., Mori, H.,et al. Hydrothermal synthesis of nanosized anatase andrutile TiO_2 using amorphous phase TiO_2. *Journal of MaterialsChemistry* 2001;**11**:1694-1703, DOI: 10.1039/B008974P

[11] Shrestha, K.M., Sorensen, C.M., Klabunde, K.J. MgO-TiO_2 mixed oxide nanoparticles: Comparision of flame synthesis versus aerogel method; characterization, and photocatalytic activities. *Journal of Materials Research* 2013;**28(3)**, pp. 431-439, DOI: 10.1557/jmr.2012.288

[12] Cullity, B.D. *Elements of X-Ray Diffraction*, AddisonWesley Publishing Company, Reading, Mass, USA, 2nd edition,1978.

Development of Cost Effective Silver Alloy Metallisation Paste for Si Solar Cells

Diarmaid Corbett[1], Alex Savidis[1], Russell Goodall[2], Jacob Corteen[2], Edwin Raj[3], Simon Johnson[3], Robin Orman[3], Gordon Kerr[4]
Solar Capture Technologies[1], PV Technology Centre, Albert Street, Blyth, Northumberland, NE241LZ.
Sheffield University[2], Western Bank, Sheffield, South Yorkshire, S102TN.
Johnson Matthey[3], Blounts Court Road, Sonning Common, Reading, RG49NH.
Phoenix Scientific Industries[4], Apex Park, Diplocks Way, Hailsham, BN27GNU.

Abstract *- Pure silver is used in conventional solar cells for front-side current collection as it exhibits high conductivity, excellent solderability and good oxidation resistance. The long term cost of silver is expected to continue to rise therefore it is important to reduce silver content in solar cells. Previous attempts to use metal additives to reduce the quantity of pure silver required have resulted in alloys with poor resistivity and/or poor oxidation resistance. This study presents a collaborative research project to develop a lower cost alternative to pure silver using a novel developmental approach to selecting, producing and testing alloys of silver. Multivariable regression techniques were used to predict the properties of binary and tertiary silver alloy systems. Samples were converted to fine powders by gas atomisation and mixed with glass frits to produce pastes. Laboratory measurements of line and contact resistance were duplicated in a solar manufacturing line and screen printed cells produced. The consortium aims to produce an 18% efficient solar cell using an optimised alloy system.*

1. INTRODUCTION

Between 2011 and 2016 the price of silver has declined, however the long term price of silver is projected to rise and over the last 6 months it has continued to do so, peaking at $17.82/Oz in April [1]. Given that around 7% [2] of the cost involved in solar cell manufacture is attributed to metallisation costs any reduction in silver content increases profitability of cell and module manufacture.

The use of silver is favoured over other metals as it has the lowest resistivity of any pure metal. Many previous attempts have used pure silver with high organic content to reduce silver content to around 75-80%. Pastes consisting of pure silver and a secondary alloy have also been used.

In order to drastically reduce silver quantity, it is suggested that a primary alloy of silver could be used in place of silver. There are significant challenges to be overcome in the reduction of silver content. Most importantly any change in chemistry requiring the addition of another process step, or unfavourable changes in process conditions (for example increased processing time or cost) is highly unlikely to be taken up by manufacturers.

In order to effectively assess a wide range of compositions and carry out sufficient testing a consortium was formed between Sheffield university (SU), Johnson Matthey (JM), Phoenix Scientific Industries (PSI) and Solar Capture Technologies (SCT); co funded by Innovate UK.

This consortium aims to produce a primary silver alloy to be used in place of pure silver, in an attempt to replicate both the desirable resistive properties of silver and produce a lower cost manufacturing process for the paste by use of novel manufacturing techniques.

2. EXPERIMENTAL

In order to develop an alloy with suitable properties for solar cell metallisation, Sheffield University screened silver alloys to predict resistivity and phase composition. This allowed suitable silver alloy systems to be selected for manufacture and test.

Samples were produced by induction and arc melting at Sheffield University and then screened based on resistivity; the most promising systems were then developed at Johnson Matthey into screen printing pastes for trial by cryo-milling and mixed with suitable frits and organics via triple roll milling. A sample grid was printed and fired in a RTP furnace. The sample was tested to obtain line and contact resistance measurements using the transmission line method [3].

Systems of interest were further investigated at Johnson Matthey using SEM, EDX and XRD to identify favourable particle size, composition and contact formation before being processed by gas atomisation for production trials.

The gas atomisation process carried out at PSI produced alloys with favourable size distribution and spherical particle geometry. The alloy powders were produced using PSI HERMIGA 100/10VI gas atomisation equipment. An advanced multi frequency classifier was used in order to isolate sub 10μm particles for paste manufacture.

The interaction between glass frit and the alloy system is particularly important in contact formation, the frit must be reactive enough to etch the antireflective coating, yet not penetrate too deeply into the cell or form too thick a glass layer during firing [4]. The organic binders should provide appropriate rheology for screen printing with minimal chemical interaction. As such, complimentary frits and organics were developed at Johnson Matthey for each paste in leaded and lead free compositions.

Test cells were previously manufactured using p-type 243mm^2 pseudosquare monocrystalline silicon wafers which were alkaline textured, diffused using $POCl_3$ to produce a 70Ω/□ emitter, had a silicon nitride antireflective coating deposited using PECVD, and then a screen printed Al back surface field was fired onto the cell.

For front contact formation initial batches of 10 cells were printed, dried and then fired with a standard profile. The firing peak was varied in steps of 10°C in order to find the optimal process window and a further 5 cells printed at this optimal peak temperature to assess the process stability.

The cell design is based on standard H-Pattern metallisation using a single print process. The finger and busbar size and spacing has been optimised to allow the widest range of paste resistivity to produce proof of concept cells; as such the maximum simulated efficiency obtainable is in the region of 19% [5].

At Solar Capture Technologies IV curve measurement, contact resistance, line resistance and peel strength testing were used to provide comparison between different experimental alloy pastes and also commercially available metallisation pastes.

3. RESULTS

Initial selection criteria have yielded many potentially suitable alloy systems, screening of these alloys has resulted in phase 1 tests of over 100 alloys. Of these two have were selected to trial in solar cell production. Four baseline silver pastes and three novel silver alloy pastes were produced; results are presented for HLS2-HLS7 (HiLoSilver Samples 2 – 7).

Baseline tests allowed comparison with standard commercially available solar cells and optimisation of frit and organic additives.

	Gintech	NSP	BigSun
Efficiency (%)	20	20	19.8
Pmax (w)	4.78	4.779	4.729
Isc (I)	9.43	9.181	9.201
Voc (mV)	653	650	641
Impp (I)	8.82	8.603	8.718
Vmpp (mV)	543	556	543
Fill Factor (ff %)	77.22	80.01	80.82
Line Resistance (Ω/cm)	0.309	0.387	0.163
Contact Resistivity (Ω/mm2)	0.021	0.248	0.08
Tab Strength (N)	1.3	1.4	1.3

Figure 1 – Commercially available solar cell comparison

Initial trials with standard silver (Ag) printing pastes confirm simulated results; efficiencies in the range of 18%-19% have been realised. Using customised frit and organics similar results have also been achieved with pure silver.

	HLS2	HLS3	HLS4
Efficiency (%)	17.5	10.4	17.45
Pmax (w)	4.243	2.52	4.236
Isc (I)	8.52	8.4	8.51
Voc (mV)	632.2	598.7	632.4
Impp (I)	7.92	5.31	7.95
Vmpp (mV)	535.6	474.5	553.2
Fill Factor (ff %)	78.8	50.06	78.8
Line Resistance (Ω/cm)	0.13	570	0.05
Contact Resistivity (Ω/mm2)	1.4	325	1.2
Tab Strength (N)	1.5	0.5	1.3

Figure 2 – Silver paste comparison

Two alloy systems have been produced by gas atomisation and manufactured into solar cells, HLS5, HLS6 and HLS7. Minor process refinement was carried out in order to produce consistent cell batches.

Both alloys were paired with suitable frit and organic binders to produce screen printing pastes before manufacturing cells. Comprehensive testing has been carried out on all cells and the best average is presented.

	HLS5	HLS6	HLS7
Efficiency (%)	16.8	5.51	1.59
Pmax (w)	3.938	1.338	0.385
Isc (I)	8.271	5.63	1.897
Voc (mV)	638.8	639.1	612.9
Impp (I)	7.544	3.344	1.051
Vmpp (mV)	522	400	366.6
Fill Factor (ff %)	74.5	37.23	33.2
Line Resistance (Ω/cm)	0.4	24.3	200
Contact Resistivity (Ω/mm2)	1.5	700	N/A
Tab Strength (N)	0.5	N/A	N/A

Figure 3 – Silver alloy paste comparison

As the Al BSF was previously fired a minor reduction in voltage for all cells produced was observed, cross sectional SEM confirms slightly overfired BSF, however cell functionality is not compromised and a single co-firing step would avoid this.

HLS5 has comparable line resistance to standard silver and as such the short circuit current is also in line with expectations. Slight oxidation caused an increase in contact resistivity and pull testing confirmed the poor adhesion. SEM analysis shows fracturing in the glass layer, causing failure of the electron tunnelling process.

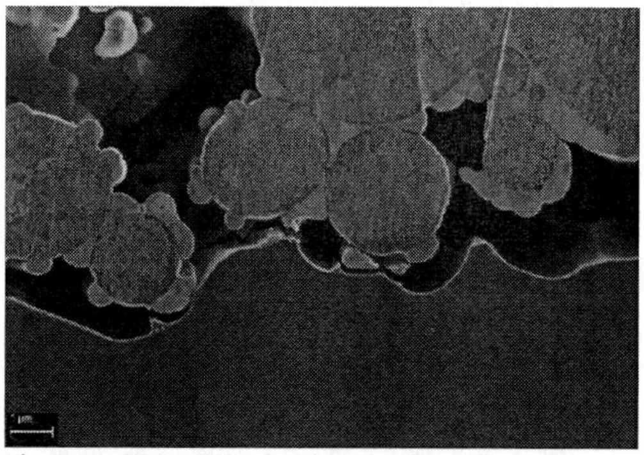

Figure 4 – HLS5 SEM showing glass layer fracture

HLS6 and HLS7 contained lead and lead free frits respectively. Both exhibit much lower current generation than standard silver, initial tests show the oxide layer is preventing contact formation. The contacts peel from the cell with ease and the high contact resistivity confirms a very poor electrical contact. This oxidation is severe enough to have also caused an increase in the line resistance. Both were fired with the same profile, only the peak temperature was adjusted; this could be a key factor in the performance variation between the leaded and lead free versions. Dark IV analysis has also been used to evaluate the alloy systems and show that the main factor causing poor performance is increased series resistance, this is the expected failure mode as all the alloys used have higher resistance than pure silver and in all cases oxidation has been present.

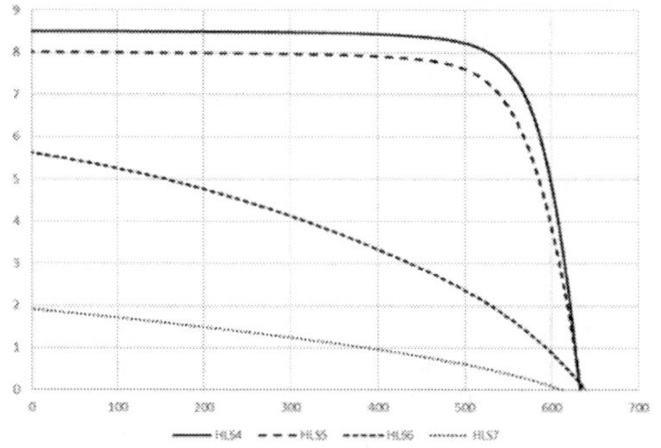

Figure 5 – Light IV curve comparison

4. CONCLUSION

With the paste formulations reported in this paper it has not been possible to create a suitable replacement for silver. HLS5 has yielded the most promising results so far, with an efficiency of 16.8%. If the minor oxidation issues seen can be overcome alongside a H-pattern optimisation for the specific paste, it is expected that an 18% efficient cell would be achieved with approximately 30% reduction in silver content over standard pastes with no major alteration to the current production methods. A number of other alloy systems are also being investigated that have the potential to meet this efficiency/cost optimisation without the oxidation issues seen in this case.

5. ACKNOWLEDGEMENTS

Many thanks to all the project partners for their support with this publication and the teams responsible for delivering these promising results. Innovate UK have been instrumental in co-funding this research and without their backing this project would not have been possible.

6. REFERENCES

[1] Kitco Silver, Silver Price Data, [Online]: Available: http://www.kitcosilver.com

[2] ITRPV Roadmap, 2016, [Online]: Available: http://www.itrpv.net/

[3] Reeves, G., Harrison, H., "Obtaining the Specific Contact Resistance from Transmission Line Model Measurements", IEEE Electron Device Letters, On page(s): 111-113, Volume: 3, Issue: 5, August 2005

[4] Tai, Y., et al "Effect of Glass Frit in Ag Paste on the Electrical Properties of Front-Side Ag Contracts for Crystalline Silicon Solar Cells", RSC Adv., 2015, 5, 92515

[5] Mette, A., "New Concepts for Front Side Metallization of Industrial Silicon Solar Cells", PhD thesis, University of Freiburg, Germany, 2007

HIGH MOBILITY AND TRANSMITTANCE TRANSPARENT CONDUCTIVE HF-DOPED IN_2O_3 THIN FILMS AND ITS APPLICATION TO SILICON HETEROJUNCTION SOLAR CELLS

W.J.Wang, G.H.Wang*, L.Zhao, H.W.Diao
Key Laboratory of Solar Thermal Energy and Photovoltaic System of Chinese Academy of Sciences, Institute of Electrical Engineering, the Chinese Academy of Sciences, Beijing 100190, China

Abstract —Hf-doped In_2O_3 transparent conductive films (IHfO) are developed at a low substrate temperature of 150 °C by radio frequency magnetron sputtering for the applications to silicon heterojunction solar cell. The maximum Hall mobility of 79.6 cm^2/Vs and the minimum resistivity of 3.76×10^{-4} Ω cm are obtained. The average transmittance of IHfO films with thickness 94 nm and 436nm in the wavelength range of 300-1100 nm is about 85%. IHfO is applied to a-Si/c-Si heterojunction solar cells, 16.77% of conversion efficiency has been obtained. We aim to improve the efficiency of solar cell in the future.
Index Terms —Hf-doped In_2O_3, Hall mobility, Transparent conductive oxide films, Silicon heterojunction solar cells.

I. INTRODUCTION

Transparent conducting oxide (TCO) films are of considerable interest in many optoelectronic applications such as optical detectors, solar cells and flat-panel displays [1,2]. Many TCO thin films with higher mobility were obtained at high temperatures or with post-thermal treatments at high temperature, which limits their applications to a-Si/c-Si heterojunction solar cells (SHJ) [3,4]. In_2O_3:W has attracted considerable attention due to their high mobility at lower than 200 °C temperature. Recently, Meng et al obtained the maximum Hall mobility of 89 cm^2/ V s for IWO films deposited by reactive plasma deposition (RPD) technique. Consequently, IWO is applied to a-Si/c-Si heterojunction solar cells, 20.8% of conversion efficiency has been obtained [5]. So far, there are few reports on high mobility and transmittance Hf-doped In_2O_3 transparent conductive films (IHfO) deposited by magnetron sputtering at low substrate temperature.

In this work, IHfO thin films were investigated for the applications to SHJ solar cells, where it serves as conductive layer on front surface of SHJ solar cells. Thus optical, structural and electrical properties were simultaneously investigated.

II.EXPERIMENTS AND MEASUREMENTS

The IHfO films were deposited on quartz or etched silicon substrate by radio frequency magnetron sputtering from a sintered ceramic Hf_2O_3-doped In_2O_3 (1wt.%) in an argon atmosphere. The substrates are kept in 150 °C during the sputtering process. The thickness of the films was measured by Veeco Dektak 200 surface profiler. The crystalline structure was monitored by X-ray diffract meter (XRD) using Cu-K_α irradiation source. The scanning electron microscopy (SEM) (ZEISS SIGMA) was employed to characterize the cross-sectional images. The optical reflection and transmission spectra of thin films were measured by Varian Excalibur HE 3100 UV-VIS-NIR Spectrophotometer.

The a-Si:H thin films were fabricated in a capacitive coupled radio frequency plasma enhanced chemical vapor deposition (RF-PECVD) system with a base vacuum of 10^{-5} Pa. The reactive gases for the deposition of thin films included silane (SiH_4), Germane (GeH_4), hydrogen (H_2), diborane (B_2H_6) and phosphine (PH_3). The pressure and flow rates were independently controlled by a downstream throttle valve controller and upstream mass flow controllers, respectively. The back ITO thin films were prepared by RPD technique. The photo current-density versus voltage (J-V) characteristics of the fabricated solar cells was measured at 25°C under 1-sun (AM1.5, 100mW/cm^2) solar simulator radiation.

III. RESULTS AND DISCUSSION

Fig.1 shows the XRD patterns of the IHfO films deposited at 16 and 50 sccm Ar flow rate. The films show typical polycrystalline structures with the weak diffraction peaks of (122), (222), (400), (411), (431), (440) and (622) together. The film is dominated by a strong peak at 2θ near 30.2° at low 16sccm flow rate. However, the intensity of (400) increases markedly at high 50 sccm flow rate. For IHfO films deposited at constant RF power, oxygen vacancies might be play a role in the changes in preferred orientations from (222) to (400) with Ar flow rate [6-8].

The columnar structure perpendicular to the substrate is shown in the cross sectional scanning electron microscope image of Fig.2 on (a) polished and (b) etched silicon substrate. The films are deposited uniformly on the two different types of substrates.

The electrical properties of the IHfO films are analyzed by Hall Effect measurements. Fig.3 illustrates the resistivity and mobility of the IHfO films as a function of the Ar flow rate. The resistivity initially shows decrease a little from

* Corresponding author. Tel.: +86 10 82547043; fax: +86 10 82547041
E-mail address: wangguanghong@mail.iee.ac.cn (G.H.Wang)

7.76×10^{-4} Ω cm to 3.76×10^{-4} Ω cm, and then increases a lot to 7.84×10^{-4} Ω cm.

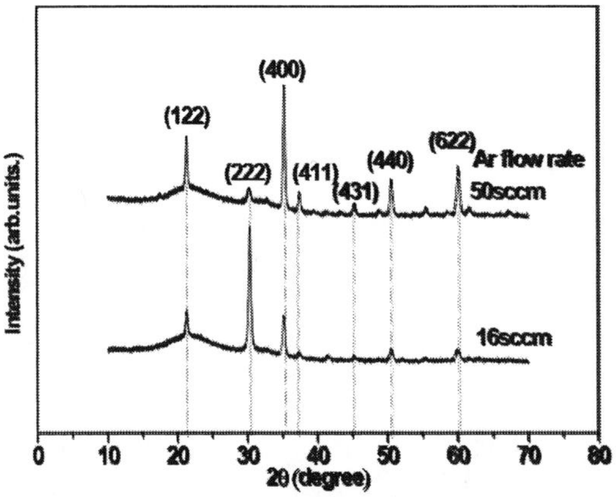

Fig.1 XRD patterns of IHfO films deposited with 16 and 50 sccm Ar flow rate

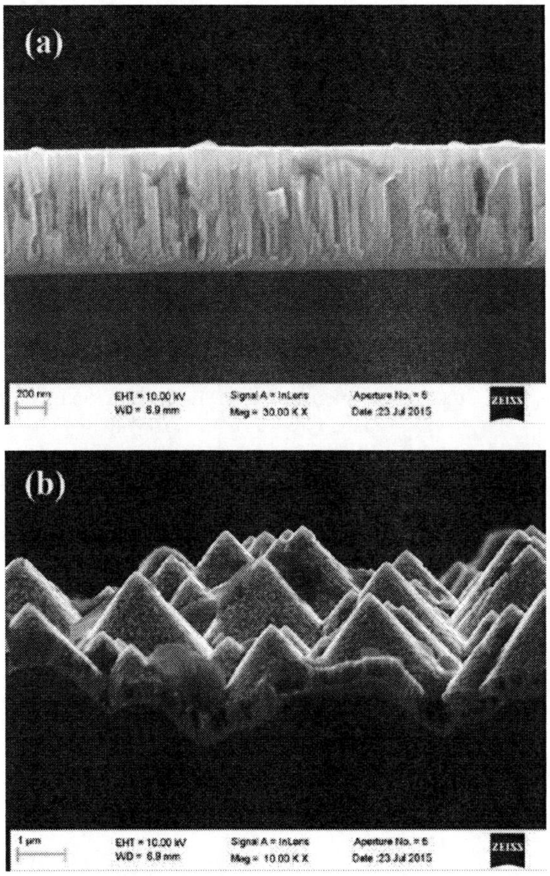

Fig.2 A cross-sectional images of the IHfO films on (a) polished and (b) etched silicon

The thick film deposited at 16 sccm Ar flow rate has post-thermal treatment for 2 hours in air at 220 °C temperature. The low temperature annealing process increased μ from 50.7 to 79.6 cm^2/Vs. Meng et al. reported that the higher oxygen partial pressure can bring high Hall mobility [5]. In this work, the oxygen is not introduced into the sputtering process. However, the annealing process in air might decrease the number of oxygen vacancies in the films, which leads to high mobility[9].

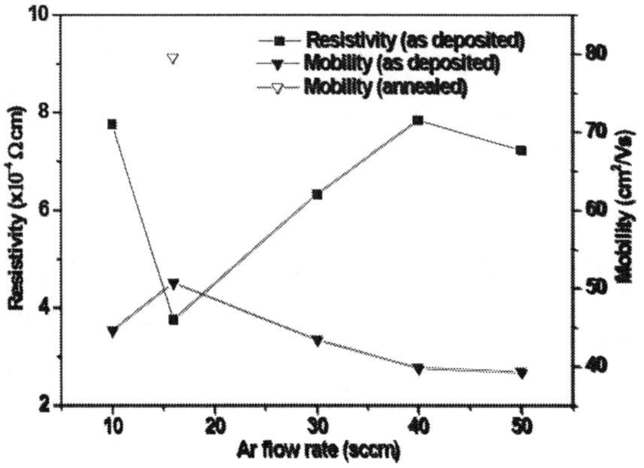

Fig.3 Resistivity and mobility of IHfO films as a function of Ar flow rate

Fig.4 shows the optical transmittance spectra of 94 nm and 436 nm thickness IHfO films without annealing. The average transmittance in the range of 300-2000 nm is over 80%. The average transmittance in the range of 300-1100 nm is about 85%.

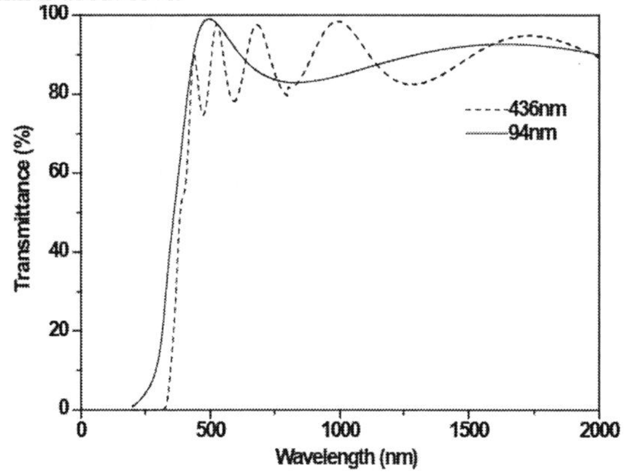

Fig.4 Transmission spectra of the IHfO films with 94 nm and 436 nm thickness

The higher transmittance and mobility of IHfO thin film makes it an excellent candidate for the applications to SHJ solar cells for enhancing the spectral response. On the basis of the inquiry of IHfO thin film characterization, we applied the IHfO film on the SHJ solar cell as the transparent electrodes. The schematic diagram of SHJ solar cells is shown in Fig.5. In this structure of SHJ solar cell, the commercial n-type Cz-Si wafer is utilized as the absorber layer. The thickness of the wafer is roughly 180μm.

The p/i/n a-Si:H layers are deposited by PECVD. The thickness of the front IHfO film is about 80 nm to minimize the sunlight reflection losses. The thickness of the back ITO film is about 100nm. Fig.6 highlights the performance of the SHJ solar cell. An efficiency of 16.77% (V_{oc}=0.64 V, J_{sc}=36.40 mA/cm^2, FF=0.72) is obtained for SHJ solar cells.

IV. SUMMARY

In this study, the structural, electrical and optical properties of the IHfO films fabricated by radio frequency magnetron sputtering are investigated. The XRD results of the films show typical polycrystalline structures. The cross sectional scanning electron microscope image shows columnar structure perpendicular to the substrate. The resistivity initially shows decrease a little from 7.76×10^{-4} to 3.76×10^{-4} Ω cm, and then increases a lot to 7.84×10^{-4} Ω cm with Ar flow rate. However, the mobility is up to the maximum value of 79.6 cm^2/Vs at 16 sccm Ar flow rate after low temperature annealing. The average transmittance of IHfO films with 94nm and 436 nm in the range of 300-2000 nm is over 80%. The average transmittance in the range of 300-1100 nm is about 85%. IHfO is applied to a-Si/c-Si heterojunction solar cells, 16.77% of conversion efficiency has been obtained.

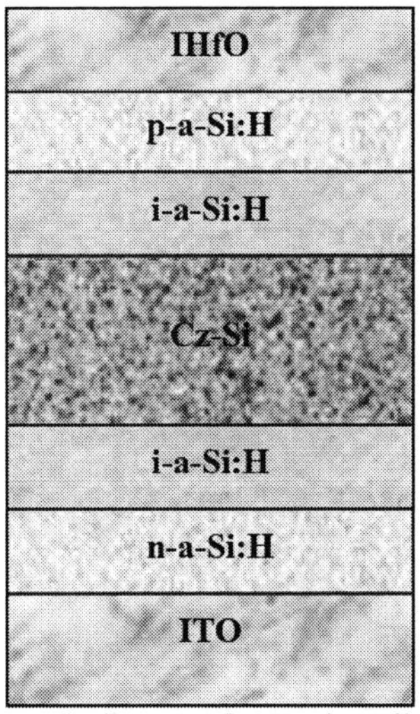

Fig.5 The schematic diagram of SHJ solar cells

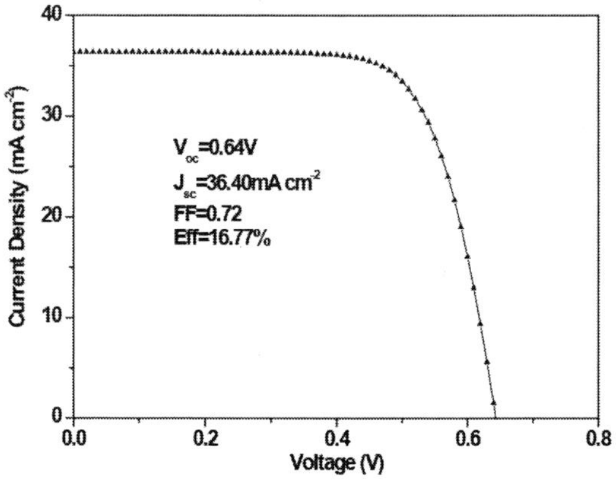

Fig.6 J-V curve and parameters of SHJ solar cells with IHfO contact layer

ACKNOWLEDGMENTS

This work is supported by National Natural Science Foundation of China (62004200 and 61274061).

REFERENCES

[1] D. S. Ginley, C. Bright. MRS Bulletin 2000, 25: 15-18.
[2] K. L. Chopra, S.Major, D.K. Pandya. Thin Solid Films 1983, 102: 1-46.
[3] T. Koida, M. Kondo. Applied Physics Letter 2006, 89: 082104-1~3.
[4] R. Hashimoto, Y.Abe, T. Nakada. Appl. Phys. Express. 2008, 1: 015002-1~3.
[5] F. Y. Meng, J. H. Shi, Z. X. Liu, Y. F. Cui, Z. D. Lu, Z. Q. Feng. Solar Energy Materials and Solar Cells 2014, 122: 70-74.
[6] O. Tuna, Y. Selamet, G. Aygun, L. Ozyuzer. Journal of Physics D: Applied Physics 2010, 43: 055402-1~7.
[7] M. Gulen, G. Yildirim, S.Bal, A. Varilci, I. Belenli, M. Oz. Journal of Materials Science: Materials in Electronics 2013, 24(2): 467-474.
[8] D. Kim, Y. Han, J.-S. Cho, S.-K. Koh. Thin Solid Films 2000, 377-378: 81-86.
[9] R. K. Gupta, K. Ghosh, S. R. Mishra, P. K. Kahol. Applied Surface Science 2008, 254: 1661-1665.

INFLUENCE OF REAR SIDE COATING ON EMITTER FORMATION DURING $POCL_3$ DIFFUSION PROCESS

Michael Steyer, Amir Dastgheib-Shirazi, Josh Engelhardt, Giso Hahn, Barbara Terheiden
Department of Physics, University of Konstanz, 78457 Konstanz, Germany

Author for correspondence: michael.steyer@uni-konstanz.de Tel.: +49 7531 882081, Fax: +49 7531 883895
Co-authors: amir.dastgheib-shirazi@uni-konstanz.de, josh.engelhardt@uni-konstanz.de, giso.hahn@uni-konstanz.de, barbara.terheiden@uni-konstanz.de

ABSTRACT: The influence of a SiN_x coating of a Si wafer on sheet resistance (R_{Sh}) of a neighbouring wafer during $POCl_3$ diffusion process is investigated. Wafers facing the SiN_x layer of the neighboring wafer in the next slot of the quartz boat show a lower R_{Sh} compared to those facing a bare Si wafer, e.g. a reduction from 61 Ω/sq to 52 Ω/sq and a thicker PSG layer are determined. The active doping profile, measured by ECV, shows a deeper plateau region while the tail region is unchanged. Accordingly, the emitter saturation current density rises from 130 fA cm^{-2} to 193 fA cm^{-2}. We propose that the thicker PSG layer originates from a lower consumption of the reactive gases ($POCl_3$-N_2 and O_2) at the SiN_x coated surface and thus a higher availability of them at the bare Si surface. On the other hand, we also investigate the influence of the thickness of the SiN_x rear coating on the emitter at the non-coated Si surface. Already a very thin layer of 20 nm SiN_x causes a significant change in R_{Sh} and the emitter profile, while there was no difference observed for the SiN_x thickness in the range from 20 nm to 160 nm. Facing a rear coated neighbor wafer during diffusion seems to improve the uniformity of R_{Sh} on 6-inch large area wafers.
Keywords: $POCl_3$, Diffusion, Emitter, PSG, PERC, Coating, Passivation

1 INTRODUCTION

Most commonly, a homogenous n-type emitter is fabricated via $POCl_3$ diffusion. Thereby the emitter is usually formed on both sides of the wafer and has to be removed chemically on the rear side for PERC (Passivated Emitter and Rear Cell) or at least an edge isolation is necessary for Al-BSF (Back Surface Field) solar cells. In order to avoid this process step, it is possible to coat the rear side before diffusion with a silicon nitride (SiN_x) capping layer, which acts as diffusion barrier. The capping layer could be already used for a texturing step before, and edge isolation after diffusion is not necessary anymore. For the manufacturing of PERC solar cells a temperature stable passivation under the SiN_x coating could be applied, which just has to be locally opened after diffusion.

During the diffusion process $POCl_3$-N_2 and O_2 forms phosphorus pentoxide (P_2O_5), which reacts with silicon to silica and the diffusing atomic phosphorus as dopant:

$$4POCl_3 + 3O_2 \rightarrow 2P_2O_5 + 6Cl_2 \quad (1)$$

$$2P_2O_5 + 5Si \rightarrow 5SiO_2 + 4P \quad (2)$$

SiO_2 and P_2O_5 forms the doping source, the PhosphoSilicate Glass (PSG) layer, which is possibly a multilayer system [1]:

$$SiO_2 + P_2O_5 \rightarrow P_2O_5 \cdot SiO_2 \quad (3)$$

In case of using rear coated wafers during diffusion, we observe a significant change in the emitter formation on the front side of the wafer facing the SiN_x coating in comparison to the emitter of an uncoated wafer. In the following, we show a detailed analysis of this effect.

2 EXPERIMENTAL SET-UP

The standard base material is p-type, 6-inch Cz silicon wafer with a bulk resistivity (R_B) of 2.7 Ωcm and 160 µm thickness. Half of the wafers received a rear side coating with SiN_x of different thicknesses (range 20 to 160 nm). After wet-chemical cleaning, a $POCl_3$ diffusion leading to a sheet resistance of 60 Ω/sq was performed in a commercial open tube diffusion furnace from Centrotherm. The wafers are arranged in the diffusion quartz boat as indicated in Fig. 1. The first and last 5 slots are used for dummy wafers. The wafers in slot 7 and 8 (as well as 34 and 35) with rear coating are arranged in back-to-back configuration. The neighboring uncoated wafers are used to characterize the emitter. On the other hand, rear coated wafers with different SiN_x layer thicknesses (indicated in Fig. 1) are arranged in the same direction (slot 10 to 21) with an uncoated wafer in between each pair. To balance small variations of R_{Sh} due to the length of the diffusion boat, the experiment is

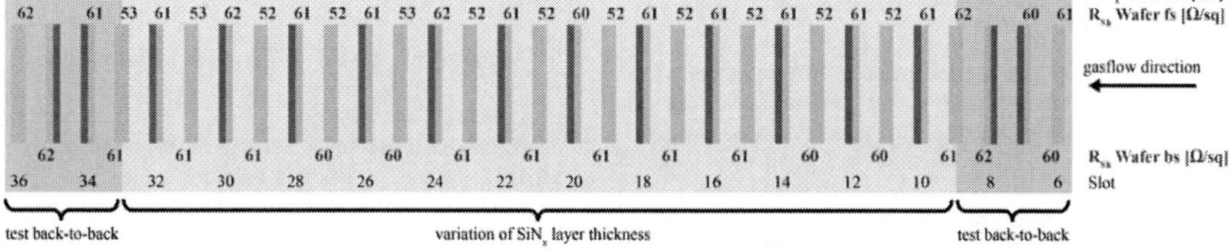

Figure 1: Wafer slot position (black bottom numbers) in the diffusion boat and resulting R_{Sh} on the wafers' front side (fs, red top numbers) and backside (bs, red bottom numbers). Some wafers have a rear coating (indicated as a blue layer) with different SiN_x layer thicknesses (black top numbers). R_{Sh} is the average value of a mapping with 25 measurements.

performed symmetrically using the slots 22 to 36.

Additionally, FZ-Si wafers with 150 mm diameter, p-type, R_B of 200 Ωcm and 250 µm thickness are placed in slots 39 and 42 (not shown in Fig. 1) for further emitter characterization. One FZ-Si wafer has uncoated neighbor wafers on both sides to create symmetrical samples, while the other FZ-Si wafer has coated neighbor wafers with SiN_x layer of 100 nm thickness. Both FZ-Si wafers are halved after diffusion. One half of each FZ-Si wafer is passivated on both sides with SiN_x by PECVD after PSG removal and co-fired in a commercial belt furnace to determine the emitter saturation current density (j_{0E}) by QSSPC [2]. After diffusion, R_{Sh} is measured on each wafer (mapping: 25 measuring points) with a Four-Point-Prober for all wafers tracking the slot position in the diffusion boat. The active doping profile of the wafers is measured by Electrochemical Capacitance Voltage (ECV). Furthermore, we determined the PSG thickness by spectroscopic ellipsometry [3].

3 RESULTS

Fig. 1 shows the measured R_{Sh} on the wafers' front side (red top numbers) in dependence of the slot position in the diffusion boat. First of all, the significant change of R_{Sh} from 61 to 52 Ω/sq on some uncoated wafers can be observed (compare slot 9 and 11). Due to the arrangement of coated and bare wafers in the diffusion boat one can also realize the reason of the change in R_{Sh}. If the coated wafers are arranged back-to-back, then there is no change in R_{Sh} (61 Ω/sq) on the bare neighbor wafers (e.g., slot 9 and 36). On the other hand, if the rear coated wafers are arranged facing one direction, R_{Sh} on the front side of the bare neighbor wafers changes to 52 Ω/sq (e.g., slot 11 and 13). So the change of R_{Sh} is caused by the opposed rear coated backside of the neighboring wafer in the next slot of the boat, independent of its own coating.

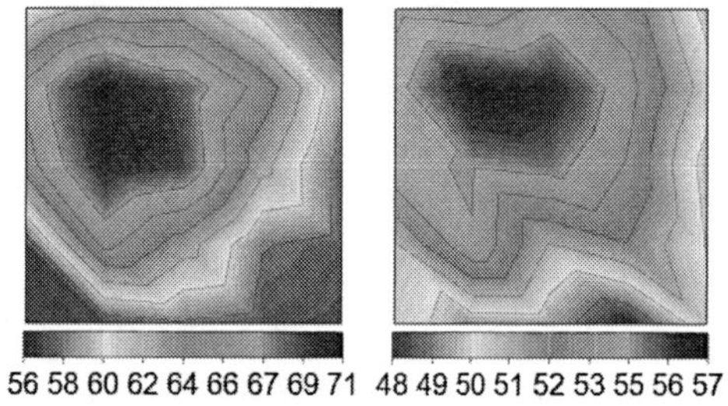

Figure 2: R_{Sh} mapping (25 measurements) of the wafers front side with uncoated neighbor wafer (left, Fig. 1 slot 9) and 20 nm SiN_x coated neighbor wafer (right, Fig. 1 slot 21) in the diffusion boat. The colored scale is in the range from minimum to maximum R_{Sh} of each wafer in the unit Ω/sq.

In Fig. 2 one can see the mapping measurement of R_{Sh} for coated and uncoated neighbor wafers. The change of R_{Sh} is also confirmed by ECV measurements, which are shown for the FZ-Si wafers (Fig. 3) and the Cz-Si wafers (Fig. 5).

The plateau depth of the emitter increases significantly by 6 nm (comparing depth values at a defined P concentration of 10^{20} cm^{-3}), while the emitter tail region is unchanged. This is a similar effect as compared to increasing the $POCl_3$-N_2 gas flow during diffusion [3].

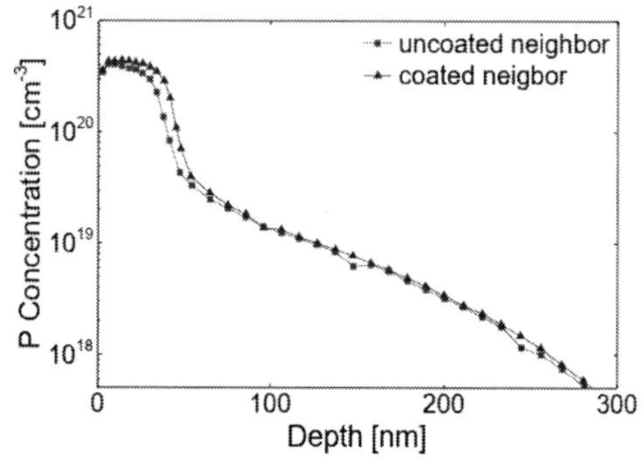

Figure 3: Emitter profiles in dependence of rear coating of neighboring wafer in the diffusion boat, measured with ECV using FZ samples. The measurement was performed in the middle of the wafer.

Furthermore, we show in Figure 4 the change in j_{0E}. If the FZ-Si wafer has uncoated neighbor wafers, j_{0E} is in average 130 fA cm^{-2} while j_{0E} rises up to 193 fA cm^{-2} for the FZ-Si wafers with coated neighbors. On the other hand, we also observe a change in thickness of the doping source, the PSG layer: The PSG thickness measured by ellipsometry changes significantly from 29 nm to 32 nm due to a coated neighbor wafer for the FZ-Si wafers (also 3 nm difference measured on the Cz-Si samples).

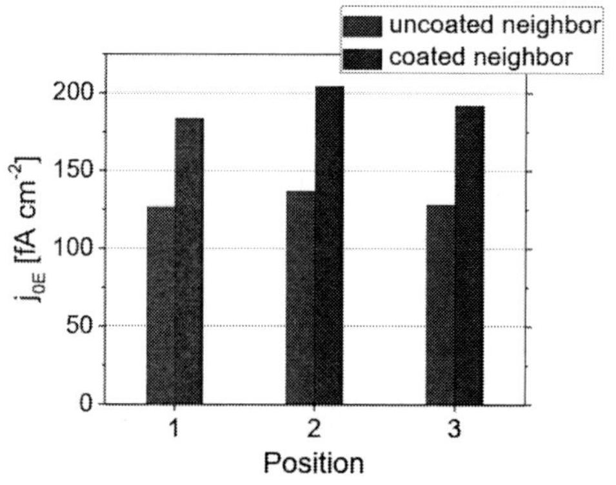

Figure 4: Emitter saturation current density measured by QSSPC. The one sample has uncoated neighbor wafers (in red), while the other one has coated neighbor wafers (in blue). Both FZ were halved after diffusion and then after PSG removal passivated by SiN_x symmetrically and co-fired. The measurement is performed in the middle (1), left (2) and right (3) position of the half 150 mm diameter FZ wafer.

We also investigated the influence of the thickness of the rear coating on the emitter properties of the neighboring wafers. As one can see in Fig. 1, there is no significant influence of the thickness in the range from 20 nm to 160 nm on R_{Sh}. Furthermore, in Fig. 5 the ECV measurements for different SiN_x thicknesses are shown. There is significant change between uncoated and 20 nm coating, but no significance for increasing the SiN_x thickness. The PSG thickness is also independent of the SiN_x thickness for the range from 20 to 160 nm.

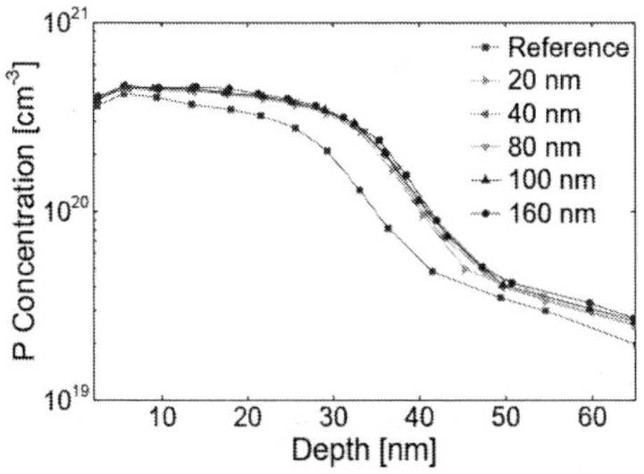

Figure 5: Emitter profiles in dependence of the SiN_x rear coating thickness of a neighboring wafer in the diffusion boat, measured with ECV. The reference sample (slot 9 in diffusion boat, see Fig. 1) has only an uncoated neighbor wafer to its front side (slot 8). The other samples have the slot position 11 (facing 160 nm SiN_x coating), 13, 15, 17, 19 and 21 in the diffusion boat. The measurement was performed in the middle of the wafer.

In addition, we show the effect for lowly doped emitters (see Tab. I) on 6-inch textured rear passivated wafers. Half of the wafers in the first part of the diffusion boat has a back-to-back orientation, while the other half of the wafers is orientated in the same direction (rear passivated backside shows in gasflow direction). For diffusion B, the rear passivated wafers with a back-to-back orientation have a sheet resistance of 108 Ω/sq (which corresponds to R_{Sh} on uncoated wafers), while the rear passivated wafers with no back-to-back orientation have a lower R_{Sh} of 88 Ω/sq. For diffusion C, R_{Sh} of the rear passivated wafers with a back-to-back orientation corresponds to 91 Ω/sq while R_{Sh} for the not back-to-back orientation decrease to 68 Ω/sq. Facing a rear coated neighbor wafer during diffusion seems to improve the uniformity of R_{Sh} on large area wafers. As one can see in Tab. I, for diffusion A, back-to-back orientation, the uniformity is the same as for diffusion B in a non back-to-back configuration, while R_{Sh} is much higher for the former.

Table I: Three diffusions, using always the same textured wafers with a SiO_2/SiN_x rear passivation. The rear passivated wafers are orientated back-to-back (b2b) or in same direction (not b2b) in the diffusion boat, which causes a dramatic change in R_{Sh} and uniformity (U).

Diffusion	Orientation	R_{Sh} [Ω/sq]	U [%]
A	b2b	63 ± 5	16
	not b2b	54 ± 3	11
B	b2b	108 ± 17	25
	not b2b	88 ± 9	16
C	b2b	91 ± 13	25
	not b2b	68 ± 5	15

4 CONCLUSIONS AND OUTLOOK

We observe a significant change in the emitter (R_{Sh}, j_{0E}, active doping profile) and the PSG thickness when the wafer side is facing a neighboring wafer in the diffusion boat which is rear coated with SiN_x. We propose that the thicker PSG originates from a lower consumption of the reactive gases ($POCl_3$-N_2 and O_2) at the SiN_x coated surface and thus a higher availability of them at the bare Si surface. Since one cannot expect an infinite doping source while increasing the $POCl_3$-N_2 gas flow [4], the change in PSG growth also changes the emitter formation.

On the other hand, we also investigated the influence of the thickness of the SiN_x rear coating. Already a very thin layer of 20 nm SiN_x causes a significant change of the emitter in R_{Sh} and the plateau depth of the emitter profile, while there was no change for increasing the SiN_x thickness up to 160 nm. In further experiments, a SiN_x layer of 10 nm and below would be of great interest to determine the minimum SiN_x thickness for a significant change in R_{Sh}.

In addition, we demonstrated the effect for highly doped emitters. Thereby one can expect a higher uniformity of R_{Sh} on 6-inch large area wafers by facing rear coated wafers. As we showed by Four-Point-Probe mapping, the highest R_{Sh} is in the middle of the wafer, where the least reactive gas molecules can be expected. Due to the SiN_x coated surface, the lower consumption of the reactive gases would be most effective in the middle of the wafer.

One has to be careful to expect always the same emitter properties, independent of a rear coating during diffusion. This is, e.g., very important for manufacturing of j_{0E} samples, which have to be symmetrical [2]. Furthermore, this systematic investigation of the rear coating during diffusion can help to transfer the emitter of a standard Al-BSF solar cell to a modern rear passivated solar cell.

5 ACKNOWLEDGEMENTS

Part of this work was financially supported by the German Federal Ministry for Economic Affairs and Energy (FKZ 0325581, 0324001 and 0325777B). The content of this publication is the responsibility of the authors.

6 REFERENCES

[1] R.N. Ghoshtagore: *Phosphorus diffusion processes in SiO_2 films*, Thin Solid Films 25(2) (1975) 501.

[2] D.E. Kane, R.M. Swanson, *Measurement of the emitter saturation current by a contactless photoconductivity decay method*, Proc. 18th IEEE PVSC, Las Vegas 1985, pp. 578-583.

[3] M. Steyer, A. Dastgheib-Shirazi, H. Wagner, G. Micard, P.P. Altermatt, G. Hahn: *A study of various methods for the analysis of the phosphosilicate glass layer*, Proc. 27th EU PVSEC, Frankfurt 2012, pp. 1325-1328.

[4] A. Dastgheib-Shirazi, M. Steyer, G. Micard, H. Wagner, P. P. Altermatt, G. Hahn: *Effects of process conditions for the n^+-emitter formation in crystalline silicon*, Proc. 38th IEEE PVSC, Austin 2012, pp. 1584-1589.

OPTICAL REFLECTION SPECTRA OF SILICON SURFACE WITH NANOWIRES PRODUCED BY SPECIAL ELECTROCHEMICAL ETCHING

Marius Treideris, Viktorija Strazdienė, Irena Šimkienė, Virginijus Bukauskas, Alfonsas Rėza, Simonas Indrišiūnas, Mindaugas Kamarauskas, Arūnas Šetkus
Center for Physical Sciences and Technology
Sauletekio ave 3, Vilnius LT-10257, Lithuania

ABSTRACT: In this work, we present promising simple method for a modification of optical response in c-Si solar cells. In this approach the spectral dependence of the optical reflectance is intentionally changed by strictly controlled electrochemical etching of the surfaces. It is demonstrated that the etching produces a layer with randomly ordered silicon nanowires (SiNW) that changes the optical response of the surface. The change in the reflectance spectrum depends on the dimensions of the SiNWs that are intentionally tuned by controlling the conditions of the etching. Our experimental results demonstrate that a maximum appears in the reflectance spectrum of c-Si surface with the SiNW in the interval of the visible wavelengths. The position of the maximum depends on the length of the SiNW and can be varied within entire interval of the visible light. We describe the relationships between the etching conditions, the properties of the SiNW layer and the optical response in this report. The technological parameters that produce the most significant effect on the dimensions of the SiNW and the reflectance are identified.
Keywords: Porous Silicon, Etching, Photoelectric Properties, Silicon Solar Cell

1 INTRODUCTION

At present silicon based solar cells are the dominant type of the cells available to various potential users on the market [1]. However an integration of this type of the cells into the buildings and other objects is frequently complicated from the aesthetic point of view due to the dominant black colour of the cells. In the urban areas, it is desired to get the solar cells with a wide optical response making it easy to choose the cells in a wide range of colours for the aesthetic view. Some possibilities to tune the optical response of Si solar cells were recently proposed by the modification of the anti-reflection coatings and the protecting glass of the solar module [2]. However the efficiency of the coloured cells is significantly lower than the standard ones because a part of the light is eliminated from the generation of the photovoltaic current due to colouring of the cell surface. It was recently introduced a very promising method for solar cell colouring [3] that is based on the optical properties of tall (>1μ) silicon nanowires (SiNW) and possibilities to choose the colouring properties by variation of the nanowire diameter. In the SiNW based approach a decrease in the amount of absorbed light and generated current due to colouring can be compensated by an increase in the efficiency due to specific properties of the SiNWs [4]. It was shown that the optical scattering of the surfaces in the SiNW containing cells depends on the geometry of the SiNWs and do not or only slightly depends on the placement order [3, 5]. However it is still a big challenge to develop a cost effective technology for the SiNW with easy tunable geometries. There is not enough information about the technological parameters acceptable to intentionally modify the selected characteristics of the cells. In this work, we propose the parameters acceptable to control the modification of the optical response of crystalline Si (c-Si) surface covered with the layer of randomly ordered SiNW.

2 MATERIALS AND METHODS

For this study the p-type Cz-grown c-Si were obtained from SIEGERT WAFER GmbH. The wafer flats were with the surface plane orientation (100) and the electrical resistivity from 1 to 5 Ω cm interval. The wafers were cut into the equal pieces for use in the separated experiments. The substrates were cleaned by the standard procedures used in the silicon technology. The substrates were mounted in the special holder adapted to the electrochemical etching in an electrolyte. A series of the etching solutions was prepared in our study by mixing HF and dimethyl sulfoxide (DMSO). The proportions of the components HF:DMSO was individual in the separated experiments and was any equal to 1:3, 1:5 and 1:8. In the separated experiments the etching current was individual depending on choice and it was set constant from the interval 6.5-16 mA/cm^2. The front side of the samples was illuminated by an incandescent lamp under the same conditions and the illumination was constant during the etching and the same for all the sequence of our experiments. The experiments of electrochemical etching were performed at constant temperature equal to about 300 K. The etching time was individual selected for separated experiments and was between 5 and 30 minutes.

The etched surface of the samples was characterised by a scanning electron microscope (SEM) JEOL JSM 6490 LV. The cut edge was analysed in the etched samples and the structure in the cross section of the surface layer was visualised. The optical characteristics of the substrates were obtained by a spectrometer SHIMADZU UV – 3600 with an integrating sphere. The spectra of the optical reflectance were measured from about 400 nm to about 1100 nm.

3 RESULTS AND DISCUSSIONS

The results of this work demonstrated that the SiNW can be obtained practically under the all conditions of the etching tested in our experiments if the the etching time was sufficiently long. However, the quality of the SiNW significantly depends on the technological variables acceptable for control of the process. Typical surfaces of the c-Si substrate after the electrochemical etching are illustrated by the SEM images in Fig. 1. The SEM images of the substrate surface in Fig. 1 are presented at two different angles of the incident beam images with respect to the surface plane.

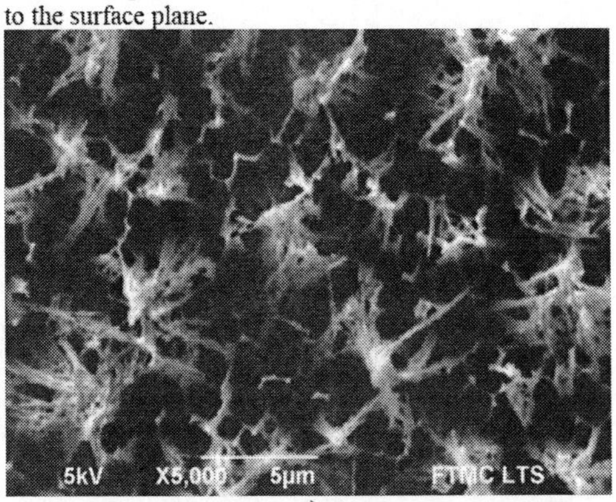

a)

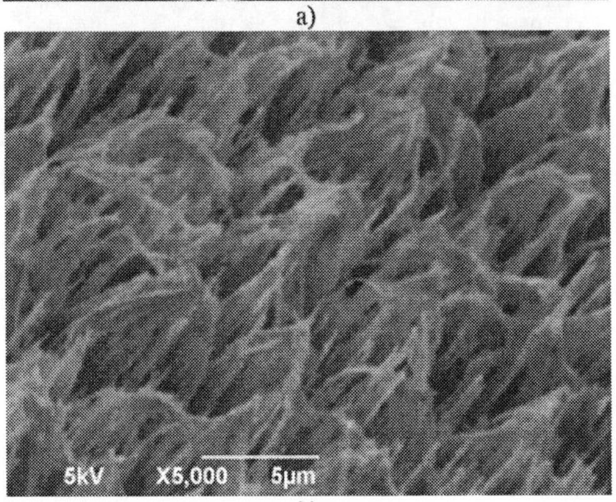

b)

Figure 1: SEM images of the c-Si surface from the vertical view (a) and the angle oriented view (b) for the sample after the electrochemical etching with t=30 min., j=11 mA/cm^2 and HF:DMSO=1:5.

It can be seen from Fig. 1b that the surface of the substrate is covered by the SiNWs and these SiNW are randomly distributed over the surface. In addition, part of the SiNWs are significantly bent and connected with each other. Therefore, in Fig. 1b the surface looks like covered by some aggregates of flakes or nonhomogeneous maze of shapes.

It followed from the analysis of the SEM images that the diameters were practically the same for all the SiNWs in the electrochemically formed layer on the c-Si surface.

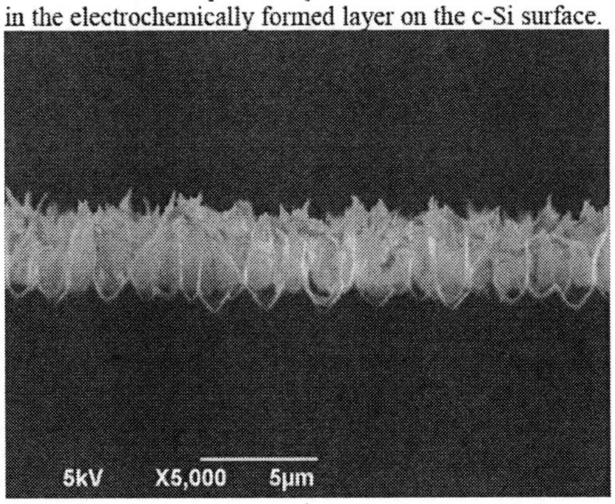

a)

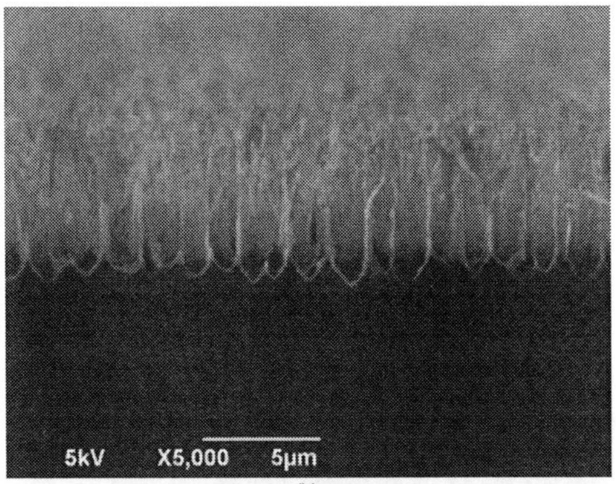

b)

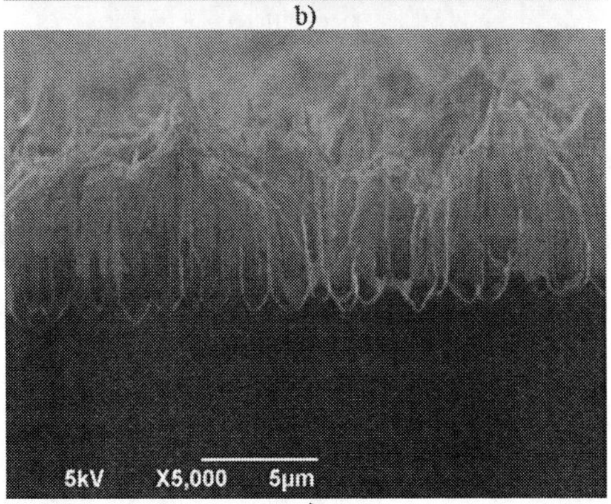

c)

Figure 2: SEM cross-section images for electrochemically etched for: a) 6.5 mA/cm^2, b) 11 mA/cm^2 and 16 mA/cm^2 samples (t=30 min., HF:DMSO=1:5)

The diameter was equal to about 150 nm and was not related to the technological variables used in this study. Typical results of these SiNWs are illustrated in Fig. 2.

However average SiNW length has strong dependence on etching current density Fig.3.

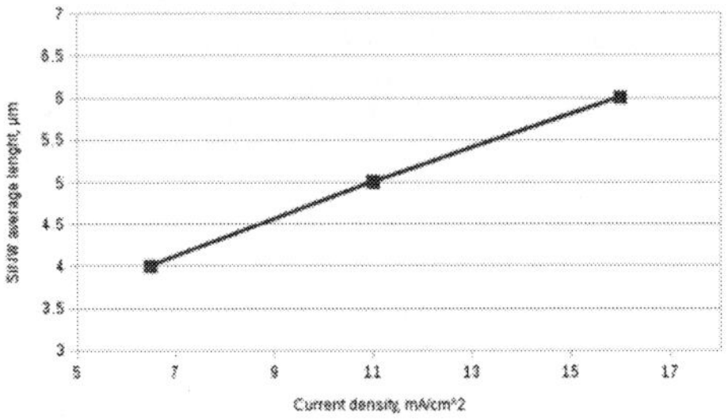

Figure 3: Average SiNW length dependence on the current density during electrochemical etching (t=30 min., HF:DMSO=1:5)

Random arrays of SiNW etched for different current density and thus a different SiNW length have optical reflectance spectra with broad peaks on different wavelengths (Fig.4).

Etching time and HF:DMSO ratio values were chosen from the series of experiments on the influence of these parameters on the etched surface morphology. It was noticed that SiNW on the surface are formed only

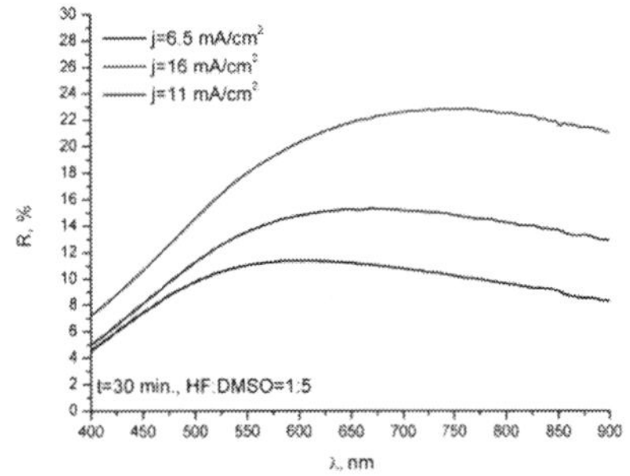

Figure 4: Optical reflectance of surfaces with randomly ordered SiNW etched under different current density

a)

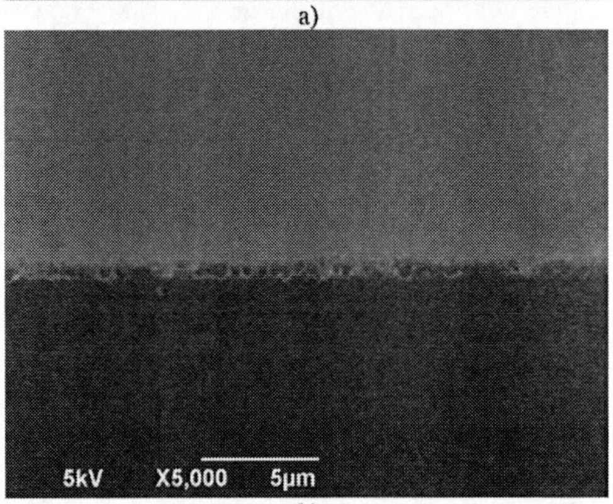

b)

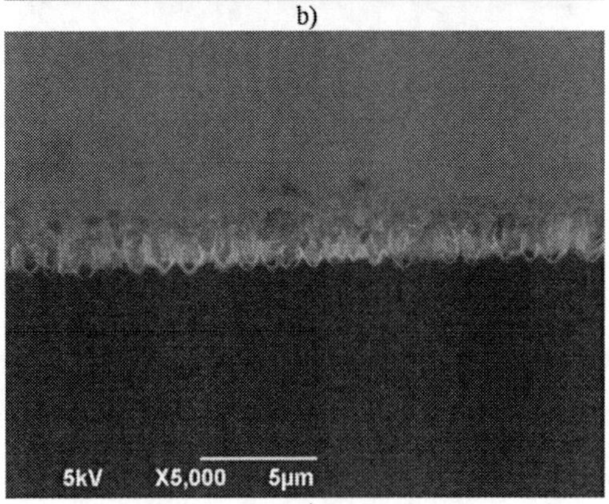

c)

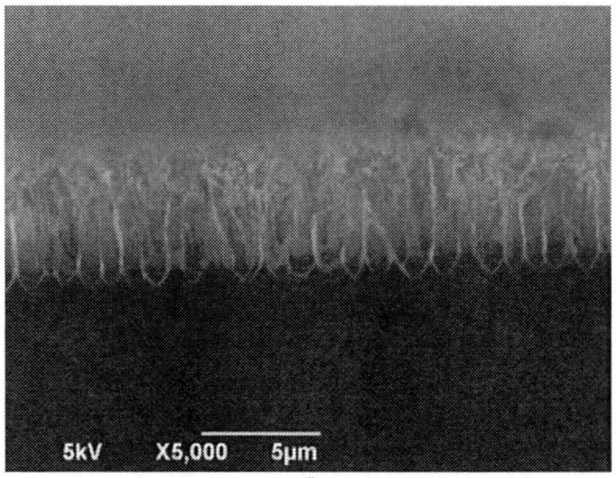

d)

Figure 5: SEM cross-section images for electrochemically etched for: a) 5 min., b) 7 min. c) 10 min. and 20 min. samples (j=11mA/cm^2, HF:DMSO=1:5)

when etching time is longer than 20 min. (for the range of current density used in this work and etchant ratio HF:DMSO=1:5) (Fig. 5).

For lower etching time rought surface with etched pores is formed (Fig. 5 a, b, c) and in samples etched for 20 min. arrays of SiNW with diameter ~150 nm and average length 4,5 µm were formed (Fig. 5d). For samples etched for 30 min. SiNW length increases to 5 µm (Fig. 2b).

Difference between these structures influences their optical response (Fig. 6).

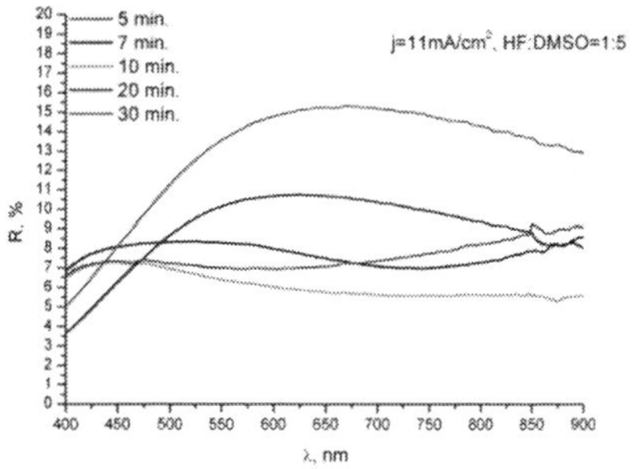

Figure 6: Optical reflectance of surfaces with randomly ordered SiNW etched under different time

For the samples etched for less than 20 min. when only rought surface is formed optical reflectance spectra is almost flat with some small variations (red, black and green lines in Fig. 6) while samples with arrays of SiNW have optical response spectra with broad peaks in different wavelengths.

The etchant composition was set as fixed technological parameter. In our experiments it was found that the change of etchant composition ratio has led to dramatical change in both the length and diameter of SiNW.

From these results we suggest that during electrochemical silicon etching pores on the surface starts to form and when the etching time increases pores depth increases and pore walls are etched to and leaves a free standing SiNW. It is known that for etching reaction there must be holes on silicon surface which is etched. So the free standing SiNW during pore etching is a possible result of a not etched silicon which is depleted because of the interaction of etching electrolyte. We suspect that depletion on the silicon surface occurs on the pore wall edges and increasing time depleted zone increases. However this model must be investigated in detail.

We demonstrate that by changing one technological parameter (current density) during electrochemical etching at fixed etching time and etchant ratio, we can control the optical response (surface color) from the surface as a result of the change in the length of SiNW (Fig. 4). From optical reflection dependence on etching time we can predict that after breaking point on the time scale were SiNW are formed optical response in visible range can be intentionally changed by changing time. Comparing results from both dependencies on technological parameter we can suggest that in general optical response of the SiNW randomly ordered arrays with similar average SiNW diameter is dependent on average SiNW length (Fig. 7).

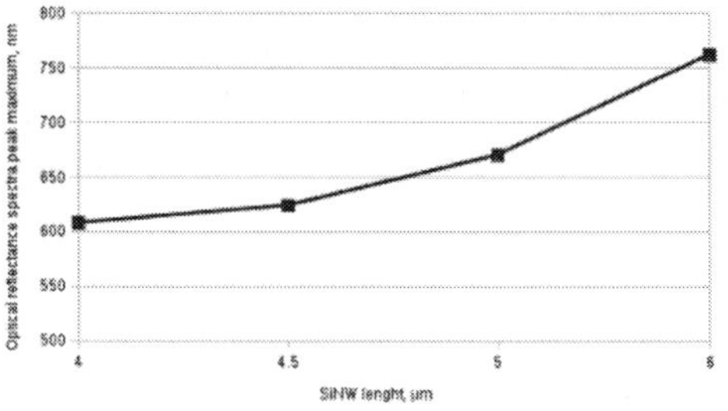

Figure 7: Optical reflectance peak maximum dependence on SiNW average length

4 CONCLUSSIONS

The results of our experimental investigation demonstrates that the optical response of the c-Si surface can be tuned by the electrochemical etching if an array of randomly distrubuted SiNWs is formed on the surface. It also follows from the results that the spectrums of the optical reflectance depend on the length of the SiNWs. The length of the SiNWs and, consequently, the color of the surface can be controlled by a change in the duration of the electrochemical etching.

However it is clear that due to the random electrochemical etching process there is some variation in SiNW diameter, lenght and alignment on the surface and it explains why we can't see sharp line in reflectance spectra. In this development phase, the absence of sharp line leads to not vivid colors of the sample surface, but optical response dependence on technological conditions is clearly seen and for some practical applications this technology can be applied. However in future investigation optimal etching conditions are expected to increase SiNW diameter uniformity and with optimized antireflection layer to lead to significant results.

5. ACKNOWLEDGEMENT

The study was partly supported by a grant (No. MIP-70/2015) from the Research Council of Lithuania.

References:

[1] T. Saga, NPG Asia Mater 2 (2010) 96-102.

[2] J. Y. Shih, S. L. Lai, H. T. Cheng, Advanced Materials Research 706 (2013) 420-425.

[3] K. Seo, M. Wober, P. Steinvurzel, E. Schonbrun, Y. Dan, T. Ellenbogen, K. B. Crozier, 11(4) (2011) 1851-1856.

[4] E. C. Garnett, M. L. Brongersma, Y. Cui, M. D. McGehee, Annual Review of Materials Research, 41 (2011) 269-295.

[5] N. Dhindsa, S. S. Saini, Optics letters, 41(9) (2016) 2045-2048.

HIGH EFFICIENT N-TYPE AND P-TYPE PERT SOLAR CELLS BY INDUSTRIALLY FEASIBLE PROCESSES

Chia-Ching Wang[1], Chia-Lung Lin[1], Yu-Ta Cheng[1], Yu-Hsiang Huang[1], Chin-Pao Tsao[1], Jung-Wu Chien[1]

I-Liang Chen[2], Chin-Huang Chen[2], Cheng-Lien Wang[2]

[1]Inventec Solar Energy Corporation, Taiwan

[2]Win Win Precision Technology Corporation (WINAICO), Taiwan

PERT crystalline silicon solar cell is getting popular in recent years due to its bifacial property which collects sunlight more effectively than monofacial solar cells. Inventec started to develop n-type PERT from 2012Q1 and published cell efficiency up to 19.7% in 2013 EUPVSEC. Mass production of n-PERT was started from 2014H2 with 40MW capacity and expanded to 80MW in 2015H2. Through continuous improvement, the average and champion efficiency of Gen1 bifacial solar cells with 3busbar are 20.12% and 20.56% in 2015Q3. Meanwhile, n-type PERT, p-type PERT and p-type PERT+ on 6" commercially available mono-Si wafers were also be studied. The cell structure, IV performance and bifaciality were compared. Surprisingly, the champion efficiency of p-PERT+ is 20.30% and average efficiency is only 0.15% lower than that of n-PERT. However, n-PERT is still the highest among these cell type in terms of both front and rear side cell efficiency. In 2016Q1, 4 busbar design and optimized process were developed and adopted to Inventec Gen2 n-PERT solar cells and the cell efficiency successfully achieved 21.15% of champion and 21.02% in average. Finally, high performance of n-PERT monofacial module achieved 300.7W by using 60 pieces of Inventec Gen2 n-PERT bifacial solar cells and Winaico advanced module technologies.

Keywords: n-type, p-type, PERT, PERT+, bifacial, module

1. INTRODUCTION

PERT (passivated emitter, rear totally diffused) solar cells became popular in recent years due to its special application and high power output which generated from both of front and rear sides [1]. N-type wafer was widely used on PERT cells since n-type wafer was not only better than p-type wafer in terms of bulk lifetime, LID, high tolerance to metal impurities, but also resist high temperature of boron diffusion [2]. However, the market share of n-type solar cells was less than 10% due to high wafer price and complicated cell processes.

In order to reduce manufacturing cost of n-PERT, a novel and industrial process was developed by Inventec and published in 2013 EUPVSEC[3]. Besides, the commercially available p-type wafers were studied on PERT. Compared with phosphorous diffusion, boron diffusion needs higher temperature to activate boron atom. Unfortunately, p-type wafer will suffer bulk lifetime degradation when process temperature is higher than 900°C [4]. Therefore, a new boron diffusion recipe was developed by Inventec to avoid thermal degradation of p-type wafer by controlling boron diffusion process time to less than 90 minutes. The advanced p-PERT and p-PERT+ cells were successfully made by implementing n-PERT process flow only changing some process parameters. Compared IV data of three types of PERT cells, n-PERT is still the best one in terms of front and rear side performance. Therefore, the advanced Gen2 n-PERT solar cell was developed and the high performance monofacial module with 300.7W power output was achieved successfully in 2016Q1.

2. DIFFERENT BIFACIAL CELL PRODUCTS AND TECHNOLOGIES

2.1 N-PERT AND MASS PRODUCTION

Inventec started n-PERT bifacial cell development in 2012Q1[5] and turned into mass production in 2014Q3. The average cell efficiency was continuously improved from 19.4% to 20.1% by process optimization and integration. In 2015H2, we expanded PERT capacity from 40MW to 80MW due to increasing demand of n-PERT.

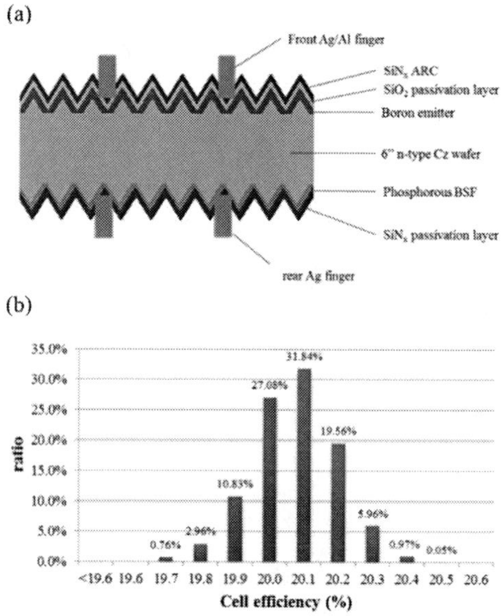

FIG.1. (a) cross-section structure of n-PERT; (b) efficiency distribution of 3 busbar n-PERT mass production in 2015Q3.

Fig.1(a) shows cross-section structure of n-PERT. BBr_3 diffusion and $POCl_3$ diffusion were applied for emitter and BSF formation. Front side was passivated by SiO_2/SiN_x stack layer while rear side was passivated by SiN_x only. Ag/Al paste and Ag paste were screen printed on front and rear surface, respectively. Fig.1(b) shows IV distribution of daily production in 2015Q3. The cell efficiency distribution is quite tight as p-type Al-BSF solar cells, the main cell efficiency ranges from 19.7% to 20.5%.

Table 1 lists average and champion IV data of Fig. 1(b). The cell efficiency of n-PERT was measured by Halm IV tester which was calibrated by AIST, Japan. The average and champion cell efficiency is 20.12% and 20.56%, respectively. Cell appearance was good on both front side and rear side just like front side of conventional p-type mono-Si Al-BSF solar cells. Compared with the other commercial n-PERT cells in the market, complicated and expensive methods were not selected in our manufacturing process, such as rear surface polishing, ion implantation, long time boron diffusion, AlOx passivation of boron emitter, wet chemical oxide passivation, emitter, selective BSF, double printing, plating, etc.

TABLE 1. IV data of GEN1 n-PERT in the Fig. 1(b).

	Eta (%)	Voc (mV)	Isc (A)	FF (%)
Average	20.12	639.5	9.29	80.95
maximum	20.56	646.1	9.34	81.41

2.2 DEVELOPMENT OF P-PERT and P-PERT+

P-type wafer was considered not suitable for PERT process due to low bulk lifetime and the potential bulk degradation under high temperature boron diffusion, however, it is still worth to be studied since p-type wafer is much cheaper than n-type wafer. Based on the n-PERT development experience, p-PERT was easily developed by using n-PERT similar processes and only changing some process parameter. Boron diffused layer is the emitter on n-type wafer but acts as BSF on p-type wafer, while phosphorous diffused layer is the BSF on n-type wafer but acts as emitter on p-type wafer. Diffusion temperature and process time (or called "thermal budget") control is important to avoid bulk quality degradation.

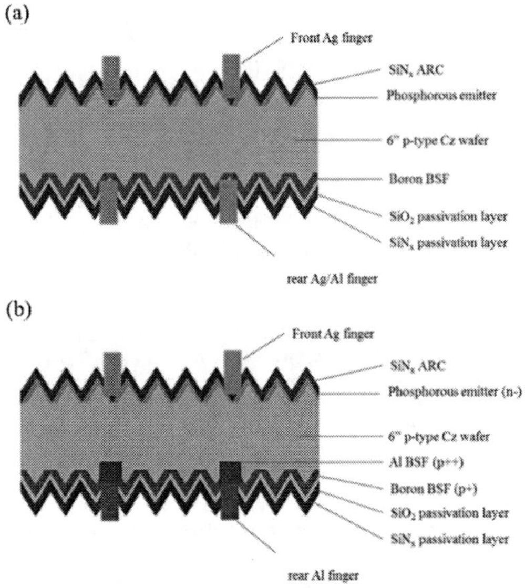

FIG.2. cross-section structure of (a) p-PERT and (b) p-PERT+

Paste cost is the second expensive material in the solar cell. Ag paste usage on PERT bifacial cell is nearly twice than conventional p-type Al-BSF solar cells and results in high manufacturing cost. Therefore, we tried to use Al paste to replace Ag/Al paste on the rear surface of p-PERT and called it "PERT+". The PERT+ is similar to the concept of PERC and other solar cells [6]. In PERT+, Al paste not only works as rear side electrode but also creates p++ BSF on the rear side. In order to have good contact, LASER was applied to open dielectric layer to assure Al paste contacts with Si wafer completely. Fig. 2 shows cross-section structure of p-PERT and p-PERT+. The only difference lies on the rear side electrode and rear local p++ BSF formation.

In order to compensate lower conductivity of Al fingers, Al fingers were printed widely to cover 25% of rear side area. Surprisingly, p-PERT+ got even higher FF than n-PERT and p-PERT despite of low conductivity of Al, as shown in Fig. 4 and Table 2. Although p-PERT+ got lower Voc and Isc than p-PERT due to larger metal contact on rear side, the cell efficiency of p-PERT+ was still higher than p-PERT by 0.26% due to high FF. In the results, Al electrode not only lowered manufacturing cost but also improved cell efficiency on p-PERT+. Fig. 3 shows the efficiency distribution of p-PERT and p-PERT+.

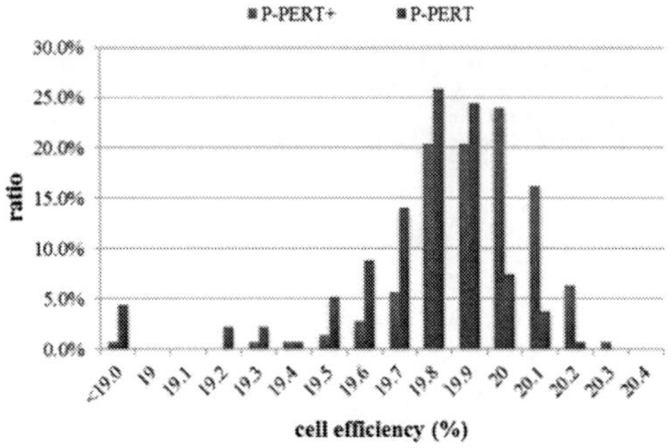

FIG.3. Efficiency distribution of 3 busbar p-PERT and p-PERT+.

2.3 COMPARISON OF N-PERT, P-PERT AND P-PERT+

Process flow and IV performance of three kinds of solar cells was listed together to compare IV performance. Besides, bifacility is the key index to evaluate performance of bifacial solar cells since albedo is normally around 20%~30% depends on the module setup.

The process flow of three type of solar cells are highly similar. Therefore, we can easily tranfer technology from n-PERT to p-PERT and p-PERT+. P-PERT saves the wafer cost and p-PERT+ saves both wafer and Ag paste cost. Despite p-PERT+ needs extra LASER opening, suitable Al paste can also be used to fire through rear dielectric layer directly to skip LASER opening process.

Fig. 4 shows IV distribution and Table 2 shows IV data of these three types of solar cell. N-PERT get higher Voc and Isc than p-PERT due to high bulk lifetime as expetced, however, FF of p-PERT and p-PERT+ are higher than n-PERT due to low bulk resistance. Therefore, efficiency of p-PERT and p-PERT+ is lower than n-PERT by only 0.41% and 0.15%, respectively. The effiency gap is much lower than our expectation.

Despite of small cell efficieny gap, however, n-PERT is still better than p-PERT and PERT+ in terms of bifaciality. From Table 3, p-PERT got lower bifaciality than n-PERT mainly due to low Isc of rear side. When measuring rear side IV data, emitter is placed back to the illumination and works as rear junction. The low bulk lifetime of p-type wafers limits minority carriers movement and results in low Isc and efficiency of rear side. On the contrary, high bulk lifetime of n-type wafer helps minority carriers move to rear junction and results in high Isc and efficiency. Therefore, we observe Isc ratio of p-PERT is only 88.4% while Isc ratio of n-PERT is 95.2%. As for p-PERT+, bifaciality is only 74% because of wide Al fingers.

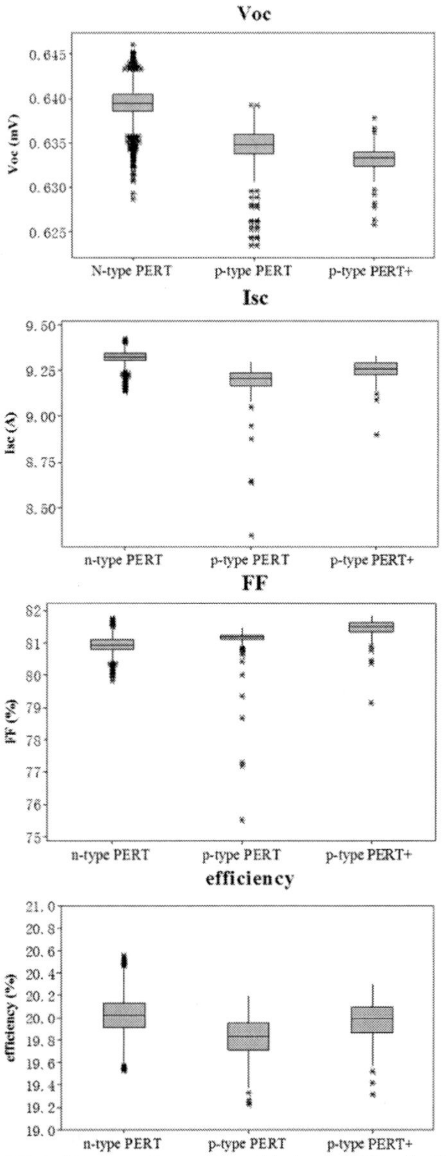

FIG.4. IV comparison of three kinds of PERT cells (3 busbar)

TABLE 2. IV data of 3 busbar n-PERT cell

Cell type		Eta (%)	Voc (mV)	Isc (A)	FF (%)
n-PERT	average	20.12	639.5	9.29	80.95
	max	20.56	646.1	9.34	81.41
p-PERT	average	19.71	634.2	9.18	80.95
	max	20.20	638.5	9.29	81.37
p-PERT+	average	19.97	633.1	9.25	81.45
	max	20.30	635.4	9.33	81.84

TABLE 3. Bifaciality of 3 busbar n-type and p-PERT

Cell type	Eta (%)	Voc (mV)	Isc (A)	FF (%)
n-PERT	95.4%	99.7%	95.2%	100.5%
p-PERT	88.7%	99.6%	88.4%	100.5%
p-PERT+	74.1%	98.7%	75.3%	99.5%

2.4 DEVELPMENT OF 4 BUSBAR N-PERT SOLAR CELLS

4 busbar cell design became main stream of crystalline solar cells due to low cell-to-module loss. By using 4 busbar design, n-PERT efficiency is obviously improved by 0.43% which average and champion cell efficiency is 20.55% and 20.76%, respectively. Table 1 shows IV data of 3 busbar n-PERT and table 4 shows IV data of 4 busbar n-PERT. By comparing average efficiency from table 1 and table 4, we found efficiency gain of 4 busbar mainly come from FF improvement. In the Fig.5, cell efficiency of 4BB is obviously better than 3BB which all cells are higher than 20%.

TABLE 4. IV data of 4 busbar n-PERT solar cells

	Eta (%)	Voc (mV)	Isc (A)	FF (%)
Average	20.55	642.6	9.34	81.82
maximum	20.76	643.8	9.40	82.01

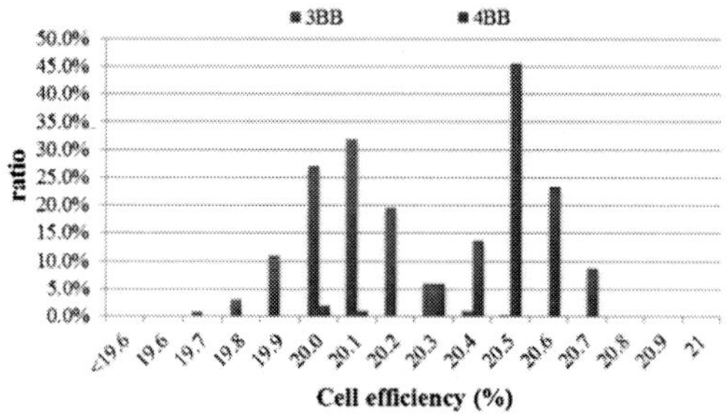

FIG.5. Efficiency distribution of 3BB and 4BB n-PERT

3. DEVELPMENT OF 21% GEN2 N-PERT SOLAR CELLS AND 300W MONOFACIAL MODULE

Besides of 4 busbar design, other parameters such as emitter, BSF profile, front and rear side electrodes can also be further optimized. Through process parameters optimization including emitter, BSF and electrode, efficiency was improved by 0.9% as Fig. 6 showed. By integrating 4 busbar design and new process parameters, we called it Gen2 n-PERT bifacial solar cells. Table 5 shows the latest IV data of Gen2 n-PERT with 21.02% of average efficiency and 21.15% of champion efficiency. Voc was also be improved to 647mV due to better emitter, BSF and passivation quality.

TABLE 5. IV data of ISEC Gen2 n-PERT solar cells by optimizing emitter, BSF, front and rear side electrode

	Eta (%)	Voc (mV)	Isc (A)	FF (%)
Average	21.02	646.5	9.71	81.32
maximum	21.15	648.5	9.72	81.44

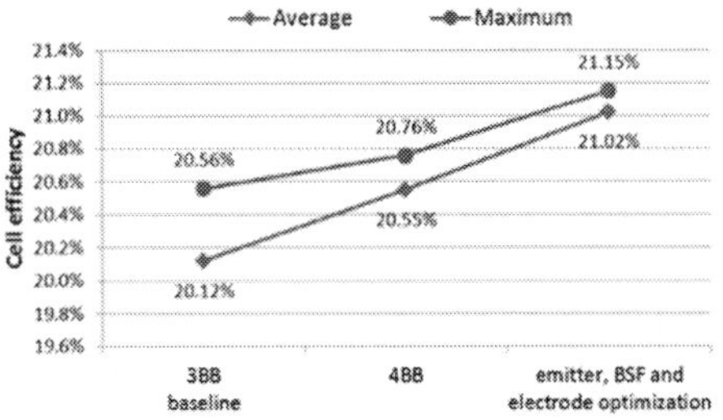

FIG.6. Efficiency improvement from 3BB and 4BB n-PERT to ISEC Gen2 n-PERT

Fraunhofer ISE had summarized the main loss factors of solar module [7] as shown in Figure 7. Suitable design can reduce optical and electrical losses. Through these principles, Winaico Corporation developed high performance p-type PERC module and used on the rooftop system in Australia. Compared with traditional p-type module, n-PERT monofacial module is more suitable for rooftop application due to high power, zero LID and zero PID. The glass/backsheet (monofacial) structure is also suitable for n-PERT bifacial solar cell since white back sheet can increase internal reflection of long wavelength sunlight and enhance module power [8]. Therefore, Inventec integrated Gen2 n-PERT bifacial solar cells with Winaico advanced module technologies to achieve high power and high quality of module.

The IV curve of n-PERT module is shown in Figure 8. The module power was up to 300.72W and CTM loss was only 1.81%. Isc was up to 10.03A by reducing light absorption of EVA/glass and increasing internal light path. 4 busbar design helped to reduce CTM electrical losses. The EL image showed no defect appeared in the 6x10 module as in the Figure 9.

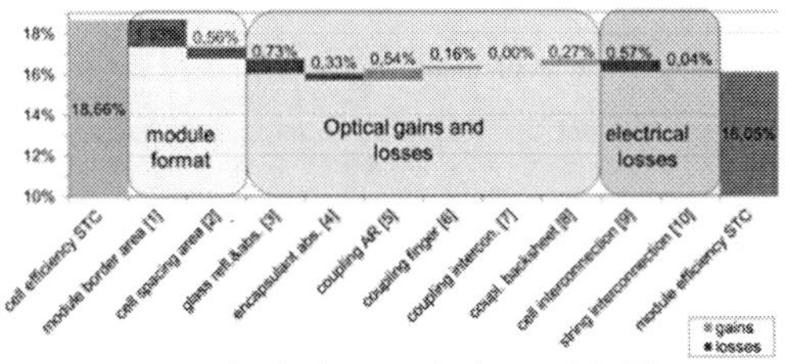

FIG.7. Loss and gain factors of solar module[7].

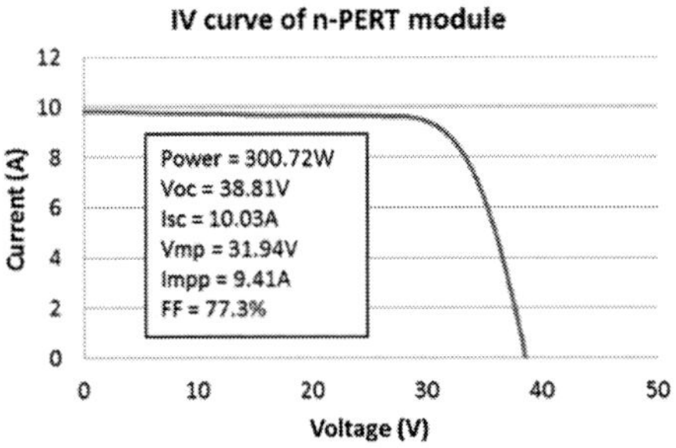

FIG.8. IV curve of 300.7W n-PERT module by using 60 pieces of Inventec 21% bifacial solar cells and Winaico advanced module technologies.

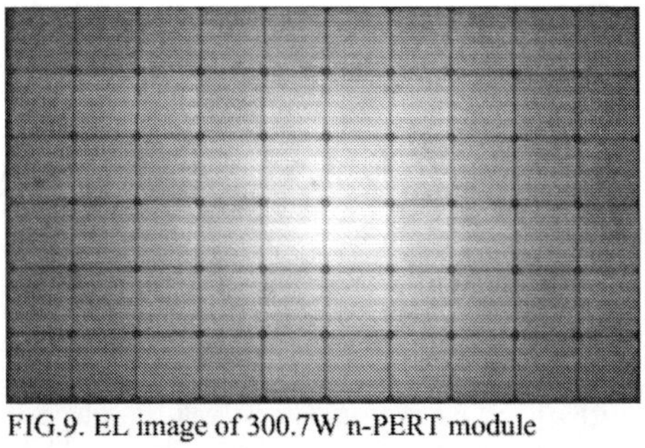

FIG.9. EL image of 300.7W n-PERT module

4. SUMMARY

N-PERT got more and more attention due to high module power than traditional p-type solar cells. In this article, we compared cell efficiency of n-PERT, p-PERT and p-PERT+. Even using cheap p-type wafer and Al paste, p-PERT+ efficiency is only 0.15% lower than n-PERT, however, bifaciality of p-PERT and p-PERT+ is much lower than n-PERT due to low bulk lifetime of p-type wafer. Therefore, n-PERT is still the best choice of bifacial solar cells. 4 busbar and process optimization helped to improved n-PERT efficiency to 21.02% and 21.15% of average and champion, respectively. Finally, 300.7W of n-PERT module is also achieved by using Winaico advanced module technologies.

ACKNOWLEDGEMENT

The authors would like to acknowledge Gigastorage Corporation for providing high quality n-type wafers to achieve 21% n-PERT solar cells and 300W monofacial module.

REFERENCE

[1] Sugibuchi K, Ishikawa N, Obara S., "Bifacial-PV power output gain in the field test using "EarthON" high bifaciality solar cells", Proc. 28th European Photovoltaic Solar Energy Conference, Paris, 2013. P.4312-7

[2] D. MacDonald and L.J. Geerligs, "Recombination activity of interstitial iron and other transition metal defects in p- and n-type crystalline silicon", Appl. Phys. Lett. 85. 4061 (2004)

[3] C.C. Chen, C.L. Lin and J.W. Chien et al., "High Efficiency N-Type Silicon Bifacial Solar Cells with Novel Industrial Process", 28th EUPVSEC (2013), 2BV.1.7, P.1053-1056

[4] A. Edler, J. Jourdan et al., "High lifetime on n-type silicon wafers obtained after boron diffusion", 25th EUPVSEC (2010), 2CV.2.54, P.1905-1907

[5] C.C. Chen, C.L. Lin and J.W. Chien et al., "High efficiency on boron emitter n-type Cz silicon bifacial cells with industrial process", Energy Procedia 38 (2013) 416 – 422

[6] A. Teppe, C. Gung, K. Zhao et al., "Progress in the industrial evaluation of the mc-Si PERCT technology based on boron diffusion", Energy Procedia 77 (2015) 208 – 214

[7] I. Haedrich, H. Wirth, M. Storz, G. Klingebiel, "PV module analysis and optimization", Fraunhofer ISE (2012)

[8] J.P. Singh, S. Guo, I.M. Peters et al., "Comparison of Glass/Glass and Glass/Backsheet PV Modules Using Bifacial Silicon Solar Cells", IEEE Journal of Photovoltaics 5 (2015) 783-791.

COMBINATION OF PLASMA-DAMAGE-LESS CAT-CVD WITH A NEW LOW TEMPERATURE IMPURITY DOPING METHOD, CAT-DOPING, FOR IMPROVEMENT OF SOLAR CELL PERFORMANCE

Huynh Thi Cam Tu, Shigeki Terashima, Koichi Koyama, Cong Thanh Nguyen, Hideki Matsumura
Japan Advanced Institute of Science and Technology (JAIST)
Asahidai, Nomi-shi, Ishikawa, 923-1211

ABSTRACT: This paper is to study on a new doping technology, Cat-doping, in which phosphorus or boron atoms can be doped into crystalline-silicon (c-Si) at temperatures as low as 80 °C by utilizing catalytically cracked species of phosphine or diborane. This paper is also to study on its application to improvement of quality of silicon-nitride (SiN_x) single passivation on c-Si, which is prepared by plasma-damage-less deposition method, catalytic chemical vapor deposition (Cat-CVD), often called Hot-Wire CVD. It is found that carrier lifetimes over 6 ms can be obtained by this SiN_x single passivation when c-Si surface is Cat-doped in advance, and the Cat-CVD SiN_x films exhibit excellent chemical resistance, which makes fabrication process of back contact solar cells easier.

Keywords: Cat-CVD, Cat-doping, Passivation, Silicon-Nitride

1 INTRODUCTION

We have discovered that phosphorous (P) or boron (B) atoms can be doped into crystalline-silicon (c-Si) at substrate temperature of 80 °C, when the c-Si is exposed to radical species generated by catalytic cracking reaction of phosphine (PH_3) or diborane (B_2H_6) with heated tungsten (W) catalyzing wire [1-4]. The doping experiments were carried out by using the conventional catalytic chemical vapor deposition (Cat-CVD) apparatus [5]. We named this new doping method as "Cat-doping".

Doping depth of Cat-doped impurities in c-Si is quiet shallow and limited to several nm [1]. Therefore, it is suitable for controlling the surface potential to make band bending of c-Si surface. The conventional doping methods, such as plasma-doping, ion-implantation or thermal diffusion are not suitable to make band bending in solar cells. When plasma-doping is utilized, the surface of c-Si suffers from the damages due to plasma attack. In cases of the ion-implantation and thermal diffusion, the doped region is not limited at the surface, which might enhance the Auger recombination [6].

On the other hand, although SiN_x/amorphous-silicon (a-Si) stacked layers exhibit excellent passivation quality on c-Si [7], it is difficult to obtain similar quality to them for SiN_x single passivation, which is more transparent than SiN_x/a-Si stacked one. However, this Cat-doping can be applied to improve the passivation quality of SiN_x single layer by introducing band bending due to Cat-doping [8].

Although the passivation quality of SiN_x single layer on c-Si is improved by Cat-doping, at the same time, it must be chemically resistant to acids such as hydrofloride acid (HF) to make fabrication process of back-contact solar cells simpler and easier.

In this paper, at first, we briefly present the main features of P Cat-doping. Then, we demonstrate the further improvement in quality of SiN_x single passivation on c-Si, in which P atoms are Cat-doped for inducing surface band-bending of c-Si. The SiN_x films are deposited by Cat-CVD to avoid destroying the delicate shallow doped region. In addition, the chemical resistance of the Cat-CVD SiN_x films is also demonstrated to show the superiority of Cat-CVD technology.

2 EXPERIMENTS

Figure 1 shows a schematic diagram of Cat-doping apparatus, which is based on Cat-CVD system. Both Cat-doping and deposition of SiN_x films are carried out by the same Cat-CVD system but different chambers. Parameters for Cat-doping and the Cat-CVD film deposition in this study are summarized in Table I.

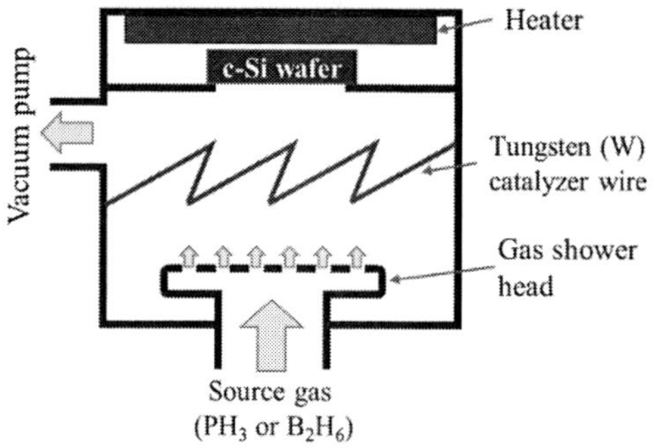

Figure 1: Schematic diagram of Cat-doping apparatus

Table I: Parameters for Cat-doping and Cat-CVD film deposition.

	P Cat doping	SiN_x	a-Si
Gas sources	PH_3	SiH_4, NH_3	SiH_4
Flow rate (FR) of gas sources (sccm)	20 (net 0.45) 2.25 % diluted by He	SiH_4: 4.2 NH_3: 250	10
Temperature of catalyzer _ T_{cat} (°C)	RT-1300	1800	1800
Temperature of substrate _ T_{sub} (°C)	80 - 350	100 - 350	90
Gas pressure _P_g (Pa)	1; 4.7	3	1
Process time (s)	60 - 600	600	45

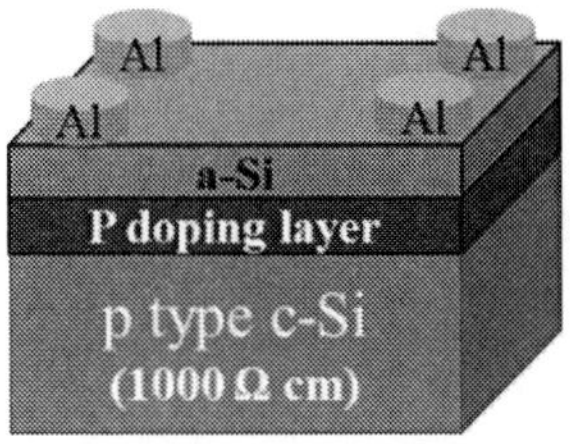

Figure 2: Schematic of sample for the Hall measurement

In order to evaluate carrier density after P Cat doping we prepared a sample with structure as shown in Fig. 2. A 10-nm-thick intrinsic (i-) a-Si film was formed on Cat-doped c-Si surface to eliminate interface defects which might pin Fermi level and disturb the Hall effect measurements [3, 9]. The sample was anealled at 350 °C for 1 min to form Ohmic contact.

Figures 3 (a) and (b) show the structures of samples for measuring the effective carrier lifetime (τ_{eff}). In this study, the 300-μm-thick-n-type (100) c-Si wafers with resistivity of about 1 - 5 Ω cm and bulk carrier lifetime around 11.5 ms were used. The Si wafers were cleaned by chemical treatment with 96 % sulfuric acid (H_2SO_4), diluted (5%) HF and deionized water [10]. In order to achieve a good surface passivation, an a-Si/SiN_x stack was deposited on the back side of the c-Si by Cat-CVD. The a-Si and SiN_x films were deposited at temperatures of substrate (T_{sub}) of 90 °C and 350 °C, respectively. The front side of the c-Si was P Cat-doped at T_{sub} = 80 °C for 5 min. Then, the Cat-CVD SiN_x single layer was deposited at various T_{sub} from 100 to 350 °C. Detailed fabrication parameters are shown in Table I. Finally, the samples were annealed at 350 °C for 30 min to improve passivation quality of SiN_x and activate the P dopants. The τ_{eff} was measured by microwave photo-conductivity decay (μ-PCD) method (KOBELCO LTA-1510EP).

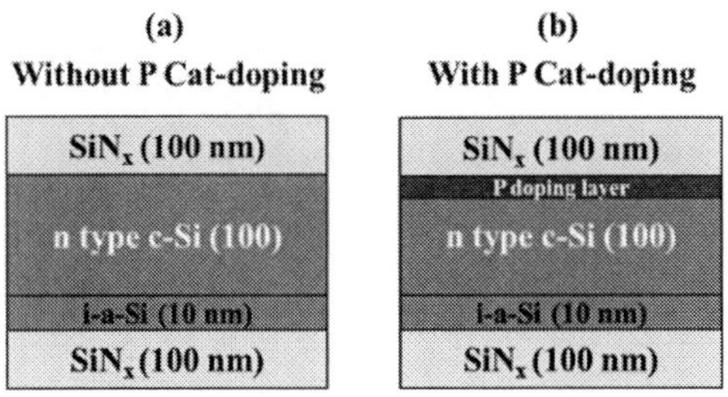

Figure 3: Cross sectional schematic of samples for measuring effective carrier lifetime: (a) without P Cat-doping and (b) with P Cat-doping

3 RESULTS AND DISCUSSIONS

3.1 Features of P Cat-doping

For both Cat-CVD and Cat-doping, temperature of catalyzer (T_{cat}) is one of the most important parameter. In the case of Cat-doping, T_{cat} is mainly fixed at 1300 °C to prevent surface etching caused by hydrogen (H) atoms generated by catalytic cracking reaction of source gas PH_3 or B_2H_6. Figure 4 shows the sheet carrier density as a function of T_{cat} for P Cat-doping into p-type c-Si. As can be seen, the sheet carrier density increases with increase

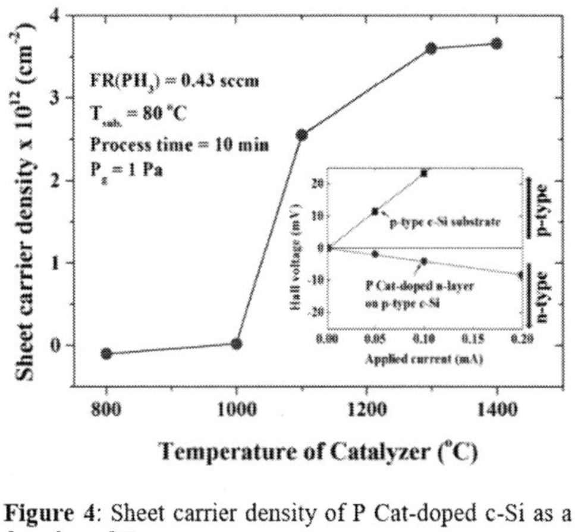

Figure 4: Sheet carrier density of P Cat-doped c-Si as a function of T_{cat}

of T_{cat}. From the inset of the figure, it is confirmed that the conduction type of P Cat-doped samples is converted from original p-type to n-type when T_{cat} exceeds 1000 °C. For T_{cat} lower than 800 °C the conduction type is kept at p-type.

Dependence of the sheet carrier density on process time for P Cat-doping at T_{cat} = 1300 °C is shown in Fig. 5. The P Cat-doping process exhibits the best effect when the c-Si are exposed to the radical species for 5 min. When exposing time is more than 5 min the sheet carrier density is likely to decrease. This reduction is supposed to be due to the etching effect by H atoms on the surface doping layer, since it is known that P doping itself is likely to saturate in initial 5 min for un-revealed reasons [1, 8].

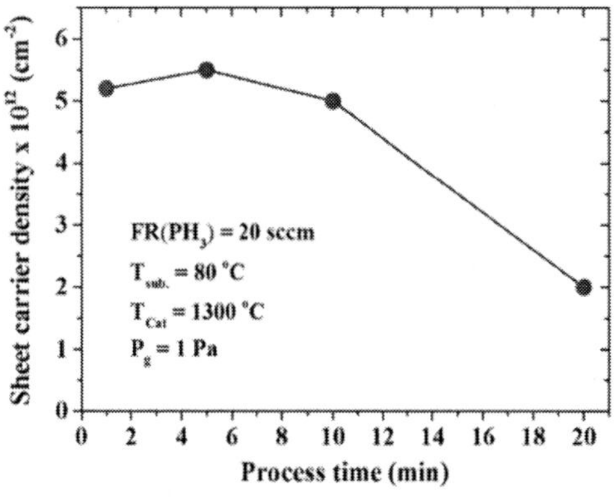

Figure 5: Sheet carrier density of P Cat-doped c-Si as a function of process time

Figure 6 shows the dependence of the sheet carrier density on T_{sub} at gas pressure (P_g) of 1 and 4.7 Pa. In our Cat-CVD system, P_g = 4.7 Pa is the maximum value obtained at FR(He containing 2.25% PH_3) = 20 sccm. As shown in the figure, the sheet carrier density increases with increase of P_g. This can be simply explained by the fact that at high gas pressure the density of P atoms increases. These atoms work as donors. Thus, the carrier density in c-Si increases as P_g increases. From the figure,

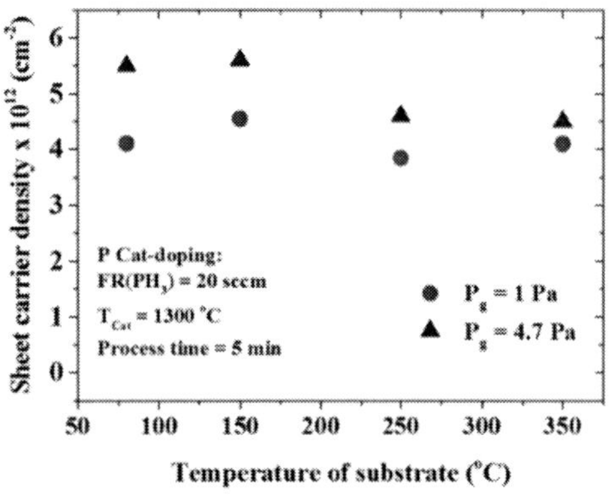

Figure 6: Sheet carrier density of P Cat-doped c-Si as a function of T_{sub} and P_g.

we can confirm that Cat-doping can perform at T_{sub} as low as 80 °C and only weakly dependent on T_{sub}.

3.2 SiN_x single passivation layer on P Cat-doped c-Si

Here, P Cat-doping is carried out at T_{cat} = 1300 °C, T_{sub} = 80 °C and P_g = 1 Pa for 5 min to improve the quality of SiN_x single passivation layer on c-Si by introducing band bending of c-Si surface.

Figure 7 shows dependence of τ_{eff} on T_{sub} for depositing SiN_x for samples with and without P Cat-doping whose structures are shown in Fig. 3. From Fig. 7, we can confirm that P Cat-doping has a positive effect in improvement of τ_{eff} especially for the SiN_x deposited at T_{sub} below 250 °C. The τ_{eff} is over 6 ms for the SiN_x deposited at 150 °C which appears enough to obtain solar cells with efficiencies over 25 %. For the SiN_x fabricated at T_{sub} = 350 °C there is no improvement in τ_{eff} . It is supposed to be due to the desorption of P atoms from c-Si surface during SiN_x deposition.

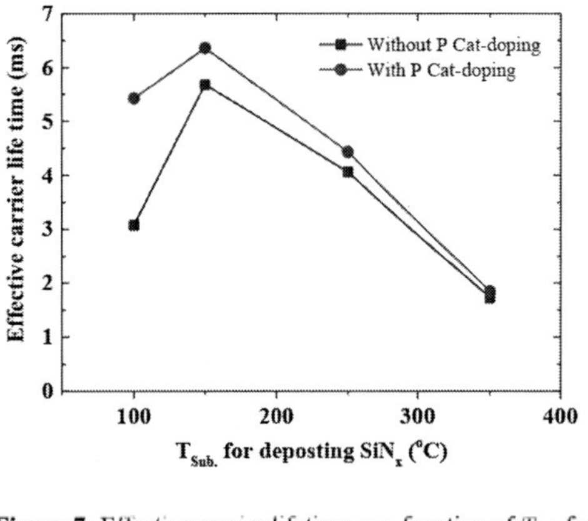

Figure 7: Effective carrier lifetime as a function of T_{sub} for depositing SiN_x

In the fabrication of back-contact a-Si/c-Si hetero-junction solar cells, the passivation layers are to be suffered from chemical etching process during the fabrication. Usually, SiN_x films are strong for alkaline process. However, they are not so strong for acid process, especially when they are prepared by PECVD [11].

Here, we investigate chemical resistance property of the Cat-CVD SiN_x by evaluating the etching rate of the

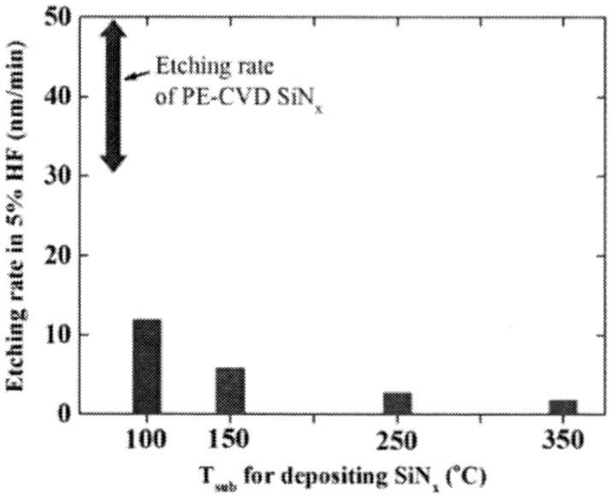

Figure 8: Etching rate of SiN_x fabricated by Cat-CVD as a function of T_{sub} for depositing SiN_x

film in 5% HF. Figure 8 shows the etching rate of SiN_x prepared by Cat-CVD plotted against substrate temperature after being annealed at 350 °C for 30 min. As can be seen from the figure, the etching rates of the SiN_x are reduced from 12 to 2 nm min^{-1} when T_{sub} increases from 100 to 350 °C, which is only about 1/10 of that of SiN_x deposited by PECVD. The etching rate is about 6 nm min^{-1} for samples of T_{sub} = 150 °C, which exhibits the best τ_{eff}. It should be noted that this etching rate is still acceptable for solar cell fabrication. If PECVD SiN_x films are used, the fabrication process will be limited due to their weakness in chemical processes.

Film formation by Cat-CVD has various advantages: 1) the surfaces of the films and substrates are free from plasma damages, 2) the efficiency of gas use is almost one order of magnitude higher than that of PECVD, thus the volume of storage of dangerous silane (SiH_4) gas in factory can be reduced, 3) large area deposition is possible, and so that, over 5,000 c-Si wafers can be processed per hour in the present mass-production system, and 4) cleaning of chambers can be avoided by installing two substrate trays surrounding catalyzing wires, since all areas looked from the catalyzing wires are covered with the trays or changeable parts. In addition to these classical advantages, the present work adds a new advantage of Cat-CVD technology in the chemical resistance which is suitable to solar cell fabrication by combining with Cat-doping.

4 CONCLUSIONS

A novel low temperature P Cat-doping is studied in this work. By P Cat-doping, *p*-type c-Si is converted to *n*-type even at the substrate temperature as low as 80 °C. In addition, by using Cat-doping the surface band-bending of c-Si is induced, which results in the improvement of τ_{eff} of SiN_x single passivation layer on c-Si, up to about 6.5 ms. Moreover, the fabricated Cat-CVD SiN_x films exhibit excellent chemical resistance that is essential requirement for fabricating high efficiency solar cells.

ACKNOWLEDGEMENTS

This work was financially supported by NEDO project, Japan.

REFERENCES

[1] H. Matsumura, T. Hayakawa, T. Ohta, Y. Nakashima, M. Miyamoto, T. C. Thi, K. Koyama, and K. Ohdaira, J. Appl. Phys. **116** (2014) 114502.
[2] T. Hayakawa, Y. Nakashima, M. Miyamoto, K. Koyama, K. Ohdaira, and H. Matsumura, Jpn. J Appl. Phys. **50** (2011) 121301.
[3] T. Hayakawa, Y. Nakashima, K. Koyama, K. Ohdaira, and H. Matsumura, Jpn. J. Appl. Phys. **51** (2012) 061301.
[4] T. Ohta, K. Koyama, K. Ohdaira, H. Matsumura, Thin Solid Films **575** (2015) 92.
[5] H. Matsumura, J. Appl. Phys. **65** (1989) 4396.
[6] S. M. Sze: Semiconductor Device: Physics and Technology (Wiley, Hoboken, NJ, (2002) 2nd ed., p. 66.
[7] K. Koyama, K. Ohdaira, and H. Matsumura, Appl. Phys. Lett. 97 (2010) 082108.
[8] T. C. Thi, K. Koyama, K. Ohdaira, and H. Matsumura, J. Appl. Phys. **116** (2014) 044510.
[9] W. Clarke, X. Zhou, A. Fuhrer, T. Reusch, and M. Simmons, Physica E **40** (2008) 1566
[10] C. T. Nguyen, K. Koyama, S. Terashima, C. Okamoto, S. Sugiyama, K. Ohdaira, and H. Matsumura, Proceedings 43rd IEEE Photovoltaic Specialists Conference (2016) (to be published).
[11] V. Gottschalch, R. Schmidt, B. Rheinlander, D. Pudis, S. Hardt, J. Kvoetkova, G. Wagner, R. Franzheld, Thin Solid Films **416** (2002) 224.

PILOT PRODUCTION OF BIFACIAL MULTICRYSTALLINE PERCT CELLS ACHIEVING 18.5% EFFICIENCY AND SINGLEFACIAL MORE THAN 19%

Andreas Teppe[1], Enyu Wang[2,3], Wei Guo[2], Chun Gong[1],
Oliver Voigt[1], Ihor Melnyk[1], Faramarz Binaie[1], Wolfgang Jooss[1], Peter Fath[1]

1) RCT Solutions GmbH, Turmstr. 20, D-78467 Konstanz, Germany
2) Shanxi Lu'an Solar Technology Co. Ltd., Changzhi, Shanxi Province, China
3) School of Science, Xi'an JiaoTong University, Xi'an, Shaanxi Province, China
e-mail: Andreas.Teppe@rct-solutions.com

ABSTRACT: In this paper we report about progress in the manufacturing of BIFACIAL and single facial multi crystalline PERCT solar cells in the pilot line of Shanxi Lu'an Solar Technology.

Compared to previous publications the PERCT solar cell process sequence and parameters are modified. These modifications increased the average cell efficiency above 19%. The new process features are a high ohmic boron diffusion on the rear, a phosphorous emitter annealing step and the changed position of the laser ablation step. Unchanged is the passivated emitter rear totally diffused (PERT) cell design, realized by a BBr_3 diffused BSF. The boron back surface field (B-BSF) is passivated by use of a PECVD SiN_x deposited onto the remaining boron glass leading to a solar cell without any use of AlO_x.
Due to the B-BSF two different cell designs can be realized, each with its distinguished screen printed rear side metallisation schemes: Standard single facial aluminium and pad printing on to a laser opened rear side passivation and alternatively, the BIFACIAL screen printed silver finger grid. For the first time a low temperature front side emitter annealing step in addition to the well passivated rear side is implemented. This combination increases the solar cell efficiency of single facial cells to an average of 19.1% and a top efficiency of 19.6%. On BIFACIAL solar cells an average of 18.6% with 87% bifaciality and a top efficiency of 18.9% (measured on a standard, not conductive black chuck) are reached. At an albedo of 20% cost savings of 15% can be achieved on module level.
To investigate the LID behaviour 6 modules were built using single facial mc PERCT cells. Three 60 cell modules reach an average power of 280.1W. Three 72 cell modules reach an average power of 335.6W. After light exposure of 5kWh /m² all 6 modules showed a low LID degradation of only 1.1% relative.

Keywords: c-Si, Multi-Crystalline, High-Efficiency, Solar Cell, Bifacial, PERT

INTRODUCTION

One effective possibility to reduce the cost of PV energy generation is the use of highly efficient BIFACIAL cells and modules. This improves the kWh/kWp power yield at none or moderate additional costs.

Today, there is a limited availability of cost effective bifacial PV panels on the market, because they are mostly based on costly n-type mono wafers and comparatively sophisticated technology.

In this paper, we present progress in the implementation of the RCT Solutions multi PERCT technology into an industrial pilot production at Shanxi Lu'an. With this technology upgrade, based on standard boron diffusion furnace in existing multi crystalline solar cell production, it was demonstrated to obtain average solar cell efficiencies above 18.5 % on the front side of BIFACIAL multi crystalline solar cells and more than 16% efficiency by illumination on the rear side.

Additionally, the single facial PERCT design, with full area Al contact is an attractive alternative to conventional PERC cell designs, which are mainly based on PECVD AlO_x/SiN_x stacks. The PERCT cells are free from AlOx, do not require any costly polishing steps and we did not observe any LID yet.

The benefit is not only due to the efficiency gain, but also because of an improvement with regard to the resulting cost of ownership due to the BIFACIAL properties of the cells. It is believed that BIFACIAL PV (BIFI-PV) has a huge market potential in many applications such as large scale power generation systems in desert regions and in regions with heavy snow fall [1], for BIPV. The multi PERCT technology [2, 3] now provides an opportunity to upgrade standard mc-Si solar cell lines to enable high efficiency single-facial and bifacial mc-Si solar cell production.

2 IMPROVED PERCT PROCESS

2.1 Cell architecture with new front side emitter

RCT's PERCT cell architecture is based on a "passivated emitter rear totally diffused" (PERT) cell design for p-type wafers. The totally diffused Back Surface Field (BSF) is realized by a cost effective and fast BBr_3 boron diffusion step in a standard diffusion tube furnace - prior to the back side passivation and protection by a PECVD SiN_x.

The boron BSF enables two options for the rear side

metallization which are shown in Figure 1: Either a metallization for single facial solar cells, done by full area aluminium print on a locally laser opened rear side. Or as an alternative the screen printing of a silver grid, which is fired through the passivation stack.

The solar cell front side is a high ohmic phosphorous emitter. This emitter is improved by a low temperature annealing step to reduce the emitter recombination activity [4], which was introduced in this work. The emitter is coated with anti-PID SiN_x and metallized by a screen print silver front grid.

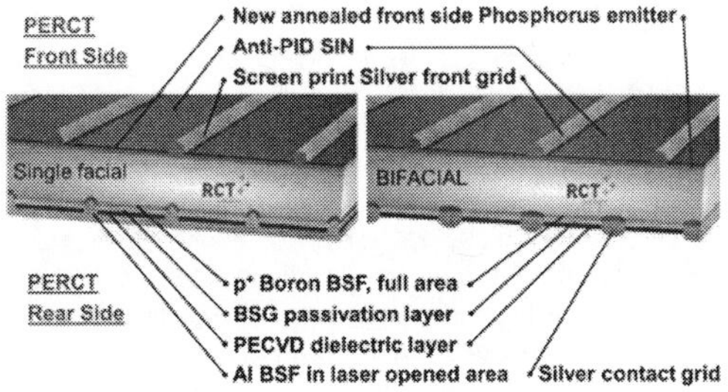

Figure 1: Schematic view of single facial (left) and BIFACIAL multi PERCT solar cell (right).

2.2 PERCT process flow

The process flow for the mc-PERCT solar cell manufacturing is described in Figure 2. It is based on a standard multi crystalline Al-BSF flow, which is upgraded by 1-3 processing steps, marked in blue in Figure 2.

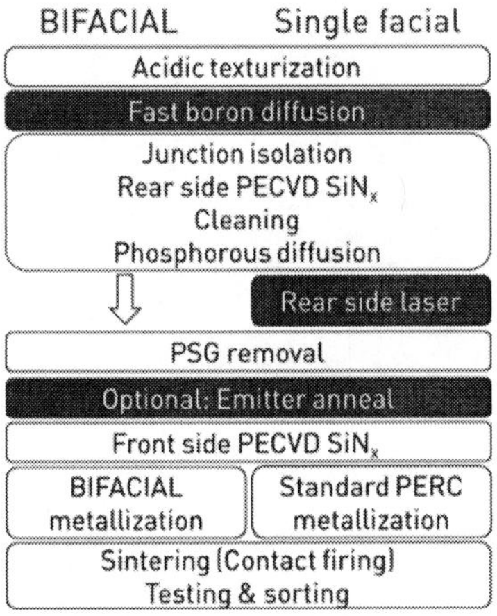

Figure 2: Optimised process sequence of single facial and bifacial Multi PERCT solar cells.

The solar cell processing starts with standard acidic texturing. By the use of modern inline equipment an additional cleaning procedure (e.g. HF/O_3) prior to the following boron diffusion can be omitted. After texturing the BBr_3 diffusion, which is not part of the Al-BSF process flow, follows, with the wafer loaded front-to-front. For the rear side passivation the boron doped BSF remains coated with the Boron-Silicate-Glass (BSG). Additionally this layer is covered by PECVD SiN_x. No aluminium oxide is needed or used, since the passivation performance of the BSG/SiN_x stack is sufficient.

After a wet chemical junction isolation and a standard HF-HCl cleaning step the $POCl_3$ emitter diffusion and PSG etch are applied. For the single facial mc-Si PERCT cell type the rear side passivation layer is opened by a laser process prior to the PSG etch. The advantage of this process sequence is a better removal of dust and parasitic oxides in the laser spot edges by HF etch compared to our previous publications, where the laser opening was done prior to screen printing.

To improve the emitter performance leading to a low J0, a low temperature anneal is introduced as an option prior to the anti-PID front side PECVD coating.

The metallization of single facial cells is standard silver paste front side grid layout in combination with aluminium paste and a silver pad layout for soldering on the rear side.

For bifacial cells there is the possibility to choose a similar approach and to use also Al paste by adjusted screen printing of an aluminium grid on to the laser opened regions [3, 5]. In this publication the laser process is avoided and commercially available Ag pastes are used which are suitable to fire through the SiN_x passivation layer. The boron BSF enables good contact resistance and good passivation of the rear side contacts. Due to the narrow silver grid lines the rear side grid causes a similar low shadowing as on the cell front side. As a future alternative, fire through aluminium pastes are a cost effective option (see section 6). Besides the significant additional power generation from the rear side in real applications, the paste consumption is reduced and cell bow is avoided.

Overall, mc-Si bifacial PERCT solar cell modules promise to achieve a highly attractive cost structure and allow new bifacial applications.

2.3 High ohmic boron diffusion

As reported in PVSEC 2015 [6] a fast BBr_3 tube diffusion was used, reaching paddle to paddle times below 90 min. Due to front to front loading, a high throughput and a competitive cost structure are reached. In several optimisation runs the sheet resistance was increased up to 350 Ohm/sq for the single facial PERCT cell. The advantage of this high resistance is a still good passivation, less consumption of BBr_3, a faster diffusion and lower Irev2 values. The local aluminium contacts in the laser opened regions ensure a low contact resistance.
For BIFACIAL applications in combination with rear side silver grid, the cells need for a low contact resistance and a good contact passivation a heavier diffusion. But also in this case R-sheet values around 200 Ohm/sq are possible.

2.4 Emitter anneal

To improve the front side emitter, an annealing step below diffusion temperature in pure N_2 gas atmosphere after PSG etch was carried out [4]. This method is a common option on monocrystalline PERC production lines. By applying this method on standard Al-BSF mc-production a small increase of +2 mV and +0.1% in efficiency was reached. Table 1 shows the results on a standard 90 Ohm/sq phosphorus emitter.

Cell process mc Al-BSF-cells	no. off cells	Average values Eff [%]	Uoc [mV]	Jsc [mA/cm²]	FF [%]	Irev2 at -12V	0.5 sun Eff [%]
without annealing	398	**18.20%**	**627.8**	**36.65**	**79.08**	**0.112**	17.73%
with emmiter annealing	394	**18.30%**	**629.7**	**36.68**	**79.23**	**0.042**	17.86%

Table 1: Al-BSF cells with and without emitter annealing

3 EXPERIMENTAL RESULTS – BIFACIAL AND SINGLE FACIAL CELLS

3.1 PERCT solar cells with single facial metallisation

To manufacture highly efficient single facial cells, the mc-PERCT- and emitter annealing processes were combined las described above. Figure 3 and 4 show the efficiency- and V_{oc}- plot of the cells, which were manufactured with 4 busbar technology. Table 2 shows the average IV data.

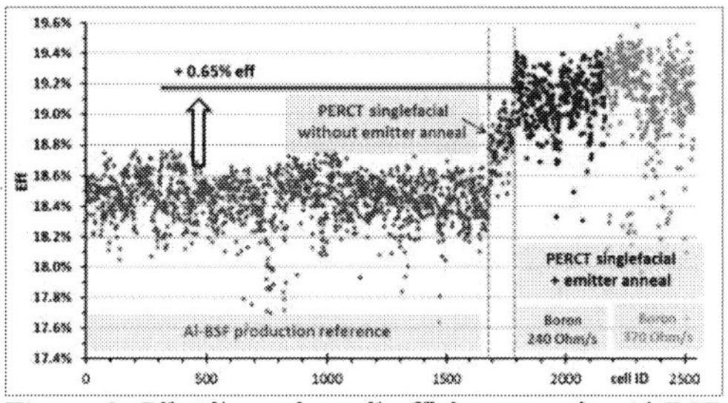

Figure 3: Pilot line solar cell efficiency results. Al-BSF references and single facial PERCT. An efficiency gain of 0.65%abs. can be observed.

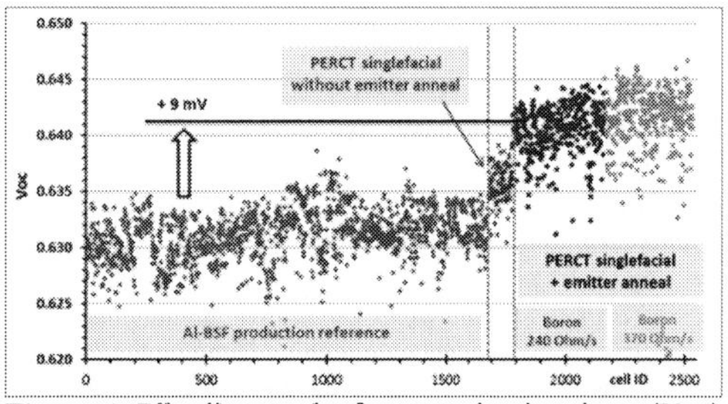

Figure 4: Pilot line results for open circuit voltage (Voc) of Al-BSF references and single facial PERCT.

Average values mc-cells	no. off cells	Eff [%]	Eff incease compared to production	Uoc [mV]	Jsc [mA/cm²]	FF [%]	Irev2 at -12V [A]	0.5 sun Eff [%]
Aluminium-BSF production	**1687**	**18.45%**		**0.631**	**36.93**	**79.14**	**0.10**	17.87%
PERCT-singlefacial Datas:								
B-BSF, 240 Ohm/sq. without Emitter anneal	98	18.81%	+ 0.36%	0.636	37.81	78.28	2.86	18.27%
B-BSF, 240 Ohm/sq. with Emitter anneal	380	19.10%	+ 0.64%	0.640	38.02	78.43	1.35	18.60%
B-BSF, 370 Ohm/sq. with Emitter anneal	369	19.11%	+ 0.66%	0.640	38.16	78.21	1.04	18.62%

Table 2: Al-BSF cells and mc-PERCT cells with and without emitter annealing and different boron sheet resistance

It can be clearly seen in table 2, that the emitter anneal delivers on PERCT structure +0.3 % in efficiency and + 4 mV on V_{oc}. This is much more effective than for Al-BSF cells (see table 1), where it delivered only 0.1% efficiency and +2mV. Al-BSF cells are limited by the rear surface, so an emitter optimization is not as effective as in PERCT cells, which are rather limited by bulk and emitter. So far this PERCT and annealing process combination leads to +0.65% efficiency gain compared to standard production. Additionally, the high sheet resistance of the B-BSF leads to lower reverse current (Irev2) values exactly like expected. Because the Irev2-breakthrough is distributed over the whole solar cell edge we accept for production usually 2.5A as a safe limit to avoid hot spots.

3.2 High efficiency PERCT solar cells with single facial and BIFACIAL metallisation

To archive a large amount of solar cells for manufacturing high power modules, a pilot production run of single facial cells and some additional BIFACIAL with GCL S4 wafer material was done. These wafers feature a little bit larger cell size of 156.75 x 156.75mm. This increase in solar cell area leads to an absolute power benefit of 2.6W, calculated for a typical 270 W module. All PERCT cells were manufactured with the emitter annealing process step. The production reference was done with standard size (156 x 156mm) GCL S3 wafer and no emitter annealing was used. All cells got the same front side screen printing. By this reason the edge distance between front side grid and solar cell edge of the PERCT cells was not perfectly optimised and the fill factor was a little bit lower than usual. We estimated an efficiency potential of 0.1%abs. with an adapted front side layout. Even with the non-optimised front side grid we reached an efficiency of 19.04%. For BIFACIAL screen printed metallisation an average of 18.55% efficiency and a bifaciality of 87% was accomplished, measured on a nonconductive black chuck. Fig 5 shows the efficiency distribution and table 3 the IV data. The measurement of all cells was done on a black and nonconductive chuck.

Due to additional light harvest on the rear side BIFACIAl cells are the most cost effective way to reduce cost of PV energy generation. The calculated cost savings on BIFACIAL module level reach 15% at 20% albedo.

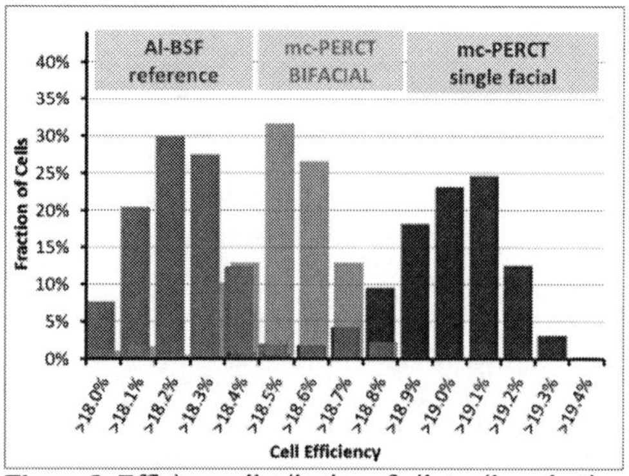

Figure 5: Efficiency distribution of pilot cell production. Al-BSF references, BIFACIAL and single facial PERCT cells, all with four busbar technology

Average values mc-cells	no. off cells	Eff [%]	bifacial-ity	Uoc [mV]	Jsc [mA/cm²]	FF [%]	Irev2 at -12V [A]	0.5 sun Eff [%]
ALBSF production	2361	18.43%	single facial	0.629	36.56	79.40	0.09	17.90%
PERCT singlefacial	2496	19.04%	single facial	0.640	38.08	78.08	0.51	18.68%
PERCT BIFACIAL	178	18.56%	87.4%	0.635	38.43	76.05	1.20	18.54%

Table 3: Pilot production IV data of. Al-BSF references, BIFACIAL and single facial PERCT, all 4BB technology.

4 EXPERIMENTAL RESULTS FOR LID ON SINGLE FACIAL CELLS AND MODULES

For module manufacturing and investigating the LID (Light Induced Degradation) effect single facial PERCT cells were used (solar cell size 156.75 x 156.75 mm^2). We used 19.2% efficient cells for the 60 cell-modules and 19.1% cells for the 72 cell modules. Due to the larger cell area the power output of a large 19.2% cell is the same like a 19.4% standard size 156.0 x 156.0mm^2 cell. Three 60 cell modules reach an average power of 280.1W before and 277.1W after light exposure of 5kWh /m². The average power of three 72 cell modules is 335.6W before and 331.9W after light exposure. The module power after light exposure was confirmed at National Centre of Supervision and Inspection on Solar Photovoltaic Product Quality (CPVT), China. For all 6 modules the LID degradation is only 1.1% relative.

Additionally some single cells were illuminated with 0.6 suns. After 60 hours no degradation was observed (see Figure 6).

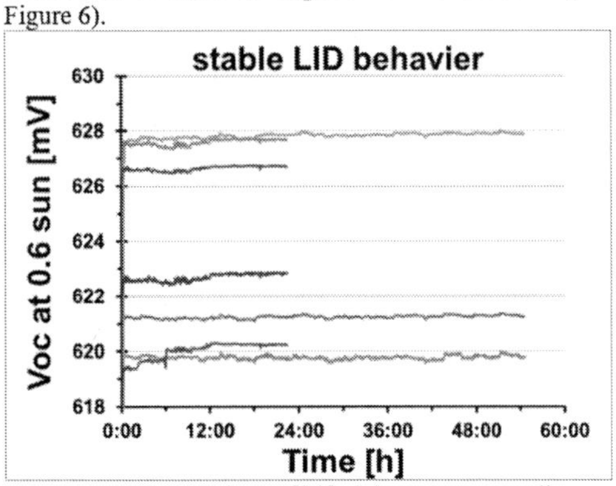

Figure 6: LID behaviour of bifacial mc PERCT cells

5 COST CALCULATION

Advantage of the mc-PERCT BIFACIAL technology is the use of cost competitive mc-wafers while yielding relatively high efficiency and bifaciality. Figure 7 shows the calculated system cost of BIFACIAL multi crystalline and n-type solar cells in relation to wafer price and efficiency. By assuming similar cell manufacturing equipment like boron and phosphorus diffusion, front- and rear side PECVD, texturing, edge isolation and printing equipment, it can clearly be seen that the wafer cost and cell efficiency dominate the system cost. At current wafer cost of $1.20 for n-type and $0.87 for multi, the multi crystalline wafer in combination with BIFACIAL technology is clearly competitive. Even when assuming that n-type wafers will reach a price level similar to current p-type Cz of around $0.90 / wafer in future, the mc-cell technology will still be cost attractive.

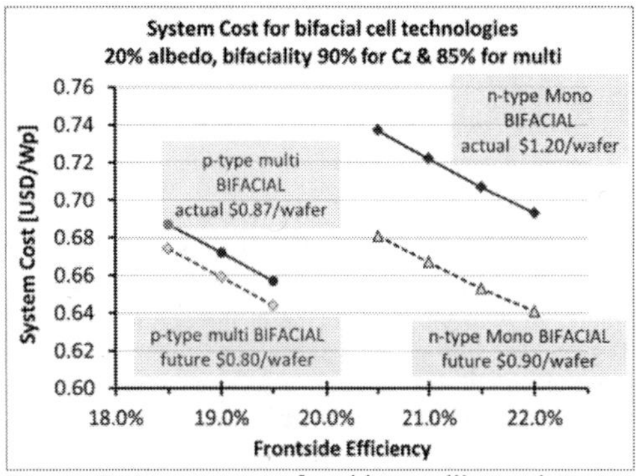

Figure 7: System cost of multi crystalline and N-type BIFACIAL solar cell technologies

A Total-Cost of Ownership calculation of different cell manufacturing technologies for multi crystalline wafers is shown in Figure 8. There is, like expected a clear advantage for BIFACIAL cells and modules due to the rear side albedo of 20%. Replacing the Ag grid by an Al grid also leads to a cost advantage. It can also be clearly seen that for single facial metallization the PERCT approach deliver the similar cost like the standard PERC-AlO_x approach, with a slight advantage for PERCT cells.

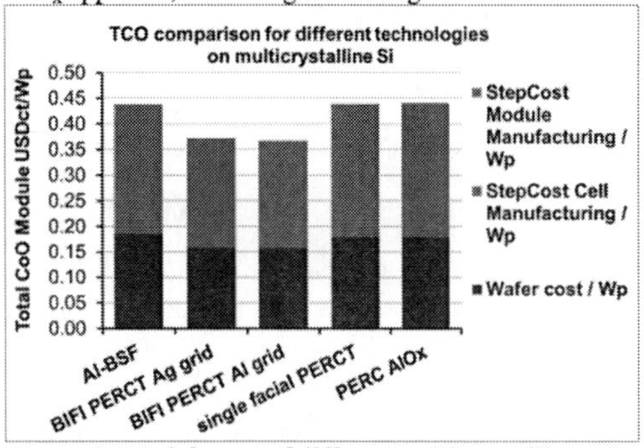

Figure 8: Module cost of different multi crystalline solar cell technologies. Assumption: 20% albedo for BIFACIAL cells.

6 OUTLOOK

For further cost reduction of the mc-PERCT BIFACIAL cell, it is intended to substitute the rear side silver grid by an aluminium grid. This can be accomplished in two ways: Either by screen-printing the aluminium grid on to a laser opened region or by applying a firing through aluminium paste.

BIFACIAL rear side metallization	Eta [%]	Jsc [mA/cm2]	Voc [V]	FF [%]
Aluminum grid on laser opening	18.4%	36.5	0.637	78.9
Aluminum grid fire though	18.2%	36.4	0.631	79.1

Table 4: IV data of laboratory mc PERCT BIFACIAL cells with rear screen printed aluminium grid. Al in 4BB

technology

The solar cell parameters of both approaches are given in Table 4. These initial lab results, which have been performed jointly with ISC Konstanz, are already promising.

7 SUMMARY

In the industrial pilot line at Shanxi Lu'an the mc-PERCT single- and BIFACIAL technology was improved with the introduction of the emitter annealing step and use of high ohmic boron BSF. We obtained for single facial PERCT cells an average efficiency increase of + 0.6% and reached an average efficiency of 19.1%. The mc-BIFACIAL cells got on average 18.56% with 87% bifaciality.

Three 60 cell modules with an average power of 280.1 W and three 72 cell modules with ab average power of 335.6 W were manufactured. All 6 modules showed after light exposure of 5 kWh/m² a low LID degradation of 1.1%rel..

Using cost competitive multi crystalline wafers and assuming a BIFACIAL gain at of 20% albedo, cost savings on module level of 15% compared to single facial Al-BSF technology will be achieved.

8 ACKNOWLEDGEMENTS

Thanks for founding some part of this work by German Federal Ministry for Economic Affairs and Energy (BMWi), Central Innovation Programme for Small and Medium-Sized Enterprises (ZIM)
Alf (Structured Aluminium Metallisation Firing of Innovative PERC/PERCT Solar Cells)
Support code : KF2818605GM4,
Consortium: ISC Konstanz and RCT Solutions

9 References

[1] K. Sugibuchi et al., "BIFACIAL-PV POWER OUTPUT GAIN IN THE FIELD TEST USING "EarthON" HIGH BIFACIALITY SOLAR CELLS", Proc. of the 28th European Photovoltaic Solar Energy Conference, Paris (2013)

[2] A. Teppe et al., Novel Technology Approach Based on Standard Quality MC Wafers Achieving Solar Cell Efficiencies Significantly above 18% in an Industrial Production Environment. Proc. of the 29th European Photovoltaic Solar Energy Conference, Amsterdam (2014)

[3] A. Teppe et al., Progress in the Industrial Evaluation of the mc-Si PERCT Technology Based on Boron Diffusion, SiliconPV 2015

[4] Amir Dastgheib-Shirazi et al, Dissolution of Electrically Inactive Phosphorus by Low Temperature Annealing, SiliconPV 2015

[5] T. Dullweber et al, 31st EUPVSEC, The PERC+ Cell: a 21%-Efficient Industrial Bifacial PERC Solar Cell, Hamburg (2015)

[6] A. Teppe et al, Bifacial Multicrystalline Solar Cells with Efficiencies above 18% Prepared in an Industrial Production Environment, Proc. of the 30th European Photovoltaic Solar Energy Conference, Hamburg (2015)

INVESTIGATION OF LASER ABLATION PROCESS FOR HIGH EFFICIENCY SOLAR CELLS

Ming Shiou Lin*, Shu Yen Liu,Yueh Lin Lee, Kuang Chieh Lai ,Yun Kuo Tsao, ,
Chia Chih Chuang, and Chi-Chun Li
Motech Industries, Xinshi Dist., Tainan City, 74145, Taiwan
(*Correspondingauthor:phone/fax/e-mail:+886-6505-0789-0617/+886-6505-8446/Joshua_lin@motech.com.tw)

ABSTRACT: Laser techniques have been applied in mass production of silicon based crystalline solar cell, such as PERC and n-PERT. The detail laser work plays an important role in the fabrication of high efficiency solar cells. On the other hand, the conventional Ag screen-printed contacts usually have high contact resistance and lead to junction shunting issue during the post firing process. The Ag screen printing can be replaced with Ni/Cu plating to reduce the production cost of solar cells. Before Ni/Cu plating, the patterned removal of SiNx is operated by 532nm green laser. However, there are still drawbacks discovered during laser process. It is observed that if the output energy and focus are not suitable, there is silicon nitride(SiNx) left on the surface or damage of wafer discovered by scanning electron microscope (SEM). By the way, the characterization of SiNx residue is identified by Energy Dispersive Spectrum(EDS). Based on the measurements, it is also believed that the low specific contact resistance(ρ_c) and high cell efficiency depend on the Ni/Cu contacts at interface strongly. The filling factor(FF) and cell efficiency both indeed correspond to variation of the opening area and metallization. According to our investigation, the optimization of laser ablation process is available and the removal of SiNx residue is under control effectively.

Keywords: Laser processing, Solar cell, Metallization

1 INTRODUCTION

Today, photovoltaic power is one of the clean energy techniques for the global environmental concerns. n-PERT solar cells are one of the advanced solar cells and have a high potential to achieve excellent performance with simple process. It is essential that the most industrial n-PERT solar cells are metallized by screen printing and PVD[1]. However, the plating process is fabricated below 100℃, which avoids the thermal effect of initial high level quality of the crystalline silicon substrate[2]. The electrode patterns are obtained by laser ablation before plating process. Besides, the loss of open-circuit voltage(V_{oc}) and filling factor(FF) may be resulted from wafer damage due to laser working[3]. In this study, the laser ablation influence and cell performance are discussed.

2 EXPERIMENTS

In this study, n-type Cz c-Si wafer was served as substrates. Afterwards, opening contact was ablated by 532 nm green laser irradiation (SL-G30W,Anderson). A plated Ni contact layer, and plated Cu conductive layer are applied to n+np+ front emitter solar cell structure consisting of boron doped emitter, intrinsic, and phosphor doped surface field. The metallization of Ni/Cu was employed as electrode side contact for n-type silicon solar cell via electrochemical solution deposition. In the end, the wafer was cut as small cells with 4 cm^2 area for the subsequent measurements as Figure 1 shown. The real small cell array and the observation of one small cell alfter Cu plating were expressed as Figure 2 (a) and (b), respectively.

The SunsV_{oc} measurements were available by Sinton. The specific contact resistance(ρ_c) was measured by contact resistance measurement (TLM,GP 4-Test Pro) and the micro-structures were observed by SEM(Hitachi,S-4800). The elements at the interface were identified by energy dispersive X-ray spectroscopy (EDS, HORIBA).

3 RESULTS AND DISSCUSSION

In this study, it is supposed that the difference of specific contact resistance(ρ_c) and filling factor between the samples is affected by laser ablation. Therefore, an experiment is designed to ensure and clarify the appearance. The energy and focus are both key parameters for the experimental design. The control of energy depends on the output intensity. The depth of focus is defined as A, B, and C. Figure 3 is the top view SEM migraphs of laser ablated areas on the SiNx surface of small cells and opening width. There are six small cells under the experimental design and they are (a),(b),(c),(d),(e),(f),and(g), respectively. Obviously, the small cells are divided as two groups. (a), (b),(c) is one group, and (d),(e),(f) is the other. The removal of nitide for the opening area was confirmed by EDS of Figure 4. According to the opening width, Focus B is clarified as the depth of focus via the verificiation of the most narrow width.

Table 1 is the SunsV_{oc} measurements of the small cells. The V_{oc} of cell (a), (b), (c) is superior to cell (d), (e), (f) because of the less damage during laser ablation. Among the samples, cell (a) has the best PFF and pseudo efficiency.

In general, the filling factor(FF) and cell performance depend on the metal contact behavior strongly. It is believed that the suitable laser ablation leads to the close metal contact for the lower specific contact resistance. In this study, the pseudo filling factor (PFF) and pseudo efficiency of cell (a) are 0.831 and 20.42%, respectively. On the other hand, the specific contact resistance of cell (a) is only 1.02Ω-cm^2. Besides, the cell efficiency of cell (a) is 20.89% and superior to the others. In summary, the the optimization of laser ablation process is available and the removal of SiNx residue is under control effectively. According to the data, the improvement of the filling factor(FF) and the benefit of cell performance are attributed to the optimization of laser irradiation woking.

4 CONCLUSIONS

In a word, the effect of laser ablation on the metal contacts of Ni/Cu plating at the interfaces between metallization layer and opening area for n-type solar cell affect the cell performance obviously. On the other hand, the lower specific contact resistance(ρ_c) corresponding the promotion of FF, which results in the enhancement of cell efficiency. In our study, the optimization of laser ablation process is available to form close metal contact and the metallization is also under control effectively.

5 ADDITIONAL COMPONENTS

5.1 Illustrations

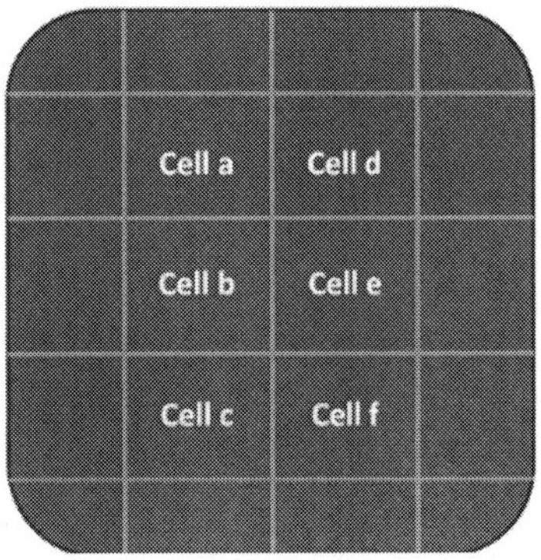

Figure 1: Schematic of the six small cells in one wafer

(a)

(b)

Figure 2: (a) Top view of small cell array with laser ablation,
(b) Observation of one small cell alfter Cu plating

Energy / Focus	Low	High
A	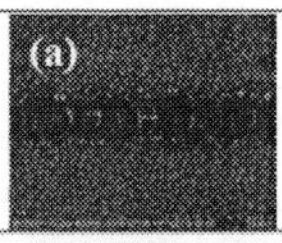	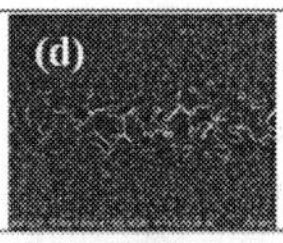
Width (μm)	26.1	29.6
B	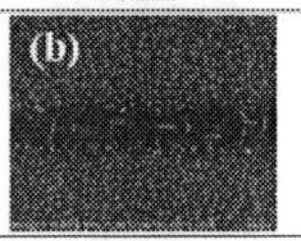	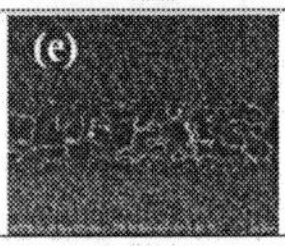
Width (μm)	22.7	29.2
C	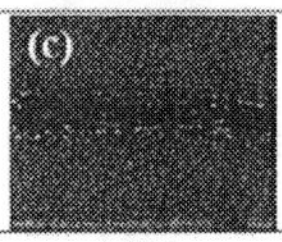	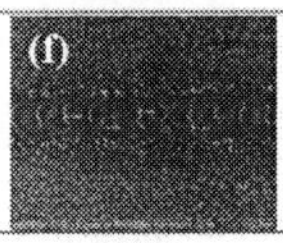
Width (μm)	21.5	33.2

Figure 3: Top view SEM migraphs of laser ablated areas on the SiNx surface of small cells and opening width. There are 6 smalls under the experimental design and they are (a),(b),(c),(d),(e),(f),and(g), respectively.

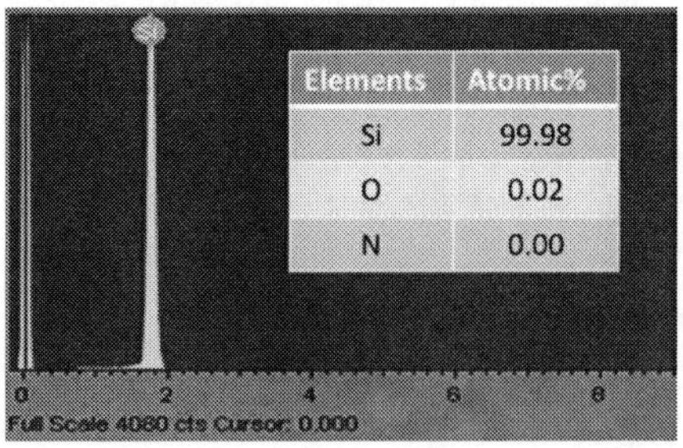

Figure 4: The EDS spectra reveals the removal of nitride.

5.2 Tables

Table I: The SunsV_{oc} measurements of the small cell

Cell	Voc (V)	PFF	Pseudo Efficiency	Ideality (n) @ 1 sun	Jo1 (A/cm^2)	Jo2 (A/cm^2)	Fitted Rsh (Ω-cm^2)
(a)	0.647	0.831	**20.42**	0.98	4.67E-13	9.47E-09	150000
(b)	0.647	0.828	20.34	1.00	4.62E-13	1.16E-08	150000
(c)	0.648	0.825	20.33	1.01	4.26E-13	1.42E-08	150000
(d)	0.636	0.824	19.91	1.04	4.67E-13	1.16E-08	150000
(e)	0.633	0.816	19.62	1.08	4.67E-13	2.00E-08	150000
(f)	0.637	0.819	19.83	1.05	4.67E-13	1.86E-08	150000

5.3 References

[1] K. Katkhouda, A. Limia, L. Bornschein, R. Koseva, T. Geppert, A. Grohe, H. Krokoszinski,and P. Schaaf, Aluminmm-Based Rear-Side PVD Metalliztion for nPERT Silicon Solar Cells, IEEE JPV, 4(1), 160-167 (2014)

[2] A. Kraft, C. Wolf, A. Lorenz, J. Bartsch, M. Glatthaar, S. W. Glunz, Long Term Stability Analysis of Copper front Side Metallization for Silicon Solar Cells, Energy Procedia 55, 478-485 (2014)

[3] G. Heinrich, M. Bahr, K. Stolberg, T. Wutherich, M. Leonhardt, A. Lawerenz, Investigation of Ablation Mechanisms for Selective Laser Ablation of Silicon Nitride Layers, Energy Procedia, 8, 592-597 (2011)

EFFECTIVE SiN_y CAPPING LAYERS ON HIGH-POWER-PLASMA PECVD AlO_x FOR HIGH EFFICIENCY (21%) INDUSTRIAL P-TYPE MONO PERC SOLAR CELLS

Chung-Jung Hung, Wu-Chun Kao, Keng-Wei Tsai, Cheng-Chieh Chen, Lu-Yi Wu, Kun-Yi Ting, Cheng-Yung Kuo, Kuo-Tung Chu, Li-Wei Cheng
Motech Industries, Inc. Taoyuan Branch
No. 1560, Sec. 1, Zhongshan Rd., Guanyin Dist., Taoyuan City 328, Taiwan (R.O.C.)
E-mail: ChungJung_Hung@motech.com.tw
Tel: +886-3-4738199 ext. 8320
Fax: +886-3-4738355

ABSTRACT: In this paper, we present the high-rate deposition (more than 3 nm/s) of aluminum oxide (AlO_x) film were deposited onto p-type monocrystalline (mono) silicon wafers by the high-power-plasma vertical plasma-enhanced chemical vapor deposition system (HPP PECVD, VCS 1200, Manz AG). High effective minority carrier lifetime (τ_{eff}) of 427.5 μs at an excess carrier density of 1×10^{15} cm^{-3} were obtained for well-optimized HPP AlO_x films exhibiting the superior passivation quality. In addition, we also provide the optimization of industrial PERC (passivated emitter rear cell) solar cells manufacturing process, in particular the effective deposition methods for silicon nitride (SiN_y) capped on HPP AlO_x and rear side polishing improvement. For the optimization procedures, we demonstrate the SiN_y capping effect by differences between direct and remote plasma excitation. Results suggest that the better cell performances would be shown by direct plasma than remote plasma attributing to the thermal budget effect. Based on the above findings, the optimization of rear side manufacturing process with HPP AlO_x/direct PECVD SiN_y passivated stacks could achieve high efficiency industrial-type PERC solar cell with 20.9% average efficiency based on 4 bus-bars metallization design. The champion cell efficiency could reach up to 21.1%.
Keywords: PERC, high power plasma PECVD, aluminum oxide, silicon nitride, passivation, capping, polishing

1 INTRODUCTION

PERC are considered to be the next generation industrial type screen printed silicon solar cells. In the PERC approach, passivation dielectric layers, such as silicon oxide (SiO_2), aluminum oxide (AlO_x), silicon nitride (SiN_y), are deposited on the rear surface, which not only improves the passivation quality of the rear surface, but also increases its internal reflectivity. Recently, AlO_x has attracted much attention as a rear surface passivation material due to its high negative fixed charge density at the AlO_x/c-Si interface providing excellent surface passivation properties [1]. In addition, more and more stack passivation schemes, such as SiO_2/AlO_x, AlO_x/a-SiN_y:H (amorphous hydrogenated silicon nitride), a-Si:H (amorphous hydrogenated silicon)/SiN_y, have also been investigated to improve the passivation effect [2].

To offer superior passivation, it will need to minimize recombination losses and increase reflection at the rear side of the PERC solar cell. We optimized several manufacturing processed to obtain effective passivation schemes, which resulted to the increase in efficiency of PERC solar cell performances.

In this study, the passivation scheme properties of the HPP AlO_x/SiN_y stacks have been investigated in terms of the manufacturing optimization strategies, electrical properties, electroluminescence (EL) measurements and IQE response, in combination with both methods (direct and remote PECVD) for rear SiN_y deposition. In addition, the polishing improvements of the rear surface flatness on PERC solar cells are also being discussed.

2 EXPERIMENTAL

The process flow of PERC solar cells made by 15.6×15.6 cm^2 p-type mono pseudo-square silicon wafers with ~190 um thicknesses is as follows: (1) texturing; (2) $POCl_3$ diffusion; (3) wet-chemical polishing; (4) bifacial passivation (consisting of rear: HPP AlO_x/PECVD SiN_y stacks & front: PECVD SiN_x); (5) rear dielectric opening; (6) metallization and firing.

For rear side SiN_y deposition methods, two different strategies have been studied. Firstly, **Cell 1**, SiN_y layers were deposited using direct PECVD. Secondly, **Cell 2**, SiN_y layers were grown by using remote PECVD. Finally, **Cell 3** is based on rear surface polishing improvement in combination with HPP AlO_x/direct PECVD SiN_y stacks without influencing other process parameters. Fig. 1 shows the process flow diagram for the PERC solar cells fabricated in this study.

The quality of the surface passivation was determined by evaluation of τ_{eff} using photoconductance decay method (M/s Sinton Instruments, USA; Model: WCT-120). Samples for lifetime measurements were coated on both sides.

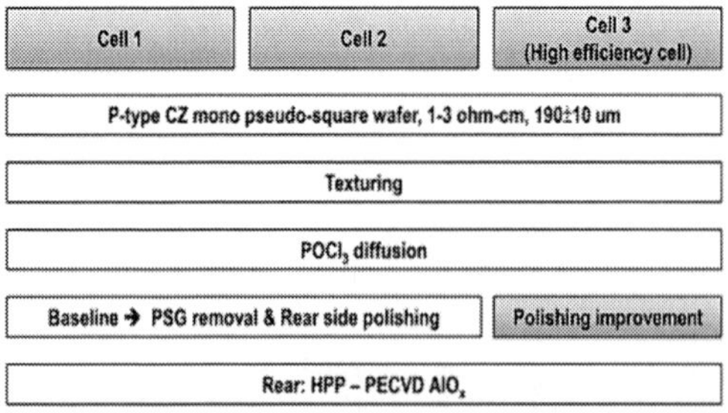

Figure 1: Process flow for the PERC solar cells fabricated in this work.

3 RESULTS AND DISCUSSION

3.1 Surface passivation quality

In order to evaluate the surface passivation quality of the HPP AlO_x films, symmetrical lifetime test structures were fabricated. τ_{eff} was measured using the transient decay photoconductance technique. The measurement is shown in Fig. 2. After firing at a measured peak temperature of ~870 °C for ~1 s, the well-optimized HPP AlO_x films provide a very good surface passivation with a median τ_{eff} of 427.5 μs at an excess carrier density of 1×10^{15} cm^{-3}. The maximum value is 509.4 μs in this study, which is corresponding to high cell efficiencies around 20.9-21.0%, as shown in the inset of Fig. 2.

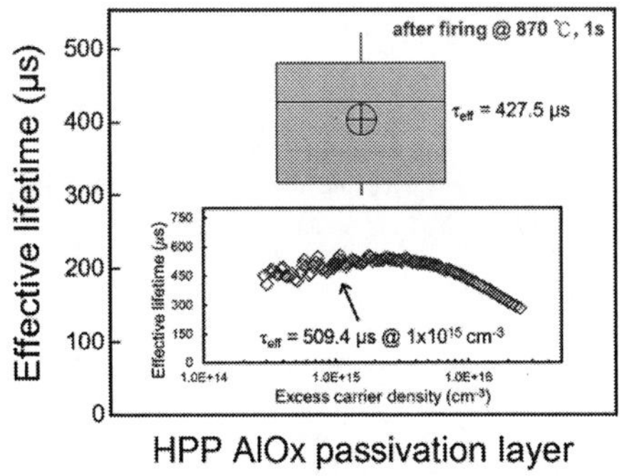

Figure 2: τ_{eff} of symmetrically passivated samples after passivation with HPP AlO_x film on both sides, measured using the Sinton WCT-120, at an excess carrier density of 1×10^{15} cm^{-3} (the inset is the best result lifetime).

3.2 Optimization of rear side manufacturing process

To achieve excellent passivation, minimize recombination losses and increase reflection at the rear side of the cell, we report that optimization of several cell processing steps led to an increase in cell efficiency of more than 0.2% absolute. The optimizations can be divided into effective rear SiN_y capping deposition methods and rear surface polishing improvement in this study. Fig. 3 shows the measured cell efficiency, open circuit voltage (V_{oc}), fill factor (FF) and short circuit current density (J_{sc}) distribution of 40 pieces cells with three cells processed parts from **Cell 1**, **2** and **3**, respectively.

For effective rear SiN_y capping deposition methods, the results obviously show that **Cell 1** average efficiency values of the HPP AlO_x/SiN_y stacks, where SiN_y has been deposited with direct plasma, are more than 0.1% higher compared to **Cell 2** efficiency values obtained after remote SiN_y capping. Furthermore, in the case of **Cell 1**, V_{oc} values and FF up to 667.2 mV and 79.5% can be achieved which are relatively higher than **Cell 2** (ca 1 mV and 0.5%), respectively. It also present clearly that the J_{sc} results obtained by the two SiN_y deposition methods are found to be the same distribution. These results show that the HPP AlO_x/direct PECVD SiN_y stacks can effectively improve the passivation properties of PERC solar cells.

To offer superior rear side reflection, **Cell 3** made by the combination of HPP AlO_x/direct PECVD SiN_y passivated stacks and improved polishing process achieves an efficiency of 20.9%, indicating a strong dependence of the rear surface smoothens. In contrast, **Cell 1**, the coverage of HPP AlO_x/direct PECVD SiN_y passivated stacks on non-optimized rear surfaces may probably be damaged at the peak and valley areas, thus the worse V_{oc} is obtained. In addition, **Cell 1** with non-optimized rear surface has slightly lower FF than that of **Cell 3** with improved surface. Also, the J_{sc} shows only a small improvement with improved polishing process.

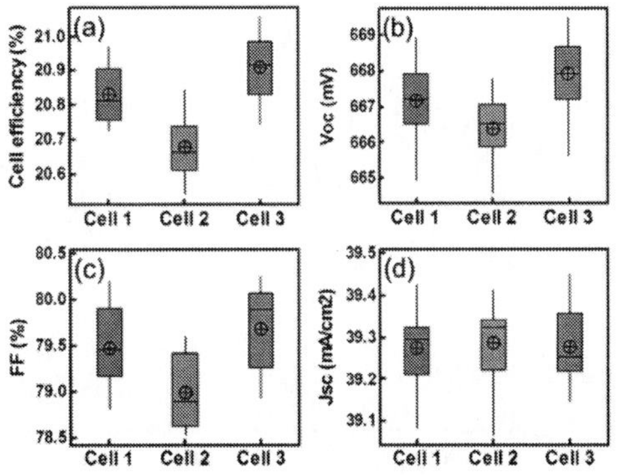

Figure 3: Distribution of (a) cell efficiency, (b) V_{oc} (c) FF and (d) J_{sc} of PERC solar cells with three types of manufacturing process. The cell performance parameters were from 40 pieces of cells.

To analyze the root cause of the lower efficiencies of the PERC solar cells, the EL measurements were carried out to study the impact of defect on cell efficiency. Table I shows the EL images sampled from **Cell 1**, **2**, and **3**, respectively, and could be found that the dark regions influence cell performance.

That makes sense since the dark regions mainly ascribe to defects and dominate the electric performance of finished cells [3]. To quantify the correlation between defect and their electric performance of finished solar cells, the area percentage of dark region corresponding to defective regions on EL images, as shown in Table I. As expected, general trends of V_{oc} and cell efficiency are inversely proportional to the percentage area of dark regions.

Therefore, by applying HPP AlO_x/direct PECVD SiN_y passivated stacks and improved polishing surface, could further decrease defect density and improve the efficiency of PERC solar cells.

Table I: EL images sampled from Cell 1, 2, and 3, respectively.

	Cell 1	Cell 2	Cell 3
EL Image	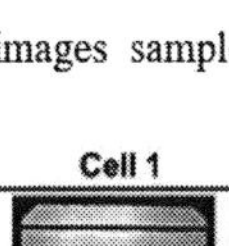	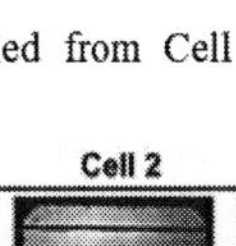	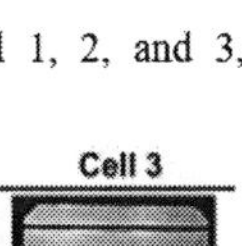
Dark Regions (% Area)	1.82	2.02	1.71

Fig. 4 displays the IQE response of PERC solar cells by applying direct and remote PECVD SiN_y capping, respectively. It could be seen that the choice of the rear side SiN_y capping (direct or remote PECVD methods) did not affect the long wavelength IQE response of PERC solar cells, indicating that these SiN_y did not affect the carrier lifetime in the base.

But it is noteworthy that the short wavelength (300-450 nm) IQE of **Cell 1** is greater than that of **Cell 2**. It appears that significantly lower interface recombination at the emitter surface with less defects and impurities [4], which attributes to direct PECVD SiN_y capping, exhibit a relatively high thermal budget (temperature × time) in comparison to remote PECVD method in this study.

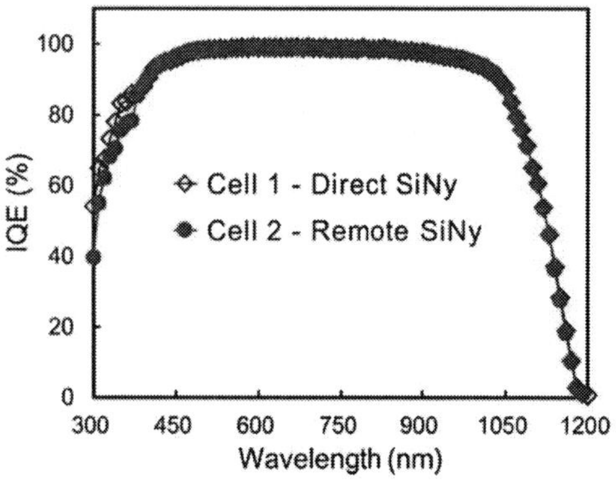

Figure 4: IQE response of PERC solar cells applying direct and remote PECVD SiN_y capping.

In order to analyze the impact of improving the flatness of rear surface, we measure the IQE and the reflectance, as shown in Fig. 5. It shows that there is a 6.7% enhancement in the IQE response for **Cell 3**, utilizing the polishing improvement in the infrared wavelength region (1000-1200 nm), compared to that in the reference of **Cell 1**. We can observe that the improved polishing rear surface, looking like a mirror, reflects more long-wavelength incident light than the amount that the non-optimized rear surfaces reflects. Also, we apply the polishing improvement to increase the emitter etch and reduce the rear surface roughness might further improve the efficiency of PERC solar cells [5], which is well agreed with I-V results, as shown in Fig. 3.

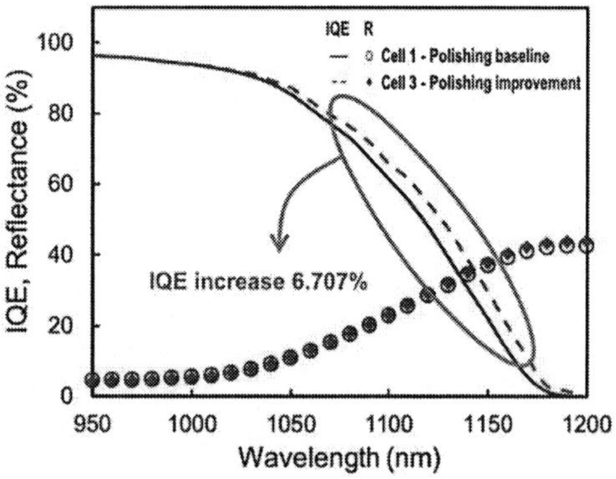

Figure 5: Comparison of IQE response and reflectance at long wavelengths between cell 1 and cell 3.

After integration of optimized processes, an average efficiency of 20.9% and an average V_{oc} of 667.9 mV were obtained, which is an obvious improvement compared to the reference PERC solar cells. In addition, the champion cell efficiency of measured under standard test conditions (AM 1.5G, 1000 W/m^2, 25 °C) has been increased to 21.1%, and the values of V_{oc}, J_{sc} and FF are 668.9 mV, 39.25 mA/cm^2, and 80.2%, respectively. The I-V curve for the champion cell is shown in Fig. 6.

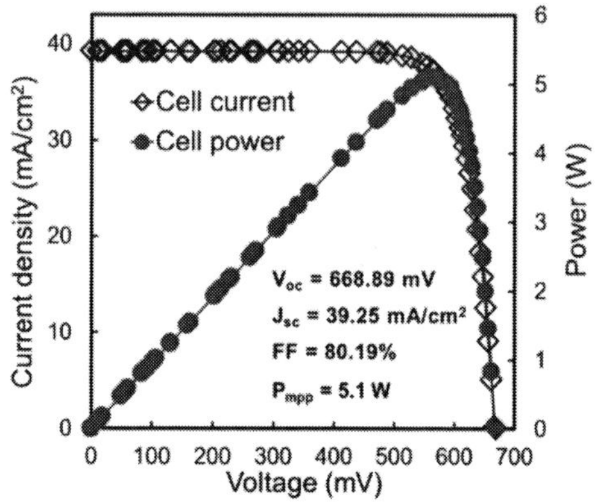

Figure 6: I-V curve of the champion cell with 21.1% efficiency.

4 CONCLUSIONS

A champion p-type PERC solar cell efficiency of 21.1% was attained on 15.6×15.6 cm^2 p-type mono pseudo-square silicon wafers utilizing commercially viable processes. Optimized cell fabrication processes consisting of the combination of effective rear HPP AlO_x/direct PECVD SiN_y passivated stacks and rear surface polishing improvement were the main contributors to this achievement. Our results suggest a possible approach for further improving the performance of PERC solar cells.

ACKNOWLEDGMENTS

The authors would like to thank the engineers of Motech and various members of Manz AG for their hard work and contributions to this paper.

REFERENCES

[1] G. Dingemans and W. M. M. Kessels, J. Vac. Sci. Technol. A 30 (2012) 040802.
[2] N. Balaji, S. Q. Hussain, C. Park, J. Raja, J. Yi and R. Jeyakumar, Trans. Electr. Electron. Mater. 16 (2015) 227.
[3] D. Mankovics, A. Klossek, Ch. Krause, T. Arguirov, W. Seifert and M. Kittler, Phys. Status Solidi A-Appl. Mat. 209 (2012) 1908.
[4] X. Gay, F. Souren, B. Dielissen, M. Bijker, R. Gortzen, D. Pysch, K. Weise, B. Sander and R. Sastrawan, 28th European Photovoltaic Solar Energy Conference, (2013) 1220.
[5] C. Kranza, S. Wyczanowski, U. Baumann, K. Weise, C. Klein, F. Delahaye, T. Dullweber and R. Brendel, Energy Procedia 38 (2013) 243.

Rear passivation and point contacts formation by laser process through stacks of a-Si:H(i) and a-Si:B for high efficiency silicon solar cell

Young-joon Han, Evan Franklin, Xinyu Zhang, Andrew Thomson and Yimao Wan
Research School of Engineering, College of Engineering and Computer Science, The Australian National University, Canberra, ACT 2601, Australia
E-mail: young.han@anu.edu.au

ABSTRACT: In this work we investigate the effectiveness of a stack of intrinsic amorphous silicon (a-Si:H(i)) and boron- (for p-type) doped amorphous silicon (a-Si:B) for producing multi-purpose films capable of providing both excellent surface passivation and effective localized doping via laser irradiation. We deposit, via a combination of PECVD and sputtering, the stack of a-Si:H(i) and a-Si:B on crystalline silicon substrates, and subsequently employ a KrF 248 nm nanosecond laser to simultaneously remove the film and dope the sub-surface region beneath the film. We evaluate and investigate the passivation capabilities of the films in non-laser processed regions from effective lifetime measurements, and also observe via PL imaging the subsequent impact of localised laser doping. We demonstrate that an intrisinc a-Si:H layer of at least 20 nm is required to sufficiently preserve surface passivation after application of sputtered a-Si:B films which have not otherwise been optimised to provide passivation. Sheet resistance measurements indicate that a high level of doping is achieved in laser processed regions for a broad range of laser parameters and dopant fractions in deposited films, with sheet resistance as low as around 25 Ω/□ being observed for a-Si:B films containing 30 vol% B and as low as 60 Ω/□ for films with 10 vol% boron. Furthermore, the qualitative characterization on the laser doped region is conducted via the low temperature photoluminescence spectroscopy.
Keywords: amorphous silicon, laser doping, passivation

1 INTRODUCTION

The ongoing drive to increase commercial solar cell efficiencies has in recent years seen renewed interest in the industrial application of PERC and PERL cell designs [1]. The key requirements for such high efficiency solar cells include excellent rear surface passivation and small, localized, typically heavily doped contact areas. Therefore, it is desirable to find a method which can further simplify the fabrication sequence for localized contact formation to reduce costs. It is also highly desirable to implement such cell designs using low-temperature only manufacturing processes, thus realising the full advantage of cheaper, lower quality materials such as Upgraded Metallurgical Grade (UMG) silicon [2]. In recent years, this concept is applied by utilizing laser processes, forming contacts by ablating passivating rear dielectric layers and/or localized doping via liquid phase diffusion. One such approach uses spin-on or spray-on dopants deposited onto the dielectric layer and irradiating via laser to form localized heavy doping. After the laser process, the deposited spin-on dopants are then required to be removed in another wet-chemistry step. Another approach is the laser-fired rear contacts, formed by laser through the dielectric layer and a metal layer on top [3]. The most advanced approach is the so-called PassDop introduced by Suwito et al [4], which provides excellent passivation layer as well as the dopant source for the laser process. This approach is very attractive since the dopant source is constitutionally part of the solar cell structure and which is not required to be removed subsequent to the laser process, therefore the cell fabrication steps could be minimized. Until today, Al_2O_3/a-SiC_x:B [5], doped silicon nitride [6] and a-$Si_{1-x}C_x$ [4] layers have been studied for PassDop application.

In this work, we are investigating the effectiveness of layer stacks of intrinsic amorphous silicon (a-Si:H(i)) and boron- (for p-type) rich amorphous silicon (a-Si:B) for producing multi-purpose films capable of providing excellent passivation and effective localized laser doping. Thin films of a-Si:H(i) are very well-known to provide outstanding surface passivation, having been successfully applied in many of the highest efficiency silicon solar cells produced to date, the so-called HIT cells and HJ-IBC cells [7,8]. Additional layer of a-Si:B acts as the dopant source for laser doping, introducing only silicon and boron atoms into the locally melted region during the laser process without any additional impurities. It has been demonstrated previously that sputtered a-Si:H films are capable of excellent surface passivation [9] and can be used to produce doped films, and likewise also that boron-doped a-Si can be deposited by PECVD. However, for reasons of equipment availability and reliability, in this work we focus on intrinsic a-Si films produced by PECVD and boron-doped a-Si films produced by sputtering. Ultimately a single machine co-optimised for both passivation and dopant source film depositions will be most attractive from an industrial processing point of view, and we see no inherent barrier to this being developed in future.

2 EXPERIMENTAL DETAILS

High resistivity (~100 Ω.cm) p-type float-zone (FZ) crystalline silicon substrates with thickness of approximately 200 μm were used for this study. All samples were first etched in 25 % tetramethylammonium hydroxide (TMAH) solution at 80 °C for 5 minutes to remove surface damages, and RCA cleaned and dipped in 1 % HF solution prior to the deposition of a-Si films.

Intrinsic amorphous silicon passivation layer (a-Si:H(i)) was deposited on both sides of substrates via Plasma-Enhanced Chemical Vapour Deposition (PECVD). The SiH_4 and He were supplied at 20 and 500 sccm flow rates respectively, the excitation power was 10 W and the deposition temperature was kept at 400 °C. To investigate the effect of thickness of a-Si:H(i), different thickness (5, 10 and 20 nm) of a-Si:H(i) were deposited by varying deposition time.

Dopant source layer (a-Si:B) was deposited on the single side of PECVD a-Si:H(i) deposited substrates by co-sputtering of pure Si and B targets. Argon (Ar) was used as the plasma source, and the sputter chamber pressure was kept at 4 mTorr. The thickness of a-Si:B was kept at 50 nm, but different percentage of B in a-Si:B were employed to study the effect of concentration of dopant atoms in a-Si on the passivation and laser doping. The deposition power for Si target was kept consistent whereas the deposition power for B target and the co-sputtering time were adjusted to provide 10 vol% and 30 vol% of B in 50 nm a-Si:B layer.

After the deposition of the films, the effective lifetime (τ_{eff}) was measured with a Sinton WCT120 photoconductance instrument [10].

We then employed a KrF 248 nm nanosecond excimer laser, with a homogenous-rectangular beam and 25 ns pulse duration, to locally heat and melt the stack of a-Si:H(i)/a-Si:B coated substrate, thus introducing dopants into the melt prior to recrystallization. Single laser pulses were applied, with a range of fluences from 1 to 4 J/cm^2. To characterize the level of the laser doping, small areas of 6 × 2 mm^2 were laser processed by overlapping 502 × 502 μm^2 beam size of laser pulses. The resultant sheet resistance of the doped region is determined by four-point probe measurement, using appropriate calibration factors for the rectangular doped areas created. In addition, qualitative characterization on the laser doped regions was conducted via the low temperature micro-photoluminescence spectroscopy (μ-PLS). [11]. We also evaluated briefly the recombination properties of locally laser processed regions via the uncalibrated photoluminescence (PL) images.

3 RESULTS AND DISCUSSION

The passivation effect of the stack of a-Si:H(i) and a-Si:B on top is observed by measuring τ_{eff}, determined at an excess minority carrier density of 1×10^{15}cm^{-3}, prior to laser processing as shown in Figure 1. Before depositing a-Si:B layer (ie. after a-Si:H only), a maximum lifetime of around 10 ms was observed for the 20 nm a-Si:H(i) coated samples in as-deposited condition. However, immediately prior to depositing the a-Si:B films, we found that the lifetime of all samples, which had been stored at room temperature for five days, approximately halved, indicating considerable degradation of surface passivation. Effective lifetime was however found to be completely restored after low temperature annealing as shown in [12], and so this same annealing condition was applied to films subsequent to a-Si:B deposition. Single side deposition of a-Si:B layer causes significant degradation of passivation quality for all intrinsic layer thicknesses, although this was much more severe in case of thin a-Si:H(i) and higher concentration of B in a-Si:B. For the 20 nm a-Si:H(i) case, the lifetime still declines due to a-Si:B layer, but reasonable lifetime of around 4.5 ms is maintained regardless of boron concentration. The corresponding thermal equilibrium recombination current pre-factor J_0 is approximately 2.5 fA/cm^2, derived by fitting the injection dependent effective lifetime measurement of the sample to full 3D device simulations using Quokka [13] as shown in Figure 2.

Doped a-Si layers have been reported previously to be able to act as a capping layer for passivating a-Si:H(i) layer [12]. However, a-Si:B capping layers used in this experiment show the opposite role, since the film is not optimised for passivation purpose. The current results lead us to conclude that a passivating amorphous layer of around 20 nm thickness is required as a minimum in order to avoid passivation degradation owing to a-Si:B top layer that has not been optimised for passivation. Further optimisation of film deposition conditions may result in thinner intrinsic silicon films being found to be resistant to degradation after application of doped layers.

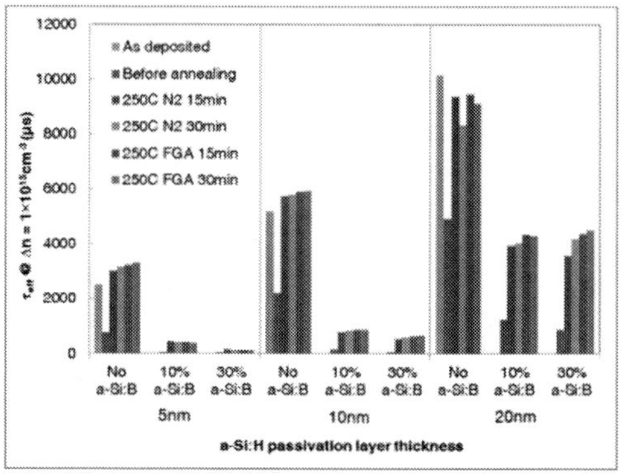

Figure 1: Measured effective lifetimes of a-Si:H(i)/a-Si:B coated samples after a-Si:H(i) deposition only, subsequent to 5 days ageing and doped layer deposition, and subsequent to various low temperature annealing conditions. All τ_{eff} values are extracted at the excess carrier density of 1×10^{15}cm^{-3}.

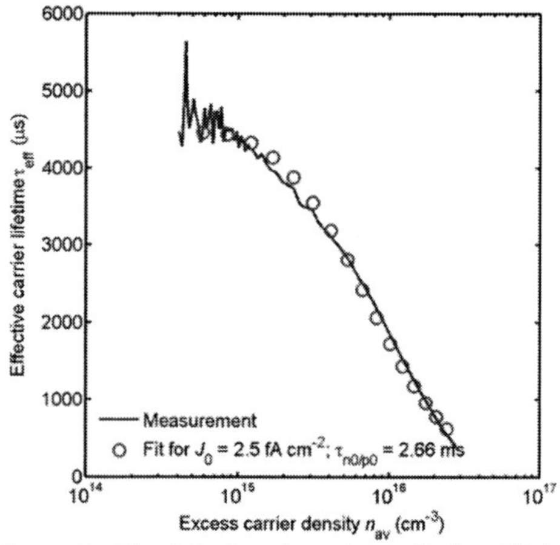

Figure 2: The injection dependent effective lifetime curve for 20 nm a-Si:H(i) / 30 vol%B a-Si:B coated sample showing J_0 of 2.5 fA/cm^2, derived from Quokka simulation.

The sheet resistance of laser processed regions under a variety of stack compositions are demonstrated in Figure 3. It is noteworthy that a wide range of doping levels are observed for the range of laser pulse fluences 1 to 4 J/cm^2 and dopant fraction in dopant source layer, with the sheet resistance of around 270 Ω/□ for maximum and around 25 Ω/□ for minimum. For a range of 5 to 20 nm thickness of a-Si:H(i) the resulting sheet resistances are impacted by the thickness only at the low fluence (1 J/cm^2). Above 2 J/cm^2, the sheet resistances are determined from the boron concentration in a-Si:B layer regardless of the thickness of a-Si:H(i). Obviously, the higher concentration of dopants enables the heavier doping. Increasing fluence shows generally lower sheet resistances, but above a fluence of 3 J/cm^2 the value of sheet resistances reach an apparent minimum, around 60 Ω/sq (10 vol% B film) or 25 Ω/sq (30 vol% B film).

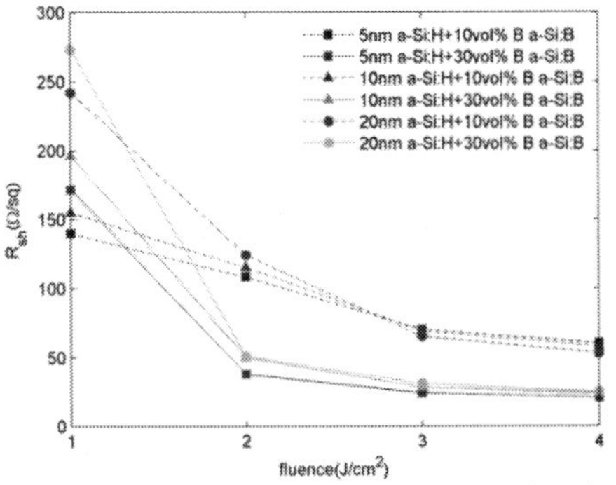

Figure 3: Measured sheet resistance as a function of fluence for regions laser doped from a-Si:H(i)/a-Si:B films. The thickness of the a-Si:H(i) layer was varied from 5 nm to 20 nm. On top of a-Si:H(i) layer, 50 nm a-Si:B was deposited with 10 vol% or 30 vol% B.

Figure 4 plots the normalized PL spectra of a variety of laser doped samples through 20 nm a-Si:H(i)/ a-Si:B, as measured at the center of the laser processed region. The presence of a-Si:H(i) is also confirmed from the PL spectra in a range of 900 ~ 1300 nm [14], measured on the passivated surface. Damage and defects caused by laser doping are represented in the PL spectra in the wavelength range of 1200 ~ 1500 nm, and the heavy doping via the local peak at around 1160 nm, owing to the band-gap narrowing effect [15]. Interestingly, the increase in the fluence shows relatively lower laser damage related PL spectra, opposite to the trend shown in [15]. This may be owing to the laser annealing effect on amorphous silicon layers, where laser pulses at low fluence (~ 1 J/cm^2) transform the amorphous silicon layers into polycrystalline layers containing a relatively high number of crystal discontinuities and defects, with little or no melting and recrystallization of the underlying substrate [16]. In contrast, higher fluences result in deeper melt zones, and likely also some direct ablation of the amorphous film, which are subsequently recrystallized from the defect-less crystalline substrate. Hence the resultant luminescent signal of laser-annealed amorphous silicon layer, and its relation to laser pulse fluence, is apparent in the longer wavelength 'damage spectrum' region shown in Figure 4 [17]. In addition, even though we have confirmed heavy doping via sheet resistance measurements, the evidence for it does not stand out via μ-PLS measurement due to the difficult to resolve local peak at around 1160 nm in the measured PL spectra. The intensity of this peak, for PL excitation at 532 nm wavelength, generally increases with doping depth [18]. We therefore speculate that doping is shallow but has higher dopant concentration at the surface.

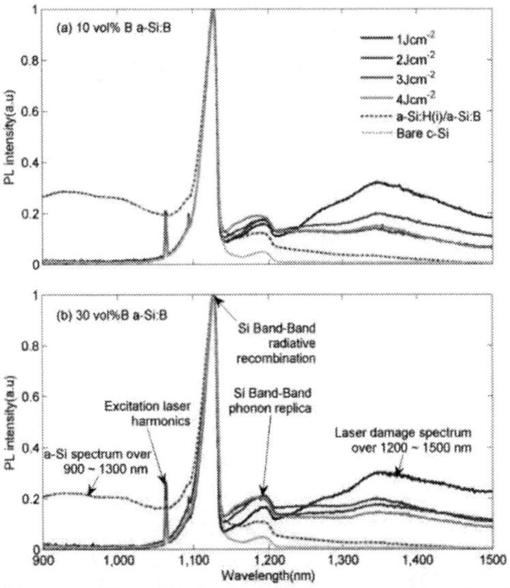

Figure 4: Normalized micro-PL spectra on laser doped samples through (a) 20 nm a-Si:H/ 50 nm 10vol% B a-Si:B or (b) 20nm a-Si:H/ 50nm 30 vol%B a-Si:B with the fluence range of 1 ~ 4 J/cm^2 at the center region. Excitation is achieved with a 532 nm laser, and substrate temperature is 79 K. The spectrum of an unprocessed crystalline Si and an 20nm a-Si:H/a-Si:B coated silicon sample is plotted for comparison.

Lastly, we perform a simple evaluation of the local recombination properties of laser doped regions. Figure 5 shows the uncalibrated PL images of laser doped samples through a stack of 20 nm a-Si:H(i) and 50 nm a-Si:B and with different fractions of doped regions. This is achieved by creating a regular rectangular array of 40 × 40 μm^2 laser doped 'spots' at varying pitch. The image was taken after forming gas annealing at 250°C for 15 minutes, since a significant improvement in effective lifetime is observed after such an annealing process. Table 1 summarizes the average PL counts for each of the fractionally laser doped arrays, with an estimated decrease in open-circuit voltage (ΔV_{oc}) that could be expected in the application of photovoltaic devices. The estimation of ΔV_{oc} is conducted based on the equation derived by Glatthar *et al.* [19]:

$$\Delta V_{oc} = V_T \ln(\frac{I_{laser}}{I_{pass}})$$

where, V_T is thermal voltage, around 26 mV at room temperature, I_{pass} is PL count in the passivated area, and I_{laser} is PL count in the laser doped region. Assuming all other factors are optimized and that device efficiency were theoretically limited by surface recombination alone, then a V_{oc} drop of about 36 mV could be expected in the case of 0.5% locally doped area. Given that J_0 of 2.5 fA/cm^2, as shown in Figure 2 in this work, is generally capable of yielding cells with very high V_{oc}, with a-Si passivated silicon solar cells having been reported with open-circuit voltages of around 750 mV [7,8], the results obtained in this work for localized laser doped structures is consistent with fabrication of cells having V_{oc} of 700 mV or over, despite relatively high recombination activity being observed at localized laser doped regions.

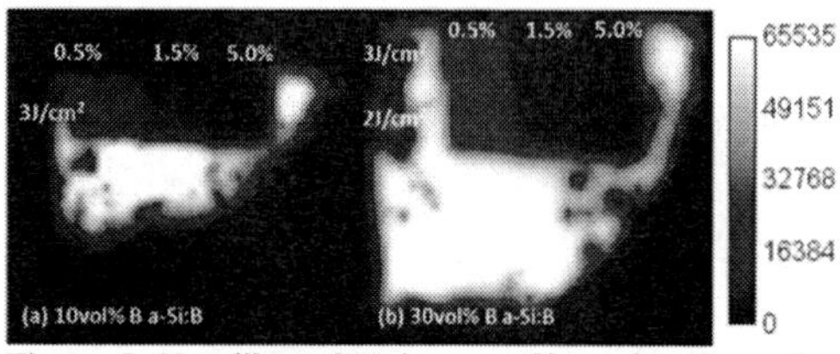

Figure 5: Uncalibrated PL images of laser doped samples through (a) 20 nm a-Si:H/ 10 vol% B a-Si:B or (b) 20 nm a-Si:H/ 30 vol%B a-Si:B after forming gas annealing at 250˚C for 15 minutes. Different fractions of the laser doped regions are employed, 0.5%, 1.5 % and 5 %

Table 1: The average of photoluminescence counts and estimated V_{oc} decrease on laser doped regions shown in Figure 5

	Doped fraction	$3J/cm^2$	ΔV_{oc} (mV)	$2J/cm^2$	ΔV_{oc} (mV)
(a)	0 %	65531	0	N/A	
	0.5 %	15944	-36.7		
	1.5 %	6028	-62.0		
	5 %	1697	-94.9		
(b)	0 %	65525	0	65525	0
	0.5 %	9350	-50.6	14996	-38.3
	1.5 %	2290	-87.2	3640	-75.1
	5 %	819	-113.9	1440	-99.2

4 CONCLUSION

A stack of a-Si:H(i) and a-Si:B provides both good surface passivation and effective localized doping via laser irradiation. The best result is found for a 20 nm a-Si:H(i) with 30 vol% B a-Si:B, showing τ_{eff} ($\Delta n = 1 \times 10^{15}$ cm^{-3}) of 4.5 ms (corresponding J_0 of 2.5 fA/cm^2) in passivated regions, and minimum R_{sh} of 25 Ω/□ in the laser doped regions. A passivating a-Si:H(i) layer of around 20 nm thickness is required as a minimum in order to avoid passivation degradation owing to a-Si:B top layer that has not been optimised for passivation. The fraction of dopant atoms in the dopant source layer determines the maximum level of doping. It is also deduced that the doped region has very shallow doping profile but high surface dopant concentration via μ-PLS measurement. Local recombination analysis shows the potential of achieving solar cells with high V_{oc}, with results for 0.5% localized doping fraction being consistent with voltages above 700 mV. We have shown via this work that multi-purpose dopant-containing a-Si passivation film stacks combined with localized laser doping offers the potential for low-temperature processing to be applied in the implementation of high-efficiency solar cell concepts.

5 REFERENCES

[1] M. A. Green, Solar Energy Materials and Solar Cells **143**, 190 (2015).

[2] P. Zheng, F. E. Rougieux, C. Samundsett, X. Yang, Y. Wan, J. Degoulange, R. Einhaus, P. Rivat, and D. Macdonald, Applied Physics Letters **108** (12), 122103 (2016).

[3] E. Schneiderlöchner, R. Preu, R. Lüdemann, and S. W. Glunz, Progress in Photovoltaics: Research and Applications **10** (1), 29 (2002).

[4] D. Suwito, Ja, x, U. ger, J. Benick, S. Janz, M. Hermle, and S. W. Glunz, Electron Devices, IEEE Transactions on **57** (8), 2032 (2010).

[5] B. Steinhauser, U. Jäger, J. Benick, and M. Hermle, Solar Energy Materials and Solar Cells **131**, 129 (2014).

[6] B. Steinhauser, M. b. Mansoor, U. Jäger, J. Benick, and M. Hermle, Solar Energy Materials and Solar Cells **126**, 96 (2014).

[7] M. Taguchi, A. Yano, S. Tohoda, K. Matsuyama, Y. Nakamura, T. Nishiwaki, K. Fujita, and E. Maruyama, Photovoltaics, IEEE Journal of **4** (1), 96 (2014).

[8] K. Masuko, M. Shigematsu, T. Hashiguchi, D. Fujishima, M. Kai, N. Yoshimura, T. Yamaguchi, Y. Ichihashi, T. Mishima, and N. Matsubara, (2014).

[9] X. Zhang, A. Cuevas, B. Demaurex, and S. De Wolf, Energy Procedia **55**, 865 (2014).

[10] R. A. Sinton and A. Cuevas, Applied Physics Letters **69** (17), 2510 (1996).

[11] R. Sauer, J. Weber, J. Stolz, E. R. Weber, K. H. Küsters, and H. Alexander, Appl. Phys. A-Mater **36** (1), 1 (1985).

[12] S. Bowden, U. Das, S. Herasimenka, and R. Birkmire, presented at the Photovoltaic Specialists Conference, 2008. PVSC '08. 33rd IEEE, 2008 (unpublished).

[13] A. Fell, IEEE Transactions on Electron Devices **60** (2), 733 (2013).

[14] J. I. Pankove and D. E. Carlson, Applied Physics Letters **31** (7), 450 (1977).

[15] Y.-J. Han, E. Franklin, A. Fell, M. Ernst, H. T. Nguyen, and D. Macdonald, Appl. Phys. A **122** (4), 1 (2016).

[16] J. Narayan, C. W. White, M. J. Aziz, B. Stritzker, and A. Walthuis, Journal of Applied Physics **57** (2), 564 (1985).

[17] R. A. Street, N. M. Johnson, and J. F. Gibbons, J. Appl. Phys. **50** (12), 8201 (1979).

[18] H. T. Nguyen, D. Yan, F. Wang, P. Zheng, Y. Han, and D. Macdonald, Phys. Status. Solidi-R **9** (4), 230 (2015).

[19] M. Glatthaar, J. Haunschild, R. Zeidler, M. Demant, J. Greulich, B. Michl, W. Warta, S. Rein, and R. Preu, Journal of Applied Physics **108** (1), 014501 (2010).

EFFECT OF LASER ABLATION ON ELECTROPLATED-METALLIZATION CRYSTALLINE SILICON SOLAR CELLS

Yueh-Lin Lee*, Ming-Shiou Lin, Shu-Yen Liu, Kuang-Chieh Lai, Chia-Chih Chuang, and Chi-Chun Li
Motech Industries, Xinshi Dist., Tainan City, 74145, Taiwan
(*Corresponding author: phone/fax/e-mail:+886-6505-0789-0608/+886-6505-8446/yuehlin_lee@motech.com.tw)

ABSTRACT: In this work, a front-side metallization based on laser ablation and electrochemical plating contacts is proposed for the improvement in shading loss and short-circuit current density. The influences of nanosecond laser ablation on finger width, surface morphology, and cell performance are investigated. The optimization of laser ablation and electroplating contacts are achieved to improve the finger width, aspect ratio, and metal shading effect. The results of Suns-Voc measurement indicate the improvement in recombination issues. A average performance of Ni/Cu plated solar cell exhibits a 0.54 mA/cm^2 increase of short-circuit current density (J_{sc}), 0.10% increase of fill factor (FF), 0.03 Ωcm^2 decrease of series resistance (R_s), and 0.28% increase of conversion efficiency (η).

Keywords: metallization, contact, performance, solar cell

1 INTRODUCTION

Recently, a simple patterned process and advanced metallization plays are interesting for silicon solar cell technology. Low surface recombination loss in silicon solar cells is achieved via the passivated surface. A method with local and low-damage opening is needed [1]. Laser process provides a feasible way to achieve pattern design, which is accurate and fast. Front-side metallization is performed by electrochemical technique consisting of selective laser ablation and electroplating deposition, thus low finger width and aspect ratio can be improved for decrease of shading [2]. The conventional Ag screen-printed contacts usually have high contact resistance and lead to junction shunting issue during the post firing process [3]. The Ag screen printing can be replaced with Ni/Cu plating to reduce the production cost of solar cells. However, some challenges still must be overcame for plating metallization technique, including the decrease of front-side shadow, increase of finger conductivity, and improvement of contact properties.

2 EXPERIMENTAL

The 156 × 156 mm^2 n-type mono-crystalline silicon wafers with a random pyramid texture, boron-doped emitter, and phosphorus-doped surface field were used as substrates, consisting of the following: A saw-damage etching, alkaline texturing, borosilicate (BSG) removal, and drive-in dopants. The plasma enhanced chemical vapour deposition (PECVD) SiNx based dielectric was used as passivation and anti-reflection coating (ARC). Prior to metallization, H-pattern of three busbars was defined in the dielectric layer using the 532 nm wavelength of nanosecond pulse laser. Both high and low laser fluences were used to form the contact opening regions. Conventional screen printing of silver paste on rear side as well as firing was performed. Ni/Cu plating metallization was employed as front side contact for n-type Si solar cells. The plated Ni contact layer and plated Cu conductive layer were applied to n$^+$np$^+$ front emitter solar cell structure. Each Ni/Cu metallization was used by forward bias plating (FBP). The process sequence of our cell with laser ablated and plated front contact is shown in Fig 1.

3 RESULTS AND DISCUSSION

Fig 2(a) and (c) show the scanning electron microscope (SEM) images of laser ablated finger using both high and low fluences. The laser ablated samples with high and low fluences are defined as L1 and L2. The finger widths of laser ablation are about 38.4 and 26.1 μm for high-fluence L1 and low-fluence L2, respectively. In order to decrease the finger width, the low fluence of laser ablation is used to achieve the narrow width of laser ablated region. The difference in surface morphology between high- and low-fluence laser ablation is observable. High-fluence laser is caused the more surface damages in the dielectric layer. Fig 2(b) and (c) show the EDX spectrum of laser ablated region using both high and low fluences. Regardless of laser fluence, no nitride signal is detected in the contact opening region. It indicates that the dielectric layer is opened completely by 532-nm laser ablation for metal contacts.

Fig 3 shows the SEM images of Ni/Cu plating metallization. For L1 condition, the finger width is about 69.3 μm after Ni/Cu plating. However, the finger width of Ni/Cu plating reduces to 50.1μm via laser condition of L2. The optimization of laser ablation and plating processes are achieved to improve the metal shading issue. Under the same condition of Ni/Cu plating, the metal finger height of L2 is higher than that of L1. The heights of Ni/Cu plated finger are about 12.5 and 15.4μm, respectively.

Table 1 shows the related parameters measured by Suns-V_{oc} for high- and low-fluence laser ablated cells. To compare L1 with L2, the loss in V_{oc} and J_{01} value is attributed to the more surface damages and higher recombination at the contacts.

Table 2 shows the performances of Ni/Cu plated cells with high- and low-fluence laser ablation. The average difference of short-circuit current density (J_{sc}) between low- and high-fluence laser ablation is about 0.54 mA/cm^2. It is due to the narrower metal finger widths, resulting in the decrease of shadow loss. Besides, the improvement of fill factor (FF) and series resistance (R_s) are attributed to the optimized width and height of metal fingers. The laser ablation removes the dielectric layer in the irradiated regions completely, so the loss in FF value is avoided. Finally, the performance of low-fluence laser ablated cell with Ni/Cu plated metallization exhibits a

0.54 mA/cm^2 increase of J_{sc}, 0.10% increase of FF, 0.03 Ωcm^2 decrease of R_s, and 0.28% increase of η.

4 CONCLUSIONS

Selective laser ablation and Ni/Cu electroplating are key processes for front side metallization silicon solar cell. The difference between high- and low-fluence laser ablation regarding the ablation micro-structure and influences on both physical and electrical properties are discussed. The finger width of Ni/Cu electroplating reduces from 69.3 to 50.1 μm by using the optimized laser ablation process. According to Suns-V_{oc} results, low contact damages and recombination current are also obtained by optimized laser ablation. For optimized laser ablation and Ni/Cu plating, average increase of J_{sc} value is about 0.54mA/cm^2. It is due to the decrease of front-side shading. The average FF and R_s values are also improved resulting from suitable finger contacts and high aspect ratio. Therefore, average enhancement of Ni/Cu electroplated cell efficiency is about 0.28%.

5 ADDITIONAL COMPONENTS

5.1 Illustrations

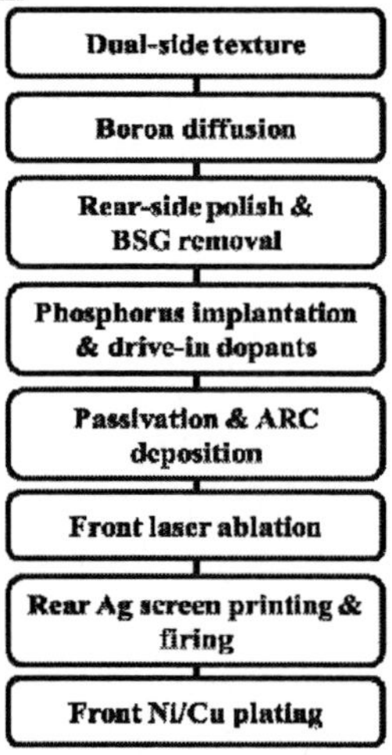

Figure 1: Process sequence for Ni/Cu electroplating cells.

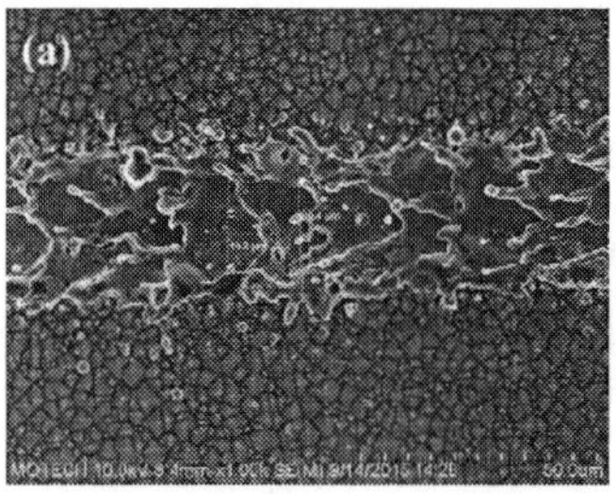

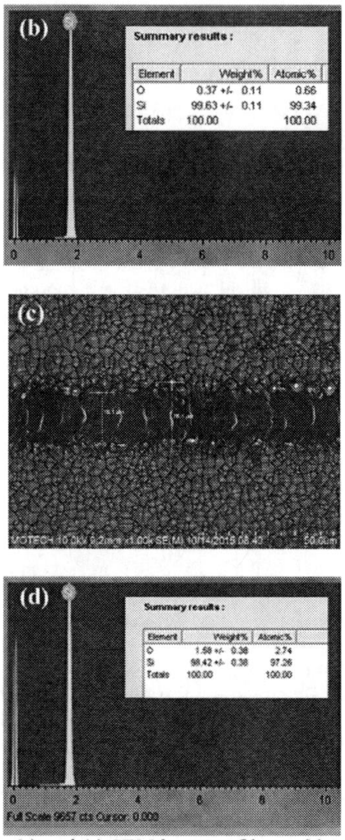

Figure 2: (a) and (c) SEM images of laser ablated finger. (b) and (d) EDX spectra of laser ablated region for high and low fluences.

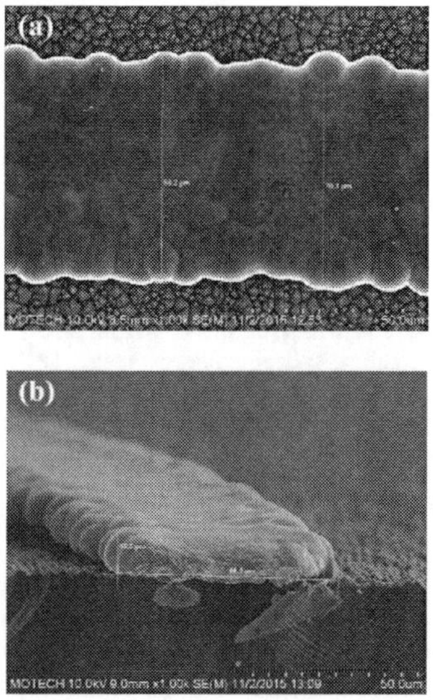

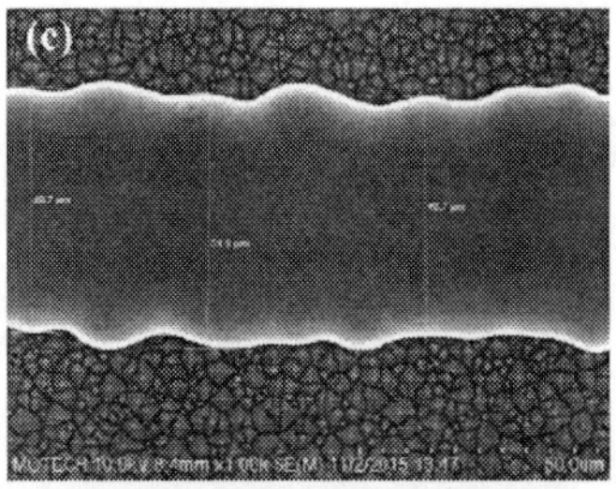

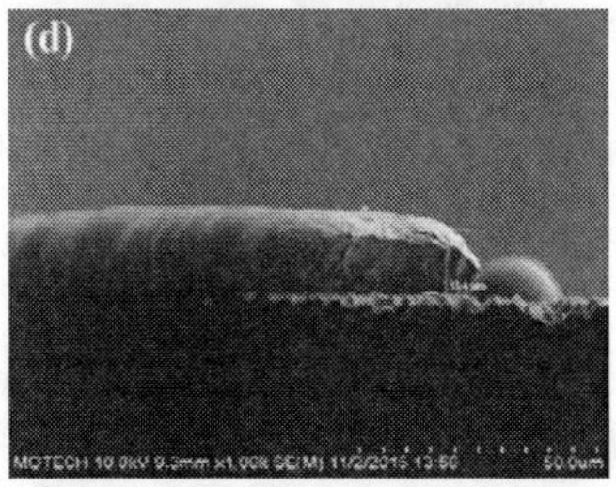

Figure 3: (a) and (c) SEM top-view images of Ni/Cu plated finger with high- and low-fluence laser ablation. (b) and (d) cross-section of Ni/Cu metallization with high- and low-fluence laser ablation.

5.2 Tables

Table 1: Suns-V_{oc} measurements for high- and low-fluence laser ablated cells.

Cell	Voc [mV]	pFF [%]	J_{01} [pA/cm^2]
Average of Ni/Cu plated cells with L1	645.7	82.9	0.53
Average of Ni/Cu plated cells with L2	646.0	83.1	0.52

Table 2: Performances of Ni/Cu plated cells with high- and low-fluence laser ablation.

Cell	η [%]	FF [%]	Jsc [mA/cm^2]	Voc [mV]	Rs [Ωcm^2]
Average of Ni/Cu plated cells with L1	Δη	ΔFF	ΔJsc	ΔVoc	ΔRs
Average of Ni/Cu plated cells with L2	Δη+0.28	ΔFF+0.10	ΔJsc+0.54	ΔVoc	ΔRs-0.03

5.3 References

[1] P. Engelhart, S. Hermann, T. Neubert, H. Plagwitz, R. Grischkel, R. Meyer, U. Klug, A. Schoonderbeek, U. Stute, and R. Brendel, "Laser Ablation of SiO_2 for Locally Contacted Si Solar Cells With Ultra-short Pulses", Prog. Photovolt. Res. Appl. 15, 521–527 (2007).

[2] A. Focsaa, D. Blanca, G. Poulaina, C. Barbosa, M. Gauthierb, N. L. Quangb, and M. Lemiti, " Copper light-induced plating contacts with Ni seed-layer for mc-Si solar cells after laser ablation of SiNx:H", Energy Procedia 38, 713 – 719 (2013).

[3] A. ur Rehman and S. H. Lee, "Crystalline Silicon Solar Cells with Nickel/Copper Contacts", published in Solar Cells - New Approaches and Reviews, (2015).

NANOSTRUCTURED SILICON NITRIDE (SI-N) ANTIREFLECTION COATING FOR C-SI SOLAR CELLS

Hemanta Ghosh[1*], Suchismita Mitra[1], Hiranmay Saha[1], Swapan Kumar Datta[1] and Chandan Banerjee[2]
[1]Centre of Excellence for Green Energy and Sensor Systems,
Indian Institute of Engineering Science & Technology, Shibpur, Howrah 711103, India
[2]National Institute of Solar Energy, Faridabad-Gurgaon Road, Gwal Pahari,
Gurgaon 122003, Haryana.
*Email: hemghosh04@gmail.com

ABSTRACT: In this paper, we present the antireflection properties of nanostructured silicon nitride layer formed by argon treatment of silicon nitride. Significant reduction of reflectance is observed after argon plasma treatment of silicon nitride layer. Different parameters like power density, gas flow rate, pressure and duration of argon plasma treatment have been varied to optimize the reflectance of the nanostructured silicon nitride layer. FESEM, FTIR and reflectance have been performed to study the morphology and optical properties of the silicon nitride layer. Nanostructures on the surface of the silicon nitride film can be observed in FESEM images. FTIR reveals a change in Si-N, Si-O and N-H bonds after plasma treatment which indicates a formation of silicon oxynitride layer. The reduction in reflection may be attributed to the formation of a double layer anti-reflection coating with silicon oxynitride layer on the top surface which is formed due to argon bombardment and silicon nitride as bottom layer. Such nanostructured silicon nitride coating can be used as anti-reflection coating on crystalline silicon solar cell, as it is compatible with present methods of fabrication
Keywords: argon plasma treatment, silicon nitride, silicon oxynitride, double layer anti-reflection coating, PECVD, solar cell

1 INTRODUCTION

The performance of conventional single junction silicon nitride coated textured crystalline silicon solar cell of 150-200 μm thickness with a <100> oriented Czochralski Si (or Cz-Si) p-type wafer is generally affected in the short wavelength range (300nm–600nm) by the reflectivity of the front surface apart from the front surface recombination velocity. In this paper, we report the fabrication of nanostructured silicon nitride on the front surface of conventional crystalline silicon solar cells which circumvents this limitation i.e. reduction of reflectivity of front surface and thereby, enhancement of photon transmission into the solar cell.

Conventional solar cells use a single antireflection coating (ARC) for reduction of reflectivity in front surface with a single layer antireflection coating designed for one specific wavelength, thereby limiting their performance. With multilayer graded-index antireflection coating [1-3] with desired value of refractive index using nanostructured dielectric materials like silica, reflectivity can be tuned further [4, 5].

However, realizing a multilayer graded index antireflection coating, especially with nanostructures is not only technologically difficult, but also an expensive method for large scale production. Also, this approach does not necessarily reduce the high reflectivity to the blue region. The present approach aims at reducing the reflectivity of single layer SiN antireflection coating, especially at the blue region of the solar spectrum. Silicon nitride is deposited on textured silicon wafer by Plasma Enhanced Chemical Vapour Deposition (PECVD) method which is further subjected to argon plasma bombardment.

This method of development of silicon nitride nanoparticles using PECVD and subsequent argon plasma treatment is a low temperature, controllable, reproducible and inexpensive one which is also compatible with the present solar cell fabrication technology.

The parameters for alteration of reflection properties by treatment of SiN layer by Argon plasma for application on crystalline silicon solar cell as an enhanced anti-reflection coating are reported in this paper. It is shown that by proper optimization of plasma power density and the time of plasma exposition, the reflectivity of the front surface of c-Si solar cells in the lower wavelength range can be decreased significantly, facilitating the injection of incident photons into the device and increasing absorption resulting into increased short circuit current and enhanced cell efficiency.

2 EXPERIMENTAL DETAILS

2.1 Deposition of Silicon Nitride:

Silicon nitride deposition on c-Si wafers was carried out in cluster PECVD tool powered by 13.56 MHz radiofrequency power supply using 5 sccm SiH_4, 25 sccm NH_3, 25 sccm H_2 at a pressure of 1 Torr and power density of 178 mW/cm^2 for 2 min. The thickness of the SiN coating was ~75-80 nm with a refractive index of 2-2.1. The structural properties of argon plasma treated silicon nitride layers were observed using FESEM from Zeiss and infrared spectroscopy was done using a Perkin Elmer Spectrum two FTIR spectrometer with a bare (no deposition) Si wafer as a reference. Reflectance for the wavelength range 300 nm to 1100 nm was measured using Bentham PVE300. These samples were then, subjected to argon bombardment.

2.2 Argon plasma treatment

The silicon nitride coated wafers were subjected to argon plasma treatment in different conditions. Different parameters like the working pressure, time duration and power density were varied keeping a constant gas flow rate of argon (100 sccm). The pressure was varied from 0.5 Torr to 1 Torr, time was varied from 30 min to 120 min and power density was varied from 250 mW/cm^2 to 1700 mW/cm^2. The IR absorption, reflectance and FESEM images have been observed for each case and have been discussed in the subsequent sections. The plasma treatment parameters which gave minimum reflectance are given in Table 1.

TABLE 1

	Argon (sccm)	Time (sec)	Pressure (Torr)	Power (mW/cm^2)
Sample 1	100	90	1	280
Sample 2	100	60	1	280

3 RESULTS AND DISCUSSIONS

3.1 Structural Characterization

In this section we shall discuss the structural properties exhibited by the argon plasma treated silicon nitride layer.

Fig. 1 shows the FESEM images of uniform distribution of silicon nitride nanoparticles on a textured silicon wafer formed after argon plasma treatment of silicon nitride for different treatment conditions which are given in Table 1.

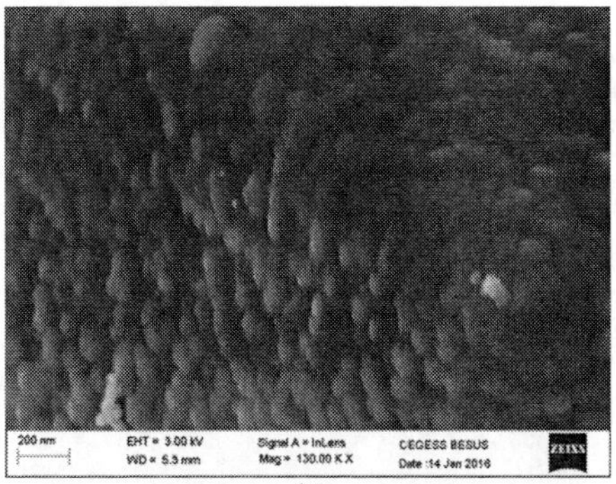

(a)

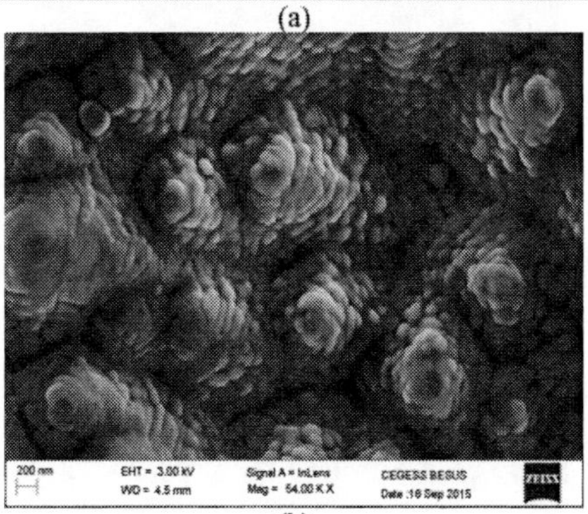

(b)

Fig.1 FESEM micrograph of silicon nitride nanoparticles formed after argon plasma treatment of silicon nitride layer for (a) Sample 1 (b) Sample 2

Fig.2 shows the FTIR spectra for as-deposited SiN and Ar treated SiN. FTIR spectra of untreated SiN show the following bonds:

a. Si-N stretching (825 cm^{-1})
b. Si-H stretching (2140 cm^{-1})
c.N-H stretching (3300 cm^{-1})
d. N-H wagging (1185 cm^{-1}) modes which is typical of silicon nitride layers [6].

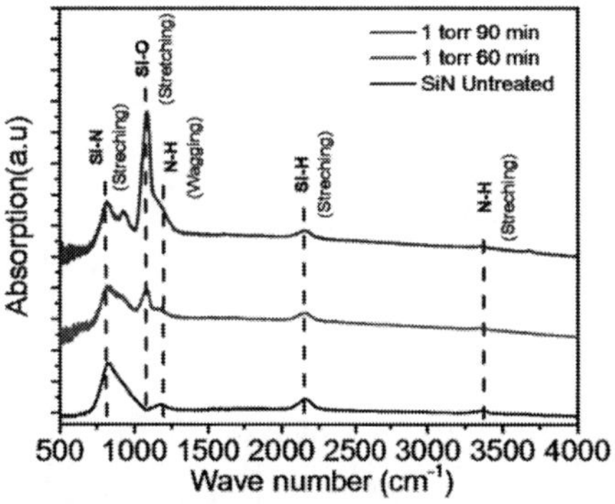

Fig. 2 FTIR spectra for as-deposited SiN and Ar treated SiN

After argon plasma treatment, Si-O stretching bonds appear at 1080 cm^{-1}. Silicon oxynitride (SiO_xN_y) is a tetrahedral amorphous alloy of silicon oxide and silicon nitride. Thus, the presence of Si-O and Si-N bonds indicate the formation of silicon oxynitride(SiO_xN_y) layer [7-9]. Absorption peak maxima and area under the peaks corresponding to Si-O and Si-N bonds depend on the (O/O+N) ratio in the silicon oxynitride layer.

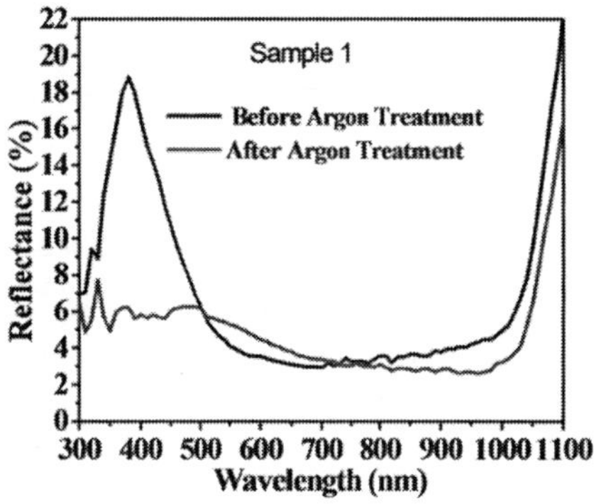

(a)

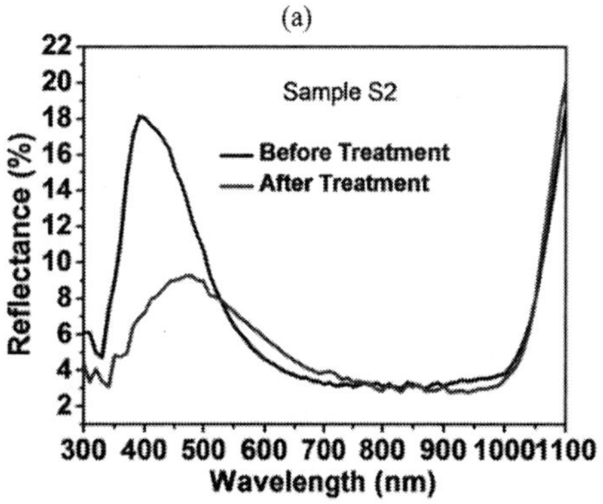

(b)

Fig.3 Reflection of ARC after argon plasma treatment for (a) Sample 1(2) Sample 2

Fig. 3 clearly shows a considerable reduction in reflection in blue region of the spectrum i.e in the wavelength region of 300-500 nm and also in the range of 800-1100nm.

The normalized injected fraction of incident photons per unit area per second before and after Argon plasma treatment has been calculated by the following relation,

$$\text{Injected Photon Fraction} = \frac{\int_{300nm}^{1100nm} N(\lambda)(1-R(\lambda))d\lambda}{\int_{300nm}^{1100nm} N(\lambda)d\lambda}$$

where N(λ) denotes the wavelength dependent photon flux density and R(λ) denotes the wavelength dependent reflectivity of the antireflection coating.

Table 2 shows the increase of photon transmission into the silicon substrate in the presence of nanostructured antireflection coating.

TABLE 2

	Total no. of injected photons ($m^{-2}s^{-2}$)	Fraction of incident photons injected (%)	Total current from injected photons (mA/cm^2)
Before Argon Treatment	$2.51x10^{21}$	94.46	40.11
After Argon Treatment	$2.54x10^{21}$	95.66	40.62

*Total no. of incident photons=$2.61x10^{21}$ $m^{-2}s^{-2}$

It can be observed from the Table 1 that an increase of ~1.5% of the incident photons due to the plasma treatment is achieved leading to a projected efficiency enhancement of the same magnitude.

C-Si solar cells using standard fabrication technique with full aluminium rear contact have been fabricated in the laboratory. The performance of the cells before and after Argon treatment has been investigated and preliminary results show an increase in the short circuit current density and efficiency of the cells by about 0.3%.

4 CONCLUSION

In this paper, antireflection property of nanostructured SiN coating by argon plasma treatment has been demonstrated. Results reveal an improvement in the anti-reflection properties compared to conventional silicon nitride coating. Rigorous experiments have been performed using different parameters of argon plasma treatment, in order to achieve the minimum reflectance and hence high short-circuit current density. FTIR spectra reveal the formation of silicon oxynitride during argon plasma treatment. The reduction in reflection after plasma treatment may be attributed to the formation of a double layer anti-reflection coating with silicon oxynitride layer on the top surface as a result of argon bombardment and silicon nitride as bottom layer.

ACKNOWLEDGEMENT

This work is supported by Department of Science and Technology, Government of India. The authors would also like to thank Prof. Subhendu Guha and Prof.Utpal Gangopadhyay for fruitful discussions.

REFERENCES

[1] Martin F. Schubert, Frank W. Mont, Sameer Chhajed, David J. Poxson, Jong Kyu Kim, and E. Fred Schubert, "Design of multilayer antireflection coatings made from co-sputtered and low-refractive-index materials by genetic algorithm", Opt. Express, 16, 5290 (2008)

[2] Daniel N. Wright, Erik S. Marstein and Awe Holt, "Double Layer Anti-Reflective Coatings for Silicon Solar Cells", Conference Record of the Thirty-first IEEE Photovoltaic Specialists Conference , 1237-1240 (2005)

[3] Zhao Y, Chen F, Shen Q and Zhang L, "Optimal design of light trapping in thin film solar cells enhanced with graded SiNx and SiOxNy structure", Opt. Express, 20, 11121–36 (2012)

[4] Sameer Chhajed, Martin F. Schubert, Jong Kyu Kim, and E. Fred Schubert, "Nanostructured multilayer graded-index antireflection coating for Si solar cells with broadband and omnidirectional characteristics", Applied Physics Letters 93, 251108 (2008)

[5] Sonali Das, Avra Kundu, Hiranmay Saha and Swapan K Datta, "Design of a high efficiency solar cell with lossless nanoentities atop and embedded in silicon substrate", J. Opt. 15, 105006 (2013)

[6] K. C. Lin, S. C. Lee, The structural and optical properties of $aSiN_x$:H prepared by plasma enhanced chemical vapor deposition, Journal of Applied Physics 72(1992) 5474-5482.

[7] F Rebib, E Tomasella, E Bêche, J Cellier, M Jacquet, FTIR and XPS investigations of a-SiO_xN_y thin films structure, Journal of Physics: Conference Series, 100 (2008) 082034

[8] S.H. Mohamed, FTIR and spectroscopic ellipsometry investigations of the electron beam evaporated silicon oxynitride thin films, Physica B, 406 (2011) 211–215

[9] D. Criado, I. Pereyra, M.I. Alayo, Study of nitrogen-rich silicon oxynitride films obtained by PECVD, Materials Characterization, 50 (2003) 167– 171

BLACK SILICON SOLAR CELLS WITH BLACK RIBBONS

Rasmus Schmidt Davidsen*[,a], Peter Torben Tang[b], Io Mizushima[b], Sune Thorsteinsson[c], Peter Behrensdorff Poulsen[c], Jesper Frausig[d], Ørnulf Nordseth[e], Ole Hansen[a]

[a] Department of Micro- and Nanotechnology, Technical University of Denmark, Ørsteds Plads building 345East, DK-2800 Kgs. Lyngby, Denmark,
[b] IPU, Nils Koppels Allé, Building 404, 2nd floor, DK-2800 Kgs. Lyngby, Denmark
[c] Department of Photonics, Technical University of Denmark, Frederiksborgvej 399 building 128, DK-4000 Roskilde, Denmark
[d] Gaia Solar A/S, Hammerholmen 9, DK-2650 Hvidovre, Denmark
[e] Institute for Energy Technology (IFE), Instituttveien 18, NO-2007 Kjeller, Norway
*rasda@nanotech.dtu.dk, phone: +45 45255848

ABSTRACT: We present the combination of mask-less reactive ion etch (RIE) texturing and blackened interconnecting ribbons as a method for obtaining all-black solar panels, while using conventional, front-contacted solar cells. Black silicon made by mask-less reactive ion etching has total, average reflectance below 0.5% across a 156x156 mm^2 silicon (Si) wafer. Black interconnecting ribbons were realized by oxidizing copper resulting in reflectance below 3% in the visible wavelength range. Screen-printed Si solar cells were realized on 156x156 mm^2 black Si substrates with resulting efficiencies in the range 15.7-16.3%. The KOH-textured reference cell had an efficiency of 17.9%. The combination of black Si and black interconnecting ribbons may result in aesthetic, all-black panels based on conventional, front-contacted silicon solar cells.
Keywords: Black Silicon, Nanostructuring, BIPV, Aesthetics, Inorganic Black Ribbons

1 INTRODUCTION

Nanoscale texturing of Si surfaces has been shown [1,2] to reduce the total optical reflectance to well below 1% averaged over a broad range of wavelengths. Compared to typical reflectance of ~2 and ~8% from conventionally textured mono- [3] and multi-crystalline [4] Si solar cells, respectively, nanoscale texturing may potentially improve the power conversion efficiency of Si solar cells due to reduced reflectance loss. Nanostructured Si surfaces suppress the reflectance of light from the surface due to the resulting graded refractive index at the Si-air interface. We use mask-less reactive ion etching (RIE) to produce nanostructured Si, also known as 'black silicon' [5-7], as texturing for front-contacted Si solar cells. In addition, there is commercial interest in obtaining aesthetic and building-integrated solar cells and panels. We propose to achieve this by combining black silicon texturing with blackened interconnection ribbons strings in order to obtain an all-black panel, while using conventional, front-contacted Si cells.

2 BLACK SILICON SOLAR CELLS

2.1 Black Silicon

Black silicon was realized on 156x156 mm^2 Czochralski (CZ) mono-crystalline Si wafers by mask-less reactive ion etching (RIE) in SF_6 and O_2 plasma at room temperature. Figure 1 shows a scanning electron microscope (SEM) image of the resulting nanostructure topology. The nanostructure topology consists of conical-like hillocks with average height of 300 nm and average spacing of 300 nm.

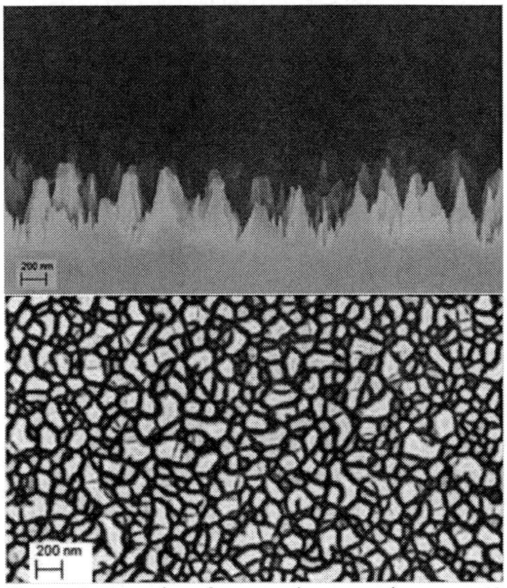

Figure 1: Scanning electron microscope image at 40 degrees (left) and 0 degrees tilt (right) of the nanostructure topology resulting from mask-less reactive ion etching.

Normal incidence reflectance measurements of the RIE-textured Si surfaces were performed using a broadband light source (Mikropack DH-2000), an integrating sphere (Mikropack ISP-30-6-R), and a spectrometer (Ocean Optics QE65000, 300-1000 nm).

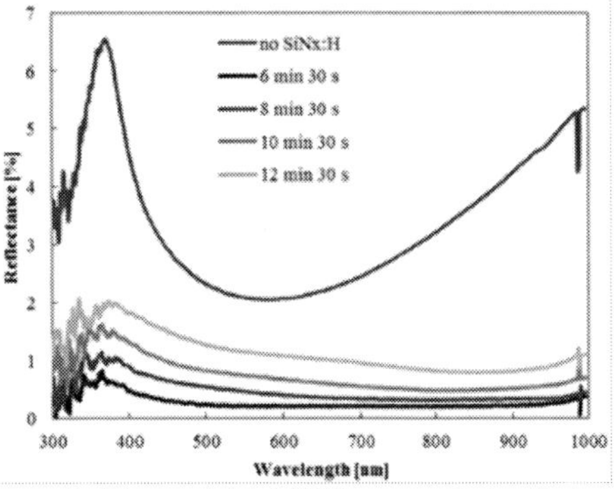

Figure 2: Measured total (diffuse+specular) reflectance of black Si with SiN_x:H AR-coating as a function of wavelength in the range 300-1000 nm for four different PECVD SiN_x:H deposition times representing four different SiN_x:H thicknesses.

For the reflectance measurements hydrogenated silicon nitride (SiN_x:H) deposited by plasma-enhanced chemical vapor deposition (PECVD) was applied. An n^+ emitter was formed on the RIE-textured Si using $POCl_3$ as dopant source prior to the deposition of SiN_x:H. Since the refractive index profile of black Si is fundamentally different from that of conventionally textured or planar Si, the optimal SiN_x:H thickness for black Si was investigated. The deposition time of the PECVD process was varied in steps of 2 minutes from the 'standard' time of 6 min and 30 s to 12 min and 30 s. The reflectance was then measured for the different SiN_x:H thicknesses on RIE-textured Si wafers. The result is shown in Figure 2 which shows that a PECVD process time of 6 min and 30 s minimizes the reflectance of black Si. In general SiN_x:H AR-coating reduces the reflectance of black Si. Deposition times are stated instead of SiN_x:H thicknesses, since the exact thickness on black Si is not fully known due to the complicated surface morphology. On conventionally textured Si a 6 min and 30 s deposition yields a SiN_x:H thickness of ~ 60 nm. Due to the broadband anti-reflective properties of black Si the SiN_x:H thickness does not need to be optimized in terms of minimized reflectance at a certain wavelength. It seems from Figure 2 that the reflectance increases monotonically with SiN_x:H thickness in the thickness range investigated.

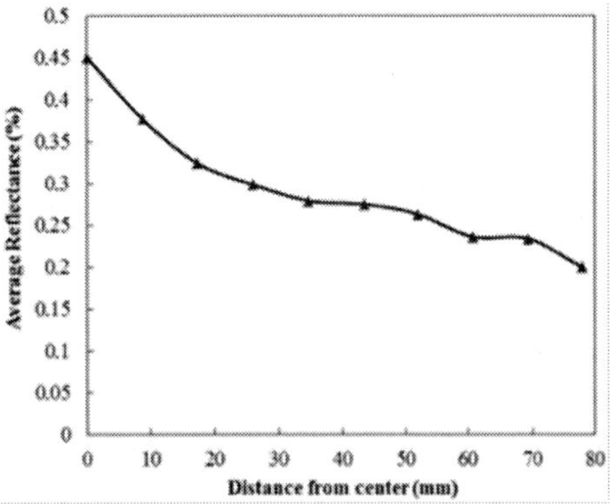

Figure 3: Measured total (diffuse + specular) reflectance of RIE-textured Si with SiN_x:H averaged over the wavelength range 300-1000 nm as function of the distance from the center of a 156x156 mm^2 CZ wafer.

In order to quantify the spatial uniformity of RIE-texturing, the reflectance was measured at different positions across a 156x156 mm^2 CZ wafer textured by mask-less RIE after $POCl_3$ emitter diffusion and PECVD of SiN_x:H (6 min and 30 s). The total (diffuse + specular) reflectance was measured in the wavelength range 300-1000 nm and the integrated average reflectance is plotted as function of distance from the wafer center in Figure 3. Figure 3 shows that the integrated average reflectance is below 0.5% across the entire 156x156 mm^2 wafer. The average reflectance decreases towards the edge of the wafer, but even though the relative difference from center to edge is significant, the absolute difference is ~ 0.2%. The deviation in reflectance across the wafer is probably due to the plasma chamber and wafer geometry.

2.2 Cell processing

Solar cells were fabricated from 156x156 mm^2 p-type CZ Si wafers using the following process steps:

1. Saw damage removal using an $HF:HNO_3:CH_3COOH$ etchant mixture
2. Maskless RIE texturing
3. Diffusion of n-type emitter in a tube furnace using a $POCl_3$ source to obtain a sheet resistance of ~120 Ω/sq.
4. Deposition of anti-reflection coating on the front surface by plasma enhanced chemical vapour deposition (PECVD)
5. Screen-printing and firing of metal contacts
6. Laser edge isolation

The formation of an n-type emitter, using $POCl_3$ as the phosphorous doping source, was carried out using a Tempress TS-8603 tube furnace at a temperature of 836-838 °C for 37 minutes, followed by a 10 minutes drive-in. Hydrogenated amorphous silicon nitride (a-SiN_x:H) was deposited using an Oxford Instruments PlasmaLab System133 PECVD tool with silane, ammonia and nitrogen at a temperature of 400 °C and a pressure of 800 mTorr for 6 minutes and 30 seconds. For the contact formation, an EKRA X5-STS semi-automatic screen printing machine was used. The contacts were printed using Ag paste (DuPont PV18A) on the front and Al paste (Monocrystal PASE-1207) on the rear. The contacted cells were subsequently dried and co-fired in a belt furnace with temperatures in the range of 780-945°C and a belt speed of 520 cm/min. Edge isolation was performed by laser ablation using a J-1030-515-343 FS System from Oxford Lasers. A wavelength of 515 nm and a repetition rate of 50 kHz were used for the laser scribing.

2.3 Characterization methods

Emitter sheet resistance was measured using a CMT-SR2000-PV unit from Advanced Instrument Technology. I-V curves were measured using a solar simulator (WACOM) to illuminate the solar cell (100 mW/cm^2) with the AM1.5 spectrum while varying the bias voltage from -1V to +1V. Electro- and photo-luminescence (EL, PL) images were measured using a LIS-R1 unit from BTImaging. The reported lifetime values are average lifetimes across the wafers at an injection level of about 10^{15} cm^{-3}.

Internal quantum efficiency (IQE) measurements using an IQE-SCAN tool from pvtools.

2.4 Results

A total of 24 full-area (156x156 mm^2) black Si solar cells were fabricated and compared with known data for KOH-textured cells. Due to reduced adhesion of screen-printed metal fingers at the edges of some of the black Si cells, the final cells were cut in 100x100 mm^2 squares in the edge isolation. The adhesion issues on RIE-textured Si will be investigated further in future studies.

Figure 4 shows the current density-voltage characteristic for 3 different black Si solar cells and the KOH-textured reference cell and quantum efficiency for one black Si cell compared to the reference cell.

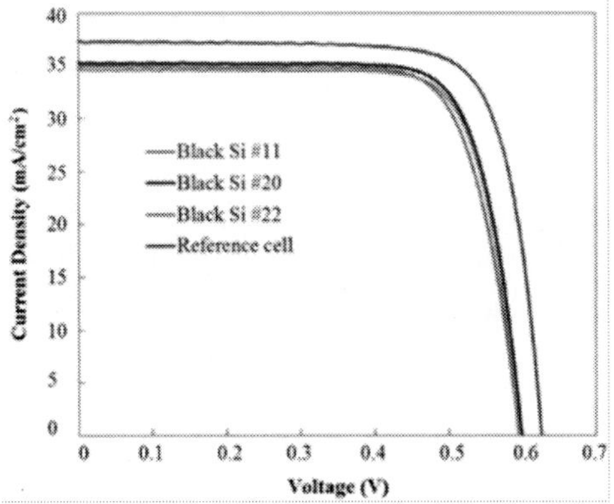

Figure 4: Current-density voltage (J-V) characteristic for 3 different black Si cells and a KOH-textured reference cell.

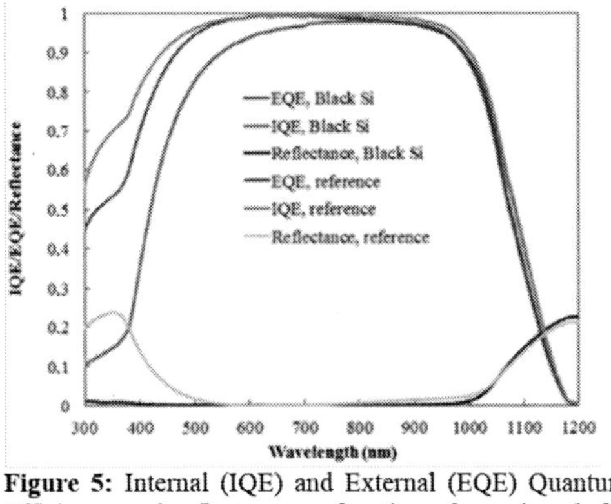

Figure 5: Internal (IQE) and External (EQE) Quantum Efficiency and reflectance as function of wavelength for the black Si cell (#11) and the KOH-textured reference cell.

Table 1: Short-circuit current (J_{sc}), open-circuit voltage (V_{oc}), fill factor (FF), power conversion efficiency and series resistance (R_s) of 3 different black Si solar cells and a KOH-textured reference. The series resistance was extracted from EL and PL images.

Cell	J_{sc} (mA/cm^2)	V_{oc} (V)	FF (%)	Eff. (%)	R_s (Ωcm^2)
BS #11	34.93	595	75.67	15.73	1.07
BS #20	35.25	598	76.91	16.22	1.03
BS #22	34.62	596	77.32	15.95	1.04
Ref cell	37.31	625	77.09	17.99	0.88

Figure 4 and Table 1 show that the black Si cells have power conversion efficiencies in the range of 15.7-16.3%. The 3 cells shown in Figure 4 and Table 1 represent the best, worst and mean of the 24 fabricated cells. The efficiency of the black Si cells is lower than that of the KOH-textured reference cell. Table 1 show that the lower efficiency is primarily due to lower J_{sc} and V_{oc}. Furthermore, the IQE measurement in Figure 5 shows that the lower efficiency of black Si cells is primarily due to increased carrier loss at wavelengths below 600 nm probably due to increased surface and emitter recombination. This was expected since previous studies [8] show similar results and since no further attempt of optimization was made in terms of improved passivation or emitter quality.

3 BLACK RIBBONS

In order to obtain an all-black appearance of cells and panels, black Si may be combined with black interconnecting ribbons. The cell results reported in this work did not involve these strings. Three different processes for blackening the ribbons were investigated. NiZnS and NiCuCo alloys, respectively, were deposited directly on Cu ribbons with soldering coating. Cu ribbons with a black oxidized surface were realized by a chemical oxidation process. Electroplating was conducted in a beaker with electrolytes containing sulfate salts and a nickel anode. Reflection measurements on the bare and coated surfaces were performed using a spectrophotometer (Shimadzu 2600).

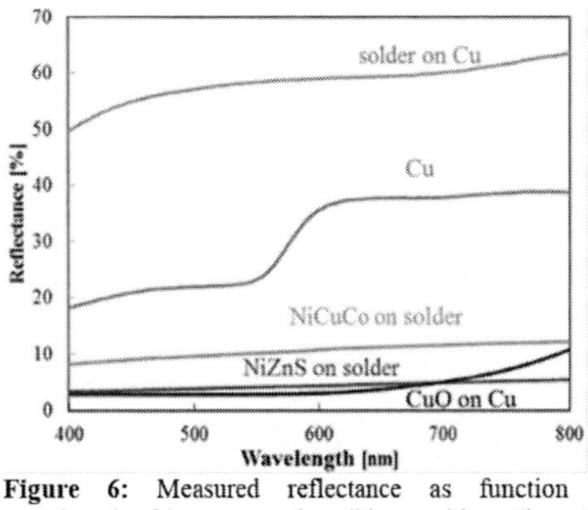

Figure 6: Measured reflectance as function of wavelength of interconnecting ribbons without (bare Cu, and Cu with solder) and with (NiCuCo, NiZnS, and CuO) black coatings.

Figure 6 shows the measured reflectance of the different ribbons. Reflectance of the strings with solder is high due to the bright metal surface, while the copper surface has a lower reflectance which increases for wavelengths above 600 nm. All of the blackened ribbons have significantly reduced reflectance. NiCuCo appears gray and has a higher reflectance compared with the other blackened ribbons, also, the reflectance of NiCuCo increases slightly towards longer wavelengths. The reflectance curve for NiZnS is quite flat and has a dark grey appearance. Oxidized Cu has the lowest reflectance, which increases slightly at longer wavelengths; indicating that the surface color has a red element due to Cu presence in the coating or due to the Cu substrate. In summary, Cu ribbons strings with CuO is the best match for black silicon with respect to the all-black appearance, because of the comparatively low reflectance.

4 CONCLUSIONS

We have presented the combination of black silicon texturing realized by mask-less RIE and blackened interconnecting ribbons as a potential method for obtaining all-black solar panels, while using conventional, front-contacted silicon solar cells. Black silicon has a total, average reflectance below 0.5% in the wavelength range 300-1000 nm across a 156x156 mm^2 silicon wafer. The absolute difference in reflectance from center to edge of the 156x156 mm^2 silicon wafer was ~ 0.2%. Black ribbons were realized by oxidized Cu resulting in reflectance below 3% in the visible wavelength range. Screen-printed Si solar cells were realized on 156x156 mm^2 black Si substrates with resulting efficiencies in the range 15.7-16.3%. The KOH-textured reference cell had an efficiency of 17.9%. The combination of black Si and black ribbons may result in aesthetic, all-black panels based on conventional, front-contacted silicon solar cells.

5 REFERENCES

[1] Priolo, F., Gregorkiewicz, T., Galli, M., Krauss, T.F., "Silicon nanostructures for photonics and photovoltaics", Nature Nanotechnology, vol. 9, pp. 19-32, (2014). DOI: 10.1038/NNANO.2013.271.

[2] Zhu, J., Yu, Z., Fan, S., Cui, Y., "Nanostructured photon management for high performance solar cells", Materials Science and Engineering R, vol. 70, pp. 330–340,(2010).

[3] Zhao, L., Zuo, Y.H., Zhou, C.L., Li, H.L., Diao, H.W., Wang, W.J., "Theoretical investigation on the absorption enhancement of the crystalline silicon solar cells by pyramid texture coated with SiNx:H layer", Solar Energy, vol. 85, pp. 530-537, (2011).

[4] Macdonald, D., Cuevas, A., Kerr, M., Samundsett, C., Ruby, D., Winderbaum, S., Leo, A., "Texturing Industrial Multicrystalline Silicon Solar Cells", Solar Energy, vol. 76, No.1, pp. 277-283, (2004).

[5] Rahman, A., Ashraf, A., Xin, H., Tong, X., Sutter, P., Eisaman, M.D., Black, C.T., "Sub-50-nm self-assembled nanotextures for enhanced broadband antireflection in silicon solar cells", Nature Communications, vol. 6, pp. 59-63, (2015). DOI: 10.1038/ncomms6963

[6] Liu, X., Coxon, P.R., Peters, M., Hoex, B., Cole, J.M., Fray, D.J., "Black Silicon: Fabrication methods, properties, and solar energy applications", Review Article, Energy & Environmental Science, Royal Society of Chemistry, vol. 7, pp. 3223-3263, (2014).

[7] Allen, T., Bullock, J., Cuevas, A., Baker-Finch, S., Karouta, F. "Reactive ion etched black silicon texturing: A comparative study", In 40th Photovoltaic Specialist Conference (PVSC), IEEE, pp. 0562-0566, IEEE, (2014). doi:10.1109/PVSC.2014.6924983

[8] Davidsen, R. S., Ormstrup, J., Ommen, M. L., Larsen, P. E., Schmidt, M. S., Boisen, A., Nordseth, Ø., Hansen, O., "Angle resolved characterization of nanostructured and conventionally textured silicon solar cells.", Solar Energy Materials and Solar Cells, 140, 134-140, (2015).

BIFACIAL P-TYPE SOLAR CELLS WITH LOW TEMPERATURE COEFFICIENTS: HETEROJUNCTION TECHNOLOGY

D. L. Bätzner, L. Andreetta, W. Frammelsberger, R. Kramer, D. Lachenal, B. Legradic,
J. Meixenberger, P. Papet, B. Strahm, G. Wahli, D. Habermann[2], S. Frigge[3], H.-P. Sperlich[3]
Meyer Burger Research AG, Rouges-Terres 61, CH-2068 Hauterive, Switzerland, Phone +41 32 566 15 23,
derk.baetzner@meyerburger.com
[2] Meyer Burger Technology, Schorenstrasse 39, CH-3645 Gwatt (Thun), Switzerland
[3] Meyer Burger (Germany) AG, An der Baumschule 6-8, D-09887 Hohenstein-Ernstthal, Germany

ABSTRACT: The large flexibility of Meyer Burger's heterojunction technology (HJT) is demonstrated by processing bifacial HJT devices not only on industrial 156mm n-type Cz wafers but more importantly on industrial 156mm p-type Cz wafers as well. The bifacial p-type HJT devices were processed into front and rear emitter devices. In case of rear emitter p-type HJT cells the contacting TCO on the rear side was varied using standard ITO as well as ZnO:Al.

A new record efficiency of 20.7% for large area p-type HJT cells was obtained as well as an impressive record Voc of over 729mV. This is to our knowledge the highest ever reported absolute Voc for a heterojunction cell on p-type substrate and likely the highest Voc for any cell technology on commercial large area (239cm2) p-type Cz wafers. This underlines the excellent surface passivation of Meyer Burger's HJT. Moreover, all devices exhibited a bifaciality of over 90% and an excellent temperature coefficient for power of -0.22%/K was determined in the temperature range of 10°C to 60°C. The p-HJT cells were rather thin with cell thicknesses ranging from 110µm to 150µm. All together this makes HJT especially from a cost point of view rather attractive giving n- and p-type HJT an advantage over PERC technology.

Keywords: Heterojunction, bifacial, p-type, temperature coefficient, solar cell production, record device

1 INTRODUCTION

Silicon heterojunction technology has been figuring prominently in research in the past years and is gathering appreciable momentum transiting from R&D to pilot production and finally to mass production state with Meyer Burger's HJT. While record efficiencies capture much attention there are other aspects of HJT of interest for production one of these being versatility of the technology. In this regards the versatility of HJT is demonstrated with a study focusing on the HJT devices processed on p-type commercial wafers as typically used in PERC production.

2 HETEROJUNCTION CELL PROCESSING

Cells were processed on p-type and as reference on a few n-type wafers. Thereby a set of standard layers of passivating and junction forming a-Si layers as well as TCO layers developed for Meyer Burger's high performance HJT cells were deposited to the different substrates. Thus devices with different emitter configuration, i.e. front side emitter (FE) and rear side emitter (RE), could be produced with great flexibility using standard recipes for passivating and contacting layers out of a model kit.

In a first experiment p-type and n-type wafers have received the same wet chemistry process and deposition of a-Si and TCO layers resulting in p-type HJT FE cells and n-type HJT RE cells with more conductive and therefore more absorbing front TCO on all cells. The cells received our standard metallisation of a front side finger print without bus bars and finer spaced rear side finger print resulting in full bifacial device structure that is adapted for integration with Meyer Burger's SWCT module technology.

The second part of the study focused on HJT devices purely on p-type wafers with FE and RE configuration and a variation of rear side TCOs namely the standard ITO layer was replaced by a lately developed TCO layer made from ZnO:Al (AZO). The TCOs were selected to demonstrate a high level of bifaciality as would be the case when integrated into glass/glass modules.

All processes were run on mass production tools using recipes developed at Meyer Burger Research in Neuchâtel that are essentially the same as used in our pilot production at Meyer Burger Germany [1].

3 P-TYPE FE vs. N-TYPE RE HJT CELLS

Two different p-type substrates and one n-type substrate with areas of 239cm^2 and resistivites of 1Ω*cm to 1.5Ω*cm were used. All devices resulted in rather thin cells with thicknesses ranging from 110µm to 150µm.

Life time measurements on cell precursors after passivation indicated the expected and previously observed strong bulk limitation on the p-type wafer due to life time tailing towards low injection [2]. This becomes evident analysing the implied FFs (iFF) for the passivated cell precursors of the champion devices. While the n-type precursor has an iFF of 85.4% at an injection density of $2*10^{15}/cm^3$ the p-type precursor exhibits a very low iFF of 79.4% at an injection density of below $9*10^{13}/cm^3$. The surface passivation quality derived from an extraction of the total J0s on both substrates is excellent and comparable, i.e. 5.2fA/cm^2 in case of the n-type wafer when extracted in the linear part of the Auger-corrected inverse life time plot at an injection density of $7*10^{15}/cm^3$ and 6.2fA/cm^2 in case of the p-type wafer when extracted at an injection density of $1.5*10^{16}/cm^3$ respectively, whereby the empirical, commonly applied parameterisation of the Auger recombination supplied with the analysis software of Sinton Instruments was used.

Therefore, it comes to no surprise that the HJT champion devices reflect above observed substrate influences in their respective IV parameters given in Table 1. It certainly is noteworthy that a new record of 20.7% efficiency for a heterojunction cell on Cz p-type industrial size wafer was obtained. Since this is to our knowledge the highest ever reported efficiency of a heterojunction cell on a p-type wafer of comparable size and properties it is very likely a world record. At the same time it is not far off the highest reported efficiency of any heterojunction cell on p-type substrate which was 21.4% on a FZ wafer for a cell size of 4cm^2 [3].

Table 1: IV-parameters of the champion devices

	Eta [%]	FF [%]	Jsc [mA/cm2]	Voc [mV]
p	20.7	75.7	37.5	728
n	22.6	81.9	37.3	740

The device quality of the p-type HJT cells shows in the excellent VOCs and other device parameters as displayed in Figure 1 for the individual wafer groups. The highest Voc of a cell from wafer group 'p-type A' was measured at 729.3mV which is – again to the best of our knowledge – the highest ever reported Voc for a heterojunction cell on a p-type wafer and might quite likely also be the highest ever reported Voc of any solar cell on an industrial size Cz p-type wafer.

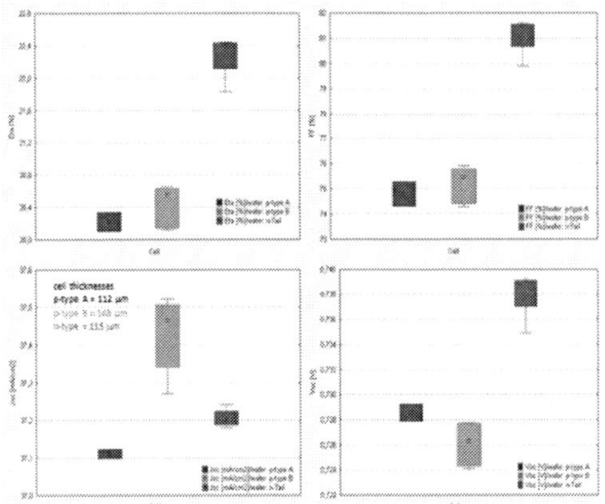

Figure 1: IV parameters of two groups of FE HJT cells on p-type wafers and a group of reference RE HJT cells on n-type wafers

The major limitation of the cell performance for HJT cells on p-type wafers is the low FF that is caused by the loss of life time in low injection. The low life time at low injection could be due stronger recombination apparent in Cz p-type wafers as compared to Cz n-type wafers but might also stem from the large asymmetry of capture the cross sections for electrons and holes at the interface of a-Si and c-Si as was also observed on p-type FZ wafers [3]. In this case it would be a surface effect that is more specific to the kind of interface passivation used.

In contrast to other cell technologies HJT will allow reducing the cell thickness without impacting cell performance negatively. This can be seen for the group of the bifacial 'p-type A' cells with thickness of 112µm that perform practically as good as the thicker 'p-type B' cell group.

4 FE vs. RE HJT CELLS ON P-TYPE WAFERS; BIFACIALITY

Adding further to the flexibility of Meyer Burger's HJT is the straight forward processing into FE or RE cell. This was investigated with the group of thin p-type wafers. The TCO of the cells in RE configuration was varied between a more standard ITO and a less common AZO the latter promising a cost advantage. The rear side TCOs were optimized to serve as an effective ARC in a bifacial glass/glass module.

Figure 2 shows the IV characteristics of p-type HJT cells groups in different configurations. All cells were measured with Meyer Burger's GridTouch contacting unit for contacting of bus bar less cells with a front side illumination (blue boxes) and a rear side illumination (orange boxes). All cells were measured against a 'black' contacting base which has as little as possible reflection. The cells' Jscs reveal a general difference between FE and RE configuration which is explained by the inferior rear emitter carrier collection due to the low diffusion length (low life time) in the p-type wafers. Likewise this is also evident when illuminating from the rear side where the 'FE/ITO' combination yields lower Jsc thatn the otherwise similar 'RE/ITO' combination that becomes a front emitter cell in the instant for rear side illumination. The RE cells with AZO on the rear side show the lowest Jsc for rear side illumination which is connected to the less favorable optical properties of AZO compared to ITO. Nevertheless, contacting the rear emitter side with AZO results in better carrier transport seen in a superior FF for the AZO contacted devices. Since the Voc is fairly similar for all configurations – omitting a couple of low flyers – it is the improved contacting with AZO that results in the best performing p-type HJT cells with a RE configuration and an AZO rear contact.

The bifaciality of these sets of HJT cell was rather high ranging from 92% to 96%. If desired a 100% bifaciality would be feasible with an adapted metallization design.

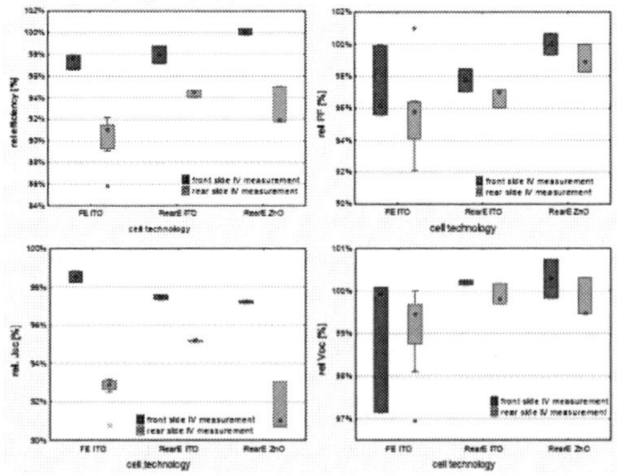

Figure 2: IV parameters of thin p-type HJT cells with FE and RE configuration and a variation of the rear side TCO

5 TEMPERATURE COEFFICIENTS

Another advantage of HJT cells are their low temperature coefficients (TCs) that are linked to the excellent surface passivation. The p-type HJT cells revealed the same favorable temperature behavior.

Table 2 shows the measured temperature coefficients on selected cells from the test groups.

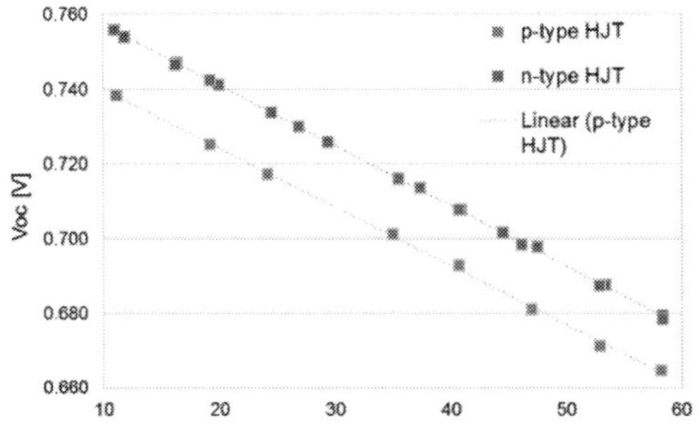

Figure 3 exhibits the measured Voc data points over the measurement temperature.

Table 2: IV-parameters of the champion devices

	TC-eta [%/K]	TC-Jsc [%/K]	TC-Voc [%/K]
p	-0.22	+0.03	-0.21
n	-0.23	+0.03	-0.21

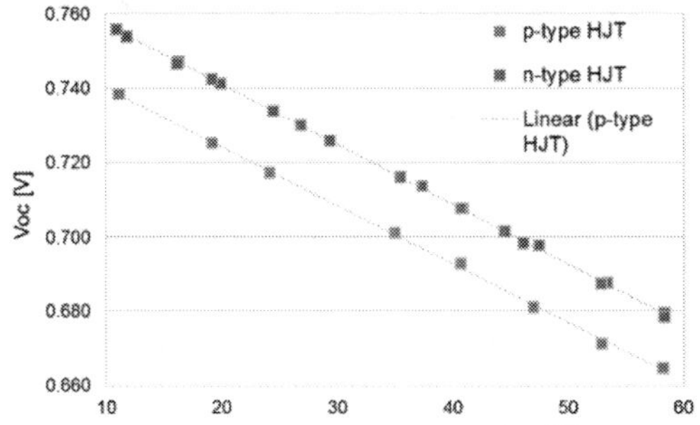

Figure 3: Temperature behavior of HJT cells on n-type and p-type wafers

The temperature behavior of the cell parameters was following a linear trend as shown in figure 3 for the Vocs of an n- and p-type HJT cell. Very similar TCs were determined for the HJT cells on n-type and on p-type wafers. The results on p-type wafer based cells might need further confirmation, but indicate that they have the desired good temperature behavior that gives an advantage in real world application and impacts LCOE calculations beneficially.

6 COST CONSIDERATIONS FOR P- AND N-TYPE HETEROJUNCTON TECHNOLOGY

With the feasibility of a 20+% p-type HJ technology demonstrated with application of standard mass production ready processes the cost structure of the cell process on Meyer Burger's HJT line are cost competitive also when using p-type CZ wafers. This indicates that the equipment can be operated with outstanding flexibility.

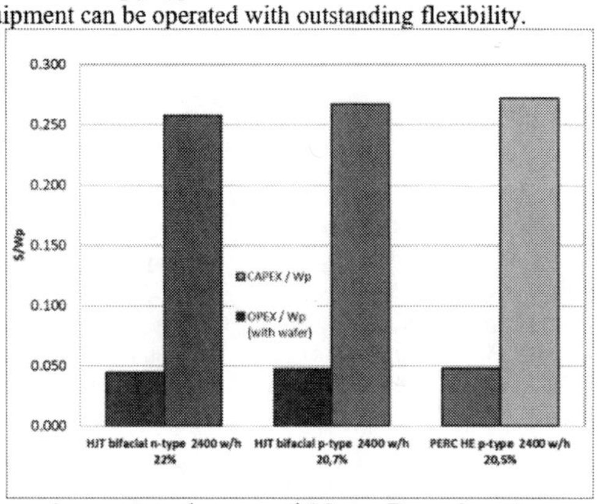

Figure 4 compares the cost of Meyer Burger's p-type HJT and n-type HJT with a typical PERC technology separated in CAPEX and OPEX. It can be seen that already the p-type HJT shows cost advantages over PERC mainly in the OPEX, however, the CAPEX beginning to be lower than for PERC. The full cost advantage of HJT will naturally be harvested using highest n-type HJT. The cost calculations include wafer prices based on typical market prices for Q2/16 and a typical depreciation period for the CAPEX.

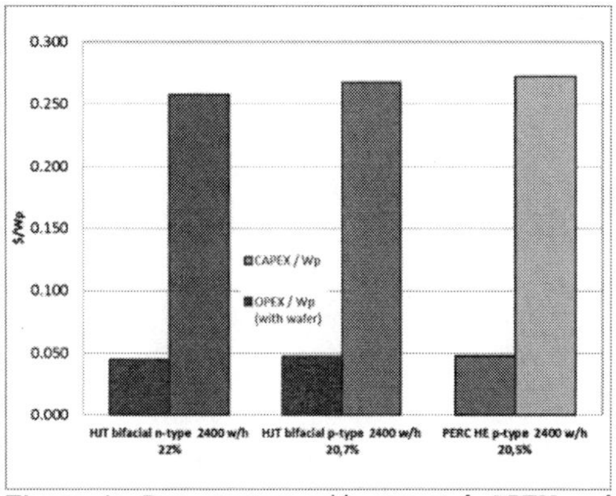

Figure 4: Cost structure with separated OPEX and CAPEX comparing a HJT line with a PERC technology line.

7 DISCUSSION AND CONCLUSIONS

The extended study of HJT cell performance on commercial 6-inch p-type CZ wafers yielded p-type HJT devices above 20.5% efficiency with Vocs just shy of 730mV, both to our knowledge record values for their category or even beyond. The flexibility in processing HJT devices with altering configuration on different type substrates into thin devices with high bifaciality and excellent temperature coefficients is unique and yielded in outstanding results in a rather sort time span. HJT devices on p-type wafers have not yet been optimized specifically, which might results in significant increases in the currently limited FFs. Thus, owing device performance which is on par or higher than established p-type PERC technology makes p-type HJT cost competitive at its current status with future upward potential.

6 REFERENCES

[1] J. Zhao et al.;"Silicon Heterojunction Solar Cells in Meyer Burger's Demo Line: Results of Pilot Production on Mass Production Tools", presented at the 32nd EU PVSEC, Munich, 2BO.9.1
[2] D.L. Bätzner, et al.; "Heterojuntion Solar Cells on p-Type Mono-Si Wafers and the Flexibility of HJT Processing", Proceedings 31st EU-PVSEC, Hamburg, 2015, 788 - 790
[3] A. Descoeudres, et al.; „>21% Efficient Silicon Heterojunction Solar Cells on n- and p-Type Wafers Compared", IEEE Journal of Photovoltaics (ISSN: 2156-3403), vol. 3, num. 1, p. 83-89

HIT CELL WITH P$^+$ EPI/POLY-SILICON INTENTIONALLY DOPED EMITTER IN CRYSTALLINE SILICON SUBSTRATE

Moustafa Ghannam, Yaser Abdulraheem, Ali Hajjiah, Jef Poortmans[@]
EE department, Kuwait University, P.O. Box 5969, 13060, Safat, Kuwait
moustafa.ghannam@ku.edu.kw
[@] IMEC, kapledreef 75, 3001, Leuven, Belgium

ABSTRACT: A modification is proposed to the standard HIT cell structure in which the interfacial inversion layer at the a-Si:H/c-Si interface is prone to collapse. It consists of substituting the p$^+$ inversion layer emitter by a very shallow p$^+$ doped layer at the surface of the c-Si prior to a-Si:H deposition. This layer could be created by p$^+$ epitaxial growth, p$^+$ recrystallized polysilicon, or simply by front p$^+$ thermal diffusion. The aim is to maintain a permanent p$^+$ emitter irrespective of the conditions of a-Si:H and cell voltage, and hence to avoid the problems encountered with the collapse of the inversion layer associated to the a-Si:H quality, TCO and operating conditions. Simulations are carried out using Silvaco tools and AFORS-HET device simulators to determine and compare the I-V characteristics of standard HIT cells suffering from inversion layer collapse, and those of similar structures implementing the proposed modification, with the characteristics of a standard well performing reference HIT cell. Restitution of the normal I-V characteristics in the dark and under illumination after applying the proposed modification is confirmed.
Keywords: HIT cell – inversion layer emitter – a-Si/c-Si hetero-interface - TCO work function

1 INTRODUCTION

It has been found that the operation of amorphous silicon/crystalline silicon heterojunction, a-Si:H(p$^+$/i)/c-Si(n) (HIT) cells strongly depends on the presence of an induced inverted p$^+$ layer heavily populated with holes at the a-Si:H/c-Si hetero-interface, in the n-type c-Si substrate side [1]. Such an inversion layer, acting as the cell emitter, results from energy band-line up in the HIT structure at equilibrium [2] and its presence has been confirmed experimentally [3-6]. For proper operation, the inversion layer should be maintained throughout the whole voltage range of operation. However, this is not always ensured since the hole population in the inversion layer and its peak concentration strongly depend on the separation between the Fermi energy E_F and the valence band edge E_V in the a-Si:H(p) layer as depicted in Figure 1. The separation E_F-E_V increases and the free hole concentration decreases when the chemical doping concentration in the bulk of a-Si:H decreases and/or when the dangling bond defect density in the a-Si:H bulk increases. In addition, if the work function of the transparent conducting oxide (TCO), used as the front contact, is significantly smaller than the work function of a-Si:H(p) the front contact exhibits a Schottky barrier

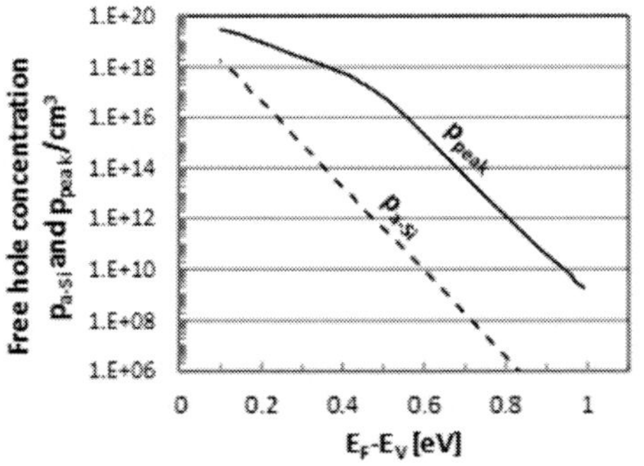

Figure 1: Free hole concentration in the bulk of a-Si:H and peak hole concentration at the a-Si/c-Si interface as a function of the separation E_F-E_V in a-Si:H.

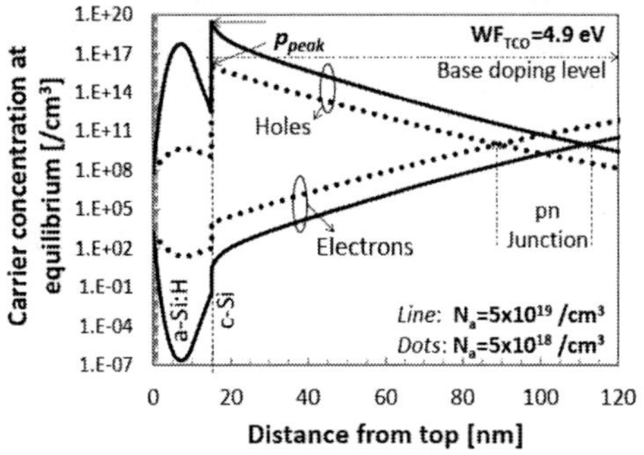

Figure 2: Free carrier profiles at equilibrium in the top 120 nm of the standard HIT structure with WF_{TCO}=4.9 eV and two different chemical doping concentrations in a-Si:H(p): N_a=5x10^{18} /cm^3 and N_a=5x10^{19}/cm^3 [7].

behavior rather than an ohmic (flat band) condition. In such a case, the interaction between the electric fields at the front contact and at the hetero-interface may cause the Fermi energy to significantly shift away from the valence band [7].

Free carrier profiles at equilibrium are displayed in Figure 2 for HIT cells with two different p-type chemical doping concentrations in a-Si:H(p): N_a=5x10^{18} /cm^3 and N_a=5x0^{19} /cm^3 and with a TCO work function WF_{TCO} of 4.9 eV. Although an inversion layer is clearly identified in Figure 2 at equilibrum, the peak hole concentration in the cell with N_a=5x10^{18} /cm^3 is in the 10^{15}-10^{16} /cm^3 range, which is apparently not high enough to maintain surface inversion. When the cell becomes forward biased at a cell voltage of 0.5 V, the inversion layer collapses as confirmed in Figure 3. It has been shown [7] that at a critical forward voltage V_{crit}, the inversion layer peak hole concentration drops sharply causing collapse of the p-type interfacial inversion layer, as depicted in Figure 4. The critical voltage for such a collapse appears to be a function of the dangling bond defect density at the hetero-interface which associates the resulting shift in the Fermi energy to the recombination at the hetero-interface or equivalently to an effective recombination velocity at

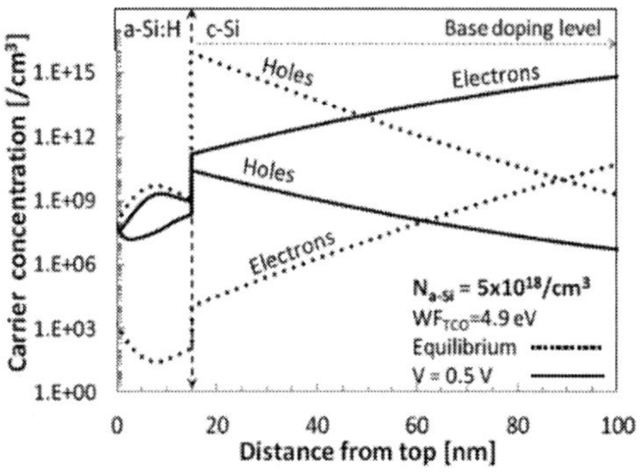

Figure 3: Free carrier profiles at equilibrium and at V = 0.5 V for a standard HIT structure with N_a=5x10^{18} /cm^3 in the a-Si:H(p) layer [7].

that interface. The collapse of the inversion layer leads to the complete failure of the HIT cell operation as the device reduces to a simple n-type resistor. Weak inversion or inversion layer collapse is probably the main reason that explains the difficulty encountered in systematically fabricating HIT cells with high efficiency close to the record efficiency of 25% achieved so far [8].

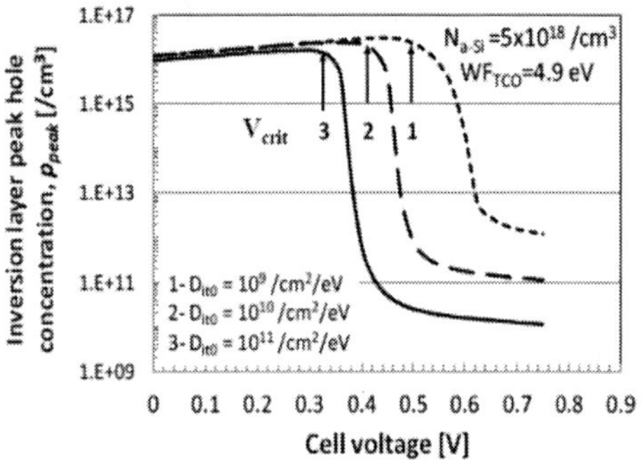

Figure 4: Peak hole concentration in the inversion layer as a function of the forward cell voltage [7].

In this work a modification to the standard HIT structure is proposed in order to avoid total device failure due to inversion layer collapse. It consists of substituting the induced p$^+$ inversion layer emitter by a shallow permanent p$^+$ doped layer. By this way the p$^+$ emitter is maintaind at equilibrium and throughout the whole voltage range of operation. Silvaco Atlas simulations are carried out to determine the I-V characteristics in the dark and under illumination of the modified structure and compare them with those of a similar standard cell suffering from inversion layer collapse. It is shown that, with such a modification, the normal I-V characteristics expected from the HIT cell in the dark and under illumination would be restituted.

2 PROPOSED MODIFICATION IN THE HIT STRUCTURE

The standard and modified HIT structures are sketched in Figure 5(a) and (b), respectively. The proposed modification consists of inserting a p$^+$ layer (emitter) at the a-Si:H/c-Si hetero-interface in the c-Si side. Such a layer is created by intentional p-type doping of a shallow region at the surface of the c-Si substrate prior to a-Si:H deposition. The emitter doping profile should be optimized to obtain the best leverage. Shallow graded profiles can be achieved by thermal diffusion while uniform doping profiles can be obtained by epitaxial growth of p-type silicon. The latter method relates to a high thermal budget process but has the advantage of producing uniform doping profiles with a wide range of doping concentrations and thickness. Alternatively, a thin layer of polycrystalline silicon deposited by LPCVD, in situ doped or doped post-deposition by ion implantation, followed by a subsequent moderate thermal budget annealing step, results in uniformly doped recrystallized polysilicon with a shallow graded doping tail [9]. Such a method has been extensively used in advanced bipolar transistor technology to reduce recombination at the emitter contact [10,11].

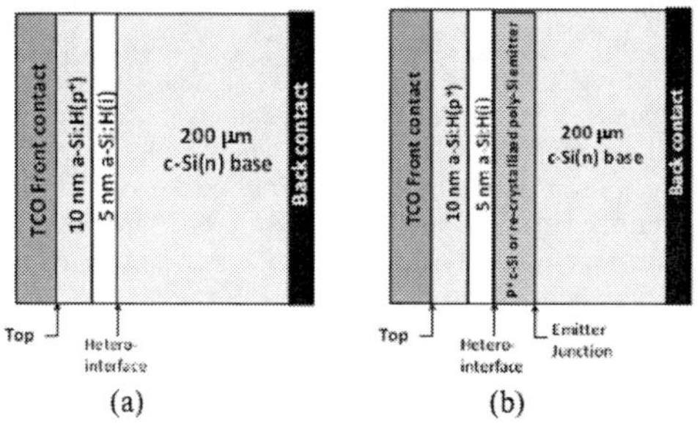

Figure 5: (a) Standard HIT cell, (b) Modified structure implementing p$^+$ doped emitter.

The degraded cell considered here for improvement is prone to inversion layer collapse as it has a TCO with WF_{TCO} = 4.9 eV and a chemical p-type doping level N_a in a-Si:H(p$^+$) of 5x10^{18} /cm^3. Such a value of WF_{TCO} is significantly smaller than the work function of a-Si:H (5.5 eV) which ensures a Schottky barrier at the front contact, and hence is appropriate for the standard HIT cell considered for improvement. As suggested in [1], the peak dangling bond defect density in the bulk of doped a-Si:H, N_{DB0}, is estimated to be 1.2 times the doping concentration, and the peak dangling bond defect density at the a-Si:H(i)/c-Si hetero-interface is fixed at 10^{11} /cm^2/eV. The latter value is also used at the interface between the a-Si:H(i) and the surface of the inserted p$^+$ doped c-Si layer.

Reference I-V characteristics in the dark and under illumination are set for assessment purposes. These reference characteristics relate to a standard HIT cell that also has WF_{TCO} = 4.9 eV but has a higher chemical doping concentration N_a=5x10^{19} /cm^3 in a-Si:H(p$^+$). Although the I-V characteristics of the reference cell under illumination do not suffer from inversion layer related problems, they exhibit current saturation and S-shape beahvior linked to the Schottky front contact.

3 ANALYSIS AND RESULTS

The benefits of the proposed modification are emphasized by comparing the I-V characteristics of the modified HIT cell in the dark and under illumination with those of the degraded cell and with the characteristics of the reference cell. These I-V characteristics are determined by means of Silvaco Atlas simulations tools and confirmed by AFORS-HET simulations. Doping dependent mobility, SRH recombination as well as Auger recombination and bandgap narrowing are taken into account.

3.1 Uniform p$^+$ emitter doping profile

In this section the proposed p$^+$ emitter to be inserted at the hetero-interface is uniformly doped with a doping concentration N_A = 5x10^{19} /cm^3 *(the subscript capital A is for doping in c-Si while N_a refers to the doping in a-Si:H(p))* and extends 100 nm in c-Si to be distinguished from The I-V characteristics in the dark and under illumination for the degraded standard HIT cell with N_a=5x10^{18} /cm^3 in a-Si:H(p) (curve 1), the reference standard HIT cell with N_a=5x10^{19} /cm^3 (curve 2), and the modified HIT cell with N_a=5x10^{18} /cm^3 (curve 3) are displayed in Figure 6 and Figure 7, respectively, As confirmed in Figure 6, the ideal exponential dark current of the modified HIT cell is one order of magnitude smaller than that of the degraded cell and is getting much closer to that of the reference cell. In the high voltage range, the inversion layer emitter collapses in the degraded cell which reduces the device to an n-type resistor with a Schottky contact. This explains the Schottky exponential forward current observed in the dark characteristics (curve 1) at high voltage. On the other hand, the emitter/base p/n junction in the modified structure is maintained throughout the whole voltage range of operation. Consequently, at high voltage and like in the reference cell, the dark current of the modified cell saturates at a level determined by the reverse saturation current of the front contact Schottky diode J_{0_Sch}.

The I-V characteristics of the modified HIT cell under illumination is displayed in Figure 7 and confirms the restitution of the normal shape and of the cell Fill Factor, FF. The open circuit voltage V_{oc} is only slightly improved as it is mainly controlled by the value of WF_{TCO}. The saturation of the dark current observed in Figure 6 at high voltage, however, is responsible for the S-shape exhibited by the I-V characteristics of both modified HIT cell and reference cell under illumination, as depicted in Figure 7.

3.2 Impact of the uniform doping level and junction depth

It appears that maintaining the uniform doping concentration of the p$^+$ layer at N_A=5x10^{19} /cm^3 while reducing the junction depth from 100 nm to 20 nm does not affect the I-V characteristics, which indicates that bulk recombination in the p$^+$ layer is negligible and that the emitter dark current is dominated by surface recombination. Since the latter is inversely proportional to the surface hole concentration, reducing the emitter concentration would raise the dark current. Indeed, the dark curent is larger when the uniform doping concentration decreases to N_A=10^{18} /cm^3 and is even larger when N_A=10^{17} /cm^3, as depicted in Figure 8.

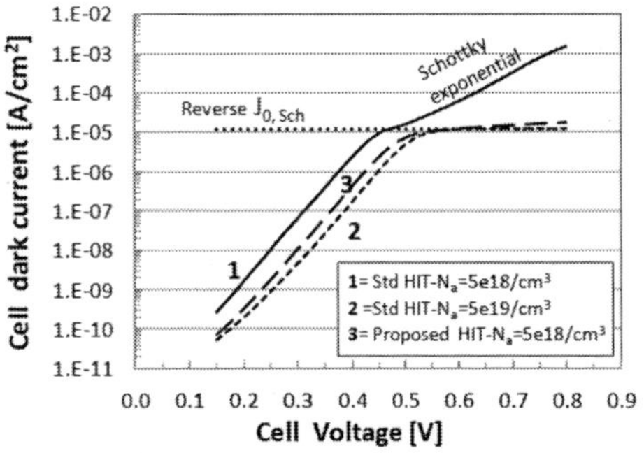

Figure 6: Dark I-V characteristics of standard and modified HIT cells.

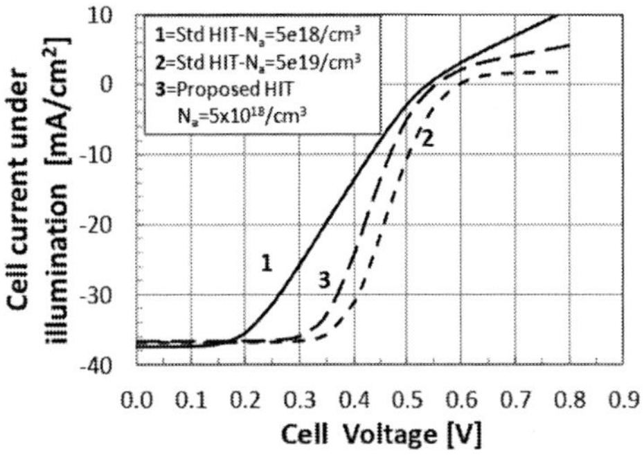

Figure 7: I-V characteristics of standard and modified HIT cells under AM1.5 illumination.

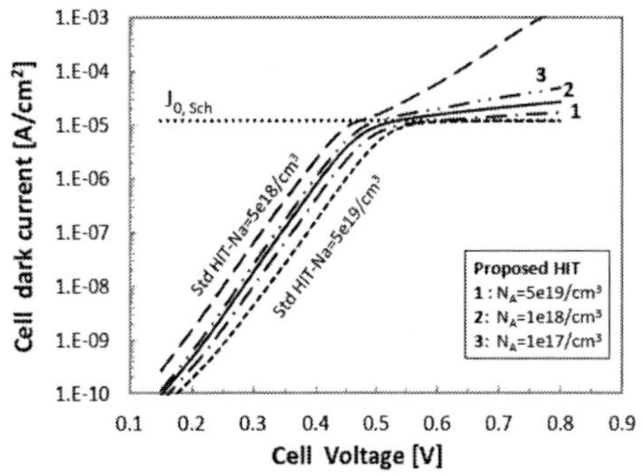

Figure 8: Dark I-V characteristics of standard and modified HIT cells having uniform p$^+$ doping profiles with N_A=10^{17}, 10^{18} and 5x10^{19}cm^3.

3.3 Graded p$^+$ emitter doping profiles

It has been shown that the exponential emitter doping profile has the merit of low surface recombination compared to uniform doping profiles due to the front surface field it generates at the emitter surface [12]. In addition, the exponential profile should exhibit smaller bandgap narrowing and smaller Auger recombination in case the latter becomes significant. Indeed, as depicted in Figure 9, the improvement in the dark current is enhanced in cells implementing a p^+ emitter with a graded doping profile (Gaussian) rather than a uniform p^+ doping profile. The improvement is identical whether the graded profile is exponential or Gaussian, both having a surface concentration N_{AS} = 5x10^{19} /cm^3 at the hetero-interface and a junction depth of 50 nm.

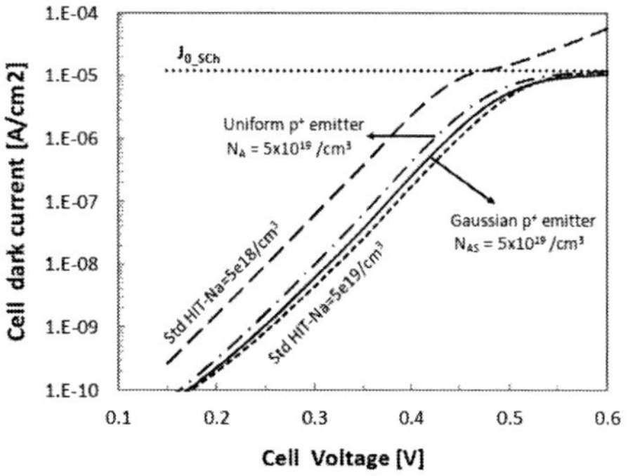

Figure 9: Dark I-V characteristics of standard and modified HIT cells with uniform and graded p^+ emitter profiles.

4 SUMMARY

The implementation of a p^+ doped shallow emitter in the n-type crystalline silicon substrate of HIT cells, prior to a-Si:H deposition, is found to be highly beneficial in cells which suffer from deficiency in the active doping in a-Si:H(p^+) and in cell using at TCO with a small work function. In such cases, the HIT cell may fail to operate properly due to the collapse of the interfacial p^+ inversion layer emitter. Simulations of the I-V characteristics using Silvaco tools and AFORS-HET confirm that the proposed modification restitutes the cell I-V characteristics under illumination to its normal shape and the FF to its expected value.

Acknowledgements: This work is supported by Kuwait University Research Administration project GE01/08 and KFAS/KU-IMEC funded project P115-15EE-01

5 REFERENCES

[1] M. Ghannam, G. Shehada, Y. Abdulraheem, J. Poortmans, Solar Energy Materials and Solar Cells 132 (2015) 320.
[2] J.P. Kleider, Physics and technology of amorphous-crystalline heterostructure silicon solar cells, W.J.H.M. Sark, L. Korte, F. Roca (Eds.), Springer-Verlag Berlin Heidelberg, 2012, Chapter 12, pp. 405-444.
[3] O.A. Maslova, J. Alvarez, E.V. Gushina, W. Favre, M.E. Gueunier-Farret, A.S. Gudovskikh, A.V. Ankudinov, E.I. Terukov, J.P. Kleider, Applied Physics Letters 97 (2010) 252110 (1-3).
[4] R. Varache, J.P. Kleider, W. Favre, L. Korte, Journal of Applied Physics 112 (2012) 123717 (1-5).
[5] J.P. Kleider, A.S. Gudovskikh, P.R.i. Cabarrocas, Applied Physics Letters 92 (2008) 162101 (1-3).
[6] J.P. Kleider, Y.M. Soro, R. Chouffot, A.S. Gudovskikh, P. Roca i Cabarrocas, J. Damon-Lacoste, D. Eon, P-J. Ribeyron, Journal of Non-Crystalline Solids 354 (2008) 2641.
[7] M. Ghannam, Y. Abdulraheem, G. Shehada, Solar Energy Materials and Solar Cells 145 (2016) 423.
[8] K.Masuko,M. M. Shigematsu, T. Hashiguchi, D. Fujishima, M. Kai, N. Yoshimura, T. Yamaguchi, Y. Ichihashi, T. Mishima, N. Matsubara, T. Yamanishi, T. Takahama, M. taguchi, E. Maruyama, S. Okamoto, IEEE Journal of Photovoltaics 4 (2014) 1433.
[9] M.Y. Ghannam and R.W. Dutton, Applied Physics letters 51 (1987) 611.
[10] T. H. Ning and R. D. Isaac, IEEE Transactions on Electron Devices 27 (1980) 2051.
[11] A. A. Eltoukhy and D. J. Roulston, IEEE Transactions on Electron Devices 29 (1982) 961.
[12] M. Ghannam and F. Alajmi, Proceedings of the 31st European Photovoltaic Solar Energy Conference EUPVSEC (2015) pp. 813-816.

SPATIALLY RESOLVED DEGRADATION AND REGENERATION KINETICS IN MC-SI

Daniel Skorka, Annika Zuschlag, Giso Hahn
University of Konstanz, Department of Physics
78457 Konstanz, Germany

ABSTRACT: The strong degradation of multicrystalline (mc) Si material upon illumination at moderate temperatures (LeTID) is an issue that might hinder the application of PERC technology to mc-Si. Interestingly, a regeneration effect can also be observed under the same conditions, starting after several hundred hours at 75°C. Studies are mainly based on time resolved investigations, and some investigations also use spatially resolved measurement techniques. Visualization of the collected data is therefore challenging, but important for a better understanding of the underlying mechanisms.
In this contribution we show that based on lifetime samples we can study the same effects as reported on solar cell level with a good spatial resolution. In ungettered samples originally good quality areas show a higher relative degradation as compared to originally poor quality areas. This behaviour changes completely for gettered samples. We also show that the regeneration process sets in earlier in good quality areas, which could be caused by the higher injection level present in these regions. The regeneration process therefore might be qualitatively similar to the known BO-related regeneration. A good point-to-point correlation of different sets of lifetime measurements allows to demonstrate the impact of local differences in material quality on degradation and regeneration behaviour.
Keywords: Degradation, Regeneration, LeTID, Multicrystalline silicon, Lifetime

1. INTRODUCTION

Recently, a phenomenon was discovered whereby mc-Si PERC (passivated emitter and rear contact) solar cells experience degradation upon illumination at elevated temperatures (LeTID) [1-3]. The loss in solar cell efficiency can be up to $10\%_{rel}$ [3] and cannot be explained by formation of BO-correlated defects or FeB pair dissociation alone [1]. This phenomenon has the potential to completely stop the application of PERC technology to mc-Si. However, a regeneration process after degradation has been observed by Kersten *et al.* [3]. The behaviour of degradation and regeneration are influenced by light intensity and temperature, but the root cause (i.e. the underlying defect) in mc-Si PERC solar cells is still unknown. Efforts to study the parameters influencing the degradation and regeneration process mainly use solar cells or even solar modules, while investigations on lifetime samples are still rare. Lifetime samples with different surface passivations were studied by Kersten *et al.* [3] and were shown to degrade, whereas Krauss *et al.* [4] employed spatially resolved minority charge carrier lifetime maps to calculate a so-called effective defect lifetime, thereby mapping effective defects which are activated upon degradation. Still missing are good methods to make best use of the spatially and temporally resolved data from, *e.g.*, lifetime mapping techniques and to correlate the information given in these sets of data.

Lifetime samples are inherently more flexible with respect to experimental design when studying the influence of specific processing steps (*e.g.*, gettering and hydrogenation as well as high temperature steps) which might influence the degree and kinetics of LeTID and regeneration. In this contribution we concentrate on three differently processed lifetime samples and study the effects visible during illumination at moderate temperature. This allows to draw conclusions and to shed more light on the underlying mechanisms for LeTID and the following regeneration in mc-Si.

2. EXPERIMENTAL

We investigate an industrially produced mc-Si PERC solar cell and several lifetime samples to compare the degradation and regeneration process on solar cell and lifetime level. Using the solar cell allows us to link and compare the results to previous studies on the LeTID effect, while the differently processed lifetime samples are intended to identify process steps that influence the degradation and regeneration behaviour of mc-Si material.

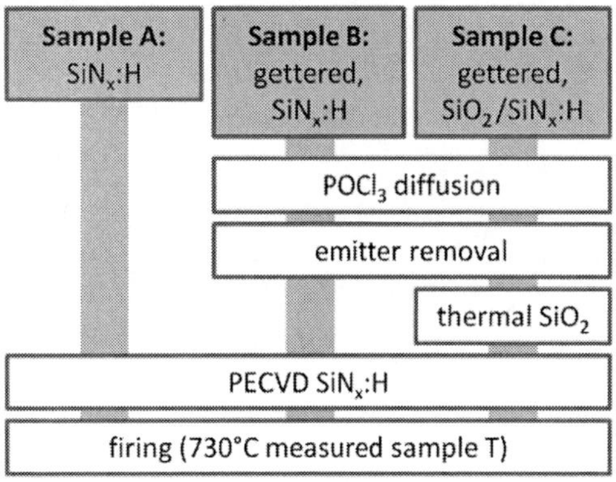

Figure 1: Process sequence of the investigated lifetime samples.

Both the wafers used for the lifetime samples (5x5 cm^2) and the investigated industrial mc-Si PERC solar cell (15.6x15.6 cm^2) originate from the same height of the same mc-Si ingot (1.5 Ωcm boron doped) and therefore have a comparable initial material quality. The lifetime samples - fabricated from sister wafers, thus having comparable grain and defect structure - were used to compare the influence of different process steps on the degradation and regeneration behaviour. A scheme of the applied process sequences is given in Fig. 1. Saw damage was removed by chemical etching. Samples B and C were gettered by a $POCl_3$ diffusion (55 Ω/□) and the emitter was subsequently removed by chemical etching. Both sample A (ungettered) and sample B (gettered) received a surface passivation by a PECVD (plasma-enhanced chemical vapour deposition) SiN_x:H layer, while sample C was passivated by a stack of thermal SiO_2 (6 nm) and a PECVD SiN_x:H layer. Firing was done in a belt furnace with a measured peak wafer temperature of 730°C.

For degradation, lifetime samples as well as the solar cell are held at a temperature of approx. 75°C under illumination with halogen lamps (0.9±0.05 suns for lifetime samples, 1±0.05 sun for the solar cell). During the degradation and regeneration process the solar cell is characterized *in situ* by automated V_{oc} measurements. For lifetime samples, the effective minority charge carrier lifetime τ_{eff} is measured repeatedly by the fast and self-calibrated TR-PLI (time resolved photoluminescence imaging) method [5, 6] at room temperature, resulting in a series of spatially resolved lifetime maps for each sample over degradation time.

3. DATA ANALYSES AND RESULTS

The *in-situ* V_{oc} measurements for the solar cell are shown relative to the starting value in Fig. 2. It is obvious that this PERC solar cell is LeTID sensitive and the V_{oc} data shows a degradation of approx. $12\%_{rel}$ after 200 h followed by regeneration as also observed previously, *e.g.*, in [3]. Data acquisition for the solar cell as well as the lifetime samples continues and we assume that the solar cell will recover further.

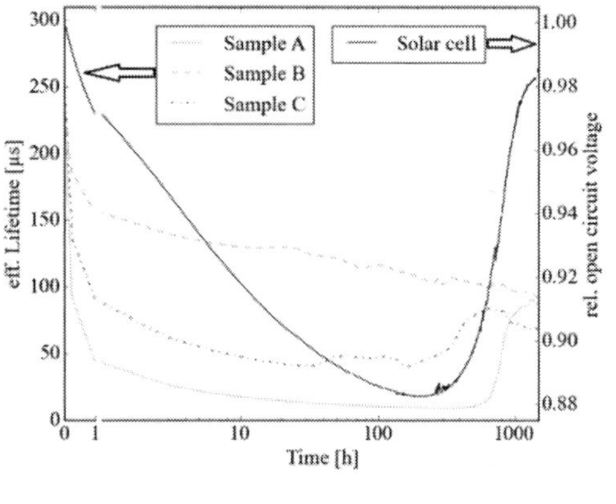

Figure 2: Left axis: harmonic average of τ_{eff} data for three differently processed mc-Si wafers (5x5 cm²). The strong influence of processing steps on the degradation and regeneration behaviour is clearly visible.
Right axis: continuously measured V_{OC} data (shown relative to starting value) of an industrial mc-Si PERC solar cell showing degradation and beginning regeneration (first 1 h: linear scale, afterwards: log scale).

Fig. 2 also shows the temporal evolution of τ_{eff} (plotted is the harmonic average of the entire effective lifetime map for each point in the time series) for the three different lifetime sister samples A, B, and C. The strong influence of processing steps on the degradation and regeneration behaviour is clearly visible. Within the first minutes a fast degradation is observed for all samples, which could be attributed to FeB- and/or BO-based degradation effects. After longer periods, significant differences are measured for samples A, B, and C regarding the strength of degradation, the range of minimum and maximum τ_{eff} in dependence of the degradation state, and the regeneration behaviour which can be clearly observed for sample A and C. Sample A (ungettered) shows the strongest degradation but also the strongest regeneration. In case of the P-gettered sample B the degradation is strongly reduced and no degradation minimum is observed up to now. In contrast to sample B, the surface passivation of sample C is realized by firing a stack of thermal SiO_2 and SiN_x:H, leading to significant differences in degradation and regeneration behaviour. Overall degradation and regeneration behaviour is closer to the ungettered sample A than to sample B. Sample C shows also a clear maximum degradation and the regeneration sets in even earlier than in case of sample A.

A comparison to the degradation behaviour of the solar cell shows that the maximum degradation level for τ_{eff} in case of lifetime samples and the V_{oc} value in case of the solar cell is reached after approximately the same range of degradation time using the same illumination level and elevated temperature. This leads to the assumption that the underlying degradation and regeneration effect can be studied either on lifetime or on solar cell level. This is an interesting finding as injection levels for a given generation should be different for lifetime samples (with very good surface passivation) and solar cells (with emitter and metalized fractions).

The material quality of mc-Si differs locally, *e.g.*, due to grain orientation, different dislocation densities or different contents and distributions of contaminations. If only the average τ_{eff} value is analyzed, local differences as visible in the TR-PLI measurements get lost.

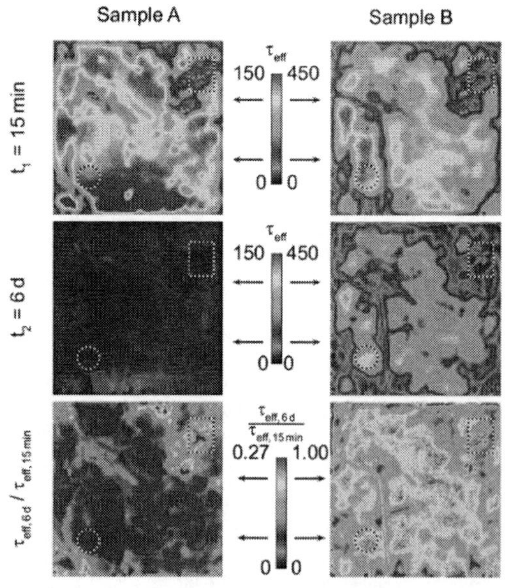

Figure 3: TR-PLI measurement of differently processed sister wafers A (ungettered) and B (gettered) after 15 min and 6 d at 75°C and under illumination (1 sun), as well as calculated maps of relative lifetime changes ($\tau_{eff,6d}$ /$\tau_{eff,15min}$), clearly indicating the different strength of degradation. Note the different scaling for sample A and B!

To demonstrate the spatially resolved information of the TR-PLI maps, Fig. 3 shows lifetime images of sample A and B at the beginning of the degradation and at the point where maximum degradation occurs for sample A, as well as calculated maps of relative lifetime changes ($\tau_{eff,degraded}/\tau_{eff,initial}$).

In case of sample A the whole wafer area degrades strongly and leads to a lifetime reduction of at least 70%. The relative lifetime change is lower in regions of lower initial lifetime as indicated by a striped box in Fig. 3 (approx. 80%) than in areas of higher initial material quality as indicated by a striped circle in Fig. 3 (approx. 90%).

If these two indicated areas of higher (circle) and lower (box) initial lifetimes are compared to the corresponding wafer regions of the gettered sample B, a different behaviour is observed. The overall degradation of the gettered sample B is much lower as already discussed for the harmonic average τ_{eff} (Fig. 2). Areas of high initial lifetime now show only a relatively weak degradation of approx. 0-15%, while areas of lower initial lifetime degrade by approx 50%.

A more detailed correlation of degradation behaviour in different sample areas and local defect structure is currently underway.

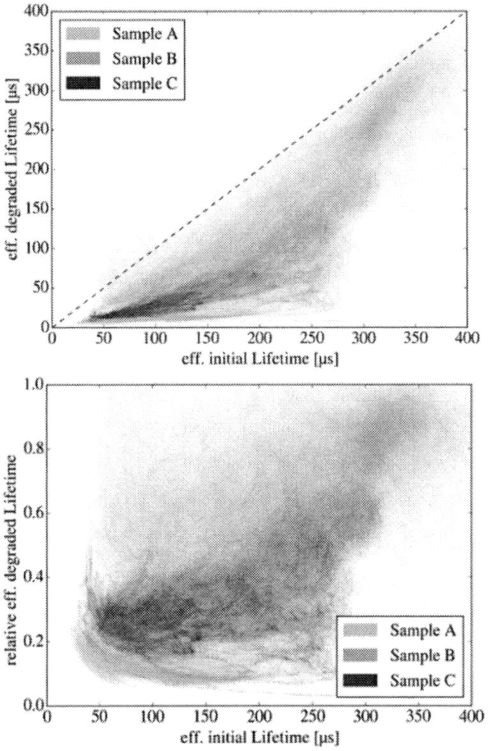

Figure 4: τ_{eff} at maximum degradation and relative τ_{eff} compared to the initial τ_{eff} for each single data point of TR-PLI maps of the different processed sister wafers A, B, and C. The strongly different influences of the applied process steps can be observed.

In Fig. 4 degraded τ_{eff} (top) as well as relative degraded τ_{eff} relative to initial τ_{eff} (bottom) for each single data point of the TR-PLI map is plotted over initial τ_{eff} for each of the differently processed sister samples A, B, and C. As also discussed for Fig. 2, sample A shows the strongest degradation over the whole sample area, while the strength of degradation is significantly reduced for all areas of the gettered sample B. An additional high temperature step (thermal oxidation after P-gettering) like in case of sample C passivated by firing of a stack of thermal SiO_2 and SiN_x:H leads to a stronger degradation over the whole sample compared to the also P-gettered sample B without thermal oxidation. The interpretation in context of Fig. 3, that the relative impact of degradation is stronger in areas of higher initial τ_{eff} in case of sample A and much lower in case of sample B, holds also true analyzing the whole sample as shown in Fig. 4 (bottom) and not only chosen sample areas as discussed for Fig. 3.

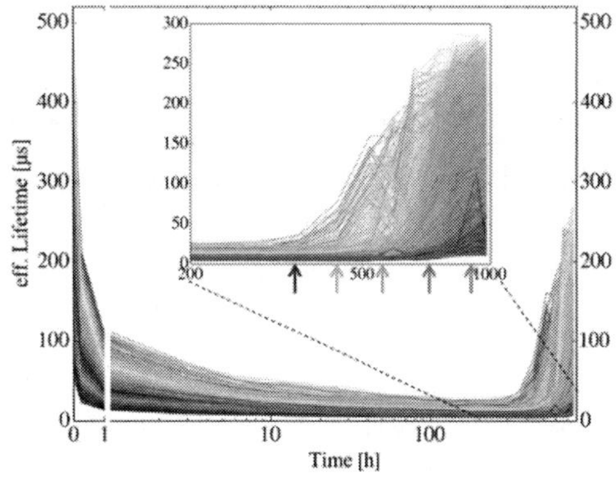

Figure 5: τ_{eff} of the ungettered sample A (first 1 h: linear scale, afterwards: log scale). Each line in the graph represents an area of 150x150 µm² in the spatially resolved TR-PLI lifetime measurements and each colour is chosen according to this areas lifetime value at the beginning of degradation. The inset highlights the different regeneration behaviour. The regeneration process starts significantly earlier in areas with higher initial τ_{eff} (red lines) than in areas of lower initial τ_{eff}, as indicated by the differently coloured arrows.

Fig. 5 shows the behaviour of τ_{eff} for sample A. Each single line represents a small area on the lifetime samples as described in the image caption. The whole ungettered sample A shows a very strong and fast degradation, as already discussed for other analysis methods above. The areas of higher initial τ_{eff} suffer more from the applied degradation conditions, and relative degradation in these areas is stronger than in sample areas with lower initial τ_{eff}. That even areas of initially high τ_{eff} exhibit such strong degradation could be explained by assuming an approximately homogenous formation of defects over the wafer area (see [7] and [8] for more details).

The effective lifetime distribution of sample A initially is very wide (approx. 20–350 µs), but shrinks down to a very narrow distribution (approx. 5–30 µs) at maximum degradation level. The maximum degradation level is reached approximately at the same illumination time for the whole sample, independently of the initial τ_{eff}. After about 300 h of illumination a regeneration process sets in, with regeneration in areas with higher initial τ_{eff} (red lines) starting significantly earlier than in areas of lower initial τ_{eff}. This behaviour is indicated in

Fig. 5 by coloured arrows representing the starting point of regeneration for different initial lifetimes. That the starting points for observable regeneration should be so different is remarkable, especially considering the very narrow τ_{eff} distribution at maximum degradation level. These differences in regeneration behaviour lead to the assumption that regeneration is not just a reversal of the first degradation reaction and that the underlying mechanism is more complex. It is not yet known whether τ_{eff} will recover its initial value. However, we currently assume that a decrease of the surface passivation quality during the degradation process of several hundred hours limits the maximum level of regeneration in case of sample A. The experiment is still running to complete the data set.

In case of sample C it is also observed that the regeneration starts earlier in sample areas of higher initial lifetime (not shown).

This indicates that the injection level might play a role in regeneration kinetics, as under homogeneous illumination injection is higher in regions of higher lifetime. This was also reported for investigations on solar cell level [3] and resembles qualitatively the behaviour also known for regeneration of BO-related defects [9].

4. SUMMARY AND OUTLOOK

Spatially resolved LeTID and regeneration was studied in mc-Si material. It could be shown that the behaviour of average τ_{eff} values in lifetime samples qualitatively resembles the behaviour of an industrial PERC solar cell processed from mc-Si of comparable quality.

Ungettered mc-Si material shows a severe degradation effect, followed by a distinct regeneration phase. In an ungettered sample, relative degradation is higher in initially good quality areas. This trend changes for gettered samples, although on a much higher lifetime level. In gettered samples relative degradation is stronger in areas of poor initial quality (most probably areas of higher density of extended defects like grain boundaries, dislocations, precipitates).

Point-to-point (or area-to-area) correlations of lifetime maps gained from TR-PLI allow for a detailed analysis of factors influencing degradation and regeneration kinetics which is not possible when only analyzing average lifetime or solar cell parameter values.

The timescale for LeTID and regeneration is rather long, and the described experiments are still running to check especially for the following regeneration behaviour.

5. ACKNOWLEDGEMENTS

Part of this work was funded by the German BMWi under contract numbers 0325581 and 0325763B. The content of the publication is the responsibility of the authors. The authors would like to thank Axel Herguth, Lisa Mahlstaedt and Barbara Rettenmaier for their assistance with this study.

6. REFERENCES

[1] K. Ramspeck, S. Zimmermann, H. Nagel, A. Metz, Y. Gassenbauer, B. Birkmann, A. Seidl, Light induced degradation of rear passivated mc-Si solar cells, Proc. 27th EU PVSEC, Frankfurt/Main, Germany, 2012, 861-865.

[2] F. Fertig, K. Krauss, S. Rein, Light-induced degradation of PECVD aluminium oxide passivated silicon solar cells, Phys. Stat. Sol. RRL 9(1) (2014), 41-46.

[3] F. Kersten, P. Engelhart, H.-C. Ploigt, A. Stekolnikov, T. Lindner, F. Stenzel, M. Bartzsch, A. Szpeth, K. Petter, J. Heitmann, J. Müller, Degradation of multicrystalline silicon solar cells and modules after illumination at elevated temperature, Sol. En. Mat. & Sol. Cells 142 (2015), 83-86.

[4] K. Krauss, F. Fertig, D. Menzel, S. Rein, Light-induced degradation of silicon solar cells with aluminium oxide passivated rear side, En. Procedia 77 (2015), 599-606.

[5] D. Kiliani, G. Micard, B. Steuer, B. Raabe, A. Herguth, G. Hahn, Minority charge carrier lifetime mapping of crystalline silicon wafers by time-resolved photoluminescence imaging, J. Appl. Phys. 110 (2011), 054508.

[6] D. Kiliani, A. Herguth, G. Micard, J. Ebser, G. Hahn, Time-resolved photoluminescence imaging with electronic shuttering using an image intensifier unit, Sol. En. Mat. Sol. Cells 106 (2012), 55-59.

[7] A. Zuschlag, D. Skorka, G. Hahn, Degradation and regeneration in mc Si after different gettering steps, this conference.

[8] A. Zuschlag, D. Skorka, G. Hahn, Degradation and regeneration analysis in mc Si, Proc. 43rd IEEE PVSC, Portland 2016, in press.

[9] A. Herguth, G. Schubert, M. Kaes, G. Hahn, A new approach to prevent the negative impact of the metastable defect in boron doped Cz silicon solar cells, Proc. 4th WCPEC, Waikoloa, USA, 2006, 940-943.

CONTACT FORMATION ON P-DOPED SI BY SCREEN-PRINTING PURE AG PASTES FOR BIFACIAL N-TYPE SI SOLAR CELLS

Josh Engelhardt, Susanne Fritz, Erkan Emre, Giso Hahn
University of Konstanz, Department of Physics, 78457 Konstanz, Germany
Author of contact: Tel. +49 7531/88-2080, Fax. +49 7531/88-3895, josh.engelhardt@uni-konstanz.de

ABSTRACT: n-type solar cell concepts increasingly utilize emitter formation by diffusion from boron doped sources. Combining the advantage of n-type silicon material and bifacial cell architecture enables high-efficiency and versatile photovoltaics. In case of boron emitters, it was standard until now to form a metal-semiconductor contact by screen-printing Al containing Ag pastes. Instead of utilizing Al to enable Ag to form a sufficient contact with the risk in loss of V_{OC} and FF, different glass compositions for pure Ag pastes were developed to form a contact with low impact on cell efficiency. In direct comparison this method in the first try already surpasses the performance of commercial Al containing Ag pastes in direct comparison. The experimental pastes show a distinctive gain in solar cell characteristics in contrast to commercial pure Ag pastes. In this case we reached an overall efficiency of 18.6% using pure Ag pastes. Contact resistivity values thereby range below 1-2 mΩcm² comparable to pure Ag pastes on n-type emitters and Ag/Al pastes on p-type emitters.
Keywords: screen-printing; Ag paste; boron emitter; contact formation; n-type

1 IINTRODUCTION

High-efficiency solar cells use n-type Si substrates due to the higher bulk lifetime and low susceptibility to degradation. With the use of n-type substrates, challenges from the common use of p-type emitters from, *e.g.*, BBr_3 diffusions arise. An example is contact formation of the screen-printed metal paste to the bulk Si substrate. Considering the widespread utilization of the method in industry, different approaches to deal with the challenge and optimize the contact quality have been done. Commonly, Al containing pastes are used to successfully form a contact to the B emitter [1] with contact resistivity values below 10 mΩcm. Yet this approach brings restrictions in firing parameters and a drop in V_{OC} and fill factor, compared to contact formation with Ag pastes on n-type emitters. Yet simply using Ag pastes on p-type emitters has so far not been successful [1-3]. B emitters lack the precipitates deemed necessary for the contact formation behavior of pure Ag pastes [4] with the standard glass composition forming shallow surface crystals [4,5]. Furthermore, B emitters usually display a surface dopant concentration below 10^{20} cm^{-3}, making it even harder to form a proper contact. The advantage of pure Ag pastes, however, is the lack of the deep crystal formation behavior that Al containing pastes display. Without it, V_{OC} and fill factor values should remain unaffected following the undisturbed space-charge region.

2 EXPERIMENTAL

Comparison and detailed investigation of the influence, physics and results of different pastes were carried out on bifacial n-type Si solar cells and TLM (transfer length method) precursors. Fully textured bifacial cells as well as TLM test structures were one-side BBr_3 diffused and afterwards obtained a selective back surface field (s-BSF) by $POCl_3$ diffusion and selective etch-back. During diffusions PECVD (plasma enhanced chemical vapor deposition) SiN_x:H was used to shield the respective backside of the wafer from parasitic in-diffusion. All samples were then passivated by double-side thermal SiO_2 and PECVD SiN_x:H stack passivation. Screen-printing and contact firing were done using mesh printing contact grids on both sides and a firing belt furnace, respectively. Finally, the cells were fully characterized using a flasher setup for IV measurements, a combination tool for reflection, thermography and additional measurements, as well as a TLM measurement setup to determine contact resistance obtained from TLM samples.

Screen-printing was done with various commercially available state-of-the-art pure and Al containing Ag pastes, as well as with experimental pure Ag pastes developed in cooperation with Johnson Matthey. Latter pastes were varied in Pb and Te content as well as glass composition to optimize contact formation on p-doped layers.

3 RESULTS AND DISCUSSION

3.1 Emitter properties

The BBr_3 emitter with a sheet resistance R_{Sheet} of 50 Ω/□ used in this study was originally developed for optimized contact formation with Al containing Ag pastes (Fig. 1).

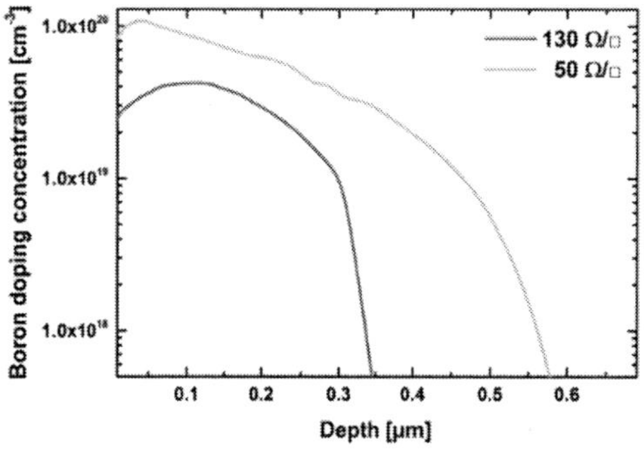

Figure 1: Profiles of the two B emitters used for bifacial solar cells and TLM test structures

In contrast, TLM studies with pure Ag pastes showed no difference in contact resistivity values on samples with different boron emitter profiles. Therefore, to show the advantage of Ag pastes on the highest level possible, we choose the emitter best suited for Al containing pastes (Tab. I) and a high ohmic emitter to show the restrictions of Ag/Al pastes. Similar emitter profiles are common to newer diffusion sources, such as CVD BSG (borosilicate glass) sources [6] also capable to be contacted with pure Ag pastes. Thereby, an emitter saturation current density of <40 fA/cm² on Float-Zone (FZ) n-type Si substrates can be reached, if passivated with Al_2O_3. Yet on the Czochralski (Cz) Si substrate passivated by SiO_2/SiN_x:H and used here for solar cell and TLM samples, the implied V_{OC} value remains 657 mV or 678 mV for the low and high ohmic emitters, respectively.

Table I: General B emitter characteristics and iV_{OC} values for SiO_2/SiN_x:H-stack passivation

R_{Sheet}	50 Ω/□	130 Ω/□
$N_{Surface}$	$8.0\cdot10^{19}$ cm^{-3}	$2.3\cdot10^{19}$ cm^{-3}
N_{Max}	$1.1\cdot10^{20}$ cm^{-3}	$4.2\cdot10^{19}$ cm^{-3}
Depth	600 nm	370 nm
$j_{0E,Al2O3}$ (FZ, n)	<40 fA/cm²	<30 fA/cm²
$iV_{OC,SiO2/SiNx}$ (Cz, n)	657 mV	678 mV

3.2 Contact investigations

Ag/Al pastes are commonly used for contacting B doped Si by screen-printed contact finger grids. Due to the high solubility of Al in Si, contact formation of metal to semiconductor is facilitated [7]. With increasing firing temperature the specific contact resistance ρ_C decreases for Ag/Al pastes on low ohmic emitters. Yet below 1-2 mΩcm², shunting of the space-charge region (SCR) in solar cells commonly occurs. On high ohmic emitters (Fig. 2) the resistance never drops so low and instead increases for higher temperatures again.

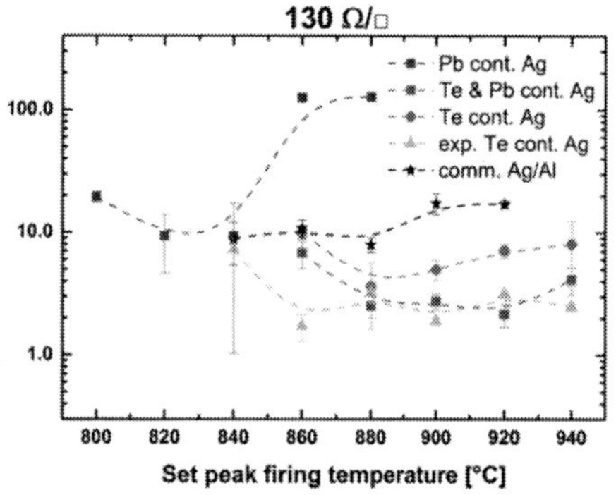

Figure 2: Contact resistance measurements from TLM test structures of different pastes dependend on the set peak firing temperature on a 130 Ω/□ B emitter

Old generation pure Ag pastes containing Pb were verified to contact at least at low set peak temperatures in contrast to older publications and at a comparable level to commercial modern pastes. Yet, the latter pure Ag pastes contact in a broader firing range and at higher temperatures as well. Adding Te, used to supplement or even substitute Pb, shifts the minimum contact resistance value to higher temperatures and outweighs the influence of Pb on contact formation behavior. By optimization of the glass composition in the experimental only Te containing paste (no Pb), the window broadens further and ρ_C values of below 2 mΩcm² can be reached. With this paste, levels only double the average value for the same paste on a far lower R_{Sheet} B emitter of 50 Ω/□ can be reached (see Fig. 3).

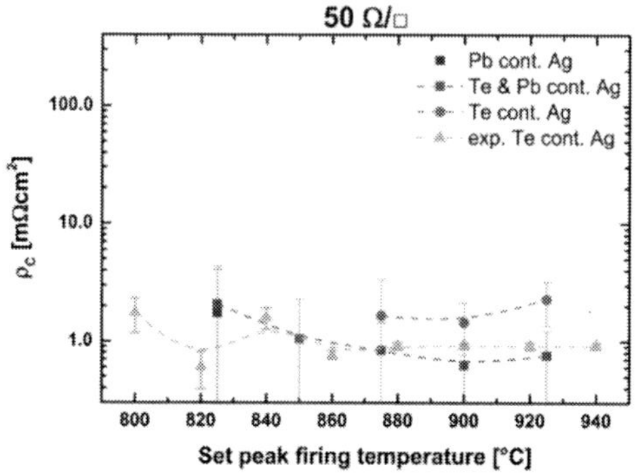

Figure 3: Contact resistance measurements from TLM test structures of different pastes dependent on the set peak firing temperature on a 50 Ω/□ B emitter

Considering the firing temperature range, the overall behavior of the pastes on a low ohmic emitter is similar to Fig. 2. It also ranges in the same resistivity range as Ag/Al do. Even the old generation Pb based Ag paste reaches a measureable resistivity value, but only for a single low temperature. Pure Ag pastes, independently of their composition, apparently do vary less with R_{Sheet} as Ag/Al pastes do. Limiting influence for the Ag/Al pastes thereby might be either the depth (area of contact with spikes) or the surface doping concentration. Since pure Ag pastes do not penetrate the space-charge-region, they do not suffer from these limiting influences and further display higher contact crystal densities. The self-limiting nature of Ag paste contact formation by glass layer growth at higher temperatures also diminishes the possibility of 'spiking'.

3.3 Solar cell results

Bifacial solar cells, processed as described before, are a straightforward and beneficial way to characterize the effect of printing the aforementioned pastes on p- and n-type Si layers. In case of the 50 Ω/□ B emitter, cells were printed with the reference group of a commercial Ag/Al paste on the B emitter and a commercial pure Ag paste on the P diffused high-low-junction. Two commercial Ag pastes, with respective lowest contact resistivity values for both sides of the cell were used as a pure Ag reference group and state-of-the-art comparison. Furthermore, a Te added old generation Pb containing pure Ag paste and a pure Te based Ag paste were printed. In direct comparison (Fig. 4) the FF is comparable on this deep and highly doped emitter except for the commercial pure Ag paste combination. The reason is an increase in series resistance on cell level despite the optimal firing temperature used for all pastes. The contact formation for the currently available pastes still lacks the optimization for B doped surfaces. Older pastes based on high Pb content perform better in contact formation, yet seem for example to in-diffuse recombination active compounds leading to a decrease in V_{OC}. The indiffusion into the SCR is also visible in a significant increase in the saturation current density of the second diode j_{02}. Standard Te based pure Ag pastes instead reach nearly the same level of Ag/Al pastes in every aspect reaching

18.6% efficiency compared to 19.1% of Ag/Al printed cells.

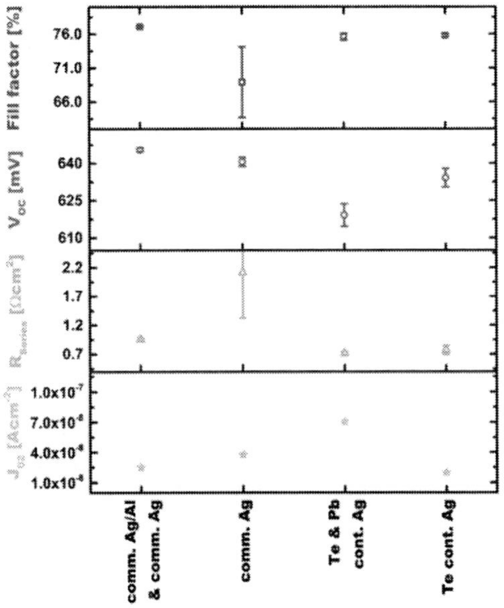

Figure 4: FF, V_{OC}, R_{Series} and j_{02} of bifacial solar cells with a 50 Ω/□ B emitter in dependency of different screen-printing pastes. From left to right: Group 1 with comm. Ag/Al on B and comm. Ag on P layer. Group 2-4 with pure Ag paste on both layers.

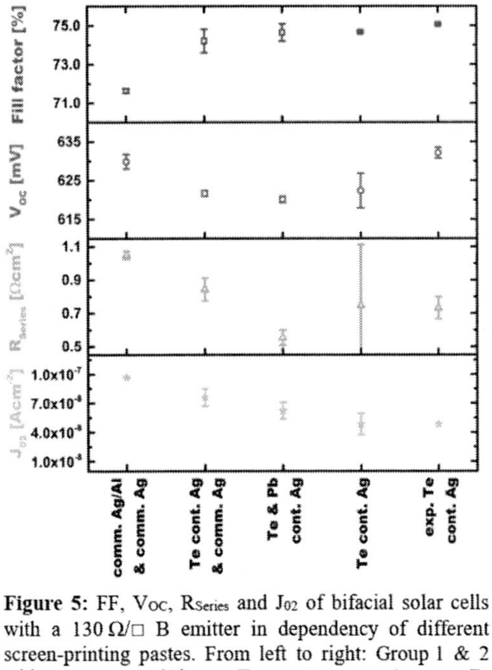

Figure 5: FF, V_{OC}, R_{Series} and J_{02} of bifacial solar cells with a 130 Ω/□ B emitter in dependency of different screen-printing pastes. From left to right: Group 1 & 2 with comm. Ag/Al or Te cont. pure Ag on B, respectively, and comm. Ag on P layer. Group 3-5 with pure Ag paste on both layers.

The optimal temperatures thereby range in the 840°C set peak firing temperature range for Ag/Al compared to 880°C for the Ag pastes. Considering the passivation of a SiO_2/SiN_x:H stack, H diffusion and passivation optimum (iV_{OC}) is at the higher temperature range, preferred by pure Ag pastes. This makes pure Ag pastes interesting for inexpensive passivation layers fired commonly at higher temperatures.

Since this 50 Ω/□ B emitter is deep enough to largely prevent the majority of negative effects Ag/Al pastes have, a higher ohmic emitter was used to demonstrate the loss of Ag/Al pastes while increasing the passivation quality of the emitter. As can be seen from Tab. I, the 130 Ω/□ B emitter can reach a far higher iV_{OC} value since the doping concentration in total as well as on the surface is reduced. Lateral conductivity is still assured, given the contact resistivity values from Fig. 2. Therefore, on cell level a significant gain should be possible.

Fig. 5 shows the respective cell results on an overall efficiency level of about 18-19%. To compare the influence of the pure Ag paste on the B emitter more explicitly, the first two groups were printed with the same commercial Ag paste on the P FSF. Only the influence of the B side is visible. Ag/Al paste now drops in FF and V_{OC}. In accordance, R_{Series} and j_{02} rise as well. Since the Te containing Ag paste is not optimized, a certain degree of contaminants now reaches the SCR, resulting in a drop in V_{OC}. There is now no difference between the Pb containing and the pure Te based Ag paste in that account. Using the latter paste for both sides, only marginally improves upon the commercial Ag paste, made for a P FSF. To regain and surpass the V_{OC} value of the Ag/Al paste and improve upon the FF further, the Te based, Pb free, pure Ag paste needs to be optimized, leading to the current experimental paste with a FF increase of 3.5%$_{abs}$ compared to commercial Ag/Al pastes.

4 CONCLUSION

We showed how pure Ag pastes, old and new ones, can contact differently diffused B emitters in a wide range of diffusion and firing parameter ranges. Furthermore, TLM measurements showed a significant influence of composition, *i.e.* Te and Pb content. Bifacial solar cells, printed with experimental pure Ag pastes, result in significant improvement of FF upon the state-of-the-art presented by commercial Ag/Al pastes. Yet commercial Ag pastes cannot reach this level due to limitations stemming from the aforementioned glass composition issues. Further work has to be done, to improve passivation and overall efficiency gain for different solar cell design types to utilize the potential of pure Ag pastes on B emitters.

5 ACKNOWLEDGEMENTS

We would like to thank T. Droste, J. Booth, N. Nowak, and B. Cela from Johnson Matthey for their paste manufacturing expertise. Furthermore, we like to thank L. Mahlstaedt, L. Kraus, and B. Rettenmaier for their technical support. Part of this work was financially supported by the German Federal Ministry for Economic Affairs and Energy (FKZ 0325581). The content is the responsibility of the authors.

6 REFERENCES

[1] R. Kopecek, T. Buck, J. Libal, R. Petres, I. Röver, K. Wambach, R. Kinderman, L.J. Geerligs, P. Fath, Proc. 14th IPSEC, Shanghai, China, 2005, pp. 892-894.

[2] R. Lago, L. Pérez, H. Kerp, I. Freire, I. Hoces, N. Azkona, F. Recart, J.C. Jimeno, Prog. Photovolt.: Res. Appl. **18**, pp. 20-27 (2010).

[3] S. Riegel, F. Mutter, T. Lauermann, B. Terheiden, G. Hahn, Energy Procedia **21**, pp. 14-23 (2012).

[4] S. Riegel, F. Mutter, G. Hahn, B. Terheiden, Energy Procedia **8**, pp. 533-539 (2011).

[5] G. Schubert, J. Horzel, R. Kopecek, F. Huster, Proc. 20th EUPVSEC, Barcelona, Spain, 2005, pp. 934-937.

[6] J. Engelhardt, A. Frey, S. Gloger, G. Hahn, B. Terheiden, Appl. Phys. Lett. **107**, p. 042102 (2015).

[7] S. Fritz, M. König, S. Riegel, A. Herguth, M. Hörteis, G. Hahn, IEEE J. Photovolt., 5(1) (2015) 145-151.

[8] J. Engelhardt, A. Frey, S. Fritz, G. Micard, S. Riegel, G. Hahn, B. Terheiden, Proc. 31st EUPVSEC, Hamburg, Germany, 2015, pp. 351-354.

OPTIMAL THERMAL ANNEALING OF A-SIOX LAYER OBTAINED BY PECVD FOR HETEROJUNCTION SOLAR CELL APPLICATION

L. Martini [2], L. Serenelli [1,2], L. Imbimbo [2], F. Menchini [1], M. Izzi [1], R. Asquini [2], M. Tucci [1]
[1] ENEA Research Centre "Casaccia" DTE-FSN-TEF, via Anguillarese 301, 00123 Rome – Italy
[2] DIET University of Rome "Sapienza", via Eudossiana 18, 00184 Rome – Italy

ABSTRACT: In a-Si:H/c-Si heterojunction solar cells the UV absorption of amorphous layers limits the current generation. The use of a more transparent material like a-SiO_x:H films is a key to further enhance the efficiency of this kind of cells. At the same time this layer must guarantee high surface passivation of the c-Si wafer to be suitable in high efficiency solar cell manufacturing. As for amorphous silicon, thermal annealing procedures are considered as valuable steps to enhance and stabilize thin film properties, when performed at opportune temperature. In order to avoid damages of the amorphous layer, each cell fabrication step should be carried at temperature lower than the emitter formation one. However, when thermal annealing is applied as one of the early fabrication step, the temperature can be high enough to obtain the best passivation properties. In this work we have analyzed the effect of several thermal treatments different for temperature, duration and surrounding atmosphere on a-SiO_x:H/c-Si/a-SiO_x:H structure. We have fully characterized the samples with the aid of effective lifetime evaluation and FTIR spectra to correlate the effect of thermal annealing to the a-SiO_x:H/c-Si interface.
Keywords: amorphous, SiOx, annealing, PECVD, passivation, FTIR

1 INTRODUCTION

The PV global market trend is to reduce costs per Wp adopting the most cost-effective technologies. Crystalline silicon based solar cells are the mainly produced, basically on p-type, even if n-type silicon is increasing its market share[1]. The reason of this rising is due to the high efficiency achievable with this type of silicon, since high efficiency means lower cost per Wp, because of lower material consumption per module area [2]. The amorphous/crystalline silicon heterojunction (HJ) solar cells largely demonstrated to be high efficiency capable, indeed the actual PV convertion world record has been recently achieved by this kind of architecture [3]. However in a-Si:H/c-Si heterojunction solar cells the UV absorption of amorphous layers limits the current generation if the amorphous emitter layer is on sun-lighted cell front side. The adoption of a more transparent material, like a-SiO_x:H films is a key to further enhance the efficiency, if it proves at the same time high passivation quality of the c-Si surface. Moreover a-SiO_x:H film already demonstrated to be effectively doped and suitable for HJ solar cell manufacturing [4], thus enhancing the photocurrent generated by short wavelength part of sunlight spectrum [5]. As for amorphous silicon, thermal annealing procedures can be considered as valuable steps to enhance and stabilize the thin film properties, when performed at opportune temperature. Within the heterojunction cell process the steps following the emitter deposition should be performed at temperature below that of the amorphous layer deposition in order to avoid damages of the amorphous layer properties. Nevertheless a effective lifetime metastability of the a-Si:H/c-Si structure is observed when the amorphous films are exposed to post-deposition treatments [6], so that deeper investigation are needed also for thermal annealing on a-SiO_x:H/c-Si. In this work we analyse the effect of several thermal treatments, different for temperature, duration and surrounding atmosphere, on a-SiO_x:H/c-Si/a-SiO_x:H structure.

2 EXPERIMENTAL

In order to compare thermal treatment effects we started producing a-SiO_x:H/c-Si/a-SiO_x:H samples on n-type CZ or FZ 1-5Ωcm, 250μm thick silicon wafers, both side polished. After standard cleaning procedure (SC1+SC2) and HF 1% dip, the wafers were entered in a 13.56MHz parallel plate PECVD system equipped with SiH_4 (5% diluted in Ar), H_2 and CO_2 gases. After an optimization procedure, related to the CO_2 flux addition to a set of parameter for a-Si:H deposition reported elsewhere [7], we fixed the following conditions:
RF: 36mW/cm2, P=1.5Torr, T=250°C, SiH_4=120sccm, CO_2=1.5sccm, H_2=177sccm. The obtained film thickness, on both wafer side, is 15nm for all samples, which is similar to the emitter and buffer cumulated thickness in HJ solar cells.

Effective lifetime (τ_{eff}) were evaluated at room temperature via transient photo conductance decay (QSSPC, WCT 120, Sinton Instruments) in low injection level ($10^{15}cm^{-3}$), while FTIR spectra were collected by Perkin Elmer Frontier spectrophotometer equipped with U-ATR accessory.

Thermal annealing were performed in ATV static furnace in nitrogen or oxygen atmosphere at different temperatures for different duration time. τ_{eff} and FTIR were collected and analyzed also after annealing.

3 RESULTS AND DISCUSSION

3.1 Increasing temperature

As a first parameters exploration, we fixed the treatment time to 15 minutes, on the base of previous experience [8]. We then performed thermal annealing on samples based on CZ wafers in nitrogen atmosphere at the lower temperature of 250°C, collecting lifetimes before and after treatment, repeating the same step until a saturation value or a decrease is registered. Then on the same sample (Sample A) we increased the temperature to 300 °C and 350 °C. Similar experiment were carried out on sample starting from 300°C and 350°C. Results are summarized in Figure 1. Firstly we noticed that just after 30 minutes the effect of a single temperature is almost saturated, so we fixed to 30 minutes the treatment duration time for the subsequent experiments. Increasing temperature from 250 to 350 °C on the same Sample A leads to a τ_{eff} increment up to a maximum of 5.5 times the initial value. 300 °C was, at the same time, sufficient to bring the lifetime of Sample B directly to the same level when applied as first treatment, with a further increment when T = 350°C was subsequently applied. Finally when a sample was directly exposed to a thermal annealing at 350 °C (Sample C) the gain reached the maximum, but only if the treatment lasted 30 minutes. We verified also that 400 °C was detrimental for the wafer surface passivation, as well known for c-Si wafers passivated by amorphous layers [8].

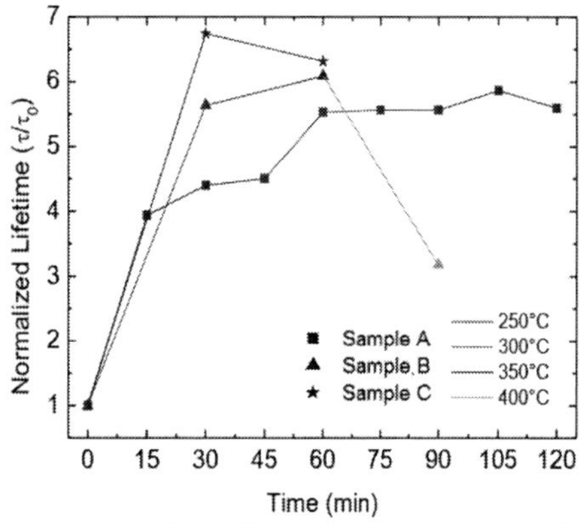

Figure 1: normalized lifetime for increasing annealing temperature. Symbols refer to the samples, while colors indicate the annealing temperature.

3.2 Different temperatures

In order to confirm the role of a single 30 minutes step at different temperature on different as-deposited samples, further experiment were carried out on higher quality (FZ) wafer, and results are illustrated in Figure 2. In this case even the stability of the resulting state was monitored. We observed a general lower lifetime enhancement up to 3.6 times, indicating that higher wafer quality leaved lower margin of improvement. A different behavior on sample treated at 350°C was evident: the normalized lifetime is less (τ/τ_0 = 3.2) than that for the sample treated at 300°C (τ/τ_0 = 3.6).

Moreover it was interesting to note the annealed state was metastable. Indeed just after a couple of hours passed from the treatment, the τ_{eff} value tended to decrease and reached a saturation, which was maintained even after weeks. In particular we observed that the 250 °C treated sample (Sample C) loses nearly 30% of its τ/τ_0 maximum value, the 300°C (Sample D) decreased of 35%, remaining at 2.4 times the initial value; Sample E (350 °C) at regime was just 20% less, that is 2.5 times τ_0.

Therefore 350 °C ensured, at regime, a global better improvement for such material. If we compare Figure 2 and Figure 1, it can be concluded that 350 °C in nitrogen atmosphere is the best annealing condition.

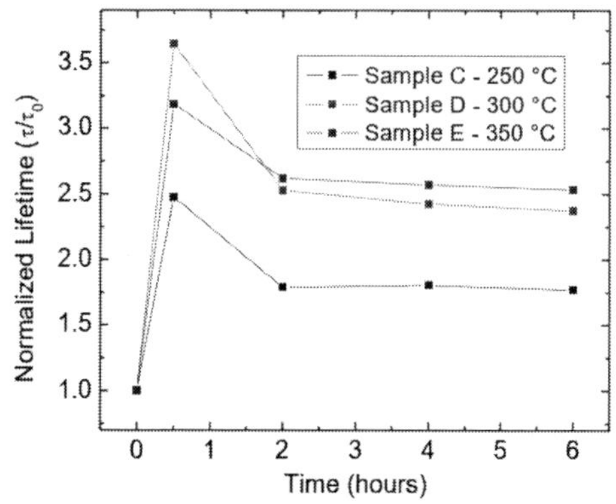

Figure 2: effect of different temperatures on lifetime enhancement and stability. 350 °C shows the best stability in time, 300 °C ensures the highest improvement after thermal annealing.

However looking at the absolute τ_{eff} values, reported in Table I, it can be noted that samples started at different values all close to 400 μs, but the best obtained lifetime was achieved on Sample B, which also kept its value at regime around 1 ms. For all these reasons we individuate 300 °C as the optimal annealing temperature for a-SiO_x:H.

Table I: effective lifetime of different samples. Sample B treated at 300°C shows the best performances.

Sample	τ_{eff} as-deposited (μs)	τ_{eff} annealed (μs)	τ_{eff} regime (μs)
A-250	430	1060	760
B-300	**400**	**1460**	**950**
C-350	350	1115	885

3.3 FTIR analysis

It is commonly addressed that SiO_x films obtained by PECVD from Silane and CO_2 discharge is a kind of amorphous material with a mixed phase of small crystals [9]. Such a kind of crystallinity fraction can be detected with the aid of Raman scattering and X-ray diffraction. However even in FTIR spectra some traces can be found to identify the mixed phase presence.

On the other hand the a-SiO_x:H film obtained by PECVD is very similar to a-Si:H film, except for a small CO_2 added quantitative, so their properties are quite similar. In particular the same model developed for a-Si:H films can be used to describe the FTIR spectra of a-SiO_x:H, examining the Si-H bonding type reported in literature. Among all peaks revealed on FTIR spectra related at Si–H bonding there are two region of particular interest that determine the main electro-optical film proprieties and thus the surface passivation. The first region is between 1980 and 2030 cm^{-1} denoted ad Low Stretching Mode (LSM) absorption, the second region starts at 2060 and ends at 2160 cm^{-1} and is caused to High Stretching Mode (HSM) absorption.

The LSM is related to the monohydrides, in particular to Si–H inside the monovacancy or at the edge of polymer-like m-vacancy. HSM instead is mainly related at dihydrites vibration, in particular when located inside divacancies and superficial voids. Monohydride peaks can also be found in the HSM region when Si–H is in center of m-vacancy or in superficial voids [10].

Generally a film with a dominant HSM absorption has inferior passivation quality and the HSM and LSM are inversely connected [11]. It is possible to make some considerations on the type of network in SiO_x:H film, amorphous or nano-crystalline, examining the Si–H bonding type detected in FTIR spectra. In general the dominance of LSM absorption indicates H in a compact amorphous film or isolated H [10].

We collected FTIR spectra of as-deposited sample. To correlate the lifetime behavior to various Si-H bonding, we repeated the measurement on the sample just thermally annealed and some hours after the treatment.

In particular we focused on the region between 1975 and 2200 cm^{-1}. We noticed some peak presence, even if partially convolute, at 1980, 2050, 2080, 2100, 2110, 2140, 2168, 2180 cm^{-1}.

From literature it is possible to assign the 1980 cm^{-1} peak to a Low Stretching Mode of Si-H in the film bulk related to mono-, di- and multi- vacancies [10]. 2050 cm^{-1} can be assigned to Si-H stretching at the amorphous-crystalline interface [12], while the presence of a narrow doublet at 2080-2100 cm^{-1} indicates [13] the presence of mono-di and trihydrides on crystalline surfaces, indicating the presence of grain boundaries in the film bulk. On the other hand 2110 and 2140 cm^{-1} are typical of $Si-H_2$ and $Si-H_3$ on the growing surface of amorphous films, that cannot be found on nc-Si films [12], unless considering this bonding as the trace of a mixed amorphous/nanocrystalline phase. 2170-2180 cm^{-1} are due to a Si-H stretching where Si is backbonded to one or two oxygen atoms. Moreover at 2100 cm^{-1} a component of Si-H LSM stretching can be found, that is a shift of 2000 cm^{-1}, occurring when oxygen is included in amorphous film [14]. Generally it can be concluded that LSM modes between 1980 and 2000 cm^{-1} are due to silicon monohydrides which are easy available to form bonding on c-Si wafer surface, enhancing the passivation properties. HSM modes, generally found between 2000 and 2150 cm^{-1}, are surface multi hydrides components convolute with other HSM between 2100 and 2150 cm^{-1}. In SiO_x, an inclusion of LSM Si-H stretching mode can be found in HSM region at 2100 cm^{-1}.

On these basis in Figure 3 we were able to compare FTIR spectra in the 1975-2200 cm^{-1} region at different stages of the experiment reported in Figure 2 for sample D, to correlate a macroscopic effect like the a-SiO_x:H/c-Si/ a-SiO_x:H effective lifetime and the evolution of hydrogen in the film and at the amorphous/crystalline interface.

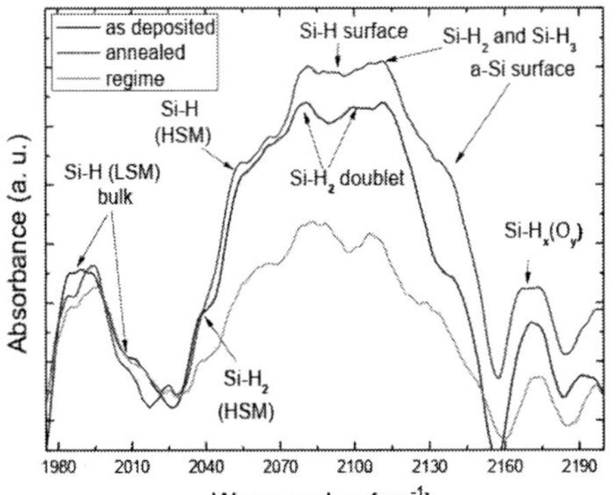

Figure 3: FTIR spectra of samples before and after a treatment at 300 °C in nitrogen.

For the as-deposited sample it was immediate to recognize the $Si-H_2$ doublet, confirming the crystalline phase presence. On the other hand LSM pronounced band was individuated, centered at 1985 cm^{-1}. Silicon hydrides can be found also on oxygen back-bonded, confirming the oxide nature of the film. Immediately after the thermal treatment at 300°C in nitrogen atmosphere three evolution manifested:

a) the LSM peak shifted its center from 1985 to 1995 cm^{-1};
b) the $Si-H_3$ component dramatically increased;
c) the whole 2070-2110 cm^{-1} increased due to the raise of 2100 cm^{-1} peak.

These considerations remarked the presence of the amorphous phase, which in turn was responsible of the improved lifetime of the sample.

As the time passed, however, it was clear that the effect of temperature was metastable: part of the peaks related to the amorphous free surface (2110-2140 cm^{-1}) dramatically decreased, and the doublet was again clearly recognizable. However the LSM peak was still present and, just near the 2080 cm^{-1} peak of the doublet, a 2085-2090 cm^{-1} was observable, which was addressable to a Si-H in HSM mode [15].

These model allowed to figure out an external surface modification and an overall hydrogen effusion which lowered some surface and bulk properties, while the most of Si-H passivating bonding still remained. This justified at regime the 30% decrease with respect to the maximum τ/τ_0 registered just after the thermal treatments (that is amorphous phase was prominent), but also that the lifetimes did not completely fall back to the starting level.

3.4 Role of the gas atmosphere on thermal annealing

On the basis of the previous considerations, various effects on the film surface were evident. To explore the possibility that the gas surrounding the samples during the thermal treatment played a role in the film annealing and consequently on the passivation properties and their stability, we performed the same annealing at 300 °C for 30 minutes in nitrogen and oxygen atmosphere on two identical samples. Normalized lifetimes in Figure 4 are illustrated.

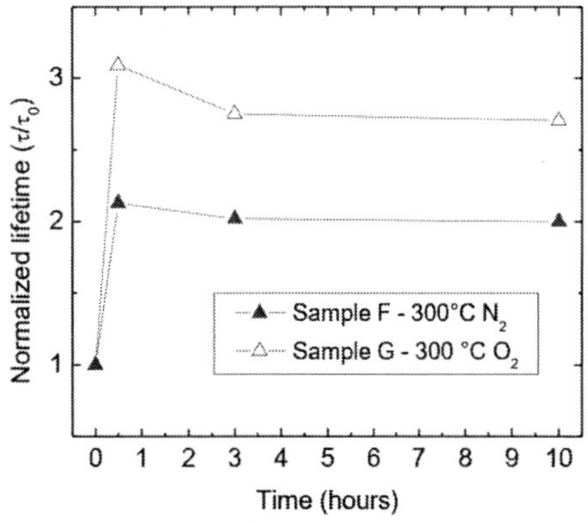

Figure 4: effect of oxygen or nitrogen surrounding the sample during the thermal annealing at 300 °C on normalized lifetime for SiO_x/c-Si/SiO_x samples.

Sample G, annealed in oxygen, showed more enhancement than Sample F, analogous at those examined before. Moreover also the stable value was higher, even if an initial larger decrease in τ/τ_0 was observed. FTIR spectra, reported in Figure 5, show almost equal curves as already examined in Figure 3 for as-deposited and annealed samples in nitrogen.

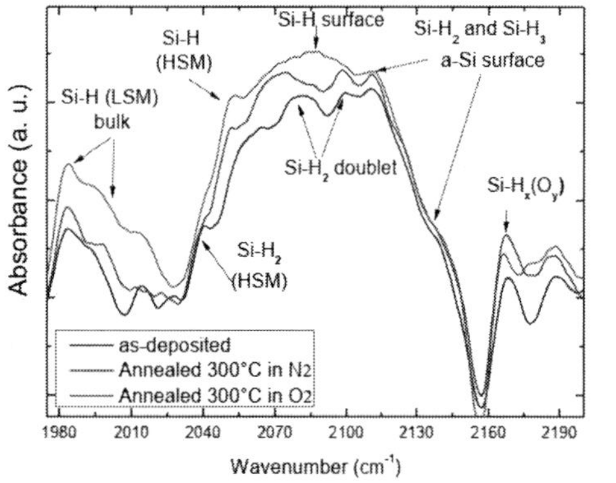

Figure 5: FTIR spectra of samples before and after a treatment at 300 °C in nitrogen and oxygen.

However a quite different behavior was noted for Sample G, annealed in presence of oxygen: firstly LSM band was broader and more intense, while Si multi hydrides were very similar to those of nitrogen-annealed sample. Secondarily was much more evident the Si-H surface peak around 2100 cm^{-1}, with the doublet that was difficult to be recognized, unless a de-convolution could show its presence. Moreover Si-H HSM bonding were further increased, while oxygen back-bonded was less present.

We can conclude that thermal treatment performed in presence of oxygen, at fixed temperature, can be suitable to maximize and stabilize the a-SiO_x:H passivation properties on c-Si surface.

Further investigations are needed for a full understanding of the mechanism that allows such large improvement, although it can be figure out a formation on the a-SiOx free surface of a very thin layer of stronger oxide, which could act as a "capping layer" avoiding the hydrogen effusion outside the film, forcing it to fill all the voids and microvoids in the amorphous bulk and at the interface with c-Si surface, enhancing the passivation properties.

4 CONCLUSIONS

In this work we investigated the effect of a thermal annealing on a-SiO_x:H obtained by PECVD on c-Si wafer surface passivation quality. We explored different temperature even increasing the temperature cumulatively on the same sample and individually on separate sample. The first conclusion is that the lifetime enhancement was related directly on the temperature, while just 30 minutes were sufficient to obtain the best condition for a specific temperature.

The second one is that thermal annealing was much more effective in improving the a-SiO_x/c-Si passivation effectiveness starting on low quality material. In turn on better wafer, the passivation offered from the as-deposited layer was already good enough, so there was less margin for improvement.

Third we individuated in 300°C the optimum temperature for the thermal annealing of a film deposited at 250 °C from SiH_4, CO_2 and H_2.

Correlating lifetime data with FTIR spectra of samples collected at different stage we individuated that the film had a mixed phase of amorphous/crystalline SiO_x, where the former become predominant after the thermal treatment. The observed metastability was related to hydrogen evolution from surface state, and from a partial regression of the amorphous phase to crystalline one.

Finally when the same thermal treatment was carried out in oxygen a higher passivation quality and stability were achieved. FTIR showed that the amorphous phase was persistent, and the crystalline phase was less pronounced, even after long time from the thermal annealing treatment.

5 ACNOWLEDGEMENT

This work is supported by the Italian Ministry of Economic Development in the framework of the Operating Agreement with ENEA for Research on the Electric System.

6 REFERENCES

[1.] PV Technology Roadmap Report – 2014, PV Technology Roadmap Report – 2014, available online at http://www.itrpv.net/Reports/Downloads/

[2.] C. Honsberg and S. Bowden, PVCDEOM, chap 5.4, available online at http://www.pveducation.org

[3.] news Panasonic Photovoltaic Module Achieves World's Highest Energy Conversion Efficiency of 23.8% at Research Level, march 2016, http://news.panasonic.com/global/press/data/2016/03/en160302-2/en160302-2.html

[4.] M. Izzi, M. Tucci, L. Serenelli, P. Mangiapane, M. Della Noce, I. Usatii, E. Esposito, L.V. Mercaldo, P. Delli Veneri, "Doped SiOx emitter layer in amorphous/crystalline silicon heterojunction solar cell", Applied Physics A 115(2):705-712, May 2014.

[5.] D. Y. Kim, E. Guijt, R. van Swaaij and M. Zeman, "Development of a-SiOx:H solar cells with very highVoc×FF product", Prog. Photovolt: Res. Appl.2015; 23:671–684.

[6.] L. Martini, L. Serenelli, R. Asquini, D. Caputo, G. de Cesare, M. Izzi, M. Tucci, "Surface photovoltage as a tool to monitor the effect of hydrogen treatment on a-Si:H/c-Si heterojunction", Proceedings of SPIE - The International Society for Optical Engineering 8823, August 2013.

[7.] L. Serenelli, R. Chierchia, M. Izzi, M. Tucci, L. Martini, D. Caputo, R. Asquini, G. de Cesare, "Hydrogen Plasma and Thermal Annealing Treatments on a-Si:H Thin Film for c-Si Surface Passivation", Energy Procedia 60 (2014) 102 – 108.

[8.] M. Tucci, L. Serenelli, "Metastability of SiNx / a-Si:H crystalline silicon surface passivation for PV application", Thin Solid Films 516(20):6939-6942,

August 2008.

[9.] H. P. Zhou, D. Y. Wei, S. Xu, S. Q. Xiao, L. X. Xu, S. Y. Huang, Y. N. Guo, S. Khan, M. Xu, "Si surface passivation by SiOx:H films deposited by a low-frequency ICP for solar cell applications", J. Phys. D: Appl. Phys. 45 (2012) 395401.

[10.] A. Smets, "Growth Related Material Properties of Hydrogenated Amorphous Silicon", ph.D thesis, Eindhoven University of Technology, 2002, chap 2.

[11.] O. Pehlivan, D. Menda, O. Yilmaz, A. O. Kodolbas, O. Ozdemir, O. Duygulu, K. Kutlu, M. Tomak, "Structural and interfacial properties of large area n-a-Si:H/i-a-Si:H/p-c-Si heterojunction solar cell", Proc. SPIE 8823, Thin Film Solar Technology V, 88230T (11 September 2013).

[12.] D.C. Marra, E. Edelberg, R. Naone, E. Aydil, "Silicon Hydride composition of plasma-deposited hydrogenated amorphous and nanocrystalline silicon films and surfaces", J. Vac.sci.technol. A 16(6) 1998 p.3199.

[13.] A. H. M. Smets, T. Matsui, and M. Kondo, "Infrared analysis of the bulk silicon-hydrogen bonds as an optimization tool for high-rate deposition of microcrystalline silicon solar cells", Applied Physics Letters, 92(3):033506, 2008.

[14.] Pavel V. Bulkin, Pieter L. Swart, and Beatrys M. Lacquet. Electron cyclotron resonance plasma enhanced chemical vapour deposition and optical properties of SiO x thin films. Journal of Non-Crystalline Solids, 226(1-2):58 – 66, 1998.

[15.] M. Z. Burrows, U. K. Das, R. L. Opila, S. De Wolf, and R. W. Birkmire. Role of hydrogen bonding environment in a-Si:H films for c-Si surface passivation. Journal of Vacuum Science and Technology A, 26:683–687, July 2008.

A NEW TYPE BACK CONTACT SOLAR CELLS BASED ON SI WAFER AND COMBINED WITH THE MULTILAYER MOOX/AG/MOOX AND CESIUM CARBONATE FILMS

Weiliang Wu[1], Jie Bao[1] and Hui Shen [1,2*]

[1] Institute for Solar Energy Systems, Sun Yat-Sen University, 5th Floor, C Block, Building of Engineering, 132 Wai Huan Dong Road, Higher Education Mega Center, Guangzhou, Guangdong, P.R. China, 510006
[2] Shunde-SYSU Institute for Solar Energy, No 1, Desheng Dong Road, Daliang,Shunde, Guangdong Province, China, 528300
*Corresponding author: Phone:+86 15521108087; Email: shenhui1956@163.com

ABSTRACT: We present a new dopant-free back contact solar cells combined with the multi-functional emitter MoO_x (20 nm)/Ag (15 nm)/MoO_x (10 nm) (MAM) and 3nm Cs_2CO_3 as back surface field (BSF). The initial efficiency of 14.06% has been achieved, featuring the short circuit current density of 36.4 mA/cm^2, open-circuit voltage of 591 mV and FF of 65.37% under STC (AM 1.5 global spectrum, 25 °C and 1000 W/m^2) on the 4 cm^2 area single side textured FZ n-Si. The front textured surface is passivated by plasma-enhanced chemical vapour deposition (PECVD) 70 nm SiN_x and thermal oxide 10 nm SiO_2. The rear-side 100 µm interdigital gap, 250 µm half-width electrode/Cs_2CO_3 and 750 µm half-width electrode/MAM are deposited by thermal evaporation with accurate control by metal masking instead of high temperature diffusion and photolithographic masking. In order to achieve higher efficiency, we analyze the power loss, due to the front surface recombination velocity can be reduced to S=2.9 cm/s, the main losses of J_{SC} are the light escape and back surface recombination. Additionally, the trend of series resistance changing with voltage is different from the homojunction back contact solar cells.
Keywords: back contact, solar cell, heterojunction

1 INTRODUCTION

The Si heterojunction structure adopted in HIT solar cell has exhibited high efficiency by depositing the intrinsic (i-type) and doped a-Si:H layers on a randomly textured c-Si wafer. The HIT solar cells can achieve high efficiency is mainly due to the excellent surface passivated properties for reducing the defects on the c-Si surface with high-quality intrinsic a-Si:H. Panasonic recently reported a new world record of 25.6% based on a combination of the interdigitated back contact (IBC) and silicon heterojunction (SHJ) technology [1].

Interdigitated back contact silicon heterojunction (IBC-SHJ) solar cells have the potential for high open circuit voltage (V_{OC}) due to the excellent surface passivation and heterojunction contacts, and high short circuit current density (J_{SC}) owing to the back contact structure [2]. However, the IBC-SHJ solar cells were prepared by the complex process for the emitter and BSF. Especially, when the a-Si was deposited by the chemical vapor deposition (PECVD) on both sides of the wafer, the silane was inflammable and explosive.

Recently, Oxide/Metal/Oxide structures have been shown to be promising electrodes and probed as anode in planar organic photovoltaic cells (OPVCs) for collecting holes. Cu or Ag layer is embedded in WO_3, V_2O_5, MoO_3 films to obtain the Oxide/Metal/Oxide multilayer structures. MoO_x/Ag/MoO_x multilayer structure has also been proven to be low sheet resistance and high work function [3]. The Cs_2CO_3 has also been proven to be a very efficient electron-injection material in Organic Light-Emitting Devices (OLEDs), including in OLEDs based on small molecules and polymers. [4]

In this paper, we propose a novel structure of silicon based solar cells combined with MoO_x (20 nm)/Ag (15 nm)/MoO_x (10 nm) multilayer films, 3 nm Cs_2CO_3 and n-Si as shown in Fig. 1. The interdigital gap was 100 µm, the half-with of back surface field (BSF) is 250 µm, and the half-width of the MAM/electrode contact is set as 750 µm. We systematically investigated the properties of elements in MAM after thermal evaporation process. And the electrical properties of MAM/n-Si and Cs_2CO_3/n-Si contacts are measured and analyzed.

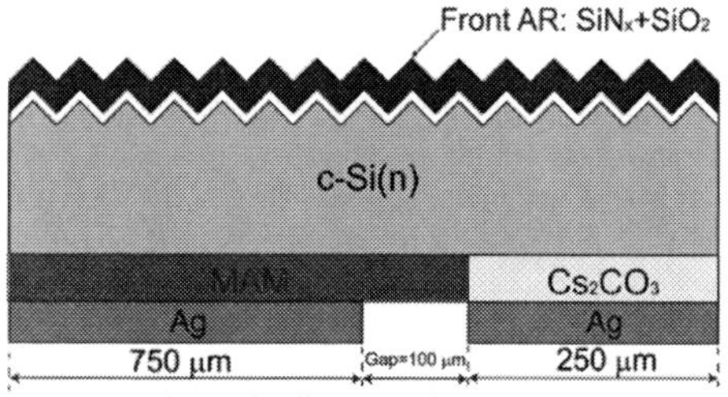

Figure 1: Schematic diagram of back contact solar cells by using MoO_x (20 nm)/Ag (15 nm)/MoO_x (10 nm) and 3 nm Cs_2CO_3 films.

2 PREPARATION AND CHARACTERIZATION OF DEVICES

The phosphorus doped and boron doped 400-µm thick (100) oriented float-zone silicon wafers were polished on one side, and both of them had resistivity of 3~5 Ω•cm. Silicon substrates were cleaned prior to the MoO_3 deposition, which consisted of standard RCA processing followed by (10%) hydrofluoric acid treatment to remove native oxide. The 99.95% pure MoO_3 powder and 99.99% Ag wire were placed into alumina-coated tungsten boat separately as source materials, the deposition rate of MoO_3 and Ag were strictly controlled to be 0.5 Å/s and 2 Å/s, especially for Ag deposition rate. The thicknesses of MoO_3 and Ag layers were monitored using a quartz crystal monitor (Model SQM200, Filtech Company, USA) positioned a short distance from the source. The pressure and temperature during the whole deposition was set 8.0×10^{-4} Pa and 25~30 °C, respectively.

3 RESULT AND DISCUSSION

3.1 Element analysis of transition metal oxide and silver in the MAM films

The bottom curve in Fig. 2a presents the X-ray photoelectron spectrum of the identify as the Mo^{6+} and Mo^{5+} valence state, corresponding to an oxygen vacancy at the vertex of an octahedron in the amorphous network. It is worth noting that the fully stoichiometric MoO_3 with only Mo^{6+} is insulating, oxygen vacancy of the MoO_x with Mo^{5+} can reduce the film resistivity to 5×10^4 Ωcm [5]. In the MAM multilayer films, the silver is not oxidized as shown in Fig. 2b, ultra low sheet resistance ~8 Ω/□ has been achieved.

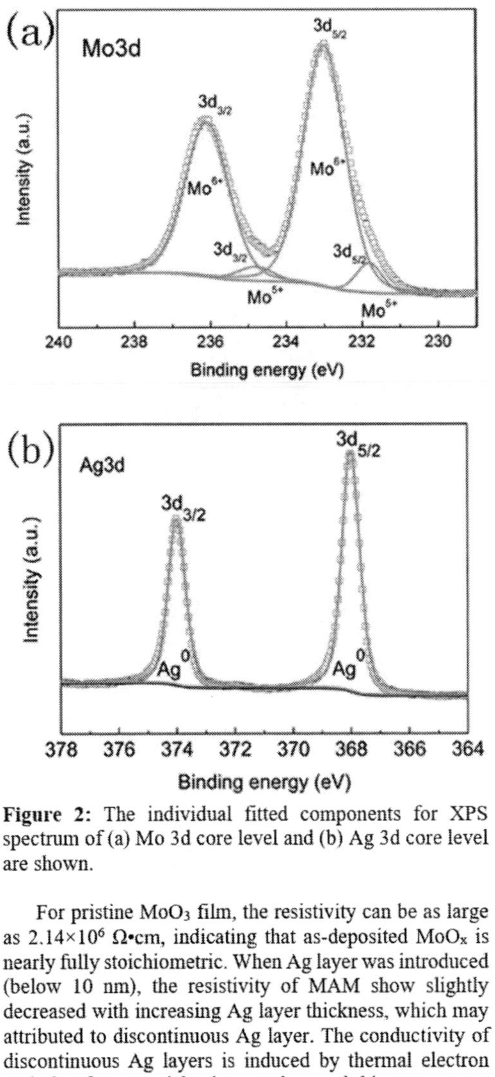

Figure 2: The individual fitted components for XPS spectrum of (a) Mo 3d core level and (b) Ag 3d core level are shown.

For pristine MoO_3 film, the resistivity can be as large as 2.14×10^6 Ω•cm, indicating that as-deposited MoO_x is nearly fully stoichiometric. When Ag layer was introduced (below 10 nm), the resistivity of MAM show slightly decreased with increasing Ag layer thickness, which may attributed to discontinuous Ag layer. The conductivity of discontinuous Ag layers is induced by thermal electron emission from one island to another, and this process can be enhanced when isolated islands are coalescence to form wormlike networks. When Ag layer thickness equals or excesses 10 nm, continuous Ag layer is formed.

On one hand, electrons can move even more easily in a complete continuous Ag layer; on the other hand, electrons inject into MoO_x film from Ag layer can improve MoO_x film's conductivity, both aspects explain the sudden resistivity values commutation from 1×10^6 to 1×10^{-5} Ω•cm. Further increasing Ag layer thickness doesn't make any significant change in resistivity. So due to different carriers transport mechanism in continuous and discontinuous Ag layer, there is a threshold value of Ag thickness (10 nm).

The morphology and electrical conductivity of the multilayer films are strongly influenced by the Ag layer thickness. We set the continuous Ag film as 15 nm in order to achieve the low sheet resistance.

3.2 J-V characteristics

Fig. 3 shows a high short-circuit photocurrent density J_{SC} of 36.4 mA/cm², an open-circuit voltage V_{OC} of 591 mV and a moderate fill factor (FF) of 65.37%, corresponding to the power conversion efficiency (PCE) of 14.06%.

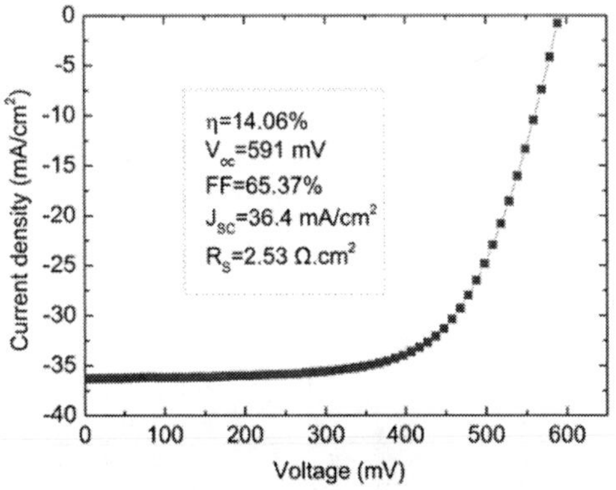

Figure 3: J–V characteristic of the back contact solar cell.

The values of recombination current corresponding to the different mechanisms are extracted by the contactless Quasi-Steady-State Photo-conductance in Table I. The term $1/\tau_{bulk}$ can be neglected owing to the long bulk lifetime of high quality wafers and the front surface recombination velocity S=2.9 cm/s can be obtained through the SiN_x/SiO_2 stack passivation. The absence of interfacial passivated layer on the back surface, characteristic recombination current for the emitter MAM is 220 fA/cm², which are close to the reported value of 230 fA/cm² for MoO_x/n-Si contact [5]. The V_{OC} is mainly limited by the recombination of back surface. And the R_S=2.53 Ω.cm² is also the dominant power loss [6]. Because of the SiN_x/SiO_2 stack passivation for front surface, the loss of J_{SC} is mainly the optical loss and also needed analyzed for further improvement [7].

Table I: Measured majority carriers lifetime and recombination current density.

Parameters	$J_{0\,emitter}$ (fA/cm²)	$J_{0\,FSF}$ (fA/cm²)	$J_{0\,BSF}$ (fA/cm²)	τ_{Bulk} (ms)
Values	220	19.5	300	3

2.3 Power loss analysis-optical loss

Based on the EQE and optical measurement, We analyze the loss of J_{SC} combined with the OPAL simulator [8] are depicted in Fig. 4a and the results are presented in Fig. 4b. The total J_{SC} loss for the back contact solar cell amounts to 7.6 mA/cm². The loss of J_{SC} causes the reduction in the V_{OC} and fill factor as well. The largest loss mechanism is the escape from front surface, which accounts for 2.57 mA/cm². Interestingly, the parasitic absorption of MAM multilayer films introduce the 1.33 mA/cm² optical loss, because the 15 nm silver film absorbs long wavelength light strongly [9, 10].

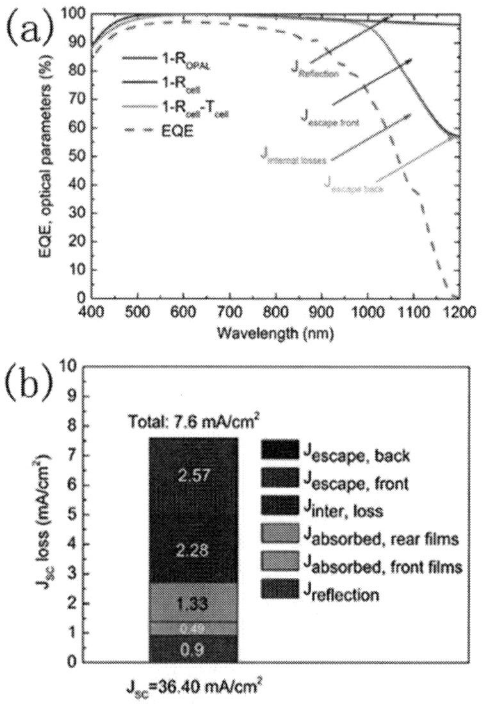

Figure 4: (a) Optical measurement (R_{cell}) and OPAL-simulated reflection of MAM/n-Si back contact solar cell. (b) The detailed loss components of short circuit current density and the J_G=44 mA/cm^2.

3.4 Power loss analysis-electrical loss

As shown in Fig. 4b, the second largest contribution to the J_{SC} losses is the inter loss including the n-Si bulk recombination, back surface defect (SRH) recombination, emitter and BSF bulk recombination. However, the R_S is also the dominant power loss resource. The Fig. 5a shows the R_S as a function of voltage measured by multi-light intensity J-V measurement, and the analyzed solar cell operates at the maximum power point (MPP) condition is remarked by purple Pentagram.

Table II: The contact resistivity ρ_c for MAM/n-Si and Cs_2CO_3/n-Si contacts.

Parameters	Cs_2CO_3/n-Si contact	MAM/n-Si contact
ρ_c (Ω.cm^2)	0.47	0.09

The trend of the R_S (V) curve is increasing under low voltage, this phenomenon contrary to the homojunction solar cells [11]. It is also found that the voltage exceeding 300 mV, the R_S is decreasing strongly, which is caused by the reduction of contact resistance with higher injection level. The higher electrical injection level could improve the carriers concentration. While the voltage exceeding 500 mV, the R_S is limited and keeps constant.

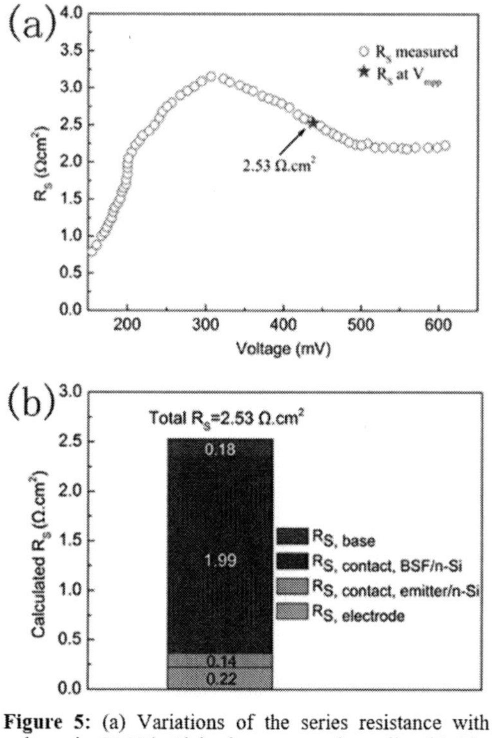

Figure 5: (a) Variations of the series resistance with voltage in TMO/n-Si back contact solar cells. (b) The detailed components of series resistance existed in devices.

According to Table II, the every part of the series resistance is measured and shown in Fig. 5b. We find that the Cs_2CO_3/n-Si contact has the largest ρ_c=0.47 Ω.cm^2 and contibute to the $R_{S,\ contact,\ BSF/n\text{-}Si}$=1.99 Ω.cm^2. The Cs_2CO_3/n-Si contact resistivity is influenced by lots of complex influences including Cs_2CO_3 thickness, process, interfacial film. The a-Si:H(i) has the relative low contact resistivity and can be used as the interfacial passivation and reduction of contact resistance [12].

4 SUMMARY

We have demonstrated that the novel structure silicon based solar cell combined with MAM and n-type Si shows the excellent performance. The resistive loss is the important loss mechanism, especially the Cs_2CO_3/n-Si contact. For the losses of J_{SC} are mainly composed of the light reflection and escape of the front surface.

Pathway to achieve higher efficiency:

- Improving the light trapping and adding the back reflector.
- Reducing the series resistance focused on the Cs_2CO_3/n-Si contact.
- Passivating the back interface between MAM, Cs_2CO_3 and n-Si.
- Simulation of device.

ACKNOWLEDGEMENTS

This work was supported by the National High-tech R&D Program of China (863 Program) (2015AA050303 and 2015AA050608), the Guangdong-Hong Kong Technology Cooperation Funding Scheme (2014B050505010), the Guangzhou collaborative innovation Major Project of producing, teaching and researching (No. 201508010011) and Jiangsu Collaborative Innovation Center of Photovoltaic Science and Engineering (Grant No. SCZ1405500002).

REFERENCES

[1] K. Masuko *et al.*, IEEE Journal of Photovoltaics **4,** 1433-1435 (2014).
[2] M. Lu *et al.*, Applied Physics Letters **91,** 63507-63507 (2007).
[3] K. Jeon *et al.*, Nanoscale research letters **7,** 1-7 (2012).
[4] Y. Zhang *et al.*, Energy & Environmental Science **8,** 297-302 (2015).
[5] L. G. Gerling *et al.*, Solar Energy Materials and Solar Cells **145,** 109-115 (2016).
[6] R. Gogolin *et al.*, IEEE Journal of Photovoltaics **4,** 1169-1176 (2014).
[7] P. J. Verlinden *et al.*, Solar Energy Materials and Solar Cells **106,** 37-41 (2012).
[8] www.pvlighthouse.com.au
[9] L. Cattin *et al.*, Thin Solid Films **518,** 4560-4563 (2010).
[10] T. Abachi *et al.*, Thin Solid Films **545,** 438-444 (2013).
[11] K. C. Fong *et al.*, Progress in Photovoltaics: Research and Applications **21,** 490-499 (2013).
[12] S.-Y. Lee *et al.*, Solar Energy Materials and Solar Cells **120,** 412-416 (2014).

WET OXIDATION EFFECTS ON THE ELECTRICAL AND INTERFACE PROPERTIES OF ALD Al_2O_3 AND ALD-AlO_x/SiN_x PASSIVATION STACKS FOR PERC SOLAR CELLS

Supawan Joonwichien*, Katsuhiko Shirasawa, Shalamujiang Simayi, Katsuto Tanahashi, and Hidetaka Takato

National Institute of Advanced Industrial Science and Technology (AIST),
Fukushima Renewable Energy Institute (FREA), 2-2-9, Machiikedai, Koriyama, Fukushima 963-0298, Japan

ABSTRACT: We investigated the impact of wet oxidation on the electrical and interface properties of aluminum oxide (Al_2O_3) using atomic layer deposition (ALD) and aluminum oxide/silicon nitride (AlO_x/SiN_x) passivation stacks on silicon. Wet oxidation has been observed to have a significant effect on the field-effect passivation, showing an increase in the flat-band voltage shift (V_{fb}), and on the chemical passivation, which resulted in a reduction of the interface trap density (D_{it}). The decrease in D_{it} resulted in an increase in the lifetime and open-circuit voltages (V_{oc}). This effect may be attributed to additional negatively charged traps originating from interstitial oxygen and hydrogen in the steam during wet oxidation. These results suggest that thermal annealing under ambient humidity is desirable to improve the quality of the passivation stacks on the rear side. This leads to high-performance passivated emitter rear cells (PERC).

Keywords: Passivation stacks, Atomic layer deposition, Silicon solar cell, Annealing, Wet oxidation

1 INTRODUCTION

An annealing process is required for aluminum oxide–silicon (Al_2O_3–Si) systems in order to obtain a high level of surface passivation. Annealing locally reconstructs Al_2O_3. Consequently, the density of negative fixed charges (Q_f) at the Al_2O_3–Si interface increases, which can augment the electron repulsions of the conduction band (p-type) [1, 2]. When accompanied by a strong hole accumulation layer in the valence band, a reduction in the surface recombination velocity can occur [1]. Negative Q_f has been reported to originate from aluminum vacancies (V_{Al}) and interstitial oxygen (O_i) [3, 4], including extrinsic defects such as interstitial hydrogen (H_i) [5]. Therefore, methods should be considered to improve the quality of surface passivation by adding additional sources of negative Q_f (V_{Al}, O_i, and H_i) to create negative charge traps in the films. A deeper understanding is still required to further improve the rear side of passivated emitter rear cells (PERC), given the underlying effects of wet oxidation on the field-effect passivation and chemical passivation.

For this purpose, we have attempted to improve the field-effect passivation using wet oxidation (bubbling N_2 though boiled deionized water). Our aim is to include additional charge traps from –OH bonds and/or O_i in the water vapor, increasing the density of negative Q_f in the films. The charge accumulates at the interface, resulting in a longer lifetime.

2 EXPERIMENTAL DETAILS

2.1 Sample preparation for characterization

The substrates used were 200-μm-thick p-type CZ-Si wafers (156 mm × 156 mm) with a resistivity in the range of 2.3 to 2.9 Ω cm. The Al_2O_3 and silicon nitride (SiN_x) films were deposited by atomic layer deposition (ALD) and plasma-enhanced chemical vapor deposition (PECVD), respectively. After saw damage removal and RCA cleaning, 20-nm-thick Al_2O_3 films were deposited on both sides of the Si substrate surface. As-deposited Al_2O_3 wafers were divided into 3 groups according to the applied processing: Group 1 (without annealing), Group 2 (dry N_2 annealing at 450 °C for 30 min), and Group 3 (wet oxidation by bubbling N_2 through boiled, deionized water 450°C for 30 min). An 80-nm-thick SiN_x capping layer with a refractive index (n) of 2.1 was then deposited on top of the as-deposited and annealed Al_2O_3 films using PECVD. After adding a SiN_x capping layer, samples were taken from Group 1 and divided into 3 additional groups: Group 4 (non-annealed aluminum oxide/silicon nitride (AlO_x/SiN_x) stacks), Group 5 (dry post-deposition annealing (PDA) of the stacks), and Group 6 (wet PDA of the stacks).

The electrical properties and surface passivation of the films and the AlO_x/SiN_x stacks were investigated by means of microwave photoconductance decay and contactless corona-voltage measurements (Semilab PV-2000, Budapest) before and after thermal annealing. The contactless corona-voltage technique used corona charging to deposit an electrical charge on a dielectric material, thus changing the electric field in both the dielectric film and the semiconductor. Corona charging changes the surface band bending in the silicon wafer. The flat-band voltage (V_{fb}) is obtained when there is no band bending observed upon corona charging [6]. Note that lifetime measurements were taken at a low injection level of 1.5×10^{13} cm^{-3} for all samples.

2.2 Large-area industrial PERC cell processing

Figure 1 shows the corresponding process flow diagram for the PERC solar cells fabricated in this work. The starting material used was the same as the samples used for characterization. After texturing and cleaning, an n^+–emitter with a sheet resistance of approximately 90 Ω/sq was formed from a $POCl_3$ source, followed by edge isolation and phosphorus silicate glass (PSG) removal. The 20 nm Al_2O_3 films were deposited by ALD on both sides of the substrates at a deposition temperature of 200 °C. At this point in the process, two groups of samples received different thermal annealing; dry annealing of Group 2 (dry annealed PERC) and wet annealing of Group 3 (wet annealed PERC); the remaining samples (Group 1) were used for PDA processing of the stacks (Group 5, and Group 6). A 190 nm SiN_x capping layer was deposited using PECVD on the back side above the Al_2O_3 films. An HF-dip was used to remove the oxide layer from the top side. An 80 nm SiN_x film was subsequently deposited on the top side. The PERC cells then received a line pattern of local contact opening using laser ablation to remove the local rear passivation layer. A chemical treatment was performed on all types of PERC cells by dipping the sample into KOH and HF solutions. Afterwards, rear side Al printing and top side Ag printing was applied and these finished PERC cells were fired.

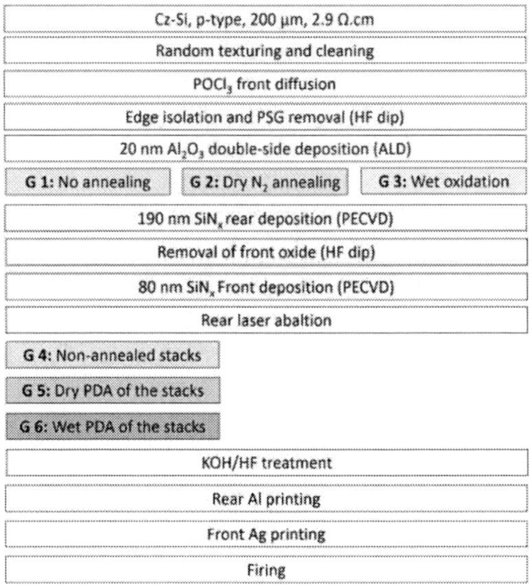

Figure 1: Process flow diagram for the PERC solar cells fabricated in this study.

3 RESULTS AND DISCUSSION

Figure 2 shows the relationship between the changes in lifetimes and the shifted V_{fb} of the annealed Al_2O_3 films after applying different thermal annealing processes of the annealed Al_2O_3 capped with a SiN_x layer and of Al_2O_3 after passivating the stacks with dry and wet PDA. An increase in the lifetimes of the annealed Al_2O_3 films with increasing V_{fb} can be clearly seen in Figs. 2a and 2b, indicating the activation of the field-effect passivation at the Al_2O_3–Si interface. These results agree with previous reports [1, 7–9] that thermal annealing under certain conditions is crucial to fully activate the field-effect passivation. This is directly related to the formation of a negative Q_f, which can assist in repelling electrons in the conduction band (p–type). When accompanied by a strong hole-accumulation layer in the valence band, a reduction in the surface-recombination velocity can occur. A hole-accumulation state can be realized by shifting V_{fb}, showing that the band bends at the interface on the energy-band diagram [7, 10].

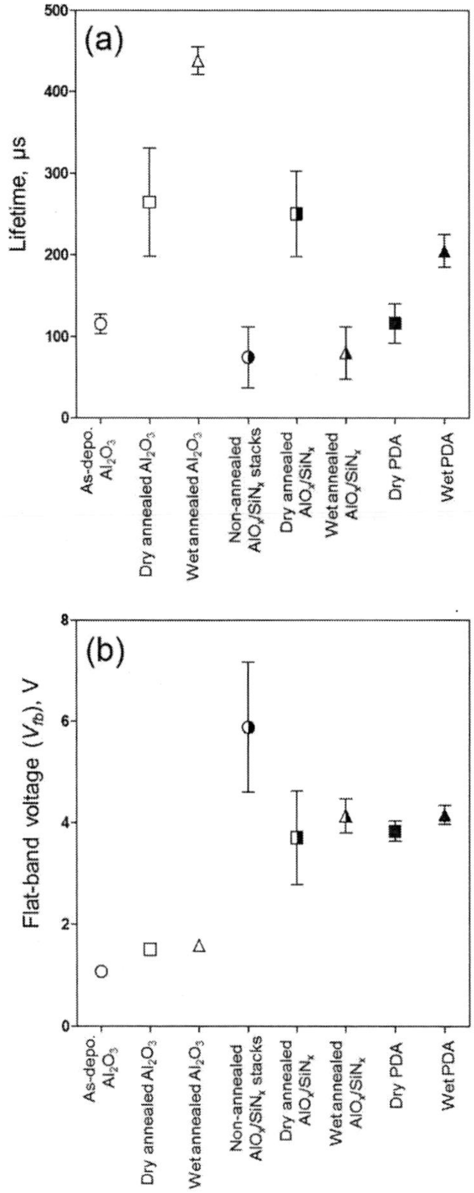

Figure 2: (a) Lifetime and (b) flat-band voltage (V_{fb}) of ALD–Al_2O_3 films and AlO_x/SiN_x stacks after applying different thermal processes.

Upon comparing dry and wet annealing of Al_2O_3 films, we see that the wet oxidation process had a greater lifetime but the V_{fb} was nearly equivalent after annealing. With respect to chemical passivation, the average interface trap density (D_{it}) values of the as-deposited, dry annealed, and wet annealed Al_2O_3 films were found to be 1.0×10^{12}, 3.1×10^{11}, and 5.3×10^{10} eV^{-1} cm^{-2}, respectively (Table 1). It is thus possible that the improved lifetime of wet annealed Al_2O_3 films compared to dry annealing could be related to the reduced D_{it} values. D_{it} is attributed to the interfacial hydrogenation of dangling bonds introduced during annealing [11–13] by hydrogen and water (atomic H, H_2, H_2O) present in the steam during wet oxidation. In the dry annealing process, the films were annealed under pure N_2 so that hydrogen is presumably not provided by the environment. This means that the hydrogen present in the as-deposited films plays an important role in achieving chemical passivation by diffusing from the hydrogen-rich bulk of the oxide films and rearranging/saturating the dangling silicon bonds at the interface.

Table I: Average interface trap densities (D_{it}) of as-deposited Al_2O_3 films, annealed films, and AlO_x/SiN_x stacks.

Sample	D_{it} (eV^{-1} cm^{-2})
As-deposited Al_2O_3 film	1.0×10^{12}
Dry annealed Al_2O_3 film	3.1×10^{11}
Wet annealed Al_2O_3 film	5.3×10^{10}
Non-annealed AlO_x/SiN_x stacks	7.5×10^{11}
Dry annealed AlO_x/SiN_x stacks	2.6×10^{11}
Wet annealed AlO_x/SiN_x stacks	8.6×10^{10}
Dry PDA of the stacks	1.3×10^{11}
Wet PDA of the stacks	9.5×10^{10}

When a SiN_x capping layer is deposited on as-deposited Al_2O_3 films, a slight decrease in the lifetime is observed, whereas the V_{fb} largely increases. The improved V_{fb} is presumably attributed to a partially activated field-effect passivation [1, 2] during the SiN_x deposition process at 450 °C. A decrease by one order of magnitude in the D_{it} value is found for non-annealed stacks, but the lifetime is not better than that of the as-deposited materials. Töfflinger *et al.* reported [14, 15] that a thermal budget during the SiN_x deposition can activate negatively charged traps in the AlO_x/SiN_x, in addition to the stable negative fixed charges formed during the annealing process. However, these negatively charged traps are unstable and are reduced when moderate voltages are applied as a result of electron detrapping and injection into the silicon.

Interestingly this study found that a large decrease in the lifetime of wet annealed Al_2O_3 films capped with a SiN_x layer was inversely correlated with an improved V_{fb}. Assuming that the field-effect passivation of wet annealed Al_2O_3 films was fully activated, a significant increase in V_{fb} after SiN_x deposition may indicate instability in the negatively charged traps originating from exposing the Al_2O_3 films to steam. Whereas a fully activated field-effect passivation is also expected to occur for dry annealed Al_2O_3 films, the improved V_{fb} was observed and the lifetime remains almost the same after SiN_x deposition. We suspected that the wet oxidation process could be acting as a source of unstable negatively charged traps in the films. The presence of a capping SiN_x layer itself and/or the deposition process is suspected to distribute unstable negatively charged traps, especially in the wet annealed stacks. These are expected to have more unstable charged traps compared to dry annealed films, reducing the lifetime with inversely improved V_{fb}.

On the basis of the results presented in this work, the SiN_x capping layer should somehow play an important role in the quality of the passivation. Not only is the field-effect passivation activated, but also the presence of a capped SiN_x layer and/or the deposition process, may cause charge instabilities and may re-distribute unstable negatively charged traps in the stacks. The D_{it} values remain of the same order of magnitude after SiN_x deposition for both dry annealed and wet annealed films. The results emphasize that the wet oxidation process acts as a source of (atomic) hydrogen in the stacks, resulting in improved chemical passivation.

The effect of different PDA processes (wet and dry PDA) on the passivation quality was also investigated by introducing wet PDA and dry PDA to the non-annealed AlO_x/SiN_x samples. The results show that wet PDA produced a higher lifetime as well as V_{fb} compared to the dry PDA. The D_{it} values of the non-annealed stacks, stacks with dry PDA, and stacks with wet PDA are 7.5×10^{11}, 1.3×10^{11}, and 9.5×10^{10} eV^{-1} cm^{-2}, respectively, showing an improvement in the chemical passivation from the wet oxidation. However, its influence remains unclear in terms of the field-effect passivation. Presumably it can be said that suppression of V_{fb} after PDA may be attributed to the coupling effects of a fully activated field-effect passivation during thermal annealing and a reduction in the unstable negatively charged traps. The latter case would somehow be related to the existence of a SiN_x layer, which usually contains a positive Q_f.

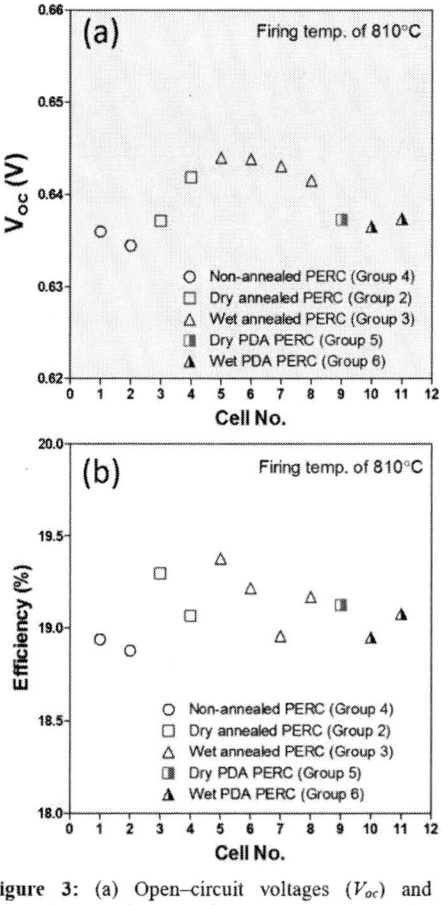

Figure 3: (a) Open–circuit voltages (V_{oc}) and (b) conversion efficiencies of large–area PERC cells with different thermal processes applied under a peak firing temperature of 810 °C.

Figure 3 presents the corresponding values for open–circuit voltages (V_{oc}) and the conversion efficiencies of PERC cells after applying different thermal processes. As can be seen in Fig. 3, the V_{oc} clearly improves from introducing the dry annealing and wet oxidation processes to the Al_2O_3 films before SiN_x deposition. This effect may be attributed to an improved field-effect passivation, as can be realized by the improved V_{fb}, accorded to an increase in the *stable* negatively charged traps in the Al_2O_3 films. The slightly higher V_{oc} of wet annealed PERC than dry annealed PERC might relate to the D_{it} values. The best conversion efficiencies in this experiment are 19.7% for the wet annealed PERC cells (Group 3) and 19.3% for the dry annealed PERC cells (Group 2).

However, the mechanism remains unclear, particularly when a SiN_x layer is involved. These results suggest that the thermal annealing conditions must be considered to improve the quality of the passivation of the stacks on the rear side to achieve high PERC cell performance.

4 SUMMARY

In this study, the impact of wet oxidation on the surface properties of Al_2O_3 films and on the passivation of AlO_x/SiN_x stacks on silicon is investigated to further improve the performance of PERC cells. As a result, wet oxidation has been observed to have a significant effect on the field-effect passivation and chemical passivation, as respectively observed from improvements in V_{fb} and a reduction in D_{it}. These resulted in an increase in the lifetime of Al_2O_3 films and the V_{oc} of PERC cells fabricated by wet annealing Al_2O_3 films as compared to the other processes. It is possible that the wet oxidation process might generate additional unstable/stable negatively charged traps from interstitial oxygen, while the Al_2O_3–Si interface may be terminated by atomic hydrogen (H_2/H_2O) in the steam during wet oxidation. The presence of a capping SiN_x layer itself and/or the deposition process is suspected to distribute unstable negatively charged traps in the stacks, resulting in reduced lifetimes and V_{oc}.

ACKNOWLEDGEMENTS

This work was supported by The New Energy and Industrial Technology Development Organization (NEDO) of Japan. The authors wish to thank Mr. Yasuhiro Kida, Mr. Satoshi Utsunomiya, and Mr. Masaaki Moriya, FREA-AIST for the technical support.

REFERENCES

[1] B. Hoex, S.B.S. Heil, E. Langereis, M.C.M. van de Sanden, and W.M.M. Kessels, Appl. Phys. Lett. **89**, 042112 (2006).
[2] B. Hoex, J. Schmit, R. Bock, P.P. Altermatt, M.C.M. van de Sanden, and W.M.M. Kessels, Appl. Phys. Lett. **91**, 112107 (2007).
[3] K. Matsunaga, T. Tanaka, T. Yamamoto, and Y. Ikuhara, Phys. Rev. B **68**, 085110 (2003).
[4] J.R. Weber, A. Janotti, and C.G. van de Walle, J. Appl. Phys. **109**, 033715 (2011).
[5] P.W. Peacock and J. Robertson, Appl. Phys. Lett. **83**, 2025 (2003).
[6] S. Duttagupata, F. Lin, M. Wilson, M.B. Boreland, B. Hoex, and A. Aberle, Prog. Photovoltaics Res. Appl. **22**, 641 (2014).
[7] S. Joonwichien, S. Simayi, K. Shirasawa, K. Tanahashi, and H. Takato, Energy Procedia (2016). *To be published.*
[8] D. Suh, and W.S. Liang, Thin Solid Films **539**, 309 (2013).
[9] T-T.A. Li, A. Cuevas, Phys. Status Solidi RRL **3**, 160-162 (2009).
[10] S. Steingrube, P.P. Altermatt, D.S. Steingrube, J. Schmidt, R. Brendel, J. Appl. Phys. **108**, 014506 (2010).
[11] G. Dingemans, N.M. Terlinden, D. Pierreux, H.B. Profijt, M.C.M. van de Sanden, and W.M.M. Kessels, Electrochem. Solid-State Lett. **14**, H1 (2011).
[12] E. Simoen, A. Rothschild, B. Vermang, J. Poortmans, and R. Mertens, Electrochem. Solid-State Lett. **14**, H362 (2011).
[13] X. Zhang, A. Thomson, A. Cuevas, ECS Solid-State Lett. **3**, N37 (2014).
[14] J.A. Töfflinger, A. Laades, C. Leendertz, L.M. Montañez, L. Korte, U. Stürzebecher, H. Sperlich, and B. Rech, Energy Procedia **55**, 845854 (2014).
[15] J.A. Töfflinger, A. Laades, L. Korte, C. Leendertz, L.M. Montañez, U. Stürzebecher, H. Sperlich, and B. Rech, Sol. Energ. Mat. Sol. Cells **135**, 49 (2015).

INVESTIGATION ON THE ANTI-PID METHOD OF MC-SI SOLAR CELL FOR MASS PRODUCTION

Junyu Lu[1], Qingzhu Wei[1], Chenyang Wu[1], Yayi Hu[1], Weifei Lian[1], Zhichun Ni[1]
1. Zhongli Talesun Solar Co., Ltd.
No. 1 Talesun Road, Shajiabang, Changshu, Suzhou, Jiangsu Province, P. R. China 215542
Junyu.lu@talesun.com, Qingzhu.wei@talesun.com

ABSTRACT: The Potential Induced Degradation (PID) is one of the most serious degradation which is known to cause power loss in c-Si PV systems. At present, there are many techniques can be applied to reduce the PID, such as modified silicon nitride anti-reflection coating (ARC), additional Silicon Oxide (SiO_2) dielectric layer, encapsulation material with improved electrical resistivity and system grounding of the negative pole. Many companies combine these methods to ensure the anti-PID ability and reliability of PV modules. In this paper, two main cell-level anti-PID methods are investigated: a. increasing the refractive index of SiNx up to 2.15; b. adding an additional SiO2 dielectric layer between emitter and ARC by inline ozone generator. The result shows that by both methods, PID resistance were obtained with the PID testing condition (-1000V, 85℃, 85%RH, 96hrs) and SiO2 dielectric layer shows more excellent PID resistivity with power degradation less than 1%. In addition, cells with method b have about 1% relative improvment in efficiency compared with method a. EQE characterization indicates that the benefit is mainly from the less light absorption of SiNx layer and a slightly better surface passivation.
Keywords: PID, Silicon Nitride, Silicon Oxide, Ozone

1 INTRODUCTION

The Potential Induced Degradation (PID) is one of the most serious degradation for crystalline silicon solar cells. It can cause critical power losses when modules work under high voltage in PV systems [1, 2]. One of the explanation of the PID effect is that the high voltage forces positive charges which come from positive sodium (Na) ions in the front glass, diffused through the encapsulation materials to the solar cell surface area and penetrated the silicon nitride (SiNx) anti-reflection coating (ARC). It is assumed that the Na ions are accumulated on certain regions of the SiNx layer. In case of p-type silicon solar cells, these charges generate local shunts in the p-n junction. [3, 4]

At present, there are several methods exist to reduce the PID on solar cell, module and system level, such as modified silicon nitride anti-reflection coating (ARC), additional Silicon Oxide (SiO2) dielectric layer, encapsulation material with improved electrical resistivity and system grounding of the negative pole. [5-9] Many companies combine these methods to ensure the anti-PID ability and reliability of PV modules.

In this work we investigated two main cell-level anti-PID methods which had been widely used for mass production: i) increasing the refractive index of SiNx up to 2.15; ii) adding an additional SiO2 dielectric layer between emitter and ARC by inline ozone generator.

2 EXPERIMENTAL

2.1 Experiment design

For the investigation, p-type multi-crystalline silicon wafers with the size of 156 x 156 mm^2, the thickness of 180 μm and the resistivity in the range 1.5–3.0Ω·cm were used as substrates.

Since the SiNx ARC and SiO_2 dielectric layer were proved to have an important impact on PID effect in cell level, so besides the anti-PID ability, we mainly compared the effect on electrical performance of different solar cells with different combinations of Refractive Index (RI) and additional SiO_2 dielectric layer.

As shown in **Figure 1**, three batches(B1,B2,B3) of solar cells were manufactured based on a conventional cell process. The additional SiO2 dielectric layer was formed by the inline ozone generator using high-frequency discharge after PSG removal and edge isolation process, and then SiNx ARC was deposited by the industrial direct type Plasma Enhanced Chemical Vapor Deposition (PECVD) process with the RI of 2.08 and 2.15 respectively.

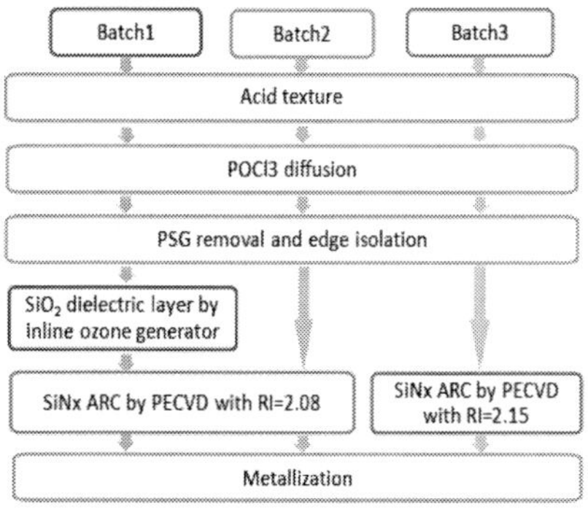

Figure 1: Process flow and information of three experimental bathes

Table 1: Grouping information of combination of different solar cell and module

Group	RI of SiNx	Additional dielectric layer	EVA with different resistivity
A	2.08	NA	3~6x10^{14}Ω·cm
B	2.08	NA	3~6x10^{15}Ω·cm
C	2.15	NA	3~6x10^{14}Ω·cm
D	2.15	NA	3~6x10^{15}Ω·cm
E	2.08	SiO_2	3~6x10^{14}Ω·cm
F	2.08	SiO_2	3~6x10^{15}Ω·cm

In order to verify the anti-PID performance, 6 module groups (group A~F) were also manufactured based on the conventional module process.. Two different encapsulation materials, standard EVA with the resistivity of 3~6x10[14]Ω·cm and PID-resistant EVA with the resistivity of 3~6x10[15]Ω·cm, were applied combining with different solar cells. The detailed grouping information is shown in **Table 1.**

2.2 Characterization

In cell level, Suns-Voc and Internal Quantum Efficiency (IQE) were carried out for the electrical performance evaluation.

There are several PID test standards with various temperature, humidity, testing time, etc. In this work, we carried out the PID test according to TS IEC 62804-1:2015 which is one of the most severity PID test methods. The scheme of PID testing was shown in Fig.2. The test conditions were as following:

- Testing time: 96h
- Chamber temperature: 85°C
- Relative Humidity: 85 %
- Potential to ground: - 1000 V

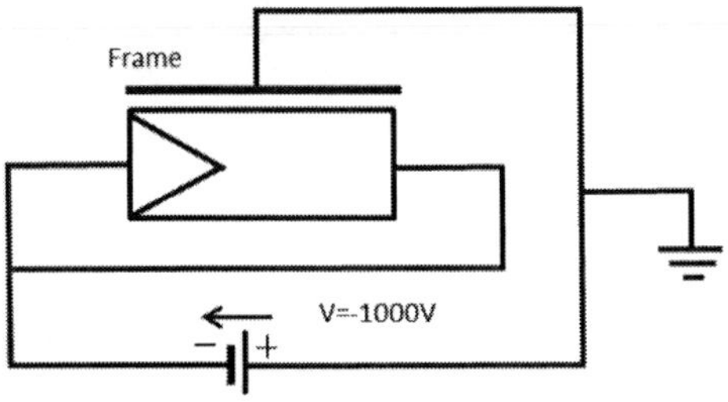

Figure 2: Scheme of PID test

For the characterization the current-voltage (IV) as well as electroluminescence (EL) was carried out before and after PID test. And all the modules were exposed under the sunlight for several days, to make sure that the LID effects had been eliminated before PID test.

3 RESULTS

3.1 Electrical performance in cell level

As shown in **Figure 3**, the electrical performance decreased when the RI was modified from 2.08 to 2.15. As the RI of SiNx increased, the extinction coefficient became larger which meant the SiNx layer would absorb more light itself and resulted in a low Isc, further led to a 0.6% decrease of efficiency relatively. On the other hand , B1 group (solar cells with SiO2 layer) showed some advantages in Voc and efficiency (0.4% relatively) compared with B2 group (solar cells without SiO2 layer). This should be contributed to SiO2 passivation for the n+ surface which could reduce the surface recombination and resulted in a better solar cell performance.

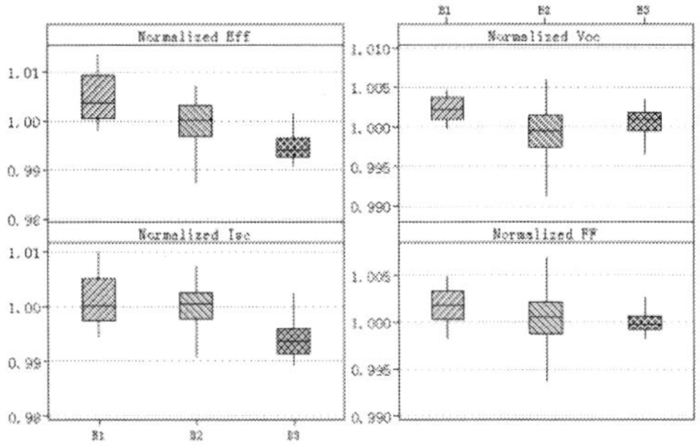

Figure 3: Comparison of different RI and Ozone oxidation treatment

In order to confirm this, we carried out the Suns-Voc and IQE measurements, see **Figure 4** and **Figure 5**. The Suns-Voc results showed a lower J_{01} and J_{02} of B1 which indicated a lower surface recombination velocity. Furthermore, the IQE results of B1 also indicated a better response at short wavelength compared with B2 which also indicated a better surface passivation.

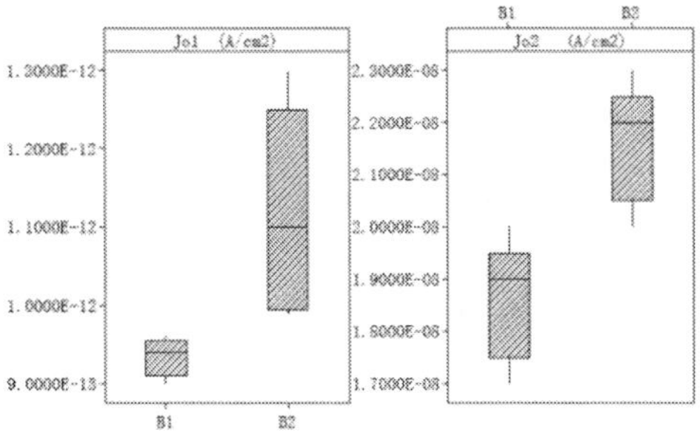

Figure 4: Suns-Voc comparison of B1 and B2

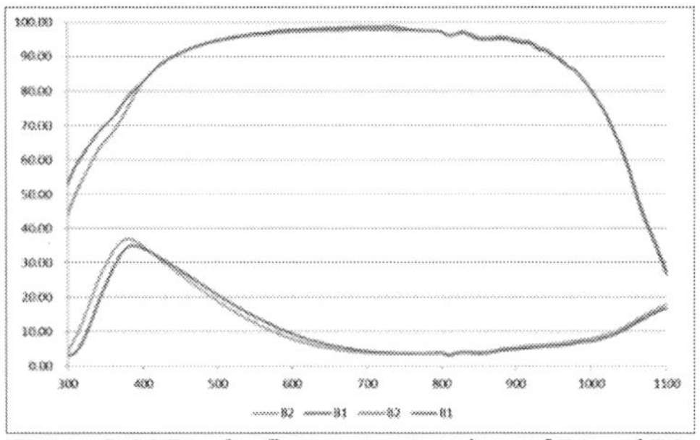

Figure 5: IQE and reflectance comparison of B1 and B2

3.2 PID behavior of the combination of different methods

Figure 6 shows the PID test results of different combination of cell and module level anti-PID methods. It indicated that both high RI and high resistivity encapsulation materials could mitigate the effect of PID, but the power degradation was variable.

Group A representing low RI and standard EVA, namely no any anti-PID treatment was applied, had the most power degradation after PID testing which was almost reached 90%.

Both using high resistivity EVA (Group B) and increasing RI (Group C) could reduce the PID degradation significantly. If we just applied one of these two methods solely, the degradation was still large than 5%. If we applied the combination (Group D), the degradation could reduce to around 5%, but t the risk of

PID still existed.

Only group E and F could be 100% sure that the power degradation was less than 5% after PID testing. That meant the additional SiO2 layer as a barrier layer between the ARC and the p-n junction was the most stable and effective method to prevent PID, and it did not rely on encapsulate material used.

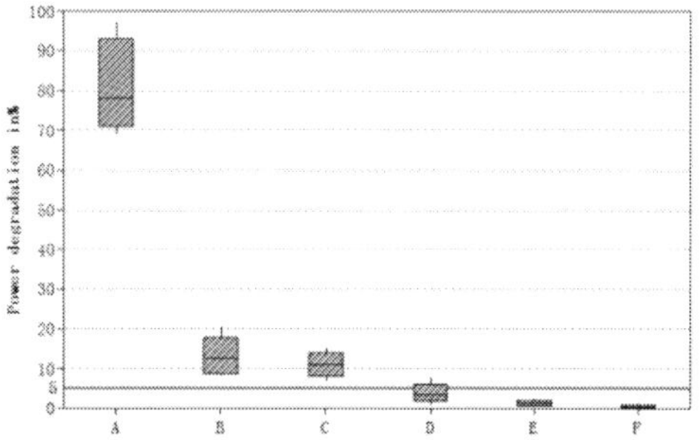

Figure 6: PID comparison of different RI and Ozone oxidation treatment

The changes of EL image before and after PID testing indicated that the degradation occurred at the region near the frame as shown in **figure 5**.

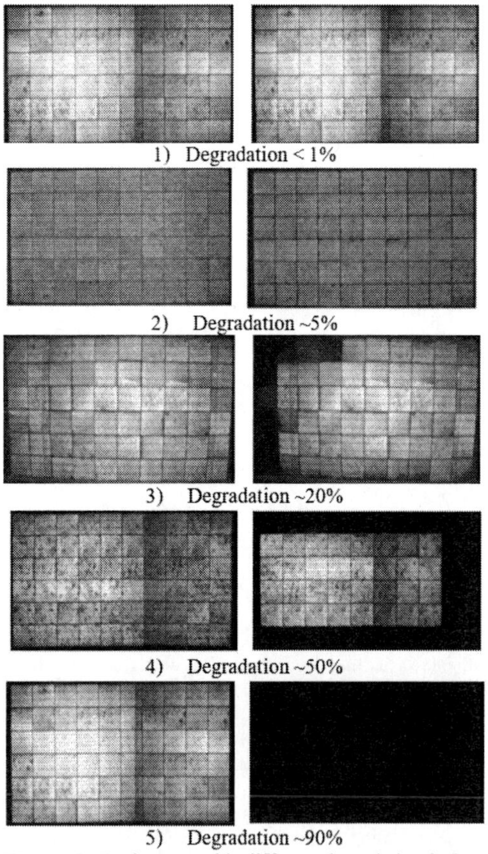

Figure 5: EL images with different degradation before and after PID testing.

4 CONCLUSION

In this paper, we demonstrated the impacts on electrical performance and anti-PID performance of two main cell-level anti-PID methods for mc-Si solar cells.

Both high RI and high resistivity encapsulation materials could mitigate the effect of PID, but still can't reach the requirements of PID.

Combined with high RI and high resistivity encapsulation materials, PID less to 5% can be relized But the solar cell efficiency decreased and the PID risk still existed.

The additional SiO2 dielectric layer formed by inline ozone generator was the most stable and effective method to prevent PID. Besides it had the advantages in solar cell efficiency.

4 REFEFENCE

[1] S. Pingel, O. Frank, M. Winkler, S. Daryan, T. Geipel, H. Hoehne and J. Berghold, "potential induced degradation of solar cells and panels", 35th IEEE Photovoltaic Specialists Conference, Honolulu, 2010.

[2] J. Berghold, O. Frank, H. Hoehne, S. Pingel, B. Richardson and M. Winkler, "potential induced degradation of solar cells and panels", 25th European Photovoltaic Solar Energy Conference and Exhibition, Valencia, 2010.

[3] J. Bauer, V. Naumann, S. Großer, C. Hagendorf, M. Schütz, O. Breitenstein, "on the mechanism of potential-induced degradation in crystalline silicon solar cells", Energy Procedia (2012).

[4] V. Naumann, C. Hagendorf, S. Grosser, M. Werner, J. Bagdahn. "Micro structural root cause analysis of potential induced degradation in c-Si solar cells", SiliconPV, Leuven, Belgium, 2012.

[5] H. Nagel, A. Metz and K. Wangemann, "crystalline Si solar cells and modules featuring excellent stability against potential-induced degradation", 26th European Photovoltaic Solar Energy Conference and Exhibition, Hamburg, 2011.

[6] S. Koch, D. Nieschalk, J. Berghold, S. Wendlandt, S. Krauter and P. Grunow, "potential induced degradation effects on crystalline silicon solar cells with various antireflective coatings", 27th European Photovoltaic Solar Energy Conference and Exhibition, Frankfurt, 2012.

[7] H. Mehlich, D. Decker, U. Scheit, M. Uhlig, S. Frigge, M. Runge, B. Heinze, H.-P. Sperrlich, J. Mai, H. Schlemm, E. Vetter, J. Hohne, S. Reichel and W. Stein, "a new method for high resistance against potential induced degradation", 27th European Photovoltaic Solar Energy Conference and Exhibition, Frankfurt, 2012.

[8] S. Koch J. Berghold, O. Okoroafo, S. Krauter, P. Grunow, "encapsulation influence on the potential induced degradation of crystalline silicon cells with selective emitter structures", 27th European Photovoltaic Solar Energy Conference and Exhibition, Frankfurt, 2012.

[9] H. Nagel, P. Saint-Cast, M. Glatthaar and S. W. Glunz, "inline process for the stabilization of p-type crystalline Si solar cells against potential induced degradation", 29th European Photovoltaic Solar Energy Conference and Exhibition, Amsterdam, 2014.

FABRICATING NANO-TEXTURE ON DIAMOND-WIRE SAW MULTI-CRYSTALLINE SILICON USING RIE-ETCH

Yadi Yang, Wei Wang, Quanhua Ye, Jianwen Dong, Jian Sheng, Chun Zhang, Qiang Huang, Jiazhen Zheng, Hua Shu

Technology development department, GCL System Integration Technology Co. ,Ltd.Solar BU, 288#, ChenFeng Road, ZhangJiaGang E&T Deve. Zone, SuZhou, JiangSu, China, 215600. yangyadi@gclsi.com

ABSTRACT: Deficiency of texturing technique impedes the upgrade of multi-diamond-wire saw silicon (MDWS) solar cell to commercial production. In this paper, Nano-textured was applied on diamond-wire saw multi-crystalline silicon by RIE etching technique. The resultant unique pyramidal nanostructures give advantages of both light-trapping and electric properties of silicon solar cells. Finally the silicon solar cells achieved an average efficiency of 18.70%, which gains an efficiency of 0.44% from silicon solar cells with micron-textured. In particular, the best efficiency of RIE 400nm-MDSS achieved 19.1%, the short circuit current achieved 290mA. In terms of the open circuit voltages, the influence of the Nano-textured size on the efficiency also has been investigated. It indicates that a larger texture size gives a higher open circuit voltage.

Keywords: Diamond-wire saw multi-crystalline silicon; Black silicon; RIE-etch;

1 INTRODUCTION

Multi–wire slurry sawing (MWSS) was a mainstream technique for slicing multi crystalline silicon ingot in the PV (Photovoltaic) industry in the past few years [1]. Recently, several research groups have demonstrated that multi-diamond wire sawing (MDWS) technology has several superiorities over MWSS, such as slicing speed, half the thickness of the saw-damage layer, and water-soluble cool without any slurry [2,3]. Importantly, MDWS technology has a great potential of cut thin wafers down to 60μm [4]. Therefore, it is expected that MDWS technology becomes a next-generation slicing technique for producing thin silicon substrates. Unfortunately, MDWS solar cells are still unacceptable in the PV industry mainly due to the lack of an effective texturing technique. An acidic solution texturing technique by using HF/HNO3 is widely used for the slurry wire sawn wafer [5]. In the slurry wire sawn wafer, damage layers exit homogeneously (uniformly), which are explained by expanding a dimple with the acidic solution [6]. By contrast, there is no homogeneous dimple shown on the diamond wire sawn wafers. However, there are deep straight grooves appearing on as-cut diamond wire sawn wafers.

Nanostructure textured black silicon technique becomes a good way to solve such problems. There are two ways of forming nanostructure textures. RIE (Reaction Ion Etch) [7] and MCCE (Metal-Catalyzed Chemical Etching) [8]. Although, MCCE is lower-cost chemical process, its stability is unsuitable for massive production. On the other hand, RIE is much more suitable for the industrial purpose, due to its stability and efficiency.

In this paper, we demonstrate that MDWS solar cells can achieve an efficiency of 18.70% after being Nano-textured by RIE-etch. The influence of texture size on the efficiency is also investigated.

2 EXPERIMENTS

In this work, there are one group of MWSS wafers and two groups of MDWS wafers (each group contains 100 pieces of p-type wafers with the dimensions of 156mm*156mm, and wafer resistivity is 1.0~3.0 Ω cm). Micron-textured surfaces were obtained on the MWSS wafers using HNO_3/HF mixture solution (Fig.1 (a)). On the other hand, after removing damage layers, two groups of MDWS wafers were textured by RIE, as shown in Fig. 1(b), first group wafers have 300nm size and another group have 400nm size. As following, consequent RIE damage were removed by BOE (Buffered Oxide Etching) solutions.

Finally, three groups of wafers (Regular -MWSS, RIE300nm-MDWS and RIE400nm-MDWS) were fabricated into solar cells in a pilot line. Their front n+ regions were formed by phosphorus diffusion and coated with SiNx anti-reflection/passivation layers, both front and back contacts were realized by screen printing and firing.

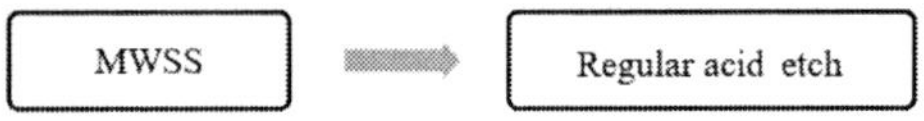

Figure 1(a): Process of Regular-MWSS

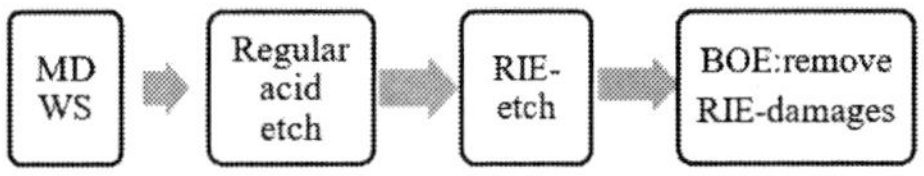

Figure 1(b): Process of RIE-MDWS

3. RESULTS AND DISCUSSION

As shown in Fig. 2 (b) and (c), uniform nanostructures were successfully fabricated on the MDWS. The size is from 200 nm to 400 nm, which is smaller than that of the MWSS wafers (3-4 μ m), as shown in Fig.2 (a). More excitingly, the size of nanostructure as produced on the MDWS wafers can be controlled by adjusting the parameters of RIE process, which can result in different size, as shown in Fig. 2(b) and Fig. 2(c).

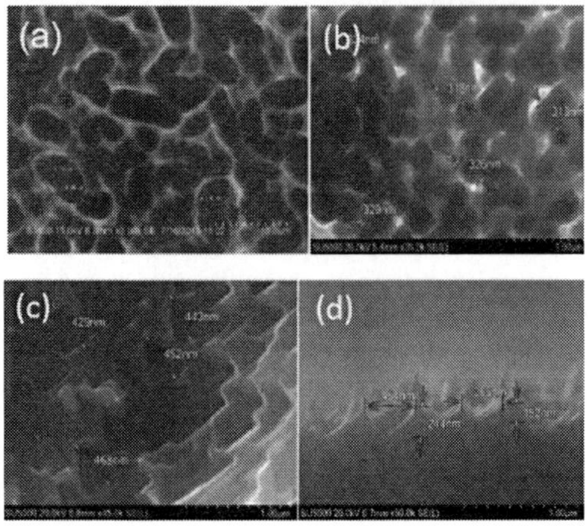

Figure 2: SEM images of (a) Regular-MWSS, (b) RIE300nm-MDWS, (c) RIE400nm-MDWS, (d) Cross-section of RIE400nm-MDWS

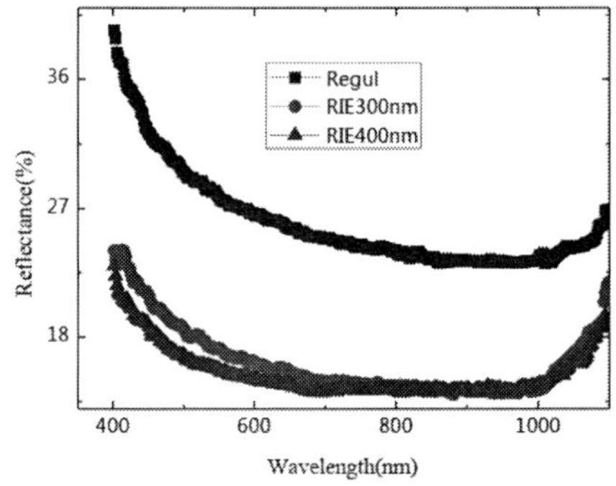

Figure 3: Reflectance curves of MWSS and RIE-MDWS wafers

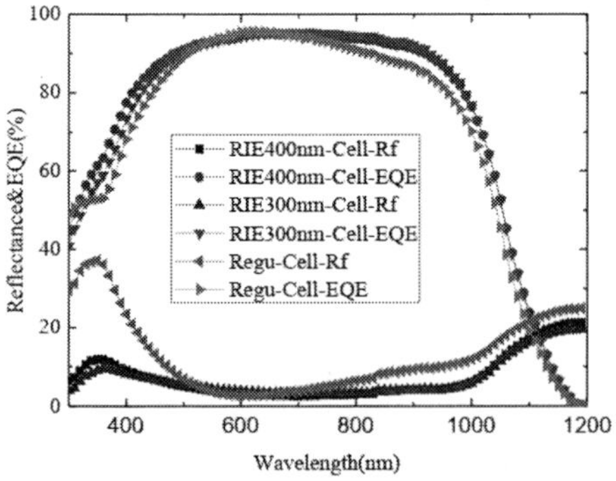

Figure 4: Reflectivity and EQE curves of MWSS-regular and RIE-MDWS solar cells

Fig. 3 shows reflectance curves (400–1050nm) of samples with three different textured surfaces. By average, the reflectivity of the RIE-MDWS wafers (14.5%) is lower than that of the MWSS wafer (22.94%). Even so, the reflectivity of MDWS-RIE wafers can be reduced to a value as low as by RIE. However, damage layer caused by ion bombardment and the high surface area of the nanostructure result in a high recombination. Thus, a post-etching step (BOE-solution) was applied to the RIE-nanostructured MDWS wafers to remove the RIE damage layer and reduce the surface area. In this work, the optimized textured size with an average reflectivity of 14.5% was used by considering the trade-off between the optical loss and the recombination loss of the silicon solar cells.

The reflectivity and external quantum efficiency (EQE) of silicon solar cells are shown in Fig.4. The reflective curves represent total reflections of the three types of cells, which include the effects of anti-reflection/passivation layers and front metal fingers. Their EQE curves overlap in the range of wavelength from 550 nm to 900 nm and then diverge at both short wavelength and long wavelength. A high reflectivity results in a low EQE, which is indicated by the Regular-MWSS cells. On the other hand, the RIE400nm-MDWS cells have the highest EQE, which do not correspond to the lowest reflectivity. These results indicate that the surface condition strongly influences the EQE value. Compared with that of the Regular-MWSS cells, the EQE of the RIE-MDWS cells (both RIE300nm-MDWS and RIE400nm-MDWS) is increased at short wavelength (350-550nm). This can be attributed to a larger surface area that short wavelength light is mostly absorbed. Nevertheless, the EQE of RIE400nm-MDWS cell is the highest value, which implies higher short circuit current values.

Table1: Main characteristic of solar cells of RIE-MDWS and Regular-MWSS

Group	Count (pcs)	Voc (V)	Isc (A)	Rs (Ω)	Rsh (Ω)	FF (%)	Efficiency (%)
RIE300nm-MDWS	98	0.6359	8.882	0.0017	115	80.00	18.57
RIE400nm-MDWS	97	0.6367	8.898	0.0016	383	80.31	18.70
Regular-MWSS	91	0.6351	8.723	0.0015	356	80.21	18.26
The best of RIE400nm-MDWS	1	0.6422	9.021	0.0019	403	80.17	19.10

Table1 shows the electrical properties of the three groups of silicon solar cells. The efficiency of the RIE400nm MDWS cells (18.70%) is 0.44% higher than that of the Regular-MWSS one (18.26%), which is mainly due to enhanced light absorption. More importantly, the open-circuit voltage (Voc) of the RIE400nm MDWS cells is increased by a value of 1.6 mV, which confirms that the problem of extra recombination introduced by the RIE Nano-structure can be neglected. In particular, the best efficiency of RIE 400nm-MDSS achieved 19.1%, the short circuit current achieved 902mA. Meanwhile, the Voc is also improved (7mV). Compare the RIE400nm MDWS-cell with the RIE300nm MDWS-cell, the efficiency is (0.13%) higher and the Voc also gain 0.8mV. Therefore, we infer that a big size of RIE Nano-structure combined with SiNx shows a better passivation parameter and a higher efficiency.

4. CONCLUSIONS

In this paper, the MDWS solar cells with an efficiency of 19.1% have been demonstrated with a unique Nano-texture fabricated by the use of RIE process after tradition acid etching on the industrial production line. Due to its enhancement of optical properties, the average Isc of RIE 400nm-MDSS solar cells is 175 mA higher than that of MWSS solar cells. The Voc is improved (7mV), which is very important for the properties of modules made from Nano-textured, the effects of SiNx coating and a front silver electrode technique, be provided by the Nano-texture MDWS solar cells. These results indicate that the RIE technique has paved the way for the mass production of MDWS solar cells.

5 REFERENCES

[1]H.J. Moller, C.Funke, M.Rinio, S.Scholz, Multi crystalline silicon for solar cells, Thin Solid Films487 (2005)179–187.

[2]B. Meinel, T.Koschwitz ,J.Acker, Textural development of Si Cand diamond wire sawed silicon wafer, Energy Procedia27(2012)330–336.

[3]N. Watanabe, Y. Kondo, D. Ide, T .Matsuki, H. Takato, I. Sakata, Characterization of polycrystalline silicon wafers for solar cells sliced with novel fixed-abrasive wire, Prog. Photo volt.; Res.Appl.18(2010)485–490.

[4]X. G. Yu, P. Wang, X.Q. Li, D. R. Yang, Thin czochralski silicon solar cells based on diamond wire sawing technology, Sol. Energy Mater.Sol.Cells 98(2012) 337–342.

[5] Wenhao Chen, Xiaomei Liu, On the nature and removal of saw marks on diamond wire sawn multi crystalline silicon wafers. Materials Science in Semi-conductor Processing 27(2014)220–227

[6] Steinert M, Acker J, Wetzig K. New aspects on the reduction of nitric acid during wet chemical etching of silicon in concentrated HF/HNO3-mixtures. J. Phys. Chem. C 2008; 112:14139-14144.

[7] J.Yoo, G.Yu, J.Yi, Large-area multi crystalline silicon solar cell fabrication using reactive ion etching (RIE), Sol. Energy Mater. Sol.Cells 95(2011)2–6

[8]X.Y.Ye,S.Zou,K.X.Chen,J.Li,J.Huang,F.Cao,X.S. Wang,L.J.Zhang,X.F.Wang,M.R. Shen,X.D.Su,18.45%-efficient multi-crystalline silicon solar cell with nano-scale pseudo pyramid texture Adv.Funct.Mater.24(2014)6708

MULTI-CRYSTALLINE DIAMOND WIRE SAWN WAFERS TEXTURED BY METAL CATALYZED CHEMICAL ETCHING

Quanhua Ye, Jianwen.Dong, Yadi.Yang, Wei Wang, Jian Sheng, Chun Zhang,Qiang.Huang and Hua.Shu

Technology development department, GCL System Integration Technology Co. ,Ltd.Solar BU, 288#, ChenFeng Road, ZhangJiaGang E&T Deve. Zone, SuZhou, JiangSu, China, 215600. yequanhua@gclsi.com

ABSTRACT: Acidic textured solution by using HF/HNO_3 is widely used for the multi-crystalline silicon solar cells. But, such textured solution is challenging for texturing the diamond wire sawn wafers, since damage layers do not homogeneously distribute on the silicon wafers. On the other hand, Metal catalyzed chemical etching (MCCE) using silver nitrate ($AgNO_3$) is an effective method for diamond wire sawn wafers. Such textured technique involves metal catalyst reaction that is spontaneously formed regardless wafer surface conditions. In this paper, fine texturing structures were formed on the diamond wire sawn wafers by using MCCE method. In order to solve the problem of crystal flower, it is necessary to add additives into MCCE solution. The efficiency of silicon solar cells with the diamond wire sawn silicon and MCCE textured surfaces is higher than that of silicon solar cells with the slurry wire sawn wafers and acidic textured surfaces. The results indicate that the MCCE method can become an alternative candidate for texturing the diamond wire sawn wafers and multi-crystalline silicon.

Key words: diamond wire sawn, MCCE, Slurry wire sawn, texture, reflectivity, cell efficiency

1 INTRODUCTION

HF/HNO_3 acidic solution is widely used for texturing the slurry wire sawn wafers. The mass production is executed with such acidic solution [1]. However, after texturing the slurry wire sawn wafer, damage layers exit homogeneously. By expanding a dimple with acidic solution, the textured structure on the silicon surface can be formed [2].By the way of contrast, there is thinner damage layer and no homogeneously dimple for the diamond wire sawn wafer. There is a deep straight groove in as-cut diamond wire sawn wafer. In particular, HF/HNO_3 acidic textured process is difficult to texture the diamond wire sawn multi-crystalline silicon wafers. Metal Catalyzed Chemical Etching (MCCE) using $HF/H_2O_2/AgNO_3$,which divided into two steps shows in Fig 1, is effective method to texture the diamond wire sawn multi-crystalline silicon wafers [3]. Nevertheless, the conventional MCCE process easily forms flower crystal on the silicon surface [3], which is not acceptable for fabricating high efficiency silicon solar cells. In this work, analyzing the flower crystal and the way of solving the flower crystal are demonstrated.

By using a reflectance test and SEM images of surface morphologies, the textured structure for the diamond wire sawn cutting multi-crystalline silicon wafers was investigated and optimized. The appearance and efficiency of MCCE textured multi-crystalline silicon solar cells are higher than that of acidic solution textured slurry wire sawn silicon solar cells.

2 METAL CATALYZED CHEMICAL ETCHING FOR DIAMOND WIRE SAWN WAFER

The processing flow of MCCE is shown in Fig. 1. Firstly, the saw damage etching (SDE) is performed by alkaline solution (KOH). The porous Si layer is formed in the following metal catalyst textured solution. The reaction system constitutes of deionized water (DI water), HF and AgNO3. 10 wt% AgNO3 was used in this work. Than immersed into DI/HF/ The detailed process conditions are written in the patent from Mitsubishi Electric Corporation [4-7].In the nitrate (AgNO3), the oxidation-reduction potential is Ag=+0.7991 [8]. As the Ag becoming adherent to silicon surfaces, the hydrogen reduced reaction is promoted by a catalyst of Ag. In order to compensate an increased part of an electronic amount of consumption, oxidation reaction occurs near the silicon surface adhering to Ag. A pit is formed in silicon wafer surfaces. In the reaction system, H2O2 enhances the oxidation of silicon surfaces and following HF solution is used to remove the oxide layer.

Fig. 2 shows the surface photographic image and SEM image of multi-crystalline silicon surfaces after the MCCE. This surface appears as black in Fig.3. From the SEM image, this surface contains pits and is coved by stains. Ag digs a pit and the etched Si chip remain on the surface. NH3/H2O/H2O2 solution is used to remove Ag. Finally, the macro-porous Si layer is formed by HF/HNO3 acid solution. The hole of pits formed by metal catalyst reaction is largely etched by HF/HNO3. Fig 4 shows the SEM image after being immersed into HF/HNO3 solution. Stains after the macro-porous step were removed with low concentration alkali solution (KOH) and Ag was removed by HF/HCL solution, as shown in Fig. 1.

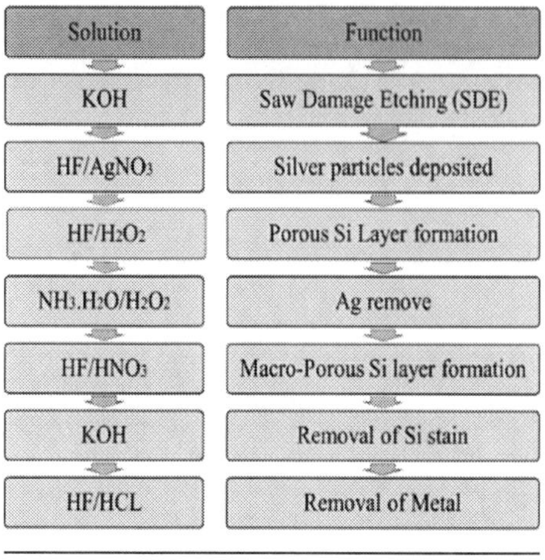

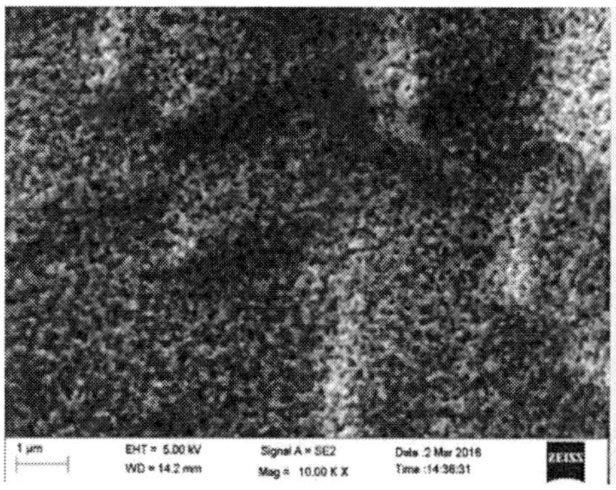

Fig 1. Process flow of metal catalyzed chemical etching

Fig 2. SEM image of after immerging into HF/AgNO3

Fig 3.Photographic imagery after MCCE

Fig 4. SEM image of after macro porous (Mag=20kx)

3. ELECTRICAL PROPERTIES OF METAL CATALYST TEXTURED SILICON SOLAR CELLS

In order to evaluate the metal catalyzed chemical etching (MCCE) for the slurry and diamond wire sawn wafer, their corresponding silicon solar cells

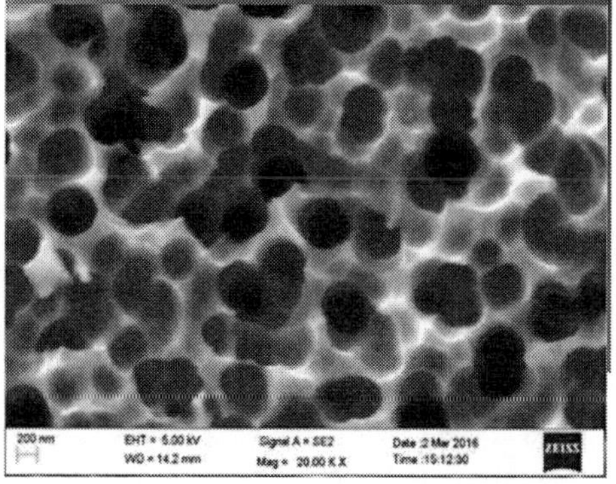

were fabricated. The fabrication and evaluation were taken place in GCL Integrated Technology FAB. The SEM image of the MCT silicon wafers is show in Fig. 4. The p-type diamond wire sawn multi crystalline silicon wafers with the dimensions of 156mm×156mm and a resistivity of 1.0~3.0 Ω.cm were used. Their thickness is around 200um +/- 10um. The fabrication process is outlined in Fig. 5. Simultaneously, reference cells by using acidic solution textured technology and slurry wire sawn wafer were produces in the same way. Textured wafers were processed by following conventional fabrication process: diffused by POCl3, edge isolation, PSG removal, anti-reflection coating (ARC) with PECVD-SiNx (80nm), rear and front metallization and co-firing.

The electrical characteristics of silicon solar cells are presented in Tabel. 1. Those solar cells were measured under AM 1.5G, 1000W/cm2 at 25 ℃. The cell area is 243.4 cm2. As indicated in Table.1, the efficiency of MCCE textured cells is 0.28% higher than that of acid solution textured cells, which is attributed to the increment of Jsc. Their corresponding EQE and reflectivity curves are show in Fig 6, which indicates that the improvement of Jsc by MCCE technology can reduce the wavelength less than 510nm reflectance.

Group	η (%)	Voc(V)	Jsc(mA/cm2)
MCCE Tex	**18.47%**	**0.6364**	**36.5**
Acid Tex	**18.19%**	**0.6363**	**35.7**

Tabel 1: Cell electrical properties of Acid texture and MCCE texture

Based on the SEM image as shown in Fig 4, we can easily obtain the MCCE textured size (around 300-500 nm), which is much smaller the acid texture size (3-5um). Due to the impact of the "dead layer" [9], the

phosphorus diffusion profile with a high sheet resistance is desirable for high efficient silicon solar cells. Consequently, the improvement of metal pastes, firing process, should also be adjusted to prevent the

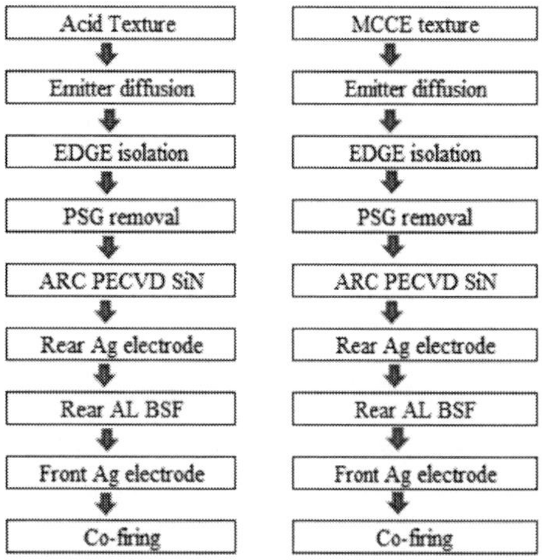

inappropriate metal contact under wiring

Fig 5. Cell fabrication procedure

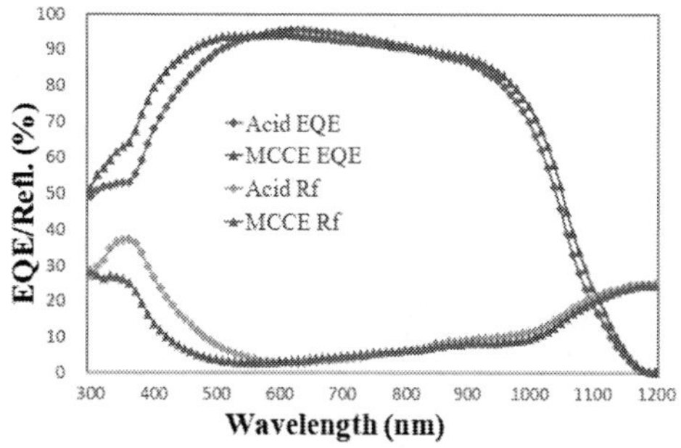

Fig 6. Acid texture EQE, MCCE texture EQE, Acid texture Rf , MCCE texture Rf

4. APPEARANCE OF METAL CATALYST TEXTURED SOLAR CELLS

The silicon solar cells without additive MCCE appears slightly shining as shown in Fig. 7(a), which is mainly due to the flower crystal. The flower crystal reasons of multi-crystal diamond wire cutting wafer, in the Fig 8 shows the SEM photograph of grain boundaries on both sides, the microstructure is obviously different, due to the anisotropic etching because of different crystal orientate between the two adjacent grains. It is obvious that appearance of both sides of the grain reflectance are inconsistent. The left side grain reflectance reached 19%, while the right side grain reflectance given 12%. Therefore, different brightness are resultant from the generation of flower crestal.

In order to find the way of solving such problem, we processed both <100> and <111> silicon wafers through MCCE. There corresponding SEM images are given in Fig 9, <100> orientated wafers give a quadrilateral morphology, and <111> wafers shows a positive hexagonal morphology, which indicates the anisotropic etching of MCCE process. In order to eliminate flower crystal, we must eliminate this anisotropic etching.

On the other hand, adding additives in HF/H2O2/AgNO3 solution allows crystalline silicon to be etched uniformly. This can be confirmed by surface morphologies shown in Fig 10. The appearance of diamond wire sawn wafers that were textured by MCCE solution with and without additives are shown in Fig 11. It is obvious that there is no flower crystal produced in the MCCE solution with additives.

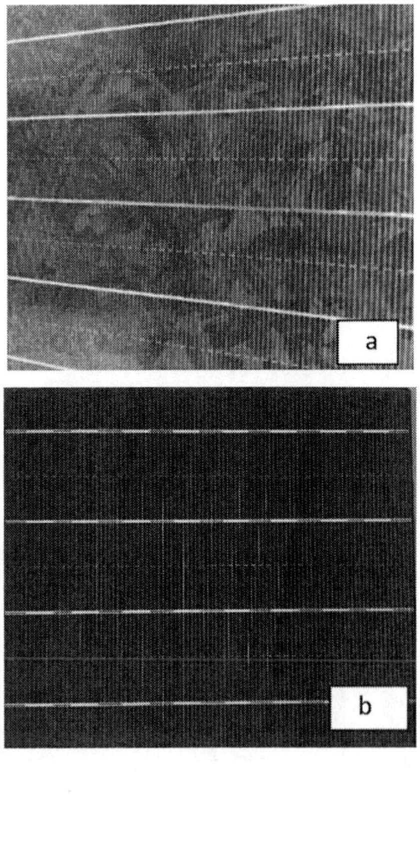

Fig 7. The photograph of cell. (a) Without additive MCCE; (b) With additive MCCE

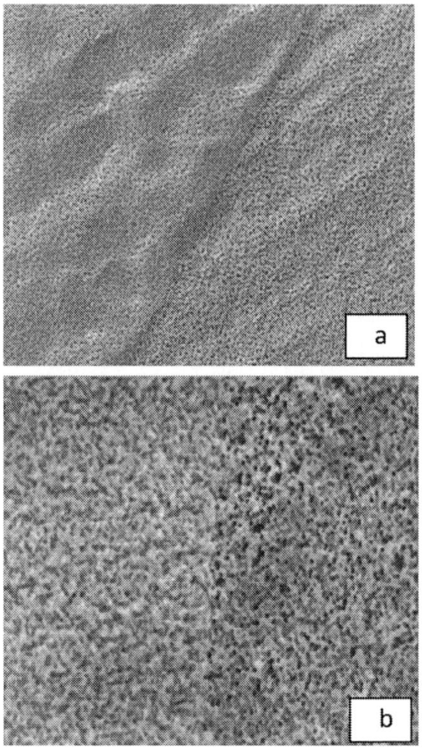

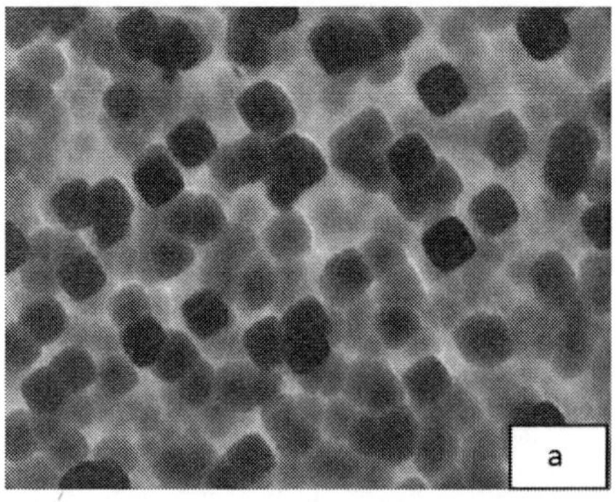

Fig 8. Grain boundaries on both sides of multi crystal wafer after MCCE, (a) Mag=5kx, (b) Mag=10k

Fig 9. Different crystal direction after MCCE <100>(a), <111>(b)

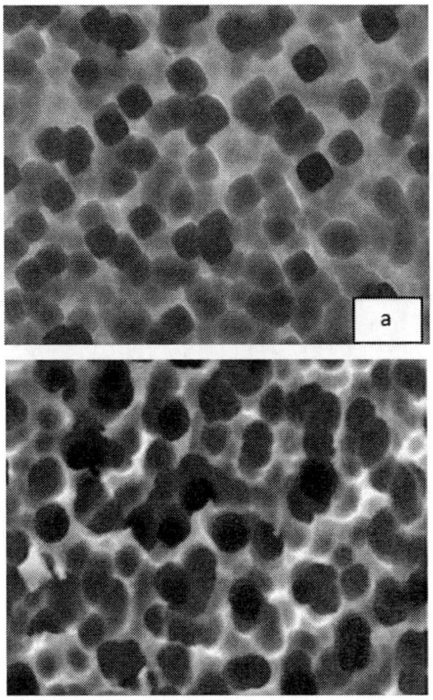

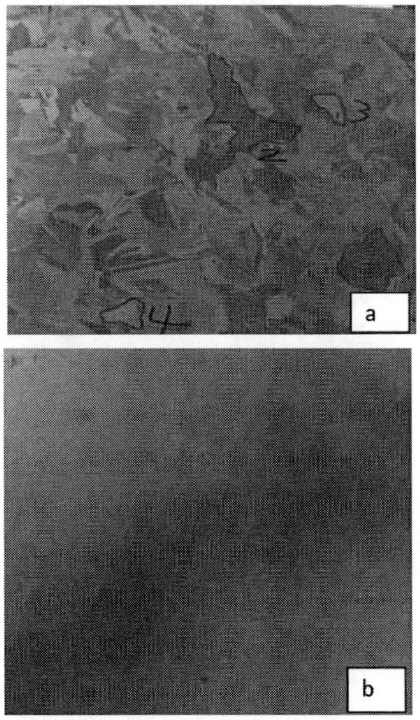

Fig 10 additive change to etch more uniform, without additive (a), with additive (b)

Fig 11 appearance without and with additive, (a) without additive; (b) with additive

4. CONCLUSIONS

Metal catalyzed chemical etching (MCCE) using silver nitrate ($AgNO_3$) is an effective method for texturing diamond wire sawn wafers. The diffusion condition (sheet resistance) and firing condition, fine-turning need to be done to meet this kind of texture structure in order to get high efficiency cell properties. The high sheet resistance is required to high efficient silicon solar cells, because the small pyramid can be easily over doped, which leads a "dead layer". The high efficient silicon cells with an improved efficiency of 0.28% could be achieved by optimizing the fabrication step. Adding additive into MCCE solution resolves flower crystal, which allows the MCCE becoming an effective textured method for multi-crystalline silicon solar cells.

5 REFERENCES

[1] De Wolf, S.,Choulat, P., Vazsony, E., Einhaus, R., Van Kerschaver, E., De Clercq, K., Szlufick, J., Towards industrial application of isotropic texturing for multi-crystalline silicon solar cells , in: Proceeding of Sixteenth European Photovoltaic Solar Energy Conference, Glasgow, UK., 2000,1521-1523.

[2] Ryosuke Watanabe, Shuuji Abe, Satoshi Haruyama, Tatsunobu Suzuki, Mitsuo Onuma, Yoji Saito, Evaluation of a new acid solution for texturization of multi crystalline silicon solar cells (Article ID 951303, Volume), Int. J. Photoenerg. 2013 (2013) 6.

[3] Texturization using metal catalyst wet chemical etching for multi crystalline diamond wire sawn wafer

[4] N. Watanabe, Y.Kondo, D.I de,T. Matsuki,H.Takato,I.Sakata , Characterization of poly crystal line silicon wafers for solar cells sliced with novel fixed-abrasive wire, Prog.Photovolt:Res.Appl.18(2010)485–490.

[5] K. Shirasawa, K. Fukui, K. Okada, Y. Inomata, H. Takahashi, Y. Fukawa, S. Fujii, Over 17% large area multi crystalline silicon solar cells, in: Fourteenth European Photovoltaic Solar Energy Conferences Barcelona, Spain, 30 JUNE 4 JULY, 1997, 384-387.

[6] Y.C. Chou, Y.C.Kuo, S.W.Yang, Y.N.Chang, Y.C.Chen, C.H.Wu, Acid treated surface morphology of fixed-abrasive sawn wafers, in: Thirty Seventh Photo- voltaic Specialist Conference, June, Seattle, USA., 2011,19-24.

[7] Method of manufacturing substrate for solar cell and solar cell, JP2012160568, US2013306148, KR20140014112, CN103339738.

[8] Marcus Lippold, Florian Buchholz, Christoph Gondek , Florian Honeit, Eckard Wefringhaus, EdwinKroke, Texturing of SiC-slurry and diamond wire sawn silicon wafers by HF–HNO3–H2SO4 mixtures, Sol.Energ.Mat.Sol.C. 127 (2014)104–110.

[9] W.ChuckHsu1, Yen-Sheng Lu2, Jung-YiChyan, J.Andrew Yeh, High-efficiency 6 multi crystal line black solar cells based on metal-nano particle-assisted chemical etching, Int.J.Photoenerg.2012(2013)7.

FIELD-EFFECT SURFACE PASSIVATION PASTE BY SCREEN-PRINTING FOR HIGH EFFICIENCY PERC

Tsuyoshi Hayasaka[1*], Syunsuke Kodama[1], Mari Shimizu[1], Mitsuyoshi Hamada[1], Naotaka Tanaka[1] and Takeshi Nojiri[1]
[1]Hitachi Chemical Co., Ltd., Core Technology Research & Innovation Center, 48, Wadai, Tsukuba, Ibaraki 300-4247, Japan
*Corresponding author. Tel.: +81-29-864-4002; E-mail address: t-hayasaka@hitachi-chem.co.jp

ABSTRACT: Passivation paste for Passivated Emitter and Rear Cell (PERC) easily formed by screen-printing is described. The layer is composed of aluminum-other metal complex oxide. Passivation layer can be formed at low cost without gas-phase process. The forming procedure is as follows: (i) Printing the passivation paste on silicon wafers. (ii) Baking the wafers and volatilizing solvent. (iii) Annealing the wafers and promoting structure change to form passivation layer. The passivation layer is hard and stuck fast to the wafer. The carrier lifetimes of samples with the passivation layer are comparable to those of the ALD-AlO_x applied sample. The negative fixed charge density of the paste is higher than -1×10^{12} cm^{-2}. Quantum efficiency of the passivation paste applied PERC is equivalent to the ALD-AlO_x applied reference cell.

Keywords: Screen Printing, Passivation, Field Effect, High Efficiency, Passivated Emitter and Rear Cell

1 INTRODUCTION

Passivated Emitter and Rear Cell (PERC) is one of the most promising structures for high conversion efficiency of silicon solar cells [1]. One of the biggest features of PERC is its electrode structure. Electrode of PERC only partially alloy with silicon substrate. Thereby, Aluminum-Silicon alloy which causes the minority carrier recombination decreases. Instead of reduced alloying part, PERC has passivation layer to reduce minority carrier recombination and improve internal reflectance of incident light. The layer composed of inorganic materials such AlO_x, SiO_2, SiN_x and a-Si:H were investigated for the passivation [2-4]. In particular, AlO_x is a candidate for high property passivation layer. Various methods to form AlO_x, including atomic layer deposition (ALD), chemical vapor deposition (CVD) were investigated [5-7]. Especially, ALD is typical method to deposit AlO_x as passivation layer. However, such deposition process needs cost to install and run the equipment. Therefore, it is needed to replace ALD with more simple method.

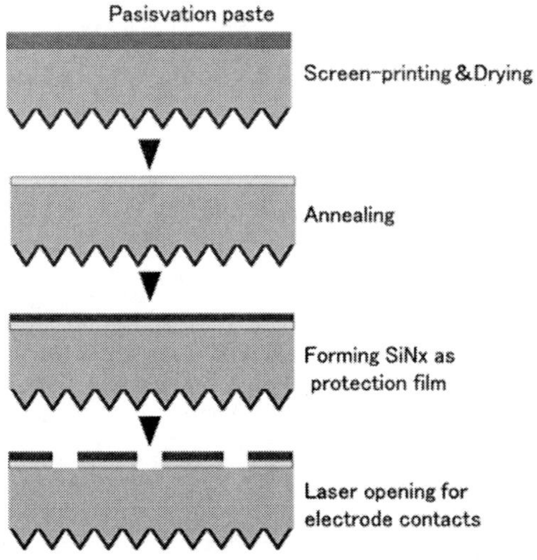

Figure 1: Printable passivation forming process flow.

We investigated easily formable passivation layer without deposition process, and developed new material having high passivation-formative ability and easy formability. Using the material, the layer can be formed only by screen-printing and thermal process. Further, the layer has good properties for PERC. Figure 1 shows forming procedures of passivation layer using the paste. In this paper, we will report the evaluation result of the firm strength of the developed passivation layer, effective carrier lifetime, negative fixed charge density, photovoltaic characteristics and quantum efficiency of the passivation paste applied cell.

2 EXPERIMENTS

2.1 Design concept of passivation material

We initially investigated AlO_x-formable material by screen-printing. However, passivation properties of AlO_x layer formed by screen-printing were deteriorated during damaging processes such CVD for coating protection SiNx and/or electrode-firing. Furthermore, these properties didn't recover with any treatment. Conceivable causes of the defects are as bellows: (i) The uniformity of the interface structure between passivation layer and silicon substrate. (ii) The unevenness due to intramembranous quality difference. The molecular simulation was performed to prove these hypotheses and find a solution. Thereby, effective additives to prevent the fatal deterioration were found out. The additives move to and repair the weak point in the layer during annealing process. In consequence, the improved passivation layer after going through cell manufacturing processes retained high properties.

2.2 Preparing the passivation material

Aluminum oxide precursors, solvents, binders and additives were mixed in the vessel. This mixture was stirred for 10 min. After stirring, stabilizing agent was added and re-stirred for 10 min. The appearance of obtained paste is pale yellow. The paste has good store stability at room temperature for several months.

2.3 Passivation layer forming

The paste was printed on silicon wafers by screen-printing. The solvents were removed by pre-baking process about several minutes. The wafers were annealed at approximately 600-800 °C. That is enough temperature for promoting structure change to the high properties having passivation layer. The annealing atmosphere was identical to the air composition. After annealing, wafers with passivation layer on the rear side were obtained. The color of the passivation layer on the wafer was uniform in the plane.

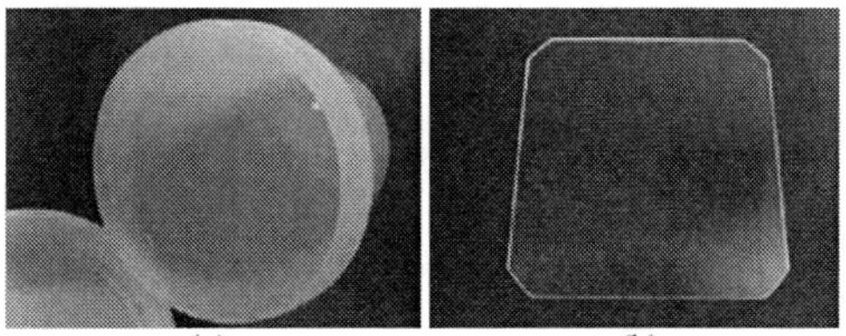

(a) (b)

Figure 2: (a) Appearance of the paste. (b) The passivation layer formed by use of the paste.

2.4 Peel test

10 mm x 10 mm area of the layer was cut at regular intervals (1 mm x 1 mm) with a laser. Poly-imide tape was stuck on the cuts. The tape was then peeled by hand at angle of 90°. After the peeling, the peel off of the layer was detected by the optical microscope.

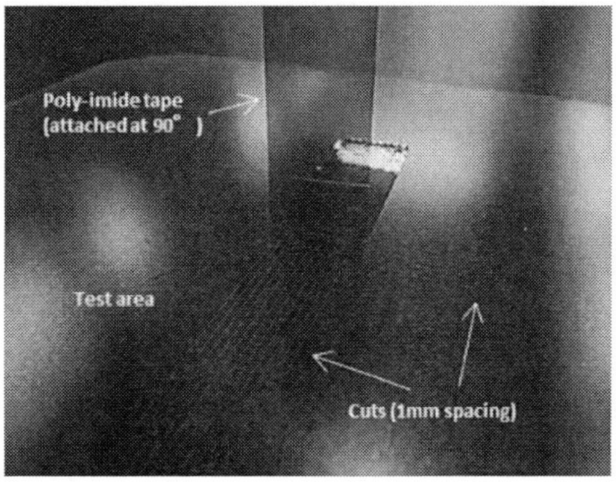

Figure 3: Peel test

2.5 Negative fixed charge density

The negative fixed charge density of the interface between wafer and the layer was measured by C-V (capacitance-voltage) method [8]. The test wafers with passivation layer on the rear side were prepared. Aluminum electrode for contact was deposited on the passivation layer by vacuum coating equipment. In this paper, we used p-type Cz mono-Si front-side-formed wafers for solar grade with a bulk resistivity of 0.7-0.9 Ωcm.

2.6 Lifetime measurements

The effective lifetime of the wafers with passivation layer on the rear side was measured by QSSPC (quasi steady state photoconductance) method [9]. The front side of the wafer was the upper side in the measuring. "Specified Minority Carrier Density" in analysis parameter was set 1.0×10^{15} cm^{-3}. The results of the continuous 10 times were averaged arithmetically.

2.7 Preparation of PERC with passivation layer

Back sides of the wafers were planarized with hot alkaline water. The passivation layer formed on back sides of wafers was coated with SiNx as protection film from aluminum electrode penetration. After laser opening for electrode contacts, aluminum and silver pastes were printed to form solar cell electrodes. Line-and-space contacts area ratio to passivation layer was approximately 4-10%. Electrodes printed on wafers were sintered by use of tunnel furnace. Reference cells applied ALD-AlO_x were obtained by same procedures.

2.8 Characteristics measurement

The photovoltaic characteristics of the obtained cell were measured by using solar simulator system under the standard test conditions. The measurement results of the continuous 10 times were averaged arithmetically. The external quantum efficiency (EQE) in the wavelength region of 300-1200 nm was measured by using quantum efficiency measurement equipment.

3 RESULTS AND DISCUSSIONS

3.1 Layer conditions

There is no peeling of the passivation layer from the wafer after the test. This result indicates that the passivation layer has excellent adhesivity with silicon. Figure 4 shows the picture of the cross section of the wafer and the layer. The uneven layer was observed in case of the germinal passivation materials. In contrast, the dense and uniform layer was observed in those of the improved material applied sample. The high quality of the layer is brought about by the effect of additives.

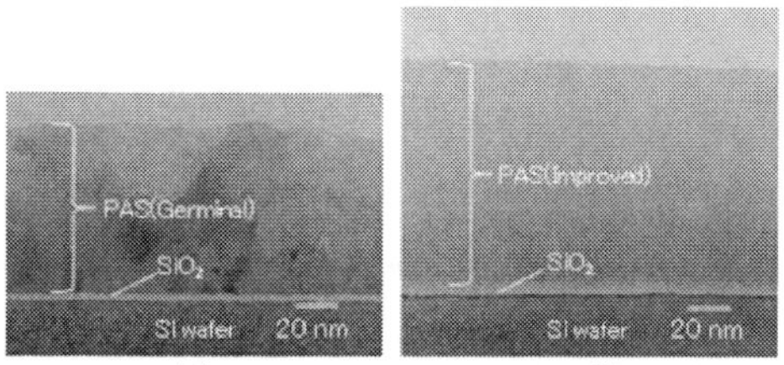

(a) (b)

Figure 4: Cross section of passivation layer (TEM).
(a) Layer formed by germinal paste. (b) Layer formed by improved paste.

3.2 Negative fixed charge density

Table I shows the measured negative fixed charge densities of the interface between each passivation layer and wafer. Typically, more than 1×10^{12} cm^{-2} negative fixed charge density is required for the passivation layer for PERC [10]. As shown in the table, the fixed charge density of the sample with the germinal passivation layer was poor. By contrast, that of the improved sample was more than 1×10^{12} cm^{-2} which is adequate value to achieve good passivation properties.

Table I: Negative fixed charge densities.

	Germinal	Improved	ALD-AlO_x
Q_f [cm^{-2}]	-1.03×10^{11}	-7.64×10^{12}	-4.22×10^{12}

3.3 Carrier lifetime retention

Figure 5 shows the measured carrier lifetime retention rates after the cell manufacturing process without electrodes printing. The carrier lifetime of the sample with the germinal paste seriously deteriorated during the process. By contrast, that of the sample with the improved passivation layer was maintained and equivalent to those with ALD-AlO_x applied samples.

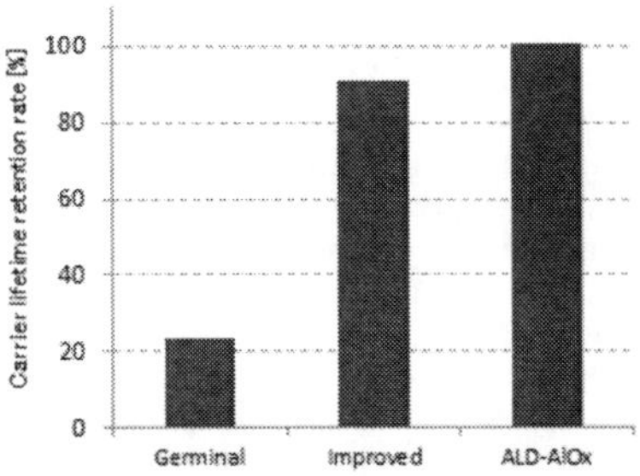

Figure 5: Carrier lifetime retention rate.

3.4 Photovoltaic properties

Table II shows the photovoltaic characteristics of the cells manufactured using improved passivation layer and ALD-AlO_x. As shown, the values of J_{sc} (short circuit current density) of both cells were comparable. The F.F. (fill factor) of the improved passivation layer applied cell (79.1) was higher than those of the ALD-AlO_x applied sample (77.5). The conversion efficiency is proportional to J_{sc}, V_{oc} and F.F. in principle. As a result, the conversion efficiencies were comparable in both of the cells. Figure 6 shows EQE of the cells. As shown, EQE curve of the cells using improved passivation layer overlapped with those of the ALD-AlO_x applied sample. Especially, the overlap in long-wavelength region of 900-1100 nm indicates the photovoltaic properties in these cells were comparable. The basis for the improvement of F.F. is still not well-defined. However, the contribution of the passivation application to the conversion efficiency and EQE was proved.

Table II: Photovoltaic characteristics.

Characteristics	Improved	ALD-AlO_x
J_{sc}	38.4	38.6
Eff. [%]	19.2	19.3

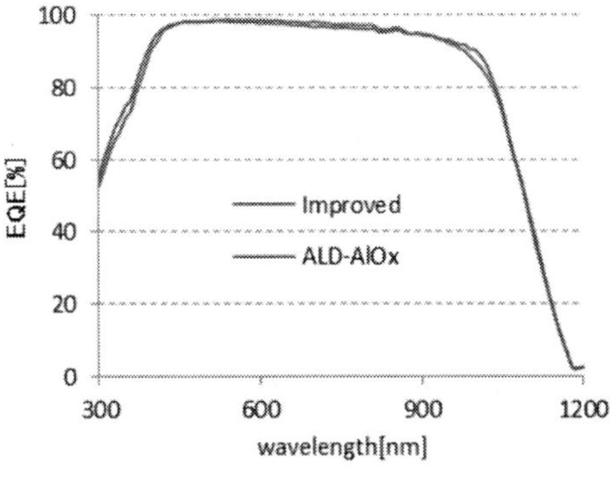

Figure 6: EQE measurements.

4 CONCLUSIONS

Screen-printable passivation material was developed. The passivation properties and endurance to the cell manufacturing process were improved by using the molecular simulation. The material changes to the passivation layer by conventional thermal process. Furthermore, the layer has high density and adhesivity. The silicon wafer with passivation layer formed by use of the improved material had excellent carrier lifetime and negative fixed charge density. Carrier lifetime is maintained after whole cell manufacturing process. PERC of applying passivation layer indicated the conversion efficiency and EQE comparable to those of the cells of applying ALD-AlO_x. This tendency was thought to thank to F.F. improvement.

ACKNOWLEDGEMENTS

I would like to thank all of the participating companies and organizations and the many people in each for their support through all of the activities.

REFERENCES

[1] M. Rauer, C. Schmiga, R. Woehl, K. Rühle, M. Hermle, M. Hörteis, S.W. Glunz, Investigation of aluminum-alloyed local contacts for rear surface-passivated silicon solar cells, IEEE Journal of Photovoltaics, 1(1) (2011) 22-28.

[2] V. Meemongkolkiat, D.S. Kim, A. Rohatgi, SiO2-Based Spin-on Dielectrics for Back Surface Passivation of p-type Si Solar Cells, In 22nd European Photovoltaic Solar Energy Conference, (2007)3-7.

[3] A. Cuevas, M. Kerr, J. Schmidt, Overview of PECVD Silicon Nitride for Solar Cells, (2003)

[4] M. Schaper, J. Schmidt, H. Plagwitz, R. Brendel, 20.1% - efficient crystalline silicon solar cell with amorphous silicon rear - surface passivation, Progress in Photovoltaics: Research and Applications, 13(5)(2005)381-386.

[5] J. Schmidt, A. Merkle, R. Brendel, B. Hoex, M. C. M. Van de Sanden, W. M. M. Kessels, Surface passivation of high - efficiency silicon solar cells by atomic - layer - deposited Al2O3, Progress in photovoltaics research and applications, 16(6) (2008) 461-466.

[6] F. Werner, W. Stals, R. Görtzen, B. Veith, R. Brendel, J. Schmidt, High-rate atomic layer deposition of Al2O3 for the surface passivation of Si solar cells, Energy Procedia, 8 (2011) 301-306.

[7] A. Morato, B. Vermang, H. Goverde, E. Cornagliotti, G. Meneghesso, J. John, J. Poortmans, Electrical characterization of ALD Al2O3-HfO2 and PECVD Al2O3 passivation layers for p-type CZ-Silicon PERC solar cells, 38th IEEE PVSEC, (2012) 1077-1082.

[8] L.M. Terman, An investigation of surface states at a silicon/silicon oxide interface employing metal-oxide-silicon-diodes, Solid State Electron, 5 (1962) 285.

[9] H. Nagel, C. Berge, A.G. Aberle, Generalized analysis of quasi-steady-state and quasi-transient measurements of carrier lifetimes in semiconductors, J. Appl. Phys., 86 (1999) 6218-6221.

[10] J. J. H. Gielis, B. Hoex, M. C. M. Van de Sanden, W. M. M. Kessels, Negative charge and charging dynamics in Al2O3 films on Si characterized by second-harmonic generation, J. Appl. Phys., 104(2008)073701.

E-TON'S PRINTED-ALOX PERC CELLS: EFFICIENCIES BEYOND 21 % WITH A NEXT-GENERATION ALOX PASTE

Tsung-Cheng Chen [1*], Jui-Yi Hung [2], Yung-Sheng Lin [1], Chun-Fu Lin [1], Chen-Hao Ku [1], Jung-Ching Wang [3], Shian-Wen Chen [3], Cheng-Shun Hu [1], and Ching-Chang Wen [1]

1. E-ton Solar Tech Co., LTD., No.498, Sec. 2, Bentian Rd., Annan Dist., Tainan City 709, Taiwan, R.O.C.
2. New E Materials Co., LTD., 2F, No. 3, Luke 1st Rd., Luzhu Dist., Kaohsiung City 82151, Taiwan, R.O.C.
3. Eternal Materials Co., LTD., No. 22, Changhsing Road, Luchu District, Kaohsiung 821, Taiwan, R.O.C.

ABSTRACT: In our previous work, we developed a special printed-AlO_x approach for our passivated emitter and rear cell (PERC) product, named ExcelTon™-III technology. With this low-cost and safer technology, our PERC product with 3-busbar (3-BB) pattern design reached an average efficiency of around 20.3 %. In this paper, we introduce a next-generation AlO_x paste, named IP1257-L, to replace the old-version paste of IP1257. One advantage of IP1257-L is that it provides a higher-passivation-quality AlO_x film, resulting in an increase in Voc of 0.4-1.3 mV and an efficiency gain up to 0.12 % by just replacing IP1257 with IP1257-L. Most importantly, the paste usage of IP1257-L is less, about 42 % saving compared to that of IP1257. This means the running cost can be further reduced. We also demonstrate that an average efficiency of 20.84 % with the best efficiency of 21.22 % was achieved in our production line by integrating IP1257-L paste together with 4-BB pattern design. With this high-performance and low-usage paste, we believe that this alternative printed-AlO_x PERC technology opens the potential for better price-performance ratio instead of other vapor-phase-deposited-AlO_x PERC technologies.
Keywords: PERC, AlO_x, paste

1 INTRODUCTION

PERC technology is currently being introduced into mass production by several solar cell manufacturers and the cell efficiency has been reported to be beyond 21 % by synergizing with selective emitter (SE) technology [1]. Nowadays, the AlO_x passivation layer for use in PERC mass production is mainly formed by vapor-phase-deposited methods, such as plasma enhanced chemical vapor deposition (PECVD) or atomic layer deposition (ALD), even though the capital investment of the vapor-phase equipment is high and the precursor of AlO_x used in PECVD or ALD, trimethyl aluminum (TMA), is quite hazardous and expensive.

In our previous work, we developed an alternative approach to form AlO_x film by screen-printing technology for our PERC product, named ExcelTon™-III technology [2]. We also demonstrated that it is feasible for mass production of cells with an average efficiency over 20.3 % and of 60-cell modules with an output power over 285 watts. Noted that this PERC product used normal-size wafers (156 mm x156 mm with diagonal of 200 mm) and adopted homogeneous emitter structure with common 3-BB pattern design.

The key feature of PERC is to improve the rear side of cells by replacing the conventional Al-back surface field with a dielectrically passivated rear surface. This means that the passivation quality at the Si/AlO_x interface is crucial for the efficiency gain of PERC cells. Three factors may affect the passivation quality at the Si/AlO_x interface. (1) The cleanness of the rear surface: Generally, a clean surface can be obtained by introducing a clean process before AlO_x deposition. (2) The flatness of the rear surface: This influences the conformity of the deposition and thus the passivation quality at the Si/AlO_x interface. We had addressed this effect in previous work [3]. (3) The quality of the AlO_x layer: This quality, i.e. negative fixed interface charge (Q_f) and interface defect density (D_{it}), mainly affects the passivation quality at the Si/AlO_x interface [4]. For further increasing the cell efficiency of our PERC product, we had focused on improving the quality of the AlO_x layer by optimizing the composition of the AlO_x paste. In brief, besides the AlO_x precursors and inducing agents, there are necessary solvent and stabilizers in the composition of the AlO_x paste for better printing properties. Our main strategy was to increase the percentage of the AlO_x precursors in the AlO_x paste. Theoretically, printing a paste with a higher percentage of the AlO_x precursors on Si surface can get a better coverage of the AlO_x precursors at the interface between Si and the printed-AlO_x paste, finally resulting in a better quality of the AlO_x layer after thermal treatment. Moreover, the wet deposit (g/wafer) of the printed paste can be further reduced by using this higher-percentage paste if the same thickness of the formed AlO_x layer is required as using a lower-percentage paste. These two advantages, i.e. higher passivation quality and lower paste consumption, can ensure a better price-performance ratio of the printed-AlO_x PERC cells compared with that of the vapor-phase-deposited-AlO_x PERC cells. After some trials for the aforementioned optimization, we developed a next-generation AlO_x paste, named IP1257-L, in which the percentage of AlO_x precursors is increased around 100 % relatively compared to that of the old-version paste (IP1257). In this paper, we report on the passivation quality of this next-generation AlO_x paste and the performance of the state-of-the-art PERC cells integrated this paste together with 4-BB pattern design.

2 EXPERIMENTAL

The PERC cells shown in this study were produced in our ExcelTon™-Cell-III mass production line and the process flow was illustrated in our previous work [2]. The minority carrier lifetime measured by WCT120 was used as the performance index to compare the passivation quality of IP1257 and IP1257-L. The cell precursors for lifetime measurements were carried out using the same PERC process flow but without rear opening and metallization steps. Lifetime after the high temperature co-firing step was also monitored since this lifetime represents the overall passivation quality and thus the cell electrical performance, especially in Voc.

To obtain the same thickness of the formed AlO_x layer as using IP1257, various screen designs, i.e. various combinations of the screen wire diameter and screen mesh number, were carried out to optimize the wet deposit needed to print for IP1257-L. Finally, a screen with the theoretical ink volume of 9.6 cm^3/m^2 was selected for IP1257-L printing.

The Multi-BB approach, which reduces the finger length between adjacent BBs and hence the series resistance losses of the front finger grid, is considered as the easiest way to further increase the cell efficiency of industrial screen-printing solar cells. Considering the market demand of high-power-output modules and the feasibility of the state-of-the-art module's interconnection technology, we further adopted larger wafers (156.75 mm x156.75 mm with diagonal of 210 mm, also known as M2 wafers) and integrated 4-BB pattern design together with IP1257-L in our ExcelTon^{TM}-Cell-III mass production line.

3 RESULTS AND DISCUSSIONS

3.1 Lower paste consumption of IP1257-L

Fig. 1 shows the monitored data of paste usage per wafer (wet deposit) for achieving the same thickness of the formed AlO_x layer by using both of IP1257 and IP1257-L pastes. The average paste usage per wafer of IP1257-L was around 0.25 g, which was 0.18 g saving i.e. approximate 42 % saving compared to that of IP1257. This is very consistent with our expectation of about 50 % saving.

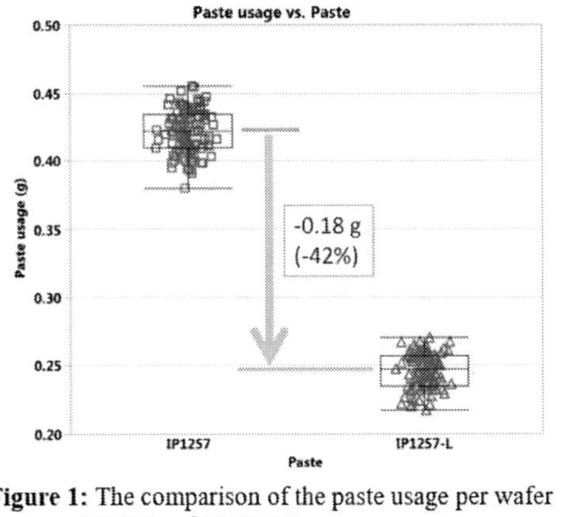

Figure 1: The comparison of the paste usage per wafer between IP1257 and IP1257-L.

3.2 Better passivation quality of IP1257-L

The minority carrier lifetime of cell precursors using IP1257-L and a reference cell precursor using IP1257 are shown in Fig.2. An average lifetime around 100 µs after SiN_x capping layer deposition on IP1257-L samples was observed, which was only 7 µs slightly higher than that of the IP1257 sample. The average lifetime of IP1257-L samples after co-firing increased to 165 µs and the corresponding average implied Voc reached 0.694 V. Moreover, the difference of the average lifetime between IP1257-L samples and IP1257 sample increased from 7 µs to about 30 µs after co-firing. All these observations indicate that the passivation quality of the AlO_x layer indeed improves by using a paste with a higher percentage of the AlO_x precursors and this benefit is more obvious after co-firing activation.

To verify the performance of PERC cells by using IP1257-L, we randomly picked three lots in our ExcelTon^{TM}-Cell-III mass production line and divided each lot into two groups. Each group contained hundreds of wafers; one group used IP1257-L and the other group used IP1257 as a reference. These three experiments confirmed that the Voc of IP1257-L groups was higher than that of IP1257 groups, which is consistent with the result of the lifetime measurement. The Voc gains and the corresponding efficiency gains of these three experiments were 0.4, 1.2, 1.3 mV and 0.02, 0.09 and 0.12 %, respectively.

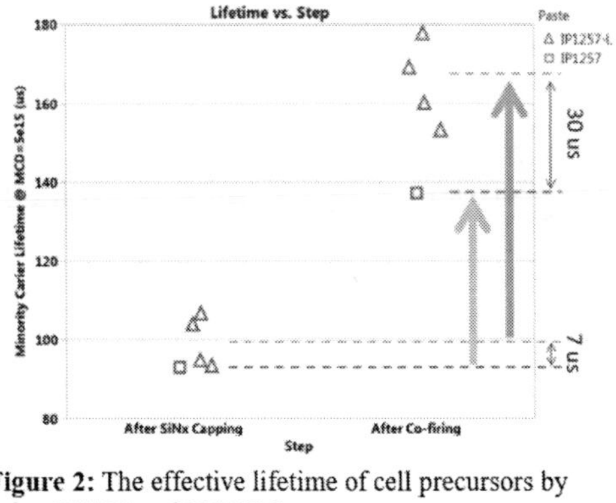

Figure 2: The effective lifetime of cell precursors by using IP1257 and IP1257-L.

3.3 The performance of the state-of-the-art PERC cells integrated with IP1257-L

To meet the market demand of high-power-output modules, some M2 wafers were purchased for development and 4-BB pattern design was implemented in our production line. Fig. 3 shows the efficiency histogram of 14,265 M2-4BB PERC cells by using IP1257-L AlO_x paste, the corresponding average and the champion cell electrical parameters are shown at Table I, respectively. At this test run, an average efficiency of 20.84 % with the champion efficiency of 21.22 % was achieved. Compared to our previous work [2], an efficiency gain of around 0.5 % was obtained, which mainly ascribes to the following two implementations. Firstly, the better-performance AlO_x paste (IP1257-L) was used as discussed in previous section, which contributes to around 0.1-0.2 % efficiency gain. Secondly, the front pattern was re-optimized. In this pattern optimization, not only 4-BB design was adopted but other factors, such as sheet resistance, finger pitch, finger width and finger height, were further optimized, which contributes to another 0.3-0.4 % efficiency gain.

From mass production point of view, we pursue not only a high average efficiency but also a narrow distribution of cell efficiencies. The distribution of the cell efficiencies in Fig. 3 is not symmetric and has a longer tail toward lower efficiencies. Especially, the efficiency below 20.1 % accounts for 0.85 % of the total cells. This abnormally long-tail distribution should come from one significant factor, such as wafers with different qualities mixed by the wafer vendor and one abnormality temporarily occurred in our production line, and together with the variances of all other materials and process steps. We are now trying to identify the root cause of this abnormally long-tail distribution and the top-variance process step in our production line, and then trying to improve them. Eliminating those low-efficiency cells (< 20.1 %) can improve the average efficiency slightly to 20.9 %. However, to further increase the efficiency, the following directions [1, 5-6] are under evaluating,

- Using 5-BB pattern design and further reducing the width of the BB to 0.5 mm. This is very easy for cell manufacturers but is very challenge for today' s module interconnection technology.
- Using fine-line printing technology. The target of the printed finger width is less than 50 um. The challenge is it needs to maintain a good aspect ratio of fingers and minimize the finger discontinuity at the same time.
- Using SE technology. This technology has potential to gain synergies with a passivated rear side and to contact at lower firing temperature that ensures a better rear-side passivation. However, the challenges are the complexity of process flow and the issue of cost.

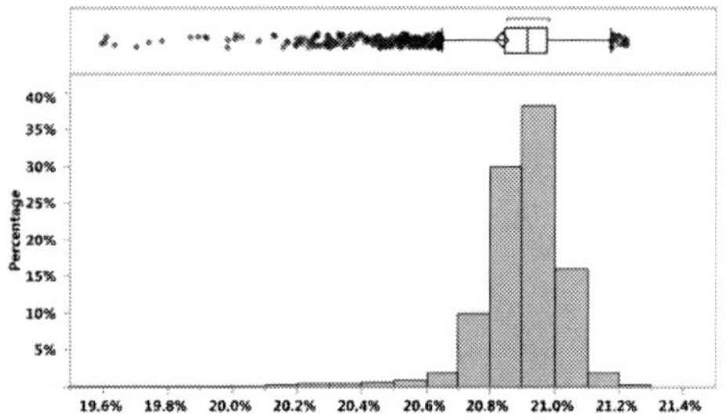

Figure 3: The cell efficiency histogram of 14,265 M2-4BB PERC cells fabricated at Eton production line.

Table I: The electrical parameters of 14,265 M2-4BB PERC cells and the champion cell.

ExcelTon™-Cell-III	Eta [%]	Voc [V]	Isc [A]	FF [%]	PCS
Average	20.84	0.656	9.78	79.31	14265
Champion Lot	21.00	0.657	9.82	79.51	997
Champion Cell	21.22	0.660	9.84	79.83	1

4 CONCLUSION

We developed a next-generation AlO_x paste of IP1257-L, which features higher passivation quality and lower paste consumption compared to those of the old-version paste (IP1257). The minority carrier lifetime of the cell precursor with IP1257-L is significantly higher than that with IP1257 after co-firing. Consequently, the cell efficiency is potential up to +0.12 % by adopting this next-generation paste.

An average efficiency of 20.84 % with the best efficiency beyond 21.2 % was demonstrated by integrating this IP1257-L paste together with 4-BB pattern design. We think that an average efficiency beyond 21.3 % of our PERC product can be expected just by further integrating 5-BB pattern design with further reducing the width of the BB or fine-line printing technology or SE technology or all in the near future.

5 REFERENCES

[1] D. Chen et al., "21.40 % efficient large area screen printed industrial PERC solar cell," in 31th European Photovoltaic Solar Energy Conference and Exhibition, 2015, pp. 334-340.

[2] T.-C. Chen et al., "R&D pilot-line production of PERC cells with printed AlO_x passivation layer," in 29th European Photovoltaic Solar Energy Conference and Exhibition, 2014, pp. 627-629.

[3] Y.-S. Lin, T.-C. Chen, C.-H. Ku, C.-S. Hu, and C.-C. Wen, "The effect of rear surface flatness on the performance of the printed-AlO_x PERC," in 31th European Photovoltaic Solar Energy Conference and Exhibition, 2015, pp. 665-668.

[4] G. Dingemans and W. M. M. Kessels, "Status and prospects of Al_2O_3-based surface passivation schemes for silicon solar cells," Journal of Vacuum Science and Technology A, vol. 30, p. 040802, 2012.

[5] T. Dullweber et al., "Fine-line printed 5 busbar PERC solar cells with conversion efficiencies beyond 21 %," in 29th European Photovoltaic Solar Energy Conference and Exhibition, 2014, pp. 621-626.

[6] T. Lauermann, B. Fröhlich, G. Hahn, and B. Terheiden, "Design Considerations for Industrial Rear Passivated Solar Cells," in 38th IEEE Photovoltaic Specialist Conference, 2012.

WET CHEMICAL METALLIZATION OF SILICON SOLAR CELLS: STATUS AND PERSPECTIVE OF INDUSTRIAL APPLICATION

A. Letize, B. Lee, D. Cullen,
MacDermid Enthone Electronics Solutions, Inc. 245 Freight St., Waterbury, Connecticut 06702, USA

ABSTRACT: Formation of electrical conductors by wet chemical metallization is well-known in the electronics industry, with decades of implementation in the fabrication of printed circuit boards and microelectronics. However, its usage in the photovoltaic industry is still in its early stages. As the solar cell fabrication process matures, wet chemical copper metallization will move from acceptable, to preferred, to required. Relative to screen printed silver paste, copper allows higher aspect ratio, self-aligned conductors, and compatibility with increasingly complex materials and processes. Many of the highest efficiency cell architectures already require, or benefit from, wet chemical metallization. Future commodity cells are predicted to require copper conductors, due to the need to continually improve efficiency while simultaneously driving costs lower. This paper is a review of the current status of wet chemical metallization in the photovoltaic space, and a prospectus for where the technology trends are leading. Also included is a brief history of plating in PV, as well as its advantages, challenges, application to specific cell types, and path to mass adoption.
Keywords: copper, wet chemical metallization, electrodeposition, cost reduction

1 INTRODUCTION

Wet chemical metallization via deposition of metals from electrolytes was initially developed in the early 1800s, and continuously refined over the past two hundred years. The technology grew further with the birth of the computer, and today enjoys mass utilization across a wide swath of electronics fabrication processes, including printed circuit boards, semiconductors, and microelectronics. The printed circuit board industry, which relies greatly on plating processes, is over $62 billion dollars today. Despite this, plating has yet to see wide-spread adoption in the photovoltaics industry, although several advantages exist. Still, PV roadmaps expect a growing percentage of cell manufacturing processes to use wet chemical metallization processes in the coming years, based on numerous factors such as efficiency advantages and cost reduction [1].

Solar cell contact formation by screen printing and firing of silver paste is by far the most common method of metallization today. Screen printing has dominated for decades because of process simplicity, high throughput and scalability of the technique. However, silver paste printing has several notable drawbacks. The efficiency of screen printed cells are limited by the capability to print narrow fingers, which is dictated by screen materials and paste rheology. Additionally, the paste firing process is limited in its ability to make low resistance contact to silicon, especially in advanced emitter concepts. Finally, silver paste is significantly more expensive than copper, the primary metal used in solar cell plating processes. Coupled with the inherent volatility in the price of silver, a precious metal, copper plated cells offer an attractive economic proposition. This paper will show how wet chemical metallization offers unique advantages over the incumbent process in each of the aforementioned aspects.

1.1 Metals and deposition methods

Historically, a wide variety of metals and alloys of metals have been utilized in the metal finishing and electronics industries. Most often the metal/alloy is chosen based on the application. Factors such as physical properties (hardness, corrosion resistance, conductivity) as well as cost considerations and compatibility with substrates/superstrates will determine the selection of metal for a particular use. The most frequently used metals in industrial applications include inexpensive transition metals such as nickel, cobalt, copper, tin and zinc, as well as more noble (and expensive) examples such as silver, gold, palladium, and platinum.

Nickel is the most common choice for metal in applications requiring direct contact to silicon, for a few reasons. First, nickel forms a low resistance contact to doped silicon [2]. The contact resistance can be further reduced by thermal annealing at low temperature, which can aid in the formation of nickel silicide intermediates [3]. This step can also aid in adhesive force, discussed later in this work. Finally, nickel is an excellent barrier to diffusion of subsequently plated layers, even at very low thickness [4]. This is especially important, since the most common primary conductor in this application is copper, which can diffuse into the space charge reason of doped silicon and increase recombination, lowering the efficiency of the device.

As mentioned, copper is the preferred metal for use as the main conductor in solar cells fabricated with wet chemical metallization. Copper is only slightly more resistive than silver (1.68 $\mu\Omega$*cm vs. 1.59 $\mu\Omega$*cm). Combined with the fact that copper is orders of magnitude cheaper than silver, the difference in required metal thickness for equivalent conductance is negligible. In addition to its economic benefit, copper electrodeposition is simple and well understood. Copper is the primary metal conductor deposited in the manufacture of printed circuit boards, as well as a significant portion of semiconductor metallization.

Because copper is susceptible to oxidation, it is normally capped with a 'final finish,' which is typically tin, silver, or an organic solderability preservative (OSP). The final finish layer serves to prevent oxidation of the underlying layer (normally copper) and maintain an oxygen-free surface suitable for bonding in the following soldering/interconnection step of module formation. The layer is normally very thin (as low as 100 nm case of silver), and doesn't add appreciably to the cell performance or cost.

The preceding metals can be applied to a solar cell in a number of different methods, depending on the specifics of the cell design and other manufacturing requirements. Traditionally, plating takes place by either electroless or electrolytic deposition. Electroless deposition is driven by differences in the galvanic potentials between two (or more) metals in the reaction. This reaction can take place directly by immersion displacement of the base metal by a more noble metal in solution, or through an intermediate reaction step whereby a catalyst drives the reduction of a metal from solution onto the substrate surface. Electrolytic deposition is effected by connecting the surface to be plated to a power source and supplying a source of electrons to the surface, which catalyzes reduction of the metal from solution onto that surface (cathode). An equal and opposite oxidation reaction occurs at the anode surface, which is normally a basket containing the same metal being deposited. By this method, the metal being plated out of solution at the cathode is replenished by dissolution from the anode Another novel method of wet chemical metallization on solar cells, described in detail by Bartsch [5], is light induced plating (LIP). The general principle of LIP is that the cell is illuminated during the deposition process and simultaneously contacted with a power source at the rear (non-plating region) of the cell. In this way, the incident light intensity on the cell surface drives the rate of metal deposition, and a uniform layer of metal is deposited on the cell in the predefined pattern. Using the proper combination of chemistry, equipment and process control, high throughput and excellent efficiency results are achievable on plated solar cells [6].

1.2 Pattern formation and plating equipment

Before wet chemical metallization can be applied to solar cells, the pattern to be plated must first be defined. In some metallization processes, the first metal contact layer is applied by a physical vapor deposition (PVD) technique such as evaporation or sputtering. After a thin, conductive layer is applied to the wafers surface, its pattern is defined by one of several lithographic techniques, such as print and etch of a screened resist, or photoresist application followed by expose/develop/etch/strip. The end result is that the metal seed layer remains only in the pattern defined for further metallization (ie. H-grid with 50 micron fingers). This patterning technique is most common for silicon heterojunction (SHJ) and interdigitated back contact (IBC) cell concepts.

For direct metallization of silicon without use of a PVD seed layer, laser patterning is the method of choice. During laser-defined patterning, the primary objective is to remove the passivation layer(s) on the surface of the cell, which may include silicon nitride antireflective coatings as well as silicon oxide dielectric layers. Once the dielectric is removed by laser ablation, the underlying silicon can be plated by the methods mentioned above. Laser patterning technology has progressed significantly over the years, from buried groove contacts [7] to laser doped selective emitters (LDSE) [8], laser chemical processing (LCP) [9], nanosecond laser ablation, and finally picosecond laser ablation, which can optionally include doping for lightly doped and/or boron emitters. The advantages and challenges of the different laser patterning techniques will be discussed below.

In order to achieve the desired results as it relates to wet chemical metallization of solar cells, it is critical to have a proven combination of chemistry and equipment. Although plating can be applied by numerous techniques, industrial production demands a simple, reliable and high throughput set of equipment. For this reason, the most common tools employed in production are conveyorized, high volume manufacturing units. Equipment manufacturer RENA offers a horizontal conveyorized unit capable of single-sided wetting and throughput in excess of 3000 wafers per hour (figure 1). Besi (Meco) offers a vertical unit for simultaneous plating of front and back of wafers, especially useful for n-type bifacial or heterojunction wafers. Innolas supplies HVM laser solutoins for patterning prior to plating.

Figure 1: (Above) high speed picosecond laser processing tool from Innolas; Inline, conveyorized high volume plating tools are available from RENA (left) and Meco (right)

2 ADVANTAGES OF PLATING

As manufacturers strive to achieve ever higher efficiency results at a competitive cost, implementing plating for both efficiency benefits and cost reductions will become the obvious choice.

2.1 Increased Jsc / reduced shading

Screen printed silver paste has a number of well known drawbacks. It is difficult to achieve finger widths below 50 microns with sufficient conductivity due to the constraints of screen printing technology. Additionally, printing such narrow contacts runs the risk of insufficient laydown, line breaks and other reliability issues. Alternative printing methods, such as double printing and use of stencils, can improve these issues, but do so at a cost of additional printing/drying steps and consumables. On the other hand, laser defined patterns can produce fully-plated contacts with finger widths of less than 30 microns (figure 2), which can significantly reduce shading and thus increase the J_{sc}. It has been shown that J_{sc} values in excess of 40 mA/cm^2 can be achieved using laser-defined and plated patterns [10]. Also, the effective optical width of plated conductors is less than paste, due to the internal reflectance of rounded copper traces [11].

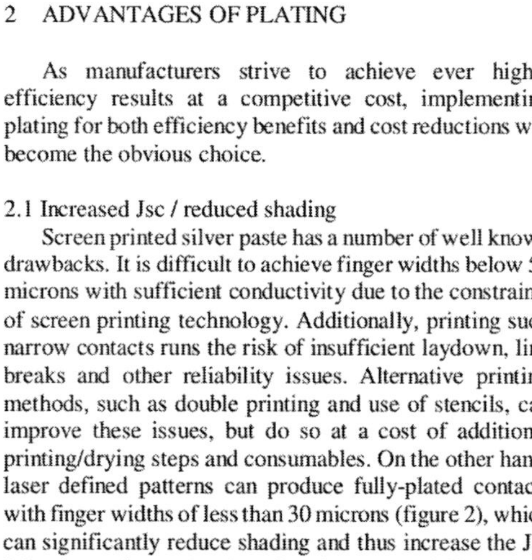

Figure 2: A fully plated Ni/Cu conductor (left), with finger width of less than 30 μm; 100% contact of plated nickel to silicon (right), leading to low R_{ser}.

2.2 Improved electrical contact

Direct plating of nickel to silicon also carries the advantage of improved contact resistance when compared to screen printed silver paste. Printed and fired paste relies on the formation of spot-contacts consisting of silver-glass crystallites, which are inherently not very conductive, and require high surface concentration of phosphorus to make acceptable contact [12]. In contrast, plated nickel forms a continuous contact to the silicon surface. Annealing of the plated deposit can result in nickel silicide formation, which further reduces contact and series resistance. It has been shown that plated nickel forms a lower contact resistance to doped silicon (figure 3) [13], and is especially advantageous in combination with lower phosphorous doping. The quality of the contact results in higher fill factors than paste, consistently higher than 80%, and a very narrow distribution of cell efficiencies within a group of plated cells, which results in less cell groups during binning [14]. There is also difficulty in contacting boron-doped emitters with silver paste, something which may be overcome by forming nickel contacts to the emitter.

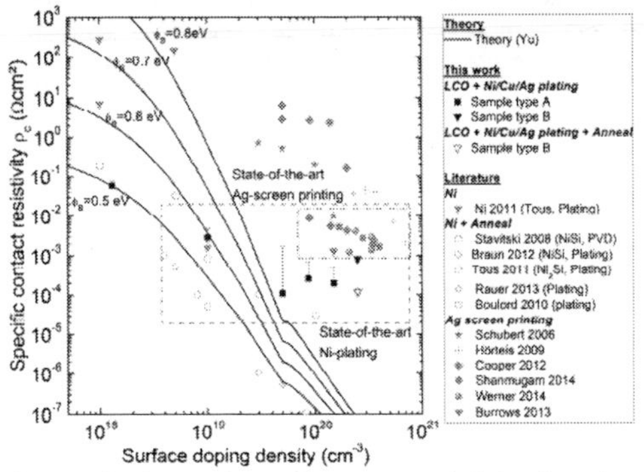

Figure 3: Comparison of contact resistivity for plated nickel and silver paste on phosphorus doped emitters, demonstrating the superior contact resistivity of nickel [13].

2.3 Reduced cost and price volatility

Finally, a major advantage of plating that can't be overstated is the cost savings when compared to paste. Spot price comparisons show that silver has averaged approximately 100X the cost of copper historically. As a result, silver paste is a significant portion (>10%) of the cell's bill of materials. Additionally, it is proposed that the cost of Ag will be materially impacted by expansion in the PV industry, even taking into account the expected reduction in usage per cell [15]. Finally, fluctuations in the price of silver, a precious metal, have a disrupting effect on manufacturers' costs, inventory, and supply contracts. All of these issues can be solved by switching to copper, a cheap, commodity metal (figure 4).

3 CELL TYPES

3.1 p-type BSF / PERC

P-type, Aluminum back surface field (BSF) crystalline silicon cells represent the current standard for industrial commodity cells, and account for nearly 90% of production as of 2015 [1]. As such, the application of plating to BSF cells has been a focus of research for years. BP Solar introduced the "Saturn" cell, featuring buried groove plated contacts, and manufactured this cell type in mass production for a decade [16]. Later, Suntech adopted a similar technology and created the "Pluto" cell, featuring LDSE patterning and plating in combination with a passivated rear [17]. At the same time, much research was going on at major institutes, manufacturers and industry suppliers to understand and progress the use of plating in solar cell manufacturing. Initially, light induced plating was used in combination with a silver paste seed layer, in order to gain some benefits of plating while still maintaining sufficient adhesion to the wafer [18]. After equipment and process improvements made it possible to plate directly to silicon after a simple laser ablation pattern step, many new wafers with high efficiency and good adhesion/reliability were created by a number of institutes and companies, detailed later in this work. Using this technology, a turnkey solution was created for patterning, plating and annealing of standard BSF or PERC cells with a simple, inline process with efficiencies above 19.5% in production on Cz silicon [19].

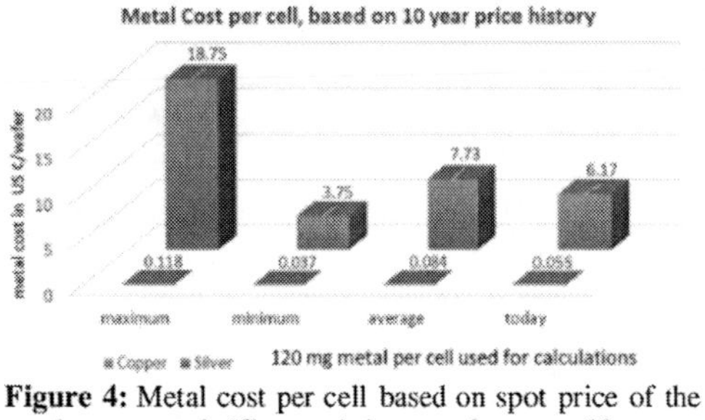

Figure 4: Metal cost per cell based on spot price of the conductor metal (Cu vs Ag) over the past 10 years, demonstrating the volatility in silver usage costs.

In recent years, standard cell production has begun shifting to the use of a rear passivation scheme to create PERC (passivated emitter and rear cell) designs. The PERC process is a relatively simple upgrade from BSF, requiring application and patterning of a rear side passivation layer, followed by standard front side metallization. Therefore, plating can be just as easily applied to the PERC cell concept. In fact, the vast reduction of rear side recombination by passivation once again directs the attention to the front side, where narrow plated conductors have the potential to realize even further gains in efficiency [20]. Plating was successfully applied in the Pluto cell mentioned above, as well as the 'i-PERC' cell concept by Russell, et al. [21], and further improved upon in subsequent testing, demonstrating efficiencies up to 20.7% [22]. Cornagliotti et. al demonstrated a 0.5% absolute efficiency improvement over screen printed paste for a plated PERC cell [23]. Additionally, the use of laser doping (selective emitter) during the front side patterning step can push the efficiency above 21% [24].

3.2 n-type PERT / bifacial

Historically, the highest efficiency cells have been fabricated using n-type silicon base material with boron-doped emitters. This is because n-type Si is not susceptible to potential induced degradation (PID) and light induced degradation (LID), which has negatively impacted p-type Si cells, although the recent practice of regeneration has resulted in efficiency recovery in these cells. N-type silicon fabrication processes can be combined with plating much like p-type cells mentioned above, with a couple key differences. First, n-type cells feature a boron emitter, as mentioned, which is more difficult to make a low-resistance contact to, especially with standard silver paste. This difficulty can be overcome by forming a selective emitter prior to patterning and plating the cell. Also, n-type cells often feature a patterned grid on both sides (bifacial design), so it is important to have a double-sided contact and plating method. It has been shown that the application of plating to an n-PERT cell (passivated emitter, rear totally-diffused) can improve efficiency by 0.4% absolute over printed paste [10]. When combined with a front side selective emitter, achieved by laser doping, the efficiency of this cell design is improved even further. Uruena et. al reported an efficiency improvement from 20.8% to 22.0% by use of laser doping, and a champion cell of 22.5% [25]. Mondon reported similar efficiencies when combined with the PassDop rear passivation process [26]. First Solar (TetraSun) has recently reported average high volume production cell efficiency in the 22% range using a unique n-type bifacial cell architecture with nickel and copper plating [27].

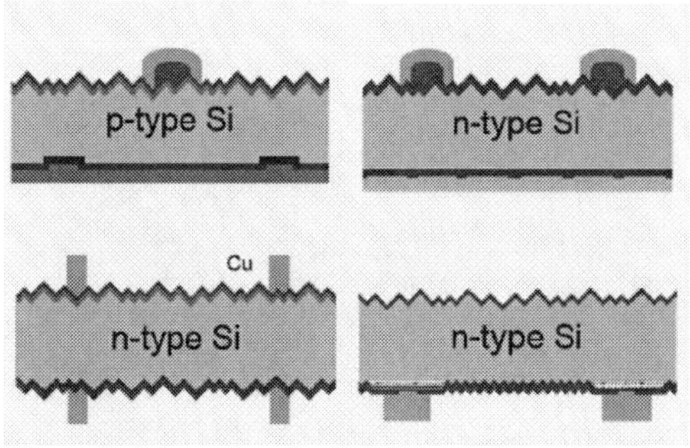

Figure 5: Schematics of p-PERC (top left), n-PERT (top right, SHJ (lower left), and IBC (lower right), featuring copper as the primary conductor.

3.3 Silicon heterojunction

Another area of intense research in recent years is the silicon heterojunction (SHJ) cell type, sometimes called heterojunction with intrinsic thin-layer (HIT). SHJ cells differ from other n-type designs in that they feature layers of amorphous silicon that serve as passivating films with a wider bandgap. These semiconducting films greatly reduce recombination at the junction interface, which leads to extremely high Voc levels for these cells [28]. The unique opportunities of heterojunction cells also come with unique drawbacks. Due to the presence of these thin amorphous silicon layers, the cell efficiency is severely damaged by high temperature (>200 °C) thermal processes. As a result, the use of silver paste as the primary conductor is not ideal; since the paste can't be fired as in a standard process, SHJ cells require much higher quantities of a thick film-like paste that is annealed at lower temperature. Cells require more than 200 mg of silver paste for monofacial designs and over 350 mg for bifacial designs in order to achieve acceptable conductivity [29]. Also, the higher quantity of paste leads to higher shading than desired. The special challenges of the heterojunction cell, therefore, make the application of wet chemical metallization especially attractive. Plating is achieved at low temperatures (typically onto a PVD seed layer), and through a resist mask, thus creating a conductor grid with high aspect ratio and excellent conductivity. Companies such as Kaneka and Panasonic have reported lab efficiencies of 25.1%, and 25.6%, respectively, by combining a heterojunction design with copper plating in a back contact configuration [30, 31]. Solar City (Silevo) has achieved efficiencies over 23% on a bifacial SHJ concept, and is currently scaling this process for mass production [32].

3.4 Interdigitated back contact

The final cell type that can benefit from wet chemical deposition is interdigitated back contact (IBC) cells. In fact, the only mass produced IBC cells have featured plating, and it can be argued that the cell design *requires* plated conductors for performance. Due to the presence of both p- and n-doped regions on the same side of the wafer, IBC cell manufacturing comprises a complex process. This process includes two separate diffusion steps, multiple lithography sequences, and copper and tin plating to form the conductors. The primary reason that wet chemical metallization is employed on IBC cells is that the conductors must be very thick (35-50 microns thick) because the effective 'finger length' spans the total length of the wafer in this cell design. Also, warpage is critical since the metallization is only on one side. IBC cells have been mass produced solely by SunPower over the past 20 years, and have consistently been the highest performing modules in the field over this time. Davis et. al have demonstrated a 25.0% efficient 4" wafer, while Trina has produced a 6" R&D cell with efficiency of 23.5% [33, 34].

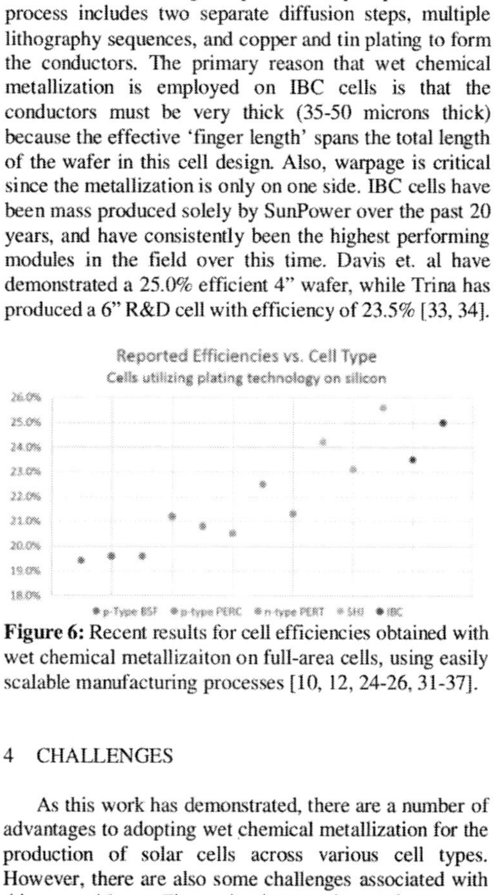

Figure 6: Recent results for cell efficiencies obtained with wet chemical metallizaiton on full-area cells, using easily scalable manufacturing processes [10, 12, 24-26, 31-37].

4 CHALLENGES

As this work has demonstrated, there are a number of advantages to adopting wet chemical metallization for the production of solar cells across various cell types. However, there are also some challenges associated with this transition. First, implementation of a new metallization process requires substantial capital expenditure, which makes it more difficult to immediately realize the economic benefit of copper conductors. Also, replacing a well known process with a new set of lesser known metallization steps may be unattractive to a risk-averse industry that prefers the 'fast-follower' approach. The slow adoption of plating in PV manufacturing can be attributed to these obstacles, as well as some of the technical challenges detailed below.

4.1 Adhesion

Ever since research has focused on replacing paste with plated contacts for solar cells, the adhesion of the plated stack to the silicon substrate has been a primary obstacle to success. For obvious reasons, adhesion failure between the seed plated layer and silicon would lead to performance and reliability failures in the field. As mentioned earlier in this paper, techniques such as laser groove buried contacts overcame this issue mechanically by metallizing a trench [7]. Also, the formation of a nickel silicide layer by thermal annealing after plating was examined for its adhesive properties [38]. The improvement of laser ablation patterning process, especially the introduction of picosecond pulse lengths, has also led to an improvement in adhesion (figure 7). This is presumably due to a rougher Si surface that serves as an anchor point [19, 26]. Some techniques actually involve intentional lasering of "anchor points" to serve as adhesion promoters [36]. The sum total of this work is that very impressive adhesion results have been presented by numerous sources detailed here, with 90' peel strengths greater than 1.5 N/mm becoming the norm. It is the author's opinion that plated adhesion will no longer be a significant obstacle for the industry going forward.

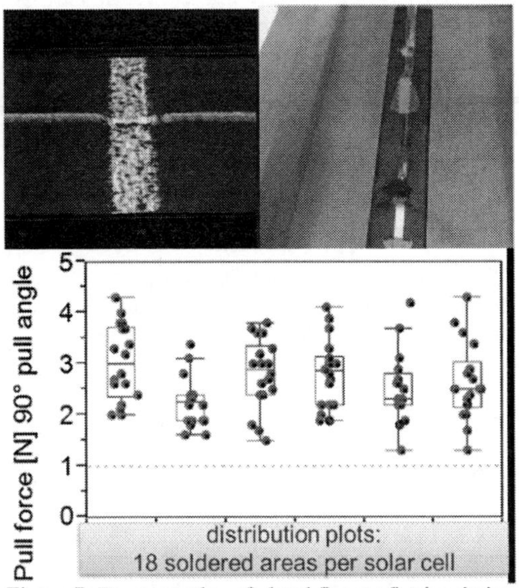

Figure 7: Demonstration of plated finger adhesion during shear testing (left); High soldered busbar adhesion leading to silicon fracture during 90° peel test (right). Box plots of peel testing also show excellent adhesion results [35].

4.2 Background plating

Another challenge to the mass adoption of wet chemical metallization has been the issue of background plating. A number of studies have focused on the causes and methods of prevention for background plating [39, 40, 41]. Background plating generally occurs when an interruption in the passivation layer covering the surface to be plated allows the silicon to plate up in undesired locations. Background plating leads to increased shading and lower J_{sc}, but generally does not cause shunting, since the barrier metal will plate in the unwanted areas as well, preventing diffusion of copper into the SCR. The primary causes of background plating occur during the wafer texturing, cleaning, and passivation steps. Impurities or other fragments are not properly cleaned from the silicon surface, and thus introduce defects in the passivated surface, which is able to plate in subsequent metallization steps. Also, micro-cracks caused by stress sometimes leave voids at the base of pyramids, which can then plate. Finally, scratching of the silicon nitride passivation layer (by excessive handling or equipment) can remove the passivation layer in regions of the cell, which will also cause background plating. The problem of background plating is an ongoing issue, but can be greatly reduced by controlling the wafer cleaning and passivation processes. In the case of direct plating on silicon, the choice of silicon activation chemistry and proper control of the pretreatment process is also critical to the incidence of background plating.

4.3 Long-term reliability

One of the greatest impediments to implementing wet chemical metallization is concerns over long-term reliability of plated cells and modules. This stems from two sources: the fear of copper penetrating into the emitter region and degrading efficiency, and the fear that, over the long term, plated conductors may delaminate and/or corrode, leading to a similar module failure. To address the first concern, numerous works have been published which demonstrate that nickel is an excellent barrier to copper diffusion. The barrier properties of nickel are also well known from the semiconductor industry. Bartsch's method of accelerated aging uses Arrhenius plots to extrapolate the time-to-failure (5% power loss) for a given metallization stack at a particular temperature [42]. Using this method, even a very thin nickel deposit should be a suitable barrier to copper diffusion for over 100 years on a rooftop [43]. In addition to this method, there are a suite of accelerated aging tests for modules specified by IEC method 61215. These include thermal cycling, damp-heat exposure, humidity freeze, and mechanical load testing. The majority of the plated cells mentioned in this paper were put through IEC testing and easily passed, some even running far beyond the required limit of cycles or time [12, 21, 32, 35]. However, perhaps the greatest proof, so far, of long-term reliability on plated modules is the ongoing study of BP Saturn modules in the field [44]. Power data gathered from Saturn modules shows remarkably high power output after 20 years of operation, demonstrating that plated modules may be just as reliable as silver paste long-term.

5 SUMMARY

There are a number of potential advantages that can be gained by switching to plated contacts. The advantages, which include lower shading, lower contact resistance, and lower cost, have been laid out in this review, and supported by the numerous works cited here. Efficiencies in excess of 20% have been achieved using applications of wet chemical metallization on various cells types. Additionally, equipment and process improvements have led to excellent adhesion, elimination of background plating, and long-term reliability in modules featuring wet chemical metallization. Plating now offers a simple and reliable solution for simultaneous efficiency improvement and cost reduction.

6 REFERENCES

[1] ITRPV 2016, International Technology Roadmap for Photovoltaic, Seventh Edition, March 2016, http://www.itrpv.net.

[2] M. Aleman, N. Bay, M. Fabritius, and S. W. Glunz, Proc. 22nd EUPVSEC, Milan, Italy, 2007.

[3] J. P. Gambino and E. G. Colgan, Materials Chemistry

and Physics., volume 52, issue 2, pp 99-189, 1998.

[4] A. Mondon, J. Bartsch, B. J. Godejohann, M. Hörteis, and S. W. Glunz, 2nd Workshop on Metallization for Crystalline Silicon Solar Cells, Konstanz, Germany, 2010.

[5] J. Bartsch, V. Radtke, C. Savio, and S. W. Glunz, Proc. 24th EUPVSEC, Hamburg, Germany, 2009.

[6] A. Letize, B. Lee, K. Crouse, D. Cullen, Proc. 28th EUPVSEC, Paris, France, 2013.

[7] M. A. Green, C. M. Chong, F. Zhang, A. Sproul, J. Zolper and S. R. Wenham, Proceeds of the 20th IEEE Photovoltaics Specialists Conference, Las Vegas, USA, 1988.

[8] B. S. Tjahjono, J. H. Guo, Z. Hameiri, L. Mai, A. Sugianto, S. Wang, and S. R. Wenham, Proc. 22nd EUPVSEC, Milan, Italy, 2007.

[9] D. Kray, N. Bay, G. Cimiotti, S. Kleinschmidt, N. Kösterke, A. Lösel, M. Sailer, A. Träger, and H. Kühnlein Proc. 35th Photovoltaic Specialists Conference, Honolulu, USA, 2010.

[10] K. Lai, S. Liu, Y. Lee, M. Lin, Y. Tsao, C. Chuang, C. Li, and C. Wang, 6th Workshop on Metallization for Crystalline Silicon Solar Cells, Konstanz, Germany, 2016.

[11] O. Schultz-Wittmann, D. De Ceuster, A. Turner, D. Crafts, R. Ong, D. Suwito, L. Pavani, and B. Eggleston, Proc. 27th EUPVSEC, Frankfurt, Germany, 2012.

[12] J. Horzel, Y. Shengzhao, N. Bay, D. Pysch, H. Kuhnlein, and P. Verlinden, Proc. 31st EUPVSEC, Hamburg, Germany, 2015.

[13] S. Kluska, J. Bartsch, A. Buchler, G. Cimiotti, A. Brand, S. Hopman, and M. Glatthaar, Energy Procedia 77 (2015), pp 733-743.

[14] N. Bay, J. Horzel, M. Passig, M. Sieber, J. Burschik, H. Kuhnlein, J. Bartsch, A. Brand, A. Mondon, D. Eberlein, C. Volker, S. Gutscher, A. Letize, B. Lee, D. Weber, and R. Bohme, Proc. 29th EUPVSEC, Amsterdam, The Netherlands, 2014.

[15] M. Redlinger, M. Woodhouse, and R. Eggert, Photovoltaics International, 31st edition, 2016.

[16] N. Mason, and D. Jordan . Proc. 10th EUPVSEC, Lisbon, Portugal, 1991.

[17] Z. Shi, S. Wenham, and J. Ji, Proc. 34th Photovoltaic Specialists Conference, Philadelphia, USA, 2009.

[18] M. Kamp, J. Bartsch, S. Nold, M. Retzlaff, M. Horteis, and S. W. Glunz, Energy Procedia 2011, volume 8, pp 558–564.

[19] A. Letize, D. Cullen, B. Lee, K. Crouse, N. Bay, J. Horzel, T. Knarozovski, H. Kuehnlein, J. Bartsch, M. Glatthaar, R. Bohme, Proc. 29th EUPVSEC, Amsterdam, The Netherlands, 2014.

[20] B. Min, H. Wagner, M. Muller, H. Neuhaus, R. Brendel, and P.P. Altermatt, Proc. 31st EUPVSEC, Hamburg, Germany, 2015.

[21] R. Russell, L. tous, H. Philipsen, J. Horzel, E. Cornagliotti, M. Ngamo, P. Choulat, R. Labie, J. Beckers, J. Bertens, M. Fujii, J. John, H. Poortmans, and R. Mertens, Proc. 27th EUPVSEC, Frankfurt, Germany, 2012.

[22] L. Tous, R. Russell, J. Beckers, J. Bertens, E. Cornagliotti, P. Choulat, J. John, F. Duerinckx, J. Szlufcik, J. Poortmans, and R. Mertens, Proc. 28th EUPVSEC, Paris, France, 2013.

[23] E. Cornagliotti, L. Tous, A. Uruena, A. Rothschild, R. Russell, V. Lu, S. Radosavjlevic, J. John, J. Toman, M. Aleman, F. Duerinckx, J. Poortmans, J. Szlufcik, B. Dielissen, F. Souren, X. Gay, and R. Gortzen, Proc. 29th EUPVSEC, Amsterdam, The Netherlands, 2014.

[24] M. Kim, D. Kyeong, K. Lee, W. Lee, and E.C. Cho, Proc. 29th EUPVSEC, Amsterdam, The Netherlands, 2014.

[25] A. Uruena, M. Aleman, E. Cornagliotti, A. Sharma, J. Deckers, M. Haslinger, L. Tous, R. Russell, J. John, Y. Yao, T. Soderstrom, F. Duerinckx, and J. Szlufcik, Proc. 31st EUPVSEC, Hamburg, Germany, 2015.

[26] A. Mondon, J. Bartsch, M. Kamp, A. Brand, B. Steinhauser, N. Bay, J. Horzel, M. Glatthaar, S. W. Glunz, Proc. 29th EUPVSEC, Amsterdam, The Netherlands, 2014.

[27] D. De Ceuster, PVCellTech Conference, Kuala Lumpur, Malaysia, 2016.

[28] S. De Wolf, A. Descoeudres, Z. C. Holman, and C. Ballif, Green, vol. 2, pp. 7-24, 2012.

[29] A. Faes, M. Despeisse, J. Levrat, J. Champliaud, A. Lachowicz, J. Geissbuhler, N. Badel, H. Watanabe, T. Soderstrom, Y. Yao, J. Ufheil, P. Papet, B. Strahm, J. Hermans, A. Tomasi, Y. Baumgartner, J. Cattin, M. Kiaee, A. Hessler-Wyser, J. Fleischer, P. V. Fleischer, A. Tsuno, C. Ballif, 6th Workshop on Metallization for Crystalline Silicon Solar Cells, Konstanz, Germany, 2016.

[30] Nikkei Technology Online. (2015). [Press release]. Retrieved from http://techon.nikkeibp.co.jp/atclen/news_en/15mk/102800130/?ST=english_PRINT

[31] Panasonic Corporation. (2014). [Press release]. Retrieved from http://news.panasonic.com/global/press/data/2014/04/en140410-4/en140410-4.html

[32] J. B. Heng, J. Fu, B. Kong, Y. Chae, W. Wang, Z. Xie, A. Reddy, K. Lam, C. Beitel, C. Liao, C. Erben, Z. Huang, and Z. Xu, Proc. 29th EUPVSEC, Amsterdam, The Netherlands, 2014.

[33] D. D. Smith, P. Cousins, S. Westerberg, R. De Jesus-Tabajonda, G. Aniero, and Y. Shen, IEEE Journal of Photovoltaics, vol. 4, No. 6, November 2014.

[34] Clean Technica. (2016). [Press release]. Retrieved from http://cleantechnica.com/2016/05/26/8-1-million-

working-global-renewable-energy-industry/8/

[35] J. Horzel, N. Bay , M. Passig , M. Sieber, J. Burschik, H. Kühnlein, A.Letize, B. Lee, D. Weber, R. Böhme, A. A. Brand, J. Bartsch, A. Mondon, D. Eberlein, and C. Völker, 5th Workshop on Metallization for Crystalline Silicon Solar Cells, Konstanz, Germany, 2014.

[36] A. Wenham, C. M. Chong, S. Wang, R. Chen, J. Ji, Z. Shi, L. Mai, A. Sugianto, S. Wenham, A. Barnett, and M. Green, Proc. 31st EUPVSEC, Hamburg, Germany, 2015.

[37] J. L. Hernández, D. Adachi, D. Schroos, N. Valckx, N. Menou, T. Uto, M. Hino, M. Kanematsu, H. Kawasaki, R. Mishima, K. Nakano, H. Uzu, T. Terashita, K.Yoshikawa, T. Kuchiyama, M. Hiraishi, N. Nakanishi, M. Yoshimi, K. Yamamoto, Proc. 28th EUPVSEC, Paris, France, 2013.

[38] A. Mondon, M. N. Jawaid, J. Bartsch, M. Glatthaar, and S. W. Glunz, Solar Energy Materials and Solar Cells, volume 117, pp 209-213, 2013.

[39] S. Braun, A. Zuschlag, B. Raabe, and G. Hahn, Energy Procedia 8, pp. 565–570, 2011.

[40] E. Lee, H. Lee, J. Choi, D. Oh, J. Shim, K. Cho, J. Kim, S. Lee, B. Hallam, S. Wenham, and H. Lee, Solar Energy Materials and Solar Cells, Volume 95, Issue 12, pp. 3592-3595, 2011.

[41] S. Wang, T. Puzzer, B. Vogl, B. Tjahjono, B. Hallam, M. Eadie, N. Borojevic, Z. Hameiri, and S. Wenham, Proc. 24th EUPVSEC, Hamburg, Germany, 2009.

[42] J. Bartsch, A. Mondon, K. Bayer, C. Schetter, M. Horteis, and S. W. Glunz, Journal of the Electrochemical Society, 157 (10), H942-H946, 2010.

[43] S. Kluska , J. Bartsch, A. Kraft, L. Ni, C. Wolf, F. Schindler, J. Schön, M. Glatthaar, and S. W. Glunz, 6th Workshop on Metallization for Crystalline Silicon Solar Cells, Konstanz, Germany, 2016

[44] M. Alonso-Abella, F. Chenlo, A. Alonso and D. González, Proc. 29th EUPVSEC, Amsterdam, The Netherlands, 2014.

Performance Enhancement of Textured and Planar Silicon Solar Cells Using Luminescent Down-Shifting Eu^{2+}-Phosphor Silica-Layer

Yu-Jie Deng, Wen-Jeng Ho*, Sheng-Kai Feng, Guan-Yi Li, and Su-Han Weng
Department of Electro-Optical Engineering, National Taipei University of Technology
No. 1, Sec. 3, Zhongxiao E. Rd., Taipei 10608, Taiwan, R.O.C.

*e-mail: wjho@ntut.edu.tw

ABSTRACT: In this study, we report the deposition of Eu-doped (3wt%) phosphors on planar and textured Si solar cells via a silicate (SiO_2) solution using spin-on film technique to enhance photovoltaic performance. The surface profile and dimensions of the phosphor particles were examined using electron scanning microscopy (SEM). Optical reflectance, photoluminescence (PL), and external quantum efficiency (EQE) were used to characterize the luminescent downshifting (LDS) of the Eu-doped silicate phosphor layer. Current density-voltage (J-V) curves under AM 1.5G simulation were used to confirm the contribution of LDS and the anti-reflectivity and shading produced by Eu-doped phosphors-particles of various dimensions. Experiment results revealed that larger phosphor particles have a more pronounced effect on the shading and reflectance of incident light. For planar solar cells, the short-circuit current density (J_{sc}) enhamcement by about 20% was obtained when deposited Eu-doped phosphors on the cells. On textured solar cells, however, the application of sub-micron Eu-doped phosphor particles increased the J_{sc} by 5.6%, far exceeding that of cells with larger Eu-doped phosphor particles (of 1.6%), compared to the reference cell.

Keywords: Eu-doped silicate phosphors, EQE response, luminescent downshifting (LDS), Si solar cell

1 INTRODUCTION

At ultraviolet (UV)-blue wavelengths (300-450 nm), the conversion efficiency of crystalline silicon (C-Si) solar cells is relatively low, due to high reflectance and low spectral response. Incident photons of higher energy (within UV-blue wavelengths) are absorbed within a very short distance from the surface, which results in high recombination loss [1-4]. Many previous studies aimed at enhancing the overall conversion efficiency have focused on the application of a down-conversion (DC) layer or down shifting (DS) layer to the top surface of the C-Si in order to improve spectral response in the UV-blue region [5-9]. However, recent studies have addressed the development of luminescent downshifting (LDS) phosphor materials for the conversion of high-energy incident photons into lower energy photons for application in photovoltaic devices [10-20].

This study characterized the LDS, shading, and reflectance of a layer of Europium-doped (Eu-doped) silicate phosphors of various particles dimensions when applied to planar-type and textured-type C-Si solar cells. Scanning electron microscopy (SEM), photoluminescence (PL), and optical reflectance were used to examine the surface morphology and optical properties of the Eu-doped silicate phosphor layer. External quantum efficiency (EQE) response and photovoltaic current density-voltage (J-V) characteristics under AM 1.5G simulation measurements were used to evaluate the improvement in photovotaic performance in both types of solar cell as a function of LDS and particle dimensions.

2 EXPERIMENT

This study employed Eu-doped silicate phosphors ($(Sr_{1-x}Ba_x)_2SiO_4:Eu^{2+}F$; O6040 ™; InteMatix Company, USA) with dimensions of 15-25 micrometers as a material for luminescent down-shifting (LDS) (referred hereafter as larger particles). Phosphor particles of smaller dimensions were obtained by grinding the larger silicate phosphor particles in an agate mortar to produce particles of 5-15 micrometers (referred to as sub-micro particles). We then produced two solutions (one with the larger phosphor particles and the other with sub-micro phosphor particles) with 0.06g Eu-doped silicate phosphor power mixed with 1.94 g silica solution. Planar and textured Si substrates were then coated with the mixed solutions using spin-on film technique at 3000 rpm for 60 s before being baked at 200℃ for 30 min under an air atmosphere, resulting in a layer of 3 wt% Eu-doped phosphors. The surface morphology of the Eu-doped phosphor layer was examined using scanning electron microscopy (SEM). The optical properties of Eu-doped silicate phosphor layer were characterized according to photoluminescence (PL) and optical reflectance using a UV/VIS/NIR spectrophotometer.

A powder of 3 wt% Eu-doped phosphors mixed with a silicate solution was deposited on the bare planar and bare textured Si solar cells using spin-on film deposition. For comparison, we also produced a Si solar cell coated with only a SiO_2 layer (using the same Silicafilm solution) without neither of the phosphor species. The resulting textured surfaces were examined using scanning electron microscopy (SEM). The optical reflectance of the bare solar cells with either a SiO_2 layer or SiO_2 layer coated with phosphor particles was characterized using a UV/VIS/NIR spectrophotometer. The external quantum efficiency (EQE) of the cells was also measured over a range of wavelengths from 350 to 1100 nm using a solar cell spectral response measurement system. The photovoltaic I–V characteristics of the proposed cells were measured using a solar simulator and Keithley 2400 source meter at 25 °C. The solar simulator was calibrated according to an NREL-certified crystalline silicon reference cell before measurements were taken. Fig. 1 shows a schematic diagram of the textured Si solar cell coated with an Eu-doped phosphor particles layer.

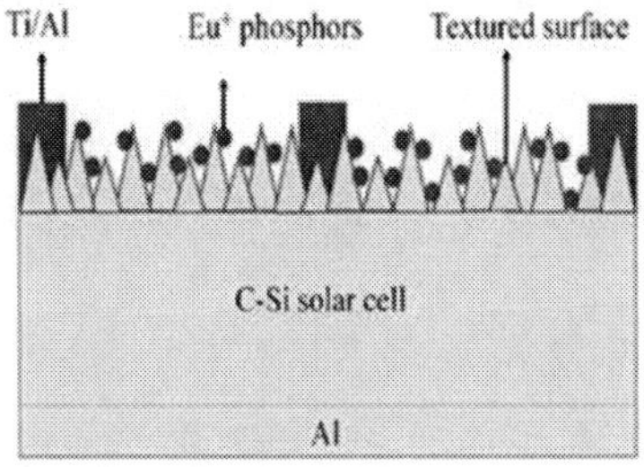

Fig. 1: schematic diagram of the textured Si solar cell coated with an Eu-doped phosphor particles layer..

3 RESULTS AND DISCUSSION

Fig. 2 presents the side-view SEM images of an Eu-doped silicate phosphor layer deposited on (a) a planar solar cell and (b) textured solar cell. The dimensions of the phosphor particles before and after grinding were approximately 15-25 μm and 5-15 μm, respectively.

(a)

(b)

Fig. 2: Side-view SEM images of an Eu-doped silicate phosphor layer deposited on (a) a planar solar cell and (b) textured solar cell.

Fig. 3 presents the PL fluorescence and PL excitation (PLE, or absorbance) spectrum of a layer comprising 3 wt% Eu-doped silicate phosphors deposited on a Si substrate. Peak PL emissions were exhibited at approximately 610 nm with full width at half maximum (FWHM) of approximately 87.5 nm. Peak absorbance was measure at approximately 375 nm with FWHM of approximately 123 nm. The Stoke shift of the Eu-doped silicate phosphor was approximately 225 nm. The measured PL emission and absorption results demonstrate that the 3wt% Eu-doped silicate phosphor layer absorbed and converted UV photons into visible photons, as indicated by the excellent LDS behaviour of the used Eu-doped silicate phosphor.

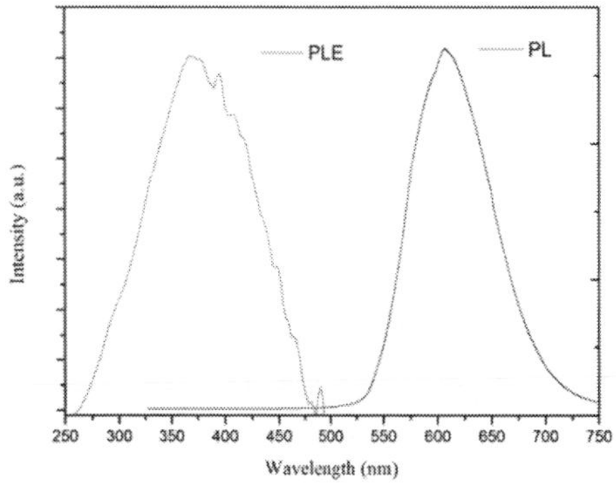

Fig. 3: PL fluorescence and PL excitation (PLE, or absorbance) of a layer comprising 3 wt% Eu-doped silicate phosphors deposited on a Si substrate.

To characterize the enhancement in absorption performance by LDS, we applied a layer of Eu-doped silicate phosphors to the top-surface of bare planar and bare textured solar cells. A comparison of optical reflectance and EQE response was used to reveal the contribution of LDS and anti-reflectivity provided by the layer of Eu-doped silicate phosphors on both types of Si solar cell. Fig. 4 presents the optical reflectance of a bare solar cell, a cell with a SiO_2 layer 250nm in thickness, and cells with either larger or sub-micron Eu-doped silicate phosphors applied over a SiO_2 layer on (a) planar-type and (b) textured-type solar cells. The reflectance of both types of solar cell was altered by the application of a SiO_2 layer, in which the lowest reflectance values were obtained from planar-type solar cells at 550 nm, due to the effects of destructive interference. The reflectance of cells with a layer of Eu-doped phosphors was monolithically reduced across the entire range of wavelengths, compared with bare-type solar cells (both planar and textured). However, compared to the cell with only a SiO_2 layer, the reflectance of the planar solar cell with Eu-doped phosphor layers dropped at wavelengths below 540 nm due to the absorption of high-energy photons by Eu-doped phosphor particles, as well as in the wavelength range 730-1200 nm due to forward light scattering caused by the phosphor particles. Furthermore, the reflectance of the textured solar cell with Eu-doped phosphor layer dropped at wavelengths below 400 nm due to the absorption of high-energy photons by Eu-doped phosphor particles and at wavelengths beyond 500 nm due to forward light scattering caused by the phosphor particles.

Fig. 5 presents the EQE response of a bare solar cell, a cell with a 250-nm-thick SiO_2 layer, and cells with Eu-doped silicate phosphors (either large or sub-micron particles) over a SiO_2 layer: (a) planar-type and (b) textured-type solar cells. As shown, the EQE values of both types of solar cell increased at wavelengths of 300-500 nm due to LDS produced by Eu-doped phosphor par-

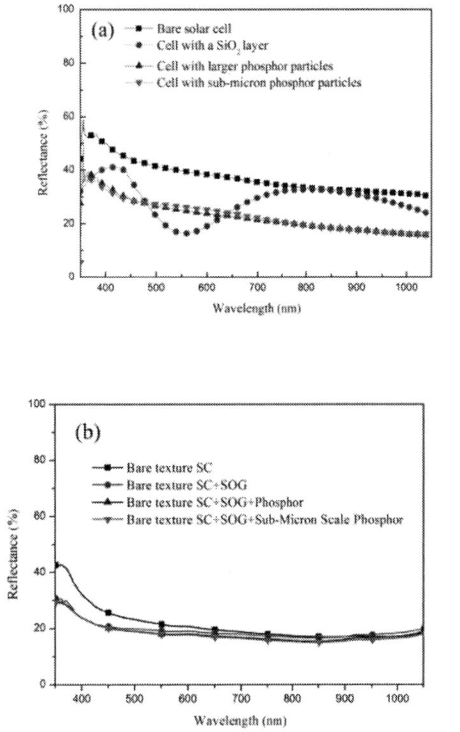

Fig. 4: Optical reflectance of a bare solar cell, a cell with a SiO_2 layer 250nm in thickness, and cells with either larger or sub-micron Eu-doped silicate phosphors applied over a SiO_2 layer on (a) planar-type and (b) textured-type solar cells.

ticles. These findings are in good agreement with PL and PLE data as well as optical reflectance at short wavelengths. Furthermore, in both types of cell, those with sub-micron phosphor particles presented EQE values higher than those fabricated using larger phosphor particles, which can be attributed to the reduced effects of shading produced by sub-micron particles.

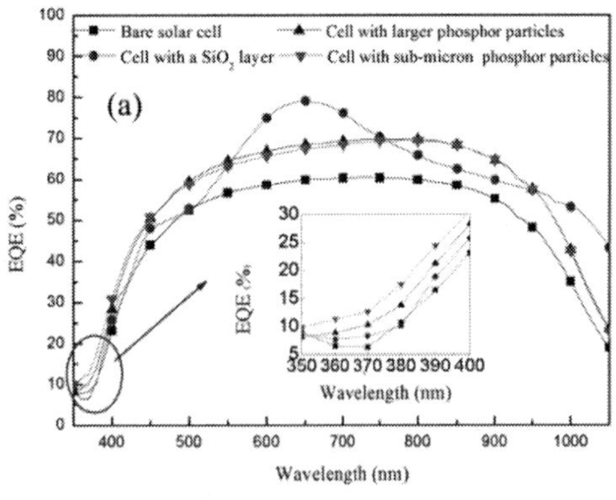

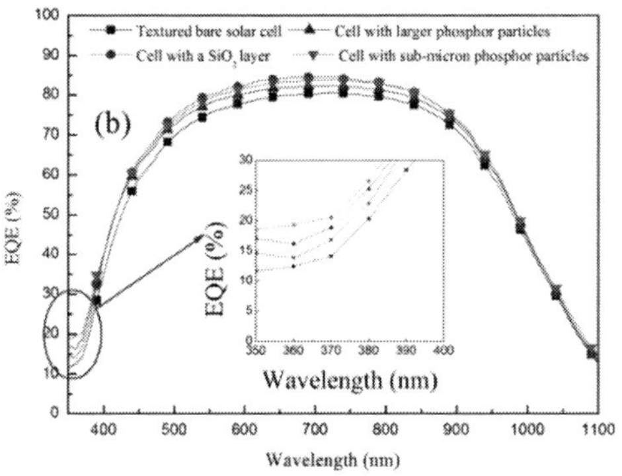

Fig. 5: EQE response of a bare solar cell, a cell with a 250-nm-thick SiO_2 layer, and cells with Eu-doped silicate phosphors (either large or sub-micron particles) over a SiO_2 layer: (a) planar-type and (b) textured-type solar cells.

Fig. 6 presents the photovoltaic current density-voltage (J-V) curves of the reference cell and cells fabricated using Eu-doped silicate phosphors (either large or sub-micron particles) on: (a) planar-type or (b) textured-type solar cells. The photovoltaic performance values of planar-type and textured-type solar cell are respectively summarized in Table 1 in Table 2. In the case of planar-type solar cells, the increase in J_{sc} in the cell with sub-micron Eu-doped phosphor particles (20.6%) was greater than that of the cell with larger Eu-doped phosphor particles (19.51%), compared to the reference cell. In the case of textured-type solar cells, the ΔJsc of the cell with sub-micron Eu-doped phosphor particles (5.6%) exceeded that of the cell with larger Eu-doped phosphor particles (1.6%). This is a clear indication that larger phosphor particles have a stronger effect on shading and reflectivity, particular on textured-type solar cells. This means that further enhancements in the photovoltaic performance of textured-type solar cells will require Eu-doped phosphor particles of smaller dimensions, such as 10-100 manometers.

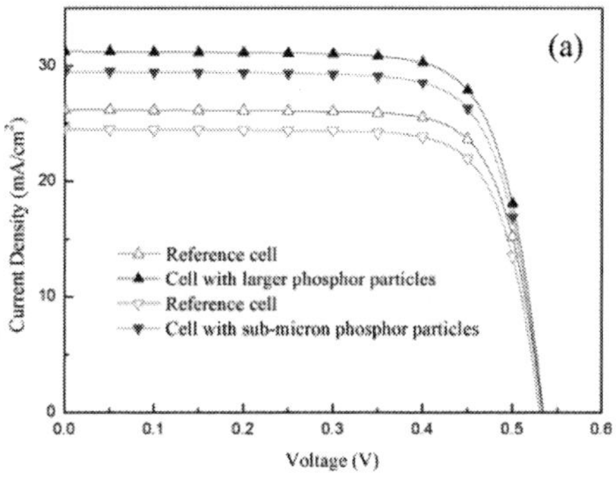

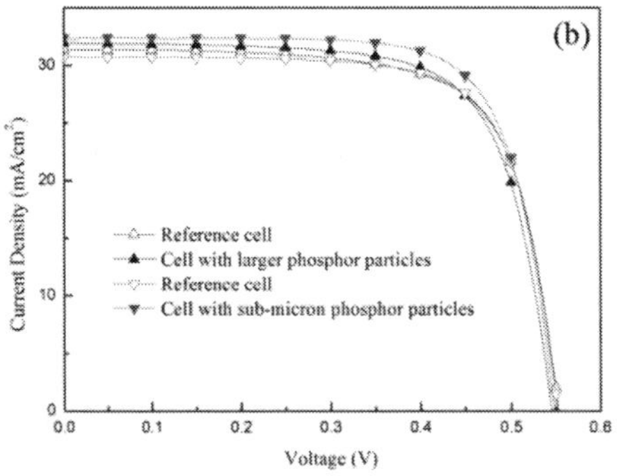

Fig. 6: Photovoltaic current density-voltage (J-V) curves of reference cell and cells with Eu-doped phosphors (either large or sub-micron particles): (a) planar-type and (b) textured-type solar cells.

Table 1: Photovoltaic performances of planar-type solar cells with and without phosphor particles.

	Reference cell	Cell with larger phosphor particles	Reference cell	Cell with sub-micron scale phosphor particles
V_{oc} (mV)	528.88	531.78	530.72	533.49
J_{sc} (mA/cm^2)	26.14	31.24	24.46	29.5
η(%)	9.79	11.74	9.92	11.89
$\triangle J_{sc}$ (%)	19.51%		20.6%	
△η(%)	19.91%		19.85%	

Table 2: Photovoltaic performances of textured-type solar cells with and without phosphor particles.

	Reference cell	Cell with larger phosphor particles	Reference cell	Cell with sub-micron scale phosphor particles
V_{oc} (mV)	552.82	545.6	552.03	548.41
J_{sc} (mA/cm^2)	31.41	31.92	30.69	32.41
η(%)	12.31	12.36	12.41	13.11
$\triangle J_{sc}$ (%)	1.6%		5.6%	
△η(%)	0.41%		5.6%	

4 CONCLUSION

This study reports the fabrication and characterization of planar-type and textured-type C-Si solar cells with a layer of LDS Eu-doped silicate phosphors. The electrical and optical properties of the cell with Eu-doped phosphors were shown to depend on the dimensions of the phosphors particles. The application of sub-micron scale Eu-doped phosphor particles to either type of solar cell led to a significant improvement in J_{sc}. The effects were particularly pronounced when sub-micron scale phosphors were applied to textured-type solar cells. Further enhancements in the photovoltaic performance of textured-type solar cells will require the application of Eu-doped phosphor particles of smaller dimensions, such as 10-100 microns. Our laboratory is currently engaged in the development of high-efficiency textured-type solar cells coated with phosphor particles of very small dimensions.

5 ACKNOWLEGEMENT

The authors would like to thank the Ministry of Science and Technology of the Republic of China for financial support under Grant MOST 103-2221-E-027-049-MY3.

6 REFERENCE

[1]. N. Chander, A. F. Khan, P. S. Chandrasekhar, E. Thouti, S. K. Swami, V. Dutta, V. K. Komarala, Appl. Phys. Lett. 105, 033904 (2014).
[2]. G. Griffini, F. Bella, F. Nisic, C. Dragonetti, D. Roberto, M. Levi, R. Bongiovanni, S. Turri, Advanced Energy Materials 5(3), 1401312 (2015).
[3]. C. K. Huang, Y. C. Chen, W. B. Hung, T. M. Chen, K. W. Sun, W. L. Chang, Progress in Photovoltaics: Research and Applications 21, 1507 (2013).
[4]. G. Shao, C. Lou, D. Xiao, Journal of Luminescence 157, 344 (2015).
[5]. M. Liu, F. Lei, N. Jiang, Q. Zheng, D. Lin, Materials Research Bulletin74, 62 (2016).
[6]. N. Chander, S. K. Sardana, P. K. Parashar, A. F. Khan, S. Chawla, V. K. Komarala, IEEE Journal of Photovoltaics, DOI 10.1109/ JPHOTOV.2015.2438633 (2015).
[7]. V. Mahalingam, J. Thirumalai, R. Krishnan, S. Mantha, Spectrochimica Acta - Part A: Molecular and Biomolecular Spectroscopy 152, 172 (2016).
[8]. S. D. Hodgson, W. S. M. Brooks, A. J. Clayton, G. Kartopu, D. A. Lamb, V. Barrioz, S. J. C. Irvine, Progress in Photovoltaics: Research and Applications 23, 150 (2015).
[9]. M. Kennedy, H. Ahmed, J. Doran, B. Norton, P. Bosch-Jimenez, M. Della Pirriera, E. Torralba-Calleja, D. Gutiérrez Tauste, L. Aubouy, S. Daren, F. Solomon-Tsvetkov, S. Galindo, C. Voz, J. Puigdollers, Physica Status Solidi (A) Applications and Materials Science 212, 203 (2015.)
[10]. F. Sgrignuoli, P. Ingenhoven, G. Pucker, V. D. Mihailetchi, E. Froner, Y. Jestin, E. Moser, G. Sànchez, L. Pavesi, Solar Energy Materials and Solar Cells 132, 267 (2015).

The application of multilayer SiN_x anti-reflection films in polycrstalline silicon solar cell production

Hongna Ma, Zhengyu Li, Long Pang, Dongsheng Zhang
Yingli Green Energy Holding Co., LTD,
State Key Laboratory of Photovoltaic Materials & Technology
3399 North Chaoyang Avenue, Baoding, 071051

ABSTRACT: In order to improve the conversion efficiency of crystalline silicon solar cells, a new multilayer PECVD process for producing SiN_x films is investigated. By using different PECVD routes, applying wafer pre-process step, utilizing optimized multilayer SiN_x deposition process, the effect of passivation and anti-reflection is enhanced, thereby the conversion efficiency of solar cells is improved. According to the result of comparative experiment, the solar cells with double layer films have 0.1% efficiency gain than single layer solar cells, solar cells with triple-layer have 0.15% efficiency gain than double layer solar cells.
Keywords: effect of passivation; effect of anti-reflection; conversion efficiency

1 INTRODUCTION

In photovoltaic fields, improvement of the conversion efficiency of crystalline silicon solar cells is an effective way to reduce the cost. Standard solar cell production process is as follows:

Texturing, phosphorus diffusion, wet etching, PECVD, metallization, testing.

The deposition of passivated anti-reflection film is one of the key process in the production of solar cells. According to the principle of thin film interference [1], single layer or more layers deposition on solar cell surface can reduce the light reflection effectively. SiN_x film as a dielectric film is widely used in microelectronics due to chemical stability, physical stability and dielectric properties advantage [2-5]. The commonly used anti-reflection film is SiN_x by using PECVD method, which has both anti-reflection and surface passivation function. During the actual SiN_x film process of solar cell production, we not only require to reduce the optical loss to show a better optical properties, but also to maximize the surface passivation. Studies have shown that higher Si-H bond density leads to better surface passivation effect [6] and higher density of SiN_x films leads to better bulk passivation effect [7]. However, this kind of structure leads to higher extinction coefficient, so it is unsuitable for anti-reflective film [8], and then the refractive index, thickness and structure of SiN_x film need to be analyzed. By means of designing rational multilayer SiN_x film process, the passivation effect can have an enhancement but less optical absorption loss at the same time, so as to improve the conversion efficiency of solar cells. Studies have shown that the most stable configuration with respect to variations in film thickness has been found to be designs with a high refractive index (n) on the substrate and a low n towards the ambient [9].

2 EXPERIMENTAL WORKS & RESULTS

2.1 Experimental materials

The experimental materials and device are shown in **Table I**.

Table I: Parameters of experimental materials

Species	Solar-grade polycrystalline silicon
Size	156mm×156mm
Thickness	160-200um
Resistivity	1-2Ω.cm
PECVD Device	Centrotherm direct plasma device

2.2 Experimental procedure

There are mainly two steps for the adjustment of SiN_x film PECVD process:

Optimizing the bilayer films based on single-layer film;
Optimizing the triple-layer films based on bilayer films.

3 RESULTS & DISCUSSION

3.1 Bilayer SiN_x process adjustment and analysis of respective layers.

After simulation and experimental verification, the final structure of bilayer is determined as follows: the thickness of first SiN_x layer is 25nm, the index of refraction is 2.15; the thickness of second SiN_x layer is 55nm, the index of refraction is 2.02. Table II shows the electrical parameter comparison of single layer and bilayer SiN_x film cell. It is obvious that the open voltage has 1 mV boost, the short circuit has 31mA boost and the conversion efficiency has 0.097% improvement. The cell parameters reflect the advantages of bilayer film process on passivation and anti-reflection.

Table II: Electrical parameter comparison between single-layer and bilayer

	Single-layer	Bilayer
U_{oc}	0	+1mv
I_{sc}	0	+31mA
FF	0	0.00%
Eff	0	0.10%

3.2 Trilayer SiN_x process adjustment and analysis of respective layers

3.2.1 Trilayer SiN_x process

Table III shows the final version of trilayer process which is determined by simulation and experimental verification.

Table Ⅲ: Trilayer structure parameters

	Thickness	Index of refraction
Layer 1	10nm	2.2
Layer 2	25nm	2.15
Layer 3	45nm	2.02

Table Ⅳ shows the detailed process parameters of respective layers:

Table Ⅳ: Trilayer process parameters

	Layer 1	Layer 2	Layer 3
NH3 (sccm)	5000	5000	7000
SIH4 (sccm)	1666	1100	538
Time (s)	96	138	500
Power (W)	6100	6100	6100
Pressure (mTor)	1300	1500	1500

3.2.2 The minority carrier lifetime test

The main steps for test is show as follows:

Select the polycrystalline wafers with consistent crystal orientation as experimental material and divided them into two groups. Then perform two groups both normal texturing, diffusion and wet etching process. But for PECVD process, one applies bilayer double-side PECVD process while the other applies trilayer double-side PECVD process. At last, anneal both groups and test minority carrier lifetimes. Figure 1 and figure 2 indicate the minority carrier lifetimes of bilayer and trilayer SiN_x films, respectively. Obviously, the trilayer group has more uniform lifetime and higher average lifetime, which means the better passivation in trilayer SiN_x film process.

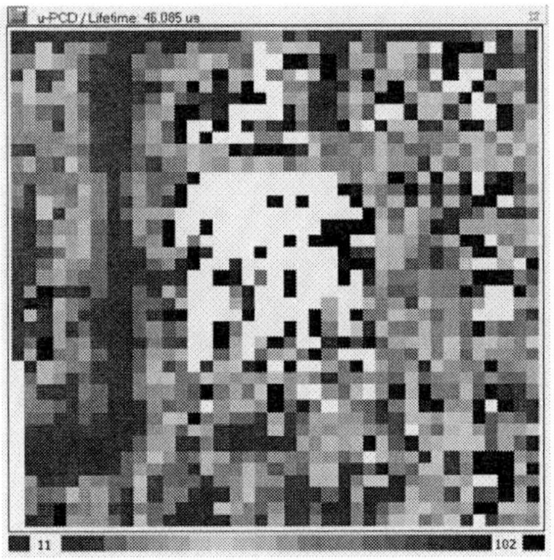

Figure 1 Lifetime of bilayer

Figure 2 Lifetime of trilayer

3.2.3 Reflectivity test

Figure 3 is the reflectivity test (wavelength range: 350nm-1200nm) result of double layer and trilayer film. It shows that the reflectivity of trilayer film is lower. In the range of shortwave, the trilayer film has obvious advantage and the lowest reflection point shift towards to the shortwave range; in the range of longwave, the reflectivity of trilayer film has a slow increase.

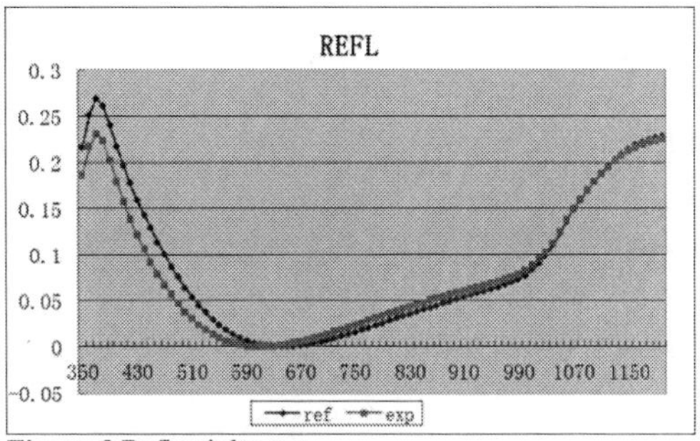

Figure 3 Reflectivity test

3.2.4 External quantum efficiency (EQE) test

The EQE test is shown in Figure 4. It can be seen that the EQE of trilayer film cell has a rapid increase (up to 99.16%) in the range of shortwave.

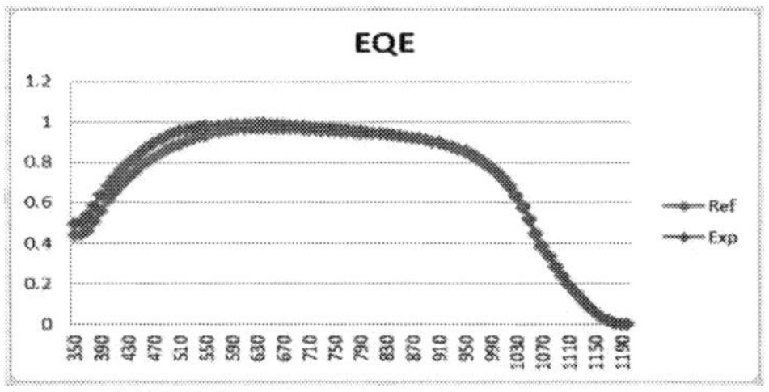

Figure 4 EQE Test

3.2.5 Cell parameters

Based on the better minority carrier lifetime, reflectivity and EQE, trilayer film cell shows better IV parameters, especially on open voltage and short circuit. Table Ⅴ is the comparison of cell parameter. The gain of the open voltage is 2mV, the gain of the short circuit is 55mA and the conversion efficiency has 0.15% enhancement.

Table V: The performance of the trilayer and bilayer cells

	Trilayer	Double layer
U_{oc}	0.633	0.631
I_{sc}	8.789	8.744
FF	78.95	78.94
Eff	18.05%	17.90%

3.3 Conclusion

In the industrialization of solar cell, the application of multi-layer SiN_x film PECVD process on polycrystalline wafers can better integrate the anti-reflection and passivation of SiN_x films. In detail, higher refractive index of SiN_x film has better passivation, which can further improve the open voltage of solar cells; the decreasing refractive index process has better anti-reflection, which can increase the short circuit of solar cell. In this research, the double layer SiN_x film and the trilayer SiN_x film can both improve the conversion efficiency while the trilayer one has better performance.

References

[1] Li Kejian.GuangxiPhysics,1995,16 (5,6) :72.
[2] LiuYu,P.Comodi.,Min.Mag.,1993;57:709.
[3] Peng Wenshi,Liu Gaokui,Acta Mineralogica Sinica,1984,(1):8.
[4] ShemeshA.Geochim.CosmochimActa.1990;54:2433.
[5] Liu kuiwu, Acta Geologica Sinica,1989,63(4):310.
[6] H.Maeckel and R. Luedemann,(2002),J.Appl. Phys.,92(5),2602.
[7]J.Hong,W.M.M.Kessels,W.J.Soppe,A.W.Weeber,W.M. Arnoldbik, and M.C.M.van de Sanden,(2003),J. Vac. Sci. Technol. B., 21(5),2123.
[8] WangXiaoquan,Wanglei,Xizhenqiang,etc.[J].Acta Energiae Solaris Sinica,2004,03.
[9] D. Bouhafs,A. Moussi, A. Chikouche,J.M. Ruiz,"Design and simulation of antireflection coating systems for optoelectronic devices:Application to silicon solar cells",Solar Energy Materials and Solar Cells,52,1998,pp.79-93.

STUDY OF ONE-STEP ANNEALING FOR PLATED NICKEL-COPPER CONTACTS ON N-PERT BIFACIAL MONOCRYSTALLINE SILICON SOLAR CELLS

Julien COUDERC[1,3,6], Hanane EL BELGHITI[4,5], Damien AUREAU[2,5,6], Julien DUPUIS[1,3,6], Pierre-Philippe GRAND[1,3,6], Élise DELBOS[4,5], Arnaud ETCHEBERRY[2,5,6] and Daniel LINCOT[2,3,6]
[1]EDF Recherche & Développement ; 6, quai Watier, 78400 Chatou, France. Phone : +33 (0)1.30.87.79.62, e-mail : julien.couderc@edf.fr
[2]Centre National de la Recherche Scientifique – Institut de Recherche et Développement sur l'Énergie Photovoltaïque (CNRS-IRDEP) ; 6, quai Watier, 78400 Chatou, France
[3]IRDEP, joint research unit EDF-CNRS-Chimie ParisTech ; 6, quai Watier, 78400 Chatou, France
[4]KMG Ultra Pure Chemicals ; Les Vieilles Hayes, 50260 Saint-Fromond, France
[5]Institut Lavoisier de Versailles ; 45, avenue des États-Unis, 78035 Versailles Cedex, France
[6]Institut Photovoltaïque d'Île-de-France (IPVF) ; 8, rue de la Renaissance, 92160 Antony, France

ABSTRACT: Chemical plating is the main alternative to screen printing for the metallization of crystalline silicon solar cells, substituting silver by copper. Yet, because the latter diffuses into silicon and drastically reduces its lifetime, a nickel layer is deposited underneath to both serve as contact for the emitter and barrier for the copper diffusion. In this work, an electrochemical metallization process is described for 180 µm-thick bifacial n-PERT silicon solar cells with 80 Ω/□ boron emitter in the textured front face and 70 Ω/□ phosphorus doping in the chemically etched rear face. After LASER dielectric opening, nickel electroless and copper electrolytic techniques are carried out with simultaneous depositions on both faces of the cell, followed by a one-step annealing of the stack. The influence of different annealing parameters is studied by looking at the nickel silicide characteristics thanks to a combination of structural and electrical characterizations. Ni(P) electroless and Cu electrolytic depositions are selective. Their growth rates are respectively around 0.1 and 1 µm/min, and they exhibit the same characteristics on both faces of the cell. On the whole, performances are quite low (particularly for FF and series resistance), probably due to contact adhesion problems in the chemically etched rear face. Efficiencies of 12.4% (16.6% for the best cell) are obtained for a 20-minute annealing at 380°C. Temperatures of 300°C and 750°C lead to shunted cells: the final one-step annealing seems to have a quite narrow process window.
Keywords: annealing, bifacial, n-type, plating.

1 INTRODUCTION

Screen printing has been the reference technique for more than twenty-five years for the metallization of crystalline silicon solar cells[1]. However, another technology has emerged for several years as the main alternative to this method: chemical plating[2]. Most of the time based on copper metallization instead of silver, this technique allows cost savings, shadowing reduction, contact quality improvement[3] and can be applied to bifacial cells. Nevertheless, the copper migration throughout the silicon wafer[4] can strongly reduce the solar cell efficiency (copper is a carrier lifetime killer in silicon). Therefore, an intermediate barrier is necessary to protect the emitter. Well known in the microelectronics domain and thanks to its properties, nickel is a good candidate to both block the copper diffusion and ensure a desirable and efficient metallic contact to silicon[5]. The formation of a silicide interlayer by annealing allows the reinforcement of the Ni/Cu stack adhesion and the reduction of the contact resistance[6]. In this study, such annealing is carried out in one step, after the nickel and copper depositions on both faces of the cell, to ensure the process compatibility with industrial developments. Fine control of its parameters is required for the nickel layer to get its good "barrier" efficiency and "seed" characteristics.

This work aims at studying the nickel-copper stack annealing thanks to structural and electrical characterizations. First of all, the electroless deposition of the nickel phosphorus layer and the electrolytic deposition of the copper layer on both faces of an n-PERT (Passivated Emitter and Rear Totally-diffused) silicon solar cell simultaneously are investigated, to characterize their growth and monitor their thickness. Then, the final one-step annealing of the stack is studied; tracking and analyses of the nickel silicide (Ni_xSi_y, noted NiSi) formation and phosphorus migration throughout the Ni layer are highlighted. In fact, only NiSi interlayer with a 1:1 stoichiometric composition gives the maximal conductivity, and the composition of the Ni layer is determined by the annealing time and temperatures. The unreacted Ni(P) ensures the role of copper diffusion barrier. Besides, it has to be efficient for the whole lifetime of the cell (approximately 25 years). That is why controlling the annealing parameters is critical in our process.

2 DESIGN OF EXPERIMENT

The experiments were carried out on 180 µm-thick 5x1 and 5x5 cm² monocrystalline n-PERT silicon solar cells with 80 Ω/□ boron emitter in the textured front face and 70 Ω/□ phosphorus doping in the chemically etched rear face. Both sides are capped with thermal silicon oxide and PECVD silicon nitride layers. First of all, the dielectric layers of both faces were opened thanks to a UV nanosecond LASER. Afterwards, the Ni(P) layer was formed using an electroless method (Solar Prep process from OMG Electrochemicals) and the Cu layer was formed using an electrolytic method (OmniCu™ Series from KMG Ultra Pure Chemicals).Their growth rates and the deposition selectivity were determined thanks to SEM and EDS measurements in order to control their thicknesses.

Then, the Ni-Cu stack was annealed in a RTA furnace under inert atmosphere (argon). The influence of different (time ; temperature) parameters was studied by looking at the nickel silicide formation — structure (NiSi, Ni_2Si, $NiSi_2$) and Ni penetration depth inside the emitter — and the P migration into the Ni layer thanks to a combination of structural (SEM, EDS, XRD, XPS) and electrical (I-V, IQE) characterizations.

A schematic view of the process flow, as well as a photograph of a final cell are displayed on Figure 1.

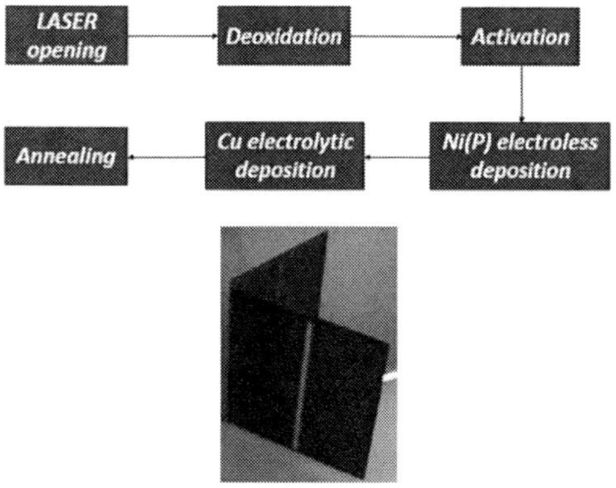

Figure 1: Schematic process flow (up) and photograph of a final cell (down)

3 RESULTS AND DISCUSSION

The silicon wafers used for these studies, based on the BiSon process[7] are represented on Figure 2.

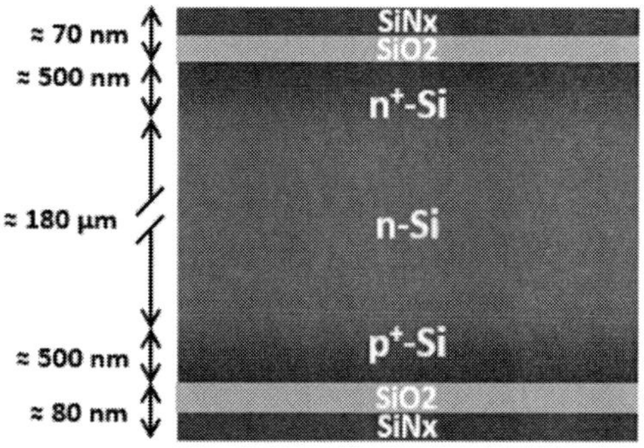

Figure 2: Schematic representation of the n-PERT bifacial monocrystalline silicon wafers. The front face is textured and the rear face is chemically etched

3.1 Ni electroless deposition

The first step has consisted in opening lines on the dielectric layers of both sides of the cell thanks to a UV nanosecond LASER. The pattern was composed of eight 25 µm-wide horizontal contact fingers and one 1.5 mm-wide vertical busbar. An image of a LASER-opened finger is available on Figure 3.

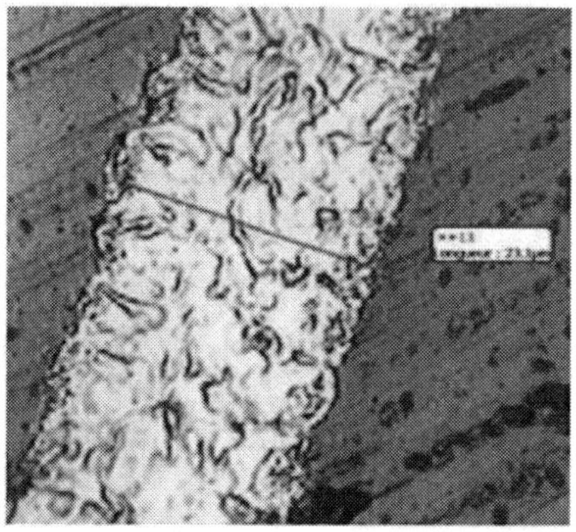

Figure 3: Image of a ns-UV LASER-opened 25-µm wide finger

The second step involved the electroless deposition of Ni(P). This contactless deposition technique, based on oxydo-reduction reactions, enables a simultaneous deposition on both faces of the cell. The samples were deoxidized in a sodium bifluoride solution, activated in a palladium chloride solution and immersed into an electroless bath composed of nickel sulphate and sodium hypophosphite.

In the bath, nickel is reduced and sodium hypophosphite is oxidized, according to the reactions (1) and (2) hereinbelow[3]:

$$Ni^{2+} + 2e^- \rightarrow Ni \quad (1)$$

$$H_2PO_2^- + H_2O \rightarrow H_2PO_3^- + 2H^+ + 2e^- \quad (2)$$

The global main reaction is consequently described by equation (3):

$$Ni^{2+} + H_2PO_2^- + H_2O \rightarrow Ni + H_2PO_3^- + 2H^+ \quad (3)$$

In addition, a side reaction takes place, resulting in the insertion of various amounts of phosphorous into the Ni matrix, noted Ni(P).

On the XPS measurements (Figure 4), the Ni and P peaks are not detected outside the lines, proving the selectivity of the Ni(P) electroless deposition, even as traces. This result is confirmed by the SEM and EDS analyses displayed on Figure 5.

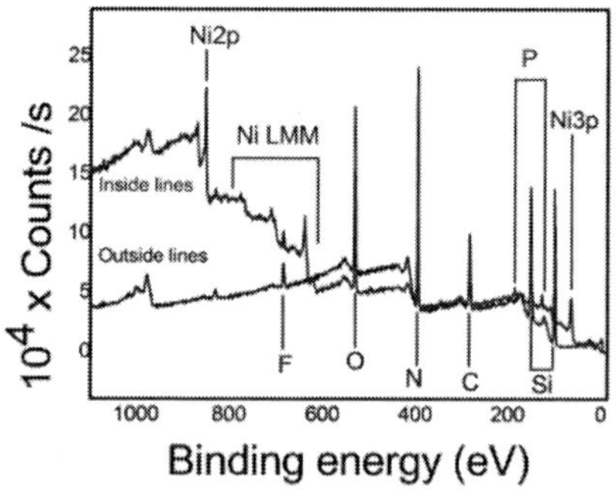

Figure 4: XPS survey spectra inside (blue line) and outside (black line) the Ni(P) lines, confirming the selectivity of the Ni(P) electroless deposition

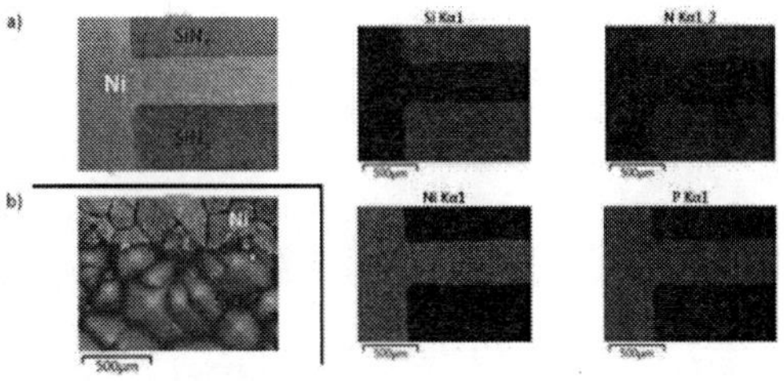

Figure 5: a) SEM micrograph (left-hand) and EDS measurements (right-hand) of an electroless Ni(P) deposition in the lines. b) Zoom on the border between the Ni(P) deposition (up) and the SiN_x layer (down) on the Si pyramids, proving the selectivity of the Ni(P) electroless deposition

Then, the Ni(P) growth rate was studied at six distinctive times: 1, 2, 5, 10, 15 and 20 minutes, in order to control the layer thickness for the future barrier formation and further Cu deposition. Figure 6 shows that, for 1 min deposition time, only palladium clusters are seen on the silicon pyramids. After 2 min, there are more clusters, but Ni(P) is not yet observed.

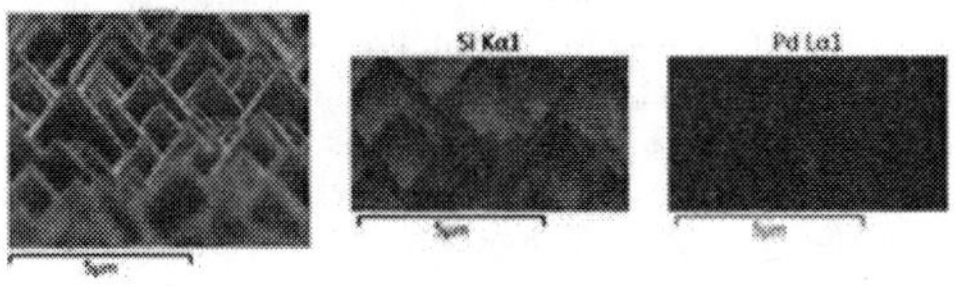

Figure 6: SEM image (left-hand) and EDX measurements (right-hand) of palladium nuclei on the Si pyramids after 1 min deposition time

After 5 min in the bath, Ni(P) is deposited under the form of spheres, the very inhomogeneous sizes of which confirm a progressive rather than an instantaneous nucleation. Moreover, the latter is preferential on the edges of the pyramids, where the bigger Ni(P) spheres are seen. For 10 min, Ni(P) layer is more homogeneous and its thickness reaches around 0.7 µm. Yet, the hollows between the pyramids are not yet filled.

For 15 min, Ni(P) is thickened (1.3 µm) and the hollows are still visible: 20 min in the bath are needed to observe a slight Ni(P) flattening. For this deposition time, the thickness is nearly unchanged (1.5 µm). The Ni(P) growth rate can be averaged around 0.1 µm/min when deposition has begun. Our experimental results confirm KMG Ultra Pure Chemicals previous studies on the optimization of this bath [2].

Figure 7 represents SEM micrographs of the Ni(P) depositions at the four characteristic times, in the same magnification (x 10000), and an EDS measurement to confirm the nature of the deposited metal.

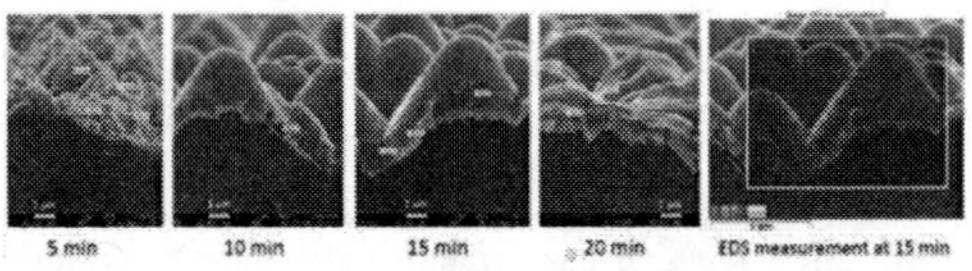

Figure 7: SEM micrographs (x 10000) of the Ni(P) electroless layer at four characteristic times and EDS measurement of the 15 min deposition

We have chosen to carry out a 15-minute Ni(P) electroless deposition in order to get a 1 µm-thick Ni layer after nickel silicide formation. With this thickness, the layer can play both roles of barrier for copper diffusion and seed for copper deposition.

3.2 Cu electrolytic deposition

We then led a copper electrolytic deposition on the Ni(P)-filled lines of the samples. This step was carried out in an electrolytic bath composed of copper sulphate and additives (a leveler, an accelerator and a suppressor) during 10 minutes at a fixed current density of 50 mA/cm². The substrate acted as the working electrode; the reference electrode was made of gold and the counter-electrode, of copper.

In the bath, copper is reduced at the working electrode (cathode) and water is oxidized at the anode, according to the reactions (4) and (5) hereinbelow[3]:

$$Cu^{2+} + 2e^- \rightarrow Cu \quad (1)$$

$$2H_2O \rightarrow 4H^+ + 4e^- + O_2 \quad (2)$$

The global main reaction is consequently described by equation (6):

$$2Cu^{2+} + 2H_2O \rightarrow 2Cu + 4H^+ + O_2 \quad (3)$$

As electrolytic Cu is deposited on conductive surfaces, its selectivity is similar as the Ni one. Besides, the layer is bright and does not peel off after deposition, and its growth rate is around 1 µm/min at this current density. Cross-section EDS analyzes after a 2 hour ionic polishing shown of Figure 8 confirm the deposition of a selective Si/Ni(P)/Cu stack in the front face.

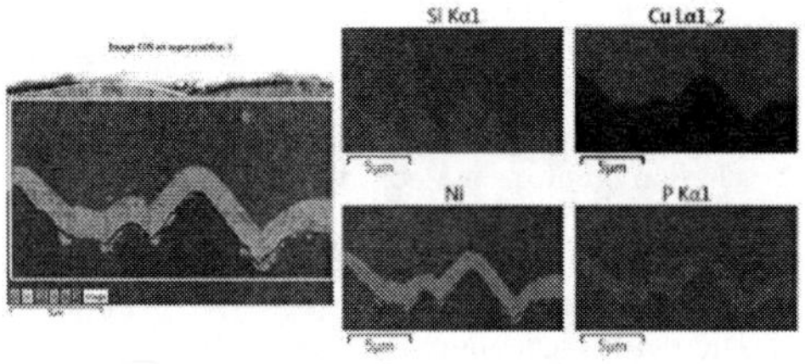

Figure 8: Cross-section EDS measurements of the Ni-Cu stack after annealing in the front face of a cell

Moreover, the atomic composition of the entire stack was determined by cross-section EDS and is displayed on Table I. The layers contain little amount carbon and oxygen; this surface contamination probably comes from the sample cutting and the bath additives and will be further investigated upon.

Table I: Atomic composition of the Ni(P)-Cu stack after annealing

Element	Atomic %
C	8.0
O	1.2
Si	37.5
P	2.2
Ni	11.3
Cu	39.8
Total	100.0

3.3 Study of the final one-step annealing

Once deposited with nickel and copper, the samples were annealed in a RTA furnace under inert atmosphere (argon). The influence of temperature and time was studied by looking at the nickel silicide formation — structure (NiSi, Ni_2Si, $NiSi_2$) and Ni penetration depth inside the emitter — and the tracking of the P content at the NiSi/Ni(P) interface. The design of experiment for this study is presented on Table II.

Table II: Design of experiment for the final one-step annealing study. 3 wafers per parameter are used to ensure reproducibility

Temperature (°C) \ Time (min)	0,5	5	10	15	20
300	-	-	-	-	3
380	3	3	3	3	3
450	3	3	3	3	3
750	3	-	-	-	-

I-V measurements were first of all carried out to draw a comparison between the different annealing parameters. The main results are displayed on Table III.

(20 min ; 380°C) are the best annealing parameters; efficiencies of 12.4% are reached. Such relatively low performances could be explained by contact adhesion problems in the chemically etched rear face of the cells, especially the low fill factor (54,5% *vs.* around 79% for standard screen-printed cells) and the high series resistance (around 100 times higher than standard screen-printed cells). These issues concern the whole nickel-copper stack and are more pronounced when the annealing temperature diverge from 380°C, which confirms the efficiency tendency observed. Nevertheless, no adhesion problems are observed in the textured front face: the nickel bath used in this study was optimized for textured surfaces, on which nucleation is quicker.

Moreover, one of the main issues for plating is the reduction of the annealing time (standard screen printing annealing only lasts 30 seconds) to lower the thermal budget and increase the process throughput. Our results show that reducing time to 15 minutes at 380°C, or increasing the temperature to 450°C with even shorter time (10 minutes) leads to an efficiency loss of 4 to 6% in absolute. The main losses come from the J_{SC} (divided by a factor two) and the fill factor, probably because the NiSi layer is not completely formed and thus does not fully play its contact and barrier role.

The best cell is obtained by performing the annealing in between nickel and copper depositions in the process flow, as done in previous works[8]. Efficiencies of 16.6% are obtained with reasonably good FF, J_{SC} and V_{OC}. Series resistance are quite high compared with standard screen-printed cells (around ten times higher), probably because LASER opening and process parameters are not optimal.

These results confirm that the final one-step annealing has a relatively narrow process window, and that its control is fundamental for the formation of the barrier layer.

Table III: I-V measurements at four representative (temperature ; time) parameters and of the best cell obtained by performing the annealing in between nickel and copper depositions

[T(°C) ; t (min)]	V_{OC} (mV)	J_{SC} (mA/cm²)	FF (%)	η (%)	R_{series} (Ω)	R_{shunt} (Ω)
[300 ; 20]	406	12.1	29.2	1.4	0.2	55
[380 ; 15]	598	29.0	51.6	8.9	0.3	346
[380 ; 20]	595	37.9	54.5	12.4	0.2	366
[450 ; 10]	601	21.6	45.1	6.9	0.2	161
Best cell	605	36.9	74.4	16.6	0.04	322

The IQE measurements, whose main results are displayed on Figure 9, confirm the best annealing parameters of (20 min ; 380°C). The collection is not optimal and explains the FF and series resistance problems previously highlighted.

Reducing the time to 15 min at the same temperature or going to 450°C for 10 minutes makes photon collection decrease for the whole wavelength range. Both curves follow the same outline and are shifted by around 5%. (15 min ; 380°C) exhibits better photon collection than (10 min ; 450°C), which corroborates the better efficiency observed for this couple of parameters.

In addition, using temperatures of 300 and 750°C lead to shunted cells (shunt resistances of 55 and 2 Ω respectively) with flat IQE curves, because no NiSi interlayer is formed (*cf.* XRD measurements below), damaging the contact conductivity and/or a too high temperature allows the Ni diffusion between two contacts, generating shunts. SIMS measurements are currently being carried out to better understand this phenomenon.

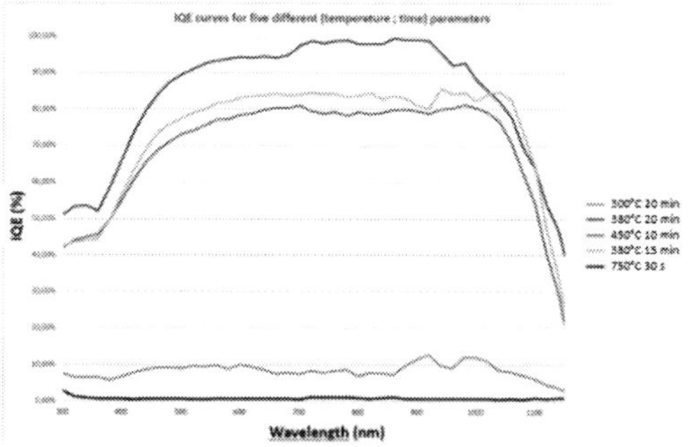

Figure 9: IQE curves of five representative (temperature ; time) parameters

Then, the barrier properties were analyzed for the best annealing parameters (20 min ; 380°C). First of all, the nickel-phosphorus layers thicknesses and stack composition were characterized by EDX profile lines, such as the one displayed on Figure 10.

Copper seems to be effectively blocked by the nickel layer, which confirms its "barrier" characteristics for the chosen annealing parameters. The evolution of its diffusion over time needs to be monitored though.

Besides, there is no detected phosphorus diffusion inside silicon, which shows that P stays in the Ni(P) layer and seems not to go inside the 800 nm nickel silicide interlayer. More precise measurements, such as SIMS profiles, will be necessary to confirm this result. Furthermore, P distribution is symmetric: no P gradient is observed in the Ni(P) layer. Phosphorus takes part neither in the contact formation, nor in the barrier characteristics.

Finally, the high carbon content on the Si layer could come from surface contamination, due to the sample preparation for the ionic polishing. Further investigations will be led to confirm its origin and understand its effect on contact formation and resistivity.

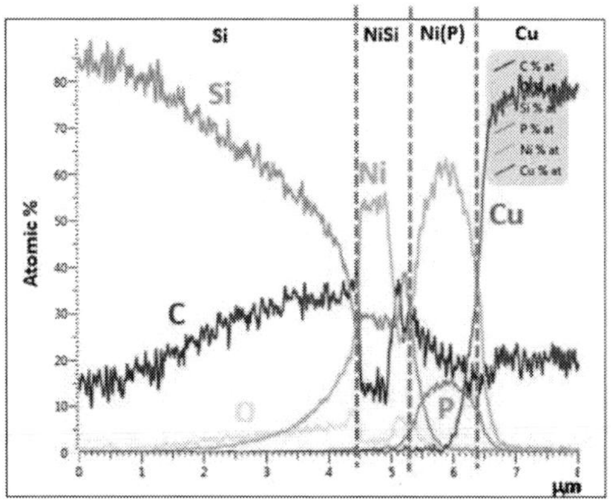

Figure 10: EDX profile line of a Si-NiSi-Ni(P)-Cu stack for a 20-minute 380°C annealing

Eventually, XRD measurements were carried out to check the stoichiometry of the nickel silicide layer. These measurements are repeatable, reproducible, and valid on both faces of the cells.

Besides, these results confirm previously published studies on nickel silicide stoichiometry[6,9]:

- Only Ni_2Si is observed at 300°C, and only $NiSi_2$ is seen at 750°C: both interlayers give a bad electric contact and lead to shunted cells.
- NiSi forms at 380°C after 5 min, but 20 min are necessary to guarantee an efficient contact.

The XRD spectrum of a 20-minute 380°C annealed sample, zoomed on the NiSi characteristic peaks, is shown on Figure 11.

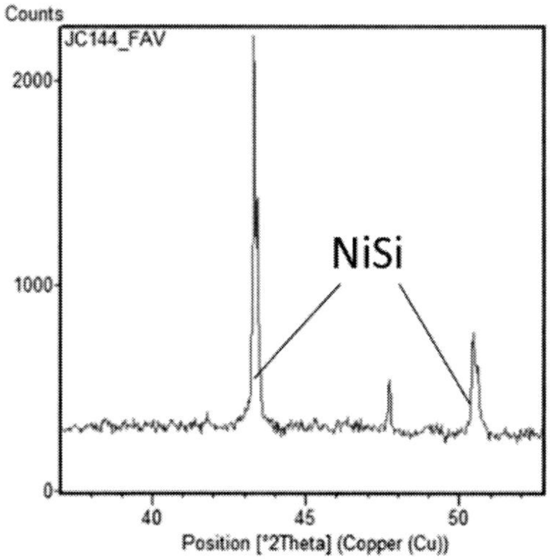

Figure 11: Zoom on the NiSi peaks of a XRD spectrum of a 20-minute 380°C annealed sample

4 CONCLUSIONS AND PERSPECTIVES

Chemical plating is the main alternative to screen printing for the metallization of silicon solar cells. Most of the time, it aims at replacing silver by copper. Yet, because the latter diffuses into silicon and drastically reduces its lifetime, a nickel layer is deposited underneath to both serve as contact to the emitter and barrier to copper diffusion.

In this work, an electrochemical metallization process for bifacial n-PERT silicon solar cells is described. Nickel-phosphorus electroless and copper electrolytic simultaneous depositions on both faces of the cell are followed by a one-step annealing under inert atmosphere. Its parameter control is critical, as it drives the stoichiometry of the resulting nickel silicide; only NiSi can play both "contact" and "barrier" roles.

Experimental results show that, once the lines are opened by LASER, Ni(P) electroless deposition is selective and begins after 5 minutes in the bath (for shorter times, we only see the activator palladium clusters). Nucleation starts on the edges of the pyramids and is progressive. For longer times, the layer thickens and its growth rate is around 0.1 µm/min. Moreover, Cu electrolytic deposition is also selective, and its growth rate is around 1 µm/min at a fixed current density of 50 mA/cm²; approximately ten times as much as the Ni(P) growth rate. Both depositions exhibit the same characteristics on both faces of the cell.

The best cell, achieved by shifting the annealing in between nickel and copper depositions, exhibits an efficiency of 16.6%. On the whole, performances are quite low (particularly for FF and series resistance), probably due to contact adhesion problems in the chemically etched rear face. Efficiencies of 12.4% are obtained for a 20-minute annealing at 380°C. Temperatures of 300°C and 750°C lead to shunted cells, because Ni_2Si and $NiSi_2$ have formed instead of NiSi.

The final one-step annealing seems to have a quite narrow process window, as reducing its time or increasing its temperature leads to efficiency losses because of the incomplete formation of the NiSi layer. Our study confirms that its control is critical for the plating process.

To understand the high series resistances, the study now focuses on the structures resistivities by TLM and the layer purities. Then, after having solved the adherence problems on the chemically etched rear face of the cells, a particular focus will be made on Ni-Cu interdiffusion, very little studied in the literature[10] to see if copper migration can explain the low performances obtained and evolves through time. Eventually, a capping layer formed by immersion into a silver bath will be added to avoid oxidation of the copper layer.

5 ACKNOWLEDGEMENTS

This project has been supported by the French Government in the frame of the program of investment for the future (Programme d'Investissement d'Avenir - ANR-IEED-002-01) and is in the frame of a collaboration between EDF R&D, the Institut Lavoisier de Versailles and KMG Ultra Pure Chemicals. The authors also want to thank ISC Konstanz for the supply of the wafers.

6 REFERENCES

[1] A. Mette, PhD thesis, Fakultät für Angewandte Wissenschaften der Albert-Ludwigs-Universität Freiburg im Breisgau (2007).

[2] É. Delbos *et al.*, Physica Status Solidi, Vol. 12 (2015) 1427-1432

[3] C. Boulord, PhD thesis, INSA de Lyon (2011).

[4] A. Focsa et al., Energy Procedia, Vol. 38 (2013) 713-719.

[5] M. Paunovic *et al.*, Journal of the Electrochemical Society, Vol. 141 (1994), 1843-1850.

[6] K. Funk *et al.*, 12th IEEE International Conference on Advanced Thermal Processing of Semiconductors (2004), 94-98.

[7] J. Libal, R. Kopecek, www.pv-tech.org, "One small step for technology, one giant leap for kWh cost reduction", article from 01-13-2015.

[8] Bifasol project (ANR-11-PRGE-0004).

[9] W. S. Yoo *et al.*, Solid State Technology, Vol. 46 (2003).

[10] S. Kaja *et al.*, Plating and Surface Finishing, Vol. 73 (1986) 42-45.

Optimize the Front-side Boron Emitter by Removing the Boron-depleted Region with Less Damage to the Texture Pyramids

Shalamujiang Simayi*[1], Yasuhiro Kida[1], Katsuhiko Shirasawa[1], Tatsunobu Suzuki[2], Hidetaka Takato[1]
[1] Fukushima Renewable Energy Institute, National Institute of Advanced Industrial Science and Technology, 2-2-9, Machiikedai, Koriyama, Fukushima 963-0298, Japan
[2] Nippon Kasei Chemical Co., Ltd., Takayama, Onahana, Iwaki, Fukushima 971-8101, Japan.

ABSTRACT: In this work, we present n-type bifacial solar cells with a BBr_3 thermally diffused emitter on the front side. Our study related mainly to the boron profiles of the front-side emitter. We optimized the front-side boron emitter by removing the boron-depleted region by means of a spin-etching technique. The boron profiles of non-etched and etched samples were examined. Passivation qualities were also studied on non-etched and etched symmetrical $p^+/n/p^+$ samples. A reduction in the emitter saturation current density J_{0e} in the sample in which the depletion region had been etched was investigated. A high cell efficiency of 19.6% was achieved by removing the depleted region on the front side.

Keywords: boron diffusion; depletion region, bifacial solar cell;

1 INTRODUCTION

Bifacial n-type solar cells having the structure shown in Figure 1(a) feature a boron emitter on the front and phosphorus emitter on the rear as a back surface field (BSF). Tube-furnace diffusion with liquid boron tribromide (BBr_3) and phosphoryl chloride ($POCl_3$), respectively, is an option for forming these emitters. This process for forming the boron emitter results in an unexpected boron-rich layer [1,2] on the top of the boron-doped region. This boron-rich layer, which has a negative effect on the performance of n-type silicon solar cells, is resistant to chemical etching, but can be removed by oxidation. In situ oxidation during the temperature ramp-down and low-temperature oxidation after deglazing are two methods that are commonly used to convert the boron-rich layer into boron surface glass. However, such oxidation steps cause depletion of boron near the surface of the wafer, because boron is more soluble in oxygen than in silicon [3,4]. This boron-depleted layer can enhance surface recombination of the minority charge carriers in the boron emitter, due to band bending within the depletion region. Consequently, elimination of the depletion region would appear to be a necessary step in optimizing the performance of the solar cells.

We have recently developed a technique for removing the unexpected boron-doped layer on the rear side of n-type passivated emitter rear totally diffused (n-PERT) solar cells that results in less damage to the pyramid textures [5]. In the current study, we applied this technique to the removal of the front-side boron-depleted emitter region from the wafer surface, and we investigated the effects of the modified emitter in solar cells.

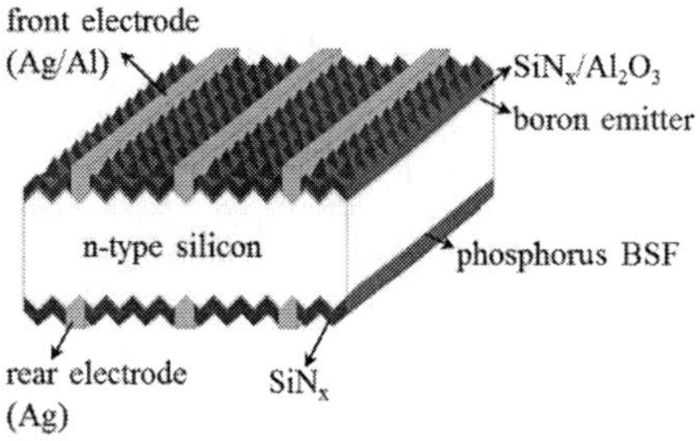

Figure 1: Cross section of n-PERT solar cell.

A high conversion efficiency requires optimization of the emitter and the presence of a suitable passivation layer. As we discussed in our previous paper, we optimized the front-side boron emitter by modulating the drive-in temperature [3]. However, the boron profiles of all the emitters showed the presence of a boron-depleted region at the surface of the wafer due to the high concentration of O_2 present during the temperature ramp-down step. As a continuation of previous research, in this study our main goal was to optimize the front-side emitter by removing this boron-depleted region of the emitter. For this purpose, we first studied the boron emitter passivation quality. A symmetrical p^+np^+ structure was employed to evaluate the passivation quality on these un-etched and etched boron emitters. Complete solar cells were then fabricated in which the depleted layer was non-etched and or etched.

2 EXPERIMENTAL

2.1 Boron emitter profile

Non-etched and etched boron emitters were prepared on 6-inch-diameter 625-μm-thick, circular n-type CZ-Si wafers. The wafer resistance was 3 Ω·cm. The boron emitter was formed by using a BBr_3 liquid source. The measured sheet resistance was 49 Ω/Sq for the non-etched boron emitter. The depleted layer was removed by etching with a solution containing HF and HNO_3 to give a sheet with a resistance of 60 Ω/Sq. The boron profiles of the non-etched and etched boron emitters were examined by secondary-ion mass spectroscopy (SIMS).

2.2 Evaluation of the passivation layer

We previously studied SiO_2 passivation on a p+ emitter [3], but the cell efficiency in this case was limited by a low open-circuit voltage (V_{oc}). Recent studies have shown that a dielectric layer of Al_2O_3 containing a high density of fixed negative charges is an ideal passivation layer for a p^+ emitter [6,7]. We examined the effect of the passivation layer on the quality of a boron emitter on a symmetrical $p^+/n/p^+$ sample. This structure was fabricated under the same conditions as those used in processing solar cells (resulting in an emitter sheet resistance of 49 Ω/sq). The saturation current was investigated on the non-etched and depleted-region-etched boron emitter. In this experiment, the saturation current was measured by the quasi-steady-state microphotoconductance decay method.

2.3 Spin-etching to remove the doped layer

The etching solution used in removing the boron-depleted layer was developed by the Nippon Kasei Chemical Company Limited. In this study, we chemically etched the silicon by a spin-etching technique in which the chemical solution was dropped onto the rapidly spinning wafer and subsequently thrown off the edge of the surface of the wafer. The solution thrown from the wafer flowed into a recycling tank and was reused until it lost its etching ability. A schematic showing the spin-etching process is given in Figure 2. Removal of the depleted region was confirmed by measuring the sheet resistance of the doped layer by the

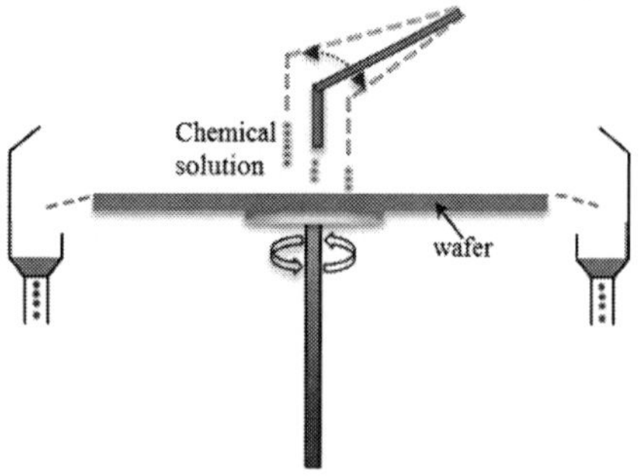

Figure 2: Schematic showing the spin-etching process

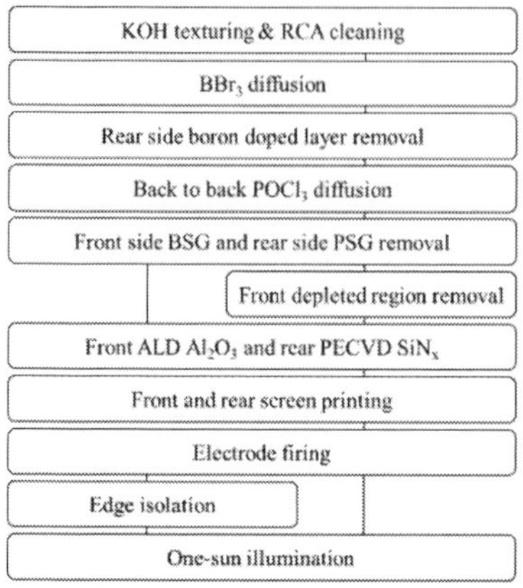

Figure 3: n-PERT solar cell fabrication process flow.

conventional four-point probe method. To determine the uniformity of the sheet resistance of the spin-etched surface, we performed measurements at 25 points, including several points 5 mm from its outer edge.

2.4 Cell fabrication

Two process flows, as presented in Figure 3, were used in the fabrication of the n-PERT cells. The solar cells were fabricated on 156 × 156 mm, 180-μm-thick, pseudo-square, n-type CZ-Si wafers, with a resistivity of about 2.1 Ω·cm. In both process flows, the p-type emitter was formed by tube-furnace diffusion with BBr_3. The resistance of the boron-diffused sheet was 49 Ω/Sq, which increased to 60 Ω/Sq after etching of the depleted region.

Both cell processes started with texturing with KOH and an RCA cleaning protocol, followed by boron (BBr_3) diffusion. The rear-side doped layer was removed with retention of the texture pyramids on the wafer surface. $POCl_3$ diffusion was followed by rear-side etching, leading to a sheet resistance of 97 Ω/Sq with a 0.4 μm diffusion length. After cleaning with HF, the front side was subjected to spin-etching to remove the boron-depleted region. Passivation stacks of Al_2O_3/SiN_x for p^+ emitter side and SiN_x for n^+ side were applied similarly in both processes. Electrodes were fabricated by screen printing of grids using Al/Ag and Ag pastes on the front and rear side, respectively, with subsequent co-firing in a belt furnace. Illuminated current–voltage (I–V; AM1.5) and characteristics was measured at 25 °C on a reflective metal-plated chuck.

3 RESULTS AND DISCUSSIONS

3.1 Boron Profiles

The boron profiles of the initially diffused 49 Ω/Sq wafer and the depleted-region-etched 60 Ω/Sq wafer are shown in Figure 4. The initially

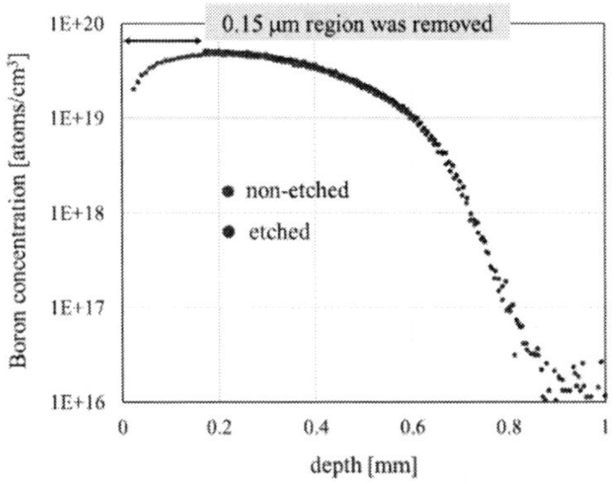

Figure 4: SIMS profiles of surfaces with non-etched and etched boron-depleted regions.

diffused boron emitter shows a boron-depleted region at a depth of 100–200 nm. The peak boron concentration was 4.8×10^{19} cm^{-3}, which decreased to 2×10^{19} cm^{-3} near the surface of the wafer. This depletion in boron is due to the high solubility of boron in oxygen. Figure 4 shows that etching to a depth of 0.15 μm produced an un-depleted boron profile. In our current experiments, the depleted region was removed after phosphorus diffusion, and passivation layers were applied at low temperatures, therefore any reformations of depleted region does not occur on the surface again.

3.2 Effects of the passivation layers on the non-etched and etched p^+ emitter

Figure 5 shows the values of the emitter saturation current density J_{0e} for surface non-etched and etched samples with SiN_x/Al_2O_3 passivation stacks applied under various conditions. The value of J_{0e} improved from 121 fA/cm² to 103 fA/cm² on removing the boron-depleted region of the emitter. Annealing and additional firing of the samples improved the implied V_{oc} by approximately 10–13 mV for both the non-etched and etched samples. However, the implied V_{oc} was similar for both etched and non-etched samples before and after the thermal treatment. The improvement in the J_{0e} might therefore be explained by the removal of the surface of the boron emitter, along with a small increase in the emitter sheet resistance.

Table I. I–V results for cells with un-etched and etched boron-depleted regions.

Cell type	J_{sc} [mA]	V_{oc} [mV]	FF [%]	η [%]
Non-etched depleted region	38.1	636.5	77.9	18.9
Etched depleted region	38.5	640.0	79.7	19.6

3.3 Solar cell approach

The measured one-sun I–V parameters for solar cells fabricated by the two processes flows are summarized in Table I; the one-sun parameters were measured on a reflective gold plated chuck. A slight increase was observed in J_{sc} in the case of sample in which the boron-depleted region was removed. This increase in J_{sc} is due to the increase in the boron concentration on the surface produced by removing the boron-depleted region. To observe how effect of the depleted region removal on the light absorption, a reflectance measurement was carried out on the depleted region etched and non-etched wafers with the same passivation scheme of SiN_x/Al_2O_3. From the Figure 6 we notice that very few reflectance increases on the depleted-region-etched surface on the short wavelength. This tiny increase implies that removing the depleted region by means of spin etching technique on the wafer surface exhibit a less damage to the pyramid textures on the wafer surface.

The other main differences observed were in the values of V_{oc} and the fill factor (FF). A 3.5 mV increment in the V_{oc} was observed in the solar cells

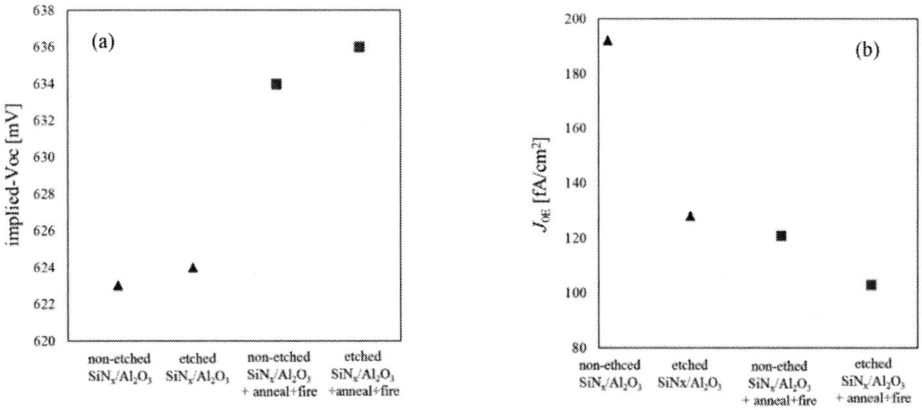

Figure 5: (a) Measured implied V_{oc} and (b) measured J_{0e} for boron-depleted-region non-etched and etched samples with as-deposed and thermally treated passivation layers

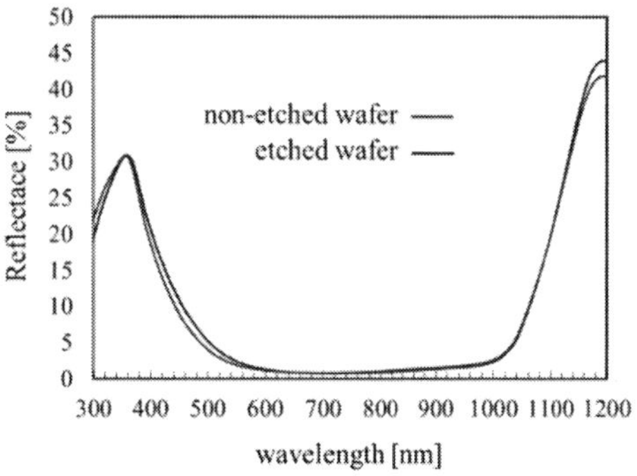

Figure 6: reflection measurement measured by internal quantum efficiency.

in which the boron-depleted region had been removed. The reduction in the J_{0e} was responsible for the increase in the V_{oc}. The FF was 1.8% lower in the case of cells in which the boron-depleted region was not etched. The dark I–V curves, series resistance, and contact resistivity are important parameters in understanding the differences in the FF.

4 SUMMARY

In this study, an important step was made toward improving the front-side emitter of n-type bifacial solar cells. An improved boron profile in the emitter was obtained by removing the boron-depleted region. J_{sc} increased on removing the boron-depleted region as a result of the increase in the boron concentration at the surface of the wafer. This increase in J_{sc} also implies that the spin-etching technique used to remove the boron-depleted region causes little damage to texture pyramids on the wafer surface. The J_{0e} values improved from 121 fA/cm^2 to 103 fA/cm^2 on removing the boron-depleted region. This improvement in J_{0e} resulted in a 3.5 mV increase in V_{oc}. The improvement in the FF is still unclear; further characterizations will be necessary to understand the difference in FF between the two processes.

5 REFERENCES

[1] Kyungsun Ryu, Ajay Upadhyaya, Hyun-Jin Song, Chel-Jong Choi, Ajeet Rohatgi, and Young-Woo Ok, APPLIED PHYSICS LETTERS, Vol. 101, 073902 (2012).

[2] Sieu Pheng Phang, Wensheng Liang, Bettina Wolpensinger, and Michael Andreas Kessler, IEEE JOURNAL OF PHOTOVOLTAICS, VOL. 3, NO. 1, 261–266 (2013).

[3] Shalamujiang Simayi, Yasuhiro Kida, Katsuhiko Shirasawa, and Hidetaka Takato, In: Proceedings of the 31th European Photovoltaic Solar Energy Conference; September 14–18th, 2015, Hamburg, Germany; pp. 947–949; DOI: 10.4229/EUPVSEC20152015-2CV.4.14

[4] E. Lohmüller, S. Werner, J. Schön, M. Thanasa, S. Mack, W. Wolf, F. Clement, and D. Biro, In: Proceedings of the 28th European Photovoltaic Solar Energy Conference; 2013, Villepinte, France; pp. 1043–1047; DOI: 10.4229/28thEUPVSEC2013-2BV.1.5

[5] Shalamujiang Simayi, Yasuhiro Kida, Katsuhiko Shirasawa, Tatsunobu Suzuki, and Hidetaka Takato, (submitted for publication).

[6] B. Hoex, J. J. H. Gielis, M. C. M. van de Sanden, and W. M. M. Kessels, JOURNAL OF APPLIED PHYSICS, Vol. 104, 113703, (2008),

[7] Supawan Joonwichien, Shalamujiang Simayi, Katsuhiko Shirasawa, Katsuto Tanahashi, Hidetaka Takato, (in press).

LOWEST SURFACE RECOMBINATION IN N-TYPE OXIDISED CRYSTALLINE SILICON BY MEANS OF EXTRINSIC FIELD EFFECT PASSIVATION

Ruy S Bonilla, Phillip G Hamer and Peter R Wilshaw
Department of Materials, University of Oxford, Parks Rd, Oxford, OX1 3PH, United Kingdom

ABSTRACT: Surface recombination remains a major factor limiting the efficiency of silicon solar cells. The post-processing of dielectric films used as surface coatings has been previously demonstrated an effective technique to improve their passivation quality. In this paper extrinsic methods are demonstrated to produce the lowest reported surface recombination velocity in solar relevant n-type silicon. Recombination velocities below 2.8 cm/s at an injection of 10^{15} cm^{-3}, are achieved using extrinsic field-effect passivation, or < 0.6 cm/s when using combined extrinsic chemical and extrinsic field effect passivation. These are equivalent to emitter saturation current densities J_{0e}<1.4 fA/cm^2 and 0.6 fA/cm^2.

Keywords: surface passivation, dielectric film, IBC silicon solar cell

1 INTRODCUTION

Recombination at surfaces and interfaces continues to be a limiting process on the path to high performance silicon solar cells. Despite their high potential efficiency, back contacted crystalline silicon cells are especially susceptible to recombination at the front surface [1]. Substantial progress has been made towards developing dielectric coatings that provide near optimal passivation to back contacted cells [2].

Their passivation quality relies on two main mechanisms. First, the reduction of defect state density at the silicon-dielectric interface by satisfying dangling bonds, and second, the presence of net charge in the dielectric which helps reduce the concentration of minority carriers at the surface, thus reducing their availability for recombination. These two mechanisms are jointly exploited in a single or double layered dielectric coating – e.g. silicon dioxide, silicon nitride or aluminium oxide.

In addition to the layer's passivation properties, minimal parasitic optical absorption and matching refractive index must also be optimised during a single dielectric deposition process. As a result, there exists a trade-off between the best chemical, field effect and optical properties in the dielectric film. An approach by which the properties of a dielectric coating can be improved after deposition has been previously suggested and referred to as extrinsic passivation as opposed to intrinsic passivation which is only achieved by tuning of the deposition process [3]. Forming gas or post-metallization anneals are well-known processes by which chemical passivation can be improved after dielectric deposition. These processes aim to introduce and/or diffuse hydrogen to the dielectric-silicon interface [4].

Another extrinsic process that can be used to improve field effect passivation is the external addition of charge onto or into the dielectric [5]. This method has been widely used at research scale, yet industrial adoption is hindered by the lack of stability of the extrinsically added charge. Addressing the stability problem is an ongoing research topic and noteworthy progress has been presented in [6,7]. Moreover, the striking effectiveness of extrinsic field effect passivation has only been highlighted recently [3]. In the present work we report the lowest recombination observed on oxidised silicon wafers using extrinsic field-effect passivation by means of corona discharge. Despite high temperature processing, effective lifetimes of up to 5 ms are possible on 1 Ωcm n-type float zone Si.

This passivation performance exceeds that reported by Kerr and Cuevas for the so called *alneal* [8], and that achieved using a double layer thermal oxide and silicon nitride [9–11]. Such surface passivation leads to outstanding performance of interdigitated back contact cells as simulated using Quokka 2.2.4 [12]. These results highlight the improvement possible via extrinsic passivation methods. The potential for increased performance with negligible cost make such methods likely to become industrially relevant once the stability issue is resolved.

2 EXPERIMENTAL METHODS

Float zone 1 Ωcm n-type 4 inch 200 μm thick silicon wafers were oxidised in a Centrotherm tube furnace at 1050 °C in a dichloroethylene (DCE) atmosphere to produce 100 nm oxide films as base passivation. The wafers were then split into two batches of wafers, and one was annealed (FGA) in forming gas (5%H_2/95%Ar) for 25 minutes at 425 °C immediately after oxidation. Effective lifetime was then recorded using Sinton photo-conductance decay (PCD) measurements as the oxide films were charged on both sides using a corona discharge apparatus optimised for uniformity, as described in [3].

Deposited charge was monitored between every step using Kelvin Probe surface potential measurements. The charge concentration was extracted assuming that all charge was situated at the surface of the dielectric, and that the potential drop inside the silicon was negligible: $Q_{corona} = \varepsilon_{SiO2}/t * [\Phi_{sm}/q - V_{KP}]$, where t is the thickness of the oxide, Φ_{sm} is the semiconductor to metal work function difference, and V_{KP} is the backing potential measured from Kelvin Probe.

A schematic of the corona discharge apparatus is shown in **Figure** 1.a. The rate of charge deposition was determined by the corona current i_c, which in turn was adjusted by setting the corona potential V_c and distance d_c. Three different conditions were used in these experiments as illustrated in **Figure** 1.b. The corona current was set to ~3-5 μA which allowed deposition of charge in reasonable time steps (~50 s). This same charging process can be used industrially in a time scale of seconds by applying a higher current using this very simple apparatus.

A photo-luminiscence (PL) lifetime measurement was performed on Wafer 4 using a BTi imaging PL system to evaluate the lifetime across the entire wafer, **Figure** 1.c.

Here, the count number was calibrated to a Sinton PC lifetime tester. Several poor lifetime spots were observed. Uniformity of the field effect passivation was demonstrated in [13], thus indicating that the recombination is induced by interface defect states, or poor bulk lifetime. Most likely the former but not further comments can be made here.

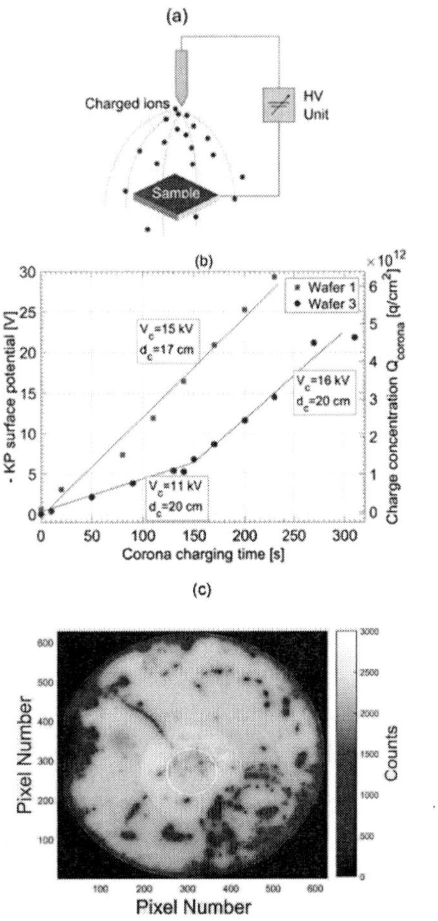

Figure 1: a. Scheme of corona discharge on top of oxidised silicon samples, b. Rate of charge deposition for different corona discharge conditions, c. PL image of Wafer 4 when corona charge has been uniformly deposited to improve surface passivation extrinsically.

3 EXTRINSIC SURFACE PASSIVATION OF C-SI

Upon corona charge deposition, the recombination rate at the silicon surface is reduced via field effect passivation. This is illustrated in **Figure 2** where PCD effective lifetime is plotted against minority carrier injection level for incrementally higher dielectric charge concentrations. Lifetimes exceeding 5 ms have been achieved using extrinsic methods to improve the passivation quality of SiO_2 films. Both chemical and field effect passivation are clearly contributing towards the overall passivation quality of these wafers. Wafers 2 and 4 were included to show repeatability. The lack of uniformity of passivation techniques, especially at the edges of the specimen [14], has recently been reported as one of the factors that limit the overall passivation quality. The substantially higher lifetimes achieved here corroborate previous studies where lifetime was limited by recombination at the edge of the specimens. The peak in lifetime observed in the shape of the lifetime curves for wafers 1 and 2 indicate that the total lifetime is limited by a bulk or near surface recombination. Bulk is likely to arise in high temperature processes as recently pointed our by Grant et al [15]. It is not possible to conclude if such a mechanism is eliminated during FGA for wafers 3 and 4 since these were oxidised using a different process.

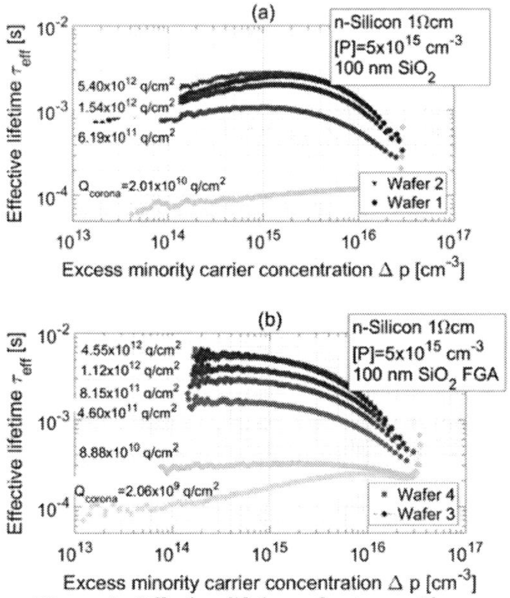

Figure 2: Effective lifetime of n-type c-Si as a function of injection level and corona deposited charge for a. Only oxidised wafers, and b. oxidised and FGA wafers.

4 SRV AND J_{0E} IN EXTRINSICALLY PASSIVATED C-SI

Upper limits to surface recombination velocity (SRV) have been extracted using Richter's [16] parametrisation of intrinsic lifetime for 1 Ωcm (5x10^{15} cm^{-3} donors) crystalline silicon, where $SRV_{UL} < (1/\tau_{eff} - 1/\tau_{int})W/2$. **Figure 3**.a and b illustrate the measured SRV as a function of injection level for different concentrations of dielectric charge. A SRV below 2.8 cm/s was calculated for the maximum charge concentration in an oxidised wafer at 10^{15} cm^{-3} carrier injection. Furthermore, a SRV < 0.6 cm/s was achieved on a wafer that underwent FGA prior to corona charge depositions. **Figure 3**.c illustrates the calculation of emitter saturation current densities using Kane and Swanson's [17] method as implemented in Sinton's lifetime software version 4.6.0 using Richter's τ_{Bulk} correction. The emitter saturation current densities were J_{0e}=0.5 fA/cm^2 and 1.36 fA/cm^2 respectively. When calculated using approximation (3) in [18], $J_{0e} = SRV * qn_i^2/(N_{dop} + \Delta p)$, the current densities increase to 1.5 fA/cm^2 and 7 fA/cm^2. To the best of our knowledge, such surface recombination parameters are the lowest reported for 1 Ωcm n-type c-Si passivated with a single layer of thermal SiO_2. Prior to the extrinsic field effect passivation achieved by the corona discharge, the SRV was ~100 cm/s and J_{0e}= 253 fA/cm^2 for oxidised wafers, and SRV was ~ 60 cm/s and J_{0e}=152 fA/cm^2 for oxidised and FGA wafers.

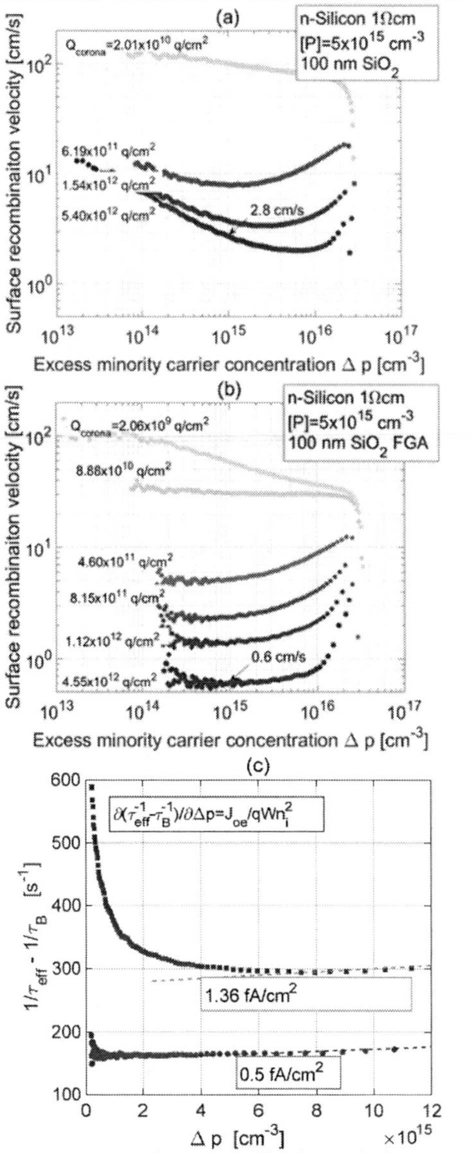

Figure 3: Surface recombination velocity as a function of injection level and corona deposited charge for a. Oxidised-only wafers, and b. Oxidised and FGA wafers, c. Extraction of emitter saturation currents for optimally passivated silicon wafers.

5 THE EFFECT OF FRONT SURFACE RECOMBINATION ON IBC CELL PERFORMANCE

Interdigitated back contact cells were simulated using Quokka 2.2.4 [12] following the parameters reported by Fell in [19]. The model was altered to evaluate the effect of the front surface passivation. The dark recombination current density at the front surface was varied in the range of values found in Section 4. **Figure 4**.a illustrates a typical meshing of the unit cell studied, including the n, p, metallised and contacted regions. J_{0e} values ranging from 0.1 to 1000 fA/cm^2 were simulated. The equivalent SRVs are displayed in the x-axis in **Figure 4**.b. Here it is evident that cell performance is strongly dependent on passivation of the front surface, when the quality of the passivation is poor (SRV>100 cm/s). A ~2% gain in efficiency is possible when reducing recombination from ~100 cm/s to below 1 cm/s (3 fA/cm^2) –i.e. from ~21 to 23% efficiency. Open circuit voltage and short circuit current have a comparable dependency with front surface recombination.

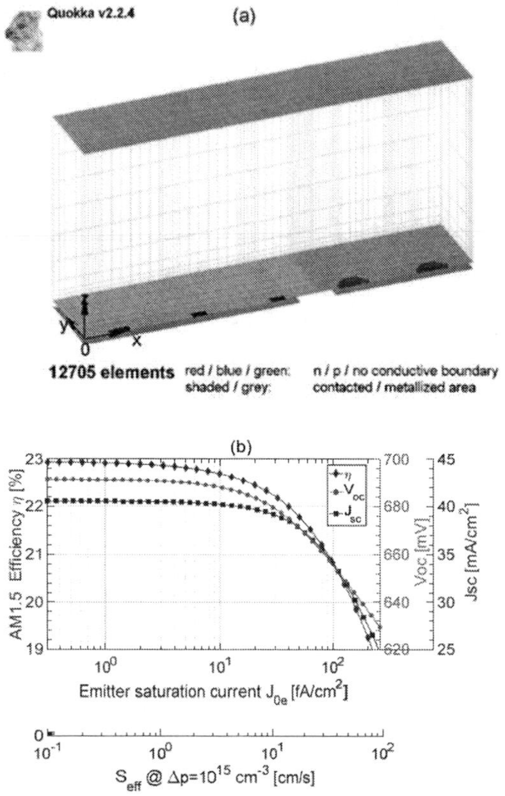

Figure 4: a. Quokka model and mesh for simulation of IBC cell performance as a function of front surface passivation, b. IBC cell performance parameters for a change in the front surface passivation quality.

6 CONCLUSIONS

This work has shown that extrinsic passivation methods are a very powerful and cost effective tool to improve the overall efficacy of surface passivation, and that they can translate into improved performance in solar cells. SRVs below 0.6 cm/s and J_{0e} of 0.5 fA/cm^2 were achieved on a 4-inch wafer by improving the passivation quality of thermally grown SiO_2 using corona charge. These are the lowest reported on 1Ωcm n-type material with a single passivation coating. The potential for these methods to be industrial exploited is hindered by the stability of the passivation obtained – which is a continuing subject of study. The effectiveness and speed of passivation, however, makes this a promising contender for the industrial manufacture of solar cells.

7 ACKNOWLEDGMENTS

RS Bonilla is the recipient of an EPSRC (UK) Postdoctoral Research Fellowship, EP/M022196/1. PR Wilshaw and PG Hamer acknowledge the support from EPSRC grant EP/M024911/1. Data published in this article can be downloaded from http://ora.ox.ac.uk. All authors are thankful to Martin Hermle and Christian Reichel at Fraunhofer ISE for provision of silicon material, and to Radka Chakalova for her extensive help in the cleanroom.

8 REFERENCES

[1] Z. Shu, U. Das, J. Allen, R. Birkmire, S. Hegedus, Experimental and simulated analysis of front versus all-back-contact silicon heterojunction solar cells: effect of interface and doped a-Si:H layer defects, Prog. Photovoltaics Res. Appl. 23 (2015) 78–93. doi:10.1002/pip.2400.

[2] M.Z. Rahman, S.I. Khan, Advances in surface passivation of c-Si solar cells, Mater. Renew. Sustain. Energy. 1 (2012) 1. doi:10.1007/s40243-012-0001-y.

[3] R.S. Bonilla, F. Woodcock, P.R. Wilshaw, Very low surface recombination velocity in n-type c-Si using extrinsic field effect passivation, J. Appl. Phys. 116 (2014) 054102.

[4] M.J. Kerr, J. Schmidt, A. Cuevas, J.H. Bultman, Surface recombination velocity of phosphorus-diffused silicon solar cell emitters passivated with plasma enhanced chemical vapor deposited silicon nitride and thermal silicon oxide, J. Appl. Phys. 89 (2001) 3821–3826.

[5] J. Schmidt, A.G. Aberle, Easy-to-use surface passivation technique for bulk carrier lifetime measurements on silicon wafers, Prog. Photovoltaics. 6 (1998) 259–263.

[6] V. Leonov, C. Hoof, M. Goedbloed, R. Schaijk, Charge injection and storage in single-layer and multilayer inorganic electrets based on SiO_2 and Si_3N_4, IEEE Trans. Dielectr. Electr. Insul. 19 (2012) 1253–1260. doi:10.1109/TDEI.2012.6259999.

[7] R.S. Bonilla, K. Collett, L. Rands, G. Martins, R. Lobo, P.R. Wilshaw, Stable, Extrinsic, Field Effect Passivation for Back Contact Silicon Solar Cells, Solid State Phenom. 242 (2015) 67–72. doi:10.4028/www.scientific.net/SSP.242.67.

[8] M.J. Kerr, A. Cuevas, Very low bulk and surface recombination in oxidized silicon wafers, Semicond. Sci. Technol. 17 (2002) 35–38. doi:10.1088/0268-1242/17/1/306.

[9] Y. Larionova, V. Mertens, N.-P. Harder, R. Brendel, Surface passivation of n-type Czochralski silicon substrates by thermal-SiO2/plasma-enhanced chemical vapor deposition SiN stacks, Appl. Phys. Lett. 96 (2010) 032105. doi:10.1063/1.3291681.

[10] F. Granek, C. Reichel, M. Hermle, D.M. Huljić, O. Schultz, S.W. Glunz, Front surface passivation of n-type high-efficiency back-junction silicon solar cells using front surface, in: 22nd Eur. Photovolt. Sol. Energy Conf., Milan, 2007.

[11] F. Granek, C. Reichel, Back-contact back-junction silicon solar cells under UV illumination, Sol. Energy Mater. Sol. Cells. 94 (2010) 1734–1740. doi:10.1016/j.solmat.2010.05.038.

[12] A. Fell, A Free and Fast Three-Dimensional/Two-Dimensional Solar Cell Simulator Featuring Conductive Boundary and Quasi-Neutrality Approximations, IEEE Trans. Electron Devices. 60 (2013) 733–738. doi:10.1109/TED.2012.2231415.

[13] R.S. Bonilla, N. Jennison, D. Clayton-Warwick, K.A. Collett, L. Rands, P.R. Wilshaw, Corona charge in SiO2: kinetics and surface passivation for high efficiency silicon solar cells, Energy Procedia. In press (2016).

[14] R.S. Bonilla, G. Martins, P.R. Wilshaw, Investigation of Parasitic Edge Recombination in High-Lifetime Oxidized n-Si, Solid State Phenom. 242 (2015) 73–79. doi:10.4028/www.scientific.net/SSP.242.73.

[15] N.E. Grant, F.E. Rougieux, D. Macdonald, Low Temperature Activation of Grown-In Defects Limiting the Lifetime of High Purity *n*-Type Float-Zone Silicon Wafers, Solid State Phenom. 242 (2015) 120–125. doi:10.4028/www.scientific.net/SSP.242.120.

[16] A. Richter, S.W. Glunz, F. Werner, J. Schmidt, A. Cuevas, Improved quantitative description of Auger recombination in crystalline silicon, Phys. Rev. B. 86 (2012) 165202. doi:10.1103/PhysRevB.86.165202.

[17] D.E. Kane, R.M. Swanson, Measurement of the Emitter Saturation Current by a Contactless Photoconductivity Decay Method, in: Proc 18th IEEE Photovolt. Spec. Conf., 1985: pp. 578–583.

[18] H. Mäckel, K. Varner, On the determination of the emitter saturation current density from lifetime measurements of silicon devices, Prog. Photovoltaics Res. Appl. (2012) n/a–n/a. doi:10.1002/pip.2167.

[19] A. Fell, K.C. Fong, K.R. McIntosh, E. Franklin, A.W. Blakers, 3-D Simulation of Interdigitated-Back-Contact Silicon Solar Cells With Quokka Including Perimeter Losses, IEEE J. Photovoltaics. 4 (2014) 1040–1045. doi:10.1109/JPHOTOV.2014.2320302.

EFFECTIVE SURFACE RECOMBINATION OF P$^+$-LAYER IN P-TYPE SILICON PERT BIFACIAL CELL

Y. Eisenberg[1,3], L. Kreinin[1], N. Bordin[1], N. Eisenberg[1], G. Grigorieva[2], M. Kagan[2], and S. Hava[3]

1- Lev Academic Center, 21 Havaad Haleumi St., Jerusalem 96110, Israel,
2 - KVANT R&P Enterprise, 16 Third Mytischinskaja St., Moscow 129626, Russia
3 - Department of E&CE, Ben-Gurion University of the Negev, Beer-Sheva, Israel
ygal@g.jct.ac.il

ABSTRACT: Boron doped BSF is a critically important factor in high efficiency p-Si bifacial solar cell design. Recombination losses represented as effective surface recombination, S_{eff}, are evaluated by simulation and experimentally. For thick heavily doped p$^+$ layers S_{eff} is determined mainly by recombination inside the layer with positive effect of top over doped region. Spectral response analysis of n$^+$-p-p$^+$ solar cells with highly doped and deep BSF (~1.6 μm) demonstrates a possibility of providing S_{eff} of 75±20 cm/s without surface passivation. For thinner p$^+$ layers surface recombination is a critical factor. n-Si wafers symmetrically doped by boron ion-implantation or by boron thermal diffusion were used for experimental validation of the simulated surface recombination contribution to S_{eff} and its dependence on the charge trapped in the passivation layer. The recombination parameters of the samples were measured using the photoconductivity decay method.

Keywords: p-Si PERT; Boron doping; Bifacial solar cell

1. INTRODUCTION

Technological difficulties of p-Si PERT solar cells fabrication resulted in attempts to substitute this cell type by others (like PERT n-type or PERC p-type cells) [1]. However, by proper managing the p$^+$-layer parameters, high efficiency and good bifaciality for p-Si PERT cells can be achieved. Retaining very high bulk minority carriers lifetime values of 0.2-1 ms was industrially proven as possible for an n$^+$-p-p$^+$ PERT structure even in Cz wafers [2].

Contribution of the back boron doped p$^+$ layer in solar cell recombination losses can be very significant and even dominating. The diode saturation current J_o is determined by the recombination inside the doped layer along with recombination on the surface. Therefore it is the effective surface recombination value (S_{eff}) that best describes the outcome parameters of a bifacial solar cell. Back spectral response, open circuit voltage, V_{oc}, as well as front and back efficiencies are controlled by the S_{eff} of the back p$^+$ layer [3-9].

Both surface and bulk of the doped layer recombination components are affected by the doping concentration and its depth [8-9]. When maximal doping concentration is below the B solubility limit and the doping is not too deep, surface passivation is required for reducing recombination losses and increasing the open circuit voltage of the cell. Different types of dielectric passivation layers, such as Al_2O_3, SiO_2 or SiN_x:H are used for silicon surfaces passivation [3-8]. These layers provide surface passivation by two different mechanisms: chemical passivation and field effect passivation. The field effect passivation is obtained by the built-in charge, which can be negative or positive. This built-in charge changes the carrier distribution at the silicon surface and suppresses or stimulates recombination.

Silicon oxide layer grown by dry thermal oxidation provides an effective chemical passivation of the n$^+$-Si. The drawback of such a layer for p-Si passivation is a positive built-in charge, which is typical for such a layer [11-12]. Aluminum oxide characterized by negative built-in charge is presently the leading passivation material for p$^+$ layers [3-5]. However, the optimal passivation method of the p$^+$ layer resulting in minimal S_{eff} for given specific properties of the layer is not developed.

In this study, the effect of p$^+$ layer doping parameters (impurity concentration and depth) on S_{eff} and on passivation efficacy in a p-type PERT cell is examined.

2. SIMULATION

PC1D program simulation was carried out for describing the influence of the p$^+$-layer on S_{eff} in the p-p$^+$ junction as well as on the limitations of V_{oc} values of n$^+$-p-p$^+$ solar cell.

One of the simulation series describes the effect of the deeply doped p$^+$ layers. Thickness of the doped layer was assumed ~1.6 μm, maximal concentration varied in the range 4.10^{18} – 3.10^{20} cm^{-3} (prototype of the profile is shown in Fig. 1). The recombination parameters of n$^+$ emitter and of the base region were assumed fairly high to not limit V_{oc} below 735 mV. The simulation was made in the frame of Boltzmann statistics and do not includes more accurate band-gap-narrowing (BGN) models. Therefore it underestimates the recombination inside of heavy doped layers [13]. The simulation results are shown in Fig. 1 as V_{oc} (and S_{eff}) functions of recombination velocity on the surface of the doped layer.

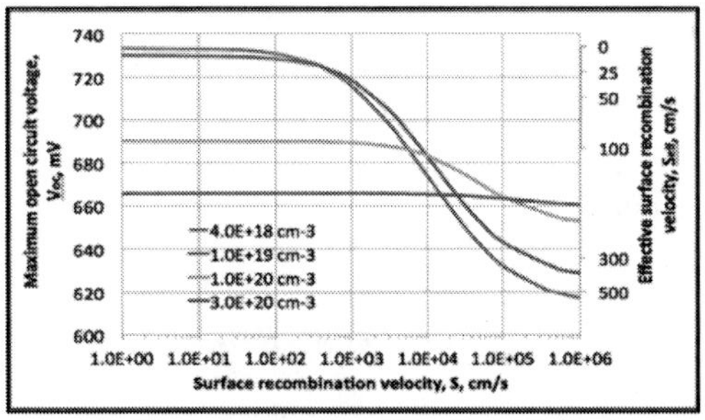

Figure 1. Maximum open circuit voltage, V_{oc}, of n$^+$-p-p$^+$ PERT cell as limited by recombination parameters of the back p$^+$ layer. Each curve refers to p$^+$ layer with mentioned maximal doping concentration.

As can be seen, S_{eff} of the highly doped p^+ layer with maximum concentration close to B solubility limit in Si (3.10^{20} cm^{-3} curve) is completely determined by bulk recombination in the doped layer and is not affected by recombination on the surface. In this case, V_{oc} doesn't depend on surface recombination that doesn't contribute in the effective surface recombination.

As doping level decreases, the contribution the surface recombination in S_{eff} value is increasing while the doped layer bulk recombination effect is decreasing. These data illustrate the expected effect of surface passivation: no real need of passivation for highly doped deep layer, and increased importance of passivation for lower doped p^+ layers. More precise evaluation of band-gap-narrowing (BGN) effect should only emphasize this result.

Simulation of the p^+ layer effect was carried out also for the structures with shallower doping. Simulation was applied for several B doping profiles in which surface concentration and position of the maximum of the concentration profile was different. Typical B profiles used for simulation are shown in Figure 2.

Some results for the structures with p^+ layers of ~0.8 μm thickness are shown in Fig. 3. In this case the figure demonstrates effects on the resulting V_{oc} of two factors: of surface recombination and of the surface charge. The open circuit voltage is plotted as a function of the recombination velocity on the cell surface. Two groups of curves describe the structures with surface boron concentration, B_s, 10^{18} and 10^{20} cm^{-3} for 3 different charge cases: positive charge, no charge or negative charge. Charge density Q (when present) is 2.10^{12} cm^{-2}.

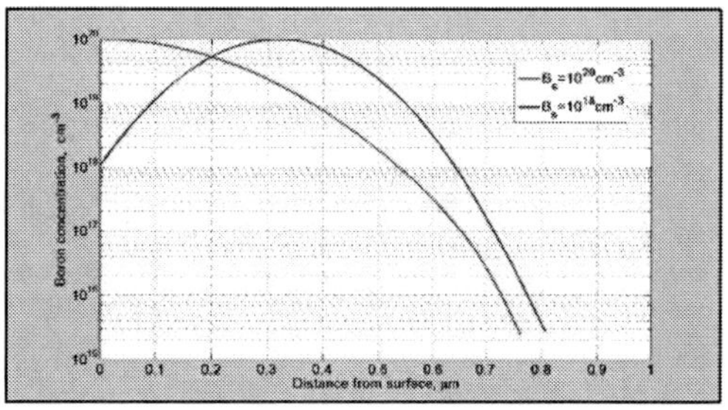

Figure 2. Typical B distribution profiles used in simulation analysis of the charge influence on the effective surface recombination. The profiles are representing the range of experimentally prepared samples.

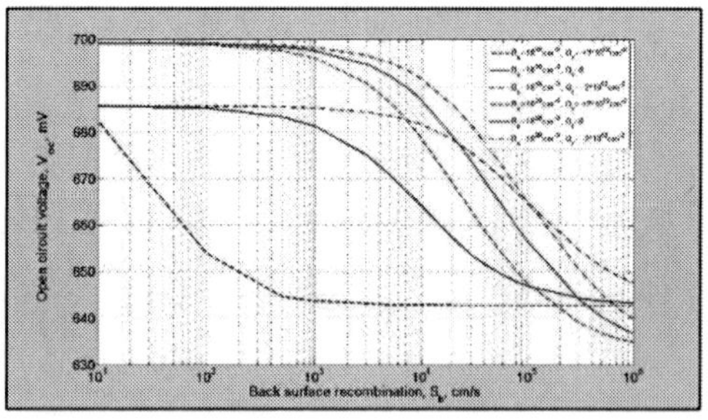

Figure 3. Effect of surface B concentration and surface charge on passivation of p^+ layer of the n^+-p-p^+ solar cell.

According to simulation, the bulk recombination in the relatively thin p^+ layer doesn't dominate as much as in the thick one (Fig. 4), and surface passivation starts to be more critical factor. However, the passivation effectiveness can be affected by the charge build in the passivation layer. The dominating parameter controlling the influence of the charge on the effective surface recombination is the B_s level. The charge in the passivation layer isn't influencing significantly when surface doping concentration is relatively high, i.e. when B_s is above ~10^{19} cm^{-3}. As can be seen, in the case of low B_s, both factors - the surface recombination and the charge - have significant influence on the V_{oc} (i.e. J_o due to effect of S_{eff}). In this case the influence of charge becomes critical causing up to 40 mV losses even for surface recombination velocity of 10^4 cm/s, which is fairly achievable with SiO_2 passivation. However, at high B_s the difference due to field effect of negative and positive fixed charge doesn't exceed ~10 mV at the same recombination velocity on the surface. So, at high B_s the charge has minor effect, and improvement of surface recombination (due to decreasing the dangling bonds density) is the controlling factor.

3. EXPERIMENTAL

The experiments were focused on the evaluation of the B doped layer parameters effect on the effective surface recombination in a p^+ layer dedicated for a bifacial solar cell. The simulation results were experimentally proven.

Wide spectrum of doping profiles with boron surface concentration of 1.8E+18 – 1.0E+22 cm^{-3} and diffusion depth of 0.45 – 1.6 μm was tested.

One series of experiments was carried out with bifacial solar cells with thermally heavily doped back. Secondary Ion Mass Spectroscopy (SIMS) data for a typical diffused bifacial solar cell sample is shown in Fig. 4a. The high B_s above the limit of solubility is distinguishing for deep doping profile. The over-doped region is a result of in-complete BSG removal.

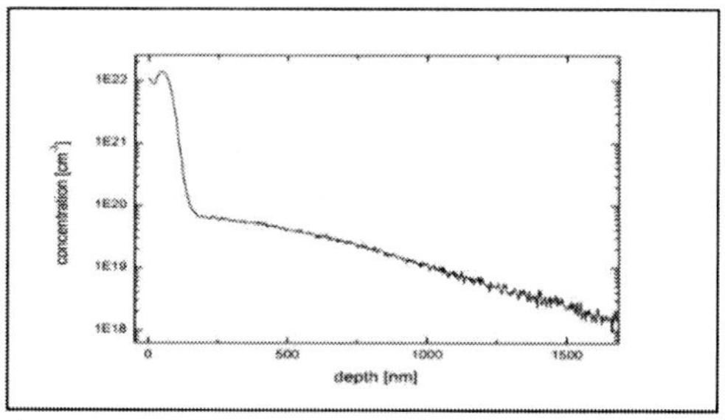

a)

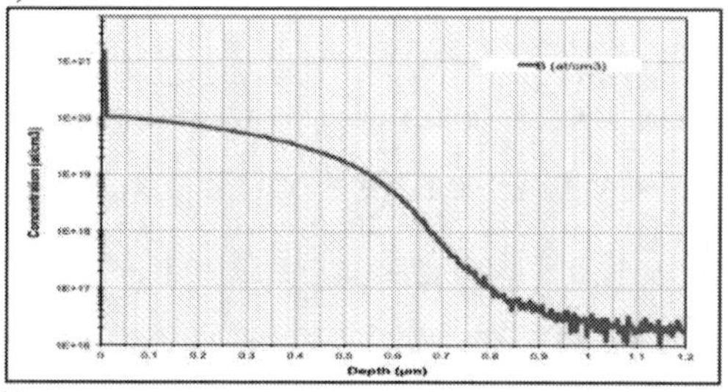

b)

Figure 4. Examples of the boron SIMS profile after high temperature diffusion.

a) Doped layer thickness ~ 1.6 μm. b) Heavy doping profile in a symmetrically doped wafer, thickness ~ 1 μm.

Another set of samples was prepared on symmetrically doped wafers. Doping was made using high temperature diffusion using solid B containing source. Secondary Ions Mass Spectroscopy (SIMS) measurements were used for the determination of boron distribution profile of very high surface concentration as shown in fig 4b.

In order to check the mutual influence of passivation and doping surface concentration. Several sets of samples were prepared. All of them are characterized by B concentrations below solubility limit and measured as electrically active impurity. Cz n-type silicon wafers were front and back symmetrically doped by boron. Doping was made using two methods. The first is by 30 keV B ion implantation with subsequent high temperature annealing. The second is the high temperature diffusion using solid B containing source. Four examples of electro-active boron concentration doping profile in the range of experiments are shown in Fig. 5. Measurements were done using Electrochemical Capacitance-Voltage (ECV) characterization apparatus.

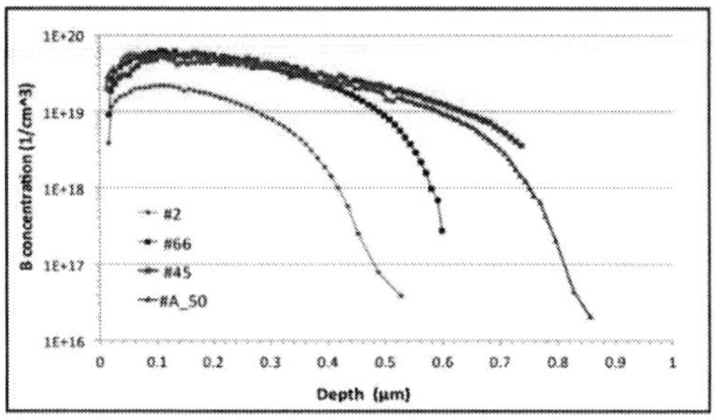

Figure 5. Boron distribution profiles in the thermally diffused p^+ layers. The data are exhibiting the range of samples being used.

Spectral quantum efficiency measurements of the back illuminated n^+-p-p^+ cells with p^+ layer heavily doped as shown in Fig. 4 were used for determination of S_{eff} of such layers.

Several passivation layers were tested: thermally grown SiO_2 alone or in combination with SiN and boron-silicon alloy, as well as quinhydrone passivation.

Lifetime measurements by Quasi Steady State Photoconductivity (QSSPC) decay technique were used for determination of J_0 and S_{eff}. The measurements were carried out on different steps of samples treatment after doping process:

1. After diffusion process followed by RCA cleaning and HF etch. (Immersed in quinhydrone solution)
2. After thermal oxidation
3. After surface RCA cleaning with subsequent SiO_2 HF etching and immersion in quinhydrone solution.

4. RESULTS AND DISCUSSIONS

4.1. S_{eff} for heavy B doped thick p^+ layer.

Back internal quantum efficiency of n^+-p-p^+ cells with over-doped p^+ layer as presented in Fig. 4. is shown in figure 6. Assuming p^+ layer with zero thickness, hence no recombination occur inside the doped layer, the shown back IQE data were used for evaluation of S_{eff} values in the range of 75±20 cm/s. (The evaluations were made using PC 1D program for the rear illuminated cell with zero thickness of the p^+ layer. Experimental data were corrected assuming p^+ layer thickness 0.57 µm and recombination losses in the layer the same as at 0.4 µm wavelength light illumination). These S_{eff} values fits to results obtained on symmetrically doped boron wafers, as shown in table 1.

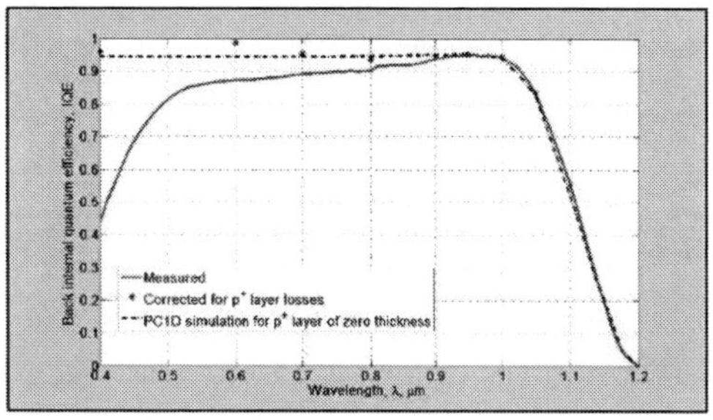

Figure 6. Spectral dependence of back internal quantum efficiency of bifacial solar cells.

High recombination rate expected for heavy doped layers according to Fig. 3 simulation results are not in agreement with relatively low S_{eff} measured for this cell. This allow assuming that the region with doping level lower than ~ 10^{20} cm^{-3} contribute in S_{eff} value. It means that heavier doped regions near the surface don't add in S_{eff} of the thick BSF layers. This explains why the deep high B doped BSF have quite low S_{eff} (which is several times lower than in Al BSF) and doesn't limit V_{oc} at level lower than ~660 – 670 mV. Moreover, this is without any surface passivation.

4.2. S_{eff} for heavy B doped thin p^+ layer.

J_0 and S_{eff} values of some representing samples are presented in table 1. After high temperature diffusion followed by RCA clean and HF etch, wafer were measured using QSSPC method with and without being immersed in quinhydrone solution. The sample wafers surfaces were hydrophilic indicating that some layer other than silicon exists on surface. A SIMS measurement was used for the determination of boron-silicon alloy on the surface. An example of such SIMS measurement is shown in figure no. 3. As can be seen in the table this boron-silicon layer provides moderate passivation. Effective surface recombination velocity values were determined using measured J_0 values. These results are demonstrating technological feasibility of p^+ layer formation enabling surface passivation without any need for a preliminary aggressive process.

Table I: Surface recombination of highly boron doped silicon

Sample #	Hydrophobic	R_{sh} before oxidation Ω/□	Surface concentration cm^{-3}	Recombination parameters without passivation		Recombination parameters in quinhydrone passivation	
				J_0, fA/cm²	S_{eff}, cm/s	J_0, fA/cm²	S_{eff}, cm/s
D_76	No	31	1.1E+22	290	115	284	113
D_45	No	33	9.9E+21	271	108	235	93.1
D_66	No	36	7.1E+21	193	76.4	136	53.8

4.3. Surface recombination depending on boron surface concentration.

J_0 values for samples with SiO_2 thermal passivation are compared in table 2 with the results of quinhydrone passivation (demonstrating the effect of the charge trapped by SiO_2). As can be seen, the experimental results confirm the simulation result. For low boron surface concentration values SiO_2 passivation layer is less effective than quinhydrone passivation (samples D_02 and D_93b). The effect of charge trapped by SiO_2 is limiting the decreasing of surface recombination: the J_0 of cell with SiO_2 passivation is ~ order of value higher than in the case of quinhydrone passivated cell. When boron concentration is higher than 1.0E+19 cm^{-3}, no significant difference is observed for quinhydrone or SiO_2 passivation.

Table II: Surface recombination parameters under passivation of SiO_2 and quinhydrone.

Sample #	R_{sh} after oxidation Ω/□	Surface concentration cm^{-3}	Junction depth µm	Recombination parameters with SiO_2 passivation		Max V_{oc}	Recombination parameters in quinhydrone solution		Max V_{oc}
				J_0, fA/cm²	S_{eff}, cm/s	V	J_0, fA/cm²	S_{eff}, cm/s	V
D_93b	330	1.8E+18	0.5	385	67	0.655	28.4	4.9	0.723
D_02	170	3.9E+18	0.45	246	131	0.667	37.8	20.3	0.715
D_06	31	1.1E+19	1.3	208	82.5	0.671	217	86.1	0.670
I_17	174	1.8E+19	0.5	324	56.4	0.66	556	104	0.646
I_50	47	3.5E+19	0.8	1000	174	0.621	1010	176	0.621

5. CONCLUSIONS

- Recombination properties of p^+ layers dedicated in n^+-p-p^+ bifacial PERT solar cells were studied experimentally and by simulation.
- Comparative measurements of thermally oxidized doped wafers and the same wafers in quinhydrone after removal the oxide layers are in the good agreement with the charge effect simulation results.
- According the simulation, the charge in the passivation layer isn't influencing significantly when B_s is relatively high, i.e. when B_s is above ~$10^{19}cm^{-3}$.
- Heavy doped layer, resulting in a Si-B alloy layer demonstrates moderate effective surface passivation.
- Back internal quantum efficiency analysis for heavily doped p^+ layers correlates with recombination data obtained on symmetrically doped wafers.

REFERENCES

[1] T. Dullweber, C. Kranz, R. Peibst, U. Baumann, H. Hannebauer, A. Fulle, S. Steckemetz, T. Weber, M. Kutzer, M. Muller, G. Fisher, P. Palinginis and H. Neuhaus. The PERC+ cell: More output power for less aluminum paste. Photovoltaic International, December 2015.

[2] L. Kreinin, N. Bordin, N. Eisenberg, P. Grabitz and G. Wahl. Industrially Fabricated Bifacial Si Solar Cells with n^+-p-p^+ Structure. Proc. of 28th European Photovoltaic Solar Energy Conf. 30 Sep - 04 Oct 2013, Paris, France.

[3] G. Dingemans, W. M. M. Kessels. Status and prospects of Al_2O_3-based surface passivation schemes for silicon solar cells. 040802-1 J. Vac. Sci. Technol. A 30(4), Jul/Aug 2012.

[4] Subhash V Ghaisas, Onkar S. Shinde, Rajiv O. Dusane, Neelkanth G Dhere. Low Temperature Surface Passivation of Silicon Solar Cells. Proc of the 43rd IEEE PVSC June 5-10, 2016 Portland, Oragon, USA

[5] J. Benic, A. Leimenstoll, and O. Schultz. Comprehensive studies of passivation quality on boron diffused silicon surfaces. Proc. 22nd European Photovoltaic Solar Energy Conf. 3-7 Sep. 2007, Milan, Italy.

[6] P.P. Altermatt, H. Plagwitz, R. Bock, J. Schmidt, R. Brendel, M.J. Kerr, A. Cuevas. In The surface recombination velocity at boron-doped emitters: comparison between various passivation techniques. Proc. of the 21st European Photovoltaic Solar Energy Conf. 4-8 Sep. 2006, Dresden, Germany.

[7] V. D. Mihailetchi, Y. Komatsu, L. Geerligs. Nitric acid pretreatment for the passivation of boron emitters for n-type base silicon solar cells. Applied Physics Letters 2008; 92, 063510- 063510-3.

[8] H. C. Chang, C. J. Huang, P. T. Hsieh, S. W. Chiu, C. C. Li. Investigation on thermal oxide thickness for passivation of n-type bifacial solar cell application. Proc. of 28th European Photovoltaic Solar Energy Conf. 30 Sep - 04 Oct 2013, Paris, France.

[9] K. R. McIntosh, L. E. Black. On effective surface recombination parameters. Journal of Applied Physics 116, 014503 (2014).

[10] L. E. Black, T. Allen, K. R. McIntosh, and A. Cuevas. Effect of boron concentration on recombination at the p-Si–Al_2O_3 interface. Journal of Applied Physics 115 , 093707 (2014); doi: 10.1063/1.4867643.

[11] J. Snel. The doped Si/SiO2 interface. Solid-State Electron. 24, 135 (1981).

[12] B.E. Deal, M. Sklar, A.S. Grove, and E.H. Snow. Characteristics of surface state charge of thermally oxidized silicon. J. Electrochem. Soc. 114, 266 (1967).

[13] D. Yan, A. Cuevas. Empirical determination of the energy band gap narrowing in p1 silicon heavily doped with boron. Journal of Applied Physics 116, 194505 (2014).

23% METAL WRAP THROUGH SILICON HETEROJUNCTION SOLAR CELLS - A SIMPLE TECHNOLOGY INTEGRATING HIGH PERFORMANCE CELL AND MODULE TECHNOLOGIES

G. Coletti[1,*], F. Ishimura[2], Y. Wu[1], E.E. Bende[1], G.J.M. Janssen[1], K. Hashimoto[2], Y. Watabe[2] and B.B. Van Aken[1]
[1]ECN – Solar Energy, P.O. Box 1, 1755 ZG Petten, the Netherlands
[2]Choshu Industry Co., Ltd., 3740, Shin-yamanoi, Sanyo Onoda, Yamaguchi 757-8511, Japan
*Corresponding author, e-mail: coletti@ecn.nl

ABSTRACT: Metal wrap through silicon heterojunction (MWT-SHJ) cells and modules combine the positive benefits of both underlying technologies: namely high Voc, higher Jsc and higher FF maintained before and after module encapsulation. We obtained 22.0% efficiency with low temperature Ag paste metallisation of front side and more than 23% on copper plated front metallisation. Voc values above 730 mV have been achieved, demonstrating that this architecture maintains the exceptional passivation of SHJ devices. The MWT cell and module structure offers even greater advantages when applied to SHJ solar cells: i) front side Ag consumption reduction up to a factor two; ii) low temperature cell interconnection concurrently with the encapsulation. Our record MWT-SHJ solar cells and modules are manufactured using industrially proven tools. The metallisation choice gives ample room to manufacturers to optimise based on internal cost structure and business strategy.
Keywords: Bifacial, PV module, energy performance, optical properties

1 INTRODUCTION

Several solar cell technologies have demonstrated the ability to reach more than 22% cell efficiency. In addition, new back contact module technologies are now available to overcome losses caused by the interconnection and due to the limited width of the tabs. High efficiencies and low materials consumption are the main drivers towards low-cost ($/Wp) silicon PV modules.

In this contribution we show the most recent development in the successful combination of metal wrap through silicon heterojunction (MWT-SHJ) cells and conductive foil module technology. The vast majority of the market implements module technologies based on interconnection of solar cells in strings by tabs soldered from the front of one cell to the rear of the adjacent one. To limit shading losses caused by these tabs, such interconnection leads to additional resistive losses in a string of cells thereby reducing the module performance. MWT technology provides a relatively small step from conventional cell technologies and has already demonstrated to increase the module power by 3%, and up to at least 5% is anticipated [1]. This is possible thanks to an integrated cell and module design in which conductive interconnection foil is used to reduce the cell-to-module power loss compared to conventional tabbing technology. Part of this gain is thanks to the reduced metal coverage on the front side, giving the solar cell performance a potential efficiency increase up to about 2.5% relative. Furthermore, thanks to the unit cell concept, i.e. the front metallisation grid of each unit cell is connected to a contact on the rear by a metal via through the wafer, the MWT cell structure decouples the wafer size from metallisation requirements allowing for lower resistive and/or shading losses.

Front and rear contacts SHJ solar cells have demonstrated more than 24% cell efficiency, e.g. 25.6% on interdigitated back contact SHJ solar cell [2], achieving excellent surface passivation with Voc exceeding 730 mV [3]. Furthermore SHJ has a lower temperature coefficient, resulting in higher module energy yield under high temperature conditions.

Nevertheless the low temperature Ag paste required for contact formation in SHJ devices is still today a challenge as it suffers from low conductivity resulting in high Ag consumption. Solutions such as multi-wire interconnection and multi-busbars have been introduced to tackle this conductivity issue [4,5]. MWT-SHJ architecture provides an elegant alternative solution to the reduced conductivity of low temperature silver pastes while maintaining the above mentioned advantages.

Moreover, the low-temperature process required by the SHJ solar cell is perfectly met by the soldering-free MWT conductive foil technology which uses conductive adhesive and single step curing for interconnection and encapsulation. MWT and SHJ cell technologies are also compatible with next generation thinner wafers resulting in a win-win situation both on cell and module level.

MWT-SHJ devices combine all the advantages of the individual concepts in a device with high open circuit voltage, high short circuit current and high module power thanks to reduced power loss.

In this paper, we show the most recent results of this MWT-SHJ solar cell architecture which resulted in a record efficiency of 23% using commercial n-type Cz 6 inch wafers. Furthermore we show potential efficiency gain of this architecture with respect to conventional front and rear contacts devices.

2 PURPOSE OF THE WORK AND APPROACH

In this contribution we describe recent development of MWT-SHJ solar cells and modules. We used a rear emitter cell structure to reduce recombination and shunt at the vias as previously reported [6]. Our approach is both theoretical and experimental. We used a 2D model to predict the experimental results and chose the best configuration and geometry of the MWT unit cells. The model has been then validated by fabricating working devices and characterisation (IV-curve and EL). In addition we demonstrated this approach with both low temperature Ag paste screen printing and Cu plating.

3 SCIENTIFIC INNOVATION AND RELEVANCE

To our knowledge, the solar cell performances obtained are record efficiency for both MWT-SHJ and MWT devices. The use of conventional low temperature Ag paste and low temperature interconnection methods, like conductive tape or multi-wire approaches, provide a costly solution that hinder the entrance in mass production of SHJ technology. MWT-SHJ addresses both issues. The MWT structure allows the reduction of the current to be transported thanks to the reduction of the unit cell size at virtually no costs and without the complicated hassle of upscaling to multi-busbar or multi-wire or smart-wire technologies. This allows a reduction of front Ag consumption by up to a factor two taking into account both the front side metallisation and the interconnection. We expect a similar gain from the rear metallisation. On module level, the cells are interconnected through the conductive rear side foil by means of conductive adhesives. During the lamination the interconnection takes place at low temperature concurrently to the encapsulation.

4 CELL STRUCTURE

MWT solar cells have the same architecture of a conventional solar cell with the addition that the front metal contact is wrapped through the wafer through metallised via holes, providing both emitter and base contacts on the rear side, see Figure 1 [6]. As the cell interconnection does not require tabbing, the busbar can be significantly slimmed down enhancing the light harvesting in comparison to conventional H-pattern cells.

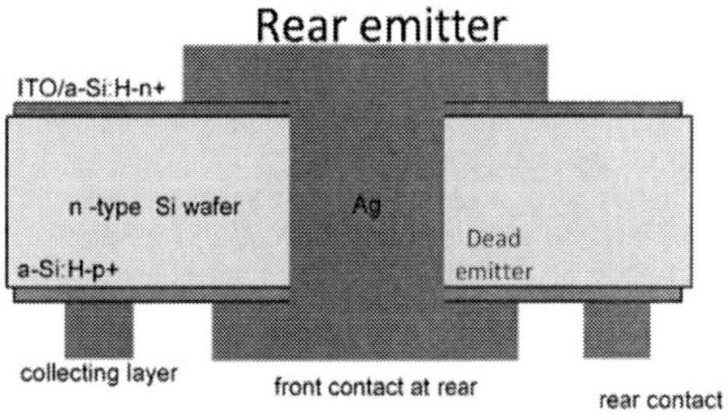

Figure 1: Schematic structure of a rear emitter MWT-SHJ solar cell. The collecting layer represents both the IWO:H and Ag rear contact

5 EXPERIMENT AND RESULTS

6" industrial n-type Cz monocrystalline full-square silicon wafers about 180 µm thick were used as substrate for MWT-SHJ solar cell fabrication. Conventional SHJ processing sequences were used for the fabrication of the solar cells: texturisation and cleaning, a-Si:H deposition by PECVD and TCO deposition. The MWT-SHJ solar cell process is the same as the SHJ process with adding a few extra steps typical for MWT cell technology [1]: vias drilling, via filling and isolation of the two metal contacts at the rear. In this experiment, the vias drilling is done before the surface preparation for the a-Si:H depositions, therefore the damage removal is done without adding any extra chemical step. In the MWT-SHJ process used, an isolation of the metal plug from the underlying Si or a-Si:H layer is not necessary. The via filling is done within the same printing step of the rear side pad.

Two types of metallisation have been used: i) low temperature silver paste by screen printing and ii) copper plating of the front metallisation. The copper plating has been entirely developed by CIC. A description of this process is will be described in a sister publication [7].

The rear side of both devices is identical and has been manufactured in two different ways: i) by screen printing of low temperature Ag paste and ii) by physical vapour deposition of a Ag blanket.

The solar cells are measured at ECN according to the ASTM-E948 norm using a Wacom class AAA solar simulator and spectral mismatch correction. IV results are shown in Table I.

Table I: IV results for MWT-SHJ solar cells with different metallisation and contacting schemes

Structure	Front metal	Jsc [mA /cm^2]	Voc [mV]	FF [%]	η [%]
H-pattern	Ag 3BB	38.9	722	77.3	21.7
MWT 4x4	Ag	39.1	726	76.6	21.7
MWT 6x6	Ag	39.2	723	77.6	22.0
MWT 6x6	Ag	39.3	719	80.0	22.6
MWT 6x6	Cu plating	39.7	731	79.8	23.1

The first three devices in Table I are manufactured in the same batch to compare the different configurations (H-pattern, 4x4 via MWT and 6x6 via MWT). Only the busbar width and number have been changed between configurations. The number and width of fingers have been kept constant (same finger metal coverage). This allowed making a direct comparison among the different configurations. The total front metal shading for the three devices is 5.4%, 3.7% and 3.4% for the 3BB, MWT 4x4 and MWT 6x6, respectively.

The fourth device in Table I is manufactured without the constraint of same finger metal coverage. Indeed by choosing for a large number of vias the unit cell area reduces and therefore the series resistance losses of the device. An optimisation of the finger width and number have been carried out to account for both metal consumption reduction and performance. This has allowed increasing the efficiency of the MWT 6x6 SHJ device by 0.6% absolute. The improved front side metallisation layout has also been combined with copper plating. Details are published elsewhere [7].

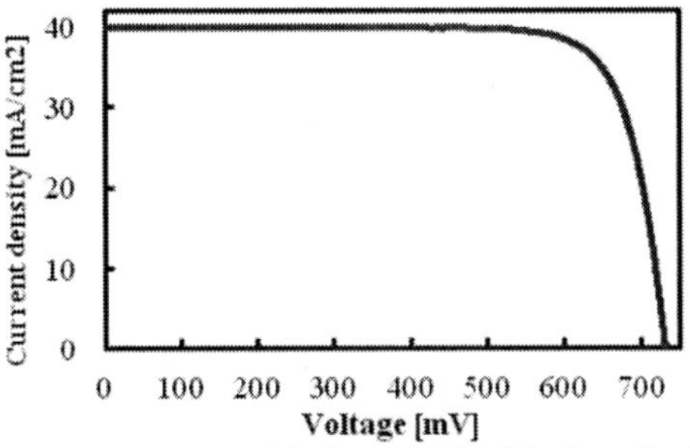

Figure 2: IV curve of the best MWT-SHJ solar cell with copper plating front side metallisation

In our MWT-SHJ, we demonstrate a Jsc that is superior to conventional H-pattern SHJ thanks to the reduced metal front coverage which results from the carrier transport being transferred to the rear side. The better cell performance is despite the fact that the measurement setups for H-pattern and MWT-SHJ solar cells are different. For the H-pattern multi-probes are usually used for the front contact of the cells. These increase the conductivity of the busbar independently of their width and thickness which results in higher FF (see table I H-pattern ref versus 4x4). This is not the case in real conditions (interconnection within the module) since the tabs used for the interconnection have a more limited conductivity than the multi-probes. MWT-SHJ measurements chucks do not use multi-probes as the current is transported to the rear side only thanks to the front metallisation. The result is that the measurements are carried out in real module conditions. In addition the FF increases on cell level since the current transport is not limited by the busbar width and by their number. The effect on the FF is even more evident on module level. Therefore a real comparison of the performance of MWT-SHJ and conventional front and rear contact cells should be carried out on module level where a reduced cell to module loss will be evident.

Figure 3 shows the dark IV-curves under negative voltage bias. Typical MWT-SHJ Irev current values are less than 0.5 A at -10 V for both the 4x4 and 6x6 configurations.

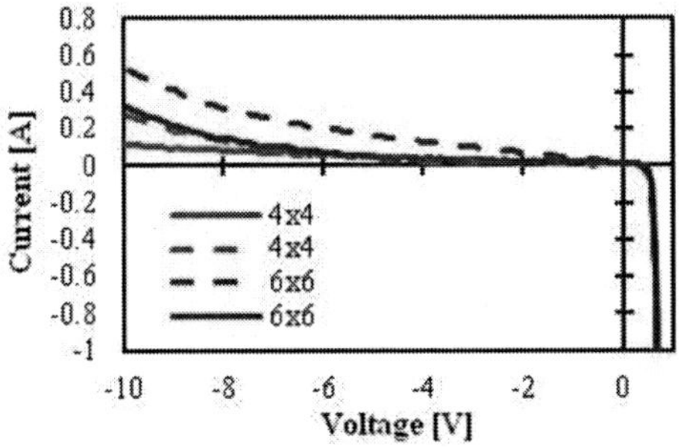

Figure 3: Reverse characteristics. of MWT-SHJ solar cells with Ag screen printing for both 4x4 and 6x6 vias configuration

In order to study the behaviour of the MWT-SHJ solar cells under low illumination conditions, we measured their performance at different illumination levels. Both H-pattern SHJ and MWT-SHJ solar cells have been measured as is shown in Figure 4. No particular difference in the illumination response for MWT-SHJ solar cells with respect to H-pattern. This indicates that the shunt and recombination of MWT-SHJ is comparable if not better than conventional front and rear contacts H-pattern SHJ devices.

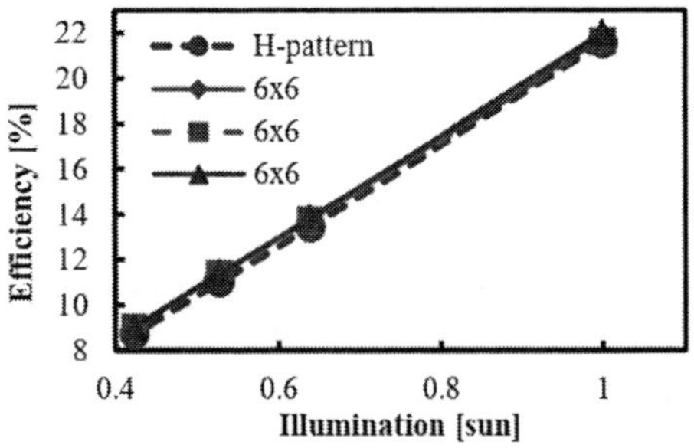

Figure 4: Efficiency as a function of the illumination level for H-pattern SHJ and 6x6 MWT-SHJ solar cells

6 CONCLUSIONS

ECN and CIC achieved 23.3% and 22.0% top efficiencies with copper plating and low temperature Ag paste, respectively. In the current configuration with LT-Ag paste and 6x6 vias we estimated a 30% cost reduction on Ag consumption with options to further reduce it to 50% with respect to 4x4 vias with no-extra manufacturing complexity. The two metallisation concepts are not yet fully optimised and there is further room for efficiency gain.

REFERENCES

[1] N. Guillevin et al., "High power n-type metal-wrap-through cells and modules using industrial processes", 28th EU PVSEC, (2013) 1304-1310.

[2] K. Masuko et al., "Achievement of more than 25% conversion efficiency with crystalline silicon heterojunction solar cell", IEEE J. Photovoltaics **4** (2014) 1433-1435.

[3] M. Taguchi et al., "24.7% Record efficiency HIT solar cell on thin silicon wafer", 39th IEEE PVSC, (2013) 96-99.

[4] S. Braun, G. Micard and G. Hahn, "Solar cell improvement by using a multi-busbar design as front electrode," Energy Procedia **27** (2012) 227–233.

[5] A. Schneider, L. Rubin and G. Rubin, "Solar cell improvement by new metallisation techniques—The DAY4TM electrode concept," in Proc. 4th World Conf. Photovoltaic Energy Conversion, 2006, pp. 1095–1098.

[6] G. Coletti et al., "20.3% MWT silicon heterojunction solar cell- A novel heterojunction integrated concept embedding low Ag consumption and high module efficiency", IEEE Journal of Photovoltaics **5** (2014) 55-60.

[7] F. Ishimure et al., "Metal wrap through heterojunction solar cell with plated electrode", 32nd EU PVSEC, (2016) 2.AV.3.14.

NOVEL LOW COST WET CHEMICAL CLEANING PROCESSES FOR INDUSTRIAL LARGE AREA N-TYPE SILICON SOLAR CELLS WITH 22% EFFICIENCY

M. Haslinger[1], J. John[1], M. Aleman[1], A. Uruena[1], E. Cornagliotti[1], L. Tous[1], R. Russell[1], A. Hajjiah[4], F. Duerinckx[1], J. Szlufcik[1], J. Poortmans[1,2,3]
[1]*imec, Kapeldreef 75, B-3001 Leuven, Belgium*
[2]*ESAT-ELECTA, KU-Leuven, Kasteelpark Arenberg 10, B-3001 Leuven, Belgium*
[3]*University Hasselt, Martelarenlaan 42, 3500 Hasselt, Belgium*
[4]*Kuwait University, College of Engineering and Petroleum, Electrical Engineering Department, Safat 13060-Kuwait*

ABSTRACT: We report on the development of industrial wet chemical cleaning processes for n-type passivated emitter, rear totally diffused (n-PERT) solar cells. PERT silicon solar cells with 22% efficiency have been manufactured comparing established silicon surface cleaning processes like sulfuric ozone mixture (SOM) with novel wet chemical processes based on DI water/ozone (DIW/O_3) and industrial cleaning processes like diluted hydrochloric acid/hydrofluoric acid mixture (HCl/HF) as used in solar cell mass production. We have tested the different wet chemical cleaning methods on prominent metal impurities like Fe and Cu. These metals typically occur on the silicon wafer surface after alkaline processing like saw damage etch and texturing. It has been shown that the impurity removal efficiency of the different cleaning methods is varying strongly for different metals. This information is used for optimizing wet chemical cleaning processes in terms of bath lifetime and chemical consumption reduction. In this paper the results of the metal removal efficiency study is applied to a n-PERT solar cell process flow, but it could be applied on other high efficiency cells concepts like IBC or HIT as well.

Keywords: Cleaning, n-type PERT, contamination.

1 Purpose of the work and relevance

Silicon solar cells based on crystalline silicon (c-Si) are the dominating technology in photovoltaics (PV) industry and have a market share of more than 85% of the modules produced for roof top installations [1]. The PV industry continues to demand even lower cost per watt solar cells so that PV electricity can ultimately compete with conventional sources of energy. One viable approach is to increase the conversion efficiency and simultaneously reduce the manufacturing cost of the device [2]. The need for cheaper levelized cost of PV-originated electricity (LCOE), i.e. cost per KWh, translates in a need for cheaper solar cells with an increase in conversion efficiency [3]. In order to reach this goal, novel cell concepts and novel fabrication methods [4] have to be developed for implementation in industrial process lines [5]. In this context, wet chemical cleaning processes should also evolve towards easy to implement, well controlled, and cost effective steps [6-8].

Process development using wet chemical processing is triggered by two main objectives: first, reducing the amount of chemicals to the minimum needed to reach the specified performance target and second, reducing the processing time to increase the throughput required in mass fabrication. In order to reach the first objective, low cost chemicals with increased chemical bath lifetimes and reduced chemical consumption by using highly diluted solutions were applied. Moreover, in order to reach the second objective, it is essential to reduce the process time for the sequential wet chemical steps by means of optimizing the wet chemical processes themselves. The purpose of this work is to provide both, reducing cost of ownership of the wet chemical cleaning process, and at the same time preserving the efficiency target.

2 Experimental approach

Currently, we still observe a large discrepancy in wet chemical cleaning methods between high efficiency cell concepts like n-PERT, HIT and IBC cells in research and development and standard cells manufactured by means of mass production lines. While lab-cells mainly use conventional silicon semiconductor cleaning sequences like RCA (SC1, SC2) and SOM (IMEC clean) to reach high efficiencies, in mass production the cleaning effort is limited to diluted HCl/HF solutions followed by a DI water rinse [9]. We could demonstrate, on one hand that the industrial cleaning process HCl/HF has its limitations concerning the removal of several metal impurities, but on the other hand, it is possible to achieve high cell efficiencies with low cost wet chemical cleaning processes, respectively.

2.1 Minority carrier lifetime (surface cleaning efficiency)

Three wet chemical cleaning processes have been tested on their efficiency to remove a well-defined amount of Fe and Cu impurities on the wafer surface. After the clean, the solar cell wafers are deposited with a thermally grown silicon oxide in order to drive in the remaining metal impurities into the bulk. Effective minority carrier lifetime is extracted by means of PL measurement (BT imaging).

2.2 Solar cell processing

The solar cell process used for the wet chemical cleaning comparison is a rear junction n-PERT flow where *Uruena et al* have reported recently conversion efficiencies above 22% [10]. The complete n-PERT process flow is depicted in fig 1.

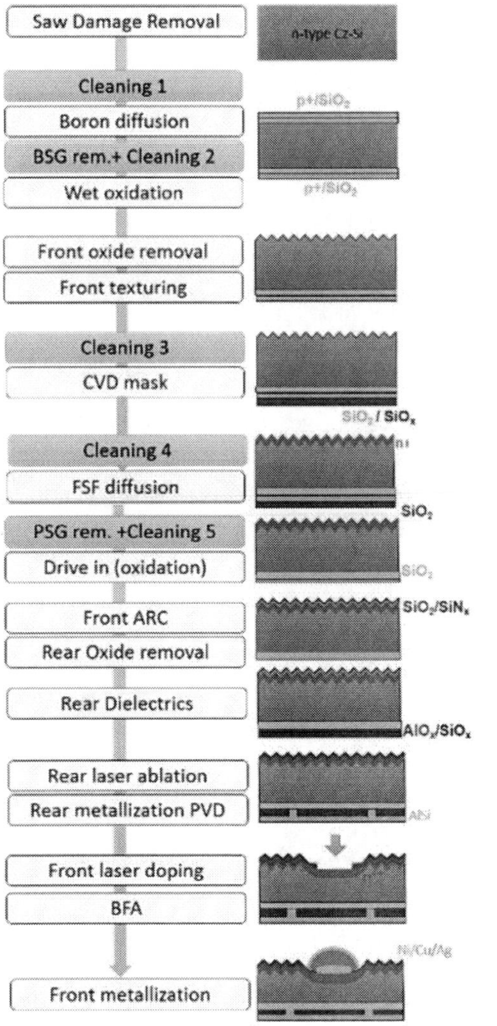

Figure 1: *Process flow for n-PERT solar cell*

We use 156x156 mm^2 commercially available Cz n-Si wafers with a base resistivity ranging from 2 to 4.5 Ωcm The rear emitter is formed by boron diffusion using BBr_3 as diffusion source and is driven-in via wet thermal oxidation, obtaining a final sheet resistance of around 100 Ω/sq. The cells have a KOH-based texturing, n+ diffused FSF with a sheet resistance in the range of 260-300 Ω/sq for the selective FSF (Laser Doping). A thin thermal oxide is used for passivation of the FSF. The anti-reflective coating (ARC) consists of a hydrogenated, non-stoichiometric silicon nitride layer (SiNx:H) obtained by plasma-enhanced chemical vapor deposition (PECVD). The rear emitter is passivated by aluminum oxide Al_2O_3 deposited by spatial atomic layer deposition (ALD) and capped with PECVD hydrogenated, non-stoichiometric silicon oxide (SiOx:H). Laser defined openings at the rear are formed by ablation of the dielectric layers creating point contacts, followed by a sputtered Al-Si (1 %) layer. Phosphorous-doped spin-on material from Filmtronics Inc. is applied on top of the front SiNx layer, with a subsequent laser treatment to form highly doped regions in the contact area for the selective FSF case. Laser treatments are performed using two —picosecond (ps) ultra-violet (UV) lasers (355 nm wavelength), one with 80 MHz and one with 200 KHz repetition rates. The wafer receives a rapid thermal anneal in a belt furnace typically used for firing of screen printed contacts. This belt furnace anneal (BFA) is applied at a moderately low temperature in the range of 400-500 C. The front metallization is based on a Light Induced Plated (LIP) Ni seed which is selectively plated on the laser-doped/ablated front pattern. This is followed by Cu LIP and electro-plating to thicken the front contact and a short Ag immersion step.

Two experimental groups for the wet chemical cleaning. Group 1: all cells are processed with the reference process (SOM+HF/HCl) and Group 2: all cells are processed with DIW/O_3+HF/HCl. The wet chemical processes are applied 5 times during the cell process. The schematic is depicted in fig 2.

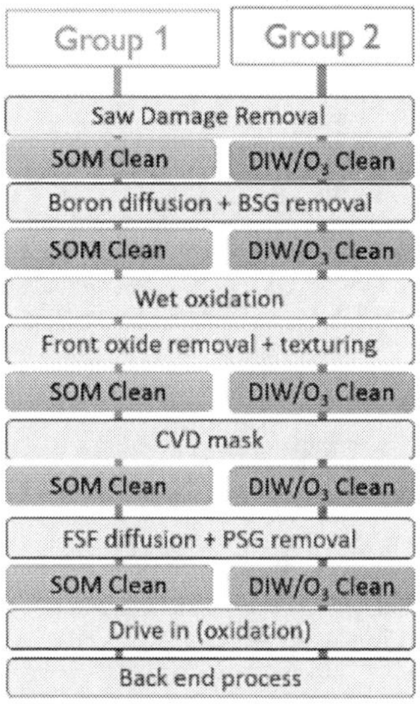

Figure 2: *Experimental matrix for cleaning*

3 Experimental results

3.1 Surface cleaning results

Surface metal impurity removal efficiency for Fe and Cu is expressed in effective lifetime vs different cleaning solutions. We can observe two striking results on these graphs. Fe surface contamination of 1e12 at/cm^2 fully degrades the minority carrier lifetime of n-type silicon after applying high temperature processing. While Fe contamination of 10^{12}/cm^2 can be removed from all the cleans quite effectively, Cu provides a problem for the industrial clean diluted HF/HCl. After this clean, Cu contamination of 5e11 at/cm^2 remains on the surface and has been driven in by the high temperature oxidation process. This results in a strongly reduced effective lifetime for the HF/HCl cleaned wafers as depicted in Fig.3.

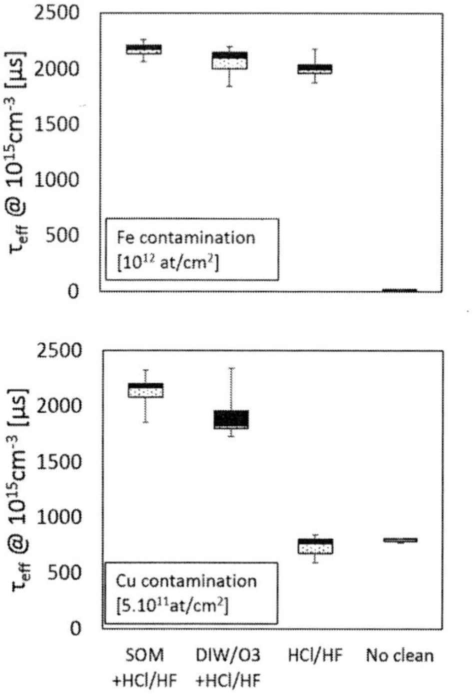

Figure 3: *Surface metal removal efficiency of different cleans for Fe and Cu. For the Fe contaminated sample, applying all cleans resulted in a recovery of the effective lifetime. On the other hand, for Cu contamination, applying only HF:HCl clean had no positive effect*

3.2 Solar cell results

During the solar cell process monitor wafer providing information on important parameters like emitter saturation current density J_{0e} and effective minority carrier lifetime τ_{eff}. Symmetric lifetime test vehicles are fabricated in parallel with the solar cells. J_{0e} and τ_{eff} are extracted from QSSPC measurements. J_{0e} after boron diffusion and wet thermal oxidation was measured to be 32 fA/cm^2 for the reference clean (SOM clean) and 35 fA/cm^2 for the novel clean (DIW/O_3 clean) on flat surfaces simulating the rear side of the n-PERT cell. Using Al_2O_3/SiOx as passivation the J_{0e} after boron diffusion is measured to be 15 A/cm^2 for both cleaning processes. After phosphorous diffusion and SiOx/SiNx passivation, emitter saturation current density J_{0e} of 10 fA/cm^2 is measured for the reference clean (SOM clean) and 12 fA/cm^2 is measured for the DIW/O_3 clean. After the fabrication, large area n-PERT cells are characterized with a solar cell simulator using a calibrated reference cell (ISE-CalLab). The measured cells show a short circuit current density Jsc of 39.6 mA/cm^2 (both Groups SOM and DIW/O_3), an open circuit voltage Voc of 691mv (SOM) and 692mV (DIW/O_3), a fill factor FF of 80.1% (SOM) and 80.2% (DIW/O_3) resulting in a conversion efficiency of 21.9% (SOM) and 22.0% (DIW/O_3). The measured solar cell parameters are depicted in figure 4. and presented in table 1.

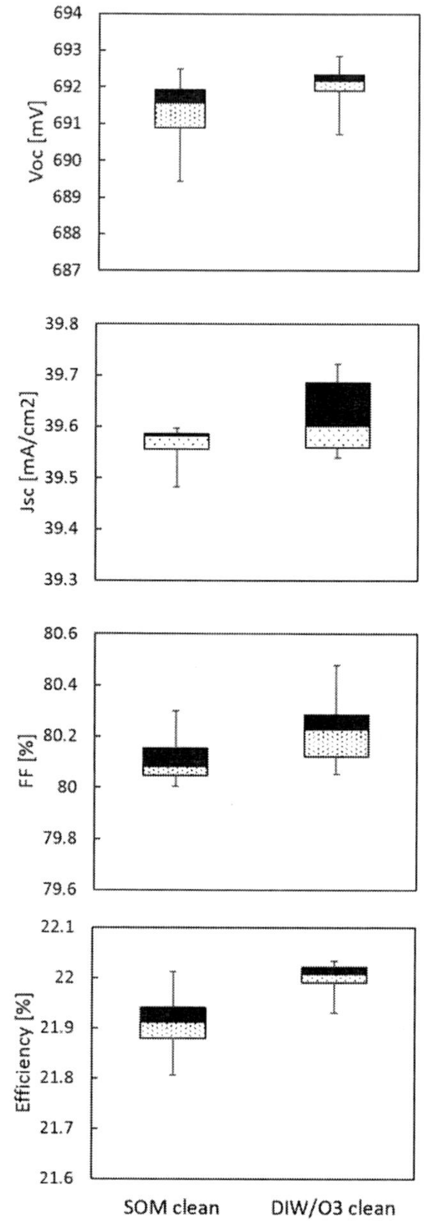

Figure 4: *solar cell parameter vs cleaning groups. All parameters were found to be indentical, hence showing a similar effeciency for the simplified DIW/O3 clean.*

Group	*SOM +HF:HCl*	*DIW/O3 +HF:HCl*
# Wafers	4	6
Voc [mV]	691.2 ± 1.3	692.0 ± 0.7
Jsc [mA/cm2]	39.6 ± 0.1	39.6 ± 0.1
FF[%]	80.1 ± 0.1	80.2 ± 0.1
Eff. [%]	21.9 ± 0.1	22.0 ± 0.1

Table 1: *Table of the solar cell parameter for the different cleaning approaches*

4 Conclusion

In this work a low cost, novel cleaning process is compared with an established reference cleaning process for high efficiency silicon solar cells. By means of controlled metal contamination of the silicon surface the metal removal efficiency of the cleaning solutions are investigated. It was demonstrated that diluted HCl/HF solutions are not effective to remove Cu contamination from the surface. Based on these experiments a two-step clean with a DI water/ozone solution followed by diluted HF/HCl solution has been developed. Applied on solar cell fabrication, the novel cleaning process provided the same cell performance than the reference process. This means, for wet chemical cleaning processes, the effort and with that, the cost could be strongly reduced without losing in performance. This has been demonstrated for large area n-PERT silicon solar cells with 22% efficiency.

Acknowledgment

"This work was funded "partially" by Kuwait Foundation for the Advancement of Sciences under project code: P115-15EE-01." The presented work is executed within the workprogram of the IIAP-SiPV.

References

[1] Jäger-Waldau A. JRC Scientific and Policy Reports-PV Status Report 2012 (2012)
[2] C. T. M. International Technology Roadmap for Photovoltaics- Results 2012. *ITRPEV. net* **2** (2012)
[3] Verlinden P. et al., presentation for the 29th EU PVSEC (2013)
[4] Recamán Payo M. et al., *Proceedings of the 30th EUPVSEC* (2015)
[5] John J. et al., *EPJ Photovolt.*, **3** (2012), p. 35005
[6] Rentsch J. et al., *Proceedings of the 25th EU PVSEC* (2009), p1113
[7] Schweckendiek J. et al., *Proceedings of the 11th UCPSS* (2012), p283
[8] Hajjiah A. et al., *Proceedings of the 31st EUPVSEC* (2015)
[9] Neuhaus H. et al., Advances in OptoElectronics, Volume 2007, Article ID 24521
[10] Uruena A et al., *Proceedings of the 31st EUPVSEC* (2015)

SAW DAMAGE REMOVAL AND TEXTURING OF CRYSTALLINE SILICON BY MASKLESS INDUCTIVELY COUPLED PLASMA (ICP) PROCESSES WITH SF_6 AND O_2

Jens Hirsch[1], Maria Gaudig[1], Marcus Gläser[2], Martina Werner[2], Stephan Großer[2], Norbert Bernhard[1], Dominik Lausch[2]
[1]Anhalt University of Applied Sciences, Faculty EMW, Bernburger Str. 55, DE-06366 Köthen
[2]Fraunhofer Center for Silicon Photovoltaics CSP, Otto-Eißfeldt-Str. 12, DE-06120 Halle (Saale)
J.Hirsch@emw.hs-anhalt.de, +49 (0)3496/67-2359

ABSTRACT: In this contribution, the wet chemical saw damage removal (SDR) and texturing of crystalline silicon wafers are replaced by maskless inductively coupled plasma (ICP) processes. For the SDR, an isotropic ICP process is applied with SF_6 as etch gas leading to semicircle surface structures with few microns width. Subsequently, an anisotropic ICP texturing is performed with SF_6/O_2 forming parabolic like surface structures with lateral extent of approx. 500 nm. A superposition of nanotexture on the SDR microtexture is performed. Optical measurements reveal an average reflection of 3.71 %, transmission of 1.31 % and absorption of 94.98 % integrated from 200-1120 nm under a normal incidence of light after PECVD SiN_x passivation. SEM images show a homogenous 70 nm SiN_x deposition on the ICP textures. A sufficient passivation is achieved by ICP SF_6 SDR with approx. 100 µs effective minority carrier lifetime at 10^{15} cm^{-3} minority carrier density (MCD). In combination with the ICP SF_6/O_2 texture, the samples have approx. 70 µs effective minority carrier lifetime at 10^{15} cm^{-3} MCD.

Keywords: Black-Silicon, Plasma, Saw damage removal, Passivation, ICP

1 INTRODUCTION

Within the last years, several groups reported about the ‘Black-Silicon-Method’ to gain enhanced light absorption of crystalline silicon solar cells [1] [2] [3]. This method was first time published by Jansen et. al. in 1995 [4] and is based on two opposite effects within a SF_6/O_2 gas discharge. On one side, F-atoms etch isotropically the silicon surface. On the other side, F-atoms and O-atoms form with the silicon surface a SiOF self-masking layer, which inhibits a sidewall etching. Ions from the glow discharge open locally the SiOF layer of the valleys, resulting in a maskless anisotropic etching [5].

However, for solar cells mass production, a saw damage removal (SDR) is required, which is usually applied by a wet chemical HF/HNO_3 or KOH/IPA solutions. For an industrial implementation of the ‘Black-Silicon-Method’, a dry SDR is needed in order to match the business case. Therefore, this contribution investigates the surface morphology, optical properties and the surface passivation of a dry plasma induced SDR and subsequent texturing.

2 EXPERIMENTAL

In this contribution, full square CZ wafer with a <100> orientation, 180 µm thickness, 12 Ω·cm boron doped resistivity are used.

The state of the art wet chemical SDR and texture are replaced by a maskless ICP SF_6 SDR and a SF_6/O_2 plasma texture. These processes are applied by means of a plasma etch chamber from Oxford Instruments (Plasmalab System 100 ICP 65). In the center of the wafer, a circular area with a diameter of approx. 5 cm is plasma etched and textured. By using inductively power coupling, no self-bias is applied on the silicon sample. Therefore, the ion acceleration is reduced to the intrinsic plasma to ground potential difference. After the plasma SDR and texturing, selected samples are passivated by an industrial standard PECVD SiN_x layer deposition. The thermal activation is done by a belt furnace.

The global reflection is measured with an integrating sphere from Perkin Elmer (Lambda 1050 UV/VIS/NIR spectrometer) under a normal incidence of light. The effective lifetimes of the minority carriers are measured by a quasi-steady-state photoconductivity (QSSPC) system BT-120 from Sinton. For the investigation of the surface morphology, scanning electron microscope (SEM) and laser scanning microscope (LSM) imaging are applied with a Hitachi SU-70 and a Carl Zeiss LSM 700, respectively.

3 RESULTS

3.1 Laser scanning microscope imaging

Fig. 1 shows the change of the surface morphology by LSM images of diamond wire sawn samples after various ICP SF_6 SDR removals (0-50 µm). The reference sample (a) shows the typical saw marks by the diamond wire saw separation in the horizontal direction. After 5 µm ICP SF_6 SDR (b), the direction of the saw marks is still visible, but is overlapped by a semicircle surface structure caused by the ICP SF_6 SDR. This semicircle surface structure has a typical width of several microns and increases with the ICP SF_6 SDR time (c-d). After 20 µm ICP SF_6 SDR, the diamond wire saw marks have almost disappeared and are replaced by the semicircle SDR structure, which are oriented along the saw mark direction. The visible accumulations at the borders of the semicircle structures will be investigated in upcoming studies. These structures are also visible in standard SiC slurry sawn wafer. Please note that a texturing of diamond sawn wafers by wet chemicals is quite sophisticate. Using plasma processes, no additional optimization has to be introduced.

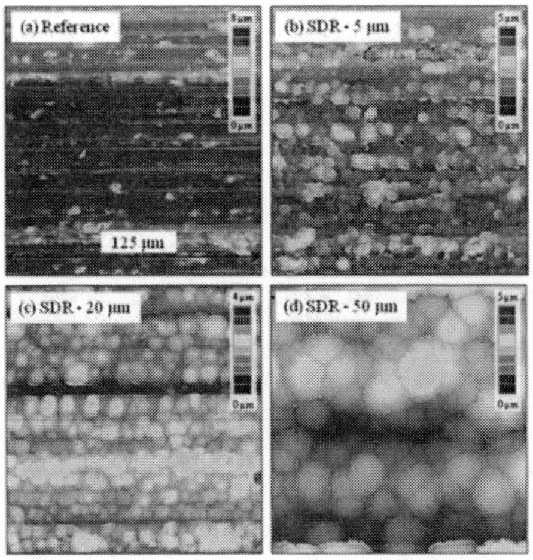

Fig. 1: Top view LSM images show the change of the surface morphology of diamond wire sawn wafers from reference (a) to 5 µm (b), 20 µm (c) and 50 µm (d) ICP SF_6 SDR.

3.2 SEM imaging

In Fig. 2, the surface morphology changes are shown after ICP SF_6 SDR (a), SF_6/O_2 plasma texturing (b) and PECVD SiN_x deposition (c-d). Fig. 2 (a) reveals the semicircle surface structure after ICP SF_6 SDR. Fig. 2 (b) shows a similar surface after ICP SF_6/O_2 plasma texturing. A superposition of the ICP SDR and plasma texture is visible. This superposition is shown in more detail in Fig. 2 (c) in cross section. The ICP SF_6 SDR bow and the parabolic like ICP texture are clearly visible. Fig. 2 (d) shows the plasma textured surface after PECVD SiN_x layer deposition in high magnification. A homogenous deposited SiN_x layer covering the whole surface without holes.

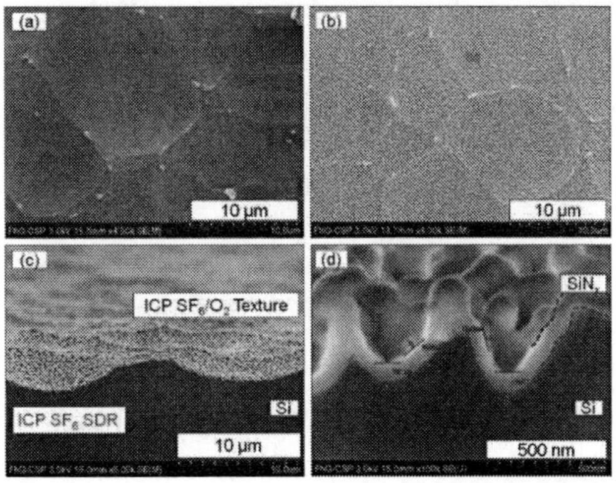

Fig. 2: Top views of crystalline silicon surfaces after ICP SF_6 SDR (a), ICP SF_6 SDR and SF_6/O_2 plasma texture (b) and cross sections of PECVD SiN_x passivated plasma textures (c-d).

3.3 Optical properties

Fig. 3 shows the average global reflection, transmission and absorption between 200-1120 nm wavelengths under a normal incidence of light after ICP SF_6 SDR, SF_6/O_2 plasma texture and PECVD SiN_x passivation. Across the shown spectrum, the graph reveals an average reflection of 3.71 %, transmission of 1.31 % and absorption of 94.98 %. Within the visible range (380-780 nm), the average reflection is below 1 % explaining why the sample appears black. In the short wavelength region (< 400 nm), the sample shows a slight increase of the reflection of approx. 10 %. In the long wavelength region (> 1000 nm), the transmission rises, caused by the low absorption coefficient of the silicon.

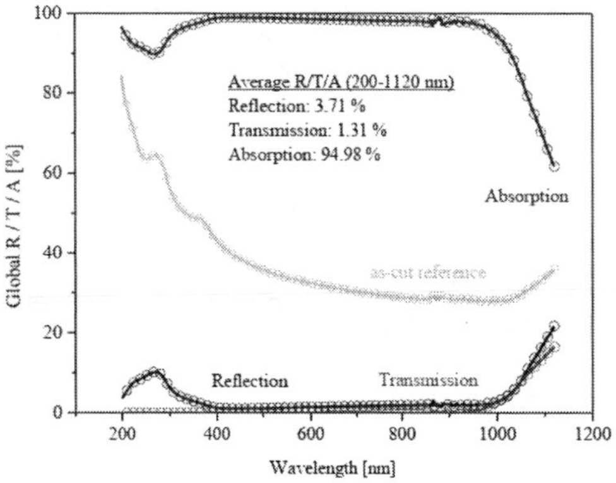

Fig. 3: Global reflection, transmission and absorption of a crystalline silicon CZ sample after ICP SF_6 SDR, ICP SF_6/O_2 plasma texturing and PECVD SiN_x surface passivation. All values are measured under a normal incidence of light.

3.4 Surface passivation

After the ICP SF_6 SDR and ICP SF_6/O_2 plasma texturing, the samples were both side passivated by an industrial standard PECVD SiN_x layer deposition and thermal activation. Prior to the SiN_x layer deposition, the samples were cleaned by an HF/HCl solution to remove metallic impurities from the plasma exposition. The surface passivation was investigated by a quasi-steady-state photoconductivity (QSSPC) measurement. These results are summarized in Fig. 4. For comparison, the result of a wet chemical HF/HNO_3 SDR and textured reference wafer is included. This reference wafer shows the highest effective lifetime of approx. 110 µs of all samples. However, the ICP SF_6 SDR sample (without texture) shows similar results and is close to the reference. Note, standard SiN layer deposition was used, hence the SiN_x layer deposition was not optimized towards plasma textured surfaces. The ICP textured sample reaches approx. 70 µs at 10^{15} cm^{-3} MCD, which is a decrease of 40 µs compared to the reference.

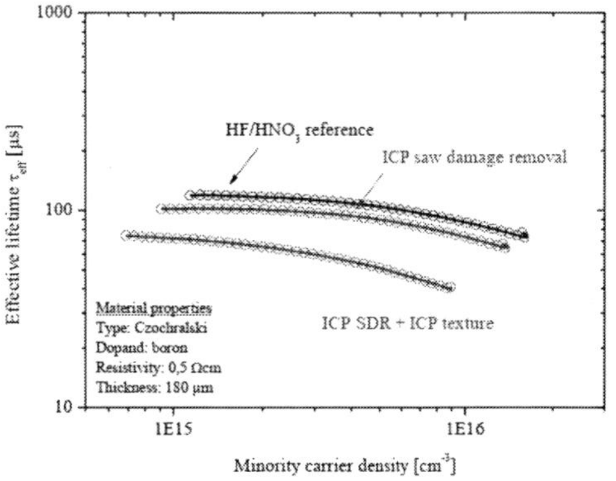

Fig. 4: QSSPC results of an HF/HNO_3 wet chemical reference wafer (black), ICP SF_6 SDR wafer (red) and an ICP SF_6/O_2 SDR and plasma textured wafer after PECVD SiN_x passivation and thermal activation.

4 DISCUSSION

The LSM images of Fig. 1 show the change of the surface morphology after various ICP SF_6 SDRs. In these figures, it can be seen (I) that the ICP SF_6 SDR forms a semicircle surface structure with a width of several microns and (II) that this structure is aligned along the diamond wire saw direction. The semicircle surface structure is caused by the exothermal silicon fluorination. Therefore, the temperature is locally enhanced, which can result in an accelerated etching in the valleys, until exposed surfaces at the edges of the semicircles will be etched back. The alignment along the saw mark direction indicates an initializing of the silicon F-atom etching by surface near damages.

The SEM images of Fig. 2 show the ICP SDR, texture and PECVD SiN_x passivation. Fig. 2 (b) shows the superposition of the µm-scaled ICP SDR and the nm-scaled ICP texture. The parabolic like plasma texture with an aspect ratio of approx. 1 and a structure width of 500 nm is a result of the pure ICP processing. Hence, no self-bias is applied on the silicon sample, which reduces the vertical acceleration of positive ions towards the surface normal [6]. These results beside a low surface damage in a lower aspect ratio compared to capacitively coupled plasma textures. Therefore, a homogenous PECVD SiN_x layer deposition is possible.

The optical measurements in Fig. 3 reveal excellent optical properties of the ICP texture after PECVD SiN_x deposition. The reflection reduction is caused by the parabolic like surface structure of the ICP texture, which works like an effective medium with an effective refraction index between air and silicon. This effect is also known as 'moth-eye-principle' [7] [8]. Furthermore, the superposition of the ICP SDR and subsequent plasma texture lead to a refraction away from the vertical direction. This further increases the optical path length within the silicon and is the reason for the high absorption, even in the long wavelengths region.

The QSSPC measurements of Fig. 4 show similar results of the HF/HNO_3 wet chemical reference and the ICP SF_6 SDR sample at approx. 100 µs effective lifetime. However, the ICP textured sample shows a decrease of the effective lifetime at 70 µs at 10^{15} cm^{-3} MCD. This decrease is caused by the surface enlargement, which increases the contribution of the surface recombination compared to the total recombination. Therefore, the PECVD SiN_x passivation must be optimized in upcoming studies.

5 CONCLUSION

In this contribution, the wet chemical saw damage removal (SDR) and texture of crystalline silicon wafers are replaced by maskless inductively coupled plasma SF_6/O_2 processes. The SDR is substituted by an SF_6 plasma step, which results in a semicircle surface structure in a scale of few microns. The semicircle structure width rises with the plasma exposure time and is likely caused by local temperature differences on the surface. The SF_6/O_2 plasma texture forms a parabolic like surface with approx. 500 nm structure width on top of the SF_6 SDR semicircles. Therefore, we have a superposition of a plasma SDR (µm-scale) and a plasma texture (nm-scale). Optical measurements revealed 3.71 % average global reflection between 200-1120 nm after PECVD SiN_x passivation. QSSPC measurements show approx. 70 µs effective minority carrier lifetime at 10^{15} cm^{-3} minority carrier density.

6 ACKNOWLEDGEMENT

This work was carried out within the joint research graduate school "StrukturSolar" of Anhalt University of Applied Sciences and Martin-Luther-University Halle-Wittenberg which was funded by the German Ministry of Education under identification code 03SF0417A/B. Ringo Köpke is acknowledge for providing the diamond sawn wafers.

7 REFERENCES

[1] Z. Zhao, P. Li, Y. Wei, C. Lu, X. Tan and A. Liu, „17.3 % efficient black silicon solar cell without dielectric antireflection coating," *Solar Energy 110 (2014) 714–719.*

[2] K. Füchsel, M. Kroll, T. Käsebier, M. Otto, T. Pertsch, E.-B. Kley, R. B. Wehrspohn, N. Kaiser and A. Tünnermann, „Black silicon photovoltaics," *Proc. of SPIE Vol. 8438 84380M-1 (2012).*

[3] R. S. Davidsen, J. Ormstrup, M. L. Ommen, P. E. Larsen, M. S. Schmidt, A. Boisen, O. Nordseth and O. Hansen, „Angle resolved characterization of nanostructured and conventionally textured silicon solar cells," *Solar Energy Materials & Solar Cells 140 (2015) 134–140.*

[4] H. Jansen, M. d. Boer, R. Legtenberg and M. Elwenspoek, „The black silicon method: a universal method for determining the parameter setting of a fluorine-based reactive ion etcher in deep silicon trench etching with profile control," *J. Micromech. Microeng. 5 (1995) 115-120.*

[5] H. Jansen, M. d. Boer, J. Burger, R. Legtenberg and M. Elwenspoek, „The black silicon method II: The effect of mask material and loading on the reactive ion etching of deep silicon trenches," *Microelectronic Engineering 27 (1995) 475-480.*

[6] J. Hirsch, M. Gaudig, N. Bernhard and D. Lausch,

„Optoelectronic properties of black-silicon generated through inductively coupled plasma (ICP) processing for crystalline silicon solar cells,“ *Applied Surface Science 374 (2016) 252–256.*

[7] P. B. Clapham and M. C. Hutley, „Reduction of lens reflexion by the 'moth-eye' principle,“ *Nauture Vol. 244 August 3 (1973).*

[8] D. G. Stavenga, S. Foletti, G. Palasantzas and K. Arikawa, „Light on the moth-eye corneal nipple array of butterflies,“ *Proc. R. Soc. B (2006) 273, 661–667.*

SURFACE PHOTOVOLTAGE STUDIES OF N-TYPE AND P-TYPE CRYSTALLINE SILICON PASSIVATED BY THERMAL-ALD ALUMINIUM OXIDE

Y. Sun[1], R. Jia[1], B.Q. Sun[2], X.M. Dou[2], K. Tao[1], J. Zhi[1], X.Y. Liu[1]

[1]Institute of Microelectronics, Chinese Academy of Science

Beijing 100029, China

[2]Institute of Semiconductors, Chinese Academy of Science,

Beijing 100083, China

ABSTRACT: In this work, we show that excellent surface passivation can be obtained on *n*-type FZ c-Si as well as *p*-type Cz c-Si by Al_2O_3 films deposited by thermal -ALD technique. The surface passivation provided by Al_2O_3 films strongly depends on thermal annealing process and the surface recombination as low as *5cm/s* can be achieved by properly annealed Al_2O_3. The electronic properties of Al_2O_3/Si interface have been investigated using contactless surface photovoltage spectroscopy technique. The experimental results indicate that the post-deposition anneal leads to significant increase in negative fixed charge density in the Al_2O_3 and suppression of interface states, both lead to great improvement of surface passivation. In addition, the measured surface photovoltage spectra are compared with a simplified theoretical model based on charge distribution method. The calculation results suggest that the strike different behavior of spectra between n-type and p-type samples can be explained in terms of SPV signals generated by two surface SCRs. The results also predict that the density of negative fixed charge in properly annealed Al_2O_3 is high enough to form inversion region close to the wafer surface which is attractive for the realization of novel high efficiency inversion solar cells.

Keywords: c-Si, Interfaces, Passivation, Surface photovoltage

1 INTRODUCTION

The imperative for continuous cost reductions for photovoltaic electricity requires both improvements in the solar cell energy conversion efficiencies and reductions in wafer thickness. These in turn lead to highly demanding surface passivation. Many surface passivation materials including thermal silicon dioxide (SiO2), silicon nitride (a-SiNx:H) have been extensively studied and are widely used for crystalline Si (c-Si) solar cells. In recent years, amorphous Al_2O_3 films synthesized by various techniques at low temperature, particularly atomic layer deposition (ALD), have demonstrated excellent passivation on highly doped p^+-emitters as well as on low resistivity *n*-type c-Si surfaces [1-2]. It has been observed that a post-deposition anneal at moderate temperatures is often required to active the surface passviation by Al_2O_3 films [3-5]. This thermal treatment induces substantial changes in the electronic properties of Al_2O_3/Si interfaces which have significant impacts on the performance of Al_2O_3-passviated solar cells. Therefore, a through understanding of thermal annealing effects on the properties of Al_2O_3/Si interface is necessary for better analysis and design of Al_2O_3-passviated solar cells.

In this work, the effect of thermal annealing on the surface passviation of Al_2O_3 synthesized by thermal ALD on both *n*- and *p*-type mono-crystalline silicon were studied. The interface properties of silicon substrate passviated by Al_2O_3 were investigated by using contactless surface photovoltage technique (SPV) and the obtained surface photovoltage spectra were analyzed using analytical models taking into account the existence of two space charge regions (SCR) contributing to the measured signals. The results show clearly striking differences between the surface photovoltage spectrum characteristics of *n*- and *p*-type samples passivated by Al_2O_3 films after anneal process. These are shown to be fundamentally due to the presence of high density of negative fixed charges located close to the interface with Si which has different effect on the *n*- and *p*- type Si electronic interface properties.

2 EXPERIMENTAL

The *p*-type (~300μm, <100>, 1~10Ωcm) Czochralski (Cz) c-Si substrates and *n*-type (280μm, <100>, 1~5Ωcm) float-zone c-Si substrates were used in this study. All substrates were double-sided polished and received a standard RCA clean with a final hydrofluoric acid dip prior to Al_2O_3 deposition to remove the native oxide and generate the hydrogen terminated surfaces. The identical Al_2O_3 films were then grown on both sides of wafers by thermal ALD technique. The thickness of the films was about 30nm as confirmed by spectroscopic ellipsometry measurements. After deposition, the samples underwent a post-deposition anneal at 400°C for various annealing time. This annealing temperature has been optimized to obtain the highest surface passviation level of Al_2O_3 as reported by us elsewhere [6].

In order to evaluate the surface passviation, the effective lifetime of all samples were measured using microwave-detected photoconductance decay (μ-PCD) system from Semilab. The interface properties of samples were investigated using non-destructive and contactless surface photovoltage (SPV) technique to avoid the change of charges in the films induced by the metal electrode required for conventionally capacitance-voltage measurements. The SPV signal defined as the illumination-induced change in the surface potential was measured using metal-insulator-semiconductor (MIS) technique as showed in Figure 1 [7]. The illumination-induced photovoltage which is proportional to the change in surface potential of whole sample were then measured between the two capacitor terminals. All SPV measurements were performed at same photo flux under low-level injection condition.

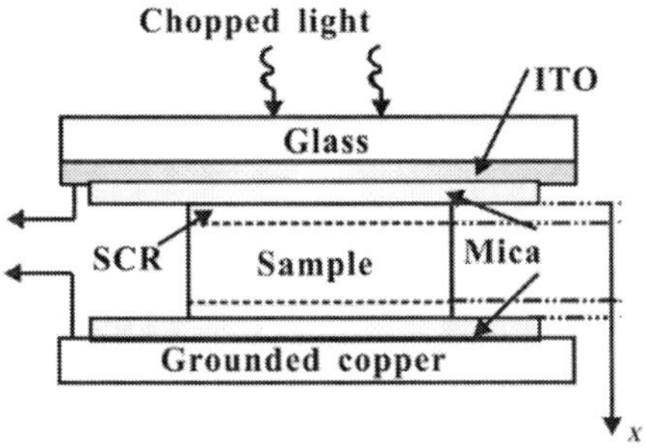

Figure 1: Schematic arrangement of the contactless surface photovoltage experiment.

3 RESULTS AND DISCUSSION

Figure 2 shows the measured effective lifetimes for investigated samples. The samples C and D are *n*-type and *p*-type control samples without passivation. The samples B and A refer to the *n*-type wafers coated by Al_2O_3 before and after 15 min annealing, respectively, while the samples E and F refer to *p*-type wafers passviated by Al_2O_3 after 2 min and 15 min annealing, respectively. The measured effective lifetime of minority carrier (6~8μs) of both *n*-type sample and *p*-type samples coated by Al_2O_3 film before annealing is close to that of control samples indicating that no passviation is provided by Al_2O_3 in the as-deposited state as commonly reported for Al_2O_3 passviation [2,8]. However, the effective lifetimes of passviated samples dramatically increase after annealing. The effective lifetime increases to 220μs after 2min annealing and reach to 340μs after 15min annealing for p-type samples. Further longer annealing time does not improve the passviation level of Al_2O_3. The maximum lifetime of 2.3ms corresponding to the effective surface recombination velocity, $S_{eff} < 5cm/s$ can be obtained on the n-type passviated sample after 15 min annealing. It should be noted that the maximum effective lifetime of *p*-type passviated sample is lower than that of n-type sample due to the lower bulk carrier lifetime limited by Sheckley-Read-Hall (SRH) recombination in the relatively low doped Cz p-type wafers, which are the main material used in the solar industry, leading to lower overall measured lifetime. Such high level surface passviation by the thermal ALD deposited Al_2O_3 films is considered as a results of the presence of large fixed negative charge in Al_2O_3 inducing strong field-effect

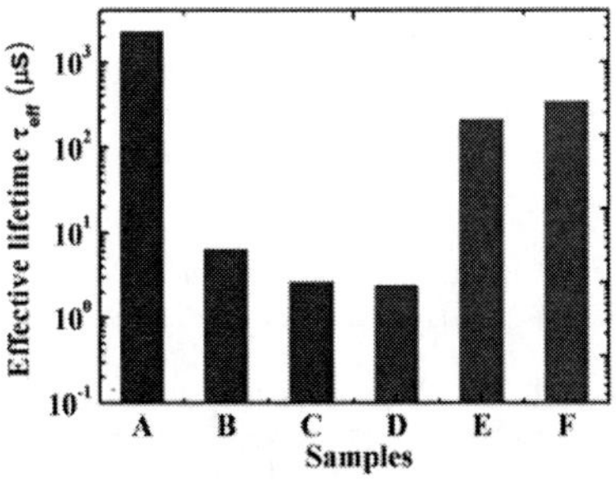

Figure 2: The measured effective minority carrier lifetime of investigated n-type and p-type samples.

passviation as dominant mechanism as proposed by Zhao *et. al.* and Hoex *et. al.* [1,9].

Figures 3 and 4 show the typical measured SPV spectra normalized to their maximum for *n*-type and *p*-type samples, respectively. For the unpassivated control

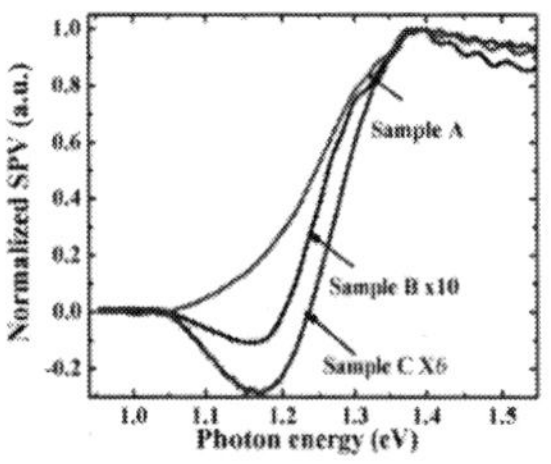

Figure 3: Typical normalized surface photovoltage spectra for *n*-type samples

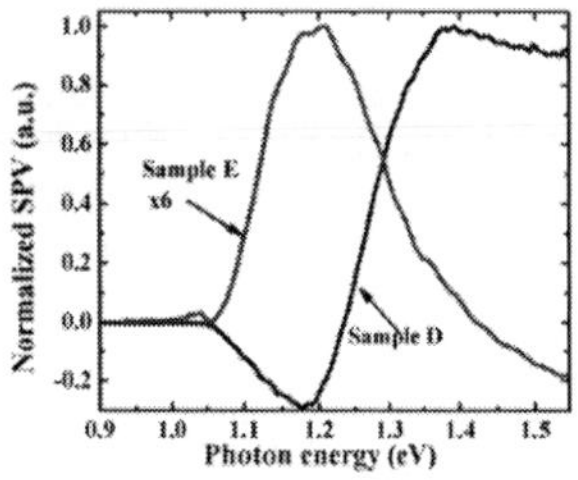

Figure 4: Typical normalized surface photovoltage spectra for *p*-type samples

samples, the inevitable presence of surface-localised states at free surface induces charge transfer between bulk and surface resulting in a surface space charge region (SCR) usually under depletion. Under illumination, the excess free carriers generated by absorbed photons via charge exchange between the valence band, surface states and conduction band are redistributed by means of the electric field within the surface SCR. Since the electric potential depends on the charge distribution via Poisson equation, the potential drops across the surface SCR changes leading to the formation of SPV. Consequently, the similar SPV spectra yielded by *n*-type and *p*-type control samples without passivation can be explained in terms of different carrier transitions. The SPV spectrum exhibits monotonic increase with increasing photon energy for the photon energy above 1.18eV and tend to be flat at high photon energy, indicating the occurrence of band-to-band transition as commonly observed in semiconductor materials [10, 11]. The SPV signal with opposite sign appears in the low photon energy range 1.047 ~ 1.17eV, strongly suggesting the electron transitions from the valence band into surface-localized states making the surface SCR more depleted instead of decrease in surface barrier caused by band-to-band transition. It can be seen from figure 3 that a large number of localized interface states present at the Al_2O_3/Si interface in sample B although the SPV signal in the low photon energy range, thus the density of gap states is reduced. However, the SPV signals due to the band-to-gap state transitions are suppressed for the sample A, and the SPV signals can be mainly attributed to the band-to-band transitions.

The spectra of n-type samples showed in figure 3 also clearly show that the maximum of SPV signal of sample A in the high photon energy range is much higher than that of sample B and C. In fact, it is almost a factor of 6 and 8 higher than maximum values of corresponding SPV in the sample C and B, respectively. In the case of surface SCR under depleted, the super-bandgap SPV signal increases with increasing initial interface band bending as reported by Munakata et. al. and Frankl et. al. [12, 13]. Therefore, the observed high SPV signal from sample A is expected as a result of enhanced interface potential barrier induced by a large fixed negative charge in Al_2O_3 located close to the interface. The lower SPV signal in sample B relative to unpassivated substrate strongly suggests that the density of fixed negative charge in thermal ALD Al_2O_3 before annealing is low and the lower initial band bending is caused by the reduced density of net negative interface charge captured by gap states.

Since the presence of fixed negative charge in Al_2O_3 tends to reduce the potential barrier of depleted interface, the SPV signal of annealed passviated p-type samples is expected to be lower than that of control sample as showed in figure 4. For passivated sample (sample F) with sufficient anneal time (15min), the density of fixed negative charge in Al_2O_3 is increased and large enough to flattening the interface potential or to make the interface in an accumulation region. This causes the SPV signal too low to be detected.

These above observations strongly suggest that the excellent passivation offered by annealed Al_2O_3 is attributed to strong field-effect passviation combining the low density of interface states. What is interesting here is that the SPV spectrum of sample E exhibits quite different behavior compared to that of sample A. It shows much high SPV signal in the low photon energy range relative to the signal in the high photon energy range. Such high long wavelength signal is unlikely induced by band-to-gap state transitions because the sub-bandgap SPV signal due to this mechanism decreases with both decreasing interface state density and surface barrier [14]. It should be noted that the absorption depth in silicon can be larger than the thickness of investigated wafer (~300μm) when the wavelength of incident light is longer than 1000nm. Therefore, the photo-generated carriers can appear at the bottom surface of sample which also contribute to the super-bandgap SPV signal via bottom SCR as supposed by Toušek et. al. [15]. As a result, the observed overall SPV signal in sample E comes from both top SCR and bottom SCR and is dominated by the signal generated at the bottom SCR as we show in the following section.

In order to evaluate the measured surface photovoltage spectra, a simplified analytical model based on charge redistribution method taking into account the super-bandgap SPV generated at both top SCR and bottom SCR was used. Considering the fact that lateral dimensions of investigated samples are much larger than the wafer thickness, one-dimensional analysis therefore was developed. To obtain the analytical solution for the photovoltage, we adopt Toušek's treatment to evaluate the SPV signal by calculating photocurrent to avoid the difficulty in calculation of the excess free carrier concentration at the surface. For simplicity, we assume that photocurrent injected into the SCR from bulk is not influenced by the recombination in the SCR in the steady-state, thus the photocurrent generated from bulk and SCR can be calculated independently. In addition, the depletion approximation is used in our model and the photocurrents generated in the SCR are neglected for simplicity. The photocurrent from the bulk can be obtained by solving the continuity equation for the minority carriers injected into the SCR. Considering a p-type sample in which the minority carrier is electron, the continuity equation for the electrons under low-level injection condition becomes [7]:

$$D\frac{d^2\delta n(x)}{dx^2} - \frac{\delta n(x)}{\tau} = -g(x) \quad (1)$$

where the subscript n is omitted for D and τ for brevity and the equation can be also used for holes in the n-type samples. Here, δn is the concentration of the excess electrons at a depth x under the illuminated surface, D and τ are the electron diffusion coefficient and lifetime of electrons, respectively. Assuming that the excess carrier density is reduced to zero by the electric field in the SCR, the continuity equation for both top and bottom SCR then can be solved with two boundary conditions:

$$\delta n(w) = 0 \text{ and } \delta n(w + d) = 0 \quad (2)$$

where w is the thickness of surface charge region as shown in figure 1.

The normalized SPV spectra have been calculated for well passivated sample A and E. In figure 5, we show the calculated SPV signals normalized to their maximum as a function of wavelength for sample E. The experimental data is also showed in figure for comparison. The calculated photovoltage fits quite well to the experimental data. The simulation results strongly suggest that the overall SPV signal caused by band-to-band transition in both sample A and E consist of two components having opposite sign. The long wavelength part comes from bottom SCR while the short wavelength part comes from top SCR. The shape of the SPV spectrum strongly depends on the excitation level of each SCR and can be dominated by bottom SCR when

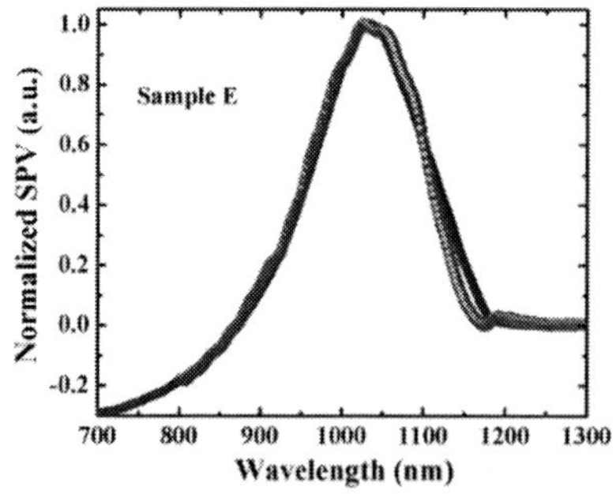

Figure 5: Calculated normalized surface photovoltage spectrum for sample E (solid line). The experimental data in solid circles are also showed in the figure for comparison.

the bottom SCR is under strongly excited as observed in sample E. The SPV signals in sample A is dominated by top SCR and the contribution from bottom SCR is small. In addition, the initial potential barrier value was taken as 0.8eV for the best fit in sample A which indicates that the high negative fixed charge density in annealed Al_2O_3 drives the top surface SCR in reversion regime. Unlike sample A, the SPV spectrum of sample E is less sensitive to the initial band bending. These results are in accord with the low SPV signals measured from sample E.

4 CONCLUSIONS

In conclusion, we have shown that excellent surface passivation can be obtained on *n*-type FZ c-Si as well as *p*-type Cz c-Si by Al_2O_3 films deposited by thermal-ALD technique. The passivation property of Al_2O_3 films strongly depends on thermal annealing process and the surface recombination as low as can be achieved by properly annealed Al_2O_3. The electronic properties of Al_2O_3/Si interface have been investigated using contactless surface photovoltage spectroscopy technique for *n*-type and *p*-type samples. The results clearly indicate that the density of negative fixed charge in the as-deposited Al2O3 film is low. Our thermal ALD as-grown Al2O3 films do not provide passviation although the density of interface states is somewhat reduced. The post-deposition anneal introduces significant changes in the electronic properties including the presence of a large number of negative fixed charges and the suppression of interface states. Therefore, the great improvement of the Al2O3 surface passivation can be mainly attributed to the field-effect passivation induced by high negative fixed charge density and to take advantage of low interface states. Furthermore, the SPV spectra of well-passivated n-type and p-type samples have been analysed by simplified analytical model based on charge redistribution method taking two surface SCRs into account. The results indicate that the density of negative fixed charge in thermal-ALD Al_2O_3 is high enough to make the *n*-type sample surface in reversion regime which can be used to develop novel inversion-layer solar cells.

ACKNOWLEDGEMENTS

This work was supported by Chinese NSF under Grant Nos. 11104319, 11274346, 512022285, 61234005 and 51172268; and by Chinese Academy of Science Solar Energy Action Plan under Grant Nos. Y1YT064001, Y1YF034001, and Y2YF014001.

References

[1] B. Hoex, J.J.H. Gielis, M.C.M. van de Sanden and W.M.M. Kessels, J. Appl. Phys. 104 (2008) 113703.
[2] J. Schmidt, B. Veith and R. Brendel, Phys. Status Solidi RRL 3 (2009) 287.
[3] G. Dingemans, N.M. Terlinden, D. Pierreux, H.B. Profijt, M.C.M. van de Sanden and W.M.M. Kessels, Electrochem. Solid-State Lett. 14 (2001) H1.
[4] F. Kersten, A. Schmid, S. Bordihn, J.W. Müller and J. Meitmann, Energy Procedia 38 (2013) 843.
[5] S. Bordihn, I. Kiesow, V. Mertens, P. Engelhart, J.W. Müller and W.M.M. Kessels, Energy Procedia 27 (2012) 396.
[6] Y. Sun, J. Rui, S. Bin, J. Zhi and X.Y. Liu (to be published).
[7] L. Kronik and Y. Shapira, Surf. Sci. Rep. 37 (1999) 1.
[8] G. Dingemans, F. Einsele, W. Beyer, M.C.M. van de Sanden and W.M.M. Kessels, J. Appl. Phys. 111 (2012) 093713.
[9] Y. Zhao, C.L. Zhou, X. Zhang, P. Zhang, Y.N. Dou, W.J. Wang, X.Z. Cao, B.Y. Wang, Y.H. Tang and S. Zhou Nanoscale Res. Lett. 8 (2013) 114.
[10] N. Kinrot, Y. Shapira and M.A.B. de Moraesces, Appl. Phys. Lett. 70 (1997) 3011.
[11] E. Fefer, Y. Shapira and I. Balberg Appl. Phys. Lett. 67 (1995) 371.
[12] C. Munakata and H. Shimizu, Semicond. Sci. Technol. 15 (2000) 40.
[13] D.R. Frankl and E.A. Ulmer Surf. Sci. 6 (1996) 115.
[14] L. Szaro, Appl. Phys. A 29 (1982) 201.
[15] J. Toušek and J. Toušková J Sol. Energ. Mat. Cells 92 1020.

SURFACE PASSIVATION BY ALO$_X$ DEPOSITED IN AN INDUSTRIAL LOW FREQUENCY PECVD EQUIPMENT

R.Monna*[1], B.Semmache[2], S. Tran[2], A.Veau[1], G.Lazzarelli[2], L.Bounaas[3] and S.Dubois[1]
[1]CEA, INES, 50 avenue du Lac Léman, BP332, F-73370, Le Bourget du Lac, France
[2] SEMCO Technologies, 625 rue de la Croix Verte, F-4196 Montpellier Cedex 5, France
[3] ECM Greentech, 109 rue Hilaire de Chardonnet, F-38100 Grenoble, France

*Corresponding author: Email: remi.monna@cea.fr, Phone: +33 479 792 931

ABSTRACT: In this study, AlO_x layers have been deposited using an industrial-type PECVD Twyn™ system from the SEMCO Technologies Company. Excellent passivation properties were obtained with i-Voc over 700 mV (SRV < 20 cm/s) demonstrated on symmetrical passivated p-type wafers covered with AlO_x/SiN_x stacks. These values are very close to those obtained with the AlO_x layers deposited by ALD and meet the quality requirements for the backside passivation of p-type PERC cells. We have demonstrated that the post annealing of the AlO_x layer has a very weak influence on the passivation quality of SiN_x capped AlO_x layer, thus the SiN_x capping layer can be deposited in the same run as the AlO_x layer leading to significant equipment investment and footprint reduction compared to ALD systems.

1 INTRODUCTION

The Al-BSF technology reached maturity with conversion efficiency between 19.0% and 19.5 % in production on Czochralski-grown p-type silicon wafers. In order to reach higher efficiencies, the rear surface of the cell needs to be optimized by applying a passivation layer between the silicon and the rear metallization. This passivation layer enhances the rear light reflection and reduces strongly the free carrier's recombination at the back surface. Applying a silicon nitride (SiN_x) layer would not be a viable option for the rear surface of p-type silicon cells. While ideally suited to n-type surfaces due to the higher density of fixed positive charges of up to $10^{13}cm^{-3}$, it creates an inversion layer when applied to p-type wafers. This phenomenon known as parasitic junction formation, in which the inversion layer touches the metal contacts on the rear surface, leads to significant losses in short-circuit current density and fill factor. In contrast aluminum oxide (AlO_x) nicely fills the gap left by the SiN_x layer. A very high fixed negative charges density up to 10^{13} cm^{-3} located near the AlO_x-silicon interface enables it to provide excellent field-effect passivation on p-type substrates [1]. AlO_x also acts as an effective hydrogen reservoir, supplying hydrogen to passivate the dangling bonds on the silicon surface during the subsequent annealing step. Nonetheless, the possibility to deposit very thin AlO_x films enables flexibility in designing passivation schemes by stacking the material with other films.

Aside from the issue of costs and quality, aluminum oxide films for rear surface passivation must be sealed, in most case with a silicon nitride layer. While PECVD (Plasma-Enhanced Chemical Vapor deposition) is a well-established technology for the deposition of silicon nitride antireflection coating, it can also be employed for depositing a capping layer on the rear side. The PECVD tools, also based on trimethylaluminum (TMA) as precursor, have a definite advantage in that both aluminum oxide capped with silicon nitride can be deposited simultaneously in one process without any need to unload and reload the substrates.

For the moment only one equipment supplier [2] has demonstrated that a direct plasma deposition system with low RF frequency (40 kHz) is able to achieve high passivation quality with AlO_x layers. The SEMCO's PECVD configuration has many advantages compared to other PECVD systems as; low gas consumption, possibility to clean externally the graphite boats which increases hugely the uptime of the equipment, passivation and capping layer deposited in one run within the same tube with a total process time within 45-50 min which leads to a low Cost-of-Ownership.

Figure 1 shows a typical PECVD Twyn™ industrial platform in which layers are deposited. An important feature of the SEMCO's PECVD tools is that the wafers lie flat on the graphite boat in the reaction chambers. This configuration allows easier wafer handling and no pin marks on the wafer edges. The wafer transport is fully automated from cassettes or inline conveyors to the tubes.

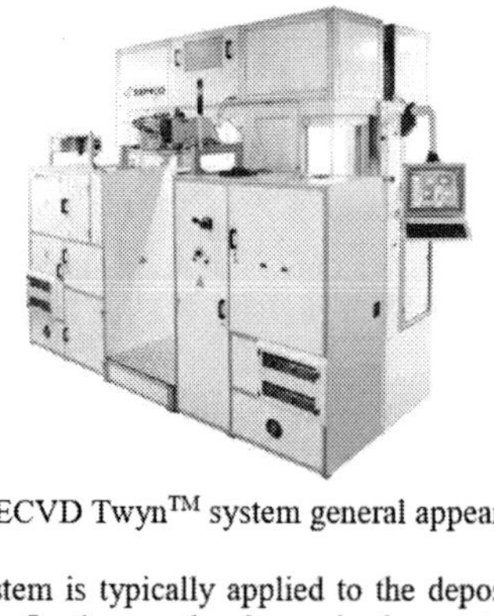

Figure 1: PECVD Twyn™ system general appearance.

This system is typically applied to the deposition of SiN_x:H antireflection coating layers in the crystalline Si PV industry with a throughput of about 2000 wafers per hour.

This paper focuses on the AlO_x deposition process using an industrial-type PECVD Twyn™ system from SEMCO. The development and the properties of these AlO_x layers are exposed and compared to AlO_x layers deposited in a lab scale ALD equipment.

2 EXPERIMENTAL APPROACH

To perform the aluminum oxide layer deposition, the wafers are placed in a vertical graphite boat as displayed in figure 2. One graphite boat consists of up to 70 graphite plates which can hold up to 4 - 156 mm ×156 mm - silicon wafers. Therefore the capacity of one boat is up to 280 wafers. For the AlO_x deposition, a gas mixture comprising nitrous oxide (N_2O) and TMA was used. For the deposition of the silicon nitride (SiN_x) capping layer, the gases ammonia (NH_3) and silane (SiH_4) are injected in the tube. Low-frequency plasma (40 kHz) is directly generated and confined to the horizontal parallel graphite plates supporting the wafers. For the determination of the surface passivation properties, symmetric test structures were processed, with passivation layers on both sides of 2- 3 Ω.cm Cz chemically polished p-type silicon wafers. The surface passivation quality of the deposited layers were compared with our standard 6 nm thick AlO_x layer deposited by ALD (BENEQ equipment) and followed either by a forming gas annealing at different temperatures or a PECVD deposition of a SiN_x capping layer needed to activate the charges.

Figure 2: PECVD Twyn™ system graphite boat vertical loading.

Then, all the test structures were fired in a conventional belt furnace which is used for contact firing. Carrier lifetime measurements were carried out on a Sinton WCT-120 equipment. All results are given for an injection level of 3×10^{15} cm^{-3}. Finally, physical characterizations were performed on the different layers with infrared spectroscopy in attenuated total reflection mode (ATR-FTIR) and ellipsometry measurements.

3 RESULTS

3.1 Thickness and refraction index measurements

Figure 3 indicates the refraction index and the extinction coefficient as a function of the incident wavelength for the AlO_x layer deposited by PECVD. We obtain a refractive index of 1.63 at 633 nm for the PECVD deposited layer and 1.65 for the ALD deposited layer. It means that only light with an incident angle lower than 26° can be transmitted through the rear layer. The deposited PECVD AlO_x layer is therefore well suited for optical applications on the rear surface of a PERC solar cell. Both layers have an extinction coefficient equal to zero for wavelengths between 300 nm and 1200 nm resulting in no (benign) light absorption.

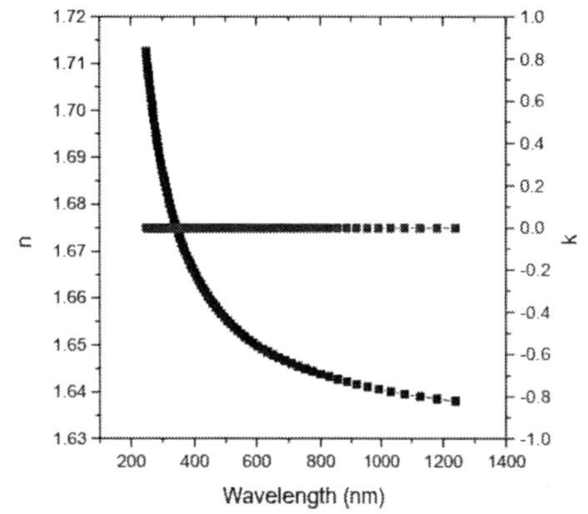

Figure 3: Refractive index and extinction coefficient as a function of the incident wavelength for the AlO_x layer deposited by PECVD.

Table I: Deposition rate, refraction index and absorption coefficient @633 nm.

Deposition rate (nm/s)	n@633nm	k@633nm
0.53	1.63	0

Depending on the deposition parameters (gas ratio and deposition temperature), the refractive index can be tuned but due to the slightly lower film density of PECVD AlO_x layers, the index is generally somewhat lower than the index of ALD deposited layers.

3.2 Development of the AlO_x/SiN_x stack elaborated by PECVD

Different AlO_x/SiNx passivation stacks were developed and studied by changing deposition parameters. AlO_x and SiN_x layers were intentionally deposited in two separated process chambers in order to make a relevant comparison with ALD AlOx deposited layers and to study the influence of the post-deposition annealing step. In the future, both layers will be deposited in the same chamber to reduce the equipment investment cost and footprint.

Figure 4 indicates the i-Voc reached on 3 different AlO_x/SiN_x stacks deposited by PECVD. The AlO_x layers were optimized by changing the deposition parameters such as gas ratio, plasma power and deposition temperature. The SiN_x capping layer has a typical thickness of 100 nm. After deposition, the wafers undergo a firing step treatment in an infrared belt furnace. This step is the same as the one used to fire the metallization of our PERC solar cells.

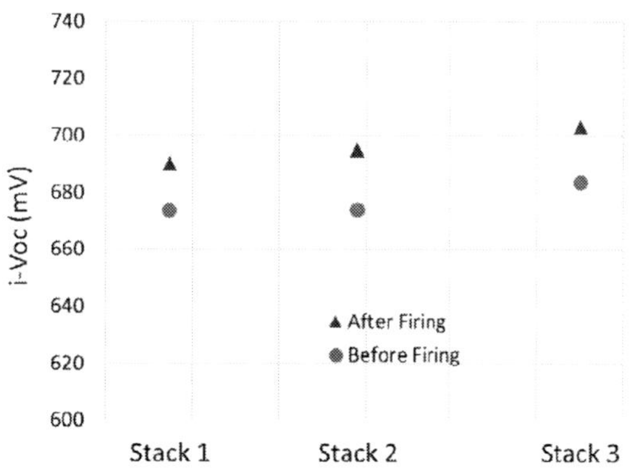

Figure 4: i-Voc obtained on different AlO_x/SiN_x passivation stacks deposited by PECVD. (Sinton measurements in transient mode)

Firstly, it can be observed that the passivation quality is increased with the firing step due to a combined effect of higher annealing step which enhances the AlO_x chemical and field effect surface passivation and the hydrogen passivation of the AlO_x – silicon interface.

Secondly, high i-Voc (> 690 mV) can be reached with all three stack layers demonstrating the possibility to use a wide range of deposition parameters.

i-Voc of 703 mV is obtained on the best passivation stack which is high enough for the backside passivation of p-type PERC cells.

3.3 Influence of the post annealing step on PECVD deposited AlO_x layer

Initial tests were performed to study the influence of the post deposition annealing temperature of the AlO_x layers on the passivation quality of AlO_x/SiNx Stacks.

We observe an improvement of the passivation quality of the PECVD and ALD deposited AlO_x layers with the annealing temperature. The interfacial oxide (SiO_x) that is formed between the c-Si and the AlO_x layer plays an essential role in the surface passivation properties [3]. The interface quality and related surface passivation properties are improved during the high temperature annealing steps. The AlO_x material properties are expected to influence both chemical and field-effect contributions to the surface passivation performance of the films. The fixed negative charge density Q_f, that increases drastically with the annealing temperature, induces the field effect passivation and is expected to be closely related to the AlO_x structural properties near the interface [4].

Figure 5 indicates the i-Voc obtained after SiN_x capping and firing of AlO_x layers deposited by PECVD and annealed at different temperatures. After SiN_x capping, we observe a slight increase of the i-Voc values mostly on the low temperature annealed AlO_x layers. An increase is observed again after the firing step due to the hydrogen coming from the AlO_x and from the SiN_x layers.

Excellent passivation properties were obtained with i-Voc over 700 mV demonstrated on symmetrical AlO_x/SiN_x passivated p-type wafers. Considering a fixed bulk carrier lifetime of 1 ms, an effective surface recombination velocity SRV below 20 cm/s could be extracted from lifetime measurements.

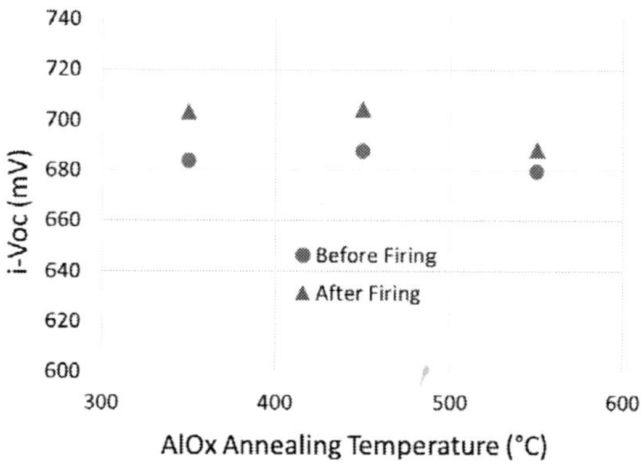

Figure 5: Influence of the post AlO_x deposition annealing steps on the i-Voc of symmetric structures with the AlO_x /SiN_x passivation stacks when the AlO_x is deposited by PECVD (Sinton measurements in transient mode)

The influence of the post AlO_x annealing step is very limited but it can be noticed that best i-Voc performances are obtained for low annealing temperatures. The temperature range required for surface passivation optimization is therefore quite wide and is in accordance with the SiNx deposition process conditions.

3.4 Influence of the post annealing step on ALD deposited AlO_x layers

Figure 6 indicates the i-Voc obtained after SiN_x capping and firing of AlO_x layers deposited by ALD and annealed at different temperatures. It can be seen that the post AlO_x deposition annealing step has no influence on the passivation quality of SiN_x capped AlO_x layers. We can also observe that the firing step has less influence on the i-Voc than on the SiN_x capped AlO_x layers deposited by PECVD.

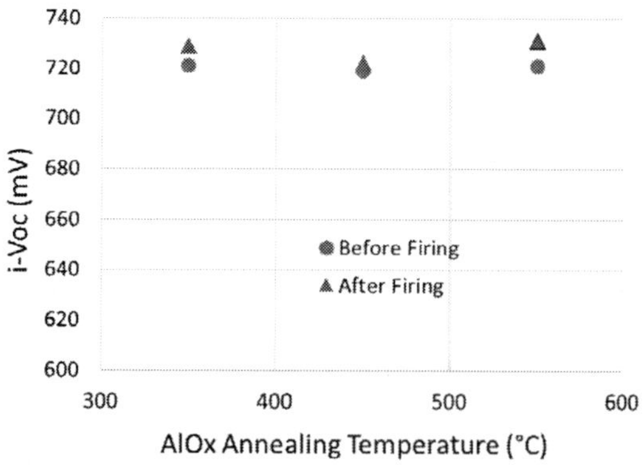

Figure 6: Influence of the post AlO_x deposition annealing steps on the i-Voc of symmetric structures with the AlO_x/SiN_x passivation stacks when the AlO_x is deposited by ALD. (Sinton measurements in transient mode)

3.5. ATR-FTIR analysis

Figure 7 shows the physical characterization that was performed on the different layers by infrared spectroscopy in attenuated total reflection mode (ATR-FTIR). FTIR analyses are usually used to obtain information on the composition and the structure of solid mater. The absorption spectra strongly depends on the interactions between the different atomic or molecular species within the solid layer.

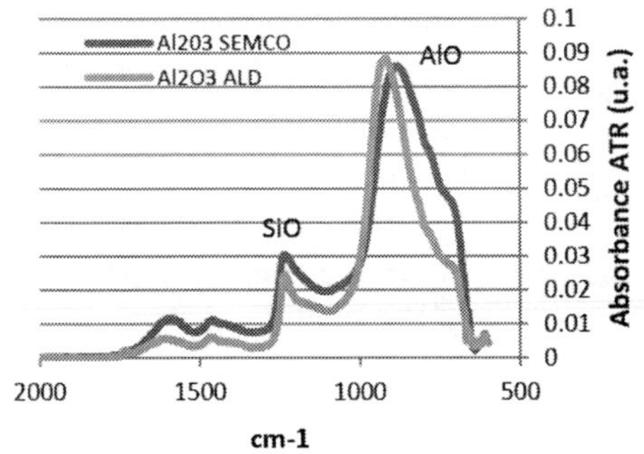

Figure 5: Physical characterization (absorbance versus the wave number) that was performed on the different layers with infrared spectroscopy in attenuated total reflection mode (ATR-FTIR).

As shown in Figure 7, the main peak is located at 900–950 cm^{-1} and corresponds to the LO mode (longitudinal optic) of the Al–O bond. The second peak located around 1240 cm^{-1} corresponds to the Si–O bond (LO mode). It can be noticed that the SEMCO's layer present a higher interfacial oxide regrowth than the layer deposited by ALD. This oxide thickness can be related to the AlO_x thickness.

Concerning the main Al-O peak, two populations can be distinguished based on the peak position and the full width half maximum (FWHM). Large ATR peaks at 900 cm^{-1} and 780 cm^{-1} for the layer deposited by PECVD and a narrower ATR peak at 920 cm^{-1} for the layer deposited by ALD can be observed. According to B. Lefez & al [5], the position of the absorption peak at 920 cm^{-1} corresponds to a α- Al_2O_3, the peaks at 930 cm^{-1} and 780 cm^{-1} correspond to a γ- Al_2O_3.

The FWHM corresponds to the order of the Al-O bonds in the aluminum oxide layer. Therefore the ALD deposited film with a narrower peak is more ordered than the PECVD deposited film suggesting an amorphous and a crystalline population.

At the moment it is not clear if the difference in passivation performances could be attributed to the difference of crystalline structure of the AlO_x layer or to a difference in terms of electric field or hydrogen content at the silicon-passivation layer interface. More investigations are underway to clarify this point.

4 CONCLUSION

The industry has been relying on the successful screen printing Al-BSF technology for the last few decades, but it is expected that PERC solar cells featuring dielectric rear side passivation will become the predominant technology in the next few years. AlO_x layer is an excellent candidate for the rear side passivation of p-type cells due to high surface passivation properties and the absence of an inversion layer. In this study, AlO_x layers have been deposited using an industrial-type PECVD Twyn™ system from the SEMCO Technologies Company. In this equipment wafers are placed horizontally on 70 graphite plates which can hold up each to 4 - 156 mm x 156 mm - silicon wafers.

Excellent passivation properties were obtained with i-Voc over 700 mV (SRV<20 cm/s) demonstrated on symmetrical passivated p-type wafers covered with AlO_x/SiN_x stacks. These values are very close to those obtained on the AlO_x layers deposited by ALD (lab-scale equipment in this study) and are high enough for the backside passivation of p-type PERC cells.

We have demonstrated that the post annealing of the AlO_x layer has a very weak influence on the passivation quality of SiN_x capped AlO_x layers, thus the SiN_x capping layer can be deposited in the same run as the AlO_x layer leading to significant equipment investment and footprint reductions compared to ALD systems.

5 ACKNOWLEDGEMENTS

The authors want to thank Dr. N. Rochat and Dr. A. Chabli for the ATR-FTIR measurements and fruitful discussions.

6 REFERENCES

[1] G.Dingemans & al., Status and prospects of Al2O3-based surface passivation schemes for silicon solar cells, J.Vac.Sci.Technol.A 30(4), Jul/Aug 2012.

[2] Tong Zhou and al, Different Dielectrics deposited by PECVD for the Industrial Manufacturing of PERC Solar Cells, 31st European Photovoltaic Solar Energy Conference and Exhibition, Hambourg, 2015

[3] B. Hoex, J.J.H. Gielis, M.C.M. van de Sanden and W.M.M. Kessels, J. Appl. Phys,. 104, 073701 (2008)

[4] G.Dingemans, R.Seguin, P. Engelhart, M.C.M. van de Sanden and W.M. Kessels, Phys. Status Solidi (RRL), 4, 10 (2010)

[5] B.Lefez & al, Fourier transform infrared reflection absorption spectroscopy and microspectroscopy, a tool to investigate thermally grown oxide scales, Thin Solid Films 516 (2007) 349–354

SURFACE PASSIVATION OF C-SI USING SILICON OXYNITRIDE - ACCENTUATING THE THERMAL STABILITY BY SILICON NITRIDE CAPPING LAYER

Anishkumar Soman, Aldrin Antony*
Department of Energy Science and Engineering, Indian Institute of Technology Bombay,
Mumbai - 400076, India.
*phone: +91 2225764400, Fax: +91 2225723480, email: aldrinantony@iitb.ac.in

ABSTRACT:
Excellent surface passivation plays a pivotal role in increasing the open circuit voltage of high efficiency c-Si solar cells. In this work we demonstrate hydrogenated Silicon Oxynitride (SiON) as a passivation layer for both p & n type crystalline silicon wafers. We have also investigated the importance of Silicon Nitride (SiN) capping layer which provides the SiON films better thermal stability and prevents degradation of SiON films and characterization were performed with the help of quasi steady state photoconductance measurement and Fourier Transform Infrared Spectroscopy (FTIR). The single layer SiON gives an implied V_{OC} of 626mV on p type and 639mV on n type CZ wafers and degrades after annealing whereas SiON/SiN stack gives an implied V_{OC} of 656 mV and improves to 673 mV for p type CZ wafers after annealing. We have obtained surface recombination velocity of < 50 cm/s for SiON/SiN layers. The degradation of SiON single layer after annealing and the improved performance of SiON/SiN bilayer passivation have been explained with the help of Si-H bond analysis from FTIR.
Keywords: Passivation, PECVD, Silicon Oxynitride

1 INTRODUCTION

Photovoltaics is now moving towards thinner wafer for crystalline silicon (c-Si) solar cells to reduce material cost; where one of the major constrain is to reduce the recombination losses at the c-Si interface which affects the open circuit voltage (V_{OC}) of the solar cell. Therefore passivation of the recombination active interface states is pivotal for high efficiency solar cells. The conventional commercial crystalline silicon solar cells use Aluminium Back Surface Field (Al-BSF) which not only increases the losses due to recombination at the interface but also causes bowing of the cell [1]. Therefore the way ahead is to replace the Al-BSF with passivating dielectric layers which reduce the rear surface recombination as well as acts as a back reflector. The current state of the art Si_3N_4 layer which passivates the emitter cannot be used as the rear surface passivation layer since it forms an inversion layer with p type silicon [2]. Alternately thermally grown silicon dioxide (SiO_2) is considered as an excellent passivation layer at the back surface for PERC solar cells [3]. However a high temperature process (900 – 1100 °C) and low deposition rates limits its application in commercial silicon solar cells. Aluminium Oxide (Al_2O_3) is considered as another option for passivation of p type silicon [4] however large scale implementation issue has limited its application.

Recently there has been an increased research interest in Silicon Oxynitride (SiON) for passivation in photovoltaics [5][6] due to multiple reasons. SiON has a low refractive index between 1.46 to 1.95, making it an effective back reflector. SiON can find application in commercial solar cells since it can be deposited by industrially viable CVD and PVD techniques [7]. However in order to find application in commercial solar cells it is very important to understand the thermal stability of these passivation layers since the conventional solar cells undergo high temperature process like firing and annealing for formation of metal contacts. Therefore, a detailed study on the passivation by SiON layers along with their thermal stability is presented in this paper. It has been observed that adding a Silicon Nitride (SiN) capping layer over SiON layer can reduce the degradation of the passivation layers during high temperature treatments. An insight of the passivation mechanism of SiON has also been explained based based on experimental results.

2 EXPERIMENTAL

2.1 Fabrication

The SiON and SiN films are fabricated using rf-PECVD system Plasmalab 100 by Oxford Instruments. The SiON film is deposited using Silane (SiH_4), Nitrous Oxide (N_2O) and Hydrogen (H_2) as the precursor gases whereas the SiN film is deposited using SiH_4 and Ammonia (NH_3). All the films are deposited at a chamber pressure of 700 mTorr and temperature of 380 °C. For SiON films the SiH_4 to H_2 gas ratio is maintained fixed at 0.75. Table I below gives the volumetric gas flow of different composition of SiON used in this work along with the deposition rates.

Sample Name	SiH_4 flow (sccm)	N_2O flow (sccm)	Deposition rate (nm/s)
SiON_0.75	15	45	0.4
SiON_0.8	15	60	0.6
SiON_0.9	15	135	0.9

Table I: Gas flow ratio of SiH_4 and N_2O for different composition SiON along with the deposition rates

For SiN the ratio of NH_3 to SiH_4 is 0.8. All the depositions are done at a power density of 60 mW/cm^2 to reduce plasma ion bombardment energy. Before PECVD process, all the Si wafers are RCA cleaned and the samples undergo a 2% HF dip before loading into the chamber. The thickness of the single SiON layer is kept fixed at 100 nm and for stack of SiON/SiN the thickness of SiON is 30 nm whereas SiN is 70 nm. The annealing experiments are done in quartz furnace with an N_2 flow of 1.5 slpm for 20 minutes at a temperature of 380 °C.

The firing of the samples is done using RTP system AccuThermo AW 610 at a temperature of 850 °C ramped up in 7 seconds from 300 °C simulating the same thermal cycles for the screen printed contact metallization of the c-Si solar cell fabrication.

2.2 Characterization

In order to study the passivation quality of the films, the minority carrier lifetime and Implied V_{OC} (iV_{OC}) of the samples are measured using Quasi steady state photoconductance (QSSPC) measurement using Sinton WCT 120. For the annealing experiments, P-type 250 μm thick, DSP Si wafers having <100> orientation, resistivity of 4-7 Ω.cm CZ wafers were used. Likewise for firing experiments N type 250 μm thick, DSP Si wafers having <100> orientation, resistivity of 4-7 Ω.cm CZ wafers were used. The surface recombination velocity (SRV) was calculated from the effective lifetime (τ_{eff}) of samples deposited on FZ wafers assuming the FZ wafers have infinite bulk lifetime using the equation below [8]:

$$SRV = W/(2.\tau_{eff}) \quad (1)$$

Where, W is the wafer thickness and τ_{eff} is measured at an injection level of 10^{15} cm^{-3}. For all the lifetime measurements SiON and SiON/SiN are deposited on both the sides for symmetrical SRV on both sides of the wafer. The bond composition of the SiON layers are understood using Fourier Transform Infrared spectroscopy (FTIR) by Perkin Elmer Spectrum 100 system with DTGS detector. The measurements are done in absorbance mode with 64 scan repetitions.

3 RESULTS AND DISCUSSION

3.1 Passivation quality of SiON and SiON/SiN stack

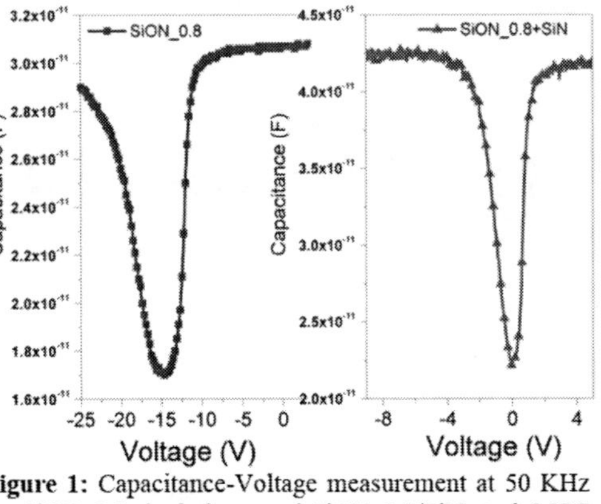

Figure 1: Capacitance-Voltage measurement at 50 KHz for SiON_0.8 single layer and SiON_0.8/SiN stack MOS capacitors

SiON layers have been used used as passivation layer for p and n type silicon [9]. Unlike SiN the passivation mechanism of SiON is relatively less explored with limited literature available. It has been reported that SiON like SiN has fixed positive charges [10]. Due to its inherent positive fixed oxide charges, it forms an inversion with p type and accumulation with n type silicon. The nature of fixed charges in the SiON layer can be verified by taking the CV measurements of MOS capacitors made with SiON and SiON/SiN dielectric layers. Fig. 1 shows the CV curve of a single SiON layer MOS cap with a considerable negative shift in the flatband voltage, which indicates the presence of positive fixed oxide charges. This infers that SiON films fabricated do exhibit field effect passivation as reported previously. However it is interesting to note that unlike single layer SiON, SiON/SiN stack has a much lesser negative flatband voltage.

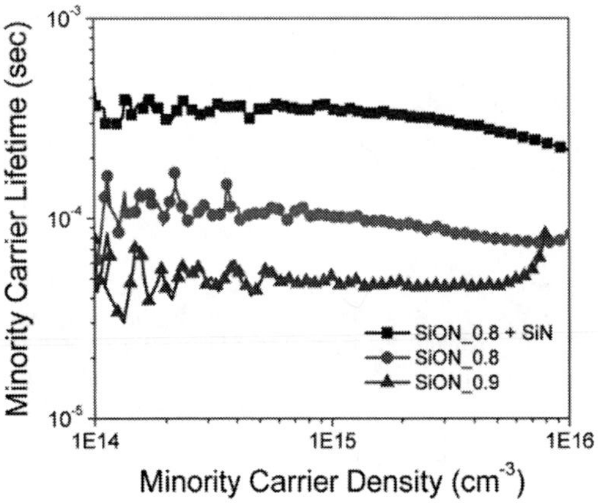

Figure 2: Minority carrier lifetime of SiON_0.8, SiON_0.9 and SiON_0.8/SiN across different injection levels

The minority carrier lifetime of the SiON was measured by depositing the films on both sides of DSP FZ p type having 1-3 Ω.cm resistivity. At an excess carrier density (Δn) 10^{15} cm^{-3}, SiON_0.8 and SiON_0.9 has an effective lifetime of 104 μs and 50 μs respectively. The SiON_0.8/SiN stack has an effective lifetime of 355 μs. Thus we can see that the stack has higher τ_{eff} as compared to single layer SiON. A detailed minority carrier lifetime results across different injection levels for single layer and stack layer is given in Fig. 2.

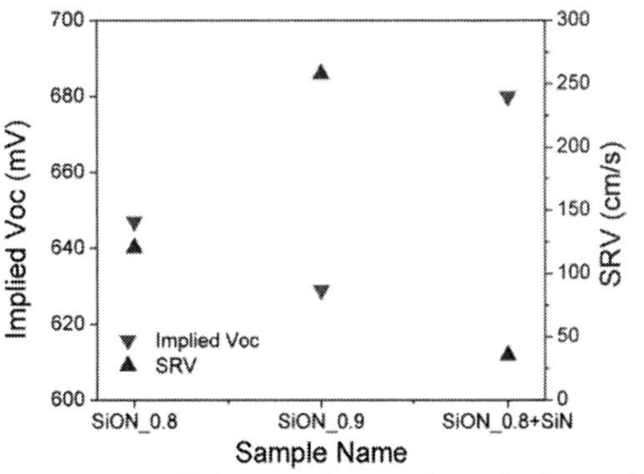

Figure 3: Implied V_{OC} and SRV values of SiON and SiON/SiN passivation layers

In order to find application in rear passivation it is important to know the SRV values of these films. The graphical illustration of SRV and Implied V_{OC} of SiON and SiON/SiN films are shown in Fig. 3. Single layer SiON_0.8 and SiON_0.9 has SRV values of 120 cm/s and 257 cm/s respectively. The 1 sun illuminated V_{OC} of these films are 647 mV and 629 mV respectively. The SRV considerably decreases for SiON_0.8/SiN to as low as 35 cm/s. These low SRV values make SiON a competent passivation structure for rear surface. The SiON/SiN stack gives a high implied V_{OC} of 680 mV.

3.2 Thermal Stability of SiON and SiON/SiN stack

a) FTIR analysis

Thermal stability of single layer SiON films has been analyzed using FTIR measurements. Fig. 4 gives the infrared spectra of SiON_0.8 and SiON_0.9 before and after firing. We observe that as deposited SiON films show characteristic Si-N peak and Si-O peak at 850 cm^{-1} and 1040 cm^{-1} respectively [11]. It also presents a $Si\text{-}H_x$ peak ~2100 cm^{-1} which shows the presence of $Si\text{-}H_2$ and $Si\text{-}H_3$ in the films stating that the films are hydrogenated in their as deposited state. However after firing the samples we observe that the $Si\text{-}H_x$ peak disappears which means that the high temperature cycle causes breakdown of Si-H bonds. It was also observed that the Si-N bond intensity decreases after firing, however the Si-O bond intensity remains the same. These observations could be explained by the fact that Si-H bonds are very weak since they have low bond energy of 3.3 eV [6]. Hence a high temperature process causes the Si-H bonds to break thereby creating defect states and increase the Si dangling bonds thereby severely affecting the passivation property of the layer. The Si-O bond is thermodynamically more stable than Si-N and Si-H [6] hence it remains unaffected. The Si-N bond has bond energy of 3.7 eV [6] hence there is only a partial breakage of the bonds. The same observations are also obtained for annealed samples; we can therefore conclude from FTIR results that the passivation quality of hydrogenated SiON films is affected by the breakage of Si-H bonds.

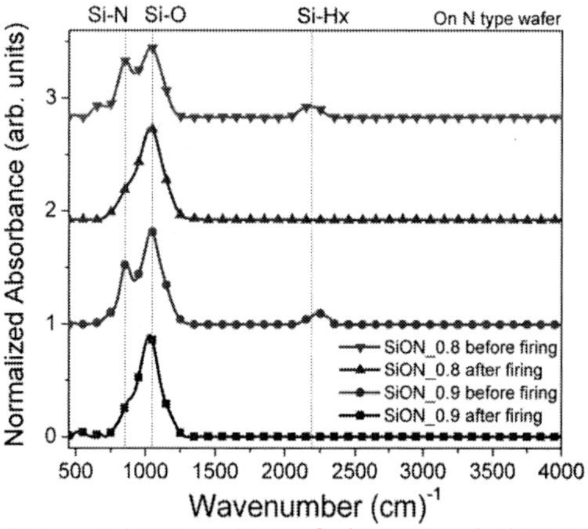

Figure 4: FTIR spectrum of SiON_0.8 and SiON_0.9 films before and after firing at 850 °C

b) QSSPC measurements

In order to understand the degradation of SiON layers at high temperature cycles we performed annealing and firing experiments on test samples having passivation layers on both the sides. From Fig. 5 we observe that the τ_{eff} of all single layer SiON layers decreases drastically after annealing. The highest lifetime by single passivation layer on p type CZ wafer before annealing was obtained for SiON_0.8; having a τ_{eff} of 93 µs and an iV_{OC} of 654 mV. SiON_0.75 and SiON_0.9 has effective lifetime of 17 µs and 75 µs respectively; corresponding to iV_{OC} of 613 mV and 640 mV. Since SiON_0.8 has the highest lifetime, a stack of SiON_0.8/SiN was annealed and found that unlike single layer the τ_{eff} of SiON_0.8/SiN increases from 104 µs before annealing to 170 µs after annealing. The corresponding iV_{OC} of SiON_0.8/SiN also increases from 656 mV to 673 mV after annealing.

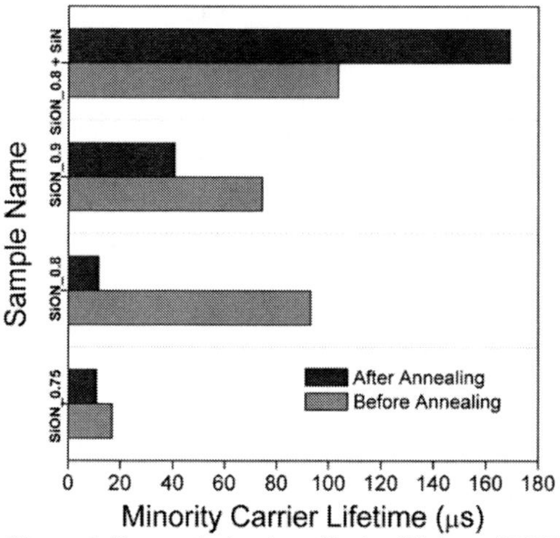

Figure 5: Bar graph showing effective lifetime of SiON and SiON/SiN before and after annealing on p-type CZ wafers

The effect of firing on the lifetime of N type CZ wafers was also analysed with SiON and SiON/SiN passivation layers on both sides, which is shown in Fig. 6. A similar trend was found in the effective lifetime values as in the previous annealing experiment. Similar to the annealing experiment, the lifetime of single layer SiON_0.8 and SiON_0.9, which are 34 µs and 247 µs respectively falls drastically after firing. Whereas for SiON_0.8/SiN, τ_{eff} increases steeply from 43 µs to 269 µs with an iV_{OC} of 639 mV.

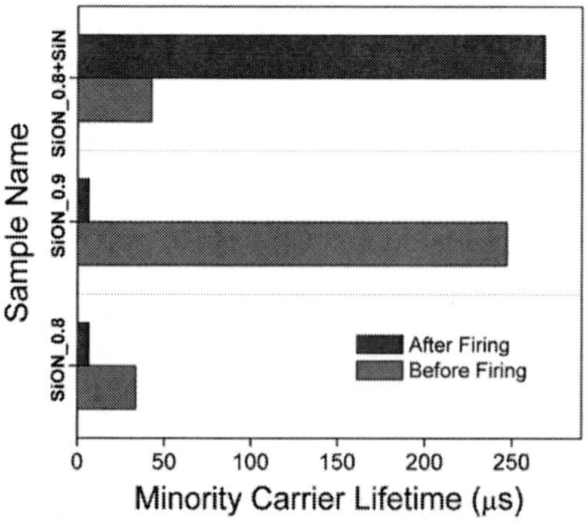

Figure 6: Bar graph showing effective lifetime of SiON and SiON/SiN before and after firing at 850 °C on n-type CZ wafers

The above results can be explained by passivation mechanism of SiON. As discussed previously passivation in SiON can arise due to the positive fixed charges resulting in field effect passivation, and Si-H bonds, which saturate the dangling bonds resulting in chemical passivation. The dominance of field or chemical passivation depends on the composition of SiON. Hence there is difference in literature regarding the passivation mechanism of SiON. From the firing and annealing experiments, it can see that for single layer SiON the lifetime decreases after high temperature process, which can be directly co-related to the FTIR results, which show the breakage of Si-H bond affecting the chemical passivation at the interface. As against this having a capping layer like SiN improves the lifetime due to different factors. Firstly the Si-H bonds break to release atomic hydrogen which effuses through the microvoids [12]; this escape of hydrogen is prevented by SiN which acts as a diffusion barrier. There is also a plausible reason that the atomic hydrogen penetrates and passivates the bulk further improving the bulk passivation. Such a mechanism has been report for SiN [13] however there are no reports for the same behavior for SiON. Secondly a high temperature process can also result in activation of charges hence the SiON/SiN stack gives a higher lifetime after a high temperature process.

SiON_0.8 presented a higher lifetime for p type wafers and lower lifetime for n type wafers. However the τ_{eff} value of SiON_0.9 for n type silicon wafer is 247 µs with an iV_{OC} of 637 mV, which is considerably high as compared to lifetime for p type wafers. This could be related to the difference in bond composition of SiON. SiON_0.8 has a higher Si-H bond density as compared to SiON_0.9 therefore the improved chemical passivation makes it a better passivating layer for p type silicon. Whereas SiON_0.9 has a higher fixed positive charge helping it result an accumulation at n type Si interface, making it a better passivation for n type silicon. Thus SiON offers the prospectus of a passivation layer for p and n type silicon.

4 CONCLUSION

In this work the passivation property and the thermal stability of SiON films are presented with the help of minority carrier lifetime and infrared spectroscopy measurements. The importance of a SiN capping layer to improve the thermal stability of SiON layers has been demonstrated. Single layer SiON films suffer degradation in passivation property at high temperature processes due to defect states created by Si-H bond breakage and diffusion of hydrogen. The SiN capping layer improves the passivation by acting as a diffusion barrier and the experimental observations where explained using the passivation mechanism of SiON. SiON/SiN stack for p-type wafers yielded an iV_{OC} of 673 mV and iV_{OC} of 639 mV for n type silicon wafers. The SRV for SiON/SiN stack was reduced to 35 cm/s. Further optimization of the composition of SiON films can help in its application as rear side passivation schemes for industrially viable solar cells.

5 ACKNOWLEDGEMENT

This work was supported by the Ministry of New and Renewable Energy (MNRE) through the project AMANSI under the National Solar Science fellowship program. We would like to thank NCPRE and CEN at IIT Bombay for the fabrication and characterization facilities.

6 REFERENCES

[1] M.M. Hilali, J.M. Gee, P. Hacke, Bow in screen-printed back-contact industrial silicon solar cells, Sol. Energy Mater. Sol. Cells. 91 (2007) 1228–1233.

[2] J. Schmidt, A.G. Aberle, Carrier recombination at silicon–silicon nitride interfaces fabricated by plasma-enhanced chemical vapor deposition, J. Appl. Phys. 85 (1999) 3626.

[3] D. Biro, S. Mack, A. Wolf, A. Lemke, U. Belledin, D. Erath, et al., Thermal oxidation as a key technology for high efficiency screen printed industrial silicon solar cells, in: 2009 34th IEEE Photovolt. Spec. Conf., IEEE, 2009: pp. 001594–001599.

[4] B. Hoex, J.J.H. Gielis, M.C.M. van de Sanden, W.M.M. Kessels, On the c-Si surface passivation mechanism by the negative-charge-dielectric Al2O3, J. Appl. Phys. 104 (2008) 113703.

[5] B. Hallam, B. Tjahjono, S. Wenham, Effect of PECVD silicon oxynitride film composition on the surface passivation of silicon wafers, Sol. Energy Mater. Sol. Cells. 96 (2012) 173–179.

[6] J. Dupuis, E. Fourmond, J.F. Lelièvre, D. Ballutaud, M. Lemiti, Impact of PECVD SiON stoichiometry and post-annealing on the silicon surface passivation, Thin Solid Films. 516 (2008) 6954–6958.

[7] P. Fath, T. Pernau, S. Keller, J. Maier, V. Mihailetchi, R. Petres, Silicon Surface Passivation by Industrial Low Frequency PECVD Films – Properties and Performance of SiCx and SiOxNy, 24th Eur. Photovolt. Sol. Energy Conf. 21-25 Sept. 2009, Hamburg, Ger. (2009) 1613–1616.

[8] S. Gatz, T. Dullweber, V. Mertens, F. Einsele, R. Brendel, Firing stability of SiNy/SiNx stacks for the surface passivation of crystalline silicon solar cells, Sol. Energy Mater. Sol. Cells. 96 (2012) 180–185.

[9] S. Wenham, B. Tjahjono, B. Hallam, Effect of PECVD Silicon Oxynitride Stoichiometry on the Surface Passivation of Silicon Wafers, 24th Eur. Photovolt. Sol. Energy Conf. 21-25 Sept. 2009, Hamburg, Ger. (2009) 1548–1551.

[10] J. Seiffe, L. Gautero, M. Hofmann, J. Rentsch, R. Preu, S. Weber, et al., Surface passivation of crystalline silicon by plasma-enhanced chemical vapor deposition double layers of silicon-rich silicon oxynitride and silicon nitride, J. Appl. Phys. 109 (2011) 034105.

[11] A. Antony, A. Soman, Fabrication of Selective and Tunable Broad Band Photonic Crystal Reflectors for Thin Film Solar Cells, 29th Eur. Photovolt. Sol. Energy Conf. Exhib. (2014) 255–258.

[12] A. Sarikov, M. Voitovych, I. Lisovskyy, V. Naseka, A. Hartel, D. Hiller, et al., Characteristics of Hydrogen Effusion from the Si-H Bonds in Si Rich Silicon Oxynitride Films for Nanocrystalline Silicon Based Photovoltaic Applications, Adv. Mater. Res. 854 (2013) 69–74.

[13] W. Soppe, H. Rieffe, A. Weeber, Bulk and surface passivation of silicon solar cells accomplished by silicon nitride deposited on industrial scale by microwave PECVD, Prog. Photovoltaics Res. Appl. 13 (2005) 551–569.

HYDROGEN PLASMA TREATMENT TO ENHANCE A-SI/C-SI INTERFACE PASSIVATION

Anishkumar Soman, Aldrin Antony *
Department of Energy Science and Engineering, Indian Institute of Technology Bombay,
Mumbai - 400076, India.
*phone: +91 2225764400, Fax: +91 2225723480, email: aldrinantony@iitb.ac.in

ABSTRACT:
The interface passivation quality of a-Si/c-Si plays the decisive role in attaining high open circuit voltage of Si heterojunction solar cells. The passivation quality of this interface can be improved by using Hydrogen plasma treatments, which helps to saturate the dangling bonds by Hydrogen termination. In this work, different plasma treatment methodologies like Pre, Post and Intermediate Hydrogen plasma treatments during the deposition of a-Si are investigated and their effect on passivation quality was studied using minority carrier lifetime measurements and Fourier Transform Infrared Spectroscopy (FTIR). The Hydrogen Plasma treatments were done using rf- PECVD at 80% Hydrogen dilution and at a low substrate temperature of 200 °C. The treatments were done for 30 seconds at a low power density of 30mW/cm^2 to avoid the plasma damage. The improved passivation quality due to hydrogen plasma treatment results in surface recombination velocity as low as ~ 24 cm/s with an implied V_{OC} of 700 mV.
Keywords: a-Si:H, Hydrogen, Passivation, PECVD

1 INTRODUCTION

Silicon Heterojunction (SHJ) solar cells is a viable industrial technology which offers lot of promise as high efficiency low cost commercial solar cells in the future. With Panasonic reaching world record efficiency of 24.7% [1] and 25.6% with interdigitated back contact [2] in silicon solar cell technology using Heterojunction solar cells for large area of 143 cm^2, the prospect of commercial SHJ solar cells which doesn't involve high temperature process penetrating the photovoltaic market, which is currently dominated by mono and multicrystalline silicon solar cells, seems feasible. This has propelled a lot of research interest to improve the efficiency of SHJ solar cells. One of the main factors which contribute to this increase in efficiency is the passivation at amorphous silicon (a-Si) and crystalline silicon (c-Si) interface. The intrinsic a-Si layer on both sides of c-Si, which acts as a buffer layer between the doped a-Si and c-Si wafer forms a critical aspect which contributes to V_{OC} over 700 mV [1] for SHJ solar cells. The reason for such a superior level of passivation at a-Si/c-Si interface is due to the hydrogen present in a-Si matrix which passivates the surface dangling bonds of silicon thereby decreasing the defect states at the interface. The device grade a-Si is fabricated using highly depleted silane plasma at the transition phase of amorphous to microcrystalline silicon [3]. To further reduce the recombination losses at the interface, hydrogen plasma treatment can form pivotal to saturate dandling bonds at the interface.

Though Hydrogen treatment has been previously used to improve passivation of a-Si [4][5][6], there has been an inconsistency in the nature of these treatment methods. Therefore in this work, different methodologies of hydrogen treatments are investigated which includes treating the a-Si films with hydrogen before (pre), after (post) and during (intermediate) deposition. The passivation quality has been determined by minority carrier lifetime and FTIR results whereas the microstructure of a-Si after treatment has been studied by XRD and HR-TEM. A detailed analysis of the passivation quality for different treatments has been done for p-type and n-type wafers in order to use the results for p and n type SHJ solar cells.

2 EXPERIMENTAL

2.1 Fabrication

The a-Si deposition and hydrogen plasma treatment is done in 13.56 MHz rf-PECVD system Plasmalab 100 by Oxford Instruments. Silane (SiH_4) and Hydrogen (H_2) were used as precursor gases for a-Si deposition. The a-Si condition has been optimized for device grade at highly diluted SiH_4 plasma. The a-Si deposition and the hydrogen treatment is done at a low temperature of 200 °C and the power density is kept as low as 30 mW/cm^2 to reduce the plasma induced damage by ion bombardment on the sensitive a-Si/c-Si interface. The a-Si deposition is done at a chamber pressure of 700mTorr whereas the Hydrogen treatment is done at 800 mTorr. The plasma treatment uses 80% diluted H_2 for a duration of 30 seconds to prevent etching of the a-Si thin film by H_2 plasma. A plasma cleaning step has also been introduced for certain samples in order to check if a short plasma cleaning before deposition improves passivation. The plasma cleaning is done using Nitrous Oxide (N_2O) and Tetrafluoromethane (CF_4) at a power of 100 Watt. All the samples are RCA cleaned before deposition and a 2% HF dip is done before loading the samples into the load lock.

2.2 Characterization

The passivation quality has been studied using minority carrier lifetime and Implied V_{OC} measurement by Lifetime measurement system WCT 120 by Sinton Instruments. The p type substrates used for experiments are double side polished FZ wafers of resistivity 4-7 Ω.cm, orientation <100> with a thickness of 250 μm. The n type substrates used are double side polished FZ wafers of resistivity 1-5 Ω.cm, with similar orientation and thickness. The surface recombination velocity (SRV) has been calculated from the effective lifetime (τ_{eff}) using the equation below [7]:

$$SRV = W/(2.\tau_{eff}) \quad (1)$$

Where W is the wafer thickness and τ_{eff} is measured at an excess carrier density (Δn) of 10^{15} cm^{-3}. Here it is assumed that the bulk lifetime of FZ wafers are infinite. The τ_{eff} values are measured using transient measurement at an injection level of 10^{15} cm^{-3}. The FTIR spectra are measured using Perkin Elmer Spectrum 100 in absorbance mode and the XRD (θ-2θ) measurement is done from 10° to 80° using Rigaku High Resolution X Ray Diffractometer (Smartlab 3KW). The HR-TEM imaging is done using FEG-TEM 300 kV-FEI Tecnai.

3 RESULTS AND DISCUSSION

3.1 Passivation using different Hydrogen treatment

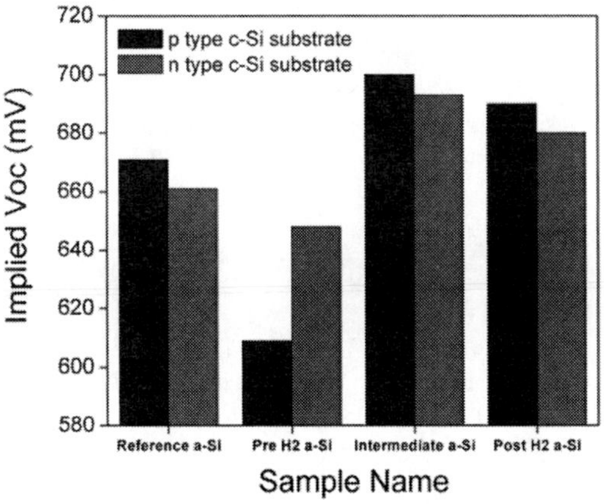

Figure 1: Implied V_{OC} (iV_{OC}) of different hydrogen plasma treated a-Si samples at 1 sun illumination for p and n type c-Si wafers.

The Implied V_{OC} of the passivation layer (intrinsic a-Si) is an important parameter to understand the recombination losses at the interface. The FZ wafers are coated with 15 nm a-Si deposited on both sides of p and n type wafers to have symmetrical surface recombination on both the surfaces. Fig. 1 shows that, the as deposited a-Si has an iV_{OC} of 671 mV and 661 mV respectively for p and n type c-Si wafers. The iV_{OC} drastically reduces to 609 mV for p type c-Si when the a-Si films are pre-treated with Hydrogen whereas it reduces to 648 mV for n type c-Si. As against this treating the a-Si with hydrogen post-deposition improves the iV_{OC} to 690 mV (for p type) and 680 mV (for n type) respectively. The highest iV_{OC} is obtained when the hydrogen treatment takes place between a-Si deposition. The iV_{OC} for p type c-Si shows an increase of 4.3% to 700 mV for intermediate H_2 treatment as compared to reference as deposited a-Si. Likewise for n type c-Si the iV_{OC} increases by 4.84% to 693 mV for intermediate treatment.

The effective lifetime (τ_{eff}) of different hydrogen treated a-Si follow similar trend and is plotted in Fig. 2. The as deposited a-Si gives a τ_{eff} of 288 μs and 397 μs for p and n type c-Si wafers. The lifetime degrades on pre-treating the samples with H_2 to 45 μs and 270 μs respectively for p and n type c-Si. The highest lifetime is obtained for intermediate treatment, where for p type it almost doubles to 594 μs and for n type it almost triples to more than 1ms. Post treated samples also show an improvement in passivation quality though less dramatic than intermediate treatment. For p type it improves to 497 μs whereas for n type it improves upto 633 μs. The same hydrogen treatment was tried over initial N_2O-CF_4 plasma cleaned sample in order to clean the c-Si surface before deposition, however the lifetime was found to degrade severely almost permanently damaging the a-Si surface, and therefore those results were not plotted in the graph.

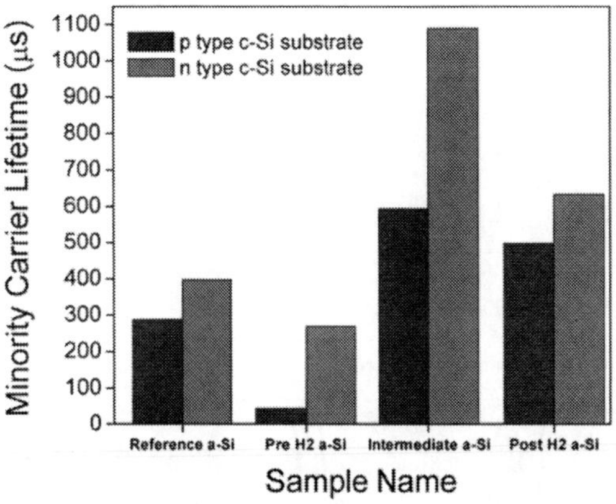

Figure 2: Minority carrier lifetime at an injection level of 10^{15} for different hydrogen plasma treated a-Si samples deposited on p and n type c-Si wafers

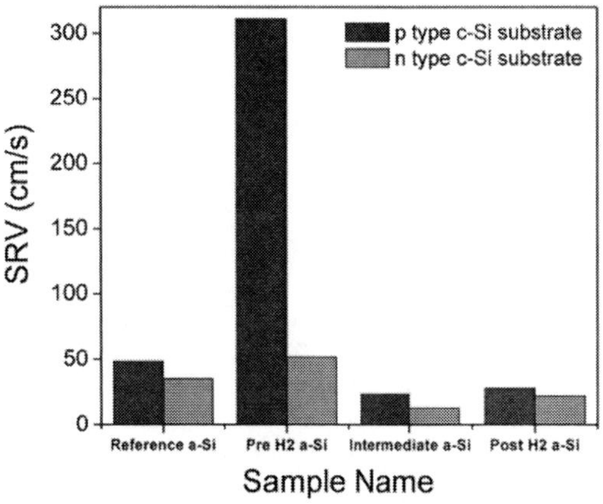

Figure 3: SRV of different hydrogen plasma treated a-Si samples for p and n type c-Si wafers.

The as deposited intrinsic a-Si inherently exhibits better passivation due to its improved chemical passivation. For p type c-Si, a-Si buffer layer presents an SRV ~ 50 cm/s, which further reduces to 35 cm/s for n type c-Si. As against this pre-treatment of H_2 shows a high SRV of ~ 300 cm/s for p type, this could be co related to the fact that ion bombardment at the pristine c-Si surface results in surface damage which acts as recombination centers. These recombination losses at the defect energy levels decrease the lifetime and Implied V_{OC} of pre-treated samples. For n type c-Si as well the SRV of pretreated a-Si is higher (52 cm/s) as compared to reference a-Si. When a-Si samples are treated after deposition, it was found that the SRV decreases to 28 cm/s for p type and 22 cm/s for n type c-Si. This shows that the interface of a-Si/c-Si is highly sensitive and an initial plasma treatment damages the interface. Once a proper interface is formed between a-Si/c-Si without any epitaxy then further hydrogen treatment saturates the remaining unsaturated bonds. Hence better results were obtained for post and intermediate hydrogen treatment. The lowest SRV obtained for p type is 24 cm/s and for n type is 13 cm/s for intermediate hydrogen treatment. From the comparative studies, it can be observed that an intermediate hydrogen treatment results in the best passivation of the recombination active defect states at the interface.

3.2 Analysis of a-Si microstructure after hydrogen treatment

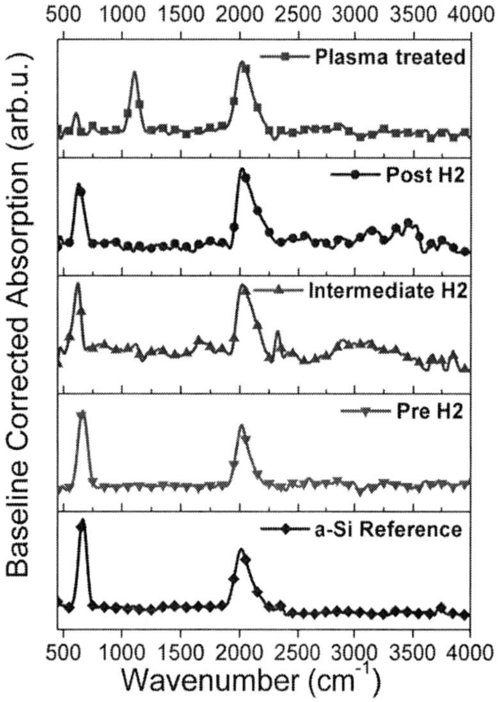

Figure 4: FTIR spectra of different hydrogen plasma treated a-Si samples

The FTIR spectra of a-Si shows 2 distinct peak, one at ~ 600 to 690 cm^{-1} for Si-H rocking bond and the other from 2000 to 2100 cm^{-1} for $Si-H_x$ bonds (SiH_2 – dihydride and $Si-H_3$ – trihydrides) [8]. For plasma cleaned sample before a-Si deposition, it was observed that the Si-H bond is highly diminished and most of the H in the film is in the presence of di or trihydrides. Also the N_2O introduces O species in the chamber which results in the formation of Si-O bonds showing peak intensity at ~1120 cm^{-1}. Thus the defects at the interface due to high energy ion bombardment and the Si-O bond formation in the film degrade the passivation quality of the film. The Intermediate and post-treated sample was also showed has a higher integrated absorbance under $Si-H_x$ peak compared to pre-treated and reference a-Si samples which is depicted by their improved passivation quality. However it is still inconclusive from the FTIR alone regarding the interface since the bond signatures are a bulk phenomenon, Hence an HRTEM imaging was performed to understand the interface using high resolution magnification of the interface.

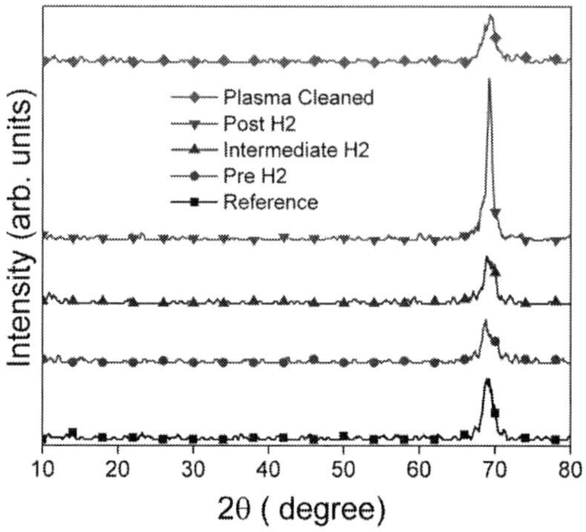

Figure 5: XRD of different hydrogen plasma treated a-Si samples

Since the a-Si:H film is being deposited near the transition from a-Si to μc-Si domain and since H_2 treatment promotes epitaxy, it is import to check if there are any crystalline fractions in the film and hence an XRD of different films were taken and are shown in Fig. 5. The XRD data only shows a single peak at ~ 69° which corresponds to the Si <100> diffraction peak from the substrate. This verifies that there is no crystalline phase in the a-Si film even after hydrogen plasma treatment.

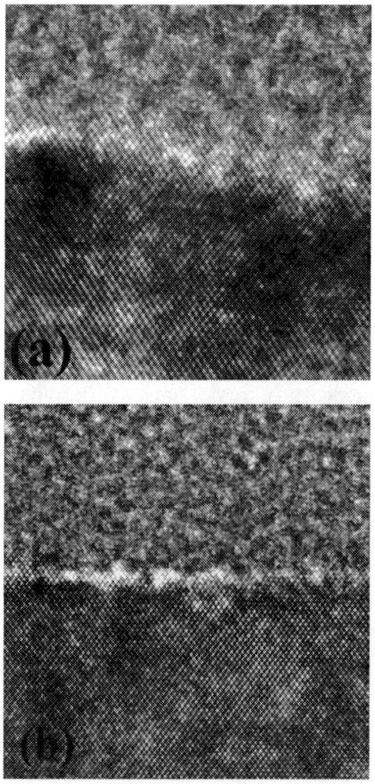

Figure 6: HR-TEM image of a) as deposited a-Si sample b) Intermediate H_2 treated a-Si sample

The FTIR data shows the bond configuration of the different hydrogen plasma treated a-Si samples and the XRD results show there are no crystalline fractions. However it still doesn't explain the superior passivation quality of Intermediate H_2 treated sample. The HR-TEM image of Intermediate H_2 treated sample and as deposited a-Si samples are shown in Fig. 6. From Fig. 6a, which is the image of as deposited a-Si, it can be observed that the interface is not sharp and there are slight extensions of the crystal plane to the a-Si film at the interface. As against this for Intermediate H_2 treated sample the interface was found to be very sharp giving least defects at the interface thereby a better passivation quality and the same can be seen in Fig. 6b. A perfect a-Si/c-Si interface coupled with saturation of surface dangling bonds can significantly improve the surface passivation.

4 CONCLUSION

In this work a comparative study of different Hydrogen plasma treatment to improve passivation of a-Si films are reported. By using hydrogen plasma treatment a lifetime of more than 1 ms and Implied V_{OC} of 700 mV were obtained. The lowest SRV obtained was 13 cm/s. The H_2 pre-treated sampled have lesser lifetime due to plasma damage of the interface. At the same time post and intermediate H_2 treated a-Si samples have improved passivation. The a-Si films show no crystalline fraction after hydrogen treatment and the high performance of the Intermediate treatment has been associated with the atomic sharp a-Si/c-Si interface with no epitaxy. This hydrogen plasma treatment to improve the life time of a-Si:H passivation will be applied for Silicon Heterojunction solar cells in the near future.

5 ACKNOWLEDGEMENT

This work was supported by the Ministry of New and Renewable Energy (MNRE) through the project AMANSI under the National Solar Science fellowship program. We would like to thank NCPRE and CEN at IIT Bombay for the fabrication and characterization facilities. The authors would also like to thank Bhagyaraj for HR-TEM imaging and Sonali. S for assistance in TEM sample preparation.

6 REFERENCES

[1] Public Relations Development Office, P. Corp., Panasonic HIT® Solar Cell Achieves World's Highest Conversion Efficiency of 24.7% at Research Level, Osaka, Japan. (2013).

[2] K. Masuko, M. Shigematsu, T. Hashiguchi, D. Fujishima, M. Kai, N. Yoshimura, et al., Achievement of More Than 25% Conversion Efficiency With Crystalline Silicon Heterojunction Solar Cell, IEEE J. Photovoltaics. 4 (2014) 1433–1435.

[3] A. Descoeudres, L. Barraud, R. Bartlome, G. Choong, S. De Wolf, F. Zicarelli, et al., The silane depletion fraction as an indicator for the amorphous/crystalline silicon interface passivation quality, Appl. Phys. Lett. 97 (2010) 183505.

[4] Y. Do Kim, S. Park, J. Song, S.J. Tark, M.G. Kang, S. Kwon, et al., Surface passivation of crystalline silicon wafer via hydrogen plasma pre-treatment for solar cells, Sol. Energy Mater. Sol. Cells. 95 (2011) 73–76.

[5] A. Descoeudres, L. Barraud, S. De Wolf, B. Strahm, D. Lachenal, C. Guérin, et al., Improved amorphous/crystalline silicon interface passivation by hydrogen plasma treatment, Appl. Phys. Lett. 99 (2011) 123506.

[6] M. Mews, T.F. Schulze, N. Mingirulli, L. Korte, Hydrogen plasma treatments for passivation of amorphous-crystalline silicon-heterojunctions on surfaces promoting epitaxy, Appl. Phys. Lett. 102 (2013) 122106.

[7] S. Gatz, T. Dullweber, V. Mertens, F. Einsele, R. Brendel, Firing stability of SiNy/SiNx stacks for the surface passivation of crystalline silicon solar cells, Sol. Energy Mater. Sol. Cells. 96 (2012) 180–185.

[8] D.J. Michalak, S.R. Amy, D. Aureau, M. Dai, A. Estève, Y.J. Chabal, Nanopatterning Si(111) surfaces as a selective surface-chemistry route, Nat. Mater. 9 (2010) 266.

Softly Doped and Deep Emitters for P/Al Solar Cell Structure

Muhammad Azam Rasool[1], Vanesa Fano[1], Ahmed. Habib[1, 2], Alona Otaegi[1], José Rubén Gutiérrez[1], Juan Carlos Jimeno[1], Nekane Azkona[1] and Eneko Cereceda[1]
[1]Technological Institute of Microelectronics (TiM), University of the Basque Country, Zamudio, Spain
[2]Faculty of Science, Department of Physics, Mansoura University, Mansoura, Egypt
Email: muhammadazam.rasool@ehu.eus Phone: +34-946018489 Fax: +34-946018490

ABSTRACT: Objective of this work is to get softly doped and deep emitters in single thermal step. Due to high temperature diffusion, dead layer (electrically inactive) is formed which produces recombination centers and increases saturation current density (Joe). For appropriate choice of selective emitter in order to decrease emitter saturation current density (Joe), it is necessary to have a passivated emitters, i.e. with silicon nitride. We have presented a process which combines with standard technology used for selective emitter formation in a single thermal step which leads to lowly doped and deep emitters. This process is feasible for industrial fabrication of P/Al solar cells. In this process the gettering is higher than conventional process. Surface concentration value ranging from $3.6x10^{19}$ cm^{-3} to $7.2x10^{19}$ cm^{-3} and depth junction values from 0.52 to 0.71 μm for sheet resistance ~100 Ω/□ has obtained. The estimated saturation current density (Joe) could be around 30-40 fA/cm² with optimization but in present work Joe is under 80 fA/cm².
KEYWORDS: soft and deep emitters, P/Al solar cell, selective emitters.

1 INTRODUCTION

Phosphorus diffusion is used in solar cells fabrication process for all kinds of crystalline-silicon (c-Si) materials to form the n^+ type emitters. Post diffusion oxidation is also common step in commercial fabrication sequence of crystalline solar cell. The quality of the emitter plays an important role for solar cell efficiency due to formation of P-N junction, which is the core of the crystalline silicon (c-Si) solar cell. If the phosphorus (P) surface concentration is high and exceeds the intrinsic charge-carrier concentration, have certain special features on electrical properties and conductivity [1-2]. Efficiency of solar cells is based on the emitter's saturation current density and passivation. The quality of emitters depends on diffusion temperature with flow of gases and doping time. At high temperature, concentration of phosphorus (P) exceeds the solubility in Si ($10^{21}/cm^3$) which forms a dead layer [1]. This dead layer is electrically inactive and produces recombination centers that increase Auger recombination which increase saturation current density (Joe) [2]. Moreover high surface P concentration reduced the passivation effect which lowers the open circuit voltage (Voc) and overall solar cell efficiency [3]. For appropriate choice of selective emitter in order to decrease emitter saturation current density (Joe), it is necessary to have a passivated front surface, i.e. with silicon nitride. Good passivated surface decreases recombination centers, as a result Joe decrease and improve the open circuit voltage [4-8].

In case of softly doped emitters (sheet resistance 90-150 Ω/□), it is difficult to make contacts between front metal and softly doped emitters, as a result contact resistivity increased which affects negatively on efficiency of solar cell. In industrial fabrication of silicon solar cells, there is a compromise between doping concentration of emitters and performance (moderate sheet resistance 50-80 Ω/□) which has sufficiently low contact resistance in order to get good efficiency. This compromise is overcome by fabrication of selective emitters. Selective emitter have high doping concentration under metal contacts grid and low doping concentration under illumination area. Selective emitter fabrication helps to reduce contacts resistance as well as lower Auger and SRH recombination and improve the open circuit voltage. There are many techniques which are mentioned in literature for fabrication of selective emitters. Röder et al, developed a process in which laser has been used to increase the doping concentration underneath the metal contacts from phosphorus glass silicate as a source of phosphorus. [9]. Lauermann et al, [10] used an etching resist (commercial available material) which is deposited by inkjet technique, with this mask, it is possible to etch back emitters between fingers area by chemical etching. Prior to this step, high doping was carried out to get high doped emitters. In this process, extra steps cleaning steps are necessary to remove the etch resist [10-12]. Fraunhofer ISE has developed an approach, which is based on simultaneous ablation of SiNx layer by laser and emitter layer underneath the ablated region by using a liquid (as P-source) laser beam is used for chemical processing (LCP) for selective emitter formation [13]. Similar work was carried out by University of New South Wales (UNSW), they have deposited phosphoric acid as doping source on the wafer prior to laser doping. Roth & Rau are working on commercialization of this technology [14].

In this work, for selective emitter's formation we have used laser as a tool to scribe on wafers followed by a heavy P diffusion. We have presented a process which combines with standard technology used for selective emitter formation in a single thermal step which leads to lowly doped and deep emitters. This process is feasible for industrial fabrication of P/Al solar cells.

2 SINGLE STEP DIFFUSION PROCESS

In the standard selective emitter's formation, it is produced by heavy phosphorus diffusion and followed by an etching process until the sheet resistance of the exposed emitter surface reaches ~60 Ω/□ [4]. Soft emitter has been made by a single thermal step process. This process runs in oversaturation conditions and it has some advantages in comparison to another processes with lower Ns target. Process starts from pre-deposition step, $POCl_3$ diffusion at 820 °C followed by a redistribution step at 950 °C. In redistribution, the junction goes deeper, aim is to the reduce of phosphorus impurity concentration. A deeper junction gives the chance of having less recombinant emitters. There is an activation of phosphorus impurities with redistribution to deeper diffusion. Thermal ramp and the oxygen flow of the proposed process is explained in the figure 1.

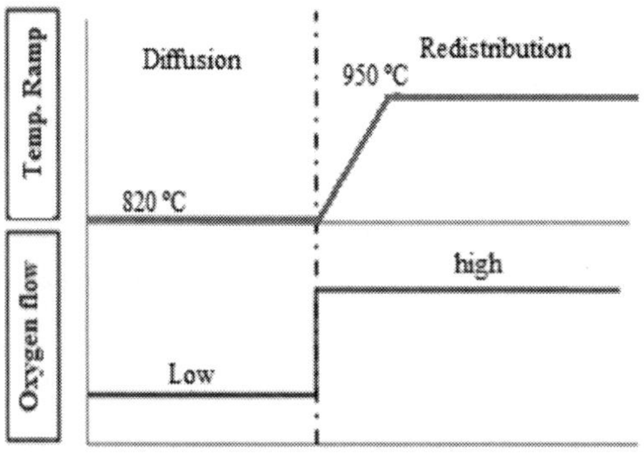

Figure 1: Thermal ramp and oxygen flow involved in the P diffusion at 820 °C and redistribution at 950°C

The influence of the passivation in the emitter reliability has been evaluated. For each emitter surface conditions, some surface recombination velocities have been simulated by PC1D. Simulation results corresponding to a conventional 0.2 and the current 0.7 µm depth junctions are plotted in figure 2, in order to highlight the importance of having deep junctions. Besides the positive influence of having deep emitters in the Joe values, we can see the more influence of a good surface passivation of emitters for lowly doped emitters, moderate sheet resistances, changing from 36 to 120 fA/cm^2, the advantages obtained from a lowly doped emitter has disappeared, as shown in figure 2.

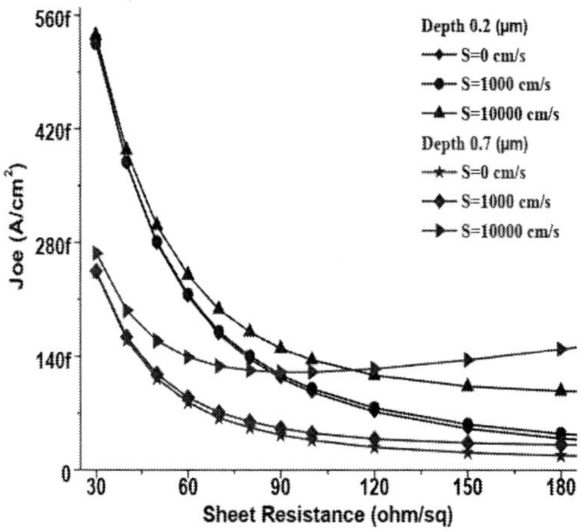

Figure 2: Saturation current of 0.2 and 0.7µm depth junction emitters resulted for several surface recombination velocities, values of S from 0 cm/s to 1000 and 10000 cm/s

After P diffusion process, a layer of SiNx (silicon nitride) is applied on front surface by plasma enhanced chemical vapor deposition (PECVD), which acts as both for masking and passivation layer. SiNx layer further decreases the emitter saturation current density, as shown in figure 3.

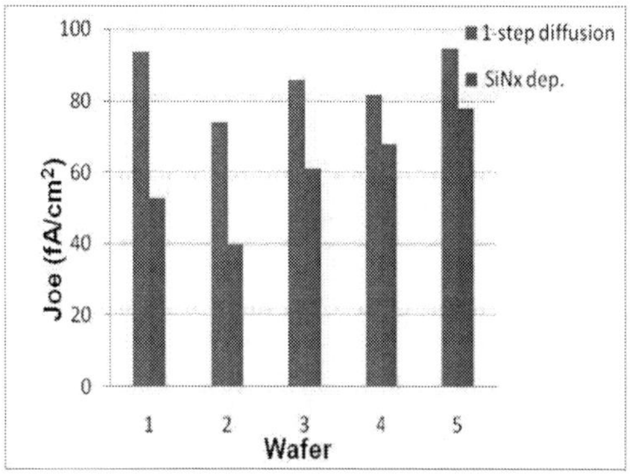

Figure 3: Saturation current density evolution after our proposed process and after the further passivation with SiNx deposition.

In addition to passivation, gettering effect in this single step diffusion process (proprosed process) is much higher than conventional process. Getterging effect is evaluated by liftime measuremts by PCD (transient photo-conductance decay). Results are given in Table I for comparison of gettering effect of our prosposed process with conventional process. For this purpose different kinds of wafers, including metallurgical grade (Metal. Grade) silicon wafers has been used.

TABLE I: Gettering in different kind of wafers

Conventional process			Proposed process		
n-type wafer	p-type wafer	Metal. Grade	n-type wafer	p-type wafer	Metal. grade
60±5 µs	25±5 µs	10 ± 3µs	500 ±20µs	200±5 µs	35±5 µs

The improvment due to gettering by our proposed process is quite high for the n and p-type wafers. In case of P-type material, we have measured lifetime around 201 µs by PCD on high resistivity (5.4Ω.cm) wafers. In case of N-type wafers, we have measured lifetime around 509µs by PCD technique on low resistivity (0.8 Ω.cm) wafers. In case of metalligical grade silicon wafers, by our proposed process improvement in lifetime is 3 times higher than conventional gettering process. Around 35µs lifetime has achieved for metallurgical grade wafers by our proposed process.

3 P/Al STRUCTURE WITH SELECTIVE EMITTERS

For P diffusion, p-type monocrystalline Cz-Silicon (($\rho = 5.4\,\Omega \cdot cm;$) with base doping N_{base} = 2.7E15 cm^{-3}and size 10 cm × 10 cm has been taken for processing after chemical etching followed by RCA1 and RCA2 cleaning. P pre-deposition was carried out at 820 ^{0}C by using $POCl_3$ as a source of phosphorus and nitrogen as a carrier gas in tube furnace under standard conditions. Redistribution process was carried in the same furnace at 950 ^{0}C. In this work we have performed P diffusion in single step, P pre-deposition together with redistribution. In between P-pre-deposition and redistribution step. This dry oxidation and redistribution process passivate the emitter's surface and converts electrically inactive P into electrically active P. The emitters obtained by above mention procedure are homogeneous and have uniform sheet resistance values (doping concentration) on whole wafer surface area. Chemical equation is given below:

$$POCl_3 + O_2 \rightarrow P_2O_5 + 3Cl_2$$
$$5Si + 2P_2O_5 \rightarrow 5SiO_2 + 4P$$

3.1 FABRICATION OF SELECTIVE EMITTERS

After soft P diffusion process, a layer of SiNx (silicon nitride) is applied on front surface by plasma enhanced chemical vapor deposition (PECVD), which acts as both for masking as well as passivation layer. SiNx layer further decreases the emitter saturation current density, as shown in figure 3. For selective emitter's formation, we have used laser as a tool to scribe on wafers in specified manner to draw superficial grooves till 2-5 µm in depth and 25-30 µm in width, on cell area of $4.2 \times 3.2\ cm^2$.

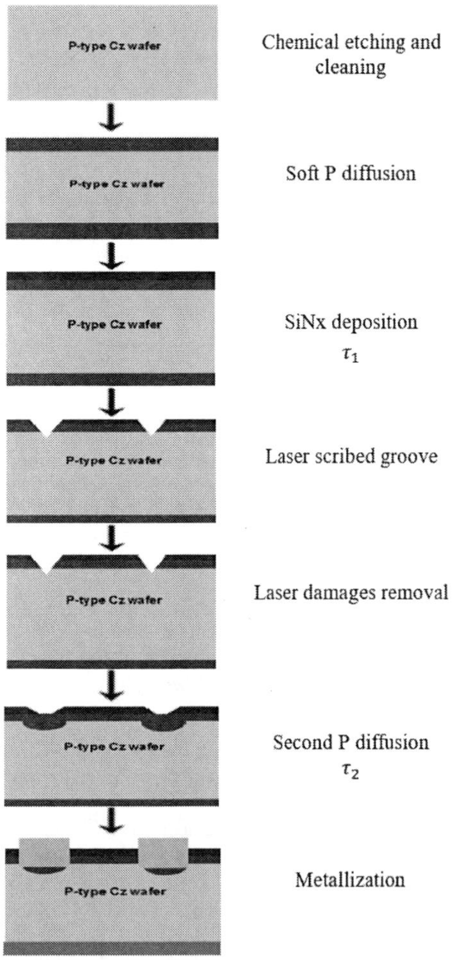

Figure 4: P/Al Solar cell fabrication scheme with selective emitters

Damages of laser were removed by alkaline selective etching, by protecting emitters under SiNx layer and rear emitters are protected by SiO_2. For this propose, after laser scribed grooves, silicon wafers were treated with alkaline solution in order to removed laser induced damages. Alkaline solution not only removed the ablated material but also etch silicon to make the grooves deeper. Fabrication process with selective emitter is shown in figure 4 [7-8].

Second P diffusion was carried out at 875 °C for 30 minutes under same gaseous flow as for soft P process. After this diffusion, we have two areas on front wafer surface with different doping concentrations, the process which is used to fabricate emitters with different concentrations. After phosphorus diffusion, we have evaluated surface of emitters through measuring the sheet resistance by removing SiNx layer. Sheet resistance values show that emitters under SiNx layer are not affected by second P diffusion or chemical attach. Low doping concentration of emitters is maintained underneath silicon nitride layer.

After Isolation of cells from wafer, metallization was carried out. A thick layer of aluminum was deposited on backside of cells by screen printing technique in order create back surface field (BSF) as well as back contact. Aluminum paste which was deposited on backside of all cells, were dried at 350°C and firing was carried at different temperature ranging from 875°C to 925°C. Optimal temperature for Al-deposition firing is 925 °C at 60inches/min belt furnace speed for good firing and p+ BSF effect formation.

3.2 CHARACTERIZATION OF SELECTIVE EMITTER

Characterization of P/Al solar cells with selective emitters was carried out by Sun-Voc and lifetime measurements by PCD (transient photo-conductance decay). After laser damage removal by NaOH and high temperature P diffusion, Lifetime of the selective emitters is decreased and emitter saturation current density is also increases which has resulted low Sun-Voc. Lifetime measurements results before and after fabrication of selective emitters are given in figure 5.

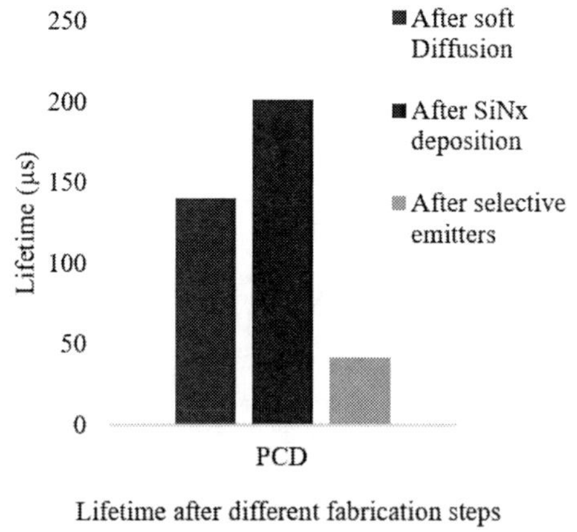

Figure 5: Lifetime measurements before and after selective emitter formation

In addition to this we have also observed that lifetime is destroyed only in laser scribed area of wafer but rest of area of wafer maintained the lifetime of minorities carriers. After laser scribtion, wafers were treated with chemical (NaOH) to remove the laser damages, chemical reacts only on the laser scribed area. There is possibility lifetime is destroyed either due to laser scribtion or by chemical attack.

As lifetime was destroyed after NaOH treatment with second Phosphorus diffusion. In order to find the reasons of lifetime destruction and to find the solution, process was repeated with NaOH as well as KOH treatment in order to remove the laser induced damages. In this batch we have used tefflon bath for NaOH and KOH treatment for extra care of process. In previous experiments, chemical etching was carried out in a metallic batch, it may be source of impurites which destroyed the lifetime. We have measured lifetime of wafers before and after chemical treatment, detail of process and results is given in table II. In addition to this some wafers were also processed without chemical treatment in order to compare the results which are given in table II.

TABLE II: Lifetime measurements

Laser damages removal process	Lifetime after SiNx deposition τ_1	Lifetime after selective emitters τ_2
No chemical treatment	139 μs	127.5 μs
NaOH metallic bath	93.1μs 162.5 μs	28 μs 29 μs
NaOH Teflon bath	137 μs	90.8μs 42.9 μs 72 μs
KOH Teflon bath	138 μs	93.5μs 83.9 μs 77 μs
Homogeneous emitters (No laser)	133 μs	118 μs

4 ALUMINUM OPTIMIZATION AND AL-BSF

The purpose of Al-optimization is to optimize the temperature and speed of belt furnace to develop Al-BSF for fabrication of industrial P/Al thin films solar cells. In order to optimize Al deposition we have taken three different kind of wafers. All the wafers were processed till selective emitters formation, as shown in figure (scheme for fabrication of selective emitters for P-type wafers). We have selected 3 different rear structures of P/Al silicon solar cells with selective emitters for Al deposition to create Al-BSF as shown in figure 6.

For this purpose, after 2nd phosphorus diffusion (selective emitter formation), about 1μm thick aluminum layer was deposited on few wafers by EBM (by eletron beam machine) in order to provide a uniform layer on backside of the wafer as shown in figure 6c. Few wafers were treated with Alkaline solution to remove rear P emittters, structure is shown in figure 6b. SiNx layer on front side of wafers protects the emitters from alkaline etching. Some wafers were processed as such, without any treatment prior to deposition of Al paste by screen printing. A thick layer of aluminum were deposited on backside of cells by screen printing technique in order create back surface field as well as back contact. For this purpose we have used Al paste Al (5540) which was dried at 350°C and was fired at different temperature ranging from 875 to 950 °C and at different speed of belt of furnace from 60 to 70 inches/minute.

Screen printed alumiunum deposition on backside to create a Al-BSF is common technique used for back surface passivation as well as to create back contacts. Formation of Al-BSF is two steps process, first deposition of Al-paste on back surface full area by screen printing and second a short time drying and annealing above the Al-Si eutectic temperature. Drying of A-paste is usually carried out at 300°C to 350°C and firing is performed at rapid thermal annealling temperature from 875°C to 950°C for short period of time (2.45 minutes). At high temperature Si is dissolve into Al-Si alloy melt. At cooling, silicon is rejected from melt and regrow on surface as an Al-doped or p+ BSF layer. In thin film solar cell when thickness of wafers decreases, back surface field (BSF) becomes more important in order to decrease the back surface recombination velocity and to increase the collection efficiency [15-16]. Structures of wafers which are shown in figure 6, represents wafers are ready for Al-deposition.

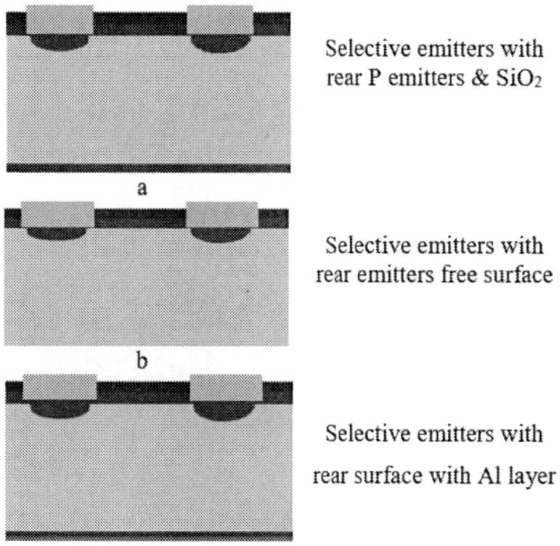

Figure 6 Structures of solar cells used for Al-optimization process.

Characterization of Al-optimization was carried out by measuring Sun-Voc, Matrix has been made in order to see the the effect of temperature, speed of belt furnace and quality of back surface field on open circuit voltage and cell performance. Thickness of BSF is measured by microscopic studies by focussing at top of surface as well as base of surface. From Sun-Voc we can extract all solar cell parameters Joe, Joz Gsh and series resistance even lifetime of bulk, which are used to characterize the performance of solar cell. Suns-Voc curve allows us to characterize lifetime of minority's carriers and its analysis also provides detailed information on the internal components of recombination in the solar cell. Sun-Voc measurements provide information of IV curves without the effect of series resistance of diode. Fitting of Sun-Voc curve is easier than illuminated curve since there is no dark current or series resistance.

After back contacts formation, cells were evaluated by Sun-Voc measurements, in order to extract diode and electrical parameters of solar cells (Joe, Joz and Gsh) as well as open circuit voltage. From Sun-Voc measurements, we have found that emitter saturation current density (Joe) increased after Al-deposition and we have Joe values around 6.4E-13 A/cm^2 and shunt conductance which were ranging from 2.0E-4 to 3.5E-3 ohm.cm^2 for p-type wafers. Effect of selective emitters on emitter saturation current density is shown in figure 7. Optimal temperature for Al-deposition firing is 925 °C at 60inches/min belt furnace speed for good firing and p+ BSF formation. In addition to this rear structure with SiO_2 surface (structure of figure 6a) of selective emitters have given the maximum voltage and good emitters saturation current density values than selective emitters with rear P emitter free surface.

Emitter's saturation current density (Joe) which we have obtained after aluminum deposition is highly increased, it limits the open circuit voltage. This Joe value is due to Al-BSF (rear emitters). We have to optimize metallization process in order to measure real Joe of P/Al solar cells. By using equation (1), we have calculated open circuit voltage around 650 mV from Joe 6.4E-13 A/cm^2. We have measured 624 mV with high shunt conductance in Sun-Voc measurements.

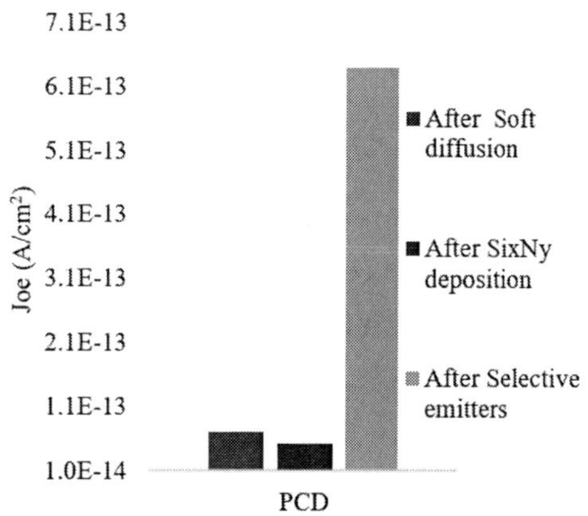

Lifetime after different fabricaton steps

Figure 7 Effect of selective emitters on Joe before and after selective emitter formation

$$V_{oc} = \frac{kT}{q} \ln \left[\frac{nN_A}{n_i^2} \right] \quad (1)$$

n, is minority carriers concentration at junction edge, ni is intrinsic carrier concentration (9E9/cm^3) at 25 °C. N_A is base doping of wafer and $\frac{kT}{q}$ is thermal voltage.

The Joe values which we have measured is due rear emitter's effect. We have to optimize some parameters in order to measure real Joe of the cells. In Sun-Voc measurements, although J_{OZ} is maintain in all wafers which is in range of 1.0E-7 A/cm^2 to 5.1E-8 A/cm^2 but Joe is varying from cell to cell and shunt conductance is also high.

5 CONCLUSION

By single thermal step, passivated emitters has been obtained with Joe is range of 40-80 fA/cm^2 with low thermal budget. Surface concentration ranging from 3.6x10E13cm^{-3} to 7.2x1019 cm^{-3} and depth junction from 0.52 to 0.71 µm has been obtained for sheet resistance ~100 Ω/□. In this process gettering is much higher than conventional process.

By deposition of SiNx layer, further improved emitters quality by passivating the surface and we have obtained Joe value as low as 40 fA/cm2. The estimated saturation current density (Joe) is around 30-35fA/cm^2 with optimization. First experiment has given Sun-Voc 624 mV, which corresponds to Joe 6.45E-13 A/cm2. This Joe is due to rear emitters (Al-BSF), it can be further improved and theoretically expected is around 650 mV.

Process which is used to fabricate selective emitter is feasible to apply for industrial fabrication P/Al silicon solar cells with expected high efficiency.

6 ACKNOWLEDGEMENTS

This work has been supported by the Spanish MINECO within the framework of the projects Crysthin TEC2011-28423-C03 and Tabaco ENE2014-56069-C4-1R. The authors would like to thank the Institute of Solar Energy (IES-UPM) for the lifetime measurements.

7 REFERENCES

[1] S. Werner, E. Lohmüller, S. Maier, A. Kimmerle, A. Spribike,S. Wasmer, F. Clement, A. Wolf. "Process optimization for the front side of the p-type Silicon". 29th European PV SolarEnergy Conference and Exhibition. Amsterdam (2014) pp 1-6.

[2] B.Bazer-Bachi, E.Fourmond, P.Papet, L.Bounass, O. Nichiporuk, N. Le Quang, M. Lemiti. "Higher Emitter quality by reducing inactive phosphorus" Solar Energy materials &Solar Cells 105, pp137-141, 2012.

[3] A. Cuevas and M. Balbuena. "Thick emitter silicon solar cells". IEEE 1988, pp 429-434.

[4] R. Khandelwal, H .Windgassen, J. VanMölken, T. Markus Pletzer and H. Kurz. "Comparison of large area high ohmic emitter silicon solar cells with standard screen – printed contacts". IEEE 2013, pp 2217-2220.

[5] R.B. Fair, J.C.C. Tsai. "A quantitative model for the diffusion in silicon and the emitter dip effect". J. ElectrochemicalSociety 124, pp1107-1118, 1977.

[6] J. Szlufcik, H.E. Elgamel, M. Ghannam, J. Nijs and R. Mertens. "Simple integral screen-printing process for the selective emitter polycrystalline silicon solar cells". Appl. Phys. Letters 59(13), pp1583-1584, 1991.

[7] V. Fano, M. A. Rasool, A. Habib, A. Otaegi, R. Gutierrez, J. C. Jimeno, N. Azkona "Lowly Doped Emitters for Crystalline Silicon Solar Cells", 42 IEEE Photovoltaic

Specialist Conference, 2015, pp.1-5.
[8] A. Otaegi, V. Fano, M. A. Rasool, R. Gutierrez, J. C. Jimeno, E. Cereceda, "Laser-Induced Damage for Crystalline Silicon Solar Cells", 42 IEEE Photovoltaic Specialist Conference, 2015, pp.1-4.
[9] T.C. Röder et al. "Add-on laser tailored selective emitter solar cells", Prog. Photovolt: Res. Appl. 18: pp505-510, 2010.
[10] T. Lauermann et al. "INSECT: An Inline selective emitter concept with high efficiencies at competitive process costs improved with inkjet masking technology", 24th EU PVSEC, Hamburg, Germany, (2009).
[11] B. Raabe, F. Book, A. Dastgheib-Shirazi and G. Hahn, "The Development of Etch-Back Processes for Industrial Silicon Solar Cells," Proceedings of the 25th European Photovoltaic Solar Energy Conference and Exhibition/5th World Conference on Photovoltaic Energy Conversion, Valencia, 6-10 September 2010, pp. 1174-1178.
[12] H. Haverkamp, A. Shirazi, B. Raabe, F. Book and G. Hahn, "Minimizing the Electrical Losses on the Front Side: Development a Selective Emitter Process from a Single Diffusion," The 33rd IEEE Photovoltaic Solar Energy Conference and Exhibition, San Diego, 11-16 May 2008, pp. 430-433.
[13] D. Kray, N. Bay, G. Cimiotti, S. Kleinschmidt, Kösterke, A. Lösel, M. Sailer, A. Tröger, H. Kühnlein, Nussbaumer, C. Fleischmann and F. Granek, "Industrial LCP Selective Emitter Solar Cells with Plated Contacts," 35th IEEE Photovoltaic Solar Energy Conference and Exhibition, Honolulu, 20-25 June 2010.
[14] B. Tjahjono, S. Wang, A. Sugianto, L. Mai, Hameiri, N. Borojevic, A. Ho-Baillie and S. R. Wenham, "Application of Laser Doped Contact Structure on Multicrystalline Solar Cells," Proceedings of the 23rd European Photovoltaic Solar Energy Conference and Exhibition, Valencia, 1-5 September 2008, pp. 1995-2000.
[15] J. D. Alamo, J. Eguren, A. Luque, Operating limits of Al alloyed high-low junctions for BSF Solar cells. Solid-State Electron, 24, pp 415-419, 1981.
[16] V. Meemongkolkiat, M. Hilali, A. Rohatgi. "Investigation of RTP belt and fired Al-BSF screen printed on textured and planar back surfaces of silicon solar cells", Proceedings of the 3rd World Conference on Photovoltaic Energy Conversion, (2003) pp1467-1470.

SURFACE PASSIVATION OF CRYSTALLINE SILICON BY HYDROGENATED AMORPHOUS SILICON\SUB-NM AL_2O_3 STACK

Asmaa S. A. Ali [a], Osama Tobail* [a,b]
[a] Center of Nanotechnology (CNT), Zewail City of Science and Technology, Sheikh Zayed District, 12588 Giza, EGYPT
[b] Egypt Nanotechnology Center, Cairo University, Shaikh Zayed Campus, 12588 Giza, EGYPT,
Email: ashaaban@zewailcity.edu.eg
Email: otobail@zewailcity.edu.eg

ABSTRACT: Atomic layer deposited aluminum oxide has been widely used as a low deposition temperature alternative material for surface passivation in the area of high efficiency silicon solar cells. Al_2O_3 thin films provide an excellent passivation quality due to chemical and field effect passivation. Fabrication of back side passivated silicon solar cells requires a complex patterning for metal contact, however, ultra-thin Al_2O_3 enables tunneling and hence avoid the complexity of metal contacts. Reducing the thickness of Al_2O_3 results in deterioration of the passivation quality. Thus, in this work, we develop doped hydrogenated amorphous silicon (a-Si:H)/sub-nm Al_2O_3 stack which compensates the passivation quality deterioration and maintains the tunneling to achieve back side passivated contact. The improvement of the surface passivation and tunneling has been observed, since the lifetime of the stack of doped a-Si:H/Al_2O_3 is from three to four times higher than Al_2O_3 reference samples. To investigate the impact of the developed stack, we incorporated the measured surface recombination velocity and contact resistance into PC-1D simulation to compare passivated contact to point contact solar cell. The simulation shows that our passivated contact has a potential to improve the solar cell performance without the need to complicated patterning.

Keywords: passivated tunneling contact, ALD, simulation

1 INTRODUCTION

High efficiency silicon solar cells (> 20 %) [1, 2] require surface passivation for silicon to suppress carrier recombination caused by the dangling bonds as a result of surface termination [3, 4]. Recently, Al_2O_3 has attracted a great interest as surface passivation material for silicon solar cells [2, 4, 5]. Al_2O_3 shows an excellent surface passivation [6] owing to the combination of both i) chemical passivation through the saturation of the dangling bonds resulting in low interface defect density (D_{it}) and ii) field-effect passivation (via repelling charges far from the interface) despite of the high negative charge density (Q_f) contained in the Al_2O_3 layer introduced by Al vacancies and O interstitials [1, 2]. For passivated back side solar cells, the local metal contact limits the cell efficiency due to the resistive losses. Thus the metal contact area needs to be minimized through complex patterning of thousands of point contacts. Resistive losses emerge owing to the limited point contact area. Therefore, the TOPCon [7] passivation scheme suggests covering the full back side area with highly doped nanocrystalline silicon on top of thin tunnel oxide. Also, thinner Al_2O_3 layer that enables tunneling of the carriers can be used as passivated tunnel contact [8]. On the other hand, the passivation quality of the Al_2O_3 layers is deteriorated by decreasing the thickness less than 5 nm [9]. Therefore, a passivation layer or stack of layers is needed to be developed for both n-type and p-type silicon, since the solar cell industry tends to use front and back side passivated solar cell concepts [10]. In addition, the passivating stack should enable tunneling to facilitate contacting the solar cell without complicated patterning of the passivating stack.

In this work, we proposed using sub-nm Al_2O_3 layers deposited by atomic layer deposition (ALD) to passivate the silicon surface followed by a layer of doped hydrogenated amorphous silicon (a-Si:H) to compensate the degraded passivation quality of the ultrathin layers. We investigate the passivation quality of both n-type and p-type a-Si:H on top of Al_2O_3 layer for both n- and p-type silicon wafers. This stack achieved high surface passivation quality which leads to measured minority carrier lifetime up to 2.4 ms for n-type wafer. PC-1D simulation was performed in order to investigate the influence of inserting the developed passivated contact stack in the solar cell compared to using local point contacts. The simulation expects an improvement of the solar cell performance in terms of efficiency and open circuit voltage (V_{oc}). The simulation results promises the fabrication of the passivated back contact cells without the complicated patterning of local point contacts.

2 EXPERIMENTAL DETAILS

Both n-type and p-type float zone (FZ) c-Si <100> double sided polished (DSP) wafers were used to investigate the surface passivation quality of a-Si:H\Al_2O_3 stack. To maximize the effect of the surface passivation on the measured effective minority carrier lifetime measurement, wafers with small thickness of 280 μm and light doping between 7 and 20 Ω.cm were selected. Prior to the deposition, the wafers were treated with diluted HF solution (5%) for 10 min. to remove the native oxide layer and achieve H-terminated surface.

Fig. 1 schematically represents the samples prepared for the lifetime measurement. Al_2O_3 ultrathin films were deposited on both sides by thermal atomic layer deposition (ALD) at 300 °C using Oxford FlexAl™ reactor. Each Al_2O_3 mono-layer were deposited by performing a cycle of a trimethyl aluminum $Al(CH_3)_3$ (TMA) and H_2O pulses with Ar purge after each pulse. In this study, we compare samples with 8 ALD cycles (~0.8 nm) and samples with 14 ALD cycles (~1.7 nm) with respect to lifetime and tunneling quality. Then 10 to 12 nm doped hydrogenated amorphous silicon (a-Si:H) was deposited on top of the Al_2O_3 films by Oxford Plasma Pro System 100 Plasma Enhanced Chemical Vapor Deposition (PECVD). For lifetime measurements, symmetrically a-Si:H/Al_2O_3/c-Si/Al_2O_3/a-Si:H samples were prepared. The a-Si:H were deposited at plasma power of 200 Watt and substrate temperature 250 °C. The n-type doped layers a-Si:H(n/n+) and p-type doped a-Si:H(p/p+) are deposited using the gas flows in table 1. The thicknesses of all deposited thin films were determined by Woollam spectroscopic ellipsometer.

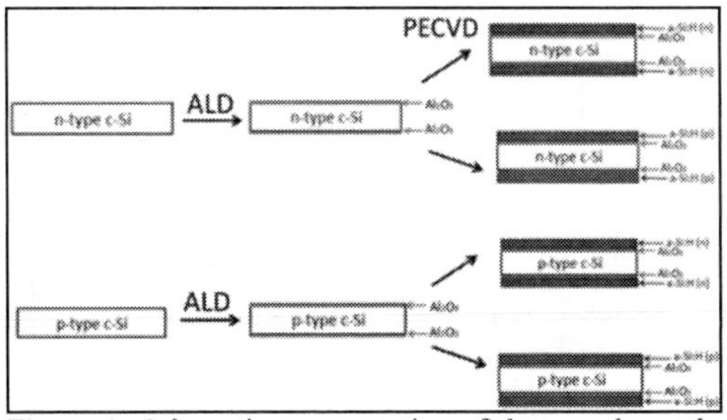

Figure 1: Schematic representation of the experimental procedures for fabrication of the passivating a-Si:H (n\p)\Al_2O_3 stack for lifetime measurements.

The effective minority carrier lifetime was measured by Quasi-Steady-State-Photoconductance (QSSPC) method using Sinton instrument (WCT-120) for as-deposited and annealed samples. The annealing was performed on a hot plate in air for 10 min at a temperature ranging from 50 °C to 540 °C. For current/voltage (*I/V*) measurements, only front side passivated samples with front and rear metallization were prepared. *I/V* characteristics were measured by Keithely 4200 parameter analyzer as a function of annealing temperature as well as as-deposited. Annealing was performed using a hot plate in air from 100 °C – 400 °C for 10 min. 1 μm aluminum contacts were deposited using Oxford Plasma Pro 400 sputtering system.

Table I: PECVD gas flows in sccm. The dopant gases are 1% diluted in H_2.

	SiH_4	H_2	PH_3/H_2(1%)	B_2H_4/H_2(1%)
a-Si:H (n)	25	180	90	
a-Si:H (n+)	5	180	90	
a-Si:H (p)	25	180		90
a-Si:H (p+)	5	180		90

3 RESULTS AND DISCUSSION

Fig. 2 shows the effective minority carrier lifetime measured at minority carrier density (MCD) of 10^{15} cm^{-3} as a function of the annealing temperature for p-type wafers with symmetrical structure of a-Si:H/Al_2O_3/c-Si/Al_2O_3/a-Si:H. The Al_2O_3 thickness of those samples is ~0.8 nm. The samples with both n- and p-type a-Si:H showed an enhancement of the surface passivation compared to the reference. In addition, the a-Si:H/Al2O3 showed an optimum annealing temperature of 200 °C for p-type a-Si:H and 250 °C for n-type a-Si:H compared to the 450 °C for reference sample with only Al_2O_3 passivation. This improvement in passivation might be ascribed to the enhanced chemical passivation due to hydrogen diffusion from the a-Si:H layer to the interface during the annealing process as a-Si:H could be considered as a hydrogen source [11]. In addition, the enhanced field effect passivation due to the increase of the negative charges from the n-type a-Si could contribute to the improvement of the measured lifetime. It is clear from Fig. 2 that n-type a-Si:H is more preferable than p-type a-Si:H for passivating p-type wafers probably due to the enhanced field-effect passivation part, as also reported in ref [12]. Beside the advantage of decreasing the needed annealing temperature from 450 °C to 250 °C for n-type a-Si:H compared to reference, also, the measured lifetime is three times higher for the n-type a-Si:H. Also, it is observed that increasing the doping of both n- and p-type a-Si:H layer deteriorates the passivation to values less than the reference sample. Fig. 3. illustrates that the samples deposited on p-type c-Si wafers show the same behaviour as n-type c-Si substrates.

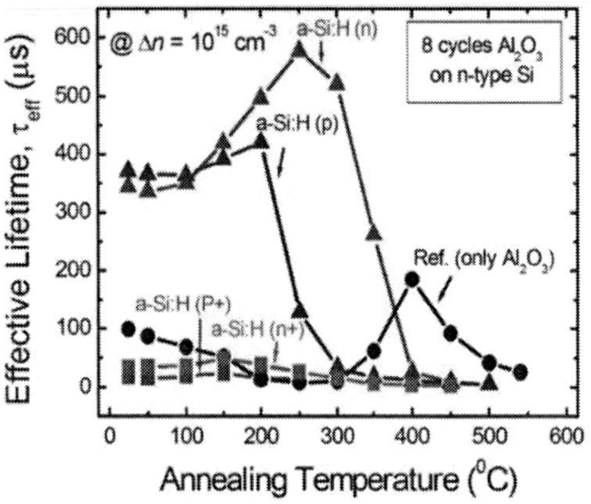

Figure 2: Measured effective minority carrier lifetime as a function of annealing temperature of n-type c-Si wafers passivated with a-Si:H (n\n+ and p\p+)\sub-nm Al_2O_3 compared with reference (only Al_2O_3).

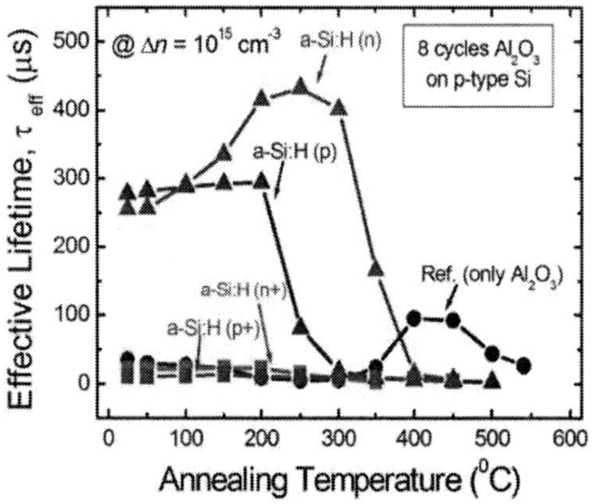

Figure 3: Measured effective minority carrier lifetime as a function of annealing temperature of p-type c-Si wafers passivated with a-Si:H (n\n+ and p\p+)\sub-nm Al_2O_3 compared with the reference (only Al_2O_3).

Fig. 4 and Fig. 5 depict the results of the effective lifetime measurements for thicker Al_2O_3 layer with a thickness of ~1.7 nm for n-type and p-type c-Si wafers respectively. Thicker Al_2O_3 films resulted in four times higher lifetimes than that of sub-nm Al_2O_3 layer (Fig. 2 and Fig. 3). Deposition of n-type a-Si:H on n-type c-Si wafer resulted in a lifetime of ~2.4 ms at 250 °C while gave a lifetime of ~2 ms at 300 °C when it deposited on p-type c-Si wafer. This could be attributed to increasing field effect passivation introduced by thicker Al_2O_3 film. The same results were shown for increasing the doping concentrations of n- and p-type a-Si:H.

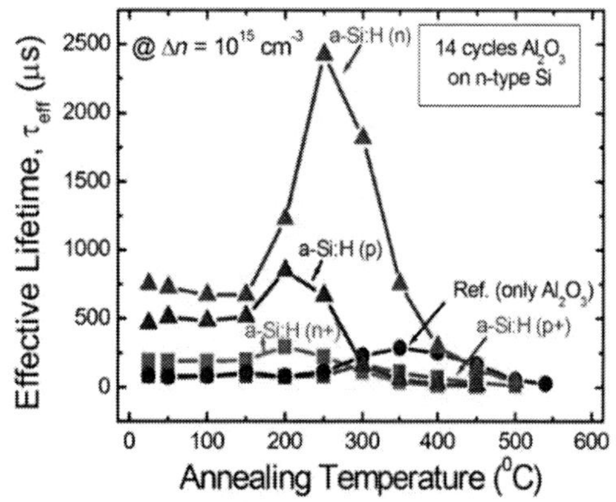

Figure 4: Measured effective minority carrier lifetime as a function of annealing temperature of n-type c-Si wafers passivated with a-Si:H (n\n+ and p\p+)\double of Al_2O_3 thickness compared with reference (only Al_2O_3).

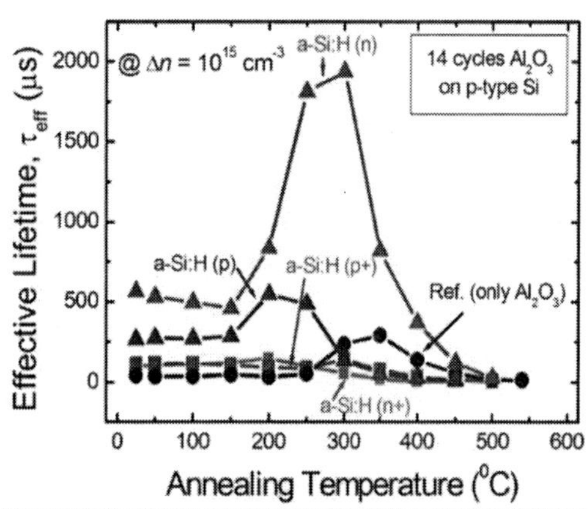

Figure 5: Measured effective minority carrier lifetime as a function of annealing temperature of p-type c-Si wafers passivated with a-Si:H (n\n+ and p\p+)\double of Al_2O_3 thickness compared with reference (only Al_2O_3).

We observed that the best surface passivation is obtained on the case of n-type a-Si:H/Al_2O_3 stack on both n-type and p-type wafers. Therefore, we measured the tunneling effect of sub-nm Al_2O_3 (~0.8 nm) in terms of current / voltage (*I/V*) characteristics of asymmetrical sample structures of Al/ a-Si:H(n)/Al_2O_3 on top of both n-type and p-type wafers with Al back contact. The test devices have an area of 0.04 cm^2 each.

Fig. 6 and Fig. 7 represent the measured *I/V* curves of the tunnel devices on n-type and p-type wafers, respectively, as a function of the annealing temperature. The contact of the devices on n-type (Fig. 6) was enhanced at annealing temperature of 200 °C, which coincides with the optimum annealing temperature range for lifetime; however, it does not show an ohmic contact. An ohmic contact is achieved for devices on p-type (Fig. 7) wafers only at an annealing temperature starting from 300 °C, which coincides with the optimum annealing temperature range for minority carrier lifetime shown in Fig. 3. Therefore, we conclude that the optimum configuration for best passivation and tunneling is a-Si:H(n)/Al_2O_3 on p-type wafers when annealed at 300 °C.

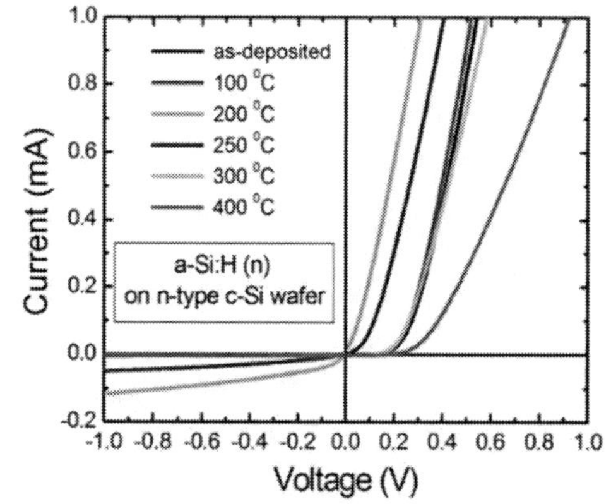

Figure 6: *I\V* curve of n-type c-Si wafer passivated with a-Si:H (n)\sub-nm Al_2O_3 at different annealing temperatures. The behavior of *I\V* curve shows non-ohmic contact with decreased barrier at 200 °C.

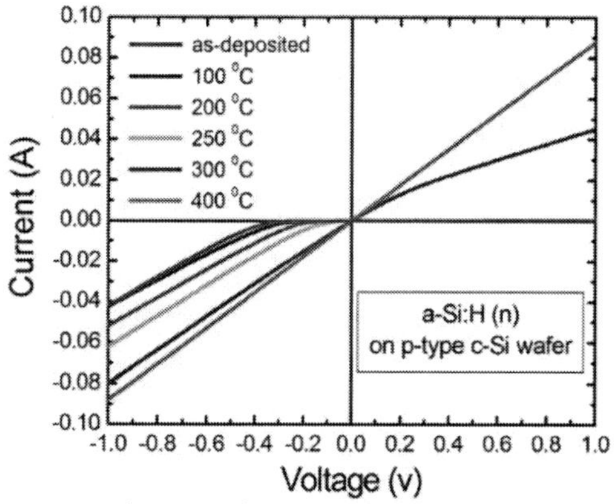

Figure 7: *I\V* curve of p-type c-Si wafer passivated with a-Si:H (n)\sub-nm Al_2O_3 at different annealing temperatures. The behavior of *I\V* curve changes with annealing temperature showing an ohmic behavior starting at 300 °C.

The observed non-ohmic behavior could be due to the re-grown native oxide after the HF dip and before transferring the sample to the ALD chamber. This grown native oxide adds to the insulator thickness and hence decreases the tunneling effect.

The best parameters using the sub-nm Al_2O_3 thickness regarding to the lifetime and tunnelling was obtained when depositing n-type a-Si:H on top of p-type wafer when annealed at 300 °C. Therefore, these parameters were incorporated into PC-1D code [13] to simulate the solar cell performance with surface passivation quality and electrical contact resistance dictated by the developed stack. We compare our developed passivated contact with the conventional passivation layer with point contacts at variable metallization fraction (α). The contact resistance of the developed passivated contact cell was extracted from the *I*/*V* measurement in Fig. 7 to be 0.95 $\Omega.cm^2$. The corresponding maximum surface recombination velocity $S_{eff\text{-}max}$ = 35 cm/s is determined from the measured carrier lifetime in Fig. 3 according to the equation $S_{eff\text{-}max} = W/(2{\cdot}\tau_{eff})$, while W is the wafer thickness and τ_{eff} is the measured minority carrier lifetime [14]. Regarding the solar cell with local point contact, the Al/Si contact resistance is taken as 3700 Ohm μm^2 [15] and the surface recombination velocity was determined from minority carrier lifetime measurement for double sided 10 nm thick Al_2O_3 coated wafer to be $S_{eff\text{-}max}$ = 20 cm/s, while $S = 10^5$ cm/s for metallized area. The effective contact resistance and effective recombination velocity were calculated as a function of the metallization fraction and used in the PC-1D simulation.

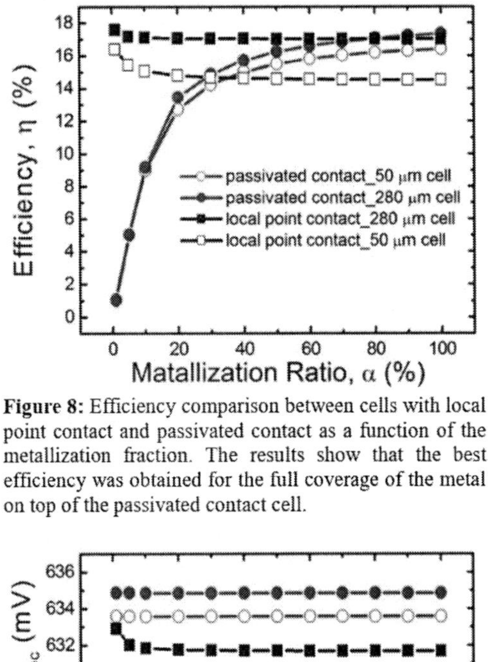

Figure 8: Efficiency comparison between cells with local point contact and passivated contact as a function of the metallization fraction. The results show that the best efficiency was obtained for the full coverage of the metal on top of the passivated contact cell.

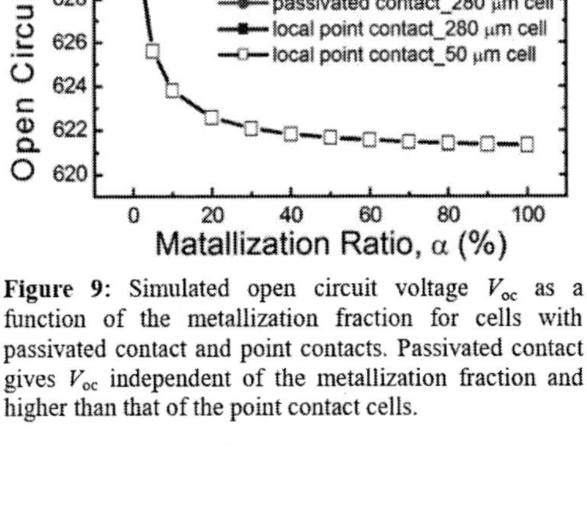

Figure 9: Simulated open circuit voltage V_{oc} as a function of the metallization fraction for cells with passivated contact and point contacts. Passivated contact gives V_{oc} independent of the metallization fraction and higher than that of the point contact cells.

Fig. 8 compares the efficiency of a cell with passivated contact and to a cell with local point contacts as a function of the metallization fraction as resulted from PC-1D. The simulation shows that efficiency increases with increasing the metallization fraction in the case of the passivated contact cell, while decreases for the local point contacts cells. Therefore, using our developed passivated contact could lead to solar cell with comparable performance of a cell with a complicated point contact with 1 to 5 % metallization fraction. Fig. 9 represents the open circuit voltage (V_{oc}) results in case of passivated contact as a function of metalization fraction compared to local point contact cell. The simulation shows that passivated contact resulted in higher V_{oc} than local point contact cells. Besides, passivated contact appears to be independent on the metalization fraction while drops in case of local point contacts cell. It is obvious from the Fig.8 and Fig. 9 that the effect of full metallization coverage on both efficiency and open circuit voltages becomes more effective with decreasing the cell thickness.

4 SUMMERY AND CONCLUSION

In this work, passivation quality of n/n+ and p/p+ type doped a-Si:H/Al_2O_3 stack was tested on both n and p-type c-Si wafers. We used two sets of Al_2O_3 thickness; ~0.8 nm and ~1.7 nm. We found that deposition of n-type a-Si:H on top of the Al_2O_3 layer shows higher lifetimes than p-type a-Si:H. Thicker Al_2O_3 (~1.7 nm) also led to higher lifetime than sub-nm Al_2O_3 (~0.8 nm). Increasing the doping of both n- and p-type a-Si:H layer deteriorates the passivation to values close to or even less than the reference sample. Current/voltage (*I*/*V*) measurements were performed as an investigation of tunnelling effect of n-type a-Si:H/sub-nm Al_2O_3 on n-type and p-type c-Si wafers. An ohmic contact was observed for sample deposited on p-type c-Si substrate starting at 300 °C of annealing temperature which works in well with achieving high passivation. Sample deposited on n-type substrate did not show an ohmic contact while it showed an enhancement of *I*/*V* current at 200 °C annealing temperature which also coincides with the range of annealing temperatures for maximum lifetimes.

In conclusion, deposition of doped a-Si:H on top of ultra-thin Al_2O_3 improves the passivation quality, since the lifetime of the stack of doped a-Si:H/Al_2O_3 is from three to four times higher than Al_2O_3 reference samples. In addition, we found that n-type a-Si:H is preferable for passivating both n- and p-type Si wafers. Using the sub-nm Al_2O_3 improves the tunneling of the carriers which facilitates the contact of the solar cell. PC-1D simulation expects an improvement in cell performance in terms of V_{oc} and efficiency by full coverage of the developed passivated contact stack.

5 REFERNCES

[1] P. Ortega, G. López, A. Orpella, I. Martín, M. Colina, C. Voz, S. Bermejo, J. Puigdollers, M. García, and R. Alcubilla, Proceedings of the 8th Spanish Conference on Electron Devices, CDE'2011 (2011) 1-4.

[2] Yan Zhao, Chunlan Zhou, Xiang Zhang, Peng Zhang, Yanan Dou, Wenjing Wang, Xingzhong Cao, Baoyi Wang, Yehua Tang, Su Zhou, Nanoscale Research Letters 8 (2013) 114.

[3] Mohammad Ziaur Rahman, Shahidul Islam Khan, Materials for Renewable and Sustainable Energy 1 (2012) 1.
[4] J. Schmidt, A. Merkle, R. Brendel, B. Hoex, M. C. M. van de Sanden, W. M. M. Kessels Progress in Photovoltaics: Research and Applications 16 (2008) 461–466.
[5] G. Dingemans, W. M. M. Kessels, Journal of Vacuum Science and Technology A 30 (2012) 040802
[6] G. Dingemans, R. Seguin, and P. Engelhart, Physica Status Solidi RRL 4 (2010) 10.
[7] S. W. Glunz, F. Feldmann, A. Richter, M. Bivour, C. Reichel, H. Steinkemper, J. Benick, M. Hermle, 27th European Photovoltaic Solar Energy Conference and Exhibition, (2015) 259 - 263.
[8] X. Bubnova, A. Petersen, F. Einsele, R. Rölver, U. Rau, C. Leendertz, T. Wagner, 31st European Photovoltaic Solar Energy Conference and Exhibition, (2012) 1538.
[9] N. M. Terlinden, G. Dingemans, M. C. M van de Sanden, and W. M. M. Kessels, Applied Physics Letters 96 (2010) 112101.
[10] Martin A. Green, Solar Energy Materials and Solar Cells 143 (2015) 190–197.
[11] Hosny Meddeb Twan Bearda, Ibrahim Abdelwahab, Valentina Ferro, Barry O'Sullivan, Yaser Abdulraheem, Hatem Ezzaouia, Ivan Gordon, Jozef Szlufcik, Jef Poortmans, Energy Procedia 55 (2014) 818 – 826.
[12] S. De Wolf and M. Kondo, Journal of Applied Physics 105 (2009) 103707.
[13] M. Belarbi, A. Benyoucef, B. Benyoucef, Advanced Energy: An International Journal (AEIJ) 1 (2014) 3.
[14] Thomas Muller, Heterojunction solar cells (a-Si/ c-Si) (2009) 14.
[15] Simon S. Cohen, Gennady Sh. Gildenblat, VLSI Electronics Microstructure science, Metal – Semiconductor Contacts and Devices Vol 13 (1986) 247

PROCESS DEVELOPMENT FOR SILICON HETEROJUNCTION SOLAR CELLS

M. Hendrichs, A. Morales-Vilches, L. Mazzarella, S. Kirner, M. Zelt, H. Rhein, S. Calnan, L. Korte, B. Stannowski, R. Schlatmann
PVcomB/Helmholtz-Zentrum Berlin für Materialien und Energie GmbH, Schwarzschildstr. 3, 12489 Berlin, Germany

ABSTRACT: This work gives an overview about the current status and the ongoing development of the silicon heterojunction (SHJ) solar cell process at HZB. Our research work focuses on the development of highly transparent nanocrystalline front surface field (FSF) layers and innovative transparent conductive oxide (TCO) layers as front electrodes for rear emitter SHJ solar cells. By introducing water vapor into the front ITO sputtering process we show an increase of 0.4 % absolute in conversion efficiency compared with our standard ITO process without water vapor. Furthermore first results for a rear emitter cell structure are presented with a conversion efficiency exceeding 21% for 4 cm² SHJ solar cells on n-type CZ-Si. Introducing further process optimizations, we expect to achieve conversion efficiencies above 22% for our baseline process sequence in the near future.
Keywords: a-Si:H, Heterojunction, ITO, Micro Crystalline Si, n-type, Silicon Solar Cell

1 INTRODUCTION

At the HZB's institute PVcomB the silicon heterojunction (SHJ) solar cell research covers a wide range of activities, such as

1) evaluation of innovative materials for transparent conductive oxide (TCO) contact layers, e.g. hydrogen doped In_2O_3:H (IOH), In_2O_3:Sn (ITO) sputtered with water vapor, and ZnO:Al (AZO) [1]
2) process optimization for nanocrystalline silicon (nc-Si:H) and nanocrystalline silicon oxide (nc-Si:O_X) layers as alternative for amorphous silicon (a-Si:H) layers [2]
3) development of high efficiency back-contact cell concepts [3]
4) benchmarking with industry and R&D partners [4]
5) bottom cells in high efficiency tandem devices, such as silicon/perovskite solar cells [5]
6) solar fuel devices [6].

A stable and reliable SHJ cell baseline fabrication process is the key to exploit synergistic effects between all research activities, aiming at continuously enhancing the solar cell performance by introducing innovations and enabling reduction of production costs.

This contribution gives an overview about the current status of the SHJ baseline process at HZB and shows selected results of recent optimization runs aimed at improving the n-doped nc-Si:H layer as well as the front ITO layer.

2 APPROACH

Fig. 1 shows the process sequence of the SHJ solar cell baseline fabrication at HZB. Generally all process steps are compatible with wafer sizes up to edge lengths of 156 mm. Recently n-type Czochralski silicon (Cz-Si) was introduced to replace n-type Floatzone silicon (Fz-Si) as the base material.

The process sequence starts with sequential wet chemical batch processes to remove the saw damage, followed by surface texturing and cleaning. Immediately after removing the surface native oxide in 1% diluted hydrofluoric acid, thin intrinsic and doped silicon layers are deposited by plasma enhanced chemical vapor deposition (PECVD). We use a three chamber Applied Materials AKT 1600 cluster tool operating at 13.56 MHz plasma excitation frequency. Plasma diagnostics enable in situ monitoring of all relevant deposition process conditions [7].

Subsequently, TCO layers are applied on both sides of the solar cell using an inline DC sputtering tool from Leybold Optics. TCO materials are In_2O_3:H (IOH), In_2O_3:Sn (ITO), and ZnO:Al (AZO), spanning the entire range from best performing material to lowest cost potential. In case of monofacial solar cells, the rear metallization consists of a full-area sputtered silver contact layer. For bifacial solar cells silver grids are applied by screen printing low-temperature silver pastes and subsequent curing. The current-voltage (*j-V*) characteristics of finished solar cells are measured in an AAA dual-source sun simulator under steady

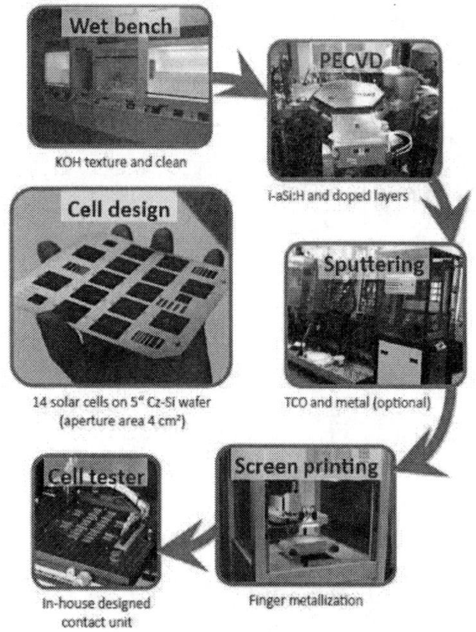

Figure 1: SHJ cell baseline fabrication process at HZB. Standard baseline cells are fabricated on 5-inch (5") n-type Cz-Si wafers carrying 14 SHJ solar cells each with an aperture area of 4 cm².

illumination. For our standard baseline cell design that consists of 14 SHJ cells (area of 4 cm^2) on 5-inch (5") Cz-Si wafer, we design a customized contact unit based on printed circuit board technology.

3 RESULTS

3.1 Optimization of nc-Si:H n-doped silicon layer

Fig. 2 shows the *j-V* parameters of 4 cm^2 n-type Cz-Si monofacial SHJ solar cells with front emitter structure and two different nc-Si:H back surface field (BSF) layers. BSF1 is nominally 15 nm thick and is deposited using a silane to hydrogen mixture of 1:200. The BSF2 process features a nominal layer thickness of 23 nm and a silane to hydrogen mixture of 1:300. For both cell groups, AZO is applied as rear TCO and ITO as front TCO. The front metal grid is screen-printed.

The cell data reveals a clear advantage for the BSF2 process. The median conversion efficiency is increased by 0.4% absolute compared with the BSF1 process. The increased layer thickness and the higher hydrogen dilution of BSF2, lead presumably to a higher crystalline volume fraction at the interface to the TCO. This leads to a lower contact resistivity between the BSF and the rear TCO layer, thus increasing the fill factor of the solar cells.

Based on the BSF process optimization for front emitter solar cells, we fabricated rear emitter solar cells, now applying the optimized BSF layer as front surface field (FSF). Table 1 shows the *j-V* characteristics of the best rear emitter solar cell. The rear emitter solar cell structure shows a higher fill factor in comparison with the front emitter cell structure. This probably originates from an improved lateral carrier collection at the grid electrode in the case of the rear emitter cell structure [8].

3.2 Introduction of water vapor to the front ITO sputtering process

Fig. 3 displays the *j-V* data of 4 cm^2 n-type Cz-Si monofacial SHJ solar cells with front emitter structure and two different front TCO layers. Both TCOs are sputtered from the same In_2O_3:SnO_2 (90/10 wt%) planar target using an argon-to-oxygen 98:2 mixture as sputter

Table 1: *j-V* characteristic of the best 4 cm^2 SHJ solar cell with rear emitter structure and nc-Si:H BSF1 process applied as FSF.

η (%)	FF (%)	j_{SC} (mA/cm²)	V_{OC} (mV)	pFF (%)	r_S (Ωcm²)
21.2	78.9	37.6	715	82.5	0.5

gas at a process pressure of $3.7x10^{-3}$ mbar. For the ITO group, the sputtering chamber is heated at T_{set} = 220 °C. In case of the ITO:H group, the sputter chamber operates without intentional heating. In addition to the standard process gases (argon and oxygen), water vapor is introduced into the process chamber [9]. The resulting water partial pressure amounts to $5.6x10^{-6}$ mbar. Before screen printing, the solar cells of the ITO:H group are pre-annealed at 180 °C for 5 minutes on a laboratory hotplate. For both groups, the standard annealing of the silver polymer paste is carried out on the same hotplate at 190 °C for 7 minutes after screen printing.

As the data clearly shows, the introduction of water vapor increases the short circuit current density without influencing the fill factor and the open circuit voltage, resulting in a gain of 0.4 % absolute in conversion efficiency. Spectral response measurements reveal lower parasitic absorption losses in the ITO:H solar cells.

Finally, the electrical contact resistivity between the front TCO layer and the screen-printed contact finger metallization is measured using the transfer length method (TLM) [10]. For both groups the contact resistivity is around 1-2 mΩcm², i.e. in an acceptably low value range. Thus, the ITO:H process is compatible with conventional screen printing metallization. This is an advantage as compared to IOH front TCO layers, where increased contact resistivities were detected in case of screen-printed silver metallization [11].

4 CONCLUSION AND OUTLOOK

The SHJ baseline process is the ideal platform for synergies between the research activities at HZB. Currently, the focus of the SHJ baseline optimization is the development of high transparency FSF silicon layers and front TCO layers for rear emitter solar cells.

In this contribution we present results of optimized nc-Si:H n-doped silicon layers. By increasing the layer

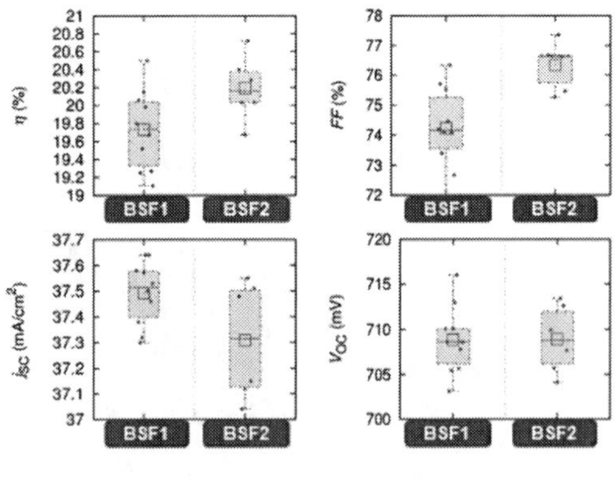

Figure 2: *j-V* parameters of front emitter SHJ solar cells with an aperture area of 4 cm² and two different BSF processes BSF1 (10 cells) and BSF2 (6 cells). Cell efficiencies are measured with an in-house AAA dual-source cell tester.

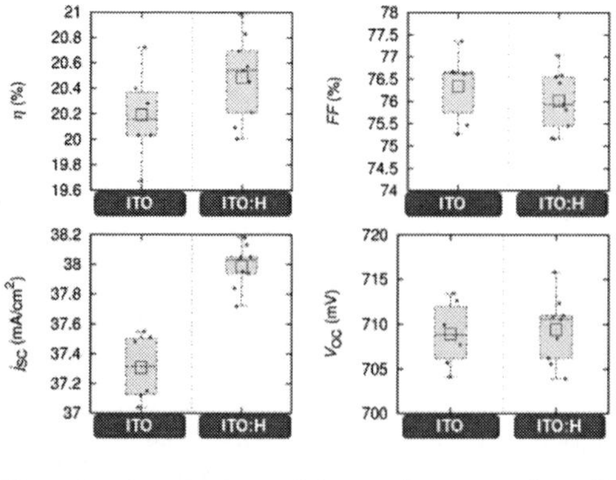

Figure 3: *j-V* parameters of front emitter SHJ solar cells with an aperture area of 4 cm² and two front TCO processes ITO (6 cells) and ITO:H (9 cells). Cell efficiencies are measured with an in-house AAA dual-source cell tester.

thickness and the hydrogen dilution the median conversion efficiency of front emitter SHJ solar cells is increased by 0.4% absolute. Applying the optimized nc-Si:H n-doped silicon layer as FSF in rear emitter solar cells we show peak conversion efficiencies over 21%. Furthermore we show that the introduction of water with a low partial pressure in the sputtering chamber reduces the parasitic absorption in front ITO:H layer. On cell level, the resulting current density gain results in a conversion efficiency gain of 0.4% absolute.

These results represent the starting point for further front layer optimization runs for rear emitter solar cells. Future experiments will focus on the reduction of the nc-Si:H FSF silicon layer thickness in order to enhance the short circuit current density of the cell concept. Additionally, we plan to implement improved IOH and ITO:H layers as front TCO.

ACKNOWLEDGEMENT

The authors gratefully thank Ana-Maria Moldovan from Fraunhofer ISE for wafer texturing and cleaning. Furthermore, the authors thank Manuel Hartig, Katja Mayer-Stillrich, Tobias Henschel, Denise Debrassine and Umut Özer for assistance in processing and characterizing the solar cells investigated here. Furthermore, we gratefully acknowledge the financial support by the German Federal Ministry of Economic Affairs and Energy within the HERA project.

REFERENCES

[1] H. Scherg-Kurmes *et al.* "High mobility In2O3:H as contact layer for a-Si:H/c-Si heterojunction and μc-Si:H thin film solar cells", Thin Solid Films Volume 594 ,pp. 316.322, 2015

[2] L. Mazzarella *et al.* "p-type microcrystalline silicon oxide emitter for silicon heterojunction solar cells allowing current densities above 40mA/cm2", APL 106, 2015.

[3] S. Ring *et al.* "Emitter patterning for back-contacted Si heterojunction solar cells using laser written mask layers for etching and self-aligned passivation (LEAP)", IEEE Journal of Photovoltaics, Volume 6, Pages 894-899, 2016.

[4] S. Calnan *et al.* "In2O3:Sn front electrodes grown from sputter targets with an SnO2 content below 10 wt.% for silicon herterojunction solar cells", Proc. EU-PVSEC, Munich, 2016.

[5] S. Albrecht *et al.* "Monolithic perovskite/silicon-heterojunction tandem solar cells processed at low temperature", Energy Environ. Sci., Pages 81-88, 2016.

[6] S. Kirner *et al.* "Architectures for scalable integrated photo driven catalytic devices - a concept study", International Journal of Hydrogen Energy, in press, 2016.

[7] O. Gabriel *et al.* "Plasma monitoring and PECVD process control in thin film silicon-based solar cell manufacturing", EPJ Photovoltaics 5, 55202, 2014.

[8] M. Bivour *et al.*, "Rear emitter silicon heterojunction solar cells: fewer restrictions on the optoelectrical properties of front side TCOs", Energy Procedia, Volume 55, Pages 229-234, 2014.

[9] D. Schroder *et al.* "Solar cell contact resistance - a review". IEEE Transactions on Electron Devices 3, Pages 637–647, 1984.

[10] T. Koida *et al.* "Hydrogen-doped In2O3as high-mobility transparent conductive oxide", Jpn. J. Appl. Phys., vol. 46, Pages L685–L687, 2007.

[11] L. Barraud *et al.*, "Hydrogen-doped indium oxide/indium tin oxide bilayers for high-efficiency silicon heterojunction solar cells", SOLMAT 115, Pages 151-156, 2013.

SILICON OXYNITRIDE–SILICON NITRIDE SURFACE PASSIVATION OF P-TYPE C-SI SOLAR CELLS WITH LASER FIRED REAR CONTACTS

Anishkumar Soman[1], Som Mondal[1], Swasti Bhatia[2], Balraj Arunachalam[3], Sandeep Kumbhar[3], S. Saravanan[3], Pradeep Nair[2], Aldrin Antony[1]*
[1]Department of Energy Science and Engineering, Indian Institute of Technology Bombay
[2]Department of Electrical Engineering, Indian Institute of Technology Bombay
[3]National Centre for Photovoltaic Research and Education (NCPRE)
Mumbai - 400076, India
*phone: +91 2225764400, Fax: +91 2225723480, email: aldrinantony@iitb.ac.in

ABSTRACT: Passivated Emitter and Rear Cell (PERC) concepts are industrially viable methods for the production of high efficiency crystalline silicon solar cells. The rear dielectric layer acts as a passivation layer as well as a back reflector. Passivation of the rear surface forms an important aspect which affects the open circuit voltage (V_{OC}), and hence the efficiency of the cell. In this work, a novel rear surface passivation scheme has been applied for CZ p-type c-Si cells with Laser Fired Contacts (LFC) at the rear side. The passivation of p-type rear surface is accomplished by using Silicon Oxynitride (SiON) single layer with a low refractive index of 1.53 and SiON/Silicon Nitride (SiN) stack layers deposited using rf-PECVD at a temperature of 380°C. The point contacts through the dielectric passivation layer is made using a 1070 nm microsecond (μs) pulsed laser. The LFC solar cells have been fabricated over an area of 156 cm^2 with rear passivation layers of SiON and SiON/SiN stacks and the importance of the Silicon Nitride capping layer has been investigated in detail. The cells have been characterized using I-V and EQE measurements and obtained a V_{OC} of 628 mV in the initial trials with single layer SiON.
Keywords: Passivation, Laser Fired Contact, crystalline silicon solar cells.

1 INTRODUCTION

Conventional Monocrystalline Silicon solar cells have screen printed Aluminium-Back Surface Field (Al-BSF) which suffers from high recombination at the rear end along with contributing to bowing the wafer [1]. In order to overcome this limitation a passivated rear surface contact cell structure called as Passivated Emitter Rear Contact (PERC) structure [2] was introduced which gave improved efficiency of over 20%. However the selective passivation removal steps by photolithography and high cost limits its application for a large scale production [3]. An alternate back contact scheme which can overcome these limitations is Laser Fired Contact (LFC) [4], where the Al deposited over the rear passivation layer is fired through the passivation layer to form contact with the silicon beneath. The laser firing is performed at room temperature thereby avoiding the high temperature firing process at ~850 °C, which is normally used for screen-printed contacts. The recombination losses at the rear surface are considerably reduced due to the passivation layer as against Al-BSF. Also the LFC process has high throughput, economically viable and doesn't contribute to bowing of the cell thereby making it an interesting candidate for commercial PERC solar cells.

In this paper p-type crystalline silicon (c-Si) solar cells were fabricated with silicon Oxynitride (SiON) rear side passivation and LFC replacing the Al-BSF. Previously Silicon Carbide (SiC), Silicon Oxide (SiO_2), Amorphous Silicon (a-Si) and Aluminium Oxide (Al_2O_3) have been used as rear passivation schemes for LFC cells [5][6][7][8]. In this work SiON is used as the rear surface passivation layer and a capping layer of Silicon Nitride (SiN) is also used [9][10] to improve surface passivation quality in LFC solar cells. The novel passivation methods using SiON/SiN layers are industrially feasible and use rf-PECVD which is currently used in the industries for depositing anti-reflection coating (ARC) for solar cells. The LFC solar cells, thus, fabricated were characterized by IV and EQE measurements and the results are presented.

2 EXPERIMENTAL

2.1 Fabrication

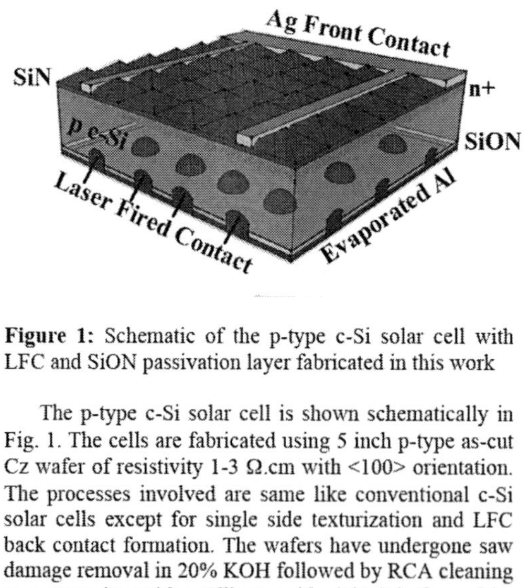

Figure 1: Schematic of the p-type c-Si solar cell with LFC and SiON passivation layer fabricated in this work

The p-type c-Si solar cell is shown schematically in Fig. 1. The cells are fabricated using 5 inch p-type as-cut Cz wafer of resistivity 1-3 Ω.cm with <100> orientation. The processes involved are same like conventional c-Si solar cells except for single side texturization and LFC back contact formation. The wafers have undergone saw damage removal in 20% KOH followed by RCA cleaning to remove impurities. Silicon oxide (SiO_2) is formed by thermal oxidation at 1000 °C followed by a single side oxide removal to prevent texturization of back surface by acting like a masking layer. After texturization in 4% KOH, the oxide is removed by 2% HF. The wafer then undergoes another RCA cleaning to remove K contaminants followed by phosphorus diffusion using $POCl_3$ at 820 °C and phosphor-silicate glass (PSG) removal. The edges are then isolated and ARC SiN is coated using rf-PECVD. The same system is used to deposit SiON and SiON/SiN passivation layers at the back surface. The SiON deposited are hydrogenated and amorphous and could be presented as a-SiON:H. The front contact is made using screen printing of Ag paste and then firing it at 850 °C. Al is then evaporated by electron beam evaporation over the back surface followed by laser firing the Al through passivation layers. The laser firing was performed at 10 kHz with 11 µs pulse duration and 100 pulses based on optimization experiments. The pitch of the contact was chosen to be 0.5 mm for 16 cm^2 cell and 0.3 mm for 156 cm^2 Annealing at 375 °C for 20 min in N_2 ambient was performed after the laser firing for the 16 cm^2 cell and annealing at 380 °C for 20 min in forming gas was performed after passivation layer deposition for 156 cm^2 cell. The process flow of the fabrication steps of the solar cell is given in Fig. 2 below for better understanding.

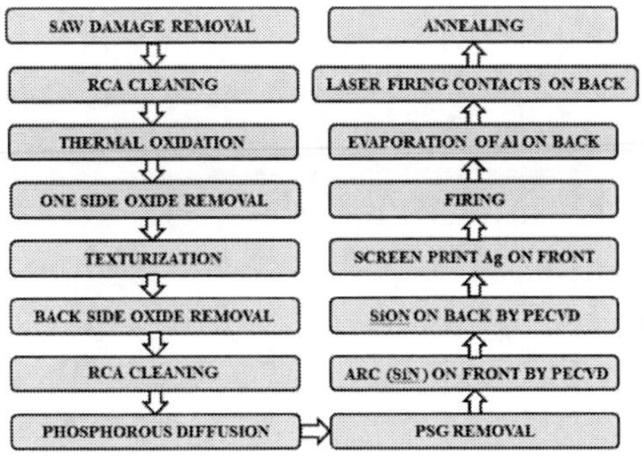

Figure 2: Flowchart for fabrication steps of solar cell

2.2 Characterization

The minority carrier lifetime (τ_{eff}) and Implied V_{OC} of the p-type diffused wafer with SiN on textured front side and passivation layer on at the rear side without any contacts have been measured using QSSPC mode in WCT 120 by Sinton Instruments. The EQE of the cells has been measured by using Bentham PV300 system and the I-V is measured using ABET technologies.

3 RESULTS AND DISCUSSION

3.1 Lifetime and Implied V_{OC} measurement of solar cell

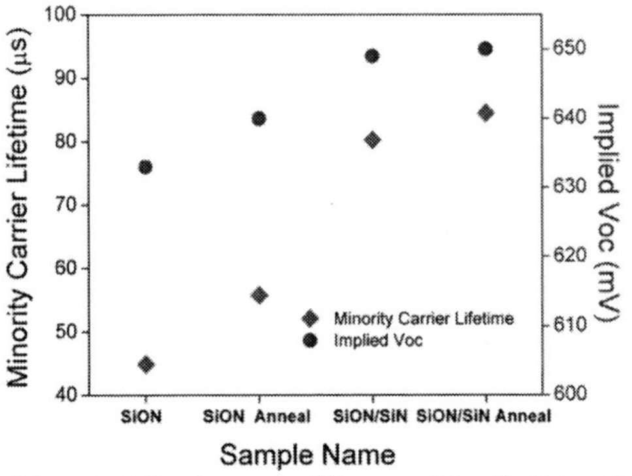

Figure 3: Minority carrier lifetime and Implied V_{OC} of p-type solar cells with SiON and SiON/SiN stack at the back surface; with and without annealing.

The QSSPC measurement at excess minority carrier density (Δn) of 10^{15} cm^{-3} of P diffused wafer with SiN ARC on the textured front and passivation layer (SiON or SiON/SiN) at the as-cut rear surfaces without metal contact is presented in Fig. 3. Lifetime of 45 µs with an Implied V_{OC} (iV_{OC}) of 633 mV was observed for sample with SiON passivation whereas 80 µs of lifetime and 649 mV of iV_{OC} were obtained with SiON/SiN stack passivation at the rear surface. It was also observed that for single layer SiON passivation, N_2 annealing and laser firing leads to degradation of the passivation quality due to H effusion [11] which can be avoided by SiN capping layer, thereby improving the passivation quality of the films [12]. As a result, improved lifetime for SiON/SiN passivation was obtained. In order to prevent the degradation of the passivation quality during annealing it is important to prevent breakage of Si-H bond and H effusion. It has been found that Forming Gas Annealing (FGA) at 380 °C for 20 minutes helps to retain the passivation quality since forming gas helps to replenish the H content in the film. After FGA, it was found that sample with SiON showed a lifetime of 56 µs and iV_{OC} of 640 mV. Whereas sample with SiON/SiN did not show any considerable improvement with FGA; giving a lifetime of 84 µs with an iV_{OC} of 650 mV. Another method to retain the passivation quality at the back surface is to anneal the solar cell after depositing Al over the passivation layer, where the Al layer could prevent the effusion of H from the passivation layer.

3.2 Solar Cell Characterization

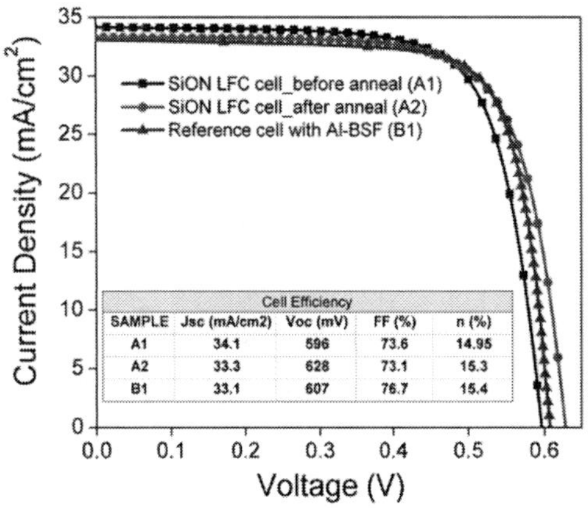

Figure 4: J-V characteristics of p type solar cells with SiON passivation and LFC at the back; with and without annealing with respect to a reference solar cell having Al-BSF.

To understand the effect of annealing on the device level small area cells of area 16 cm^2 has been fabricated using the structure given in Fig.1. The cell has a single layer SiON as a passivation layer and reflector. It has been found that before anneal the cells have a V_{OC} of 596 mV and an efficiency of 14.95%. After annealing the V_{OC} improves by 32 mV to 628 mV and the efficiency to 15.3%. This increase in V_{OC} can be attributed to the activation of charges as well as "al-annealing" wherein the Al layer prevents the degradation of the SiON layer beneath by H effusion. The performance of our passivation layer and LFC at the rear can be better understood by comparing it with a reference solar cell made in the same batch with Al-BSF at the back side on 156 cm^2. The reference cell has a V_{OC} of 607 mV with an efficiency of 15.4 %. Thus it can be clearly see that the SiON passivation helps to improve the V_{OC} by 21 mV. However there is significant difference in efficiency due to the fact that the reference cell has a higher fill factor of 76 % while the cell with SiON rear passivation exhibit lower fill factor of 73 % which could be further improved by optimizing the LFC to smaller pitch along with improving the shunt resistance. The J-V characteristics of the cells before and after annealing along with the reference cell are presented in Fig. 4.

The wavelength dependent absorption of the solar cell using External Quantum Efficiency (EQE) measurement has been investigated. The EQE can also help us understand the performance of the solar cell especially at the longer wavelength which corresponds to the effectiveness of the rear side passivation. A comparative study has been made with a reference cell made in the same batch with an Al-BSF as done previously and is presented in Fig. 5. As compared to the reference solar cell, the cell with SiON rear passivation and LFC has lower external quantum efficiency at the longer wavelength (> 900 nm). This could be attributed to the fact that the cell with SiON at rear surface is single side textured so the reflection losses are high at the rear surface. In order to understand the actual working of the cell under lighted condition, the EQE was measured with bias light and it was observed that under bias the cell performs better than the reference cell at longer wavelength. The bias light helps to saturate the defect states as well as helps in charge activation thereby resulting in higher carrier collection at longer wavelength. We also observe improved performance in the visible range with and without bias for cell with SiON passivation as compared to reference cell with Al-BSF. This could be attributed to the impurity gettering, which takes place during oxidation and removal.

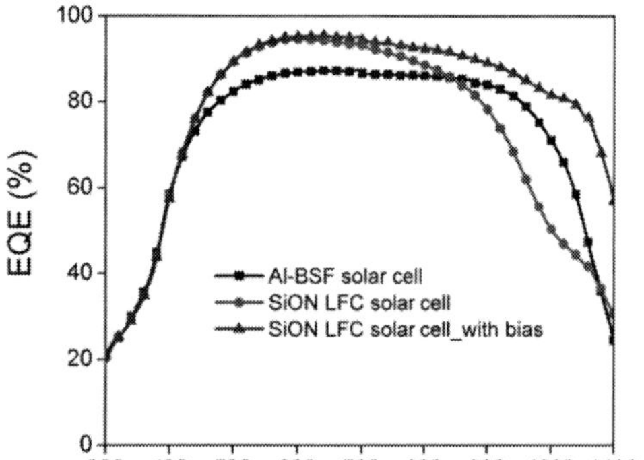

Figure 5: External Quantum Efficiency of p type solar cells with SiON passivation and LFC at the back; with and without bias light with respect to a reference solar cell having Al-BSF.

After studying the passivation quality of SiON and annealing effects on solar cell using I-V and EQE measurements, the final cells were fabricated on 156 cm^2 pseudo-square 5 inch ascut p type CZ wafers with 1-3 Ω.cm resistivity and <100> orientation. The cells with SiON single layer and SiON/SiN stack layer passivation schemes were fabricated. The thickness of SiON single layer was 100 nm and for the stack the SiON/SiN thickness was 30 nm and 70 nm respectively. The J-V characteristics of the cells are given in Fig. 6. The cells show lower V_{OC} of 603 mV for single layer SiON and 612 mV for SiON/SiN stack. This is due to the experimental limitation of annealing 5 inch wafers, hence these results are without annealing the solar cells and expect the V_{OC} to improve after annealing. Also for large area cells the edges are not fired resulting in high series resistance and low fill factor. Therefore to improve the fill factor the pitch has been reduced along with an increase in the back contact area. This helped us to improve the fill factor to ~ 78%. Thus it can be stated that with optimizing the LFC condition we can use this technology to replace Al-BSF for commercial c-Si solar cells

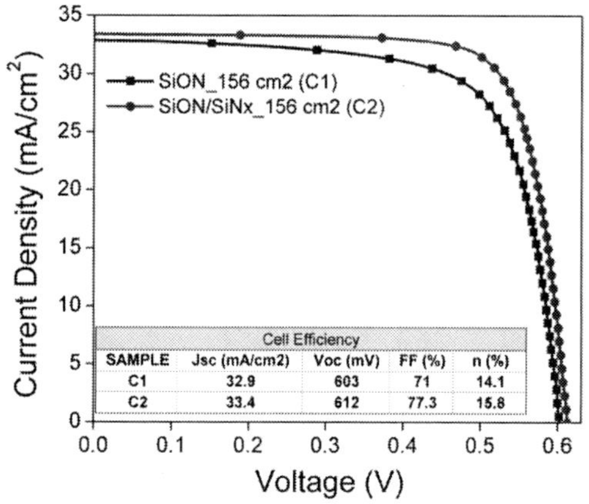

Figure 6: J-V characteristics of 156 cm^2 p type solar cells with SiON and SiON/SiN passivation with LFC at the back

4 CONCLUSION

In this work p-type c-Si solar cells were fabricated with SiON and SiON/SiN as the rear surface passivation layers and the back contacts were made using laser fired point contacts. QSSPC measurements showed an implied V_{OC} of 633 mV for single layer SiON and 649 mV for SiON/SiN stack on solar cells without metal contacts. The effect of annealing to improve the V_{OC} without degrading the passivation layers was performed and solar cells with SiON rear passivation layer presented the highest V_{OC} of 628mV. A highest efficiency of 15.8 % was obtained using SiON/SiN stack passivation layer with reference to 15.4% for cells made in the same batch with Al-BSF. Thus we have demonstrated a viable back contact method, which can replace Al-BSF on large area solar cells. A further optimization in LFC can give higher efficiency by improved fill factor.

5 ACKNOWLEDGEMENT

This work was supported by the Ministry of New and

Renewable Energy (MNRE) through the project AMANSI under the National Solar Science fellowship program. We would like to thank NCPRE and CEN at IIT Bombay for the fabrication and characterization facilities.

6 REFERENCES

[1] M.M. Hilali, J.M. Gee, P. Hacke, Bow in screen-printed back-contact industrial silicon solar cells, Sol. Energy Mater. Sol. Cells. 91 (2007) 1228–1233.

[2] A. Metz, D. Adler, S. Bagus, H. Blanke, M. Bothar, E. Brouwer, et al., Industrial high performance crystalline silicon solar cells and modules based on rear surface passivation technology, Sol. Energy Mater. Sol. Cells. 120 (2014) 417–425.

[3] A.W. Blakers, A. Wang, A.M. Milne, J. Zhao, M.A. Green, 22.8% efficient silicon solar cell, Appl. Phys. Lett. 55 (1989) 1363.

[4] E. Schneiderlöchner, R. Preu, R. Lüdemann, S.W. Glunz, Laser-fired rear contacts for crystalline silicon solar cells, Prog. Photovoltaics Res. Appl. 10 (2002) 29–34.

[5] S.W. Glunz, M. Hermle, S. Janz, D. Suwito, High-Efficiency Silicon Solar Cells with Intrinsic and Doped A-SiCx Rear Side Passivation, 24th Eur. Photovolt. Sol. Energy Conf. 21-25 Sept. 2009, Hamburg, Ger. (2009) 2197–2201.

[6] C. Molpeceres, F. Perales, M. Colina Brito, I. Sánchez Aniorte, R. Alcubilla González, C. Voz Sánchez, et al., Optimization of the Rear Point Contact Scheme of Crystalline Solar Cells Using Laser-Fired Contacts, 25th Eur. Photovolt. Sol. Energy Conf. Exhib. / 5th World Conf. Photovolt. Energy Conversion, 6-10 Sept. 2010, Val. Spain. (2010) 2126–2129.

[7] J. Cárabe, J.J. Gandía, C. Molpeceres, D. Munoz-Martin, J.P. González, S. Fernández, et al., Effect of the Passivating a-Si:H Layer on Laser-Fired Contacts in p-CZ Silicon-Heterojunction Solar Cells, 28th Eur. Photovolt. Sol. Energy Conf. Exhib. (2013) 2024–2028.

[8] R. Preu, J. Rentsch, D. Biro, N. Kohn, L. Gautero, M. Hofmann, et al., p-Type Cz-Si PERC-Type Solar Cells Applying PECVD Aluminium Oxide Rear Surface Passivation, 25th Eur. Photovolt. Sol. Energy Conf. Exhib. / 5th World Conf. Photovolt. Energy Conversion, 6-10 Sept. 2010, Val. Spain. (2010) 2516–2518.

[9] R. Preu, J. Rentsch, D. Biro, A. Grohe, J.F. Nekarda, L. Gautero, et al., Industrial Type CZ Silicon Solar Cells with Screen-Printed Fine Line Front Contacts and Passivated Rear Contacted by Laser Firing, 23rd Eur. Photovolt. Sol. Energy Conf. Exhib. 1-5 Sept. 2008, Val. Spain. (2008) 1704–1707.

[10] O. Schultz, M. Hofmann, S.W. Glunz, G.P. Willeke, Silicon oxide/silicon nitride stack system for 20% efficient silicon solar cells, in: Conf. Rec. Thirty-First IEEE Photovolt. Spec. Conf. 2005., IEEE, 2005: pp. 872–876.

[11] A. Sarikov, M. Voitovych, I. Lisovskyy, V. Naseka, A. Hartel, D. Hiller, et al., Characteristics of Hydrogen Effusion from the Si-H Bonds in Si Rich Silicon Oxynitride Films for Nanocrystalline Silicon Based Photovoltaic Applications, Adv. Mater. Res. 854 (2013) 69–74.

[12] C. Schwab, M. Hofmann, R. Heller, J. Seiffe, J. Rentsch, R. Preu, Effects of high-temperature treatment on the hydrogen distribution in silicon oxynitride/silicon nitride stacks for crystalline silicon surface passivation, Phys. Status Solidi. 210 (2013) 2399–2403.

EMITTER AND CONTACT OPTIMIZATION FOR HIGH-EFFICIENCY IBC MERCURY CELLS

[1]Agnes Mewe, [1]Pierpaolo Spinelli, [1]Antonius Burgers, [2]Ard Vlooswijk, [1]Nicolas Guillevin, [1]Eric Kossen, [1]Ilkay Cesar
[1]ECN Solar Energy, Petten, NL-1755 LE, The Netherlands
Tel: +31 88 515 4323; Fax: +31 88 515 8214; E-mail: mewe@ecn.nl
[2]Tempress Systems, Vaassen, NL-8171 MD, The Netherlands

ABSTRACT: In this paper we present the results of a study of the emitter contact properties of the Mercury cell, which is an n-type IBC cell with a front floating emitter. We found that contact recombination becomes higher and contact resistance becomes lower with higher firing temperature. Additionally, the contacts to light and heavy emitters were evaluated and the results showed that in case of a heavier emitter, the emitter contact recombination is reduced. The emitter contact resistance to both the light and heavy emitter appeared to only depend on the sheet resistance, but not on the exact shape of the doping profile. The contact recombination, calculated from measured V_{oc} values of test structures, appeared to be dependent on which side of the sample was illuminated: extracted V_{oc} values from front side illumination lead to much lower fitted $J_{0,contact}$ values than from rear side illumination. Quokka simulations indicate that the lower front side illumination values are closer to the real values, which can be calculated from the results of the measurements with either orientation.
Keywords: Back Contact, c-Si, Characterisation, Contact, Metallization, n-type, Recombination, Screen Printing

1 LOSSES IN MERCURY IBC CELLS

The Mercury cell which we developed at ECN is an interdigitated back contact (IBC) cell which features a front floating emitter (FFE). This FFE enhances lateral transport of holes and limits the effect of electrical shading [1], enabling easy module integration. In Figure 1, a schematic cross-section of the Mercury cell is shown.

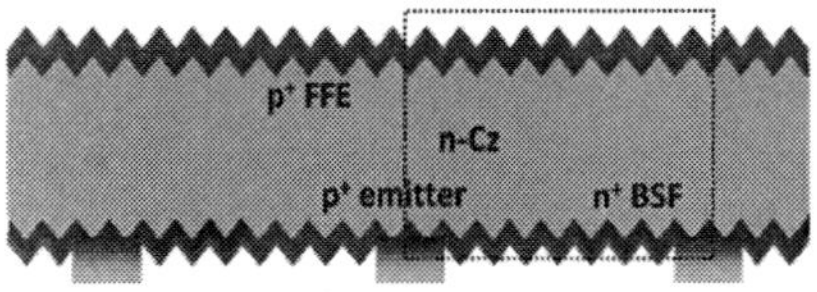

Figure 1: Schematic cross-section of the Mercury cell, with the unit cell, as used in simulations, indicated

Recently, we published improvement of our best Mercury cell to 21.1% [2], still following an industrial process flow that uses the same equipment as we use for our industrial n-Pasha cell.

1.1 Loss analysis

We performed a loss analysis of the Mercury cell. The bulk and surface recombination and transport losses are based on 2D device physics with Quokka [3] on a cross-section of the unit cell. The ohmic losses related to metal patterns are based on the calculated contact and line resistance of the fingers and busbars. Finally, circuit simulations were used to include several edge (non-unit-cell) effects.

The loss analysis that we performed on these cells, revealed that we suffer from important losses in our emitter contact, both recombination as well as ohmic losses. The loss analysis breakdown is shown in Figure 2.

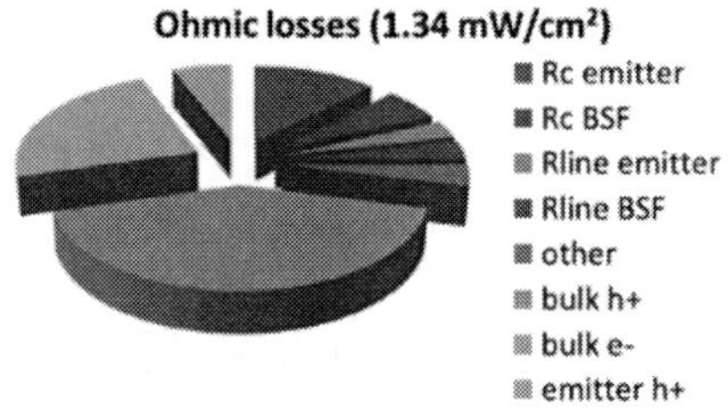

a

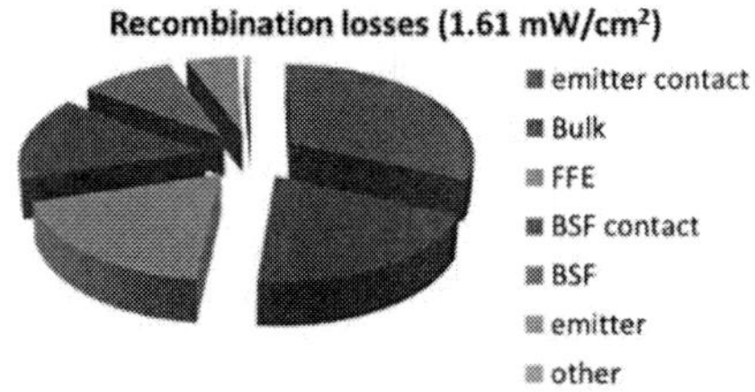

b

Figure 2: Loss analysis breakdown of the Mercury IBC cell, split in a) ohmic and b) recombination losses

1.2 Emitter contact

The loss breakdown shows us that, besides the emitter contact, the other large contributor to the losses is the n-Cz bulk. The bulk shows high ohmic losses as high-ohmic material was considered, and these could be partly mitigated when thinner and/or medium resistivity wafers are used. Considering the metal contact losses, the emitter contact recombination has the largest contribution and needs to be reduced. The losses in the BSF contacts are much smaller. The recombination in the FFE is considerable and can be further reduced by application of a lighter boron doping of 150 Ω/sq, which will be presented in this paper as well.

Based on these findings, we investigated the effects of different pastes and contact firing on the contact properties (contact resistance and contact recombination). Additionally, we made emitters with different R_{sheet} and doping profiles by varying the thermal diffusion process and by different etch-back of the as-diffused emitters.

In this paper we present the results of both routes to improve the emitter contact.

2 MEASUREMENT METHODS

2.1 Test structures

To evaluate the losses on a device which is processed

like an IBC cell, a special test structure was designed. Several key parameters can be evaluated, like contact recombination, as proposed in [4], and contact resistance, which we do from the combination of different structures on one wafer. This wafer is processed like an IBC cell, and differs only in the design of the doped areas and contacts.

2.2 J_0 fit from V_{oc} measurement

Structures as depicted in Figure 3 can be used to fit the contact recombination parameter $J_{0,c}$ by varying the contact area and evaluating the fitted J_{01} (from a Suns-V_{oc} measurement) as a function of metal contact area.

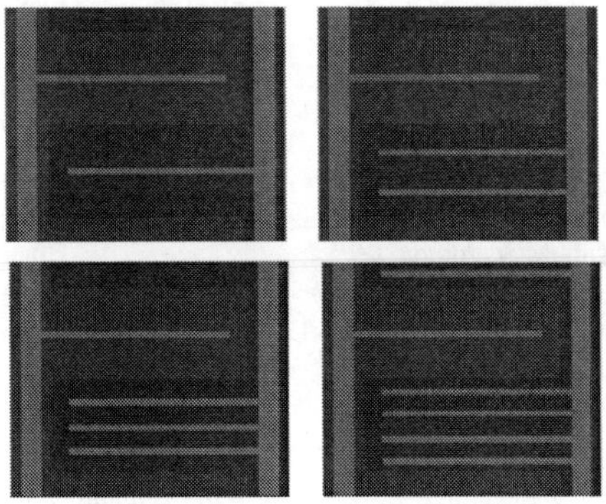

Figure 3: Test structures for measuring the emitter contact recombination. Blue and red indicate the BSF and emitter diffusions, grey the metallization

2.3 $R_{contact}$ using Transfer Length Method (TLM)

The transfer length method is used to determine the contact resistance of the emitter contact to the emitter. At ECN we use an automated TLM tool (PV-tools) that enables us to map the contact resistance over a complete wafer and to improve statistics, as the measurement is fast enough to evaluate several wafers per case.

3 INFLUENCE OF PASTE AND FIRING

We investigated the effect of the metal paste and the firing temperature on the contact properties. The contact recombination depends on how well the metal contact is shielded by the emitter, which depends on the doping profile (i.e. doping concentration as a function of depth) in combination with the etch depth of the metal paste into the doped layer (as schematically shown in Figure 4). The latter is related to paste chemistry and contact firing temperature, which are the parameters in this experiment. The same parameters will also influence the contact resistance.

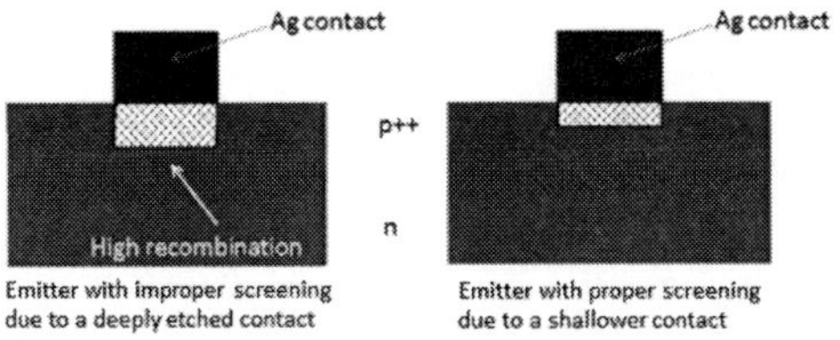

Figure 4: Schematic view of contact formation by a deep or shallow contact etch into the emitter profil

3.1 Set-up of experiment

Two different Ag/Al metal pastes were applied to a boron emitter with an industrial profile and a sheet resistance of 70 Ω/sq, and fired through the passivation layer using different firing temperatures. The set temperature differences in the firing furnace varied 40 to 60 °C for the two pastes, divided over 4 or 5 settings. Paste 1 is a commercially available Ag/Al paste, and paste 2 is a paste that was designed for improved contact properties, for which results on our n-Pasha concept were presented earlier this year [5].

The result of paste 3, also a commercially available paste, was added at one firing temperature setting to relate it to the emitter variation results in section 4.

3.2 Results

As shown in Figure 5, the $J_{0,c}$ clearly increases with higher firing temperature. For the lowest temperatures, the $J_{0,c}$ of the emitter contact can decrease down to below 1500 fA/cm^2, which is much lower compared to the J_0 for a more standard firing setting (2000-2500 fA/cm^2).

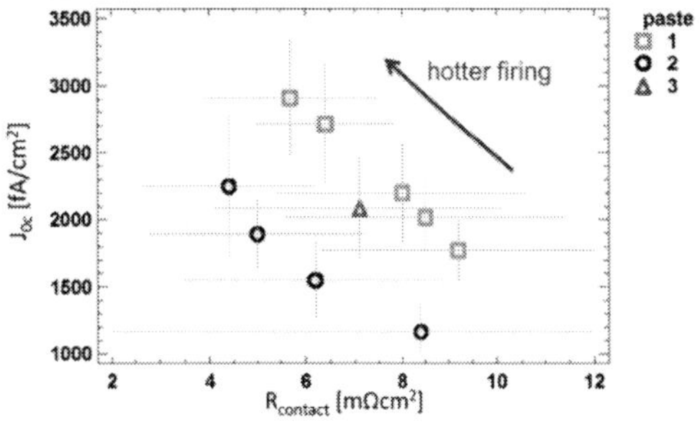

Figure 5: Fitted emitter contact recombination (vertical) against contact resistance (horizontal) for three different pastes for different firing settings (values towards higher firing temperature indicated)

The contact resistance of the fired metal contact to the emitter shows an opposite trend: it becomes higher for the lower temperatures. This means that in a cell, a balance should be found between contact resistance and recombination, to get the lowest cumulative losses and the highest performance. This will depend on the metallization geometry as well.

Paste 2 clearly performs better than paste 1 (and paste 3), which means that the combined ohmic and recombination losses in the cell will be smaller at the optimal firing setting.

3.3 Cells

The improved paste was used to make Mercury IBC cells, applying a slightly lighter doped emitter than tested in the contact properties experiment, as this is more optimal for the complete solar cell.

We applied the paste by screen printing and by stencil printing. Using stencil printing, we can achieve narrower fingers with better aspect ratio, which led to 2 mV higher V_{oc} due to the reduced emitter contact area, resulting in 0.1% absolute higher efficiency than the cells with screen printed metallization. The efficiency results are shown in Figure 6.

The average efficiency of the cells with stencil printed metallization was 21.0%, with the best cells reaching 21.1% (internal measurement, corrected for spectral mismatch), which equals our best cell result.

This is especially a promising result, because there are some processing factors that we already improved for the previously achieved best result and have not been included in this cell run. It means that there is definitely room for further improvement.

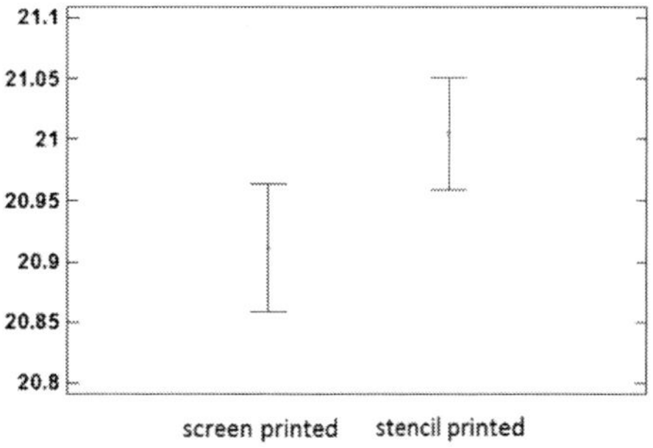

Figure 6: Efficiency results for Mercury cells with the improved emitter metallization paste (paste 2) using screen and stencil printing. Group sizes are 11 and 14 cells respectively

4 INFLUENCE OF EMITTER PROFILE

In this experiment we investigated the effect of the doping profiles on contact recombination and contact resistance. As already mentioned, a deeper profile is associated with better contact shielding and therefore low contact recombination, as sketched in Figure 7.

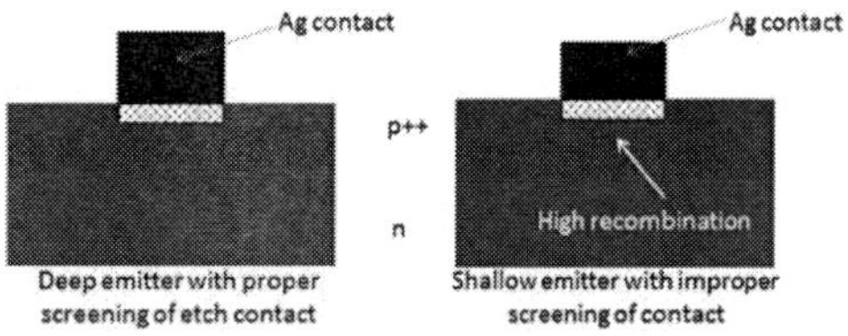

Figure 7: Schematic view of contact formation into a deep (heavy) or shallow doping profile

4.1 Set-up of experiment

The different profiles were realized by changing the diffusion parameters (a standard light emitter and a heavy emitter) and additionally etching back to increase the sheet resistance to 50 (only for the heavy emitter), 70 and 85 Ω/sq. We used the previously mentioned paste 3 for the emitter contacts, and a standard firing setting.

4.2 Results

In Figure 8 it is shown that the heavier emitter results in lower emitter contact recombination compared to the light emitter. Counterintuitively, etching back the same emitter profile does not lead to higher recombination values. For both emitter profiles, a clear trend of contact recombination with sheet resistance is absent.

As expected, for the contact resistance we observe a trend of higher $R_{contact}$ with higher emitter R_{sheet}. Although the margin of error is quite large, the trend seems to be linear (not visible in the graph). There are no contact resistance differences between the heavy and light profiles with the same R_{sheet}.

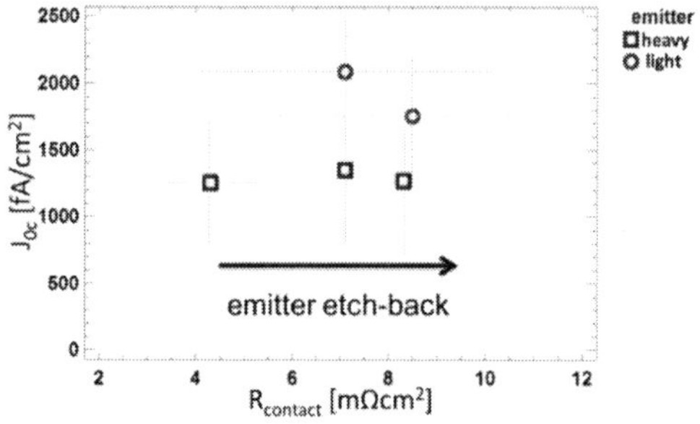

Figure 8: Fitted emitter contact recombination against contact resistance for a heavy emitter (blue) and more lightly doped emitter (red) after different etch-back (direction towards higher sheet resistance indicated)

4.3 Outlook for cells

We conclude that we can further improve the cell results if we combine paste 2 with the contact properties of the heavy emitter.

Additionally, to avoid excessive recombination in the FFE, we can apply an etch-back on the front side, which is a good way to reduce losses in the cell further. The front floating emitter needs to remain conductive to allow lateral electron transport in an efficient way. From device simulations, we calculated that a FFE up to 150 Ω/sq is conductive enough and does not lead to additional ohmic losses.

In a test experiment on symmetrically diffused samples, we reduced the doping of the front floating emitter, to obtain sheet resistance values ranging from 70 to 150 Ω/sq, and evaluated the surface passivation on symmetric n-Cz samples with the boron FFE on each side. In Figure 9 the results of the J_0 for one side are presented for five sample groups, which shows a 40% reduction of passivated surface J_0 if we decrease the emitter R_{sheet} from 70 to almost 150 Ω/sq.

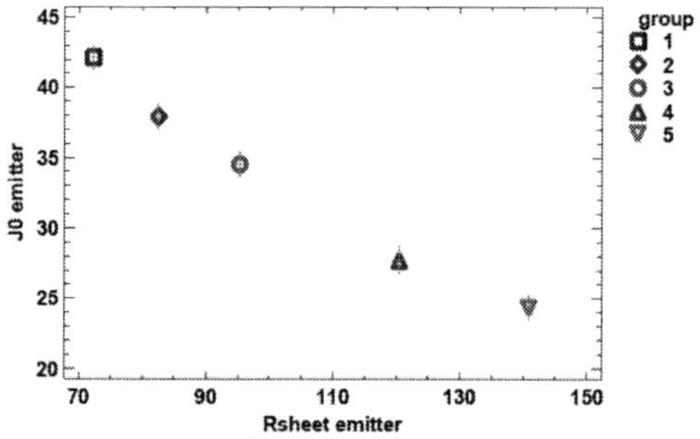

Figure 9: Emitter recombination decrease (single side evaluation of J_0, in fA/cm^2, with increasing FFE sheet resistance, in Ω/sq

The combined effects of paste and emitter choice, and a lighter FFE could result in 0.3-0.4% absolute efficiency gain, according to our simulations. This would bring our current record Mercury IBC cell of 21.1% (in-house measurement, corrected for spectral mismatch) to about 21.5% with only minor process changes.

5 DISCUSSION

5.1 Measurement method for $J_{0,c}$ fit

We fit the metal contact recombination parameter ($J_{0,c}$) from the slope of the J_0 with the metal fraction. The J_0 is calculated from the V_{oc} that we measure. We executed the V_{oc} measurements of the test structures in this paper with the metallized side of the wafer facing the flash of the Suns-V_{oc} set-up. This was done because of practical reasons, as we had to contact different locations of the test structures on the wafer, and therefore we need to put the contact probes on many different locations.

However, this procedure is not completely representative for the IBC cell, as the illuminated side of the test structure (metallized side) is different from the illuminated side of an IBC cell (non-metallized side). In addition to that, the varying metal coverage of the test structures may lead to errors in the measurements due to different light exposure of the cell.

To avoid this, we designed a printed circuit board that exactly matches with the contact locations of the test structures, so we can use it as a contacting chuck. All test structures can be individually contacted by correct placement and the measurement can be done with the non-metallized side up.

We compared the values that we obtained for the fitted contact recombination parameter using both orientations of the wafer (rear side up and front side up). In this case, the data came from samples with a lightly doped emitter from a different experiment. We found that the $J_{0,c}$ values were quite different for the two orientations, as shown in Table I.

Table I: Fit of $J_{0,c}$ values from V_{oc} measurements obtained with different sample orientations

Orientation	$J_{0,c\_emitter}$ [fA/cm^2]
Front illumination	1217
Rear illumination	1834

We looked into this issue in more detail using Quokka simulations.

5.2 Quokka simulation results

We simulated the influence of the test structure orientation in the simulation software package Quokka, using an input value for the emitter contact recombination of 1210 fA/cm^2, which corresponds to the fitted value $J_{0,c}$ from experimentally measured V_{oc} values. For front illuminated samples, Quokka calculates V_{oc} values that lead to a fitted $J_{0,c}$ of 1386 fA/cm^2 for the emitter, while for rear illumination and the same input value, the fitted $J_{0,c}$ value becomes much higher, 1910 fA/cm^2. This is shown in Table II.

Table II: Fit of $J_{0,c}$ values in fA/cm^2, from V_{oc} measurements and from V_{oc} values resulting from test structure simulations in Quokka, obtained with different sample orientations. The Quokka input value for $J_{0,c}$ was 1210 fA/cm^2

Orientation	Measurement	Simulation
Front illumination	1217	1386
Rear illumination	1834	1901

The fits from the simulated and measured V_{oc} values show the same trend for front and rear illuminated conditions. However, both situations appear not to lead to the right value. Even for front illuminated samples, which is the condition closest to the IBC cell measurement, a small correction of the fitted $J_{0,c}$ value will be necessary. At the moment, we are investigating and validating this further.

6 CONCLUSION

We showed that the firing temperature, metallization paste and emitter profile cause a significant change in $J_{0,c}$ and $R_{contact}$ of the emitter contact.

Based on the fitted $J_{0,c}$ and $R_{contact}$ values for the different emitter cases, and the introduction of a lighter FFE, we foresee that we can improve the performance of the Mercury IBC cell to at least 21.5% with only small process modifications like different emitter and FFE profiles and contacting paste.

Although more validation is needed to obtain the correct absolute numbers for $J_{0,c}$, we can roughly estimate from the raw V_{oc} data what the gain could be in case of a process change.

7 ACKNOWLEDGEMENTS

This work was supported by the Dutch Ministry of Economic Affairs, within the TKI framework, project IBCense.

REFERENCES

[1] I. Cesar, N. Guillevin, A.R. Burgers, A.A. Mewe, M. Koppes, J. Anker, L.J. Geerligs, and A.W. Weeber, "Mercury: a back junction back contact front floating emitter cell with novel design for high efficiency and simplified processing" *Energy Procedia,* vol 55 2014, p. 633.

[2] P. Spinelli, P. Danzl, N. Guillevin, A. Mewe, S. Sawallich, A. Vlooswijk, B. van de Loo, E. Kessels, M. Nagel, and I. Cesar, "High-resolution sheet resistance mapping to unveil edge effects in industrial IBC solar cells", *SiliconPV,* 2016, Chambéry, to be published.

[3] A. Fell, "A free and fast 3D/2D solar cell simulator featuring conductive boundary and quasi-neutrality approximations," IEEE Transactions on Electron Devices, Vol 60 (2), pp. 733–738, 2012.

[4] T. Fellmeth, A. Born, A. Kimmerle, F. Clement, D. Biro, and R. Preu, "Recombination at Metal-Emitter Interfaces of Front Contact Technologies for Highly Efficient Silicon Solar Cells, *Energy Procedia* vol 8, 2011, p. 115

[5] E. Kossen, "Contacting high ohmic boron emitters with a screen print paste on n-Pasha", *Metallization Workshop*, 2016, Constance

LIFETIME OF NANO-STRUCTURED BLACK SILICON FOR PHOTOVOLTAIC APPLICATIONS

Maksym Plakhotnyuk*[a], Rasmus Schmidt Davidsen[a], Michael Stenbæk Schmidt[a],
Radu Malureanu[b], Eugen Stamate[c], Ole Hansen[a]
[a] Department of Micro- and Nanotechnology, Technical University of Denmark, Ørsteds Plads building 345East, DK-2800 Kgs. Lyngby, Denmark
[b] Department of Photonics, Technical University of Denmark, Ørsteds Plads Building 345West, DK-2800 Kgs. Lyngby, Denmark
[c] Department of Energy Conversion and Storage, Technical University of Denmark, Frederiksborgvej 399, DK-4000 Roskilde, Denmark
*Corresponding author: makpl@nanotech.dtu.dk, phone: +45 27575092

ABSTRACT: In this work, we present recent results of lifetime optimization for nano-structured black silicon and its photovoltaic applications. Black silicon nano-structures provide significant reduction of silicon surface reflection due to highly corrugated nanostructures with excellent light trapping properties. We applied reactive ion etching technology at -20°C to create nano-structures on silicon samples and obtained an average reflectance below 0.5%. For passivation purposes, we used 37 nm ALD Al_2O_3 films. Lifetime measurements resulted in 1220 μs and to 4170 μs for p- and n-type CZ silicon wafers, respectively. This is promising for use of black silicon RIE nano-structuring in a solar cell process flow.

Keywords: c-Si, black silicon, ALD Al_2O_3, passivation, lifetime, light trapping, nano-structured silicon

1 INTRODUCTION

Nano-textured silicon, known as black silicon, is currently a subject of great interest in photovoltaics [1–4, 5]. It has extremely low surface reflectance, in some cases below 1%, in a broad range of wavelengths and incident angles even with a simple antireflective coating [6, 7]. Reactive plasma etching is a most promising alternative to any other silicon texturing method like wet etching [8–11], plasma immersion ion implantation etching, metal nanoparticles assisted etching and laser induced etching [12–15]. Key advantages of dry plasma silicon structuring are no usage of toxic chemicals, easy morphology control of structure size and shape, less silicon waste and mask-less processing. Solar cells with improved light trapping nano-structures show improved characteristics, such as high open circuit voltage and large short circuit current. However, surface-texturing methods like RIE also induce surface damage and potentially contamination, and therefore increased surface recombination velocity resulting in poor performance of nano-structured solar cells [3, 5, 16, 17]. Thus, effective surface passivation of nano-textured surfaces and optimization of the texturing process towards reduction of surface defect damage is required for further black silicon application in photovoltaics [18–22]. Atomic layer deposition (ALD) of Al_2O_3 is one of the best methods for black silicon passivation, since it has conformal coverage of plane and corrugated surfaces. Thermal ALD of Al_2O_3 leads to high chemical interface quality on crystalline silicon, low density of defect states, and the strongest known field effect due to fixed negative charges [18, 20, 21, 23–32].

2 EXPERIMENT

2.1 Sample Details and Black Silicon Etching

Black silicon nanostructuring was realized on Czochralski (CZ) mono-crystalline Si 4" (100) n- and p-type (1-20 Ωcm) wafers with thickness 350 μm by mask-less reactive ion etching (RIE) in SF_6 and O_2 inductively coupled plasma at -20°C in DRIE SPTS Pegasus (Fig.1). Prior to final sample preparation, the RIE process was optimized: platen power was reduced from 50 to 10 W to reduce kinetic energy of ions, directed towards the substrate surface; coil power was increased from 2500 to 3000 W to increase homogenous ion plasma density and maintain a stable etch rate; aspect ratio of nanostructures was controlled via the low gas pressure (38 mTorr) which determined the directionality of the ions and the physical etching components by reducing the mean free path of the plasma species; average flow of SF_6 gas (70 sccm) allowed to a stable Si etch rate and high O_2 flow (100 sccm) allowed to passivate Si surface and reduce etch rate. The first samples had shown low lifetime after ALD Al_2O_3 passivation and are not referred in this report. For conducting the experiment we used six silicon wafers of n- and p-type: two wafers had single side RIE processing for 16 min, another two wafers had double sided RIE processing for 16 min on each side and the last two wafer were kept as plain polished references.

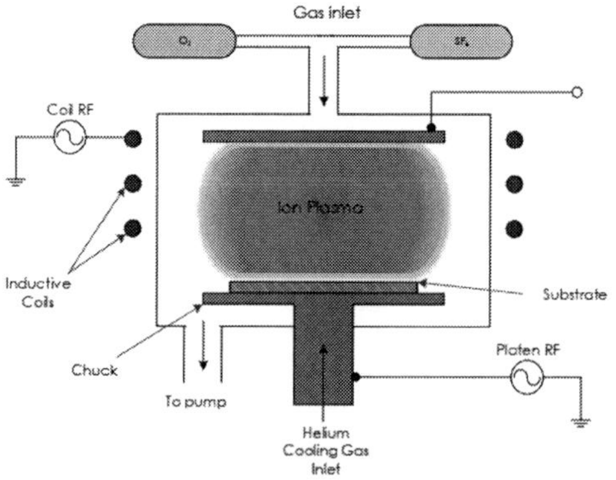

Figure 1: Schematic of inductive coupled plasma system used for RIE

The black silicon nano-structuring process was conducted in the following order as shown on Fig. 2. A sample wafer was loaded in to the ICP RIE chamber and cooled to -20°C. SF_6 gas was supplied to the chamber and fluorine radicals rapidly attacked the silicon and destroyed native oxide on top of the wafer forming volatile SiF_4. In the next step, oxygen was supplied to the chamber. Oxygen radicals form silicon-oxyfluorine $SiF_4+O^*=SiO_xF_y$ at low temperature (-20°C). SiO_xF_y acts as an etch stop for F* and passivated sample surface. Horizontal planes were more intensively bombarded by ions from plasma while on the vertical sidewalls the ion bombardment was weaker due to directionality of the plasma ions and the sidewalls therefore remained protected from chemical etching by fluorine radical [1-2].

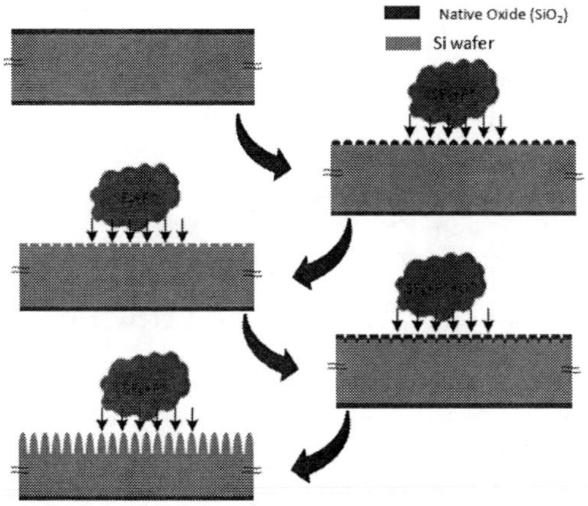

Figure 2: Schematic view of black silicon nanostructures formation

2.2 Surface cleaning and ALD Al_2O_3 passivation

After RIE all samples were cleaned in standard RCA cleaning solutions, RCA1 and RCA2, each with a subsequent HF-dip, rinsed in deionized water and spin dried. Subsequently, wafers were coated with 380 cycles of ALD Al_2O_3 synthesized from trimethylaluminium (TMA) and H_2O. For reference purposes two polished wafers (p- and n-type) were also included in ALD Al_2O_3 passivation procedure. The passivation layers were activated by post-deposition in-situ annealing in N_2 ambient at 375-390 °C for 30 min. The resulting Al_2O_3 thickness of 37 nm was measured from polished reference samples using ellipsometry. Charge carrier lifetime was measured after annealing with the microwave detected photoconductivity (MDP) method in transient as well as in injection dependent single point modes, and lifetime mapping mode using a MDPmap setup from Freiburg Instruments.

3 RESULTS AND DISCUSSIONS

3.1 Black Silicon SEM Morphology Study

In Fig. 3 scanning electron microscope (SEM) images of the resulting black silicon nano-structures are shown without Al_2O_3 coating. The nanostructure topology consists of slightly rounded conical-like hillocks with average height of 500 nm and average spacing of 250 nm. There is a slight variation in the dimensions of the surface structures. The height of the nanostructures was controlled by etching time with an average etch rate of 30 nm/min.

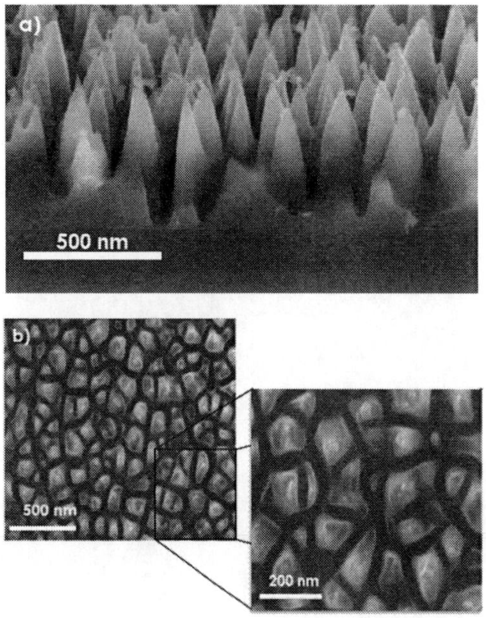

Figure 3: SEM image of the black silicon nano-structure topology processed with plasma assisted mask-less reactive ion etching: a) side view with scale 500 nm, b) top views with scale 500 nm and 200 nm

3.2 Reflectance

The main purpose of nanostructuring is to reduce reflectance and improve light trapping properties of the Si surface [3, 33]. The spectral reflectance of black silicon and polished samples were measured in the wavelength range from 500 to 1100 nm. Fig. 4 shows the reflectance of the samples as a function of wavelength. Photographic images are inserted in the reflectance figure to illustrate the difference between polished and nanostructured samples. It can be seen that polished silicon wafer in average has reflectance of 30%, while the black silicon sample has reflectance below 1% up to 1000 nm. Above 1000 nm the reflectance still remains low but shows a slight increase up to 5% which is still much lower than that of the polished wafer, and this long wavelength reflectance may partly be due to reflectance of the chuck below the silicon sample.

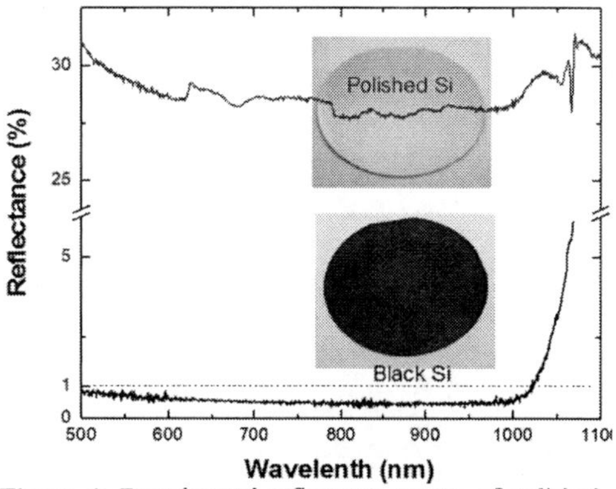

Figure 4: Experimental reflectance spectra of polished and black silicon without Al_2O_3 coating

3.3 Minority Carrier Lifetime

Fig. 5 shows measured effective minority carrier lifetime for p- and n-type wafers for polished as well as for nano-textured samples. MDP measurements were conducted with a MDPmap setup from Freiberg Instruments to determine the effective carrier lifetimes as a function of injection level. The effective minority lifetime is a figure of merit to estimate surface recombination and surface damage [18, 21, 25, 34, 35].

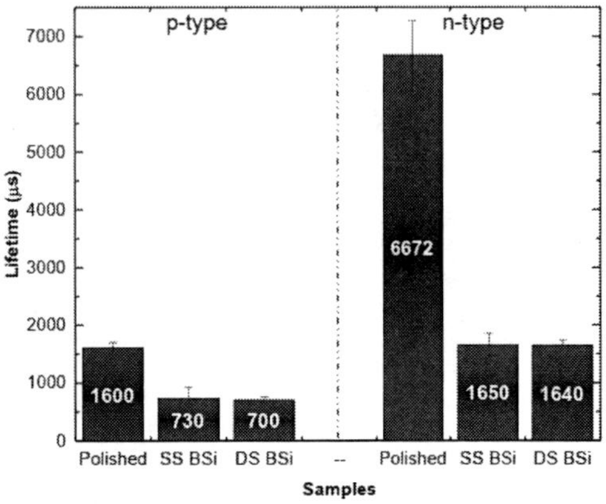

Figure 5: Comparative graph of average lifetime wafer mapping values with standard deviation for p- and n-type silicon samples with polished, single side nanostructured and double side nanostructured black silicon, passivated with ALD Al_2O_3 37 nm film. Legend: SS BSi - single side nanostructured black silicon, DS BSi - double side nano-structured black silicon

For p-type Si wafers, the lifetime difference between polished and textured samples is a factor of two. For some samples we have recorded a lifetime approaching 1 ms and higher, however to report these results additional testing is required. Injection dependent lifetime for p-type samples shown in Fig. 6 confirms stability of above results. A similar tendency is present for n-type samples, however lifetime difference between polished and nanostructured samples is in that case a factor of four.

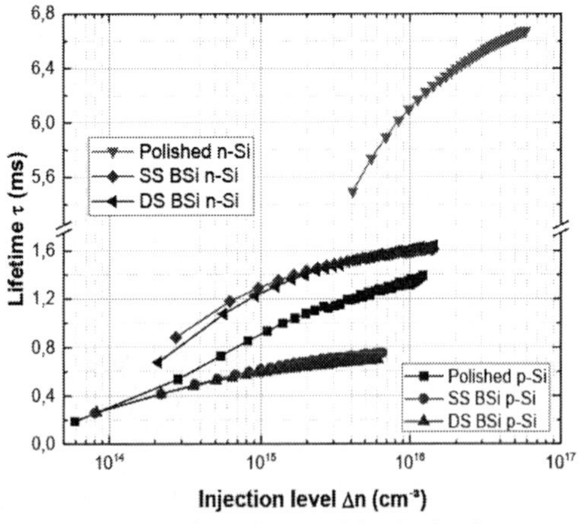

Figure 6: Lifetime dependency on injection level

The surface recombination velocity was calculated from effective minority carrier lifetime from the equation shown in Fig. 6 [25].

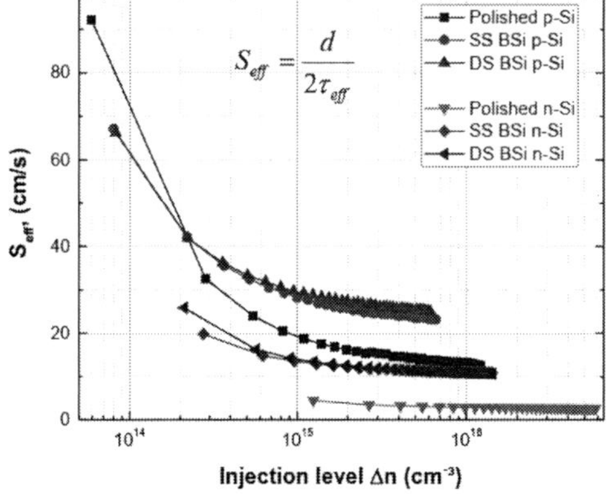

Figure 7: Surface recombination velocity dependencies on injection level

where d is the wafer thickness, τ_{eff} is the measured effective lifetime. Fig. 7 shows the calculated surface recombination velocity as a function of injection level.

4 CONCLUSION

We have presented recent results of effect of reactive ion process optimization for improvement of minority carrier lifetime. The optimized RIE recipe with reduced ion damage of the surface allows us to achieve comparatively high effective lifetime for n- and p-type wafers and for single and double side nano-textured samples. For further investigation of electro-optical properties all samples were passivated with Al_2O_3 and annealed at 375-390°C for 30 min. The surface morphology of textured samples was studied using SEM, and the optical reflectance was measured with an integrated sphere in the range of 500-1100 nm. Reflectance measurements allow us to evaluate the antireflective properties of nanostructures. With the plasma-assisted mask-less etching method we achieved broad spectral reflectance below 0.5%. Finally, the effective lifetime was measured on all textured and reference polished samples to evaluate the passivation effect and to estimate surface recombination velocity. Our overall results are promising for black silicon fabrication and lifetime improvement to facilitate incorporation in solar cell fabrication.

5 REFERENCES

[1] Koynov, S., Brandt, M. S. & Stutzmann, M. Black thin film silicon. J. Appl. Phys. 110, 043537 (2011).

[2] Kroll, M. et al. Black silicon for solar cell applications. Proc. SPIE 8438, 843817 (2012).

[3] Koynov, S., Brandt, M. S. & Stutzmann, M. Black nonreflecting silicon surfaces for solar cells. Appl. Phys. Lett. 88, 203107 (2006).

[4] Otto, M. et al. Black Silicon Photovoltaics. Adv. Opt. Mater. 3, 147–164 (2015).

[5] Davidsen, R. S. et al. Black silicon laser-doped selective emitter solar cell with 18.1% efficiency. Sol. Energy Mater. Sol. Cells 144, 740–747

(2016).
[6] Koynov, S., Brandt, M. S. & Stutzmann, M. Black thin film silicon. J. Appl. Phys. 110, 043537 (2011).
[7] Raghunathan, D. Black silicon for higher efficiency in solar cells. Appl. Mech. Mater. 787, 92–96 (2015).
[8] Huang, Z., Geyer, N., Werner, P., De Boor, J. & Gösele, U. Metal-assisted chemical etching of silicon: A review. Advanced Materials 23, 285–308 (2011).
[9] Koynov, S., Brandt, M. S. & Stutzmann, M. Black nonreflecting silicon surfaces for solar cells. Appl. Phys. Lett. 88, 203107 (2006).
[10] Liu, Y. et al. Nanostructure formation and passivation of large-area black silicon for solar cell applications. Small 8, 1392–1397 (2012).
[11] Lu, Y.-T. & Barron, A. R. Nanopore-type black silicon anti-reflection layers fabricated by a one-step silver-assisted chemical etching. Phys. Chem. Chem. Phys. 15, 9862–70 (2013).
[12] Iyengar, V. V. et al. Properties of ultrafast laser textured silicon for photovoltaics. Sol. Energy Mater. Sol. Cells 95, 2745–2751 (2011).
[13] Kontermann, S., Gimpel, T., Baumann, A. L. L., Guenther, K.-M. M. & Schade, W. Laser processed black silicon for photovoltaic applications. Energy Procedia 27, 390–395 (2012).
[14] Sarnet, T., Carey, J. E. & Mazur, E. From black silicon to photovoltaic cells, using short pulse lasers. AIP Conference Proceedings 1464, 219–228 (2012).
[15] Schütz, V., Horn, A., Nagel, H. & Stute, U. Black silicon solar cell processing with high repetitive laser systems. 27th European Photovoltaic Solar Energy Conference and Exhibition 1864–1866 (2012).
[16] Oh, J., Yuan, H.-C. & Branz, H. M. An 18.2%-efficient black-silicon solar cell achieved through control of carrier recombination in nanostructures. Nat. Nanotechnol. 7, 743–748 (2012).
[17] Savin, H. et al. Black silicon solar cells with interdigitated back-contacts achieve 22.1% efficiency. Nat. Nanotechnol. 10, 1–6 (2015).
[18] Dou, B. et al. Surface passivation of nanotextured silicon solar cells by atomic layer deposited Al_2O_3 films. J. Appl. Phys. 114, 174301 (2013).
[19] Hsu, C.-H. et al. Fabrication and characteristics of black silicon for solar cell applications: An overview. Mater. Sci. Semicond. Process. (2014). doi:10.1016/j.mssp.2014.02.005
[20] Otto, M. et al. Passivation of different black silicon surfaces by ALD deposited Al_2O_3. 2013 IEEE 39th Photovoltaic Specialists Conference (PVSC) 1277–1279 (IEEE, 2013). doi:10.1109/PVSC.2013.6744374
[21] Otto, M. et al. Passivation of optically black silicon by atomic layer deposited Al_2O_3. Energy Procedia 38, 862–865 (2013).
[22] Zhao, Z., Zhang, B., Li, P., Guo, W. & Liu, A. Effective passivation of large area black silicon solar cells by SiO_2/SiN_x:H stacks. Int. J. Photoenergy 2014, 1–6 (2014).
[23] Otto, M. et al. Extremely low surface recombination velocities in black silicon passivated by atomic layer deposition. Appl. Phys. Lett. 100, 191603 (2012).
[24] Repo, P. et al. Passivation of black silicon boron emitters with atomic layer deposited aluminium oxide. Phys. status solidi - Rapid Res. Lett. 7, 950–954 (2013).
[25] Repo, P. et al. Effective passivation of black silicon surfaces by atomic layer deposition. IEEE J. Photovoltaics 3, 90–94 (2013).
[26] Wang, W.-C. et al. Surface passivation of efficient nanotextured black silicon solar cells using thermal atomic layer deposition. ACS Appl. Mater. Interfaces 5, 9752–9 (2013).
[27] Benick, J. et al. Effect of a post-deposition anneal on Al_2O_3/Si interface properties. Photovolt. Spec. Conf. (PVSC), 2010 35th IEEE 891–896 (2010). doi:10.1109/PVSC.2010.5614148
[28] Bordihn, S. et al. Impact of the deposition and annealing temperature on the silicon surface passivation of ALD Al_2O_3 films. Energy Procedia 27, 396–401 (2012).
[29] Dingemans, G. & Kessels, W. M. M. Status and prospects of Al_2O_3-based surface passivation schemes for silicon solar cells. J. Vac. Sci. Technol. A Vacuum, Surfaces, Film. 30, 040802 (2012).
[30] Dingemans, G. & Kessels, W. M. M. Aluminum oxide and other ALD materials for Si surface passivation. ECS Trans. 41, 293–301 (2011).
[31] Kotipalli, R. et al. Passivation effects of atomic-layer-deposited aluminum oxide. EPJ Photovoltaics 4, (2013).
[32] Schmidt, J., Werner, F., Veith, B. & Zielke, D. Industrially relevant Al_2O_3 deposition techniques for the surface passivation of Si solar cells. Photovolt. Sol. 4–7 (2010). doi:10.4229/25thEUPVSEC2010-2AO.1.6
[33] Zhu, J., Yu, Z., Fan, S. & Cui, Y. Nanostructured photon management for high performance solar cells. Materials Science and Engineering R: Reports 70, 330–340 (2010).
[34] Sinton, R. A. & Cuevas, A. Contactless determination of current–voltage characteristics and minority-carrier lifetimes in semiconductors from quasi-steady-state photoconductance data. Appl. Phys. Lett. 69, 2510 (1996).
[35] Stephens, A. W. & Green, M. A. Novel method for minority-carrier mobility measurement using photoconductance decay with chemically passivated and plasma damaged surfaces. J. Appl. Phys. 80, 3897 (1996).

MATCHING THE CONTACT ADHESION STRENGTH OF SCREEN PRINTED SOLAR CELLS WITH COPPER-PLATED CONTACTS

R. Chen, S. Wang, A.Wenham, Z.Shi, T. Young, J. Ji, M.Edwards, A. Sugianto,
L. Mai, S.Wenham and C. Chong

School of Photovoltaic and Renewable Energy Engineering, University of New South Wales, Sydney NSW 2052, Australia

ABSTRACT: Copper-plated contacts have the potential to dominate metallisation for silicon solar cells if durability concerns relating to the metal's adhesion strength can be addressed. In this work, a range of approaches were used for forming the plated metal/Si interface but with no discernable difference resulting in the corresponding metal adhesion strength, even though this metal/Si interface was shown to be the weakness in each case. Importantly, the volume of plated metal gave a much stronger correlation with adhesion strength, suggesting that internal stresses within the plated metal, possible partly linked to the thermal expansion mismatch between the metal and the silicon, was having significant impact on the resulting Ni/Si interfacial strength. To overcome the impact of such variability, a new approach capable of ensuring good adhesion strength for plated contacts was developed. It involves the formation of laser-ablated anchor points in the heavily-doped silicon surface which when plated greatly strengthen the metal's mechanical adhesion strength.

keywords: Cu-plated contacts, contact adhesion, light-induced plating, anchor points, silicon solar cells

1. INTRODUCTION

Experts predict that copper is the most likely replacement for most silver-based metal contact schemes for silicon wafer-based technologies in the future [2]. Its high conductivity that is comparable to that of silver [3,6], its suitability for fine-line metallisation designs [4,5,6], its compatibility with simple plating techniques and its relative abundance and low cost [7], are some of its important attributes. Never the less, many large-scale cell manufacturers still appear cautious about using the plating of metals directly onto the silicon. Adhesion strength and hence durability of such plated contacts appear to be the major concerns.

A new approach involving the formation and use of anchor points in the silicon surface has been developed in this work for improving the adhesion and durability of plated contacts [18,19]. Such anchor points emulate the "Saturn technology" developed and commercialised by BP Solar into large-scale manufacturing [6], which essentially provides a continuous anchor point by plating metal into grooves buried beneath the surface. This "anchoring" of the plated metal has demonstrated superior adhesion strengths, even when compared to screen-printed contacts [6]. This has motivated the present work on evaluating and simplifying the formation of such anchor points while simultaneously facilitating significantly higher efficiencies than were achieved by Saturn. The formation of such anchor points is easily combined with laser doping processes [11]. Laser doping for forming selective emitters in conjunction with plated contacts has been used for many years for achieving high performance cells [8,9] and has been successfully used in large-scale manufacturing [4]. The formation of "anchor points" in the silicon surface is shown in Figure 1 [19]. This can be achieved without complicating the laser-doping process by increasing the energy per laser pulse in the location where each anchor point is required without varying the scanning speed or frequency of the laser. The increased pulse energy allows a small amount of the molten silicon to be ablated. This results in the formation of a hole with the remaining molten silicon in the walls of the hole regrowing via liquid phase epitaxial growth during cooling while incorporating active phosphorus dopants to make the walls of such holes heavily n-type doped as shown in Figure 1 [19].

This paper investigates in detail the solutions for solving the adhesion challenges identified in ITRPV [2] for plated metallisation. In terms of improving adhesion of metal plated contacts, anchor points were formed by using localised laser ablation to produce holes in the heavily laser-doped lines. The new plated contacts with anchor points show superior adhesion strength, not only to alternative approaches for forming the plated metal contacts, but even to screen-printed cells.

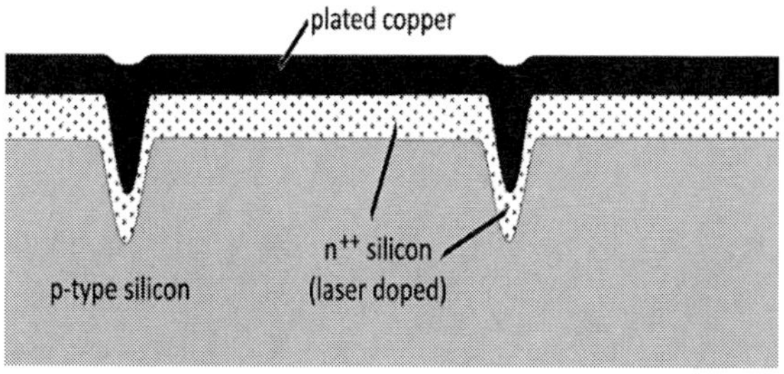

Figure 1: Laser-doped line with 10μm anchor points every 100 μm to improve copper adhesion.

Peel tests [12-14] are used in this work for adhesion testing of the metallised contacts at busbar level. The peel tester, typical of those used industrially, was designed based on European Adhesion Standard EN50461 and used to characterise the adhesion of metal contacts in this work. In each case, interconnect wire for the peel test was attached via soldering or conductive adhesives, depending on the strength and hence ability of the contacts to withstand the thermal stresses associated with soldering. In this study, a 90° peel test angle was used. For plated contacts, the solderable surfaces were created through the formation of parallel fingers spaced 30 μm apart. This in turn was achieved through the use of 10 μm wide laser lines spaced 30 μm apart, with the space between such lines still coated with SiNx and therefore unable to be plated with metal. The actual metal/silicon interface in these tests for the plated contacts was therefore only 30-40% of the test area.

2. EXPERIMENTAL METHODS

The preferred approach for forming plated contacts to silicon solar cell surfaces in recent years appears to be through the use of light-induced plating (LIP) of Ni and Cu [15,16]. An initial thin layer of Ni is used to make ohmic contact to the silicon and to provide a barrier to prevent the penetration of copper into the silicon [7]. Unfortunately, the conductivity of Ni is about a factor of ten worse than the copper, necessitating the use of the latter as the main conducting metal. Such copper plating is then usually capped with Ag or tin to prevent the copper surface from oxidising and to provide a solderable surface [7]. There are more than a 1 GW of solar cells using such a combination of plated metals already operational in the field with published field performance data [17] confirming the reliability and durability of such a metalisation scheme. Initial experimental work to study adhesion aimed to firstly ascertain the point of weakness in such a metalisation scheme and secondly, to evaluate different approaches for forming metal contacts in this way to look at the variability or improvement in adhesion able to be achieved.

In all approaches used for forming the plated contacts, the point of failure of the metal adhesion to the silicon was always the Ni/Si interface as shown by the example in Figure 3. To be more specific, four common approaches for plating the metal were used and evaluated, namely:

1. plating of Ni to a heavily n-type laser-doped Si surface followed by Ni sintering at 350 °C for 2 mins, Cu plating and then Ag capping layer;
2. plating of Ni to heavily n-type laser-doped Si followed by Ni at 350 °C for 2 minutes, followed by chemically removing any residual nickel, followed by Cu plating and Ag plating of cap;
3. same as 1. but without a Ni sinter;
4. the same as in 3. above but with the addition of a post-processing sinter for 2 minutes at 350 °C.

In the above approaches, the purpose of the 350 °C sinter in each case is two-fold, firstly to enhance the Ni to Si bond strength and secondly to form a nickel silicide layer that has been widely reported [7] to act as an effective diffusion barrier to prevent copper from diffusing to deep in the silicon material. Five groups of cells were prepared corresponding to the above four approaches for forming the plated Ni/Cu/Ag contacts plus a fifth group based on industrially fabricated solar cells with standard screen-printed Ag contacts. Groups 1-4 were LDSE solar cells as shown in Figure 2, utilizing light-induced plated Ni/Cu contacts with metal fingers spaced 1mm apart in conjunction with p-type, 1-3 Ω.cm, Cz-Si (100) wafers and 100-120 Ω/□ emitters. These were all fabricated using identical industrial tool sets for processes in common. The processing sequence used for device fabrication to form the structure of Figure 2 was as follows:

1. Wafer texturing
2. Phosphorus 100-120 Ω/□ emitter diffusion
3. Rear etch
4. Front surface PECVD silicon nitride deposition
5. Rear aluminium screen-printing and firing
6. Phosphoric acid coating
7. Front localised laser-doping
8. Ni light-induced plating (LIP)
9. Optional belt-furnace Ni sintering at 350 °C and/or surplus Ni removal
10. Cu LIP + Ag capping
11. Testing and sorting.

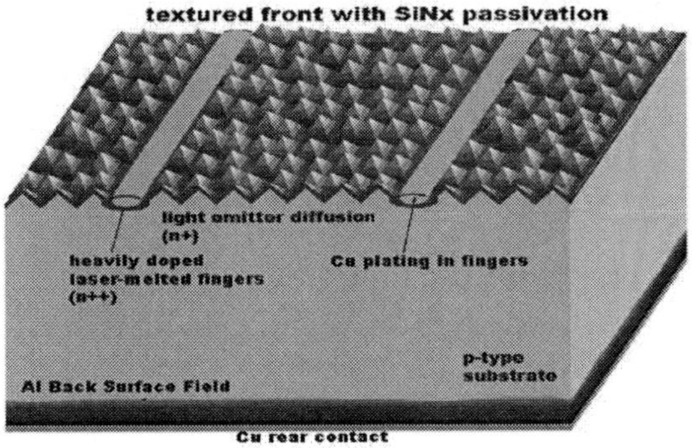

Figure 2: Schematic of the laser-doped LDSE cell

3. Results And Discussion

As expected, the Ni/Si interface was the weakness for all LDSE cells in groups 1 to 4 using the LIP metal contacts. Figure 3 shows the underneath side of the metal following dislodgement from the Si. The clear pattern of the pyramids can be seen impregnated in the metal layer. The metal does not plate to the dielectric coated silicon, with minimal contribution to the adhesion being contributed by these regions as evidenced by the lack of observable damage to the metal in Figure 3 (a) in the regions adjacent to the laser-doped region.

To identify and study the peeling interface for these plated LDSE cells shown in Fig 3, scanning electron microscopy (SEM) with energy dispersive x-ray spectrometry (EDS) was used to analyse the underside of a peeled finger. As seen in Fig 3b, Ni remained on the underside finger surface after testing where the Si surface was exposed for nucleation of the plating, suggesting that the Si-Ni interface was disrupted by the test and therefore represented the weakness in the adhesion strength. In comparison, there was no Ni evident in the textured regions away from where the plating nucleated at the exposed Si surface, confirming the masking ability of the SiN_x layer to prevent plating to the regions coated by the SiN_x. This however exacerbates the adhesion problem since it implies there is minimal adhesion between the metal and Si in the textured regions of Figure 3a which are SiN_x coated.

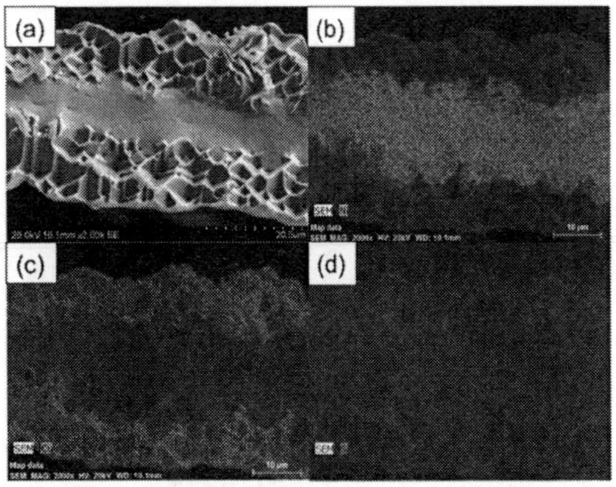

Figure 3: SEM images with EDS analysis of an underside peeled plated finger: (a) original image (b) Ni elementary detection (c) Cu detection (d) Si elementary detection

Surprisingly however, despite the Ni/Si interface being the weakness for each of the four groups of cells, there was no statistically significant difference in adhesion strength between the cells in each group despite using the significantly different approaches for establishing the Ni/Si interface. This was even the case regardless of whether the Ni was sintered or even subsequently removed prior to Cu plating. This appears to be because the metal adhesion strength is relatively unaffected by changes in the Ni processing in comparison to variations in the volume of Cu plated that seems to cause changes in internal stress that weakens the Ni/Si interface. This important aspect contributing to the metal adhesion strength is evaluated in Figure 4 while new innovative approaches for forming the Ni/Cu/Ag plated contacts that remove the adhesion strength's dependence on the volume of plated metal are evaluated in Part B.

In studying the impact that the plated volume of metal has on the adhesion strength of the plated contacts, the LDSE solar cell design was used and with the same wafer type and processing sequence as used above. The only variable in this case is the duration of the plating process for the Cu, which in turn leads to different volumes of plated metal and associated metal heights. Five groups of cells therefore provide five data points for the evaluation of the relationship between the plated metal volume and the adhesion strength. In each group, the Ni is sintered using the standard conditions of 350 °C for 2 minutes. A 6th group is also identically processed but with no metal sinter to further check the apparent finding that the formation conditions for the metal/Si interface are relatively unimportant compared to the volume of metal plated when determining the metal's adhesion strength to the Si. For each group of cells, the interface between the dislodged metal and the exposed Si surface was studied to better understand the failure of the metal adhesion and the importance of the internal stresses within the plated metal.

Figure 4 summarises the relationship between the adhesion strength and the volume of plated Cu indicated by its height. It is clear that increased metal height corresponds to reduced adhesion strength for the metal, probably as a result of increased stresses within the plated Cu that arise in part from the thermal expansion coefficient mismatch between the Cu and the Si. Although thinner Cu has the same thermal expansion coefficient mismatch with the Si, thinner Cu is less able to exert significant resulting force on the interface with the Si and therefore has less impact in reducing the adhesion strength. To further study the impact of the thermal expansion coefficient mismatch, cells from each group were deliberately thermally cycled between room temperature and 350 °C. Interestingly, the cells with greatest plated volume of Cu (14 μm height in Figure 4) could not even tolerate a single thermal cycle before most metal fingers peeled from the Si surface. Figure 5 shows a typical plated cell where the plated metal is peeling following the thermal cycling, even without application of the peel test.

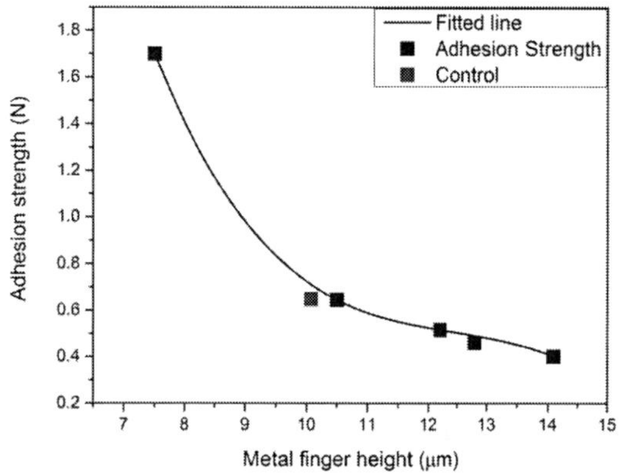

Figure 4: Tested metal adhesion strength for Ni/Cu/Ag contacts with different metal heights

Figure 5: Example of a plated contact peeling prior to the application of the peel test.

Those plated contacts with intermediate height in the range 10-13 μm gave quite variable responses to the thermal cycle, some peeling on the first cycle, some on latter cycles, while some still remained in tact after the maximum of 10 thermal cycles. In all cases though, the metal adhesion strength was significantly damaged by the thermal cycling, with even the contacts that remained in tact suffering from reduced adhesion strengths, typically retaining only half the original values prior to thermal cycling.

Interestingly, the group of cells with significantly reduced copper (7.5 μm height), not only achieved much greater adhesion strength prior to thermal cycling, but in general did not appear to suffer any loss in adhesion strength for the metal as a result of 10 thermal cycles between room temperature and 350 °C. There was however a small degradation evident in the electrical performance of the latter cells following thermal cycling, but this appeared to be the result of increasing ideality factor rather than any apparent change in the contact resistance for the metal. This increase in ideality factor is consistent with previous work [6] where this type of metalisation scheme is subjected to elevated temperatures in this way and it is concluded that the degradation arises from penetration of the metal into active regions of the Si, particularly in the vicinity of the junction.

It is perhaps somewhat surprising that the cells with metal height of 7.5 μm have so much greater adhesion strength compared to the other groups of cells. The fitted line in Figure 4 is only intended to be a guide to the eye and not a quantification of the relationship between metal height and adhesion strength.

Regarding the processing of the metal/Si interface and its importance to the adhesion strength, the 6th group of cells, fabricated identically but without any sintering of the Ni to the Si, is also shown in Figure 4 as the control indicated by the red dot. It is clear that the significant difference in processing is having minimal impact on the adhesion strength in comparison to variations in the volume of plated Cu. Cells with and without Ni sintering have the same adhesion strength for the metal provided the metal heights are the same.

4. PART B - Experimental Methods

The focus in PART B of this work is to develop and demonstrate innovative approaches for forming the plated contacts that remove the adhesion problems that result from variability in plated metal volumes. Five groups of cells were prepared. Group 1, as a control, was industrially produced solar cells with standard screen-printed Ag contacts. Groups 2-5 were solar cells utilizing LIP Ni/Cu contacts with metal fingers spaced 1mm apart in conjunction with p-type, 1-3 Ω.cm, Cz-Si (100) wafers and 100-120 Ω/□ emitters. These were all fabricated using identical industrial tool sets for processes in common. Group 2 was LDSE cells, and group 3 was commercialised LDSE cells with significantly reduced amounts of plating that gave much lower metallised contact heights. Groups 4 & 5 were the same as group 3 except for the inclusion of the process for simultaneously forming the anchor points while carrying out the laser-doping process (process #7 below). In the case of group 5, the density of the anchor points was so high as to cause them to overlap and therefore create a continuous groove equivalent to the buried contact solar cell (BCSC).

The processing sequence used for device fabrication in this work is only modified slightly from that used in Part A whereby the front localised laser-doping process includes the anchor point formation. Each of the processes was implemented through the use of standard commercial manufacturing tools except for the laser system as described above. More specifically, group 4 wafers were fabricated by laser-ablating anchor points uniformly distributed along laser-doped line with spacing of about 0.2 mm. This was done for simplified testing despite earlier analysis and evaluation showing that the adhesion strength towards the ends of each finger is most important and sufficient in determining the reliability and durability of such a metalisation scheme. Following processing, cells in all five groups were adhesion strength tested and analysed.

5. PART B - Results And Discussion

Table 1: Peel adhesion results for different metallisation schemes normalised to the average adhesion strength for standard screen-printed metal fingers (group 1).

Metallisation type	Peel test	Metal height μm (σ)
SP (Group 1)	80-120%	20.6 (1.7)
LDSE (Group 2)	5-7%	12.8 (2.7)
Pluto (Group 3)	30-50%	7.5 (0.5)
LAASE (Group 4)	100-150%	8.0 (1.5)
BCSC (Group 5)	100-130%	10.0 (1.7)

σ represents standard deviation

The work in Part A showed that for all the plated cells, Ni and its interface with the silicon appears to be the key factor that affects the adhesion strength in a Ni/Cu contact LDSE cell. A clear conclusion from Table 1 however is that anchor points can increase the adhesion strength for the plated contacts by more than an order of magnitude compared to conventional plated contacts onto planar silicon. In fact in this work, the adhesion strength measurements did not even determine the strength of the metal/silicon interface, but rather the force necessary to remove large chunks of silicon that pulled away with the plated metal as shown in Figure 6. The use of such anchor points appears to more than compensate for the potential problems created by increased volumes of plated copper, poor interfaces between the Ni and silicon, and the lack of adhesion between the SiN_x and the overlying Cu that plates across the surface. With regard to the latter, the EDS result from Part A in Figure 3 (c) confirms the low interfacial adhesion at the SiN_x/Cu interface with the evidence of Cu remaining on the underside of peeled fingers adjacent to the remaining nickel but with no evidence of residual Cu on the SiNx juxtaposed to the laser-doped line. Without the use of anchor points, it appears that the area covered by such regions as well as the height of the plated metal, both need to be minimized to gain adhesion strength approaching that of SP cells.

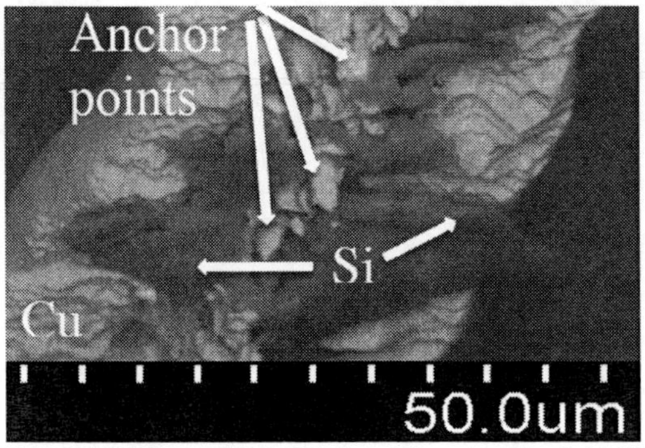

Figure 6: Underneath of plated metal contact for LAASE cells after tape-pull test showing chunks of silicon removed rather than failure of the metal/Si interface.

6. Conclusion

Adhesion strength and durability of plated contacts are probably the main issues causing many cell manufacturers to be cautious about considering such metallisation for large-scale manufacturing. A range of different approaches for forming the plated metal/Si interface have been evaluated in this work but with no discernable difference resulting in the corresponding metal adhesion strength, even though this metal/Si interface appeared to be the weakness in each case. In comparison, the volume of plated metal gave a much stronger correlation with adhesion strength, suggesting that internal stresses within the plated metal are probably having significant impact on the resulting strength of the interface between the plated metal and the silicon. Reducing the plated metal height to about 7 µm or less appears to achieve metal adhesion strength comparable to that of screen-printed cells acting as controls.

For contacts requiring metal heights greater than 7 µm for increased metal conductivity, a new approach capable of ensuring good adhesion strength for plated contacts of almost any height was developed. It involves the formation of laser-ablated anchor points and grooves in the heavily doped silicon surface which when plated greatly strengthen the mechanical adhesion strength for the metallisation. The adhesion strengths for this and various other metal contacting schemes have been evaluated and compared including: screen-printed (SP) silver contacts, nickel/copper (Ni/Cu) light induced plated (LIP) contacts for laser-doped selective emitter (LDSE) cells, commercialised LDSE cells which are also called PLUTO cells, buried contact solar cells (BCSCs) with LIP contacts as well as Ni/Cu LIP laser-ablated anchoring points in selective emitter (LAASE) cells. The latter LAASE cells showed superior adhesion strength to all the other types of cells evaluated including the SP cells, although the BCSCs also demonstrated metal adhesion strengths superior to those of SP cells and only slightly worse than for LAASE cells.

Acknowledgements

The work presented in this paper has been supported by the Australian Government via the Australian Renewable Energy Agency (ARENA). This support included direct grant funding as well as provision of Postdoctoral Fellowship for Dr M. Edwards and PhD scholarship for S. Wang. All these contributions are gratefully acknowledged.

References

[1] R. Jacobsson, "*Measurement of the adhesion of thin films*", Thin Solid Films, vol. 34, pp. 191, 5/17/ 1976.

[2] ITRPV Working Group, *International technology roadmap for photovoltaics (itrpv.net): Results* 2015.

[3] Kray D. and McIntosh K.R., IEEE Trans. on El. Devices, Vol. 8 (56), 2009, p1645

[4] Z. Shi, S. Wenham and J. Ji, *"Mass Production of the Innovative PLUTO Solar Cell Technology"* Invited Plenary Session Paper, 34th IEEE PV Specialists Conference, Philadelphia, June 2009.

[5] S. R. Wenham and M.A. Green, *"Self Aligning Method for Forming A Selective Emitter and Metallization in a Solar Cell"*. US Patent No 6429037, August 2002.

[6] S. Wenham, *"Laser Grooved Silicon Solar Cells"*, PhD thesis, The University of New South Wales, 1986.

[7] A. Lennon, Y. Yao and S. Wenham, *"Evolution of Metal Plating Techniques for Silicon Solar Cell Metallisation"*, accepted for publication, Wiley Journal Progress in PV, March 2012

[8] B. Hallam, et al, *"Record Efficiency on Large Area P-Type Czochralski Silicon Substrates"*, Japanese Journal of Applied Physics, Vol. 51, No. 10, p. 10NA08, 2012.

[9] Moonseok Kim et al, *"Copper Metallization Of Silicon PERL Solar Cells"*, 29th EuPVSEC, Sept 2014.

[10] T.L.Young, et al, *"Design and characterization of an adhesion strength tester for evaluating metal contacts on solar cells"*, 40th IEEE Photovoltaics Specialist Conference, Denver, CO, USA, June 2014

[11] A. Wenham, *"High Efficiency Solar Cells"*, Undergrad thesis, The University of New South Wales, 2009.

[12] J. Wendt et al, *"Improved quality test method for solder ribbon interconnects on silicon solar cells,"* in ITherm, 2010 12th IEEE Intersociety Conference on, 2010, pp. 1-4.

[13] A. Mondon et al, *"Microstructure analysis of the interface situation and adhesion of thermally formed nickel silicide for plated nickel–copper contacts on silicon solar cells"*. Solar Energy Materials and Solar Cells, vol. 117, pp. 209-213, 2013.

[14] J. Horzel et al., *"Low cost metallisation based on Ni/Cu plating enabling high efficiency industrial solar cells"* in 29th European Photovoltaic Solar Energy Conference, Amsterdam, NL, 2014

[15] L. Mai, et al, *"Photoplating of metal electrodes for solar cells"*, U.S. Patent 9,040,128, May 2015.

[16] J. Ji, S. Wenham and Z. Shi, "Method for electrochemical depositing solar cell metallic electrode", Patent CN101257059A, Nov 30, 2007.

[17] M. Alonso et al, *"Toledo PV plant 1MWp 20years of operation"*, EU PVSEC, 2014

[18] C. M. Chong et al, "Improved adhesion for plated Solar cell metallisation" In 42nd IEEE PVSC , 2015.

[19] A. Wenham, et al, *"Overcoming Issues with Copper Plated Contacts for Large-Scale Manufacturing"*, EU PVSEC, Hamburg, Germany 2015

AUTHOR INDEX

AUTHOR INDEX

AUTHOR INDEX

AUTHOR INDEX

AUTHOR INDEX

AUTHOR INDEX

AUTHOR INDEX

AUTHOR INDEX

AUTHOR INDEX

AUTHOR INDEX

AUTHOR INDEX

AUTHOR INDEX

AUTHOR INDEX

AUTHOR INDEX

AUTHOR INDEX

AUTHOR INDEX

AUTHOR INDEX

AUTHOR INDEX

AUTHOR INDEX

AUTHOR INDEX

AUTHOR INDEX

AUTHOR INDEX

AUTHOR INDEX

AUTHOR INDEX

AUTHOR INDEX

32nd European Photovoltaic Solar Energy Conference and Exhibition (EU PVSEC 2016)

Munich, Germany
20-24 June 2016

Volume 2 of 4

Editors:

M. Topic
P. Helm
N. Taylor

ISBN: 978-1-5108-3651-8

Printed from e-media with permission by:

Curran Associates, Inc.
57 Morehouse Lane
Red Hook, NY 12571

Some format issues inherent in the e-media version may also appear in this print version.

Printed by Curran Associates, Inc. (2017)

WIP – Renewable Energies
Sylvensteinstr. 2
81369 Munchen
Germany

Phone: +49 89 72012735
Fax: +49 89 72012791

wip@wip-munich.de

Additional copies of this publication are available from:

Curran Associates, Inc.
57 Morehouse Lane
Red Hook, NY 12571 USA
Phone: 845-758-0400
Fax: 845-758-2633
Email: curran@proceedings.com
Web: www.proceedings.com

32nd European Photovoltaic Solar Energy Conference and Exhibition

Proceedings of the International Conference held in Munich, Germany

20 June - 24 June 2016

Edited by:

M. TOPIČ
University of Ljubljana
Chairman of European Technology & Innovation Platform Photovoltaics
Ljubljana, Slovenia

N. TAYLOR
European Commission - DG Joint Research Centre
Ispra, Italy

P. HELM
WIP Renewable Energies
Munich, Germany

Coordination of the Technical Programme:

European Commission - DG Joint Research Centre
Via E. Fermi 1
21020 Ispra (VA)
Italy

Institutional Support:

European Commission

UNESCO – United Nations Educational, Scientific and Cultural Organization, Natural Sciences Sector

WCRE – World Council for Renewable Energy

Supporting Organisations:

ARE – Alliance for Rural Electrification

EREF – European Renewable Energies Federation

ESA – European Space Agency

EUFORES –The European Forum for Renewable Energy Sources

EUREC – The Association of European Renewable Energy Research Centres

European Photovoltaic Technology Platform

IEA PVPS – International Energy Agency Photovoltaic Power Systems Programme

REN21 – Renewable Energy Policy Network for the 21st Century

Institutional PV Industry Cooperation:

SolarPower Europe

SOLARUNITED

Supporting Associations:

AIE – European Association of Electrical Contractors

AREA – African Renewable Energy Alliance

CABA – Continental Automated Buildings Association

EPIC – European Photonics Industry Consortium

SASIA - Saudi Arabia Solar Industry Association

EU PVSEC realised by:

WIP
Sylvensteinstr. 2, 81369 München, Germany
Tel: +49 89 720 12 735, Fax: +49 89 720 12 791
Email: pv.conference@wip-munich.de
www.photovoltaic-conference.com
www.wip-munich.de

Proceedings produced and published by:

WIP
Sylvensteinstr. 2, 81369 München, Germany
Tel: +49 89 720 12 735, Fax: +49 89 720 12 791
Email: pv.conference@wip-munich.de
www.photovoltaic-conference.com
www.wip-munich.de

ISSN 2196-0992

32nd EUROPEAN PHOTOVOLTAIC SOLAR ENERGY CONFERENCE AND EXHIBITION
20 JUNE – 24 JUNE 2016
MUNICH, GERMANY

EU PVSEC COMMITTEES

INTERNATIONAL SCIENTIFIC ADVISORY COMMITTEE (ISAC)

Chair
G.F. De Santi, European Commission, DG JRC, Director of the Institute for Energy and Transport, Ispra, Italy

Secretary
N. Taylor, European Commission, DG JRC, Institute for Energy and Transport, Ispra, Italy

Committee Members
P. Frankl, Head of the Renewable Energy Division, International Energy Agency, France
P. Hofer-Noser, CTO, Meyer Burger Technology, Switzerland
P. Malbranche, CEA, Le Bourget du Lac, France
S. Nowak, Managing Director of NET Nowak Energy & Technology, Switzerland
W. Sinke, Program Development Manager, ECN Solar Energy, The Netherlands
P. Verlinden, Chief Scientific and Vice Chair, State Key Lab. TRINA Solar, China
E. Weber, Director, Fraunhofer ISE, Germany
P. Helm, WIP, Munich, Germany
A. Grassi, ETA-Florence Renewable Energies, Florence, Italy

CONFERENCE EXECUTIVE COMMITTEE

Conference General Chairman
M. Topič, University of Ljubljana, Chairman of European Technology & Innovation Platform Photovoltaics, Ljubljana, Slovenia

Technical Programme Chairman
N. Taylor, European Commission, DG JRC, Institute for Energy and Transport, Ispra, Italy

Committee Members
H. Ossenbrink, European Commission, DG JRC, Institute for Energy and Transport, Ispra, Italy
S. Nowak, Managing Director of NET Nowak Energy & Technology, St. Ursen, Switzerland
W. Sinke, Program Development Manager, ECN Solar Energy, The Netherlands
P. Helm, WIP, Munich, Germany
A. Grassi, ETA-Florence Renewable Energies, Florence, Italy

2016 SCIENTIFIC COMMITTEE
Topic Organisers and Paper Review Experts

Topic 1 – New Materials and Concepts for Solar Cells and Modules

G. Beaucarne, Dow Corning, Seneffe, Belgium
N.J. Ekins-Daukes, Imperial College London, United Kingdom
A. Martí Vega, UPM, Madrid, Spain
J. Poortmans, imec, Leuven, Belgium
M. Rusu, HZB, Berlin, Germany
J.A.M. Van Roosmalen, ECN, Petten, Netherlands

V. Bukauskas, Center for Physical Sciences and Technology, Vilnius, Lithuania
F. Falk, IPHT, Jena, Germany
J. Govaerts, imec, Leuven, Belgium
J.F. Guillemoles, CNRS, Chatou Cedex, France
M.A. Hernandez-Fenollosa, UPV, Valencia, Spain
J.C. Hummelen, University of Groningen, Netherlands
D. König, UNSW, Sydney, Australia
I. Konovalov, Ernst Abbe University of Applied Science Jena, Jena, Germany
C. Lévy-Clément, CNRS, Thiais, France
Y. Okada, University of Tokyo, Japan
B. Rech, HZB, Berlin, Germany
F. Roca, ENEA, Portici, Italy
V. Saly, Slovak University of Technology, Bratislava, Slovakia
H. Shirai, Saitama University, Japan
W. Warta, Fraunhofer ISE, Freiburg, Germany

Topic 2 – Wafer-Based Silicon Solar Cells and Materials Technology

O. Anspach, PV Crystalox Solar, Erfurt, Germany
D.L. Bätzner, Meyer Burger Research, Hauterive, Switzerland
R. Brendel, ISFH, Emmerthal, Germany
F. Ferrazza, Eni, San Donato Milanese, Italy
S.W. Glunz, Fraunhofer ISE, Freiburg, Germany
G. Hahn, University of Konstanz, Germany
D. Muñoz, CEA, Le Bourget du Lac, France
S. Riepe, Fraunhofer ISE, Freiburg, Germany
A.W. Weeber, ECN, Petten, Netherlands
P. Wohlfart, Singulus Technologies, Kahl am Main, Germany

A.G. Aberle, SERIS, Singapore
N. Azkona, UPV/EHU, Zamudio, Spain
J. Bagdahn, Fraunhofer CSP, Halle (Saale), Germany
R. Barinka, Fill Factory, Rožnov pod Radhoštm, Czech Republic
P.A. Basore, PV Specialist Services, Solana Beach, USA

D. Biro, Fraunhofer ISE, Freiburg, Germany
S. Bowden, Arizona State University, Tempe, USA
T.M. Bruton, TMB Consulting, Woking, United Kingdom
B. Dauksher, Arizona State University, Tempe, USA
M. Debucquoy, imec, Leuven, Belgium
T. Desrues, CEA, Le Bourget du Lac, France
D. Dimova-Malinovska, Bulgarian Academy of Sciences, Sofia, Bulgaria
S. Dubois, CEA, Le Bourget du Lac, France
T. Dullweber, ISFH, Emmerthal, Germany
P. Fath, RCT-Solutions, Konstanz, Germany
A.J. Galdikas, Modernios E-Technologijos, Vilnius, Lithuania
M. Hermle, Fraunhofer ISE, Freiburg, Germany
K. Hesse, Wacker Chemie, Burghausen, Germany
J. Horzel, CSEM, Neuchâtel, Switzerland
J. John, imec, Leuven, Belgium
S.D. Kim, SunPower, San Jose, USA
J. Libal, ISC Konstanz, Germany
A. Metz, h.a.l.m. elektronik, Frankfurt am Main, Germany
J.W. Müller, Hanwha Q CELLS, Bitterfeld-Wolfen, Germany
Y. Ohshita,TTI, Nagoya, Japan
S. Peters, Hanwha Q CELLS, Bitterfeld-Wolfen, Germany
K. Petter, Hanwha Q CELLS, Bitterfeld-Wolfen, Germany
R. Preu, Fraunhofer ISE, Freiburg, Germany
J. Rentsch, Fraunhofer ISE, Freiburg, Germany
E. Sauar, Brighterlite, Oslo, Norway
H. Schlemm, Meyer Burger, Hohenstein-Ernstthal, Germany
M.C. Schubert, Fraunhofer ISE, Freiburg, Germany
B. Terheiden, University of Konstanz, Germany
Y. Veschetti, CEA, Le Bourget du Lac, France
G.P. Willeke, Fraunhofer ISE, Freiburg, Germany
D. Wood, Hemlock Semiconductor, Hemlock, USA
A. Yamada, Tokyo Institute of Technology, Japan
D.L. Young, NREL, Golden, USA

Topic 3 – Thin Film Solar Cells and Modules

C.J. Brabec, University of Erlangen-Nuremberg, Germany
J. Cárabe, CIEMAT, Madrid, Spain
A. Meeder, Flisom, Dübendorf, Switzerland
J. Meier, PV Consultant, Berlingen, Switzerland
S. Siebentritt, University of Luxembourg, Belvaux, Luxembourg
M. Topic, University of Ljubljana, Slovenia
S.C. Veenstra, ECN, Eindhoven, Netherlands
W. Wischmann, ZSW, Stuttgart, Germany

C. Ballif, EPFL&CSEM, Neuchâtel, Switzerland
R. Birkmire, University of Delaware, Newark, USA
D. Cheyns, imec, Leuven, Belgium
S. Cros, CEA, Le Bourget du Lac, France
T. Dalibor, AVANCIS, Munich, Germany
P. Delli Veneri, ENEA, Portici, Italy
B. Dimmler, Manz CIGS Technology, Schwäbisch Hall, Germany
K. Durose, Stephenson Institute for Renewable Energy, Liverpool, United Kingdom
S. Gall, HZB, Berlin, Germany
I. Gordon, imec, Leuven, Belgium
S. Huang, UNSW, Sydney, Australia
S.J.C. Irvine, Swansea University, St. Asaph, United Kingdom
R. Klenk, HZB, Berlin, Germany
C.-F. Lin, NTU, Taipei, Taiwan
M.C. Lux-Steiner, HZB, Berlin, Germany
E. Mellikov, Tallinn University of Technology, Tallinn, Estonia

M. Meuris, imec/IMOMEC - ATFPV, Leuven, Belgium
C. Molpeceres, UPM, Madrid, Spain
J. Parisi, University of Oldenburg, Oldenburg, Germany
R. Po, Eni, Novara, Italy
M. Powalla, ZSW, Stuttgart, Germany
A. Romeo, University of Verona, Verona, Italy
V. Sittinger, Fraunhofer IST, Braunschweig, Germany
A. Slaoui, InESS, Strasbourg, France
A.N. Tiwari, EMPA, Dübendorf, Switzerland
A. Virtuani, SUPSI, Canobbio, Switzerland
M. Zeman, Delft University of Technology, Netherlands

Topic 4 – Concentrator and Space Applications

R. Campesato, CESI, Milan, Italy
G. Flamand, imec, Leuven, Belgium
F. Rubio Berenguel, Freiburg, Germany
C. Signorini, ESA-ESTEC, Noordwijk, Netherlands

I. Anton Hernandez, UPM, Madrid, Spain
A.W. Bett, Fraunhofer ISE, Freiburg, Germany
V. Khorenko, AZUR SPACE, Heilbronn, Germany
R. King, Arizona State University, Tempe, USA
G. Siefer, Fraunhofer ISE, Freiburg, Germany
T. Takamoto, SHARP, Nara, Japan
G. Timò, RSE, Piacenza, Italy

Topic 5 – Operations, Performance, Reliability and Sustainability of PV

J. Binder, ZSW, Stuttgart, Germany
M. de Wild-Scholten, SmartGreenScans, Groet, Netherlands
R. Gottschalg, Loughborough University, United Kingdom
U. Jahn, TÜV Rheinland Energy, Köln, Germany
N.M. Pearsall, Northumbria University, Newcastle upon Tyne, United Kingdom
M. Perrin, CEA, Le Bourget du Lac, France
C. Protogeropoulos, Phoenix Solar, Athens, Greece
K. Radouane, EDF EN, Paris La Defense, France
T. Sample, European Commission DG JRC, Ispra, Italy
W.G.J.H.M. van Sark, Utrecht University, Netherlands
K. Wambach, Wambach-Consulting, Aindling, Germany

C. Alonso-Tristán UBU, Burgos, Spain
M. Bächler, PerVorm, Ulm, Germany
T.R. Betts, Loughborough University, United Kingdom
P. Blanc, MINES ParisTech, Paris, France
R. Einhaus, Apollon Solar, Lyon, France
G. Friesen, SUPSI, Canobbio, Switzerland
V. Fthenakis, Columbia University, New York, USA
W.J. Gambogi, DuPont, Wilmington, USA
L. Gomes de Freitas, Federal University of Uberlândia, Brazil
G. Graditi, ENEA, Portici, Italy
M. Grottke, WIP - Renewable Energies, Munich, Germany
Y. Hishikawa, AIST, Tsukuba, Japan
T. Huld, European Commission DG JRC, Ispra, Italy
A.R. Lagunas, CENER, Sarriguren, Spain
O. Mayer, GE Global Research, Garching, Germany
M.A. Muñoz, UPM, Madrid, Spain
H. Nagel, Fraunhofer ISE, Freiburg, Germany
H. Nussbaumer, Zurich University of Applied Sciences, Winterthur, Switzerland
C. Nyman, Soleco, Porvoo, Finland
S. Oktik, Sisecam, Istanbul, Turkey
W. Palitzsch, Loser Chemie, Zwickau, Germany
A.H.M.E. Reinders, University of Twente, Enschede, Netherlands

J. Remund, Meteotest, Bern, Switzerland
J.N. Roy, IIT Kharagpur, Kharagpur, India
M. Schroedter-Homscheidt, DLR, Wessling, Germany
M. Sengupta, NREL, Golden, USA
R. Sinton, Sinton Instruments, Boulder, USA
J. Sutterlueti, Gantner Instruments, Schruns, Austria
S. Tselepis, CRES, Pikermi, Athens, Greece
M. Van Iseghem, EDF R&D, Moret-sur-Loing, France
E. Voroshazi, imec, Leuven, Belgium
A. Wade, First Solar, Berlin, Germany
J. Wohlgemuth, NREL, Golden, USA
M. Zehner, Rosenheim University of Applied Sciences, Germany

Topic 6 – PV Applications and Integration

F.P. Baumgartner, Zurich University of Applied Sciences, Winterthur, Switzerland
F. Bonemazzi, ENEL, Rome, Italy
F. Frontini, SUPSI, Canobbio, Switzerland
P. Malbranche, CEA, Le Bourget du Lac, France
A. Scognamiglio, ENEA, Portici, Italy
X. Vallvé, Trama TecnoAmbiental, Barcelona, Spain
I. Weiss, WIP - Renewable Energies, München, Germany

R. Burkhardt, Energy Depot, Kreuzlingen, Switzerland
S. Caneva, WIP - Renewable Energies, Munich, Germany
I. Da Silva, Strathmore University, Nairobi, Kenya
J.P. de Barros Duarte Fonseca, University of Cape Verde, Mindelo, Cape Verde
B. de Boer, TNO, Eindhoven, Netherlands
H. Fechner, University of Applied Sciences Technikum Wien, Breitenfurt, Austria
I.B. Hagemann, Architekturbüro Hagemann, Aachen, Germany
C. Helmke, Management Consultancy Research, Erfurt, Germany
E. Kaijuka-Okwenje, Makerere University, Kampala, Uganda
G. Mütter, Alternative Energy Solutions, Vienna, Austria
K. Peter, ISC Konstanz, Germany
T. Reijenga, BEAR-iD Sustainable Urban Planners + Architects, Gouda, Netherlands
S. Wittkopf, HSLU, Horw, Switzerland

Topic 7 – PV Economics, Markets and Policies

C. Breyer, Lappeenranta University of Technology, Finland
M. Getsiou, European Commission DG RTD, Brussels, Belgium
G. Masson, Becquerel Institute, Brussels, Belgium
T. Nordmann, TNC Consulting, Feldmeilen, Switzerland
S. Nowak, NET Nowak Energy & Technology, St. Ursen, Switzerland
E. Perezagua, Consultores de Energía Fotovoltaica, Madrid, Spain

I. Kaizuka, RTS, Tokyo, Japan
D. Mayer, MINES ParisTech,Paris, France
P. Mazzucchelli, EUREC,Brussels, Belgium
D. Stickelberger, Swissolar, Zürich, Switzerland
S. Szabó, European Commission DG JRC, Ispra, Italy

SUBJECT INDEX

Topic Code

1 New Materials and Concepts for Solar Cells and Modules
2 Wafer-Based Silicon Solar Cells and Materials Technology
3 Thin Film Solar Cells and Modules
4 Concentrator and Space Applications
5 Operations, Performance, Reliability and Sustainability of PV
6 PV Applications and Integration
7 PV Economics, Markets and Policies

Session Type

P = Plenary Session
O = Oral Session
V = Visual Session

Day Codes

A = Monday, 20 June 2016
B = Tuesday, 21 June 2016
C = Wednesday, 22 June 2016
D = Thursday, 23 June 2016
E = Friday, 24 June 2016

e.g. 3CP.1 ⇒ 3=Thin Film Solar Cells and Modules, C=Wednesday, P=Plenary Session, 1=Session 1

TABLE OF CONTENT
of EU PVSEC 2016 Proceedings Papers

New Materials and Concepts for Solar Cells and Modules

Visual PRESENTATIONS 1BV.6 New Materials and Concepts for Cells

Wafer-Based Silicon Solar Cells and Materials Technology

Visual PRESENTATIONS 2AV.3 Silicon Solar Cell Improvements and Innovation (III)

Thin Film Solar Cells and Modules

Oral PRESENTATIONS 3AO.4 Special Session on CdTe and Kesterites

Oral PRESENTATIONS 3AO.5 Buffer and Contacts for Thin Film Devices

Oral PRESENTATIONS 3AO.6 Interfaces for Thin Film Devices

Oral PRESENTATIONS 3BO.5 Amorphous Silicon-Based Thin-Film PV Devices

Oral PRESENTATIONS 3BO.6 Silicon-Based Thin-Film Materials and Devices

Concentrator and Space Applications

Oral PRESENTATIONS 4CO.6 III-V-based Devices for Terrestrial and Space Applications / Concentrator and Space Systems

Operations, Performance, Reliability and Sustainability of PV

Visual PRESENTATIONS 5BV.2 Operation of PV Systems

Visual PRESENTATIONS 5BV.3 Balance of System Components

Visual PRESENTATIONS 5CV.3 Solar Resource and Forecasting / Sustainability and Recycling

PV Applications and Integration

Plenary SESSION 6DP.2 PV as Part of Our Built Environment: Solutions for Integration into Building Envelops and Energy Systems

Oral PRESENTATIONS 6CO.12 Grid and Energy System Integration (III) - Technology Solutions

Visual PRESENTATIONS 6AV.5 PV in Buildings and the Environment

Visual PRESENTATIONS 6AV.6 Utility-Scale PV / PV Applications without a Centralised Grid

PV Economics, Markets and Policies

Visual PRESENTATIONS 7DV.4 PV Economics and Markets / PV Global Issues, Policies and Strategies

FOREWORD

The EU PVSEC 2016 in Munich concluded with the strong message that photovoltaic solar energy is here to stay. Research and innovation across the sector, together with its deployment and endless application option, are paving the way to putting PV at the center-stage of sustainable energy systems. 1,742 conference participants from 73 countries participated in this most inspiring platform. It provided a vital forum for information exchange in the field of PV solar energy research, demonstration and applications.

The EU PVSEC 2016 Proceedings give a comprehensive overview of the photovoltaic solar sector, its current status and future prospects in science, research, innovation, development and deployment on more than 5,000 pages. Selection for inclusion in the conference was made by paper review experts and topic organisers (see the listing on pages III-V), to whom we express our sincere gratitude for their comprehensive review work and overall contribution to the success of the conference. In addition to the 658 submitted papers, the proceedings includes 82 presentations (slides) shown during the plenary and oral presentations as well as 269 poster files of the visual presentations. In total this amounts to 1,009 publications!

The Proceedings constitute a powerful tool for targeted and quick information search and retrieval, enabling you to search by keyword, paper title and authors. We are sure that these features will help to simplify the use and exploit the full potential of this extensive source of information.

The Conference Proceedings are also published online. **A DOI code (Digital Object Identifier) has been assigned to each paper. This ensures unequivocal and permanent identification and full citability.** The papers can be viewed and downloaded from the EU PVSEC Proceedings website www.eupvsec-proceedings.com.

We are confident that these Proceedings will play an important role in providing a comprehensive overview of the current actors and activities in the global PV sector and that they will disseminate information on the state-of-the-art of technologies and applications. This can generate further research, add momentum to innovation and promote interest in PV worldwide.

We would like to cordially thank all authors and participants of the EU PVSEC 2016 for their contributions and look forward to welcoming you in 2017 at the next EU PVSEC.

The Editors

Joint Research Centre

the European Commission's in-house science service

Serving society
Stimulating innovation
Supporting legislation

EU PVSEC 2016 Highlights

Nigel Taylor
Technical Programme Chair

European Commission

First some stats ...

Participants: 1742

Presentations: 1053 (334 oral, 719 posters)

Parallel Events: 6

European Commission

European
Commission

Silicon: strength in depth

European
Commission

Q.ANTUM
Champion R&D module results

Champion module **features**

- Use current production tools
- Standard soldering technique
- Same BOM as in standard modules (high quality ARC glass, back sheet)
- Standard 60 cells module size
- No additional light-trapping schemes

- Usage of champion cell technology with module optimized ARC coating
- Half cell technology and splitted junction box
- Minimized intercell distance

- η = 19.5 % (ap)*
- V_{oc} = 41.5 V
- J_{sc} = 9.30 A
- FF = 77.4 %

- P = 301 W (ta)*

Classification	Efficiency (%)	Area (cm²)	V_{oc} (V)	I_{sc} (A)	FF (%)	Test centre (date)	Description
Si (crystalline)	23.8±0.5	11562 (ap)	53.4	6.32*	81.6	AIST (1/16)	Panasonic (72 cells) [11]
Si (multicrystalline)	19.5±0.4	15349 (ap)	41.53	9.299*	77.4	FhG-ISE (12/15)	Hanwha Q Cells (120 cells) [12]

M. Green et al, Prog. Photovolt: Res. Appl. 2016; 24:905–913

Current Status of High-Efficiency Q.Antum Technology with New World Record Module Efficiency of 19.5%
M. Scherff, P. Kowalzik, C. Gerber, K. Duncker, M. Junghänel, C. Fahrland, S. Kunath, S. Hörnlein, M. Schütze, L. Niebergall, B. Klöter & J.W. Müller
Hanwha Q CELLS, Bitterfeld-Wolfen, Germany

European Commission

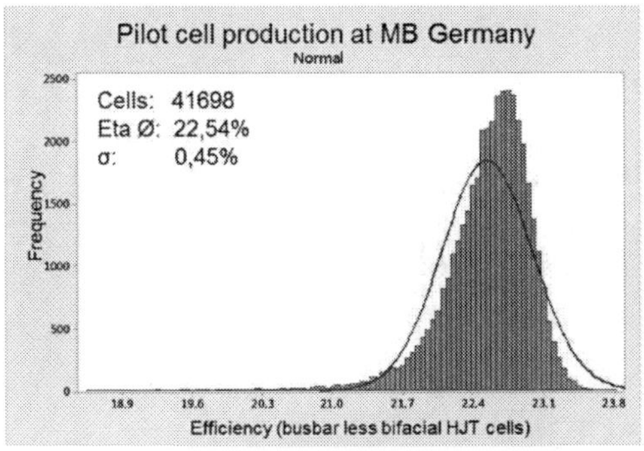

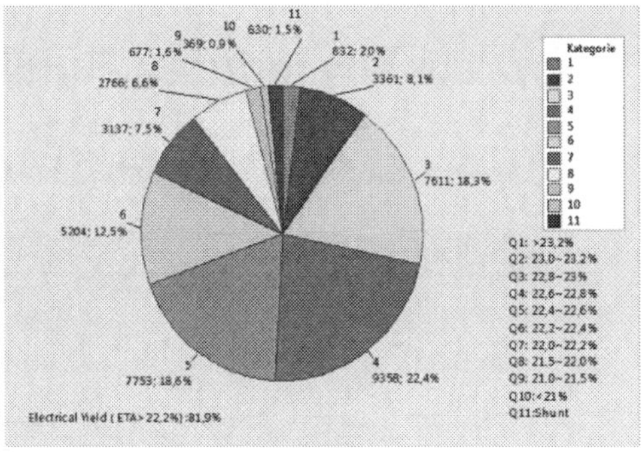

Golden Cell

Status	Cell	Area (cm²)	Eff. (%)	V_{oc} (mV)	FF (%)	J_{sc} (mA/cm²)
09/2015	Cz	239,5	23,3	742,5	81,3	38,64

- In this pilot production, total 41698 SWCT(busbar less) bifacial cells were processed
- Average ETA reached 22,54%
- 34617 cells (yield of 82%) are over 22,2%
- Record cell reached 23,3% cell efficiency, which is from its high V_{oc} and FF.

Silicon Heterojunction Solar Cells in Meyer Burger's Demo line: Results of Pilot Production on Mass Production Tools
J. Zhao et al

European Commission

Tipping point:

2BO.3 Heterojunction Solar Cell Concepts

Speaker gender balance for the first time!

European Commission

Vibrationally Induced Fluidization

1. CapEx, Full Cost, and Cash Cost will be lower than legacy processes

China: $/Kg on a Polysilicon Plant Basis(1)			
	Genesis™	H-FBR	HC-TCS/Siemens
CapEx	43	65	68
Full Cost	10.25	14.00	18.45
Cash Cost(2)	7.35	9.65	13.90

2. Advantages include
   - Only 1% yield loss to dust
   - Only 1 KwHr/Kg for V-FBR process unit – only 8 KwHr/Kg for entire plant
   - 10X higher reactor productivity
   - Self-seeding and in-situ particle size classification
   - On/off fluidization and "Ok" to temperature cycle
   - Continuous self-regulating product withdrawal

2BP.1.2 New Monosilane Fluid Bed Decomposition Technology for the Production of Solar Quality Silicon Feedstock, M. Dassel, SiTec, Augsburg, Germany

European Commission

Material Limits of Si from State-of-the-Art PLI Techniques

WAFER ELBA

Exact specification of loss mechanisms by ELBA

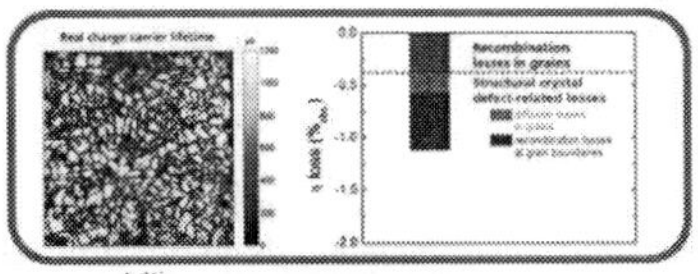

INGOT

Precise PL imaging of τ and [Fe_i] on ingot level

modulum

harmonically modulated luminescence

Fraunhofer ISE

ELBA

CUSTOM APP

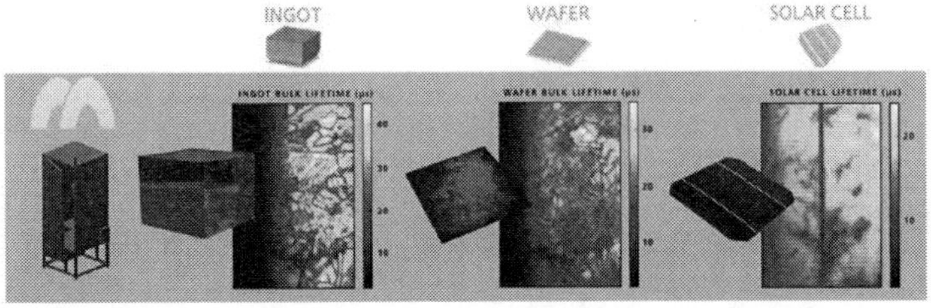

2CO.1.2 Material Limits of Silicon from State-of-the-Art Photoluminescence Imaging Techniques, F. Schindler, J. Giesecke, B. Michl, W. Warta & M.C. Schubert
Fraunhofer ISE, Freiburg, Germany

European Commission

Thin film competes on efficiency

European Commission

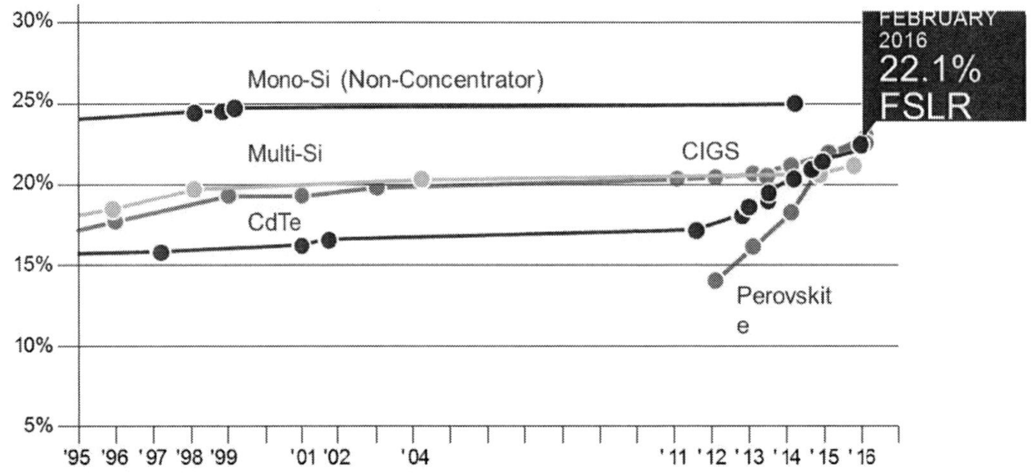

First Solar: current production module average 16.2%, top 18.2% and set to increase

European Commission

AVANCIS' Track Record of Certified Champion Efficiencies

LATEST 2016 EXTERNALLY CERTIFIED CHAMPION MODULE

STC measurement under natural sunlight according to IEC 60904-1/ IEC 60891

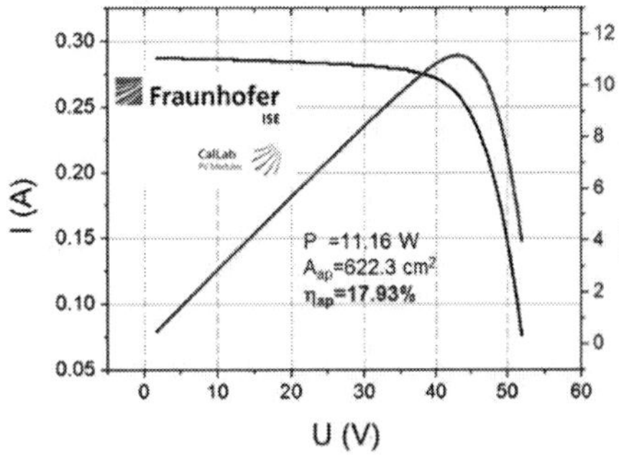

Comparison of champion modules:

	η_ap (%)	Jsc (mA/cm2)	Voc (mV)	FF (%)
2014 Champ (NREL)	16.6	38.5	621	69.4
2016 Champ (ISE)	**17.9**	39.7	627	72.2
Improvement	+7.8%	+3.1%	+1%	+4%

▶ Key improvements 2016 at a glance:

- Optimized cell width and dead area
- Improved TCO layer

▶ Major improvements on FF and Jsc

3BO.8.2 Efficiency Improvement of CIGSSe/Cd-Free Solar Module by Optimized Cell and Interconnect Design

P. Eraerds, C. Schubbert, T. Kwast, M. Grave, F. Braun, M. Algasinger, R. Lechner, T. Dalibor & J. Palm
AVANCIS, Munich, Germany

European Commission

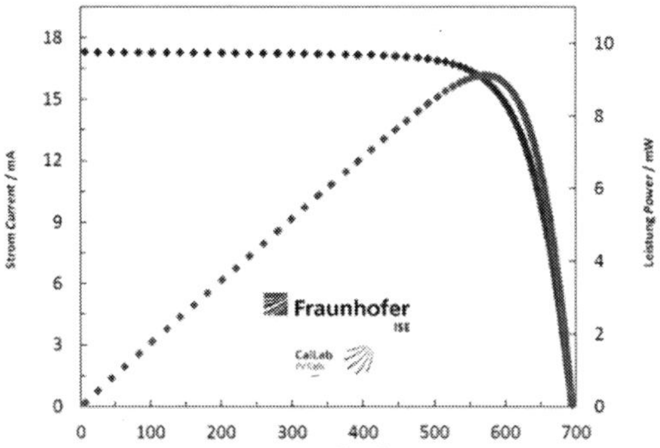

V_{OC}	= (	694.5	±	2.4	) mV
I_{SC} (Ed.2 - 2008)/3/	= (	17.28	±	0.33	) mA
J_{SC}	= (	33.58	±	0.64	) mA/cm²
I_{MPP}	=	15.94			mA
V_{MPP}	=	572.0			mV
P_{MPP}	= (	9.12	±	0.18	) mW
FF	= (	75.98	±	0.49	) %
η	= (	17.72	±	0.37	) %

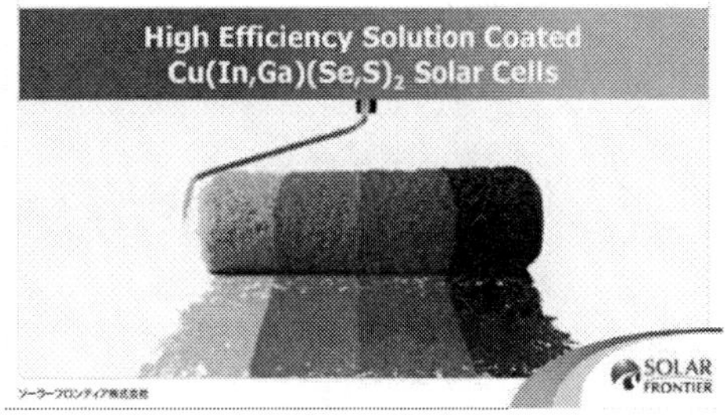

- **Eff 17.72% has been approved by Fraunhofer Institute**
- **Best result for the solution coated CIGSSe cell**
  (Nano solar : Eff 17.1%)

Photo conversion efficiency of 18.7% was achieved with solution coated $Cu(In,Ga)(S,Se)_2$.

3BO.8.3 High Efficiency Solution Coated Cu(In,Ga)(Se,S)2 Thin Film Solar Cells

T. Aramoto & Y. Kawaguchi
Solar Frontier, Atsugi, Japan
Y.-C. Liao, Y. Kikuchi, T. Ohashi, H. Iida & A. Nakamura
Tokyo Ohka Kogyo, Koza-Gun, Japan

European Commission

Lab scale PSC baseline development: mini-modules

- **Perovskite module development: lab scale**
  - Glass/sputtered ITO/P1 laser scribe /e-beam TiO_2/spin coat/Perovskite/spin coat OMeTAD/P2 mechanical scribe/evaporate Au/P3 mechanical scribe
  - **Status Solliance**: **16 cm²** mini-module performance of **14,2%**

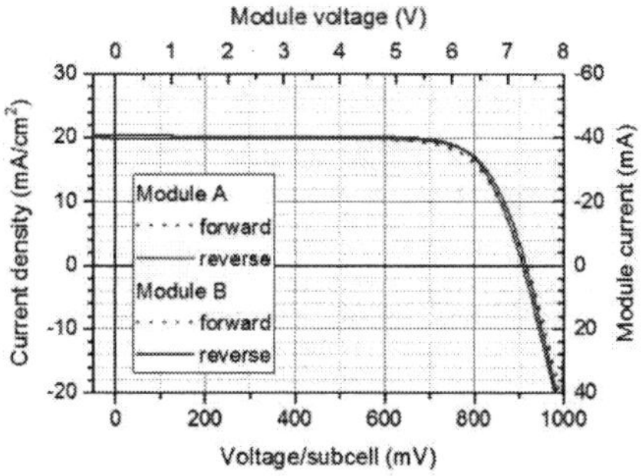

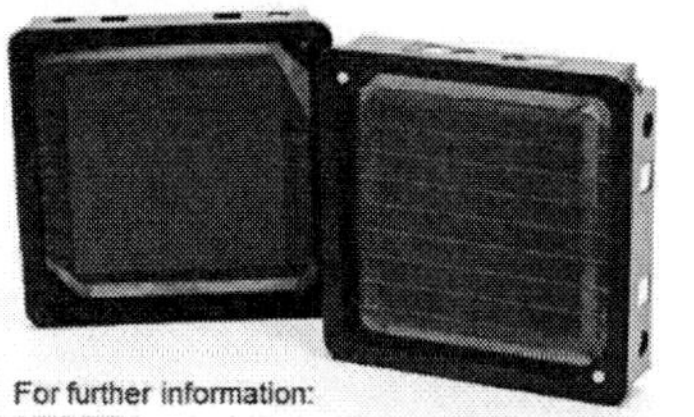

For further information:
3CO.7.5
Loss Analysis and Optimization for High Efficiency Perovskite Photovoltaic Modules
R. Gehlhaar, T. Merckx, C. Masse de la Huerta, L. Rakocevic, W. Qiu & M. Jaysankar
imec, Leuven, Belgium

Towards Roll-to-Roll Manufacturing of Perovskite Based PV Modules

R. Andriessen & P. Poodt, TNO, Eindhoven, Netherlands, T. Aernouts, imec, Leuven, Belgium. S. Veenstra, ECN, Eindhoven, Netherlands, R. Janssen & A. Createore, TU/e, Eindhoven, Netherlands, D. Vanderzande, University of Hasselt, Diepenbeek, Belgium T. Kirchartz Forschungszentrum Jülich, Germany

European Commission

Computational Screening of A_2MX_6 Compounds

Lead-free Perovskite Materials for Photovoltaic Applications

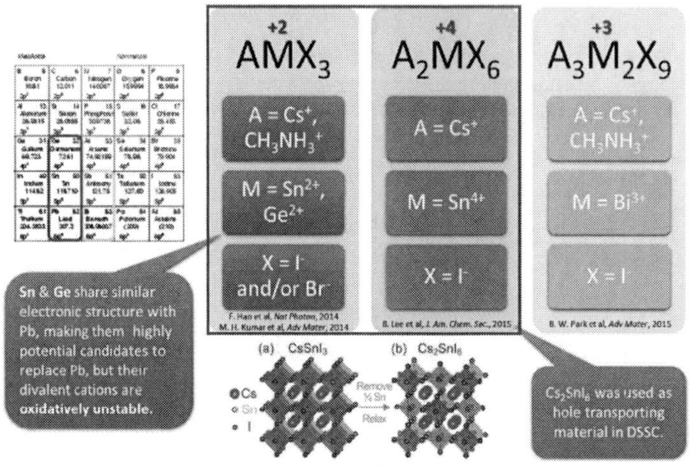

Tin(IV)-based Iodide Perovskites A2SnI6 (A = Rb+, Cs+, MA+, FA+)

Y. Chen, T. Krishnamoorthy, C. Yao, T. Baikie, S. Dharani, J.H. Fu, S.T. Chien, M. Asta, N. Mathews, L.H. Wong & S.G. Mhaisalkar
Nanyang Technological University, Singapore, Singapore & University of California, Berkeley, USA

European Commission

Concentrators:

new paradigms

European Commission

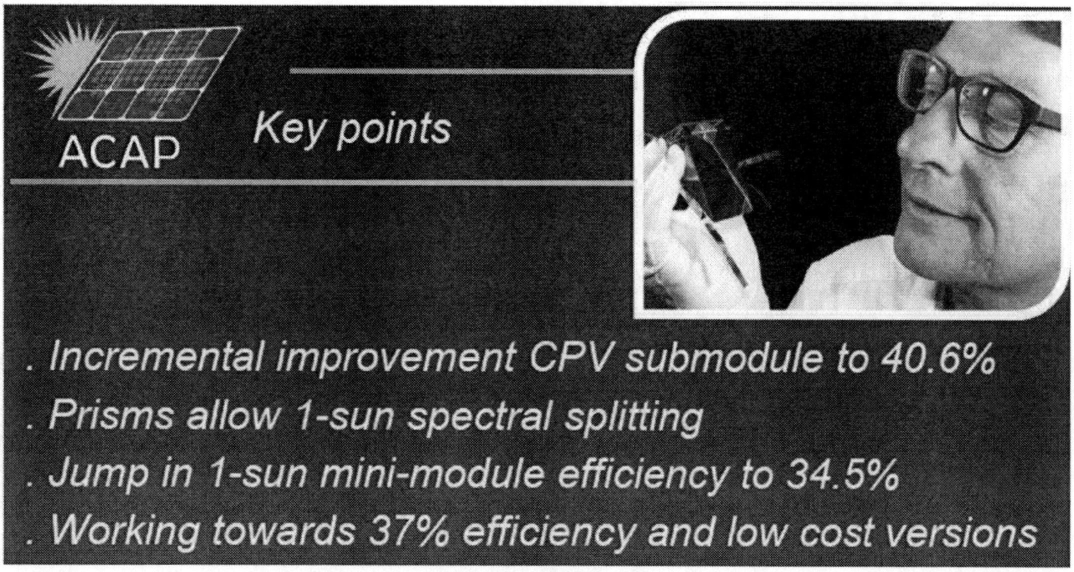

1AP.1.1 37% Efficient One-Sun Minimodule and over 40% Efficient Concentrator Submodules, M.A. Green, M.J. Keevers, B. Concha-Ramon & J. Jiang, UNSW, Sydney, Australia, P.J. Verlinden, Y. Yang & X. Zhang, Trina Solar, Changzhou, China

European Commission

4CO.6.4 Luminescent Solar Noise Barrier – Large Scale Testing and Modeling L.H. Slooff ECN, Petten, Netherlands S. Verkuilen Heijmans Wegen, Rosmalen, Netherlands M.M. de Jong & M.N. van den Donker SEAC, Eindhoven, Netherlands M. Kanellis & M.G. Debije Eindhoven University of Technology, Netherlands

European Commission

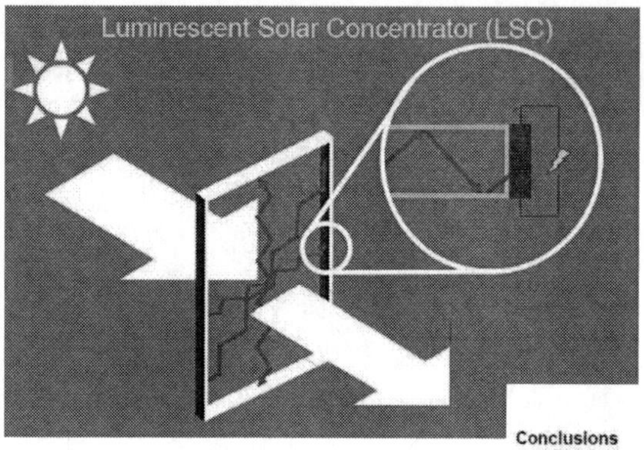

Conclusions

- Tm^{2+} doped halides discovered
  - Colorless
  - No self-absorption
  - >60% sunlight absorption
  - Expected ~15% power efficiency
- Made as both powders and thin-films
- Scattering still a challenge for BIPV

Outlook

- Not a lot known/understood (only ~4 major studies)*
  - Fundamental luminescence properties
  - Optimizing thin film growth
- Realize a lab-scale Tm^{2+} LSC prototype

1AO.2.4 Efficient Luminescent Solar Concentrators Based on Self-Absorption Free, Tm2+ Doped Halides

O.M. ten Kate, M. ten Kate, M. De Jong, O.M. ten Kate & E. van der Kolk
Delft University of Technology, Netherlands
K.W. Krämer
University of Berne, Switzerland

European Commission

Operation

European Commission

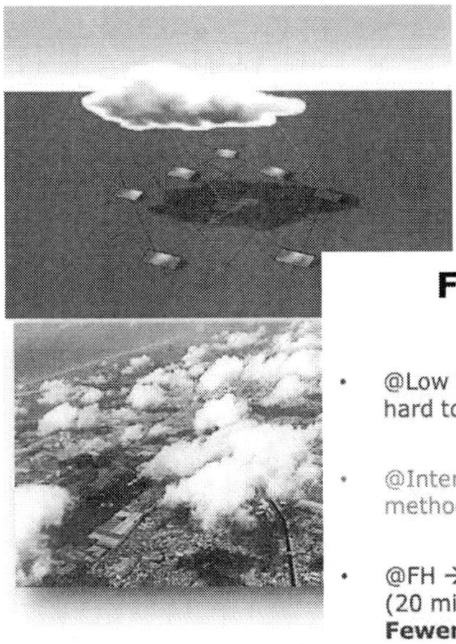

Peer-to-Peer (P2P) Forecasting:

Forecast Skill vs. Forecast Horizon

- @Low FH: **persistence** hard to beat
- @Intermediate FH: P2P method shows its **value**
- @FH → T_{Obs} (20 min. in this example.): **Fewer** to **no** available periph. $k_j^*(t-\tau_{ij})$ timeseries for P2P FC

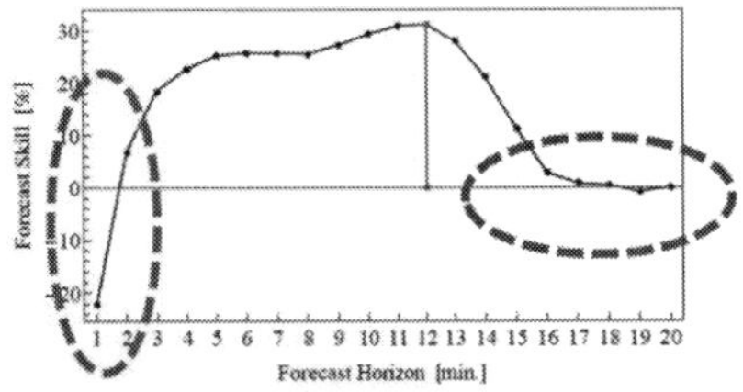

Dependence of Peer-to-Peer Solar Forecast Skill on Irradiance Variability

B. Elsinga & W.G.J.H.M. van Sark
Utrecht University, Netherlands

European Commission

Analysis of Failure Occurrence

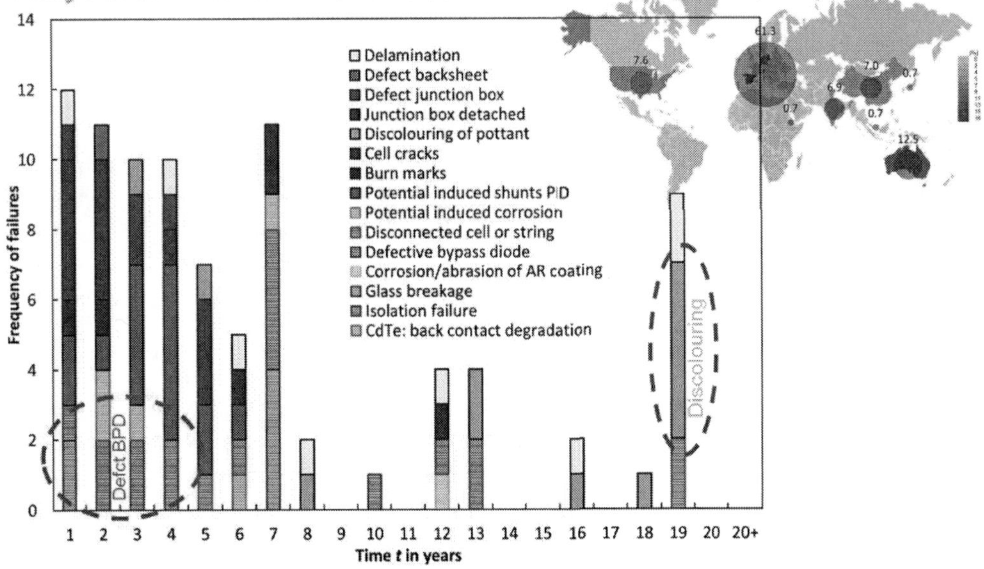

5DP.1.2 Mean Degradation Rates in PV Systems for Various Kinds of PV Module Failures

M. Köntges & S. Altmann
ISFH, Emmerthal, Germany
U. Jahn
TÜV Rheinland, Cologne, Germany

European Commission

Implementation and Monitoring

2015/16

Detailed measurement

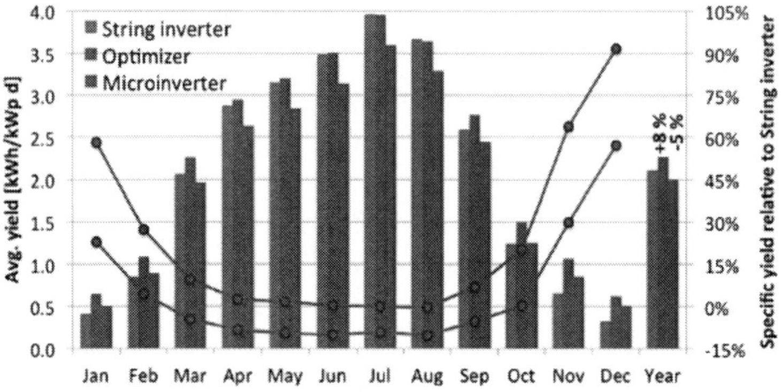

TNC

(4.6.2015 – 3.6.2016)

12

5AO.9.4 Performance of Recent Inverter Systems under Partial Shading Conditions

R. Lingel, T. Nordmann & T. Vontobel
TNC Consulting, Feldmeilen, Switzerland

European Commission

Energy Solutions

European Commission

IS "BIPV" ALREADY "VINTAGE"...?

SO... BIPV "MAINSTREAM POTENTIAL"

.TECHNOLOGICAL INTEGRATION | PV AS A PART OF A WHOLE ENERGY SOLUTION SET
.POLICY DRIVERS ARE CRUCIAL

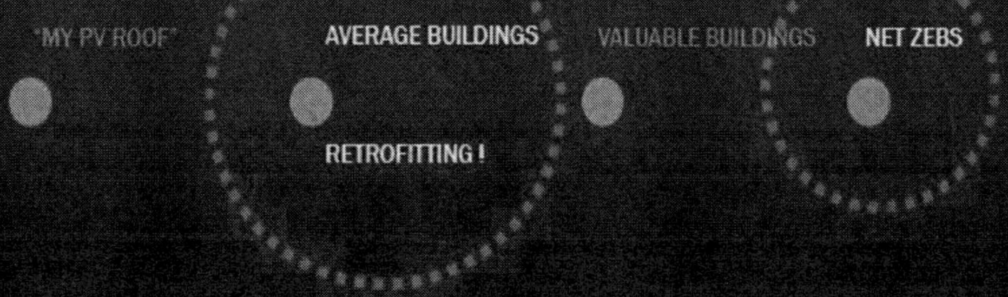

Kinetic envelope for heating cooling and light comfort + PV generation

Case Study

- Single Zone Office
- Zurich Weather Data
- 50 Panels in a Single Cluster
- Heating COP: 4
- Cooling COP: 3
- Lighting Load: 11.74 W/m^2

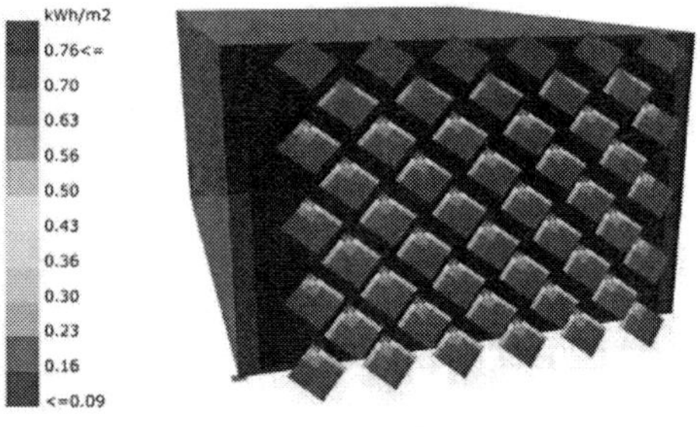

6DO.8.5 Energy Performance of PV Modules as Adaptive Building Shading Systems
J. Jayathissa, J. Schmidli, J. Hofer & A. Schlueter
ETH Zurich, Switzerland

European Commission

New metrics for PV + storage performance!

Definition of the Storage Performance Index (SPI)

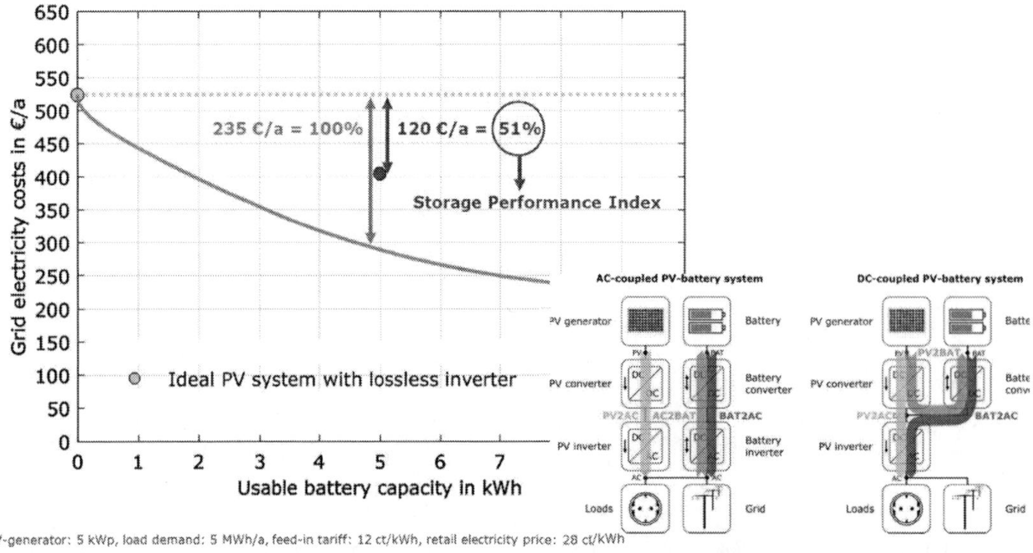

PV-generator: 5 kWp, load demand: 5 MWh/a, feed-in tariff: 12 ct/kWh, retail electricity price: 28 ct/kWh

Emerging Performance Issues of Photovoltaic Battery Systems, Johannes Weniger, Tjarko Tjaden, Joseph Bergner, Volker Quaschning, HTW Berlin - University of Applied Sciences, Germany

European Commission

Thanks!

European Commission

Start working on excellent abstracts for EU PVSEC 2017!

European Commission

We thank the EU PVSEC 2016 Sponsors

BERGER Lichttechnik

RENA

SINGULUS

VITRONIC
the machine vision people

BIFACIAL SOLAR CELLS FABRICATED BY PERC PROCESS FOR MASS PRODUCTION

Sung-Yu Chen[1], Han-Chang Liu[1], Hung-Yi Chang[2], Ying-Yen Chiu[2], Yu-Hui Wang[2], Yu-Hsuan Lin[1], Sin-Han Yu[1], Wen-Jong Lih[1], Chen-Hsun Du[1]
[1] Green Energy & Environment Research Lab., Industrial Technology Research Institute,
No.195, Sec. 4, Chung Hsing Rd., Chutung, Hsinchu, Taiwan 31040, R.O.C.
[2] Big Sun Energy Technology Inc.,
No.458-9, Sinsing Rd., Hukou, Hsinchu, Taiwan 303, R.O.C.

ABSTRACT: Passivated emitter and rear cells (PERC), which have already been proved to be a technology of high-efficiency solar cell theoretically and experimentally, become the one of main structures for silicon crystalline solar cells market in recently years. In this paper, p-type four-busbars bifacial solar cells (BSC) were fabricated base on industrial PERC process. The rear metallization is locally screen printing so that the cost was decreased by using fewer aluminum pastes but the front cell efficiency was kept as high as passivated emitter and rear cells (PERC). It was also investigated that the influence of rear passivated stack layers for bifacial solar cells (BSC). There is a standing-wave phenomenon on rear external quantum efficiency according to rear thickness of SiNx layers; therefore, the rear performance of bifacial solar cells can be improved by optimizing the SiNx layers. The best cell efficiency of 20.71% and 13.92% for the front side and rear side of bifacial solar cells (BSC), respectively, were demonstrated in this study with an area of 156×156 mm^2. It was concluded that p-type bifacial solar cell (BSC) with passivated emitter and rear cells (PERC) approach is useful and feasible for mass production.
Keywords: Silicon solar cell, Bifacial, High-Efficiency

1 INTRODUCTION

It is well known that PERC technology is useful for improving the cell performance and p-type PERC has already been mass-production for silicon crystalline solar cells market in recently years [1-2].

Because of the similar procedure between PERC and BSC (as shown in Figure 1), p-type BSC could be fabricated according to industrial PREC process by local printing instead of full printing on rear surface [3]. On the other hand, there are some advantages for this structure so that it is unavoidable tendency to replace PERC:

(1) Cost was down by using just local but not full aluminum paste on rear surface. The weight of aluminum decreases from 1g to 0.13g for six-inch wafers roughly.

(2) The power loss of module is lower due to the additional contribution from back electrical power.

(3) The local back metallization allows the infrared radiation to penetrate so that the cells can be work under a lower temperature due to higher infrared transparency.

In spite of the fact that more and more institutes focus on the n-type BSC because there is no light induced degradation compare to p-type [4-7]. The formation of emitter and back-surface field for n-type is more difficult and complicated compare to p-type and the cost is more expensive due to silver paste on both sides so that p-type BSC has superiority and the most possibility to become the one of main structures in future years [8].

In this paper, we present an approach of p-type four-busbars BSC for mass production industrially based on PERC fabricated methods without any additional process and equipment. In addition, the lifetime variation and external quantum efficiency for different rear passivated stack layers were also investigated how to influence the photovoltaic characteristics.

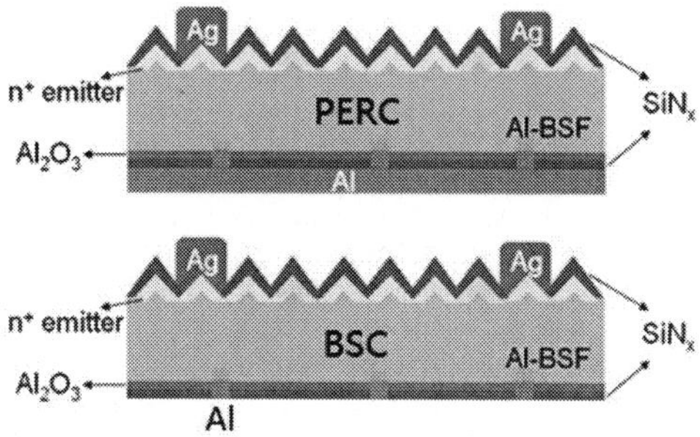

Figure 1: the comparison of schematic cross section for PERC and BSC

2 EXPERIMENTAL

Figure 2 shows the process flow of BSC which is the same with PERC procedure but adjusting conditions. In this study, the area of 156 mm × 156 mm p-type single-crystalline silicon CZ wafer with 200µm thickness was used. The size of wafer is M1, which means that the diameter is 205 mm.

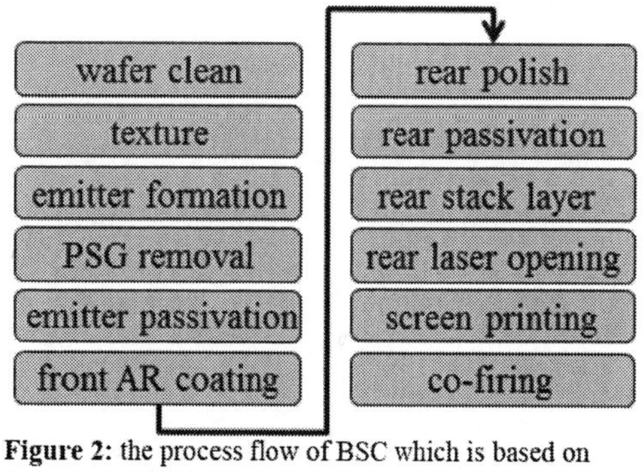

Figure 2: the process flow of BSC which is based on PERC procedure

In order to clean contamination and oxidation layer

on a silicon surface, a wafer clean procedure is necessary before texturization by using KOH solution. After a emitter with sheet resistance of 95 ohm/sq. was formed on the surface by $POCl_3$ diffusion, phosphor-silicate glass (PSG) was removed by HF dipping. The immediately following step is emitter passivation by thermal oxidation process at 800℃ for 20 min and then the SiNx anti-reflective layer was coated on the front surface by plasma-enhanced chemical vapor deposition (PECVD). The rear oxidation layer, n-layer removal and surface polish were complete at the same time by single-step etching in NaOH solution. After that, an ultra-thin oxidation layer was formed at 550℃ and then the aluminum oxide (Al_2O_3) layer was deposited on the rear flat surface by atomic layer deposition (ALD) for rear passivation. After annealing, a SiNx layer was deposited onto Al_2O_3 for forming an Al_2O_3 /SiNx stack layers. In the following, the Al_2O_3/ SiNx passivated stack layers was locally opened by a laser ablating. The pitch of line opening is half of front pattern. Finally, the silver paste and aluminum paste were printed on the front and back side with four-busbars patterns respectively and then the metallization was complete after co-firing process.

3 DISSCUSSIONS

Compare to PERC, the only one difference of BSC is rear structure so that we focus on the relation between rear passivated stack layers and photovoltaic characteristics on BSC. Figure 3(a) shows a semi-finished BSC (before laser opening and metallization processes) for lifetime measurement. The passivation effect is not sensitive as changing the thickness of rear stack layers, including Al_2O_3 and SiNx (shown as Figure 3(b)).

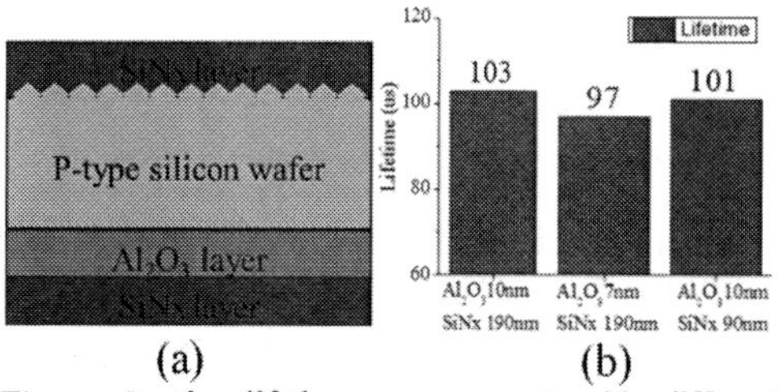

Figure 3: the lifetime measurement with different passivated stack layers: (a) cell structure for lifetime measurement (b) lifetime results

In addition, the lifetime variations in thickness and annealing temperature of Al_2O_3 were also investigated. The Al_2O_3 /SiNx stack layers were formed on both side of the wafer for lifetime measuring (shown as Figure 4(a)). Figure 4(b) shows lifetime results and it is indicated that the influence of annealing temperature on passivation effect is critical. Thicker Al_2O_3 layer anneal at 600℃ shows higher lifetime..

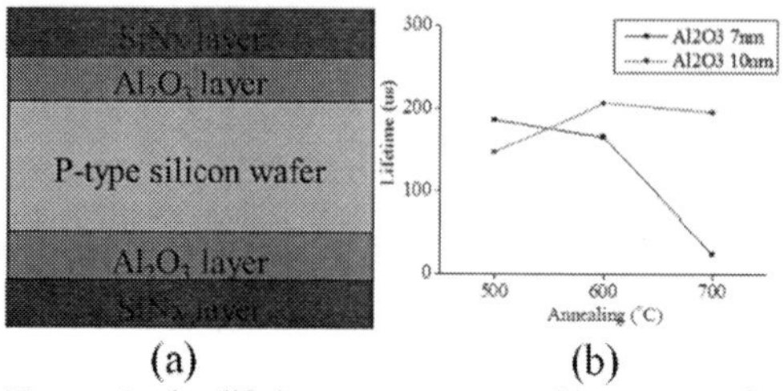

Figure 4: the lifetime measurement: (a) structure for measuring (b) lifetime results

Figure 5 shows results of front external quantum efficiency (EQE) of BSC with different passivated stack layers. As the thickness of SiNx is 190 nm, thicker Al_2O_3 shows higher EQE in wavelengths from 300 nm to 1200 nm. Isc and Voc increase due to reduce surface recombination. As a result, BSC with thickness of 10 nm Al_2O_3 and 190 nm SiNx shows the best cell efficiency on 20.71 %.

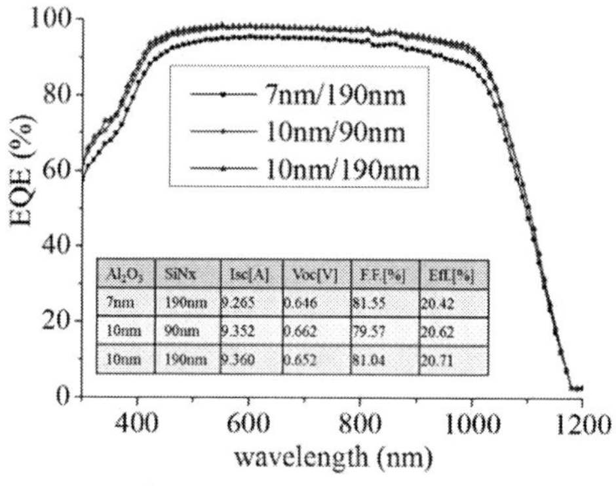

Figure 5: the front external quantum efficiency of BSC

Figure 6 shows the rear quantum efficiency for BSC with different passivated stack layers. It obviously shows a standing wave phenomenon and the period is dependent on the thickness of SiNx. According to the comparison of Isc, we found that the thinner SiNx layer the higher cell efficiency. As a result, the best cell efficiency of 13.92% on rear surface of BSC with 10 nm Al_2O_3 and 90 nm SiNx was achieved. The rear performance of BSC still can be improved by optimizing the SiNx layer.

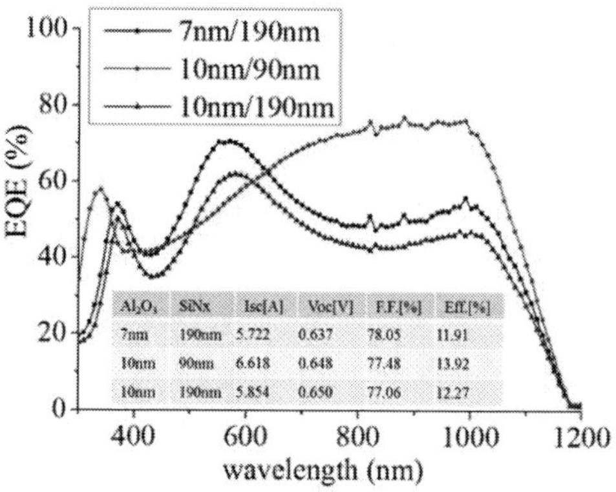

Figure 6: the external quantum efficiency on rear surface of BSC

Table I shows the average photovoltaic characteristics for PERC and both sides of BSC. The thickness of rear Al_2O_3 and SiNx are 10 nm and 190 nm. Compare to PERC, the front photovoltaic properties of BSC was kept as high as PERC and even the cost was reduced by using fewer Al paste. It shows that BSC is useful and feasible for mass production.

Table I: the IV characteristics for PERC and both sides of BSC

Cell	Voc[V]	Jsc[A]	F.F.[%]	Eff.[%]
PERC	0.653	9.394	80.89	20.67
Bif-front	0.652	9.346	81.06	20.66
Bif-rear	0.650	5.867	76.86	12.25

4 CONCLUSIONS

The thickness of SiNx on rear surface of BSC is sensitive to the cell performance of the rear side but not front side. Therefore, the rear efficiency still has a potential to be improved by optimizing the rear SiNx layer. The best cell efficiency of 20.71% and 13.92% for the front side and rear side respectively of p-type four-busbars BSC were demonstrated in this study. Compare to PERC, the front efficiency of BSC was kept almost 20.7% as high as PERC. It was concluded that p-type BSC is the most possible for mass production.

ACKNOWLEDGMENT

The financial support provided by Bureau of Energy, Ministry of Economic Affairs in Taiwan is gratefully acknowledged.

REFERENCES

[1] Martin A.Green, Solar Energy Materials & Solar Cells 143 (2015) 190.
[2] Gerd Fischer, Energy Procedia 77 (2015) 515.
[3] T. Dullweber et al., 31st European Photovoltaic Solar Energy Conference, 2015, 341.
[4] Andreas Hu¨bner, Armin G. Aberle, and Rudolf Hezel, Appl. Phys. Lett. 70, 1008 (1997)
[5] H. Ohtsuka et al., Progress in Photovoltaics Research and Applications 8 (2000) 385.
[6] Hung-Chih Chang et al., Energy Procedia 55 (2014) 643.
[7] Chuan-Chi Chen et al., Energy Procedia 38 (2013) 416.
[8] P. Rothhardt et al., 29th European Photovoltaic Solar Energy Conference, 2014, 653.

INFLUENCE OF THE BOTTOM WO_3 LAYER ON THE SERIES RESISTANCE IN SILICON BASED SOLAR CELLS WITH WO_3/AG/WO_3 EMITTER

Jie Bao,[1,2] Weiliang Wu,[1] Zongtao Liu,[2] Lun Cai[1] & Hui Shen [1,2,3*]

1. Institute for Solar Energy Systems, School of Physics, Sun Yat-sen University (SYSU), Guangzhou University Town, Guangzhou, Guangdong Province, 510006, China
2. Shunde-SYSU Institute for Solar Energy, No 1, Desheng Dong Road, Daliang, Shunde, Guangdong Province, 528300, China
3. Jiangsu Collaborative Innovation Center of Photovoltaic Science and Engineering, Changzhou University, No.1 Gehu Road, Changzhou, Jiangsu Province, 213164, China

* Corresponding author: Phone: +86 15521108087; Email: shenhui1956@163.com

ABSTRACT: The silicon based solar cells was fabricated on n-type CZ silicon wafer, employing triple-layer WO_3/Ag/WO_3 films as emitter, cesium carbonate (Cs_2CO_3) as back surface field (BSF), referred to as WAW/n-Si solar cells. All of those films were deposited by physical vapor deposition (PVD) at room temperature. In this work, series resistance (R_s) in WAW/n-Si solar cells, mainly comprising all the film's series and contact resistances, were quantitatively analyzed and discussed. The results indicated that contact resistivity (ρ_c) for n-Si/WO_3 contact is strongly dependent on bottom WO_3 layer thickness. 10~20 nm-thick bottom WO_3 layer is needed for WAW/n-Si solar cells with favorable R_s. When optical performance was taken into consideration, WAW/n-Si solar cells of 9.47% best efficiency was achieved on polished silicon substrate under STC (AM 1.5 global spectrum, 25 °C and 1000 W/cm^2). In absence of the effect of R_s, efficiency of 11.52% can be reached.

Keywords: Heterojunction, Silicon Solar Cell, Contact resistivity, WO_3/Ag/WO_3 multilayer, Cs_2CO_3

1 INTRODUCTION

In the last few years, transition metal oxides (TMOs), compromising MoO_3, WO_3 and V_2O_5, have been extensively explored as emitter in n-type silicon heterojunction (SHJ) solar cells due to wide band-gap, high-work-function and facile deposition. [1-4] Power conversion efficiency (PCE) of 14.3% have been reached for MoO_x/n-Si heterojunction solar cells. [5] Moreover, when inserting an intrinsic a-Si:H layer into MoO_x/n-Si heterojunctions for passivation, PCE as high as 22.5% has been recorded for this novel structure (MoO_x/a-Si:H/n-Si) solar cells, [3] indicating the extraordinary hole-selective property of TMOs.

Since the nm-thick TMOs exhibit ultra-high sheet resistance on the order of 10^9 Ω/sq, [4] a TCO (e.g. ITO, IO:H) layer covered on TMOs is essential for charge-carriers lateral transportation and collection. However, the sputter deposition of ITO would severely degrade surface passivation as well as PCE. [3, 6, 7] Here, we solve this problem by inserting one thin metal film into TMOs to achieve charge-carriers lateral transportation and collection, that is, forming oxide/metal/oxide (OMO) multilayer structure instead of utilizing TCO films. WO_3/Ag/WO_3 (WAW) multilayer [8, 9], possessing the merits of high-work-function and low-resistivity, is therefore incorporated into n-type SHJ solar cells as emitter in this work.

The Cs_2CO_3 and LiF have been shown to be very efficient electron-injection (electron-selective) materials in organic electronics. [10, 11] Since Cs_2CO_3 exhibits better electrical properties with Al, Ag or Au electrode than the devices using LiF with Al as electrode. [12] What's more, thermally evaporated Cs_2CO_3 would decompose into CsO_2 and cesium suboxides, yielding a heavily doped n-type semiconductor with an intrinsically low work function, [11, 13] which is critical for electron-selectivity. Additionally, the Cs_2CO_3 interfacial layer can decrease the contact resistance between silicon and electrode. [14] In this work, we develop a novel dopant-free SHJ solar cells employing triple-layer WO_3/Ag/WO_3 as emitter, Cs_2CO_3 as back surface field (BSF) film via a simple PVD method, referred to as WAW/n-Si solar cells. The series resistance (R_s) in WAW/n-Si solar cells, comprising all the film's bulk and contact resistances, were quantitatively analyzed and discussed, the results may pave a new way for analyzing the resistance component in dopant-free SHJ solar cells.

2 EXPERIMENTAL DETAILS AND METHOD

2.1 WAW/n-Si solar cells fabrication

2×2 cm^2 in area, alkali polished n-type (100) oriented CZ silicon wafers (0.3~1 Ω•cm^2, 180 μm) were used for all the device fabrications in this study. Silicon wafers were cleaned prior to the WO_3 deposition, which consisted of standard RCA processing followed by (10 %) hydrofluoric acid treatment to remove native oxide. After cleaning and dewatered, the substrates were immediately loaded into the evaporating chamber. 99.99% pure WO_3 powder, 99.999% silver wire and 99.99% Cs_2CO_3 powder were placed into alumina-coated tungsten boat separately as source materials. WO_3 films, Ag and WO_3 were thermally evaporated sequentially onto the front side of wafer, Cs_2CO_3 layer was evaporated onto the rear side subsequently. For intermediate Ag layer and Cs_2CO_3 layer, the deposition pressure was kept at 8.0×10^{-4} Pa. For both WO_3 layers, O_2 gas was introduced to ensure that the vacuum pressure was 1.5~2.5×10^{-2} Pa, detailed procedure could be found in reference [15]. The thicknesses of WO_3 and silver films were monitored using a quartz crystal monitor (Model SQM 200, Filtech) positioned a short distance from the source. The solar cells were metalized via depositing 500-nm-thick Ag at the front and rear as electrode.

2.2 Contact resistivity of n-Si/WO_3 contact

Contact resistivity (ρ_c) of n-Si/WO_3 contact was investigated using transfer-length-method (TLM). [16-18] A typical arrangement for a TLM test pattern involves a number of identical rectangular ohmic contacts of length L and width W placed at variable intervals, as shown in Fig.1a. If a current flows through two such contacts spaced by L, then the total resistance, R_T, is a sum of two components:

$$R_T = \frac{R_S}{W}L + 2R_C \quad (1)$$

$$L_T = \sqrt{\frac{\rho_C}{R_S}} \quad (2)$$

$$R_C = \frac{\rho_C}{L_T W} \quad (3)$$

$$R_T = \frac{R_S}{W}(L + 2L_T) \quad (4)$$

Where R_C is the contact resistance comprises the resistance of the WO_3/c-Si and WO_3/metal interfaces as well as the WO_3 bulk resistivity; L_T is transfer length; R_S is the surface resistivity of the inversion layer (p^+-Si) under the WO_3 layer. Equations 2, 3 are put into equation 1, we can get equation 4. Plotting the R_T vs spacing length L and linear fitting, the R_S and L_T can be obtained, then we can get ρ_c from equation 2. It should be noted that the extracted ρ_c comprises the interfacial resistivities and bulk resistivities of all layers in between the c-Si and the outer metal electrode.

2.3 Contact resistivity of WO_3/Ag contact

ρ_c of WO_3/Ag contact was investigated using transfer-length-method (TLM). Silver pad of length L and width W are placed at variable intervals on WAW multilayer, as shown in Fig.2a. Supposing the current was confined to the intermediate Ag layer, the total resistance, R_T, between two silver pads with L spacing is:

$$R_T = \frac{R_S}{W}L + 4R_C \quad (5)$$

$$R_T = \frac{R_S}{W}(L + 4L_T) \quad (6)$$

Where R_C is the contact resistance comprises the resistance of the WO_3/Ag contact and WO_3 bulk resistivity; L_T is transfer length; R_S is the surface resistivity of the intermediate silver layer. Equations 2, 3 are put into equation 5, we can get equation 6. In the same way, ρ_c of WO_3/Ag contact can be extracted.

2.4 Contact resistivity of n-Si/Cs_2CO_3 contact

Since Cs_2CO_3 layer was sensitive to humility and couldn't be exposed to the air directly, ρ_c of n-Si/Cs_2CO_3 contact was measured using Cox and Strack method. In the Cox-Strack method, [19] a structure comprises an array of different diameter front contacts and a full-area rear contact, the rear side of n-Si was heavily n^+ doped with phosphorus (> 1e19 cm^{-3}). The total resistance can now be presented as:

$$R_T = \frac{\rho}{\pi d}\arctan\frac{4t}{d} + \frac{4\rho_C}{\pi d^2} + R_0 \quad (7)$$

Where ρ and t are the resistivity and thickness of the front metal electrode separately; d is the contact diameter; ρ_c is the contact resistivity comprises the resistance of the n-Si/Cs_2CO_3 and Cs_2CO_3/metal interfaces as well as the Cs_2CO_3 bulk resistivity. Plotting the R_T vs contact diameter d and fitting with the equation 7, the ρ_c can be obtained.

2.5 Simulation of the photo-generated current

The photo-generated current of WAW/n-Si solar cells were simulated by *Wafer Ray Tracer* [20] developed by *PV Lighthouse*, refractive index (RI) data of amorphous WO_3 layer [21] and other materials required for simulation were sourced from the RI Library.

3 RESULT AND DISCUSSION

Fig. 1a shows the typical I-V measurements and ρ_c extractions for n-Si/WO_3 contact. The test device for TLM measurement is schematic presented in inset graph, silver layer with 1 μm-thick is utilized as electrode. Obviously, the current-voltage curves are linear, indicating the Ohmic characteristics of n-Si/WO_3 contact. After performing a linear fit (bottom inset graph), ρ_c of 0.314Ω•cm^2 is obtained for n-Si/WO_3 contact with 10 nm WO_3 layer.

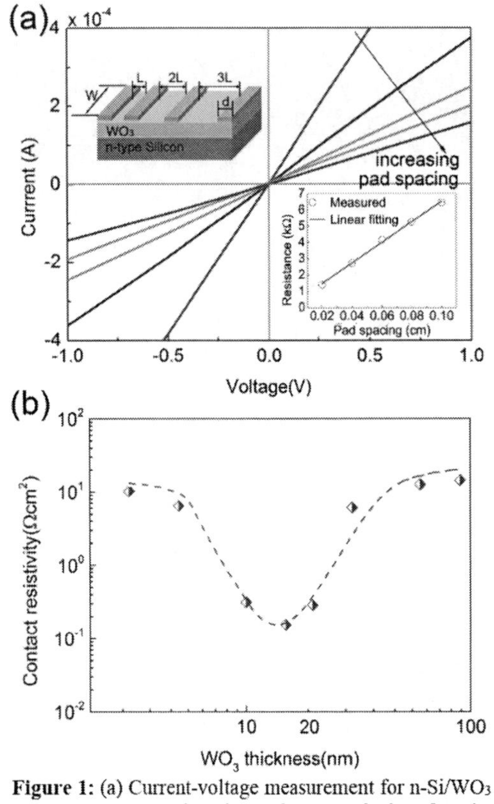

Figure 1: (a) Current-voltage measurement for n-Si/WO_3 contact. Inset graphs show the test device for the measurement of n-Si/WO_3 contact resistivity by TLM (top) and the fitting results (bottom). (b) Dependence of contact resistivity on WO_3 layer thickness for the n-Si/WO_3 contact.

In the same way, the dependence of ρ_c on WO_3 layer thickness for the n-Si/WO_3 contact are extracted and presented in Fig. 2a. It can be noted that the ρ_c for n-Si/WO_3 contact is strongly dependent on WO_3 layer thickness. Initially, the ρ_c drops rapidly from 10.25 Ω•cm^2 to 0.31 Ω•cm^2 while the thickness of WO_3 film is increased from 3 nm to 10 nm, and demonstrates the lowest value in the 10~20 nm range. After that, ρ_c climbs to 6.09 Ω•cm^2 for 30 nm-thickness WO_3 layer and keeps on increasing for thicker WO_3 layers. Therefore, thickness in the range of 10~20 nm for bottom WO_3 layer is needed for WAW/n-Si solar cells with favorable series resistance (R_s).

The WAW/n-Si contact was characterized by cross sectional TEM. The analyzed structure consists of two WO_3 layers (10 nm bottom layer, 55 nm top layer) and an intermediate Ag layer (12 nm) deposited on silicon wafer (covered with ~1 nm of native SiO_2), as shown in Fig. 2a. The surface oxidation of n-Si is invoked by the reaction between WO_3 and silicon substrate. [22] It can be seen that 10 nm-thickness WO_3 layer is continuous and the n-Si/WO_3 interface is clearly resolved. As films deposited via thermal evaporation always display island growth (Volmer-Weber nucleation) [23], WO_3 layer within 10 nm is potentially uncontinuous and with rough surface, which may hinder charge carriers transportation and contribute to enhanced resistive loss. While WO_3 layer exhibits good insulating properties, the followed rising of ρ_c is probably a result of the bulk resistance of WO_3 film.

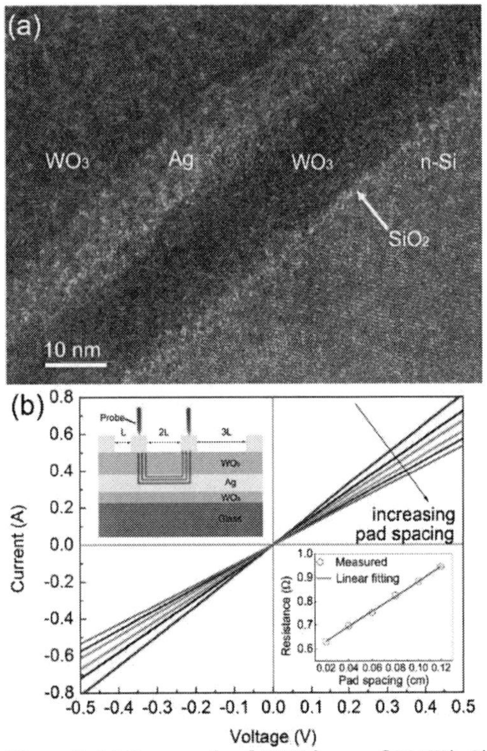

Figure 2: (a) Cross sectional TEM image of WAW/n-Si contact, (b) Current-voltage measurement for WO_3/Ag contact. Inset graphs show the test device for the measurement of WO_3/Ag contact resistivity by TLM (top) and the fitting results (bottom).

Since the ρ_c of WO_3/metal contact can be overestimated utilizing TLM with metal electrode/WO_3/glass substrate structure, [4] we develop metal electrode/WO_3/ metal/WO_3/glass substrate structure for the ρ_c measurement. In this work, the ρ_c of WO_3/Ag contact was measured using TLM with Ag/WO_3/Ag/WO_3/glass (1 μm/55 nm/ 12 nm/10 nm/1 mm) structure, and assuming that the current flow is confined to the intermediate Ag layer, as depicted in inset graphs of Fig. 2b. We find the ρ_c of WO_3/Ag contact shows minor differences with the WO_3 layer thickness. Representative I-V measurements are shown in Fig. 2b. The ρ_c value of 11 mΩ•cm² was obtained for WO_3/metal contact after a linear fitting.

Even though Cs_2CO_3 could be vacuum evaporated without difficulty, the film was extremely sensitive to humility and can't be exposed to ambient air directly. Therefore, the ρ_c of n-Si/Cs_2CO_3 contact was measured using Cox and Strack method. After fitting the total resistance RT with contact diameter d, the ρ_c value of 0.47 Ω•cm² was obtained for of n-Si/Cs_2CO_3 contact (detailed results were not shown here).

The dependence of photo-generation current density (J_g) for WAW/n-Si (polished) solar cells on bottom and top WO_3 layer with various thicknesses is simulated by *Wafer Ray Tracer* hosted by *PV Lighthouse* [20], the results are shown in Fig. 3.

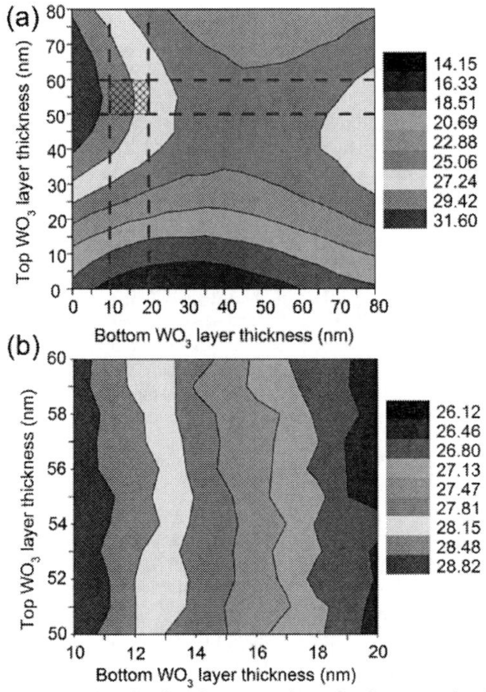

Figure 3: (a) Simulated contour plots of J_g for WAW/n-Si (polished) solar cells upon both WO_3 layers with variation thicknesses, the intermediate Ag layer with 12 nm thickness. (b) Magnified view of the corresponding highlighted regions in (a).

The dependence of photo-generation current density (J_g) for WAW/n-Si (polished) solar cells on bottom and top WO_3 layer with various thicknesses is simulated by *Wafer Ray Tracer* hosted by *PV Lighthouse* [20], the results are shown in Fig. 3. Since the bottom WO_3 layer in 10~20 nm thickness range is critical for WO_3/n-Si contact, it can be noted that the top WO_3 layer in 50~60 nm thickness range is favorable for higher J_g (Fig. 3a). The magnified view of the corresponding highlighted regions is shown in Fig. 3b. Obviously, the J_g decreases gradually with increasing the thickness of bottom WO_3 layer, whereas it doesn't show any differences with top WO_3 layer thickness (in 50~60 nm range). Therefore, the top WO_3 layer thickness is set as 55 nm and will be used in the subsequent investigation.

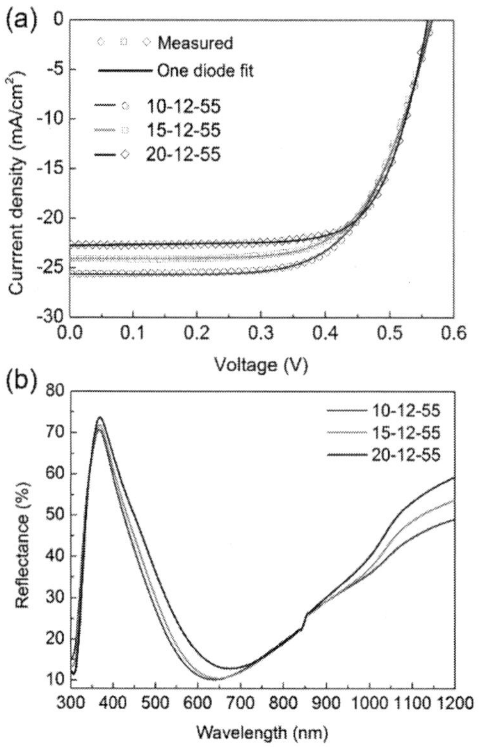

Figure 4: (a) Measured (data points) and one diode fit (lines) J–V curves for WO_3/n-Si solar cells with various bottom WO_3 thicknesses. (b) The corresponding reflectance of WO_3/n-Si solar cells

Fig. 4a shows the J-V curves of WAW/n-Si solar cells with different thicknesses of bottom WO_3 layer, while the thickness of intermediate Ag layer and top WO_3 are fixed at 12 and 55 nm, respectively. The detailed parameters of the devices are fitted by one-diode model and presented in Table I. it can be noted that the short-circuit current density (J_{sc}) values decrease gradually with increasing the thickness of bottom WO_3 layer (10~20 nm), whereas the differences in V_{oc} values are slightly. The decreased J_{sc} values can be attributed to the fact that a thicker bottom WO_3 layer results in higher reflection loss, as shown in Fig. 4b, which is in accordance with the optical simulation in Fig. 3. WAW/n-Si solar cell with 10 nm-thick bottom WO_3 layer possesses the highest PCE of 9.46%, featuring J_{sc} of 25.6 mA/cm^2, open-circuit voltage (V_{oc}) of 567 mV and 65% fill factor (FF).

It has been proven that the electrode resistance, bulk resistance of active materials and contact resistance contribute to the R_s of one solar cell. In WAW/n-Si solar cells, the normalized R_s (1 cm^2 unit area) comprises the silver electrode resistance ($R_{electrode}$), the ρ_c of electrode/WO_3 contact ($\rho_{c,\ electrode/WO3}$), the ρ_c of WO_3/Ag contact ($\rho_{c,\ WO3/Ag}$), the ρ_c of n-Si/WO_3 contact ($\rho_{c,\ Si/WO3}$), the ρ_c of n-Si/Cs_2CO_3 contact ($\rho_{c,\ Si/Cs2CO3}$), the bulk resistivity of WO_3 ($R_{bulk,\ WO3}$)and the bulk resistivity of n-Si ($R_{bulk,\ Si}$). Due to the low resistivity of n-Si (0.3~1 $\Omega\cdot$cm^2), the $R_{bulk,\ Si}$ was omitted. The $R_{electrode}$ and $\rho_{c,\ electrode/WO3}$ were calculated by *Grid* developed by *PV Lighthouse*. The resistivity of as-deposited WO_3 layer we measured was $\sigma\sim1.85\times10^{-5}$ S/cm, which is close to the reported value of 1.4×10^{-5} S/cm. [7] The $R_{b,\ w}$ could be measured by $\sigma\cdot d$ (d-the thickness of the film). Therefore, the series resistive components in WAW/n-Si solar cells are obtained and are summarized in Table II. For WAW/n-Si solar cells with 10 nm and 15 nm bottom WO_3 layer, the calculated R_s (sum of each component) matches well with the R_s fitted by one diode model. However, for 20 nm bottom WO_3 layer, the calculated R_s is inconsistent with the one-diode fitting R_s, which is probably the errors inherent to the TLM itself.

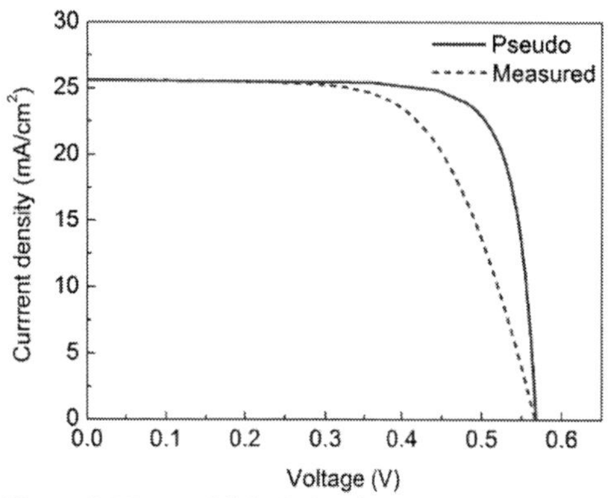

Figure 5: Measured light J-V and Suns-V_{oc} curves of the WO_3/n-Si solar cells with 10 nm-thick bottom WO_3 layer.

Fig. 5 shows the measured light J-V and Suns-V_{oc} curves of the device with 10 nm-thick bottom WO_3 layer. The Suns-V_{oc} curves are measured by WCT-120 lifetime test instrument using Quasi-Steady-State-Photo-Conductance (QSSPC) method. A notable pseudo FF (pFF) of 79.0% is obtained for the WAW/n-Si solar cells, without considering the R_s effect, the PCE of 11.52% can be reached.

Table I: List of the WAW/n-Si solar cells parameters with various thicknesses of bottom WO_3 layer

NO:	V_{oc} (V)	J_{sc} (mA/cm^2)	FF (%)	Eff. (%)	R_s ($\Omega\cdot$cm^2)	R_{sh} ($\Omega\cdot$cm^2)
10-12-55	0.567	25.64	65	9.46	1.95	3334
15-12-55	0.563	24.07	67	9.08	1.59	8370
20-12-55	0.559	22.71	71	9.04	1.24	1142

Table II: List of the series resistive components in WAW/n-Si solar cells (unit: $\Omega\cdot$cm^2)

NO:	$R_{electrode}$	$\rho_{c,\ WO3/Ag}$	$\rho_{c,\ Si/Cs2CO3}$	$\rho_{c,\ electrode/WO3}$	$\rho_{c,\ Si/WO3}$	$R_{bulk,\ WO3}$	R_T
10-12-55					0.314	0.351	1.804
15-12-55	0.358	0.011 × 2	0.470	0.289	0.154	0.378	1.671
20-12-55					0.287	0.405	1.831

4 CONCLUSION

In summary, the silicon based solar cells using $WO_3/Ag/WO_3$ as emitter and Cs_2CO_3 as BSF (WAW/n-Si solar cells) were developed via simple PVD. The effect of the thicknesses of the bottom WO_3 layer on R_s was investigated, as well as other series resistive components. Results have shown that ρ_c for n-Si/WO_3 contact was strongly dependent on bottom WO_3 layer thickness. The thickness of bottom WO_3 layer in the region of 10~20 nm is favorable for the device performance. Moreover, considering the light absorption, the top WO_3 layer was set 55 nm. WAW/n-Si solar cell with 10 nm-thick bottom WO_3 layer possesses the highest PCE of 9.46%, featuring J_{sc} of 25.6 mA/cm^2, V_{oc} of 567 mV, FF of 65% and R_s of 1.95 Ω•cm^2. Additionally, the efficiency can increase from 9.46% to 11.52% when omit the effect of the R_s.

5 ACKNOWLEDGEMENTS

This work was supported by the Guangdong-Hong Kong Technology Cooperation Funding Scheme (2014B050505010), the National High-tech R&D Program of China (863 Program) (2015AA050303 and 2015AA050608), the Guangzhou collaborative innovation Major Project of producing, teaching and researching (No. 201508010011) and Jiangsu Collaborative Innovation Center of Photovoltaic Science and Engineering (Grant No. SCZ1405500002).

REFERENCE

[1] C. Battaglia *et al.*, Applied Physics Letters **104,** 113902 (2014).
[2] M. Bivour *et al.*, Solar Energy Materials and Solar Cells **142,** 34-41 (2015).
[3] J. Geissbühler *et al.*, Applied Physics Letters **107,** 081601 (2015).
[4] L. G. Gerling *et al.*, Applied Sciences **5,** 695-705 (2015).
[5] C. Battaglia *et al.*, Nano letters **14,** 967-971 (2014).
[6] J. Ziegler *et al.*, Applied Physics A **120,** 811-816 (2015).
[7] L. G. Gerling *et al.*, Solar Energy Materials and Solar Cells **145,** 109-115 (2016).
[8] S. Y. Ryu *et al.*, Applied Physics Letters **92,** 23306-23306 (2008).
[9] K. S. Yook *et al.*, Applied Physics Letters **93,** 3301 (2008).
[10] J. Huang *et al.*, Advanced Materials **18,** 114-117 (2006).
[11] J. Huang *et al.*, Advanced Functional Materials **17,** 1966-1973 (2007).
[12] T. Hasegawa *et al.* 11.3: Novel electron-injection layers for top-emission oleds. In: Proceedings of the SID Symposium Digest of Technical Papers, 2004. Wiley Online Library.
[13] H.-H. Liao *et al.*, Applied physics letters **92,** 173303-173303 (2008).
[14] Y. Zhang *et al.*, Energy & Environmental Science **8,** 297-302 (2015).
[15] C. Song *et al.*, Applied Physics Express **5,** 041102 (2012).
[16] S. S. Cohen, Thin Solid Films **104,** 361-379 (1983).
[17] D. L. Meier *et al.*, IEEE transactions on electron devices **31,** 647-653 (1984).
[18] J. Bullock *et al.*, Applied Physics Letters **105,** 232109 (2014).
[19] R. Cox *et al.*, Solid-State Electronics **10,** 1213IN71215-71214IN81218 (1967).
[20] *Wafer Ray Tracer* simulation tool, Version 1.6.4, PV Lighthouse Pty. Ltd., Australia. Available online: http://www.pvlighthouse.com.au (accessed on 1 May 2016).
[21] M. Hutchins *et al.*, Materials chemistry and physics **98,** 401-405 (2006).
[22] Z. Liang *et al.*, Physical Chemistry Chemical Physics **17,** 27409-27413 (2015).
[23] L. Ottaviano *et al.*, Thin solid films **490,** 59-67 (2005).

SIN_X/SIO_XN_Y STACK PASSIVATION FOR N-TYPE SI

Junjie Zhu[a], Chunlan Zhou[b], Rune Søndenå[a], Erik Stensrud Marstein[a], Sean Erik Foss[a]
[a]Institute for Energy Technology, Instituttveien 18, 2007 Kjeller, Norway
[b]Institute of Electrical Engineering, Chinese Academy of Sciences, No.6 Beiertiao, Zhongguancun, Beijing, China
Email: Junjie.Zhu@ife.no

ABSTRACT: The passivation properties of a SiNx/SiOxNy stack on n-type Czochralski (Cz) Si wafer and n-type emitter with phosphorus diffusion are determined. The SiNx/SiOxNy stack shows considerably better passivation properties on both n-type Si wafer and n-type phosphorus diffused emitters relative to the single SiNx layer after a firing process. The improvement is more distinct for emitters with higher sheet resistance. In addition, the stability of the SiNx/SiOxNy stack passivation is also investiga-ted. The passivation stability during extended illumination reveals similar behavior as that of a single SiNx layer. Illumination at one-sun starts to degrade the passivation quality after an initial increase. Heating at 250°C reduces the lifetime as well, but only on fired samples. The passivation quality of the as-deposited stack, however, can be improved during the heating process, most likely due to hydrogen in-diffusion. Moreover, the degradation can be largely recovered by further thermal activation. Minimal degradation was observed for samples stored in the dark. largely be recovered by further thermal activation. Minimal degradation was observed for samples stored in the dark.
Keywords: Passivation, SiN_x/SiO_xN_y stack, Silicon, Stability

1 INTRODUCTION

Thin amorphous layers of SiN_x are widely used as antireflection coatings (ARC) on crystalline silicon (Si) solar cells due to both their optical and electrical properties. In addition, the ARC electrically passivates defect sites on the wafer surface reducing the electron-hole recombination, thus increasing the performance of the cell. More advanced layered structures of thin films, such as SiN_x/a-Si, SiN_x/SiO_x, SiN_x/AlO_x, are interesting for use in devices due to their improved passivating properties as well as the optical properties. However, there are still challenges in using the stack structure: passivation quality, stability as well as fabrication cost. The SiN_x/SiO_xN_y stack has been reported as a promising passivation material for both p- and n-type Si with good thermal stability [1-3]. Such stacks have been applied in p-type multi-crystalline Si solar cells as an ARC resulting in improved conversion efficiency and resistance to potential induced degradation (PID) [4]. However, in order to apply this in mass production, more knowledge is required. In the present work, the passivation properties of a SiN_x/SiO_xN_y stack on n- and p-type Czochralski (Cz) Si wafers with phosphorus diffused emitters are determined. In addition, the stability of the passivation is investigated.

2 EXPERIMENTAL DETAILS

Both n- and p-type Cz Si wafers with resistivity of 1~3 Ω-cm were used in this work. The wafers were chemically polished in a mixture of HF, HNO_3 and CH_3COOH solution followed by piranha (a mixture of sulfuric acid and hydrogen peroxide) and HF cleaning. Phosphorus diffusion was carried out on p-type wafers using a Tempress tube furnace. Emitters with different sheet resistances were prepared by the etch-back approach. A Roth & Rau SiNA XS PECVD reactor was employed for both SiO_xN_y and SiN_x deposition. A SiN_x/SiO_xN_y stack structure was symmetrically deposited on both sides of the Si wafers with an optimized process. Here, the thickness of the SiO_xN_y and SiN_x layers were kept at 10 nm and 70 nm, respectively. For comparison, Si wafers coated with a single SiN_x layer were also prepared. A firing process corresponding to a standard Si solar cell metallization process was carried out in a belt furnace with a peak temperature of ~800°C for 3 seconds after deposition. Three different storage conditions were utilized in order to investigate the passivation stability (see Fig. 1): Illumination by a Xenon lamp at one-sun intensity; heating at 250°C on a hot plate in the dark; storage in the dark. The effective minority carrier lifetime was obtained by quasi-steady-state photo-conductance (QSSPC) measurement. The heated samples were cooled to room temperature before each lifetime measurement.

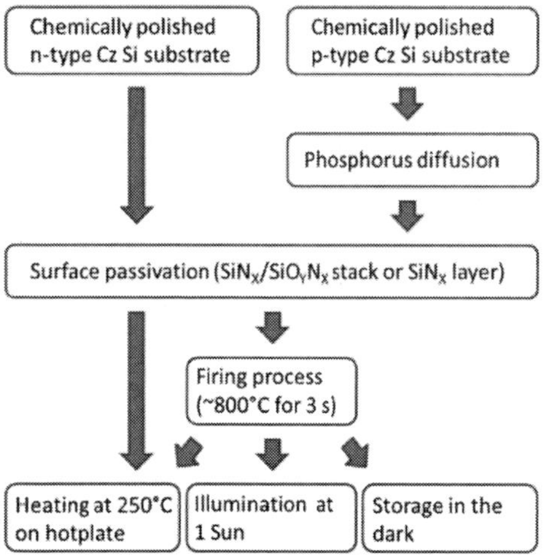

Figure 1. Samples, processing and stability testing performed in the work.

3 RESULTS AND DISCUSSION

Minority carrier lifetimes of n-type Cz-Si samples passivated using a SiN_x/SiO_xN_y stack are compared to a reference passivated with a single SiN_x layer in Figure 2. The lifetime for both as-deposited samples is around 10 us at an injection level of 10^{15} cm^{-3}. However, the lifetime is increased considerably for both samples after firing due to the release of hydrogen from the deposited dielectric layers, which passivate the defects well in the bulk and surface. In addition, the advantage of the stack on the passivation quality is obvious as the minority carrier lifetime reaches 946 µs at an injection level of 10^{15} cm^{-3} compared to that of single SiN_x passivated sample which is only 300 µs. This improvement on n-type Si passivation with the SiO_xN_y/SiN_x stack is almost the same as we applied the stacks for p-type Si wafer passivation [3].

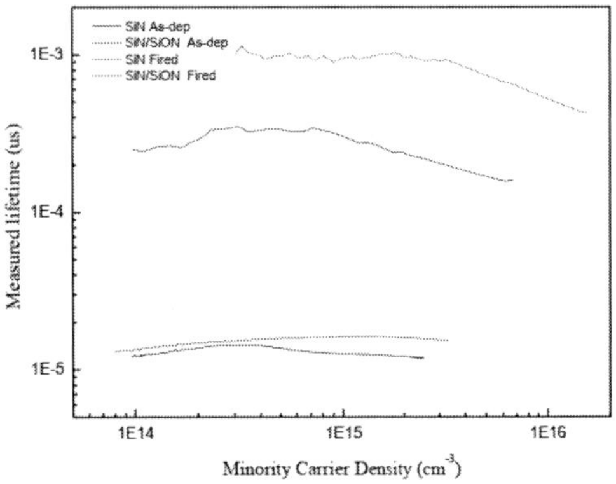

Figure 2: Minority carrier lifetime of Cz n-type Si passivated with SiN_x, SiN_x/SiO_xN_y stacks (as-deposited and fired).

N-type emitter with phosphorus diffusion on p-type Cz-Si wafers was also passivated with the SiN_x/SiO_xN_y stack and single SiN_x layers. Figure 3 shows a comparison of the effective minority carrier lifetime of SiN_x and SiN_x/SiO_xN_y passivated emitters with different sheet resistance. The stack passivated samples show higher lifetime compared to those of the single SiN_x layer passivated samples at all sheet resistances. When the sheet resistance is increased, the lifetime of SiN_x passivated samples almost stabilize at 80 us. However, the lifetime of the SiN_x/SiO_xN_y stack passivated samples start to increase significantly. It achieved 220 us as the sheet resistance of the emitter is increased to 330 Ω/sq. Thus, when the phosphors emitter has high sheet resistance, the advantage of using the SiN_x/SiO_xN_y stack as the passivation layer is more obvious.

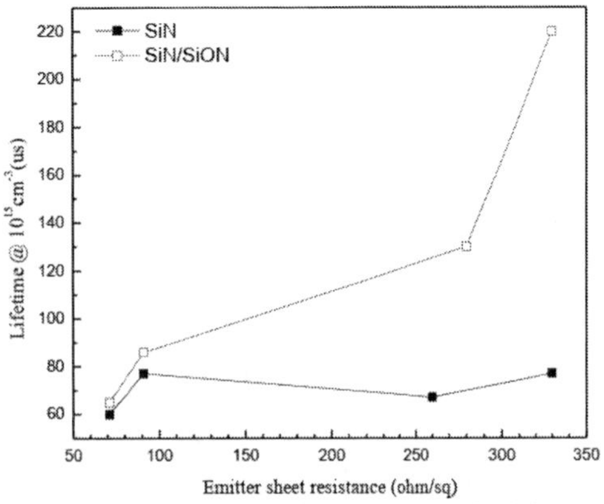

Figure 3: Minority carrier lifetime of SiN_x, SiN_x/SiO_xN_y passivated emitter with different sheet resistances.

Thermal activation is an important step for improving the passivation quality with SiN_x/SiO_xN_y. While, stability of the dielectric layer at high temperature is also important for the application especially for solar cells. Fig. 4 shows the lifetime of stack passivated samples as a function of time while heated at 250°C in the dark. After heating the fired samples for 120 h, the lifetime degraded from 686 µs to 498 µs, which is 72% of the initial value. By repeating the firing process, the lifetime comes back to 800 µs, even higher than the initial lifetime. On the contrary, the as-deposited samples demonstrate an increase in the lifetime with heating reaching a maximum value after 10 h. Minimal subsequent degradation was observed until 120 h, when the fired samples already degraded a lot. All those indicate that the chemical passivation is an important mechanism for the lifetime improvement.

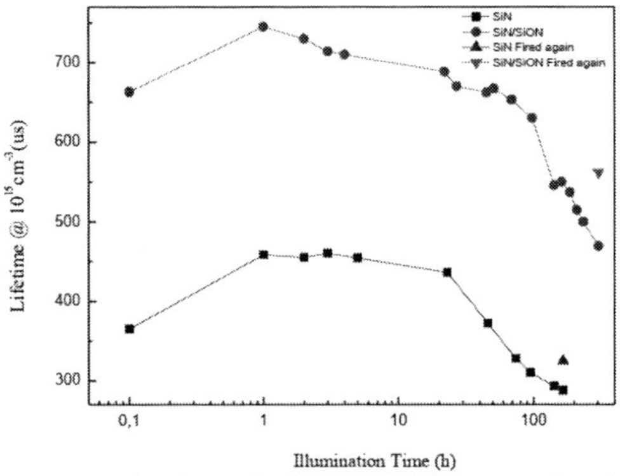

Figure 4: Minority carrier lifetime of different passivated samples as a function of illumination time at 1 Sun.

The minority carrier lifetime as a function of the illumination time is shown in Fig. 5. The stack passivated sample shows almost the same behavior as the sample passivated by a single SiN_x layer when it is illuminated at 1 sun. In the beginning, the lifetime slightly increased. However, after ~1 hour the minority carrier lifetime starts to decrease and reaches ~70% of the initial value after 300 hours. However, the passivation can be reactivated by repeated firing, indicating that this is related to the passivating properties of the thin films and not changes in bulk properties.

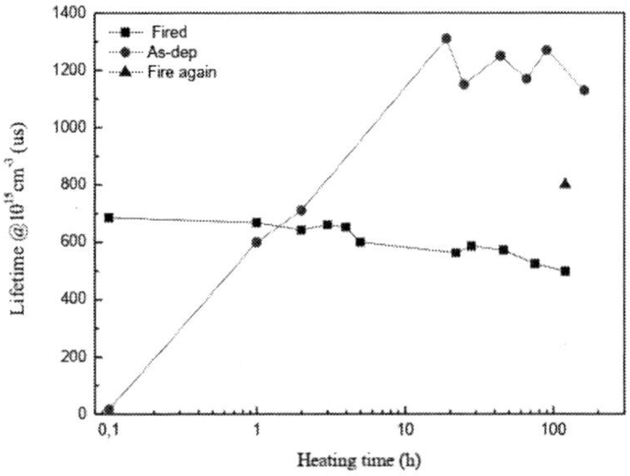

Figure 5: Minority carrier lifetime of SiN/SiO_xN_y passivated samples (as-dep and fired) as a function of heating time.

Finally, the lifetime of the passivated samples stored in the dark was investigated. Fig. 6 shows the storage time dependent of the minority carrier lifetime for the samples stored in the dark. Both the stack and single SiN_x passivated samples show a stable lifetime without significant degradation.

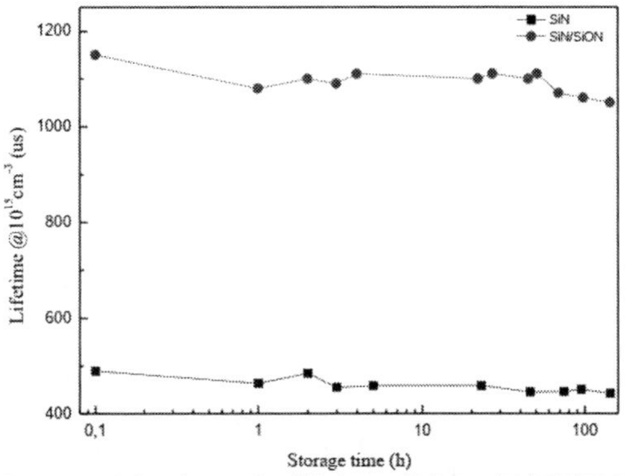

Figure 6: Minority carrier lifetime of SiN and SiN/SiON passivated samples as a function of storage time in the dark.

4 SUMMARY

SiN_x/SiO_xN_y stacks have shown considerably better passivation properties for n-type Si relative to the single SiN_x layer after a firing process. They also improve the passivation on phosphorus diffused n-type emitters on p-type wafers. The improvement is more distinct for emitters with higher sheet resistance.

Lifetime measurements during extended illumination reveal similar behavior for both single layer SiN_x and SiN_x/SiO_xN_y stacks. Illumination at one sun intensity can degrade the passivation quality after an initial increase in the lifetime. Heating at 250°C will reduce the lifetime as well, but only on fired samples. The passivation quality on as-deposited stacks, however, can be improved during the heating process, most likely due to hydrogen in-diffusion. Moreover, the degradation can largely be recovered by further thermal activation. Minimal degradation was observed for samples stored in the dark.

ACKNOWLEDGEMENT

Funding was partly provided by the Norwegian Research Council and industry partners in Norway through the EnergiX programme and the Centre for Environment-friendly Energy Research.

REFERENCES:

[1] J. Dupuis, E. Founnond, J.F. Lelievre, D. Ballutaud and M. Lemiti, Impact of PECVD SiON stoichiometry and post-annealing on the silicon surface passivation, Thin Solid Films 516 (2008) 6954-6958.

[2] A. Laades, M. Blech, H. C. Biank, C. Maier, M. Roczen, C. Leschinski, H. Strutzberg , A. Lawerenz, Fudamental study of silicon oxynitrides for photovoltaic applications, in 26th European Photovoltaic Solar Energy Conference and Exhibition, Hamburg, Germany, 2011.

[3] J. Zhu, S. Zhou, H. Haug, E.S. Marstein, S.E. Foss, WJ. Wan, A firing stable passivation for p-type Si substrate, in 29th European Photovoltaic Solar Energy Conference and Exhibition, Amsterdam, The Netherlands, 2014.

[4] CL Zhou, J. Zhu, S.E. Foss, H. Haug, Ø. Nordseth, E. S. Marstein, WJ. Wang, SiO_xN_y/SiN_x stack anti-reflection coating with PID-resistance for crystalline silicon solar cells, Energy Procedia 77 (2015) 434–439.

Gap in pagination due to unavailable paper.

Pages 783-785

NEW PROMISING C-SI SOLAR CELL AND BUSBAR CONCEPTS FOR INDUSTRY APPLICATION

W. Mühleisen[1], L. Neumaier[1], C. Hirschl[1], S. Seufzer[2], M. Trobej[3],
W. Pranger[4], J. Scheurer[5], R. Lorenz[6], M. Schwark[7]
[1] CTR Carinthian Tech Research AG, Europastr.12, 9524 Villach/St.Magdalen, Austria
(Author and Contact: wolfgang.muehleisen@ctr.at; T: +43 4242 56300 238)
[2] KIOTO Photovoltaics GmbH, Solarstr.1, 9300 St.Veit/Glan, Austria, (Siegfried.Seufzer@kioto.com)
[3] Energetica Energietechnik GmbH, Adi-Dassler Gasse 6, 9073 Klagenfurt, Austria, (M.Trobej@energetica-pv.com)
[4] Ulbrich of Austria GmbH, Industriestr.1, 7052 Müllendorf, Austria, (wpranger@ULBRICH.com)
[5] Polytec PT GmbH, Polytecplatz 1-7, 76337 Waldbronn, Germany, (j.scheurer@polytec-pt.de)
[6] Teamtechnik GmbH, Planckstr.40, 71691 Freiberg a.N., Germany, (Robert.Lorenz@teamtechnik.com)
[7] AIT Austrian Institute of Technology GmbH, Giefinggasse 2, 1210 Vienna, Austria, (Michael.Schwark@ait.ac.at)

ABSTRACT: Different busbar concepts with three, four, five and six busbars (BB) on standard silicon solar cells on both full cell format (156 mm x 156 mm) and half-cell format (156 mm x 78 mm) were investigated. 40 prepared single-cell mini module specimens, five of each per variation, were tested with a flash cell tester to determine series resistance and maximum power for a relative comparison. Simulations with original measured parameters were compared to the determined values of the experiment. A dependency was found between the number of busbars and the finger grid design, all together affecting series resistance and efficiency. An optimized busbar concept with at least 4-BB technology is recommended as a first step due to performance improvement and economic issues.
Keywords: PV Module, Silicon Solar Cell, Design, Ribbons, Electrical Properties

1 INTRODUCTION

In the PV module production, crystalline silicon solar cells with 2-BB and later 3-BB have been used over the last few years. In 2016, while a silicon PV module manufacturer has free choice in using solar cells with different cell design, there are fewer and fewer options to order solar cells with an old 2-BB or 3-BB concept. Therefore manufacturers are forced to retool consisting machines to be competitive.

The basic understanding of how a busbar and finger grid concept can affect the solar cell's and module's efficiency is of high interest. Since the step from scientific innovation to industry has been completed successfully for some time, there are new trends in the variable market to use four, five, six or multi wire busbar technologies [1-5].

There are many advantages of new silicon solar cells with innovative cell grid concepts - which have sporadically found their way into industry. But which concept should a module manufacturer with existing machines choose? Three concepts are ranked as feasible solutions.

One of the three concepts is to achieve less shading by fine printing pastes coupled with a thinner finger grid and a better aspect ratio. The second concept is the introduction of more busbars to reduce series resistance. The third concept proposes dividing solar cells to reduce current and ohmic losses. It is worth mentioning that these three "new" concepts have been known for some time but the technological possibilities and good price conditions were not available.

To optimize a solar cell, different factors such as contact resistance and shading grade by pastes should be considered. On one hand: the more the electrical contacts, the less the series resistance. On the other hand: the more the electrical contacts, the less the useable solar cell area. In the past, a grid design with 2-BB was the best decision for pastes because of paste quality and expensive machine expenditure. Later, with better pastes, the 3-BB technology was favored despite the retooling investment step, leading to a better cell to module efficiency. Due to new high tech fine line printing pastes are reliable, narrow and high grid fingers and busbars are possible. This development leads to high aspect ratios and thus opens the door for new cell grid concepts. With more busbars, a solar cell with micro cracks is more protected against loosing big cell fragments which could lead to hot spots. Such hot spots were the case with PV modules that used the 2-BB concept in the past.

With the introduction of sophisticated laser cutting processes, cheaper process costs and better solar cells with higher currents above 9 amps, the idea to divide solar cells for a better performance was born, however, has not been fully realized in industry due the costs.

The investigations presented in this paper are meant to assist module manufacturers in choosing a suitable cell concept which is coupled with economic and technical aspects.

2 EXPERIMENTAL SET-UP AND SIMULATIONS

2.1 Experimental set-up

Multi-crystalline silicon solar cells with different amount of busbars for full- and half-cell format were prepared with corresponding soldering ribbons (Fig. 1).

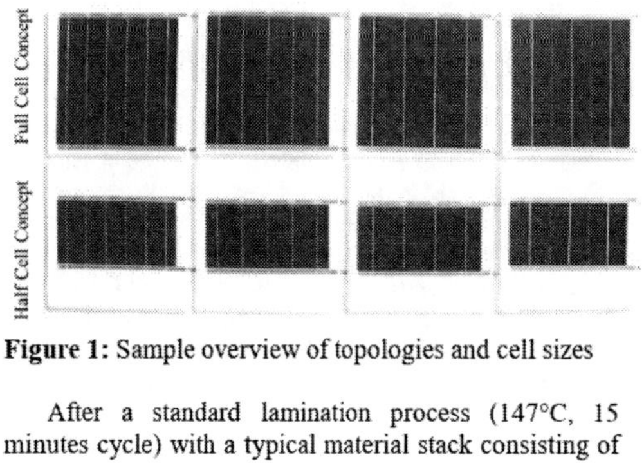

Figure 1: Sample overview of topologies and cell sizes

After a standard lamination process (147°C, 15 minutes cycle) with a typical material stack consisting of

3.2 mm glass, 400 µm EVA foil, solar cell, 400 µm EVA foil and 350 µm backsheet foil, single-cell mini modules were produced. To do a statistical evaluation, every equal specimen sort is available five times resulting in a total amount of 40 specimens in eight groups. There are full cell (156 mm x 156 mm) and half cell (78 mm x 156 mm) specimens with three, four, five and six busbars. Every specimen was measured four times with a class AAA cell flash tester (Sinton FCT-350) to determine series resistance (Rs) and maximum power (Pmpp) (Fig. 2).

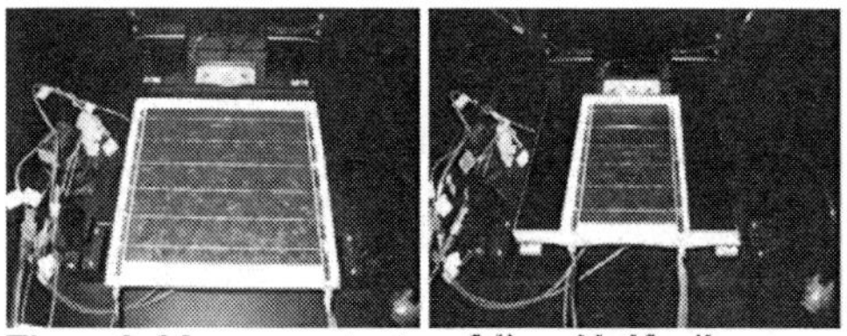

Figure 2: Measurement set-up full- and half cell

The wafer raw material quality and cell efficiency class is unknown but it is assumed as similar. While this circumstance may present an uncertainty factor, it is still considered sufficient to allow for a comparison with the simulation and to see a tendency. For a real qualitative comparison of measured and simulated values, solar cells with the same raw material would have been required. To be able to simulate with original data, the dimensions of every cell grid were measured (Fig. 3).

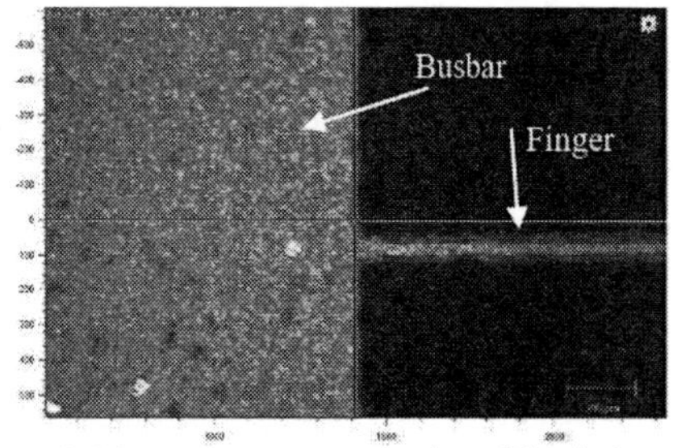

Figure 3: Measurement geometry data of the design

2.2 Simulations

To prove the design concept of different busbar variations, finite element analysis (FEA) simulation software was used. The software "Griddler2" [7] was written by the Solar Energy Research Institute of Singapore (SERIS). This powerful software has many input fields to describe effects of the material properties and loss processes. For the calculations, realistic parameters were chosen. To copy the real cell- and grid dimensions into the simulation, every cell concept was measured out with a microscope to generate the drawing, saved as own DXF file (Fig. 4). It was not necessary to create completely new files because an existing similar file delivered with the software could be modified and saved with transferred real cell dimensions (Table I and II). For the modification process only a CAD program is necessary. In this case the open source program "QCAD" was used [8]. In order to change the drawings, it is necessary to work with existing layers and "Griddler2" rules since otherwise the program cannot read the file again. The entered values were the same for all design variations. As a result, the material and loss effects are the same for every cell. Only the finger and busbar grid design is the variable.

Table I: Geometry data full cells

No of BB	No of Fing.	Finger width [µm]	Pad Front [mm]	Pad Back [mm]	Ribb. [mm]
3	82	60	1.4	1.6	1.3
4	100	53	0.95	1.67	1
5	92	63	0.8	1	0.7
6	99	48	0.6	0.7	0.6

Table II: Geometry data half cells

No of BB	No of Fing.	Finger width [µm]	Pad Front [mm]	Pad Back [mm]	Ribb. [mm]
3	41	73	1.41	2.17	1.3
4	45	53	0.93	1.34	1
5	44	53	0.54	0.77	0.7
6	50	48	0.43	0.69	0.65

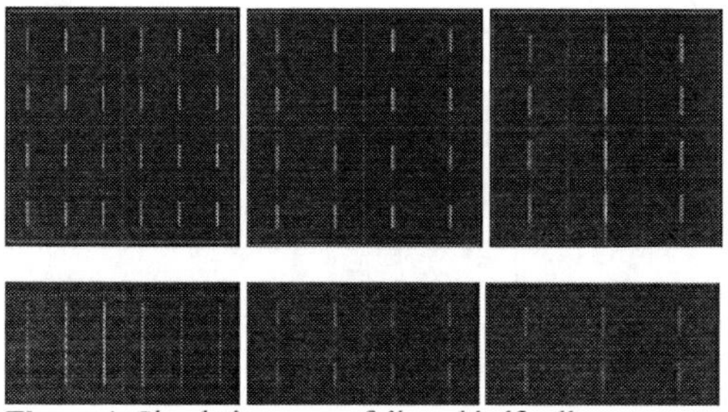

Figure 4: Simulation set-up full- and half cells

3 RESULTS

The presentation of the results is organized in a measurements- and a simulations section. Both, measurements and simulations were compared to evaluate the simulation tool and to evaluate the effect of the cell design for full and half cells.

3.1 Measurements Results

Full cell concept: The measured series resistance including wiring is the highest for the 3-BB design and in the region of 70 mOhm. For all other concepts (4-BB, 5-BB and 6-BB) Rs is in the region of 50-55 mOhm (Fig. 5). The measured efficiency with the cell flash tester shows a gain for the 4-BB, 5-BB and 6-BB concept compared to the 3-BB one. The 5-BB cell concept shows a drop in efficiency in comparison to 4-BB and 6-BB. The variation among five identical samples is around 0.5 % (Fig. 6).

Half cell concept: The series resistance including wiring is the highest for the 3-BB concept and in the region of 90 mOhm. Both 4-BB and 6-BB concepts show values in the region of 70 mOhm while the 5-BB concept shows 80 mOhm (Fig. 7). There is an ascending trend in efficiency from the 3-BB to the 6-BB concept (Fig. 8). The large variation in the region of 1.2 % for the 4-BB measurements are indicative of a not properly working cell sorting.

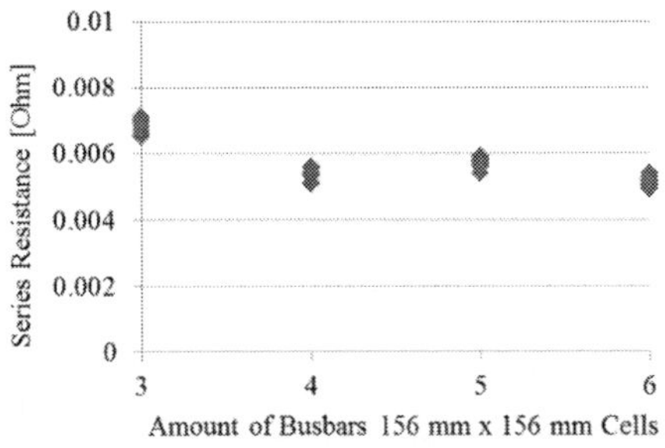

Figure 5: Series resistance measurement full cell

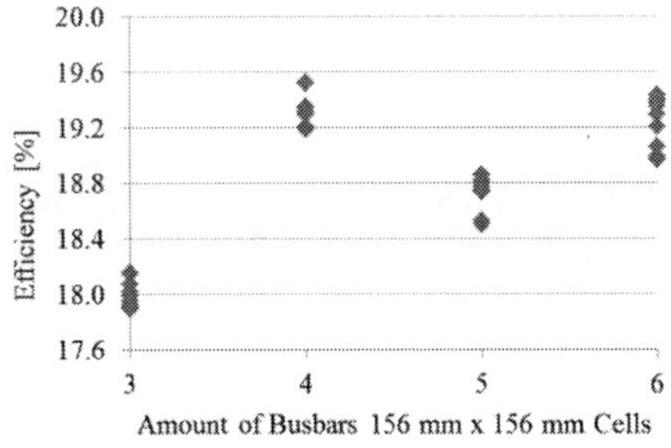

Figure 6: Efficiency measurement full cell

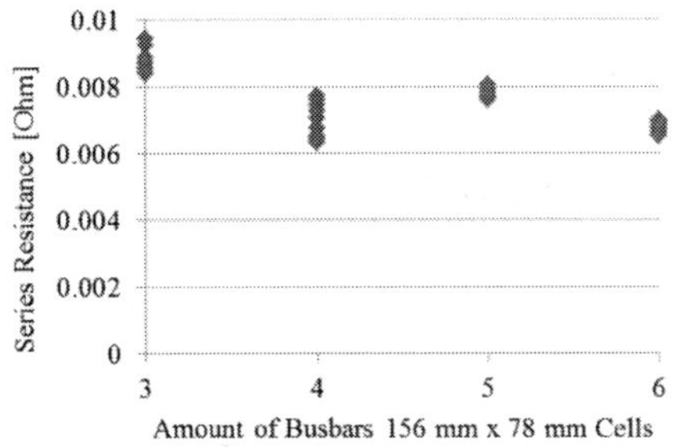

Figure 7: Series resistance measurement half cell

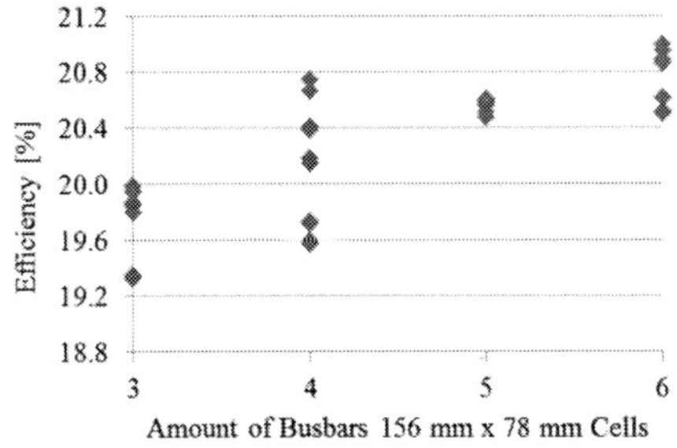

Figure 8: Efficiency measurement half cell

3.2 Simulations Results

Full cell concept: It has been shown that the front shading for the full cell format for the 3-BB and 4-BB cell is the same at 5.7 % while the 5-BB cell is at 6 %. The best 156 mm x 156 mm cell with regard to shadowing due to grid design is the 6 BB cell with 5.3 % shaded cell area (Fig. 9).

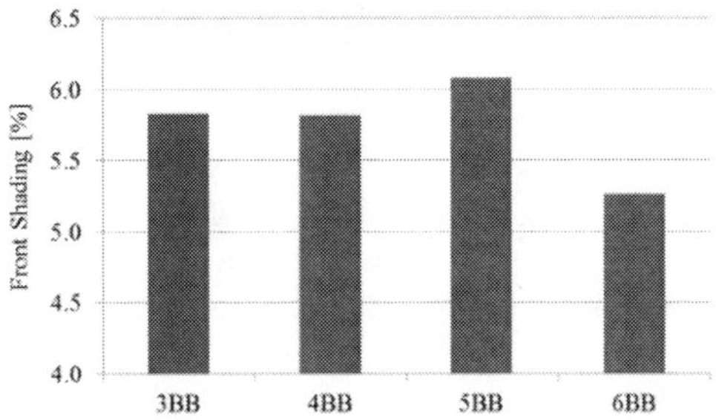

Figure 9: Front shading full cell

Based on the calculations of the grid design, standard cell parameter and ribbon dimensions, an efficiency was calculated. The 4-BB and 6-BB design are similarly good at 18 %, the 3-BB design is the worst. The efficiency grade of the 5-BB design is between the 3-BB and 4-BB/6-BB one. This is a strong indication that the 5-BB grid design is not optimized. The simulated efficiency gain among 3-BB and 4-BB cell grid design is 0.25 % and 0.3 % among 3-BB and 6-BB type (Fig. 10).

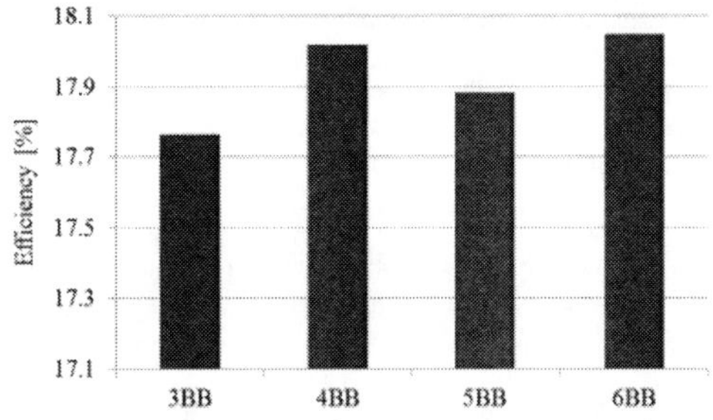

Figure 10: Efficiency simulation full cell

Half cell concept: For the simulated different half cells, the front shading descends from the 3-BB to the 5-BB concept. The 6-BB design concept shows no improvement in comparison to the 5-BB concept (Fig. 11).

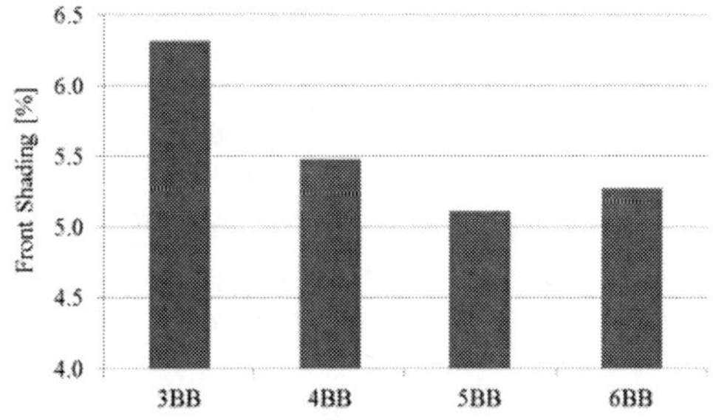

Figure 11: Front shading half cell

The simulated efficiency ascends from 3-BB to 6-BB with the half cell grid design concept. The gain between 3-BB and 6-BB design is 0.5 % for the same material (Fig. 12).

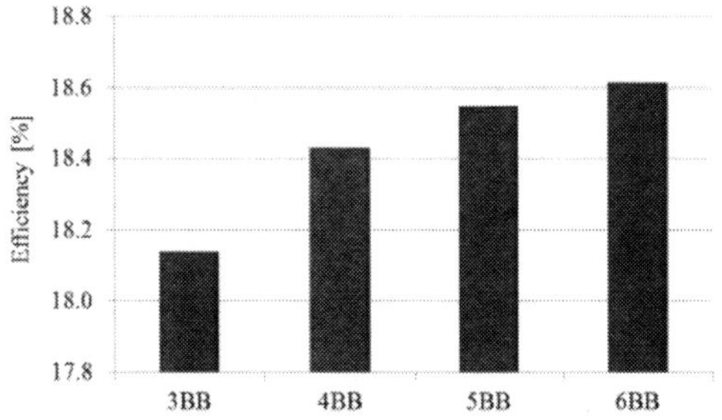

Figure 12: Efficiency simulation half cell

4 DISCUSSION

Full cell concept: Fig. 9 shows that there is front shading as a percentage of measured full cell design and it is obvious that the 5 BB cell has the highest shading grade. The expectation for an optimized design is a descending trend from 3 BB to 6 BB concept or at least a consistent trend. The poor 5-BB grid design is visible in in the simulated efficiency as well as in the measurements. The gain between 4-BB and 6-BB design is small although the 6-BB grid design has one percent less shading grade. But, in comparison with the minimal additional gain in efficiency, the effort to connect 6-BB cells in series is very high compared to 4-BB technology (Figs. 10, 6). Therefore, the advice for a module manufacturer is to choose the 4-BB technology when using 156 mm x 156 mm cells.

Half cell concept: The 78 mm x 156 mm half cell format shows a gain in efficiency compared to the 156 mm x 156 mm cell format (Figs 10, 12). This is due to the fact that the ohmic power loss (P_{LOSS}) is proportional to the square of the maximum power current (I_{MPP}^2) multiplied by the series resistance (R_{SERIES}). By reducing the current to the half, the power loss is only one fourth. Therefore the half cell concept is more advantageous than the full cell concept. This fact pushes the efficiency in the range of 0.4 % for a 3-BB or 4-BB concept. Daßler et al. experimentally found similar results for a produced half cell and full cell module [6]. The yield for a half cell module was on average 3 % higher than for the full cell version.

All in all it seems that a 3-BB concept is underprivileged compared to the 4-BB concept. This is valid for both full and half cell format. The manufacturers of solar cells are using new fine line printing pastes and module manufacturers must pay an extra fee when ordering 3-BB concept solar cells. So the next step for a module manufacturer is to invest in new stringer technology. But with more busbars at cell level the manufacturer need more ribbon meters. The ribbon manufacturers reel the copper ribbon to a reel. But the copper coil weight is the same for a 3-BB cell design with 1.3 mm ribbon and a 4-BB cell design with a 1.0 mm ribbon. Only the wiring time is longer because of more meters of copper ribbons. So the introduction of more busbars for solar cells could lead to a price not only measured per kilogram but also per kilogram and meter.

The step from 3-BB to 4-BB technology is very lucrative because of more efficiency. Another positive effect of this design development is a lower failure probability. This is due to the fourth or fifths busbar ribbon. In case of an induced cell crack the cell fragment is small compared to a 2-BB or 3-BB concept. But is a 4-BB design the best one for a module manufacturer? The economic factors for a possible change in the cell concepts are easy to address. For a cell manufacturer it is more lucrative to sell cells with 4-BB or more because of more generated efficiency than producing the same cells with 3-BB concept. A module manufacturer is in the situation to calculate new: The choice is between using powerful 4-BB cells with a better failure probability and as an argument to convince customers, or keeping the 3-BB concept and paying an extra charge. According to the results of experiment and simulation, it is not necessary to use more than four busbars for a full cell concept. For a half cell concept a 5-BB or 6-BB cell is the best choice but the dividing process is cost intensive.

The best decision for a manufacturer is to retool consisting machines to 4-BB technology with the option to expand to 5-BB or 6-BB technology. The cell cutting process could be integrated whenever wanted.

5 CONCLUSION

Silicon solar cells with different busbar concepts 3-BB, 4-BB, 5-BB and 6-BB in the format 156 mm x 156 mm (full cell) and 156 mm x 78 mm (half-cell) were investigated. 40 produced single-cell mini module specimens, five of each per variation, showed in a relative comparison that 4-BB technology is both for full cell and half cell format the best economic and technical decision at the moment. Simulations with original measured parameters were compared to the determined values of the experiment. The simulated efficiency output gain between 3-BB and 4-BB technology is in the range of 0.3 %. One 5-BB cell was found with a non-optimized cell grid design showing a poorer performance than a 4-BB or 6-BB cell design.

6 ACKNOWLEDGEMENTS

This work was conducted as part of the Austrian "Energy Research Program" project INFINITY, using results from the predecessor project InnoModu. Funded by the Austrian Climate and Energy Fund and the Austrian Research Promotion Agency (FFG), are both gratefully acknowledged.

7 REFERENCES

[1] H. Oh, D. Kyeong, I. Chung, H.-Y. Son, I.-S. Moon, W. Lee, E.-C. Cho, "Multi-Wire Ribbon Technology Evaluation and Contact Design Optimization for Commercial Application" Proc. 29th EU PVSEC, Amsterdam, The Netherlands, 2014; 218 – 220

[2] A. Faes, M. Despeisse, J. Levrat, J. Champliaud, N. Badel, M. Kiaee, T. Söderström, Y. Yao, R. Grischke, M. Gragert, J. Ufheil, P. Papet, B. Strahm, B. Cattaneo, J. Cattin, Y. Baumgartner, A. Hessler-Wyser, C. Ballif, "SmartWire Solar Cell Interconnection Technology" Proc. 29th EU PVSEC, Amsterdam, The Netherlands, 2014; 2555 - 2561

[3] H. Chen, C. Chen, M. Chang, C.H. Hsueh, E. Yen, K.L. Ho, "The Influence of Cell Busbare Pattern on PV Module Reliability" Proc. 29th EU PVSEC, Amsterdam, The Netherlands, 2014; 2562 – 2565

[4] A.F. Dethlefsen, M. Volk, K. Ramspeck, C. Buchner, "Reliability Testing of Multi-Busbar Modules" Proc. 29th EU PVSEC, Amsterdam, The Netherlands, 2014; 3323 – 3326

[5] T. Hammam, "An Optimisation Study of the Copper Ribbon for Crystalline Silicon Solar Cell Interconnections" Proc. 26th EU PVSEC, Hamburg, Germany, 2011; 3329 – 3335

[6] David Daßler, "Leistungsanalyse von Halb- und Vollzellenmodulen basierend auf mehrjährigen Freifeldmessungen" Proc. 32. Symposium Photovoltaische Solarenergie, Bad Staffelstein, Germany, 2016; C11

[7] Software "Griddler2" – The handy solar cell simulator http://www.seris.sg/Seris/OurServices/Griddler.html

[8] Software "QCAD" – Open Source CAD http://www.qcad.org/en/

ANALYSIS ON EMITTER OF N-TYPE MONOCRYSTALLINE SILICON PERT PHOTOVOLTAIC CELL

Takayuki Morioka, Tatsuro Watahiki, Shinya Nishimura, Kunihiko Nishimura, Daisuke Niinobe, Yumiko Kobayashi, Hidetada Tokioka, and Mikio Yamamka
Advanced Technology R&D Center, Mitsubishi Electric Corporation
8-1-1 Tsukaguchihonmachi Amagasaki, Hyogo 661-8661, Japan

ABSTRACT: We have developed the boron emitter for n-type monocrystalline silicon photovoltaic cells and achieved the conversion efficiency of 21.53% by passivated emitter and rear totally diffused (PERT) structure with a total area of 239 cm^2. The center area of the cell shows 22.08% (222.0 cm^2). The developed cells contain a more deeply diffused and lightly doped boron emitter to obtain higher efficiency than the previous cell we have reported.

We have also studied a rear emitter structure in the n-type monocrystalline silicon PERT photovoltaic cell by a 2D device simulation and the experimental results; especially the correlation between substrate resistivity and the photovoltaic property. The V_{oc} of rear emitter cell is nearly constant regardless of the substrate resistivity. On the other hand, the J_{sc} is decreased sharply by lowering substrate resistivity. The J_{sc} of 0.55 Ω·cm substrate is 5 % lower than that of 5.6 Ω·cm one. These results are attributed to front surface field (FSF) effect which shows a dependence on the substrate resistivity. To achieve higher J_{sc} for the cells with FSF, the barrier height for the minority carriers are important.
Keywords: n-type, c-Si, High-Efficiency, Silicon Solar Cell, Modeling

1 INTRODUCTION

We have achieved the conversion efficiency of 19.7% for the p-type monocrystalline silicon photovoltaic cell [1]. On the other hand, the n-type monocrystalline silicon photovoltaic cell has a potential for higher conversion efficiency. The n-type Si wafers possess a high durability against a high temperature process and metal impurities which cause the quality degradation of the wafers. A longer minority carrier lifetime of the n-type silicon wafer compared to that of the p-type wafer is also profitable. Many researchers reported their research results related to the n-type cell [2], and the conversion efficiency has been improved continuously. Previously we have reported our PERT cell with an efficiency of 21.3% [3]. For a PERT cell, a boron-doped emitter, a rear side structure and passivation layers for each surface have a great influence on the cell property. In this paper, we report the theoretical and experimental details of the boron emitter of our n-PERT cell. We have fabricated n-type Si photovoltaic cell with a Passivated Emitter and Rear Totally diffused (PERT) structure.

We have also studied rear emitter (RE) structure of the n-type monocrystalline silicon PERT photovoltaic cell; especially the correlation between the substrate resistivity and the photovoltaic property. We reported rear emitter Hetero Junction Cell with an efficiency of over 23% [4]. Therefore, the rear emitter structure would be better than the front emitter structure even for Homo Junction Cell to achieve higher efficiency. On the other hand, resistivity variation of n-type Si wafers through the identical ingot is wider than that of p-type Si wafers. That is because a segregation coefficient of phosphorous atoms is lower than that of boron atoms. Therefore, we investigated the correlation between the substrate resistivity and the photovoltaic property of the rear emitter cells.

2 EXPERIMENTAL

Figure 1 shows the schematic illustration of the developed n-type PERT cell. The n-type CZ silicon wafers were used for the cell fabrication. Front surfaces of the cell were textured with alkaline solution, resulted in random pyramids. For the investigation of rear emitter cells, the substrate resistivity was varied from 0.55 to 5.6 Ω·cm. The emitter layer was fabricated by a deposition of a boron silicate glass (BSG) layer on the substrate and by the thermal diffusion process. The BSG layers were deposited with atmospheric pressure chemical vapor deposition (APCVD). The phosphorus doped regions with two different peak concentrations were fabricated on the rear side of the cell using a thermal diffusion process. The highly concentrated phosphorus layer was designed for low resistance ohmic contacts between the silicon wafer and a rear electrode. Sheet resistance of lowly concentrated phosphorus area is designed to show more than 100 ohms per square. The SiN/Al_2O_3 stacks were applied to the cell as a passivation layer for the surface of the boron layer. Screen-printed electrodes were fabricated on both surfaces, followed by a firing process.

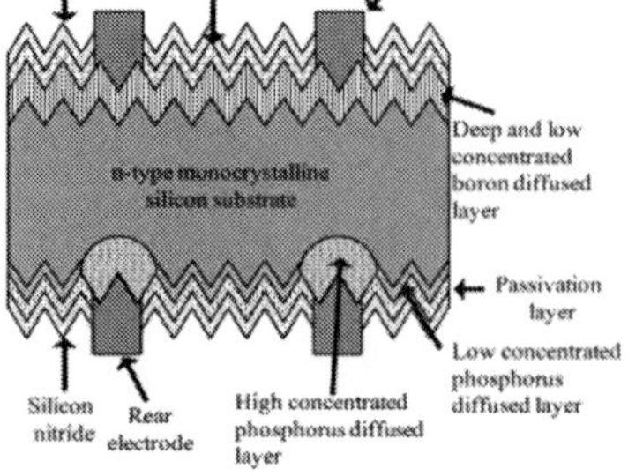

Figure 1: Schematic cross-sectional diagram of the developed photovoltaic cell.

3 IMPROVING THE CHARACTERISTICS OF BORON EMITTER

We calculated a correlation between the cell structure and the cell performance with PC1D version 5.9 to investigate on the boron layer. The parameters used in PC1D are listed in Table I. A peak boron concentration and a diffusion depth were varied to clarify the effect of boron profiles on the cell performance. The boron peak concentration of the front surface was varied from 8×10^{18} cm^{-3} to 5×10^{20} cm^{-3}. The depth of the boron layer was controlled by setting a depth factor which is varied from 0.020 µm to 0.58 µm. Figure 2 (a) and (b) shows V_{oc} and J_{sc}, respectively, as functions of the peak doping concentration and the depth factor. Each value improved as the boron peak concentration is reduced. To achieve both high V_{oc} and J_{sc}, the peak concentration should be lower so that the depth factor can be chosen from a wide range. Although the shallow junction may also improve the V_{oc} and J_{sc}, the shallowly diffused boron emitter is difficult for the fabrication in mass production due to less process margin. Therefore, making a low concentrated and deeper boron layer is a better way to improve the efficiency in the mass production.

(a)

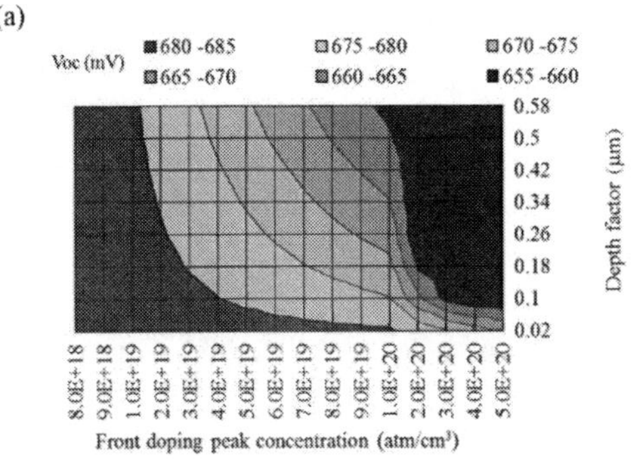

(b)

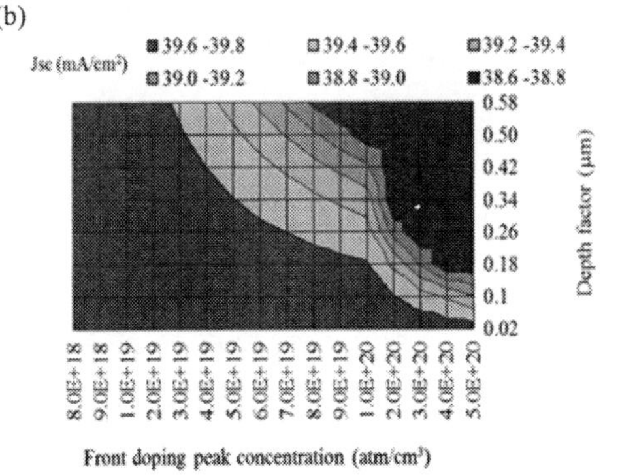

Figure 2: Calculated results of (a) V_{oc} and (b) J_{sc}, depending on the peak boron concentration and depth factor.

Taking these simulation results into account, we have developed the emitter with the less boron concentration than the previously reported cell. In order to reduce boron peak concentration and to increase the depth of the layer, we have controlled the boron concentration in BSG and extended thermal diffusing time than the previous condition. Table II shows the photovoltaic property we have developed. The V_{oc} was successfully increased from 678 mV to 683 mV, without any degradation of the Fill Factor (FF). As a result, the cell efficiency was increased from 21.30% to 21.53% (Certificated by National Institute of Advanced Industrial Science and Technology (AIST) Japan.).

Table I: Parameters of the PC1D simulation

Parameter	Value
Device area	$1cm^2$
Front and rear surface	Textured
Internal front and rear surface	Optically rough
Wafer thickness	200µm
Wafer background doping	$2.4\times10^{15}/cm^2$
Rear phosphorus peak concentration	$7\times10^{19}/cm^2$
Depth factor of the rear diffused layer	0.15µm
Peak position of the rear diffused layer	0µm
Bulk lifetime	$\tau_n=\tau_p=1ms$
Front and rear surface recombination velocity	100cm/s
Fill Factor	0.83

TABLE II: Photovoltaic property of n-type monocrystalline Silicon PERT photovoltaic cell (certificated by AIST Japan.)

	V_{oc} / V	J_{sc} / mA/cm²	F.F. / %	Eff / %
Previous	0.678	39.6	79.4	21.30
This study	0.683	39.5	79.7	21.53

4 ANALYSIS OF SHORT CIRCUIT CURRENT LOSS IN REAR EMITTER CRYSTALLINE SI SOLAR CELL

4.1 Photovoltaic property dependence on substrate resistivity

Figure 3 plots the J_{sc} and the V_{oc} over the substrate resistivity. The V_{oc} is approximately 680 mV and nearly constant in this range of the substrate resistivity. On the other hand, the J_{sc} and the substrate resistivity have a strong positive correlation. The J_{sc} was decreased sharply in lower substrate resistivity cells. The J_{sc} with 0.55 Ω·cm resistivity substrate dropped 5 % from that of 5.6 Ω·cm one. This strong correlation resulted in the variation of the efficiency between 20.0 and 20.7%.

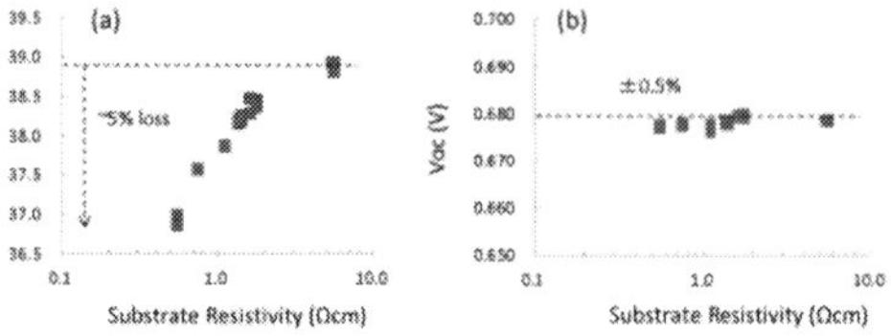

Figure 3: Photovoltaic property dependence on substrate Resistivity

Figure 4 shows internal quantum efficiency (IQE) curves of the bifacial solar cell that we have developed. J_{sc} loss was mainly due to the loss in the short to medium wavelength region. It suggests that J_{sc} loss was most probably due to the bulk recombination and/or recombination in n^+/n^{++} region where the Auger recombination and surface recombination occurred.

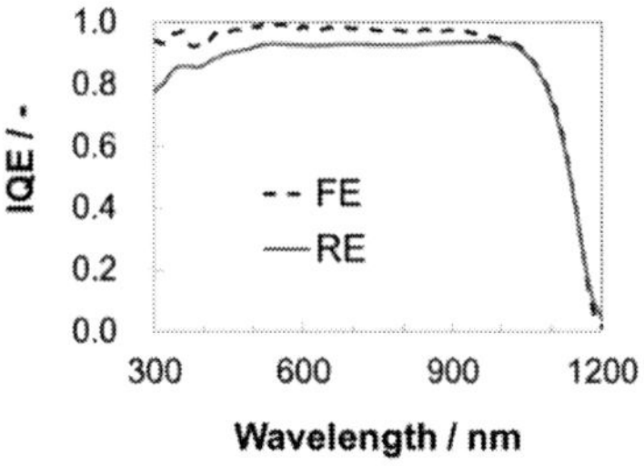

Figure 4: IQE characteristics of bifacial solar cell.

4.2 Comparison between experiment and calculation for rear emitter Si solar cell

For the calculation of the I-V characteristics of the solar cells, we have used 2D device simulator ATLAS (Silvaco, Inc.). Schematic structures of the RE cell for the calculation models are shown in Figure 6. The details of the model are summarized in Table III [5].

To explain the decrease of J_{sc}, we first check the effect of the carrier lifetime of the substrate, assuming the variation of the lifetime is the main cause of the decrease in J_{sc}. Then the bulk lifetime of the substrate with a different resistivity used for the calculation is chosen to match with the experimental J_{sc}. The similar trends of J_{sc} can be obtained by decreasing the lifetime from 1 ms down to 0.25 ms as a minimum for the lowest resistivity (dotted line in Figure 6). In case the bulk lifetime is low, J_{sc} would be strongly affected by the recombination loss in the substrate. However, the calculated V_{oc} shows the noticeable decrease in contrast to the experimental results in our case. This data shows that the recombination in the bulk substrate may not be the major cause of the J_{sc} loss for the RE cell with a low resistivity substrate.

Then, we calculate the I-V characteristics using a structure model with n^+ and n^{++} doping regions on the front side (FSF model: dashed line in Figure 6). In this case, the bulk lifetime is chosen to match V_{oc} (from 1 ms to 0.65 ms). If the bulk life time is constant regardless of the resistivity, the calculated V_{oc} increases with the decreasing resistivity, and shows slight gap from the experimental V_{oc}. The both calculated V_{oc} and J_{sc} match better with the experimental data by using FSF model.

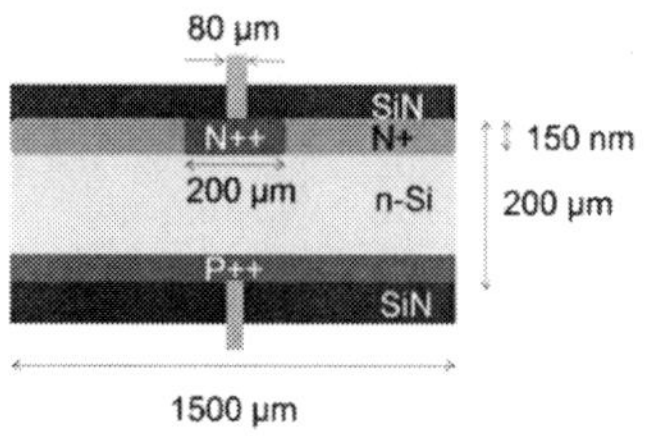

Figure 6: The schematic structure of the rear emitter cell used for the calculation.

TABLE III: Basic parameters used for the calculation [5].

Region	Width /μm	Doping concentration / cm^{-3}	Recombination velocity cm/s	Thickness / μm
n^+region	1300	1×10^{19}	200	0.15
n^{++}region	200	5×10^{19}	200	0.15
Substrate	1500	8×10^{14}-9×10^{15}	...	200

(a)

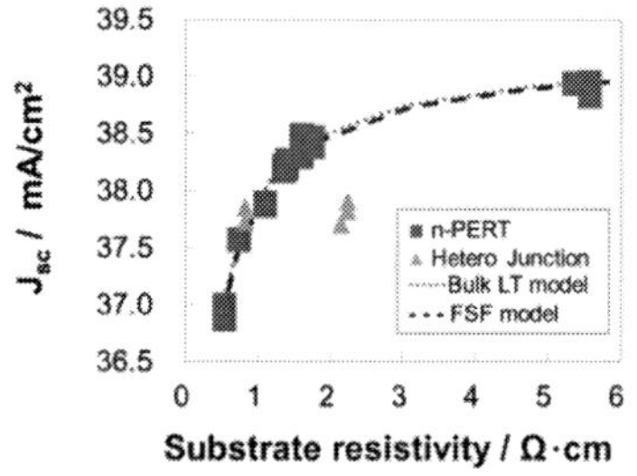

(b)

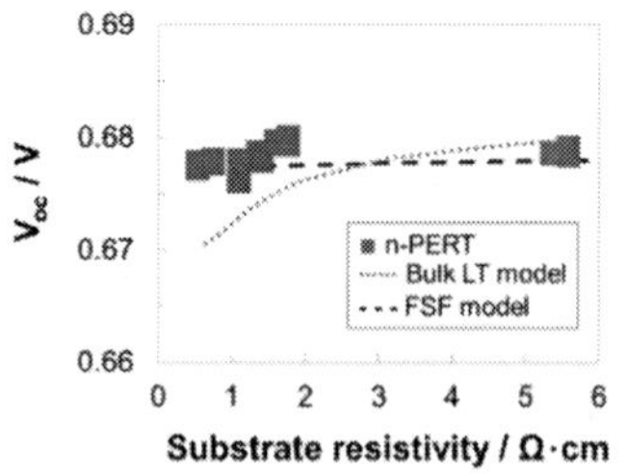

Figure 6: Measured and calculated (a) J_{sc} (b) V_{oc} as a function of substrate resistivity.

The increase in the J_{sc} loss in the RE cell denotes the decrease in the hole concentration. Therefore, it seems reasonable that a change in the substrate resistivity leads to the large drop of hole concentration and non-negligible decrease of V_{oc} at the same time. However, both the experimental and the calculated results show almost no change in V_{oc}. The V_{oc} is equal to a difference between the quasi-fermi level of electron and hole when J = 0; therefore, it can be calculated using Δp and N_{dsub} as shown in the following equation:

$$V_{OC} = \frac{kT}{q}\left(\frac{(\Delta p + N_{d_{sub}})\Delta p}{n_i^2}\right)$$

where n_i is the intrinsic carrier concentration. The minority carrier concentration in the substrate decreases from around 4×10^{15} cm^{-3} to 2×10^{15} cm^{-3} by decreasing the substrate resistivity from 5.6 Ω·cm to 0.55 Ω·cm, including the effect of the change in the bulk lifetime (Figure 7). The reduction of the hole concentration itself leads to around 15 mV reduction of V_{oc}. On the other hand, the increase of N_d causes the fermi level of electron to be 13–14 mV higher. (The electron concentration increases from around 5×10^{15} cm^{-3} to 1×10^{16} cm^{-3}, by decreasing the resistivity, including the electron generated by the light.) This increment cancels the effect of the reduction of the minority carrier concentration in the substrate. As a result, the reduction of V_{oc} is limited to a few mV.

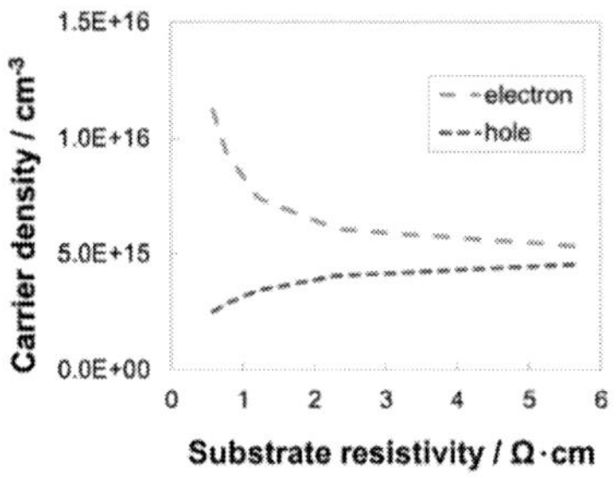

Figure 7: Calculated carrier density of the substrate near the FSF at V_{oc}.

4.3 Calculated J_{sc} as a function of the barrier height for the minority carriers

Figure 8 shows the dependence of the calculated J_{sc} on the barrier height. In this calculation, the energy level of the valence band for FSF layer is varied. The substrate thickness is 200 μm and the doping concentration at FSF is kept as 1×10^{19} cm^{-3}. To confirm only the effect of the barrier height, the bulk lifetime is kept high to eliminate the effect of the bulk recombination loss. The optical bandgap is kept the same as Si, so the generation rate in the cell is the same in all conditions. As shown in the figure, J_{sc} is not the function of the resistivity anymore but the function of the barrier height. The barrier height lower than 0.1 eV drastically decrease J_{sc}. When the barrier height is above 0.2 eV, then J_{sc} saturates and no resistivity dependence is practically observed. This result also indicates that the resistivity itself is not the point but the barrier height is a key point to explain the J_{sc} loss shown in Figure 3.

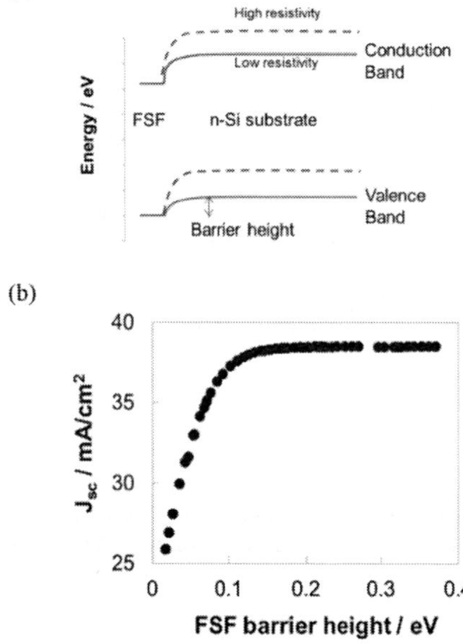

Figure 8: (a) Energy band diagram for FSF/n-Si substrate and (b) calculated J_{sc} as a function of the barrier height for the minority carriers.

It is difficult to achieve the barrier height higher than 0.2 eV for the homojunction cell, especially for a high V_{oc} cell. However the band gap engineering is possible for the heterojunction cell to reduce the resistivity dependence. The triangle shape plots in figure 6 show the experimental results for the heterojunction rear emitter cell. Details of the cell structure are given in Ref. 4. The J_{sc} trend clearly indicates that J_{sc} does not depend on the resistivity of the substrate for the heterojunction solar cell. The Auger recombination as well as the surface recombination is not the problem because an intrinsic layer has no doping and the barrier height more than 0.5 eV. V_{oc} as high as 0.740 V shows the very low recombination rate near at FSF. The absorption loss for the short wavelength at the n-layer also helps the distribution of the generation rate within the substrate to shift toward the emitter, results in the transport of minority carriers to the emitter more effectively. Therefore, the heterojunction solar cell possesses all the features listed previously to reduce the J_{sc} loss caused by the resistivity variation.

5 CONCLUSION

We have investigated the boron diffused layers of the n-type monocrystalline silicon photovoltaic cell. According to the calculation, the boron layer should be diffused deeply, and its peak concentration should also be decreased to achieve a high efficiency. The deeply diffused and low concentrated boron layer was fabricated by APCVD process. As a result, we achieved the conversion efficiency of 21.53% by passivated emitter and rear totally diffused (PERT) structure with a total area of 239 cm^2. The center area of the cell shows 22.08% (222.0 cm^2).

The dependence of J_{sc} loss on the base resistivity of the substrate is analyzed by 2D device simulator and compared with the experimental results. V_{oc} does not show the clear dependence on the resistivity in the range of 0.55–5.6 Ω·cm. On the other hand, J_{sc} strongly depends on the barrier height between FSF and the substrate for the minority carriers. The substrate with a higher resistivity produce a higher surface field effect, thus it brings higher J_{sc}. The lower substrate resistivity leads to lower barrier height, thus the J_{sc} loss is high because of the Auger recombination and the surface recombination. The control of the doping concentration in FSF is very important to minimize the J_{sc} loss in FSF. Furthermore, the passivation layer with a wide bandgap, such as amorphous Si layer, enables to reduce the J_{sc} loss at FSF drastically.

ACKNOWLEDGEMENT

A part of this work was supported by New Energy and Industrial Technology Development Organization (NEDO).

REFERENCES

[1] D. Niinobe, S. Hiza, T. Sato, S. Matsuno, T. Toyoshima, T. Katsura, T. Okamoto, T. Ishihara, S. Hamamoto, H. Morikawa and S. Arimoto, in technical digest of 21st International Photovoltaic Science and Engineering Conference, 2011, 5a-1o-11.

[2] D. Kania, T. S. Böscke, M. Braun, P. Sadler, C. Schöllhorn, M. Dupke, D. Stichtenoth,A. Helbig, R. Carl, K. Meyer, A. Grohe, J. Lossen, and H.-J. Krokoszinski, 28th European Photovoltaic Solar Energy Conference, 2014, 2BV.2.54.

[3] S. Nishimura et al., IEEE Journal of Photovoltaics Volume:6, Issue:4 (2016) 846-851.

[4] T. Watahiki et al., Appl. Phys. Express, v8 (2015) 021402.

[5] T. Watahiki et al., Journal of Applied Physics, 119, (2016) 204501.

ADVANTAGES OF TRANSITION TO FOUR AND FIVE BUSBAR FRONT CONTACT GRID DESIGNS FOR NI/CU/AG PLATED SILICON SOLAR CELLS

D. Pysch[a], J. Burschik[a], Y. Shengzhao[b], N. Bay[a], K. Crouse[c], A. De Rose[a], A. Hoffmann[a], H. Kühnlein[a], B. Lee[c], A. Letize[c], M. Passig[a], M. Sieber[a], P. Verlinden[b], K. Vosteen[a]
a) RENA Technologies GmbH, Freiburg Solar Technology Center, Hans-Bunte Str. 19, 79108 Freiburg, Germany
b) Trina Solar, 333 North Caoixi Rd, CCIG International Plaza Building B, Shanghai 200030, PR China
c) MacDermid Enthone Inc., 245 Freight street, Waterbury, CT06702, USA
eMail damian.pysch@rena.com

ABSTRACT: The intention of this work is to show the advantages of the Ni/Cu/Ag plating process when the number of busbars on the solar cell is increased. Cells with four and five busbars were processed and compared to those with three in respect of efficiency, finger and solder adhesion. Simulations to calculate the metal reduction and the experimental values are compared. Estimations how much the production costs may be reduced are shown.

Keywords: solar cells; plating; busbars; adhesion; Ni/Cu/Ag; cost reduction

1 INTRODUCTION

The aim to produce more efficient and cost effective solar cells and modules has led to an increased number of busbars. The current standard of three busbars is expected to be replaced by four to five or multiwire concepts within the next years (1). Some manufacturers have already installed such lines.

An increased number of busbars leads to shorter effective finger length (2) (3). This can result in a lower series resistance R_s keeping the finger cross section constant. It is also possible to reduce the finger cross section with constant R_s leading to a cost reduction for metallisation.

The most common technology for metallisation is screen printing of silver paste. A large proportion of the total metallisation costs is taken by the price of silver, which is also subject to fluctuation of market price. Substituting copper for silver and plating for printing promises considerable savings.

Replacing an established technology requires at least equal technical performance and a financial benefit. Both aspects will be compared and discussed in the following.

2 SIMULATION

In order to estimate the required finger cross section Gridsim (4) calculations for three, four and five busbar grid designs were performed. It is assumed that the finger cross section is defined by the width of the laser contact opening w and the thickness of the plated layer h (Figure 1). The growth of the layer is isotropic so that the extensions to the sides of the opening are equal to the height h. Properties of the busbars were assumed as constant in respect to width, height and conductivity. The difference in the number of busbars was accounted for in the length of the fingers unity cell.

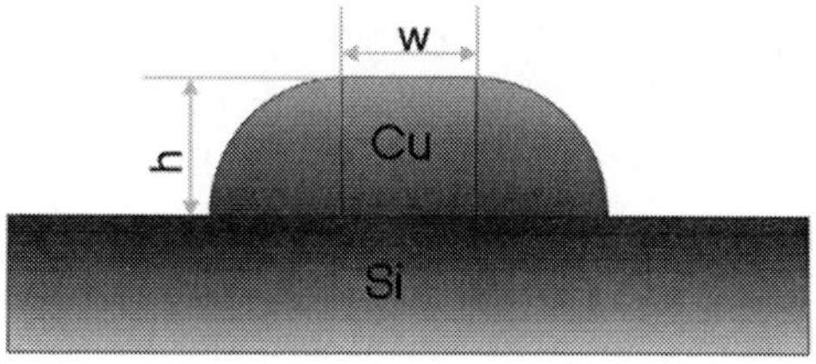

Figure 1: Sketch of the cross section of a plated copper finger for front side metallization.

The grid design for three busbars was optimised in regards to power loss. A higher finger cross section improves conductivity but increases shading loss and decreases the short circuit current I_{sc}. From the minimum in power loss for the three busbar case, $R_s = 0.36\ \Omega cm^2$ is obtained. The cross sections for the other busbar numbers are calculated to give the same value of R_s (Table 1).

Table 1: Results of simulation. Minimum power loss for three busbars of R_s =0.36 Ωcm^2 defines the finger cross section to yield the same R_s for four and five busbars.

No. of busbars	Finger length (mm)	Rs (Ωcm^2)	Finger height (µm)	Finger cross section (mm^2)
3	25	0.36	11	0.041
4	19	0.36	6	0.018
5	15	0.36	4	0.011

3 PROCESS SEQUENCE

The processing of plated cells is quite different to the manufacturing of screen printed cells (5). The process consists of the following individual steps:

- Laser contact opening: To ablate the silicon nitride layer, a pico second laser with 355 nm wavelength is used. The choice of laser has a big impact on the adhesion of the metallisation.

- Pretreatment: To remove the native oxide, a buffered diluted HF solution is used.
- Ni LIP: A thin nickel layer is deposited using light induced plating (LIP) (4) (6). This layer forms the contact to the silicon and acts as a diffusion barrier against copper diffusion into silicon.
- Cu LIP: The copper layer is also grown by a light induced plating step. This layer forms the bulk of the contact finger and acts as the conducting layer.
- Ag: A very thin (~100 nm) silver layer is added in an immersion bath. Without this layer the Cu surface would tarnish and the solderability would suffer.
- Anneal: To improve contact formation and adhesion, the cells are annealed for a short time around 300°C in a nitrogen atmosphere.

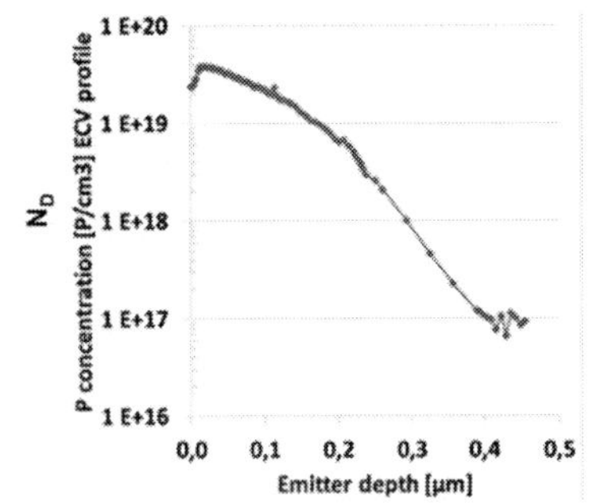

Figure 2: Modified emitter doping profile for plating with reduced surface concentration and deeper diffusion.

All process steps from pretreatment to the deposition of the silver layer are single side processes so that the rear side of the solar cells is kept dry. As precursors, 156x156 mm² p-type Cz wafers with a modified doping profile for plating were used. A reduced surface concentration and deeper diffusion results in a 120 Ω/sq emitter (Figure 2). Compared to screen printing, the nickel layer can contact lower doped surfaces. On the rear side of the wafers a full Al BSF had been formed.

150 fingers were lasered and three busbar layouts were used: 3 x 1.5 mm, 4 x 1 mm and 5 x 0.9 mm width. For the 3 and 5 busbar case the shading is equal for four busbars lower.

4 EXPERIMENTAL RESULTS

4.1 Metal mass variation

To find the required deposited amount of metal, several cells were processed with varying metal mass. Considering that a low mass would show itself in a reduction of the fill factor (FF), 80.2% was chosen as a criterion for FF (Figure 3). For each busbar design there is a range of masses that mostly give fill factors of 80.2% or above.

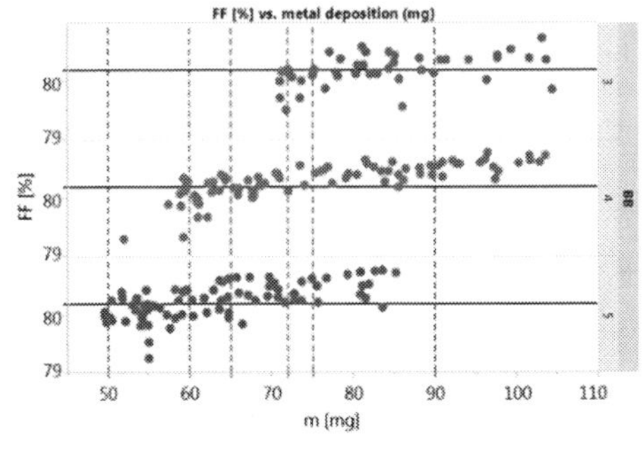

Figure 3: FF plotted against the deposited metal mass for three (red), four (green) and five (blue) busbars. The black solid lines mark the chosen criterion of 80.2% respectively

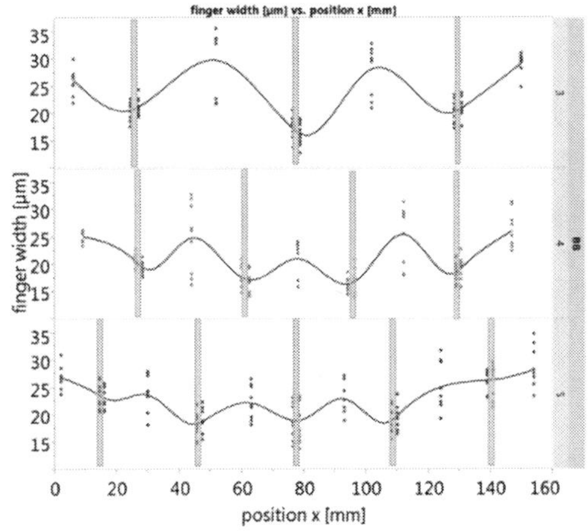

Figure 4: Finger width vs. position on the wafer for three, four and five busbar cells. The vertical grey bars mark the position of the busbars on the cell. Solid line are guides to the eye for the assumed finger width.

Table 2: Metal deposition from simulation and experiment and resulting metal reduction.

	Metal deposition (mg)		Metal reduction(%)	
No. of busbars	Simulation	Experimental	Simulation	Experimental
3	130	75-90	0	0
4	65	60-72	50	20
5	46	50-65	65	30

The comparison of simulation and experiment reveals significant deviations for all busbar designs (Table 2). One possible explanation is that the finger width is not uniform across the wafer. The current distribution seems not to be uniform across the wafer surface. Another explanation is that next to the busbars the finger width is narrower than halfway between the busbars (Figure 4).

Furthermore close to the busbar the layers grown on the individual laser pulses can still be seen like beads on a string (Figure 5).

In addition to the reduced metal laydown, lesser shading increases the cell efficiency (Figure 6). The four busbar layout has less shading from the busbars than the other two and therefore performs better. Best cell efficiencies achieved are up to 19.5%.

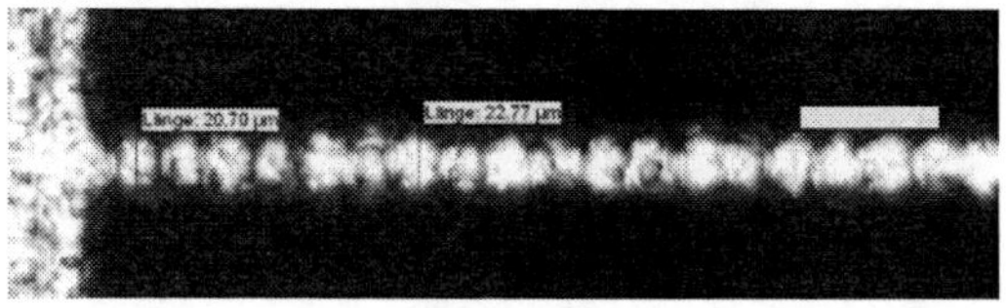

Figure 5: Microscope image of a finger close to a busbar.

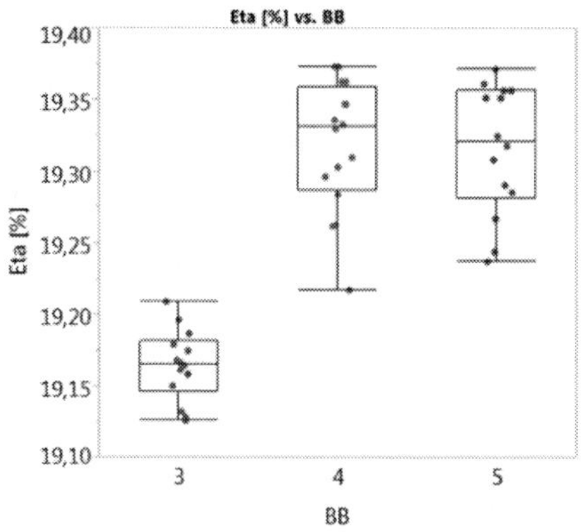

Figure 6: Efficiency gain with increased no. of busbars.

4.2 Finger and busbar adhesion testing

In the past, plating often did not show adequate adhesion compared to other metallisation methods (5). With progress in processing however it has been shown that plating can achieve adhesions high enough to enable mass production (7). Adhesion is tested on the fingers and after soldering on the busbars. The method used for the fingers is based on the IPC-TM-650 standard used in printed circuit board testing, the difference being that the adhesive tape used is from a different manufacturer.

The number of fingers delaminated by the test divided by the total number of fingers tested gives the pull off ratio (Figure 7). Wafers with less mass perform better in the tape test due to a lower finger pull of ratio. Primarily this can be explained by reduced mechanical stress in the metal stack. Another reason is the increased adhesive force of the tape to broader fingers (higher mass). Compared to a second method of testing using a stylus (8), the results differ (Figure 8). In this case there is no difference between the two investigated groups apparent so further investigation is necessary.

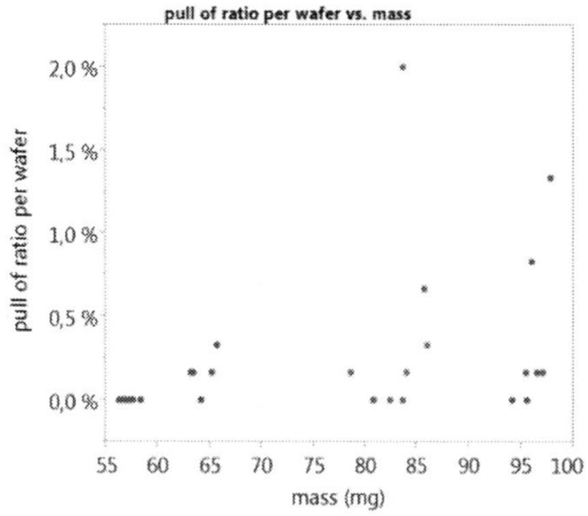

Figure 7: Finger pull off ratio vs. metal mass for three (red), four (green) and five (blue) busbar layouts.

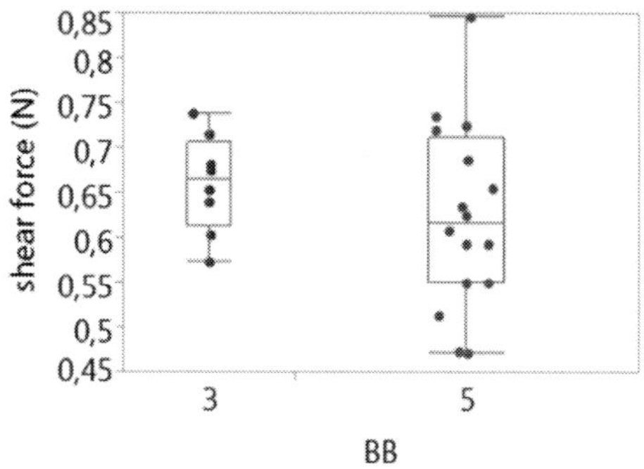

Figure 8: Shear force on fingers, tested for a three busbar (red) and five busbar cell (blue).

To solder the wafers, a semiautomatic soldering station was employed. The temperature at the solder tips is about 300°C, 1.5 mm x 0.2 mm Cu ribbons with SnPb 60/40 solder were used. Solder adhesion is tested in a 90° peel test according to IEC 61189-3 (9). The results in Figure 9 show that the peel force changes with the number of busbars. However the peel force per busbar width is mostly above the threshold value of 1 N/mm which is considered sufficient for module assembly. Manual assembly is responsible for the majority of the lower values, on the other hand the peel force is partly high enough to cause wafer fracture (Figure 10).

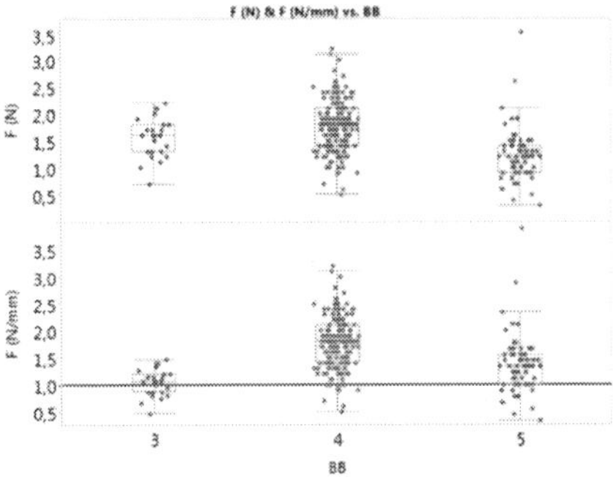

Figure 9: Busbar adhesion measured with 90° peel tests after soldering.

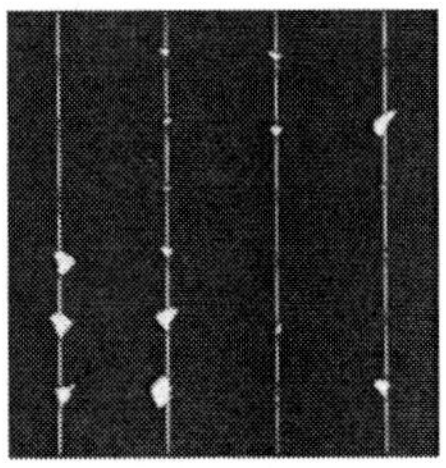

Figure 10: Wafer after peel test showing wafer fracture.

7 COST OF OWNERSHIP

If plating wants to successfully compete against screen printing not only the performance of the cells has to be on equal level but the technology has to be financially attractive. For the three busbar design, the cost of ownership is compared for both technologies. To estimate the costs of screen printing, a silver price of 480 €/kg was taken and a silver paste consumption of 90 mg (1) with double print was assumed. In this scenario no efficiency gains are accounted for. So in total plating adds up to 9 €ct/wafer versus 12.4 €ct/wafer for screen printing (Figure 11). The largest share for screen printing is the cost for the silver paste which raises the part of the consumables to over 70% (Table 3). On the other hand, tool costs are higher for the plating cluster and are almost the same as the cost for consumables. All other factors such as labour, floor space, facility, waste treatment etc. are summed up in the category others. Although waste water treatment is expected to draw some attention in the near future, the expenses associated with it are only estimated to be 1.5% of the total cost.

Taking the metal deposit from the experimental data, the cost of metallisation for plating is expected to diminish to 8.6 €ct/wafer going from three to five busbars. With the lower metal deposit, the process time in the plating bath is reduced and therefore also the consumption of chemicals. Additionally the plating bath can also be shortened leading to a shorter and cheaper plating tool. The reduction in consumables makes up about 75% of the savings, the other 25% going to the tool. The efficiency gain from lesser shading for the increased number of busbars constitutes an additional benefit.

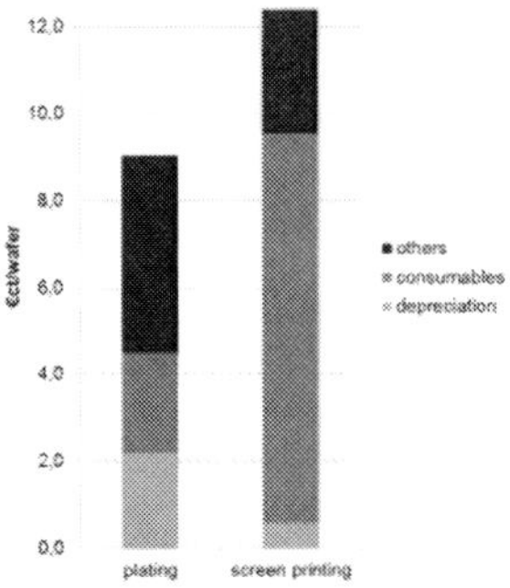

Figure 11: Comparison of cost of ownership for plating and screen printing. The category others includes factors as labour, floor space, facility and waste treatment.

Table 3: Current distribution of the costs for plating and screen printing technology.

	Consumables (%)	Depreciation (%)	Other (%)
plating	25.5	24.4	50.1
screen printing	72.2	4.8	23

8 CONCLUSION

It is possible to process wafers for four and five busbar designs with equal or better performance than wafers with the prevalent three busbar grid design. Improvements in efficiency through lesser shading and finger adhesion have been seen. Additionally the required metallisation can be reduced which improves the manufacturing costs. Not only is the amount of consumables is decreased but also the production tool can be shortened.

9 REFERENCES

[1] International Technology Roadmap for Photovoltaic (ITRPV). [Online] 2015. [Cited: 10.05.2016.] http://www.itrpv.net/Reports/Downloads/2016/.

[2] Braun, Stefan, et al., The multi-busbar design: an overview. *Energy Procedia.* 43, 2013.

[3] *Assessing the impact of multi-busbars on metallization cost and efficiency of solar cells with digital inkjet-printed gridlines.* Chen, Nian, et al [ed.] IEEE. 2013. High Capacity Optical Networks & Enabling Technologies (HONET-CNS)

[4] Mette, Ansgar. New concepts for front side metallization of industrial solar cells. Freiburg: Fraunhofer ISE, 2007.

[5] Cimiotti, G., et al., Design Rules for Solar Cells with Plated Metallization. *Energy Procedia.* 2015, Bd. 67, S. 84-92.
[6] Lennon, AlisonYao, Yu and Wenham, Stuart. Evolution of metal plating for silicon solar cell metallisation. *PROGRESS IN PHOTOVOLTAICS: RESEARCH AND APPLICATIONS.* 2012.
[7] L. Tous et al., Large area copper plated silicon solar cell exceeding 19.5% efficiency. Elsevier. *Energy Procedia.* 2012, Vol.21, pp. 58-65.
[8] *Design and Characterization of an Adhesion Strength Tester for Evaluating Metal Contacts on Silicon Solar Cells.* Young, T., Hee K., Lennon A., Egan R., Wilkie O., Yao Y.. Denver USA: s.n., 2014. 40th IEEE Photovoltaic Specialists Conference .
[9] *Improved quality test method for solder ribbon interconnects on silicon solar cells.* Wendt, J. et al., Las Vegas: IEEE, 2010. Thermal and Thermomechanical Phenomena in Electronic Systems (ITherm), 2010 12th IEEE Intersociety Conference.

ACKNOWLEDGMENT
The authors kindly thank all industrial partners for their support.

Metal Wrap Through Heterojunction Solar Cell with Plated Electrode

°Fumiharu Ishimura[1], Wenjun Li[1], Eiji Kobayashi[1], Kimikazu Hashimoto[1], Seiji Sato[1], Yoshimi Watabe[1], Evert Bende[2], and Gianluca Coletti[2],

[1]Choshu Industry Co., Ltd., 3740, Shin-yamanoi, Sanyo Onoda, Yamaguchi 757-8511, Japan

[2]ECN Solar Energy, Petten, NL-1755 LE, the Netherlands

E-mail : ishimura.fumiharu@choshu.co.jp

ABSTRACT: We report our recent efforts to improve cell and module efficiency by Metal Wrap Through technology. We implemented this work in the cooperation with Energy Research Centre of the Netherlands, ECN. Plating technologies were applied in a MWT-SHJ cell, which achieved a cell efficiency of 23.3%. A single cell module was fabricated and achieved a module efficiency of 21.9%. Current collecting electrodes were formed by electrolytic copper plating and electrolytic tin plating. A significantly low contact resistance of 10^{-5} $\Omega \cdot cm^2$ on TCO was achieved. Line width of 23 μm and line height of 21 μm of plated electrode was formed for MWT-SHJ cell.

Keywords: Heterojunction, Back contact, Electroplating

1. Introduction

In order to improve the efficiency of solar cell, it is essential to reduce the resistance loss and shadow loss. Usually, the low temperature silver paste with high resistivity has been used for silicon heterojunction cells. An applying the electroplated electrode with low resistivity would be expected to improve the cell efficiency. Metal wrap through (MWT) technology as shown in figure 1 is one of the effective method to improve the cell and module's efficiency. The via holes created inside the cell contribute to reduce the distance across electric conduction. Therefore, integration of electroplating technology and MWT technology for silicon heterojunction cells would be expected to achieve high efficiency cells and modules.

In this paper, we report our recent achievements of plating technology and MWT technology for silicon heterojunction cell.

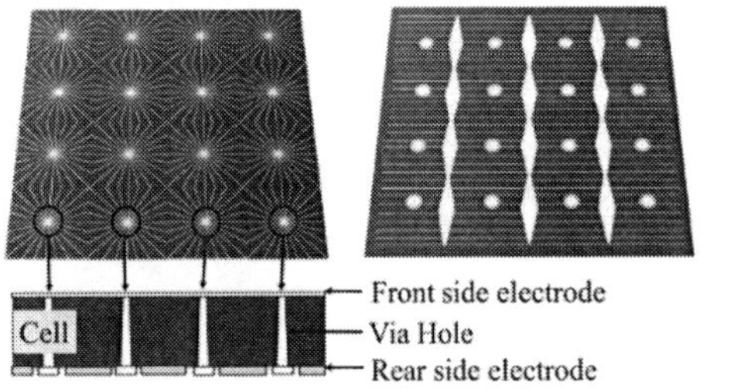

Figure 1 Schematic of metal wrap through cell: Via holes were created inside of cell. The front side electrode is connected the rear side electrode through the via holes.

2. Experimental

Figure 2 shows the MWT cell process applying the electroplating technology. These MWT processes were advanced based on published data[1]. After the TCO deposition, APC was deposited on the TCO as electrical contact layer. APC is a kind of silver based alloy consisted of silver, palladium and copper. Mask pattern for plated electrode was formed by photolithography of dry film resist. Copper plating was followed by tin plating. Width and height of plated electrode were almost 20μm, 20μm, respectively. APC layer was removed on the wet etching without under the plated electrode. Rear side electrode pattern was formed by screen printing of silver paste. In this experimental, rear emitter silicon heterojunction cells were prepared and applied 4x4 structure of metal wrap through as shown figure 3.

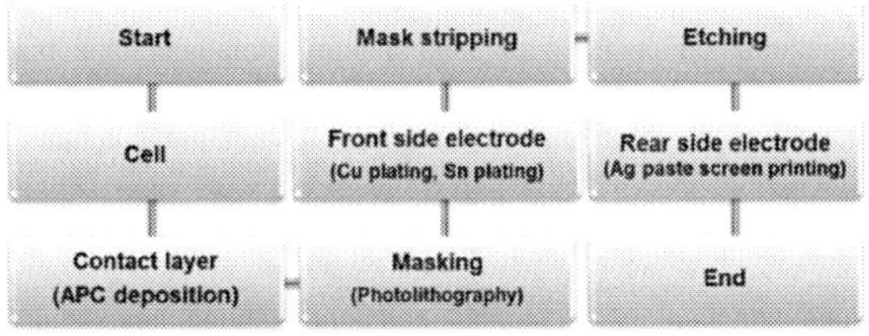

Figure 2 Plating processes for MWT-SHJ cell.

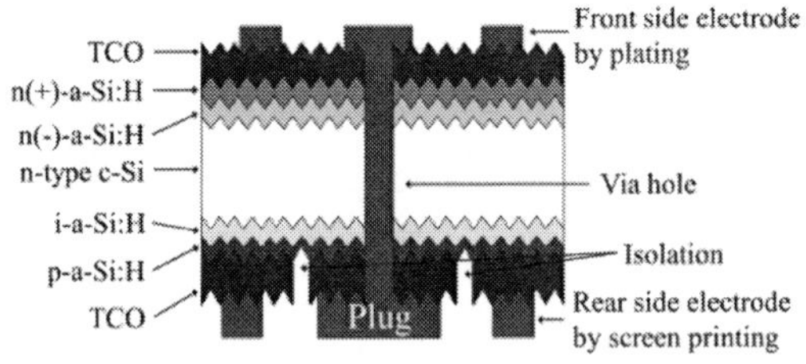

Figure 3 Schematic drawings of the rear emitter silicon heterojunction cell applying the MWT

3. Results

3.1 Contact resistance

Contact resistance of plated electrode was evaluated by TLM method. Table 1 shows contact resistance of Sn / Cu / APC / TCO structure and low temperature Ag paste / TCO structure. The contact resistance of plated electrode fabricated on the process of figure 2 was reduced than that of low temperature Ag paste / TCO structure. Contact resistance of Sn / Cu / APC / TCO doesn't contribute to decrease the FF for less than 30μm width of finger electrode. Additionally, cost impact for cell mass production is definitive because consumption of APC was low.

Table 1 Contact resistance of Sn / Cu / APC / TCO structure and low temperature Ag paste / TCO structure: The contact resistance of plated electrode was reduced than that of low temperature Ag paste / TCO structure.

Structure	Contact resistance / Ω·cm^2
Sn / Cu / APC / TCO	1.1×10^{-5}
LT Ag paste / TCO	6.0×10^{-4}

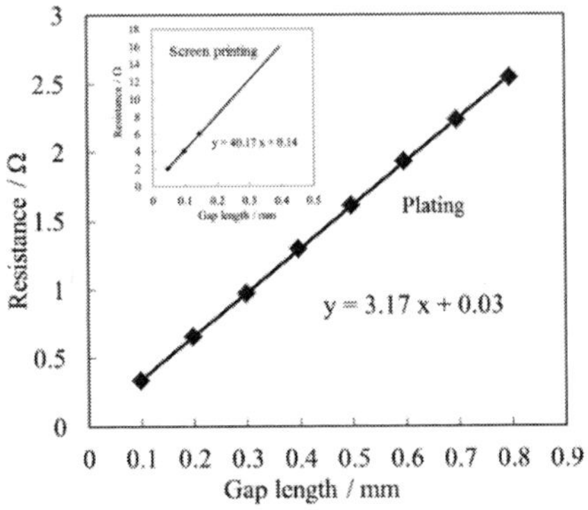

Figure 4 Contact resistance of TCO / electrode was measured by TLM method. Measurement pattern was formed by photolithography and printing, respectively.

3.2 Cell and module characteristics

Figure 5 (a) shows the front side of metal wrap through cell. Fine line finger electrodes were formed by electroplating. Figure 5 (b) shows the rear side of MWT cell. Fingers, 3 busburs and plugs were printed Ag paste. Figure 6 shows microscope image of plated finger electrode. Line width of 23 μm and line height of 21 μm was observed. High aspect ratio of finger shape was obtained because vertically dike was formed by dry film resist of photolithography.

(a) (b)

Figure 5 (a) Front side of MWT cell: Fine line finger electrodes were formed by electroplating. (b) Rear side of MWT cell: Fingers, 3 busburs and plug were printed Ag paste.

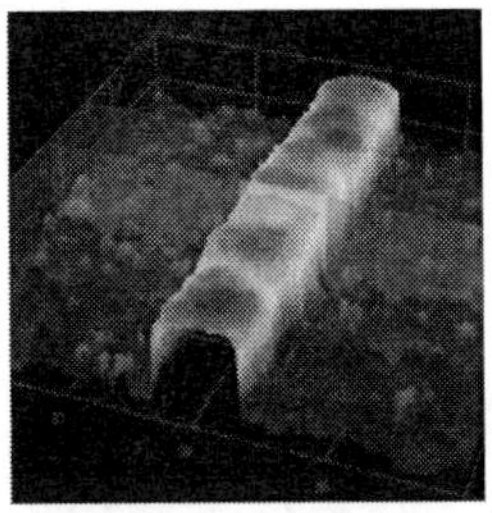

Figure 6 Microscope image of finger electrode: Line width was 23 μm and line height was 21 μm. high aspect ratio of finger was observed because vertically dike was formed by dry film resist of photolithography.

Figure 7 shows the output characteristics of our best MWT cell applying the plated electrode measured by Pasan spot light cell tester with PCB touch system and a single cell module. The module was consisted of as shown in figure 8. We obtained a high *I*sc and a high FF in the cell's IV characteristics because of fine line electrode reducing the resistive losses by plated electrode technology and MWT technology. As a result, the MWT cell with plated electrode had high efficiency, 23.3 %. However, Module efficiency was lower than we expected because FF was decreased drastically. FF might be maintained the cell's FF in MWT module because high conductivity Cu foil is located as instead of interconnector. The optimization of module fabrication such as contact resistance between conductive adhesive and Cu foil backsheet would be necessary.

Figure 9 shows the EL image of single cell module. There was no dark area and no crack in EL image.

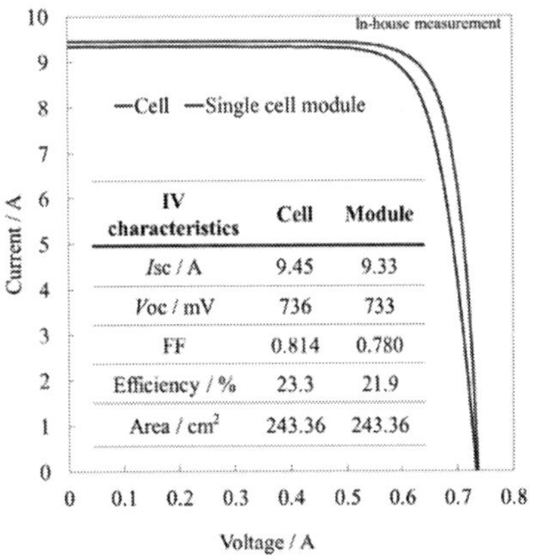

Figure 7 Cell efficiency of 23.3 % and single cell module efficiency of 21.9% were achieved with 6-inch the rear-emitter SHJ structure with MWT technology and plating electrode. (In-house measurement).

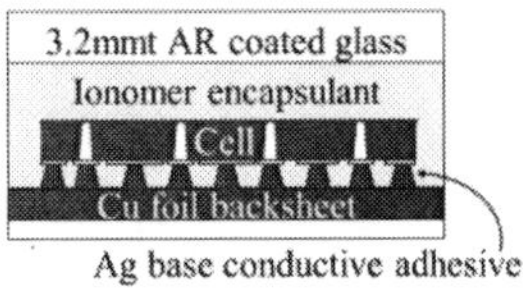

Figure 8 Module structure of MWT: It consists of AR coated glass, ionomer encapsulant, MWT cell, Ag base conductive adhesive and Cu foil backsheet. Standard Cu foil backsheet for MWT was used. Conductive adhesive assume the connection between cell's rear side and Cu foil backsheet.

Figure 9 EL image of single cell module. There was no dark area and no crack in EL image

4. Conclusion

The MWT cell applying the plated electrode for heterojunction cells has been investigated to find the possibility of high efficiency cell and module.

A best MWT cell was obtained 23.3% of cell efficiency. However, optimization of module fabrication would be necessary because the large Cell-To-Module loss was observed.

REFERENCE

1) Coletti, G. *et al.*, IEEE JOURNAL OF PHOTOVOLTAICS, **5** (1) 55-60 (2015)

IMPLANTATION OF PHOSPHORUS INTO PYRAMIDAL TEXTURE IN SILICON SOLAR CELL

*Katsuto Tanahashi, Masaaki Moriya, Yasuhiro Kida, Tetsuo Fukuda, Katsuhiko Shirasawa, and Hidetaka Takato
Renewable Energy Research Center, National Institute of Advanced Industrial Science and Technology (AIST)
2-2-9, Machiikedai, Koriyama, Fukushima 963-0298, JAPAN

ABSTRACT: We investigate the impacts of ion beam irradiation methods in processes of phosphorus (P) implantation on P-emitter silicon solar cell performance. P implanted emitter was fabricated on *p*-type CZ silicon wafer. Following the wafer texture, P^+ ions were implanted with tilt and step-by-step wafer rotation. The textured solar cell has a three-dimensional shape of tetrahedral pyramids in surface. Incident angle of ion beam on the textured surface depends on tilt and wafer rotation. Thus, tilt and wafer rotation are the key factors of as-implantation depth profile, resulting in the improvement of cell efficiency. We found cell efficiency increased with increase of tilt angle within a range of 7° to 35°, or decreased of rotation angle within a range of 15° to 90°. Cell efficiency is 19.15% as implantation of P is performed with an acceleration voltage of 3 keV, a total dose of 4×10^{15} atoms/cm^2, a tilt angle of 35°, and a rotation angle of 15°.

1 INTRODUCTION

Ion-implantation-based silicon solar cells have been proposed, such as selective emitters [1], passivated emitters, and rear totally-diffused (PERT) cells [2-4], passivated emitter and rear cell (PERC) [5], and back-contacted cells [6]. These papers reported that the number of processing steps during the fabrication of solar cells decreased by replacing diffusion processes with ion implantation. A simplification of the fabrication processes reduces the cost of solar cells.

Ion implantation technology has been widely used in ultra-large scale integrated circuit (ULSI) processing. Effective technologies of ion implantation are implemented to form various structure of advanced transistors such as shallow junction, deep well, and embedded stressor; pocket implantation, multiple implantation, co-implantation and so on. From the point of view of ion implantation technology, silicon solar cells have two characteristics which are different from LSIs. One is surface morphology, and the other is post-implantation annealing (PIA).

Silicon wafers for LSIs are mirror polished flat surface. However, solar cells have a textured surface composed of upright tetrahedral pyramids. They consist of four (111) planes with an inclination of 54.7° to the (100) plane. Ion implantation involves several factors that affect the dopant distribution, such as the acceleration energy, dose, tilt, and wafer rotation. Fuse et al. investigated the depth profiles of boron (B) with large tilt-angle implantation in order to control the narrow width effect of n-channel MOSFETs by sidewall implantation dose [7]. They concluded that as-implantation B profiles with large tilt-angle implantation up to 82° were different from the conventional ones below 60° tilt-angle implantations in the surface regions due to the forward scattering effect. Krügener et al. performed structural investigation of ion implantation of B on random pyramidal textured silicon wafer for photovoltaics [8]. They found the impacts of tilt angle and screening oxide on local doping profiles, resulting in the secondary defect distribution after post-implantation-annealing (PIA). Thus, surface morphology also has an influence on as-implantation profile and residual defects.

In fabrication of LSI, Rapid Thermal Processing (RTP) is employed for PIA in order to keep as-implantation profiles. Recently, we observed two-dimensional carrier distribution of P-implanted emitter in silicon solar cell with Al-BSF by Scanning Capacitance Microscopy [9] and Super-higher-order Scanning Nonlinear Dielectric Microscopy [10]. According to SCM observation, junction depth of sidewall of texture is 0.59 μm [9]. Junction depth of solar cell is an order of magnitude deeper than as-implantation depth profile. Therefore, furnace annealing is generally used as PIA in order to make a deep junction. Acceleration voltage of ion implantation into flat surface is important factor on cell efficiency. However, not only acceleration voltage but also tilt and rotation are also ones to determine as-implantation profile in textured surface. Moreover, their impacts on cell efficiency are unknown.

In this paper, local doping profiles of P-implanted single pyramidal texture with tilt and rotation are discussed. And we investigate their impacts on silicon solar cell performance of P-implanted emitter with Al-BSF.

2 EXPERIMENTAL

Figure 1 shows the processing steps during the fabrication of a solar cell. P-implanted emitter was fabricated on a *p*-type (boron) CZ silicon wafer with a size of 156 × 156 mm^2, a resistivity of 3 Ω·cm, and a thickness of 180 μm. Following the wafer texture, the *n*–type layer was fabricated by implantation of P. Implantation of P was performed with an acceleration voltage of 3-10 keV, a dose of 4×10^{15} atoms/cm^2, and a tilt of 7-35°. The sample was subjected to isothermal annealing in a nitrogen atmosphere at 900 °C for 10 min in order to recover the damage, activate the P atoms, and make *p-n* junction. Then, a SiNx layer was deposited on the implanted textured surface by using plasma-enhanced chemical vapor deposition (PE-CVD).

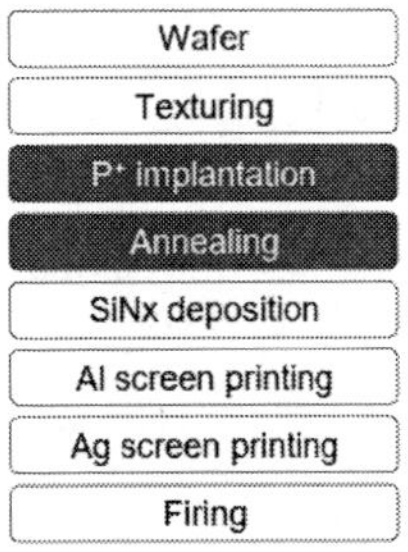

Fig. 1 Processing steps during the fabrication of a solar cell.

In this paper, ion implantation with step rotation is implemented, which is step-by-step wafer rotation during ion implantation processes as shown in Fig. 2. Table 1 summarizes the conditions of rotational ion implantation. Rotation angles are (a) 15° a step, (b) 35° a step and (c) 90° a step, respectively. The case of rotation angle of 15° is explained as follows. Implantation of P is performed with a dose of 1.67×10^{14} atoms/cm^2 as the 1st step. Following the 1st step, wafer was rotated by 15°. In the 2nd step, implantation of P is also performed with the same condition as 1st step. Total number of steps depends on the rotation angle; 24 steps for rotation angle of 15°. Wafer also makes one rotation during the process of implantation. Total dose for whole wafer was 4×10^{15} atoms/cm^2, which is the 24 times as dose as 1.67×10^{14} atoms/cm^2.

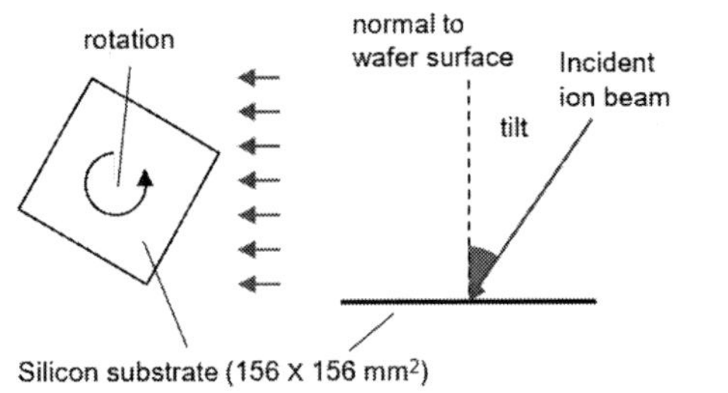

Fig. 2 Schematic illustration of ion implantation into silicon substrate. (a) plane view and (b) cross section.

Table 1 Rotation angle, number of steps, and dose in ion implantation with step rotation.

	(a)	(b)	(c)
Rotation angle (degree)	15	35	90
Number of steps (times)	24	12	4
Dose/step (x10^{14} /cm^2)	1.67	3.33	10
Total dose (x10^{15} /cm^2)	4	4	4

3 RESULTS AND DISCUSSION

3.1 As-implantation local P profile of in texture

First, local profiles of P in as-implanted single pyramidal texture are discussed. Figure 3 shows schematic illustration of a single textured pyramid during implantation of P under a certain tilt angle and a certain rotation angle. Upright tetrahedral pyramids consist of four (111) planes with an inclination of 54.7° to the (100) silicon surface. Incident angle of ion beam into the textured (111) surface, α, depends on not only tilt, but also wafer rotation. Incident ion beam angle to (111) plane, α, is changed following wafer rotation, i.e. rotation of texture.

For simplicity, single texture is assumed to be rotated around the perpendicular line from the top of pyramid to the bottom, and α is calculated in each step (not shown). Figure 4 shows the simulated as-implantation local P profile of single texture in each rotation step by transport of ions in matter (TRIM) calculation [11]. Horizontal axis, depth, corresponds to the distance perpendicular to the (111) surface of side wall of texture. Projection range decreases and concentration of P increases in each step with decrease of α as denoted by arrow in Fig. 4. Total P Profile consists of each local profile. It becomes homogeneous with increase of the total number of rotation steps.

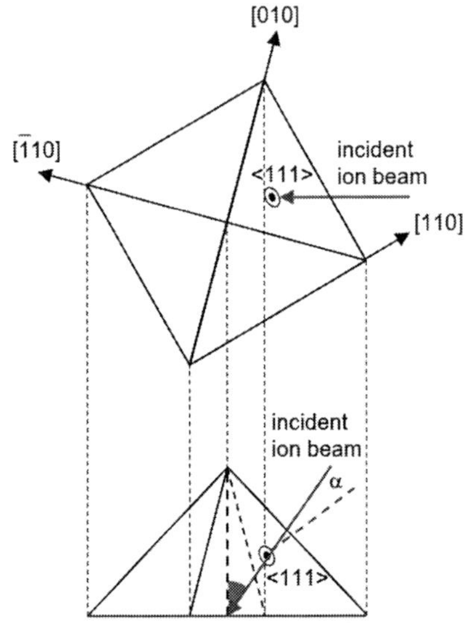

Fig. 3 Schematic illustration of a single textured pyramid during phosphorus implantation under a certain tilt angle and a certain rotation angle. (a) plane view and (b) cross section.

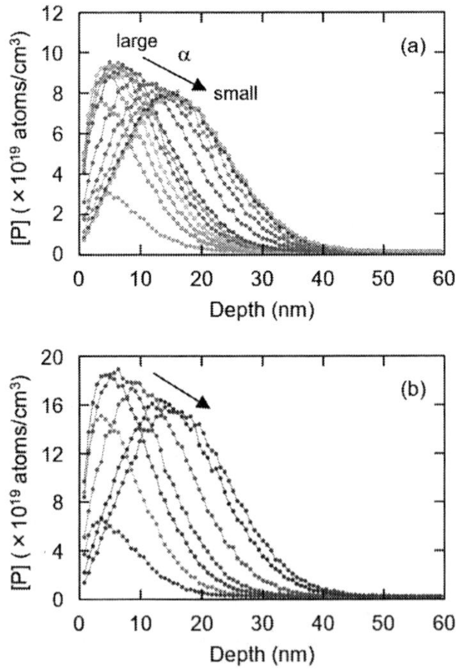

Fig. 4 Simulated as-implantation local P profiles of single texture in each rotation step with an acceleration voltage of 10 keV, and a tilt of 35°, a rotation angle, a dose of (a) 15°, 1.67×10^{14} atoms/cm^2, (b) 30°, 3.33×10^{14} atoms/cm^2.

3.2 Characterization of P-implanted region

Figure 5 shows Transmission Electron Microscopy (TEM) images of a textured surface implanted with an acceleration voltage of 3 keV, a dose of 4×10^{15} atoms/cm^2, a tilt of 35°, and a rotation angle of 15°. Amorphous layer with constant depth of about 10 nm in thickness is clearly observed along the textured surface as shown in Fig. 5(a). As dose exceeds critical value, amorphous layer is formed. Amorphous layer with constant depth reflects the P-implanted layer with constant depth along the textured surface. After PIA at 900°C for 10 min in nitrogen atmosphere, atomic arrangement is observed just under the SiNx layer. This means above annealing condition is enough to recover the damage due to implantation of P under the condition used in this experiment.

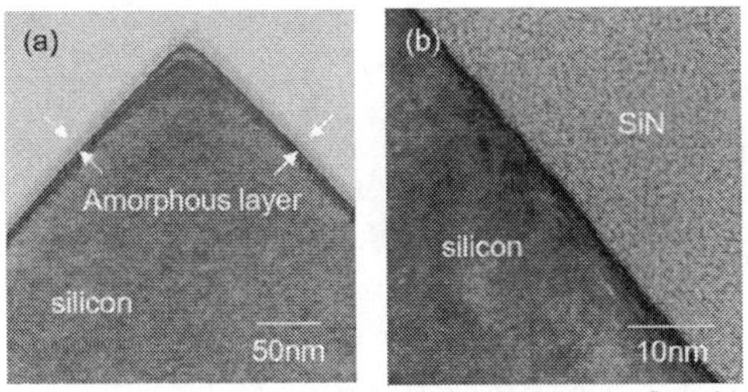

Fig. 5 TEM images of a P-implanted texture surface (a) before and (b) after annealing at 900°C for 10 min in nitrogen atmosphere. Dark contrast just under (a) amorphous layer and (b) SiNx layer are due to not damages, but surface roughness of silicon. Image (b) corresponds to the side wall of upright tetrahedral pyramid.

3.3 Solar cell performance of P-implanted emitter

Fig. 6 shows the simulated as-implanted local P profiles of single texture with a dose of 1.67×10^{14} atoms/cm^2, α of 17.24°, acceleration voltages of 3 keV, 5 keV, and 10 keV. α =17.24° is incident angle with a tilt of 35°and total rotation angle of 30°, or a tilt of 35° and total rotation angle of 60°. Roughly speaking, change of as-implantation depth profile due to one rotation of texture corresponds to that of acceleration voltages of 3-10 keV from the comparison between Fig. 4(a) and Fig. 6. Therefore, effect of tilt and rotation on cell efficiency should be considered.

Figure 7 shows the dependence of tilt and rotation angles on the cell efficiency of P-implanted emitter with an acceleration voltage of 3 keV, a total dose of 4×10^{15} atoms/cm^2. In Fig. 7(a), horizontal axis is the tilt angle, which corresponds to an angle of intersection between incident ion beam and normal angle to wafer (100) surface. Vertical axis in Fig. 7(a) is cell efficiency. In Fig. 7(b), horizontal axis is the rotation angle of wafer in each step during ion implantation with step rotation processes, which corresponds to rotation angle in Table 1. Cell efficiency increases with increase of tilt angle within a range of 7° to 35°, or decrease of rotation angle within a range of 15° to 90°. According to Fig. 4, 24 local profiles, and 12 local profiles are overlapped in rotation angles of 15°, and 35°, respectively. Resulting depth profile of P have a tendency to become constant in depth direction with increase of the number of rotation step. This may has a positive effect on the quality of *p-n* junction. We further proceed to investigate the above while optimizing ion implantation as well as annealing conditions. Conversion efficiency of 19.15% has been achieved as implantation of P is performed with an acceleration voltage of 3 keV, a total dose of 4×10^{15} atoms/cm^2, a tilt of 35°, and a rotation angle of 15° as listed in Table 2.

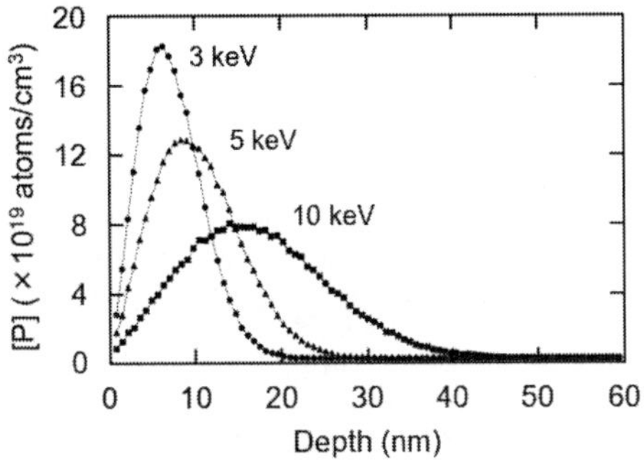

Fig. 6 Simulated as-implanted local P profiles of single texture with a dose of 1.67×10^{14} atoms/cm^2, α of 17.24°, acceleration voltages of 3 keV, 5 keV, and 10 keV.

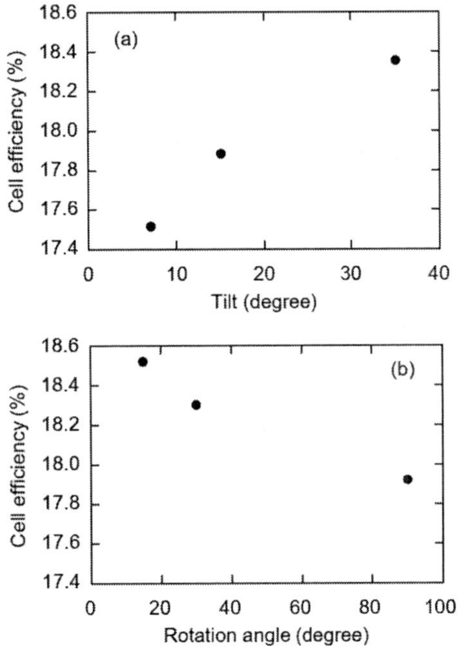

Fig. 7 Cell efficiency as a function of (a) tilt and (b) rotation angles.

Table 2 Electrical characterization of phosphorus-implanted solar cell.

R_{sheet} (ohm/sq.)	57.7
J_{sc} (mA/cm^2)	37.76
V_{oc} (mV)	631.8
FF (%)	80.28
η (%)	19.15

4 CONCLUSIONS

In conclusion, we investigated the impacts of ion beam irradiation methods in processes of P implantation on cell characteristics. Ion implantation with step rotation was implemented, which was step-by-step wafer rotation during ion implantation processes. Local profiles of P in as-implantation single pyramidal texture were discussed based on the simulated profiles.

We found cell efficiency increased with increase of tilt angle within a range of 7° to 35°, or decreased of rotation angle within a range of 15° to 90°. Tilt and wafer rotation are the key factors of as-implanted profile in depth direction, resulting in the improvement of cell efficiency. Conversion efficiency of 19.15% was achieved as implantation of P was performed with an acceleration voltage of 3 keV, a total dose of 4×10^{15} atoms/cm^2, a tilt of 35°, and a rotation angle of 15°.

4 ACKNOWLEDGEMENT

This work was supported by New Energy and Industrial Technology Development Organization.

REFERENCES

1. Su S-P, Chantarat N, Huang Y-H, Kang C-M, Chen Sean H T, Cheng L-W. N-TYPE AL-ALLOYED REAR JUNCTION SILICON SOLAR CELLS WITH IMPLANTED FRONT SURFACE FIELD, in *38th IEEE Photovoltaic Specialist Conference (PVSC)* 2012; pp. 1010-1013.
2. Ok YW, Upadhyaya AD, Tao Y, Zimbardi F, Ning S, Rohatgi A. Ion-implanted and Screen-Printed Large Area 19.6% Efficiency N-type Bifacial Si Solar Cell, in *38th IEEE Photovoltaic Specialist Conference (PVSC)* 2012; pp. 2240-2243.
3. Chen C-C, Lin C-L, Chien J-W, Chen Y-T, Lin Y-J, Huang F-M, Wu H-C. High efficiency on boron emitter n-type Cz silicon bifacial cells with industrial process, *Energy Procedia* 2013; **38**, pp. 416-422.
4. Benick J, Muller R, Bateman N, Hermle M. FULLY IMPLANTED N-TYPE PERT SOLAR CELLS, in *27th European Photovoltaic Specialist Conference and Exhibition (EuPVSC)* 2012; pp. 676-679.
5. Tao Y, Ok Y-W, Zimbardi F, Upadhyaya AD, Upadhyaya VD, Rohatgi A. Fully Ion-Implanted and Screen-Printed 20.2% Efficient Front Junction Silicon Cells on 239 cm^2 n-Type CZ Substrate, *IEEE Journal of Photovoltaics* 2014; **4**(1) pp. 58-63.
6. Young D-L, Nemeth W, LaSalvia V, Reedy R, Essig S, Bateman N, Stradins P, Interdigitated Back passivated Contact (IBPC) Solar Cells Formed by Ion Implantation, *IEEE Journal of Photovoltaics* 2016; **6**(1) pp. 41-47.
7. Fuse G, Umimoto H, Odanaka S, Wakabayashi M, Fukumoto M, Ohzono T. Depth Profiles of Boron Atoms with Large Tilt-Angle Implantations, *Journal of the Electrochemical Society* 1986; **133**(5) pp. 996-998.
8. Krügener J, Bugiel E, Osten H-J, Peibst R, Kiefer F, Ohrdes T, Brendel R. Structural Investigation of Ion Implantation of Boron on Random Pyramid Textured Si(100) for Photovoltaic Application, in *20th Ion Implantation Technology (IIT) 2014*; pp. 1-4.
9. Tanahashi K, Moriya M, Kida Y, Suzuki N, Fukuda T, Shirasawa K, Takato H. TWO-DIMENSIONAL CARRIER DISTRIBUTION IN PHOSPHORUS-IMPLANTED EMITTER OBSERVED BY SCANNING CAPACITANCE MICROSCOPY, in *31th European Photovoltaic Specialist Conference and Exhibition (EuPVSC)* 2015; pp. 35-38.
10. Hirose K, Tanahashi K, Takato H, Chinone N, Cho Y. Two-dimensional Analysis of Carrier Distribution in Phosphorus-Implanted Emitter and Phosphorus-Diffused Emitter using Super-Higher-Order Scanning Nonlinear Dielectric Microscopy, in *43th IEEE Photovoltaic Specialist Conference (PVSC)* 2016.
11. Ziegler J-F, Biersack J-P, Littmark U. The Stopping and Range of Ions in Solids, New York, Pergamon, 1985.

PASSIVATION OF SILICON SOLAR CELLS VIA LOW TEMPERATURE WET CHEMICAL OXIDATION

Gamze Kökbudak[1,2], Emine Hande Çiftpınar[1,2,3], Olgu Demircioğlu[1,2], Rasit Turan[1,2,3]

[1] The Center for Solar Energy Research and Applications (GÜNAM), Middle East Technical University, Dumlupinar Bulvari no:1, 06800 , Ankara Turkey

[2] Micro and Nanotechnology Graduate Program, Middle East Technical University, Dumlupinar Bulvari no:1, 06800 , Ankara Turkey

[3] Department of Physics, Middle East Technical University, Dumlupinar Bulvari no :1 , 06800 , Ankara Turkey

ABSTRACT: In the development of high efficiency crystalline Si solar cells, decreasing bulk and surface recombination velocities of the minority carriers is vital. As the bulk recombination could be suppressed by enhancing the material quality, the effect of surface recombination on cell performance becomes more dominant. Also, recent studies have revealed that the area under the metal contacted region needs to be passivated to minimize the carrier recombination. The passivation of front and back surface of the cell can be achieved using different techniques and materials. Dry oxidation is a well-known process by the industry and requires high temperature treatment that increases thermal budget of the whole fabrication sequence and also degrades bulk lifetime. For this reason, a low temperature technique like wet chemical oxidation is highly desirable for the growth of thin oxide layers. In this work, 4 different wet chemical oxidation techniques based on nitric acid (HNO_3), hydrogen peroxide (H_2O_2), RCA II (HCl: H_2O_2) and hydrochloric acid (HCl) solutions were studied and corresponding cell performances was compared with that of dry oxidation.

Keywords: c-Si, Passivation, Wet Chemical Oxide, Solar Cell

1 INTRODUCTION

In the development of high efficiency crystalline Si solar cells, decreasing bulk and surface recombination of carriers is vital. Surface recombination effect on cell performance becomes more dominant when the bulk recombination could be moderated by enhancing the material quality. The use of SiO_2 becomes an efficient way of surface passivation with development of new device architectures like PERC, PERT and IBC cells. Also, recent studies have revealed that the area under the metal contacted region needs to be passivated to minimize the carrier recombination. In all these new approaches, the use of SiO_2 appears to be an efficient way of surface passivation for both n and p-type regions. In a recent work, very high efficiency values have been achieved through passivation of the whole back side of the wafer with a thin SiO_2 layer [1]. In the fabrication of this solar cell called TOPCon (Tunnel Oxide Passivated Contact), on n type Si substrate, a thin SiO_2 layer is grown by wet chemical oxidation using a solution containing nitric acid. Similar cell architecture has also been proposed for p-type substrates.

To have efficient carrier collection at the metal contacts, transport of one type of carrier is needed while preventing the transport of the other type. In this way, reducing recombination in the final cell structure is achieved by passivating under the metal contacted region. While designing high-efficiency solar cells, covering full surface with a carrier-selective material using a suitable interface-passivation technique is the most desirable way to achieve effective passivation with less effort.

In this paper, results of a comparison study on different oxidation techniques for the surface passivation of p-type silicon is presented. For this purpose, methods of dry and wet chemical oxidation are employed to form a thin SiO_2 layer onto Si wafer. Dry oxidation is a well-known process by the industry and requires high temperature treatment that increases thermal budget of fabrication and also degrades bulk lifetime. For this reason, a low temperature technique like wet chemical oxidation is highly desirable for the growth of thin oxide layers. This was indeed shown to be effective in the above mentioned study on high efficiency solar cell. In this work, 4 different wet chemical oxidation techniques based on nitric acid (HNO_3), hydrogen peroxide (H_2O_2), RCA II (HCl: H_2O_2) and hydrochloric acid (HCl) solutions were studied and compared with the dry oxidation. Passivation quality of the grown oxide layers was characterized via lifetime and ellipsometric measurements on symmetrical lifetime samples while current-voltage, external quantum efficiency, electroluminescence and SunsVoc measurements were carried out for the characterization of fabricated solar cells.

2 EXPERIMENT

2.1 Cell structure

p-type, solar grade, Cz Si wafers with thickness of 180 μm and dimensions of 156 mm x 156 mm were used as substrate and 7 cells were fabricated for each set of samples. Samples were pyramid textured by alkaline solutions and cleaned using RCA I & II solutions. After texturing, a 50 Ω/□ emitter was formed by Phosphorous diffusion which was followed by PSG removal and additional RCA I & II cleanings. Oxide passivation processes were applied separately for each set of samples. Dry oxidation was conducted at a temperature less than 900°C in a vacuum furnace under $O_2(g)$ flow, whereas wet chemical oxide have been grown at approximately 70°C via immersing the wafers into corresponding solutions of HCl, H_2O_2, RCA II and HNO_3. Oxidation step was followed by deposition of 80 nm thick anti-reflective SiN_x. Screen printing and firing steps were applied as the last steps of cell fabrication procedure which is summarized in Figure 1 with the corresponding cell structure shown in Figure 2.

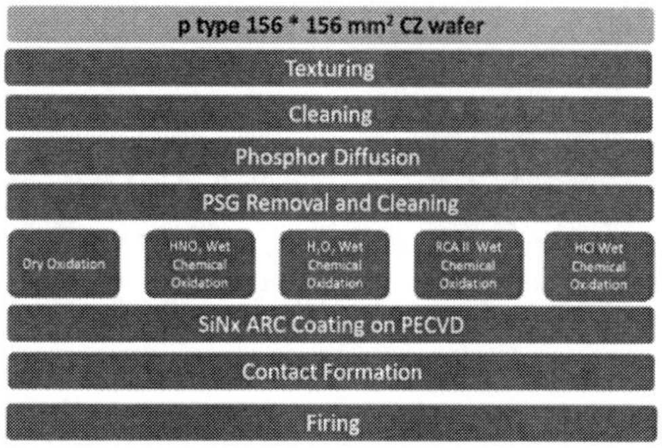

Figure 1: Process sequence of dry and wet chemical oxidized solar cells

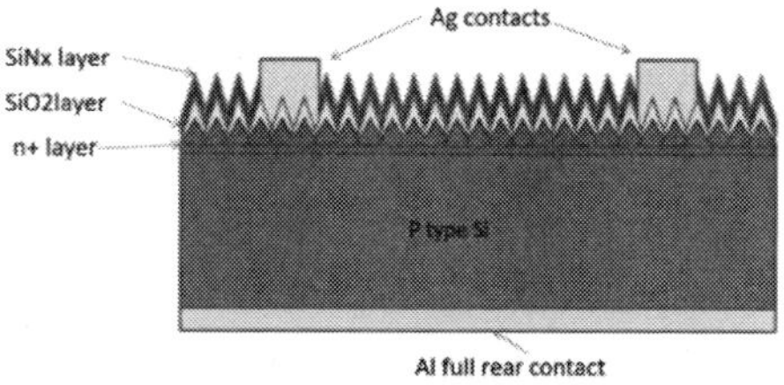

Figure 2: Cross-sectional view of corresponding cell structure

2.2 Lifetime samples structure

Parallel to production of solar cells, effect of oxidation method on passivation quality was investigated via lifetime measurements on symmetrical test structures. For symmetrical test structures, RCA I&II cleanings followed by HF dipping prior to oxidation with dry and wet chemical oxidation techniques. Following the oxidation, nitride deposition and firing steps were carried out to observe the effect of different process steps on passivation quality. Structure of symmetrical lifetime samples is shown in Figure 3.

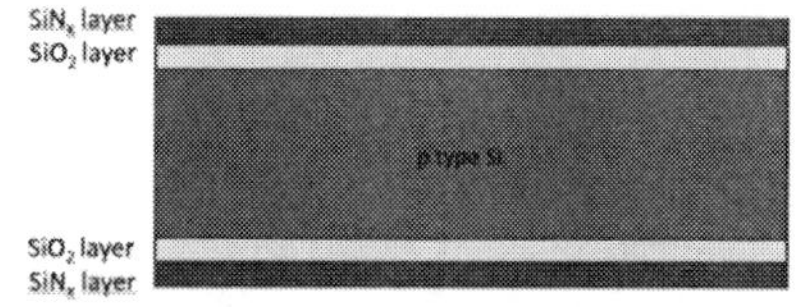

Figure 3: Cross-sectional view of corresponding lifetime sample structure

3 RESULTS

3.1 Lifetime measurements

Figure 4 shows the comparison of lifetime values obtained by dry oxidation and different types of wet chemical oxidation method. Although dry oxidation provided the highest lifetime values, it is apparent that HNO_3 and RCA solution based oxidation provided a comparable level of passivation with a greater performance compared to H_2O_2 and HCl solutions.

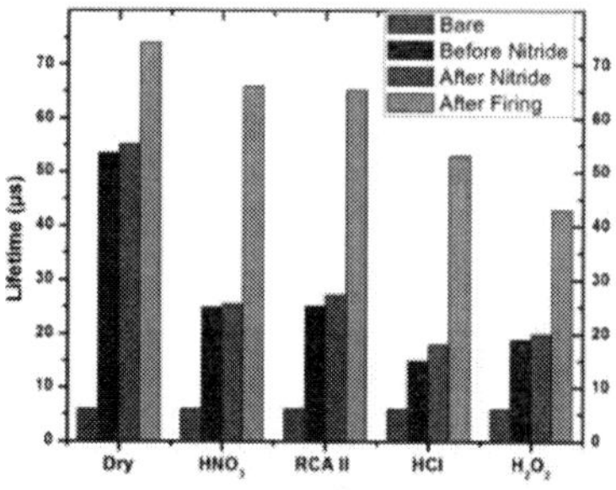

Figure 4: Lifetime values for sequential process steps

Therefore, further cell processes were carried out for 3 sets including dry oxidation and HNO_3 & RCA solution based wet oxidation. Comparable results were obtained for corresponding cell efficiency values for these 3 sets of samples.

3.2 Cell Performances

Although the best performance belongs to dry oxidation set as expected, in the wet oxidation sets, the efficiency results were more stable with less deviation in between different cells of the same group (Figure 5).

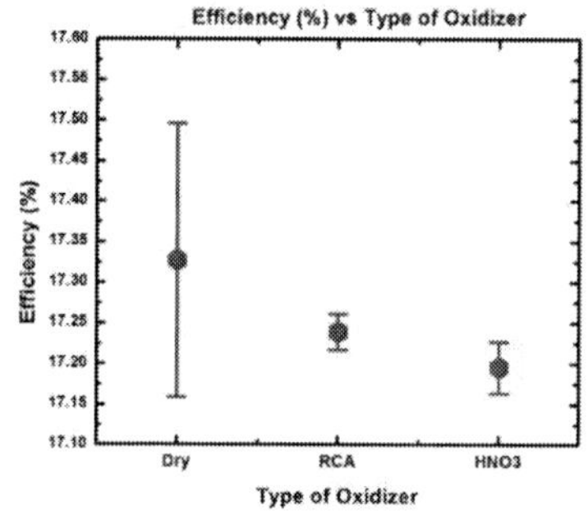

Figure 5: Effect of oxidizer on efficiency

For fill factor, dry oxidation has the best performance as shown in Figure 6.

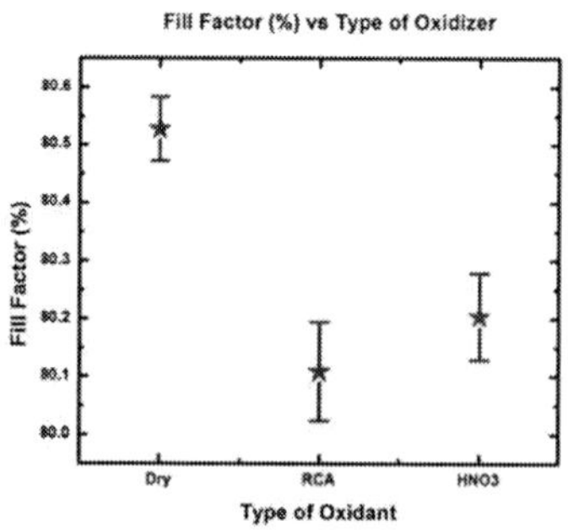

Figure 6: Effect of oxidizer on fill factor

Although the metallization parameters were optimized for dry oxidation sets, wet oxidation sets have comparable fill factor values which could be further enhanced by optimizations for firing temperature and conveyor belt speed. Corresponding Voc values can be seen in Figure 7.

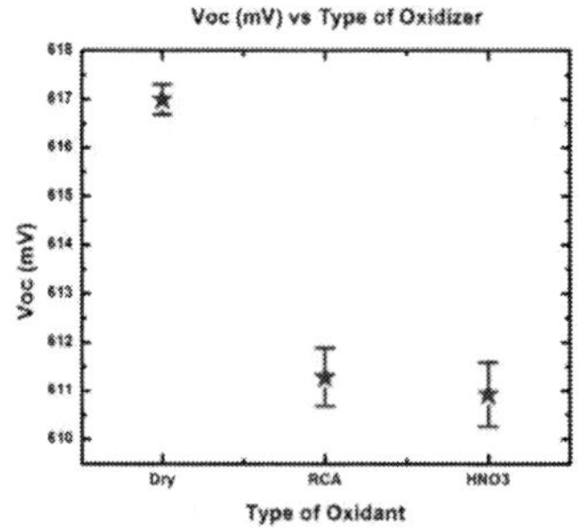

Figure 7: Effect of oxidizer on open circuit voltage

Voc values of wet chemical oxide samples are relatively lower than that of dry oxide samples as expected which is related to poorer film quality of wet chemical oxide films but still providing comparable passivation quality. Considering Figure 8, enhancement in short circuit current density values of wet chemical oxidation samples was observed which could be an indication of reduced series resistance losses for thinner wet chemical oxide layers.

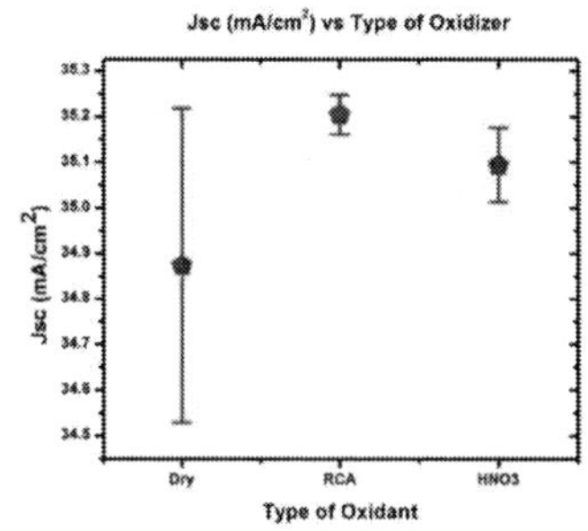

Figure 8: Effect of oxidizer on short circuit current density

In order to be sure about the reason of this enhancement, series resistance analysis and E.Q.E measurements were also carried out and corresponding results are demonstrated in Figure 9 and Figure 10 respectively. Despite an increase in the series resistance, wet chemical oxide samples represent an enhancement in the blue response compared to that of dry oxide passivated samples which could result in the observed enhancement in short circuit current density shown in Figure 8.

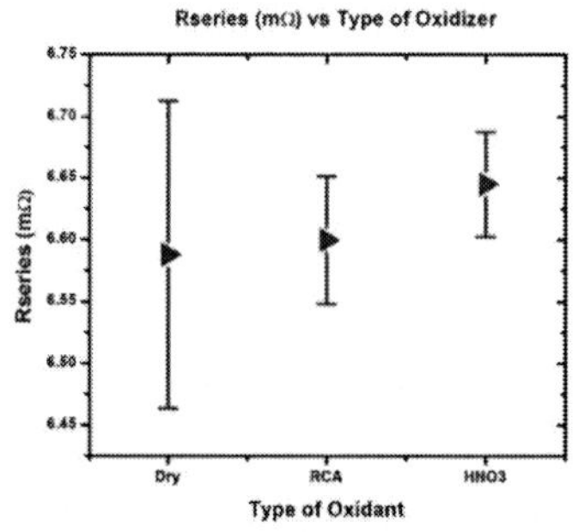

Figure 9: Effect of oxidizer on series resistance

Based on the enhancement in UV regime of E.Q.E plots shown in Figure 10, it was deduced that RCA II based wet chemical oxidation provided slightly better performance in terms of passivation which could be related to lack of high temperature oxidation step promoting the formation of boron-oxide compounds in p-type Si wafers.

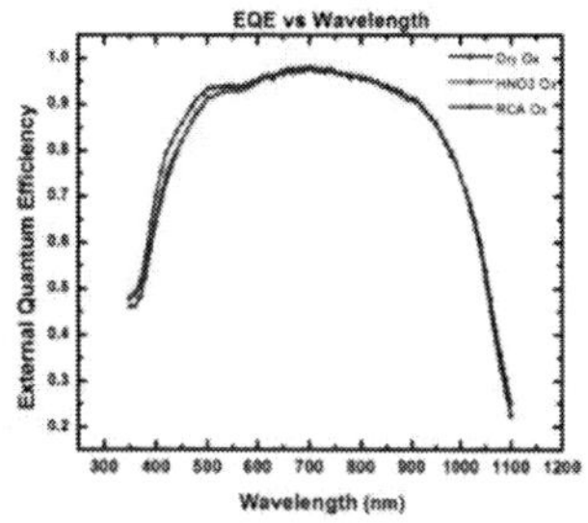

Figure 10: Effect of oxidizer on external quantum efficiency

4 CONCLUSIONS

Considering cell performances and lifetime values, all wet chemical oxide samples showed promising performance in terms of efficiency and open circuit voltages. Voc values of wet chemical oxide samples are relatively lower than that of dry oxide samples as expected which is related to poorer film quality of wet chemical oxide films. Considering Figure 8, enhancement in short circuit current density values of wet chemical oxidation samples was observed. Considering these comparable performance, it could be concluded that wet chemical oxidation would be an alternative passivation method.

5 REFERENCES

[1] Feldmann, F., Bivour, M., Reichel, C., Steinkemper, H., Hermle, M., & Glunz, S. (2014). Tunnel oxide passivated contacts as an alternative to partial rear contacts. Solar Energy Materials and Solar Cells, 46-50.

SURFACE PASSIVATION PROVIDED BY AN ALNEAL THROUGH SiO_2/TiO_2 BILAYER

K.A. Collett, M. Cyrson, R.S. Bonilla and P.R. Wilshaw
Department of Materials, University of Oxford
Department of Materials, University of Oxford, Parks Rd, Oxford, OX1 3PH

The process of annealing a SiO_2 dielectric layer coated in aluminium, termed the alneal, is known to produce some of the most effective surface passivation for silicon. However, it is traditionally performed on a SiO_2 dielectric coating which has sub-optimal anti-reflection properties. In this work it is shown that it is possible to achieve alneal passivation through a double layer SiO_2/TiO_2 stack. TiO_2 was investigated as it is one of the most effective anti-reflection coatings available and its deposition technology is advanced and cost effective. Here, the alneal was carried out on n-type ~40 Ωcm Cz-Si coated with a SiO_2/TiO_2 dielectric stack. In the best case, the alneal produced a lifetime increase from ~15 μs to 3084 μs, which equates to a SRV ≤ 10 cm/s or a J_{0e} of 32 fA/cm^2. This increase in lifetime was comparable to that achieved by a conventional alneal on a single layer SiO_2 specimen. It was also found that the thicker the TiO_2, the less effective the alneal was at passivating. Lastly, the passivation achieved after the alneal on the SiO_2/TiO_2 bilayer was found to be more stable than that achieved on the conventional single SiO_2 layer.
Keywords: surface passivation, silicon solar cell, dielectric coating.

1 INTRODUCTION

The silicon solar cell industry is beginning to adopt high efficiency back contact cell designs which are able to produce higher efficiencies [1], [2]. However, one major issue for back contact cells is that minority carriers generated near the silicon surface must diffuse a greater distance before reaching the p-n junction. Thus, the need for high quality surface passivation becomes far more significant. Recombination is prevalent at the silicon surface due to the abundance of bandgap states that exist as a result of lattice discontinuities. There are two aspects to passivation. Firstly, the number of inter-bandgap states can be reduced by chemically passivating the surface. Secondly, carrier access to the inter-bandgap states can be limited by means of an electric field, known as field effect passivation (FEP). Both of these passivation schemes can be provided by a dielectric capping layer. However, this is not the only job of this capping layer; it must also act as an anti-reflection coating (ARC) to improve the optical absorption efficiency of the cell. For this reason, the choice of capping layer, and the parameters used to deposit or grow it, must be optimised to achieve effective passivation and anti-reflection.

One of the most effective techniques to improve the passivation quality of the dielectric once it is in place is the so called *alneal* process. It was first developed by Eades [3] and later used by Kerr and Cuevas [4] to minimise surface recombination and characterise bulk recombination processes. The alneal is conventionally performed on SiO_2 [4] and, according to Kerr and Cuevas, is able to produce SRVs as low as 2.4 cm/s on 1.5 Ωcm n-type Si [4]. However, SiO_2 does not provide effective anti-reflection characteristics since the refractive index is too low [5] ($n = 1.46$ at $\lambda = 632$ nm [6]). If the high quality passivation provided by an alneal could be used on a superior ARC, the efficiency of the entire cell design would be improved. TiO_2 is a top quality anti-reflection coating for solar cells due to its high refractive index and low absorption [7]. Despite this, TiO_2 is no longer used in the silicon PV industry due to its poor passivation properties [5], [8]. One method to overcome this poor passivation is by using an intermediate SiO_2 layer that provides better chemical passivation. It is known that the passivation provided by a SiO_2 layer can be further improved by an alneal and thus the effectiveness of the alneal on a SiO_2/TiO_2 double layer dielectric is investigated in this work.

2 EXPERIMENTAL PROCEEDURE

Czochralski (Cz), ~ 40 Ωcm, n-type silicon wafers were dry oxidised at 900 °C to grow a 30 nm thermal SiO_2 layer. A SiO_2 layer of this thickness was required for certain characterisation techniques and further work will be done to investigate the impact of using a thinner SiO_2 layer. The wafer was then diced into 3 x 3 cm^2 samples, and batches of these were then deposited with TiO_2. Specimens with a single SiO_2 layer were kept in order to contrast the conventional alneal with that of the double layer. The TiO_2 was deposited in a thermal evaporator on both sides of the sample. The thickness of the resulting films was measured using a Vecco DekTak 6M stylus profiler.

The alneal was performed on both the single SiO_2 layer and double layer (SiO_2/TiO_2) specimens. In order to do this, the samples were coated, on both sides, with ~100 nm aluminium using a thermal evaporator. Controls without an aluminium coating accompanied both the SiO_2 and SiO_2/TiO_2 specimens for all processing except the thermal evaporation of the aluminium. All samples were annealed in a tube furnace in an argon atmosphere for 30 minutes at 400 °C. After annealing, the Al was etched off in HCl at 80 °C. The controls, without the Al coating, were also placed in the etchant. The process of etching the Al also removed the TiO_2 on the double layer specimens. This is believed to be due to the low temperature at which the TiO_2 was deposited, which resulted in poor adhesion of the TiO_2 film to the SiO_2 surface. Future work will investigate using APCVD deposited TiO_2. Sinton photo-conductance measurements were taken to calculate the lifetime of the specimens. All lifetime values are quoted at an injection level of 10^{15} cm^{-3}. The TiO_2 was deposited in several different batches in order to vary the thickness of the deposition. Each deposition batch was assigned a batch number and the characteristics of each are shown in Table I. The refractive index of one of the samples from batch 1 was measured using an optical reflectometer and it was found to be 1.96 at 632 nm.

Table I Batch characteristics of TiO_2 depositions

Batch Number	TiO_2 thickness (nm)	
	Front	Rear
1	66 ± 5.6	57 ± 2.5
2	133 ± 3.5	158 ± 22
3	89 ± 8.8	77 ± 19
4	109 ± 4.9	123 ± 5.7
5	82 ± 5	84 ± 10
6	105 ± 7	101 ± 9

3 PASSIVATION DUE TO TiO_2 DEPOSITION

After depositing the TiO_2 in the thermal evaporator the lifetime of each sample was recorded. These are shown in Table II along with the batch number for the TiO_2 deposition as described in Table I.

Table II: Lifetime of samples before and after TiO_2 deposition

SiO_2/TiO_2		SiO_2
Batch no.	Lifetime (μs)	Lifetime (μs)
1	15.72	12.56
1	16.04	12.80
2	18.00	14.49
2	17.83	12.94
3	15.13	12.91
4	13.23	12.27
4	16.34	13.02

The data in Table II shows that the addition of the TiO_2 layer on top of the SiO_2 causes a slight increase in lifetime. On average, the SiO_2/TiO_2 double layer samples have a lifetime ~3 μs greater than the SiO_2 single layer samples; this is an insignificant increase in lifetime.

4 ALNEAL ON SiO_2/TiO_2 stack

In order to alneal the samples, 100 nm of aluminium was thermally evaporated on top of the dielectric(s). As controls, both a single layer SiO_2 sample and a double layer SiO_2/TiO_2 sample were annealed without the aluminum layer.

Figure 1 shows the injection dependent minority carrier lifetimes for four samples after the anneal. The alnealed SiO_2/TiO_2 sample shown in Figure 1 was the one that demonstrated the best improvement from the alneal process. In this case the TiO_2 deposited was from batch 1. The other samples shown in Figure 1 were all annealed in the same process. The lifetimes achieved by other alnealed SiO_2/TiO_2 and SiO_2 samples are shown in Table III.

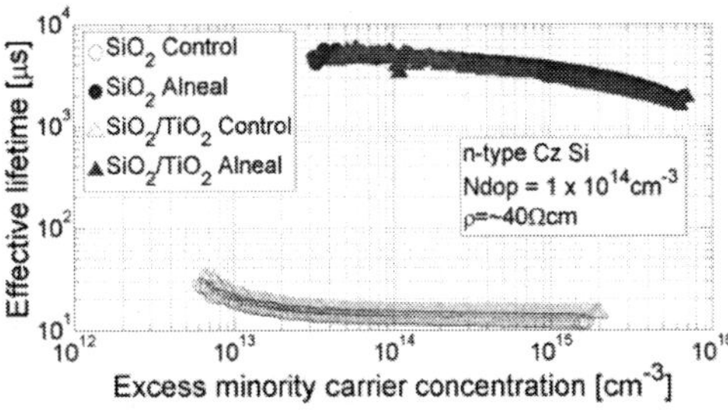

Figure 1 Effective lifetime vs minority carrier concentration curves for Cz ~40 Ωcm n-type Si with both alnealed (full symbols) and control (hollow symbols) SiO_2 (blue) and SiO_2/TiO_2 dielectrics (green). The TiO_2 on these samples was from batch 1 and ~ 62 nm thick.

From Table II and Figure 1 it is evident that both the controls, which were annealed in the tube furnace but were not coated in aluminium, achieved negligible change in lifetime. Both remained at their initial lifetimes of 12 μs and 15 μs for the SiO_2 single layer and SiO_2/TiO_2 double layer respectively. In contrast, the alnealed single layer SiO_2 sample reached a lifetime of 3691 μs, and the double layer SiO_2/TiO_2 specimen achieved a lifetime of 3084 μs.

Effective surface recombination velocity (SRV) was calculated as SRV $\leq (1/\tau_{eff} - 1/\tau_{bulk})W/2$ where W is the thickness of the silicon. It is assumed that the bulk lifetime is only limited by the Auger recombination (as calculated by Richter's parameterisation [9]) and the radiative recombination as modelled by Altermatt [10]. This leads to a SRV ≤ 8.9 cm/s and ≤ 10.7 cm/s at injection level 10^{15} cm^{-3} for the SiO_2 single layer and SiO_2/TiO_2 double layer alneal specimens respectively. This model assumes that recombination in the bulk due to defects is zero and therefore the calculated SRV must be considered to be an upper limit.

Emitter saturation current, J_{0e}, was calculated using Kane and Swanson's technique [9] for both alnealed samples. Effective lifetime was corrected using Richter's bulk lifetime model and values for J_{0e} were extracted directly from Sinton's lifetime software. Figure 2 shows the data used for this calculation. The saturation currents for the SiO_2 single layer and the SiO_2/TiO_2 double layer were calculated to be 32 fA/cm^2 and 33 fA/cm^2 respectively.

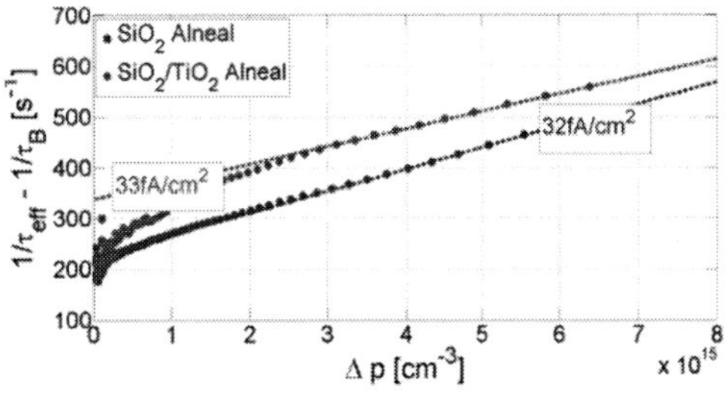

Figure 2 J_{0e} calculation using Kane and Swanson's method implemented in the Sinton lifetime software.

The alneal was repeated several times on both the single SiO_2 layer and the double SiO_2/TiO_2 layer samples. The lifetimes of these samples after alnealing are presented in Table III and the lifetime of the controls in Table IV.

Table III: Lifetime of samples after the alneal process at 400 °C for 30 minutes.

	SiO_2/TiO_2 Alneal		SiO_2 Alneal	
Batch no.	τ (μs)	SRV	τ (μs)	SRV
1	3084	10.7	3691	8.9
5	1506	22.2	4276	7.6
5	706	47.6	3593	9.1
6	312	108	3218	10.2

Table IV: Lifetime of samples after an anneal (with no aluminium) at 400 °C for 30 minutes.

	SiO_2/TiO_2 No Al		SiO_2 No Al	
Batch no.	τ (μs)	SRV	τ (μs)	SRV
1	15.5	2177	12.2	2766

5 EFFECT OF TiO_2 THICKNESS

It is clear from Table III that the minority carrier lifetime of the samples that underwent an alneal is significantly greater than the control samples that were simply annealed without any aluminium on the dielectric surface. Thus, one can conclude that the increase in lifetime is due to the presence of the aluminium during the anneal, and not the anneal itself.

When comparing the increase in lifetime between the double SiO_2/TiO_2 layer and the single SiO_2 layer, two points are evident. Firstly, the lifetimes achieved by the SiO_2/TiO_2 double layer samples are not as high as those achieved on the alnealed SiO_2 single layer samples. This could indicate that the TiO_2 presents a partial barrier to the mechanism by which the alneal process passivates the silicon surface. Secondly, the range in lifetimes achieved after the alneal is much greater for the SiO_2/TiO_2 double layer samples than for the single layer SiO_2 samples.

Previous reports by Zhao [11], Reed [12] and Cuevas [13] have suggested that the mechanism by which the alneal works is via hydrogen passivation. Hydrogen is said to be released due to aluminium interaction with hydroxyl ions present in the silicon dioxide [11]. This hydrogen is then free to diffuse and passivate dangling bonds at the silicon surface.

If this model is applied to the SiO_2/TiO_2 bilayer then there are two possible locations where the Al may react with hydroxyl ions, either within the SiO_2 or within the TiO_2. If the Al reacts with hydroxyl ions within the SiO_2, then the Al must diffuse through the TiO_2. If the reaction takes place within the TiO_2, any hydrogen produced in the TiO_2 would need to diffuse through both the TiO_2 and the SiO_2 (including the TiO_2/SiO_2 interface) in order to reach the silicon and passivate the surface. It is also possible that the TiO_2 layer may not be as rich in hydroxyl ions as the SiO_2 layer.

By varying the thickness of the TiO_2 layer, it is possible to assess whether the TiO_2 layer is acting as a diffusion barrier to the source of passivation or if a different mechanism limits the passivation. The average thickness of the TiO_2 layer against lifetime is shown in Figure 3.

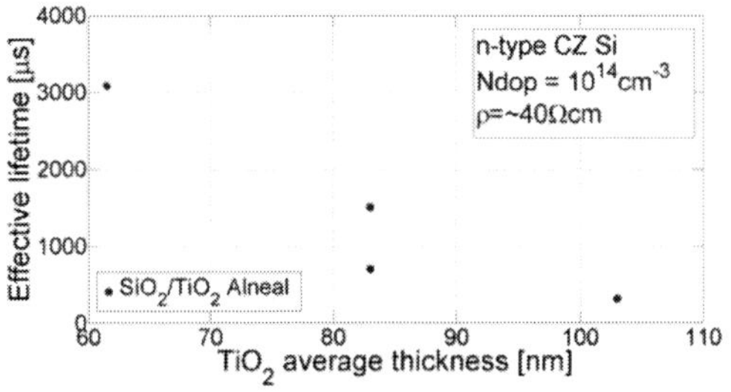

Figure 3 Effective lifetime at injection level of 1 x10^{15} cm^{-3} after alneal vs average TiO_2 thickness on SiO_2/TiO_2 double layer dielectric sample.

The results displayed in Figure 3 show that the thickness of the TiO_2 layer does have an effect on the effective lifetime after the alneal: the thicker the TiO_2 layer, the lower the effective lifetime. This supports the theory that the TiO_2 acts as a diffusion barrier to the source of passivation.

The variation between the two samples with a TiO_2 of 83 nm indicates that it is not only the TiO_2 thickness that affects the change in lifetime, there is another factor. This is corroborated by the variation seen in the results for the alnealed SiO_2 samples, where there is a difference of 1 ms for identically processed samples. One likely cause of this variation is a temperature profile within the tube furnace.

6 STABILITY OF LIFETIMES

The lifetimes of one of the alnealed single layer SiO_2 and double layer SiO_2/TiO_2 samples were monitored after the alneal. The TiO_2 deposition on the monitored sample was from batch 5 and ~ 83 nm. The stability of the effective lifetime curve with time is shown for each sample in Figure 4 (a) and (b).

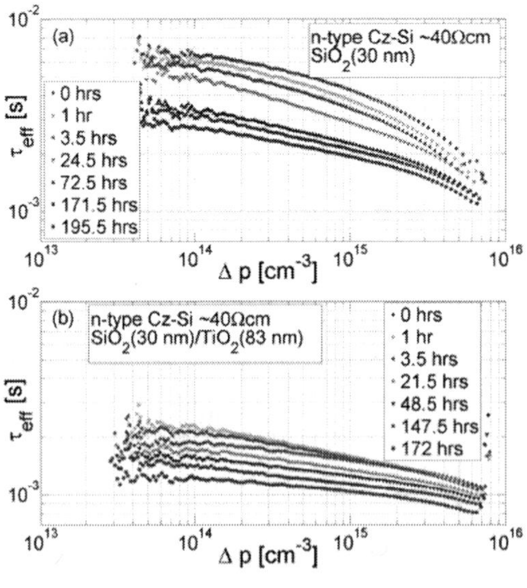

Figure 4 Effective lifetime vs minority carrier concentration at a given time after the alneal process on ~40 Ωcm Cz silicon with (a) a 30 nm SiO_2 dielectric and (b) a 30 nm SiO_2/ 83 nm TiO_2 dielectric stack

For clarity, the effective lifetime at a minority carrier concentration of 10^{15} cm^{-3} for both samples is shown against time in Figure 5. This data was used to extract the decay time constant and better define the lifetime stability.

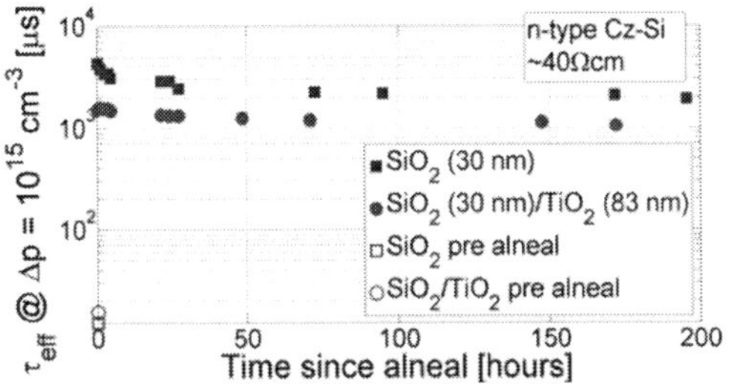

Figure 5 Effective lifetime at 10^{15} cm^{-3} against time after alneal of ~40 Ωcm Cz Si with a single SiO_2 layer (blue) and with a SiO_2/TiO_2 double layer dielectric (red).

From Figure 5 it is evident that the initial rate of decay in lifetime of the single SiO_2 dielectric layer is greater than of the SiO_2/TiO_2 bilayer. The decay time constants for the two curves (after the initial rapid decay) were calculated and found to be 20.4 days and 38.2 days for the SiO_2 and SiO_2/TiO_2 double layers respectively.

7 IMPACT OF ETCHING

It was observed that the decay in lifetime began only after the aluminium and TiO_2 had been removed in HCl at 80 °C and the samples rinsed in DI water. This was found by alnealing two samples, from the same TiO_2 deposition batch, together. One of these samples was etched immediately after the alneal. The other was stored, with the aluminium and TiO_2 intact, for a week before it was also etched. The lifetimes of both samples were monitored from immediately after etching. The results of this are shown in Figure 6. The samples used were from batch 6 with an average of 103 nm of TiO_2.

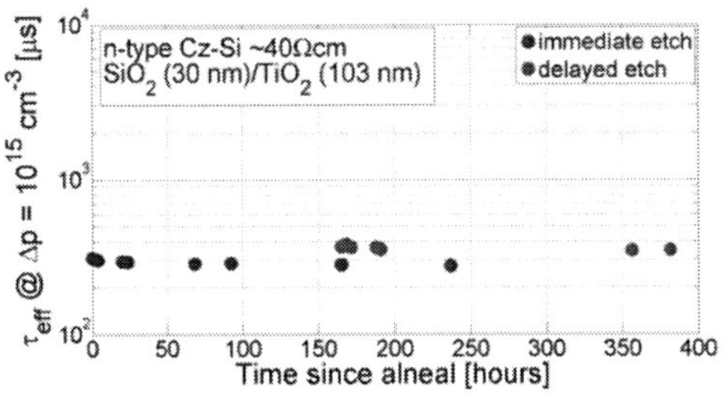

Figure 6 Lifetime vs time after alneal for a sample where the Al and TiO_2 were etched immediately after the alneal (blue) and another sample where the Al and TiO_2 were not etched until a week after the alneal (green).

A pair of single layer SiO_2 samples underwent the same procedure. The results of this are shown in Figure 7.

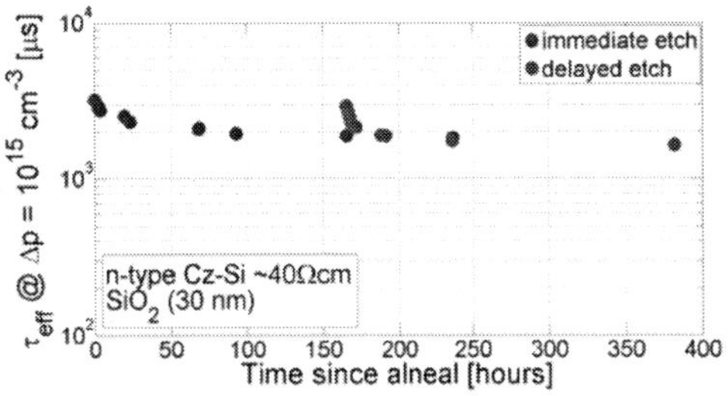

Figure 7 Lifetime vs time after alneal for a sample where the Al was etched immediately after the alneal (blue) and another sample where the Al was not etched until a week after the alneal (green).

From Figure 6 and Figure 7 it is evident that the lifetime decay is stimulated when the aluminium covering the samples is etched and the samples rinsed in DI water. The difference in the starting lifetime values for the samples in both Figure 6 and Figure 7 are most likely due to temperature variation within the furnace.

The possibility that the lifetime decay is caused by light irradiation was investigated. Twin samples, with a single layer SiO_2 dielectric, were processed identically. They were both etched immediately after the alneal. After etching, one was kept in the dark between lifetime measurements; the other was kept on the benchtop in a transparent plastic bag. The lifetimes of both were monitored with time and are shown in Figure 8.

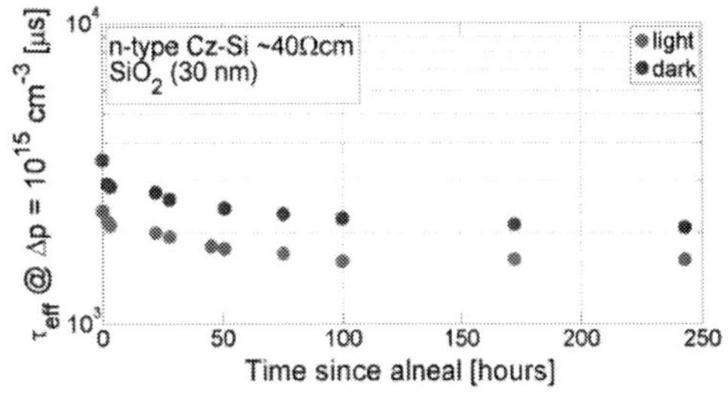

Figure 8 Lifetimes with time since alneal for SiO_2 coated ~40 Ωcm Cz silicon samples kept in the dark (purple) and kept in a transparent bag on the benchtop (teal)

As the decay rates of the two curves are very similar, it can be concluded that light is not the cause of the lifetime decay.

The trigger for the decay is not yet understood, however, two possibilities are that either the decay is stimulated by the wet processing itself or that the wet processing introduces a further source of passivation that decays with a shorter time constant.

8 CONCLUSIONS

It has been demonstrated that the effective passivation process of annealing SiO_2 passivation layers coated with aluminium, known as the alneal, is also possible on SiO_2/TiO_2 double layers. In this work the most effective alneal was seen to improve the lifetime of the n-type ~40 Ωcm Cz-Si coated in a SiO_2/TiO_2 double layer dielectric from ~15 μs to 3084 μs at 10^{15} cm^{-3}. This equates to a SRV ≤ 10 cm/s or a J_{0e} of 32 fA/cm^2. The thickness of the TiO_2 dielectric layer was seen to affect the post-alneal lifetime of the sample indicating that the

TiO_2 is likely acting as a diffusion barrier to the source of passivation. The stability of the passivation was also monitored. It was seen that the alneal passivation on the SiO_2 dielectric decayed at a greater rate than the passivation on the SiO_2/TiO_2 dielectric layer. The decay rates were calculated to be 20.4 days and 38.2 days on the single and double layer samples respectively. The major advantage of this finding to the solar cell industry is that this provides a method by which the excellent anti-reflection properties achieved by TiO_2 coatings could once again be exploited.

ACKNOWLEDGMENTS

K A Collett would like to thank EPSRC for funding her doctoral studies. Thanks also go to the Armourers and Brasiers Gauntlet Trust and Supergen Super Solar for travel grants to attend EUPVSEC. R S Bonilla is the recipient of an EPSRC (UK) Postdoctoral Research Fellowship, EP/M022196/1. P R Wilshaw acknowledges the support from EPSRC grant EP/M024911/1. Data published in this article can be downloaded from http://ora.ox.ac.uk. All authors are thankful to Martin Hermle and Christian Reichel at Fraunhofer ISE for provision of silicon material, and to Radka Chakalova for her extensive help in the cleanroom.

REFERENCES

[1] S. Pukhrem, "Concepts In Photovoltaic Technology," *Solar Love,* 2014. [Online]. Available: http://solarlove.org/concepts-photovoltaic-technology/. [Accessed: 08-Apr-2015].

[2] "SunPower Sets Solar Cell Efficiency Record at 24.2% - Renewable Energy World." [Online]. Available: http://www.renewableenergyworld.com/articles/2010/06/sunpower-sets-solar-cell-efficiency-record-at-24-2.html. [Accessed: 10-Sep-2015].

[3] W. D. Eades, "PhD Thesis," Stanford University, 1985.

[4] M. J. Kerr and A. Cuevas, "Very low bulk and surface recombination in oxidized silicon wafers," vol. 35, pp. 13–17, 2002.

[5] B. S. Richards, "Comparison of TiO2 and other dielectric coatings for buried-contact solar cells: A review," *Prog. Photovoltaics Res. Appl.*, vol. 12, no. 4, pp. 253–281, 2004.

[6] "Refractive Index of SiO2, Fused Silica, Silica, Silicon Dioxide, Thermal Oxide, ThermalOxide for Thin Film Thickness Measurement." [Online]. Available: http://www.filmetrics.com/refractive-index-database/SiO2/Fused-Silica-Silicon-Dioxide-Thermal-Oxide-ThermalOxide. [Accessed: 04-Feb-2016].

[7] K. Ali, S. A. Khan, and M. Z. M. Jafri, "Effect of Double Layer (SiO 2 / TiO 2) Anti-reflective Coating on Silicon Solar Cells," vol. 9, pp. 7865–7874, 2014.

[8] B. S. Richards, J. E. Cotter, and C. B. Honsberg, "Enhancing the surface passivation of TiO2 coated silicon wafers," *Appl. Phys. Lett.*, vol. 80, no. 7, pp. 1123–1125, 2002.

[9] A. Richter, S. W. Glunz, F. Werner, J. Schmidt, and A. Cuevas, "Improved quantitative description of Auger recombination in crystalline silicon," *Phys. Rev. B - Condens. Matter Mater. Phys.*, vol. 86, no. 16, pp. 1–14, 2012.

[10] P. P. Altermatt, F. Geelhaar, T. Trupke, X. Dai, a. Neisser, and E. Daub, "Injection dependence of spontaneous radiative recombination in crystalline silicon: Experimental verification and theoretical analysis," *Appl. Phys. Lett.*, vol. 88, no. 26, 2006.

[11] J. Zhao, A. Wang, P. P. Altermatt, S. R. Wenham, and M. A. Green, "24% Efficient Silicon Solar Cells," pp. 1477–1480, 1994.

[12] M. L. Reed and J. D. Plummer, "Chemistry of Si-SiO2 interface trap annealing," *J. Appl. Phys.*, vol. 63, no. 12, pp. 5776–5793, 1988.

[13] A. Cuevas, P. A. Basore, G. Giroult-Matlakowski, and C. Dubois, "Surface recombination velocity of highly doped n-type silicon," *J. Appl. Phys.*, vol. 80, no. 6, pp. 3370–3375, 1996.

19.31%-efficient Multi-crystalline Silicon Solar Cell with MCCE black silicon technology

Shuai Zou, Xusheng Wang*, Fang Cao, Xiaoya Ye and Guoqiang Xing

Canadian Solar Inc., 199 Lushan Road, SND, Suzhou, China, 215129

**Corresponding Author: Xusheng WANG, xusheng.wang@canadiansolar.com*

ABSTRACT A novel nano-scale pseudo-pit texture can be formed on the surface of *mc*-Si wafer by using a MCCE technique plus an additional chemical treatment. We have made a desired nanoscale inverted pyramid texture by optimizing the recipe of MCCE solution and a self-made chemical post-treatment, and, the depth and the width of the inverted pyramid can be adjusted in 100~900nm range. Average 18.90%-efficient MCCE black *mc*-Si solar cells have been fabricated on CSI's industrial production line, which equated to an efficiency gain of ~0.4% (absolute) at the cell level, and the efficiency of champion cell reaches to 19.31%.

Keyword: Black silicon; Metal-catalyzed chemical etching; Nanostructure textured; Chemical post-treatment

I. Introduction

The power conversion efficiency η is over 19.5% and 18.5% for single-crystalline silicon (*sc*-Si) and multi-crystalline silicon (*mc*-Si) solar cells in most of industrial production lines, respectively. [1] The cell's performance was improved by elevating the both electrical and optical properties, i.e. the quality of wafer, the better passivation in the front and the rear surfaces, the enhance light trapping via the interdigitated back contact and metal wrap through, and etc. [2, 3, 4, 5]

Recently years, *mc*-Si solar cells are holding absolute leading position in the photovoltaic (PV) market. However, the higher light reflection at the front textured surface make its' η about 1-1.5% lower than that of *sc*-Si one. In fact, the random pits or honeycomb texture formed on the surface of *mc*-Si via isotropic acidic etching, traps light weaker than that the pyramid texture formed on *sc*-Si wafer based on anisotropic alkali etching. [6, 7] Therefore, it gives a potentiality to raise η, even breaking 19%, if more effective light-trapping texture can be fabricated into *mc*-Si cells.

It was well-proven that the nanostructure textured black silicon can be more effective to trap light, thus has great potential for applications in silicon-based solar cells. [8, 9, 10] Several advantages of black silicon solar cells can be concluded as: the excellent light trapping in wide spectrum arrange from 300-2000nm; [11] the possibility to remove the expensive vacuum processing of plasma-enhanced-chemical-vapor-deposition (PECVD); wider acceptance angle of light. [12] Mainly, there are three kinds of techniques to fabricate black silicon: laser texturing, [8, 13] reactive ion etching (RIE), [10, 14, 15] and metal-catalyzed chemical etching (MCCE). [9, 16, 17, 18] For industrial production, the RIE and MCCE techniques have obtained high expectations.

II. Process flow and experiment

In **Fig.1**, the process flow simplifications of the MCCE, the RIE black *mc*-Si solar cell and typical solar cell are compared. Obviously, MCCE is much more suitable for the current production line, where the conventional texturing process is also based on wet chemical etching.

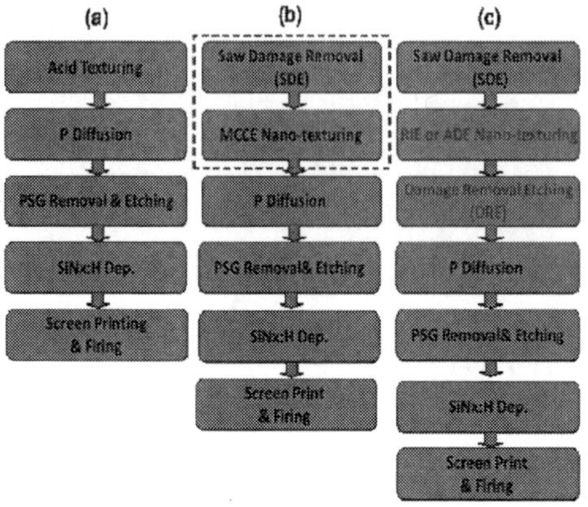

Figure 1. Process flow simplifications: (a) Typical *mc*-Si, (b) MCCE black *mc*-Si and (c) RIE black *mc*-Si solar cell.

Oh *et al.* reported 18.2%-efficiency in black single-crystalline silicon (B*sc*-Si) solar cells. [19] Although the efficiency is still lower than that of industry level, it demonstrates a great progress in MCCE technique. However, the reported efficiencies for black multi-crystalline silicon (B*mc*-Si) solar cells based on either RIE or MCCE are still very low, i.e. 12-16.6%. **Table I** lists the state-of-the-art of black Si solar cells. The presented techniques are still in laboratory level but not in industry scale. Certainly, the nanostructured silicon can absorb sunlight over a broad range of wavelengths and incidence angle, but also introduces more recombination centers and nonuniform doping into cell. [19, 20] Therefore, a tradeoff between optical gain and recombination loss must be considered to achieve high efficiency.

Table 1. State-of-the-art of nanostructured Si solar cells

Cell (type, size)	Method	Passivation and ARC	R (%)	η (%)
mc-Si (P, 15.6×15.6cm^2)	RIE	PECVD-SiN$_x$	13.7	16.1 [10]
mc-Si (P,15×15cm^2)	RIE	PECVD-SiN$_x$	N/A	17.1 [21]
sc-Si (N, 4cm^2)	RIE	ALD-Al_2O_3	1.9	18.7 [22]
sc-Si (P, 0.8081cm^2)	MCCE($AgNO_3$)	Thermal oxide SiO_2	4.6	18.2 [19]
sc-Si (P, N/A)	MCCE($AgNO_3$)	Al_2O_3	2.5	18.2 [23]
mc-Si (P, 4cm^2)	MCCE ($AgClO_4$)	PECVD-SiN$_x$	22	16.6 [24]

In this work, we report that a novel nano-scale pseudo-pit texture can be formed on the surface of *mc*-Si wafer by using a MCCE technique plus an additional chemical treatment. **Fig.2** is an illustration of the main steps of MCCE nano-texturing. In the first step of MCCE nano-texturing, a noble metal, such as Au, Ag, and Pt is deposited on the Si surface usually as nanoparticles (NPs).In the second step, a porous layer on the surface is formed by dipping the silicon substrate into a mixed aqueous solution of an oxidizing reagent (such as H_2O_2)and HF acid. Finally, to reduce the recombination due to high surface area, the third process of chemical post-treatment is very critical. An average efficiency of 18.90% has been achieved in CSI's industry product line. The champion cell reached 19.31%, which is highest reported value in black silicon up to now.

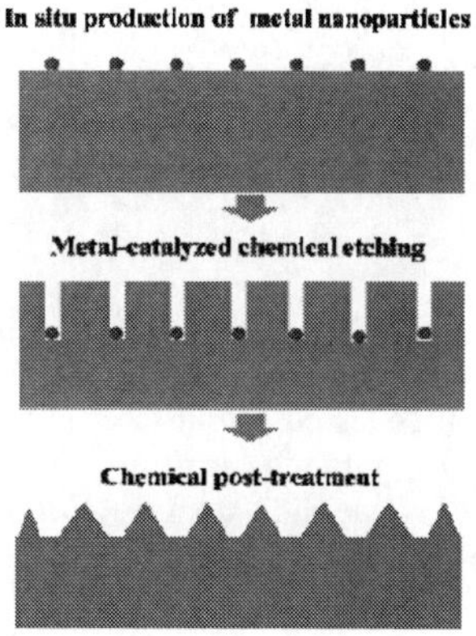

Figure 2. Main steps of MCCE nano-texturing process.

III. Results and discussions

The nanoscale structures of different size and shape are shown in **Fig.3**. We found that the morphology of nanoscale structures possesses high dependence on the concentration of metal ion, the volume ratio of H_2O_2:HF:DIW, the chemicals of post-treatment, and etching time of above process steps. In our experiment, by adjusting the recipe of MCCE solution and a proper alkali post-treatment, a nanoscale pyramid or inverted-pyramid texture can be obtained. For instance, a high Ag ion concentration can result in nanoscale pyramid texture, on the contrary, a low Ag ion concentration will lead to an inverted-pyramid texture. By adjusting the recipe of MCCE solution and a proper acid post-treatment, a nanoscale pseudo-pit texture can be formed. If adding a special processing after metal NPs deposition, an orderly nanoscale wormhole texture will be formed.

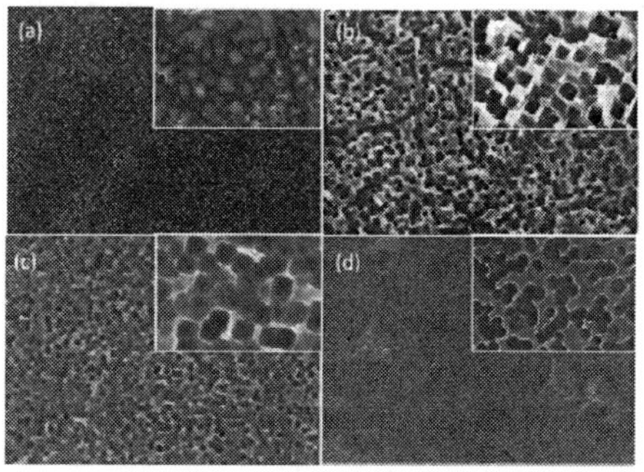

Figure 3. SEM surface images of (a) nanoscale pyramid texture, (b) nanoscale inverted- pyramid texture, (c) nanoscale pseudo-pit texture, and (d) nanoscale wormhole texture.

We have made a desired nanoscale inverted pyramid texture by optimizing the recipe of MCCE solution and a self-made chemical post-treatment, as shown in **Fig.4**. And, the depth and the width of the inverted pyramid can be adjusted in 100-900nm range. A proper size of inverted pyramid texture can ensure the reduced surface recombination while keeping low reflectance.

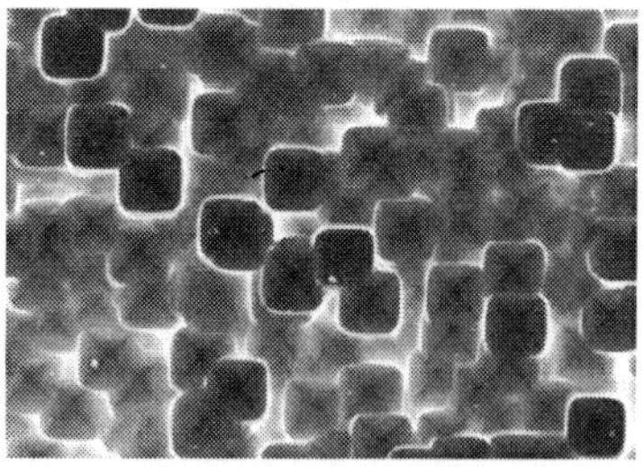

Figure 4. SEM image of desired nanoscale inverted pyramid texture.

After above optimization of nanoscale structure texture, the average efficiency of mass-produced MCCE black *mc*-Si solar reaches to 18.90%, while the baseline yielded 18.51%, this equated to an efficiency gain of ~0.4% at the cell level. The mean parameters of the black cells are concluded in **Table 2**. As expected, the short-circuit current has an increase about 140mA, which can be mainly due to its enhanced light absorption. The average V_{oc} value of MCCE black *mc*-Si solar cells is about 1.0mV higher than that of typical *mc*-Si solar cells (Baseline), which thanks to the low J_{02} value of MCCE black *mc*-Si solar cells. Based on a two-diode model, the J_{02} accounts for the injection dependent Shockley-Read-Hall recombination in the space charge region, moreover, such low injection recombination can be attributed to a flat pn junction after MCCE texturing. Certainly, the average FF of MCCE black *mc*-Si solar cells is about 0.3% higher than baseline.

Compared with blue appearance of typical *mc*-Si solar cell, the appearance of MCCE *mc*-Si solar cell is dark blue, as shown in **Fig.5**. However, the grains of MCCE *mc*-Si solar cell are more legible than typical acid texturing solar cell, because that the reaction rates of MCCE in different grains of random crystalline-orientation are different.

Table2. Main characteristics for mass-produced MCCE black *mc*-Si and typical *mc*-Si solar cells (Baseline).

	Uoc(mV)	Isc(A)	Rs(mΩ)	Rsh(Ω)	FF(%)	EFF(%)
Baseline-Avg	637.6	8.842	1.80	506	79.90	18.51
MCCE-Avg	638.6	8.981	1.70	607	80.19	18.90
Champion cell	642.4	9.109	1.68	599	80.31	19.31

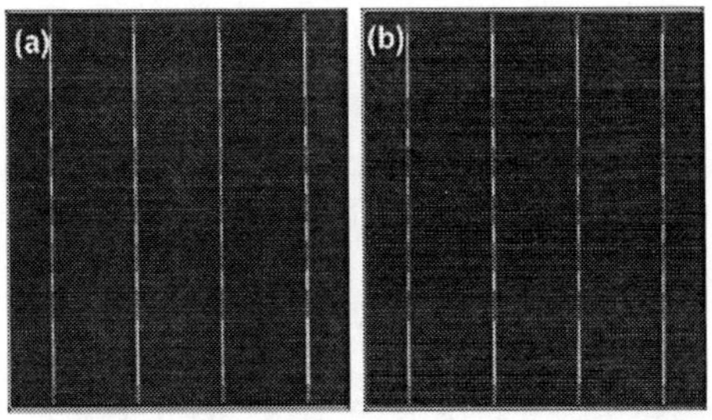

Figure 5. Cell photos of (a) Typical *mc*-Si and, and (b) MCCE black *mc*-Si solar cell.

IV. Conclusions

Metal-catalyzed chemical etching is a more economical and efficient approach for industrial black *mc*-Si manufacturing. A desired nanoscale structure can be made by adjusting the recipe of MCCE solution and a self-made chemical post-treatment processing, which will ensure a tradeoff between optical gain and recombination to achieve high efficiency. The conversion efficiency of mass-produced MCCE black *mc*-Si solar cells reaches to ~18.90% in CSI, which equated to an efficiency gain of ~0.4% (absolute) at the cell level, and the efficiency of the champion cell reaches to 19.31%. There still are huge potential to improve the efficiency of MCCE black *mc*-Si cells through an effective passivation technique and optimizing the MCCE process. Moreover, we believe that the MCCE is an ideal technique to solve the texture problem for diamond wire sawn *mc*-Si wafers and direct wafers.

Reference

[1] E. Lee, H. Lee, J. Choi, D. Oh, J. Shim, K. Cho, J. Kim, S. Lee, B. Hallam, S. R. Wenham, H. Lee, *Solar Energy Materials and Solar Cells* **2011**, *95*, 3592–3595.

[2] S. J. Eisele, T. C. Röder, J. R. Köhler, J. H. Werner, *Appl. Phys. Lett.* **2009**, *95*, 133501.

[3] S. Gatz, K. Bothe, J. Müller, T. Dullweber, R. Brendel, *Energy Procedia* **2011**, *8*, 318.

[4] N. Bateman, P. Sullivan, C. Reichel, J. Benick, M. Hermle, *Energy Procedia* **2011**, *8*, 509.

[5] W. W. Yin , X. S. Wang, F. Zhang, L. J. Zhang, *IEEE Journal of Photovoltaics* **2013**, *3*, 697.

[6] P. Panek, M. Lipinski, J. Dutkiewicz, *J. Mater. Sci.* **2005**, *40*, 1459.

[7] H. Seidel, L. Csepregi, A. Heuberger, H. Baumgärtel, *J. Electrochem. Soc.* **1990**, *137*, 3626.

[8] T. H. Her, R. J. Finlay, C. Wu, S. Deliwala, E. Mazur, *Appl. Phys. Lett.* **1998**, *73*, 1673.

[9] F. Toor, H. M. Branz, M. R. Page, K. M. Jones, H. C. Yuan, *Appl. Phys. Lett.* **2011**, *99*, 103501.

[10] J. Yoo, G. Yu, J. Yi, *Solar Energy Materials & Solar Cells* **2011**, *95*, 2-6 (2011).

[11] M. Halbwax, T. Sarnet, Ph. Delaporte, M. Sentis, H. Etienne, F. Torregrosa, V. Vervisch, I. Perichaud, S. Martinuzzi, *Thin Solid Films* **2008**, *516*, 6791.

[12] H. C. Yuan, V. E. Yost, M. R. Page, P. Stradins, D. L. Meier, H. M. Branz, *Appl. Phys. Lett.* **2009**, *95*, 123501.

[13] C. H. Crouch, J. E. Carey, J. M. Warrender, M. J. Aziz, E. Mazur, *Appl. Phys. Lett.* **2004**, *84*, 1850.

[14] P. Choulat, G. Agostinelli, D. Dehertoghe, G. Beaucarne, *Proceedings 22nd European Photovoltaic Solar Energy Conference*, **2007**.

[15] H. Jansen, M. de Boer, R. Legtenberg, M. Elwenspoek, J. Micromech. Microeng. 1995, 5, 115; D. S. Ruby, S. H. Zaidi, S. Narayanan, B. M. Damiani, A. Rohatgi, *Sol. Energy Mater. Sol. Cells* **2002**, *74*, 133.

[16] S. Koynov, M. S. Brandt, M. Stutzmann, *Phys. Stat. Sol. (RRL)* **2007**, *1*, R53.

[17] S. Koynov, M. S. Brandt, M. Stutzmann, *Appl. Phys. Lett.* **2006**, *88*, 203107.

[18] S. Bastide, N. Quang, R. Monna, C. Lévy-Clément, *Phys. Status Solidi C*, **2009**, *6*, 1536.

[19] J. Oh, H. C. Yuan, H. M. Branz, *Nature Nanotechnology* **2012**, *7*, 743.

[20] M. D. Kelzenberg, S. W. Boettcher, J. A. Petykiewicz, D. B. Turner-Evans, M. C. Putnam, E. L. Warren, J. M. Spurgeon, R. M. Briggs, N. S. Lewis, H. A. Atwater, *Nature Materials* **2010**, *9*, 239.

[21] K. Fukui, K. Okada, Y. Inomata, H. Takahashi, S. Fujii, Y. Fukawa, K. Shirasawa, *Solar Energy Materials and Solar Cells* **1997**, *48*, 219.

[22] P. Repo, J. Benick, V. Vähänissi, J. Schön, G. von Gastrow, B. Steinhauser, M. C. Schubert, M. Hermle, H. Savin, *Energy Procedia* **2013**, *38*, 866.

[23] W.C. Wang, C.W. Lin, H.J. Chen, C.W. Chang, J.J. Huang, M.J. Yang, B. Tjahjono, J.J. Huang, W.C. Hsu, M.J. Chen, *ACS Appl. Mater. Interfaces* **2013**, *5*, 9752.

[24] K. Tsujino, M. Matsumura, Y. Nishimoto, *Sol. Energy Mater. Sol. Cells* **2006**, 90, 100.

NOVEL VACUUM-FREE TECHNIQUE AND TECHNOLOGIES FOR HIGH EFFICIENT AND LOW-COST PHOTOVOLTAICS

Zhavnerko G.K., Shiripiov V.Ya., Sergeev O.V.*
Izovac Technologies Ltd., Bogdanovicha 155, Minsk, Belarus
* NEXT ENERGY · EWE Research Centre for Energy Technology at the University of Oldenburg, Carl-von-Ossietzky-Str. 15, 26129 Oldenburg, Germany
Tel.: +375 172931842, Fax: + 375 17 2931845, E-Mail: zhavnerko@izovac.com

ABSTRACT: Unique production facility with advanced environmental-friendly vacuum-free processing is developed in order to efficiently exchange some of existing technologies of solar cells fabrication. We report successful implementation of our approaches in the engineering of *h*eterojunction with intrinsic thin layer solar cells, although application of them is very prospective for all other types of solar cells and modules. Innovative technological solutions that concern formation of front contact, improvement of light management, solar cell protection from environment influence, the formation of tandem solar cells, etc. are explained in details. Our technology is suitable for application on large-scale and flexible photovoltaic devices. The principle of minimizing the capital cost of equipment and the technologies have been taken into account for new technique construction.

Keywords: Silicon Heterojunction, Nanostructure formation, Wear-resistant coatings

1 INTRODUCTION

There are two obvious problems must be solved to reduce the cost of energy generated by the solar module – to increase the efficiency of solar module and to reduce its prime cost. First way is to reduce optical losses with coating the front cover glass of the module with antireflective layer based on nanostructures. Secondly, we can improve the utilization of broader sun spectrum with modification of cover glass or individual solar cells (SC) with down converters or with complementary layers forming the tandem solar cell, which can enhance solar power device efficiency dramatically. Thirdly, we can give dirt resistance properties to the module and thus to eliminate losses of transparency due to soiling and abrasion by the influence of wind and dust particles, that also will decrease maintenance costs of installed modules. Moreover, application of such functional coatings directly on solar cells at the same time can substitute the lamination layers. These techniques are compatible with existing industrial units of production technologies, that do not require high temperatures and special conditions, easy up scalable, allow reducing of capital costs, working expenses and material consumption. Developed technologies can be utilized on broad variety of materials of different dimensions and shapes (glass, wafers, foils, polymers, etc.).

The method is fascinating with the easiness of its process-control and reproducibility, with the ability to use the wide range of different materials and substrates, their dimensions and variety of structures that can be deposited and with highest degree of useful material utilization. In comparison to vacuum coatings, where less than 20% of used material are only efficiently spent for layer formation, which is highly crucial parameter for expensive materials, this method provides close to 100% of useful material consumption. The process take place at room temperature and for most applications does not require any thermal post-processing, but could be easily combined if necessary. Moreover, there is no need to use and to neutralize after that any expensive, dangerous and harmful process gases, like it mostly to be in vacuum processes, that is why our technology is very environmental friendly.

The goal of the work is to demonstrate the advanced capabilities of the new equipment and related technologies in the production of solar cells and modules based on heterojunction with intrinsic thin layer (HIT) technology.

2 DESCRIPTION

Our developments in Langmuir-Blodgett (LB) film technology [1,2] have led to the construction of the coater (Fig.1) that allows both defect-free and continuous monolayer formation on rigid (glass, wafers, ceramics) and flexible substrates.

Figure 1: The Coater (conveyer type) for continuous surface modification with monolayer films by Roll-to-Roll-compatible deposition method.

We suggest to use additional layers on the front ITO layer and on the cover glass (Fig.2) that can be formed by molecular R2R method.

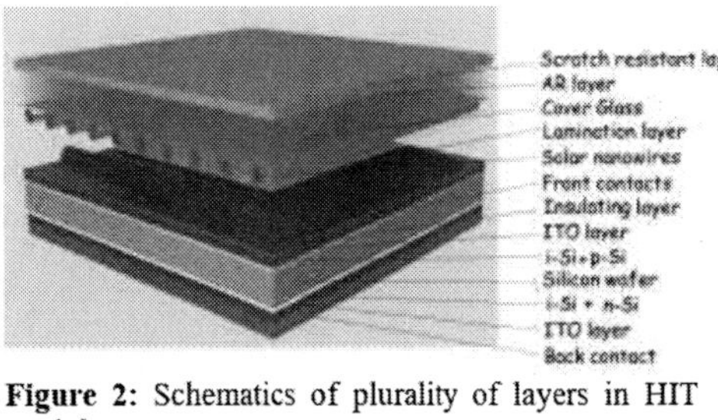

Figure 2: Schematics of plurality of layers in HIT module

Technological solutions will allow increase of the efficiency and durability by the formation of the antireflective layer (AR layer in Fig. 1) on the cover glass of a module, by giving the surface the anti-abrasive and environmental protection properties (scratch resistant layer in Fig. 2), by the formation of additional down converter or complementary layers (solar nanowires in Fig. 2) forming the tandem solar cell. In particular, the formation of antireflective layer on front cover glass of the module will increase module efficiency by more than 1%. The formation of wear-resistant oleofobic coating reduces losses due to the scattering of light from matt and soiled surfaces, that is an actual case for non-protected PV-modules affected for the long time with influence of dust particles, sand and ice. Some of those applications we already reported previously [2].

3 RESULTS

Functional coating can be used to protect the individual solar cell surface and structure from destructive external impact. In particular, environmental tests confirm that the deposition of additional special organic monolayer coating on the front side of HIT solar cell keeps the efficiency of the cell unchanged as opposed to unprotected cell that degrades rapidly enough (up to one absolute percentage of efficiency per day at 25 °C, 85 % humidity, and constant illumination of 1100 W/m^2), see Table 1.

Table 1: The data of climatic test for HIT solar cell both unprotected and covered with transparent hydro- and oleophobic monolayer film

Sample	Pseudo "Effi-ciency"	Voc, V	Jsc, A/cm^2	Pseudo -FF
Initial SC	16.03	0.694	0.031	75.32
Initial SC_climate	15.13	0.687	0.029	75.92
SC with protective layer	17.50	0.706	0.031	78.88
SC with protective layer_climate	17.80	0.705	0.032	78.92

R2R coater can be extremely useful for additional layers formation in the field of solar cell technology, for example, to prepare extremely scratch-resistant, transparent, self-cleaning (oleophobic) layer on the top of the cover glass for module structures. For example, protective glass remains hydrophobic (105-110 degree for water contact angle) after 8000-10000 cycles of abrasive wearing with stainless-steel wool by applying of 1 kg to 1 cm^2 wearing area.

It should be emphasized that the protective coating may also serve as a mask during formation of the front contact at solar cell surface through plating of metals on the regions activated by Femtosecond laser. Narrow (10-20 μm) fingers consisting of nickel, copper and tin are deposited by subsequent plated metallization. Another possibility is to prepare both nanostructured conductive transparent metal layer and busbars by plating on the nanostructured template providing low contact resistivity with highly conductive bulk metals. High performance of metallization process, high adhesion and low material cost allow consider that technology is quite promising for HIT SC production in nearest future.

Our technology can also form a tandem device, for example, by deposition of additional layer of a GaAs nanowire array with varying their packing density (10-30% of the total surface area) that does not reduce the transparency of cover glass significantly. Hence, III-V-nanowires in combination with today's silicon photovoltaic technology have very good potential for the development very high performance devices, with efficient use of materials and at low cost.

4 CONCLUSION

We demonstrated capabilities of new equipment, the principles of its operation, and the results of climatic test at HIT solar cell coated with protective layer. Proposed approaches can increase both the efficiency and the durability of solar cells and modules. The very wide commercial potential of proposed technology will also give the industry direct tool for efficient, ecology friendly and high quality fabrication of complex protective, photonic and nanoelectronic structures, contacts and transparent electrodes on different kinds of rigid and flexible solar cells and modules.

REFERENCES

[1] G.K. Zhavnerko and G. Marletta. Developing Langmuir–Blodgett strategies towards practical devices. Materials Science and Engineering B. 169 (2010) 43–48.

[2] G.K. Zhavnerko, O.V. Sergeev, A. Neumüller, V.Ya. Shiripiov. Vacuum-free technology of nanostructured coatings for enhanced light conversion and for module protection. 31st European Photovoltaic Solar Energy Conference and Exhibition. 2015, Hannover. 220-222.

PHOSPHOROUS DOPING FROM APCVD DEPOSITED PSG

Felix Book[1], Holger Knauss[2], Carsten Demberger[2], Florian Mutter[1], Giso Hahn[1]
[1]University of Konstanz, Department of Physics, 78457 Konstanz, Germany
[2]Gebr. Schmid GmbH, Robert-Bosch-Str. 32-36, 72250 Freudenstadt, Germany
Author for correspondence: felix.book@uni-konstanz.de, Tel.: +49 7531 88 3386

ABSTRACT: The phosphorous diffusion from atmospheric pressure chemical vapor deposition (APCVD) deposited phosphorus silicate glass (PSG) promises reduced process costs compared to the standard $POCl_3$ diffusion process, since no $POCl_3$ gas flow is necessary during the diffusion process. Therefore, much smaller or no spacing between the wafers is necessary and the throughput of the diffusion process can be significantly increased. Furthermore, it allows a structuring of the doping source prior to diffusion. We investigate the effect of basic process parameters concerning the deposition of the PSG and the capping layer on sheet resistance and uniformity. On standard aluminum back surface field (Al-BSF) solar cells, cell efficiencies of up to 19.6 % were achieved. In a high temperature co-diffusion process with reduced P content, the APCVD-PSG emitter passivated with fired PECVD-SiN_X features low j_{0E} of 100 fA/cm² at 50 Ω/sq. This results in a high cell V_{OC} of 639 mV while leading to a j_{SC} loss due to increased Auger recombination in the deep emitter profile. This loss can partly be compensated by a selective emitter etch-back. It would not occur, when the doping profile is located at the rear side of a bifacial or back contact solar cell as a BSF.
Keywords: Diffusion, Doping, CVD Based Deposition

1 INTRODUCTION

The electrical quality of the n-type emitter plays a major role for the efficiency of today's industrial type solar cells. It is typically formed by a $POCl_3$ diffusion process (see, *e.g.*, [1]) which is well-established and optimized, however still a topic of current research. The PSG is grown during the high temperature diffusion from the $POCl_3$ gas phase. It is therefore deposited on all sides of the wafer, and requires an edge isolation later in the process to prevent shunting by the rear side metallization.

The APCVD technology represents an alternative way to deposit the PSG as dopant source before the actual diffusion process [2,3]. This potentially also allows structuring the PSG for selective doping. The diffusion can be carried out either in an inline furnace [3] or, like in this work, in a tube furnace as used for $POCl_3$ diffusion. Since no homogenous gas flow is required between the wafers, a smaller or no spacing between the wafers could be used. This can drastically increase the throughput and reduce the cost of the diffusion process. Another advantage is the one-sided deposition, which can make edge isolation obsolete, if the PSG wrap-around is removed by a single side HF clean before the diffusion.

A common application of P surface doping is a front or back surface field (FSF/BSF) for high efficiency n-type solar cells. Most of these cell concepts feature a B emitter, which can be formed in a co-diffusion process with P, but requires a higher diffusion temperature due to the lower diffusion constant of B compared to P. This results in a very low n-type sheet resistance, if an infinite P source is used. The sheet resistance therefore has to be controlled by the P content in the PSG or an additional diffusion barrier.

In this work we determine the main effects of relevant process parameters on the sheet resistance and uniformity. Furthermore, we investigate the possibility of controlling the sheet resistance of a high temperature diffusion (900 °C) by reducing the P content in the PSG. The electrical quality of the doping profiles is characterized in terms of emitter saturation current density j_{0E} and IV parameters of standard p-type Al-BSF solar cells.

2 EXPERIMENTAL

The solar cells presented in this work are processed on 2 Ωcm p-type Cz-Si according to the well-known process flow for industrial p-type Al-BSF solar cells:

- Alkaline texture
- PSG deposition (except $POCl_3$ reference)
- Diffusion
- PSG removal / edge isolation
- SiN_X:H deposition
- Screen printing
- Co-firing

The PSG was deposited at the University of Konstanz in a lab-type APCVD system (Schmid 5K6) with single lane ceramic roller wafer transport. The system features two chambers to deposit PSG and an undoped SiO_X capping layer in one process step. The first chamber can also be used to deposit undoped SiO_X, BSG or AlO_X. When process gas flows are varied, the total gas flow through the injector is kept constant by adapting the N_2 dilution. The P weight percentage of the PSG layer (P [$\%_{wt.}$]) is calculated from the gas flows and can be set as an input process parameter.

The diffusion was carried out in an oxidation tube furnace or a $POCl_3$ tube furnace. For solar cells with a selective emitter, an etch-back process was added after the diffusion, which consists of printing a mask, etch-back and mask removal [4].

An HF clean and, if applied, a single side etching of the rear side was carried out before SiN_X:H deposition. The metallization was performed by screen printing of industrially available Ag and Al paste with subsequent co-firing.

For some emitter profiles, symmetrical samples were processed to determine j_{0E} and the doping profile. These samples were processed on chemically polished 200 Ωcm p-type FZ-Si wafers. The effective lifetime was measured with a Sinton WCT-120 tester, the fit of j_{0E} was performed at MCD = $3 \cdot 10^{15}$ cm^{-3}.

3 RESULTS

3.1 Homogeneous emitter for p-type solar cells

If the APCVD-PSG is used instead of a $POCl_3$ PSG for the formation of an emitter for standard Al-BSF or PERC solar cells, no additional boundary conditions cocerning the diffusion process have to be considered. It was observed in previous experiments that for the APCVD-PSG a capping layer of undoped SiO_x is necessary to improve the process stability. As a starting point, the standard $POCl_3$ diffusion program without $POCl_3$ flow and a drive-in temperature of 840 °C for 45 min was chosen. For a $POCl_3$ emitter, this program typically results in a sheet resistance of 50–55 Ω/sq.

A full factorial design of experiment (DoE) was performed to investigate the effect of the following factors on the emitter sheet resistance and uniformity:

- PSG P concentration (C_PSG = 20/25/30 $\%_{wt.}$)
- PSG thickness (d_PSG = 10/20/30 nm)
- Capping thickness (d_Cap = 25/50/75 nm)
- Diffusion in $POCl_3$ or oxidation tube furnace (effect of P from the tube wall)

The main effects and interaction of the factors on R_{sheet} and uniformity are presented in Fig. 1 and 2. The different PSG and capping layers result in a wide range of R_{sheet} (41–650 Ω/sq) and a uniformity of 3–64 %. Each data point represents the mean value of all samples with the indicated factor combination.

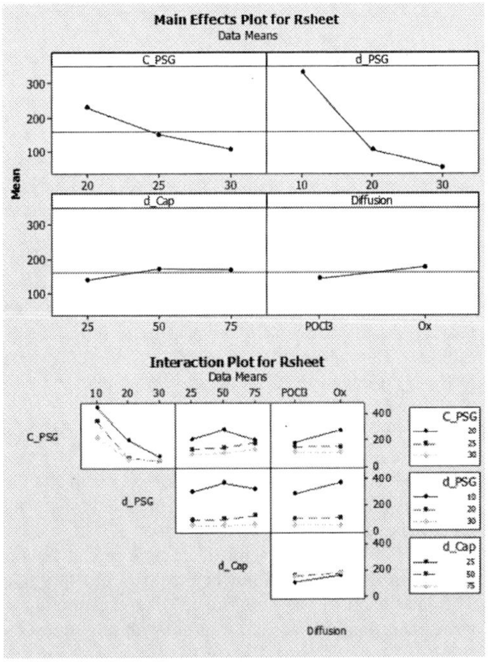

Figure 1: Main effects and interactions of the investigated factors on R_{sheet} of the P diffusion from APCVD-PSG.

A strong effect of C_PSG can be observed, especially for the thin PSG layers. R_{sheet} values below 80 Ω/sq could only be achieved with d_PSG of at least 20 nm. On these samples, the uniformity improves with increasing d_PSG. A thicker capping layer leads to an increase of R_{sheet}, which can be attributed to a diffusion of P in the capping layer. In terms of R_{sheet} uniformity, a thick PSG layer with a thin to medium capping layer seems desirable. This combination leads to an R_{sheet} of 70 Ω/sq for the lowest C_PSG of 20 $\%_{wt.}$. Higher C_PSG result in R_{sheet} below the typical range of Al-BSF solar cell emitters. The diffusion of P from the $POCl_3$ furnace tube wall shows a small effect, which only becomes relevant for thin d_PSG and d_Cap.

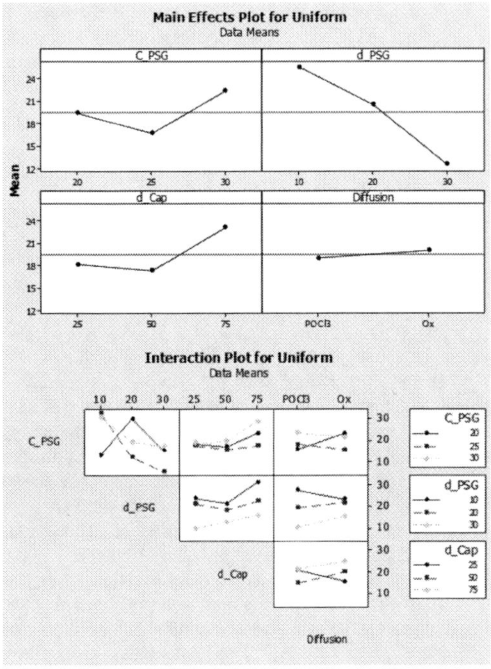

Figure 2: Main effects and interactions of the investigated factors on R_{sheet} uniformity of the P diffusion from APCVD-PSG.

The samples with a sheet resistance below 80 Ω/sq were used to process Al-BSF solar cells without edge isolation. This results in a shunt resistance of 200-1400 Ω/sq and shows, that despite the one-sided deposition, at least a single side HF-Dip after the PSG and capping deposition is necessary. However, it has almost no effect on V_{OC} and j_{SC}, which are show in Figure 3. The expected increase in both values with increasing R_{sheet} can be observed. In the 50–65 Ω/sq range, some solar cells show a reduced j_{SC}. These cells also exhibit a poor R_{sheet} uniformity and can be assigned to the PSG combination C_PSG = 30 $\%_{wt.}$ and d_PSG = 20 nm.

On selected solar cells, an additional edge isolation with a dicing saw was carried out. This results in an efficiency of 19.6 % for the best cell of which the IV data is shown in Tab. I.

Table I: IV data of the best Al-BSF solar cell from the APCVD-PSG and $POCl_3$ emitter group.

PSG	V_{OC} [mV]	j_{SC} [mA/cm²]	FF [%]	Eta [%]
APCVD	640	38.1	80.4	19.6
$POCl_3$	639	38.3	79.7	19.5

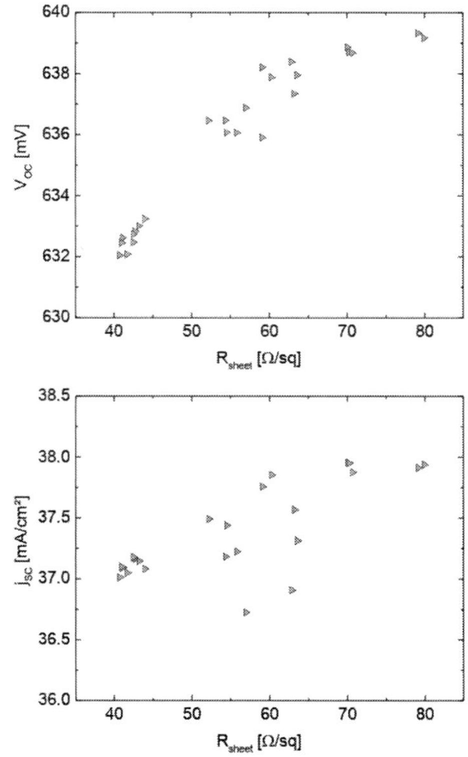

Figure 3: V_{OC} and j_{SC} vs. R_{sheet} of 6" standard p-type solar cells with an emitter diffused from various APCVD layers of the DoE shown in Figure 1.

The high fillfactor of this cell with an R_{sheet} of 79 Ω/sq demonstrates, that the APCVD-PSG emitter can be well contacted with conventional screen printing paste. The reference solar cells with a 65 Ω/sq $POCl_3$ emitter were processed equally but with edge isolation by single side etching after the diffusion. The high and similar cell efficiency compared to the well optimized $POCl_3$ emitter proves, that the APCVD-PSG allows for at least equally high emitter quality. Since only one cell per parameter set was processed, this experiment cannot be regarded as an optimization in terms of solar cell efficiency. This will be performed in an upcoming experiment with selected deposition parameter sets, that have resulted in good R_{sheet} uniformity.

3.2 High temperature diffusion

For this study, a 900 °C diffusion with high temperature plateau of 45 min was chosen. This temperature profile is sufficient to obtain an R_{sheet} below 100 Ω/sq for a B diffusion from APCVD BSG. In order to achieve sufficiently high R_{sheet}, the P content in the PSG was reduced and varied between 6-15 $\%_{wt.}$ in a 50 nm PSG layer capped with 50 nm of undoped SiO_X, that was deposited in the second APCVD chamber.

j_{0E} was measured after diffusion with the PSG as a passivation, after removing the PSG (unpassivated) and with a 75 nm SiN_X passivation before and after firing (see Figure 4).

The unpassivated samples show an increase in j_{0E} for high R_{sheet} emitters due to a reduced field effect passivation. The PSG passivated samples feature j_{0E} values below the SiN_X passivated ones before firing. However, the PSG is hygroscopic and dissolves to phosphoric acid in air, which impedes its application for solar cell passivation. If the time between deposition and diffusion exceeds several hours, the wafers therefore should be stored in a dry cabinet.

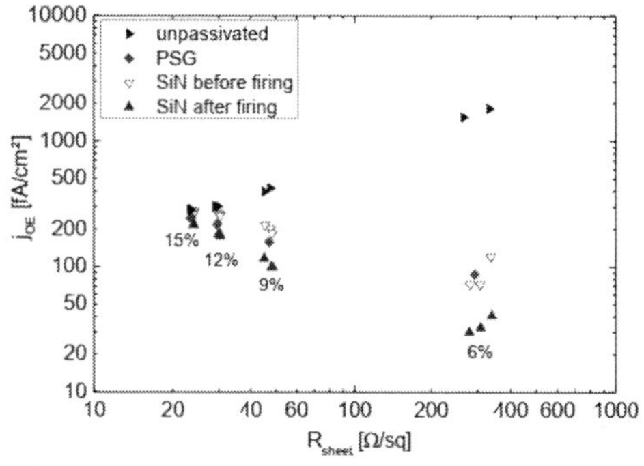

Figure 4: j_{0E} of emitters diffused from APCVD-PSG with varied P content. Data shown for unpassivated surfaces and passivation with the PSG or SiN_X before and after firing.

The APCVD-PSG emitter passivated with a fired SiN_X shows the lowest j_{0E} values. For 9 $\%_{wt.}$ PSG, 100 fA/cm² are achieved at an R_{sheet} of 50 Ω/sq. At the same R_{sheet}, a $POCl_3$ emitter typically features a value of approx. 220 fA/cm² [5].

Doping profiles measured by ECV are shown in Figure 5. The high temperature diffusion leads to a very deep profile. For a reduced P content in the PSG, the profile is "shifted" towards the surface, leading to a shallow kink with very similar surface P concentration for P > 9 $\%_{wt.}$. At 6 $\%_{wt.}$ the kink disappears and surface concentration drops, resulting in a drastic increase in R_{sheet}. In direct comparison of the 50 Ohm/sq emitter with a reference $POCl_3$ emitter profile of similar sheet resistance, a deeper profile with lower doping concentration (below the solubility limit) within the kink is observed for the APCVD-PSG, resulting in reduced recombination losses and j_{0E}.

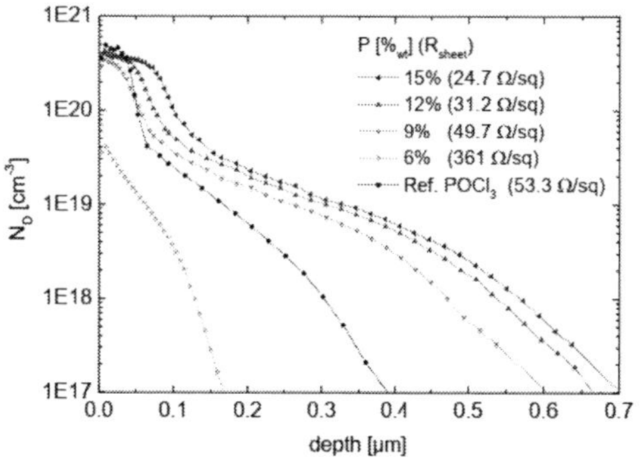

Figure 5: Doping profiles from ECV measurements of emitter samples shown in Figure 4.

The emitter from the high temperature diffusion was further investigated by processing Al-BSF solar cells from 2 Ωcm 6" p-type Cz-Si. The optimal P content that allows a good contact with screen printed Ag paste was determined in an experiment not shown here to 10 $\%_{wt.}$. On some solar cells with an APCVD-PSG emitter, an additional selective emitter etch-back with a variation of the etch depth was applied. All solar cells were processed in parallel, V_{OC} and j_{SC} are shown in Figure 6.

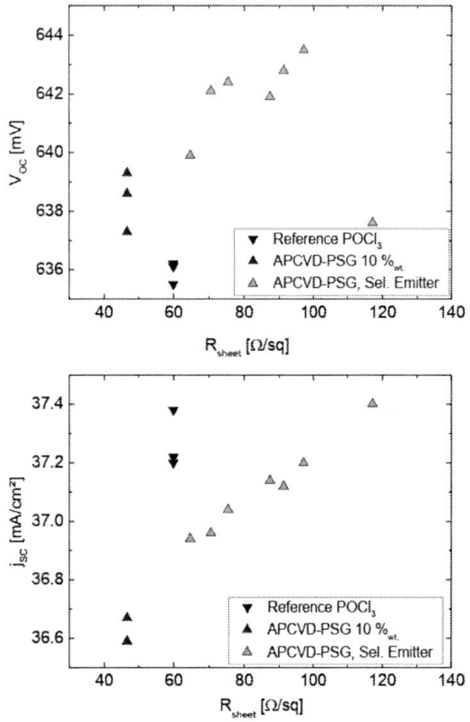

Figure 6: V_{OC} and j_{SC} of 6" standard p-type solar cells with a reference $POCl_3$ emitter and emitter diffused from APCVD PSG and selectively etched back to various R_{sheet}.

In good agreement with the j_{0E} results, the homogeneous APCVD-PSG emitter features higher V_{OC} than the reference $POCl_3$ emitter, even with a lower R_{sheet}. However, a significant loss in j_{SC} is observed. In combination with the higher contact resistance of the APCVD-PSG emitter solar cells, this leads to an average loss in cell efficiency of 0.4 % and a highest cell efficiency of 18.7 % vs. 19.1 % for the reference cells. The generally lower level compared the solar cells presented in section 3.1. is due to a higher area fraction of the front side metallization.

The selective etch-back of the APCVD-PSG emitter leads to an increase in V_{OC} and j_{SC}. The latter rises to the level of the reference cells for an etch-back to approx. 100-120 Ω/sq. The increase in R_{sheet} reduces the fillfactor (not shown here) of the selective emitter solar cells. The highest cell efficiency of 18.9 % was achieved at 75 Ω/sq.

The R_{sheet} at which the kink is completely removed can be calculated from ECV profiles to approx. 80 Ohm/sq. At an etch-back to this value, the losses resulting from the reduced blue response are still higher than for the reference cells. This indicates, that a large contribution of the emitter recombination arises from the deep tail of the profile. In contrast, the low surface P concentration leads to comparably small losses in the kink. The efficiency gain, that can be achieved by selectively etching the emitter surface, is therefore smaller than for typical $POCl_3$ emitters with high surface P concentration.

4 SUMMARY

The similarly high cell efficiency of 19.6/19.5 % indicates, that the electrical quality of the APCVD-PSG emitter can compete with the well-established $POCl_3$ emitter while promising low process costs, improved process control and compatibility to advanced cell concepts.

If R_{sheet} of the high temperature co-diffusion process is controlled by the P content in the PSG, the deep doping profile leads to significant j_{SC} loss when located at the illuminated side of the cell. However, this is not relevant for the application as a BSF on n-type solar cells, which is therefore the preferred cell structure for co-diffusion from APCVD deposited doping sources on both sides of the solar cell. By applying an additional selective etch-back, the j_{SC} loss could be reduced, but the loss in efficiency compared to the reference $POCl_3$ emitter solar cells was not fully compensated.

6 ACKNOWLEDGEMENTS

Part of this work was funded by the German BMWi (0325632, 0325581). The content is the responsibility of the authors.

7 REFERENCES

[1] G.S. May, S.M. Sze, Fundamentals of Semiconductor Fabrication (Wiley-VCH, 2004).
[2] A. Slaoui, B. Hartiti, J.C. Muller, R. Stuck, M. Loghmarti, P. Siffert, in 22nd IEEE Photovolt. Spec. Conf. (Las Vegas, USA, 1991) 445-449.
[3] K.O. Davis, K. Jiang, C. Demberger, H. Zunft, H. Haverkamp, D. Habermann, W.V. Schoenfeld, Phys. Stat. Sol. RRL 7 (2013) 319.
[4] H. Haverkamp, A. Dastgheib-Shirazi, B. Raabe, F. Book, G. Hahn, in 33rd IEEE Photovolt. Spec. Conf. (San Diego, 2008), 430-433.
[5] F. Book, Hocheffiziente Industriesolarzellen mit selektiver Oberflächendotierung, Dissertation, University of Konstanz, 2014.

FORWARD-BIAS-PLATED NI/CU FRONT CONTACTS FOR 20.5% EFFICIENCY N-TYPE BIFACIAL SOLAR CELL

Shu-Yen Liu*, Yueh-Lin Lee, Ming-Shiou Lin, Chih-Ming Wei, Kuang-Chieh Lai, and Chia-Chih Chuang
Motech Industries, Inc. Science Park Branch, Chief Technology Officer Office
No.96, Dashun 7th Rd., Xinshi Dist., Tainan City74145, Taiwan

ABSTRACT: N-type silicon bifacial solar cells with forward-bias-plated Ni/Cu front contacts are presented in this work. Two different front-side metallization processes were compared in this study, including conventional Ag/Al paste printing and Ni/Cu forward bias plating. We used laser to ablate the passivation layers on front side of the cells and thus the boron-doped silicon was exposed for Ni/Cu forward-bias plating. For the Ag/Al paste printed cells, 3 bus bars/90 fingers front electrode pattern design was used. For the Ni/Cu plated cells, 3 bus bars/90 fingers and 4 bus bars/110 fingers electrode pattern designs were used to observe the effect of electrode patterns on cell performances. We found that increasing bus bar number and finger number of the plated cells can obviously improve the fill factor (FF). Besides, the relationship between the electrode pattern designs and FF of the n-type bifacial cells were predicted by the simulation. The resistivity of the plated metal was about 1.9 μΩcm which was evaluated from the line resistance and cross-section area of the plated finger electrodes.
Keywords: Bifacial, Solar Cell, Metallization

1 INTRODUCTION

Nowadays, most of the global market share for crystalline Si solar cell is p-type solar cells. It is well known that an obvious reduction of cell performances occurs when the p-type cells are exposed to light. This is called light induced degradation (LID) effect. For p-type solar cells, under illumination B-O complexes were formed that degrade the bulk lifetime of Si wafer [1]. N-type Si solar cells have been developed to avoid LID effect, because there is no boron doping for the n-type Si substrates. Some products using n-type Si wafers are available, including interdigitated back-contact (IBC) cells and heterojunction with intrinsic thin layer (HIT) cells. However, the cost of these products is high due to the additional processing steps. Bifacial n-type cell structure is a potential candidate for achieving high efficiency and keeping low cost by using industrial production processes. To further lower the cost, Ni/Cu plating is a cost effective approach to replace the high material cost Ag paste and Ag/Al paste printing metallization process. Besides the low cost, there are many advantages of Ni/Cu plating metallization process, including the good ohmic contact property by NiSi forming, the needless high temperature firing process, and fine-line electrode structure for the plated metal, which is benefit to fill factor (FF), open-circuit voltage (Voc), and short-circuit current density (Jsc) of the cells, respectively.

In this work, the electrical properties of the conventional printed and forward bias plated n-type bifacial solar cells had been analyzed. In addition, the effects of front-metal-electrode designs on the plated n-PERT cells performances were discussed with the simulation and the experimental results. The resistivity of the plated metal was also evaluated from the line resistance and cross-section area of the plated finger electrodes.

2 EXPERIMENTS

The wafers we used were 156 x 156 mm^2 n-type Cz silicon wafers. After the alkaline texturing process, front emitter was formed by boron diffusion and back surface field was formed by phosphorous implantation. The passivation layers and SiNx anti-reflection layers were deposited on the doped area. The process described in detail for the n-type bifacial cell precursors was available from the previous report [2]. The rear-side metallization was completed by screen-printing commercially available Ag pastes. Two different front-side metallization processes were compared in this study, including conventional Ag/Al paste printing and Ni/Cu forward-bias plating. For conventional printing metallization process, it needs high temperature firing to form ohmic contact. In contrast, it does not need high temperature firing for Ni/Cu plating metallization to form ohmic contact. We used 532 nm green ns laser to ablate the passivation layers on front side of the cells and thus the boron-doped emitter silicon was exposed for Ni/Cu plating. Light-induced plating is a major method of plating metallization for silicon solar cells [3]. In this report, forward-bias plating was used to deposit the Ni/Cu front contacts because front p-type emitter structure was used.

For the Ag/Al paste printed cells, 3 bus bars/90 fingers front electrode pattern design was used. For the Ni/Cu plated cells, 3 bus bars/90 fingers and 4 bus bars/110 fingers electrode pattern designs were used to observe the effect of electrode patterns on cell performances. Besides, the relationship between the electrode pattern designs and FF of the n-type bifacial cells were predicted by the simulation.

3 RESULTS AND DISCUSSION

Fig. 1 shows the schematic cross-section of the n-type bifacial cell structure with Ni/Cu plated front electrode. Table I shows the cell performances for the n-type bifacial cells with printing and plating metallization processes. With the electrode design of 3 bus bars/90 fingers, Ni/Cu plated cells had higher Jsc, lower FF, and higher series resistance (Rs) than those of the Ag/Al paste printed cells. Because the front-plated Ni/Cu finger-type electrode had smaller line width than that of the front-printed Ag/Al finger-type electrode, higher Jsc were obtained for the plated cells. Even though the line resistance is lower for the plated Ni/Cu electrode than that of the printed Ag/Al electrode, the laser ablation process was not optimized that the worse contact property was obtained for the plated cells resulting in higher Rs and lower FF. Back to Table I, increasing bus bar number and finger number of the plated cells can obviously improve the Rs and FF. This result was consistent with the simulated result as shown in Fig. 2. Increasing the finger number means the shorter distance between two finger-type electrodes, so the photo-generated carriers can be collected more effectively. In addition, the effective line resistance can be reduced by increasing the bus bur number from 3 bus bars to 4 bus bars. Therefore, increasing bus bar number and finger number of the plated cells can obviously improve the Rs and FF. Fig. 3 shows the I-V curve of the peak cell with 4 bus bars/110 fingers electrode pattern, and the insert shows the cell performances. The highest efficiency of the Ni/Cu plated cell is 20.55 %. The SEM images of the plated finger-type electrode are shown in Fig. 4. The plated Ni/Cu finger width and finger height are 38.4 μm and 16.4 μm, respectively. The resistivity of the plated metal is about 1.9 μΩcm which was evaluated from the line resistance and cross-section area of the plated electrode.

4 CONCLUSION

In this study, the rear-side metallization was completed by screen-printing commercially available Ag pastes. Two different front-side metallization processes were compared in this study, including conventional Ag/Al paste printing and Ni/Cu forward bias plating. We used laser to ablate the passivation layers on front side of the cells and thus the boron-doped emitter silicon was exposed for Ni/Cu plating. In this work, forward bias plating was used to deposit the Ni/Cu front contacts. For the Ag/Al paste printed cells, 3 bus bars/90 fingers front electrode pattern design was used. For the Ni/Cu plated cells, 3 bus bars/90 fingers and 4 bus bars/110 fingers electrode pattern designs were used to observe the effect of electrode patterns on cell performances. Besides, the relationship between the electrode pattern designs and FF of the n-type bifacial cells were predicted by the simulation. The resistivity of the plated metal was also evaluated from the line resistance and cross-section area of the plated finger electrodes.

5 ADDITIONAL COMPONENTS

5.1 FIGURES

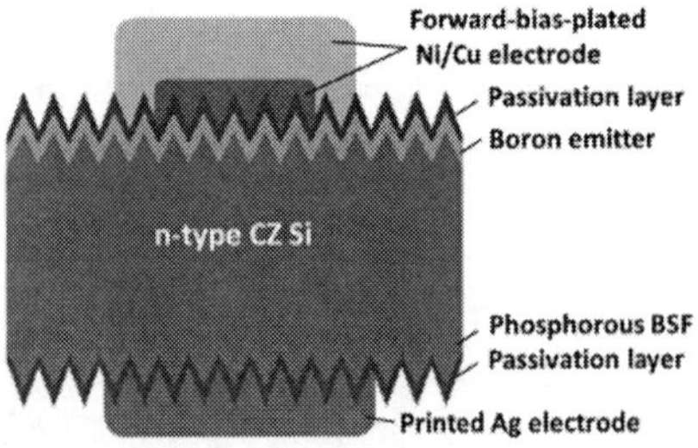

Figure 1: Schematic cross-section of the n-type bifacial cell structure

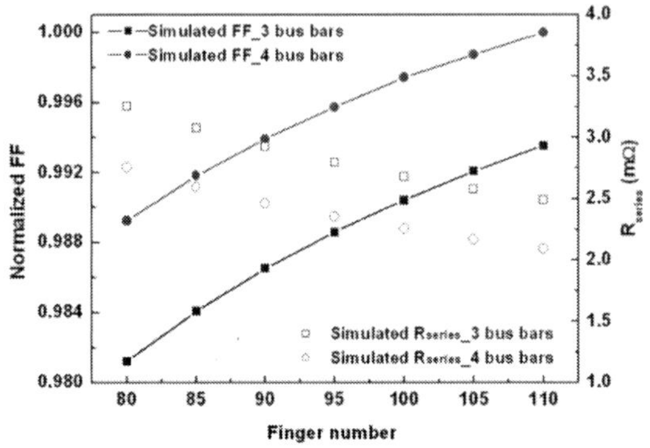

Figure 2: Simulated results of FF and R_{series} for the cells with 3 and 4 bus bars and different finger numbers

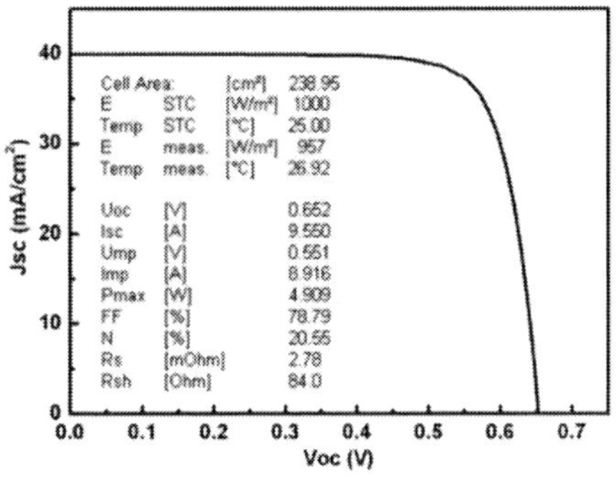

Figure 3: I-V curve of the peak cell with 4 bus bars/110 fingers, the insert shows the cell performances

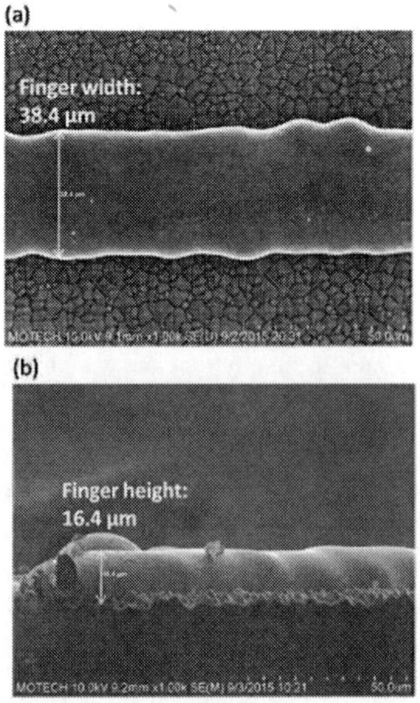

Figure 4: SEM images of the forward bias plated Ni/Cu contact (a) top view, (b) lateral view

5.2 TABLES

Table I: Cell performances for the n-type bifacial cells with printing and plating metallization

Cell types	V_{oc} (mV)	J_{sc} (mA/cm^2)	*FF* (%)	*Efficiency* (%)	R_{series} (mΩ)
With printed-front contacts (3 bus bars/90 fingers)	649	39.88	78.2	20.25	3.42
With plated-front contacts (3 bus bars/90 fingers)	648	40.55	76.6	20.13	4.31
With plated-front contacts (4 bus bars/110 fingers)	652	39.93	78.8	20.52	2.71

5.3 REFERENCES

[1] S. W. Glunz, S. Rein, W. Warta, J. Knobloch, W. Wettling, "Degradation of carrier lifetime in Cz silicon solar cells", Sol. Energy Mater. Sol. Cells, vol. 65, pp. 219 – 229 (2001).

[2] P. T. Hsieh, C. J. Huang, H. C. Chang, S. W. Chiu, C. C. Li, "20.2% efficiency n-type bifacial solar cell by industrial fabrication processes", 28th European Photovoltaic Solar Energy Conference, pp. 1174 – 1176 (2013).

[3] S. Kluska, J. Bartsch, A. Büchler, G. Cimiotti, A. Brand, S. Hopman, M. Glatthaar, " Electrical and mechanical properties of plated Ni/Cu contacts for Si solar cells", Energy Procedia, vol. 77, pp. 733 – 743 (2015).

THE EFFECT OF SURFACE PASSIVATION AT LOW-INJECTION LEVEL ON FILL FACTOR OF SILICON HETEROJUNCTION SOLAR CELLS

Lin Zhang[1,2], Mingchong Ren[1,2], Jin Wang[1,2], Rong Yang[1,2], Liwei Li*[1,2], Yuan Meng[1,2], Ted Guo[1,2]
[1]ENN Solar Energy Co., Ltd., [2]State Key Laboratory of Coal-based Low Carbon Energy
No.106 Huaxiang Road, Langfang, Hebei, P. R. China 065001
Email: liliwei@enn.cn, zhanglinl@enn.cn

ABSTRACT: In this paper, we present our study of fill factor (FF) loss mechanism for silicon heterojunction solar cells (SHJ) in relation to the quality of surface passivation. The effective minority carrier lifetime (τ_{eff}) as a function of carrier injection density (Δn) has been modeled to identify the limiting factor of FF in our research. We have found that defects at a-Si:H/c-Si interface can cause low-quality surface passivation, which can be characterized by minority carrier lifetime at low-injection level, leading to reduction of fill factor. We propose a new mechanism that the presence of shallow defects with positive charges at the interface contributes to degradation of fill factor of SHJ cells. This hypothesis was verified using an AFORS-HET simulation tool in our study. Based on this study, we have developed an anti-contamination tray for wafer handling at the PECVD processing step. Our results show improvement of fill factor by an absolute value of 2%. SHJ solar cells with high efficiency of greater than 23% have been successfully fabricated using this technique.
Keywords: Heterojunction, silicon solar cell, simulation.

1 INTRODUCTION

To achieve high efficiency in silicon heterojunction (SHJ) solar cells, it is crucial to obtain high fill factor (FF) which allows maximum output of power generation. The mechanisms limiting FF of SHJ solar cells are mainly resistive losses and recombination of electron-hole pairs. There have been many studies and significant efforts put into addressing the resistive losses of SHJ solar cells, from both PV academia and industry. However, few researches have been conducted on the impact of carrier recombination on FF. In this work, we focus on the loss mechanism of FF caused by carrier recombination at low-injection level due to insufficient surface passivation.

The carrier recombination can be studied by a simple but effective method, e.g. the measurement of τ_{eff} as a function of carrier injection level (Δn). The shape of the τ_{eff} curve reveals information about the nature of recombination [1-2]. It was demonstrated that poor passivation of the c-Si/a-Si:H interface strongly affects the field effect passivation at low-injection level ($\Delta n<10^{15}$cm^{-3}) and the pseudo fill factor [3-5]. In 2015, Adachi et al. reported that an impressive FF improvement of 2.7% was obtained by increasing the τ_{eff} at low-injection level [6].

Up to now, several mechanisms have been proposed to interpret the decrease of τ_{eff} at low Δn. Tomasi *et al.* have found that the presence of TCO could decrease the τ_{eff} at low injection [7]. Steingrube *et al.* have proposed a model where the interface recombination is caused by surface damage [8]. Reusch *et al.* have investigated the field effect passivation by simulating the charges of doped a-Si:H films [9]. The local-ideal factor and wafer doping concentration have also been proved to induce the injection-level-dependent τ_{eff} [10-11]. However, it is difficult to determine from above mentioned studies which mechanism limits the low-injection passivation of specific type of silicon wafers. In addition, the physical nature of the deterioration of τ_{eff} at low-injection level is seldom involved.

In this study, we demonstrate the influences of surface passivation at low-injection level on fill factor and that the interface defects result in poor passivation at $\Delta n<10^{15}$cm^{-3}, by modeling of τ_{eff} data. Based on our study, we propose a surface charge mechanism that explains the reduction of τ_{eff} at low Δn. Lastly, using the simulation results, we have developed an anti-contamination tray for wafer handling at PECVD step, and achieved a remarkable increase of SHJ fill factor and high cell efficiency of 23.27%.

2 EXPERIMENTAL

The mono-facial SHJ solar cells were prepared on commercially available monocrystalline silicon wafers (5-inch n-type). The wafers were textured by a NaOH solution and cleaned by a wet chemical process. After stripping off the chemical oxide using a diluted hydrofluoric solution, the front i/p and rear i/n amorphous silicon layers were deposited in cluster plasma enhanced chemical vapor deposition (PECVD) reactors. Subsequently, TCO and the rear metal contact were deposited in a magnetron-sputtering reactor. Ag grid electrodes were printed on the front side followed by a baking step at around 200°C. The I-V parameters and carrier lifetimes were characterized using a solar simulator and Sinton WCT-120 tester, respectively.

3 RESULTS AND DISCUSSION

3.1 Effects of injection-dependent τ_{eff} on the FF

In our early study, we found that the variation of efficiency was mainly caused by the fluctuation of FF. Table 1 shows the parameters of two typical solar cells in a same batch. The open-circuit voltage (Voc) and short-circuit current density (Jsc) of the two cells are similar but FF difference reaches an absolute value of 2.9%. No difference of series resistance is extracted, which is also verified by the resistance measurement of Ag finger and TCO. Thus, we speculated that the difference resulted from the p-n junction and measured the curve of injection-level-dependent τ_{eff}, as shown in Figure 1. The solar cell with low FF features of low τ_{eff} at low-injection level, resulting in decreased Pseudo Fill Factor (PFF). It is demonstrated that at high-injection level, τ_{eff} of different solar cells are similar, however, at low-injection level, poor surface passivation likely results in low τ_{eff} and low FF, which is attributed to high recombination rate. Therefore, τ_{eff} at low-injection level can be an effective indicator for surface passivation quality which is strongly correlated to FF. A factor γ, denoted as the ratio of $Voc_{0.1SUN}/Voc_{1SUN}$, is employed to characterize the degree of the reduction of τ_{eff} from high to low-injection level, under 1 and 0.1 SUN illumination, respectively. The FF and EFF as a function of γ are plotted, as shown in Figure 2. A clear dependence of both FF and EFF on γ is observed. The results show that γ can be used as a reflection of solar cell quality and guideline to further improve solar cell performance.

Table I: The performance parameters of SHJ solar cells

Sample No.	EFF (%)	Voc (mV)	Jsc (mA/cm2)	FF (%)
Cell 1	21.6	725	38.5	77.5
Cell 2	20.9	726	38.6	74.6

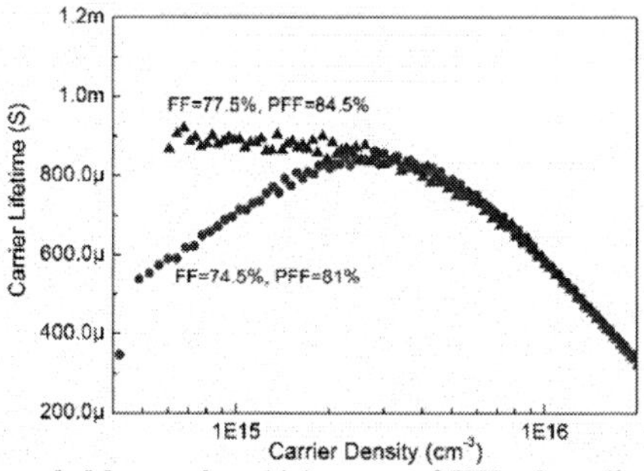

Figure 1: Measured $\tau_{eff}(\Delta n)$ curves of SHJ solar cells

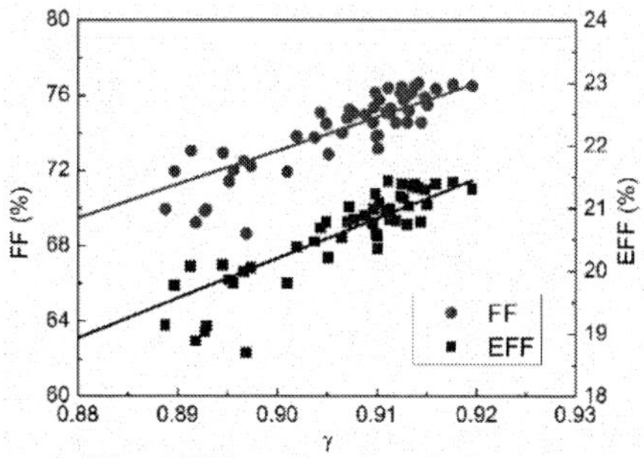

Figure 2: FF and EFF as a function of γ factor

3.2 Modeling of injection-dependent τ_{eff}

We model τ_{eff} to quantitatively extract the dominated carrier recombination from the measured τ_{eff} data. The τ_{eff} is expressed as:

$$\tau_{eff} = \frac{\Delta n}{(U_{Auger}+U_{Rad}+U_{bulk,SRH}+U_{li}+U_{hi})} \tag{1}$$

where U_{Auger} is the Auger recombination rate, U_{Rad} the radiative recombination rate, and $U_{bulk,SRH}$ the Shockley-Read-Hall (SRH) recombination rate of silicon wafer, as defined in references [12-13]. U_{li} and U_{hi} are the surface recombination rate of low injection and high injection, respectively. Note that the U_{li} and U_{hi} are similar with the saturation current densities J_{01} and J_{02}, which are related to the intrinsic recombination at high-injection level and a-Si:H/c-Si interface recombination, respectively. Thus, the corresponding recombination rates are described as follows:

$$U_{hi} = \frac{J_{ohi}exp\left(\frac{qV}{n_{hi}k_bT}\right)}{qW} \text{ and } U_{li} = \frac{J_{oli}exp\left(\frac{qV}{n_{li}k_bT}\right)}{qW} \tag{2}$$

$$\Delta n = \frac{-N_d + \sqrt{N_d^2 + 4\left(n_{i,eff}^2 exp\left(\frac{V}{k_bT}\right)\right)}}{2} \tag{3}$$

Where N_d is the doping concentration, $n_{i,eff}$ effective intrinsic carrier concentration, J_{0hi} (J_{0li}) and n_{hi} (n_{li}) recombination current and ideal factor at high (low) injection level. Figure 3 shows the decomposition of the effective lifetime measured from a pi/c-Si/in configuration. This model predicts that the surface recombination dominates the shape of the effective carrier lifetime curve at low injection and it is probably limiting the fill factor because of its high surface recombination current (J_{0li}).

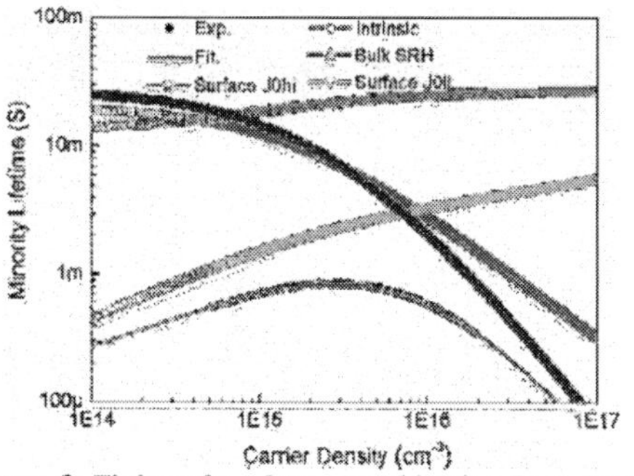

Figure 3: Fitting of τ_{eff} data: recombination from Auger, bulk SRH, and surface recombination

A reverse calculation of IV curve is carried out by the components of $\tau_{eff}(\Delta n)$ curves to investigate how J_{0li} affects FF. The equation of the cell current is expressed as:

$$J = J_{gen}(1-shading) - qW(U_{Auger}+U_{Rad}) - qWU_{SRH} - J_{0hi}\left(exp\left(\frac{qV}{n_{hi}k_bT}\right)-1\right) - J_{0li}\left(exp\left(\frac{qV}{n_{li}k_bT}\right)-1\right) \tag{4}$$

The FF loss due to surface recombination J_{0li} is calculated under this approximation. Figure 4 shows that increase of J_{0li} dramatically shifts the maximum power point (MPP). With J_{0li} decreasing from 1.6×10^{-4} to 1.6×10^{-6} mA/cm^2, calculated FF increases from 78% to 85%. However, FF tends to saturate with further J_{0li} increase.

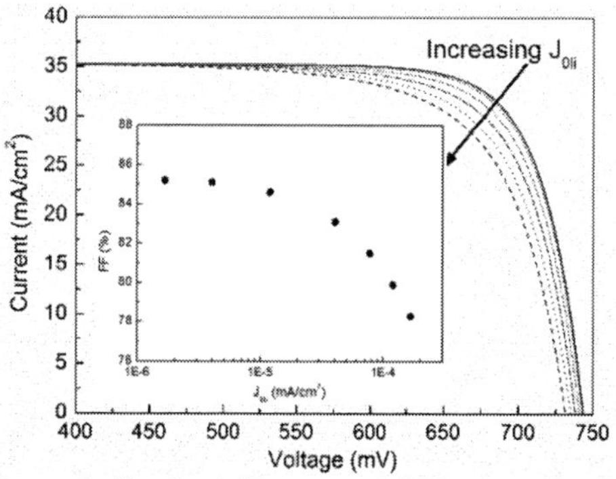

Figure 4: Simulated IV curves of SHJ solar cells with different combination current J_{0li} at low-injection level (inset: extracted FF versus J_{0li}).

A mechanism is proposed to interpret the FF loss caused by the surface defects. We speculate that the presence of shallow energy level defects with positive charges on the surface decreases the effective charges (Q_c) of the inversion layer, leading to the decrease of FF. As shown in Figure 5, charge neutrality results from the balance between the charges in p-type a-Si:H (Q_g), the charges in the interface states (Q_i), and the charges in c-Si (Q_c). When positive defects (Q_d) exist on the interface, they can trap negative charges, thereby possibly cancelling part of the field-effect passivation provided by the negative charges in the passivation film. The effects of negative charges (Q_g plus Q_i) on the performance of SHJ solar cell are investigated by the AFORS-HET tool [14]. Figure 6 (a) shows the simulation results that the hole concentration of the inversion layer as a function of the distance from the intrinsic a-Si:H/c-Si interface. With the decrease of negative charges from -9.0×10^{12} to $-1.0\times10^{11}cm^{-2}$, the hole concentration decreases dramatically. No more hole concentration decrease is observed when negative charges are less than $-1\times10^{11}cm^{-2}$. Increasing the hole density in the inversion layer can reduce the concentration of electrons, leading to improvement of field effect passivation and thus increase of fill factor [4]. As shown in Figure 6(b), with the reduction of the negative charges, the FF decreases from 83.0% to 79.5%. When the charge turns from negative to positive, an intense decrease of fill factor is observed. On the other hand, no obvious Voc changes observed, as shown in Figure 6(b). It is well known that the shallower the defect level, the less the carrier recombination. Thus, it is reasonable to speculate that the state of defects is shallow and near the valence band. Such defects can possibly capture the charges in the inversion layer, leading to the decrease of fill factor.

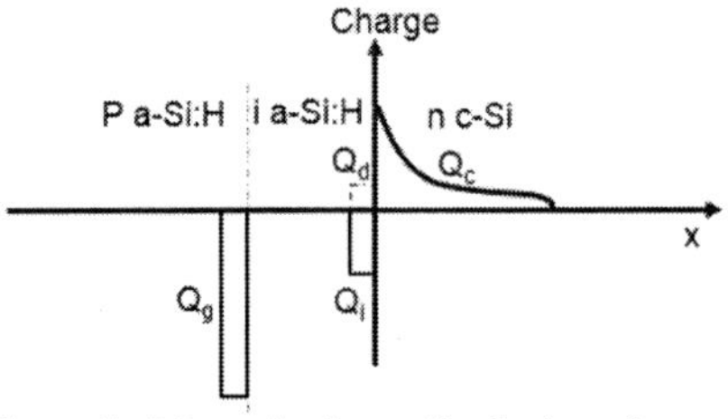

Figure 5: Schematic charge distribution of n-type doped c-Si passivated by i/p a-Si:H layers

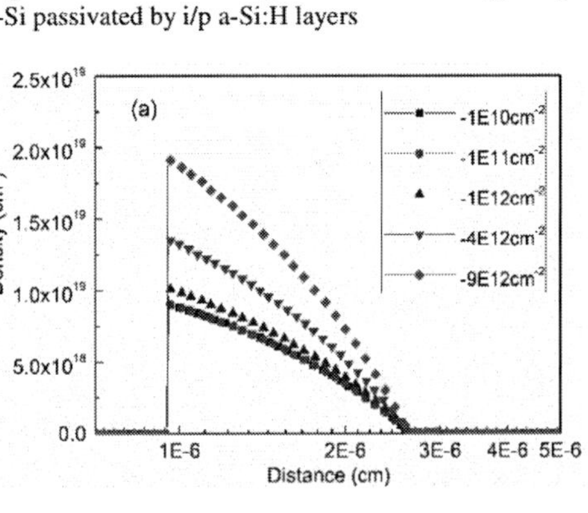

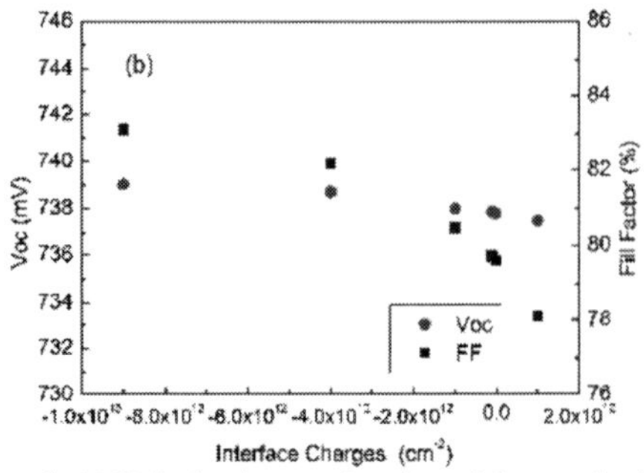

Figure 6: (a) Hole density as a function of distance in the near surface of silicon, (b) Voc and FF as a function of interface charges

3.3 Solar cell process optimization

According to the simulated results and passivation mechanism, the fabrication process of SHJ solar cell can be improved. As mentioned above, the defects on the interface are shallow. It is well known that dangling bonds on the surface can cause deep-level defects. Thus, the shallow-level defects are unlikely caused by the intrinsic a-Si:H layer. We speculate that the defects could be caused by contamination during wafer handling. Thus, we introduce an anti-contamination tray to eliminate contamination in a-Si:H/c-Si interface at PECVD step. Figure 7 shows the effects of using anti-contamination tray on SHJ solar cell performance. By using the anti-contamination tray, the median of FF increases from 75% to 77% and EFF increases by an absolute value of 0.6%. However, there is no obvious Voc and Jsc increase. The result is consistent with AFORS-HET simulation. The FF improvement exactly comes from PFF increase (from 80.5% to 82.5%), as shown in Figure 8. We also extracted the series resistance from the IV curves and no difference is observed between different trays. Thus, the FF improvement of is attributed to PFF increase, resulting from low surface combination rate at low-injection level.

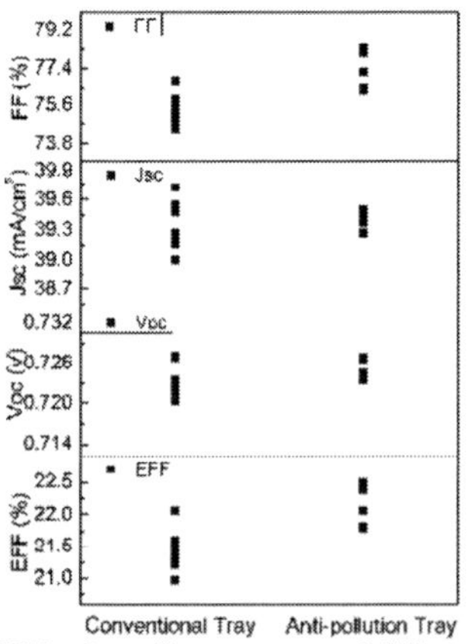

Figure 7: SHJ solar cells parameters of different trays

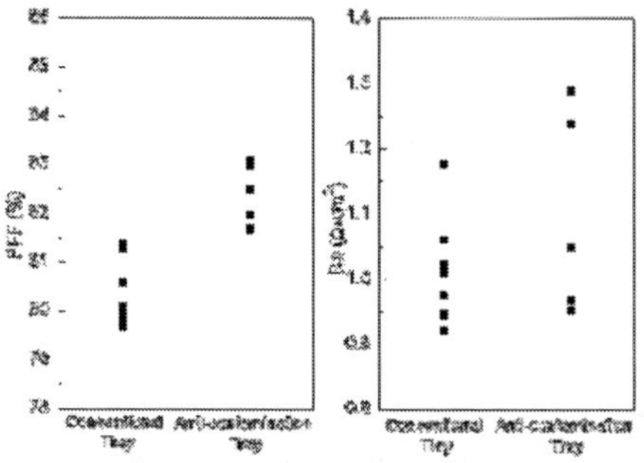

Figure 8: PFF and Rs of SHJ solar cells using conventional tray and anti-contamination tray

By using the anti-contamination tray, we have achieved SHJ solar cell efficiency of 23.27%. Figure 9 shows the solar cell I-V characteristics, with a high fill factor up to 82.05%.

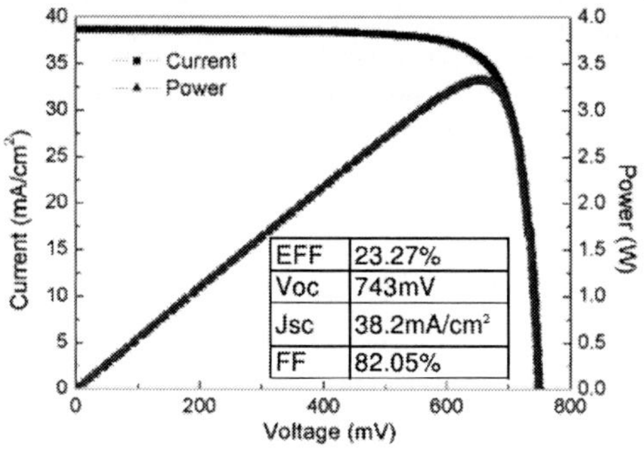

Figure 9: SHJ solar cell IV curve

4 CONCLUSIONS

In this work, we have found that silicon heterojunction solar cells with low fill factor (FF) showed low effective carrier life time (τ_{eff}) at low-injection level. By modeling recombination sources of τ_{eff}, we speculated that poor passivation, i.e., defects at a-Si:H/c-Si interface, are dominating contributor to more recombination and thus low τ_{eff}, at low-injection level, which further results in low FF. The defects mentioned above are shallow-level defects that trap positive charges and lead to decreased FF, this is consistent with AFORS-HET simulation of carrier transportation. Based on this FF loss mechanism, we have developed an anti-contamination tray for wafer handling at PECVD step to improve passivation quality which is confirmed by increased τ_{eff} at low-injection level. A significant increase of SHJ FF and high cell efficiency of 23.27% are achieved.

5 REFERENCES

[1] A. Descoeudres et al, IEEE Journal of Photovoltaics, Vol. 3 (2013) 83.

[2] C. Leendertz et al, Applied Physics Letters, Vol. 98 (2011) 202108.

[3] M. Ghannam et al, Solar Energy Materials & Solar Cells, Vol.132 (2015) 320.

[4] A. Tomasi et al, In the proceedings of the 42nd Photovoltaic Specialist Conference, Vol.3 (2015) 1.

[5] M. Filipic et al, Journal of Applied Physics, VOL.114 (2013), pp.074504-1-7.

[6] D. Adachi et al, Applied Physics Letters, Vol.107 (2015) 233506.

[7] B. Demaurex et al, IEEE Journal of Photovoltaics, Vol. 4 (2014) 1387.

[8] S. Steingrube et al, Journal of Applied Physics, Vol. 108 (2010) 014506.

[9] M. Reusch et al, Energy Procedia, Vol. 38 (2013) 297.

[10] Mark J. Kerr et al, Journal of Applied Physics, Vol. (2002) 2473.

[11] M. Kessler et al, Journal of Applied Physics, Vol. 111 (2012) 054508.

[12] A. L. Blum et al, In proceedings of 28th European Photovoltaic Solar Energy Conference and Exhibition (2013) 1521.

[13] S. Y. Herasimenka, A Dissertation for the Degree Doctor of Philosophy, Arizona State University (2013) 134.

[14] A. Froitzheim et al, Proceedings of the 3rd World Conference on Photovoltaic Energy Conversion (2003), 1P-D3-34.

DOPED A-SI:H/μC-SI:H HYBRID LAYERS USED TO IMPROVE THE PERFORMANCE OF TOP-CON SILICON SOLAR CELLS

Ke Tao[1], Jin Wang[2], Rui Jia[1], Yun Sun[1], Zhi Jin[1], Xinyu Liu[1]

1, Institute of Microelectronics, Chinese Academy of Sciences, 3# Bei-Tu-Cheng West Road, Beijing100029, China

2, College of Electronic Information and Optical Engineering, Nankai University, #94 Weijin Road, Tianjin300071, China

ABSTRACT: Carrier-selective passivated contact with low minority carrier recombination and efficient majority carrier transport has become an active area of investigation for high-efficiency n-type Si solar cells. This paper presents our research on the TOP-Con silicon solar cells, which focuses on the passivated contacts based on a thin tunneling oxide layer and a silicon thin film. The ultra-thin (< 20Å) tunnel oxide layers are grown chemically on silicon wafer, and the p-type and n-type silicon thin films deposited by PECVD system are applied for the front emitter and BSF. A post-annealing is used to facilitate the crystallization of the as-deposited Si layers and activation of dopants. The influence of the microstructure of silicon thin film on the passivation property was studied. The quality of contact passivation was estimated by means of QSSPC measurement on the symmetric structures composed of tunnel oxide and the n+ poly-Si layer. The experimental results indicated that the tunnel oxide combined with a-Si:H featured a good passivation property compared to the μc-Si:H. whereas, the annealed μc-Si:H led to a high fraction of crystallization, and no blistering was observed. A hybrid structure containing a-Si/μc-Si:H thin films was suggested to improve both the passivation and contact properties. The effective carrier lifetime of 3.2ms, implied Voc of 716 mV and implied fill factor (FF) of >85% were achieved with symmetric structure on n-type Cz substrate (3~8 Ωcm, 300 μm), indicating that our tunnel oxide/n+ hybrid-Si provides excellent passivation.

1 INTRODUCTION

The surface passivation always takes the priority for the design and optimization of crystalline silicon solar cell. PERC and PERL solar cells, which are the hotspot of industrialization, alleviate the problem in surface passivation by introducing insulators such as silicon oxide, aluminum oxide and silicon nitride [1,2] . Recently, the named TOP-Con (Tunnel Oxide Passivated Contact) solar cell technology has been a study focus in the world due to the excellent passivation ability without opening [3~6]. The TOP-Con structure consists of a doped Si layer and an ultra-thin tunneling SiO_2 layer which is sufficiently thin to enable the tunneling of majority carriers while blocking the transport of minority carriers at the same time. The main feature of TOP-Con solar cells is the surface-passivation by tunnel oxide and realized selective contact by highly doped silicon thin films, resulting in a very low recombination rate. Owing to its excellent passivation and junction properties, remarkable cell efficiency up to 25.1% has been reached, which is the world record for both sides-contacted silicon solar cell [7].

For the fabrication of TOP-Con structure, after the deposition of Si thin film on the tunneling SiO_2 layer, an annealing process is usually needed to increase the crystallization fraction [8]. Highly crystallized Si layer allows more efficient doping, leading to the excellent field passivation and junction property. In this work, the passivation quality of doped Si thin film was investigated by vary the hydrogen dilution rate. It was found that doped a-Si layer exhibited good passivation quality due to the large content of hydrogen atoms in the film. However, with further increasing the hydrogen content in the a-Si layer, blistering was observed on the TOP-Con samples, leading to the degradation of passivation quality. Compared to a-Si layer, μc-Si layer showed a lower passivation quality due to the smaller content of hydrogen. However, a lower thermal budget was required to re-crystallize and better crystalline quality was achieved. Simultaneously, no blistering can be observed by μc-Si layer after the thermal annealing. Based on this, new hybrid structure consisting of a thin a-Si layer and a thin μc-Si layer was designed for TOP-Con samples. No blistering was observed while a high passivation quality was reserved.

2 EXPERIMENTAL

Passivated contacts were fabricated on n type CZ wafers with resistivity of 3~8Ω-cm on symmetric structures poly-Si(n+)/SiOx/c-Si(n)/SiOx/poly-Si(n+), as shown in Fig. 1. All the wafers are double-side polished. The wafers were firstly chemically cleaned by a RCA process, followed by a HF-dip to remove the surface oxide. Then the tunnel oxide layer was grown in 70 wt% HNO3 acid at 110°C for 10 min. The resulted tunnel oxide thickness was about 1.5 nm, which is thinner than the maximum allowed oxide thickness of 2 nm for efficient tunneling [9]. Next, a thin phosphorus-doped Si thin film was deposited on both sides of the wafer using a single-chamber PECVD system. A subsequent anneal at 800°C crystallized Si thin film into n+ poly-Si. Forming Gas Annealing (5% H_2 diluted in N_2) at 450°C was performed to further improve the contact passivation. In order to investigate the microstructure of Si thin films when deposited at different conditions, quartz glass was used as substrate.

Film thickness was evaluated by spectroscopic ellipsometry (SE400, Sentech, Germany), and micro-structure was characterized by Raman spectra. The doping profiles were measured by electrochemical capacitance-voltage (ECV) profiling (CVP21, WEP, Germany). The implied Voc and effective minority carrier lifetime τeff were obtained by quasi-steady-state photoconductance (QSSPC) method (WCT-120, Sinton Instruments, USA).

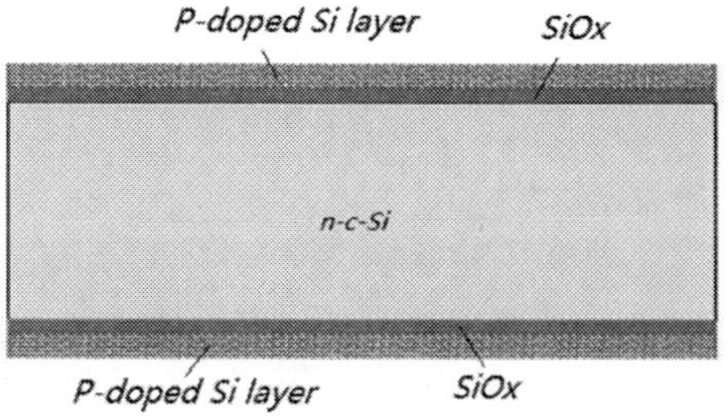

Fig.1 the cross-sectional structure for passivated contact samples.

3 RESULTS AND DISCUSSION

It is well known that the atomic hydrogen can improve the interface passivation quality. Therefore, forming gas anneal [10] and remote plasma hydrogen passivation [11] are often used after the re-crystallization of a-Si:H thin films. During the deposition of Si thin films, hydrogen dilution rate can also affect the interface passivation quality. A group of samples were fabricated by varying the Hydrogen dilution rate ($R=H_2/SiH_4$) from 3 to 209, and the passivation quality was studied. As shown in Fig. 2, the effective minority carrier lifetime was drastically increased, when the hydrogen dilution rate was augmented from 3 to 33. This suggested that increasing hydrogen dilution rate helped to improve the passivation quality. However, With further increasing the hydrogen dilution rate, the effective minority carrier lifetime degraded rapidly. The implied Voc exhibited the similar tendency when changing the hydrogen dilution rate. The maximum of effective minority carrier lifetime and implied Voc were 3.6 ms and 719 mV respectively.

In fact, the increase in hydrogen dilution rate during the plasma enhanced chemical vapor deposition of a-Si:H results in the increase in the atomic hydrogen content. During the re-crystallization of a-Si:H, atomic hydrogen will in-diffuse from a-Si layer into Si/SiO_x interface. A mass of atomic hydrogen can saturates the dangling bonds, thus improve the passivation quality. When the hydrogen dilution rate is further increased, Si thin film will transform from amorphous phase into crystalline phase, leading to a decrease in the hydrogen content in the Si thin films. This will degrade the passivation ability.

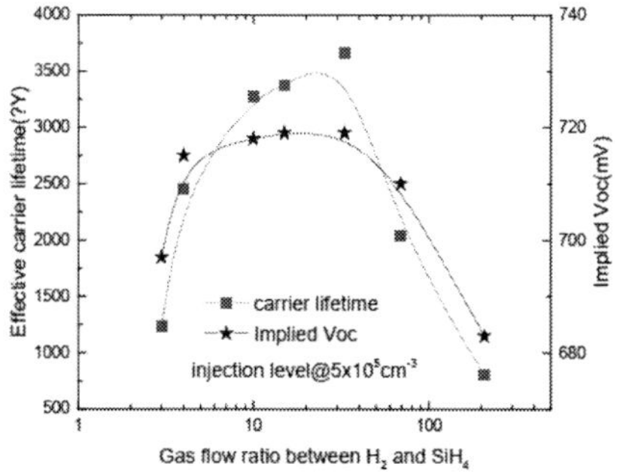

Fig.2 The effective carrier lifetime and implied Voc as a function of gas flow ratio between H_2 and SiH_4.

A post-deposition thermal treatment is beneficial to activate of dopants and improve the micro-structure. High annealing temperature results in a decrease of Si-H bonds corresponding to a reduction of the amorphous matrix in the layer promoting the formation of covalent Si-Si bonds [12]. The crystallization of the Si layer in TOP-Con structure brings advantages. For instance, the μc-Si has higher doping efficiency than a-Si, which means better field passivation and junction property [13]. Moreover, a more ordered structure of the Si layer improves the carrier transport property due to the increased crystallization fraction [14]. In addition, ITO (Indium Tin Oxide), the most commonly used transparent conductive coating, is n-type doped degenerate semiconductor. A depletion region is formed at the interface of ITO and the TOP-Con emitter. A high doping concentration in the Si layer prevents the complete depletion induced by the ITO, which improves the Voc and the FF of the TOP-Con solar cells.

Although μc-Si thin films exhibit week passivation quality compared with a-Si thin films, the above mentioned advantages of μc-Si thin films make it valuable to be applied into solar cells devices. Besides, only a low budget is needed for the μc-Si thin films to crystallize. As shown in Fig. 3(a), the a-Si thin films are still amorphous phase after annealing at 600°C for 1 hour. Whereas, μc-Si thin films already show a big crystalline fraction when annealed at 600°C for 1 hour, as shown in Fig. 3(b). Both a-Si thin films and μc-Si thin films exhibit high crystalline fraction when they are annealed at a higher temperature, i.e. ≥700°C. However, a higher crystalline fraction and better crystal quality are obtained by μc-Si.

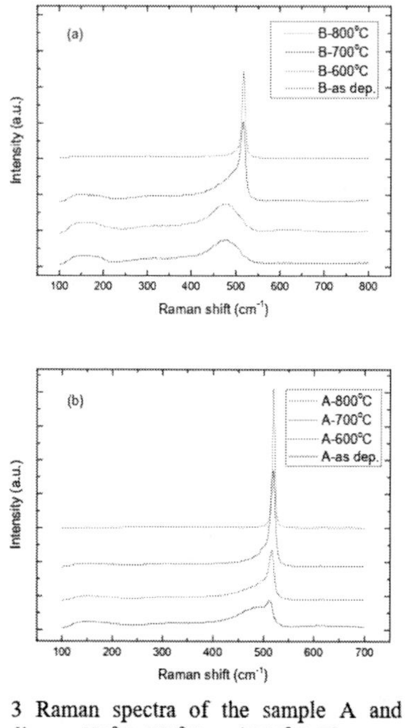

Fig. 3 Raman spectra of the sample A and B after annealing at 600°C, 700°C, and 800°C. The amorphous Si layer is characterized by a broad peak around 480 cm^{-1} and the μc-Si phase is described by a superposition of the

amorphous and crystalline Si peak (520.7 cm^{-1})

In addition, blistering is observed for a-Si thin films after they were annealed at 700°C, and it is aggravated after annealing at 800°C, as shown in Fig.4(a) and (b). Interestingly, μc-si thin films never blister even annealed at 800°C, as shown in Fig.4 (c). The blistering of Si layer was caused by hydrogen effusion which was provided from the breaking of weakly bonded hydrogen coming from SiH_2-bonding and SiH_3-bonding, as well as strongly bonded hydrogen of isolated Si-H bonds [15]. The hydrogen can accumulate at the interface between Si substrate and SiO_2 layer, leading to the increase of the hydrogen pressure at the interface. When the hydrogen pressure exceeds some critical value, the explosion of the blistering occurs [14, 16]. The elimination of blistering for μc-Si layer is due to its lower hydrogen incorporation in the silicon network compared with a-Si layer [17].

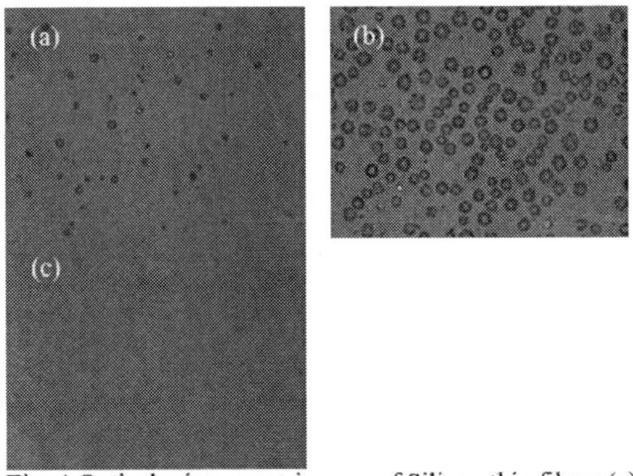

Fig. 4 Optical microscope images of Silicon thin films: (a) a-Si annealed at 700°C, (b) a-Si annealed at 800°C and (c) μc-Si annealed at 800°C. Blistering can be observed in (a) and (b).

The observed blistering can undoubtedly weaken field passivation due to loss of doped Si layer. In addition, the in-diffusion of dopants into Si substrate is accelerated though the blistering area, as shown in Fig. 5. Both a-Si and μc-Si thin films deposited on c-Si substrate were annealed at 800°C for 1hour. A abrupt drop in dopant concentration at Si/SiO_x interface was observed for μc-Si samples, whereas, an obvious in-diffusion of dopant into c-Si substrate was observed for a-Si samples. The in-diffusion of dopants into the Si substrate also deteriorated the interface quality.

In order to combine the advantage of a-Si and μc-Si thin films, a hybrid structure was designed, in which a thin a-Si layer was firstly deposited on tunnel oxide, followed by another μc-Si layer. The thickness for each layer was 10 nm and 15 nm, respectively. In fact, the a-Si can work as the incubation layer for μc-Si thin films, thus, promote the crystallization of μc-Si. Fig. 6 shows the injection dependent effective minority carrier lifetime for the symmetrically passivated samples with a-Si, μc-Si and hybrid layer. An effective minority lifetime of 3.2 ms was achieved, which is comparable to that of a-Si samples, but far better than that of μc-Si samples. No blistering was observed again for the TOP-Con samples when using hybrid layer.

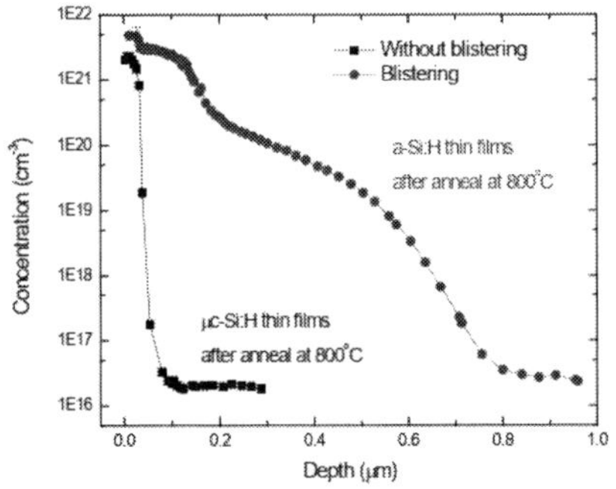

Fig. 5. The doping profile of TOP-Con samples passivated by μc-Si and a-Si after annealed at 800°C for 1 hour.

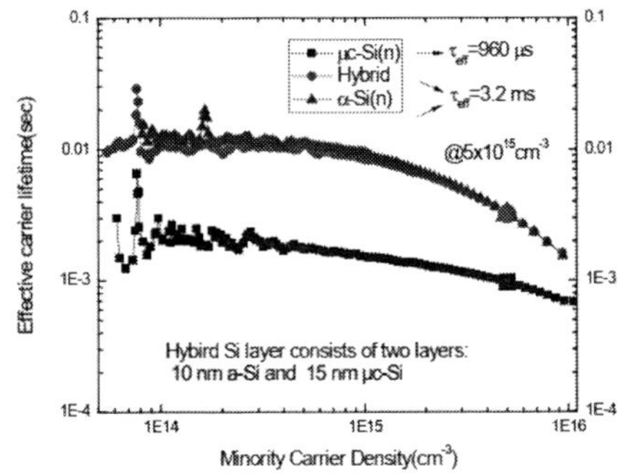

Fig. 6 Comparison of injection dependent effective minority carrier lifetime for the symmetrically passivated samples with a-Si, μc-Si and hybrid layer. The figure also depicts the lifetime at injection level of 5×10^{15} cm^{-3}.

4 CONCLUSION

In this work, the passivation quality of doped Si thin film was investigated by vary the hydrogen dilution rate. The impact of doped a-Si and μc-Si thin films on the passivation quality has been compared. A hybrid structure containing a-Si/μc-Si:H thin films was suggested to improve both the passivation and contact properties. The effective carrier lifetime of 3.2 ms, implied Voc of 716 mV and implied fill factor (FF) of >85% were achieved with symmetric structure on n-type Cz substrate (3~8 Ωcm, 300 μm), indicating that our tunnel oxide/n+ hybrid-Si provides excellent passivation.

ACKNOWLEDGE

This work was supported by the Project of Beijing Municipal Science and Technology Commission (Grant No. 39Z151100003515003) and National Natural Science Foundation of China (Grant No.110751402347).

REFERENCES

[1] D. Weiwei, C. Daming, X. Zhen, P. J. Verlinden, D. Jianwen, Y. Feng, IEEE Journal of Photovoltaics. 2016;6:3-9.
[2] J. Zhao, A. Wang, and M. A. Green. Progress in Photovoltaics: Research and Applications, 7 (1999): 471–474.
[3] Feldmann F, Bivour M, Reichel C, Hermle M, Glunz SW. Sol Energy Mater Sol Cells. 2014;120, Part A:270-4.
[4] P. Stradins; S. Essig; W. Nemeth; B. G. Lee; D. Young; A. Norman; Y. Liu; J.-W. Luo; E. Warren; A. Dameron; V. LaSalvia; M. Page; A. Rohatgi; A. Upadhyaya; B. Rounsaville; Y.-W. Ok; S. Glunz; J. Benick; F. Feldmann; M. Hermle, Proceedings in 6th World Conference on Photovoltaic Energy Conversion, Kyoto, Japan November 23–27, 2014
[5] F. Feldmann, M. Simon, M. Bivour, C. Reichel, M. Hermle, S. W. Glunz, Applied Physics Letters 104 (2014) 181105.
[6] Y. Tao, V. Upadhyaya, C. Chen, A. Payne, E. L. Chang, A. Upadhyaya, A. Rohatgi, Prog. Photovolt: Res. Appl. (2016).
[7] https://www.ise.fraunhofer.de/en/press-and-media/press-releases/press-releases-2015/fraunhofer-ise-achieves-new-world-record-for-both-sides-contacted-silicon-solar-cells.
[8] Feldmann F, Simon M, Bivour M, Reichel C, Hermle M, Glunz SW. Appl Phys Lett. 2014;104:181105.
[9] J. Shewchun J, R. Singh, M. Green. Journal of Applied Physics 48 (1977) 765.
[10] B. Nemeth, B. Nemeth, D. L. Young, H. Yuan, V. Lasalvia, A.G. Norman, M. Page, B.G. Lee, P. Stradins, 40th IEEE Photovoltaic Specialist Conference, 2014, pp. 3488-3452.
[11] F. Feldmann, M. Bivour, C. Reichel, H. Steinkemper, M. Hermle, and S.W. Glunz, Solar Energy Materials and Solar Cells 131 (2014): 46-50.
[12] S. B. Amor, H. Meddeb, R. Daik, A. B. Othman, S. B. Slama, W. Dimassi W, et al. Appl. Surf. Sci. 2016;360, Part B:572-8.
[13] M. Bailly, J. V. Carpenter, Z. Holman, S. Bowden. Photovoltaic Specialist Conference (PVSC), 2014 IEEE 40th2014. p. 1201-5.
[14] L. Hamui, B. M. Monroy, K. H. Kim, A. López-Suárez, J. Santoyo-Salazar, M. López-López, et al. Mater. Sci. Semicond. Process. 2016;41:390-7.
[15] J. Kakalios, R. A. Street, W. B. Jackson. Phys. Rev. Lett. 1987;59:1037-40.
[16] U. Kroll, J. Meier, A. Shah, S. Mikhailov, J. Weber, J. Appl. Phys. 1996; 80:4971-5.
[17] S. Acco, D. L. Williamson, P.A. Stolk, F. W. Saris, M. J. van den Boogaard, W. C. Sinke, et al. Physical Review B. 1996;53:4415-27.

A COMPARISON OF DIFFERENT LASER SEPARATION METHODS FOR HALF CELL PRODUCTION

J. Röth[1], K. Kaufmann[1], S. Eiternick[3], C. Belgardt[2], M. Grimm[2], N. Bernhard[1]
[1]Anhalt University of Applied Sciences, Bernburger Str. 55, 06366 Köthen, Germany
[2]3D-Micromac AG, Technologie-Campus 8, 09126 Chemnitz, Germany
[3]Fraunhofer Center for Silicon Photovoltaics CSP, Otto-Eißfeldt-Straße 12, 06120 Halle (Saale), Germany
j.roeth@emw.hs-anhalt.de

ABSTRACT: Thermal Laser Separation (TLS) and Laser Induced Cutting (LIC) for manufacturing of half-cell solar modules are compared with respect to crack deviation and the current–voltage-characteristics, especially the shunt resistance. First the process window of each process was investigated. It turns out, that the TLS process has a wider process window compared to the LIC method. This results in a better process stability of TLS. A glance at the crack deviations shows a similar situation. The TLS process is more accurate than LIC. We measured a 52% increase of the mean deviation range. In addition we observed a wider distribution by using the LIC technique. This fact again emphasizes the higher process stability using TLS. The shunt resistance measurement shows a non-significant difference between the two separation methods. In conclusion, criteria in favor of TLS are the higher accuracy, a larger process window, which acts as an indicator for a high process stability, and a high shunt resistance.
Keywords: Thermal laser separation, half cells, laser induced cutting,

1 INTRODUCTION

The PV industry is a very cost driven industry. Smart improvements in the solar module efficiency can cause a reduced price per Wp. One of these smart improvements is the use of half cells. Half cells, connected in series, increase the voltage and decrease the current. This causes a higher energy yield based on a reduced resistance loss. The resistance loss is given as the sum of all electrical resistances multiplied by the squared current. [1][2][3]

$$P_{loss} = \sum R * I^2 \qquad (1)$$

This formula shows that the current has a major influence on the resistivity loss and explains the half cell advantage. Studies show that the total energy yield of a test module increases up to 3.9 % rel. [1] Three different separation methods are described in the literature: Scribe and break, Thermal Laser Separation (TLS) and Laser Induced Cutting (LIC). [2][4][5] These methods work as follows: The scribe and break technique is the routinely used process in the PV industry, i.e. the nowadays available half cell modules are usually produced with the scribe and break technique. It uses a pulsed laser beam to create a predetermined breaking line. Afterwards the cell will be divided by mechanical breakage. (Figure 1, left) The other two methods are still under development. [3][4]

LIC is a two-step process. First a pulsed laser beam or a diamond needle produces an initial crack. This very small initial damage is the starting point of the cleaving process. Then a continuous laser beam heats up the cell along the section line. During this process the material next to the initial crack expands and a compressive stress arises. (Fig. 1, middle, force no. 1) In the rear of the laser, where the temperature is not that high anymore, the material is not extended so much. Hence there is a tensile stress. (Same fig., force no. 2) This drives the crack to propagate.

TLS also needs an initial crack. After this a continuous laser beam heats up the cell and expands the material next to the crack. Immediately afterwards a directed stream of sprayed water cools down the material behind the laser. This cooling process reduces the volume again and causes tensile stress (Fig. 1, right, force no. 2). Due to this the cell is separated into two halves. Thus, TLS is in principle an LIC process with additional active cooling by a directed aerosol stream.

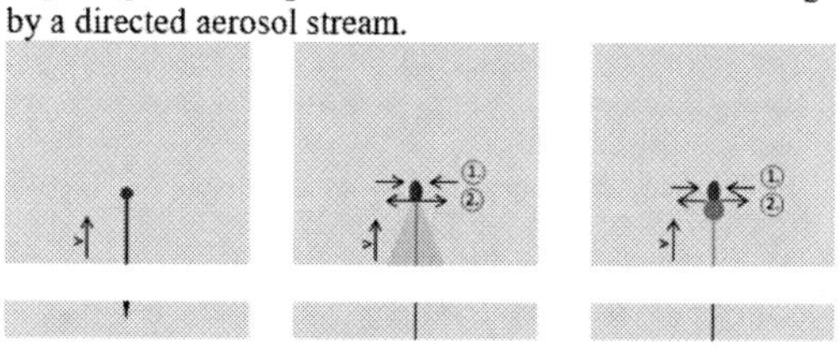

Fig. 1. Schematic representation of the different separation methods. (From the left to the right: scribe and break, LIC, TLS)

This study compares the advantages and disadvantages of TLS and LIC with respect to crack deviation and shunt resistance. For the deviation test we have used blank monocrystalline <100> Cz photovoltaics wafers. The deviations have been measured with the TLS tool's internal microscope camera system.

The I-V-curve was recorded before and after the separation of standard polycrystalline solar cells, so that it is possible to observe a changing shunt resistance after the separation.

2 EXPERIMENTAL

2.1 Preparation

For the TLS and LIC separation we used a 3D-Micromac AG TLS tool with full TLS functionality. If the cooling nozzle is disabled, the tool has still an LIC functionality. This tool uses a 200 Watt 1070 nm CW laser with an elliptical laser beam. The beam heats up the wafer along the section line. The initial crack was completed with a diamond needle. The cooling nozzle uses a DI-Water aerosol to cool the wafer behind the laser.

The TLS and LIC technologies are complex with respect to the impact of the various process parameters and therefore need a comprehensive process development. The influence of the process parameters of the TLS process was investigated in a former study [3]. This study focuses on the process window of TLS in comparison to LIC. The investigation of the process window was done on <100> Cz standard PV wafers. The separations were performed with a speed of 150 mm/s and we modified the laser power and the focus position. For each cut we used a new wafer to ensure that the cut is always done in the middle of the wafer and with the same dimensions.

The crack deviations were measured with the internal microscope camera (wide camera) from the 3D-Micromac AG MicroDice TLS tool. This microscope allows the measurement of the deviations down to 5µm.

To measure the shunt resistance, we used a fully automated flasher tool. (PV-Tools LOANA) With this tool we measured the complete current-voltage-characteristics of each cell before and after the separation. This procedure enables to detect shunting during the cleaving process. For this measurement we used commercial standard polycrystalline PV cells without special opening gaps in the metallization for the laser separation.

3 RESULTS

3.1 Process window and parameter setting.

At the focus position of -3 mm we noticed an optimum due to a shorter laser spot. Because of the cylindrical lenses the laser spot is not symmetric around the focus. The length of the long elliptical half axis extends steadily from -10 to +10 mm focus offset. The focus position is defined as the focal length of the x-axis (shorter half axis). Due to this the shorter half axis is symmetric around the focus. The graph shows that a minimum spot width is necessary. Thus a cut above +1.5 mm and below -1.5 mm focus is not possible. The tensile stress behind the laser arises due to a temperature gradient and the temperature gradient of the LIC technology arises due to the heat conduction. At the focus position of -3 mm we found the optimal spot length for the chosen velocity. With this setting less laser power is needed. That is why we especially chose this caustic. High mechanical stress levels due to a high laser power favors crack deviations. [6] Because of this we have chosen the lowest possible laser power.

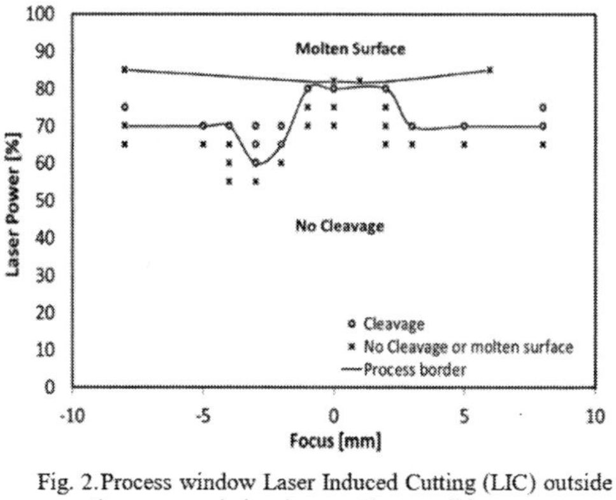

Fig. 2.Process window Laser Induced Cutting (LIC) outside the process window is no cutting possible.

The TLS process has a larger process window. The cooling nozzle makes it possible to cut the wafer on every focus position in the investigated range between +/- 10 mm. To compare the results of each technology we have chosen for the further investigations the same laser parameter setup at 60% of the maximum laser power and a focus position of -3 mm.

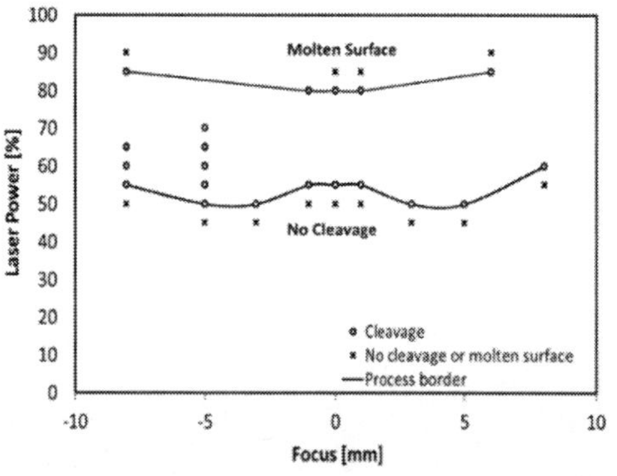

Fig. 3.Process window Thermal Laser Separation (TLS) outside the process window is no cutting possible.

3.2 Crack deviation

During the sample preparation the crack deviation was measured. The maximum deviation on the left and on the right side of the desired cutting line was recorded. After this the two deviations were combined to one range. The following bar chart shows the distribution of the deviation range for the LIC and TLS method. The scribe and break technology, with a very small deviation range of about 10 µm according to own observations, is the most accurate technique and was therefore not further investigated here. Its high accuracy is caused by an about 60 µm deep permitted section line and the crack is exactly following this laser scribe, so that there aren't any major deviations from this line. Any deviations are dependent on the chosen focus and only within the section line.

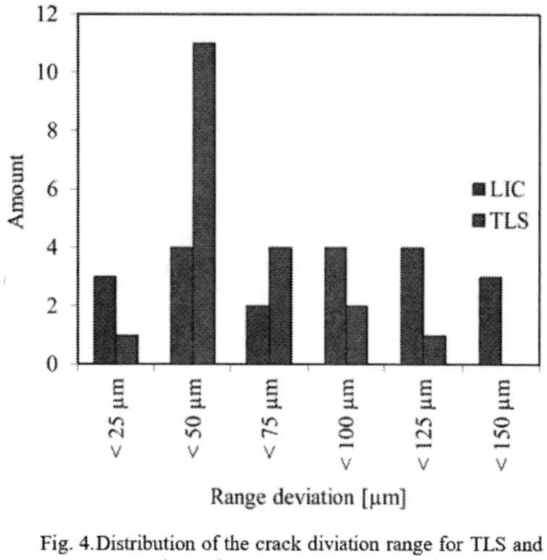

Fig. 4.Distribution of the crack diviation range for TLS and LIC cutting. The amount scale shows how often a defined range was measured.

The histogram shows that the TLS technology is more accurate than the LIC method. The mean of the deviation range is 54 µm by the use of TLS and 82 µm by the use of LIC. Also the spread around this mean value is narrower for TLS. This result can be understood by thermographic imaging. The pictures below show the relative temperature distribution, but not the exact temperature due to the temperature-dependent emissivity.

All the same, they can be used to describe the conditions during the process. The left picture has a considerable spreading temperature profile. This image shows the LIC process. The tensile stress arises as a result of the temperature gradient behind the laser. The temperature variation on the left takes place on a larger area compared to the one on the right, showing the TLS technology. We suppose this difference, mainly caused by the active aerosol cooling, is responsible for the higher accuracy of TLS.

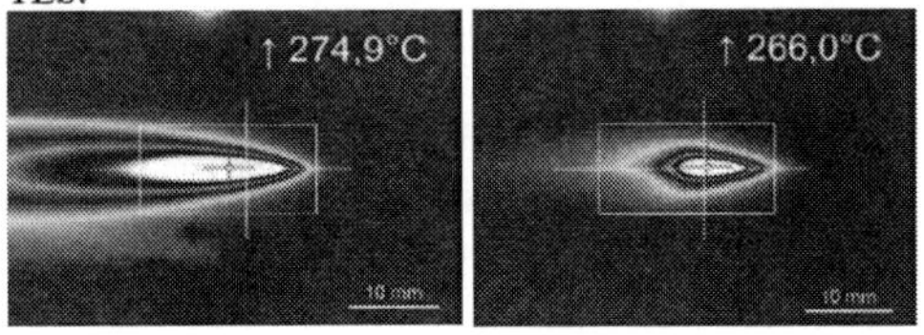

Fig. 5. Thermographic image of the LIC (left) and the TLS (right) technology. The indicated temperatures are not the real ones, because no emissivity correction was applied.

3.3 Shunt resistance

Finally, we measured the I-V-curve. A low shunt resistance, mainly caused by defects, has a major influence on the whole cell efficiency. Therefore the measurement of the shunt resistance before and after the separation can be used to evaluate the quality of the cell separation method. In this investigation we separated 40 cells into 80 halves. The first half of the sample was separated with the LIC and the second with the TLS technology. The following box plot shows that the shunt resistance does not significantly decline during the treatment, which in favor of both methods. Also the other data of the I-V- curve, for example FF, Rseries, Uoc, Isc, do not show any significant change relative to the values measured before separation. Both separation methods use mechanical stress, which produces very smooth edges in comparison to the scribe and break technique. Du to this the TLS and LIC technologies have to be assessed as equivalent with respect to the shunt resistance and other solar cell parameters. TLS and LIC are suitable for the half cell production in respect to electrical parameters, but TLS has the larger process window with respect to a successful separation, as shown above.

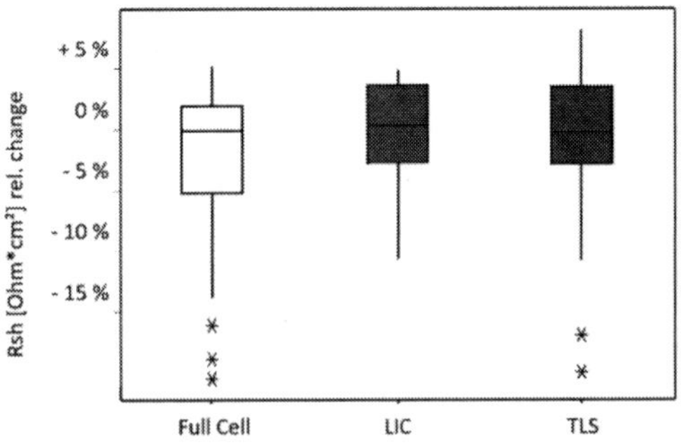

Fig. 6. Shunt resistance before the treatment (Full Cell) and after the treatment with the respective technology.

4 CONCLUSION AND OUTLOOK

In this study we compared Thermal Laser Separation (TLS) and Laser Induced Cutting (LIC) with respect to crack deviation and current–voltage-characteristics. First, we expect that the processing of half cells using the TLS technology is more production-worthy. The process window is wider and therefore the process stability should be higher. The separation using the LIC method is more challenging, because the process window is smaller and it is more dependent on the used caustic.
Secondly, the TLS technology is more accurate in relation to the LIC separation. The mean deviation range is significantly reduced by the use of a cooling nozzle (TLS). It is reduced from 82 µm (LIC) to 54 µm (TLS).
All the same, both separation methods have no significant influence on the shunt resistance, at least to the extent examined in this study.
With respect to use of a water aerosol for cooling, it should be mentioned, that in today's photovoltaics manufacturing a direct contact of water to the cell surface with metallization is undesirable. To maintain the advantages of additional active cooling, further investigations will concentrate on replacing the water spray cooling by a dry cooling.

5 ACKNOWLEDGEMENTS

We thank A. Facchini for engineering assistance. This work was funded by the German ministry of education and research (BMBF) under contract no. 03FH005IX4.

6 REFERENCES

[1] Stefan Eiternick, Kai Kaufmann, Jens Schneider, Marko Turek, „Loss analysis for laser separated solar cells", 4th International Conference on Silicon Photovoltaics 2014, p. 326 - 330
[2] Stefan Eiternick, Felix Kaule, Hans-Ulrich Zühlke, Thomas Kießling, Michael Grimm, Stephan Schoenfelder, Marko Turek, „High Quality Half-Cell Processing Using Thermal Laser Separation", 5th International Conference, on Silicon Photovoltaics 2015, p. 340 - 345
[3] J. Röth, N. Bernhard, C. Belgardt, M. Grimm, F. Kaule, „Thermal Laser Separation (TLS) Dicing Process Study – a New Technology for Cutting Silicon Solar Cells for High-Efficiency Half-Cell Modules", 31st European Photovoltaic Solar Energy Conference and Exhibition, p.716 - 718
[4] S. Weinhold, A. Gruner, R. Ebert, J. Schille, H. Exner, „Study of fast laser induced cutting of silicon materials", Laser Applications in Microelectronic and Optoelectronic Manufacturing, March 2014, p. 53 - 60
[5] H.-U. Zühlke, G. Eberhardt, R. Ullmann; TLS-Dicing – an innovative alternative to known technologies; Advanced Semiconductor Manufacturing Conference 2009, p. 28 – 32
[6] D. Gross, P. Seelig, "Bruchmechanik mit Einführung in die Mikromechanik", Springer Berlin, p. 233

THE BIFACIAL N-PERT SOLAR CELL COUPLING BORON SPIN-ON WITH $POCL_3$ DIFFUSION AND IT'S GLASS-GLASS MODULE PERFORMANCE

Chenyang WU *, Paul NI, Qingzhu WEI, Junyu LU, Weifei LIAN, Yayi HU, Xiaorui LIU, Sanyang ZHANG
Zhongli Talesun Solar Co.,Ltd.
No.1 Talesun Road, Shajiabang, Changshu, Suzhou, Jiangsu Province, PR China , 211542
Chengyang.wu@talesun.com,paul@talesun.com

ABSTRACT: N-type silicon solar cell is being more and more focused because of the Si substrate feathering with long minority lifetime and no B-O complex related degradation. Bifacial Passivation Emitter and Rear Total-diffused (PERT) n-type solar cells is a high efficiency solution which can be realized by industrialization easily due to the process simplification. One of the key nPERT process was the P^+ emitter formation. BBr_3 diffusion has been proved for mass production for years, but the dopant uniformity and the process integration still limited the industrialization of n-type solar cells. In this work, nPERT solar cell combined boron ink spin coating and $POCl_3$ diffusion was investigated. The results showed that solar cell efficiency greater than 20.2% was obtained with bifaciality more than 90%. The outdoor performance of glass-glass module with nPERT solar cells showed that the power output gained about 7-10% with grass background, and increased higher with higher light intensity.
Keywords: n-type, PERT, bifacial, outdoor performance

1 INTRODUCTION

Advanced solar cell structure with higher efficiency was trend to introduce in PV industry because of the market desire of high power modules. Currently, commercial Aluminum back-surface-field solar cell made from p-type material is update to PERC structure by many PV manufactures [1]. On the other side, marketing share of n-type base material solar cell is growing. The best known IBC (Integrated Back-Contact) and HIT (Heterojunction with Intrinsic Thin-film) solar cells with conversation efficiency greater than 25% were manufactured by Sunpower and Panosonic respectively [2],[3]. F. Feldman reported a selective contact n-type structure, so called TOPCon (Tunnel Oxide Passivation Contact), and the efficiency is also higher than 25%[4].

n-type silicon has several advantages over p-type. Firstly, n-type Si has longer lifetime over micro-seconds, because minority carrier hole of n-type material is less sensitive to many common metallic impurities. This is the reason why most of the high efficiency solar cell made from n-type material. Secondly, there is no Light Induced Degradation (LID) of n-type silicon for the absence of boron-oxygen complex in phosphorous doped substrate. However, to benefit from these advantages of the bulk material, advanced cell concept like either Back Contact or heterojunction based on n-type c-Si still have some issues to be solved in mass production with acceptable cost. As a transition from BSF solar cell to Back Contact or Heterojunction solar cells, bifacial n-PERT solar cell caused more and more interesting.

Especially, the n-type bifacial solar cell shows higher performance because of rear total diffused and good front and rear passivation with low surface recombination rate. To realize nPERT structure, a lot technology can be chosen such as diffusion [7], implantation [5-7], CVD deposition [8], and advanced metallization technologies. More importantly, nPERT can be retrofitted from the existing traditional p-type solar cells line.

In this paper, we developed industrial bifacial n-type PERT solar cell based on boron ink spin coating combined with $POCl_3$ diffusion with screen printing. The bifacial glass-glass module outdoor power output performance was investigated.

2 EXPERIMENTS

The n-type PERT solar cells were fabricated on 6 inch CZ phosphorous-doped silicon wafers with resistivity of 0.8–4 Ωcm and thickness of 180μm. The PERT solar cells were fabricated in industrialized equipments. The solar cell scheme was shown in Figure1. Both sides were textured with pyramids, front side spin-on boron ink and then baking, boron diffusion in a quartz tube furnace to form homogeneous pn junction. Besides, the BSF was fabricated in the industrial $POCl_3$ tube diffusion system. Thermal Oxide silicon and direct PECVD silicon nitride were deposited on both sides as stack passivation, and both sides were screen printed with H-pattern fingers.

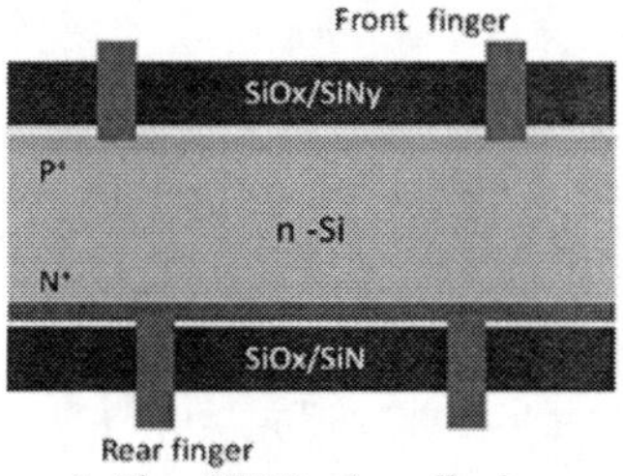

Figure 1. The nPERT solar cell scheme

2.1 Formation of Boron doped Emitter

Liquid boron ink was spin coating on the surface and then baked at 150℃about 20s. The coating substrates were delivered into a quartz tube furnace immediately. The full doping process took about three steps.

- Firstly, the boron source was oxidized to forming boron silica glass (BSG) at 700-900℃.
- Secondly, the furnace was heat up to 950°C and stabilized 15min, and the Boron in the BSG diffused into Si substrate.
- Finally, an oxidation process was applied to form a thick BSG on the c-Si surface, and redundant B dopant was removed by SiO_x.

Representative as-diffused Sheet resistance mapping measured by 49-points was shown in figure 2. A standard deviation as low as 3Ω/□ was obtained, which was quite good for industrial P^+ type emitter application. The uniformity of dopant can be improved by spin coating and backing process before thermal treatment, and the surface concentration and junction depth can be optimized during the thermal process , which means the boron profile tuning is unique during the thermal processing.

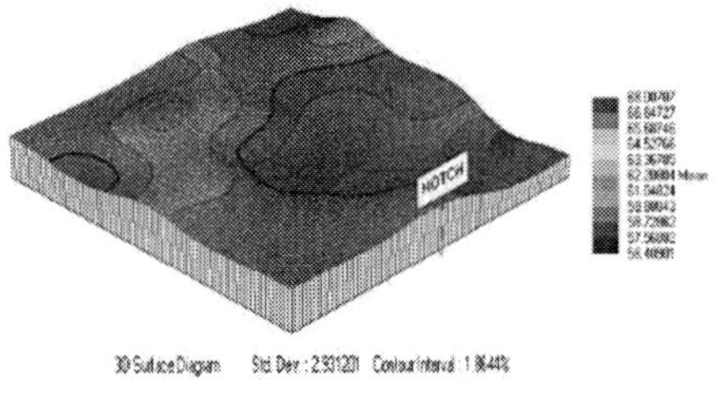

Figure 2. Sheet resistance mappignof 7*7 points test (Target 60Ω/□, mean 62Ω/□)

The wafer to wafer uniformity was evaluated by sheet resistance measurement of different heating zone. The samples location was shown in figure 3.

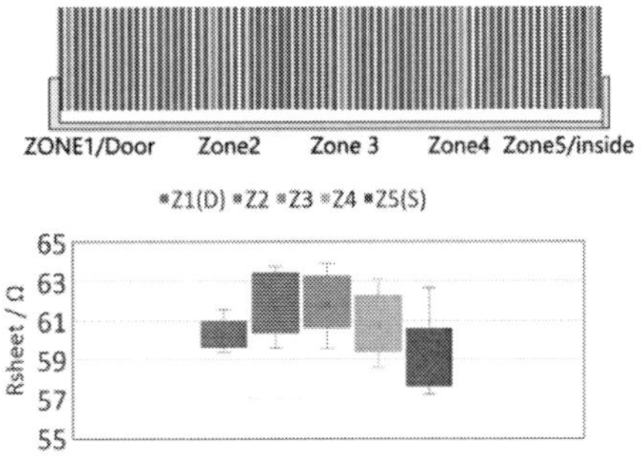

Figure 3: Wafer to wafer sheet resistance uniformity,

2.2 Optimization of phosphorous back surface filed

The rear side phosphorous doped layer not only provided a Metal-Silicon contact agent, but also acted as light absorption layer of. The resistivity distribution of n type wafers was very wide for n-PERT solar cells. In order to investigate the effective of BSF, 3 groups of phosphorus doping profile have been applied to coupling the P+ emitter:

- G1: Shallow phosphorus doping with high surface concentration
- G2: Deep phosphorus doping with high surface concentration
- G3: Deep phosphorus doping with low surface concentration

Table 1: Relationship of phosphorus doping BSF and solar cell performance

	BSF Depth	BSF C_B	Voc	FF	Eta
G1	Middle	Middle	1	1	1
G2	Deep	Heavy	1.015	1.006	1.002
G3	Deep	light	1.017	1.005	1.003

Table 1 listed the relative IV parameters with different BSF. The dopant profile of G1 was similar to typical n type emitter of p type Al-BSF solar cell. Unlike the n-p junction, the homologous n^+n high-low junction is not a complete diode device. No doubt the surface P-concentration of G1 was good enough to form contact with Ag paste, but the fill factor of the cell was lower. Deeper phosphorous doped region like G2 or G3 demonstrated a better fill factor, meanwhile Voc got better over reference G1. The surface P-concentration of G3 was reduced and the doping depth was almost similar to G2. The Voc gains of G3 came from the lighter dopant.

3 SOLAR CELL performance

The emitter and BSF was formed in quartz tube furnace step by step, the full solar cell was passivated by thermal SiO_x while front and rear contact with Ag-Al and Ag paste respectively. Solar cells were measured by commercial solar simulator under STC 25℃ @1sun.

Table 2. IV parameters of nPERT solar cell

	Eff	Voc/ mV	Isc / mA·cm^2	FF
Mean	20.11	652	39	79.0
Mean$_{rear}$	18.55	637	37.5	77.6
Best	20.23	654	39.3	78.8

Table 2 showed the optimized 30pcs batch of solar cell results. The mean front efficiency was 20.11%, while rear average efficiency was 18.55% for the, which means an average bifaciality greater than 90%. The best front efficiency is 20.23%. Due to the single side diffusion, no more diffusion etching for back side was required. Furthermore, better efficiency should be obtained by optimizing doping profiles for Boron and Phosphorous dopant respectively to suit the surface passivation. The IQE, EQE and reflectance of front side 20% efficiency bifacial solar cell was shown as Fig. 4.

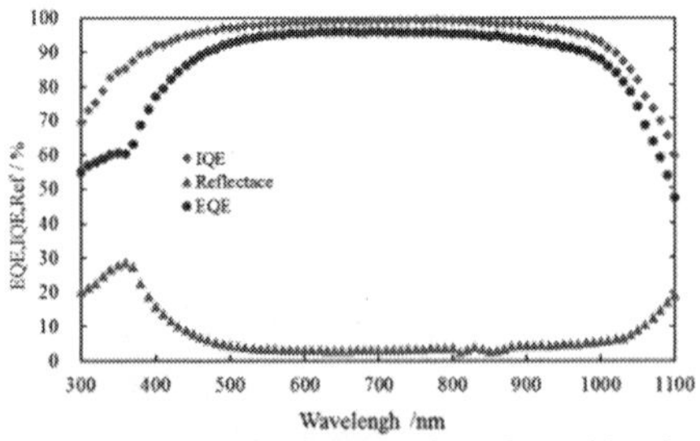

Figure 4: IQE, EQE and reflectance from front side of 20% bifacial solar cell

4 OUTDOOR EXPERIMENTS

To demonstrate the bifacial benefited from the bifacial PERT cell, we have fabricated full-size 660 (serial 6*10 cells) glass-glass modules in Suzhou. As reference, commercial momo and multi c-Si back sheet modules were installed at the same place independently. The system was set up with four panels in series with MPPT and dynamic load. The output data such as IV parameters and total power output were recorded every 2 hours.

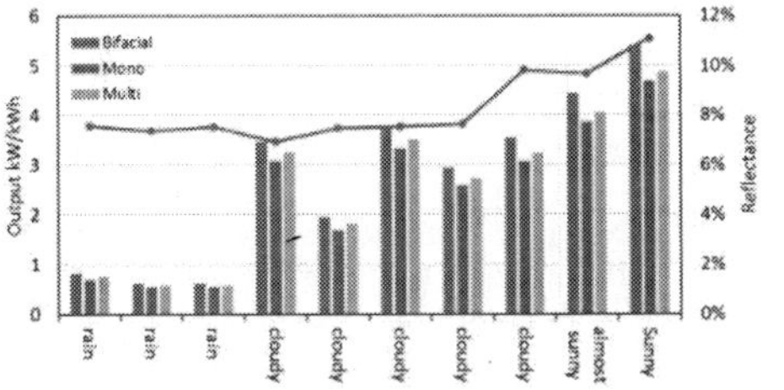

Figure 5. Daily power output comparison

It's clear that bifacial module power gains, which have been report by Talesun and others [9][10], were watched on different backgrounds. Figure 5 showed daily power outputs of three kinds of modules installed on grass background. Average of 9% power gains was recorded of bifacial modules, and the power gains increased higher with higher light intensity. Due to the higher Voc, n-type modules works longer in a whole day.

5 CONCLUSION

In this work, combination of Spin coating Boron ink and $POCl_3$ gas source was used to form a p-type emitter and back surface field for bifacial n-type silicon solar cells. Efficiency greater than 20.2% was obtained on nPERT cells by using industrial screen printing metallization. Bifacial glass-glass module of the cells showed good low-irradiance performance. And more power output was observed on nPERT bifacial glass-glass modules with different weather conditions.

6 ACKNOWLEGEMENTS

This work is supported by the Project of Major Science and Technology Achievement Transformation from Jiangsu Province (No. SBA2015030129).

7 REFERENCE

[1] S.Engelhart, et al. "3 years of high quality mc Si Q-Antum production experience-approaches for efficient cell and module development " , 31st European Photovoltaic Solar Energy Conference (2015) , 2BP.1.3

[2] D.Smith et al. " Toward the practical limits of silicon solar cells", IEEE Journal of Photovoltaics 6, pp. 1465-1469 (2014)

[3] K. Masuko et al. "Achievement of more than 25% conversion efficiency with crystalline silicon heterojunction solar cell" IEEE Journal of Photovoltaics 4, pp. 1433-1435 (2014).

[4] S.W.Glunz, F.Feldmann, "The irresistible charm of a simple current flow pattern-25% with a solar cell featuring a full-area back contact", 31st European Photovoltaic Solar Energy Conference (2015) , 2BP.1.1

[5] Y.W. Ok et al. "Ion implanted and screen printed large area 20% efficient N-type front junction Si solar cells", Solar Energy Material Solar Cells 123, 92-96 (2014)

[6] I.G. Romijn et al. "Industrial n-type solar cells with >20% cell efficiency". China PV Technology International Conference (CPTIC), 2013

[7] D. Kania "Pilot line production of industrial high-Efficient bifacial n-type silicon solar cells with efficiencies exceeding 20.6 %", 5th SiliconPV 2015

[8] J.Cho et al. "21% efficiency n-type rear-junction PERT solar cell with thin industrial 156mm CZ single crystline silicon wafer", 5th SiliconPV 2015

[9] C. Comparottoa et al. "Bifacial n-type solar modules: indoor and outdoor evaluation", 5CV.2.11. 29th European Photovoltaic Solar Energy Conference (2014)

[10] S.Gonsui et al. "n-typt high efficiency bifacial solar cell with the extremely high bifaciality of 90% in average fabricated by using conventional diffusion method", 28th European Photovoltaic Solar Energy Conference (2013)

THE IBC STRUCTURE AS SUPPORT FOR THREE BAND-GAPS TANDEM DEVICES

J. C. Jimeno[1], J. R. Gutierrez[1], V. Fano[1], A. Habib[1,2], C. del Cañizo[3], M. A. Rasool[1] & A. Otaegi[1]
[1] Technolgical Institute of Microelectronics (TiM), UPV/EHU, Bilbao, Spain. email: jc.jimeno@ehu.eus
[2] Faculty of Science, Department of Physics, Mansoura University, Mansoura, Egypt
[3] Instituto de Energía Solar (UPM), Universidad Politécnica de Madrid, Madrid, Spain

ABSTRACT: The IBC structure is one of the champions in the efficiency competition of silicon devices. Industrial cells are produced with efficiency values surpassing 24%, what is due to an excellent design in terms of separation from the region where carriers are energized and emitters where these carriers are selectively separated. Their position in the back of the cell allows the access to both quasi-Fermi-levels from this surface, and enables the access to their corresponding levels of top or bottom cells from this place, far away from the entrance of light and simplifying the design of complex multi-band-gap structures. This paper will shown several of these structures, many of them achievable with well known materials with technology developed for the microelectronic sector. These solutions would be to cells with efficiencies in the 30-32 %. Using optimum materials, not clear at this moment, we estimate a technological limit of 39.4 % for these new structures.
Keywords: Tandem, Back Contact, Silicon Solar Cell

1 INTRODUCTION

Since the early 90's it was shown that the technological limit to efficiency for crystalline silicon technology is close to 26% [1]. This limit was early reached [2] getting to laboratory cells with 25% of conversion efficiency. Even in mass production there are technologies producing silicon cells with efficiencies of 24% [3] and only very recently has been surpassed this efficiency world record with a 25.6% for a silicon solar cell has been announced by Panasonic utilizing the IBC solar cell design [4] over the active area or 25.2 % over the full cell area [5]. This paper shows structures of silicon solar cells capable of produce efficiencies values in the 35-40% range. Obviously these structures uses tandem configurations mixing several materials but being a silicon solar cell the support for all of those. Technical approaches followed so far have been monolithically interconnected or mechanically stacked [6] tandem solar cell structures, getting at the date to devices with efficiencies close to that of single silicon solar cell and recently a mechanical stacked tandem cell [7] has reached a superb 29.8% efficiency but in a 4 terminal configuration. There is a great challenge to exceed 25% of efficiency with tandem cells being silicon as one of their components.

2 CAPTURING THE INFRARED SPECTRUM

Single silicon solar cells are capable of recovering up to 42 mA/cm^2 within the upper part of the solar spectrum (≤1100 nm). In the lower part another 26 mA/cm^2 can be achieved and obviously there is not any other material capable of the improvement in efficiency to a silicon cell when it is used as low band-gap material in a series-connected 2-junction structure. This work suggest to work connecting in parallel the silicon cell with two cells of lower band-gap working at half voltage than the silicon, as it is shown in figure 1. Ge is able to capture more than 17 mA/cm^2 and because of that it has been used in many attempts of tandem cells on silicon. However the V_{oc} of grown Ge cells is not enough to compliment a good Si cell and finally the tandem efficiency would be lower than the only silicon one.

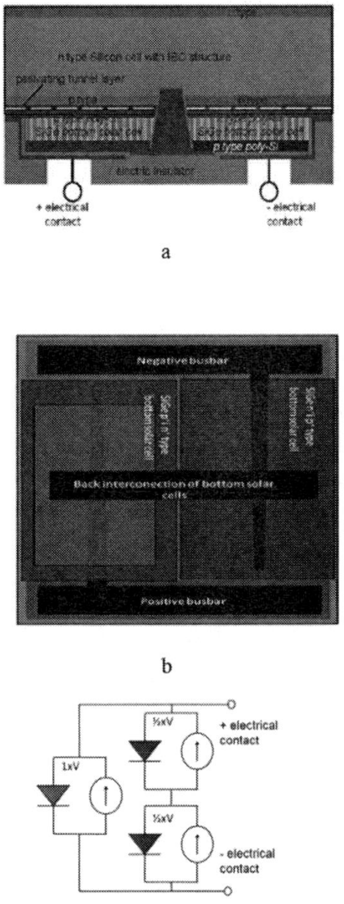

Figure 1: a) transversal section, b) bottom view and c) equivalent circuit of the IBC structure with the 2 infrared SiGe cells

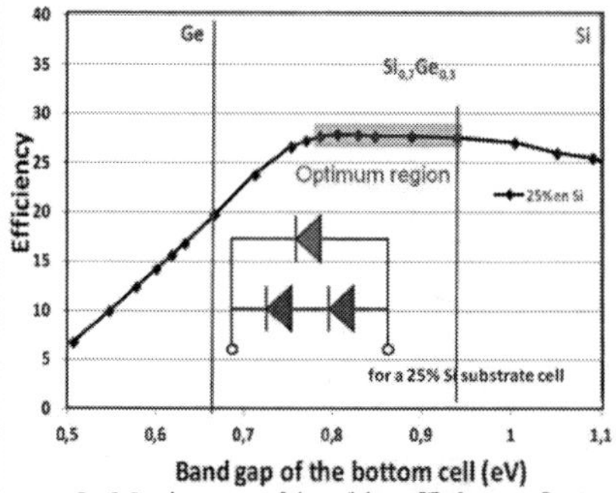

Figure 2: Maximum achievable efficiency for a voltage matched tandem configuration composed by a 25 % efficient silicon cell and 2 infrared bottom cells series interconnected

According to figure 2 there is a near independent band-gap maximum of efficiency using a tandem composed by a 25 % efficient top silicon cell and two bottom cells with band-gap value within the 0.75-0.95 eV range. For the lower E_G values the tandem device is limited by the IR cell but in the higher region this limitation is due to the silicon cell. Because the silicon technology is well developed and understood it sounds more convenient to work in this higher band-gap region and SiGe with a high Si content could be a good option.

From the 2 rear SiGe cells and according with results from figure 2 an extra efficiency of 2.7 absolute points would be accounted. Several materials would be used but in our understanding the crystalline silicon-germanium is the best option wit band gaps in the 0.8 to 0.9 eV range. Silicon-germanium can be easily deposited by PECVD epitaxially on silicon at low temperatures (under 600 °C). Regarding figure 1 it could be shown that this not needed to dope n or p type this material because it was found in a sandwich of higher bang-gap materials, what are the p and n type surrounding polycrystalline silicon layers.

3 EXPLOITING ULTRA VIOLET PHOTONS

3.1 The 3 terminals approach

To improve the exploitation of UV photons their energized electrons must be extracted from the tandem cell at a higher voltage than of the silicon cell. An extra third terminal is needed and it is placed on the top of the cell. To maintain these energized electrons enough time before they can be extracted a larger band-gap semiconductor is needed on the top of the structure, and a classical pn junction made on the top is sufficient for the separation of carriers. The resulting structure is shown in figure 3. The n type of top cell is in contact with the n type front emitter of the silicon cell. In this design holes from the top cell are directed to the front contact and electrons to the terminal 3 on the back of the silicon cell. A doping grading in the n type region of the top cell will helps to avoid the injections of holes from this top cell into the silicon. The dual structure, having the Si cell a front p-type floating junction and being the top cell n-p with the p type region in contact with the silicon works in a similar manner than the previously explained. Being the transit between materials of the same type, n-n or p-p, the relevance of interface defects is lowered.

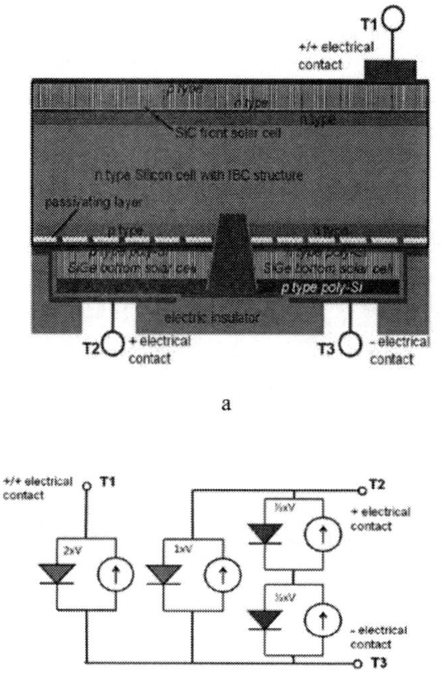

Figure 3: a) transversal section and b) equivalent circuit of a three terminals IBC structure with the 2 infrared SiGe cells and a high voltage top cell

3.2 Choosing the top semiconductor

The three terminals device already described and shown in figure 3 can be include in two terminal modules if their operating voltage is a multiple of the voltage from the silicon cell. Next section will explain that with more detail. In the simplest case that this voltage is double for this top cell there is a dependence of the theoretically achievable efficiency regarding the band-gap of the top semiconductor, as it is shown in figure 4, with a maximum of efficiency for this top semiconductor of 1.7 eV.

Regarding figure 4 it can be shown that for the top cell and depending of their band gap there is a short range of efficiencies from this top cell in which the total tandem efficiency benefits for the inclusion of this top cell. That is due to the fact that not all the cells of 1.7 eV can work at double voltage than the silicon cell. However, the total efficiency of the tandem cell is reduced if the top cell doesn't reach this minimum efficiency. For example, for a material with the optimum band-gap of 1.7 eV we can get the maximum efficiency of 36.7 % if the individual top cell presents 24.6 % efficiency. However for efficiencies lower than 16.3 % of this top cell the final device always will have efficiency under 25%, degrading the electrical behavior of the single silicon cell. Figure 4 represents also the current efficiency of different technologies and curiously none of the current amorphous technologies is capable of supplement the efficiency of the silicon cell. The situation is the opposite if a crystalline top cell is used.

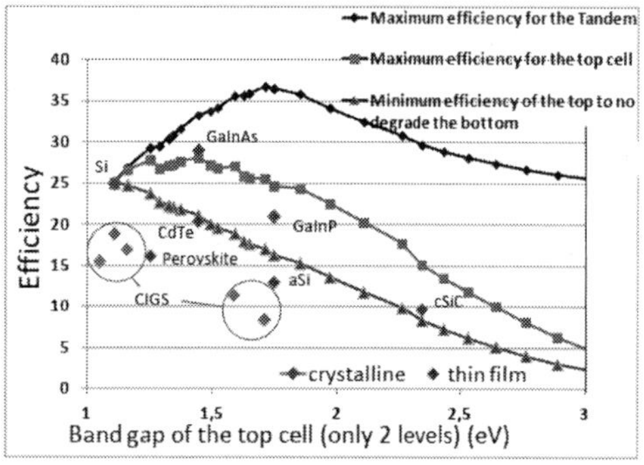

Figure 4: Maximum achievable efficiency for a 2 levels parallel connected solar cell, being the bottom one a 25% efficient silicon solar cell and indicating the range of efficiencies of the top cell to reach an ensemble with more than 25%

The final efficiency depends on the relation $\Delta V_{oc}/\Delta E_G$ of the top cell, what is a factor depending of physical and technological facts. The red line on figure 4 shows the frontier from an improving or degrading top cell. Several values of efficiency vs. band-gap from different technologies are represented in figure 4 and none of the amorphous materials analyzed and represented by blue triangles fulfill the requirement of the minimum efficiency.However that is not the case of crystalline materials represented by the red diamonds. We suspect that polycrystalline materials would also be in the region of a net efficiency gain but we don't have evidence of that.

In order to understand the limitations of each material, figure 5 represents the open circuit voltage of the top cell and their operating voltage in the case of being parallel connected to two 25 % silicon solar cells. As in figure 4 green lines represents the ideal case and red ones the limit for the degradation of the silicon behavior of 25 %. For the ideal case it can be saw that the operating voltage follows to the open circuit voltage from the lower E_G values to the maximum 1.7 eV and saturates for larger values.

We can say that the total structure is limited by the top cell for E_G values under 1.7 eV and from the silicon cell for larger E_G values. Also, the green line has a behavior of $\Delta V_{oc}/\Delta E_G$=1, while the red line, corresponding to the case of degrading the full structure presents this tendency $\Delta V_{oc}/\Delta E_G$=0.5. The region under 1.7 eV presents a low range in terms of efficiency of acceptable top materials, because as previously explained the device is limited from this, but there is more tolerance to the quality of that in the upper region and sounds more convenient to work with materials with wider band gap. Figure 8 also includes results of record cells. In blue they are represented the V_{oc} values from different CIGS cells with different band gaps [9]. Surprisingly all these devices accommodate to the lower red line and cannot upgrade any good silicon solar cell. But the good accommodation to this $\Delta V_{oc}/\Delta E_G$=0.5 tendency seems to indicate the existence of any fundamental limitation, probably extendable to other amorphous materials. For crystalline record materials [10], in contrary, there is near to perfect accommodation to the ideal behavior represented in the green line in figure 5.

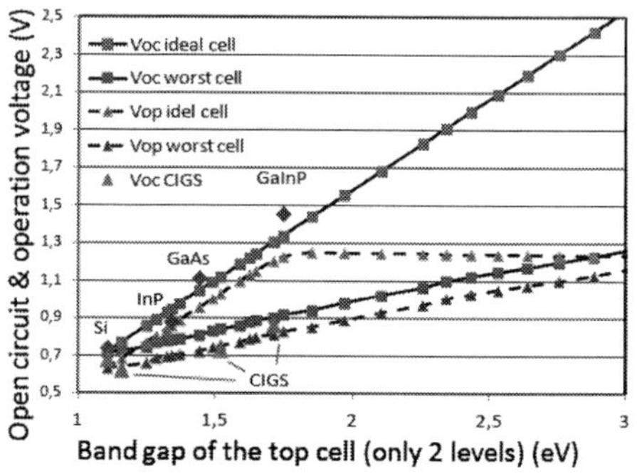

Figure 5: Open circuit and operating voltages of the top cell under the same assumptions than figure 4

Regarding the optimum materials it appears that III-V compounds, as GaP, give a clear opportunity for high efficiency devices, however until now mixing silicon and III-V materials is a challenge topic and always solved by not large throughput techniques like MOCVD. Other materials as $ZnSnN_2$, $ZnSiAs_2$ and others sound very promising but their technology is not currently available. However, using materials of higher band gap, like beta-SiC (2.4 eV), would give to lower increase on efficiency but also to a more secure behavior in the silicon solar cell, which efficiency is not affected by the addition of this top cell. Silicon carbide is a well known material, used in their amorphous state as passivant for silicon surfaces; it can be epitaxially deposited on silicon substrates at relatively low temperatures [8] (around 900 °C) and used in microelectronics for large temperature devices. It can be doped in the same manner than silicon, using the same dopants, and also oxidized and with full technology developed to work with that. The only problem is that their band-gap of 2.4 eV is far from the optimum material, but even in this case a final efficiency of 29.6 % would be obtained for a 2 materials device and 32.5 % for a 3 layers device. In the case of being the voltage at the top cell thriple than the silicon one, as shown in figure 6, it appears a wide region of band-gaps of the top layer, from 1.85 to 2.35 eV within the theoretical efficiency can be over 34 %. In this case SiC is one of the best candidates.

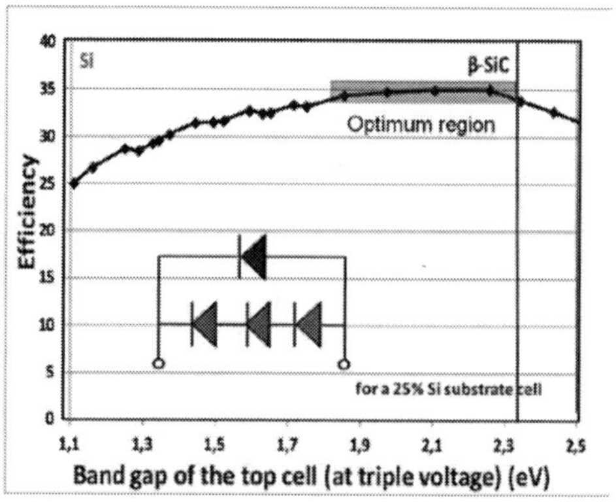

Figure 6: Maximum achievable efficiency for a voltage matched tandem configuration composed by three 25 % efficient silicon cells and a ultra-violet top cell

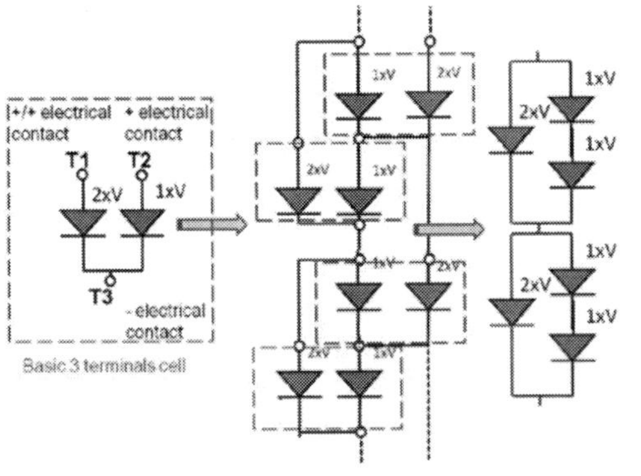

Figure 7: Connection of the basic three terminals cell into 2 terminals PV module and its simplified circuit

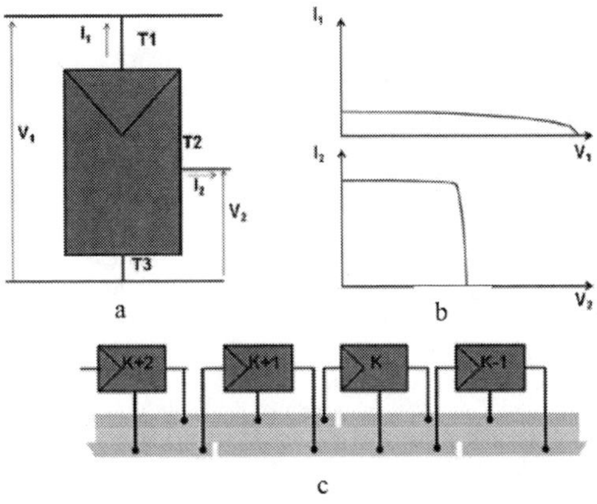

Figure 8: a) symbolic representation of the three terminals solar cell, b) their corresponding I-V curves and c) example of interconnection in a double layer back-sheet laminated circuit

4 MODULE INTERCONNECTION

The interconnection in a two terminals PV-module is simple if the operating voltage of each individual solar cell component is a multiple of the lower voltage cell. Some approaches to that would be found in [11] and also different tandem approaches for solar cells are suggested in [11]-[15], however the approach from this work is unique and new [16] (currently in a patenting process).

For the special case -and also others- of each sub-cell from a given band-gap works at double or half voltage than the previous or the next band-gap sub-cells it is possible to find simpler structures able to be interconnected in 2 wire modules. Figure 7 shows examples of that, being the starting device a 3 terminal tandem cell, what can be interconnected in 2 terminal modules. Finally, their implementation using a 2 layers copper bach-sheet can be done easily, as it is shown in figure 8.

5 CONCLUSIONS

A new structure of tandem solar cell has been presented. This structure gives to devices with two or three terminals but easily connected into two terminal modules. Although the concept presented is independent from their realization this structure would implement materials deposited in silicon by standard industrial techniques, like LPCVD. The new approach of connecting the cells in parallel is less restrictive than the conventional series connected giving to a continuous improving on efficiency from the full region of 0.75 to 3 eV for the extra cells added. This relaxation of this configuration from the top and bottom materials used gives to being less critical to manufacturing tolerances, degradations or exposition to different light spectrums. In addition the parallel connection avoids the use of tunnel diodes for the interconnecting different materials, resulting in simpler structures.

From the top cell we shown that an extra of 11.7 of absolute efficiency would be added to the tandem device, however not all materials and cells with adequate band gap can do that. We demonstrate that crystalline solutions, like that made on InP, GaAs, GaInP or even β-SiC can improve the final results over the starting silicon cell efficiency, however those cells based on amorphous materials, like CIGs cells, aSi or maybe perovskites seem to not be able for this purpose. We have found that any fundamental phenomenon is under this limitation and more research seems necessary in order to overcome that.

5 ACKNOWLEDGEMENT

This work has been supported by the Spanish MINECO within the framework of the projects Crysthin TEC2011-28423-C03 and Tabaco ENE2014-56069-C4-1R.

6 REFERENCES

[1] R. Mertens, "Silicon solar cells". 12th EPSEC, pp. 1-6, 1994.

[2] J. Zhao, A. Wang & M. A. Green, 24.5% "Efficiency Silicon PERT Cells on MCZ Substrates and 24.7% Efficiency PERL Cells on FZ Substrates". Prog. Photovolt: Res. Appl. 7, 471-474 (1999)

[3] D.D. Smith, et al. "Generation III high efficiency lower cost technology: transition to full scale manufacturing". 38th IEEE Photovoltaic Specialists Conference, 2012.

[4] K. Masuko, M. Shigematsu, T. Hashiguchi, D. Fujishima, M. Kai, N. Yoshimura, T. Yamaguchi, Y. Ichihashi, T. Yamanishi, T. Takahama, M. Taguchi, E. Maruyama, S. Okamoto, Achievement of more than 25% conversion efficiency with crystalline silicon heterojunction solar cell, 40th IEEE

Photovoltaic Specialist Conference, Denver, Colorado, USA, 2014.
[5] D.Smith et all., "Silicon Solar Cellas with total area efficiency above 25%". 43rd IEEE PVSC, Portland 2016.
[6] J. Yang, et al. "III-V on Silicon Multi-Junction Solar Cell with 25% 1-Sun Efficiency via Direct Metal Interconnect and Areal Current Matching". 27th EPSEC, 2012.
[7] S. Essig, "Si-based tandem solar cells with 29.8% record one-sun efficiency". 6th International Conference on Crystalline Silicon Photovoltaics, 2016.
[8] C.W. Liu, J.C. Sturm, "Low temperature chemical vapor deposition growth of β-SiC on (100) Si using methylsilane and device characteristics" J. Appl. Phys. 82, 4558,1997
[9] A.Luque & S. Hegedus, "Handbook of Photovoltaic Science and Engineering". John Wiley & Sons, pp. 601, 2003.
[10] M.A. Green et al. "Solar cell efficiency tables (version 44)". Prog. Photovolt: Res. Appl. 2014; 22:701–710.
[11] J.M. Gee, "A comparison of different module configurations for multi-band-gap solar cells". Solar Cells, Volume 24, Issues 1–2, May–June 1988, Pages 147-155.
[12] T. Trupke, P. Würfel, "Improved spectral robustness of triple tandem solar cell by combined series/parallel interconnection" J. Appl. Phys. 2004, v.96, N.4, pp.2347-2351.
[13] B.J. Stanbery, "Monolithic voltage-matched tandem photovoltaic cell and method for making same". Patent number: 5,261,969. Nov. 1993.
[14] K. Wada, "Silicon solar cell with germanium backside solar cell". Patent number: WO 02/091482 A2. 2002.
[15] S. Werner, E. Lohmüller, S. Maier, A. Kimmerle, A. Spribike, S. Wasmer, F. Clement, A. Wolf. "Process optimization for the front side of the p-type Silicon". *29th European PV Solar Energy Conference and Exhibition*. Amsterdam 2014
[16] JC. Jimeno, R. Gutierrez, V. Fano and C. Del Cañizo, "Célula fotovoltaica, panel fotovoltaico y método de fabricación de células fotovoltaicas", Spanish patent requested P160293ES

ITO/N-SI BASED SOLAR CELLS: THE INFLUENCE OF INTERFACES ON SOLAR CELL EFFICIENCY

A. Simashkevich[1], L.Bruc[1], N. Curmei[1], D. Serban[1], M. Rusu[2], A. Thogersen[3], A. Ulyashin[3]

[1]Institute of Applied Physics, Chisinau, Moldova
[2]Helmholtz Zentrum Berlin Germany
[3]SINTEF Materials and Chemistry, Oslo, Norway

ABSTRACT: The goal of this work is to demonstrate that heterojunction ITO/Si based solar cells fabricated by non-vacuum spray pyrolysis technique for deposition of indium-tin-oxide (ITO) based layers, can be considered as a promising approach for the cost effective processing of ITO/n-type Si based solar cells. It is established that preparation of an intermediate layer between ITO and Si is a crucial step, which determines efficiencies of ITO/Si based solar cells. Two approaches for the formation of such buffer layer have been considered: (i) chemical etching, which results in a formation of a thin Si porous/sponge-like ITO/Si interface layer and (ii) low- temperature (~450 °C) anneals, which result in a formation of a thin SiOx layer at the ITO/Si interface. A set of analytical characterization methods (SEM, TEM, EELS, EDS) have been used for the analysis of individual ITO layers as well as ITO/Si interfaces. It is established that formation of an ultra-thin SiOx layer at the ITO/Si interface is the most promising route to fabricate heterojunction ITO/Si solar cells with the reasonable (above 14%) efficiencies at a low-cost. General technological trends, which lead to an improvement of the conversion efficiency of ITO/Si solar cells, as well as further steps for optimization and improvement of the low-cost processing routes for ITO/Si solar cells are discussed.

1 INTRODUCTION

It is well established that transparent conductive (TCO) layers can be used for fabrication of Si based TCO/Si heterojunction (HJ) solar cells. The well known approach is based on an introduction of a-Si:H intermediate layer at the ITO/Si interface, as it has been as it is demonstrated by SANYO [1]. It should be noted that investigations of Si based heterojunction solar cells, which consist of transparent conductive oxide (TCO)/Si structures started a long time ago [2-7] and still attract some attention recently [8-12]. Such solar cells can be considered as low-cost photovoltaic devices, since the cost reduction is provided by the low-temperature hetero junction-formation step and by the unique properties of the TCO layer, which serves as an emitter, a conductive top electrode and as an antireflection coating.

Nevertheless, a number of problems in this field still remain unsolved. In particular, it is established, that fabrication of reasonable efficiency HJ ITO/Si solar cells can be done exclusively on n-type silicon substrates. It has been established also, that peculiarities of the interfaces between such TCO as indium-tin oxide, and n-type Si substrate, are crucial for fabrication of ITO/n-Si solar cells with reasonable efficiencies. In particular, positive charges, which appear upon formation of ITO/Si interface, decrease the surface barrier height in the case of n-type Si, while increase the barrier in the case of p-type Si. Therefore, ITO/n-type Si solar cells exhibit better PV properties compared to those based on ITO/p-Si structures.

The goal of this work is to demonstrate that heterojunction ITO/n-Si based solar cells fabricated by non-vacuum spray pyrolysis technique for deposition of ITO layers can be considered as a promising approach for the cost effective processing of ITO/Si based solar cells.

2 EXPERIMENTAL

Two types of preparation steps have been applied to 1-3 Ohmcm polished n-type Si substrates:

(i) chemical etching: NH_3: H_2O_2: H_2O (1:1:8) : 70-80°C; 15-20 minutes; HNO_3: HF (3:1) 75 sec; HF dip, DI water;
(ii) low-temperature oxidation: HF, 2 minutes, DI water, 450°C 10 min in air.

Spray pyrolysis of ITO has been performed under following conditions: 1M ($InCl_3$+$3H_2O$): 1M($SnCl_4$+$5H_2O$): C_2H_5OH (1:0,1:20); substrate temperature ~450 °C.
Deposition rate: ~100 nm/min; solution flow: 1.2 ml/min, oxygen flux ~pressure ~1.1 atm.

A sketch of spray system, proposed in [13,14], which has been used in experiments, is shown in **Figure 1.**

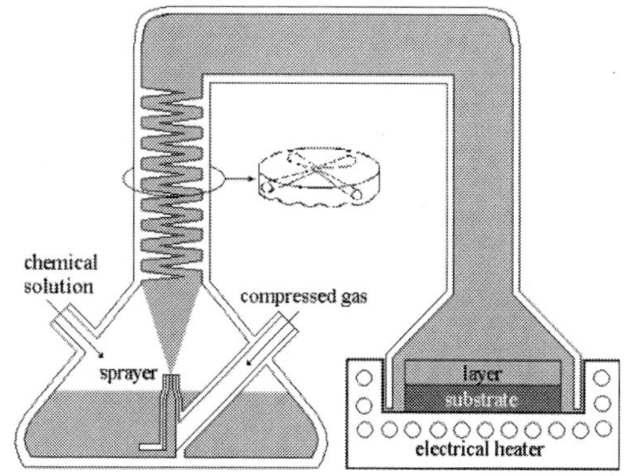

Figure 1: Sketch of a spray system for deposition of ITO layers.

Front side metallization was performed using evaporation of Ag and back side has been performed using evaporation of Cu/Ag layers for fabrication of 2cm^2 solar cells.

Atomic force measurements (AFM) have been used to establish thickness of ITO layers deposited on polished Si surface. Scanning electron spectroscopy (SEM), high-resolution transmission electron microscopy (TEM), electron energy loss spectroscopy (EELS) and Energy-dispersive X-ray spectroscopy (EDS) analyses have been used to characterise structural properties of ITO/Si interfaces.

3 RESULTS

Figure 2 shows results of ITO layers thicknesses, obtained after different deposition times and measured at the steps between ITO/Si and uncovered Si structures. These results demonstrate that deposition rate of ITO layers by non-vacuum spray pyrolysis method is very high - about 100 nm per minute. Thus, ITO layers can be deposited extremely fast on properly prepared Si surfaces and thus, ITO/Si structures can be fabricated at relatively low temperatures (~450 °C) at a low-cost.

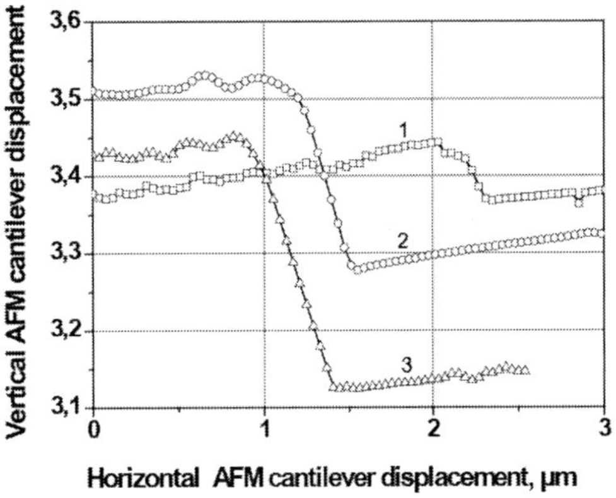

Figure 2: AFM linear scanning measurements at the steps between spray deposited ITO and uncovered Si surface: 1,2.3-depositions for 1,2,3 minutes, respectively.

ITO layers have resistivity about $4\cdot10^3$ $Ohm^{-1}cm^{-1}$, electron concentration $8.8\cdot10^{20}cm^{-3}$ and mobility about $28cm^2V^{-1}s^{-1}$ (Hall measurements). As an example, SEM images of the ITO film surface as well as cross section can be seen from **Figure 3**.

Figure 3 shows that ITO films grown by spray pyrolysis, consist of vertical poly-crystals with pyramidal shapes in the subsurface regions.

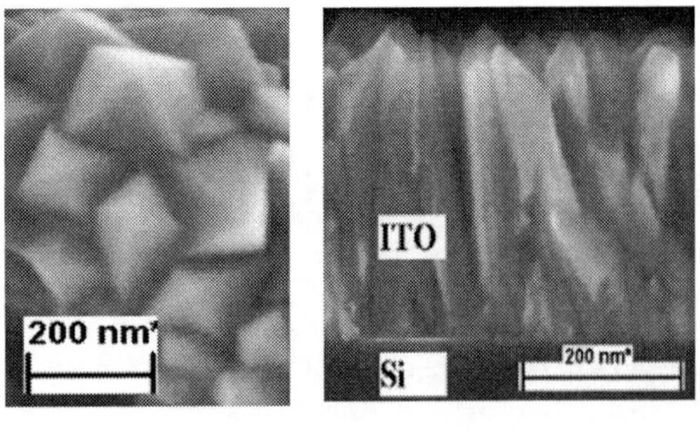

Figure 3: SEM image of spray deposited ITO surface (left) and ITO/Si interface cross section (right).

The light IV measurements of the ITO/Si solar cells fabricated on Si substrates subjected to different treatments prior ITO deposition is given in **Table I.** From **Table I** it can be concluded that low-temperature oxidation of Si substrate provides better parameters of ITO/Si solar cell structure compared to that based on chemically etched Si, and, as a results, higher final conversion efficiency for this type of ITO/Si based solar cells.

Table I: Light IV measurements of ITO/n-Si cells processed using different tretaments of Si substrates.

Type of Si treatment	J_{sc}, mA/cm^2	V_{oc}, V	FF	η, %
Chemical etching	36,2	0,53	0,69	13,4
low- temperature oxidation	36,6	0,55	0,70	14,2

To clarify peculiarities of both solar cell structures under investigation, dark I-V measurements have been performed.

Figure 4 shows results of dark I-V measurements performed at different temperatures for ITO/n-Si structure, in which Si substrate was chemically etched.

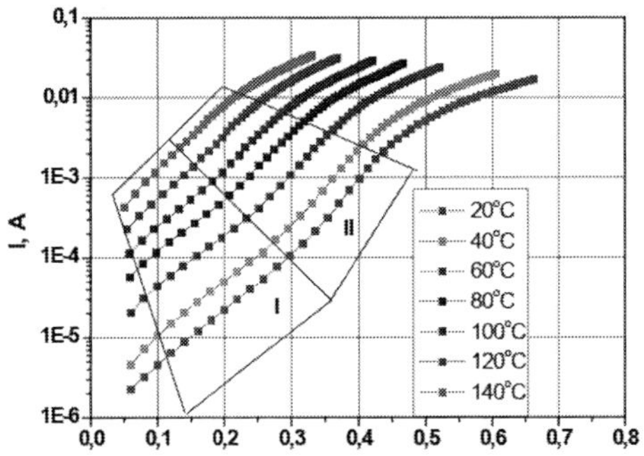

Figure 4: Results of dark I-V measurements for ITO/etched Si solar cell structure

Figure 5 shows similar results of dark I-V measurements for ITO/n-Si structure, in which Si substrate was subjected to a low-temperature oxidation at ~450 °C for 10 minutes.

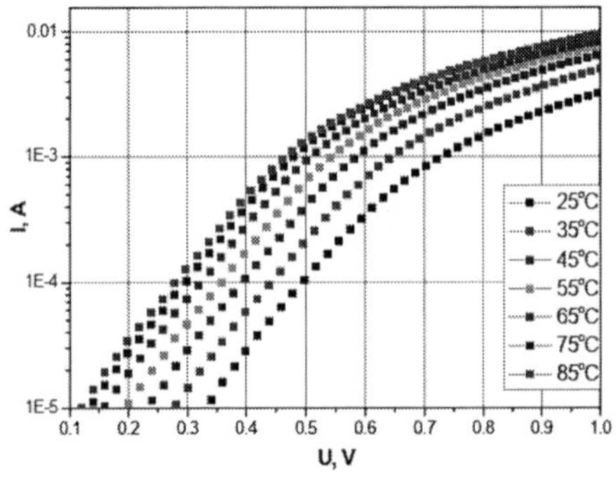

Figure 5: Results of dark I-V measurements for ITO/oxidised Si solar cell structure

From **Figures 4,5** it can be concluded that the mechanism of carrier transport is different structures under consideration is different. Following methodology for the analysis of the carrier transport through potential barriers, developed in [15], it can be concluded that in case of ITO/etched Si solar cell structure an emission based mechanism occurs, while in case of ITO/oxidised Si solar cell structure, mechanism of a tunnelling through a barrier takes place. It can be assumed, that in a later case, thin SiOx layer is formed at the ITO/Si interface [16,17].

To clarify peculiarities of ITO/Si interfaces, high-resolution TEM measurements have been performed for both ITO/Si structures under consideration.

Figure 6 shows TEM cross section image of ITO/Si interface of ITO/etched Si solar cell structure.

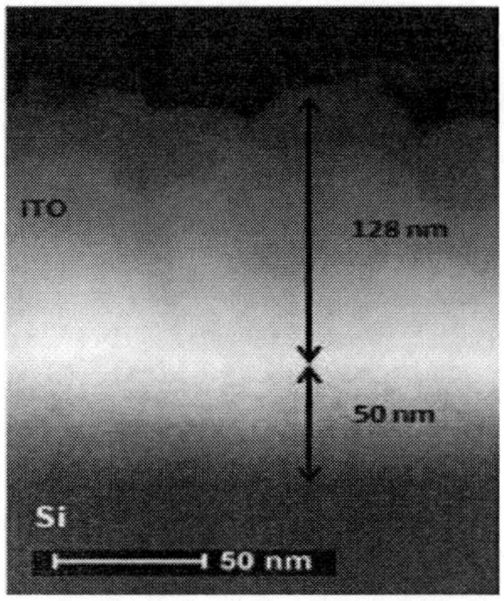

Figure 6: TEM cross section image of ITO/etched Si interface.

Formation of an interface layer ~50 nm thick clearly can be seen from **Figure 6.**
Higher resolution TEM image (**Figure 7**) shows that the interface layer consists of Si, which probaly has porous or sponge-like structure.

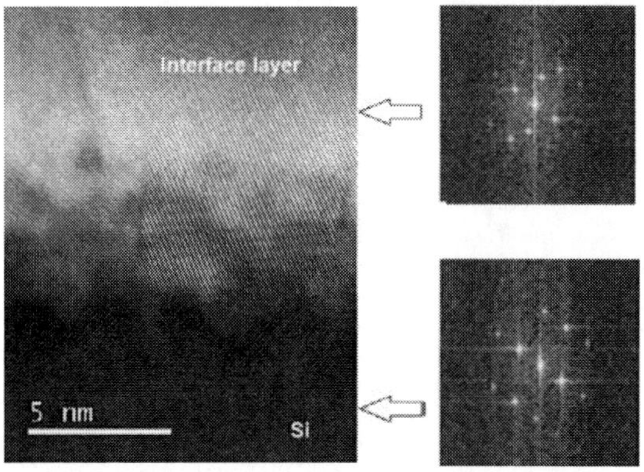

Figure 7: High resolution TEM cross section image of ITO/etched Si interface.

EDS line scanning measurements have been performed to clarify chemical composition of porous/sponge like interface layer, formed by chemical etching of n-type Si substrate and modified by ITO deposition (**Figure 8**).

From **Figure 8** it can be concluded that Indium penetrates into interface layer upon ITO deposition, which indicates that this layer is not dense enough and most probably have porous/sponge structure, as it has been assumed above.

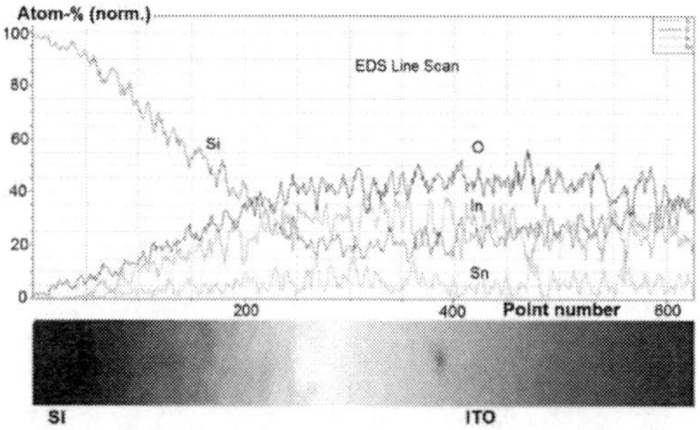

Figure 8: EDS line scanning measurements for interface layer in ITO/Si structure processed by chemical etching of Si substrate followed by deposition of ITO layer.

In case of ITO/ ITO/oxidised Si solar cell structure, situation is different – at the ITO/Si interface a thin (~1 nm) SiOx layer is formed, which can be seen from **Figure 9** (TEM) and **Figure 10** (EDS line scan).

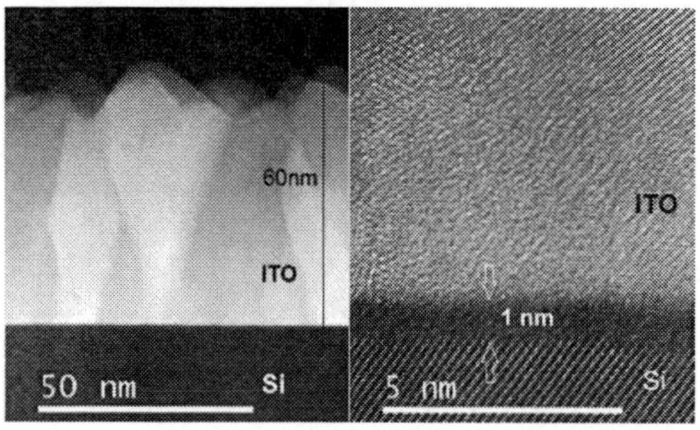

Figure 9: TEM cross section images of ITO/ oxidised Si interface measured at different magnifications.

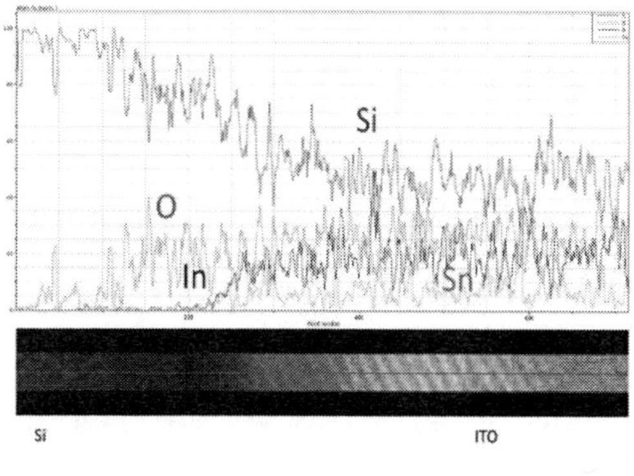

Figure 10: EDS line scanning measurements for interface layer in ITO/ oxidised Si structure followed by deposition of ITO layer.

In this case, a sharp decrease of Indium signal at the ITO/Si interface occurs (**Figure 11**), which indicated that penetration of Indium into Si substrates is not that pronounced as in case of ITO/etched Si solar cell structure (**Figure 8**).

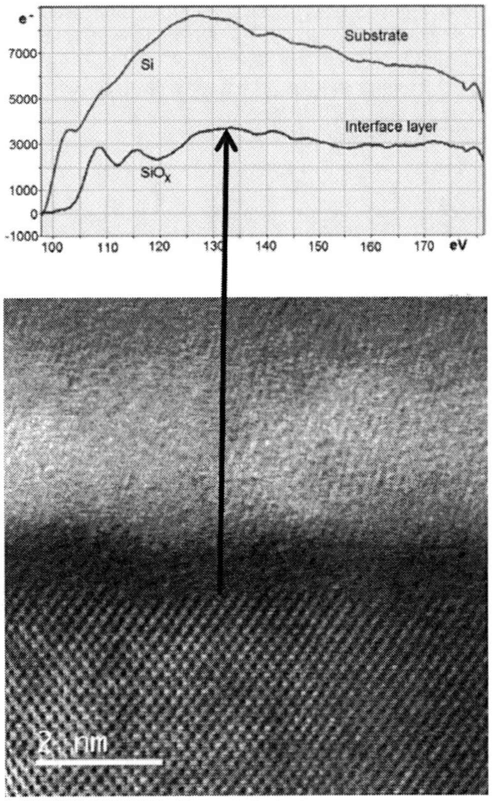

Figure 11: Results of EELS measurements for ITO/ oxidised Si interface.

5 CONCLUSIONS

Based on the experiments obtained so far, the following conclusions concerning ITO/n-type solar cell structures processed by spray pyrolysis of ITO layer on Si substrates with different interfaces can be drawn:

- Heterojunction ITO/n-type Si cells can be fabricated using simple spray pyrolysis deposition of ITO on n-type Si substrates
- In case of **"chemical etching"** of n-Si substrates, best efficiency for Ag/ITO/etched Si/n-Si/Cu/Ag structure obtained so far, is about 13.4%.
- In case of a **"low-temperature oxidation"** of n-type Si substrates, best efficiency for Ag/ITO/SiO_x/n-Si/Cu/Ag structures is 14.2%.
- It can be concluded, that properties of a buffer/intermediate layer at the ITO/Si interface influences essentially efficiency of this type of Si based heterojunction cells.
- It can be assumed, that optimization of interface properties of ITO/Si structures can provide higher efficiencies for ITO/Si HJ solar cells at a low processing cost.
- It can be expected that implementation of other than SiOx oxides or their combination with SiOx in a form of a stack (as it has been demonstrated for a-Si:H/Si HJ solar cells recently [18,19]) can be considered as a further possible step for the efficiency enhancement of low-cost ITO/Si solar cells.

6 ACKNOWLEDGEMENTS

Part of this work is funded by the H2020 EU project CABRISS (grant agreement N° 641972).

REFERENCES

[1] M. Taguchi, K. Kawamoto, S. Tsuge, T. Baba, H.Sakata, M. Morizane, K. Uchihashi, N. Nakamura, S. Kiyama, O. Octa, Progr. Photovolt. 8 (2000) 503.

[2] K. Kajiyama, Y. Furukawa, Jpn. J. Appl. Phys. 6 (1967) 905.

[3] T. Nishino, Y. Hamakawa, Jpn. J. Appl. Phys. 9 (1970) 1085.

[4] R.L. Andersen, Appl. Phys. Lett. 27 (1975) 691.

[5] S.J. Fonash, J. Appl. Phys. 46 (1975) 1286.

[6] A.K. Gnosh, C. Fishman, T. Feng, J. Appl. Phys. 49 (1978) 3490.

[7] T. Feng, A. Gnosh, C. Fishman, J. Appl. Phys. 50 (1979) 4972.

[8] F. Roca, J. Carabe, A. Jager-Waldau, in: W. Hoffmann, J.-L. Bal, H. Ossenbrink, W. Palz, P. Helm (Eds), Nineteenth European Photovoltaic Solar Energy Conference, WIP Munich, ETA Florence, Vol.II, 2004, p. 1321.

[9] A. Chen, K. Zhu, Solar Energy 107 (2014) 195–201.

[10] Wen-Jeng Ho, Min-Chun Huang, Yi-Yu Lee, Zhong-Fu Hou and Changn-Jyun Liao, Nanoscale Research Letters 2014, 9:658.

[11] A. Simaschevici, S. Serban and L. Bruc L. Solar cells on the base of semiconductor-insulator-semiconductor structures. Book Chapter in: "Solar cells – silicon wafer-based technologies". Ed. L. Kosyachenco, InTech, (2011), 299–332.

[12] Z.Q. Ma and B. He TCO-Si Based Heterojunction Photovoltaic Devices, Book Chapter in: "Solar Cells – Thin-Film Technologies", Ed. L. Kosyachenko, InTech, (2011), 111-136

[13] W. Badawy, F. Deckert and K. Doblhofer, Solar Energy Materials. (1983), **8**, 363-369.

[14] H.H. Afify, R.S. Momtaz, W.A. Badawy, S.A Nasser Journal of Materials Science: Materials in Electronics, (1991), **2**, 40-45

[15] A. R. Riben & D. L. Feucht, Int.J. Electronics 1966, v.20, pp.583-587

[16] Spyros Diplas, Andriy Romanyuk, Annett Thøgersen, and Alexander Ulyashin, Phys. Status Solidi A, (2015), **212**, 1, 47–50.

[17] O. M. Løvvik, S. Diplas, A. Romanyuk and A. Ulyashin, J. Appl. Phys. (2014); 115, 083705.

[18] M. Bivour, J. Temmler, H. Steinkemper, M. Hermle, Solar Energy Materials & SolarCells 142 (2015) 34–41.

[19] J. Geissbuhler, J. Werner, S. Nicolas, L. Barraud, A. Hessler-Wyser, M. Despeisse, S. Nicolay, A. Tomasi, B. Niesen, S. De Wolf, and Christophe Ballif, Appl. Phys. Lett. 107, 081601 (2015).

LEAD FREE OHMIC CONNECTIONS ON HIGH EFFICIENCY SILICON SOLAR CELLS

E. Skuras[1], G. Sempros, H. Zoubos, E. Mantzopoulou, T. Giousis, D.Anagnostopoulos
Department of Materials Engineering, University of Ioannina, Ioannina 45110, Greece

Th. Makris, P. Fleming, A. Santamaria
IPSol Energy, Unit 12 Easter Park, Lenton Lane, Nottingham NG7 2PX, United Kingdom.

ABSTRACT: In this paper, we study lead-free ohmic connections made on highly efficient mono-crystalline Si solar cells. Flat copper wires coated with either Sn(96.5%)/Ag(3.5%) or Bi(58%)/Sn(42%) eutectic alloys were soldered to the bus bars of industrially-manufactured Si cells to form environmentally friendly ohmic connections. Different combinations of preheat and soldering temperatures were tested to establish the optimum conditions for achieving low resistance ohmic contacts. Temperature uniformity along the 156x156 mm^2 front surface of the preheated cells was monitored using low-cost, self-adhesive irreversible temperature labels. The electrical performance of the Si cells and their ohmic connections were evaluated from the analysis of the recorded I-V characteristic curves. Efficiencies up to 18.7% and fill factors up to 78% were deduced. Micro-XRF measurements were performed after the completion of each individual step in the process used for making the ohmic connections. Analysis of the X-ray fluorescent spectra has shown compliance with the RoHS directive.
Keywords: mono-crystalline Si solar cells, Pb-free, eutectic solder alloys, micro XRF

1 INTRODUCTION

In this paper, we study lead-free ohmic connections made on industrially-manufactured mono-crystalline Si solar cells. Flat copper wires coated with Sn(96.5%)/ Ag(3.5%) or Bi(58%)/Sn(42%) eutectic alloys are soldered to the bus bars of the high efficiency Si cells. Preheat and soldering temperatures in different combinations, depending on the solder alloy used, are tested to establish the optimum conditions for achieving low resistance ohmic connections [1,2]. Temperature uniformity along the 156x156 mm^2 front surface of the preheated cells is monitored using low-cost, self-adhesive irreversible temperature labels. The electrical performance of the ohmic connections and the Si solar cells were evaluated from the analysis of the recorded I-V characteristic curves under physical light. Efficiencies up to 18.7% and fill factors up to 78% were deduced. After the completion of each individual step in the process used for making the ohmic connections micro X-ray fluorescence measurements were performed for testing compliance with RoHS directive.

2 LEAD FREE OHMIC CONNECTIONS

Our lead-free ohmic connections were made on some of the best-performing commercially-produced standard mono-crystalline Si solar cells. Manufactured by Sunways AG, they have a surface area of 156 x 156 mm^2, a thickness of 180 ± 20 μm and efficiencies up to 18.7% under Standard Test Conditions. Other manufacturers produce equally-efficient Si cells of the same surface area and thickness. A four-wire configuration for eliminating lead resistance was used on two-bus-bar (2BB) and three-bus-bar (3BB) cells. The flat copper wires used were 1.5 or 2.0 mm wide and 0.15 or 0.18 mm thick and coated in either Sn(96.5%)/ Ag(3.5%) or Bi(58%)/Sn(42%) eutectic alloys with melting temperatures 221 ^{0}C and 138 ^{0}C respectively. Solder alloy thickness was varied between 10 and 15 micrometers. The wires were manufactured by Ulbrich Solar Technologies and Indium Corporation. Tabbing ribbons were soldered to the bus bars to form environmentally friendly ohmic connections. Four-wire connections were employed to eliminate lead resistance, as were double ribbons to extract the high currents generated when concentrated sunlight was used. Prior to soldering, a lead-free tin-silver paste is applied to the bus bars of the preheated cells.

Different combinations of preheat and soldering temperatures, depending on the eutectic solder alloy used, are tested to establish the optimum conditions for achieving low resistance ohmic connections [1,2]. Temperature uniformity along the 156x156 mm^2 front surface of the preheated cells is monitored using low-cost, self-adhesive irreversible temperature labels. A gradual decrease in temperature, after soldering, is critical for slowing the cooling and avoiding cell breakages [3].

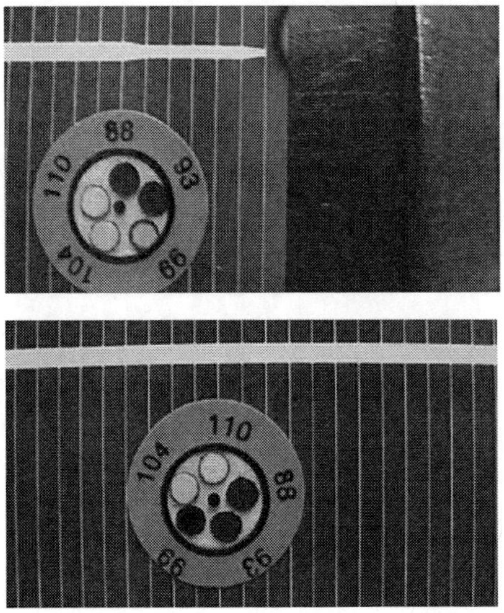

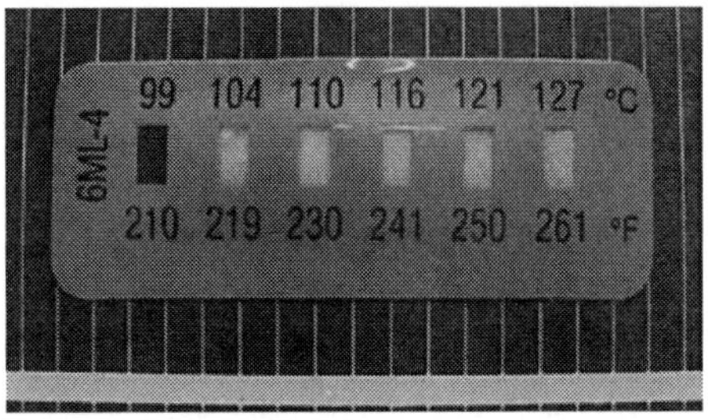

Figure 1: Different parts of the front surface of a three-bus-bar mono-crystalline Si solar cell are shown. The cell rests on a hot plate to control the preheat temperature prior to and during the making of the ohmic connections. At this stage of the process, low-cost, self-adhesive irreversible temperature labels are used for monitoring the uniformity of the temperature along the 156 x 156 mm^2 cell surface.

3 ELECTRICAL MEASUREMENTS

The electrical performance of the Si cells with the Pb-free ohmic connections was evaluated from the analysis of the I-V characteristic curves. A four-wire configuration for eliminating lead resistance was used. The I-V curves were recorded with a 2611A Keithly source-measure unit equipped with a 10 A pulsed electric current source. Efficiencies up to 18.7% and fill factors up to 78% were deduced. In Figure 2 the characteristic I-V curve of a high efficiency mono-crystalline Si solar cell is shown. The intensity of the physical light was 1030 Watt/m^2. The flat copper wires were coated with a lead-free Sn(96.5%)/Ag(3.5%) solder alloy. The preheat and soldering temperatures used were relatively high i.e. 130 ^{0}C and 310 ^{0}C respectively.

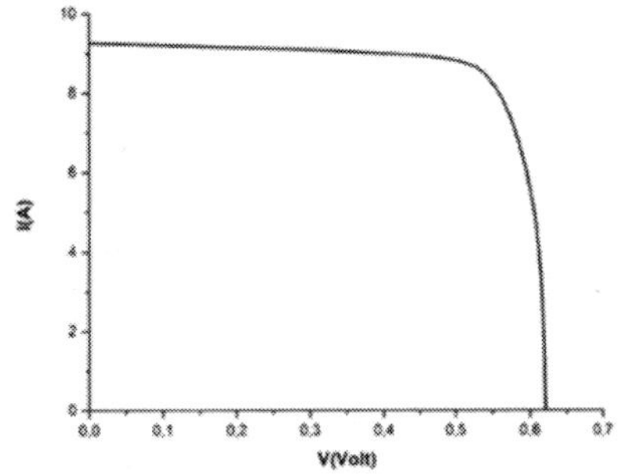

Figure 2: I-V characteristic curve recorded at 1030 Watt/m^2 of physical light. The lead-free ohmic connections were made using preheat and soldering temperatures of 130 ^{0}C and 310 ^{0}C respectively. The solder alloy coating the flat copper wires was Sn(96.5%)/Ag(3.5%).

The reliability of Bi(58%)/Sn(42%) coated flat copper wires as lead-free ohmic connections to industrially-manufactured Si solar cells was tested under different configurations such as a four-wire connection or a double-ribbon geometry combined with a four wire configuration. A temperature of only 200 ^{0}C was used for soldering the copper ribbons to the bus bars due to the low (138 ^{0}C) melting temperature of the Bi(58%)/ Sn(42%) eutectic alloy. Prior to soldering the Si cells were preheated to 120 ^{0}C. An environmentally-friendly solder paste was also used.

Due to the high electric currents generated by the 156 x 156 mm^2 cells, only circles or squares located at the center of the front side of each cell were irradiated with concentrated sunlight. Outside these areas the cells were blanketed. The exposed areas correspond to up approximately 20% of the total cell surface. All ohmic connections were soldered to the middle of the three bus bars. To extract the high currents generated, double Bi(58%)/Sn(42%) coated flat copper wires were used for the current connections. Using Fresnel lenses, concentrations of solar radiation up to 5 Xs were employed and high currents up to 10 A were measured. The electrical performance of the Si cells with Bi(58%)/Sn(42%) coated flat copper wires as lead-free ohmic connections was evaluated from the analysis of the I-V characteristic curves recorded at solar radiation concentrations from 1 X to 5 Xs.

In Figure 3 I-V characteristic curves were instantly recorded from an irradiated circle with 40 mm radius for 1 X and 5 Xs respectively. The irradiated circle area corresponds to ~23% of the total 156x156 mm^2 surface of the cell. The intensity of the physical light was 950 Watt/m^2. In the case of 5 Xs a high current of 10 A was generated.

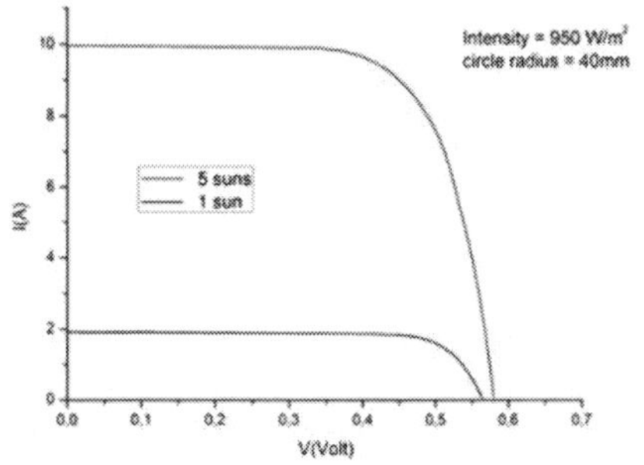

Figure 3: I-V characteristic curves recorded from an irradiated circle with 40 mm radius, immediately after its exposure, for 1 X (blue curve) and ~5 Xs (red curve) respectively. The irradiated circle area corresponds to 23.64% of the total 156x156 mm^2 surface of the cell. The intensity of the physical light was 950 Watt/m^2. In the case of 5 Xs a high current of 10 A was generated and extracted through double lead-free flat copper wires from the middle bus bar.

4 MICRO-XRF MEASUREMENTS

After the completion of each individual step in the process used for making the ohmic connections micro X-ray fluorescence measurements were performed. Elemental concentrations deduced from the analysis of the micro-XRF data were used for testing compliance with RoHS directive. A relatively high content of bromine (Br) was detected in the lead-free solder paste applied to the bus bars of the preheated Si solar cells prior to the soldering of the flat copper wires. Lead (Pb) was present in a minute quantity compliant with the RoHS directive.

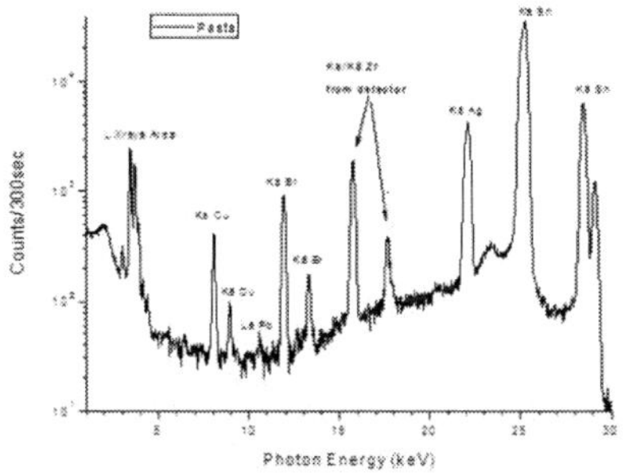

Figure 4: XRF spectum recorded from the lead-free solder paste applied to the bus bars of the preheated Si solar cells prior to the soldering of the flat copper wires. A relatively high content of Br is detected. Lead is present in a minute quantity compliant with the RoHS directive.

The flat copper wires used for making the ohmic connections were coated either with Sn(96.5%)/Ag(3.5%) or Bi(52%)/Sn(48%) solder alloys. XRF spectra of the tabbing wires were taken prior to removal of the oxides. Lead is present in a concentration below the limit of 0.1% by mass (1000 ppm) of homogeneous material specified by the RoHS directive.

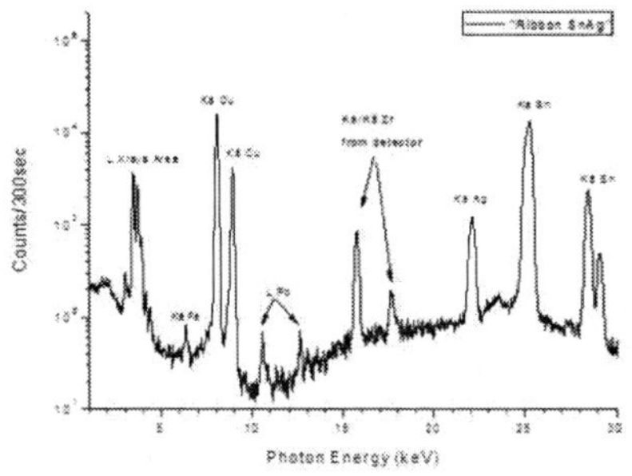

Figure 5: XRF spectrum recorded from a flat copper wire coated with a 10-15 μm thick Sn(96.5%)/Ag(3.5%) solder alloy prior to removal of the oxides. Lead is present in a concentration below the limit of 0.1% by mass (1000 ppm) of homogeneous material specified by the RoHS directive.

Micro X-ray fluorescence spectrometry provides highest spatial resolution with spot sizes a few tenths of a micrometer. Due to the small size of its beam micro X-ray fluorescence is suitable for performing elemental analysis on tiny sample areas. In Figure 6 the beam is placed in the middle of a front side bus bar of an industrially-manufactured mono-crystalline Si solar cell used in this study. The characteristic X-ray fluorescence emission from the bus bar allows for elemental mapping along its 156 mm length and across its 1.5 mm width. The relatively high content of Pb and Bi detected is due to the Pb- and Bi-oxide glass frits contained in the Ag paste. The μ-XRF technology used allows the production of beams with sufficient intensities even at sizes down to 100 μm.

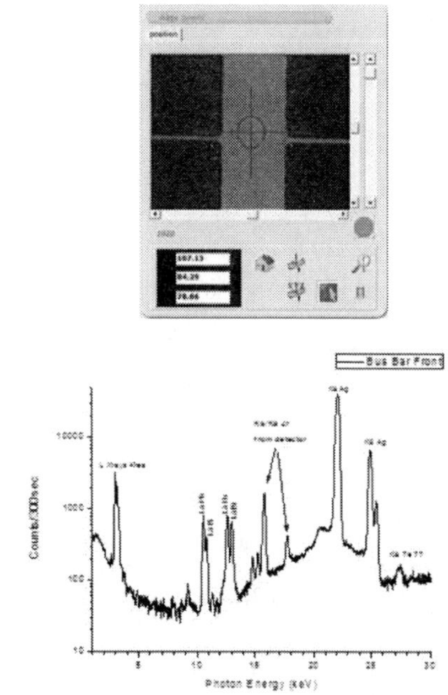

Figure 6: XRF spectrum recorded in the middle of a front side bus bar of an industrially-manufactured mono-crystalline Si solar cell used in this study. The relatively high content of Pb and Bi detected is due to the Pb- and Bi-oxide glass frits contained in the Ag paste. The μ-XRF technology used allows the production of beams with sufficient intensities even at sizes down to 100 μm.

Elemental analysis based on the micro-XRF spectra shows that the small quantities of lead detected in the solder paste, the Sn(96.5%)/Ag(3.5%) and Bi(58%)/Sn(42%) coated flat copper wires (before and after the technique applied for removal of the oxides), the solder and the tinned soldering tip used all comply with the RoHS directive. In contrast, the bromine content found in the lead-free soldering paste is relatively high. Although no lead is introduced at any of steps in the process or materials used, when ohmic contacts are made, lead and bismuth are detected in the spectra recorded from the bus bars of the industrially-manufactured cells. They are attributed to the Pb and Bi-oxides of the glass frits in the Ag paste used during the screen printing fabrication of the Si cells [4, 5].

6 CONCLUSIONS

Lead-free ohmic connections were made on industrially-manufactured mono-crystalline Si solar cells. Flat copper wires coated with either Sn(96.5%)/ Ag(3.5%) or Bi(58%)/Sn(42%) eutectic alloys were soldered to the bus bars of the Si cells. Different combinations of preheat and soldering temperatures were tested to establish the optimum conditions for achieving low resistance Pb-free connections. For the first time, temperature uniformity along the 156x156 mm^2 front surface of the preheated cells was monitored using low-cost, self-adhesive irreversible temperature labels. The electrical performance of the Si cells and their connections was evaluated from the data analysis of the recorded I-V characteristic curves. Efficiencies up to 18.7% and fill factors up to 78% were deduced. Micro-

XRF measurements were performed after the completion of each individual step in the process used for making the connections. Analysis of the X-ray fluorescent spectra has shown compliance with the RoHS directive.

1) E. Skuras, Corresponding author, e-mail address : eskuras@cc.uoi.gr

6 REFERENCES

[1] P. Schmitt, P. Kaiser, C. Sario, M. Tranitz, U. Eitner, Energy Procedia 27 (2012) 664-669.
[2] S. Schindler, S. Wiese, 18th European Microelectronics and Packaging Conference (EMPC) 2011, ISBN: 978-1-4673-0694-2.
[3] P.Grunow, Proceedings of the 2nd Workshop on Metallization for Crystalline Silicon Solar Cells, 2010, pp.59-62.
[4] Yu-Chou Shih, Yeong-Her Lin, Jiun-Pyng You, Frank G. Shi, Journal of Electronic Materials, 42(3) (2013) 410-416.
[5] Ji-Hyun Jeong, Sung-Jin Kim, Chang-Rok Son, Soon-Chul Ur, Soon-Yong Kweon, Materials Science Forum, 724 (2012) 393-396.

INVESTIGATIONS ON LASER FIRED CONTACTING AND ANNEALING OF RST SILICON PERC-TYPE SOLAR CELLS

Benjamin Albrecht[a], Yves Patrick Botchak[a], Fabrice De Moro[b], Philipp Keller[a], Giso Hahn[a]
[a]Department of Physics, University of Konstanz, 78457 Konstanz, Germany
[b]Solarforce SA, 1 rue du Dauphin, 38300 Bourgoin-Jallieu, France

Author for correspondence: benjamin.albrecht@uni-konstanz.de Tel.: +49 7531 882082, Fax: +49 7531 883895

ABSTRACT: The Ribbon on Sacrificial Template (RST) casting process is a kerf-loss free direct wafer casting method, which enables the production of silicon ribbon wafers below 100 µm thickness for thin solar cells. But the thinner the cell, backside passivation is all the more important. In our lab-type PERC solar cell process, the back contact is realized by laser fired contacts (LFC). In this study the influence of different LFC parameters on back side passivation is shown, based on lifetime samples made of standard 1 Ohmcm FZ wafers. With the best parameter set, surface recombination velocity $S_{eff,LFC}$ of 34.1 cm/s is possible with reasonable contacting. Beside backside passivation, an improvement of bulk lifetime of RST material was achieved by different surface passivation and annealing steps. These improvements resulted in a new effective lifetime record for RST material of 39.6 µs (measured by QSSPC).
Keywords: Annealing, Laser Fired Contacts, Lifetime, Passivation, Ribbon Silicon

1 INTRODUCTION

The Ribbon on Sacrificial Template (RST) casting process [1] enables the production of silicon ribbon wafers down to 50 µm thickness for thin solar cells. It is a kerf-loss free direct wafer casting method which means that energy-intensive silicon material can be saved. But the thinner the cell, backside passivation is all the more important. In our lab-type PERC solar cell process [2], the back contact is realized by laser fired contacts (LFC). Single side metalized and LFC contacted FZ lifetime samples showed that the standard LFC parameters, used by us up to now, damage backside passivation significantly.

In this study new LFC parameter sets are studied on single side metalized and LFC contacted 1 Ωcm p-type FZ lifetime samples (sec. 2.1, 3.1). Minority carrier lifetimes are measured before metallization and after LFC by TR-PLI [3] (time resolved photoluminescence imaging). Double side metalized and LFC contacted contact resistance samples are produced in parallel to measure contact resistances of the LFCs.

Beside backside passivation, improvement of bulk quality is very important for RST material to reach higher cell efficiencies. P-gettered and one-time H-passivated 1 - 3 Ωcm, 50 – 80 µm thick, p-type RST wafers from different batches, produced by the company Solarforce SA (France) [4], show typical bulk lifetimes of 15 – 30 µs. These wafers are used in this study for further treatments.

In the first RST experiment (sec. 2.2, 3.2), RST wafers are surface passivated with either SiN_x:H via PECVD (plasma-enhanced chemical vapor deposition) or Al_2O_3 via ALD (atomic layer deposition) and are annealed up to 3 h at 400 °C in N_2 atmosphere. Lifetimes are measured several times in between via QSSPC (quasi-steady-state photo conductance) measurements. Standard 1 Ωcm FZ wafers are processed in parallel to determine S_{eff} and use it to calculate τ_{bulk} for the RST wafers.

The second RST experiment (sec. 2.3, 3.3) is about multi-passivation of RST material with Al_2O_3. One passivation step includes Al_2O_3 deposition (ALD) and 1 h annealing at 400 °C in N_2 atmosphere. Multi-passivation comprises three passivation steps with etching off the passivation layer in between. Lifetimes are measured after each step via QSSPC.

The third RST experiment (sec. 2.4, 3.4) comprises all annealing steps which are used in our solar cell process [2] to establish electrical contacting on front and rear side of the solar cell. But only rear side contacting was realized, so that lifetime measurements with TR-PLI [3] could be applied after each step. Compared to the first (annealing) experiment, also low temperature annealing (70 – 170 °C) and annealing with Al present is applied here.

2 EXPERIMENT

2.1 Optimisation of LFC parameters

To find out LFC parameter sets which damage surface passivation as little as possible but still enable reasonable electrical contact, single side LFC contacted lifetime samples and double side LFC contact resistance samples are produced in parallel.

For this investigation 250 µm thick, 5x5 cm² 1 Ωcm p-type FZ wafers are used. At first, all wafers receive a wet chemical cleaning, are symmetrically coated with 37.5 nm Al_2O_3 by ALD and are annealed for 60 min at 400 °C in N_2 atmosphere. TR-PLI [3] is used to determine $\tau_{eff,Al2O3}$. τ_{bulk} was assumed to 3,800 µs for all 1 Ωcm FZ wafers and the surface recombination velocity $S_{eff,Al2O3}$ was calculated.

The LFC lifetime samples are then coated on the backside with 2 µm Al via EBPVD (e-beam evaporation), are LFC contacted with a 1064 nm laser (pitch 0.5 mm and different pump current/frequency sets) and annealed for 40 min at 350 °C in N_2 atmosphere. TR-PLI [3] is used to determine τ_{eff} of LFC lifetime samples.

The LFC contact resistance samples are coated on both sides with 2 µm Al by EBPVD and LFC contacted with the same LFC parameter sets as the LFC lifetime samples. 4 point probe resistivity measurements are used to calculate the resistance R_{LFC} of one LFC point. An optical microscope is used to measure diameters of the

LFC points additionally.

2.2 Annealing experiment on RST material

This experiment (400 °C for all annealing steps, τ_{eff} measured with PCD) examines the effect of annealing after passivation of RST wafers with Al_2O_3 (ALD) compared to SiN_x:H (PECVD) on τ_{bulk}. Standard FZ wafers are processed in parallel to determine S_{eff}, so that τ_{bulk} of RST wafers can be calculated from τ_{eff} (assuming that S_{eff} for FZ and RST wafers is equal).

All wafers receive a $POCl_3$ diffusion, are coated with 75 nm SiN_x:H via PECVD on the front side, and are fired in a belt furnace to enhance τ_{bulk} before SiN_x:H and emitter are etched off. A standard wet chemical cleaning followed before the wafers are separated into two groups.

Both groups contain three subgroups: FZ samples, RST batch 1 and RST batch 2. Each subgroup contains four wafers. RST batches vary in resistivity (batch 1: 1 – 2 Ωcm; batch 2: 2 – 3 Ωcm) and some crystallization parameters. The first group is coated with 37.5 nm Al_2O_3 (ALD) on both sides and is annealed for a total of 3 h. Four τ_{eff} measurements are performed in between. The second group is passivated with 37.5 nm SiN_x:H (PECVD) on both sides, annealed for 3 h, fired and annealed a total of 2 h more. Four τ_{eff} measurements are performed after firing. Then, SiN_x:H is etched off and the samples are coated also with 37.5 nm Al_2O_3 (ALD) and annealed for 30 min. τ_{eff} is measured again to compare both groups with the same passivation layer.

2.3 Multi-passivation of RST material

In this investigation some samples of the annealing experiment are processed further. The other samples have received only the standard treatment before (P-gettering, PECVD SiN_x:H-coating, firing, Al_2O_3 surface passivation (ALD) and 1 h annealing at 400 °C). All samples have exactly one Al_2O_3 surface passivation before this experiment.

In this experiment all samples receive three Al_2O_3 surface passivation steps in addition, one after the other. One passivation step includes etching off the old Al_2O_3 passivation, new Al_2O_3 deposition (ALD) and 1 h annealing at 400 °C in N_2 atmosphere. Lifetimes are measured after each step via QSSPC.

2.4 Low temperature annealing, Al metallization and LFC on RST material

This experiment comprises all annealing steps which are used in our solar cell process [2] to establish electrical contacting on front and rear side of the solar cell. But only rear side contacting was realized, so that lifetime measurements with TR-PLI [3] can be applied after each step.

RST samples with different history are used. Some have been used already for the experiments 2.1 and/or 2.2, the others received only the standard treatment before (P-gettering, PECVD SiN_x:H-coating, firing, Al_2O_3 surface passivation (ALD) and 1 h annealing at 400 °C).

To establish front side contacting, a photolithography process is used in our solar cell process [2]. This includes different annealing steps: 30 min at 170 °C, 1 h at 70 °C and a ramp from 70 °C to 105 °C subsequently (~20 min) and 1 h at 100 °C. In this experiment all annealing steps of the photolithography process are applied directly one after another.

To establish rear side contacting, the samples are coated with 2 µm Al on the rear side by EBPVD. During this process (~30 min) the samples are heated up to 220 °C. Then LFC is applied (pitch 0.5 mm and different pump current/frequency sets). An annealing step (40 min, 350 °C) with Al on the rear side follows.

3 RESULTS

3.1 Optimisation of LFC parameters

All stated lifetime values τ_{eff} in this experiment are arithmetic mean values from TR-PL images [3]. τ_{eff} = 1,340 µs is the mean value of seven 1 Ωcm FZ wafers after Al_2O_3 passivation. This corresponds to $S_{eff,Al2O3}$ = 6.0 cm/s. This value was used to calculate $S_{eff,LFC}$ of LFC contacted back sides of the wafers from $\tau_{eff,LFC}$.

Our LFC standard parameters are: pump current 34 A, frequency 1,000 Hz, pitch 0.5 mm. Higher frequency (up to 20,000 Hz) means reduction of pulse energy, smaller LFCs and less damage of the passivation layer. More than 20,000 Hz means that the LFC points become inhomogeneous. That is why LFC parameter sets with frequency 20,000 Hz were used primarily in this examination. The pump current was varied between 27 A and 34 A. Below 27 A, LFC points become also more inhomogeneous. A comparison of results between standard parameter set and the weakest parameter set (27°A, 20,000 Hz) is shown in Tab. I. A pitch of 0.5 mm was used in all parameter sets.

Table I: Comparison of LFC results between standard parameter set and the weakest parameter set.

Parameter set:	34 A, 1 kHz	27 A, 20 kHz
$\tau_{eff,LFC}$ [µs]	168	545
$S_{eff,LFC}$ [cm/s]	142	33.6
R_{LFC} [Ω]	33.5	93.1
$\varnothing_{LFC}$ [µm]	130	45

Fig. 1 shows all results for frequency 20,000 Hz and pump currents from 34 A down to 27 A.

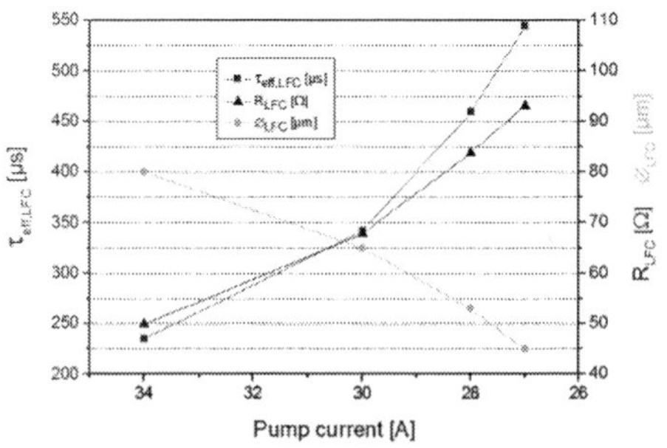

Figure 1: $\tau_{eff,LFC}$, R_{LFC} and $\varnothing_{LFC}$ depending on pump current. Frequency is 20,000 Hz and pitch is 0.5 mm for all measurement points.

3.2 Annealing experiment on RST material

This experiment was carried out to investigate the effect of annealing at 400 °C on τ_{bulk} of RST wafers. Results of this experiment are shown in Fig. 2 and Fig. 3

(QSSPC lifetimes). FZ wafers have been used to determine S_{eff}. With these values, τ_{bulk} was calculated from τ_{eff} for RST wafers. Fig. 2 shows the results of group 1 (passivated and annealed with Al_2O_3). The green curves belong to RST batch 1, the red curves belong to RST batch 2. Triangles show τ_{eff}, squares show τ_{bulk}. Each point in the graph is a mean value calculated from four samples. The graph shows that τ_{bulk} increases for both batches up to an annealing time of 60 min, although only by approx. 0.5 µs. Surface passivation (blue curve) improves up to an annealing time of 120 min.

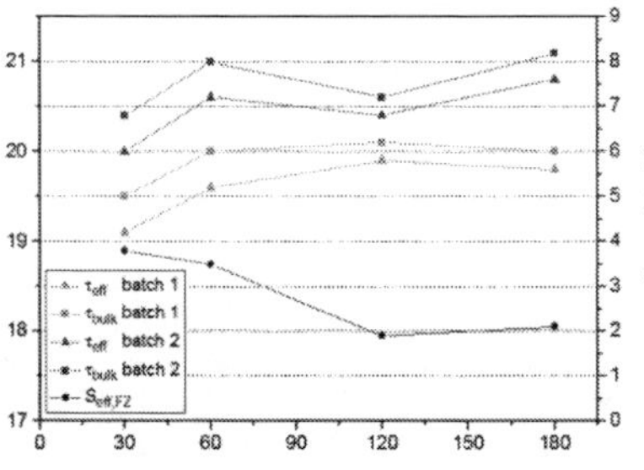

Figure 2: QSSPC lifetime results of group 1. τ_{bulk} was calculated from τ_{eff} with S_{eff} from FZ wafers.

Fig. 3 shows the results of group 2. Each point of the green and blue curves is a mean value calculated from four samples. The red curve depicts only one sample. After deposition of SiN_x:H, the samples were annealed for 3 h in total and were measured several times, but nothing really interesting happened. So, the first point in the graph includes SiN_x:H deposition and 3 h annealing. Then the samples were fired and annealed again. τ_{bulk} of batch 1 improves slightly by approx. 1.5 µs during the first 60 min of annealing. The only sample of batch 2 was annealed only for 30 min. The last step (etching off the SiN_x:H, passivation and annealing with Al_2O_3) shows that $S_{eff,FZ}$ and τ_{eff} of batch 1 reach the same level as in group 1 (Fig. 2) with the same surface passivation layer. τ_{bulk} of batch 1 improves by approx. 1 µs and τ_{bulk} of the sample from batch 2 rises strongly from 24.1 µs to 36.8 µs in this last step.

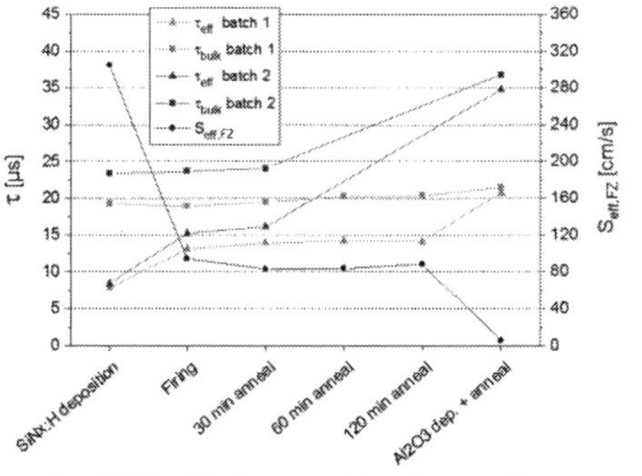

Figure 3: QSSPC lifetime results of group 2. τ_{bulk} was calculated from τ_{eff} with S_{eff} from FZ wafers.

3.3 Multi-passivation, low temperature annealing, Al metallization and LFC on RST material

Multi-passivation induces lifetime improvement of 2 – 5 µs for most RST wafers. But after three passivation steps there is no further improvement visible. Primarily good regions improve by multi-passivation. τ_{eff} of the sample from batch 2 in Fig. 3 could be improved further by multi-passivation up to 39.6 µs (QSSPC measurement). This means a new record for RST material.

In Fig. 4 the effects of multi-passivation, low temperature annealing, Al deposition and annealing on lifetime are depicted by the example of three samples. All three samples are from batch 1 (see 2.2 and 3.2). SF_1 is from group 1 (Fig. 2). SF_2 and SF_3 are from group 2 (Fig. 3). These samples were processed further in this experiment. SF_1 and SF_3 got all process steps (steps 1-3 in Fig. 4). SF_2 got no multi-passivation (step 1 in Fig. 4). Arithmetic and harmonic mean effective lifetime values of TR-PLI measurements are shown in Fig. 4. If primarily good regions improve, the arithmetic mean value increases stronger than the harmonic mean value. If primarily poor regions improve, the harmonic mean value increases stronger than the arithmetic mean value. The three process steps (abscissa in Fig 4) are depicted in Table II (see also 2.3 and 2.4).

Table II: Starting situation and three process steps in Fig. 4

Step	Description
0	Starting situation: One Al_2O_3 passivation, at least 1 h annealing at 400 °C
1	Multi-passivation
2	Low temperature annealing (70 °C – 170 °C)
3	Al deposition and 20 min annealing at 350 °C

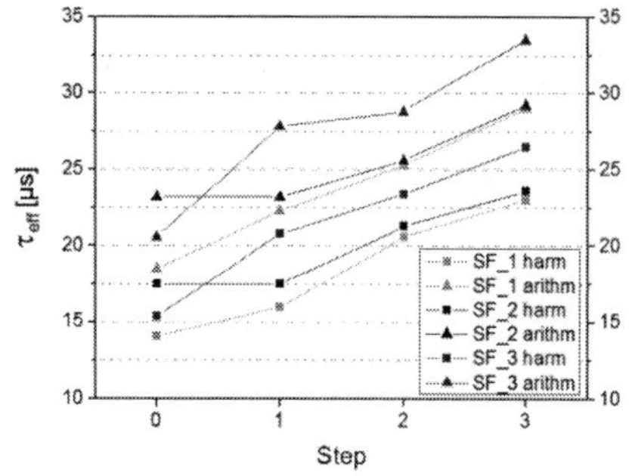

Figure 4: TR-PLI lifetimes after starting situation (step 0) and three process steps (see Table 2).

Fig. 4 shows that the total improvement of the three samples is in the same range (9 – 12 µs in total for all steps). This result corresponds to all other samples which were processed in this way, regardless of the fact that the other samples were not part of the annealing experiment (sections 2.2, 3.2) before and were only annealed for 1 h with Al_2O_3 passivation before this experiment. Furthermore, the increase of lifetime of sample SF_2, which got no multi-passivation (3 times etching off the old Al_2O_3 passivation, new Al_2O_3 deposition (ALD) and 1 h annealing at 400 °C in N_2 atmosphere), behaves similar to SF_1 and SF_3 after step 1.

By comparing the development of arithmetic and harmonic mean values, we can state that the harmonic mean values increase stronger than the arithmetic mean values for all samples during low temperature annealing. For all other process steps the arithmetic mean values increase stronger than the harmonic mean values. This means that low temperature annealing primarily increases the lifetimes of poor quality regions. The other process steps increase primarily the lifetimes of good regions.

Fig. 5 shows TR-PLI images before and after low temperature annealing (left and right map with histogram) of sample SF_3. The left image and histogram reveals that the poorest regions have lifetimes between 3 µs and 7 µs before low temperature annealing. The right image and histogram shows that there is no point with lifetime less than 11 µs after low temperature annealing.

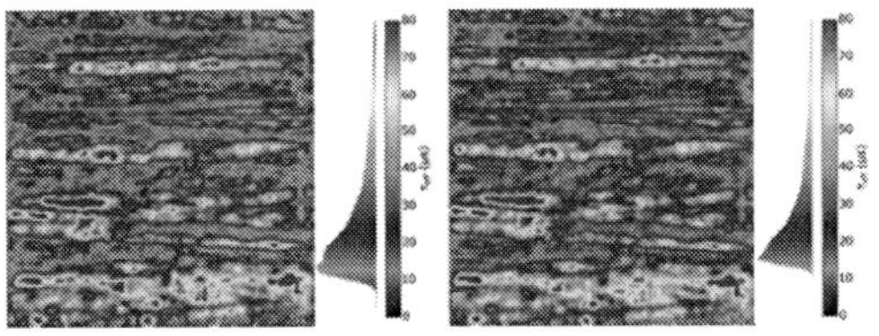

Figure 5: TR-PLI images of sample SF_3. Left: before low temperature annealing. Right: after low temperature annealing. Minimum lifetime increases from 3 µs (left) to 11 µs (right).

Fig. 6 shows TR-PLI images before and after LFC and 40 min annealing at 350 °C. The four marked regions were contacted with different LFC parameter sets. Tab. 3 lists the parameter sets and harmonic mean values of τ_{eff} before and after LFC. Pitch 0.5 mm was used in all regions.

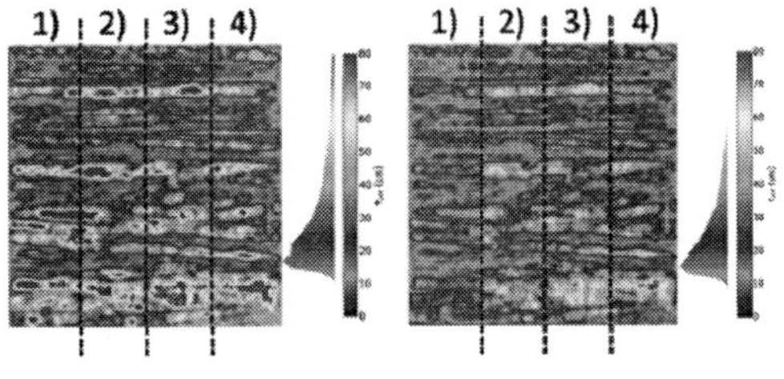

Figure 6: TR-PLI images of sample SF_3. Left: before LFC. Right: after LFC. Four marked regions contacted with different LFC parameter sets (Tab. 3).

Table III: Parameter sets and harmonic mean values of τ_{eff} before and after LFC. Pitch 0.5 mm was used in all regions.

Region	LFC parameter set	$\tau_{eff,harm}$ [µs] before LFC	$\tau_{eff,harm}$ [µs] after LFC
1	34 A, 1 kHz	25.5	18.5
2	30 A, 20 kHz	26.5	22.3
3	28 A, 20 kHz	28.0	24.3
4	27 A, 20 kHz	25.9	22.9

Fig. 6 and Tab. III reveal that also for RST material higher τ_{eff} values are possible with the new LFC parameter sets, which were tested first on FZ wafers (sec. 2.1, 3.1).

4 DISCUSSION

The optimization of LFC parameter sets (sec. 2.1, 3.1) revealed that $\tau_{eff,LFC}$ up to 545 µs and $\varnothing_{LFC}$ down to 45 µm are possible with a frequency of 20,000 Hz and pump currents down to 27 A. Further investigation is necessary to determine the lower limit more precisely. R_{LFC} = 93 Ω and a pitch of 0.5 mm result in a contribution to series resistance of 0.24 Ωcm², compared to 0.09 Ωcm² with standard parameters. This means a decrease of fill factor by ~ 0.7 %. But the higher τ_{eff} should easily overcompensate this disadvantage.

The annealing experiment of RST shows that one time passivation with either Al_2O_3 or SiN_x:H passivation layer and annealing for several hours at 400 °C has only little effect on bulk lifetime. But Fig. 3 reveals that the combination of SiN_x:H passivation, annealing, etching off the passivation layer and afterwards Al_2O_3 passivation and annealing seems to be an effective way to increase bulk lifetime for some kinds of RST material. This effect has to be investigated further.

Multi-passivation with Al_2O_3 is a possibility to increase bulk lifetime of RST material. But improvement is limited to ~ 5 µs and 3 additional passivation steps. Nevertheless, the new record of τ_{eff} = 39.6 µs was reached in this experiment.

The results of low temperature annealing (70 - 170 °C) are promising, because it enables to improve primarily poor regions of RST wafers. Fig. 5 shows that the lifetime of the poorest region of sample SF_3 could be increased from ~ 3 µs to over 11 µs. This effect has to be investigated further and the optimal annealing temperature and time have to be found.

The deposition of Al and subsequent annealing at 350 °C (with Al) leads to further lifetime improvement. This is another promising result which has to be investigated further.

Fig. 6 depicts that the new LFC parameter sets are also beneficial for RST material. Because of the elongated grain structure with closely packed grain boundaries, even smaller or asymmetrical pitches could be beneficial for RST material.

5 CONCLUSION

The results with new LFC parameter sets are very promising and should enable significantly higher V_{oc} in future solar cell experiments.

Different passivation and annealing processes were examined to improve τ_{bulk} of RST material. If applied one after another, a total improvement of more than 10 µs for τ_{eff} compared to a standard passivation treatment was demonstrated in this study. This is a good starting point for further investigation. Particular attention should be paid to the effect of low temperature annealing, because it is a process which primarily improves the bulk lifetime of poor regions.

Finally the new record τ_{eff} = 39.6 µs was set in this study.

6 ACKNOWLEDGEMENTS

Part of this work was funded by the German BMWi (0325581). The content is the responsibility of the authors.

7 REFERENCES

[1] C. Belouet et al., Journal of Crystal Growth 82 (1987) 110.
[2] J. Junge et al., Proceedings 25th European Photovoltaic Solar Energy Conference (2010) 1722.
[3] D. Kiliani et al., Solar Energy Materials and Solar Cells 106 (2012) 55.
[4] F. De Moro et al., Phys. Status Solidi C9(10-11) (2012) 2092.

TEMPERATURE DEPENDENCE OF THE MAIN CHARACTERISTICS OF HIT ELEMENTS

A.V. Sachenko[a], Yu.V. Kryuchenko[a], V.P. Kostylyov[a], A.V. Bobyl[b], E.I. Terukov[b,c], S.N. Abolmasov[c], A.S. Abramov[c], D.A. Andronikov[c], M.Z. Shvarts[b], I.O. Sokolovskyi[a], and M. Evstigneev[d*]
[a]V. Lashkaryov Institute of Semiconductor Physics, NAS of Ukraine,
41 prospect Nauky, 03028 Kyiv, Ukraine
[b]Ioffe Institute, 194021 St.-Petersburg, Russian Federation
[c]TFTC Ioffe R&D Center, 194021 St.-Petersburg, Russian Federation
[d]Department of Physics and Physical Oceanography, Memorial University of Newfoundland,
St. John's, NL, A1B 3X7 Canada
[*]Author to whom correspondence should be addressed. Electronic mail: mevstigneev@mun.ca

Silicon heterojunction solar cells (HJSC) with the efficiency of about 20 % are manufactured. Their short-circuit current, open-circuit voltage, photoconversion efficiency, and fill factor of the current-voltage curve are measured in a broad temperature range from 80 to 420 K. It is established that the open-circuit voltage, the fill factor, and the photoconversion efficiency are non-monotonic functions of temperature, having a maximum in the vicinity of 200 K. A new approach to modeling of HJSCs is proposed, which allows one to obtain quantitative agreement with the experimental results at temperatures above 200 K, as well as to describe the results published in the literature on the solar cells under AM1.5 conditions. The temperature coefficient of photoconversion efficiency in HJSCs is discussed and its low value is shown to be related to the low surface and volume recombination rates.
Keywords: Solar Cell Efficiencies, Silicon, Heterojunction, High-Efficiency

1 INTRODUCTION

Recently, there has been substantial progress in increasing the photoconversion efficiency of the silicon heterojunction solar cells (HJSC, also known as heterojunction with intrinsic thin layer, HIT), see Refs 1 and 2 for a state-of-the-art review. For instance, Masuko *et al.*[3] report the efficiency of 25.6 % under standard conditions (AM1.5G spectrum, 1000 W/m^2 incident power density). According to the estimates performed in several works, such as Refs. 4 and 5, the highest theoretical efficiency of silicon-based photoconverters under unfocused solar radiation is about 30 %. In order to approach the limit efficiency, it is necessary to intensify the research aiming at optimizing the parameters of these structures. This requires deeper understanding of the peculiarities of the processes that take place in silicon HJSCs as compared to the standard ones.

The principal difference between the HJSCs and the p-n junction-based ones is that in the former, in order to contribute to the photoinduced current, the photogenerated electrons and holes need to travel through the wide-bandgap layers. Typically, these are the layers of a-Si:H[2]. However, in a view papers, other materials are used, such as a-SiC:H[6] and a-SiO$_x$:H[7]. The band diagram of a heterojunction depends both on the bandgaps of the semiconductors in contact and on the respective electron affinities. The absence of a barrier on the heterojunction boundary for at least one band is not a rule, but an exception. In the case considered, the very thin layers of a-Si:H, whose bandgap exceeds the bandgap of Si by at least 0.5 eV, behave as dielectrics. The current transport through such layers proceeds via either direct or multistep trap-assisted tunneling[8]. As estimates show, the former mechanism cannot produce the transfer of photogenerated electrons and holes for dielectric thicknesses of the order of 10 nm, the typical thickness of the a-Si:H layers in HJSCs. Therefore, the photocurrent through a-Si:H layers is most likely due to the trap-assisted tunneling.

This work reports the results of our experimental and theoretical investigation of HJSCs' principal characteristics – short-circuit current density, J_{SC}, open-circuit voltage, V_{OC}, current-voltage curve fill factor, FF, and photoconversion efficiency, η, – obtained in a broad temperature range from 80 to 420 K. We are aware of only a few studies published in the literature, where the temperature dependence of these parameters in HJSCs is reported. However, the temperature interval probed in those works is much narrower than in our experiment, from the room temperature to T not exceeding 150 K above the room temperature[7,9,10]. In Ref. 11, dark $I-V$ curves were measured in an interval from 100 to 400 K. Finally, Schulze et al.[12] performed these measurements under illumination conditions, probing the interval from 240 K to 360 K. In the present work, we found that at lower temperatures, the output power, open-circuit voltage, and fill factor develop a maximum at about 200 K. We introduce a semianalytical model of the temperature dependences measured, allowing one to obtain quantitative agreement with the experiment at temperatures above 200 K.

We note that the analogous theoretical studies that we are aware of usually employ numerical simulations of, e.g., the package AFORS-HET, and do not contain direct comparison with the experiment[13-15]. In particular, their findings cannot be employed to explain our experimental results, because they were obtained for a fixed temperature $T = 298$ K under AM1.5G irradiation conditions.

In our theoretical treatment, we derive the conditions, under which the determination of the photoconversion efficiency η in the HJSCs can be performed using the same procedure as in usual p-n junction-based solar cells. The model presented here takes into account the distinct features of the HJSCs as compared to the conventional solar cells. In the first place, these features are related to the high bulk Shockley-Hall-Read lifetime $\tau_{SHR} \geq 1$ ms. Due to this, the inequalities $\Delta p \geq N_d$ and $L >> d$ hold in a broad temperature range, where Δp is the excess electron-hole density in the base region (Si single crystal), N_d the doping level, L the diffusion length of the minority carriers in the base, and d the base thickness. We show that for $\Delta p \geq N_d$, the open-circuit voltage V_{OC} is higher than in the standard case $\Delta p < N_d$ not only because of the higher Δp-value, but also because of the additional contribution due to the back surface of the solar cell.

We take into account several recombination mechanisms. We establish that in HJSC solar cells at T > 200 K, Shockley-Hall-Read and surface recombination dominate at relatively low base doping level, whereas interband Auger recombination dominates at high base doping level. We explain the notable insensitivity of the HJSCs' photoconversion efficiency to temperature as compared to the commercial solar cells[16,17], first and foremost, by the essential reduction of the total bulk and surface recombination rates in the HJSCs.

In order to explain the experimental temperature dependence of the fill factor and photoconversion power, we take into account the increase of the series resistance in the low-temperature range, which is due to the increase of the contact resistance. Also, experimental measurement of the shunt resistance yielded the rather high value of about $7 \cdot 10^4$ Ω, allowing us to neglect its effect in the modeling.

TABLE I. Parameters of our HJSC. J_{SC}, V_{OC}, and η are given for $T = 300$ K.

N_d, cm^{-3}	d, μm	τ_b, ms	J_{SC}, mA/cm^2	V_{OC}, V	S, cm/s	η, %	R_{S0}, Ohm	A_t, μOhm	φ_c, V	E_{00}, meV
10^{15}	150	1	36.54	0.682	25	18.4	0.2	0.4	0.39	23

2 FABRICATION AND CHARACTERIZATION OF HETEROJUNCTION SOLAR CELLS

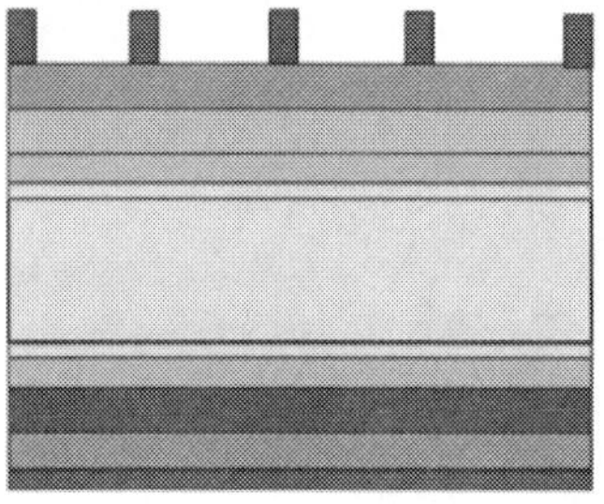

Figure 1: Schematic illustration of Si-HJSC used in this study; surface texture is not shown.

The structure of the Si-HJSC used in this study is shown in Fig. 1. It contains a textured single-crystal n-type silicon substrate of (100) orientation, produced by Czochralski method (Cz). It has the doping level $N_d \approx 10^{15}$ cm^{-3}, thickness $d \approx 150$ μm, and Shockley-Hall-Read lifetime of charge carriers $\tau_{SHR} \approx 1$ ms. On its front side we deposited a buffer α-SiC layer followed by layers of intrinsic amorphous silicon α-Si, p-type α-Si, transparent conducting indium tin oxide (ITO), and a silver grid that collects current. On the rear side are a buffer layer, followed by intrinsic α-Si, n-type α-Si, ITO, and silver layers. All α-Si:H layers were produced in a Gen5 (110x140 cm^2) plasma enhanced chemical vapor deposition (PECVD) reactor, whereas ITO and Ag layers were deposited using a magnetron sputtering system[18]. The buffer films ensured suppression of the epitaxial film growth at the α-Si:H/c-Si inteface. The thicknesses of these layers is about 2-3 nm. The parameters of our Si-HJ solar cells are indicated in Table 1. The last fabrication step was deposition of a solid contact film on the back side and contact grid on the front side of the solar cell. The photoactive surface area of the cell, including the contact grid, was $A = 4.34$ cm^2.

In order to facilitate absorption of the incident radiation by the semiconductor bulk on both sides of the substrate, we created microtextures in the form of randomly located vertical pyramids with characteristic size of a few micrometers. In Ref. 3 it was shown that such surface morphology allows one to obtain the short-circuit current density J_{SC} of 41.8 mA/cm^2 under AM1.5 conditions. Estimates that account for shadowing of the solar cell by the contact grid indicate that, in this case, the external quantum yield is close to 96 %, whereas the internal quantum yield of the photogenerated current is close to 100 %.

3 THEORETICAL FORMULATION

The mathematical description of photoconversion processes in HJSCs is similar to the already solved problem of photoconversion efficiency in structures consisting of a conducting layer, a tunneling layer of SiO_2, and a layer of Si, where the majority carriers (electrons) enter the metal via thermionic emission, and the minority carriers (holes) tunnel through the dielectric layer[8]. In HJSCs, both electrons and holes tunnel through the thin dielectric layer. In order to be able to treat the nanometer layers on the front and the rear sides of the HJSC as dielectrics, light absorption and electron-hole pair generation in these layers must be weak. This condition holds very well for the rear layer, because the energy of the photons reaching it is much smaller than the bandgap energy. With respect to the front one, before reaching it, light has to travel through the relatively thick (about 100 nm) ITO layer, where the most energetic photons are absorbed. For the remaining lower-energy photons, the product of the absorption coefficient and the α-Si:H layer thickness is much smaller than 1, so that they are practically not absorbed by α-Si:H. In that case, by introducing the phenomenological tunneling coefficients for electrons and holes, it is possible to write down the criteria, under which the transport through these layers occurs without losses, and the open-circuit voltage is practically unaffected by α-Si:H. These coefficients can be calculated, e.g., using the AFORS-HET package.

Taking into account tunneling through the wide-bandgap layers and the fact that the density of surface states is extremely small, the boundary conditions for the electron and hole current densities, J_n and J_p, in the case of non-degenerate semiconductor read:

$$J_p = q\frac{V_p}{4}\theta_p\left[p(x) - p_0(x)\right]_{x=0},$$

$$J_n = q\frac{V_n}{4}\theta_n\left[n(x) - n_0(x)\right]_{x=d}. \quad (1)$$

Here, q is the elementary charge, V_p and V_n are the mean thermal velocities of holes and electrons, respectively, θ_p and θ_n are their tunneling coefficients, $p(x)$ and $n(x)$ are the densities of electrons and holes, and $p_0(x)$ and $n_0(x)$

are the respective equilibrium densities. In the second expression, d is the thickness of Si layer.

As shown in Ref. 8, the minority carriers' current (for definiteness, we assume these to be holes) is collected without recombination losses if the condition

$$V_{pe} = \frac{V_p \theta_p}{4} e^{-y(0)} >> V_r + S , \tag{2}$$

is fulfilled, where V_{pe} is the effective hole emission rate from the semiconductor into the metal, $y(0)$ is the nonequilibrium electrostatic potential (band bending) normalized to k_BT/q in the $x = 0$ plane, V_r and S are, respectively, bulk and surface recombination rates. The inequality (2) should hold not only in the short-circuit regime, but also in the regime of maximal power. Estimates obtained for the typical parameter values of the HJSCs indicate that (2) is fulfilled for $\theta_p \geq 10^{-5}$, which is the case in our HJSCs.

The dependence of the short-circuit current on temperature is due to the temperature dependence of the bandgap and absorption coefficient[19]. It can be expressed as

$$J_{SC}(T) = q\int_{\lambda_0}^{\lambda_m(T)} d\lambda Q_e(\lambda, T)\Phi(\lambda)\frac{\lambda}{hc}, \tag{3}$$

where $\lambda_m(T) = hc/E_g(T) \approx 1.24\ \mu\text{m eV}/E_g(T)$ is the photoelectric threshold in Si, λ_0 is the absorption edge, $Q_e(\lambda, T)$ is the external quantum efficiency of the cell, and $\Phi(\lambda)$ is the spectral density of the incident radiation. The temperature dependence of the bandgap in Si can be found, e.g., in Ref. 20. The increase of $J_{SC}(T)$ with temperature is due to the variations of both $\lambda_m(T)$ and $Q_e(\lambda, T)$ with T. Because the latter dependence is difficult to determine theoretically, we use, instead of Eq. (3), an empirical expression, which approximates the product $Q_e\Phi$ in the integrand of (3) by a black-body spectrum:

$$J_{SC}(T) = J_{SC}(300K)\frac{F(T)}{F(300K)}, \tag{4}$$

Where

$$F(T) = \int_0^{\lambda_m(T)} \frac{d\lambda}{\lambda^4\left(e^{hc/(kT_{eff}\lambda)} - 1\right)}. \tag{5}$$

In general, the effective temperature T_{eff} is smaller than the radiation temperature T_L. However, under the AM0 conditions, or when an incandescent lamp is used, the two temperatures are practically the same, provided that the quantum efficiency is close to 1.

For the open-circuit voltage in HJSCs, the following general expression applies:

$$V_{OC} = \frac{k_BT}{q}\cdot \left(\left(\Delta y_{OC}(0) - \Delta y_{SC}(0)\right) - \left(\Delta y_{OC}(d) - \Delta y_{SC}(d)\right)\right), \tag{6}$$

where $\Delta y_{OC}(0)$ and $\Delta y_{SC}(0)$ are the dimensionless variations of the band bending in the open-circuit and short-circuit regimes, respectively, at $x = 0$, and $\Delta y_{OC}(d)$ and $\Delta y_{SC}(d)$ are the respective values at $x = d$.

Our estimates indicate that in HJSCs that completely conduct the current generated in the short-circuit regime, the following conditions hold: $\Delta y_{OC}(0) << 1$ and $\Delta y_{SC}(d) << 1$. Therefore, in this case, the value of V_{OC} is practically the same as in the standard p-n junction-based solar cells. Taking into account that $p_0n_0 = n_i^2$, where n_i is the intrinsic charge carrier density, the open-circuit voltage is

$$V_{OC} \approx V_{OC1} + V_{OC2},\ V_{OC1} = \frac{kT}{q}\ln\frac{\Delta p_0 n_0}{n_i^2}$$

$$V_{OC2} = \frac{kT}{q}\ln\left(1 + \frac{\Delta p_0}{n_0}\right), \tag{7}$$

where Δp_0 is the excess electron-hole pair density in the Si bulk in the open-circuit regime. The expression (7) for V_{OC} is valid when the moduli of band bending at $x = 0$ and $x = d$ in the absence of illumination exceed the respective values in the open-circuit regime,

$$y_0(0) > qV_{OC1}/kT,\ y_0(d) > qV_{OC2}/kT. \tag{8}$$

Usually, these inequalities are well satisfied in HJSCs, at least at the operating temperatures.

Eq. (7) is a quadratic equation for Δp_0 with the solution

$$\Delta p_0 = -\frac{n_0}{2} + \sqrt{\frac{n_0^2}{4} + n_i^2 e^{qV_{OC}/k_BT}}. \tag{9}$$

The open-circuit voltage can be found from the balance equation relating the short-circuit current and the recombination currents due to the Shockley-Hall-Read mechanism with the characteristic time τ_{SHR}, radiative recombination with the time τ_r, Auger recombination with the rate R_{Auger}, surface recombination with the rate S, and recombination in the space-charge region with rate R_{SC}:

$$J_{SC} = q\left[d\left(\tau_{SRH}^{-1} + \tau_r^{-1}\right) + R_{Auger} + S + R_{SC}\right]\Delta p_0 \tag{10}$$

where[21,22]

$$R_{Auger} = C_p\left(n_0 + \Delta p_0\right)\Delta p_0 + C_n\left(n_0 + \Delta p_0\right)^2,$$

$C_p = 10^{-31}$ cm^6/s,

$$C_n = \left(2.8\cdot 10^{-31} + \frac{2.5\cdot 10^{-22}}{\left(n_0 + \Delta p_0\right)^{0.5}}\right)\ \text{cm}^6/\text{s}. \tag{11}$$

The space-charge region recombination rate is given by[23]

$$R_{SC}(\Delta p_0) \approx \frac{L_D}{\tau_{SRH}}\int_{y_{pn}}^{-0.1} dy \frac{n_0 + \Delta p_0}{\sqrt{e^y - y - 1}} \times \left[\left(n_0 + \Delta p_0\right)e^y + n_i(T)e^{\varepsilon_r} + b\left(\frac{n_i(T)^2}{n_0 + \Delta p_0} + \Delta p_0\right)e^{-y} + n_i(T)e^{-\varepsilon_r}\right]^{-1}, \tag{12}$$

where $L_D = \sqrt{\varepsilon_0\varepsilon_S kT/(2q^2 n_0)}$ is Debye screening length, ε_0 the relative dielectric constant of the semiconductor, $b = \sigma_p/\sigma_n$ the ratio of the capture cross-sections of a hole and an electron by a deep recombination center, $\varepsilon_r = E_r/kT$ the normalized energy of the deep recombination level measured from the middle of the band gap, y the dimensionless potential, and y_{pn} the dimensionless potential at the p-n junction boundary.

The intrinsic charge carrier density in Si is given by the empirical expression[20]

$$n_i(T) = 5.71\cdot 10^{19}\left(\frac{T}{300K}\right)^{2.365} e^{-6773K/T}\ \text{cm}^{-3}, \tag{13}$$

and the equilibrium electron density can be obtained from the relations $n_0 = p_0 + N_d$, $p_0n_0 = n_i^2$:

$$n_0 = \frac{N_d}{2} + \sqrt{\frac{N_d^2}{4} + n_i^2}, \tag{14}$$

where N_d is donor density.

As our estimates using the parameters from Table I have shown, in the temperature range from 200 K to 300 K, Δp_0 is about $5\cdot 10^{15}$ cm^{-3}, and R_{SC} does not exceed 0.1 cm/s. Thus, the space-charge region recombination rate is insignificant in comparison to the total bulk and surface recombination.

The fill factor, FF, can be written as

$$FF = \frac{J_m V_m}{J_{SC} V_{OC}}\left(1 - \frac{R_S J_m A}{V_m}\right), \qquad (15)$$

where J_m and V_m are the photoinduced current density and voltage in the maximum-power regime, A is the surface area, and R_S the series resistance of the solar cell.

To determine J_m and V_m of a HJSC, one has to consider the current-voltage relation [cf. Eq. (10)],

$$J(V) = J_{SC} - q\left(d\left(\tau_{SRH}^{-1} + \tau_r^{-1}\right) + R_{Auger} + S\right)\Delta p(V), \qquad (16)$$

where

$$\Delta p(V) = -\frac{n_0}{2} + \sqrt{\frac{n_0^2}{4} + n_i^2 e^{qV/k_B T}}\,. \qquad (17)$$

The maximum power condition $d(VI(V))/dV = 0$ yields the value of V_m, whose substitution into (16) allows one to determine J_m.

Another approach to this problem can be based on the two-diode model[12]. We do not use this model, because it describes the dark current-voltage curves. In our work, we use the light $I-V$ measurements, neglecting tunneling current and current in the space-charge region. The reason for this neglect is that, under the maximum-power conditions (V_m = 580 mV), and especially in the open-circuit regime (V_{OC} = 682 mV), their contribution is small. At the same time, the salient feature of our approach is that, in the current-voltage characteristics, we take into account not only the front, but also the rear side of the cell. In the language of the two-diode model this means that, in the limiting case when the excess electron-hole pair density Δn is much greater than the equilibrium density n_0 in the base region, the $I-V$ curve with the ideality factor 2 is realized, whereas in the opposite limit the ideality factor is 1, see Ref. 24.

The total series resistance R_S consists of the bulk resistance of the material, R_{bS}, sheet resistance R_{SS} of the ITO film used as a transparent conductor, and contact resistance R_{CS}. The first two contributions to series resistance are usually temperature-independent. The temperature dependence of the contact resistance is determined by the nature of the current transport mechanism through the contact. If the current is due to thermionic or field emission, R_{CS} increases on cooling. The resistance R_{CS} is presumably related to the resistance of ITO-metal contact. Because the ITO film is degenerate, we can expect that the current transport through the contact is due to the field emission mechanism. Then, the expression for the series resistance can be written as[25]

$$R_S = R_{S0} + A_t(T)\exp\left(\frac{q\phi_C}{E_{00}\coth(E_{00}/k_B T)}\right), \qquad (18)$$

where R_{S0} is the temperature-independent component of the series resistance, $A_t(T)$ the prefactor that depends on temperature according to a power law, ϕ_C is the contact barrier height, E_{00} is the characteristic electron tunneling energy in the semiconductor. If the barrier ϕ_C is sufficiently high, then one can neglect the temperature dependence of the prefactor.

The photogenerated power can be found from the standard expression

$$P = J_{SC} A V_{OC} FF \qquad (19)$$

And the photoconversion efficiency is given by

$$\eta = \frac{P}{AP_S}, \qquad (20)$$

where P_S = 1000 W/m^2 is the surface power density of the incident radiation and A = 4.34 cm^2 is the surface area of the sample.

4 COMPARISON BETWEEN THE THEORETICAL AND EXPERIMENTAL CHARACTERISTICS OF HETEROJUNCTION SOLAR CELLS

In the measurements of the HJSC current-voltage curves, we used a simulator of solar radiation (Yamashita Denso Corp., Model SS-80AA, class AAA) that produces illumination of 1 kW/m^2 with the AM1.5G spectrum for the irradiation region of 30 mm x 30 mm. The light coming from this source was directed by an optical collimation system to a quartz window (diameter 50 mm) of a vacuum chamber, in which the SC was placed. The chamber contained a vertical cryostat (model VPF-100, Janis Research) with a cryogenic temperature controller (Model 325, Lake Shore Cryotronics, Inc.), that allowed us to perform temperature measurements in the range 80-400 K and enabled temperature stability of 1 K. The chamber was evacuated with a HiCube 80 Eco pump (Pfeiffer Vacuum) down to 10^{-4} mBar. The current-voltage recording system consisted of a bipolar voltage source Kepco BOP 36-12, a reference resistor of 0.1 Ohm (Burster GmbH, model 1282-0.1, precision 0.02 %), and two multimeters Keithley 2000, which measured the voltage across the HJSC and across the reference resistor. To determine and control the illuminance in the sample plane, we used a standard World Photovoltaic Scale SC calibrated at the Fraunhofer Institute for Solar Energy Systems as a reference[26]. The calibration of photoinduced current in our samples at T = 298 K was performed under direct illumination from the solar simulator using a thermally stabilized base with a Peltier element, a temperature controller based on the Eurotherm 2216e unit (Eurotherm Co.) with a temperature detector RTD-100 (Omega, U.S.A.).

For the sample placed inside the vacuum chamber, the photoinduced current measured at the calibration stage was used as an indicator of the correct value of the incident radiation power to compensate for a small loss of irradiation energy in the collimation block and in the quartz window. Outside of the chamber, a control photocollector was placed in order to monitor the stability of the irradiation intensity during the measurements of the temperature dependence of HJSC's characteristics.

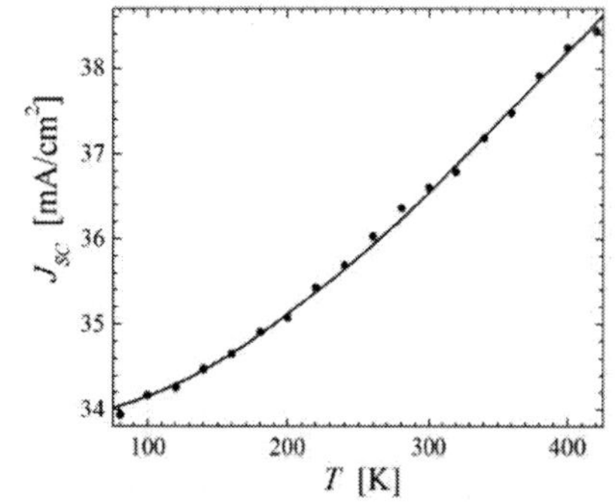

Figure 2: Experimental (circles) and theoretical (line) temperature dependence of the short-circuit current.

After having set the required photocurrent in the SC,

corresponding to the irradiation power density of 1000 W/m^2, the sample was cooled down to 80 K under constant irradiation from the solar imitator. In this way, we could compensate for the small warming up of the sample due to irradiation, and to ensure a good agreement between the real temperature of the SC and its measured value. The $I-V$ curve and the photoelectric parameters (J_{SC}, V_{OC}, FF, and P_{max}) were measured on heating the sample.

Fig. 2 shows the short-circuit current vs. temperature curve. The theoretical curve was produced by fitting the experimental data points with Eq. (4); the best fit was achieved for the effective temperature $T_{eff} = 4030$ K. The agreement of the experimental and the theoretical curve in the whole temperature range is obvious.

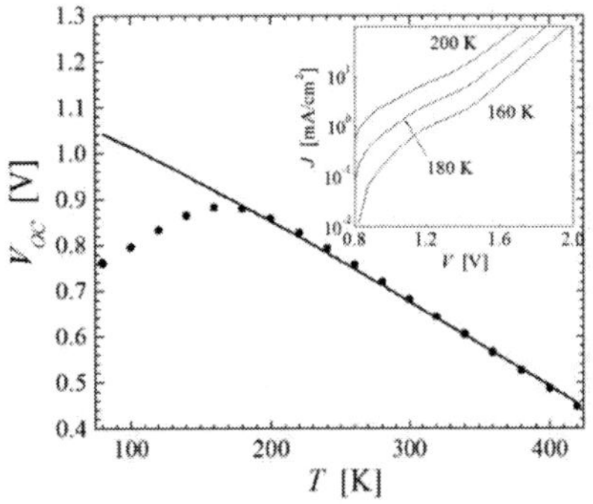

Figure 3: Experimental (circles) and theoretical (lines) temperature dependence of the open-circuit voltage obtained using the general expression (10) with parameters shown in Table I. The inset shows the current-voltage curves at three different temperatures in the low-temperature range.

Shown in Fig. 3 is the temperature dependence of the open-circuit voltage, $V_{OC}(T)$. The theoretical curves are obtained by numerically solving Eq. (10) supplemented by Eqs. (11) and (4) with parameters from Table I. When building these curves, we used the temperature-dependent $J_{SC}(T)$, and the intrinsic carrier density $n_i(T)$ given by Eq. (13). We neglected the temperature dependence of the Shockley-Hall-Read, radiative, surface, and Auger recombination rates, because they are much weaker than the $n_i(T)$ dependence. It is noteworthy that for the equilibrium electron densities $n_0 \cong 10^{15}$ cm^{-3}, which correspond to the doping level, interband Auger recombination and radiative recombination can be neglected in comparison to the bulk Shockley-Hall-Read and surface recombination. Auger recombination becomes essential when $n_0 \geq 10^{16}$ cm^{-3}.

As can be seen in Fig. 3, the theoretical curve agrees well with the experimental results at $T \geq 180$ K. The temperature coefficient of $V_{OC}(T)$ at $T > 200$ K is about 1.8 mV/K, which is close to the typical value of $|dV_{OC}/dT| \approx 2$ mV/K in the industrial silicon-based solar cells with diffused p-n junctions. At temperatures below 180 K, the experimental and theoretical values of $V_{OC}(T)$ deviate from each other: the theoretical open-circuit voltage increases, whereas the experimental one decreases on cooling. Finding possible reasons for experimentally observed non-monotonic $V_{OC}(T)$ behavior is a complicated problem. First of all, as follows from the calculations based on the spectrum of localized donor-like and acceptor-like states in α-Si:H, this material is a heavily compensated semiconductor[27], whose degree of compensation increases on reducing the temperature. Therefore, the charge in the acceptor-doped α-Si:H decreases on cooling, leading to the reduction of $y(x = 0)$. This implies that, although the criterion (2) at $T \leq 200$ K is fulfilled in the short-circuit regime, it breaks down at a relatively weaker bias. This point is illustrated by the experimental current-voltage curves of our HJSCs measured at $T = 200$, 180, and 160 K, see inset in Fig. 3. Similar to Ref. 28, these curves exhibit sections due to the hole tunneling from the narrow-band single-crystal Si into α-Si:H for relatively high direct bias (≥ 1.4 V), whereas for weaker bias (≤ 0.82 V), the dark current-voltage curves are adequately described by the expressions of the form (16) and (17) (with $J_{SC} = 0$). Thus, the reduction of $V_{OC}(T)$ on cooling is due to several reasons. It is the decrease of $y(x = 0)$ and failure of the criterion (2), i.e. the reduction of the photogenerated current at low temperatures. As the value of $y(x = 0)$ goes down on cooling, the inequality $kTy_0(x = 0) > qV_{OC}$ breaks down, where $y_0(x = 0)$ is the dimensionless band bending in the absence of illumination. Then, the expression (7) becomes inadequate, and the open-circuit voltage becomes smaller than the value predicted by (7). Finding a better formula for V_{OC} in the low-temperature range is a complicated task, which requires the knowledge of the distribution of the localized states in the doped and undoped layers of α-Si:H. Additional research is needed to attack this problem.

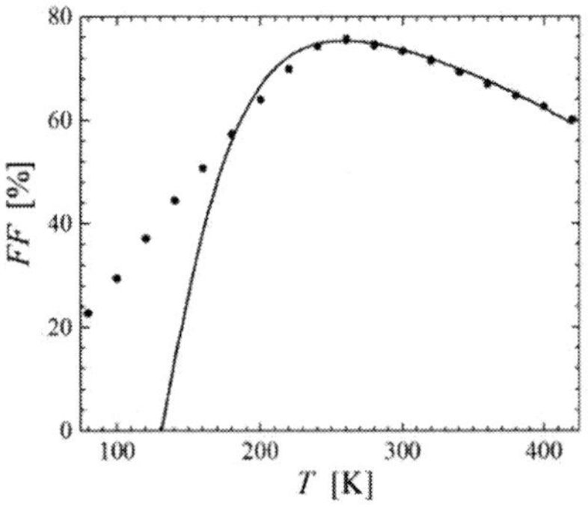

Figure 4: Experimental (circles) and theoretical (lines) temperature dependence of the current-voltage curve fill factor.

Fig. 4 shows the temperature dependence of the fill factor. It is seen that FF at low temperatures becomes anomalously low, signaling a strong increase of the series resistance R_S on cooling. The theoretical curve was obtained from Eqs. (15) and (18) for the same parameters as the theoretical curve $V_{OC}(T)$ from Fig. 3. We took $A_t = 4 \cdot 10^{-7}$ Ohm, $\varphi_c = 0.387$ V, $E_{00} = 0.0225$ eV, $R_{S0} = 0.2$ Ohm. The agreement between the theoretical and the experiment curves is very good at $T \geq 200$ K, whereas at $T < 200$ K, the discrepancy between theory and experiment is observed for the reasons discussed above.

We note that the non-monotonic temperature dependence of the fill factor in HJSCs was observed in the work by Taguchi et al[9] for the double and triple thicknesses of a-Si:H layers, and in the work by Seif et al[7] for the case of a-SiO$_x$:H layers. This non-monotonicity was explained in Ref. 7, based on the work by Anderson[30], by the fact that, at low temperatures, not all photogenerated current can pass through the wide-bandgap layer due to the band offset at the interface between c-Si and a-Si layers. In our case, the maximum of the $FF(T)$ curve is observed at a much lower temperature. For this reason, we believe it to be principally due to the temperature dependence of the contact resistance.

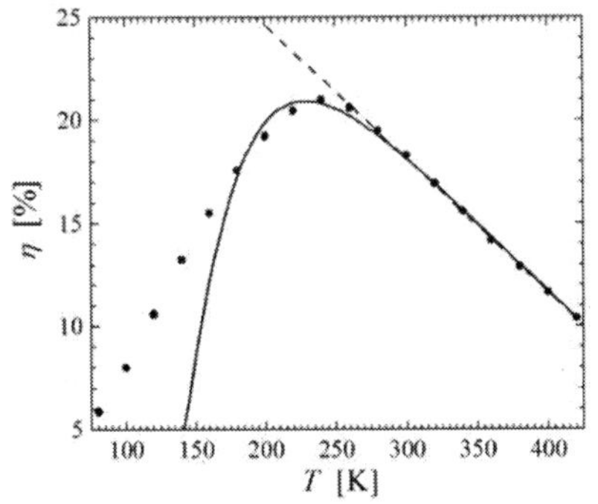

Figure 5: Experimental (circles) and theoretical (lines) temperature dependence of the photoconversion efficiency.

The temperature dependence of the photoconversion efficiency η, see Eq. (20), is shown in Fig. 5. The agreement between theory and experiment at $T \geq 200$ K, including its non-monotonic behavior, is obvious. Above about 280 K, the efficiency decreases approximately linearly with temperature. This linear decay can be described by an expression[16]

$$\eta(T) = \eta(300K)(1 - \beta(T - 300K)), \qquad (21)$$

which defines the temperature coefficient β. A linear fit to the experimental data at $T \geq 280$ K yields $\beta_{exp} = 0.355$ %/K, see Fig. 5, dashed line. Theoretically, the photoconversion efficiency vs. temperature curve slightly deviates from the linearity (21), see the solid line in Fig. 6 showing the theoretical derivative of the efficiency with respect to temperature normalized to η(300 K), as well as the respective experimental estimates. To obtain the experimental curve based on the set of experimental support points (T_n, η_n), n = 1, 2, ..., we used the third-order accurate approximation for the derivative at the midpoint $T_{n+1/2} = (T_n + T_{n+1})/2$:

$$\frac{1}{\eta(300K)}\frac{d\eta}{dT}(T_{n+1/2}) \approx \frac{1}{\eta(300K)}\frac{\eta_{n+1} - \eta_n}{T_{n+1} - T_n}. \qquad (22)$$

The overall trend towards saturation of both experimental and theoretical curves is clear, and the discrepancy between the two sets of data is due to the amplification of the experimental uncertainty by the derivative expression (22).

In the work 17, the efficiency temperature coefficient in HJSCs with $\eta = 23$ % was measured. Its value at the largest open-circuit voltage V_{OC} of 0.745 V, which was realized at the smallest solar cell thickness (see Fig. 3 of Ref. 17), was 0.23 %/K. Our theoretical estimates at $T = 298$ K and parameter values from Ref. 17, in the thickness range of the Si substrate from 80 to 165 μm, gives $\beta \approx 0.25$ %/K, quite close to the experimental values.

We note that the low values of $\beta(T)$ are directly related to the high excitation level in the open-circuit regime. This owes, in particular, to the minimal surface recombination rate in HJSCs. An important role here is played by atomic hydrogen, which leads to a strong reduction of the number of surface centers responsible for recombination. As shown in Ref. 31, it was precisely the use of atomic hydrogen that allowed obtaining minimal surface recombination rate of about 0.25 cm/s on Si.

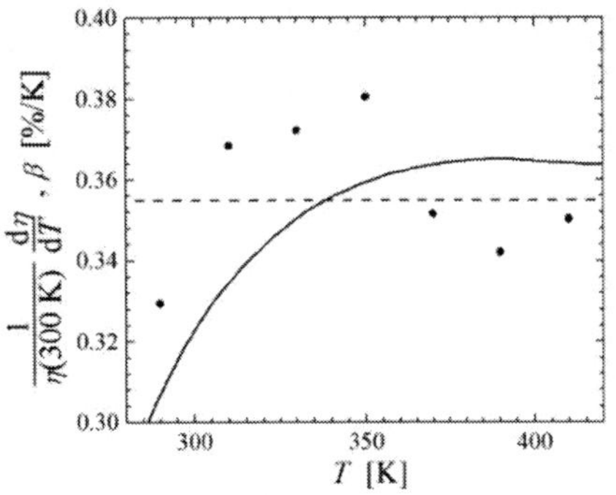

Figure 6: Experimental (dashed line, circles) and theoretical (solid line) temperature dependence of the photoconversion power reduction coefficient. The dashed line was obtained by fitting the $P(T)$ curve, Fig. 5, at $T >$ 280 K. The circles are obtained by numerically differentiating the experimental data using Eq. (22); the solid line is obtained from the derivative of the respective theoretical curve.

The bulk recombination rate, $V_r = d/\tau_b$, where τ_b is the bulk lifetime, is also much smaller in HJSCs than in the diffused p-n junction-based solar cells. This is so, because, as shown above, HJSCs employ silicon with long Shockley-Hall-Read lifetimes exceeding 1 ms. Secondly, d in HJSCs is also smaller than in the usual solar cells. For instance, for $d = 100$ μm and $\tau_{SHR} = 1$ ms, $V_r = 10$ cm/s. These features make it possible to maintain a high excitation level $\Delta p_0 >> n_0$ in HJSCs, and thus to maximally reduce the temperature dependence of V_{OC} and η, and to minimize β.

Table II: Theoretical determination of V_{OC}, FF , and η for the systems studied in Refs. 3, 17, and 29. In all calculations, we have used the doping level $N_d = 4.9 \cdot 10^{15}$ cm^{-3} and the Shockley-Hall-Read lifetime $\tau_{SHR} \approx 3$ ms. The thickness d and short-circuit current density J_{SC} are taken from the respective publications. Experimental values from Refs. 3, 17, and 29 are indicated in brackets.

Ref.	d, µm	S, cm/s	J_{SC}, mA/cm^2	V_{OC}, V	FF, %	η, %	R_S, mΩ
3	150	1.5 (1.5)	41.8	0.74 (0.74)	82.8 (82.7)	25.6 (25.6)	1.4
17	98	6	38.8	0.743 (0.743)	79.2 (79.1)	22.8 (22.8)	10

29	98	1.5 (<2)	39.5	0.75 (0.75)	83.4 (83.2)	24.7 (24.7)	1

The approach introduced in this work allows one to obtain the main characteristics of HJSCs under standard test conditions (AM1.5G, 1000 W/m^2). Table II shows the results of our calculations of V_{OC}, FF, and η for the parameters from Refs. 3, 17, and 29. These parameters include the doping level of $4.9 \cdot 10^{15}$ cm^{-3}, the lifetime in the base region equal to 3 ms, the SC thickness, and the experimental values of the short-circuit current. They were used to calculate the open-circuit voltage under AM1.5 conditions at the temperature 25 °C, using the expressions (9)-(11). In these equations, the surface recombination rate was treated as a fit parameter. As seen in Table II, in two cases out of three, the fit and the experimental values for the surface recombination rate are in very good agreement; in one case, the respective experimental value was not given in the original publication 17.

Next, from the condition of maximum power, using Eqs. (16) and (17), the values of J_m and V_m were found, and then the fill factor was determined from Eq. (15). In this calculation, another fit parameter appears, namely, the series resistance. It can be determined uniquely by comparing the theoretical FF with the experimental value. Finally, the photoconversion efficiency is found using Eq. (20). As seen from Table II, the theoretical and experimental values agree with each other up to the fourth significant figure for all three SCs. With respect to the fill factor, the agreement between theory and experiment is slightly worse; however, the two sets of data do agree in three significant figures. In view of the empirical character of the short-current expression (4), we have used the experimental values of J_{SC} as the input parameters. The agreement between the theoretical and experimental results for V_{OC}, FF, and η demonstrates the correctness of the relations (10)-(20).

5 CONCLUSIONS

Peculiarities of HJSC as opposed to the standard solar cells include, on the one hand, tunneling of electrons and holes through the α-Si:H layers; on the other hand, large number densities of the excess electron-hole pairs and large diffusion length of the minority charge carriers. In this work, we have derived the criteria, under which tunneling processes do not lead to the deterioration of the HJSCs' characteristics as compared to those of the solar cells based on the diffused p-n junctions, and obtained the expressions for the main parameters that characterize the performance of HJSCs.

These expressions are compared to the experimental results. We have obtained good agreement between the theoretical and experimental temperature dependences of the HJSC parameters in the temperature range $T \geq 200$ K. At low temperatures, the series resistance, R_S, grows dramatically on cooling due to the contribution of the contact resistance. The reduction of the open-circuit voltage, V_{OC}, fill factor, FF, and photoconversion efficiency, η, at low temperatures is interesting from the scientific point of view. The reduction of FF and η is explained by the reduction of V_{OC}, and by the increase of R_S. Of main practical interest are the results obtained for $T > 0$ °C. They are completely explained by the specifics of the HJSCs studies, especially the high excitation levels under AM1.5 conditions.

We have shown that the rather low value of the calculated temperature coefficient of photoconversion efficiency, $\beta(T) \approx 0.3$ %/K, near room temperature is in agreement with the data from Ref. 17. The physical reason of its low values in HJSC solar cells is low surface and bulk recombination rate as compared to the silicon diffused p-n junction-based solar cells.

The practical importance of these findings lies in the fact of their applicability in the analysis of HJSCs' performance under extreme temperature conditions realized, e.g., in the Earth polar regions, as well as in the outer space. In particular, high temperatures are realized, e.g., on the orbits of Venus, whereas low temperatures are found on the orbits of those planets which are far away from the Sun, such as Jupiter.

ACKNOWLEDGMENTS

This work was performed under the project of Minobrnauka RF No. 14.607.21.0075 (RFMEFI60714X0075). M.E. is grateful to the Natural Sciences and Engineering Research Council of Canada (NSERC) and to the Research and Development Corporation of Newfoundland and Labrador (RDC) for financial support.

REFERENCES

[1] W.G.J.H.M. van Sark, L. Korte, and F. Roca, Physics and Technology of Amorphous Crystalline Heterostructure Silicon Solar Cells (Springer, Berlin, 2012).
[2] S. De Wolf, A. Descoeudres, Z.C. Holman, and C. Balif, Green 2, 7 (2012).
[3] K. Masuko, M. Shigematsu, T. Hashiguchi, D. Fujishima, M. Kai, N. Yoshimura, T. Yamaguchi, Y. Ichihashi, T. Mishima, N. Matsubara, T. Yamanishi, T. Takahama, M. Taguchi, E. Maruyama, and S. Okamoto, IEEE J. Photovolt. 4, 1433 (2014).
[4] W. Shockley and H.J. Queisser, J. Appl. Phys. 32, 510 (1961).
[5] A. Richter, M. Hermle, and S.W. Glunz, IEEE J. Photovoltaics 3, 1184 (2013).
[6] D. Suwito, U. J□ ager, J. Benick, S. Janz, M. Hermle, and S.W. Glunz, IEEE Trans. Electron. Devices 57, 2032 (2010).
[7] J.P. Seif, A. Descoeudres, M. Filipič, F. Smole, M. Topič, Z.C. Holman, S. De Wolf, and C. Ballif, J. Appl. Phys. 115, 024502 (2014).
[8] A.Ya. Vul' and A.V. Sachenko, Sov. Phys. Semicond. 17, 865 (1983).
[9] M. Taguchi, E. Maruyama, and M. Tanaka, Jap. J. Appl. Phys. 47, 814 (2008).
[10] J. P. Seif, G. Krishnamani, B. Demaurex, Ch. Ballif, and S. De Wolf, IEEE J. Photovoltaics 5, 718 (2015).
[11] M. W. M. van Cleef, M.W.H. Philippens, F.A. Rubinellia, M. Koltera, and R.E.I. Schroppa, Mater. Res. Soc. Symp. Proc. 420, 239 (1996).
[12] T.F. Schulze, L. Korte, E. Conrad, M. Schmidt, and

B. Rech, J. Appl. Phys. 107, 023711 (2010).
[13] N. Dwivedi, S. Kumar, A. Bisht, K. Patel, and S. Sudhakar, Solar Energy 88, 31 (2013).
[14] X. Wen, X. Zeng, W. Liao, Q. Lei, and S. Yin, Solar Energy 96, 168 (2013).
[15] M.H. Vishkasougheh and B. Tunaboylu, Energy Conversion and Management 72, 141 (2013).
[16] E. Skoplaki and J.A. Palyvos, Sol. Energy 83, 614 (2009).
[17] T. Mishima, M. Taguchi, H. Sakata, and E. Maruyama, Sol. Energy Mater. Sol. Cells 95, 18 (2011).
[18] S. Abolmasov, A. Abramov, D. Andronikov, K. Emtsev, G. Ivanov, I. Nyapshaev, D. Orekhov, A. Semenov, G. Shelopin, E. Terukov, B. Strahm, G. Wahli, P. Papet, T. Söderström, Y. Yao, T. Hengst, and G. Kekelidze, Proceedings of the 31st European Photovoltaic Solar Energy Conference and Exhibition (Hamburg, Germany), 1046 (2015).
[19] D.W. Spaderna and D.H. Navon, IEEE Trans. El. Dev. 25, 1290 (1978).
[20] M.A. Green, J. Appl. Phys. 67, 2944 (1990).
[21] A. Hangleiter and R. H□ acker, Phys. Rev. Lett. 65, 215 (1990).
[22] A. V. Sachenko, A. P. Gorban, V. P. Kostylev, and I. O. Sokolovskii, Semiconductors 41, 281 (2007).
[23] A.V. Sachenko, A.I. Shkrebtii, R.M. Korkishko, V.P. Kostylyov, M.R. Kulish, I.O. Sokolovskyi, and M. Evstigneev, J. Phys. D: Appl. Phys. 48, 455101 (2015).
[24] A.V. Sachenko, A.I. Shkrebtii, R.M. Korkishko, V.P. Kostylyov, N.R. Kulish, and I.O. Sokolovskyi, Semiconductors 49, 264 (2015).
[25] S.M. Sze and K.K. Ng, Physics of Semiconductor Devices, 3rd Ed. (John Wiley and Sons, 2007).
[26] C.R. Osterwald, S. Anevsky, K. B□ ucher, A.K. Barua, P. Chaudhuri, J. Dubard, K. Emery, B. Hansen, D. King6, J. Metzdorf, F. Nagamine, R. Shimokawa, Y. X. Wang, T. Wittchen, W. Zaaiman, A. Zastrow, and J. Zhang, Prog. Photovolt.: Res. Appl. 7, 287 (1999).
[27] Yu.V. Kryuchenko, A.V. Sachenko, A.V. Bobyl, V.P. Kostylyov, E.I. Terukov, A.S. Abramov, E.V. Malchukov, and I.O. Sokolovskyi, Semiconductors 49, 683 (2015).
[28] J.P. Donnelly and A.G. Milnes, Int. J. Electron. 20, 295 (1966); Proc. IEE 113,1468 (1966).
[29] A. Jano, S. Tohoda, K. Matsuyama, Y. Nakamura, T. Nishiwaki, K. Fujita, M. Taguchi, and E. Maruyama, Proc. 28th European Photovoltaic Solar Energy Conference and Exhibition, pp. 748-751 (2013).
[30] R.L. Anderson, Appl. Phys. Lett. 27, 691 (1975).
[31] E. Yablonovitch, D.L. Allara, C.C. Cheng, and T.B. Bright, Phys. Rev. Lett. 57, 249 (1986).

A SIMULATION STUDY OF DEPLETION EFFECT OF NEGATIVELY CHARGED PASSIVATION LAYER ON *n*-TYPE BACK-CONTACT BACK-JUNCTION SILICON SOLAR CELL

Chih-Ming Wei*, Chi-Chun Li, and Chia-Chih Chuang
Motech Industries, Inc., Science Park Branch
No. 96, Dashun 7th Rd., Xinshi Dist., Tainan 74145, Taiwan
*E-mail: alex_wei@motech.com.tw

ABSTRACT: In this study, the effect of fixed charges contained in the passivation films that are applied to the gap region of back-contact back-junction (BCBJ) solar cells has been demonstrated by numerical device simulation. Usually, the BCBJ solar cells are fabricated with *n*-type Si wafers. It is found that the negatively charged passivation film results in an additional depletion region near the surface of the gap region of the BCBJ solar cell while the positively charged passivation film does not. This additional depletion region would also lead to a larger carrier generation rate under reverse bias, which might in turn which might in turn makes larger the current under reverse bias and smaller the shunt resistance. Based on the observed results, experiments can be planned to effectively improve the cell performance with reduced R_{SHUNT} by optimizing the charge state of the passivation film.
Keywords: back contact, passivation, simulation, silicon, solar cell

1 INTRODUCTION

The development of high-efficiency silicon solar cells is one of the research topics in the field of silicon solar cells. Several different novel cell concepts were developed to make cell efficiencies greater than 20%. [1] However, the required techniques make some of these high-efficiency solar cells not able to be fabricated in a cost-effective way. Sunpower Corp. showed that it is possible to reach average cell efficiencies up to 22% [2] in mass production with a back-contact back-junction silicon solar cell structure. In this kind of solar cell, both metallization grids are located at the back side of the cell. This design results in free shading losses at the front side and wide emitter and base metallization. Free shading losses at the front side increase short circuit current density. Wide emitter and base metallization fingers at the back side of the cell reduce the series resistance.

For high efficiency crystalline silicon solar cells, the surface passivation which is utilized to minimize the surface recombination rate of carriers is very important. a-SiN_X:H, SiO_2, and Al_2O_3 are popular candidates that are used to provide good passivation and to reduce the surface recombination rate of carriers. Several studies have shown that not all the passivation schemes are applicable to *n*- and *p*-type Si surfaces. [3-5] Generally, films containing positive fixed charges, such as a-SiN_X:H and SiO_2, provides good passivation quality when applied to the *n*-type Si surfaces. On the other hand, the *p*-type Si surfaces are passivated by the films containing negative fixed charges, e.g. Al_2O_3. The previous study has shown that a-SiN_X:H films can lead to compromised solar cell performance when applied on the *p*-type Si base at the rear side. [5] This is related to the presence of an inversion layer below the surface induced by the fixed positive charge density, which can lead to current flow between the rear metal contacts. This effect is known as parasitic shunting. [5]

In this study, the effect of fixed charges contained in the passivation films that are applied to the gap region of back-contact back-junction (BCBJ) solar cells is demonstrated by numerical device simulation. The BCBJ solar cells are usually fabricated with *n*-type Si wafers. It is found that the negatively charged passivation film results in an additional depletion region in the gap region of the BCBJ solar cell while the positively charged passivation film does not. This additional depletion region would also lead to carrier generation under reverse bias, which might in turn makes larger the current under reverse bias and the shunt resistance smaller.

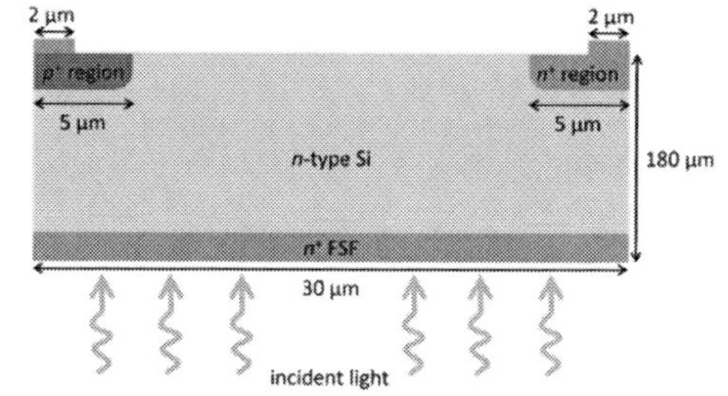

Figure 1: The schematic diagram of the structure used in the simulations.

Table I: Some parameters used in the simulation.

Parameter	Value
Bulk lifetime	1 ms
Wafer resistivity	2 Ωcm
Doping profile (all)	Gaussian
Surface doping concentration (all)	$1x10^{20}$ cm^{-3}
Doping profile depth (all)	500 nm

2 SIMULATION

The technology computer-aided design (TCAD) tool ATLAS from Silvaco was used for device simulations. TCAD simulations are based on physical models resolved at each node of the meshed structure and able to provide physics-based predictive results. In this study, the simulations are evaluated on BCBJ solar cells made of *n*-type wafers. The presented work focuses the effect of the fixed charges contained in the passivation films on the current density-voltage (J-V) characteristics under reverse bias and the shunt resistance at dark state. In order to focus on the effect that we are interested in, the two-dimensional (2D) structure for numerical simulations is designed with a reduced and simplified structure as shown in Fig. 1. Some related parameters used in the simulations are listed in Table I. Different kinds and different amounts of fixed charges are placed on the surface of gap region to simulate the effect of coating positively or negatively charged passivation films onto the surface of gap region.

3 RESULTS AND ANALYSIS

The simulated dark J-V characteristic curves under reverse bias are displayed in Fig. 2. Interesting phenomena are found as different amounts and types of fixed charges are placed on the surface of the gap region. When positive fixed charges are placed on the surface, the current density remains almost unchanged as increasing the amount of positive fixed charges, as shown in Fig. 2(a). Contrarily, in the case of negative fixed charges, the current density increases when increasing the amount of negative fixed charges as displayed in Fig. 2(b). The shunt resistance can be estimated via the following equation:

$$\frac{1}{R_{SHUNT}} = \left|\frac{dJ}{dV}\right|_{V=0} \quad (1)$$

where J is the current density, V is the bias voltage, and R_{SHUNT} stands for the shunt resistance in unit of Ωcm^2. The estimated shunt resistances as a function of fixed charge density are plotted in Fig. 3. It is found that the shunt resistance remains almost constant for the case of positive fixed charges. On the other hand, the shunt resistance decreases remarkably as increasing the negative fixed charge density. The contour maps of the carrier concentration in the vicinity of the gap region as placing different types of fixed charge densities are shown in Fig. 4. Figures 4(a) and 4(b) show the electron distribution when the fixed charge densities are 10^{12} cm^{-2} and -10^{12} cm^{-2}, respectively. Figures 4(c) and 4(d) are the hole distribution when the fixed charge densities are 10^{12} cm^{-2} and -10^{12} cm^{-2}, respectively. It can be easily found that the holes are attracted toward the surface and the electrons are repelled away in the presence of negative fixed charges. When placing positive fixed charges, the electrons and the holes behave in an opposite way.

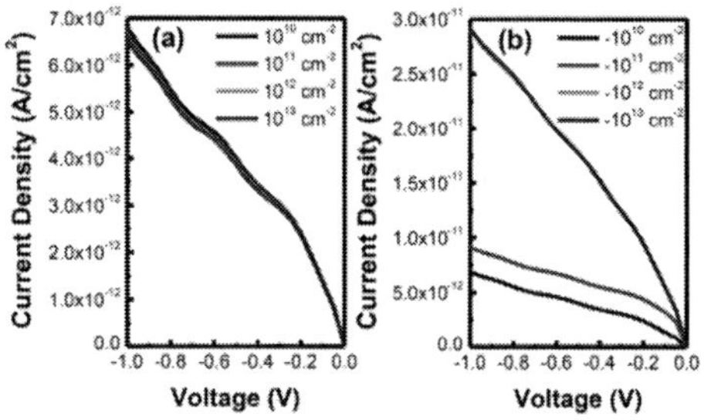

Figure 2: The dark current density-voltage characteristics under reverse bias when different amount of (a) positive and (b) negative fixed charges are deposited.

Figures 5(a) and 5(b) display the contour maps of the generation rate of the tested structure at the bias voltage of -1 V as the fixed charge densities are 10^{12} cm^{-2} and -10^{12} cm^{-2}, respectively. The boundaries of depletion region are also marked. There are two features that can be easily observed in Figs. 5(a) and 5(b). One is the size of the depletion region; the other is the magnitude of the integrated generation rate. In the case of negative fixed charges, there is an additional depletion region in the gap region. Besides, there is also a larger generation rate in this additional depletion region. The obtained results imply that the negatively charged passivation films lead to higher current under reverse bias and smaller shunt resistance than the positively charged passivation films do. Based on these results, it is implied that there exists a strong correlation among the polarity of the fixed charges, the size of the depletion region, the magnitude of the generation rate, the magnitude of the reverse current density, and the shunt resistance.

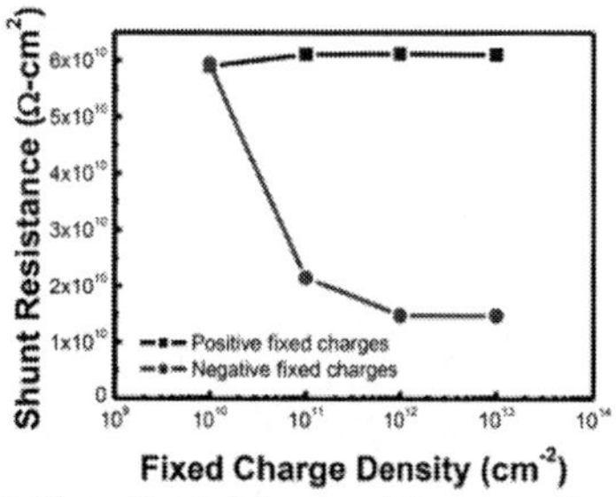

Figure 3: The estimated shunt resistance as a function of positive (square) and negative (dot) fixed charges.

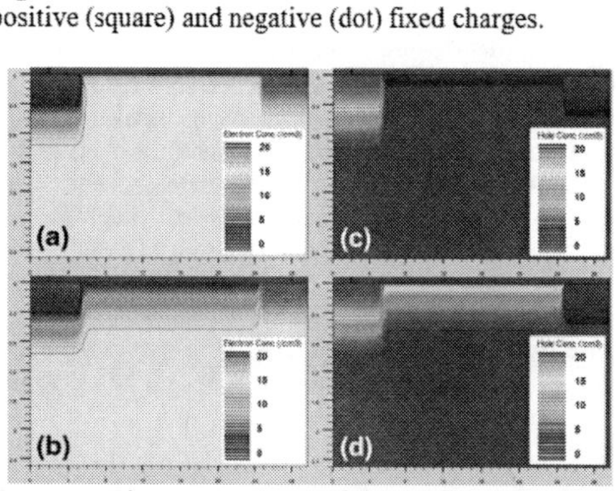

Figure 4: The contour maps of the carrier concentration in the vicinity of the gap region as different types of fixed charge densities are placed. Theses contour maps are plotted in logarithmic scale.

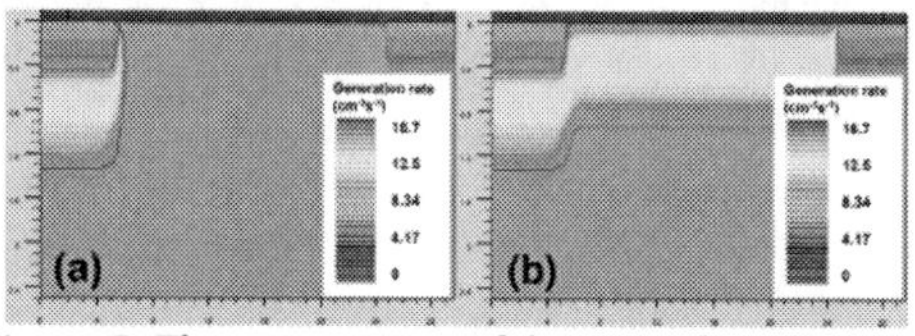

Figure 5: The contour maps of the generation rate of the tested structure at the bias voltage of -1 V as the placed fixed charge densities are 10^{12} cm^{-2} and -10^{12} cm^{-2}, respectively. Theses contour maps are plotted in logarithmic scale.

Christensen has reported that the surface charges would result in the bending of energy band near the surface and a surface conduction channel. [6] McWhorter and Kingston have also found that there is excess current under reverse bias in grown germanium *p-n* junction diodes due to surface charges. [7] According to these reports, the observed phenomena can be well explained as follows. The negative fixed charges placed on the surface of the gap region would induce an electric field which attracts the holes toward the surface and repels the electrons, as shown in Fig. 4. In other words, the induced electric field depletes carriers near the surface of the gap region. The drift of the carriers near the surface of the gap region would result in an additional depletion region. It is well-known that the current of the silicon diode under reverse bias mainly comes from the carrier generation in the depletion region. [8] The larger the depletion region is, the larger the current under reverse bias is. Therefore, the additional depletion region induced by the negative fixed charges causes additional carrier generation and current under reverse bias, as shown in Fig. 2(b). Compared to the case of negative fixed charges, the electric field induced by the positive fixed charges makes carriers behave in an opposite way. In this case, the depletion effect does not occur near the surface of the gap region under this circumstance. Thus, the current under reverse bias is smaller and remains almost unchanged, as shown in Fig. 2(a).

4 CONCLUSION

In conclusion, 2D TCAD numerical device simulations have been used to address the effect of fixed charges contained in the passivation films that are applied to the gap region of the BCBJ solar cells. It is found that the negatively charged passivation film results in an additional depletion region near the surface of the gap region of the BCBJ solar cell while the positively charged passivation film does not. This additional depletion region would also lead to a larger carrier generation rate under reverse bias, which in turn which might in turn makes larger the current under reverse bias and the shunt resistance smaller. Based on these results, experiments can be planned to effectively improve the cell performance with reduced R_{SHUNT} by optimizing the charge state of the passivation film.

REFERENCES

[1] M. A. Green, Prog. Photovoltaics: Res. Appl. **17** (2009) 183.
[2] D. De Ceuster, P. Cousins, D. Rose, D. Vicente, P. Tipones, and W. Mulligan, Proceedings of the 22nd European Photovoltaic Solar Energy Conference (2007) 816.
[3] M. J. Kerr and A. Cuevas, Semicond. Sci. Technol. **17**, (2002) 166.
[4] S. De Wolf, G. Agostinelli, G. Beaucarne, and P. Vitanov, J. Appl. Phys. **97** (2005) 063303.
[5] S. Dauwe, L. Mittelstädt, A. Metz, and R. Hezel, Prog. Photovoltaics: Res. Appl. **10** (2002) 271.
[6] H. Christensen, Proceedings of the IRE **42** (1954) 1371.
[7] A. L. McWhorter and R. H. Kingston, Proceedings of the IRE **42** (1954) 1376.
[8] S. M. Sze and M. K. Lee, Semiconductor Devices: Physics and Technology, 3rd edition (John Wiley & Sons, Inc., New York, 2012).

TCAD MODELING OF TLM CONTACT RESISTANCE STRUCTURES

Geoffrey Gregory[1], Siyu Guo[1], Andrew M. Gabor[2], Adam M. Payne[3], Kristopher O. Davis[1], Rob Janoch[2], Andrew Anselmo[2]
[1]Florida Solar Energy Center – University of Central Florida
[2]BrightSpot Automation LLC
[3]Suniva Inc.
[1]1679 Clearlake Road, Cocoa, FL, USA 32922-5703, Tel: +1-321-638-1000, Fax: +1-321-638-1010
Email: ggregory@fsec.ucf.edu
[2]5 Abbot Mill Lane 10F, Westford, MA, USA 01886, Tel: +1-508-633-5413
[3]5765 Peachtree Industrial Boulevard, Norcross, GA, USA 30092, Tel: +1-404-477-2700

ABSTRACT: The TLM approach to measuring contact resistivity is commonly performed to help optimize cell designs, materials, and processing recipes, and for use in quality control at the factory. Despite the popularity of the technique, there are no standards on how to perform the measurement or how to prepare the structures. In order to better understand the method and its sensitivities to the physical characteristics of its test structures, we have constructed TCAD and circuit simulator models for the measurement sample. The models correctly predict trends in the measured contact resistivity as the sample width is varied. The models also show how the TLM is affected by the presence of unprobed contacts when measuring solar cells. The results are used to create correction factors that can be incorporated into the TLM measurement.
Keywords: contact resistance, TLM, photovoltaic cells

1 INTRODUCTION

As metallization processes improve in the crystalline silicon solar cell industry, it is important that the study of contact resistivity is well developed. The traditional technique for measuring contact resistivity is the TLM approach, which was first introduced by Berger [1]. To prepare test structures, either a special structure is fabricated with variable spacing between the contacts, or a standard solar cell is cut into strips parallel to the busbars. Resistance values are then measured between pairs of contacts on the strips. When using strips cut from cells, contacts can be skipped over to achieve the variable spacing.

We previously [2] explored the dependence of these measurements on sample preparation methods and found a dependence of the measured contact resistivity on the sample width. When sample widths are high (~ 25 *mm*) there is a tendency for the measured contact resistivity to be overestimated. When samples are large and/or the contacts are narrow, the voltage drop down the length of the contact can be significant such that error is introduced to the measurement. Similarly, when sample widths are low (~ 5 *mm*) there is a tendency for the measured contact resistivity to be overestimated. This effect may be due to the mechanism of edge shunting, which is most likely to occur when the ratio of strip perimeter length to strip area is high.

Additionally, the idea of skipping over contacts to achieve variable spacings was not fully captured within the original TLM theory, and it is possible that significant error could result from current entering unprobed contacts as the current flows across the strip. Finally, the firing process for forming silver paste contacts on solar cells might damage the emitter underneath the contacts, thus increasing the sheet resistance there, or selective emitter structures could be implemented that lower the sheet resistance in the contact regions. These factors are also not captured by the original TLM theory.

There is a need to quantify the trends that contact resistivity measurements exhibit as sample preparation methods vary, and to modify the calculations in order to take into account the effects mentioned above. In the present study, we first describe the fundamentals of the TLM method and provide an overview of a useful semi-automatic tool, the *ContactSpot*, used to perform this measurement. We present a numerical, physically based three dimensional model along with a circuit analysis detailing the effects of strip width on contact resistivity measurements. We compare these two models to experimental data acquired using the *ContactSpot* and make conclusions regarding optimal sample conditions for the TLM method. The models are also used to create to create correction factors that can be incorporated into the TLM measurement.

2 TLM TECHNIQUE

Consider the fitted curve shown below in Fig. 1. When performing the TLM technique, the total resistance R_T between two contacts of length W and width f is measured and plotted as a function of contact spacing d. Three parameters can be extracted from this plot: the contact resistance R_C, the sheet resistance ρ_{sh}, and the transfer length L_T. The value of the total resistance at the y-intercept of the plot is 2 R_C. Given this value, the effective contact resistivity $\rho_{c\text{-}eff}$ can be calculated using the area of the contact segments $(W*f)$. As contact spacing d increases, the effect of the sheet resistance on the total resistance measurement increases, thus creating the slope of the TLM plot. Therefore, the value of sheet resistance can be extracted using:

$$R_T/d = \rho_{sh}/W \qquad (2.1)$$

The value of the contact spacing at the intercept where R_T=0 yields 2 L_T. The transfer length found here indicates how much of the contact width is actually being used to insert current into the semiconductor. This can be used in place of the contact width f to calculate an actual contact resistivity.

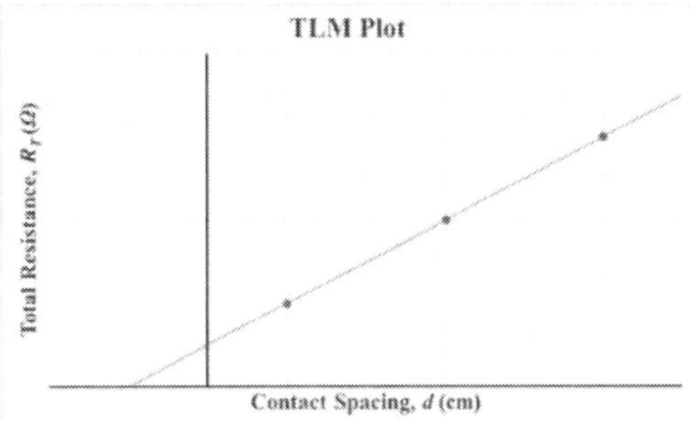

Figure 1: Fitted curve of total resistance versus contact spacing.

Table I: Definitions

Parameter	Definition	Units
ρ_c	Contact resistivity	$m\Omega$-cm^2
$\rho_{c\text{-}eff}$	Effective contact resistivity	$m\Omega$-cm^2
ρ_{sh}	Sheet resistance	$\Omega/\square$
R_c	Contact resistance	Ω
W	Sample width	*mm*
F	Contact width	*mm*
L_T	Transfer length	*mm*
R_T	Total resistance	Ω
D	Contact spacing	*cm*

The *ContactSpot* is a semi-automatic tool for performing contact resistivity measurements. Fig. 2 shows a photo of two *ContactSpot* units. This unit includes a sample platform and microscope with translation and rotation controls for easy contact alignment. Each probe head contains 10 sets of current and voltage probes allowing the measurement of 10 contacts at a time.

This tool can also be used to measure the line resistance of contact segments. During this measurement current is probed along the length of a contact and the resultant voltage is then measured. The resistance of the metal can then be calculated and used to find a value of resistance per unit length.

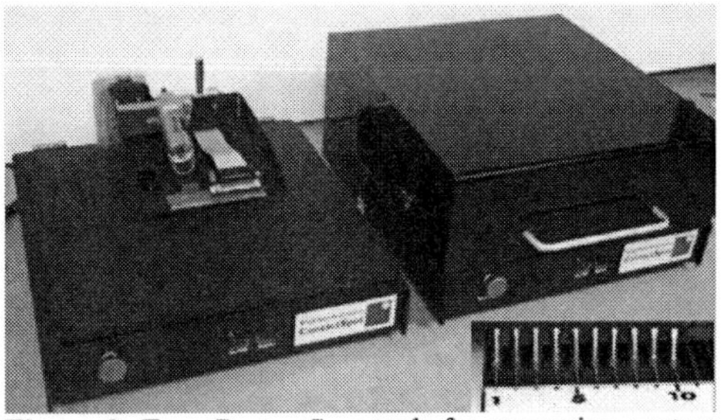

Figure 2: Two *ContactSpot* tools for measuring contact resistivity

As an example of distortions seen measuring strips from solar cells, Fig. 3 shows the results of an experiment where we measured $\rho_{c\text{-}eff}$ versus strip width on the *ContactSpot* for several Suniva solar cells. High values of $\rho_{c\text{-}eff}$ were seen when the strip widths went above 15 *mm* and below 10 *mm*. When strip widths are low (~ 5 *mm*) the ratio of the sample's perimeter length to area is high. In this case, effects such as edge shunting and recombination may cause the measured $\rho_{c\text{-}eff}$ to be artificially high. Issues with the conductivity of contacts may also skew the measured values, especially when the TLM structure's strip width is high (~ 25 *mm*). We explore these and other effects below by modeling the system.

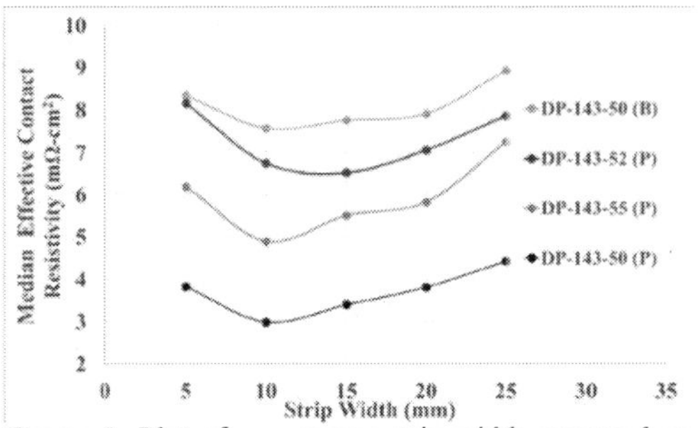

Figure 3: Plot of $\rho_{c\text{-}eff}$ versus strip width measured on strips cut from Suniva cells.

3 PHYSICAL MODEL AND CIRCUIT SIMULATOR

We used the physically based device simulator, Silvaco Atlas®, to model of the TLM test samples. This TCAD software allows the user to define the physical structure to be simulated, the physical models to be used, and the bias conditions for which the electrical characteristics are to be simulated [4]. Table 2 shows a list of the parameters we used to create the TLM structure model. These parameters are typical values measured on silicon solar cells.

Table II: TCAD and Circuit Model Base Parameters

Parameter	Value	Units
Contact resistivity, ρ_c	2	$m\Omega$-cm^2
Sheet resistance, ρ_{sh}	85.0	$\Omega/\square$
Sample width, W	5.0 - 25.0	*mm*
Contact width, f	60.0	μm
Contact spacing, d	0.175	*cm*
Current level, I	0.01	*A*
Emitter thickness, t	0.3	μm

Silvaco Atlas® requires that each structure be defined on a mesh that covers the physical simulation domain. In order for the model to yield an accurate solution, it is important that the mesh is defined properly. Therefore, a very fine mesh is defined near the metal-semiconductor interface and then made coarse away from the contacts for the sake of computational efficiency.

We created the circuit simulation using LTspice IV software. The parameters used for simulation are the same as those used in the TCAD model, shown in Table 2. The two models are compared quantitativley in order to verify the accuracy of the solutions.

It's possible that edge shunting effects may be introducing significant amounts of error in the TLM results. When current is injected into a contact, it can find a path of low resistance through the structure along the edge of the sample. In order to determine whether edge shunting introduces error into the TLM measurement, we incorporated the mechanism into the models. Fig. 4 shows a plot of $\rho_{c\text{-}eff}$ versus strip width for structures containing shunt paths. We also implemented structures without the shunt in order to quantify the impact of this phenomenon on the measured $\rho_{c\text{-}eff}$. Both models show that the edge shunt modifies the measured value of $\rho_{c\text{-}eff}$

at all strip widths. The most notable effect is at a low strip width of 5 *mm*.

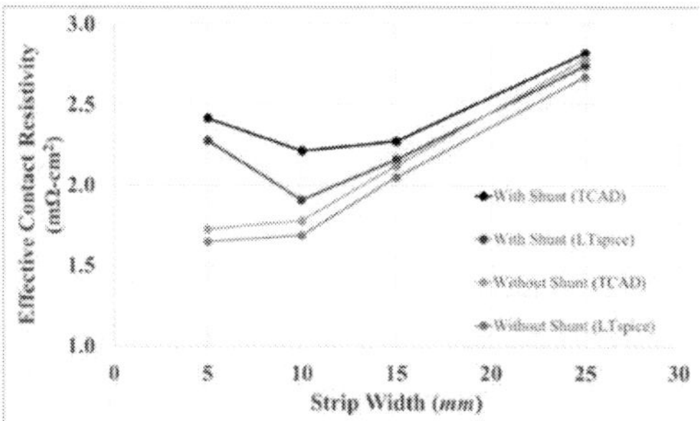

Figure 4: Plot of $\rho_{c\text{-}eff}$ versus strip width for TLM structures with and without shunt paths.

It can be seen that for both the structure with an edge shunt and the one without an edge shunt, the $\rho_{c\text{-}eff}$ increases with increasing strip width. In order to investigate the cause of this phenomenon, we created three structures using the base parameters given in Table 2. We then defined different line resistance values for each structure's metal contacts. This was done to study the effect that the line resistance of the contacts has on the measured $\rho_{c\text{-}eff}$. Fig. 5 shows the value of $\rho_{c\text{-}eff}$ measured for each combination of input parameters.

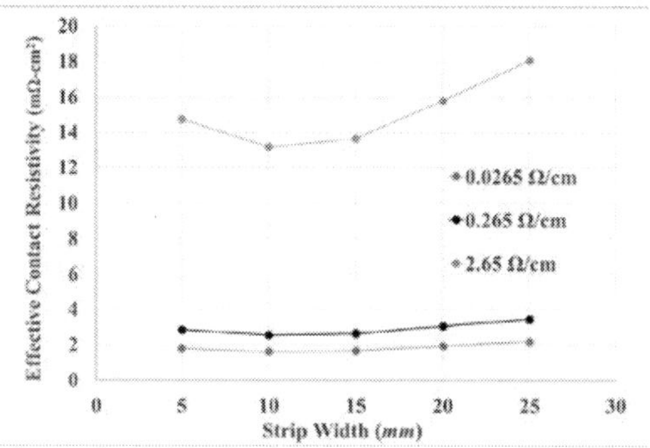

Figure 5: TCAD simulation of $\rho_{c\text{-}eff}$ versus strip width for varying line resistance values of the contacts.

It can be seen that the trends in increasing $\rho_{c\text{-}eff}$ are most pronounced for the structures with high line resistance (2.65 *Ω/cm*), and the measured value approaches the input contact resistivity (2 *mΩ-cm*^{2}) as the line resistance of the contacts decreases.

When standard solar cells are used for contact resistivity measurements it is necessary to skip over contacts in order to measure the total resistance R_T at different contact spacings d. In the past, groups [3] have reported seeing consistently higher values of ρ_c when using TLM structures with large amounts of unprobed contacts (> 5). There is a concern for the effect of the unprobed contacts that are skipped over during the resistance measurements of the TLM. In order to study this effect, we modeled TLM structures using the base parameters from Table 2, and each structure was then used to measure $\rho_{c\text{-}eff}$ with and without the unprobed contacts. We also modeled structures with varying amounts of contacts in order to see how the quantity of unprobed contacts would affect the measurement. The results of this simulation can be seen below in Fig. 6.

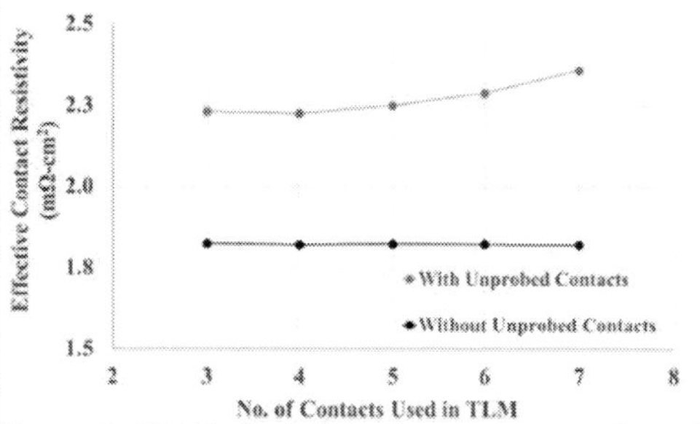

Figure 6: TCAD simulation of $\rho_{c\text{-}eff}$ versus number of contacts on the TLM structure.

Fig. 6 shows that $\rho_{c\text{-}eff}$ remains relatively stable for the structures that did not have unprobed contacts. For the structures that did have unprobed contacts, the $\rho_{c\text{-}eff}$ does increase as the number of contacts used to create the TLM plot increases. The simulated $\rho_{c\text{-}eff}$ is always higher for the structures containing unprobed contacts than the structures that did not contain them.

4 RESULTS AND DISCUSSION

The plot in Fig. 4 shows a high value of $\rho_{c\text{-}eff}$ at the lowest strip width (5 *mm*) for the structure that has a shunt path incorporated into the models. This plot indicates that edge shunting does skew the measured $\rho_{c\text{-}eff}$ most notably when the ratio of strip perimeter length to strip area is high. The measured value is skewed since the shunt paths decrease the value of R_T at each point in the TLM plot. This will decrease the slope of the plot and increase the value at the y-intercept, $2R_c$.

The same plot also shows that the simulated value of $\rho_{c\text{-}eff}$ increases as the width of the sample increases. This increase is a result of an additional voltage drop along the probed contacts that the TLM does not account for. We used the circuit simulator to show the voltage drop along a probed contact (Fig. 7). As the contacts become longer, the voltage drop gets larger, and a higher measured $\rho_{c\text{-}eff}$ is obtained. This is verified in Fig. 5, where the effect is amplified as the line resistance of the contacts increases. If the contact is unable to effectively conduct the injected current, then voltage will have a significant drop along the length of the contact, making the current flow more concentrated towards its center.

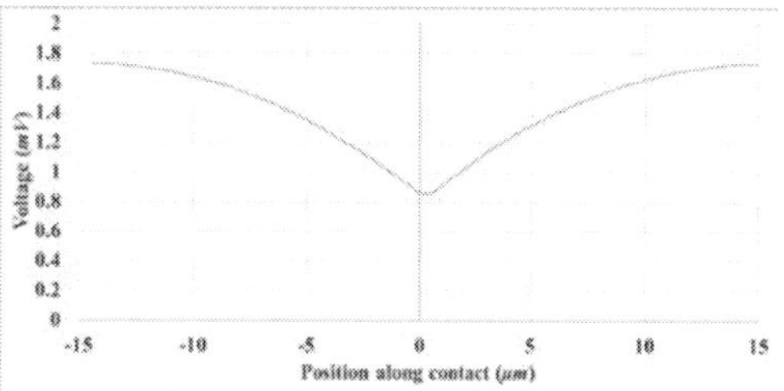

Figure 7: Plot of voltage along the length of a contact from circuit simulation.

As trends in metallization steadily move toward narrower, less conductive contacts, it is necessary to correct for the line resistance of the metal in order to obtain more accurate $\rho_{c\text{-}eff}$ measurements. From simulation results, we observed that current flow through the contacts always has a linear decrease/increase (shown in Fig. 8).

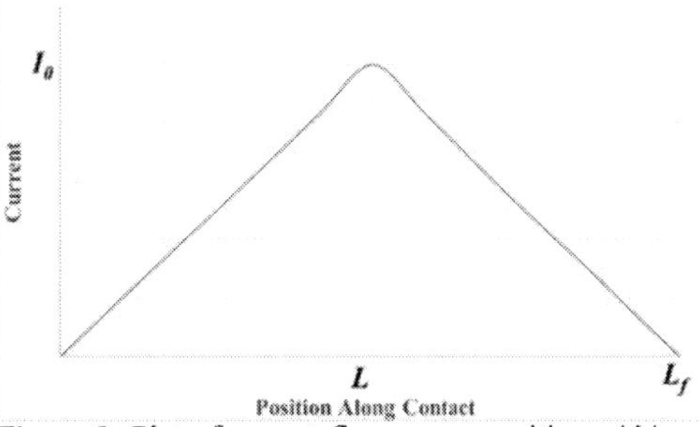

Figure 8: Plot of current flow versus position within a contact.

Assuming the total injection current is I_0, the current flow to each direction is $I_0/2$. The voltage drop from the edge of a contact to its center can be calculated by integration:

$$V_{drop} = \int_0^L \left(\frac{I_0}{2} \times \frac{x}{L}\right) \rho_l \, dx = \frac{\rho_l L I_0}{4} \qquad (4.1)$$

where ρ_L is the metal line resistance in Ω/cm, I_0 is injection current in A, and L is half of the full contact length L_f in cm. Assuming the voltage drop to be equal along both halves of the contact, we calculate the total voltage drop to be:

$$V_{total} = \int_0^L \left(\frac{I_0}{2} \times \frac{x}{L}\right) \rho_l \, dx = \frac{\rho_l L I_0}{4} \times 2 = \frac{\rho_l I_0 L_f}{4} \qquad (4.2)$$

From this, the additional resistance caused by the current flow through the contact can be calculated:

$$R = \frac{V_{total}}{I_0} = \frac{\rho L_f}{4} \qquad (4.3)$$

This value can then be subtracted from the measured contact resistance R_c in order to obtain a more accurate value of $\rho_{c\text{-}eff}$.

Another factor which influences the measured $\rho_{c\text{-}eff}$ is the current flow through the unprobed contact, which is not accounted for in the original TLM technique. As is shown in Fig. 6, the existence of unprobed contacts in a total resistance measurement will increase the final value of $\rho_{c\text{-}eff}$. As each unprobed contact is added to a TLM structure, the current path becomes more conductive. If the sheet resistance ρ_{sh} of the sample is low, then current will have a propensity to enter into the intermediate contacts as it travels through the sample. We used the circuit model to determine the percentage of current that flows through the unprobed contacts as a function of sheet resistance for a given value of $\rho_{c\text{-}eff}$ (Fig. 9).

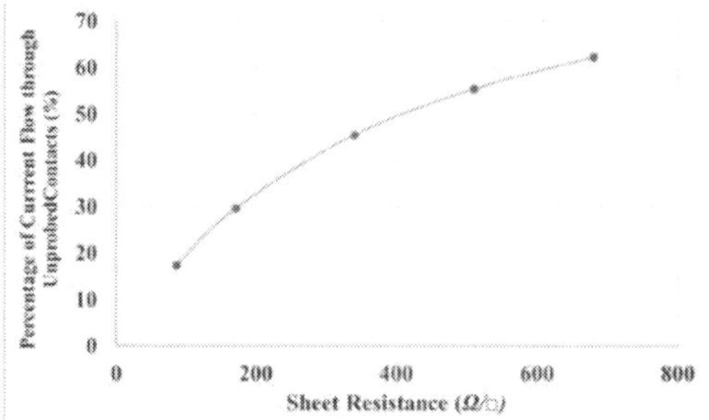

Figure 9: Circuit simulation of percentage of current flow through the unprobed contacts as a function of sheet resistance.

By knowing the percentage of current that flows through the unprobed contacts, we can determine the true voltge drop between two contacts. Assuming 20% of the total injected current flows through the unprobed contacts, the true voltage drop will be 80% of the measured voltage drop. We then adjust the contact width f to $0.8f$, which corrects the value of $\rho_{c\text{-}eff}$ to a more accurate result. Fig. 10 shows a plot of $\rho_{c\text{-}eff}$ at different strip widths before and after correction.

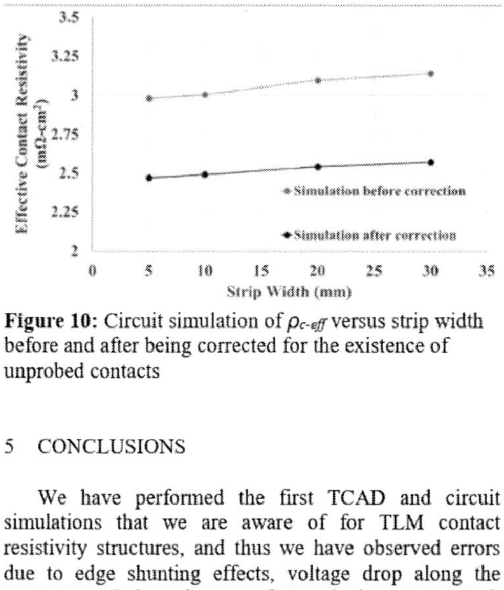

Figure 10: Circuit simulation of $\rho_{c\text{-}eff}$ versus strip width before and after being corrected for the existence of unprobed contacts

5 CONCLUSIONS

We have performed the first TCAD and circuit simulations that we are aware of for TLM contact resistivity structures, and thus we have observed errors due to edge shunting effects, voltage drop along the contacts, and the existence of unprobed contacts. Both the correction for the voltage drop along the contact and the correction for current entering unprobed contacts can be incorporated into tools like the *ContactSpot*. These corrections will increase the viability of the TLM technique as contacts become narrower and less conductive and as emitter or BSF sheet resistance increases.

We have shown with our models that edge shunting introduces a considerable error in the measured $\rho_{c\text{-}eff}$ at low strip widths. For our structures, the strip width should be kept higher than 5mm in order to avoid this effect, but in general this will likely depend on the cell structure and the strip preparation method.

REFERENCES

[1] Berger, H. H. (1972). "Models for contacts to planar devices." Solid-State Electronics 15: 145-158.

[2] A.M. Gabor, G. Gregory, A. Payne, R. Janoch, A. Anselmo, V. Yelundur, and K. O Davis, *Dependence of Solar Cell Contact Resistivity Measurements on Sample Preparation Methods,* in 43rd *IEEE Photovoltaics Specialist Conference.* 2016, IEEE: Portland, OR, USA. (Not yet published)

[3] R. Janoch, A.M. Gabor, A. Anselmo, and C.E. Dube, et al., *Contact Resistance Measurement-Observations on Technique and Test Parameters,* in *42nd IEEE Photovo,ltaic Specialist Conference.* 2015, IEEE: New Orleans, LA, USA.

[4] Silvaco International (2015), "Atlas User's Manual, Device Simulation Software." Pg. 25.

ULTRA-THIN SILICON SOLAR CELL: FLEXIBILITY, MODELING AND PREDICTION

Jianshu Han[1], Malcom Abbott[1], Phillip Hamer[2], Bram Hoex[1], Anthony Lochtefeld[3] and Allen Barnett[1]

[1]School of Photovoltaic & Renewable Energy Engineering, UNSW Australia, UNSW Sydney, NSW 2052, Australia
[2]Department of Materials, University of Oxford, Oxford, OX1 2JD, UK
[3]AmberWave, Inc., Salem, 03079-2435, USA

ABSTRACT: Silicon solar cells with reduced thickness have the potential to achieve higher open circuit voltages (V_{OC}) due to reduction in intrinsic recombination in the bulk region. At UNSW Australia, the ultra-thin silicon on steel (UTSi) solar cell developed is based on epitaxial growth on porous silicon. Efficiencies of 16.8% on 4 cm^2 devices and 15.9% on 90 cm^2 devices have been demonstrated, both with an active layer thickness of less than 20 μm.

In this paper, a three-dimensional cell model is developed to accurately simulate an UTSi cell with a confirmed efficiency of 15.9%. This model determines the unknown $J_{0,rear}$ component of the device. The cause and magnitude of various losses in the device is then calculated, followed by performance predictions when these losses are minimized using available technologies. The model predicts this device can reach 20.4% efficiency with readily available technologies, and is capable of achieving 22.2% efficiency.

The UTSi device also demonstrates excellent flexibility, and is very robust against mechanical movements. The degree of damage caused by such physical impairment was quantitatively investigated in this paper.

Keywords: Flexible substrate, modeling.

1 INTRODUCTION

Thin silicon solar cells have the potential to achieve higher open circuit voltages compared to conventional solar cell designs due to reduction in intrinsic recombination in the bulk region. This design can be achieved by epitaxially growing a layer of crystalline silicon on porous silicon. Solar cell fabricated with this method has recently demonstrated conversion efficiency of 22.5%, suggesting epitaxial material could achieve quality as good as conventional CZ wafers [1].

The ultra-thin silicon (UTSi) solar cell is epitaxially grown with a thickness of just 21 μm. The device is bonded to a steel substrate which provides physical support as well as a conductive rear surface to the solar cell. The structure of UTSi solar cell is demonstrated in Figure 1. The 21 μm cell is an *n*-type monocrystalline device with a rear junction. The front and rear surfaces were passivated by $SiON_x$ and SiO_2 respectively. A stack of Ni/Cu fingers and busbars contacts the heavily doped n^+ openings on the front of the cell. Al point contacts were formed covering approximately 0.5% of the rear surface.

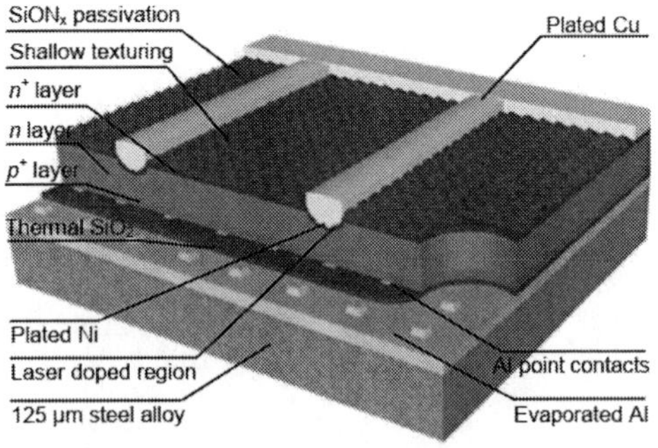

Figure 1: The UTSi solar cell structure [2].

Light trapping is critical for thin devices. The front surface of the UTSi cell is textured with pyramid sizes of approximately 2 μm. The thickness and refractive index of the $SiON_x$ layer is tuned to minimize the total reflection over the solar spectrum. The rear evaporated Al layer reflects long wavelength photons which have passed through the thickness of the device. This increases the probability that they will generate electron-hole pairs in the device.

The device fabrication procedure was described by Wang et al [3] and Han et al [2].

2 MODELING WITH QUOKKA

One problem with the unique steel substrate is related to its conductive nature. It makes this design incompatible with standard photoconductance decay measurements of lifetime [4, 5], leaving bulk lifetime, front and rear surface dark saturation current ($J_{0,rear}$) unknown. This problem can be studied by simulation. In this paper, a 3-D UTSi solar cell model is developed to evaluate the rear surface recombination. The modeled result will be presented with a comparison to real cell measurements.

2.1 Simulation Method and Results

The solar cell simulator "Quokka" [6-8] developed by Andreas Fell from Australia National University was used to model a fabricated 90 cm^2 UTSi solar cell with confirmed efficiency of 15.9%.

In order to accurately evaluate $J_{0,rear}$ by simulation, all input device parameters for Quokka will be independently measured or calculated. Other simulators such as EDNA [9], OPAL2 [10, 11] and "Wafer Ray Tracer" [12] developed by PV Lighthouse were used in this work to determine the various optical and electrical aspects of this cell. $J_{0,rear}$ will then be determined by fitting simulated V_{OC} to measured V_{OC}. This method can only produce an explicit result when there is no more than one unknown parameter in the model.

The simulated device parameters are listed in Table 1.

Table I. Device parameters used in Quokka simulation.

Category	Parameter	Value	Method
Bulk	Resistivity	0.99 Ω.cm^2	
	Background lifetime	50 µs	Test structure
Front non-contact	Junction depth	2 µm	
	Collection efficiency	100%	
	$J_{0,front}$	20 fA/cm^2	Kane-Swanson [13]
	Sheet Resistance	274 Ω/sq	
Front laser doped Si-Ni contacts	Junction depth	3 µm	
	Line contact half width	8 µm	
	Contact $J_{0,front}$	153 fA/cm^2	SRV = 1×10^{7}cm/s at 10 Ω/sq [9]
	Sheet resistance	10 Ω/sq	
	Contact resistance	0.02 Ω.cm^2	From Ref. [14]
Rear emitter & Si-Al contact	Junction depth	1 µm	
	$J_{0,rear}$	**578 fA/cm^2**	***V_{OC}* fitting**
	Sheet resistance	651 Ω/sq	
	Collection efficiency	92%	
	Point contact radius	8 µm	
	Contact $J_{0,rear}$	9100 fA/cm^2	SRV = 1×10^{7}cm/s at 651 Ω/sq [9]
	Sheet resistance	651 Ω/sq	
	Contact resistance	0.0001 Ω.cm^2	From Ref. [14]
External circuit	*External R_S*	0.122 Ω.cm^2	Grid calculator
	External R_{SH}	1004 Ω.cm^2	J-V slope at V=0
Generation	Shading	11%	
	Transmission, reflection and absorption of front SiN_x	λ dependent	OPAL 2 [11]
	Pathlength enhancement factor	λ dependent	Long wavelength fitting for normalized external quantum efficiency (EQE)

By V_{OC} fitting, the UTSi Quokka model finds a $J_{0,rear}$ of 578 fA/cm^2 for the champion solar cell. The J-V curve simulated by this model agrees well with measured data of the device, as shown in Table II.

Table II. Comparison of the measured and simulated key parameters.

	Measured	Simulated
V_{OC}	632 mV	632 mV
J_{SC}	33.7 mA/cm^2	33.7 mA/cm^2
FF	75.0%	74.5%
Efficiency	15.9%	15.9%

2.2 Simulation Result Validation

Quokka assumes carrier diffusion at the surface as a recombinative boundary condition and applies quasi-neutrality outside of the space-charge region [6]. It is therefore important to evaluate whether these assumptions still hold for thin solar cell devices. A comparison between Quokka and PC1D was performed based on a simplified structure, with the result shown in Figure 2. For these simulations a simplified model with 0% contact area and no light trapping was used. The front and rear layers in both models have the same doping level as shown in Table I. The bulk lifetime was assumed to be 50 µs while J_0 for all surfaces were set to be 90 fA/cm^2. Quokka agrees well with PC1D when the device thickness is more than 120 µm. Even at 20 µm there is a net difference of only 4 mV, representing an equivalent error of 0.6% in V_{OC}.

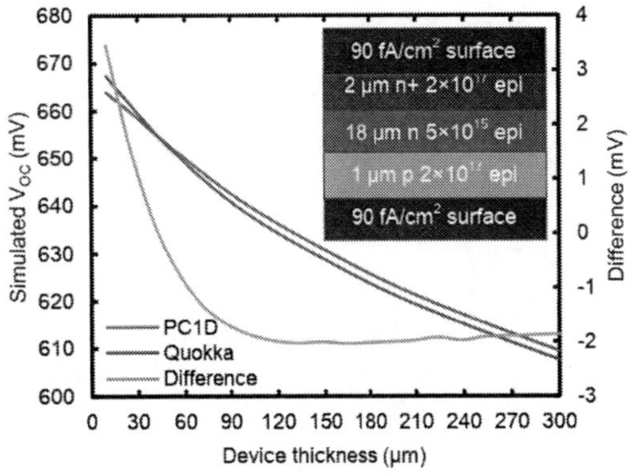

Figure 2: Comparison of Quokka and PC1D simulation results as a function of the device thickness.

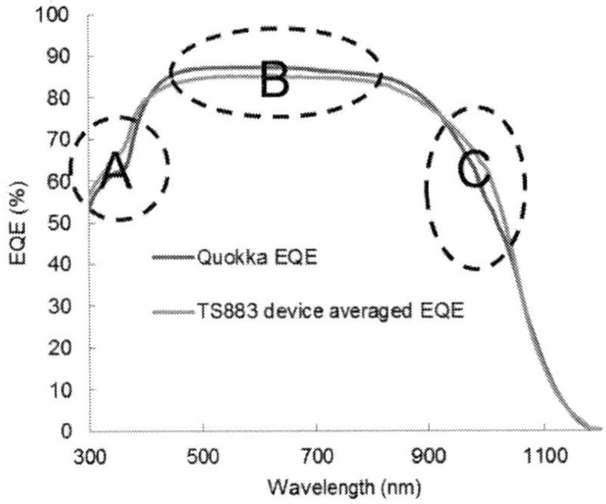

Figure 3. Comparison of the simulated and measured EQE curves for the champion UTSi device.

A comparison of simulated and measured EQE curves is shown in Figure 3. The modelled curve closely fits real device EQE measurement. There are three regions of misfit, as labelled in "A", "B" and "C" in the diagram. Mismatch "A" is related to the spectral response near the front surface. This is most likely caused by the uneven etching rate during shallow texturing. As a result of this process, part of the n^+ front surface field was completely removed while part of it remained on the sample surface. Quokka is programmed to simulate only one of the situations, and for simplicity reason, the n+ front surface field was assumed to be fully remained on the surface in the UTSi model. EQE mismatch in region "B" does not reflect inaccuracy in the broadband shading estimation, since the curves fit well in the long wavelength range. Rather, it is caused by J_{02} recombination losses that weren't included in this model. These losses are attributed to the junction second diode recombination, edge recombination, and local recombination such as shunting in laser doped lines. Mismatch in region "C" can be explained by perturbation and inaccuracy of the ray tracing simulation.

2.3 Performance Prediction

The established UTSi model can be used to assess impacts of any design and fabrication changes to device performance. The model suggests that a better rear passivation (by replacing SiO_2 with AlO_x/SiN_x stack), a higher emitter doping and a higher bulk resistivity should immediately improve the conversion efficiency to 17.9%. A key advantage of the epitaxial growth process is the simplicity of transitioning into a new emitter and bulk design. In long term, process improvement in laser doping (reduced shading, better Ni-Si contacts), edge isolation and surface passivation can potentially improve UTSi conversion efficiency to over 22%, as shown in Figure 4.

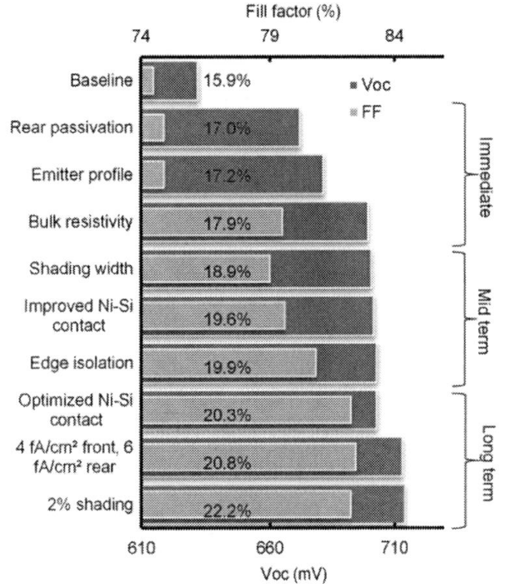

Figure 4: Cumulative performance improvement predicted by Quokka model. Predicted efficiencies are labeled next to each column.

4 UTSI DEVICE FLEXIBILITY

In order to demonstrate the flexible and robust nature of the UTSi design, a 7 cm x 7 cm UTSi solar cell was bent, pivoting alone the center of the device with a vertical displacement of 2 cm at the ends (Figure 6). This process was repeated, while IV measurement was taken after the following steps:

#1: Before bending
#2: 1x bend, hold for 5 sec each
#3: 3x bend, hold for 5 sec each
#4: 10x bend, hold for 5 sec each
#5: 2x bend, hold for 1 min each

The physical impairment to this solar cell is assessed based on IV measurements. There has been a total of 3 mV drop (-0.5%) after the 16 bends, while no loss in fill factor is observed, as shown in Figure 7.

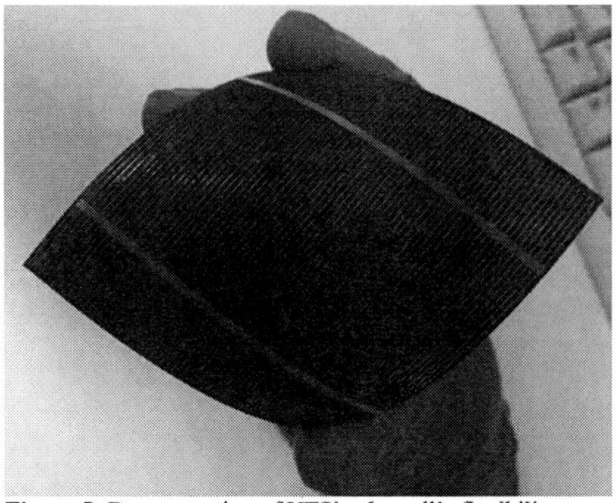

Figure 5: Demonstration of UTSi solar cell's flexibility and robustness.

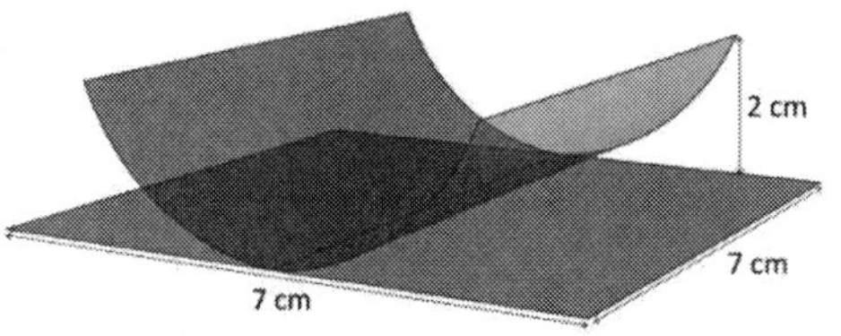

Figure 6: Geometry illustration of the bending process.

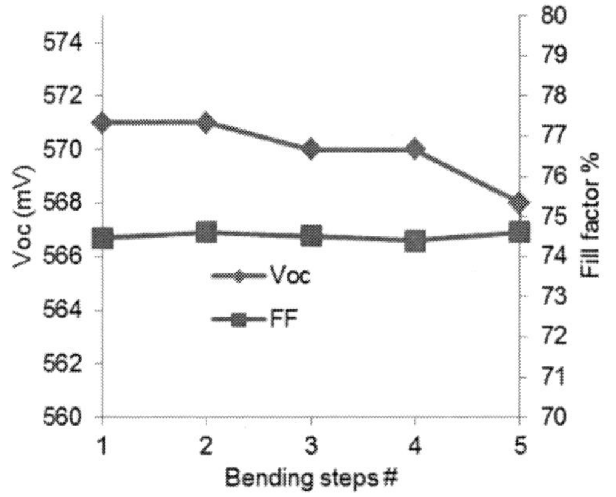

Figure 7: V_{OC} and FF measured after each of the five bending steps.

5 SUMMARY

In this paper, a 15.9% efficient large area UTSi solar cell is modeled with three-dimensional simulator Quokka. The simulation determines a $J_{0,rear}$ of 578 fA/cm^2, which was not compatible with direct measurement. The model predicts a conversion efficiency of over 22% with the appropriate design and fabrication improvements.

UTSi solar cell uniquely offers a robust and flexible device structure compared to conventional crystalline solar cell technologies, making it suitable for BIPV and other portable PV applications.

REFERENCES

[1] R. Hao, T. S. Ravi, V. S. Vatus, I. Kuzma-Filipek, F. Duerinckx, M. Recaman-Payo, *et al.*, "Kerfless Epitaxial Mono Crystalline Si Wafers With Built-in Junction And

From Reused Substrates For High Efficiency PERx Cells," *43rd IEEE PVSC Proceedings,* 2016.
[2] J. Han, L. Wang, A. Lochtefeld, A. Gerger, H. Li, M. Carroll, *et al.*, "Fabrication of large area ultra-thin silicon solar cells," presented at the EUPVSEC, Amsterdam, 2014.
[3] L. Wang, A. Lochtefeld, J. Han, A. P. Gerger, M. Carroll, J. Ji, *et al.*, "Development of a 16.8% Efficient 18-μm Silicon Solar Cell on Steel," *Photovoltaics, IEEE Journal of,* vol. 4, pp. 1397-1404, 2014.
[4] R. A. Sinton and A. Cuevas, "Contactless determination of current - voltage characteristics and minority - carrier lifetimes in semiconductors from quasi - steady - state photoconductance data," *Applied Physics Letters,* vol. 69, pp. 2510-2512, 1996.
[5] R. A. Sinton, A. Cuevas, and M. Stuckings, "Quasi-steady-state photoconductance, a new method for solar cell material and device characterization," in *Photovoltaic Specialists Conference, 1996., Conference Record of the Twenty Fifth IEEE,* 1996, pp. 457-460.
[6] A. Fell, "A free and fast three-dimensional/two-dimensional solar cell simulator featuring conductive boundary and quasi-neutrality approximations," *Electron Devices, IEEE Transactions on,* vol. 60, pp. 733-738, 2013.
[7] A. Fell, K. C. Fong, K. R. McIntosh, E. Franklin, and A. W. Blakers, "3-D simulation of interdigitated-back-contact silicon solar cells with quokka including perimeter losses," *Photovoltaics, IEEE Journal of,* vol. 4, pp. 1040-1045, 2014.
[8] A. Fell, K. R. McIntosh, M. Abbott, and D. Walter, "Quokka version 2: selective surface doping, luminescence modeling and data fitting," in *23rd Photovoltaic Science and Engineering Conference, Taipei,* 2013.
[9] K. R. McIntosh and P. P. Altermatt, "A freeware 1D emitter model for silicon solar cells," in *Photovoltaic Specialists Conference (PVSC), 2010 35th IEEE,* 2010, pp. 002188-002193.
[10] S. C. Baker-Finch and K. R. McIntosh, "A freeware program for precise optical analysis of the front surface of a solar cell," in *Photovoltaic Specialists Conference (PVSC), 2010 35th IEEE,* 2010, pp. 002184-002187.
[11] K. R. McIntosh and S. C. Baker-Finch, "OPAL 2: Rapid optical simulation of silicon solar cells," in *Photovoltaic Specialists Conference (PVSC), 2012 38th IEEE,* 2012, pp. 000265-000271.
[12] PVLighthouse. (2015, 3rd September). *Wafer Ray Tracer.* Available: https://www.pvlighthouse.com.au/calculators/wafer%20ray%20tracer/wafer%20ray%20tracer.html
[13] D. Kane and R. Swanson, "Measurement of the emitter saturation current by a contactless photoconductivity decay method," in *IEEE photovoltaic specialists conference. 18,* 1985, pp. 578-583.
[14] L. Wang, H. Li, C. Shen, J. Han, P. Teng, M. Abbott, *et al.*, "Photoluminescence Analysis of a 16.8% Efficient 18-um Silicon Solar Cell," *International Journal of Emerging Technology and Advanced Engineering,* vol. 5, 2015.

LIGHT INDUCED DEGRADATION OF SOLAR CELLS FROM HIGH-PERFORMANCE MULTICRYSTALLINE SILICON

Kai Sporleder[1], Tabea Luka[1], Stefan Eiternick[1], Kati Hübener[2], Kai Petter[2], Roland Scheer[3], Marko Turek[1]*
[1]Fraunhofer Center for Silicon Photovoltaics CSP, Otto-Eissfeld-Str. 12, 06120 Halle (Saale), Germany
[2]Hanwha Q CELLS GmbH, Sonnenallee 17-21, 06766 Bitterfeld – Wolfen, Germany
[3]Martin Luther-Universität Halle-Wittenberg, von-Danckelmann-Platz 3, 06120 Halle
*Corresponding Author: Tel: +49-345-5589-5121, E-mail: marko.turek@csp.fraunhofer.de

ABSTRACT: The diversified c-Si materials market is dominated by high-performance multicrystalline (HPmc) wafer. This material is characterized by a large number of small grains while these grains exhibit a strongly reduced density of structural intra-grain defects. Thus, solar cells made of HPmc-wafer can show a significantly increased performance. However, it has been shown, that depending on the wafer production process, light induced degradation (LID) might occur, which results in an efficiency loss of up to 10 $\%_{rel}$. This degradation has been termed "sponge-LID" due to its characteristic appearance in luminescence imaging. It cannot be explained by the well-known mechanisms due to the formation of boron-oxygen-pairs or the separation of iron-boron complexes and is also unrelated to the mc-LID effect, observed on mc-Si PERC cells. Sponge-LIDs harmful impact on efficiency is related to a decrease in all cell parameters. However, strongest decay is observed in short circuit current which mainly due to a change in bulk recombination, while emitter and rear recombination are mostly unaffected. Kinetics of defect activation is strongly influenced by intensity of illumination. On the other hand, a variation of the degradation temperature between 25 °C and 75 °C does not influence the sponge-LID kinetics which is in contrast to other degradation mechanisms. Spatially resolved light beam induced current measurements of sponge-LID on single grains show that the defects are located within a small area around grain boundaries.
Keywords: LID, Sponge-LID, multi crystalline silicon, high performance multi crystalline silicon

1 INTRODUCTION AND MOTIVATION

Light induced degradation of solar cells can cause major efficiency losses already during the first hours of operation. The most prominent degradation mechanisms are due to the formation of boron-oxygen complexes [1] or the separation of iron-boron complexes [2]. Mc-Si PERC cells can show degradation under illumination and elevated temperatures (mc-LID) [3]. Recently, a new form of light induced degradation termed "sponge-LID" has been discovered. Sponge-LIDs detrimental impact on cells from HPmc wafers cannot be explained by any of these degradation mechanisms [4], [5]. It has been shown, that up to 20 % - 30 % of an ingot might be affected by "sponge-LID" inducing an efficiency loss of up to 10 $\%_{rel}$. Typically, cells from the lower third of an ingot are "sponge-LID" sensitive. Furthermore, it has been shown that an adapted cell process leads to solar cells that are stable regarding "sponge-LID" [4, 5].

In this work, we will present results on localizing the root-cause for the "sponge-LID" effect. We give an overview of the cell parameter degradation based on current-voltage-measurements (I-V-measurements). We show that sponge-LID mostly affects the short circuit current I_{sc}, while open circuit voltage V_{oc} and fill factor FF are less affected. The resulting cell power loss of about 7 $\%_{rel}$ compares to losses smaller than 1 $\%_{rel}$ of a sponge-LID prone cell from a sister wafer processed by an adapted sponge-LID resistant process. We investigate the change of degradation behavior with constant illumination and variation in temperature as well as the differences in degradation kinetics at different illumination levels at fixed temperature. Afterwards, we study to what extend the front and rear side as well as the bulk is affected by "sponge-LID" by applying a loss analysis model based on quantum efficiency measurements [7]. In analogy to the degradation impact of mc-LID, the current loss is due to recombination in the bulk material [8]. Spatially resolved light beam induced current (LBIC) measurements show that the Sponge-LID defects are located nearby grain boundaries, which differs from what is known about mc-LID [9]. However, the here presented results indicate that during crystallization or cell process there are different ways in how inactive defect centers spread through the grains before activation

2 EXPERIMENTAL APPROACH

Three batches of industrially produced solar cells (boron doped and aluminum back surface field) are investigated. Batch A and B are processed from "sponge-LID" sensitive sister wafers, taken out of the same brick. Batch A cells from the lower third of the ingot show "sponge-LID". Batch B cells are processed using an optimized "no-sponge-LID" cell process. Batch C cells show "sponge-LID" but stem from a different ingot.

To evaluate the adapted low sponge process, the impact of sponge LID on a batch A cell and a batch B sponge-LID resistant cell is quantified. The degradation conditions are given by an illumination intensity of 1 sun (1000 Wm^{-2}) and temperature of 25 °C. The cells were exposed to light for about 20 hours, split on 4 days. The cells were stored in the dark at room temperature until the next light treatment. During degradation, I-V-measurements were performed to quantify the degradation rate and amount of the cell parameters. In addition, light beam induced current (LBIC) images at a wavelength of 960 nm and a lateral resolution of 100 µm of a 2 cm x 2 cm section were taken before and after illumination to evaluate lateral effect of sponge-LID.

In a further experiment, the impact of illumination level and temperature on sponge-LID was studied. For investigation of illumination dependence three batch A cells from sister wafers were degraded using different intensities: 0.2 suns, 0.4 suns, and 0.6 suns. Cell temperature during light treatment was held at 25 °C. IV measurements with 1 sun intensity and temperature 25 °C were performed to quantify degradation. To investigate the impact of temperature, three batch C cells from sister wafers were exposed to light for 20 hours at intensity of

0.6 suns. During illumination, cells were held at different temperatures of 25 °C, 50 °C, and 75 °C.

Furthermore, we analyzed spectrally and spatially resolved measurements. This allows an assessment whether the degradation occurs in the emitter, volume, or rear side as well as an assessment whether the "sponge-LID" is lateral homogeneous or dominated by local loss centers.

3 RESULTS AND CONCLUSIONS

The impact of sponge-LID on cell parameters of a sponge-LID sensitive cell (cell A) and a cell from the adapted process (cell B) is shown in Figure 1. For both cells, relative decrease of I_{sc}, V_{oc}, FF, and η are compared. Cell A shows 6.9 $\%_{rel}$ loss in efficiency which is dominated by a 4.4 $\%_{rel}$ decrease in short circuit current while open circuit voltage and fill factor are less affected. The 0.7 $\%_{rel}$ loss in efficiency of Cell B is mainly due to a loss in I_{sc}. However, an adaption of the cell process can reduce impact of LID by almost 90 %. Measured values for short circuit current (circle) and open circuit voltage (cross) during degradation are shown. Both, short circuit current and circuit voltage, measured values have been fitted to

$$P_t = P_0 [\, 1 - \beta\, (1\text{-}e^{-kt})\,], \qquad (1)$$

where P_t is the measured value after illumination time t. Parameter β is given by $1\text{-}P_\infty/P_0$ and thus determines the amount of degradation. Rate k quantifies the kinetics of the degradation with 1/k being the typical time scale on which the sponge-LID occurs. For the sponge-LID sensitive cell, about 50 % of degradation losses appear within the first three hours of illumination. The degradation saturates after about 15 hours of 1 sun illumination at 25 °C. The corresponding degradation rate k = 0.15 h^{-1} was determined by fitting Eq. (1) to the experimental data and implies a typical sponge-LID time scale of about 7 hours.

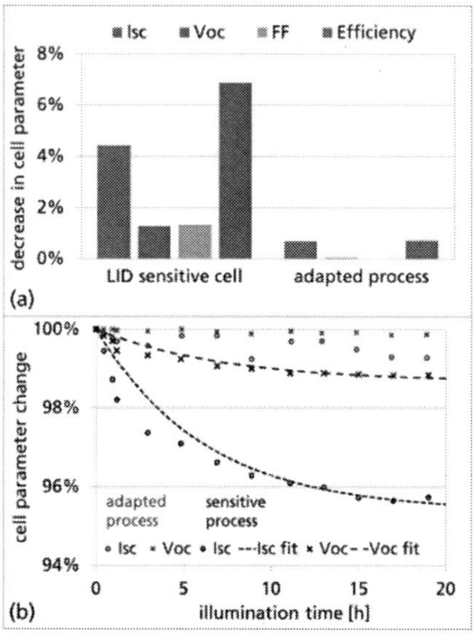

Figure 1: (a) Relative change of I_{sc}, V_{oc}, FF, and efficiency after 20 hours of illumination. Cell A is prone to sponge-LID, cell B from sister wafer is resistant to sponge-LID due to an adapted cell process. (b) Decrease of I_{sc}, V_{oc} plotted against illumination time.

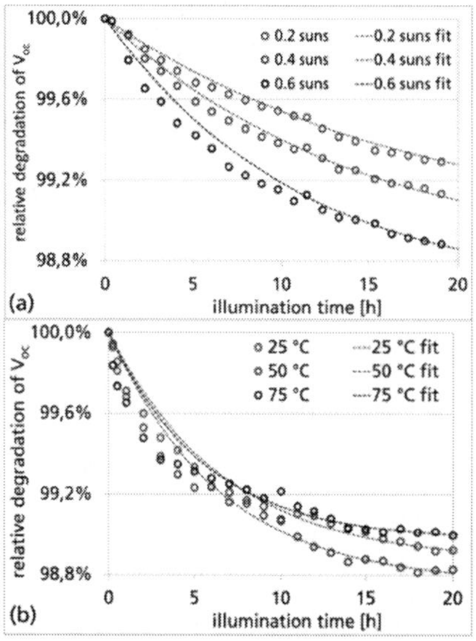

Figure 2: (a) V_{oc} plotted against the time of illumination. Three batch A cells from sister wafers are degraded at fixed cell temperature of 25 °C. Intensity of illumination is varied from 0.2 suns to 0.6 suns. (b) V_{oc} during 20 hours light exposure from batch C cells. All cells were illuminated with 0.6 suns while temperature during degradation was changed.

it can be observed that stronger illumination during the degradation leads to faster decay of the cell parameters. It is found that intensities up to 0.6 suns are not sufficient to reach a saturated degradation state of sponge-LID within 20 hours of light exposure.

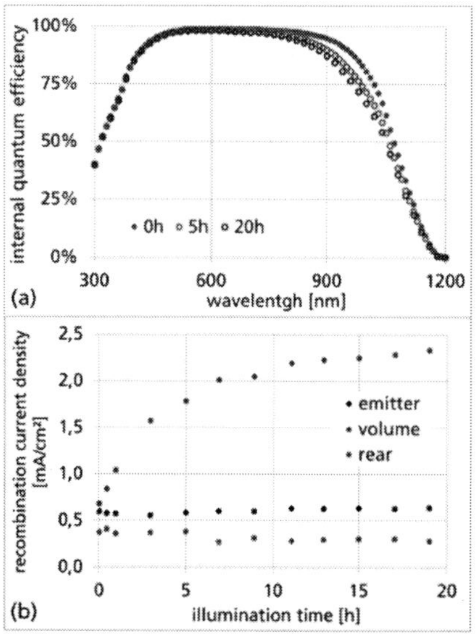

Figure 3: (a) Internal quantum efficiency as a function of wavelength at different times of illumination. (b) Recombination current versus degradation time. Data from batch A cell.

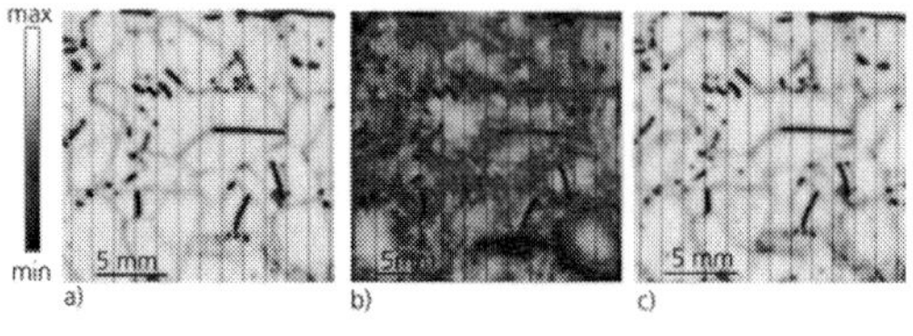

Figure 4: 2 cm x 2 cm LBIC using 960 nm with lateral resolution of 100 µm. All images at same scale. a) Cell A before degradation. b) Cell A after 17 hours light c) Cell B after 17 hours illumination

A comparison of defect activation times $1/k$, see Table 1, highlights a strong dependence on illumination. In contrast to the illumination dependence, sponge-LID is rather insensitive to an increased temperature during degradation, see Fig. 2 (b) and Table 1. A higher temperature leads only to a slightly decreased defect activation time. This is in clear contrast to other known degradation mechanisms as B-O-LID [1], mc-LID [3], and Fe-B-LID [6].

Table 1: Defect activation time $1/k$ and relative sponge-LID losses β in V_{oc}

Intensity [Wm^{-2}]	200	400	600
$1/k$ [h]	17.9	14.3	11.1
β [%]	1.1	1.2	1.4

Cells from batch A. Cell temperature fixed at 25 °C.

Temperature	25 °C	50 °C	75 °C
$1/k$ [h]	6.7	6.4	5.5
β [%]	1.1	1.3	1.0

Cells from batch C. Illumination intensity fixed at 0.6 suns.

Based on the strong decay in the short circuit current, the change in the cells internal quantum efficiency due to sponge-LID was investigated next, see Figure 3. It becomes evident, that sponge-LID is characterized by decay in quantum efficiency for wavelengths above 700 nm. A current loss analysis, based on the evaluation tool LASSIE [7], shows an increase of recombination current in the volume while the emitter and rear recombination remains mostly unchanged.

Spatially resolved light beam induced current measurements have been performed to assess sponge-LID on single grains and grain boundaries as can be seen on Figure 4. While images a) and b) show cell A before and after degradation, image c) is a measurement of cell B from sponge-LID free process, but showing the very same grain structure after illumination. Before degradation, lower values of IQE due to higher recombination are directly located in grain boundaries as well as in decorated grain boundaries, while intra grain regions exhibit a high IQE signal, see Figure 4 a). When exposed to light, sponge-LID causes a decrease of internal quantum efficiency within all grains of the section, see Figure 4 b). While in most cases regions within a 1 mm – 2 mm distance from grain boundaries are affected by sponge-LID, some of the grains completely suffer from the degradation. Sponge-LID is not necessarily related to decorated grain boundaries, as neighbored areas do not show stronger sponge-LID. Examinations of regions nearby grain boundaries show that sponge-LID may appear in a blurred structure or lead to a rather severe degradation. LBIC images of earlier degradation stages (not presented here) show that areas do not change their size which is seen as an indication that sponge-LID is caused by activation of fixed defect centers rather than by diffusion of impurities. The very same grains on cell B, with the adapted process, Figure 4 c), do not show any of the sponge-LID features

4 SUMMARY

We have investigated the light induced degradation effect termed "sponge-LID" which can affect cells made of high performance mc-Si wafers. We found that the degradation leads to a significant current loss while other cell parameters are less affected. This is reflected in a decrease of quantum efficiency for wavelengths above 700 nm and can be associated to an increased recombination in the cell bulk. The degradation strongly depends on the intensity of illumination. However, unlikely B-O-LID, Fe-B-LID, and mc-LID, it is rather insensitive to a variation in cell temperature during degradation. Furthermore, our results indicate that "sponge-LID" is caused by activation of localized defect centers rather than by diffusion of impurities during the degradation phase. The grain boundaries appear as a source of the degradation as regions nearby are mostly affected by the degradation. In addition to our investigation of "sponge-LID" we have furthermore analyzed solar cells produced in an adapted "no-sponge-LID" process. We found that these cells show no significant "sponge-LID".

5 REFERENCES

[1] Bothe, K. and Schmidt, J. "Electronically activated boron-oxygen-related recombination centers in crystalline silicon." Journal of Applied Physics 99.1, pp 013701, (2006).

[2] Geerligs, L. J. and Macdonald, D. "Dynamics of light Induced FeB pair dissociation in crystalline silicon", Applied Physics Letter 85, (2004).

[3] Ramspeck, K. et al. 2012, "Light induced degradation of rear passivated mc-Si cells", Proc. 27th EU PVSEC, Frankfurt, Germany, pp 861.

[4] Petter, K. et al. "Light induced degradation of solar cells made from high performance multi wafers. Proceedings of the 29th EUPVSEC, p 551 – 554, Paris, (2013).

[5] Fahrland, Christian, et al. "Sponge LID-A new degradation mechanism?." Photovoltaic Specialist Conference (PVSC), 2014 IEEE 40th. IEEE, 2014.

[6] Istratov, A. A., Hieslmair, H. and Weber, E. R. "Iron and its complexes in silicon." Applied Physics A 69.1 (1999): pp 13-44.

[7] Fischer B. Loss analysis of crystalline silicon solar cells using photoconductance and quantum efficiency measurements. Cuvillier Verlag, Göttingen, 2003.

[8] Luka, T. et al. 2015, "Investigation of light induced degradation of multi-crystalline PERC cells",Proc. 31st EU PVSEC, Hamburg, Germany, pp. 826–828.

[9] Luka, T. et al. [forthcoming], "Intra-grain versus grain boundary degradation due to illumination and annealing behavior of multi-crystalline solar cells".

PERFORMANCE OF C-SI PHOTOVOLTAIC DEVICES BASED ON OPTICAL MEASUREMENTS AND SPECTRAL IRRADIANCE IN THE ATACAMA DESERT

Pablo Ferrada[1,2], Aitor Marzo[1,2], Enrique Cabera[3], Haifeng Chu[3], Jorge Rabanal[1,3], Andreas Schneider[3]

[1]Centro de Desarrollo Energético Antofagasta (CDEA), Av. Angamos 601, Antofagasta, Chile
[2]Solar Energy Research Center (SERC Chile), Av. Tupper 2007 4th floor, Santiago, Chile
[3]International Solar Energy Research Center, Rudolf Diesel-Str. 15, Konstanz, Germany
Phone: +56 55 2 513530; Email: pablo.ferrada@uantof.cl

ABSTRACT: The potential of laminated solar cells, namely a Standard p-type, PERC, Bifacial and IBC, was calculated considering a Solar Spectrum of Atacama Desert, the Transmittance of glass-encapsulant-glass structures and External Quantum Efficiency. Using these data, the maximum theoretical value of the current density, which can be photo-generated, was determined. Regarding the solar spectrum in Atacama, an average air mass (AM) at noon for this location in Atacama Desert averaged 1.17 and the PV modules tilt angle 20°. When studying the impact of using glass and encapsulants combined with a solar cell under the same solar spectrum, EVA with low UV encapsulant led to the higher current density values, up to 2% higher with the IBC solar cell compared to the other solar cells. The highest current gain, when studying the impact of the two spectra in the 300-1200 nm wavelength range, was 7.4% obtained with a standard 3.2 mm glass, a thermoplastic material as encapsulant for the IBC solar cell. Considering the UV part of the spectrum, the current gain was maximized with a glass with an anti reflection coating combined with the TM encapsulant for the IBC solar cell (25%).
Keywords: c-Si, Solar Cells; PV materials, Encapsulation

1 INTRODUCTION

The Atacama Desert (Chile) exhibits the highest surface irradiation values on Earth, a high mean altitude, a large number of days with clear sky condition and low absorption ozone and water vapor columns. Thus, global horizontal irradiation (GHI) can reach more than 8 kWh/m^2 per day resulting in more than 2500 kWh/m^2 per year [1]. This desert is located along the Pacific coast in South America with a length close to 1000 km between latitudes 20°S and 30°S and a surface of approximately 105.000 km^2. It is characterized to be hyper arid with annual precipitations lower than 50 mm [2]. Mean temperature values are 10°C to 20°C in winter and 20°C to 30°C in summer, with an air temperature below 38°C [3]. With regard to chemical composition, nitrates accounting for 28% of the soil and water-soluble salts such as perchlorates and iodides are found [4].

The sunlight and environmental conditions of the Atacama Desert can strongly differ to those of North America or Europe for which standards of fabrication and certification for PV modules were defined. Usually PV modules are produced and certified according to international norms such as IEC and UL. It is intended that modules fulfill environmental test requirements, safety standards and reach long lifetimes. These standards are referred to data from long-term analysis of existing PV installations mainly located in the US and Europe. Recently, Chile entered to the PV market with large-scale PV plants. The cumulative installed PV capacity reached more than 1 GW in March 2016 with 2 GW of approved PV projects to be added in the following years [5]. Specific investigations on the module materials and modules itself are necessary considering the specific environmental conditions of a location, like Atacama Desert. With this regard, a first step can be the study of the PV potential when using different module materials with selected solar cell concepts. A second step can be the study of the material degradation due to specific conditions of a location.

2 CURRENT INVESTIGATION

2.1 Goal

In order to identify the best module materials for best performing solar cell concepts, calculations of the current generation considering local conditions of the Atacama Desert are required. For this purpose, 4 selected crystalline silicon (c-Si) solar cells were selected. These technologies can be commercially relevant and potentially implemented in PV modules in Atacama. The conditions under which PV modules are exposed must be taken into consideration since Atacama solar spectrum can be different compared to ASTM G173-03 Reference used for standard testing conditions (STC).

Consequently, the main idea of this work is to evaluate the potential with regard to photo current generation of different encapsulated crystalline silicon (c-Si) solar cells considering the conditions of the Atacama Desert in Chile. The c-Si technologies consisted of (1) a Standard p-type mono c-Si, (2) a Passivated Emitter and Rear Contact (PERC), (3) a bifacial (BiSoN) n-type mono c-Si and (4) an Interdigitated Back Contact (IBC) solar cell.

2.2 Approach

The performance potential with regard to the photo current generation of laminated solar cells was calculated based on the following data: (A) The Global Solar Spectrum of the Sunlight in the Atacama Desert and the ASTM G173-03 Reference Spectrum, (B) The transmittance of glass-encapsulant-glass structures, (C) the External Quantum Efficiency (EQE) of solar cells

For (A), the Simple Model of the Atmospheric Radiative Transfer of Sunshine (SMARTS) v. 2.9.2 code was used to calculate the Global Tilted Spectral Irradiance (GTI_λ) in the Atacama Desert. For (B), a set of PV glass types of thicknesses 1.5 mm and 3.2 mm were used in combination with three different encapsulants, namely, Ethylene Vinyl Acetate (EVA), a Thermoplastic Material (TM) and a low UV light cut-off EVA (EVA-U). Thus, the transmittance of glass-encapsulant-glass was measured with a spectrophotometer. For (C), the solar cells of type (1) and (2) were measured indoor at ISC Konstanz. Data of solar cells of type (3) and (4) were simulated from PV lighthouse [6].

Integrating the product of the spectral irradiance, the transmittance, EQE and considering physical constants, it is possible to compute the maximum theoretical current density to be generated by each PV device [7]. The details are described in section 4.

3 EXPERIMENTAL

3.1 Sample preparation

PV glass of 3.2 mm and 1.5 mm thickness and size 5×5 cm^2 were used. Float-glass as standard (Std), float glass with antireflection coating (ARC) and casted glass (Text). Table I indicates the types of glass for this study.

Table I: PV glass types.

	Thickness (mm)	Surface
Glass 1	1.5	Standard
Glass 2	3.2	Standard
Glass 3	3.2	ARC
Glass 4	3.2	Textured

The encapsulants tested were a 450 μm thick EVA (E), a 450 μm thick thermoplastic material (TM) and 400 μm thick EVA with a low light UV cutoff (U). The PV glass and encapsulants were combined to achieve glass-encapsulant-glass structures. While the front glass was either 3.2 mm or 1.5 mm thick, the glass of the rear side was always 1.5 mm thick. The rear side glass was used as support plate for preparation. Samples were laminated at temperatures of 145 ± 5 °C in 16 min. Table II shows the glass-encapsulant combinations for measuring the transmittance and reflectance.

Table II: Glass-encapsulant combinations.

Glass type	Encapsulant		
G1 (1.5Std)	E	TM	U
G2 (3.2Std)	E	TM	U
G3 (3.2ARC)	E	TM	
G4 (3.2Text)	E	TM	U

3.2 Measurement devices

The transmittance (T) as well as the reflectance (R) of glass, encapsulants and glass-encapsulants structures was measured as a function of the wavelength with a Perkin Elmer Spectrophotometer. The Current-Voltage (IV) characteristics were measured for the standard and bifacial solar cells using a Flasher device produced by h.a.l.m. elektronik. The Quantum Efficiency and reflection of metallized solar cells (standard and bifacial) were determined with an IQE Scan of PV Tools GmbH.

4 METHODOLODY

4.1 The photo current

A maximum value of the photo-generated current density (J_{max}) was calculated according to Eq. 1.

$$J_{max,front} = \frac{q}{h \cdot c} \cdot \int_{\lambda_1}^{\lambda_2} \Phi_{in}(\lambda) \cdot T(\lambda) \cdot EQE(\lambda) \cdot \lambda \cdot d\lambda \qquad (2)$$

The constants are q the elementary carrier charge, h the Planck's constant, c the speed of light, $\Phi_{in}(\lambda)$ the spectral irradiance given in W/m^{-2}nm^{-1}, $T(\lambda)$ the transmittance of glass-encapsulant-glass structures and EQE(λ) the external quantum efficiency of a solar cell. In [8], the optical performance of encapsulants is compared in terms of photo current density.

4.2 The global solar spectrum in the Atacama Desert

The photogenerated current density calculation was carried out by using the Global Tilted Spectral Irradiance (GTI$_\lambda$) for a specific location in Atacama (the Solar Platform of the Atacama Desert or PSDA). The PSDA is located at coordinates 24° 05' 14" S. 69° 54' 47" W at 963 m.a.s.l, and belongs to the University of Antofagasta, Chile. Photo current density was calculated also with the reference spectrum under which solar cells and modules are usually rated.

The GTI$_\lambda$ at the PSDA was obtained by using SMARTS v. 2.9.2 code [90,10]. This spectrum was compared with the ASTM G173-03 Reference Spectra also derived from SMARTS v. 2.9.2. [9,11,012,130]. The current standard G173-03 was issued by the North American PV industry and the American Society for Testing and Materials (ASTM) and the Research and Development Laboratories of the US government. The standard spectral distribution of solar irradiance at ground level is calculated according to the North American atmospheric and geographic conditions. An air mass (AM) equal to 1.5 was considered, a receiving surface inclined at 37° and mean values of atmospheric conditions from US.

The Authors want to highlight that this solar spectrum calculation was performed without information about atmospheric parameters, which are not available yet for Atacama. Because of this, atmospheric parameters were not modified from the reference spectrum but only the astronomical inputs. Note that the AM is the main parameter, which affects strongest solar spectrum along the year (day and hour). AM is derived from equations related to time and location.

5 RESULTS

5.1 The solar spectrum

An AM mean value was calculated at noon along the year for the PSDA. The obtained result was 1.17. A tilted surface of 20° was also considered, which corresponds to the tilt angle of installation for PV modules at this place.

In the absence of measurement data for the atmospheric parameters to calculate the solar spectrum, this input data was set equal to the values of the ASTM G173-03.

5.2 T and R of glass and encapsulant samples

5.2.1 PV glass types

A comparison of the wavelength range from 400 nm to 1100 nm between G1, G2 and G3 resulted in an average and standard deviation (StdDev) of 91.0% ± 0.2%, 91.0% ± 0.3% and 94.4% ± 0.8% for the transmittance, respectively. Likewise, same calculation for the reflectance resulted in 8.5% ± 0.4%, 8.3% ± 0.4%

and 5.1% ± 0.8%, respectively. The ARC led to an increase of 4% in transmittance and a reduction of 3% in reflection. A measurement for G4 is not available.

5.2.2 Encapsulants

The transmittance of E, TM and U encapsulants showed averaged values of 88.9% ± 0.6%, 87.3% ± 0.6% and 88.6% ± 0.4%, respectively, in the 400-1100 nm spectral range. Finally, the reflectance in the same spectral range was 9.9% ± 1.0%, 12.1% ± 1.5% and 10.1% ± 0.9%. Each encapsulant exhibits a different wavelength from which light is transmitted. This cutoff is 360 nm for E, 305 nm for TM and 330 nm for U encapsulant.

5.2.3 Glass-encapsulant-glass structures

The calculation of J_{max} was performed using the transmittance of glass-encapsulant-glass structures. The J_{max} values obtained are an approximation since light will travel through a glass-encapsulant layer before reaching the solar cell. This approximation is valid within the framework of this work due to the high transmittance of the used materials. The transmittance of glass-encapsulant structures is predictable to be between the transmittance of glass and the transmittance glass-encapsulant-glass structures. According to the sample configuration, taking the worst case into account, which is a 7.3 % difference in transmittance between those of the G3-TM-G1 structure and G3, the maximum absolute error is 3.6 %. This error is close to (E_T) the measurement standard deviation 1.8%, calculated according to Eq. 2:

$$E_T = \sqrt{StdDev(T_{G3\text{-}TM\text{-}G1})^2 + StdDev(T_{G3})^2 + StdDev(T_{TM})^2} \quad (2)$$

Table III summarizes the mean and standard deviation values of the transmittance of glass-encapsulant-glass structures for the 400-1100 nm spectral range. While the combination G3-E-G1 achieved the highest transmittance due to the ARC, the combination G2-TM-G1 led to the lowest transmittance. In order to differentiate how materials respond to short wavelengths, the spectral transmittance and reflectance curves are required (see Fig. 1 and Fig. 2).

Table III: Mean and standard deviation values of the transmittance of glass-encapsulant-glass structures.

Structure	Mean (%)	Std. Dev. (%)
G1-E-G1	88.0	1.1
G1-TM-G1	86.8	0.5
G1-U-G1	88.2	0.9
G2-E-G1	87.7	1.3
G2-TM-G1	84.5	0.8
G2-U-G1	87.7	0.9
G3-E-G1	90.7	1.8
G3-TM-G1	87.1	1.5
G4-E-G1	87.2	1.5
G4-TM-G1	84.9	0.8
G4-E-G1	87.3	1.1

In the case of the reflectance, a similar calculation can be given. The worst case is found when R of G3-TM-G1 is compared to that of E resulting in 6.4%. Thus, the maximum absolute error for R is 3.2% contrasting with the instrumental error E_T. Using Eq. 2 for R, E_T=1.9%.

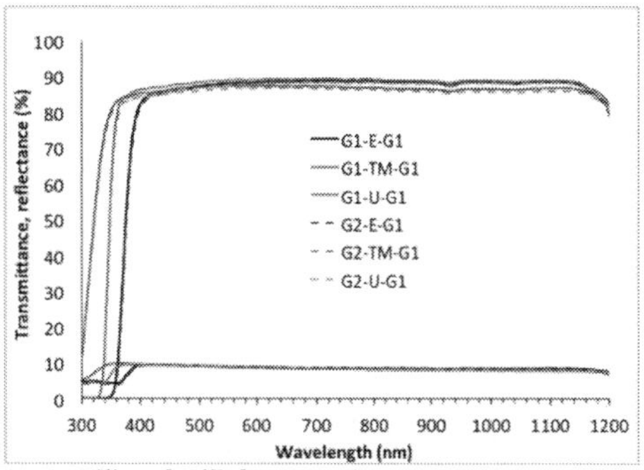

Fig. 1: $T(\lambda)$ and $R(\lambda)$ for G1 and G2 structures.

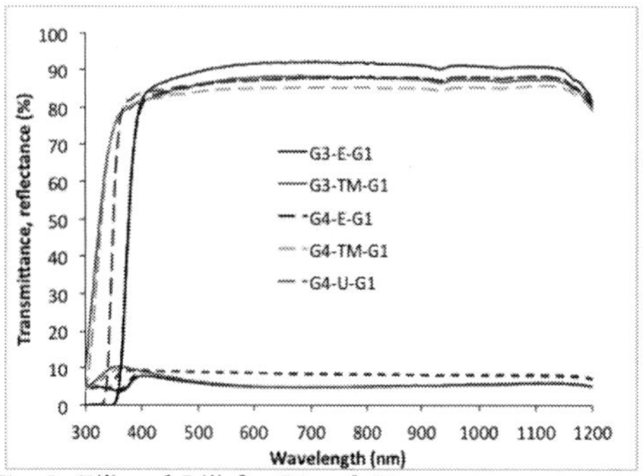

Fig. 2: $T(\lambda)$ and $R(\lambda)$ for G3 and G4 structures.

5.3 Short circuit current density (J_{sc})

J_{sc} values for each solar cell are shown in Table IV, where Std stands for standard p-type, PERC for passivated emitter and rear contact, Bif-f for bifacial front side, and IBC for the interdigitated back contact solar cell. The PERC, Bifacial and IBC cells are compared with the Std solar cell, defined as reference.

Table IV: J_{sc} values from IV measurement.

Sample	J_{sc} (mA/cm^2)
Std	36.9
PERC	39.7
Bif-front (f)	39.2
IBC	42.0

It is expected to achieve the same correspondence after calculating J_{max}, since the total EQE, including the non-shaded and shaded regions of the cell were considered. Thus, J_{max} from the integral of Eq. 1 can be compared with J_{sc} from IV measurements.

5.4 Calculated J_{max}

5.4.1 Verification of Eq. 1

First, the integrated values of J_{max} using Eq. 1 were obtained without considering the glass-encapsulant structures, i.e. $T(\lambda)$=1. In that way an error with respect to the J_{sc} values can be calculated. The comparison between J_{max} and J_{sc} is possible since the total EQE was measured as mentioned above. Table V shows the integrated J_{max} values and the deviation defined as $(J_{max}-J_{sc})/J_{sc}$. This value keeps below 1% validating Eq. 1.

Table V: J_{max} without glass/encapsulant structures.

Sample	J_{max} (mA/cm^2)	$(J_{max}-J_{sc})/J_{sc}$ $/J_{sc}$ (%)
Std	36.6	0.8
PERC	39.4	0.7
Bif-f	39.9	0.9
IBC	42.9	0.3

5.4.2 J_{max} for laminated solar cells

The final calculation was performed taking the transmittance of the glass-encapsulant-glass structures into account. The integral of Eq. 1 was calculated for the wavelength range 300 to 1200 nm with the solar spectrum AM1.17 of the Atacama location (red columns) and the ASTM AM1.5 (blue columns).

While Fig. 3 summarizes the J_{max} values obtained for the standard and PERC solar cells, Fig. 4 shows J_{max} values for the bifacial and IBC solar cells. These results are further discussed within section 6.

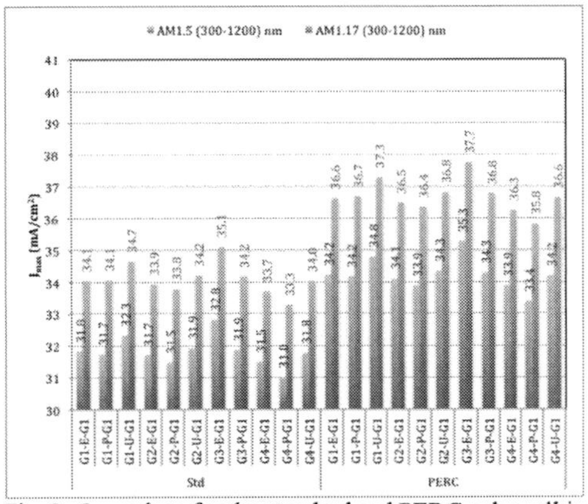

Fig. 3. J_{max} values for the standard and PERC solar cell in the 300-1200 nm spectral range.

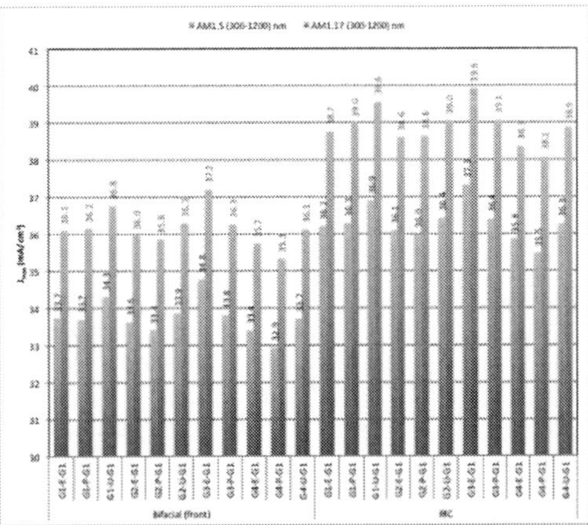

Fig. 4. J_{max} values for the bifacial (front side) and the PERC solar cell in the 300-1200 spectral range.

6 DISCUSSION

6.1 Glass type

First observations regarding the glass indicate that G1, G2, G3 and G4 transmitt light for wavelengths from 300 nm on and therefore, based on the UV light cutoff, glass type is expected to have a less impact than encapsulants on J_{max} values.

6.1.1 Thickness

G1 and G2 differentiate each other only by thickness. However, it was found that $T_{G1} \approx T_{G2}$ (section 5.2.1) and the transmittance of the combination G1-encapsulant-G1 is higher than those of G2-encapsulant-G1. This result led to higher J_{max} values obtained with G1-encapsulant-G1 instead of with G2-encapsulant-G1 (each comparison considers the same encapsulant E, TM or U). Note that the standard deviation in Table III suggests that T of the glass-encapsulant-glass structures for G1 and G2 as front glass exhibits no significant difference. In this framework, higher J_{max} were determined when using G1 instead of G2 as front glass, holding for all solar cell types. A quantification of the relative differences $(J_{maxG1}-J_{maxG2})/J_{maxG1}$ led to 0.4% for E, 0.8% for TM and 1.3% for U encapsulants. These differences were equal for each solar cell concept, both under AM1.5 and AM1.17. It follows that the thickness of the glass had only a small impact on the photo generated current.

6.1.2 Surface

G3 exhibits an ARC coating and G4 is a textured glass. Results obtained with G3 and G4 can be compared with those with G2 as it has the same thickness and standard surface. Due to the ARC, J_{max} with G3 as front glass reached the highest values compared to all other cases. Indeed, calculating a percentage difference as $(J_{maxG3}-J_x)/J_{maxG3}$ where x can be G1, G2 and G4 as front glass, following results are obtained (each comparison considering the same encapsulant). First, with respect to G1, 3% for E and 0.4% for TM was obtained. Second, with respect to G2, which can be considered a reference for G3 since it has the same thickness, 3.4% for E and 1.2% for TM was calculated. Third, with respect to G3, 4% for E and 2.6% was determined. These results mean that ARC produced a gain in J_{max} of at least 3% compared to its reference (G2). Solar spectra produced same J_{max} gain.

6.2 Encapsulant

When comparing for each solar cell the J_{max} produced by using one encapsulant with respect to a J_{max} obtained with a reference encapsulant (encapsulant E) keeping the same front glass and the same solar spectrum, a percentage difference was defined as the following $(J_{max,G1\text{-}ENC\text{-}G1}-J_{max,G1\text{-}ENC\text{-}G1})/J_{max,G1\text{-}ENC\text{-}G1}$. For instance, if G1 is considered and TM is compared with E, calculation is given by $(J_{max,G1\text{-}TM\text{-}G1}-J_{max,G1\text{-}E\text{-}G1})/J_{max,G1\text{-}E\text{-}G1}$. Likewise, if G2 is considered and U is compared with E, the calculation is $(J_{max,G2\text{-}U\text{-}G1}-J_{max,G2\text{-}E\text{-}G1})/J_{max,G2\text{-}E\text{-}G1}$.

6.2.1 Standard solar cell

The TM encapsulant led to the worst J_{max} compared to that obtained with E. Thus, the J_{max} loss under AM1.5 was 2.9% when using G3 as front glass, 1.5% with G4, 0.7% with G2 and 0.2% loss with G1. Note that the transmittance of the TM encapsulant for the wavelength range 400-1100 nm was the lowest producing the highest loss when combined with G3 (glass with ARC).

On the contrary, the higher T of U in the wavelength range 400-1100 nm correlates with the J_{max} gain, which was 1.6% when using G1 (T values in subsection 5.2.1), 0.9% when using G4 and 0.6% when using G2 as front glass under AM1.5. Note that there is no U encapsulant combined with G3 as front glass (Table II).

6.2.2 PERC solar cell

As for the standard solar cell under AM1.5, TM and U encapsulants led to the similar behavior for the PERC solar cell. Using G1, the TM and U encapsulants exactly led to the same differences obtained with the standard solar cell. However, for G2, G3 and G4 it is observed that the loss with TM was reduced by 0.1% and the gain with U is increased by 0.1%.

6.2.3 Bifacial and IBC solar cells

Comparing the bifacial to the standard solar cell under AM1.5, a reduction of the J_{max} loss with TM of 0.1% and a gain with U of 0.1% was obtained for all front glass types.

The IBC solar cell under AM1.5 together with the TM encapsulant produced a J_{max} gain within G1 of 0.2% (instead of the 0.2 % loss with the standard solar cell) and reduced the loss within G2, G3 and G4 by 0.3 %, 2.5% and 1.1%. The use of U encapsulant led to an increased J_{max} gain with respect to the values obtained with E of 1.9 % for G1, 0.9 % for G2, and 1.1% for G4 all as front glass (contrasting with the lower gain values of 1.6%, 0.6% and 0.9% for the standard solar cell, respectively for same glass types).

6.3 The solar spectra and quantum efficiency

The effect of solar spectra and quantum efficiency of the solar cells can explain the behavior of J_{max} values. First of all, it is pointed out that ASTM AM1.5 and the obtained AM1.17 for the Atacama location can exhibit differences. In fact, calculating the fraction of photons $f(\lambda)$ with wavelength below λ, both spectra differentiate more from each other as λ increases. Values of f at 350 nm are 1.7% and 1.8% for AM1.5 and AM1.17 respectively. At 400 nm, 800 and 1100 nm, f values are 5.7%, 64% and 88.3% under AM1.5, and 6.2%, 70.3% and 96.2% under AM1.17. In [14] the influence of solar spectra on optical performance of encapsulants is reported.

With regard to the quantum efficiency, the average values for the 300-400 nm, 400-800 nm and 800-1200 wavelength ranges are highly dependent on the solar cell concept. The IBC solar cell is the most sensitive to UV with an EQE of 87% followed by that of the bifacial cell with an EQE of 67%, the PERC with an EQE of 66% and the standard solar cell with an EQE of 58%. These differences explain the possibility or non-possibility to take advantage of using an encapsulant with a low UV cutoff such as TM or U. The overall EQE in the 300-1200 nm range resulted in average values of 87% for IBC, 80% for the PERC, 79% for the bifacial and 72% for the standard solar cell. The same order is found for the VIS and IR spectral ranges.

As a result, the gain values given in subsection 6.2 are in correspondence with the differences in the solar spectra and solar cell technology. The J_{max} loss is reduced and J_{max} gain is increased as more sensitiveness is incorporated in the UV range together with a better overall efficiency. The change of the solar spectrum from AM1.5 to AM1.17 means that more UV photons will be available. This light can be conveniently used when more UV light is transmitted by the glass-encapsulant structure together with a higher EQE. This effect explains the tendencies in the loss/gain analysis of subsection 6.2.

6.4 The encapsulants in the UV

Considering the Eq. 1 with integral limits of 300 nm and 400 nm, J_{max} values obtained for G1, G2, G3 and G4 combined with the TM encapsulant were between 2 up to 2.1 times higher than those reached with the E encapsulant. Likewise, J_{max} values with glass types 1 to 4 combined with the U encapsulant were 1.7 times higher with respect to those with E encapsulant. These values are valid for the standard, PERC and bifacial solar cells. For the IBC solar cell, glass types 1 to 4 combined with the TM encapsulant produced a 2.1 to 2.2 times higher J_{max} compared to those with E encapsulant. Similarly, the U encapsulant led to a 1.8 times higher J_{max} compared to those with the E encapsulant. These calculations resulted to be the same under AM1.5 and AM1.17.

7 CONCLUSIONS

The potential with regard to photo current generation (J_{max}) under an average air mass at noon (AM1.17) of a specific location in Atacama Desert and the ASTM G173-03 AM1.5 reference spectrum was calculated for laminated crystalline silicon solar cell concepts. The photo current was calculated in order to determine how PV glass of 1.5 mm and 3.2 mm thickness with standard, antireflection coating and textured surfaces combined with EVA (E), a Thermoplastic Material (TM) and an EVA with a low UV cutoff (U) perform.

On one hand, the glass types having same flat surface but differentiating each other by their thickness showed not to have an impact on final J_{max} values. On the other hand, glass with ARC led to higher J_{max} values compared to those with textured and flat surface glass.

Encapsulants E, TM and U exhibited a respective UV light cutoff at 360 nm, 305 nm and 330 nm. The TM encapsulant, however, averaged a lower transmittance and higher reflectance in the 400-1100 nm spectral range.

When glass and encapsulants were combined, following generalizations can be concluded:

A. Comparing for each solar cell the J_{max} obtained with one encapsulant with respect to a J_{max} value obtained with another encapsulant for same glass and same solar spectrum.

B. Comparing for each solar cell the J_{max} obtained with AM1.17 with respect to a J_{max} value obtained with AM1.5 for the same glass & encapsulant

For (A) the main findings can be summarized as follows:

- Encapsulant U produced higher J_{max} (with respect to those J_{max} values obtained with E and TM) since U combines high light transmission (88.6%) and a UV cutoff in between of those E and TM (305 nm and 360 nm, respectively).
- The J_{max} loss when comparing encapsulants for the same glass (and same AM) is reduced, as well as the J_{max} gain is increased as more sensitiveness is incorporated in the UV together with a better overall quantum efficiency.
- When considering only the UV spectral range, i.e. J_{max} integrated over 300-400 nm, the TM and U encapsulants produced up to a 2.2 and 1.8 times higher J_{max} compared to that with E encapsulant under AM1.17.

For (B), main conclusions are the following:

- The impact of the two solar spectra on J_{max} was 7.4% gain for a 3.2 mm standard glass combined with the TM encapsulant in the IBC solar cell.
- The more ability of transmitting more UV photons and using them is clearly seen when observing J_{max} values integrated between 300 nm and 400 nm under AM1.5, and stronger pronounced for AM1.17.
- If only UV is considered, the J_{max} gain reached 25% with ARC glass combined with TM encapsulant for the IBC solar cell.

APPENDIX A

The fraction of photons $f(\lambda)$ with wavelength below λ of section 6 was obtained according to Eq. A1.

$$f(\lambda)=\frac{\int_{300}^{\lambda}\Phi_{in}(\lambda')d\lambda'}{\int_{300}^{1200}\Phi_{in}(\lambda')d\lambda'} \qquad \text{A.1}$$

where $\Phi_{in}(\lambda)$ is the spectral irradiance in $W/(m^2nm)$.

ACKNOWLEDGMENTS

We acknowledge Christopth Peter and Jayaprasad Arumughan for the measurement of the quantum efficiency on the bifacial and standard cells, respectively.

The authors acknowledge the support provided by the FONDECYT Project Number 3160190, as well as CONICYT/FONDAP/15110019 "Solar Energy Research Center" SERC-Chile, and the BMBF Solar Collaboration between Chile and Deutschland (Solar ChilD) project number 01DN14005.

REFERENCES

[1] R. Escobar et al. Renewable Energy 71 (2014), 324-332.
[2] T. Larraín, R. Escobar R. Renewable Energy 41 (2014), 123–133.
[3] C. P. McKay et al. Astrobiology 3 (2003), 393–406.
[4] R. Navarro-González. Science. 302:5647 (2003), 1018–1021.
[5] CIFES Report. Renewable Energy in the Chilean Electricity Market. March 2016.
[6] A. Fell et al. IEEE Journal of Photovoltaics 5:4(2015), 1250-1263.
[7] B. Fischer. PhD Thesis. University of Konstanz. 2003.
[8] K. R. McIntosh et al. Proc. Photovoltaic Specialists Conference (PVSC), 2009 34th IEEE, 544-549.
[9] C. A. Gueymard. Solar Energy 71 (2001), 325-46.
[10] C. A. Gueymard. SMARTS. A Simple Model of the Atmospheric Radiative Transfer of Sunshine: Algorithms and Performance Assessment. SMARTS 1995.
[11] ASTM International. ASTM G173-03 (2012). Standard Tables for Reference Solar Spectral Irradiances: Direct Normal and Hemispherical. 2012.
[12] C. A. Gueymard et al. Solar Energy 73 (2002), 443-467.
[13] C. A. Gueymard. Solar Energy 76 (2004), 423-453.
[14] K. R. McIntosh et al. Proc. Photovoltaic Specialists Conference (PVSC), 2010 35th IEEE, 269-274.

Rapid calculation of series and shunt resistance values for a crystalline silicon solar cell

F. Ghani[1,2], T. S. O'Donovan[1], and M. Zaglio[2]
[1]Engineering & Physical Sciences, Heriot-Watt University, Edinburgh, United Kingdom
f.ghani@hw.ac.uk
[2]Sunamp Ltd, Macmerry, East Lothian, United Kingdom

ABSTRACT: The series and shunt resistance values play an influential role in the output of a photovoltaic device. Additionally, they may also be useful parameters to assess the cell or module quality. Calculation of these values however is non-trivial, consequently a number of experimental and non-experimental techniques currently exist in order to estimate their values. In previous work [1], a method was proposed to calculate the values of series and shunt resistances at maximum power primarily using the experimental data typically published by the manufacturer. This numerical method based on the multi-dimensional Newton-Raphson method and current voltage equations expressed using the Lambert *W*-function is however computationally burdensome and therefore unsuitable for implementation in an environment where speed is needed such as manufacturing or field testing. Here we present a rapid method of calculation based on a feed forward type multi-layer perceptron artificial neural network (ANN). Current-voltage data were experimentally acquired for a single multi-crystalline silicon cell under varying levels of outdoor illumination from which the short circuit current, open circuit voltage, and location of the maximum power point values were obtained as typically published by the manufacturer, forming the training set for the ANN. The corresponding values of series and shunt resistances were then calculated using the method of Ghani and Duke [1] to form the network target set. It was found in this study that the network could be successfully trained for this application providing a method to rapidly calculate these loss values. It may therefore be a beneficial quality assessment tool for photovoltaic device manufacturers, installers, and users.

Keywords: Series resistance, shunt resistance, characterisation, photovoltaic, artificial neural network, single-diode model

1 INTRODUCTION

The series and shunt resistance values, henceforth referred to as R_s and R_{sh} respectively, are two lumped parameter values unique to the device (solar cell, module, etc.) under study. These values play an important part in modelling photovoltaic devices when using characterisation models such as the single and double diode types. The single-diode model is extensively used in the literature based on its ability to strike a good balance between modelling accuracy and complexity [2]. Within this model type are several variants such as the four and five parameter model. Research work based on experimental results has previously revealed that the five parameter model was overall able to reflect the actual behavior of a crystalline silicon solar cell [3]. This equivalent circuit diagram for this model is shown in Figure 1 below.

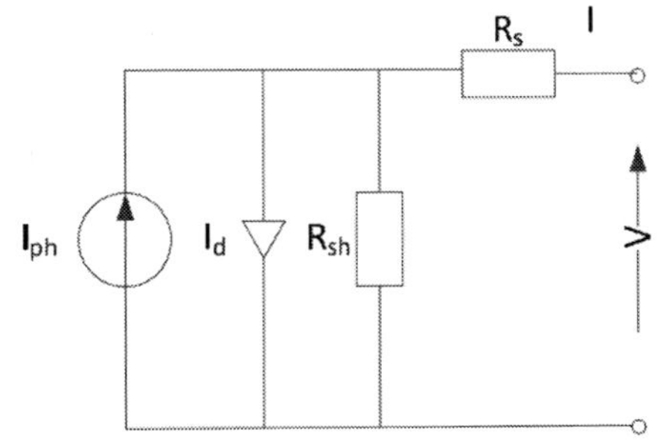

Figure 1: Equivalent circuit diagram for the single-diode, five parameter model.

Applying Kirchhoff's laws to Figure 1, we can obtain current (I), as an implicit non-linear function of current and voltage (V). This equation is shown by Eq. (1).

$$I = I_{ph} - \frac{V + IR_s}{R_{sh}} - I_0 \left[e^{\left(\frac{V+IR_s}{nV_{th}}\right)} - 1 \right] \quad (1)$$

Due to the implicit, non-linear nature of Eq, (1) it has recently become popular to express current as an explicit function of voltage using the Lambert *W*-function [4]. This new equation is shown by Eq. (2) [5].

$$i = \frac{V}{R_s + R_{sh}} - \frac{LambertW\left(\frac{R_{sh}(R_s I_{ph} + R_s I_0 + V)}{nV_{th}(R_s + R_{sh})}\right) nV_{th}}{R_s} + \frac{R_{sh}(I_0 + I_{ph})}{R_s + R_{sh}} \quad (2)$$

By being able to express current as an explicit function of voltage, improving computational efficiency, a number of studies have been conducted examining the effects of mismatch losses due to non-homogenous shading [6, 7] and temperature [8, 9].

Regardless of which equation is used, Eq. (1) or Eq. (2), it is required to calculate the values of five lumped parameter values or characterisation parameters; R_s, R_{sh}, I_{ph}, I_0, and n. These values are unique to the device under study and therefore to accurately model the device, they must be individually sought as their values are typically not published by photovoltaic manufacturers. It is therefore up to the modeler to calculate these values, however, their calculation is by no means a trivial task. This is indicated by the large number of studies devoted to the topic, which can be classified into three main groups; analytical [10], numerical [11], and advanced methods utilizing soft computing techniques.

In previous work by Ghani and Duke [1], a hybrid analytical/numerical approach was employed where the values for Iph, I0, and n where determined via analytical methods, while the values for Rs and Rsh which have a strong impact on modelling accuracy where determined via a more computationally demanding numerical approach. For conciseness the details of this approach are not provided in this paper, but it is useful to note that the objective of this method is to calculate the values of the series and shunt resistances by attempting match the output of Eq. (2) to the maximum power point value published by the manufacturer of the photovoltaic device using the Newton-Raphson method. This method was later experimentally validated, with one of the results from this study shown in Figure 2. Inspecting this figure, we note that the experimental and simulated data sets are closely matched at maximum power point, the point at which a photovoltaic system is targeted to operate at.

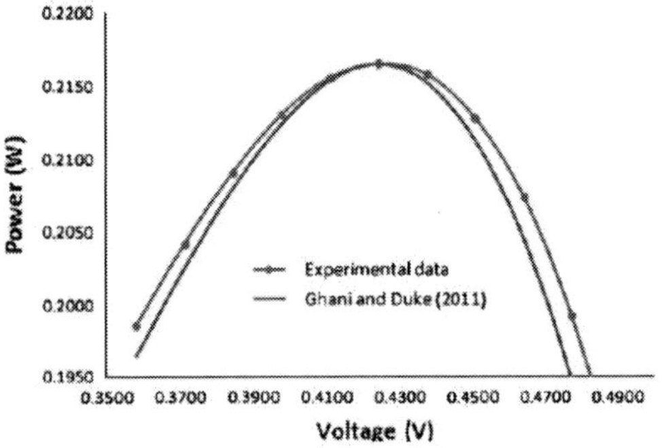

Figure 2: Comparing simulated and experimental data sets when modelling with values for R_s and R_{sh} using the method of Ghani and Duke [1, 12].

An issue with the method presented in [1] is the computational burden due to the iterative nature of the numerical approach. Estimates for values must first be specified to initiate the algorithm, these values must be substituted into the partial differential equations which make the Jacobian matrix, and finally matrix operations must be carried out to determine the updated vectors of values (i.e. R_s and R_{sh}). For research applications, the issue of computational burden is less of an issue, but if fast calculation is required, to suit the manufacturing environment for example this is a problem. To overcome this issue, it is investigated in this study to implement an artificial neural network (ANN).

Artificial neural networks are computational models which attempt to mimic the function of a biological network. They have been used for a number of applications such as control, data compression, foecasting, optimsation, pattern recognition, classification, speech, and vision [13]. In recent times there has been strong interest in tis application specifically in the discipline of renewable energy systems [14-18].

A number of neural network architectures exist and have been investigated. The feed-forward, multi-layer perceptron (MLP) is common used, particularly in the field of solar energy [19, 20]. The architecture of this type of model typically consists of three layers; an input layer, a hidden layer, and an output layer. As the name suggests, inputs to the network are fed to the network via the input layer were they are then processed by the hidden layer and finally the outputs of the network are presented at the output layer.

In addition to the architecture of the neural network, it is the value of the neural network weights that is essential to the operation of the ANN. These weights are continually adjusted during the training phase of the network which are then held constant was this phase has been completed. Before the training phase is initiated, the weights and bias values are randomly selected values. The training cycle is carried out by providing the input and output values to the network and allowing the neural network to adjust the weight values such that the output from the neural network matches the output from the training data.

The proceeding section will discuss the method applied to train a neural network with the objective to calculate the values of R_s and R_{sh} using only the data provided by the manufacturer (i.e. I_{sc}, V_{oc}, I_{mp}, and V_{mp}).

2 METHOD

In this section, the method applied to train an ANN such that when presented with the values of I_{sc}, V_{oc}, I_{mp}, and V_{mp}, values typically provided by the manufacturer, the neural network will output the values for the series and shunt resistances. The architecture investigated is shown in Figure 3 below.

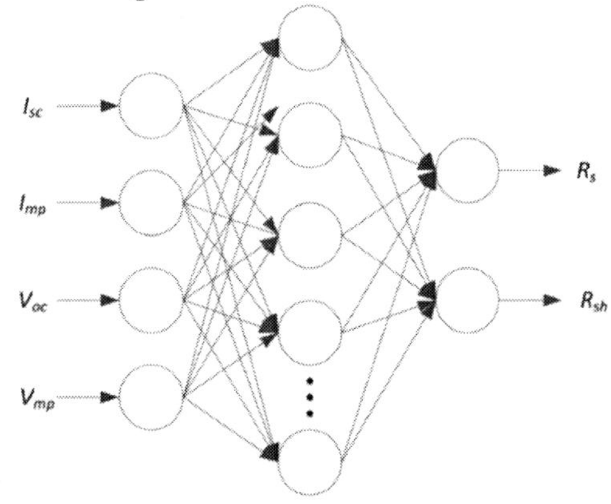

Figure 3: Architecture of the feed forward MLP neural network used in this study.

As already discussed in the introduction section, a critical step is the training phase where the hidden layer weights are adjusted based on the training data presented to the network. To carry out this step, training data must first be obtained. This was achieved by carrying out the following steps:

1. Collecting current-voltage data from a mc-Si solar cell experimentally using a source measuring unit (SMU).
2. Calculating the values of Rs and Rsh using the method outlined in [1].
3. Compiling an input training matrix consisting of the values for Isc, Voc, Imp, Vmp.
4. Compiling an output training matrix of Rs and Rsh calculated from step 2.

5. Training the neural network using the Levenberg-Marquardt back-propagation learning algorithm.

The current-voltage data was collected from a small mc-Si type solar cell manufactured by Silicon Solar (USA). A four-wire approach was used similar to the method applied by the authors in previous studies [9, 21, 22] in order to eliminate resistance in measuring probes. Schematic of the approach applied is shown in Figure 4. A precise source measuring unit manufactured by National Instruments (model: NI PXI-4130) was used within a PXI-1033 chassis. The hardware was software controlled by a program developed using LabVIEW software.

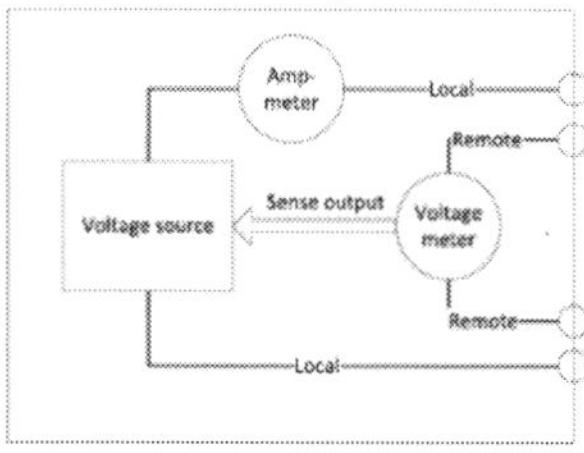

Figure 4: Schematic of the four-wire approach used to measure the current-voltage data from the solar cell.

The solar cell was illuminated in natural sunlight under varying levels of irradiance and temperature. A program was compiled to periodically measure the I-V data from the cell over an extended period of time such that 500 data sets were obtained for the cell.

Once the current-voltage data had been acquired, the values for I_{sc}, V_{oc}, I_{mp}, and V_{mp} were extracted and then used to calculate the values of R_s and R_{sh} using a Matlab program based on [1]. This data was used as the training data for the neural network. Training of the neural network was carried out using the Matlab Neural Network Toolbox where the training data was portioned so that 80% of the data collected was used for training, 20% was used for validation, and the final 20% was used for testing. The results from the neural network training are presented in the proceeding section.

3 RESULTS & DISCUSSION

Here the results from the neural network training and testing phases are presented. Inspecting Figure 5 where the results from the training, testing, and validation steps are presented in the format of regression plots, we can see that a high R-value was achieved for all steps carried out. This indicates that the neural network was able to successfully adjust weights such that a relationship between the input and output vectors was successfully determined.

Once the computationally intensive training phase has been carried out, and the hidden layer weights/bias values have been finalized and fixed in value, the calculation time of the neural network is only a fraction of a second. This is a significant improvement over the computational time required to numerically solve the I-V data for R_s and R_{sh} as originally conducted.

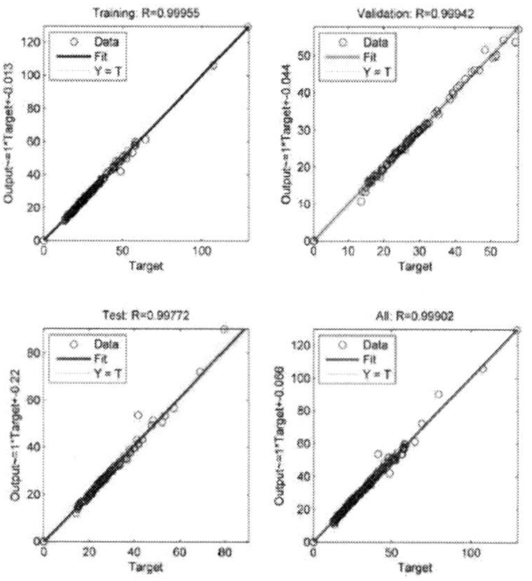

Figure 5: Results from the neural network training. High R-values indicate the neural network was able to successfully 'learn' the relationship between the input and output vectors.

4 CONCLUSION

In this study, a new approach is presented to determine the values of the series and shunt resistances using an artificial neural network. Current-voltage data was experimentally acquired for a mc-Si type solar cell using a precise source measuring unit. This data was then used to calculate the values of Rs and Rsh offline using the numerical approach outlined in previous work by the authors. Using the I-V data and values for Rs and Rsh as the training data, a neural network based on the feed forward multi-layer perceptron architecture was trained using the back-propagation learning algorithm. Results indicated via a regression analysis that the ANN was able to successfully map the relationship between the input and output vectors making it suitable for rapid calculation of these values.

A limitation of this work is the neural network training phase which requires first the data for training and the computational burden in its execution.

5 REFERENCES

[1] F. Ghani, M. Duke, Numerical determination of parasitic resistances of a solar cell using the Lambert W-function, Solar Energy 85(9) (2011) 2386-2394.
[2] M.G. Villalva, J.R. Gazoli, E.R. Filho, Comprehensive Approach to Modeling and Simulation of Photovoltaic Arrays, IEEE Transactions on Power Electronics 24(5) (2009) 1198-1208.
[3] A.N. Celik, N. Acikgoz, Modelling and experimental verification of the operating current of mono-crystalline photovoltaic modules using four-and five-parameter models, Applied Energy 84 (2007) 1-15.
[4] J.H. Lambert, Observationes variae in mathesin puram, Acta Helvetica, physico-mathematico-anatomico-botanico-medica 3 (1758) 128-168.

[5] A. Jain, A. Kapoor, Exact analytical solutions of the parameters of real solar cells using Lambert W-function, Solar Energy & Materials 81 (2004) 269-277.
[6] D. Picault, B. Raison, S. Bacha, J. de la Casa, J. Aguilera, Forecasting photovoltaic array power production subject to mismatch losses, Solar Energy 84 (2010) 1301-1309.
[7] G. Petrone, G. Spagnuolo, M. Vitelli, Analytical model of mismatched photovoltaic fields by means of Lambert W-function, Solar Energy Materials and Solar Cells 91 (2007) 1652-1657.
[8] F. Ghani, M. Duke, J.K. Carson, Effect of flow distribution on the photovoltaic performance of a building integrated photovolatic/thermal (BIPV/T) collector, Solar Energy 86(11) (2012) 1518-1530.
[9] F. Ghani, G. Rosengarten, M. Duke, J.K. Carson, On the influence of temperature on crystalline silicon solar cell characterisation parameters, Solar Energy 112(0) (2015) 437-445.
[10] J.C.H. Phang, D.S. Chan, J.R. Phillips, Accurate analytical method for the extraction of solar cell model parameters, Electron letters 20(10) (1984) 406-408.
[11] J.P. Charles, M. Abdelkrim, M.Y. H., P. Mialhe, A practical method of analysis of the current-voltage characteristics of solar cells, Solar Cells 4 (1981) 169-178.
[12] F. Ghani, M. Duke, Numerical determination of series and shunt resistances using the Lambert W-function, Conference Record of the IEEE Photovoltaic Specialists Conference, 2011, pp. 002861-002865.
[13] A. Sözen, T. Menlik, S. Ünvar, Determination of efficiency of flat-plate solar collectors using neural network approach, Expert Systems with Applications 35 (2008) 1533-1539.
[14] S.A. Kalogirou, Applications of artificial neural networks in energy systems A review, Energy Conversion and Management 40 (1999) 1073-1087.
[15] S.A. Kalogirou, Artificial neural networks in renewable energy systems applications: a review, Renewable and Sustainable Energy Reviews 5(4) (2001) 373-401.
[16] A. Mellit, S.A. Kalogirou, Artificial intelligence techniques for photovoltaics: A review, Progress in Energy and Combustion Science 34 (2008) 574-632.
[17] A.N. Celik, Artificial neural network modelling and experimental verification of the operating current of mono-crystalline photovoltaic modules, Solar Energy 85(10) (2011) 2507-2517.
[18] E. Karatepe, M. Boztepe, M. Colak, Neural network based solar cell model, Energy Conversion and Management 47(9–10) (2006) 1159-1178.
[19] I. Farkas, P. Geczy-Vig, Neural network modelling of flat-plate collectors, Computers and Electronics in Agriculture 40 (2003) 87-102.
[20] M.S.S. Ashhab, Optimization and modeling of a photovoltaic solar integrated system by neural networks, Energy Conversion and Management 49 (2008) 3349-3355.
[21] F. Ghani, M. Duke, J. Carson, Numerical calculation of series and shunt resistance of a photovoltaic cell using the Lambert W-function: Experimental evaluation, Solar Energy 87(0) (2013) 246-253.
[22] F. Ghani, G. Rosengarten, M. Duke, J. Carson, The numerical calculation of single-diode solar-cell modelling parameters, Renewable Energy 72 (2014) 105-112.

Influence of Thermal Dry Oxidation Process on the Silicon Solar Cell Emitter Profiling and Performance

A. Habib[1,2], V. Fano[1], M. A. Rasool[1], A. Otaegi[1], J. R. Gutierrez[1], J. C. Jimeno[1], M.T.Ahmed[2]

[1]Technological Institute of Microelectronics (TiM), University of the Basque Country, Zamudio, Spain
[2]Faculty of Science, Department of Physics, Mansoura University, Mansoura, Egypt
Email: ahmedhabib@mans.edu.eg

ABSTRACT:In this work our target is to obtain low surface emitter concentration with Gaussian profilesdue to that have a strongly effect on solar cell efficiency. Two kinds of phosphorus profiles were proceed experimentally, the first emitter profile is obtained with only pre-deposition process for various diffusion temperature and the second emitter profile by adding a drive-in after introducing thermal dry oxidation step after diffusion process for various diffusion temperature. Diffusion profiles after Dry oxidation process predict a reduction in surface recombination due to removing the inactive phosphorus layer that is play as recombination centers. Both V_{oc} and J_{sc} simulation results increased after the emitter surface concentration has been reduced with the thermal oxidation process.The V_{oc} and J_{sc}valuescalculated by PC1D, using a Gaussian model with values of the emitter surface concentration N_s and sheet resistance for the corresponding values of depth junction. The first emitters show maximum efficiencies using Pc1D simulation program about ($\eta \approx$16-19%) with emitter surface concentration N_s = (0.8-3) x10^{20}cm^{-3} and junction depth range (0.35-0.6)μm. For second emitters the surface concentration is reduced with the same diffusion parameters and show improvement of the maximum efficiencies ($\eta \approx$ 20-22.5%) with surface doping concentration N_s = (0.6-2) ×10^{19} (cm^{-3}) and junction depth range of (1.3-2.3)μm. Silicon solar cell performance and parameters are improved after the thermal dry oxidation process and become more efficient.

Keywords: Thermal oxidation, Phosphorus profiles, Sheet resistance, PC1D.

1 INTRODUCTION

In commercial silicon solar cell technologies, moreover for the fabrication process in case of p-type silicon solar cells, the emitter is generally formed by thermal diffusion of $POCl_3$ to have the p-n junction and build-in voltage. Diffusion profiles obtained have the typical kink and tail shape, with a highly doped region near the cell surface followed by a deeper tail at lower surface concentration [1-3]. For highly phosphorus doped emitters (>10^{20} cm^{-3}), near the surface, the concentration can exceed the solid solubility limit of phosphorus in silicon, result phosphorus precipitation which form a dead layer [4]. This layer create recombination centres, which enhance Auger recombination, band gap narrowing and a high surface recombination velocity (FSRV) that decrease emitter lifetime and affect negatively on silicon solar cell efficiency [5]. Also, at high dopant concentration [6], the contact regime reduces the series resistance losses almost due to decreasing of contact resistance and for lowly P concentration at non-contact regime reduces the recombination losses optimizing the spectral response of the silicon solar cell. The amount of inactive phosphorus (dead layer) on the wafer surface at highly doped emitters can be modified by changing the diffusion parameters [7].

Thermal oxidation process is well known process from the semiconductor industry and research and it is important to keep bulk lifetime of the wafer [8].

A lot of papers shows that the improvement of silicon solar cell efficiency can be achieved when the sheet resistance is increased and the emitter surface concentration reduced on a narrow and thin emitter junction depth [9]. The electrical properties of silicon solar cell devices have been established as an essential tool for both the improvement of existing those ones and development a new device. For several years many modelling tools and computer simulating programs have been developed such asTSUPREM-4, PC1D and AMPS that it is specific to PV devices.

The aim of this work is to study the effect of thermal oxidation process on the different emitter characteristics (Emitter surface concentration about (1.8-3) x10^{20} cm^{-3}) after pre-deposition process. Our experimental results show the changing in emitter profiles diagram after thermal oxidation step and the highly doped region is near to the wafer surface. It's also clear the effect of thermal oxidation process in the simulation results using TSUPREM4 program, a 1-dimensional process simulator by Silvaco International Inc. TSUPREM4 simulates the changes in a semiconductor's structure that result from the various processing steps used in its manufacture [10-12].

Our important results obtained are related to (1) Emitter profiles for different temperatures with various surface concentration and junction depth (2) How the shape of profiles and surface concentration of emitters can change after thermal oxidation step (3) Obtaining sheet resistance of diffused regions in silicon layers where some Phosphorus precipitates can appear because the concentration can be over the solid solubility limit of phosphorus in silicon. After that, we Pc1D [13] device simulator program that can calculate the cell parameters and see how thermal oxidation process can improve the silicon solar cell parameters. Pc1D is more or less a specified in this field; Pc1D was developed by Basore and co-workers at Sandia National Lab [14] and was further developed at UNSW, Australia.

2 PROFILES AND ELECTRICAL CELL PARAMETERS

2.1 Experimental details

The experiment was carried out on p-type monocrystalline Czochralski Silicon wafers≈ (ρ2 Ω·cm), <100> oriented, 10.5x10.5 cm2 silicon wafers (thickness = 20 0 μm). A two-step process has been proposed, Firstly, the pre-deposition step in a gas tube furnace for several temperatures, from 800 to 875°C. In this process, the emitter was formed by diffusing phosphorous tri-chloride ($POCl_3$) dopant into p-type silicon substrates; the emitter resultant has a high concentration very near to surface. The dopant gas reacts with silicon surface at high temperature in the presence of O_2. The following reactions are taking place:

$$Si+O_2SiO_2 \rightarrow$$

$$4POCl_3+3O_2 \rightarrow 2P_2O_5 +6Cl_2$$

Later on, a Phosphosilicate Glass (PSG) layer was removed using an HF acid dip. The sheet resistance is measured using four-point probe technique, and the amount of electrically active phosphorus concentration is calculated by using Tsai model [15] for different diffusion temperatures. Afterwards, a dry thermal oxidation process is carried out for different temperatures and times, in order to see suitable parameters which we can use, with the corresponding of oxide growth. In our research we see that the dry thermal oxidation process was carried out at 950 °C for 60 min, that is more suitable for our work, then, the oxide layer was removed in HF and the sheet resistance was able to measure showing what is happened in emitter surface concentration after thermal process. The Profile measurement is carried for both kinds of emitters, Emitter 1(without dry oxidation process) and Emitter 2 (with dry oxidation process) see Figure1. For this two kinds of wafers we made profile measurements is the same, using a 2 wt % NaOH solution at 60 °C ~50 nm of material is removed. The sheet resistance of the exposed surface is measured by 4-point probe, obtaining the amount of electrically active phosphorus Ns in at/cm^{-3}. We follow with this chemical etching and 4-point measurement until the sheet resistance shows that a total removal of the P emitter happened.

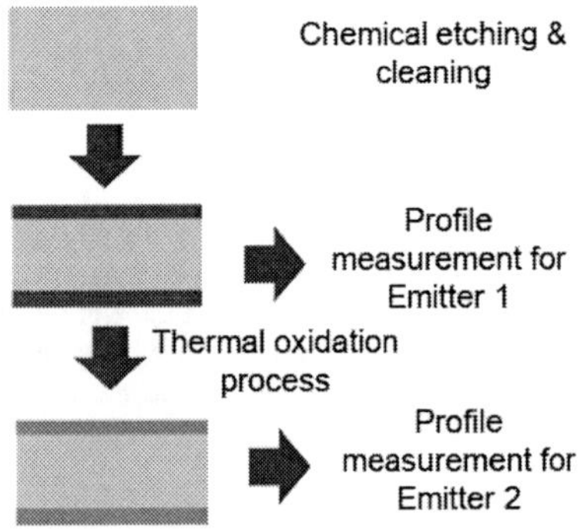

Figure1: Profile measurement Process for two kind of emitters.

2.2 Phosphorus profiles results

a. Phosphorus profiles For Emitter 1. (Bare pre-deposition)

To measure phosphorus doping profiles in silicon solar cells after diffusion etched samples are processed. Figure 2 shows the phosphorus diffused profiles obtained directly from the sheet resistance measurement. Regions of phosphorus profiles in silicon cell that have phosphorus concentrations exceeding the equilibrium concentration, these regions contain electrically inactive phosphorus. The inactive phosphorus can be mobile and immobile, this depends on the saturation concentration N_{sat} [16]. Above this saturation, the concentration of phosphorus exists as immobile SiP precipitates [16, 17]. These can cause reduction in the lifetime and efficiency of solar cell and also leading to higher saturation current densities [18]. Our emitter profiles have a fraction of electrically inactive phosphorus which exceeds the saturation concentration of N_{sat}~ (0.9-1.2) $\times 10^{20}$cm^{-3}.

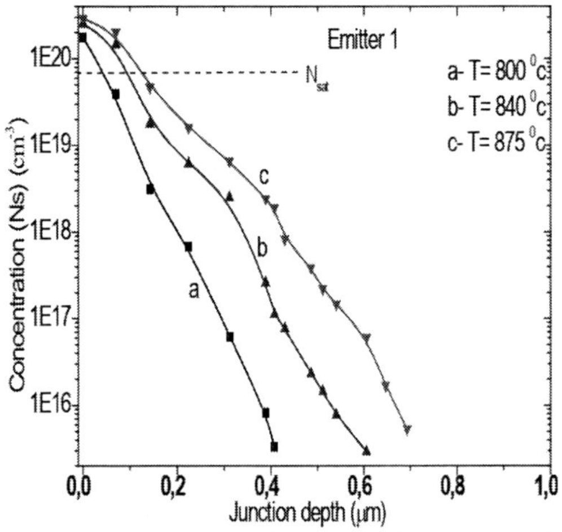

Figure2: Phosphorus doping profiles measured experimentally for different diffusion temperature without thermal Dry oxidation process (Emitter 1).

b. Phosphorus profiles For Emitter 2.(adding an oxidation and drive-in steps)

As is drawn in Figure 3, the Phosphorus emitter profiles after wet oxidation step, that reduces the P surface concentration and drives in the emitter more inside in the silicon wafer. Also, in the cleaning process carried before the oxidation step, few nanometres of highly doped silicon are removed [19] increasing the sheet resistance. This effect is indemnity by the oxidation step that depends both on the diffused emitter concentrations and the oxidation process parameters. Thus, the electrically inactive phosphorus near to the surface becomes electrically active. Our results show the more doped emitters the higher oxide thickness. The growth rate of phosphorous silicate glass in silicon oxide is given by [20]

$$\frac{x_m^2}{t_m} = 1.7 \times 10^{-7} e^{-1.46/kt} cm^2/sec$$

X_m, is the silicon oxide thickness and t_m is the diffusion time.

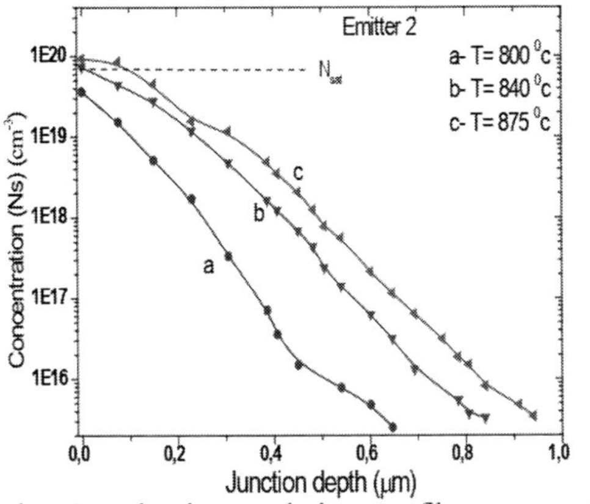

Figure3: Phosphorus doping profiles measured experimentally for different diffusion temperature after thermal Dry oxidation process (Emitter 2).

The diffusion profiles after oxidation process predict a reduction in surface recombination due to removing the inactive phosphorus layer that play as recombination centres. Furthermore, all phosphorus become active and the cell performance increases. In the next section, we will start to simulate this process using PC1D and calculate the changes in cell parameters (open circuit voltage Voc, emitter saturation current density Joe and cell efficiency) before and after oxidation step.

2.3 Solar Cell parameters using PC1D program

The numerical simulation program PC1D is very used in silicon solar cells simulation; study the effect of process condition and modification on the cell parameters [21, 22]. In our work we have several simulations to see the influence of thermal dry oxidation process on the efficiency of solar cell according to our profiles. The typical physics physical parameters that used as input for computer simulation program is two simulation data listed in Table I. The average I-V data for several dopant concentrations of the Emitter 1 and Emitter 2 and the improvement that is observed on the solar cell parameters were are listed in Table II and Table III.

Table I: Simulation parameters input for PC1D

No.	Base ResistivityΩ·cm	Carrier Life time τ	Surface recombination velocity cm/sec
Simulation 1	2.3	600	100
Simulation 2	2.3	200	10000

Table II: I-V data for several dopant concentrations of the Simulation parameters 1

Diffusion Temperature (°C)	Cell parameters For Emitter 1	Cell parameters For Emitter 2
800 °C	V_{oc}=0.693 V I_{sc}=39.5mA/cm^2 Efficiency=20.1%	V_{oc}=0.7281V I_{sc}=39.89mA/cm^2 Efficiency=22.3%
840 °C	V_{oc}=0.677 V I_{sc}=39.2 mA/cm^2 Efficiency=19.5%	V_{oc}=0.733 V I_{sc}=39.6 mA/cm^2 Efficiency=21.3%
875 °C	V_{oc}=0.635 V I_{sc}=35.7 mA/cm^2 Efficiency=16.4%	V_{oc}=0.680 V I_{sc}=37.8mA/cm^2 Efficiency=20.9%

Table III: I-V data for several dopant concentrations of the Simulation parameters 2

Diffusion Temperature (°C)	Cell parameters For Emitter 1	Cell parameters For Emitter 2
800 °C	V_{oc}=0.665 V I_{sc}=37.43mA/cm^2 Efficiency=18.5%	V_{oc}=0.695 V I_{sc}=38.84mA/cm^2 Efficiency=19.9%
840 °C	V_{oc}=0.663 V I_{sc}=37.31 mA/cm^2 Efficiency=17.9%	V_{oc}=0.691 V I_{sc}=37.93 mA/cm^2 Efficiency=19.4%
875 °C	V_{oc}=0.646 V I_{sc}=36.3 mA/cm^2 Efficiency=16.8%	V_{oc}=0.676 V I_{sc}=36.8mA/cm^2 Efficiency=18.9%

Both V_{oc} and J_{sc}values result are increased when diffusion temperature decrease emitter surface concentration decreasing after Dry oxidation process. Our simulated profiles and calculated parameters calculation show a good agreement with our previous experimental results [23].

The I-V curve of simulated results for the different cells give us an explanation about how can the efficiency and short circuit voltage changes with the changing of emitter surface concentration and junction depth of the emitters as shown in Figure 4 and 5 for two simulation parameters. When the emitter surface concentration N_s decreases from $3x10^{20}$ to $2x10^{19}$ cm^{-3}, the cell efficiency increases from 16.3 to 20.7 for 875°C diffusion temperature that is indicate the dry oxidation process is greatly affected on solar cell performance.

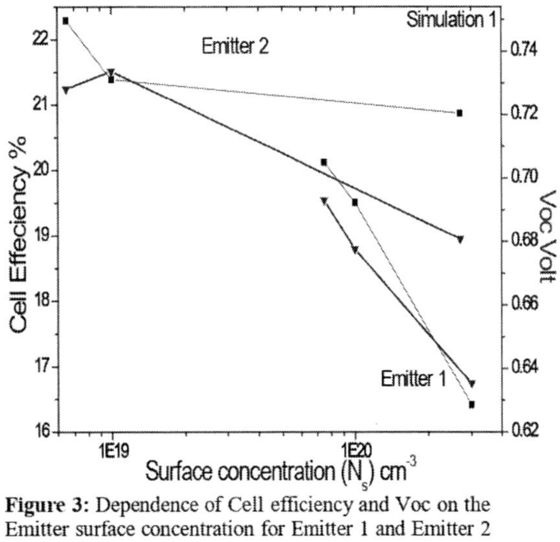

Figure 3: Dependence of Cell efficiency and Voc on the Emitter surface concentration for Emitter 1 and Emitter 2 with simulation data 1

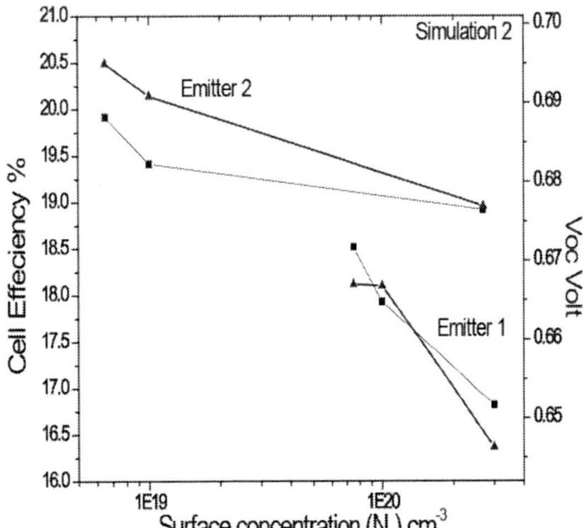

Figure 4: Dependence of Cell efficiency and Voc on the Emitter surface concentration for Emitter 1 and Emitter 2 with simulation data 1

Due to the diffusivity of phosphorus in silicon oxide is much lower than in silicon [24] so the little part of P at the surface that is highly concentrated is diffused. The intrinsic and extrinsic diffusivity (D_i, D_x) of phosphorus in silicon is given by [25]

$$D_i = 3.85e^{-\frac{3.66}{kt}}\frac{cm^2}{s}$$

$$D_x = D^0 + D^*[\frac{n}{n_i}]^2$$

Where n_i is the intrinsic carrier concentration and D^0, D^* is given by

$$D_0 = 3.85e^{-\frac{3.66}{kt}}\frac{cm^2}{s}$$

$$D^* = 44.2e^{-\frac{4.37}{kt}}cm^2/s$$

The sheet conductance C measured experimentally by the four-point probe technique, giving the concentration of phosphorus at the silicon oxide/silicon interface [20].

$$C = 4.53\int_0^x q\mu(C)[C_2(x) - C_3]dx$$

Where q is the electronic charge, μ(c) is the majority carrier mobility, $C_2(x)$ is the concentration of phosphorus and C_3 is the impurity concentration originally present in the bulk silicon.

The Emitter surface concentration results show high doping concentrations that increase the surface recombination and lead to lower cell performance [26]. The dependence of these parameters was changed after applying thermal dry oxidation process. Concentration of the P impurities is a decrease so much in Emitter 2 and junction depth also increase, this means that the oxide layer push emitter more inside into p-type silicon substrate.

3 SUMMARY AND CONCLUSIONS OF THE WORK

The phosphorus surface concentration is play very important rule to improve silicon solar cell efficiency. The electrically inactive phosphorus that appears in case of highly doped emitters near to the surface becomes electrically active after the dry thermal oxidation process and contributes in solar cell connection.The dry thermal oxidation process was carried out at 950 °C for 60 min that is more suitable for our work.The silicon oxide layer thicknesses become thicker for highly surface doped emitters.Decreasing of emitter surface concentration after the Dry oxidation process showing high improvement in I-V results using PC1D program.

4 REFERENCES

[1]R.B.Fair,J.C.C.Tsai,A quantitative model for the diffusion of phosphorusin silicon andtheemitterdipeffect,JournaloftheElectrochemicalSociety124 (1977) 1107–1118.

[2] A.F.W.Willoughby, Interactions between sequential dopant diffusions in silicon — a review, Journal of Physics D: Applied Physics 10(1977)455–480.

[3] B. Bazer-Bachi, E.Fourmond, P.Papet, L.Bounaas, O.Nichiporuk, N.LeQuang, M.Lemiti , Solar Energy Materials & Solar Cells 105 (2012) 137–141

[4] M. Finetti, P. Negrini, S. Solmi, D. Nobili, Electrical properties and stability of supersaturated phosphorus-doped silicon layers, Journal of the Electrochemical Society 128 (1981) 1313–1317.

[5]J.Ostoja,F.Gueri,P.Negrini,S.Solmi,Theeffects of phosphorus precipitation on theopencircuitvoltageinn+/p siliconsolarcells,SolarCells11(1984) 1–12.

[6]A. Fallisch, D. Wagenmann, R. Keding, D. Trogus, M. Hofmann, J. Rentsch, H. Reinecke, D. Biro, IEEE JOURNAL OF PHOTOVOLTAICS · october 2012.2200455

[7] A. Cuevas, P.A. Basore, G. Giroult-Matlakowski, C. Dubois, Surface recombina- tion velocity of highly doped n-type silicon, Journal of Applied Physics 80 (1996) 3370–3375.

[8] Nicollian, E.H. and J.R. Brews, MOS (Metal OxidSemiconductor) Physics and Technology. 1982: BellTelephone Laboratories.

[9] I. KuzmaFilipek, M.R. Payo, E.V. Kerschaver, R. Mertens, E. Schmich, J. V. Hoeymissen, K. V. Nieuwenhuysen, G. Beaucarne, J. Poortmans, S. Lindekugel, S. Reber. Efficiency (>15%) for thin film epitaxial silicon solar cells on 70cm^2 area offspec silicon substrate using porous silicon segmented mirrors. Progress in Photovoltaics: Research and Applications 143(2010) 18:137.

[10]C. P. Ho, S. E. Hansen, and P. M. Fahey. "SUPREM III—A Program for Integrated Circuit Process Modelling and Simulation," Technical Report No. SEL84-001, Stanford Electronics Laboratories, Stanford University, Stanford, California, July 1984.

[11] R. E. Bank, W. M. Coughran Jr., W. Fichtner, E. H. Grosse, D. J. Rose, and R. K. Smith. "Transient Simulation of Silicon Devices and Circuits," IEEE Trans. Elec. Dev., 32(10) (1985) 1992-2007

[12] P. B. Griffin and J. D. Plummer. "Process Physics Determining 2-D Impurity Profiles in VLSI Devices," IEDM Technical Digest, (1986) 522-525

[13]D.Clugston, P.Basore, Pc1D Version 5: 32-bit Solar Cell Modeling on Personal Computers. Proceeding of the 26th IEEE Photovoltaic Specialists Conference, Anaheim

CA, (1997) 207–210. Also, see http://www.pv.unsw.edu.au/pc1d/
[14] P.Basore. Numerical modeling of textured silicon solar cells using Pc1D. IEEE Transactions on Electron Devices, 37(2) (1990) 337–345.
[15]R.B. Fair, J.C.C. Tsai, A quantitative model for the diffusion in silicon and the emitter dip effect // J. Electroch. Soc., 124, p. 1107 (1977).
[16] S. Solmi, A. Parisini, R. Angelucci, A. Armigliato, D. Nobili and L. Moro, "Dopant and carrier concentration in Si in equilibrium with monoclinic SiP precipitates", Physical Review B, 53(1996) 7836-7841.
[17] M. Finetti, P. Negrini, S. Solmi and D. Nobili, "Electrical properties and stability of supersaturated phosphorus-doped silicon layers", Journal of the Electrochemical Society: Solid-State Science and Technology, 128 (1981) 1313-1317.
[18] P. Ostoja, S. Guerri, P. Negrini and S. Solmi, "The effects of phosphorus precipitation on the open-circuit voltage in N+/P silicon solar cells", Solar Cells, 11(1984) 1-12.
[19] A. Kimmerle, A. Wolf, U. Belledin and D. Biro, "Modelling carrier recombination in highly phosphorus-doped industrial emitters", Energy Procedia, 8(2011) 275-281.
[20] C. T. SAH, H. SELLO and D. A. TREMJZRE, J. Phys. Chem. Solids Pergamon Press, 11 (1959) 288-298.
[21]P.A. Basore, "Numerical modeling of textured silicon solar cells using PC-1D," IEEE Transactions on Electron Devices, 37(2) (1990) 337–343.
[22]D.T. Rover, P.A. Basore and G.M. Thorson, "Solar cell modelling on personal computers," Proc. 18th IEEE Photovoltiac Specialists Conference, Las Vegas, (1985) 703–709.
[23] Fano, V.; Rasool, M.A.; Habib, A.; Otaegi, A.; Gutierrez, J.R.; Jimeno, J.C.; Azkona, N.Photovoltaic Specialist Conference (PVSC), 2015 IEEE 42ndYear: (2015)1 – 5.
[24]K. A. Ellis and R. A. Buhrman, Electrochemical and Solid-State Letters, 2 (10) (1999) 516-518
[25] R.B. Fair, "concentration profiles of diffuse dopants in silicon," in F.Y.Y Yang, Ed. "Impurity dopant process in silicon" North Holland (1988)
[26]V. Yelundur, K. Nakayashiki, M. Hilali and A. Rohatgi, Implementation of a Homogeneous High-Sheet-Resistance Emitter in Multicrystalline Silicon Solar Cells. Conference record of the Thirty-First IEEE Photovoltaic Specialists Conference (2005) 959-962, IEEE Electron Devices Society.

INVERTED RANDOM PYRAMIDS: GENERATION IN HF-HCL-CL_2-SOLUTIONS AND SIMULATION OF THE INFLUENCE OF SURFACE TEXTURE ON LIGHT ABSORPTION IN PERC SOLAR CELLS

André Stapf, Christoph Gondek, Peter Nattrodt, Edwin Kroke
Institute of Inorganic Chemistry, Technische Universität Bergakademie Freiberg
Leipziger Str. 29, D-09599 Freiberg, Germany

andre.stapf@chemie.tu-freiberg.de, +49 3731 39-3583
edwin.kroke@chemie.tu-freiberg.de

ABSTRACT: The Texturization of crystalline silicon wafers by wet-chemical treatment is commonly used for the enhancement of light absorption in solar cells. Recent studies of anisotropic acidic etchants based on HF-HCl-Cl_2 led to an innovative random inverted pyramid surface structure. By using the 'wafer ray tracer' of pvlighthouse.com.au the theoretical light absorption properties of the random inverted pyramid structure were compared to the well-known textures for PERC solar cells. We simulated pyramidal and inverted pyramidal structures with monomodal sizes and with various average pyramid widths. Besides, simulation of a PERC solar cell on the roof top (30°, south) of our building in Freiberg (N 50°55'32.16", E 13°19'58.43") on June 21st, 2016 (solstitium) was performed. The results encourage further development of new textures on monocrystalline silicon wafers by maskless acidic wet chemical treatment with HF-HCl-oxidant mixtures.

Keywords: Silicon, Texturization, Etching, Modelling, Light trapping

1 INTRODUCTION

Theoretical discussions on the texture of monocrystalline silicon wafers have been made since the 1980s [1]. These studies led to powerful Monte-Carlo simulation tools like OPAL 2 [2] and the 'wafer ray tracer' [3], which are based on assumptions of various ray paths and give the share of reflection, transmission and absorption of light for photogeneration of charge carriers [4].

The goal of the present study was to identify ideal textures for a PERC solar cell. We examined four different types of surface structures on monocrystalline silicon wafers. Random inverted pyramid structures, recently found by HF-HCl-Cl_2-treatment [5], are compared to the well-known types of surface structures on monocrystalline silicon wafers: regular upright, random upright and regular inverted (Figure 1).

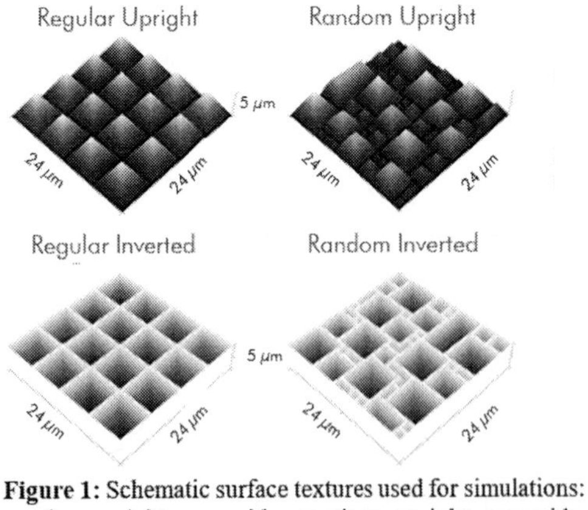

Figure 1: Schematic surface textures used for simulations: regular upright pyramids, random upright pyramids, regular inverted pyramids, and random inverted pyramids.

2 EXPERIMENTAL AND CALCULATIONS

For preparing the etching mixtures hydrochloric acid (37 wt%, analytical grade, VWR), hydrofluoric acid (48 wt%, analytical grade, Sigma-Aldrich) and deionized (DI) water were mixed in a PFA beaker. Dried gaseous chlorine was bubbled into the etching solution. Silicon wafer pieces of approximately 1 cm x 1 cm ((100) orientation, boron doped, thickness of 190 µm, resistivity of 0.5 – 2.0 Ω cm^{-1}, Deutsche Solar GmbH, Freiberg, Germany) were placed with tweezers for 20 minutes in 30 mL of the etching mixture.

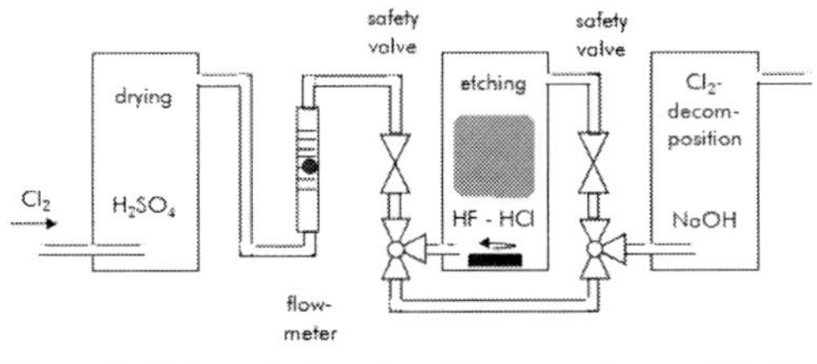

Figure 2: Schematic drawing of the apparatus used for the etching of silicon in HF-HCl-Cl_2 mixtures.

Experiments were performed at room temperature with as-cut diamond wire-sawn silicon (100) wafers. The etching reaction was quenched by rinsing the wafer piece with DI water. Each experiment was performed using freshly prepared solutions.

The etch rate is the quotient of the etched silicon layer per side and the immersion time. The thickness of the removed silicon layer per side (Δd) is calculated from the mass loss obtained by differential weighing.

The simulations were performed with the 'wafer ray tracer' of pvlighthouse.com.au [3]. Parameters like wafer thickness, front and rear films were kept constant. Surface morphologies were varied as shown in Figure 1. Parameters for the Zenith and Azimuth were used from www.suncalc.org [6]. The given values for absorption and reflectivity are average values for the wavelengths between 300 nm and 1200 nm.

2 MASKLESS GENERATION OF RANDOM INVERTED PYRAMIDS IN HF-HCl-CL_2 MIXTURES

Using HF-HCl-Cl_2 mixtures for treatment of monocrystalline silicon wafers, random inverted pyramidal structures are generated without masking or lithographic processes. Silicon etching rates up to 0.35 µm min^{-1} are observed during etching of monocrystalline, diamond wire-sawn wafers at room temperature. SEM pictures show pyramid edge widths between 2 and 20 µm (Figure 3). On the Inside of the inverted pyramids a microstructure is observed.

Figure 3: SEM pictures of the random inverted structure generated on (100) silicon surface by HF-HCl-Cl_2 treatment.

Chlorine is acting as oxidant for silicon surface atoms [7]. A higher bubbling rate of chlorine gas increases the rate of dissolution of silicon. A higher concentration of hydrofluoric acid increases the complexation of the oxidized silicon surface species. A higher concentration of hydrochloric acid decreases the etching rate but improves the uniformity of the surface texture. Also stirring has a strong influence on the etching rate.

3 CALCULATIONS ON PYRAMID WIDTH DEPENDENCE

For calculations on the dependence of light absorption on the pyramid width, simulations were performed without any front or rear film, e.g. anti-reflex-coating. By varying the average pyramid width from 2 µm to 10 µm, no significant dependence of the light absorption was observed (Figure 4). The regular upright pyramid texture shows the lowest absorption. Surprisingly, the random inverted structure shows a significant increase of the absorbed part of the incident light, also independent of the pyramidal width.

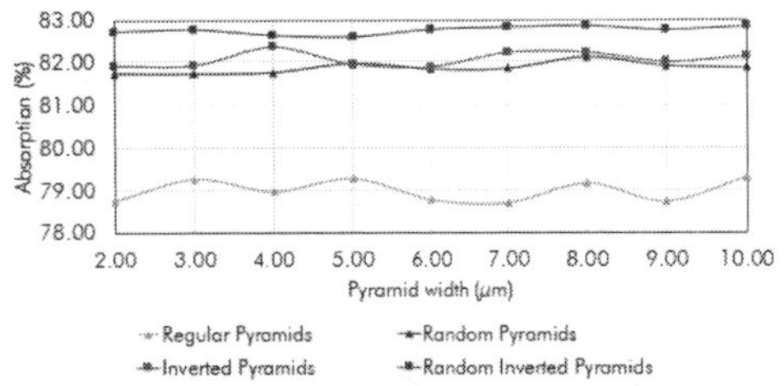

Figure 4: Dependence of light absorption in silicon wafers on the average pyramid width. Textured silicon wafers without any front or rear coating (e.g. ARC).

4 SIMULATIONS ON INFLUENCE OF INCIDENT LIGHT ANGLE

Simulation using the 'wafer ray tracer' was also performed for PERC solar cells on the roof top (30°, south) of the Clemens-Winkler-Bau, TU Bergakademie Freiberg (N 50°55'32.16", E 13°19'58.43") on June 21st, 2016 (solstitium). Figure 5 shows an example for simulation of losses and absorption for photogeneration of charge carriers as share of incident light for PERC solar cell with random inverted pyramid texture.

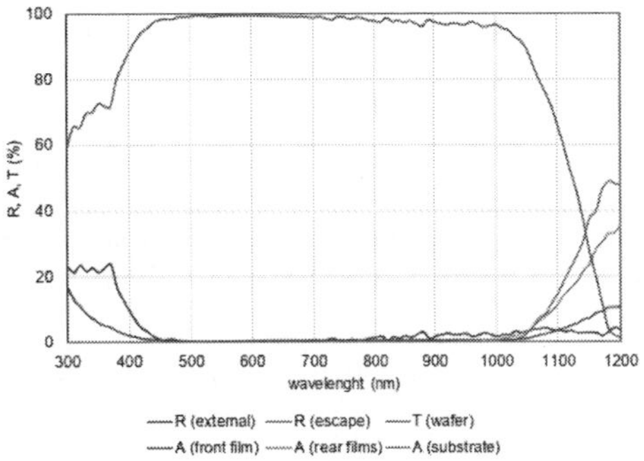

Figure 5: Example for simulation of losses and absorption for photogeneration of charge carriers as share of incident light with the 'wafer ray tracer' (PERC solar cell with random inverted pyramid texture, June 21st, 2016, 13:00 at Clemens-Winkler-Bau, TU Bergakademie Freiberg).

PERC solar cells with a regular texture, whether it is upright or inverted, show a significant dependence on the angle of the incident light, even in lower deviation of the zenith ($\Delta\theta < 2$ °). Cells with an upright pyramid texture show lower absorption then inverted pyramid structures (Figure 6). Interestingly, random inverted pyramids show the highest absorption of light for photogeneration of charge carriers.

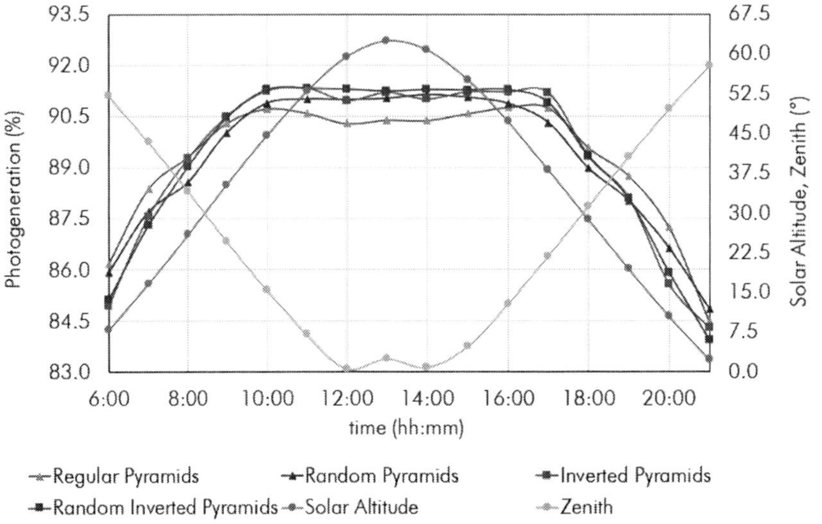

Figure 6: Simulation of the light absorption for photogeneration of charge carriers as share of incident light, dependent on the sun path on June 21st, 2016 at Clemens-Winkler-Bau, TU Bergakademie Freiberg (Germany).

5 CONCLUSIONS

HF-HCl-Cl_2 solutions can be used for generation of random inverted pyramid structures on monocrystalline silicon wafers. The dissolution of silicon depends on the concentrations of hydrochloric acid and hydrofluoric acid, the bubbling rate of chlorine and the stirring rate. Etching rates up to 0.35 µm s^{-1} are observed for the texturization of monocrystalline silicon wafers. Simulations indicate that a silicon wafer surface with random inverted pyramids give the highest light absorption, independent of the pyramid edge width. According to the calculations, random structures should be more independent of incident light angle then regular structures. These results encourage us in the further development of new structures by maskless wet chemical treatment with HF-HCl-oxidant mixtures.

5 ACKNOWLEDMENTS

We kindly acknowledge the Solar World Innovations GmbH for the supply of the silicon wafer material. Financial support was in part provided within the Cluster of Excellence "Structure Design of Novel High-Performance Materials via Atomic Design and Defect Engineering (ADDE)" that is financially supported by the European Union (European regional development fund) and by the Ministry of Science and Art of Saxony.

6 REFERENCES

[1] P. Campbell, M. A. Green, Light trapping properties of pyramidally textured surfaces, J. Appl. Phys, 62 (1987) 243.

[2] K. R. McIntosh, S. C. Baker-Finch, Proceedings of the 38th IEEE Photovoltaic Specialists Conference, Austin (2012).

[3] S. C. Baker-Finch, K. R. McIntosh, M. Abbott, and B. Sudbury, *Wafer Ray Tracer*. Available: www.pvlighthouse.com.au/calculators/wafer%20 ray%20tracer/wafer%20ray%20tracer.html. 12.05.2016.

[4] S. C. Baker-Finch, K. R. McIntosh, Reflection of normally incident light from silicon solar cells with pyramidal texture, Prog. Photovolt: Res. Appl, 19 (2011) 406.

[5] A. Stapf, F. Honeit, C. Gondek, E. Kroke, Siliziumwafer, Verfahren zum Strukturieren eines Siliziumwafers und Solarzelle, DE 10 2016 105 866.1(31.03.2016).

[6] T. Hoffmann, *Suncalc*. Available: www.suncalc.org. 12.05.2016.

[7] A. Stapf, C. Gondek, M. Lippold, E. Kroke, HF-$(NH_4)_2S_2O_8$-HCl mixtures for HNO_3- and NO_x-free etching of diamond wire- and SiC-slurry-sawn silicon wafers: Reactivity studies, surface chemistry, and unexpected pyramidal surface morphologies, ACS Appl. Mater. Interfaces, 7 (2015) 8733.

A RIGOROUS TESTING ON REGENERATED PERC SOLAR CELL

Gaofei Li*, Jilei Wang, Jin Huang, Shaojian Fu, Juan Zhang, Yanhui Bai, Liyou Yang
Jinneng Clean Energy Technology Co., Ltd
Jinergy, No.1 Wenshui Economic Development Zone, Lvliang, Shanxi 032100, China

ABSTRACT: P-type PERC solar cell containing boron as well as oxygen suffers from light induced degradation (LID) due to the formation of boron-oxygen related recombination active complexes. The so-called 'regeneration' recovered from degraded stage is proven to be less recombination active with high carrier injection by illumination or bias voltage. In this work, various feasible methods of anti-LID were explored. Rigorous testing as well as calculated simulation was performed on regenerated PERC solar cells, which were treated with carrier injection by bias voltage. Meanwhile, the stability and reliability were mainly tested. Upon process optimization, commonly-occurred LID could be controlled effectively to be lower than 1%, on average, for PERC+ (PERC solar cell after anti-LID treatment from Jinergy) in mass production with an average 21.2% efficiency.

Keywords: PERC, LID, Boron-oxygen complexes, annealing process

1 INTRODUCTION

The boron-doped PERC solar cells are well known for their significant efficiency, while undesirable light induced degradation (LID) results in a considerable loss thanks to the formation of boron-oxygen complexes with high recombination activity [1-3]. Especially, highly boron doped (N_D>10^{16} cm^{-3}) oxygen rich materials are generally susceptible to get a serious loss in efficiency under the typical working conditions. It has been shown that the total number of B-O complexes scale linearly with the boron concentration [4–5] and correspondingly scales about quadratically with the interstitial oxygen concentration [6–7], suggesting a stoichiometric composition of BO_2. The feasibility methods of anti-LID is numbered, albeit more and more treating processes have been reported.

According to previous reports [8–10], using of p-type dopants and reducing of interstitial oxygen concentration can preclude the negative impact of defect formation, but it is difficult to completely eliminate LID. A straightforward approach has been put forward by Herguth et al [11-15] to suppress LID as described in Fig 1. The introduction of carrier injection (by light or bias voltage) or slightly elevated temperatures between 60 and 200 ℃ may produce a so called regeneration procedure. This regenerated state can effectively recover its bulk lifetime from degraded stage. Therefore, to reach regeneration step is a practical and effective way to permanently improve the minority carrier lifetime as well as the efficiency of the boron-doped PERC solar cells. In this paper, we compare the various method to suppress light induced degradation and to verify reliability and repeatability though a large amount of data.

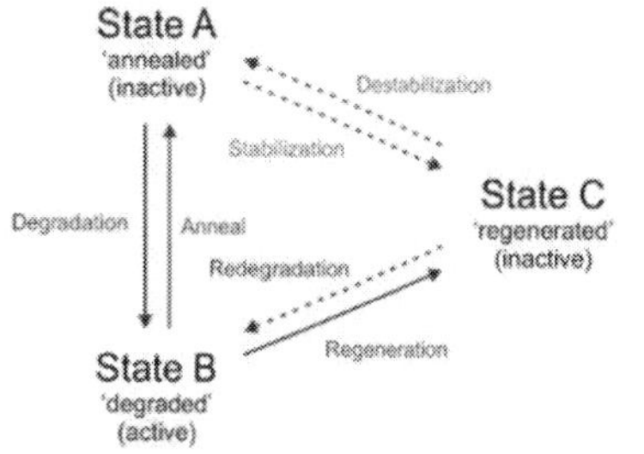

Fig. 1 Three state reaction scheme of degradation and regeneration [11].

2 EXPERIMENT

For our experiment, all initial samples were made from commercially available boron-doped Czo-chralski (Cz) silicon solar wafers and gallium-doped Cz silicon wafers, with the characteristics of 156.75×156.75 mm^2 and a base resistivity in the range of 1-3 Ωcm. A standard industrial mono-crystalline solar cell process was applied including KOH saw damage etch, phosphorus emitter from a $POCl_3$ diffusion, a SiNx:H antireflection coating (ARC) on the front side by plasma enhanced chemical vapor deposition (PECVD), passivation with special atomic layer deposition (ALD) Al_2O_3 layer on the back side, subsequently capped by SiNx layer, picosecond green laser was adopted to make opening for aluminum-silicon contact, then deposited a screen printed silver front and aluminum back contact, co-firing of aluminum back and silver front contact. The structural representation of PERC solar cells was displayed as Fig.2 and the IV characteristics were measured under standard test condition. An average 21.2% solar cell efficiency has been achieved in mass production, from which all samples were selected for this work.

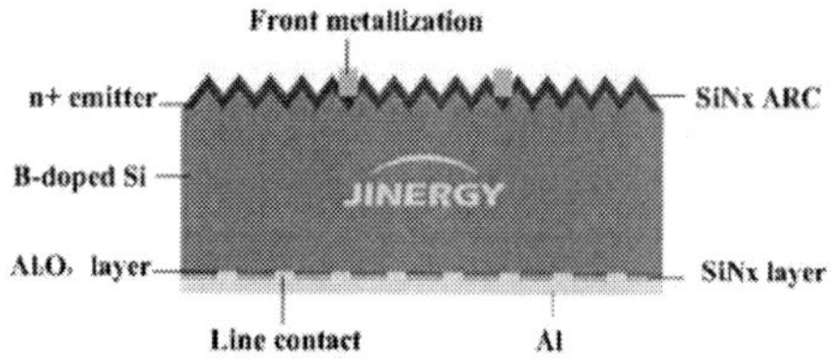

Fig. 2 The structural representation of PERC solar cells.

Three different typical anti-LID methods were applied to PERC solar cells during experiment (as shown in Fig. 3). After been processed by different methods, several groups of PERC cells were obtained correspondingly, including the B-doped, Ga-doped, B/Ga-codoped PERC cells, PERC+ cells (B-doped but treated by bias voltage), Regl and Reg2, by the way, Regl and Reg2 represented regenerated B-doped PERC cells treated by the injection of illumination (treated with equipments from different suppliers).

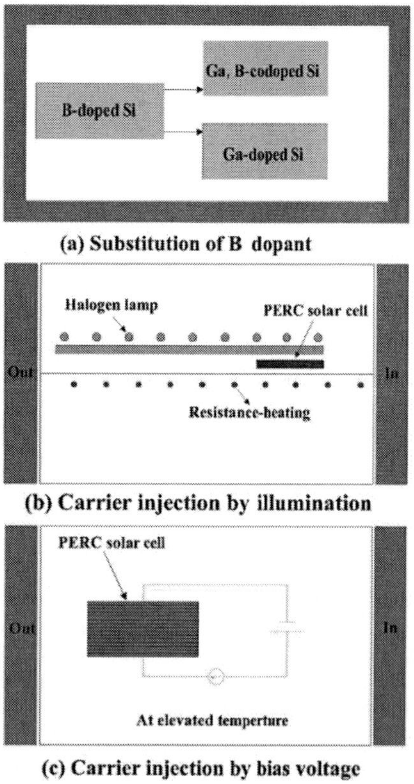

Fig. 3 Three typically feasible strategies to suppress LID. (a) B dopant was substituted by Ga, B-codoped and Ga-doped, (b) Carrier injection by illumination and (c) Carrier injection by bias voltage.

3 RESULTS AND CONCLUSION

3.1 Investigation of anti-LID performance

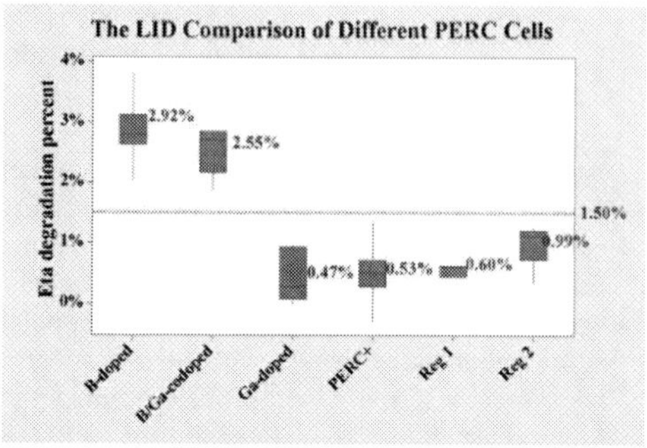

Fig. 4 LID comparison of different PERC cells with illumination of 8 kWh.

To compare the LID performance, all these PERC solar cells were exposed to illuminations of 8 kWh and the results of LID are shown in Fig. 4. Obviously, compared with B-doped, the others showed lower LID to different degrees. The LID value of PERC+ reached a very low level, as well as that of Regl and Reg2, although Ga-doped PERC cell showed the lowest LID.

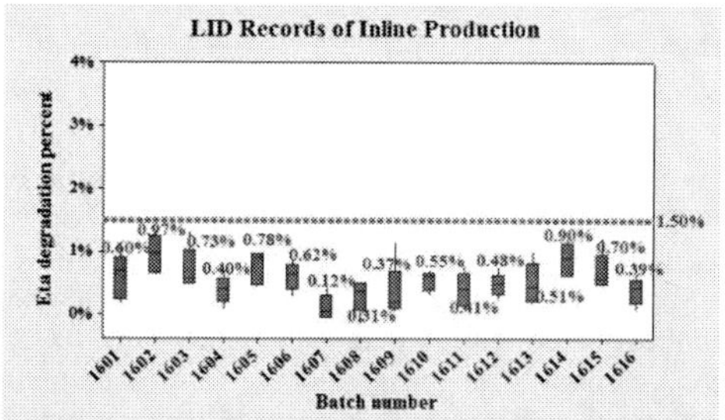

Fig. 5 Spot check records of LID for inline production PERC+ with illumination of 8 kWh.

The repeatability and reproducibility of anti-LID characteristics of $PERC^{+}$ were monitored in the production line. LID of different batches from continuous production were shown in Fig. 5. All of the LID values are less than 1.5%, and median value is less than 1%. PERC+ was demonstrated to be a promising candidate as anti-LID material.

3.2 Investigations of LID trends on PERC and PERC+

The LID behaviors of PERC and PERC+ were both studied. The test results of LID level with illumination duration of 1, 10, 100 and 1000 minutes were shown in Fig. 6. The LID trend of both PERC and PERC+ tend to be depend approximately linearly on the natural logarithm of illumination time. Furthermore, the slop of PERC is 6 times higher than PERC+, in turn, the LID of PERC+ is rather lower than PERC in the long run. It could provide some academic bases and references to forecast the LID of PERC and PERC+. According to our calculation, the LID of PERC+ products will be 0.92% in the first year. In addition, LID will be lower than 0.1% in each next year by assuming 1200 kWh illumination per year.

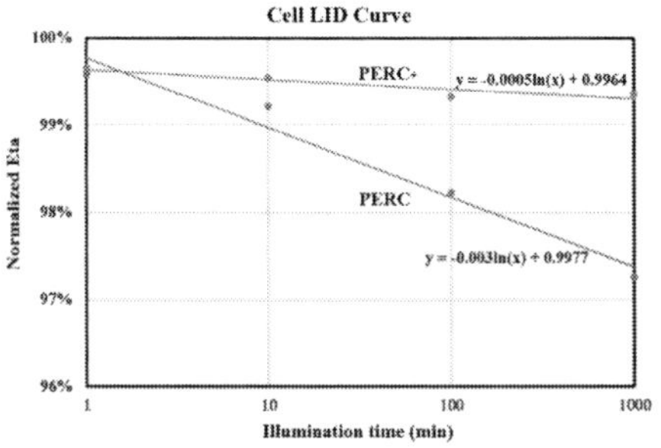

Fig. 6 The LID trends versus illumination time of normal PERC and PERC+ product

3.3 Analysis of module LID via long-term tracking

Meanwhile, the potential reliability and stability of PERC+ have been verified by further outdoor testing. Five manufactured photovoltaic modules that composed of PERC+ were exposed outdoors in this experiment. Data were collected and analyzed as shown in Fig. 7. Referring to the monitored data, LID of PERC+ maintained a relative steady level in 75 days, and the total Pmax degradation was less than 1.5%, which is in the same level of traditional multi-crystalline Si solar products, demonstrating their superior performance of excellent reliability and stability.

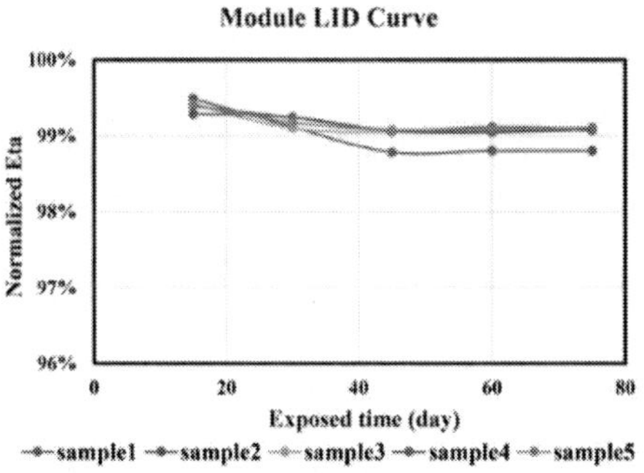

Fig. 7 Analysis of LID via long-term tracking data.

4 CONCLUSION

Various anti-LID process methods on p-type Cz PERC solar cells have been investigated in this paper. Except B/Ga-codoped PERC cells, the others mentioned above all exhibit relatively low LID to different degrees. More importantly, the method of carrier injection by bias voltage is proved to be an efficient mode, which is not only suitable for mass production but also reaches remarkable effects. All of the LID values in the production line are less than 1.5%, and the median value is less than 1%. The reliability and stability of PERC+ were checked both on cell and module level, all results show PERC+ product is as stable as widely used traditional BSF products.

5 REFERENCES

[1] K. Graff, H. Pieper, Physica Status Solidi A 49 (1978) 137.
[2] J. Schmidt, A. G. Aberle, Journal of Applied Physics 81 (1997) 6186.
[3] S. Dubois, N. Enjalbert., J. P. Garandet, Applied Physics Letters 93 (2008) 103510.
[4] SW. Glunz, S. Rein, W .Warta, Solar Energy Materials and Solar Cells 65 (2001) 219.
[5] J. Schmidt, A Cuevas, Journal of Applied Physics 86 (1999) 3175.
[6] K. Bothe, J. Schmidt, Journal of Applied Physics 99 (2006) 013701.
[7] J, Schmidt, K, Bothe, Physical Review B 69 (2004) 024107.
[8] S. W. Glunz, S. Rein, J. Y. Lee , W. Warta, Journal of Applied Physics 90 (2001) 2397.
[9] J. Schmidt , A. Cuevas, Journal of Applied Physics 86 (1999) 3175.
[10] D. Macdonald1, F. Rougieux1, A. Cuevas1, B. Lim, J. Schmidt, M. Di Sabatino, L. J. Geerligs, Solar Energy Materials and Solar Cells 65 (2001) 219.
[11] A. Herguth, G. Schubert, M. Kaes, G. Hahn, Progress in Photovoltaics: Reserch and Applications 16 (2008) 135.
[12] A. Herguth, G. Schubert, M. Kaes, G. Hahn, in Proceedings of the 32nd IEEE PVSC (4th WCPEC), Waikoloa USA (2006) 940.
[13] A. Herguth, G. Schubert, M. Kaes, G. Hahn, Prog. Photovoltaics 16 (2007) 135.
[14] S. Wilking, A. Herguth, G. Hahn, Journal of Applied Physics 113 (2013) 194503.
[15] S. Wilking, S. Ebert, A. Herguth, G, Hahn, Journal of Applied Physics 114 (2013) 194512.

The design and industry road of a low cost and high efficient multi busbar technology

Songbo Wan, Xusheng Wang*,Dengzhi Wang, Yanfen Wu, Zhengyue Xia, Guoqiang Xing
Canadian Solar Inc., 199 Lushan Road, SND, Suzhou, Jiangsu, China
* Corresponding author: Phone: +86 512 66908088; Fax: +86 512 66908089; Email: Xusheng.wang@canadiansolar.com

ABSTRACT: Multi busbar becomes more and more hot nowadays as it can increase the module power output and save the silver consumption on the cell level at the same time. We found the saved silver, the power gain are highly correlated with the number of multi busbar, the width of multi busbar and the spec of the ribbon by theoretical modeling. Then the modeling results are well proved in the following experiment. The optimized design saves at least 50% silver, and increases the module power output 2% above, which is very attractive for the industry. Multi busbar can be the future trend in PV.
Keywords: Multi busbar technology, high efficiency module, low cost

1 INTRODUCTION

As a strategic emerging industry, solar Photovoltaic has become a kind of necessary clean energy of human development, and obtains national recognition and praise in recent years as its safety, less pollution and renewable. However, the cost of photovoltaic power generation is one of the important reasons restricting photovoltaic products wider application. Future solar cell designs not only have to take into account an increase of efficiency but also reduced production costs to be competitive on the growing solar market. A lot of new technologies can increase the photoelectric conversion efficiency of photovoltaic products. However, they will increase the cost of production at the mean time. But the multi busbar technology can not only increase the photoelectric conversion efficiency of photovoltaic module, but also reduce the production cost of photovoltaic module.

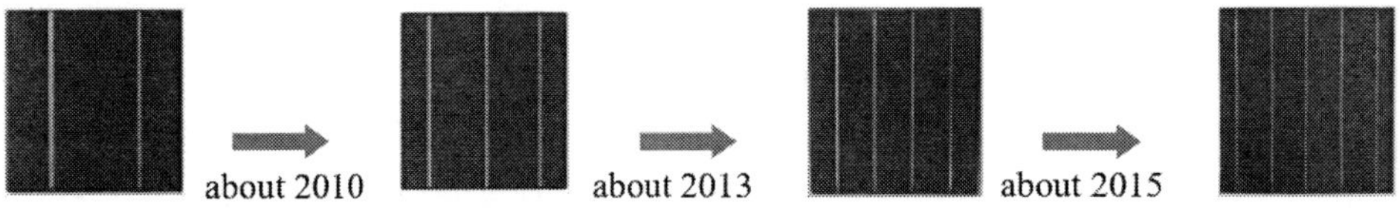

Figure 1: The road map of of solar cell busbar number

2 THE HISTORICAL CHANGES OF SOLAR CELL BUSBAR NUMBER

In the early years, the solar cell and module usually adopted two busbar design. In order to get higher power output, since 2009, the PV industry had begun to import three busbar gradually. Since then the multi busbar design has been on the application way in PV industry (Figure 1). CSI has taken the lead ship in four busbar cell mass production since 2013. At present, some PV manufacturers are taking an active part in advancing the five busbar solar cell and module. Three busbar, four busbar and five busbar solar cell and module, has little difference compared with the traditional two busbar solar cell and module on production process, is just a gradual changed process. The multi busbar solar cell and module in this paper, refer to more than five busbar solar cell and module technology.

3 INTRODUCTION OF THE MULTI-BUSBAR TECHNOLOGY

When solar cell busbar number is increased to more than five, the traditional solar cell and module manufacturing technology will be unsustainable. New metallization project will be needed so that we can realize the industrialization of the multi busbar solar cell and module.

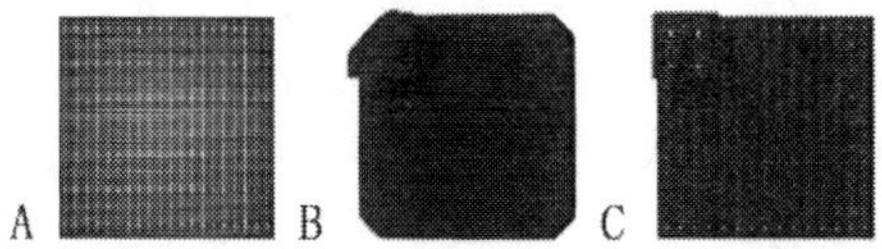

Figure 2 The multi busbar cell

The solar cells used in traditional photovoltaic modules usually have busbar design, but the multi busbar technology is not quite like that. The multi busbar technology can use many kinds of solar cell. As shown in Figure 2A, a kind of multi busbar cell developed by CSI, has more busbars with narrower width. And some other solar cells with no busbar (Figure 2B) or special shape busbar (Figure 2C) [1] can also be used in the multi busbar technology field.

Figure 3 Cells connection conducted by CSI

The multi busbar cells need more ribbons. Figure 3 shows a type of multi busbar cell developed by CSI. As the busbar number is more , the ribbon must be narrower in order to maintain a reasonable shading area of the ribbon. So many and so narrow ribbons will make difficulty of the soldering increased greatly. In this case, the traditional cell interconnection technology would be very difficult to meet our requirements of making the modules, so a new kind of interconnection technology is necessary.

There are many kinds of multi busbar interconnection technology. According to their characteristics, we generally divide them into four categories: soldering, low temperature alloy lamination, conductive adhesive laminating and the prefabricated interconnected grid method (particularly refer to the "Merlin" technology of GTAT) (Table 1).

Among them, the multi busbar technology of soldering is the mostly close to the traditional cell interconnection technology. Schmid company is the earliest one who developed the multi busbar machine for soldering. But now, few domestic equipment manufacturers are also trying to develop this machine. The cell structure adopted by soldering is similar with the conventional cell which retains the busbar design. The cell need to be printed with busbar and finger in front side, while the back electrode and back surface field in rear side (busbar and finger need to be printed on the back side of bifacial cell). The multi busbar technology of soldering still adopt tin-coated ribbon. The electrical connection can be realized between cell and ribbon under the condition of hot soldering. As large number of busbar, the available ribbons are only the ones with small cross section. The ribbon section shap is usually round. The advantage of this multi busbar technology is that its interconnection technology is similar to that of the conventional cells, and related equipment manufacturers also have many years of development experience. In addition, the current equipment can be upgraded and rebuilt to produce the multi busbar module. However, keeping the yield will face the challenge for this technology when the number of soldering point continually increases. Besides, this technology can't be applied in heterojunction (HJT) cell as it's the high temperature process. This technology also faces a bigger challenge as the future solar cell will become more and more thinner.

The typical case of low temperature alloy laminated method is that the SmartWire technology developed by MeyerBurger company. The busbar design can be eliminated for the cell of SmartWire technology. It is a kind of no busbar solar cell. All the cells connect with each other with eighteen or more pieces of round copper wires coated with low temperature alloy. At first, these copper wires are arranged on the polymer film. Then, the copper wires will be laid on the cell together with the polymer film. In the subsequent laminating process, the cell and the metal wire will be connected together as the low temperature alloy coated on the wire surface will melt. As both sides of the SmartWire cell do not need busbar, this technology can save a lot of silver paste. But as this technology introduces more expensive low temperature alloy and other matched encapsulating material, its manufacturing cost is relatively high. The advantage of SmartWire technology is that it has no high temperature soldering process, so it can also be applied to HJT cell and thinner cell.

Conductive adhesive laminating is a kind of multi busbar technology connecting ribbon and cell with conductive adhesive, which also has no high temperature process. The cells adopted by the multi busbar technology of conductive adhesive laminated method have busbar design or no busbar dsign, which mainly depends on the performance of the conductive adhesive. An example of this technology is CF technology of the Hitachi Company. The cell of CF technology has multi busbar design on the rear and no busbar design on the front. The interconnection process of CF technology is as following. Firstly, conductive adhesive film is stick on the busbar location of the both sides of the cell to replace the busbar of the traditional cell. Then the ribbon is placed in the position of conductive adhesive of cell. Next the ribbon and cell will be laminated and connected together through a about 200℃ hot laminating process. The gelatinizing way of conductive adhesive multi busbar technology is not only limited to conductive film way. It also include the way of printing glue on the cell, the way of gluing to the cell and the way of gluing to the ribbon. All conductive adhesive laminating technologies undergo no high temperature process, so this multi busbar technology can be compatible with HJT cell and thinner cell. As conductive adhesive can work only if it maintain certain contact area of ribbon and cell, the ribbon adopted by this multi busbar technology is usually flat shap. As the shading area of flat ribbon is larger, the allowed busbar number design by this multi busbar technology of conductive adhesive laminated method is limited.

Another kind of multi busbar technology is prefabricated interconnected grid, particularly refer to the "Merlin" technology of GTAT. The most obvious characteristic of this technology is to use metal grid to replace the traditional strip shape ribbon. The metal grid is made up of main copper wire and floating cables. The main copper wire plays a role in traditional confluence and the floating cables can maintain the shape of metal grid. The interconnection methods between the metal grid and the cell of "Merlin" contain hot Soldering, low temperature alloy laminating or conductive adhesive laminating

Table 1 Main realization methods of the multi busbar technology

	Soldering	Low temperature alloy laminating	Conductive adhesive laminating	Merlin

Interconnection method	Cells are connected to copper wires by the tin alloy coated on copper wires	Cells are connected to copper wire in lamination by the low temperature alloy coated on copper wires	Cells are connected to copper strips during lamination by conductive adhesive	Soldering, low temperature alloy laminating or conductive adhesive laminating
Printing pattern	Multi busbar	No busbar	Multi busbar/no busbar	No busbar
Printing station	3-4	2	2-4	2-3
Ribbon shap	Round	Round	Rectangle	Copper net
Paste consumption saved	Medium	High	Medium	Medium
Power gains	High	High	High	High
Advantage	Low cost, similar with current production line	Compatible for HJT and thinner wafer	Compatible for HJT and thinner wafer	-
Disadvantage	Unadoptable for HJT and thinner wafer	Expensive for typical cell	Technology readiness level is low	-
Adoptable cells	Typical, bifacial	Bifacial, HJT, thinner wafer	Typical, bifacial, HUT, thinner wafer	-

4 THEORY SIMULATIONS OF MULTI BUSBAR CELL DESIGN AND MODULE INTERCONNECTION

The radiation of sunlight on solar cells, part transmits into the cell interior, and part is reflected by the cell surface. Only the sunlight successfully radiated into solar cells inside, motivating electron and hole pairs which can be successful divided by a p-n junction, can be converted into effective electricity. In the process of charge collection from p-n junction to electrode, the electricity will be wastage by all kinds of resistance loss like the diffusion layer resistance at the top of the cell, contact resistance of fingers and silicon material, busbar resistance and so on.

After cell encapsulated into module, current is collected by the busbar and then transmit mainly along the ribbon at the top of busbar. In addition, as the encapsulation material added, the sunlight reflection behavior will also change, which will affect the conversion efficiency of sunlight into electric energy. In this paper, we consider the influencing factors of the photoelectric conversion efficiency from the module side, including the busbar number and finger number, finger width and height (silver paste consumption), shape and cross-sectional area of ribbon.

Martin a. Green put forward 6 kinds of power losses mechanism associated with the top electrodes on cell side 【2】.In this paper, we also consider the corresponding 6 power loss items (see table 2) related to the busbar and finger design from the module side. Among them, Prf, Pcf and Ptl are the same as Martin a. Green's. For Psf finger shading power loss, we propose a correction coefficient in our calculation, as the light reflected by the fingers can also be used again in encapsulation material due to the module Internal reflection effect. The cell busbar shading power loss is replaced by the ribbon shading power loss Psb. For the similar reason, a correction factor is proposed in our calculation. After collected by busbar, current mainly transmit along the ribbon. So we use the resistance loss Prb which is caused by both sides ribbon, to replace that which is caused by one side busbar.

Table 2 The power loss related to busbar and finger

P_{rf}	The finger resistance power loss
P_{sf}	The finger shading power loss*
P_{cf}	The contact resistance of finger and silicon material power loss
P_{tl}	The diffusion layer transverse current power loss
P_{rb}	The ribbon resistance power loss
P_{sb}	The ribbon shading power loss*

*: Contain internal reflection correction coefficient

The total power loss P-Loss, is equal to the sum of the above 6 items, that is:

P-Loss = Prf + Psf +Pcf + Ptl + Prb + Psb

Others loss, such as the contact resistance between the cell busbar and ribbon, cell BSF transmission resistance and so on, are ignored in the calculation of this paper for their small value. In addition, incident light reflection loss on the surface of the glass and other parameters which are not related to busbar number or finger shape are not included in this paper too.

4.1 Simulation conditions

In this paper, the calculation model is conventional P type cell. The input parameters required in calculation are shown in table 3.

The finger height is one of the main variables. The basic value is 12 um. This height will directly affect the resistance loss of finger and silver paste consumption.

Table 3 All input parameters involved in calculation

parameter	value
The length of the cell	156 mm
The finger shading width	60 um
The finger cross section average width	30 um
The finger height	12 um
The ribbon thickness	0.2 mm

The finger volume resistivity after sintering	3.40*10-6 Ω·cm2
The ribbon volume resistivity	1.75*10-6 Ω·cm2
The contact resistivity of finger and silicon	4.0*10-3 Ω·cm2
The diffusion square resistance	80 Ω/□
The current density at the module maximum operating point	0.035 A·cm2
The single cell voltage at the module maximum operating point	0.512 V
The finger internal reflection correction coefficient	56%
The ribbon internal reflection correction coefficient	67%
The round wire internal reflection correction coefficient	50%

4.2 Effect to power loss

Figure 4 shows the relationship between the busbar number design and module power loss when keeping the finger shape (silver paste consumption) and the module total ribbon width (3.08% of the cell area) the same. Power loss is inversely proportional to the square of the transmission distance. With the busbar number increasing, the current transmission distance along the finger will decrease. Based on this principle, the module power loss will decrease with the increase of the number of the busbar. As the present mainstream 60 pieces of polycrystalline module on the market for example, when the busbar number increased from 3 to 5, P - Loss relatively reduced by 1.3% (as shown in Figure 4). The increase of the corresponding module power watt is 260 x1. 3% = 3.4 watt. When the busbar number continually increases, the finger resistance power loss will further reduce. But the range of reduction will gradually become smaller.

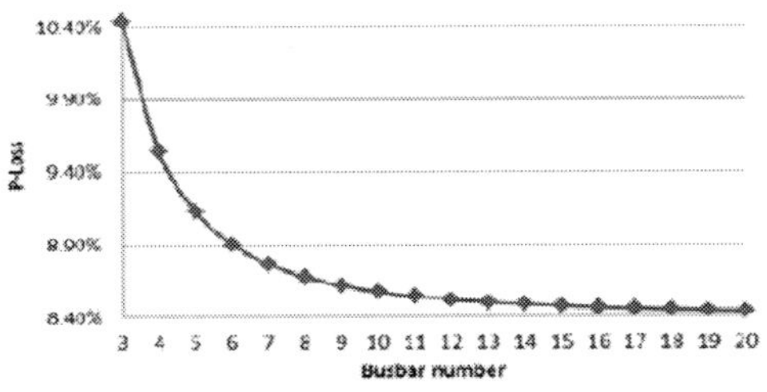

Figure 4 For certain silver consumption, finger resistance power loss plotted over the number of busbars.

4.3 Effect to silver consumption

Silver paste is the most important composition of cell conversion cost. For the MBB module technology, one of the most charm is greatly reduced cell silver paste consumption under the precondition of maintaining or improving the module power. This is a revolutionary significance for the development of the entire photovoltaic technology. Figure 5 shows that, take four busbar as baseline, the relationship between the busbar number and silver paste consumption under the precondition of keeping the same finger resistance power loss. As seen in the figure, when the busbar number increase to 12, the silver paste consumption can be reduced more than 85%.

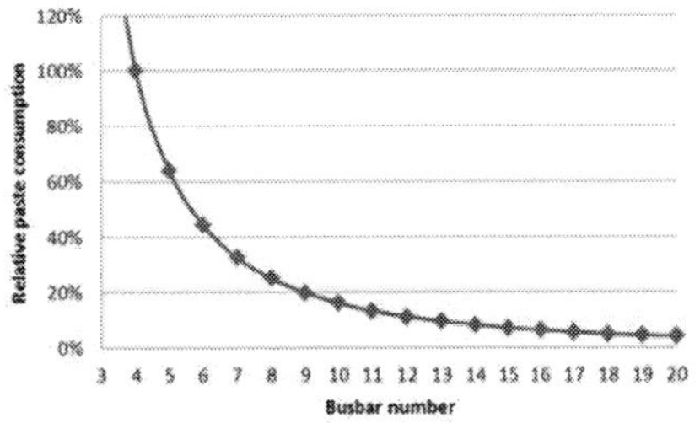

Figure 5 To maintain the same finger resistance, the required silver paste consumption plotted over busbar number.

4.4 Best solution

Is there a design, which need low silver paste consumption while achieve the lower finger resistance power loss? Figure 6 shows that the finger resistance power loss distribution with different busbar number and silver paste consumption. In the right lower area of the picture, we can see that the silver paste consumption is less than that of conventional four busbar which is 50%. The finger resistance power loss is relatively small, which is far lower than that of the conventional four busbar.

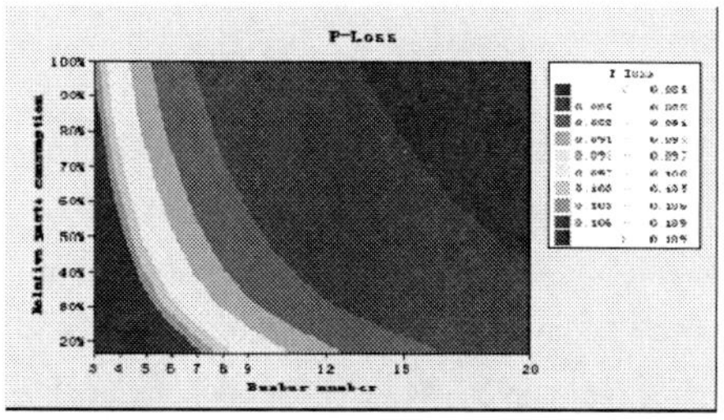

Figure 6 Power loss of finger resistance for different busbar number and different silver consumption

We also do multi dimensionally theoretical simulation on some main parameters, such as diffusion square resistance, finger number, busbar number, ribbon shape and size and so on, to provide the BOM optimization direction for cell and module.

5 THE MULTI BUSBAR TECHNOLOGY EXPERIMENTAL RESULTS

We fabricated many multi busbar cells and modules. Here is a representative experiment :

We prepare 1000 pieces of polycrystalline wafers, and then texture, diffusion, etching and SiNx deposition together. After that, we divide all the above wafers into four groups at the ratio of 1/10, 3/10, 3/10 and 3/10, and then the wafers were printed and sintered into four busbar cells, nine busbar cells, twelve busbar cells and fifteen busbar cells. The four group's cell electrode graphic and morphology parameters are shown in table 4. After test, each group take 60% of the highest efficiency cells to fabricate into modules. Then the above nine busbar cells, twelve busbar cells and fifteen busbar cells are respectively divided into three parts with corresponding size ribbon (Φ0.32 mm, Φ0.37 mm and Φ0.40 mm) and fabricated into three modules for each group.

Table 4 The cell electrode graphic and morphology parameters

	4BB	9BB	12BB	15BB
Busbar number	4	9	12	15
Busbar width	1.1mm	0.1mm	0.1mm	0.1mm
Finger number	101	101	101	101
The finger shading width	60 um	60 um	60 um	60 um
The average width of the finger cross section	30 um	30 um	30 um	30 um
The height of the finger	12 um	10 um	10 um	10 um

Take four busbar cell front silver paste consumption as a baseline, nine busbar, twelve busbar, fifteen busbar cell front silver paste consumption is shown in Figure 7. All kinds of multi busbar cell silver paste consumption are lower than that of the conventional four busbar cell.

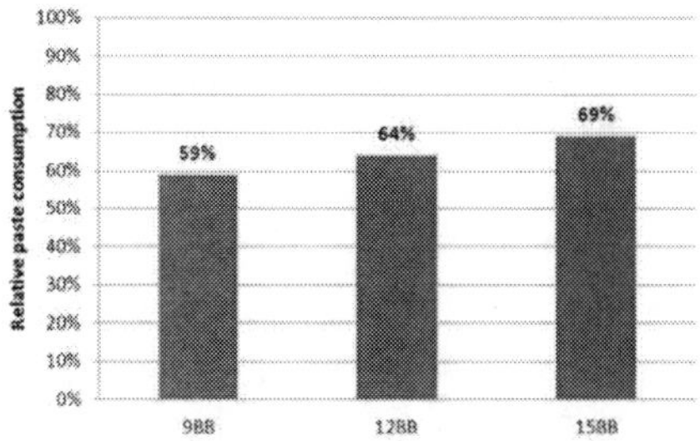

Figure 7 The multi busbar cell relative silver paste consumption

Take four busbar module output power as a baseline, the rest nine module output power relative gains are shown in Figure 8. As shown in the figure, the multi busbar module has more 1% power gain than the conventional four busbar module. How much the power gain of the multi busbar module can bring will depend on the busbar number design and ribbon size choice. The Figure 8 also shows that each group module power gain of theoretical simulation. The theoretical simulation result and experimental result match very well.

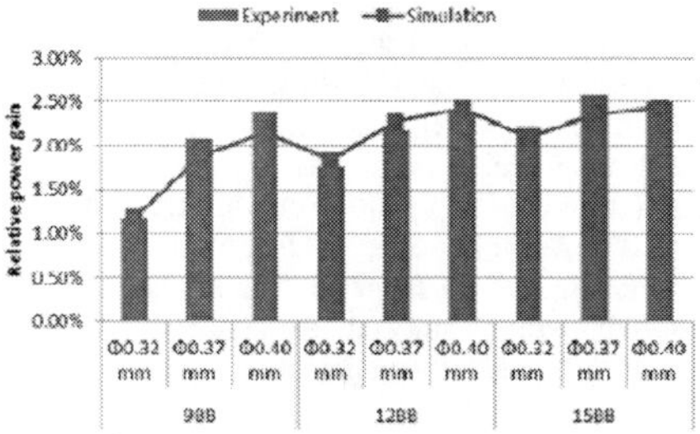

Figure 8 The experimental and theoretical simulation power gain of the multi busbar module

6 CONCLUSIONS

The multi busbar technology can greatly reduce the silver paste consumption, increase the output power of module at the same time, is the inevitable trend of the future technology development. CSI has built the multi busbar simulation calculation method, provides the reliable design optimization direction for the multi busbar cell and module, and the theoretical simulation result and experimental result match very well.

7 REFERENCES

[1] Johann Walter, Marco Tranitz, Michael Volk, Christian Ebert, Ulrich Eitner, Energy Procedia 55 (2014) 380 - 388;

[2] Martin A. Green, Solar cell work principle, technology and system application;

MONOFACIAL IV MEASUREMENTS OF BIFACIAL SILICON SOLAR CELLS IN AN INTER-LABORATORY COMPARISON

Michael Rauer[1,*], Karsten Bothe[2], Corrado Comparotto[3], Pi Danzl[4], Maarten Debucquoy[5], Nicolas Enjalbert[6], Jochen Hohl-Ebinger[1], Petra Manshanden[4], Yannik Veschetti[6], Johnson Kai Chi Wong[7]

[1]Fraunhofer Institute for Solar Energy Systems (ISE), Heidenhofstraße 2, 79110 Freiburg, Germany
[2]Institute for Solar Energy Research Hamelin (ISFH), Am Ohrberg 1, 31860 Emmerthal, Germany
[3]International Solar Energy Research Center (ISC), Rudolph-Diesel-Straße 15, 78467 Konstanz, Germany
[4]ECN Solar, Westerduinweg 3, 1755 LE Petten, The Netherlands
[5]imec, Kapeldreef 75, 3001 Heverlee, Belgium
[6]CEA, LITEN, INES, 50 av. du Lac Léman, 73370 Le Bourget du Lac, France
[7]Solar Energy Research Institute of Singapore (SERIS), 7 Engineering Drive 1, Singapore 117574, Singapore

[*]Corresponding author: phone: +49(0)761/4588-5564, e-mail: michael.rauer@ise.fraunhofer.de

ABSTRACT: Standardizing illuminated IV measurements of bifacial solar cells and modules is a central objective for the introduction of bifacial products into the market. In this paper, the application of monofacial IV measurement conditions to bifacial solar cells is evaluated in an inter-laboratory comparison among seven research institutes in Europe. Bifacial silicon solar cells manufactured at five different sites with five different fabrication technologies were used in this investigation. We demonstrate that several characteristics of the measurement setups which are of minor importance for the measurement of monofacial solar cells can significantly affect bifacial solar cell measurements: (i) the reflectance of the measurement chuck and (ii) the electrical conductance of the chuck implying specific contacting schemes. When dividing the measurement results of this round robin into two groups according to chuck reflectance and conductance, the deviations in the IV parameters among the different partners are mostly within the uncertainty limits commonly reported for monofacial solar cell measurements. For standardization of bifacial solar cell measurements, it is therefore important to define admissible ranges for the chuck reflectance and to specify the contacting scheme in the standard.
Keywords: Bifacial Solar Cells, Standardization, IV Measurement, Round Robin

1 INTRODUCTION

In today's solar cell market, bifacial solar cells and modules are becoming more and more important. Because of their ability to convert light from both front and rear surface, bifacial modules can produce 5 to 25 % more energy than standard monofacial products [1]. Therefore, the market share of bifacial silicon solar cells is estimated to increase to up to 20 % in the next 20 years [2]. Although the current development of bifacial solar cell and module technologies is promising, an essential issue hindering the introduction into the market is the missing IV measurement standardization [3].

The standards of the International Electrotechnical Commission (IEC) describe the conditions for the measurement of solar cells to ensure inter-laboratory comparability of the solar cell characteristics. However, there is as yet no standard defining the measurement of bifacial solar cells. The most common method to determine the illuminated IV characteristics of bifacial solar cells is the application of monofacial measurement conditions including single-sided illumination. This, however, can lead to additional uncertainties: light that is transmitted through the bifacial solar cell can be reflected at the measurement chuck and reenter the solar cell through its rear surface. Therefore, the chuck reflectivity influences the measured current significantly [4]. In addition, the rear electrical contacting scheme depends on the measurement chuck design as well: whereas reflective chucks are normally made from copper or brass blocks, which electrically contact the entire rear metal grid of the solar cells, non-reflective chucks are commonly covered by non-conducting black foils, so that only the rear busbars are contacted. In general, this leads to the measurement of different fill factors [5]. However, reliable and accurate characterization procedures are essential for solar cell manufactures in order to optimize the solar cell structures for high energy yield.

To clarify the impact of the measurement setup on the illuminated IV characteristics of bifacial solar cells under monofacial conditions, a round robin among seven different research institutes in Europe was carried out. Bifacial solar cells of five different fabrication techniques were investigated. The aim was to evaluate inter-laboratory comparability and to elaborate the reasons for measurement deviations.

2 EXPERIMENTAL

Twenty-seven industrial bifacial *n*-type silicon solar cells with edge lengths of 156 mm and three busbars were provided by five participants of this round robin. The investigated solar cells thus comprise five different fabrication technologies. The illuminated IV characteristics of the solar cells were independently measured by the different partners under standard testing conditions (25°C, 1000 W/m^2, AM1.5G spectrum) using single-sided illumination. Separate measurements of front and rear characteristics were carried out. The measurement chucks used by the partners can be grouped into (i) "reflective and conductive chucks", which are generally realized by metal chucks with reflective surface and electrical contact to the entire rear metal grid of the solar cells, and (ii) "non-reflective and non-conductive chucks", which are chucks with black foils and electrical contacts only to the rear busbars of the solar cells (see

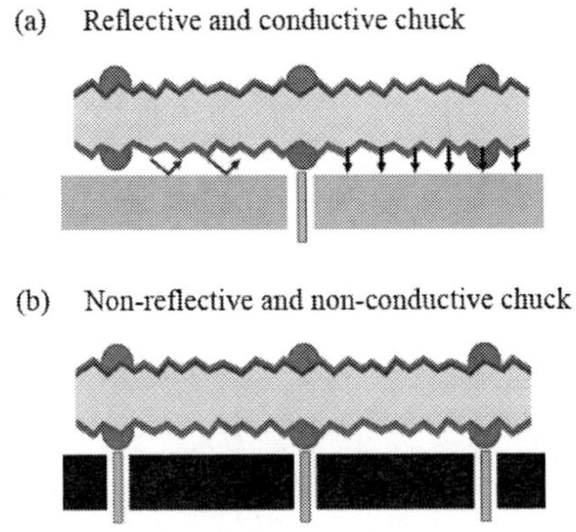

Figure 1: Schematics of the two different measurement setups applied in this study. The figures are adapted from [5].

Figure 1).

Table 1 gives an overview of the chucks applied by the partners. For simplicity, reflective/conductive chucks are referred to as *"reflective chucks"*, non-reflective/non-conductive ones as *"non-reflective chucks"* in the following.

Partners 1, 5 and 6 carried out additional spectral response measurements for the spectral mismatch correction, thereby using corresponding identical measurement chucks. Partner 5 measured the IV characteristics with a non-reflective chuck at the beginning and at the end of the round robin. The accordance of the measured IV parameters was within the uncertainty limits given by the partner.

In order to assess the inter-laboratory comparability, the non-area-related IV characteristics, namely the short-circuit current I_{sc}, the open-circuit voltage V_{oc}, the fill factor FF and the absolute power P_{mpp}, were evaluated. To investigate the influence of the series resistance of the metal grids, the busbar-to-busbar resistances were additionally measured.

3 DEVIATIONS IN THE MEASURED IV PARAMETERS

In order to compare the IV parameters measured by the different partners of this round robin, median values were calculated for each cell, thereby distinguishing between reflective and non-reflective chucks. The relative differences of the parameters measured by the partners to the respective cell medians were determined. To eliminate the effect of outliers, median instead of average values were used for the evaluation.

Figure 2 shows the deviations of the short-circuit current, the open-circuit voltage and the fill factor from the respective medians. For reasons of clarity, the deviations in power are not depicted. The dashed lines indicate the expanded measurement uncertainties (coverage factor 2) for the measurement of monofacial solar cells as given by partners 1 and 5.

The distributions can be used to investigate the random and the systematic deviations between the partners for each IV parameter. Random variations are quantified by the standard deviations of the distributions, *i.e.* by the widths of the distributions, systematic variations by the absolute deviations from the median, *i.e.* by the offsets of the distribution averages [6].

Table 1: Measurement setups used by the partners of this round robin for the determination of the illuminated IV characteristics. "x x" means that two different, independent setups of the respective configuration were applied.

Partner	Reflective and conductive chuck	Non-reflective and non-conductive chuck
1	x	x
2	x	
3	x	
4		x
5	x x	x
6	x x	x
7	x	

3.1 Short-circuit current deviations

The I_{sc} deviations observed for bifacial solar cells between the partners in this round robin are within the uncertainty limits commonly reported for monofacial solar cell measurements (see dashed lines in Figure 2), if the measurements are differentiated by the two chuck types. No significant differences in the random and systematic variations were thereby found between measurements on reflective and on non-reflective chucks: The distributions of the I_{sc} values show that, for both chuck types, the random variations at each partner's setup are in the same order or larger than the systematic deviations among the different partners.

In this study, the random deviations are more pronounced than in other round robin evaluations published before [6]. We attribute this discrepancy to the use of solar cells of different technologies together with the partial omission of the spectral mismatch correction in the present round robin. As the solar cells investigated here exhibit diverse spectral responses, using the same reference solar cell for all the measurements leads to variations in the spectral mismatches between the tested solar cells and the reference cell. If no mismatch correction is carried out, this in turn can result in a broadening of the I_{sc} distribution.

In addition, the non-application of a monitor setup to correct the measurement for fluctuations in the lamp intensity of the sun simulator can also contribute to the random deviations [7].

These two aspects would occur for the measurement of monofacial and bifacial solar cells alike. There are also effects characteristic for bifacial solar cells, though, which lead to systematic deviations between the partners. These are discussed in more detail in section 4.1.

3.2 Open-circuit voltage deviations

Figure 2 shows that the deviations in V_{oc} among the different partners for reflective and non-reflective chucks are mostly within the uncertainty limits specified by partners 1 and 5 for the measurement of monofacial solar cells. The random variations in V_{oc} given by the widths of the distributions are thereby significantly smaller than the random variations in I_{sc}

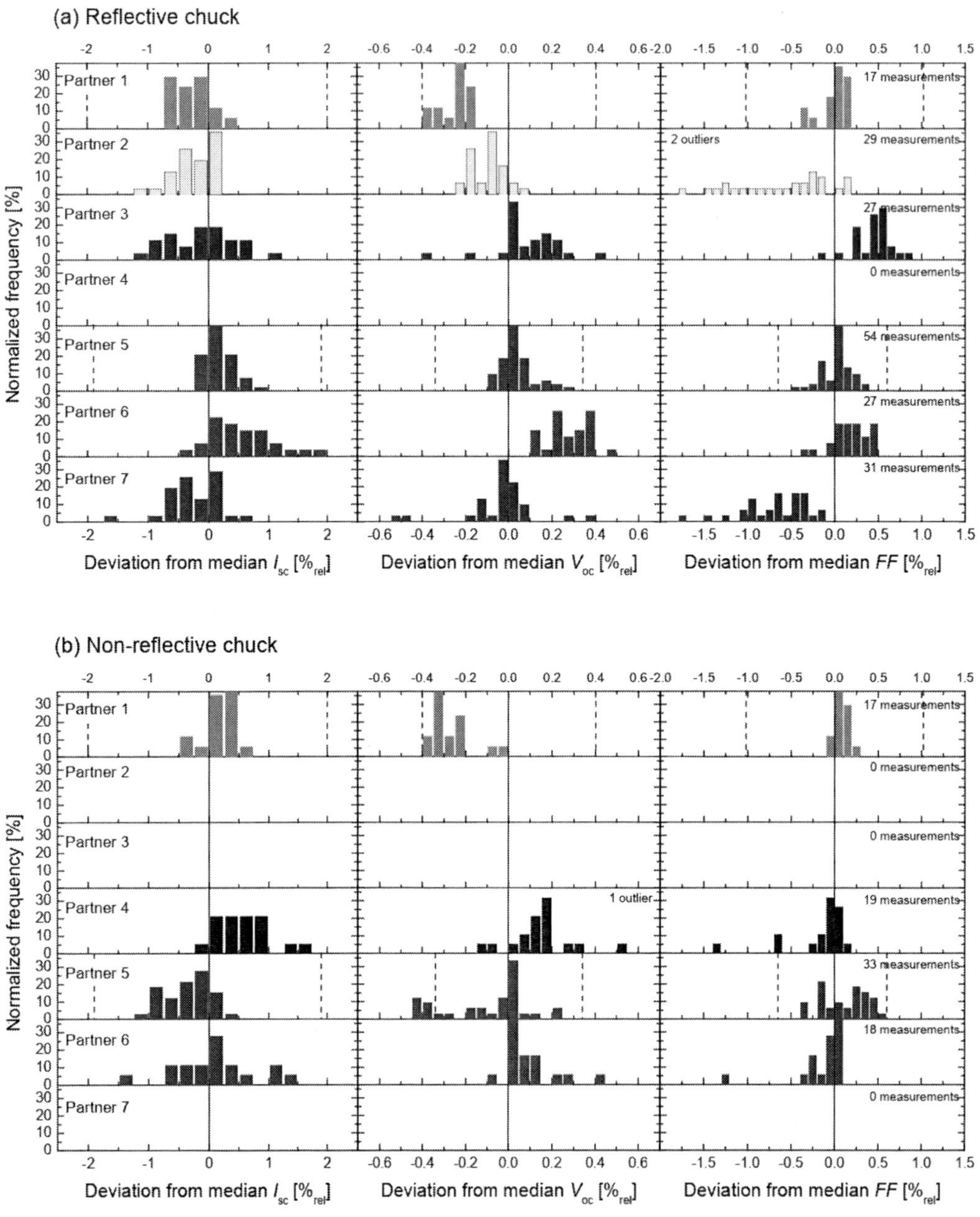

Figure 2: Relative deviation of the short-circuit current I_{sc}, the open-circuit voltage V_{oc} and the fill factor FF measured by each partner from the respective median value for measurements on (a) reflective chucks and (b) non-reflective chucks. Please note that not all the partners measured each solar cell; only solar cells measured by at least four different institutes were taken into account in the evaluation. The numbers in the upper right corners of the graphs indicate the amount of cells measured by the partners contributing to the evaluation. The dashed lines in the graphs of partners 1 and 5 demonstrate the respective expanded measurement uncertainties determined for monofacial solar cell measurements (coverage factor of 2). For the sake of clarity, outliers are not shown but indicated in the graphs.

discussed above, which has also been reported for earlier round robin evaluations [6]. Within the uncertainty limits, systematic deviations in V_{oc} among the partners occur. In relation to the widths of the respective distributions, the systematic deviations are larger for V_{oc} than for I_{sc} measurements.

It is well-known that the major uncertainty in the V_{oc} measurement comes from the control of the solar cell temperature. The small random variations in V_{oc} observed here show that the solar cell temperature is generally controlled accurately at the partners' setups, so that the V_{oc} determination is highly reproducible.

The larger systematic differences in V_{oc}, however, indicate that the absolute temperature may vary among the different partners due to different positioning of the temperature sensor (*e.g.* directly on the solar cell or indirectly on the measurement chuck). This has also been reported in earlier round robin evaluations [6, 7]. Figure 2 shows that this issue occurs for reflective and non-reflective measurement chucks alike.

For partner 5, the deviations from the median V_{oc} exceed the respective measurement uncertainty for measurements on non-reflective chucks. Provided that the medians approximate the respective real values adequately, this implies that the uncertainty budget for non-reflective chuck measurements needs to be adapted.

In section 4.2, the differences in temperature control on reflective and on non-reflective chucks are investigated in more detail.

3.3 Fill factor deviations

For most partners, the deviations in *FF* are within the uncertainty limits of monofacial solar cell measurements for both reflective and non-reflective chucks. Only the fill factor values measured by partners 2 and 7 exceed these limits and are broadly distributed towards smaller values. It was found that the *FF* deviations of partner 2 thereby correlate with the series resistances of the metal grids of the tested solar cells, *i.e.* the higher the grid series resistance, the higher the *FF* deviation from the respective median. This suggests that a systematic error is the cause for the measurement deviations of partner 2. No equivalent correlations were found for the measurements of partner 7. Further investigations on this issue are still pending.

Apart from the measurement results of partners 2 and 7, no significant differences in the systematic and random *FF* variations were found between measurements on reflective and on non-reflective chucks.

In section 4.3, reasons for the systematic deviations are investigated and discussed in more detail.

4 SYSTEMATIC DEVIATIONS SPECIFIC TO THE MEASUREMENT OF BIFACIAL SOLAR CELLS

In order to investigate the systematic deviations further, the differences in IV parameters between reflective and non-reflective chuck measurements were examined as observed directly by the different partners. This allows for a higher precision in the assessment of deviations between the two chuck types, since the variations measured by one partner are generally smaller than the variations among the different partners [7, 8].

4.1 Chuck reflectance

The differences ΔI_{sc} in the short-circuit current from reflective to non-reflective chuck measurements as detected by partners 1, 5 and 6 are shown in Figure 3. ΔI_{sc} could thereby only be determined for the partners that carried out measurements on both reflective and non-reflective chucks.

A significant offset between the ΔI_{sc} distribution of partner 1 and the distributions of partners 5 or 6 can be seen. Although the average ΔI_{sc} values of partners 5 and 6 agree, the widths of the respective distributions differ.

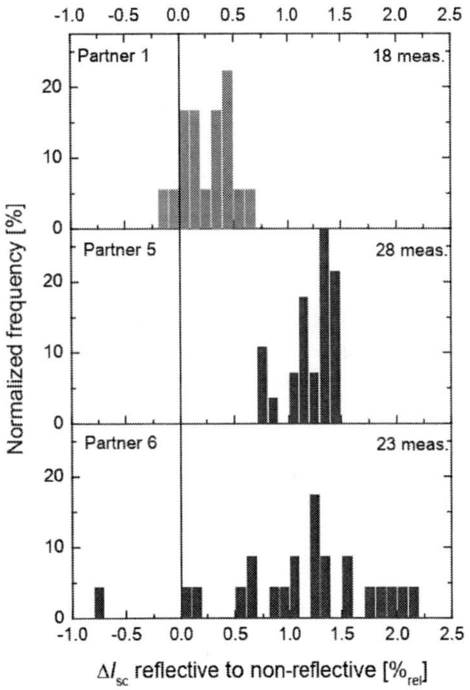

Figure 3: Relative difference ΔI_{sc} between the short-circuit currents measured on reflective and non-reflective chucks by the different partners. The difference could only be determined for the partners that carried out measurements on both reflective and non-reflective chucks.

In order to investigate the offset in ΔI_{sc} further, the reflectances of the measurement chucks used by the partners were measured, as shown in Figure 4. It can be observed that (i) the reflectances corresponding to one chuck type vary significantly among the partners, and (ii) the differences in reflectance between reflective and non-reflective chucks vary as well. Partner 1 intentionally uses a non-reflective chuck with slightly higher reflectance. The focus of the chuck design is to enable the accurate thermal control of the tested solar cells.

Differences in chuck reflectance lead to different contributions of the transmitted and reflected light to the respective short-circuit current. Thus, the larger the difference in the chuck reflectance between reflective and non-reflective chuck, the larger the short-circuit current difference ΔI_{sc}.

It becomes obvious that the classification into reflective and non-reflective chucks actually has to be differentiated further as there are no *typical* measurement chucks. The chuck reflectance is an important parameter for the measurement of bifacial solar cells because it can significantly affect the measured I_{sc} values. For standardization of the IV measurements, it is therefore important to either define an admissible reflectance range for reflective and non-reflective measurement chucks or to explicitly specify the chuck reflectance in the measurement report.

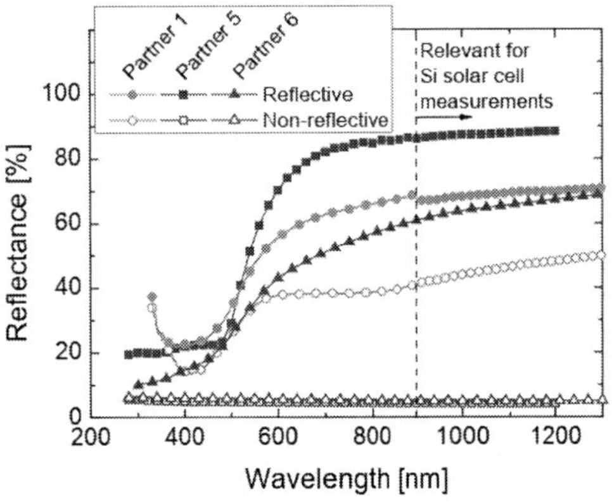

Figure 4: Reflectance of the measurement chucks applied by partners 1, 5 and 6. Full symbols represent measurements on reflective chucks, empty symbols on non-reflective chucks.

4.2 Temperature control

In section 3.2, systematic differences in the V_{oc} distributions among the different partners have been identified. These deviations can most likely be attributed to differences in the solar cell temperature control.

It has yet to be investigated if the temperature control is different for the two chuck types. Figure 5 shows the differences in V_{oc} between reflective and non-reflective chuck measurements for partners 1, 5 and 6. All ΔV_{oc} distributions are systematically shifted to positive voltages, *i.e.* V_{oc} is measured higher on reflective chucks. This shift is thereby more pronounced for partners 5 and 6, who also exhibit broader distributions.

We have considered the implications of physical origins for the V_{oc} differences: assuming similar measurement conditions, the only physical reason for the measurement of different V_{oc} values on reflective chucks is the increased injection resulting from an increased chuck reflectance. To assess the relevance of the increased injection, we calculated the corresponding V_{oc} increase using the ideal diode equation. We calculated values well below 0.1 $\%_{rel}$, which shows that this effect is negligible. Measurable differences in V_{oc} on reflective and non-reflective chucks are therefore measurement-related and are attributed to inaccuracy in temperature control. The comparison of the V_{oc} values measured with the two different chuck types thus provides a good measure to check consistency in temperature control.

Reflective chucks exhibit good thermal conductance. For these chucks, there is an adequate thermal contact between the solar cells under test and the chuck. For non-reflective chucks, though, the insulating layer generally covering the surface is of poor thermal conductance. The temperature adjustment is therefore more critical for non-reflective chucks [9]. Underestimating the heating of the solar cell under test during measurement on non-reflective chucks results in the determination of too small V_{oc} values and thus to $\Delta V_{oc} > 0$. This can be seen for the measurements of partners 5 and 6. Partner 1 applies a non-reflective measurement chuck which is optimized for improved thermal contact to the tested solar cells. Figure 5 shows that this leads to a smaller difference in V_{oc} between the two chuck types.

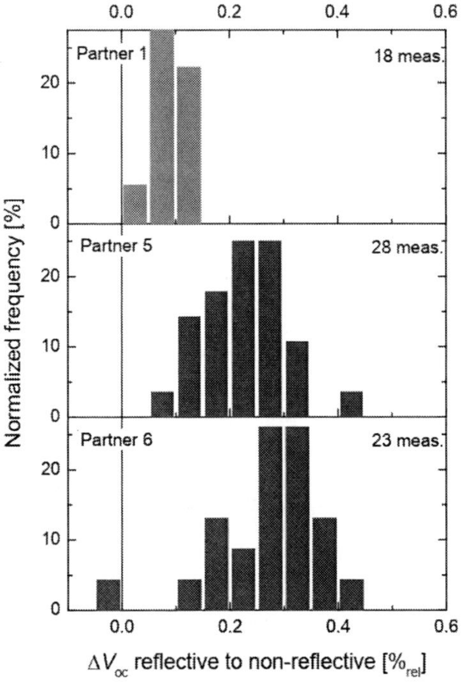

Figure 5: Relative difference ΔV_{oc} between the open-circuit voltages measured on reflective and non-reflective chucks by partners 1, 5 and 6.

In conclusion, the temperature of bifacial solar cells has to be adjusted carefully to standard testing conditions. This round robin showed that there are not only general differences in the temperature control among the different partners, but also between the different chucks used by the individual partners.

4.3 Influence of chuck conductivity

Figure 6 shows the differences in fill factor ΔFF between reflective and non-reflective chuck measurements as observed by partners 1, 5 and 6. A systematic shift to positive ΔFF values can be seen for all three partners, *i.e.* the fill factors are measured higher on reflective as compared with non-reflective chucks.

A physical reason for the measurement of different fill factors with the two chuck types is the different injection conditions leading to different series resistance losses. Using Quokka simulations of a typical industrial bifacial Si solar cell [10], this effect was estimated to be below 0.1 $\%_{rel}$, though, and thus to be negligible.

A measurement-related reason for $\Delta FF > 0$ is the difference in the rear contacting scheme of the two chuck types. Whereas reflective chucks contact the entire rear metal grid of the solar cells, non-reflective chucks contact only the rear busbars. Therefore, the lateral resistance of the rear metal finger grid contributes to the series resistance of the tested solar cells for non-reflective but not for reflective chuck

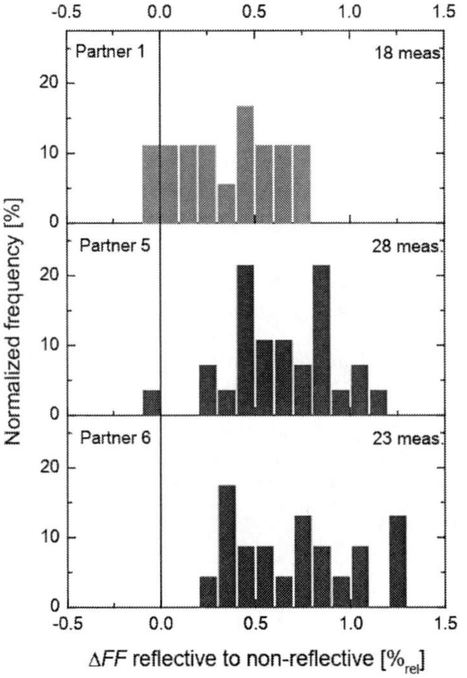

Figure 6: Relative difference ΔFF between the fill factors measured on reflective and non-reflective chucks by partners 1, 5 and 6.

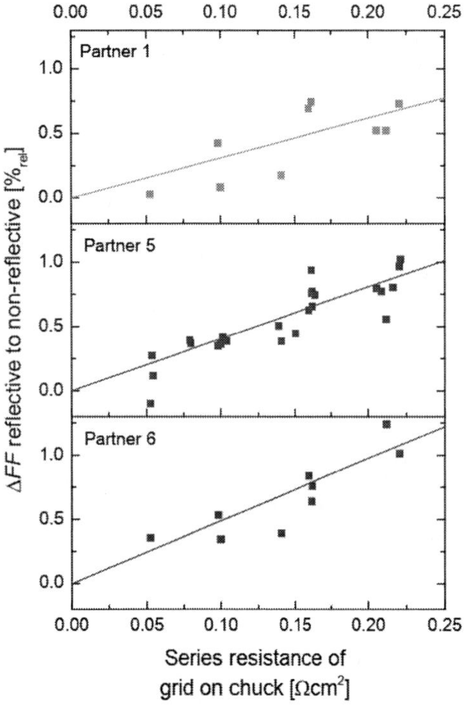

Figure 7: Relative difference ΔFF between the fill factors measured on reflective and non-reflective chucks as a function of series resistance of the metal grid contacting the chucks. The lines are guides to the eyes.

measurements. To evaluate this effect, the series resistance of the front and rear metal grid was determined for each solar cell of this round robin by measuring the respective average busbar-to-busbar resistances. The fill factor differences ΔFF between reflective and non-reflective chucks were then correlated with the series resistances of the grids contacting the chucks during the measurements, *i.e.* ΔFF corresponding to front side measurements of a specific solar cell was related to the grid resistance of its rear side and *vice versa*. Figure 7 shows that there is a clear correlation between ΔFF and the grid series resistance for all three partners.

This demonstrates that the rear contacting scheme significantly affects the measured *FF* for bifacial solar cells. This effect also occurs in the measurement of conventional monofacial solar cells, however to a much lesser degree. Since conventional solar cells feature a full-area rear metal contact, the series resistance due to lateral current flow to the rear contacting unit is typically one order of magnitude smaller (0.02 to 0.03 Ωcm^2 [11]). Therefore, the influence of the contacting scheme associated with the chuck type on the measured *FF* is significantly smaller.

Although all partners measure higher fill factors with reflective as compared with non-reflective chucks for the solar cells of this round robin, the *extent* how much higher the *FF* is measured is different for partner 1 and partners 5 and 6, *i.e.* there is an offset in the ΔFF distributions between these partners. This can partially be explained by the differences in the control of the solar cell temperature, which has been discussed in the previous section. Underestimating the temperature on non-reflective chucks leads to the measurement of too small fill factors and a slight overrating of ΔFF. Further contributions to ΔFF differences possibly result from differences in placement and distance of current and voltage pins among the partners. Since the contact pin geometry affects reflective and non-reflective chucks differently (front contact for reflective chucks, both front and rear contacts for non-reflective chucks), different contact pin geometries of the partners could result in differences in ΔFF.

In conclusion, it has been shown that the contacting scheme is more critical for the measurement of bifacial as compared to the measurement of monofacial solar cells due to the high series resistance of the metal grids. For standardization of the IV measurements, it is therefore crucial to explicitly specify the contacting scheme used for the measurements.

5 SUMMARY

A round robin among seven different research institute partners in Europe has been carried out to investigate the application of monofacial IV measurement conditions to bifacial silicon solar cells. Twenty-seven solar cells of five different fabrication techniques were used for the measurements. The measurement chucks applied by the partners were divided into two categories: (i) reflective and conductive chucks and (ii) non-reflective and non-conductive chucks.

It has been shown that, for both chuck configurations, the deviations in the short-circuit currents, open-circuit voltages and fill factors measured for the bifacial solar cells among the different partners are mostly within the uncertainty limits commonly reported for monofacial solar cell measurements. Within these limits, systematic differences in the IV parameters among the partners larger than random variations were found for measurements on both chuck types. The measurements which exceeded the uncertainty limits were examined further and possible causes for these discrepancies were discussed.

In order to investigate the systematic deviations among the partners in more detail, the differences in the IV parameters between reflective/conductive and non-reflective/non-conductive chuck measurements as observed directly by the different institutes were examined. It was thus demonstrated that there are several effects, which play a minor role in the measurement of monofacial solar cells, but are strongly relevant for bifacial solar cell measurements.

The comparison of the chuck reflectances has shown that the reflectances differ significantly for both reflective/conductive and non-reflective/non-conductive chucks among the partners. It is demonstrated that this has a significant effect on the measured short-circuit current of bifacial solar cells.

Significant differences in the open-circuit voltages among the different partners were observed, which is likely caused by the temperature control of the solar cells. In addition to differences in the positioning of the temperature sensor among the partners, which is expected to affect both types of measurement chucks alike, the heating of the bifacial solar cells on non-reflective/non-conductive chucks seems to result in an additional underestimation of V_{oc} on these chucks for some partners.

The contacting scheme, which differs for the two measurement chuck types (contact to entire rear grid for reflective/conductive chucks, contact to the busbars only for non-reflective/non-conductive chucks), significantly affects the measured fill factor due to the different contribution of the rear metal grid. Although this effect also occurs for monofacial solar cells, it is significantly more pronounced for bifacial solar cells due the higher series resistance of the rear contact grid compared to a conventional full-area rear contact.

For standardization of bifacial solar cell measurements, it is therefore important to either define admissible ranges for the chuck reflectance and to specify the contacting scheme in the standard or to explicitly state these parameters in the measurement report.

ACKNOWLEDGEMENTS

This work has partially been carried out in the framework of the project "SOPHIA – Solar Photovoltaic European Research Infrastructure". SOPHIA is a project supported by the European Commission in the Seventh Framework Programme under grant agreement no 262533.

REFERENCES

1 C. Reise, A. Schmid, *Realistic yield expectations for bifacial PV systems – An assessment of announced, predicted and observed benefits*, Proc. 31st EUPVSEC, Hamburg, Germany, (2015), 1175.

2 International Technology Roadmap for Photovoltaic (ITRPV) Results 2014. 2015: Berlin.

3 R. Kopecek, Y. Veschetti, E. Gerritsen *et al.*, *Bifaciality: One small step for technology, one giant leap for kWh cost reduction*, Photovoltaics International, 26, 2014.

4 J. Hohl-Ebinger, W. Warta, *Bifacial solar cells in STC measurement*, Proc. 25th EUPVSEC, Valencia, Spain, (2010), 1358.

5 J. Lossen, T. Böscke, H.-J. Krokoszinski, *Increased Energy Yield for Bosch's n-Type Solar Cells in Bifacial Application*, presented at 1st bifi PV workshop, Constance, Germany, 2012.

6 P. Manshanden, N.J.C.M van der Borg, M. Bliss *et al.*, *Round robins of solar cells to evaluate measurement systems of different European research institutes*, Proc. 28th EUPVSEC, Paris, France, (2013), 861.

7 G. Hahn, A. Herguth, A. Helfricht, *et al.*, *IV measurements of mc-Si solar cells: Comparison of results from insitute and industry partners within the CrystalClear project*, Proc. 23rd EUPVSEC, Valencia, Spain, (2008), 1182.

8 L.M. Popescu, A. Schneider, E. Wefringhaus, *et al.*, *Round robin analysis of IV data for CZ and mc-Si solar cells: Comparison of results from institute and industry partners within FutureFab project*, Proc. 29th EUPVSEC, Amsterdam, The Netherlands, (2014), 3374.

9 A. Herguth, S. Riegel, *Challenges occuring during the electrical characterization of (bifacial) solar cells*, presented at 1st bifi PV workshop, Constance, Germany, 2012.

10 A. Fell, K.R. McIntosh, P.P. Altermatt, *et al.*, *Input Parameters for Simulation of Silicon Solar Cells in 2014*, IEEE Journal of Photovoltaics 2015; 5(4): 1250.

11 T. Fellmeth, *Silicon Solar Cells for the Application in Low Concentrator Systems - Development and Characterization*, Dissertation, University of Tübingen, Tübingen, 2014.

A FIRST STUDY OF TERAHERTZ EMISSION FOR A-SI:H/C-SI PASSIVATED INTERFACE IN HIT SOLAR CELLS

Jonathon Mitchell[1], Akira Ito[2], Toshimitsu Mochizuki[1], Hidetoshi Nakanishi[2], Hidetaka Takato[1]
[1]Fukishima Renewable Energy Institute, The National Institute of Advance Industrial Sciences and Technology, 2-2-9 Machiikedai, Koriyama, Fukushima 963-0298, Japan
[2]SCREEN Holdings Co. Ltd. 322 Furukawa-cho, Fushimi-ku, Kyoto, 612-8486, Japan

ABSTRACT: In this paper, we present the first measurement and analysis of terahertz (THz) generation from the hydrogenated amorphous silicon (a-Si:H) passivated c-Si surface in the absence of an external magnetic field. Using laser terahertz emission microscopy (LTEM), the terahertz emission from the surface of silicon wafers with an intrinsic a-Si:H passivation layer and those with a doped a-Si:H Hit-like stack have been measured, revealing hidden information relating to surface photocharges and band-bending not detectable by other methods without metallization. We can directly observe the polarity of the charges at the interface, as well as the strength of the inherent field-effect influencing band-bending, as a result of deposition conditions and a-Si:H quality. In addition, we are able to apply LTEM to image mapping of the deposited intrinsic a-Si:H only interface, and that of the a-Si:H(i/n) stacked layers, revealing electronic features associated with surface preparation undefined from other methods such as photoluminescence.

Keywords: a-Si:H, Heterojunction, Terahertz, Band-bending

1 INTRODUCTION

In understanding the surface boundary conditions for the a-Si:H/c-Si interface region, as with any solar cell, has typically only been possible through MIS-based structures and capacitance-voltage (C-V) analysis. However, the incorporation of a metallised contact strongly influences the internal electric field of both the crystalline silicon and the a-Si:H layer; and in the case of the latter, the polarization state. Therefore, measurement of these properties in the absence of these contacts by a non-destructive method can provide insight into the intrinsic field which modifies the interface potential. By applying Laser Terahertz Emission Microscopy (LTEM)[1,2,3], we are able to study surface photocharges, electric field distribution and band-bending conditions with sub-picosecond resolution, at any stage of fabrication, from bare wafer to a completed solar cell.

THz wavelets emitted from the surface of suitable materials are typically low-energy in terms of the excitation light source compared to other common techniques in photovoltaic characterisation[4]. While this does limit the spatial resolution to the laser spot diameter, probing of the transient photogenerated carriers at the c-Si surface is made available. The excitation of photocarriers by femtosecond laser pulses, leads to ultrafast charges being transport driven by the intrinsic surface depletion field of the c-Si, or via the photo-Dember effect which occurs from a difference in diffusion velocities between electrons and holes. Therefore, THz radiation, due to the photocurrent, is proportional to both the difference in mobility of photocharges, and the gradient of the carrier density (Fig. 1). Furthermore, modulation of the photocurrent occurs due to the local electric field extrinsically or intrinsically induced at the laser illumination site.

As part of this work, we present a first analysis in the use of LTEM for the evaluation of the a-Si:H/c-Si interface, whereby, the THz emission waveform from the c-Si silicon interface region with passivating intrinsic a-Si:H and the Hit-like stacked layer (a-Si:H(i)/a-Si:H(n)) during photoexcitation by Ti:Sapphire laser pulse is measured over suitable time-domains.

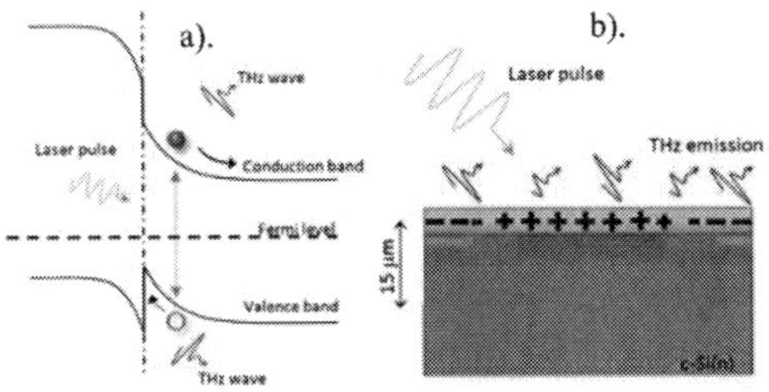

Figure 1: a). Illustration on the source of the THz waveform orginating from the presence of positive (open) or negative (closed) photocharges within the surface region, b). Sketch of the c-Si(n) surface passivated with thin a-Si:H(i) showing regions of both postive and negative photocharges dependent on surface conditions. Indicated charges (-,+) occur at the c-Si surface region and are offset in the diagram for clarity.

2 EXPERIMENTAL DETAILS

2.1 Preparation of a-Si:H/c-Si

Float-zone crystalline silicon wafers were etched in HNO_3/HF solution, followed by standard RCA clean and native oxide removal in 10% HF solution, prior to being loaded into a PECVD system which is then evacuated.

Deposition of a-Si:H(i) and a-Si:H(n) layers was in a 13.56MHz parallel-plate PECVD system with inverted growth surface. Deposition conditions were varied, however, all occurred at temperatures of 225°C and *rf*-power of 4W (35mW/cm^2). All deposited samples underwent stepwise post-deposition thermal annealing at 240°C to improve minority carrier lifetime, measured by QSSPC. The composition of hydrides in the deposited layers was confirmed by FTIR, with the amorphous nature confirmed by RAMAN spectroscopy.

2.2 THz measurement

Non-contact, non-destructive THz measurement was achieved by Laser Terahertz Emission Microscopy (LTEM), using a spiral-net antenna, with THz imaging also being performed for a spot size of 10 mm (pitch: 1 mm).

Figure 2 illustrates the experimental setup of the THz spectroscopy technique, where femtosecond pulses from the 800 nm mode-locked Ti:Sapphire laser were focused onto a sample, mounted on 3-axis stage, before passing to the GaAs LT detector and also by way of a delay-path to the detector. The photoexcitation rate is at 80MHz with an incident angle of 45° leading to a spot size of 10 mm with a power density of 127 mW/cm^2 (averaged for 100 fs pulses). Measurement of the THz signal is done by sweeping the delay stage relatively slowly compared to the pulse time (via the lock-in amplifier). THz emission from the sample surface region radiate into free-space, and are focused by a pair of off-axis parabolic mirrors onto a spiral-type photoconductive antenna made of low-temperature grown GaAs.

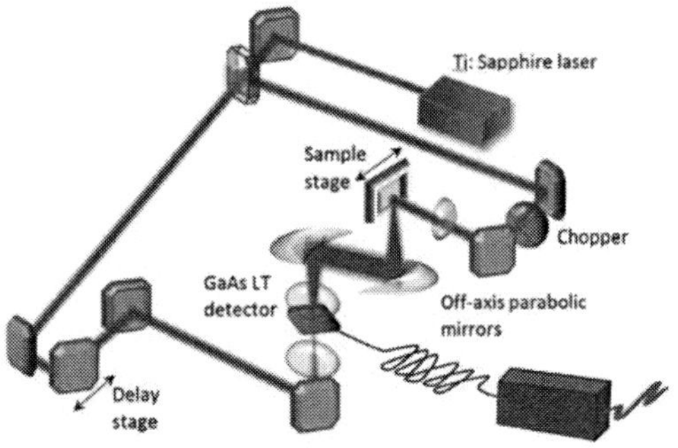

Figure 2: Schematic of LTEM apparatus

For silicon wafers, the THz emission is small compared with other materials, so a larger spot-size of 10 mm in diameter is needed to provide sufficient S/N ratio. The polarity of the surface photocharge can be detected using LTEM, and is determined by the value at approximately 10 ps. At this time, the terahertz signal has stabilised and the initial 'twitching' signal (occurring approx. 6 ps) has occurred.

3 RESULTS AND DISCUSSION

3.1 Analysis of Terahertz waveform for a-Si:H(i)/ c-Si(n)

For a-Si:H(i) layers deposited onto 1.4Ωcm c-Si(n) FZ, the surface photocharge are observed to be inverted compared to normal surface charge on bare (no native oxide) c-Si(n) (Fig. 3). The thickness of the a-Si:H(i) layer appears to influence the severity of this inversion. Thinner layers most significantly altering photocharge, while thicker layers (up to 50nm) having reduced affect. Strangely, very thick 500nm a-Si:H(i) has significantly great influence compared to all layers, and likely caused by the enhanced external field of the now active a-Si:H layer.

3.2 Terahertz waveform for a-Si:H(i/n) stack on c-Si(n)

The deposition of stacked a-Si:H(n)/a-Si:H(i) layers onto similar c-Si(n) surfaces (Fig 4), sees the polarity of photocharges at the surface reverting to the same THz-emission polarity as measured on the bare c-Si(n) surface. Very slight variance can be observed with differing thicknesses of a-Si:H(n). Plotting peak height at 10ps, the influence of the a-Si:H(i) was negligible, with 20nm ≤ a-Si:H(n) ≤ 80nm having minimal influence on the strength of the THz emission.

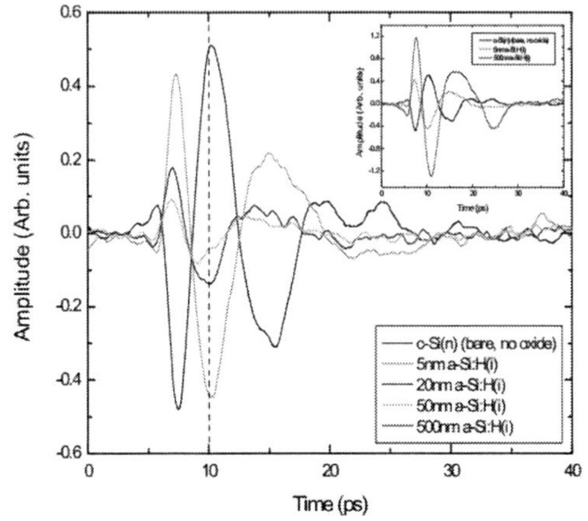

Figure 3: Thin a-Si:H(i) layers on 1.4Ωcm c-Si(n) *FZ*

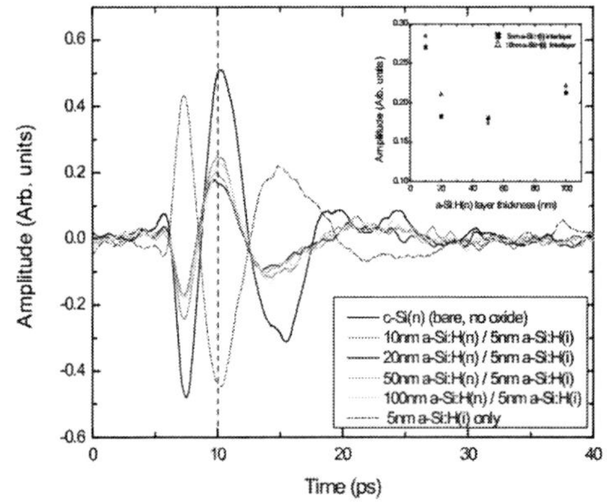

Figure 4: Thin a-Si:H(i/n) heterostructures on 1.4Ωcm c-Si(n) *FZ*

3.3 Analysis of Terahertz waveform on c-Si(p)

Deposition of a-Si:H(i) layers onto 2.5Ωcm c-Si(p) *FZ* (Fig. 5) does not cause inversion of the THz emission at the interface.

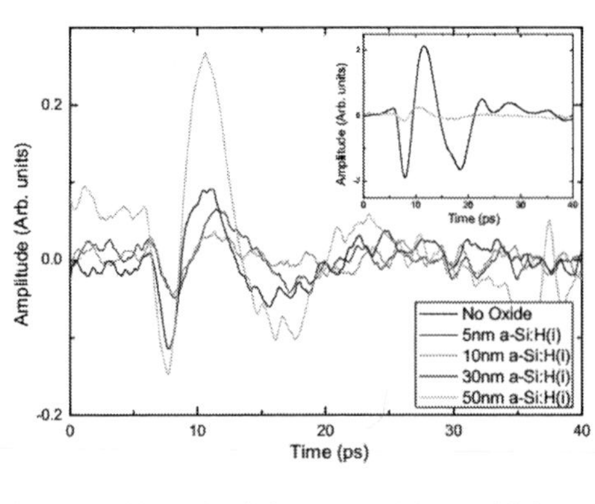

Figure 5: Thin a-Si:H(i) layers on 1.4Ωcm c-Si(n) *FZ*

Although the deposited layer thickness does appear to play a role in the mitigating the THz signal, we can still observe the expected surface polarity despite shifts towards flat-band conditions in the absence of a doped a-Si:H layer or junction. Additional information may be extracted from the phase shift, however, further analysis is required.

3.2 Terahertz imaging vs photoluminescence

In Figure 6a, the normalized, contrast enhanced imaging by photoluminescence (top image surface) is shown compared to the THz imaged c-Si surface (lower image surface). Given that PL imaging by spontaneous emission are mostly responsive to spatially resolved minority carrier lifetimes and are susceptible to background doping density, for simple a-Si:H(i) passivating layer on c-Si(n) limited information can be obtained. Despite, normalized contrast enhancement, the PL image appears to suggest some passivation occurring in part of the c-Si surface (indicated by red), this is despite τ_{eff} = 1.6 ms being measured by QSSPC.

Comparatively, THz imaging was more sensitive to spatial information relating to recombination at the surface interface for samples with minimal PL emission. A (+) THz signal (red) relating to upward band bending can be observed with the THz imaging (lower surface), neutral or approaching flat-band conditions (green), and (-) polarity (blue) can be extracted, even from samples with poor PL emission. THz signals are stronger at defected or high recombination regions, which is not apparent from the PL image.

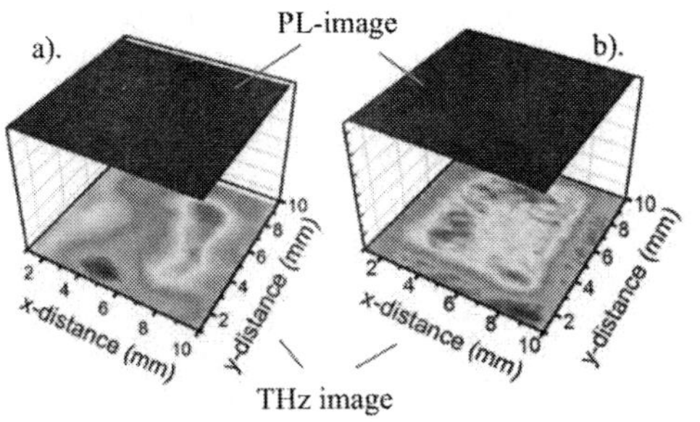

Figure 6: THz imaging (lower surfaces), and Photoluminescence mapping (upper surfaces) with contrast enhanced. a). for the a-Si:H(i)/c-Si(n) interface, b). for the a-Si:H(n)/a-Si:H(i)/c-Si(n) stacked interface.

Similarly, for the a-Si:H(n)/a-Si:H(i) stack deposited onto c-Si(n) (see Figure 6b), the PL image displays fractional information concerning possible recombination regions without indicating the cause of the recombination within the surface region. Whereas, THz imaging reveals much greater detail concerning band bending occurring at the surface as it relates to photocharges within the conduction and valence bands. By further analysis, we are able to reveal information from this previously unseen band-bending relating to the interface between the deposited a-Si:H and the c-Si wafer for partially completed devices.

Notably, we see in both THz images a distinct region of negative photocharges (blue) occurring at the x.y coordinates (4, 2) in Figure 6a, and two edge regions in Figure 6. In the first case, this is where the tweezer held the wafer during etching of the silicon surface. For the stacked a-Si:H/c-Si THz image, the negative photocharge region corresponds to the sections of the c-Si wafer which were supported during the Si-etch step. It is apparent, that THz imaging is suitable for identifying hidden information relating to a fundamental, early-stage processing step for c-Si solar cells and may assist in identifying improvements to optimise solar cell fabrication methodologies. Interestingly, the only possible response in the PL images to the surface in this way appears in the form of regions where no a-Si:H deposition took place, which occurs where the c-Si was supported (and masked) during deposition. In this way we can see the usefulness of the THz imaging technique for understanding the importance of surface preparation on the overall device over other methods.

4 CONCLUSIONS

In this work, we have demonstrated THz spectroscopy by LTEM to be suitable in measuring the dynamics of photocharges at the a-Si:H/c-Si interface, showing polarity inversion and shifts towards flat-band conditions arising from the deposition of relatively thin a-Si:H. The THz emission waveform from the a-Si:H(i)/c-Si interface was shown to be influenced by the quality and thickness of the a-Si:H(i) or a-Si:H(doped) layer(s). THz imaging, although lower in resolution, has revealed spatial information concerning recombination regions for photocharges on the c-Si surface which are not apparent for PL imaging or other techniques. Time-resolved THz-spectroscopy provides a non-contact method for probing the surface-charge potential with sub-picosecond resolution. For solar cells, THz imaging provides both quantitative and qualitative analysis of device quality and represents a highly useful new tool during fabrication stages of solar cells.

5 REFERENCES

[1] M. Tonouchi. Nature Photonics, 1(2): 97-105, February 2007

[2] H. Nakanishi, S. Fujiwara, K. Takayama, I. Kawayama, H. Murakami, and M. Tonouchi. Applied Physics Express, 5(11): 112301, October, 2012

[3] H. Nakanishi, A. Ito, K. Takayama, I. Kawayama, H. Murakami, and M. Tonouchi. AIP Advances, 5(11): 117129, November 2015

[4] T. Trupke, B. Mitchell, J. W. Weber, W. McMillan, R. A. Bardos and R. Kroeze, *Energy Procedia* (2012), Vol. 15, pp. 135-146.

6 ACKNOWLEDGEMENTS

This work was made possible with support from NEDO, and SCREEN Holdings Co. Ltd

AUTOMATED STATISTICAL ALGORITHMS TO INTERPRET ROOT CAUSE VARIANCE IN PHOTOVOLTAIC CELL MANUFACTURING

Rhett Evans[1,2], Matthew Boreland[3]
[1] Australian Centre for Advanced Photovoltaics, UNSW, Sydney, NSW, 2052, Australia.
E: evansrhett@gmail.com, P: +61 411 418004
[2] 3Solinno Pty Ltd, Bulli, NSW 2516, Australia
[3]School of Photovoltaic and Renewable Energy Engineering, UNSW, Sydney, NSW, 2052, Australia

ABSTRACT: This study presents the use of a multivariate statistical approach known as Structural Equation Modelling (SEM) to help interpret the sources of variance in end-of-line performance data. Using only the measured relationships between the J_{sc}, V_{oc}, FF and R_s, it is shown that through the use of a path model, it is possible to infer the variance associated with a change in the front screen printed grid. This calculation is shown to closely match the variance contribution estimated directly from the actual measured finger width of the front screen. These models are very powerful and can be run automatically to detect changes in the sources of variance in production over time, and to trigger action. Depending on the available data, these models can become detailed and complex, and they can also be used to determine the most valuable and cost effective metrology for running a stable production process.

1 INTRODUCTION

Cost effective options exist for the measurement of most aspects of a cell's performance throughout the photovoltaic (PV) manufacturing sequence. Progressively larger amounts of data is being generated in the PV manufacturing sector due the rapidly expanding scale and the increasing implementation of low cost, high value, in-line metrology on production lines. This data is highly valued during times of poor cell performance (*process deviation* or *loss-of-process*), but often remains unexploited during times when production is operating within its expected performance range (*in-process*).

In particular, metrology data generated when the line is *in-process* can be a valuable tool for reducing production variance. It should be noted that production variance is not only a measure of production quality, but also represents a significant cost factor. For example, a long tail on the distribution of solar cell efficiencies equates to lost revenue as the cells need to be sold at less than premium price, or in worse case become yield-loss or reject cells that earn no revenue. The underutilization of this data is in part due to the lack of automated statistical algorithms specifically tailored to PV manufacturing, which can provide a timely and efficient method for analysis of the mountain of metrology data.

Some manufacturers also have the ability to track wafers, throughout the manufacturing sequence enabling all the collected data to be joined for each cell. The Q CELLS, Tra.Q system is an example of such a tracking system, using a laser barcode [1]. The "joined-up" data provides the potential for even greater insight into process variance via the use of advanced multivariable statistical models. Models that provide information on the sources of variance in production can be very sensitive indicators of the quality of the manufacturing system.

Multivariate statistical techniques are commonplace in other manufacturing sectors [2], but there is very little evidence of their use in photovoltaic manufacturing [3, 4, 5]. There remains a significant opportunity to improve and optimise PV production by using more advanced statistical analysis.

This study presents the use of a multivariate statistical approach known as *Structural Equation Modelling (SEM)* [6] to help interpret the sources of variance in end-of-line performance data. Using only the measured relationships between the J_{sc}, V_{oc}, FF and R_s, it is shown that, via a *Path Model* [7], it is possible to infer the variance associated with a change in the front screen printed grid. This calculation is shown to closely match the variance contribution estimated directly from the actual measured finger width of the front screen. These models are very powerful and can be run automatically to detect changes in the sources of variance in production over time, and to trigger action. Depending on the available data, these models can become detailed and complex, and they can also be used to determine the most valuable and cost effective metrology for running a stable production process.

2 PATH MODELLING

The *Path Model* [from *Structural Equation Modeling* [8] is ideally suited to interpreting PV manufacturing variance because of the strength of the known physical models that can also be used to help interpret the relationships. SEM is a framework for statistical modelling, encompassing many statistical techniques, which focuses on analysising causal processes [9]. SEM moves beyond purely descriptive statistical models of variance to enable models that include a hypothesis for causation (cause & effect) to link variance to root causes. The Path Model in particular provides a graphical method to describe the cause and effect relationships, which is then used to formulate the model to be solved.

An example one possible path model to describe the cell manufacturing system is shown in figure 1. There are many variants of this model and many ways to express or measure different aspects of a cell's performance. Figure 1 is in no way definitive, but serves the purpose of introducing the path model concept. Some path models are more tractable than others so it is entirely legitimate to try to find the best representation for any given situation

The underlying purpose of the path diagram is to provide a hypothesis for the sources of variance in end of line performance. Whenever a correlation exists between two or more variables, there is often a latent or measured variable that explains the root cause of the correlation, and this root cause may be a more sensitive indicator of process variance. The term latent variable is used to describe a variable that is implied but not directly measured. The path diagram is written such that there is no repetition in the causation path amongst the arrows. For example, in figure 1, the emitter sheet resistivity will influence the FF, but there is no direct path between the two. It will directly influence the R_s, and directly influence the recombination, both of which will influence the FF in turn. The use of two boxes representing lifetime / recombination effects is just a matter of convenience and tractability. There are several simple and effective parameterizations of the "enhanced recombination" effect, most commonly with some sort of "J_{02}" type of parameter modelled from an IV curve. This type of modelled parameter is useful even where there is no direct proof the recombination can be strictly described by an n=2 mechanism.

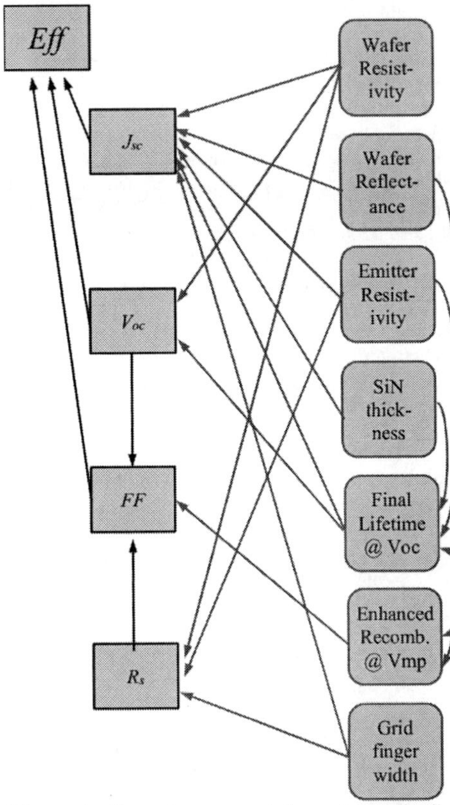

Figure 1: Example representation of a path diagram describing some common sources of variance in PV manufacturing. Measured variables are typically in boxes with latent variables in circles, but the rounded boxes in this case represent variables that can sometimes be measured and can sometimes be treated as latent variables when solving the model. The arrows indicate directions of causality, and double ended arrows imply correlation. The formalized rule structure of path diagrams can be found in [7]

Intuitively, it would not be possible to solve a path diagram like the one shown in figure 1 with only the data in the five blue boxes (i.e. too many relationship with too little measured data). This is what is known as an underdetermined model. The ideal situation is an over determined model because fitting algorithms like Maximum Likelihood Estimation (MLE) can be used with the degrees of freedom available for determining the statistical confidence in the results [6, 7]. In the case of photovoltaic manufacturing, physical models of the cell can be used to help solve for the relationships amongst some of the paths. A convenient open source software for solving these models is the commonly known software R [10], running the *lavaan* package [11].

2 APPLICATION TO GRID FINGER WIDTH VARIANCE

A simplified subset of this path model is considered here for further analysis as a case study for the potential of the approach. Figure 2 shows this simplified path model.

This simplified model has five measured variables, two different expressions are used for the FF to improve model tractability (FF_0 is the FF expected from V_{oc} only [12] and ΔFF is the difference between measured FF and FF_0). The remaining latent parameters are the hypothesis root causes for the variance in the measurable parameters. This subset model is chosen based on broad interest to a range of production lines, particularly those where only limited end-of-line data is available. It is also based on some a-priori knowledge of where the significant source of variance are in the data and so which paths can be excluded and amalgamated in the path model of figure 2 compared to figure 1, without introducing significant error.

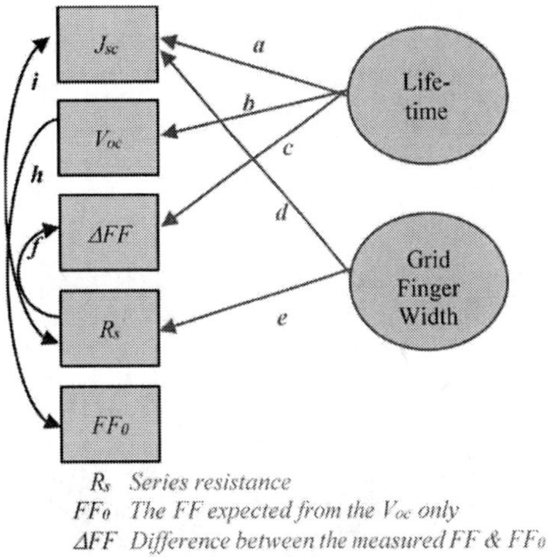

Figure 2: The path model as applied to the production data in this study, where the measured variables are in boxes and the latent variables are in circles. Straight lines indicated paths to the latent variable and curved lines indicate correlations between measured variables.

The paths are labelled in figure 2 with path coefficients, which are weightings for the relationships between variables that are used later for solving the model. In this model the relationships are assumed to be linear, which is reasonable over small parameter ranges. As noted earlier, the Path Model provides a graphical representation of the cause and effect relationships between variables. By following all the paths connecting measurable variables (direct and indirect), the path diagram is solved for the path coefficients using a correlation matrix.. So for example, the correlation between J_{sc} and R_s is described by the following connecting paths: j is the direct correlation path and de is the indirect correlation path via the latent variable, giving an overall correlation of $j + de$

The same approach is continued to make a complete system of equations in terms of the correlation values, as

Table 1: Correlation matrix expressed in terms of the coefficients from the path model

	J_{sc}	V_{oc}	ΔFF	R_s	FF_0
J_{sc}	1	ab	$ac + def + jf$	$j + de$	$j + de$
V_{oc}	ab	1	$bc + bajf$	abj	h
ΔFF	$ac + def + jf$	$bc + bajf$	1	$f + acj$	$cbh + fjabh$
R_s	$j + de$	abj	$f + acj$	1	$abhj$
FF_0	abh	h	$cbh + fjabh$	$abhj$	1

described in Table 1.

Solving the model involves calculating the correlation matrix for an actual data set of measured variables and equating with the correlation matrix expressed in terms of the path coefficients. The subsequent system of equations is numerically solved to extract values for the coefficients. In the case of an undetermined model, additional constraints can be imposed on the coefficients, such as a physical model of the cell, to assist in converging to a solution.

The data used in the study is from over 9,000 Q.antum cells made in the Hanwha QCELLS production facility in Bitterfeld-Wolfen [13]. There are over 70 pieces of data for each cell, which are tracked throughout the production system by the unique Tra.Q system [1] that uses a laser barcode on each cell. Data measured from the cell throughout the production sequence can all be joined to the end-of-line data.

This subset model was chosen because the joined data from Q CELLS Tra.Q system also includes directly measured finger width data. This information can be used to check the efficacy of the path model. In the general case, a large joined data set offers the opportunity to build considerably more sophisticated models, beyond the scope of this introductory treatment

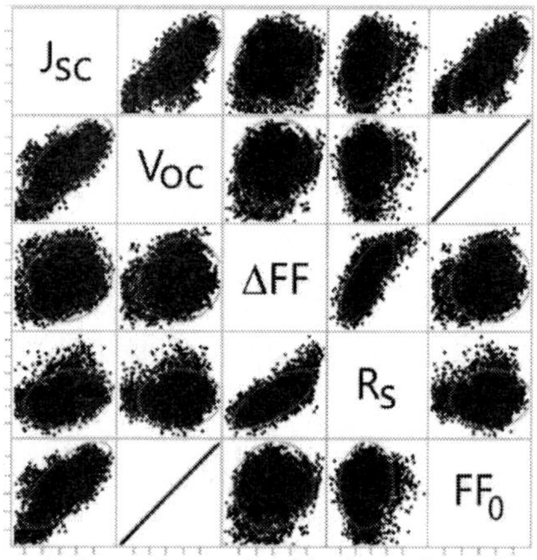

Figure 3: Scatterplot matrix of the data under study.

4 RESULTS

Figure 3 shows the scatterplot matrix of all the measured variables plotted in pairs and table 2 gives the associated correlation matrix. By equating table 1 and table 2 it is possible to solve for thepath coefficients.

The solution is under-determined, and further physical models are needed to solve for all path coefficients. The path equations resolve an estimate for the product ***de***, which is related to the correlation between J_{sc}, R_s and grid finger width. A simple physical model can be used to estimate the ratio of these coefficients. For example, suppose the standard grid caused 3% shading and 50% of the series resistance contributions. A 10% change in finger width would cause a 0.3% change in J_{sc} and a 10% change in R_s (at a constant finger aspect ratio). These changes in J_{sc} and R_s, normalised to their respective standard deviations, will give the ratio of the path coefficients. In the case of this data, a 0.3% J_{sc} / 10% R_s change gives a $d{:}e$ ratio of 0.65. As the coefficients are between 0 and 1, it is reasonably robust to the assumptions in the physical model. In practical cases, these physical models should be tuned to the precise conditions of the line under examination.

The solutions for the coefficients *a-j* are shown in table 3. The value for ***b***=1 comes from the assumption of the wafer quality being responsible for all of the variance in the V_{oc}. With the values now available for all the path coefficient, the model can be used to estimate the contribution of the latent variables to the variance in the measured variables. The calculated variance impact for the finger width is shown in table 3.

Table 2: The correlation matrix for the model of figure 3. The model is under-determined from the path equations and a physical model is required

	Jsc	Voc	ΔFF	Rs	FF0
Jsc	1.0000	0.6950	0.1596	0.2534	0.6950
Voc	0.6950	1.0000	0.0710	0.0503	1.0000
ΔFF	0.1596	0.0710	1.0000	0.7079	0.0720
Rs	0.2534	0.0503	0.7079	1.0000	0.0498
FF0	0.6950	1.0000	0.0720	0.0498	1.0000

The Hanwha Q CELLS' data set also includes measured finger width data, so the variance associated with this variable can be directly estimated by the R-sq of a linear regression model of J_{sc} and R_s against finger width. Table 2 also compares the linear regression model to the latent variable model. As can be seen the results are in close agreement and sufficient for practical use.

Table 3: Estimates for the path coefficients

a	*b*	*c*	*d*	*e*	*f*	*h*	*j*
0.69	1.00	-0.029	-0.34	-0.52	0.71	1.00	0.072

Table 4: Variance in J_{sc} and R_s related to the finger width, calculated by path model or by linear regression.

	Variance in J_{sc} (due to finger width)	**Variance in R_s** (due to finger width)
Path Model	12%	27%
Linear Regression	8%	25%

Although there is enough information to get an estimate for all path coefficients, there is not enough information to have degrees of freedom available for estimating the goodness of fit. One informal solution to this is to check the consistency of the model across a number of data sets. When the data sets are sufficiently large (the solution here is for 9,000 cells), the path coefficients are reasonably consistent across different days of data. When the data is broken down into smaller groups, the results are less consistent. This is mostly related to the influence of data autocorrelation on the validity of two-variable correlations calculated by the standard pairwise approach (as shown in figure 4). Improved calculations of correlation are required for more consistency in the model. Examples of this problem can also be found in [4].

6 DISCUSSION

The results clearly demonstrate that useful information can be extractd from a relatively small set of commonly measured end-of-line data. The multivariabel basis of the approach can also potentially provide a more sensitive indicator of process variance compared to traditional single parameter run plots.

However of more fundamental importance here is that running a high quality production involves understanding as much as possible about its sources of variance. Variance models do not often perfectly explain all the variance in the manufacturing system. The aim should always be to understand as much as is possible with all the information that is available, and then to improve that understanding over time with improved models and more experience of the behaviour of the manufacturing system. The path model presented here is an example of one method that can be used to help improve the understanding of variance - the latent variables, with the help of a physical model, have improved the understanding of variance with only a very small amount of measured data. In some instances, variance models are a more sensitive indicator of consistency and quality than the absolute performance of a production line. These models can be run automatically on a set of data over an hour, a day, or a week to check the consistency of the components of variance.

A latent variable model does not invalidate the need for directly measured metrology, by contrast the more metrology that is available, the more detailed the variance model can become. For example, if the known finger width data is included in the subset path model, then other components of the full path model in figure 1 can be included as latent variables. A good candidate for this would be the emitter resistivity, particularly as this is only directly known in the case of wafer data being fully tracked. The wafer data in this study was fully tracked, but that is not always the case in all production facilities. As another variant, if finger resistances were available, say with a busbar to busbar resistance measurement [14], it would be possible to infer more details regarding the resistance, even right down to estimating the accuracy of the R_s parameterisation itself. The influence of all data in a path model can be easily assessed and the most valuable metrology steps can be determined. In some cases, latent variables may even be superior to measured ones because they are not subject to the same levels of measurement error and other measurement challenges such as speed of measurement. The applications of these types of models is potentially very broad.

6 CONCLUSIONS

One of the big opportunities to further optimise PV manufacturing in coming years will be through the intelligent use of the data already being generated. An understanding of variance is a critical part of running a high quality process and the types of statistical models explained here offer an excellent opportunity to use existing data sets to do this.

ACKNOWLEDGEMENTS

The authors would like to thank Hanwha Q CELLS for their support of this work. The Australian Center for Advanced Photovoltaics is supported by the Australian Government through the Australian Renewable Energy Agency (ARENA). Responsibility for the views expressed herein is not accepted by the Australian Government.

REFERENCES

[1] Wanka, S., Rychtarik, D., Müller, J., Geissler, S., Kappe, P., Thormann, S., Spallek, M., vom Bauer, U., Ludwig, C., Wawer, P. 2011, "Tra.Q - Laser Marking for Single Wafer Identification - Production Experience from 100 Million Wafers", 26th EUPVSEC, 2011

[2] Köksal, G. et al 2011, "A review of data mining applications for quality improvement in manufacturing industry", Expert Systems with Applications, vol. 38, no. 10, pp. 13448-13467.

[3] Coleman, J.L. & Nickerson, J. 2005, "A multivariate exponentially weighted moving average control chart for photovoltaic processess", 31st IEEE PVSEC

[4] Evans, R., Dore, J., Van Voorthuysen, E., Zhu, J., Green, M.A. 2014, "Interpreting manufacturing variance using a data mining approach", 29th EUPVSEC, 2014.

[5] Evans, R., Stegemann, E., Dinkel, T., Klöter, B., Stoye, P., Petter, K., Boreland, M., 2015 "Mulitvariate Analysis of Wafer Process Data", 30th EUPVSC, 2015

[6] Raykov, T., Marcoulides, G. A., 2006 "A first course in structural equation modeling", 2nd Edition.

[7] Loehlin, J. C., 2004 "Latent Variable Models", 4th Ed.,

[8] Raykov, T., Marcoulides, G. A., 2006 "A first course in structural equation modeling", 2nd Edition.

[9] J. B. Grace, D. R. Schoolmaster, G. R. Guntenspergen, A. M. Little, B. R. Mitchell, K. M. Miller, and E. W. Schweiger, "Guidelines for a graph-theoretic implementation of structural equation modeling," Ecosphere, vol. 3, no. 8, pp. 1–44, Aug. 2012.

[10] https://www.r-project.org/

[11] http://lavaan.ugent.be/

[12] Green, M.A. 1982, "Accuracy of analytical expressions for solar cell fill factors", Solar Cells, vol. 7, no. 3, pp. 337-340

[13] Klöter B., Kowalzik, P., Ploigt, H.C., Buss, D., Hussack, A., Schütze, Stegemann, E., Stolze, L., Müller, J.W., 2013 "Current Status of High-Efficiency Q.antum Technology at Hanwha Q.Cells", 28th EUPVSEC, 2013

[14] Meier, D. L., Chandrasekaran, V., Gupta, A., Yelundur, V., Rohatgi, A., 2013 "Silver Contact Grid: Inferred Contact Resistivity and Cost Minimization in 19% Silicon Solar Cells", IEEE Journal of Photovoltaics, Vol 3, No. 1, Jan 2013.

ALTERNATIVE INLINE ANALYSIS OF ACIDIC ETCHING BATHS

L. Mohr, T. Dannenberg, M. Zimmer, J. Rentsch
Fraunhofer Institute for Solar Energy Systems ISE
Heidenhofstraße 2, 79110 Freiburg, Germany
Phone: +49 761 4588 5657. E-Mail: lena.mohr@ise.fraunhofer.de

ABSTRACT: A process control system was developed which continuously monitors the composition of acidic etching baths. The simple physical parameters, sound velocity, conductivity, and refractive index are verified for suitability to determine the concentrations of hydrofluoric acid, nitric acid, and hexafluorosilicic acid. In pre-tests the characteristics of the individual concentrations were observed with the measuring devices and the effect of the formation of nitrous gases on the physical measuring devices is observed. In contrast to the conductivity electrode, the sound velocity and refractive index sensors show highly reproducible results. The formation of nitrous gases has no significant effect on the physical measurement methods. The influence of different concentrations of a mixture on the physical parameters is determined building up a design of experiment. With aid of multiple linear regression model equations were created which describe the quantitative relationship between the concentration and the physical variables. The model equations which lead to a quantitative correlation between the concentration of the acids as a function of the parameters sound velocity, conductivity, refractive index and temperature are compared and finally validated. The model equation predicts the concentration of HF with a recovery rate of 96.40%, HNO_3 with 96.38% and Si in H_2SiF_6 with 78.74%.
Keywords: Silicon Solar Cells, Manufacturing and Processing, Etching, Design of Experiments, Modelling

1 INTRODUCTION

Acidic based etching baths are frequently used for industrially manufacturing of multicrystalline silicon solar cells, mainly for texturization and rear side polishing before diffusion [1,2]. The resulting surface morphology after etching particularly relies on the composition of the solution, consisting of hydrofluoric acid (HF) and nitric acid (HNO_3) [3,4]. During the processes different reaction products are formed, especially hexafluorosilicic acid (H_2SiF_6) and the concentrations of the reactants decrease. This means that monitoring of the composition is indispensable, to achieve stable process results. Currently, analytical methods such as titration, near infrared spectroscopy (NIRS) and ion chromatography (IC) with post column derivatization and UV-Vis detection for determination of solved silicon, are used [5–7]. These methods have been successfully implemented but the analysis instruments have high costs of ownership and trained manpower is required. Inline analytics allow, unlike the offline and atline processes, a continuous correlation between the obtained information and the properties of the process [8]. Inline monitoring is commonly used for classical measured variables, such as temperature, pressure or flow. However, further research is needed in the process control of mixtures and their compositions [9]. Due to calibration relationships a concentration can be determined by means of physical dimensions (as electrical, acoustical or optical values). For example, ultrasonic measurement is gaining more and more importance, because it provides a reproducible, non-invasive measurement of flow rate and in some cases concentration, with short time response [8]. However, other measuring devices have similar advantages and in a mixture of components, more variables are needed, such as refractive index (n) or conductivity (κ) which can be linked to the ultrasonic velocity (c). The reference methods used were titration as well as ion chromatography.

2 EXPERIMENTAL

2.1 Preliminary tests

A series of measurements of each acid – HF, HNO_3, H_2SiF_6 - was taken with the physical devices to observe the physical behavior change within the different concentrations and to gain knowledge about the handling and the operating range. In addition two measuring points of each curve were taken on different days in order to check the reproducibility of the analytical equipment. Furthermore a mixture (reference solution) of $c_{HF} = 60$ g/l, $c_{HNO_3} = 600$ g/l and $c_{Si\ in\ H_2SiF_6} = 5$ g/l was prepared to perform long-term measurements. This mixture was in the range of the composition of a polishing bath.

2.2 NO_x test

During texturing nitrogen oxides are formed which lead to a yellowing of the solution due to the reaction with HNO_3 and Air [10,11]. To rule out interference on the measuring devices due to this reaction, a NO_x experiment was built up. Various solutions of different acid concentrations (HNO_3, HF, H_2SiF_6) and the reference solution were produced and their exact concentrations recorded using the reference methods. Furthermore, the physical parameter of the test solutions and a blank value were recorded before the NO_x experiment was carried out. Nitrogen dioxide (NO_2) was produced in a reaction chamber, filled with 10% HNO_3 and a diluted sodium nitrite solution ($NaNO_2$), according to equation (1), so that a strong yellow coloration was achieved.

$$2\,NaNO_2(l) + 2\,HNO_3(l) \rightarrow 2\,NaNO_3(l) + NO_2(g) + NO(g) + H_2O(l) \quad (1)$$

The NO_2 is passed through a gas washing bottle filled with solution, thereby turning it yellow. Immediately, the solutions were measured again with the physical devices and the reference method.

2.3 Design of Experiments (DoE)

The influence of the different concentrations of a

mixture on the physical parameters was determined building up a Design of Experiments (DoE). An orthogonal and rotatable central composite design (CCD) was created, with the response values n, κ as well as c and the factors HF, HNO_3, H_2SiF_6 and temperature (t). The factor levels were: $c_{HF}\,(g/l) = \{50, 70\}$, $c_{HNO_3}\,(g/l) = \{500, 700\}$, $c_{Si\ in\ H_2SiF_6}\,(g/l) = \{0.5, 9.5\}$ and $t\,(°C) = \{10, 30\}$. To draw conclusions about the amount of Si which is brought in during the process basin the factor H_2SiF_6 is given as Si in H_2SiF_6. The experimental design, consisting of 36 solutions, was randomized and measured in an experimental setup (fig. 1). The conductivity sensor was placed in the solution. The solution was circulated through a temperature basin to achieve specific temperatures, followed by the refractometer and sound velocity sensor.

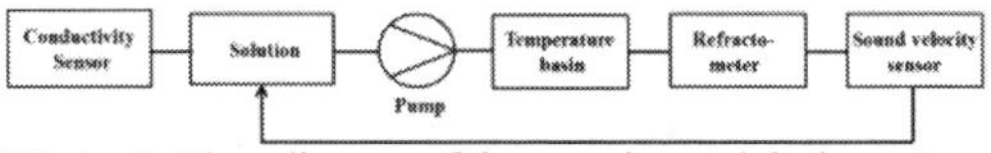

Figure 1: Flow diagram of the experimental design setup

The aim was to obtain an empirical model, which quantitatively describes the relationship between the studied factors (concentration of the acids) and the response values (physical variables). With the aid of multiple regression, model equations were created [12]. A general regression model with $k = 4$ factors based on a linear combination of basic functions, was used.

$$y_i = \beta_0 + \sum_{i=1}^{k} \beta_i x_i + \sum_{i=1}^{k} \beta_{ii} {x_i}^2 + \sum_{i=1}^{k} \sum_{j=i+1}^{k} \beta_{ij} x_i x_j \quad (2)$$

With the response value (dependent variable) y_i, the regression coefficients β_i and linear x_i, squared x_i^2 as well as interactive $x_i x_j$ effects (explanatory or independent value). The modeling was done by progressive elimination of factors and interactions which did not show a significant effect on the response value. The analysis of variance (ANOVA) has been used to investigate the significance of true effects. ANOVA uses variance estimates to compare response mean differences [13]. The variance of response values is explained by the influence of various factors. In order to assess if the factors have a significant impact on the acid concentration, the coefficients are compared with the width of the confidence interval. The beta coefficients ($Beta_j$), in contrast to the regression coefficients (β_i) allow the comparison with each other. Due to the different units within the regression coefficients it is not possible to get information about the strength of changing on the response value by marginal changes of the coefficients. Standardized beta coefficients

$$Beta_j = \beta_j \frac{s_{xj}}{s_y} \quad (3)$$

allow a relative comparison of the explanatory values [14], with the standard deviation s_{xj} of the independent variable x_j and the standard deviation of the dependent variable s_y. Thus, the $Beta_j$ indicates the change in the response value with a change of the independent variables by one standard deviation (of x_j) while keeping all other independent variables constant. During the evaluation high beta values could indicate multicollinearity. This occurs, if there is a linear relationship between two or more independent variables. This problem of regression analysis may affect the accuracy of the model equation. A relatively simple identification is the variance inflation factor (VIF).

$$VIF_j = \frac{1}{1 - R_j^2} \quad (4)$$

Where R_j^2 is the coefficient of determination obtained by regressing the jth independent variable on the remaining independent variables. A VIF of 1 corresponds to the ideal state, a $VIF \geq 10$ indicated already a high multiple correlation [15]. For the model validation a confirmatory test run was carried out. Six different solutions were prepared and detected with the experimental setup. The solutions contain random compositions of acids in the concentration range of the experimental design.

3 RESULTS AND DISCUSSION

3.1 Preliminary tests

In order to identify trends, two solutions at each concentration series were analyzed repeatedly at different days. Sound velocity and conductivity were taken tenfold, refractive index threefold and the reference solution 16-fold. The reproducibility of the sensoric devices was expressed via the relative standard deviation (υ) and is listed for each repeated measurement as well as for the reference solution in table I. Figure 2 (i) shows the results of the preliminary tests sound velocity as a function of weight percent for different acids. A good reproducibly, which can be seen on the replicate points and the υ results of the reference solution (table 1) was observed. The curve of HF in contrast to the other acids decreases with increasing concentration. To explain this curve progression, equation (5) was used and the bulk modulus plotted as a function of weight percent [16].

$$c^2 = \frac{1}{\rho \cdot \delta} \quad (5)$$

With sound velocity c, density ρ and the bulk modulus δ. Figure 3 (ii) shows that the bulk modulus of $\delta_{H_2SiF_6}$ decreases with a greater gradient than δ_{HF} ($\delta_{H_2SiF_6} > \delta_{HNO_3} > \delta_{HF}$). In addition, the density is plotted against the concentration in weight percent (fig. 2 (iii)). The results show that the bulk modulus of H_2SiF_6 in contrast to the HF falls steeply, thus the denominator in equation (5) becomes smaller, whereby the sound velocity becomes larger. The bulk modulus of HF changes only slightly with increasing concentration. This means that the density of HF has the greater influence in equation (5), the denominator becomes larger and the sound velocity decreases. It should be noted that the non-linearity of the HNO_3 for high concentrations occurs in the standard measuring range which may lead to misinterpretations during the equation modeling (fig. 2 (i)).

Table I: Mean relative standard deviation (υ) for the replicates and the reference solution.

Relative Standard deviation	c [%]	κ [%]	n [%]
HNO_3	0.06	13.64	0.11
HF	0.05	10.67	0.004
Si in H_2SiF_6	0.09	9.64	0.01
Reference solution	0.14	1.97	0.03

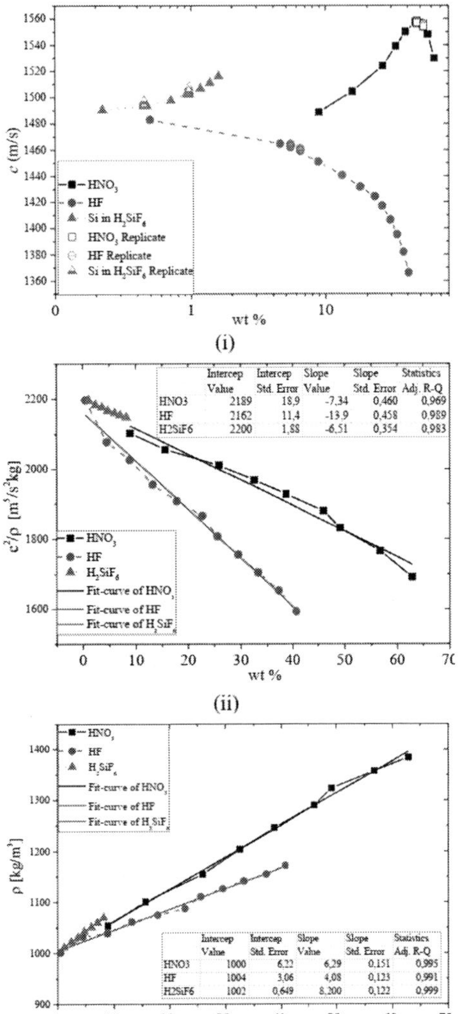

Figure 2: Sound velocity (i), bulk modulus (ii) and density (iii) of individual acids as a function of the concentration in weight percent.

Figure 3 (i) shows the difference between strong and weak electrolytes. Strong electrolytes, such as HNO_3 differ in high concentrations due to the relaxation and electrophoretic effects from linear behavior. The HNO_3 curve reaches up to a conductivity of almost 900 mS/cm. It should be noted that the non-linearity of the HNO_3 curve could lead to misinterpretation during the equation modeling. Furthermore the conductivity electrode shows in this region a high relative standard deviation (table 1). In comparison, the standard deviation of the standard solution shows a low value of 1.97%. This is due to the lower conductivity of the standard solution and the associated proximity to the calibration range. The reproducibility of the refractometer is very good, as demonstrated by the low relative standard deviations in table 1 column n and the replicate points on the curve in figure 3 (ii).

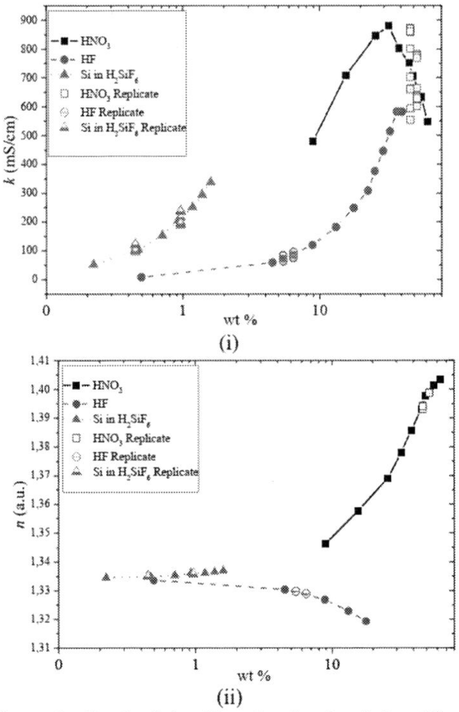

Figure 3: Conductivity (i) and refractive index (ii) of individual acids as a function of the concentration in weight percent.

3.2 NO_x test

To examine the effect of the nitrous gases the relative standard deviation is estimated with the values of the physical measurement devices before and after the experiment as well as with ion chromatography. The results of the ion chromatography show that during the experiment NO_2 has increased with an average of $m_{NO_2} = 345.2$ mg. The mean relative standard deviation of all samples per each physical device was calculated. The highest value was found for the conductivity sensor $\upsilon_\kappa = 1.581\%$, the least for the refractometer $\upsilon_n = 0.003\%$. The mean relative standard deviation of the sound velocity was $\upsilon_c = 0.009\%$. Despite a low increase in the amount of nitrite no or negligible effects on the physical measurement methods are expected during a HF-HNO_3 process.

3.3 DoE

For the statistical evaluation, especially the modelling the factors and response values of the CCD are exchanged to obtain an empirical model, which quantitatively describes the relationship between the concentration of the acids (c_{HF}, c_{HNO3}, $c_{Si\ in\ H2SiF6}$) and the physical variables (n, κ, c, t). The following pareto charts (fig. 5) of standardized (*Beta*) coefficients show the significant variables of the respective response value out of all effects (eq. (2)). During the evaluation the significance level (p-value) was set to 0.05 which means that the results are representative with a 95% level of confidence. Due to the use of beta coefficients information is obtained how a change of a factor affects the response value. These factors are listed in figure 4 in descending magnitude.

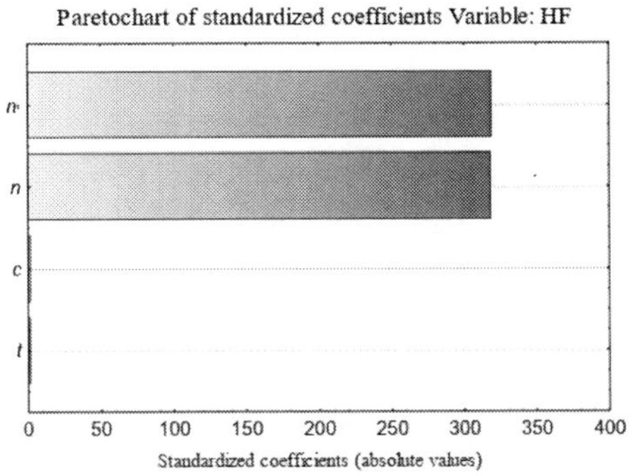

(i)

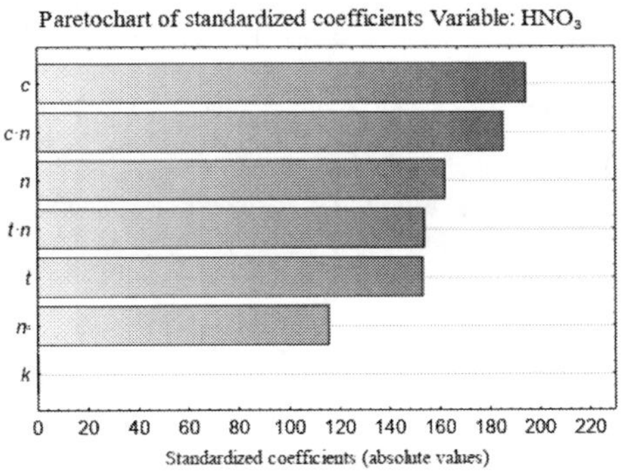

(ii)

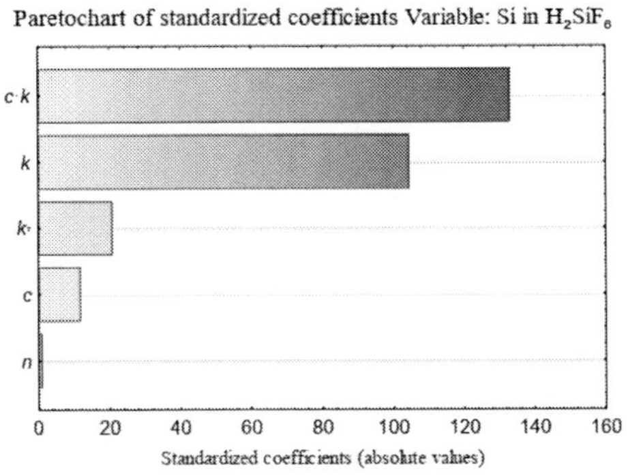

(iii)

Figure 4: Pareto charts of standardized coefficients of the response values HF (i), HNO_3 (ii) and Si in H_2SiF_6 (iii).

The concentration of HF is determined with the use of the physical devices refractive index, sound velocity and the aid of temperature as shown in figure 4 (i). A change in refractive index value – squared (n^2) and linear (n) - has the greatest impact on the concentration of HF. A small change in sound velocity and in the interaction of sound velocity with refractive index indicates a big change in the concentration of HNO_3 (fig. 4 (ii)). Compared to HF (four terms) the equation of HNO_3 involves more terms (seven terms), including linear (c, n, t, κ), interactive ($c{\cdot}n$, $t{\cdot}n$) and a squared term (n^2). To determine the concentration of Si in H_2SiF_6 five terms are needed, including the linear values of κ, c and n the interaction of c and κ and κ^2 (fig. 4 (iii)). With these pareto charts it is shown, that a change in refractive index mainly indicates a change in the concentration of HF, a change in sound velocity mainly indicates a change in HNO_3 and a small change in the conductivity mainly indicates a big change in Si in H_2SiF_6. The presumption of multicollinearity, indicated by very high beta values was confirmed by calculating the *VIF* (table II). It was found that in the linear terms of the equation the temperature ($VIF_t \approx 5$) correlates with the physical variables. Moderate correlation is not necessarily problematic; the variance of the regression coefficient increases, which could lead to a less accurate prediction of acid concentrations. The tolerance in table II is defined as $1\text{-}R^2$. The smaller a tolerance of a variable, the redundant is their contribution to the regression.

Table II: *VIF* of the linear factors c, n, t, κ.

	Tolerance	*VIF*	R^2
t	0.1973	5.07	0.8027
c	0.2210	4.52	0.7790
κ	0.2607	3.84	0.7393
n	0.3793	2.64	0.6207

The results graphically shown in the pareto charts can also be described as model equations:

$$y_{HF} = \beta_1 + \beta_{1,1}x_c + \beta_{1,3}x_n + \beta_{1,4}x_t + \beta_{1,3}{x_n}^2 \quad (6)$$

$$y_{HNO_3} = \beta_2 + \beta_{2,1}x_c + \beta_{2,2}x_\kappa + \beta_{2,3}x_n + \beta_{2,4}x_t + \beta_{2,5}x_c x_n + \beta_{2,6}x_t x_n + \beta_{2,8}{x_n}^2 \quad (7)$$

$$y_{Si\ in\ H_2SiF_6} = \beta_3 + \beta_{3,1}x_c + \beta_{3,2}x_\kappa + \beta_{3,3}x_n + \beta_{3,7}x_c x_\kappa + \beta_{3,9}{x_\kappa}^2 \quad (8)$$

With β_{ij} where i = {HF, HNO_3, Si in H_2SiF_6} and j = $\{c, \kappa, n, t, c{\cdot}n, t{\cdot}n, c{\cdot}\kappa, n^2, \kappa^2\}$. The adequacy of the model equation is expressed by the adjusted coefficient of determination. The value y_{HNO3} with $R^2_{ad} = 0.9737$ is close to 1 and indicates a true description of the actual concentrations with the model. The results of y_{HF} ($R^2_{ad} = 0.5045$) and $y_{Si\ in\ H2SiF6}$ ($R^2_{ad} = 0.4157$) are very low. In figure 5 (i), (ii) and (iii) the observed versus the predicted values of each acid are shown. Especially in figure 5 (i) and (iii) the distribution of the predicted values is high, which reflects the low R^2_{ad} values of HF and Si in H_2SiF_6. Especially the moderate adjusted coefficient of determination of Si in H_2SiF_6 could be caused by the high relative standard deviation of the conductivity, since this value is present in equation (8) in three out of five effects. That the observed values do not scatter, is due to the consistent concentrations of the acids, which have been used for the DoE.

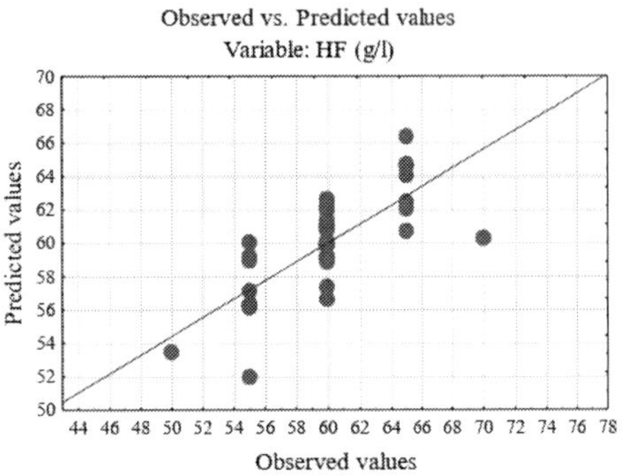

(i)

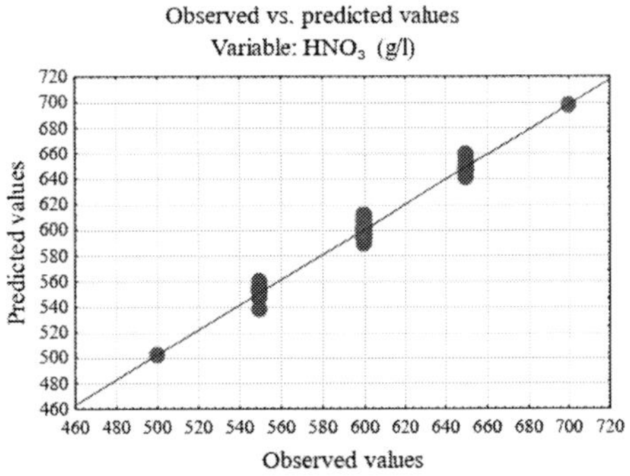

(ii)

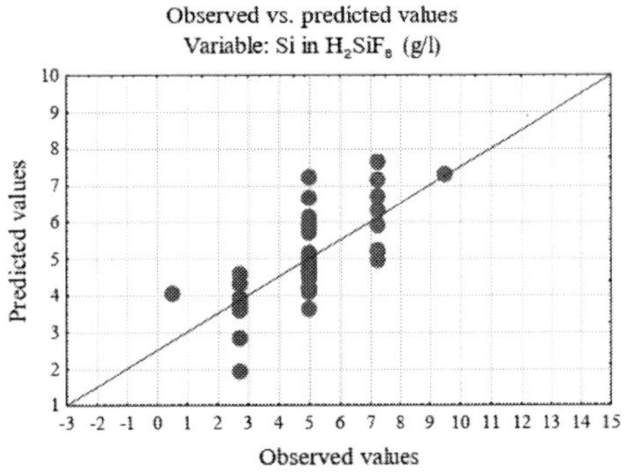

(iii)

Figure 5: Observed versus predicted values and the adjusted determination of coefficient of the response value HF (i), HNO_3 (ii) and Si in H_2SiF_6 (iii).

3.4 Validation

The model equations have been validated with a test run. Therefore six solutions with different concentrations of HF/HNO_3/Si in H_2SiF_6 have been used. The values of c, n, t, κ were obtained and the intermediate recovery rate calculated. Although model y_{HF} shows a less R^2_{ad} the result of the intermediate recovery rate is the highest with 96.40%. The concentration of HNO_3 was found with an accuracy of 96.38%. The model equation for Si in H_2SiF_6 with an accuracy of 78.74% leads to a big uncertainty of results. The model equations are valid for the concentration range seen in figure 5 and should not be extrapolated.

4 SUMMARY

This work demonstrates as proof of concept that through the combination of sound velocity, conductivity and refractive index a system can be developed, which can be used for the inline analysis of all main components of etching baths. In preliminary tests was shown that the measuring results of sound velocity and refractive index are very reproducible compared to conductivity. In addition it was observed that NO_x gases have no or negligible effects on the physical devices. The influence of different acid concentrations with respect to the physical parameters could be determined using DoE. With the aid of multiple linear regression model equations were created which describe the relationship between the concentration of HF, HNO_3 and Si in H_2SiF_6 and the physical values c, n, t, κ. The concentration range used were HF 50 - 70 g/l, HNO_3 500 - 700 g/l and Si in H_2SiF_6 0.5 - 9.5 g/l. The measuring temperature should be between 10 and 30 °C. Furthermore it should be noted that the presumption of multicollinearity indicated by high beta-values was confirmed. The model equation predicts the concentration of HF with a recovery rate of 96.40%, HNO_3 with 96.38% and Si in H_2SiF_6 with 78.74%.

5 ACKNOWLEDGEMENTS

This work was funded by the German Federal Ministry for the Environment, Nature Conservation and Nuclear Safety under the frame of the project "NAPOLI" (0325654B).

6 REFERENCES

[1] D.-H. Neuhaus, A. Münzer, "Industrial Silicon Wafer Solar Cells", *Advances in OptoElectronics*, 2007.

[2] X. Liu *et al.*, "Black silicon: fabrication methods, properties and solar energy applications", *Energy & Environmental Science*, vol. 7, no. 10, pp. 3223–3263, 2014.

[3] H. Robbins, B. Schwartz, "Chemical Etching of Silicon", *J. Electrochem. Soc.*, vol. 106, no. 6, p. 505, 1959.

[4] H. Robbins, B. Schwartz, "Chemical Etching of Silicon", *Journal of The Electrochemical Society*, vol. 107, p. 108, 1960.

[5] M. Zimmer, A. Oltersdorf, J. Rentsch, "Online process control of acidic texturisation baths with ion chromatography", *Talanta*, vol. 80, no. 2, pp. 499–503, 2009.

[6] A. Henßge, J. Acker, "Chemical analysis of acidic silicon etch solutions: I. Titrimetric determination of HNO, HF, and H SiF", *Talanta*, vol. 73, no. 2, pp. 220–226, 2007.

[7] M. Zimmer *et al.*, "Towards the automation of the acidic texturisation process", *European Commission: 25th European Photovoltaic Solar Energy Conference and Exhibition, EU PVSEC 2010. Proceedings: 5th World Conference on Photovoltaic Energy Conversion, 6-10, September 2010, Valencia, Spain. München: WIP-Renewable Energies, 2010, pp. 1719-1721*, 2010.

[8] Kessler, Rudolf W., "Prozessanalytik", WILEY-VCH Verlag GmbH & Co. KGaA, 2006.
[9] J. Sacher, Rg, "Ein Ohr für PAT", *Chemie Technik, Heidelberg*, vol. 36, no. 7, pp. 72–74, 2007.
[10] K. Mayer, "Chemische Ansätze zur Neuordnung des Solarzellenprozesses ausgehend vom Wafering bis hin zur Emitterdiffusion, 2009.
[11] Kästner, Gero, "Gas phase characterization of acidic etching mixtures for industrial single side solar cell prodction sequences, Master Thesis, 2013.
[12] A. González, "Two level factorial experimental designs based on multiple linear regression models", *Analytica Chimica Acta*, vol. 360, no. 1-3, pp. 227–241, 1998.
[13] Allen, Theodore T., "Introduction to Engineering Statistics and Lean Sigma: Statistical Quality Control and Design of Experiments and Systems", Springer London, 2010.
[14] Ernst, Huib, "Angewandte Statistik in Geografie und Umweltwissenschaften", vdf Hochschulverlag AG an der ETH Zürich, 2011.
[15] F. Y. HSIEH *et al.*, "An Overview of Variance Inflation Factors for Sample-Size Calculation", *Evaluation & the Health Professions*, vol. 26, no. 3, pp. 239–257, 2003.
[16] U. Hempel, S. Wöckel, J. Auge, "Ultraschallbasierte Informationsgewinnung in der Verfahrenstechnik", *Chemie Ingenieur Technik*, vol. 82, no. 4, pp. 491–502, 2010.

DRY PLASMA TEXTURING OF MONO-SI FOR SILICON HETEROJUNCTION SOLAR CELLS APPLICATION

M.L. Addonizio(*), L. Fusco, A. Antonaia, A. Spadoni
ENEA, Portici Research Centre, P.le E. Fermi,1, 80055 Portici, Napoli, Italy
(*)Phone: +39-081-7723255, Fax: +39-081-7723344, E-mail: marialuisa.addonizio@enea.it

ABSTRACT: Dry and maskless texturing process of mono-crystalline silicon wafers using CF_4/O_2 plasma in a reactive ion etching (RIE) system has been studied for silicon heterojunction solar cells application. The main RIE process parameters have been changed and optimized in order to obtain a surface texture suitable to reduce the reflection of the silicon surface. Depending on gaseous mixture composition and process pressure, the etching mechanism can be modified from isotropic to anisotropic giving different shapes and sizes of surface features. At this purpose, effects induced by different plasma etching conditions on wafer surface morphology were systematically analysed in order to optimize the scattering properties. We verified that some typical texture parameters such as average Haze and RMS are not able to well explain reflection reduction. Differently, some specific texture functions such as angular distribution of scattered light and inclination angle distribution of the textured surface proved to be very effective in defining the most appropriate geometric surface parameters suitable to obtain large reflection reduction. In particular, with the aim of obtaining a textured surface with the lowest average reflectance in the wavelength range of interest, we have related the geometric parameters of the morphological features with the angular distribution of the scattered light obtained by ARS technique. Consequently, in this study we showed that, if shape and size of the characteristic surface features were appropriately modified by changing RIE process parameters, ARS curves very close to Lambertian behaviour can be obtained for etched silicon surface giving the highest reflection reduction.

Keywords: Texturing; reactive ion etching; reflectance.

1 INTRODUCTION

The surface of mono-crystalline silicon (mono-Si) wafers, utilized for silicon heterojunction solar cells, usually reflects around 39% of the incoming light in the wavelength range 300-1200 nm. Wafer surface texturing process has been acquiring more and more appeal in order to increase the photovoltaic conversion efficiency by means of both reduced surface reflectance and increased light trapping [1, 2]. Conventional wet texturing, obtained by using an alkali solution with anisotropic etching characteristics, is normally applied to mono-Si wafer [3, 4]. This alkali wet etching gives texturing of both wafer sides with an increase of its fragility. Furthermore, several problems concerning its industrial application such as expensive chemical waste treatment, stocking and manipulation with large volume dangerous chemicals, increase of DI water consumption have to be solved [5].

Within the last years, also in order to further improve the PV device performances, reactive ion etching (RIE) treatment has been investigated as alternative approach to develop low-cost, large-area and maskless texturing method, taking into account that RIE process can give anisotropic etching characteristic and short process time, with notable advantages for PV industries [6, 7]. Furthermore, plasma texturing is a low temperature process with reduced environmental impact and it also allows single side texturing of Si wafer. In particular, RIE process carried out with fluorinated gases, such as CF_4, SF_6 etc. mixed with O_2, appears highly attractive due to self-masking effects with enhanced anisotropic etching behaviour [8]. Effectively, it is well known that, during the etching process, the fluorine radicals (F*) play a role as active etchant species both resulting in volatile silicon fluoride (SiF_4) species and depositing passivation layers of silicon oxyfluoride (SiO_xF_y).

Depending on the process RIE parameters selected to carry out effective Si surface texturing, different morphologies can be obtained, from needle-like surface structure (very high aspect ratio) to sponge-like surface structure (moderately high aspect ratio). However, plasma texturing of c-Si generally increases surface recombination drastically due to both increased surface area and increased defects introduced by plasma etching [9, 10]. Consequently, highly textured needle-like structure (black silicon), also being more attractive in terms of optical gain, does not lend itself to fabricate highly efficient PV devices, whereas less textured Si surface can better fit PV device requirements. A reasonable aim related to plasma texturing could be represented by a reduction of the average Si surface reflection from the initial value of 39% down to about 15%, where this average reflection value corresponds to that normally obtained by conventional wet etching.

In this study we have carried out a maskless surface texturing of mono-Si wafers by means of RIE process as an effective method in order to modify the wafer surface by using gaseous mixture with different CF_4/O_2 ratios, for application in silicon heterojunction solar cells. At this purpose, the effects induced by different plasma etching conditions on wafer surface morphology were systematically analysed in order to optimize the scattering properties. We worked towards obtaining sponge-like surface structures taking into account that this surface texture, also not giving the best optical gain, could well represent a good compromise between proper scattering properties and sufficiently low surface recombination.

Another crucial aspect related to plasma etching of Si surface consists in the different behaviour shown by n-type and p-type Si wafers, especially in terms of etching rate and morphology features formed on the etched surface. Specifically, plasma etching rate is less effectiveness on p-type Si due to unfavourable polarity of the space charge present in the depletion layer [11, 12]. At the same time, a less pronounced anisotropic etching occurs on p-type Si [13] giving less effective scattering properties and, consequently, textured Si surface with less effective reflection reduction. As in this study we chose to texture p-type Si surface, we had to take into account this further serious problem in pursuing our experimental aim.

Although several previous studies showed some feasibility for improving efficiency of silicon solar cells by means of surface plasma treatments, it is not yet well clear how surface features form by maskless RIE process and, among them, what feature shape is more effective to increase solar cell efficiency. In particular, with the aim of obtaining a textured surface with the lowest average reflectance in the wavelength range of interest, we considered highly strategic to relate the geometric parameters of the morphological features (AFM and SEM analysis) with the angular distribution of the scattered light (ARS technique). As a consequence, in this study we pursued the goal of appropriately modifying shape and size of the characteristic surface features, by changing RIE process parameters, for obtaining ARS curves very close to Lambertian behaviour.

2 EXPERIMENTAL

Texturing of mono-crystalline FZ silicon wafers has been carried out by reactive ion etching (RIE). c-Si substrates (100) p-type, 270 μm thick and with a resistivity in the range 1-5 &·cm have been used. They were cleaned with the standard RCA method before being loaded into RIE system.

Plasma treatments were carried out in a glow-discharge radio frequency (13.56 MHz) reactor with high purity carbon tetrafluoride–oxygen (CF_4/O_2) gas mixtures. RF power was supplied to a circular metallic electrode in a vacuum chamber through a capacitively coupled matching network. Different RIE set-up parameters such as gaseous mixture composition, amount of total gas flow, working pressure, RF power and process time have been explored.

Ultraviolet–Visible–near Infrared (UV–VIS–NIR) reflectance measurements were carried out in the range 300–1000 nm utilizing a Perkin Elmer Lambda 900 double beam spectrophotometer. Average spectral haze H_R, defined as diffuse to total reflectance ratio $[R_T - R_{spec}]/R_T$ (where R_T and R_{spec} are total and specular reflectance respectively), was determined.

A detailed analysis was performed on surface roughness, profile of structures produced and size/aspect ratio of RIE-textured surface features. Morphology and surface roughness properties were studied using a LEO mod. S360 scanning electron microscope (SEM) and a Veeco mod. NSIV atomic force microscope (AFM), respectively.

In order to better understand the light scattering capability of the textured surfaces, the inclination angle distribution of the surface features was determined using the experimental AFM images. The AFM images were processed using MATLAB programming. Angular Resolved Scattering (ARS) measurements in reflectance mode were carried out for obtaining information about the angular distribution of the scattered light. The ARS measurements have been performed with laser illumination at 630 nm. The resulting intensity profiles were normalized respect to the maximum intensity of the scattered light and subsequently integrated over the azimuth angle (under assumption of isotropic scattering).

3 RESULT AND DISCUSSION

3.1 Plasma texturing process

As the characteristics of the surface features were dependent on the RIE set-up parameters such as gaseous mixture composition, amount of total gas flow, working pressure, RF power and process time, an accurate optimization of the RIE parameters has been carried out.

Removal reaction on the silicon surface mainly occurs by means of F* radicals via the following reaction:

$$4F^* + Si\,(s) \rightarrow SiF_4\,(g) \qquad (1)$$

In our study, primarily we have investigated the influence of O_2 concentration in the gas reactant mixture on the weight loss and on the texturing during the chemical etching step.

Currently, gas mixtures of CF_4 and O_2 are used because addition of O_2 to a CF_4 glow discharge increases the amount of free F* radicals and accelerates the etching process via:

$$CF_4 + O^* \rightarrow COF_2 + 2F^* \qquad (2)$$

Depending on gaseous mixture composition and RF power utilized, the etching mechanism can be modified from isotropic to anisotropic due to the formation of a SiO_xF_y film. The good selectivity of fluorine, preferentially removing silicon respect to silicon oxide [14], gives rise to a self-masking effect that leads to anisotropic etching with different shapes and sizes of surface features.

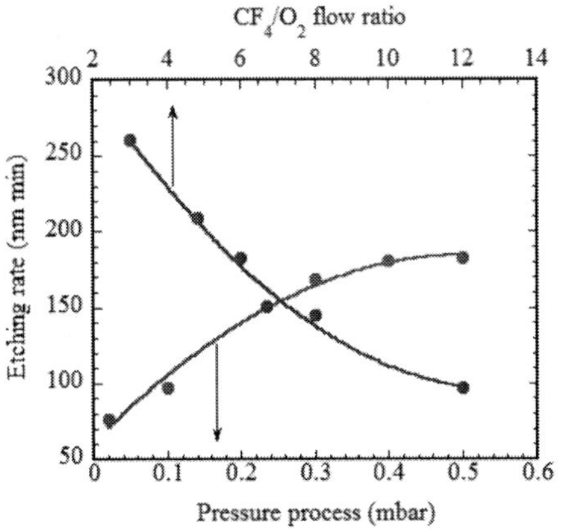

Figure 1: Etching rate of c-Si substrate as a function of CF_4/O_2 flow ratio and process pressure.

At this purpose, the effects induced by different plasma etching conditions on wafer surface morphology were systematically analysed in order to optimize the scattering properties. Etching rates, evaluated by weight loss of Si wafer after 25 min of plasma treatment using different CF_4/O_2 flow ratios, are shown in Fig. 1. We can observe that etching rate increases quite monotonically at the increase of O_2 content and this behaviour can be well explained taking into account the reaction (2) where further free F* radicals become available to accelerate the removal reaction (1). In the same Fig. 1 the etching rate is also shown as a function of the process pressure and we can observe how it increases at the increase of the pressure. As the process pressure increase corresponds to an increase of the residence time spent by the free active radicals inside the process chamber, we can well suppose that free F* radicals have more time to spend for reaching wafer surface and for giving the removal reaction (1). However, for process pressure higher than 0.4 mBar the etching rate settles down at a plateau value and this behaviour can be explained in terms of lower F* radical diffusivity at very high chamber pressure.

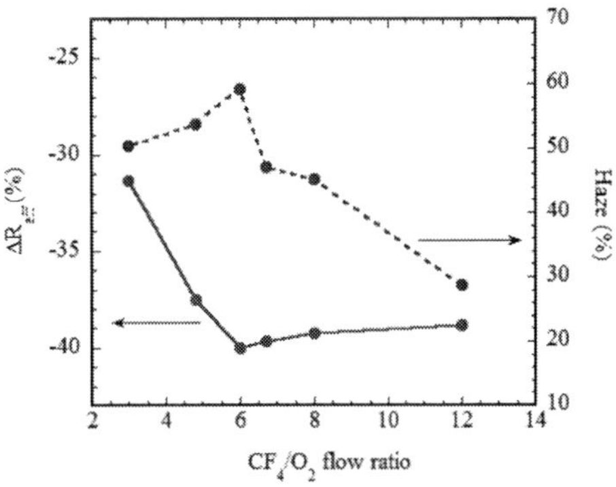

Figure 2: Reflection reduction percentage and reflective haze factor of dry etched c-Si substrate as a function of CF_4/O_2 flow ratio.

In Fig. 2 the average Haze evaluated in reflection mode (H_R) and the average reflection reduction, evaluated as relative change of the reflectance measured before and after dry etching process ($R_R=[R_{fin} - R_{in}]/R_{in}$), are shown as a function of the CF_4/O_2 flow ratio. We can observe that, for a flow ratio value of about 6, the maximum peak of H_R corresponds to the maximum beneficial effect in decreasing Si surface reflectance (minimum peak of R_R). However, for CF_4/O_2 ratio values between 6 and 12, at a strong variation of H_R (from about 60 % down to about 30 %) does not correspond a strong variation of R_R (the behaviour is almost flat). In fact, the average H_R value takes account of the scattered reflected light in a large range of scattering angles and it may contain large amount of photons scattered at very low scattering angles. This portion of scattered light is not able to give rise to effective light trapping inside the textured surface and, consequently, it does not give appreciable contribution to reduce the Si surface reflection. We found a similar lack of perfect correlation between variation of RMS values and corresponding variation of R_R values. For this reason, we concluded that H_R and RMS cannot represent the best texture parameters suitable both to perform the optimization procedure in reducing Si surface reflection and to understand what roughness type should be more appropriate for the aforementioned aim.

3.2 Texture surface assessment

Starting from the above experimental observation, great attention has been dedicated to two specific texture functions such as angular distribution of scattered light Ps(,) and inclination angle distribution $S(\Phi)$ of the textured surface. Both these texture functions are more specific in properly describing the light scattering phenomenon coupled with the partial light trapping occurring on the textured Si surface.

According to this aim, four plasma etched samples, obtained at different RIE process conditions, have been selected for the large difference existing in terms of surface structure and spectral Haze. In Fig. 3 AFM images of dry etched c-Si wafers clearly show different surface morphologies where sponge-like structures are characterized by cavities with different average size (in the range 0.1 - 1 µm) and different size distribution function. These different surface morphologies give rise to large difference in spectral Haze behaviours (see Fig. 4 and Tab. I) where surfaces showing larger cavities (samples S3 and S4) clear give Haze values higher than surfaces characterized by well distributed small cavities (samples S1 and S2). If we would exclusively take into account the evidence of the spectral Haze curves shown in Fig. 4, we could erroneously conclude that the surface structure of the sample S4 represents the most promising surface texture to reduce the optical reflection of etched Si wafers.

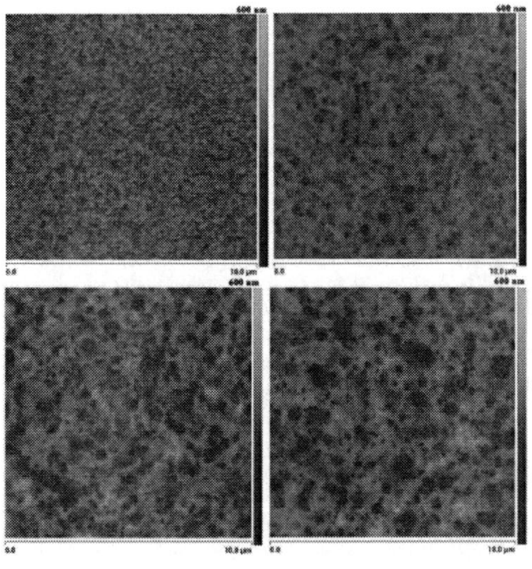

Figure 3: AFM images of dry etched c-Si wafer with different surface morphologies.

Actually, experimental spectral reflectance curves, related to the same four samples and shown in Fig. 5, give different results from those previously advanced, where also the surface structure of the sample S1 (with the lowest Haze value) represents a very promising surface texture in reducing the optical reflection.

Table I: Haze factor, RMS roughness and average reflectance (in the wavelength range 300-1000 nm) for the samples of Fig. 3.

Sample	Haze(%)	RMS(nm)	R(%)
S1	19	22	17.7
S2	28.7	30	23.9
S3	59.2	50	23.4
S4	76.2	74	19.5

At this point, it is fundamental to understand how this discrepancy arises. For the same four samples above reported, related AFM images were processed using MATLAB programming to obtain distribution histograms of surface inclination angles. In Fig. 6 we can observe that the inclination angle distribution having the largest shift towards high angles is represented by the sample S1, whereas also the sample S4 shows the maximum peak at a comparatively high angle. We have to observe that both these samples show low spectral reflectance (see Fig. 5) and, consequently, it is well conceivable a good correlation between inclination angle distribution shifted towards high angles and effective reflection reduction.

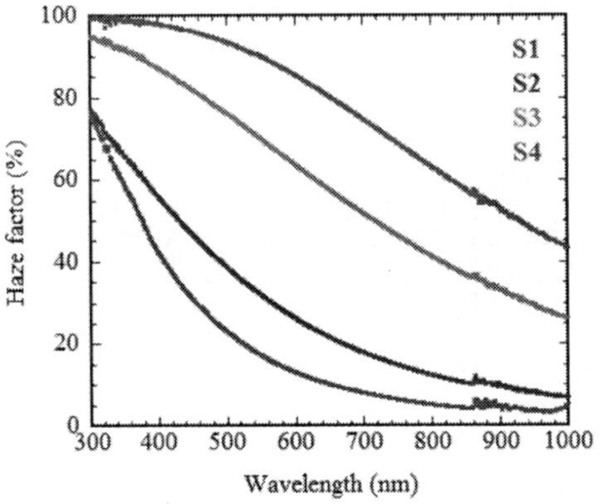

Figure 4: Spectral Haze curves, obtained in reflective mode, related to dry etched c-Si substrates with different surface texture.

However, well distributed small cavities with very sloping walls (sample S1) seem to give the maximum beneficial effect both in favouring light trapping inside textured surface and, consequently, in reducing reflectance. Low spectral Haze and low RMS value of this sample proved to be not effective in evaluating proper scattering properties in terms of large reflection reduction.

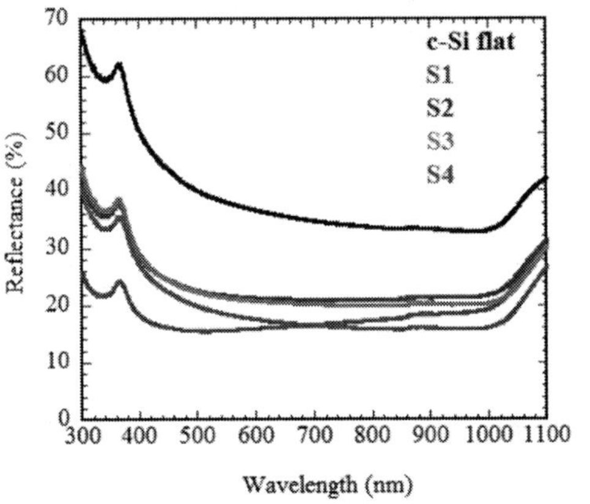

Figure 5: Spectral global reflectance of plasma etched c-Si surfaces, having different surface texture, related to the same samples of Fig. 3. Reflectance of flat c-Si wafer not exposed to plasma is also included as reference.

Differently, Angular Resolved Scattering (ARS) measurements in reflectance mode can represent a powerful tool to better understand what type of scattering is strongly necessary for reducing optical reflection of textured Si surface. In Fig. 7 the scattered power Ps(,) behaviour of reflected intensity, derived from sin-weighted ARS measurements, is reported for the same aforementioned four samples. The Ps(,) intensities are very different for the four textured surfaces and this evidence well agrees with spectral Haze curves reported in Fig. 4, where samples with high H_R values (S4 and S3) clearly show the highest Ps(,) values. However, the different angular distribution behaviours of the Ps(,) intensity represent the most meaningful characteristic suitable to well explain the effectiveness of a textured scattering surface in reducing the optical reflection.

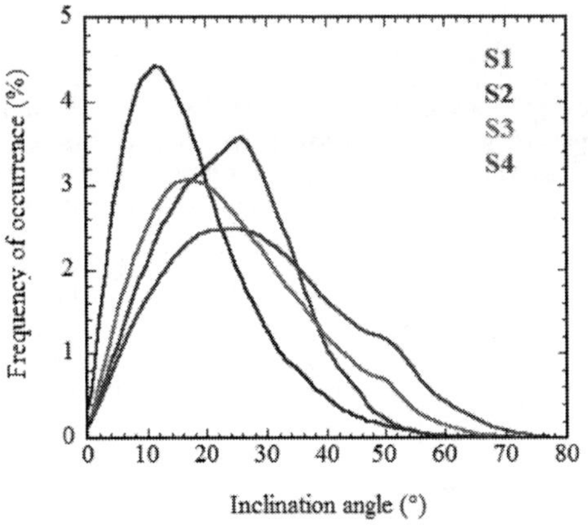

Figure 6: Distribution histograms of inclination angles, derived from AFM images, related to the dry etched c-Si substrates shown in Fig. 3.

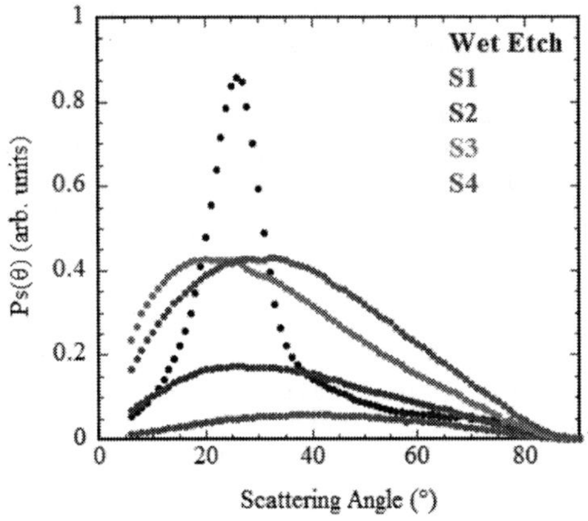

Figure 7: Scattered power of reflected intensity, derived from sin-weighted ARS measurements, as a function of the scattering angle for all the plasma etched substrates shown in Fig. 3. For comparison, P_s(,) distribution of wet alkaline chemical etched c-Si is also reported.

Indeed, ARS curves quite close to Lambertian behaviour, with a symmetric distribution around a maximum peak positioned between 38° and 45°, represent quasi-ideal reflective diffuser. This is the case of the samples S1 and S4 (see Fig. 7) whose textured surfaces also give the largest reflection reduction (see fig.5). An average reflection value of about 17% has been obtained, very close to that generally obtained by conventional wet etching.

In conclusion, we can affirm that angular distribution of scattered light Ps(,) and inclination angle distribution of the textured surface represent two powerful tools suitable to properly perform an optimization procedure for obtaining the best surface texture in terms of maximum reflection reduction.

4 CONCLUSIONS

RIE technique was used to texture Si wafer and effects induced by different plasma etching conditions on surface morphology were systematically analysed in order to identify a proper optimization procedure to pursue the aim of a strong reduction of the optical reflection. Using CF_4/O_2 gaseous mixtures, we observed that etching rate increases quite monotonically at the increase of O_2 content and it also increases at the increase of the process pressure. The average Haze H_R and the average reflection reduction R_R were analysed and we observed that, for a CF_4/O_2 flow ratio value of about 6, the maximum peak of H_R corresponds to the maximum beneficial effect in decreasing Si surface reflectance (the lowest value of R_R). However, for CF_4/O_2 ratio values between 6 and 12, at a strong variation of H_R (from about 60 % down to about 30 %) does not correspond a strong variation of R_R (the behaviour was almost flat). Starting from this observed discrepancy, great attention has been dedicated to two specific texture functions such as angular distribution of scattered light Ps(,) and inclination angle distribution S(Φ) of the textured surface. AFM images of dry etched c-Si wafers clearly showed different surface morphologies where sponge-like structures were characterized by cavities with different average size and different size distribution function. Both S(Φ) and Ps(,) functions proved to be power tools in properly describing these different surface morphologies and the related light scattering phenomenon coupled with the partial light trapping occurring on the textured Si surface. Surface texture, characterized both by well distributed small cavities and by high average angle inclination, proved to be more effective in giving reflection reduction respect to other surface textures characterized by inhomogeneously distributed large cavities and/or low average angle inclination. An average reflection value of about 17% has been obtained, very close to that generally obtained by conventional wet etching.

Furthermore, we verified that ARS curves very close to Lambertian behaviour (ideal reflective diffuser) corresponded to Si textured surfaces with the largest reflection reduction.

5 ACKNOWLEDGEMENT

This work was supported by the Italian Ministry of Economic Development in the framework of the Operating Agreement with ENEA for the Research on the Electric System.

6 REFERENCES

[1] E. Vazsonyi, K. De Clercq, R. Einhaus, E. Van Kerschaver, K. Said, J. Poortmans, J. Szlufcik, J. Nijs, Sol. Energy Mater. Sol. Cells 57 (1999) 179–188.

[2] D. Iencinella, E. Centurioni, R. Rizzoli, F. Zignani, Sol. Energy Mater. Sol. Cells 87 (2005) 725–732.

[3] A.K. Chu, J.S. Wang, Z.Y. Tsai, C.K. Lee, Sol. Energy Mater. Sol. Cells 93 (8) (2009) 1276–1280.

[4] Jung M. Kim, Young K. Kim, Sol. Energy Mater. Sol.Cells 81 (2004) 239–247.

[5] G. Agostinelli, P. Choulat, H.F.W. Dekkers, S. De Wolf, G. Beaucarne, Proceedings of the 31st IEEE PVSC, Orlando, 2005, p.1149.

[6] J. Yoo, K. Kim, M. Thamilselvan, N. Lakshminarayn, Y.K. Kim, J. Lee, K.J. Yoo, J. Yi, J. of Phys. D: Appl. Phys. 41 (2008) 125205-125211.

[7] M. Moreno, D. Daineka, P. Roca i Cabarrocas, Sol. Energy Mater. Sol.Cells 94 (2010) 733-737.

[8] Henri Jansen, Meint de Boer, Henk Wensink, Ben Kloeck, Miko Elwenspoek, J. Microelectron. 32 (2001) 769–777.

[9] Jan Kegel, Heike Angermann, Uta Stürzebecher, Erhard Conrad, Mathias Mews, Lars Korte, Bert Stegemann, Applied Surface Science 301 (2014) 56–62.

[10] K. Ma, X. Zeng, Q. Lei, J. Xue, Y. Wang, C. Zhao, Front. Optoelectron. 7(1) (2014) 46-52.

[11] A. Tachibana, S. Kawauchi, T. Yamabe, J. of Physical Chemistry 95 (1991) 24 71-2476.

[12] S. Kawauchi, A. Tachibana, T. Yamabe, J. of Physical Chemistry 95 (1991) 6303-6308.

[13] Elena Dilonardo, Giuseppe V. Bianco, Maria M. Giangregorio, Giovanni Bruno, Pio Capezzuto, Maria Losurdo, Energy Procedia, 10 (2011) 1-7.

[14] Alfred J van Roosmalen, Vacuum 34(3-4) (1984) 429-436.

A STUDY OF IMPROVING WAFER QUALITY WITH THE PHOSPHORUS GETTERING PROCESS ON SILICON HETEROJUNCTION SOLAR CELLS

Zih-Yu Shih, Wen-Chi Hsieh, Hsiang-Wei Yin, Jay Chang, Ming-Yu Chen
AU Optronics Corporation, Advanced Technology Dev. Dept. No. 1 JhongKe Rd.
Central Taiwan Science Park Taichung 40763, Taiwan, R.O.C.

ABSTRACT: Silicon heterojunction (SHJ) solar cells have potential to achieve high efficiencies. To improve the efficiency, minimizing the recombination at the bulk region is important. Metal impurities in the bulk wafers make recombination sites and they are the major reason for recombination. Phosphorus (P) gettering was introduced to reduce metal impurities in the bulk region of Si wafers and then to improve the efficiency of SHJ solar cells fabricated on the wafers. P gettering would be feasible to improve the efficiency of SHJ solar cells fabricated on P-doped Si wafers.

Keywords: Heterojunction solar cell, metal impurity, gettering

1 INTRODUCTION

Many researchers point out that wafer quality affects performance of SHJ solar cells. High-efficiency solar cells are fabricated from wafers in most cases because of their high material quality [1, 2]. Especially when using Czochralski silicon (Cz-Si) wafers for high efficiency solar cell concepts exceeding 20 % efficiency, minor defects can limit cell performance [3]. Metal impurities incorporated in Si wafers, such as Fe, Ni, and Cu, are highly mobile, diffuse over long distances, and act as a recombination center [4-6]. Furthermore, solar cell performance is largely limited by the regions of poorest material quality [6]. Such recombination center in the wafers deteriorates carrier collection efficiency. P gettering of Si wafers is known to increase bulk recombination lifetimes through the removal of mobile impurities, resulting in improved cell performance. In this paper, we study the possible benefits of gettering impurities using P diffusion to improve the wafer quality on SHJ solar cells.

2 EXPERIMENTAL

SHJ solar cells were processed on 6 inch as-cut n-type Cz-Si wafers. As-cut wafers were cut from two different ingots, named as Q1 wafers and Q2 wafers, respectively. The wafers in the top, middle, and bottom parts of each ingot were taken and tested. Q1 and Q2 wafers were P gettered at 750 °C for 30 minutes in a quartz tube furnace, named as GQ1 and GQ2, respectively. While a reference sample was not gettered. To remove the gettered layers where impurities were assumed to be gathered, GQ1 and GQ2 wafers were immersed in potassium hydroxide solution for 10 minutes to etch back. The reference underwent the same etch-back process. The metal impurities in bulk wafers with texturing were measured by inductively coupled plasma mass spectrometry (ICP-MS). The analytic solutions were prepared in electronic-grade hydrofluoric acid and nitric acid.

A cleaning procedure of the textured wafers prior to deposition was the immersion of the wafers in an oxide-removed solution, followed by rinsing in overflow DI water. A PECVD system was used to deposite a-Si:H passivation layer. The front side of n c-Si wafer was deposited by i a-Si:H and p a-Si:H passivation layer. The rear side of n c-Si wafer was deposited by i a-Si:H and n a-Si:H passivation layer. Then the transparent conductive oxide (TCO) layer was deposited on both sides of n c-Si wafer. Ag grid on the front side and rear side were formed by screen printing. Figure 1 shows that the solar cell scheme in this study consists of Ag grid/ TCO/ p a-Si:H /i a-Si:H /n c-Si/ i a-Si:H / n a-Si:H/ TCO/ Ag. Detail process information can be found from the article [7, 8]. The conversion efficiencies and J-V characteristic of SHJ solar cell were measured at 25 °C at an illumination of AM1.5, 100 mW/cm^2.

Figure 1: The structure of SHJ solar cell.

3 RESULTS AND DISCUSSION

P gettering has been reported to occur at an optimum temperature as it relies on the transportation of released impurities towards a segregation region near the surface of the samples [9]. In Figure 2, the higher concentrations at about 2.21 x 10^{12} atom/cm^3 up to 1.08 x 10^{14} atom/cm^3 of metal impurities existed in the bulk of Q1 wafers indicates that the wafer quality of Q1 is worse than one of Q2. After running at the same bath of the gettering process, most of metal concentrations of GQ1 and GQ2 were relatively lower than ones of Q1 and Q2, respectively. Through P gettering, that generates stress silicon lattice which acts as a sink by external means. Besides, generated carriers move through the bulk region to electrodes, less recombination sites in bulk would get better efficiency of solar cells [10-14].

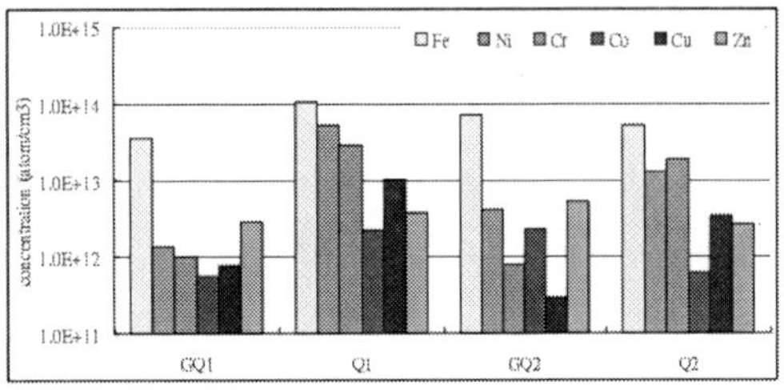

Figure 2: Accumulated elemental concentrations in each ingot.

The minority carrier lifetime (MCLT) of GQ1 is shown in Figure 3A. It is clearly shown that the values of the GQ1 wafers are increased at ca. 2200 µs in comparison to the reference wafers at ca. 1500 µs. It indicates that the wafer quality of Q1 after P gettering process would be improved. In Figure 3B, it is shown that the values of the GQ2 wafers are slightly reduced at ca. 900 µs in comparison to the reference wafers at ca. 950 µs. This indicates that there was no big enhancement of MCLT for Q2 wafers. That would be probably due to decreasing the less concentrations of metal impurities for Q2 wafers after the P gettering process.

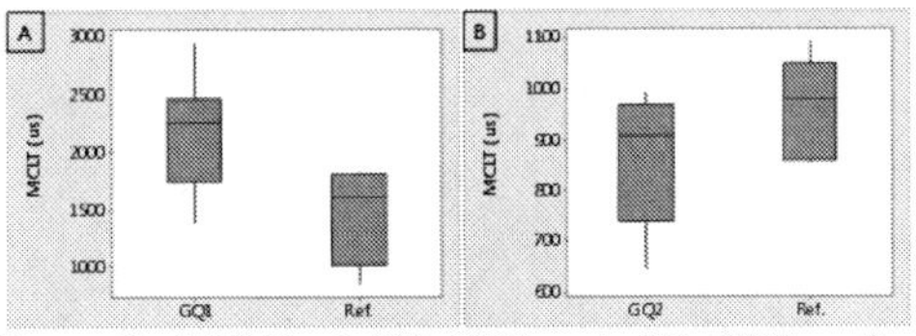

Figure 3: MCLT of (A) GQ1 and (B) GQ2 wafers compared with reference wafers.

In Figure 4A, the normalized efficiency value for Q1 ingot via the P gettering process was averagely gained from 100.0 % to 102.3 %. From the normalized Voc values in Figure 4B, the average increment of GQ1 ingot in comparison with reference was also seen from 100.0 % to 100.4 %. In Figure 4D of the normalized FF values, the average of GQ1 ingot in comparison with reference was raised from 100.0 % to 101.2 %. This results show that the wafer quality of Q1 ingot could be effectively improved through P gettering process.

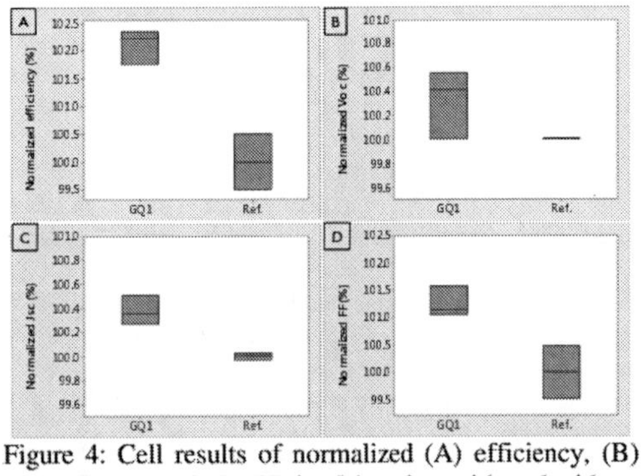

Figure 4: Cell results of normalized (A) efficiency, (B) Voc, (C) Jsc, and (D) FF for Q1 wafers with and without gettering.

In Figure 5A, there was no obvious increment of normalized efficiency value of Q2 ingot with gettering. The increased variation of normalized efficiencies were probably due to different batches of cell processing causing about 1.1 % variation of FF values in Figure 5D. From the normalized Voc values in Figure 5B and Jsc values in Figure 5C, there seems almost no affection for gettering process. The averaged Voc value of GQ2 was 99.7 % in the comparison of averaged reference value (99.9 %). The averaged Jsc value of GQ2 was 100.0 % in the comparison of averaged reference value (99.9 %).

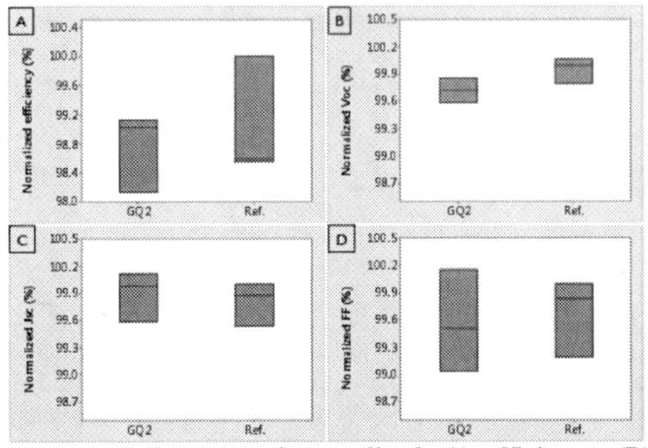

Figure 5: Cell results of normalized (A) efficiency, (B) Voc, (C) Jsc, and (D) FF for Q2 wafers with and without gettering.

4 SUMMARY

The MCLT of GQ1 was increased at ca. 2200 µs in comparison to the reference wafers at ca. 1500 µs. The wafer quality of Q1 after P gettering process would be improved. On the other hand, there was no big enhancement of MCLT for Q2 wafers. That would be probably due to decreasing the less concentrations of metal impurities for Q2 wafers after the P gettering process. Through this study, we found that P gettering makes improved wafer quality of silicon substrates for SHJ solar cells.

5 REFERENCES

[1] A. Lawerenz, K. Lauer, M. Blech, et al., Proceedings of the 25th EU-PVSEC, Valencia, Spain, 2486, 2010.
[2] J. Haunschild, J. Geilker, I.E. Reis, et al., Physica Status Solidi RRL, 5, 199, 2011.
[3] P. J. Cousins, D. D. Smith, H.-C. Luan, et al., Proceedings of the 35th IEEE PVSC, Honolulu, Hawaii, USA, 000275, 2010.
[4] A. Luque and S. Hegedus, "chapter 5. Solar grade silicon feedstock," in *Handbook of Photovoltaic Science and Engineering*, pp. 188–189, JohnWiley & Sons, 2011.
[5] S. M. Myers, M. Seibt, and W. Schroter, *Journal of Applied Physics*, 88, 7, 3795–3819, 2000.
[6] A. Luque and S. Hegedus, *Handbook of Photovoltaic Science and Engineering*, JohnWiley & Sons, 2011.
[7] Y.-J. Chien, Proc. 39th IEEE Photovoltaic Specialist Conference, 2013.
[8] M.-Y. Chen, J. Chang, F.-S. Chen, et al., Proc. 28th European Photovoltaic Solar Energy Conference and Exhibition, 2013.
[9] L. J. Caballero, C. del Canizo, P. Sanchez-Friera, et al., *Solar Energy Materials and Solar Cells*, 88, 3, 247–256, 2005.

[10] H. Nagel, A. G. Aberle, and S. Narayanan, *Solid State Phenomena*, 67-68, 503–508, 1999.
[11] A. Bentzen, H. Tathgar, R. Kopecek, et al., *Proceedings of the 31st IEEE Photovoltaic Specialists*, pp. 1074–1077, January 2005.
[12] D. Macdonald, A. Cuevas, A. Kinomura, et al., *Proceedings of the 29th IEEE Photovoltaic Specialists Conference*, pp. 285–288, May 2002.
[13] A. Cuevas, M. Stocks, S. Armand, et al., *Applied Physics Letters*, 70, 8, 1017–1019, 1997.
[14] J. S. Kang and D. K. Schroder, *Journal of Applied Physics*, 65, 8, 2974–2985, 1989.

EVALUATION OF SPATIAL ALD OF AL_2O_3 FOR REAR SURFACE PASSIVATION OF MC-SI PERC SOLAR CELLS

Friederike Kersten, Ina Förster, Stefan Peters
Hanwha Q CELLS GmbH
Sonnenallee 17-21, 06766 Bitterfeld-Wolfen, Germany

ABSTRACT: In this work, the evaluation results on industrial *p*-type mc-Si PERC cells using spatial atomic layer deposition for aluminum oxide from SoLayTec are shown. By optimizing the process flow and integrating a post-deposition anneal into the SiN_x capping process we achieve 0.15% higher cell efficiency compared to remote microwave plasma-enhanced chemical vapor deposition. Furthermore, it is shown that the improved passivation quality resulting in V_{oc} gain remains constant during light and elevated temperature induced degradation measurements.
Keywords: Multicrystalline Silicon, Passivation, Degradation.

1 INTRODUCTION

Deposition of aluminum oxide (Al_2O_3) by an atomic layer deposition (ALD) process has been evaluated on a high throughput deposition tool from SoLayTec ("InPassion ALD"). The key features of this new deposition technology are high flexibility in layer thickness, high throughput due to a deposition rate of up to 1 nm/s with excellent uniformity and single sided deposition on standard 6" solar wafers while maintaining proven superior surface passivation known from batch ALD.

We evaluate the InPassion ultrafast spatial ALD tool from SoLayTec for industrial production of mc-Si solar cells based on Hanwha Q CELLS Q.ANTUM technology [4]. Only the rear side passivation layer was varied by using spatial ALD for Al_2O_3 deposition in comparison to remote microwave plasma-enhanced chemical vapor deposition (MW-PECVD) of AlOx.

Based on recently published severe light-induced degradations levels of mc-Si PERC cells [1-3] the long-time performance stability of these ALD passivated mc-Si PERC cells during light and elevated temperature (LeTID) was investigated.

2 EXPERIMENTAL

The in-line spatial ALD Al_2O_3 system from SoLayTec works with a gas mixture comprising of trimethylaluminium (TMAl) and water (H_2O) vapor. The two process gases are separated by nitrogen (N_2) gas curtains. In a high throughput ALD process both half-reactions are spatially separated, as shown by the sketch in Fig. 1. The wafers move contactless on nitrogen gas bearings and pass TMAl and H_2O vapor inlets sealed off by a flow of pressurized N_2, forming isolated reaction zones. The wafers move back and forth underneath the reactor head, each passage under the injector head results in two complete ALD cycles. The deposition rate is determined by the cycle rate of the wafer in the reactor, enabling deposition rates of up to 1 nm/s. In a conventional ALD reactor operating under vacuum conditions, pump and vent times can severely limit the wafer throughput. As the spatial ALD concept works under atmospheric pressure, any pump or vent times are avoided [5].

Subsequently to the Al_2O_3 deposition a post-deposition anneal (PDA) integrated into the deposition of the SiN_x capping layer in a conventional tube furnace direct plasma PECVD was used.

For performance stability measurements highly LeTID susceptible mc-Si wafer material was processed to 6" PERC solar cells. Part of the cells were passivated with MW-PECVD Al_2O_3 as reference batch. The other cells were passivated with SoLayTec InPassion ALD. The layer thickness was varied between 4.7, 10 and 20 nm ALD Al_2O_3. All cells were processed with a LeTID susceptible process flow. Only the PECVD batches and the 4.7 nm ALD Al_2O_3 cells were additional processed by using the LeTID controlled solar cell process sequence of Hanwha Q CELLS [3]. The LeTID treatments are carried out in V_{oc} mode at 75°C. The excess carriers are injected by illumination at 1000 W/m². After 800 h exposure time all cells were regenerated without illumination at 200°C for 10 min in muffle furnace.

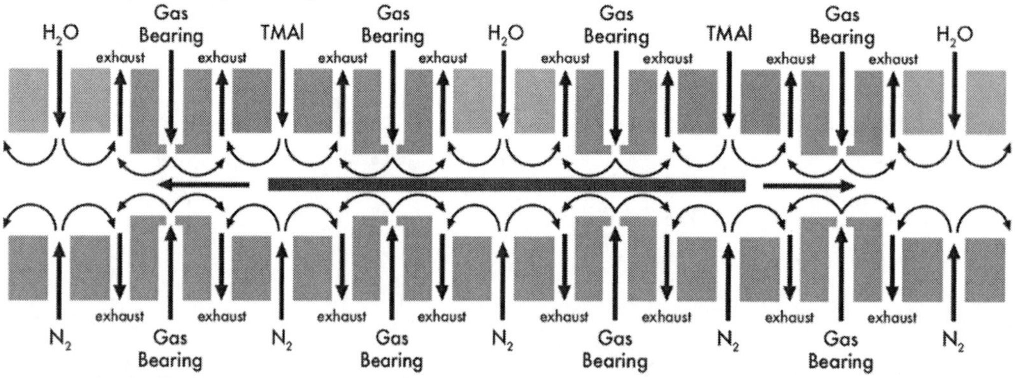

Figure 1: Sketch of in-line spatial ALD concept [6]. The TMAl and H_2O half-reaction zones are separated by N_2 curtains. Contactless transport of the wafer is facilitated by N_2 gas bearings below the wafer.

3 RESULTS

The extent of uncoated and wrap-around borders due to the ALD gas inlet design are shown in Fig. 2. ALD Al_2O_3 single sided deposition in steps of 10 nm up to 180 nm shows a stripe of less thickness at the top and bottom wafer of less than edge < 1 mm (Fig. 2 a) and average wraparound of 0.3 mm and maximum of 1 mm (Fig. 2 b).

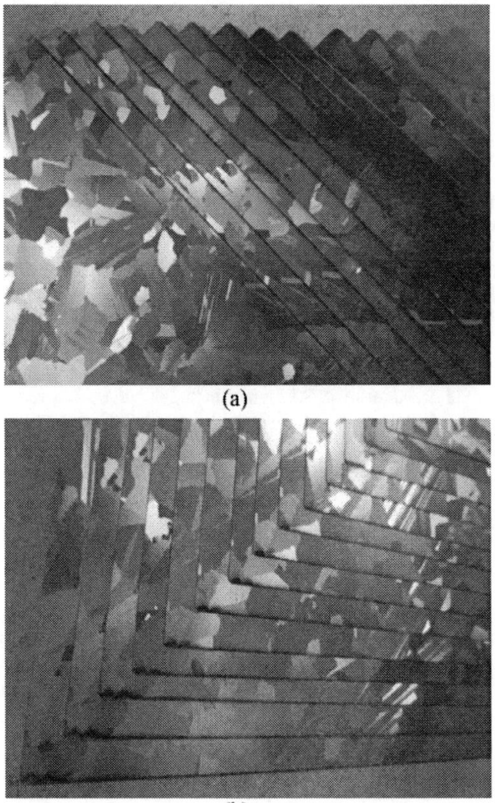

Figure 2: Wafers with Al_2O_3 single sided deposition in steps of 10 nm up to 180 nm (a) front side and (b) rear side.

3.1 Evaluation of the InPassion ALD tool from SoLayTec

In the upper part of Fig. 3 an efficiency gain of ~ 0.15% of spatial ALD passivation versus MW-PECVD during different calendar weeks is shown. This efficiency gain was reached by depositing only 4.7 nm Al_2O_3 and using an integrated PDA into the deposition of the SiN_x capping layer in a conventional tube furnace direct plasma PECVD. During the evaluation period, the amount of solar cells produced varied between 300 and 700 wafers per week this results in throughput of 1000 w/h. The electrical data for ΔV_{oc}, ΔJ_{sc} and ΔFF are shown in the lower graph of Fig. 3. The good layer uniformity in spite of low layer thickness results in V_{oc} improvement of $\geq$ 3 mV. The *FF* of the ALD cells was not optimized yet. If the *FF* would be improved for ALD PERC cells, than this cells can reach 0.25% efficiency gain with respect to MW-PECVD.

As shown in Fig. 4 when leaving out the PDA we achieved the same eta as the reference process which clearly shows that spatial ALD needs proper PDA treatment for achieving best passivation properties. Without the PDA a throughput of 1800 w/h was reached in the tube furnace PECVD. In order to achieve the best passivation properties the spatial ALD Al_2O_3 needs proper PDA treatment which is implemented in a direct plasma PECVD capping layer.

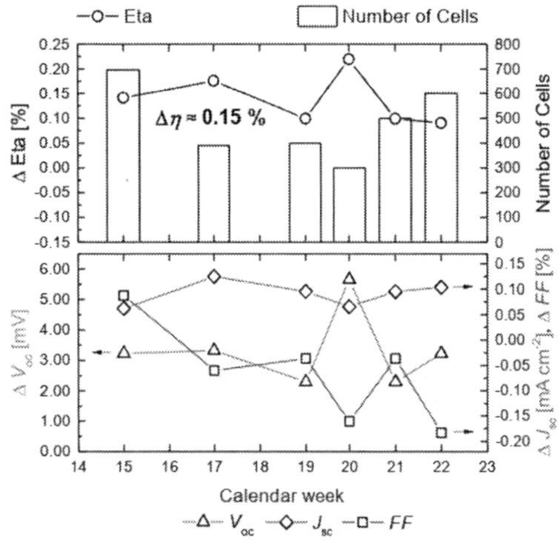

Figure 3: Efficiency gain, amount of solar cells and electrical data delta of ALD versus MW-PECVD.

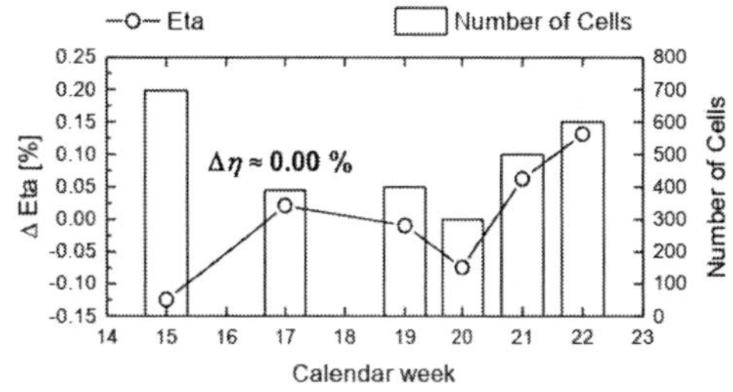

Figure 4: Efficiency gain and amount of solar cells of ALD versus MW-PECVD without PDA.

3.2 Long-time performance stability

The relative cell efficiency degradation for MW-PECVD and ALD Al_2O_3 passivated mc-Si PERC cells with susceptible (w/ LeTID) and LeTID controlled (w/o LeTID) process sequence are shown in Fig. 5. The degradation behavior is independent of Al_2O_3 layer thickness. All ALD passivated mc-PERC cells degrade up to ~ 3% after 400 h exposure time. The ALD cells processed with the LeTID controlled process sequence show a slightly lower degradation level, but also higher than the MW-PECVD cells with defect optimized process sequence. Fig. 5 shows only the relative loss in cell efficiency. It should be noted that there are significant losses in *FF* during degradation, handling and measurement procedure. This results in a higher eta degradation in Fig. 5. The real open-circuit voltage measurement are shown in Fig. 6. A V_{oc} improvement of 5 mV were determined for LeTID controlled ALD cells with an Al_2O_3 layer thickness of 4.7 nm. The V_{oc} gain is constant during the whole degradation measurement. After regeneration occurring at 800 h exposure time, all cells nearly recover to their initial V_{oc} value.

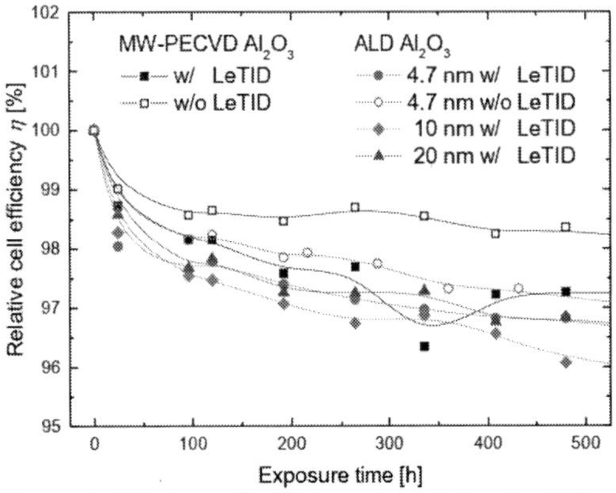

Figure 5: Relative cell efficiency degradation for MW-PECVD and ALD Al_2O_3 passivated mc-Si PERC cells with susceptible (w/ LeTID) and LeTID controlled (w/o LeTID) process sequence.

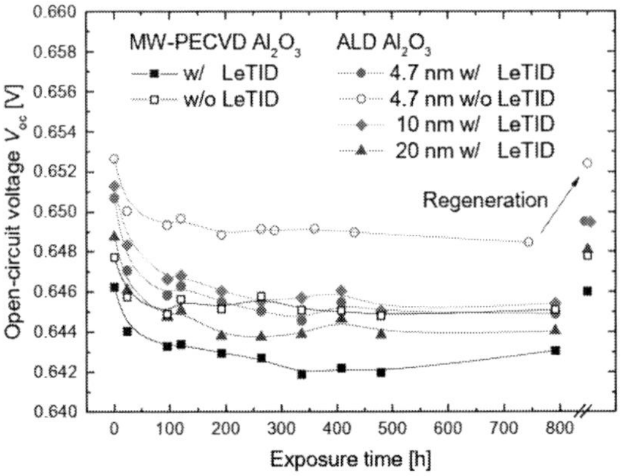

Figure 6: Real open-circuit voltage measurement during degradation for MW-PECVD and ALD Al_2O_3 passivated mc-Si PERC cells with susceptible (w/ LeTID) and LeTID controlled (w/o LeTID) process sequence.

4 CONCLUSION

In this work we show an efficiency gain of spatial ALD Al_2O_3 for the passivation of mc-Si PERC cells compared to MW-PECVD. The biggest advantages of the ALD tool compared to MW-PECVD are a better step coverage, a good layer uniformity and a layer thickness requirement of only 4.7 nm ALD Al_2O_3. An integrated PDA in the SiN_x capping process is needed for improved solar cell performance. By using the LeTID controlled solar cell process sequence of Hanwha Q CELLS [3] the V_{oc} values are continuously higher than for remote MW-PECVD passivated cells during the whole degradation measurements, which shows a better passivation performance for ALD Al_2O_3 compared to MW-PECVD also during degradation. We conclude that spatial ALD provides a promising tool for the integration of Al_2O_3 passivation layers into industrial solar cells.

5 ACKNOWLEDGEMENTS

The authors would like to thank Xavier Gay, Floor Souren, Roger Görtzen from SoLayTec and the entire team of Reiner Lemoine Research Center and pilot line at Hanwha Q CELLS for their contribution to that work.

6 REFERENCES

[1] K. Ramspeck, S. Zimmermann, H. Nagel, A. Metz, Y. Gassenbauer, B. Brikmann, A. Seidl, Light induced degradation of rear passivated mc-Si solar cells, in: Proceedings of the 27th EUPVSEC, Frankfurt, Germany 2012, pp. 861-865.

[2] F. Fertig, K. Krauß, S. Rein, Light-induced degradation of PECVD aluminium oxide passivated silicon solar cells, in: Physica Status Solidi RRL (2014) 1-6.

[3] F. Kersten, P. Engelhart, H.-C. Ploigt, A. Stekolnikov, T. Lindner, F. Stenzel, M. Bartzsch, A. Szpeth, K. Petter, J. Heitmann and J.W. Müller, Degradation of multicrystalline silicon solar cells and modules after illumination at elevated temperature, in: Solar Energy Materials and Solar Cells 142 (2015) 83-86.

[4] P. Engelhart, D. Manger, B. Klöter, S. Hermann, A. Stekolnikov, S. Peters, H.-C. Ploigt, A. Eifler, C. Klenke, A. Mohr, G. Zimmermann, B. Barkenfelt, K. Suva, J. Wendt, T. Kaden, S. Rupp, D. Rycharik, M. Fischer, J. W. Müller, P. Wawer, Q.ANTUM – Q-Cells next generation high-power silicon cell & module concept, in: Proceedings of the 26th EUPVSEC, Hamburg, Germany, 2011, pp. 821-826.

[5] F. Werner, W. Stals, R. Görtzen, B- Veith, R. Brendel, J. Schmidt, High-rate atomic layer deposition of Al_2O_3 for the surface passivation of Si solar cells, in: Energy Procedia 8 (2011) 301-306.

[6] SoLayTec InPassion ALD High Throughput ALD, http://www.solaytec.com/products/inpassion-ald.

UPGRADE OF AN INDUSTRIAL Al-BSF SOLAR CELL LINE INTO PERC USING SPATIAL ALD Al_2O_3

Floor Souren, Xavier Gay, Bas Dielissen and Roger Görtzen
SoLayTec, Dillenburgstraat 9G, 5652 AM, Eindhoven, The Netherlands
e-mail address: Floor.Souren@solaytec.com
Telephone number: +31 40 2380508

ABSTRACT: In this paper, we report the results of an upgrade from an Al-BSF solar cell line to a PERC (passivated emitter and rear cell) solar cell line, based on p-type mono-crystalline silicon (mono c-Si) material. For the rear side Al_2O_3 passivation, an InPassion ALD system of SoLayTec was used. The PERC solar cell optimization has been done in three main steps: first we optimized the rear side SiN_x capping layer, leading to a solar cell efficiency increase of $+0.4\%_{abs}$ with respect to the mono c-Si Al-BSF solar cell baseline. Secondly, we optimized the rear side polishing depth from 1-2 µm to 3-4 µm, which results in a solar cell efficiency gain of $+0.5\%_{abs}$ with respect to the standard Al-BSF solar cell baseline. For the third optimization, the emitter sheet resistance has been increased from (78±3) Ω/sq to (90±5) Ω/sq, resulting in a solar cell efficiency of (20.44±0.17)% and a gain of $>+0.8\%_{abs}$ with respect to the Al-BSF baseline. After several repeat runs, where the number of PERC solar cells has been increased to >1000 wafers per run with $>1.0\%_{abs}$ solar cell efficiency gain, >50000 PERC solar cells have been processed over two days with a stable solar cell efficiency of (20.5±0.3)% and best cell with efficiency of 21.1%. The reliability of the InPassion ALD system has been improved by decreasing the breakage rate from $0.19\%_{abs}$ to $0.05\%_{abs}$. Finally, it has been proven that with the latest configuration of the system, the breakage rate can be reduced even more to $\leq 0.05\%_{abs}$.
Keywords: PERC, ALD Al_2O_3, Manufacturing and Processing

1 INTRODUCTION

Leading PV manufacturers have started mass production of the passivated emitter and rear cell (PERC) concept within the last five years [1-3]. According to the ITRPV roadmap, the PERC solar cell concept has the largest market share of ~10% of all advanced solar cell concepts [4]. In addition, the ITRPV roadmap predicts that the PERC solar cell concept will also have the largest increase in market share: to up to 20% in 2017 and to up to >30% in 2019 [4]. One of the main reasons for this is that the PERC solar cell design leads to a solar cell efficiency increase of $0.5\%_{abs}$-$0.8\%_{abs}$ based on p-type multi-crystalline Si (mc-Si) material and $0.6\%_{abs}$-$1.0\%_{abs}$ based on p-type mono crystalline Si (mono c-Si) material with respect to the standard Al-BSF solar cell, respectively. An additional benefit of the PERC solar cell concept is that it can be relatively easily implemented on a conventional screen printing Al-BSF Si solar cell manufacturing line. Additional tools for PERC are a rear side passivation system to deposit the Al_2O_3, a post-deposition annealing tool, a tool for rear side SiN_x capping and a laser. For the rear side SiN_x capping layer, the tool for SiN_x anti reflection coating, which is available in the standard Al-BSF solar cell line, can be used. The post-deposition annealing step can be integrated in the SiN_x capping process. It is optional to upgrade the edge isolation tool to increase the polishing depth from 1-2 µm to 3-4 µm [5,6,7].

In this paper we report on an upgrade of an Al-BSF solar cell line into PERC, based on mono c-Si material. Several runs have been performed to optimize the PERC solar cell process and after several repeat runs, for the final PERC run >50000 PERC solar cells have been produced.

2 EXPERIMENTAL SETUP

2.1 Solar cell processing flow

For the upgrade of the standard Al-BSF solar cell into a mono c-Si PERC solar cell line, based on p-type Czochralski Si wafers (156 mm·156 mm, 1-3 Ω·cm), the following equipment is added: an InPassion ALD of SoLayTec for rear side ALD Al_2O_3 passivation, a laser for rear side contact opening, and an upgrade of the edge isolation to increase the polishing depth from 1-2 µm to 3-4 µm. The system for the SiN_x anti-reflection coating was also used for the deposition of rear side SiN_x capping layer on the ALD Al_2O_3 layer. The deposition time of the SiN_x anti reflection coating, which is at a thickness between 70 nm and 80 nm, has been increased to deposit a thickness of the SiN_x capping layer with a value between 100 nm and 150 nm. The post-deposition annealing has been done external, but can be integrated in the rear side SiN_x capping process. Originally, the standard Al-BSF solar cell line was based on mc-Si material and, therefore, for the upgrade into mono c-Si PERC solar cells, the saw damage removal and texturing step has been changed from acidic etching to alkaline etching. In order, to have a proper comparison between PERC and Al-BSF, the mc-Si Al-BSF baseline was converted to mono c-Si Al-BSF and all results presented in this paper are based on mono c-Si material. An overview of the processing sequences for the mono c-Si PERC solar cells versus the mono c-Si Al-BSF solar cells is presented in Figure 1.

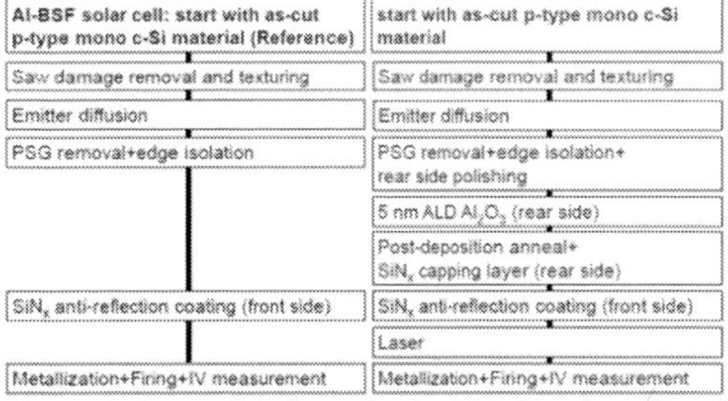

Figure 1: Processing flows for the Al-BSF solar cell and PERC solar cell.

2.2 ALD Al_2O_3 equipment

The ALD equipment of SoLayTec is based on six identical modules with a gross throughput of 3600 wafers/hour at an ALD Al_2O_3 layer thickness of 4.7 nm. An image of the six modules ALD system is presented in Figure 2.

Figure 2: InPassion ALD equipment of SoLayTec, which consists of six modules, with loader and unloader.

The wafers are placed on the main conveyor belt via a loader system. The main belt transports the wafers to one of the six deposition modules. In a module, the wafer is first in a pre-heating stage where it is heated to the desired processing temperature of 200°C and successively the wafer starts floating on air. In the next step, the wafer is transported into the reactor to deposit Al_2O_3 layer via spatial ALD [8]. Floating on air, the wafer is moving back and forward, and for each pass, the wafer is going through the different reactor zones. The reactor zones consist of a consecutive exposure to: H_2O, TMA (=trimethylalumium), H_2O, TMA and, again H_2O, separated by N_2 to prevent the mixing of precursors. Schematic overview of the injector head is presented in Figure 3.

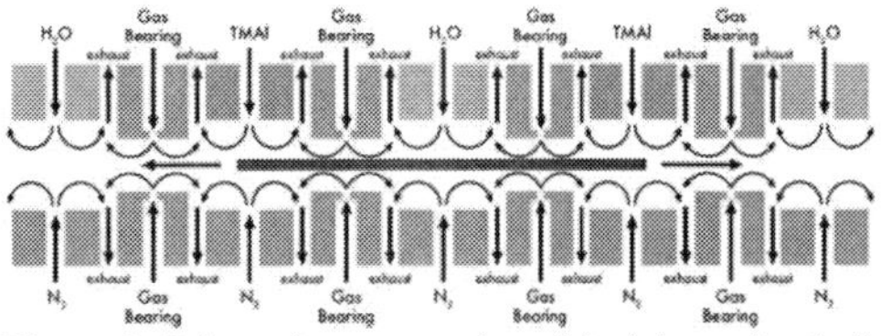

Figure 3: Schematic cross section of the injector head of the spatial ALD reactor.

Per pass through the injector head, 2 layers of ALD Al_2O_3 are deposited leading to a typical deposition rate of 1 nm/s. The deposition thickness can be controlled by setting the number of passes. After deposition of the requested layer thickness, the wafer is cooled down and placed back on the main conveyor belt and finally to the unloader system. The modules can process independently from each other to ensure the maximum up-time.

3 RESULTS

3.1 PERC optimization runs

The first PERC processing run resulted in a solar cell efficiency of 0.4%$_{abs}$ lower than the Al-BSF solar cell. We have optimized the solar cell efficiency of the PERC cell in three main optimization steps, SiN_x capping optimization, polish depth optimization, and emitter sheet resistance optimization. First, we optimized the SiN_x capping layer deposited on the ALD Al_2O_3, which has led to a solar cell efficiency increase of 0.4%$_{abs}$ with respect to the standard Al-BSF solar cell. The results are presented in Table I. The PERC solar cell efficiency increase is caused by an improvement in rear side passivation quality and optimized rear side reflection, resulting in an 3 mV increase of the open circuit voltage and an 1.2 mA/cm^2 increase of the short circuit current density, respectively.

Table I: Overview of solar cell efficiency parameters for the standard Al-BSF solar cell and for the PERC solar cell with rear side SiN_x capping recipe A and B, respectively. The open circuit voltage (V_{oc}), short circuit current density (j_{sc}), series resistance (R_s), shunt resistance (R_{sh}), fill factor (FF) and solar cell efficiency (η) show a clear improvement with optimized SiN_x capping recipe.

	Standard Al-BSF	PERC SiN_x capping recipe A	PERC SiN_x capping recipe B
V_{oc} (mV)	631±1	632±2	634±2
j_{sc} (mA/cm^2)	38.1±0.2	39.2±0.2	39.3±0.2
R_s (mΩ)	3.0±0.3	3.9±0.2	4.0±0.3
R_{sh} (Ω)	164±47	47±11	75±34
FF (%)	79.2±0.4	77.7±0.5	78.0±0.6
η (%)	19.06±0.17	19.27±0.20	19.44±0.23
Number of solar cells	42	27	31

In the second optimization step, we increased the rear side polishing depth from 1-2 μm to 3-4 μm and thereby increased the solar cell efficiency gain to 0.5%$_{abs}$ with respect to the standard Al-BSF solar cell. The additional efficiency gain was related to an improved open circuit voltage as well as to an improved fill factor difference between PERC and the Al-BSF solar cell. A smaller surface area leads to less dangling bonds which results in less surface recombination and, therefore, to a lower surface recombination velocity and to an increase of the open circuit voltage. The increase of polishing depth from 1-2 μm to 3-4 μm leads to a smaller silicon surface area and, therefore, to an increase of the open circuit voltage. The results are presented in Table II.

Table II: Overview of solar cell efficiency parameters for the standard Al-BSF and PERC solar cell with increased etch depth of 3-4 μm.

	Standard Al-BSF	PERC 3-4 μm polishing
V_{oc} (mV)	632±1	636±2
j_{sc} (mA/cm^2)	39.0±0.1	40.2±0.1
R_s (mΩ)	3.4±0.3	4.1±0.4
R_{sh} (Ω)	103±48	148±42
FF (%)	78.3±0.4	77.7±0.5
η (%)	19.32±0.11	19.85±0.16
Number of solar cells	39	39

In the third optimization step, we increased the sheet resistance of the emitter from (78±3) Ω/sq to (90±5) Ω/sq by decreasing the diffusion temperature of the diffusion process. The results are presented in Table III. The PERC solar cell with optimized emitter sheet resistance of (90±5) Ω/sq showed a higher solar cell efficiency of

0.2%$_{abs}$ with respect to the PERC solar cell with standard emitter sheet resistance of (78±3) Ω/sq. This increase is caused by a higher open circuit voltage of 3 mV while maintaining the fill factor. Based on this result, we concluded that the limiting factor of the PERC solar cell was recombination losses in the emitter. By increasing the sheet resistance of the emitter, the losses of minority carriers in the emitter were reduced, which led to an increase of the open circuit voltage. Because of a similar fill factor of the PERC group with increased sheet resistance of (90±5) Ω/sq compared to the PERC group with standard sheet resistance (78±3) Ω/sq, we can conclude that the PERC group with higher sheet resistance can be contacted properly. The emitter optimization led to a total increase in solar cell efficiency for PERC of 0.8%$_{abs}$ compared to the standard Al-BSF solar cell. The obtained gain in open circuit voltage and short circuit current density are 12 mV, and 1.5 mA/cm^2 respectively. Note that for the optimized emitter sheet resistance, a loss in fill factor of 1.1%$_{abs}$ with respect to the standard Al-BSF solar cell was obtained.

Table III: Overview of solar cell efficiency parameters for the standard Al-BSF solar cell and for the PERC solar cell with standard and optimized emitter sheet resistance of (78±3) Ω/sq and (90±5) Ω/sq, respectively.

	Standard Al-BSF	PERC with standard emitter sheet resistance $R_{standard\_sheet}$ =(78±3)Ω/sq	PERC with optimized emitter sheet resistance $R_{optimized\_sheet}$ =(90±5)Ω/sq
V_{oc} (mV)	635±2	644±3	647±3
j_{sc} (mA/cm^2)	38.5±0.1	39.9±0.1	40.0±0.1
R_s (mΩ)	2.1±0.1	2.6±0.2	2.8±0.2
R_{sh} (Ω)	162±74	94±29	96±30
FF (%)	80.1±0.1	79.0±0.3	79.0±0.3
η (%)	19.61±0.10	20.28±0.18	20.44±0.17
Number of solar cells	40	37	39

In Figure 4a, the solar cell efficiency of the PERC and Al-BSF baseline is presented for the different optimization steps as discussed before, followed by three repeat runs. In Figure 4b, the same data are presented, but the solar cell efficiency of PERC and the Al-BSF solar cell is presented relative to each other. For the optimization runs as well as for the first repeat run, the number of wafers per group were relatively small at <100 wafers. For the second and third repeat run, the number of PERC solar cells was increased to 1100 and 1340 wafers, respectively. The open circuit voltage (V_{oc}), short circuit current (j_{sc}) and fill factor (FF) of the optimization and repeat runs are presented in Figure 5a, 5b and 5c, respectively. The increase of the solar cell efficiency for PERC and the Al-BSF solar cell for the repeat runs 2 and 3 compared to repeat run1 are caused by an increase of the open circuit voltage as well as by an increase of the fill factor. Different and better wafer material leads to an increase of the open circuit voltage and an optimized firing condition results in an improved fill factor.

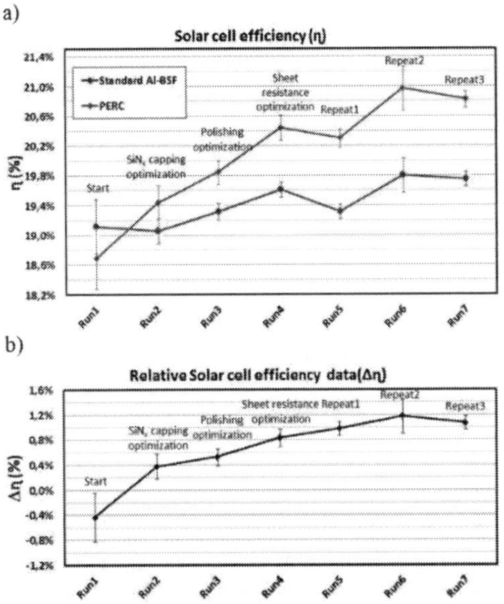

Figure 4: Absolute solar cell efficiency (a) and the solar cell efficiency of PERC and Al-BSF solar cell relative to each other (b).

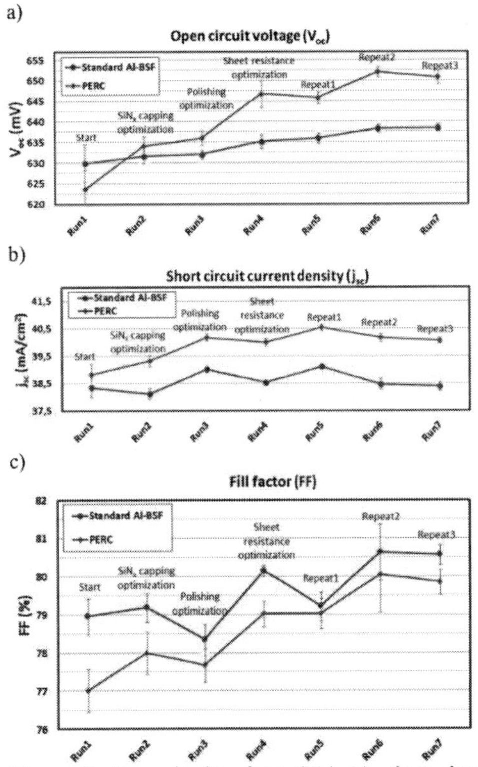

Figure 5: Open circuit voltage (V_{oc}) (a), short circuit current density (j_{sc}) (b), and fill factor (FF) (c) for PERC solar cell and standard Al-BSF solar cell for the different optimization runs and repeat runs.

3.2 PERC production ramp-up

Finally, production ramp-up was initiated and over 2 days >50000 PERC solar cells have been manufactured. Figure 6 shows the efficiency of the solar cells with average solar cell efficiency of (20.5±0.3)% and the best PERC cell reaches 21.1%. The cell distributions of day 1 and day 2 are fitted to a normal distribution, as presented in Figure 7a and Figure 7b, respectively. The solar cell efficiency distribution is as expected for solar cells: a faster drop off for the higher efficiency cells and a slower drop off for the lower efficiencies, which means that there are relatively more bad cells than would be expected from a perfect normal distribution. From the fitted normal distribution, the average solar cell efficiency (μ_{normal}), the standard deviation (σ_{normal}) as well as the full width at half maximum ($FWHM_{normal}$) are determined. For day 1 and day 2, the average efficiency is equal to (20.6±0.2)%, which is slightly higher than the average value of the solar cell efficiency data and can be explained by the fact that for the normal distribution there are less low fliers compared to the solar cell efficiency data. For day 1 and day 2, the standard deviation and full width at half maximum are determined to be at $0.2\%_{abs}$, and $0.4\%_{abs}$, respectively. The FWHM of the normal distribution is similar to the FWHM of mono c-Si PERC solar cells as published by Trina, showing the relatively good quality of the solar cell line [9]. In this paper, we show the results of the successful upgrade into mono c-Si PERC solar cells originally starting from a mc-Si Al-BSF solar cell line instead of a mono c-Si Al-BSF line. Where the results presented in this paper for PERC and Al-BSF solar cells are all based on mono c-Si material to make a correct comparison.

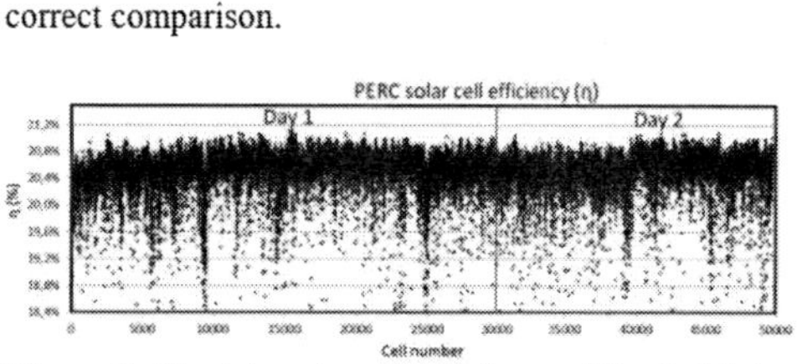

Figure 6: Absolute solar cell efficiency of the first 2 days of the production ramp-up of PERC. A stable efficiency of (20.5±0.3)% has been achieved.

a)

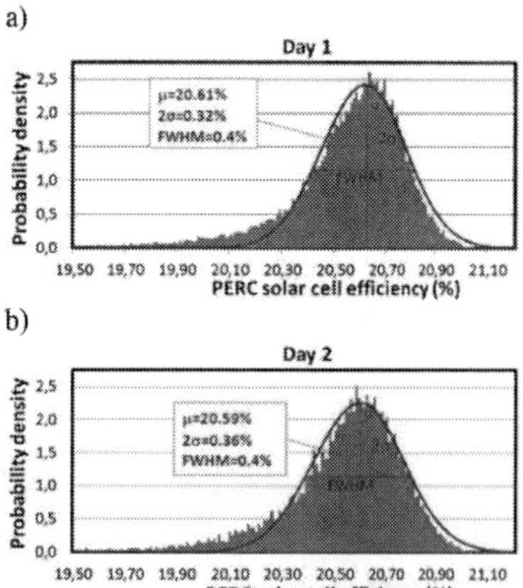

Figure 7: Probability density of PERC solar cell efficiency for day 1 (a) and day 2 (b), where for both days based on the normal distribution the average solar cell efficiency is at 20.6% and FWHM is determined to be on 0.4%. FWHM is the width of the distribution where the distribution is equal to half of the maximum.

The corresponding open circuit voltage (V_{oc}), short circuit current (j_{sc}), and fill factor (FF) of the PERC solar cells, are presented in Figure 8a, 8b and 8c, respectively.

a)

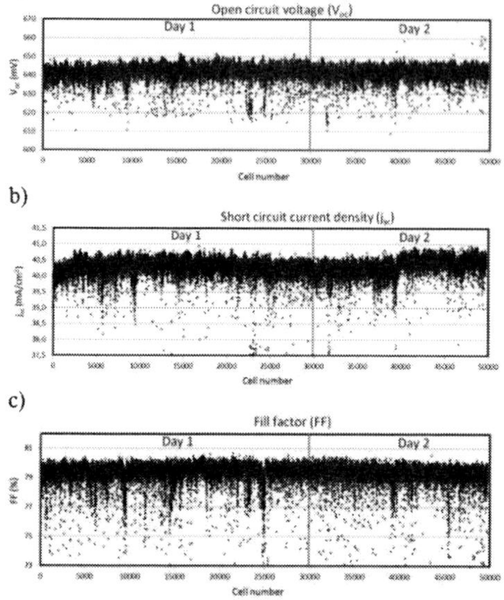

Figure 8: Open circuit voltage (V_{oc}) (a), short circuit current density (j_{sc}) (b), and fill factor (FF) (c) of the first 2 days of the production ramp-up of PERC.

Of all the results of open circuit voltage, short circuit current density and fill factor, the fill factor has the widest distribution, with a lot of low fliers (e.g. around cell numbers: 10000, 15000, 25000, 40000 and 45000) leading also to low fliers in the solar cell efficiency. This indicates that for further increase of the solar cell efficiency, the fill factor needs to be improved, which shows also the further optimization potential. Moreover, the PERC solar cell efficiency increases with ca $0.1\%_{abs}$ for the first couple of 1000 solar cells of day 1 and at day 2 after ~40.000 cells, which corresponds to an increase of the short circuit current density. Open circuit voltage has the smallest distribution, which shows that the passivation with ALD Al_2O_3-SiN_x stack is performing well. Overview of the solar cell efficiency parameters and parameters which characterize the normal distribution are presented in Table IV.

Table IV: Overview of solar cell efficiency parameters for the PERC production of day 1 and day 2, respectively. The solar cell efficiency distribution has been fitted according normal distribution and from this distribution, the average solar cell efficiency ($\eta_{average\_normal}$), standard deviation (σ_{normal}) and full width at half maximum ($FWHM_{normal}$) have been determined.

	PERC production day 1	PERC production day 2
V_{oc} (mV)	642±4	641±3
j_{sc} (mA/cm^2)	40.3±0.3	40.3±0.3
FF (%)	79±1	79±1
η (%)	20.5±0.4	20.5±0.3
Number of solar cells	29918	19830
	Fit with Normal distribution:	Fit with Normal distribution:
η_{normal} (%)	20.6	20.6
σ_{normal} (%)	0.2	0.2
$FWHM_{normal}$ (%)	0.4	0.4

3.3 Reliability improvement of InPassion ALD

Over the last few months, the reliability of the InPassion ALD of SoLayTec has been significantly improved, which is presented by the breakage rate in Figure 9. The breakage rate is monitored on a production level of more than 75000 wafers per day. The Module-type 2015 reached an average wafer breakage rate of 0.19%$_{abs}$, while the module-type 2015+ shows a reduced breakage rate at an average value of 0.05%$_{abs}$. Preliminary results present that for the Module-type 2016, the average breakage rate has been decreased even more to ≤0.05%$_{abs}$ which is based on 5000 wafers per day for a few consecutive days.

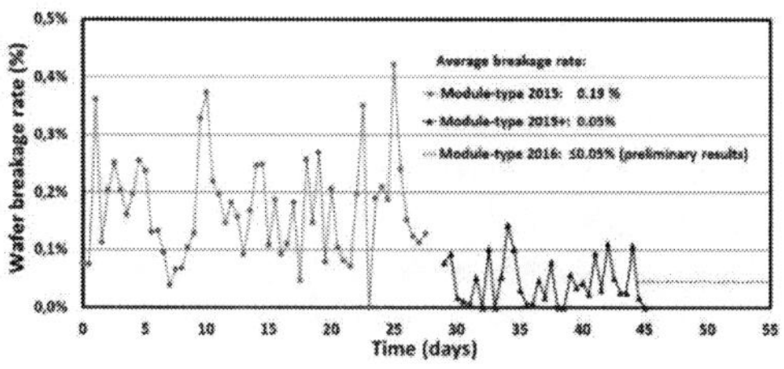

Figure 9: Wafer breakage rate as function of days for 3 different module types: Module-type 2015, Module-type 2015+ and Module-type 2016.

4 CONCLUSIONS

Ramp-up of PERC solar cell line based on p-type mono c-Si material has been done in three optimization steps: optimization of rear side SiN_x capping process, the rear side polishing depth, and optimization of the emitter sheet resistance. After the third optimization, the solar cell efficiency for PERC was determined at (20.44±0.17)%, which was >0.8%$_{abs}$ higher with respect to the standard Al-BSF solar cell reference. Successfully, for two repeat runs, even >1.0%$_{abs}$ solar cell efficiency gain has been obtained caused by an increase of wafer quality and optimized firing condition. Finally, >50000 PERC solar cells have been processed at average solar cell efficiency of (20.5±0.3)% where the solar cell efficiency distribution has been fitted to a normal distribution, showing an average value of (20.6±0.2)% and $FWHM_{normal}$ of 0.4%$_{abs}$, and this value is similar as presented by Trina for mono c-Si PERC, and shows that the ramp-up was successful. These results show also that implementation of mono c-Si PERC solar cell can be done successful starting from a mc-Si Al-BSF solar cell line. All results of Al-BSF solar cells and PERC solar cells presented in this paper, are based on mono c-Si material. Based on these >50000 PERC solar cells, the parameter with widest distribution and most low fliers is the fill factor and shows further optimization potential which can increase solar cell efficiency. The parameter with the smallest distribution is the open circuit voltage, which is caused by the good rear side passivation of the ALD Al_2O_3-SiN_x stack. Together with an improved breakage rate down to ≤0.05%$_{abs}$ of the InPassion ALD, SoLayTec offers an interesting solution for mass production of ALD Al_2O_3 for PERC solar cells.

5 REFERENCES

[1] D. Chen et al., 21.40% Efficient Large Area Screen Printed Industrial PERC Solar Cells, 30th European Photovoltaic Solar Energy Conference and Exhibition, Hamburg, Germany.

[2] J.W. Müller et al., Current Status of Q-Cells' High-Efficiency Q.ANTUM Technology with New World Record Module Results, 27th European Photovoltaic Solar Energy Conference and Exhibition, Frankfurt, Germany.

[3] B. Tjahjono et al., Optimizing Celco Cell Technology in one Year of Mass Production, 28th European Photovoltaic Solar Energy Conference and Exhibition, Paris, France.

[4] International Technology Roadmap for Photovoltaic (ITRPV) 2015, 6th edition, April 2015.

[5] R. Sastrawan et al., Implementation of a Multicrystalline ALD-Al_2O_3 PERC-Technology into an Industrial Pilot Production, 28th European Photovoltaic Solar Energy Conference and Exhibition, Paris, France.

[6] D. Pysch et al., Implementation of an ALD-Al_2O_3 PERC-Technology into a Multi- and Monocrystalline Industrial Pilot Production, 29th European Photovoltaic Solar Energy Conference and Exhibition, Amsterdam, The Netherlands.

[7] Xavier Gay et al., Post-Deposition Thermal Treatment of Ultrafast Spatial ALD Al_2O_3 for the Rear Side Passivation of p-type PERC Solar Cells, 28th European Photovoltaic Solar Energy Conference and Exhibition, Paris, France.

[8] Poodt et al. High-Speed Spatial Atomic-Layer Deposition of Aluminum Oxide Layers for Solar Cell Passivation, Advanced Materials, 22(32), 3564-3567, 2010.

[9] D. Chen et al., 21.40% Efficient Large Area Screen Printed Industrial PERC Solar Cell, 31st European Photovoltaic Solar Energy Conference and Exhibition, Hamburg, Germany.

THE OPTIMIZATION OF LASER CONTACT OPENING PROCESS FOR N-TYPE REAR JUNCTION PRINTING PERT SOLAR CELLS

Jongchul LEE, Yoonseok CHOI, Jieun LEE, Hoon OH, Dohyeon Kyeong,
Taejun KIM*, Myungick HWANG, and Eun-Chel CHO
Hyundai Heavy Industries Co., Ltd., Yongin 446-912, South Korea

ABSTRACT: In this paper, we present n-type rear junction (RJ) PERT (Passivated Emitter Rear Totally-diffused) solar cell with the process improvement of Laser Contact Opening (LCO) process and rear pattern opened by laser. In detail, effect of laser process parameter such as laser power and pulse duration was evaluated in terms of impact on the cell efficiency and opening shape. Also LCO pattern such as line, dash and dot was optimized to reduce rear electrode contact area to minimize the recombination characteristics of n-RJ-PERT cells. Through this, we achieved 21.77% efficiency, and 686 mV open-circuit voltage(in-house measured) of n-RJ-PERT solar cell with plated Ni/Ag front contacts, Al_2O_3 rear passivation, and screen-printed local Al emitter on 110μm-thickness 6-inch n-type Czochralski (Cz) single crystalline.
Keywords: n-type, passivation, recombination, laser processing

1 INTRODUCTION

N-type silicon wafer solar cells are receiving increasing attention for industrial application in recent years, such as the n-type PERT solar cells. N-type PERT cells have been studied and under production in many groups, at least 4 of them showed cell efficiency over 21% so far [1-5]. Due to the high bulk minority carrier diffusion length, the p-n junction of an n-PERT cell can be placed either at front or rear side [6].

We chose the back junction PERT structure because of main two reasons. Back junction structure is suitable for high V_{OC} than the front-junction cell, probably because the back-junction cell is less sensitive to metal-induced emitter recombination. Also its structure and the process sequence is very similar to that of industrial p-type PERC (Passivated Emitter Rear Contact), which enables cell makers to produce highly-efficient n-type solar cells with minimal upgrades into p-type PERC production lines. Therefore, n-RJ-PERT structure is very attractive candidate as a structure of highly efficient n-type solar cell with reasonable cost [7].

Open-circuit voltage (Voc) of n-type rear junction PERT printed back Al in almost groups including us is about 670mV.[8] It is necessary to improve the process that affects high carrier recombination to achieve Voc over 680mV. As a result of investigating recombination losses on each area of rear junction cells, for example, front/rear passivation and electrodes, recombination of rear electrode contact area has to be improved preferentially to obtain high Voc. This paper is focusing on study about optimizing laser process condition and rear dielectric layer ablation pattern to reduce the recombination of rear electrode contact area.

2 N-TYPE REAR JUNCTION PERT SOLAR CELL FABRICATION PROCESS

N-RJ-PERT cell depicted in Fig. 1 was fabricated with large-area (6", 239 cm^2), commercial-grade, phosphorous-doped Czochralski silicon wafers. Process sequence is shown in Fig.2. After a saw-damage etch, we deposited a BSG on the rear side by APCVD and it was annealed to fabricate boron emitter. BSG and BRL (Boron Rich Layer) were removed by HF and HNO_3 solution. Then we prepare a dielectric on the rear side by PECVD. The dielectric served as a mask for the subsequent texturing and $POCl_3$ diffusion. The rear mask was dissolved during the PSG (phosphosilicate glass) removal step. After formation of phosphorus doped FSF, we stacked dielectric layers on the front and back with a combination of dry oxidation, PECVD, and ALD processes. Wet cleaning procedures were employed in between some deposition steps as needed. After opening up the back dielectric stack locally with a laser contact opening(LCO), we use the laser-doped selective emitter (LDSE) process [9] to simultaneously open the front dielectric and form the selectively heavy-doped region. Rear side metallization was performed with screen-printing and firing Al electrode. Light-induced plating (LIP) method is used to create nickel–silver (Ni/Ag) front contacts. [10-12]

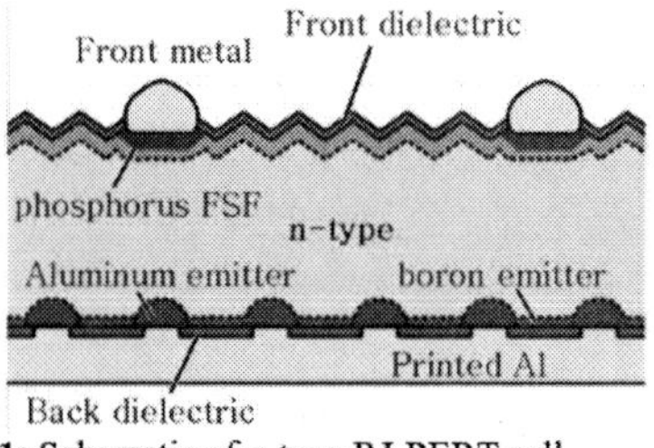

Figure 1: Schematic of n-type RJ-PERT cell

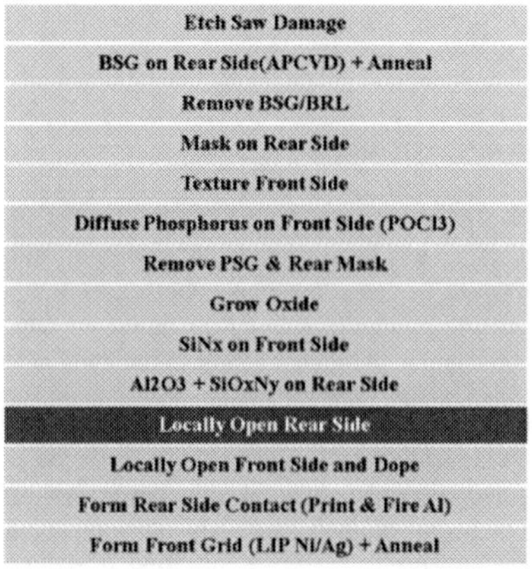

Figure 2: The process sequence of n-RJ-PERT cell

Recombination characteristic of boron emitter and phosphorus doped FSF were measured by Quasi-Steady-State Photo Conductance (QSSPC) in symmetric test structure to confirm the recombination velocity on each area of cell.

When ablating the back dielectric stack locally with laser, Opening shape formed by laser ablating at different laser condition was measure by the microscope. Recombination characteristics of cell were investigated by measuring I-V and PL intensity.

3 OPIMIZATION OF LASER PROCESS

3.1 Measure of recombination losses

To improve recombination characteristics of n-RJ-PERT cell with Voc of 670mV, recombination losses about each part were measured. [13][14] In an experiment, sample has symmetric structure with a double side doping and passivation layer. Recombination losses are measured by QSSPC with samples by varying laser aperture ratio of the front and rear side for electrodes as shown in Fig. 3.and table I.[8]

Photoluminescence and implied V_{OC} characteristics by QSSPC were measured for each sample and J_0 was calculated from i-V_{OC} data. $J_{0,\,open}$ in the opened area and $J_{0,\,passi}$ in the passivated area can be extracted from $J_{0,\,tatal}$ versus the fraction of open area (C_{open}) irrespective of the relative magnitude of bulk recombination and recombination at the passivated surface (Eq. (1)).[13] In other words, slope and y-intercept in Fig. 4 stands for recombination characteristics (J_0) of electrodes and passivated area that includes bulk recombination, respectively. As a result, Recombination loss in rear electrode represent the largest portion at 44% of the total, LCO process which form that opening area has to be modified for high Voc and efficiency.

$$J_{0,\,total} = 2J_{0,\,passi} + C_{open}\,(J_{0,\,open} - J_{0,\,passi}) \quad (1)$$

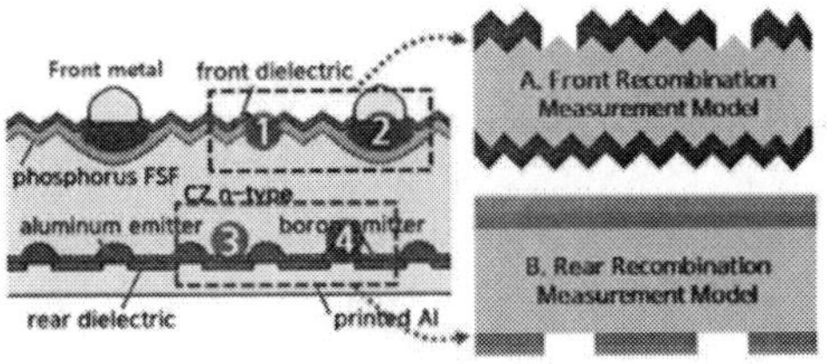

Figure 3: Sample structure for measuring recombination characteristics

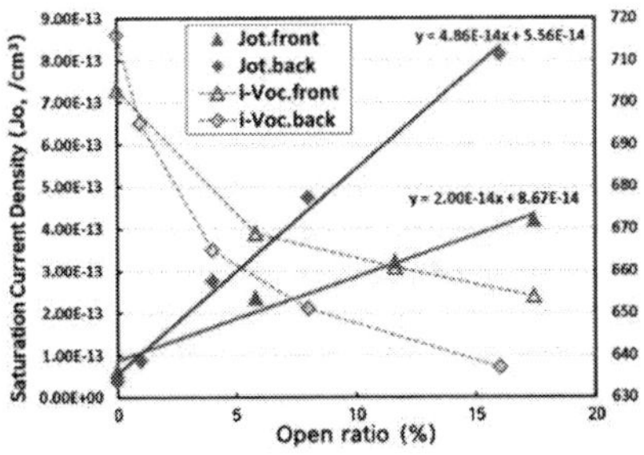

Figure 4: Total saturation current density and i-Voc vs open ratio.[8]

Table I: J_0 in each area and J_0 distribution in the entire area of our n-RJ-PERT cell with Voc of 670mV [8]

Recombination Area	J_0 [fA/cm^2]	Area [cm^2]	J_0 distribution in the cell [%]
Front Passivation	43	23422	22
Front electrodes	2000	478	20
Rear Passivation	28	22944	14
Rear Electrode	2175	956	44

3.2 Optimization of laser process condition for LCO process

LCO process was performed using by nano-second laser, which has infrared wavelength (1054nm). The samples before LCO process were coated by different dielectric films by PECVD. Rear side reflectance of dielectric films with different capping layer was measured by reflectometer. LCO process by Laser ablation was performed by laser power and duration at the given laser system. Opening shape by LCO process was characterized by microscope. Dielectric films by recipe A and B had reflectance of 30% and 10%, respectively, at IR range as shown in Figure 6. Dielectric opening with dielectric film by recipe A and the given laser ablation condition was irregular, where opening shape was not circular shape and residual dielectric film existed. In the case of dielectric film with lower reflectance deposited by recipe B in Figure 6, the opening shape was circular shape. As a conclusion, LCO condition of dielectric film should be considered reflectance of the sample after capping dielectric layer deposition process.

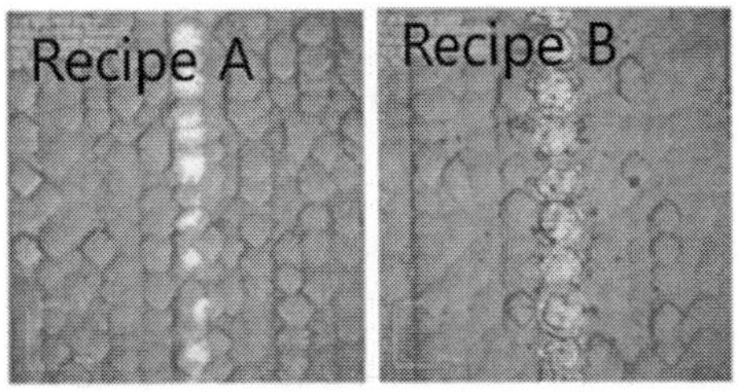

Figure 5: Opening shape of sample with different recipe of capping layer.

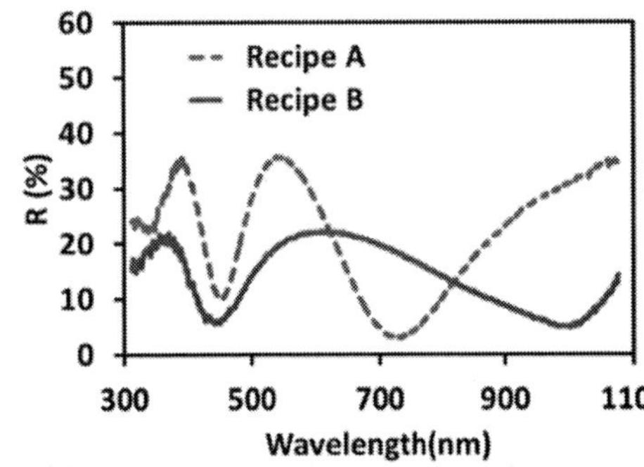

Figure 6: Reflectance of sample with different recipe of capping layer.

Laser power and pulse duration was varied to optimize LCO process condition with capping laser recipe fixed. When laser power is lower than threshold power, dielectric ablation was not performed clearly and spot size was small. When laser power was too high, cell efficiency was decreased because of thermal damage caused by the laser. If the pulse duration of laser is long, the thermal damage could be increased. So the LCO condition should be optimized considering both the ablation pattern shape and the thermal damage of the laser.

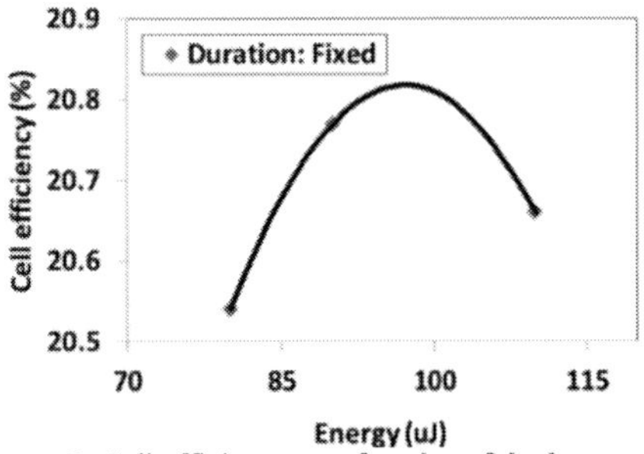

Figure 7: Cell efficiency as a function of the laser energy at fixed pulse duration

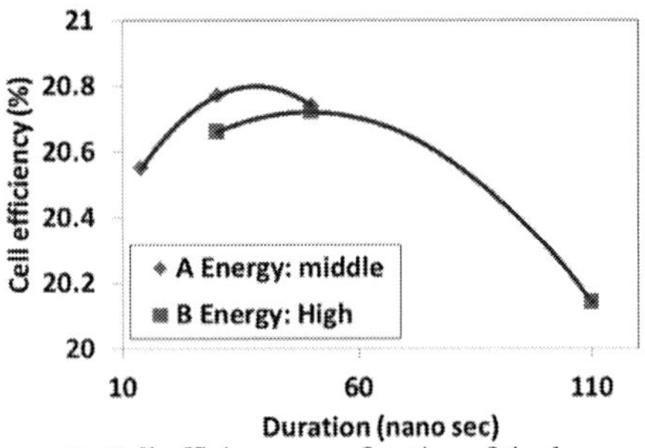

Figure 8: Cell efficiency as a function of the laser pulse duration

3.3 Optimization of opening pattern for LCO process

Rear electrode contact area showed the higher recombination characteristics than that of the passivated area, so the simplest way to reduce the total recombination is decreasing the LCO area. Normally LCO pattern is the line type, it is necessary to change dash or dot pattern for minimizing recombination as shown figure 9. Dash pattern was performed by spacing between lines, finally dot pattern was formed when spacing length is long and line length is very short.

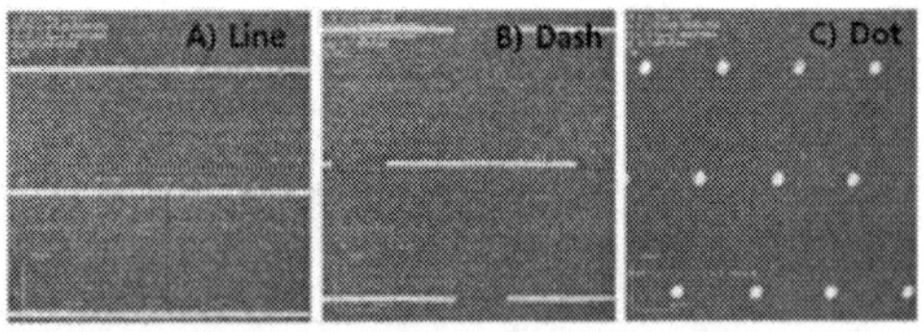

Figure 9: Microscope image of LCO patterns

Laser opening is necessary for metal contact, but a passivation property of the LCO patterned dielectric film on rear surface have to be optimized to better Voc and FF, thereby highest Voc. PL intensity was measured by laser opening ratio. As shown in figure 10, PL intensity was inversely proportional to contact opening ratio because opening area can be defined as high recombination site. But it occur the side-effect such as void formation and/or series resistance increase when a pitch of LCO pattern is very long or dot size is very small. So dot size and distance have to be optimized for obtain high voltage without decreasing FF. Three LCO patterns shown in figure 9 were tested to get higher efficiency. Dot pattern with contact opening of 1% give best result shown in figure 11.

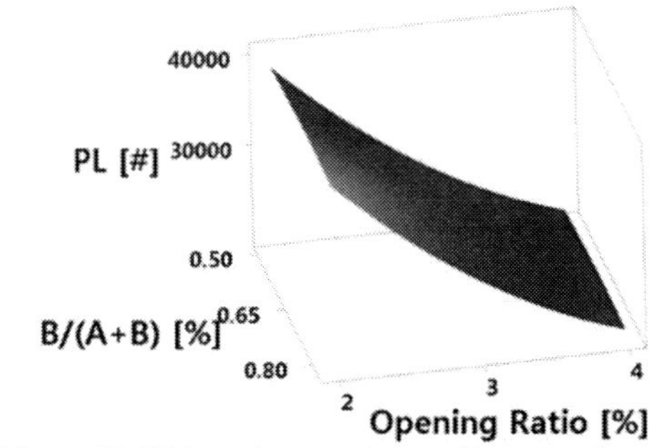

Figure 10: PL intensity as variation of laser opening ratio [A: line length, B: not opening length between lines]

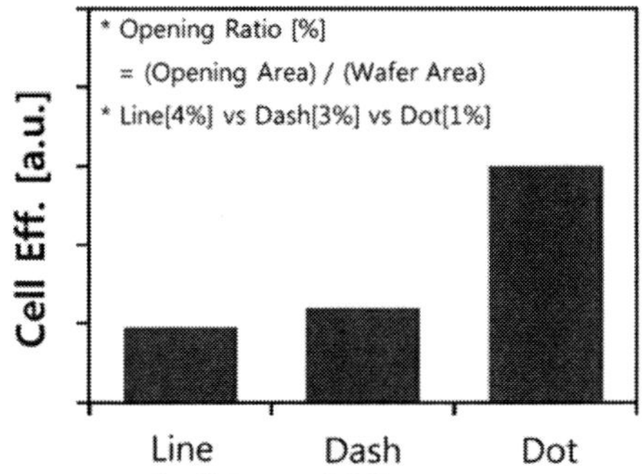

Figure 11: Cell efficiency of different LCO pattern type

4 SUMMARY

We have been working on the LCO process development for n-type Rear Junction PERT cells. Laser power and pulse duration were optimized considering the rear surface reflectance, opening shape and thermal damage of the laser. LCO pattern has been improving to reduce the recombination of rear electrode contact area. So far we achieved 21.77% efficiency of n-RJ-PERT solar cell with plated electrodes on 156mm-large industrial Czocharalski-grown wafer, we are trying to optimize LCO pattern to improve the open circuit voltage and efficiency further more than 22.0%.

Table II: I-V of n-RJ-PERT cell measured by SERIS calibration center

	Thickness [μm]	Voc [mV]	Isc [A]	FF [%]	η [%]
N-PERT cell with plated front contact	110	686.1	9.35	81.12	21.77
N-PERT cell with screen printed front contact	110	679.2	9.27	79.59	21.05

ACKNOWLEDGMENTS

This work was partly supported by the New & Renewable energy Core Technology Program of the Korea Institute of Energy Technology Evaluation and Planning (KETEP), with the financial grant provided by the Ministry of Trade, Industry & Energy, Republic of Korea (No. 20133010011780).

REFERENCES

[1] Y.-W. Ok, Solar Energy Material Solar Cells 123, 92–96 (2014)
[2] P. Rothhardt, Energy Procedia 55, 287 – 294 (2014)
[3] H.C. Chang, Energy Procedia 55, 643-648 (2014)
[4] F. Duerinckx, Proceeding of 29th European Photovoltaic Solar Energy Conference and Exhibition (2014)
[5] V. Mertens, 28th European Photovoltaic Solar Enrgy Conference and Exhibition (2013)
[6] Jia Chen, Prog. Photovolt: Res. Appl. (2015)
[7] J. Cho, Proceeding of the 5th International Conference on Silicon Photovoltaics, SiliconPV 2015
[8] Jieun LEE, Proceeding of the32th European Photovoltaic Solar Energy Conference and Exhibition (2015)
[9] Hallam B, Wenham S, Sugianto A, Mai L, Chong CM, Edwards M, Jordan D, Fath P, IEEE J Photovolt 1:43-48 (2011)
[10] J. Cho, Proceeding of the 23rd Photovoltaic science and engineering conference (2013)
[11] T. Kim, Energy Procedia 55, 431-438, (2014)
[12] M. Kim, Proceeding of the 29th European Photovoltaic Solar Energy Conference and Exhibition (2014)
[13] Chi Danga,b, Riet Labiea, Energy Procedia 55 (2014) 649 – 655
[14] Dechers J, Proceeding of the 28th European Photovoltaic Solar Energy Conference and Exhibition (2013)

A SIMPLE ROUTE TO FABRICATION OF LOCAL BACK CONTACTS TO SILICON SOLAR CELLS

C.K. Hsu[1], J.H. Yang[1], C.W. Kuo[2], T.M. Kuan[2], C.Y. Yu[2] and I.C. Chen[1]

[1] Institute of Materials Science and Engineering, National Central University, Zhongli 320, Taiwan
[2] TSEC Corporation, Hsinchu 303, Taiwan

ABSTRACT: A new and simple patterning technique has been developed to fabricate local back contacts (LBCs) of rear-surface-passivated Si solar cells. Randomly hole-patterned polystyrene (PS) films were fabricated on p-type Si wafers by spin-coating of PS films followed by soaking in an acetone (ACE) solution. The hole-patterned PS films were then served as wet-etch masks of SiO_2/SiN_x stacked passivation layers, and LBCs were formed by wet-etch process followed by screen-printing of Al paste. The hole area fraction on the PS film can be adjusted by change of ACE soaking time. Passivated emitter and rear cells (PERCs) have been demonstrated using this patterning process.

Keywords: Back-Surface-Field; Passivation; Silicon Solar Cell

1 INTRODUCTION

Solar industry grows rapidly in recent years. The global photovoltaic market is currently dominated by crystalline silicon (Si) technologies. Si solar cells with a full area screen-printed Aluminum back surface field (Al-BSF) have been widely used over the last decades. However, the full Al-BSF structure possesses several major issues hindering the device to reach better electrical performances [1], such as high rear surface recombination velocities. The high rear surface recombination velocities prevent the full Al-BSF solar cells from being fabricated thinner to take advantage of cost reduction and reduced bulk recombination [2]. Therefore, passivated emitter and rear cell (PERC) structure was proposed for improving conventional Al-BSF solar cells.

Much research has been devoted to develop technologies for fabriation of local back contacts (LBCs), including laser ablation, laser firing, fire-through contacts and photolithography. Photolithography is the most precise approach for forming LBC patterns. It has been demonstrated to fabricate high efficiency Si solar cells [3]. However the technique is costly and time-comsuming. Another method of forming aluminium LBCs is via fire-through contact technology. This method uses a fire-through metal paste, which is locally applied onto the rear passivation layer, and then local penetration of the dielectric passivation layer occurs during contact firing [4].

Currently, the back contact patterns of PERCs are majorly fabricated via laser technology, which are mainly divided into two categories: (1) laser ablation process, where the dielectric layers are formed by laser scribing of passivated dielectric layers, and (2) laser firing process, where the contacts are created by creating openings in dielectric and making contacts to the Si substrate in one single step [5].

Compared to other methods for LBC fabrication, the laser technology is suitable for commercial solar cell production. However, the high equipment cost is the disadvantage of this technology. In this study, we propose a simple approach to fabricate LBC patterns. By dewetting of the polymer film on the dielectric film, holes can be randomly formed on the surface and served as contact patterns. Random local Al-BSF solar cells with a hole-patterned rear surface passivation layer have been prepared.

2 EXPERIMENTAL

2.1 Sample preparation

200 μm-thick p-type Si wafers with a resistivity range between 1 and 5 Ωcm were used for the cell fabrication. The wafers were cleaned by acetone, isopropyl alcohol and deionized water. After surface texturation by KOH etching, phosphorous diffusion was carried out by spin-on-dopant method. A SiO_2 diffusion barrier layer was deposited on the back-side surface of the wafer by plasma-enhanced chemical vapor deposition (PECVD). A highly doped n-type emitter layer was made on the front-side surface by spin-on H_3PO_4 solution followed by annealing at 900 °C for 30min. Then, the wafer was immersed into hydrofluoric acid to remove the phosphorus silicate glass (PSG) and the diffusion barrier layers. The antireflection coating (ARC) and passivation layer were deposited on the n^+-emitter and the back-side surface using PECVD, repsectively. The stacked passivation layers was composed of a SiO_2 and SiN_x thin films.

The polystyrene (PS) powders were dissolved in toluene solution. The PS film was formed on the passivation layer by spin-coating of the PS solution and the baking process. Hole-patterned PS films, which serve as wet-etch masks, could be obtained after soaking PS-coated wafers into acetone. The holes were formed randomly on the polymer films. The soaking time was adjusted to control the hole size and hole fraction. The local rear surface passivation layer can be formed by etching away exposed SiO_2/SiN_x stacked passivation layers, and then the Ag paste and Al paste were scrren-printed on the ARC and the local passivation layers, respectively. The sample was annealed at 800 °C for the metallization.

2.2 Sample characterization

For the characterization, hole formation in the PS films is visualized and measured with a scanning electron microscope (SEM) and an optical microscope (OM). The carrier lifetime was measured by Quasi Steady State Photo Conductance (QSSPC) using WCT-120 silicon wafer lifetime detector (Sinton Consulting Inc.).

The photovoltaic properties of the fabricated solar cells were evaluated at room temperature based on the illuminated current density versus voltage (J-V) and the internal quantum efficiency (IQE) characteristics under 1-sun air mass 1.5 global spectrum.

3 RESULTS AND DISCUSSION

3.1 Observation of PS thin films with different hole fractions

Hole-patterned PS thin films were formed by dewetting of PS films on Si substrates. The different hole fractions were changed by controlling the soaking time in acetone. Figure 1 shows the SEM images of the hole-patterned PS films with different hole fractions. The hole size and hole fraction were estimated from the SEM and OM images using software ImageJ. Figure 2 shows hole diameter and hole fraction as a function of soaking time. As shown in Figure 2, the hole size and fraction increase with increasing the soaking time. Initially, the hole density also increases as the soaking time increases, but the hole density would decrease at certain time due to the merge of small holes.

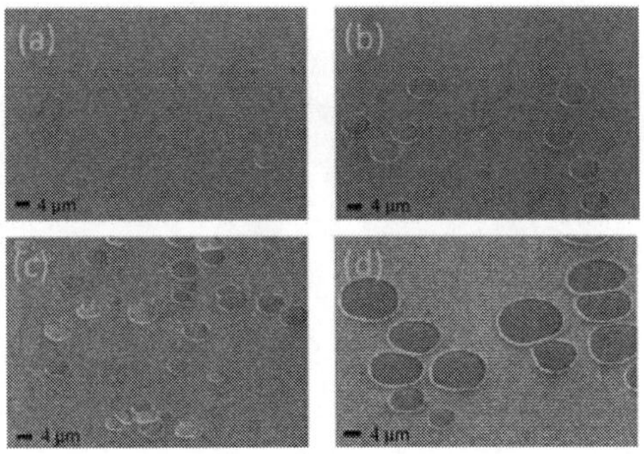

Figure 1: SEM images of the hole-patterned PS films with a hole fraction of (a) 2.7%, (b) 7.3%, (c) 18.7% and (d) 29.5%.

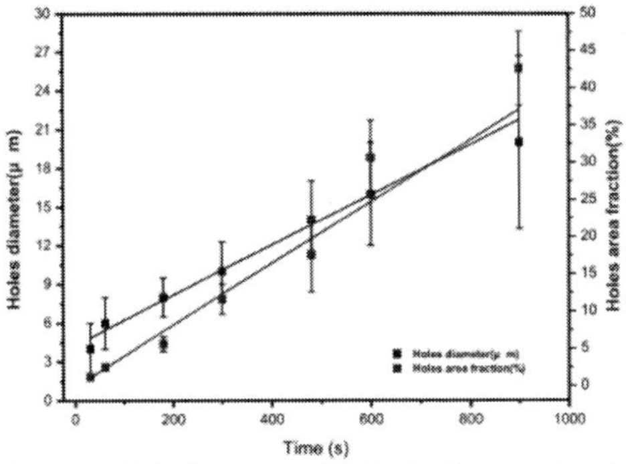

Figure 2: Hole diameter and hole fraction as a function of soaking time.

3.2 Lifetime and implied V_{oc} measurements

Figure 3 shows effective carrier lifetime and implied V_{oc} as a function of hole fraction. The lifetime is about 40 μs without hole formation. Both lifetime and implied V_{oc} decrease with increasing the hole fraction up to 45%, and then the values of lifetime and implied V_{oc} remain nearly the same when the hole fraction is above 45%.

3.3 Performance of random Al BSF Si solar cells

Table 1 shows cell parameters with various hole fractions. The illuminated J-V curves corresponding to each sample are displayed in Figure 4, and their open-circuit voltage (V_{oc}), short-circuit current density (J_{sc}), fill factor (FF), and efficiency are shown in Table 1. The cell with smallest hole fraction possesses the highest V_{oc}, but the value of FF is lowest due to high series resistance. The sample with a hole fraction of 7.3% could offer a best cell performance since it possesses a good passivation compared with a full Al-BSF cell and keeps a relatively low series resistance compared with the cell with a hole fraction of 1.5%.

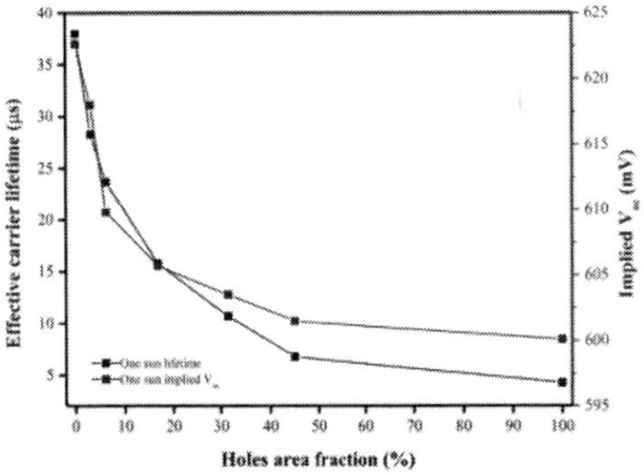

Figure 3: effective carrier lifetime and implied V_{oc} as a function of hole fraction.

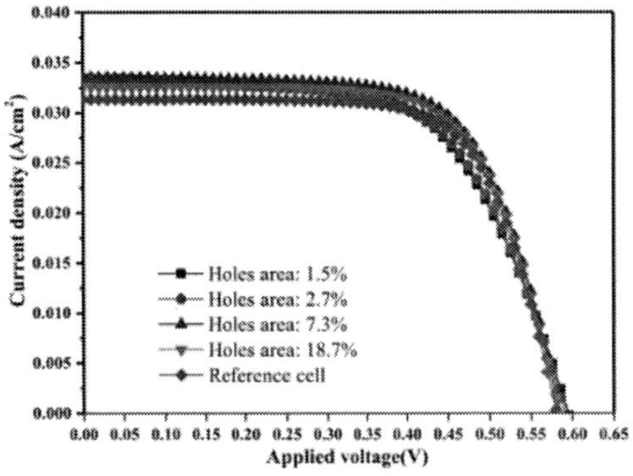

Figure 4: Illuminated J-V characteristics of solar cells with various hole fractions.

Table I: Cell parameters with various hole fractions

Holes area fraction (%)	V_{oc} (mV)	J_{sc} (mA/cm²)	Fill factor (F.F, %)	Efficiency (%)
1.5	595.1	33.42	62.1	12.35
2.7	591.1	33.17	63.23	12.4
7.3	587.6	33.6	68	13.43
18.7	584.2	32.55	69.4	13.19
100	580.8	31.38	70.3	12.81

4 REFERENCES

[1] A.W.Y. Ho, S.R. Wenham, Solar Energy Materials & Solar Cells 91 (2007) 1234.

[2] C.B. Honsberg, et al., Proceedings of First World Photovoltaic Specialists Conference, Vol.2 (1994) 1473.

[3] A. Wang, J. Zhao, M.A. Green, Appl. Phys. Lett. 57 (1990) 602.

[4] B. Thaidigsmann, et al., Solar Energy Materials & Solar Cells 108 (2013) 164.
[5] E. Schneiderlochner, et al., Prog. Photovolt: Res. Appl. 10 (2002) 29.

ACKNOWLEDGEMENTS
This research was supported by Ministry of Science and Technology under Grant Nos. 105-2622-E-008-003-CC2 and 103-2221-E-008-041.

FINE LINE DOUBLE PRINTING + FOR TODAY AND TOMORROW CELL METALLIZATION AND INTERCONNECTION

Marco Galiazzo, Oscar Borsato, Emiliano Bortoletto

Applied Materials Italia, Via Postumia Ovest, 244 Olmi di S. Biagio di Callalta (TV), Italy

Corresponding author: marco_galiazzo@amat.com

ABSTRACT: One common requirement for all frontside contacted cells is to utilize fine finger lines with good conductivity. An established technique already proven in mass production is Fine Line Double Printing, able to deliver finger widths of 40-45 µm and an Aspect Ratio of 0.5. In this paper an evolution of this approach is described, where: 1) busbars are printed separately from fingers and 2) fingers are printed twice with the same screen. This patented concept allows for increased printing consistency and a greatly simplified process in terms of screens and tool.

Keywords: c-Si, metallization, screen printing

1 INTRODUCTION

According to ITRPV roadmap [1] the reduction rate of frontside finger widths has been of 10 µm in the last 5 years. Clearly this rate will reduce but it will continue, with predicted finger widths of 25 µm in 2026. The direct consequence of finger width decrease is a reduction of Ag paste consumption per cell, with clear economic benefit for cell manufacturers. Although alternative metallization approaches are constantly being improved in laboratory and pilot scale [2], the mainstream production technology is screen printing. The critical mass in terms of paste, screen and equipment development is such that there is high confidence that this technique will be able to meet also future requirements by evolutionary steps.

In order to preserve suitable line conductivity and avoid interruptions at narrow linewidths of 45 µm and below it is common to perform Fine Line Double Printing (FLDP), as first proposed by our group in 2009 [3]. Another double step printing approach that has been introduced in the last years is Dual Printing (DuP), with separate printing steps for busbars and fingers [4]. Both processes are applicable to all frontside cell designs, with FLDP typically delivering 0.2% efficiency gain over reference Single Printing (SP) and DuP being optimized for cost reduction with around 10 mg paste saving, but marginal efficiency benefit.

Motivation of this work is to introduce an alternative printing concept able to combine the benefits of FLDP and DuP in terms of efficiency gain and paste savings and at the same time to greatly simplify the tool and screen requirements needed to run it in mass production.

2 FLDP+ DESCRIPTION

DuP and FLDP processes have been widely presented in literature and schematized in Fig.1 together with FLDP+, which is basically their combination.

In fact, in FLDP+ process, the busbars are normally printed in the first step as in DuP allowing to use dedicated Ag paste with low laydown, suitable for: 1) high peel strength for soldering and 2) limited etching of SiN layer and low recombination. The first point allows to reduce paste consumption, the second increases the cell Voc especially for Cz-Si in the range of 1-2 mV depending on busbar layout.

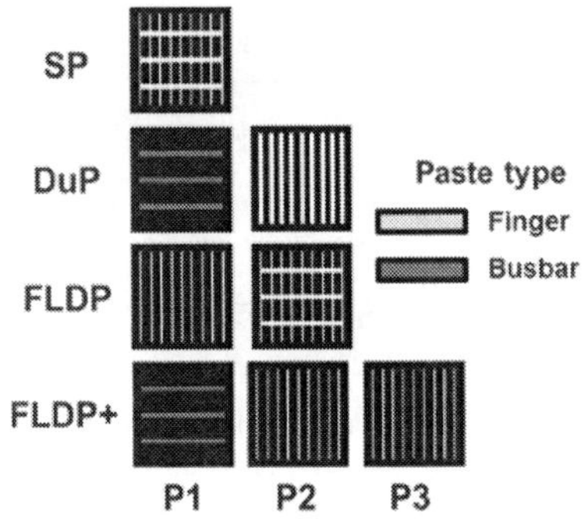

Figure 1: schematic representation of SP, Dual Printing (DuP), Fine Line Double Printing (FLDP) and Fine Line Double Printing + (FLDP+) processes.

With respect to fingers, the target is to have high aspect ratio as in the FLDP by double printing them. But since the busbar has already been printed there is the possibility to use the same screen layout with fingers only.

In terms of tool, this means that cells can be re-circulated and the same screen can be used twice in a print-dry-print sequence, thus reducing the constraints on screen matching and deformation control. With this concept the tool has to be modified in order to preserve the full throughput of the line.

This approach would greatly simplify the process to produce very narrow fingers in a consistent way in mass production and has been patented by our group.

3 FLDP+ PROFILE OPTIMIZATION

An important parameter for FLDP+ process is the printing offset in the finger direction. Since the same screen is used, if this offset is set to zero between the two printing steps there could be two issues: 1) an interruption would be directly replicated and 2) the mesh mark pattern of peaks and valley would also be replicated and leading to twice the height of the peaks and depth of the valleys.

In order to understand the optimal offset a dedicated

printing test has been performed on two different screen meshes: 326-16 with 23 µm opening and 500-16 with 25 µm opening, both screens having 15 µm emulsion over mesh.

The finger profile of a single printed line for each screen has been measured by laser profilometer in order to find the characteristic period of finger profile variation. For the printing test we tested three different offsets: zero offset, offset equal to a quarter period and offset equal to half period. All data in this section refer to dry metallization as after firing a thickness reduction of around 20% for all conditions is normally observed.

In order to understand the effect on the finger morphology the optical image of each screen has been analysed and correlated to the 3d map of the double printed finger image. Additionally, morphology data have been acquired and summarized in Tab.1.

Table I: Morphology data summary for different mesh and offset applied to FLDP+ process. All data refer to finger thickness.

Screen type	Offset [µm]	Avg T [µm]	Max T [µm]	Min T [µm]	Rz [µm]
325-16	0	24.1	36.5	6.5	24.9
	P/2	20.6	31.3	11.7	15.0
	P/4	18.9	27.2	10.8	12.8
500-16	0	22.4	28.8	13.5	11.9
	P/2	22.3	27.4	17.0	9.4
	P/4	24.0	30.4	18.9	9.1

It can be seen that the optimal offset is a quarter period, leading both to highest average thickness, lower minima points and lowest Rz values.

This is explained by the mesh orientation with respect to the screen openings, with a fixed blocking period, where the mesh is almost completely filling the opening, and a secondary period which is half the main period and is related to single mesh wires crossing the opening and leading to reduced deposit as well, as shown in Fig.2 for 500-16 mesh.

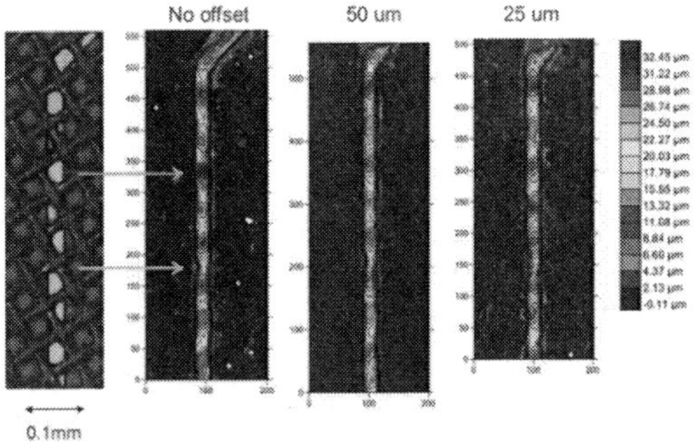

Figure 2: Optical image (left) of the screen mesh and opening and 3d profiles of FLDP+ fingers with respectively 0, 50 µm and 25 µm.

4 EXPERIMENTAL DETAILS

In order to assess if FLDP+ is suitable for future metallization requirements we performed a laboratory test with aggressive printing setup, with single printing fingers at 30 µm and double printing at 23 µm (squeegee side). For all lots a 4 busbars layout has been adopted, with continuous busbars and 125 fingers for all conditions.

The only variation introduced in the test has been the printing process, as summarized in Tab.2. The processed wafers are full square 6" Cz-Si with 90 Ω/sq emitter, alkali textured, POCl diffused, PECVD ARC and with full BSF on the rear side.

Table 2: Summary of test conditions. Printing sequence for DuP and FLDP lots is based on busbars first to reduce the deposit

LOT	Screen mesh	Opening [µm]	Paste Type
DuP	325-16	30	B+F
FLDP	380-14	23+23	F+F
FLDP+	325-16	23+23	B+F+F

Cells have been IV tested and then further characterized for peel strength.

5 CELL RESULTS

Finger morphology images are reported for all conditions in Fig.3. Due to suitable paste printability through narrow openings, DuP finger width is in the range of 40 µm, with an aspect ratio of 0.28, while for FLDP a final width of 35 µm is achieved, with aspect ratio of 0.5.

When implementing FLDP+, there is a benefit in terms of finger width, reduced by 2 µm if compared to FLDP.

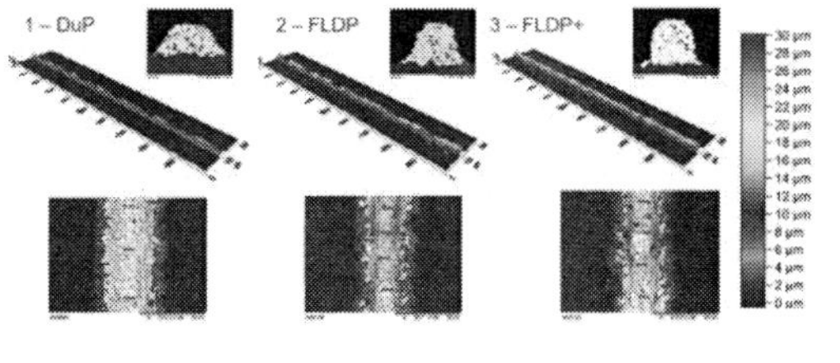

Figure 3: Morphology images for DuP, FLDP and FLDP+ lots.

Morphology and electrical data are summarized in Tab.3 and data distributions reported in Fig.4.

Table 3: Morphology and electrical data summary

	Unit	DuP	FLDP	FLDP+
Fing. width avg	µm	40	35	33
Fing. thick. avg	µm	12	17	17
Fing. thick. min	µm	7	7	8
Aspect Ratio		0.28	0.5	0.51
Isc	A	9.23	9.32	9.29
Voc	V	0.647	0.646	0.647
Fill Factor	%	79.59	79.91	79.7
R_series	mΩ	1.6	1.5	1.6
Efficiency	%	19.53	19.74	19.68
Paste weight	mg	106	130	108
Busbar thick.	µm	5	8	4
Peel Force	N	2.5	1.8	1.8

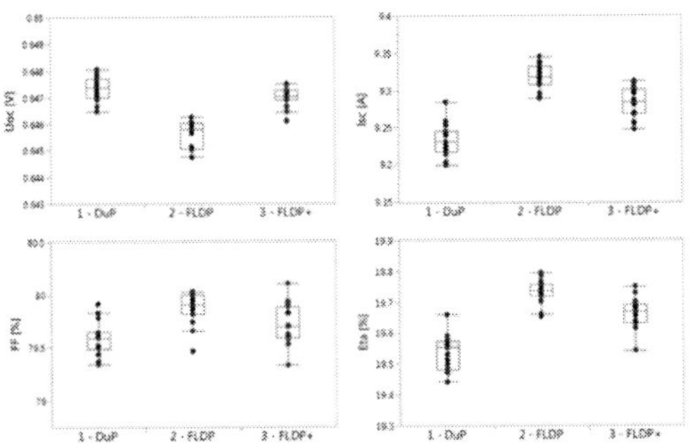

Figure 4: Electrical data distribution. Each lot is composed of 15 cells.

As expected, due to non-contacting paste on the busbars both DuP and FLDP+ benefit from higher Voc if compared to FLDP, where contacting paste is used. In terms of Isc gain, due to reduced finger width there is an advantage for FLDP and FLDP+ over DuP.

Due to the high FF and Isc values, FLDP shows the highest efficiency, outperforming DuP by 0.18%. The efficiency gain of FLDP+ is at 0.12% and limited by FF and Isc.

In order to test peel strength, 1.7 mm wide and 0.2 mm thick ribbon with SnPb coating has been manually soldered to one cell per lot and pulled by an automatic system at 180° angle.

During the printing process the busbar thickness has been controlled for both DuP and FLDP+ to around 5 µm in order to reduce paste consumption. Since this paste is designed for high adhesion it is possible to work under these conditions and preserve the peel force to above the minimum value of 1N/mm defined by the standard.

CONCLUSION

A novel approach for solar cell metallization is proposed, based on the combination of DuP and FLDP. This allows to benefit from both the reduced paste consumption on the busbars and increased efficiency due to narrow finger lines.

Additionally, with this process the same screen can be used for first and second printing allowing a great simplification of the process tolerances in mass production.

REFERENCES

[1] www.itrpv.net

[2] G. Hahn and S. Joos, 'State-of-the-Art Industrial Crystalline Silicon Solar Cells', Semiconductors and Semimetals 90(2014):1-72 · December 2014

[3] M. Galiazzo et al, 'New technologies for improvement of metallization line'. 24th EUPVSEC, Hamburg, 2009, pp. 1502–1505

[4] E. Kossen et al, 'Comparison of two step printing methods for front side metallization', 25th EUPVSEC, Valencia, 2010, pp. 2099–2104

HIGH-PRODUCTIVE ALUMINUM DEPOSITION OF BACK CONTACTS FOR HETERO-JUNCTION SOLAR CELLS BY ELECTRON BEAM EVAPORATION

Jens-Peter Heinß[1], Hermann Schlemm[2], Frank Wünsch[3]
[1]Fraunhofer Institute for Organic Electronics, Electron Beam and Plasma Technology (FEP)
Winterbergstrasse 28, 01277 Dresden, Germany
[2]Meyer Burger (Germany) AG, [3]Roth & Rau AG, An der Baumschule 6-8, 09337 Hohenstein-Ernstthal, Germany

ABSTRACT: Currently back contacts of Silicon HJT (Hetero Junction Technology) cells are sputtered or screen printed Ag-layers or structures. The present work aims to replace these cost-intensive manufacturing steps by electron beam evaporation of aluminum using axial beam guns.
Al-layers of the necessary thickness have been evaporated with a dynamic deposition rate of up to 3000 nm·m/min while keeping the solar cell below 200°C to avoid any thermal damage of underlying layers and junctions.
The deposited aluminum layers exhibit specific electrical resistance of $2.8 \cdot 10^{-6}$ Ω·cm, specific electrical contact resistance of 10^{-3} Ω·cm^2 and more than 92 % averaged reflection in the NIR.
By exchanging only the metallization step in a standard HJT cell processing without further process-optimization lead to cell efficiencies of 20.6 %.
It can be stated that the electron beam evaporation can meet the technical and economical demands of the HJT production process. The high deposition rate can ensure the rear-side metallization of a 100 MW production line in one single deposition equipment.
Keywords: Back Contact, Metallization, Cost Reduction, High Deposition Rate

1 INTRODUCTION

Aluminum evaporated backside contacts had been investigated and successfully implemented for standard silicon solar cells in the past. For multi c-Si solar cells with diffused Al-BSF (back surface field) as well as for the better passivated PERC [1] cells maintaining the same contact interface Al/BSF/p-type c-Si. For both Solar cell types additional metal-pads have to be deposited on parts of the rear side for solderable interconnections [2].

Compared to PERC cells the application of evaporated Al-layers to Heterojunction solar cells (HJT) as a low temperature contact system on TCO creates a totally different contact interface, since now Al is no longer a diffusion source for a local BSF in p-type c-Si. Instead an Al/TCO interface has to replace the standard sputtered or screen printed Ag/TCO interface at the HJT cells rear for further reduction of production costs. With this standard Ag/TCO rear contact HJT efficiencies of up to 23 % have been obtained in an industrial environment in a monofacial configuration [3].

This contribution demonstrates in a first approach, that the high rate electron beam evaporation process can be applied to Heterojunction solar cells while keeping their efficiency above 20%, if optimum electron beam evaporation conditions are applied. High dynamic deposition rates are achieved which are well suited for economic industrial production.

The investigations include three main steps:

1. Adaptation of high-rate electron beam evaporation with adherence of evident temperature limits during deposition
2. Prove of non-influence of electron beam evaporation onto the electronic wafer and interface quality with passivated cell precursor samples via excess charge carrier lifetime measurements.
3. Metallization of HJT solar cells back contact by using the high-rate electron beam evaporation and measurement of the solar cell efficiency

2 EXPERIMENTAL

The aluminum evaporation experiments have been carried out in the eight chamber in-line vacuum coating equipment MAXI (Fig. 1). For the high-rate evaporation an axial electron beam gun was installed, which was designed and manufactured at FEP (P_{ES}=160 kW, beam deflection angle: 30°, deflection frequency: 3.5 kHz).

Two different types of evaporation crucibles were used: water-cooled copper crucible with an orifice of 200 mm (method A) and boron nitride crucible (method B) with 160 mm. The wafer located in a carrier were moved in a height of about 300 mm above the evaporation pool.

Figure 1: Fraunhofer FEP's in-line deposition equipment for sheets and metal strips - "MAXI".

Ion-polished cross sections of the aluminum layers were prepared using an argon ion-beam preparation technique (Cross Section Polisher, SM-0910, Jeol). The cross sections were analyzed by field emission scanning electron microscopy (FE-SEM, SU 8000, Hitachi) using the channelling contrast for imaging the microstructure.

Chemical-depth profiles of the aluminum layers were determined by rf glow discharge optical emission spectrometry (GD-OES, GD Profiler 2, HORIBA Jobin Yvon).

The electrical sheet resistance was measured with the four point probe measurement device VEECO FPP-5000.

The electrical contact resistance was analyzed according to the transfer length method [4]. The measurement pads (3 x 20 mm^2) with ten different distances between 0.2 and 2 mm were deposited through a steel mask with a thickness of 0.2 mm.

Determination of the excess-charge carrier lifetime by Sinton-calibrated Photoluminescence measurements were carried out on passivated test samples. For methodic reasons the passivated samples were covered by glass during the Al deposition. So the samples were exposed to process-induced X-ray and heat during the evaporation cycle without being metallized.

The diffuse reflection was measured with a spectral photometer (Lambda 900, Perkin Elmer) using an integrating sphere between 700 and 1200 nm.

To provide solderable solar cell rear sides, nickel vanadium layers were sputtered without vacuum interruption onto the aluminum layers. Test wires were manually soldered to these layer stacks and cross sections were prepared from the solder joints to prove the microstructure and to evaluate the wetting.

3 RESULTS AND DISCUSSION

By increasing the electron beam power up to 36 kW a dynamic deposition rate (ddr) of 3000 nm·m/min was attested for the copper crucible. The application of higher beam powers allows a further increase of the deposition rate (Fig. 2).

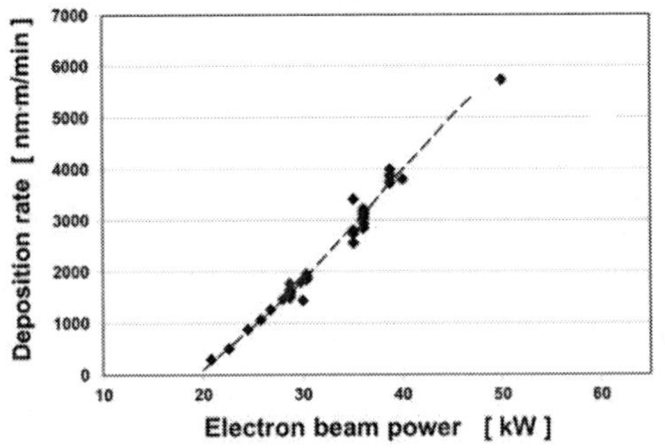

Figure 2: Dynamic deposition rate in dependency on the electron beam power by using a water-cooled copper crucible.

With the boron nitride crucible a dynamic deposition rate of 3000 nm·m/min was obtained with a beam power of 12 kW only. This reduction by two third of the applied power in comparison to the copper crucible is caused by the lower heat conductivity of the crucible wall material.

The deposition of a 1 µm thick aluminum layer onto a 180 µm thick c-Si wafer can be carried out without exceeding the actual predetermined temperature limit of 200 °C for the Hetero Junction cell (Fig. 3).

It should be noted that an increase of the deposition rate did not lead to higher wafer temperatures. This is caused by lower proportion between the heat from layer condensation and parasitic process heat quantity at higher deposition rate values. This means that the aluminum deposition can be adapted to even higher production demands.

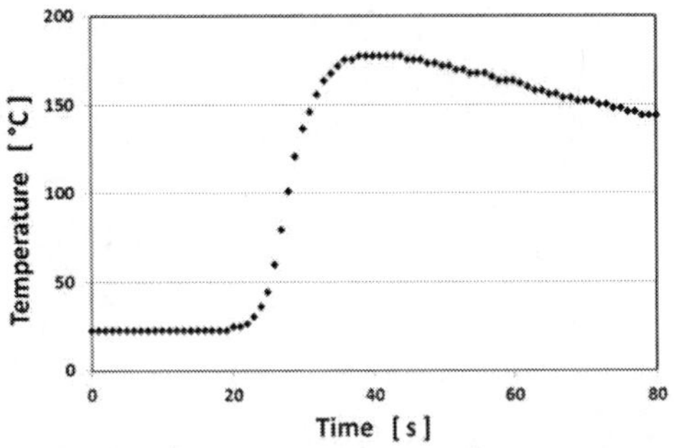

Figure 3: Temperature in the wafer centre during the aluminum deposition (method A, ddr: 3000 nm·m/min).

Additional temperature measurements were carried out at several positions on the wafer (Fig. 4). In this example the dynamic deposition rate was about 1500 nm·m/min. In comparison to the deposition according Fig. 3 the transportation speed was halved. The amount of parasitic heat in relation to the heat of condensation is increased for lower deposition rates. Therefore an increase has to be also considered for the total heat input. The temperature on the wafer was increased up to 250 °C. So the following fact can be demonstrated quantitatively a little bit more clearly: The rear wafer position show higher temperatures. The difference of the maximum temperatures in the transportation direction amounted up to 15 K. This dependency is induced by the heat conduction inside the wafer and was confirmed in consistent FEM simulation calculations.

The structure of the deposited aluminum layers were characterized basically by SEM imaging. The cross sections (Figure 5 and 6) show block-shaped, very dense layer structures for both evaporation methods A and B. The lateral size of the aluminum crystallites amounts about one micrometer and is in the same dimension like the layer thickness. Such dense and compact structures are known for the evaporation of water-cooled crucibles. Fortunately we achieved the same layer quality by the evaporation from a ceramic crucible.

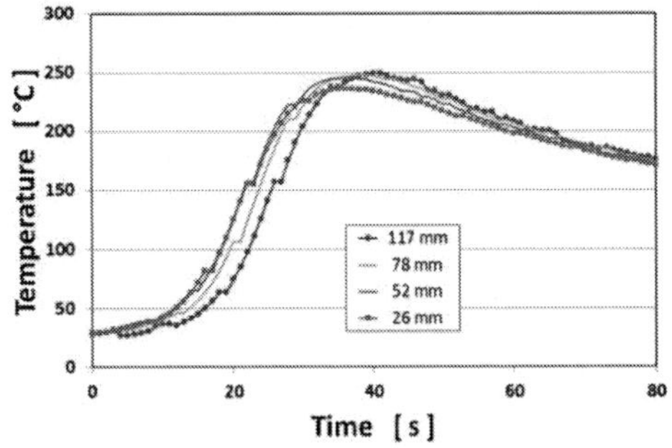

Figure 4: Temperatures at four different wafer positions during the aluminum deposition (method A, ddr: 1500 nm·m/min). The description in the legend represents the distance of the thermocouple for the single measuring position to the front edge in transportation direction

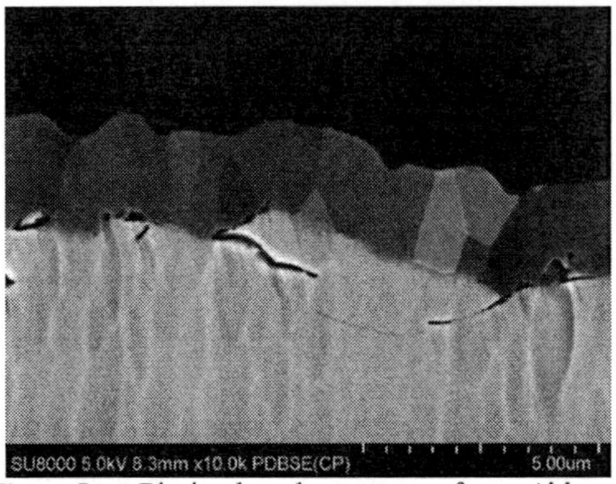

Figure 5: Block shaped structure of an Al-layer, evaporated from copper crucible onto as-cut c-Si wafer.

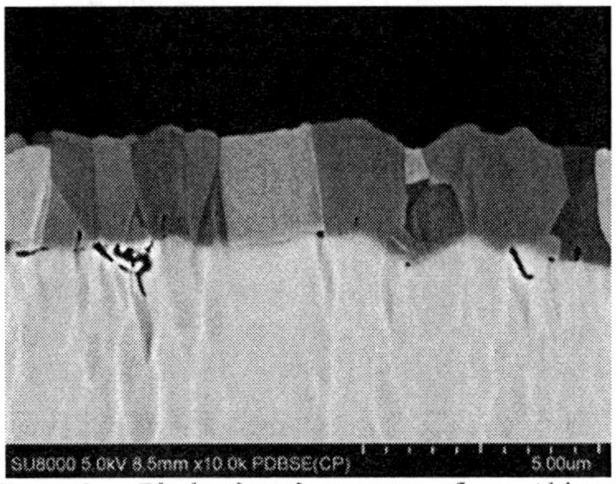

Figure 6: Block shaped structure of an Al-layer, evaporated from boron nitride crucible onto as-cut c-Si wafer.

The result of the chemical analysis by GD-OES of an aluminum layer deposited with method A is given in Fig. 7. The concentration depth profiles for the elements aluminum and silicon in the interface region are broadened and influenced by the surface geometry of the used as-cut wafer. The concentration values for hydrogen were multiplied by a factor of 10 to illustrate them in the same scale. The aluminum layer shows maximum hydrogen contents of 0.3 wt.-% and is caused by the electron beam evaporation process.

The oxygen in the near surface region of the aluminum layer results from the storage period between deposition and analyzing the aluminum layers. The element carbon could be detected in the near-surface and the interface region in maximum with about 0.04 wt.-%. For the layer itself no oxygen, nitrogen or carbon could be detected in the 0.01 wt.-% detection limit. It can be concluded that the high-rate evaporation enables clean aluminum layers.

The chemical analysis of the aluminum layers according to deposition method B show similar results for the depth profile and for the amounts of pollution (Fig. 8). Again this confirms the high quality for aluminum layers which were evaporated from the ceramic crucible.

Relating to the application as contact layers for Hetero-Junction solar cells the following properties of the aluminum layers were analyzed:

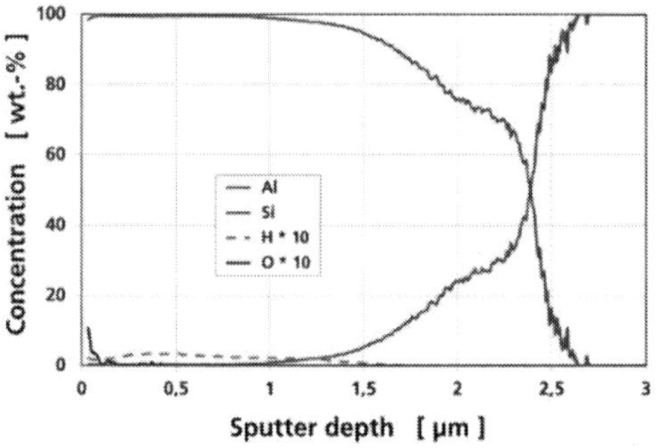

Figure 7: GD-OES depth profile for aluminum layers evaporated from a copper crucible.

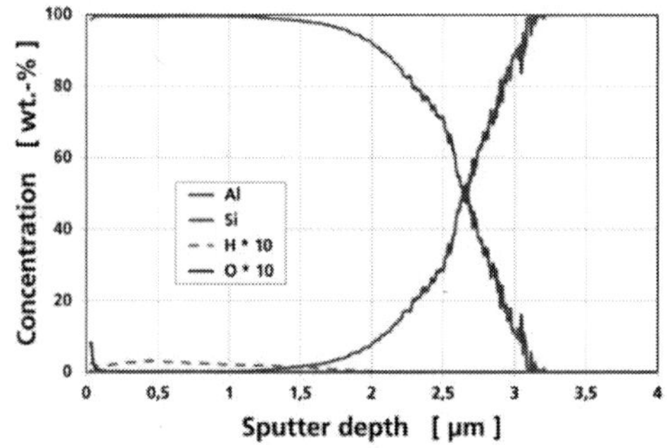

Figure 8: GD-OES depth profile for aluminum layers evaporated from a boron nitride crucible.

After a few optimization steps for both evaporation methods A and B no electronic degradation of the HJT-cell precursors were observed. The comparison of carrier lifetime between their initial state and after exposure to the Al-deposition process cycle confirmed that the electronic quality of the silicon wafers and especially the passivated c-Si surface was not deteriorated by electron beam-induced radiations (Fig. 9). Surprisingly sample areas with an initially lower excess charge carrier lifetime (lower right corner and lower part in left images of Fig. 9) are electronically improved. Eventually the described deposition test procedure causes a thermal annealing of the interface.

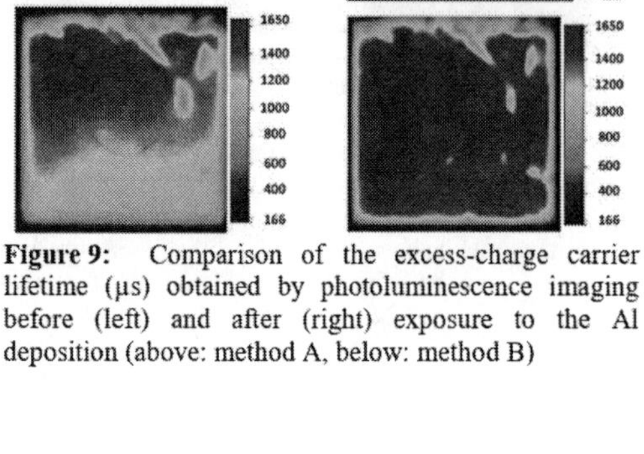

Figure 9: Comparison of the excess-charge carrier lifetime (µs) obtained by photoluminescence imaging before (left) and after (right) exposure to the Al deposition (above: method A, below: method B)

The specific electrical resistance of the deposited Al-layers was determined to be $2.7 \cdot 10^{-6}$ Ωcm (method A) and $2.8 \cdot 10^{-6}$ Ωcm (method B) respectively. A maximal deviation of only 4 % from the literature value for aluminum bulk material confirms the high quality of the deposited layers.

The specific electrical contact resistance to two different transparent conducting oxides (TCO, indium tin oxide and indium zinc oxide) was estimated within the range 10^{-3} ... 10^{-2} Ωcm². This value leads to a contact resistance of less than 0,1 mΩ on 156x156 mm² pseudo-square wafers and ensures low serial resistances for the back contact of a solar cell.

For deposited aluminum layers the index of refraction and the absorbance of separately coated c-Si samples were measured. From these data the reflection at the aluminum layers from air was calculated (Fig. 10). The comparison with data for aluminum bulk material from Palik [5] shows a high conformity. The average over the NIR (700-1200 nm) range was estimated to more than 92 %. The high reflection of aluminum can contribute to an efficient light management inside the solar cell.

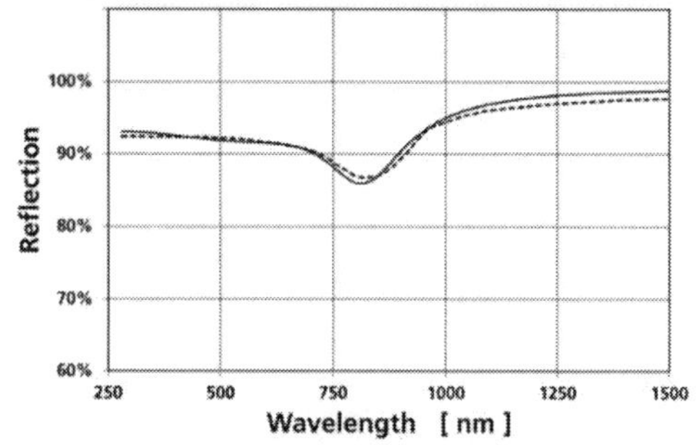

Figure 10: Reflection of deposited aluminum layers on a c-Si substrate (dasehd line: reference curve for aluminum bulk calculated with data from Palik [5]).

After a first optimization manual soldering tests on a NiV/Al/Si-stack showed that the soldering flux wet the metallized wafer almost completely (Fig. 11). The additional NiV sputter layer enabled the close contact between the solder and the aluminum layer. Delamination was not observed. Various NiV layer thicknesses prepared between 50 and 150 nm showed the same wetting and adhesion behaviour after soldering. Even with preliminary manual soldering procedures, the wafer temperature could be kept below the 200°C limit.

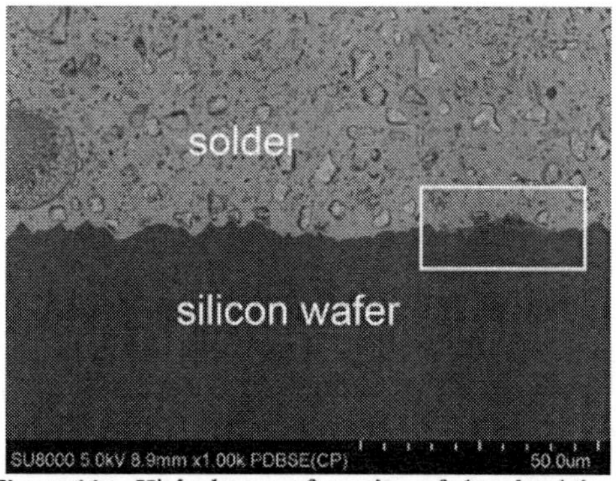

Figure 11: High degree of wetting of the aluminized silicon (dark gray below) by a low melting soldering flux (bright gray top) after sputtering an additional NiV layer.

In the frame in Fig. 11 which shows a small region with non-successful wetting the 1 µm thick aluminum layer on the top of the silicon wafer is still visible. Elsewhere between the solder and the silicon wafer the NiV and the Al layers are not visible because of the magnification of the picture.

Encouraged by the positive pretests a first series of Hetero-Junction cells was metallized by the electron beam evaporation of aluminum. All other process steps were performed according to the standard pilot production at that time.

Table I: Data for the Al metallized and the reference Ag metallized HJT solar cells

Type of back metallization		Al	Ag
U_{OC}	(mV)	731	730
j_{sc}	(mA)	37.2	37.4
fillfactor	(%)	75.6	77.3
effiency	(%)	20.5	21.1

Cell processing without further process-optimization lead to cell efficiencies up to 20.6 %. The open circuit voltage remains the same as for the cells which were metallized in the standard procedure. This confirmed that no degradation of the silicon during the electron beam evaporation and deposition occurred. In total the fill factor and the efficiency was slightly lower than for cells receiving the baseline rear metallization. With respect to the small number of processed cells the reasons for this performance difference cannot be related to the metallization method and should therefore be a task of further studies.

A cross section of the back contact of a full processed solar cell is shown in Fig. 12. On the textured surface the intermediate and the 1 µm thick aluminum layer are visible. The aluminum layer is very dense and compact and features crystallites with size up to 2 µm.

As a last point technological relevant properties were studied. The bow induced upon 180 µm thick 156 x 156 mm² pseudo-square silicon wafers for aluminum thicknesses up to 2 µm remains below 1 mm. Such values are well accepted for common wafer handling systems.

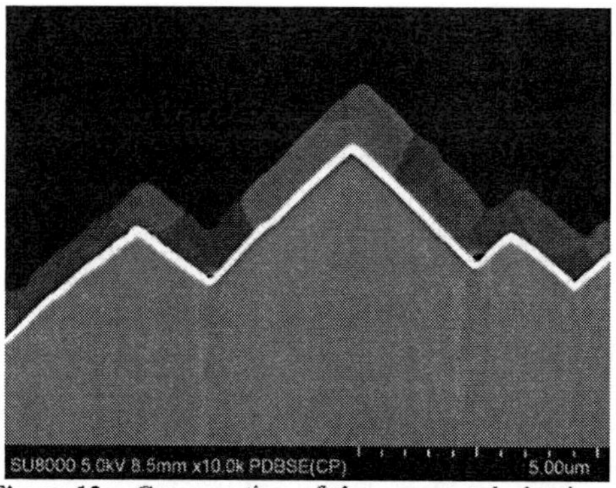

Figure 12: Cross section of the evaporated aluminum back contact of a full processed Hetero Junction solar cell.

The extension of the coating width in the equipment was realized by the superposition of the evaporation zones of two boron nitride crucibles which were located side by side. Over a width of 340 mm a homogeneity of ±4 % could be achieved (Fig. 13). This is a confirmation that the needed precision for the back contact metallization can be implemented by the high-rate electron beam evaporation also for larger coating widths with a larger number of installed evaporation crucibles.

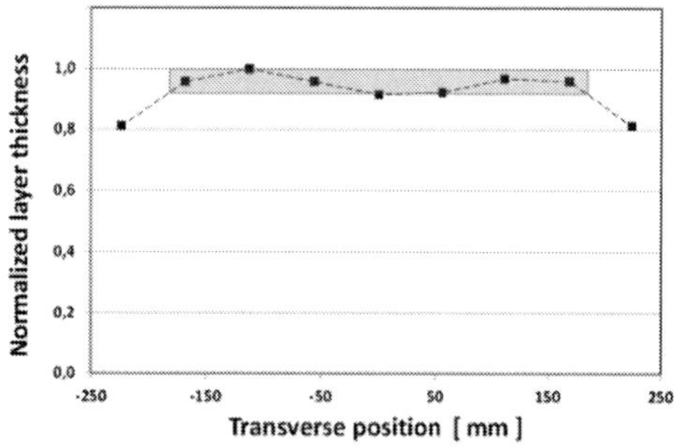

Figure 13: Layer thickness distribution in dependency on the transverse position for an Al evaporation from two crucibles.

4 CONCLUSIONS

The achieved results confirm that the high-rate electron beam evaporation for metallizing the back contact of Hetero-Junction solar cells is a very promising new technology.

High-rate electron beam evaporation can realize valuable aluminum contact layers. The low specific electrical sheet and contact resistance fulfill the demands for the metallization of Hetero Junction solar cells. By using the electron beam evaporation for metallization step cell processing without further process-optimization lead to cell efficiencies of 20.6 %.

The high deposition rate in combination with the adaptable deposition width ensures the quantity of coated solar cells in production lines which are planed nowadays. A cost estimation for a production volume of 100 MW gives an attractive number of below 0.02 € per W_P for the metallization of Hetero Junction Cells.

5 REFERENCES

[1] D. Zielke, J.H. Petermann, F. Werner, B. Veith, R. Brendel, J. Schmidt, Proceedings 26th European Photovoltaic Solar Energy Conference (2011) 1115

[2] H. Schulte-Huxel, R. Witteck, P. van Laak, T. Brendemühl, D. Hinken, K. Bothe, R. Brendel, 6th International Conference on Silicon Photovoltaics, SiliconPV 2016

[3] B. Strahm et al., Proceedings 29th European Photovoltaic Solar Energy Conference (2014) 467

[4] D.K. Schroder, Semiconductor Material and Device Characterization, John Wiley and Sons, 1990

[5] E. D. Palik (ed.), Handbook of Optical Constants of Solids, Academic Press San Diego, 1998

6 ACKNOWLEDGEMENTS

The authors thank M. Oberst for carrying the contact resistance measurements, L. Decker for the soldering experiments, O. Zywitzki and T. Modes for the analytical investigations and especially L. Klose for his assistance and commitment during the deposition experiments

The project was funded by the European Union and the Free State of Saxony (funding reference 100095132).

Europa fördert Sachsen.
EFRE
Europäischer Fonds für regionale Entwicklung

ULTRAHIGH PID-RESISTANCE FOR MONO SILICON PERC SOLAR CELLS BY USING INDUSTRIAL MASS-PRODUCTION TECHNOLOGY

Cheng-Wen Kuo, Ta-Ming Kuan, Li-Guo Wu, Chih-Chiang Huang, Hsin-Ying Peng, and Cheng-Yeh Yu
TSEC Corporation, No.85, Gaungfu N. Rd., Hsin-Chu 30351, Taiwan.

ABSTRACT: In this paper, using the industrial mass-production technology on commercially available 156 mm × 156 mm p-type Czochralski-grown crystalline mono silicon wafers, the best cell efficiencies more than 21% are demonstrated. We improved the solar cell devices by improving front side anti-reflection coating and back side process parameter which results in higher open circuit voltages and short circuit current. Besides, the mono PERC solar cell process pass the UL/ITRI potential induced degradation (PID) test which condition was indicated to be 85 degree Celsius and 85% RH in climatic chamber apply -1500 V for 300 hours.
Keywords: High Efficiency, PERC, PID-Resistance,Mono

1 INTRODUCTION

Photovoltaic industry had been developed for decades and had been growing rapidly in recent years, due to awareness of global warming effect and shortage of natural resources such as fossil energy. Hence, there are many research institutes, universities and companies reported on the fabrication of high efficient p-type passivated emitter and rear cell (PERC) structures recently [1-7]. The PERC applies dielectric passivation layers to the front and rear side in the solar cell [1], which is a significantly better passivation performance in the rear side of solar cell, can be achieved by dielectric layers compared with the full-area aluminum back surface field (Al BSF). The additional dielectric layer also improves the internal rear reflectance that reduces absorption losses in the Al back contact.

Crystalline silicon wafers are the main stream substrates for current solar cells photovoltaic, the increased crystalline silicon conversion efficiency and market attention which become more important. It is indicated the industrial PERC technology will be employed in order to achieve target. However, in the PERC process, the rear contacts are formed locally after opening the dielectric layer. The opening of the dielectric may be done by laser, by using an etching paste or by masking of the dielectric and wet chemical etching. Afterwards the rear side contact metallization usually using Al is carried out. The Al may be screen printed, evaporated or sputtered on the full area of the rear side or locally as lines or dots. Finally, the contacts are fired [2].

In this work, we proposed the improving front side anti-reflection coating and back side process parameter process parameter applied to screen-printed PERC structure. It could be found that employing optimizing PERC process is better than without optimizing PERC process. Besides, the PERC solar cell process pass the UL/ITRI potential induced degradation (PID) test which condition was indicated to be 85 degree Celsius and 85% RH in climatic chamber apply -1500 V for 300 hours.

2 EXPERIMENTAL

Figure 1 showed the PERC structure in this work. The high efficiency 6-inch (156 mm x 156 mm) PERC solar cells were fabricated on 180 μm thick p-type boron doped Czochralski (Cz)-silicon wafers with a resistivity of 1~3Ω-cm.

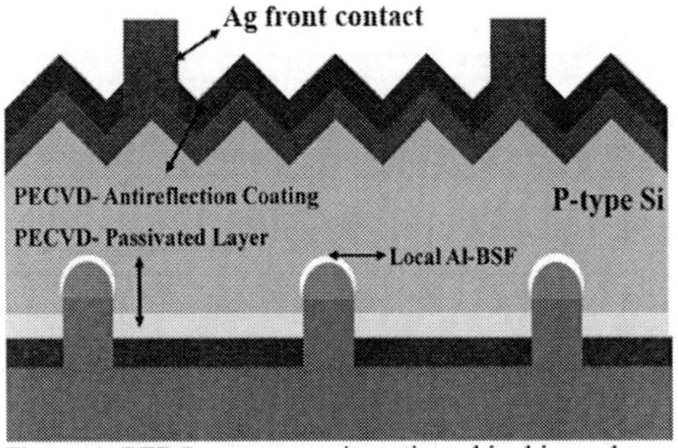

Figure 1: PERC structure as investigated in this work.

The process sequence in this work was shown in Figure 2. Chemical wet etching methods were used for saw damage removal and surface texturing, which include both acidic etching for multi-crystalline silicon wafers and alkaline etching for mono-crystalline silicon wafers. The solution composition of acidic etching was nitric acid (HNO3) : hydrofluoric acid (HF) : deionized water (DI) = 1 : 2 : 1.5, and the solution composition of alkaline etching was potassium hydroxide (KOH) : deionized water (DI) = 1 : 50. Then, the rear surface is polished in order to reduce the roughness and its impact on the effective charge carrier lifetime [5]. Conventional phosphorous diffusion was employed by furnace to form the n+ emitters with a sheet resistance of 60 – 80 Ω/□. The diffusion process also improved to uniformity for this work.

The phosphosilicate glass (PSG) was then removed by HF-contained solutions, formed edge isolation in the meantime. In order to decrease PID effect and increase efficiency, we optimized the CVD process parameter to improve. An antireflection coating (ARC) layer of silicon nitride (SiN_x: H) and/or SiN_x/SiO_x dual ARC was deposited by plasma enhanced chemical vapor deposition (PECVD) with thickness of 80 nm. The passivation was carried out on rear side also using a PECVD. Afterward, the rear side laser opening ratio is also optimized [7]. All PERC solar cells have the same laser-ablated opening width in the rear-side passivation stacks which was about 45 μm, as shown in Figure 3. The front side silver and backside silver and aluminum metallization was formed by screen printing technique and followed by co-firing process, which resulted in formation of front silver grid contact and local back-surface-field. Finally, the current-voltage measurements of solar cells under AM 1.5 illumination were performed.

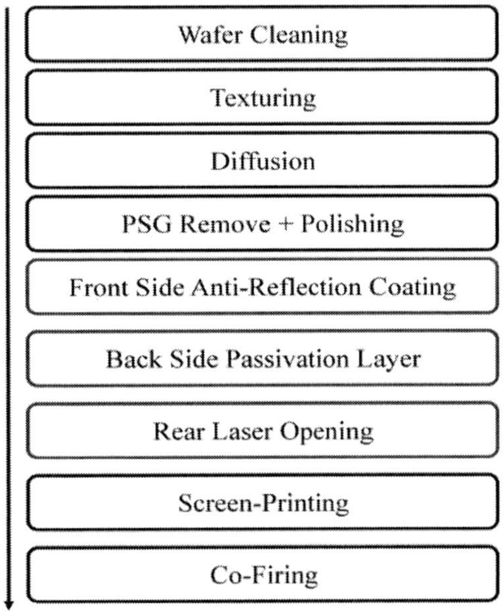

Figure 2: Process flow for the fabrication of a PERC solar cell as developed in this work.

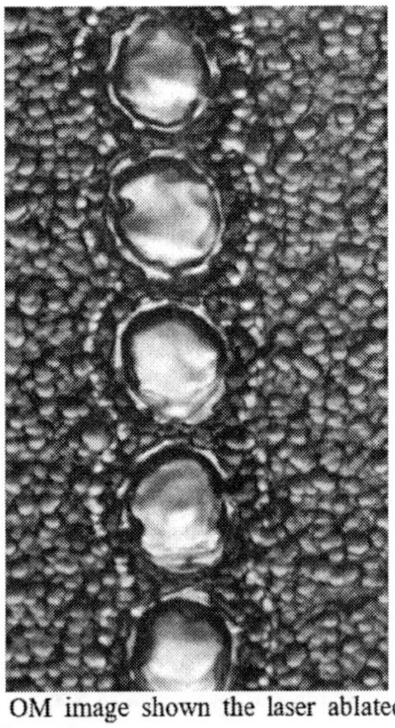

Figure 3: OM image shown the laser ablated opening width in this work.

3 RESULTS AND DISCUSSION

Figure 4 showed the measured current and power as function of the voltage for the optimum PERC solar cell. Table I summarized the average and best results of conversion efficiency (EFF), open circuit voltage (V_{OC}), short- circuit current (I_{SC}), fill factor (FF), series resistance (R_S) and shunt resistance (R_{SH}). The processing of the wafers is developed at a standard industrial level, and no special equipment or processing was required for achieving these efficiencies.

It could be found that a strong dependence of the measured series resistance RS on the laser opening ratio with improved parameter of PERC solar cell. It was observed to the laser opening ratio decreased with the R_S increased which could be determination by the rear contact resistance [1]. On the other hand, the PERC solar cells induced the influence on void formation, as shown in Figure 5. It can be found dark region in EL image which might be the existence of voids. The number of voids increased with the contact spacing increased [3]. These illustrations are improved with this work.

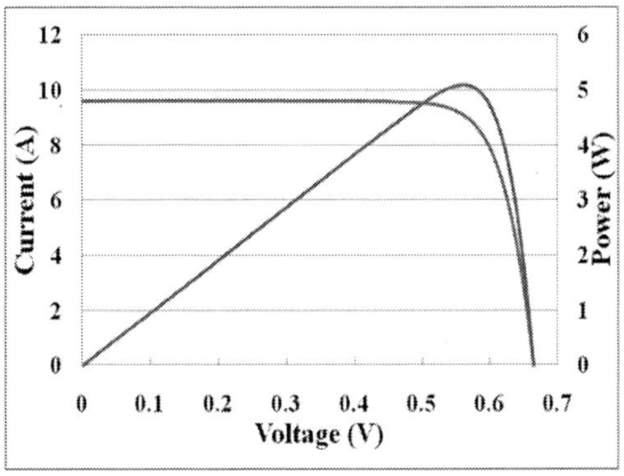

Figure 4: Measured current and power vs. voltage of the PERC solar cell with the 5% opening ratio.

Table I: Average electrical results for more than 100,000 mass production PERC solar cells with different laser opening ratio. * is the best solar cell parameter

Cell	PERC	Improved PERC	Improved PERC*
EFF (%)	20.15	20.77	21.05
V_{OC} (V)	0.654	0.665	0.667
I_{SC} (A)	9.487	9.578	9.636
FF (%)	79.25	79.72	80
R_S (Ω)	2.29	2.13	2.06
R_{SH} (Ω)	167.58	169.59	192.22

In order to reduce the series resistance, the contribution of the rear contact resistance had to be minimized, one could increase the laser opening ratio. However, it would result in an increased rear surface recombination velocity and a decreased reflectance. Therefore, designing rear side laser pattern would take account of the laser opening ratio.

On the other hand, it could be found that a strong dependence of the measured open circuit voltage Voc on the back side process with improved parameter of PERC solar cell. It was observed to the efficiency increased with the Voc increased which could be determination by optimized back side process.

Besides, the PERC solar cells also passed the the mono PERC solar cell process pass the UL/ITRI PID test which condition was indicated to be 85 degree Celsius and 85% RH in climatic chamber apply -1500 V for 300 hours as shown in Figure 6. Observed the power loss < 5% and no EL darkened area were found after PID test for the PERC solar cells.

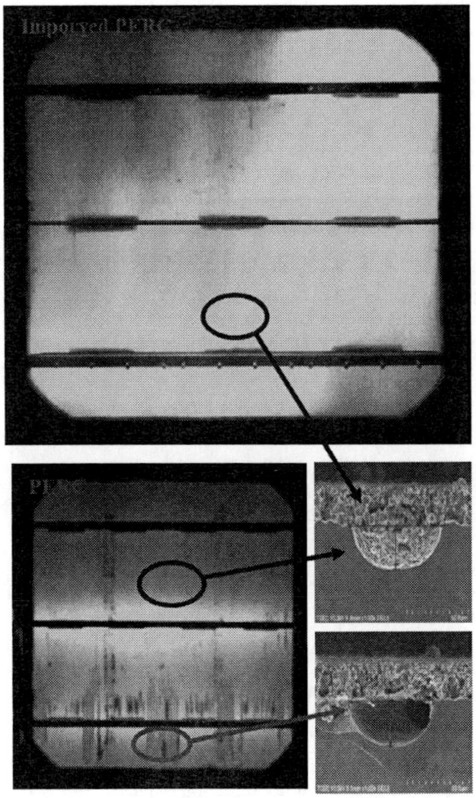

Figure 5: EL measurements and SEM micrographs of PERC and improved solar cell.

ITRI	Decay Rate	EL	
	Power Loss	Pre-PID	Post-PID
Sample A	**4.45 %**		
Sample B	**1.89 %**		

Figure 6: Summary of power loss and EL images of Pre-PID and post-PID of mono-crystalline silicon PERC solar cells. PID test condition is 85 degree Celsius / 85% RH biased -1000V for 300 hours.

4 CONCLUSIONS

In this work, the industrial process has been proposed for the fabrication of PERC solar cells, leading to the best efficiency of 21.05 % on large area (156 mm × 156 mm) p-type mono crystalline Cz silicon wafer. Morever, It could be demonstrated that the improved the solar cell devices by improving front side anti-reflection coating and back side process parameter which results in higher open circuit voltages and short circuit current. On the other hand, the PERC solar cell process pass the UL/ITRI potential induced degradation (PID) test which condition was indicated to be 85 degree Celsius and 85% RH in climatic chamber apply -1500 V for 300 hours.

5 REFERENCES

[1] S. Gatz, T. Dullweber, and R. Brendel, "Evaluation of Series Resistance Losses in Screen-Printed Solar Cells With Local Rear Contacts" IEEE J. Photovolt, vol. 1, pp.37, 2011.

[2] S. Riegel, F. Mutter, T. Lauermann, B. Terheiden, G. Hahn, "Review on Screen Printed Metallization on P-type Silicon" Energy Procedia, vol. 21 pp.14, 2012.

[3] E. Urrejola, K. Peter, H. Plagwitz, and G. Schubert, "Distribution of Silicon in the Aluminum Matrix for Rear Passivated Solar Cells" Energy Procedia, vol. 8, pp.331, 2011.

[4] E. Urrejola, K. Peter, H. Plagwitz, and G. Schubert. "Al–Si Alloy Formation in Narrow P-type Si Contact Areas for Rear Passivated Solar Cells" J. Appl. Phys. vol. 107, pp. 124516, 2010.

[5] E. Urrejola, R.Petres, J. Glatz-Reichenbach, K. Peter, E. Wefringhaus, H. Plagwitz, G. Schubert., "High Efficiency Industrial PERC Solar Cells with all PECVD-Based Rear Surface Passivation" 26th EUPVSEC, pp. 2233, 2011.

[6] S. Mack, A. Wolf, B. Thaidigsmann, E. Lohmüller, U. Jäger, M. Pospischil, F. Clement, D. Eberlein, R. Preu, and D. Biro, "Technology for Mass Production of > 20% Efficient P-type Silicon Solar Cells", 28th EUPVSEC, pp. 761, 2013

[7] C.-W. Kuo, T.-M. Kuan, C.-C. Liu, C.-C. Huang, L.-G. Wu, C.-Y. Yu, "Impact of Laser Opening Ratio for Mass Production High Efficiency Mono-Crystalline Silicon Solar Cells" 29th EUPVSEC, pp. 1392, 2014

COMPARISON OF INFLUENCE ON MC-SI SOLAR CELL PERFORMANCE OF DISLOCATION CLUSTERS AND GRAIN BOUNDARIES BY USING PHOTOLUMINESCENCE IMAGING

Xiaolong Niu[1], Song Qiao[1], Limo Zhang[1], Mingcui Pan[1], Yunfeng Zhang[1], WenKuan Gao[1,2], Bo Yu[1,2], Dengyuan Song[1,2]
[1] Yingli Green Energy Co., LTD,
[2] State Key Laboratory of Photovoltaic Materials & Technology
3399 North Chaoyang Street, Baoding, China

ABSTRACT: Inline photoluminescence (PL) imaging has been applied to identify impurity-decorated dislocation clusters and grain boundaries in as-cut wafers. The ratio of impuritiy-decorated dislocation clusters area and grain boundaries area to the total area in wafer can be calculated separately by software. Ten wafers are selected from bottom and top region of a central block with a defect area ratio of 7.5% to fabricate solar cells respectively. The result shows that for the samples with main grain boundaries defect, there is a 0.11% absolute efficiency higher than that of samples with dislocation clusters defect. It can be concluded that grain boundaries contribute a relatively less negative effect on the performance of solar cells after the gettering process. Meanwhile, impurities prefer to interact with dislocation clusters in top area of ingot.
Keywords: multicrystalline silicon, dislocation cluster, grain boundary, impurity, PL

1 INTRODUCTION

In the current PV market, multicrystalline silicon (mc-Si) solar cells have occupied a major share of the market due to their reliable performance and low cost. Recently, high performance multicrystalline silicon (HPMC Si) based on seed-assisted growth is being widely used in multicrystalline silicon industrial field because of its excellent crystal quality. In this method, proper mc-Si such as silicon beads or chips with diameter of several millimeters are used as external seed for nucleation and grain growth, and crystals with much lower dislocation defects are obtained [1-4]. However, compared with monocrystalline Si (m-Si), there are enormous crystal defects in HPMC Si, such as grain boundaries (GBs), dislocation clusters (DCs), impurities, and precipitates. They can heavily decrease the material quality by drastically reducing the diffusion length of minority carriers, and then limit the solar cell efficiency.

Transition metal impurities play a large role in decreasing mc-Si device performance relative to m-Si [5,6]. These impurities strongly interact with existing crystal defects to form complexes, accumulate at dislocations or grain boundaries in different forms, or even form silicide precipitates which simultaneously contain several metal impurities [7-9]. Fortunately, the negative impact of these impurities can often be reduced by phosphorus or aluminum gettering [10] which are contained in traditional mc-Si solar cell process. These gettering techniques are usually effective at removing interstitial impurities, and since these are often the dominant lifetime-killers, large improvements in lifetime can result [11]. However, as internal getting centers for metallic impurities, GBs and DCs may contributed different effects on the performance of mc-Si solar cells. Our aim in this study is to compare and discuss the effects of impuritiy-decorated dislocation clusters and grain boundaries in as-cut wafers on the performance of mc-Si solar cells after gettering process.

Inline photoluminescence (PL) imaging has been applied to identify impuritiy-decorated dislocation clusters and GBs in as-cut wafers [12,13]. Solar cell performance can be predicted quantitatively based on defects index of as-cut wafer from PL imaging [14]. In this work, the ratio of impuritiy-decorated dislocation clusters area and GBs area to the total area in wafer can be calculated separately by software.

2 EXPERIMENT AND MEASUREMENT

A central block with the dimension of 156×156×340mm^3 from an industrially produced P-type HPMC Si ingot is used in this experiment. Top and bottom parts of the block are cut off due to their low minority carrier lifetime. As-cut wafers with a thickness of 180μm are obtained after the wire sawing process, and the surfaces of wafers are perpendicular to the ingot solidification direction. Inline PL module (Hennecke, PL-2.0) is employed to identify regions of high recombination and calculate the area ratio of grain boundaries and dislocation clusters in as-cut, non-passivated wafers from the whole block, as shown in Fig.1.

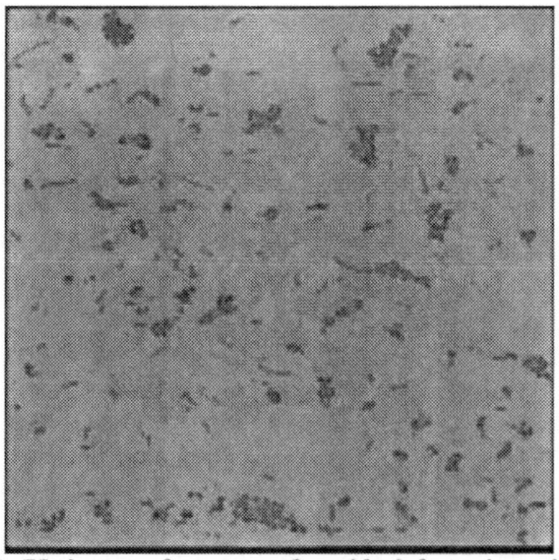

Fig.1. PL image of as-cut wafer with defects: impurity-decorated dislocation clusters (red line surrounded) and grain boundaries (purple lines).

Main defects of bottom region wafers are grain boundaries , while dislocation clusters are main defect of wafers from top region. Ten wafers are selected from each part (i.e., bottom and top region) with a defect area ratio of 7.5% to fabricate solar cells respectively, as shown in Fig.2.

Fig.2 Schematic diagram of wafers from top and bottom region in a silicon block.

PL images of as-cut wafer with equivalent defect area ratio of GBs and DCs, as shown in Fig.3. Solar cells are fabricated by a conventional manufacturing process including texture, phosphorous diffusion, PSG removal, PECVD SiN, screen printing and firing.

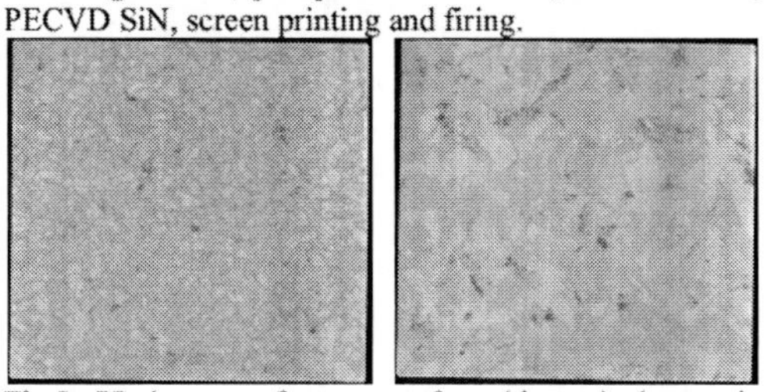

Fig.3. PL images of as-cut wafer with equivalent ratio (7.5%) of GBs (left) and dislocation clusters (right) to the total area.

3 RESULTS AND DISCUSSIONS

3.1 Performance of solar cells with different defects

Efficiency of Solar cells fabricated of as-cut wafers with different type of defects is shown in Fig.4. The result shows that for the samples with main GBs defect, there is a 0.11% absolute efficiency higher than that of samples with DCs defect.

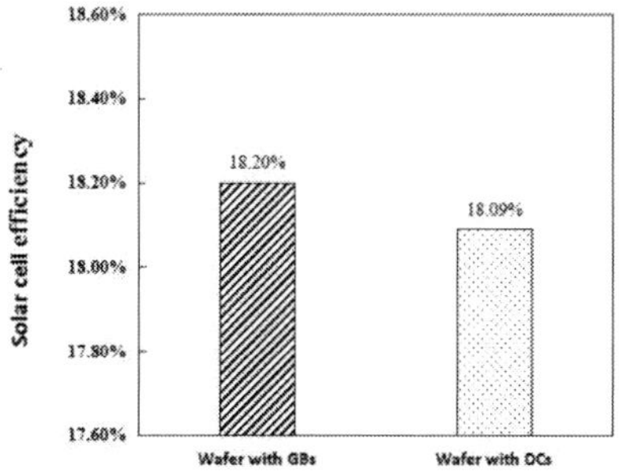

Figure 4. Efficiency of Solar cells fabricated by as-cut wafers with different defects

3.2 Discussion

The source of impurities that interact with GBs and DCs are diverse during the casting process of mc-Si, as shown in Fig.5[15].

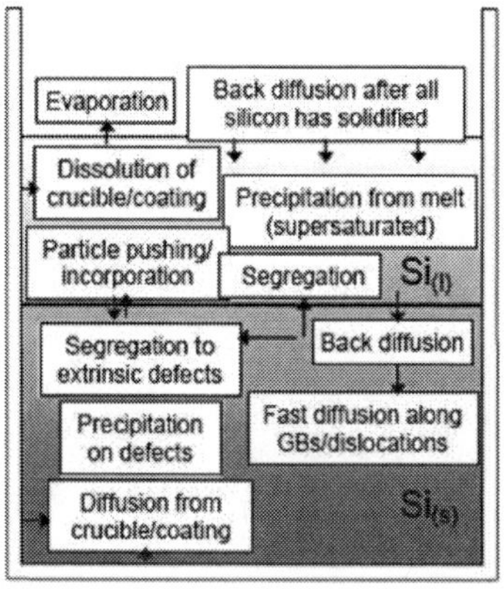

Fig. 5. Overview of some of the most important mechanisms affecting concentration, chemical state, and localization of impurities in mc-Si [15].

The impurities presenting in as-grown ingot are localized mainly by segregation and diffusion. Under the effect of segregation, most impurities gather at the top of the ingot and precipitate in the defects, even in grains due to overhigh concentration. At the bottom of ingot, impurities in crucible, coating, and the impurity-rich area (top of ingot) diffuse to the solidified ingot and a certain amount of interstitial impurities are generated. As previous paragraph mentioned, GBs and DCs are the main defects at the bottom and top of the ingot respectively, the former are decorated mainly by interstitial impurities while precipitations interact with the latter. The difference of gettering effect indicate that gettering techniques are effective at removing interstitial impurities in GBs but not effective for precipitation in DCs. It can be concluded that GBs contribute a relatively less negative effect on the performance of solar cells after the gettering process. Meanwhile, right PL image in Fig.3 shows almost no impurity interact with the GBs in top wafers where DCs are the main defect, which demonstrate impurity gettering ability of DCs is higher than that of GBs in top area.

4 CONCLUSION

The result shows that for the samples with grain boundaries defect, there is a 0.11% absolute efficiency higher than that of samples with dislocation clusters defect. The difference of gettering effect indicate that gettering techniques are effective at removing interstitial impurities in grain boundaries but not effective for precipitation in dislocation clusters. It can be concluded that grain boundaries contribute a relatively less negative effect on the performance of solar cells after the gettering process. Meanwhile, impurities prefer to interact with dislocation clusters in top area of ingot.

5 REFERENCES

[1] D. Zhu, L. Ming, M. Huang, et al. Seed-assisted growth of high-quality multicrystalline silicon in directional solidification, *Journal of Crystal Growth*, 386 (2014) , pp. 52-56.

[2] Y.T. Wong, C. Hsu, C.W. Lan, Development of grain structures of multicrystalline silicon from randomly

orientated seeds in directional solidification, *Journal of Crystal Growth*, 387 (2014) , pp. 10-15.
[3] C.C. Hsieh, Y.C. Wu, A . Lan, et al. Comparison of defect formations in solar silicon growth from small random and large oriented seeds, *Journal of Crystal Growth*, 419 (2015) , pp. 1-6 .
[4] H. Zhang, D. You, C. Huang, et al. Growth of multicrystalline silicon ingot with both enhanced quality and yield through quartz seeded method. *Journal of Crystal Growth*, 435 (2016) , pp. 91-97.
[5] T. Buonassisi, A. A. Istratov, M. D. Pickett, et al. Metal precipitation at grain boundaries in silicon: Dependence on grain boundary character and dislocation decoration. *Applied Physics Letters*, 2006, 89(4): 042102.
[6] M. Seibt, D. Abdelbarey, V. Kveder, et al. Interaction of metal impurities with extended defects in crystalline silicon and its implications for gettering techniques used in photovoltaics. *Materials Science and Engineering: B*, 159(2009), pp. 264-268.
[7] T. Buonassisi, M. Heuer, A.A. Istratov, et al. Transition Metal Co-Precipitation Mechanisms in Silicon. *Acta Mater.* 55 (2007), pp.6119-6126.
[8] C. Rudolf, P. Saring, L. Stolze, et al. Co-precipitation of copper and nickel in crystalline silicon. *Materials Science and Engineering: B*, 159 – 160(2009) pp. 365-368.
[9] P. Saring, C. Rudolf, L. Stolze, et al. Light-beam-induced current measurements on copper-nickel co-contaminated Cz-silicon bicrystals. *Materials Science and Engineering: B*, 159 – 160(2009) pp. 216-218.
[10] D. Macdonald, A . Cuevas, A . Kinomura, et al. Phosphorus gettering in multicrystalline silicon studied by neutron activation analysis, *Proceed-ings of the 29th IEEE Photovoltaic Specialist Conference*, IEEE, New Orleans, LA, USA, 2002, pp. 285-288.
[11] A. Cuevas, et al. High minority carrier lifetime in phosphorus-gettered multicrystalline silicon, *Appl. Phys. Lett.* 70(1997), pp. 1017-1019.
[12] T. Trupke, J. Nyhus, J. Haunschild. Luminescence imaging for inline characterization in silicon photovoltaics. *Phys. Status Solidi (RRL)*, 5(2011), pp. 131-137.
[13] S. Johnston, F. Yan, K. Zaunbrecher, et al. Imaging study of multi-crystalline silicon wafers throughout the manufacturing process. *Proc. 37th IEEE Photovoltaic Specialists Conf, Seattle*, USA, 2011: pp. 002885-002890.
[14] S. Fu, Z. Xiong, Z. Feng, et al. Cell Performance Prediction based on the Wafer Quality, *Energy Procedia*, 2013, pp. 43-48.
[15] R. Kvande, L. J. Geerligs, G. Coletti, et al. Distribution of iron in multicrystalline silicon ingots. *Journal of Applied Physics*, 104(2008), pp. 064905.

FABRICATION AND ELECTRICAL CHARACTERIZATION OF SEMI-TRANSPARENT SILICON SOLAR CELLS

Th. Makris[1], E. Skuras[2]
Department of Materials Engineering, University of Ioannina, Ioannina 45110, Greece

A.R. Long
School of Physics and Astronomy, University of Glasgow, Glasgow G12 8QQ, United Kingdom.

A. Santamaria, P. Fleming
IPSol Energy, Unit 12 Easter Park, Lenton Lane, Nottingham NG7 2PX, United Kingdom.

Humayra Chowdhury
Department of Electrical Engineering and Computer Technology,
University of Patras, Patra 26504, Greece

ABSTRACT: Experimental results from the fabrication and electrical characterization of semi-transparent mono-crystalline Si solar cells, with transparencies exceeding 10%, are presented. The semi-transparency is achieved by laser-cutting ellipses through industrially-manufactured mono-crystalline Si cells. The ellipses are designed to be distributed uniformly over the whole surface area of the highly efficient cells and form two-dimensional periodic lattices. Depending on their size they are either confined between successive grid lines or intersect them. Incident sunlight is transmitted through the ellipses. The large number of cutting elements results in a uniform distribution of the solar radiation transmitted. High efficiencies of 14.2% and 13.2% have already been achieved with 156x156 mm^2 semi-transparent mono-crystalline Si solar cells consisting of 91 and 106 ellipses respectively. In our newly-designed semi-transparent Si cells presented in this paper, ellipses with considerably larger sizes are used. In these cells both transparencies and the corresponding efficiencies exceed 10%.
Keywords: Fabrication, Silicon Solar Cells, Semi-transparent, Laser Processing

1 INTRODUCTION

In this paper we present experimental results from the fabrication and electrical characterization of semi-transparent mono-crystalline Si solar cells, with transparencies exceeding 10%. The semi-transparency is achieved by laser-cutting ellipses through industrially-produced mono-crystalline Si cells [1, 2]. The ellipses are designed to be distributed uniformly over the whole surface area of the highly efficient cells and form two-dimensional periodic lattices. Incident sunlight is transmitted through the ellipses. A novel two-level metal base [3, 4] was invented to minimize breakages of the thin and large surface area cells by creating conditions of mechanical stability during the laser cutting of the pattern elements. A uniform distribution of the solar radiation transmitted is achieved due to the large number of cutting elements. A high efficiency of 14.2% has already been achieved with an 156x156 mm^2 semi-transparent mono-crystalline Si solar cell consisting of 91 ellipses uniformly distributed on its surface [4]. The transparency of the cell was 4.6%. All its ellipses were confined within successive metal grid lines. In a similar Si cell with 106 ellipses - the largest number reported in the literature - the efficiency was 13.2%. In our newly-designed semi-transparent Si cells, presented in this paper, ellipses with considerably larger sizes are used. As a consequence the laser beam intersects successive grid lines as it moves to cut them and form a periodic pattern. In these cells both transparencies and efficiencies exceed 10%.

2 FABRICATION OF SEMI-TRANSPARENT Si SOLAR CELLS

The semi-transparency of our large surface area cells was achieved by creating periodic patterns of ellipses on industrially-manufactured and highly efficient Si cells. The ellipses were designed to be distributed periodically and almost uniformly over the entire 156x 156 mm^2 surface of the commercial cells. To create the periodic patterns of ellipses, the Bystar 4020 laser beam equipped with a 6kW CO_2 Bystronic laser cutting system [5] was used. Bystar 4020 has a position accuracy of ±0.1 mm, a repeatability of ±0.05 mm, an edge detection precision of ±0.5mm and a 0.3mm cross section of the laser beam. Bystar's 4020 coaxial jet helps in removing melted material and avoiding the formation of microcracks due to the rapid re-deposition and solidification of the melted material. Control of the flow rate of the inert gas coaxial to the laser beam can lead to even cleaner cuts [6].

Semi-transparent cells have very recently been used in active bio-climatic architecture and energy self-sufficient solar houses where they are integrated into glass roofs, atriums or facades and as fixed or movable solar shading devices. They were first designed for industrial production and fabricated by Takeoka et al [7] using either photolithography and chemical etching or laser patterning techniques, then by Willeke and Fath using mechanical texturing [8, 9].

Figure 1: A novel two-level metal base used for converting a three-bus-bar standard Si cell into a semi-

transparent one. It consists of two aligned and welded parallel iron plates with a square surface of 230 x 230 mm^2. A two-dimensional periodic pattern of ellipses, located at the center of the lower metal plate is created using the Bystar 4020.

A novel two-level metal base, shown in Figure 1, was invented to minimize breakages of the thin and large surface area cells by creating conditions of mechanical stability during the laser cutting of the pattern elements. It consists of two aligned and welded parallel iron plates with a square surface of 230 x 230 mm^2 [2]. A two-dimensional, periodic pattern of ellipses, located at the center of the lower metal plate, is created using the Bystar 4020. The lengths of their minor and major axes are substantially larger than those of the ellipses designed to be cut through standard Si cells to convert them into semi-transparent ones. All bases used in this study were constructed by the engineering team of Metallemporiki S.A. in Greece. In Figure 2, a schematic view is shown of the cross section of the laser system and the two-level metal base during the cutting of the ellipses. Bystar's 4020 coaxial jet helps in removing melted material and avoiding the formation of micro-cracks due to the rapid re-deposition and solidification of the melted material. Control of the flow rate of the inert gas coaxial to the laser beam can lead to even cleaner cuts [6].

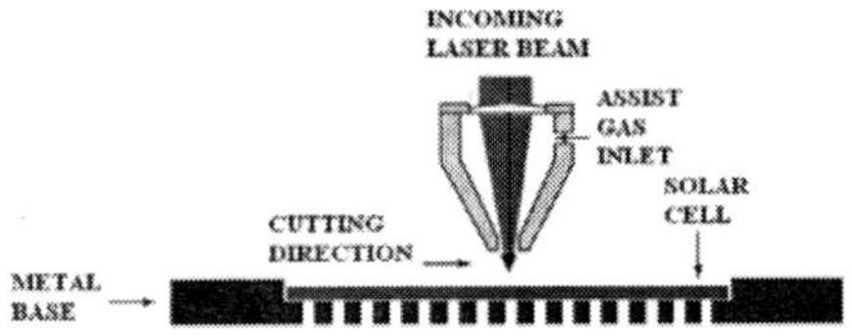

Figure 2: Cross section of the laser system, the metal base and a Si solar cell placed on it during the fabrication of a semi-transparent silicon solar cell.

Fabricated 156x156 mm^2 semi-transparent silicon solar cells with periodic pattern of ellipses are shown in Figures 3 and 4. The 106 ellipses of the semi-transparent Si cell shown in Figure 3 were designed to be located symmetrically within the area confined by successive grid fingers of the front side ohmic contacts of the cells and orientated with its minor axis perpendicular to the direction of the fingers. The ellipses were designed so as not to intersect with bus bars or grid fingers to ensure that the paths of the conduction electrons created by the absorbed photons to the metal grid lines, and through them to the bus bars, are not obstructed.

Figure 3: A 156 x 156 mm^2 two-bus bar semi-transparent mono-crystalline Si solar cell with a two-dimensional periodic pattern of 106 ellipses.

For our recently fabricated semi-transparent Si cells, shown in Figure 4, ellipses with considerably larger sizes are used. As a consequence the laser beam intersects successive grid lines as it moves to cut them and form a periodic pattern. The more the cutting elements and the larger their size, the longer the length of their perimeter and the laser beam travel length, and consequently the less the efficiency of the cell [4] given the same reduction of the sun-active area.

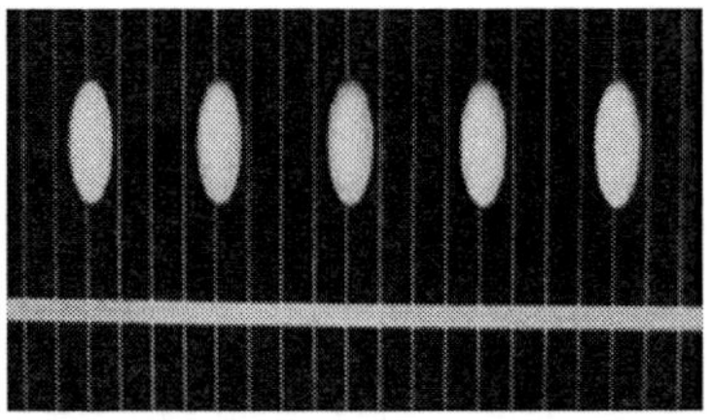

Figure 4: A close-up view of a newly designed and fabricated semi-transparent Si solar cell. Its transparency is achieved through large ellipses. As a consequence, the laser beam intersects successive grid lines as it moves to cut them and form a periodic pattern.

3 ELECTRICAL MEASUREMENTS AND ANALYSIS

Flat copper wires uniformly coated with either Sn(96.5%)/Ag(3.5%) or Bi(58%)/ Sn(42%) eutectic alloys were soldered to the fabricated semi-transparent Si cells to form Pb-free ohmic connections. The I-V characteristic curves were recorded with a 2611A Keithley source-measure unit equipped with a 10 A pulsed electric current source. In Figure 5, the I-V and P-V curves of a fabricated semi-transparent silicon cell with a two-dimensional periodic pattern consisting of 91 ellipses are shown. The curves were recorded at 930 Watt/m^2 of physical light. The transparency of the cell is 4.6% and its efficiency has decreased from 17.3% of the standard two-bus-bar cell to 14.2% after laser cutting. The 4.6% reduction of the sun-active area causes a decrease in its efficiency to 16.5%. The rest of the decrease, i.e. 2.3%, is attributed to the damage induced by the laser beam as it successively cuts the 91 ellipses by moving along their perimeters. Their total perimeter length is 1650 mm [4].

For the fabricated semi-transparent silicon cell with a two-dimensional periodic pattern consisting of 106 ellipses the transparency is 5.4% and its efficiency decreases from 17.5% of the three-bus-bar standard cell to 13.2% after laser cutting. For both cells the dimensions of the ellipses after laser cutting are 1.8 mm (minor axis) and 8.6 mm (major axis).

There are two contributing factors responsible for the decrease in the efficiency when fabricating a semi-transparent mono-crystalline silicon solar cell from a standard one. The first is the material damage induced by the laser beam as it successively cuts its pattern elements Ideally this damage should be proportional to the total length the laser beam travels during cutting. The second is the reduction of the sun-active cell area due to the removal of the pattern elements. To distinguish between

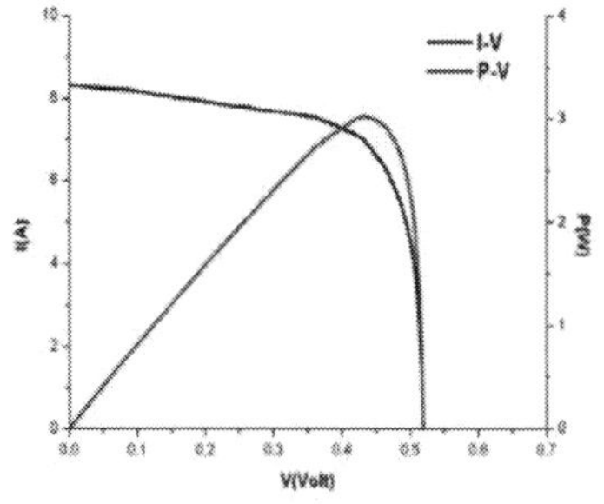

Figure 5: I-V and P-V curves recorded at 930 Watt/m^2 for a 156 x 156 mm^2 two-bus bar semi-transparent mono-crystalline Si solar cell with 91 ellipses.

the effects each factor has on the cell's efficiency, special samples have been designed and fabricated. They consist of standard 156x156 mm^2 three-bus-bar mono-crystalline Si solar cells with two circles, each with a 20 mm radius, having been removed from their central area by laser cutting. The circles are symmetrically cut either side of the central bus bar of each cell. This causes a reduction in its sun-active area of 10.5% i.e. twice that of the semi-transparent cell with the periodic pattern of the 91 ellipses shown in Figure 3.

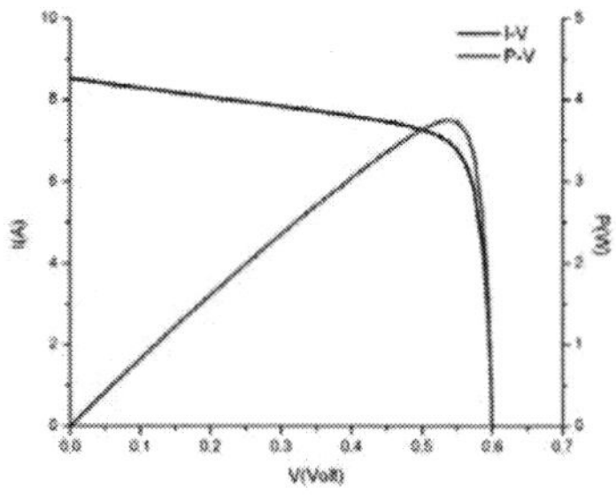

Figure 6: I-V and P-V curves, for 1030 Watt/m^2 for a 156x156 mm^2 three-bus bar mono-crystalline Si solar cell with two circles with 20 mm radius each placed between the bus bars.

Figure 6 shows the I-V and P-V curves of the designed sample recorded at a sunlight radiation of 1030 Watt/m^2. An efficiency of 15.2% is deduced from the analysis of the electrical data. The 10.5% reduction of its sun-active area explains the decrease in its efficiency from the initial value of 18.6%, prior to the circles being cut, to 16.8%. The total distance the laser beam travels along their perimeters is 250 mm. This is significantly less than the length of 1650 mm the laser beam moves when cutting the semi-transparent cell's 91 ellipses.

For our recently fabricated semi-transparent Si cells presented in this paper, ellipses with considerably larger sizes are used. As a consequence the laser beam intersects successive grid lines as it moves to cut them and form a periodic pattern. The more the cutting elements and the larger their size, the longer the length of their perimeter and the laser beam travel length, and consequently the less the efficiency of the cell [4] given the same reduction of the sun-active area. Despite the larger number of ellipses and the longer length of their minor and major axes, compared with our previous designs our cells exhibit efficiencies just over 10% for transmittances exceeding 10%.

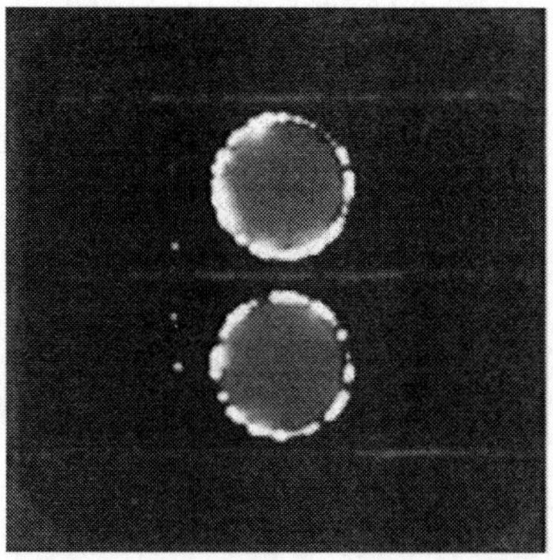

Figure 7: An image of reverse biased electro-luminescence measurements performed to detect localized shunts caused by the laser beam as it moves to successively cut the two circles from the central area of a specially-designed sample.

Detailed electroluminescence measurements [10-12] are used to detect any micro-cracks or localized shunts caused by the laser beam as it moves to successively cut the ellipses of a periodic lattice. In Figure 7 localized shunts caused by the laser beam as it moves to successively cut the two circles are shown. The image was recorded using reverse biased electroluminescence measurements and is used to explain an observed gradient at the intersection of the I-V curve with its corresponding current axis.

4 CONCLUSIONS

In this paper the fabrication and electrical characterization of semi-transparent mono-crystalline Si solar cells have been presented. The semi-transparency was achieved by laser-cutting ellipses through industrially-manufactured mono-crystalline Si cells. The ellipses were designed to be distributed uniformly over the whole surface area of the highly efficient cells and form two-dimensional periodic lattices. Depending on their size they are either confined between successive grid lines or intersect them. Incident sunlight is transmitted through the ellipses. The large number of cutting elements results in a uniform distribution of the solar radiation transmitted. High efficiencies of 14.2% and 13.2% have already been achieved with 156x156 mm^2 semi-transparent mono-crystalline Si solar cells consisting of 91 and 106 ellipses respectively. In our newly-fabricated semi-transparent Si cells presented in this paper, ellipses with considerably larger sizes were used. Their efficiencies deduced from the analysis of their corresponding I-V characteristic curves were just over 10% for transparencies exceeding 10%.

1) Present address : IPSol Energy, Unit 12 Easter Park, Lenton Lane, Nottingham NG7 2PX, United Kingdom.
2) E. Skuras, Corresponding author, e-mail address : eskuras@cc.uoi.gr.

REFERENCES

[1] Th. Makris, K. Tsevas, D. Anagnostopoulos and E. Skuras, Fabrication of transparent monocrystalline silicon solar cells, Proceedings of the 24th European Photovoltaic Solar Energy Conference, Hamburg, Germany, 2009, pp. 1323-1325, ISBN:3-936338-25-6.

[2] Th. Makris and E. Skuras, Electrical Characterization of Large Surface Area semi-transparent Silicon solar cells, Proceedings of the 27th European Photovoltaic Solar Energy Conference, Frankfurt, Germany, 2012, pp. 2072-2074, ISBN:3-936338-28-0.

[3] Th. Makris and E.Skuras, Electrical Characterization of Large surface Area Semi-transparent Si Solar cells, 31st International Conference on the Physics of Semi-conductors, 29th July – 3rd August 2012, ETH - Swiss Federal Institute of Technology, Zurich, Switzerland, AIP Conference Proceedings 1566 (2013) 397-398.

[4] Th.Makris, K. Tsevas, L. Kadylis, L. Mprechas, E. Skuras, Journal of Solar Energy Engineering, 135 (2012) 34503-1 – 34503-4.

[5] www.bystronic.com/cutting_and_bendings/us/en

[6] R. Hendel, Laser Applications in Solar Cell Manufacturing, Laser Technik Journal, 5 (2008) 32-35.

[7] A. Takeoka, S. Kouzuma, H. Tanaka, H, Inoue, K. Murata, M. Morizane, N. Nakamura, H. Nishiwaki, M. Ohnishi, S. Nakano and Y. Kuwano, Development and application of see-through a-Si solar cells, Solar Energy Materials and Solar Cells 29 (1993) 243-252.

[8] G. Willeke, P. Fath, Mechanical wafer engineering for semitransparent polycrystalline silicon solar cells, Applied Physics Letters 64 (1994) 1274-1276.

[9] A. Boueke, R. Kühn, P. Fath, G. Willeke and E. Bucher, Latest results on semitransparent POWER silicon solar cells, Solar Energy Materials and Solar Cells 65 (2001) 549-553.

[10] Otwin Breitenstein, Jan Bauer, Thorsten Trupke and Robert A. Bardos, Progress in Photovoltaics Research and Applications, 16 (2008) 325-330.

[11] Otwin Breitenstein, Materials Science Forum, Volume 725 (2012) 115-122.

[12] T. Trupke, B. Mitchell, J.W. Weber, W. McMillan, R.A. Bardos and R. Kroeze, Energy Procedia 15 (2012) 135-146.

OPTICAL LOSS ANALYSIS OF PV MODULES

Malcolm Abbott, Keith McIntosh and Ben Sudbury

PV Lighthouse, Coledale, NSW 2515, Australia

ABSTRACT: Photovoltaic modules present a complex system of interactions between various optical materials and the solar cells. Understanding how this system operates optically is an important part of designing modules with high output power. This paper demonstrates how simulations based on a combination of Monte Carlo ray tracing and thin film optics can be used to determine the optical loses in photovoltaic cells and modules. It finds that the performance of commercial cells and modules can be reproduced allowing a detailed loss analysis of the optics to be determined. Various examples are given on the use of such a simulation approach, including an assessment of the module performance as a function of cell spacing, rear surface lambertian fraction and optimization of front surfaces under encapsulation.

1 INTRODUCTION

Minimizing cell-to-module losses is vital to achieving modules with high output power. This task is made very difficult due to the complex combination of different materials, thin film optics, surface geometries and the various interactions with the active device. Simulation software can play an important role in helping to characterize these interactions, assess the loss mechanisms and ultimately optimize the module design. There are many well developed software tools to simulate the performance of photovoltaic cells, however for modules there are many less options and those that are available tend to be slow and difficult to use.

Recently, PV Lighthouse [1] introduced a new software tool specifically designed to simulate the optical performance of photovoltaic modules. Through optimisation of the algorithms, combined with the scalable processing power of cloud computing, the solving is extremely fast allowing many variations of the module layers to be rapidly simulated. This paper discusses some of the fundamental principles of this approach and demonstrates how it may be used to assess the optical losses for a particular module design as well as to understand the sensitivity of different aspects to the overall performance.

2. RAY TRACING BASELINE MODULES

The simulation method used in this work is based on a combination of ray tracing and thin film optics. A Monte Carlo approach is employed in which random numbers are used to determine both the initialization and progression of rays of a particular wavelength through the system of optical materials. As they interact with interfaces where thin films are present the transfer matrix method is used to solve the reflection, absorption and transmission (RAT) of the beam [2]. Interfaces can also be defined more simply as having particular RAT properties. In both cases a Lambertian scattering model can be employed. Surface texturing is accounted for using a variety of geometric surface morphologies implemented as unit cells. The random nature of certain textures is handled through random shifts in the beam position during entry and exit to the unit cell. A more detailed description of this approach, as well as a comparison to experimental results, has recently been presented [3].

The impact of metal contacts (fingers, busbars and coatings) was not included in the version of the MRT used in this work. Where it was relevant the final value of Jsc presented was scaled based on the percentage of shading, taking into account the effective shading width [10]. It should be noted that more recent versions of the MRT include the ability to simulate all types of common metal contact patterns, including special busbar alternatives such as the so called MultiWire approach. Results on some of these types of simulation were presented at this conference [4].

2.1 Baseline module performance

There are many different flavours of multicrystalline cells and modules currently on the market. To get an estimate of the most common performance and design parameters we conducted a survey of publically available datasheets (cells and modules) from a selection of commercial manufacturers. Here we focus on the data related to standard screen print cells fabricated on multi-crystalline substrates. Rather than quote every exact source of this data, we keep the manufacturer names anonymous and refer the reader instead to the ENF database where all datasheets were sourced [5]. Roughly half of the data was from tier 1 manufacturers, the rest were tier 2 and 3. The summary of the I_{sc} (Figure 1) reveals an average value of 8.75 A for the cells and 9.05 A (per cell) for modules. This gain in current for the modules is expected and is due to several factors including collection of carriers from outside the cell area and enhanced trapping of light initially reflected from metal fingers and isotexture.

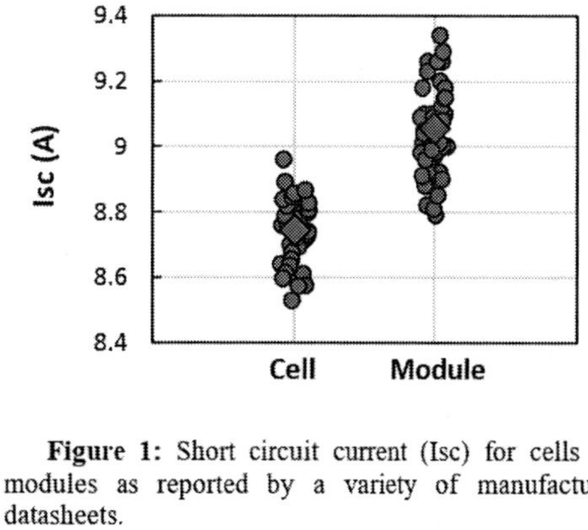

Figure 1: Short circuit current (Isc) for cells and modules as reported by a variety of manufacturers datasheets.

The basic inputs for the material and geometric parameters of the cells and modules were also extracted from the data sheets. In this case there was significantly more error since the exact material properties cannot be known and would be expected to vary between products. Further work is required to establish common values for these more accurately; however for the purposes of this work the values used (Table 1) were a reasonable approximation. Note the use of a front-side AR coating on the glass which we found to be very common on the manufacturers datasheets.

Table I. Simulation inputs for baseline module

Location	Property	Value
Front glass	Thickness	3.2 mm
	Material	Borofloat33
	AR coat	Enki cleanARC
EVA	Thickness	450 μm
	Material	[6]
Cell bulk	Thickness	180 μm
	Material	[8]
Cell texture	Type	Inverted caps
	Angle	60°
	Height	4 μm
Cell ARC	Thickness	75 nm
	Material	SiNx (2.03) [7]
Cell rear	Reflection	65%
	Absorption	35%
	Lamb. fraction	70%
Cell layout	spacing	0.2 mm
Backsheet	Reflection	85%
	Absorption	15%
	Lamb. fraction	50%

Conversion of an optical generation current into collected cell current required wavelength dependent collection efficiency data (CE). This was generated using PC1D with common front, bulk and rear recombination inputs [9] and 100% internal reflection (to remove impact of light trapping). The MRT software allows this CE to be entered as an input, resulting in an EQE curve and cell Jsc as outputs.

The results for the baseline module under standard test conditions are shown in Table 2 and Figure 2. The optical losses have been grouped by material where EVA includes the surrounding area of the cell. Weighting these losses via the AM1.5G spectrum allows them to be expressed in terms of a generation current (Table 2). The error in the current is the calculated 95% confidence interval based on the set of results from 200 different simulation runs of 5000 rays.

Table 2. Optical losses simulated for the baseline module

	Mean (mA/cm^2)	Fraction of Jinc (%)
Incident	46.71	100
Reflected – external	1.13 ±0.017	2.41
Reflected – escape	1.82 ±0.015	3.89
Absorption glass	0.57 ±0.001	1.23
Absorption EVA	1.05 ±0.003	2.26
Absorption in SiNx	0.11 ±0.000	0.25
Absorption in Si	37.71 ±0.023	80.7
Absorption in Al rear	4.12 ±0.007	8.81
Absorption in backsheet	0.19 ±0.002	0.42

The wavelength dependent reflection and absorption for this baseline module reveals which parts of the spectrum are lost in the different layers. At very short wavelengths over 80% of the light was absorbed in the EVA. For longer wavelengths the absorption at the rear of the solar cell was significant.

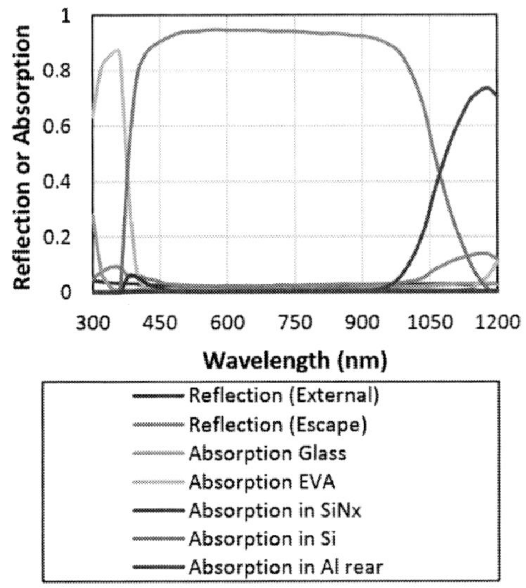

Figure 2: Wavelength dependent reflection and absorption losses simulated for the baseline module without metalisation.

After applying the collection efficiency and accounting for the metal shading, the simulated current of the cell was 8.8 A with no encapsulation and 9.0 A when incorporated into a module unit cell. This agrees well with the data shown in Figure 1 for the commercially available cells and modules. With so many free variables it is hardly surprising that we were able to reproduce these numbers. However, it is encouraging that the output of the simulator is in the correct range and for the purposes of this paper our baseline module can be used to demonstrate the power and versatility of the simulation approach.

2.2 Improving the confidence interval

Inherent to the Monte Carlo approach of ray tracing is the need to define the number of rays to trace as well as the maximum number of interaction with interfaces. Lower numbers lead to faster simulations but with less accurate results. To demonstrate this effect we simulated the baseline module several times and employed the sweeping feature of the MRT to vary the maximum number of rays for each simulation. The results are plotted in Figure 3 in terms of absolute current for each run as well as the 95% confidence interval. From the data it is clear that increasing the number of simulated rays improves the accuracy of the result. However, it is interesting to note that with even relatively few rays (here the minimum used was 30,000) the final value of current from the solar cell was within 30 mA (0.12 mA/cm^2) of the more accurate value achieved with 1 million rays. If speed of simulation is required then this level of error may be acceptable.

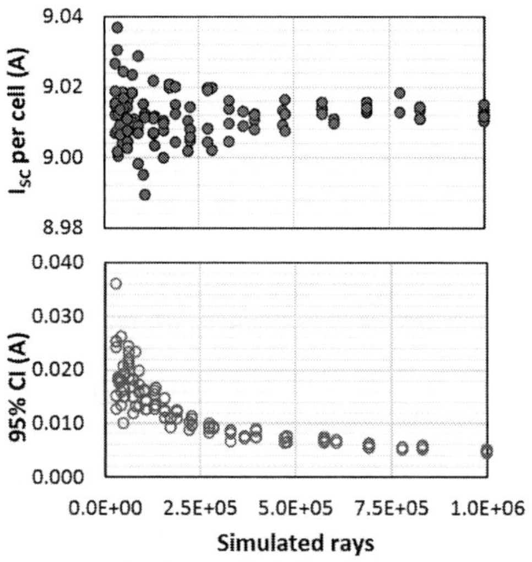

Figure 3: Impact of the total number of rays simulated on (top) the total module collected current per cell and (bottom) on the 95% confidence interval.

2.3 Angular dependences of losses

Modules in the field experience a variety of incident angles of the incoming light. To study the performance of a module design as the sun moves across the sky we simply sweep the zenith angle. In doing so the overall incident intensity of the light is reduced, as such it is instructive to present the results both in absolute terms and as a percentage of the incoming light intensity.

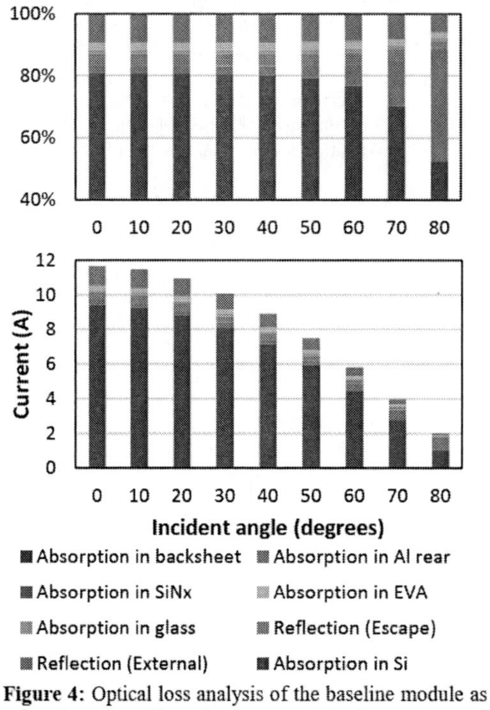

Figure 4: Optical loss analysis of the baseline module as a function of zenith angle shown as a (top) percentage of total available current and (bottom) absolute cell current.

The angular dependent losses showed that, as expected, the most significant impact of the zenith angle was an increase in the percentage of loss due to reflection from the front surface of the glass.

3. FURTHER APPLICATIONS

The previous section described the simulation of baseline modules. This section demonstrates how the simulation input parameters can be varied to assess the impact of different properties.

3.1 Spacing between cells

The space between the cells in a module is an important region to optimize. This area does not contain a solar cell absorber, however, it is well known that light incident on these areas can be collected via reflection from the backsheet and internal light trapping beneath the glass. The effectiveness of this process is very dependent on the properties of the backsheet. There are many options for material to use, some designed for high efficiency and others intended to create an appealing look (i.e. all black modules).

An extremely important property of a backsheet is the ability to scatter light. Here we model this using a simple Lambertian function in which the fraction of light scattered in a Lambertian fashion is defined by a value that does not change with wavelength. The impact of this on the collection of current as a function of the spacing between the cells is shown in Figure 5. In this experiment the spacing between cells for the baseline module was varied along with the Lambertian fraction of the rear interface (i.e. the backsheet).

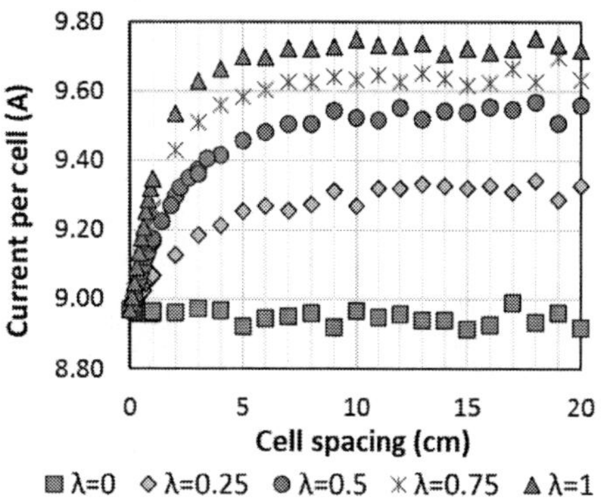

Figure 5: Current produced under normally incident light by a module unit cell as a function of cell separation, simulated for the baseline module using a backsheet with various values of Lambertian fraction.

The simulated values of module current for different white space properties (Figure 5) demonstrate the strong impact on module performance of these regions. With no lambertian scattering and normally incident light the module current sees no change as the total area is increased. This occurs as no light incident on those areas is able to spread sideways onto the cell. By comparison the samples with lambertian reflection show increasing total current due to the collection of light from outside the cell area. As the spacing is further increased the current saturates as the longer path to the cell results in other losses dominating. Interestingly the model predicts that for the inputs used in this work, collection of light can occur from as far away as 10cm.

3.2 Optimisation of front surfaces

It is common for device engineers to optimize the cell design based on electrical performance measured using an IV tester in which the layer above the cell is air. However, the end module product includes layers of glass and EVA above the cell that create the possibility of internal light trapping. To demonstrate this impact we simulated the case of front surface AR coating thickness optimization. Figure 6 plots the normalized module current, under normally incident light, for three different surface morphologies as a function of the thickness of the front AR coating (SiNx).

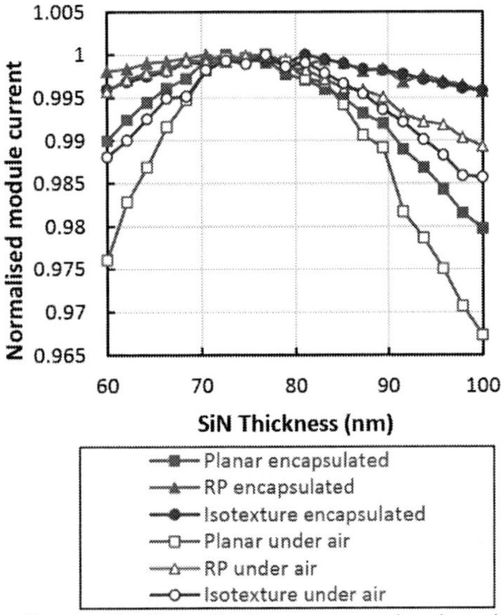

Figure 6: Normalised current produced under normally incident light as a function of solar cell front AR coating thickness, shown for different surface morphologies under air and under encapsulation.

The impact of the front side AR coating film thickness on the Isc of a cell is significantly reduced when that cell is encapsulated. This effect is greatest for samples with either random pyramid (Figure 6 triangles) or isotexture (Figure 6 circles) morphologies on the surface. In those cases the current varies less than 5%rel for a range of thicknesses from 60 nm to 100 nm. In all cases the optimal thickness was found to be 75 nm. This provides just one example of how encapsulation should be considered in cell design. In this case the optimal point was the same, but the sensitivity to the value was reduced under encapsulation.

4. CONCLUSIONS

This paper investigated the application of ray tracing to the optical analysis of solar modules. Specifically it presented a new tool from PV Lighthouse capable of rapidly solving many thousands of rays through a series of optical layers joined with textured, thin film interfaces. As an example a baseline module was described and compared to the performance of current commercial modules and cells as reported in manufacturer datasheets.

It was shown that the simulation approach is capable of reproducing the performance of such modules and that this then allows a detailed loss analysis of the optics to be determined. Furthermore the impact of the number of rays used within the simulation on the errors in the final result was presented.

Finally the paper demonstrated a number of example applications of module simulations. These included determining the module performance as a function of cell spacing, rear surface lambertian fraction and optimization of front surfaces under encapsulation.

Future work on the module ray tracer will seek to advance the approach by incorporating metal contacts as well as presenting more detailed validation through comparisons with experimentally measured datasets.

5 REFERENCES

[1] https://www.pvlighthouse.com.au

[2] K.R. McIntosh and S.C. Baker-Finch, "OPAL 2: Rapid optical simulation of silicon solar cells," in *38th IEEE PVSC*, Austin, 2012, pp. 265–271.

[3] Keith R. McIntosh, Thomas G. Allen, Simeon C. Baker-Finch and Malcolm D. Abbott, "Light trapping in isotextured silicon wafers", *43rd IEEE PVSC*, Portland, 2016

[4] Keith R. McIntosh, Malcolm D. Abbott, Matt Edwards, Rhett Evans and Yu Yao, "Optical evaluation of multi-wire modules", *32nd EU-PVSEC*, Hamburg, 2016

[5] http://www.enfsolar.com/pv/panel

[6] K.R. McIntosh, J. Cotsell, J. Cumpston, A. Norris, N. Powell and B. Ketola, "An optical comparison of silicone and EVA encapsulants for conventional silicon PV modules: A ray-tracing study," *34th IEEE PVSC*, Philadelphia, pp. 544–549, 2009.

[7] S. Duttagupta, F. Ma, B. Hoex, T. Mueller and A. Aberle, "Optimised antireflection coatings using silicon nitride on textured silicon surfaces based on measurements and multidimensional modeling," *Energy Procedia* 15(0), pp. 78–83, 2012.

[8] M.A. Green, "Self-consistent optical parameters of intrinsic silicon at 300K including temperature coefficients," *Solar Energy Materials and Solar Cells*, vol. 92 (11), pp. 1305–1310, 2008.

[9] A. Fell *et al.*, "Input Parameters for the Simulation of Silicon Solar Cells in 2014," in *IEEE Journal of Photovoltaics*, vol. 5, no. 4, pp. 1250-1263, July 2015.

[10] Pietro P. Altermatt, "Models for numerical device simulations of crystalline silicon solar cells—a review", *Journal of Computational Electronics*, September 2011, 10:314

LARGE AREA IBC ZEBRA SOLAR CELLS IN PILOT PRODUCTION: THE RESULTS OF FP7 HERCULES PROJECT

Giuseppe Galbiati*, V. D. Mihailetchi, A. Halm, H. Chu, and Radovan Kopecek
International Solar Energy Research Center (ISC) Konstanz, Rudolf-Diesel-Straße 15, D-78467 Konstanz, Germany
** Corresponding author. Tel.: +49-7531-361-83-361; E-mail address:giuseppe.galbiati@isc-konstanz.de*

ABSTRACT: In this work we present the latest results on large area interdigitated back contact (IBC) Zebra cells. Within the framework of the European FP7 HERCULES project, we demonstrated the transfer and scaling-up of our IBC ZEBRA cell process to the pilot line production and fabricate more than 1100 solar cells for the industrial feasibility of the processes and for modules demonstration. The solar cells were fabricated in several batches at the ISC-Konstanz R&D laboratory. The n-type Cz Si wafers were provided by the project partner NorSun AS, with additional commercial reference n-type Cz wafers added as process control. It was found that the efficiency results depend on wafers supplier, with median of 21.1 % for NorSun material and 21.4 % for the reference material.
In order to maximize the efficiency of ZEBRA cells in a pilot line production we investigated the effect that silicon material quality and resistivity variation occurring along the entire n-type Cz ingot have on cell performance. In addition, we studied the stability of the process on different wafer thickness.

Keywords: IBC solar cell, BC-BJ solar cells, n-type, modeling

1. INTRODUCTION

The evolution of the photovoltaic market devices toward solar cells with higher conversion efficiency enhances the industrial relevance of back-contact back-junction (BC-BJ) technologies. In agreement with the ISC Konstanz's roadmap mission, we have developed the Zebra technology as a low-cost high-efficiency industrial feasible process which implies manufacturing of interdigitated-back-contact (IBC) solar cells with front floating emitter (FFE) and screen-printed firing-through metal contacts [1]. The Zebra cells (see Fig.1) are fabricated on standard 156 x 156 mm^2 n-type Cz-Si wafers where the emitter and back-surface-field regions are formed using industrial-size quartz tube furnace diffusion using BBr_3 and $POCl_3$ source respectively. The entire process requires only one masking step while the interdigitated pattern on the back side is obtained by laser processing.

In this work we present the latest Zebra solar cells results for the process optimization conducted at the ISC Konstanz laboratory within the EU-founded HERCUES project [2]. The concept proposed by the HERCULES is to develop innovative n-type mono-crystalline silicon (c-Si) device-structures based on back-contact solar cells with alternative junction formation, as well as related structures including hybrid concepts (homo-heterojunction). These are the most promising silicon-based technologies to reach high efficiencies with industrially relevant processes. The project started in 2014 and in now in its last year. Within the framework of the project, ISC-Konstanz has the role to provide IBC-Zebra solar cells with homo-junction emitter obtained with high temperature processes (i.e., diffusion process) and high efficiency, in order to fabricate back contacted modules with more than 320 Wp. The work herewith presented was needed to ensure a close connection between cell processing and the specification for the Cz-Si wafers used for the fabrication of these solar cells. We investigate the variation in cell performance of our process over large variation of base resistivity as a consequence of the distribution of the dopant concentration. Moreover we tested the stability of the ZEBRA process and the yield losses arising by using thinner wafers material.

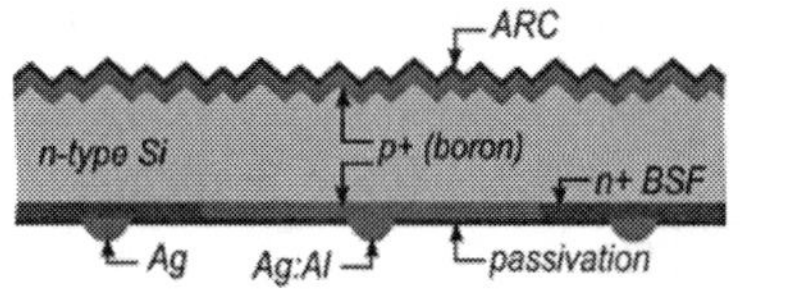

Figure 1: Sketch of the Zebra cell. The main process are i) high temperature BBr3 and $POCl_3$ diffusions; ii) laser patterning and thermal SiO_2/SiN_x passivation stacks; iii) screen printed and firing through metallization contacts.

2. APPROCH

In our first test we processed as-cut Norsun Si-wafers with thickness from 140 to 180 µm. The cells performance and the process breakage are reported. In a second analysis, the quality of the silicon wafers was tested from different part of the ingot as presented in Fig.2. At the latest we fabricated approximately 1150 IBC-Zebra cells as a pilot production run. The cells were fabricated on industrial equipment with only manual handling capability. Due to the limited availability of automatization, the entire cell-production was divided in batches of approx. 144 wafers each, spanned over eight weeks of lab-processing. As a starting material we received 6-inch n-type Cz-wafers from the top part of the ingot (from 40 to 800 mm along the ingot direction). For reference purposes, as well as for process control, we added several commercial n-type Cz wafers in each fabricated batch.

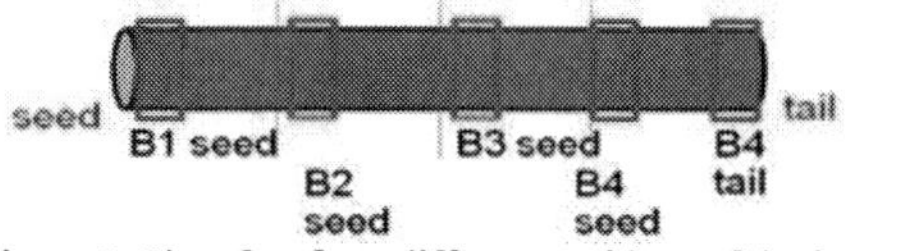

Figure 2: Si wafers from different positions of the ingot results variation of base resistivity due to the segregation effect of the dopant.

3. RESULTS

3.1 Wafer thickness variation

In Fig.3 the performance of the ZEBRA cells are reported as function of the wafer thickness. Notable are the opposite trends of the fill-factor (FF) versus current (Jsc) and voltage (Voc). Indeed, it is shown that increasing the thickness the Jsc and Voc increase due to better light absorption and the reduce surface recombination whereas FF drops due to larger series resistance. The resulted efficiencies are maximized for thinner wafers (~145 μm). Zebra solar cells made with wafers of 195 μm show a drop in average efficiency of 0.43 $\%_{abs}$.
The disadvantages of using very thin Si wafers arise from the handling of this material. Thin wafers are in general more fragile that results in higher breakage rate and have the tendency for larger bowing, which contribute for the low performance of the solar cell. For instance, we experienced several problems during wet-chemical batch-processes due to the larger bowing of the thin wafers placed in the standard process carrier: in this case wafers placed in adjacent positions touch and bound together leaving solar cells not completely textured or half cleaned. Nevertheless, the breakage rate remain lower that 5 %.

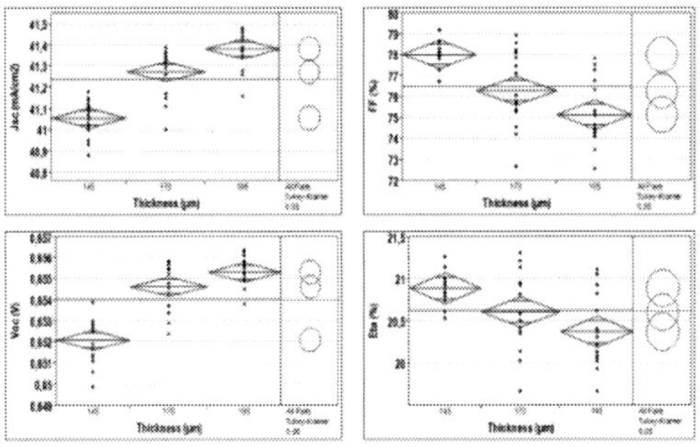

Figure 3: Zebra cell efficiency as a function of as-cut wafer thickness. Standard-one-way ANOVA reveals a significant difference within the groups. Best efficiency of 21.3 % was obtained with wafer thickness of 145 μm. From the total batch of 65 wafers, the breakage counts for 5 wafers with 145 μm thickness and 1 wafer with 195 μm thickness. The final thickness of the cells was ~120 μm; 150 μm and 170 μm respectively.

3.2 Base resistivity variation

Fig. 4 shows the cell results on wafers taken from different parts of the n-type ingot that have different base resistivity. Low values of resistivity correspond to the tail part of the ingot. The data show that the efficiencies reach an optimum for base resistivity range between 5-7 Ω·cm for the given lifetime material.

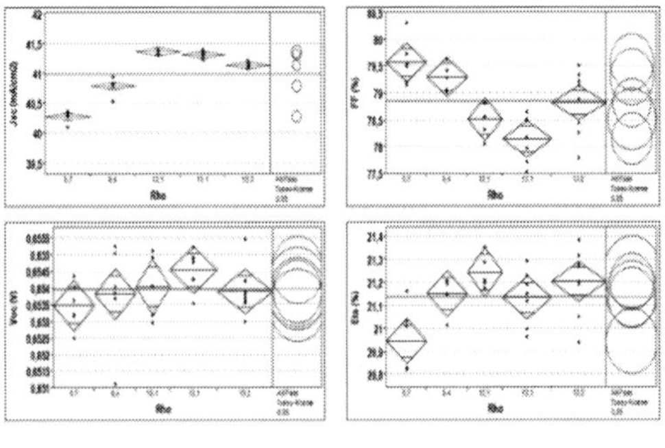

Figure 4: The data represent solar cells obtained from wafers located in different parts of the n-type ingot. Each group corresponds to a different base resistivity. Low values of resistivity correspond to the tail part of the ingot. The data show that the efficiency reaches a plateau for resistivity range of 5-7 Ω·cm.

3.3 ZEBRA pilot line results and module assembling

Here we present the results for our first ZEBRA pilot line production. We started with 1150 wafers and we finalized 1065 cells, with less than 6% yield loss despite of manual handling and batch-like process. This is in agreement with the general yield for batch process lines using manual handling.
From the IV data we excluded the tail of production of about 5 %. Fig.5 and Fig.6 shows the efficiencies distribution of the ZEBRA cells for the different material As can be seen, the median efficiency obtained is >21.4% on reference material and only 21.15% on NorSun material. The reason for the overall lower performance using NorSun material is caused by particular lower wafers quality of one of the three wafer lots used in the fabrication (see Fig.7). This particular wafer-lot originates from the top part of the ingot which has high concentration of thermal donors and oxygen induced stacking faults. Nevertheless, similar best cell efficiencies of about 22% were obtain on both, NorSun and reference wafers (see Table I), respectively. The ZEBRA cells used for the production have a final thickness of 150 μm.

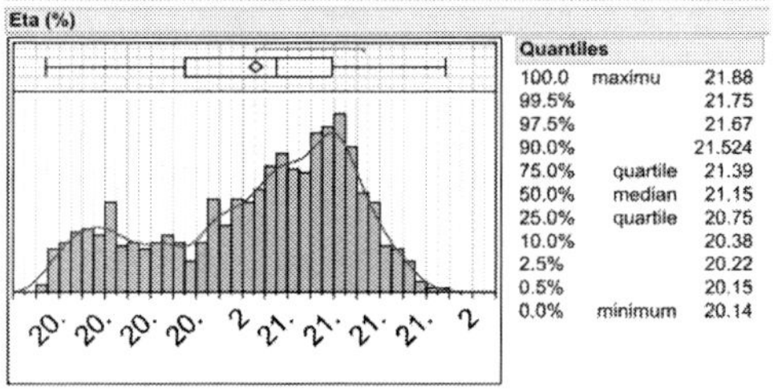

Figure 5: Efficiency distribution of the ZEBRA cell production using NorSun wafers.

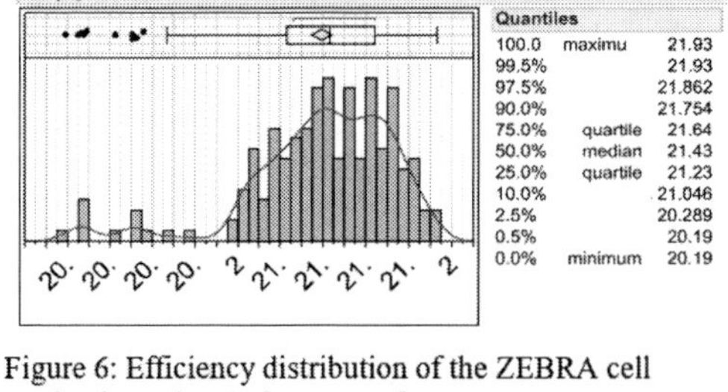

Figure 6: Efficiency distribution of the ZEBRA cell production using Reference wafers.

The resulted ZEBRA cells have been use to manufacture 60 cell modules in the Eurotron BV (HERCULES partner) back contact module assembly line. Fig.8 shows the certified IV data measurements of the first 300 Wp module, while in Table II the results are reported.

The cell to module power drop is with 1.8% relative extremely low for a n-IBC solar module. The module's fill factor is 76.5%. The dimension is 1647x985 mm^2 including frame making it smaller than standard 60 cells.

The consumables utilized during production were commercially available back-sheet, 200 micron thin encapsulant for front encapsulation, low Ag electric conductive adhesive which is used in standard MWT module manufacturing and 3.2 mm ARC coated solar glass.

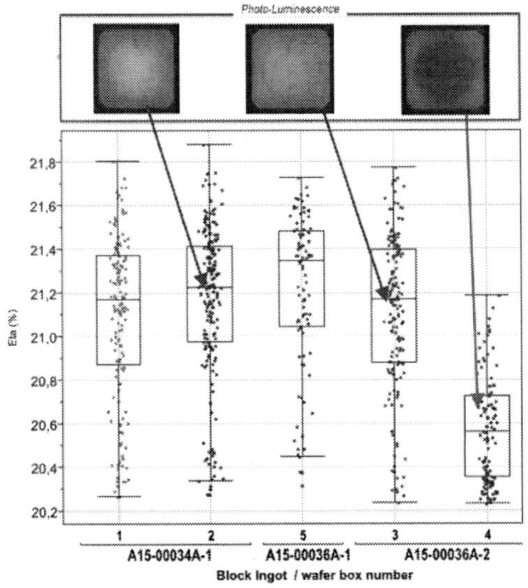

Figure 7: Zebra cell efficiency versus wafer block. The PL-images clearly show the influence of the oxygen content on the performance of the cells.

Table I: Champion Zebra cells: IV data for cells processed on different wafers supplier.

Zebra Cell [material]	*area* [cm²]	J_{SC} [mA/cm²]	V_{OC} [mV]	*FF* [%]	η [%]
NorSun	239	41.6	664.1	79.3	**21.9***
Reference	239	41.5	654.8	80.6	**21.9***

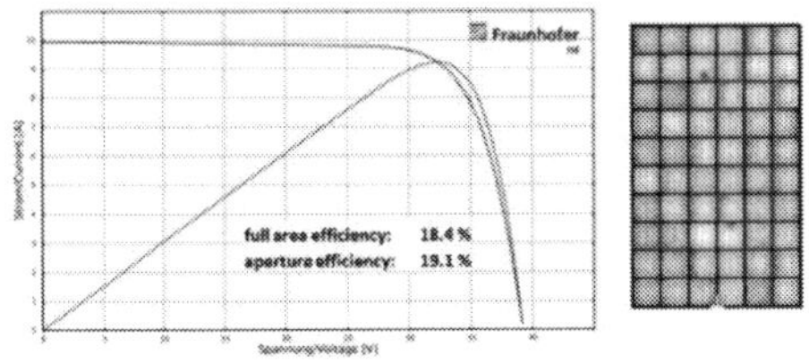

Figure 8: IV curve of a 300 Wp ZEBRA 60 cell module, as measured by Fraunhofer ISE Callab. The PL image of the module is shown on the right side.

Table II: IV average data for the n-IBC module.

Zebra Module	I_{SC} [A]	V_{OC} [V]	*Impp* [A]	*Vmpp* [V]	*Pmpp* [W]	*FF* [%]	η [%]
Average	9.94	39.24	9.23	**32.32**	**298.34**	**75.5**	**18.4**

4 CONCLUSION

The fabrication of more than 1000 ZEBRA cells was completed at ISC Konstanz. The best cell efficiency of approx. 22% was obtained. However the resulted median efficiency is about 21.4%. This is due to lower material quality and process variations as a result of manual handling and batch processing. The ZEBRA cells fabricated were used to assemble the first back-contact module with power output of approx. 300 Wp.

5 REFERENCES

[1] G.Galbiati et al., Large Area Back-Contact Back-Junction Solar Cell with Efficiency Exceeding 21%, IEEE Journal of Photovoltaics, Jan 2013. (DOI:10.1109/JPHOTOV.2012.2224848).

[2] http://www.helmholtz-berlin.de/projects/hercules/

[3]..www.eurotron.nl/public/img/ZEBRA%20press%20release %20-%2030-09-2015.pdf

IMPACT OF OPERATING TEMPERATURE AND ABSORPTION-LAYER THICKNESS ON ALL-BACK-CONTACT SOLAR CELL EFFICIENCY

Joseph E. O'Connor and Sherif Michael
Naval Postgraduate School, Department of Electrical and Computer Engineering / Space Systems Academic Group
1 University Circle, Monterey, CA 93940, USA
jeoconno1@nps.edu & michael@nps.edu
+1(831)656-2252 Office
+1(831)656-2760 Fax

ABSTRACT: The efficiency of all-back-contact (ABC), single-crystalline, silicon solar cells has increased steadily since Lammert and Schwartz introduced their novel interdigitated-back-contact design in 1975 [1]. This work explores the impact of high-temperature operation and absorption-layer thickness on performance by means of device simulations in Silvaco® Atlas™. We begin by establishing a benchmark model that matches the experimentally derived parameters of a physical solar cell. Then, we alter the model to determine the optimum absorption-layer thickness as well as the cell's performance at various temperatures. Finally, we combine both variables to simulate an optimally thin silicon ABC cell at elevated temperature in order to derive realistic performance characteristics for comparison.
Keywords: Back Contact, Solar Cell, Thermal Performance

1 INTRODUCTION

The development of high-efficiency silicon solar cell technology has advanced rapidly in recent years, with SunPower® manufacturing large-area cells that approach the generally accepted "maximum manufacturable efficiency" of 26% [2][3]. This trend is primarily due to improved fabrication processes and crystal growth techniques developed to meet demand for high-efficiency solar cells.

Among the many solar cell designs that show promise, the all-back-contact (ABC) design has achieved the highest efficiency for silicon solar cells under 1-sun illumination [2]. The key advantage of the ABC design is the placement of the emitter with its associated electrical contacts on the back of the cell, which improves optical and electrical performance. In traditional cell design, emitter contacts on the front of the cell produce shade and reduce efficiency. Consequently, designers are forced to balance a tradeoff between electrical grid coverage and serial-resistance power loss.

This work explores the impacts of varying operating temperature and absorption-layer thickness on an ABC silicon cell through modeling and simulation. Chapter 2 begins with a brief primer on solar cell modeling in Silvaco® Atlas™ (hereafter "Atlas"). Chapter 3 examines the impact of varying absorption layer thickness. Chapter 4 explores the relationship between operating temperature and open-circuit voltage (V_{OC}), and Chapter 5 discusses results and conclusions.

2 MODELING A SILICON ABC CELL

"Atlas is a [semiconductor] device simulator that predicts the electrical characteristics associated with [device] structures and bias conditions". The simulator is capable of modeling devices in two or three dimensions as shown in Figure 1 and Figure 2, respectively. Simulation provides insight into what is happening in a semiconductor device and displays theoretical knowledge in a visual platform [4].

At the core of the Atlas model are fundamental

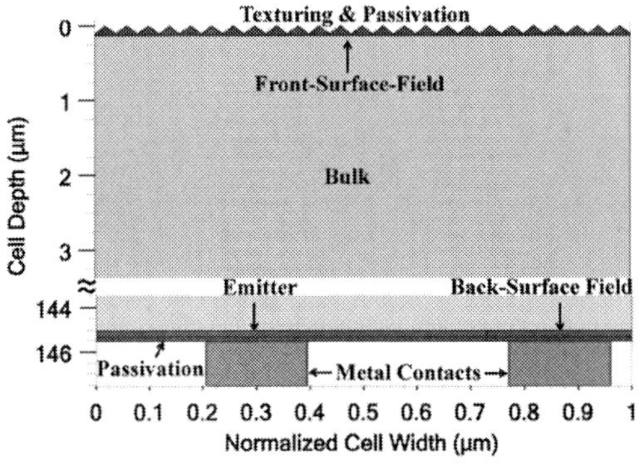

Figure 1: Two-dimensional Atlas model of an ABC solar cell. The emitter and back-surface field (BSF) contacts alternate periodically, and a single section is shown here.

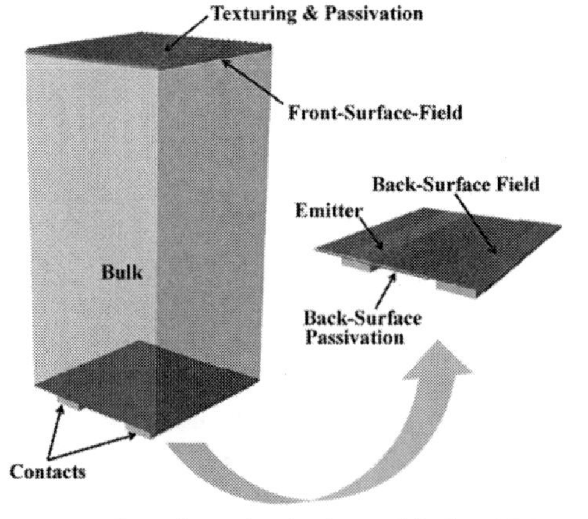

Figure 2: Three-dimensional Atlas model of an ABC solar cell with alternating emitter and BSF.

equations that account for carrier movement in semiconductor materials. During a simulation, Atlas solves differential equations at distributed nodes within the device structure to model the transport of charge carriers through a grid. Parameters such as the electric-field can be easily measured as shown in Figure 3.

A step-by-step approach to modeling a silicon ABC cell in Atlas is presented in [5]. The authors explore many design factors important for optimization and conclude that high-efficiency in ABC cells depends primarily upon device geometry, front and back-surface passivation quality, doping levels, semiconductor material quality, and minority carrier lifetime.

The silicon ABC design developed and optimized in [3],[6]-[8] has successfully addressed each of the design factors listed above to achieve very high efficiency for full-production cells. A recent manufacturing run of a large-area "generation III" solar cell produced a median efficiency of 23.6%. Details of a cell that achieved 25% efficiency were published in [3] and independently verified by the Fraunhofer Institute for Solar Energy Systems. Cumulative information from the references enabled an accurate two-dimensional model to be developed in Atlas as shown in Figure 1. Since all design specifications were not reported, we make no claim that our model is an exact representation; however, the simulation matched output parameters well as shown in Figure 4.

3 OPTIMAL ABSORPTION LAYER THICKNESS

The geometry of the ABC cell has many parameters that must be optimized. One of the most important parameters is absorption layer thickness. As bulk thickness is increased, minority carrier lifetime must increase proportionally; otherwise, bulk recombination increases and performance suffers.

Minority carrier lifetime (in seconds) due to Shockley-Read-Hall (SRH) and auger recombination in the solar cell bulk layer is defined as

$$\frac{1}{\tau_{bulk}} = \frac{1}{\tau_{SRH}} + \frac{1}{\tau_{auger}}. \tag{1}$$

SRH recombination due to the presence of impurities doping elements or unintentional manufacturing defects dominates below a certain doping concentration, while auger recombination dominates above a certain doping concentration. In high quality substrates, higher doping concentration may yield a higher V_{OC}; however, higher doping also increases auger recombination losses which decrease short-circuit current density (J_{SC}), therefore optimization is required to obtain maximum efficiency.

ABC cells collect carriers at the back of the cell; therefore, minority carriers must exist long enough to diffuse to the emitter without recombining; setting the base thickness equal to (or less than) the minority carrier diffusion length for a given lifetime is good design practice for ABC cells. This can be problematic though, since photogeneration occurs predominantly at the front of the cell as shown in Figure 5. To overcome this dillema, high quality substrates are utilized to increase carrier lifetime and diffusion length in order to keep recombination loss low. Although electrons have higher mobility and longer diffusion length than holes, it is easier to manufacture high-quality n-type wafers, which explains the trend towards n-type silicon ABC cells [5].

To determine the optimum absorption layer thickness for the cell depicted in Figure 1, we vary the thickness and evaluate the impact on J_{SC}. A direct correlation is established as shown in Figure 6, and we note that the

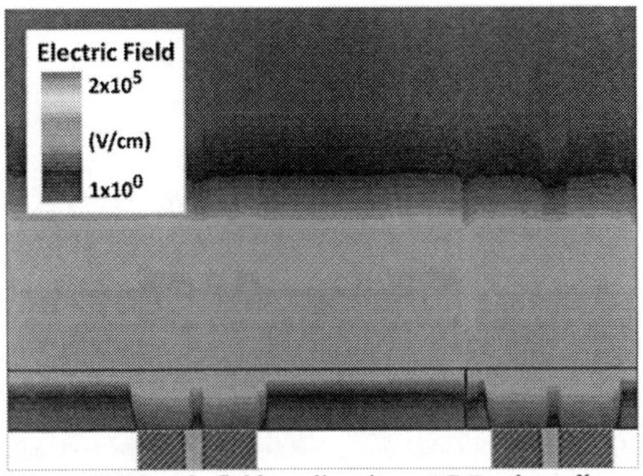

Figure 3: Electric field gradient in an ABC solar cell

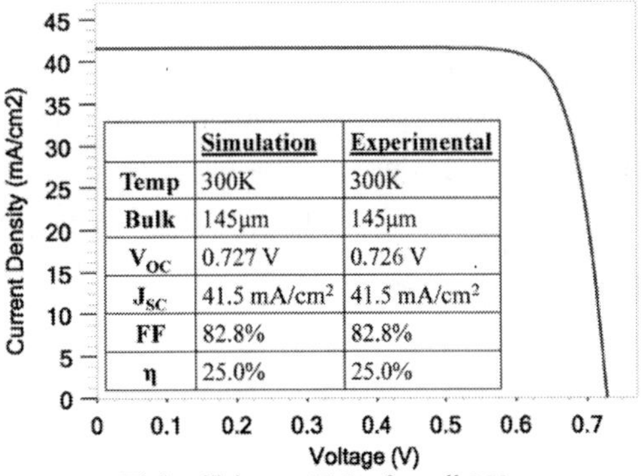

	Simulation	Experimental
Temp	300K	300K
Bulk	145μm	145μm
V_{OC}	0.727 V	0.726 V
J_{SC}	41.5 mA/cm²	41.5 mA/cm²
FF	82.8%	82.8%
η	25.0%	25.0%

Figure 4: High-efficiency ABC solar cell J-V curve

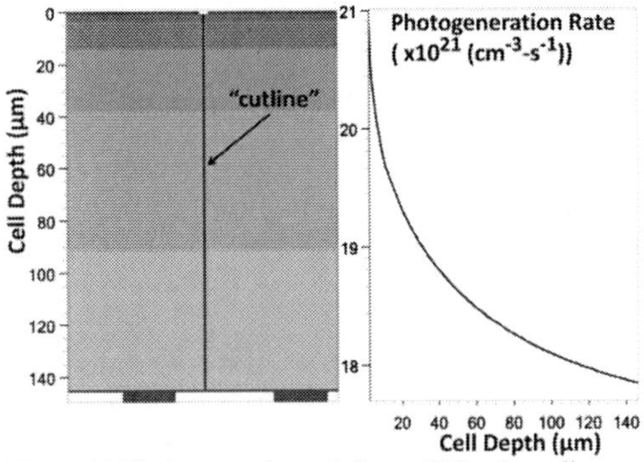

Figure 5: Photogeneration rate in an ABC solar cell

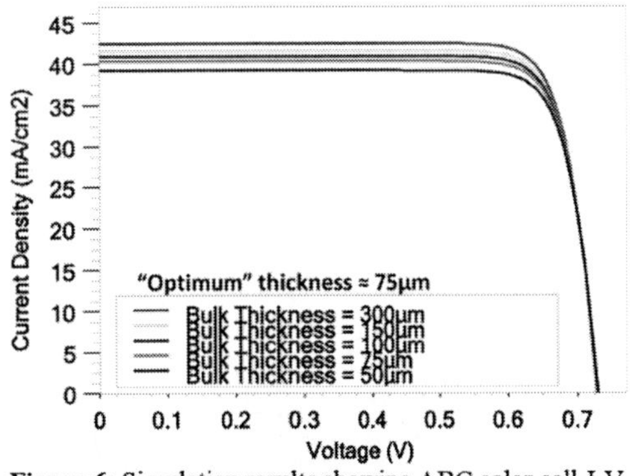

Figure 6: Simulation results showing ABC solar cell J-V curves as a function of bulk thickness. Optimum thickness is about 75μm (green line). The simulation was run at 300K under a simulated AM1.5 spectrum.

optimum thickness for this particular cell is about 75µm (green line), after which J_{SC} drops below 40mA/cm^2.

4 TEMPERATURE DEPENDENCE

Operating temperature is an important design consideration for solar cells due to its negative correlation to V_{OC}. Of the three primary metrics of a solar cell – V_{OC} , J_{SC}, and fill-factor (FF) – voltage is impacted most because of its inverse correlation to reverse saturation (dark) current, which increases with temperature [9-10]. Reverse saturation current is the current across a P-N junction in reverse bias, and is often referred to as recombination or "leakage" current, since it is the combined effect of carriers recombining on either side of a P-N junction.

Increasing the temperature of a solar cell reduces the bandgap of the semiconductor which in turn lowers V_{OC} on the order of a few millivolts per degree K. This is an important consideration since solar cells often operate well above room temperature due to the absorption of heat in the crystal lattice.

The temperature dependence of the bandgap E_G is described in [9] as

$$E_G(T) = E_G(0) - \frac{\alpha T^2}{(T+\beta)}, \tag{2}$$

where α and β are both material-dependent table constants. From (2) we note that E_G is lowered when the temperature increases. V_{OC} is dependent upon reverse saturation current density J_0 as

$$V_{OC} = \frac{kT}{q} \ln\left(\frac{J_{SC}}{J_0}+1\right). \tag{3}$$

Taking the derivative with respect to temperature we get

$$\frac{dV_{OC}}{dT} = \frac{V_{OC}}{T} + \frac{kT}{q}\left(\frac{1}{J_{SC}}\frac{dJ_{SC}}{dT} - \frac{1}{J_0}\frac{dJ_0}{dT}\right), \tag{4}$$

where

$$\frac{kT}{q} = V_{th}. \tag{5}$$

Combining (4) and (5), we obtain

$$\frac{dV_{OC}}{dT} = \frac{V_{OC}}{T} + V_{th}\left(\frac{1}{J_{SC}}\frac{dJ_{SC}}{dT} - \frac{1}{J_0}\frac{dJ_0}{dT}\right), \tag{6}$$

which demonstrates that V_{OC} is dependent on J_0, and J_0 is dependent on temperature. J_0 may then be calculated as

$$J_0 = qn_i^2\left[\frac{D_n}{L_n N_A} + \frac{D_p}{L_p N_D}\right], \tag{7}$$

where D_n and D_p are the diffusion coefficients for electrons and holes; L_n and L_p minority carrier diffusion lengths for electrons and holes; and N_D and N_A are the doping concentrations for donors and acceptors, respectively. Equations (6) and (7) demonstrate that diffusion coefficients and intrinsic concentration n_i show temperature dependence; however, n_i dominates temperature dependent parameters due to the squared term. n_i is defined as

$$n_i = 4\left(\frac{2\pi kT}{h^2}\right)^3 (m_e^* m_h^*)^{\frac{3}{2}} \exp\left(-\frac{E_G}{kT}\right), \tag{8}$$

where m_e^* and m_h^* are the effective masses of electrons and holes, respectively; and E_G is the bandgap. Simplifying (8), we get

$$n_i = BT^3 \exp\left(-\frac{E_G}{kT}\right), \tag{9}$$

where B is a constant of collected terms independent of temperature.

Substituting (9) into (7), we get

$$J_0 = CT^3 \exp\left(-\frac{E_G}{kT}\right), \tag{10}$$

where C is a constant of grouped terms. Now it is apparent that J_0 increases with temperature, which in turn lowers V_{OC} as we see from (3).

To summarize, increasing the temperature of a semiconductor increases the reverse saturation current density, which decreases V_{OC}. This phenomenon is related to a decrease in the bandgap of the semiconductor at higher temperatures. It can also be shown that increasing the temperature has an adverse effect on FF which further impacts the efficiency of a solar cell [10]. Finally, we note that the short circuit current density J_{SC} slightly increases with higher temperature, but the increase is small in comparison to the large decrease in V_{OC} [12]. Further information may be found in [9]-[12].

A visual representation will further clarify concepts described in this chapter. Figure 7 depicts normalized power of the AM1.5 solar spectrum as a function of wavelength. Shown also in Figure 7 is the spectrum response of the high-efficiency silicon cell model presented in Chapter 2. The bounded area between the functions represents the portion of spectrum power that contributes to heat in the crystalline lattice. The excess heat lowers the bandgap of the semiconductor which in turn reduces V_{OC}.

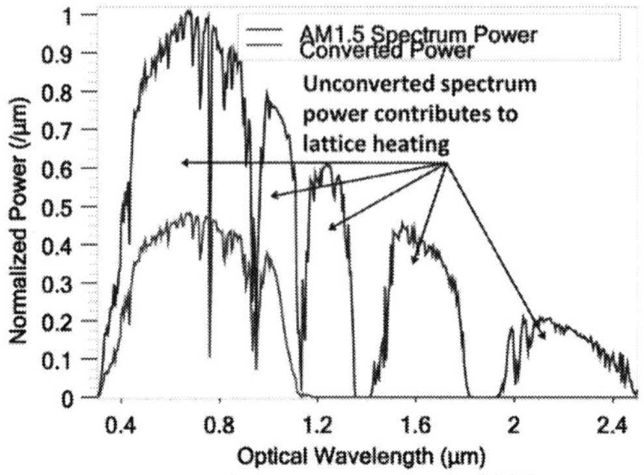

Figure 7: AM1.5 solar spectrum power and high-efficiency ABC silicon solar cell power conversion shown as a function of optical wavelength. Integrating each function and taking the difference gives the portion of spectrum power that contributes to lattice heating.

Now we apply the concepts to our model to show the adverse impact of varying temperature from 300K to 360K as shown in Figure 8. The temperature range correlates with experimental results obtained by the U.S. National Renewable Energy Laboratory (NREL) in [12]; however, peak temperature is extended from 340K to 360K to simulate performance in hotter climates. As expected, a direct correlation is established between temperature and V_{OC}. We note that the voltage drops about 1.8mV per degree Kelvin.

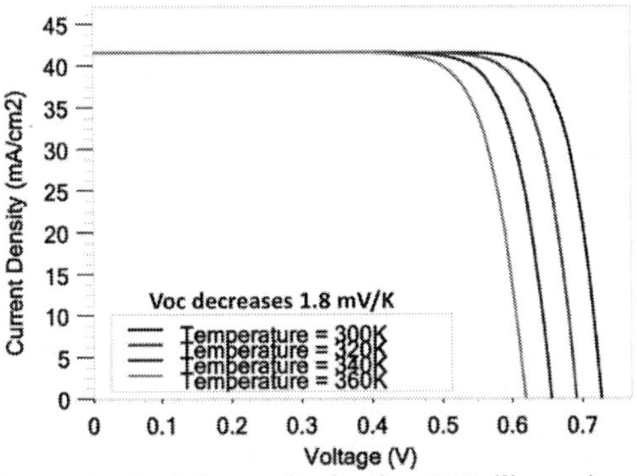

Figure 8: Simulation results showing ABC silicon solar cell J-V curves as temperature is increased through a typical operating range for terrestrial operation. Results indicate that *Voc* decreases by approximately 1.8mV per degree Kelvin under a simulated AM1.5 spectrum.

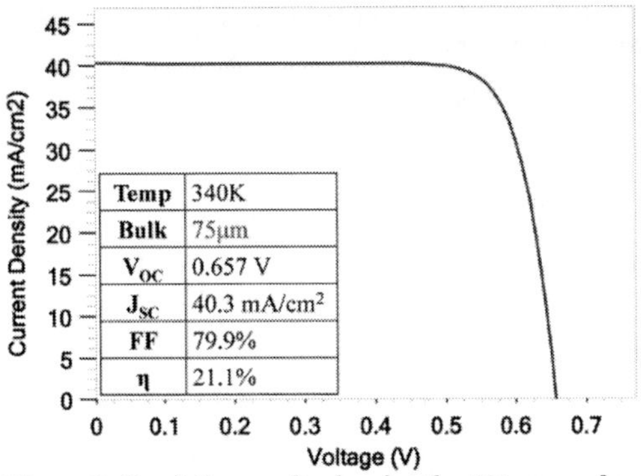

Figure 9: Simulation results showing the J-V curve of an ABC solar cell with "optimal" bulk thickness (75μm) at typical operating temperature under an AM1.5 spectrum.

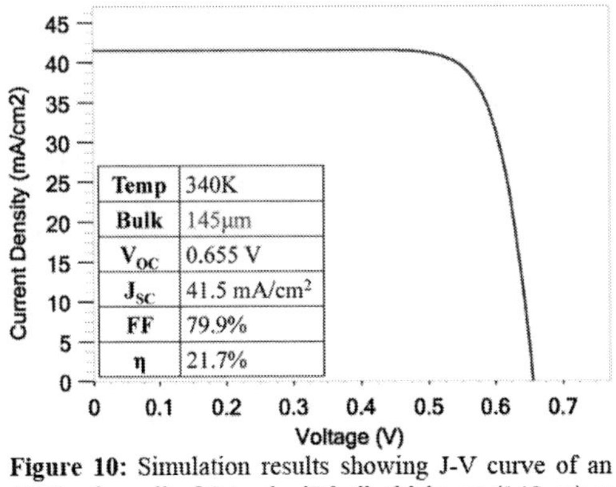

Figure 10: Simulation results showing J-V curve of an ABC solar cell of "standard" bulk thickness (145μm) at typical operating temperature under an AM1.5 spectrum.

5 RESULTS AND CONCLUSION

Now we incorporate the work from sections 3 and 4 to simulate an "optimally thick" high-efficiency silicon ABC cell operating to 340K. This temperature represents the highest recorded temperature of silicon cells in [12] where peak $T_{ambient}$ = 307K. We note from Figure 9 that V_{OC} and J_{SC} drop by 9.6% and 2.9%, respectively; while conversion efficiency decreases from 25% to 21.1%. Considering this scenario to be a worst-case scenario for cells operating at solar noon in a moderate climate, the performance is impressive. Running the same simulation for the 145μm bulk thickness reported in [3], Figure 10 shows that V_{OC} and J_{SC} drop by 9.9% and 0%, respectively; while efficiency decreases from 25% to 21.7%.

We have shown the effects of high temperature operation on silicon ABC cell performance, and evaluated the impact of reducing absorption layer thickness. Our simulation of an "optimally thin" silicon ABC cell at typical operating temperature provides realistic performance characteristics for comparison.

6 REFERENCES

[1] M. Lammert and R. Schwarz, "Silicon solar cells for high concentration applications," in *Electron Devices Meeting*, vol. 21, 1975, pp. 350-352.

[2] M. Green, K. Emery, Y. Hishikawa, W. Warta, and E. Dunlop, "Solar cell efficiency tables (version 46)," in *Progress in photovoltaics research and applications,"* vol. 23, no. 7, pp. 805-812, 2015.

[3] D. Smith, P. Cousins, S. Westerberg, R. Jesus-Tabajonda, G. Aniero, and Y. Shen, "Toward the practical limits of silicon solar cells," in *IEEE Journal of Photovoltaics*, vol. 4, no. 6, pp. 1465-1469, 2014.

[4] ATLAS™ User's Manual, Device Simulation Software, SILVACO® Inc., February 19, 2016.

[5] D. Diouf, J. Kleider, and C. Longeaud, "Two-dimensional simulations of interdigitated back contact silicon heterojunctions solar cells," in *Physics and Technology of Amorphous-Crystalline Heterojunction Silicon Solar Cells*, 1st ed. Chennai, India: Springer, 2012, ch. 15, pp. 483-519.

[6] R. Swanson, "The role of modeling in SunPower's commercialization efforts," in *Challenges in PV Science, Technology, and Manufacturing: A workshop on the role of theory, modeling and simulation*, Purdue University, West Lafayette, IN, 2012.

[7] D. Smith et al, "Generation III high efficiency lower cost technology: transition to full scale manufacturing," in *IEEE Photovoltaic Specialists Conference*, 2012, pp. 1594-1597.

[8] P. Cousins, D. Smith, H. Luan, J. Manning, T. Dennis, A. Waldhauer, K. Wilson, G. Harley, and W. Mulligan, "Generation 3: Improved performance at lower cost," in *IEEE Photovoltaic Specialist Conference*, 2010, pp. 275-278.

[9] S. Grover and J. Li, "Theory and analysis of temperature coefficient of open-circuit voltage in heterojunction solar cells," in *Proc. IEEE Photovoltaic Specialists Conference*, 2015, pp. 1-3.

[10]Photovoltaic Education. [Online]. Available: www.pveducation.org. [Accessed 13 March 2015].

[11] P. Singh and N. Ravindra, "Temperature dependence of solar cell performance – an analysis," in *Solar Energy Materials & Solar Cells*, pp. 36-45, 2012.

[12] T. Silverman, M. Deceglie, B. Marion, S. Cowley, B. Kayes, and S. Kurtz, "Outdoor performance of a thin-film gallium-arsenide photovoltaic module," in *Proc. IEEE Photovoltaic Specialist Conference*, 2013, pp. 103-108.

SCRAP RECYCLING IN AN ELECTROMAGNETIC CASTING FURNACE

Authors: José Manuel Míguez[1] (jmmiguez@ferroatlantica.es), Nicolas Pourade[2] (npourade@emix.fr), Florine Boulle[2] (fboulle@emix.fr), Gonzalo Varela[1] (gonzalo.varela@ferroatlantica.es), Ramón Ordas[1] (ramon.ordas@ferroatlantica.es)

[1]SILICIO FERROSOLAR. GRUPO FERROATLÁNTICA
Polígono Industrial de Sabón s/n, 15142 Arteixo (A Coruña) Spain.
Telephone: +34 981 647010; Fax: +34 981 602354

[2]EMIX. GRUPO FERROATLÁNTICA
Parc d'activités de la Croisière 23300 St Maurice. La Souterraine. France
Telephone: +33 (0)555 633 030; Fax: +33 (0)555 639 331

ABSTRACT: In this work we show the results of development of recycling silicon scrap in an Electromagnetic Casting Furnace. This EMC furnace has a number of advantages in terms of product costs: high growth rate, eliminating the need to use expensive quartz crucibles and graphite components but it also has advantages in terms of product quality due to the efficiency in carbon removal. Carbon and oxygen are measured along the column and in several cross sections of the column showing that these impurities concentrate in the sidewalls of the ingots. The efficiency in carbon removal is compared with some reported results on Directional Solidification furnaces. This study confirms the promising use of EMC technology for recycling scrap material.
Keywords: EMC Furnace, Scrap, Recycling, Polysilicon, SiC

1 BACKGROUND

The top of MC-Si ingot, approximately 10% in mass of the whole ingot, is highly contaminated by impurities such as iron, aluminum, Si_3N_4, and especially SiC, thus must be cut off as scrap. This scrap silicon should not be used as a charge for the next ingot because it will reduce the performance of the solar wafers. If inclusions and metallic impurities in the top-cut scraps can be removed, these top-cut scraps can become good feedstock for silicon PV cells. The current practice is to use the Directional Solidification furnaces to recycle this silicon scrap but this is an expensive process because it requires a quartz crucible which is a costly consumable.

2 DESCRIPTION OF THE MACHINE

The technology of producing multicrystalline silicon ingots by EMC method consists in induction heating and melting of silicon held in a water-cooled (cold) segmented copper crucible. Molten silicon is stirred due to the electromagnetic pressure difference and hence contaminating inclusions with higher electric resistance in the molten silicon are ejected to the molten surface and sidewalls where they are trapped. Then, by cutting 20 mm of sidewalls it is possible to remove up to 80% of SiC of the incoming material.

Figure 1: EMC furnace

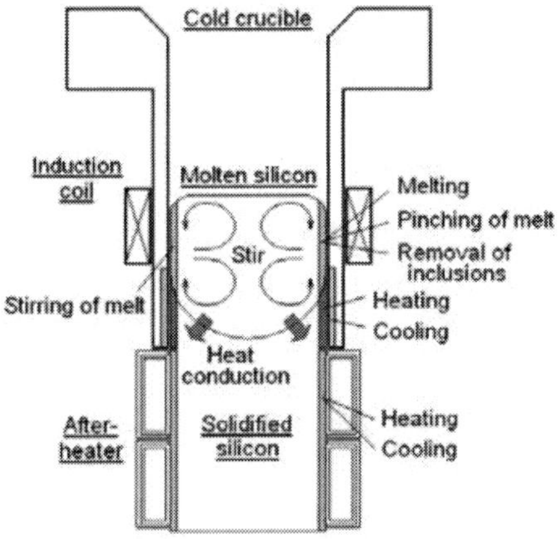

Figure 2: Effect of the electromagnetic in the liquid silicon[1]

3 RESULTS

3.1 Silicon scrap characterization (top cuts)

Silicon scrap surface is quite rough due to the presence of impurity inclusions, mainly in the form of fibers. The length of these fibers could be between 0,5 and 1,0 mm.

The elemental analysis of these fibers detects nitrogen. In order to separate these fibers, we used FH/NO_3H to dissolve the silicon matrix.

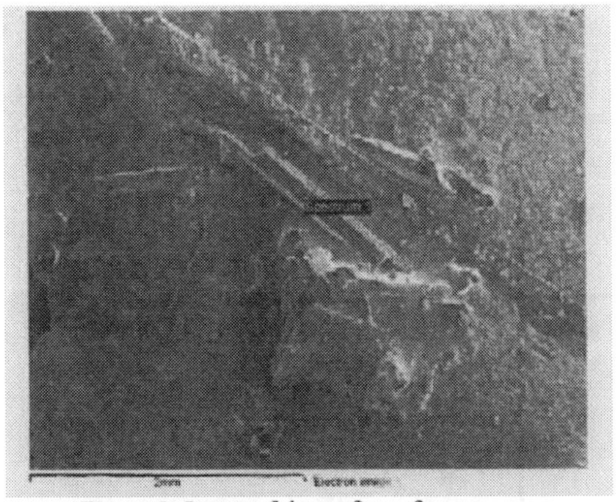

Figure 3: Image of the surface of top scrap

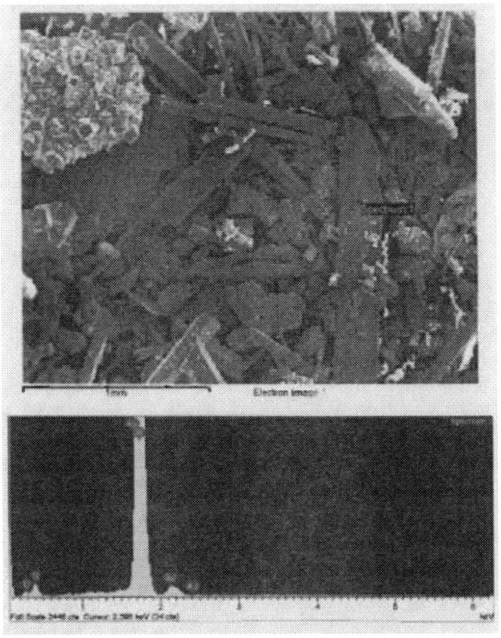

Figure 4: Image of fibers of silicon nitride

In table I, we show some chemical analysis of the dissolutions we did to obtain the prismatic particles of SiC and the fibers of silicon nitride. Top scrap discs normally have 15 mm thickness. Top A was taken from the top 7 mm of the slice and Top B comes from the other 8 mm. B13 is a pure center brick and B25 is the worst corner.

Table I: Analysis of acid dissolutions of top scrap

		Al	Ca	Fe	Ti	Cu
B13	Top A	4,30	2,60	5,60	0,20	1,97
B13	Top B	2,10	1,20	1,00	0,10	0,05
B25	Top A	110,80	20,10	67,00	3,70	12,93
B25	Top B	1,20	1,70	2,00	0,20	0,51

Figure 5: Image of the prismatic particles of SiC

3.2 Solidification in the EMC furnace

A total amount of 1750 kg of silicon scrap (tops from directional solidification of polysilicon) were grown in the EMC furnace with a copper crucible cross section of 340mmx340mm. This manner we obtained a column of 6,6 m in length that was cut in 11 ingots of 0,60 m. The last ingot is where all impurities are concentrated so that it should be excluded.

A representative sample of the charged silicon scrap was taken before it was melted and then we analyzed the carbon and oxygen content by Inert Gas Analysis. Due to the electromagnetic forces, we expected SiC inclusions to accumulate in the sidewalls (20 mm) so that the rest of material could be used as silicon feedstock to produce solar wafers. In order to check this hypothesis, we will take three silicon samples in a transversal cut of ingots 5, 7 and 9. These samples were cubic silicon pieces of 2 cm on each edge and were taken one in the border, one in the center of the ingot and the last one in the middle between the center and the border. In figure 6, we could see the position of the samples and the resistivity of the column.

In Table II, we could see the results of the carbon analysis. It seems clear that carbon concentrates in the sidewalls so it could be removed by cutting this part of the ingot. The final concentration in the good material could be estimated between 15-20 ppmw.

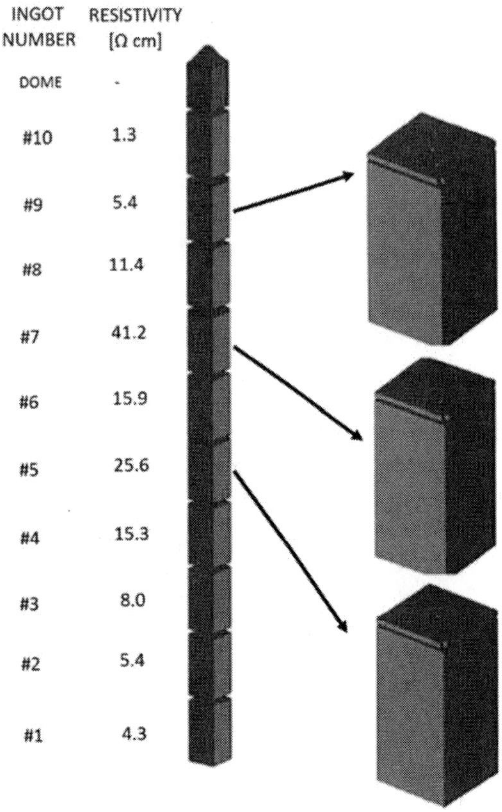

Figure 6: Representation of the silicon column made in the EMC furnace. It is also shown the position of the samples for analysis and the resistivity of each ingot.

Table II: Carbon distribution in the column

	Carbon (ppmw)		
	Center	**Intermediate**	**Sidewalls**
Ingot 5	15,0	17,0	34,0
Ingot 7		29,0	93,0
Ingot 9	14,0		43,0

As far as the oxygen is concerned, the concentration in ingots 5 and 7 are quite similar but there is an increase in the concentration at the end of the column (see Table III).

Table III: Oxygen distribution in the column

	Oxygen (ppmw)		
	Center	**Intermediate**	**Sidewalls**
Ingot 5	5,4	5,1	5,2
Ingot 7		6,9	6,9
Ingot 9	21,0		46,0

Table IV: Calculation of carbon distribution between the side cuts and usable silicon

	Si mass	**C [ppmw]**	**C distribution**
side cuts EMC	22%	56,7	46,0%
usable column	78%	18,8	54,0%

In order to have reference values to provide context for this information, we will see the level of concentration of C and O after growing polysilicon in a directional solidification furnace.

	Carbon	Oxygen	Nitrogen
7N Si feedstock	<0.5	<1.0	0.03
5N Si feedstock	37	200	0.55
7N corner	55	11	1.3
7N center	71	8.0	1.1
7N edge	54	7.8	0.96
7N:5N corner	41	16	1.1
7N:5N center	49	18	1.4
7N:5N edge	49	13	1.2
Typical SIMS detection limits	0.5	1.0	0.02
SEMI PV Si wafer specification	100	80	no spec

Figure 7: Carbon and oxygen distribution after processing different silicon qualities in the Directional Solidification furnace[2]

In figure 7, it is shown a study carried out by EAG in which they used polysilicon (7N material) and UMG silicon (5N material) to compare the C and O concentrations before growing and after growing. The result of the study was that there was a pollution of C and O inside the furnace. The mechanism of such contamination starts with the quartz crucible that adds oxygen to the liquid silicon. Therefore, there is an increase in the oxygen concentration so that it will eventually generate SiO which reacts with the carbon parts inside the furnace. The chemical product of this reaction is CO that goes back to silicon increasing the C level in the liquid.

In relation to the metallic impurities content, in table V we show the improvement from the raw material (polysilicon top scrap) and the purified silicon after growing in the EMC furnace.

	Al	**Ca**	**Fe**	**Ti**
Raw ICPMS	1,75	1,40	4,50	0,20
Purif GDMS	0,08	<0,1	<0,05	<0,005

Table V: Chemical analysis of raw material and final silicon. Samples from final silicon were taken from ingot 5

4 SUMMARY

As a conclusion of this study and according to the exposed results, the EMC furnace is a promising technology for the purification of scrap silicon due to its lower operation cost and higher productivity in comparison with the current procedure in the Directional Solidification furnaces. It shows a purification efficiency of around 50% in the carbon content but there is also purification in the metallic content which are necessary to reuse the material as feedstock for solar wafers.

5 REFERENCES

[1] K. Kaneko, K. Nakagawa, T. Onizuka, K. Sasatani, N. Kubo, M. Kida, SUMCO SOLAR Corporation, Proceedings 16th Workshop on Crystalline Silicon Solar Cells and Modules: Materials and Processes, (2006), 8

[2] R. S. Hockett, Evans Analytical Group, Proceedings 18th Workshop on Crystalline Silicon Solar Cells and Modules: Materials and Processes, (2008), 55

RECENT SOLAR CELL RESULTS USING UMG-SILICON

Eugenia Zugasti, Jon Armentia, Mikel Ezquer, Miguel Murillo, María Jesús Rodriguez, Ana Rosa Lagunas
NATIONAL RENEWABLE ENERGY CENTRE (CENER)
C/ Ciudad de la Innovación 7, 31621 Sarriguren (Navarra), Spain
Telephone: +34 948 25 28 00; Fax: +34 948 27 07 74

ABSTRACT: The developments set out on the fabrication route to produce upgraded metallurgical silicon (UMG-Si) have led to an optimized material, whose results are promising when used to process standard Al-BSF solar cells. Therefore, this UMG-Si material can be considered as a real and feasible alternative to produce industrial wafer-based silicon solar cells at lower cost and energy consumption. The objective of this study is to compare the performance of different solar cells starting from UMG-Si material, in order to establish the factors influencing their behavior. For this purpose, solar cells based on UMG-multicrystalline wafers with different resistivity values, height in the ingot and compensation levels have been processed. At last, the analysis of the results obtained from the characterization of the material properties (lifetime and resistivity) and from the complete characterization of the final solar cells (EL, IV-Curve and SR) is presented.
Keywords: metallurgical silicon, multi-crystalline, silicon solar cell

1 INTRODUCTION

UMG-Si is obtained by direct purification of metallurgical grade silicon without chlorine in contrast to the complex gas phase conventional Siemens route. Therefore, in comparison with standard polysilicon, this material is influenced by dopants naturally present in silicon, like boron and phosphorus, and metals [1]. A purification method to minimize the contaminants quantity in UMG material has been developed in latest years. This improved UMG silicon is able to be used in efficient crystalline silicon solar cells.

In this study, wafers from ingots obtained in three stages during the development of UMG silicon purification route have been considered. Apart from this, wafers come from different ingot heights. In this way, the variation within the ingot height as well as among ingots can be analyzed by means of IV, EL and SR measurements.

2 EXPERIMENTAL

2.1 UMG Silicon substrates

The substrates used in this study are 156 mm x 156 mm multicrystalline wafers which come from three ingots with different nominal resistivities. The ingots have been grown in a directional solidification system using various types of UMG Silicon from a supplier, as their fabrication route has progressed.

In Table I, resistivity and UMG Silicon GEN is shown for each ingot.

Table I: P-type multicrystalline silicon ingots selected for this study

Ingot Name	Nominal Resistivity (Ω·cm)	Generation
A	1	Gen1
B	0.6	Gen2
C	1.5	Gen3

The wafers used are extracted from 5 heights of each ingot labeled as follows: top, middle-top, middle, middle-bottom and bottom.

2.2 Solar Cell Processing

A standard Al BSF solar cell structure and screen printing technology was applied to the p-type multicrystalline wafers from UMG-Si feedstock, as it is depicted in Figure 1. This standard process sequence has been performed in a pilot fabrication line with efficiency ranging between 16.5% and 17% in standard multicrystalline silicon solar cells.

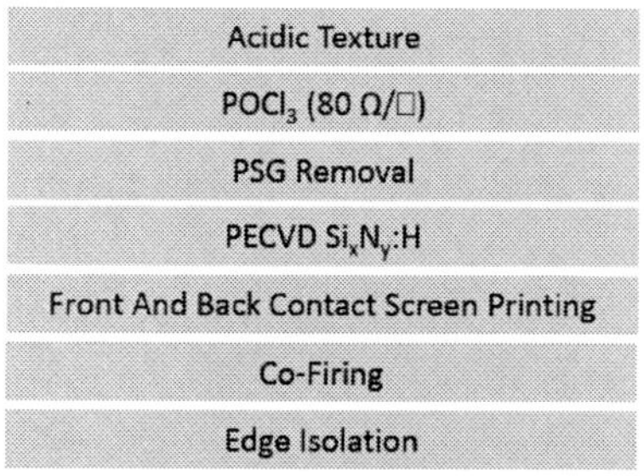

Figure 1: Solar cell processing sequence

Wafers are textured with an isotropic acidic solution and cleaned. Then, an 80 Ω/□ emitter layer is formed by a $POCl_3$ diffusion process. After phosphorus glass removal and cleaning, hydrogen-rich silicon nitride (SiN_x:H) is deposited by plasma enhanced chemical vapour deposition (PECVD). The metallization is performed by screen printing of a full Al rear contact and a Ag front grid. Finally, the contact is formed by firing and edge isolation is obtained through laser scribing.

3 RESULTS AND DISCUSSION

3.1 Substrate Characterization

In a first step, the resistivity of all the wafers involved in the study was measured and the obtained results are depicted in Figure 2.

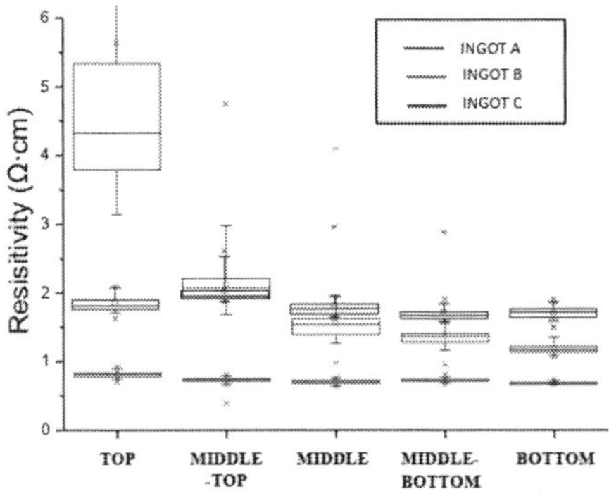

Figure 2: Resistivity of the different substrates used in this study

In ingot A, wafer average resistivity, ρ, increases from 1.0 Ω·cm, at the bottom, to 4.3 Ω·cm, at the top, showing the largest variation among the three ingots. This effect can be explained by the different segregation coefficients of dopant contaminants, boron K_B=0.8 and phosphorus K_P=0.35 [2], which leads to an increased concentration of phosphorus at the top of the ingot. The consequence is that motilities, net doping and net charge carrier concentration decrease with ingot height, and so the resistivity [3]. Nevertheless, in ingots B and C, this variation has been reduced to values below 0.3 Ω·cm after improvements of UMG-Si purification and compensation procedure, as shown in Table II.

Table II: Average resistivity and its variation along the three ingots

Ingot Name	Average ρ (Ω·cm)	(Max. ρ –Min. ρ) along ingot height (Ω·cm)
A	2.0	3.3
B	0.7	0.1
C	1.7	0.3

In a second step, lifetime spatial distribution of the wafers after texturization, see Figure 3, is also measured using a SEMILAB WT2000.

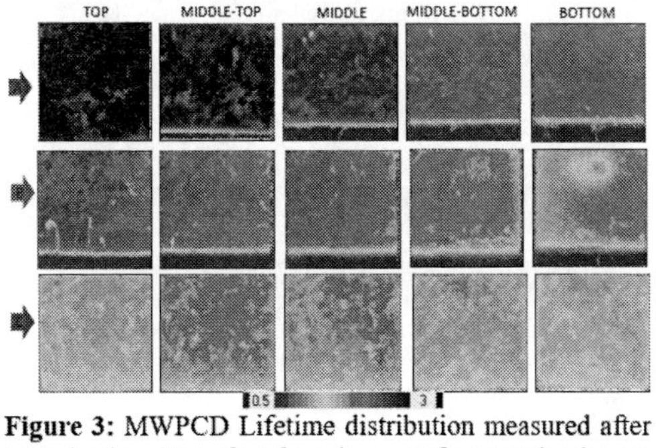

Figure 3: MWPCD Lifetime distribution measured after texturization. Note that there is no surface passivation on the samples

In ingots A and B, there are some "red zones", located at the lower parts of the maps, which correspond to lower lifetime regions. These reduced lifetime can be atributed to metalic contaminations diffused from the crucible walls during crystallization process. Other variations of lifetime are mainly related to crystallization defects and segregation of impurities. Apart from this, and in general, higher lifetimes are observed in the more compensated regions due to reduction in the free carrier concentrations so, in these wafers, there is a correspondence between higher resistivity values and lifetime values [4].

3.2 Final solar cell characterization

After lifetime and resistivity measurements of the wafers, the solar cells have been processed and characterized. In addition to standard I-V curves, electroluminescence (EL) and spectral response (SR) techniques have been used.

3.2.1 IV Curve

Solar cells from the three ingots have been characterized by means of I-V curve and their main electrical parameters, V_{oc}, J_{sc} and efficiency, extracted. In addition, and, as a reference, a multicrystalline solar cell, starting with conventional polysilicon and fabricated with the same process sequence has been included. The electrical parameters of this multicrystalline solar cell are included in Figure 4, Figure 5 and Figure 6, below, as a baseline. From there, the differences are calculated as a Δ of each parameter, defined as the value of each solar cell minus the value of the reference divided by the value of the reference, in percentage. In this way, the influence of the manufacturing processes will be minimized and the relative differences between ingots and the reference are highlighted.

In Figure 4, the open circuit voltage difference between each cell and the multicrystalline reference is shown for the three ingots. In the case of ingot A, the open circuit voltage V_{oc} decreases from bottom to top ingot height, reaching the value obtained by the reference solar cell at the bottom part. This trend might be due to the decreasing material quality and to the higher structural defect density at the top of the ingot, originating active recombination centers [5]. Also, as seen in Table III, ingot A suffers some non-homogeneity. For example, ingot B has a much more stable open circuit voltage along the ingot height, sign of a better material and, Ingot C, apart from its homogeneity, has higher values of V_{oc} than the others, demonstrating the improvement of the material quality too. In addition, the performance in terms of V_{oc} is very similar, only 0.4% below, respect to the results obtained in the reference multicrystalline solar cell.

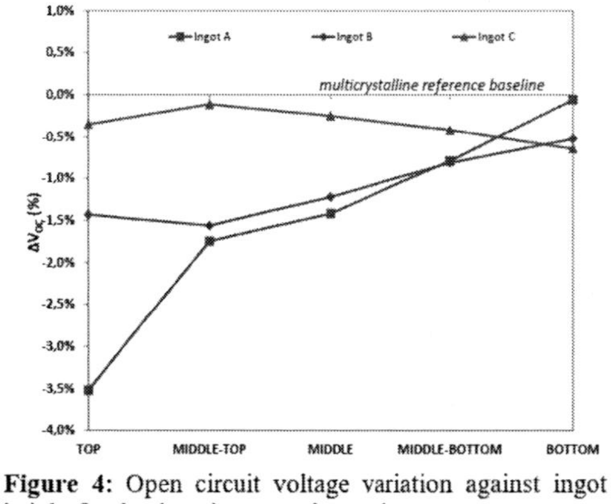

Figure 4: Open circuit voltage variation against ingot height for the three ingots under study

Table III: Average variation of open circuit voltage along the three ingots respect to the reference

Ingot Name	Average ΔV_{oc} along ingot height (%)	(Max. ΔV_{oc} – Min. ΔV_{oc}) along ingot height (%)
A	1.5	3.5
B	1.1	1.0
C	0.4	0.5

Regarding short circuit current density, Figure 5 shows that this parameter increases when going to the bottom area, especially for ingot A and B, while ingot C has a more homogeneous behavior along the ingot height, see Table IV. Furthermore, the difference between the reference and UMG-Si short circuit current density has been minimized, from 7.9% and 9.6%, for ingots A and B, up to around 5.2% in ingot C, with a deviation from top to bottom up to 0.8% in this case.

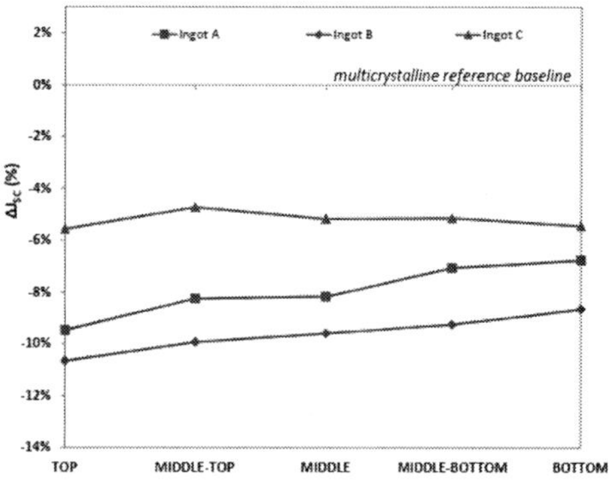

Figure 5: Short circuit current density variation against ingot height for the three ingots under study

Table IV: Average variation of short circuit current density along the three ingots respect to the reference

Ingot Name	Average ΔJ_{sc} along ingot height (%)	(Max. ΔJ_{sc} – Min. ΔJ_{sc}) along ingot height (%)
A	7.9	2.7
B	9.6	2.0
C	5.2	0.8

The efficiency shows the same tendency observed in open circuit voltage and short circuit current density, as depicted in Figure 6. An efficiency improvement in ingot C of around 5%, averaged along the ingot height, is obtained with regard to ingots A and B and its homogeneity is improved see Table V.

The relative difference in efficiency between the reference multicrystalline solar cell and the solar cells from ingot C has been reduced to around 5.6%. Note that the UMG solar cells have been manufactured with the standard process developed for the reference multicrystalline solar cell starting from conventional polysilicon (not optimized for the new material).

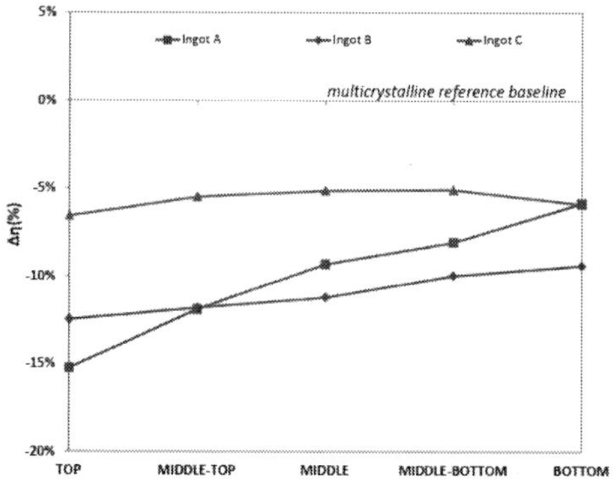

Figure 6: Relative efficiency variation against ingot height for the three ingots under study

Table V: Average variation of efficiency along the three ingots respect to the reference

Ingot Name	Average $\Delta\eta$ along ingot height (%)	(Max. $\Delta\eta$ – Min. $\Delta\eta$) along ingot height (%)
A	10.1	9.4
B	11.0	3.1
C	5.6	1.5

3.2.2 EL measurement

The electroluminescence (EL) images show the spatially resolved intensity of photo-electrons per second which are proportional to the emitted photons.

In Figure 7, reverse EL images (IBIAS = -3.0A), shown in red, have been superposed to direct EL images, in grey scale, for several solar cells in order to localize the breakdown voltage points.

In direct EL images it has been observed that ingot A has more dark zones than ingot B, and ingot C is the one with fewer dark zones. This outcome can potentially mean a better material quality as it is also translated into less severe breakdown points in reverse EL imaging. This effect will be discussed later on.

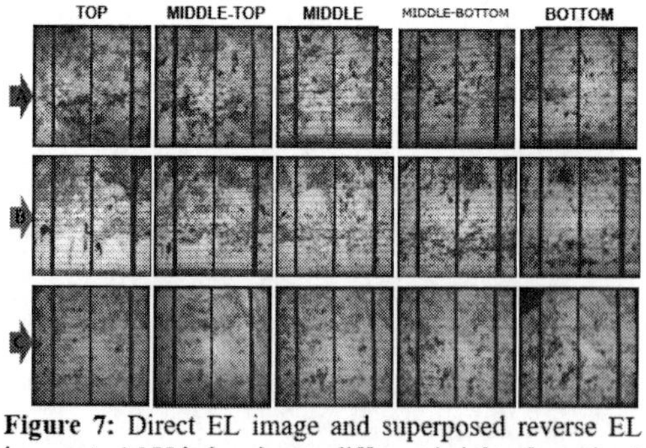

Figure 7: Direct EL image and superposed reverse EL image at -16 V belonging to different heights from the 3 ingots

There are different breakdown mechanisms which affect solar cells. In UMG material two can mainly be highlighted: defect-induced and avalanche breakdown mechanisms [6].

Defect-induced type correlates with recombination-active crystal defects and they showed up at around -12 V. Net-doping concentration of B and P, Fe and other metal precipitates are associated to this type of breakdown [6] thatis related with the red points depicted in reverse-bias EL at that voltage and corresponds to dark zones shown in forward bias EL. Figure 8 shows a zoom of a defective area of a cell from the middle-bottom height of ingot C.

The avalanche breakdown mechanism are caused by etch pits and dislocations and they occur at bias voltage beyond -12 V [6].

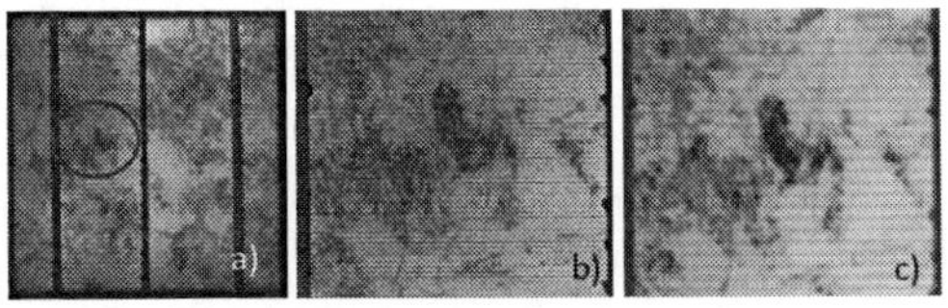

Figure 8: a) direct EL of the whole cell at 7 A, b) direct EL zoom, c) direct and reverse EL zoom superposed; in red are defect-induced breakdown mechanism points at -14 V. Both zooms correspond to the red circle zone depicted in a).

In figure 9, a plot of the reverse IV curves of ingot A, B and C solar cells is shown. As it can be observed, each solar cell breaksdown at different voltage. The mean values for each type of material are tabulated in Table VI. ingot C arrives to higher breakdown voltage (-17 V) than ingot A and B, which means a consistent improvement of the UMG material.

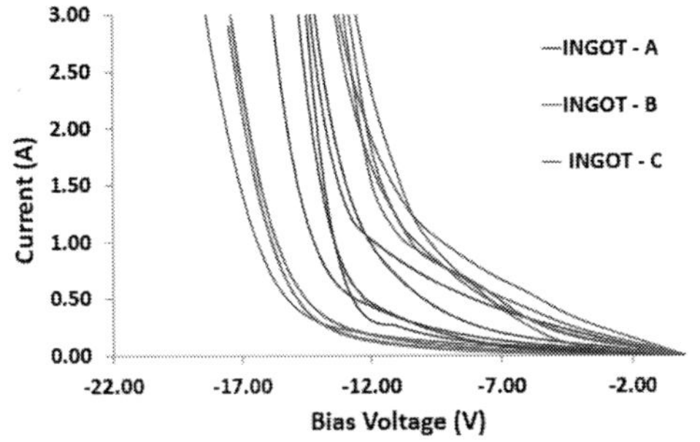

Figure 9: Reverse I-V of the cells from different heights of the three ingots.

Table VI: Average breakdown voltage in the three ingots

Ingot Name	Average $V_{breakdown}$ along ingot height (V)	(Max. $V_{breakdown}$ – Min. $V_{breakdown}$) along ingot height (V)
A	-14.2	1.7
B	-12.1	1.0
C	-17.0	0.9

3.2.3 Spectral Response measurement

Local measurements of the spectral response have been carried out in several areas of each solar cell where performance is known to be different deduced from EL images. Two areas of each solar cell, one brighter and another darker (EL), have been chosen to perform the analysis. In Figure 10, Figure 11 and Figure 12, the spectral response for ingots A, B and C including the 5 heights are shown respectively. Dashed curves correspond to worse areas and solid curves the better areas.

This analysis has been performed to identify the worst zones existing in the substrates and their influence in the final performance of the solar cell.

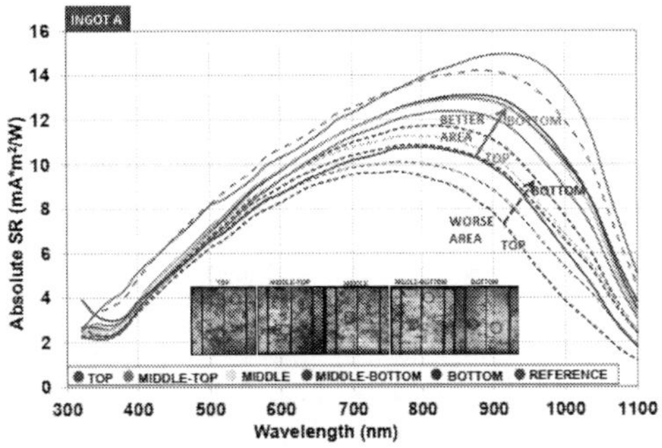

Figure 10: Spectral response obtained in different areas of the solar cells from ingot A

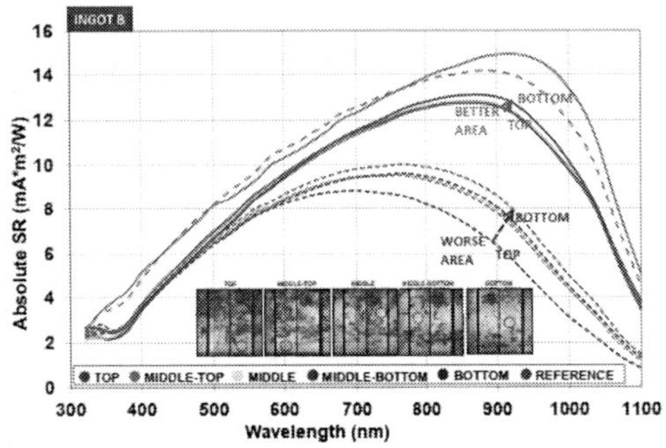

Figure 11: Spectral response obtained in different areas of the solar cells from ingot B

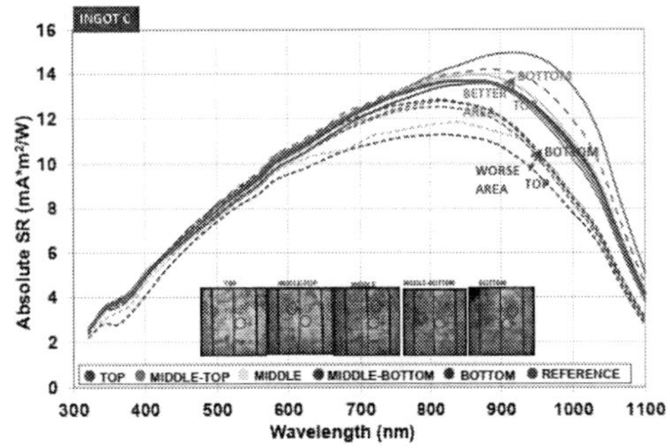

Figure 12: Spectral response obtained in different areas of the solar cells from ingot C

The areas with higher defects density and/or contaminants (darker areas) have a significantly lower spectral response in the long wavelength's range. The measurements show that, in general, at the ingot top, there is a lower spectral response in both, better and worse areas, in the wavelength range between 800 nm to 1100 nm. Besides, the spectral response improves in this wavelength range from ingot A to C.

The differences of the photogenerated current have been quantified and integrated in this wavelength range for the three ingots and the results are summarized in Table VII.

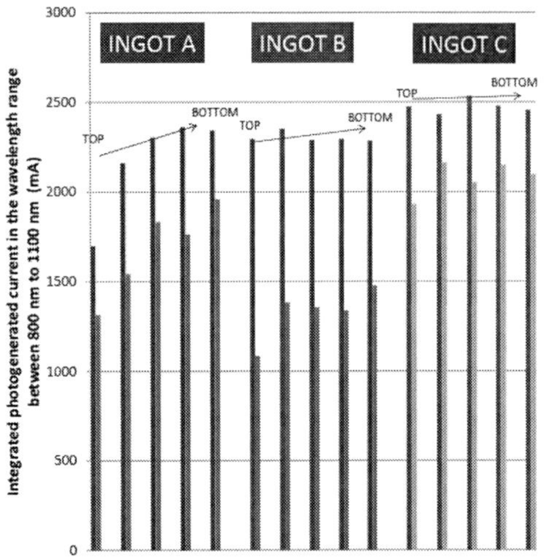

Figure 13: Total photogenerated current in the wavelength range between 800 nm to 1100 nm, taking into account better and worse areas localized in each solar cell

Table VII: Averaged photogenerated current, I_{ph}, in the wavelength range between 800 nm to 1100 nm, taking into account better and worse areas localized in each solar cell

Ingot Name	Average I_{ph} along ingot height (mA)	(Max. I_{ph} – Min. I_{ph}) along ingot height (mA)
A	1929	1047
B	1816	1266
C	2277	603

Comparing all the measurements, ingot C exhibits an improved spectral response compared to ingots A and B and its results are much more homogeneous along the ingot height.

4 CONCLUSIONS

The different GEN of UMG silicon feedstock under study has improved in quality as a consequence of the development of its purification technology and compensation procedure.

The main electrical parameters, V_{oc}, J_{sc} and η, extracted from the IV curves of UMG-Si multicrystalline solar cells have dramatically reduced their gap compared to conventional polysilicon multicrystalline solar cells manufactured with the same process. These results are congruent with the results obtained in electroluminiscence and spectral response, which show and improvement of the material.

Regarding homogeneity, initially, in ingot A, the performance and electrical characteristics varied obviously from top to bottom, being the bottom solar cells the ones which exhibited better results. However, uniformity has increased in ingot B and clearly more on ingot C. This improvement has an important impact on the usable fraction of the ingot produced.

At last, as a conclusion, and according to the exposed results, UMG-Si material improvement is, nowadays, a real alternative to process efficient silicon solar cells with a lower cost.

5 REFERENCES

[1] D. Kohler, B. Raabe, S. Braun, S. Seren, G. Hahn, in 24th European Photovoltaic Solar Energy Conference, Hamburg, pp. 1758-1761 (2009)

[2] F. A. Trumbore, The Bell System Technical Journal, 39, 1, pp. 205-233 (1960)

[3] F. Schindler, M. C. Schubert, A. Kimmerle, J. Broisch, S. Rein, W. Kwapil, W. Warta, Solar Energy Materials & Solar Cells, 106, pp. 31-36 (2012)

[4] D. Macdonald, A. Cuevas, Journal of Applied Physics, 109, 043704 (2011)

[5] R. Bakowskie, G. Kesser, R. Richter, D. Lausch, A. Eidner, P. Clemens, K. Petter, Energy Procedia. 27, pp. 179-184 (2012)

[6] O. Breitenstein, J.Bauer, K. Bothe, W. Kwapil, D. Lausch, U. Rau, J. Schmidt, M. Schneemann, M. Schubert, J. M. Wagner, W. Warta, J. Appl. Phys. 109, 071101 (2011)

NEUTRON ACTIVATION ANALYSES (NAA) INVESTIGATION OF TRANSITION-METAL IMPURITIES CONTENTS IN SOLAR-GRADE SILICON FEEDSTOCK FOR DIRECTIONAL SOLIDIFICATION OF PHOTOVOLTAIC HEM SILICON INGOT

Y. Chettat[1], L. Hamidatou[2], M. Salhi[2], H. Slamene[2], A. LAMI[1], A. Benmounah[3]
[1]Centre de Recherche en Technologie des Semi-Conducteurs pour l'Energétique (CRTSE). Division CCPM.
02 Bd, Frantz Fanon-Alger, B.P. N°140, Alger, Sept Merveilles-16200, Algérie,
[2]Centre de Recherche Nucléaire de Birine (C.R.N.B). B.P. : 180 Ain Oussera - Djelfa, Algérie,
[3]Unité de Recherche en Matériaux, Procédés et Environnement (UR–MPE). Université M'hamed Bougara, Avenue de l'Indépendance, 35 000, Boumerdès, Algérie.

ABSTRACT:

The degradation of multicrystalline silicon (mc-Si) Lifetime is mainly caused by defects such as: precipitates, inclusions, dislocations and transition-metals impurities simply known by traps or lifetime killers. The main source of theses contaminants are diffusion from solar silica crucible, silicon nitride coating and solar grade silicon feedstock starting material. Some impurities such Ta, W, V, Nb and Fe are very dangerous and can degrade the efficiency of the silicon solar cell at very low concentrations. Therefore, a strict control of the silicon feedstock material specifications via using very high sensitive analytical technique for traces and ultra-traces analyses is more than necessary. In this work we have used an advanced high sensitivity and multi elementary analytical technique such Neutron Activation Analyses (NAA) which require no chemicals treatment for sample preparation in attempt to determine exactly which transition-metal impurities and with which concentrations they are present in three different batch samples of solar grade silicon feedstock supplied by Silicon Recycling Services (SRS) supplier, and three other different standards mc-Si wafers from the edge of the silicon ingot obtained by HEM crystallization of thereof solar grade silicon feedstock. We have used also a lifetime tester type Sinton WCT-120 in mode QSSPC to investigate the electrical quality of the heart of the silicon ingot. The obtained results show that silicon feedstock material and mc-Si ingot (wafers) are heavily contaminated by transition and alkaline metals impurities. The main dominates metallic impurities found in the silicon feedstock material are : Mo, Nd, Zr, W, Cr, Fe, Co, Zn, Au, Sc, Ce, La, Th, Rb, and Ba, Ca, Na, K with concentrations exceeding thousand-folds their-sholds concentrations in silicon feedstock material, and the dominates metallic impurities found in mc-Si wafers are: W, Fe, Cr, Co, Ag, Sc, Ce, La, Hf with concentrations exceeding hundred-folds their-sholds in mc-Si type P for photovoltaic application, and that the lifetime seems to be remarkably degraded with only 4,8 µs max in the heart of the ingot. This degradation is mainly caused by the intense transition metal-contamination of silicon feedstock material.
Keywords: Solar-grade silicon feedstock, HEM growth, silicon ingot, Multicrystalline silicon wafer, transition-metal, alkaline-metal, impurity, Neutron activation Analyses (NAA), diffusion and segregation.

1 INTRODUCTION

Solar grade silicon feedstock material is used in photovoltaics (PV) as the starting material for manufacturing both monocrystalline and multicrystalline as well as for thin film silicon modules. More than 98% of the annual solar cells production is based on crystalline silicon particularly mc-Si wafers doped type P. Therefore, silicon is the most important material for PV industry today. The challenge which the PV-industry is currently facing to is to decrease the manufacturing costs and at the same time increase the efficiency of solar cell to reach the grid parity. To achieve this goal many venues are followed among them the development and using sufficiently cheap ultra-pure solar grade silicon feedstock material for manufacturing silicon ingots via using different routes and technologies permitting to achieve the desired criteria for solar grade silicon feedstock in terms of cost and purity. The principal quality of the solidified silicon ingot is minority carrier lifetime. The degradation of mc-Si silicon lifetime is mainly caused by defect precipitates, inclusions, dislocations, and essentially transition-metals impurities simply known by traps or lifetime killers such as: Ta, Mo, Nd, Zr, W, Cr, Fe, Co, Zn, Ag, Au, Al, Ti, Cu, etc. which cause additional recombination or by scattering-induced mobility loss [1]. The main source of these contaminants is often diffusion from the silica solar crucible and the silicon nitride coating during the crystallisation and essentially solar-grade silicon feedstock starting material. Some impurities such Ta, W, V, Nb and Fe are very dangerous in such way that, the presence of each one of them in the bulk as-cut silicon wafer in concentration as small as 1 ppb ($\sim 10^{11}$ atm/cm^3) could lead to a reduction of 50 % of its nominal efficiency[02]. For these reasons the maximum concentrations of most of these impurities in the solar-grade silicon feedstock must not exceed the value of 10^{13} atm/cm^3 (1 ppb) each [03]. Such very low concentrations cannot be detected by physical technique such SIMS, FTIR and Deep Level transient Spectrometry (DLTS) [1]. In the last few years many scientific works have been reported in this topic, all them are monitoring that using low purity solar grade silicon feedstock intensely contaminated by transition metallic impurities such as: Fe, V, W etc. lead to a significant degradation of the Lifetime quality among the hearth of the ingot and that using both higher purity solar grade silicon feedstock and higher purity silicon nitride coating-crucible system lead to a significance improvement of the whole silicon ingot both the center and the edge of the ingot [4, 5, 6]. Therefore, a strict control of the silicon feedstock material specifications using very high sensitive analytical technique is more than necessary.

In this work we have used a powerful advanced high

sensitivity and multi elementary analytical technique such Neutron Activation Analyses (NAA) which require no chemicals for sample preparation in attempt to determine exactly which metallic impurities of the above mentioned contaminants, and with which concentrations, they are present in three different batch samples of solar grade silicon feedstock supplied by Silicon Recycling Services (SRS) supplier and three different standard multicrystalline silicon (mc-Si) wafers doped type P obtained from the edge of HEM directionally solidified mc-Si ingot of thereof solar grade silicon feedstock in CRTSE. We have used also Lifetime tester type Sinton WCT-120 in mode QSSPC to investigate the impact of the purity of solar grade silicon feedstock starting material on the electrical quality of the mc-Si wafers at the heart of the HEM silicon ingot.

One of the objectives of this NAA investigation is to compare between the purities concentrations in solar grade silicon feedstock material and the HEM mc-Si ingot or wafer and the solar grade silicon and wafer specifications for photovoltaic application from literature, and to correlate them with the electrical quality of the grown HEM mc-Si ingot.

2 EXPERIMENTAL

2.1 Crystallization

The studied ingot was grown in an industrial fully automated HEM growth furnace (manufactured by GT-Solar Tech.) at the Silicon Technology Development Unit (UDTS) producing 80 Kg ingot each batch in standard fused silica solar crucible, the crucible is G3 with dimensions 44 x 44 cm^2 square cross-section and about 34 cm height. Prior to crystallization SoG-Si chunks was P-doped with boron to a base resistivity of ~ 1 Ω.cm and melted at 1530 °C in fused silica crucible, this last was lined by a release coating composed of a deionized water as solvent, coating powder (95 % α Si_3N_4) and an organic binder. The raw coating was then fired at high temperature to get rid of carbone residue. The molten silicon is directionally solidified from below by extracting of heat from the crucible bottom resulting in a convex solide/liquid interface.

2.2 Sawing and wafering of the ingot

The edges of the as-grown silicon ingot were removed by cropping 2 cm on the top and the bottom as well as on the walls sides that these regions are very rich in impurities due to diffusion from silica crucible and coating materials, and because of segregation phenomena. The ingot is cut into sixteen vertical blocks (bricks) of 10 x 10 cm^2 square cross section by using an automated diamond wire saw which are subsequently horizontally cut into standards wafers of 10 x 10 cm^2 with 350 µm in thickness for each by using a multi wire saw machine (HCT). The studied mini bricks samples for NAA analysis are approximately 12 x 12 mm^2 each. They were obtained by the successive sectioning the standards mc-Si wafers by the same diamond wire saw machine. In total three samples were selected: Mc-Si1 (bottom), Mc-Si2 (middle), and Mc-Si3 (top) respectively. They were all cut 30 mm distance from the edge of the silicon ingot.

2.3 Samples Preparation for NAA Analyses

2.3.1 Solar grade silicon feedstock samples

Three samples of solar grade silicon feedstock chunk supplied by Silicon Recycling Service (SRS) were selected in aleatory manner from three different batch samples for NAA Analyses. Each sample of them is represented by another three samples, weighing about 90 mg each.

2.3.2 Multicrystalline silicon wafers samples.

The selected as-cut mini silicon wafers for NAA analysis testing were chemically degreased and cleaned in electronic grade trichloroethylene, acetone and deionized water. Then, they were etched in a HF: HNO_3: CH_3COOH solution to depth of approximately 6 µm per side to remove saw damage and surface contamination. Each sample is weighing about 90 mg. The selected standard mc-Si wafers for Lifetime testing by QSSPC were also chemically degreased and cleaned in electronic grade trichloroethylene, acetone and deionized water solution. Then, they were etched in basic solution based of NaOH: HCL: H_2O to depth approximatively 6 µm per side to remove any saw damage and surface contamination. Then, they were passivated in Ethanol: Iode solution prior to lifetime measurements.

2.4 Neutron Activation Analysis (NAA)

All solar grade silicon feedstock samples and as-cut mc-Si wafers were activated under the same conditions by a thermal–neutron flux of about 10^{13} n/cm^2 s for eight hours at the Es-Salam Multi Purpose Heavy Water Research Reactor (MHWRR), Birine Nuclear Research Centre (CRNB)-Algeria. The resulting gamma rays emitted by the decaying nuclei were monitored by a high purity Ge Detector (HPGe) cooled with liquid nitrogen at temperature (-196°C) firstly for 3600 seconds, then again for 7200 seconds. The measurements were based on the monostandard method using an AIEA certified synthesized standards.

2.5 Minority carrier lifetime tests

Three standard passivated mc-Si wafers were selected from the heart of the ingot for lifetime tests. The measurements of lifetime were carried out on the above mentioned standards mc-Si samples using Lifetime time tester type Sinton WCT-120 in mode QSSPC under standard illuminating conditions at 10^{15}/cm^{-3} at ambient temperature.

3 RESULTS AND DISCUSSIONS

3.1 NAA Analysis of solar grade silicon feedstock

In Figure.1 are shown the impurities detected by NAA in the investigated solar grade silicon feedstock samples. This investigation reveals the presence of large number of contaminants and that the dominants impurities are found to be: Fe (35 000 ppbw), Cr (12 683 ppbw), W (35,1 ppbw), Mo (45 ppbw), Zn (3 300,00 ppbw), Zr (4 112,00 ppbw), Nd (397 ppbw), Au (545,00 ppbw), Ce

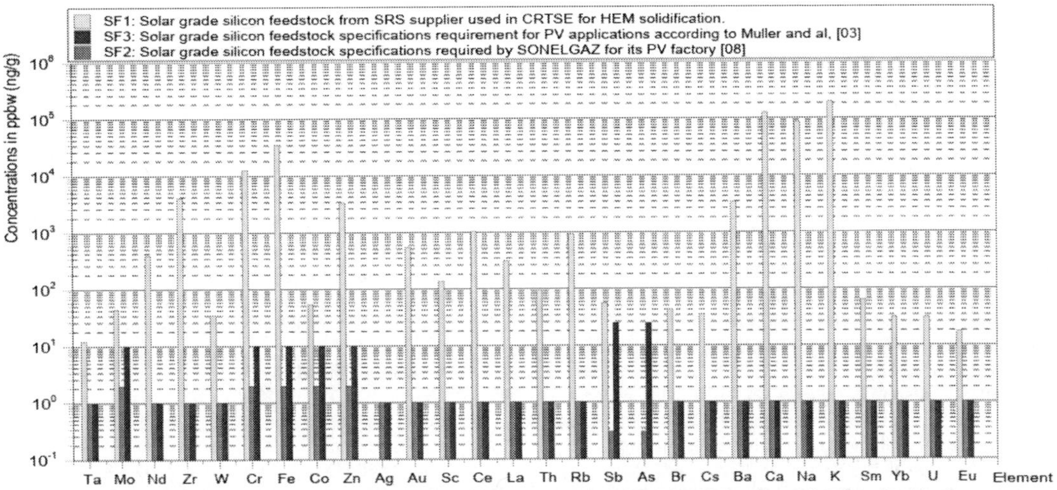

Figure. 1: a) Transition metal impurities conentrations as determined by means of NAA in SRS solar grade silicon feedstock.
b) Specifications of solar grade silicon feedstock required by SONELGAZ for its photovoltaic solar factory.
c) Specifications of a type solar grade silicon feedstock for PV applications according to Muller and al,

(1 038,00 ppbw), Rb (935,00 ppbw), La (313,40 ppbw), Th (92,60 ppbw) and Ca (123 529,00 ppbw). Most of these metallic impurities such as: Fe, Cr, Ca are exceeded several thousand-fold their-sholds concentrations in the silicon feedstock material for photovoltaic application [3]. This strong contamination of silicon feedstock material by transition-metal impurities can lead to the contamination and the degradation of the electrical quality of the as-grown silicon ingot. NAA investigation has revealed that the studied silicon feedstock material is also intensely contaminated by alkaline metallic impurities such: Na (85 395,00 ppbw) and K (192 353,00 ppbw) and Ba (3 339,00 ppbw). These Alkaline-metal impurities are exceeded hundred thousand-folds their-sholds in the silicon feedstock material [3]. This intense contamination of solar grade silicon feedstock material by alkaline metal-impurities is very revealing about the origin of the feedstock material which is very probably electronic or photovoltaïc industry wastes, and it is also very revealing about the techniques may be being used in the past for its treatments which could be based of KOH and NaOH chemicals leaching solutions widely used in such cases for silicon cleaning. These high contents of alkaline contaminants could also explain the adhering of the silicon ingot to the silica crucible silicon nitride coating during the crystallization resulting in bad releasing properties. It is well known that the presence of sodium in higher concentrations in silicon wafer could cause later the Potential Induced Degradation (PID) of the manufactured silicon Device.

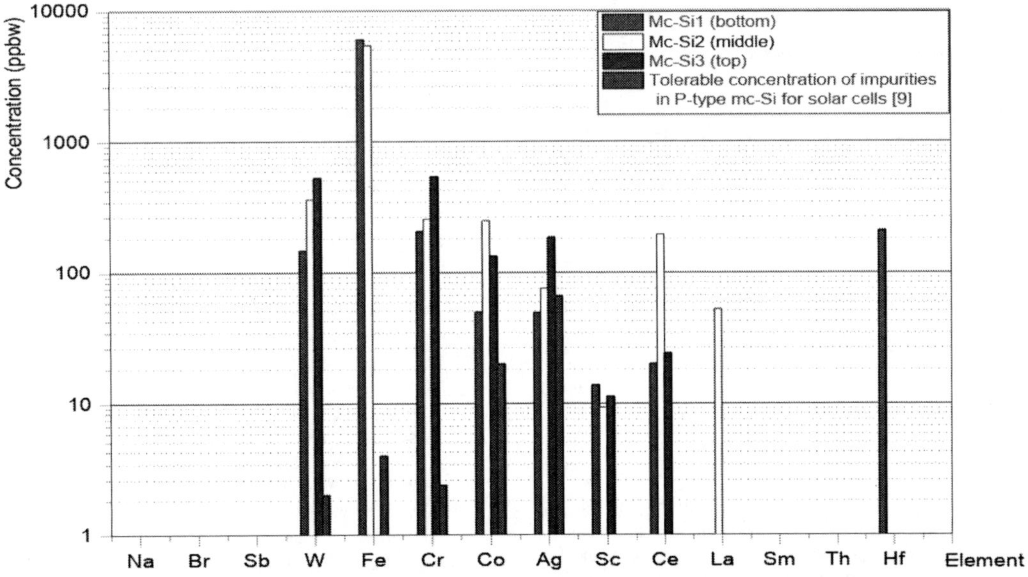

Figure.2 : Transition-metal impurities concentrations as determined by means of NAA in the edge of the silicon ingot and theresholds in P-type Mc-Si for photovoltaic applications.

3.2 NAA Analysis of mc-Si silicon wafers

The Figure.2 shows the impurities revealed by NAA in three investigated different mc-Si wafers from the edge of the silicon ingot. This NAA investigation shows that the dominants metallic impurities in the edge of the silicon ingot are found to be: Fe (6040 ppbw), Cr (540 ppbw), W (584 ppb), Ag (200 ppbw), Co (248 ppbw), Ce (200 ppbw), Hf (208 ppb) and La (52,54 ppbw). Most of these impurities concentrations are above the level of 10^{15} at/cm^3 which mean that it is very difficult to get rid of them even by using an expensive high temperature aggressive and long gettering [10]. The concentrations of W, Fe, Cr, Co, and Ag impurities in mc-Si wafers are 26 333, 1 508, 225, 6,67 and 2,8 folds their-sholds respectively in mc-Si doped type P for photovoltaic applications. This high level of transition-metal impurities in the as-grown silicon ingot is due to the intense contamination of the solar grade silicon starting and/or silica crucible and silicon nitride coating materials because some impurities such W, Ag and Co have very low concentrations in the silicon feedstock starting material and very low effective coefficient of segregation (K_{eff}).

3.3 Minority carrier lifetime measurements

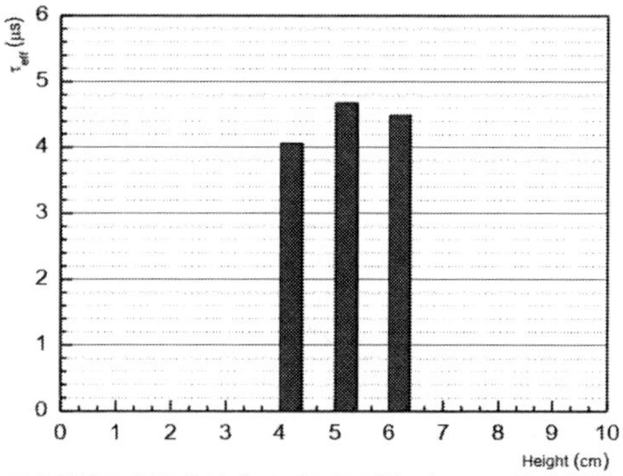

Figure. 3: Lifetime distribution in the center of mc-Si ingot measured by QSSPC

The Figure.3 gives an overview of the lifetime distribution of three representative wafers from the heart of the silicon ingot measured by WCT-120 in mode QSSPC. The average maximum lifetime values in the heart of the grown HEM silicon ingot is about 4,5 µs. According to the reference [11] this degradation of lifetime in the center of the silicon ingot is mainly caused by the intense contamination of the solar grade silicon feedstock by transition metal-impurities because the electrical quality in the center of the silicon ingot is strongly governed in its magnitude by feedstock purity and not by crucible and silicon nitride coating purity which are so far from the heart of the silicon ingot.

4 CONCLUSION

NAA investigation has revealed that the analysed solar grade silicon feedstock from SRS supplier is strongly contaminated by transition-metal impurities particularly Fe, Cr, Zn, Zr, Ce, Au, Ca and Rb and by alkaline metals impurities such Na, K and Ba. Their concentrations are largely exceeding several thousand-fold their-sholds in the solar grade silicon feedstock. The NAA Investigation shows also that the edge of the mc-Si ingot is also intensely contaminated by transition metal-impurities such Fe, Cr, W, Ag, Co, Ce, Hf and La. Their concentrations are exceeding several hundred-folds their –sholds in the mc-Si type P for PV applications. The lifetime measurement by QSSPC has monitored the degradation of electrical quality in the heart of the silicon ingot which is mainly caused by an intense contamination of the starting silicon feedstock material by transition-and alkaline metallic impurities. We recommend using an ultra-pure cheap solar grade silicon feedstock such as Wacker Chemie, and/or Hemlock Semiconductors quality to improve the electrical properties of the silicon ingot.

5. AKNOWLEDGEMENTS

The authors gratefully acknowledge the co-financing support of this study by the National Fond of Research (FNR) of DGRSDT/MESRS of Algeria. The authors would also warmly thanks *Mr. Rahab Hamid and Mr. Meribai Nadjib* for their friendly technical support in cutting the studied mc-Si wafers and for Mr. Khelifati Nabil for its assistance in Lifetime measurements.

6. REFERENCES

[1] MacDonald D, Cuevas A, Kinomura A, Nakano Y, Geerlings LJ, Transitions-metal profiles in a multicrystalline silicon ingot, journal of Applied Physics, 97(3) : 033521-033523 : (2005).

[2] Sergio Pizzini, Towards solar grade silicon: Challenges and benefits for low cost photovoltaics Solar Energy Materials & Solar Cells 94 (2010) 1528–1533

[3] A. Müller, M. Ghosh, R. Sonnenschein and P. Woditsch, Silicon for Photovoltaic applications, Mat Sc & Eng B134 (2006) 257-262.

[4] Sylke Meyer , Stefanie Wahl, Alexander Molchanov , Kristin Neckermann , Christian Möller , Kevin Lauer, Christian Hagendorf, Influence of the feedstock purity on the solar cell efficiency, Solar Energy Materials & Solar Cells 130 (2014) 668–672

[5] Tingting Jiang, Xuegong Yu, Lei Wang, Xin Gu, and Deren Yang, On the low carrier lifetime edge zone in multicrystalline silicon ingots, JOURNAL OF APPLIED PHYSICS 115, 012007 (2014)

[6] J. Hofstetter, J.F. Lelièvre, C. del Cañizo, A. Luque, Acceptable contamination levels in solar grade silicon: From feedstock to solar cell, Materials Science and Engineering B 159–160 (2009) 299–304

[7] W, Kwapil and al, Influence of crucible and coating on the contamination of directionally solidified silicon: First results of the German Researches network "Solar Wins" 27th EU PVSEC, 627-635, Frankfurt, Germany (2012)

[8] Catalogue des fiches techniques de composants de modules et systèmes photovoltaïques, Salon de la sous-traitance photovoltaïque- 10 &11 juillet 2011

[9] Popov, V. G., 2000, Solar cells based on multicrystalline silicon, *Semiconductor physics, Quantum Electronics & Optoelectronics*, **V. 3. N4,** 479-488.

[10] Daniel Macdonald, Andres Cuevas, Atsushi Kinomura and Yukihiro Nakano, Phosphorus gettering in multicrystalline silicon studied by neutron activation analysis, 29th IEEE PVSC, New Orleans, May 2002.

[11] Bulk lifetime Testing: BLS-I, BCT-400, SBS-150, Sinton Instruments, Inc, (2010).

MATHEMATICAL MODELING OF METALLURGICAL-GRADE SILICON PLASMA-CHEMICAL PURIFICATION PROCESS

Sergey M. Karabanov[1], Victor I. Yasevich[2], Dmitry V. Suvorov[1], Evgeny V. Slivkin[1], Andrey S. Karabanov[2], Dmitry Yu. Tarabrin[1]
1- Ryazan State Radio Engineering University
59/1 Gagarina St., Ryazan 390005, Russia
2- Energy Ltd.
51 B Novaya St., Ryazan 390027, Russia
e-mail: pvs.solar@gmail.com

ABSTRACT: One of the major lines of solar energy development is the creation of environmentally friendly, wasteless and cheap solar-grade silicon production technology.
Currently the main silicon production technologies are based on reduction of silicon from hydrogen and chlorides: trichlorosilane, tetrachlorosilane, monosilane. These technologies use harmful and potentially dangerous compounds in quantity. These productions can be profitable only at big volumes.
One of the research directions in the field of chlorine-free technology for solar-grade silicon production is the method of plasma-chemical purification of metallurgical-grade silicon. This environmentally friendly, power efficient, scalable method allows to avoid application of chlorine-containing compounds. However, it is difficult to describe the processes in silicon melt by simple heat- and mass exchange equations.
The paper presents the results of mathematical modeling of silicon purification process using the method of melt plasma-chemical treatment by directed arc discharge in argon atmosphere, containing water vapor impurities.

Keywords: feedstock; silicon; purification, mathematical modeling

1. INTRODUCTION

Currently the main solar-grade silicon production technologies are based on reduction of silicon from trichlorosilane (TCS) and monosilane [1]. These processes are connected with the use of toxic and explosive substances (TCS, SiH_4, HCl and H_2. The production is profitable only at big volumes, more that 10000 tons per year.

There are a number of other technological approaches to chlorine free solar-grade silicon production [2, 3].

One of the directions in the chlorine-free silicon production is the plasma chemical technology [4-9]. This technology has a number of advantages: lack of toxic and explosive substances, low energy consumption, high productivity. The paper presents the research results of mathematical modeling and experimental research of metallurgical-grade silicon purification by the method of plasma chemical treatment of silicon melt up to solar-grade level with further direct crystallization in one reactor.

2. THE MATHEMATICAL MODEL

The mathematical model is based on axially symmetrical approximation of plasma flux impact on the melt surface.

The mathematical model of the installation for silicon purification from impurities is based on axially symmetrical approximation of the crucible geometry in which silicon melt is located. According to the published research plasma-chemical effect is made by a plasma flux directed to the melt center [5-9]. The impurities removal from silicon melt takes place on the liquid-gas border due to chemical and plasma-chemical reactions with formation of volatile compounds which are effectively evaporate from the surface and also due to sublimation process.

In the course of purification continuous updating of the purified near-surface layer with the rate determined by melt convective mass transfer takes place. The flow velocity of travel is non-uniform along the melt surface and is determined by operating conditions of the plant: the crucible geometry, input power, etc. As impurities removal takes place on the heterogeneous liquid-gas border, the purification intensity is proportional to the plasma flux rate on the melt surface. With increase of the input power the plasma flux rate increases; however at the same time the melt and crucible temperature increases that can result in its melting. Besides, the influence of crucible geometrical parameters on the surface plasma flux rate is not clarified.

The purpose of mathematical modeling is optimization of geometrical parameters of the crucible and the input power level according to criteria of providing the maximum plasma flux rate of melt on the surface and in case if the crucible material is not melted.

The mathematical modeling of the melt convective mass transfer is expressed by the following equation:

$$\frac{\partial T}{\partial t}+w_x\frac{\partial T}{\partial x}+w_y\frac{\partial T}{\partial y}+w_z\frac{\partial T}{\partial_z}=\alpha\left[\frac{\partial^2 T}{\partial x^2}+\frac{\partial^2 T}{\partial y^2}+\frac{\partial^2 T}{\partial z^2}\right] \quad (1),$$

where w_x, w_y, w_z are the components of convective mass transfer, α – the heat transfer coefficient.

At a first approximation the purification rate and depth are determined by mass transfer rate with environment. The mass transfer rate is determined y the expression:

$$\theta(r)=v(r)m_s(r) \quad (2),$$

where v is the radial speed of melt movement, m_s – a local coefficient numerically equal to effective mass of the near-surface layer of the purified substance.

According to this system the mathematical model of silicon purification has been developed. Figure 1 shows the axially symmetrical geometry of the numerical model used for calculation. The local influence area of plasma flux is presented in the form of a hemisphere with the diameter of 40 mm in which all input power is concentrated. The velocity vector of a gas-dynamic flux is directed to the crucible bottom. The plasma flux rate is used in calculations as a boundary condition. Figure 2 shows the triangulation computational grid.

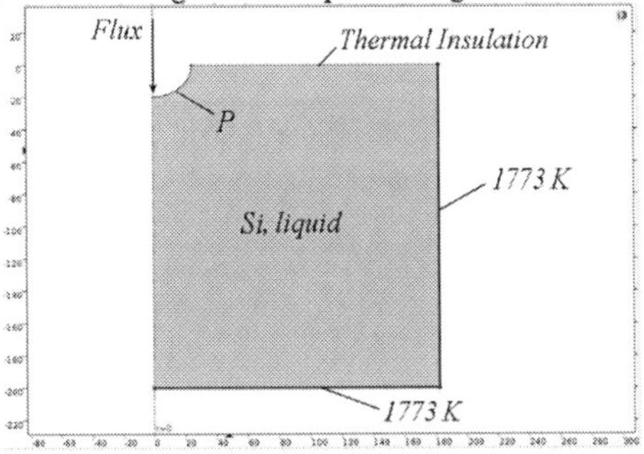

Figure 1: Axially symmetrical geometry of the numerical model.

On the lower and side boundaries heat transfer to provide constant temperature occurs. The arrow shows the flux movement direction with constant speed and constant heat power. The thermal insulation condition is applied to the upper boundary.

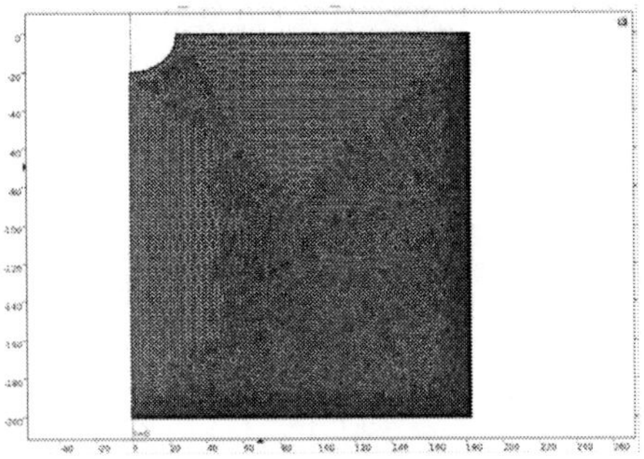

Figure 2: Computational grid of the numerical model.

3. INITIAL CONDITIONS OF THE NUMERICAL EXPERIMENT

By means of numerical model the calculation for a crucible with the height of 50-400 mm and diameter of 240-710 mm at constant temperature of the crucible walls of 1773 K for the plasma flux rate of 0.1 m/s, and the input power of 2.5-20 kW is made. The purpose of the performed modeling is optimization of the reactor parameters and technological process for providing maximum surface mass exchange rate.

4. THE MODELING RESULTS

Figures 3 and 4 show distribution of mass transfer rate and temperature inside the melt (the input power is 10 kW, flux rate is 0.1 m/s). One can see that the main initiating flux inside the melt volume is located under the area of plasma flux influence and is directed to the crucible bottom. The initiating plasma flux together with convective mass transfer results in turbulent flows in the volume and on the melt surface. The analysis of the melt thermal field shows that the area of the highest temperature is located along the crucible axis and corresponds to the main mass transfer area.

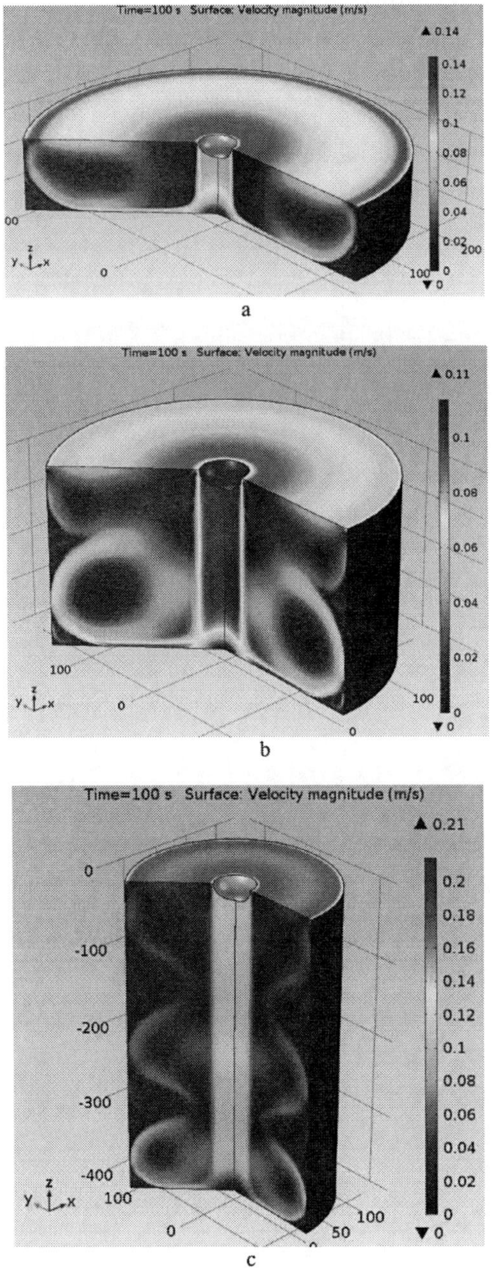

Figure 3: Distriburtion of the flow rate in steady-state conditions: input power -10 kW diameter – 360 mm; the crucible height – 100 mm (a), 200 mm (b), 400 mm (c)

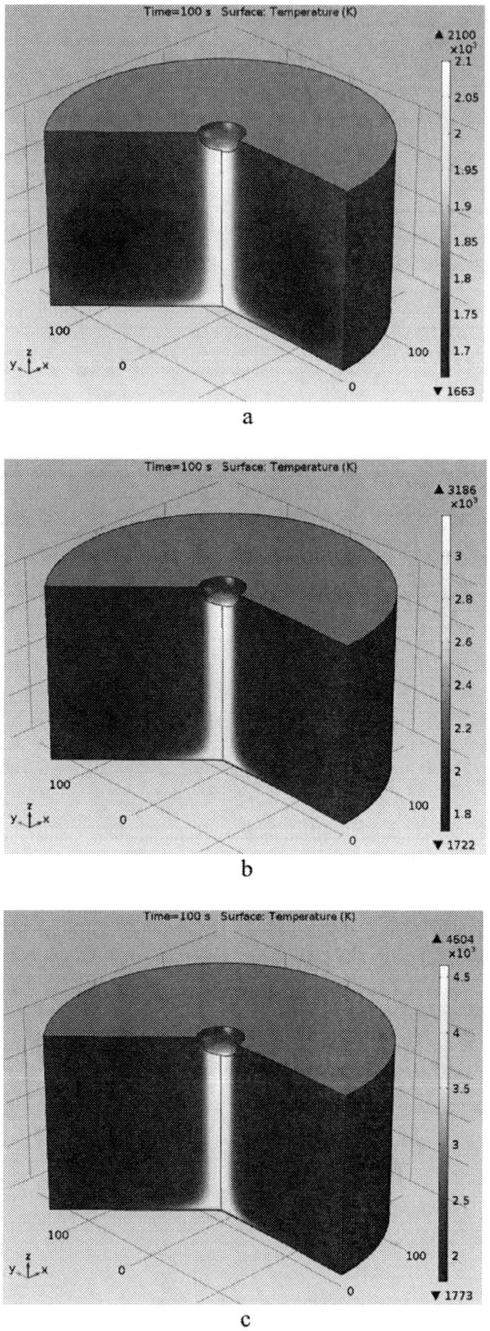

Figure 4: Temperature distriburtion in steady-state conditions: the crucible height – 200 mm, diameter – 360 mm, input power -2.5 kW (a), 10 kW (b), 20 kW (c).

The research of influence of geometrical parameters of the crucible – height and diameter (Fig. 5), has shown that with height reduction the heat exchange rate increases, however the temperature in the area of "a thermal column" rises sharply, and with the height increase the main mass transfer happens in melt volume, and to a greater extent the melt interacts with crucible walls than with the surface. The following optimum parameters of the crucible have been chosen: height – 200 mm, diameter – 360 mm.

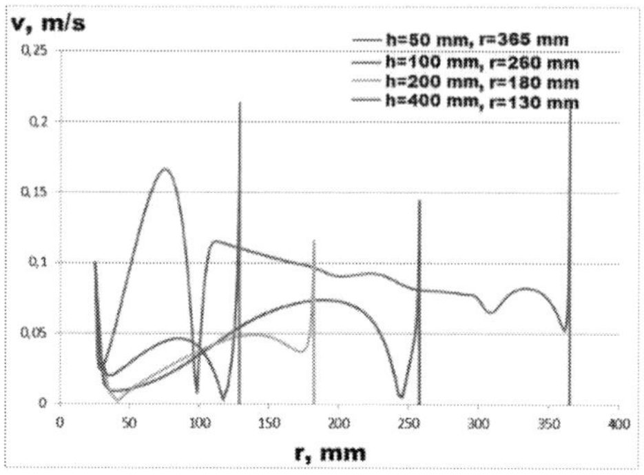

Figure 5: Distribution of the flow rate of liquid silicon on the surface from r coordinate at different height and crucible diameter (plasma input power – 10 kW).

Figure 6 shows the distribution of the flow rate of liquid silicon on the surface from r coordinate at different levels of input power.

It is clear that with increase of the input power the surface flow rate increases, however the degree of the rate increase is slowed down with the power increase. On the basis of the performed calculations the recommended plasma power is 10-15 kW.

The fulfilled modeling shows that plasma chemical purification process is scalable, i.e., one and the same surface flow rate of silicon can be obtained at various crucible parameters and plasma power.

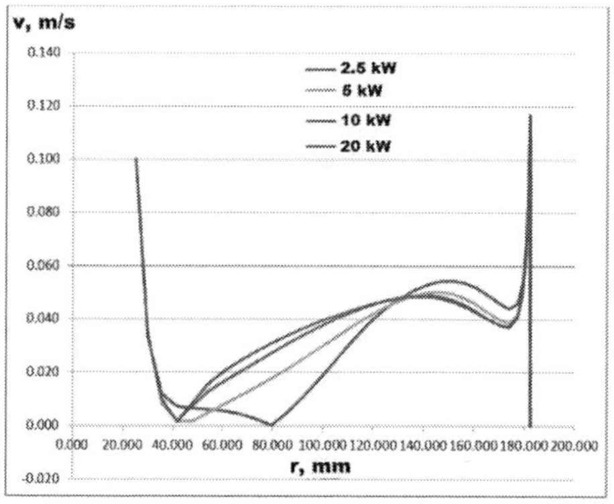

Figure 6: Distribution of the flow rate of liquid silicon on the surface from r coordinate at different levels of input power. (the crucible height – 200 mm, diameter – 360 mm).

In order to increase the silicon purification rate by increase of silicon surface flow rate and surface area widening with maximum flow rate it is possible to use several plasmatrons.

The obtained data shows that the modeling method demonstrates the results close to the ones received experimentally [5-9], and can be used in further works on improvement of plasma-chemical purification process of silicon melt

5. CONCLUSION

The mathematical model of the processes happening during metallurgical-grade silicon plasma-chemical purification is developed.

The work has the following results:

- distribution diagrams of the flow rate of silicon melt masses in a crucible and temperature inside silicon melt are obtained;
- it is determined that plasma-chemical treatment of silicon melt surface results in effective purification from impurities;
- regularities of the silicon purification process by the method of plasma-chemical treatment of the melt surface with arc discharge plasma are determined;
- optimum conditions of silicon purification are determined: discharge power, treatment duration, crucible geometry.

The obtained data shows that the modeling method demonstrates the results close to the ones received experimentally, and can be used in further works on improvement of plasma-chemical purification process of silicon melt.

5 REFERENCES

[1] www.silicontimes.com/ru/production
[2] S.M. Karabanov, New technologies for solar-grade silicon production, MRS Proceedings, Vol. 716 (2002), pp. 495-499.
[3] A.S. Skryabin, Patent 2434683 RU, MPK B03B 7/00 (2008.01), Bulletin of inventions, 20111, No. 33.
[4] B.M. Moon and G. Kim, Boron refining and continuous casting of solar-grade polysilicon using steam-hydrogen plasma with electromagnetic induction melting, 28th European Photovoltaic Solar Energy Conference, Poster Presentation.
[5] S.M. Karabanov, V.L. Dshkhunyan, V.I. Yasevich and M. Hoshino, Method for manufacturing ingots of polycrystalline silicon, Patent 2465201 RU, December 13, 2010.
[6] S.M. Karabanov, V.L. Dshkhunyan and V.I. Yasevich, Method for purification of metallurgical-grade silicon by moisturized AC plasma in vacuum, Patent 2465202 RU, November 17, 2010.
[7] S.M. Karabanov, V.L. Dshkhunyan, V.I. Yasevich and M. Hoshino, Method for refinement of metallurgical-grade silicon, Patent 2465200 RU, February 14, 2011.
[8] S.M. Karabanov, V.L. Dshkhunyan, V.I. Yasevich and M. Hoshino, Method for refinement of metallurgical-grade silicon with dry argon plasma with water injection on the surface of melted silicon with subsequent directional crystallization, Patent 2465199 RU, November 17, 2010.
[9] S.M. Karabanov, V.L. Dshkhunyan, V.I. Yasevich and M. Hoshino, The study of the method of plasma-chemical purification of metallurgical-grade silicon, Proceedings of 31st European Photovoltaic Solar Energy Conference, 14-18 September, 2015, Hamburg, Germany, pp. 555 – 558.

PERFORMANCE OF FBR BLENDED CZ WAFERS

Ørnulf Nordseth[1], Yacine Boulfrad[2], Gene Garrett[3], Rune Søndenå[1], Chang Chuan You[1], Marie Syre Wiig[1], Junjie Zhu[1], Bent Thomassen[1], Sean Erik Foss[1]
[1] Institute for Energy Technology, Instituttveien 18, 2007 Kjeller, Norway
[2] Norwegian Crystals AS, 8161 Glomfjord, Norway
[3] REC Silicon, 3700 Buffalo Speedway, Suite 620, Houston, TX 77098, USA

ABSTRACT: In this study a comparison of the properties and performance of p-type Czochralski (Cz) silicon wafers produced with different feedstocks has been performed. Two Cz ingots were produced, where one ingot was made from virgin polycrystalline feedstock (Poly) and a second ingot was made from a blend of 80 % Poly and 20 % granular silicon made by the fluidized bed reactor (FBR) technology. The positional dependence of the cell performance and light induced degradation were measured showing little or no correlation between the measurements and the feedstock choice. A 0.1 % (abs.) higher cell efficiency for FBR blended wafers was observed.
Keywords: Czochralski, feedstock, solar cell efficiencies

1 INTRODUCTION

The need for high performance silicon wafers is increasing together with the demand for higher efficiency solar cells. At the same time there is a constant push toward lower production costs for these wafers. This can be achieved through several means, from using crucibles that allow for longer, continuous feed ingot pulling, to using different types of feedstock. The use of fluidized bed reactor (FBR) produced granular silicon has grown in recent years. Utilizing FBR material gives several advantages: lower cost compared to virgin polycrystalline silicon (Poly), simpler handling and denser packing in the crucible, as well as the possibility of continuous feeding during Czochralski (Cz) ingot pulling.

In this study, the performance of 6-inch p-type monocrystalline Si solar cells produced from one ingot made from virgin polycrystalline feedstock (Poly) material and one ingot made from a blend of 80 % Poly and 20 % FBR material has been compared.

2 EXPERIMENTAL DETAILS

2.1 Si material properties

Monocrystalline 6-inch semi-square p-type slurry-cut Si wafers from two different ingots were used in this investigation. The minority carrier lifetime was comparable for the Poly and the 20 % FBR ingot. The wafers had a nominal thickness of 200 µm. For the Poly ingot the resistivity was measured with a four-point-probe in the range from 2.93 Ohm-cm near the seed end of the ingot to 1.70 Ohm-cm at the tail end. For the 20 % FBR ingot, the resistivity was measured in the range from 2.89 Ohm-cm near the seed end of the ingot to 1.65 Ohm-cm at the tail end of the crystal. Both oxygen and carbon were measured on the Poly ingot using Fourier transform infrared spectroscopy (FTIR). The results are shown in Table I. Commercially available 6-inch Cz wafers were also included in this study in order to serve as reference wafers.

2.2 Solar cell fabrication and testing

Solar cells from each ingot at different positions were fabricated and characterized. The cells were made using a conventional aluminum back-surface-field process. The wafers went through a five step cell processing sequence, which includes (1) saw damage removal and surface texturing, (2) diffusion of n-type emitter in a tube furnace using a $POCl_3$ source, (3) deposition of SiN_x anti-reflection coating on the front surface by plasma enhanced chemical vapour deposition (PECVD), (4) screen-printing and firing of metal contacts, and (5) laser edge isolation. Current-voltage (I-V) measurements of the solar cells were performed under standard conditions (1 Sun illumination, AM1.5g spectrum, 25 °C) using a solar simulator (WACOM) with the bias voltage varying from -1 V to +1 V. A LIS-R1 unit from BTImaging was used for measuring the minority carrier lifetime for surface passivated wafers, as well as the series resistance of solar cells [1]. Internal quantum efficiency (IQE) measurements were carried out using an IQE-SCAN tool from pv-tools. I-V-light induced degradation (I-V-LID) measurements were performed to investigate the effect of light-soaking on the I-V response of the solar cells. The cells were illuminated with 1 Sun for 48 hours at 25°C and the I-V-LID sampling interval was 1 hour.

Table I: Resistivity and concentration of oxygen, carbon and boron for the seed and tail end of the Poly and 20 % FBR ingot.

Parameter	Ingot	Seed	Tail
Resistivity (Ohm-cm),	Poly	2.93	1.70
Resistivity (Ohm-cm),	20 % FBR	2.89	1.65
$[O_i]$ (ppma)	Poly	16.87	12.78
$[O_i]$ (ppma)	20 % FBR	16.14	13.74
$[C_s]$ (ppma)	Poly	0.03	0.27
$[C_s]$ (ppma)	20 % FBR	0.00	0.39

3 RESULTS

3.1 I-V measurements

The average I-V data for the cells made from Poly, 20 % FBR, and reference wafers are summarized in Table II. The data shows that the I-V parameters for all three wafer sets are quite similar, with the 20 % FBR cells showing a 0.1 % abs. higher efficiency than the Poly and reference cells.

The cell efficiency as a function of ingot height for the two ingots is shown in Figure 1. The data suggests that the cell efficiency is about 0.1 % abs. higher throughout the ingot for the cells made from the 20 % FBR ingot compared to the cells made from the Poly ingot. For both ingots, the cell efficiency decreases towards the seed and the tail end of the ingot.

Table II: Average I-V data for cells made from Poly, 20 % FBR, and reference wafers, based on fit to the recorded I-V curve [2].

Parameter	Poly	20 % FBR	Ref.
J_{sc} (mA/cm^2)	36.83	36.96	36.81
V_{oc} (mV)	630	630	631
Fill factor (%)	78.87	78.87	78.85
Efficiency (%)	18.31	18.41	18.32
Series resistance (Ω·cm^2)	1.22	1.22	1.24

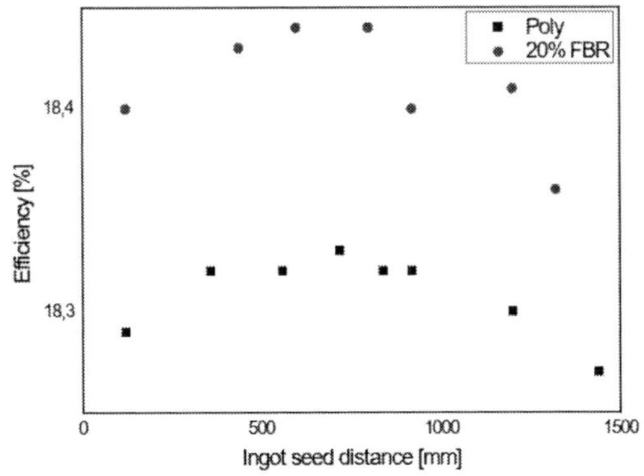

Figure 1: Efficiency vs. ingot seed distance for cells made from Poly and 20 % FBR wafers.

3.2 Power loss analysis

An analysis of the power loss mechanisms for the solar cells was carried out. The loss analysis is similar to the method proposed by Aberle et al. [3], and is based on a set of electrical measurements, including current-voltage curves, luminescence images, and IQE. Table III summarizes the results of the power loss analysis, for which seven different power loss mechanisms were defined. The optical and carrier recombination losses were extracted from IQE measurements.

Table III: Power loss (mW/cm^2) for cells made from Poly, 20 % FBR, and reference wafers (MPP = maximum power point).

Loss mechanism	Poly	20 % FBR	Ref.
Shadowing/metal shading	1.60	1.53	1.59
Front surface reflectance	0.36	0.38	0.37
Other optical losses	0.11	0.14	0.12
Series resistance	1.47	1.44	1.47
Shunt resistance	0.05	0.04	0.04
Forward-bias current at MMP	1.29	1.26	1.32
Recombination losses	0.76	0.71	0.76
Total losses	5.64	5.50	5.67

Figure 2(a) and (b) show the corresponding percentage power loss for the Poly and 20 % FBR cell, respectively. Figure 2 along with the data in Table III suggest that there are no major differences in the power loss mechanisms for cells made from the Poly and 20 % FBR ingot.

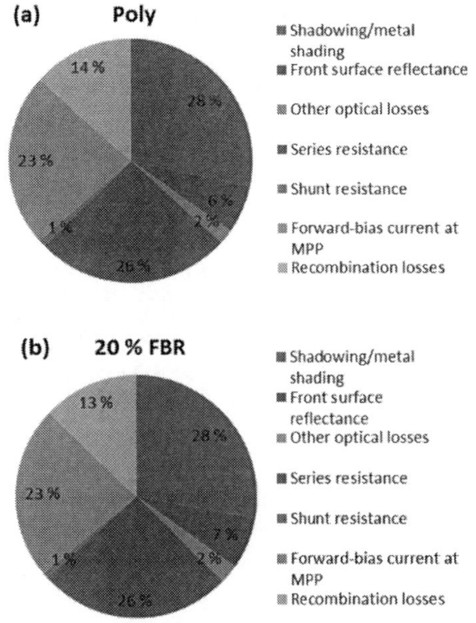

Figure 2: Percentage distribution of seven different power loss mechanisms for cells made from (a) Poly and (b) 20% FBR (MPP = maximum power point).

3.3 I-V-LID measurements

Figure 3 shows the normalized change in V_{oc} for five different cells during an I-V-LID measurement series, for which the first measurement was taken after one hour of illumination. For the "seed" cells, the distance from the seed was 120 mm, whereas for the "tail" cells, the distances from the seed were 1440 mm and 1320 mm for Poly and 20 % FBR, respectively. For both the "seed" and the "tail" cells, the degradation rate is slightly lower for the cell made from the 20 % FBR ingot compared to that for the cell made from the Poly ingot.

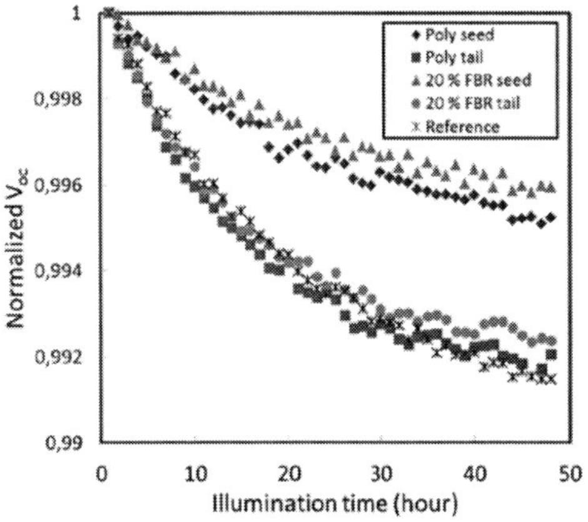

Figure 3: Normalized V_{oc} as a function of light-soaking time (1 Sun) for cells made from the seed and tail end of the Poly and 20 % FBR ingot.

Figure 3 also shows that for both Poly and 20 % FBR, the cells made from wafers near the seed degrade less than the cells made from wafers near the tail. The larger reduction in V_{oc} for cells near the tail may be correlated with the larger boron concentration near the tail. For the Poly ingot the boron concentration was measured to $\sim 4\times10^{15}$ and $\sim 9\times10^{15}$ at respectively the seed and tail end of the ingot, using FTIR. The higher concentration of boron may result in higher concentrations of boron-oxygen complexes which have been attributed to causing LID [4].

4 CONCLUSION

The aim of this investigation was to compare p-type monocrystalline solar cells made from two different Cz ingots: one ingot made from virgin polycrystalline feedstock (Poly) and a second ingot made from a blend of 80 % Poly and 20 % granular FBR silicon, in order to identify possible effects of the ingot production method on the performance of solar cells. I-V measurements showed that the conversion efficiency was slightly higher (~0.1 % abs.) for cells made from 20 % FBR wafers compared to that for cells made from Poly wafers. There was no significant difference in the power loss mechanisms for cells made from Poly and 20 % FBR wafers. I-V-LID measurements for 48 hours revealed a near identical degradation of the overall electrical performance of the cells made from Poly and 20 % FBR wafers. In conclusion, measurement of the positional dependence of the performance and LID for the fabricated solar cells showed little or no correlation between the measurements and the feedstock choice.

ACKNOWLEDGEMENTS

This work has been funded by the Research Council of Norway through "The Norwegian Research Centre for Solar Cell Technology", NFR Project No. 192839.

REFERENCES

[1] H. Kampwerth, T. Trupke, J. W. Weber, Y. Augarten. Advanced luminescence based effective series resistance imaging of silicon solar cells, Appl. Phys. Lett. 93 (2008) 202102-4.

[2] N. Enebish, D. Agchbayar, S. Dorjkhand, D. Baatar, I. Ylemj, Numerical analysis of solar cell current-voltage characteristics, Sol. Energy Mat. Sol. Cells 29 (1993) 201-208.

[3] A. G. Aberle, W. Zhang, B. Hoex, Advanced loss analysis method for silicon wafer solar cells, Energy Proc. 8 (2011) 244-249.

[4] K. Bothe and J. Schmidt, Electronically activated boron-oxygen-related recombination centers in crystalline silicon, J. Appl. Phys. Vol. 99 (2006) 013701.

SILICON PRODUCTION BY CENTRIFUGE CVD REACTOR ON THE WAY TO INDUSTRIAL VERIFICATION

Werner O. Filtvedt[1*,2], Hallgeir Klette[2], Sverre Sørensen[1], Josef Filtvedt[1]
[1]Dynatec Engineering, Rakkestadvn. 1, 1814 Askim, Norway
[2]Institute For Energy technology Instituttvn. 2027, Kjeller Norway

ABSTRACT: A novel patented design of a rotating CVD chamber for production of polysilicon by decomposition of monosilane is demonstrated. The method has been proven through a first and second generation. It has been demonstrated that first grade polysilicon can be produced at a cash costnlower than all methods being utilized by the industry today. In this presentation we will give an update on the progress of the industrial verification project which will demonstrate the technology in full industrial scale.
Keywords: Polysilicon, Feedstock

1 Introduction

The are only two methods being utilized by the industry today for production of high purity silicon feedstock to electronics and solar cell applications. In regards to high purity we refer to SEMI PV17 1 or 2nd grade. The two methods used are the Siemens Bell Jar reactor and the Fluidized Bed reactor, first proposed by H.W. Ling of E.I. du Pont [1, 2]. Both designs was proposed in the late 1950s, and traditionally the Siemens reactor has facilitated high quality at the expense of a high cost while the fluidized bed has given lower cost at the expense of lower quality. The quality challenges of the FBR have both included purity, fines, porosity and crystallinity [3]. The motivation for proposing a new design has been to combine the benefits of the Siemens reactor with the cost structure of the FBR. Especially being able to produce low cost and high purity without suffering from high production of dust has been a key motivation for the new technology.
The concept has been through two stages. A prototype reactor Dynatec 1.0 and later the upscaled Dynatec 2.0 has been and successfully tested. In this article the team discloses some of the principles being used in the industrial pilot reactor Dynatec 3.0 which is an industrial demonstration unit.

Theory

Decomposition of silane
In the early days of silicon CVD, several gases were explored for the purpose, but in today's reactors monosilane SiH_4 (silane) and trichlorosilane $SiHCl_3$ (TCS) are the ones being commercially utilized. Upon being heated to the decomposition temperature, the gases undergo the following reactions:

$$SiH_4 \rightarrow Si + 2H_2 \qquad (1)$$

The silane decomposition sequence initiation is about 420°C.
However at this temperature, pressure and the present hydrogen concentration, the silicon will form amorphous structures with dispersed hydrogen. One needs temperatures of about 700°C to avoid this problem or even higher depending on the design of the process [4 - 6].

Silicon growth

When the monosilane molecule decomposes, a chain reaction is initiated [6]. The chain reaction involves a large number of intermediate reactions and species and will depend on silane concentration, temperature, pressure, temperature of confining surfaces and gradient of these variables.
In general the decomposition process involves the development of higher order silanes moving through silicon hydrogen complexes into a silicon-hydride structure that gradually desorbs the hydrogen moving into crystalline silicon [5 - 9]. There has historically been a discussion on whether the growth rate is limited by the transport of gaseous species to the surface or the surface chemistry that forms the incoming silicon into a crystalline structure.
In the earlier proposed designs the transport of gaseous species to the surface has been depending on gas diffusion and concentration gradients. The limiting factor controlling the surface chemistry seems to be connected directly to the rate of hydrogen desorption from the surface, which is inversely dependent of the hydrogen partial pressure outside the surface. High hydrogen partial pressure outside the surface seems to limit the hydrogen desorption rate and thus the silicon growth at the surface.

Influence of reactor layout

Since there are a number of stages in silane decomposition and silicon growth, the history of the gas flow in terms of concentrations, temperature, pressure and time will all contribute to the probability of competing events and products.

In general there are two competing macro mechanisms, gas phase nucleation and growth and surface initiated growth. To complicate this picture the produced silicon particles may be scavenged by the growing surface, thus also contributing to the bulk growth [3, 5, 6].

The more developed the intermediate species are in terms of complete dehydrogenated particles, the more complicated it will be to fully incorporate them into a lattice without defects. It is therefore crucial to capture and incorporate the species as early as possible in their development.

When optimizing a silane decomposition reactor the key factor is control of the reaction sequence and ultimately the growth.

Centrifuge CVD

Fundamentally there are two challenges in terms of bulk growth that needs to be addressed to increase the growth speed, limit the fines production, obtain high silane utilization and grow a dense high quality material.

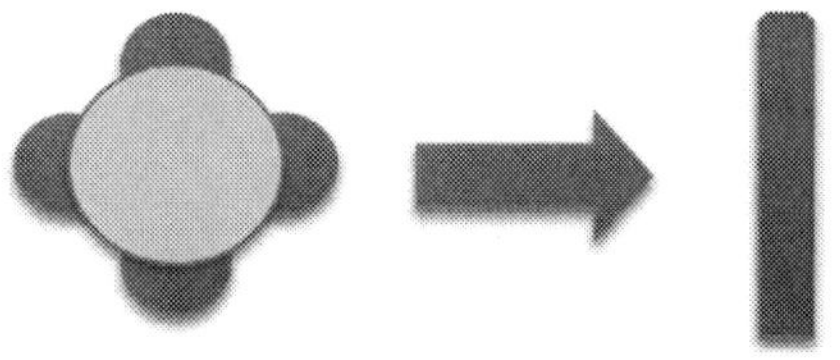

Figure 1: Effective transport of all silicon containing species to the growth surface

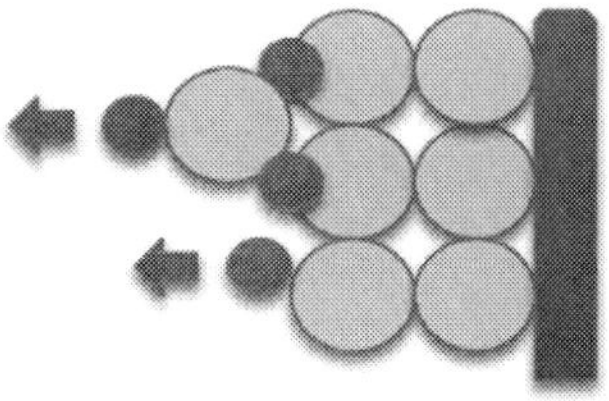

Figure 2: Low hydrogen partial pressure at the growth surface to increase hydrogen desorption and thus increase the silicon lattice incorporation efficiency

These two fundamental mechanisms controlling the silicon growth are both addressed when designing the whole reactor as a centrifuge, see figure 1 and 2. The silane and all other higher order species, complexes and particles will experience a force that actively moves them to the deposition surface. At the same time the hydrogen will experience the opposite effect and be actively moved towards the centre of reactor. This effect will be the same as when releasing hydrogen into air and that it then moves directly upwards. In a centrifuge up is toward the centre and the g force is substantially higher than earth gravity.

At the growth surface the hydrogen will be released faster as the hydrogen partial pressure outside the surface is reduced due to this effective transport of hydrogen away from the surface. This release will both directly supply more vacant bonds and increase the capture of incoming silicon atoms, but also indirectly enhance the hydrogen transport to the surface from the bulk material since the amount of hydrogen inn the close to surface lattice is reduced. The team has been able to grow over 150 micron pr minute with the method which is substantially faster than all presently used commercial methods.

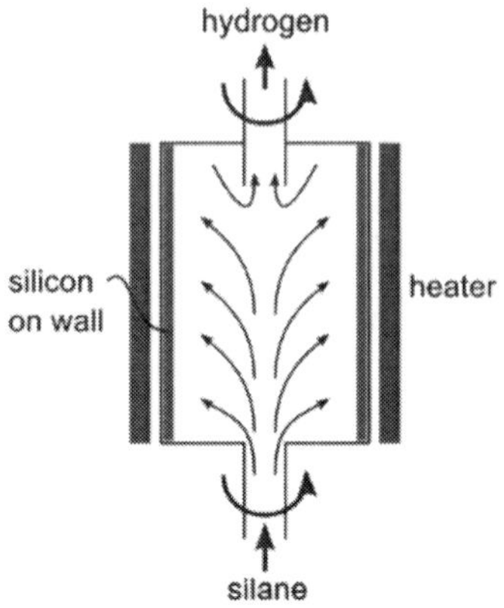

Figure 3: Centrifuge CVD reactor (CCVD), heavy silane is forced to the wall and decomposes, light hydrogen goes to the centre and out

Figure 4: The Dynatec reactor 2.0

The material

The material produced in the Dynatec reactor is a dense crystalline material of high purity. The nature of the material is polycrystalline, but differs in appearance from other polysilicon sources in that there is very high homogeny in the structure of the material. The material consist of very similar crystalline columns growing out from the seed layer on the wall towards the centre of the reactor.

Figure 5: Material produced in the Dynatec reactor

The material has been tested by several institutions among them Evans Aanalytical Group (EAG) in the United States. The result of this investigation is presented in table 1. The team see that we are able to produce according to SEMI PV17 1st grade and that it may be possible to go beyond this quality with minor modifications to the reactor and if the gas feed, environment and product handling is dealt with appropriately.

Silicon product quality PPB			
Purity	Max.	Min.	Average
Ti	14	13	13
Fe	< 0,4	< 0,4	< 0,4
Cr	< 0,2	< 0,2	< 0,2
Mn	< 0,1	< 0,1	< 0,1

Table 1. Purity of the material from Dynatec 2.0, EAG

The industrial reactor

The industrial reactor will like Dynatec 2.0 be a semi continuous reactor where harvest and seeding is done automatically hot without cooling down and heating the reactor.

Conclusions

The Dynatec reactor is the world's first method for production of high quality bulk polysilicon at a cost lower than fluidized bed. The gas utilization, low energy consumption, high degree of automation and low fines production makes the cost to price ratio very beneficial and the CCVD method is likely to play a major role in the polysilicon market of tomorrow.

The largest difference to the FBR method is however the low production of fines which facilitates the possibility of a high price to cost ratio for the produced material.

Hot harvest is demonstrated and the Dynatec reactor is the world's first reactor to demonstrate harvest of bulk polysilicon and initialization without cooling down the reactor between runs. This is crucial to make use of the high growth rate of the method, which is demonstrated at over 150 micron pr minute.

The upscaling of the reactor to version 3.0 is in the engineering phase and the reactor is planned to be finalized in 2017.

Acknowledgements

The authors would like to thank the Norwegian Research Council and the EnergiX program.

References

[1] H. Gutsche Attn. Siemens-Schuckertwerke, Method for production of highest-purity silicon for electronic semiconductor devices US Pat 3.042.494

[2] Harry W. Ling Attn. E. I. du Pont de Nemours and Co. Production of Silicon, US Pat 3.012.861

[3] W. O. Filtvedt, M. Javidi, A. Holt, M. C. Melaaen, E. Marstein, H. Tathgar, P. A. Ramachandran, Development of fluidized bed reactors for silicon production, Solar Energy Materials & Solar Cells 94 (2010) 1980 – 1995.

[4] Robert H. Allen, James E. Boone, Process for preparing polysilicon with diminished hydrogen content by using a fluidized bed with a two-step heating process, Assigned to Ethyl Corporation, US Pat 5.242.671. (1993)

[5] J. O. Odden, P. K. Egeberg, A. Kjekshus, From monosilane to crystalline silicon. Part III. Characterization of amorphous, hydrogen containing silicon products, Journal of Non-Crystalline Solids 351 (2005) 1317–1327.

[6] W.O. Filtvedt, A. Holt, P.A. Ramachandran, M.C. Melaaen, Chemical Vapor Deposition of Silicon from Silane: Review of Growth mechanisms and Modeling/Scaleup of Fluidized Bed Reactors, Solar Energy Materials and Solar Cells 107, 188–200 (2012)

[7] A. A. Onischuk, V. P. Strunin, M. A. Ushakova, V. N. Panfilov, On the path ways of aerosol formation by thermal decomposition of silane, Journal of Aerosol Science 28 (2) (1997) 207–222.

[8] S. P. Walch, C. E. Dateo, Thermal decomposition pathways and rates for silane, chlorosilane, dichlorosilane and trichlorosilane, Journal of Physical Chemistry 105 2015–2022 (2001)

[9] C. S. Herrick, D. W. Woodruff, The Homogeneous Nucleation of Condensed Silicon in the Gaseous Si-H-Cl System, J. Electrochem. Soc.: Solid-state Science and Technology, 2417-2422, (1984)

[10] T. Mongstad, H. Klette, Ø. Nordseth, T. J. Preston, G. M. Wyller, W. O. Filtvedt New technologies for silicon production from monosilane; Pilot testing and on-line process monitoring aiding the development. Silicon for the Chemical and Solar industry XIII

SILICON PURIFICATION THROUGH MAGNESIUM ADDITION AND ACID LEACHING

Jafar Safarian, Gabriella Tranell
Department of Materials Science and Engineering, Norwegian University of Science and Technology (NTNU), No-7491 Trondheim, Norway

ABSTRACT: Silicon is the dominant material in photovoltaic devices and with regard to the PV market growth, the demand for solar grade silicon feedstock is expected to be high in the future. Although solar silicon is currently produced through both chemical and metallurgical routes, the chemical route is still the main production method, which is expensive and have environmental challenges. Acid leaching of silicon, which is a metallurgical refining process is applied for silicon purification in the present research work. In this case, magnesium in different portions is added to liquid silicon containing B, P, Fe, Al, and other trace impurities. The Si-Mg melts are solidified under a controlled cooling rate and then they are treated by acid leaching through using dilute HCl solutions. The effect of magnesium concentration on the segregation and removal of the impurities, in particular P and B, is studied by electron microscopy and chemical analysis techniques. It is indicated that the process is effective for removing the impurities, and the purification is depending on the amount of Mg added into the silicon.

Keywords: Silicon, magnesium, solar silicon, solidification, segregation, leaching, magnesium silicide

1 INTRODUCTION

Silicon is produced through carbothermic reduction of quartz in the electric arc furnace; the product being metallurgical grade (MG-Si), which is the basis of many metallurgical, chemical and electrical applications. MG-Si contains about 99wt% Si with impurities such as Fe, Al, Ti, Ca, B, and P [1]. Solar Grade Silicon (SoG-Si) with more than 6N (+99.9999) purity is produced from MG-Si and it is the feedstock for the production of silicon solar cells. Ultra-high pure silicon with around 11N purity, which is called Electronic Grade Silicon (EG-Si), is used for the fabrication of electronic devices [1]. The majority of SoG-Si and all the EG-Si existing in the market are currently produced from MG-Si through the *Siemens process* or newly developed *fluid bed reactor (FBR) technology*. In these chemical processes, pure silicon is deposited on rods or seeds from a gas phase, which is produced from MG-Si and contains gaseous compounds of silicon such as $SiHCl_3$ or SiH_4. The Siemens process in particular is an expensive process with regard to high energy consumption [2, 3]. The production of SoG-Si through *metallurgical refining processes* is more energy efficient and environmentally friendly than chemical route which in turn may encourage a faster growth of the global PV market. This has been the motivation of the development of several refining processes such as the *ELKEM Solar Silicon (ESS) process* in Norway, *NS Solar process* in Japan, *Chinese metallurgical routes* such as those of *Shanghai Propower* and *Ningxia,* and the *Photosil process* in France [3]. Solvent refining of silicon using Al as the refiner is another method as reported by *Silicor Materials* [4]. Almost all the present impurities in MG-Si except B and P can be removed by directional solidification due to large segregation is solidification[5], which is usually a final key process step in metallurgical approach. An effective process for silicon purification is acid leaching in which a metal e.g. Ca is added to silicon melt and the solidified silicon is hydrometallurgically treated by acid. The impurities are segregated during solidification and concentrated on the grain boundaries in association with a silicide phase such as $CaSi_2$ soluble in the acid. Therefore, the impurities are separated physically or dissolved chemically into the solution from the silicon grains and highly pure silicon particles are produced.

Acid leaching process for silicon purification have been studied through several studies [6,7,8,9,10,1112] and the process is effective for the removal of many impurities, in particular the elements with large segregation in solidification such as Fe, Al, Mg, Sb, Sn, Zn, Cu, Au, Ni and Ti. The process is also effective for partial removal of P. The purification of silicon is depending on the particle size of the silicon particles subject to acid leaching and better purification is obtained when smaller silicon particles are treated [6, 10]. The removal of impurities in acid leaching is higher at higher temperatures [8, 10]. And it has been observed that the purification is more effective when HCl and HF acids are used in comparison with other types of acids [6, 8, 11]. Inspecting these studies reveals that the silicon has been treated without the addition of any element to liquid silicon or through the addition of Ca into liquid silicon before solidification.

In the present study, magnesium is added into molten silicon and the effect of this element addition in silicon purification by acid leaching is investigated. Electron microscopic examination of Mg-doped silicon up to 10wt%Mg is studied using. Moreover, the purification of the samples is studied through chemical analysis study before and after leaching.

2 EXPERIMENTAL PROCEDURE

2.1 Silicon doping by magnesium

A metallurgical grade silicon (MG-Si) was used in the present study. , and its chemical composition is shown in Table I.

Table I: Chemical composition of used silicon (wt%).

Mg	Al	Fe	Ca	P	B
0.005	0.16	0.17	0.014	0.0027	0.0060

Magnesium silicide (Mg_2Si) with purity above 99% was mixed with silicon in different portions and Mg-containing silicon alloys with 0.2, 0.6, 2.1, 3.6 and 4.9 wt% Mg were obtained. The mixtures were then heated and smelted in high density graphite crucibles at 1500°C in an induction furnace. The materials were then slowly cooled down and solidified with average of 10°C/min.

2.2 Leaching

The solidificed silicon samples were crushed and particles with 1-3 mm particle size were leached by hydrochloric acid solutions. The leaching was performed in two steps; first by using 15%HCl at 80°C for 3 hours, and then by 2.5%HCl at 80°C for 3 hours.

Silicon samples before and after leaching were analyzed by high-resolution Inductively Coupled Plasma-Mass Spectroscopy technique. Moreover, metallography samples were prepared and their microstructures was studied through scanning electron microscope.

3 RESULTS AND DISCUSSION

The results of the present work are presented and discussed as follows.

3.1 Microstructural analysis

The microstructure of polished solidified Mg-containing silicon samples were studied using secondary electron (SE) and backscattered electron (BSE) modes of electron microscope supported by Energy dispersive mass spectroscopy (EDS). Figure 1 shows the microstructure of the samples with different Mg concentrations. As seen in all the samples we have three phases; a grey matrix phase, a dark grey secondary phase and a bright phase. Based on the EDS results shown in Figure 2, the grey phase is silicon matrix (Fig. 2.a). In addition, some Si particles distributed in the dark grey phase. The dark grey phase is Mg_2Si phase based on the quantified EDS spectrum (Figure 3.b). Figure 1 shows clearly that the amount of Mg_2Si is increased with increasing Mg. Moreover, the shape of this phase becomes different at high Mg concentrations. According to Figure 2, the bright phase contains mainly Si, Fe, and Al and small concentrations of Ca and Mg. This phase is observed bright compared to the other phases in backscattered mode due to the existence of heavier elements in it. Since Si, Fe and Al are the main elements in the bright phase with variations in the chemical composition, we call it here Si-Fe-Al phase.

The microstructural images show that for low Mg concentrations such as images (a), (b) and (c) in Figure 1 the Mg_2Si and the Si-Fe-Al are located between the silicon grains. It was observed that for Mg concentrations lower than 2.1 wt% there is not significant Si particles together with Mg_2Si. This may be due to the growth of secondary Si phase due to a eutectic reaction over the primary Si grains. This may be due to short diffusion distances available for Si precipitation. For higher Mg concentration, in which significant amount of liquid is transformed to Si+Mg_2Si, based on the Si-Mg phase diagram, secondary silicon phase is formed through the eutectic reaction between the grains which causes the existence of Si particles together with Mg_2Si between the silicon grains as we see in image (d) in Figure 1. The Si-Fe-Al phase is obviously formed at the end of solidification and for low Mg containing melt it is located in contact with the primary silicon grains with more less the same distribution of Mg_2Si phase. In high Mg containing melts, however, the Si-Fe-Al phase is also observed in the eutectic Si+Mg_2Si phase as observed in image (d) in Figure 1.

Microscopic study shows also that there is some cracks in Si matrix and in addition significant amount of cracks in Mg_2Si phase.

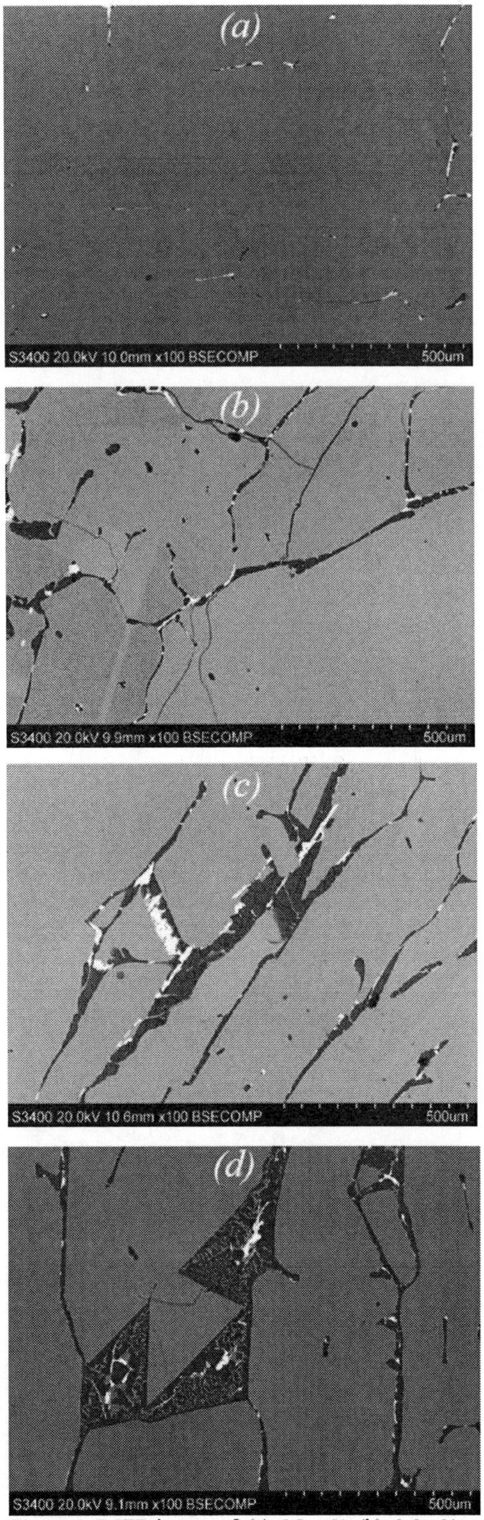

Figure 1: DSED images of (a): 0.2 wt%, (b): 0.6wt%, (c): 2.1wt%, (d): 3.6% Mg-containing silicon.

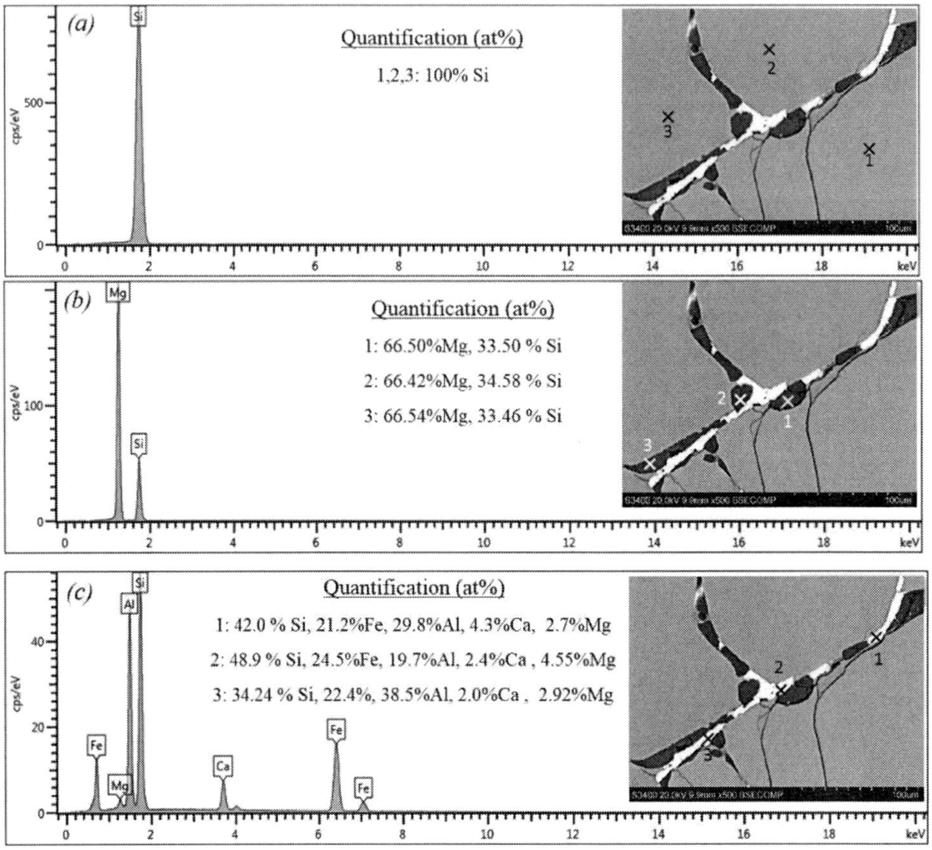

Figure 2: EDS analysis of coexisting phases in samples.

3.2 Leaching effect

Electron microscopy of specific locations of some samples after leaching in different durations indicated the effect of leaching on the removal of Mg_2Si phase adjacent to the silicon grains (Figure 3).

The solution of Mg_2Si causes the physical separation of intermetallic phases as observed in Figure 3. All the impurities, in particular trace elements such as P and B in the both phases are therefore removed from the silicon.

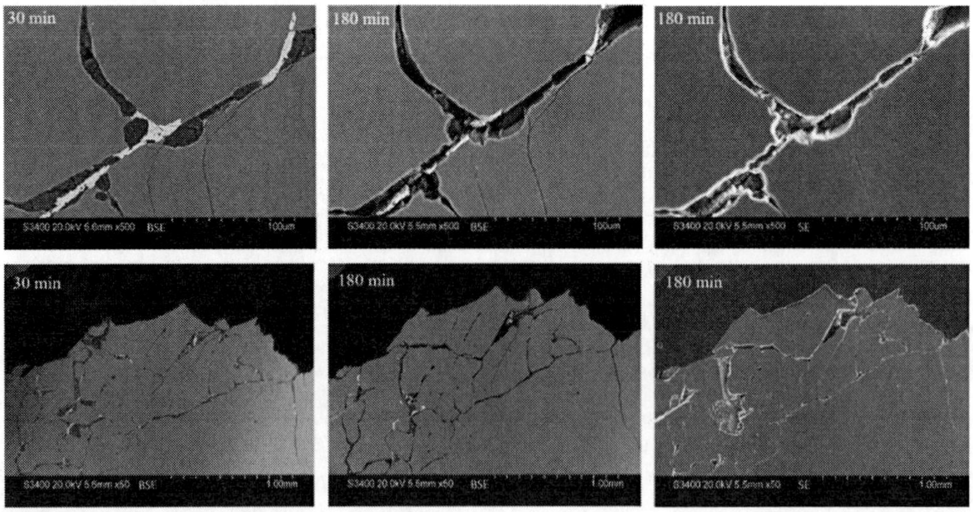

Figure 3: SE and BSE images of 2.1%Mg-containing sample during different durations of leaching.

The chemical analysis of the samples before and after leaching shows that there is significant decrease in the concentrations of the impurities in silicon. Figure 4 shows the removal degree of the impurities calculated based on the chemical composition analysis by the following equation:

$$Removal\ degree = \frac{C_i - C_f}{C_i} \times 100$$

Where C_i and C_f are concentrations of the impurity before and after leaching, respectively. Figure 4 shows that all the impurities are removed to some degree through acid leaching. The removal degree is increased with increasing the amount of the added Mg into silicon for the majority of the impurities. However, the addition of Mg in higher than 2.1% does not show a significant effect in terms of B removal. It may however be generally concluded that the segregation of the impurities and their separation through acid leaching is improved with increasing the amount of Mg in silicon.

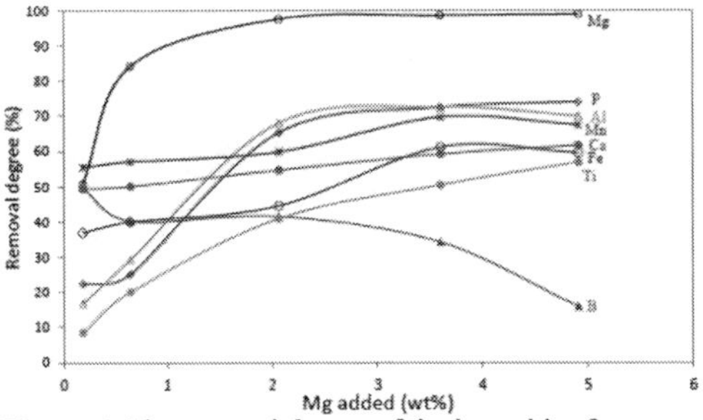

Figure 4: The removal degree of the impurities for different concentrations of Mg in silicon.

The chemical analysis measurements show that the removal of P is significantly improved with increasing the concentration of Mg in silicon as clearly shown in Figure 5. This may indicate that the segregation of P is better when more Mg exist in the silicon melt.

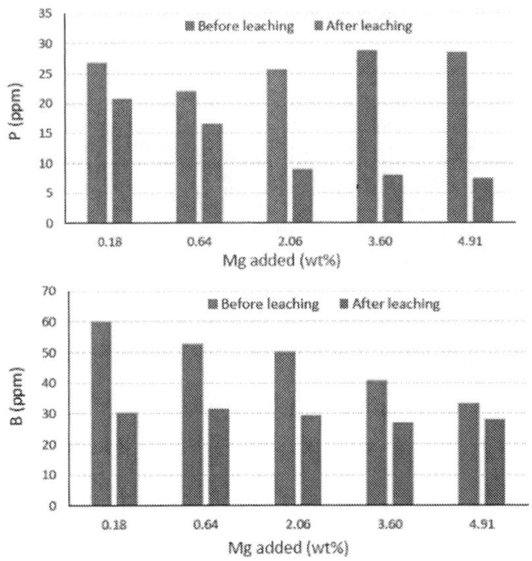

Figure 5: The concentrations of B and P before and after acid leaching of different Mg-containing silicon samples.

The B removal results in Figure 5 show that the B segregation and separation is minimally increased with Mg addition in small amount to 2.1wt%, and higher Mg concentrations do not affect B removal significantly. Figure 5 shows that B is removed around 50% in low Mg concentrations, and this may indicate that the segregation of B during solidification is improved.

The presented data in this research are based on selected leaching conditions and optimal purification conditions may be obtained through more experimental work.

4 CONCLUSIONS

- The alloying of molten silicon with magnesium followed by slow cooling and solidification shows the precipitation of Mg_2Si and Fe-Al-Si-Ca intermetallic phases between silicon grains.
- The leaching of silicon samples by 15%HCl solution is effective for dissolving Mg_2Si phase.
- The removal of Mg_2Si phase causes significant removal of the Fe-Al-Si-Ca intermetallic particles due to their sizes and distribution along Mg_2Si particles.
- The ICP-MS results show that all types of the impurities are removed from silicon through Mg addition followed by acid leaching.
- Phosphorous in particular shows better segregation with increasing the amount of the added Mg to silicon.

ACKNOWLEDGEMENT

Authors would like to appreciate the funding from research domain RD3 in SFI[13]-metal production and the Research Council of Norway.

REFERENCES

[1] A. Schei, J.Kr.Tuset, H. Tveit: Production of high silicon alloys. Trondheim, Tapir Forlag; 1998.
[2] A.F.B. Braga, P.R. Zampieri, J.M. Bacchin, P.R. Mei: Solar Energy Materials & Solar Cells, 92 (2008), 418.
[3] J. Safarian, G. Tranell, M. Tangstad: Energy Procedia, 20 (2012), 88.
[4] M. Heuer, M. Kaes, A. Turenne, T. Jester, Silicon for the chemical and solar industry XII, pp. 253.
[5] R.H. Hopkins, A. Rohatgi: J. Cryst. Growth, 1986, vol. 75, pp. 67-79.
[6] J. Dietl, Solar Cells, 10 (1983), 145.
[7] J.M. Juneja, T.K. Mukherjee, Hydrometallurgy, 16 (1986), 69.
[8] I.C. Santos, A.P. Goncalves, C. Silva Santos, M. Almeida, M.H. Afonso, M. Joaquina Cruz, Hydrometallurgy, 23 (1990), 237.
[9] T. Shimpo, T. Yoshikawa, K. Morita, Metall. Mat. Trans. B, 33B(2004), 277.
[10] Y. Zhan-liang, M. Wen-hiu, D. Yong-nian, Y. Bin, L. Da-chun, D. Wie-ping W. Jin-xian, Trans. Nonferrous Met. Soc. China, 17 (2007), s1030.
[11] Z. Jian, L. Tingju, M. Xiaodong, L. Dawei, L. Ning, L. Dehua, Journal of Semiconductors, 30 (2009), 053002-1.
[12] F. He, S. Zheng, C. Chen, Metall. Mat. Trans. B, 43B (2012), 1011.
[13] SFI Norwegian acronym for Centre for Research driven Innovation (CRI).

CONTAMINATION OF SILICON DURING ELECTRON BEAM MELTING

Alexey Kravtsov, Anatoly Kravtsov
SIA "KEPP EU"
LV-1034, Carnikavas 5, Riga, Latvia

ABSTRACT: In our previous works the possibility of growing silicon rods suitable for FZ process was shown, however, the use of the high-energy process such as electron beam melting (EBM) may lead to generation of impurities which substantially reduce the efficiency of solar cells. EBM grown rods were recrystallized with FZ. Due to FZ process purity all impurities are relied to EBM process and rod growth. FTIR was used to conrtol levels of impurities. Slight removal of phosphorus and the generation of boron and aluminium were noticed. The level of contamination is hundreds of ppta and is not crucial for the quality of the single crystals used in solar energy. The attempt to reliably examine the level of other impurities with GDMS analysis has failed due to insignificant quantities of impurities and the high resistance of the monitored sample. High lifetime at the end of the grown single crystal indirectly evidences of the absence of significant contaminations. Thus, single crystals obtained utilising the described technical processes are a promising material for the production of highly-efficient solar cells.
Keywords: Electron beam, c-Si, float-zone, purification, pulling

1 INTRODUCTION

In our previous works [1-3] the possibility of growing silicon rods suitable for FZ process was shown. Technical advantages of using FZ for solar power were shown repeatedly [4-5], however, the use of the high-energy process such as electron beam melting (EBM) may lead to generation of impurities which substantially reduce the efficiency of solar cells. The aim of this paper is to study the impurity composition of rods grown using electron-beam heating.

2 EXPERIMENTS

To analyse the purity of the process of growing silicon rods using EB heating, rods [2] have been grown from melt and were used as the source rods for obtaining dislocation-free single crystals in the FZ process. Obatined rods have been mechanically and chemically treated as usual, then the dislocation-free single crystals were grown in argon atmosphere. Due to the expected concentration of impurities of not more than 10^{13}-10^{16} at/ccm, it was decided to analyse the content of impurities by the FTIR method in samples cut from single crystals.

In this case, the reliability of estimates is determined by well-known regularities of impurity distribution in FZ process [6] and the purity of the process itself. The generation of impurities in FZ does not exceed 10^{11} at/ccm. The disadvantages of FTIR method include the absence of information base for monitoring of a number of impurities in silicon, which we formerly notices in the EBM process [7]. To detect metal impurities, it is more convenient to use GDMS if the impurity level is higher than or comparable to the method's detection limit. All metal impurities in silicon have distribution coefficients of 10^{-2} – 10^{-6}. Thus it is natural to expect maximum metal concentration in the end of the FZ single crystal, namely in the area where the residual melt crystallisation occurs at the end of the process. Samples for measurements were taken from these parts of crystals.

The rod No. 52 was grown in the VU2PCZ furnace (Fig. 1). Polycrystalline silicon with the properties listed in Table 1 were used as a raw material. Properties of grown rod shown in Table 2.

Table I:

Impurity	Content
Donors (P, As,Sb), max	300 ppta
Acceptors (B), max	100 ppta
Total of bulk metals (Fe, Cu,Ni, Cr, Zn, Na), max	2 ppbw
Surface Metals	
Fe, max	5 ppbw
Cu, max	0,5 ppbw
Ni, max	0,5 ppbw
Cr, max	0,5 ppbw
Zn, max	1 ppbw
Na, max	3 ppbw

Figure 1: Sample rod No. 52 grown from polysilicon in EBM process

Table II: Parameters of sample rod No. 52.

Resistivity, Ω•cm		Diameter	
min	max	min	max
400	1700	85	99
Conductivity	Length	Length of cylinder	
P-type	305,7	276,8	

Before growing single crystals, the rod was mechanically and chemically treated. A single crystal of 60-70 mm diameter was grown out of it in the VZA1505 furnace, in IKZ (Germany). Figure 2 shows the marked places where samples for FTIR were taken.

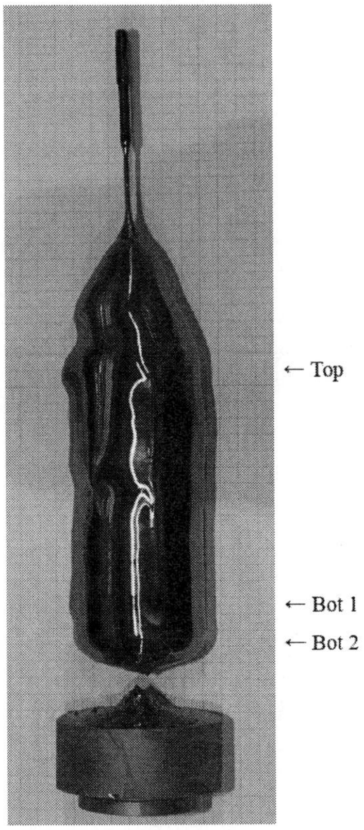

Figure 2: FZ single crystal grown from rod No. 52 with FTIR samples position indicated

3 RESULTS AND DISCUSSION

3.1 Results of analysis

Results of FTIR analysis, resisistivity and lifetime measurements for grown FZ single crystal are shown in Table 3.

Table III: Results of FTIR analysis, resisistivity and lifetime measurements for FZ single crystal grown from rod No. 52.

Sample	52Top	52Bot1	52Bot2
Impurity content, ppta:			
B	264	392	466
P	46	240	258
Al	34	94	96
Conductivity	P-type	P	P-type
Resistivity	2000	2600	3800
Lifetime, us	Not mea-sured	PCD: 1320 Sinton: 667,9	Not mea-sured

It is shown [6], that no significant fluctuations of the impurity concentration should occur in the cylindrical part of the single crystal grown in the FZ process when the uniform source rod is used.

However, Table 3 shows significant differences. At the TOP point, concentrations differ significantly from the concentration of impurities in the original polysilicon. Next, the concentration of all impurities grows along the length of the crystal. Noting that the generation of impurities during FZ process does not occur, we should relate this change of the concentration to an increase of the impurity concentrations in the source rod. Such increase should have originated during the growth of the source rod in the EBM process.

It is obvious that during the EBM process of source rod growth aluminium was introduced. Increase of impurity content along the crystall indicates that the concentration of such impurity increased during EBM process. Difference between determined content of impurities and their content in source polysilicon is shown at Fig. 3.

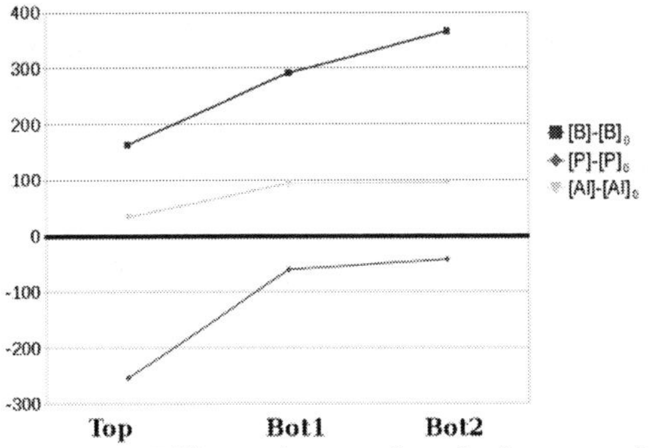

Figure 3: Difference between determined content of impurities in single crystal grown from rod No. 52 and their content in source polysilicon (ppta)

The phosphorus content in the crystal is lower than stated in the polycrystalline silicon specifications and boron and aluminium concentrations are higher.

However, when supply of fresh material into the zone stops, segregation takes place and impurity concentrations vary along the crystal axis in accordance with [6]. To analyse the presence of other impurities, the samples were taken from the last crystallised zone of the single crystal (Fig. 4). In that part concentration of introduced contaminants increases significantly. and can show the presence of introduced contaminants Thus monitoring of the sample cut from such part can show contaminants presence if their concentration exceeds the detection limit of the chosen monitoring method. The results of the GDMS analysis are given in Table 4.

Table IV: Results of GDMS analysis.

Impurity	[B]	[Na]	[Al]	[P]	[Ca]	[Ti]	[Cr]
Content, ppbw	60	20	<20	40	<500	<5	<5
Impurity	[Mn]	[Fe]	[Ni]	[Cu]	[As]	[Sb]	
Content, ppbw	<5	<150	<40	<25	<100	<20	

Figure 4: End of the crystal with highlighted zone for GDMS samples taking.

As the Table 4 shows, besides impurities of boron and phosphorus, only sodium has the concentration that exceeds the detection limit. The results obtained are weakly correlated with those obtained in FTIR measurements. The most probable cause is some instability or non-homogeneousity during glow discharge caused by sample high resistivity.

SIMS analysis was used to check P and B content in 52Top sample, result was 600 and 86 ppt respectively, which agrees with FTIR results.

3.2 Discussion

The conducted process of growing of rods with EBM is a type of floating-zone refining. For the formation of the zone when heating from the top, it is required to hold melt at constant power in order to heat up and thermally stabilise the system. As the rod is pulled (material withdrawal), the zone moves deeper supplying the melt with the source material. Sample 52 revealed that the purification of silicon from phosphorus and it's contamination with boron and aluminium occur during the melt holding for thermal stabilisation of the system. After pulling of a certain part of the melt, the supply of the zone with the source silicon becomes significant and phosphorus concentrations are rising to the level of the source material. At the same time despite supplying the source polysilicon zone, the boron and aluminium concentrations increase evidencing the generation of impurities in the melt during the entire process. However, the level of introduced impurities does not exceed hundreds of ppta and at end of the crystal reaches a level of 100-300 ppta, which meets the single crystal quality requirements for solar energy. The minor carriers lifetime obtained at the end of the FZ single crystal shows a good potential for solar cell manufacturing.

4 CONCLUSIONS

1) The introduction of impurities controlled by FTIR was studied during the pulling of the source rods with the EBM. It has been established that a slight removal of phosphorus impurities and the generation of boron and aluminium impurities occur during the process.
2) The level of contamination is hundreds of ppta and is not crucial for the quality of the single crystals used in solar energy
3) The attempt to reliably examine the level of generation of a wider range of impurities has failed due to their insignificant quantities and the high resistance of the monitored sample, which hampers GDMS analysis; high lifetime at the end of the grown single crystal indirectly evidences of the absence of significant contaminations.
4) The single crystals obtained utilising the described technical processes are a promising material for the production of highly-efficient solar cells.

5 ACKNOWLEDGEMENTS

Part of work covering minimization of background impurities level and creating a technology for obtaining silicon feedstock applicable in power electronics is conducted by SIA "KEPP EU" and CFI LU, Riga, Latvia and supported by EEA grants program "National climate policy" supervised by VRAA on a project Nr.2/EEZLV02/14/AK/010/005 „Energoefektīvas tehnoloģijas izstrāde polikristāliskā silīcija ražošanai" – Create the energy-efficient technology to produce polycrystalline silicon.

6 REFERENCES

1. A. Kravtsov, "Ingots Pulled with Electron Beam Heating from Skull - a New Feedstock for FZ Crystals Applicable for Solar Cells", in IEEE PVSC40, Denver, CO, USA, 2014
2. An. Kravtsov, "New Feedstock for FZ Crystals Applicable for Solar Cells", in 29th European Photovoltaic Solar Energy Conference and Exhibition, Amsterdam, Netherlands, 2014, pp. 703-705
3. Kravtsov, Al; Kravtsov, An. ; Fuksis, R. ; Pudzs, M., Ingot pulling with electron beam heating: Process enhancements, in IEEE PVSC40, New Orleans, LA, USA, 2015
4. J. Vedde, T. Clausen, J. Borregaard and P. Kringhøj, "Float-Zone Crystal Growth For Pv – Where Is The Future?", 23th EU PVSEC 2011, pp. 703-705
5. J. Vedde, T. Clausen and L. Jensen, "Float-Zone Silicon For High Volume Production Of Solar Cells", in 3rd World Conference on Photovoltaic Energy Conversion, Osaka, Japan, 2003
6. William G. Pfann, «Zone melting», 2nd Edition, John Wiley & Sons, 1966
7. A. Kravtsov, A. Kravtsov, "Electron Beam Silicon Purification", 26th EU PVSEC 2011, pp. 1814-1816

USE OF THE CZOCHRALSKI GROWTH TECHNIQUE TO REMOVE DEFECTS OF POLYCRYSTALLINE UPGRADED METALLURGICAL GRADE SILICON

Francisco C. Marques[1], Andressa D. S. Côrtes[2], Rafael B. Merlo[1], Douglas S. Silva[1], Gustavo A. Viana[1], and Paulo R. Mei[2]

[1]Institute of Physics "Gleb Wataghin", State University of Campinas, Campinas, SP, 13083-859, Brazil

[2]Faculty of Mechanical Engineer, State University of Campinas, Campinas, SP, 13083-970, Brazil

ABSTRACT: Upgraded metallurgical grade silicon (UMG-Si) was developed using two different metallurgical routes and used in the fabrication of solar cells. For that purpose, metallurgical grade silicon (MG-Si) was purified by the vacuum degassing (VD) technique. However, solar cells made with this polycrystalline material have poor efficiency due to grain boundary defects. Thus, monocrystalline ingots were growth by the Czochralski (CZ) technique to get monocrystalline wafers, free of grain boundaries. The combined VD with CZ growth routes provided UMG-Si with 99.999% purity and conventional p-n silicon solar of 13% efficiency.
Keywords: silicon, Czochralski, metallurgical-grade.

1 INTRUDUCTION

Current commercial photovoltaic (PV) panels are mainly made using solar cells fabricates with electronic grade crystalline silicon (EG-Si), where the silicon feedstock is purified by the Siemens process. In order to reduce cost in the fabrication of the solar cells, the Siemens process has been replaced by simple processes, but able to produce silicon with purity called "solar grade" (SG-Si), in which high efficient solar cells can be processed, without losing much in the efficiency compared to solar cells made with EG-Si [1, 2]. Some of these approaches adopts a modified Siemens process [3,4].

In this work we adopted two metallurgical routes, vacuum degassing (VD) and Czochralski (CZ) growth, to purify MG-Si to a level that makes possible the fabrication of solar cells of reasonable good efficiency. The reduction in the impurities is achieved mainly by the VD technique. However, this process is used to prepare wafers of polycrystalline silicon. Solar cells made with this material are poor due to a combination of low purity and grain defects. Thus, to eliminate the grain defects we adopted the CZ technique to supply monocrystalline ingots. The CZ technique reduces only slightly the concentration of impurities, but eliminates the grain defects, since one can get monocrystalline silicon using this technique. Thus, using these two processes we could fabricate solar cell of 13 % with potential for higher efficiency using better fabrication process.

2 EXPERIMENTAL

Initially 500g of MG-Si was melted in an electron-beam system (EMO-LEW model) to degas high vapour pressure contaminants. In this process, the melted silicon was held in a vacuum chamber for 30 minutes at a base pressure of 10^{-2} Pa. The process was performed several times to supply few kilograms of purified silicon, which was used in a CZ system to fabricate a monocrystalline silicon ingot. The impurity concentration of the silicon feedstock and the UMG-Si were determined by glow discharge mass spectroscopy (GDMS).

Solar cells were fabricated using a conventional p-n silicon junction, where phosphorous (P) was thermally diffused using $POCl_3$ precursor. To improve que quality of the junction, a gettering process was developed, using P diffusion. Aluminium (Al) back surface field was applied at 800°C. Thermally evaporated Ti/Palladium (Pd)/Silver (Ag) was used for the top and bottom contacts. A tin dioxide (SnO_2) antireflective coating of 75nm thick was deposited adopting the spray pyrolysis technique.

3 RESULTS AND DISCUSSION

The main impurities found in the silicon feedstock, determined by GDMS, are displayed in Fig. 1 together with the impurities of the UMG-Si obtained only by the application of the VD technique (MG-Si+VD) and UMG-Si obtained with the combination of VD and CZ growth (MG-Si+VD+CZ). One can observe that the main reduction in the impurity concentration is achieved by using the VD technique, which was effective for P, Ti, Fe, Al and many other impurities not displayed in the figure. This technique is not good to reduce boron (B) since the vapour pressure of boron is lower than the operating pressure of the electron-bean system (10^{-2} Pa). Adopting the VD only we obtained silicon with 99.9986% purity.

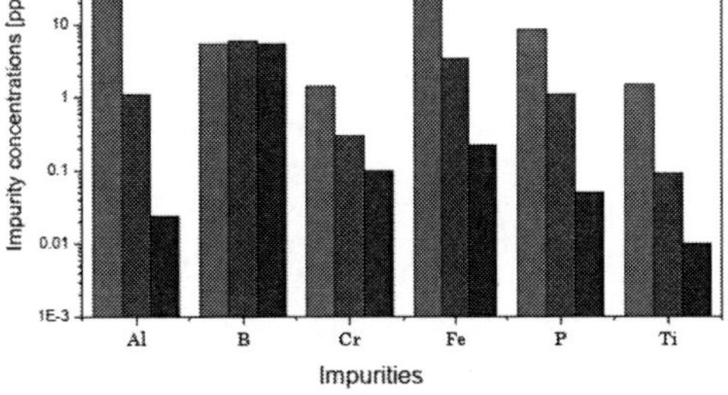

Figure 1: Impurity concentration of the silicon feedstock (MG-Si), UMG-Si obtained by vacuum degassing and UMG-Si obtained by a combination o vacuum degassing and CZ

Using a few kilograms of silicon purified by the VD technique we prepare an ingot using a CZ system. Fig. 1 shows that the impurities concentrations (MG-Si+VD+CZ) also reduce during the CZ process due to segregation phenomena. As can be observed, CZ is not effective in reducing the concentration of B because it has a segregation coefficient close to 1. However, the combination of VD and CZ growth techniques produced material with 99.999% purity.

Using the combination of VD and CZ we obtained monocrystalline silicon with reasonable high purity and free of grain boundary defects. Using these materials we fabricated solar cells in wafers of about 300 micron thick adopting a conventional process as described in the experimental section.

In order to improve the cell efficiency we developed gettering processes based in other experiments using polycrystalline EG-Si reported in literature [5-7]. The idea here was to investigate the differences in the gettering process adopted for poly-EG-Si versus mono-UMG-Si, i.e., high purity/poly versus low purity/mono silicon wafers.

Three different gettering processes (1, 2 and 3) were adopted with the following steps: 1) P diffusion at 900°C, 40 min followed by chemical etch, followed by P diffusion at 900°C, 18 min as the final n^+-layer of the n-p junction; 2) P diffusion at 900°C, 40 min followed by annealing at 900°C, 60 min, followed by chemical etch and the final P diffusion at 900°C, 18 min; 3) P diffusion at 850°C, 50 min, followed by annealing at 850°C, 120 min, chemical etch and final P diffusion at 850°C, 50 min. The final diffusion of each process provided an n^+-layer with sheet resistance of about 20-30 Ohm/square. The solar cells were fabricated using mono-EG-Si and mono-UMG-Si. The basic differences between the three gettering processes were the temperature (850°C or 900°C) and the annealing times.

The effect of the gettering process can be evaluated by comparing the evolution of the diffusion length, Fig. 3. The gettering 1 provided the highest diffusion length for electronic grade silicon (EG-Si), while gettering 3 (850°C) provided the highest value for UMG-Si explaining why those gettering are the best for each material.

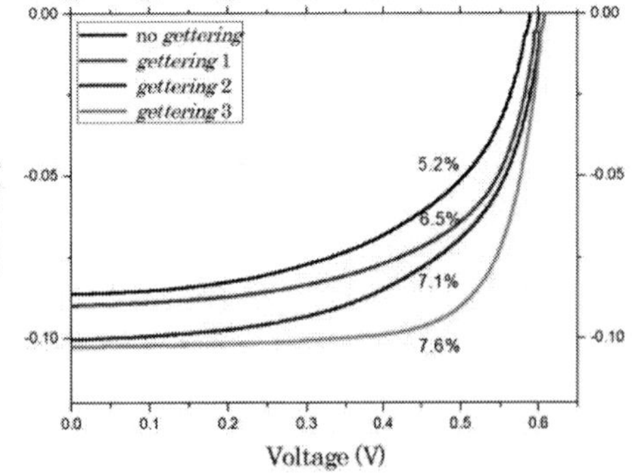

Figure 2: Current vs. voltage curves of the best solar cells fabricated with UMG-Si fabricated with different gettering process

The photovoltaic parameters were obtained by I x V curves, under 100 mW/cm^2 illumination using a solar simulator. Fig. 2 shows the best cell for each of the three gettering process and a the conventional process (without gettering). One can observe that the gettering 1 (900°C) was the best for monocrystalline EG-Si, but for the UMG-Si, the gettering 3 (850°C) provided higher efficiency. At the higher temperature (900°C), some trapped impurities are redistributed within the UMG-Si, which degrades the final efficiency. This effect is partially reduced by adopting the lower temperature of 850°C.

The highest efficiency displayed in Fig. 3 was 7.6 %. However, optimizing the fabrication process we could prepare solar cell on UMG-Si with efficiency as high as 13 %. Using the same processes we got efficiency of 15.5% in monocrystalline EG-Si. It is well know that much higher efficiency can be achieved in EG-Si. Commercial solar cell are fabricated today with efficiency of about 18%. Thus, much higher efficiency is expected for our UMG-Si.

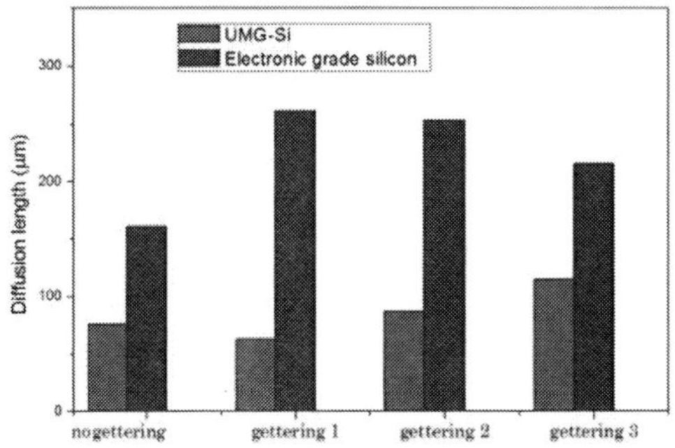

Figure 3: Carrier diffusion length of monocrystalline UMG-Si compared to electronic grade silicon (EG-Si)

Fig. 4 shows the difference in efficiencies of two groups of solar cells, i.e, polycrystalline silicon obtained from the VD process and monocrystalline silicon obtained from the combination of the VD process and CZ.

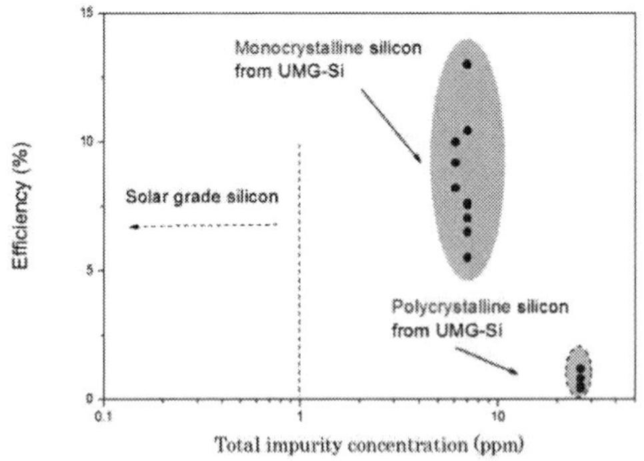

Figure 4: Efficiency vs. total impurity concentration of solar cells made with UMG-Si

The highest efficiency obtained with poly-UMG-Si was about ~1 %, while solar cells of 13% were obtained with mono-UMG-Si. One can notice that the improvement in the material quality is not significant after the CZ growth. The VD wafers (polycrystalline) has about 15ppm impurities, while the VD+CZ (monocrystalline) has about 7 ppm. However, the improvement in efficiency is significant. Thus, the reduction of the impurity concentration obtained by the CZ growth cannot explain the substantial increase in the efficiency. It might be related to the reduction of grain boundary defects that kills the diffusion length of poly-UMG-Si, since the impurity concentration is much higher than the accepted values for solar grade silicon.

4 CONCLUSIONS

We adopted a combination of two techniques to purify metallurgical grade silicon to a level of 99.999% purity and to fabricate solar cell of 13 %, with potential to achieve efficiency higher than 15%. The vacuum degassing technique was effective in reducing the impurity concentration to a reasonable level, while the CZ was efficient in eliminating the grain boundaries defects by supplying a monocrystalline ingot. The difference in efficiency of the solar cells made with materials of about the same concentration of impurity but one been polycrystalline and the other monocrystalline were significant. That improvement justified the extra cost due to the CZ process, since it is also used for the fabrication of commercial silicon wafers using silicon feedstock purified by the Siemen process

ACKNOWLEDGMENT

This work was founded by FAPESP, CNPq, INES/INCT and CAPES. We are also in debt to RIMA Industrial S/A for supplying MG-Si and for performing the CZ growth. We are also grateful for A. A. Coelho for his support in the use of the laboratory CZ system.

REFERENCES

[1] S. Pizzini, "Solar grade silicon as a potential candidate material for low-cost terrestrial solar-cells," Sol. Energy Mater. 6, 253–297, 1982.
[2] P. Woditsch and W. Koch, "Solar grade silicon feedstock supply for PV industry," Sol. Energy Mater. Sol. Cells. 72, 11–26, 2002.
[3] C. P. Khattak, D. B. Joyce, and F. Schmid, "A simple process to remove boron from metallurgical grade silicon," Sol. Energy Mater. Sol. Cells. 74, 77–89, 2002.
[4] S. Pizzini, "Towards solar grade silicon: Challenges and benefits for low cost photovoltaics," Sol. Energy Mater. Sol. Cells 94, 1528–1533, 2010.
[5] D. Macdonald and A. Cuevas, "Understanding carrier trapping in multicrystalline silicon," Sol. Energy Mater. Sol. Cells. 65, 509–516, 2001.
[6] S. Ponce-Alcantara, C. Del Canizo, and A. Luque, "Adaptation of monocrystalline solar cell process to multicrystalline materials," Sol. Energy Mater. Sol. Cells 87, 411–421, 2005.
[7] T. Saitoh, T. Warabisako, E. Kuroda, H. Itoh, S. Matsubara, and T. Tokuyama, "Impurity gettering of polycrystalline solar-cells fabricated from refined metallurgical-grade silicon," IEEE Trans. Electron Devices 27, 671–677, 1980.

BEHAVIOUR OF THE SLIP-CAST CRUCIBLE AS A CONTAMINATION SOURCE DURING SILICON DIRECTIONAL SOLIDIFICATION

Hanna Vaksvik Skarstad[1], Antoine Autruffe[1], Gaute Stokkan[2], Marisa Di Sabatino[1]
[1]Department of Materials Science and Engineering, NTNU NO-7491 Trondheim, Norway. E-mail: hannavak@stud.ntnu.no, antoinea@material.ntnu.no, marisa.di.sabatino@material.ntnu.no.
[2]SINTEF Materials and Chemistry, Department of Metallurgy, P.O.Box 4760 Sluppen, NO-7465Trondheim, NORWAY: E-mail: gaute.stokkan@sintef.no

ABSTRACT: The present work investigates the silica slip cast crucible's role as a contamination source of multicrystalline silicon during directional solidification. The diffusivity of iron in the crucible material and quartz glass was quantified at three different temperatures (1100, 1200 and 1300 °C). Samples of quartz glass and samples of a silica slip cast crucible were heat treated in contact with solid steel. The diffusion profile of iron in quartz glass was obtained with line scans using an Electron Probe Micro Analyser. The diffusion profile of iron in the silica crucible was obtained with Semi-quantitative X-ray mapping and image analysis. The concentration profiles of iron were fit to the solution of the diffusion equation and the temperature independent diffusion coefficients could also be obtained. The diffusivity of iron in both the crucible and quartz glass increased with temperature and the diffusivity was higher in the porous crucible material than in the denser quartz glass. The temperature independent diffusion constants and activation energies were estimated to be $1{,}12*10^{-4}$ cm²/s and 1,52 eV for the crucible, respectively, and $8{,}1*10^{-5}$ cm²/s and 1,8 eV for the quartz glass, respectively. These results indicate that the impurity transport is strongly dependent on the crucible structure. The estimated transport parameters of iron through the crucible material were applied in a numerical model that predicts the contamination of silicon during directional solidification.
Keywords: SiO_2, Multicrystalline Silicon, Photovoltaic

1 INTRODUCTION

Defects and impurities in silicon interact during the crystallization process to degrade the solar cell efficiency. The slip cast solidification crucible is generally considered a significant source of contamination of multicrystalline silicon ingots during directional solidification [1-3]. Little is known about transport of impurities through the slip-cast crucible. Quantification of impurity transport parameters as well as knowledge about the impurity behaviour is necessary in order to optimize the crucible material and reduce the silicon contamination.

The slip cast crucible consists of sintered quartz glass particles, processed from natural quartz, which contains impurities such as metal elements. The crucible structure is porous, contains pores and grains, and the material has lower density than amorphous silica glass. During directional solidification the crucible is coated with a silicon nitride layer. The impurities in both the crucible and coating can diffuse through the material and be incorporated into the silicon. Research has shown that the diffusivity measured in amorphous silica is considerably slower than the one measured in the slip cast crucible material [2,4-6]. This indicates that the structure of the crucible affects the degree of silicon contamination.

The present work investigates the diffusivity and solubility of iron in fused quartz glass and in a silica slip-cast crucible, which are representative of the crucible structures used for Czochralski and multicrystalline silicon, respectively. Samples of an actual slip-cast crucible and samples of quartz glass were cut and heat treated in contact with iron at 1100, 1200 and 1300 °C. Electron Probe Micro Analyzer (EPMA) line scans were performed on the quartz glass and semi-quantitative X-ray mapping on the crucible material. The mapping intensities were averaged and converted into concentration profiles by Matrix Laboratory (MatLab). The diffusivity and solubility of iron in both materials were quantified by fitting the concentration profiles of iron to the solution of the diffusion equation. The temperature independent diffusion coefficient and the activation energy were estimated by applying the Arrhenius equation. The transport parameters of iron through the crucible material were applied in a numerical model that predicts the contamination of silicon during directional solidification.

2 EXPERIMENTAL

2.1 Diffusion experiments

Samples of steel were used as the contamination source of iron in the experiments. The contact surfaces were polished prior to the heat treatment in order to ensure good contact between the samples. The samples were secured with Kanthal wire to ensure that the contact remained throughout the heat treatments. The Kanthal wire was not touching the contact surface and therefore not considered a contamination source. Figure 1 shows the set up of the diffusion couples with a) a sample of slip cast crucible material and b) a sample of quartz glass. All of the crucible samples were taken from the same part of the crucible to ensure small structure variation between the samples.

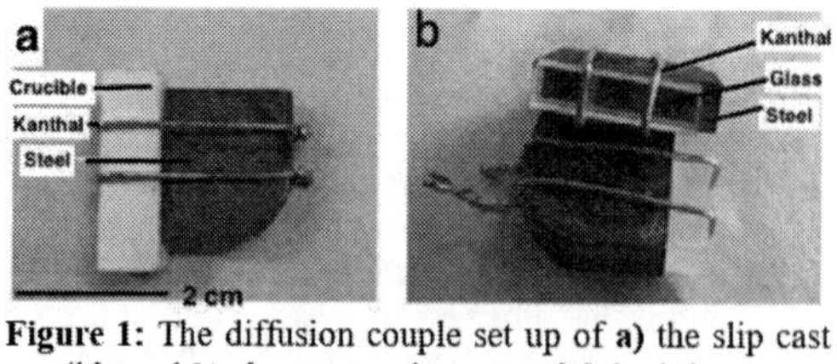

Figure 1: The diffusion couple set up of **a)** the slip cast crucible and **b)** the quartz glass material, both in contact with steel, and tightened with Kanthal wire.

The diffusion couples were heat treated at 1100, 1200 and 1300 °C with different holding times.

2.2 Characterization method

In order to enable quantitative analyses of the silica samples, the cross sections were cut on a Conrad D-38678 Claustal-Zellerfeld diamond saw and embedded in epoxy resin. Line scans were performed on the quartz glass sample cross sections using a JEOL JXA-8500F EPMA. Figure 2a shows a backscatter image of a quartz glass cross section with an EPMA line scan from a sample heat treated at 1100 °C. The line scan starts at the sample edge and can be seen as a red line, going from a white iron oxide surface layer and approximately 100 µm along the sample cross section. The white iron oxide surface layer is formed within 10 minutes of the heat treatment and behaves as an infinite contamination source throughout the heat treatment. The time of surface layer formation is considered negligible compared to the duration of the heat treatment. The boundary concentration, C_0, is given in Equation 1 and is constant throughout the heat treatment. C_0 is the concentration at the interface between iron oxide layer and SiO_2. This concentration is assumed to correspond to the solubility of iron in amorphous quartz.

The slip-cast crucible material is very porous and line scans unfortunately cannot be performed on the surface. A secondary electron micrograph is shown in Figure 2b. The micrograph shows the porous structure of the silica crucible with grains up to 60 µm wide.

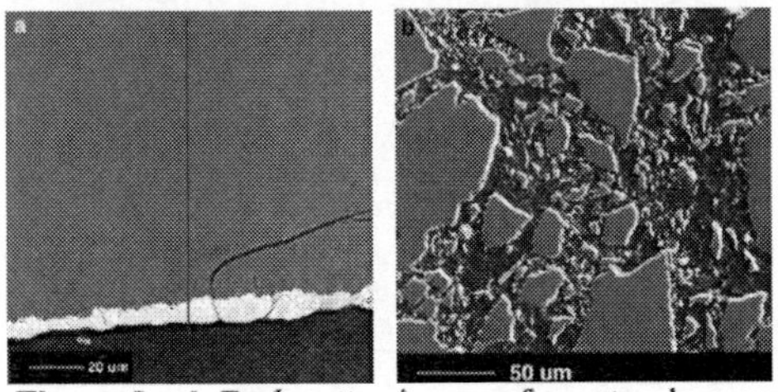

Figure 2: a) Backscatter image of quartz glass cross section with EPMA line scan of a sample heat treated at 1100 °C **b)** Secondary electron micrograph of the sintered quartz glass particles in the slip-cast crucible structure before heat treatment.

Semi-quantitative X-ray mapping was used to obtain the iron distribution in the crucible material. The X-ray mapping of the sample cross section, heat treated at 1100 °C holding temperature, is shown in Figure 3. Four micrographs were connected in Adobe Photoshop to obtain several areas, in order to avoid that large grains dominated the area investigated.

Figure 3: X-ray mapping of a crucible sample heat treated at 1100 °C holding temperature .

2.3 Analysis

The concentration of iron obtained from the quartz glass cross section, was fit to the solution of diffusion equation for a semi-infinite medium under the boundary condition $C(x,t) = C_0$, when $x = 0$ and $t>0$, and the initial condition $C(x,t) = 0$ when $x>0$ and $t=0$. The solution is:

$$C(x,t)=C_0(1\text{-erf}(x/2(Dt)^{0.5})) \quad (1)$$

Where C is concentration, C_0 is the boundary concentration, x is distance, t is holding time, D is the temperature dependent diffusion coefficient and erf is the error function [8].

The X-ray mapping micrographs were imported in to MatLab and a script was run to extract the diffusion profiles. This script converted the image into a matrix of grey scale intensities ranging from 1 to 256, where the intensity of every pixel was obtained. The maximum intensity was equivalent to 70 mass % of iron. The script converted the intensities into equivalent concentrations of iron. In the case of the crucible material, there is no clear interface between the surface layer and silica and the value of C_0 becomes less accurate value. From previous work [7] an approximate solubility of iron in quartz glass was found to be 0,5 mass%. It was considered a reasonable assumption that this solubility also applies to the crucible material. Only values below this concentration were considered dissolved and may be fit to Equation 1, which is the solution of the transport equation with no sink term (as precipitates would be). A diffusion profile for every column of the image was obtained and the concentrations of the columns were averaged to form one diffusion profile. The diffusion profile was fit to the solution of the diffusion equation the same way as quartz glass.

3 RESULTS

3.1 Diffusion profiles

The diffusion profile of quartz glass and the crucible material are shown in Figure 4 and Figure 5, respectively. The curve fit to the solution of the diffusion equation is shown as a red line. In Figure 4, the high concentration at the beginning of the line scan is due to the iron oxide phase that behaves as the infinite iron source. The diffusion profile starts at the interface, at C_0, where the continuous points of data begin. The scattered values are not included as it is likely that these values are approaching the detection limit of the EPMA instrument.

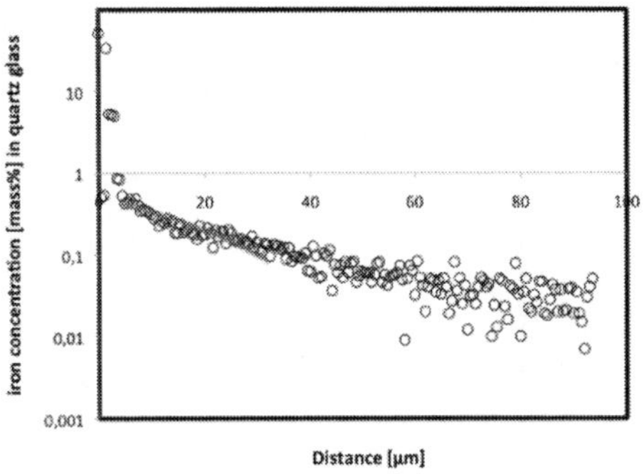

Figure 4: Concentration of iron as a function of distance [µm] in quartz glass, obtained from a EPMA line scan. A curve fit to the diffusion equation is shown as a red line.

In Figure 5 the first data points are not part of the diffusion profile as the MatLab script make sure the first columns does not contain many zeros, and are unnaturally higher than the rest of the data points. The diffusion profile therefore starts by the red fit curve.

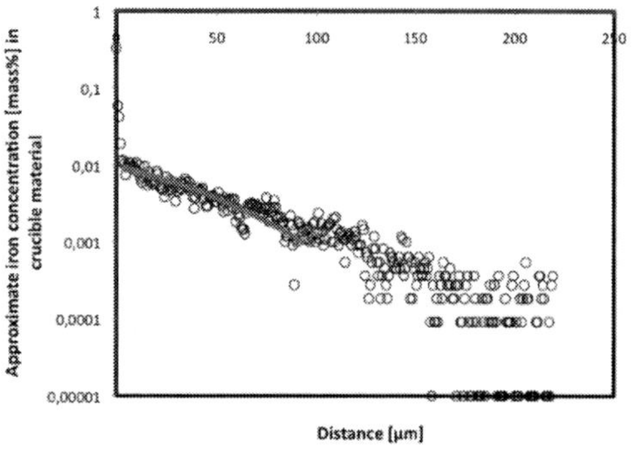

Figure 5: Concentration of iron as a function of distance [µm] in slip-cast crucible, obtained from X-ray mapping. A curve fit to the diffusion equation is shown as a red line

3.2 Average diffusivities

The average diffusivities at 1100, 1200 and 1300 °C in the quartz glass and the slip cast crucible are given in Figure 6 and Figure 7, respectively.

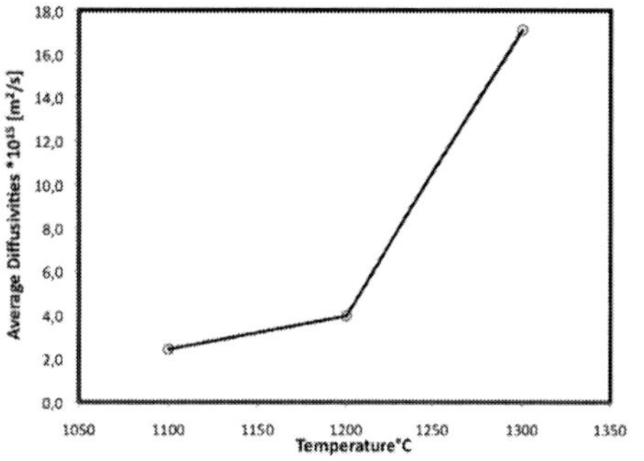

Figure 6: The average diffusivities at 1100, 1200 and 1300 °C in quartz glass.

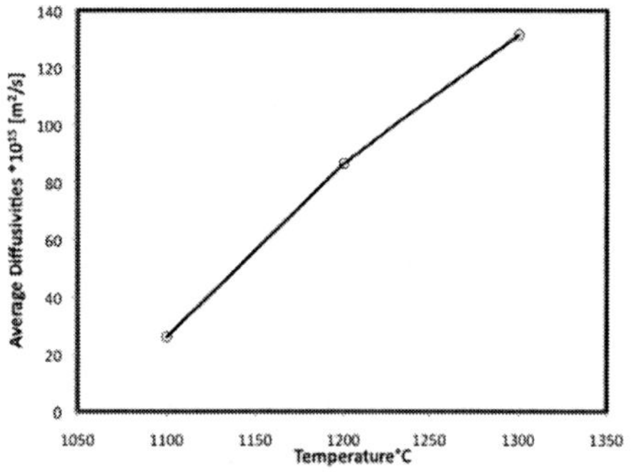

Figure 7: The average diffusivities at 1100, 1200 and 1300 °C in the slip cast crucible.

The temperature independent diffusion constants and activation energies were estimated to be $1{,}12 \cdot 10^{-4}$ cm²/s and 1,52 eV for the crucible, respectively, and $8{,}10 \cdot 10^{-5}$ cm²/s and 1,81 eV for the quartz glass, respectively.

4 DISCUSSION

4.1 Diffusivities

The transport diffusivity of both the crucible and quartz glass increased with temperature. As expected the diffusivity was higher in the porous structure of the crucible material than in the denser quartz glass. The difference in transport parameters of iron in the two materials indicates that the structure of the silica crucible is important in determining to which degree it behaves as a contamination source of silicon.

4.2 Arrhenius plot

Arrhenius plots of the diffusivities in quartz glass and crucible material with a linear regression are given in Figure 8 and Figure 9, respectively. From the linear regression and the Arrhenius equation, the temperature independent diffusion coefficient, D_0 and activation energy of the diffusion, E_a, could be estimated. The Arrhenius equation can be expressed as:

$$D=D_0\exp(-E_a/RT) \qquad (2)$$

Where T is the temperature and R is the universal gas constant.

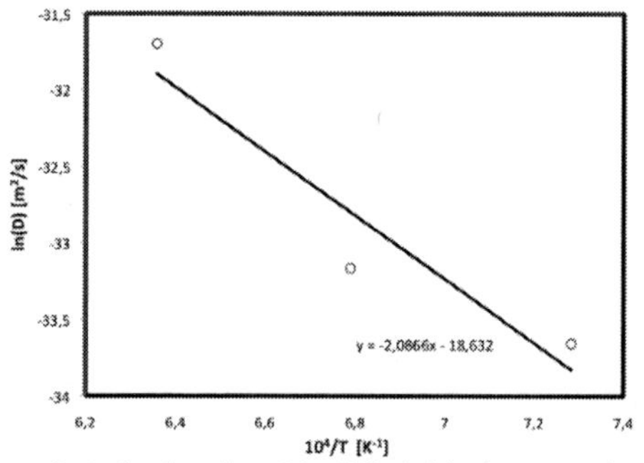

Figure 8: Arrhenius plot of the diffusivities in quartz glass with a linear regression.

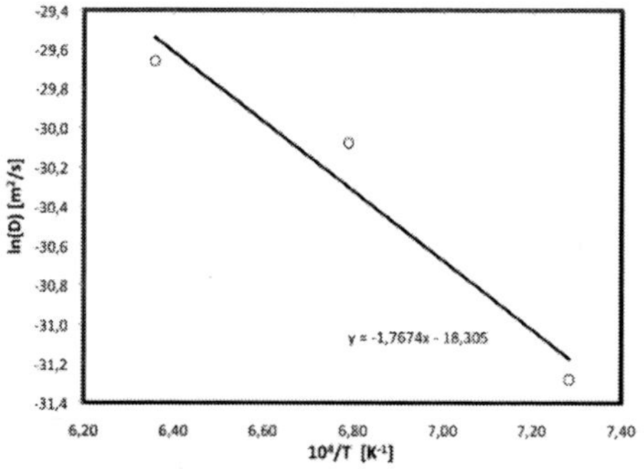

Figure 9: Arrhenius plot of the diffusivities in the slip cast crucible with a linear regression.

4.3 Comparison of diffusivities

Using the obtained values of D_0 and E_a of iron in the quartz glass and the crucible, diffusivities were plotted as a function of inverse temperature in Figure 10. Diffusivities from Ramappa et al, Atkinson, Kononchuc and Schubert [2,4-6] are also included in the Figure for comparison; Ramappa et al has performed a quantitative analysis of iron diffusion (700-1100 °C) in a silicon dioxide layer made from thermal oxidation on a silicon wafer. Kononchuc has investigated the diffusivity (900-1150 °C) of iron in the silica layer of silicon-on-insulators (SIO) structures. Atkinson has measured the diffusivity (500-1000 °C) of iron in synthetic silica glass. Schubert's results are based on fitting a diffusion model to actual impurity measurements in silicon solidified in a slip cast crucible. It can be seen from the graph that the values for the diffusivity in the crucible match well with Schubert's diffusivities although the slope is steeper in Schubert's estimations. The quartz glass diffusivities in this work are considerably higher than Atkinson, Kononchuc and Ramappa's estimations for amorphous silica but lower than for the slip cast crucible. This Figure shows it is crucial to account for the structure of the amorphous silica in order to estimate the diffusivity in a particular system. Crucibles made of quartz glass should experience considerably faster diffusion than expected from literature values, and the porous structure of the slip cast crucible makes this system an even more potent source of contamination by diffusion into the coating, melt and solidified crystal.

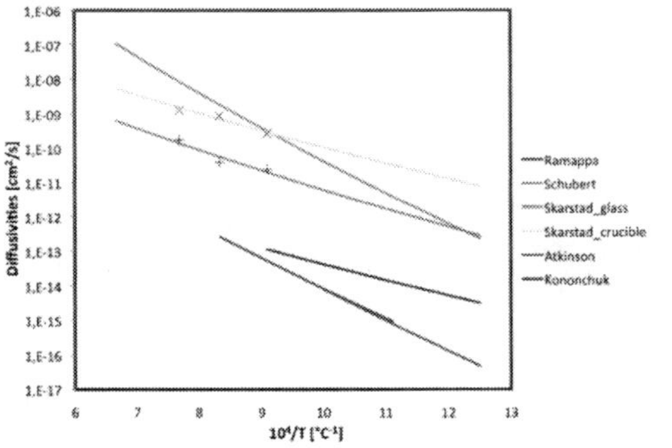

Figure 10: Comparison of the diffusivities obtained from Ramappa, Atkinson, Kononchuc, Schubert and this work (Skarstad).

4.4 Simulation

The temperature independent diffusion coefficient and activation energy of the iron diffusion in the slip-cast crucible were applied in a numerical model that predicts the contamination of silicon during directional solidification. The model is one-dimensional and uses COMSOL multiphysics as a solver. Schubert's estimations of impurity and diffusivity in the coating were applied in the simulation. The time was set to 12 hours and the temperature to 1414 °C. The model quantified the approximate iron concentration that diffuses from the crucible and coating into the melt during the melting stage. By applying the Scheil equation, the iron concentration in the ingot could be obtained. The concentration of iron in the ingot centre amounted to $1{,}65 \cdot 10^{10}$ cm^{-3}. Schubert's estimations for the iron concentration estimated in a G4 ingot was $3{,}2 \cdot 10^{10}$ cm^{-3} [2], but this model also accounts for contamination of the melt during solidification, and a factor of two in difference between these two models can be expected.

5 CONCLUSIONS

The diffusivities of quartz glass and a silica slip cast crucible were quantified at three different temperatures (1100, 1200 and 1300 °C). Samples of quartz glass and samples of a silica slip cast crucible were heat treated in contact with solid steel. The quartz glass was analyzed with EPMA line scans and the silica crucible was analyzed with Semi-quantitative X-ray mapping. The concentration profiles of iron were fit to the solution of the diffusion equation. The diffusivity in the crucible and quartz glass increased with temperature, and the temperature independent diffusion coefficient and activation energy on the diffusion could also be obtained. The diffusivity was also higher in the porous crucible material than in the denser quartz glass. The diffusivity of the crucible material supports Schubert's estimations of diffusivity in the silica slip cast crucible. Crucibles made of quartz glass should experience considerably faster diffusion than expected from literature values of diffusion. The porous structure of the slip cast crucible makes this system an even more potent source of contamination by diffusion into the coating, melts and solidified crystal.

6 ACKNOWLEDGEMENTS

The work has been a part of a collaboration project called "Impurity Control in High Performance Multicrystalline Silicon" where NTNU, IFE and SINTEF are research partners. The project is sponsored by the Research Council of Norway, REC Solar, REC Silicon, Steuler Solar and The Quartz Corp.

7 REFERENCES

[1] E. Olsen and E. J. Øvrelid, Silicon Nitride Coating and Crucible- Effects of Using Upgraded Materials in the Casting of Multicrystalline Silicon Ingots. Progress in Photovoltaics: Research and Applications, 16(2):93–100, mar 2008.

[2] M. C. Schubert, J. Schon, F. Schindler, W. Kwapil, A. Abdollahinia, Bernhard. Michl, Stephan. Riepe, C. Schmid, M. Schumann, S. Meyer, and W. Warta. Impact of Impurities From Crucible and Coating on mc-Silicon Quality-the Example of Iron and Cobalt. IEEE Journal of Photovoltaics, 3(4):1250–1258, October 2013

[3] R. Kvande, L. Arnberg, and C. Martin. Influence of crucible and coating quality on the properties of multicrystalline silicon for solar cells. Journal of Crystal Growth, 311(3):765–768, January 2009

[4] O. Kononchuk, K. G. Korablev, N. Yarykin, and G. A. Rozgonyi. Diffusion of iron in the silicon dioxide layer of silicon-on-insulator structures. Applied Physics Letters, 73(9):1206, 1998.

[5] A. Atkinson and J.W.Gardner.The diffusion of Fe^{3+}in amorphous SiO_2 and the protective properties of SiO_2 layers. Corrosion Science, 21(1):49–58, jan 1981.

[6] D. A. Ramappa. Diffusion of Iron in Silicon Dioxide. Journal of The Electrochemical Society, 146(10):3773–3777, October 1999.

[7] H. V. Skarstad. Diffusvity and Soulubility of iron in fused quartz crucibles. Project thesis, Norwegian University of Science and Technology, 2015.

[8] J. Crank. The Mathematics of Diffusion. Clarendon Press, 1979.

IMPURITIES AND DEFECTS DISTRIBUTION DURING THE GROWTH OF PV SILICON: INFLUENCE OF MELT CONVECTION AND GRAVITY

A. Le Donne[1], M. Acciarri[1], C. Reimann[2], J. Friedrich[2], T. Jauss[3], A. Cröll[3], T. Sorgenfrei[3] and S. Binetti[1]*
[1]University of Milano-Bicocca, Department of Materials Science and MIBSOLAR center, Milano, Italy
[2]Fraunhofer IISB, Erlangen, Germany
[3]University Freiburg, Institute for Geosciences, Freiburg, Germany
* email: simona.binetti@unimib.it

ABSTRACT: During the growth of silicon for photovoltaic applications, the melt can be deeply contaminated by carbon, transported via gas atmosphere to the melt surface or already present in the initial feedstock. Preliminary investigations have shown that gravity and melt convection are most likely determining the particles transport. However, the interaction of such particles with the moving crystallization front and their incorporation into solid silicon are not yet fully understood. Aim of this work is to study the interaction of impurities and defects like foreign phase particles during silicon solidification in terrestrial experiments and in microgravity conditions during the TEXUS 51 flight mission. All the samples were analyzed by infrared transmission microscopy, infrared spectroscopy and photoluminescence. The concentration and spatial distribution of impurities and defects were determined and related to the growth process and the gravity effect.

Keywords: PV materials, Silicon, Silicon Carbide, Characterization, Space.

1 INTRODUCTION

It is widely known that during the crystallization of silicon for PV applications the melt can be deeply contaminated by carbon transported via the gas atmosphere to the melt surface, or already present in the initial feedstock [1]. It is also well known that the presence of SiC particles can influence the grain growth and the formation of dislocations [2], which all act as recombination centers and cause problems during the sawing process [3,4]. Moreover, SiC precipitates can also induce shunts reducing the performance of Si solar cells [4]. Nevertheless, the interaction of such particles with the moving crystallization front and their incorporation into solid silicon are not yet fully understood. One of the major goals of the present work is therefore to get new insights on the interaction of foreign phase particles with the moving solid-liquid interface during directional solidification of silicon. To do that, silicon samples grown in terrestrial experiments and in microgravity conditions during the TEXUS 51 flight mission [5] were considered.

The focus of this study is to understand the particle interaction with a moving solid-liquid interface and their incorporation into solid silicon. As a matter of fact, under the assumption that the particle would form in the melt, a critical growth velocity V_{cr} should exist depending on the density, size and morphology of the particles. Below V_{cr} a particle will be pushed in front of the growth interface, whereas above V_{cr}, the particle will be captured and frozen into the growing crystal. This particle engulfment depends on the so-called interface, drag, lift and gravity forces acting on the particle in the vicinity of the interface.

Preliminary investigations where SiC particles were added to the melt have shown that gravity and lift forces (due to melt convection) are most likely determining the transport of the particles under terrestrial conditions. Depending on the size and density of the particles and on the convection intensity they can either sink to the bottom of the melt or float on the melt surface. Since gravity affects melt movement, solute distribution, and sedimentation of the particles, microgravity experiments offer the unique opportunity to study particle nucleation, transport and incorporation under purely diffusive conditions. Aim of this work is therefore the investigation of the interaction of impurities and defects (like foreign phase particles) during silicon solidification in terrestrial experiments and in microgravity conditions during the TEXUS 51 flight mission. Infrared transmission microscopy, infrared spectroscopy and photoluminescence were used to determine the concentration and spatial distribution of impurities and defects as a function of both the growth process parameters and the gravity effect.

2 EXPERIMENTAL DETAILS

In order to study pushing and engulfment phenomena of SiC particles during silicon solidification, experiments were performed in the well-known ELLI mirror furnace [6]. During all the experiments, <100>-oriented single crystalline silicon rods 8 mm in diameter were used. In the preliminary terrestrial experiments reservoirs of SiC particles with particle size varying from 7 to 60 µm were introduced into the silicon rods. In order to avoid Marangoni convection the rods were oxidized at 1050 °C for up to 140h in order to grow a 5 µm thick oxide skin on the silicon surface. After oxidation, silicon rods were mounted in a silica ampoule, which is evacuated and then filled with 1.5 bar oxygen to ensure the oxide skin stability during the whole solidification experiment. Finally, the silicon rods were zone melted at different growth rates, both constant (0.2 mm/min) and variable during the growth (2.5 to 10 mm/min). For comparison purposes, silicon samples without SiC particles added to the melt were also prepared. Furthermore, terrestrial experiments were also carried out both without oxide skin (i.e. in presence of Marangoni convention [7]) and under the influence of static axial magnetic fields up to 5 T (i.e. with reduced buoyancy convention [8]).

First of all, the samples were analyzed by infrared transmission microscopy (IR-TM) to determine the distribution of the SiC particles in the silicon rod after processing and to gain information about the particle incorporation depending on the growth rate. More details about the experimental procedure are reported in [9].

Then, the samples were analyzed by Fourier transform infrared spectroscopy (FTIR) in the 400–4000 cm^{-1} range to determine the distribution of interstitial oxygen (indicated in the following as O_i) and substitutional carbon (indicated in the following as C_s) in the silicon rods. All FTIR spectra were performed in air at 300 K by a Jasco 6200 spectrometer. The calibration factors and the calculation procedure for the determination of interstitial oxygen and substitutional carbon are reported in [10,11], respectively.

Photoluminescence (PL) analyses at low temperature (15 K) were also carried out in order to get additional information on impurities and possible defects present in the Si matrix. All PL measurements were performed with a spectral resolution of 6.6 nm using a standard lock-in technique in conjunction with a single grating monochromator and a short wavelength enhanced InGaAs detector with maximum responsivity at 1540 nm. A quantum well laser (λ_{exc}=805 nm) with a spot size of 0.3x0.3 mm^2 and a power density of 15.5 W cm^{-2} was used as excitation source. A cooling system consisting of rotary pump, turbomolecular pump and He closed circuit cryostat was used to perform PL spectra at low temperature.

3 RESULTS AND DISCUSSION

Silicon samples grown in terrestrial conditions showed that, when no SiC particles were added to the melt, the presence of the oxide skin avoid carbon contamination, as expected, while increasing the oxygen content in the Si matrix (see Fig. 1). Incidentally, the O_i concentration is strongly higher than the solubility limit of oxygen in silicon, however the Si:O solid solution is stable since no SiO_X precipitates were detected by FTIR at 12 K (not shown here).

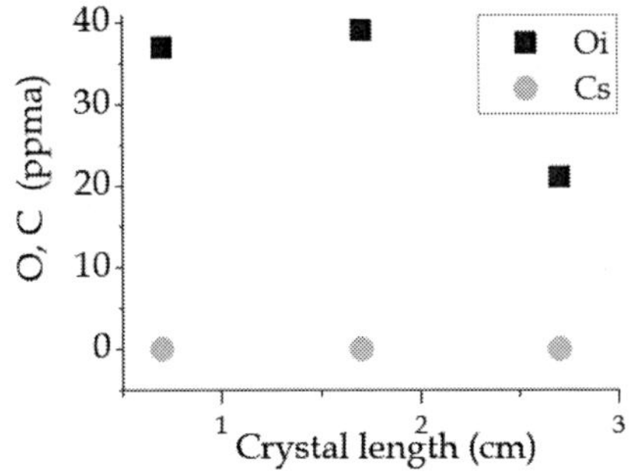

Figure 1: O_i and C_s content of a Si sample with oxide skin (no SiC particles added, 2.5 to 10 mm/min growth rate)

Accordingly, samples grown without oxide skin showed the presence of SiC precipitates (see Fig. 2) since the melt was supersaturated with carbon, whose concentration in the melt increased with the solidified length due to the increased growth rate (see Fig. 3).

Figure 2: IR-TM image of a Si sample without oxide skin (no SiC particles added, 2.5 to 10 mm/min growth rate).

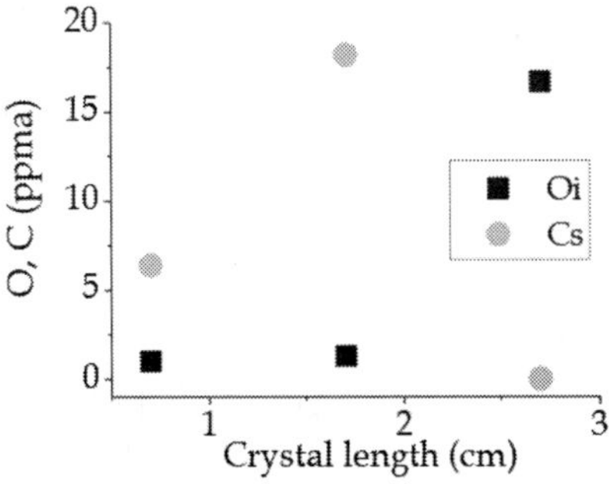

Figure 3: O_i and C_s content of a Si sample without oxide skin (no SiC particles added, 2.5 to 10 mm/min growth rate).

Conversely, samples grown with oxide skin showed an increase of the oxygen content with decreasing growth rates (see Fig. 4) since the increased time of liquid silicon in contact with the oxide skin for smaller growth rates dissolves more oxygen into the melt. Despite also in this case very high oxygen content were determined (up to 47 ppma), neither FTIR nor PL analyses at low temperature (not shown here) seem to suggest the presence of oxide related defects in this sample set.

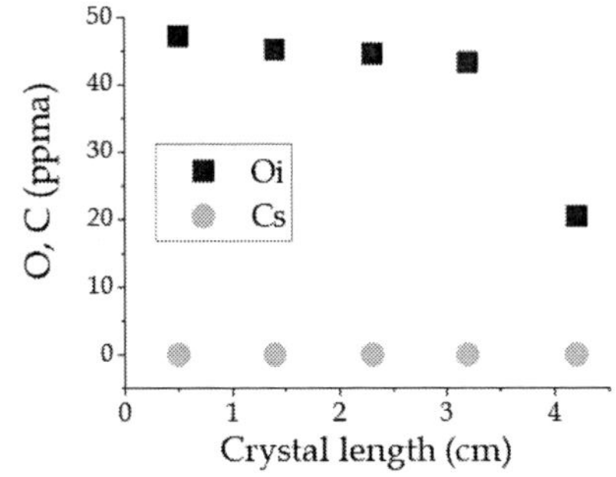

Figure 4: O_i and C_s content of a Si sample with oxide skin (no SiC particles added, 0.2 mm/min growth rate).

Samples grown in terrestrial conditions with the same reservoir of SiC particles and oxide skin but different applied magnetic fields (0 to 5 T) showed that SiC precipitates are present. In the case of 5 T applied magnetic field, SiC precipitates were found at an early stage of solidification due to the fact that the carbon solubility was already reached at this stage of solidification (see Fig. 5).

Figure 5: IR-TM image of a Si sample with 5 T applied static magnetic field (oxide skin, SiC particles added, 2.5 to 10 mm/min growth rate).

In spite the influence of static magnetic field on the convection conditions is still not clear, a higher SiC particles dissolution was observed. Fig. 6 clearly demonstrates that a higher C incorporation in the Si matrix occurs, which is confirmed by the PL spectrum reported in Fig. 7 showing the presence of C related defects. Most probably, with reduced buoyancy convection, a nearly stagnant melt would lead to a higher C concentration in the vicinity of the SiC particles because of less mixing. Also in this case, despite the very high oxygen content (11 to 46 ppma), neither FTIR (not shown here) nor PL analyses (see Fig. 7) at low temperature seem to suggest the presence of oxide related defects.

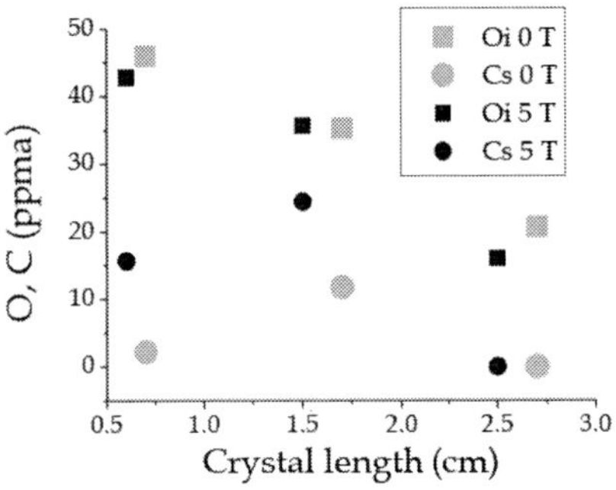

Figure 6: O_i and C_s content of a Si sample as a function of the applied static magnetic field (oxide skin, SiC particles added, 2.5 to 10 mm/min growth rate).

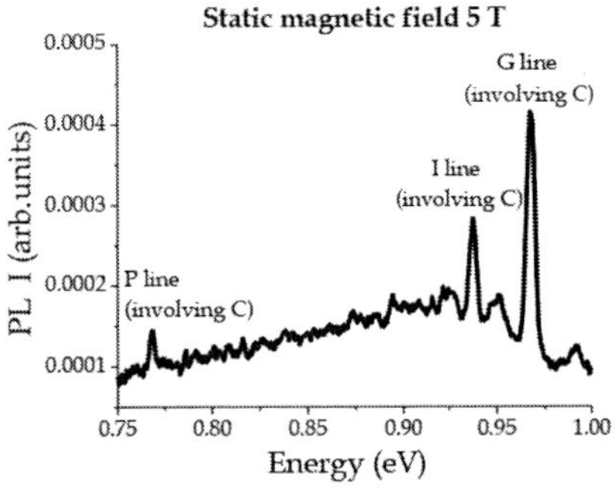

Figure 7: PL spectrum at 15 K of a Si sample with 5 T applied static magnetic field (oxide skin, SiC particles added, 2.5 to 10 mm/min growth rate).

For the microgravity experiment onboard TEXUS-51 and its reference on the ground (i.e. 1 g conditions), the silicon rods were prepared as described in the Experimetal Details section, using a mixture of 7 µm and 60 µm SiC particles. The process-time diagrams for both the TEXUS-51 experiment and for its reference under terrestrial conditions are reported in [12]. Figures 8 and 9 show the actual growth velocity V_{gr} as a function of the crystal length for the reference and TEXUS-51 experiments, respectively [12]. Obviously, the behavior of V_{gr} for both experiments is nearly the same, with the exception of grown crystal lengths larger than 14 mm.

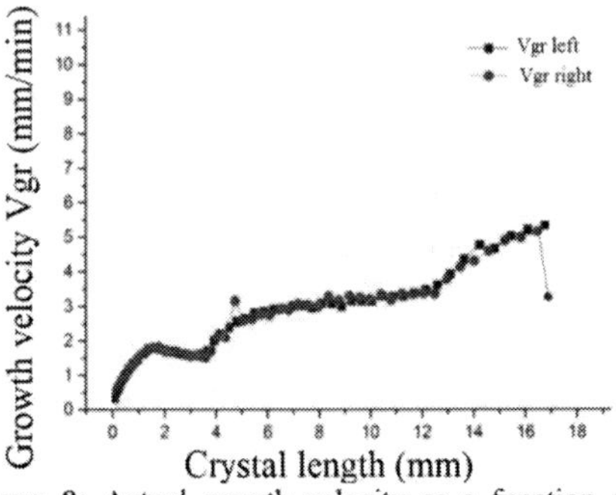

Figure 8: Actual growth velocity as a function of the crystal length for the reference experiment (i.e. 1 g conditions).

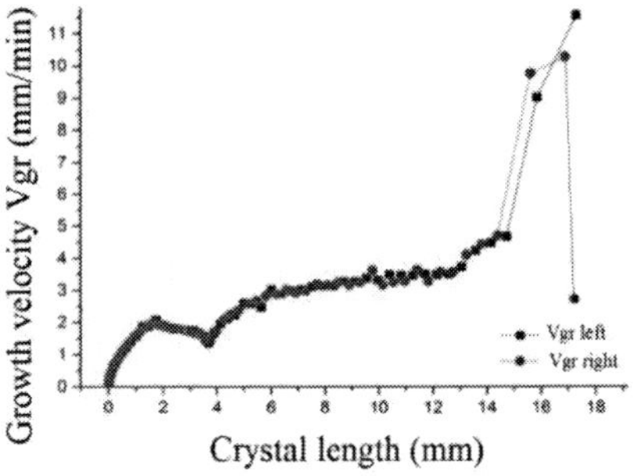

Figure 9: Actual growth velocity as a function of the crystal length for the TEXUS-51 experiment (i.e. µg conditions).

As it is apparent from the comparison of Figs. 8 and 9, more pronounced deviations are present between the two profiles after this crystal length value, due to the different heat loss mechanisms under µg and 1 g conditions [12].

IR-TM analyses (not shown here) [12] showed that under microgravity conditions SiC particles are incorporated with a lower critical growth velocity of the moving solid-liquid interface, due to the absence of the lift-force (3-4 mm/min for 1 g and 2.2 mm/min for µg). Furthermore, FT-IR measurements (see Fig. 10) showed a lower oxygen incorporation from oxide skin in the TEXUS-51 experiment.

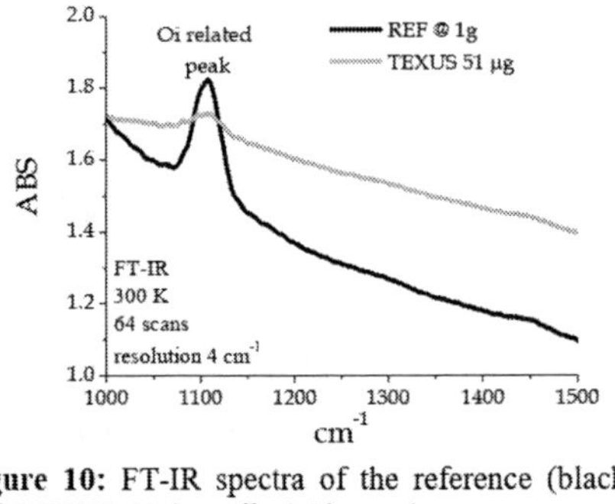

Figure 10: FT-IR spectra of the reference (black line) and TEXUS-51 (grey line) Si samples.

4 CONCLUSIONS

In this work, silicon solidification experiments were carried out both under terrestrial and microgravity conditions to study the incorporation of SiC particles in the Si melt. The influence of both Marangoni and buoyancy convections as well as the effects of lift forces were investigated. Microgravity experiments established that SiC particles are incorporated with a lower critical growth velocity of the moving solid-liquid interface, due to the absence of the lift-force.

5 ACKNOWLEDGEMENTS

Funding is provided by Italian Space Agency (ASI) through the project n° 2013-052 (SiSSi project) and by DLR Space Administration with funds from the Federal Ministry of Economics and Technology (ParSiWal project).

6 REFERENCES

[1] C. Reimann, M. Trempa, T. Jung, J. Friedrich, G. Müller, J.Cryst.Growth 312 (2010) 878
[2] T. Duffar, Recent Res. Dev. Cryst. Growth 5 (2010) 61.
[3] G. Du, N. Chen, P. Rossetto, Sol. Energy Mater. Sol. Cells 92 (2008) 1059
[4] O. Breitenstein, J.P. Rakotoniaina, M.H. Al Rifai, M. Werner, Progress Photovolt.: Res. Appl. 12 (2004) 529
[5] http://www.sscspace.com/texus-51
[6] A. Eyer, A. Cröll, R. Nitsche, J. Cryst. Growth 71 (1985) 173
[7] A. Eyer et al., J.Crys.Growth 71 (1985) 249
[8] Cröll et al., Prog. Crystal Growth and Charact. 38 (1999) 133
[9] J. Friedrich, H.J. Blankenburg, R. Colditz, M. Azizi, C. Reimann, D. Gruss, German Patent DE10 2012 017 183 A1, 2012.
[10] ASTM Standard, F1188-88, Standard test method for interstitial atomic oxygen content of silicon by infrared absorption, Annual Book of ASTM Standards, Electronics (II), 1996
[11] ASTM Standard, F1391-93, Standard test method for substitutional atomic carbon content of silicon by infrared absorption, Annual Book of ASTM Standards, Electronics(II), 1998
[12] J. Friedrich, C. Reimann, T. Jauss, A. Cröll, T. Sorgenfrei, J.Crys.Growth 447 (2016) 18

INVESTIGATION OF DEEP-LEVEL DEFECTS IN THE ACTIVE LAYER OF MULTICRYSTALLINE SILICON SOLAR CELLS

V.G. Litvinov[1*], N.V. Vishnyakov[1], V.V. Gudzev[1], A.V. Ermachikhin[1], A.D. Maslov[1], A.S. Karabanov[2], S.M. Karabanov[1], S.P. Vikhrov[1]
[1]Ryazan State Radio Engineering University, 59/1 Gagarin Str., Ryazan, 390005, Russian Federation
[2]Helios-Resource Ltd., Saransk, Mordovia, 1 Proletarskaya Str., 430001, Russian Federation
[*]tel. +7 4912 460299, fax +7 4912 922215, e-mail: vglit@yandex.ru

ABSTRACT: Three solar cells (SCs) based on multicrystalline silicon (mc-Si) with different efficiencies 10.1, 16.8 and 20.4% were investigated by current deep level transient spectroscopy (I-DLTS) and by the method of capacitance-voltage (C-V) characteristics at different temperatures and frequencies. From various places along the diagonal of SC's substrate five pieces with an area ~20 mm^2 were cleaved and studied. I-DLTS spectra of each of the five pieces from three SCs were measured. The features of deep levels concentration distribution along the SCs surface were studied. The correlation between the total concentration of deep traps and the values of the SCs conversion efficiency is found.
Keywords: Solar Cell, Multicrystalline Silicon, Electrical Properties, Defects

1 INTRODUCTION

Solar cells (SCs) based on multicrystalline silicon (mc-Si) have a lower production cost and comparable efficiency of solar energy conversion into electricity to SCs based on monocrystalline silicon. Mc-Si is a highly defective material. Defects or deep centers (DCs) form deep energy levels (DLs) in the semiconductor band gap. Currently actual questions of increasing the conversion efficiency of solar energy into electricity and reasons of degradation of SCs are continuously studied. The conversion efficiency of SCs depends on the structure of the solar module, the type of semiconductor materials and etc. Certain structural defects with deep energy levels or deep traps in the band gap in active semiconductor layer affect the stability of SCs electrical properties. DLs are responsible for the generation-recombination processes and form the recombination component of direct current through the p-n junction [1, 2].

Therefore DLs affect the efficiency of solar energy conversion. The recombination current leads to reduction of conversion efficiency of solar energy into electricity up to 25% [1]. It is important to study the DLs parameters in order to optimize the structure and technology of SCs and to obtain advanced parameters.

Investigation of the DL's energy spectrum of defects in SCs active layers with different conversion efficiencies is necessary to find the relationship between DL's parameters and efficiency of solar energy conversion. The subject of this work is the investigation of deep level defects influencing on conversion efficiency of SCs based on Multicrystalline silicon (mc-Si) p-n junction. It is also interesting to investigate how the defects are distributed along the SCs surface. Mc-Si is a very promising material for photovoltaic applications and its investigation is a very actual task.

2 EXPERIMENTS

For investigation of DLs in SCs well known methods I-DLTS [3] and C-V were used. SCs have a large area of the test sample, a high leakage current. The traditional DLTS method [4] based on measuring the electrical capacitance relaxation is not quite suitable for investigation of DLs in SCs. The I-DLTS method of studying the electric current relaxation is more suitable. A specially designed measuring system that allow to measure I-DLTS-spectra of barrier structures and SCs characterized by the high leakage current, large area of the p-n-junction and relatively large electrical capacitance was used for investigation of the DLs parameters [5]. The measuring complex provides the compensation of the leakage current and the measurement of samples with high electrical capacitance. C-V characteristics of SCs were measured in order to clarify I-DLTS-spectra measurement. The data obtained from C-V characteristics were used to calculate the DL's concentration.

C-V characteristics at temperatures 100 – 350K and at fixed frequencies in the range 1 kHz – 1 MHz were obtained using Janis CCS400/204N cryostat and Agilent E4980A LCR-meter with option 001.

Boron doped p-type mc-Si 156x156 mm^2 wafers of 0.5 – 2 Ω·cm resistivity were used to fabricate SCs by n-type 1 μm thick layer forming by phosphorus gas diffusion in Helios-Resource Ltd. The growing method is the directional solidification. Oxygen content $<1.0\times10^{18}$ at/cm^3 was measured by ASTM1188. The antireflection films, front and backside silver contacts were formed also [6]. Standard SC with efficiency coefficient η = 16.8% (sample #2), defective SC with η = 10.1% (sample #1) and SC with extremely high η = 20.4 % (sample #3) were selected. From various places along the diagonal of SC's substrate five pieces with an area ~20 mm^2 were cleaved (Fig. 1).

Free hole distribution profiles in the p-type base of the diode structure were estimated from C-V characteristics and used for I-DLTS analysis. The data about the DLs energy spectra and its parameters such as activation energy, capture cross section, concentration are obtained.

3 RESULTS AND DISCUSSION

The C-V characteristics were measured at bias voltage –5 to 1 V and amplitude of test signal 30 mV. In Fig. 2 typical C-V characteristics of the sample with η = 16.8 % in coordinates 1/C^2 (C is the capacitance per unit squire) versus applied voltage are shown. It should be noted that for a reverse bias (it is not shown in the Fig. 2) up to –2V the linear character of dependencies was observed.

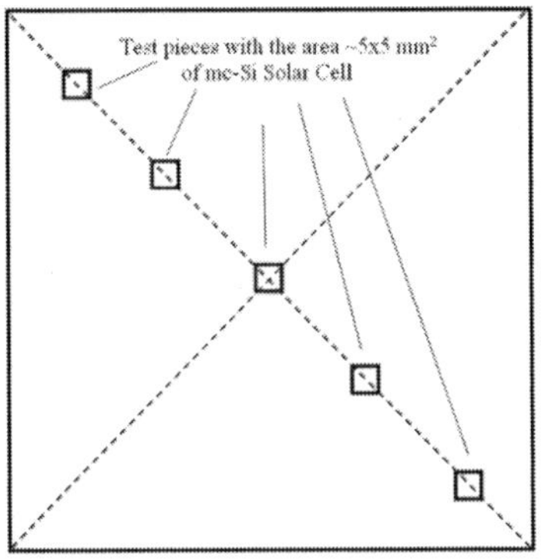

Figure 1: The scheme of test five pieces with an area ~20 mm^2 which were cleaved from places along the diagonal of SC's substrate

The free hole concentration was estimated from the slope of linear part of C-V curves in the Fig. 1. It is changed from $1.1\cdot10^{15}$ cm^{-3} to $1.2\cdot10^{15}$ cm^{-3} with changing the test signal frequency from 1 kHz to 1 MHz. From the C-V characteristics of all test samples from the SCs with η = 16.8% and 20.4% the homogeneous distribution of free hole concentration along the SCs surface was found.

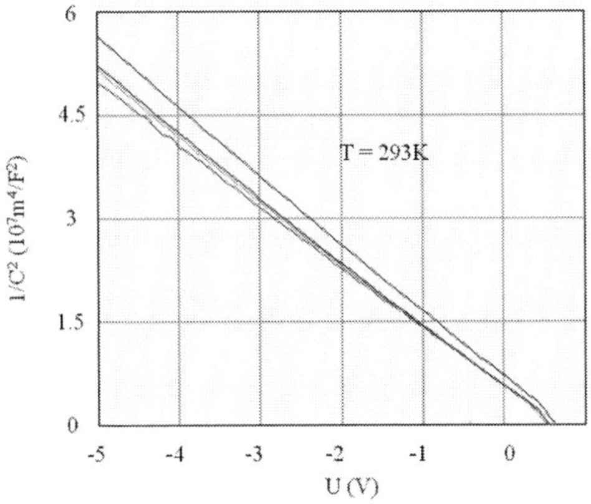

Figure 2: C-V characteristics of mc-Si based SC with η = 16.8 % measured at room temperature and frequencies of 1 kHz, 10 kHz, 100 kHz, 1 MHz (curves from down (red) to up (violet) respectively)

Typical I-DLTS spectra of the pieces cleaved from the central part of SLs are presented in Fig. 3-5. I-DLTS spectra were measured in DL's recharge mode by applying a pulse voltage. The amplitude 0.6 V of the filling pulse voltage practically corresponds to open-circuit voltage of the SC. At this condition DLs in the active layer where electron-hole pairs are generated under illumination during SC operation are recharged. The amplitude of the reverse bias pulse was -1 V. Detected DLs are hole traps with the activation energies E_t of 0.17, 0.24, 0.31 and 0.56 eV.

The numbering the DLs peaks were made in the sequence of their appearance in the I-DLTS spectrum. I-DLTS spectra of the pieces from sample #1 (SC with η = 10.1 %) contained two peaks predominantly, H1 and H2 (see Fig. 3). The peak H1 was on the shoulder of the peak H2, so its activation energy may be estimated only by simulation. The possible nature of the DLs was identified with the DLs, previously observed in monocrystalline silicon. The activation energy of H2 peak was 0.31 eV. DL H2 is broadened and may be explained by meta-stable B-O defect [7]. In the pieces from the sample #2 (SC with η = 16.8 %) H1 peak with the activation energy 0.24 eV was observed at the same conditions that H1 in the sample #1. The hole trap with the activation energy 0.23 eV is attributed to the well known divacancy [8]. Parameters of obtained DLs are presented in table I. Capture cross sections were calculated from the Arrhenius plots.

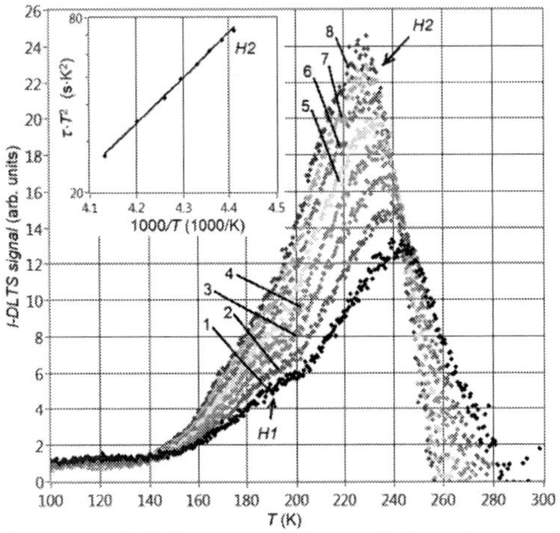

Figure 3: I-DLTS spectra of the sample from the central part of SC with η = 10% at different relaxation time constant τ: 1 – 455 μs; 2 – 621 μs; 3 – 771 μs; 4 – 910 μs; 5 – 1043 μs; 6 – 1170 μs; 7 – 1292 μs; 8 – 1412 μs. In the insertion the Arrhenius plot is presented

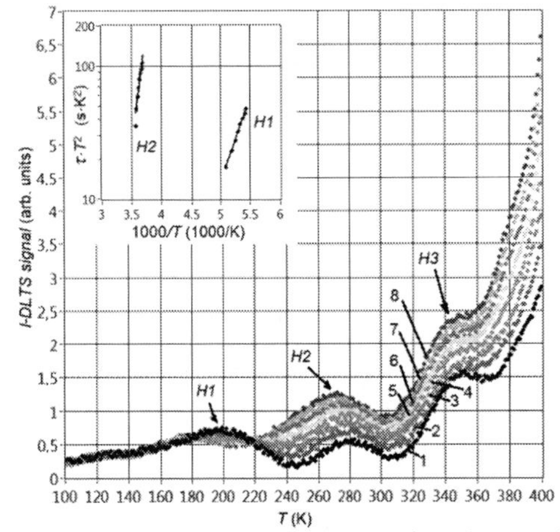

Figure 4: I-DLTS spectra of the sample from the central part of SC with η = 16.8% at different relaxation time constant τ: 1 – 455 μs; 2 – 621 μs; 3 – 771 μs; 4 – 910 μs; 5 – 1043 μs; 6 – 1170 μs; 7 – 1292 μs; 8 – 1412 μs. In the insertion the Arrhenius plots are presented

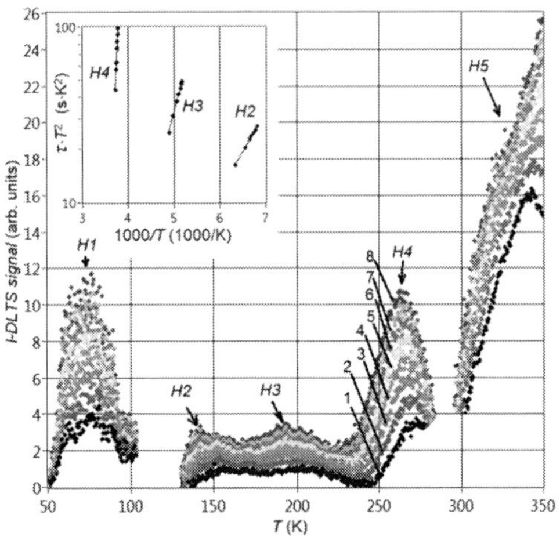

Figure 5: I-DLTS spectra of the sample from the central part of SC with η = 20.4% at different relaxation time constant τ: 1 – 455 μs; 2 – 621 μs; 3 – 771 μs; 4 – 910 μs; 5 – 1043 μs; 6 – 1170 μs; 7 – 1292 μs; 8 – 1412 μs. In the insertion the Arrhenius plots are presented

Table I: Parameters of DLs obtained from I-DLTS measurements, where E_t is the activation energy, σ_p is the capture cross section, N_t is the DL's concentration range obtained for five pieces of each SC, $N_{t\Sigma}$ is the total average DL's concentration

Sample #	1	2	3
η (%)	10.1	16.8	20.4
Deep level	H1 H2	H1 H2 H3	H1 H2 H3 H4 H5
E_t (eV)	0.25±0.05 (simulation) 0.31±0.02	0.24±0.02 0.56±0.02 –	– 0.17±0.02 0.21±0.02 0.56±0.02 –
σ_p (cm^2)	– $6.3\cdot10^{-17}$	$8.7\cdot10^{-17}$ $3.9\cdot10^{-11}$ –	– $1.3\cdot10^{-17}$ $2.3\cdot10^{-17}$ $8.5\cdot10^{-11}$ –
N_t (cm^{-3})	$(2.7\text{-}5.9)\cdot10^{15}$ $(0.9\text{-}1.3)\cdot10^{16}$	$(6.3\text{-}10.7)\cdot10^{14}$ $(3.3\text{-}15.1)\cdot10^{14}$ $(0.9\text{-}2.5)\cdot10^{15}$	– $(0.7\text{-}1.5)\cdot10^{14}$ $(0.7\text{-}1.7)\cdot10^{14}$ $(2.5\text{-}5.9)\cdot10^{14}$ $(1.1\text{-}2.5)\cdot10^{15}$
$N_{t\Sigma}$ (cm^{-3})	$1.5\cdot10^{16}$	$3.4\cdot10^{15}$	$2.5\cdot10^{15}$

I-DLTS-spectra of the pieces from the sample #2 (SC with η = 16.8 %) contained mainly three intensive peaks. The activation energy of H2 was 0.56 eV. The H2 and H3 peaks positions in the I-DLTS-spectra of sample #2 (Fig. 2) were similar to the H4 and H5 peaks positions on the I-DLTS-spectra of the pieces from the sample #3 (Fig. 3).

The DL H3 concentrations for the sample #2 and DL H5 for the sample #3 were estimated only because these peaks are hidden behind a sharp increase in the I-DLTS-signal due to the abrupt increase of leakage current at high temperatures. Due to this reason, the nature of H3 and H5 DLs is not completely clear. The temperature peak H1 position in the sample #3 did not depend on temperature at different relaxation time constant. A possible explanation of such observation is the recharge of surface states.

The activation energies of H2 and H3 peaks were found as 0.17 and 0.21 eV respectively. The DL H2 with the activation energy 0.17 eV may be explained by different ways in different literature references. DL 0.21 eV probably has the same nature as the H1 in the sample #2. The total DLs concentration $N_{t\Sigma}$ averaged by the I-DLTS data obtained from five test sample pieces of each of three SCs was also calculated. The highest total average DLs concentration observed in the sample with the lowest efficiency η. The DL H1 concentration in the sample #3 was not considered in calculation of the $N_{t\Sigma}$. Further analysis of the DLs influence on the conversion efficiency will be associated with the improvement of defects parameters characterization techniques and modeling of the DLs influence on the SC's stability characteristics and reliability.

4 CONCLUSION

Thus, the energy and concentration of deep level defects localized in mc-Si are measured. The correlation between the total DL's concentration in the active layer of SCs based on mc-Si and conversion efficiency was found. The possible nature of detected DLs is analyzed. The total concentration of DLs was $1.5\cdot10^{16}$, $3.4\cdot10^{15}$ and $2.5\cdot10^{15}$ cm^{-3} for the SCs with an efficiency of 10.1, 16.8 and 20.4% respectively. The total DLs concentration of the sample with the lowest efficiency 10.1 % is higher in six times than the total DLs concentration in the sample with the efficiency of 20.4%.

A correlation between the total DCs concentration in the active layer of SCs based on mc-Si and conversion efficiency is experimentally observed. The measurement system and the method for the SC's I-DLTS-spectra measuring are developed.

5 ACKNOWLEDGMENTS

This work is executed at financial support of the Ministry of Education and Science in the Regional Center of Probe Microscopy of Ryazan State Radio Engineering University.

6 References

[1] J.J. Wysocki and P. Rappaport, J. Appl. Phys., Vol. 31 (1960) 571.
[2] S.M. Sze and Kwok K. Ng, Physics of semiconductor devices, John Wiley&Sons, New York, (2006).
[3] J.A. Borsuk and R.M. Swanson, IEEE Transactions on electron devices, ED-27 (1980) 2217.
[4] D.V. Lang, J. Appl. Phys., Vol. 45 (1974) 3023.
[5] V.G. Litvinov et al., Power Electronics and Renewable Energy Conversion USB Proc. IEEE International Conference on Industrial Technology, (2015) 1071.

[6] V.G. Litvinov et al., MRS Advances, Vol. 1, Issue 14 (2016) 911.
[7] J. Schmidt and A. Cuevas, J. Appl. Phys., Vol. 86 (1999) 3175.
[8] M. Mamor and etc., Physical Review B, Vol. 63 (2000) 045201.

Bulk Lifetime Improvement of N-type Czochralski Silicon Crystals Grown from The Melt in "Liquinert" Quartz Crucible

[1]Tetsuo Fukuda, [2]Yukichi Horioka, [3]Shiro Sakuragi,
[4]Katsuto Tanahashi, [5]Shalamujiang Simayi, [6]Katsuhiko Shirasawa, and [7]Hidetaka Takato

[1, 4, 5, 6, 7]Renewable Energy Research Center, National Institute of Advanced Industrial Science and Technology (AIST), 2-2-9 Machiike-dai, Koriyama, Fukushima 963-0215, Japan
[2]Frontier Technology Business Research Institute Co. LTD., 11-9 Yamazakiumenodai, Noda, Chiba 278-0024, Japan
[3]Union Materials Inc., 1640 Oshido, Tone-machi, Ibaraki 300-1602, Japan

Corresponding Author: Phone/Fax/Email: +81-29-849-1547/+81-24-963-0824/tetsuo-fukuda@aist.go.jp

ABSTRACT

In order to improve the quality of monocrystalline PV silicon, we have developed LCZ (Liquinert-crucible-applied CZ) crystal growth method using vitreous silica crucibles whose inner surfaces have non-wettable (melt-phobic) property. We previously reported that the bulk lifetime of p-type LCZ silicon improved twice as large as that of commercially available p-type silicon due to carbon reduction. In this study, we showed that the lifetime of n-type LCZ silicon is equal to or longer than that of commercially available n-type MCZ silicon, and that also metal reduction has the high possibility to improve the lifetime, considering the incorporation mechanism from vitreous silica crucibles.

1. Introduction

In photovoltaic (PV) industry, conversion efficiency has been recognized as a key metric of solar cell technologies. Since crystalline silicon currently accounts for over 90% of all PV cell substrates, the improvement in silicon quality has had the large impact on the development and production of silicon-based cells.

Multicrystalline silicon growth techniques have made dramatic progress[1), 2)], resulting in higher conversion efficiency of multicrystalline-based PV cells. On the other hand, the Czochralski (CZ) monocrystalline silicon growth technique has been considered to be a field with mature technologies. Currently, large-diameter CZ silicon crystals with the highest quality (such as having long lifetime) are obtained by employing the Magnetic-field-applied CZ (MCZ) technique.

Sakuragi[3), 4)] proposed the concept of "liquinert (liquid is an inert state)" technique in crystal growth to improve the crystal quality. They have successfully grown many kinds of excellent crystals based on their liquinert technique putting RAP[3)] gas like silicon tetrachloride ($SiCl_4$) into the crystal growth environment (RAP: Reactive-gas Atmosphere Processing).

In the previous report[5)], we showed the lifetime improvement in monocrystalline p-type CZ silicon with an alternative liquinert technique, the employment of liquinert silica crucibles[6)], whose inner-surfaces have melt-phobic (non-wettable) property against molten silicon not in $SiCl_4$ ambience but argon (Ar) ambience. In this report, we describe Liquinert-crucible-applied CZ technique as LCZ technique. Using the LCZ growth method, we have succeeded to grow lower-carbon content p-type silicon crystals than conventional ones[5)]. Lifetime values of the p-type wafers cut from our LCZ silicon exceed 700 μs[5)] which is more than twice those of commercially available p-type silicon wafers (since our purpose is to reveal the bulk lifetime of silicon, no Phosphorus Diffusion Gettering (PDG) treatment was performed as has been well done to improve lifetime values).

In this report, by comparing the bulk lifetime of both LCZ and commercially available MCZ silicon, we show that our LCZ technique has the advantage also for the growth of n-type silicon.

2. Experiments

We prepared two kinds of monocrystalline CZ silicon crystals doped with phosphorus (P). The one is LCZ silicon using a 24 inch diameter liquinert crucible and the other is a commercially available silicon ingot grown by a silicon vendor with their MCZ technique. The diameter of both crystals was approximately 200 mm.

As is shown in Figure 1, we prepared square-prism-bar samples (20mm×20mm×80mm) and 2 mm thick round wafers from seed-side or tail-side of the ingots. The square-prism-bar samples, finished with lapping, are for the bulk lifetime measurement and round wafers, finished with lapping and subsequent etching with hydrofluoric-nitric acid to reveal the mirror surfaces, are for the measurement of resistivity, oxygen, carbon and metals (Fe, Cu and Ni).

The bulk lifetime was measured by microwave photoconductive decay method according to JIS H0604, measuring of minority-carrier lifetime in silicon single crystal by photoconductive decay method[7]. The resistivity was determined using the four-point probe measurement method. The oxygen content was measured by Fourier Transform Infrared (FTIR) spectroscopy with the conversion factor of $3.14 \times 10^{17} cm^{-1}$ standardized in ASTM-F1188. The carbon content was measured by secondary ion mass spectroscopy (SIMS) and the metal contents were determined using Induction Coupled Plasma Mass Spectroscopy (ICP-MS) method.

We also observed the cross-sectional views at the initial melt surfaces both in conventional and liquinert crucibles after the completion of growth.

3. Results

Figure 2 shows the dependence of oxygen (1) and carbon (2) concentrations on solidification fraction in LCZ and commercially available MCZ silicon crystals. Except for a limited portion (at the low fraction less than 20%), LCZ silicon has higher oxygen content than MCZ silicon. Figure 3 shows that the carbon content of LCZ silicon is higher in all fractions than that of MCZ silicon although the carbon content of p-type LCZ was lower than that of conventional p-type CZ silicon[5]. The reason why higher carbon content in LCZ silicon is probably because we finally succeeded to grow a monocrystalline silicon after three-times failures from the same melt in the same liquinert crucible. It is well known that the longer the molten silicon is kept in the same crucible in a furnace ambience, the heavier the melt becomes contaminated with carbon.

Figure 4 shows the dependence of bulk lifetime on resistivity for LCZ and commercially available MCZ silicon. Generally speaking, bulk lifetime proportionally increases to the resistivity. However, the lifetime of LCZ silicon at the resistivity of 0.8 – 0.9Ωcm is equal to those of MCZ silicon at the resistivity of 1.5 – 2.5Ωcm despite that fact that the carbon of LCZ silicon content is higher than that of MCZ silicon.

The MCZ sample at the resistivity of 2.9Ωcm has lower lifetime despite of the higher resistivity. We will consider the reason in the next chapter.

Bulk concentrations of metal impurities were below the background level (5ng/g) in all LCZ and MCZ samples.

Figure 5 shows the position (framed blue) where post-growth samples were taken (1) and the comparison of the cross-sectional views in conventional (a) and liquinert (b) crucibles after the completion of growth. Figure 5 (2) (b) clearly shows that the dissolution of liquinert layer was extremely suppressed because the contact surface below the initial melt curve was almost unchanged.

4. Discussion

Using Figure 3, we previously concluded that the longer lifetime of p-type LCZ silicon was due to carbon reduction. Although n-type LCZ silicon has high possibility of longer lifetime than that of n-type MCZ silicon from Figure 4, the excellent lifetime is not caused by carbon reduction but other improvement mechanism.

Figures 5 (2) (a) and (b) shows that the dissolution of liquinert-crucible wall was suppressed less than one-tenth of conventional crucible wall. This means that the metal incorporation from the dissolved crucible to molten silicon will be lower than one-tenth of conventional crucible. Since vitreous silica crucibles for CZ growth contain metal impurities at sub-ppm order, shown in Table 1[9], such as Fe: 0.3ppm, K: 0.7ppm, Li: 0.5ppm[9], and so on, it is highly expected that the use of liquinert crucibles makes the metal contamination lower than that of conventional vitreous-silica crucibles. Kitagawara[8] reported the dependence of bulk lifetime on Fe concentration, showing that the lifetime increases as Fe concentration decreases from 10^{12} down to $10^{9}cm^{-3}$. Considering the results of lifetime improvement both in p-type and n-type LCZ silicon, we conclude that the improvement results not only from carbon reduction but also from metal reduction.

The MCZ sample with lower lifetime at the resistivity of 2.9Ωcm in Figure 4 was taken from the position of low fraction of 5% and from the data in Figure 2 (1), the oxygen content was the highest ($1.1 \times 10^{18} cm^{-3}$) of all samples. Since it was reported that lifetime becomes lower as the oxygen content increases[8], the lowest lifetime in the MCZ silicon is due to the highest oxygen-content.

5. Conclusion

LCZ growth technique using a liquinert crucible improves the bulk lifetime of n-type silicon longer than that of commercially available n-type MCZ silicon. We speculate that the lifetime improvement results not only from carbon reduction but also from metal reduction.

References

1. A-K. Soiland, G. Halvorsen, K. Friestad, A. H. Amundsen, J. O. Odden, and T. Ulset, Proceedings of EUPVSEC (2014), pp. 536–541.
2. Z. Xia, K. Guo, H. Sun, H. Pan, J. Li, R. Liu, H. Cui, R. Sidhu, B. Zhang, and G. Xing, Proceedings of 29th EUPVSEC (2014), pp. 542–547.
3. S. Sakuragi, T. Shimasaki, G. Sakuragi, and H. Nanba, Proceedings of 19th EUPVSEC (2004), pp. 1197–1200.
4. S. Sakuragi, Proceedings of 19th EUPVSEC (2004), pp. 1201–1204.
5. T. Fukuda, Y. Horioka, N. Suzuki, M. Moriya, K. Tanahashi, S. Simayi, K. Shirasawa, and H. Takato, J. Cryst. Growth **438** (2016), pp. 76 – 80.
6. Y. Horioka and S. Sakuragi, US Patent: US 2014/0150714 A1.
7. JIS H0604.
8. Y. Kitagawara, T. Yoshida, T. Hamaguchi, and T. Takenaka, J. Electrochem. Soc., Vol. 142, No. 10, (1995) pp. 3505 - 3509.
9. http://www.iotaquartz.com/product-range.cfm

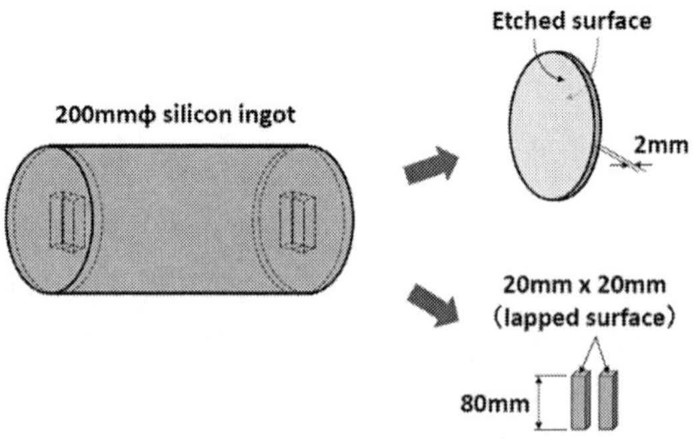

Figure 1. Sample preparation for the measurement of impurities (oxygen, carbon and metals) concentration and bulk lifetime.

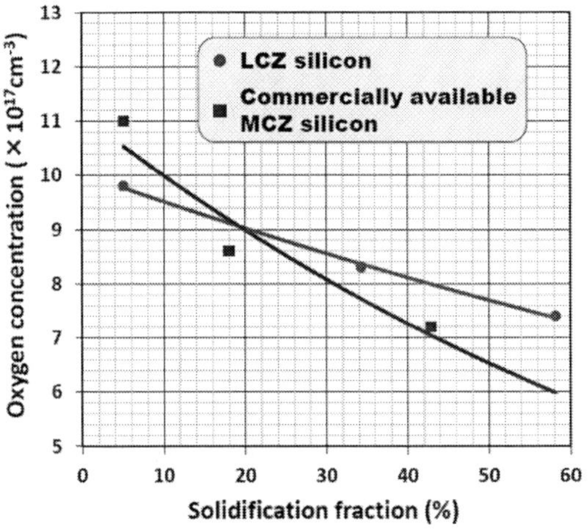

(1)

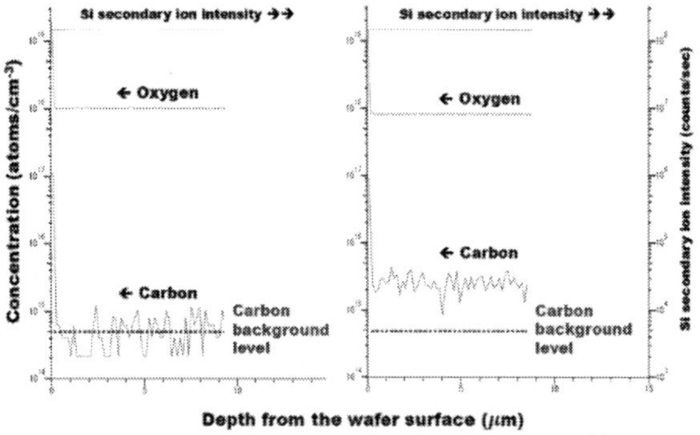

(1) (2)

Figure 3. The impurity concentrations (oxygen and carbon) of the p-type LCZ (1) and conventional CZ silicon (2)[5].

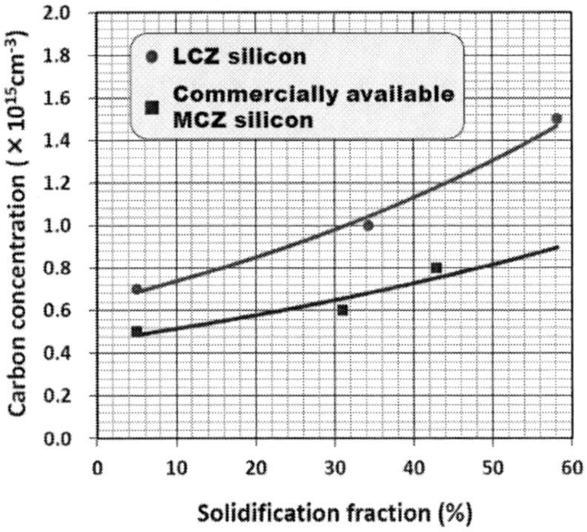

(2)

Figure 2. The dependence of oxygen (1) and carbon (2) concentrations on solidification fraction in LCZ and commercially available MCZ silicon.

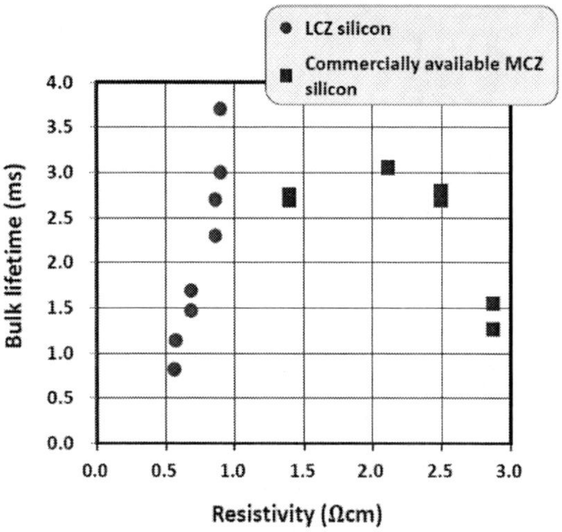

Figure 4. The dependence of bulk lifetime on resistivity for LCZ and commercially available MCZ silicon.

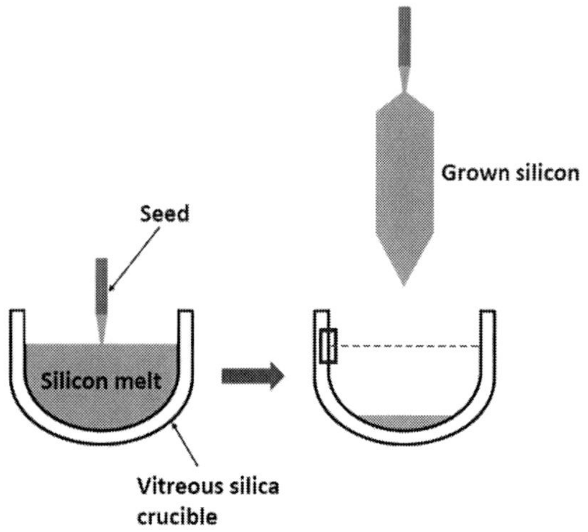

(1)

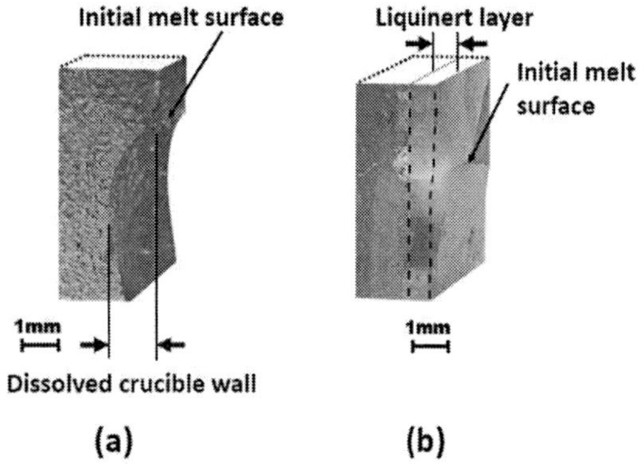

(2)

Figure 5. The position (framed blue) where post-growth samples were taken (1) and the comparison of the cross-sectional views in conventional (a) and liquinert (b) crucibles after the completion of growth.

Table 1. Typical Values (ppm) of Impurities in Vitreous Silica Uploaded by IOTA[9)]

	Al	B	Ca	Cr	Cu	Fe	K	Li	Mg	Mn	Na	Ni	P	Ti	Zr
IOTA 4	8	<0.05	0.7	0.007	0.034	0.3	0.4	0.2	0.07	0.013	1.0	0.002	0.1	1.4	0.01
IOTA 6	8	<0.05	0.7	0.003	0.001	0.2	0.1	0.2	0.07	0.008	<0.1	0.002	0.1	1.4	0.01
IOTA 6-SV	8	<0.05	0.5	0.002	0.001	0.1	<0.1	0.2	0.02	0.004	<0.05	0.001	<0.05	1.3	0.01
IOTA 8	8	<0.05	0.4	0.001	<0.001	<0.05	<0.05	<0.05	0.01	0.001	<0.05	<0.001	<0.05	1.3	0.01
IOTA STD	14	<0.10	0.5	0.006	0.026	0.3	0.7	0.5	0.04	0.039	1.0	0.001	0.1	1.2	0.01
IOTA CG	14	<0.10	0.6	0.007	0.019	0.3	0.7	0.5	0.04	0.029	1.0	0.001	0.1	1.2	0.01
IOTA STD-SV	14	<0.10	0.3	0.004	0.003	0.1	<0.1	0.5	0.01	0.007	<0.05	0.001	0.1	1.1	0.01

CZ SILICON BENCHMARK FOR P-TYPE PERC SOLAR CELLS

Pierre Saint-Cast, Isolde Reis, Johannes Greulich, Sabrina Werner, Elmar Lohmüller, Hannes Höffler, Jonas Haunschild and Ralf Preu
Fraunhofer Institute for Solar Energy Systems ISE, Heidenhofstraße 2, 79110 Freiburg, Germany
Telephone: +49 761 4588 5480, fax: +49 761 4588 9000, e-mail: pierre.saint-cast@ise.fraunhofer.de

ABSTRACT: In this paper, seven Czochralski (Cz) and one Float Zone *p*-type silicon materials are processed as PERC solar cells in the Fraunhofer ISE PVTEC pilot line. In order to compare materials with different bulk resistivities, the rear contact distance is varied in order to keep the spreading resistance in the same range. As-processed and regenerated, energy conversion efficiencies between 20.5% and 21.0% are achieved. Within processing and measurement accuracy, all materials performed on the same level when comparing the solar cell conversion efficiencies. Implied open-circuit voltage samples revealed to be more sensitive to the material quality than the solar cells and give some more details. Due to the small differences in the results, we will not provide a ranking of the material tested in this paper. However, all the materials tested here are capable of achieving at least an efficiency of 20.7%, when applied in a PERC solar cell process.

Keywords: PERC, Cz material, light induced degradation

1 INTRODUCTION

During recent years, impressive progress was made in increasing the solar cell conversion efficiency as well as in decreasing the cost of crystalline silicon solar cell fabrication. This was made possible thanks to great efforts along the whole value chain of crystalline silicon solar cells, including silicon feedstock, solar cell processing and module fabrication. Today, industry is rapidly moving towards the "Passivated Emitter and Rear Cell" (PERC) structure which allows conversion efficiencies well above 20% and has been in the focus of research and development for more than 15 years [1,2,3]. An increased spreading of high efficiency solar cells brings up and emphasizes the need for high quality silicon material.

This paper presents a quality comparison of commercially available p-type monocrystalline Cz-grown silicon material. Fraunhofer ISE organized a benchmark to determine the best Cz-Si material for p-type PERC solar cells. The results show that the material tested in this study is capable of yielding high efficiencies with the potential for further process optimization.

2 MATERIAL PRETEST

2.1 Eligibility and wafer specification

In order to take part in this benchmark, the participants had to provide *p*-type monocrystalline silicon wafers based on the Czochralski growth method which are produced for the photovoltaic industry, participants who are manufacturing wafers for microelectronics applications and retailers were not accepted.

Originally, in order to be accepted for this test, the wafer had to meet the specifications listed in Table I. However all the wafers received were used in the test. A pretest was carried out in order to verify whether the wafers meet the specification or not.

2.2 Participation and pretest

Cz material manufacturers all over the world were invited to participate to this benchmark. Seven suppliers answered the call and were capable of supplying wafers according to the specification. Reference material used as standard material in our pilot was also added to the test. The reference material used was *p*-type magnetic-Czochralski-grown with a bulk resistivity $\rho_{bulk} \approx 1.4\ \Omega cm$.

The suppliers were numbered in order to anonymize the participant of the benchmark. The material numbers 1 to 7 correspond to the suppliers which took part in the benchmark, the material number 8 corresponds to the reference material. The numbers are consistent throughout the paper.

Before processing the wafers to solar cells, a pretest was applied to all of the provided as-cut wafers in order to verify if the wafers meet the specified criteria and to find correlations to solar cell results. The following values were measured inline: thickness (on three tracks), resistivity (one track), saw marks, effective lifetime (MDP-method [4]). In addition, the following methods were also applied: photoluminescence imaging (PL), infrared imaging for the detection of micro-cracks, and optical inspection in order to detect: stains, edge defects and geometry.

Table I: Wafer specific requirement in order to participate to the test.

Item / Measurement	Specification	Unit
Material	Crystalline silicon	-
Method	Czochralski-grown	-
Doping Type	*p*-type Si Boron or Gallium doped	-
Orientation	<100>	
Resistivity	0.5-3.0	Ωcm
Resistivity Variation	± 0.5	Ωcm
Oxygen Content	as low as possible	atoms/cm^3
Carbon Content	< 5.0×10^{16}	atoms/cm^3
Iron Content	< 2.0×10^{10}	atoms/cm^3
Lifetime	as high as possible (> 20)	µs
Size	156.0 ± 0.5	mm
Diameter	Pseudo square : 200.0 ± 0.5 Full square : 219.0 ± 0.5	mm
Etch Pit Density (EPD)	≤ 3000	cm^{-2}
Average Wafer Thickness	180 ± 20	µm
Total Thickness Variation (TTV)	≤ 30	µm
Saw Mark Depth	≤ 15	µm
Cracks/Holes	no	-
Edge Defects/Chips	no defects > 0.5	mm
Surface Condition	as cut, cleaned, no stains	-

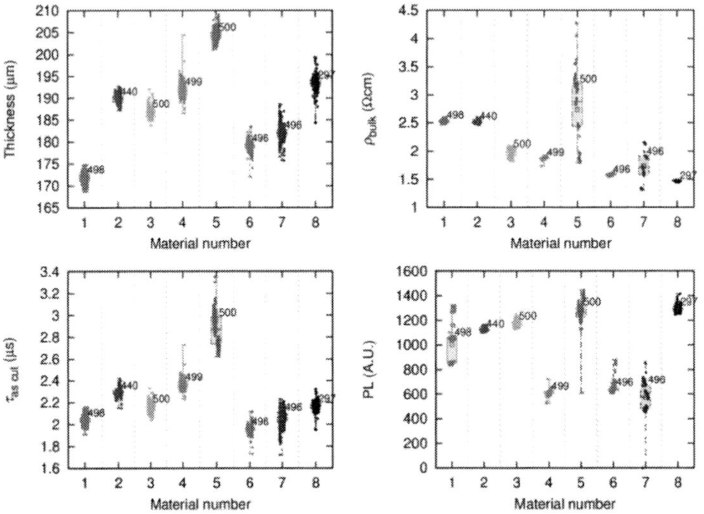

Figure 1 Selected results of the pretest applied to the wafers of all suppliers (material number 1-7) and the reference material (number 8). All results are within the expected variations.

A part of the results of the pretest is displayed in Figure 1. The thickness, the resistivity ρ_{bulk}, the effective minority carrier lifetime of the as-cut wafers $\tau_{as\text{-}cut}$, and the intensity of the PL signal during PL imaging are shown. The results from the resistivity measurement are required later in order to adapt the distance (or pitch) between the rear line contacts. The value of $\tau_{as\text{-}cut}$ and of PL are limited due to surface recombination. Despite that, these values might be correlated with the electrical quality of the material and are given as an indication.

At this stage, several suppliers did not meet exactly all of our criteria:

- Material number 2 is Float Zone (FZ) material and not Czochralski grown.
- Material number 5 has a resistivity between $2\ \Omega cm \leq \rho_{bulk} \leq 3.5\ \Omega cm$, however the accepted variation of the resistivity was maximal 0.5 Ωcm (see Table I). Selected wafers were used for further processing.
- Material number 7 is pseudo square and has a diameter of 205 mm. Our specification for pseudo square material was a diameter of 200 mm. No screen printing mask for the front and the rear side was available for this format. Therefore, these wafers were processed further as if they were 200 mm diameter.

Due to the resonable number of participants, no supplier was excluded from the benchmark and all materials were fully processed and characterized.

2 SOLAR CELL PROCESSING AND CHARACTERISATION

2.1 Processing

The solar cells are completely processed in the industry-related PV-TEC pilot line at Fraunhofer ISE [5]. Commercially available industrial tools are used for each of the solar cell fabrication processes. The process flow is presented in Figure 2.

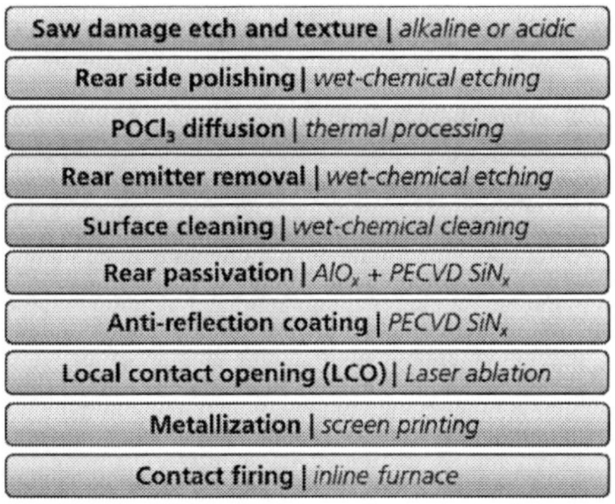

Figure 2 Process flow used for the fabrication of the PERC solar cells.

The wafers are first labeled with laser-marking for tracing and randomized in groups of 10 wafers in order to minimize the influence of process fluctuations on the out coming results. After an alkaline saw-damage removal, the wafers are alkaline-textured and the rear side is chemically polished resulting in a thickness reduction of about 40 µm in total. The homogeneous emitter is formed in a tube furnace diffusion process using $POCl_3$ with in-situ oxidation [6]. Then, a wet-chemical edge isolation (CEI) is carried out in order to remove the emitter on the rear side. Prior to passivation, a wet-chemical cleaning (SC1/SC2 [7]) is carried out. The passivation of the rear surface is obtained by a 6 nm-thick aluminum oxide (Al_2O_3) layer which is deposited by fast atomic layer deposition (ALD). The ALD deposition is followed by an annealing step performed in a tube furnace under N_2 atmosphere at 550°C for 10 min. A silicon oxide (SiO_x, 100 nm thick) and silicon nitride (SiN_y, 100 nm thick) layer stack serves as capping layers. The capping layers are deposited by plasma enhanced chemical vapor deposition (PECVD). Concerning the front surface passivation, a PECVD SiN_z layer is used. All the processes applied up to this stage will be later referred to as the "front-end".

Then, the formation of the metal contacts is carried

out. The rear layers are opened using a laser process in order to obtain a line-shaped local contact opening (LCO). The front and rear side metallization is formed by screen printing. The front grid features five busbars and a finger distance of 1.56 mm with a printed finger width $w_f \approx$ 55-70 µm. The rear electrode does not feature Ag/Al soldering pads. Finally, the contact firing is performed in an industrial conveyor belt furnace and the current voltage characteristics of the solar cells are measured with an industrial cell tester.

In addition to the solar cells, special test samples are made. Implied open-circuit voltage (iV_{OC}) samples correspond to wafers, with only the front-end processes being applied. These samples are then directly fired without laser opening or screen printing processes. These samples allow determining the V_{OC} potential independent of the metal recombination losses. Emitter saturation current density (j_{0e}) samples mirror symmetrically on both front and back side the front surface of the iV_{OC} samples. The j_{0e} samples are also fired without metal. iV_{OC} samples and j_{0e} samples are measured using the quasi-steady-state photoconductance (QSSPC [8]) method. For the determination of j_{0e}, a recently presented refinement [9] of the Kane-Swanson method [10] is applied.

2.1 Pitch variation

Materials with different bulk resistivity need to be processed with different contact distances on the rear (pitch) in order to fully exploit the respective material's potential. Our approach to meet the processing condition is a variation of the pitch. First, the materials were separated in three different categories:

- Low resistivity: 1.25 Ωcm $\leq \rho_{bulk} \leq$ 1.75 Ωcm. The materials 6, 7 and 8 belong to this category.
- Middle resistivity: 1.75 Ωcm $\leq \rho_{bulk} \leq$ 2.25 Ωcm. The materials 3 and 4 belong to this category.
- High resistivity: 2.25 Ωcm $\leq \rho_{bulk} \leq$ 3 Ωcm. The materials 1, 2 and 5 belong to this category.

Each category has its own pitch variation. Each pitch has been chosen in order to keep the spreading resistance within a range of 0.15 Ωcm² $\leq R_{Spread} \leq$ 0.25 Ωcm². In Fig. 3, the spreading resistance is given as a function of wafer resistivity and the pitch, for an effective contact width of 55 µm and a wafer thickness of 150 µm. The spreading resistances presented in Fig. 3 are calculated using an analytical model [11]. In Table II, the pitch variation for each category is given.

In addition to the pitch variation, a firing temperature variation was performed, so that from the initial 63 wafers per material processed, only 7-8 wafers are processed identically (the size of each group might vary slightly due to wafer breakage).

4 RESULTS AND DISCUSSION

4.1 Other measurements

The measurement of the emitter saturation current density j_{0e} shows a very good repeatability between the materials. The median values of j_{0e} varies between 79 fA/cm² $\leq j_{0e} \leq$ 85 fA/cm² and the standard deviation is about 8 fA/cm² for all materials. It can be expected that the emitter behaves similarly for the different materials. In addition to the j_{0e} values, these samples also deliver information on the bulk lifetime by fitting the invert

Table II: Distance between the rear contacts lines (pitch) used for each resistivity category.

Resistivity category	low	middle	high
Resistivity range [Ωcm]	1.25-1.75	1.75-2.25	> 2.25
Pitch 1 [µm]	1000	800	600
Pitch 2 [µm]	1200	1000	800
Pitch 3 [µm]	1400	1200	1000

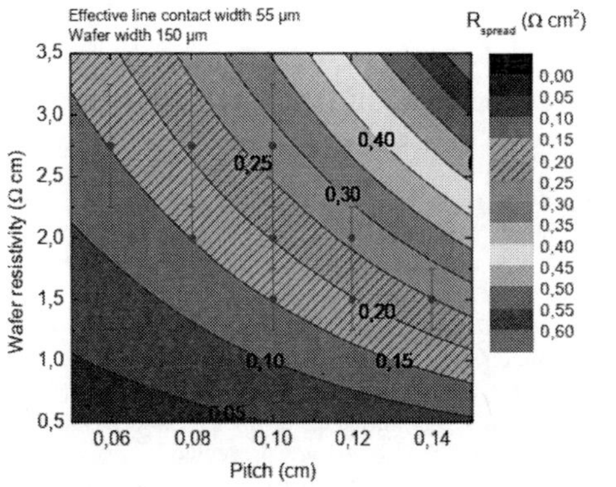

Figure 3 Spreading resistance (R_{spread}) as a function of the wafer resistivity and the distance between the rear contacts (pitch). The calculations are done supposing a wafer thickness of 150 µm and an effective contact width of 55 µm.

effective carrier lifetime. The median minority carrier lifetime in the bulk extracted varies between 300 µs and 1000 µs. Although this lifetime difference appears to be significant, it can be explained by the large bulk doping variation between the materials. Thus it cannot be directly correlated with the V_{OC} potential of the solar cells within this experiment. The maximal implied V_{OC} is a more appropriate value. This value corresponds to the maximal V_{OC} allowed by the front half of the solar cell which includes the recombination at the emitter and half of the recombination in the bulk. This value corresponds to the iV_{OC} at an irradiance equivalent to 2 suns of the j_{0e} samples. The results are presented in Fig. 4.

The implied open circuit voltage (iV_{OC}) is measured on the iV_{OC} samples (see section 2.2). The iV_{OC} value measured corresponds to the potential of the front-end without the recombination due to the metallization. More specifically the iV_{OC} sample includes the recombination

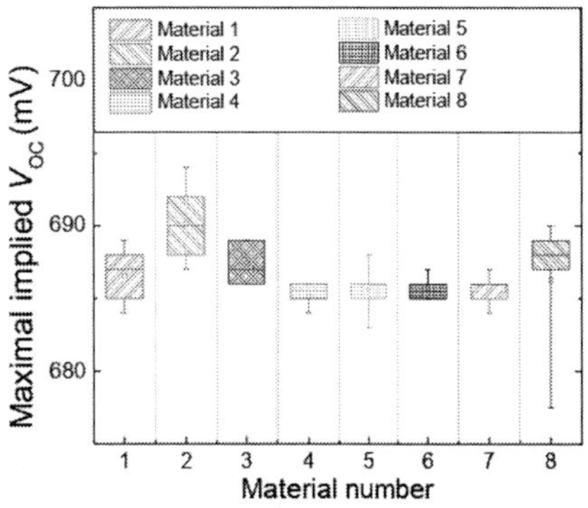

Figure 4 Implied V_{OC} at an irradiance equivalent to 2 suns of j_{0e} samples, which corresponds to the maximal implied V_{OC} allowed by the front of the solar cell and half of the bulk.

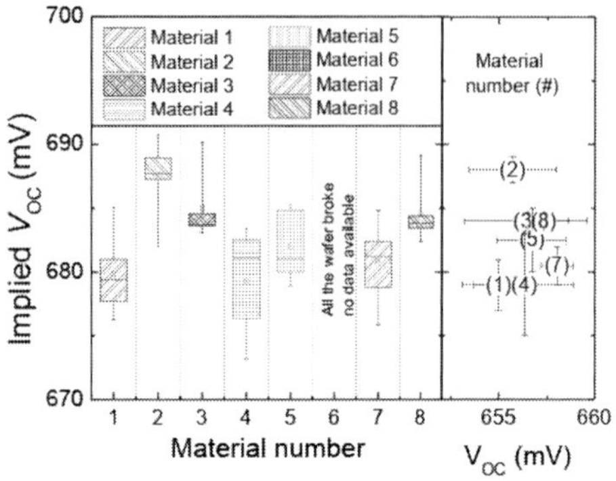

Figure 5 On the left hand side, the implied V_{OC} at irradiation equivalent to 1 suns of iV_{OC} samples, which corresponds to the V_{OC} allowed by the front-end process of the solar cell. On the right hand side, the implied V_{OC} as a function of the measured V_{OC} of the solar cells from the same material.

at the emitter in the bulk and on the passivated rear surface. The iV_{OC} results are presented in Fig. 5. These results correlate very well with the ones presented in Fig. 4. As mentioned earlier, the j_{0e} values are very similar between the materials. Therefore, the difference in iV_{OC} between the groups can be mainly attributed to the material quality. A variation up to 10 mV is observed for the iV_{OC} samples, the variations in V_{OC} are much smaller (see Fig. 6). These variations do not correlate with the V_{OC} variation of the solar cells. Therefore it seems that for this experiment, the performance before metallization does not correlate with the performances after metallization. This could be explained by instabilities in the metallization process influencing the recombination parameters.

4.2 Influence of the pitch

From the three firing temperatures applied, two led to a noticeable decrease in the solar cell performance, indicating that the process window was relatively small. Thus, only the results of the solar cells fired at the lowest temperature will be presented. In Fig. 6, the results of the illuminated current-voltage measurement are given. The measurement was performed directly after processing (as-processed). The energy conversion efficiency η, the fill factor FF, the short-circuit current density j_{SC} and the open-circuit voltage V_{OC} are given for the material 1-8 and for each pitch.

The efficiencies obtained are ranging from $20.5\% \leq \eta \leq 21.0\%$. For each material, the variation of the pitch has a noticeable impact on the efficiency of more than 0.2%. The expected impact of the pitch is the following:

- For a large pitch, the series resistance increases and therefore the FF decreases. However, the rear recombination decreases and therefore the V_{OC} increases. A small increase of the j_{sc} can also be expected due to the increase of the internal reflection.
- For a small pitch, the effect is opposite, the FF increases, the V_{OC} decreases and the j_{sc} decreases slightly.
- The overall impact cannot be estimated as they have opposite impacts on the efficiency and we do not know the details of each individual effect quatitatively.

The expected behavior can be observed for most of the materials. For materials 1, 2 and 3 the V_{OC} clearly increases with the pitch, for the materials 4, 5 and 6 the largest pitch has the highest V_{OC} values, however, the trend is unclear for the lower pitch. For the materials 7 and 8 no clear correlation to V_{OC} can be found. For the materials 1-5 and 7 the FF clearly decreases with the pitch as expected. For the materials 6 and 8, no clear difference can be observed between the pitch 2 and 3. Concerning the j_{sc}, no clear trend can be observed in the results and its variation might be related to process

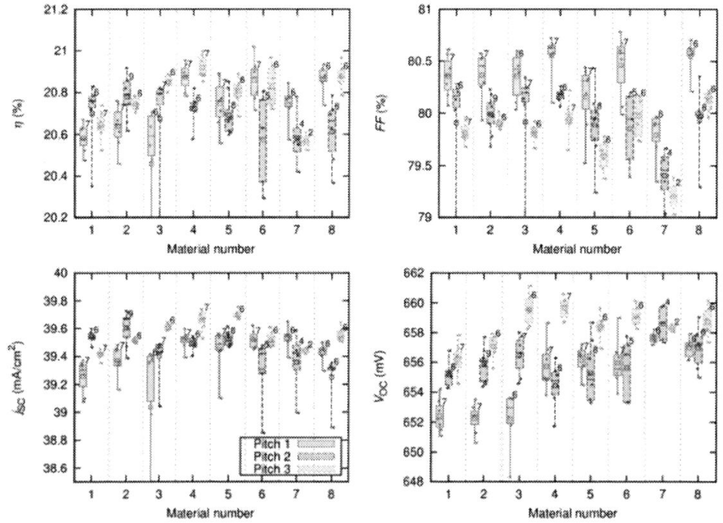

Figure 6 Results of the illuminated current voltage measurement for the lowest firing temperature applied. The material 1 to 8 and all the pitches are displayed. The measurement was performed directly after processing. Between the end of the process and the measurement, the cells were kept in a dark environment.

instabilities. In general the pitch related behavior of V_{OC} and *FF* are not consistent with the expectation in all cases and are probably also related to process fluctuations. As all materials seem to have a different optimal pitch, the results of each material will be treated independently from the pitch.

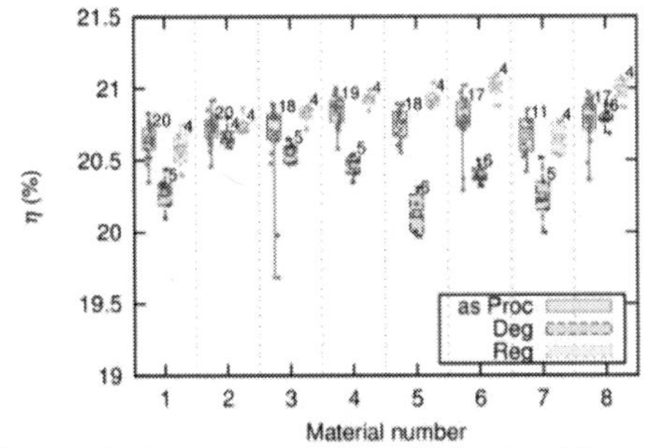

Figure 7 Energy convention efficiencies (η) as the function of the material for the as process (as Proc), the degraded (Deg) and regenerated (Reg) state. All the pitchs are presented together.

4.3 Light induced degradation and regeneration

Light induced degradation (LID) is very important to consider when studying the quality and performance of *p*-type Cz-Si material. Five wafers representative of each material (close the median efficiency of the material) are selected in order to study the effect of LID on the material. After processing (as Proc), the selected solar cells are degraded (Deg) during 96 h. Then, the solar cells are regenerated (Reg) at 140°C under 1 sun irradiation for 2 hours. For each group, a test sample is regenerated for an additional hour in order to check the completeness of the regeneration. No substantial improvement has been observed for this sample, therefore after 2 hours the regeneration can be considered being completed. After regeneration the solar cells are degraded again for 96 h to verify the stability of the regenerated results and no significant change in efficiency could be measured. The efficiency in the as-processed, degraded and regenerated state are presented in Fig. 7.

As it could have been expected, the materials 2 and 8 did not show any significant degradation, which is explained by the low oxygen concentration of Float Zone and magnetic Cz-Si material. The other materials degraded from 0.5%$_{abs}$ until 0.9%$_{abs}$ in efficiency. This value did not correlate with the doping concentration, which tends to show that the oxygen concentration also varies among the materials. It is difficult to predict which state is the most representative of the field condition. Recent studies tend to show that a complete degradation is too pessimistic and that a kind of regeneration process takes place naturally in the field. In addition, fast stabilization processes are now under development for industrial production, which could greatly reduce the impact of LID [13]. In a future benchmark, the use of fast regeneration would allow to obtain better statistical results after regeneration.

Although there are noticeable differences for the materials after regeneration, these results are based on only 4 samples per group. Due to the lack of statistical significance, we will abstain here from giving a ranking of the suppliers. A larger experiment should be conducted to verify the results including a regeneration of all the samples. The efficiencies obtained are representative of actual efficiencies reach in production [14], the solar cell process present a remarkable flexibility allowing to process different material up to a similar efficiency level. In fact, all the materials tested in this experiment allowed for similar efficiencies (as-Pros or Reg) compared to the high quality magnetic Cz-Si (Material 8), which tend to show that all the materials have also a high quality. The higher the material quality, the higher are the requirements on the solar cell processing quality. In order to increase the sensitivity on solar cell level, a more stable PERC solar cell process with higher efficiency potential should be applied [15].

It is interesting to notice that difference in efficiencies between the materials is 0.4%$_{abs}$ in the regenerated state and 1.0%$_{abs}$ in the degraded state, therefore fast regeneration could be used in a production line, in order to level the difference in the incoming material quality.

5 CONCLUSION

Wafers from seven different *p*-type monocrystalline silicon material suppliers, in majority Cz-Si, are processed as PERC solar cells. Except from the rear contact pitch, all materials are used without adapting the process. Energy conversion efficiencies from 20.5% until 21.0% are achieved in the as-processed state and confirmed via regeneration. This result shows the high flexibility of the PERC process used in this experiment.

The material performance observed on solar cell level is on a very similar level within measurement accuracy. However, more sensitive iV_{OC} and j_{0e} samples have shown differences between the materials have been observed. Two deductions follow, first the quality of all the materials used in this experiment allowed for after-regeneration efficiencies close to the high quality magnetic Cz-Si reference, second the cell process applied here, which is representative of the actual industry quality level, is not sensitive enough to resolve clearly the differences in the material performance for such high material quality.

We will therefore abstain from giving a ranking of the suppliers. However, it can be stated that all the materials tested here are capable to reach at least an efficiency of 20.7%, when applied in a PERC solar cells process.

ACKNOWLEDGEMENTS

The authors would like to thank all colleagues at the Fraunhofer ISE Photovoltaic Technology Evaluation Center (PV-TEC), the companies which send material and make this benchmark possible: Ubiquity Solar, Norwegian Crystals, Tianjin Zhonghuan Semiconductor Co. Ltd, AUO Crystal Corp. and other companies. This work was funded by the German Federal Ministry for Economic Affairs and Energy within the project "CUT-A" under contract no. 0325823.

REFERENCES

[1] A. W. Blakers, A. Wang, A. M. Milne, J. Zhao and M. A. Green, "22.8% efficient silicon solar cell", Applied

Physics Letters, 55(13), 1363–1365 (1989).
[2] Press realise PV magazine Trina 16th Dec. 2015, Solar World 14th Jan. 2016, Gintech 5th Jan. 2016
[3] T. Dullweber, H. Hannebauer, U. Baumann, T. Falcon, K. Bothel, S. Steckemetz, R. Brendel, "Fine-Line Printed 5 Busbar PERC Solar Cells With Conversion Efficiencies Beyond 21%" 30th EU PVSEC, Paris, France, Sep. 2014.
[4] K. Dornich *et al.*, "New spatially resolved inline lifetime metrology on multicrystalline silicon for pv" 24th EUPVSEC, Hamburg, Germany, pp. 1106–1108.
[5] D. Biro, R. Preu, S. W. Glunz, S. Rein, J. Rentsch, G. Emanuel et al., "PV-Tec: Photovoltaic technology evaluation center - design and implementation of a production research unit", Proceedings 21st EUPVSEC 2006, pp. 621–624.
[6] S. Werner et al. Procedings 29th EU PVSEC, 2014, pp. 1342.
[7] Kern, W. et al., RCA Review, Vol. 31, pp. 187-205, 1970.
[8] R. A. Sinton and A. Cuevas, Appl. Phys. Lett, vol. 69, no. 17, pp. 2510–2512, 1996.
[9] A. Kimmerle, J. Greulich, and A. Wolf. "Carrier-diffusion corrected J_0-analysis of charge carrier lifetime measurements for increased consistency." Solar Energy Materials and Solar Cells 142 (2015): 116-122.
[10] D.E. Kane, and R. M. Swanson. "Measurement of the emitter saturation current by a contactless photoconductivity decay method." IEEE photovoltaic specialists conference. 18. 1985.
[11] P. Saint-Cast, Ph.D. Thesis, Physic, Constance University, pp. 61, 2012.
[12] P. Saint-Cast *et al.*, "Analysis of the losses of industrial-type PERC solar cells". to be published.
[13] A. Herguth *et al.*, "Accelerating Boron-Oxygen Related Regeneration: Lessons Learned from the BORNEO Project" 30th EUPVSEC, Hamburg, Germany, 2015, pp. 804-807
[14] B. Tjahjono et al., Proceedings 28th EUPVSEC, Paris, France (2013), pp. 775
[15] H. Neuhaus, PVCellTech Conference, Kuala Lumpur, Malaysia (2016)

THE BENEFIT OF ULTRA-HIGH MINORITY CARRIER LIFETIME SILICON WAFERS FOR HIGH-EFFICIENCY AND INNOVATIVE SOLAR CELLS

Ian MacLellan[U], Siva Sivoththaman[W], Sjouke Zijlstra[U], Jan Vedde[S], Rafael Kleiman[M], Peter Dold[F], Jerry Olson[J], Tor Hartmann[U], Zhen Gao[W], KC Chang[U], Tom Cadwell[U], Jørgen Bødker[B], Frank Faller[G]

(U) Ubiquity Solar Inc., Sarnia, Ontario, Canada; (W) University of Waterloo, Canada; (S) SiCon Consulting, Denmark; (M) McMaster University, Hamilton, Ontario, Canada; (F) Fraunhofer CSP, Halle, Germany; (J) Jerry Olson Consulting, Boulder, Colorado, USA; (B) SmarterEnergy, Copenhagen, Denmark; (G) FSC Solar Consulting, Germany, corresponding author, Email ffaller@f-s-c.eu

ABSTRACT: Over the past 10 years, average PV cell efficiencies (multi/mono) have improved by ~ 3-5% absolute from about 15-18% on volume weighted average basis to about 18-23%, and costs at the cell level have been reduced by over a factor of 10. This has been driven by improvements in wafer quality, cell design and technologies, economies of scale and manufacturing equipment and processes. The need for high minority carrier lifetime in wafers historically has been limited due to the quality of the cell processing steps and the success of gettering techniques developed to improve poor quality material. However, with recent trends in moving advanced PV cell concepts (PERC-family, hetero- or tunnel-junction, etc.) into production, and continued growth of current high-efficiency solar cell production (IBC, HIT), wafer quality and specifically minority carrier lifetime may now be a limiting factor for further improvement of the cell efficiency. In the near future, wafer quality will become more important to fully exploit the current state-of-the-art technologies, and implementing even more advanced solar cell designs. Mono-crystalline wafers produced using the Float Zone (FZ) process are known to provide the highest lifetimes, but have so far been too expensive to produce. An innovative technology to produce cost-effective FZ wafers is being introduced in this paper. A pilot production line is currently being set up by Ubiquity Solar, USI, in Canada, to produce n- and p-type ingots and wafers.

1. INTRODUCTION

The impact of minority carrier lifetime (MCL) on the efficiency of photovoltaic cells is well known [1]. What is not as well understood is to what degree the minority carrier lifetime of the as-received wafer controls the *effective* minority carrier lifetime of the finished product, and to what degree the lifetime limits the efficiency of advanced solar cell concepts. The previous question is relevant since the cell fabrication processes can change the minority carrier lifetime through the addition or removal of impurities via gettering, the introduction or passivation of deep levels and local changes in the position of the quasi-Fermi level associated with p-n junction formation and surface passivation [1] [2] [3]. Minority carrier lifetime also changes after cell formation in response to thermal cycles and exposure to light, voltage and current flow.

With solar cell architectures and process technologies advancing to higher theoretical efficiencies, the ultimate question now is to what extent the effective wafer MCL limits the exploitation of the theoretical efficiency potential.

High quality starting materials allow for the development of more complex cell structures and higher improvements in efficiency than just the evolution of PV cell processes and structures [4].

This paper discusses the influence of the minority carrier lifetime of the starting wafer on the PV cell conversion efficiency and an innovative Float-Zone (FZ) ingot technology, which will allow to manufacture ultra-high lifetime wafers at competitive cost.

2. CURRENT MARKET SITUATION AND NEED

After years of global production overcapacity for solar components several market research firms are now predicting a possible bottleneck for certain products, for example silicon wafers. While this possible shortage might mainly affect mainstream state-of-the-art technology, it is worth noting that there are a number of advanced solar cell technologies, which are currently pushing on the market. Most notably are there new players like Solarcity/Silevo and FirstSolar/Tetrasun, which both have innovative solar cell concepts as well as robust backing from their potent mother companies. Companies like that, but also most of the incumbent large players, will drive a technology development as is predicted by the SEMI International Technology Roadmap for PV (ITRPV) technology roadmap [5], shown in Figure 1. In its March 2016 7th Edition, ITRPV predicts the possible market entry of FZ wafers for production of highest efficiency solar cells.

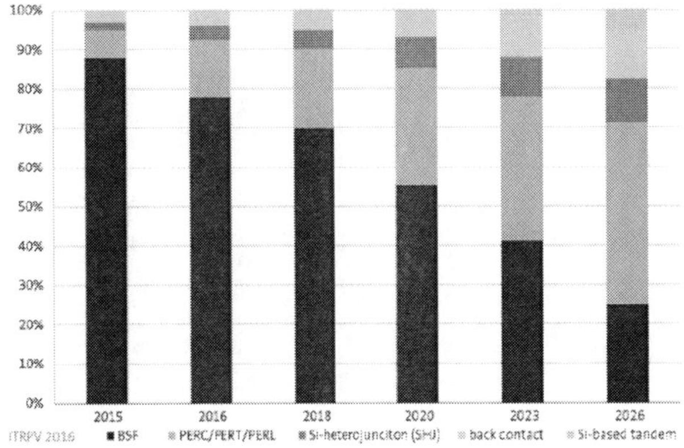

Figure 1: ITRPV Roadmap

Over the next years the PERC-based solar cell concepts will become the dominant mainstream technology, with other advanced technologies continuously increasing their market share. During the same time, established high-efficiency product companies like Sunpower and Panasonic will further push their solar cell performances towards the optimum as well.

Advanced solar cell concepts, like for example back-side IBC cells, require high-lifetime wafers. Today, manufacturers have to use lightly doped substrates to maintain a sufficiently high wafer lifetime. If high or ultra-high lifetime wafers could be realized with relatively high conductivity, the series resistance loss of IBC cells will be reduced. Furthermore, this can reduce the necessity of a frontside-surface-field, FSF, which may lead to one less high temperature process for IBC cell fabrication.

More generally, surface passivation techniques have been improved in the past years by the introduction and fine-tuning of several novel approaches: ALD Al_2O_3 passivation, intrinsic a-Si:H heterojunction passivation, tunnel oxide passivation, etc. To take full advantage of better surface passivation, wafers with higher bulk lifetime are required.

The entire development to higher efficiencies will push the requirements for the materials and processes to higher standards. In the past, the majority of the steady increase in average cell efficiencies year upon year has been the result of fine-tuning cell fabrication processes and adopting new device structures [6]. This method of efficiency improvement has been required since previously the photovoltaics industry has not had access to low cost, very high quality starting materials. This has been changing, with the increase in the size of the photovoltaics industry.

Today about 90% of polysilicon is used by the PV industry instead of ten years ago when 90% of the polysilicon was used by the semiconductor industry. This change in customer negotiating power is allowing access to high quality starting materials that have higher minority carrier lifetimes. High quality starting materials allow for the development of more complex cell structures and higher improvements in efficiency than just the evolution of PV cell processes and structures [4].

On the side of the silicon wafers, however, there is today a lack of availability of ultra-high lifetime products, which would enable these advanced technologies to better exploit their theoretical potential.

The ITRPV Roadmap also clearly forecasts a trend towards high-performance wafers, and favours mono- over multi-crystalline wafer technology, see Figure 2.

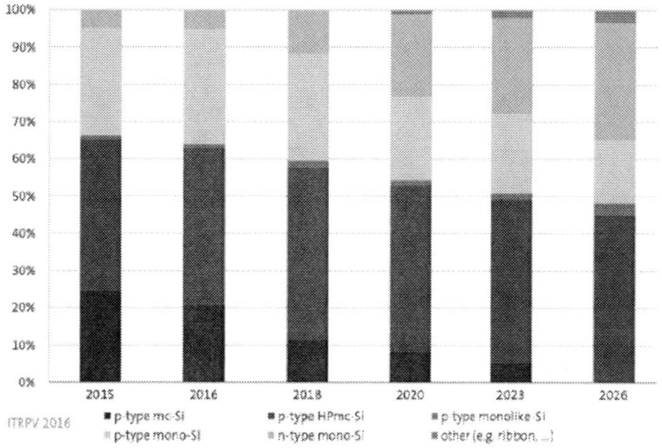

Figure 2: ITRPV Roadmap forecast on wafer types

While multi wafers are predicted to fully shift to the high-performance end of their quality range, the roadmap even forecasts an increasing demand for ultra high-quality FZ wafers of near 20% share of the mono market by the year 2025.

The demand for very-high and ultra-high lifetime wafers will have a significant increase over the next years.

3. IMPACT OF HIGH-MCL WAFERS ON CURRENT PV CELL

High minority carrier lifetime is important to achieving high cell efficiencies. Minority carrier lifetime measures the mean time between formation and recombination of an electron-hole pair [7]. Recombination losses decrease as the minority carrier lifetime increases.

The ITRPV roadmap has tracked the recombination current as a way to describe recombination losses and is forecasting its future development as seen in Figure 3.

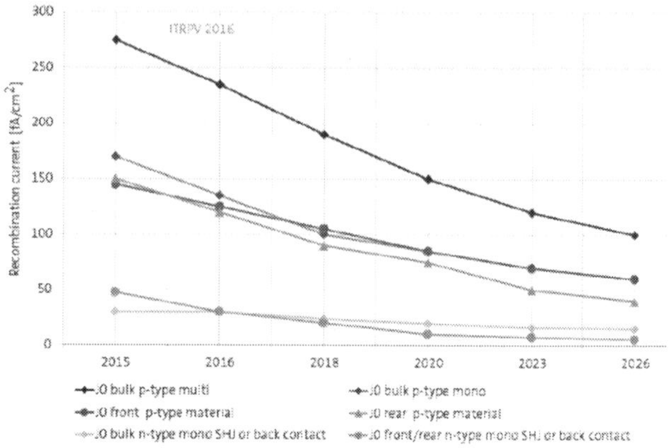

Figure 3: ITRPV Roadmap showing predictions for the reduction of loss currents in solar cells

It is noteworthy that this aspect of technology development has indeed happened much faster than predicted in the past. The bulk recombination current J_{0bulk} for p-type multi, for example, was predicted in the 2013 ITRPV roadmap to reach 200 fA/cm^2 in the year 2023. However, in the current 2015 version of the same roadmap this value is forecasted to be reached six years earlier in 2017.

Along with high minority carrier lifetime, enhanced back-surface passivation can improve efficiency. As the back-surface passivation improves, the cell efficiency benefits more from improved minority carrier lifetime. Conversely, if the minority carrier lifetime is not high enough then the effect on efficiency is minimal [1]. The trend currently underway to improve surface passivation provides a growing opportunity for higher wafer quality to strongly contribute to efficiency improvements. Thus, wafer and cell manufacturers are closely linked, as cell manufacturers depend on high wafer quality to achieve the best results from their cells.

4. SIMULATION OF SOLAR CELL EFFICIENCY DEPENDING ON WAFER LIFETIME

In order to demonstrate the benefit of higher lifetime wafers, solar cell simulations were carried out by the University of Waterloo using the 2-dimensional MEDICI software.

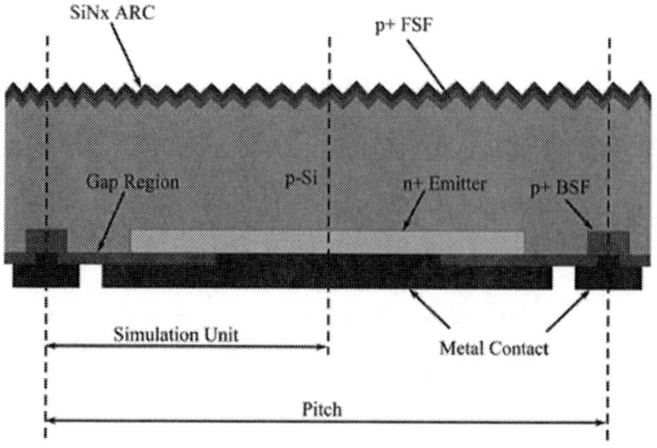

Figure 4: IBC back-side contact solar cell architecture used for simulation

In this first calculation, a back contact IBC-type solar cell structure was used, as shown in Figure 4.

More simulations considering other advanced solar cell architectures, like PERC/PERT cells, hetero-junction/HIT, tunnel junction, etc. are being performed currently, and will be presented when available.

The key solar cell parameters used for the IBC cell simulation were chosen to resemble a true manufacturing (not laboratory) situation as close as possible, and are shown in Figure 5.

The calculations were performed for variations of the p-type wafer minority lifetime from 100 µs up to 3000 µs and wafer thicknesses from 140 µm to 200 µm. Bulk wafer lifetimes of even greater than 3000 µs are achievable using the Floatzone (FZ) technology presented in this paper.

Cell Structure	IBC Solar Cell
Substrate MCL	100 - 3000 µs
Substrate Thickness	140 - 200 µm
Pitch Size	200 µm
Simulation Unit Size	100 µm
Emitter Coverage	70 µm
Back Surface Field Coverage	15 µm
Gap between n+ and p+	15 µm
Light Confinement	Random Pyramid Texturing + SiN_x ARC, Input Based-on Experimental Data
Substrate Doping Level	5E15 cm^{-3}
Front Passivation	Front Surface Field & SiN_x Passivation
Front Surface Recombination Velocity	10 cm/s
Gap Passivation	SiN_x Passivation

Figure 5: Key parameters used for Medici simulation of IBC solar cell structure

The results of the simulation show an expected increase of solar cell efficiency with increasing lifetime, as well as an expected increase of efficiency with decreasing wafer thickness. The summary of all curves is shown in Figure 6.

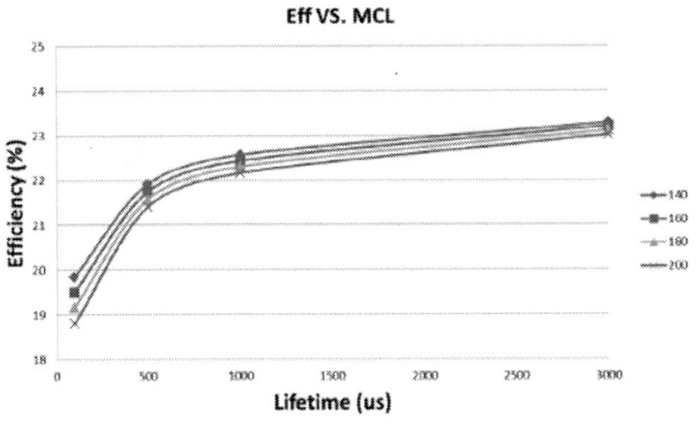

Figure 6: Simulation results: Solar efficiency dependent on wafer bulk lifetime for various wafer thicknesses

The best efficiency of 23.3% was achieved for a solar cell made from a 140 µm thick wafer with 3000 µm lifetime. A reduction of the wafer lifetime to 1000/500/100 µs results in reduced efficiencies of 22.6/21.9/19.9%, respectively. The detailed results of the simulation are shown in the table in Figure 7.

MCL (µs)	Thickness (µm)	Isc (mA/cm^2)	Voc (V)	FF (%)	Efficiency (%)
100	140	37.8729	0.645	81.24	19.846
	160	37.489	0.640	81.30	19.507
	180	37.0353	0.635	81.50	19.166
	200	36.5258	0.630	81.74	18.809
500	140	38.9615	0.695	80.95	21.919
	160	38.8903	0.691	80.93	21.748
	180	38.7812	0.688	80.87	21.579
	200	38.6453	0.685	80.89	21.412
1000	140	39.1013	0.706	81.77	22.572
	160	39.0717	0.704	81.58	22.440
	180	39.0091	0.702	81.46	22.306
	200	38.9246	0.700	81.34	22.162
3000	140	39.195	0.714	83.22	23.290
	160	39.1934	0.713	83.07	23.214
	180	39.1623	0.712	82.92	23.121
	200	39.1127	0.710	82.90	23.020

Figure 7: Full results of the IBC solar cell simulation

For current industry typical high-quality mono wafers of 180 µm thickness and approx. 500 µs lifetime, the efficiency benefit from using ultra-high lifetime wafers of 1000 µs or 3000 µs is 0.7% or 1.5% absolute, showing a very significant benefit of such wafers.

5. UBUIQUITY SOLAR'S FLOATZONE CONCEPT

FZ crystallization process typically yields higher quality wafers, including higher MCL, than other crystallization methods. As it has been presented in this paper, a higher MCL in the base material results in higher performance solar cells in terms of conversion efficiency. However, the higher cost associated with the wafers from FZ process has been a barrier for a wider adoption in the fabrication of solar cells. Polysilicon feedstock for the FZ process requires very high quality polysilicon rods that are completely crack-free with excellent morphology. These high quality polysilicon rods for the FZ process requires a relatively slower process in the chemical vapor deposition (CVD) stage compared to the rods required for more widely used Czochralski (Cz) crystallization process for monocrystalline wafers resulting in a higher manufacturing cost. As a result, the overall cost of manufacturing FZ wafers is higher than Cz wafers. Even with the performance increase in the solar cells using FZ wafers, approximately 1% - 2% absolute increase in cell efficiency over conventional monocrystalline wafers using Cz process, the premium price of the solar cells associated with the increased performance is insufficient in compensating for the current higher manufacturing cost.

Recognizing this challenge associated with the high performance FZ wafers, Ubiquity Solar Inc. (USI) is currently developing an innovative manufacturing process to significantly lower the cost of manufacturing FZ wafers. The unique process being developed by USI allows a lower cost, high quality, crack-free polysilicon rods to be used in the FZ process through integrated Continuous Chlorination Trichlorosilane (CC-TCS™) generation and CVD processes. The use of the integrated processes results in simpler purification steps reducing overall cost of manufacturing polysilicon rods while producing a polysilicon rods with a better morphology. Furthermore, the use of a silicon alloy as the feedstock for the TCS generation process provides an unique method of purifying silicon material in a cost-effective manner.

A simplified block diagram of the high quality polysilicon rod manufacturing process is shown in Figure 8.

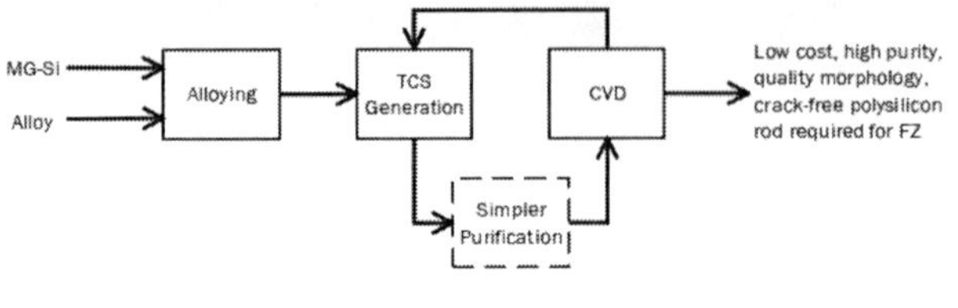

Figure 8: Simplified block diagram of Ubiquity's poly-silicon rod manufacturing process

As a result of these innovative approaches in polysilicon rod manufacturing, the overall cost of high quality rods for FZ process is substantially lower than using the conventional Siemens method along with the standard CVD process of producing high quality rods including hydro-chlorination (HC) and direct chlorination (DC) for a similar volume of polysilicon production. USI expects the overall manufacturing cost of the Company's CC-TCS™ process to be approximately 25 – 30% less than the conventional HC and DC Siemens-based production process for the same quality of material.

In addition to the benefits of cost reduction in polysilicon production, the FZ monocrystalline wafers from USI's innovative processes are expected to have highly desirable wafer properties to support increased solar cell performance – higher conversion efficiency due to higher MCL and higher reliability by minimizing light-induced degradation caused by Oxygen. 5" and 6" wafers with various thickness options are targeted to have minority carrier lifetime of $\geq$ 3,000 µs for n-type and $\geq$ 1,000 µs for p-type as well as a narrow resistivity range. Furthermore, the wafers are targeted to have dislocation density $\leq 300/cm^2$ as well as an average Oxygen concentration of $\leq 2.0 \times 10^{16}$ $atoms/cm^3$ and average Carbon concentration $\leq 2.0 \times 10^{16}$ $atoms/cm^3$.

In combination of higher quality monocrystalline wafers using FZ process and a lower cost manufacturing process, USI products are expected to enable higher performance solar cells without a significant increase in overall cost.

6. Ubiquity Material Handling

In order to further optimize the benefits of the above-mentioned FZ process, USI is also developing proprietary methods to handle the materials left from the conversion of Floatzone ingots into bricks, and a subsequent optimized Cz process which turns these left-over materials into mono-wafers with superior carrier lifetime. A first batch of Cz wafers has been produced and underwent thorough investigation and characterization within the framework of a Benchmark Study conducted by Fraunhofer ISE [8], in which 8 different Cz materials from 7 suppliers were characterized.

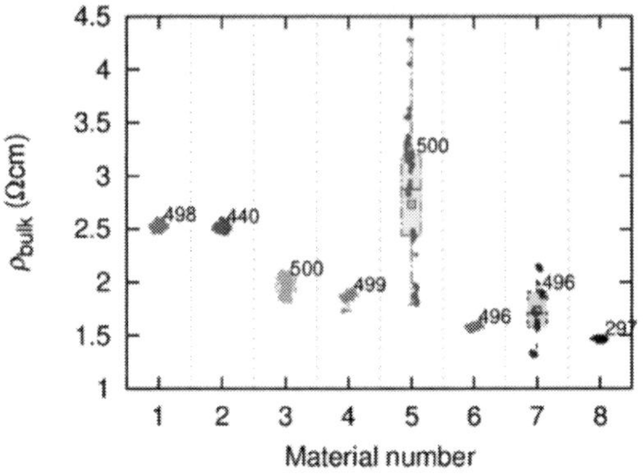

Figure 9: Fraunhofer ISE benchmark: Resistivity distribution (USI is material 1)

The wafers supplied by the USI, denominated "Material 1" in the Benchmark Study, were the thinnest wafers of the benchmark, with thickness between 170-175 µm (compared to 180-205 µm). The results of the benchmark show overall good and partly above average features of the Ubiquity material, see Figure 9, 10. Thickness, resistivity and as-cut lifetime are highly consistent, promising low variation for finished cells.

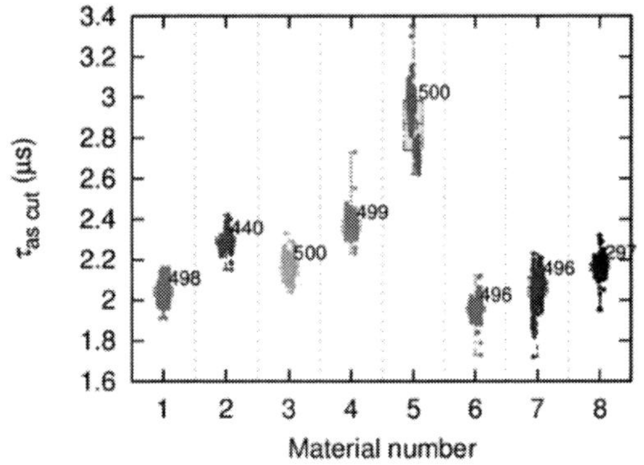

Figure 10: Fraunhofer ISE benchmark: As-cut lifetime

Despite much lower wafer thickness USI material exhibits similar or higher as-cut lifetime and good PL signals, which indicate high bulk lifetime.

Subsequent to characterization, solar cells were fabricated within the benchmark study, based on a PERC solar cell process involving a thermal diffusion of emitter, AlO_x+SiN_x passivation on the rear-surface, laser opening of local contact, and co-firing in an inline furnace.

After processing (as Proc), the solar cells were degraded for 96 hours under controlled irradiation. Then, the solar cells were regenerated at 140°C under 1 sun irradiation for 2 hours.

Figure 11 shows the solar cell efficiencies as a function of the material for the as process (as Proc), the degraded (Deg) and regenerated (Reg) state.

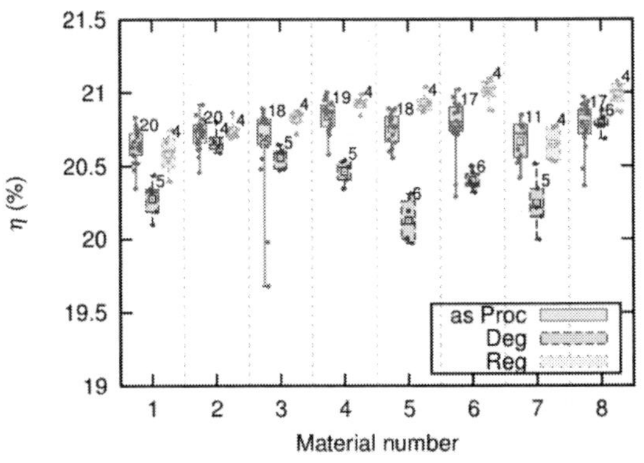

Figure 11: Fraunhofer ISE benchmark: PERC solar cell efficiencies

The wafer provided by Ubiquity show good finishing cell performance with efficiencies up to 20.8% (average above 20.5% after regeneration) and of FF ~80%. Remarkably, an average short circuit current of ~39.5 mA/cm^2 has been achieved, almost identical to that of thicker sample wafers.

7. CONCLUSIONS

Increased solar cell efficiency remains a key factor to further reduce the cost of electricity from solar PV. As the PV industry has grown, it has evolved from using scrap silicon from the semiconductor industry ten years ago to near-semiconductor grade today. In parallel, cell designs have become increasingly complex to increase cell efficiencies, and have done so while also driving down manufacturing costs. Together, these advances have brought solar PV to grid parity in some jurisdictions and substantially closer in others. As this trend continues, higher efficiency silicon solar cells will require higher quality wafers, due to the close link between cell efficiency and minority carrier lifetime. With the increasing use of complex cell structures, wafer minority carrier lifetime is much more of a limiting factor now as compared to ten years ago, and advanced cell designs require and will benefit from improved wafer quality, as is shown by the simulation results presented in this paper.

Of all the options for improving wafer quality, FZ wafers provide the highest performance, but have historically been prohibitively expensive to produce. An innovative polysilicon rod process has been presented, which enables cost-competitive manufacturing of FZ wafers with ultra-high minority carrier lifetimes.

First encouraging benchmark characterization results were obtained and presented of Cz wafers, which will further improve the economics of the USI FZ process.

USI is currently setting up a pilot plant in Sarnia, Ontario, Canada to produce high quality FZ and Cz wafers

FZ wafers are the best choice for the high performance part of the market and are likely to be adopted for a rapidly growing part of the market as FZ wafers become increasingly cost competitive.

REFERENCES

[1] A. Goodrich, M. Woodhouse and P. Hacke, "The Value Proposition for High Lifetime (p-type) and Thin Silicon Materials in Solar PV Applications," in 2012 IEEE Photovoltaic Specialists Conference, 2012.

[2] S. K. Pang, A. Rohatgi, B. L. Sopori and G. Fiegl, "A Comparison of Minority-Carrier Lifetime in As-Grown and Oxidized Float-Zone, Magnetic Czochralski, and Czochralski Silicon," Journal of The Electrochemical Society, vol. 137, no. 6, pp. 1977-1981, 1990.

[3] G. Coletti, "Sensitivity of State-of-the-Art and High Efficiency Crystalline Silicon Solar Cells to Metal Impurities," Progress in Photovoltaics: Research and Applications, 2012.

[4] K. A. Munzer, J. Schone, A. Teppe, M. Hein, R. E. Schlosser, M. Hanke, J. Maier, A. Yodyunyong, J. Isenberg, T. Friess, C. Ehling, K. Vamer, S. Keller and P. Fath, "Towards 19.5% Industrial Crystalline Silicon Solar Cells," in European Photovoltaic Solar Energy Conference and Exhibition, 2011.

[5] International Technology Roadmap for Photovoltaic (ITRPV), "Results 2015," International Technology Roadmap for Photovoltaic, 7th Edition, March 2016.

[6] T. Saga, "Advances in crystalline silicon solar cell technology for industrial mass production," NPG Asia Materials, vol. 2, pp. 96-102, 2010.

[7] D. K. Schroder, "Carrier Lifetimes in Silicon," in Handbook of Semiconductor Silicon Technology, Norwich, Noyes Publications, 1990, p. 550639.

[8] P. Saint-Cast, "CZ SILICON BENCHMARK FOR P-TYPE PERC SOLAR CELLS", 32 EU-PVSEC, 2DV.3.22

A COMPARISON OF RESIDUAL STRESS INDUCED BY FIXED ABRASIVE DIAMOND WIRE SAWING AND LOOSE ABRASIVE SLURRY WIRE SAWING IN MULTICRYSTALLINE SILICON WAFERS

Vanessa Pogue, Shreyes Melkote, RGR Prasath, Steven Danyluk
The George W. Woodruff School of Mechanical Engineering
Georgia Institute of Technology
Atlanta, Georgia 30332-0405
U.S.A.
Ph: 404.894.3832
vpogue@gatech.edu

ABSTRACT: The breakage of thin Silicon (Si) wafers is a significant problem that leads to lower production yields in the manufacture of Si-based solar cells. Residual stress of Si wafers is therefore a major concern due to its detrimental effect on wafer mechanical integrity. This work investigates the spatial distribution of residual stress in multicrystalline silicon (mc-Si) wafers produced by the two prevalent wire sawing methods used in industry: loose abrasive slurry wire sawing (LAWS) and fixed abrasive diamond wire sawing (DWS). The saw damage layer of as-cut industrially sawn DWS and LAWS wafers was removed at 2 μm steps using wet chemical isotropic etching and residual stress was measured as a function of etch depth using a near-infrared birefringence polariscope. Results show that in as-cut wafers, DWS creates a larger number of high maximum residual shear stress locations compared to LAWS. In DWS, points of highest residual stress were aligned with wire striation lines in the direction of wire travel. In contrast, LAWS resulted in a random pattern of locations of high residual stress. After 12 μm of material removal by etching, the points of highest residual stress were localized at grain boundaries in both DWS and LAWS wafers.

Keywords: Manufacturing and Processing, Multicrystalline Silicon, Grain

1 INTRODUCTION

Residual stress in materials is studied due to its influence on mechanical and structural properties. In multicrystalline silicon (mc-Si) wafers used as solar cell substrates, residual stress may contribute to fracture and breakage of the wafers during processing. Silicon (Si) is a very brittle material, and therefore breakage due to residual stress is a significant issue, which causes lower production yields and is responsible for a large proportion of the overall solar cell manufacturing cost. The understanding of manufacturing-induced residual stress is therefore critical to improve wafer mechanical integrity and increase yields in production.

The manufacturing process of a Si wafer first comprises of a high temperature Si heating, melting, and solidification process to produce a Si ingot from polycrystalline silicon, which is then cut into bricks and subsequently sawn into wafers using wire sawing. Residual stresses are produced in the wafer from the thermal gradient induced by solidification and from the wire sawing process. In addition to residual stress, wire sawing produces a saw damage layer on the surface of the finished wafers comprised of micro-cracks, pits, amorphous regions, and areas of brittle fracture.

The traditional loose abrasive slurry wire sawing process is shown schematically in Figure 1. In wire sawing, a thin, long stainless steel wire, approximately 120-160 μm in diameter is wrapped multiple times around spools to create a web of cutting wires. The Si brick is then fed into the fast moving web of wires to yield wafers. There are two wire sawing methods used in industry to cut Si wafers, loose abrasive slurry wire sawing (LAWS) and fixed abrasive diamond wire sawing (DWS).

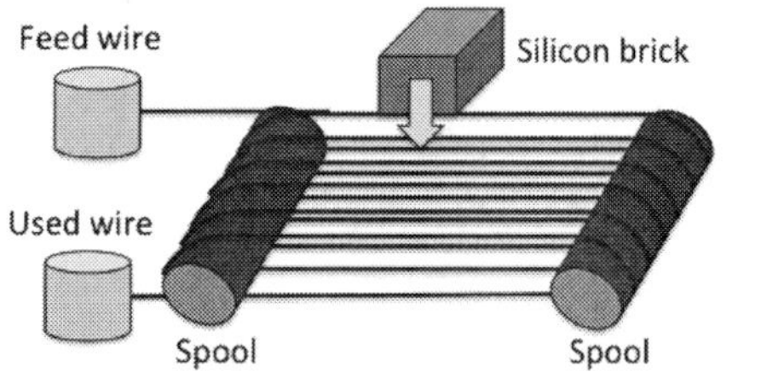

Figure 1. Schematic of wire sawingwire sawing process

In the LAWS process, a loose abrasive slurry consisting of silicon carbide (SiC) particles entrained in polyethylene glycol is continually poured onto the wire web during sawing. The abrasive grits in the slurry cut the Si via a rolling-indenting material removal process.

Recently, manufacturers have begun to replace the LAWS process with the more efficient fixed abrasive DWS process, especially for mono-Si wafers [1]. In DWS, instead of using a slurry with loose SiC abrasives, diamond grits are fixed onto the wire by nickel electroplating. The DWS process has been shown to yield a two-fold increase in the cutting speeds, reduce the thickness of the saw damage layer, reduce kerf loss, and to enable easier kerf recycling [2-4].

Because the grits are fixed to the wire, the dominant cutting mechanism in DWS is a scratching-indenting process compared to the rolling-indenting process in LAWS. It is known through previous studies [5] that the fracture strength of DWS wafers is highly dependent on the direction the stress is applied relative to the sawing direction. Yang et al. [5] used four point bending tests to determine the wafer fracture strength. The results showed that while DWS had lower fracture strength than LAWS when bending in a direction perpendicular to wire travel, DWS exhibited similar fracture strength to LAWS in bending direction parallel to the direction of wire travel.

In addition, Anspach et al. [6] investigated the fracture stength of LAWS wafers and found a dependence of fracture strength to the wire feed direction and wafer location.

While the depth of saw damage, surface morphology, and fracture strength of Si wafers has been investigated for the two wire sawing processes, the residual stress produced in mc-Si wafers by the wire sawing process, namely DWS and LAWS, as a function of etch depth, has not been reported.

The purpose of this investigation is to understand the spatial distribution of residual stress produced in the surface layers of mc-Si wafers by the two wire sawing processes, namely, LAWS and DWS. The approach utilizes wet chemical etching to remove layers of saw damage at intervals and near-infrared birefringence polariscopy to measure the residual stress.

2 METHODOLOGY

2.1 Overview

A process flow diagram demonstrating the experimental approach is shown in Figure 2.

Figure 2. Process flow diagram of experimental approach

Multicrystalline silicon "sister wafers" of each sawing process type, DWS and LAWS, were first diced into quarters to facilitate sample handling. Sister wafers are wafers located side-by-side in the same Si brick drawn from the same Si ingot. Therefore, they exhibit the same crystallographic structure and can be assumed to have similar manufacturing-induced residual stress distributions and magnitudes.

A wet chemical isotropic etching solution of HNA ($HF:HNO_3:CH_3COOH$) was used to remove saw damage from both sides of the mc-Si wafer. Each sister wafer was etched to a progressively increasing etch depth from 0 μm to 12 μm at 2 μm intervals to result in seven sister wafers at 0 μm, 2 μm, 4 μm, 6 μm, 8 μm, 10 μm, and 12 μm etch depths. Subsequently, the residual stresses in the wafers were measured using a near-infrared birefringence polariscope.

2.2 Wet chemical Isotropic Etching

Isotropic etchants such as HNA result in uniform Si removal rates, independent of crystalline orientation. Because near-infrared birefringence polariscopy is dependent on the sample thickness for the calculation of the maximum residual shear stress, τ_{max}, the use of an isotropic etchant enabled even Si removal and therefore uniform sample thickness.

An HNA ($HF:HNO_3:CH_3COOH$) solution with a ratio of (9:75:17) was used to yield an etch rate of 5 μm/minute, producing a planar etch [7]. The chemical compositions were 49% Hydrofluoric acid (HF), 69-70% Nitric acid (HNO_3) and >99% Acetic acid (CH_3COOH).

In the etching process, the wafers were placed individually in a sample holder and submerged horizontally in the HNA bath. The sample holder prevented contact of the wafer with the bottom of the bath. Slight manual agitation was applied during etching to produce uniform Si removal. After etching, the wafers were immediately placed into two successive deionized water (DI H_2O) baths for 5 minutes each, rinsed with DI H_2O, and dried completely using nitrogen (N_2) gas.

The etch stop time was adjusted to control the depth of etching. To verify the etch depth accuracy, prior to etching, a small (<1 cm^2) area on each Si wafer was masked using Apiezon Black W. After the etching process, the mask was removed, and a stylus profilometer (KLA Tencor Alpha-Step IQ Surface Profiler) was used to measure the step height between the masked and unmasked areas to provide a measurement of etch depth.

2.3 Near infra-red birefringence polarisceope

The near infra-red birefringence polariscope is a through-thickness measurement method, which produces an average full-field maximum residual shear stress map of the mc-Si wafer. The experimental set-up is shown in Figure 3.

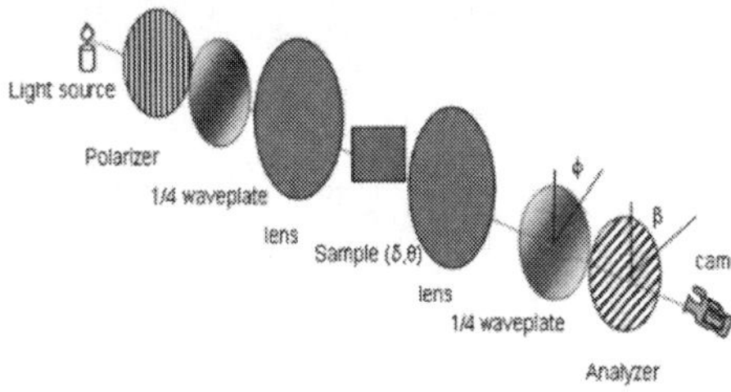

Figure 3. Experimental set-up of near-infrared birefringence polariscope [8]

The polariscope consists of a series of optics comprised of polarizers, quarter wave-plates, lenses to collimate the light through the sample, and a tungsten 1150 nm lamp light source. A digital camera is used to record the resulting grey-scale intensity images. The polariscope uses near-infrared light to which Si is transparent to, and the principles of photoelasticity to determine the phase retardation of light through the wafer and thereby provide a measurement of the residual stress.

The polariscope set-up and procedure for the residual stress measurement in Si wafers was previously developed by Zheng et al. [9, 10], Danyluk et al. [11], He et al. [12], Li [13], Skenes[14] Skenes et al. [8, 15], and Prasath et al. [16]. Details of the ten-step phase stepping procedure implemented to determine the photoelastic parameters is given in Prasath et al. [16]. The ten-step phase stepping procedure analyzes the differences in the grey scale intensities of the images to determine phase retardation, δ and the isoclinic angle, θ for each point imaged in the wafer sample.

The polariscope measurements are used to compute the maximum shear stress, τ_{max}, using the stress optic law,

$$\tau_{max} = \frac{|\sigma_1 - \sigma_2|}{2} = \frac{\lambda\delta}{2\pi t C}$$

where σ_1 and σ_2 are principal stresses, λ is the wavelength of light used, δ is the retardation of light through the sample, t is the thickness of the sample, and C is the relative stress optic coefficient. The difference $|\sigma_1 - \sigma_2|$ is the absolute value of the difference between the largest and smallest principal stresses, equal to twice the maximum shear stress, $2\tau_{max}$. For each point in the wafer, τ_{max} is calculated to generate a full-field residual stress map.

3 RESULTS

DWS and LAWS produced very different τ_{max} distributions in as-cut sawn wafers. Figure 4 shows near-infrared transmission images of two regions of interest selected from the DWS and LAWS wafers with grains and grain boundaries in each region. The regions of interest are representative of the larger diced wafer quarters and demonstrate the effect of wire sawing.

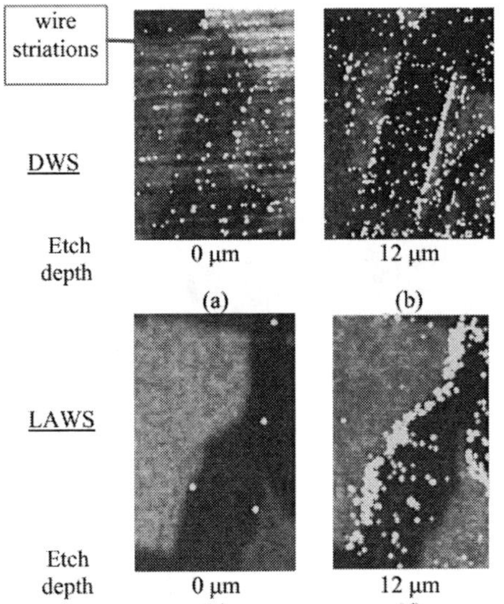

Figure 4. (a) and (b): Near-infrared transmission images of DWS region of interest. Superimposed yellow data points show locations of $\tau_{max} > 25$ MPa at (a) 0 µm, and (b) 12 µm etch depths. (c) and (d): Near-infrared transmission images of LAWS region with superimposed yellow data points showing locations of $\tau_{max} > 25$ MPa at (c) 0 µm and (d)12 µm etch depths.

In the results shown in Figure 4, the threshold of 25 MPa was selected to show the locations of points of highest τ_{max}. The threshold of 25 MPa was selected to represent the "tail" of the residual stress distribution. An example residual stress distribution is shown in Figure 5.

In Figure 4, the points of τ_{max} greater than the 25 MPa threshold were superimposed on the near-infrared transmission images of the region of interest to show the spatial distribution of high τ_{max}. The results compared in Figure 4a and Figure 4c, clearly show that DWS produced more points of τ_{max} greater than 25 MPa compared to LAWS. In the DWS region, all points of highest τ_{max} were predominantly aligned with the wire striation lines in the direction of wire travel. In comparison, LAWS resulted in random locations of τ_{max} greater than 25 MPa; the locations were not aligned with a particular direction.

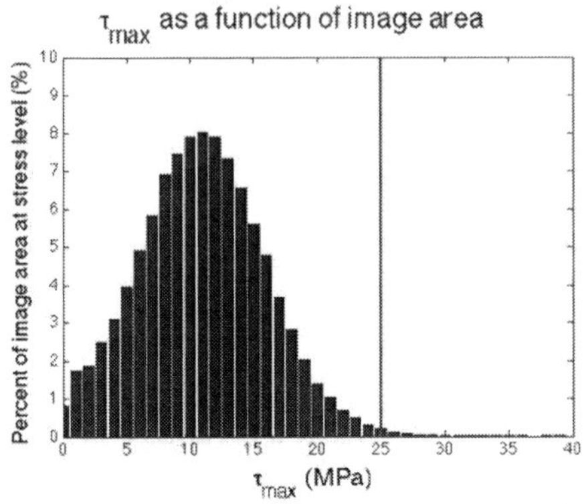

Figure 5. τ_{max} as a function of percent image area for the same region shown in Figure 4a. The distribution shows the 25 MPa threshold representing the tail of the distribution, or the points of highest τ_{max}.

Shown in Figure 4b and Figure 4d, at 12 µm etch depth, the points of $\tau_{max} > 25$ MPa were localized along the grain boundaries in both the DWS and LAWS wafers. At 12µm of Si material removal, it can be deduced that the majority of the residual stress due to saw damage is removed based on the location of high τ_{max} concentrated at the grain boundaries, which is likely caused by residual stress due to thermal gradients present during solidification of the Si ingot, and is manifested in the wafers.

4 DISCUSSION

As-cut DWS wafers were found to have a larger number of points of $\tau_{max} > 25$ MPa when compared to the LAWS wafers. Depending on the type of stress (tensile versus compressive), the greater number of locations of high τ_{max} seen in the as-cut DWS wafers could lead to increased fracture during manufacturing and therefore outweigh the benefits of decreased kerf loss when compared to LAWS. A detailed wafer breakage and cost analysis would be required to confirm this possibility.

It is also interesting to note the localization of high τ_{max} at grain boundaries for the 12 µm etch depth cases, implying that the majority of high residual stress (τ_{max}) caused by wire sawing is removed at 12 µm in wafers sawn by either LAWS or DWS. The results suggest that the stress induced by the thermal gradients generated in the solidification process during casting dominates the through-thickness residual stress state at this etch depth. The residual stress maps confirmed the non-directional fracture strength results of Yang et al. [5].

5 CONCLUSIONS

The objective of the experiments described in this work was to understand the spatial distribution of residual stress produced by DWS and LAWS in mc-Si wafers. The conclusions can be summarized as follows:

- Diamond Wire Sawing (DWS) produces high residual maximum shear stresses, ($\tau_{max} > 25$ MPa) that are spatially aligned with wire striations.

- Loose Abrasive Wire Sawing (LAWS) produces high τ_{max} stresses (> 25 MPa) that are randomly distributed, with no directional preference.
- At 12 µm of material removal, DWS and LAWS wafers demonstrated similar residual stress spatial distribution maps showing a high number of points concentrated at grain boundaries. This suggests:
    - The majority of residual stress produced by wire sawing was removed at 12 µm for wafers processed using either LAWS or DWS.
    - Casting and solidification induced residual stresses tend to accumulate at the grain boundaries in mc-Si wafers.

Further investigations are needed to understand the principal stress types (tensile vs. compressive), as tensile and compressive stresses are known to have different effects on the mechanical strength and fracture of materials. This work will be presented in a future paper.

6 ACKNOWLEDGEMENT

The authors would like to thank the Silicon Solar Consortium (SiSoC), a National Science Foundation (NSF)—Industry/ University Cooperative Research Center (I/UCRC), for financial support. The authors would also like to thank SunEdison, Inc. for providing the samples used in the study.

7 REFERENCES

[1] Hali Forstner, S. Bandil, M. Zwegers, R. Bollen, Gianluca Coletti, Wim Sinke, *et al.* (2014). *International Technology Roadmap for Photovoltaic Results (ITRPV) (5th ed.).* Available: http://www.itrpv.net/

[2] Y. Kondo, N. Watanabe, D. Ide, T. Matsuki, H. Takato, and I. Sakata, "Characterization of multicrystalline silicon wafers for solar cell applications sliced with a fixed abrasive wire," in *23rd European Photovoltaic Solar Energy Conference and Exhibition*, Valencia, Spain, 2008, pp. 1297-1301.

[3] K. Tomono, S. Miyamoto, T. Ogawa, H. Furuya, Y. Okamura, M. Yoshimoto, *et al.*, "Recycling of kerf loss silicon derived from diamond-wire saw cutting process by chemical approach," *Separation and Purification Technology,* vol. 120, pp. 304-309, 2013.

[4] N. Watanabe, Y. Kondo, D. Ide, T. Matsuki, H. Takato, and I. Sakata, "Characterization of polycrystalline silicon wafers for solar cells sliced with novel fixed-abrasive wire," *Progress in Photovoltaics: Research and Applications,* vol. 18, pp. 485-490, 2010.

[5] C. Yang, H. Wu, S. Melkote, and S. Danyluk, "Comparative analysis of fracture strength of slurry and diamond wire sawn multicrystalline silicon solar wafers," *Advanced Engineering Materials,* vol. 15, pp. 358-365, 2013.

[6] A. Grün, A. Lawerenz, R. Porytskyy, and O. Anspach, "Investigation of wafer surfaces with space-resolved breaking strength tests and corresponding analysis of the crack depth," in *26th European Photovoltaic Solar Energy Conference and Exhibition*, Hamburg, Germany, 2011, pp. 961-965.

[7] K. E. Bean, "Anisotropic etching of silicon," *IEEE Trans. Electron Devices* vol. 25, pp. 1185-1193, 1978.

[8] K. Skenes, F. Li, and S. Danyluk, "Analysis of residual stress in thin silicon wafers with an NIR polariscope," in *Proceedings of the 26th European Photovoltaic Solar Energy Conference and Exhibition*, Hamburg, Germany, 2011.

[9] T. Zheng and S. Danyluk, "Study of stresses in thin silicon wafers with near-infrared phase stepping photoelasticity," *Journal of Materials Research,* vol. 17, pp. 36-42, 2002.

[10] T. Zheng and S. Danyluk, "Determination of stresses in thin silicon plates with near-infrared phase stepping photoelasticity," in *Proceedings of the 7th ASME NDE Tropical Conference*, San Antonio, TX, 2001, pp. 195-199.

[11] S. Danyluk and S. Ostapenko, "Full field birefringence measurement of grown-in stresses in thin silicon sheet," National Renewable Energy Laboratory, Golden, Co. NREL/SR-520-44237, November 2008.

[12] S. He, T. Zheng, and S. Danyluk, "Analysis and determination of the stress-optic coefficients of thin single crystal silicon samples," *Journal of Applied Physics,* vol. 96, pp. 3103-3109, 2004.

[13] F. Li, "Study of stress measurement using polariscope," Ph.D. Dissertation, Mech. Eng., Georgia Inst. of Techn., 2010.

[14] K. Skenes, "Characterization of residual stresses in birefringent materials applied to multicrystalline silicon wafers," Ph.D. Dissertation, Mechanical Engineering, Georgia Institute of Technology, Georgia Institute of Technology, 2014.

[15] K. Skenes, R. G. R. Prasath, and S. Danyluk, "Polariscopy measurement of residual stress in thin silicon wafers," in *Conference Proceedings of the Society for Experimental Mechanics Series*, Lombard, IL., 2014, pp. 79-85.

[16] R. G. R. Prasath, K. Skenes, and S. Danyluk, "Comparison of phase shifting techniques for measuring in-plane residual stress in thin, flat silicon wafers," *Journal of Electronic Materials,* vol. 42, pp. 2478-2485, 2013.

EFFECT OF ABRASIVE GRIT SHAPE ON SURFACE MORPHOLOGY, SUBSURFACE DAMAGE AND FRACTURE STRENGTH OF DIAMOND WIRE SAWN SILICON WAFERS

Arkadeep Kumar [1*], Shreyes N. Melkote[1] Steffi Kaminski [2], and Chris Arcona [2]
[1] George W. Woodruff School of Mechanical Engineering, Georgia Institute of Technology, Atlanta, GA 30332
[2]Saint-Gobain Northborough R&D Center, Northborough, MA 01532
*Corresponding author: arkadeepkumar@gatech.edu

ABSTRACT: The surface morphology of and subsurface damage in silicon wafers produced by fixed abrasive diamond wire sawing influence the mechanical strength of photovoltaic silicon solar cells. One approach to cutting wafers with favorable surface and subsurface characteristics is using diamond wires made with different grit shapes. In this paper, we study mono-crystalline silicon wafers cut by wires with two nominally different grit shapes (denoted A and B), to determine the effect of grit shape on the wafer surface morphology, roughness, subsurface damage, and fracture strength. It is found that the surface morphology of wafers cut by wire A exhibits more ductile grooves while the wafer surface cut by wire B exhibits more brittle fracture. The surface roughness data show that wafers cut by wire A have lower surface roughness than those cut by wire B. However, biaxial flexure tests on laser-cut wafer coupons show that the wafers cut by wire A have lower fracture strength than those cut by wire B. The subsurface damage in wafers cut by wire A consists of fewer but longer median cracks than those cut by wire B, which has shallow median and lateral cracks. We hypothesize that the observed variation in fracture strength results from the dominance of the subsurface condition over the surface condition.

Keywords: Manufacturing and Processing, Reliability, Silicon, Diamond Wire Sawing, Grit Shape, Surface Damage, Subsurface Damage, Fracture Strength

1 INTRODUCTION

High cost is a significant barrier to the adoption of renewable sources of energy such as photovoltaic solar cells to augment power generation from traditional sources. Most photovoltaic solar cells are manufactured using crystalline silicon wafers. These wafers are produced by slicing silicon ingots by wire sawing processes that utilize either a loose abrasive slurry or a fixed abrasive diamond wire. Recent trends indicate a shift from loose abrasive slurry sawing (LAS) to fixed abrasive diamond wire sawing (DWS) [1]. Cost can be reduced by saving material and by producing thinner wafers with higher mechanical strength. The mechanical strength of wafers depends on the surface morphology generated by DWS and associated subsurface damage (e.g. micro cracks). A smoother wafer surface with fewer micro-cracks is expected to yield superior mechanical strength and reliability of solar cell substrates.

The dominant material removal mechanisms in DWS include ductile mode material deformation coupled with brittle fracture [2]. Ductile mode removal leads to features indicative of scratching on the wafer surface while brittle fracture leads to intermittent chipping and cracks on the wafer surface/subsurface. A critical wire-related parameter affecting the mode of material removal in DWS is the abrasive shape/geometry [3]. The geometry of abrasive grit in contact with the silicon affects the shape, size, and extent of micro-cracks generated by brittle fracture, or the surface condition if the material is plastically removed by ductile mode cutting. Silicon is known to undergo pressure-induced phase transformations and become plastically deformed [4, 5], enabling ductile mode material removal.

This paper focuses on the effect of grit shape on the cut-surface morphology, subsurface damage, and fracture strength of mono-crystalline wafers cut by two wires with nominally different grit shapes.

2 EXPERIMENTS

The two diamond wires used in this study are labeled A and B. Figure 1 shows representative diamond grit shapes on the two wires. Table 1 gives the wire sawing conditions used in the experiments.

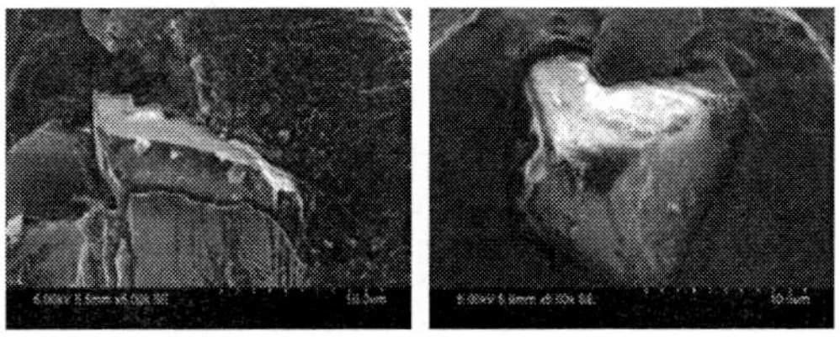

(a) (b)

Figure 1. Representative diamond grit shapes (a) grit A (b) grit B.

Table 1. Wire sawing conditions.

Wire speed	10 m/s
Feed speed	1 mm/min
Total cut length	130 mm
Number of wafers	9
Wire supply	1.5 m/min
Cycle time	3 sec / 15 sec
Tension	20 N
Total cutting time	135 min

Wafers cut by the two wires were examined in a scanning electron microscope (SEM, FEI Nova) to evaluate the surface morphology. Subsurface damage was evaluated using an edge polishing method that involved embedding the wafer specimen and polishing them smooth to obtain statistics of crack length and crack density per length of edge of the wafer. Wafers were diced into small coupons and embedded in epoxy resin with steel springs such that the wafer thickness was perpendicular to the plane viewed in an optical microscope. First, the embedded wafer samples were ground using successively finer silicon carbide paper of grit sizes 280, 500, 1200, 2400 and 4000. Next, samples were polished to mirror finish using Buehler Texmet cloth with 1 μm diamond suspension and Buehler Microcloth with 0.05μm diamond suspension. The polished edges were observed under an optical microscope to get detailed statistics of the micro-cracks. A second method of analyzing subsurface damage consisted of focused ion beam (FIB) cross sectioning of the wafer coupon surface at selected locations.

Surface roughness of the wafers was measured using confocal microscopy (Olympus LEXT microscope). Three dimensional roughness measurements were taken at 10 locations on the wafer surface to quantify the average surface roughness of the wafer. Fracture strength of the wafers was measured using a biaxial flexure method [6, 7] to minimize the effect of wafer surface anisotropy (due to ductile grooves) on the fracture strength results. Circular coupons of 35mm diameter were laser-cut from the wafers. Care was taken to ensure that there was minimal edge damage due to laser-cutting. Fifteen (15) samples each from wafers sliced by wires A and B were tested using biaxial flexure, to obtain the fracture strength statistics reported in the paper.

3 RESULTS

The surface morphology results (Figures 2 and 3) indicate that wafers cut by the diamond wire with grit shape A exhibit greater evidence of ductile material removal than the wafer surface produced by grit shape B. Wafers cut by grit shape A exhibit long grooves with intermittent brittle fracture. Wafers cut by grit shape B show more random pits and chipping on the surface, which are characteristic of brittle fracture.

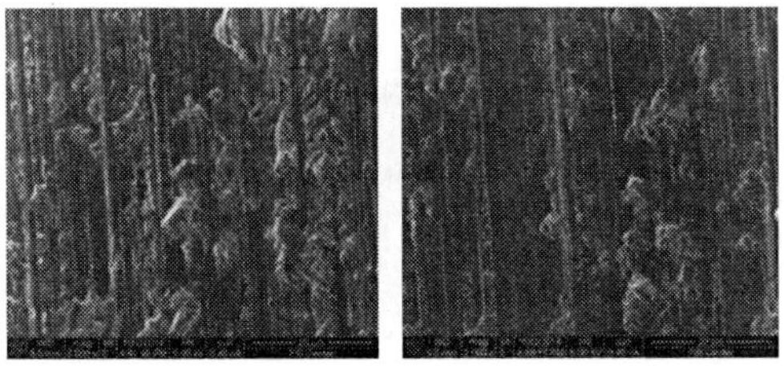

Figure 2. Surface morphology of wafers cut by wire with grit shape A.

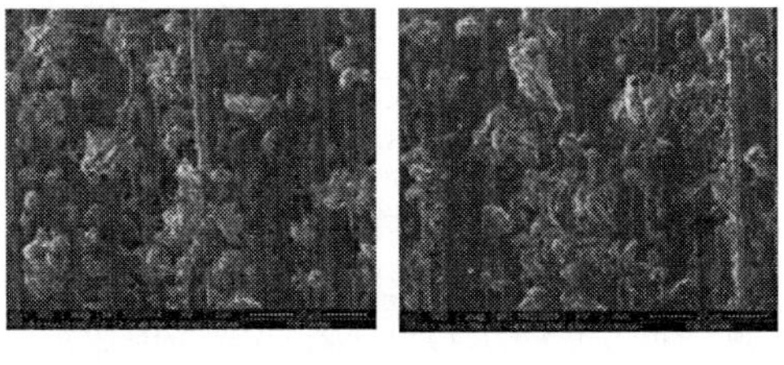

Figure 3. Surface morphology of wafers cut by wire with grit shape B.

Figure 4 shows a comparison of the average surface roughness of the wafers cut by grit shapes A and B. It is clear that grit B yields a rougher surface than grit A. This result is consistent with the earlier observation of greater brittle fracture on the wafer surface generated by grit B compared to the surface generated by grit A (see Figures 3 and 2, respectively).

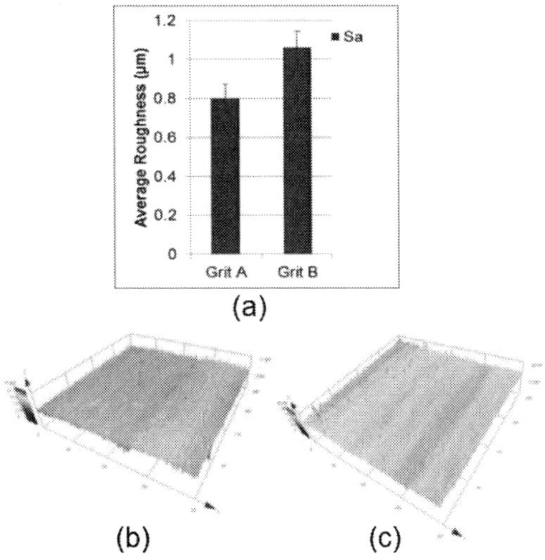

Figure 4. (a) Average surface roughness of wafers. 3D surface topography generated by (b) wire A (c) wire B.

Figure 5 shows two representative images of the subsurface damage for the two cases. Figure 6 shows the distribution of number of cracks and depth of cracks measured using the edge polishing method described earlier. It can be seen that wafers cut by grit shape A had, on average, fewer micro-cracks (33) than wafers cut by grit shape B (40), though some were longer *inclined median cracks*. Wafers cut by grit shape B had more *lateral cracks*, most of which were shallow. The difference in the mean number of cracks for the two wires was found to be statistically significant at a 95% confidence level. The average crack depth for grit A was measured to be 4.7 ± 1.9 μm versus 4.3 ± 1.6 μm for grit B. In contrast to the number of cracks, the difference in average crack depths was not found to be statistically significant. However, since wafer breakage depends on the maximum crack length rather than the average crack length, a comparison of the maximum subsurface crack lengths for the two cases indicates that the maximum crack length in the wafers cut by grit shape A is higher (10.1 μm) than in the wafers cut by grit shape B (7.7 μm). In general, wafers cut by grit shape A had 20% fewer cracks than wafers cut by grit shape B, but the cracks were longer.

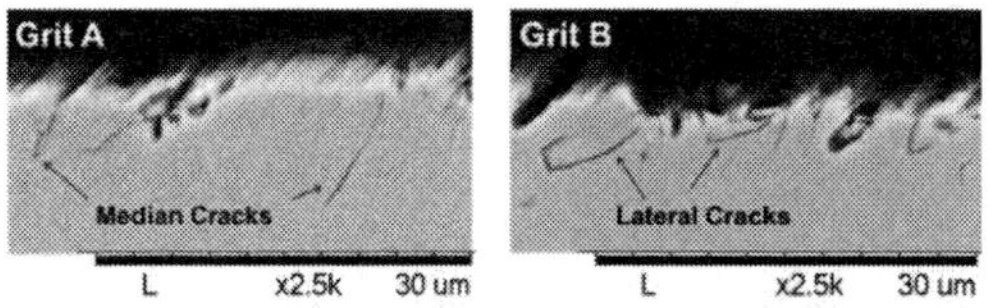

Figure 5. Micrographs of subsurface damage for grit A (left) and grit B (right).

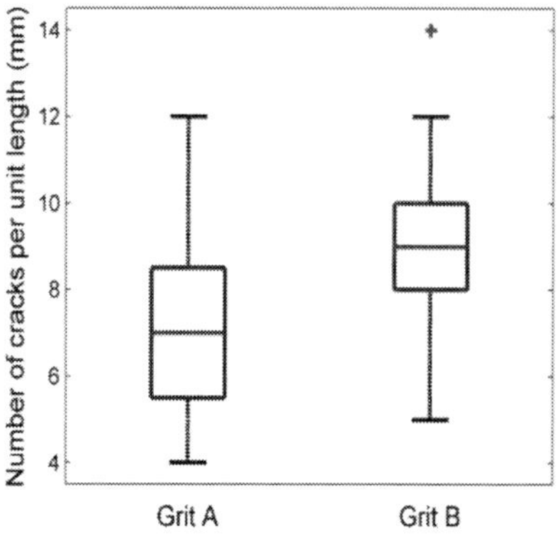

(a)

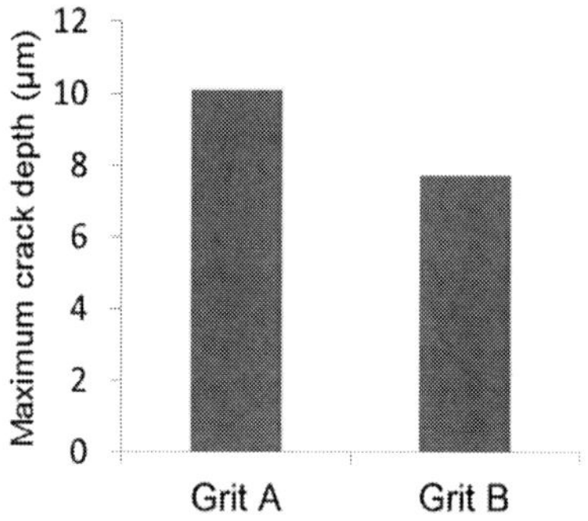

(b)

Figure 6. Boxplots showing distribution of (a) number of cracks per unit length of wafer edge, and (b) maximum depth of cracks for the two grit shapes.

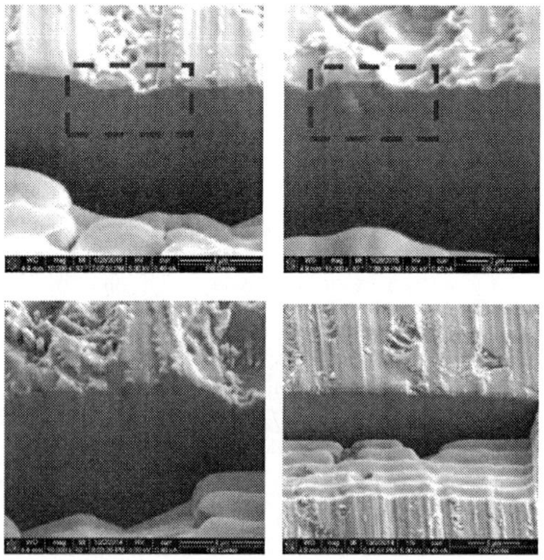

Figure 7. FIB images of wafer subsurface for grit shape A.

FIB sections of the wafer subsurface for the two grit shape cases are shown in Figures 7 and 8. They show that most of the subsurface cracks are lateral cracks, with evidence of networks of cracks in some cases. It should be noted that, unlike the edge polishing method, the FIB method is limited to a localized region of the subsurface. Nevertheless, the FIB results are in general agreement with those obtained from edge polishing. Specifically, the FIB images reveal that there are fewer subsurface cracks in the wafers cut by grit A (Figure 7) compared to the wafers cut by grit B (Figure 8). The crack depths for grit A and B appear to be similar in locations where the FIB sections were made. This is also in agreement with the conclusion obtained from the statistical analysis of crack depths derived from the edge polishing method.

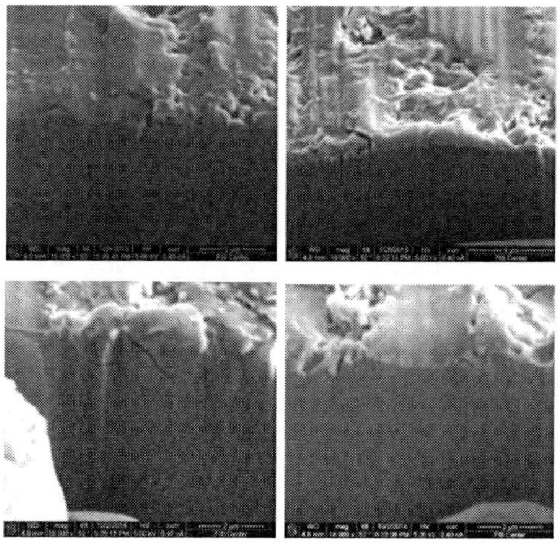

Figure 8. FIB images of wafer subsurface for grit shape B.

Figure 9 shows the biaxial fracture strength comparison for wafers cut by the two grit shapes. Interestingly, the mean fracture strength of wafers cut by grit A is found to be lower than the mean fracture strength of wafers cut by grit B. This is non-intuitive since the wafers cut by grit A show more ductile material removal characteristics. However, according to Bidiville et al. [8], long grooves in diamond wire sawn wafers act as edge cracks and can lower the stress level at which the wafer fails. In this study, wafers cut by grit shape A exhibited more long ductile grooves and less brittle fracture when compared to wafers cut by grit shape B. Furthermore, the subsurface damage results suggest that the wafers cut by grit shape A have longer median cracks than the shallow lateral cracks found in the subsurface of wafers cut by grit shape B. Since, for a given applied stress, crack length determines wafer fracture, and the maximum crack depth for grit A is 3 µm more for than grit B, the observed fracture strength results are supported by the other observations of this study. A two-sample t-test gave a P-value of 0.017 which is smaller than 0.05, hence the difference in fracture strength is statistically significant at a 95% confidence interval.

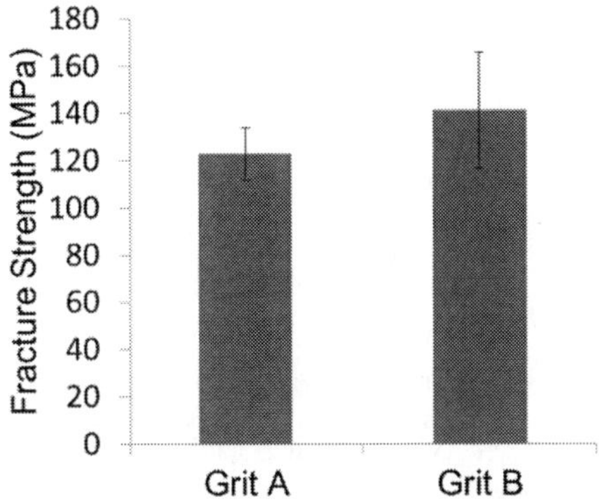

Type	Number of Samples	Mean (MPa)	Standard Deviation (MPa)
Grit A	15	122.8	11.1
Grit B	15	141.1	24.5

Figure 9. Biaxial fracture strength results.

4 CONCLUSIONS

The paper analyzed the effects of abrasive grit shape in fixed abrasive diamond wire sawing on the surface morphology, roughness, subsurface damage, and fracture strength of sliced mono-crystalline wafers. Two wires with nominally different abrasives shapes (A and B) generate observable differences in surface morphology and surface roughness of the silicon wafers. Wafers sliced by wire A showed more ductile material removal compared to those cut by wire B. Average surface roughness was lower for wafers cut by wire A compared to those cut by wire B. Subsurface damage studies showed that wafers cut by wire A had fewer micro-cracks per edge length than wafers cut by wire B. The average crack depth was not found to be statistically different for wires A and B. However, the maximum length of crack was greater for wafers cut by wire A than wire B. The biaxial fracture strength results showed that wafers cut by wire A are weaker than those cut by wire B. The reason for this non-intuitive observation is attributed to a combination of the long grooves acting as stress raisers and the presence of longer inclined median cracks or higher maximum depth of damage for the wafers cut by wire A, compared to the mostly random areas of brittle fracture (pits) and shallow subsurface lateral cracks found in wafers cut by wire B.

5 ACKNOWLEDGEMENT

The authors are grateful for support of this work by the Saint-Gobain Northboro Research and Development Center.

6 REFERENCES

[1] ITRPV, "International Technology Roadmap for Photovoltaics " *http://www.itrpv.net/Reports/Downloads/2014/,* 2014.

[2] E. Cai, B. Tang, W. R. Fahrner, and L. Zhou, "Characterization of the surfaces generated by diamond cutting of crystalline silicon," *26th European Photovoltaic Solar Energy Conference and Exhibition,* pp. 1884 - 1886, 2011.

[3] A. Kumar, A. Kovalchenko, V. Pogue, E. Pashchenko, and S. N. Melkote, "Ductile Mode Behavior of Silicon During Scribing by Spherical Abrasive Particles," *Procedia CIRP,* vol. 45, pp. 147-150, // 2016.

[4] T. G. Bifano, T. A. Dow, and R. O. Scattergood, "Ductile-regime grinding of brittle materials: experimental results and the development of a model," in *32nd Annual Technical Symposium. International Society for Optics and Photonics*, 1989, pp. 108-115.

[5] M. B. Cai, X. P. Li, and M. Rahman, "High-pressure phase transformation as the mechanism of ductile chip formation in nanoscale cutting of silicon wafer," *Proceedings of the Institution of Mechanical Engineers Part B-Journal of Engineering Manufacture,* vol. 221, pp. 1511-1519, Oct 2007.

[6] "ASTM F394-74T ''Tentative Test Method for Biaxial Flexure Strength (Modulus of Rupture) of Ceramic Substrates''," *American Society for Testing and Materials, Philadelphia, PA,* 1974.

[7] C. Funke, E. Kullig, M. Kuna, and H. J. Möller, "Biaxial fracture test of silicon wafers," *Advanced Engineering Materials,* vol. 6, pp. 594-598, 2004.

[8] A. Bidiville, Heiber, J., Wasmer, K., Habegger, S., Assi, F., "Diamond wire wafering: Wafer morphology in comparison to slurry sawn wafers," *Proceedings of 25th European Photovoltaic Solar Energy Conference and Exhibition,* pp. 1673 - 1676, 2010.

CRACK DISTRUBUTION AND STRENGTH OF SILICON WAFERS CONSIDERING THE CRYSTALLOGRAPHIC ORIENTATION OF THE SILICON INGOT IN DIAMOND WIRE SAWING PROCESS

Carola Klute[1], Ringo Koepge*[1] and Stephan Schoenfelder[1,2]
[1]Fraunhofer Center for Silicon Photovoltaics CSP, Otto-Eissfeldt-Strasse 12, 06120 Halle (Saale), GERMANY
[2]Leipzig University of Applied Science, Karl-Liebknecht-Str. 132, 04277 Leipzig, GERMANY
*corresponding author: ringo.koepge@csp.fraunhofer.de

ABSTRACT: The mechanical properties and the failure of silicon wafers are defined by surface defects, like cracks. In the wire sawing process different wire types and particle sizes as well as the machine settings affects the quality of the wafer surface. In order to optimize wire sawing technologies the influence of the crystallographic orientation on the strength and crack distribution of silicon wafers is investigated in this work. By ball-on-ring-tests the fracture strength of different orientations was determined and compared to the crack distribution, which was investigated by bevel polished method. The results show that a rotation of the silicon crystal by 45° (<110> direction) leads to an increase of the crack density in the surface of the samples and a decrease in the fracture strength of the wafers in comparisons to the standard sawing orientation (<100> direction).
Keywords: silicon, crystallographic orientation, wafer strength, crack length distribution, diamond wire sawing

1 INTRODUCTION

Surface defects define the mechanical properties and the failure of silicon wafers. The quality of the wafer surface is affected by the wire sawing process, e.g. by different wire types and diamond particle sizes as well as the machine settings [1, 2]. Investigations of the strength behavior of silicon wafers show that in general the strength depends on the orientation of the wafer referred to the loading direction [3, 4]. For a diamond wire sawing process the diamond particles are fixed on the sawing wire, which leads to another abrasion [5] and subsurface damage on the wafer surface compared to wafers sawn with loose abrasives in the slurry sawing process [6-8]. Furthermore, the crystal orientation could influence the subsurface cracks due to the different cleavage planes in silicon. In order to understand the sawing process the knowledge of crack distribution and crack length caused by the sawing process in correlation to the crystallographic orientation is helpful to identify critical loading directions in wire sawing and to improve the wafer quality.

In this work, two batches of as-cut monocrystalline silicon wafers that were manufactured by two different sawing directions (standard crystal orientation and a rotation of the crystal by 45°, cf. Fig. 1) are compared to each other. The mechanical strength as well as the crack distribution are analyzed. Due to crystal and cleavage plane orientation there might be an influence of the sawing process to the crack formation and strength distribution.

2 MATERIALS AND METHODS

2.1 Sample Preparation

In order to analyze the influence of the crystallographic orientation of the silicon brick on the properties of the resulting wafers a diamond wire based wafering process was performed on a Meyer Burger DS265 Multi Wire Saw. A diamond wire was used with a core diameter of 120 µm and electro plated diamond particles with a particle diameter of 10 µm to 20 µm. A feed rate of 860 µm/min was chosen with a wire speed of 15 m/s. The diameter of two {100} silicon crystals was about 80 mm and the crystal heights about 50 mm. Figure 1 shows a schematic view of these two silicon samples that were cut into wafers with a thickness of approx. 150 µm to 160 µm within in the same sawing run. The major difference for that experiment was the crystal orientation during the wafering process. Sample 1 was cut into wafers by running the diamond wire parallel to the <100> direction whereas sample 2 was cut parallel to the <110> direction, the main cleavage direction on silicon [9].

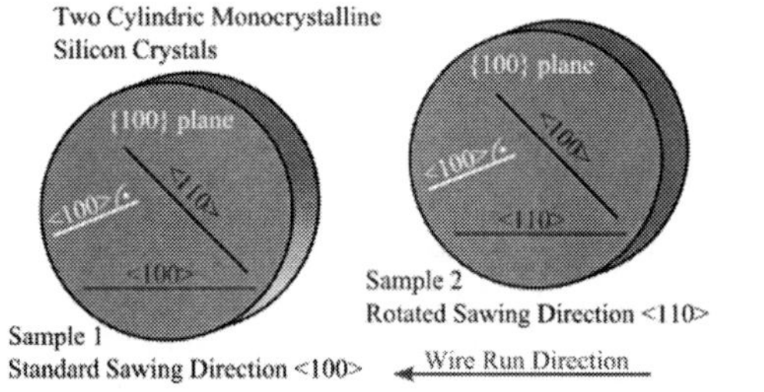

Figure 1: Schematic view of the orientation of two silicon crystals during the wafering process. For sample 1 the diamond wire is running parallel to the <100> direction (common for wafering of silicon). Sample 2 was rotated by 45° around the <100> axis (normal to the {100} plane) that the wire run direction was parallel to the <110> direction, the main cleavage direction in silicon.

2.2 Determination of Wafer Strength

The mechanical strength of the different orientated samples was determined by the ball-on-ring-test (see Figure 2). This test primarily loads the wafer surface instead of the edges. Thus, the mechanical behavior of the surface of the samples is characterized. For this test, the wafers were cut by laser into chips of 10x10 mm^2. The loading ball had a diameter of 6 mm and the ring of the test setup was 7.5 mm diameter.

During the experiment, a force and resulting bending moment was applied, which causes biaxial tensile stresses on the bottom side of the sample. In this axi-symmetrical biaxial stress state all defect directions can be stressed assuming a representative defect distribution in the loaded region. The deflection and the force were recorded until the sample breaks. In order to consider nonlinearities during the experiment, Finite-Element models were used to calculate numerically the exact fracture stress.

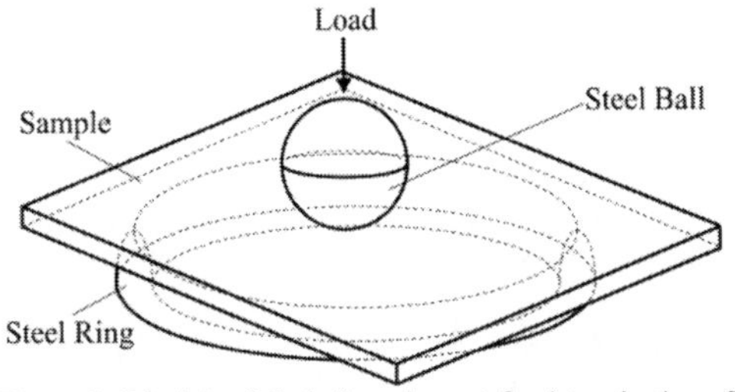

Figure 2: Principle of the ball-on-ring test for determination of the mechanical strength.

Brittle materials, like silicon show a large scattering of the fracture stresses due to their defect distribution. Hence, the Weibull distribution was used for the statistical evaluation [10], where the statistical parameters, the characteristic fracture stress σ_θ and Weibull modulus m were determined.

A sample overview is listed in Table 1, with the thickness of the samples and the number of investigated samples.

Table 1: Sample overview for strength investigation.

Sample No.	Sawing Direction	Sample Thickness [µm]	Sample Number Ball-on-Ring Test
1	<100>	150.1	35
2	<110>	156.0	30

2.3 Crack Distribution - Bevel Polished Method

The subsurface damage on the wafer surface was investigated by the bevel polished method. A bevel polished surface is used to determine the crack depth and their distribution. A schematic view is shown in Figure 3.

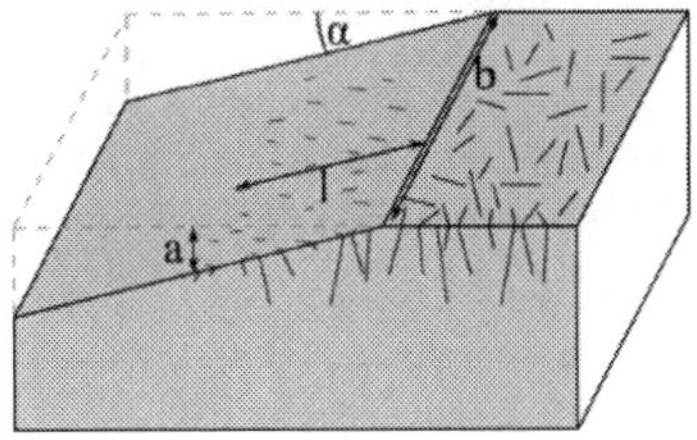

Figure 3: Schematic view of a bevel polished sample. Blue area denotes the bevel, the crack depth a that will be calculated with the measured length l and the bevel angle α. Length b is the analyzed length of the sample.

The blue area denotes the bevel polished surface and the grey area the untouched wafer surface. Red dashes mark the investigated cracks, which become visible after the polishing and Secco etching [11]. The crack depth a was calculated by the measured length l on the bevel surface and the angle α of the bevel. This crack depth is a minimum crack depth because cracks need to reach deeper in the material to be visible on the beveled surface.

In this experiment the crack distribution of the sample batch 1 and batch 2 was determined. From each batch two samples at a length of 2.8 mm per sample were analyzed. Table 2 summarizes the samples of the microstructure analysis.

Table 2: Summary of samples analyzed by the bevel polished method.

Sample No.	Sample ID	Sawing Wire Direction	Full Analyzed Length [mm]
1	1.1 / 1.2	<100>	5.6
2	2.1 / 2.2	<110>	5.6

3 RESULTS

3.1 Mechanical Evaluation

The two sample batches of different crystallographic orientations show a significant difference regarding their fracture strength. The fracture stresses of both batches are shown in histograms in Figure 4. The frequency of fracture stress depends on the crystallographic orientation of silicon material during the wafering process. The <110> sample shows higher frequency at lower fracture stresses.

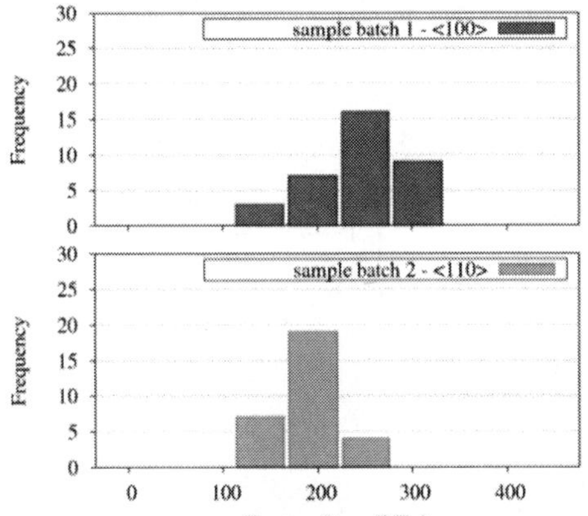

Figure 4: Histogram of all fracture stresses of the analyzed sample batches 1 and 2.

In terms of Weibull parameters the wafers of the <100> sawing direction show a higher characteristic fracture strength than the wafers of the <110> sawing direction. The results are visualized in a Weibull diagram in Figure 5. Two strength distributions are shown with the same slope with a shifted distribution to higher fracture stresses for the wafers out of the <100> sawing run. The Weibull parameters of the investigated samples are summarized in **Fehler! Verweisquelle konnte nicht gefunden werden.**.

Table 3: Weibull parameters with 90% confidence intervals for the ball-on-ring test.

Sample No.	Sawing Direction	m [-]	σ_θ [MPa]
1	<100>	6.3 (5.0…7.8)	266 (254…279)
2	<110>	6.4 (5.0…7.9)	198 (188…208)

The strength measurement shows a significant higher fracture stress for the standard sawn wafers with the <100> crystallographic orientation parallel to the wire run direction. The different characteristic fracture strength but same Weibull modulus (scattering of strength) indicate a different crack size but similar defect distribution in comparison to the <110> sample batch. Thus, for better understanding the determination of the crack distribution of the wafers is necessary.

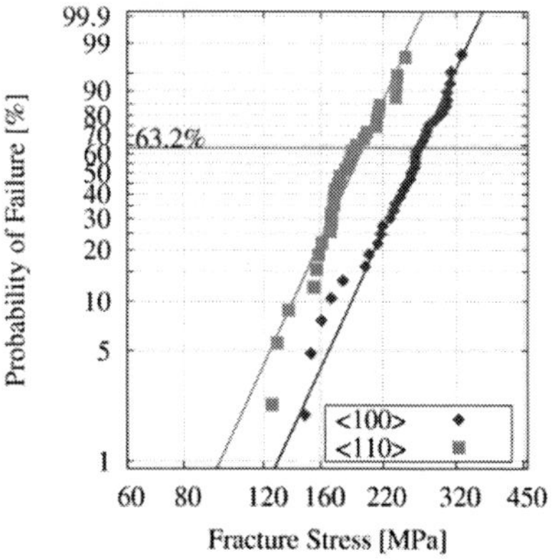

Figure 5: Results of mechanical strength testing shown as Weibull plots for the wafers out of the standard <100> sawing direction and <110> sawing direction.

3.2 Crack Distributions

The influence of the crystallographic orientation on the microstructure in the near-surface area of the wafers was investigated by the bevel polished method. Figure 6 show an optical microscopy image of the bevel polished surface.

Figure 6: Image of bevel polished surface, the dark grey area at the bottom of the image is the as-cut wafer surface. Tiny lines and curves at the light grey bevel polished surface are the cracks starting at the silicon surface into the material. Orange box indicates the area of crack depths smaller 3μm.

Two samples of each batch 1 and 2 were examined. Figure 7 shows the histograms of the minimum crack depth of the investigated samples starting at a crack depth of 3 μm. Smaller cracks are out of detection capability because of a noisy transition area between the wafer surface and the bevel surface (represented by orange box in Figure 6).

As expected, the crack density of all samples is higher for smaller cracks and the maximum crack depths of the wafer samples are in a similar range. The wafer surfaces of the standard sawing direction <100> show a reduced crack density in a range up to 5 μm crack depth compared to the parallel orientation of the <110> direction samples. Furthermore, the maximum crack depth is higher for the sample batch 2 (<110> direction), cf. Table 4.

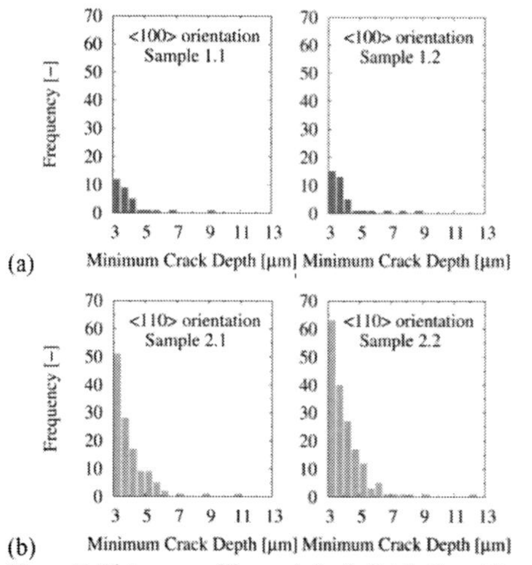

Figure 7: Histograms of the crack depth distribution at the wafer surface, sawn with (a) standard silicon material orientation (<100>) and (b) rotated silicon material by 45° (<110>).

Table 4: Maximum crack depths and the number of cracks for the investigated samples.

	1.1	1.2	2.1	2.2
Total Number of Cracks	31	39	124	173
Max. Crack Depth [μm]	9.2	8.6	10.7	12.1

The clearly different number of cracks and the existence of larger crack depths that were determined for <110> sawing direction correlates well to lower fracture stresses in the ball-on-ring-test.

4 DISCUSSION

The investigation shows that a rotation of the silicon crystal at the standard diamond wire sawing process leads to significant changes in the fracture strength of the as-cut wafers. Furthermore, the crack density as well as the crack depth is higher for the parallel orientation of the <110> direction samples, which correlates to the lower fractures stresses in the mechanical tests. A configuration of diamond scratches parallel to the <110> direction causes a crack propagation along the weak cleavage {110} and {111} planes and the <110> direction [9]. Thus, larger defects arise and a lower strength occurs. Aside from the correlation between the strength and crack depth distribution it can be seen that the scattering of strength distribution that is characterized by the Weibull modulus does not change by changing the wire run direction in the silicon crystal. That could mean that the amount and length of cracks is defined by the crystal orientation (i.e. cleavage planes) regarding the abrasion process in wire sawing. But the crack shape and defect structure formed by the sawing process are independent of the crystal orientation.

5 SUMMARY AND CONCLUSION

The understanding of the influence of the crystallographic orientation on the crack length and crack distribution in the surface of silicon wafers is an important topic in order to optimize wire sawing technologies. The experiments have shown that a rotation of the silicon material in another crystallographic orientation has a strong influence on the mechanical strength and the crack distribution in the silicon wafer. The <110> orientated wafers (rotation of the standard brick by 45° about normal axes to {100} plane) show more and larger cracks in the near surface area within and above 5 µm compared to <100> oriented wafers. This leads to a higher probability of breakage that is equal to the detected lower fracture strength of the wafers. The microstructural investigations strongly correlate to the strength measurements.

In conclusion the mechanical characterization can be used to analyze sawing process parameters in order to improve the wafer quality. Furthermore, the crystal orientation is an important parameter for high quality wafers, whereas the standard orientation shows fairly good results. In further investigations intermediate crystal orientations will be analyzed as well as detail investigations of crack formation and propagation regarding more efficient sawing directions and optimization of the diamond wire sawing process.

ACKNOWLEDGEMENT

This work has been financially supported by the Investitionsbank Sachsen-Anhalt with the research project "Diadem" (contract no. 6060240201) and the Federal Ministry for Economic Affairs and Energy (BMWI) in "ENOWA-II" (contract no. 0325805I), which is gratefully acknowledged.

REFERENCES

[1] R. Koepge, C. Klute, F. Kaule, K. Sunder, O. Anspach, S. Schoenfelder: "Mechanical and Microstructural Wire Sawn Wire Sawn Wafers Considering the Wire Wear in Sawing Process." 31st EUPVSEC, 2015

[2] E. Chai, B. Tang, W. R. Fahrner, L. Zhou: "Characterization of the Surfaces Generated by Diamond Cutting of Crystalline Silicon.", 26th EUPVSEC, p. 1884-1886, 2011

[3] A. Bidville, J. Heiber, K. Wasmer, S. Habegger, F. Assi: "Diamond Wire Wafering: Wafer Morphology in Comparison to Slurry Sawn Wafers.", 25th EUPVSEC, p. 1673-1676, 2010

[4] A. Bohne, S. Schoenfelder, J. Bagdahn: "The Influence of Wire Sawing Process on Mono- and Multicrystalline Silicon Wafers.", 23rd EUPVSEC, p. 1780-1784, 2008

[5] H. J. Moeller: "Wafering of Silicon", Chapter Two of Semiconductors and Semimetals, Vol. 92, p. 63-109, 2015

[6] A. Bidville, K. Wasmer, R. Kraft, C. Ballif: "Diamond Wire-Sawn Silicon Wafers - From the Lab to the Cell Production.", 24th EUPVSEC, p. 1400-1405, 2009

[7] J.-I. Bye, L. Norheim, B. Holme, O. Nielsen, S. Steinsvik, S. A. Jensen, G. Fragiacomo, I. Lombardi: "Industrialised Diamond Wire Wafer Slicing for High Efficiency Solar Cells.", 26th EUPVSEC, p. 956-960, 2011

[8] F. Coustier, J.-D. Penot: "Diamond wire sawing: State of the art and perspectives.", 15th Photovoltaics International, p. 60-65, 2012

[9] R. Pérez, P. Gumbsch: Directional Anisotropy in the Cleavage Fracture of Silicon. Phys. Rev. Lett., American Physical Society 84, 5347-5350, 2000

[10] W. Weibull, A Statistical Theory of the Strength of Materials. Generalstabens Litografiska Anstalts Foerlag, 1939

[11] F. Secco d'Aragona: Dislocation Etch for (100) Planes in Silicon. J. Electrochem. Soc. 119, 948-951, 1972

ECONOMIC EVALUATION OF ADVANCED COOLANT SUPPLY TO DIAMOND WIRE SAWS IN SILICON WAFER PRODUCTION

Jochen Ruth (1), Gerd Heser (1), Rolf Berndt (2)
(1) Pall GmbH, Philipp-Reis-Str. 6, D-63303 Dreieich, Germany (correspondence)
Phone +49-6103-307-292, FAX +49-6103-307-552, jochen_ruth@europe.pall.com
(2) RBFM Consulting, Dresden, Germany, rbfm_consulting@t-online.de

ABSTRACT: Wafer slicing by diamond wire saws will substantially contribute to cost reduction in mono c-Si (mcSi) wafer production provided the coolant supply technology is economically viable. Aqueous (water based) coolants appear to be generally advantageous. Meanwhile new coolants with more attractive prices can be found on the market. The ways how to apply a coolant and how to handle spent coolants need to be optimized to achieve further improvements. "LOW-FLOW" and "HIGH- FLOW" coolant mode, both with or without coolant regeneration and reuse, should be distinguished. A detailed economic comparison clearly shows that "HIGH-FLOW" coolant usage, imperatively combined with reclaim of spent coolant, lowers the impact of coolant supply on wafer production costs significantly. Generally, "HIGH-FLOW" application of aqueous coolants in collaboration with coolant regeneration/ reuse is preferable to any "LOW-FLOW" mode if disadvantages due to continuously increasing solid load during one cut are tolerable, and if spent coolant is regenerated and reused. Coolant regeneration and reuse are essential prerequisites for cost-efficient coolant supply regardless of the mode of coolant application. Without regeneration, even "high-flow" mode is inefficient.
Keywords: c-Si, Manufacturing and Processing, Cost reduction, Wafering, Coolant, Recycling

1 BACKGROUND AND MOTIVATION

Wafer slicing by diamond wire saws may substantially cut costs in mcSi wafer production. Among the decisions to be taken in a process design, two aspects are significant. The first one refers to the kind of coolant. Previous analysis highlighted advantages of aqueous (water-based) coolants over alcohol-based coolants [1]. Inexpensive coolant formulations meanwhile available have supported this tendency. A second decision applies to the way how coolants are used, regenerated, and reused. Improving these procedures makes supply of coolants economically viable and gives access to an impressive optimization potential.

As far as usage of coolants during sawing is concerned, "LOW-FLOW" and "HIGH- FLOW" coolant mode should be distinguished. Both of them can be combined with coolant regeneration and reuse. Benchmarking of coolant supply technologies necessarily includes cost analysis for those technologies, considering Total Cost of Ownership (TCO) and Investment Costs (IC) in typical scenarios.

2 VERSIONS OF COOLANT APPLICATION

Aqueous coolants may be applied in three viable versions. They are illustrated in Fig. 1.

2.1 "LOW-FLOW" mode with coolant regeneration and reuse (Version A)

Particle-free coolant is fed from a CLEAR COOLANT TANK to wire saws at moderate flow-rate (typically 0.5 to 2 m³/h per saw), and discharged to a SPENT COOLANT TANK after <u>single use</u>. The spent coolant is clarified, typically by CROSSFLOW MICROFILTRATION [2] in two stages. A particle-containing concentrate is sent to waste-water treatment whereas the particle-free permeate enters the CLEAR COOLANT TANK where it is complemented by coolant concentrate and DI water (COOLANT MAKE-UP). Now it is available to be supplied to the saws again.

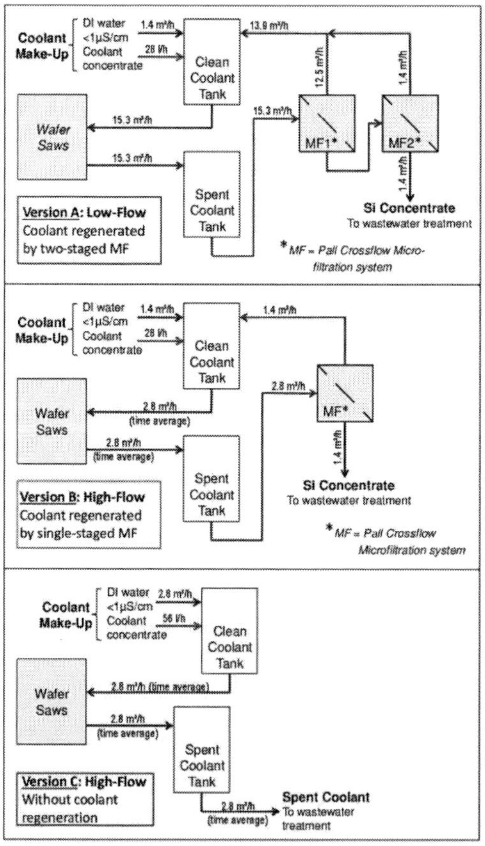

Figure 1: Three viable versions of coolant application technology

Since the saws are always fed with particle-free coolant, the conditions during one cut are invariable as far as particle load is concerned. No additional recirculation within the saw takes place.

2.2 "HIGH-FLOW" mode with coolant regeneration and reuse (Version B)

Coolant intensely circulates in a saw temporarily (during one cut) under ongoing increase of its particle load. High flow-rate (typically 3 to 6 m^3/h) of the internally circulating coolant prevents sedimentation of accumulated kerf material. Only after the cut, the contaminated coolant is discharged into a SPENT COOLANT TANK after multiple use and regenerated by filtration, typically single-staged crossflow microfiltration. Again a particle-containing concentrate is sent to waste-water treatment whereas the clear permeate enters the CLEAR COOLANT TANK for COOLANT MAKE-UP and is reused in subsequent sawing cuts.

Since the coolant circulates in the saw during one cut, it collects more and more kerf material. Thus the conditions during one cut are variable and worsened as far as particle load is concerned.

2.3 "HIGH-FLOW" mode without coolant regeneration and reuse (Version C)

This version resembles version B as far as coolant application is concerned. Different from version B, spent coolant is neither regenerated nor reused after one cut; it is completely discarded after single use.

2.4 Differences and similarities

Version A has solid-liquid separation in common with Version B. As the silicon particles are less than 1 µm in size only microfiltration applying crossflow membrane modules is economically viable [2]. Fig. 2 shows a typical crossflow microfiltration system for complete removal of kerf particles from spent coolant. It is meant to concentrate these particles up to 100…120 g/l solids in the concentrate stream to be treated as waste-water.

Version C works without any coolant regeneration so that more fresh coolant is needed and more spent coolant is sent to the wastewater treatment plant. To proceed without filtration certainly lowers investment costs. On the other hand, it would increase coolant procurement costs.

Hence any economic ranking of Versions, A, B and C needs an analysis of overall costs.

3 MODEL SCENARIO FOR COST ANALYSIS

3.1 Basics

Concrete analysis of overall costs needs a concrete scenario with practical relevance that enables cost calculation. The scenario applied in this paper is characterized by

- 165 million wafers per year sliced on 24 saws
- Spent coolant discharged at 10 g/l Si particle concentration in version A, and at 56 g/l in versions B and C
- Si particles are removed from spent coolant by crossflow microfiltration in versions A and B, and collected in a concentrate at 112 g/l solids load in both cases.

Internal and external flow rates result from common balancing of material flow as indicated in Fig. 1.

3.2 Unit cost figures

Unit costs are based on typical market prices for consumables, materials, energy, waste discharge and labor. Table 1 indicates essentials.

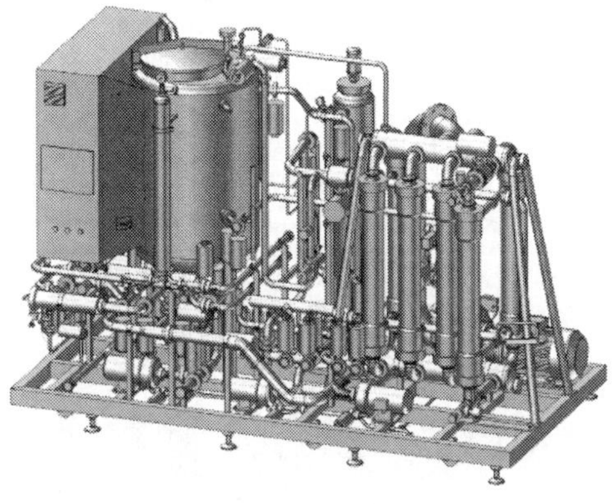

Figure 2: Crossflow microfiltration system type Microflow HS-6A (Pall Corp.)

Table 1: Essential unit costs

Item	USD	per
Electric energy	0.10	kWh
DI Water	5.00	m^3
Compressed air	0.02	Nm^3
Coolant concentrate	7.00	kg
Labor cost filtration, file management	50k	apiece p.a.
Labor cost administration	80k	apiece p.a.
Cleaning w/water discharge	2.00	m^3
Si sludge discharge	5.00	m^3

3.3 Further assumptions

Assets are depreciated over 10 years on a straight-line basis. Degradation of coolant components during usage is neglected. Effects of that kind might appear with certain coolant types, or might be completely negligible for others. Degradation may be compensated by disproportionate make-up which would raise coolant procurement costs slightly. The general economic picture will remain unchanged.

4 COST ANALYSIS AND RESULTS

4.1 Bases of calculation

Cumulated *Total Costs* C_T for X years were calculated by

$$C_T = C_I + X \cdot C_O \tag{1}$$

with C_I – *investment costs* and C_O – *Annual Total Cost of Ownership*

$$C_O = C_{DR} + C_P \tag{2}$$

incorporating *Annual Total Cost of Ownership of Coolant Delivery and Regeneration* C_{DR} as well as *Annual Coolant Procurement Costs* C_P. Individual cost figures for Versions A, B and C are summarized in Table 2.

4.2 Discussion and interpretation

As to sheer *investment costs* (95,000 USD for coolant handling), Version C (no regeneration) might be economical. But *Annual Total Costs of Ownership* C_O unmasks **Version C** as an inefficient technology.

Fig. 3 indicates *Total Costs* C_T over time. It gives a clear picture on cost efficiency in all three analyzed versions. **Version B** ("HIGH-FLOW" with filtration) outperforms Version C after less than five months, and saves 4.3 million USD in three years. Even **Version A** ("LOW-FLOW" with filtration) is superior to **Version C** at long sight, despite higher initial investment costs.

Table 2: Analysis and comparison of costs

COST CATEGORIES		Unit	Version A	Version B	Version C
TCO Coolant Delivery & Re-generation C_{DR}	per year	kUSD per year	628	250	157
	per wafer	*USD per wafer*	*0.0038*	*0.0015*	*0.0009*
Coolant Procurement Costs C_P	per year	kUSD per year	1,646	1,646	3,373
	per wafer	*USD per wafer*	*0.0100*	*0.0100*	*0.0204*
Investment Costs C_I	totally	kUSD	1,921	718	95
Total Costs C_T in First Three Years	totally	kUSD	8,742	6,404	10,684

Version B cuts coolant supply costs C_{DR} + C_P per wafer to 0.012 USD compared to 0.014 USD for Version A and to unacceptably high 0.021 USD for Version C. From this point of view, Version B ("HIGH-FLOW" with filtration) should be the preferred way how coolants are used, regenerated, and reused. This judgment, however, does not consider the difference in coolant quality during one cut. Version A provides a constantly good coolant quality free from particles throughout a cut. The coolant provided to the kerf by Version B is different. Its solids concentration starts at Zero but increases during one cut to a considerable level (about 40…80 g/l, depending on kerf size and recirculating coolant volume). Possibly this deteriorating quality affects the sawing process, e.g. concerning surface quality, wire lifetime, energy and kerf width. Such effects have not been investigated by the authors and are not subject of this paper. If they demand attention, Version A might be a better choice, the more so as its economic figures in the long term are nearly as good as those ones of Version B.

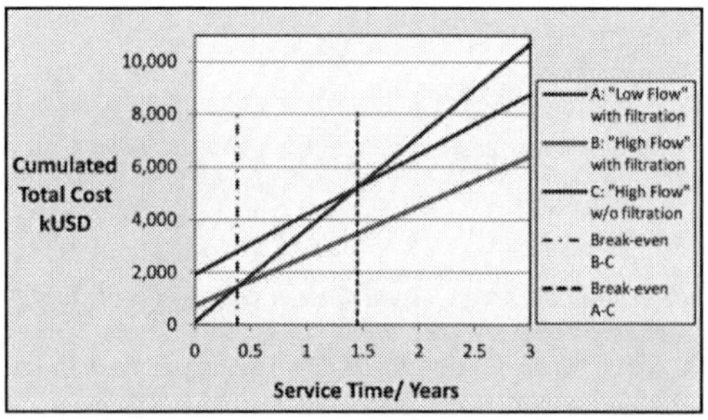

Figure 3: Comparison of cumulated total costs over time for three versions of coolant supply

4. SUMMARY AND RECOMMENDATIONS

1) "High-flow" application of aqueous coolants on diamond wire saws facilitates coolant supply at low cost. It is preferable to any "low-flow" mode if disadvantages due to temporarily increasing solid load during one cut are tolerable, and if spent coolant is regenerated and reused.

2) Coolant regeneration and reuse are essential pre-requisites for cost-efficient coolant supply regardless of the mode of coolant application. Without regeneration, even "high-flow" mode is inefficient.

3) Crossflow micro-filtration is a powerful tool to cut cost of coolant supply. Up to 40 % cost reduction in coolant supply per wafer is achieved through coolant reclamation.

REFERENCES

[1] R. Berndt, J. Ruth, S. Waters, Proceedings 26th European Photovoltaic Solar Energy Conference, (2011) 1951

[2] J. Ruth, S. Waters, R. Berndt, Proceedings 27th European Photovoltaic Solar Energy Conference, (2012) 939

COST EFFECTIVE GROWTH OF SILICON MONO INGOTS BY THE APPLICATION OF A MOBILE RECHARGE SYSTEM IN CZ-PULLER

F. Mosel[1], A.V. Denisov[1], B. Klipp[1], B. Spill[1], R. Sharma[2], R. Kunert[2], P. Dold[2]
[1] PVA Crystal Growing Systems GmbH, Im Westpark 10-12, 35435 Wettenberg, Germany
[2] Fraunhofer CSP, Otto-Eissfeldt-Str. 12, 06120 Halle/Saale, Germany
frank.mosel@pvatepla.com, phone: +49 64168690-125, fax: +49 64168690-822

ABSTRACT: The permanent demand on the improvement of solar cell efficiency and reduction of production costs has placed much focus on the multiple Cz-crystal pulling technique from one crucible. The use of lower quality silicon as feedstock for monocrystalline material is not an alternative track because it is in contradiction to the quest for high cell efficiencies.
PVA Crystal Growing Systems Gmbh has developed a novel mobile recharge system for Czochralski pullers, the SiCharger MRS. This systems is capable of handling silicon feedstock material up to a chip size of 15 mm and was extensively tested under industrial conditions. The assignment for a chip size of 40 mm is currently under series of tests, the results are promising. One MRS-system can serve several pullers, presuming a suitable production organization. Investment cost and thus amortization period are minimized.
In this paper the MRS-system is presented in detail as well as its operation during the recharging of the hot crucible. The influence of the different types of feedstock on the recharge process is presented. The segregation effect of impurities on the quality of multiple ingots grown in one crucible is shown.

Keywords: Czochralski, Melt Recharge, Economic Analysis

1 INTRODUCTION

About 90% of the solar cells produced in the world are based on crystalline silicon (c-Si). Currently, the wafer market is dominated by casted materials. It is expected that in the next 10 years the market share of mono-Si will increase up to 50%, coupled with a strong trend towards high efficient cells on n-type wafers. Fig.1 shows the expected proportions of the different crystal growing technologies and wafer types for the next ten years.

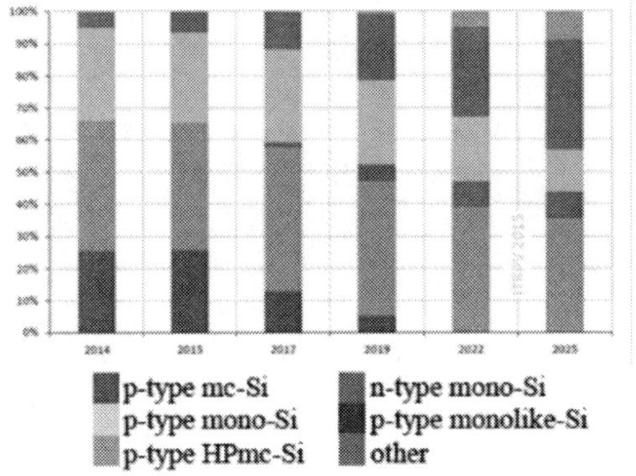

Figure 1: World market shares for wafer / ingot types. (International Technology Roadmap for Photovoltaic ITRPV 2014 Results, www.itrpv.net).

The casting process of mc-silicon might be scaled rather easily, today, G6 is already standard and G7 or G8 are in use, too. Up-scaling in the case of Czochralski mono-ingot growth is more difficult, the diameter and the length of the ingots are more or less fixed. Multipulling, i.e. the growth of several ingots out of the same crucible is a powerful method to reduce production costs: in a standard Czochralski process, costs for the crucible amount to approximately one quarter of the total consumable expenses. Furthermore, multipulling reduces the process time (no lengthy cool-down period between single runs) and extends the life-time of the graphite hot-zone. The challenge is the design and operation of a sophisticated but still economic feeding system, which allows the refill of the crucibles in a well-controlled way.

2 EXPERIMENTAL

2.1 Multipulling of Silicon Monocrystals

At present, silicon monocrystals are still produced mostly according to the standard Cz-process. In this batch process the quartz crucible is filled to top with silicon feedstock and a crystal is pulled by the well-known Cz-procedure. After the process the entire Cz-puller is cooled down, which inevitably leads to the destruction of the quartz crucible. Besides the loss of the expensive crucible, hotzone components may deteriorate by the temperature changes of the heat-up and cool-down cycles. In order to reduce the production costs, in particular the crucible costs, and to economize the Cz-process by saving unproductive down times during the heat-up and cool-down cycles, multiple batch recharging is a favorable track. This multiple batch recharging is also called "multipulling Czochralski" and is defined here as the growing of several ingots in one crucible with melt replenishment after each growth process. During the cool down of the crystal in the gate chamber, which is separated from growth chamber by means of a gate valve, the hotzone is maintained under melting conditions. Thus, the crucible can be recharged during the cool down cycle of the ingot. Several concepts of replenishing the hot crucible are well known since the early days of silicon mass production [1].
Simple, but effective recharge systems are the "slab-feeder", where the side slabs of squared ingots are attached to the central pulling drive and are slowly lowered into the hot crucible, or the "central feeder", where a quartz tube filled with silicon chunks also attached to the central pulling device is lowered into the crucible and emptied into the restmelt [2]. The

disadvantage of both systems is that the previous ingot had to be cooled down and taken out of the gate. This procedure is quite time consuming. Fig. 2 shows another concept of a "stationary feeding unit" from PVA Crystal Growing Systems GmbH. The feeder is permanently

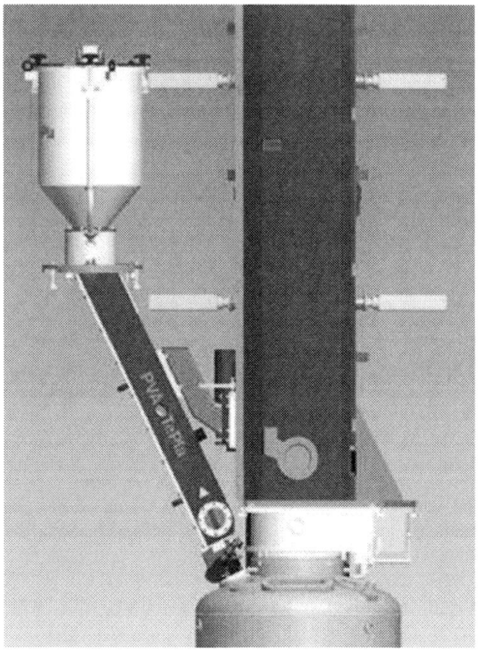

Figure 2: Stationary feeding unit mounted on Cz-puller

mounted to the puller. The storage vessel (hopper) is removable for charging with silicon feedstock material up to a chip size of 10mm. Inside the connection piece between the hopper and the furnace vessel a system of movable quartz tubes is installed, which can be moved back and forth, depending whether the feeder is in operation for feeding or separated from the growth chamber by means of a gate valve for refilling the hopper. This kind of feeding system is commonly used when multiple Czochralski growth is performed. However for each Cz-puller a separate feeding unit is required which implies high installation costs.

2.2 Mobile Recharge System

The PVA Crystal Growing Systems GmbH has developed a mobile recharge system, which offers the unique possibility to serve several Cz-pullers time-delayed. To make the mobile feeding unit compact and easy to move, the hopper is separated from the passive conveyor, the quartz tube system. The connection of the hopper with the quartz tube is realized by an active vibratory conveyor. The advantage of the vibratory conveyor is that no moving parts of the conveyor have contact to the silicon charge and thus contamination is avoided. Therefore the conveyor drain is lined with silicon plates. The silicon chips are drawn off from the hopper by the active vibratory conveyor and are transported to the quartz tube and thus feeded into the crucible.

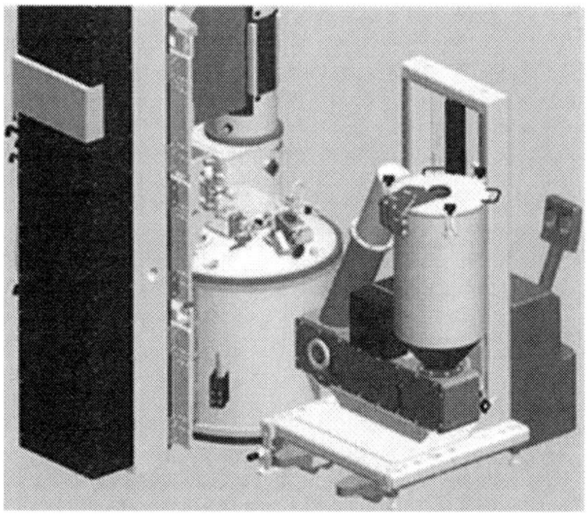

Figure 3: Mobile recharge system MRS undocked on transportation cart

The active conveyor is a small-parts conveyor drive, which is a linear vibration conveyor excited by means of an oscillating electromagnet at a fixed frequency of 100Hz. The working unit is a V-shaped drain which transmits the oscillation of the vibration unit to the transported silicon chunks. The mass flow is controlled by the magnitude of the oscillating amplitude and can be adjusted in a wide range up to several kg/min. It is also possible to control the oscillation of the vibratory unit by a closed control loop. This ensures the compatibility of different mobile SiChargers.

Figure 4: Mobile Recharge System docked on Cz-puller

The MRS-system can be docked in any phase of the growth process, for instance in the melting or in the endcone phase. The docking valve is closed and after docking, the MRS-system is evacuated and leak tested. When the melting phase or the ingot growth is completed and the ingot is separated from the growth chamber, the feeding of the hot crucible can be initiated. Therefore the whole feeding system is pressurized to the growth chamber condition. The docking valve is opened and the quartz tube is lowered to its feeding position in the hot crucible. Feeding can be executed under argon gas flow. The feeder system is attached to the Czochralski puller only for the time of topping up the crucible or refilling after crystal growth. Once the crucible is filled up, the quartz-tube is retracted in its chamber and the docking valve is closed. The tightness of the docking valve is proven by means of pressure drop test. The MRS-system is disconnected and can be recharged for the next refill process.

The MRS-system is controlled by the PLC of the puller, which is the case for Cz-pullers from PVA CGS Gmbh. For other manufactures it is also possible to control the mobile feeder by its own PLC, assuming a flange for the docking valve is available.

2.3 Influence of particle size and distribution

Well controlled feeding rates are crucial for a successful operation of the system. If the amount of recharged silicon is too high, the residual melt might freeze and the crucible would be damaged. If the recharge rate is less than the amount, which the system could handle, economics are affected negatively. We performed measurements of feed rate for different types of feedstock using a stationary feeder. Not surprising, applying similar parameters, the granular material from Fluidized Bed Reactors (FBR) showed the highest feeding rate[1]. Crushed silicon feedstock from Siemens rods, even if they had the same size than the granules, resulted in significantly lower feed rates (Fig. 3). Compared to the spherical granules, the crushed silicon pieces are strongly elongated, an aspect ratio of 1:4 or higher is found quite often. This results in strong interdigitating and increased friction. Even clogging might be seen, if the transport system is not several times larger in diameter than the maximum length of the silicon pieces. In any case, it makes clear that a sophisticated control system is essential, which is capable to measure reliably the amount of transported silicon and which can control the feeding process as part of the overall control loop. In our presentation, we will demonstrate the correlation between feed rate, feedstock type and feeder design.

Particle release during the feeding process is a rather critical issue. Although most of the particles are in the lower micrometer or in the sub-micrometer range, as our measurements have shown, they might accumulate in certain areas of the growth chamber and might be released at a later point, in particular, when the gas flow or the flow configuration changes. Typically, this is the case, when the crystal body enters the area of the gate valve: the cross-section for the argon flow is reduced and the overall flow pattern changes. We identified two sources for particle formation: (I) extrinsic reasons, the material is already delivered with a certain dust load and (II) intrinsic reasons, particles are formed during the feeding process due to abrasion and friction. The dust load of the virgin material might be removed by proper pre-conditioning (inert gas blowing, washing), but it will increase cost. Dust formation during the process itself might be minimized by short transport length, by a low-angel conveyer system, and by a sophisticated gas flow system. We will present corresponding measurements and design solutions.

[1] Although FBR silicon shows the best flow behavior, other aspects like interaction with the crucible or splashing during melting have to be considered, too, when discussing the use of FBR granules.

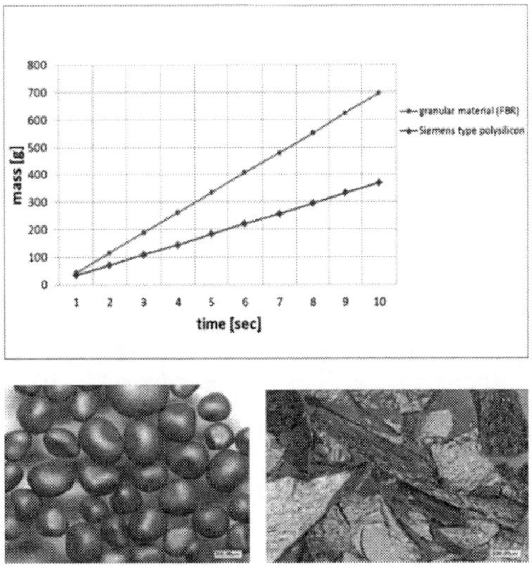

Figure 5: Feed rate for granular material (blue) and crushed Siemens rods. Size and feeding parameters had been identical, but the flow and transport properties vary significantly. Below: Magnified image of granules from FBR process (left) and crushed Siemens silicon (right).

particle size	below the crucible	above the crucible
1 µm	388656	361637
5 µm	10513	9655
10 µm	169	334

Table I: Measurement of particle load inside the growth chamber: The measurements have been performed at room temperature and under normal pressure. Within 4 minutes, about 6.7 kg of silicon had been filled into the crucible. A steady argon flow of 20 slm was used.

2.4 Impurity redistribution in Cz-ingots grown from a replenished melt

An estimation of impurity accumulation in the melt due to the sequential melt replenishment has been published by Hopkins et al. [3]. Using the well-known "normalfreeze" relation (1) given by Pfann [4]

$$(1)\quad C_S = k_{eff} C_0 (1-g)^{k_{eff}-1}$$

and the concept of the effective segregation coefficient k_{eff} [5], which is the ratio of the solute concentration in the growing crystal C_S to the solute concentration in the bulk liquid, the axial concentration profiles in Cz-ingots grown from replenished melts can be calculated as a function of the solidified fraction g, with C_0 is the initial concentration of the impurity in the liquid. The initial impurity concentration C_L^i (n) in the melt at the start of the n-th pull (2) can be calculated according to Hopkins with $p = (1-g)^{k_{eff}-1}$ [3].

$$(2)\ C_L^i(n) = C_0\left[p^{n-1} + g\left(\frac{p^{n-1}-1}{p-1}\right)\right]$$

Assuming a working life of a standard quartz crucible of ~150 h and a mean process duration of 40 hours for an 1.5 m long 8 inch ingot without heat up and cool down times it should be possible to grow three ingots out of one crucible.

Fig 6 shows the accumulation of impurities with a small segregation coefficient in the melt calculated for the sequential replenishment of the crucible. The refill is calculated after solidified fraction of 90% (g=0.9). For the calculation an effective distribution coefficient $k_{eff} = 10^{-3}$ was applied, which is in the range for metal impurities, having a strong impact on the solar cell performance [6]. Tab. II shows measured segregation coefficients of main metal impurities in Cz-silicon [3].The main origins of the metal impurities have been identified by Harada et al [7] as the silicon feedstock and the quartz crucible.

metal	k_{eff}	metal	k_{eff}
Co	1.0×10^{-5}	Mo	4.5×10^{-8}
Cr	1.1×10^{-5}	Ni	3.2×10^{-5}
Cu	8.0×10^{-4}	Ti	2.0×10^{-6}
Fe	6.4×10^{-6}	V	4.0×10^{-6}
Mn	1.3×10^{-5}	Zn	1.0×10^{-5}

Table II: Effective distribution coefficient k_{eff} of metals derived by Hopkins et al. [3]

No remarkable difference is evident for segregation coefficients below this value. It can be estimated that the maximum accumulation of impurities in the melt with small segregation coefficients is less than a factor of 3. This holds also for the axial impurity distribution in the grown crystals, which can be estimated using equation (1) and (2).

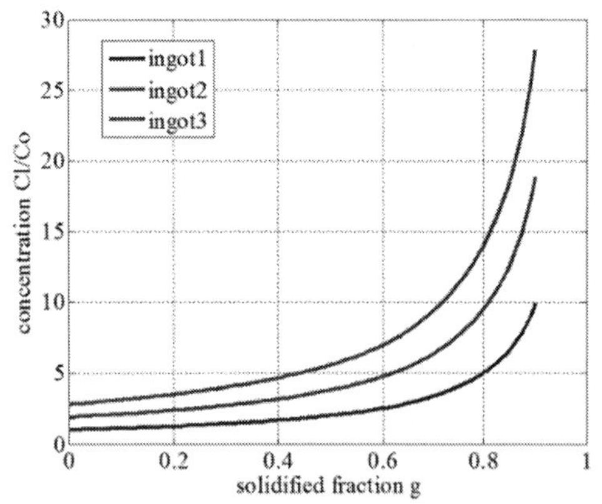

Figure 6: Impurity build-up curves for a replenished melt for the growth of 3 ingots in one crucible calculated according to Hopkins et al. [3].

Several ingots have been grown with the application of the SiCharger in the Crystal Growth department of the PVA CGS GmbH. The concentrations of the O_i and C_S have been measured by FTIR and compared to crystals grown under standard Cz-conditions. The results are shown in table: III. It can be seen that the process conditions of the MRS-system have no remarkable impact on the crystal quality.

Cz/MPCz	$O_i \times 10^{17}$ atoms/cm^3		$C_s \times 10^{16}$ atoms/cm^3	
	top	tail	top	tail
Cz 1	9.55	5.86	0.38	2.01
Cz 2	10.05	5.97	2.1	2.93
MPCz A1	9.97	6.43	0.01	1.62
MPCz A2	10.29	8.02	0.13	0.88
MPCz B1	10.42	5.19	0.24	2.05
MPCz B2	9.93	6.39	0.53	1.59
MPCz C1	10.97	5.4	0.18	2.41
MPCZ C2	9.81	6.08	0.36	1.73
MPCz D1	9.02	4.85	0.16	3.7
MPCz D2	8.42	5.71	2.26	3.92

Table III: O_i and C_S-concentration measured by FTIR at top and tail of crystals grown at PVA crystal growth department. Cz is standard grown ingot, MPCz X1 and MPCz X2 is 1st and 2nd crystal of growth run A,B,C,D grown by multipulling Cz-method.

3 ECONOMIC ANALYSIS OF THE MULTIPULLING Cz-PROCESS

An assessment of the economic potential cannot be presented quantitatively, because many factors depend on the technological condition of the process development as well as on the training level of the operational employees. Furthermore production costs may have different values depending on the location of the production plant and are subject to substantial fluctuations in the market.

The main economic potentials are the cost-saving of crucible consumption and the reduction of the down-time between the standard batch processes. The crucible costs are indirectly proportional to the number of refill processes. Thus 50% of crucible costs can be reduced even with one refill process. The non-productive down-time between the growths of 2 crystals can be reduced effectively. This shown in fig. 7, where the heater power of a standard one batch process is sketched together with a refill process. The curves were derived from real crystal growth experiments performed at the crystal growth department of PVA CGS GmbH. The results are summarized in table IV. The standard charge weight is 140 kg for an 8 inch ingot with a body length of 1600 mm. The blue curve represents a complete standard process followed by the cool down and heat-up sequence for the second ingot ending at the dipping point. The red curve shows the course of the heater power for the crystal growth and the subsequent refill process of the hot crucible, also ending at the dipping point. The red curve has a time offset to the blue one due to an additional feeding procedure after the initial melting for a better fill level of the crucible. It is evident, that in the case of melt replenishment the down-time can be reduced in the range of 50-60%. The energy consumption for the refill process is in the same range as the standard Cz-process, as shown in figure 7.

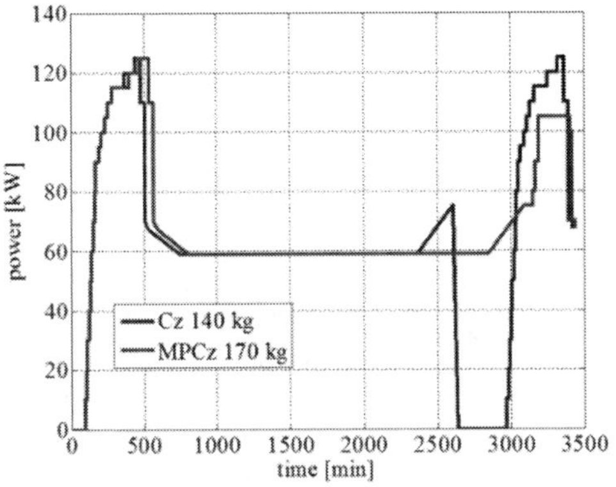

Figure 7: Power consumption of the heater versus process time for a standard Cz- (blue curve) and a recharge MPCz-process (red curve).

of the puller. This fact on the other hand takes care on the hotzone, especially the graphite felt, which degrades very rapidly, when the puller is opened in a hot state under atmosphere. A reduced temperature cycling may also be beneficial for the lifetime of the other rigid parts of the hotzone, due to reduced thermal stresses.
An uncomplex comparison of the throughput of a standard one batch Cz-process with a multiple recharge process and a simple cost analysis shows the economic potential of the MRS system.

process stage	time [h] Cz	time [h] MPCz
pump down (incl. leak test)	1.5	1.5
melting to dip temperature	7.5	8.5
stabilization, neck, crown	3.5	3.5
body	27	33
endcone	4	4
time to next dip temperature	13.5	6

Table IV: comparison of a standard Cz-process with a MPCz-process.

50 weeks of production per year with an output of 150 ingots per puller should result in an annual output of 100%. Assuming that 3 ingots can be grown out of one crucible, it is possible to optimize the fillevel of the crucible of about 20-30 % per ingot resulting in an equal increase of the throughput (20-30 ingots/year). The longer duration of the body growth due to the topping up of the crucible is compensated by reducing the downtime (see fig.7). An increase of productivity of 25 % (25 ingots/year) and an economization of 67 % of the crucible cost should be possible using the MRS system. One MRS-system should serve for 5-7 pullers.

4 CONCLUSIONS

High quality crystals can be grown by the application of the MPCz-technique. With this MRS-system capital costs can be minimized. It is shown that the additional investment for the recharge system is quickly amortized by the economization of crucibles and other operation costs.

5 ACKNOWLEDGEMENTS

This work was supported by the German Ministry of Economy and Energy under contract number 0325883A.

6 REFERENCES

[1] W.Zulehner, D. Huber, Crystals 8: Silicon Chemical Etching, J. Grabmeier ed., Springer Verlag, Berlin (1982)
[2] H.Kato et al.: U.S. Pat. Pub. No.: US 2012/0160156 A1, Pub. Date: Jun.28,2012
[3] R.H. Hopkins et al., Journal of Crystal Growth **42** (1977), 493-498
[4] W.G. Pfann, Zone melting, John Wiley, New York (1966)
[5] J.A. Burton, R.C. Prim, W.P. Slichter, J.Chem. Phys. **21** (1953) 1987-1991
[6] G. Coletti, Impurities in silicon and their impact on solar cell performance, thesis
[7] K. Harada et al., Journal of Crystal Growth **154** (1995), 47-53

An Approach to High Efficient CdTe Solar Cells with Wide Spectral Response

Lili Wu, Lianghuan Feng, Jinquan Zhang, Wenwu Wang, Wei Li, Hang Xu, Cai Liu, Bing Li, Guanggen Zeng
College of Materials Science & Engineering, Sichuan University
South Section 1, Yihuan Road, 610065 Chengdu, China
Tel: +86 (28) 8541 2542, Fax: +86 (28) 8541 0130, Email: wulily77@163.com

ABSTRACT: How to improve short circuit current of CdTe solar cells while maintaining high open circuit voltage and fill factor is a key point to achieve high efficiency. In this paper, we used homely designed vapor transport deposition system to prepare CdTe films and proposed a solar cell structure which can lead to 16.28% efficiency solar cell with wide spectral response. The property of CdTe films were optimized by VTD parameters in details. Appropriate high substrate temperature is very important to prepare CdTe films with larger grain size, higher hole density, lower resistivity and better crystallinity. The solar cell deposited at 610℃ shows the best conversion efficiency. CdS thickness has obvious influence on the spectral response of CdTe solar cells in short wavelength region. The quadra like spectral response was obtained by using 30nm CdS window layer, while the device efficiency is less than 10%. After a high resistive buffer layer is inserted between TCO and CdS, the solar efficiency apparently increased up to 16.28%. This CdTe solar cell has a better spectral response than that from GE and NREL and nearly the same response with that from First Solar in short wavelength region. This suggests that CdTe solar cell with the structure of very thin CdS and buffer layer can reach a high efficiency with wide spectral response. A higher efficiency for CdTe solar cell can be expected by using this structure.
Keywords: cadmium telluride solar cell, thin film, spectral response.

1 INTRODUCTION

Although the efficiency of CdTe Solar cells has been improved to 21.5% [1], it is still unknown how to achieve this high efficiency for public researchers. The most remarkable improvement in the performance of CdTe Solar cells reported by First Solar lies in the obvious increase of short circuit current [2]. From the reported literature [3-4], we know that high current density originates from a very wide and quadra like spectral response. But how can we do that? This work tries to propose a possible way to obtain high efficient CdTe solar cell with wide spectral response.

Vapor transport deposition is an advanced deposition technology for CdTe thin films with high speed, low cost, high quality advantages. However, the details of this technology are seldom reported[5-7]. We designed a vapor transport deposition system by ourselves and further studied the effects of parameters such as carrier gas, pressure and substrate temperature etc. on film property and device performance.

Based on the optimized CdTe films by VTD technique, a new CdTe structure with different buffer layer between TCO and CdS was constructed. The best CdTe solar cell showed a quadra like spectral response and the efficiency reached 16.28% for small area CdTe solar cells. These results demonstrated that we possibly have found a right way to obtain higher CdTe efficiency toward 20%. We hope that the hint from the result may inspire other scientific researchers to improve CdTe solar cell efficiency quickly.

2 EXPERIMENTAL

2.1 Fabrication of CdTe solar cells

CdTe solar cells were fabricated as following steps. First, a high resistive buffer layer was deposited on Pilkington tec 10 substrates. Then 30~100nm CdS film was prepared by RF sputtering and 4 μm CdTe film was grown at high substrate temperature by vapor transport deposition(VTD)[8]. Back contact (if exist) and gold electrode were deposited by vacuum evaporation. For the best CdTe solar cell in this paper, 100nm MgF_2 as an anti-reflection layer was coated on the glass.

2.2 Characterization of CdTe films and solar cells

SEM images were obtained by field emission scanning electron microscope (Hitachi S-4300). Photoluminescence spectra were measured using Edinburgh Instruments FLS 980 fluorescence spectrometer. The external quantum efficiency (EQE) was measured by QEX10 solar cell quantum efficiency measurement system (PV Measurements, Inc.), and was calibrated with a Si diode sourced to NIST before measurement. The light current−voltage measurement was performed at National PV Product Quality Supervision and Inspection Center in Chengdu.

3 RESULTS AND DISCUSSION

3.1 Optimization of VTD parameters

Substrate temperature in VTD has significant effect on film property and performance of CdTe solar cells. Fig. 1 shows the surface morphology of CdTe films deposited at different substrate temperatures. It is clearly seen that high temperature is helpful to grow CdTe films with larger grain size and better crystallinity. Table I gives electrical parameters of these CdTe films. The data shows that high substrate temperature will also enhance hole carrier density in CdTe film and decrease the resistivity of films. These are due to the decrease of activation energy for carrier density, which is helpful to form a strong built-in electric field at CdS/CdTe interface.

Both oxygen partial pressure in carrier gas and the working pressure are very important to deposit high quality CdTe films. The details will be stated in other paper.

Figure 1: Surface morphology of CdTe films deposited at different temperatures by VTD

Table I: Electrical parameters of CdTe films deposited at different substrate temperatures

Substrate temperature	Electrical parameters (300K)		Activation energy		
	ρ	p	E_a	E_{ap}	$E_{a\mu}$
	Ω·cm	cm^{-3}	eV	eV	eV
520℃	1.8×10^6	1.9×10^{10}	0.58	0.61	-
580℃	2.9×10^6	1.9×10^{12}	0.55	0.37	0.18
620℃	5.8×10^5	2.4×10^{13}	0.50	0.28	0.19

Fig.2 gives the surface photoluminescence spectra of CdTe films deposited at different temperatures. The previous research has shown that ~1.55eV luminescence in the free surface excitation spectrum is attributed to bound exciton recombination. For polycrystalline films, this peak becomes broad and asymmetric due to a large number of small grains and grain boundaries. The non-radiative recombination via boundaries decreases with grain growth and fewer boundaries, which follows that the peak intensity in this position will increase obviously. In Fig.2, this luminescence becomes more apparent in sample at higher temperatures, especially at 640℃. For all samples, ~1.42eV luminescence band is the broadest and strongest one and it varies in shape and intensity in samples. The peak intensity increases with substrate temperature while the peak width has not changed. It is noteworthy that the 640℃ sample shows a series of structures in both sides of the peak position. The spacing of the structure is about 21meV, which demonstrates the involvement of LO phonon because 21meV is the characteristic LO phonon energy in CdTe. The more obvious the structure is, the better crystalline quality CdTe has. It is to be noted that the sample with obvious structure shows high exciton emission, which implies that the non-radiative recombination declines with grain growth and the reduction of grain boundaries. These results are consistent with SEM images.

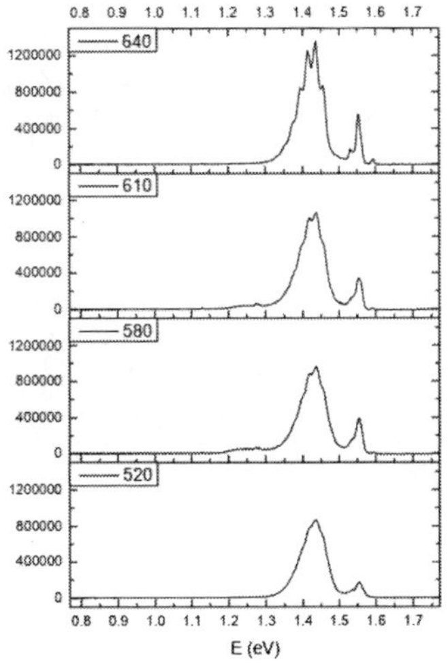

Figure 2:Surface photoluminescence of CdTe films at different substrate temperatures

The performance of CdTe solar cells without window buffer layer and back contact at different substrate temperatures are shown in Fig. 3. The results suggest that high temperature(600℃~610℃) has positive effect on device performance, but solar cells would degrade when the temperature is higher than 620℃. The underlying reason is considered that an ideal interdiffusion between CdS and CdTe will form at an appropriate temperature.

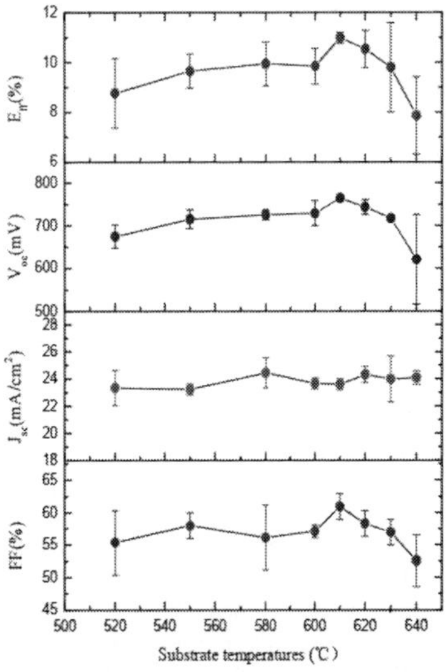

Figure 3: Light I-V parameters of CdTe solar cells at different substrate temperatures

3.2 CdTe Solar cell with optimized window layer

Based on the optimized VTD parameters, we decreased the thickness of CdS layer from 100nm to 30nm, the corresponding spectral response are shown in figure 4. The tendency is apparent that the spectral response in short wavelength increase with the decrease of CdS thicknesses. The current density calculated from spectral response spectra increases clearly with the reduced CdS thickness. When CdS is reduced to 30nm, a quadra like spectra response was obtained. However, the efficiency of CdTe solar cell with this spectrum is less than 10% because there is a weak diode between TCO and CdTe film when CdS is very thin.

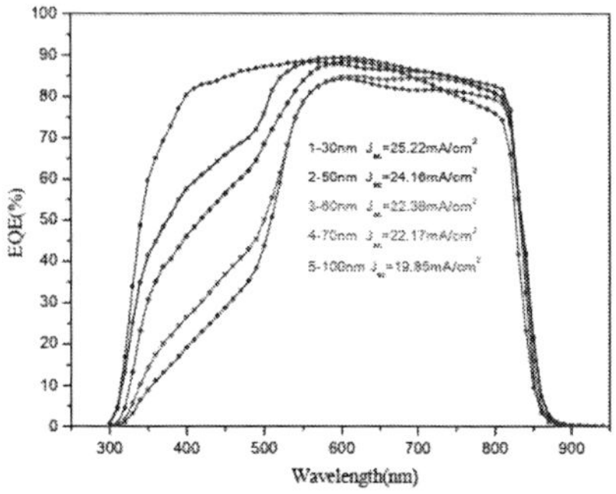

Figure 4: Spectral response of CdTe solar cells with different CdS thicknesses

In order to improve the efficiency of CdTe solar cell while maintaining the wide spectral response, a high resistive buffer layer was inserted between TCO and CdS. The device performance was greatly enhanced after both the resistivity and the thickness of buffer layer were optimized. The best solar cell has shown 16.28% efficiency and the light I-V curve is shown in Fig.5. Although the efficiency is not so attractive, it is worthy to note that the current density has reached 27.92mA/cm^2 with commercial Tec15 substrate.

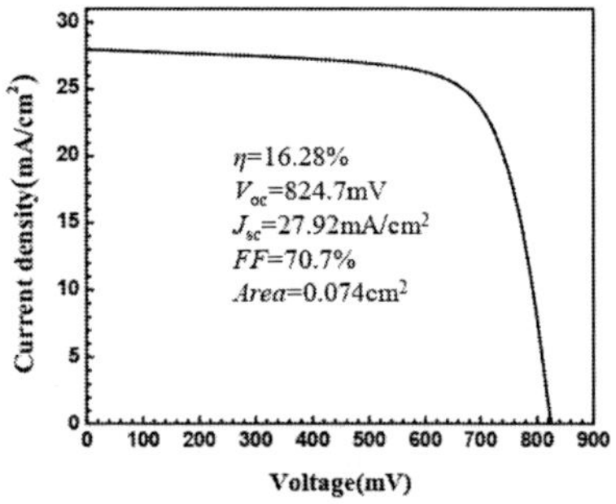

Figure 5: Light I-V curve of 16.28% CdTe solar cell

The corresponding spectral response is shown in Fig.6 compared with CdTe solar cells from First Solar, GE and NREL. From Fig.6, we know that our CdTe solar cell has a better short wavelength response than that from GE and NREL and nearly the same response with that from First Solar. This suggests that CdTe solar cell with the structure of very thin CdS and buffer layer can reach a high efficiency with wide spectral response. We believe that CdTe solar cells with higher efficiency will be fabricated in near future in our lab.

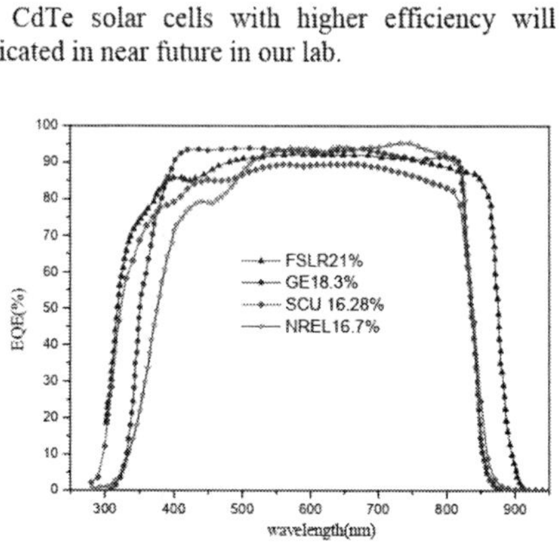

Figure 6: Comparison of spectral response of CdTe solar cells from First Solar, GE, SCU and NREL

4 CONCLUSIONS

CdTe solar cells with wide spectral response were fabricated with the combination of 30nm CdS and high resistive buffer layer between TCO and CdS. High quality CdTe films were prepared by Vapor Transport Deposition technology. Appropriate high substrate temperature is very important to prepare CdTe films with larger grain size, higher hole density, lower resistivity and better crystallinity. The best CdTe solar cell has achieved 16.28% conversion efficiency with a better spectral response than that from GE and NREL and nearly the same response with that from First Solar in short wavelength region. A higher efficiency for CdTe solar cell can be expected by using this structure since TCO substrate was tec 15 from Pilkington Company.

ACKNOWLEDGMENT

This work was supported by the National High Technology Research and Development Program of China (Grant no.2015AA050610).

REFERENCES

[1] First Solar press release. February 23, 2016, http://investor.firstsolar.com/releases.cfm.
[2] Martin A. Green, Keith Emery, Yoshihiro Hishikawa, Wilhelm Warta and Ewan D. Dunlop, Solar cell efficiency tables (version 47), Prog. Photovolt: Res. Appl. 23(2015)805.
[3] Bülent M. Başol and Brian McCandless, Journal of Photonics for Energy, 4(2014) 040996.
[4] Naba R. Paudel, Jonathan D. Poplawsky, Karren L. Moore and Yanfa Yan, IEEE Journal of Photovoltaics, 5(2015)1492.
[5] James M. Kestner, Sarah McElvain, Stephen Kelly et al., Solar Energy Materials & Solar Cells 83 (2004) 55.
[6] James M. Kestner, Thesis (Ph.D.), Colorado School of Mines, 2003.
[7] Keedong Yang*, Bokyung Cha, Duchang Heo, and Sungchae Jeon, Phys. Status Solidi C 11, No.7–8, (2014) 1341.
[8] Cao sheng, Wu Lili, Feng Lianghuan et al. Journal of Inorganic Materials, 31(2016)135.

THE IMPACT OF OXYGEN INLET DURING CLOSE-SPACED SUBLIMATION PROCESS ON THE AS-DEPOSITED AND CHLORINE TREATED MICROSTRUCTURE OF CDTE LAYERS

D. HIRSCH[1)], O. ZYWITZKI[1)], T. MODES[1)], H. MORGNER[1)], B. SPÄTH[2)], B. SIEPCHEN[2)], C. DROST[2)], C. KRAFT[2)], V. KRISHNAKUMAR[2)] and CHR. METZNER[1)]
[1)] Fraunhofer Institute for Organic Electronics, Electron Beam and Plasma Technology (FEP), Winterbergstraße 28, 01277 Dresden, Germany
[2)] CTF SOLAR GmbH, Zur Wetterwarte 50, 01109 Dresden, Germany

ABSTRACT: In this study CdTe thin films were deposited by close-spaced sublimation (CSS) in an oxygen ambient and the influence of oxygen on the properties of as-deposited and chlorine activated CdTe microstructures were investigated. FE-SEM imaging of ion-polished cross sections of the as-deposited CdTe-layers revealed a significant change from a block-like to a columnar microstructure with increasing oxygen concentration in the gas flux. After the chlorine activation step at 440 °C the oxygen free sample shows a strong sulfur diffusion whereas samples with low oxygen inlet show a suppressed diffusion. Due to the enlarged amount of grain boundaries in the as-deposited microstructure the sulfur diffusion became stronger again with increasing oxygen concentration.
Keywords: CdTe, close-spaced sublimation, sulfur diffusion, oxygen, GD-OES

1 INTRODUCTION

Cadmium telluride is a promising candidate for thin film solar cells to reach high efficiencies. Its direct Bandgap of about 1.49 eV [1] leads to a high degree of absorbed photons.

In this study we deposited CdTe and CdS layers by the close-spaced sublimation process (CSS) [2, 3, 4, 5] which is shown in Figure 1. The CSS technology is described in a patent of Gegenwart and Richter [3]. For the CSS process a flat crucible is used so the distance (h) between glass and sublimation material is kept small and constant during the deposition. This small distance leads to a high yield of material and a dense microstructure which are the main advantages of this technology. Another characteristic feature of this technology is the heated crucible walls to suppress deposition.

The CSS technology is applied in an industrial scale by a series connection of several crucibles. Thus, a CdTe thickness of 4 to 5 μm can be achieved. However, for research purposes the substrate is oscillating above one CdTe crucible.

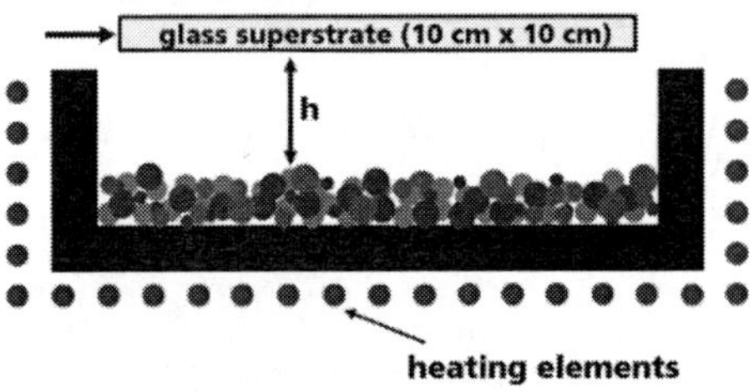

Figure 1: Schematic depiction of the close-spaced sublimation process.

It is already known that the presence of oxygen during the deposition of the absorber material increases the performance of a CdTe solar cell [5]. Oxygen causes a decrease in grain size [6, 7, 8]. Furthermore, there is no detectable amount of oxygen found in as-deposited CdTe layers with EDS analysis [9].

2 EXPERIMENTAL

The superstrate configuration used in this study attains the best efficiencies. As a substrate we used a Pilkington TEC C15D with the size of 10 cm times 10 cm. This includes a float glass coated with a FTO- and an intrinsic SnO_2 buffer layer. We coated this substrate with CdS by CSS. The CdS layer has a thickness of about 90 to 120 nm. The deposition of CdS remained constant for all samples. The used CSS pilot plant "CATE" is shown in Figure 2.

Figure 2: The CSS pilot plant "CATE" at the Fraunhofer FEP.

The next process step is the deposition of the CdTe which acts as the absorber layer. For all the samples in this study the CdTe crucible temperature of 685 °C and the substrate temperature of 510 °C have been kept constant. During the CSS deposition of CdTe an oxygen - argon gas mixture intake took place.

The oxygen proportion of the gas mixture is varied from 0 % to 20 % of the total gas flow. The net gas flow rate of the oxygen-argon mixture remained constant to avoid changing effects caused by pressure variation. We used a total gas flux rate of about 500 sccm which leads to a total pressure of 0.15 to 0.18 mbar. A MKS Baratron 626B gauge was used to determine the pressure independently from gas components. The reference cells were made after a complete cleaning of the chamber to exclude any influence of oxygen on the CdTe growth.

All those samples were activated with $CdCl_2$ in a methanolic solution with an activation temperature of 440 °C. The back contact consisting of a copper-containing primary contact and a secondary contact made of molybdenum is equally processed for all samples.

For structure investigations ion-polished cross-sections of the solar cells were prepared using an argon ion-beam preparation technique (Cross Section Polisher,

SM-0910, Jeol). The cross-sections were analyzed by field emission scanning electron microscopy (FE-SEM, SU 8000, Hitachi) using the channeling contrast for imaging the CdTe microstructure.

Chemical-depth profiles of the CdTe solar cells were determined by RF glow discharge optical emission spectrometry (GD-OES, GD Profiler 2, HORIBA Jobin Yvon). Additionally, the texture of the CdTe layers in dependence of oxygen content was analyzed by X-ray diffraction (XRD, D8 Advance, Bruker) in Bragg-Brentano geometry using Cu-Kα radiation and calculating the texture coefficient from six diffraction peaks.

The electrical characterization of the solar cells was done by measurement of the external quantum efficiencies (Oriel IQE-200, Newport).

3 RESULTS

Figure 3 shows FE-SEM cross-sections of all samples. A small decrease in thickness of the CdTe layers can be seen with increasing oxygen concentration. This is caused in a higher material loss due to chemical reaction of the oxygen. The most significant effect of oxygen is a drastic change of microstructure with increasing oxygen concentration. Without oxygen and with 1 % oxygen inlet the layers have a typical block-like microstructure with large lateral grain size of about 1 µm. In contrast layers with oxygen inlet between 5 % and 20 % exhibit a columnar microstructure with small lateral grain size of about 0.5 µm and a rougher surface.

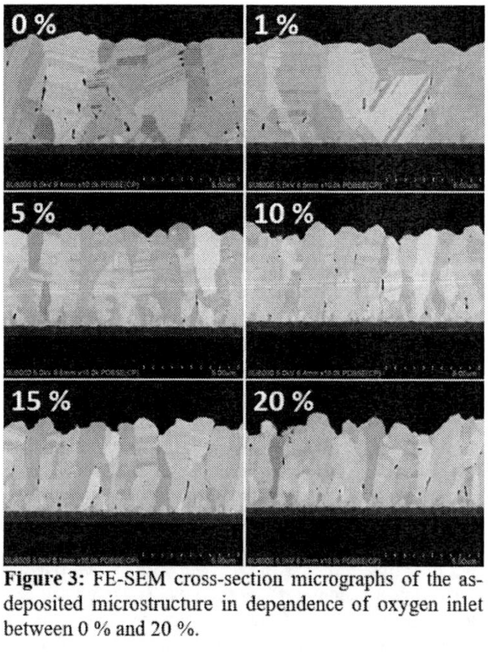

Figure 3: FE-SEM cross-section micrographs of the as-deposited microstructure in dependence of oxygen inlet between 0 % and 20 %.

Twin boundaries reveal the distribution of the crystal orientation. By imaging in channeling contrast the twin boundaries are identifiable as parallel lines within the grains. At 0 % and 1 % oxygen concentration a random orientation can be seen. However, at higher oxygen concentrations the twin boundaries are mainly arranged parallel to the substrate surface. This indicates that there is a preferred crystal orientation.

This result was confirmed by XRD measurements. The CdTe layers deposited with 0 % and 1 % oxygen show a statistically orientation distribution. On the other hand, the samples with higher oxygen content show a preferred (111)-crystal orientation. The texture coefficient of these samples with columnar microstructure amounts to about 5 independent of the oxygen content.

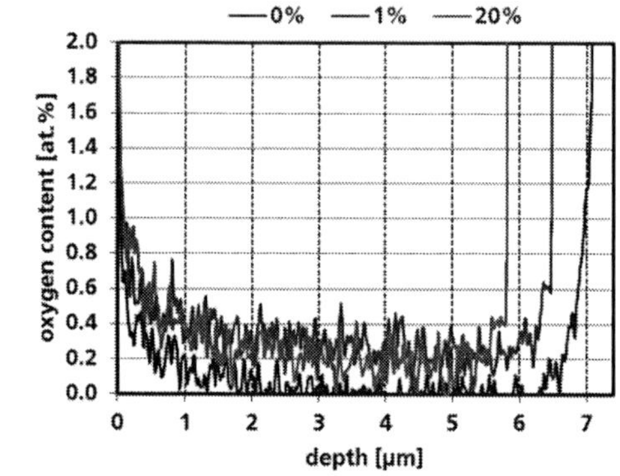

Figure 4: Depth profile of as-deposited CdTe layers grown in an oxygen ambient investigated with GD-OES.

The oxygen content in the CdTe layers was measured with glow discharge optical emission spectrometry (GD-OES). The GD-OES signal which is shown in Figure 4 detects an oxygen enrichment at the CdTe surface. The increasing oxygen signal at the substrate side is due to sputtering of the TCO layer stack.

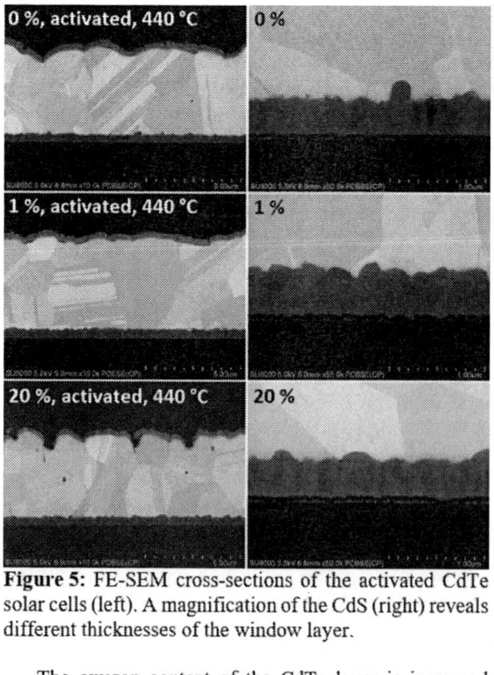

Figure 5: FE-SEM cross-sections of the activated CdTe solar cells (left). A magnification of the CdS (right) reveals different thicknesses of the window layer.

The oxygen content of the CdTe layer is increased from about 0.05 at. % without oxygen inlet to 0.2 at. % with oxygen inlet. But we cannot find any significant difference in oxygen content for the samples treated with different concentrations of oxygen during CdTe deposition.

Figure 5 (left column) shows the effect of the chlorine

treatment at 440 °C on the microstructure of oxygen treated samples. Activated samples exhibit a recrystallized microstructure with lower porosity. The 0 % and 1 % samples show a block-like microstructure with larger lateral grains in comparison to the as-deposited state shown in Figure 3. Samples with higher oxygen content show a more drastic change in microstructure: The columnar microstructure changes to a globulitic one by recrystallization. The lateral grain size increases to 1 to 2 µm and the porosity decreases. But the rougher surface of this sample remains preserved.

The right column of Figure 5 shows a magnification of the CdS layer. For the sample without oxygen the CdS layer is almost completely consumed and the CdTe is in direct contact with the FTO layer. When the oxygen content is increased to 1 % a complete different result can be seen. The CdS layer is still a nearly closed layer. With 20 % oxygen a strong agglomeration and dissolution of the CdS layer is visible.

CdS forms only some isolated grains. The most of CdS is dissolved within the CdTe layer. The CdTe is mainly in direct contact with the FTO layer. To investigate the sulfur diffusion in more detail chemical depth profiles are determined by GD-OES measurements which are shown in Figure 6.

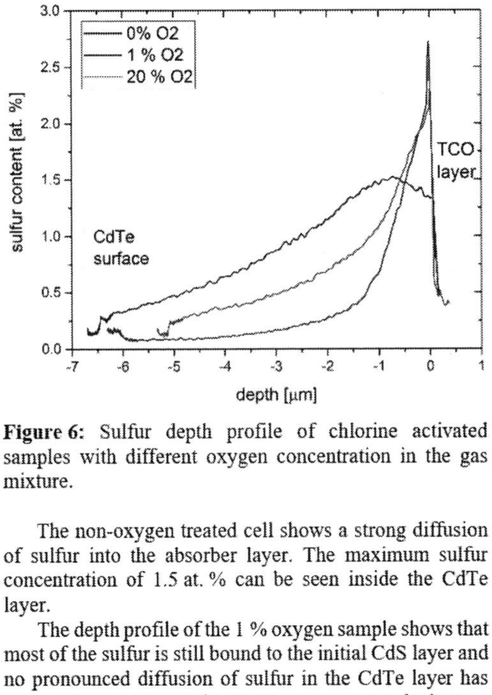

Figure 6: Sulfur depth profile of chlorine activated samples with different oxygen concentration in the gas mixture.

The non-oxygen treated cell shows a strong diffusion of sulfur into the absorber layer. The maximum sulfur concentration of 1.5 at. % can be seen inside the CdTe layer.

The depth profile of the 1 % oxygen sample shows that most of the sulfur is still bound to the initial CdS layer and no pronounced diffusion of sulfur in the CdTe layer has occurred. In contrast the 20 % oxygen sample have a diffusion of sulfur through the whole CdTe layer up to the surface.

The sulfur depth profile can be adjusted with the oxygen flow in combination with the activation temperature. The consequences of this phenomenon can be seen in the EQE curves in Figure 7. There are significant changes in the EQE trend. In the wavelength range of 320 to 500 nm a higher quantum efficiency is achieved by lower CdS layer thickness. While the 1 % oxygen sample shows a typical CdS absorption shoulder the EQE-curves of a non-oxygen treated sample and 20 % oxygen sample is almost rectangular. There is also a small shift in the absorption edge of CdTe by sulfur incorporation above 850 nm detectable

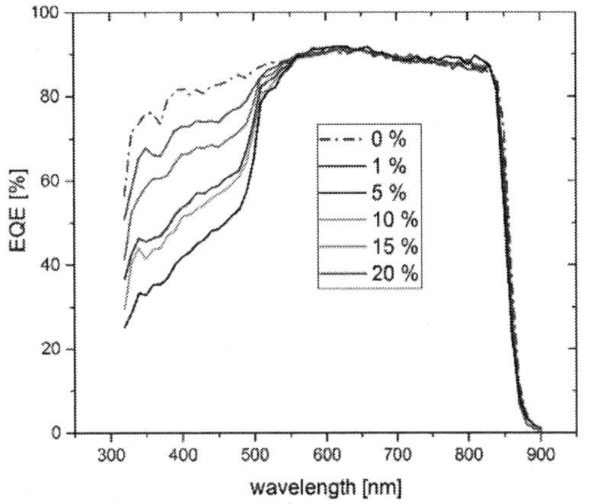

Figure 7: EQE-curve of activated CdTe solar cells with different oxygen concentration during the deposition of CdTe.

4 DISCUSSION

CdTe layers deposited without oxygen inlet have a block-like microstructure in the as-deposited state due to the used high substrate temperature of about 510 °C. We assumed that the mobility of these grain boundaries is decreased with increasing oxygen flux. Thus, oxygen is pinning the grain boundaries. The lower grain boundary mobility of CdTe grains leads to a formation of a columnar microstructure.

Because of relative high activation temperature of 440 °C a strong sulfur diffusion can be expected [10]. Therefore it is understandable, that the CdS layer of the non-oxygen treated sample is almost consumed. Incidental photons with a wavelength in the range of 320 to 500 nm can easily travel through the overlying layers without being absorbed in the CdS. This increased transparency causes the higher EQE signal. The range of 500 to 550 nm is characterized by the formation of a $CdS_{1-x}Te_x$ alloy [12] due to tellurium diffusion into the CdS which lowers the bandgap of the window layer. So the EQE signal in this range of wavelength increases with higher sulfur diffusion.

Furthermore, the strong sulfur diffusion into the CdTe layer of non-oxygen treated samples causes the formation of $CdTe_{1-x}S_x$. More sulfur incorporation in the absorber material is minimizing the bandgap of CdTe. Hence a small shift of the absorption edge can be seen.

However, with just a small amount of 1 % oxygen the sulfur diffusion can probably be suppressed effectively due to incorporation of oxygen at CdTe grain boundaries. The passivation effect of oxygen at grain boundaries in CdTe layers is already studied [11]. By suppressing of sulfur diffusion the CdS is preserved also at relative high activation temperature of 440 °C. Consequently, the absorption of photons in the range of 300 to 500 nm increases which leads to a decreasing EQE signal.

With increasing oxygen concentration the CdS consumption is getting stronger again. The columnar microstructure of the as-deposited state leads to an enlarged amount of grain boundaries. So we assumed that this effect enhances the sulfur diffusion at grain boundaries during the activation process. This effect can be seen in the FE-SEM cross-section by consuming the CdS layer. Furthermore, the GD-OES depth profile shows an increasing sulfur diffusion compared to the 1 % oxygen solar cell.

5 CONCLUSION

The CSS-technology is a common method to deposit CdS and CdTe on a flat substrate. By adding oxygen to the CdTe sublimation process the as-deposited microstructure changes from block-like to a more columnar type. We assumed that this effect is due to lower grain boundary mobility during the CdTe deposition. XRD measurements revealed that the oxygen inlet prefer the formation of a (111)-texture. A slight difference in oxygen concentration between non-oxygen treated and oxygen treated samples was found by GD-OES.

After a chlorine activation at 440 °C the grain size of samples processed with 20 % oxygen concentration in the gas flux increases due to recrystallization. They also change their microstructure from columnar to globulitic after activation. A stronger sulfur diffusion with 20 % oxygen flux is explained by larger amount of grain boundary diffusion.

So by using oxygen during the CSS process a new opportunity to control the sulfur depth profile arises. Sulfur diffusion has a strong impact on the device performance remarkably on the short circuit current density.

6 REFERENCES

[1] Mathew, X. and Pantoja Enriquez, J. 2000. Temperature dependence of the optical transitions in CdTe thin film – Investigation with photoresponse spectra. Solar Energy Materials and Solar Cells 63, 4, 347–354.

[2] Bonnet, D. 2000. Manufacturing of CSS CdTe solar cells. Thin Solid Films 361-362, 547–552.

[3] Gegenwart, R. and Richter, H. 2002. Apparatus for depositing CdS and CdTe layers on substrates by means of a CSS process. Google Patents.

[4] Siepchen, B., Drost, C., Späth, B., Krishnakumar, V., Richter, H., Harr, M., Bossert, S., Grimm, M., Häfner, K., Modes, T., Zywitzki, O., and Morgner, H. 2013. Thin film CdTe solar cells by close spaced sublimation: Recent results from pilot line. Thin Solid Films 535, 224–228.

[5] Tyan, Y.-s. and Perez-Albuerne, E. A., Eds. 1982. EFFICIENT THIN-FILM CdS/CdTe SOLAR CELLS.

[6] Falcão, V. D., Pinheiro, W. A., Ferreira, C. L., and Cruz, Leila Rosa de Oliveira. 2006. Influence of deposition parameters on the properties of CdTe films deposited by close spaced sublimation. Mat. Res. 9, 1, 29–32.

[7] Flores Mendoza, M. A., Castanedo Pérez, R., Torres Delgado, G., Márquez Marín, J., Cruz Orea, A., and Zelaya Angel, O. 2011. Structural, morphological, optical and electrical properties of CdTe films deposited by CSS under an argon–oxygen mixture and vacuum. Solar Energy Materials and Solar Cells 95, 8, 2023–2027.

[8] Rose, D. H., Levi, D. H., Matson, R. J., Albin, D. S., Dhere, R. G., and Sheldon, P. The role of oxygen in CdS/CdTe solar cells deposited by close-spaced sublimation. In Conference Record of the Twenty Fifth IEEE Photovoltaic Specialists Conference - 1996, 777–780. DOI=10.1109/PVSC.1996.564243.

[9] Hernández-Torres, M. E., Silva-González, R., Casarrubias-Segura, G., and Gracia-Jiménez, J. M. 2006. SEM, EDS, PL and absorbance study of CdTe thin films grown by CSS method. Solar Energy Materials and Solar Cells 90, 15, 2241–2247.

[10] Zywitzki, O., Modes, T., Morgner, H., Metzner, C., Siepchen, B., Späth, B., Drost, C., Krishnakumar, V., and Frauenstein, S. 2013. Effect of chlorine activation treatment on electron beam induced current signal distribution of cadmium telluride thin film solar cells. J. Appl. Phys. 114, 16, 163518.

[11] Yan, Y., Albin, D., and Al-Jassim, M. M. 2001. Do grain boundaries assist S diffusion in polycrystalline CdS/CdTe heterojunctions? Appl. Phys. Lett. 78, 2, 171.

[12] Ohata, K., Saraie, J., and Tanaka, T. 1973. Optical Energy Gap of the Mixed Crystal CdS x Te 1- x. Jpn. J. Appl. Phys. 12, 10, 1641–1642.

7 ACKNOWLEDGEMENT

The project is founded by "Europäischer Fonds für regionale Entwicklung" and the free state of Saxony (funding reference: 100206775/3005).

Europäische Union

Europa fördert Sachsen.
EFRE
Europäischer Fonds für regionale Entwicklung

SODIUM INDUCED MICROSTRUCTURAL CHANGES IN MOCVD-GROWN CDTE THIN FILMS

Arezoo Amikhalili[1], Vincent Barrioz[1], Stuart J. C. Irvine[2], Neil S. Beattie[1] and Guillaume Zoppi[1*]
[1]Department of Physics & Electrical Engineering, Ellison Building, Northumbria University, Newcastle upon Tyne, NE1 8ST, UK
[2]Centre for Solar Energy Research, Swansea University, St Asaph, LL17 0JD, UK
arezoo.amirkhalili@northumbria.ac.uk, vincent.barrioz@northumbria.ac.uk, s.j.c.irvine@swansea.ac.uk
neil.beattie@northumbria.ac.uk, guillaume.zoppi@northumbria.ac.uk
*corresponding author

ABSTRACT: The properties of CdTe thin films deposited by metal-organic vapour deposition (MOCVD) on Mo-coated glass substrates are studied with particular interest in the effect of Na. It is shown that for radio frequency sputtered Mo layers on soda-lime glass substrates the Na content in the metal layer is regulated by the deposition pressure of the Mo layer itself due to increased porosity of the metal layer. Na diffusion from the glass into the CdTe layer results in grain enlargements to such extend that film continuity is broken following $CdCl_2$ treatment for both low and high Na content. When Na is added post-Mo deposition on quartz substrates, the growth of large faceted grains is observed in the as-deposited stage, increasing by an order of magnitude compared with Na-free layers. However, following the $CdCl_2$ activation step the average grain size is reduced but still observed to be larger than in the Na-free case. The CdTe crystal structure is analysed for the different sample combinations and initial device data suggest a small improvement in open circuit voltage when Na is present.
Keywords: Sodium, Grain Structure, Magnetron Sputtering, CdTe Thin Film, MOCVD

1 INTRODUCTION

The production of photovoltaic (PV) modules based on thin film technologies is currently dominated by those based on CdTe/CdS structure with manufacturing costs as low as $0.6 /Wp in the superstrate configuration [1] and efficiencies as high as 22.1 % [2] for small area cells. In order to further reduce production costs, the use of alternative substrates may be desirable [3] with the solar cell structure grown in the substrate configuration. Substrate configuration advantages include possible reduction in optical losses at the front surface and potential for better control of the junction formation following the absorber heat treatment. However the maximum efficiency reported to date in the substrate configuration is limited to 13.6 % [4] and further improvement are required.

Diffusion of impurities and dopants play a major role in PV devices and a prime example is sodium (Na) in $CuInGaSe_2$ (CIGS) solar cells. The properties of both the back contact *e.g.* molybdenum (Mo) and the substrate control this diffusion. In the case of Mo, growth parameters such as pressure and power have major effects not only on the film growth [5, 6] but also on the Na diffusion from the glass [7]. Those are also influenced by the type of deposition technique, *e.g.* direct current (DC) or radio frequency (RF) magnetron sputtering or even electron beam evaporation. DC films usually lead to larger grains compared with RF deposited Mo films but a reduced Na diffusion from the glass due to a reduction in grain boundaries density. On the other hand, RF films provide better adhesion to the substrate but with a slightly reduced planar conductivity [8].

The presence of Na in CdTe solar cells has received little interest to date but can result in increased grain size of the absorber layer, increased acceptor density and open circuit voltage (V_{oc}) as Na acts as a shallow acceptor in CdTe with low activation energy [9-12]. The morphological and structural properties of CdTe grown by metal-organic chemical vapour deposition (MOCVD) where Na is introduced using an evaporated sodium fluoride (NaF) thin layer are investigated. The MOCVD process is capable of depositing a variety of materials and it has been shown to be an effective technique for the deposition of polycrystalline CdTe PV cells [13]. The ability to fabricate the entire device structure using one method at atmospheric pressure and the potential to have an inline roll-to-roll process also provide further benefits of this technique [14].

2 EXPERIMENTAL

Mo thin films were deposited as back contact on soda-lime glass (SLG) and quartz substrates using either DC or RF magnetron sputtering at room temperature. The process parameters are detailed in Table I. The Mo films are referred to as RF-Mo and DC-Mo for those deposited by RF and DC magnetron sputtering, respectively.

Table I: Process parameters for the deposition of the Mo films and film thicknesses.

	DC	RF
Target size (cm^2)	330	181
Substrate	Quartz	SLG
Base pressure (Torr)	$7x10^{-6}$	$6x10^{-7}$
Power (W)	900	300
Ar pressure (mTorr)	3;7	1.5;4.5;7;13;20
Deposition time (min)	60	120
Mo thickness (nm)	1000	900;850;760;670;530
Resistivity (μΩcm)	23	31-91
CdTe thickness (nm)	2750	1750
NaF thickness (nm)	10	-

NaF (~ 10 nm thick) was deposited using thermal evaporation after the metal deposition on the DC-Mo sputtered films. Arsenic doped CdTe thin films were grown by MOCVD in a horizontal configuration quartz tube, using purified H_2 as the carrier gas and with a growth temperature of 390 °C. The metal-organic precursors used were dimethylcadmium (DMCd), diisopropyltelluride (DiPTe) and trisdimethylaminoarsine (tDMAAs) for Cd, Te, and As, respectively. The cadmium chloride ($CdCl_2$) activation treatment was carried out in-situ after the growth of the CdTe film by depositing 1 μm thick $CdCl_2$ layer annealed under H_2 at 420 °C for 10 min. Further details can be obtained in [13].

All sets of samples were characterised by scanning electron microscopy (SEM), X-ray diffraction (XRD) and secondary ion mass spectroscopy (SIMS).

3 RESULTS

3.1 Na diffusion through SLG

CdTe is normally deposited on borosilicate glass where a barrier layer is included to minimize impurity diffusion from the glass substrate. Significant amounts of Na are present in SLG (typically 12-14 wt%) and Na is known to diffuse during thermal processes above 300 °C. The presence of Na promotes uncontrolled grain enlargement in CdTe, deposited using vacuum evaporation [12]. For MOCVD films, this is also the case as shown in Fig. 1. The as-deposited films showed grain size in the range 100 – 200 nm while the $CdCl_2$-treated films deposited on SLG and quartz had a mean grain size of 1.8 μm and 0.9 μm, respectively. In extreme cases (Fig. 1 d), Na causes widening of grain boundaries, creating shunt paths in the solar cell reducing the overall efficiency. It is therefore advantageous to adequately control grain growth while maintaining film continuity. Fig. 1 already demonstrates that the combination of Na and Cl can effectively double the average grain size of MOCVD-grown CdTe layers at relatively low process temperature. CdTe films deposited on RF-Mo/quartz had identical grain sizes (not shown here) to the films deposited on bare quartz indicating that the surface state of the substrate plays no part in grain growth.

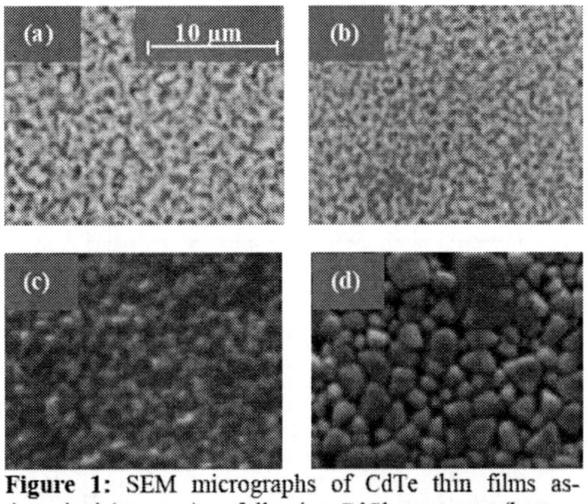

Figure 1: SEM micrographs of CdTe thin films as-deposited (top row) or following $CdCl_2$ treatment (bottom row) deposited on quartz substrate (left column) or SLG (right column) without Mo layer.

Qualitative SIMS measurements were performed in order to assess the in-depth Na diffusion through the RF-Mo layers before and after annealing at 530 °C for 30 min as a function of deposition pressure (Fig. 2). In general, as-deposited films showed lower Na levels at higher sputtering pressure with a graded depth profile (Fig. 2 dashed lines). After annealing, significant Na diffusion occurs throughout the depth of the Mo layer and this concentration increased with sputtering pressure as demonstrated on Fig. 3. This compares well with the work of Scofield [7] albeit for DC sputtered Mo films.

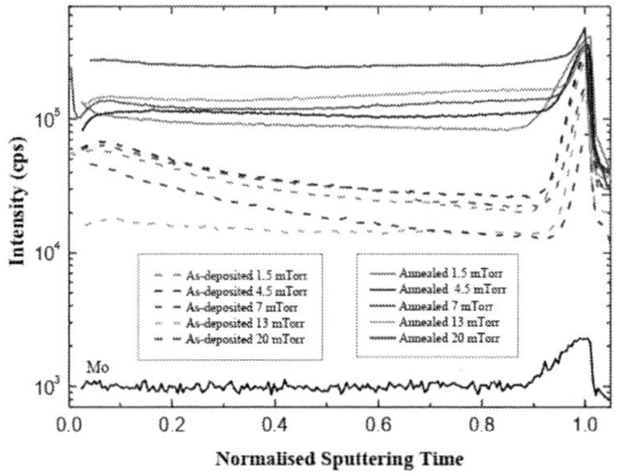

Figure 2: SIMS depth profiles showing the Na content in the as-deposited (dashed lines) and Ar annealed (530 °C for 30 min) (solid lines) RF-Mo thin films on SLG at varying sputtering pressures of 1.5, 4.5, 7, 13, and 20 mTorr.

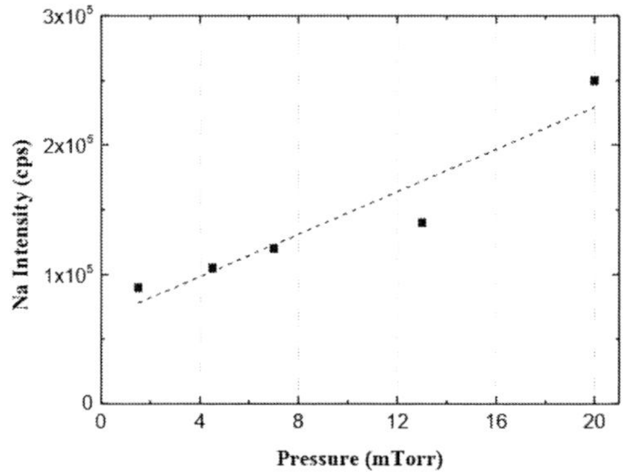

Figure 3: SIMS Na intensity as a function of sputtering pressure for RF-Mo films annealed in Ar at 530 °C for 30 min.

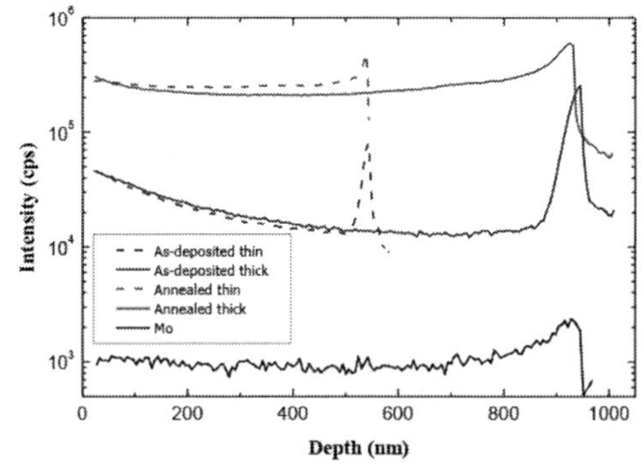

Figure 4: SIMS depth profiles showing the Na content in RF-Mo films deposited at 20 mTorr on SLG with thicknesses of 530 and 910 nm.

The RF-Mo presented large variations in thickness due to the change in sputtering pressure for a fixed deposition time (see Table I). However, the thickness had no impact on the diffusion rate of Na from the SLG as shown in Fig. 4 where films of thicknesses 530 and 910 nm deposited at 20 mTorr are compared before and after annealing, respectively.

The effect of varying the RF-Mo sputtering pressure on the microstructure of CdTe films post-$CdCl_2$ treatment is shown in Fig. 5. There are substantial morphological changes when the sputtering pressure increases. The mean grain size was measured to be 1.4, 1.2, 1.3, 1.8, and 2.0 μm with respect to increasing pressure. At the lowest pressure the film presents a mixture of very large and small grains with wide gap between the grains. As pressure increases the grain size distribution becomes more homogeneous (narrower) and the films are more compact. From 4.5 mTorr upwards, the average grain size increases with sputtering pressure however at the highest pressure the film again presents a widening of the grain boundaries. It is confirmed that the presence of Na in $CdCl_2$-treated MOCVD-grown CdTe can promote the formation of very large grains; however, this must be balanced with maintaining film continuity and therefore a carefully controlled Na dose is required.

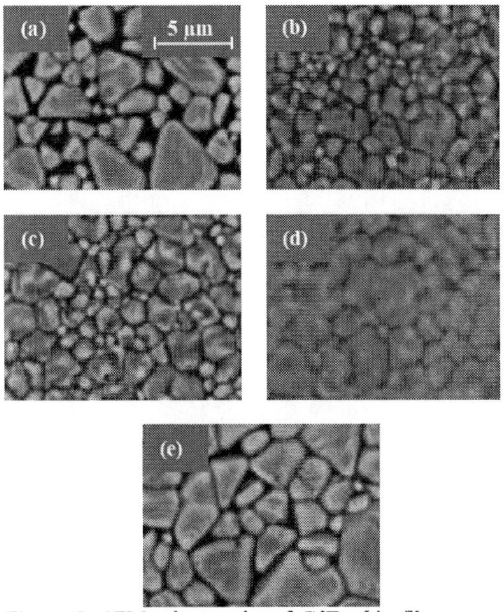

Figure 5: SEM micrographs of CdTe thin films post-$CdCl_2$ treatment on SLG at varying RF-Mo sputtering pressures of: (a) 1.5 mTorr, (b) 4.5 mTorr, (c) 7 mTorr, (d) 13 mTorr and (e) 20 mTorr.

3.2 CdTe morphology with intentionally added Na

In the previous section, it was shown that diffusion of Na through the SLG promotes grain growth but can be difficult to control. In order to further control the effects of Na on the CdTe grain size and shape, a fixed amount of Na was added to structures deposited on quartz substrates. The NaF layer was deposited after the DC-Mo back contact followed by CdTe and $CdCl_2$ treatment. The surface morphologies of the as-deposited and heat treated films are shown in Fig. 6 a and b. The micrographs were analysed to determine the grain size distribution and were fitted to a Rayleigh function to obtain average values and standard deviations. In the as-deposited state and for films processed without Na, the grains show very small (< 0.2 μm) pyramid shaped structures (Fig. 6 a). Following the $CdCl_2$ treatment (Fig. 6 c), the grains become more rounded with less sharp faceted edges and a grain growth (~ 0.5 μm) is observed as is usually observed in MOCVD-grown CdTe thin films. Fig. 6 c shows the effect of NaF on the as-deposited CdTe films. Comparing samples with and without Na, there is a clear morphological change: large faceted CdTe grains are formed in the presence of Na (Fig. 6 b) in the as-deposited state with an average grain size of 1.9 μm. The $CdCl_2$ annealing step yields more rounded grains and smoother films (Fig. 6 d) with an average grain size of ~ 1 μm showing a reduction in grain size following heat treatment. On the other hand, the $CdCl_2$ treated grains have become uniformly distributed compared to the as-deposited layers where the grain size distribution is broader (Fig. 7).

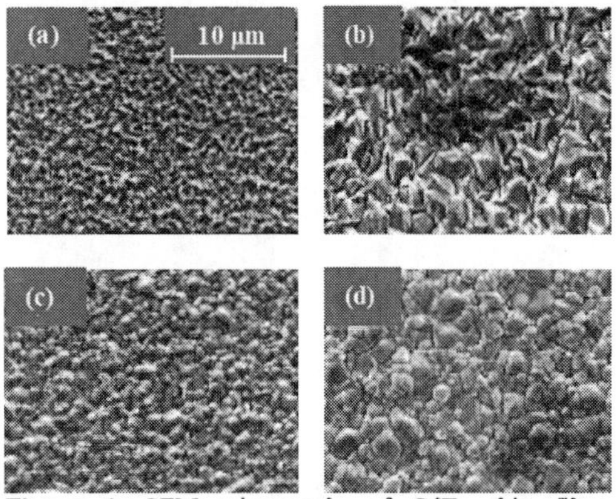

Figure 6: SEM micrographs of CdTe thin films deposited on DC-Mo on quartz substrate: as-deposited (a-b), $CdCl_2$-treated (c-d) and Na-free (a-c) or with a NaF layer evaporated between the DC-Mo and CdTe (b-d).

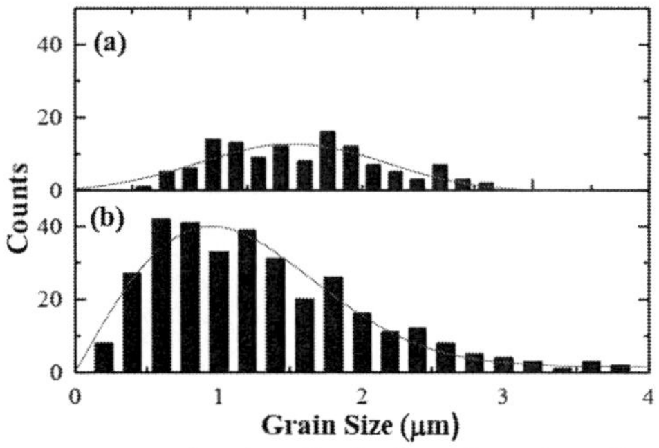

Figure 7: Grain size distributions for CdTe films deposited on NaF/DC-Mo/quartz. (a) as-deposited and (b) $CdCl_2$-treated.

X-ray diffraction was used to analyse the crystallographic properties of the deposited layers for CdTe films deposited on DC-Mo. Peaks corresponding to (111), (220), (311), (400), (331), (422) and (511) planes were observed for all samples (Fig. 8) along with the Mo substrate peaks. The effects of growth conditions on the texture and preferred orientation of the samples were investigated by calculating the texture coefficient C_{hkl}. The preferred orientation of each layer, as a whole, was analysed from the standard deviation σ of all C_{hkl} values as compared with the reference data. The calculated σ values post-$CdCl_2$ treatment for the Na-free and SLG/RF-Mo samples ranged between 1.6 – 1.7 and 1.5 – 1.7 respectively, indicating less randomly oriented samples compared to the samples with intentionally added Na that had a σ value ranging between 0.6 – 1.2. This is reflected in an increased texturing along the (111) direction (Fig. 8). Fig. 8 b i also indicates that the intensity of the C_{111} peak increases post-$CdCl_2$ treatment for samples prepared without Na; however, the opposite was observed when Na is present (Fig. 8 b ii and c). It is also noteworthy that Na-free and SLG/RF-Mo samples that were treated with $CdCl_2$ resulted in the disappearance of the (331) peak (Fig. 8 b i and iii).

Fig. 9 b and c shows the texture coefficient of (400) and (220) peaks respectively, where it is clear that as the intensity of the C_{111} increases (Fig. 9 a), the intensity of the C_{400} and C_{220} decreases in accordance with the variation of preferred orientation. The Na-free sample shows an increase for C_{111} post-$CdCl_2$ treatment.

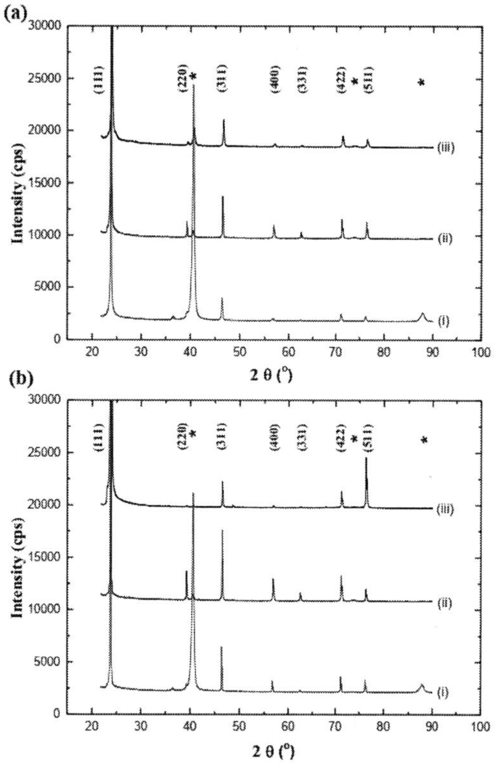

Figure 8: XRD pattern of CdTe thin films (a) as-deposited and (b) post $CdCl_2$ treatment with the following configurations: (i) SLG/RF-Mo, (ii) quartz/DC-Mo with Na (iii) quartz/Mo Na-free. The DC-Mo films were deposited at a sputtering pressure of 7 mTorr. The peaks corresponding to Mo are marked by asterisks.

Some samples were further processed into PV devices. A cadmium sulfide (CdS) window layer was deposited using chemical bath deposition process onto the CdTe layer. Thereafter, a 60 nm thick intrinsic transparent zinc oxide (ZnO) and a 200 nm thick indium tin oxide (ITO) layers were deposited by magnetron sputtering. The devices were finished by evaporating a front contact grid consisting of 50 nm nickel and 1 μm aluminium, deposited by electron-beam evaporation through a shadow mask. Each substrate was then defined by mechanical scribing into 0.16 cm^2 devices. The devices had low efficiency (<3 %) arising from weak current-voltage parameters. Some devices showed clear roll over in forward bias and weak diode characteristics, indicating non-ohmic back contact and possible lack of interdiffusion at the junction arising from the low temperature deposition of the CdS layer and the absence of post-CdS heat or second $CdCl_2$ treatment. Despite this the results showed that the addition of Na improved the open circuit voltage from 300 mV to >400 mV; however further device optimization is required.

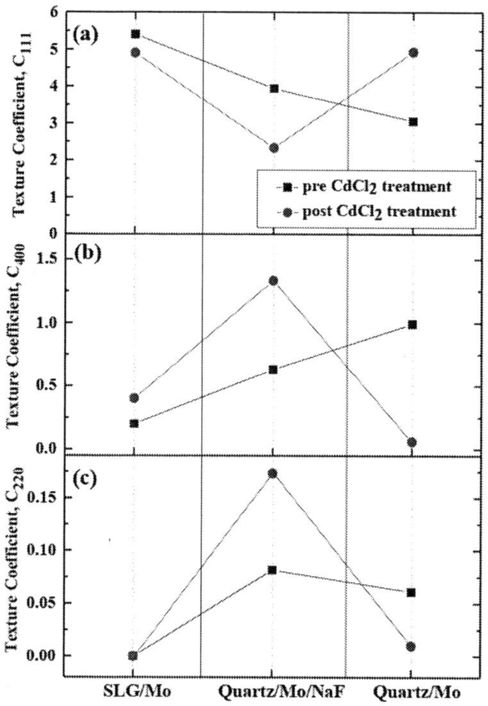

Figure 9: Texture coefficient values for (a) (111), (b) (400) and (c) (220) peaks plotted to show the variation of intensity with different deposition layers. The DC-Mo films were deposited at a sputtering pressure of 7 mTorr.

4 CONCLUSION

Preliminary results relating to the effect of Na on the CdTe microstructure have been investigated. RF-Mo film can control the amount of Na delivered to the absorber layer via diffusion from the glass however the effect on CdTe grains are less predictable. The microstructural investigation shows that the presence of Na after the back-contact layer increases the grain size of the as-grown films (DC-Mo), where the average grain size reach 1.9 μm compared to < 0.2 μm for as-deposited Na-free samples. Post-$CdCl_2$ treatment the samples showed smaller rounded grains (1.0 μm) but more uniformly distributed. Even though the devices showed low efficiency (<3 %) arising from weak current-voltage parameters, there was an improvement in the open circuit voltage. Combination of Na and Cl treatment provides a new avenue for the development of MOCVD CdTe thin films in substrate configuration and should be further exploited.

5 REFERENCES

[1] Empa-Technology-Transfer, "Production method for CdTe solar cells in substrate configuration," 2014.

[2] M. A. Green, K. Emery, Y. Hishikawa, W. Warta, and E. D. Dunlop, *Progress in Photovoltaics: Research and Applications,* vol. 24, pp. 905-913, 2016.

[3] A. Romeo, M. Terheggen, D. Abou-Ras, D. L. Bätzner, F. J. Haug, M. Kälin, *et al., Progress In Photovoltaics: Research and Applications* vol. 12, pp. 93-111, 2004.

[4] L. Kranz, C. Gretener, J. Perrenoud, R. Schmitt, F. Pianezzi, F. La Mattina, *et al., Nature Communications,* vol. 4, pp. 1-7, 2013.

[5] G. Zoppi, N. S. Beattie, J. D. Major, R. W. Miles, and I. Forbes, *Journal of Materials Science,* vol. 46, pp. 4913-4921, 2011.

[6] M. Jubault, L. Ribeaucourt, E. Chassaing, G. Renou, D. Lincot, and F. Donsanti, *Solar Energy Materials and Solar Cells,* vol. 95, pp. S26–S31, 2011.

[7] J. H. Scofield, S. Asher, D. Albin, J. Tuttle, M. Contreras, D. Niles, *et al., Conference Record of the Twenty-fourth. IEEE Photovoltaic Specialists Conference,* pp. 164-167, 1994.

[8] K. Aryal, H. Khatri, R.W. Collins, and S.Marsillac, *International Journal of Photoenergy,* vol. 2012, pp. 7, 2012.

[9] E. Molva, J. P. Chamonal, and J. L. Pautrat, *Physica Status Solidi B-Basic Research,* vol. 109, pp. 635-644, 1982.

[10] L. Kranz, J. Perrenoud, F. Pianezzi, C. Gretener, P. Rossbach, S. Buecheler, *et al., Solar Energy Materials and Solar Cells,* vol. 105, pp. 213-219, 2012.

[11] J. Sites and J. Pan, *Thin Solid Films,* vol. 515, pp. 6099-6102, 2007.

[12] R. Dhere, K. Ramanathan, J. Keane, Z. Jie, H. Moutinho, S. Asher, *et al., Conference Record of the Thirty-first. IEEE Photovoltaic Specialists Conference,* pp. 279-282, 2005.

[13] G. Kartopu, A. J. Clayton, W. S. M. Brooks, S. D. Hodgson, V. Barrioz, A. Maertens, *et al., Progress in Photovoltaics: Research and Applications,* vol. 22, pp. 18-23, 2014.

[14] V. Barrioz, G. Kartopu, S. J. C. Irvine, S. Monir, and X. Yang, *Journal of Crystal Growth,* vol. 354, pp. 81-85, 2012.

CHEMICAL BATH DEPOSITED ZINC OXIDE AS TRANSPARENT CONDUCTIVE CONTACT FOR CIGS SOLAR CELLS

Jérôme Steinhauser*, Peter Fuchs*, Dimitrios Hariskos**, Wiltraud Wischmann**, David Bremaud ***, Yaroslav E. Romanyuk*, Ayodhya N. Tiwari*
* Empa, Swiss Federal Laboratories for Materials Science and Technology
Überlandstrasse 129, CH-8600 Dübendorf, Switzerland
** ZSW, Zentrum für Sonnenenergie- und Wasserstoff-Forschung Baden-Württemberg
Industriestrasse 6, DE-70565 Stuttgart
*** Flisom AG
Überlandstrasse 129, CH-8600 Dübendorf, Switzerland Author(s)

ABSTRACT:
Chemical bath deposited (CBD) zinc oxide (ZnO) as a front contact in $Cu(In,Ga)Se_2$ (CIGS) based solar cells was recently demonstrated as a possible non-vacuum alternative to the standard transparent conductive oxide (TCO) deposited by dry processes. In this contribution, our latest CIGS solar cells results with CBD ZnO layers as TCO are presented. Using non-vacuum CBD ZnO TCO an efficiency of 18.3 % was achieved, approaching the efficiencies obtained with sputtered bilayer window consisting of undoped and aluminum doped zinc oxide (ZnO/ZnO:Al) TCO. Details on the implementation of CBD ZnO TCOs on cells grown on glass substrates with cadmium sulfide (CdS) and zinc sulfide oxide (Zn(O,S)) buffers as well as on cells deposited on flexible polyimide substrates are presented.
Keywords: TCO Transparent Conducting Oxides, ZnO, CIGS

1 INTRODUCTION

Transparent conductive oxides (TCO), a material class that combines optical transparency and electrical conductivity, serve as transparent contacts in solar cells [1]. Indium tin oxide (ITO) and zinc oxide (ZnO) deposited by sputtering as well as tin oxide (SnO_2) and ZnO deposited by chemical vapor deposition are commonly used materials for this application [1,2]. Expensive vacuum equipment is needed for these techniques, therefore low cost, non-vacuum processes are explored as alternatives. Dense, conductive ZnO layers can be deposited from aqueous solutions [3-6]. The use of chemical bath deposited (CBD) ZnO as a front contact in $Cu(In,Ga)Se_2$ (CIGS) solar cells was recently demonstrated as a possible low temperature alternative to standard TCO deposited by dry processes [6-9]. This work presents the implementation of CBD ZnO TCO in CIGS solar cells and a comparison of cells performances with cells having a bilayer window consisting of undoped and aluminum doped zinc oxide (ZnO/ZnO:Al) TCO. CBD ZnO was implemented in cells with cadmium sulfide (CdS) as well as zinc sulfide oxide (Zn(O,S)) buffers. Furthermore, results of CBD ZnO implemented on CIGS cells grown on flexible polyimide (PI) substrate are also presented.

2 EXPERIMENTAL

2.1 CBD of ZnO

The deposition procedures for CBD ZnO were based on previously published procedures. Experimental details regarding the deposition can be found in [8]. The substrates were first seeded by ~30 nm ZnO obtained by three successive dips in a ZnO (50 mM) ammonium hydroxide (5 M) solution followed by drying with a nitrogen air gun. The samples were then heated in air at 110°C for 10 minutes. Samples were then immersed in the CBD bath heated at around 90 °C. The CBD solution was composed of 50 mM of ZnO, 50 mM of NH_4NO_3, 0.97 M NH_4OH, and 3 mM ammonium citrate dissolved in deionized water. The samples were then exposed to 70 $mW.cm^{-2}$ UVA light from a mercury lamp for 12 minutes.

2.2 AZO RF magnetron sputtering

For cells with sputtered TCO, prior to the ZnO:Al TCO coating, the cell was coated with 80 nm of sputtered undoped zinc oxide in the case of cells with CdS buffer or with zinc magnesium oxide in the case of cells with Zn(O,S) buffers. RF magnetron sputtered AZO layers were deposited without intentional substrate heating in a vacuum chamber with a direct facing target. Ceramic $ZnO:Al_2O_3$ (98:2 wt%, purity 99.995%) targets with a diameter of 10.16 cm (4 in.) were sputtered in a mixed Ar/O_2 atmosphere with a total pressure of 0.15 Pa and a power density of 2.5 W cm^2.

2.4 Solar cells fabrications

CIGS absorber layer on molybdenum coated glass was deposited by an evaporation process combined with an alkali post-deposition treatment (PDT) [10, 11] or was inline deposited by a dynamic multistep process without an additional PDT [12]. CdS and Zn(O,S) buffer layer growth was performed by CBD [13]. CIGS absorber layers were deposited on flexible PI by elemental co-evaporation [14]. After TCO deposition, 50/4000/50 nm thick Ni/Al/Ni grids were deposited by e-beam evaporation. The cells were defined to a square of ~36 mm^2 by mechanical scribing. Selected cells were coated with evaporated MgF_2 antireflection coating (ARC). Figure 1 shows a picture of a 5x5 cm^2 sample with the cell geometry used in this study.

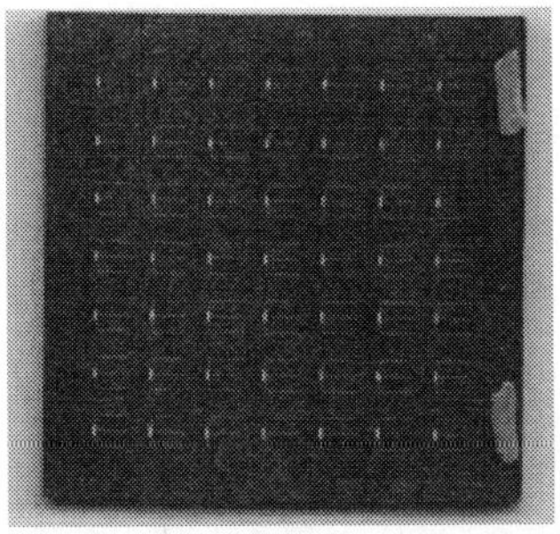

Figure 1: Picture of a 5x5 cm^2 sample with the cell geometry used in this study. The cells are about 36 mm^2

2.3 Characterization

The sheet resistance of the ZnO thin films was measured by a four-probe method with a Nagy meter SD-600. Hall measurements were conducted with an Ecopia HMS 3000 setup at room temperature in the dark using a van der Pauw contact geometry. Transmittance measurements were performed with a Shimadzu UV-3600 spectrophotometer with an integrating sphere. Transmittance values were measured with the glass substrate stack against air reference. External quantum efficiency (EQE) of solar cells was measured with a lock-in amplifier. A chopped white light source (900 W, halogen lamp, 360 Hz) and a dual grating monochromator generated the probing beam. The beam size was adjusted such that the illumination area was smaller than the device area. A certified monocrystalline silicon solar cell from Fraunhofer ISE was used as the reference cell. Cell temperature was controlled at 25°C with Peltier cooling and white light bias was applied. J–V characteristics of solar cells were measured under simulated standard-test conditions (25°C, 1,000 $W.m^{-2}$, AM 1.5 G illumination) in a sun-simulator with a HMI light source. A Keithley 2400 source meter with four-terminal sensing was used to acquire J–V characteristics.

3 CBD ZNO PROPERTIES

After optimization of the solution temperature, chemistry and deposition time, CBD ZnO layers with a sheet resistance of 20 Ω_{square} for a thickness of 2.7 µm leading to a resistivity of 5.2 10^{-3} Ω.cm with a mobility of 10 $cm^{-2}.V^{-1}.s^{-1}$ and a carrier density of 1.2 10^{20} cm^{-3} were obtained on glass substrates. The conductivity of these layers slowly degrades when exposed to humid atmosphere [8], the sample should therefore be encapsulated for long term usage. Figure 2 shows the transmittance and the properties of such CBD ZnO layer deposited on glass compared to a typical sputtered ZnO:Al TCO. Figure 3 shows the SEM cross section of a typical CBD ZnO layer. The compact growth could be achieved by controlling the quantity of ammonium citrate in the deposition solution [7]. Layers with the same deposition condition were implemented as front TCO in the different type of solar cells presented in the following sections.

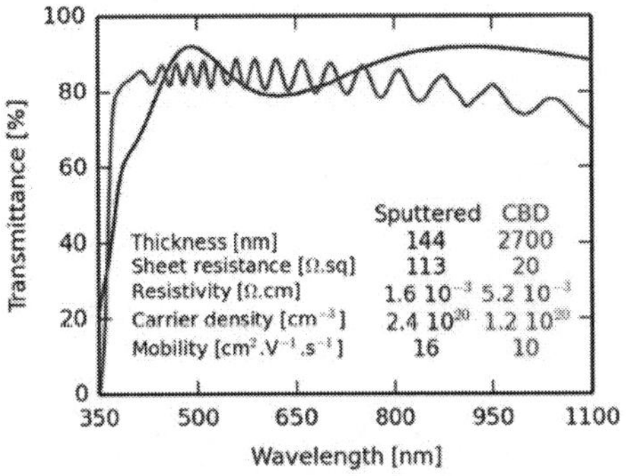

Figure 2: Comparison of the transmittance and electrical properties of CBD ZnO with sputtered ZnO:Al deposited on glass

Figure 3: SEM micrographs cross section of a CBD ZnO layer deposited on glass

4 SOLAR CELLS RESULTS

4.1 Cells with CdS buffer

CBD ZnO layers were implemented as front contact in CIGS solar cells with CdS buffer grown on glass substrates. Figure 4 shows SEM cross section of such a cell compared to a similar cell with a sputtered ZnO/ZnO:Al TCO.

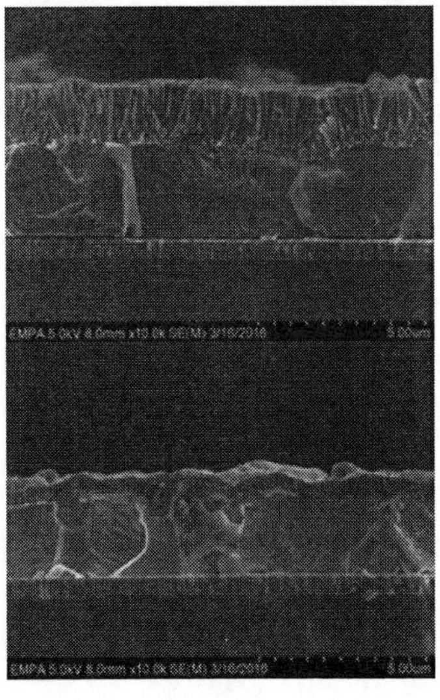

Figure 4: SEM of cross section of cells with CBD ZnO layer (up) and with sputtered ZnO/ZnO:Al (down)

Figure 5 shows the current-voltage characteristics of those cells and figure 6 their external quantum efficiency. With a CBD ZnO TCO, we achieved an efficiency of 16.6 % with a current density of 33.5 $mA.cm^{-2}$, an open circuit voltage of 674 mV and a fill factor of 73.8 % for a 0.36 cm^2 cell area. This efficiency is close to 17.5 %, the efficiency achieved with sputtered ZnO:Al TCO on the same batch of CIGS cells. The lower open circuit voltage and fill factor for cells with CBD ZnO TCO is probably due to a more difficult cell scribing process in the case of the thicker CBD ZnO which can introduce shunts.

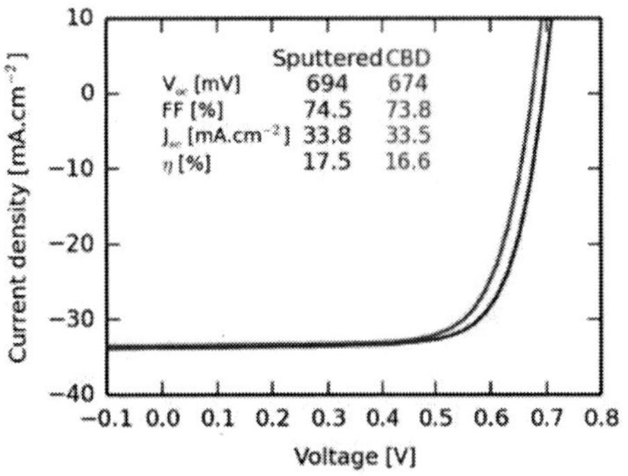

Figure 5: Comparison of J-V characteristics of the best CIGS/CdS solar cells with CBD ZnO and with sputtered ZnO/ZnO:Al as front TCO

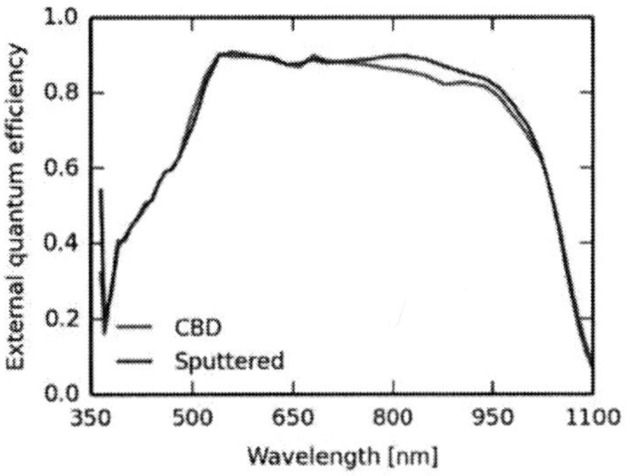

Figure 6: Comparison of the EQE of the best CIGS/CdS solar cells with CBD ZnO (red) and with sputtered ZnO/ZnO:Al (blue) as front TCO

4.2 Cells with CdS buffer, high quality absorber and AR coating

CBD ZnO layers were then implemented as front contact on high quality CIGS absorber with CdS buffer on glass substrate. MgF_2 antireflection coating was also added. Figure 3 shows the current-voltage characteristics of the best cells achieved: 18.3 % efficiency with a Voc of 703 mV, a FF of 70.4 % and a Jsc of 37 mA.

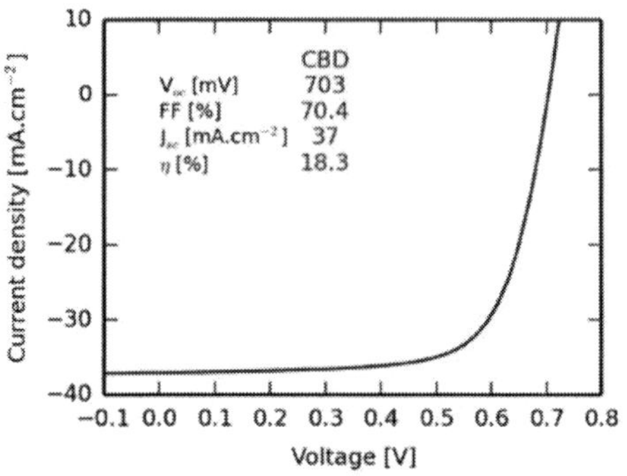

Figure 7: J-V characteristics of the best 36 mm^2 CIGS/CdS solar cell with CBD ZnO and AR coating

4.3 Cells with CdS buffer on flexible substrate

CBD ZnO layers were also implemented as front contact in CIGS solar cells with CdS buffer grown on flexible polyimide absorber. In this case an efficiency of 13.8 % was achieved with a Voc of 644 mV, a FF of 67.6 % and a Jsc of 31.8 mA as shown on figure 5. These values are comparable with results of solar cells obtained with similar absorber with sputtered ZnO TCO and show that CBD ZnO TCO are compatible with flexible polyimide substrates.

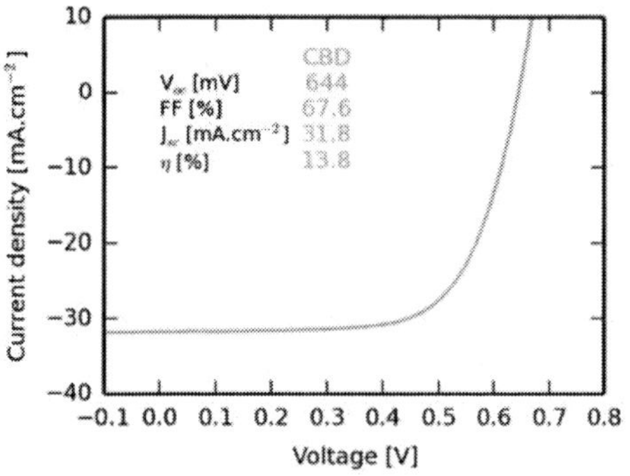

Figure 8 : J-V characteristics of the best 36 mm^2 CIGS/CdS solar cells with CBD ZnO on a flexible PI substrate

4.3 Cells with Zn(O,S) buffer

Finally we implement CBD ZnO as TCO in cells with Zn(O,S) buffer layer on glass substrates. The absorber is comparable with the one used in section 4.1 with CdS buffers. Figure 9 shows the current-voltage characteristics of the best cell with CBD TCO. We achieved an efficiency of 16 % with a current density of 35.2 $mA.cm^{-2}$, an open circuit voltage of 642 mV and a fill factor of 70.6 % for a 0.36 cm^2 cell area. With this result we realized a cell with cadmium-free fully solution processed contact (buffer and TCO).

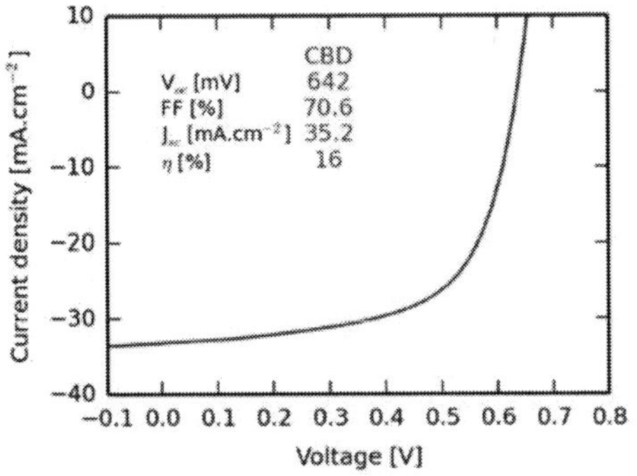

Figure 9 J-V characteristics of the best 36 mm^2 CIGS/Zn(O,S) solar cell with CBD ZnO

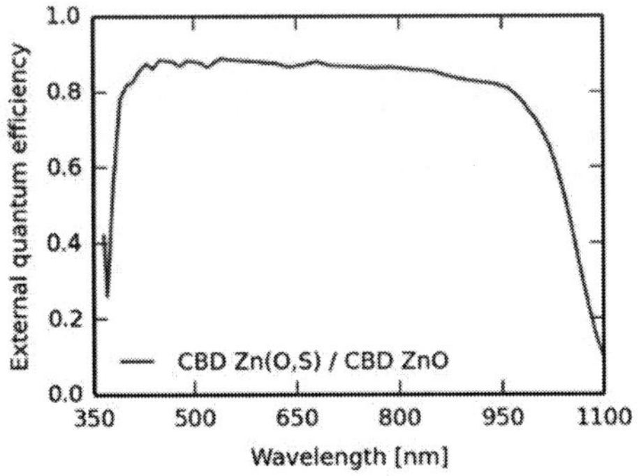

Figure 10: External quantum efficiency of the best CIGS/Zn(O,S) solar cells with CBD ZnO TCO

5 CONCLUSIONS

Non-vacuum, solution processed, ZnO layers used as TCO in CIGS solar cells were presented. With these CBD ZnO layers implemented as TCO in CIGS solar cells with CdS buffers, an efficiency of 18.3 % was achieved, value comparable with cells obtained with sputtered ZnO:Al TCO. The CBD ZnO layers were also grown on Zn(O,S) buffers achieving an efficiency of 16 % and on flexible cells grown on PI substrates up to an efficiency of 13.8 %. Table I summarize the best cells characteristics achieved.

Table I: Summary of the best cells characteristics

Substrate	Buffer	TCO	Voc (mV)	Jsc (mA.cm^{-2})	FF (%)	Eff (%)
Glass /CIGS	CdS	sputt i-ZnO/ZnO:Al	694	33.8	74.5	17.5
Glass /CIGS	CdS	CBD ZnO	674	33.5	73.8	16.6
Glass / High quality CIGS	CdS	CBD ZnO + ARC	703	37	70.4	18.3
PI / CIGS	CdS	CBD ZnO	644	31.8	67.6	13.8
Glass / CIGS	Zn(O,S)	CBD ZnO	642	35.2	70.6	16

With this study we demonstrate that CBD ZnO TCO could be considered as an non vacuum alternative to sputtered ZnO:Al in CIGS cells. Large area uniformity as well as long term stability of CBD ZnO layers has to be addressed in order to prove industrial relevance.

6 ACKNOWLEDGEMENTS

The authors acknowledge the support of the solar-era.net project "NovaZOlar" financed by the Swiss Office of Energy (BFE), under project no. 8100073 and the German Federal Ministry for Economic Affairs and Energy (BMWi) under project no. 0325749.

7 REFERENCES

[1] D. Ginley, H. Hosono, and D.C. Paine, Handbook of Transparent Conductors (Springer, New York/Heidelberg/Dordrecht/London, 2010)
[2] K. Ellmer, A. Klein, and B. Rech, Transparent Conductive Zinc Oxide (Springer, Berlin/Heidelberg, 2008)
[3] J. Rousset, E. Saucedo, K. Herz, and D. Lincot, Prog. Photovolt. Res. Appl. 19, 537 (2011).
[4] M. Podlogar et al.,Adv. Funct. Mater., 22, 3136-3145 (2012)
[5] M. Miyake, H. Fukui, and T. Hirato, Phys. Status Solidi A 209, 945 (2012).
[6] M. Kevin, G. H. Lee, and G. W. Ho, Energy Environ. Sci. 5, 7196 (2012).
[7] H. Hagendorfer, et al, Adv. Mater., 26: 632–636. doi: 10.1002/adma. (2014).
[8] P. Fuchs , et al., RSC Adv.,6, 53768-53776, (2016).
[9] Y. E. Romanyuk, et al. (2015), Adv. Funct. Mater., 25: 12–27. doi: 10.1002/adfm.201402288.
[10] P. Jackson, D. Hariskos, R. Wuerz, O. Kiowski, A. Bauer, T. M. Friedlmeier, and M. Powalla, Phys. Status Solidi RRL 9, No. 1, 28-31, 2015.
[11] P. Jackson, D. Hariskos, R. Wuerz, W. Wischmann, and M. Powalla, Phys. Status Solidi RRL 8, No. 3, 219-222, 2014.
[12] G. Voorwinden, R. Kniese, P. Jackson, M. Powalla, Proceedings of the 22nd European Photovoltaic Solar Energy Conference EU-VSEC, Milan, Italy, September 3–7, 2007, pp.2115–2118.
[13] D. Hariskos, P. Jackson, W. Hempel, S. Paetel, S. Spiering, R. Menner, W. Wischmann, and M. Powalla, Proceedings of the 43rd IEEE Photovoltaic Specialists Conference, Portland USA, June 5-10, 2016
[14] A. Chirilă, P. Bloesch, S. Seyrling, A. Uhl, S. Buecheler, F. Pianezzi, C. Fella, J. Perrenoud, L. Kranz, R. Verm, D. Guettler, S. Nishiwaki, Y.E. Romanyuk, G. Bilger, D. Brémaud and A.N. Tiwari, Prog. Photovolt: Res. Appl., 19: 560–564. doi:10.1002/pip.1077, 2011

CHARACTERIZATION OF THE BACK CONTACT OF CIGS SOLAR CELL AS THE ORIGIN OF "ROLLOVER" EFFECT

Takuya Kato, Kimihiko Kitani, Kong Fai Tai, Rui Kamada, Homare Hiroi, and Hiroki Sugimoto
Atsugi Research Center, Solar Frontier K.K.
123-1 Shimo-kawairi, Atsugi, Kanagawa 243-0206

ABSTRACT: In this study, the back contact in $Cu(In,Ga)(S,Se)_2$ (CIGS) solar cell was investigated as the origin of the rollover effect, which is characterized as a distortion of current-voltage (*J–V*) characteristics under illumination at forward bias with saturating current at higher voltage. Presence of Schottky barrier between CIGS and $Mo(S,Se)_2$/Mo back contact in the solar cell with rollover *J–V* was elucidated by the composition and the valence band energy measured on the CIGS backside exposed by exfoliation. The device simulation using the obtained values confirmed that the barrier at the back contact is responsible for the rollover. Furthermore, the impedance analysis clearly showed the existence of the back diode in the equivalent circuit of the device. The circuit simulation illuminated that the rollover is described as a distortion of *J–V* curve due to the voltage distribution between the *pn* junction diode and the back diode. By using *J–V* characteristics of the back contact, a potential *J–V* curve intrinsic to the *pn* junction was derived. As a result, the cell with rollover *J–V* was found to have similar or even higher potential performance as compared with the one without rollover. It was also found that although the rollover was less apparent in the *J–V* curve, the device possibly holds a performance loss due to the back contact barrier.
Keywords: Thin Film Solar Cell, Chalcopyrite, CIGS, Rollover, Back Contact, Back Diode, Schottky Barrier

1 INTRODUCTION

Recently Solar Frontier has achieved energy conversion efficiencies of 22.3 and 22.0% on $Cu(In,Ga)(S,Se)_2$ (CIGS) solar cells with CdS and Cd-free buffer layers, respectively [1]. These achievements demonstrate great potential of CIGS as an absorber material for high-efficiency and low-cost thin-film photovoltaic modules. In Solar Frontier, CIGS film is formed by selenization and sulfurization of sputter-deposited metal precursors [2,3]. This absorber layer holds two major advantages for the solar cell operation: i) efficient carrier collection due to the Ga gradient, and ii) reduction of buffer/absorber interface recombination rate resulting from the lowered valence band energy due to high sulfur composition at the surface of CIGS film. On the other hand, CIGS solar cell occasionally experiences rollover effect, which is characterized as a distortion of current-voltage (*J–V*) characteristics under illumination at forward bias with saturating current at higher voltage [4,5]. The rollover effect can be explained by the existence of a barrier preventing carrier transport in the solar cell device. In the CIGS solar cell having well optimized buffer and window layers, this barrier most likely originates from the back contact, i.e., CIGS/$Mo(S,Se)_2$/Mo interface, acting as Schottky diode. In particular, the CIGS absorber film containing sulfur is subject to suffer from the rollover effect because of the lower valence band energy at not only surface but also backside. In this study, the CIGS/$Mo(S,Se)_2$/Mo back contact interface was investigated as the origin of the rollover effect by comparing the cell with and without rollover behavior. The material and device analyses with simulations consistently indicated that the back contact is responsible for the rollover. The circuit simulation study provided insight into the mechanism of the appearance of the rollover *J–V* curve. Finally, it was demonstrated that a potential *J–V* curve intrinsic to the *pn* junction was derived by using the *J–V* characteristics of the back contact junction.

2 EXPERIMENTAL

The CIGS solar cells investigated in this study were fabricated by our standard process [3]. The CIGS absorber layer was formed by selenization and sulfurization of metal precursors deposited on Mo-coated glass substrate followed by chemical bath deposition of Cd-free buffer and metal-organic chemical-vapor deposition of boron-doped/nondoped bilayer ZnO. The *J–V* characteristics were measured under the standard test conditions of AM 1.5 illumination. Two types of CIGS solar cells were chosen to study the rollover effect; the cell exhibiting normal (rollover) *J–V*, referred to as the sample N (R), respectively. The capacitance-voltage (*C–V*) measurement and impedance spectroscopy were performed using KEITHLEY 4200-SCS semiconductor characterization system and Solartron 1260A impedance analyzer, respectively. To obtain *J–V* characteristics of the back contact junction, the window/buffer layers were etched off by using dilute hydrochloric acid solution followed by sputter deposition of gold contact on the exposed CIGS surface to form Au/CIGS/$Mo(S,Se)_2$/Mo device. The Ohmicity at Au/CIGS contact was confirmed separately (not shown in this paper). All the device measurements mentioned above were performed under relaxed state (Rlx) and light-soaked state (LS). To investigate the material properties at the back contact

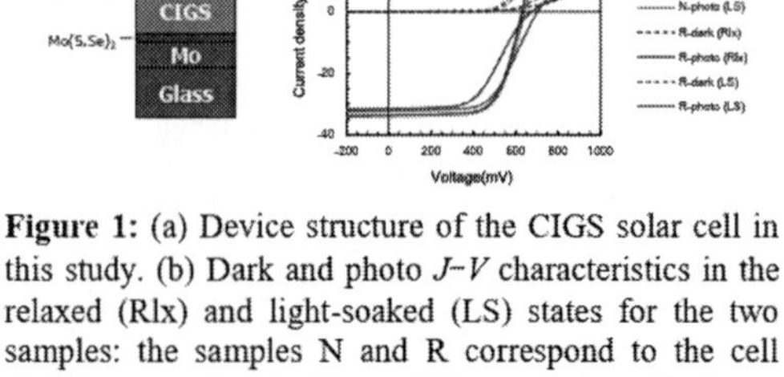

Figure 1: (a) Device structure of the CIGS solar cell in this study. (b) Dark and photo *J–V* characteristics in the relaxed (Rlx) and light-soaked (LS) states for the two samples: the samples N and R correspond to the cell showing normal and rollover *J–V* curves, respectively.

Table I: Solar cell parameters and acceptor density for the samples shown in Fig. 1.

Sample	Condition	η (%)	J_{sc} (mA/cm^2)	V_{oc} (mV)	FF (%)	N_A (10^{15} cm^{-3})
N	Rlx	14.4	33.7	630	67.8	3.3
	LS	14.7	34.4	642	66.3	5.1
R	Rlx	11.4	31.6	675	53.3	6.0
	LS	13.6	32.3	705	59.5	8.0

interface, CIGS and superjacent layers were forced to be exfoliated from $Mo(S,Se)_2$/Mo back contact using a quartz second substrate and epoxy adhesive. The sister devices of the samples N and R were used for the exfoliation study. Secondary ion mass spectrometry (SIMS) and ultraviolet photoelectron spectroscopy (UPS) were performed on the exposed CIGS back surface and back contact surface to analyze the compositions and the valence band energy. Device and circuit simulations were performed by using SCAPS-1D [6] and LTspice [7], respectively.

3 RESULTS AND DISCUSSION

Figure 1b shows the dark and photo $J-V$ characteristics of the samples N and R in the relaxed and light-soaked states. The solar cell parameters are listed in Table I with acceptor density (N_A) measured by $C-V$. Sample R showed apparent rollover, which was weakened by the light soaking treatment probably due to the increase of N_A.

The material properties at the back contact were investigated by SIMS and UPS on the CIGS backside and $Mo(S,Se)_2$ surface exposed by the exfoliation as shown in Fig. 2a. The sulfur content at the back contact was higher in the sample R, which leads to the lower E_V of CIGS backside (E_V^{CIGS}). While the sulfur content of $Mo(S,Se)_2$ was similar to that of CIGS backside in both samples, E_V of $Mo(S,Se)_2$ (E_V^{MoSSe}) did not change significantly. As a result, the valence band offset (Schottky barrier) at the back contact interface becomes larger in the sample R, as shown in Fig. 2c. Here, the band gap of CIGS backside was obtained by using the well-known relation between

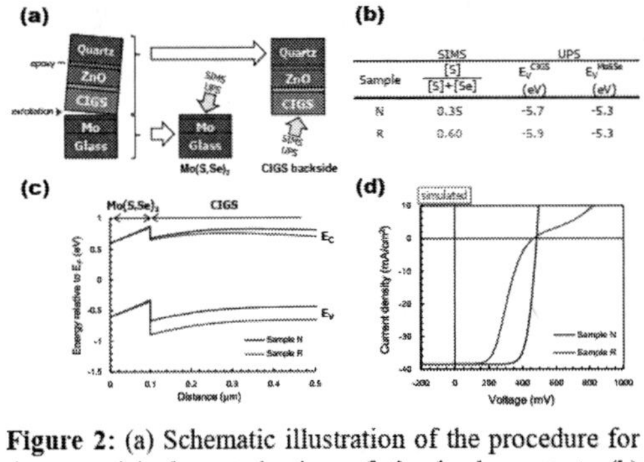

Figure 2: (a) Schematic illustration of the procedure for the material characterization of the back contact. (b) Summary of the results of SIMS and UPS measured at CIGS backside and $Mo(S,Se)_2$ surface. The sulfur composition at CIGS backside and $Mo(S,Se)_2$ were almost the same. E_V is defined relative to the vacuum level. (c) Simulated band structure at CIGS/$Mo(S,Se)_2$/Mo interface using obtained values. (d) Simulated $J-V$ characteristics using the band structure in (c).

Table II: Results of impedance analysis obtained by the fitting shown in Fig. 3b.

Sample	Condition	R_s (Ωcm^2)	CPE^{pn}-T (nF/cm^2)	R_p^{pn} (kΩcm^2)	CPE^{bc}-T (nF/cm^2)	R_p^{bc} (Ωcm^2)
N	Rlx	5.6	28.1	10.2	–	–
	LS	5.0	28.0	6.7	–	–
R	Rlx	5.3	35.6	24.5	56.1	68
	LS	6.0	37.2	15.2	48.0	23

the bandgap and compositions [8]. By using the obtained values, the normal and rollover $J-V$ behaviors were successfully reproduced by the device simulation, as shown in Fig. 2d. These results clearly suggest that higher Schottky barrier at the back contact due to higher sulfur incorporation in sample R causes the distortion in the $J-V$ curve.

To elucidate the existence of the back contact barrier in the device, impedance analysis was carried out. Figure 3a represents AC equivalent circuit model for the CIGS solar cell with the back contact barrier. When the contribution of Schottky junction at the back contact becomes significant, in addition to the main *pn* junction, the second junction component is expected to appear. The results of the impedance analysis are summarized in Fig. 3b and Table II. The Nyquist plot clearly presented the apparent contribution of the back contact barrier in the sample R, while the second junction component was negligibly small in the sample N. In addition, the reduced rollover by light soaking (Fig. 1b) can be explained by the decrease of the parallel resistance of the back contact barrier (R_p^{bc}) within this equivalent circuit model. Therefore, the rollover becomes stronger (weaker) when the back contact junction shows better (worse) diode characteristics with higher (lower) Schottky barrier, respectively.

To understand how the rollover effect appears in the solar cell circuit under operation, a circuit simulation study using the DC equivalent circuit model shown in Fig. 4a was performed. The Schottky junction at the back contact was implemented as the reversely connected diode with parallel resistance in accordance with the AC circuit shown in Fig. 3a. The applied external voltage (V_{bias}) is divided into the voltage at *pn* junction (V_{pn}) and at the back contact junction (V_{bc}) when the series resistance R_s was involved in each junction component. As seen in Fig. 4b, the sign of V_{bc} is changed at $V_{bias} = V_{oc}$. V_{pn} is modified from V_{bias} by V_{bc}, whose behavior is

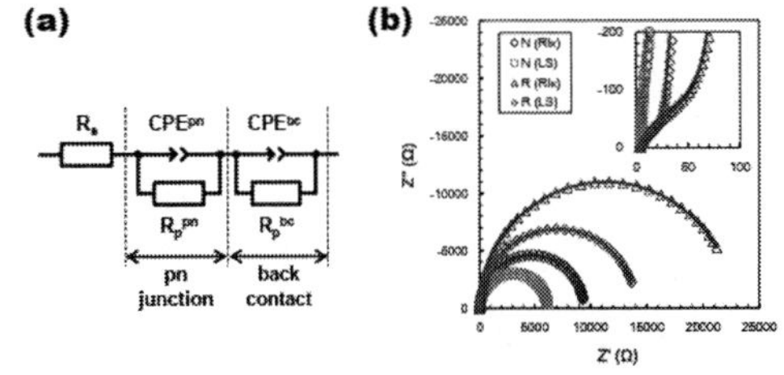

Figure 3: (a) AC equivalent circuit model for the CIGS solar cell having two junctions – the main *pn* junction (denoted as pn) and Schottky junction at the back contact (bc). CPE stands for constant phase element. (b) Nyquist plots for the samples shown in Fig. 1. The solid lines represent fitted curves to the equivalent circuit shown in (a). The inset shows enlarged plots in the high-frequency region indicating the existence of the second impedance arc in the sample R.

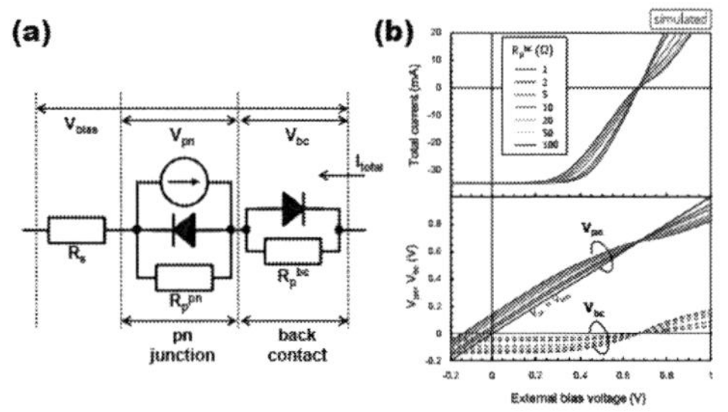

Figure 4: (a) DC equivalent circuit model for the CIGS solar cell under illumination with a back diode (corresponding to the AC model shown in Fig. 3a). (b) Circuit simulation results of the total current (I_{total}), the voltage at *pn* junction (V_{pn}), and the voltage at the back contact junction (V_{bc}) versus the external bias voltage (V_{bias}) with varying the parallel resistance of the back diode (R_p^{bc}).

defined by the characteristics of the back diode. Hence the rollover effect can be described as a distortion of *J–V* curve due to the voltage distribution between the *pn* diode and the back diode [4].

The characteristics of the back diode were directly measured on the actual devices as shown in Fig. 5. The devices only consist of the back contact junction (referred to as the back contact device) were fabricated by etching off the window and buffer layers of the original solar cell (the full device). The obtained *J–V* curve exhibited weak diode behavior with high leakage current and significantly low breakdown voltage. The facts that the back diode in the sample R has lower leakage and that the light soaking enhances the leakage agree well with the above-mentioned findings that the higher Schottky barrier and higher R_p^{bc} are responsible for the rollover (Fig. 2 and Table II).

Finally, a correction of *J–V* curves to extract intrinsic performance of *pn* junction using the back diode characteristics was demonstrated. Since the current flowing the *pn* junction and the back diode is equal, by knowing *J–V* characteristics of the back diode, V_{bc} at any J (therefore at any V_{bias}) in the original *J–V* curve (Fig. 1b) can be estimated. Then, V_{pn} can be calculated by subtracting V_{bc} from V_{bias} as shown in Fig. 6a and 6b for the samples N and R, respectively. Lastly, the corrected *J–V* curves intrinsic to the *pn* junction were obtained by plotting J versus V_{pn} as shown in Fig. 6c and 6d. The resulted *J–V* curves give conversion efficiency of 15.4 and 15.7% for the samples N and R, respectively, though

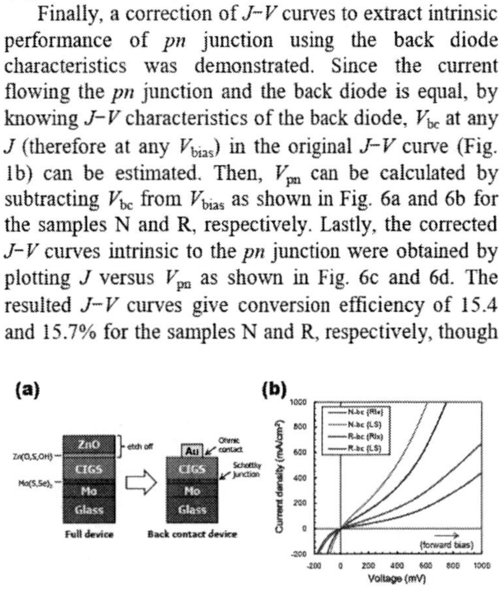

Figure 5: (a) Schematics for the fabrication procedure of the back contact device. (b) *J–V* characteristics measured on the back contact device fabricated from the samples N and R shown in Fig. 1b. The voltage axis is taken so that the positive side corresponds to the forward direction of the Schottky diode.

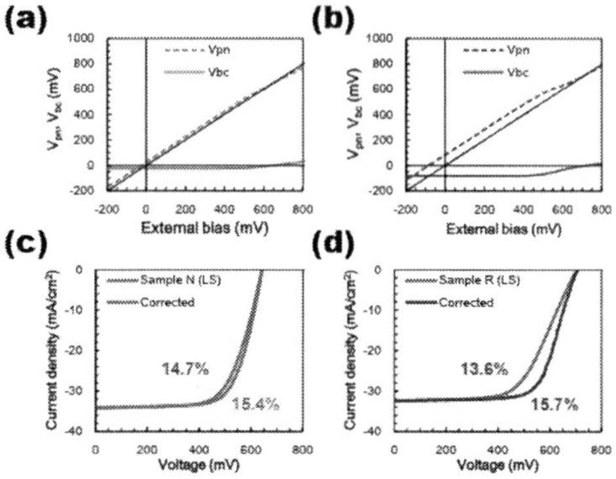

Figure 6: Demonstration of the derivation of potential *J–V* curve intrinsic to the *pn* junction from the original *J–V* curve including the contribution of back diode. (a,b) V_{pn} and V_{bc} versus external bias voltage (corresponding to the simulated curves in the bottom panel of Fig. 4b) and (c,d) corrected *J – V* curve for the samples N and R, respectively. The gray curves in (c) and (d) represent the original *J–V* curve as in Fig. 1b.

some distortion in the *J–V* curve still remains in the sample R probably due to non-identical conditions between the full device and the back contact device. The result suggests that the sample R has equivalent performance to the sample N except for the significance of the Schottky barrier at the back contact, which leads the strong rollover effect. It is also indicated that even in the sample N, where the rollover is less apparent, there still exists a performance loss to some extent due to the back contact barrier.

4 CONCLUSION

The back contact of CIGS solar cell was studied to investigate the rollover effect. Both material analysis and device analysis including simulations consistently showed that the back diode caused by the Schottky barrier at the back contact is responsible for the rollover. The current-voltage characteristics of the back contact Schottky junction were directory observed on the back contact device, where the main *pn* junction was excluded. By using the back diode characteristics, which also evidenced the Schottky junction of the back contact, the correction of the current-voltage characteristics was demonstrated to reveal the potential performance intrinsic to the *pn* junction. The rollover sample was found to have equivalent potential performance to the non-rollover sample. It was also found that there can be a performance loss due to the back contact barrier even if the rollover is less apparent. The *J–V* correction technique described in this study will be helpful to diagnose whether the device suffers from the back contact barrier and to evaluate the potential performance of the *pn* junction in the device.

ACKNOWLEDGEMENT

This work was partly consigned from the photovoltaics research and development programs of New Energy and Industrial Technology Development Organization (NEDO).

REFERENCES

[1] Solar Frontier Press Release, http://www.solar-frontier.com/eng/news/2015/C051171.html, (2015); R. Kamada, T. Yagioka, S. Adachi, A. Handa, K.F. Tai, T. Kato, H. Sugimoto, Photovoltaic Specialists Conference (PVSC) IEEE 43rd (2016) to be published.

[2] K. Kushiya, Sol. Energ. Mat. Sol. Cells 122 (2014) 309.

[3] M. Nakamura, N. Yoneyama, K. Horiguchi, Y. Iwata, K. Yamaguchi, H. Sugimoto, T. Kato, Photovoltaic Specialists Conference (PVSC) IEEE 40th (2014) 0170.

[4] S. H. Demtsu, J. R. Sites, Thin Solid Films 510 (2006) 320.

[5] T. Eisenbarth, T. Unold, R. Caballero, C. A. Kaufmann, H. W. Schock, J. Appl. Phys. 107 (2010) 034509.

[6] http://scaps.elis.ugent.be/; M. Burgelman, R. Nollet, S. Degrave, Thin Solid Films 361-361 (2000) 527.

[7] http://www.linear.com/solutions/ltspice

[8] M. Bär, W. Bohne, J. Röhrich, E. Strub, S. Lindner, M. C. Lux-Steiner, Ch.-H. Fischer, T. P. Niesen, F. Karg, J. Appl. Phys. 96 (2004) 3857.

PUNCH-THROUGH EFFECT IN CIGS THIN FILM SOLAR CELLS

Thomas Ott[1]*, Thomas Walter[2] Raymund Schäffler[3] and Hans-Jörg Fecht[1]
[1]University Ulm, Albert-Einstein-Allee 47, 89081 Ulm, Germany
[2]University of Applied Sciences Ulm, Albert-Einstein-Allee 55, 89081 Ulm, Germany
[3]Manz CIGS Technology GmbH, Alfred-Leikam-Str. 25, 74523 Schwäbisch Hall, Germany
*Author for correspondence: (thomas.ott@uni-ulm.de)

ABSTRACT: $Cu(In,Ga)Se_2$ (CIGS) continuously demonstrates more of its potential with cell efficiencies exceeding 22% on laboratory scale. A goal for the future is the reduction of the absorber thickness. In the past a phototransistor model was introduced to explain the cell behavior for low temperatures or after long term endurance test. This phototransistor model depends on a back contact barrier which becomes more important for a reduced thickness or low doping densities of the absorber. In this contribution the occurrence of a punch-through effect during long term endurance test with an applied negative bias is discussed. In the field negative voltages could occur due to partial shading. From transistor theory the punch-through effect is well known. If the space charge region of the Schottky barrier and the main junction touch, a punch-through occurs resulting in a strong increase of the current and a significantly reduced open circuit voltage. A reduced doping density of the absorber or/and an increased barrier at the back contact enhance this behavior. This assumption is verified with simulations based on the phototransistor model. Finally, the influence on highly efficient cells with reduced absorber thickness is discussed. For thick absorbers this phenomena should play no role under normal operating conditions. But for a reduced absorber thickness the punch-through effect can occur in the initial state and have a negative influence on the cell performance for cells with thinner absorbers. If the trend towards reduced absorber thicknesses in the future continues, further investigations regarding the back contact barrier including the adjustment of the doping density are necessary.
Keywords: Punch-through Effect, Back Contact, CIGS, Thin Film Solar Cell.

1 INTRODUCTION

Cu(In, Ga) Se_2 is the leading thin-film solar cell technology with efficiencies exceeding 22% on laboratory scale and 16% on module size [1, 2]. Beside cell efficiency also the long term stability is a crucial point to all solar cell technologies. The behaviour of CIGS solar cells at low temperatures or after an endurance test at elevated temperatures can be explained best by the phototransistor model [3]. The key part of the model is a barrier at the back contact. One of the future goals for CIGS thin film solar cells is the reduction of the absorber thickness to reduce the cost of material. With respect to the back contact the influence of a Schottky barrier will increase for thinner absorber. This also applies for lower doped absorbers. A low negative bias during acceleration tests leads to lower doping of the CIGS thin film solar cells absorbers [4]. In the field negative bias can occur at partially shaded cells in a module. For higher negative voltages this phenomena should increase. A widening of the space charge region is the consequence. In certain cases the space charge region of the heterojunction and the Schottky barrier will touch. From transistor theory this phenomena is called the punch-through effect [5, 6]. The consequences for CIGS solar cells of reduced thickness or lower doped absorber regarding the phototransistor model has still not been investigated.

First the phototransistor model is introduced. Also the punch-through effect is explained. The results of endurance tests with negative applied bias will be presented. The evolution of JV-characteristics and CV-characteristics are discussed. Simulations of the band diagram with different doping densities of the absorber were performed to support the occurrence of a punch-through effect. The corresponding JV-characteristics and CV-characteristics are shown and compared with the measurement results. Finally the influence of the punch-through effect to thinner absorbers and different barrier heights at the back contact is verified with simulations and presented.

2 EXPERIMENTAL AND RESULTS

2.1 Theory

Figure 1 shows the simulated band diagram based on the phototransistor model in the thermal equilibrium. The simulations were performed using SCAPS1D. The main part is a Schottky barrier at the back contact. In the underlying parameter set a barrier height of 400meV was chosen. Together with the heterojunction this results in a band diagram comparable to that of a transistor. The absorber can be seen as the base of the transistor, the ZnO as the emitter and the Schottky barrier as the collector. For a phototransistor the base current is replaced by photo generated electron hole pairs. In the dark no base current is present, thus no current can flow. Under illumination a base current is present and a current can flow.

In the semiconductor theory the punch through effect is well known in different fields [5]. For diodes this effect can exhibit for one-sided abrupt junctions. In that case one part of the diode is much stronger doped than the other. If the depletion-layer width is bigger than the width of the less doped semiconductor the depletion layer touches the back contact. A breakdown occurs. An increase of the doping density (higher doped semiconductor) will lead to no change. From the bipolar transistor theory the punch-through effect can occur for a very narrow base width or a very low doped base. The neutral base width is reduced to zero due to sufficiently high voltages between the collector and the base. The depletion regions of the emitter and collector touch each other and a high current will flow [6]. A breakdown is visible in the output characteristics of the transistor [6].

2.2 Measurement

To investigate the punch-through effect an endurance test in the dark at a high temperature of 105°C was performed. During the accelerated ageing a high negative

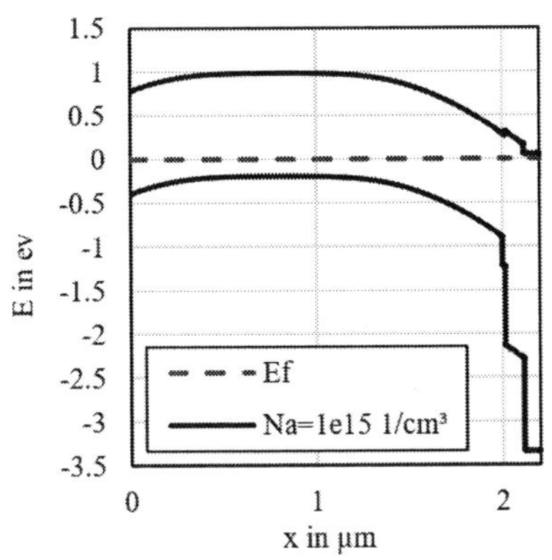

Figure 1: Simulated band diagram of a CIGS solar cell based on the phototransistor model, in the thermal equilibrium with a Schottky barrier of 400meV.

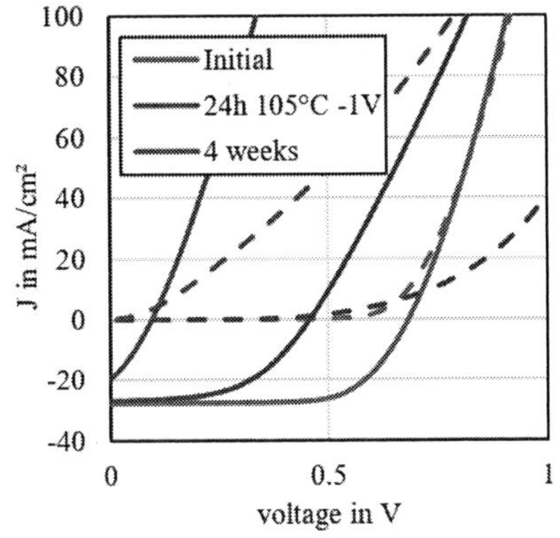

Figure 2: Evolution of the measured JV-characteristics of a CIGS thin film solar cell during an endurance test at 105°C with an applied bias of -1V.

bias of -1V was applied to a cell. The junction was measured in certain time intervals. JV-characteristics in the dark and under illumination of 1 sun were recorded, also the CV-characteristic in the dark. Figure 2 shows the evolution of the dark (dashed line) and illuminated (solid line) JV-characteristics. In the initial state (green) a "normal" solar cell behaviour is obvious. After 24h (red) a strong cross over occurs accompanied by a blocking of the current in the dark. Under illumination a significant decrease of V_{OC} is visible. Based on previous findings one would expect an increase of these anomalies with increasing test time [4]. But after 4 weeks (blue) the blocking of the dark current is reduced.

Figure 3 shows the corresponding CV-measurements to the IV-characteristics of the endurance test at 105°C with -1V presented in figure 2. In the initial state (green) a common CV-characteristic of a CIGS solar cell is visible. For higher negative applied biases the capacitance C_D of the solar cell decreases. From theory the capacitance of the solar cell can be compared with the one of a capacitor and depends strongly on the width of the space charge region w_D respectively the width between the plates.

$$C_D = \frac{\varepsilon_S}{w_D} \qquad (1)$$

where ε_S is the permittivity of the semiconductor.

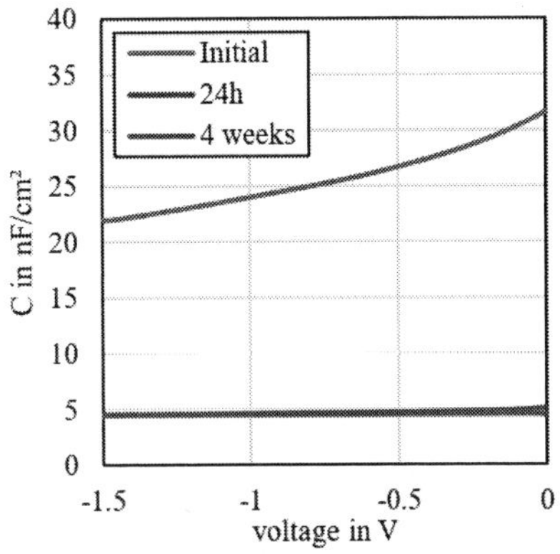

Figure 3: Corresponding CV-characteristics to the JV-characteristics of the endurance test presented in figure 2.

Considering a one-sided abrupt junction (stronger doping of the window layer) and other simplifications, equation (1) can be rewritten to:

$$C_D = \sqrt{\frac{q\varepsilon_S N}{2}}\left(V_{bi} - V - \frac{2kT}{q}\right)^{-\frac{1}{2}} \qquad (2)$$

where q is the unit electronic charge, V_{bi} the build-in voltage, N the doping density of the absorber, V the applied bias, k the Boltzmann constant and T the temperature. For an increasing negative bias the capacitance of the solar cell decreases due to an increasing width of the space charge region. After the first 24h a common CV-characteristic of a CIGS solar cell is no longer visible. The measurement exhibits a strong decrease of the capacitance accompanied by a flat characteristic in the sense that the capacitance does not depend on the applied bias. A constant value independent to the applied value is measured. A possible interpretation could be that the space charge region of the hetero interface reaches the space charge region of the Schottky contact [5, 6]. As mentioned in the first chapter this phenomena is also known as the Punch-through effect. A change of the bias has no influence on the output characteristic of the junction. The very low value of the capacitance indicates a very low doping density of the absorber. As a result the geometric capacitance of the solar cell is measured. In the next step simulations are performed to verify this assumption.

2.3 Simulation

For the simulations the solar cell capacitance simulator (SCAPS1D) was used [7]. A parameter which has a strong influence to the width of the space charge region is the acceptor density N_a of the absorber. For the simulation a standard parameter set of a CIGS solar cell was used. This model was extended by a Schottky barrier at the back contact (phototransistor model). The barrier height ϕ_B was set to 400meV. This value corresponds to measured values in the literature for CIGS thin film solar cells after an endurance test at elevated temperatures [8]. The simulations were performed for three different acceptor densities, 1e14 1/cm³, 1e15 1/cm³ and 1e16 1/cm³. All other parameters were not changed. Different simulations have been performed. In this contribution the results for the band diagram, the JV-characteristics in the dark and

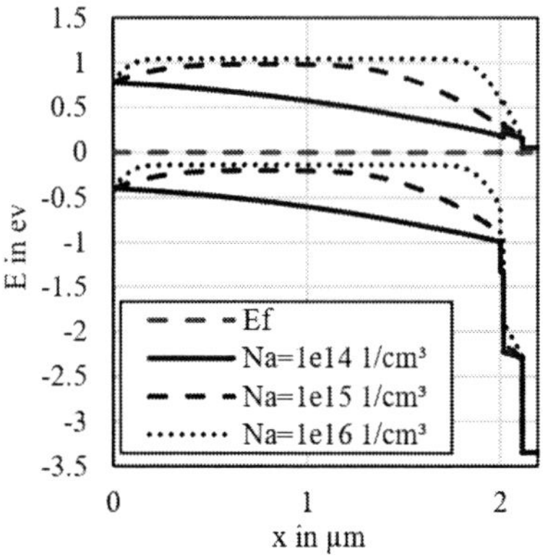

Figure 4: Simulated band diagram with a Schottky barrier of 400meV, for an acceptor density of 1e14 $1/cm^3$ (solid), 1e15 $1/cm^3$ (dashed) and 1e16 $1/cm^3$ (dotted).

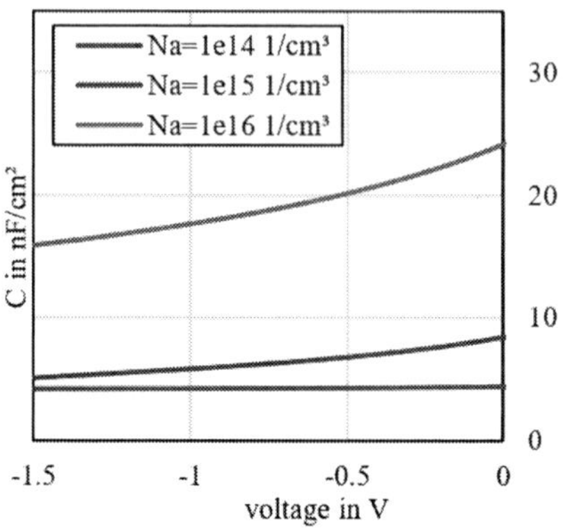

Figure 6: Corresponding CV-characteristics for the band diagrams of figure 4. Simulated in the dark for an acceptor density of 1e15 $1/cm^3$ (blue), 1e14 $1/cm^3$ (red) and 1e16 $1/cm^3$ (green).

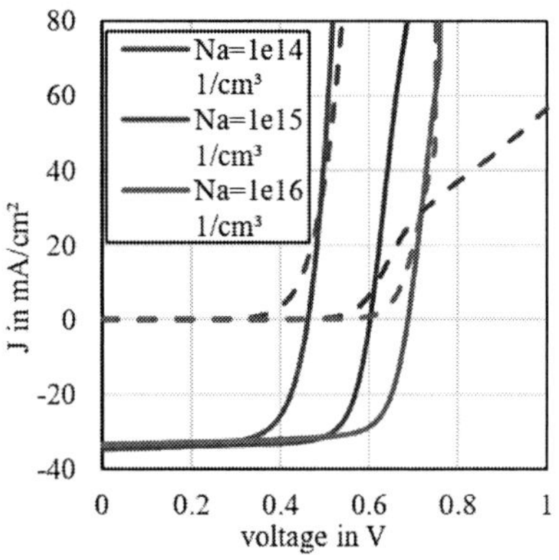

Figure 5: Corresponding JV-characteristics for the band diagrams of figure 4. Simulated in the dark (dashed) and under illumination (dotted) for an acceptor density of 1e14 $1/cm^3$ (blue), 1e15 $1/cm^3$ (red) and 1e16 $1/cm^3$ (green).

under illumination and the CV-characteristics will be discussed. The thickness of the absorber was set to 2μm. Figure 4 shows the simulated band diagram for an acceptor density of 1e14 $1/cm^3$ (solid line), 1e15 $1/cm^3$ (dashed line) and 1e16 $1/cm^3$ (dotted line). For an acceptor density of 1e16 $1/cm^3$ there is no touching of the two space charge regions. For a lower doping density of 1e15 $1/cm^3$ the two space charge regions still do not touch each other. But there are changes in the band diagram caused by the lower doping density. The space charge region of the main junction is wider compared to the higher density. Also the space charge region of the Schottky barrier goes deeper through the absorber. For a lower doping density this trend continuous and the two space charge regions touch each other. So the space charge region stretches through the complete absorber. There is only a small barrier left for the injection of electrons. This leads to an increased current and a reduced open circuit voltage.

Figure 5 shows the corresponding simulated JV-characteristics for the band diagrams of figure 4. For the highest acceptor density of 1e16 $1/cm^3$ (green) in the dark a normal diode behaviour of the CIGS solar cell is visible. The super position principle of the characteristic in the dark and under illumination is valid. No cross over is detected. For the acceptor density of 1e15 $1/cm^3$ (dashed red) the situation changes. The JV-characteristic in the dark exhibits a blocking of the current. A strong cross over of the JV-characteristic in the dark and under illumination is visible. The blocking of the dark current can be explained with the phototransistor model due to a barrier at the back contact [3, 9]. Another decrease of one order of magnitude in the doping density of the absorber to 1e14 $1/cm^3$ leads again to a change of the characteristic. The blocking of the dark current almost disappeared accompanied by a strong decrease of the open circuit voltage (blue).

The corresponding CV-characteristics for the band diagrams shown in figure 4 were also simulated. The results are shown in figure 6. For the highest acceptor density of 1e16 $1/cm^3$ (green) the capacitance exhibits a "normal" behaviour and increases to 0V. A strong decrease of the capacitance is detected for a lower acceptor density of 1e15 $1/cm^3$ (red). Still a gradient to 0V is observed. For the lowest acceptor density of 1e14 $1/cm^3$ (blue) the simulation show no change of the capacitance. A very low value is achieved. This simulated CV-characteristic corresponds to the measurement results with a negative bias of 1V. In figure 4 the corresponding band diagram shows that in this case the space charge region stretches to the back contact.

Figure 7 shows simulated JV-characteristics in the dark (dashed) and under illumination (solid) for two absorbers thicknesses, 2μm standard (green and yellow) and 0.5μm thin (blue and red). Additionally a Schottky barrier of 300meV was added (red and yellow). For a thick absorber the Schottky barrier at the back contact has no influence on the JV-characteristics. Only a small decrease of the open circuit voltage is visible (yellow and green). This changes for the thin absorber. The Schottky barrier has a stronger influence on the JV-characteristics. The JV-characteristics with a Schottky barrier (red) exhibits a strong decrease of the open circuit voltage compared to the simulation without a Schottky barrier (blue). The decrease of the short circuit current density is due to a weaker

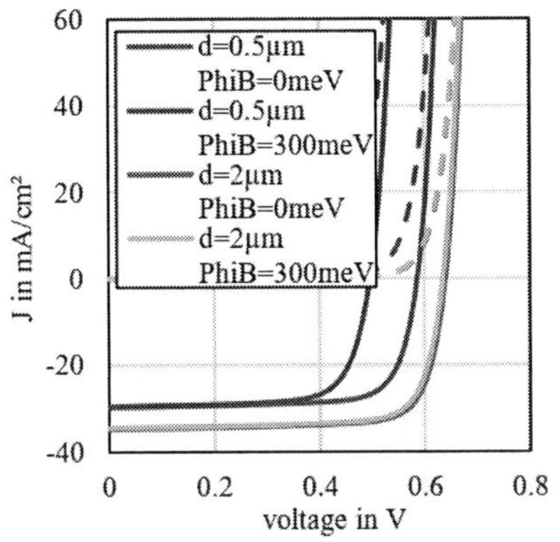

Figure 7: Simulated JV-characteristic for a CIGS thin film solar cell in the dark (dashed) and under illumination (solid). With (red and yellow) and without (blue and green) a Schottky barrier at the back contact for an absorber thickness of 0.5 (blue and red) and 2µm (green and yellow).

collection in the thinner absorber respectively a lower external quantum efficiency.

3 DISCUSSION

The phototransistor model explains the cell behavior for low temperatures or after long term endurance test. The phenomena that the space charge region goes through the whole absorber is well known from the transistor theory as the punch-through effect. CIGS thin film solar cells after accelerated ageing in the dark with a negative bias during the endurance test show an untypical behavior of the JV- and CV-characteristics. With increasing test time a blocking of the dark current occurs which disappears after longer test times. The capacity decreases and gets finally independent of the applied bias. A possible interpretation of the observed chances is the punch-through effect known from transistor theory. A strong hint is the constant capacitance and the change in the dark JV-characteristic. For higher acceptor densities the Schottky barrier is not visible. For a lower doping the quasi neutral region decreases and electrons can reach the back contact. Under illumination the barrier decreases and the current increases, like for a phototransistor. For very low doping densities the quasi neutral region disappears and the space charge regions of the heterojunction and the Schottky barrier touch each other. The current increases due to the fact there is only a small barrier left for the electrons.

The main factors which are influencing the punch-through effect are the doping density of the absorber, the thickness of the absorber and the height of the Schottky barrier. For modules in the filed a decrease of the acceptor density can occur due to partial shading.

4 CONCLUSION

During long term endurance test with an applied negative bias an uncommon cell behavior was observed. Based on the phototransistor model the occurrence of a punch-through effect in CIGS solar cells was observed. For standard cell this phenomena should have no influence to the cell performance. But regarding the topics thinner absorber and partial shading the phenomena should be paid more attention. The CV-characteristics seems to be an appropriate measurement to identify the punch-through effect in CIGS thin film solar cells.

5 REFERENCES

[1] Press release Solar Frontier K.K.: "Solar Frontier Achieves World Record Thin-Film Solar Cell Efficency: 22,3%. (08.12.2015). [Online] Available: http://www.solar-frontier.com/eng/news/2015/C051171.html

[2] Press release Manz AG: 16 Percent: Manz achieves new world record for efficiency of CIGS thin-film solar modules. (27.04.2015). [Online] Available: http://www.manz.com/media/news/16-prozent-manz-erzielt-neuen-weltrekord-bei-wirkungsgrad-von-cigs-duennschicht-solarmodulen-615/

[3] T. Ott, T. Walter and T. Unold, "Phototransistor Effects in $Cu(In,Ga)Se_2$ Solar Cells", Thin Solid Films 535 (2013).

[4] P. Mack, T. Ott, T. Walter, D. Hariskos and R. Schäffler, "Optimization of Reliability and Metastability of CIGS Solar Cell Parameters", Proc. 25th EUPVSEC (2010).

[5] H. Dejun, L. Guohui, Y.F. Zhang und E.-J. Zhu, "Ultrahigh-Sensitive AlGaAs-GaAs Punchthrough Heterojunction Phototransistor", IEEE Photonics Technology Letters, 1997.

[6] S.M. Sze and K.K. Ng, "Physics of Semiconductor Devices", John Wiley & Sons, New York, 3.edition, 2007.

[7] M. Burgelman, P. Nollet and S. Degrave, "Modelling polycrystalline semiconductor solar cells", Thin Solid Films, 361-362: 527-532, (2000).

[8] T. Ott, T. Walter and R. Schäffler, "On the Interpretation of Admittance and IV(T) Measurements of CIGS Thin Film Solar Cells", Proc. 31th EUPVSEC (2015).

[9] T. Ott, et al., "Verification of Phototransistor Model for $Cu(In,Ga)Se_2$ Solar Cells", Thin Solid Films (2014).

6 ACKNOWLEDGMENTS

This work has been supported by the German Federal Ministry for Economic Affairs and Energy.

MONOLITHIC INTERCONNECTION OF MICROMORPH SILICON TANDEM THIN FILM SOLAR CELLS ON FLEXIBLE AND OPAQUE SUBSTRATES USING LASER ABLATION

Kristina Borzutzki[1], Stefan Geißendörfer[1], Ortwin Siepmann[1], Jörg Ohland[2], Oleg Sergeev[1], Martin Vehse[1], Carsten Agert[1]
[1]NEXT ENERGY. EWE Research Centre for Energy Technology at the University of Oldenburg,
Carl-von-Ossietzky-Str. 15, 26129 Oldenburg, Germany
[2] Carl von Ossietzky University of Oldenburg, Carl-von-Ossietzky-Str. 9 – 11, 26129 Oldenburg, Germany

ABSTRACT: Monolithic interconnection of thin film solar cells is an essential step of PV module development. So far, monolithic interconnection using laser scribes has been established as a favoured method for silicon thin film solar cells deposited on glass substrates. Transferring this method to flexible substrates is an important key for industrial realisation of an efficient module concept. In this work, we demonstrate the feasibility of monolithic interconnection of micromorph silicon tandem thin film solar cells deposited on flexible aluminium substrates by applying the interconnection on one cell stripe using laser scribes. Fabrication problems concerning the adhesion of the back contact and the microcrystalline bottom cell were solved by plasma treatment. For the P2 and P3 scribe the laser parameters were adjusted in order to ablate the thick layer stack successfully. In case of the P3 scribe, a modified ablation method was used to avoid melting of the layers which predominantly occurred. As final result we demonstrate a well working monolithic interconnection concept of micromorph tandem solar cells on aluminium substrates without significant electrical losses compared to a non interconnected reference cell. Furthermore, our optimised parameters for all scribes are presented.
Keywords: a-Si/µc-Si, Silicon tandem solar cell, Thin film solar cell, Laser processing, PV module

1 INTRODUCTION

Flexible and light weight substrates have become of high interest for silicon thin film solar cells due to new fields of mobile and stationary applications as well as the possibility to use a low cost roll-to-roll production process. So far, their series connection was performed using methods like shingle construction [1] or laser scribe applications which use an additional grid that involves losses due to shading and/or isolating material to fill the scribes [2] which is a rather complex process. On glass substrates on the other hand, a monolithic series interconnection using three laser scribes (P1, P2, and P3) and no further support techniques is an already well established and preferred method to realize their series connection [3]. Know-how transfer of the monolithic series interconnection from glass to flexible substrates is a great step forward to enable the low cost and efficient industrial roll-to-roll production of thin film photovoltaic on flexible substrates.

For samples deposited on glass substrates laser ablation is done from the substrate-side. Applying it on flexible and opaque substrates it is done from the substrate averted side. This allows for a depth selective ablation but the realisation is not that trivial as other physical processes for instance melting instead of ablation take place.

A previous study demonstrates the monolithic series interconnection of amorphous silicon thin film solar cells on aluminium substrates [4]. In this study, we firstly apply this concept on a more efficient and more complex cell stack which is a micromorph tandem thin film solar cell.

2 EXPERIMENTAL

2.1 Cell production

The solar cells used for laser ablation were fabricated on lacquered aluminium substrates developed by the company Alanod in a joint project [5]. The silver/TCO back contact was deposited using electron beam evaporation and magnetron sputtering, respectively. Microcrystalline silicon layers (µc-Si) for the bottom cell and amorphous silicon layers (a-Si) for the top cell were applied by plasma enhanced chemical vapour deposition (PECVD). For the 110 nm thick ITO (indium tin oxide) front contact magnetron sputtering was used. More details of the solar cell stack can be found in the study of Neumüller et al. [6]. In between the layer deposition of back contact, active layer (including all layers of µc-Si and a-Si) and front contact the laser scribes, known as P1, P2 and P3, were performed. The scribes are shown in Figure 1.

The P1 scribe is supposed to isolate two adjacent cell stripes from each another, meaning that the back contact needs to be ablated along the scribe and no connection of the back contact across the scribe can be tolerated. The P2 scribe enables the series connection. It ablates the active layer and the front contact fills the scribe when it is deposited in the following step. This way a connection between the front contact of one cell to the back contact of the adjacent cell is created. Finally, the P3 scribe is performed for isolation of adjacent cell stripes.

In Figure 1 the layout used for the development of series interconnection is sketched as a cross sectional view. Figure 2 shows the top view. As shown in these sketches, one cell stripe was created using the three laser scribes P1, P2 and P3. For isolation of the interconnected cell against the remaining substrate, it is surrounded by a P1 as well as a P3 scribe placed on top of each other.

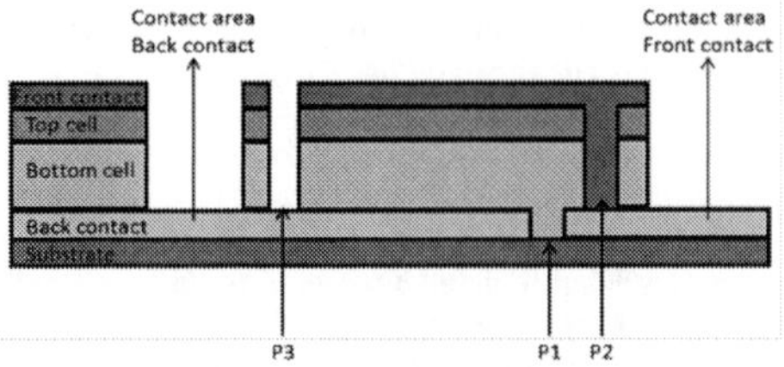

Figure 1: Cross section of the tandem solar cell stack which includes the layout used for an interconnection of one cell stripe.

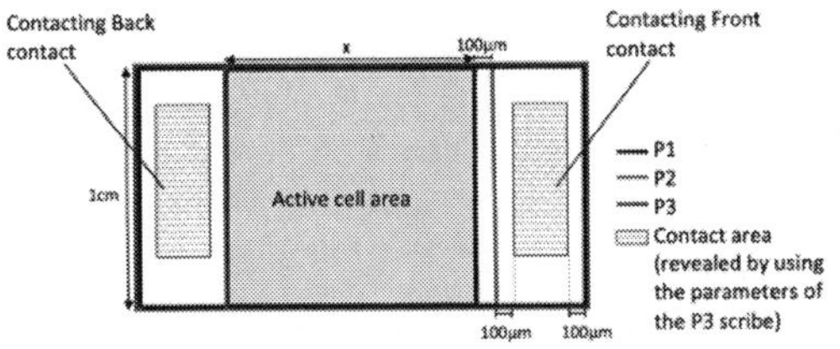

Figure 2 Layout for the monolithic series interconnection of one cell stripe using the three laser scribes P1, P2 and P3.

In addition to the laser scribes, two small areas of the back contact are revealed to contact back and front contact. This way, the contacting of the front contact is performed by relocating it to the higher conductive back contact by the P2 scribe.

Optimal width of the cell stripe, denoted as x in Figure 2, was determined to be 0.5 cm for this cell stack. This result was calculated theoretically by evaluating omic losses due to the resistance of the front contact and area losses of the inactive area [7]. Further, this value was confirmed experimentally (not shown in this work). For comparison of the monolithically interconnected cell a reference cell of area 8 x 8 mm^2 was built using the P3 scribe to separate the cell area from the rest of the substrate.

2.2. Laser scribing

Laser scribes were performed using the micromachining system microSTRUCT Vario fabricated by 3D-MICROMAC. The laser installed in the system is a diode pumped Yttrium-Vanadat ($Nd{:}YVO_4$) laser by Lumera having a pulse duration of 8,5 ps. Its wavelength is 1064 nm and can be frequency doubled resulting in a wavelength of 532 nm, respectively. The repetition rate of the laser was 100 kHz. Parameters that were varied in order to receive optimal ablation results are: the wavelength, laser power, the speed of the galvo-head scanner defining the path of the laser beam and therefore the laser spot overlap and the amount of passes (single vs. multi pass scribes).

2.3 Analysis of scribes and cell stripes

To evaluate the quality of the laser scribes confocal microscopy as well as scanning electron microscopy (SEM) and energy dispersive X-ray spectroscopy (EDX) were performed. For analysis of the cell stripe the current-voltage-characteristic (IV) under standard test conditions and dark lock-in thermography (D-LIT) were measured.

3 RESULTS AND DISCUSSION

3.1 P1 laser ablation

The P1 scribe was carried out using optimized parameters known from previous works on series interconnection of a-Si single cells [4]. In contrast to the deposition of the a-Si cell stack on the P1 structured TCO back contact [4] here we examined delamination of the µc-Si cell directly after deposition (Figure 3a) even though our cells have been treated equally to the a-Si cells in [4]. The delamination starts at the previously performed P1 scribe and the substrate edges and propagates on the cell surface. Different process modifications were carried out to avoid the delamination. Slowing down the cooling process did not show a significantly better behaviour. Deposition of a thin germanium layer (acting as an adhesion agent) in between back contact and µc-Si active layer led to a counter diode behaviour in most cases. A satisfactory result, presented in Figure 3b, was achieved for samples that have been treated with plasma of an argon/ oxygen mixture prior to the deposition of the microcrystalline bottom cell. Probably the plasma treatment activates the surface as it breaks weak atom bonds and removes contaminants whereby the adhesion of back contact and µc-Si increases.

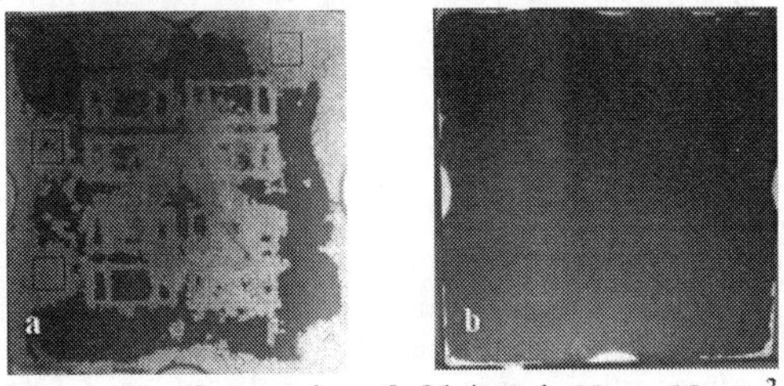

Figure 3: Photographs of fabricated 10 x 10 cm^2 substrates presenting the resulting adhesion of the microcrystalline bottom cell on: (a) an untreated substrate and (b) a plasma (Ar/O) treated substrate.

3.2 P2 laser ablation

Schweitzer et al. [4] found that for a front contact consisting of 10 nm highly conductive silver and 70 nm of low conductive AZO the P2 scribe has to be performed using a multi-pass scribe with shifted scribes in order to achieve a homogenous silver layer on the scribe walls. Based on this finding, we also carried out a multi pass shifted scribe as well as a multi pass scribe. Here, the former one was realized placing multiple scribes in a distance of 10 µm next to each other increasing the number of passes with each scribe. In Figure 4 the profile of such a multi pass shifted scribe is presented.

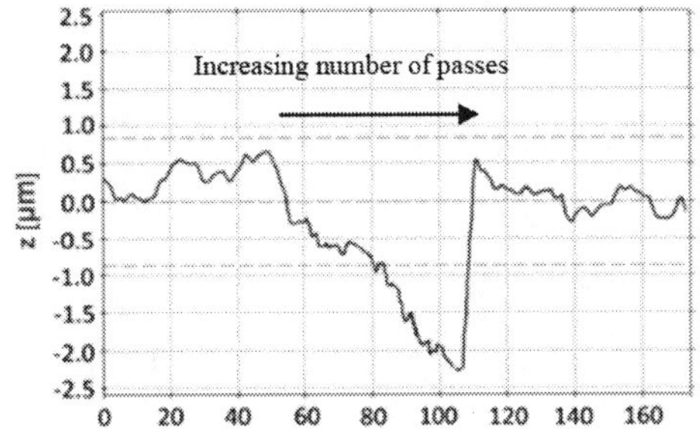

Figure 4: Profile of the multi pass shifted P2 scribe performed with a wavelength of 1064 nm, mean laser power of 0,82 W, velocity 1700 mm/s, 1-6 passes, 6 scribes shifted by 10 µm.

A satisfactory result for the P2 scribe is achieved if the back contact is revealed along the whole scribe length. In order to evaluate and develop optimal parameters for both scribe types optical microscopic measurements were performed.

3.3 P3 laser ablation

Working with glass substrates or a-Si single cells on flexible substrates, the P3 scribe is performed by a single or multi pass laser scribe. Here, these scribe types were also investigated but a short circuit was caused in all samples. Figure 5 shows a confocal microscopy image of a multi pass scribe. It can be seen that spike building in the P3 scribe occurs. The SEM image in Figure 6a shows that the active layer melts and quick solidification of the molten material causes this peak building. Further, EDX analysis of the scribe was done in order to analyse the scribe. The results, presented in Figure 6 b-d, show that some indium (Figure 6c) that can be ascribed to the front contact as well as silicon (Figure 6b) is detected in the scribe. Thus, the short circuit observed can be explained by a transverse conductivity of the molten material in the scribe. It was found that sufficient cell separation can be achieved if all active material is ablated along the scribe. Therefore, a multi pass shifted scribe was investigated because a single or multi pass scribe results in inhomogeneous ablation. Figure 7 shows a multi pass shifted P3 scribe as it was realised for the final interconnection. It was performed using 15 shifted laser scribes with a scribe distance of 25 µm. The EDX analysis shows that almost no silicon is left in the scribe (Figure 7b) but silver and indium (referring to the ITO back contact) are detected. This result is contradictory to the result found for monolithic interconnection of a-Si single cells. In that case best results were achieved if half of the active layer is ablated [4]. Further studies will optimise these P3 laser parameters in order to reduce the scribe width.

Figure 5: Image of the multi pass P3 scribe analysed by confocal microscopy.

3.4 Interconnected cell

The optimized parameters developed for the different cell scribes are listed in Table I. Applying these parameters and the layout shown in Figure 1 interconnected cell stripes were produced. The IV characteristics of the monolithically interconnected cell stripes for the two different P2 scribe realisations (multi pass shifted and multi pass) are averaged over 12 samples and compared to a reference cell without the monolithic interconnection concept. The results are presented in Figure 8. One can see that there is no significant difference in cell performance for the cells fabricated with different P2 scribes. Therefore, if an ITO front contact is used it is not necessary to perform a multi pass shifted scribe. Nevertheless, the result demonstrates the feasibility to perform such a scribe on micromorph tandem cells too, in case that other front contacts with a thin metal layer are used. The interconnected cells show a high parallel resistance which is identical to the one of the reference cell indicating an ideal cell separation. Their series resistance is slightly increased compared to the reference cell which also leads to a lower fill factor (*FF*) than the reference cell. Further analysis of the monolithically interconnected cell with multi pass P2 scribe by dark lock-in thermography (D-LIT), presented in Figure 9, indicates that the laser scribes are performed satisfactory. The active cell area is isolated well against the surrounding cell area as the heating just takes place inside the active cell area and not at the P3 scribe and the surrounding cell area. Additionally, the intensity profile illustrates a low temperature increase at the P2 scribe which is still lower than the one in the active cell area. The increased series resistance as well as the lower *FF* of the interconnected cells occurring in the IV-characteristic can therefore be explained by the sheet resistance of the front contact but not with high resistance due to the P2 scribe.

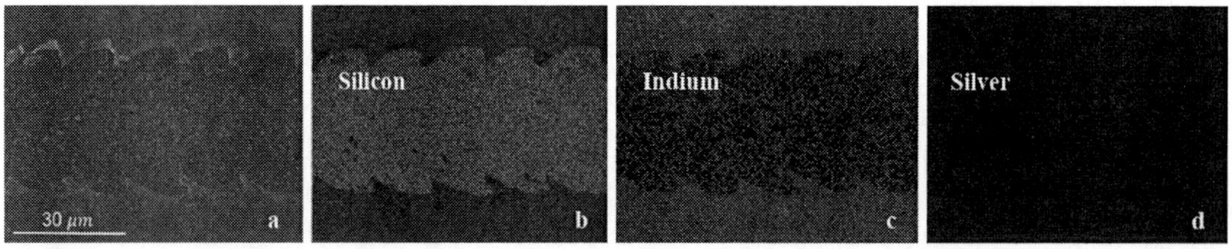

Figure 6: EDX analysis of a multi pass P3 scribe having a horizontal scribe direction: (a) SEM image, (b) silicon detection, (c) indium detection and (d) silver detection.

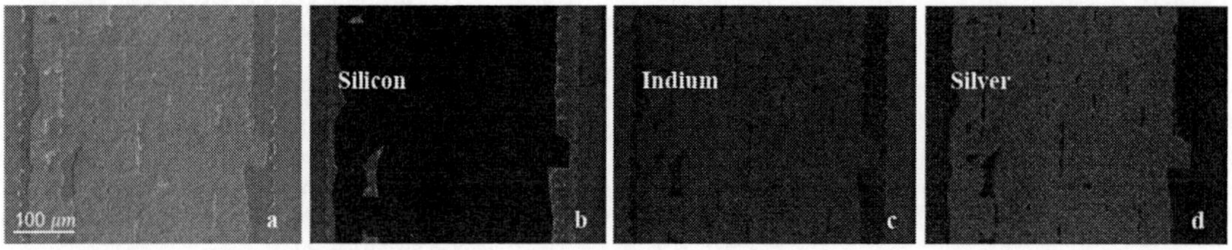

Figure 7: EDX analysis of a multi pass shifted P3 scribe having a vertical scribe direction: (a) SEM image, (b) silicon detection, (c) indium detection and (d) silver detection.

Table I: Optimised parameters developed for the different scribes. The parameters are the laser wavelength λ, mean laser power P, speed v and number of passes n.

Scribe	λ [nm]	P [W]	v [mm/s]	n	Remarks
P1	532	0,73	1500	1	-
P2 multi pass	1064	2,76	1000	2	-
P2 multi pass shifted	1064	0,82	1700	1-6	6 lines, Line distance 10 μm
P3	532	1,5	2500	1	15 lines, Line distance 25 μm

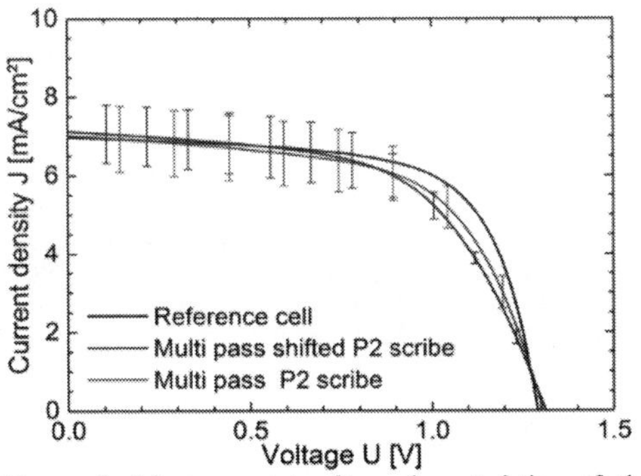

Figure 8: Mean current-voltage-characteristics of the interconnected cell stripes for the two different P2 scribes and current-voltage-characteristic of a reference solar cell.

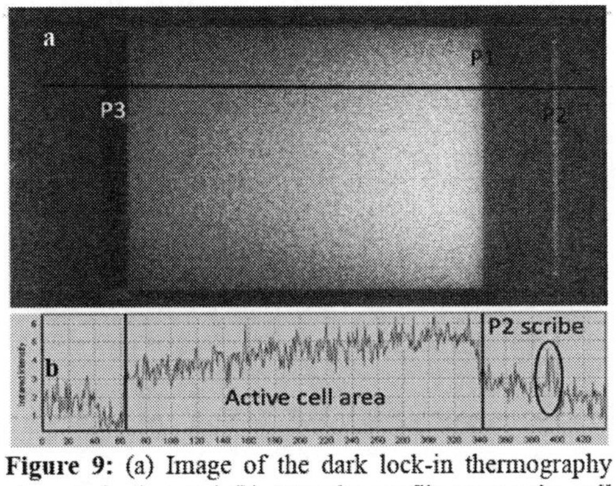

Figure 9: (a) Image of the dark lock-in thermography characterisation and (b) Intensity profile across the cell area

4 CONCLUSION

In this work we performed a proof of concept showing the feasibility of monolithic series interconnection of micromorph tandem thin film solar cells deposited on flexible aluminium substrates using three laser scribes without the established shingle technique or an additional grid. This interconnection method which is well established for glass substrates was firstly transferred to a micromorph cell stack deposited on a flexible substrate. Based on our results and the optimised laser parameters which were presented monolithic series interconnection can be applied on the complete module of this layer stack. Therefore, more efficient modules can be fabricated by integrating the monolithic series interconnection in a cost efficient roll-to-roll process.

5 ACKNOWLEDGEMENTS

We would like to thank Ulf Kirstein, Martin Kellermann, Antje Schweitzer, Volker Steenhoff, Jens Sürig-Morieng and Alex Neumüller for cell depositions, plasma treatment and helpful discussions. Financial support from the German Federal Ministry of Education and Research in the framework of the SiSoFlex project (FKZ 03SF0418) is gratefully acknowledged.

6. REFERENCES

[1] Kessler F., *4. Workshop: Photovoltaik-Modultechnik, flexible PV-Dünnschichtmodule, Übersicht weltweit*, Köln, 2007

[2] Löffler J. et al., *Depth selective laser ablation for monolithic series interconnection of solar cells,* 24th European Photovoltaic Solar Energy, 2009

[3] Bovatsek, J., et al., *Thin film removal mechanisms in ns-laser processing of photovoltaic materials*, Thin Solid Films, 2010. 518(10): p. 2897-2904.

[4] Schweitzer, A., et al., *Monolithic Interconnection of Silicon Based Thin Film Solar Cells on Aluminium Substrates*, 31st European Photovoltaic Solar Energy Conference and Exhibition, 2015.

[5] Geißendörfer, S., et al., *The SiSoFlex Project: Silicon Based Thin-Film Solar Cells on Flexible Aluminium Substrates,* 29th European Photovoltaic Solar Energy Conference and Exhibition, 2014: p. 1667 - 1670.

[6] Neumüller, A., et al., *Influence of Argon Plasma Treatment at i-/p- Interfaces on Silicon Thin-Film n-i-p Solar Cells,* 31st European Photovoltaic Solar Energy Conference and Exhibition, 2015.

[7] Haas, S., *Untersuchung und Optimierung der Serienverschaltung von Silizium-Dünnschicht-Solarmodulen.* Vol. 91. Forschungszentrum Jülich, 2010, p. 164 ff.

EXTERNAL QUANTUM EFFICIENCY AS FUNCTION OF APPLIED VOLTAGE OF MULTI-JUNCTION HYDROGENATED AMORPHOUS SI BASED CELL: PERFORMANCE OPTIMIZATION AFTER STABILIZATION

Canino A.[1,a], Condorelli G.[1], Battaglia A.[1] and Gerardi C.[2]
[1]3SUN, Contrada Blocco Torrazze Zona Industriale 95121, Catania, Italy
[2]Enel Green Power; Contrada Blocco Torrazze Zona Industriale 95121, Catania, Italy
[a]Corresponding author: andrea.canino@3sun.com

ABSTRACT: A multiple junction device based on hydrogenated amorphous silicon (a-Si) and microcrystalline-amorphous (μ-Si) mixed phase material has the advantage to reach an high Voc. One of the most important point for the development of a multiple junctions device is to find the correct balance between the sub-cells current because this is the limiting factor to the total device output. Because of the a-Si Staebler-Wronski degradation the sub-cell currents can vary hence, for better cell performance, it is fundamental to balance currents after the light exposure. We developed a method to measure the intrinsic performance of mj-device to define the correct balance among the cells leading to a higher device stability in time. In the specific case of a tandem cell, the correct balance determination allows to avoid also to waste material thanks to the possibility to well define the thickness of each layer. In this work we studied the impact of current density of μ-Si sub-cell with the aim of reaching the higher efficiency with lower light induced degradation (LID) factor. Actually we measured an initial efficiency between 12% and 13% on small area samples and after LID we found values around 11.5% reaching a LID factor of about 9%.
Keywords: a-Si/μ-Si, Fabrication, Degradation, Durability

1 INTRODUCTION

Thin-film Si solar cells have been for several years a good choice for big PV installation due to their potentials for large areas and low-cost manufacturing. For high-efficiency thin film tandem solar cells, optimized light trapping and device structure are mandatory to obtain the higher current density minimizing the production total time. The deposition of silicon active layers, i.e. the two p-i-n junctions, the amorphous (a-Si) one and the microcrystalline (μ-Si) one, is extremely long and significantly impacts the cost. The device architecture must be the best compromise among process time and performance optimization. In this optic the determination of the correct current matching among junctions and, as consequence, the determination of deposition time for the active layers can give a fundamental gain in the development of the technology. From the industrial point of view, to reduce production cost and at same time to increase the efficiency are both key elements of thin film silicon technology.

We studied as particular case the micromorph tandem cell that is a sequence of two p-i-n junctions, top cell of a-Si:H and bottom cell of c-Si:H, which allows solar energy conversion in a wide range of the sunlight spectrum (from ~ 300nm to 1100nm). The same concept can be applied to every multiple junctions device.

As the cells behave like two current generators connected in series, the overall cell current is determined by the lower current between the two cells. Consequently, a large mismatch between the cell currents is not ideal. Moreover the difference in fill factor between sub-cells makes complicated to find the correct match between them [1]. Typically, the best condition is to have a top cell to bottom cell current ratio of ~ 1 during operating conditions. The Top/Btm current ratio is a fundamental parameter in regulating the impact of light induced degradation (LID) of a tandem cell [2]. In fact, when the initial current of the Top cell is lower than that of the bottom cell, i.e. the tandem cell is top limited, the tandem cell LID factor is mainly determined by the Staebler-Wronsky (SW) degradation [3] of the a-Si:H top cell. This implies a higher degradation of the tandem cell performances caused by significant short circuit current reduction. In Figure 1 we summarized the theoretical extrapolation for a tandem photovoltaic device.

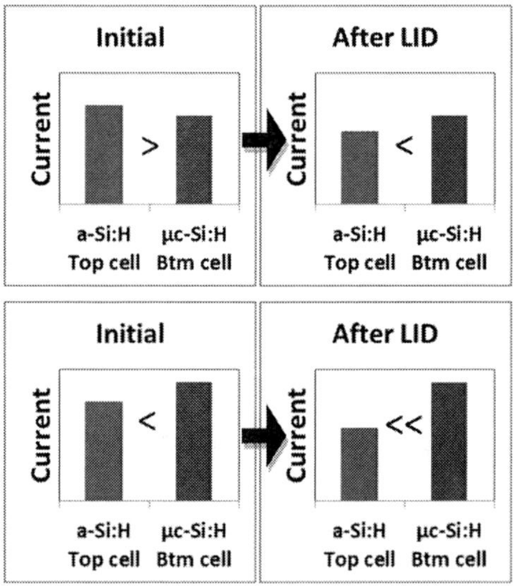

Figure 1. Currents balance general concept before and after LID in the general case of a tandem cell.

The upper graph of Figure 1 reports the case of an initial state for a tandem cell with Top/Btm currents ratio > 1. In these conditions the reduction of current of Top amorphous cell due to SW effect, is partially mitigated by the initial current mismatch in favor of the Top cell. In the lower panel the Top/Btm < 1 case as initial state is reported. By increasing the Btm junction current (i.e. for example by increasing the thickness of intrinsic layer) above the current of Top cell, there is a current excess in the Btm cell that is wasted. This condition leads to an extra degradation due to a decrease of Isc after light soaking and to the predominance of the top amorphous cell degradation.

2 EXPERIMENTAL DETAILS

2.1 Tandem cell deposition

Doped and undoped silicon-based films have been prepared using an industrial 13.56 MHz multi-chamber plasma-enhanced chemical vapor deposition (PECVD) system. The reactor has been developed to ensure high throughput and low cost of ownership. For this reason the process is carried out simultaneously on 8 glasses each one of 1.4m x 1m.

In Figure 2 we reported the scheme of the tandem cell device. Despite the conceptual simplicity of the device, the deposition of all layers requires about 100 process steps and last about 4 hours. The a-Si intrinsic layer is about 250 nm while the m-Si intrinsic layer is about 1450 nm.

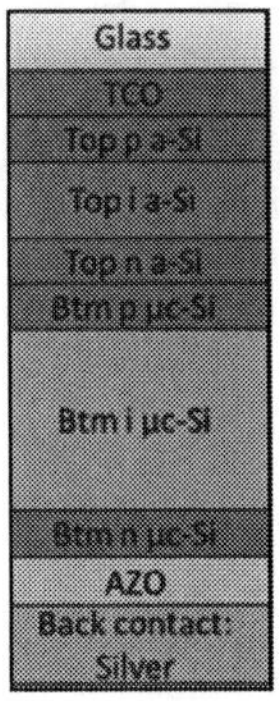

Figure 2. Tandem cell device architecture.

2.2 Quantum efficiency measurements

The external quantum efficiency (EQE) allows to determine the current of sub-cells. We used a Bunko Keiki customized tool that allows to measure separately the junctions in stack. The tool has a monochromatic lamp that is the probe for the junction that must be measured and an AM1.5G light source and different filters to switch on the other cells. It is possible to measure the EQE applying different external voltage bias.

The system allows to maintain constant the applied voltage during the measure. With this method we were able to calculate the maximum current that can flow through the device at each operating voltage point up to the maximum power point. Over this voltage the tandem cell with an external imposed direct bias starts to conduct in the opposite way thus avoiding to reach the Voc limit, i.e. the point at which the device total current is zero. Never the less, up to the Pmax point we can follow the trend of calculated I-V curve and we can compare the results with I-V curves measured directly under 1 sun at AM1.5G (not shown here). In Figure 3 the experimental data show that after 80h of LID one of the amorphous junction clearly is limiting the current flow at maximum power point while at zero voltage the currents are almost aligned.

The data reported in Figure 3 clearly show that at maximum power point the limiting junction after degradation is the amorphous one. Actually even if the two junction are both able to sustain the same Jsc, there is a mismatch during operating condition. Moreover it is clear the series shunt resistance is higher for the microcrystalline junction than for the amorphous one as expected, i.e. the internal losses are higher in the amorphous layer.

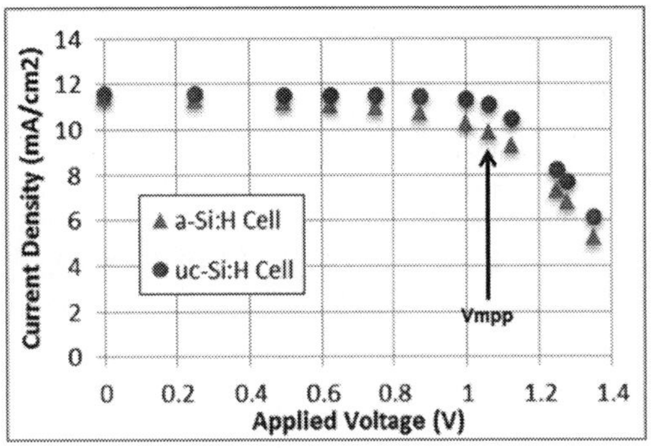

Figure 3. Currents as function of an external bias calculated from EQE.

2.2 Electrical parameters trend after LID

In order to verify how the tandem device behaves under illumination after LID, we produced several modules changing the thickness of the bottom layer. This way was possible to extract small area samples and to measure the performances as function of the Top to Btm current ratio. The samples have been made by isolating with an offline laser 3 cells monolithically integrated. Each cell has an area of about 2.05 cm^2 so the total area is around 6.15 cm^2.

Despite the initial Pmax value is higher, as expected, for Top to Btm close to 1, after 500h of exposure in open circuit condition at 50°C and under 1 sun AM1.5G of illumination, the best performances have been reached with an initial ratio close to 1.06.

In Figure 4 we reported separately the initial and final value of efficiency on small area measured on our samples as function of Top to Btm currents ratio. The values are referred to the aperture area of the cells. If we scale the numbers to full module the higher efficiency at 11.6% corresponds to a module efficiency of about 10.5% considering all the dead areas (lasers, metallization and trimming).

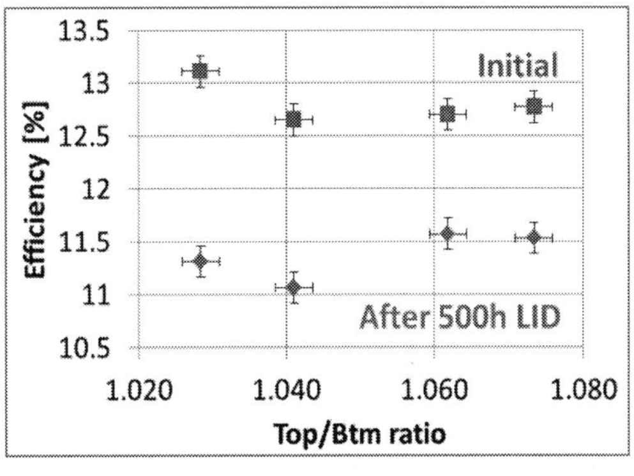

Figure 4. Efficiency before and after 500h of LID as function of Top to Btm currents ratio.

In Figure 5 we reported the relative variation of electrical parameters. As it is possible to see in the graphs the main gain is related to almost zero loss of short circuit current (Jsc) and open circuit voltage (Voc).

The reduction to zero of Jsc loss is an indirect parameter to evaluate the mismatch of current between the cell. The zero reduction means for sure that the current of the top cell has not been reduced below the bottom cell current in short circuit. We can say that between the average ratio 1.06 and 1.08 we are in a condition that minimize the loss of current. Considering that below the 1.06 limit the Pmax degradation start to be very high while above the 1.08 the initial Pmax start to be too low. Considering the best compromise between the currents, we were able to reduce the relative Pmax loss to about 9%.

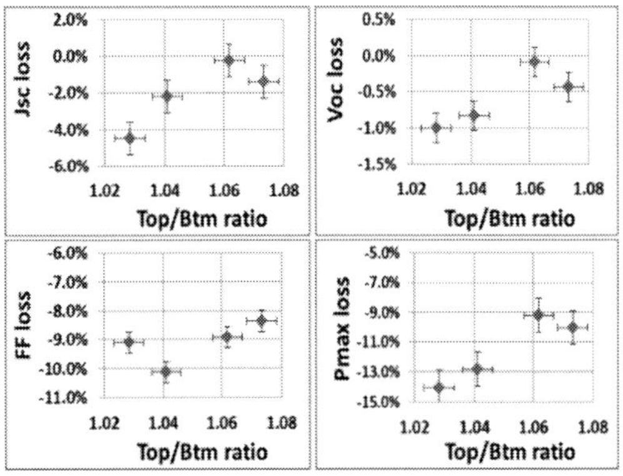

Figure 5. Electrical parameters relative losses as function of the Top to Btm current ratio.

3 OUTDOOR PERFORMANCES EVALUATION

The last evaluation step in this phase of the development program was to evaluate the stabilization of performances during outdoor exposure. We studied different modules with an average Top to Btm current ratio of 1, 1.03 and 1.06. We refer to average values because on a very large scale (1.4m x 1.0m) we cannot speak of perfectly defined ratio over the whole area. Actually we have a spread of ratio values over the deposited area. Actually in our case, i.e. in the case of a large area PECVD reactor, we must consider several different causes as well as the usual measurements indetermination. Actually we must take into account the uniformity of all the production steps of the modules within the fab in standard production configuration. As an example the deposition non-uniformity is well below the 10% on a whole module (1m * 1.4m) and to apply the results to a large production scale we must consider the observed results as lower limit of a distribution for all the points of the module.

We know from previous evaluations that our modules reaches the stability during outdoor exposure in about 4 weeks. In Figure 6 we reported the normalized Pmax variation for 3 different modules exposed at the same time and working at maximum power point. The results showed that by reducing the Btm intrinsic layer thickness it was possible to obtain a better results also from the LID point of view. Actually we are able to decrease the LID factor from about 11% to slightly less than 10%.

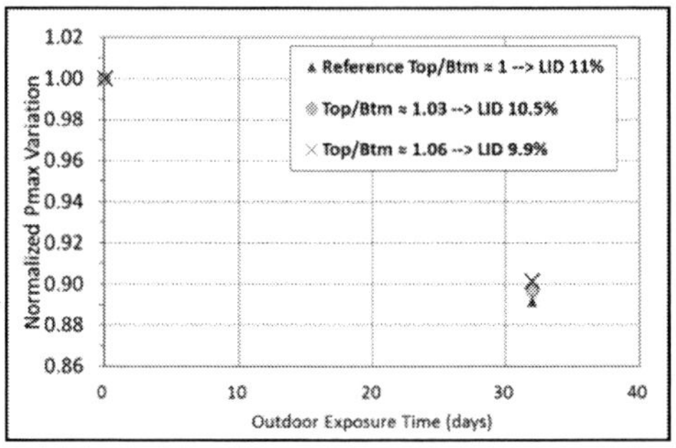

Figure 6. Outdoor exposure results. Normalized Pmax variation as function of exposure days.

4 SUMMARY

We used an effective method to evaluate how the two sub-cells of a tandem device work together. By calculating the current flow through the device by using the EQE data at various applied voltage we were able to determine that during the degradation of the cell due to the SW effect, the main loss can be reduced to an Impp reduction and, as consequence, to a FF reduction by correct matching the initial currents of the sub-cells.

Actually this method can be also used to determine the correct balance among different cells connected in series in a module or to cells connected in tandem in a different way. We are referring for example to crystalline silicon modules and to the criteria used for the sorting of cells. Or for example to the studies on tandem cell architecture based on different materials, e.g. peroskite/c-Si tandem cells. In all these cases the determination of the working point will help to optimize the output of the modules.

This study allowed to demonstrate that one of the key parameters for the minimization of the LID factor of tandem cell is the correct balance between Top and Btm cell current. The main problem on large area scale is to find a compromise among the initial Pmax value and LID factor. We found that it is possible to reach a good working point for Top/Btm ratio around 1.06 or slightly higher that allows, with our actual technology, to save deposition time and, as consequence, costs without losing performances. On the other hand we were also able to reach a LID factor on modules around the 10%. On the basis of this results a more deep study on the intrinsic properties of materials has started with the aim of a further quality improvement of mj-devices based on silicon.

From industrial point of view the reduction of thickness to a value that allows a ratio of 1.06 leads to an increase of throughput of about ~1÷2% that means for 3SUN line in full production about 2÷4MW per year. This means an overall reduction of the production cost.

5 REFERENCES

[1] Baojie Yan, Guozhen Yue, Jeffrey Yang and Subhendu Guha, Photovoltaic Specialists Conference, 2008. PVSC '08. 33rd IEEE.

[2] M. Bonnet-Eymard n, M.Boccard, G.Bugnon, F.Sculati-Meillaud, M.Despeisse, C.Ballif, Solar Energy Materials & Solar Cells 117 (2013) 120–125

[3] D. L. Staebler and C. R. Wronski, Appl. Phys. Lett. 31, 292 (1977).

ANALYSIS OF CARRIER LIFETIME IN LIQUID-PHASE CRYSTALLIZED SILICON ON GLASS

Michael Vetter, Annett Gawlik, Jonathan Plentz, Gudrun Andrä
Leibniz-Institute of Photonic Technology (IPHT), Dept. Functional Interfaces
Albert-Einstein-Str. 9, 07745 Jena, Germany

ABSTRACT: Liquid-phase crystallized silicon on glass (LPCSG) presents a promising material to fabricate high quality silicon thin films on glass, e.g. for solar cells and modules. To achieve a high electronic material quality after the crystallization process on glass, first barrier layers (BL) and then an amorphous silicon layer are deposited on the glass followed by crystallization with a line focus laser beam. In this paper we use injection level dependent lifetime data generated by the quasi steady-state photoconductance decay method (QSSPC) to characterize LPCSG absorbers. This contactless method allows a determination of the LPCSG absorber quality already at an early stage of solar cell fabrication, and provides a monitoring of the absorber quality during solar cell fabrication steps. So far, we found minority carrier lifetimes higher than 500 ns (e.g. n-type absorber with $N_D = 4x10^{16}cm^{-3}$) indicating a surface recombination velocity $S_{BL} < 1600$ cm/s at the BL-Si interface and an effective diffusion length $L_{eff} > 20$ µm in an 8 µm thick absorber.

Keywords: multi-crystalline silicon, thin film, laser crystallization, carrier lifetime, quasi steady-state photoconductance

1 INTRODUCTION

Liquid-phase crystallized silicon on glass (LPCSG) is a interesting material to fabricate solar cells and modules. Recently, solar cells with LPCSG absorbers produced by line focus laser beam or electron beam (e-beam) crystallization of amorphous silicon (a-Si) layers on glass, showed open circuit voltages (Voc) of 656 mV [1, 2, 3] similar to highest efficient multi-crystalline (mc) wafer solar cells [4]. LPCSG could merge the advantages of crystalline silicon (c-Si) wafer technology with its high efficiency potential and thin film technology with low Si consumption and low cost monolithic integration for module fabrication. Therefore, LPCSG technology could be an approach to overcome the emerging limits for further cost reduction in standard wafer-based module technology as e.g. reduction of wafer thickness or increasing wafer size. Tandem and bifacial configurations are under investigation to reach higher efficiencies.

An important role to achieve a high electronic material quality after the crystallization process on glass play barrier layers (BL) introduced between the glass and the a-Si layer. The BLs should ensure good adhesion of Si (wettability), facilitating crystallization of high-quality Si, blocking diffusion of impurities from the glass, passivating Si interface defects and acting as a transparent antireflection coating for the superstrate configuration, furthermore it could serve as a dopant source for the absorber. Recent investigations were focused on combination of SiO_2, SiN_x, SiO_xN_y and SiC_x layers [5, 6], however to find the optimal combination fulfilling all the requirements is still objective of investigation.

Most developments of the LPCSG absorbers have been made by fabricating solar cells, measuring and analyzing current-voltage (IV) characteristics and quantum efficiency (QE) or on quasi-solar cells measuring the Suns-Voc characteristic.

In literature [1] the effective diffusion length (L_{eff}) of high-quality LPCSG absorbers is estimated from QE data to lie around 10-30 µm of solar cells with absorber thickness (W) in the range of 5-10 µm and with efficiency in the range of 10%. Carrier lifetime can be calculated from the basic relation $\tau = L^2/D$ and diffusion constant $D = \mu kT/q$ where μ is the carrier mobility. We have calculated τ in function of L for c-Si [8] where μ, D and τ are calculated by the "mobility calculator" on the pvlighthouse web page [9]. The μ in LPCSG is lower than in c-Si [2] resulting in an underestimation of τ_{eff} for LPCSG in this calculation, nevertheless for effective $L_{eff} \approx$ 10-30 µm [1] we suggest that effective lifetime, τ_{eff} of n-type LPCSG absorbers (W $\approx$ 5-10 µm) will be in the range from approx. 10 ns to 1 µs at an injection level of 10^{15} cm^{-3}.

In this work, we measure and analyze injection level dependent lifetime (τ_{eff}) data generated by the quasi steady-state photoconductance decay (QSSPC) method [7]. This contactless method allows a determination of the LPCSG absorber quality already at an early stage of solar cell fabrication, and provides a monitoring of the absorber quality during the solar cell fabrication steps.

2 EXPERIMENTAL

To measure τ_{eff} in function of the injection level (Δn) in LPCSG absorbers we modified a Sinton WCT-120 Photoconductance tool [8]. LPCSG absorbers are prepared on 3 mm thick Borofloat 33 (Schott) glasses. First BLs are deposited by sputtering on the Borofloat glass then the deposition of the a-Si layer is done by depositing non-doped a-Si layers (5-10 µm) by e-beam evaporation. Doping of absorbers is achieved by depositing a thin highly doped a-Si film on the intrinsic film before crystallization. Latter is performed by continuous wave line focus laser irradiation at 808 nm. The typical preparation steps at the beginning of the hetero emitter solar cell fabrication process then are removal of a SiO_2 layer formed in the laser crystallization process in air. Next follows hydrogen plasma passivation of bulk defects at about 500 °C and a surface etch to remove surface damage produced in the plasma process. Then, we perform RCA cleaning and deposit thin intrinsic amorphous silicon (a-Si) and a p-type a-Si layer and a transparent conductive oxide (TCO) layer for hetero emitter formation.

3 RESULTS / DISCUSSION

An important prerequisite to be able to measure the lifetime in thin Si layers in the thickness range of a few µm is a very good surface passivation. For insufficient surface passivation (e.g. surface recombination velocity (S) reaching the thermal electron/hole velocity (v_{th}=2x10^7 cm/s)) the upper limit of τ_{eff} is given by $1/\tau_{eff} = 1/\tau_{bulk} + 12*D/W^2$ [10] where τ_{bulk} is the lifetime of the Si absorber bulk material. From this equation, one calculates an upper limit of τ_{eff} in the range of a few ns ($N_D \approx 10^{15}$-10^{17}cm^{-3}), in this case bulk lifetime will not be visible. Therefore, the first is to find a barrier layer which provides good passivation at the glass/Si interface, in addition the top surface of the absorber should be well passivated. In this work we use a combination of a 300 nm thick SiO_2 layer which serves as a diffusion barrier and a thin 20 nm thick SiON layer which provides good surface passivation and good surface wetting in the crystallization process.

Figure 1 shows lifetime measurements of the initial preparation steps of the solar cell fabrication process and its impact on τ_{eff} in a non-doped absorber with initial thickness of 11 µm. A non-doped absorber has been prepared to see the possible diffusion of contaminants from the glass into the absorber during the laser crystallization process. The specific resistance resulted in about 800 Ωcm indicating that no active doping atoms are incorporated in the absorber during the laser process suggesting that the SiO_2/SiON BL stack is an effective barrier. In addition, two different etch solutions, perchlorate etch (HNO_3/CH_3COOH/HF/ $HClO_4$, etch rate ≈ 1µm/min) and purple etch ($KMnO_4$/HF, etch rate ≈ 100 nm/min) both known as polishing etches, are used to remove surface damage and surface contamination.

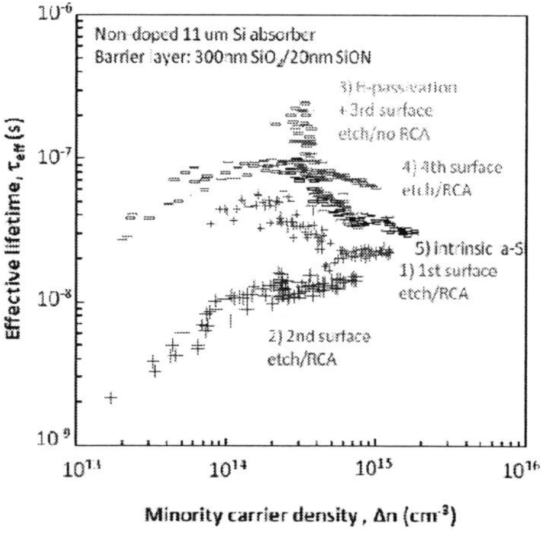

Figure 1: Lifetime measurements of an initially 11 µm LPCSG absorber after different preparation steps, top surface passivated by HF-dip after RCA cleaning (steps 1-4).

After laser crystallization in air, first the SiO_2 layer formed in this process and about 500 nm of silicon has been etched away by perchlorate etch to remove any surface contamination produced in the laser process. Then an RCA cleaning with final HF-dip is performed and lifetime is measured immediately. The H-termination of the silicon surface of the HF-dip provides a very good surface passivation for about 30 min. The τ-data is shown in figure 1 by purple plus symbols indicating a τ_{eff} of about 20 ns in high injection (marked by "1) 1st surface etch/RCA"). However, τ increase in low injection which we attribute to carrier trapping effects, typically found in mc-Si [11]. Next, we have performed a surface polishing etch using purple etch removing about 100-200 nm of silicon [12]. RCA cleaning with final HF-dip was performed again followed by immediate τ measurement. The τ-data is presented in figure 1 by big red plus symbols (marked with "2) 2nd surface etch/RCA"). τ_{eff} in high injection lies at about 15 ns and now decreases in low injection, indicating that a large number of trapping centers have been removed by the purple etch. In the following process step the sample was heated to at about 500 °C and has been exposed to hydrogen plasma with the objective to passivate bulk and grain boundary defects in the multi-crystalline material. Afterwards again about 500 nm of silicon was etched from the surface with the perchlorate etch and an HF-dip has been performed. τ was measured which is shown in figure 1 by the green minus (marked with "3) H-passivation+3rd surface etch/noRCA"). The τ-curve again increases in low injection indicating again trapping effects. Next, another purple etch and RCA cleaning with final HF-dip has been performed with the sample and τ was measured again shown by the blue minus symbols in figure 1 (marked with "4) 4th surface etch/RCA"). The τ has increased to about 60-90 ns due to the hydrogen plasma treatment and trapping effects are not visible. Finally, a thin (5 nm thick) intrinsic a-Si layer was deposited on the sample and τ was measured again, the τ-curve is presented in figure 1 by black minus symbols (marked with "5) intrinsic a-Si"). The τ-curve coincides again with the τ-curve directly after hydrogen passivation indicating that the very thin a-Si layer does not provide such good surface passivation as the HF-dip. This also did not change by depositing an additional p-type a-Si layer, the next step for hetero emitter formation (not shown).

Interestingly, non-doped Si absorbers present a p-type conductivity (indicated by +-symbols in figure 1) which is determined by the change of the contact voltage when heating of two electrical measurement probes on the absorber surface (Seebeck-effect). The conductivity type changes to n-type (- symbols in figure 1 after a hydrogen plasma passivation step of electronic defects.

Figure 2 shows τ-data of a similar fabrication sequences with an intentionally n-type doped 8 µm thick LPCSG absorber with a specific resistivity of 0.16 Ωcm (N_D approx. 4x10^{16} cm^{-3}). Figure 3 presents the corresponding Suns-Voc curves calculated from the photoconductance data and Table I the values of the data fitting of Suns-Voc curves in the high injection range.

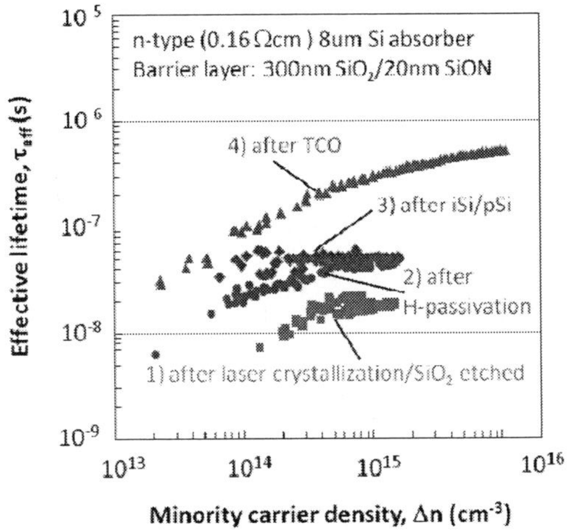

Figure 2: Lifetime measurements of an 8 μm LPCSG absorber after different preparation steps, in process steps 1 and 2 top surface is passivated by HF-dip after RCA cleaning.

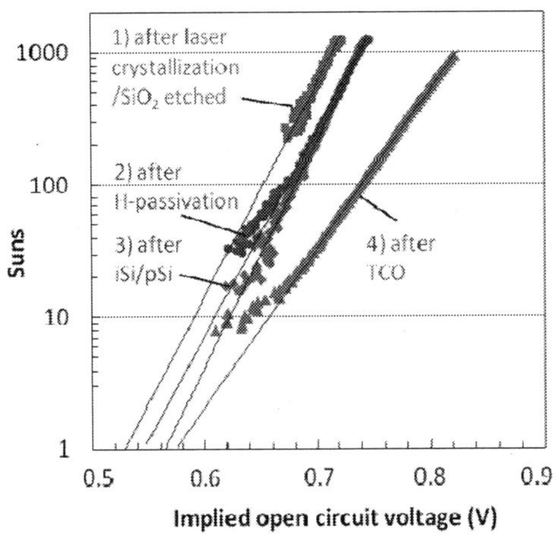

Figure 3: Suns-Voc curves calculated from conductance data of the 8 μm LPCSG absorber of figure 2. Lines present data fitting in the high injection region with fitting values in Table I.

For this sample after laser crystallization the SiO_2 layer formed in the process was removed by HF solution and only purple etch has been performed since it provides better surface passivation than the perchlorate etch. Then, an RCA cleaning with final HF was made and τ was measured immediately (marked with "1) after laser crystallization/SiO_2 etch"). The τ-curve and Suns-Voc curve are shown by the green squares in Fig. 2 and Fig 3. Next, hydrogen plasma treatment has been performed followed by a purple etch and RCA cleaning with final HF-dip. The curves are presented by purple dots in Figs. 2 and 3 (marked with "2) after H-passivation"). τ_{eff} has increased from about 20 ns to about 50 ns in high injection and the implied Voc at 1 Sun (extrapolated from the data fitting in high injection) has increased by 20 mV (see Table I). The slopes of the Suns-Voc curves are about 75 mV/dec indicating recombination processes in the absorber. Next, intrinsic a-Si layer and p-type a-Si layer were deposited in one vaccum process and τ was measured again (presented by red rhombus in Figs. 2 and 3, marked as "3) after iSi/pSi"). The implied Voc at 1 Sun increases by 15 mV to 559 mV and the slope of the Suns-Voc curve has decreased below 70 mV/dec which let us suggest that in this case surface recombination is slightly less than with H-termination from HF-dip.

Finally, a TCO layer consisting in a thin 5 nm thick Al_2O_3 layer and a 200 nm thick Al-doped ZnO (AZO) layer was deposited on the absorber and τ was measured again. τ increases about one order of magnitude in high injection but decreases in low injection. The Suns-Voc curve presents with 563 mV still slightly higher Voc at 1 sun than with the a-Si layers however, the slope of the curve increased to 99 mV/dec.

Table I: Values of fitting Suns-Voc curves in Fig. 3 in high injection.

Process step	V_{oc}(@1Sun) (mV)	Slope (mV/dec)
1) after laser crystal.	524	74
2) after H-passivation	544	77
3) after iSi/pSi	559	69
4) after TCO	563	99

Considering the lifetime of about 500 ns at high injection with TCO layer suggests a nearly perfect surface passivation at the top surface (e.g. with S_{top}<<100 cm/s). Then the effective surface recombination velocity (S_{BL}) at the BL-Si interface can be estimated from the relation $1/\tau_{eff} = 1/\tau_{bulk}+S_{BL}/W$. Attributing all recombination losses to the bulk, then the τ_{eff} of 500 ns indicates an upper limit of the S_{BL} of 1300 cm/s. On the other hand attributing all recombination to the BL/Si interface the τ_{eff} of 500 ns indicates a lower limit of L_{eff} > 20 μm.

4 CONCLUSION

Effective lifetime data in function of the injection level of LPCSG absorbers has been generated by the QSS-PC method at various fabrication steps of a hetero emitter solar cell process. Low resistivity of non-doped LPCSG absorbers after laser crystallization indicates that a 300 nm SiO_2 / 20 nm SiON barrier layer stack provides a good barrier to prevent diffusion of impurities from the Borofloat 33 glass to the absorber during melting of the silicon layer. Surface preparation of the absorber top surface after hydrogen plasma treatment is important to achieve a low surface recombination velocity and to prevent trapping effects in a hetero emitter structure. A LPCSG absorber with a hetero emitter consisting of very thin intrinsic and p-type doped silicon layers in combination with an Al_2O_3/AZO stack presents a τ_{eff} of about 500 ns at an injection level of $5x10^{15}$ cm^{-3} (100 Suns) indicating an upper limit of the surface recombination velocity at the BL/Si interface $S_{BL}< 1600$ cm/s and a effective diffusion length of L_{eff} >20 µm in the 8 µm thick absorber.

ACKNOWLEDGEMENTS

The authors like to thank H.-P. Linke and U. Brückner for technical assistance. This project has received funding from the European Union's Horizon 2020 research and innovation programme under the Marie Sklodowska-Curie grant agreement No. 657115 (cSiOnGlass, M.V.). We like to thank the Thüringer Aufbaubank (TAB) for funding the project Bi-PV (FKZ 2015 FGR 0078) and the European Commission (ESF) for co-financing the project Bi-PV (ERDF).

References

[1] J. Haschke, D. Amkreutz, L. Korte, F. Ruske, B. Rech B, "Towards wafer quality c-Si thin-film solar cells on glass", Sol. Energy Mater. Solar Cells 128, (2014) 190-197.

[2] J. Dore, D. Ong, S. Varlamov, R. Egan, M.A.Green, "Progress in laser-crystallized thin-film polycrystalline silicon solar cells: intermediate layers, light trapping, and metallization", IEEE J. Photovolt.4 (1) (2014) 33-39.

[3] A. Gawlik, I. Höger, J. Bergmann, J. Plentz, T. Schmidt, F. Falk, and G. Andrä, "Optimized emitter contacting on multicrystalline silicon thin film solar cells", Phys. Status Solidi RRL 9 (7) (2015) 397–400.

[4] M. A. Green, K. Emery, Y. Hishikawa, W. Warta, E. D. Dunlop, "Solar cell efficiency tables (version 46)", Progr. Photovol.Res. Appl. 23 (2015) 805-812.

[5] O. Gabriel, T. Frijnts, S. Calnan, S. Ring, S. Kirner, A. Opitz, I. Rothert, H. Rhein, M. Zelt, K. Bhatti, J-H Zollondz, A. Heidelberg, J Haschke, D. Amkreutz, S. Gall, F. Friedrich, B. Stannowski, B. Rech, R. Schlatmann, "PECVD Intermediate and absorber layers applied in liquid-phase crystallized Si solar cells on glass substrates", IEEE J. Photovolt. 4 (6) (2014) 1343-1348.

[6] J. Dore, S. Varlamov, M.A. Green, "Intermediate layer development for laser-crystallized thin-film Si solar cells on glass", IEEE J Photovolt.5 (1) (2015) 9-16

[7] R. Sinton, A. Cuevas "Contact-less determination of current-voltage characteristics and minority carrier lifetimes in semiconductors from quasi steady-state photoconductance data", Appl. Phys. Lett. 69, (1996) 2510.

[8] M. Vetter, A. Gawlik, J. Plentz, G. Andrä, "Carrier Lifetime in Liquid-Phase Crystallized Silicon on Glass ", SiliconPV2016, Chambery (France) 7.-9.3.2016, to be published in Energy Procedia 2016.

[9]https://www.pvlighthouse.com.au/calculators/mobility%20calculator/mobility%20calculator.aspx

[10] I. Martin, Thesis (2003, Universitat Politècnica de Catalunya, Barcelona, Spain) p. 45.

[11] A. Cuevas, M. Stocks, D Macdonald; M. Kerr; C. Samundsett, "Recombination and trapping in multicrystalline silicon", IEEE Trans.Electr. 46 (1999) 2026.

[12] I. Höger, M. Himmerlich, A. Gawlik, U. Bueckner, S. Krischok, G. Andrä,, "Influence of intermediate layers on the surface condition of laser crystallized silicon thin films and solar cell performance", J. Appl. Phys 119, (2016) 04530.

DEPENDENCE OF THE TRANSPORT LENGTH IN $CH_3NH_3PbI_3$ POWDERS ON LIGHT SOAKING: A SURFACE PHOTOVOLTAGE STUDY

Th. Dittrich*[1,2], O. Shargaieva[1], F. Lang[1], N. H. Nickel[1], B. Rech[1], J. Rappich[1],
[1]Helmholtz-Zentrum Berlin für Materialien und Energie GmbH, Kekuléstr. 5, 12489 Berlin, Germany
[2]Helmholtz-Zentrum Berlin für Materialien und Energie GmbH, Hahn-Meitner-Platz 1, 14109 Berlin, Germany
*E-mail: dittrich@helmholtz-berlin.de

ABSTRACT: The diffusion or transport length (L) of photo-generated charge carriers has been measured directly in $CH_3NH_3PbI_3$ powders by applying the surface photovoltage (SPV) method after Goodman. The average size of crystallites in the powder was changed between tens of µm and about 1 mm. The values of L ranged between 3 and 30 µm and decreased to 0.6 µm in some cases after light soaking. It was shown that light soaking does not necessarily lead to a degradation of L. An instability of L was observed under certain conditions depending on conditioning of $CH_3NH_3PbI_3$ crystallites, the value of the constant SPV signal and light soaking. The results have consequences for the further development of solar cells based on $CH_3NH_3PbI_3$.

Keywords: perovskite, diffusion length, light soaking, surface photovoltage

1 INTRODUCTION

Despite high reported solar energy conversion efficiencies [1] of solar cells based on organic-inorganic hybrid lead halide perovskites such as $CH_3NH_3PbI_3$ numerous physical properties and their origin of this class of materials remain still unclear. In particular, light soaking, altering efficiency and/or stability due to, for example, photo-induced degradation has been reported [2] but the mechanisms behind are not really well understood yet. One of the most important properties of any absorber is its diffusion or transport length. Diffusion lengths of up to 10 [3] or even 175 [4] µm have been reported for $CH_3NH_3PbI_3$ single crystals. However, these indirect measurements were based on the lifetime and there was no information about the influence of light soaking on the diffusion length.

The surface photovoltage (SPV) technique after Goodman is well suitable for the direct measurement of a diffusion length (L) [5]. In this method, the SPV signal is kept constant and the intersection point of the intensity dependence on the absorption length gives the value of L [5]. The very recent and accurate measurement of the optical parameters of $CH_3NH_3PbI_3$ by ellipsometry [6] makes the application of SPV after Goodman possible. Further, powders of $CH_3NH_3PbI_3$ crystallites with sizes larger than 10 µm are well suitable for studying the influence of light soaking on L in crystallites of $CH_3NH_3PbI_3$.

In this work, we started a systematic study of the dependence of the diffusion length in powders of $CH_3NH_3PbI_3$ (size fractions of tens of µm up to mm) on light soaking under red or blue light. Further it will be shown for a certain sample that the value of the constant SPV signal can have strong influence on the measured values of L.

2 EXPERIMENTAL DETAILS

2.1 Sample preparation

Powders of $CH_3NH_3PbI_3$ were prepared by precipitation from a saturated solution. For this purpose, methyl amine iodide (CH_3NH_3I) and lead iodide (PbI_2) were dissolved in γ-butyrolactone (Sigma Aldrich, $CH_3NH_3I:PbI_2 = 1:1$). The solution was kept at 80°C and permanent evaporation of the solvent was possible so that particles were able to precipitate. The size of precipitated $CH_3NH_3PbI_3$ crystallites and particles increased with increasing time and increasing depth in the precipitate at the bottom of the vial. The fraction of $CH_3NH_3PbI_3$ crystallites has been changed. For the series shown in this work, three samples were prepared from the same solution:

Sample A – extraction at high depth 17 days after preparation of the solution (large grain fraction with order of 1 mm)

Sample B – extraction at low depth 10 days after preparation of the solution (small grain fraction with order of 10 µm)

Sample C – extraction at higher depth 11 days after preparation of the solution (larger grain fraction with order of hundreds of µm)

Carbon pads were placed on glass substrates coated with a SnO_2:F contact layer. The extracted powder was dried in nitrogen. After that, some powder was crumbled onto a carbon pad and fixed by slight pressing with a clean glass sheet. Unfixed powder was removed by hitting the sample on a glass plate. The surface coverage of a carbon pad with $CH_3NH_3PbI_3$ crystallites was on the order of 80%.

2.2 Set-up for SPV measurement and light soaking

Figure 1 shows the scheme of the experimental set-up used in this work. Modulated surface photovoltage spectra were measured in the fixed capacitor arrangement with a mica layer placed between the sample and the reference (SnO_2:F on a quartz cylinder) electrodes [7]. A fresh mica layer was used for each powder sample in order to avoid contamination.

Modulated SPV signals were excited with a halogen lamp and a quartz prism monochromator (SPM2). The modulation frequency was 10 Hz. The modulated light was directed with a parabolic mirror through the quartz window of the vacuum chamber to the electrode with the sample. The measurement resistance was 50 GΩ. The SPV signals were detected with a high impedance buffer and measured with a double phase lock-in amplifier (EG&G 7260). A part of the modulated light was coupled out with a beam splitter and directed to a detector with a silicon photodiode (Elektronik Manufaktur Mahlsdorf). The intensity of the modulated light was measured with a second lock-in amplifier (EG&G 7260).

For measurements after Goodman, the analog output of the in-phase signal was connected to the input of a power supply with an integrated feedback unit (Elektronik Manufaktur Mahlsdorf) at which the desired SPV signal was set. The in-phase SPV signal was kept constant by adjusting the power of the halogen lamp.

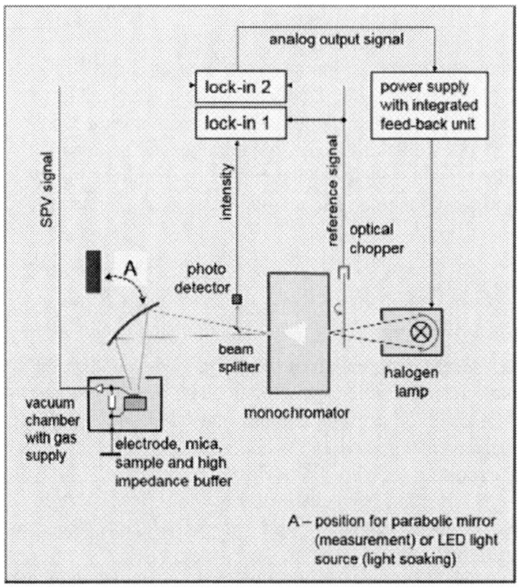

Figure 1: Schematic of the setup used for the measurement of the diffusion length by keeping the modulated surface photovoltage signal constant. The quartz cylinder carrying the SnO_2:F front electrode is not shown for keeping the scheme as simple as possible. For light soaking the parabolic mirror was replaced by the LED light source.

After preparation in a glove box, the samples were immediately placed into the vacuum chamber and the chamber was pumped down and filled with nitrogen to a pressure of 700 mbar in order to avoid uncontrolled oxidation during the SPV measurements.

Light soaking was performed with ultra-bright red and blue LEDs without removing the sample from the electrode. During light soaking, the light of the LEDs was directed to the sample with an optical fiber which was placed above the quartz window in front of the electrode.

2.3 Analysis after Goodman

The analysis after Goodman makes use of the dependence of the absorption length on the wavelength. For this purpose the data of Löper et al. [6] were interpolated with a cubic spline function (figure 2).

Figure 3 shows the overview spectra of the in-phase and phase-shifted by 90° SPV signals measured for sample A. The signs of the in-phase and phase-shifted by 90° signals were negative and positive, respectively. This means that photo-generated electrons were separated towards the external surface of $CH_3NH_3PbI_3$ crystallites.

For measurements after Goodman, the in-phase SPV signal (X) was kept constant and the light intensity was plotted as a function of the absorption length (see figure 4 for sample A). The value of L was obtained at the intersection point of the linear function. The analysis was performed over a wide range of constant X signals. For the given sample and measurement conditions, L was 5.7 µm which was much shorter than the size of $CH_3NH_3PbI_3$ crystallites in sample A.

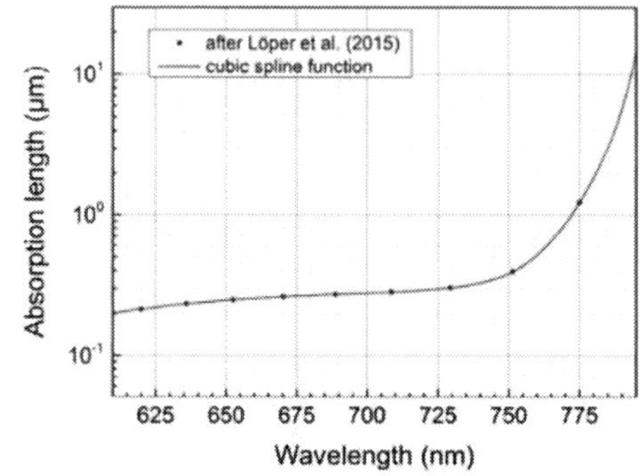

Figure 2: Spectrum of the absorption length of $CH_3NH_3PbI_3$ used for the analysis in this work. The data points are from Löper et al. [6].

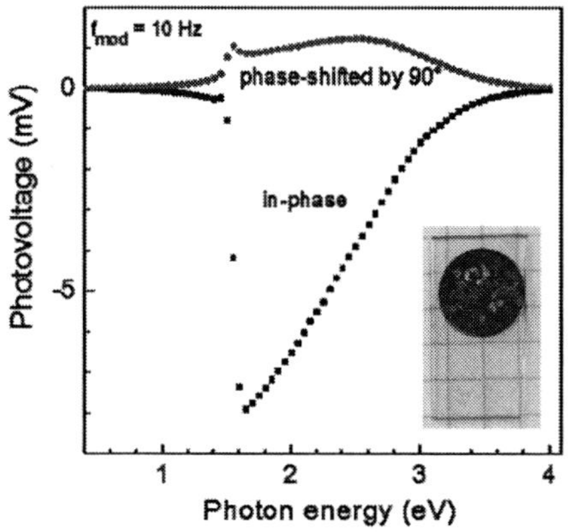

Figure 3: Overview spectra of the in-phase (squares) and phase-shifted by 90° (circles) surface photovoltage signals measured for sample A. The insert shows a photo of sample A (distance between lines 5 mm).

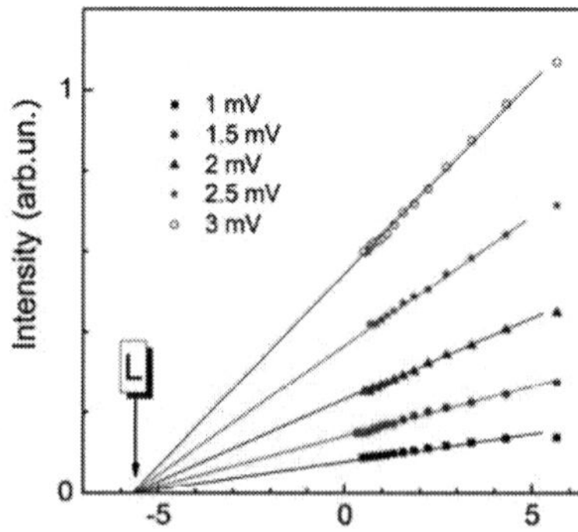

Figure 4: So-called Goodman plots or dependencies of the intensity required for keeping the SPV signals constant at 1, 1.5, 2, 2.5, and 3 mV (squares, circles, triangles, stars and open circles, respectively) on the absorption length of $CH_3NH_3PbI_3$ for sample A. L marks the value of the diffusion or transport length of photo-generated charge carriers.

For common semiconductors, there is one common intersection point independent of X. However, we found for $CH_3NH_3PbI_3$ powders that there can be a strong dependence of L on X depending on the sample and light soaking. Therefore, X was varied over a wide range.

3 RESULTS AND DISCUSSIONS

3.1 Sample A

Figure 5 shows the dependence of L on X for sample A. Before light soaking, L varied between 6.7 µm at lower values of X (X between 10 and 100 µV) and 5.7 µm at higher values of X (X between 2 and 4 mV). The change of L was weak after light soaking with the red LED for 10 and 20 min. The values of L tended to increase to up to 8…9 µm for lower values of X and 6…7 µm for X between 1 and 4 mV. Prolonged light soaking for 293 min led to a reduction of L to values of the order of those before light soaking. Therefore, sample A showed only a weak dependence on the value of the SPV and on light soaking.

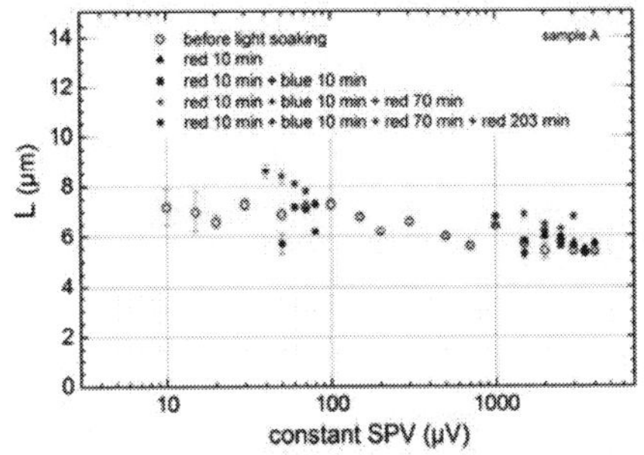

Figure 5: Dependence of L on the value of the constant in-phase SPV signal for sample A before light soaking (open circles) and after light soaking with the red LED for 10, 20, 90 and 293 min (triangles, squares, stars and circles, respectively).

3.2 Sample B

Figure 6 shows the dependence of L on X for sample B. Before light soaking, L varied between 3 µm (X 115 µV) and 2 µm (X 260 µV). After light soaking for 10 min with the red LED, L decreased to about 1.5 µm and decreased further to about 1 µm after light soaking for 33 min. For longer times of light soaking, L decreased to values up to 500…600 nm (164 min). After the strong decrease of L due to light soaking with the red LED for 164 min, L remained practically constant after additional light soaking with the blue LED. As remark, there was practically no dependence of L on X after light soaking.

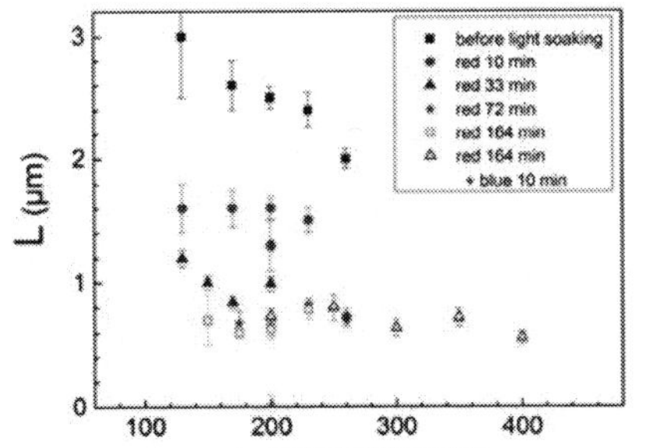

Figure 6: Dependence of L on the value of the constant in-phase SPV signal for sample B before light soaking (filled squares) and after light soaking with the red LED for 10, 33, 72 and 164 min (circles, triangles, stars and open squares, and respectively).

3.3 Sample C

Figure 7 depicts some Goodman plots for sample 3 after light soaking with red light for 10 min. Sample C behaved completely different in comparison to samples A and B. First, the values of L changed over a range between 3 and 30 µm. Second, two values of L could be extracted for Goodman plots measured under certain conditions whereas these conditions were not stable and could not be fixed in terms of a defined value of X of a certain procedure of measurement or light soaking.

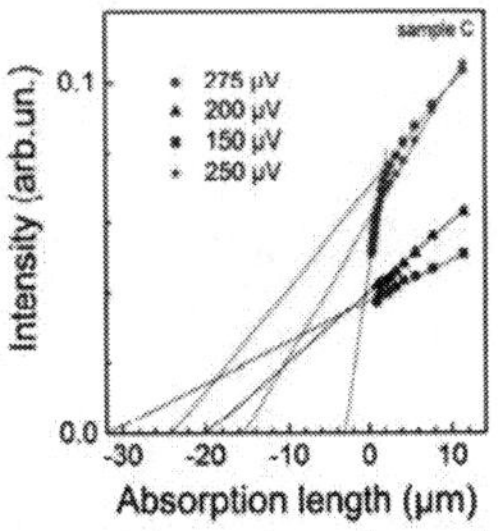

Figure 7: Some Goodman plots of sample C after light soaking with the red LED for 10 min for constant X values of 275, 200, 150 and 250 µV (circles, triangles, squares and stars, respectively). Note the wide range of resulting values of L and the possible appearance of two values of L.

For getting a better impression about instabilities in sample C, the intensity is plotted as a function of the absorption length on a log scale after light soaking for 20 min in figure 8.

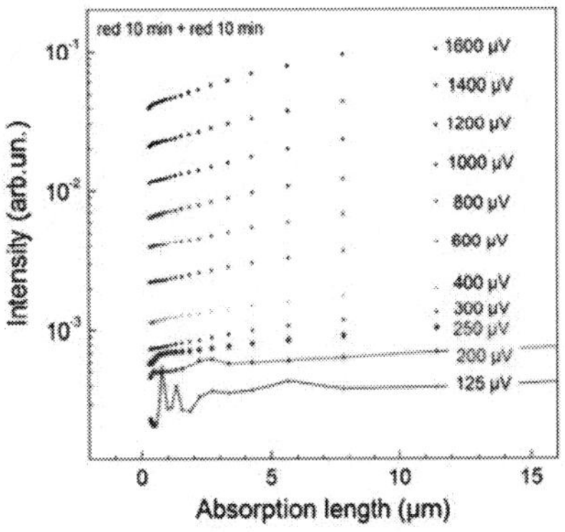

Figure 8: Dependence of the intensity on the absorption length on a log scale for a wide range of constant X signals.

At the higher values of X, the dependencies of L on the absorption length were stable. However, for the given sample and measurement conditions, deviations and instabilities appeared in the dependencies for low values of X (X below 250 µV) so that the determination of L from the corresponding Goodman plots became impossible. Therefore, the region of photo-generation and the values of the quasi Fermi-levels were crucial for the investigation of L for sample C. It seems that an inhomogeneous distribution of defects plays an important role. It can be speculated that sample C was rather sensitive to the conditions during extraction of the fraction of $CH_3NH_3PbI_3$ particles from the saturated solution and drying.

Figure 9 summarizes the dependencies of L on X for sample C before and after light soaking. The higher L (L_1) decreased with increasing X at the higher values of X (X above 400...500 µV). For example, L_1 was 7 and 13 µm at 1.6 and 0.6 mV, respectively, for sample C before light soaking. After light soaking with the red LED for 20 min, L_1 decreased to 4.8 and 11.5 µm at 1.6 and 0.6 mV, respectively.

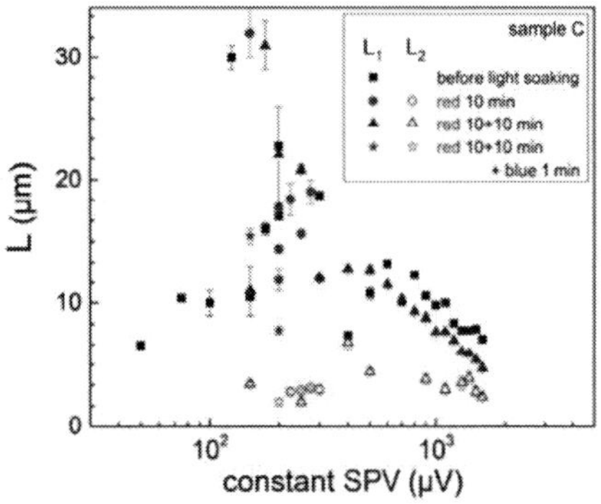

Figure 9: Dependence of L on the value of the constant in-phase SPV signal for sample C before light soaking (squares) and after light soaking with the red LED for 10 and 20 min (circles and triangles, respectively) and with the red LED for 20 min and blue LED for 1 min (stars). The filled and open symbols correspond to the higher and lower values of L, respectively.

For very low values of X (X below 150 µV), L could be measured only for sample C before light soaking and amounted to 10 µm at 100 µV.

The values of L_1 were characterized by large fluctuation for values of X between 120...150 and 400...500 µV. For example, L_1 could be as high as 30 µm at 125 µV and about 10 µm at 100 and 150 µV before light soaking. The large fluctuations appeared also after light soaking. It should be pointed out that the fluctuations were much larger than the experimental error and therefore caused by changes in the sample. The value of L_2 varied mainly between 2 and 4 µm and showed less fluctuations in comparison to L_1.

4 Conclusions

The diffusion length of photo-generated charge carriers was directly measured in powders of $CH_3NH_3PbI_3$ for a wide range of constant values of in-phase SPV signals before and after light soaking. It was shown that degradation of L under light soaking is not a general property of powders of $CH_3NH_3PbI_3$. For the given powders, L was lower than the average grain size. In some cases, L was as high as 30 µm whereas defect states could lead to strong fluctuations of measured values of L.

The appearance of two values of L in one Goodman plot and the observed instabilities under certain conditions gave evidence for the importance of the distributions of defects in space and in energy. More detailed investigations are necessary for getting a deeper understanding about the nature of defects involved and the role of surface passivation for the stabilization of L under light soaking. For example, the specific influence of the formation of PbI_2 on L will be of interest since the interface between PbI_2 and $CH_3NH_3PbI_3$ (i) is passivated [8], (ii) enables transfer of electrons from $CH_3NH_3PbI_3$ into PbI_2 [9], and (iii), indicates a change in the type of conductivity in $CH_3NH_3PbI_3$ [10].

The precise knowledge of the absorption spectrum is crucial for the analysis after Goodman. However, the absorption spectra are not well known for other metal halide and mixed perovskites. It would be really useful to have data about the dependence of the absorption spectra on the stoichiometry.

The method after Goodman is rather simple, accurate and relatively fast. Therefore, it can be well applied not only to the investigation of basic properties of organic-inorganic hybrid metal halide perovskites but also to the monitoring of the diffusion or transport length in relation to technological processes.

5 References

[1] W. S. Yang, J. H. Noh, N. J. Jeon, Y. C. Kim, S. Ryu, J. Seo, S. I. Seok, Science 348, 1234 (2015).

[2] C. Zhao, B. Chen, X. Qiao, L. Luan, K. Lu, B. Hu, Adv. Energy Mater. (2015) 1500279.

[3] D. Shi, V. Adinolfi, R. Comin, M. Yuan, E. Alarousu, A. Buin, Y. Chen, S. Hoogland, A. Rothenberger, K. Katsiev, Y. Losovyi, X. Zhang, P. A. Dowben, O. F. Mohammed, E. H. Sargent, O. M. Bakr, Science 347 (2015) 519.

[4] Q. Dong, Y. Fang, Y. Shao, P. Mulligan, J. Qiu, L. Cao, J. Huang, Science 347 (2015) 967.

[5] A. M. Goodman, J. Appl. Phys. 32 (1961) 2550.

[6] P. Löper, M. Stuckelberger, B. Niesen, J. Werner, M. Filipič, S.-J. Moon, J.-H. Yum, M. Topič, S. de Wolf, C. Ballif, J. Phys. Chem. Lett. 6 (2015) 66.

[7] V. Duzhko, V. Yu. Timoshenko, F. Koch, Th. Dittrich, Phys. Rev. B. 64 (2001) 075204.

[8] T. Supasai, N. Rujisamphan, K. Ullrich, A. Chemseddine, Th. Dittrich, Appl. Phys. Lett. 103 (2013) 183906.

[9] V. Somsongkul, F. Lang, A. Jeong, M. Rusu, M. Arunchaiya, Th. Dittrich, Phys. Stat. Sol. RRL 8 (2014) 763.

[10] A. Naikeaw, P. Prajongtat, M. Ch. Lux-Steiner, M. Arunchaiya, Th. Dittrich, Appl. Phys. Lett. 106 (2015) 232104.

HIGH EFFICIENCY SOLUTION COATED $Cu(In,Ga)(Se,S)_2$ THIN FILM SOLAR CELLS

Tetsuya Aramoto[1], Yuka Kawaguchi[1]
Yueh-Chun Liao[2], Yubun Kikuchi[2], Takuya Ohhashi[2], Hiroyuki Iida[2], Akimasa Nakamura[2]
[1]Atsugi Research Center, Technology Organization, Solar Frontier K.K.,
123-1 Shimokawairi, Atsugi, Kanagawa 243-0206, Japan
Tel: +81-46-245-1750, Fax: +81-46-245-8761
*corresponding author e-mail: tetsuya.aramoto@solar-frontier.com
[2]Advanced Material Development Division 4, Research & Development Department,
Tokyo Ohka Kogyo Co., Ltd.
1590 Tabata, Samukawa-machi Koza-gun, Kanagawa 253-0114, Japan

ABSTRACT: Solar Frontier developed a new type of solution coating method for $Cu(In,Ga)(Se,S)_2$ solar cells. Carrier density, grain growth, recombination velocity are controlled with newly developed absorber formation ink. Photo conversion efficiency (PCE) of 17.72% was achieved with new process and certified by Fraunhofer Institutes, Germany. Internally PCE of 18.7% was measured these results are the highest reported efficiency for a solar cell with solution coated $CuInGa(S,Se)_2$. EBIC, Suns Voc, SSRM, C-V measurement are employed to analyze the difference of physical properties for the cells with doped layer and un-doped layer.

Keywords: CIGS, solution, coating, print, slit coat, 18.7%, EBIC, Suns Voc

1 INTRODUCTION

Solar Frontier has over 20 years of history for research and development of CIS solar cells since 1993. In 2005 Showa Shell K.K. decided the investment for PV business and in 2007 we started our first factory for commercial production with manufacturing capacity of 20MW at Miyazaki, Japan. And now we have gigawatt-scale manufacturing capacity, including one of the world's largest plant, the Kunitomi Plant, which opened in 2011 at Miyazaki, Japan.

With this long history of research and development, Solar Frontier has exploit sputtering and vapor phase sulfurization after selenization (SAS) process for absorber formation technique consistently and we announced small size CIS cell with conversion efficiency 22.3 % the highest performance of all thin film PV in 2015[1]. Meanwhile in 2013 we started to develop solution coating method for CIS manufacturing as one of a next generation manufacturing method. We aim for achieving higher efficiency than vacuum processed CIS solar cell such as sputtering method or co-evaporation method with solution coated CIS, not only to achieve better manufacturability nor lower manufacturing cost.

In this paper we discuss what kind of merit and demerit lies on solution coating method compared with existing conventional manufacturing method and its potential as a future manufacturing process for thin film PV.

Three stage method, all sputtering method, and Surfurization after Selenization (SAS) method are usually employed for the absorber formation on the CIGS field. And several groups are working on solution coating method for low cost manufacturing method of CIGS solar cells. They usually mention the reason of necessity to study solution coating method is because we can reduce manufacturing cost with employing the solution coating method, we don't need expensive vacuum manufacturing equipment. Therefore the capital expense will be so much suppressed. But this is only a partly right explanation. Actually depreciation of manufacturing equipment for absorber layer occupies only a small part of manufacturing cost. Therefore if this part of cost will be reduced to a half it won't make strong impact to the manufacturing cost. Currency exchange rate or metal related material cost has far higher impact.

Many research groups have been developing solution coating CIGS utilizing various row materials and solvents. [4] [5]

IBM group reported lowtemperature processed CIGS solar cell with hydrazine as its solvent[2]. On the other hand, nano-solar reported 17.1% efficiency with nano particle ink technology[3].

We can adduce several reasons to employ solution coating method for thin film PV manufacturing.

1) High material utilization: recycle and maintenance of the reaction chambers are needed with sputtering or co-evaporation method

2) Short reaction time (High through put deposition): Crystal can be grown in a short time with solution coating method, since our ink solution contains VI group element.

3) Atmospheric pressure formation: High speed in-line manufacturing is possible since all coating, drying and annealing process are carried under atmospheric pressure condition.

4) Lower cost for annealing atmosphere: Manufacturing cost can be reduced since filtering facility for treatment gas and safety equipment are unnecessary.

5) Easy formation for Band gap Profile: band gap and physical properties can be easily controlled for depth direction with stacking different composition layers.

We need to pay attention for several issues about composition of the "Ink" to make high quality absorber layer and to realize practical production method. The first point is safety matter. Inflammable agent like hydrazine requires special safety design for manufacturing apparatus. It is also highly reducible, highly toxic and dangerously unstable unless handled in solution. The second point is facileness of handling. When we try to use nano particle, we face difficulty of synthesis and manage the condensation of nano particles. The third point is a problem of organic residue. If we try to use ink with a certain amount of organic constituents, a lot of carbon containing residuals are formed on a bottom of the absorber layer and degrade device performance.

2 EXPERIMENTS

In this study, we developed a new generation CIS ink with Tokyo Ohka Kogyo Co., Ltd. (TOK). This new generation ink solved the problems mentioned above, which contains metal chalcogenide and dimethyl sulfoxide (DMSO) is used for solvent. Characteristics of the material and device will be discussed in this paper.

We coated the new generation ink on the sputtered Mo glass substrate by spin coater or slit coater. Spin coater was used mostly for the experiment to test the various ink formuraltion. After the coating the coated film is heated in inert atmosphere and annealed in the Se containing atmosphere. After annealing, Buffer layer, TCO layer, Al grid electrode and MgF_2 AR coat was formed on it.

Photo conversion efficiency of 17.72% was achieved with this process and certified by Fraunhofer Institutes, Germany. The band gap of the cell estimated from EQE was 1.18eV and Voc deficit was 485mV. This represents the highest reported efficiency for a solar cell with solution coated $CuInGa(S,Se)_2$ and its open circuit voltage is approaching 700mV. We plan to approach 20% conversion efficiency by optimizing the sulfur profile for the front side of the device in the near future.
In this study, we used two types of top layer ink, same formulation ink was used for the bottom layer for all samples. We tried a lot of combination of the material and we got better performance with a limited combination samples.

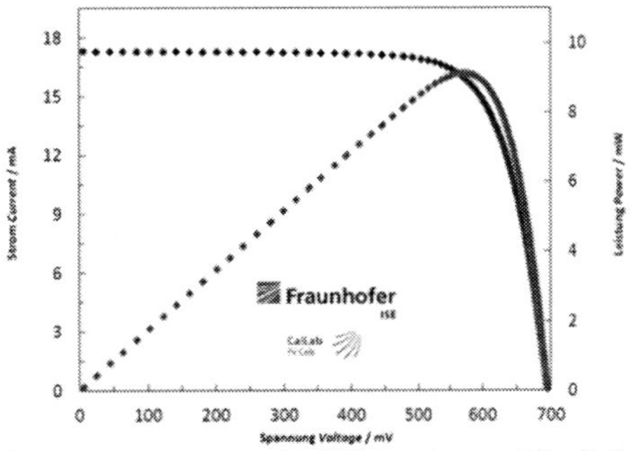

V_{OC}	=	(	694.5	±	2.4	)	mV
I_{SC} (Ed.2 - 2008)/3/	=	(	17.28	±	0.33	)	mA
J_{SC}	=	(	33.58	±	0.64	)	mA/cm^2
I_{MPP}	=		15.94				mA
V_{MPP}	=		572.0				mV
P_{MPP}	=	(	9.12	±	0.18	)	mW
FF	=	(	75.98	±	0.49	)	%
η	=	(	17.72	±	0.37	)	%

Figure 1: I-V curve of the Fraunhofer Institute approved high efficiency cell

3 RESULTS

In 2015 we are obtained close to 18% efficiency with simple CIGSSe formulation ink. Figure 1 shows I-V curve of the ink coated CIGSSe cell. This cell was approved by Fraunhofer Institute as 17.72% Photo conversion efficiency. We got this result only with optimizing annealing temperature, annealing time and the amount of Se vapor. Mo back electrode, TCO, buffer layer, Al grid, and AR coating was formed exactly same with conventional Solar Frontier's small area high efficiency cell condition. The open circuit voltage is 694.5mV, Short circuit current is 33.58mA/cm2, Fill factor is 76.0 and Voc deficit is 474mV.

Figure 2 shows the comparison of I-V curves for two types of top layer ink. Red line shows the I-V curve for the cell with doped top layer, blue line shows conventional top layer cell. Table 1 shows the comparison for the I-V characteristics of these cells. We can see each Voc and Jsc is better with doped top layer cell.

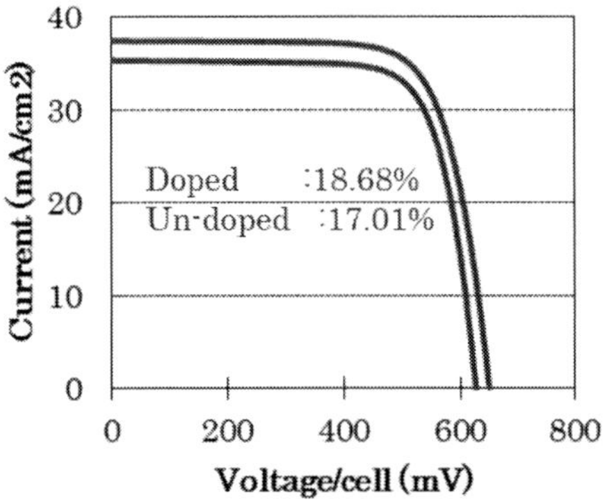

Figure 2: comparison of I-V curves for two types of top layer ink.

Top layer	Eff (%)	Jsc (mA/cm^2)	FF	Voc/cell (mV)
Un-doped	17.01	35.3	0.760	633
doped	18.68	37.2	0.760	660

Table 1: I-V characteristics of doped and un-doped top layer cell

We examined EBIC, SSRM, C-V, Suns Voc to understand the difference of these two types of cell.
Figure 3 shows EBIC image of the cell without doping for the top layer and Figure 4 shows EBIC image of the cell with doped top layer.
We can see many differences for these samples. Firstly we can obviously see the thickness of absorber layer. With doped top layer, total thickness of the absorber has approximately 2 micron thickness, on the other hand, the absorber thickness without doped top layer is around 1.5 microns. Both of the samples have voids in between top layer and bottom layer and we can see the EBIC signal is observed only in the top layer part, even on the area without voids the EBIC signal seems cannot spread across the interface of top layer and bottom layer. This may be due to the lattice mismatch related defect between top layer and bottom layer. Inter diffusion and relaxation of this lattice mismatch must be improved in the future.
The most important information from EBIC is the behavior of the depletion layer. The depletion layer width of un-doped top layer cell (reference cell) is approximately 0.5 microns. And the depletion layer width of the doped top layer cell looks approximately 1micron at its thicker part and its thickness looks not so uniform across the cross section. Depletion layer width estimated from C-V measurement is 0.46 micron and 0.59 micron

respectively. These results match with EBIC measurement.

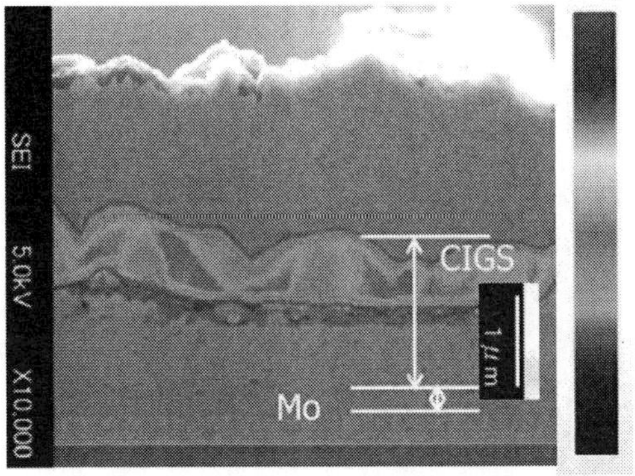

Figure 3: EBIC image of the reference cell

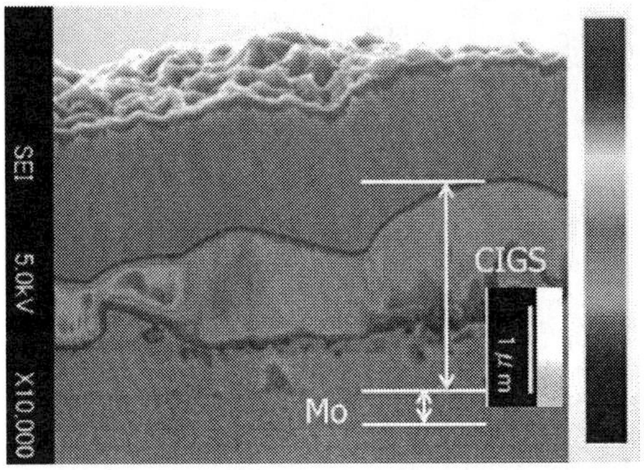

Figure 4: EBIC image of the doped top layer cell

Figure 5 and 6 shows SSRM mapping for these two types of top layer cells. The resistance is electrically measured by touching the micro probe on the surface of the sample for this mapping. On figure 5 the top layer of the sample shows 1e+9 ohm and the bottom layer is 1e+7 ohm. The bottom layer has two digit lower resistivity than that of the top layer. On the other hand, figure 6 shows SSRM image of the doped top layer sample. Most of the absorber layer has an order of 1e+8 ohm resistance. Only the interface layer to the back contact has an order of 1e+7 ohm. This structure may help to make better back contact to the Mo electrode.

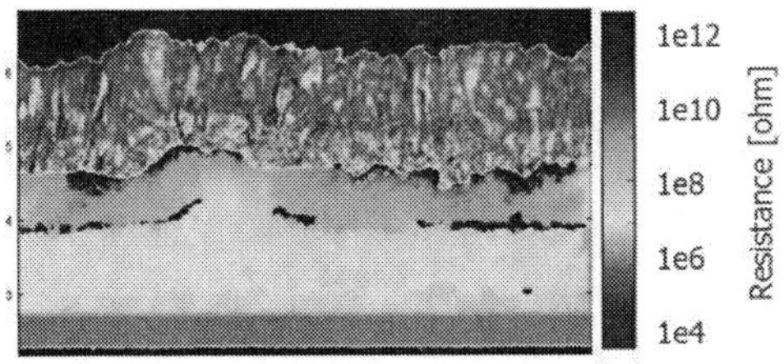

Figure 5: SSRM image of reference cell

These results show good match with the data estimated from C-V measurement. Table 2 shows carrier density and depletion layer width estimated from C-V measurement. Reference cell has acceptor density of 1e+15 on its surface region and 1e+16 on its bulk region. On the other hand, the cell with doped top layer has acceptor density of 1e+15 on for both its surface region and bulk region.

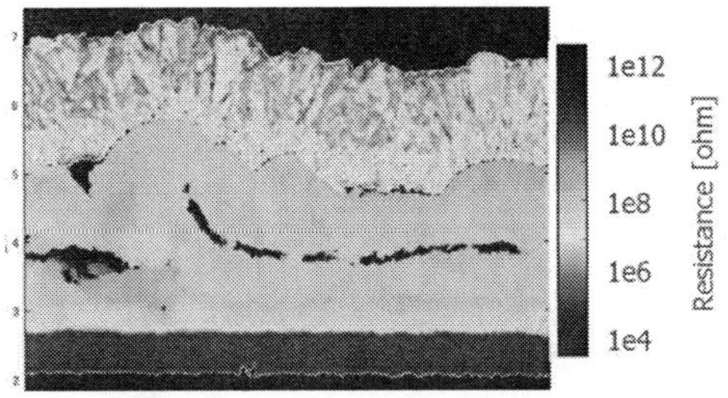

Figure 6: SSRM image of doped top layer cell

Top layer	Carrier density bulk(cm^{-3})	Carrier density surface(cm^{-3})	Depletion width (um)
Un-doped	1.48E+16	6.26E+15	0.455
Doped	6.07E+15	3.48E+15	0.585

Table 2: Carrier density and depletion layer width estimated from C-V measurement

Top layer	R^d_0 ($cm^{-2}s^{-1}$)	$R^i_0+R^b_0$ ($cm^{-2}s^{-1}$)	Dark J_0
Un-doped	6.42E+11	1.01E+07	6.9E-6
Doped	5.02E+11	7.52E+06	2.4E-6

Table 3: Recombination velocity from Suns Voc and dark J-V characteristics

Table 3 shows the recombination velocity of the depletion layer (R^d_0) and recombination velocity of the interface and bulk ($R^i_0+R^b_0$) calculated from Suns Voc measurement and dark I-V characteristics for each samples. Recombination velocity for both depletion layer and interface + bulk layer shows lower at doped top layer cell. Dark current also shows lower value for doped top layer cell. Each data indicate better quality with doped top layer. This means top layer doping gives better diode quality for the absorber layer.

4 DISCUSSION

We successfully developed the new generation solution coated CIGSeS technology with doped top layer ink.

According to the above mentioned analysis, the top layer ink has three functions for the CIGSeS absorber formation. The first function is better grain growth of the top layer observed by SEM image which is done with EBIC measurement.

The second function is carrier density of the absorber layer can be controlled by the doped top layer. Which is observed by C-V measurement and SSRM. We can get better short circuit current by changing width of depletion layer. Depletion layer width is appropriately controlled by this doping technology. Figure 7 shows the comparison of these two cells. Photo current with the wave length range of over 700nm is clearly enhanced with doped top layer. This is due to the depletion layer is controlled appropriate thickness to absorb this range of wave length light. Usually, wave length range of the band edge response is easily lost by band grading and narrow SCR width.

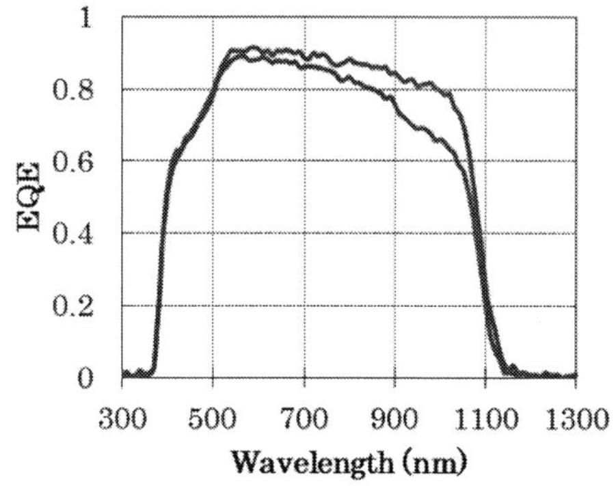

Figure 7: Comparison of EQE for doped and un-doped top layer cell

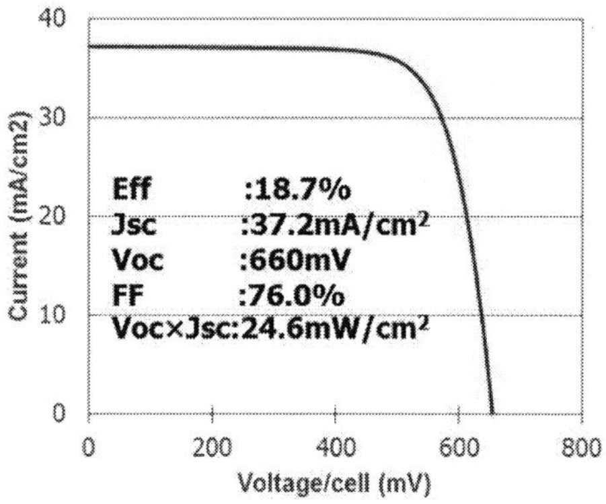

Figure 8: I-V curve of the internally measured latest cell

The third function of the doped top layer is suppression of recombination on the depletion layer. Recombination velosity calurated from Suns Voc data and dark J_0 shows lower recombination at depletion layer with doped top layer, these results explain higher Voc with doped top layer. Furthermore, according to the SCAPS simulation with changing the carrier density of the surface. In this simulation, carrier density of the surface part of CIGS layer is changed from 4.0e+15 to 1.0e+17. The simulated I-V curves is showed on Figure 10. I-V curves, espetially fill factor is decreasing when carrier density of the surface part is increased more than 4.0e+16. This means fill factor also affected by the physical property of the depletion layer.

When we use this solution coating technology, these all kind of important characteristics can be controled by changing the composition of the toplayer and bottom layer ink. We can still see some non-uniformity of the doping from EBIC measurement. But once these points are solved, solution coating method is really suitable for making high efficiency solar cells. Because physical characteristics are easilly controlled with this technology.

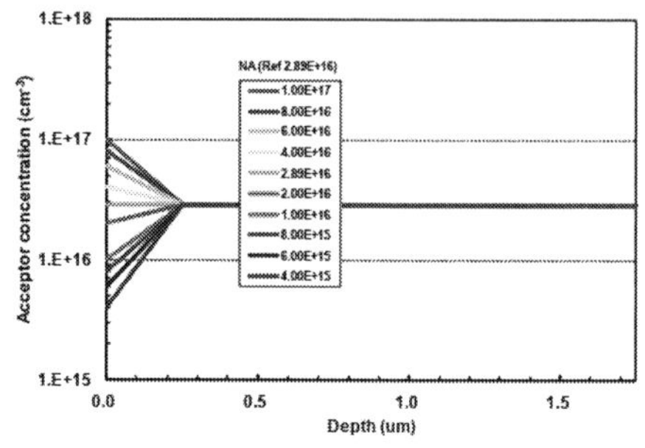

Figure 9: Acceptor concentration condition on the surface

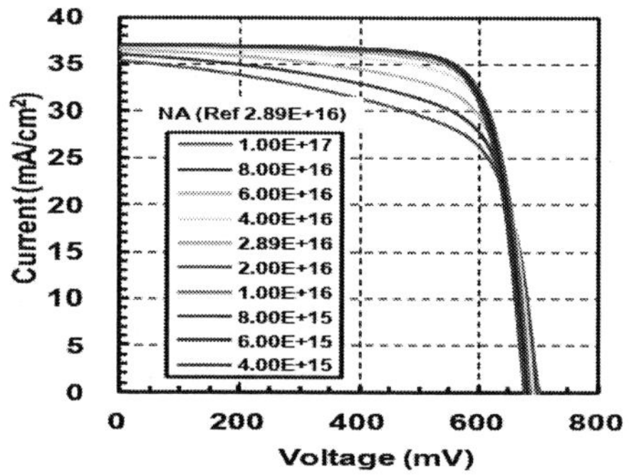

Figure 10: I-V curves from SCAPS simulation

5 SUMMARY

1. Better grain growth on CIGSeS absorber is possible with doped top layer ink.
2. Carrier density for the depletion layer can be controlled appropriate for light absorption with doped top layer ink, so we can control the thickness of depletion layer from 0.2 micron to 1.0 micron.
3. With doped top layer ink, absorber layer can be lower recombination velocity and lower dark current.

Photo conversion efficiency of 17.72% was achieved with this process and certified by Fraunhofer Institutes, Germany. And internally Photo conversion efficiency of 18.7% was achieved these results represent the highest reported efficiency for a solar cell with solution coated $CuInGa(S,Se)_2$.

REFERENCES

[1] R. Kamada, Proceeding 43rd IEEE PVSC, (2016)
[2] M. Yuan, D. Mitzi, Thin Solid Films 519 (2010) 852
[3] G. Brown, P. Stone, Proceeding 38th IEEE PVSC, (2012) 3230.
[4] P. Arnou, C. Cooper, Proceeding 42nd IEEE PVSC, (2015)
[5] Q. Guo, H. Hillhouse, Progress in Photovoltaics, 21(2013) 64

REVEALING LASER-INDUCED DAMAGES IN CIGSE BASED SOLAR CELLS BY MEANS OF PHOTOLUMINESCENCE AND THERMOGRAPHY

Christof Schultz[a], Guillermo Farias[a,b], Christian Wolf[c], Björn Rau[c], Rutger Schlatmann[a,c], Bert Stegemann[a]

[a] University of Applied Sciences – HTW Berlin, Wilhelminenhofstr. 75a, D-12459 Berlin, Germany
[b] Technical University of Berlin, Straße des 17. Juni 135, D-10623 Berlin, Germany
[c] PVcomB / Helmholtz-Zentrum Berlin für Materialien und Energie GmbH, Schwarzschildstr. 3, D-12489 Berlin, Germany

ABSTRACT: Laser technology has been successfully applied for P2 patterning of CIGSe solar cells either for modifying the layer composition (using ns pulses) or by layer ablation (using ps pulses). In this study, complementary information on laser-induced damages of the CIGSe layer in the vicinity of the P2 laser scribe are revealed by spatially and spectrally resolved photoluminescence (PL) and infrared illuminated lock-in thermography (LIT). The results consistently show, that laser-induced damages deplete radiative recombination (detected by PL), while defect-related recombination and, thus, heat emission (detected by LIT) is enhanced. The derived heat-affected areas are significantly larger than expected from inspections by optical microscopy. For the ps treatment the affected area is even larger than for the ns treatment, due to a modification of the band structure and to thermal decomposition. From this data, distinct insights into laser based material modifications and essential information for further minimization of the dead area is obtained.
Keywords: Ablation; $Cu(InGa)Se_2$; Laser Processing; Phase Transformation; Photoluminescence; Serial Integration; Thermography

1 INTRODUCTION

Laser processing has become a key technology for the industrial production of crystalline as well as thin film solar cells, due to its prospects of reducing manufacturing costs and improving solar cell efficiencies. In thin film photovoltaics laser scribing technologies for the first (P1), second (P2), and third (P3) patterning step for monolithical serial interconnection are currently developed and successively introduced to the market. Main challenges are still the tolerance of the laser processes to spatial inhomogeneities particularly in the absorber layer, the selectivity of the process and the high heat sensitivity of CIGSe. However, as compared to mechanical scribing, laser scribing has several advantages like well-defined scribing lines, higher scribing speed, better process control and less down-time. Thus, complete laser scribing of CIGSe modules promises smaller dead areas and therefore higher efficiencies [1].

Chalcopyrite solar cells (i.e., $Cu(In_x,Ga_{1-x})Se_2$ or CIGSe) are monolithically integrated utilizing different methods for the P2 patterning step, see Fig. 1. The P2 can be done mechanically, i.e. by using a needle to remove the material. Alternatively, laser-based patterning is done either by ablating the material with picosecond (ps) laser pulses (Fig. 1a) or by transforming the material into a metallic phase using nanosecond (ns) laser pulses (Fig. 1b) [2]. Regardless of the utilized method, the electronic and physical properties of the CIGSe layer suffer from deterioration in the vicinity of the P2 scribes, either due to chipping caused by the needle or due to the impact of the ns and ps laser pulses [3, 4]. Therefore, the aim of this work is to reveal the material degradation and defect creation in the vicinity of the patterned lines for the different patterning methods, to determine the extent of the affected areas and to identify the underlying mechanisms.

A common defect characterization method is photoluminescence (PL) [5], since it may provide information on changes of a band structure in CIGSe due to material degradation. Spectral resolved PL allows identification of defect levels involved in the optical transitions [6].

Thermal, infrared imaging was applied for analysis of material degradation within the laser-affected area by identification of power dissipating areas featuring non-radiative recombination pathways. To overcome limitations in the signal-to-noise ratio and the localization of a heat producing area, illuminated lock-in thermography (ILIT) was chosen [7].

(a)

(b)

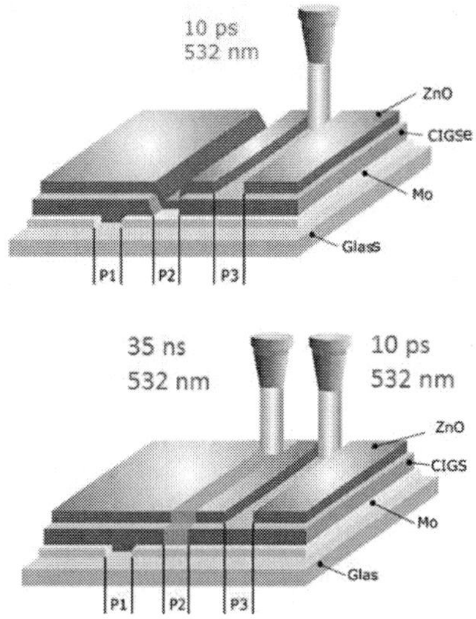

Figure 1: Schematic illustration of the scribing configurations: (a) P2 / P3 laser ablation by 10 ps laser pulses, (b) P2 scribe using 35 ns laser pulses for local phase transformation and simultaneous P3 scribe for selective TCO (ZnO) layer ablation using 10 ps laser pulses.

2 EXPERIMENTAL

2.1 Sample preparation and laser patterning

All solar cells were prepared using the baseline process of the Competence Center for Thin-Film and Nanotechnology for Photovoltaics in Berlin (PVcomB) by sequential growth of the functional cell layers on soda lime glass [8]. The samples typically consist of a SiO_xN_y barrier layer, a 800 nm back contact Mo layer, a 1.6 µm CIGSe absorber layer, a 40-50 nm CdS buffer layer, and a 1 µm transparent conductive oxide (TCO) front contact (i-ZnO ~150 nm / ZnO:Al ~850 nm) layer.

For laser patterning a sophisticated motion control system (Aerotech, Rofin Baasel Lasertech) is used that combines high speed (up to 1.2 m/s) and high accuracy (< 5 µm). This patterning tool is equipped with several pulsed laser sources, enabling the selection of the proper laser parameters to match the specific properties of the material being processed. The available laser sources include a picosecond laser source λ = 1064 / 532 / 355 nm, τ_{pulse} = 10 ps) and a nanosecond laser source λ = 532 nm, τ_{pulse} = 35 ns). Both laser beams possess Gaussian spatial intensity distributions. As a reference, mechanical patterning was done with a tungsten carbide needle with a nominal tip-width of 40 µm.

2.2 Characterization

Optical microscopy (OM) with a Carl Zeiss AxioScope A1 was used to characterize the surface morphology.

Illuminated lock-in thermography (ILIT) measurements were done using a Thermosensorik TDL 640 M system with a maximum resolution of 640 × 512 pixels in the mid-wave infrared window. The camera head consists of an InSb focal plane array which is operated at 77 K to enhance the thermal sensitivity.

Photoluminescence was excited by a cw laser at λ = 532 nm with a diameter ($2\omega_0$) of approximately 40 µm and analyzed with an S&I TriVista spectrometer. To avoid compositional changes induced by the measurement itself, pre-test were performed and the incident laser power on the sample surface was attenuated with optical filters to 0.15 mW.

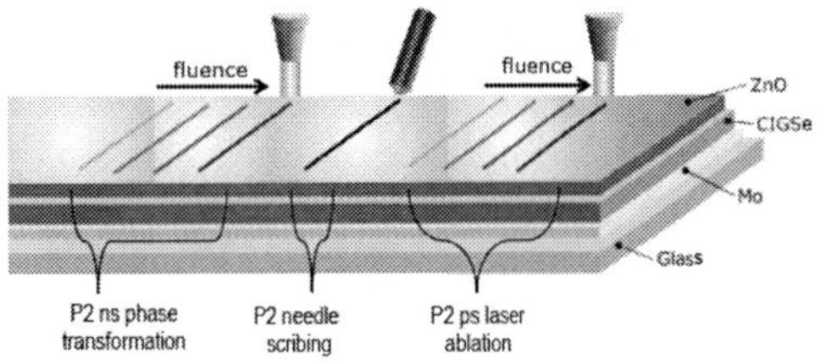

Figure 2: Schematic illustration of the experimental approach. Multiple lines were patterned in the CIGSe layer by using ns and ps laser pulses at systematically varied energy densities and by mechanical needle scribing.

2.3 Approach

Multiple lines were patterned in the CIGSe layers by using ns- and ps laser pulses at systematically varied pulse energies (1, 2, 4, 12 µJ). As reference process conventional needle based patterning was utilized. After inspection of the scribed areas by OM, spectrally and spatially resolved PL and ILIT measurements were performed at the scribe and its vicinity to obtain complementary information about the dominant recombination mechanism and to determine the extent of the laser-affected areas

3 RESULTS

3.1 P2 ps laser structuring

P2 laser structuring of the absorber layer with 10 ps pulses is successfully achieved by direct multi-spot ablation at 1064 nm, 532 nm and 355 nm at rather high scribing speeds of up to 1 m/s. In all cases high laser spot overlaps of > 99% and rather low laser fluences of < 0.5 J/cm^2 are necessary to produce homogeneous trenches of constant width (< 30 µm), though slight melt rims may appear at the scribe edges. As recently shown by electrical characterization of complete solar cells, P2 ps laser patterning is competitive to conventional needle scribing with the best efficiency obtained at wavelength of 532 nm [4]. With respect to the laser pulse energy, a wide process window was established that ensures robust processes.

In Fig. 3a an optical micrograph of such a P2 scribe obtained by ps laser ablation with pulse energy 4 µJ is shown. PL and ILIT measurements were done across the scribe with a spacing of 10 µm and 6 µm, respectively, to investigate the laser impact in the vicinity of P2 scribes and to determine the range of the laser-affected zone. The intensities of the main PL emission peak (band-band transition) and of the IR radiation detected by ILIT are also given in the figure and plotted as a function of distance from the center of the P2 scribe.

Noticeably, the PL intensity is depleted not only within the ablated area but also in its vicinity. The ILIT measurement is well in accord with this observation. The apparent asymmetry of the data (also seen in Figs. 3b and 3c) is considered as a systematic artifact due to an asymmetric illumination of the sample). However, the observed temperature behavior clearly points toward a heat-induced modification of the material resulting in suppression of radiative recombination.

3.2 P2 ns laser structuring

In contrast to monolithic serial integration by ps laser pulse ablation, the heat input of the ns laser pulses was well-controlled and beneficially used by selecting the proper laser fluence to re-melt the absorber layer locally into a highly conductive, metal-like compound which serves as P2 interconnect.

To identify optimal P2 phase transforming scribing conditions the laser pulse energies were systematically varied between 1 µJ and 12 µJ, covering a range of slightly surface disturbance till ablation of the whole layer stack. Homogenous phase transformation requires a high spot overlap (< 99%) and, thus, the accumulation of the incident energy. As demonstrated previously, intermediate pulse energies lead to optimal phase transformation, i.e. formation of a clearly confined and highly conductive line with lowest series resistances [3, 9]. The locally enhanced conductivity is due to an increase of the concentration of the metallic constituents, i.e. Zn, Ga and Cu, and the loss of In in the phase transformed zone, as was revealed by energy dispersive

X-ray analysis (EDX) [9]. If a too low laser fluence is chosen, incomplete phase transformation of the CIGSe occurs, while a too high laser fluence leads to a broad conductive zone, but at the same time material from the center of the scribe is removed and cracking of the molybdenum back contact layer occurs.

In the same way, as described for P2 ps scribing, the impact of the ns laser pulses on the CIGSe layer was systematically investigated by PL and ILIT as a function of the pulse energy and as a function of the distance from the P2 scribes. In this graph, the PL data points represent the normalized maxima of the PL signal (band-band transition), ILIT the normalized temperature maxima of the corresponding line scan. The results for a pulse energy of 4 µJ are shown in Fig. 3b. Again, the PL intensity is depleted within the scribed area and its vicinity. The ILIT measurement is well in accord with this observation. It presents a clear increase of the temperature in the vicinity of the scribe representing an ohmic contact behavior of the affected material along the scribe indicating a large compositionally changed, conductive area beside the trench.

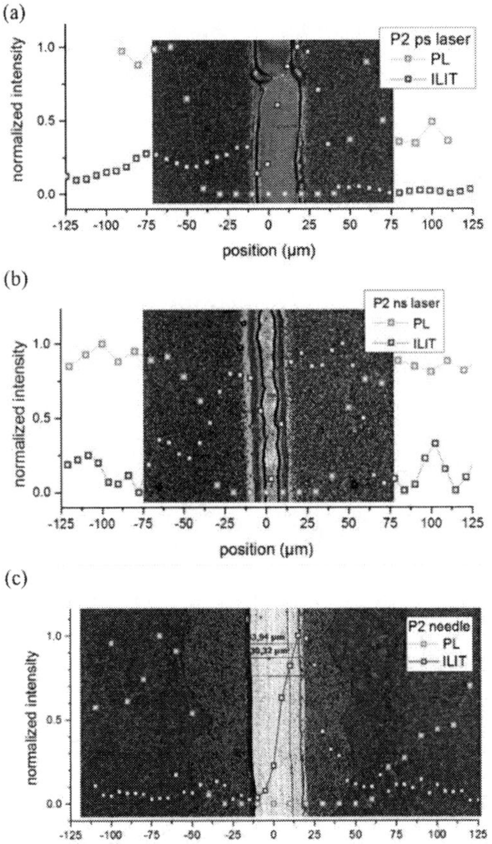

Figure 3: Optical micrographs of a P2 scribe obtained by (a) ps laser ablation (pulse energy 4 µJ), (b) ns phase transformation (pulse energy 4 µJ), (c) needle scribing. PL and ILIT data were acquired across the scribe with a spacing of 10 µm and 6 µm, respectively. PL data represent the normalized maxima of the PL signal (band-band transition), ILIT the normalized maxima of the corresponding line scan.

3.3 P2 needle structuring

As reference process, P2 mechanical needle scribing was included in the investigation. Needle scribing is known to yield various distinct effects, which are visible by OM (see Fig. 3c), such as delamination, corrosion of the Mo back contact layer and chipping which increases the connection width. Again a depletion of the PL signal as well as an increase of the ILIT signal found, which in this case nearly corresponds to the affected range visible by OM.

4 DISCUSSION

A quantitative evaluation of the extent of the area affected by the laser or needle treatment is summarized in Fig. 4. This representation uses concentric circles to visualize the range of the affected range detected by OM (black), ILIT (red) and PL (blue). Moreover, in green the diameter of the laser beam ($2\omega_0$) is given, i.e. 59 µm for the ps laser pulses and 73 µm for the ns laser pulses, as well as the nominal width of the needle tip (40 µm).

The results demonstrate that for both ps and ns laser treatment the extension of the affected zone visible by OM is smaller than the beam diameter, due to the Gaussian line shape. Moreover the extension of the affected zone visible by PL and ILIT is larger than is expected from optical inspection. Though the beam diameter of the ps pulses is smaller than for the ns pulses, the affected area for the ps treatment exceeds the affected area for the ns treatment. This behavior points towards a minor influence of thermal input, but rather an intensity related effect, which might induce mechanical stress [10]. A further explanation could be heat accumulation within the layer that the athermal nature of the short ps laser pulses vanishes. As a result, elemental decomposition or the formation of binary phases might occur [11].

As expected, in the absence of thermal input (i.e. in the case needle scribing) the altered range can nearly be determined based on optical inspection. Just a slight increase is detected by PL and ILIT, mainly caused by inevitable occurrence of chipping. However, even by taking into account the widths of the heat-affected areas for laser-based patterning processes, it is concluded that needle-based pattering is outperformed by laser-patterning.

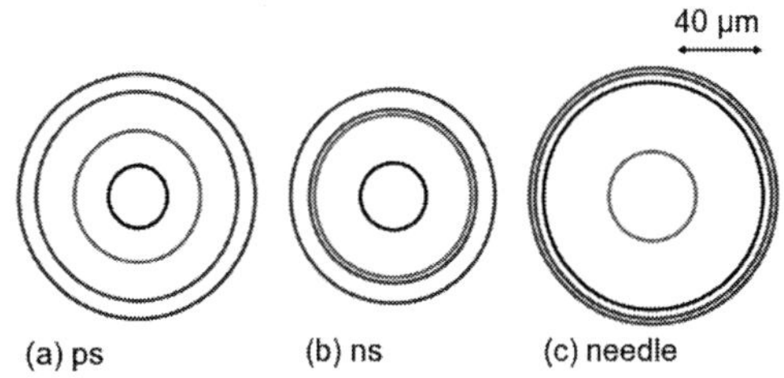

Figure 4: Circular representation of the affected zones for (a) P2 ps laser ablation (pulse energy 4 µJ), (b) P2 ns phase transformation (pulse energy 4 µJ), (c) needle scribing as determined by different techniques: optical microscopy (black), ILIT (red), PL (blue). Green: (a,b) diameter of the laser beam $2\omega_0$ (ps laser: 59 µm, ns laser: 73 µm, (c) nominal needle tip width (40 µm).

Figure 5 gives an overview of the width of the affected zones as a function of the applied laser fluence, as determined from the pulse energies and the respective beam diameter, for all three patterning approaches. All measured data are normalized to the laser beam diameter $2\omega_0$, indicated by a straight line, or the nominal tip width (in case of needle scribing). The visual diameters of the patterned trenches, as obtained by OM, are compared with the widths derived from the PL and ILIT intensities across the trenches. Values from needle patterning obtained by OM, PL and ILIT are added as straight line. Irrespective of the characterization technique, there is distinct increase of the laser-affected range with increasing laser fluence. The asymptotic black fit-line represents the Gaussian shape of the laser beam which fits well with the visual spot diameters, and which is assumed as well for the propagation of the laser-affected zone detected by PL and ILIT (dashed curves in Fig. 5).

As already assumed from Fig. 4, this graph suggests that the affected zone is significantly larger (~ 50 %) in the case of ps patterning as compared to ns patterning, but in both cases lower than for mechanical needle scribing. Thus, further minimization of the 'real' dead area is feasible by P2 ps and ns scribing, even by taking into account the widths of the heat-affected areas for laser-based patterning processes.

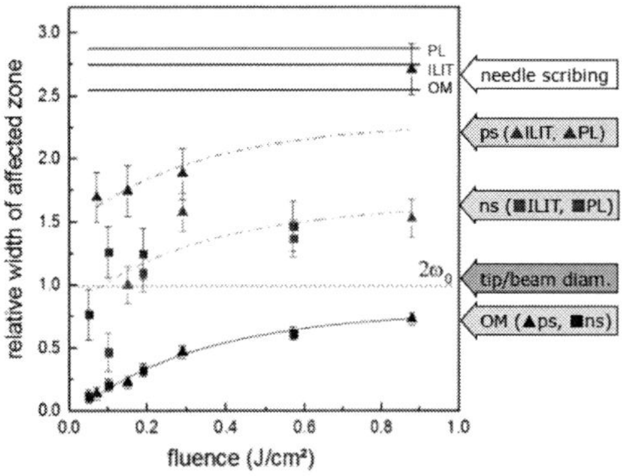

Figure 5: Comparison of the relative widths of the affected zones as a function of the distance to the scribe center and of the applied fluence. Scribing lines were obtained by ns pulses, ps pulses and needle patterning. The diameter of the laser beam ($2\omega_0$) is marked with a dotted line.

5 CONCLUSIONS

PL and ILIT have been demonstrated to reveal complementary but consistent results on the range of the laser-induced material modification in the vicinity of laser-based P2 scribing lines. Both techniques provide distinct insights into laser based material modifications and are, thus, essential for further minimization of the 'real' dead area.

The laser-affected areas turned out to be significantly larger than expected from visual inspection by OM. P2 ps scribing is found to generate larger affected area than ns scribing, possibly to to intensity-related effects rather than thermal effects. Even by taking into account the widths of the heat-affected areas for laser-based patterning processes, it is shown that P2 needle pattering is outperformed by P2 ns or ps laser patterning.

ACKNOWLEDGEMENTS

The authors thank Karolina Mack and Iris Dorbandt (PVcomB) for experimental support. This work was supported by the Federal Ministry of Education and Research (BMBF) and the state government of Berlin (SENBWF) in the frame work of the program "Spitzen-forschung und Innovation in den Neuen Ländern" (grant no. 03IS2151A,B, C).

REFERENCES

[1] G. Heise, A. Börner, M. Dickmann, M. Englmaier, A. Heiss, M. Kemnitzer, J. Konrad, R. Moser, J. Palm, H. Vogt, H.P. Huber, Progress in Photovoltaics: Research and Applications, 23 (2015) 1291-1304.

[2] P.O. Westin, U. Zimmermann, M. Ruth, M. Edoff, Solar Energy Materials and Solar Cells, 95 (2011) 1062-1068.

[3] C. Schultz, M. Schule, K. Stelmaszczyk, M. Weizman, O. Gref, F. Friedrich, C. Wolf, C.A. Kaufmann, B. Rau, R. Schlatmann, Proc. 42nd IEEE Photovoltaic Specialist Conference (PVSC), 2015, 4 pp.

[4] B. Stegemann, M. Schule, C. Schultz, K. Stelmaszczyk, M. Weizman, C. Wolf, C.A. Kaufmann, B. Rau, R. Schlatmann, F. Fink, Proc. 42nd IEEE Photovoltaic Specialist Conference (PVSC), 2015, 4 pp .

[5] O. Breitenstein, J. Bauer, T. Trupke, R.A. Bardos, Progress in Photovoltaics: Research and Applications, 16 (2008) 325-330.

[6] K. Stelmaszczyk, C. Schultz, M. Schuele, M. Weizman, C.A. Kaufmann, R. Schlatmann, B. Rau, V. Quaschning, B. Stegemann, F. Fink, Proc. 29th European Photovoltaic Solar Energy Conference, 2014, pp. 1763-1769.

[7] O. Breitenstein, W. Warta, M.n.E. Langenkamp, Lock-In Thermography - Basics and Use for Evaluating Electronic Devices and Materials, Springer, 2012.

[8] S.S. Schmidt, C. Wolf, H. Rodriguez-Alvarez, C.A. Kaufmann, J.-P. Bäcker, M. Hartig, S. Merdes, F. Ziem, I. Dorbandt, C. Köble, S. Cinque, R. Mainz, D.Abou-Ras, R. Schlatmann, submitted to Progress in Photovoltaics: Research and Application, (2016).

[9] C. Schultz, M. Schuele, K. Stelmaszczyk, M. Weizman, O. Gref, F. Friedrich, C. Wolf, N. Papathanasiou, C.A. Kaufmann, B. Rau, R. Schlatmann, V. Quaschning, F. Fink, B. Stegemann, Sol. Energ. Mat. Sol. Cells, accepted for publication (2016).

[10] J. Bovatsek, A. Tamhankar, R.S. Patel, N.M. Bulgakova, J. Bonse, Thin Solid Films, 518 (2010) 2897.

[11] M. Ruckh, J. Kessler, T.A. Oberacker, H.W. Schock, Jpn. J. Appl. Phys. 32 (1993) 65.

REVERSE-VOLTAGE BIAS INDUCED SHUNT FORMATION IN $Cu(In,Ga)Se_2$ THIN FILM SOLAR CELLS: AN APPROACH WITH THREE-DIMENSIONAL ELECTRO-THERMAL SIMULATIONS

Michael Richter*, Janet Neerken, Jürgen Parisi

Laboratory for Chalcogenide Photovoltaics (LCP), Energy and Semiconductor Research Laboratory, Department of Physics, University of Oldenburg, 26111 Oldenburg, Germany, *email: michael.richter@uni-oldenburg.de

ABSTRACT: By partial shading of thin-film solar panels a high reverse voltage drops across shaded solar cells. These solar cells may be driven into breakdown, which causes in the worst case severe damage to the cell. With dark lock-in thermography the damaged sites are then visible as shunt signals observed. An important issue to prevent cells from damaging is to understand the reasons why a specific site tends to get damaged at reverse bias and how it might be identified before the damage occurred. We developed a three-dimensional simulation structure of a single cell to perform numerical electro-thermal simulations and calibrated the current-voltage and thermal characteristics to a sample we extracted from a module. We studied the thermal signal and corresponding efficiency loss of the device when different types of defects are present. Furthermore, we simulated at the conditions at which real solar cells get damaged. We found that these conditions are not sufficient to generate enough heat at already existing shunts to cause additional damage. However, local recombination sites cause high electric fields at reverse bias, which increase the probability of breakdown processes as avalanche breakdown or band-to-band tunneling.
Keywords: CIGS, shunts, simulation, heat, breakdown, reverse bias

1 INTRODUCTION

Under normal operating conditions, photovoltaic panels could be exposed to shading or certain solar cells in a module could underperform. In both cases, a reverse voltage bias stresses single cells that need to withstand this without irreversible damage. Especially in thin-film modules as those based on $Cu(In,Ga)Se_2$ (CIGS), the reverse voltage can reach high values. Understanding the defect or hot spot formation under reverse bias is a key issue to improve the stability and lifetime of the modules. Local material failures or small defects might be a source of destructive defect formation. Defects could be distinguished in ohmic and diode-like type (or a mixture of both) which reflects their voltage dependence of the dark lock-in thermography (D-LIT) signal [1]. With this study we want to answer the question if reverse-voltage bias induced defects were caused by already existing (weak) defects that might be observable beforehand and which mechanisms might be responsible for the damaging. The latter is investigated by means of numerical three-dimensional electro-thermal simulations.

2 EXPERIMENTAL DETAILS

Samples of solar cells with $Cu(In,Ga)Se_2$ (CIGS) absorbers were extracted from a module produced up to the step of cell separation by patterning front contact and absorber (P3 scribe). The sample's front contact – the cathode – was connected via the molybdenum (Mo) of the adjacent cell. The molybdenum was uncovered and indium patches were used to get reliable contacts. The back contact of the cell – the anode – was connected via the aluminum-doped zinc-oxid (ZnO:Al) front contact of the opposite adjacent cell. The electrical contacting through two P2 scribes deteriorates the performance by a relatively high series resistance that has to be considered in the simulation.

This structure was implemented in a three-dimensional simulation structure as shown in Fig. 1, consisting of a layer stack of ZnO:Al/i-ZnO/CdS/CIGS/Mo while the cathode and anode electrode connections are localized at the outer sides of the ZnO (Fig. 1, blue line) and Molybdenum (Fig. 1, red line), respectively. Here, the current density distribution is shown when a forward voltage of 1 V is applied. The simulations were performed with the commercial TCAD platform Sentaurus by Synopsys [2]. The main simulation parameters are provided in Tab. 1. The CIGS absorber bandgap is assumed to be linearly graded from top to bottom.

Figure 1: 3D-simulation structure of one cell out of a monolithically connected module at a voltage bias of +1 V. The dimension of the X:Y:Z scale is 1000:1000:1. The local current density is color-coded and the anode (at ZnO:Al) and cathode electrode (at Mo) is indicated with blue and red lines respectively. The white line indicates the position of the pn-junction.

Table I: Main simulation parameters of the applied solar cell.

parameter	ZnO:Al	i-ZnO	CdS	CIGS
thickness [µm]	0.87	0.07	0.06	1.92
doping [1/cm³]	$1x10^{20}$(n)	$5x10^{17}$ (n)	$1x10^{17}$ (n)	$2x10^{16}$ (p)
bandgap [eV]	3.5	3.5	2.5	1.18 – 1.3
electron affinity [eV]	4.6	4.6	4.3	4.37 – 4.26
mobility e/h [cm²/Vs]	100/25	100/25	100/25	100/25
permittivity	10	10	10	10

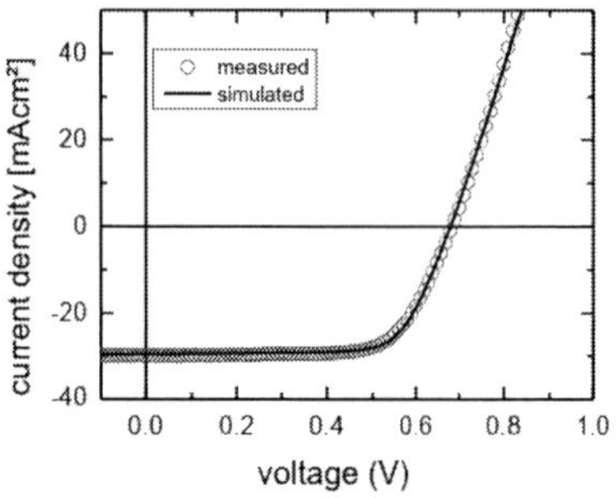

Figure 2: Measured and simulated IV-curve at standard test conditions.

The excellent match of simulated and measured current-voltage (IV) characteristics as shown in Fig. 2 ensures reliable predictions from the simulation.

The thermal simulation includes several thermal processes as illustrated in Fig. 3. The most relevant self-heating mechanisms are Joule heat generation due to currents of electrons and holes, thermalization (in case of illuminating the sample), heat generation by recombination, and Peltier heating and cooling at the pn-junction [3]. The equations given in Fig. 3 are only estimations to get an idea of their dependencies. The different processes are coupled in a system of differential equations. The counterpart of heat generation is heat flux to the ambient. Here we simplify the heat flow as a function of the temperature difference between cell and ambient and a thermal resistor which is defined for the surface and backside of the cell stack. At the sides of the cell the heat flux is set to zero. The thermal resistivity of the surface and backside depends on many factors such as emissivity or heat conduction through the glass. This makes it difficult to measure the thermal resistivity directly. We therefore assume them equal for front and backside and adjusted the value in order to match the temperature profile with a difference of about 50 mK across the cell which we measured with D-LIT at a voltage of +1 V – compare Fig. 4 left and Fig. 5 top. This temperature profile is caused by non-uniform heat generation with about 50 mK higher temperature at the P1 side which is also reproduced by the simulation. Due to the different electrical conductivity of molybdenum and ZnO:Al the voltage drop across the ZnO:Al layer leads to an increased current density near the anode, so at the P1 scribe. Here the full bias voltage drops across the junction whereas at P3 side the applied voltage is reduced by the drop across ZnO:Al. Towards the P1 scribe the Joule heat generation is increased especially within the ZnO:Al layer but also within the i-ZnO because the low conductivity resists the vertical current flow. Furthermore, the increased forward junction voltage at this position enhances the recombination current and thus recombination heat in the CIGS layer.

In order to study the influence of different types of defects on the performance and temperature profile of the solar cell, we implemented on the one hand a region with Gaussian-shape increased bulk mid-gap trap density in the CIGS layer with a peak value of 150 times the baseline value, and on the other hand a region with metallic conduction properties. By these configurations, we can reproduce either diode-like defects or ohmic shunts [1]. The lateral extension of these defects influences the loss in efficiency of the solar cell. As the solar cell is much more sensitive to ohmic shunts their size has to be assumed about 100 times smaller than the size of a diode-like shunt to get at least a diode behavior of the solar cell. A simulation example of a diode-like shunt is shown in Fig. 5 bottom. Compared to the shunt-free sample (Fig. 5 top) a spot of increased heat generation can be observed in the middle of the cell when a voltage of +1 V is applied.

3 RESULTS AND DISCUSSION

3.1 Simulation of defects

The question to be answered by the simulation is whether there are any conditions at which a local inhomogeneity (here two different defect types) could cause temperatures that might destruct the material. In contrast to the ohmic shunt type, a diode-like defect does not open a direct current path at reverse bias, which in that case excludes additional heat generation by these defects at this condition. However, it might be interesting to investigate the temperature development at forward bias. At least to gain a better understanding of forward bias D-LIT measurement results. The highest impact of local defects on the local temperature is at high reverse voltages for ohmic shunts and at forward voltages for diode-like defects. The simulation of the different shunt types with increasing shunt size gives the maximum temperature in the shunt as shown in Fig. 6. For cells

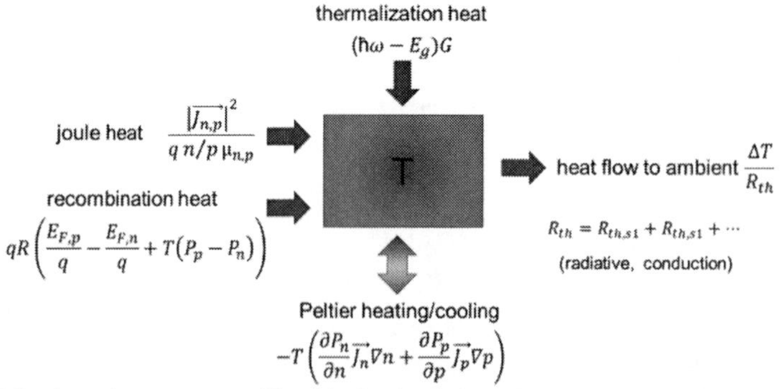

Figure 3: Scheme of the thermal processes considered in the simulation. The equations are only estimations. Symbols are: J – current density, n/p – electron-/hole density, E_F – quasi-Fermi energy, μ - mobility, R – recombination rate, P – thermoelectric power, T – temperature, Rth – thermal resistivity, G – generation rate, Eg – bandgap, ħω – photon energy.

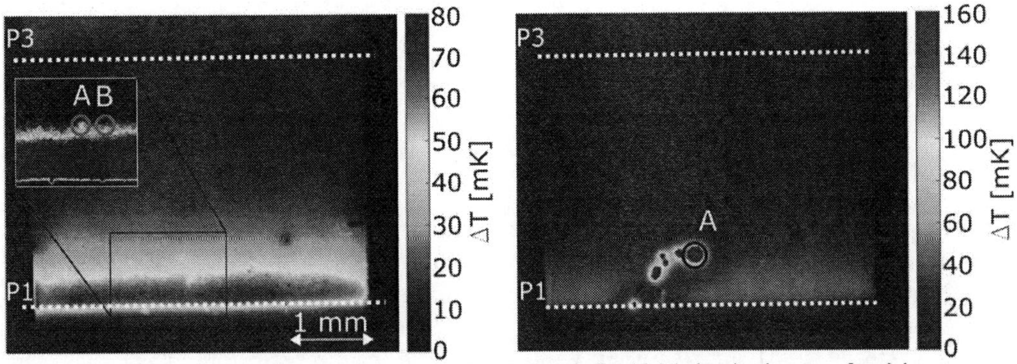

Figure 4: D-LIT image of a CIGS solar cell at +1 V before (left) and after (right) shunting because of applying a reverse bias voltage of -6 V. For reasons of clarity the scaling of the right image was limited to 160 mK the peak value is 1600 mK. The inlet in the left image shows one location (A) at which the cell is shunted and a location next to it where no shunt occurred (B).

with an ohmic shunt (Fig. 6, top) the local temperature at this location which establishes at -6 V and -10 V shows a monotonic increase with increasing shunt size and reaches up to 1.6 K at -10 V. The shunting reduces the open circuit voltage (Voc) and especially the fill factor (FF) dramatically. For a diode-like shunt with a FWHM diameter of around 100 µm (Fig. 6, bottom) the temperature at +1 V has a maximum of 0.42 K above room temperature and decreases at higher shunt sizes due to broader distribution of recombination heat.

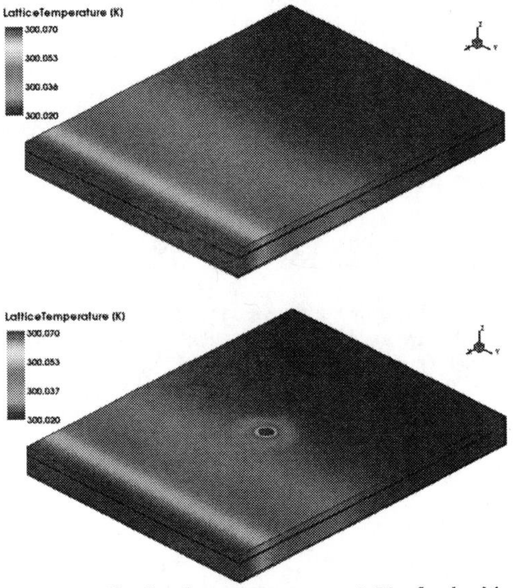

Figure 5: Simulated temperature at +1 V of a healthy (top) and a shunted cell (bottom). The dimension of the X:Y:Z scale is 100:100:1. The shunt at the bottom is realized by a local increase of CIGS mid-gap trap density with a Gaussian-shape area of 20 µm and a peak density of 150 times the baseline value. The color code gives the temperature of the materials while the scale is limited to 300.07 K. The peak temperature of the shunt is 300.26 K.

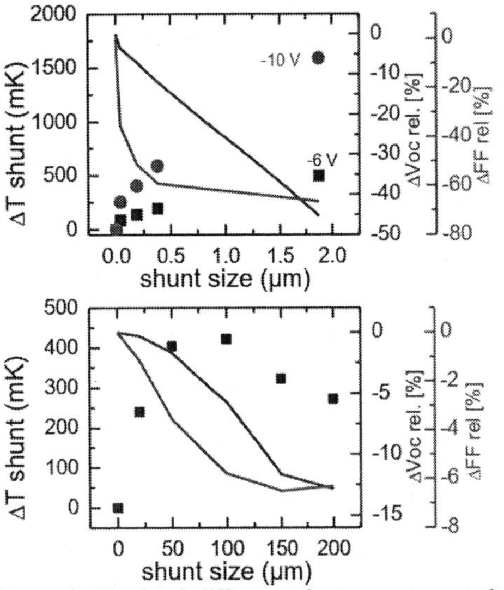

Figure 6: Simulated differences in temperature at the shunt location and electrical losses of the solar cell with varying shunt sizes for ohmic (top) and diode-like (bottom) shunt types. The ohmic shunted cell is operated at -6 V (squares) and -10 V (dots) whereas the diode-like at +1 V.

A major conclusion of this study is that both ohmic and diode-like defects at high reverse or forward bias do not generate sufficient heat to reach temperatures that could cause irreversible damage to the layers. Hence, the shunt formation at reverse bias condition cannot be explained by this model.

3.2 Reverse bias induced damage

After applying a severe reverse bias of -6 V to -12 V, depending on the sample, a shunt pattern – also referred to as "worms" – is formed on a very short time scale (milliseconds), see Fig. 4 right. It should be noted that for the actual module this bias only occurs at very harsh shading conditions. In Fig. 7 (top) the IV characteristic before and after shunting is shown. It can be seen that not only the parallel resistance drops significantly, but also the reverse breakdown behavior changes dramatically.

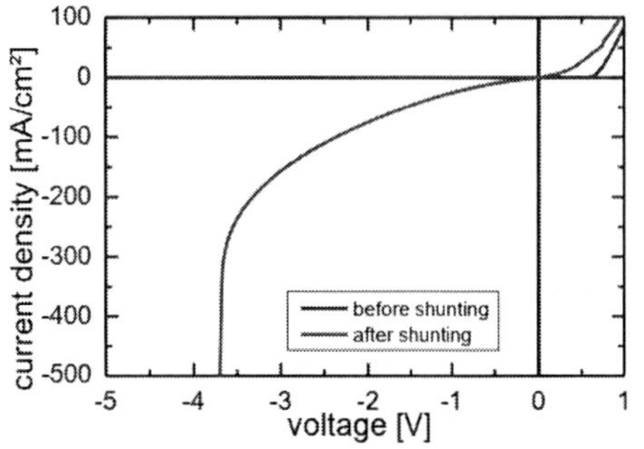

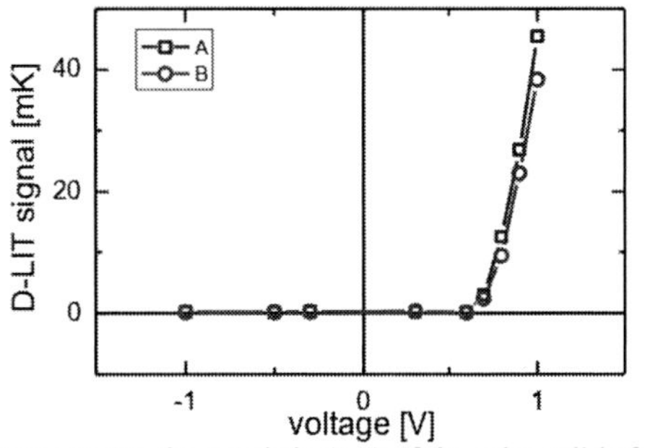

Figure 7: IV-characteristic (top) of the solar cell before and after applying -6 V. The bottom figure shows the voltage dependence of the D-LIT signal of the shunt location A before shunting and a location next to it (B).

The location of the highest shunt signal (Fig. 4, right A) could be identified before shunting as a location of slightly increased heat generation (Fig. 4, left, inlet A). Fig. 7 (bottom) shows the voltage dependence of this location which could be assigned to a "weak diode" or a weak diode-like defect. However, a clear indication of a destructive spot is not given since the pre-damage characteristic of position A is not significantly different from other locations.

Simulations of diode-like defects show that there is no excessive heat generation at the recombination site, because there is a lack of charge carriers at reverse bias. However, the electric field differs at the shunt, which can be seen in the local band diagram shown in Fig. 8. A high electric field promotes any breakdown mechanism such as avalanche ionization, band-to-band tunneling or tunneling-assisted Poole-Frenkel conduction [4, 5]. It is likely to assume that the formation of shunts at reverse bias is due to a highly localized heat generation caused by high currents that are flowing at local breakdown sites. A "weak-diode" enhances the probability of a breakdown that produce currents that could damage the solar cell in a way that the breakdown is more enhanced at adjacent regions and "worms" are formed. Scanning electron microscope images identify these "worms" as areas with a lift-off of the ZnO/CdS from the CIGS which was also observed at laser (heat)-induced shunts [6]. Furthermore, the electric field is higher near the P1 scribe, which enforces these breakdown mechanisms at this location. A defect is therefore more detrimental if it is located near to the P1 scribe.

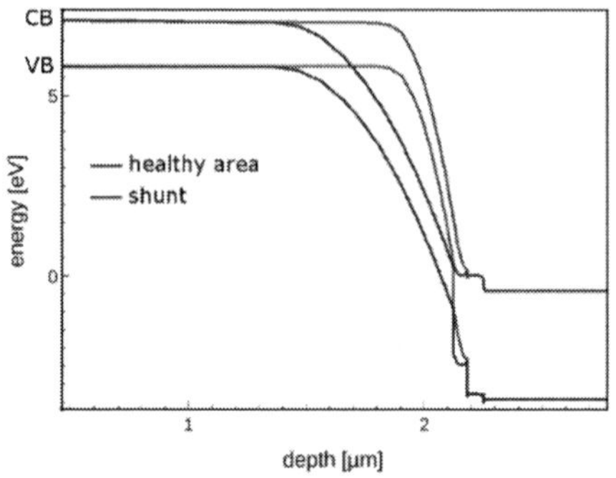

Figure 8: Band diagram of the solar cell at -6 V bias at a healthy area and at the shunt location.

4 CONCLUSION

We investigated the temperature development in CIGS solar cells by means of three-dimensional electro-thermal simulation. The heat conduction properties were adjusted by the temperature profile across the cell stripe at forward bias voltage measured with D-LIT. We found that neither diode-like defects at forward bias nor ohmic shunts at high reverse bias cause hot spots with sufficient heat generation to cause irreversible damage to the material. We conclude that heat generation by additional current generation due to breakdown mechanisms are responsible for the bias-induced defect formation that we observed at high reverse voltages. These breakdown mechanisms are enhanced by high electric fields that occurs at diode-like defects and close to the P1 scribe. Indeed, a small "weak-diode" was observed at the location where the damage occurred prior to the application of a severe reverse bias. However, a clear characteristic that allows for damage prediction was not found, yet.

ACKNOWLEDGMENT

We acknowledge the German Ministry of Economy and Energy for funding (OptiCIGS, 0325724C) and R. Schaeffler and J. P. Theisen (MANZ AG, Germany) for providing the solar cell samples. The LCP group gratefully acknowledge financial support by the EWE AG Oldenburg, Germany.

REFERENCES

[1] M. Richter, F.-S. Babbe, A. Vetter, S. J. Heise, J. Ohland, I. Riedel, J. Parisi, Proceedings 31th European Photovoltaic Solar Energy Conference, Hamburg (2015), p. 1069-1073.
[2] Synopsys, Sentaurus Device User Guide, 2014.
[3] O. Breitenstein, W. Warta, M. Langenkamp, 2nd ed. Springer, Berlin (2010).
[4] X. Sun, M. A. Alam, J. Raguse, R. Garris, C. Deline, T. Silverman, Proceedings 42nd IEEE PVSC, New Orleans (2015).
[5] P. Szaniawski J. Lindhal, T. Törndahl, U. Zimmermann, M. Edoff, Thin Solid Films 535 (2013), p. 326–330.
[6] A. Vetter, F. S. Babbe, B. Hofbeck, P. Kubis, M. Richter, S. J. Heise, J. Ohland, I. Riedel, C. J. Brabec, Prog. Photovolt: Res. Appl. (2016), DOI: 10.1002/pip.2749

PEDOT:PSS/rGO/CuNWs BASED COUNTER ELECTRODE FOR USE IN DSSCs

A.S. Shikoh[1], Z. Ahmed[1], Z. Zhu[2], T. S. Mankowski[2], F. Touati[1], R.A. Shakoor[3], M.A. Benammar[1], M.A. Mansuripur[2], C.M. Falco[2].

[1]Department of Electrical Engineering, College of Engineering, Qatar University, P. O. Box 2713, Doha, Qatar
[2]College of Optical Sciences, The University of Arizona, Tucson, Arizona 85721, USA
[3]Center of Advanced Materials (CAM), Qatar University, P.O. 2713, Doha, Qatar

ABSTRACT: High aspect ratio CuNWs were synthesized using a one step solution based process, yielding an average wire length and diameter of 75 µm and 45 nm, respectively. In order to fabricate hybrid Transparent Conducting Electrodes (TCEs), layer of CuNWs was deposited onto glass and PET substrates using spray coating, along with the subsequent deposition of rGO and PEDOT:PSS layers, used to prevent NW oxidation and control the RMS roughness. The electrodes were later characterized for their morphological, optical and electrical properties using various sets of equipment. The fabricated hybrid films were eventually integrated within Dye Sensitized Solar Cells (DSSCs) as a counter electrode, which were later characterized for their performance using a solar simulator. According to the initial results gathered, an efficiency of 0.21% was noted in case of CuNW counter electrode based DSSCs.
Keywords: Dye-Sensitised, Nanowires, Transparent Conducting Electrodes, Hybrid thin films.

1 INTRODUCTION

Dye sensitized solar cells (DSSCs) are electrochemical devices that have the potential of solving problems like energy depletion, enviornmental pollution and ecological destruction [1]. They are believed to be a promising replacement of conventional crystalline silicon solar cells because of their high efficiency and potential low fabrication cost.

A typical DSSC mainly consists of four parts i.e. photo-anode, a dye sensitized semiconductor layer, electrolyte and a catalyst coated counter electrode. The counter electrode is of great importance since it is responsible for charge transfer to electrolyte ions. The counter electrode is typically based on metal oxide i.e. ITO, FTO etc. with a thin coating of platinum catalyst, used to accelerate the regeneration of electrolyte ion species. Unfortunately, there are a couple of issues related to the use of Metal Oxide/Pt assembly, that need to be addressed. Metal oxides like Indium Tin Oxide (ITO) are brittle in nature, limiting their applicability to non-flexible devices, only. Furthermore Indium and Platinium are rare metals, making the price of ITO/Pt counter electrodes ever increasing. In the past researchers have investigated various transparent conducting electrode technologies that are cheap, versatile and easily integerateable within various opto-electronic devices, including DSSCs [2].

Currently, graphene based materials, Carbon nanotubes (CNTs), conducting polymers, nanowires (NWs) and metal grids are amoungst the forefront technologies considered as a promising substitute to ITO [3, 4]. Metal based nanostrutures have gained a lot of interest oweing to their high optical transparency and low sheet resistance. Bulk metals cannot be used in transparent conducting electrodes since they have high visible region reflection. If the metals are shrunk to nanometers range, as done in case of randomly scattered nanowires, high optical transparency can be achieved. Amoungst the two major types, randomely scattered networks that utilize copper nanowires (CuNWs) are considered to be more practicle as compared to the ones made from silver, since copper is 100 times cheaper but has a bulk conductivity equal to silver [5]. However issues like oxidation and tarnishing that eventually result in the degradtion of CuNW network hurdel the wide spread of this technology [6]. Further the root mean squre (RMS) roughness of the CuNW layer also needs to be controlled to achieve the proper functioning of the devices it is incorporated in.

The main purpose of this investigation is to come up with an alternative counter electrode that addresses the drawbacks of Metal Oxide/Pt assembly and thus act as a suitable replacement, for use in DSSC applications. In order to do so we propose an alternative electrode based on CuNWs, rGO and PEDOT:PSS which offers key benefits such as low cost and flexible nature.

2 EXPERIMENTAL

2.1 Fabrication of Copper Nanowires

The copper nanowires were synthesized using a one step solution based procedure mentioned by Guo et al [7]. In the first step, 1.6 mmol Copper chloride ($CuCl_2$) and 0.8 mmol $Ni(acac)_2$ were dissolved in Oleylamine and the mixture was kept at 80°C for 30 minutes, with constant stirring to help with the dissolution. Later, to begin the growth of the nanowires the temperature was ramped up to 175°C. The temperature of the flask containing the solution was kept constant at 175°C for 10 hours to allow for the synthesis to complete, Eq. (1).

$$CuCl_2 \xrightarrow[Ni(acac)_2\ (Catalyst)]{\substack{Oleylamine\ (Solvent) \\ 175°C}} CuNWs \quad (1)$$

At the completion of the reaction the solution was allowed to cool down and excess hexane was poured into the flask, to precipitate the nanowires. The CuNWs were separated from the solution by centrifuging at 6000 rpm, for 15 minutes. The centrifuged NWs were submerged into Toluene, at the completion of the synthesis process.

2.2 Fabrication of Transparent Conducting Electrode:

The synthesized copper nanowires were detached from Toluene and transferred to Isoporpyl Alcohol (IPA) containing 1wt% of PVP (Polyvinylpyrrolidone). Later, the nanowires were centrifuged to remove excess PVP, and were stored in IPA.

Spray coating technique was utilized to deposit a layer of nanowires, since it is fast and is known for its ability of fabricating large scale homogeneous coatings and specific patters [2]. Cleaned substrates were laid down on a hot plate with the temperature maintained at 60°C, for the prevention of coffee-stain effect that could have compromised the homogeneous deposition of films.

To deposit rGO layer over the CuNWs, GO platelets were sprayed onto the substrate. The samples containing GO layer were annealed in forming gas environment (95% Nitrogen – 5% Hydrogen) at varied temperatures, ranging from 100°C to 300°C, in order to reduce GO. The reduction of GO to rGO helped in restacking of Graphene sheets and improving the structure [8, 9]. Annealing also had a profound effect on the conductivity of the CuNWs network by burning excess organics off the surface of CuNWs and reducing the resistance at the NW intersections.

A final PEDOT:PSS layer was deposited, with the intension of reducing the RMS roughness of the TCE. Briefly, the polymer was heated to 70°C and spun coated onto the electrode samples, while being spun at 6000 rpm.

The fabricated TCEs were characterized using SEM, TEM, AFM, spectrophotometer and sheet resistance measurement equipment.

2.3 Photo anode preparation and integration

In the start, a layer of P25 TiO_2 nano-porous semiconductor was deposited on ITO coated glass substrates, using electrophoretic deposition (EPD) process mentioned by Liou et al [10]. The samples were later sintered in a tube furnace at 450°C, to establish low resistance interconnection between the TiO_2 nanoparticles. The sintered photo anodes were kept in the solution containing Ru based N719 dye for 24 hours, in order to maximize the absorption of dye in the nano-porous TiO_2 layer. The photo-anodes were later taken out of the dye solution, rinsed with ethanol and dried in ambient enviornment. Few drops of electrolyte containing I^- and I_3^- redox species were dropped onto the photo-electrode. In the final step, CuNWs/rGO/PEDOT:PSS based counter electrode was attached to the photo-anode with the sealent between them, to form a compete DSSC, as depicted in Figure 1.

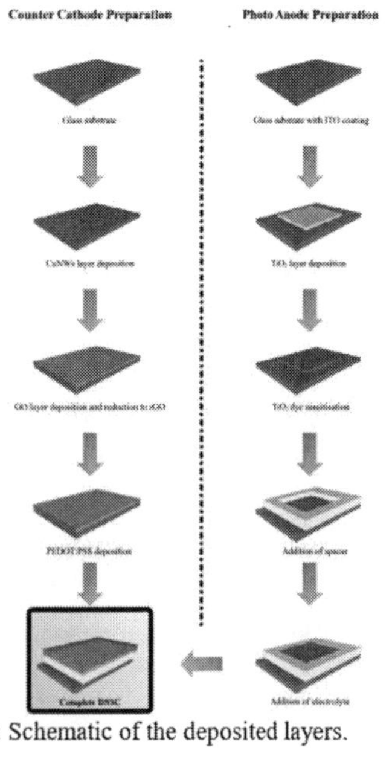

Figure 1: Schematic of the deposited layers.

3 RESULTS AND DISCUSSION

According to the perculating theory, for every doubling of the aspect ratio (L/D) the sheet resistance drops by a factor of ~4, in dense nanowire networks [11]. Hence it is of great important to synthesize NWs as thin and as long as possible, to lower down the sheet resistance of facricated TCEs without any comprommise on the transmittance. The CuNWs synthesized in the current study yielded an aspect ratio of 72.9 μm/44.5 nm, that surpassed the figures quoted in the previous studies [7, 12]. Spray coating the nanowires on to the the glass substrate, produced a network of CuNWs that was homogenously distribuited and had frequent nanowire overlappings, thus resutling in lower sheet resistance, Figure 2.

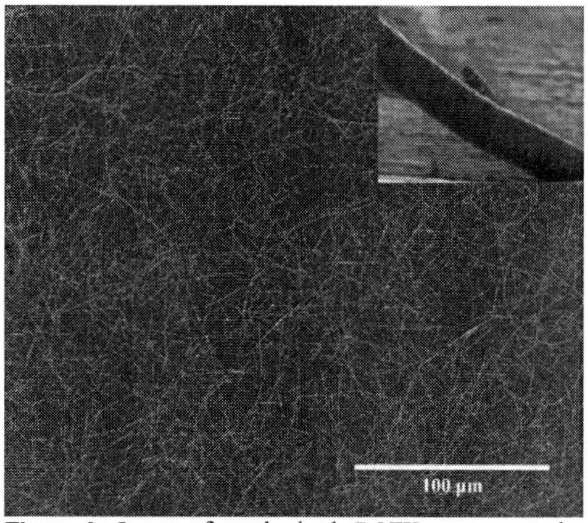

Figure 2: Image of synthesized CuNWs spray-coated onto a glass substrate, gathered by SEM; the nanowires are seen to be homogeneously distributed, inset: The surface of synthesized CuNWs.

CuNWs are susceptable to oxidation as a consiquence of continuous contact with the enviornment. The rGO coating provied an effective way to prevent the oxidation by offering impenetrability against oxygen and water molecules. The RMS roughness was effectively reduced using PEDOT:PSS, as illustrated in Figure 3, thus preventing the CuNWs from shortening subsequent layers of device, the electrode is integerated in.

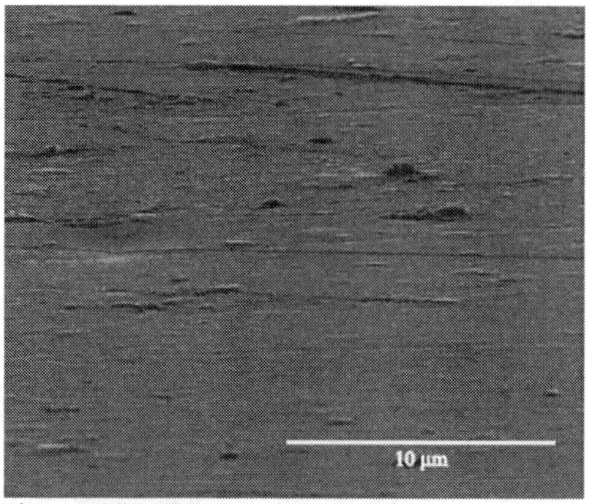

Figure 3: PEDOT:PSS coverage over the

The fabricated hybrid electrodes were integerated within the DSSC, which were later characterized for their performance based on various parameters, using Keithley 2400 source meter and SunLite Solar Simulator, manufactured by Abet Technologies.

Figure 3, shows the I-V curve of the CuNWs/rGO/PEDOT:PSS TCE integerated DSSCs. The Isc and Voc turned out to be 0.95 mA and 0.58V, respectively.

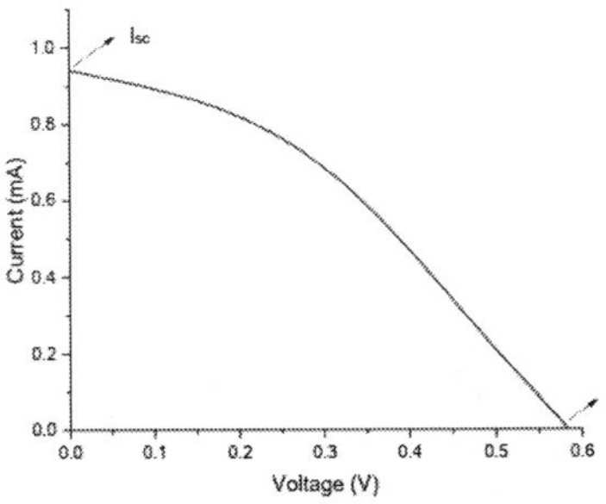

Figure 2: I-V Curve of hybrid TCE based DSSC.

The Fill Factor (FF) for the fabricated hybrid TCEs based DSSCs turned out be 37.67%. The overall efficiency achieved for the CuNWs integerated DSSCs was 0.21%. One of the main reasons behind the lower efficiency could be due to the fact that the graphene produced by the reduction of GO exhibits inheriently low aspect ratio of graphene flakes and high contact resistance between sheets of graphene [13]. In addition to that, the electrolyte used contained Iodine and Lithium iodide salts, which are known to corrode majority of the metals including copper, thus severely effecting the performance by destroying the CuNW network. Further analysis needs to be done in order to find out the main casues behind the lower efficiency.

4 CONCLUSION

Transparent conducting electrodes (TCEs) based on solution processed high aspect ratio copper nanowires were fabricated. Later rGO and PEDOT: PSS layers were deposited onto the electrode that effectively prevented CuNWs from oxidation and reduced RMS roughness, repectively. When integerated within DSSCs as a counter electrode, the TCEs produced an efficiency of 0.21%. Further investigation needs to be carried out in order to find out the reason behind the lower efficiency.

5 ACKNOWLEDGEMENT

This publication was made possible by NPRP grant # 5-546-2-222 from Qatar National Research Fund (a member of Qatar Foundation). The statements made herein are solely the responsibility of the authors.

6 REFERENCE

[1] B. He, Q. Tang, M. Wang, C. Ma, S. Yuan, Complexation of polyaniline and graphene for efficient counter electrodes in dye-sensitized solar cells: Enhanced charge transfer ability, Journal of Power Sources, 256 (2014) 8-13.

[2] Q. Zheng, Z. Li, J. Yang, J.-K. Kim, Graphene oxide-based transparent conductive films, Progress in Materials Science, 64 (2014) 200-247.

[3] D.-S. Leem, A. Edwards, M. Faist, J. Nelson, D.D.C. Bradley, J.C. de Mello, Efficient Organic Solar Cells with Solution-Processed Silver Nanowire Electrodes, Advanced Materials, 23 (2011) 4371-4375.

[4] T. Akter, W.S. Kim, Reversibly Stretchable Transparent Conductive Coatings of Spray-Deposited Silver Nanowires, ACS Applied Materials & Interfaces, 4 (2012) 1855-1859.

[5] Z. Zhu, T. Mankowski, K. Balakrishnan, A.S. Shikoh, F. Touati, M.A. Benammar, M. Mansuripur, C.M. Falco, Ultrahigh Aspect Ratio Copper-Nanowire-Based Hybrid Transparent Conductive Electrodes with PEDOT:PSS and Reduced Graphene Oxide Exhibiting Reduced Surface Roughness and Improved Stability, ACS Applied Materials & Interfaces, 7 (2015) 16223-16230.

[6] A.S. Shikoh, Z. Ahmad, F. Touati, R.A. Shakoor, N.J. Al-Thani, Z. Zhu, T. Mankowski, M.A. Benammar, M. Mansuripur, C.M. Falco, Integration of the inexpensive CuNWs based transparent counter electrode with dye sensitized photo sensors, RSC Advances, 6 (2016) 53123-53129.

[7] H. Guo, N. Lin, Y. Chen, Z. Wang, Q. Xie, T. Zheng, N. Gao, S. Li, J. Kang, D. Cai, D.-L. Peng, Copper Nanowires as Fully Transparent Conductive Electrodes, Scientific Reports, 3 (2013) 2323.

[8] B.-T. Liu, H.-L. Kuo, Graphene/silver nanowire sandwich structures for transparent conductive films, Carbon, 63 (2013) 390-396.

[9] M.S. Miller, J.C. O'Kane, A. Niec, R.S. Carmichael, T.B. Carmichael, Silver Nanowire/Optical Adhesive Coatings as Transparent Electrodes for Flexible Electronics, ACS Applied Materials & Interfaces, 5 (2013) 10165-10172.

[10] Y.-J. Liou, P.-T. Hsiao, L.-C. Chen, Y.-Y. Chu, H. Teng, Structure and Electron-Conducting Ability of TiO2 Films from Electrophoretic Deposition and Paste-Coating for Dye-Sensitized Solar Cells, The Journal of Physical Chemistry C, 115 (2011) 25580-25589.

[11] S. Ye, A.R. Rathmell, Z. Chen, I.E. Stewart, B.J. Wiley, Metal Nanowire Networks: The Next Generation of Transparent Conductors, Advanced Materials, 26 (2014) 6670-6687.

[12] A.R. Rathmell, S.M. Bergin, Y.-L. Hua, Z.-Y. Li, B.J. Wiley, The Growth Mechanism of Copper Nanowires and Their Properties in Flexible, Transparent Conducting Films, Advanced Materials, 22 (2010) 3558-3563.

[13] J.H. Lee, D.W. Shin, V.G. Makotchenko, A.S. Nazarov, V.E. Fedorov, Y.H. Kim, J.Y. Choi, J.M. Kim, J.B. Yoo, One-Step Exfoliation Synthesis of Easily Soluble Graphite and Transparent Conducting Graphene Sheets, Advanced materials, 21 (2009) 4383-4387.

Organic Photovoltaics for energy harvester of wireless sensor network

Yoichi Aoki

ROHM Co.,Ltd, 21 Saiin Mizosaki-cho, Ukyo-ku Kyoto 615-8585 Japan

TEL:+81-075-321-1481, FAX:+81-075-311-0383 E-mail:Yoichi.Aoki@dsn.rohm.co.jp

ABSTRACT:

Organic photovoltaic (OPV) module has great photovoltaic performance under room light because OPV has good spectral matching with UV-Vis light owing to its organic absorber. In this study, we investigate the electric characteristics of OPV and conventional amorphous silicon solar cells (a-Si) used for energy harvester of wireless sensor network. The photovoltaic performance of OPV module is $7.9\mu m/cm^{-2}$ and $43.4\mu m/cm^{-2}$ at fluorescent light 200lux and 1000lux respectively. And the data transmission rate of the wireless sensor node embedded OPV is 30-40% improved under the dim light condition compared to a-Si.

Keyword: organic photovoltaics, energy harvester, wireless sensor network

1. INTRODUCTION

The wireless sensor network (WSN) is expected to enrich our lives automatically because wireless sensor nodes with energy harvesters detect ambient environment (temperature, humidity, room light intensity, motion, and so on) and feedback the sensing data to the hosting system in order to maintain the comfort of the home and save the energy.

The energy harvesters like electro-dynamic and thermoelectric generator is one of key technologies for WSN to operate the wireless sensor nodes stably. In case of home and office, however, indoor solar cells are the most appropriate and convenient method to provide the μW to sub-mW class power.

Generally, solar cells have wide range of absorption against the sunlight for outdoor application [1]. For the energy harvester, however, solar cells should have strong absorption against the UV-Vis spectrum of indoor lighting because the wireless sensor nodes are used in the house and buildings [2].

Organic photovoltaic (OPV) has the ultra-thin 0.2μm organic absorber. This layer has the strong absorption and high external quantum efficiency in UV-Visible 350-750nm spectrum [3,4]. So, the single junction OPV cell has enough thickness to absorb the weak room light source $>0.1mW/cm^2$ effectively in spite of its thinner absorber. And it has good spectral matching with the room light like the fluorescent light and warm & cool colored LEDs which have 350-700nm emitting spectrum (Fig1).

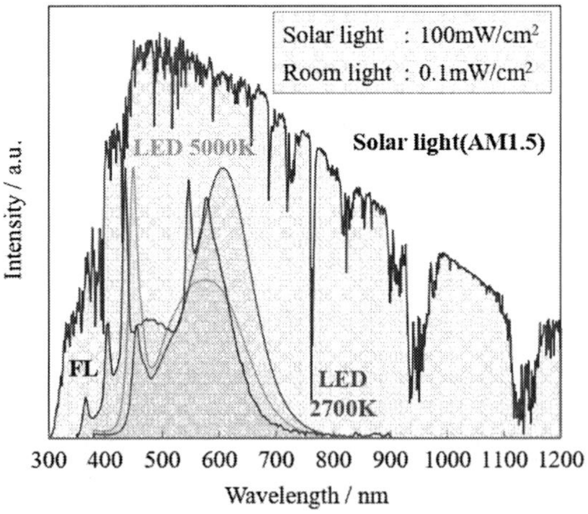

Fig1. The normalized spectra of the room light [full white fluorescent light (FL), LED 5000K, LED 2700K] and the AM1.5G standard solar light

2. EXPERIMENTAL

2-1. The electrical characteristics of OPV module

The OPV cell consists of a solution-processed organic layer sandwiched with ITO and metal electrode on the glass substrate and covered with passivation film in order to avoid exposure from oxygen and moisture. Ultra-thin cover glass is combined with glass substrate with OPV cells by adhesive resin to ensure the mechanical scratch resistance.

The OPV module has 8 cells in series and the module dimension is 35.0mm x 12.8mm x t0.46 mm. The OPV cell color is black which is familiar with the enclosure of wireless sensor nodes. The a-Si has same specification and external form as OPV module. And the photovoltaic performance of OPV and a-Si is measured under the full-white (5000K, Ra70) fluorescent light at each light intensity. 50lux (0.005mW/cm^2) and 1000lux (0.1 mW/cm^2) are the same brightness at an emergency staircase and desktop in the office. The energy harvester is mainly used at less than 200lux.

Fig2. Organic photovoltaic module for the energy harvester of wireless sensor network

OPV has more than 30 - 40% output power compared to a-Si at illumination of fluorescent light(Fig3). The OPV module harvests 7.9μW/cm^{-2} and 43.4μW/cm^{-2} at fluorescent light 200lux and 1000lux. a-Si module harvests 5.6μW/cm^{-2} and 28.5μW/cm^{-2} at fluorescent light 200lux and 1000lux. Notably, OPV's performance goes up with the incident light intensity. OPV module provides 27.3μW, 150μW, and 766μW at 200lux, 1000lux, and 5000lux respectively.

Fig4 shows the typical J-V characteristics of each solar cell under the 50 -1000 lux. Because organic absorber is well matched with the emission spectrum of room light, OPV module has 20% much more short circuit current density compared to a-Si module. And OPV has over 0.7 fill factor which gives higher operating voltage and current range under the dim light condition.

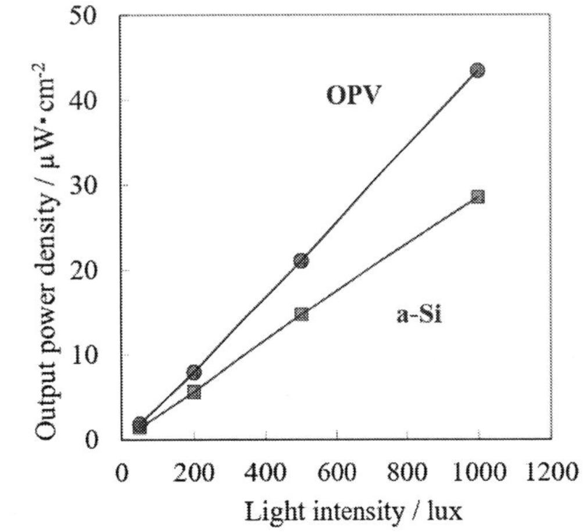

Fig3. Maximum output power density of OPV and a-Si modules under the low light condition

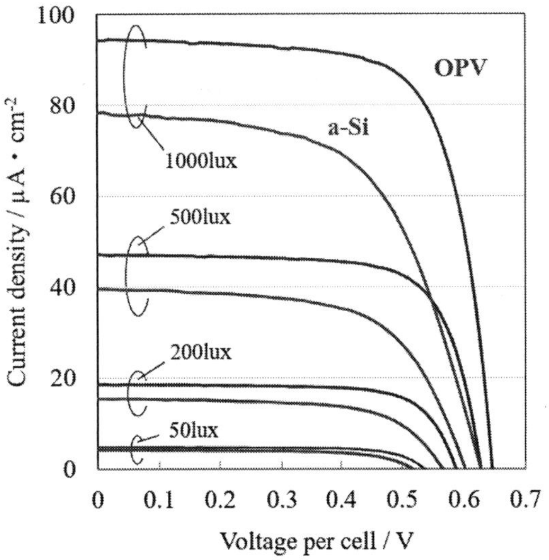

Fig4. The J-V Characteristics of OPV and a-Si module for energy harvester of WSN under the fluorescent light 50, 200, 500, and 1000lux.

2-2. Demonstration of wireless sensor node with OPV

In order to investigate the practical application, the OPV or a-Si module is embedded the wireless sensor node (EnOcean STM 431J) with temperature sensor widely used for home and office application.

This finger-sized sensor node [2cm×7cm×t1.5cm] detects ambient room temperature and transfers the temperature data wirelessly to the receiver every 10 seconds. The energy harvester generates and provides the electric power to the capacitor. And this capacitor drives sensor, IC controller and RF transceiver. For continuous operation, this node requires more than 13.5μA and 2.1V.

Fig5 shows the transmission state from wireless sensor node for room temperature with a-Si solar cell at fluorescent light 200lux. "No data transmitted" means charging period from solar cells to capacitor until the threshold current is charged.

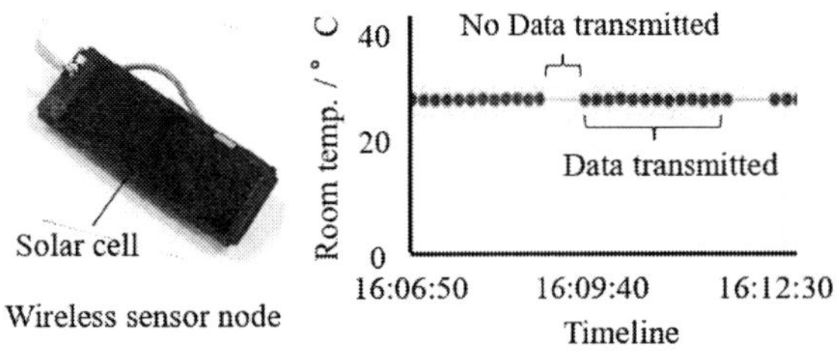

Fig5. The transmission state from wireless sensor node with a-Si solar cell at fluorescent light 200lux

The data transmission rate depends on the output power of energy harvester. So, High performance harvester gives high transmission rate. The score is calculated by "Data transmission rate = Transmitted point / All point". The wireless sensor nodes with OPV and a-Si are measured under the each light intensity 50-500lux (Fig6). Over 500lux, OPV and a-Si provide enough output power to drive the nodes at short interval of time. The wireless sensor node with OPV has 30-40% higher performance because OPV has great photovoltaic performance.

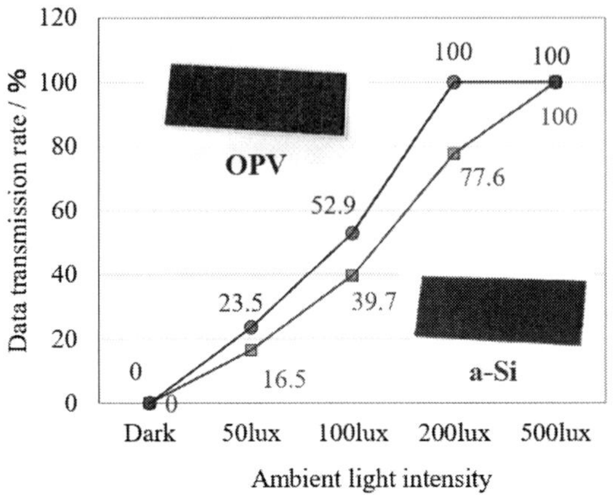

Fig6. Data transmission rate from the wireless sensor node driven by OPV or a-Si at each ambient illuminance

3. CONCLUSION

OPV is great energy harvester for the wireless sensor network μW to sub-mW class power compared to conventional a-Si. OPV enables us to decrease the size of solar cell panel area or stabilize continuous operation of wireless sensor network under the low light condition if OPV panel size is same as a-Si.

4. ACKNOWLEDGMENTS

The author wishes to thank Hironori Kashiwa and Ryosuke Ishii for continued support.

REFERENCE

[1] M.A.Green, K.Emery, Y.Hishikawa, W.Warta, E.D.Dunlop, Prog. Photovolt,Vol.23,7(2015)805-812

[2] N.Sridhar, D.Freeman,10.4229/26th EUPVSEC 2011-1CO.10

[3] Y.Aoki, 10.4229/EUPVSEC2015 2015-3BV.5.48

[4] B. Minnaert ; P. Veelaert, Proc. SPIE 7722, Organic Photonics IV, 77221P (May 18, 2010)

This is a pre-peer review version of the paper submitted to the journal Progress in Photovoltaics.
It has been selected by the Executive Committee of the EU PVSEC 2016 for submission to Progress in Photovoltaics.

AN INVESTIGATION OF PRE-CONDITIONING ROUTINES FOR ACCURATE EFFICIENCY MEASUREMENTS OF PEROVSKITE SOLAR CELLS

Ricky B. Dunbar[1]*, Walied Moustafa[1,2], Noel Duffy[3], Ben Duck[1], Kenrick F. Anderson[1], Timothy W. Jones[1], Christopher J. Fell[1], Gregory J. Wilson[1]

[1] CSIRO Energy, Mayfield West, NSW, 2304 Australia
[2] Hamburg University of Applied Sciences, 20099, Hamburg, Germany.
[3] CSIRO Energy, Clayton, VIC, 3800 Australia

*email: ricky.dunbar@csiro.au

ABSTRACT: Despite the rapid rise in efficiency reported for research-scale perovskite solar cells, significant measurement challenges remain. A key challenge is the tendency for perovskite solar cells to respond slowly (on timescales of seconds to minutes) to changes in voltage bias and irradiance, which typically leads to a dependence of measurement results on cell history and measurement approach. Here, we investigate pre-conditioning treatments, which can be applied to prevent transient effects being mistaken for voltage-dependence during measurement of current-voltage curves. We find the measurement history of a device can significantly influence transient current characteristics when the device is transitioned from being in the dark to being illuminated. For the device structure investigated in this work, light-soaking periods of 5-10 minutes can be sufficient to return the device to the illuminated steady-state provided the device was light-soaked for 10 minutes no more than 45 minutes previously. Otherwise, longer additional light-soaking periods are required. This is important information for efficiency measurement of devices subject to performance degradation on these timescales. If sufficiently long dark rest periods are allowed prior to measurement, distinct dips are readily observable in current transients between 0 and 100 seconds after illumination commences.
Keywords: Perovskite Solar Cells, Performance, Characterization

1 INTRODUCTION

Many commercially established photovoltaic technologies, such as CIS- and CdTe-based devices, are sensitive to measurement and exposure history prior to measurement of the device efficiency. This is usually addressed using pre-conditioning treatments [1], [2] which can prevent a transient current response to illumination being misinterpreted as a voltage dependence during voltage sweeps. The high sensitivity to measurement history that has been observed for perovskite solar cells [3] suggests that the identification of appropriate pre-conditioning treatments will be a critical step towards determining robust measurement guidelines for assessing the performance of perovskite solar cells.

Pre-conditioning is typically performed by illuminating the device or applying a voltage bias for an extended period, or a combination of both. In this work, we investigate the influence of pre-conditioning light level and voltage bias for p-i-n lead-acetate perovskite cells.

An improved understanding of perovskite cell stabilisation will assist in the identification of pre-conditioning treatments that bring devices to the required state prior to measurement, and hence improve the accuracy of the measured efficiency. There are two device states that are of particular interest:

1) The dark state, where the device behaves equivalently to the case where it has been kept in the dark for an indefinitely long period of time; and

2) The illuminated steady-state, where the device behaves equivalently to the case where it has been exposed to constant irradiation for an indefinitely long period of time.

By definition, once the device has been pre-conditioned to reach either of these states, the influence of measurement history on subsequent behaviour will have been removed, and will depend only on the starting state.

Pre-conditioning treatments that reliably bring the cell to the illuminated steady-state are important for determining a value of efficiency indicative of continuous operation at standard test conditions. This is of particular importance for laboratories, such as the CSIRO PV Performance Laboratory (PVPL), that hold accreditation for PV performance testing. As reported STC efficiencies for research prototype perovskite solar cells have already reached a similar level to records set by commercialised thin-film technologies, and further improvements may be possible, it is critical that the influence of pre-conditioning on measured efficiency is well understood.

2 METHOD

A standard p-i-n perovskite solar cell with structure ITO | PEDOT:PSS | Perovskite | PCBM | Ag was used for our study. PEDOT:PSS, perovskite and PCBM layers were deposited by spincoating. The perovskite was deposited from a solution of 40 wt.% 3:1 $CH_3NH_3I:Pb(OAc)_2$ in DMF, and annealed under glovebox conditions at 90 °C for 10 min. Devices were encapsulated with a coverslip and a UV-curable epoxy used as an edge seal. The PEDOT:PSS used was Clevious P VP.Al 4083 from Heraeus. The PCBM was purchased from Lumtec. Precursor materials for the perovskite film, MeAI and lead acetate, were purchased from Dyesol Ltd. and Sigma-Aldrich, respectively. The cell active area was 0.3 cm^2.

Current-voltage (I-V) measurements were performed using a Keithley 2400 source-measure unit with a 150 W Oriel solar simulator. Irradiance was limited to the cell area using a shadow mask.

Device stabilisation is investigated using transient current measurements following changes in bias or light level. These measurements are performed using a multi-function measurement platform (Paios 3.0, Fluxim), with a white LED light source. The cell temperature is maintained at 25 °C for all transient measurements using a thermoelectric stage. Three different irradiance levels, corresponding to approximately 0.4, 0.8 and 1.9 suns, are used in the study.

3 RESULTS AND DISCUSSION

Current-voltage curves of the cell used in this study, performed with a scan rate of approximately 50 mV/s were performed prior to the transient measurements. The scans indicated a power conversion efficiency of approximately 6.5% in both scan directions. The similarity of the results for forward- and reverse- scans has been widely observed for this cell architecture. The efficiency measured for this cell is lower than the current record efficiency [4] for this architecture, 18.9% for a device area of 0.09 cm^2; we attribute this to a lack of optimisation in our fabrication process and to the larger area of our cell compared to the record device. We note that it is likely that the loss mechanisms limiting lower performance devices will result in transient characteristics that differ from high performance devices. The influence of cell efficiency on the transient processes that we investigate will be the subject of future work.

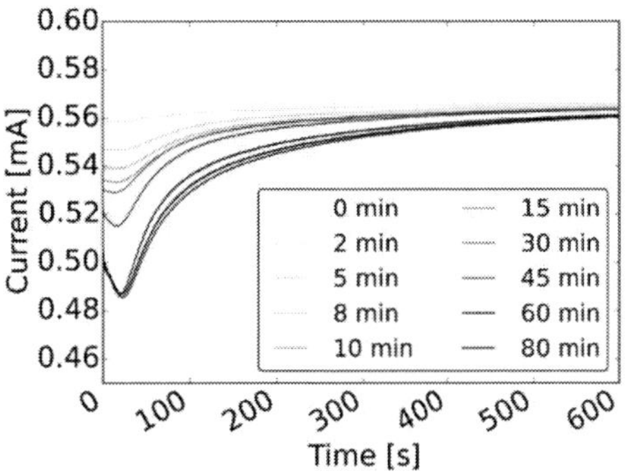

Figure 1: Transient current response to an irradiance step of zero to 0.8 suns following rest periods of 0-80 minutes in the dark. The cell is held at short-circuit throughout.

Figure 1 shows current transients over 10 minutes of illumination. Immediately prior to this measurement, the cell undergoes a fixed light soaking treatment of 10 minutes followed by a rest period of variable duration in the dark. The device is held at short-circuit throughout, and the measurements are carried out in order of increasing rest period. The shape of the current transient is clearly influenced by the duration of the preceding rest period. For longer rest periods, dips in the transient characteristic indicate that when transitioning from the starting state to the illuminated steady-state, processes occur which collectively initially reduce the current until a minimum value is reached at around 30 seconds, after which the current increases. For shorter rest periods, dips are reduced or absent, indicating the reversal of these processes during the rest period was incomplete before the transient measurement. This behaviour, characterised by transient processes with timescales of seconds to minutes, is consistent with the presence of mobile ions in the perovskite film[5], [6] that can be liberated from their lattice positions to influence charge-carrier dynamics in the cell. Evidently for this cell, rest periods of at least 45 minutes in the dark are necessary to achieve a current transient similar to that observed for the longest wait period investigated here. This as an indication that the dark state can be re-established after approximately 45 minutes in the dark at short-circuit.

It can be inferred from Figure 1 that if the cell has been in the dark for less than 45 minutes immediately following a previous 10 minute light soaking treatment, then 5- 10 minutes of further light-soaking is sufficient to return the device to its illuminated steady-state current. Otherwise, longer additional light-soaking periods are required. This information is of high importance when testing devices that are subject to performance degradation,

In order to assess whether the measurement sequence has an influence on the trend observed in Figure 1, this experiment was repeated with three repeat measurements for each rest period duration with a random measurement sequence (Figures 2a)-c)). The randomisation is used to attempt to isolate the influence of the rest period duration from the additional prior history of the cell. For each sequence, each curve is normalised to the maximum current value obtained in the sequence. This was then repeated for three irradiance levels: 0.4, 0.8 and 1.9 suns, respectively.

Figures 2a)-c) show the same overall trend as Figure 1, but display a considerable spread in the transient curves for a given value of rest period duration, even for the longest rest period investigated. This likely indicates that the rest periods investigated here do not completely permit complete relaxation to the dark state, but some dependence on previous measurement history remains.

The relative change in current in the transient curves following long rest period is observed to increase with irradiance level, whereas the shape of the curves, determined by the decay time of the transient processes, remain the same. This suggests that the mechanism governing the transient behaviour is the same for the three irradiance levels investigated here, and it is only the magnitude of the relative change in current that changes with irradiance. This is confirmed quantitatively by applying mono-exponential fits to the curves between 100 and 600 s. As expected, these fits yield time constants that are relatively invariant with rest time and exponential coefficients that increase with rest period duration, the latter of which is shown in Figure 2d.

The application of a voltage bias during pre-conditioning can also strongly influence the stabilisation time of the cell. Figure 3a)-b) shows transient current measurements at 1.9 suns at short-circuit, prior to which the cell was held under bias in the dark for 60s. Reverse biases are observed to yield increasing transients (Figure 3a)). For the forward biases investigated, a distinct change in the current transient curve from an increasing to a decreasing single exponential function was observed as the pre-conditioning bias is increased above a threshold value of approximately 0.5 V.

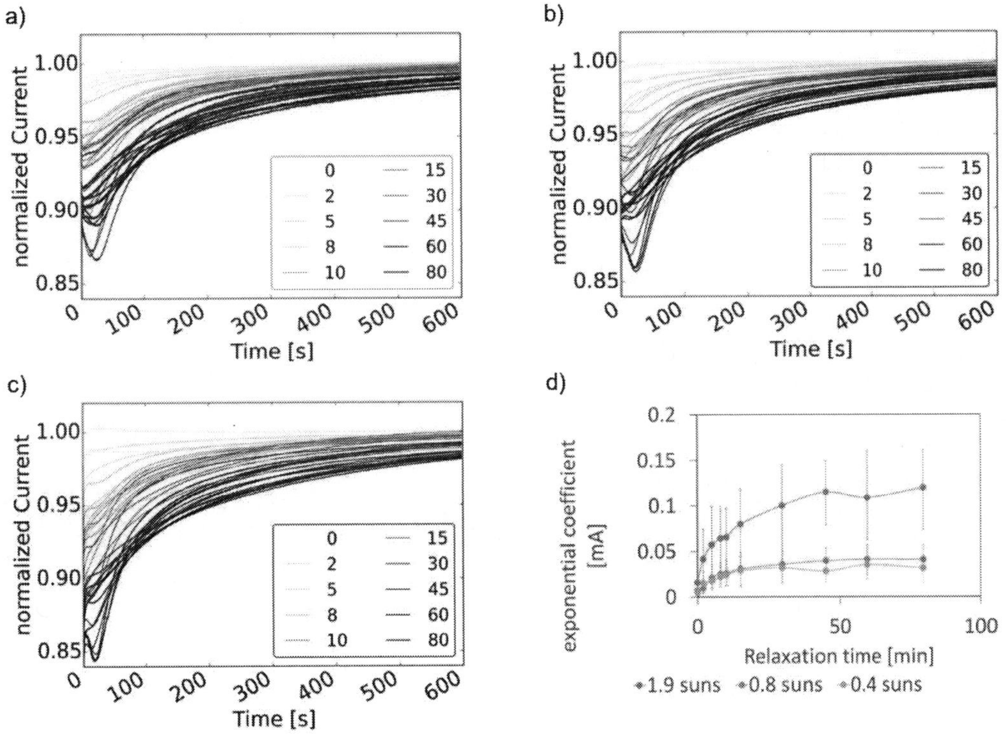

Figure 2: a), b) and c) Normalised transient current curves for rest periods of 0-80 minutes at irradiance values of 0.4, 0.8 and 1.9 suns, respectively; d) time-constant obtained from a single-exponential fit to data in for the time period 100-600 s. Error bars constructed using the range of three repeat measurements for each rest period.

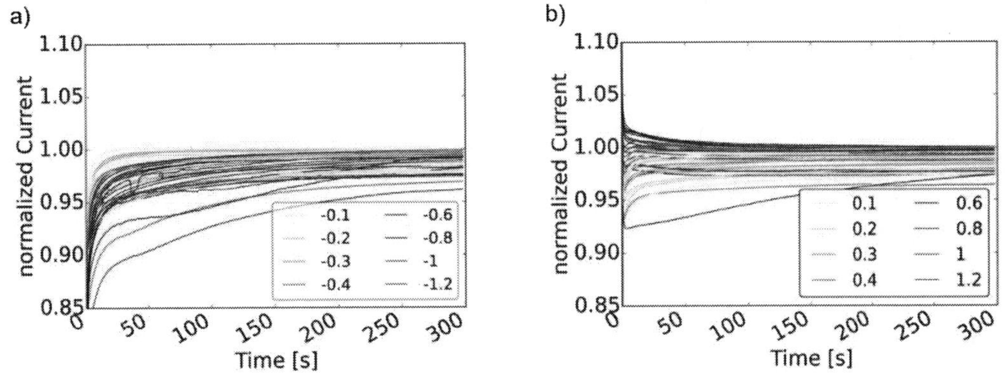

Figure 3: Transient current behaviour at 25 °C at 1.9 suns following a 60 second period in the dark under applied bias: a) reverse bias voltages in the range -0.1 to -1.2 V; b) forward bias voltages in the range 0.1 to 1.2 V.

4 CONCLUSION

This work has shown that measurement and exposure history of the lead-acetate perovskite cell investigated can significantly influence the device state, which can persist even after ten minutes of light-soaking. The effect of dark treatments and the application of voltage bias on cell stabilisation have been investigated. Following rest periods at short-circuit in the dark for a sufficient duration, states that limit current generation, resulting in dips in current transients, can be readily observed.

4 REFERENCES

[1] "Characterisation of Performance of Thin-film Photovoltaic Technologies," *13, IEA PVPS Task, T13-022014*, 2014.

[2] "Guidelines for PV Power Measurement in Industry," *Performance FP6 Integrated Project*. 2010.

[3] E. L. Unger, E. T. Hoke, C. D. Bailie, W. H. Nguyen, A. R. Bowring, T. Heumuller, M. G. Christoforo, and M. D. McGehee, "Hysteresis and transient behavior in current-voltage measurements of hybrid-perovskite absorber solar cells," *Energy Environ. Sci.*, vol. 7, no. 7,

pp. 3690–3698, 2014.

[4] Q. Dong, Y. Yuan, Y. Shao, Y. Fang, Q. Wang, and J. Huang, "Abnormal crystal growth in $CH_3NH_3PbI_{3-x}Cl_x$ using a multi-cycle solution coating process," *Energy Environ. Sci.*, vol. 8, pp. 2464–2470, 2015.

[5] W. Tress, N. Marinova, T. Moehl, S. M. Zakeeruddin, M. K. Nazeeruddin, and M. Grätzel, "Understanding the rate-dependent J–V hysteresis, slow time component, and aging in $CH_3NH_3PbI_3$ perovskite solar cells: the role of a compensated electric field," *Energy Environ. Sci.*, vol. 8, no. 3, pp. 995–1004, 2015.

[6] M. De Bastiani, G. Dell'Erba, M. Gandini, V. D'Innocenzo, S. Neutzner, A. R. S. Kandada, G. Grancini, M. Binda, M. Prato, J. M. Ball, M. Caironi, and A. Petrozza, "Ion Migration and the Role of Preconditioning Cycles in the Stabilization of the J-V Characteristics of Inverted Hybrid Perovskite Solar Cells," *Adv. Energy Mater.*, 1501453, 2015.

CALIBRATION PROCEDURE FOR THE ACCURATE POWER MEASUREMENTS OF SLOW RESPONDING PV DEVICES (HETERO-JUNCTION, DYE-SENSITIZED AND PEROVSKITE SOLAR CELLS)

Giorgio Bardizza*, Diego Pavanello, Roberto Galleano, Tony Sample and Harald Müllejans
European Commission, DG JRC, Institute for Energy and Transport, Renewables and Energy Efficiency Unit
Via E. Fermi 2749, 21027 Ispra (VA), Italy
*Tel: +39 0332785529, e-mail: giorgio.bardizza@ec.europa.eu

ABSTRACT: In this work we propose a procedure for the calibration of slow responding PV devices which allows for an accurate evaluation of their response time and, consequently, for a reliable calibration fulfilling all requirements specified in the standards (IEC 60891 and IEC 60904 series). The application of this procedure to the case of three dye-sensitized solar devices (one single cell and two mini-modules) is presented. The spectral responsivity (SR) measurements were performed at 1 Hz chopping frequency. The low chopping frequency resulted from an analysis of the device time response. To ensure reliable electrical characterization, current-voltage measurements were performed using different sweep-times in both the forward and reverse direction. By comparing these measurement results, a reliable evaluation of the appropriate sweep-time for I-V characterization was determined. Combining appropriate SR and I-V measurements yielded a calibration of these devices according to the IEC 60891 and IEC 60904 series of standards.
Keywords: Photovoltaic, calibration, DSSC, hetero-junction, perovskite, slow response, PV devices

1 INTRODUCTION

The procedures for the power characterization of PV devices are described in IEC 60891 and IEC 60904 series of standards. For c-Si PV devices the procedures to perform reliable measurements are well established among reference laboratories like ESTI (European Solar Test Installation). However, in case of emerging PV technologies like back-contact and hetero-junction solar cells, dye-sensitized solar cells (DSSC) or perovskite solar cells (PSC) new procedures need to be specifically developed taking into account their characteristics, i.e. the slow response time. IEC 60904-1 contains a remark about the measurements of devices with high capacitance, in which an analysis of the effect of the sweep-time and sweep-direction is required when performing a calibration of these devices.
The slow response time of DSSC devices is well known in the literature and has been previously studied by Sommeling [1, 2], Hishikawa [3] and Takagi [4]. They presented examples of characterization of DSSCs and described methods for the evaluation of their response time. Snaith [5] firstly reported about an anomalous hysteresis in perovskite solar cells (PSC), but no procedure has been proposed yet for their calibration. The need for a general method and procedure to reliably measure and characterize PV devices possessing slow or unknown response times for photocurrent generation by incident light is evident. In this work we propose a procedure for the calibration of these devices which allows for an appropriate evaluation of their response time and, consequently, for a reliable calibration fulfilling all requirements specified in the standards (IEC 60891 and IEC 60904 series). The application of this procedure to the case of different DSSC devices is presented and a calibration of one of them is finally shown.

2 CALIBRATION PROCEDURE

The procedure proposed here consists of an initial response time analysis needed to estimate the chopping frequency suitable for SR measurements and the required sweep time to perform I-V measurements. Consequently a set of multiple I-V measurements performed at various sweep time and changing of the sweep direction are part of the evaluation. The goal of this part of the time response evaluation is to obtain constant and reproducible I-V curves and to find an overlap between measurements performed in both the forward and reverse sweep directions. Finally SR and I-V measurements can be performed reliably following the procedures in IEC 60891 and IEC 60904. A schematic of the procedure is shown in figure 1.

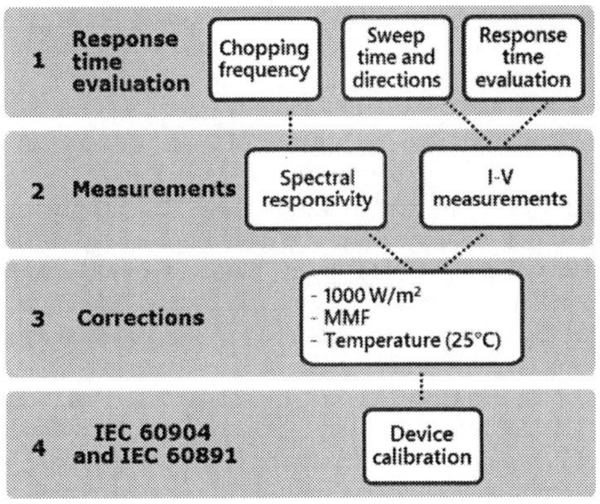

Figure 1: Schematic of the procedure.

3 RESULTS

3.1 Devices

Three DSSC devices were utilized in the measurements presented here: a single DSSC in a glass-glass configuration having an active area of 77.4 mm^2 (device area 550 mm^2, ESTI code PL84), one DSSC masterplate in a glass-glass configuration with five serially connected cells (cell area 360 mm^2, module area: 89.7 cm^2, ESTI code PP81) and one DSSC masterplate in a glass-glass configuration with five cells each with external connectors (cell area 360 mm^2, module area: 89.7 cm^2, ESTI code PP85). A liquid electrolyte was incorporated in all the devices. The devices were provided by two external laboratories under sample transfer agreements and two of them are shown in figure 2.

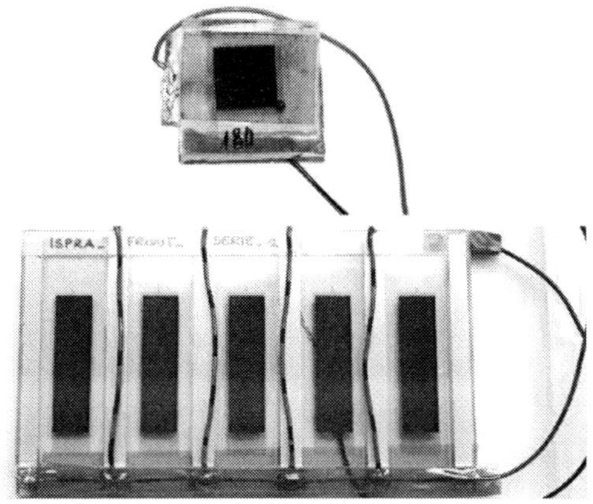

Figure 2: Single DSSC cell (PL84, top) and one of the DSSC masterplates (PP81, bottom).

3.2 Response time evaluation

A reliable calibration of a PV device with an unknown response time requires the evaluation of its capability to respond to light, which could be performed in many different ways. In this work we present a method based on the SR setup available at ESTI previously described in detail [7]. Practically, a reference c-Si device (REF) and a DSSC device under test (DUT) are simultaneously illuminated by a chopped broadband light generated by a steady-state Xenon light source passing through a mechanical chopper. The voltage signals generated by DUT and REF across calibrated shunt resistors are simultaneously acquired by a digital storage oscilloscope. An example of the signals of the DUT and REF acquired at three different representative chopping frequencies are shown in figure 3 a, b and c. In this example a data processing of the signals was performed: their voltage offset was removed and the amplitude was normalized to the initial value. The variation of the peak-to-peak signals with chopping frequency is shown in the plots on the right. Looking at figure 3a it is evident that at the chopping frequency of 1.5 Hz the shape of DUT and REF signals are comparable and their peak-to-peak amplitude within the 0.5 to 1.5 Hz range is almost constant. In figure 3b the same signals acquired when the chopping was about 10 Hz are shown. Here the DUT signal in blue is clearly distorted and has a different shape when compared to the REF signal and to the shape of the DUT below 1.5 Hz. The peak-to-peak amplitude of the REF signal remained constant also up to 10 Hz as expected, on the contrary the amplitude of the DUT signal strongly decreased to almost 30% of its initial value. The same trend continued on increasing the chopping frequency up to 80 Hz, the amplitude of the REF signal remained constant while the signal of DUT continued to reduce in amplitude and to be distorted as shown in figure 3c. A phase shift between REF and DUT signals was also noticeable. The results of the response time analysis applied to the DSSC devices involved in this study are reported in figures 4 and 5. They correspond to the measurements performed on the single cell PL84 and on the masterplate PP81. In both cases the amplitude variation of the DUT signal below 1.5 Hz is comparable to the variation of the REF signal along the full frequency range (standard deviation of about 1%) which is acceptable. These results suggested that a chopping frequency below 1.5 Hz has to be chosen when measuring devices of this type to ensure full signal amplitude. Measuring above that chopping frequency will result in an underestimation of the DUT spectral responsivity.

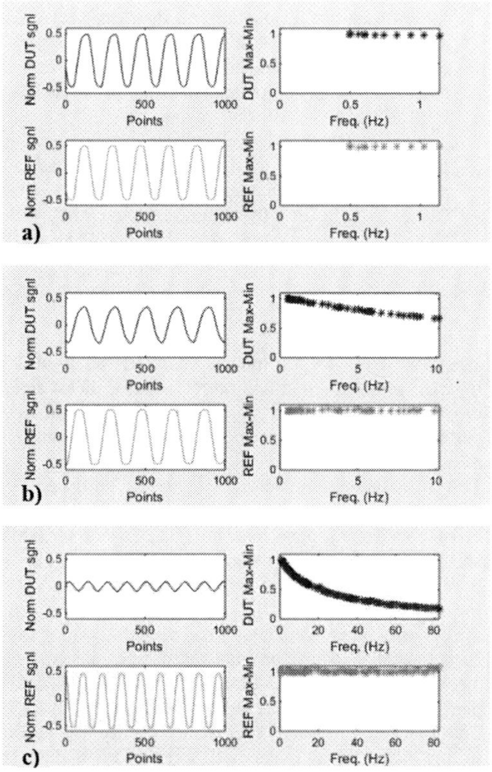

Figure 3: Photocurrent signals of the DUT and REF illuminated by a chopped broadband light at three different stages of the response time analysis: 1.5 Hz (a), 10 Hz (b) and 80 Hz (c). The variation of the peak-to-peak signals with changing frequency is shown on the right.

The results of the response time analysis applied to the single cell and masterplate under study yield the same conclusion. Although this was expected since the devices are of the same technology (dye-sensitized solar cell with liquid electrolyte). We selected a chopping frequency of 1 Hz to be appropriate for a reliable measurement of the spectral responsivities of these devices.

3.3 Spectral responsivity measurements

At ESTI the spectral responsivity of PV devices is measured on a dedicated setup for cells and mini-modules [7, 8]. The existing SR apparatus typically operates at a chopping frequency of around 70 Hz. This chopping frequency has been found to be optimal for standard c-Si PV devices, while it is certainly too high for slow responding PV devices. A chopping frequency of

1 Hz was found to be appropriate for these devices to ensure reliable SR measurements (see section 3.2). A validation of the ESTI setup operating at a 1 Hz chopping frequency was performed and results can be found in previous publication [6].

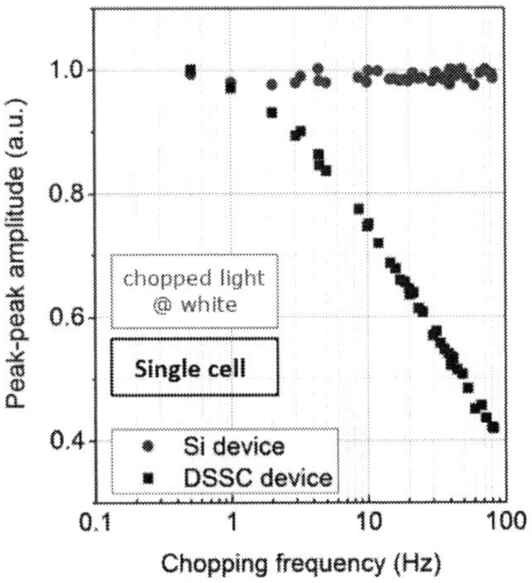

Figure 4: Normalized (to low-frequency) amplitude variation of the peak-to-peak photocurrent signals of a c-Si reference cell and the DSSC device (single cell) illuminated by chopped broadband light at different chopping frequencies.

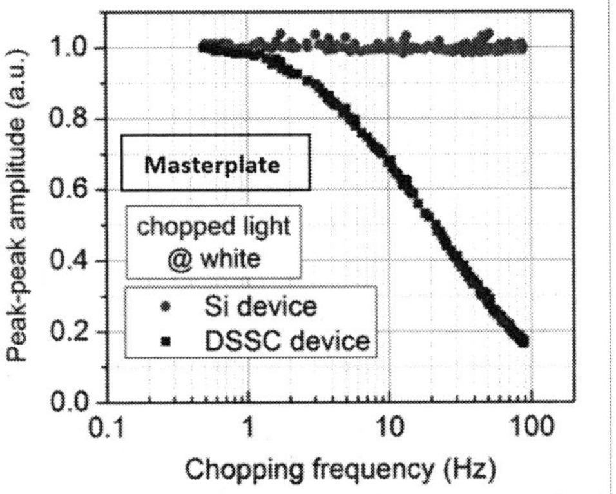

Figure 5: Normalized (to low-frequency) amplitude variation of the peak-to-peak photocurrent signals of a c-Si reference cell and the DSSC device (serially connected masterplate) illuminated by chopped broadband light at different chopping frequencies.

In figure 6, 7 and 8 the absolute SR measurements of the single cell, of the masterplate with serially connected cells and of two cells of the masterplate with separated cells are shown respectively. Measured data and interpolated fitting curves are shown in the figures. Piecewise Hermite polynomials were used to fit the data and interpolate between the measured points in the wavelength range from 300 to 1200 nm with a step of 1 nm. Figure 9 shows all SR curves in one graph for comparison purposes.

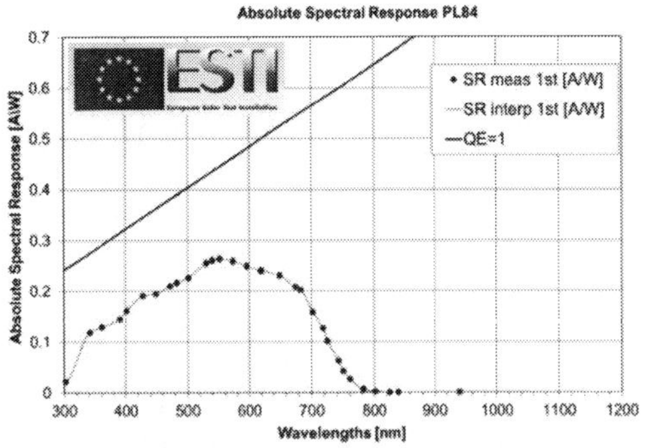

Figure 6: Absolute spectral responsivity (measured and interpolated data) of the single cell (PL84) obtained at a 1Hz chopping frequency.

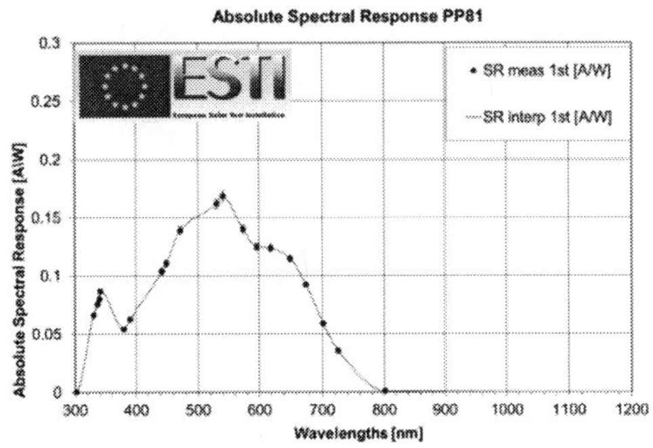

Figure 7: Absolute spectral responsivity (measured and interpolated data) of the masterplate with serially connected cells (PP81) obtained at a 1Hz chopping frequency.

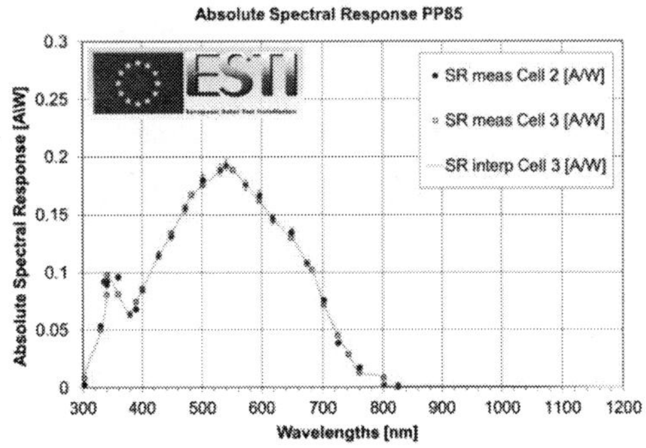

Figure 8: Absolute spectral responsivity (measured and interpolated data) of two cells (cell 2 and 3) of the masterplate with separated cells (PP85) obtained at a 1Hz chopping frequency.

Once the SRs of the three devices were measured, we calculated the MMF values according to IEC 60904-7 [9]. The results of the calculation are reported in Table 1.

Table 1: MMF values calculated from SR measurements for the devices under study.

	PL84	PP81	PP85 cell 2	PP85 cell 3
MMF	1.0025	1.0208	1.0196	1.0163

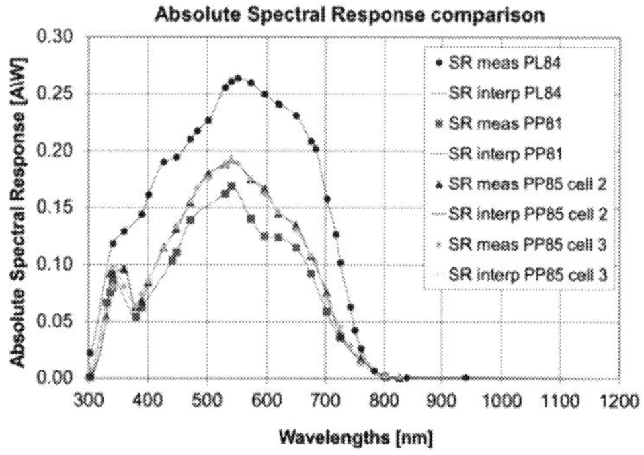

Figure 9: Absolute spectral responsivity comparison

3.4 I-V measurements

A WACOM dual lamp steady-state solar simulator WXS 140 Super with an illuminated area of 14 cm by 14 cm and of class AAA (IEC 60904-9 [10]) was used for measuring the I-V characteristics. The procedure for measuring current-voltage characteristics of PV devices is described in IEC 60904-1 [11]. Following the remark included in the standard; relating to the case of highly capacitive devices (or with unknown response times to light) the effect of the sweep-time on the power measurements of the devices involved in this work has been studied. The I-V measurements were performed after stabilization of the device temperature (in the dark). The devices were illuminated at ~1000 Wm^{-2} and a linear forward voltage ramp was applied going from voltage values slightly below zero to values slightly above V_{oc} to ensure the crossing of both axes.

The light was kept on only during the sweep to avoid heating the DUT. I-V measurements in the reverse sweep direction (from voltage values slightly above V_{oc} to voltage values slightly below zero) were also performed. The sweep time was varied from ~0.5 seconds to about 20 seconds to try and obtain a good overlap between measurements performed in both directions.

Figure 10 shows some selected I-V measurements performed on the single cell PL84. When the sweep time was short (432 ms) a significant hysteresis between measurements performed in opposite sweep directions was observed, resulting in a relative difference in P_{max} of about 9%. Increasing the sweep time, the hysteresis decreased and a good overlap between measurements performed in both directions could be found at a sweep time of about 5 seconds, where the corresponding relative difference in P_{max} was 0.5 %. It was noticeable that as the sweep time was increased both I-V curves (forward and reverse) converged to an intermediate I-V curve. Even though the intermediate I-V curve was closer to those measured in the forward direction. A sweep time of about 5 s appeared to be sufficiently long to guarantee reliable I-V measurements of this device.

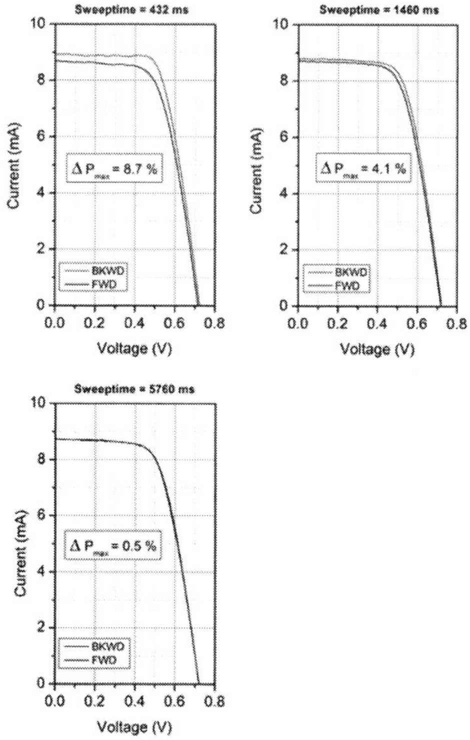

Figure 10: I-V measurements performed on the single cell (PL84) in both the forward sweep direction (from 0 V to V_{oc}) and the reverse sweep (from V_{oc} to 0V) using different sweep-times.

The I-V measurements performed on the masterplates PP81 and PP85 (cell 3) are shown in figures 11 and 12 respectively. The hysteresis between measurements performed in each sweep direction was evident in both cases and was still noticeable when the sweep time was increased up to 10 seconds for PP81 and 18 seconds for PP85. Longer sweep times could be tested, however additional problems could arise, e.g. device instability under light and particularly temperature increases above accepted limits. Looking in details at the measured curves and at the electrical parameters extracted from them reported in table 2 (only the parameters relative to the measurements performed on PP85 are reported as example) the main parameter influenced by the sweep time was the FF in this case. Moreover the I-V curves measured with a reverse voltage sweep appeared to be more constant and less influenced by the sweep time. This may suggested that applying a reverse voltage sweep could be preferable when measuring devices of this type as supposed to c-Si devices where forward sweep is generally preferred.

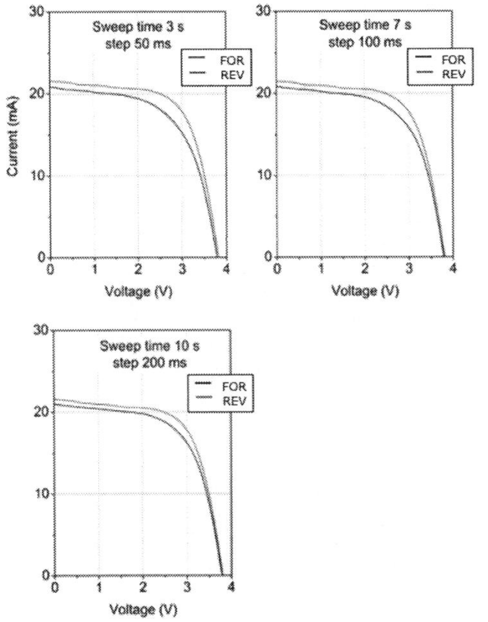

Figure 11: I-V measurements performed on the masterplate PP81 in both sweep directions: from 0 V to V_{oc} (forward sweep) and from V_{oc} to 0 V (reverse sweep) using different sweep-times.

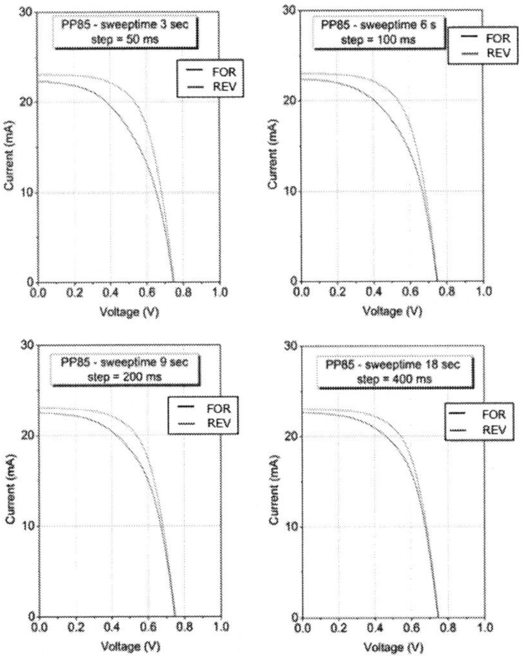

Figure 12: I-V measurements performed on the cell 3 of the masterplate PP85 in both sweep directions: from 0 V to V_{oc} (forward sweep) and from V_{oc} to 0 V (reverse sweep) using different sweep-times.

Table 2: Electrical parameters extracted from I-V curves measured on PP85 in both sweep directions at different sweep times.

PP85 - REVERSE				
Sweep Time (sec)	Isc (mA)	Voc (V)	Pmax (mW)	FF (%)
2.8	23.08	0.749	10.788	62.44
5.6	23.06	0.749	10.796	62.52
8.6	23.06	0.749	10.808	62.55
17.6	23.04	0.748	10.906	63.25
MAX-MIN / AVG %	**0.16**	**0.11**	**1.09**	**1.29**
PP85 - FORWARD				
Sweep Time (sec)	Isc (mA)	Voc (V)	Pmax (mW)	FF (%)
2.8	22.35	0.745	8.546	51.34
5.6	22.44	0.746	9.025	53.90
8.6	22.52	0.746	9.399	55.96
17.6	22.65	0.744	9.896	58.69
MAX-MIN / AVG %	**1.33**	**0.23**	**14.65**	**13.38**

3.4 Corrections and calibration

In order to conclude with the final calibration of the devices additional I-V curves need to be measured for statistics and corrections to standard test conditions (STC) must be applied. The example of the single cell PL84 is reported. Three I-V measurements of DUT performed close to STC conditions (and subsequently corrected to STC) using a sweep time of 5760 ms were selected for the calibration of the device. In figure 13 the I-V characteristics are plotted (solid lines) together with the power at each point (dashed lines). The electrical parameters (I_{sc}, V_{oc}, P_{max} and FF) calculated from the I-V curves shown in figure 13 are reported in table 3, as average values and their standard deviations due to repeatability. The standard deviations are well below the overall measurement uncertainties of the ESTI apparatus.

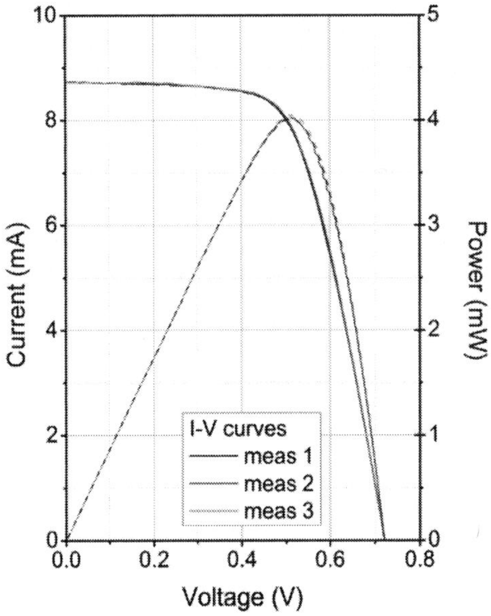

Figure 13: Three I-V measurements performed using a sweep time of 5760 ms and forward sweep direction for final calibration of DUT PL84. I-V (solid lines) and P-V (dashed lines) curves are reported in the graph. Incident irradiance was about 990 W/m^2 and correction to 1000 W/m^2 was applied. The temperature was monitored and controlled to be around 25°C ± 0.3°C during the sweep.

Table 3: Electrical parameters extracted the from three I-V curves reported in figure 13.

Electrical Performance Data				
	I_{sc} [mA]	V_{oc} [V]	P_{max} [mW]	FF [%]
Measurement 1	8.723	0.7207	4.012	63.82
Measurement 2	8.722	0.7204	4.025	64.05
Measurement 3	8.725	0.7214	4.046	64.28
Average	**8.723**	**0.7208**	**4.028**	**64.05**
Standard dev.	**0.002**	**0.0005**	**0.017**	**0.23**

In order to complete the calibration of PL84 and obtain final electrical values corrected to STC, MMF correction was applied. The MMF value as calculated in section 3.3 for the PL84 was 1.0025 and the corrected parameters are reported in table 4 together with their uncertainties. The calculated combined expanded uncertainties (k = 2) of the ESTI apparatus come from the analysis presented by Müllejans [12].

Table 4: Final electrical parameters values after MMF correction (MMF = 1.0025) and corresponding expanded uncertainties of the measurements in absolute and relative units.

Final values corrected to STC				
1000 Wm^{-2}, 25°C and AM1.5 Global				
		UC	Units	k = 2 [%]
I_{sc}	8.75	± 0.10	[mA]	± 1.2
V_{oc}	0.7208	± 0.0017	[V]	± 0.24
FF	64.05	± 0.58	[%]	± 0.90
P_{max}	4.04	± 0.061	[mW]	± 1.5

4 CONCLUSIONS

In this work a general procedure for the calibration of slow responding PV devices is described and discussed. It allows for an evaluation of their response time and, consequently, for a reliable calibration fulfilling all requirements specified in the standards (IEC 60891 and IEC 60904 series). The procedure consists of an initial response time analysis needed to estimate the chopping frequency suitable for SR measurements. This is followed by SR measurements and of a set of multiple I-V measurements performed at various sweep times and in both the forward and reverse sweep directions. Three examples have been shown in this work as a demonstration of its practical applicability: the procedure was applied to a single dye-sensitized solar cell incorporating a liquid electrolyte and to two masterplates of dye-sensitized solar cells incorporating a liquid electrolyte. However, the method is not limited to this type of device and can be generally applied, in particular to photovoltaic devices which are known to have a long response time or whose response is unknown.

REFERENCES

[1] Sommeling P M, Rieffe H C, Kroon J M, van Roosmalen J A M, Schönecker A and Sinke W C Spectral response and response time of nanocrystalline dye-sensitized TiO_2 solar cells 1997 *Proceeding of the 14th EUPVSEC Barcelona (Spain)* 1816.

[2] Sommeling P M, Rieffe H C, Van Roosmalen H C, Schönecker A, Kroon J M, Wienke J A, Hinsch A Spectral response and IV-characterization of dye-sensitized nanocrystalline TiO2 solar cells 2000 *Solar Energy Materials and Solar Cells* **62**, 4 399–410

[3] Hishikawa Y, Yanagida M, Koide N. Performance Characterization of the Dye-Sensitized Solar Cells 2005 *Conference record of the 31st IEEE Photovoltaic Specialists Conference Orlando (Florida)* 67-70.

[4] Takagi K, Magaino S, Saito H, Aoki T, Aoki D Measurements and evaluation of dye-sensitized solar cell performance 2013 *Journal of Photochemistry and Photobiology C: Photochemistry Reviews* **14** 1-12

[5] Snaith H J, Abate A, Ball J M, Eperon G E, Leijtens T, Noel N K, Stranks S D, Tse-Wei Wang J, Wojciechowski K and Zhang W Anomalous Hysteresis in Perovskite Solar Cells 2014 *Journal of Physical Chemistry Letters* **5** 1511-1515.

[6] Bardizza G, Pavanello D, Müllejans H and Sample T Spectral responsivity measurements of DSSC devices at low chopping frequency (1Hz) 2016 *Progress in Photovoltaics: Research and Applications* **24** 428-435

[7] Ebner B, Agostinelli G and Dunlop E Automated Absolute Spectral Response Characterisation for Calibration of Secondary Standards 2000 *Proceedings of the 16th EUPVSEC Glasgow (UK)* 2202–2205.

[8] Pravettoni M., Komlan A., Galleano R., Müllejans H. and Dunlop E An alternative method for spectral response measurements of large-area thin-film photovoltaic modules 2011 *Progress in Photovoltaics: Research and Applications* **20** 416-422.

[9] IEC 60904-7 (Ed. 3) Photovoltaic devices – Part 7: Computation of the spectral mismatch correction for measurements of photovoltaic devices 2008

[10] IEC 60904-9 (Ed. 2) Photovoltaic devices – Part 9: Solar simulator performance requirements 2007

[11] IEC 60904-1 (Ed. 2) Photovoltaic devices – Part 1: Measurement of photovoltaic current-voltage characteristics 2006

[12] Müllejans H, Zaaiman W and Galleano R Analysis and mitigation of measurement uncertainties in the traceability chain for the calibration of photovoltaic devices 2009 *Measurement Science and Technology*, **20** 7 075101

ELECTRICAL PROPERTIES OF CZTS THIN FILMS GROWN BY COEVAPORATION AND ITS RELATION WITH SECONDARY PHASE FORMATION

G. Gordillo[1], R. Moreno[1], A.A. Ramirez[2] F.E. Guzman[1] and J.S. Oyola[3]
[1] Departamento de Física, Universidad Nacional de Colombia, Bogotá, Colombia
[2] Departamento de Ingeniería Química, Universidad Nacional, Bogotá, Colombia
[3] Grupo de Materiales Semiconductores y Energía Solar, Universidad Nacional de Colombia, Bogotá, Colombia

The relation of the electrical transport properties of Cu_2ZnSnS_4 (CZTS) thin films prepared by co-evaporation with secondary phases formation was investigated trough measurements of photoconductivity and conductivity as a function of the temperature. From temperature dependent electrical conductivity measurements it was found that at temperatures greater than 600K, the electrical conductivity is dominated by band conduction and at lower temperatures, Mott variable-range hopping (VRH) is observed. On the other hand, photocurrent studies revealed that successive slow photoresponse transient following after a fast rise is due to the establishment of equilibrium between the charge carriers in the conduction band and the trapping centers generated by native defects in the CZTS films.
By using the real part of Fourier transform, the experimental curves of photocurrent rise have been analyzed and time constants and amplitudes were estimated.

Keywords: Thin films, Cu_2ZnSnS_4, electrical transport properties, recombination mechanisms.

1 INTRODUCTION

Cu_2ZnSnS_4 (CZTS) and $Cu_2ZnSn(S,Se)_4$ (CZTSSe) have been studied as p-type semiconductors for absorber layers of thin film solar cells. In particular, CZTS has a suitable optical bandgap energy of 1.4–1.5 eV [1,2]; in addition, all the constituents of CZTS are abundant and non-toxic. Solar cells based on CZTSSe absorber layers have been fabricated using both solution based processes [3,4] and vacuum based processes [5,6]; a 12,6% efficient $Cu_2ZnSn(Se,S)_4$ device was fabricated using a method that combines advantages of both solution processing and particle based deposition [4]. The absorber layers of high-efficiency CZTS solar cells are predominantly fabricated using vacuum based deposition processes [7-9].

In this study, we present results regarding the relation of the electrical transport properties of Cu_2ZnSnS_4 films prepared by co-evaporation using a novel procedure, with secondary phases formation. The influence of secondary phases formation on mechanisms affecting the electrical transport as well as on processes of recombination was studied through measurements of photoconductivity and conductivity as a function of temperature.

2 EXPERIMENTAL

The Cu_2ZnSnS_4 thin films were prepared by simultaneous evaporation of their precursors (Cu, Zn, Sn, S) on a substrate heated at 250°C, followed by annealing at temperatures greater than 500°C during 30 minutes in presence of elemental sulphur; Cu and Sn are evaporated from the same molybdenum crucible and Zn from two sources type Knudsen cell placed concentrically to either side of the crucible of molybdenum, with the purpose of improving the homogeneity in chemical composition of the CZTS films both laterally and in depth. X-ray diffraction (XRD) and Raman spectroscopy measurements made it possible to verify that thin films of CZTS prepared by co-evaporation at substrate temperatures equal to or greater than 500°C grow on the Cu_2ZnSnS_4 phase without formation of secondary phases, while in CZTS films deposited at temperatures below 500°C was found evidence of secondary phases formation.

The XRD measurements were performed with a Shimadzu-6000 diffractometer and Raman spectroscopy with a Thermo Scientific DXR Raman spectrometer with a DPSS laser of 785 nm, 15 mW focused with a 50X objective.

For photocurrent and conductivity vs T measurements, electrical contacts of Mo were deposited by DC magnetron sputtering on the sample surface. The electrodes are placed at two opposite edges of the front surface of the sample with 1cm of distance between them. The photoresponse was measured under vacuum ambient ($2x10^{-1}$ mBar) using a 250−W Halogen lamp and applying a stabilized dc field (10 V/cm to 12 V/cm). The σ *vs* T measurements were done under vacuum ambient ($6x10^{-2}$ mBar) varying the sample temperature between 100 and 700 K. Both, the transient photocurrent and σ *vs* T curves were recorded continuously by a program developed in LabVIEW.

3 RESULTS

3.1 XRD and Raman analysis

In Fig. 1 and 2 are compared XRD and Raman spectra of a typical Cu_2ZnSnS_4 thin films prepared by co-evaporation at 500°C under optimized deposition parameters with those of a CZTS film prepared at 450°C.

It is observed that the CZTS films prepared under optimized conditions (using a growth temperature of 500°C) present Bragg peaks of (101), (112), (200), (220), (116) and (224) which belong only to the kesterite type Cu_2ZnSnS_4 phase (JCPDS data file # 26-0575), whereas the CZTS sample prepared at 450°C present an additional peak of (101) corresponding to the secondary Cu_3SnS_4 phase (JCPDS 33–0501). Similar results were also found through analysis with Raman spectroscopy (see Fig. 2).

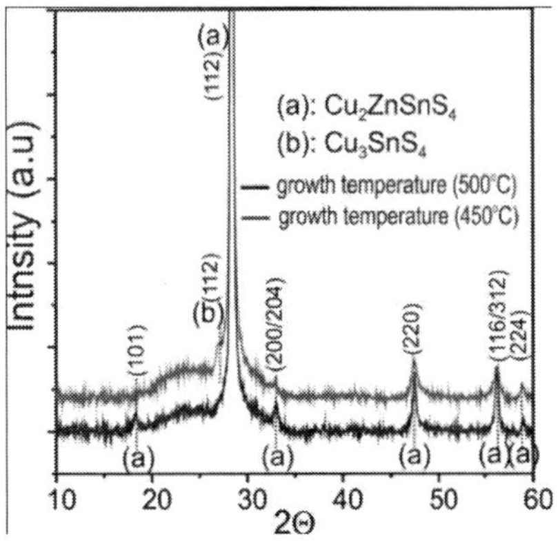

Figure 1: Comparison of the XRD pattern of a typical CZTS film prepared by co-evaporation at a growth temperature of 500°C with that of CZTS films prepared at a growth temperature of 450°C.

Fig. 2 shows the Raman spectra of the same two CZTS samples whose XRD spectra are depicted in Fig.1.

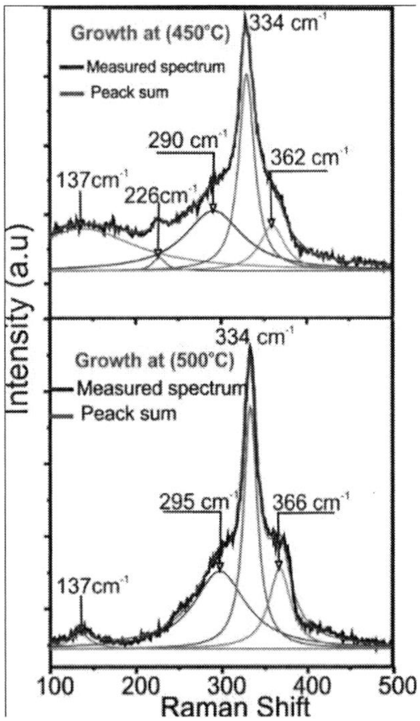

Figure 2: Raman spectra of CZTS film prepared by co-evaporation using growth temperatures of 500°C and 450°C respectively.

The strongest peak in the two samples is located at 334-330 cm^{-1}, which is identified as the main vibrational A symmetry mode from kesterite CZTS [10-14]. The weaker peaks at about 287, and 365 cm^{-1} were also found by fitting the spectra with Lorentzian curves, and can be attributed to kesterite CZTS [10-14]. The Raman spectrum of the sample grown at 450°C show additionally a weaker broad peak at 225 cm^{-1}, which have been assigned to the phase of orthorombic Cu_3SnS_4 [15] and to the phase of monoclinic Cu_2SnS_3 [10,16].

3.2 Study of electrical transport and recombination processes

The temperature dependence of the electrical conductivity of CZTS films prepared by co-evaporation using growth temperatures of 500°C and 450°C respectively, are shown in figure 3.

It is observed that the conductivity varies very slowly with the temperature in the region of low temperatures (T<600K) while in the region of high temperatures, the conductivity varies abruptly, indicating that electrical transport at low temperatures is dominated by a mechanism different from the transport at high temperatures.

In an attempt for determining which mechanism dominates in the regions of low and high temperatures, curves of $ln\sigma$ *vs* $1000/T$ and $ln(\sigma \cdot T^{0,5})$ *vs* $1/T^{1/4}$ were done (see Fig. 4 and 5); The linear behavior at high temperatures ($T > 600$ K) of the curve of $ln(\sigma)$ *vs* $1000/$T indicates that the conductivity can be expressed by the relation:

$$\sigma(T) = \sigma_O Exp[-\frac{\Delta E}{kT}] \quad (1)$$

Giving evidence that the electrical conductivity in both types of CZTS samples is predominantly dominated by transport of holes in extended states of the valence band. The curve of σ vs T of the sample prepared at 450°C shows an inflection around 650°C probably caused by a contribution of the phase Cu_2SnS_3 to the total conductivity.

The graphics displayed in Fig. 4 reveal that the CZTS sample prepared at 500°C presents an energy of activation of 0.33 EV (determined from the slope of the curve of $ln\sigma$ *vs* $1000/T$), while the sample prepared to 450°C presents two energies of activation (0.55 eV and 1.2 eV), apparently attributed to deep states generated by two different types of native defects.

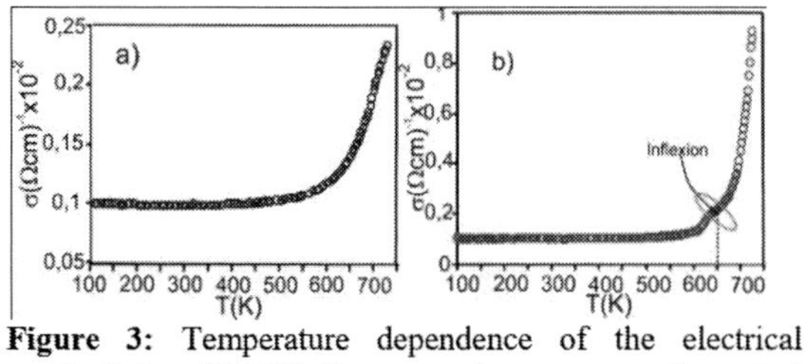

Figure 3: Temperature dependence of the electrical conductivity of CZTS film grown by co-evaporation, at a) 500°C and b) 450°C.

At lower temperatures, the conductivity shows a typical Mott variable-range hopping (VRH) behavior [17, 18]. VRH conduction was originally developed by Mott [17], pointing out that if there is a constant density of states $N_O(E_F)$ near the Fermi level the conductivity is given by [17, 19].

$$\sigma_M(T) = \sigma_{OM} T^{-1/2}[-\left(\frac{T_M}{T}\right)^{\frac{1}{4}}] \text{ and}$$
$$T_M = \frac{\alpha}{k_B \varepsilon^3 N_O(E_F)} \quad (2)$$

Where σ_{0M} is a constant, T_M measures the degree of disorder in the film, α = 18.1 is a constant, and ε is the localization length which characterizes the hopping probability between sites. This $T^{1/4}$ Mott's law has been observed in various classes of lightly doped semiconductors [20,21].

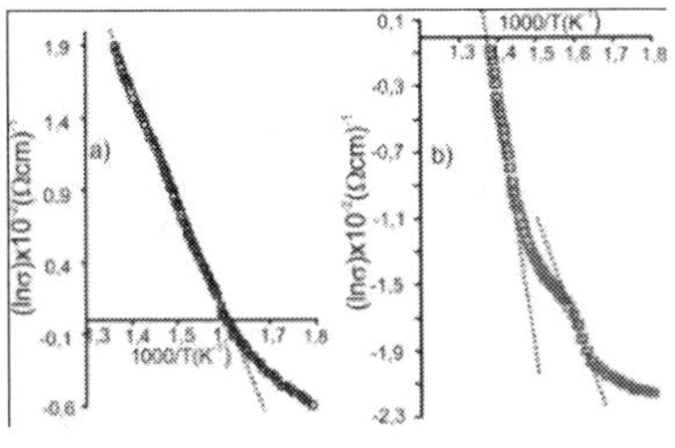

Figure 4: Curves of *lnσ vs 1000/T* corresponding to CZTS film grown at a) 500°C and b) 450°C

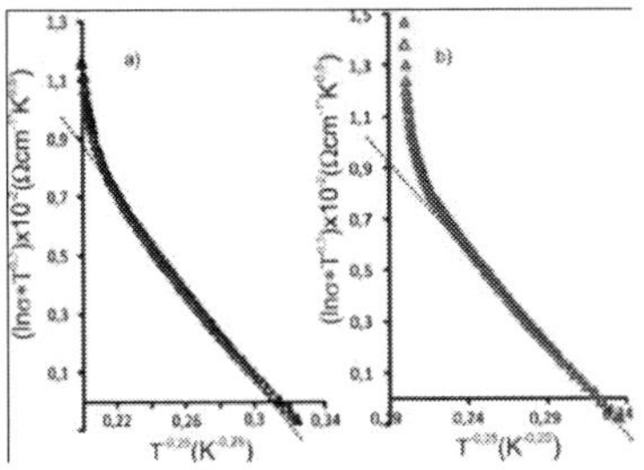

Figure 5: Curves of $ln(\sigma*T^{0.5})$ *vs* $T^{-0.25}$ corresponding to CZTS film grown at a) 500°C and b) 450°C

In addition to the study of mechanisms affecting the electric transport of CZTS thin films, transient photoconductivity measurements were performed to get information about trapping and recombination processes. The amplitude, time constants, and delayed time (escape time from the traps) have been determined simulating the relaxation curves using the method proposed by N.V. Joshi et al., [22], that assumes that if recombination process is other than linear or the time constant varies during the process, then the form of the relaxation curves can be expressed by multiexponential terms. In this case the rise and decay curves can be expressed by [22] :

$$I_{ph} = \begin{cases} \sum_{i=1}^{M} I_i\left(1-e^{-t/\tau_i}\right) + u(\lambda)\sum_{j=1}^{K} I_j\left(1-e^{-(t-\tau_{oj})/\tau_j}\right), \textbf{\textit{Rise}} \\ \sum_{i=1}^{M} I_i e^{-t/\tau_i} + u(\lambda)\sum_{j=1}^{K} I_j e^{-(t-\tau_{oj})/\tau_j}, \ \textbf{\textit{Decay}} \end{cases}$$
(3)

Where u (λ) is a step function defined as: $u(\lambda)=0$ for $t < \tau_{Oj}$ and $u(\lambda)=1$ for $t > \tau_{Oj}$

Here, the summation in the first term of Ec. 3 goes up to M, where M is the number of contributing components and the summation for delayed terms goes from 1 to K, where K is the number of types of traps which hold charge carriers for a certain time, thus, contributing in a delayed signal

Equations corresponding to the real and imaginary parts of the Fourier transform of Eqs. (3) (see details in Ref. 13) provides sufficient information in order to obtain the values of I_i, I_j, τ_i, τ_j and τ_{oj} uniquely and precisely without ambiguity, in contrast to numerical or curve fitting approaches.

We have found that the rise und decay curves of the two studied CZTS films can be simulated with two non delayed and one (or two) delayed time constants. Therefore the time-dependent behavior of photocurrent for the sample prepared at 500°C becomes:

$$I_C(t) = I_1\left(1-e^{\frac{-t}{\tau_1}}\right) + I_2\left(1-e^{\frac{-t}{\tau_2}}\right) - u(\lambda)I_3\left(1-e^{\frac{-(t-\tau_0)}{\tau_3}}\right) \quad (4a)$$

And for the sample prepared at 450°C becomes:

$$I_C(t) = I_1\left(1-e^{\frac{-t}{\tau_1}}\right) + I_2\left(1-e^{\frac{-t}{\tau_2}}\right) - u(\lambda)\left[I_3\left(1-e^{\frac{-(t-\tau_0)}{\tau_3}}\right) + I_4\left(1-e^{\frac{-(t-\tau_0)}{\tau_4}}\right)\right] \quad (4b)$$

The Fourier transform of the experimentally obtained data was performed with the help of a conventional Cooley-Tukey algorithm. In Table 1 are listed values of I_i, I_j, τ_i, τ_j and τ_{oj} obtained for the two types of studied CZTS samples, from the solution of the equations resulting for the real part of the fast Fourier transform.

Table 1. Parameter values used to obtain the simulated curves of the rise photocurrent for the two types of studied CZTS samples.

Parameters	Sample prepared at 500°C	Sample prepared at 450°C
$I_1\,[\mu A]$	0.25	1.80
τ_1 [s]	8.29	5.98
$I_2\,[\mu A]$	0.086	0.98
τ_2 [s]	612.9	573.30
$I_3\,[\mu A]$	0.19	1.20
τ_3 [s]	9.51	3.38
$I_4\,[\mu A]$	---------	0.0043
τ_4 [s]	--------	3.18
τ_0 [s]	0.03847	0.005

Using data of table 1, the relaxation curves were constructed and compared with the experimental one. In Figure 6 are compared rise photocurrent curves carried out at a constant bias of 10 V and excitation power density of 500 W m^{-2}, with curves simulated with the help of Eq.4, using the parameter values listed in table 1. The results of Fig. 6 reveal that the rise curve of the single phase CZTS film (prepared at 500°C) , can be simulated with two non delayed and one delayed time constants, while the CZTS sample prepared at 450°C, which present secondary phase formation, was simulated with two non delayed and two delayed time constants. In both types of samples, the major contribution I_1 (0.25 and 1.8 μA) comes from the direct-band-gap process and the trap-assisted contribution I_2, is less than the previous contribution. The contribution I_3 and I_4 from the traps whose activation process is delayed is significantly less (about 55% and 44.6%) of the contribution from the band-to-band process.

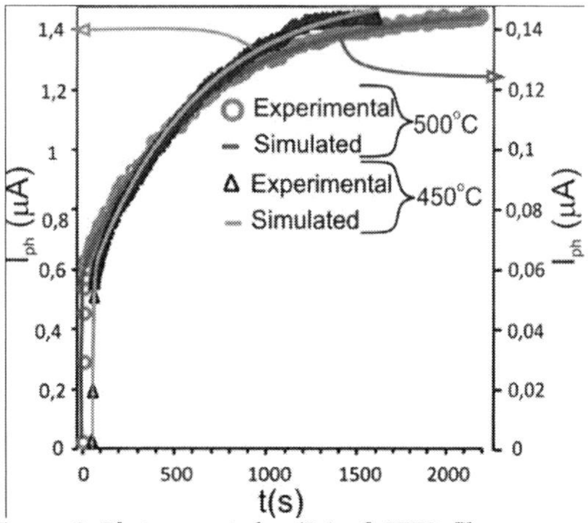

Figure 6: Photocurrent rise (I_{ph}) of CZTS film grown at 450°C and 500°C.

It is also observed that CZTS film prepared at 500°C present only a trapping center (with time constants of 9.51),called single trapping center or β- type center [23]; in this case the establishment of equilibrium between trapping levels and the conduction band requires a time υ which is much larger that the life time τ. The CZTS film prepared at 450°C present two trapping centers (with time constants of 3.38 and 3.18 s) that could be attributed to native defects formed in the Cu_2ZnSnS_4 phase as well as in the secondary Cu_2SnS_3 phase. The traps corresponding to these levels are called multiple trapping centers or α-type centers [23], characterized by having a time $\upsilon << \tau$. These results agrees with the one obtained from the curves of *lnσ vs 1000/T* (see Fig.4).

4 CONCLUSIONS

The relation of the electrical transport properties of CZTS films prepared by co-evaporation with secondary phases formation was investigated trough measurements of photoconductivity and conductivity as a function of the temperature. From the analysis of the temperature dependence of the electrical conductivity it was found that at temperatures higher than 600 K, the electrical conductivity of the CZTS films is dominated by band conduction and at lower temperatures , the conductivity shows a typical Mott variable-range hopping behavior.

By using the real part of Fourier transform, the experimental curves of photocurrent rise have been analyzed and time constants and amplitudes were estimated. The analysis revealed that single phase CZTS film prepared at 500°C present a single β- type trapping center, whereas the CZTS film prepared at 450°C present two trapping centers that could be attributed to native defects formed in the Cu_2ZnSnS_4 phase as well as in the secondary Cu_2SnS_3 phase. The traps corresponding to these levels are α-type trapping centers .

ACKNOWLEDGEMENTS: This work was supported by Colciencias (Contract #038/2013) and Universidad Nacional de Colombia, Sede Bogotá , Facultad de Ciencias, Grupo GMS&ES, Bogotá DC, Colombia (Proy. 20287 supported by DIB).

5 REFERENCES

[1] H. Katagiri: Thin Solid Films 480–481 (2005) 426.
[2] K. Jimbo, R. Kimura, T. Kamimura, S. Yamada, W. S. Maw, H. Araki, K. Oishi, and H. Katagiri: Thin Solid Films 515 (2007) 5997.
[3] T. K. Todorov, J. Tang, S. Bag, O. Gunawan, T. Gokmen, Y. Zhu, and D. B. Mitzi: Adv. Energy Mater. (in press) [DOI: 10.1002/aenm. 201200348].
[4], W. Wang, M.T. Winkler, O. Gunawan, T. Gokmen , T.K. Todorov, Y. Zhu and D.B. Mitzi, Device Characteristics of CZTSSe Thin-Film Solar Cells with 12.6% Efficiency. Advanced Energy Materials 2013, doi:10.1002/aenm.201301465.
[5] B. Shin, Y. Zhu, N. A. Bojarczuk, S. Jay Chey, and S. Guha: Appl. Phys. Lett. 101 (2012) 053903.
[6] G. Brammertz, M. Buffie` re, Y. Mevel, Y. Ren, A. E. Zaghi, N. Lenaers, Y. Mols, C. Koeble, J. Vleugels, M. Meuris, and J. Poortmans: Appl. Phys. Lett. 102 (2013) 013902.
[7] H. Katagiri, K. Jimbo, S. Yamada, T. Kamimura, W. S. Maw, T. Fukano, T. Ito, and T. Motohiro, Enhanced Conversion Efficiencies of Cu2ZnSnS4-Based Thin Film Solar Cells by Using Preferential Etching Technique: Appl. Phys. Express 1 (2008) 041201.
[8] B. Shin, O. Gunawan, Y. Zhu. N. A. Bojarczuk, S. J. Chey and S. Guha, Thin fi lm solar cell with 8.4% power conversion efficiency using an earth-abundant Cu2ZnSnS4 absorber, Prog. Photovolt: Res. Appl. 2013; 21 :72–76
[10] Xavier Fontané, " Caracterización por espectroscopía Raman de semiconductores, Cu2ZnSnS4 para nuevas tecnologías fotovoltaicas", PhD Tesis, Universidad de Barcelona, 2013
[11] K. Wang, B. Shin, K. B. Reuter, T. Todorov, D. B. Mitzi, and S. Guha, Structural and elemental characterization of high efficient Cu2ZnSnS4 solar cells, Appl. Phys. Lett. 98, 051912, 2011
[12] F. Biccari, Fabrication of Cu2ZnSnS4 solar cells by sulfurization of evaporated precursors, Energy Procedia,10 (2011) 187-191
[13] P. Fernandes, P.M.P. Saloméa, A.F.D. Cunha, Study of polycrystalline Cu2ZnSnS4 films by Raman scattering, J. Alloys Compound, 509 (2011) 7600.
[14] X. Fontané, L. Calvo-Barrio, V. Izquierdo-Roca, E. Saucedo, A. Pérez-Rodriguez, J.R. Morante, D.M. Berg, P.J. Dale, S. Siebentritt, In-depth resolved Raman scattering analysis for the identification of secondary phases: charactrization of Cu2ZnSnS4 layers for solar cell applications, Appl. Phys. Lett. 98 (2011) 181905.
[15] P. A. Fernandes, P.M.P. Salome′ and A. F. da Cunha, A study of ternary Cu2SnS3 and Cu3SnS4 thin films prepared by sulfurizing stacked metal precursors, J. Phys. D: Appl. Phys. 43 (2010) 215403
[16] X. Fontané et al., selective detection of secondary phases in CuZnSn(S,Se)4 based absorbers by pre-resonant

Raman Specttoscopy, Proc. 39th IEEE PVSC , Tampa Fl, (2013) 2581-2584

[17] Mott N F and Davis E A 1971 *Electronic Processes in Noncrystalline Materials* (Belfast, Oxford: Clarendon)

[18] Efros A L and Shklovskii B I 1985 *Electronic Properties of Doped Semiconductors* (Berlin: Springer)

[19] Paul D K and Mitra S S 1973 *Phys. Rev. Lett.* 31 1000

[20] Viana E R, Gonzalez J C, Ribeir´ o G M and Oliveira A G 2012 *Phys. Status Solidi RRL* 6 262

[21] Kosyak V, Karmarkar M A and Scarpulla M A 2012 *Appl. Phys. Lett.* 100 263903

[22] Kosyak V, Karmarkar M A and Scarpulla M A 2012 *Appl. Phys. Lett.* 100 263903

[23] A. Serfaty and N. V. Joshi, Fourier-transform analysis of multiexponential curves observed in photoconductivity studies, Physical Review B Vol. 41, No 12 ,1990, 8545-8548

CADMIUM SULFIDE FILMS GROWN BY PHOTOCHEMICAL DEPOSITION AND THEIR APPLICATION IN CIGS SOLAR CELLS

Jun Zhang , Xiaojie Yuan, Lexi Shao*
School of Physical Science and Technology, Lingnan Normal University, Zhanjiang 524048, P. R. China,
Tel./fax: 86 759 3183406. Email: zhangjun@lingnan.edu.cn

ABSTRACT: Uniform Cadmium Sulfide (CdS) thin films have been deposited on CIGS thin films as buffer layers by photochemical deposition (PCD) technique. The CdS film is formed and deposited only in the illuminated region of the substrate where UV light is focused. Therefore, the thickness of deposited film is easily controlled by turning on/off the illumination or by changing the illumination intensity. The films were characterized by X-ray diffraction(XRD), Raman spectroscopy and optical transmission spectroscopy. The performances of the CIGS-based solar cells fabricated using PCD-CdS as buffer layers were compared to the standard CdS solar cells deposited by chemical bath deposition (CBD). Our results showed an efficiency of 7.5% for the CIGS solar cells with PCD-CdS buffer layers.
Keywords: Photochemical deposition; Cadmium selenide; Solar cells;Thin films

1 INTRODUCTION

Several techniques have been used to fabricate CdS thin films buffer layer, such as electrodeposition [1], chemical bath deposition (CBD) [2], screen printing (SP) [3] and physical vapour deposition [4]. Among those methods, CBD is applied intensively for deposition of CdS thin layers on different substrates in manufacture due to its low cost, simplicity and high device efficiency. However, the CBD method leads to the formation of a large amount of hazardous solvent waste. Besides, the utilization of precursor materials is very low: typically, only 2% of the cadmium source is deposited into the CdS thin film during an intermittent process[5]. Therefore, it would be interesting to use another similar chemical solution deposition technique that requires less material and could be continuous operated, which would lower the waste material as well as processing cost and thus arouse the manufacture revolution of CIGS thin film solar cell. Photochemical deposition (PCD) has been developed to deposit metal sulfide semiconductor thin films from an aqueous solution[6]. This method is accessed to be very facile, cheap, much easier to scale up and the deposition can be carefully controlled by controlling the illumination of light on the solution. More importantly, it manages to operate the deposition process continuously with minimized material waste since the deposited film is merely generated in the UV light illuminated area, which drives it to become a quite promising technique for producing CdS thin films for large-scale solar cell applications. On the other hand, The growth mechanism of PCD-CdS thin film is still uncertain. To achieve the CdS thin film manufacturing on thin film solar cell, comprehensive research on the formation of CdS thin film are in urgent requirement.

In this paper, we report in detail the photochemical deposition technique of CdS using a commercial low-pressure mercury arc-lamp with the main line of 254nm. In addition, a CIGS solar cell was fabricated with a typical PCD CdS buffer layer to verify the potential application of PCD method in CIGS solar cell fabrication.

2 EXPERIMENT

Schematic sketch of the photochemical deposition apparatus is shown in Fig. 1. A CIGS\Mo\soda-lime glass and ITO substrate were used as a substrate, which provides the nucleation centers needed for the film growth. A degreased substrate was immersed at less than 1mm below the solution surface and illuminated by a commercial low-pressure mercury arc lamp of 24 W. Thus, the compound is formed on the illuminated region of the substrate. The diameter of the illuminated region was about 15 mm. The growth solution was mildly stirred during the deposition.

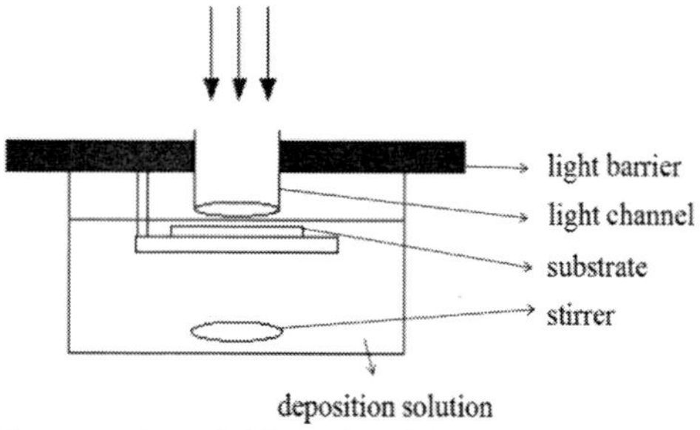

Figure 1 Schematic illustration of the apparatus for PCD

In a typical deposition, CdS thin films were deposited from aqueous solutions containing 2 mmol/l $CdSO_4$, 100 mmol/l $Na_2S_2O_3$, 0.5mmol/l Na_2SO_3, and a small amount of H_2SO_4 for adjusting pH to 3.5. Before the deposition, the CIGS substrate was successively cleaned by ultrasonic in acetone, deionized water and alcohol for 10 min, respectively, followed by drying with a blow dryer. Then, a small amount of the solution was dropped on the CIGS substrate and irradiated right under the lamp for 5 min. This process is to make nuclei on the substrate and improves the reproducibility. Subsequently, the CIGS substrate was immersed in the deposition solution and irradiated for 60min. The lamp was 5cm above the light barrier. The temperature of the solution was room temperature and considered to be static during the illumination due to the low irradiation and calorific power of low-pressure mercury arc lamp. When the deposition finished, the as-prepared sample was washed with deionized water and dried in air at 50℃.

To compare the PCD CdS buffer layer (PCD-CdS) with the conventional CdS buffer layer fabricated by CBD method (CBD-CdS), CIGS solar cells of PCD-CdS and CBD-CdS were fabricated with an optimal PCD condition and a referred CBD process elsewhere, respectively[8]. For both CIGS absorber layers, the thickness was about 2.3 μm, the Cu/[In+Ga] ratio was 85–90 at%, and the Ga/[In+Ga] ratio was 30–40 at%, which was reported elsewhere[8]. Solar cell devices were completed by depositing a radio-frequency sputtered i-ZnO/ZnO:Al top-window layer right after the CdS layer growth. Ni/Al/Ni electrode grid was deposited on top of the ZnO:Al window layer by e-beam evaporation for better current collection. No anti-reflection coating is deposited for these solar cells.

3 CHARACTERIZATION

The morphology and microstructure of the CdS thin films and the fabricated CIGS solar cells were observed using a field emission scanning electron microscope (SEM, MERLIN). The structure and phases of the deposited CdS thin films were analyzed by X-ray diffraction (XRD, PHLIPS, X' Pert-MPD System) using Cu Kα radiation. The bonding status in CdS thin films were studied by X-ray Photoelectron Spectroscopy (XPS, ESCALAB 250, Thermo Fisher Scientific) with Al Kα X-ray at 15kV and 150W. The films chemical compositions were investigated by the electron probe micro analyzer (EPMA, Shimadzu, EPMA 1600). The transmittance spectra of the prepared CdS-ITO films were measured by a UV-vis spectrophotometer. The power density of UV light illumination around 254nm line was measured by an LS126C irradiatometer. The J-V measurements of the CIGS solar cells were performed under the standard AM1.5 spectrum with 100 mW/cm^2 at 25 ℃. The light source intensity of the solar simulator was calibrated by a standard single crystal Si solar cell certified by NIST.

4 RESULTS AND DISCUSSIONS

While CdS has successfully grown into nano-film, it is necessary to investigate the deposition process of CdS thin film in our PCD system. In our study, CdS thin films of various deposition times from 40 min to 70 min with 5 min interval were obtain by PCD. The SEM images of the samples with different deposition times are shown in Fig.2. It is clearly depicted that CdS particles became more and more compact when the deposition time prolonged and completely covered the substrate after a 60 min deposition. In fact, when the deposition started forth, the PCD process of CdS suffered from an induction period, which could account for the small amount of CdS nuclei when the deposition time was 40min. After this period, CdS nuclei was generated rapidly and then grew into larger particles to form a compact film. However, the decrease of CdS particle size was also notable as the deposition went on. This should be caused by the photo-induced decomposition of CdS since CdS was probable to suffer from photo-corrosion under UV light illumination:

$$CdS + h\nu \rightarrow Cd^{2+} + S + 2e^- \quad (1)$$

With the coexistence of decomposition and regeneration, CdS possiblly grew into a plat compact film, which is beneficial to a construct a non-crack buffer layer film onto CIGS absorber layer with good adhesion. This indicates that the growth of CdS thin film in PCD is mostly conducted by an ion-by-ion mechanism with cluster decomposition rather than a cluster-by-cluster mechanism according to the similar CBD theory of CdS thin film. The optical transmission spectra for the compact films with different deposition times of 60min, 65min and 70min are also given in Fig. 2H. Since the CdS films are sufficiently thin, the transmission of the samples was high and close to that of ITO substrate. And the transmission in the visible range decreased as the deposition time was prolonged because of the compacter construction and increased thickness. However, besides the absorption edge of 300nm for ITO appears, the absorption edge of CdS is not shown for all deposited samples. This should be accounted by the thinness of CdS film and the disturbance of fluctuant spectrum derived from the light interference phenomenon of ITO thin film on glass.

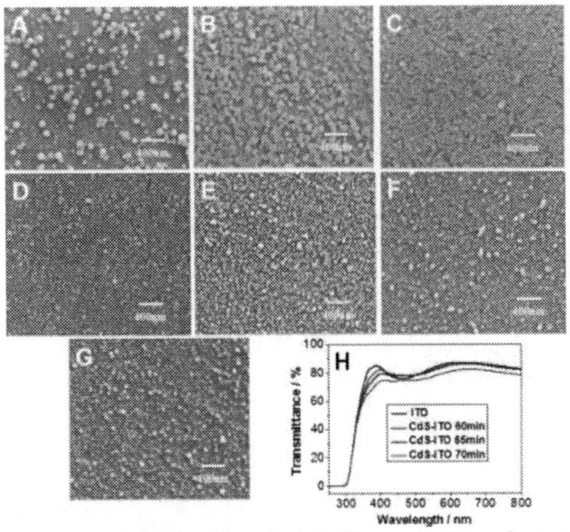

Figure 2 SEM images of CdS films on ITO substrates deposited within various time: (A) 40min, (B) 45min, (C) 50min, (D) 55min, (E) 60min, (F) 65min and (G) 70min, (H) Optical transmission spectra for (a) ITO substrate and the films deposited in (b)60min, (c)65min and (d)70min.

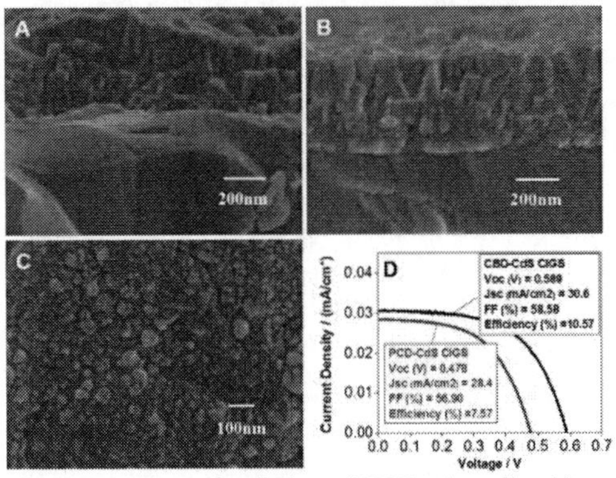

Figure 3 Cross sectional view of CIGS solar cells with different CdS layer: (A) CBD-CdS and (B) PCD-CdS, (C) image of PCD-CdS on CIGS in a typical process, (D) I-V curves of CBD-CdS CIGS and PCD-CdS CIGS solar cells.

The cross sectional images of CBD-CdS CIGS and PCD-CdS CIGS solar cells are shown in Fig. 3A and 3B, respectively. In a CBD process, the fabricated CdS buffer layer displays a compact and uniform appearance on

CIGS grains. In comparison, CdS buffer layer prepared in PCD process shows an ununiform thickness between 20~80nm, which is correspondent to the top-view appearance of the PCD-CdS with some vacant areas on CIGS layer in Fig. 3C. Figure 3D shows the current–voltage (J–V) data of both CBD-CdS CIGS and PCD-CdS CIGS solar cells measured under standard AM1.5 illumination. We could notice that the CBD-CdS CIGS solar cell displays higher efficiency, open circuit voltage (Voc), short circuit current density (Jsc) and fill factor (FF) than PCD-CdS CIGS solar cell. The inferior Voc, Jsc and FF of the PCD-CdS CIGS solar cell should be reduced by a lower coverage of CdS layer and the fluctuant thickness. These features should be attributed to two main reasons. Firstly, during our PCD process, the solution fluxion on the CIGS substrate was considered to be slightly unstable to achieve the uniform dispersion of Cd2+ and sulfur and to fabricate CdS film with fixed thickness. Besides, the diffusion of Cd^{2+} ions into the surface of CIGS films could supply activated sites for the nucleation[5]. This process would shorten the induction time in PCD process and facilitate a faster precipitation of CdS films on CIGS compared to deposition on an ITO substrate. And therefore, much more research should be carried on to fabricate a uniform CdS buffer layer on CIGS by PCD method. And these would be discussed in our future reports.

5 CONCLUSIONS

We have deposited CdS thin films by PCD using a commercial low-pressure mercury arc-lamp, and investigated the influence factors in CdS film fabrication. The fabricating process of CdS thin film in PCD and the relative mechanism was discussed. The optimized PCD-CdS thin film was utilized to construct the buffer layer of CIGS solar cell. Although the efficiency of the PCD-CdS CIGS solar cell was slightly lower than the CBD-CdS CIGS solar cell, the efficiency could be further optimized by the adjusting the deposition process on CIGS in stead of ITO substrate. We will report CIGS solar cells with higher efficiency by further optimized PCD-CdS buffer layer.

ACKNOWLEDGEMENTS

This work was supported by Project of International as well as Hongkong , Macao&Taiwan Science and Technology Cooperation Innovation Platform in Universities in Guangdong Province No. 2015KGJHZ028,

REFERENCES

[1] G.Sasikala, R. Dhnasekaran and C. Sbramanian, Electrodeposition and optical characterisation of CdS thin films on ITO-coated glass, Thin Solid Films 30 (1997) 71

[2] M. A. Martinez, C. Guillen and J. Herrenro, Cadmium sulphide growth investigations on different SnO_2 substrates, Applied Surface Science 140 (1999) 182-189.

[3] S. A. Kuhaimi, Influence of preparation technique on the structural, optical and electrical properties of polycrystalline CdS films, Vaccum 51 (1998) 349.

[4] R. W. Birkmire, B. E. McCandless, S. S. Hegedus. Optimization of Vapor Post-Deposition Processing for Evaporated CdS/CdTe Solar Cells, Solar Energy 145-154 (1992) 9.

[5] A. Bayer, D. S. Boyle, M. R. Heinrich, D. J. Otway, O. Robbe, P. O'Brien, Developing environmentally benign routes for semiconductor synthesis: improved approaches to the solution deposition of cadmium sulfide for solar cell applications, Green Chemistry 2 (2000) 79-86

[6] F. Goto, M. Ichimura, E. Arai. A New Technique of Compound Semiconductor Deposition from an Aqueous Solution by Photochemical Reactions, Japanese Journal of Applied Physics 36 (1997) 1146-1149.

[7] F. Goto, M. Ichimura, Y. Ono, E. Arai, Deposition of CdS and ZnS from aqueous solutions by a new photochemical technique, Journal of Crystal Growth 198 (1998) 308-312.

[8] T. Nakada, A. Kunioka, Direct evidence of Cd diffusion into Cu(In, Ga)Se_2 thin films during chemical-bath deposition process of CdS films, Applied Physics Letters 74 (1999) 2444.

[9] K. Zhang, C. L. Yang, L. Yin, Z. Liu, Q. M. Song, H. L. Luo, Z. Y. Xiong, M. M. Xu, X. D. Xiao, Fabricating highly efficient Cu(In,Ga)Se_2 solar cells at low glass-substrate temperature by active gallium grading control, Solar Energy Materials and Solar Cells 120 (2014) 253-258.

EFFECT OF Cd AND Te_2 VAPOR PHASE MIXTURE IN CMBD ON GROWTH RATE AND MORPHOLOGY OF CdTe FILMS FOR USE IN THIN-FILM SOLAR CELLS

T.M.Razykov[1,3], A.Bosio[2], B.Ergashev[1], C.S.Ferekides[3], D. Y.Goswam[3], K.M.Kouchkarov[1], A.Romeo[4], N.Romeo[2], H.S.Ullal[5], H.M.Upadhyaya[6], R.Yuldashev[1].

[1]Physical-Technical Institute, Scientific Association "Physics-Sun"
Phone: +998-71-235-4103, Fax: +998-71-235-4291, E-mail: razykov@uzsci.net
[2]University of Parma, G.P Usberti 7/A - 43124 Parma, Italy
[E3]University of South Florida, 4202 E. Fowler Ave., ENB 0118 Tampa, FL, 33620, USA
[4]Universita' di Verona, Ca' Vignal 2- Strada Le Grazie 15, 37134 Verona, Italy
[5]National Center for Photovoltaics, National Renewable Energy Laboratory, Golden, CO, 80401 USA
[6]Institute of Materials and Manufacturing, Brunel University London, UB8 3PH, UK

ABSTRACT: We have developed revolutionary novel and nonvacuum and low cost chemical molecular beam deposition (CMBD) method in the atmospheric pressure gas flow for fabrication of II-VI binary, ternary and multinary films. CdTe films with different compositions were fabricated by this CMBD method in the atmospheric pressure hydrogen and argon flows using Cd and Te granules of 99.999% purity as precursors. The films were deposited at the substrate temperature of 500 °C. The growth rate varied in the wide range of 0.2-30 Å / sec. The composition of CdTe films were controlled by changing the molecular beam intensities (MBI) ratio of Cd and Te (evaporated amount). Samples were deposited on glass substrates. Under atmospheric pressure hydrogen flow we succeeded in obtaining stoichiometric (Cd / Te = 1.0) as well as p (Te rich)- and n-CdTe (Cd rich) films. The film thickness was 1.5 μm. Surface morphology of CdTe films was examined by SEM. The as deposited films contain high density and very well oriented cubic polycrystalline grains with the average grain size of 8 – 10 μm at high growth rates of 20- 30 Å / sec and nanocrystalline films at low growth rates of 0.2- 1 Å / sec.
Keywords: CdTe, fabrication, thin film.

1 INTRODUCTION

Thin film solar cells using CdTe semiconductor absorber layers are one of the primary contenders for large-scale commercialization of photovoltaics [1]. The two key properties of this material are its near ideal band gap of 1.45 eV for photovoltaic conversion efficiency, and its high optical absorption coefficient. A thin film of CdTe with thickness of approximately 2μm will absorb nearly 100% of the incident solar radiation. Substantial progress has been made thus far in the area of material research, device fabrication and technology development and numerous applications based on CdTe have been deployed worldwide. Thin film CdS – CdTe solar cells with efficiency of 22.1 % have been obtained [2].

Thin film CdS-CdTe solar cells have potential to increase the efficiency to 25 %. In order to achieve this, it is necessary to develop a better understanding of the basic properties of the materials and processes involved in fabricating the photovoltaic device structure.

One of critical stages in thin film solar cell fabrication is the deposition of CdTe layer with controllable composition and stoichiometry. It has been pointed out that high efficient solar cells possess Te-rich CdTe surfaces with smooth interfaces of p-CdTe/n- CdS [3].

Another issue is fabrication of low resistivity p-CdTe film. This can be solved by intrinsic or extrinsic point defectss [4].ilm Intrinsic point defects strongly depend on the composition of CdTe films. Therefore, fabrication of CdTe films with controllable composition in growth process is very important.

Earlier, we had reported about revolutionary novel and low cost chemical molecular beam deposition (CMBD) method in atmospheric pressure gas flow for fabrication of II – VI binary, ternary and multinary films from separate metallic and chalcogenic precursors [5]. Some characteristics of CdTe films with different compositions fabricated by CMBD method were discussed in [6-10].

The influence of Cd and Te vapor phase mixture on growth rate of CdTe films is discussed in this report.

2 EXPERIMENTAL

The CdTe films with different compositions were fabricated by revolutionary novel and low cost CMBD method in atmospheric pressure hydrogen as well as argon flows [5]. Cd and Te granules of 99.999 % purity were used as precursors. The composition of CdTe films were controlled by changing the molecular beam intensities (MBI) ratio of Cd and Te. Films were fabricated in the wide range of MBI ratio $10^{-2} \leq Cd / Te \leq 10.0$ on glass substrates at 500 °C. SEM data were obtained with a Focused Ion Beam, Quanta 3D microscope.

3 RESULTS AND DISCUSSION

At the evaporation temperature the metallic Cd and Te components transfer into the vapor phase

$$Cd_s + 2Te_s + H_2 = Cd_g + Te_{2g} + H_2 \quad (1)$$

Cd and Te atoms reach the surface of the substrate and as a result of their interaction the CdTe film is formed

$$2Cdg + Te_{2g} + H_2 = 2CdTe_s + H_2 \quad (2)$$

Here, the formation reaction of H_2Te (hydrogen telluride) does not occur, since for the temperature range used in our experiment the vapor pressures are such that $P_{Te2} \gg P_{Te} \gg P_{H2Te}$.

The kinetics of the growth process of films by CMBD in a atmospheric pressure gas flow depends strongly on the

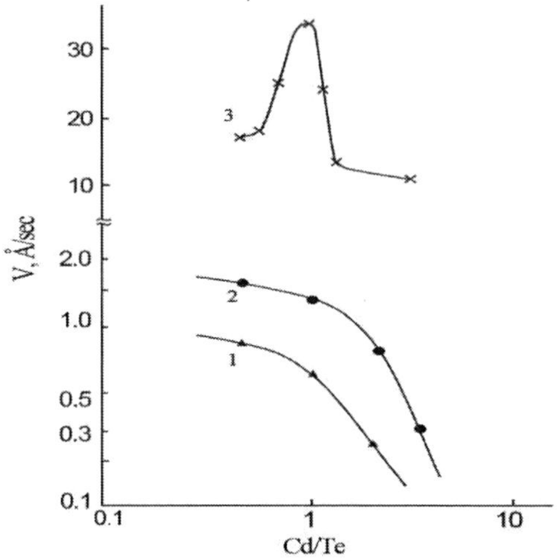

Figure 1: Dependence of the growth rate on molecular beam intensity ratio of Cd / Te in argon (1) and hydrogen (2, 3) flows at MBI $10^{10 \div 11}$ cm^{-2} (curves 1 and 2) and $10^{13 \div 14}$ cm-2 (curve 3).

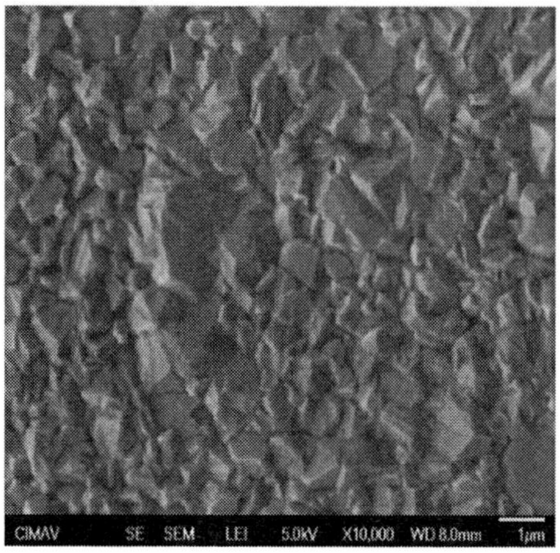

Figure 2: SEM image of the morphology of as deposited CdTe film fabricated at substrate temperature of 500 C and Cd / Te = 1.0.

MBI ratio of the metal and the chalcogen and on the nature of the gas medium. Films did not condense for MBI ratios Cd / Te « 1 and Cd / Te » 1. The dependence of the growth rate V of CdTe films on the MBI ratio is shown in Fig. 1. For CdTe films this dependence was studied in H_2 and Ar flows. Films were condensed at $10^{-2} \leq$ Cd / Te ≤ 10. V increased from 0.2 to 0.3 Å /s with equalization of MBI ratio beginning when the ratios Cd / Te $\leq$ 5–7. The value of V was saturated and equal 1–2 Å /s at the congruent evaporation Cd / Te = 1–1.5. The value of V did not change when Cd / Te $\leq$ 0.2–0.5. The results of this study of the growth process of CdTe films in H_2 and Ar flows showed that V was 2–3 times higher in the first case than in the second. The results of our investigation of the dependence of V on the MBI ratio (Fig. 1) showed that formation of films requires the interaction of Cd and Te atoms on the surface of the substrate and that the kinetics of the growth process of the film depends strongly on the MBI ratio. The absence of film formation for Cd / Te « 1 is caused by the fact that Te atoms do not stick to the surface of the substrate unless Cd atoms are present since their condensation temperature is higher than the evaporation temperature of the sources. Consequently, for Cd / Te $\leq$ 1–1.5, the Te excess is desorbed from the substrate causing saturation of V. The value of V is limited by the arrival rate of Cd atoms on the substrate. V is saturated when there is congruent evaporation of Cd and Te sources. Cluster formation of Cd atoms was observed for Cd / Te » 1. The fact that for CdTe films V was 2–3 times higher in a H_2 atmosphere than in an Ar atmosphere is attributed to the high heat conductivity, lower molecular weight and the dimensions of the H_2 molecule in comparison to Ar.

Surface morphology of all CdTe films was examined by SEM.. We did not observe significant difference on SEM depending on MBI ratio at high growth rates (Fig.1, curve 3). Typical SEM image is shown in Fig. 2. It is seen that the films have high density and very well oriented cubic polycrystalline grains. The average grain size was 8–10 μm.

4 CONCLUSIONS

By the precise control of the MBI we were able to regulate the Cd / Te ratio in a revolutionary novel and low cost CMBD technique. Under atmospheric pressure hydrogen as well as argon flows we succeeded in obtaining stoichiometric (Cd / Te = 1.0) CdTe films. Also, we can control the growth rate by deliberately changing the stoichiometry. The kinetics of the growth process of films by CMBD in a gas flow depends strongly on MBI of the metal and the chalcogen and on the nature of the gas medium. This CMBD method can successfully be used for fabrication of high efficiency and low cost thin film CdS – CdTe solar cells.

5 REFERENCES

[1] T.M.Razykov, C.S.Ferekides, D.Morel, E.Stefanakos, H.S.Ullal, H.M.Upadhyaya, Solar photovoltaic electricity: current status and future prospects. Solar Energy 85 (2011), 1580-1608.

[2] http://www.pvmagazine.com/news/details/beitrag/first-solar-sets-new-cadmium-telluride-thin-film-cell-efficiency-record-at-22.1_100023341/#ixzz415kIS7YW

[3] Z.C. Feng, H.C. Chou, A. Rohatgi, G.K. Lim, A.T.S. Wee, K.L. Tan, Correlations between CdTe/CdS/SnO2/glass solar cell performance and the interface/surface properties *J. Appl. Phys.*79 (1996) 2151-2153.

[4] X.Mathew, Development of CdTe thin films on flexible substrates – a review, Solar Energ.Mater. & Sol. Cells 76 (2003) 293-303.

[5] **T.M.Razykov**, Physical properties of II-VI binary and multinary compound films and heterostructures fabricated by chemical vapor deposition, Thin Solid Films, 163-164, (1988) 301-308.

[6] T.M.Razykov, N.Khusainova, K.Kouchkarov, K.Sato, T.Shimizu, A.Troushin, X-Ray diffraction analysis of CdTe films with different compositions fabricated by CMBD in hydrogen flow, Tech.Digest, PVSEC-12, Korea (2001) 539-540.

[7] T.M.Razykov, A.N.Georgobiani, K.M.Kouchkarov, K.Sato, M.A.Zufarov, Photoluminescence of CdTe Films With Different Compositions Fabricated by CMBD, Technical Digest, PVSEC-14, Thailand (2004) 713-714.

[8] T.M.Razykov, T.Anderson, V.Craciun, O.Crisalle, Y.Goswami, K.Kouchkarov, S.Li, Characteristics of CdTe films of different compositions fabricated by CMBD, Proc. Intern. IEEE PVSC-31, Orlando, FL, USA (2005) 484-485.

[9] T.M.Razykov, G.Contreras-Puente, G.Chornokur, M.Dybjec, Y.Emirov, B.Ergashev, C.Ferekides, Y.Goswami, A.Hubbimov, B.Ikramov, K.Kouchkarov, X.Mathew, D.Morel, S.Ostapenko, E.Stefanakos, O.Vigil-Galan, Y.Vorobiev, H.Zhao, Effect of CdCl2 treatment on structure and photoluminescence of CdTe films with different compositions fabricated by CMBD. Technical Digest of the International PVSEC-17,Fukuoka, Japan (2007) 5P-P3-43, 855–856.

[10] T.M. Razykov, G. Contreras-Puente, G.C. Chornokur, M. Dybjec, Yu. Emirov, B. Ergashev, C.S. Ferekides, A. Hubbimov, B. Ikramov, K.M. Kouchkarov, X. Mathew, D. Morel, S. Ostapenko, E. Sanchez-Meza, E. Stefanakos, H.M. Upadhyaya, O. Vigil-Galan, Yu.V. Vorobiev. Structural, photoluminescent and electrical properties of CdTe films with different compositions fabricated by CMBD, Solar Energy 83 (2009) 90–93.

SURFACE PHOTOVOLTAGE STUDY OF $Cu_{1.95}Zn_{1.1}Sn_{0.96}Se_4$ SINGLE PHASE POWDER

Th. Dittrich[1], L. E. Valle Rios[2], G. Gurieva[1], S. Kapil[1], N. Rujisamphan[3], S. Schorr[1,2],
[1]Helmholtz-Zentrum Berlin für Materialien und Energie GmbH, Hahn-Meitner-Platz 1, 14109 Berlin, Germany
[2]Freie Universität Berlin, Institute of Geological Sciences, Malteserstr. 74-100, 12249 Berlin, Germany
[3]Faculty of Science, King Monkut's University of Technology Thonburi, 10140, Bangkok, Thailand
*E-mail: dittrich@helmholtz-berlin.de

ABSTRACT: Modulated surface photovoltage (SPV) spectra, frequency dependent SPV signals and SPV transients were measured on a $Cu_{1.95}Zn_{1.1}Sn_{0.96}Se_4$ (CZTSe) single phase powder at different modulation frequencies, temperatures and excitation wavelengths. The sign and the onset energies of the modulated SPV signals depended on temperature and modulation frequency. A band of acceptor states with an energy of 65 meV above the valence band edge was observed and the disorder in the bulk of $Cu_{1.95}Zn_{1.1}Sn_{0.96}Se_4$ crystallites is described by band tails with a tail parameter of 30 meV. It was shown that the band gap of $Cu_{1.95}Zn_{1.1}Sn_{0.96}Se_4$ crystallites can be obtained from the spectra of the parameters of the fits of SPV transients by stretched exponentials (1.00 eV at -80°C).

Keywords: CZTS, photoelectric properties, defects, surface photovoltage

1 INTRODUCTION

Kesterites based on $Cu_2ZnSn(S_xSe_{1-x})_4$ (CZTSSe) have been successfully applied as absorber layers in thin film solar cells and solar energy conversion efficiencies of up to 12.7 % [1] have been achieved. However, this technology is based on a highly toxic hydrazine dissolvent. Therefore research started in directions such as low waste ink-jet printing resulting in a solar energy conversion efficiency of 6.4 % [2].

A deeper understanding of defects in CZTSSe is important for the further improvement of related solar cells. For example, the open circuit voltage is limited mainly by tail states in CZTSSe kesterite solar cells [3] and order-disorder effects in the Cu-Zn layers have a strong influence on the band gap of $Cu_2ZnSnSe_4$ (CZTSe) [4]. Further, the highest solar energy conversion efficiency has been reached for Zn-rich absorbers [5] whereas single phase kesterite powder has been discovered close to the B-type Cu-poor/Sn-poor line of CZTSe (see figure 1, [6]). According to neutron diffraction analysis on the investigated sample, antisite defects $Zn_{Sn}+2Zn_{Cu}$ have been observed together with the Cu-Zn disorder defect.

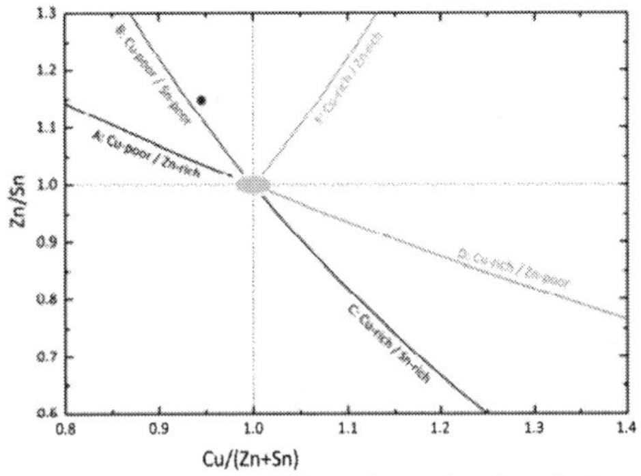

Figure 1: Schematic of the cation ratio plot of CZTSe (after [6]). The lines of A-, B-, C-, D- and F-type mark ratios at which certain substitution processes occur (see [6]). The red point marks the cation ratios for the single phase $Cu_{1.95}Zn_{1.1}Sn_{0.96}Se_4$ kesterite powder investigated in this work.

Single phase powders have the advantage that an influence of secondary phases and interfaces with other semiconductors on measurements can be excluded. Surface photovoltage (SPV, [7]) measurements have the advantage to be sensitive only to a region within the charge separation length and/or diffusion length, i.e. the morphology of a powder sample does not play an important role.

In pre-studies, we investigated the SPV response for numerous CZTSe powders and found an SPV response for some single phase powders along the B-line whereas the highest response was obtained for a single phase $Cu_{1.95}Zn_{1.1}Sn_{0.96}Se_4$ kesterite powder. Therefore, in this work, a single phase $Cu_{1.95}Zn_{1.1}Sn_{0.96}Se_4$ kesterite powder was investigated by spectral dependent modulated and transient SPV measurements at different temperatures in order to obtain information about the specific influence of defects on charge separation.

2 EXPERIMENTAL DETAILS

2.1 Sample preparation

The single phase $Cu_{1.95}Zn_{1.1}Sn_{0.96}Se_4$ kesterite powder was prepared by reaction in a closed quartz tube as described recently by Valle Rios et al. [6]. For SPV measurements, a carbon pad was placed onto a glass substrate coated with a molybdenum layer. Some $Cu_{1.95}Zn_{1.1}Sn_{0.96}Se_4$ kesterite powder was crumbled onto the carbon pad and fixed by slight pressing with a clean glass plate. Unfixed powder was removed by hitting the sample on a glass plate. The surface coverage of the carbon pad with $Cu_{1.95}Zn_{1.1}Sn_{0.96}Se_4$ crystallites was on the order of 80%.

2.2 Modulated and transient SPV measurements

Modulated surface photovoltage spectra and SPV transients were measured in the fixed capacitor arrangement with a mica spacer (thickness about 30 µm) between the sample and the reference (SnO_2:F on a quartz cylinder) electrodes [8] (see also the scheme in figure 2).

Modulated SPV signals and SPV transients were excited with a halogen lamp and a quartz prism monochromator (SPM2) and pulses of a tunable Nd:YAG laser (EKSPLA, NT 342/1/UVE, duration time 5 ns, intensity about 1 mJ/cm²), respectively. The modulation frequency was varied between 3 and 3120 Hz with a mechanical chopper. Modulated SPV signals were excited with a red LED in some experiments allowing to increase the modulation frequency to up to 120 kHz. The repetition rate of laser pulses was 1 Hz for transient measurement. Measurements were performed at temperatures between -160 and 120°C.

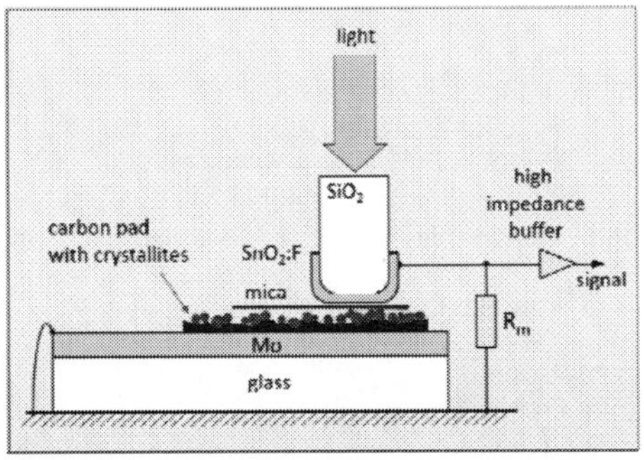

Figure 2: Schematic of the electrode configuration with the sample, the mica layer, the reference electrode, the measurement resistance and the high impedance buffer.

3 RESULTS AND DISCUSSIONS

3.1 Modulated SPV spectra

Figure 3 shows in-phase and phase-shifted by 90° modulated SPV spectra measured at low (-160°C) and higher (36°C) temperatures at low modulation frequency (10 Hz). In-phase signals correspond to the response following rapidly the modulation period. The sign of the in-phase signals is related to the preferential direction of modulated charge separation (positive sign: towards the internal interface, negative sign towards the external surface).

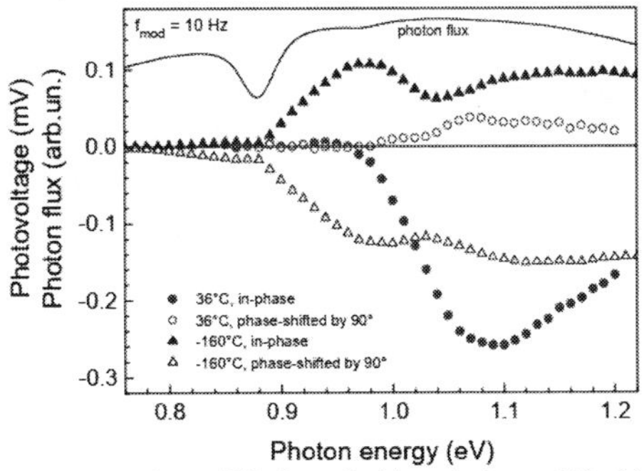

Figure 3: In-phase (filled symbols) and phase-shifted by 90° (open symbols) modulated SPV spectra measured at 36 (circles) and -160 (triangles) °C. The solid line shows the spectrum of the photon flux.

The sign of the in-phase signals was negative or positive for measurements at 36 or -160°C, respectively. The in-phase signals set on at photon energies of 0.98 eV (36°C) or 0.88 eV (-160°C) whereas the onset region was dominated by a peak for the measurement at -160°C. Phase-shifted by 90° signals correspond to the response strongly retarded in relation to the modulation period. The onset energies of the phase-shifted by 90° signals were 1.02 and 0.87 eV for measurements at 36 and -160°C, respectively. The change of the sign of the in-phase SPV signals gave evidence that the modulated charge separation was dominated by the change of the surface charge due to filling of trap states.

The onset energy of in-phase signals can often be related to energy close to the band gap of a semiconductor. However, electronic transitions including defect states can strongly affect SPV spectra near the band edge. This is illustrated in figure 4 showing the spectra of the SPV amplitudes, i.e. the square roots of the sum of the squared in-phase and phase-shifted by 90° signals, measured at different modulation frequencies. The spectra of the SPV amplitude measured at 36°C were practically identical. This means that the SPV signals decayed much faster than the modulation period at higher temperature. In contrast, the spectra of the SPV amplitude measured at -160°C changed strongly with changing modulation frequency.

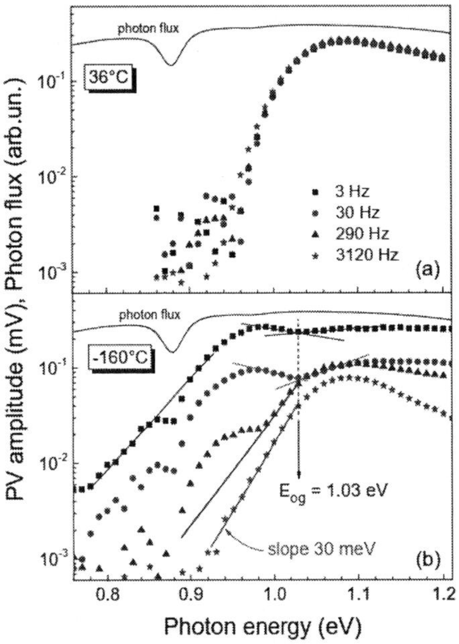

Figure 4: Spectra of the surface photovoltage amplitude measured at 36°C (a) and -160°C (b) with modulation frequencies of 3, 30, 290 and 3120 Hz (squares, triangles, circles and stars, respectively). The solid line shows the spectrum of the photon flux.

For the measurement at -160°C at 3120 Hz, the spectrum of the SPV amplitude was dominated by an exponential increase (slope 30 meV) with increasing photon energy up to about 1.03 eV (optical band gap) and a decrease of the signal occurred at photon energies larger than about 1.1 eV. For the measurement at -160°C at 200 Hz, an additional pronounced shoulder appeared at photon energies below 0.98 eV, the slope of the exponential increase changed to 35 meV and the decrease of the signal at photon energies above 1.1 eV became weaker than for the measurement at 3120 Hz. For the measurement at -160°C at 30 and 3 Hz, the shoulder became a peak (maximum at 0.99 eV) setting on at photon energies below 1.03 eV, the slope of the exponential increase increased to 45 meV and the decrease of the signal at photon energies above 1.1 eV disappeared.

3.2 Modulated SPV signals at higher frequencies

Figure 5 shows the dependencies of the in-phase SPV signals excited with a red LED on the modulation frequency for different temperatures. At low frequencies, the in-phase SPV signals were about 3.8 mV at -160°C and decreased to about 1.5 mV at -100°C. The sign of the in-phase SPV signals at low frequencies changed to negative at higher temperatures (-1.6 mV at -40°C) and increased to up to -5.2 mV at 20°C. A further increase of the temperature led to a decrease of the in-phase SPV signals.

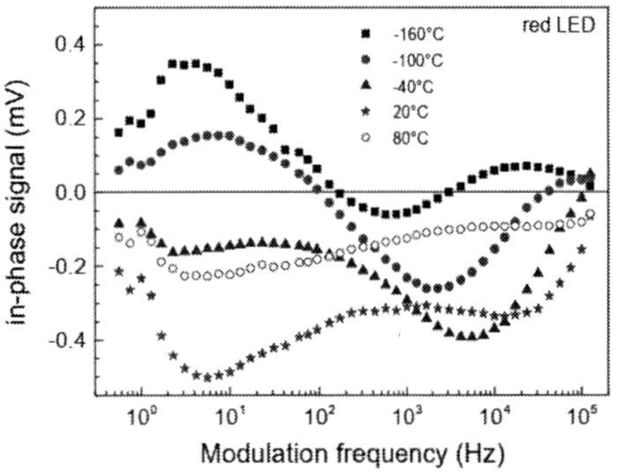

Figure 5: Dependencies of the in-phase SPV signals excited with a red LED on the modulation frequency measured at -160, -100, -40, 20 and 80°C (squares, circles, triangles, stars and open circles, respectively).

At low temperatures, the in-phase SPV signals decreased with increasing modulation frequency and changed from positive to negative around 100 Hz. Further the in-phase SPV signals became more negative and started to change to more positive at frequencies between 600 Hz (-160°C) and 2 kHz (20°C). The in-phase SPV signals changed the sign from negative to positive at about 3 kHz (-160°C) and at higher frequencies for higher temperatures.

The change of the sign of the in-phase SPV signals gave evidence for a change of the dominating mechanism of charge separation setting on at frequencies at which the in-phase SPV signals started to change to more positive. Therefore the corresponding frequency can be treated as a transition frequency at which the dominating mechanism of modulated charge separation started to change due to changes in the modulated population of trap states. The Arrhenius plot of the transition frequency is plotted in figure 6. An activation energy of about 65 meV has been deduced from the Arrhenius plot.

It is important to note that the activation energy of the transition frequency is close to the difference between the optical band gap and the defect peak in the modulated SPV spectra. It was shown by theoretical analysis of stoichiometric CZTSe that Zn_{Sn} forms acceptor states with energy levels of 0.09 (-/0) and 0.18 (2-/-) eV above the valence band edge [9]. Further, V_{Cu} forms shallow acceptors with an energy of about 0.04 (-/0) eV above the valence band edge [9]. Variations in the stoichiometry can lead to variations in the energy levels. Therefore, more sophisticated experiments will be needed to judge whether the observed shallow acceptor band in single phase $Cu_{1.95}Zn_{1.1}Sn_{0.96}Se_4$ kesterite powder is related to Zn_{Sn} or V_{Cu}.

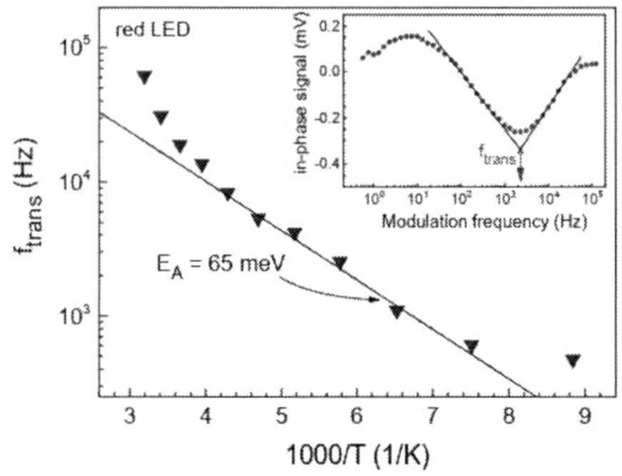

Figure 6: Arrhenius plot of the transition frequency at which the in-phase SPV signals started to become more positive (see also the insert).

3.3 SPV transients

The measurement of SPV transients over a range from ns to hundreds of ms allowed a more detailed study of the specific role of the defect states. Figure 7 shows SPV transients measured at -80, 0 and 80°C and excited with photon energies below (wavelength 1300 nm), around (wavelength 1200 nm) and above (wavelength 1100 nm) the band gap of CZTSe.

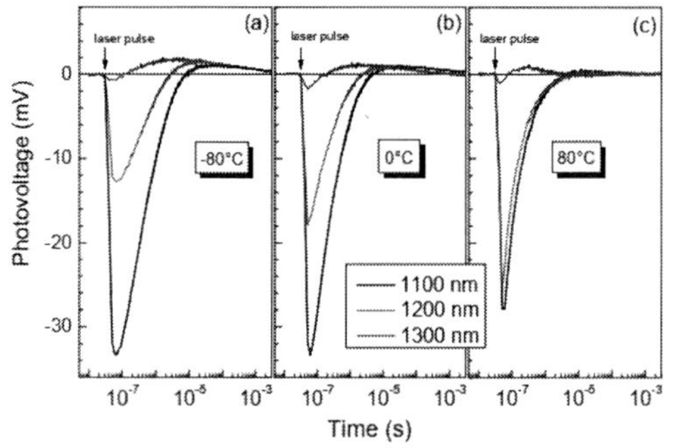

Figure 7: Surface photovoltage transients measured at -80, 0 and 80°C (a-c, respectively) and excited at wavelengths of 1100, 1200 and 1300 nm (thick black, medium red and thin blue lines, respectively). The arrow marks the onset of the laser pulse.

The SPV signals excited with 1100 nm set on within the laser pulse and reached -33 or -28 mV within 20 ns after switching on the laser pulse for measurements at -80 and 0 or 80°C, respectively. In the following the SPV transients decreased to -10 mV within 1 µs, 500 ns and 200 ns for measurements at -80, 0 and 80°C, respectively, i.e. the decays became faster with increasing temperature. At times between 7 and 4 µs, the SPV signals changed the sign to positive whereas the positive signal was very weak for the measurement at 80°C. The SPV signals excited with 1200 nm reached -13, -18 and -26 mV for the measurements at -80, 0 and 80°C, respectively, whereas the qualitative behavior was similar to excitation with 1100 nm. For excitation with 1300 nm, the fast negative signals were very low (-0.6, -1.5 and -0.9 mV for measurements at -80, 0 and 80°C, respectively), the sign changed at 100, 140 and 40 ns for measurements at -80, 0 and 80°C, respectively, and the positive signals reached 1.8, 1.1 and 0.8 mV at 4, 2 and 0.2 μs for measurements at -80, 0 and 80°C, respectively.

The large and fast negative SPV signals can be related to charge separation within the space charge region of a depleted p-type doped semiconductor, the positive signals can be interpreted as a capture of holes at deep states in the near surface region.

The detailed analysis of the SPV transients is not straight forward since the transients shall be fitted with several stretched exponentials which leads to significant uncertainties if comparing transients measured at different temperatures. Figure 8 shows a SPV transient excited with 1150 nm and measured at -80°C as an example. This transient could be well fitted with three stretched exponentials (large negative signal with a time constant of 370 ns and a stretching parameter of 0.45, small positive signal with a time constant of 1.6 ms and a stretching parameter of 0.4, one relatively large fast positive signal with a time constant of 16 ns and no stretching).

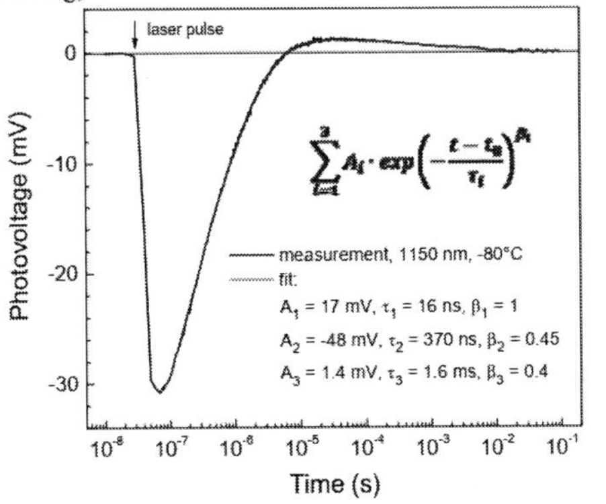

Figure 8: Surface photovoltage transient measured at -80°C and excited at 1150 nm (black line) and transient fitted with three stretched exponentials (red line). The arrow marks the onset of the laser pulse.

For a given temperature, the amplitudes, time constants and stretching parameters can be plotted as a function of the photon energy. This is shown in figure 9 for the measurements at -80°C. The amplitude of the negative signal part started to increase continuously at about 1 eV. The amplitudes of the fast or slow signal parts increased or decreased abruptly, the time constants increased from 6 to 14 ns (fast positive signal part) and from 30 to 500 μs (slow positive signal part) and the stretching parameters increased from 0.2 to 0.34 (slow positive signal part) and from 0.29 to 0.39 (negative signal part) in the same range of photon energies around 0.99...1.00 eV. These strong qualitative changes around 1 eV have to be attributed to a qualitative change of the mechanisms of charge separation and relaxation by the onset of photo-generation of free electron hole pairs. In this sense the band gap of the investigated single phase $Cu_{1.95}Zn_{1.1}Sn_{0.96}Se_4$ kesterite powder at -80°C can be determined as 1 eV. This value is about 20 meV above the band gap of CZTSe [10].

The stretching parameters saturated at 0.4 (slow positive signal part) and 0.45 (negative signal part) for photon energies larger than 1.02 eV. Therefore, the mechanisms of charge separation and relaxation still changed between 0.99 and 1.02 eV, probably due to some influence of structural disorder.

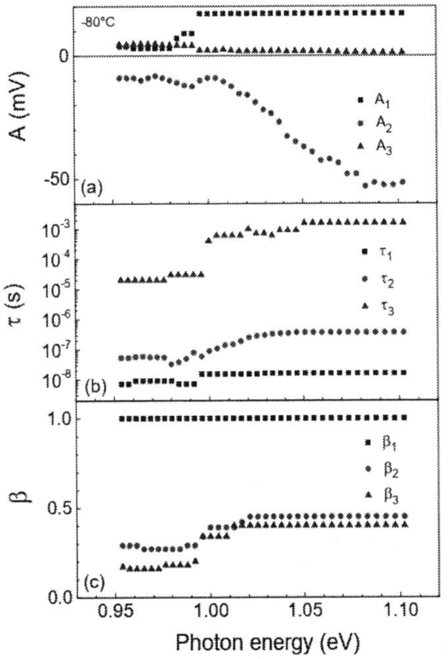

Figure 9: Dependence of the amplitude (a), the time constant (b) and the stretching parameter (c) for the three stretched exponentials (squares, circles and triangles for indices 1,2 and 3, respectively) used for fitting of SPV transients measured at -80°C on the photon energy of the exciting laser pulse. The two dashed lines mark photon energies at which pronounced changes occur.

The SPV transients measured at high temperatures could be fitted with two negative stretched exponentials, i.e. the positive signal parts disappeared and the sign of the slow signal part changed from positive to negative in comparison to the measurement at low temperature. The time constant of the large negative signal part decreased to 40 ns which was one order of magnitude faster than for the SPV transient measured at -80°C. At the same time, the stretching parameter of the large negative signal part did not change significantly at the high temperature. The time constant of the slow negative signal part was about 1μs which was three orders of magnitude faster than for the slow positive signal part measured at -80°C.

The existence of stretched exponentials in the decay of SPV transients gives evidence for limitation of the relaxation by transport via localized trap states. For a model system with a single CdSe quantum dot (QD) layer deposited on ITO, SPV transients could be fitted with only one stretched exponential and the single QD approximation model has been developed by Fengler [11] resulting in a broad distribution of trap states for holes at the surface of a single QD recombining with an electron at the ITO side. The situation is more complicated for the single phase $Cu_{1.95}Zn_{1.1}Sn_{0.96}Se_4$ kesterite powder for which a superposition of three processes shall be considered. It seems that stretching parameters around 0.4...0.5 obtained for excitation with photon energies above the band gap of CZTSe are related to defect states distributed relatively narrow in energy and homogeneously in space in the bulk of CZTSe crystallites. The much lower stretching parameters obtained for excitation with photon energies below the band gap of CZTSe give evidence for a broad distribution of the density of states in energy which is usual for surface states or states close to a surface due to increased disorder.

4 CONCLUSIONS

Spectral dependent modulated and transient SPV signals close to the band gap of a single phase $Cu_{1.95}Zn_{1.1}Sn_{0.96}Se_4$ kesterite powder were investigated in a broad range of temperature, modulation frequency and response time. A band of acceptor states with an energy of about 65 meV above the valence band edge was observed and had strong influence on modulated SPV signals. Disorder in the bulk of $Cu_{1.95}Zn_{1.1}Sn_{0.96}Se_4$ kesterite crystallites was characterized by a tail parameter of about 30 meV.

SPV transients were strongly affected by disorder and could be fitted at lower temperatures by three stretched exponentials. Surface states caused trapping of holes, stretching parameters of the order of 0.2 and time constants of the order of hundreds of µs. Bulk tail states limited the relaxation of the charge in the space charge region and resulted in stretching parameters of the order of 0.4..0.5 and time constants between tens and hundreds of ns depending on temperature. The abrupt change of parameters obtained from the fits by stretched exponentials as a function of excitation wavelength served as a criterion for the determination of the band gap of the $Cu_{1.95}Zn_{1.1}Sn_{0.96}Se_4$ crystallites (1.00 eV at -80°C). However, one has to be a bit careful with interpretation of precise values without applying a model for simulation and fitting of SPV transients since disorder in space and energy has to be considered.

For future, a correlation between electronic states and well defined order-disorder control in the Cu-Zn layers and parameters of off-stoichiometry in single phase CZTSe powders and solar cells would be interesting.

5 ACKNOWLEDGMENTS

N.R. is grateful to the Higher Education Research Promotion and National Research University Project of Thailand, office of the Higher Education Commission.

6 REFERENCES

[1] J. Kim, H. Hiroi, T. K. Todorov, O. Gunawan, M. Kuwahara, T. Gokmen, D. Nair, M. Hopstaken, B, Shin, Y. S. Lee, W. Wang, H. Sugimoto, D. B. Mitzi, Advanced Materials 26 (2014) 7427.

[2] J. Lin, J. Kavalakkatt, M. Ch. Lux-Steiner, A. Ennaoui, Adv. Sci. 2 (2015) 1500028.

[3] T. Gokmen, O. Gunawan, T. K. Todorov, D. B. Mitzi, Appl. Phys. Lett. 103 (2013) 103506.

[4] G. Rey, A. Redinger, J. Sendler, T. P. Weiss, M. Thevenin, M. Guennou, B. El Adib, S. Siebentritt, Appl. Phys. Lett. 105 (2014) 112106.

[5] W. Wang, M. T. Winkler, O. Gunawan, T. Gokmen, T. K. Todorov, Y. Zhu, D. B. Mitzi, Adv. Energy Mater. 4 (2014) 1301465.

[6] L. E. Valle Rios, K. Neldner, G. Gurieva, S. Schorr, J. Alloys and Compounds 657 (2016) 408.

[7] L. Kronik, Y. Shapira, Surface Science Reports 37 (1999) 1.

[8] V. Duzhko, V. Yu. Timoshenko, F. Koch, Th. Dittrich, Phys. Rev. B. 64 (2001) 075204.

[9] S. Chen, A. Walsh, X.-G. Gong, S.-H. Wei, Adv. Mater. 25 (2013) 1522.

[10] S. G. Choi, T. J. Kim, S. Y. Hwang, J. Li, C. Persson, Y. D. Kim, J.-H. Wie, I. L. Repins, Solar Energy Mats. & Solar Cells 130 (2014) 375.

[11] S. Fengler, E. Zillner, Th. Dittrich, J. Phys. Chem. C 117 (2013) 6462.

CHARACTERIZATION AND POST-PROCESSING OF CADMIUM SULFIDE POLYCRYSTALLINE THIN FILMS

Hang Xu, Lili Wu*, Cai Liu, Wenwu Wang, Guanggen Zeng, Wei Li, Bing Li, Jingquan Zhang and Lianghuan Feng
*Corresponding author: Lili Wu E-mail: wulily77@163.com
College of Materials Science and Engineering, Sichuan University, Chengdu 610064, China

ABSTRACT: Cadmium Sulfide (CdS) thin films were grown by chemical bath deposition (CBD) method on glass substrates. The as-deposited films experienced heat treatment in N_2, N_2-O_2 and H_2. Characterization methods such as XRD, SEM, XPS, UV-Visible transmission test and dark electrical conductivity-temperature test were employed to study the properties of the CdS thin films. The analysis shows that preferential orientation in (220), (400) direction and grain size of the CdS thin films depend on the treatment method. Whether $CdCO_3$, $CdNO_3$, CdO and $CdSO_4$ exist on surface of the CdS film depend on the post-processing as well. The band gap of as-deposited CdS films is 2.44 eV and decreased to 2.27 eV, 2.28 eV and 2.34 eV after treatment in N_2, N_2-O_2 and H_2, respectively. The results show that electrical conductivity activation energy of CdS thin film decreased from 0.17 eV to 0.07 eV when treated in H_2. CdTe solar cells were also fabricated to show the influence of N_2-O_2 treatment on device performance.
Keywords: CdS thin film, CdTe solar cell, heat treatment

1 INTRODUCTION

It is well known that the short circuit current density loss in CdS/CdTe solar cells is optical absorption in the cadmium sulfide (CdS) layer [1,2]. Reduction in thickness of CdS layer below a critical value about 30 nm improves device quantum efficiency but also results in the degradation of open circuit voltage and fill factor. Thus, it is important to find methods to keep CdS thin films work well or enhance the properties of CdS films with suitable thickness [3]. Post-processing is an efficient method to approach CdS film with high quality, such as improve optical properties, n-type doping level, reduce defect and improve crystalline quality [4-8].

Previous work reveals that the films are described by cubic CdS layer for the as-grown and TAT=208 ℃ samples, while the samples with TAT=260 ℃ are best described by hexagonal CdS layers [9]. CdS thin film was partially oxidized to CdO and $CdSO_4$ on the cluster surface when annealed in air [10,11]. Large CdS grains with good crystalline quality formed through recrystallization during the $CdCl_2$ heat treatment [12-14]. Long-term study on the thermal treatment of CdS thin films has been conducted. However, it seems few methods with much effectivity for improving CdTe solar cell performance.

In this study, CdS thin films were experienced thermal treatment in N_2, N_2-O_2 and H_2. A thorough analysis of the film structural, morphological, surface chemical composition, optical and electrical properties has been carried out. The CdS thin films were treated in N_2-O_2 for different time to study the effect of annealing on CdTe solar cell performance.

2 EXPERIMENTAL

2.1 Sample preparation

CdS thin films were grown by CBD method on silica glass substrates and commercial FTO substrates. The thickness of the CdS films was about 60 nm. Cadmium chloride ($CdCl_2$) solution of 0.01 mol/L concentration was sprayed on the surface of CdS thin films directly by ultrasonic method. The surface of CdS samples was maintained at 120 ℃ during the spray process. Then the samples were experienced thermal treatment at 380 ℃ in different atmosphere. The temperature of treatment for CdS thin film in hydrogen was 300 ℃. Because higher temperature usually results in reduction reaction, making damage to the CdS films and increase pinholes in CdTe thin films [3]. The details of post-processing conditions are listed in **Table I**.

Cadmium telluride (CdTe) thin films were deposited on the CdS substrates by close space sublimation (CSS) method. The CdTe thin films were covered with a thin layer of $CdCl_2$ by ultrasonic method and then were annealed at 385 ℃ for 35 min in N_2-O_2 atmosphere. After heat treatment, the surface of CdTe film was etched by Br_2-methanol solution of 0.2% volume for 4 s. Zinc telluride (ZnTe) thin film doped with copper was deposited by vacuum evaporation as back contact layer. The devices were experienced quick treatment at 190 ℃ for 15 minutes and were finished using gold as electrode.

Table I: The details of post-processing for CdS thin films: as deposited, treated in N_2, N_2-O_2 and H_2.

NO.	Atmosphere (pressure)	T (℃)	Time (min)	Surface treatment
CdS-None	None	None	None	As deposited
CdS-N_2	N_2	380	30	0.01mol/L $CdCl_2$ solution
CdS-N_2-O_2	N_2/O_2	380	30	
CdS-H_2	H_2(2 kPa)	300	5	As deposited

2.2 Characterization

The surface morphology of the films was determined by Scanning Electron Microscope (SEM, Hitachi-S4800 and Japan). The thickness of CdS thin film was measured with a step-profile system (Ambios XP-2, America). The physical structure and phase of the CdS thin films were studied with an X-ray diffractometer (XRD) (Dandong Fangyuan DX-260, China). XPS measurements were conducted with a Kratos surface-science instrument and an Al Ka (1486.7 eV) x-ray source under high vacuum (1×10^{-9} Torr). The transmittance of the films was characterized by an UV-Visible spectrophotometer (Perkin Elmer Lambda 950, America). The light *J-V* characteristics were measured at AM 1.5 illumination condition under 1 Sun using a Class-A Sun Simulator.

3 RESULTS AND DISCUSSION

3.1 Crystal structure of CdS thin films

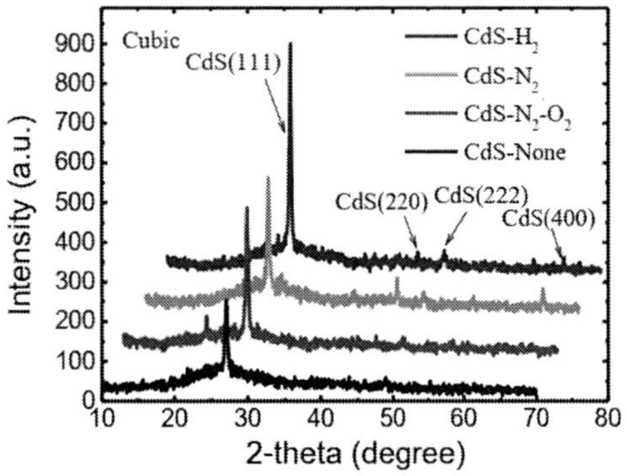

Figure 1: XRD patterns of CdS thin films, as deposited, with post-processing in N_2, N_2-O_2 and H_2.

Table II: The interplanar distance of (111) direction and grain size of CdS thin films treated in different atmosphere.

	Grain size (nm)	2-Theta of (111) (deg)	Value of d_{111} (Å)
CdS-H_2	24.8	26.79	3.32
CdS-N_2	20.6	26.75	3.33
CdS-N_2-O_2	19.5	26.87	3.31
CdS-None	17.2	26.97	3.30
PDF	-	26.97	3.30

The structure exhibits cubic phase with preferential orientation in (111) direction for all the CdS thin films (PDF 65-8873) (as shown in **Figure 1**). The analysis shows that orientation in (220), (400) direction was also observed when the CdS film has been treated in N_2. Thermal treatment increased the grain size of CdS thin films obviously (as listed in **Table II**), and the grain size of CdS thin film increased from 17.2 nm to 24.8 nm when treated in H_2. This indicates that the treatment at lower pressure is to the benefit of larger grain size. The as deposited CdS thin film has the same value of d with CdS cubic structure in PDF 65-8873. Increased interplanar distance in (111) direction of the CdS thin films was observed after the annealing step. The distance of crystal face in CdS thin film treated in N_2 is larger than that of others CdS thin films. It was found that treatment in N_2 results in more defects, such as dislocation in CdS thin film.

3.2 The morphology of CdS thin films

The CdS thin films treated in N_2 and N_2-O_2 atmosphere show different roughness, surface flatness and crack distribution in comparison with the as grown CdS thin film or the film treated in H_2 obviously. The surface morphology of CdS thin film treated in N_2 was up heaved which is not good for the performance of CdTe solar cells.

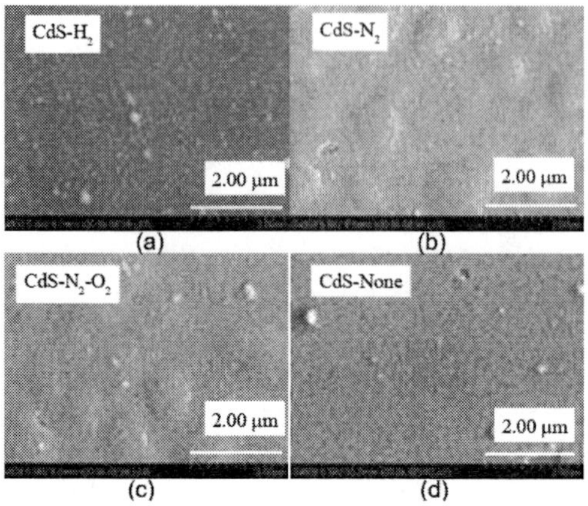

Figure 2: SEM shows morphology of the film, such as surface flatness and crack distribution: (a) Treated in H_2, (b) Treated in N_2, (c) Treated in N_2-O_2 and (d) The as deposited film.

3.3 Chemical composition on surface of CdS thin films

Figure 3 shows the XPS spectra of CdS thin films. The results indicate that element Cd, S and O can be found on surface of the CdS thin films. Element Cd possibly exists as cadmium, CdS, halides, CaO, CaO_2, $Cd(OH)_2$ and $CdCO_3$. Element S mainly exists as sulfide. Element O possibly exists as several kinds of compounds, such as hydroxides, sulfates and carbonates. To our discussion, higher content of oxygen not only results in CdO and $CdSO_4$ but also $CdCO_3$ and $CdNO_3$ on the surface of CdS thin film. The compounds including CdO and $CdSO_4$ were also reported by other works [15]. As well-known that oxidation reaction also happens during the CdTe thin film deposition step. The extent of oxidation reaction depends on heating rate, temperature and content of oxygen. The effect of those compounds on CdS thin film is complicated for CdTe solar cells. Additional surface treatment for the N_2-O_2 annealed CdS thin films should be conducted before CdTe deposition to remove those compounds. It was considered that the H_2-treatment at appropriate temperature is an efficient method and with no-damage to clean the surface of oxidized CdS films.

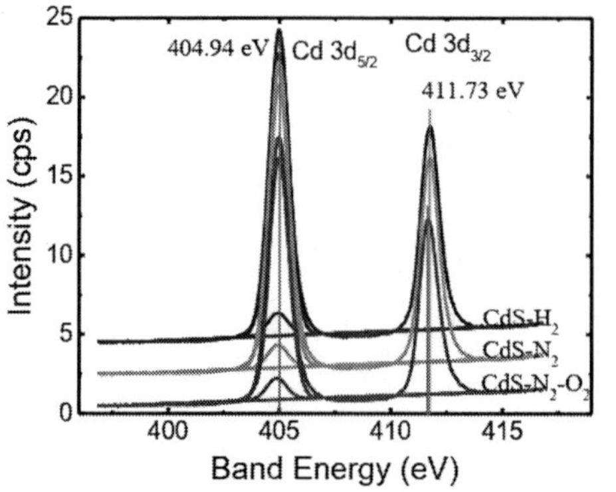

(a)

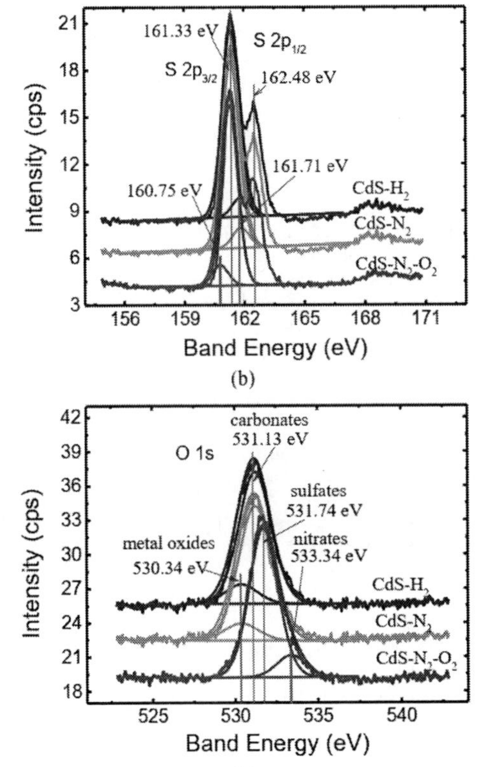

Figure 3: The XPS spectra of the treated CdS films: (a) the atom state details of Cd 3d, (b) S 2p and (c) the atom state details of O 1s.

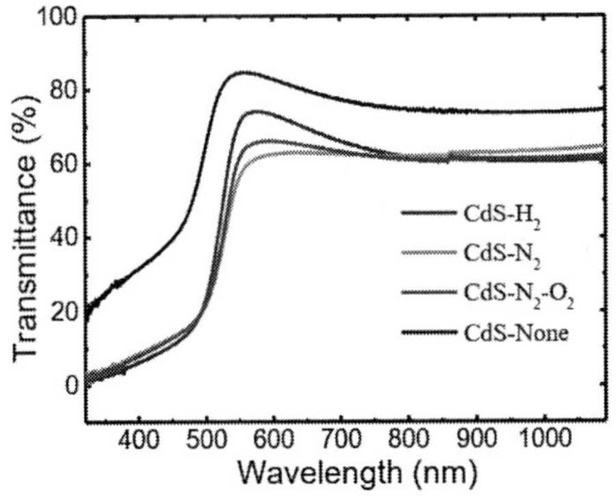

Figure 4: Transmittance of various CdS films: as deposited, treated in N_2, treated in N_2-O_2 and H_2.

3.4 The optical properties of CdS thin films

Figure 4 describes the optical properties of CdS films treated in different atmosphere at 300 K in the wavelength ranges from 320 nm to 1200 nm. The as grown CdS thin film shows a much higher transmittance in the wavelength range of 320-1200 nm. And the film treated in N_2 shows the lowest transmittance in the wavelength of 550-600 nm. This possibly due to the increase of grain size, defects and new compounds on the surface of CdS thin film. The surface roughness of CdS thin film can also lead to lower transmittance in the wavelength range from 550 nm to 1200 nm.

The band edge steepness value was obtained by plotting $(\alpha h\nu)^2$ versus $h\nu$ and taking the slope of the graph at the beginning of band-to-band absorption. The band gap of as-deposited CdS film is 2.44 eV. The optical energy gap decreased to 2.27 eV, 2.28 eV and 2.34 eV after treatment in N_2, N_2-O_2 and H_2, respectively (as listed in **Table III**).

Table III: The energy band gap and activation energy of conduction of CdS thin film treated in different atmosphere.

	E_g (eV)	E_a (eV)
CdS-H_2	2.34	0.07
CdS-N_2	2.27	0.13
CdS-N_2-O_2	2.28	0.14
CdS-None	2.44	0.17

3.5 The electrical properties of CdS thin films

The conduction activation energy of CdS thin films decreased when experienced heat-treatment, especially treated in hydrogen. This is agree with others work which showed that electrical properties of CdS film could be controlled by temperature, time and ambient of heat-treatment [5]. In the annealing process the oxide barriers on crystals are removed, the crystals are grown together and the electrical contacts between crystals are improved. The lowest activation energy of conduction was obtained by treatment in hydrogen of 0.07 eV (as listed in **Table III**).

In conclusion, the CdS thin films treated in N_2 usually with too much defects, such as dislocation and cracks for CdTe solar cell fabrication. The treatment in N_2-O_2 mixture is conductive for CdS thin films to increase crystallinity, electrical properties and with fewer defects, which seems with potential to improve performance of CdTe solar cells. The hydrogen treatment during CdTe thin film is conductive to improve crystalline and electrical properties of CdS thin film as well.

3.6 Performance of CdTe solar cells

CdTe solar cells were fabricated on CdS thin films treated in N_2-O_2 atmosphere. The CdS thin films were treated for different time: as deposited, 35 min, 40 min and 45 min. Hydrogen treatment was down for all the CdS substrate at 300 ℃ before the sublimation stage of CdTe thin film growth process. **Figure 6** shows the device performance of the CdTe solar cells. The box charts are based on 15 unit solar cells with area of 0.0707 cm^2 on each 4.5 cm×3 cm sample. A unit CdTe solar cell with area of 0.48 cm^2 on each sample was used to test the QE curve.

In **Figure 5** (a), the efficiency of the treated samples decreased in comparison with the CdTe solar cells fabricated on CdS substrate without annealing. This is due to the decrease of *Voc* and *FF*. However, increased *Jsc* about 3.9% was obtained when the CdS thin film was treated for 40 min as shown in **Figure 5** (b) and (c). This possibly because of that CdS treatment in N_2-O_2 for appropriate time can decrease trapping-centre and is good for carrier-collection in CdS thin film. The *Voc* and *FF* possibly decreased by the inter-diffusion effect at the interface of CdS/CdTe, which was decreased by the N_2-O_2 annealing. Another interesting phenomenon is that the treatment for CdS substrates results in fewer pinholes during higher temperature deposition of CdTe thin films obviously. Higher temperature should be employed to grow or anneal the CdTe thin film on the treated CdS substrates. These indicate that the heat-treatment could protect the CdS thin film from consumption during the CdTe film deposition and post-treatment process. Thus the enhancement for CdTe solar cells through CdS treatment is possible rely on the optimization of CdTe layer.

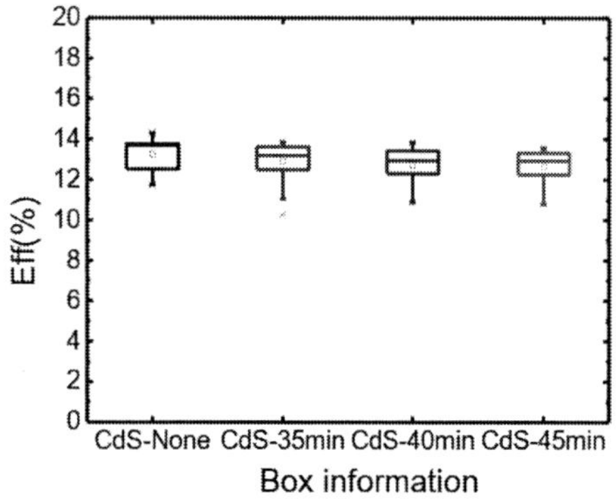

(a)

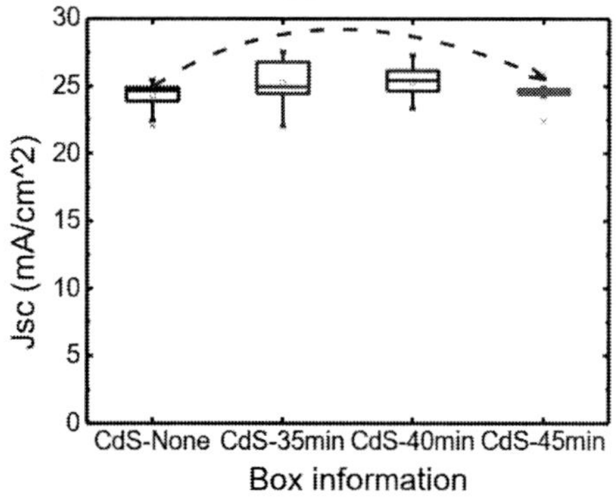

(b)

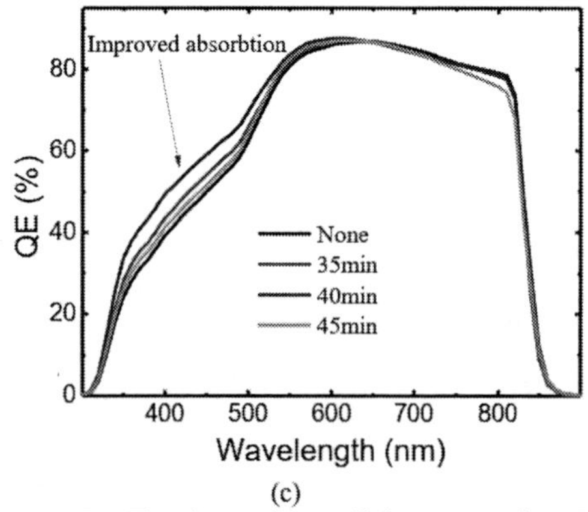

(c)

Figure 5: The box charts, light *I-V* and QE characteristics of CdTe solar cells with CdS annealed in N_2-O_2.

4 CONCLUSION

The heat treatment for CdS thin films leads to larger grain size, lower optical gap and higher conductivity in different atmosphere. Several compounds including CdO, $CdNO_3$, $CdSO_4$ and $CdCO_3$ were generated after N_2-O_2 treatment. CdS thin film treated in N_2-O_2 showed better structure properties than that treated in N_2 atmosphere. Larger grain size and lower E_a of the post-treated CdS thin films decreased the carrier-recombination and leads to higher QE at the wavelength ranged from 350 nm to 500 nm. The device performance suggests that higher deposition temperature and higher temperature rising rate should be engaged to grown CdTe thin films on treated CdS substrates.

ACKNOWLEDGEMENTS

This work is financially supported by the National High Technology Research and Development Program of China No. 2015AA050610. And we would like to thank the Analytical and Testing Center of Sichuan University f or XPS measurement.

REFERENCES

[1] T. Potlog, Solar Energy Materials and Solar Cells 80 (2003) 327
[2] S. Shibutani S. Kumazawa, T. Nishio, T. Aramoto, H. Higuchi, T. Arita, A. Hanafusa, K. Omura, M. Murozono, H. Takakura, Solar Energy Materials and Solar Cells 49 (1997) 8
[3] Donghwan Kim Hyeongnam Kim, Solar Energy Materials & Solar Cells 67 (2001) 8
[4] A. Cortes, Solar Energy Materials and Solar Cells 82 (2004) 21
[5] J. Hiie, K. Muska, V. Valdna, V. Mikli, A. Taklaja, and A. Gavrilov, Thin Solid Films 516 (2008) 7008
[6] H. Metin, S. Erat, F. M. Emen, V. Kafadar, A. N. Yazici, M. Ari, and N. Kulcu, Journal of Luminescence 130 (2010) 1531
[7] E. Yücel, S. Kahraman, and H. S. Güder, Mater Res Bull 68 (2015) 227
[8] M. T. S. Nair, P. K. Nair, R. A. Zingaro, and E. A. Meyers, Journal of Applied Physics 75 (1994) 1557
[9] A. Mendoza-Galván, G. Martínez, and R. Lozada-Morales, Journal of Applied Physics 80 (1996) 3333
[10] N. Maticiuc, N. Spalatu, V. Mikli, and J. Hiie, Appl Surf Sci 350 (2015) 14
[11] J. C. Bernede H. El Maliki, S. Marsillac, J. Pinel, X. Castel, J. Pouzet, Appl Surf Sci 205 (2003) 10
[12] Lei Wan, Zhizhong Bai, Zerong Hou, Deliang Wang, Hao Sun, and Limin Xiong, Thin Solid Films 518 (2010) 6858
[13] Xavier Mathew, Jose S. Cruz, David R. Coronado, Aduljay R. Millán, Gildardo C. Segura, Erik R. Morales, Omar S. Martínez, Christian C. Garcia, and Eduardo P. Landa, Solar Energy 86 (2012) 1023
[14] Faisal R. Ahmad, Aharon Yakimov, Robert J. Davis, Jae-Hyuk Her, James R. Cournoyer, and Nana Menya Ayensu, Thin Solid Films 535 (2013) 166
[15] J.C. Bernede, S. Marsillac H. El Maliki, J. Pinel, X. Castel, J. Pouzet, Appl Surf Sci 205 (2003) 15

IDENTIFICATION OF LOSS MECHANISMS IN CIGS MICRO-CELLS FOR CONCENTRATOR APPLICATIONS

E. Lotter, P. Jackson, S. Paetel, W. Wischmann

Zentrum für Sonnenenergie- und Wasserstoff-Forschung Baden-Württemberg (ZSW), Industriestr. 6, D-70565 Stuttgart, Germany

Corresponding author: Erwin Lotter, Email: erwin.lotter@zsw-bw.de, Phone: +49 711 7870 261

ABSTRACT: Recent results have proven the efficiency gain of micrometer-sized CIGS thin-film cells under medium to high concentration. In this work, co-evaporated CIGS cells with sizes in the range of 0.04 mm^2 to 1 mm^2 were investigated under white light illumination and concentrations up to 100 suns. The cells were isolated from finished standard test cells from two different CIGS sources: the inline deposition system and the high-efficiency cell line at ZSW.
High-efficiency cells with alkali post-deposition treatment were found to exhibit an internal series resistance with a temperature behavior which indicates a built-in barrier. This barrier can limit the efficiency of micro-cells already at moderate concentration factors.
The origin of shunting at the cell edges and its dependency on illumination intensity was studied for different preparation techniques such as mechanical scribing, pico-second laser ablation, and chemical etching of the ZAO layer. With the latter, very good edge passivation and efficiencies of up to 22.5% at 77 suns were achieved.

Keywords: CIGS, Thin Film Solar Cell, Concentrator Cells, Shunts, Laser Processing

1 INTRODUCTION

Recently, concentrating thin-film solar cells based on $Cu(In,Ga)Se_2$ (CIGS) have attracted attention because of the high efficiency gains under concentrated light which were demonstrated for micrometer-sized cells fabricated by different methods [1, 2].

Advantages of this micro-cell approach with low to medium concentration compared to conventional high concentrating systems are lower costs for lenses and tracking systems and a compact module design. For large-area CIGS modules with integrated micro-lens arrays, some constraints exist. First, the cell interval has to be small enough to permit the dissipation of the generated heat without a significant temperature rise within the cells. On the other hand, adjusting the micro-lens array to the cell array on a square-meter-sized glass substrate requires to oversize the cell area to compensate for position mismatch. In addition, there is some excess area needed to compensate for imperfect tracking and optical collection. The total excess area, however, should be kept small to limit parasitic currents in dark parts of the cells. From these constraints a pitch in the range of 1 mm or slightly less seems plausible. At a concentration factor of 100 the core cell size will then amount to 100 µm and an additional excess area of 100 percent will fit into an only 20 µm wide frame around the core. From a practical point of view this already seems to be difficult to manage. Therefore, a further miniaturization doesn't appear desirable. Instead, a somewhat lower concentration level in the range of 30x to 50x might help to reduce precision demands while still offering a substantial efficiency gain. Additionally, at lower concentration the voltage difference between the illuminated and dark parts of the cells is smaller and thus drives exponentially less parasitic dark current.

In this framework we started to examine how the reduction of cell sizes of a few millimeters, where cells behave more or less comparable to standard test cells, towards 100 µm impairs the performance of the micro-cells together with increasing concentration. For the preparation of the cells we tried to use techniques which should be applicable in module manufacturing. Our approach to cut micro-cells from completely finished standard test cells allowed us to directly compare the performance of micro-cells and standard cells and helped to identify problems related to miniaturization. Additionally, this approach enabled access to a large pool of high-quality cells fabricated with a variety of process modifications.

Discussions of the performance of solar cells under concentrated illumination and the influence of device properties have been published repeatedly and will not be replicated here. A compact illustration can e.g. be found in [3] together with a discussion of the influence of cell geometry and contacting.

2 EXPERIMENTAL

2.1 Sample preparation

$Cu(In,Ga)Se_2$ samples were grown in two different deposition systems [4]. In the high-efficiency cell line, CIGS absorbers (2.5–3.0 µm) were grown onto 1.2 mm thick alkali-aluminosilicate glass substrates covered by 500–900 nm sputtered molybdenum as a back contact. Some of these cells received an alkali post-deposition treatment [5]. The cells were completed with chemical-bath-deposited CdS as a buffer layer (30–50 nm), sputtered undoped ZnO (50–100 nm), sputtered Al-doped ZnO (ZAO, 150–200 nm) as conductive window material, and a nickel/aluminum grid.
The ZSW module line implements an in-line multistep process. Here, high-quality CIGS (2 µm) was evaporated onto Mo-coated soda lime glass (0.5 µm / 3 mm) followed by the standard CBD CdS (50 nm) and RF sputtered i-ZnO (75 nm) buffer layers. DC-sputtered ZAO (400 nm) and a Ni/Al contact grid were used as front contact.
Anti-reflective coatings (ARC) were not applied to the micro-cells.

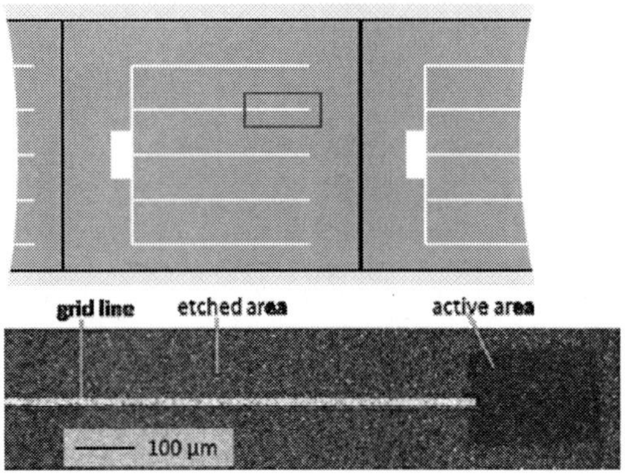

Figure 1: Illustration of the utilization of the standard cell grid for mechanically scribed cells (top).
Bottom: Etched micro-cell with an active area of about 0.04 mm². An evaporated Ni-Al grid line is used to contact etched cells as well.

Micro-cells were located around some part of the fingers of the Ni/Al contact of the standard cells. First, the metallic grid finger was cut to the desired length with a mechanical scribing tool (Figure 1). The borders of the micro-cells were defined by either mechanical scribing, laser ablation, or chemical etching of the window layer. In the beginning when cell areas in the 1 mm² range were created, the cells were formed symmetrically around the grid fingers and contacted in the center. With smaller cells the shading by the contact tips became more critical, so cells were located at an end of a grid line. The window layer was etched along the remaining part of the grid line, which is used for contacting.

2.2 Characterization

A class AAA solar simulator with a continuous AM 1.5 spectrum was used to measure I–V curves. In order to boost the basic light intensity of 100 mW/cm², a lens system was attached which can be adjusted to deliver concentration factors between 3 and 100 at the sample position (Figure 2).

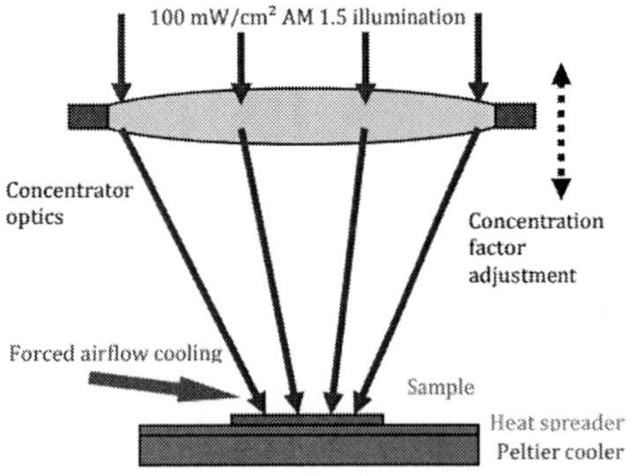

Figure 2: Schematic sketch of the concentrator setup used for I–V characterization.

Due to the simple design of this setup, the light intensity hitting the sample cannot be measured directly with sufficient precision. To overcome this problem, the short-circuit current density (j_{sc}) of the cell under investigation was used to determine the concentration factor. The linear relation between j_{sc} and the light intensity - which is assumed in this approach - was verified by use of attenuation filters for concentration factors up to 100 in agreement with measurements reported in [6]. Furthermore, for a better comparability between micro-cells and the parent standard test cells, the area of the micro-cells was determined from j_{sc} at STC which was assumed to be identical to the j_{sc} measured for the whole parent cell. In doing so, the shading by the contact grid of the full cell which covers about 3% of the total area is transferred to micro-cells, neglecting that micro-cells suffer much more from shading by contacts. Efficiencies calculated on this basis are therefore slightly less than ‘designated area efficiencies’. Cell areas calculated this way were additionally checked by microscopical size measurement.

During I–V measurements, the samples were cooled by a forced airflow across the surface. Otherwise, significant heating of the cells occurs already at moderate concentration factors of 20 to 30 and results in a notable voltage loss. This effect is caused by the provisional concentrator setup which does not allow proper masking of the illuminated area and therefore the vicinity of the active area also heats up. As a consequence, heat generated within the active area cannot be distributed sufficiently. Additionally, the integrated thermoelectric cooler is not designed to cool adequately under concentrated illumination. Even airflow cooling cannot prevent sample heating thoroughly when light concentration advances towards 100x. Therefore, the cell performance begins to suffer from heating at concentration factors beyond about 50. This effect becomes noticeable mainly as a deviation of the voltage (and efficiency) growth from the expected logarithmic behavior. Additionally, it should be noted that fill factors appear to be higher when the scan direction during I–V sampling is from low to high voltage as in our setup because the sample heats up during the scan.

3 RESULTS AND DISCUSSION

3.1 Edge Passivation

In the case of mechanically scribed cells, already for relatively large samples with areas in the square-millimeter range we observed an intensified shunting of the micro-cells. A closer look revealed that the parallel resistance R_P of the cells did not scale with the cell area, but rather with the edge length. Lock-in thermography (DLIT) was used to confirm that the shunts are in fact located directly at the scribed grooves.

Unfortunately, the impact of this shunting on the cell performance is not reduced for increasing illumination intensity since the shunt conductance of the cells scales in a way very similar to the intensity and I_{SC} as shown in Figure 3. In order to verify that this light-dependent component of the shunting is related to the cells edges as well, the cell borders were covered with black tape. Although the coverage was not perfect, it is clearly seen that now the shunting is much less for increasing illumination intensity. This result is a strong indication that the shunting is not caused by flakes of ZAO or other rubbed-off particles, but rather is caused by the CIGS absorber whose high photo-conductivity could explain the strong light dependence of the shunting. Since I–V curves of those cells show a linear increase of the shunt current over a wide voltage range, recombination is also ruled out as a cause. Instead, small imperfections at the edges introduced by the scribing process seem to act as leakage paths.

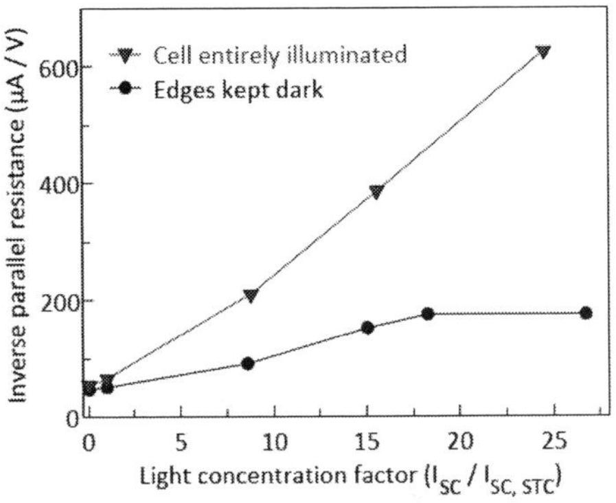

Figure 3: Parallel conductance of a mechanically scribed micro-cell with an area of 2 mm² and a total edge length of 7 mm. The shunting behavior of small cells strongly worsens not only with increasing edge-to-area ratio, but also with illumination intensity (red). The black dots show the clearly less pronounced shunting of the same cell after the edges were covered with black tape.

To avoid shunts at the cell edges, laser ablation was used for cell definition in a next step. Picosecond laser pulses with 12 ps duration at a wavelength of 532 nm and a pulse energy of 1 - 2 µJ were focused to spots of about 30 µm in diameter in order to lift off the ZAO. This process had been developed for P3 patterning of modules with good results. Compared to mechanically scribed cells, the parallel resistance is much higher and also decreases less with increased illumination, as shown in Figure 4. At higher concentration factors, however, a different type of deterioration appears. The cell efficiency starts to drop due to an earlier bending of the I–V curve near the maximum power point. In Figure 4 this is illustrated for an I–V curve at 20 suns. To emphasize the effect, the measured curve is supplemented with a modelled single diode curve, which does not consider the measured kind of imperfection. The fill factor loss increases strongly with further increase in illumination intensity and is also enhanced by light-soaking.

Usually, this kind of non-ideality is attributed to recombination processes. In our case, only a very narrow region near the edges, which accounts for a few percent of the cell area only, is locally damaged. Therefore, a more plausible explanation of this observation is that the thermally induced damage of the p-n junction near the edges leads to a reduction of the cell voltage in that region.

Thermally induced damage becomes more prominent when the thickness of the ZnO window is reduced as is the case for high-efficiency cells. Here the ZnO thickness is trimmed to maximize the impact of an antireflective coating. The elevated damage is most probably caused by the reduced mechanical strength of thinner window layers, which results in a narrower ablation width. Regions of the cell which have come closer to the laser beam and are therefore damaged more seriously by heating thus remain active.

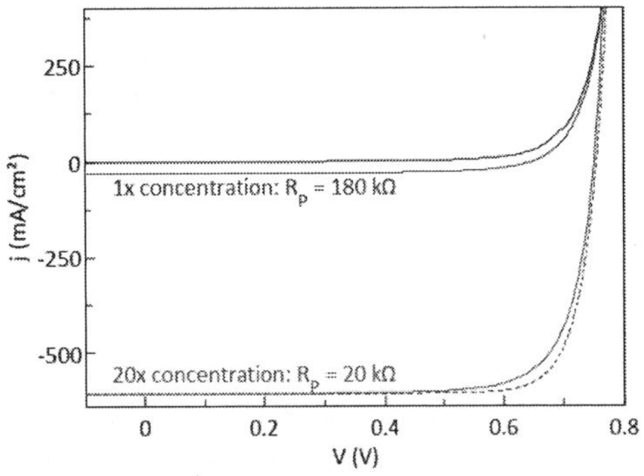

Figure 4: I–V characteristics of a laser-scribed cell with an area of 3 mm². Shunting is very low compared to mechanically scribed cells. At higher illumination levels, however, cells show a different kind of defect at the edges which manifests itself as a fill factor loss. The dotted blue line is a single diode fit to the 20x I–V curve in which the effect of this defect is omitted.

There is a way to circumvent this kind of damage as one can predetermine a ZnO breaking line with a preceding ultra-violet laser scribe. Thereby, the ablation of the ZAO can be extended beyond the heat-affected zone [7]. For practical reasons, this method has not yet been applied to micro-cells.

Instead, etching the window layer with diluted hydrochloric acid was applied, an approach which has already been reported for CIGS concentrator cells earlier [8].

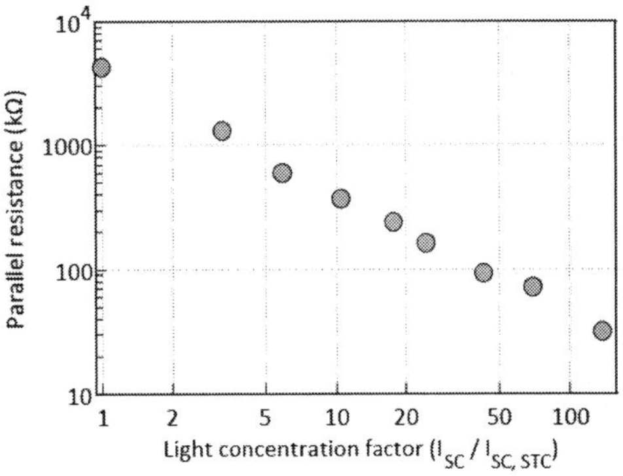

Figure 5: Parallel resistance of an etched cell with an area of 0.044 mm² as a function of the illumination level.

In Figure 5 the parallel resistance of an etched cell with an area of 0.044 mm² is plotted versus the illumination level. R_P still behaves inversely proportional to the light intensity, quite similar to the behavior observed for mechanical scribing in Figure 3, however on a much higher level. Based on the areas relation of 1:45, R_P is increased by a factor of about 5 only. The total length of the edges, which strongly influences R_P, differ much less, however, because the edges not only include the active area but also the grid line which serves as a contact pad. While the edges of the 2 mm² sized cell amount to a total length of 7 mm, the borders of the 0.044 mm² sized cell total 3.4 mm = 0.8 mm + 2.6 mm for active area and contact pad, respectively. Based on the edge length, R_P of the etched cell is therefore 100 times higher than R_P of the scribed cell. The data are not sufficient to distinguish how much R_P of the etched cell originates from contributions of surface area and edges, but it seems very clear that the R_P values determined for scribed standard cells predominantly result from the edges if no defect-related localized shunts are involved.

For the preparation of micro-cells, this result means that cell definition and isolation can be conducted with a quite simple lithographical masking and etching process.

3.2 Series resistance

After having adapted the etching process to micro-cell preparation and thereby having satisfactorily solved the shunting problem, we started to investigate the influence of the second CIGS cell property which is especially important for application under concentrated light, the series resistivity. Mainly three different types of CIGS absorber materials were used, CIGS from the ZSW module line without alkali-PDT and from the high-efficiency cell line with and without alkali-PDT. Additionally, the influence of other layers' properties in the cell stack, like the thickness of the i-ZnO and the buffer layer and also buffer materials (CdS, ZnOS) have been examined. Although some of these latter variations had an impact on cell properties, the main influence was found to come from the CIGS absorber itself. It turned out that cell und module line, which operate on slightly different efficiency levels, produce cells that do not differ very much from each other with respect to concentrated light performance, in particular series resistivity. The big difference comes into play with alkali-PDT. Already under STC operation of standard test cells, one can observe that even the best cells' I–V curves exhibit series resistances high enough to have a noticeable negative impact on the cell efficiency. Under concentrated illumination, this negative impact grows quickly, resulting in fill factor loss and saturating efficiency already at very moderate concentration levels of 10 to 20 suns as for the 0.05 mm² cell shown in Figure 6. The fill factor is on a very high level already at low illumination, indicating very good edge isolation, but starts to drop already at 20 suns. Consequently, the cell efficiency does not continue to increase although the V_{OC} still follows the expected growth.

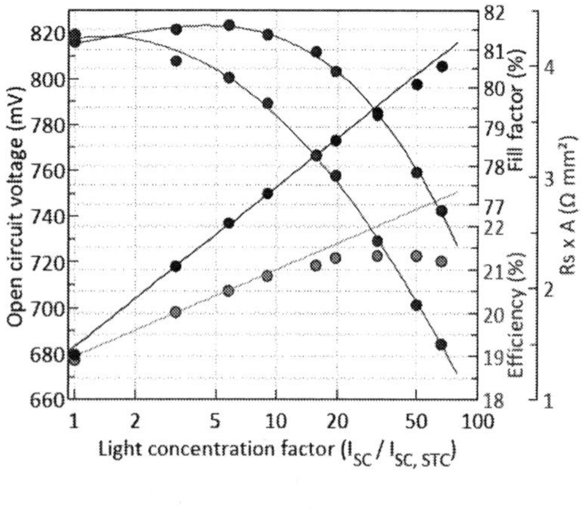

Figure 6: Performance of an alkali-PDT cell (0.05 mm²) under concentrated light. Due to its high series resistance, the fill factor starts to drop already at 20 suns.

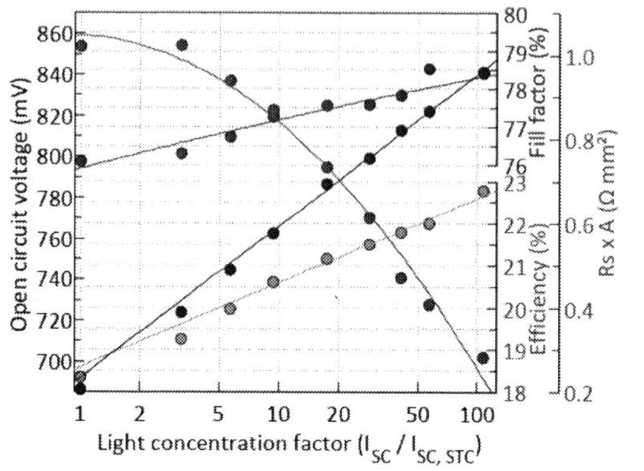

Figure 7: Performance of a 0.06 mm² cell without alkali-PDT under concentrated light.

The reason for this unfavorable behavior becomes apparent when the series resistance is compared to R_S of a similar cell without alkali-PDT (Figure 7). For the non-PDT cell, the fill factor is quite low due to a weak shunt, but increases continuously with light intensity. This behavior indicates that the additional shunt is not light sensitive or may be located in the dark below the contact pad. Thereby, the cell efficiency increases logarithmically up to the highest light intensities available, exceeding 22.5% at 100 suns. The series resistance of this cell is about five times lower compared to that of the alkali-PDT cell and also drops faster with increasing light intensity. The origin of the increased series resistance becomes apparent when the temperature behavior of the cell's I–V characteristics is examined. The I–V curves shown in Figure 8 were taken at increasing cell temperatures due to increasing duration of illumination at 70 suns without cooling. The heating not only leads to an observable decrease of V_{OC}, but also steepens the I–V curves due to a pronounced reduction of the series resistance.

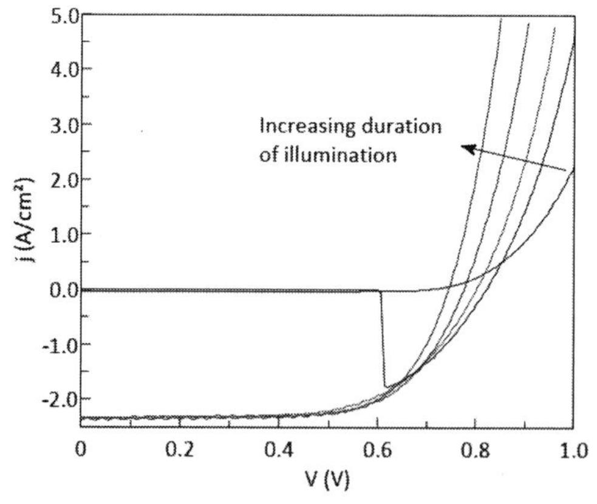

Figure 8: Effect of a temperature rise on the I–V curves of a cell with alkali-PDT. At 70 suns without cooling, the cell heats up with increasing duration of illumination. This situation not only leads to an observable decrease of V_{OC}, but also to a pronounced reduction of the series resistance.

Depending on the strength of the effect this behavior, which is fully reversible and not observed for non-PDT cells, can even result in a positive temperature coefficient of the efficiency as in the example shown. In some samples with still higher series resistance, even rollover signatures have been observed in the I–V characteristics, which shift to higher currents with increasing temperature. This behavior can be explained by the presence of a built-in electrical barrier, probably induced by the surface modification seen e.g. by photoemission spectroscopy [9].

4 CONCLUSIONS

The origin of the dominating part of the parallel resistance of micro-cells was found to be located at the cell edges. A strong dependency on the illumination intensity indicates that the photo-conductivity of the CIGS absorber is involved in the shunt mechanisms. Mechanically scribed edges exhibit a purely ohmic shunt characteristic, while pico-second laser ablation and chemical etching tend to suffer from a damage of the p-n junction near the edges, which becomes apparent as a fill factor loss particularly at higher light intensities. By adjusting the etch process we have been able to almost completely transfer the STC performance of large area cells to micro-cells as small as 0.04 mm^2.

Low shunting is also important for adequately dimensioned excess area to compensate for misalignments of the concentrator optics and imperfect tracking.

High efficiency cells with alkali post-deposition treatment turned out to be limited to low concentration factors due to their high internal series resistance. The temperature behavior of this resistance indicates a built-in barrier. At higher concentration levels, CIGS absorbers without alkali-PDT proved to be superior. A 0.06 mm^2 sized cell reached an efficiency of 22.5% at 77 suns, a gain of 4% absolute compared to its parent cell.

Other cell properties besides the series resistance turned out to be of minor importance. However, cells with deficiencies e.g. due to imperfections of their buffer layer, always operated worse also under concentrated light and often didn't benefit thereof at all. Summarized, the best cells at STC clearly perform best under concentrated light with the exception of the series resistance problem of cells with alkali-PDT.

The details of this deficit will be examined further with the aim to make the enhancement due to the alkali-PDT usable also for concentrator cells.

5 ACKNOWLEDGEMENTS

This research is part of the EU CHEETAH project and has received funding from the European Union's Seventh Programme for research, technological development and demonstration under grant agreement No 609788.

6 REFERENCES

[1] M. Paire, L. Lombez, N. Péré-Laperne, S. Collin, J.-L. Pelouard, D. Lincot, and J.-F. Guillemoles, Appl. Phys. Lett. 98, 264102 (2011).

[2] M. Paire, C. Jean, L. Lombez, S. Collin, J.-L. Pelouard, I. Gérard, J.-F. Guillemoles, and D. Lincot, Thin Solid Films 582, 258 (2015).

[3] B. Reinhold, M. Schmid, D. Greiner, M. Schüle, D. Kieven, A. Ennaoui, and M.C. Lux-Steiner, Prog. Photovoltaics Res. Appl. 23, 1929 (2015).

[4] M. Powalla, P. Jackson, W. Witte, D. Hariskos, S. Paetel, C. Tschamber, and W. Wischmann, Solar Energy Materials and Solar Cells, 119, 51-58 (2013).

[5] P. Jackson, D. Hariskos, R. Wuerz, O. Kiowski, A. Bauer, T. M. Friedlmeier and M. Powalla, Phys. Status Solidi RRL, 2015, 9, 28–31.

[6] M. Paire, A. Shams, L. Lombez, N. Péré -Laperne, S. Collin, J.-L. Pelouard, J.-F. Guillemoles, D. Lincot, J. Energy Environ. Sci. 2011; 4: 4972–4977.

[7] European patent EP2177302 (B1), „Method of removing layered material of a layered construction with a laser beam, with a preliminary grooving step and a removing step".

[8] J. S. Ward, K. Ramanathan, F. S. Hasoon, T. J. Coutts, J. Keane, M. A. Contreras, T. Moriarty, and R. Noufi, Prog. Photovoltaics 10 (2002) 41.

[9] E. Handick, P. Reinhard, J.-H. Alsmeier, L. Köhler, F. Pianezzi, S. Krause, M. Gorgoi, E. Ikenaga, N. Koch, R. G. Wilks, S. Buecheler, A. N. Tiwari, and M. Bär, ACS Appl. Mater. Interfaces 2015, 7, 27414.

ELABORATION OF ZNO: GA THIN FILMS BY SPRAY PYROLYSIS FOR PHOTOVOLTAIC APPLICATIONS

[1]Z.El khalidi, [*,1]B.Hartiti, [1]S.Fadili ,[2]A.Lfakir, [3]P.Thevenin
[1]MAC & PM Laboratory, ANEPMAER Group FSTM, Hassan II Casablanca University Casablanca, Morocco
[*]Senior Assoc, ICTP,UNESCO, Trieste, Italy
[2]University Moulay Ismail, FST ERRACHIDIA, Morocco
[3]LMOPS Laboratory, Department of Physics University of Lorraine Metz, France

ABSTRACT:
Zinc oxide Semiconductor is a promising material for optoelectronic applications. It is studied for the advantageous properties such as good transparency, optical and electrical characteristics that makes it used as window layer in photovoltaic cells. It can be elaborated using several preparation methods, besides introducing dopants to enhance its optical and physical properties. Therefore, in the present work we have been interested in the synthesis of ZnO doped gallium with different concentrations, using spray pyrolysis at 400°C. The characterization using XRD indicates that the films exhibit polycrystalline structure, and a preferential orientation along (002) perpendicular to the substrate. The electrical data using 4 point probe shows a resistivity of ($25.10^{-4}\Omega$.cm) for ZnO layer. The optical study showed a transparent behavior of the films with a transmittance up to 90%. The refractive index n(λ), extinction coefficient, and dielectric constants were also calculated. The lowest band gap energy is about (3.28 eV). The Urbach energy in function of Ga % was studied to determine the disorder in the structure which is confirmed by the structural data. These results indicate the conformity of the films to be potentially applicable as transparent layer in photovoltaic devices.
Keywords: ZnO:Ga, Spray Pyrolysis, Urbach energy, refractive index , window layer.

1 INTRODUCTION

Solar cells technology has known a wide use due to its role in harvesting clean energy with a remarkable rate of efficiency [1-4]. The window layer materials such as zinc oxide, which is a semiconductor material, characterized with high performance and low cost, and the other transparent conductor oxides like CuO, In2O3, NiO, SnO2 [5-13] were studied for the advantageous properties they provide such as good transparency, optical and electrical characteristics. Zinc oxide is a binary semiconductor n type material with a direct bandgap of 3.37 eV and several outstanding properties, besides its abundance and non-toxicity [14]. Zinc oxide thin films are appropriate in optoelectronic systems, solar cells, Coating-protective layers or smart, Gas sensors and ultrasonic oscillators, transparent electrodes [15-21]. They are also found in many applications such as acoustic surface devices, optical waveguides and micro actuators. They can be used as an alternative material of tin oxide (TO) and Indium tin oxide (ITO) which are most used nowadays [22-24]. There are many methods to prepare Zinc oxide thin films such as Sol-Gel [25-34], sputtering [35-37],Chemical Bath Deposition [38], Pulsed-laser Deposition [39] and Spray Pyrolysis [40-41] which is the method adopted in our process. We elaborated in the previous works, many layers of intrinsic zinc oxide using several conditions. In the present work, the preparation conditions of the best intrinsic zinc oxide layer has been chosen to be doped with gallium at different ratios, by spray pyrolysis method. We investigated the effect of doping on structural optical and electrical properties of zinc oxide, to be used as window layer in photovoltaic solar cells.

2 EXPERIMENTAL DETAILS

The elaboration of Zinc oxide thin films was considered in our process, which includes spray pyrolysis as method of synthesis. The starting solution contains zinc acetate ($C_4H_6O_4Zn.2H_2O$) as precursor with the concentration of 0.05 M in distilled water, and prepared at 400°C. Then, different percentages of gallium nitrate hydrate were added to the ZnO, with the ratios [Ga]/ [Zn] from 1 to 4 %. The rate of the Spray was constant and equal to1 mL.min-1, the distance between the substrate and the nozzle is about 30 cm, and the pressure was maintained at two Bar. All these parameters were kept constant during the experiment. X-Ray Diffraction (XRD) gave the structural properties, the transmission spectra was recorded at room temperature and near to the normal incidence, where we used a Lambda 900 UV/VIS/NIR spectrometer. The electrical data were given by 4 point probe.

3 RESULTS AND DISCUSSION

3.1 Structural Data

To understand the material's behavior it is necessary to report the structural data, to extract structural parameters from the X-ray diffraction pattern of the studied samples. Fig.1 shows that the doped and undoped ZnO are polycrystalline and exhibit multiple peaks (as indicated in the JCPDS 36-1451) related to the wurzite structure.

THE GRAPHIC ON THIS PAGE WAS NOT AVAILABLE ON THE E-MEDIA VERSION OF THIS PAPER.

Figure 1: Diffraction spectra of ZnO:Ga thin films

deposited on glass substrates

Table I: The structural parameters of ZnO:Ga thin films. JPCDS card N°: 36-1451 (a = 3,250 Å and c = 5,207Å).

Dopant %	d Spacing (hkl) $d_{hkl}(100)$	$d_{hkl}(002)$	D (nm)	$\delta*10^{-5}(nm^{-2})$
ZnO	2.8084	2.6031	31.7	99.513
ZnO:Ga 1%	2.8037	2.6025	32.96	92.05
ZnO:Ga 2%	2.8130	2.6035	27.46	132.617
ZnO:Ga 3%	2.8082	2.6001	26.36	143.9
ZnO:Ga 4%	2.8083	2.6004	21.95	207.1

Dopant %	Lattice constants a(Å)	c(Å)	c/a
ZnO	3.24	5.21	1.608
ZnO:Ga 1%	3.237	5.205	1.607
ZnO:Ga 2%	3.248	5.207	1.603
ZnO:Ga 3%	3.242	5.206	1.605
ZnO:Ga 4%	3.242	5.200	1.603

The intensity of the plan (002) decreases when we increase the gallium concentration, similar results were found by Muchuweni et al [42]. To understand more this phenomenon, we calculated several parameters set in Table 1, where we regrouped the lattice constants a and c [43], the crystallite size D_{hkl} [44], the dislocation δ [45]:

$$\frac{1}{d^2}=\frac{4}{3}\left(\frac{h^2+hk+k^2}{a^2}\right)+\frac{l^2}{c^2} \ (1) \quad D=0{,}9\frac{\lambda}{\beta\cos\theta} \ (2) \quad \delta=\frac{1}{D^2} \ (3)$$

where D_{hkl} is the crystallite diameter, λ_{hkl} is the wavelength of the incident radiation (λ = 0.154nm), β_{hkl} is the full width at half maximum (FWHM) of the peak corrected by the instrumental broadening, expressed in radian, θ_{hkl} is the Bragg half-angle diffraction [46]. The lattice constants of the ZnO films are in agreement with standard values. As the indices are known from reference data, the lattice parameters a (a=b) and c were calculated from the peaks of reflections (002) and (100). The parameters calculated including lattice parameters, dislocation increase in function of doping, the crystallite size decreases. This indicates that the doping in this case alters the structure of the films by degrading their crystallinity. The injection of the impurities causes also a mechanical stress due to the size difference of ionic radii by substitution of Zn^{2+} with smaller Ga^{3+} ions [47-48].

3.2 Optical Data

Fig.2 shows the variation of transmittance in function of the wavelength of GZO films.

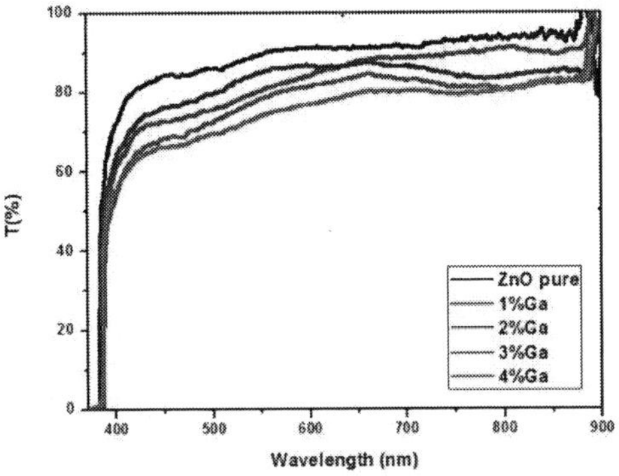

Figure 2: Transmittance spectra of ZnO:Ga films

The doped and undoped samples are transparent in the visible range up to 90 %. The transparency decreases with higher doping level. This could be explained by the increase of photons scattering by defects of crystal through doping [49]. The transmission curves as a function of the wavelength, allow the calculation of the absorption coefficient for each wavelength [50-51]:

$$\alpha=\frac{1}{d}\ln\left(\frac{1}{T}\right) \quad (4)$$

α is the absorption coefficient, T the transmittance and d is the thickness of the films which is between 350 and 400 nm. We plot the variation of $(\alpha h\upsilon)^2$ as a function of (hυ). The width of the band gap Eg of these films can be determined from the following equation [50]:

$$(\alpha h\upsilon)^2 = B\,(h\upsilon - Eg) = f\,(h\upsilon) \quad (5)$$

where hυ is the energy of the incident photons. Eg is the optical gap and B is a constant. As shown in Fig.3, the values of the gap energy Eg of GZO samples were obtained by extrapolating the linear part of the curves to the zero absorption Table 2.

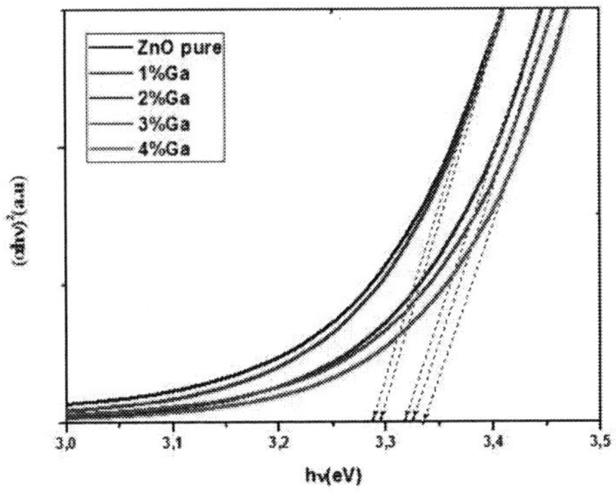

Figure 3: Determination of the energy gap of ZnO:Ga by extrapolation method

The optical band gap increases from 3.28 to 3.33 eV with increasing the doping concentration. This shifting corresponding to the gallium concentration altering and refers to the Burstein-Moss effect [52]. For semiconductor materials, the Fermi level increasing in the conduction band results a broadening in the band gap energy with carrier concentration increasing. Fermi level is within the conduction band, its position depends on the free electrons density. When we increase the gallium doping, an excitation of the electrons from the top of the valance band to the bottom of the conduction band occurs. The lifting of the Fermi level into the conduction band of the degenerate semiconductor due to the increase in the carrier density leads to the energy band broadening (shifting) [53]. We can deduce the disorder from the changes of the absorption coefficient. The following relation [54] relates the absorption coefficient to the disorder:

$$\alpha = \alpha_0 \exp(\frac{h\nu}{E_u}) \quad (6)$$

where α_0 is a constant, and E_u is the Urbach energy or the disorder, that determines the slope of the exponential edge, and calculated by fitting the linear portions of $\ln(\alpha)$ versus the photon energy $h\upsilon$. The increasing in this case is clearly related to the doping levels of gallium, also in accordance with the literature [55] and confirmed by the previous data given by XRD.

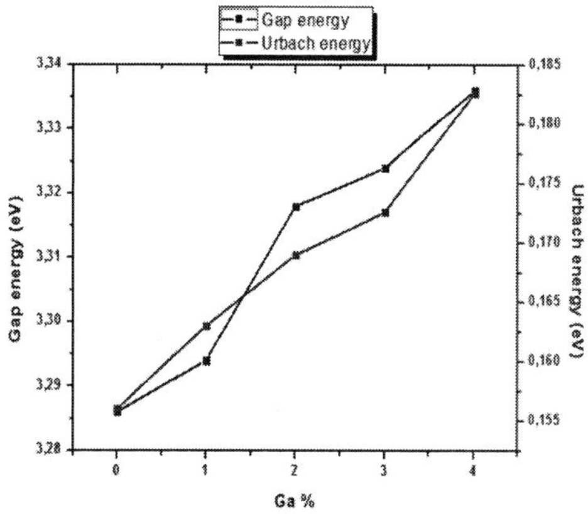

Figure 4: Variation of optical band gap and Urbach energy of ZnO:Ga films.

The refractive index has been calculated according to the relation [56]:

$$n = \left(\frac{1+R}{1-R}\right) + \sqrt{\left(\frac{4R}{(1-R)^2}\right) - k^2} \quad (7)$$

where k is the extinction coefficient ($k = \alpha\lambda/4\pi$), R is the reflectance. Through these equations, one can calculate the values of n and k. The variations of refractive index n and extinction coefficient k with wavelength are represented in Figs. 5 and 6. The optical constants n, k exhibit a decrease in function of the wavelength and increase in function of doping to 2.3. This could be related to the porosity of the film, roughness of the surface, which affects light scattering on the surface that changes the refractive index in consequence.

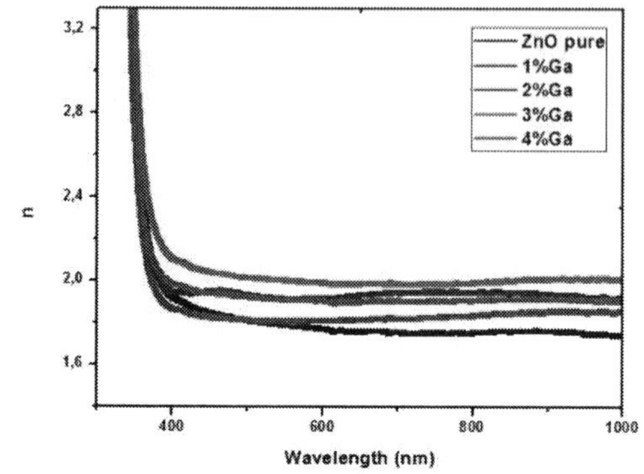

Figure 5: The variation of the Refractive Index corresponding to the wavelength of ZnO:Ga samples

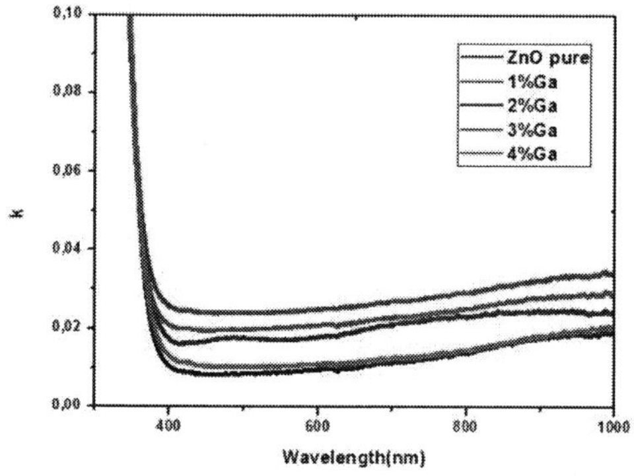

Figure 6: Variation of the Extinction Coefficient corresponding to the wavelength of ZnO:Ga samples

The electron excitation spectrum of the film was described by the complex electronic dielectric constant. The dielectric constant is defined as [57]:

$$\varepsilon = \varepsilon_1 + i\varepsilon_2 \quad (8)$$

$\varepsilon 1$ and $\varepsilon 2$ are real and imaginary parts of the dielectric constant, respectively and these values were calculated using the formulas [58]:

$$\varepsilon_1 = n_2 - k_2 \quad (9) \qquad \varepsilon_2 = 2nk \quad (10)$$

Figs. 7 and 8 representing the real and imaginary dielectric constants. It is observed from these figures, that ε_1 and ε_2 decrease corresponding to the wavelength. The ε_1 values are higher than of ε_2 values for all samples. These optical parameters indicate that our films are conform to be potentially applicable as transparent layer in photovoltaic devices.

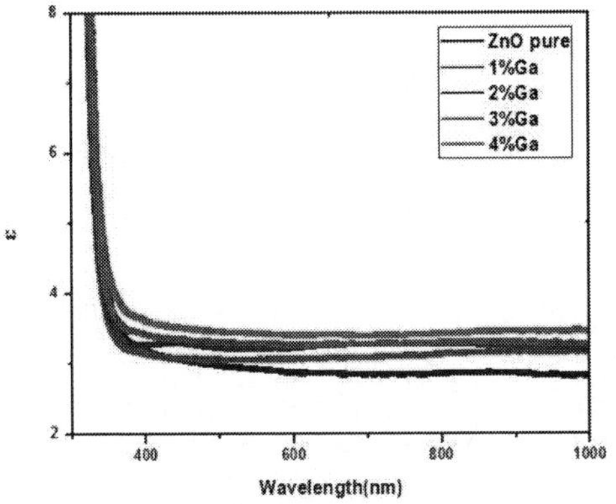

Figure 7: Variation of the real part of ZnO:Ga dielectric constant

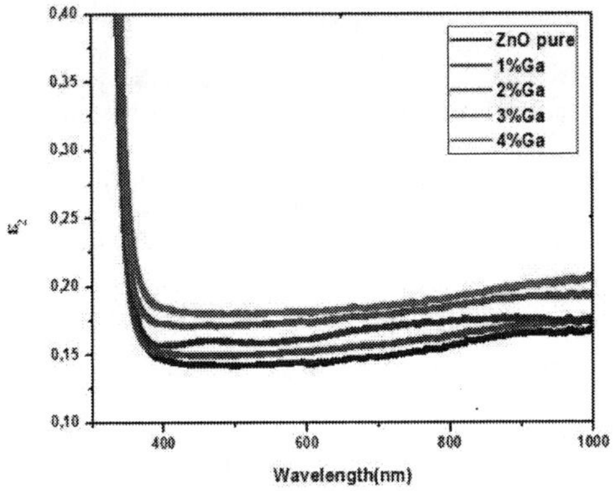

Figure 8: Variation of the imaginary part of dielectric constant of ZnO:Ga films

3.3 Electrical Data

To measure the resistivity using 4-point probe, we passed a current through two outer probes and reported the voltage through the inner probes. The electrical resistivity is then calculated using the following equation [59]:

$$\rho = Bt(\frac{V}{I}) \quad (11)$$

where B= π /ln(2) , V is the voltage across inner probes, I is the current in outer probes, and t is the thickness. From Fig.9 the lowest resistivity is about 0.25×10^{-2} Ω.cm better than the values found by Muchuweni [42] and Prasada [55]. The resistivity decreases for the first doping percentage and increases for the heaviest doping levels. This phenomenon is well explained by the previous authors [42],[55], the low resistivity is related to the shrinkage of the grain boundary potential barrier, and the substitution of Zn^{2+} by Ga^{2+} introduces more electrons causing an increase in carrier concentration as result. Otherwise, the increase in the resistivity above 1% doping could be explained by the massive electron density in the structural network, which leads to a difficulty of electrons circulation and the built of charge carrier trapping sites at grain boundaries [60]. Another reason is related to the structural aspect, explained by the disorder of the structure at higher level of doping by replacing Zn^{2+} by Ga^{2+} that increases the mechanism of phonon scattering and ionized impurity scattering that increases the resistivity [61]. These results are in good agreement with structural data given by XRD and the calculation of Urbach energy, regarding the disorder created at higher doping levels.

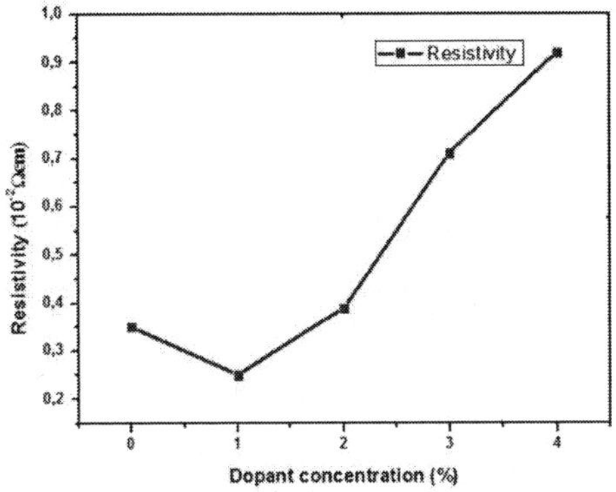

Figure 9: variation of resistivity of GZO thin films

3 RESULTS AND DISCUSSION

In this work we deposited ZnO doped Ga (GZO) on glass substrates using spray pyrolysis technique. We reported different variations and changes in structural, optical and electrical properties of the deposited films. XRD data revealed that ZnO doped and undoped films are polycrystalline with an hexagonal wurtzite structure, and a preferential orientation along (002). The calculation of structural parameters extracted from XRD data show the deterioration and defects created by increasing the gallium doping, which is confirmed by the calculation of Urbach energy and the analysis of the shift resistivity. The films exhibit a high transmittance above 90%, the band gap varies from 3.28 to 3.33 eV and explained by Burstein-Moss effect. The best electrical resistivity compared to the previous research is about 0.25×10^{-2} Ω.cm, at 1% of gallium doping. These results imply that our films can be used in many applications, particularly as window layer in photovoltaic devices.

References

[1] B. Oregan and M. Gratzel, Nature 353, 737 (1991).
[2] H. Liu, Superlattices Microstruct. 48, 458 (2010).
[3] O. Lupan, J. Cryst. Growth 312, 2454 (2010).
[4] C. Y. Jiang, Appl. Phys. Lett. 92, 143101 (2008).
[5] K. Mageshwari, Materials Science in Semiconductor Processing .16(2013)337-343
[6] G.G. Untila, Thin solid films 518 (2009) 1345–1349.
[7] Y.-M. Sung, Vacuum 83 (2008)161–165.
[8] K. Iwata, Thin Solid films 480–481 (2005) 199–203.
[9] J. Muller, Solar energy 77 (2004) 917–930.
[10] S.-M. Park, T. Ikegami, K. Ebihara, P.-K. Shin, Appl. Surf. Sci.253 (2006) 1522–1527
[11] J. Montero, C. Herrero, J. Guillen, Sol. Energy Mater.Sol. Cells 94 (2010) 612–616.
[12] Y. Huang, G. Li, J. Feng, Q. Zhang, Thin solid films 518 (2010) 1982-1896.
[13] Lingxiangyu.Lia, Science of Total Environment 505 (2015)253-260.

[14] I. Stambolova, Lanthanum doped SnO2 and ZnO thin films sensitive to ethanol and humidity, Materials Chemistry and Physics, 2000, 63, p. 104-108.
[15] C. Jie, G. Xin-shi, Solar Energy Materials & Solar Cells, 1998, 55, p. 323-329.
[16] P. Jin, L. Miao, Formation and characterization of TiO2 thin films with application to a multifunctional heat mirror, Applied Surface Science, 2003, 212-213, p. 775-781.
[17] J.Y. Lee, A study on the origin of emission of the annealed n-ZnO/p-GaN heterostructure LED, Thin Solid Films, 2009, 517(17), p. 5157-5160.
[18] J. Nishino, J. Mater. Sci. Lett.1 (1997) 629.
[19] M. Ritala, Mater. Res. Soc. Symp. Proc.426 (1996) 513.
[20] R. Wang, L.L.H. King, W.W. Sleight, J. Mater. Res.11 (1996) 1659.
[21] V. Gupta, A. Mansingh, J. Appl. Phys. 80 (1996) 1063.
[22] M.M. Bertolotti, J. Mater. Res.5 (1990) 1929.
[23] T. Schuler, M.A. Aegerter, Thin Solid Films 351 (1999) 125
[24] T. Schuler, M.A. Aegerter, Thin Solid Films 351 (1999) 125.
[25] M.S. Tokumoto, J. Non-Cryst. Solids 247 (1999) 176.
[26] M.N. Kamalasanan, S. Chandra, Thin Solid Films 288 (1996) 112.
[27] J.-H. Lee, Thin Solid Films 426 (2003) 94.
[28] A.E. Jiménez González, J.A. Soto Urueta, Sol. Energy Mater. Sol. Cells 52 (1998) 345.
[29] G.G. Valle, P. Hammer, S.H. Pulcinelli, C.V. Santilli, J. Eur. Ceram. Soc. 24 (2004) 1009.
[30] R.F. Silva, M.E.D. Zaniquelli, J. Non-Cryst. Solids 247 (1999) 248.
[31] M. Ohyama, H. Kozuka, T. Yoko, J. Am. Ceram. Soc. 81 (1998) 1622.
[32] M. Ohyama, H. Kozuka, T. Yoko, Thin Solid Films 306 (1997) 78.
[33] Y. Talahashi, M. Kanamori, A. Kondoh, H. Minoura, Y. Ohya, Jpn. J. Appl. Phys. 33 (1994) 6611.
[34] J.A. Anna Selvan, H. Keppner, A. Shah, Mater. Res. Soc. Symp. Proc. 426 (1996) 497.
[35] T. Nakada, N. Murakami, A. Kunioka, Mater. Res. Soc. Symp. Proc. 426 (1996) 411.
[36] B. Szyszka, S. Jäger, J. Non-Cryst. Solids 218 (1997) 74
[37] P. O'Brien, J. Mater. Chem. 6 (1996) 1135
[38] S. Hayamizu, H. Tabata, H. Tanaka, T. Kawai, J. Appl. Phys. 80 (1996) 787.
[39] C. Messaoudi, D. Sayah, M. Abd-Lefdil, Phys. Status Solidi 151 (1995) 93.
[40] M.S. Tokumoto, A. Smith, C.V. Santilli, S.H. Pulcinelli, E. Elkaim, V. Briois, J. Non-Cryst. Solids 273 (2000) 302.
[41] M.S. Tokumoto, A. Smith, C.V. Santilli, S.H. Pulcinelli, A.F. Craievich, E. Elkaim, A. Traverse, V. Briois,Thin Solid Films 416 (2002) 284
[42] E. Muchuweni et al. effect of gallium doping on the structural, optical and electrical properties of zinc oxide thin films by spray pyrolysis. Ceramic International, (2016).
[43] Bunn, C.W. (1935) Proceedings of the Physical Society, 47, 836.
[44] RANI S., SURI P., SHISHODIA P.K., MEHRA R.M., Sol. Energ. Mat. Sol. C., 92 (2008), 1639.
[45] WANG X.S., WU Z.C., WEBB J.F., LIU Z.G., Appl. Phys. A-Mater., 77 (2003), 561.
[46] B. Marí, M. Mollar, D. Soro, R.Henríquez, R. Schrebler, H. Gómez. Int. J. Electrochem. Sci., 8 (2013) 3510- 3523.
[47] A. Kumar Srivastava, J. Kumar, Effect of zinc addition and vacuum annealing time on the properties of spin-coated low-cost transparent conducting 1 at% Ga–ZnO thin films, Sci. Technol. Adv. Mater. 14 (2013) 065002 15pp
[48] I. Winer, G.E. Shter, M. Mann-Lahav, G.S. Grader, Effect of solvents and stabilizers on sol–gel deposition of Ga-doped zinc oxide TCOfilms, J. Mater. Res.26 (2011) 1309–1315.
[49] S.S. Shinde, P.S. Shinde, Y.W. Oh, D. Haranath, C.H. Bhosale, K.Y. Rajpure,Structural, optoelctronic, luminescence and thermal properties of Ga-doped zinc oxide thinfilms, Appl. Surf. Sci. 258 (2012) 9969–9976.
[50] Yoichi Yasaki, Noriyuki Sonoyama, Tadayoshi Sakata. Journal of Electroanalytical Chemistry 469 (1999) 116–122.
[51] R.A. Smith, Semiconductors, Academic Publishers, Calcutta, 1989, pp. 461–463.
[52] Natsume Y, Sakata H. Thin Solid Films 2000; 372: 30–6.
[53] R. Ferro, J.A. Rodr´ıguez, Thin Solid Films 347 (1999) 295–298.
[54] Mott NF, Davis EA. Electronic process in non-crystalline materials. Oxford: Calendron Press; 1979
[55] T.P. Rao, M.C.S. Kumar, Physical properties of Ga-doped ZnO thinfilms by spray pyrolysis, J. Alloy. Compd. 506 (2010) 788–793.
[56] S. Aksay, B. Altiokka, phys. Stat. sol. (c) 4, No. 2, 585-588 (2007) /DIO 10. 1002/ pssc. 200673236.
[57] J. N. Hodgson, Optical absorption and dispersion in Solids. Chapman and hall LTD, 11 New fetter Lane London EC4, 1970.
[58] E. Abd El-Wahabb, A. F. Mott, Philos. Mag. 22 (1970) 903
[59] K.L. Chopra, Thin Film Phenomena, MC Graw Hill, New York, USA 1969, p. 255.
[60] C. Tsay, K. Fan, C. Lei, Synthesis and characterization of sol–gel derived gallium-doped zinc thin films oxide, J. Alloy. Compd. 512 (2012) 216–222.
[61] B. Joseph, P.K. Manoj, V.K. Vaidyan, Ceram. Int. 32 (2006) 487–493.

MECHANISM OF EARLY-STAGE DEGRADATION OF CIGS SOLAR CELLS INDUCED BY AIR EXPOSURE

Jiro Nishinaga[1,2], Yukiko Kamikawa[1,2], Takashi Koida[1,2], Hajime Shibata[1,2], and Shigeru Niki[2]
[1]Research Center for Photovoltaics, National Institute of Advanced Industrial Science and Technology (AIST)
[2]Renewable Energy Research Center, National Institute of Advanced Industrial Science and Technology (AIST)
[1]Central 2, 1-1-1, Umezono, Tsukuba, Ibaraki 305-8568, Japan
[2]2-2-9 Machiikedai, Koriyama, Fukushima 963-0298, Japan

ABSTRACT: The degradation mechanism of unencapsulated Cu(In, Ga)Se_2 (CIGS) solar cells on exposure to air has been investigated. Exposure to air at room temperature slightly reduces the conversion efficiency of CIGS solar cells. However, this conversion efficiency decreases significantly under damp heat testing at 85 °C and 85% relative humidity for several hours. The shunt resistance and conversion efficiency are completely recovered after removing the side edges of the CIGS solar cells by mechanical scribing. This result suggests that low-resistive layers are formed on the sidewalls of the solar cells during damp heat testing. The low-resistive layers on the sidewalls are identified to be molybdenum oxides and sodium molybdate by Auger electron spectroscopy. After etching the oxides on the sidewalls by alkaline solution, the saturation current density and ideality factor are confirmed to be improved. These results suggest that metal oxides on the sidewalls of CIGS solar cells may act as recombination centers.
Keywords: Cu(InGa)Se_2, Degradation, Shunts

1 INTRODUCTION

The chalcopyrite compound Cu(In,Ga)Se_2 (CIGS) has potential to be used as a semiconductor material for thin-film photovoltaic devices because of its high absorption coefficient and appropriate band gap [1-6], and CIGS solar cells have recently shown record conversion efficiency over 22% [7]. However, there have been few reports investigating the long-term stability of CIGS solar cells compared with Si solar cells, and long-term operation for 20 years more is considered to be essential for decreasing the cost of solar cells. Therefore, investigation of all degradation mechanism to promote CIGS solar cells is extremely important.

Pern et al. investigated the stability of encapsulated CIGS solar cells under accelerated degradation tests [8,9]. Using several kinds of backsheets and moisture barriers, they reported that some CIGS cells exhibited degradation of open-circuit voltage (V_{oc}), fill factor (FF), and conversion efficiency because of increased series resistance (R_s) after damp heat (DH) testing. Lee et al. reported the mechanism for increasing R_s during exposure to damp heat [10,11], i.e., during damp heat testing, water has penetrated into CIGS modules, and Al-doped ZnO (AZO) layers are transformed into $Zn(OH)_2$ at high temperatures, thereby increasing R_s. Yamaguchi et al. and Fjällström et al. reported potential-induced degradation (PID) of CIGS solar cells [12,13]. CIGS modules, such as crystalline Si solar cells, are affected by PID, which is mainly influenced by the enhancement of recombination centers at the pn junctions.

These reports are very important for long-term reliability because these degradations may occur after a relatively long time at a high temperatures. In contrast, degradation of conversion efficiency frequently occurs for unencapsulated CIGS solar cells kept in atmosphere even at room temperature. This degradation mechanism should differ from the models reported in literatures [8-13]. In this study, we investigate the degradation mechanisms of unencapsulated CIGS solar cells with air exposure and DH testing, and discuss the origin of degradation on cell performance.

2 EXPERIMENTAL PROCEDURE

CIGS solar cells are fabricated on soda lime glass (SLG) substrates. Figure 1 shows the schematic layer structure of the CIGS solar cells. A 1.0-μm-thick Mo back contact is deposited by in-line type direct current (DC) sputtering on a SLG substrate. A 2.0-μm-thick CIGS layer is deposited by evaporation using a three-stage process. The growth temperature of the first stage is 350 °C, and that of the second and third stages is 550 °C. The beam equivalent pressure (BEP) ratio of selenium to group III elements (P_{Se} / (P_{Ga} + P_{In})) is set at 7 [14,15]. The atomic ratio of Ga to group III elements is fixed at 0.4, and that of Cu to group III elements is 0.9. After KCN etching of the CIGS layers, a 50-nm-thick CdS layer is formed by chemical bath deposition. A 60-nm-thick intrinsic ZnO layer and a 350-nm-thick n-type AZO layer are deposited by facing-targets radio frequency and DC magnetron sputtering depositions. An aluminum electrode is fabricated on the AZO by evaporator, and the devices are separated by mechanical scribing. No anti-reflection coating is used in this study.

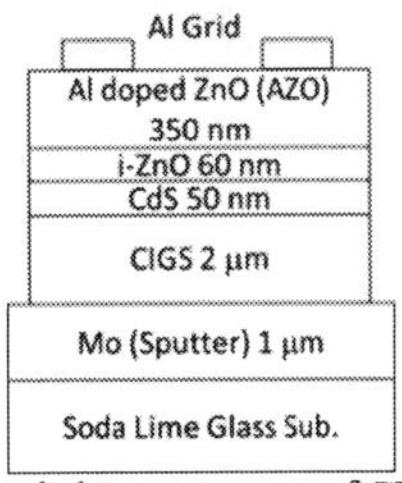

Figure 1: Schematic layer structure of CIGS solar cell.

The photocurrent-voltage (*I-V*) characteristics are measured with the solar radiation of an air-mass (AM) 1.5 spectrum at 100 mW/cm^2 at 25 °C with an active area of 0.514 cm^2. DH tests are performed under conditions at 85 °C and 85% relative humidity for 15 hours. After DH testing, the surface conditions are observed by secondary electron microscopy (SEM), and the surface element compositions are measured by Auger electron spectroscopy (AES).

3 RESULTS AND DISCUSSION

To clarify the degradation mechanism for CIGS solar cells with exposure to air, we first investigate the *I-V* curves for unencapsulated CIGS solar cells under both exposure to air and DH testing. Figure 2 shows semi-logarithmic plots of dark *I-V* curves for the CIGS solar cell exposed to air at room temperature for 90 days. Initially, the conversion efficiency (η) is 18.5% and the R_{sh} is confirmed to be 5.0 kΩcm^2. In contrast, η for the solar cell with air exposure for 90 d is 17.6% and R_{sh} is 1.5 kΩcm^2. These results indicate that low-resistive layers are formed during exposure to air, and concentration of recombination centers should be increased. And, the V_{oc} and the short-circuit current density (J_{sc}) are confirmed to be decreased.

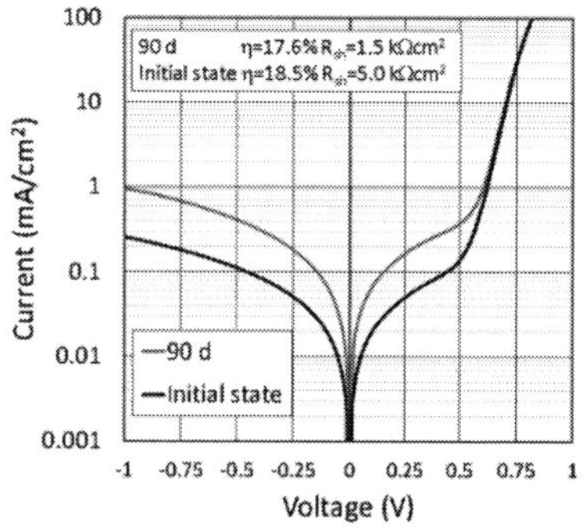

Figure 2: Semilogarithmic plots of dark *I-V* curves for the CIGS solar cells exposed to air at room temperature.

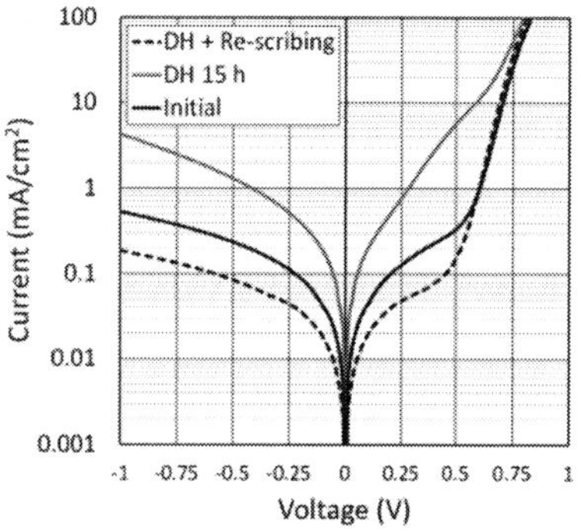

Figure 3: Semilogarithmic plots of dark *I-V* curves for the CIGS solar cells subjected to DH testing for 15h.

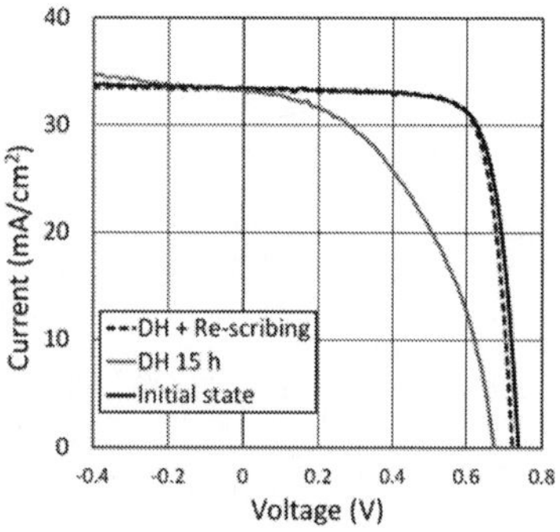

Figure 4: *I-V* curves under solar radiation for CIGS solar cells exposed to DH testing and re-mechanical scribing.

Figure 3 shows semilogarithmic plots of dark *I-V* curves for the CIGS solar cell subjected to DH testing for 15 h, and figure 4 shows *I-V* curves under solar radiation with an AM 1.5 spectrum at 100 mW/cm^2. After DH testing, some of the CIGS solar cells show low R_{sh} and conversion efficiency. In particular, the CIGS solar cells without intrinsic ZnO layers are frequently deteriorated by DH testing for several hours. Intrinsic ZnO layers are reported to prevent leak currents at interface of CdS and AZO [16]. These results indicate that intrinsic ZnO layers should be effective for long-term reliability. Table I shows the solar cell parameters for a CIGS solar cell deteriorated by DH testing. After DH testing, R_{sh} becomes less than 500 Ωcm^2, and η is evaluated to be 10.6%. The V_{oc} and FF are significantly decreased due to the low shunt resistance.

Table I: Parameters of the CIGS solar cells with DH testing and re-mechanical scribing.

	Eff. %	V_{oc} V	J_{sc} mA/cm^2	FF
Initial stage	18.9	0.737	33.4	0.766
DH 15 h	10.6	0.673	33.2	0.473
DH + Re-scribing	18.7	0.721	33.3	0.776

Surface AZO layers are relatively stable under exposure to air and humidity. The series resistances are confirmed to be identical after DH testing. Therefore, the degradation mechanism for unencapsulated CIGS solar cells is different from the model of increasing R_s reported in literatures [8-11]. During air exposure, the affected area should be the side edges of CIGS solar cells because of exposure of pn junction to oxygen and humidity. After removing the side edges of the deteriorated CIGS solar cells by mechanical scribing, the R_{sh} and conversion efficiency are completely recovered, as shown in Figs. 3 and 4. These results show the low-resistive layers should be formed on the side edges of CIGS solar cells during exposure to air and DH testing.

Mechanical scribing is undesirable from a practical viewpoint because it reduces the active area of the CIGS solar cell. Therefore, only the low-resistive layers should be removed by wet chemical etching. Table II shows the solar cell parameters for the CIGS solar cell with DH testing. As with Figs. 3 and 4, R_{sh} are confirmed to be low after DH testing, and V_{oc} and FF are significantly decreased. After the deteriorated CIGS solar cells are etched by 3 wt% NH_3 solution for 30 seconds, the R_{sh} and conversion efficiency are confirmed to recover, as shown in Table II. These results suggest that the low-resistive layers and recombination centers can be removed by NH_3 etching. The R_{sh} value after NH_3 etching is confirmed to be larger than the initial R_{sh} because the alkali metal oxides on the sidewalls should be removed by such etching. In contrast, V_{oc} and FF after NH_3 etching are slightly deteriorated because the NH_3 solution may damage surface AZO layers. Therefore, the AZO should be protected from NH_3 etching.

Table II: Parameters of the CIGS solar cells with DH testing and NH_3 etching.

	Eff. %	V_{oc} V	J_{sc} mA/cm^2	FF
Initial stage	18.9	0.737	33.6	0.762
DH 15 h	14.9	0.719	33.5	0.620
DH + NH_3 etching	18.6	0.729	33.9	0.753

Next, we discuss the low-resistive layers formed after DH testing. Figure 5 shows a bird's eye view of SEM images of the sidewalls initially (a), after DH testing (b), and after NH_3 etching (c). Initially, the solar cell structures (AZO/CIGS/Mo) are clearly observed. In contrast, additional materials are formed on AZO/CIGS interface and Mo electrode after DH testing in Fig. 5(b). After NH_3 etching, all the additional materials are removed; however, the NH_3 solution also etches the surface AZO layers as shown in Fig. (c). Therefore, V_{oc} may be slightly decreased after NH_3 etching, as shown in Table II.

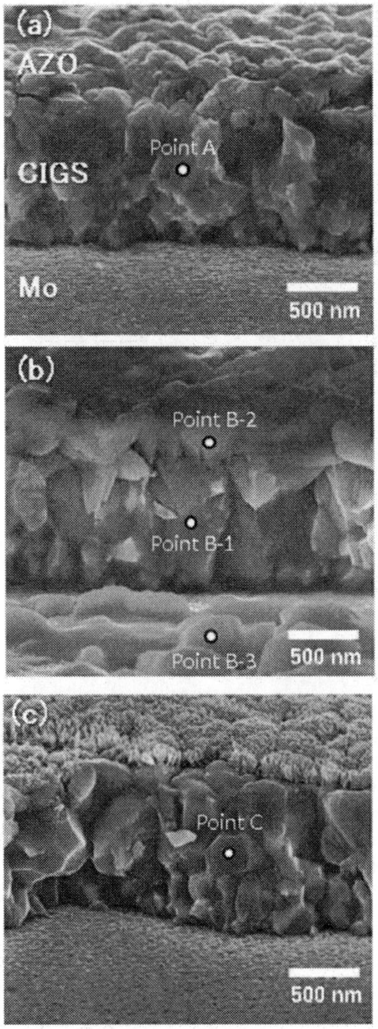

Figure 5: Bird's eye view of SEM images of the sidewalls of the CIGS solar cells.

Table III: Surface elemental composition of the CIGS solar cells evaluated by AES.

Atomic %	Cu	In	Ga	Se	O	Na	Mo	C
Point A	3	4	2	59	5	4	2	21
Point B-1	5	6	3	56	12	3	3	11
Point B-2	-	-	-	-	63	3	24	10
Point B-3	-	-	-	-	64	3	25	9
Point C	4	3	2	70	6	3	2	9

AES measurements are performed for analyzing the additional materials. Table III shows surface elemental compositions at the points shown in Fig. 5. The spatial resolution is 30 nm, and the depth resolution is 5 nm. At point A, Cu, In, Ga, and Se signals are detected, and the layers are confirmed to be CIGS layers. The atomic percent of selenium is overestimated because of the weak signal of selenium photoelectrons. After DH testing, the atomic percent of oxygen is increased at point B-1, and the additional materials are identified to be molybdenum oxides and sodium molybdate at point B-2 and B-3. During DH testing, molybdenum electrodes are oxidized, and molybdenum oxides react with sodium metals from SLG substrates. Sodium molybdate can be easily dissolved in water, and molybdate ions can diffuse to the sidewalls of CIGS solar cells. Molybdenum oxides are easily dissolved in alkaline solutions. Therefore, after NH_3 etching, all the additional materials are removed, and the atomic percent of oxygen is decreased at point C.

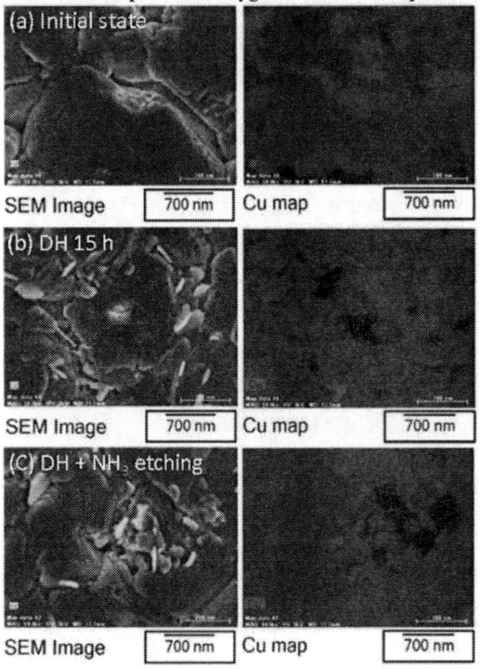

Figure 6: SEM images and cupper mapping for CIGS surfaces evaluated by EDX.

Table IV: Surface elemental composition of CIGS solar cells evaluated by XPS.

Atomic %	Cu	In	Ga	Se	O	Na	Cu/(In+Ga)
Initial state	8	15	10	39	18	9	0.33
DH 15 h	8	12	7	21	40	12	0.42
DH + NH_3	13	16	10	25	35	-	0.5

In order to precisely investigate the surface element compositions of CIGS surfaces after DH testing, X-ray photoelectron spectroscopy (XPS) and elemental mapping evaluated by energy dispersive X-ray spectroscopy (EDX) are performed on the CIGS layers on Mo/SLG substrates. Figure 6 shows SEM images and Cu mapping for CIGS surfaces initially (a), after DH testing (b), and after NH_3 etching (c). Table IV shows the surface elemental compositions on CIGS layers evaluated by XPS. After exposure to air, the surfaces of the CIGS layers are oxidized, and the presence of oxygen and sodium are confirmed even at the initial state. The atomic ratio of Cu to group III elements is evaluated to be 0.33 initially, indicating the formation of $Cu(In,Ga)_3Se_5$ layers on the surface [17]. However, figure 6 shows Cu rich area appears near grain boundary. Therefore, Cu compositions on the CIGS surfaces should be inhomogenious. After DH testing, the atomic percents of oxygen and sodium are increased, and additional flat materials appear on the Cu rich area in figure 6(b). Cu oxides are relatively stable with NH_3 solution compared with In and Ga oxides. Therefore, after NH_3 etching, the ratio of Cu to group III elements is increased.

Table V: Parameters of the CIGS solar cells with alkaline solution etching.

	Eff. %	R_{sh} Ωcm^2	J_0 mA/cm^2	n
Initial state	18.5	5000	1.7×10^{-8}	1.36
Air for 90d	17.6	1500	1.2×10^{-8}	1.33
DH + NH_3 etching	18.5	10000	6.5×10^{-9}	1.30

Table V shows the parameters of the CIGS solar cell with exposure to air and alkaline solution etching. After exposure to air for 90 d, J_{sc} and V_{oc} are decreased because of recombination centers on the sidewalls of the CIGS solar cells. In order to remove the low-resistive layers while protecting AZO layers, tetramethylammonium hydroxide $(CH_3)_4N(OH)$ solution with a pH 12 is used for etching. Thereafter, the shunt resistance and conversion efficiency are completely recovered, and the saturation current density J_0 and ideality factor n are confirmed to be improved. These results suggest that the metal oxides on the sidewalls of CIGS solar cells may act as recombination centers, and the saturation current density and ideality factor could be improved by means of etching the oxides with weak alkaline solutions.

4 CONCLUSIONS

The mechanism behind the degradation of unencapsulated CIGS solar cells induced by air exposure has been investigated. After damp heat testing, some CIGS solar cells show low shunt resistance and conversion efficiency. In particular, the CIGS solar cells without intrinsic ZnO layers are frequently deteriorated by damp heat testing after several hours. After removing the side edges of the deteriorated CIGS solar cells by mechanical scribing, the conversion efficiency is completely recovered. In addition, alkaline solution etching is confirmed to be an effective way of removing the low-resistive layers. The low-resistive layers on the sidewalls are identified to be molybdenum oxides and sodium molybdate by AES. Molybdate ions can diffuse from Mo electrodes to the sidewalls during air exposure. Na, In, and Ga oxides on the sidewalls may act as recombination centers. Molybdenum and sodium oxides could be etched by alkaline solutions, and the saturation current density and ideality factor are confirmed to be improved after etching the oxides on the sidewalls.

ACKNOWLEDGEMENTS

The authors would like to thank Dr. S. W. Choi, H. Takahashi, M. Iioka, H. Higuchi, and A. Kurokawa for help with experiments. This work is partly supported by the New Energy and Industrial Technology Development Organization (NEDO) under the ministry of Economy, Trade and Industry (METI), and Grants-in-Aid for Scientific Research (C) (25420298) from the Japan Society for the Promotion of Science (JSPS).

REFERENCES

[1] S. Wagner, J.L. Shay, P. Migliorato, and H.M. Kasper, Appl. Phys. Lett. 25 (1974) 434.
[2] R.A. Mickelsen and W.S. Chen, Appl. Phys. Lett. 36 (1980) 371.
[3] S. Niki, M. Contreras, I. Repins, M. Powalla, K. Kushiya, S. Ishizuka, and K. Matubara, Prog. Photovoltaics 18 (2010) 453.
[4] I. Repins, M.A. Contreras, B. Egaas, C. DeHart, J. Scharf, C.L. Perkins, B. To, and R. Noufi, Prog. Photovoltaics 16 (2008) 235.
[5] T. Nakada, D. Iga, H. Ohbo, and A. Kunioka, Jpn. J. Appl. Phys. 36 (1997) 732.
[6] T. Minemoto, T. Matsui, H. Takakura, Y. Hamakawa, T. Negami, Y. Hashimoto, T. Uenoyama, and M. Kitagawa, Sol. Energy Mater. Sol. Cells 67 (2001) 83.
[7] Solar Frontier Press Release 8 December 2015: http://www.solar-frontier.com/eng/news/2015/C051171.html
[8] F.J. Pern, S.H. Glick, R. Sundaramoorthy, B. To, X. Li, C. DeHart, S. Glynn, T. Gennett, R. Noufi, and T. Gessert, Proc. 35th IEEE PVSC (2010) p. 001166.
[9] F.J. Pern, L. Mansfield, C. DeHart, S.H. Glick, F. Yan, and R. Noufi, Proc. 37th IEEE PVSC (2011) p. 002798.
[10] D. Lee, O. Kwon, J. Song, C. Park, K. Park, S. Nam, and Y. Kim, Sol. Energy Mater. Sol. Cells 105 (2012) 15.
[11] D. Lee, W. Cho, C. Jang, J. Song, C. Park, K. Park, J. Ryu, H. Lee, and Y. Kim, Curr. Appl. Phys. 15 (2015) 285.
[12] S. Yamaguchi, S. Jonai, K. Hara, H. Komaki, Y. Shimizu-Kamikawa, H, Shibata, S. Niki, Y. Kawakami, and A. Masuda, Jpn. J. Appl. Phys. 54 (2015) 08KC13.
[13] V. Fjällström, P.M.P. Salomé, A. Hultqvist, M. Edoff, T. Jarmar, B.G. Aitken, K. Zhang, K. Fuller, and C.K. Willeams, IEEE J. Photovol. 3 (2013) 1090.
[14] S. Ishizuka, A. Yamada, H. Shibata, P. Fons, and S. Niki, Thin Solid Films 519 (2011) 7216.
[15] S. Ishizuka, A. Yamada, P. Fons, and S. Niki, Prog. Photovoltaics 21 (2013) 544.
[16] S. Ishizuka, K. Sakurai, A. Yamada, K. Matsubara, P. Fons, K. Iwata, S. Nakamura, Y. Kimura, T. Baba, H. Nakanishi, T. Kojima, and S. Niki, Sol. Energy Mater. Sol. Cells 87 (2005) 541.
[17] Y. Hashimoto, N. Kohara, T. Negami, M. Nishitani, T. Wada, Jpn. J. Appl. Phys. 35 (1996) 4760.

EFFECTS OF STACKING SEQUENCES IN THE FORMATION OF CZTS THIN FILM USING ELECTRON BEAM EVAPORATION

Kannan P.K.*, Sushmita Chaudhari, Suhash R. Dey
Department of Materials Science and Metallurgical Engineering, Indian Institute of Technology Hyderabad, Kandi, Sangareddy-502285, Telangana, INDIA
E-mail: kannan.pkk@gmail.com, Telephone: +91 (0)40 23016096, Fax: +91 (0)40 23016032

ABSTRACT: This study investigated the influence of stacking sequences in the formation of deleterious secondary phases which has degrading effects on the performance of Cu_2ZnSnS_4 (CZTS) thin film solar cells. Various stacking of absorber layer precursors for e.g. Zn/Sn/Cu, Sn/Zn/Cu, Cu/Zn/Sn/Cu and Cu/Sn/Zn/Cu are prepared on glass substrates without heating using electron beam evaporation of elemental metallic precursors. Interdiffusion of metals is confirmed from the XRD results. Further, the precursor films are sulfurized at 500°C for 1 hr in the presence of elemental sulfur powder having N_2 as the carrier gas. The effect of ramping rates on the formation of secondary phases are investigated using XRD and Raman spectroscopy. It is found that precursor stacking sequence of Sn/Zn/Cu sulfurized at a rate of 1°C/min and Cu/Sn/Zn/Cu at a sulfurization ramping rate of 1°C/min are the optimum condition for the formation of phase pure CZTS thin films.
Keywords: CZTS, Thin Film Solar cells, Electron Beam Evaporation, Ramping rate

1 INTRODUCTION

Cu_2ZnSnS_4 (CZTS) is widely viewed as a possible replacement for CIGS thin film solar cells, due to non-toxicity and relative abundance of its constituents [1]. CZTS is a direct band gap semiconductor with a high absorption coefficient (>10^4 cm^{-1}) and band gap identical of 1.5 eV which is very close to the ideal band gap of 1.45 eV according to the Schockley-Queisser limit [2]. Unfortunately, formation of secondary phases such as Cu_2SnS_3, Cu_xS, ZnS and SnS during the growth of the CZTS absorber layer makes the preparation of phase pure CZTS film very difficult [3].

Secondary phases formed in the CZTS absorber layer affects the band gap and electrical conductivity thus limiting the open circuit voltage of the CZTS solar cell. Determining the means to avoid the formation of secondary phases is crucial to enhance the electrical properties and quality of the CZTS thin films. Controlling the composition of the metallic precursors and phase transformation during sulfurization are the major ways to achieve phase pure CZTS. Hence, in this study, the effects of precursor stacking sequences and sulfurization ramping rate on the formation of phase pure CZTS has been investigated.

2 EXPERIMENTAL DETAILS

CZTS thin films are prepared by sulfurization of multilayer precursors, made of Zn, Sn and Cu deposited by e-beam evaporation on soda lime glass substrates. The CZT metallic precursor layer is deposited from the sequential evaporation of Zn, Sn and Cu layers using electron beam evaporation with an acceleration voltage of 3.3kV and beam current of 78 mA, 41 mA and 100 mA respectively, at a working pressure of 6 x 10^{-6} mbar with a substrate rotation of 50 rpm. The thickness of the deposited Zn, Sn and Cu layers are 160 nm, 230 nm and 180 nm respectively. Prior to the deposition, glass substrates are cleaned with acetone, ethanol, methanol and deionized water respectively in an ultrasonicator bath for 10 minutes each. Four samples with stacking sequences Zn/Sn/Cu (ZTC), Sn/Zn/Cu (TZC), Cu (90 nm)/Zn/Sn/Cu (90 nm) (CZTC) and Cu (90 nm)/Sn/Zn/Cu (90 nm) (CTZC) are prepared in this study. The major reason behind the choice of Cu as the last layer is to prevent the loss of Sn from the stack during sulfurization and oxidation of Zn.

The precursor film is sulfurized at 500°C for 1 hr in the presence of elemental sulfur powder using N_2 as carrier gas in a tubular furnace at various ramping rates of 1, 5 and 10°C/min respectively. The effect of stacking sequence and ramping rate on the formation of CZTS is investigated using X-ray diffraction (XRD Bruker, Discover D8) and Raman spectroscopy (Bruker SENTERRA). The morphology and the composition of the films are obtained using Field Emission Gun-Scanning Electron Microscopy (Carl Zeiss Germany, Model Supra-40) equipped with Energy Dispersive X-ray Spectroscopy (EDS) (Oxford Instrument, UK). The band gap of the films are calculated using UV–Vis spectroscopy (Perkin Elmer UV-Vis spectrophotometer with integrating sphere).

3 RESULTS AND DISCUSSION

3.1 Effect of precursor stacking sequence

The XRD of the as deposited films of various stacking sequence is shown in fig. 1. Without any further treatment at room temperature itself, formation of Cu_6Sn_5 (JCPDF card No. 00-045-1488), Cu_5Zn_8 (JCPDF card No. 03-065-6566) and $Sn_{0.85}Zn_{0.15}$ (JCPDF card No. 01-072-3273) phases confirm the inter-diffusion of the metallic layers. XRD results indicate that peak intensities of Cu_6Sn_5 and Cu_5Zn_8 phases in stacking sequence ZTC and TZC films are more when compared to CZTC and CTZC which indicates that films deposited with Cu at both the ends have comparatively less precursor alloy formation. From the XRD of ZTC film, it can be inferred that formation of Cu_5Zn_8 is very low and Cu_6Sn_5 is high which can be due to the poor diffusion of Zn to the Cu layer through the Sn layer. Similarly, formation of Cu_6Sn_5 is very low and Cu_5Zn_8 is high in case of TZC indicating poor diffusion of Sn to the Cu layer through the Zn layer. The same phenomena is not observed in case of CZTC and CTZC films where the formation of Cu_6Sn_5 and Cu_5Zn_8 is in equal amount due to the easy diffusion of elements with the adjacent Cu layers itself. From the XRD results of as deposited metallic stacks, it can be concluded that the diffusion and phase formation of elements among the adjacent layers is very high.

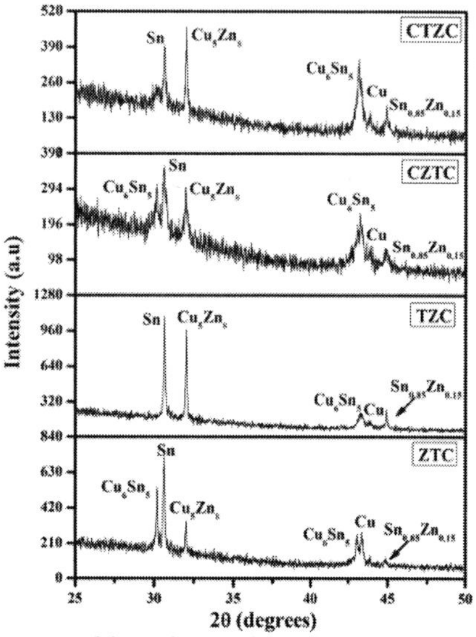

Fig. 1: XRD of the as deposited metallic precursor film for various stacking orders

3.2 Effect of sulfurization ramping rate

For the formation of CZTS thin films, the precursor films are sulfurized at 500°C for 1 hr in the presence of elemental sulfur powder with N_2 as a carrier gas at various ramping rates. For ZTC stack, the effect of ramping rates on the formation of CZTS is studied using XRD as shown in fig. 2.

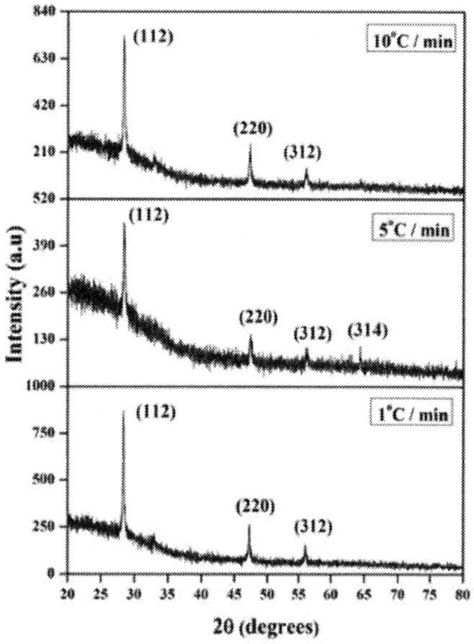

Fig. 2: XRD of CZTS films prepared at various ramping rate with ZTC stack

From the XRD results, formation of CZTS is confirmed (JCPDF Card No. 00-026-0575). It is found that film deposited at a lower ramping rate of 1°C/min is showing maximum intensity when compared to others. Since, the 2θ values of the secondary phases such as Cu_2SnS_3 and ZnS lies in the same values as of CZTS[4], Raman spectroscopy of the films is needed to confirm the phase purity of the films.

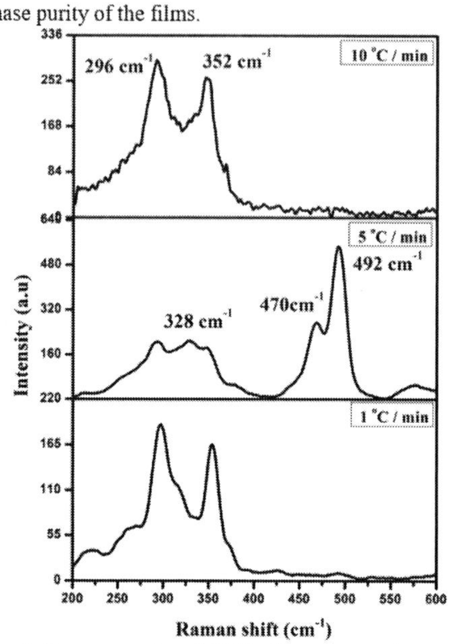

Fig. 3: Raman spectra of CZTS films prepared at various ramping rate with ZTC stack.

Raman spectra of the CZTS films with ZTC stack is given in fig. 3. Formation of tetragonal Cu_2SnS_3 (296 and 352 cm^{-1}) is observed in all the films clears that the prepared film is not having CZTS. Also for the films sulfurized at a rate of 5°C/min, presence of $Cu_{2-x}S$ is also observed (470 and 492 cm^{-1}). Thus, Raman results clearly suggests that ZTC is not the optimum stacking sequence for the formation of CZTS films.

The XRD of the CZTS films prepared with TZC stack is shown in fig. 3. No significant change in the crystallinity of the films is observed.

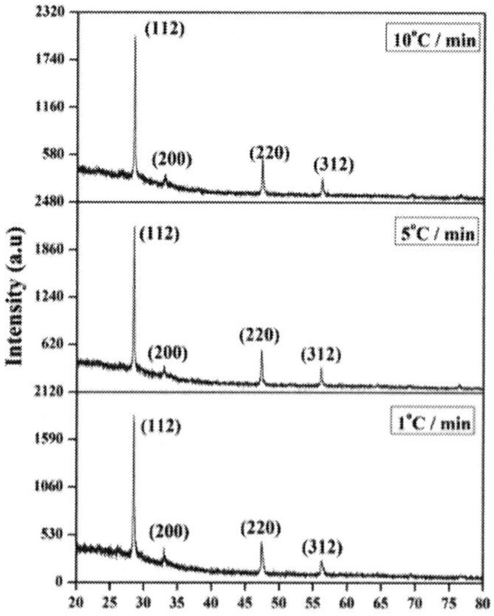

Fig. 4: XRD of CZTS films prepared at various ramping rate with TZC stack

But from the Raman measurements, it is clear that film sulfurized at 1°C/min is showing maximum intensity confirming that films sulfurized at lower ramping rate is more crystalline when compared to others as shown in fig. 5.

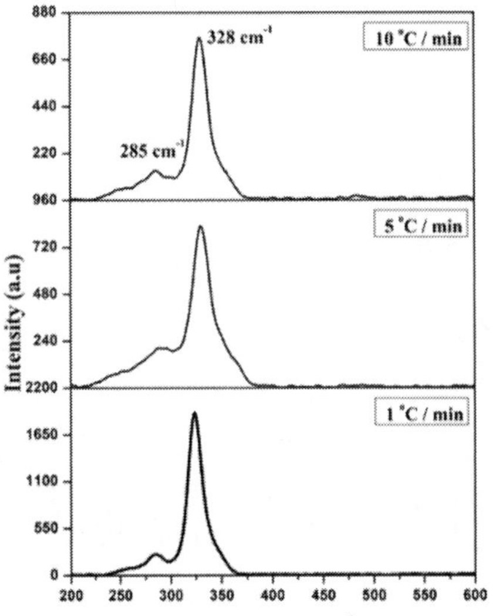

Fig. 5: Raman spectra of CZTS films prepared at various ramping rate with TZC stack

In case of CZTS films made using CZTC stack also, no significant change is observed in the XRD for films annealed in various ramping rates which is shown in fig. 6.

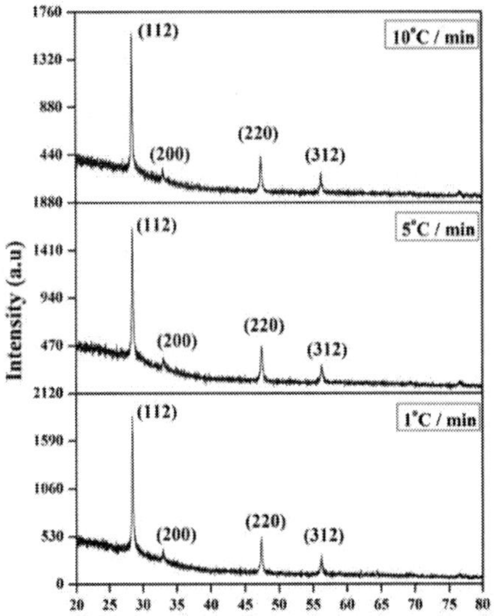

Fig. 6: XRD of CZTS films prepared at various ramping rate with CZTC stack

No significant change is observed from the intensity of the Raman spectra of CZTS films prepared with CZTC stack is shown in fig. 7. These results suggests that the ramping rate is not having a significant effect on the crystallinity of CZTS films prepared using CTZC stack.

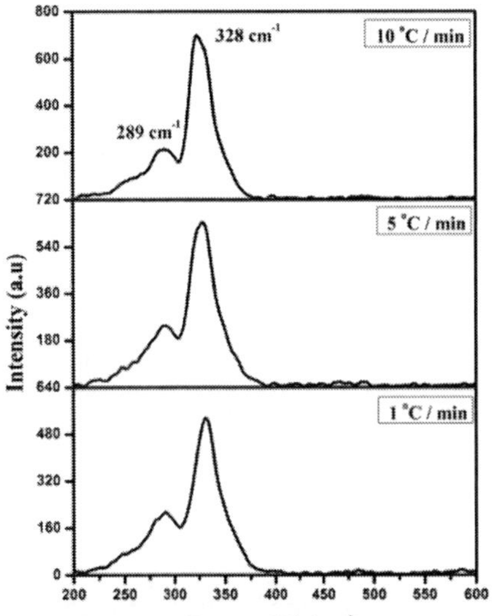

Fig. 7: Raman spectra of CZTS films prepared at various ramping rate with CZTC stack

The XRD of CZTS films made using CTZC stack is shown in fig. 8. From this, it can be clearly seen that films deposited with a low ramping rate is having maximum intensity and hence more crystalline.

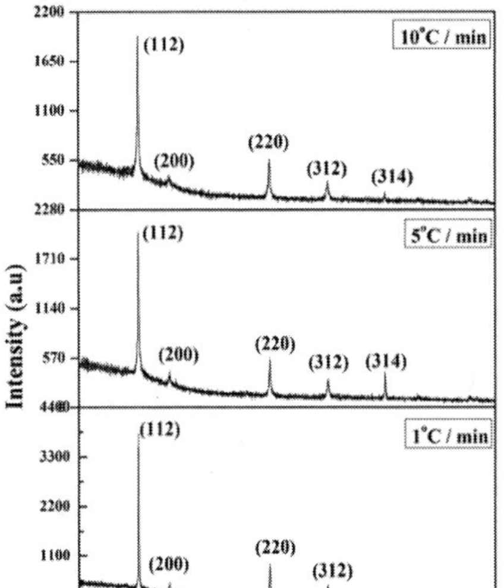

Fig. 8: XRD of CZTS films prepared at various ramping rate with CTZC stack

From the Raman spectroscopy results of CZTS films prepared using CTZC stack, it is observed that films prepared at lower ramping rate is having $Cu_{2-x}S$ impurity phase (470 cm^{-1}). The intensity of the CZTS is almost similar for all the cases. Thus for CZTS films prepared using CZTC stack, higher ramping rate of 10°C/min yields maximum crystalline CZTS thin films.

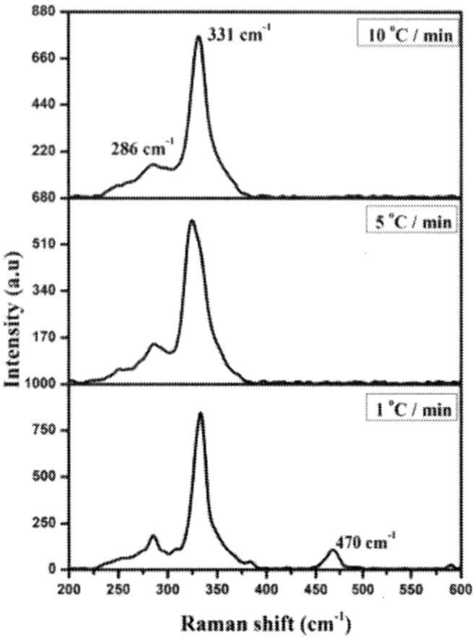

Fig. 9: Raman spectra of CZTS films prepared at various ramping rate with CTZC stack

Thus, from the XRD and Raman results, it can be concluded that for the films deposited TZC stack, a lower ramping rate of 1°C/min yields good crystalline CZTS thin films and for the films with CZTC and CTZC stack, a higher ramping rate of 10°C/min yields better phase pure crystalline CZTS thin films.

SEM micrograph of the as deposited metallic stack films is shown in fig. 10. Films deposited with ZTC stacking sequence is having a smoother morphology. Also formation of granular structures is observed in other stacking sequences and it can be seen that CTZC precursor film is coarse granular and CZTC film is fine granular and a dispersed structure of coarse and fine grains is observed in TZC films.

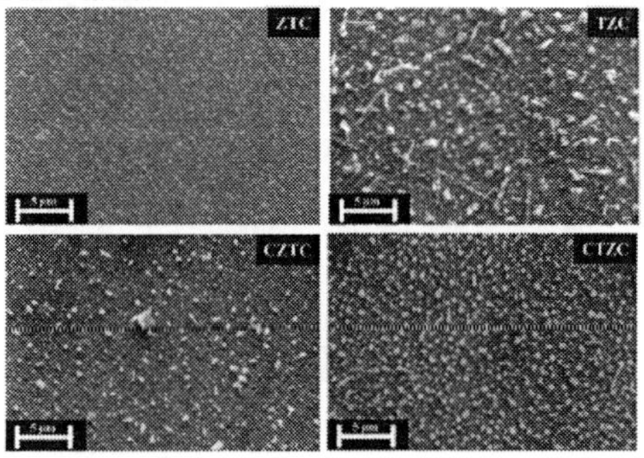

Fig. 10: SEM micrograph of the as deposited metallic precursor film for various stacking orders

SEM images of the CZTS films prepared using sulfurization of different stacked precursor film at different ramping rates is provided in fig. 11. It can be seen that irrespective of the ramping rate, the CZTS films prepared using CZTC and CTZC stack is having fine granular structure when compared to that of ZTC and TZC stacks. The surface of the film is not uniform and therefore it needs further optimization of annealing conditions for the formation of uniform, smooth CZTS films.

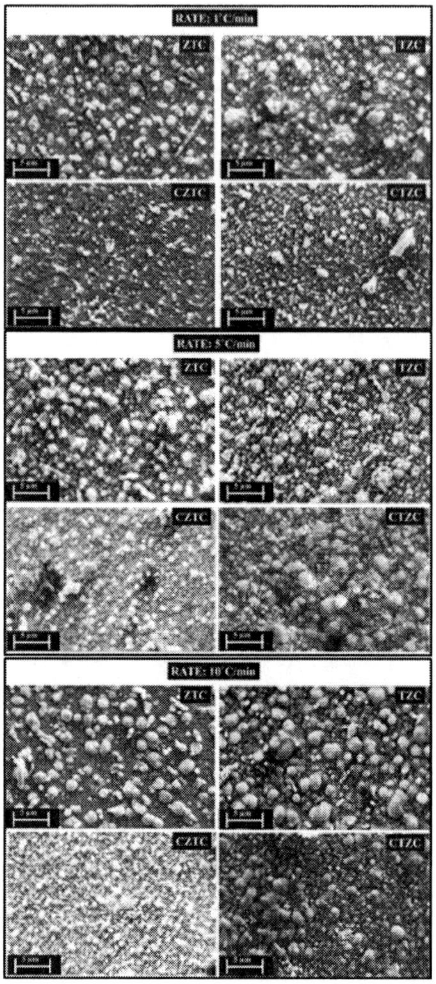

Fig. 11: SEM micrograph of the CZTS films prepared at various ramping rate with various stacking sequences

STACK	Cu/Zn+Sn			Zn/Sn			S/metal		
	1°C/ min	5°C/ min	10°C /min	1°C/ min	5°C/ min	10°C /min	1°C/ min	5°C/ min	10°C /min
ZTC	2.07	1.98	1.96	0.11	0.31	0.07	0.73	0.79	0.71
TZC	1.04	1.36	1.23	1.15	0.92	0.97	0.78	0.78	0.81
CZTC	1.62	1.54	1.39	0.56	0.55	0.64	0.81	0.81	0.81
CTZC	1.25	1.25	1.26	1.15	1.21	0.87	0.78	0.81	0.80

Table I: EDS results of the CZTS films prepared at various ramping rate with various stacking sequences

The EDS results of the CZTS films are given in Table I. The results clearly reveal that loss of Zn from the film during sulfurization in case of CZTS films prepared with ZTC and CZTC stacks is more when compared to CZTS films prepared with TZC and CTZC stacks. In both the CTZC and TZC stacked film, the loss of Zn is higher from the film as the sulfurization ramping rate is increased from 1 to 10°C/min. It is inferred that for stacking order of TZC a sulfurization ramping rate of 1°C/min and for CTZC, a rate of 10°C/min is more suitable for the fabrication of phase pure Zn rich CZTS thin film.

From the UV-Vis absorption spectra, the bandgap of the best CZTS films prepared by the sulfurization of TZC and CTZC stack are calculated using Tauc plot which is shown in fig. 12.

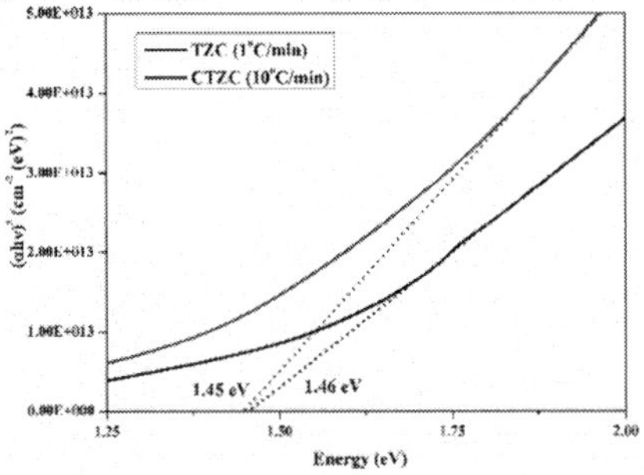

Fig. 12: Tauc plot of the best CZTS films prepared using TZC and CTZC stack

From the above results, it can be seen that the bandgap of the CZTS films prepared using TZC stack at a ramping rate of 1°C/min and CTZC stack at a ramping rate of 1°C/min is estimated as 1.46 eV and 1. 45 eV which is identical to the optimal bandgap required for maximum efficiency solar absorber layer for single junction according to the SQ limit [5].

Thus, from the above study it is clearly evident that precursor stacking sequence and sulfurization ramping rate have a significant effect on the formation of phase pure CZTS films. It is found from this study that TZC and CTZC stacked precursor films, sulfurized at 500°C for 1 hour in a ramping rate of 1°C/min and 10°C/min in elemental S powder environment using N_2 as a carrier gas is the optimum condition for the fabrication of highly crystalline and phase pure CZTS thin films.

4 CONCLUSIONS

The effect of precursor stacking sequence and sulfurization ramping rate on the formation of CZTS films is investigated and it is concluded from this study that Sn/Zn/Cu films sulfurized at a ramping rate of 1°C/min and Cu/Sn/Zn/Cu films sulfurized at ramping rate of 10°C/min lead to formation of phase pure, Zn rich CZTS films with a bandgap of 1.46 eV and 1.45 eV respectively.

REFERENCE

1. H. Yoo, J. Kim, Solar Energy Materials & Solar Cells 95 (2011) 239.
2. Hironori Katagiri, Nobuyuki Sasaguchi, Shima Hando, Suguru Hoshino, Jiro Ohashi, Takaharu Yokota, Solar Energy Materials and Solar Cells 49 (1997) 407.
3. Sushmita Chaudhari, Srinivas Palli, Kannan P.K., Suhash R. Dey, Thin Solid Films 600 (2016) 169.
4. P. A. Fernandes, P.M.P. Salom'e, A.F. da Cunha, Journal of Alloys & Compound 509 (2011) 7600.
5. Lee, Myoung-Bok, Song,Kyu-Ho, Park, Kwang-Mook; Jung, Ji-Hee; Bae, So-Ik, Journal of Nanoscience and Nanotechnology 12 (2012) 5691.

EFFECTS OF ANNEALING ATMOSPHERE AND STABILIZING AGENT ON THE FORMATION OF CZTS FILM USING A SIMPLE DIP COATING TECHNIQUE

Sushmita Chaudhari*, Kannan P.K., Suhash R. Dey
Department of Materials Science and Metallurgical Engineering, Indian Institute of Technology Hyderabad
Kandi, Sangareddy-502285, Telangana, INDIA
Email: sushmita_uit@yahoo.co.in, Telephone: +91 (0)40 23016096, Fax: +91 (0)40 23016032

ABSTRACT: Fabrication of CZTS film is proposed through non-hydrazine based dip coating technique. To obtain uniform, phase pure CZTS film, effects of annealing atmosphere and stabilizing agent Triethanolamine (TEA) on synthesis of CZTS thin film are investigated. Phase formation of CZTS films as well as mechanism behind reduction of voids and porosity in the film in presence of TEA is investigated. The complex formation of metal chlorides with methanol and the decomposition of Thiourea is discussed. X-Ray diffraction pattern and Raman spectra revealed the formation of kesterite structure of CZTS. SEM micrograph of CZTS film using TEA confirms the formation of less porous and more compact film. No significant change has been observed in morphology and composition of films annealed in N_2 atmosphere and in presence of elemental sulfur with N_2 as carrier gas which inferred that a simple nitrogen annealing itself is sufficient instead of a complex and harmful sulfurization process for the synthesis of uniform CZTS thin film.
Keywords: Cu_2ZnSnS_4 (CZTS), Thin film solar cell, dip coating, Triethanolamine, Kesterite structure.

1 INTRODUCTION

Due to its high absorption coefficient and ecofriendly nature of CZTS, several solar photovoltaics researchers are focusing towards CZTS as an absorber layer in thin film solar cells. CZTS can be deposited through vacuum and non-vacuum based techniques. Among all techniques, dip coating is technically and economically viable technique. In dip coating method all cations and anions are present in the precursor solution and substrate has to be dipped into the solution followed by heat treatment to form CZTS [1]. Through chemical route method, Wei Wang et al. achieved 12.6% efficiency using hydrazine hydrate [2]. The major disadvantage of hydrazine hydrate is that it is toxic and explosive in nature and hence, an alternative route is required. Generally, CZTS films are synthesized by dipping the substrate into the precursor solution followed by heat treatment at elevated temperature. Heat treatment can be done in various atmosphere such as N_2, Ar, H_2S or in presence of elemental powder of S with an inert carrier gas. In the synthesis of CZTS absorber layer, sulfurization is a tedious and environmentally hazardous process and hence hazardless approach is needed to avoid this step.

The purpose of this work is to develop a cost effective, simple, non-vacuum based method to synthesis CZTS films through a non-hydrazine solution based dip coating technique followed by heat treatment in inert atmosphere. In order to achieve this, effects of various annealing atmosphere and stabilizing agent Triethanolamine (TEA) on the structural and morphological properties of CZTS are investigated.

2 EXPERIMENTAL DETAILS

For the synthesis of Cu_2ZnSnS_4 (CZTS), first glass substrates are cleaned with acetone, ethanol, methanol and deionized water respectively in ultrasonicator bath for 20 min each. Precursor solution is prepared by dissolving 0.1M $CuCl_2$, 0.05M $SnCl_2$, 0.05M $ZnCl_2$, 0.5M thiourea and 2mM triethanolamine (TEA) using methanol as a solvent and then cleaned glass substrates are dipped into precursor solution for 5 minutes. As dipped films are annealed in various atmosphere such as Air, Argon, Nitrogen and in presence of elemental sulfur powder using N_2 as carrier gas. The effects of using stabilizing agent (triethanolamine) and effects of various annealing atmosphere on the formation of CZTS film are investigated. X-Ray diffraction (XRD) (Bruker Discover D8) with CuKα radiation (λ = 1.5406 Å) from 2θ angle 20° to 80° and Field Emission Gun Scanning Electron Microscopy (FEG SEM) (Carl Zeiss Germany, Model Supra-40) equipped with Energy Dispersive X-ray Spectroscopy (EDS) (Oxford Instrument, UK) at an acceleration voltage of 10 KV, are used to investigate structural characteristics, surface morphology and chemical composition of synthesized films. Structural information and band gap calculation of CZTS films are assessed respectively by Raman Spectrometer (Bruker SENTERRA) with the incident laser light wavelength of 532 nm and UV-Vis Spectrophotometer (Perkin Elmer UV-Vis spectrophotometer with integrating sphere) in reflectance and transmittance mode from 600 nm to 1000 nm wavelength range at room temperature.

3 RESULTS AND DISCUSSION

3.1 Effects of annealing atmosphere without TEA

Structural analysis of CZTS films, prepared without using stabilizing agent (TEA) and annealed in different atmosphere (air, argon, nitrogen and in presence of elemental sulfur using nitrogen as carrier gas) at 300°C for 1 hr with 10°C/min ramping rate is studied using X-ray diffractometer and shown in Fig. 1. XRD peaks of 2θ at 28.6°, 47.4° and 56.1° correspond to (112), (220) and (312) planes of CZTS [JCPDS: 26-0575] respectively and confirm the formation of kesterite structure without the presence of any other secondary phases in all the films annealed at different atmosphere. From Fig. 1, it is noticed that there is a variation in the intensity of the peaks for films annealed at different atmosphere and maximum intensity is achieved for the film annealed in nitrogen atmosphere which suggest that most optimum atmosphere for the formation of CZTS is nitrogen atmosphere. The trouble with CZTS is that, all probable secondary phases such as ZnS and CTS have quite similar unit cells and hence their 2θ values also lies in the same place. For further confirmation of phase pure CZTS formation, all the films annealed at different atmosphere is characterized by Raman spectroscopy as shown in Fig. 2.

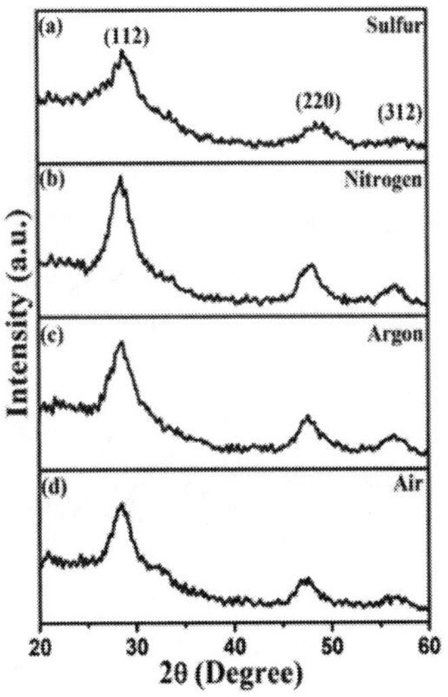

Fig. 1: XRD of CZTS films annealed at different atmosphere without TEA.

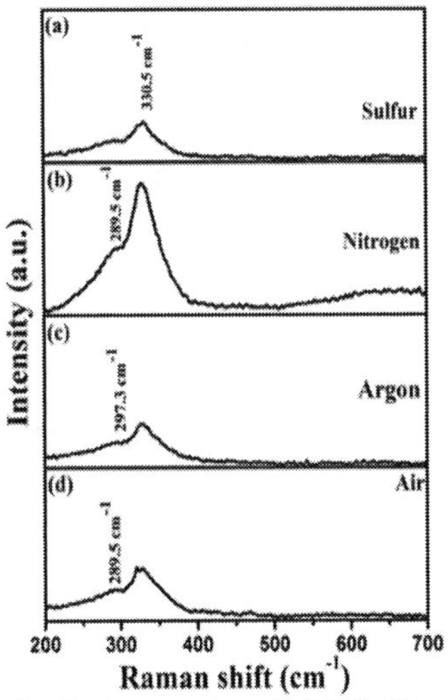

Fig. 2: Raman spectra of CZTS films annealed at different atmosphere without TEA.

Presence of broad peak around 330.5 cm^{-1} indicate the formation of CZTS phase in all the films [3]. Together with peak at 330.5 cm^{-1} one more extra peak at 297.3 cm^{-1} in the film annealed in argon atmosphere shows the presence of secondary phase CTS [4] which is shifted to 289.5 cm^{-1} (corresponding to CZTS) [5] in case of the film annealed in air and nitrogen atmosphere and confirm the formation of phase pure CZTS. In comparison of the film annealed in air atmosphere, film in nitrogen atmosphere have more intense peak at 330.5 cm^{-1} and confirms that nitrogen is optimal atmosphere for the formation of phase pure CZTS.

EDAX result of CZTS films annealed at different atmosphere is shown in Fig. 3. The EDS results disclosed that most appropriate atmosphere for the formation of CZTS is nitrogen atmosphere where atomic percentage of Cu, Zn, Sn and S are 27 at.%, 15 at.%, 13 at.% and 45 at.% (near stoichiometric) respectively. Still as per SEM micrograph as shown in Fig. 4, film annealed at nitrogen atmosphere is not compact and having voids which is not desirable since presence of voids will reduce the efficiency of the photovoltaic cells as it can acts as recombination center. Hence, in order to reduce the voids and ensure uniform conformal coating of the film over the substrate, Triethanolamine is used as a stabilizing agent.

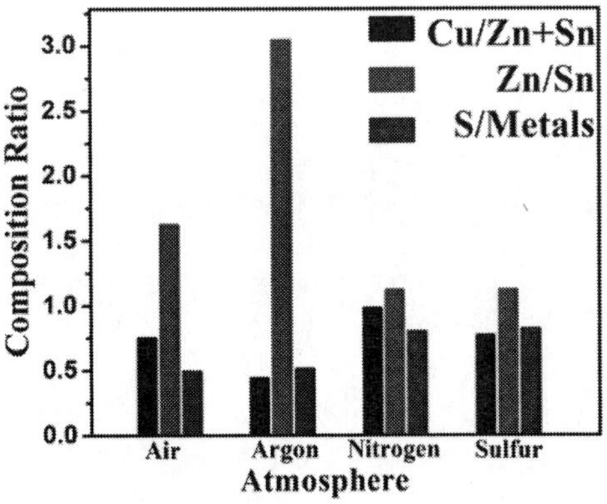

Fig. 3: EDS of CZTS films annealed at different atmosphere.

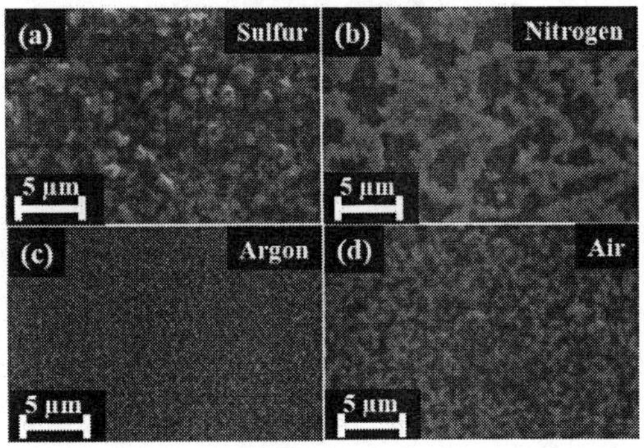

Fig. 4: SEM micrograph of CZTS films annealed at different atmosphere without TEA.

3.2 Effects of stabilizing agent (Triethanolamine):

To achieve voids free film, Triethanolamine (TEA) is added into the precursor solution as stabilizing agent. Addition of TEA enhances the solubility of metal salts into the methanol due to the presence of both hydrophilic and hydrophobic groups in TEA, which can be endorsed to the strong coordination of -OH and $-NH_2$ groups with metal ions [1]. And also, uniform spreading distribution of solution is noticed probably due to the high viscosity of TEA. Due to the alkaline nature of TEA, it favors the decomposition of Thiourea too into S^{2-} ions.

$(NH_2)_2CS_{(s)} \rightarrow NH_4SCN_{(s)} \rightarrow CS_{2(g)} + H_2S_{(g)} + NH_{3(g)}$ [6,7]

Thiourea has C=S and $-NH_2$ functional group which makes strong coordination with metal cations. When combining tin (II), zinc (II) and copper (II) precursors altogether in methanol, Sn^{2+} is oxidized to Sn^{4+} and Cu^{2+} is reduced to Cu^{+} due to the reaction between $Cu^{2+}$and Sn^{2+}. When mixing thiourea into it, Cu^{+}, Sn^{4+} and Zn^{2+} can coordinate with thiourea and form precursor complex. Precursor complex is represented here as $[Cu((NH_2)_2CS)_n]^{+}$, $[Sn((NH_2)_2CS)_n]^{4+}$, $[Zn((NH_2)_2CS)_n]^{2+}$ respectively [1].

The possible formation mechanism of methanol can be formulated as follows [1]:

$$Cu^{2+} + Sn^{2+} \xrightarrow{Redox} Cu^{+} + Sn^{4+}$$

$$Cu^{+} + n[(NH_2)_2CS] \xrightarrow{Coordination} [Cu((NH_2)_2CS)_n]^{+}$$

$$Sn^{4+} + n[(NH_2)_2CS] \xrightarrow{Coordination} [Sn((NH_2)_2CS)_n]^{4+}$$

$$Zn^{2+} + n[(NH_2)_2CS] \xrightarrow{Coordination} [Zn((NH_2)_2CS)_n]^{2+}$$

$$[Cu((NH_2)_2CS)_n]^{+} + [Sn((NH_2)_2CS)_n]^{4+} + [Zn((NH_2)_2CS)_n]^{2+} \rightarrow [CuZnSn((NH_2)_2CS)_n]^{m+}$$

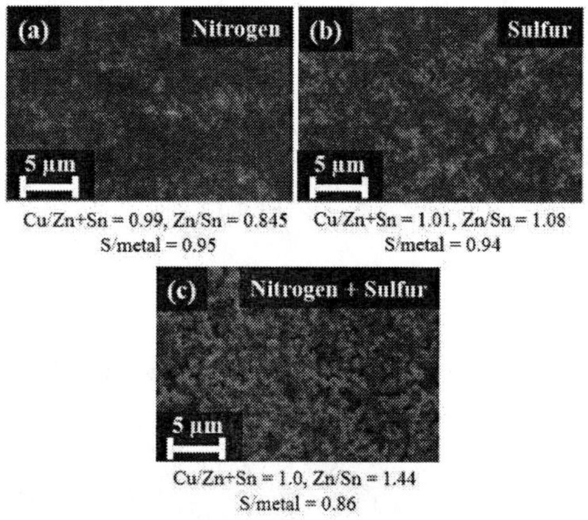

Fig. 5: SEM-EDS of CZTS films with TEA and annealed in (a) N_2 atm (b) N_2 in presence of elemental S powder (c) N_2 atm followed by sulfurization.

Film prepared using TEA followed by annealing at 300°C in nitrogen atmosphere is more compact and dense when compared to without TEA film as per SEM micrographs of film shown in fig. 5(a). Formation of phase pure CZTS film (with TEA) is also confirmed by X-ray diffraction pattern and Raman spectra as shown in fig. 6(a) and 6(d) respectively.

X-Ray diffraction patterns of CZTS film annealed at 300°C for 1 hr in N_2 atmosphere (a), N_2 atmosphere in presence of elemental sulfur (b) and N_2 atmosphere followed by annealing in N_2 atmosphere in presence of elemental sulfur (c) confirm the formation of kesterite structure of CZTS as shown in Fig. 6 but as per SEM micrograph films annealed in N_2 atmosphere and sulfur atmosphere is more compact and voids free compare to the film annealed in N_2 atmosphere followed by annealing in N_2 atmosphere in presence of elemental sulfur.

Crystallinity, morphology and composition of the films annealed in N_2 atmosphere and annealed in sulfur atmosphere are quite similar and there is no significant change has been noticed. Tauc plot of the film prepared using TEA and annealed in N_2 atmosphere is shown in Fig. 7 and found that band gap of the film is 1.47 eV which is close to the optimal value for single juction thin film solar cell. Hence, from the above results it is concluded that uniform, crystalline CZTS thin film can be obtained using TEA as the stabilizing agent without sulfurization and by annealing in N_2 atmosphere.

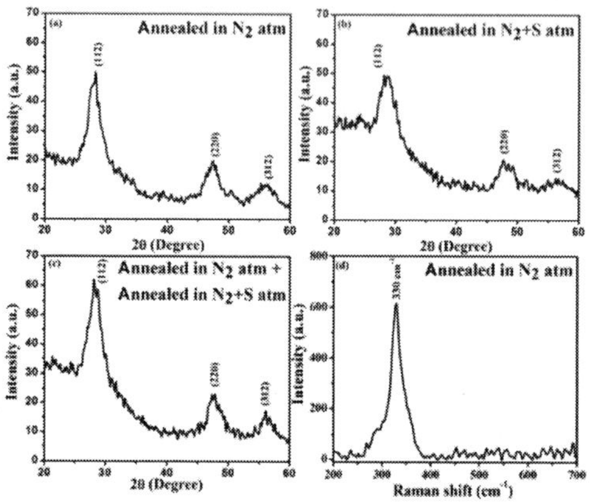

Fig. 6. X-Ray diffraction pattern of CZTS films with TEA and annealed in (a) N_2 atm (b) N_2 in presence of elemental S powder (c) N_2 atm followed by sulfurization (d) Raman spectra of CZTS film with TEA and aneealed in N_2 atm [8].

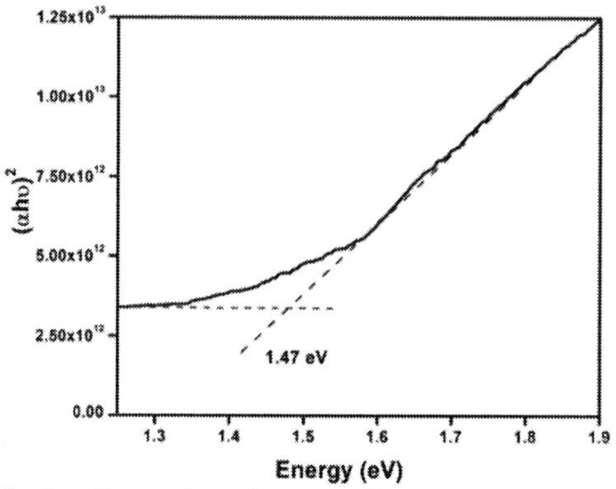

Fig. 7: Tauc plot of CZTS films with TEA and annealed in N_2 atm [6].

4 CONCLUSION

A chemical route to fabricate CZTS film is proposed through non-hydrazine based dip coating technique using triethanolamine (TEA) as a stabilizing agent. In this study, conventional ways of heat treatment for the synthesis of CZTS are investigated and found that a simple annealing in N_2 atmosphere at 300°C for 1 hr with 10°C/min ramping rate is sufficient for the formation of phase pure CZTS films. This study excluded the need of other harmful sulfurization processes that were used traditionally for the synthesis of CZTS film. Voids free, Zn rich Cu poor uniformly distributed CZTS film having a band gap of 1.47 eV can be synthesized using TEA through annealing in simple N_2 atmosphere without the need of further sulfurization.

REFERENCE

1. Yuxiu Sun, Yongzheng Zhang, Hao Wang, Ming Xie, Kai Zong, Huijuan Zheng, Yingqi Shu, Jingbing Liu, Hui Yan, Mankang Zhua and WoonMing Lau, Journal of Materials Chemsitry A 1 (2013) 6880.

2. Wei Wang, Mark T. Winkler, Oki Gunawan, Tayfun Gokmen, Teodor K. Todorov, Yu Zhu, David B. Mitzi, Advanced. Energy Materials 4 (2013) 1.

3. Yuxiu Sun, Kai Zong, Huijuan Zheng, Hao Wang, Jingbing Liu, Hui Yan, Mankang Zhu, Materials Letters 92 (2013) 195.

4. Vincent Tiing Tiong, John Bell, Hongxia Wang, Beilstein J. Nanotechnol. 5 (2014) 438.

5. P. A. Fernandes, P.M.P. Salom'e, A.F. da Cunha, Journal of Alloys & Compound 509 (2011) 7600.

6. Shun Wang, Qingyu Gao, Jichang Wang, The Journal of Physical Chemistry B 109 (2005) 17281.

7. Michihiro Ohta, Shinji Hirai, Hisanaga Kato, Vladimir V. Sokolov Vladimir V. Bakovets, Materials Transactions 50 (2009) 1885.

8. Sushmita Chaudhari, P.K. Kannan, Suhash R. Dey, 43rd IEEE Photovoltaic Specialists Conference (2016).

OPTIMIZATION OF SULPHURIZATION TEMPERATURE FOR OBTAINING DENSE Cu_2ZnSnS_4 FILMS WITH PHASE PURITY AND PREFERRED COMPOSITION

Amrut Agasti, Sudhanshu Mallick, Parag Bhargava
Department of Metallurgical Engineering and Materials Science, IIT Bombay, Mumbai, India.

ABSTRACT: This study focusses on the effect of sulphurization temperature on properties of Cu_2ZnSnS_4 (CZTS). Cu-Zn-Sn precursor films were co-electrodeposited using electrolyte containing salts of Cu, Zn and Sn and a complexing agent (tri-sodium citrate) followed by sulphurization in inert atmosphere (Ar) at different temperatures ranging from 250°C to 600°C for 30 minutes. Samples sulphurized at these temperatures were characterized for structure, morphology and composition. XRD and Raman spectroscopy suggested that, sample sulphurized at 500°C has a phase purity unlike samples sulphurized at 450°C, 550°C and 600°C which showed peaks corresponding to CZTS phase along with some impurity phases. SEM images for samples sulphurized at 450°C, 500°C, 550°C and 600°C indicated compact morphology but, only the sample sulphurized at 500°C showed preferred composition for CZTS absorber material. Structural, morphological, compositional studies suggested that, sample sulphurized at 500°C resulted in CZTS film with characteristics for an absorber layer for thin film solar cell.
Keywords: thin film solar cell; co-electrodeposition; sulphurization temperature.

1 INTRODUCTION

Cu_2ZnSnS_4 (CZTS) is an attractive p-type absorber material to be used for thin film solar cells as it is a direct band gap material, has high absorption coefficient ($>10^4$ cm^{-1}) and band gap value of 1.5eV falling within the range of optimum band gap for solar cells [1]. There are several vacuum and non-vacuum based techniques for deposition of CZTS, of which electrodeposition is a promising technique due to its advantages like low in cost, faster deposition times, and potential for large area application. Deposition of CZTS can be done by sequential electrodeposition (Cu, Zn, Sn deposition is done sequentially) or co-electrodeposition (all metal salts contained in same electrolyte).

Electrodeposited films are usually subjected to sulphurization in inert atmosphere in order to obtain crystalline CZTS. Table I shows sulphurization conditions (temperature and time) used in different studies. It can be observed that, sulphurization conditions vary drastically in these studies with sulphurization temperature and time varying from 500°C to 590°C and 15 minutes to 2 hours respectively. Thus, it is important to find out the optimum sulphurization conditions in order to get dense CZTS films with phase purity, preferred composition and optimum band gap.

Table I: Sulphurization conditions used in various studies

Type of electrodeposition	Sulphurization conditions	Reference
Co-electrodeposition	550°C, 1hr	[2]
Sequential electrodeposition	550°C, 2hrs	[3]
Co-electrodeposition	350°C 1hr followed by 580°C 2hrs	[4]
Co-electrodeposition	500°C, 1hr	[5]
Co-electrodeposition	590°C, 15 min	[6]
Co-electrodeposition	530°C, 15 min	[7]

2 EXPERIMENTAL

During this study, co-electrodeposition of Cu-Zn-Sn is performed using electrolyte (pH=6) containing $CuSO_4$ (0.01M), $ZnSO_4$ (0.015M), $SnSO_4$ (0.006M), tri-sodium citrate (0.2M) in aqueous medium at deposition potential of -1.15V for 20 minutes at room temperature without stirring.

As electrodeposited samples were sulphurized at temperatures 250°C, 350°C, 450°C, 500°C, 550°C, 600°C in Argon atmosphere in the presence of Sulphur in tube furnace. All sulphurized samples were characterized for structural properties using X-ray diffraction (PANalytical X-ray diffractometer) and Raman Spectroscopy (HORIBA HR800). Compositional analysis for all these samples was performed using Inductively coupled plasma-Atomic emission spectroscopy (ICP-AES). FEG-SEM was used to study the morphology of the sulphurized films. In order to obtain optical band gap for sulphurised films, optical absorption studies were performed by UV visible spectroscopy (Jasco V-650) in the wavelength range of 350 – 900 nm.

3 RESULTS AND DISCUSSION

Figure 1 show XRD plots for as deposited sample and samples sulphurized at 250°C, 350°C, 450°C, 500°C, 550°C, 600°C. All samples show diffraction peak corresponding to the substrate which is Mo (JCPDS # 42-1120). As electrodeposited and sample sulphurized at 250°C show peaks corresponding to intermetallic phases of Cu, Zn and Cu, Sn [Cu_5Zn_8,(JCPDS # 65-6566), Cu_6Sn_5(JCPDS # 45-1488)]. With further increase in annealing temperature, diffraction peaks corresponding to CZTS starts to appear in XRD. For sample sulphurized at 350°C, small diffraction peaks corresponding to (112) and (220) crystallographic planes of CZTS with kesterite crystal structure (JCPDS # 26-0575) along with a diffraction peak for Cu_5Zn_8 were observed. With further increase in annealing temperature, diffraction peaks corresponding to above mentioned intermettallic phases vanished and peaks corresponding to CZTS became stronger. For samples annealed at 450°C, 500°C, 550°C, 600°C diffraction peaks corresponding to (112), (200), (220), (312) crystallographic planes of CZTS were present. For samples annealed at 450°C, 550°C and 600°C, diffraction peaks corresponding to Cu_4SnS_4 (JCPDS # 27-0196) and Cu_2S (JCPDS # 53-0522) were also present. XRD results suggested that, temperatures lower than 450°C are not sufficient to fully convert intermetallic phases into CZTS, while temperatures higher than 500°C are causing CZTS to decompose into secondary phases.

Raman spectroscopy was performed to confirm the existence of phase pure CZTS in the sample sulphurized at 500°C, 550°C and 600°C as ZnS with blende structure also exhibits X-ray diffraction peaks close to that of CZTS (112) and (200) planes. Figure 2 shows that, two main vibrational modes (339 cm^{-1} and 288 cm^{-1}) of CZTS (kesterite phase) with A1 symmetry were present in all samples [5] [8]. No other impurity peaks were found in all these samples.

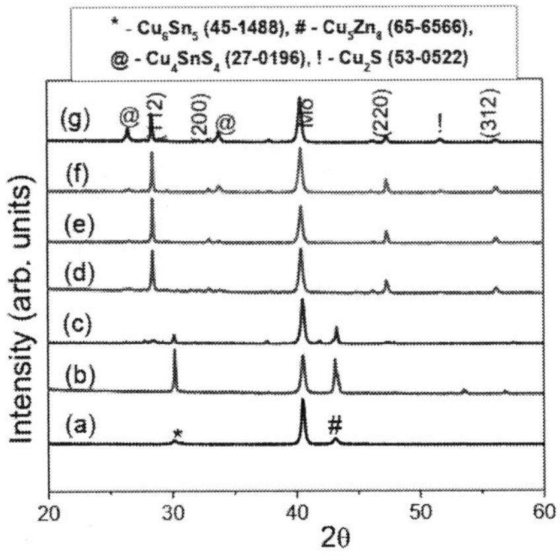

Figure 1: XRD plots for samples (a) as electrodeposited (b) sulphurized at 250°C (c) sulphurized at 350°C (d) sulphurized at 450°C (e) sulphurized at 500°C (f) sulphurized at 550°C (g) sulphurized at 600°C

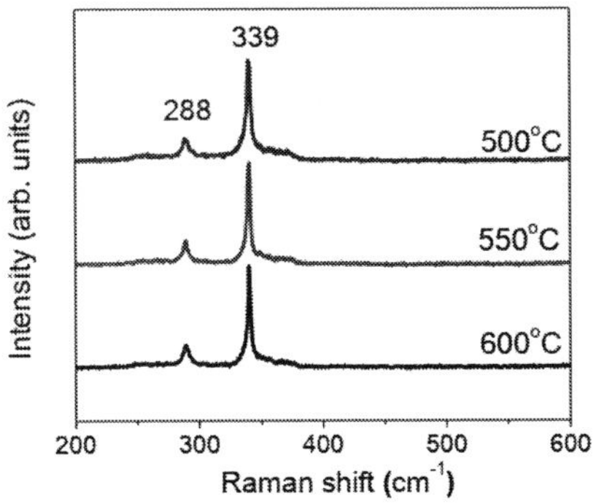

Figure 2: Raman spectra for samples sulphurized at 500°C, 550°C and 600°C

Figure 3 shows FEG-SEM images for as electrodeposited (pH=6, -1.15V, 20 min.) sample and samples sulphurized at 250°C, 350°C, 450°C, 500°C, 550°C, 600°C. Sample sulphurized at 250°C did not show any significant change in the morphology as compared to that of as electrodeposited sample. With further increase in sulphurization temperature, morphology changes starts to appear with the presence of some Sn rich platelet kind of structure at 350°C. At sulphurization temperature of 450°C, CZTS phase had appeared causing further changes in morphology of the sample. Micrographs for samples sulphurized at 450°C, 500°C, 550°C, show dense films with slightly smaller grain size observed in case of the sample with sulphurization temperature of 550°C. The reduction in grain size could be due to recrystallization at this temperature.

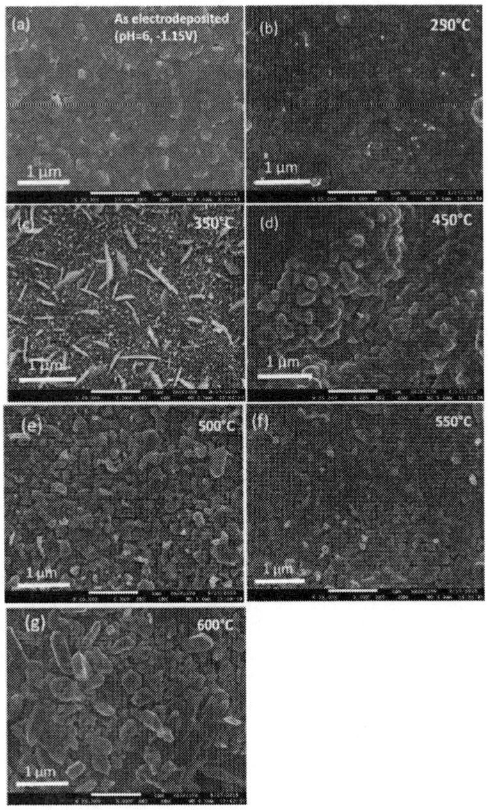

Figure: 3 FEG-SEM images for samples (a) as electrodeposited with electrolyte having pH=6 and deposition potential of -1.15V (b) sulphurized at 250°C (c) sulphurized at 350°C (d) sulphurized at 450°C (e) sulphurized at 500°C (f) sulphurized at 550°C (g) sulphurized at 600°C

Table II: ICP-AES results for samples sulphurized at various temperatures

Sample	Cu (at%)	Zn(at%)	Sn(at%)	S (at%)
250°C	45.6	23.7	25.1	5.6
350°C	40.7	22.7	26.1	10.5
450°C	29.3	14.3	16.2	40.2
500°C	25.1	12.8	12.6	49.5
550°C	25.1	12.6	11.1	51.2
600°C	27.9	10.4	9.1	52.6
Stoichiometric	25	12.5	12.5	50

Table II shows ICP-AES results for samples sulphurized at temperatures 250°C, 350°C, 450°C, 500°C, 550°C, 600°C. As the annealing temperature is increased, the S content increases which is in accordance with the XRD results where CZTS phase can be seen at 450°C. With increasing suphurization temperature, Sn content had decreased. It could be due to vaporization of SnS at higher temperatures [6]. It is well known that, Cu poor and Zn rich conditions in CZTS absorber helps in improving the efficiency of CZTS thin film solar cells [9]. Thus, Cu/(Zn+Sn), Zn/Sn ratios become vital in order to comment on the quality CZTS film. Among these samples, sample sulphurized at 500°C shows Cu poor and Zn rich composition [Cu/(Zn+Sn) = 0.98, Zn/Sn = 1.01] with S/metal ratio close to 1.

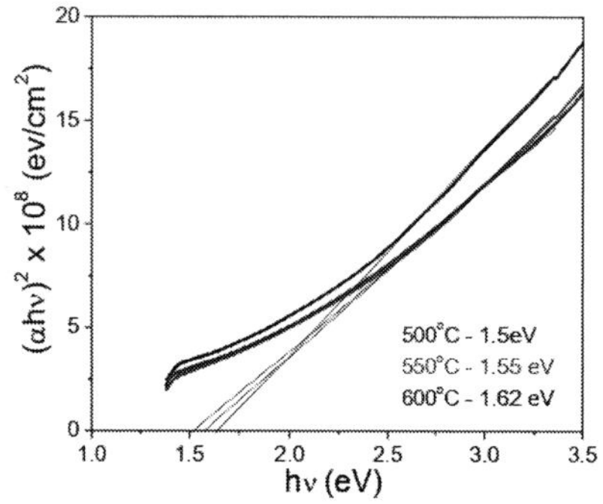

Figure 4: (αhν)2 vs hν plot for samples sulphurized at 500°C, 550°C and 600°C

Figure 4 shows (αhν)2 vs hν plots for sulphurised samples. Band gap values for samples sulphurised at 500°C, 550°C, 600°C, determined from the given plot, are 1.5eV, 1.55eV and 1.62eV respectively. Band gap for these CZTS sample sulphurized at 500°C, 30 minutes shows band gap value of 1.5eV which is in good agreement with those reported in the literature [2] [4] [6].

4 CONCLUSIONS

Cu-Zn-Sn thin films were co-electrodeposited at -1.15V and sulphurized at different temperatures from 250°C to 600°C for 30 minutes in Argon atmosphere. Structural characterization by XRD and Raman spectroscopy suggested that, sample sulphurized at 500°C showed phase purity. Morphology and compositional characterization also confirmed that, sample with sulphurization temperature of 500°C has compact morpholgy with preferred composition (Cu poor and Zn rich). Moreover, sample with sulphurization temperature of 500°C showed a band gap of 1.5eV which is in good agreement with the values reported in the literature. Thus, films sulphurized at 500°C, 30 minutes can be used as an absorber material for CZTS based thin film solar cells.

REFERENCES

[1] K. Ito and T. Nakazawa, Japanese Journal of Applied Physics, Vol. 27 (1988) 2094
[2] S. M. Pawar, B. S. Pawar, A. V. Moholkar, D. S. Choi, J. H. Yun, J. H. Moon, S. S. Kolekar, and J. H. Kim, Electrochimica Acta, Vol. 55 (2010) 4057.
[3] J. J. Scragg, P. J. Dale, and L. M. Peter, Thin Solid Films, Vol. 517 (2009) 2481.
[4] K. V. Gurav, S. M. Pawar, S. W. Shin, M. P. Suryawanshi, G. L. Agawane, P. S. Patil, J.-H. Moon, J. H. Kim, J. H. Yun, and J. H. Kim, Applied Surface Science, Vol. 283 (2013) 74.
[5] M. Valdes, M. Modibedi, M. Mathe, T. Hillie, and M. Vazquez, Electrochimica Acta, Vol. 128 (2014) 393.
[6] J. Ge, J. Jiang, P. Yang, C. Peng, Z. Huang, S. Zuo, L. Yang, and J. Chu, Solar Energy Materials and Solar Cells, Vol. 125 (2014) 20.
[7] C. Gougaud, D. Rai, E. Chassaing, and D. Lincot, Journal of The Electrochemical Society, Vol. 160 (2013) D485.
[8] B. Ananthoju, F. J. Sonia, A. Kushwaha, D. Bahadur, N. V. Medhekar, and M. Aslam, Electrochimica Acta, Vol. 137 (2014) 154.
[9] S. Chen, X. G. Gong, A. Walsh, and S.-H. Wei, Applied Physics Letters, Vol. 96 (2010) 021902.

ACKNOWLEDGEMENTS

We acknowledge SAIF, IIT Bombay for providing access to characterization facilities. We also acknowledge Solar Energy Institute for India and the United States (SERIIUS) and National Centre for Photovoltaic Research and Education (NCPRE) for their financial assistance. The authors acknowledge the receipt of Mo coated glass from NREL, USA.

INDIRECT LASER ABLATION OF Cu(In, Ga)Se2-LAYERS BY ns PULSES WITH A WAVELENGTH OF 1342 nm

Kai Kaufmann[1], Christian Hagendorf[2]
[1] Hochschule Anhalt University of Applied Sciences, Bernburger Straße 55, 06966 Köthen
[2] Fraunhofer Center for Silicon Photovoltaics CSP, Otto-Eißfeldt-Straße 12, 06120 Halle

ABSTRACT: Indirect laser ablation of thin layers from a substrate (lift-off) is a commonly known alternative to direct ablation. Since energy deposition preferably takes place at the interface between thin film and substrate, the thin film layer and especially the resulting scribe edges are less affected by laser irradiation. This reduces the probability of thermally induced damage. For indirect laser ablation of Cu(In, Ga)Se_2 layers the wavelength of 1342 nm is well suited, because CIGS layers are transparent regardless of their composition. We show that 10 ns pulses at 1342 nm are capable to remove CIGS layers from the underlying Mo substrate. Prior to ablation experiments the mechanical interface strength of two different sample systems is analyzed with a scratch test method. It is shown that low adhesion strength promotes an indirect ablation driven by mechanical effects without melting the CIGS material. High interfacial adhesion strength leads to an evaporation driven lift-off process resulting in molten residuals at the crater bottom.
Keywords: CIGS, laser scribing, monolithical series interconnection

1 INTRODUCTION

Lift-Off processes to remove CIGS layers from Mo substrates by using ns laser pulses with a wavelength of 1064 nm, have been reported to be a promising way to achieve the monolithical series interconnection via laser structuring [1,2,3]. However, irradiation with a wavelength of 1064 nm cannot be used for CIGS layers with an [Ga]/([In]+[Ga]) stoichiometric ratio below 0.25. In this case the band gap is below 1.16 eV [4] which leads to an optical absorption of irradiation inside the CIGS layer. The resulting heating of the CIGS layer is likely to cause melting and makes the desired lift-off impossible.
An alternative is the use of 1342 nm laser pulses. At this wavelength CIGS is transparent independently of its composition. Laser energy with a wavelength of 1342 nm is able to penetrate the CIGS layer and is absorbed at the CIGS/Mo interface which is required for lift-off processes. In this work we show successful lift-off processes on two different CIGS layers of different thickness and mechanical adhesion strength.

2 EXPERIMENTAL

The structuring experiments were conducted on two different CIGS samples. The layer stack was in both cases glass/Mo/CIGS. The layer thicknesses and the minimal ratio [Ga]/([In]+[Ga]) are given in Table 1. The CIGS layers were deposited using a three stage coevaporation process [5]. Analysis of the [Ga]/([In]+[Ga]) ratio was done using a ToF-SIMS MCs^+-cluster approach [6].
Sample A reveals a minimum ratio of [Ga]/([In]+[Ga]) of 0.3 with an minimum band gap of 1.2 eV. For sample B the minimum ratio of [Ga]/([In]+[Ga]) is 0.18 so the minimum band gap is 1.12 eV which is below the photon energy of laser irradiation with a wavelength of 1064 nm.
The interface strength of the samples was measured using a self-built scratch test setup which allows force measurements in all three spatial directions during scratching (see Figure 1). The applied force normal to the sample (max. 15 N) surface is kept constant during the scratch test. Scribing speeds are varied between 80 nm/s and 500 mm/s.

For interface characterization a number of mechanical scribe processes were performed while measuring the occurring force in scribing direction with high temporal resolution. The scribing was done with a tungsten needle with a flat tip (tip diameter 50 µm). The scribe area which is needed to calculate the maximum shear stress was analyzed using optical microscopy with subsequent image analysis.
The laser used for structuring was a Xiton Photonics IMPRESS-1.34-Q Nd:YVO4 laser with a wavelength of 1342 nm. The pulse duration was about 10 ns with pulse energies up to 200 µJ. The pulse shape was Gaussian. The laser was directly focused onto the sample surface. So the creation of scribes was then realized by moving the sample which was mounted to a X-Y table.
The SEM analysis of the scribe edges was carried out using a Hitachi SU-70 instrument.

Table 1, Layer thickness, composition and minimum band gap of the samples.

property	Sample A	Sample B
Thickness CIGS [nm]	200	2000
Thickness Mo [nm]	400	660
$[Ga]/([Ga]+[In])_{min}$	0.3	0.18
$E_{gap,min}$ [eV]	1. 2	1.12

Figure 1, Experimental setup for scratch testing, the sample is moved with an X-Y-table.

3 RESULTS

3.1 CHARACTERIZATION OF THE CIGS/Mo INTERFACE

Preliminary scratch tests indicated that samples A and B have significantly different adhesion strength at the CIGS/Mo interface. For a more quantitative estimation of the interface strength a scratch tests were performed in our advanced scratch tester with a constant normal load of 50 mN and a scribing speed of 0.5 mm/s. During scratch testing the force in scribing direction was recorded The resulting scribe features chippings similar to those known from mechanical scribing of CIGS thin films.. Figure 2 shows the force measurement for sample A as an example. The measured force in scribing direction shows an oscillating behavior. During scribing the force does not drop below a minimum of about 20 mN. This can be attributed to the friction force F_{fric} between the scribing needle and the Mo back contact layer.

Further, the visible spikes in scribing force strongly correlate with the positions of the characteristic chippings and reach maximum values just before a chip is removed rapidly. A maximum shear stress is calculated from these maximum values. First, the friction force F_{fric} is subtracted. Then the resulting maximum scribing force is divided by the area of the corresponding chip.

The maximum shear stress for sample A is 6.39±0.25 MPa and for sample B 1.10±0.17 MPa calculated from the above scratch tests. Although these are no quantitative values for interface adhesion it can be concluded that sample B exhibits a significantly weaker Mo/CIGS interface than sample A.

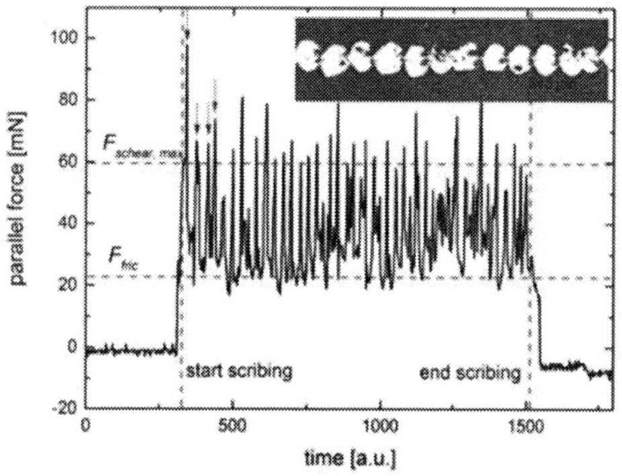

Figure 2: Force measurement during scratch test on sample B shows an oscillation of the scribing force above a minimal value F_{fric} which can be attributed to the friction of the scribing needle on the Mo substrate. The peaks of the scribing force strongly correlation with the positions of the characteristic chippings of the resulting scribe.

3.2 LASER STRUCTURING

Laser scribing of sample A and B results in significantly different behavior with respect to residuals on the scribe bottom. Figure 3 shows optical micrographs of the achieved scribing results.. Sample A showed a clean Mo surface, sample B exhibits a layer of molten material on the bottom. The fluence given represents the maximum fluence in the pulse center which has been calculated from the pulse diameter and the known pulse energy.

On sample A (Fig. 3a) the thin CIGS layer was removed completely at a maximum fluence of 0.47 J/cm². In this case no residuals remain at the scribe bottom. An increased maximum fluence leads to visible residuals at the scribe bottom and finally to Mo damage at a fluence of 0.74 J/cm² (not shown here).

On sample B with (Fig. 3b) a thicker CIGS layer the lift-off was possible at higher fluences of about 0.91 J/cm². The lift-off was incomplete for lower fluences. For a fluence of 0.82 J/cm² the CIGS layer was erected conically around the pulse center. It is concluded that the CIGS layer was pushed away from the CIGS/Mo interface due to evaporation of material and the resulting gas pressure. However, when the lift-off was complete on sample B residuals were always found on the scribe bottom with a small gap of Mo without residual next to the scribe edge.

Despite the fact that the material was removed indirectly by a layer lift-off this result has to be considered as non-ideal, because the clearly visible droplets cause the risk of shunting when dropped on the scribe edge. For sample B the phenomenology of the laser scribe did not change for fluences up to 1.7 J/cm². Even at high fluences Mo damage was not observed.

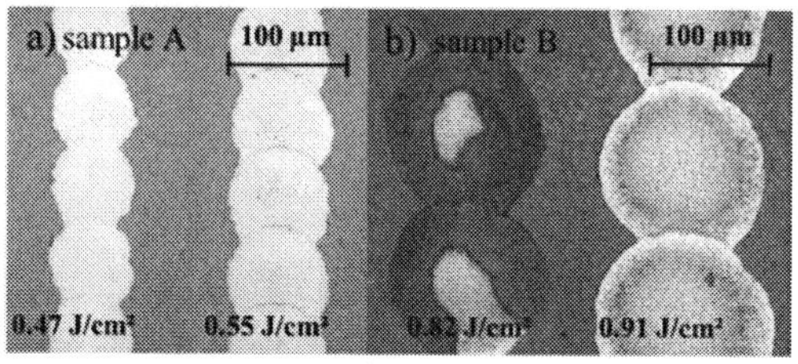

Figure 3: Optical micrographs of scribes after lift-off in CIGS, in a) the lift-off leaves no residuals on the crater bottom, in b) molten CIGS is visible in the center region of the scribe

For a more detailed investigation of the scribe edges SEM images were acquired for both of the samples. Figure 4a) shows the scribe wall of sample A after successful lift-off. There are clearly no droplets or molten residuals visible. The Mo surface does not show any damage or the formation of visible cracks. In Fig. 4b) the scribing edge of sample B is shown. The edge itself is free of molten and redeposited CIGS material. On the scribe bottom droplets are visible. Although the spatial density of these droplets declines towards the scribe edge, it has to be investigated if single droplets on the scribe wall may lead to local shunting of the solar cell.

It can be concluded that the underlying mechanisms of film removal are different for both cases. On sample A molten material was not observed at the crater bottom. Accordingly the ablation is expected to be driven predominantly by mechanical effects due to the different thermal expansion coefficients of CIGS and Mo. On sample B melting was observed at the scribe bottom when the lift-off was achieved. Since the used fluences on sample B are approximately twice as high compared to sample A it is expected that the maximum temperature on sample B was also significantly higher than in sample A.

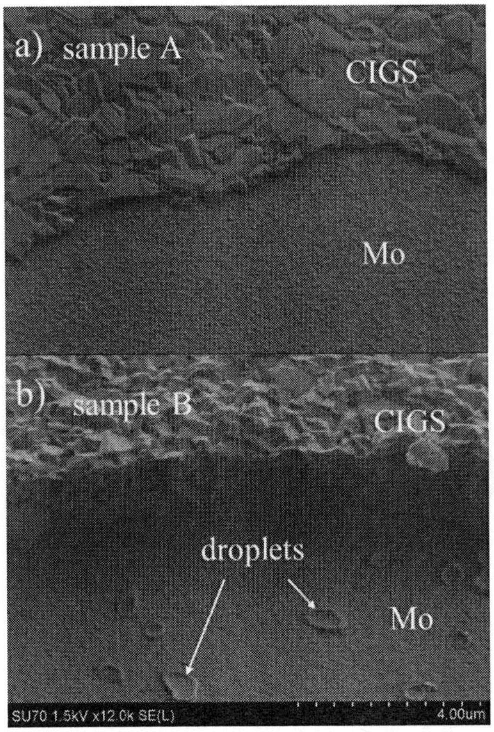

Figure 4: SEM micrographs of scribe walls after successful lift-off on the thin CIGS layer a) and a thicker CIGS layer b). In a) the Mo surface is completely free of residuals. In b) molten residuals are found in the crater center and as droplets near the crater edge as well. The edge itself remains residual free.

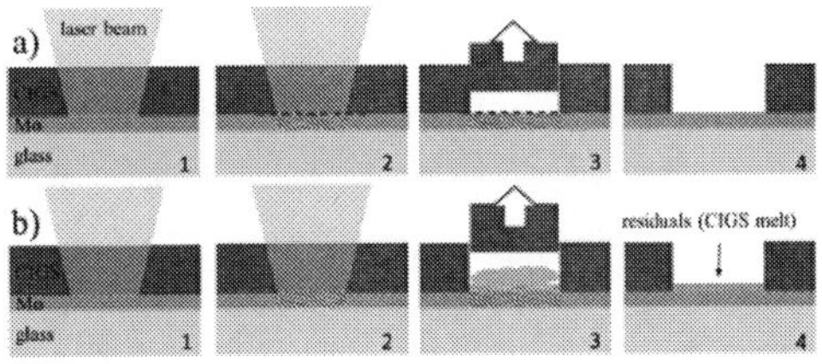

Figure 5: Mechanisms of the lift-off process. The mechanically driven lift-off in a) results in residual free scribes: The laser is also absorbed at the Mo/CIGS interface (1)+(2). The layer is removed by mechanical stress (3), melting temperatures of the CIGS are not reached. The resulting scribe is free from molten residuals (droplets)

In b) the lift-off is driven by evaporation of CIGS material: The laser beam passes the CIGS absorber (1) and is absorbed at the Mo/CIGS interface (2). Evaporation causes the lift-off of the CIGS layer (3). The resulting scribe features molten residuals on the scribe bottom (4).

Based on the described experimental observations two mechanisms for the lift-off are proposed. Figure 5 shows these two mechanisms schematically. In Fig. 5a) the Mo is heated through the absorbed laser irradiation. Prior to the CIGS melting a thermo-mechanical delamination of the CIGS layer from the underlying Mo layer causes the CIGS to be removed. Since the temperatures are below the CIGS melting point there are no molten residuals on the scribe bottom. In Fig. 5b) the Mo layer is also heated through laser irradiation. Due to the elevated mechanical strength of the CIGS layer (thickness, adhesion strength) melting and evaporation occurs at the Mo/CIGS interface. During evaporation the gas pressure forces an explosive ablation of the CIGS layer.

4 CONCLUSIONS

We have shown that pulses with a length of 10 ns at 1342 nm are capable of removing CIGS layers from a underlying Mo substrate based on lift-off-processes. Due to the transparency of the CIGS layer this process is independent of the [Ga]/([In]+[Ga]) stoichiometric ration. Two different lift-off mechanisms could be identified. The evaporation driven lift-off leads to residuals on the scribe bottom and is considered as non-ideal. The mechanically driven process leads to residual free scribes without any Mo damage. It is assumed that interfacial adhesion is a key factor for the lift-off processes. We have developed an in-situ measurement setup in order to quantitatively evaluate the interfacial adhesion strength. Maximum shear stresses for sample A at 6.39 ± 0.25 MPa and for sample B at 1.10 ± 0.17 MPa have been calculated. These values are used to explain the lift-off behavior of the corresponding samples within the framework of a basic mechanistic model. Shorter pulses in the ps range might be an option to improve the lift-off by reducing thermal effects on the Mo/CIGS interface. Despite being subject of current research[7] [8], up to now there is no high energy ps laser with a wavelength of 1342 nm system available commercially.

6 ACKNOWLEDGEMENTS

This work was carried out within the joint research graduate school ''StrukturSolar'' of Anhalt University of Applied Sciences and Martin-Luther-University Halle-Wittenberg which is funded by the German Ministry of Education and Research under identification code 03SF0417A. We also would like to thank Dr. Martin Nittmann of XITON Photonics for performing the laser experiments.

7 REFERENCES

1 R. Murison, et al., CIGS P1, P2, P3 laser scribing with an innovative fiber laser, 35th Photovoltaic Specialists Conference (PVSC), 2010

2 A. Buzás, Z. Geretovszky: Nanosecond laser-induced selective removal of the active layer of CuInGaSe2 solar cells by stress-assisted ablation, Physical Review B 85 (2012), Nr. 24, S. 245304

3 S. H. Lee, C. K. Kim, J. H. In, H. S. Shim, S. H. Jeong: Selective removal of CuIn1− xGaxSe2 absorber layer with no edge melting using a nanosecond Nd:YAG laser,

Journal of Physics D: Applied Physics 46 (2013), Nr. 10, S. 105502. – ISSN 0022–3727

4 M. Bar, W. Bohne, J. Rohrich, E. Strub, S. Lindner, M. C. Lux-Steiner, C. H. Fischer, T. Niesen, F. Karg, Determination of the band gap depth profile of the penternary Cu (In (1-X) Ga X)(SYSe(1-Y))2 chalcopyrite from its composition gradient, J. Appl. Phys. 96 (2004), Nr. 7, S. 3857-3860

5 S. Niki, et al., CIGS absorbers and processes, Progress in Photovoltaics: Research and Applications 18.6 (2010): 453-466

6 K. Kaufmann, et. al., Quantitative elemental analysis of photovoltaic Cu(In, Ga)Se2 thin films using MCs+ clusters, Surface and Interface Analysis 45.1 (2013): 434-436.

7 A. Rodin et. al., High average power picosecond laser for selective material processing at 1342 nm wavelength, International Society for Optics and Photonics, (2015)

8 Y. Chen et. al., High Energy, High Peak Power 1342-nm Picosecond Nd: YVO4 Regenerative Amplifier, In: Quantum Electronics, IEEE Journal of 51 (2015), Nr. 9, S. 1–6

VARIABLE-RANGE HOPPING VERSUS INTER-GRAIN TUNNELING IN $Cu_2ZnSn(S_xSe_{1-x})_4$ THIN-FILMS PREPARED BY SPRAY PYROLYSIS

K.G. Lisunov[1,*], L. Bruc[1], M. Rusu[2,3], G. Gurieva[2], M. Guc[4], S. Levcenko[2], L. Dermenji[1], N. Curmei[1], D.A. Sherban[1], A.V. Simashkevich[1], S. Schorr[2,5], E. Arushanov[1]

[1]*Institute of Applied Physics, str. Academiei 5, MD 2028, Chisinau, Republic of Moldova*
[2]*Helmholtz-Zentrum Berlin für Materialien und Energie GmbH, Hahn-Meitner-Platz 1, D-14109 Berlin, Germany*
[3]*Moldova State University, 60 A. Mateevici str., 2009 Chisinau, Republic of Moldova*
[4]*IREC, Catalonia Institute for Energy Research, Sant Adrià del Besòs (Barcelona), 08930, Spain*
[5]*Institute of Geological Sciences, Free University Berlin, Malteserstr. 74-100, 12249 Berlin, Germany*
[*]*E-mail: kgl-official@hotmail.com*

ABSTRACT: Thin $Cu_2ZnSn(S_xSe_{1-x})_4$ films with x = 1.0 and 0.85 were prepared by the spray pyrolysis method with subsequent annealing in the presence of elemental sulfur or selenium. The surface sensitive Raman measurements indicated a high crystalline quality of all samples and absence of secondary phases. Small amount of $Cu_{2-x}S$ secondary phase was detected from grazing incidence X-ray diffraction patterns (information depth of about 500 nm). The resistivity, ρ (T), is governed by the Mott variable-range hopping conduction at high temperatures, when the mean hopping length R_h (T) exceeds the mean distance between the acceptors, R_A, but is still smaller than the typical inter-grain distance, a_{ig}. However, with lowering the temperature the resistivity becomes dominated by inter-grain tunneling, provided that R_h (T) > a_{ig}. The analysis of the ρ (T) function has yielded the values of the acceptor concentration and the localization radius, with the evidence that all the samples are characterized by a state far different from the metal-insulator transition. The radius of the mean size grain as well as values of a_{ig} were in addition determined.

Keywords: Kesterites, Thin Films, Spray Pyrolysis, Electronic Properties

I INTRODUCTION

A solar energy conversion efficiency of 12.6 % has been reached by using the $Cu_2ZnSn(S_xSe_{1-x})_4$, or CZTSSe, solid solutions as absorber layers in thin film solar cells [1]. The CZTSSe thin films can be obtained with different methods. The spray deposition method is particularly attractive since it is a relatively fast and vacuum-free technique [2].

The resistivity ρ (T) of the CZTSSe single crystals is governed by the Mott variable-range hopping (VRH) transport within a wide temperature range with the onset already close to the room temperature. This is connected to a large value of the acceptor activation energy $E_A \sim 140$ meV [3], accompanied by a high intrinsic lattice disorder [4]. However, in CZTSSe thin-films another mechanism of the charge transfer related to the inter-grain tunneling (IGT) of the holes becomes important [5]. On the other hand, the investigated CZTSSe films have exhibited only one out of the transport mechanisms above [5].

In this work, we apply for the preparation of CZTSSe thin-films the spray pyrolysis method performed at ambient pressure from aqueous solutions and subsequent selenization or sulfurization. We investigate the CZTSSe transport properties as function of material stoichiometry by performing temperature dependent resistivity measurements in a wide temperature range by using the Van der Pauw method. The analysis of the resistivity data is carried out by taking into account a possible competition between the VRH and IGT transport, depending on the typical grain size and the inter-grain distance.

II EXPERIMENTAL DETAILS

In our experiments, first, the Cu_2ZnSnS_4 (CZTS) thin films were prepared on glass substrates by spray pyrolysis at ambient atmosphere from aqueous solutions of metal salts (CuCl, $Zn(O_2CCH_3)_2$, $SnCl_4$) and thiourea ($SC(NH_2)_2$) [6,7] at temperatures between 350 and 410 °C. Then, the CZTS films were annealed in presence of elemental S or in presence of elemental Se in evacuated quartz ampoules at 525 °C for 30 min. This allowed to prepare polycrystalline samples with high crystallites quality, while the presence of Se allowed to prepare CZTSSe solid solutions with x = 1.0, 1.0 and 0.85 cited in the following as #1, #2 and #3, respectively. The elemental composition of the layers was determined using the energy dispersive X-ray (EDX) analysis. In addition to the anion ratio, the EDX analysis showed that all the samples are Cu-poor, while sample #2 is Zn-rich and #1 and #3 are in addition Zn-poor.

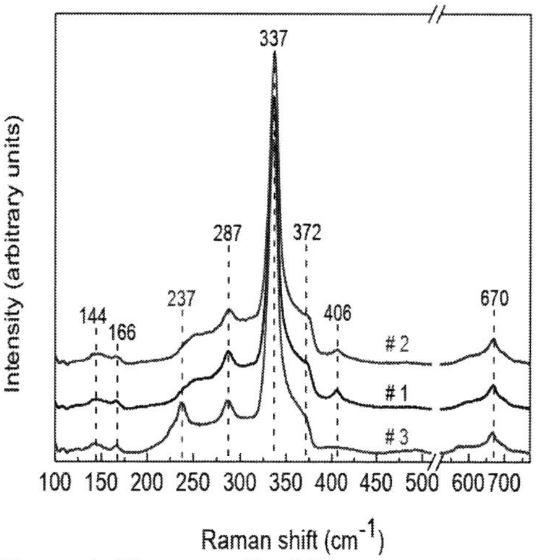

Figure 1: The normalized Raman spectra.

The Raman spectra were excited by a solid state lasers (lines 532 nm and 785 nm) and by a gas laser (line 325 nm) and were recorded using a *i*HR-320 Horiba Jobin Yvon spectrometer in conjunction with a CCD detector. All spectra were measured in backscattering configuration through an Olympus metallographic microscope. The lasers power was selected to avoid any thermal effects at the samples surface. Spectra excited with a green laser are presented in Fig. 1. Here, all the peaks correspond to a pure CZTS or CZTSSe phase [8,9]. The narrow width of all peaks as well as observation of low intensity peaks (like peaks at 144 and 166 cm^{-1}) indicate the high crystalline quality of the analyzed samples. In case of #3, appearance of a peak at 237 cm^{-1} denotes the formation of CZTSSe phase, and its relative intensity comparing to the peak at 337 cm^{-1} yields the Se/(Se + S) ratio ~ 0.15 [9]. This is in good agreement with the EDX measurements. Additional Raman spectra obtained with 785 nm and 325 nm laser lines also do not show presence of any secondary phases, like SnS or ZnS [10].

The structural characterization of the thin-films was carried out by grazing incidence X-ray diffraction (GIXRD). The samples have been measured at incidence angle ω = 1°. Considering the thickness of the films, the information obtained corresponds to the middle of their thickness, i.e., to about 500 nm information depth. The GIXRD patterns in Fig. 2 show distinguishable Bragg reflexes, allowing to assign most of the peaks to kesterite type CZTS (#1 and #2) and CZTSSe (#3) phases (space group $I\bar{4}$). The $Cu_{2-x}S$ phase is present in all of the thin-films, however, its amount should not be high since no evidences of this phase was found in Raman spectra. Using LeBail refinement, lattice constants of the main kesterite type phase were derived (Table I). Using the Vegard law, the Se/(Se + S) ratio ~ 0.20 for #3 was obtained, in satisfactory agreement with that found from the analyses of the Raman and EDX data.

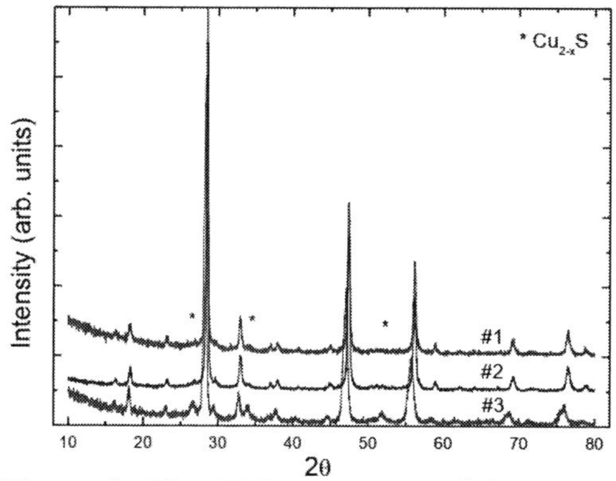

Figure 2: The GIXRD patterns of the CZTS and CZTSSe thin films.

The hot probe method showed p-type conductivity for all the samples. The resistivity was measured with the Van der Pauw method.

III RESULTS AND DISCUSSION

As can be seen in Fig. 3, all samples exhibit a strongly activated resistivity within the whole temperature range.

The Mott VRH conduction, $\rho(T) = AT^{1/4} \exp[(T_{0M}/T)^{1/4}]$, where A is a constant, $T_{0M} = \beta_4/[k_B\, g(\mu)\, a^3]$, $\beta_4 = 21$, $g(\mu)$ is the density of states (DOS) at the Fermi level, μ, and a is the localization radius [11],

Table I: The values of the lattice constants (a, c), the temperature interval (ΔT_M) and the characteristic temperature (T_{0M}) of the Mott VRH conduction, the corresponding data (ΔT_g and T_{0g}) of the IGT conduction, the width of the acceptor band (W), the relative values of the acceptor concentration (N_A/N_c) and of the localization radius (a/a_B), the typical inter-grain distance (a_{ig}), and the radius of a typical grain, a_g.

Sample	a Å	c Å	ΔT_M K	T_{0M} 10^6 K	ΔT_g K	T_{0g} 10^3 K	W meV	N_A/N_c	a/a_B	a_{ig} nm	a_g nm
# 1	5.435	10.833	181-84	3.19	86-53	3.32	90	0.30	1.4	7	40
# 2	5.433	10.831	233-181	13.0	188-152	9.28	150	0.19	1.2	7	90
#3	5.497	10.899	165-90	5.66	93-80	4.21	100	0.24	1.3	7	50

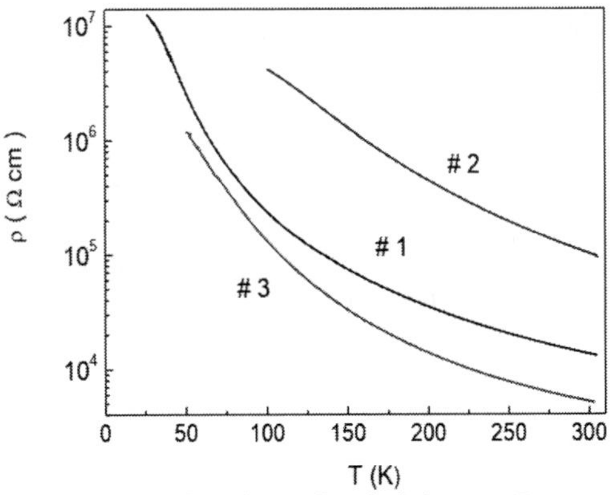

Figure 3: The plots of resistivity vs. T.

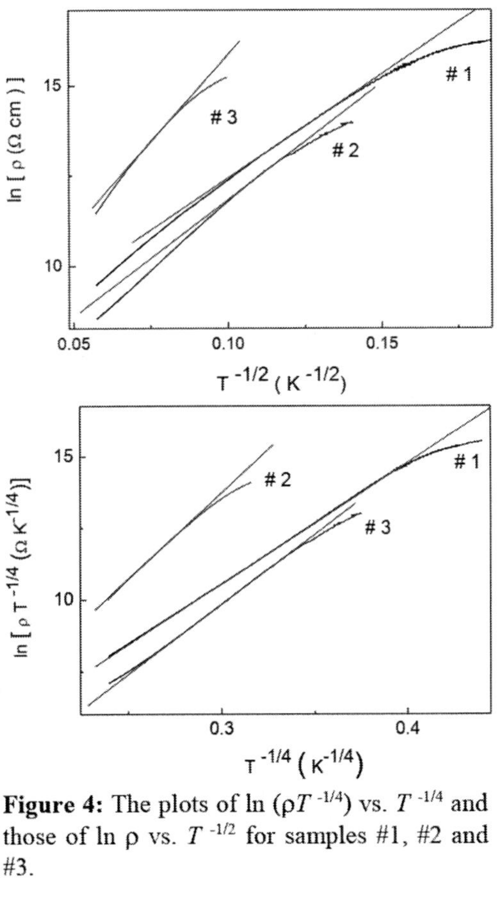

Figure 4: The plots of ln ($\rho T^{-1/4}$) vs. $T^{-1/4}$ and those of ln ρ vs. $T^{-1/2}$ for samples #1, #2 and #3.

dominates at high temperatures (bottom panel of Fig. 4). The values of this interval, ΔT_M, and those of T_{0M} are collected in Table I.

On the other hand, intervals ΔT_g of the IGT conduction, $\rho(T) = \rho_0 \exp [(T_{0g}/T)^{1/2}]$, where ρ_0 is the prefactor and T_{0g} is the IGT characteristic temperature [12], can be seen at lower temperatures (top panel of Fig. 4). The values of ΔT_g and T_{0g} are shown in Table I. The reason to such behavior of $\rho(T)$ is that the hopping length, R_h (T), increases with decreasing T. Therefore, at high T one can expect that R_h (T) does not exceed the inter-grain distance, a_{ig}. Hence, in the interval of $R_A < R_h (T) < a_{ig}$, where $R_A = 2(4\pi N_A/3)^{-1/3}$ is the mean distance between acceptors with the concentration N_A, the majority of hops lie inside the grains, and the Mott VRH is predominant. However, when R_h (T) with lowering T becomes comparable with a_{ig}, the hops beyond a single grain become more important, leading eventually to the transition to the IGT transport mechanism.

Analysis of the Mott VRH conduction allows to determine the values of the acceptor band (AB) width, $W \approx 0.5 k_B (T_{VM}^3 T_{0M})^{1/4}$ [13], where T_{VM} is the Mott VRH onset temperature (the data of W are collected in Table I). In addition, the relative values of N_A/N_c, where N_c is the critical concentration of the metal-insulator transition (MIT), as well as those of a/a_B, where a_B is the Bohr radius, can be found with the method of Ref. 13 (see Table I). As follows from Table I, the ratio of N_A/N_c is small and that of a/a_B is close to unity, indicating that all investigated samples are far from the MIT. With the equations $a_B = \hbar^2\kappa_0/(me^2)$ (where $\kappa_0 \approx 7$ is the dielectric permittivity of an ideal lattice [14] and $m \approx 0.47\ m_0$ is the hole effective mass [13]), and $N_c^{1/3} a_B \approx 0.25$ (the Mott criterion [15]) one finds the values of $a_B \approx 0.8$ nm and $N_c \approx$

3.1×10^{19} cm^{-3}. Then, the absolute values of $N_A \approx 9.3$, 6.0 and 7.3×10^{18} cm^{-3} and those of $a \approx 1.1$, 1.0 and 1.05 nm, can be found for #1, #2 and #3, respectively. The latter, together with the data of T_{0M}, yields the values of g (μ) $\approx g_{av} \equiv N_A/(2W) \approx 5.1$, 1.9 and 3.7×10^{16} meV^{-1} cm^{-3}.

The mean hopping length can be expressed as R_h $(T) = Ca$ $(T_{0M}/T)^{1/4}$. The condition of R_h $(T_{VM}) \approx R_A$, or the onset of the Mott VRH conduction, yields the value of $C \approx 0.45$, which deviates somewhat from that of $C \approx 0.37$ evaluated for a case of μ lying in the center of the AB [16]. This means, that the position of μ is near one of the band edges. Then, the typical inter-grain distance can be found with the equation R_h $(T_{vg}) \approx a_{ig}$, where T_{vg} is the IGT onset temperature (Table I), yielding the values of a_{ig} collected in Table I. In turn, $a_{ig} \approx e/(\beta k_B T_{vg} \kappa_0)^{1/2}$, where $\beta = \kappa_{0g} k_B T_{0g}/(4e^2)$, and κ_{0g} is the dielectric permittivity of the inter-grain medium [12]. The values of $\kappa_{0g} \approx 8.8$, 3.5 and 7.3 for #1, #2 and #3, respectively, leading to $\kappa_{0g} = 7 \pm 2$, have been obtained by fitting a_{ig} in Table I with the expression of a_{ig} above. The data of κ_{0g}, lying close to $\kappa_0 \approx 7$ of CZTS [14], suggest that the inter-grain medium represents a similar material as in grains, differing from the latter mainly by a reduced concentration of defects (acceptors). This is supported by the values of the inter-grain potential barrier [12], $U \approx (\kappa_{0g} k_B T_{0g} \hbar)^2/ (128me^4) \approx 3.9$, 4.8 and 4.3 meV for #1, #2 and #3, respectively. Indeed, these data are by an order of the magnitude lower than those of metallic granular films, where the difference between the metallic grains and dielectric medium is much higher [12]. Finally, with the equation $a_g \approx 128me^6/[\kappa_{0g}{}^3(\hbar k_B T_{0g})^2]$, following from Ref. 12, one can estimate the radii of a typical grain, which are collected in Table I.

CONCLUSIONS

The resistivity of CZTSSe thin films, obtained by the spray pyrolysis method, exhibits broad intervals of the Mott VRH conduction, followed by the IGT mechanism when temperature is decreased. Such behavior is connected to the granular film structure and the hops only inside the grains at high temperatures, leading to the overall Mott picture of the conductivity. The transition to the IGT behavior with decreasing the temperature corresponds to the hopping beyond separate grains due to the increase of the hopping length.

ACKNOWLEDGEMENTS

The support of this research by the project STSU 5985 and by Institutional Project CSSDT 15.817.02.04A is acknowledged.

REFERENCES

[1] W. Wang *et al.*, Advanced Energy Materials 4 (2014) 1301465.

[2] M. P. Suryawanshi *et al.*, Materials Technology 28 (2013) 98.

[3] S. Levcenko *et al.*, Phys. Rev. B 86 (2012) 045206

[4] A. Lafond *et al.*, Acta Crystallographica B 70 (2014) 309.

[5] L. Dermenji *et al.*, Proc. of 29th EU PVSEC (2014) 1801; M. Guc *et al.*, Proc. of 28th EU PVSEC (2013) 2449.

[6] L. I. Bruc *et. al.*, Proc. of 27th EU PVSEC (2012) 2763.

[7] G. Gurieva *et. al.*, Physica Status Solidi C 10 (2013) 1082.

[8] M. Guc *et. al.*, Scientific Reports 6 (2016) 19414.

[9] M. Dimitrievska *et. al.*, Journal of Alloys and Compounds 628 (2015) 464.

[10] P.A. Fernandes *et. al.*, Journal of Alloys and Compounds 509 (2011) 7600.

[11] B. I. Shklovskii and A. L. Efros, Electronic Properties of Doped Semiconductors, Springer, Berlin, 1984.

[12] V. F. Gantmakher, Electrons and Disorder in Solids, University Press, Oxford, 2005.

[13] K. G. Lisunov *et al.*, Sol. Energ. Mater. Sol. Cells 112 (2013) 127.

[14] O. Gunavan *et al.*, Appl. Phys. Lett. 100 (2012) 253905.

[15] N. F. Mott, Metal-Insulator Transitios, Taylor and Francis, London, 1990).

[16] K. G. Lisunov *et al.*, J. Appl. Phys. 94 (2003) 5912.

ON THE INTERPRETATION OF PHOTOLUMINESCENCE AND VIBRATING KELVIN PROBE METHOD FOR QUALITY CONTROL OF CU(IN,GA)SE2 THIN FILMS

Tetiana Lavrenko*, Thomas Walter
Ulm University of Applied Sciences, Albert-Einstein-Allee 55, 89081 Ulm, Germany
(lavrenko@hs-ulm.de)
Balázs Plesz
Budapest University of Technology and Economics, Magyar tudósok körútja 2, H-1117, Budapest, Hungary

ABSTRACT: Photoluminescence imaging is considered as a valuable tool for the quality control of $Cu(In,Ga)Se_2$ (CIGS) thin films due to a fast data acquisition and processing. However, the PL intensity can be affected by minority carrier lifetimes and additionally by electric fields leading to a charge separation. Therefore, the interpretation of PL measurements is not always straightforward. In this contribution, the interpretation of PL measurements is discussed using complementary vibrating Kelvin probe measurements. It will be demonstrated that thermal ageing of CIGS thin films (with a CdS buffer layer) leads to a reduction of the PL intensity and a change in a surface band-bending as deduced from vibrating Kelvin probe measurements under illumination. It will be shown that 2D-maps of the work function correlate well with images of the PL intensity. Furthermore, it has been found that etching away the CdS buffer layer reduces the previously measured band-bending significantly pointing out the importance of the buffer layer deposition process.
Keywords: $Cu(In,Ga)Se_2$ solar cells, photoluminescence, vibrating Kelvin probe measurement, band-bending.

1 INTRODUCTION

The quality control of CIGS thin films is a key requirement for solar module production with a high yield. Especially, photoluminescence (PL) imaging is considered as a promising tool for process monitoring at different stages of a production chain. Due to its contactless nature, the in-line application of PL already on the absorber level or after buffer layer deposition is of particular interest. Since PL intensity is determined by the rate of spontaneous emission via band-to-band transitions, imaging results can be directly correlated to the final cell performance by assessing the minority carrier lifetimes, the quasi-Fermi levels splitting, and the diode voltage. [1] However, the interpretation of PL measurement data is not always unambiguous. It has been discussed in the literature that PL transients are significantly affected by electric fields which sweep photogenerated charge carriers apart and inhibit radiative recombination. [2], [3] Electric fields are inherent in hetero-junctions but also appear as a result of surface band-bending in absorber thin films due to the presence of charges close to the surface. In this work, the impact of surface band-bending on the PL intensity is investigated by using the vibrating Kelvin probe measurement: in the dark – to map surface potentials which will be correlated to spatially-resolved PL images – and under illumination – to scan the surface photovoltage which is, in other words, a photoinduced shift of the work function originating from the surface band-bending. It will be demonstrated that thermal ageing of CIGS/CdS thin films leads to the degradation of the PL intensity and the change in the local band-bending. Furthermore, it will be shown that the CdS buffer layer plays an important role in inducing the observed band-bending leading to the decrease of the PL intensity.

2 EXPERIMENTAL

2.1 Samples under investigation

For the discussed measurements, Cu(In,Ga)Se2 absorbers were grown by a coevaporation process and coated with the CdS buffer layer deposited by a wet chemical bath process. Photoluminescence imaging and vibrating Kelvin probe measurements have been performed on the $Cu(In,Ga)Se_2$/CdS heterostructure at room temperature in the initial state and after local ageing. The local ageing test comprises of keeping a sample in the climate chamber at elevated temperatures and enforcing strong degradation effects only on one side of the sample.

2.2 Measurement setups

Photoluminescence is a phenomena based on the spontaneous light emission from a sample in response to the optical excitation. In this study, photoluminescence imaging has been done by using a near infrared camera based on the InGaAs detector. Such a camera offers a good spatial resolution of 640 x 512 pixels with a pixel size of 20 μm and a frame rate of 25 Hz. The luminescence signal was induced by illuminating the sample with two red LED arrays with the emission wavelength of 625 nm.

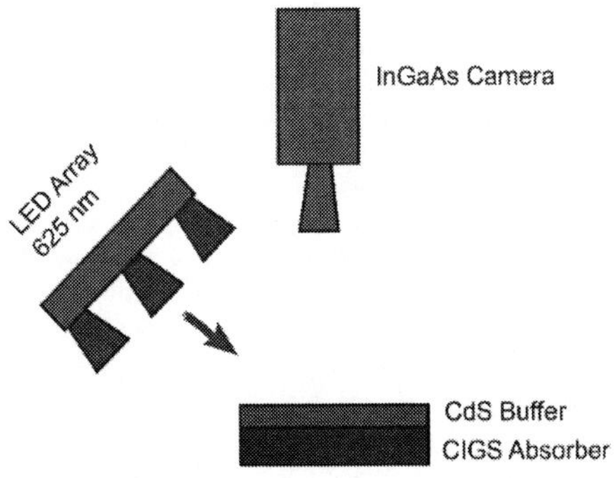

Fig. 1. Photoluminescence imaging setup.

Vibrating Kelvin probe arrangement is a simple and accurate method to measure the work function at the surface of materials. This measurement is based on the compensation of the electric field (work function difference) between a vibrating electrode and the surface to be investigated. [4] At the compensated state, when alternating current induced by the vibrating capacitance approaches zero, the compensating voltage equals to the work function difference or contact potential difference. Therefore, surface potential variations can be detected by using this method.

In this measurement setup, the probe comprises of a graphite electrode with a tip diameter of 100 µm. A grounded graphite ring is placed around the vibrating electrode to provide additional shielding. Minimum step size of the measurement is 64 µm. The maximum resolution of the setup is equal to the tip diameter, therefore variations in surface potentials on small areas can be successfully detected.

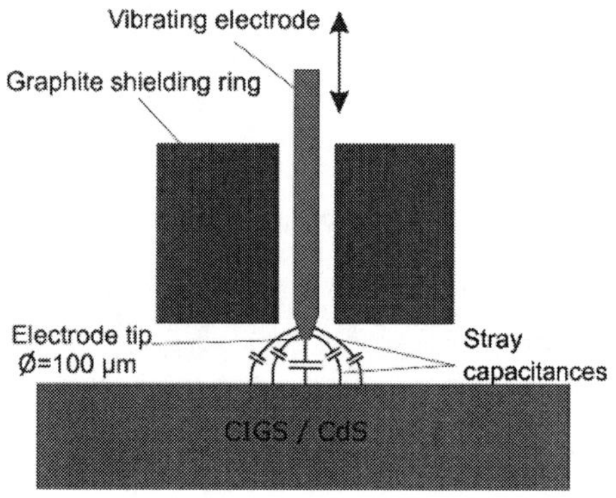

Fig. 2. Vibrating Kelvin probe measurement setup.

Both, photoluminescence and vibrating Kelvin probe measurements were performed in a light blocking box with blackened interior to exclude an impact of ambient light and interference with reflected light inside the box.

3 RESULTS AND DISCUSSION

The main objective of this work is to establish a correlation between steady-state photoluminescence intensity and vibrating Kelvin probe measurements in order to build an appropriate model to explain the PL intensity degradation after thermal ageing of CIGS thin films.

3.1 Photoluminescence imaging

Fig. 3 reproduces PL images of the absorber/buffer heterostructure in the initial state (a) and after additional local ageing (b). Capacitance-voltage measurements on the completed devices aged in the same way (not shown here) exhibited a decrease of the acceptor density which can be associated to the well-known metastabilities in CIGS-based devices. [5]

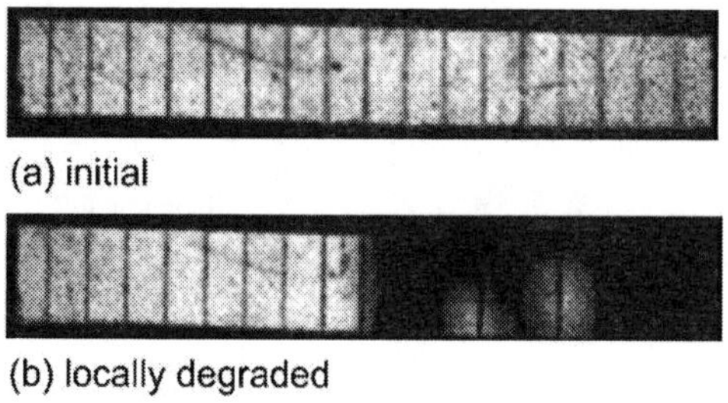

Fig. 3. Steady-state photoluminescence images of the CIGS/CdS sample in the initial state (a) and after local degradation (b).

A change of the doping concentration in the absorber surface layer will directly affect the dynamics of the photogenerated carriers due to an induced band-bending close to the absorber/buffer interface.

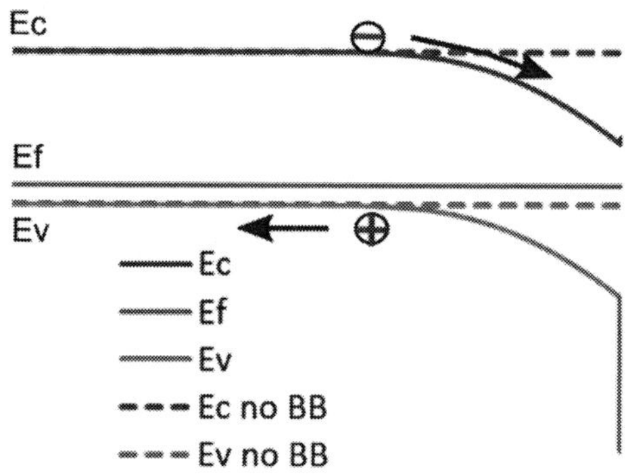

Fig. 4. Sketch of the energy bands in the CIGS layer: dashed lines – without band-bending; solid lines – with band-bending. Photogenerated carriers are swept away by electric field induced by band-bending at the CIGS/CdS interface.

The occurrence of this band-bending after thermal aging may be caused by defect states at the interface or in the CdS layer. These states can act as traps for holes and get positively charged if holes are captured. The accumulation of the positive charges at the interface would move the Fermi level close to the mid-gap position.

Due to band-bending the photogenerated electrons and holes are separated as shown in Fig. 4, and therefore cannot recombine radiatively, resulting in a decrease of the PL intensity. Similar behaviour was reported in [3] where transient PL measurements of completed devices demonstrated shorter decay times compared to the measurements on the CIGS or CIGS/CdS films.

3.2 Surface potential mapping

In order to obtain a deeper insight into these mechanisms vibrating Kelvin probe measurements were performed which allow the contactless determination of the local surface potential, and in combination with illumination the determination of the local band bending. Therefore, inhomogeneities in the material doping and the surface band bending can be detected by this technique.

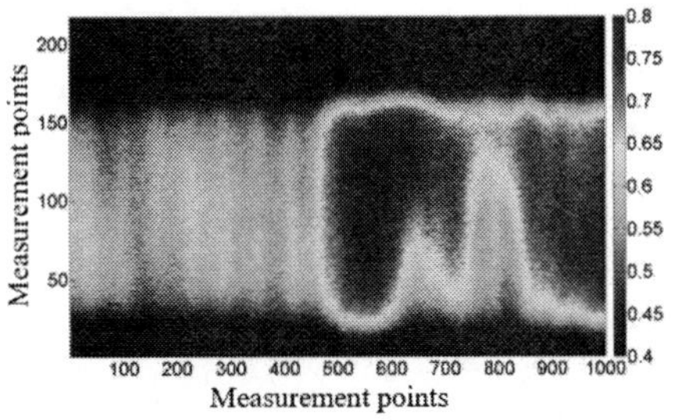

Fig. 5. Mapping of surface potential variations on the locally degraded sample (stepsize of 64 microns).

Fig. 5 reproduces the Kelvin probe mapping of the locally aged sample shown in Fig. 3(b). The lateral inhomogeneities of the PL-intensity were also observed as a difference in the surface potential. It can be seen that areas with higher PL values show up as areas with lower potentials on the surface potential map. Moreover, the two areas with the higher PL intensity on the aged part and P1 scribes can be also distinguished on the surface potential map.

3.3 Surface photovoltage

The Kelvin probe measurements in combination with optical excitation allow testing the spatial homogeneity of the surface photovoltage. In this study, the measurements have been done on the same locally aged sample in the dark and under illumination. The illumination was chosen to ensure absorption only in the absorber layer (red light) as well as in the whole structure (white light).

The surface photovoltage measurement is sensitive to band-bending and recombination rates thus can be used to investigate the electronic properties of the absorber-buffer interfaces. The surface photovoltage was determined as a difference between the surface potentials in the dark and under illumination.

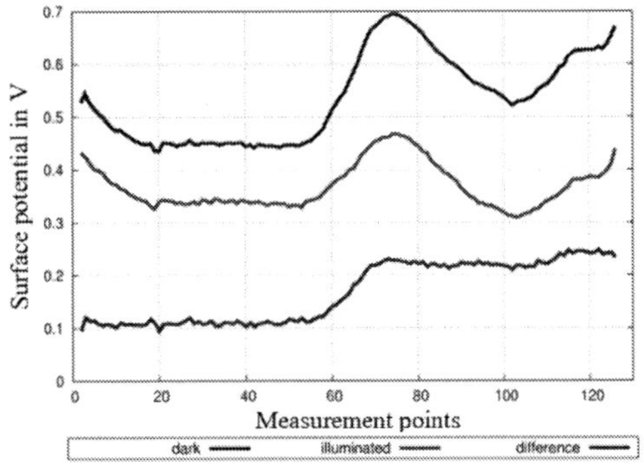

Fig. 6. Line scan of the surface potential in the dark (black), under white light illumination (red) and the difference between the two (blue). The stepsize of the measurement is 0.5 mm.

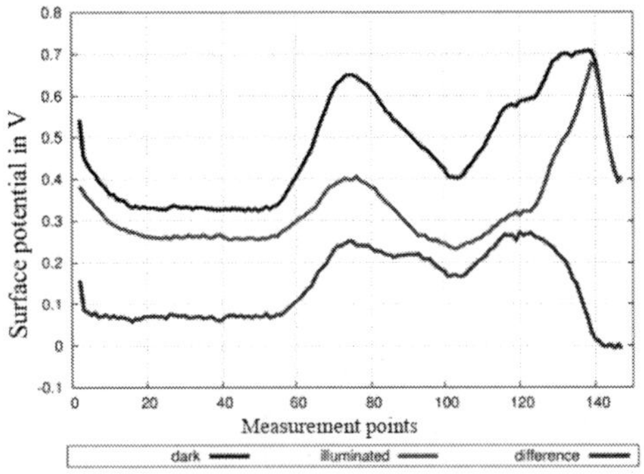

Fig. 7. Line scan of the surface potential in the dark (black), under red light illumination (red) and the difference between the two (blue). The stepsize of the measurement is 0.5 mm.

Looking at the surface photovoltage in Fig. 6 (blue) clearly shows that the band bending in the aged part is about 120 meV higher. Thus, this ageing induced band bending can explain the observed quenching of the PL-intensity after local ageing which is obviously not related to a decrease in the minority carrier lifetime. This indicates that these metastabilities are related to a change of the charge distribution in the layer rather than a decrease of the minority carrier lifetimes. A comparison with Fig. 7 (no absorption in the CdS) further implies that charges in the CdS or at the interface to the CIGS play an important role for these ageing effects.

3.3 Impact of a CdS buffer layer

In order to find out which role the CdS buffer layer plays in the band-bending phenomena, the investigated sample was dipped into a 5 w% - diluted solution of hydrochloric acid for 3 mins. Such a chemical bath is expected to etch away the CdS layer without a damaging the CIGS layer.

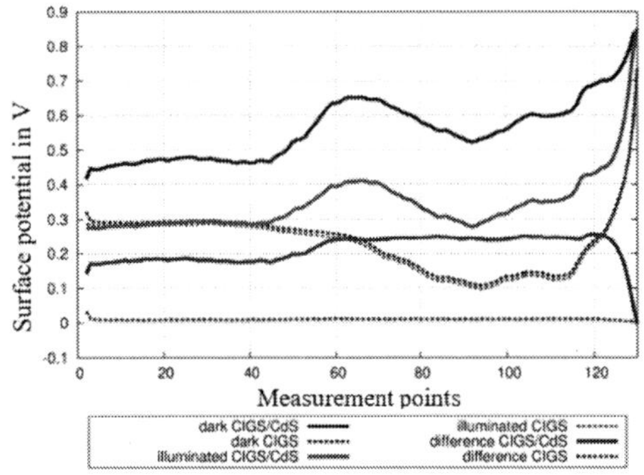

Fig. 8. Line scan of the surface potential in the dark (black), under white light illumination (red) and the difference between the two (blue). Solid lines correspond to the measurements on the CIGS/CdS heterostructure, whereas dotted lines correspond to the measurements on the bare CIGS absorber. The stepsize of the measurement is 0.5 mm.

As can be seen in Fig. 8 etching away the buffer layer eliminates the surface photo voltage, indicating a reduction of the band-bending measured previously. The surface potential scans in the dark and under illumination overlap pointing out once again that the charge accumulation at the CIGS/CdS interface or in the CdS layer is responsible for the observed degradation effects.

4 CONCLUSION

With the high resolution and short measurement time, PL imaging can be used as an in-line characterization tool to enhance production of highly-efficient CIGS solar cells. Since PL intensity is directly linked to the quasi-Fermi level splitting, PL images can serve as a quality metrics of the CIGS absorbers if the imaging results are interpreted correctly.

In this work, it has been shown that the degradation effects in CIGS/CdS films after thermal ageing result in the quenching of PL intensity. The vibrating Kelvin probe measurements applied complementary on the same sample demonstrated that the band-bending at the CIGS/CdS interface induced by the accumulation of charges at the interface or in the CdS layer can be responsible for the observed PL intensity degradation. Etching away the buffer layer confirmed the statement. The Kelvin probe measurements on the bare absorber showed significantly reduced band-bending as compared to the measurements with the CdS layer. This finding indicates that PL intensity may additionally be affected by electric fields invoked at grain boundaries or by charge accumulation close to the surface. Such electric fields lead to a charge separation inhibiting radiative band-to-band recombination and increasing the rates of the non-radiative recombination at the interface.

5 ACKNOWLEDGEMENT

This work has been funded by German Federal Ministry of Economic Affairs and Energy within the OptiCIGS project and by the Donauhochschulen – Forschung am Wissensfluss.

6 REFERENCES

[1] T. Trupke, B. Mitchell, J. W. Weber, W. McMillan, R. A. Bardos und R. Kroeze, „Photoluminescence imaging for photovoltaic applications,“ *Energy Procedia,* pp. 135-146, 2012.

[2] W. K. Metzger, R. K. Ahrenkiel, J. Dashdorj und D. J. Friedman, „Analysis of charge separation dynamics in a semiconductor junction,“ *Physical review B,* p. 035301, 2005.

[3] M. Maiberg, C. Spindler, E. Jarzembowski und R. Scheer, „Electrical characterization of Cu(In,Ga)Se2 - solar cells by voltage-dependent time-resolved photoluminescence,“ *Thin Solid Films,* pp. 379-382, 1 May 2015.

[4] J. Mizsej, „Surface potential transients of ultrathin SiO2/Si structures,“ *Solid state electronics,* pp. 951-957, 1991.

[5] P. Mack, T. Ott und T. Walter, „Optimization of reliability and metastability of CIGS solar cell parameters,“ in *Proceedings of the 25th EU PVSEC,* Valencia, Spain, 2010.

THE NEGATIVE INFLUENCES OF EXCESSIVE OXYGEN GAS ON THE ELECTRICAL PROPERTIES OF ITO FILMS DEPOSITED BY MAGNETRON SPUTTERING

Xuehai Tan[1], Shou Peng[2,3], Xin Cao[1,2], Jingong Pan[4], Xinwei Wang[5], Alan E. Delahoy[1] and Ken K. Chin[1]
1. CNBM New Energy Materials Research Center, New Jersey Institute of Technology, Newark, NJ, USA.
2. Bengbu Design & Research Institute for Glass Industry, China.
3. China Triumph International Engineering Co. Ltd, Shanghai, China.
4. CNBM Optoelectronic Materials Co. Ltd., Chengdu, China.
5. Evans Analytical Group, Liverpool, NY, USA.

ABSTRACT: Indium tin oxide (ITO) films deposited on glass substrate can be used as the transparent electrode for thin film photovoltaics, such as CdTe-based thin film solar cells. The most widely used fabrication technique for ITO films is magnetron sputtering directly from the ceramic target. While the preparation of ITO films is known for decade, there is still lacking of consensus whether adding oxygen to the sputtering gas of Ar is beneficial to the film electrical properties. Herein, we present a systematic experimental study comparing the properties of ITO films deposited under pure Ar versus under Ar-O_2 mixture with increasing O_2 partial pressure. Our results of Hall measurements show that adding more oxygen in the sputtering gas has negative influences on the electrical conductivity of ITO films. The origin of these negative effects are discussed with a focus on the concentration of ionized impurities and the scattering of grain barriers. We employed extensive XPS analysis to provide insight into the variation of oxygen vacancy concentration as a function of sputtering O_2 partial pressure. The ITO film microstructure was characterized using XRD, STEM and high-magnification SEM analyses. Our results point to the oxygen vacancy concentration that dictates the carrier concentration and thus the resistivity of magnetron sputtered ITO films.
Keywords: ITO; magnetron sputtering; XPS; oxygen vacancy concentration; carrier concentration

1 INTRODUCTION

In the past decades, transparent conducting oxide (TCO) films, which transmit light and conduct electrical current simultaneously, are of increasing importance for various applications, including optoelectronic and photovoltaic devices [1,2]. Recently, the world record of thin film solar cell conversion efficiency has been increased significantly [3]. Through the spectral quantum efficiency measurement and data analysis, it was estimated that the most noticeable efficiency improvements have been originated from the development of better superstrate - including a high performance TCO layer [4-6], a high resistant transparent (HRT) layer [7,8] and a wide band gap buffer layer [9-11]. For TCO layer that is used as the front contact for thin film solar cells, a higher electrical conductivity presumes a thinner film needed. This improves both short circuit current density and light transmission, and thus enhances the conversion efficiency of solar cells.

Among the large list of different TCO materials [12-16], indium tin oxide (ITO) is one of the common materials because of its low resistivity, high transparency and good etchability [17,18]. While ITO films can be prepared from a wide range of deposition methods [19-21], the most widely used fabrication technique for ITO films is magnetron sputtering directly from the ceramic target [22,23]. The electrical properties of ITO films depend on the sputtering conditions, such as the oxygen partial pressure [24,25]. Choi et al. showed that the increase of oxygen partial pressure results in a reduction of both carrier mobility and concentration, leading to an increase of resistivity in ITO thin films [26]. Kim et al. on the other hand showed that carrier mobility increases with oxygen partial pressure, leading to a decrease of resistivity in ITO thin films [27]. A more recent study by Chen et al. also visited such issues and showed that at higher oxygen partial pressure the carrier concentration increases but the carrier mobility decreases [28]. Apparently, despite having been subject to extensive research, there is still lacking of consensus whether adding oxygen to the sputtering gas of Ar is beneficial to the electrical properties of ITO films. We believe more work can be useful to clarify such point, maybe supporting one argument over another.

Herein, we intend to address the aforementioned issue by presenting an experimental study comparing the properties of ITO films deposited under pure Ar *versus* under Ar-O_2 mixture with increasing O_2 partial pressure. The comparisons were performed systematically - using identical sputtering condition (except oxygen flow rates), a simple sputtering process at 250 °C without additional postdeposition annealing treatment, and identical film thickness. To avoid variation in target composition, single ITO target has been used to deposit all films investigated in this study. We used Hall effect measurement to evaluate the carrier mobility, concentration, and the film resistivity. We utilized a variety of analytical techniques including X-ray diffraction (XRD), high-magnification scanning electron microscopy (SEM) and scanning transmission electron microscopy (STEM) to characterize the film microstructure, obtaining information such as grain size distribution. We employed X-ray photoelectron spectroscopy (XPS) as our primary tool to analyze the ionized impurities as a function of oxygen partial pressure in the sputtering gas mixture. In a typical TCO materials, the ionized impurities could serve as the doping agents but at the same time as the scattering centers for carriers [17]. Combining the results of these analyses demonstrates the interrelationship between the oxygen vacancies and the carrier density in ITO films. The diminishing oxygen vacancy concentration, observed for ITO films deposited with increasing oxygen partial pressure, accounts for the negative effects on the electrical properties of ITO films.

2 EXPERIMENTAL

All ITO thin films were deposited on glass substrates by dc magnetron sputtering using a sputtering power of 80W. The substrate temperature was maintained at 250 °C for all depositions. The sputtering target was purchased containing 90 wt.% of In_2O_3 and 10 wt.% of SnO_2, according to the information provided by the manufacturer. To avoid variation in target composition, single ITO target has been used to deposit all films investigated in this study. The glass substrates were consistently cleaned before placing into the deposition chamber. Argon and oxygen gases with purity of 99.999% were used at a working pressure of 0.4 Pa, with a maximum base pressure of 8×10^{-4} Par. While Ar flow rate was kept constant at 80 sccm, various oxygen flow rates, i.e. 0, 2, 3, 4, 5 and 6 sccm were used for different sputtering gas mixtures, corresponding to an increasing ratio of O_2 flow rate over the total Ar-O_2 flow rates [O_2/(Ar+O_2) %] of 0, 2.44, 3.61, 4.76, 5.88 and 6.98%, respectively. For simplicity, the ITO films deposited under these conditions were denoted as ITO-1 to ITO-6, respectively. Considering the sputtering gas mixture as ideal gas, the flow rate of O_2 is proportional to its partial pressure given $P\ (dV/dt) = (dn/dt)\ RT$. Therefore, the increasing O_2 flow rate at a fixed Ar flow rate can be described more generally as an increasing oxygen partial pressure. The sputtering rate was accurately measured in-situ using a crystal monitor at the substrate plane. Additional ex-situ calibrations were performed using a DekTak thin film thickness profiler (Veeco) to confirm the sputter rates employed and the resultant thicknesses. The thicknesses of all ITO films were fixed at 100 nm, unless otherwise stated.

The Hall effect measurements were performed in van der Pauw geometry at room temperature using the Ecopia HMS-3000 Hall measurement system. The film sheet resistance was measured using a 4-point probe (Guardian) for quick check of film conductivity. X-ray photoelectron spectroscopy (XPS) measurements were performed on an ULTRA (Kratos Analytical) spectrometer under ultrahigh vacuum (10^{-9} Torr) using monochromated Al Kα radiation ($h\nu = 1486.6$ eV) operated at 210 W. All XPS spectra were calibrated using the universal hydrocarbon contamination C 1s peak at 284.8 eV. X-ray diffraction (XRD) analysis was performed on a Bruker AXS diffractometer (Bruker Discover 8) using Cu Kα radiation ($\lambda = 1.5406$ Å) that was monochromated using a single Gobel mirror. The diffractometer was equipped with a Histar general area two-dimensional detection system (GADDs) with a sample–detector distance of 15 cm. For XRD grain size analysis, the instrumental broadening was carefully determined by using LaB_6 standard material and subtracted from the measured line broadening. Scanning electron microscopy (SEM) analysis was performed with a Zeiss Sigma field emission SEM operated at 10 kV accelerating voltage. Focused ion beam (FIB) lift-out was performed on an FEI Helios 660 Dual Beam FIB/SEM to prepare the cross-section specimen for scanning transmission electron microscopy (STEM) analysis. Prior to introduction into the dual beam, a thin layer of ink was applied to the surface along with a thin sputter-coated Pt layer. Before milling an FIB deposited protective Pt-C layer was applied. The electron transparent cross-section specimen was imaged in a Hitachi HD2700 STEM operated at 200 kV.

3 RESULTS AND DISCUSSIONS

Figure 1 shows the results of Hall measurements on 100 nm thick ITO films. It is found that adding more oxygen in the sputtering gas has negative influences on the electrical conductivity of ITO films. The measured resistivity of ITO film rises almost exponentially from 7.9×10^{-4} to 4.1×10^{-2} Ω.cm with increasing oxygen partial pressure. Specifically, the carrier concentration of ITO films decreases from 4.8×10^{20} to 5.4×10^{18} cm^{-3} for O_2/(O_2+Ar)% from 0 to 6.98%. The Hall mobility shows a general trend of increasing from 17 to 45 $cm^2V^{-1}s^{-1}$ for oxygen addition up to O_2/(O_2+Ar)% of 4.76%; beyond that the Hall mobility decreases and drops to a value of 28 $cm^2V^{-1}s^{-1}$ at O_2/(O_2+Ar)% of 6.98%. It should be mentioned that our results actually do not agree with any of the three works cited above. Instead, our results are consistent with the findings of another work, in which the dependencies of resistivity, carrier density and Hall mobility on oxygen deposition pressure were investigated for ITO films grown by pulsed laser deposition [29]. It is clear that the variation of carrier mobility as a function of oxygen partial pressure is not monotonic. As will be elaborated later, this is due to the combined influences of several charge carrier scattering mechanisms.

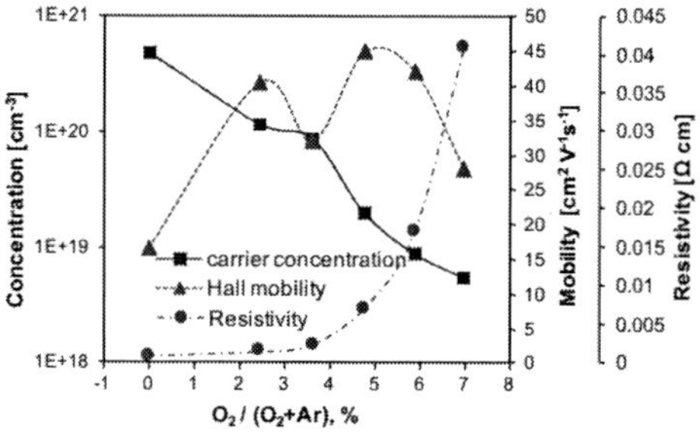

Figure 1: The measured carrier concentration, Hall mobility and resistivity of ITO films as a function of increasing O_2/(Ar+O_2)% in the sputtering gas mixture.

Figure 2 summarizes the results of microstructural characterizations. The STEM micrographs for the cross-section of a typical ITO thin film deposited on glass substrate are shown in Figure 2(a). For the purpose of STEM analysis, the thickness of this ITO film was selected to be 25 nm. The deposited film is polycrystalline with no or low evidence of amorphous phases. As shown by the high resolution STEM micrograph (see bottom inserted micrograph), each ITO crystallite may actually occupy the entire film thickness. On the basis of this observation, we gestimated the columnar grain structure from the faceted top surface, as sketched in the inserted micrograph near the top. We acknowledge that the columnar grain structure observed for 25 nm thick film may not represent the microstructure of thicker films. As described in a typical competitive van der Drift type crystallite growth that occurs during film deposition [30,31], the grain probability of survival diminishes with increasing film thickness.

Figure 2(b) shows the results of XRD analysis on 100 nm films of ITO-1 to ITO-6. All films are polycrystalline single-phase SnO_2-doped In_2O_3 solid solution with a body-centered cubic (bcc) crystal structure (Ia-3 (206)). There is no detectable variation in terms of the diffraction peak position. This is consistent with the fact that all six ITO films should in principle contain the same amount of SnO_2. As oxygen partial pressure increases, the resulted ITO films become slightly more textured with [222] preferred

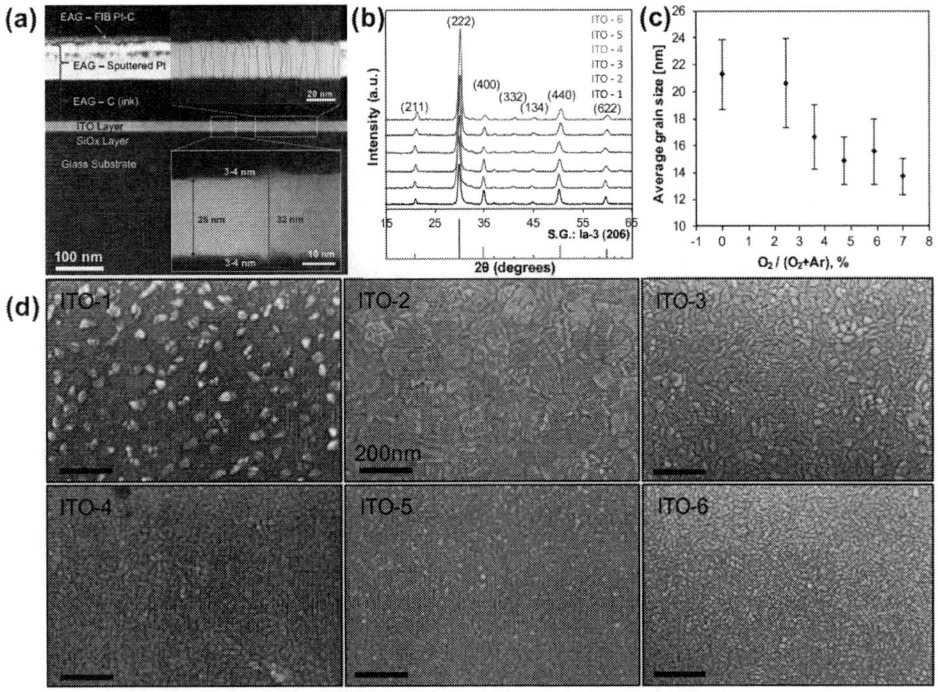

Figure 2: (a) STEM micrographs for the cross-section of a typical ITO thin film deposited on glass substrate. (b) XRD of ITO films deposited under different oxygen partial pressures in the Ar-O_2 sputtering gas mixture. (c) The results of grain size analysis. The error bars show the deviation from the mean. (d) High-magnification SEM micrographs of ITO films. ITO-1 to 6 represent films deposited with $O_2/(Ar+O_2)$% of 0, 2.44, 3.61, 4.76, 5.88 and 6.98%, respectively.

growth orientation. Close inspection of the XRD patterns reveals that there is diffraction peak broadening for ITO films deposited at higher oxygen partial pressure. Accordingly, we performed grain size analysis using Scherrer equation to estimate the average grain size of each film. As depicted in Figure 2(c), the average grain size of ITO film becomes gradually smaller as a function of increasing $O_2/(Ar+O_2)$% in the sputtering gas mixture. This trend of decreasing grain size can also be confirmed from the direct observation via SEM imaging. Figure 2(d) shows the high magnification SEM micrographs of the six ITO films.

Considering the carrier mobility in polycrystalline degenerately doped semiconducting material is limited by scattering mechanisms such as ionized impurity, lattice vibration, crystallographic defects and grain barrier scattering [32,33]. A decrease in grain size should in principle result in higher probabilities of grain barrier scattering, leading us to expect a decrease in carrier mobility as the oxygen partial pressure increases in the sputtering gas. However, as shown in Figure 1, the measured Hall mobility does not follow such expectation, at least for $O_2/(O_2+Ar)$% up to 4.76%. This suggests that for high carrier concentrations the change in grain size is not the predominant factor, or plays only a secondary role behind the observed trend of carrier mobilities.

In addition, we compare the mean free path of electrons (l_e) with the average grain size. The mean free path for electrons in n-type degenerately doped semiconductor can be calculated using [13,34]:

$$l_e = \left(\frac{h}{2q}\right)\left(\frac{3n}{\pi}\right)^{1/3}\mu$$

where h is the Planck's constant, q is the electron charge, n is the measured carrier concentration and μ is the measured Hall mobility. For ITO-1 to 4, the calculated values of l_e are 2.6, 4.0, 2.9 and 2.5 nm, respectively, all being considerably smaller than the corresponding grain size. These comparisons support our argument qualitatively that the grain barrier scattering only has minor/negligible effect on carrier transport for ITO films that have carrier concentration above 10^{19} cm^{-3}. Our argument is in accordance with the conclusions of an earlier study on carrier transport in various polycrystalline TCO films: While grain barriers were found to limit the carrier mobility in ZnO for carrier concentration as high as 2×10^{20} cm^{-3}, the influences due to grain barriers were not observed for ITO down to a carrier concentration of ~ 10^{18} cm^{-3} [33].

As the measured carrier concentrations of ITO-1 and ITO-6 differ by almost two orders of magnitude, the electrical conductivities of ITO-1 to 6 may be predominantly determined by their decreasing carrier concentrations. Herein, we used x-ray photoelectron spectroscopy (XPS) as our primary technique to study the ITO films' chemical bonding environment, in order to find evidences correlating the ionized impurities and the varied carrier concentrations. Figure 3(a) shows the XPS O 1s spectra of the ITO films that were deposited at different oxygen partial pressures. The O 1s spectra can be fitted with two components at ~ 529.8 eV and ~ 531.4 eV, respectively [35]. The former component, labeled as O_I, represents oxygen ions that are surrounded by In or the substituted Sn atoms with their full complement of nearest neighbor O^{2-} ions; the latter component, labeled as O_{II},

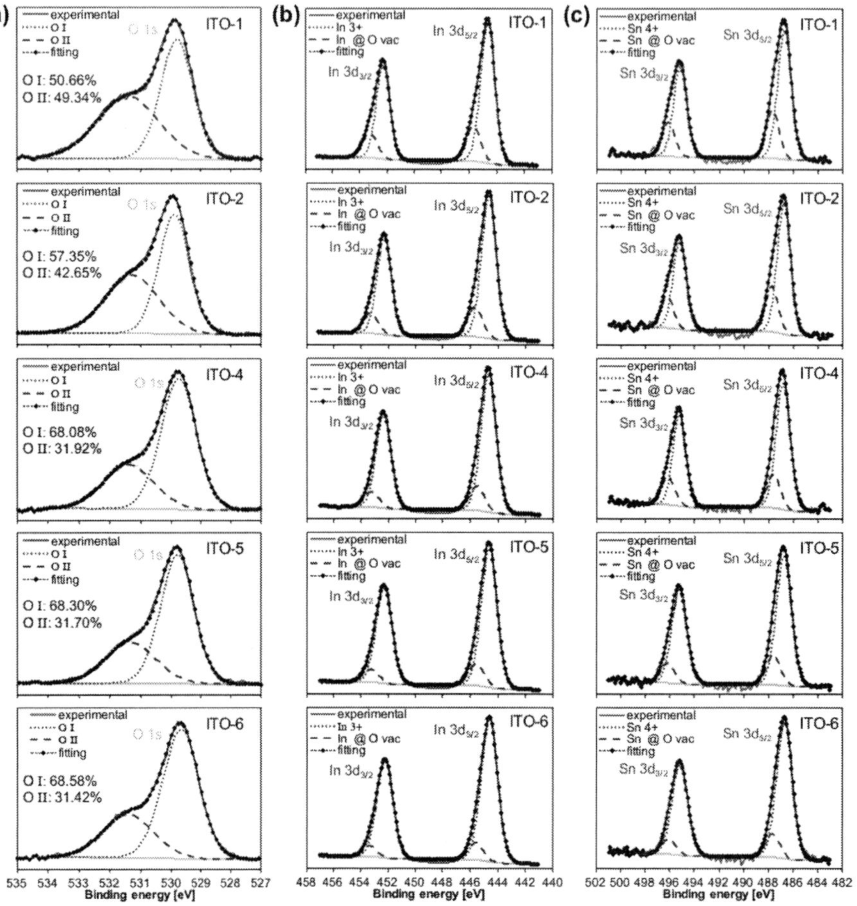

Figure 3: Results of XPS analysis for ITO films: (a) O 1s, (b) In 3d, and (c) Sn 3d. ITO-1 to 6 represent films deposited with $O_2/(Ar+O_2)$% of 0, 2.44, 3.61, 4.76, 5.88 and 6.98%, respectively.

represents O^{2-} ions arising from the oxygen-deficient region, which may be connected in part to the oxygen vacancies [36-38]. As a result, comparing the ratio of integrated areas of O_{II} and O_I for different ITO films enables us to obtain a sensitive indicator of the variation of oxygen vacancy concentration [36,38]. It must be kept in mind that such ratio does not represent the actual level of oxygen vacancies in ITO. The XPS O 1s spectrum should also contain components arising from the sample surface adsorption/contamination, but the amount of surface adsorption should be similar across all ITO films that were examined here.

Our results demonstrate a decreasing trend of oxygen vacancy concentration as a function of increasing $O_2/(O_2+Ar)$% in the sputtering gas. At present, it is accepted by many researchers that oxygen vacancy is one of the fundamental sources acting as donors in ITO and some other TCOs [39,40]. An oxygen vacancy in the In_2O_3 lattice is known to generate two free electrons [17]. By plotting the ratio of two integrated areas O_{II}/O_I together with our measured data of carrier concentration, not in a log scale (see Figure 4(a)), two sets of data actually have analogous variation trends as a function of $O_2/(O_2+Ar)$%. This comparison reveals a clear evidence demonstrating the interrelationship between the oxygen vacancies and the carrier density of the ITO films.

Figure 3(b) and (c) show the XPS spectra of In 3d and Sn 3d, respectively, for the ITO films that were deposited at different oxygen partial pressures. For ITO, the substitution of In^{3+} by Sn^{4+} makes Sn the extrinsic ionized impurity, contributing to the free electron population. Table 1 shows the results of quantitative analysis of the XPS spectra. The near-surface chemical compositions of all ITO films are quite consistent, showing In wt.% of 91.2±0.2% and Sn wt.% of 8.8±0.2%. Additionally, there is no evidence of chemical state Sn^{0+} or Sn^{2+} for any ITO films examined here. Therefore, it may be safe for us to argue that the ionized impurity Sn is not responsible for the decreasing carrier concentration of ITO films deposited at higher oxygen partial pressures.

Table I: Quantitative analysis of the XPS results, summarizing the In and Sn wt.% at the near-surface of ITO films.

	ITO-1	ITO-2	ITO-4	ITO-5	ITO-6
In wt%	90.9	91.4	91.2	91.4	91.1
Sn wt%	9.1	8.6	8.8	8.7	8.9

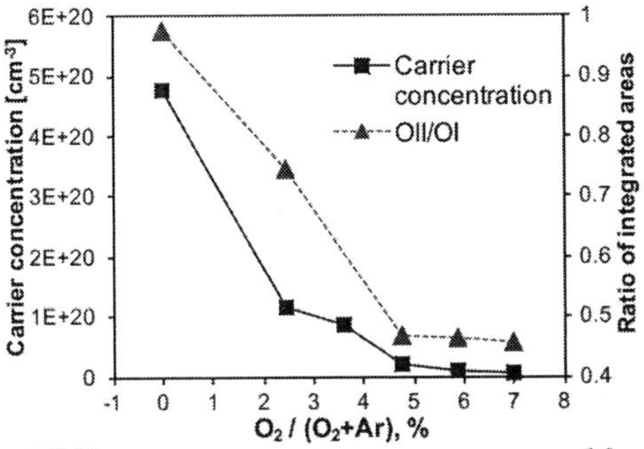

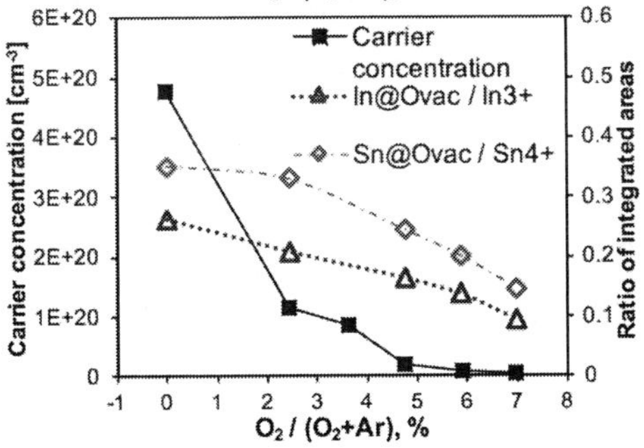

Figure 4: Comparisons of (a) O_{II}/O_I, and (b) $In@O_{vac}/In^{3+}$ and $Sn@O_{vac}/Sn^{4+}$ with our measured data of carrier concentration shown in Figure 1.

Interestingly, for both In and Sn each of their spectra can be fitted with two components, one corresponding to the valance state of In^{3+} in In-O bond (or Sn^{4+} in Sn-O bond), and one corresponding to a core level (CL) that actually has shifted to an even higher binding energy. Similar CL shift to a higher binding energy has been previously reported for Zn 2p XPS spectra of ZnO, which have been assigned to Zn^{2+} in Zn-O bond but being surrounded by oxygen vacancies [41]. It is clear from Figure 3(b) and (c) that the latter component of In or Sn also diminishes in its integrated area as a function of increasing oxygen partial pressure. This observation indicates that the latter component of In or Sn spectrum can be related to oxygen vacancies, and thus being labeled as $In@O_{vac}$ or $Sn@O_{vac}$, respectively. By forming an oxygen vacancy in the oxide lattice, the remaining valance electrons, which were originally shared by In-O or Sn-O in the covalent bond, are now bonded much weakly. During the XPS analysis, these electrons can be excited into the conduction band, and consequently the CL of In or Sn can shift towards a higher binding energy comparing to that of In^{3+} or Sn^{4+}.

As shown in Figure 4(b), we have plotted the ratio of integrated areas for $In@O_{vac}/In^{3+}$ and for $Sn@O_{vac}/Sn^{4+}$ as a function of $O_2/(O_2+Ar)\%$. Both ratios are decreasing, but do not exactly follow the variation trend of carrier concentration or that of oxygen vacancy concentration (see Figure 4(a)). Probably, the arguments on oxygen vacancy concentration on the basis of In and Sn XPS spectra can be only taken with a grain of salt. However, from an overall viewpoint, our XPS results on O, In and Sn provide a consistent picture regarding the dependence of oxygen vacancy concentration on the oxygen partial pressure in the sputtering gas. This in turn predominantly determines the carrier concentration of the ITO films.

Finally, it is important to discuss that while oxygen vacancy and Sn are the doping agents, they also serve as scattering centers for carriers. At high dopant levels, the charge carrier scattering processes due to ionized impurities should be the dominant mechanisms limiting the carrier mobilities of our ITO films. Since oxygen vacancies are doubly charged, its scattering power is higher than that of singly ionized dopant Sn in ITO. As oxygen vacancy concentration decreases, the population of oxygen vacancy scattering centers reduces accordingly. This could explain our observation that the carrier mobility increases from ITO-1 to ITO-4, until the carrier concentration of ITO film reduces to a certain threshold. It is noted that further lowering of carrier concentration does not lead to higher mobility. Earlier study on ITO has shown that the onset of grain barrier limited transport, transiting from the ionized impurity limited transport, is in the order of 10^{18} cm^{-3} [33]. As carrier concentrations of ITO-5 and ITO-6 drop to the order of 10^{18} cm^{-3}, their mobilities start to show positive correlation with the measured film grain size. In this region, the carrier mobility exhibits strong dependence on grain barriers and other crystallographic defects.

4 CONCLUSION

In this study we systematically clarified the influences of higher oxygen partial pressure on the electrical properties of ITO films deposited by magnetron sputtering. The film conductivity is negatively affected, because the carrier concentration of ITO films is decreasing with increasing oxygen partial pressure in the $Ar\text{-}O_2$ sputtering gas. The negative influences on electrical properties are discussed based on the results of extensive microstructural characterizations. XPS was used as the primary technique to elucidate oxygen-related defects. On the basis of XPS analysis, it is safe for us to conclude that: It is the oxygen vacancy concentration, which decreases as oxygen partial pressure increases in the sputtering gas, that dictates the carrier concentration and thus the conductivity of magnetron sputtered ITO films.

5 ACKNOWLEDGEMENTS

The authors acknowledge the China Triumph International Engineering Co. Ltd. (CTIEC), Shanghai, China, which offers generous financial support for this work. Authors thank Evans Analytical Group materials characterization division for the FIB-STEM analysis.

6 REFERENCES

[1] Minami T, *Thin Solid Films* **2008**, 516: 5822-5828.
[2] Fortunato E, Ginley D, Hosono H, Paine DC, *MRS Bulletin* **2007**, 32: 242-247.
[3] http://www.firstsolar.com/en/About-Us/Press-Center/Blog/2015/June/First-Solar-Sets-Another-Module-Efficiency-Record.aspx
[4] Delahoy AE, Guo S, "Transparent conducting oxides for photovoltaics". Chapter 17 in *Handbook of Photovoltaic Science and Engineering*, 2nd Edition, John Wiley & Sons, Ltd, Chichester, pp. 716–796, 2011.
[5] Jeyadheepan K, Thamilselvan M, Kim K, Yi J, Sanjeeviraja C, *J Alloys Compds* **2015**, 620: 185-191.

[6] Ghorannevis Z, Akbarnejad E, Salar Elahi A, Ghoranneviss M, *J Inorg Organomet Polym* **2015**, 25: 1486-1489.
[7] Wang JJ, Ling T, Qiao SZ, Du XW, *ACS Appl Mater Interfaces* **2014**, 6: 14718-14723.
[8] Pena JL, Hernandez-Rodriguez E, Rejon V, Mis-Fernandez R, Riech I, "Thermal Behavior of ITO/TiO_2/CdS/CdTe Solar Cells" in *42nd IEEE Photovoltaic Specialists Conference*, **2015**.
[9] Meysing DM, Wolden CA, Griffith MM, Mahabaduge H, Pankow J, Reese MO, Burst JM, Rance WL, Barnes TM, *J Vac Sci Technol A* **2015**, 33(2): 021203.
[10] Duncan DA, Kephart JM, Horsley K, Blum M, Mezher M, Weinhardt L, Haming M, Wilks RG, Hofmann T, Yang W, Bar M, Sampath WS, Heske C, *ACS Appl. Mater. Interfaces* **2015**, 7: 16382-16386.
[11] Kephart JM, Geisthardt RM, Sampath WS, *Prog Photovolt: Res Appl* **2015**, 23: 1484-1492.
[12] Ellmer K, *Nat Photonics* **2012**, 6: 809-817.
[13] Gao Q, Li M, Li X, Liu Y, Song CL, Wang JX, Liu QY, Liu JB, Han GR, *J Alloys Compds* **2013**, 550: 144-149.
[14] Zhu Y, Mendelsberg RJ, Zhu J, Han J, Anders A, *J Mater Sci* **2013**, 48: 3789-3797.
[15] Panda D, Tseng TY, *J Mater Sci* **2013**, 48: 6849-6877.
[16] Lu Y, Khan S, Song CL, Wang KK, Yuan GZ, Li W, Han GR, Liu Y, *J Alloys Compds* **2016**, 663: 413-418.
[17] Minami T, *Semicond Sci Technol* **2005**, 20: S35–S44.
[18] Tsai TH, Wu YF, *Microelectronic Engineering* **2006**, 83: 536–541.
[19] Linss V, *Surface & Coatings Technology* **2016**, 290: 43–57.
[20] Joshi SM, Gerhardt RA, *J Mater Sci* **2013**, 48: 1465-1473.
[21] Wegener M, Kato M, Kakimoto K, Spallek S, Spiecker E, Roosen A, *J Mater Sci* **2015**, 50: 6124-6133.
[22] Hu Y, Diao X, Wang C, Hao W, Wang T, *Vacuum* **2004**, 75: 183-188.
[23] Terzini E, Thilakan P, Minarini C, *Materials Science and Engineering B* **2000**, 77: 110-114.
[24] Kurdesau F, Khripunov G, da Cunha AF, Kaelin M, Tiwari AN, *J Non-Crystalline Solids* **2006**, 352: 1466-1470.
[25] Tseng KS, Lo YL, *Appl Surf Sci* **2013**, 285P: 157–166.
[26] Choi CG, No K, Lee WJ, Kim HG, Jung SO, Lee WJ, Kim WS, Kim SJ, Yoon C, *Thin Solid Films* **1995**, 258: 274-278.
[27] Kim YJ, Jin SB, Kim SI, Choi YS, Choi IS, Han JG, *Thin Solid Films* **2010**, 518: 6241-6244.
[28] Chen A, Zhu K, Zhong H, Shao Q, Ge G, *Sol Energy Mater Sol Cells* **2014**, 120: 157-162.
[29] Kim H, Gilmore CM, Piqué A, Horwitz JS, Mattoussi H, Murata H, Kafafi ZH, Chrisey DB, *J Appl Phys* **1999**, 86: 6451-6461.
[30] Ophus C, Luber E, Mitlin D, *Acta Maters* **2009**, 57: 1327-1336.
[31] Ophus C, Ewalds T, Luber EJ, Mitlin D, *Acta Maters* **2010**, 58: 5150-5159.
[32] Stapinski T, Leja E, Pisarkiewicz T, *J Phys D: Appl Phys* **1984**, 17: 407-413.
[33] Ellmer K, Mientus R, *Thin Solid Films* **2008**, 516: 4620–4627.
[34] Wohlmuth W, Adesida I, *Thin Solid Films* **2005**, 479: 223-231.
[35] Pammi SV, Chanda A, Ahn JK, Park JH, Cho CR, Lee WJ, Yoon SG, *J Electrochem Soc* **2010**, 157: H937-H941.
[36] Simanjuntak FM, Panda D, Tsai TL, Lin CA, Wei KH, Tseng TY, *J Mater Sci* **2015**, 50: 6961-6969.
[37] Chen M, Wang X, Yu YH, Pei ZL, Bai XD, Sun C, Huang RF, Wen LS, *Appl Surf Sci* **2000**, 158: 134-140.
[38] Kim JS, Ho PKH, Thomas DS, Friend RH, Cacialli F, Bao GW, Li SFY, *Chem Phys Lett* **1999**, 315: 307-312.
[39] Gwamuri J, Marikkannan M, Mayandi J, Bowen PK, Pearce JM, *Materials* **2016**, 9: 63.
[40] Stadler A, *Materials* **2012**, 5: 661-683.
[41] Chen YY, Hsu JC, Lee CY, Wang PW, *J Mater Sci* **2013**, 48: 1225-1230.

$CuInSe_2$ NANOSTRUCTURES PREPARED BY METAL ORGANIC CHEMICAL VAPOUR DEPOSITION FOR HYBRID PHOTOVOLTAIC DEVICES

S. Vatavu[1,2,*], N. von Morze[1], J. Albert[1], S. Wiesner[1], V. Hinrichs[1], M.-Ch. Lux-Steiner[1], M. Rusu[1]
[1]Helmholtz-Zentrum Berlin für Materialien und Energie GmbH,
Hahn-Meitner-Platz 1, 14109 Berlin, Germany
[2]Moldova State University, 60 A. Mateevici str., 2009 Chisinau, Moldova
(*e-mail: svatavu@usm.md)

ABSTRACT: $CuInSe_2$ nanostructures for application in organic/inorganic hybrid solar cells have been prepared by the use of Metal Organic Chemical Vapour Deposition (MOCVD). We demonstrate stoichiometric $CuInSe_2$ nanostructures with controllable nano-crystallites' size and spacing suitable for application in hybrid photovoltaics. The control of the crystallites' stoichiometry, size and spacing has been carried out by the fine adjustment of the MOCVD preparation parameters such as precursor flows, substrate temperature and deposition time. 10 and 20 nm high $CuInSe_2$ nanostructures have been prepared on Mo/Na-barrier/glass substrates at 500°C. The structures' physical properties have been studied by grazing incidence x-ray diffraction and Laser Ablation Inductively Coupled Plasma Mass Spectrometry.
Keywords: $CuInSe_2$, nanostructure, MOCVD, hybrid.

1 INTRODUCTION

The search of a better route in achieving high efficiency photovoltaic (PV) converters brought into light the possibility of combining inorganic and organic materials. By using the later ones one is benefiting such advantages as low temperature preparation as well as the easiness of flexible substrate use. The main challenges whenever organics is used in PV are related to short excitonic diffusion length, low mobility of charge carriers and morphology of the resulting structure [1]. Doping in organic semiconductors can help in overcoming some of the problems [2]. The donor or acceptor organic materials choice is influencing the open circuit voltage [3].

A combination of organic and inorganic materials in hybrid solar cells might result in higher efficiencies compared to that of "pure" organic cells, which is determined by increase of the photon absorption and the choice of the inorganic semiconductor. Hybrid solar cells are predicted to have high conversion efficiencies [4]. Among the inorganic semiconductors, $CuInSe_2$ has already proven its use in PV as conventional thin film. Being prepared by RF sputtering [5] as a film, one can also prepare it in the nanoscale range [6]. The influence of [Cu]/[In] composition ratio on the properties of $CuInSe_2$ by Pulsed Laser Deposition followed by selenization has been investigated for thin films [7]. The road to nanoscale can be reached, as colloidal nanocrystals [8] or, for a more practical use in hybrid PV, nanoparticle $CuInSe_2$ can be prepared from printed CuO/In_2O_3 nanoparticle-based inks under Se ambient [9].

$CuInSe_2$ nanocrystals' bandgap decreases from 3.3 eV down to 1.3e V as their size increases from 1 nm to 6 nm [10] clearly demonstrating the possibilities of main semiconductor parameter tuning. There are still insufficient experiments aimed at studying the compositional variations of $CuInSe_2$ nanostructures by other preparation methods. Thus, MOCVD being a relatively "low-rate" deposition method might be a suitable choice.

2 EXPERIMENTAL

$CuInSe_2$ nanostructures have been prepared by MOCVD (Aixtron system) on Mo/Na-barrier/glass substrates by using the following precursors: Cyclopentadienyl (triethylphosphine) Copper (I) $(C_5H_5)CuP(C_2H_5)_3$, Trimethylindium & N, N Dimethyldodecylamine $C_{14}H_{31}N(CH_3)_3In$, and Ditertiarybutylselenide $Se(C_4H_9)_2$. A steady substrate temperature of 500°C has been kept for all processes. The [Cu]/[In] precursor flow ratio has been varied in the range of 0.612 to 1.708 for reaching the optimum for nano-$CuInSe_2$ growth conditions. The variation of the process time allowed the preparation of different nanostructures consisting of nanoparticles (NPs) (nanostructures) with controlable height of 10 and 20 nm (growth rate - 1 nm/min).

The X-ray diffraction based characterization setup has been used for phase analysis, namely: PANalytical X'Pert Pro MRD diffractometer (Cu X-ray tube) in grazing incidence (GI-XRD) configuration.

The elemental composition of the as-prepared $CuInSe_2$-nanostructures were analyzed by Laser Ablation Inductively Coupled Plasma Mass-Spectrometry (LA-ICP-MS) (Teledyne CETAC LSX-213 Laser Ablation System). The vaporization of the nanostructured thin films and nanostructures ocurred from the entire thickness.

Surface morphology and cross-section views have been analyzed by use of a Zeiss GEMINI LEO Supra 35 SEM system.

3 RESULTS AND DISCUSSION

Structural characterization included the analysis of X-ray diffraction in grazing incidence configuration (grazing incidence angle varied between 0.1° and 1.1° for choosing the optimum measurement conditions). A detailed phase analysis of the prepared samples has been performed for 0.3° grazing incidence angle for the whole range of precursor flow ratio and depostion time (Fig. 1). The typical morpholgy of the obtained $CuInSe_2$ nanostructures depending on the precursor flow ratio is given in Fig. 2.

The XRD pattern is clearly dominated by the presence of $CuInSe_2$ peaks, namely (112), (204)/(220), (116)/(332) (73351-ICSD). It can be seen (Fig. 1) that

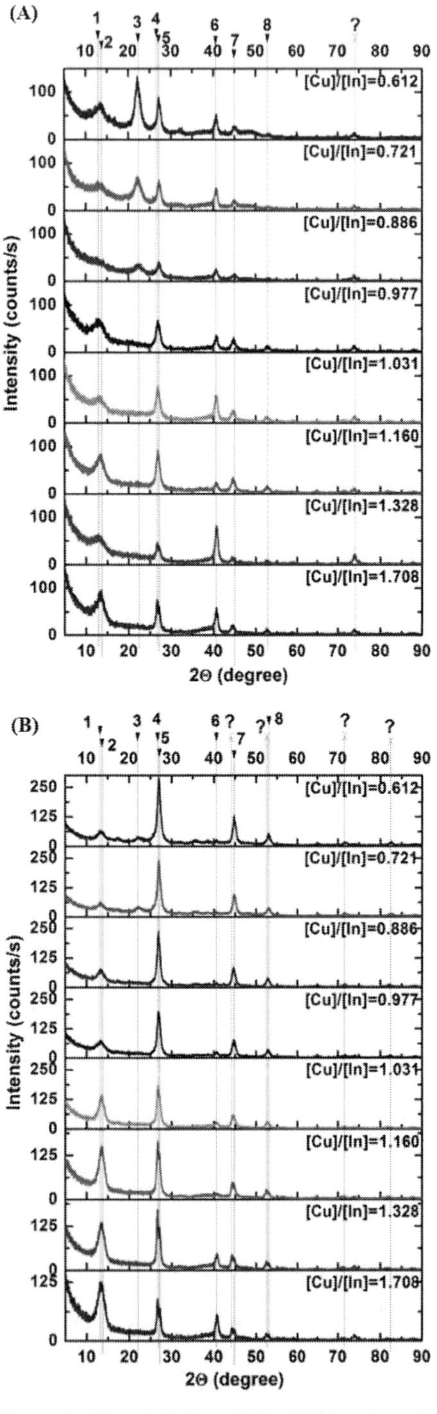

Figure 1: The evolution of the XRD pattern for as-deposited $CuInSe_2$ nanostructures related to the change of the [Cu]/[In] precursor flow ratio (0.3° grazing incidence angle, 10 min (A) and 20 min (B) deposition time). The peaks are marked as: 1 - $MoSe_2$; 2 - $Cu_{2-x}Se$; 3 - (006) InSe (023002-ICSD or In_6Se_7 (9135-ICSD) and In_2Se_3 (1376-ICSD); 4 - (112) $CuInSe_2$; 5 - (111) $Cu_{2-x}Se$; 6 - (110) Mo (173127-ICSD); 7 - (204)/(220) $CuInSe_2$; 8 - (116)/(332) $CuInSe_2$.

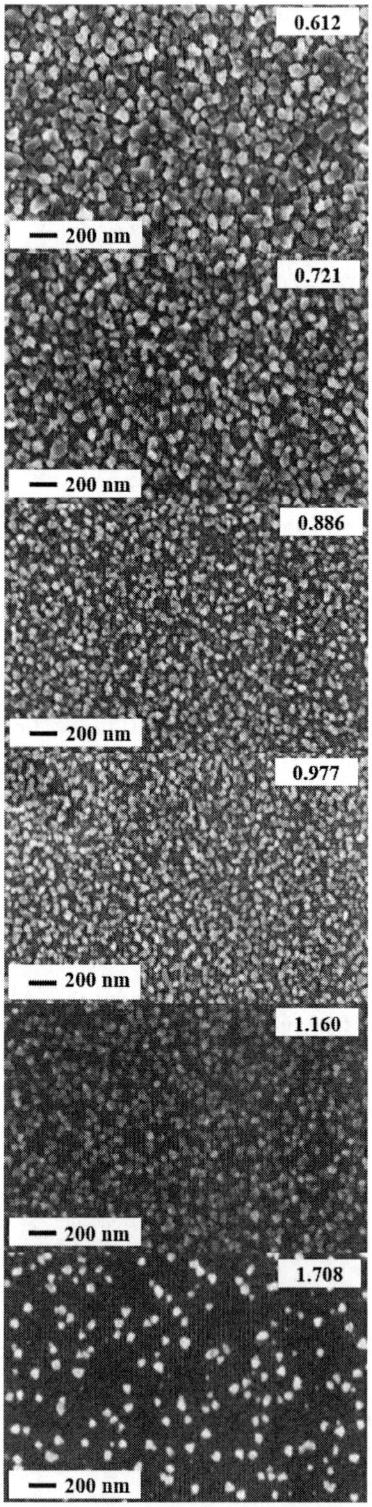

Figure 2: SEM images showing the dynamics of the $CuInSe_2$ nanostructure growth morphology depending on the precursor [Cu]/[In] flow ratio.

secondary phase peaks are still detected, their contribuition to the XRD pattern higly depending on precursor flow ratio.

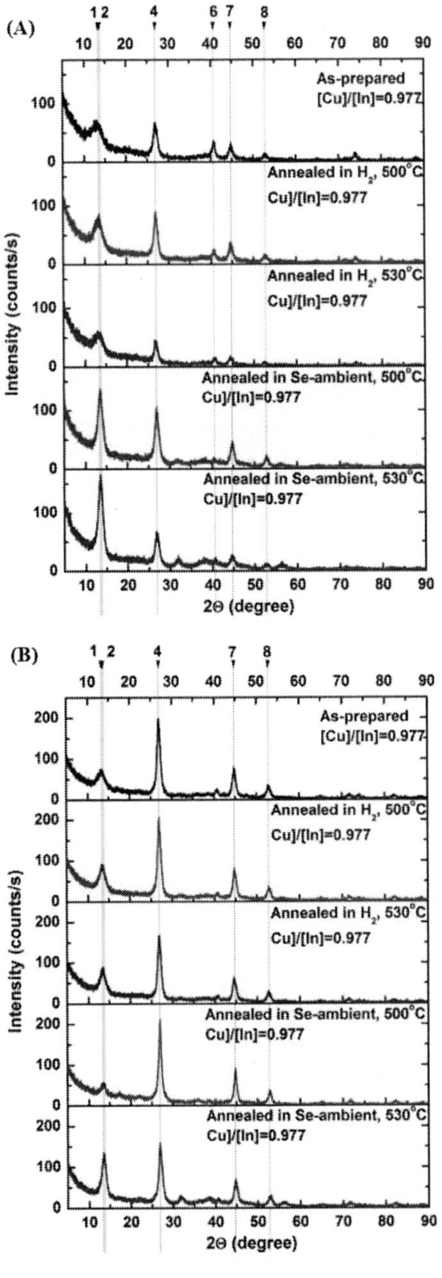

Figure 3: The influence of annealing (15 min) in different ambients on the evolution of the XRD pattern for $CuInSe_2$ nanostructures ([Cu]/[In] precursor flow ratio is 0.977, 0.3° grazing incidence angle, 10 min (A) and 20 min (B) deposition time). The peaks are marked as: 1 - $MoSe_2$; 2 - $Cu_{2-x}Se$; 3 - (006) InSe (023002-ICSD or In_6Se_7 (9135-ICSD) and In_2Se_3 (1376-ICSD); 4 - (112) $CuInSe_2$; 5 - (111) $Cu_{2-x}Se$; 6 - (110) Mo; 7 - (204)/(220) $CuInSe_2$; 8 - (116)/(332) $CuInSe_2$

One can trace out a double peak structure at about 26° which is highly evidentiated at the highest value of [Cu]/[In] precursor ratio (1.708). This peculiarity completely vanishes for [Cu]/[In] ratio of 0.977.

The results of the structural investigations by GI-XRD point at 0.977 [Cu]/[In] precursor flow ratio as the most appropriate for the preparation of single phase $CuInSe_2$ nanostructures (the lowest content of secondary phases). LA-ICP-MS investigations have revealed a constant excess of Se in samples in comparison to the one necessary for a stoichiometric film. Also, the peak at 13.10° (of complex structure) needes futher investigations for to conclude on its nature.

An attempt to understand the nature of several problematic peaks resulted in annealing experiments (different ambients and temperature) - Fig. 3. The elimination of the secondary phases has been achieved by annealing in H_2 and Se containing ambient at temperatures up to 530°C.

<u>10 nm high $CuInSe_2$ NPs:</u> The increase of the annealing temperature results in decrease of the (112) $CuInSe_2$ peak intensity regardless of the ambient; annealing in Se ambient causes the increase of the 1 and 2 peaks (probably due to formation of the Copper Selenide or Molybdenum Selenide).

<u>20 nm high $CuInSe_2$ NPs:</u> Annealing in H_2 decreases the overall pattern' intensity. Se-ambient annealing at 500°C diminishes the secondary phase presence; temperature increase up to 530°C increases the signal of the peak at 13.10°.

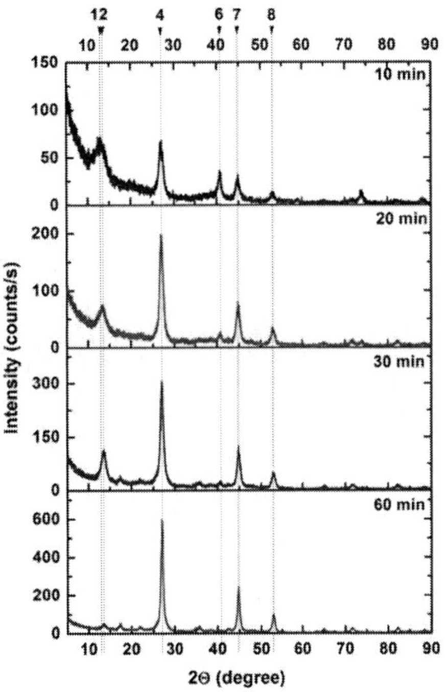

Figure 4: The influence growth time on the evolution of the XRD pattern for $CuInSe_2$ nanostructures ([Cu]/[In] (precursor flow ratio is 0.977, 0.3° grazing incidence angle, deposition time as in Figure). The peaks are marked as: 1 - $MoSe_2$; 2 - $Cu_{2-x}Se$; 3 - (006) InSe (023002-ICSD or In_6Se_7 (9135-ICSD) and In_2Se_3 (1376-ICSD); 4 - (112) $CuInSe_2$; 5 - (111) $Cu_{2-x}Se$; 6 - (110) Mo; 7 - (204)/(220) $CuInSe_2$; 8 - (116)/(332) $CuInSe_2$

The persistance of Se excess in the samples, point at a possible formation of secondary phases. The experiments being carried out have proven that the peak at 13.10° can be attibuted to molybdenum selenide phase.

4 CONCLUSIONS

$CuInSe_2$ nanostructures have been successfully prepared by MOCVD method. The results of GI-XRD and LA-ICPMS investigations point at 0.977 [Cu]/[In] precursor flow ratio as the most appropriate for the preparation of single phase $CuInSe_2$ nanostructures. Complete removal of secondary phases is achieved by a fine tuning of the MOCVD recipe and post preparation annealing.

5 ACKNOWLEDGEMENTS

The authors gratefully acknowledge the support of the Helmholtz-Gesellschaft Deutscher Forschungszentren e.V. (HGF) under the project "Hybrid-PV".

The authors would like to thank T. Mete, A. Ioakimides and K. Foustiropulous for assistance and discussions on technical issues.

6 REFERENCES

[1] H. Hoppe, N. Serdar Sariciftci, J. Mater. Res. 19 (2004) 1924.
[2] B. A. Gregg, S.-G. Chen, R. A. Cormier, Chem. Mater. 16 (2004) 4586.
[3] B. P. Rand, D. P. Burk, Stephen R. Forrest, Phys. Rev. B 75 (2007) 115327.
[4] M. Wright, A. Uddin, Sol. Energy Mater. Sol. Cells 107 (2012) 87.
[5] J.-W. Lim, J.-H. Choi, I.-H. Choi, J. Korean Phys. Soc. 30 (1997) 293.
[6] S. Huang, L. Zhang, Z. Dai, X. Zhu, F. Qu, S. Fu, J. Vac. Sci. Technol. B (20)5 (2002) 2043.
[7] K. H. K. Fianti, Chalcogenide Letters 11 (2014) 605.
[8] O. Yarema, D. Bozyigit, I. Tousseau, L. Nowack, M. Yarema, W. Heiss, V. Wood, Chemistry of Materials 25 (2013) 3753.
[9] F. A. Pulgarin-Agudelo, S. Lopez-Marino, A. Fairbrother, M. Placidi, V. Izquierdo-Roca, P. J. Sebastian, F. Ramos, B. Pina, A. Perez-Rodriguez, E. Saucedo, J. Renew. and Sustain. Energy 5 (2013) 053140.
[10] T. Oamata, K. Nose, S. Otsuka-Yao-Matsuo, J. Appl. Phys. 105 (2009) 073106.

Two-Stage Synthesis of CZTS thin Films and the Influence of Geometry and Sulphur and Tin Sulphide Supply

S. Mazzamuto[1*], Z. Wei[2], G. Kissling[3], T.M. Watson[2], L.M. Peter[3], N.M. Pearsall[1], I. Forbes[1]

[1]NPAG, Department of Physics and Electrical Engineering, Northumbria University, Newcastle NE1 8ST, UK
[2]SPECIFIC, College of Engineering, Swansea University, Engineering East, Bay Campus, Swansea, SA1 8EN, UK
[3]Department of Chemistry, University of Bath, Bath, BA2 7AY, UK
* Corresponding email: samantha.mazzamuto@northumbria.ac.uk

ABSTRACT: This study reports the synthesis and characterization of Cu-poor/Zn-rich CZTS as a function of sulphur content in order to improve the homogeneity and the quality in CZTS film formation. The absorber layers were produced using a two-stage process. The first stage involved the sputter deposition of metallic Cu-Zn-Sn precursors followed by the evaporation of a sulphur capping film. The second stage involved the conversion of the precursor stacks to the kesterite absorber layer by heating the samples in a tube furnace under argon ambient. Scanning Electron Microscopy together with Energy Dispersive X-ray Spectroscopy was used to determine composition and morphology, whilst phase characterization was carried out using Raman spectroscopy and X-ray Diffraction. Finally, photo-electrochemical External Quantum Efficiency measurements were used to characterize the optoelectronic response of the samples.The paper will show that the kesterite formation, its composition and the uniformity of the conversion are strongly dependent on the sulphur location, sulphur quantity and the internal geometry of the graphite box used to contain the samples during conversion.
Keywords: CZTS, Annealing, Sputtering, Crystallization.

1 INTRODUCTION

Kesterite compounds are very promising materials for PV applications because they are p-type semiconductors, with a direct band gap, that is 1.5 eV for the Cu_2ZnSnS_4 and 1.0 eV for the $Cu_2ZnSnSe_4$ [1], and a very high absorption coefficient ~10^4 cm^{-1}. Among kesterite compounds, Cu_2ZnSnS_4 has reached a record efficiency of 9.2 % [2]. Although, Kesterite materials are considered a promising alternative to CIGS and CdTe solar cells because they are made from earth abundant and non-toxic elements, CZTS efficiency is still below the theoretical limit.

The poorer performance of CZTS solar cells can be attributed to several factors, such as the lower formation energy of secondary phases among which the binary phase, Cu_2S, and the ternary phase, Cu_2SnS_3 (CTS), are the most detrimental because they are both p-type semiconductors but Cu_2S is highly conductive which may short the solar cell and CTS has an energy-gap (0.98-1.35 eV) very close to that of CZTS and could affect the carrier collection efficiency [3].

Moreover, it has been reported by several authors the blistering effect [4] and the formation of voids during the conversion both at the interface with the molybdenum [5] and also on the surface [6-8] in CZTS films. Another challenge is related with the uniformity of the material in terms of composition. It has been proved that the use of a gas, such as H_2S, during the conversion, allows having a better uniformity of the material across the whole area of the sample because the use of this gas can form large grained CZTS films with a uniform diffusion of the elements [9]. The drawback concerning the H_2S is its very high toxicity that it makes this gas not very suitable to be used at industrial level because in case of accident in the factory this gas could be released in the atmosphere causing serious problems for the human health and the environment.

In this paper, a study on the geometry of the graphite box and on the supply used both during the conversion is shown. In particular, it is shown how kesterite formation, composition, uniformity of the conversion and the density of voids is strongly dependent on the sulphur location and amount of the sulphur source and the geometry of the box.

2. EXPERIMENTAL

2.1 Processing

Metallic precursors Cu, Zn and Sn were deposited by magnetron sputtering onto Mo-coated soda lime glass with an area of 25 mm x 50 mm. All targets had 5N purity. The target powers for Cu, Zn, and Sn were respectively 140 W, 65 W and 150 W. A working Argon pressure of 1.3x10^{-3} Torr and a deposition time of 25 min were used for all samples. The precursors thickness used for this study was 450 nm and their atomic composition was a Cu-poor and slightly Zn-rich composition of Cu/(Zn+Sn) ≈ 0.7, Zn/Sn ≈ 1.1.

In this study, only two samples of the six used have a sulphur cap thermally evaporated onto the precursors with a final thickness of 10 μm (corresponding to around 25mg). This high S thickness ensured an approximately 10-fold excess of sulphur during the conversion.

All the six samples studied here were converted adding some sulphur (S – Aldrich, 99.998%) and tin sulphide (SnS – Alfa Aesar, 99.5%) pellets, as a supply, directly in the box during the conversion. All the details about the supply and the geometric conditions are summarized in Table I.

All samples were placed in a graphite box and heated to 550 °C in a tube furnace and kept at temperature for 30 minutes at 5 mbar Argon pressure. Afterwards, samples were allowed to cool naturally to room temperature. The conversion has been studied also from the point of view of the box geometry used. In particular, with the exclusion of sample A, all the samples used for this study has been converted using a box with a lid with one hole (Ø = 0.5mm) in the centre of the lid (see Table I).

Table I: Summary of the supply used during the conversion for all the samples used for this study.

Sample	S Cap	S Supply	SnS	Lid
A	10 μm	100 mg	15 mg	No Hole
B	10 μm	100 mg	15 mg	1 Hole
C	-	100 mg	-	1 Hole
D	-	100 mg	30 mg	1 Hole
E	-	500 mg	-	1 Hole
F	-	500 mg	30 mg	1 Hole

2.2 Characterization

All the absorber layers were characterized by Scanning Electron Microscopy (SEM) on Tescan combined with an Oxford Instruments Energy Dispersive X-ray (EDS) Spectroscopy System in order to measure the elemental compositions of the samples. EDS was carried out using an accelerating voltage of 20 kV.

The structural and phase characterization was carried out using X-ray diffraction (XRD) and Raman spectroscopy. The XRD measurements were carried out with a Siemens using a CuKa radiation source (λ = 1.5418 Å) and the Raman measurements were carried out using a Renishaw inVia Raman spectrometer with 532 nm excitation light.

Finally, in order to characterize the optoelectronic response of the samples, electrochemical External Quantum Efficiency (EQE) measurements were performed using a home-built system. Before the EQE measurement the samples were etched for 30 s in 5 % aqueous KCN solution and then coated with an 80 nm thick CdS layer deposited by CBD. Measurements were performed in a standard three electrode cell using a Pt gauze counter electrode and a Ag/AgCl (3 M KCl) reference electrode, with an applied potential of -0.2 V versus an Ag/AgCl reference electrode. Acetate buffer (pH 4.7) was used as the electrolyte. The illuminated area was circular with an approximate area of 0.5 cm^2.

3. RESULTS AND DISCUSSION:

3.1 Sulphur Cap

At first sight, several concentric rings could be observed at the surface on the converted samples when a lid without holes is used. Two main areas could be distinguished: central area and edge area.

Figure 1 shows SEM images in the centre and at the edge of samples A and B after the conversion. In particular, it can be seen how the morphology of the films can be affected by the nature of the box. In fact, while in the first case (sample A – graphite box with a lid without holes), the film shows a very high density of voids and a not uniform behaviour from the centre to the edge; in the second case (sample B - graphite box with a lid with one hole), the sample looks to be uniform across the whole area of the sample and the density of voids is reduced.

The atomic composition measured by EDS (see Table II) and the Raman (see Fig 2) spectra prove the uniformity of the sample B with respect to sample A also in terms of composition and phases. In fact, the central area of sample A is totally decomposed in binary phases and a weak signal coming from the CZTS phase. In particular from Figure 2a, Raman spectra shows SnS peaks (located at 160 cm^{-1}, 191 cm^{-1}, 220 cm^{-1}), a $Cu_{2-x}S$ peak located at 475 cm^{-1} and a weak signal coming from kesterite. At the contrary, the film at the edge is mainly kesterite (peaks located at 289 cm^{-1}, 338 cm^{-1} and 348 cm^{-1}) and $Cu_{2-x}S$. Sample B is only CZTS plus a signal coming from the $Cu_{2-x}S$ phase.[10, 11]

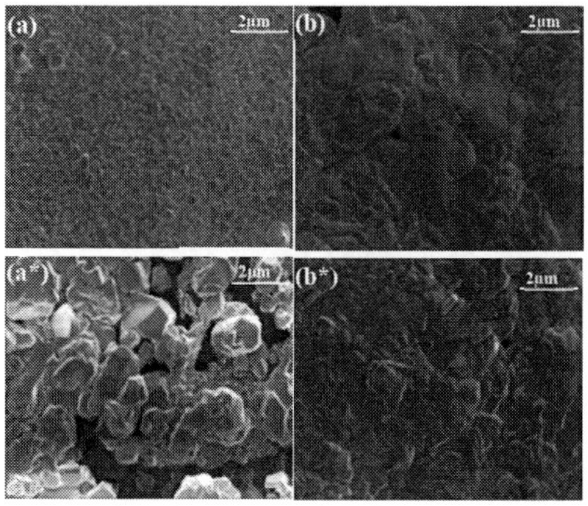

Figure 1 Plan-view SEM images of the centre and edge (*) of thin films converted using two different box configurations (a) No hole, Supply, (b) 1 hole, Supply.

Table II: Atomic compositions determined by EDS of precursor and converted films using Sulphur cap.

		Cu / (Zn + Sn)	Zn / Sn	Cu / Sn	S / Metals
Precursor		0.7	1.1	1.5	-
Sample A	Centre	0.2	2.7	0.9	0.7
	Edge	1.0	1.2	2.3	0.8
Sample B	Centre	1.0	1.3	2.3	0.9
	Edge	1.0	1.4	2.3	0.8

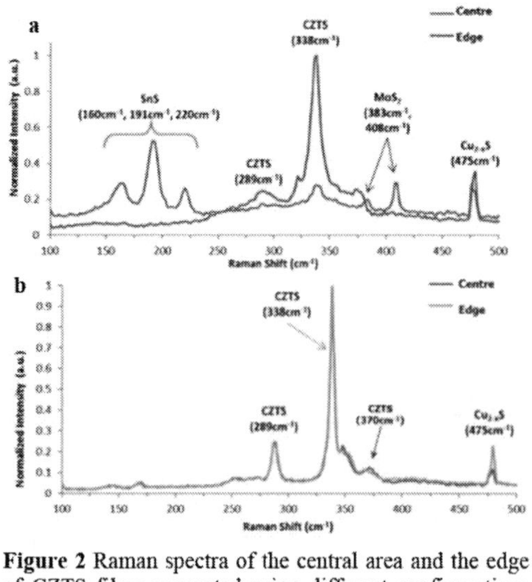

Figure 2 Raman spectra of the central area and the edge of CZTS films converted using different configurations (a) No hole, Supply, (b) 1 hole, Supply (see Table I).

It must be also observed that the MoS_2 [12] phase is detected by Raman in the central area of sample A. Since

Raman is a surface measurement, MoS_2 detection indicates that the voids density and the consumption of material in sample A are very high compared with the sample B.

The composition in sample B is uniform across the whole area of the sample, with only a tiny difference in terms of S and Sn distribution across the area of the sample.

Comparing these two samples, a possible explanation is that the kesterite formation seems to take place much more easily at the edges of the sample, that is, in those areas in which there is more free volume available inside the box and where the volatile species can lie for a longer time. From this perspective, the consumption of material observed in the centre of sample A is probably due to a vapour phase escape path towards the edges of the box that is sealed by 4 screws in the corners. At the contrary, the use of a box with the lid with a hole in the centre allows to have a better control of the vapour phase escape path forcing the volatile species (such as S and SnS) to collect much more in the centre of the sample. This enhances the CZTS growth because it allows having the minimum in order the reaction (1) is pushed towards the kesterite formation.[13]

$$Cu_2ZnSnS_4 \Leftrightarrow Cu_2S(s) + ZnS(s) + SnS(s) + \frac{1}{2}S_2(g) \quad (1)$$

This first part of the study highlighted that the use of a graphite box with one hole in the centre of the lid is beneficial for the CZTS growth because it allows having uniform films in term of atomic composition and phase formation across the whole area of the sample. In particular, the uniformity has been reached using S and SnS pellets and not a toxic gas such as H_2S that is not easily scalable at industrial level because of its high toxicity.

In sample B, many voids are still present and the atomic composition measured is not the best one for photovoltaic applications. In fact, the high density of voids can cause short-circuits in the device, while the "not ideal" composition can cause a high density of mid-bandgap levels in the CZTS that can act as recombination centres and/or traps.

With the composition of sample B (Table II), Copper could replace Zinc or Tin and this creates two deep levels in the CZTS gap replacing the most important shallow and acceptor level that is V_{Cu} (that determines the population of the majority carriers) and clusters such as $[V_{Cu} + Zn_{Cu}]$ that can facilitate the separation of photo-generated electron-hole pairs [14].

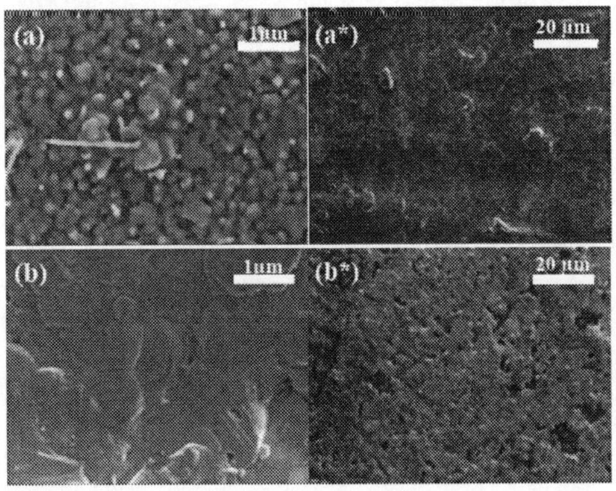

Figure 3 Plan-view SEM images with respectively 5μm and 100 μm (*) view field of the centre of (a) Sample C, (b) Sample D.

3.2 Sulphur cap versus Sulphur pellets

With the aim to improve the quality of the converted films in terms of composition and in order to reduce the density of voids and material consumption, another comparison was made between a sample converted with a Sulphur cap (sample B) and a sample converted by adding some sulphur supply directly in the graphite box during the conversion (sample C) or adding sulphur plus tin sulphide (sample D). The lid, used for the graphite box, had one hole in the centre for all the three samples.

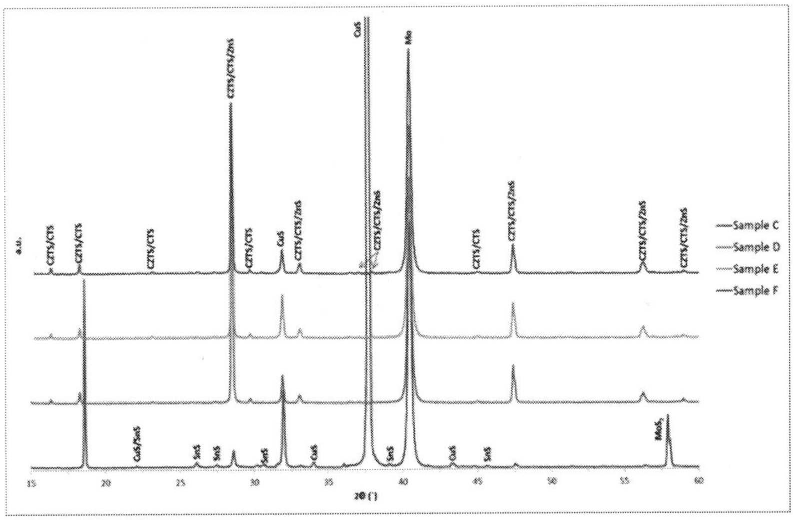

Figure 4 X-ray diffraction patterns using a Cu K_α radiation from films converted using a narrow box with a hole in the centre of the lid of respectively sample C (blue line) – 100mg of S, D (orange line) – 100mg of S + 30mg of SnS, E (green line) – 500mg of S and F (violet line) – 500mg of S + 30mg of SnS.

This experiment is made in order to see whether or not there are differences in the CZTS quality when the supply source is changed.
Figure 3 shows sample C and D with respectively high and low magnification in order to highlight the density of voids formed after the conversion. In particular, from the SEM-image (Fig 3-(a)) sample C is decomposed in the binary phases. In fact, some SnS crystals (pipe form) and ZnS crystals (squared and bright) can be seen. This is also proved by RAMAN and XRD measurements. From Figure 4 and 5 (blue line), sample C is decomposed in SnS [JCPDS 73-1859 (SnS/Orth.)] and CuS [JCPDS 78-2122 (CuS/Orth.)], so that also ZnS is probably present as well following the reaction (1). Also the MoS_2 phase is visible and this is an evidence of the very high density of voids still present in this sample C. At the contrary, sample D shows (Figure 3 (b)) uniform grains, with a dimension of 1-2μm that are pure CZTS [JCPDS 26-0575 (CZTS/Tet.)] plus a weak signal (observed only in the XRD measurement) coming from CuS. No binary phases are detected by Raman (Fig 5 – red line).
The main difference between these two samples is that some SnS, in addition with S pellets, was added in sample D during the conversion. So it can be concluded that the sulphur supply is, of course, important in order to let the reaction (1) happen, as well as working in Sn excess ambient. On this sense, it must be noted that the starting composition for the precursors used for this study is Cu-poor and Zn-rich, meaning that the Sn already present in the bulk should be enough in order to Kesterite from reaction (1).
Looking at the compositions, in particular for the two samples B (S cap – S+SnS pellets) and D (S+SnS pellets) it seems that the S cap favours a huge tin loss from the substrate independently from the SnS supply added during the conversion and, as a consequence of this, the formation of binary phases such as SnS and $Cu_{2-x}S$ (Figure 2 b) due to the fact that the reaction (1) of the kesterite phase formation is pushed towards the formation of binary sulphides.
The composition of sample D is Cu-poor, but it is Zn-poor as well (Zn/Sn=0.9). The not "ideal" composition, reported by many authors to give the best performance solar cells (Cu/(Zn+Sn) = 0.8 and Zn/Sn = 1.2), of Sample D is related with its optoelectronic response. Figure 7 shows the EQE spectrum in wavelengths ranging from 350nm to 1000nm. The EQE spectrum for sample D (red line) increases gradually from 350 nm to ca. 450 nm, steeply rises to ca. 525 nm, when the maximum value of EQE is measured corresponding to 13%, and then declines gradually to the onset wavelength of ca. 800nm, corresponding to the bandgap energy of CZTS of 1.55eV which is calculated by extrapolating the plot of $[h\nu \times \ln(1\text{-EQE})]^2$ vs. hν [15] (inset plot in Fig 7).

Table III Atomic compositions of precursor and converted films using sulphur pellets determined by EDS in the central area of the samples.

	$\frac{Cu}{(Zn + Sn)}$	$\frac{Zn}{Sn}$	$\frac{Cu}{Sn}$	$\frac{S}{Metals}$
Precursors	0.7	1.1	1.5	-
Sample B	1.0	1.3	2.3	0.9
Sample C	0.3	1.9	0.9	0.5
Sample D	0.9	0.9	1.8	1.0
Sample E	0.8	1.0	1.6	1.0
Sample F	0.8	1.0	1.6	1.0

The reduction of EQE response in the short wavelength region is due to absorption losses in the buffer (CdS) layer. On the other hand, the gradual weakening of the EQE response in the long wavelength region indicates loss of deeply absorbed photons due to poor minority carrier diffusion length compared with the depletion width and/or insufficient penetration of depletion width

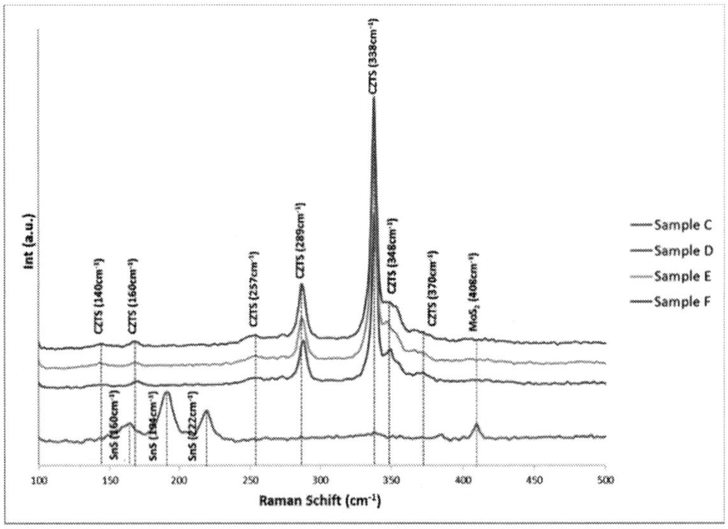

Figure 5 Raman spectra from films converted using a narrow box with a hole in the centre of the lid of respectively sample C (blue line) – 100mg of S, D (red line) – 100mg of S + 30mg of SnS, E (green line) – 500mg of S and F (violet line) – 500mg of S + 30mg of SnS.

into the absorber [16].Such a failure would be caused by high majority carrier concentration of the CZTS film, i.e., defect density in the CZTS film is still too high to achieve an optimum. It must be noted that the EQE measurements were carried out using a reverse bias of -0.2V and that, when higher reverse bias were applied, the resulting EQE was lower indicating a not very long carrier diffusion length compared with the depletion width. Thus, further investigation should be carried out in order to improve the material quality by efficient production of a well-defined CZTS film with a small amount of defects and majority carrier density.

3.3 Sulphur pellets

Although sample D is uniform in terms of composition and in terms of phase formation, it has a not ideal composition therefore a low EQE. From Table III, the ratios Cu/(Zn+Sn) and Zn/Sn are not respectively 0.8 and 1.2. The higher value of the ratio Cu/(Zn+Sn) and the lower Zn/Sn can introduce a high density of mid-gap defects acting as recombination centres causing a lower EQE response as already said in the previous paragraph.

In order to get samples with the optimal composition other two experiments has been made: sample E converted using 500mg of S and sample F converted using 500mg of S plus 30mg of SnS.

In these last two cases the samples are very uniform and much more compact across the whole area of the sample with a very low density of voids (Figure 6). Moreover, from XRD in Figure 4 (lines green and violet), they show to be mainly CZTS plus very weak signals coming from SnS and CuS phases, while from Raman spectra (Figure 5) the only phase present is the kesterite phase. The composition of both the two samples is closer to the ideal one with a Cu/(Zn+Sn) = 0.8, but with a lower Zn/Sn ratio than the ideal one (see Table III).

Figure 6 Plan-view SEM images with respectively 5μm and 100 μm (*) view field of the centre of (a) Sample E, (b) Sample F.

The EQE and the energy gap of the last two samples are respectively 22% and 1.61 eV for sample E and 19% and 1.62 eV for sample F (see Figure 7, its inset).

Even if the EQE has improved respect the sample D, it is still quite low. The main cause of this is related again with the composition of the samples and with the high density of defects introduced by the not "ideal" composition.

During the measurement, we could not introduce a strong band bending (we carried out the measurements applying a -0.2 V bias) and, as said in the previous paragraph, this could be due to mid-gap defects states probably due to the composition of the converted samples.

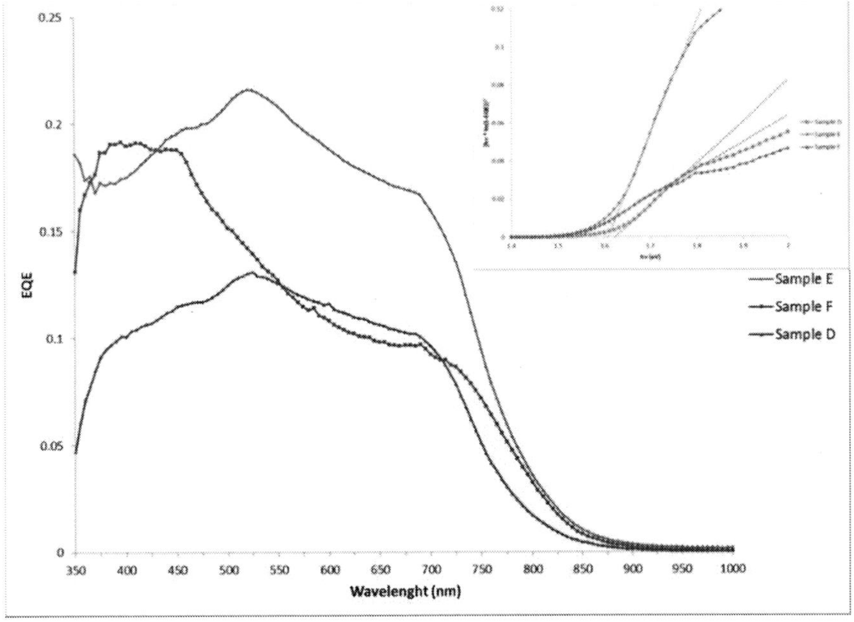

Figure 7 EQE measurements of the CZTS samples D (red line), E (green line) and F (violet line); inset: the band gaps of the CZTS absorber layers calculated by extrapolating the plots of $[h\nu \times \ln(1\text{-EQE})]^2$ vs. hν.

Comparing the EQE curves for samples E and F, sample F shows a strong inflexion between 450 nm and 700 nm respect to sample E. This could be due to a high doping of the sample F respect to the sample E, even if further investigations should be carried out.

CONCLUSIONS

In this work, different configurations of the graphite box, used during the conversion, were studied.
In particular, the first two samples were converted using a box with a lid (one without any holes and the other one with a hole). From this first part, we can conclude that it is possible to obtain uniform films starting from metallic precursors converted with some supply such as S+SnS by using a graphite box during the conversion that must have a lid with one hole. In this case the uniformity of the sample is comparable to the use of the H_2S that has many drawbacks related with its high toxicity.
In order to improve our results, a following comparison was made between the kinds of Sulphur supply: in particular between sulphur deposited by Physical Vapour Deposition (PVD) and Sulphur supplied directly into the graphite box.
From this second comparison we could conclude that, even if the Sulphur quantities are pretty much the same, the conversion is dependent from the supply. In particular, samples converted using Sulphur pellets exhibited a better control of the composition compared with the ones converted with a sulphur cap.
Finally, a study on the quantity of supply is done. Converting sample with a very high quantity of supply allows having uniform films in term of composition and phase formation and a very much reduced density of voids. For these last samples an EQE of 21% was achieved. This poor value depends from several factors and thus, further investigation should be carried out in order to improve the material quality.

ACKNOWLEDGEMENTS

This work has been undertaken as part of the EPSRC-funded PVTeam project (EPSRC Ref.:EP/L017792/1).
Dr S. Mazzamuto would like to thank Mako Ng of the University of Bath for his help in CdS deposition and Dr G. Kissling for the EQE measurements.

REFERENCES

[1] Siebentritt, S. and S. Schorr. Progress in Photovoltaics, 2012. **20**(5): p. 512-519.
[2] T. Kato, H.H., N. Sakai, S. Muraoka, H. Sugimoto.. Proceedings of the 27th EU-PVSEC, 2012: p. 2236-2239.
[3] Vanalakar, S.A., et al. Solar Energy Materials and Solar Cells, 2015. **138**: p. 1-8.
[4] Bras, P., J. Sterner, and C. Platzer-Björkman. Journal of Vacuum Science & Technology A, 2015. **33**(6): p. 061201.
[5] Liu, F., et al. ACS Applied Materials & Interfaces, 2015. **7**(26): p. 14376-14383.
[6] Jung, H.R., et al. Ceramics International, 2015. **41**(10, Part A): p. 13006-13011.
[7] Johnson, M.C., et al. Chemistry of Materials, 2015. **27**(7): p. 2507-2514.
[8] Fernandes, P.A., P.M.P. Salomé, and A.F. da Cunha. Thin Solid Films, 2009. **517**(7): p. 2519-2523.
[9] Khalkar, A., et al. International Journal of Photoenergy, 2015. **2015**: p. 7.
[10] Cheng, A.J., et al. Journal of Vacuum Science and Technology A: Vacuum, Surfaces and Films, 2011. **29**(5).
[11] Dumcenco, D. and Y.-S. Huang. Optical Materials, 2013. **35**(3): p. 419-425.
[12] Li, H., et al. Advanced Functional Materials, 2012. **22**(7): p. 1385-1390.
[13] Weber, A., R. Mainz, and H.W. Schock. Journal of Applied Physics, 2010. **107**(1): p. 013516.
[14] Chen, S., et al. Adv Mater, 2013. **25**(11): p. 1522-39.
[15] Zoppi, G., et al. Progress in Photovoltaics: Research and Applications, 2009. **17**(5): p. 315-319.
[16] Todorov, T.K., et al. Advanced Energy Materials, 2013. **3**(1): p. 34-38.

INVESTIGATION OF LIGHT INDUCED METASTABILITIES THROUGH COLORED FILTERS ON THE CZTSSe MONOGRAIN PV DEVICES

A. Mittal [a,b], G. Ujvari'[a], T. Dimopoulos [a], M. Rennhofer [a], S. Lopez [c], L. Plessing[c], V.Schlosser [b]

(a) Austrian Institute of Technology, Donau-City-Strasse 1, 1220 Vienna, Austria.
Tel: +43(0) 50550-6348, Fax: +43(0) 50550-6390
Email: ankit.mittal.fl@ait.ac.at

(b) Faculty of Physics, University of Vienna, Boltzmanngasse 5, Wien, 1090, Austria.

(c) Crystalsol, Simmeringer Hauptstrasse 24, 1110 Wien, Austria.

ABSTRACT: Measurements of the photovoltaic (PV) activity of solar cells under low-light conditions using different wavelength-selective filters can provide useful information about nature, location and activation of defects, and possible recombination mechanisms inside-of CZTSSe solar cell bulk and interfaces. In this work, we use this approach for the characterization of flexible monograin membrane kesterite (CZTSSe) PV devices. Time dependence of the open circuit voltage, short circuit current, fill factor and maximum power, under low intensity of white, red and blue light is studied and related to carrier recombination processes in the bulk of the absorber and at the interface with the CdS buffer and the electrodes. The different illumination conditions cause moderate metastabilities in the electrical parameters, which are reversible when the cell is stored in dark for some hours. Reversible and irreversible contributions to the light-soaking effects are further analyzed.

Keywords: Kesterite, Light Soaking, Defects.

1 INTRODUCTION

The kesterite CZTSSe offers to be a good substitute material for replacing the CIGS based PV devices. Since the light induced metastable effect is quite known in the CIGS solar cells, which are reversible in nature. Therefore, it is quite likely that might be these metastable effects in the kesterite based solar cells as they follow the same cell architecture. In a kesterite solar cell light enters through the TCO contact constituting: silver nanowires and aluminium doped zinc oxide (Ag NWs/AZO), transverses the window layer of zinc oxide (ZnO) and buffer layer of cadmium sulphide (CdS) before being absorbed in the kesterite (CZTSSe), generating charge carriers. As a result, light of shorter wavelengths is significantly absorbed in the top layers, while light of longer wavelengths is mainly absorbed in the CZTSSe. Exploiting this property can help in studying the inherent metastabilities within the cells and devices.

The purpose of the proposed wavelength selective based light soaking is to determine the influence of defects on the amplitude of change in the performance parameters of the solar cells and devices. A series of measurements under the white and colored filtered light conditions were carried on the monograin CZTSSe PV devices. The work presented here will focus on a mini - module. Time dependence of PV performance under white or color-filtered low-light conditions of solar cells can give information about the direction and origination of the metastability inside the solar cell configuration. In the following section 2, the measurement procedure will be explained and in section 3 the results for the measurements done on the mini-module will be discussed.

2 METHODOLOGY

In this work, a mini-module was used for the characterization. The mini-module consists of 12 cells in series of a total area of 65.7 cm^2. It was irradiated with approximately **80 W/m^2** and IV (current as a function of voltage)-curves were measured under the same illumination conditions in a Static Solar Simulator Class AAA with a shutter mechanism using an electronic load and a reference cell for monitoring irradiance. To quantify the effects of specific wavelengths two sets of filters were used: blue (450nm) and red (650 nm) for light-soaking. For the white light soaking, two sheets of paper were used to attenuate the light to 80 W/m^2 level.

Before performing the light soaking measurement, the mini-module was kept overnight in dark to avoid parasitic light and ensure stabilization. Each light-soaking measurement was carried for approximately 250-300 minutes. After each light-soaking measurement it was again kept overnight in dark to ensure stabilization before the next light-soaking was performed.

An IV curve measurement at the standard test conditions (STC) was also carried out to determine the real performance parameters of the device. The standard test condition is defined as: 1000 W/m^2, 25 °C and Air Mass 1.5. The STC measurement was carried out before starting with the light-soaking, immediately after completing the light soaking and after storing the device overnight in dark.

3 RESULTS AND DISCUSSION

3.1 Effects of light soaking at low irradiances

All the electrical parameters under white, blue and red light showed an interesting behavior under different light conditions. The results shown in figure 1 are of V_{OC}

(open circuit voltage), P_{MPP} (Maximum power point), FF (Fill factor) and I_{SC} (short circuit current). Light soaking under white and blue (450nm) light showed a similar behavior for all the electrical parameters but I_{SC}. The white and blue light-soaking showed a strong decrease in the V_{OC}, P_{MPP} and FF. Under red light a slight decrease in the V_{OC}, P_{MPP} and FF can be observed, whereas I_{SC} suffers a slight increase.

All electrical parameters under red (650 nm) light showed a similar behavior for the first initial minutes, as observed under white and blue light. Nevertheless, after 300 minutes they recover to their initial values. V_{OC} decreases by 2.7 % after 3 minutes, and comes back to the initial value after 300 minutes. This is in contrast with the results of white and blue (450 nm) light, which causes a decrease of 4.4 % and 6.2 %.

An interesting trend was seen for I_{SC}, using white, blue and red light. All cases showed an increase in I_{SC} during the first 10 minutes. Thereafter, I_{SC} continued to increase under white light by approximately 2%, but showed a decrease under blue and red light reaching its initial starting point. Under white and blue light conditions, P_{MPP} showed a similar decrease of 8-9 % and in FF a decrease of 4-5 % from their initial values.

This similar decreasing trend under blue and white light conditions gives an indication to the majority of the metastabilities might be originating from the blue light component. Since, blue light contributes to an absorbance in the buffer and red light is absorbed in the bulk of the absorber of CZTSSe device. When comparing the negative slope under white light with the slope under blue light it can be seen that trapping defects are located at the proximity of the heterojunction, causing these transient metastabilities.

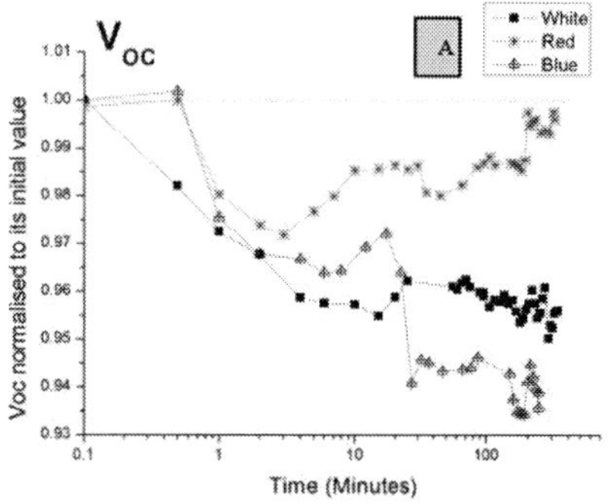

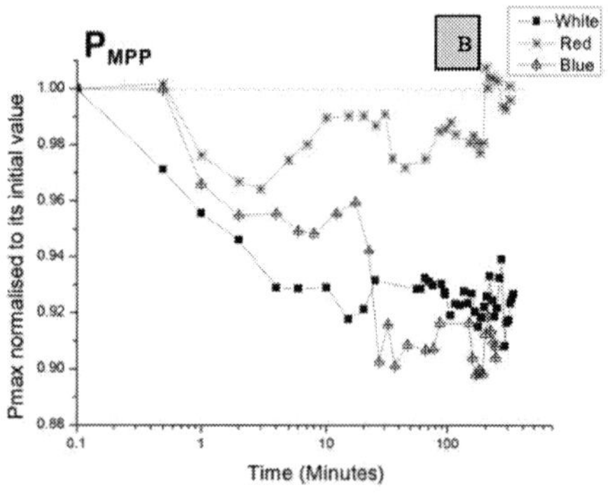

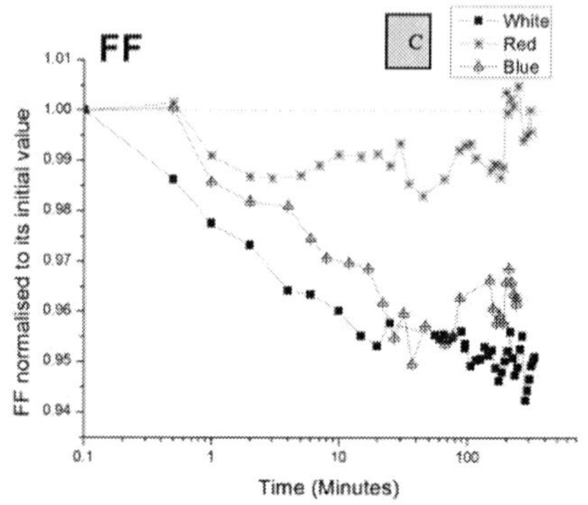

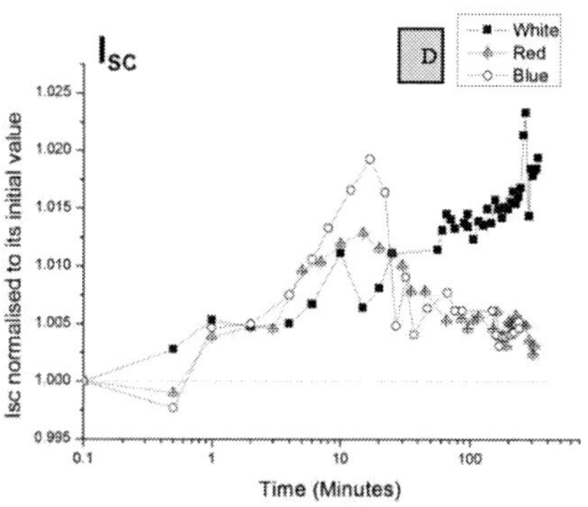

Figure 1: *The evolution of the normalized electrical parameters : (a) V_{OC} (open circuit voltage),(b) P_{MPP} (maximum power point), (c) FF (Fill factor) and (d) I_{SC} (short circuit current) under white, red and blue color light soaking of CZTSSe mini-module.*

3.2 Comparison of IV curves at the standard test conditions

All the STC IV curves performance parameters which were measured before and after performing the light-soaking is tabulated in table 1. As shown in the STC curves in figure 2 and 3, during the white and blue light-soaking the mini-module showed a decrease in power after the light soaking, whereas from the red color light soaking the power increased afterwards as shown in figure 4.

Table I: *Effect of white and colored light soaking (LS) on the performance parameters values at the standard test conditions prior and after light-soaking of the CZTSSe mini-module.*

Parameters	White		Red		Blue	
	Initial	Final	Initial	Final	Initial	Final
Voc	8.010 V	8.014 V	7.96 V	7.80 V	8.013 V	7.89 V
Pmpp	0.342 W	0.336 W	0.331 W	0.333 W	0.341 W	0.335 W
Isc	67.03 mA	67.7 mA	66.1 mA	66.1 mA	66.8 mA	67.4 mA
FF	63.84 %	61.95 %	62.95 %	64.57 %	63.90 %	63.05 %
Efficiency	5.218 %	5.119 %	5.045 %	5.073 %	5.195 %	5.11 %

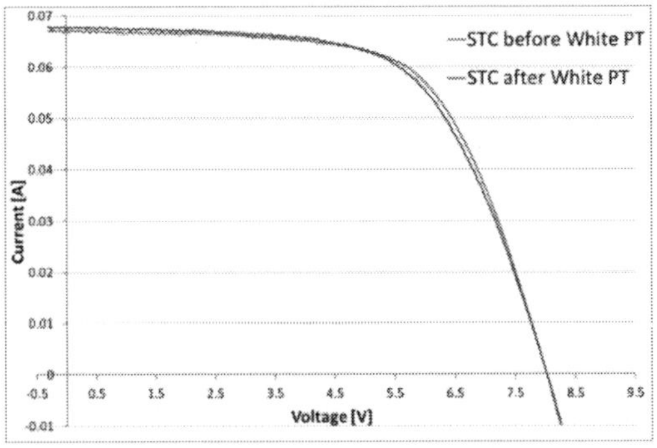

Figure 2: *Comparison of IV curves at the standard test condition before and after the white light soaking.*

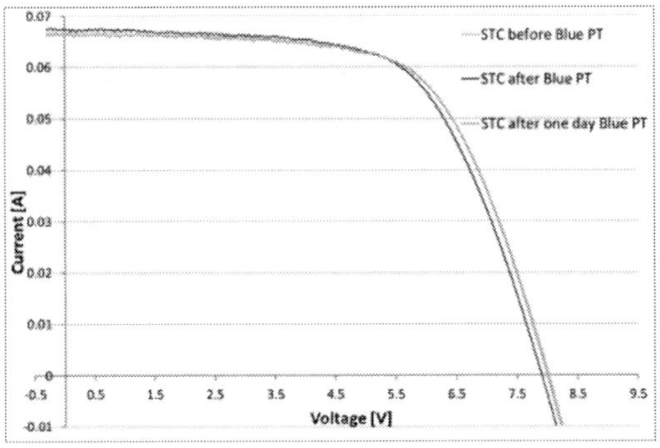

Figure 3: *Comparison of pre and post characterization after blue light-soaking of the mini-module at the standard test conditions. Please note that the IV curves taken before blue light-soaking and after one day dark relaxation are superimposing over each other.*

Under the white and blue color light soaking the IV curves at the STC conditions revealed an increase in series resistance, thus causing loss in performance of the device. Under white light it was clearly seen a decrease in the maximum power point, figure 2, which is reflected in the decreasing value of the FF and efficiency by 1.8 % and 0.1 % . While under blue light it caused a complete shift in IV curve to the left side after the completion of the light soaking, see figure 3. It reduced the overall FF by 0.8 % and efficiency by 1 %.

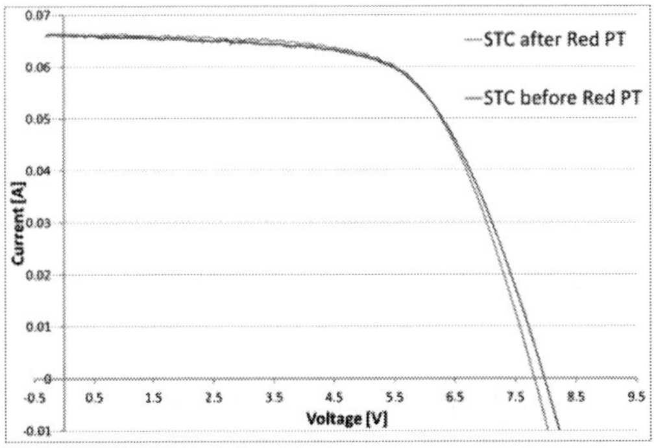

Figure 4: *Comparison of pre and post characterization after red light-soaking of the mini-module at the standard test conditions.*

An increase in performance is revealed after the red light soaking, along with a loss in series resistance. There was a slight decrease in V_{OC}, which might have been caused by an increase in barrier height of the CdS layer. But the overall performance was increased.

An interesting effect of reversibility of the metastable state was observed after overnight dark relaxation of the CZTSSe device. It is shown in figure 3, but since the initial and final IV curves at STC conditions are superimposing over each other therefore it is difficult to observe it in the graph.

These measurements help in revealing the defect locations in relation to the change in electrical state of the cell heterojunction or interfaces. This study can help in determining the location of defects which lead to the metastable state and also the quality of the thin-film PV device for further use.

For industrial standards the power rating plays an important role, but if the device under test (solar cells) is changing its electrical behavior over time under the influence of irradiation then it can lead to the measurement of an erroneous power. Therefore, it is most important to take into account the light induced metastability prior to the measurement routine of the cell and modules to ensure a reliable power rating.

4 CONCLUSIONS

The present work shows how low-light conditions can introduce moderate changes in the electrical parameters of the monograin-CZTSSe technology within a short interval of time. It was found out that the observed metastabilities might be most likely to be originated majorly near the heterojunction in the CZTSSe solar cells. This was shown via colored light soaking. It was also found that the light-induced metastabilities are reversible. All these changes are reflected (can be followed) in the electrical parameters of the solar cells.

THE INFLUENCE OF SODIUM IN HIGH GA-CONTENT $CU(IN_{1-x},GA_x)SE_2$ (CIGS) SOLAR CELLS

Xia Hao[1*], Kazi Taslimuddin Chowdhury[1], Takeaki Sakurai[1], Yukiko Kamikawa[2], Shogo Ishizuka[2], Akimasa Yamada[2], Hajime Shibata[2], and Katsuhiro Akimoto[1]
[1]Institute of Applied Physics, University of Tsukuba, Tsukuba, Ibaraki 305-8573, Japan
[2]National Institute of Advanced Industrial Science and Technology (AIST), Tsukuba, Ibaraki 305-8568, Japan

ABSTRACT: This work compares a pair of CIGS solar cells with Ga content of 0.75 deposited on Na-incorporated (soda lime glass, SLG) and Na-free (ceramic) substrates, respectively. The power conversion efficiency of the SLG device is remarkably higher (11.06 %) than that of the ceramic one (4.44 %). The degraded open circuit voltage and fill fact are the main reason of the poor performance for the ceramic device. The XRD results suggest a presence of $MoSe_2$ diffraction peak in SLG device and the formation of $MoSe_2$ has already been proved to form an ohmic contact at the CIGS/Mo interface and this might be a possible reason for the decreased series resistance for the SLG device. Besides, the defect spectra indicate that the incorporation of Na will also affect the defect level and distributions in CIGS thin films.
Keywords: effect of sodium, CIGS solar cells, defects

1 INTRODUCTION

Renewable energy has got considerable interest in the past a few decades since it can be served as an alternative to the traditional fossil fuels. A representative device that utilizes renewable energy is photovoltaic device, which directly converts light into electricity. Among the photovoltaic materials, $Cu(In_{1-x},Ga_x)Se_2$ (CIGS) is quite fitful in realizing low-cost thin-film photovoltaic technology since it has achieved the power conversion efficiency of 22.6%.[1] As is well known, the device performance could be greatly enhanced if the CIGS thin films were fabricated on glass substrates containing sodium. However, the effect of sodium is still unclear to date. In this study, we investigated the effect of Na on the performance of CIGS solar cells with high Ga.

2 EXPERIMENT

A pair of CIGS devices were examined to illustrate the effect of Na in device performance. One was deposited on ordinary sodium lime glass and with a standard concentration of Na. And the other was fabricated on ceramic substrate that is free of Na. The CIGS solar cells were fabricated by a 3-stage process using a molecular beam epitaxy (MBE) system. The Ga content was controlled to be 0.75 and the copper content of the CIGS films was roughly fixed at approximately Cu/(In+Ga)=0.9.[2] The current density-voltage (*J-V*) characteristics were carried out with a *J-V* source meter (Advantest R6245) under illumination at intensity of 100 mW/cm^2, AM 1.5 G spectrum. The X-Ray Diffraction (XRD) was taken on CIGS bare samples deposited on Mo-coated ceramic and SLG substrates at room temperature in the range of 15°~70° by Rigaku X'pert diffractometer at θ~2θ mode with Cu-Kα (α=1.541837Å) radiation operated at 45 kV and 40 mA.

3 RESULTS AND DISCUSSIONS

3.1 Photovoltaic performances of the solar cells

The *J-V* curves of the devices were shown in Fig. 1. For device free of Na, the deterioration in short circuit current (J_{SC}), open circuit voltage (V_{OC}) and fill factor (*FF*) resulted in a total decrease in power conversion efficiency (*PCE*). The photovoltaic parameters are shown in Table Ⅰ. The device fabricated on ceramic substrate shows poor performance, with J_{SC} of 12.29 mA/cm^2, V_{OC} of 0.70 V and *FF* of 0.40. And the poor performance of the device free of Na mainly due to the drastic in J_{SC} and *FF,* namely, the series resistance (R_S). In the present work, the difference between the resistivity of CIGS thin films in pairs of devices could be ignored. We then deduced that the huge difference in the R_S may be mainly due to the different contact resistance of the devices.

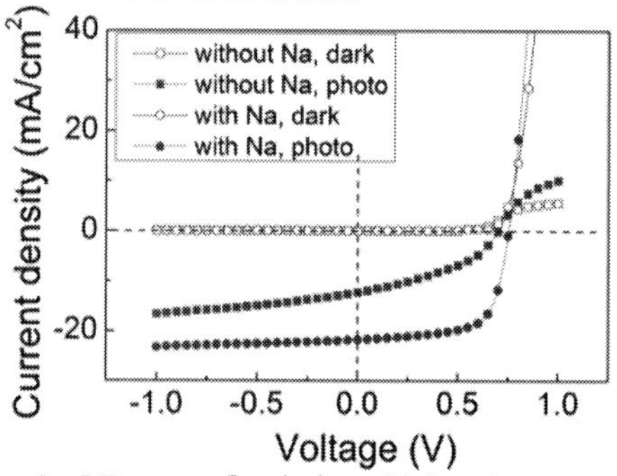

Fig. 1. *J-V* curves for devices fabricated on ceramic (without Na, empty and filled squares) and SLG substrates (with Na, empty and filled circles).

Table Ⅰ. Photovoltaic parameters of device fabricated on ceramic and SLG substrates.

Substrate	J_{SC} (mA/cm^2)	V_{OC} (V)	*FF*	*PCE* (%)	R_S (Ω-cm^2)
Ceramic	12.29	0.70	0.40	4.44	3.52
Glass	21.69	0.75	0.68	11.06	0.39

3.1 Structural properties

XRD patterns of the CIGS absorbers are shown in Fig. 2. An obvious difference is the formation of $MoSe_2$ phase in CIGS/Mo thin film deposited on SLG substrate, which is indicated by peaks around 32.53 ° and 58.73 ° corresponding to the (012) and (113) planes. It has been reported that the formation of $MoSe_2$ at CIGS/Mo interface makes it a favorable ohmic-type heterojunction, rather than Schottky-type.[3] Therefore, the dramatically decreased R_S in device with standard structure is mainly due to the formation of $MoSe_2$.

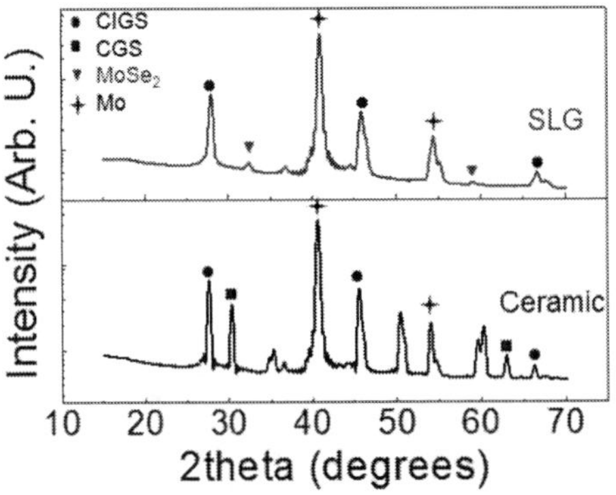

Fig. 2. XRD patterns corresponding to $CuIn_{0.25}Ga_{0.75}Se_2$ thin films fabricated on Mo coated ceramic substrate (without Na, black) and on Mo coated SLG substrate (with Na, red)

3.4 The defect spectra

In order to get further insight into the effects of Na on CIGS devices, the admittance spectroscopy measurements were carried out to determine the density of states and the activation energy of traps by varying frequency and temperature. Fig. 3 depicts the admittance spectroscopy of the devices fabricated on ceramic and SLG glass. The activation energy of the designated defect could be obtained by recording the frequency and temperature of the steps in capacitance by using the following equation:

$$\frac{E_a}{T} = kTln\frac{2\xi_0}{\omega}, \tag{1}$$

simply as:

$$\ln\left(\frac{\omega}{T^2}\right) = -\frac{Ea}{k}\frac{1}{T} + ln(2\xi_0). \tag{2}$$

Then one can calculate the activation energy Ea and emission factor ξ_0 if plotted inflection frequency as a function of inverse temperature.

Two defects with Ea of 46 meV and 250 meV were detected in device fabricated without Na and marked as α and ζ, as shown in Fig. 4(a). While in case of the SLG device, three defects were observed, with Ea of 10 meV, 95 meV and 230 meV, marked as α, β and ζ, respectively. It is worth noting that the values of activation energies for α- and ζ-defects of the standard SLG device are smaller than those of the ceramic device. The difference in the defect level of the pair of devices indicates that the compositional or structural variation may be caused by the incorporation of Na. Besides, in device free of Na, the β-defect was disappeared. The defects located around 100 meV on valence band may be originated by the interface defects, bulk defects V_{Se} donors or the $(In,Ga)_{Cu}$ defects.[4] However, we cannot clarify the origination of this phenomenon in the present devices because the origin of β-defect is still unclear.

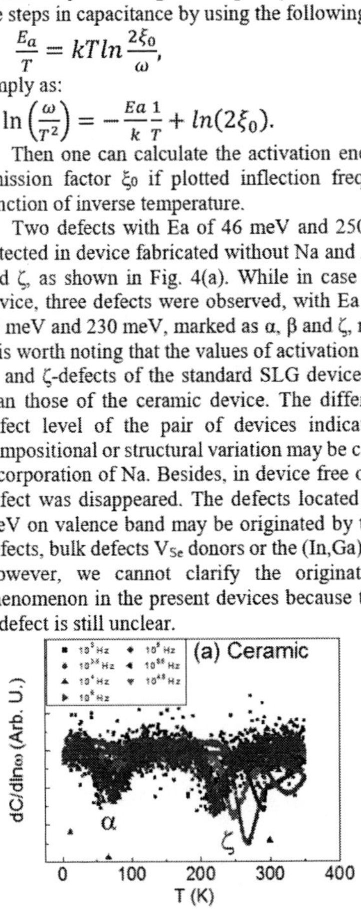

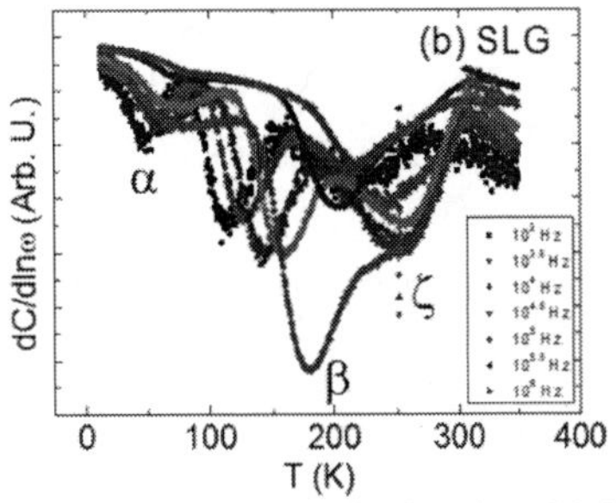

Fig. 3. Admittance spectra of devices fabricated on (a) ceramic and (b) standard SLG substrates.

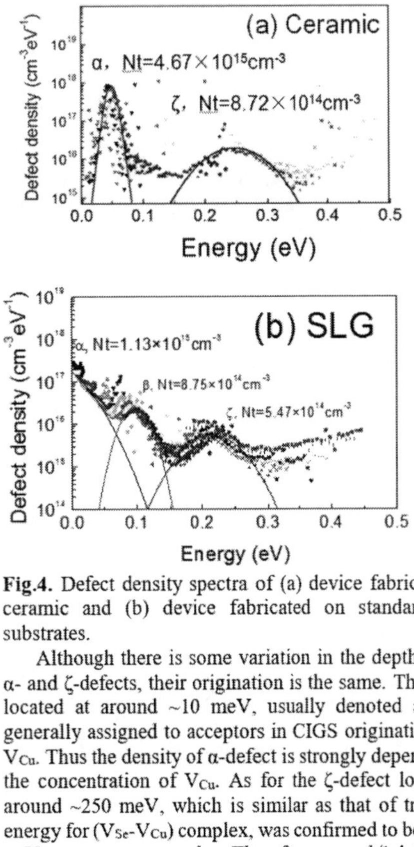

Fig.4. Defect density spectra of (a) device fabricated on ceramic and (b) device fabricated on standard SLG substrates.

Although there is some variation in the depths of the α- and ζ-defects, their origination is the same. The defect located at around ~10 meV, usually denoted as α, is generally assigned to acceptors in CIGS originating from V_{Cu}. Thus the density of α-defect is strongly dependent on the concentration of V_{Cu}. As for the ζ-defect located at around ~250 meV, which is similar as that of transition energy for $(V_{Se}\text{-}V_{Cu})$ complex, was confirmed to be related to $V_{Cu\text{-}Se}$ vacancy complex. Therefore, α- and ζ-defects are both sensitive to the V_{Cu}, since in case of ζ-defect, the variation of V_{Cu} will consequently affect the $(V_{Se}\text{-}V_{Cu})$ complex. It was reported that the incorporation of Na will reduce the concentration of V_{Cu}, therefore, the defect densities of α- and ζ-defects in SLG devices should be lower than that of the ceramic one.

To verify this, the measured the defect density of both the device by utilizing the model developed by Walter et al.[5] which allows one to directly determine the energetic distribution of defects from experimentally obtained capacitance data C (f,T). The spectra of the defect density for these two devices are shown in Fig. 4. It is clearly observed that the defects densities for ceramic devices, 4.67×10^{15} and 8.72×10^{14} cm^{-3}, for α- and ζ-defects while those for the SLG device are 1.13×10^{15}, 8.75×10^{14} and $5.47\times 10^{14} cm^{-3}$ for α-, β- and ζ-defects, respectively.

In summary, the effects of Na on CIGS solar cells with Ga content x=0.75 were researched. The improvement of device performance in Cell with Na is mainly ascribed to be the formation of $MoSe_2$ which favors the carrier transportation and decreases the series resistance. Besides, presence of Na, will also affect the properties of CIGS absorbers, as well as the device performances by varying the defect level and the trap density.

REFERENCES:

[1] https://www.zsw-bw.de/en/about-us/history.html.
[2] X. Hu, et al, J. Appl. Phys., **116**, (2014) 163703.
[3] T. Wata, et al., Thin Solid Films, **387**(2001)118.
[4] Q. Gao, et al, Adv, Energy Mater., **1** (2011) 845.
[5] R. Wuerz, et al, Sol. Energy Mater. Sol. Cells, **100** (2012) 132.

ACKNOWLEDEMENT
This work was supported by the Incorporated Administrative Agency, New Energy and Industrial Technology Development Organization (NEDO) under the Ministry of Economy, Trade, and Industry (METI).

LIGHT MANAGEMENT IN THIN FILM SOLAR CELLS

W. Soppe[1], D. Zhang[1], K. van der Werf[1], R. van Swaaij[2], M. Creatore[3], B. Williams[3],
Z. Vroon[1], J. van Deelen[1], B. Crombach[4], and R. van Erven[5]

1) Solliance, High Tech Campus 21, Eindhoven, the Netherlands;
2) Photovoltaic Materials and Devices, Delft University of Technology, P.O. Box 5031, 2600 GA Delft, The Netherlands
3) Eindhoven University of Technology, PO Box 513, 5600 MB Eindhoven, The Netherlands
4) C-Coatings B.V. Garnizoenstraat 6, 5363 VX Velp, The Netherlands
5) Morphotonics BV. De Run 4281, 5503 LM Veldhoven, The Netherlands

ABSTRACT: Application of structured back reflectors, made by Nano-Imprint Lithography is shown to result in excellent light trapping in a-Si and nc-Si cells on steel foil. Applying these concepts we managed to achieve back reflectors with a Haze factor of more than 95% for the entire visible wavelength regime. On these back reflectors, nc-Si and a-Si cells were fabricated and J_{sc}s of more than 24 mA/cm^2 for nc-Si cells with an absorber layer thickness of 1000 nm were obtained. We applied the anti-reflection texture on the front glass of a hybrid tandem (in which a c-Si bottom cell was combined with a semi-transparent a-Si top cell) and reached an improvement of the J_{sc} of the bottom cell of 1.7 mA/cm^2. The lacquer that has been developed for textured back reflectors in CIGS solar cells has been applied to make periodic and random textures. First experiments to implement these back reflectors in thin CIGS solar cells were carried out indicating that the growth of CIGS on textured substrates is non-conformal on textures with high aspect ratios. Further modeling, in which this non-conformity is taken into account, is required to determine the optimum textures for CIGS
Keywords: Thin Film Solar Cell, a-Si/nc-Si; CIGS; Light Trapping; Nano Imprint Lithography

1 INTRODUCTION

Improvement of the light management is one of the key routes to increase the efficiency of thin film solar cells. The first step here is obviously to minimize the reflection losses at the front side of the solar cell. The common way to realize this is to implement an anti-reflection (AR) layer at the front and/or to introduce a scattering texture which further minimizes the reflection. Main issues here are the optical absorption losses in these layers and the production costs [1]. In this paper we will demonstrate a low cost approach to fabricate AR textures on glass by Nano-Imprint Lithography (NIL).

The next step is to increase the optical path length in the absorber material by creating textured interfaces in the solar cell. Yablonivitch in 1980 determined an upper boundary for the optical enhancement factor that can be obtained by this approach: the famous $4n^2$ limit [2]. Already in 1986, Derrick et al presented optical modeling results of the light trapping effect of periodic grating textures in amorphous silicon solar cells [3]. It took however decades before such grating textures could be realized in an affordable way. It is through NIL that the fabrication of light trapping textures has become practically relevant [4].

2 APPROACH

Within the project LIMANIL we investigated the implementation of textures made by NIL in thin film solar cells, to improve the light trapping in these cells. The first step in this approach was to perform optical modeling to determine the ideal textures for the different solar cells. The next step was to implement these textures in small area solar cells. Finally, we up-scaled the NIL process to larger areas, to demonstrate the industrial applicability of the technology.

2.1 Modeling

We used Finite Difference Time Domain FDTD software (MEEP and Comsol) for optical simulations of periodic textures in a-Si/nc-Si tandem cells and in CIGS solar cells.

2.2 Implementation in solar cells

In the project we implemented AR textures on hybrid tandems, consisting of wide bandgap amorphous silicon top cells and c-Si bottom cells; we implemented Distributed Bragg Reflectors (DBR) in a-Si/nc-Si tandem cells and we implemented textured back reflectors in CIGS cells. All textures have been made by NIL.

2.3 Upscaling of NIL

In collaboration with the Dutch equipment manufacturer, Morphotonics and Solliance designed and built a roll-2-plate pilot system for semi-automatic nano-imprinting on glass sheets with sizes up to 0.6x1.2m^2.

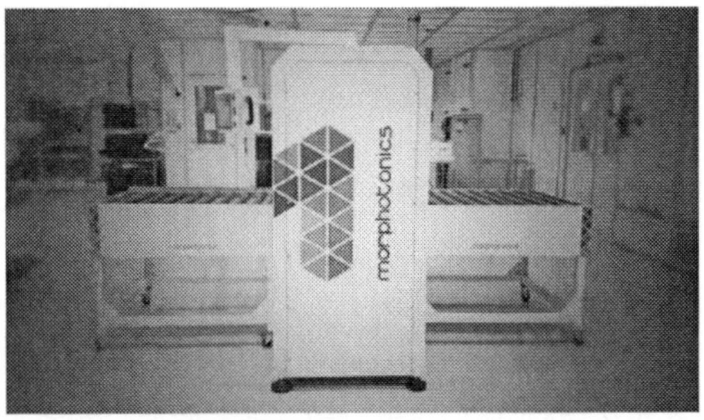

Figure 1: Roll2sheet NIL tool in the clean room of Solliance.

3 RESULTS

3.1 Modeling

a-Si/nc-Si tandems

These simulations have been carried out on thin-film silicon based tandem solar cells on glass in which a grating structure was implemented and using a three-layer DBR as intermediate reflector between the a-Si:H top cell and nc-Si:H bottom cell.

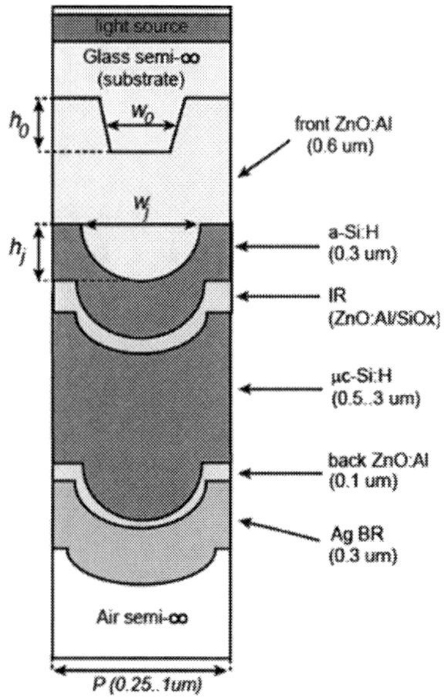

Figure 2: Schematic lay-out of a thin-film silicon tandem solar cell on a regular grating. The grating is characterized by the width, w0, and the height, h0, of the trapezoid. A leveling parameter, not shown here, indicates the degree of leveling of the layers deposited on top of this grating.

The important parameters in the simulations are: the width and the height of the trapezoid making up the grating; the leveling parameter that indicates the degree of conformity of the grating structure upon film growth, and the optical properties of the individual layers (in particular the complex refractive index). As intermediate reflector a DBR was used with the following structure: ZnO/a-Si:H/ZnO. The major results of these simulations are presented in the figure below.

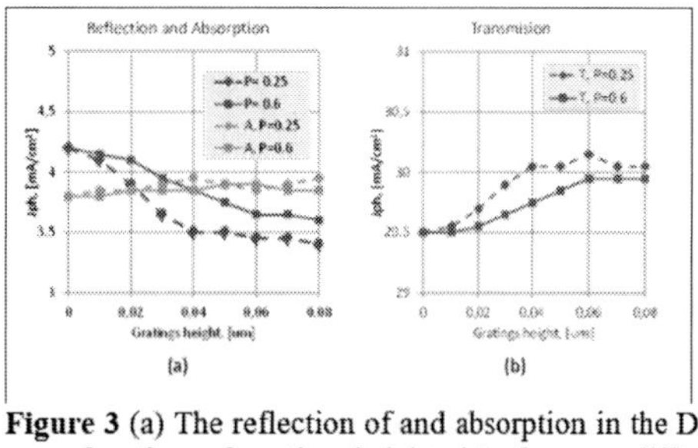

Figure 3 (a) The reflection of and absorption in the DBR as a function of grating height, h0, for two different grating periods. (b) transmission of the DBR as a function of grating height, h0, for two different grating periods. Note that all properties are expressed in the photocurrent under AM1.5

the absorption in the DBR is not affected by the grating. The reflection and transmission of the DBR are influenced by the grating, although operation of the DBR can still be observed; b) the degree of levelling of the layers has an effect on the current matching of tandem solar cell (not shown here).

CIGS solar cells

We modelled the effect of various periodic textures of the back reflector in CIGS solar cells with 2D simulations. We varied the geometry (triangles, rectangles, sinusoidals) and the pitch and height of these textures and determined the absorption in the CIGS. In the figure below some typical results are shown for triangular textures.

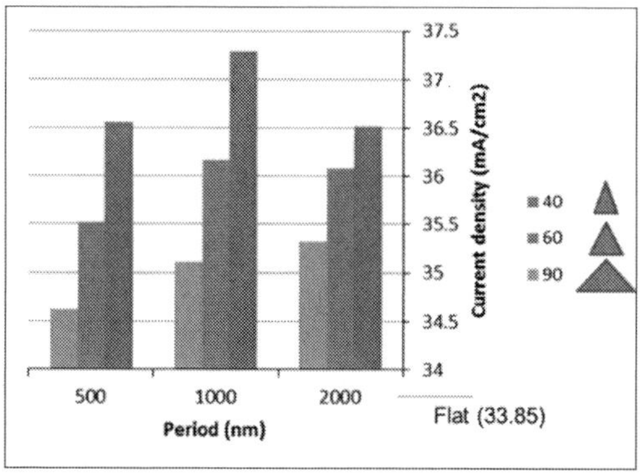

Figure 4: Enhancement of J_{sc} for 2000 nm absorbers through triangular textures with different top angles and periods.

A general trend is that textures with higher aspect ratios theoretically yield the highest light trapping enhancement. The practice however is less ideal, and side-effects like increased shunting of the solar cells on textured back contacts, play also a role.

3.2 Implementation in solar cells

Hybrid tandems

In Figure 5 the effect of an AR texture (a replica of a random pyramid structure of silicon) on the current of the bottom cell in an a-SiOx/c-Si tandem is shown. Note that the AR texture is applied on the front glass of the device, e.g. in front of the PSC top cell. The AR texture virtually has no effect on the light trapping of the PSC top cell.

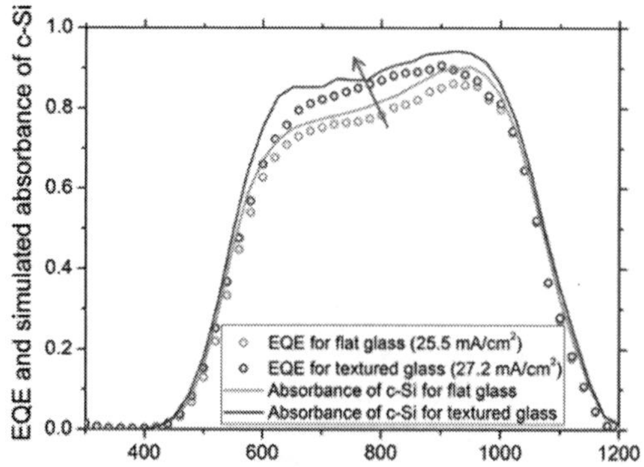

Figure 5: EQE of the c-Si bottom cell of a hybrid tandem, for which the front glass was flat (green) or coated with an AR texture (red)

Application of NIL structured back reflectors results in excellent light trapping in a-Si and nc-Si cells on steel foil. Applying these concepts we managed to achieve back reflectors with a Haze factor of more than 95% for the entire visible wavelength regime. We made nc-Si and a-Si cells on these back reflectors and J_{sc}s of more than 24 mA/cm^2 for nc-Si cells with an absorber layer thickness of 1000 nm.

CIGS cells

New UV curing lacquers were developed for the NIL process, which can withstand the higher processing temperatures of CIGS. This development was a challenge because of crack-formation and shrinkage of the material. Some shrinkage must be accepted and can be accommodated by anticipation when selecting the feature sizes of the stamp. In the figure below a typical example of an imprint is shown.

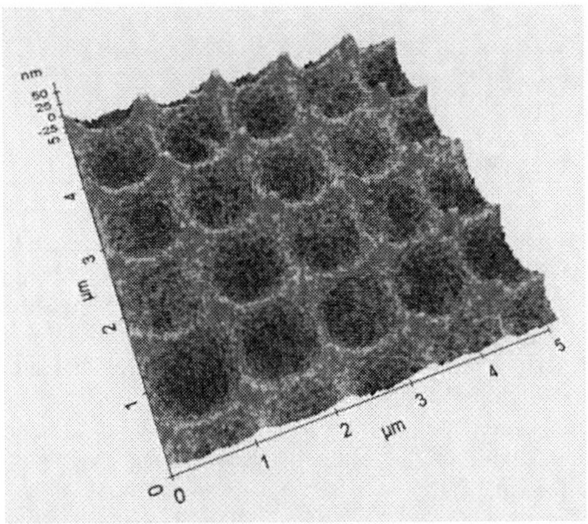

Figure 6: 2D periodic texture imprinted in a UV curable lacquer, after annealing at 550 ^{0}C. The R_{pv} is about 80 nm; i.e. about 50% of the R_{pv} of the master.

The growth of CIGS material on a textured substrate, however appears to develop differently than on a flat substrate. In the figure below one can see that on a textured surface the grains are smaller than on a flat surface, and that the overall morphology is different.

This is probably the reason that the cells that we made so far on textured substrates suffer from lower *FF* and V_{oc} than for cells made on flat substrates. Next experiments therefore will focus on textures with larger periods.

Figure 7: SEM cross section of CIGS solar cells on flat surface (left) and testured surface(right).

3.3 Upscaling of NIL

Morphotonics has developed a pilot system for semi-automatic nano-imprinting on glass sheets with sizes up to 0.6x1.2 m^2 and together with NTS a fully automatic industrial production system. Furthermore, an in-house capability was developed for nano-imprinting of substrates up to 1.6x2.0 m^2 which was used for the implementation of NIL on substrates for CIGS samples. For these samples a texture was selected based on sputter-etched AZO (aluminum doped zinc oxide) which provides U-shaped features that are known to prevent defects compared with V-shaped features..

These samples were manufactured with high temperature resistant lacquer provided by C-Coatings and tested with CIGS deposition conditions demonstrating the scalability of the concept. Next to the LIMANIL application of light trapping for thin -film PV these systems are also used for the display and lighting markets which can be considered as spin-off activities.

Figure 8: Large area shim (left) and imprint of a replica of sputtered-etched AZO (right)

4 CONCLUSIONS

The project LIMANIL has resulted in various new concepts for light management in thin film solar cells. The key technology used: NIL, has shown to be a very viable, and industrially applicable technique to implement flexible light management strategies.

ACKNOWLEDGEMENTS

This work is financially supported by the Dutch Ministry of Economic Affairs through TKI projects TKIZ01010 and TEZ0114003

REFERENCES

[1] A. Gombert et al., Solar Energy 68 (2000) 257.
[2] E. Yablonovitch, Proc. IEEE, 1982, p. 501.
[3] G. Derrick et al, Applied Optics 26 (1986) 3690.
[4] W. Soppe et al, Phys. Status Solidi A 210 (2013) 707.

TEMPERATURE DEPENDENCE OF EXTREMELY BRIGHT EL INHOMOGENEITIES IN CDTE PV DEVICES

Matevž Bokalič [a], Rok Kimovec [a], James R. Sites [b], Marko Topič [a]
[a] University of Ljubljana, Faculty of Electrical Engineering, Tržaška 25, SI-1000 Ljubljana, Slovenia
[b] Colorado State University, Fort Collins, CO 90521, USA

ABSTRACT: Extremely bright electroluminescence (EL) inhomogeneities are observed in cadmium telluride (CdTe) photovoltaic (PV) devices. Small circular laboratory CdTe cells and commercially available CdTe PV modules are inspected with EL at different temperatures. The phenomenon appears as bright ring around the cells, and as bright lines along and bright dots at the end of the monolithic contacts of the modules. The areas of the devices that exhibit the phenomenon correspond to the areas where the local applied voltage is the highest. The exact shape of the phenomenon appears grainy and may be defined by the grainy structure of the CdTe. The effect is present on devices when the temperature of the devices is below a certain threshold (around 30 °C in our case), and when the applied forward bias is high enough (around one half of the short circuit current). The results suggest that the phenomenon may be correlated with either temperature or voltage break down of the back contact barrier which exists between the CdTe and the metal back contact.

Keywords: CdTe, Electroluminescence, Solar Cell, PV Module, Back Contact

1 INTRODUCTION

The cadmium-telluride (CdTe) based photovoltaics (PV) production has increased to almost 3 GW/a in the last year, ensuring its top position in thin-film market share [1]. The recently achieved 21.0% record cell efficiency, and 18.6% record module efficiency [2], make the CdTe the most efficient conventional thin-film technology, which is the result of intense scientific effort.

Our research of CdTe PV devices was focused on spatially resolved characterization [3], where we often observed different forms of extremely bright inhomogeneities in electroluminescence (EL) images of various CdTe devices. They appeared either as bright rings around the small circular laboratory cells [4,5] or bright lines in commercial CdTe PV modules [3]. Other groups have also observed similar inhomogeneities, such as bright ring around the cell in LiT image [6], bright ring around the cell in EL image and short bright line along the monolithic junction of the CdTe sub-module [7]. Although other authors have noticed the phenomenon, they have only provided a brief description.

In this work we report on the existing results in more detail and focus on temperature dependence of extremely bright inhomogeneities in small circular laboratory CdTe cells and commercially available CdTe PV modules using EL.

2 EXPERIMENTAL

2.1 Small circular laboratory CdTe cells

Small circular (1 cm diameter) laboratory CdTe solar cells were manufactured by Materials Engineering Laboratory at Colorado State University (CSU) following their standard manufacturing procedure: Tec10 TCO glass superstrate, close-spaced sublimation (CSS) of a thin cadmium sulfide (CdS) layer and a thicker CdTe layer, cadmium chloride ($CdCl_2$) heat treatment, copper chloride deposition and annealing, and finally application of a graphite/nickel back contact [8]. The cells were characterized in Photovoltaic Laboratory at CSU [9]. Characterization includes electroluminescence imaging with a 9 mega pixel silicon camera and temperature dependent EL imaging. The latter was conducted by installing the cells into transparent mini climatic chambers, where the temperature was controlled by either ventilating the chamber with evaporated liquid nitrogen or heating using hot air. The temperature was measured at a single point at the center of the back contact of the cell. Correction of the images includes background subtraction, bright outliers removal and vignetting correction.

2.2 Commercially available CdTe modules

The CdTe modules presented herein are commercially available products and were independently purchased on the market. General EL imaging, magnified module EL and selective cooling was carried out in the Laboratory of Photovoltaics and Optoelectronics (LPVO) at the University of Ljubljana. A 16 mega pixel silicon camera was used. The selective cooling was realized by placing ice on the designated area of the module (area labeled in Fig. 6). After approximately 30 minutes the ice was removed, module was wiped dry and immediately imaged by EL. According to the thermal camera, the temperature of the designated area was around 6° C at the beginning of the EL measurements. Magnified EL images were acquired at selected points of interest in the module using extension tubes to achieve magnification. Image correction for general EL and selective cooling includes background subtraction, bright outliers removal and vignetting correction, while for magnified EL vignetting correction was not performed.

The temperature dependent measurements were performed at Forschungszentrum Jülich where the temperature controlled chamber of the sun simulator was used to acquire the temperature dependent EL images with a 320k pixel InGaAs camera. Image correction includes temperature corrected background subtraction and outliers removal. The thermal background (i.e. black body radiation of the imaged objects) becomes noticeable with the InGaAs camera at temperatures above room temperature.

3 RESULTS AND DISCUSSION

3.1 Extremely bright EL phenomenon

Extremely bright EL phenomenon appears as bright ring around the small circular CdTe cells (Fig. 1 a)) and as bright lines along and dots at the end of the monolithic contacts (MC) in the CdTe modules (Fig. 2). In the cell, the location of the bright ring corresponds to the highest local voltage, which occurs along the edge of the cell due the resistance of the TCO [5]. Zoomed areas (labeled with A and B in Fig. 2) of three cells and two MCs in the PV modules is shown in Fig. 3. Figures a) and b) display the same EL image with different color scales to emphasize the magnitude of the phenomenon. A visual image of the same area A with backlight illumination is added in figure c) to allow the determination of the exact location where the phenomenon occurs. Comparison of figures a) to b) in Fig. 3 reveals that the bright lines appear along and immediately above the P3 cut in the active area of the cell, where the local junction voltage is the highest due to typical current distribution in thin film cells governed mainly by the TCO's resistance [10]. Furthermore, very thin bright lines are observed at the opposite side of the MC just below the P1 cut. In the downward direction, the bright EL line is immediately followed by a much weaker EL signal. Such a strong fluctuation of the voltage without an obvious reason seems highly unlikely and suggests that this may be an optical effect. Additionally the zoomed in EL images of the bright dots at the end of the MC have also been taken. The same EL image of area B is presented in different scales in Fig. 3 d) and e) to emphasize the magnitude of the phenomenon. Figures d) and e) are aligned to figure c) for determination of the MC's location. Here the bright dots no longer appear as dots but rather as bright areas. Similar to the bright lines, they appear near the MC, but in contrast to the previous case, the former explanation may only be applied to the phenomenon around the lower MC, while the upper MC shows the phenomenon only near the P1 cut but not near the P3 cut. Explanation of this behavior requires additional research.

When inspected on the micrometer scale (Figs. 1 and 3), the ring, the lines and the dots do not appear homogeneous. Their shape and intensity appears grainy, which may indicate that on the micrometer level the area occupied by the phenomenon is defined by the grainy structure of the CdTe.

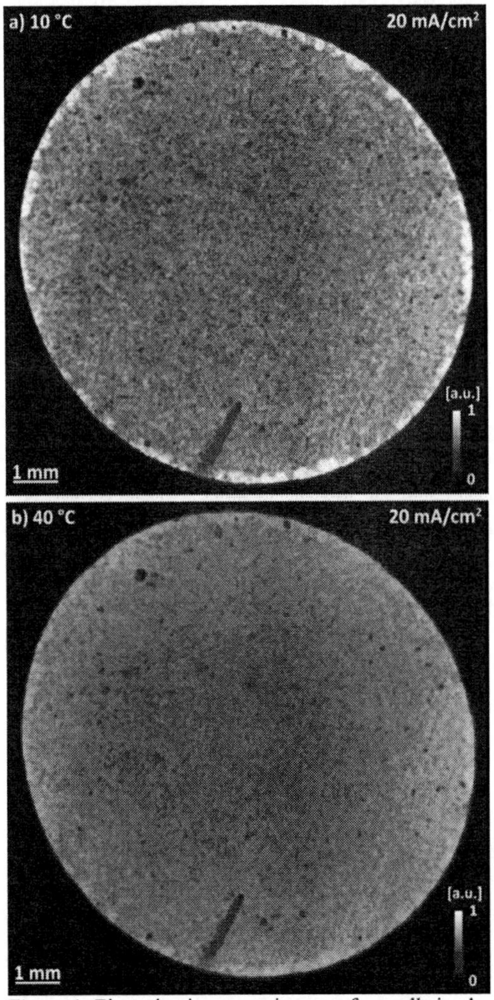

Figure 1: Electroluminescence images of a small circular laboratory CdTe solar cell acquired under 20 mA/cm^2 forward bias current density at a) 10 °C and b) 40 °C.

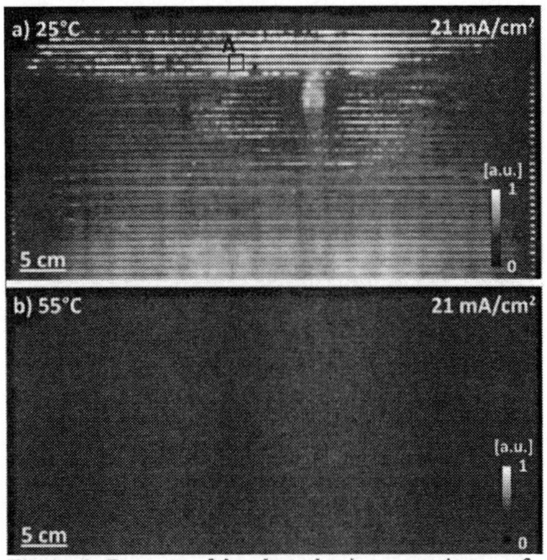

Figure 2: Excerpts of the electroluminescence image of a commercial CdTe PV module acquired under 21 mA/cm^2 forward bias current density at a) 25 °C and b) 55 °C. Additional ROIs A and B are labeled.

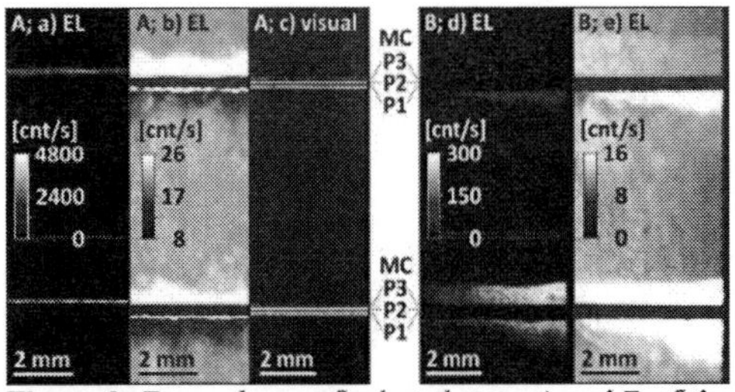

Figure 3: Zoomed area of selected areas A and B of the commercial CdTe PV module. a) and b) the same EL image showing extremely bright EL lines in different color scale, c) visual image of the same area with backlight illumination, d) and e) the same EL image showing extremely bright EL lines in different color scale.

3.2 Temperature dependence

The EL images of the CdTe cell in Fig. 1 a) and b) were taken at 10 and 40 °C, respectively, while the EL images of the CdTe module in Fig. 2 a) and b) were taken at 25 and 55 °C, respectively. The figures clearly show that the bright phenomenon disappears when either the CdTe cell or the CdTe module is heated.

3.3 Selective cooling (s.c.)

EL images acquired at different temperatures reveal temperature dependence of the bright EL phenomenon. In order to confirm the temperature activation of the phenomenon we selectively cooled one third of the upper half of the module. EL images acquired before and after selective cooling are presented in Fig. 4 a) and b), respectively. The bright phenomenon (bright lines and dots in case of the module) is not exhibited by the selected area when the device is at room temperature, but when the same area is cooled it exhibits the phenomenon.

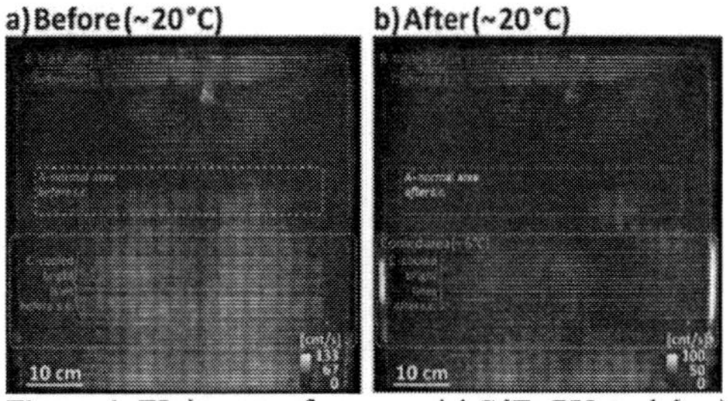

Figure 4: EL images of commercial CdTe PV module a) before and b) after selective cooling.

Quantitative analysis of EL intensity at varying forward bias current density before and after selective cooling is presented in Fig. 5 for A – normal area, where no bright lines are observed, B – region of interest (ROI) limited to the top bright lines, and C – ROI limited to the bright lines in the cooled area. The ROI A exhibits potential growth expected for the EL intensity [5,11]. The growth of the ROI B also resembles the potential growth and similarly does the ROI C before selective cooling. The interesting observation can be made on the ROI C after selective cooling. Here, at low injection current densities (J_F), the curve at first follows the uncooled one, afterwards it exhibits rapid growth at middle J_F and finally it starts to follow the bright line's curve at high J_F. Such a behavior indicates also forward injection current or forward voltage bias activation of the phenomenon.

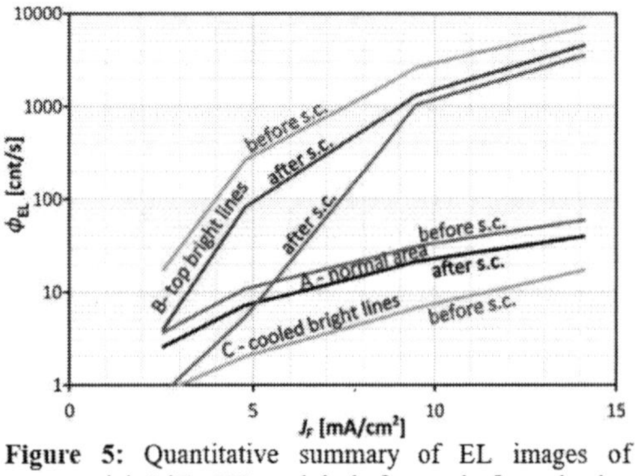

Figure 5: Quantitative summary of EL images of commercial CdTe PV module before and after selective cooling.

4 CONCLUSION

The extremely bright EL inhomogeneities are reported as bright rings in the case of small circular laboratory CdTe solar cells and as bright lines along and dark dots at the end of the monolithic contacts of the commercial CdTe PV module. In both cases the phenomenon is exhibited in the areas where the local voltage across the device is the highest according to the current and voltage distribution in such thin-film devices.

The phenomenon also exhibits a temperature dependence which manifests in "turning on" the phenomenon below a certain temperature threshold.

Combining the temperature and injection current dependence reveals that both temperature and injection conditions must be met to excite the phenomenon: the temperature must be low enough and the injection current density (and simultaneously also the applied voltage) must be high enough.

The results suggest that an energy barrier governs the extremely bright EL phenomenon: at lower temperatures the energy barrier is not thermally activated, however by forcing the forward injection current through the device, the local voltage becomes high enough to overcome the barrier and locally inject high current through the main junction resulting in the extremely bright EL phenomenon. An appropriate barrier in CdTe devices exists between the CdTe and the metal back junction, however, to prove the connection, further research is underway.

5 Acknowledgments

The authors acknowledge the financial support from the Slovenian Research Agency (Research Programme P2-0197, Project J2-7105, Bilateral project Bi-US/15-16/008). M. Bokalič acknowledges the Slovenian Ministry of Education, Science and Sport for providing post-doc funding through the European Social and Cohesion Fund (UL FE-664). Material Engineering Laboratory at Colorado State University and Forschungszentrum Jülich are acknowledged for cell fabrication and temperature dependent module measurements, respectively.

6 References

[1] Fraunhofer Institute for Solar Energy Systems, *Photovoltaics Report*, Freiburg, **2016**.

[2] M. A. Green, K. Emery, Y. Hishikawa, W. Warta, E. D. Dunlop, *Prog. Photovolt: Res. Appl.* **2016**, *24*, 3.

[3] M. Bokalič, M. Topič, *Spatially Resolved Characterization in Thin-Film Photovoltaics*, Springer International Publishing, Cham, **2015**.

[4] K. N. Zaunbrecher, S. W. Johnston, J. R. Sites, in *2012 38th IEEE Photovoltaic Specialists Conference (PVSC)*, **2012**, pp. 001716–001720.

[5] M. Bokalič, J. Raguse, J. R. Sites, M. Topič, *J Appl Phys* **2013**, *114*, 123102.

[6] D. Shvydka, J. P. Rakotoniaina, O. Breitenstein, *Applied Physics Letters* **2004**, *84*, 729.

[7] V. I. Kaydanov, T. R. Ohno, *Studies of Basic Electronic Properties of CdTe-Based Solar Cells and Their Evolution during Processing and Stress*,

National Renewable Energy Laboratory, Golden, Colorado, US, **2007**.

[8] R. A. Enzenroth, K. L. Barth, W. S. Sampath, V. Manivannan, A. T. Kirkpatrick, P. Noronha, *J. Sol. Energy Eng.* **2009**, *131*, 21012.

[9] J. Raguse, J. T. McGoffin, J. R. Sites, in *2012 38th IEEE Photovoltaic Specialists Conference (PVSC)*, **2012**, pp. 448–451.

[10] M. Bokalič, K. Brecl, J. Raguse, K. Zaunbrecher, J. R. Sites, M. Topič, in *27th EUPVSEC Proceedings*, Frankfurt, Germany, **2012**, pp. 2222–2225.

[11] K. J. Price, A. Vasko, L. Gorrelland, A. D. Compaan, *MRS Online Proceedings Library* **2003**, *763*, 195.

Structural, Morphological, and Optical Properties of Single Step Electrodeposited Cu_2ZnSnS_4 (CZTS) Thin Films for PV Applications

H. Kirou, L. Atourki, E.H. Ihalane, A. Elfanaoui, K. Bouabid, M. Nya & A. Ihlal
Materials and Renewable Energy Laboratory.Physics department, Ibn Zohr University, Agadir, Morocco.
*Corresponding author :hassan.kirou@edu.uiz.ac.ma

ABSTRACT: Cu_2ZnSnS_4 (CZTS) is a promising compound for use as absorber in thin film solar cells. In this paper we report on the results of CZTS thin film grown on SnO_2/F coated glass substrates by co-electrodeposition at different applied potentials. The obtained thin films were sulfurized in N_2 atmosphere at 550°C for 1h. The structural, morphological and optical properties were investigated using X-ray diffraction (XRD), scanning electron microscopy (SEM) and optical absorption techniques. SEM analysis shows a very dense and uniform surface with large particles. X-ray diffraction patterns exhibit the kesterite structure of CZTS with the preferred orientation of the (112) plan and other kesterite peaks assigned to (002), (101), (110), (103),(200), (220), (224)and (400) planes. Raman spectroscopy measurements confirm the XRD results with clear peaks at 253, 289, 339 and 368 cm^{-1}. The Raman shift was attributed to CZTS kesterite mode vibration. Finally, the bandgap of CZTS film is found to be close to 1,5eV.
Keywords: CZTS, Cu_2ZnSnS_4, Electrodeposition, annealing, kesterite,

1. INTRODUCTION

High Efficiency and low cost fabrication are the majors persisting problems which restraint the industrialization of solar cells in the large scale. Actually, the highest efficiency in the commercial photovoltaic modules is 25% achieved by monocrystalline silicon (mono-Si) cells [1]. Mass production of this type of solar cells is limited by the high cost of processing [2]. Thin film materials like CIGS are emerging as promising candidates to compete with mono-Si and recorded recently an efficiency of 22.3 % by Japan's Solar Frontier [3]. However, CIGS contains the relatively scarce elements indium and gallium, as well as toxic selenium, which limit its viability for mass production in the mid-to long- term. Another quaternary absorber material has attracted recently the greatest attention is $Cu_2ZnSnSe(S)_4$ (CZTSe(S)) due to its tailorable band gap from 1 to 1.5 eV and its high absorbance coefficient higher than 10^4 cm^{-1}[4]. Moreover, they are based on earth-abundant, inexpensive and nontoxic elements [5]. Record device efficiencies are currently around12.6% achieved by Wei Wang et al. [6]. In this work, we present some results of CZTS thin films synthesized in single bath deposition.

2. Experimental details

CZTS thin films were deposited from aqueous solution containing 20mM of copper sulfate pentahydrate ($CuSO_4.5H_2O$), 20mM of Zinc sulfate heptahydrate ($ZnSO_4.7H_2O$) and 13 mM of tin chloride ($SnCl_2$) and 20 mM of sodium thiosulfate ($Na_2S_2O_3$) as source of sulfur. Two complexing agents were used to bring the reduction potentials of metals ions closer together and facilitate the solubilization of the various species (citric acid and tartaric acid with concentration of 20mM for each one). The pH of the bath solution was kept up to 3.5 by NaOH and Na_2SO_4 was used as supporting electrode.

The electrodepositions thin films were carried outin a conventional tree electrodes system controlled by PGZ 301 Voltalab Potensiostat. Platinum (Pt) grid was used as counter electrode, saturated calomel electrode (SCE) as a reference electrode and FTO substrate as working electrode.

The deposited thin films were dried and annealed in the tube of a furnace for 1h at 550°C under a small flow of nitrogen gas. Then, the annealed CZTS thin films were characterized with several techniques. X-Ray diffraction spectrum was obtained by Bruker Diffractometer employing Cu Kα1 radiation wavelength 1.54060 Å at 45kV. Raman spectroscopy accompanying with 514 nm Argon ion laser was carried out to distinguish between pure CZTS compound and their secondary phases. Morphological and optical proprieties were performed using Bruker Scanning electron microscope (SEM) and JASCO V-730 UV–vis spectrophotometer, respectively.

3. Results and Discussion

Cyclic voltammetry (CV) scanning was investigated to determine reduction potential region of ions (Cu^{2+}, Zn^{2+}, Sn^{2+} and S^{2-}) existing in the electrolyte bath. The result of CV curve shows a large cathodic peak between -1 V and -1.5 V (vs SCE) indicating the reduction of all elements in this region (figure 1). Then we limited our study to this region. CZTS thin films deposition was carried out at -1 V, -1.1 V and -1.2 V for 10 minutes. We have avoided going into the high negative potentials where the hydrogen reduction is dominant.

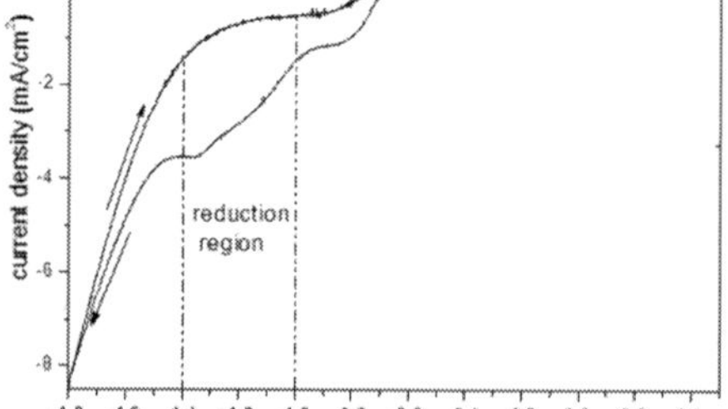

Figure 1: Cyclic voltammetry (CV) scanning from -1.8V to 0.4V (vs. SCE) of the aqueous solution

The chronoamperometry curves of deposited CZTS thin films for different chosen potentials are shown in figure 2. It can be shown that the current density is constant and increasing more negatively with decreasing the imposed potential what means that deposing element becomes very important in the more negative potential.

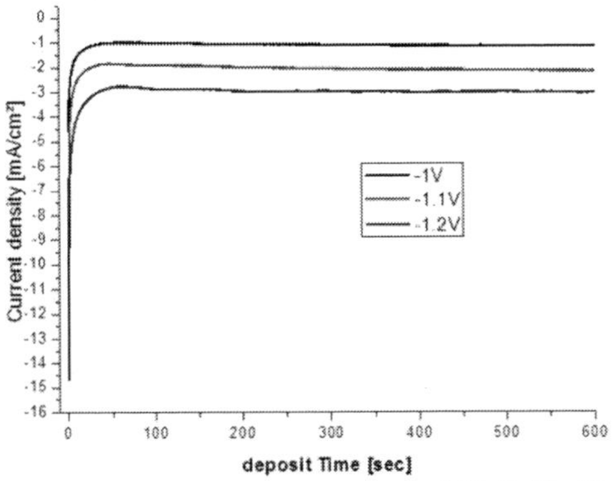

Figure 2: chronoamperometry curves of CZTS thin films deposited at -1V, -1.1V and -1.2V

Figure 3 shows the XDR patterns of annealed CTZS thin films deposited at different potentials. In addition to FTO peaks, XRD patterns exhibit more than ten peaks (002, 101, 110, 112, 103, 200, 220, 312, 224 and 400) plans assigned to CZTS in the kesterite structure according to **PDF JCPDS card #26-0575.** Particularly, the CZTS layer prepared at -1.1V potential, exhibit intense and sharp peaks indicating a good crystallinity of the prepared CZTS thin films. Lattice parameters were calculated using inter-reticular distance d(h,k,l) formula of tetragonal crystal system with XRD peaks (112) and (220) plans. The average values of a and c parameters were find to be 5.433 Å and 10.942 Å, are very close to lattice parameters calculated from data card 26-0575 (a=5.427Å, c=10.780 Å)

Secondary phase were detected in XRD spectra in two positions with small peaks at 32.2° and 42.25° attributed to $Cu_{2-x}S$. This impurity can be eliminated by immersing the sample in the potassium cyanide (KCN)[7].

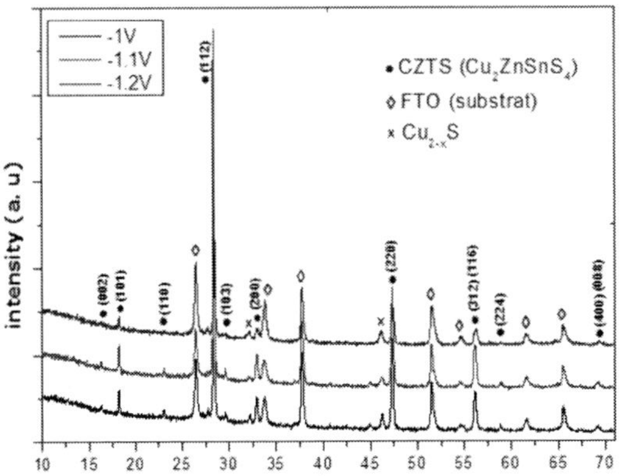

Figure 3: XR-diffraction of deposited thin films.

As the XRD analysis cannot distinguish between pure CZTS phase and two others secondary phases (ZnS and Cu_2SnS_3) due the overlap of the peaks of these two alloys. Therefore, Raman measurements were carried out to distinguish between these phases. The Raman spectrum shows the presence of peaks at 253, 289, 339 and 368 cm-1 attributed to CZTS vibrational mode. Moreover, additional peak at 674 cm^{-1} of CZTS in the second order is shown in the Figure 4 indicating a high quality and crystallinity of CZTS thin layer. $Cu_{2-x}S$ trace is detected at 480 cm^{-1}. These results are on good agreement with XRD measurements and with other works in the literature [8,9].

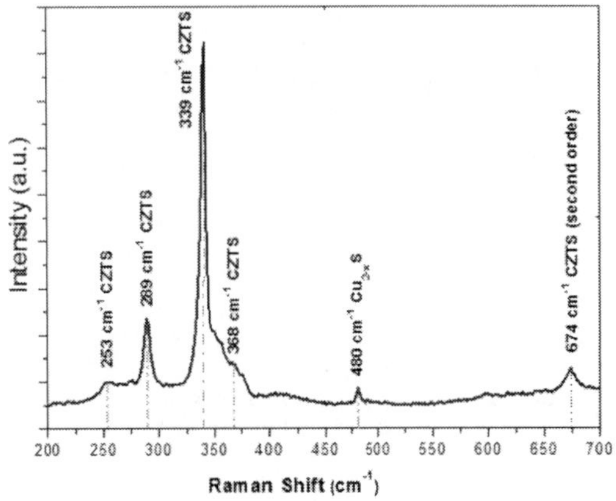

Figure 4: Raman spectrum of sample deposited at -1.1V potential (vs. SCE) for 10 minutes.

Surface morphology and cross section of sample deposited at -1.1V are presented in Figure 5. SEM micrograph shows a high density and compact morphology with defined agglomerations of particles. The grain size was estimated to be around 1-2 µm. The layer thickness was determinate by cross section image; and found to be in the range 1.6- 1.8µm.

Figure 5: SEM image and cross section of layer deposited at -1.1V

Transmittance and absorbance measurements were recorded in visible and near infrared region using JASCO V-730 UV–vis spectrophotometer in the wavelength range 400-1100 nm (Figure 6). The transmittance of the film is around 0% in the visible region and doesn't exceed 15% in the near infrared region. Moreover, the absorption coefficient of the film is higher than 10^4 cm^{-1}.

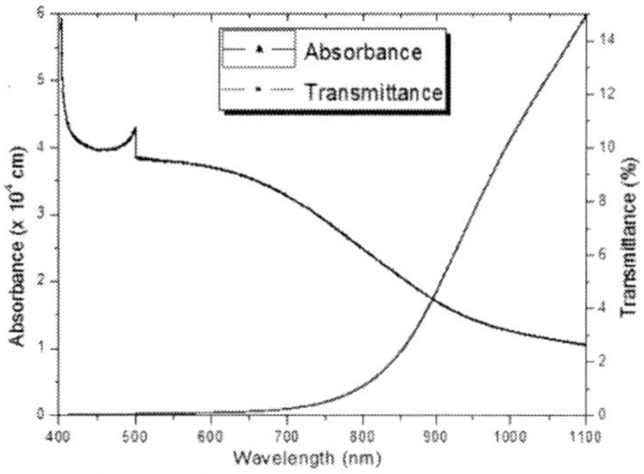

Figure 6: Absorbance and transmittance measurements

The band gap energy of the CZTS thin film was determent by extrapolating to the zero line, the curve plotting (αhϑ) versus hϑ in the liner region (Figure 7). The band gap energy is around 1.52 eV, this result is in good agreement with several reported studies [10,11].

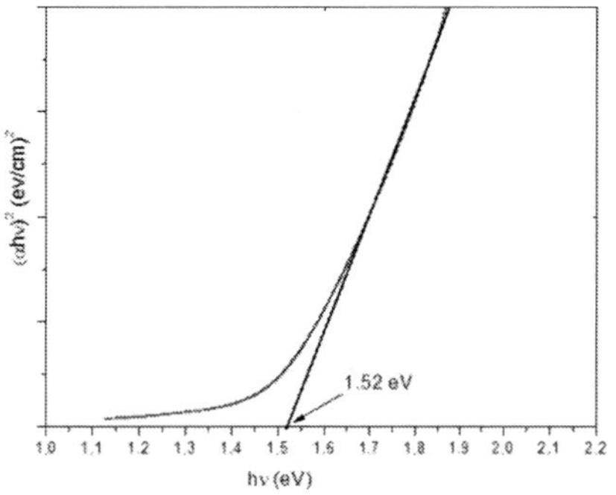

Figure 7: plotting curve of (αhϑ) versus hϑ.

4. CONCLUSION

A cost effective and facile technique was deployed to synthesis a CZTS quaternary compound in the single bath deposition. Reduction potential zone was detected by cyclic voltammetry technique. X-ray diffraction shows a near stoichiometry Cu_2ZnSnS_4 with applied potential of -1.1V vs SCE. Raman spectroscopy measurement and XRD analysis demonstrate the good quality of our kesterite CZTS thin films. SEM observations show that the synthesized films exhibit a compact and uniform surface morphology with agglomeration of particles. The absorption coefficient is higher than 10^4 cm^{-1} and the band gap energy was found to be close to 1.5 eV.

[1] Zhao J, Wang A, Green MA. Progress in Photovoltaics: Research and Applications 4 (1999) 7:471.
[2]K.L.Chopra, P.D.Paulson, V.Dutta,Progress in Photovoltaics: ResearchandApplications12(2004)69–92.
[3] Solar Frontier Achieves World Record Thin-Film Solar Cell Efficiency: 22.3%, 2015. <http://www.solar-frontier.com/eng/news/2015/C051171.html> (accessed 12.02.16).
[4] K. Ito and T. Nakazawa, Japanese Journal of Applied Physics, 27 (1988) 2094–2097.
[5] H. Katagiri, N. Sasaguchi, S. Hando, S. Hoshino, J. Ohashi, and T. Yokota, Solar Energy Materials and Solar Cells, 49 (1997) 407–414.
[6]Wei Wang, Mark T.Winkler, Oki Gunawan, Tayfun Gokmen, Teodor K. Todorov, YuZhu, et al. ,Advanced Energy Mater 4 (2014) 1301465
[7] M. Bär, B. A. Schubert, B.Marsen et al. AppliedPhysicsLetters. 99 (2011) 15-152111.
[8] Md Ibrahim Khalil, Roberto Bernasconi, Luca Magagnin, ElectrochimicaActa 145 (2014) 154–158.
[9]XiancongHe,HonglieShen,WeiWang,JinhongPi,YuHao,Xiaobo Shi, Applied Surface Science,(282), (2013) 765–769
[10]Yanfeng Cui, ShaohuaZuo, Jinchun Jiang, Shengzhao Yuan, Junhao Chu Solar, Energy Materials & Solar Cells 95 (2011) 2136–2140
[11]SeungWook Shin, Jun Hee Han, Chan Yeong Park, Sae-Rok Kim, Yeon Chan Park, G.L. Agawane, A.V. Moholkar, Jae Ho Yun, Chae Hwan Jeong, Jeong Yong Lee, JinHyeok Kim, Journal of Alloys and Compounds 541 (2012) 192–197.

CHARACTERIZATION OF (AG, CU)$_2$ZNSN(S,SE)$_4$ KESTERITE SOLAR CELL FABRICATED BY SPRAY PYROLYSIS OF AQUEOUS PRECURSOR SOLUTION.

Wei-Chih Huang, Shih-Yuan Wei, Chung-Hao Cai, Tzu-Ying Lin, and Chih-Huang Lai
Department of Materials Science and Engineering, National Tsing Hua University
No. 101, Section 2, Kuang-Fu Road, Hsinchu, Taiwan 30013, R.O.C.

ABSTRACT: The effect of Ag incorporation into CZTSSe kesterite solar cell by a non-vacuum environmental-friendly process has been studied. We first demonstrated the fabrication of (Ag, Cu)$_2$ZnSn(S,Se)$_4$ kesterite solar cells using chemical spray pyrolysis. Ag content in the ACZTSSe thin film can be controlled by precursor solution concentration. For kesterite with high Ag content, aggregation of grain and formation of second phases occurred at elevated temperature, but can be solved when being annealed at relatively low temperature. Enhancement of grain size at lower process temperature is fulfilled by Ag incorporation. With 20% Ag addition, efficiency of the kesterite increased from 5.3% to 6.8% with the improvement of V_{OC} from 388 mV to 444 mV. Bandgap of ACZTSSe can be controlled from 1.06 to 1.2 eV with Ag content varied from 0~50% and the tailing energy was reduced by Ag.
Keywords: ACZTSSe, kesterite, spray pyrolysis, phase stability

1 Introduction

Kesterite Cu$_2$ZnSn(S, Se)$_4$ (CZTSSe) has been regarded as a promising absorber candidate for solar cell application. Cu$_2$ZnSn(S, Se)$_4$ is composed of earth abundant elements and possesses high absorption coefficient, direct and adjustable bandgap, which is suitable for photovoltaics. The record power conversion efficiency (PCE) of CZTSSe is 12.7%[1], but still far away from that of CIGS, which is 21.7%[2]. First principle calculation showed that Cu_{Zn}, Zn_{Cu} are majority defects in kesterite solar cells[3]. These defects may reduce V_{OC} of CZTSSe by two mechanisms. First, Cu_{Zn} defect at the surface pins the Fermi level at the middle of the bandgap, and therefore type inversion of CZTSSe is limited[3]. Second, [Cu_{Zn}^- + Zn_{Cu}^+] defect complex may cause electrostatic fluctuation in CZTSSe, and result in the band tailing states in CZTSSe, proposed to be one of the origin of V_{OC} deficient in kesterite solar cells[4]. Ag incorporation in kesterite is one of the possible way to reduce the electronic defects. Recently, Yuan et al. showed that the formation energy of the Ag_{Zn} in AZTS is far larger than Cu_{Zn}. Therefore, by incorporation of Ag in kesterite solar cell, detrimental defect Cu_{Zn} can be suppressed and efficiency might be further improved. A strategy to increase the V_{OC} by means of (Ag, Cu)$_2$ZnSn(S, Se)$_4$ (ACZTSSe) solid solution or Ag graded ACZTSSe was also proposed[5]. Recently, over 10% efficiency was reached with 10% Ag substitution.[6] However, the improvement of efficicncy of kesterite solar cells by Ag substitutuin can only be fulfilled at low Ag content[6][7]. The effect of Ag content to kesterite solar cell is still unclear. In this work, We fabricated ACZTSSe kesterite solar cells with various Ag content by environment-friendly spray pyrolysis process. The effect of Ag content and selenization temperature on the microstructure, crystal structure, and device properties of ACZTSSe solar cells was further investigated.

2 Experimental

2.1 Preparation of precursor solution and precursor films

CuCl$_2$·2H$_2$O (98%, Acros Organics), AgNO$_3$ (99%, Sigma Aldrich), 12 mM SnCl$_2$ (98%, Alfa Aesar), 14.7 mM Zn(NO$_3$)$_2$· 6H$_2$O (99%, JT Baker), and thiourea (99%, Sigma Aldrich) were dissolved in 150 ml DI water to form aqueous precursor solution. A series of ACZTSSe solar cells with various of Ag/(Cu+Ag) ratio of 0, 0.20, 0.35, 0.5 and 1 and denoted as CZTSSe, Ag 20, Ag 35, Ag 50 and AZTSSe respectively. The corresponding chemical concentration is shown in Table I. The precursor solution was then sprayed on the hot plate heated at 300 °C to form chalcogenide precursor films. The spray rate is 2.5 ml/min with 5 bar N$_2$ as carrier gas.

Table I: Chemical concentration for ACZTSSe precursor solution.

Sample No.	CuCl$_2$·2H$_2$O	AgNO$_3$
CZTSSe	22.6 mM	0 mM
Ag 20	18.1 mM	4.5 mM
Ag 35	14.7 mM	7.9 mM
Ag 50	11.3 mM	11.3 mM
CZTSSe	0 mM	22.6 mM

2.2 Device fabrication and characterization

The as sprayed film was selenized with 200 mg selenium and 4 mg SnSe powder under different temperature. The temperature used for the selenization the different Ag content is shown in Table II. The devices were completed by chemical bath deposition of 80 nm CdS, sputtering of 50nm i-ZnO/ 200 nm ZnO:Al as window layers and screen printing Ag grids as front contacts.

Table II: Selenization temperature used for ACZTSSe.

Sample No.	Annealing T (°C)		
CZTSSe	-	-	550
Ag 20	-	-	550
Ag 35	-	470	-
Ag 50	-	470	550
CZTSSe	400	470	550

The crystal structure was determined by the X-ray diffraction (XRD, D2 phaser, Bruker) with monochromatic Cu-Kα1 radiation (λ= 1.5418 Å). The morphology was observed by field emission scanning electron microscope (SEM, SU8010, Hitachi). Current-voltage (I-V) characteristics were implemented by the Keithley 4200SCS four-point probe system under AM 1.5G spectrum, generated by an AAA solar simulator, which is calibrated by a standard Si solar cell certified by

Fraunhofer. External quantum efficiency (EQE) was measured from 350 nm to 1500 nm with calibrated silicon and germanium photodiodes form National Institute of Standards and Technology (NIST).

3 Results and discussion

3.1 Phase stability of ACZTSSe kesterite films

Fig 1 shows the XRD patterns of AZTSSe and Ag 50 with differnet annealing temperature. For samples annealed at 400˚C, characteristic peaks of the AZTSe kesterite is demonstrated. ZnSe was also observed because the composition of the precursor solution was designed to be Zn-rich to achieve high efficiency. However, as the anneal temperature raised, the ZnSe peak intensity became stronger. Other second phases such as Ag_2Se and $SnSe_2$ were also observed and were shown in Fig 1(b)(c). The second phases may originate from decomposition of AZTSSe at elevated temperature. This observation is in agreement with the observation from powder specimen [8]. For Ag 50, the XRD pattern demonstrates that Ag 50 is more stable at slightly higher temperature. However, as the temperature reaches 550˚C, which is often used for CZTSSe, peak intensity of ZnSe increases again with small amount of Ag_2Se. From SEM EDX analysis of the Ag 50 annealed at 550˚C, the aggregated large grain shows the normal composition of ACZTSSe. However, the composition of small grains is Zn rich, indicating the ZnSe second phases, which is in agreement with the observation from XRD.

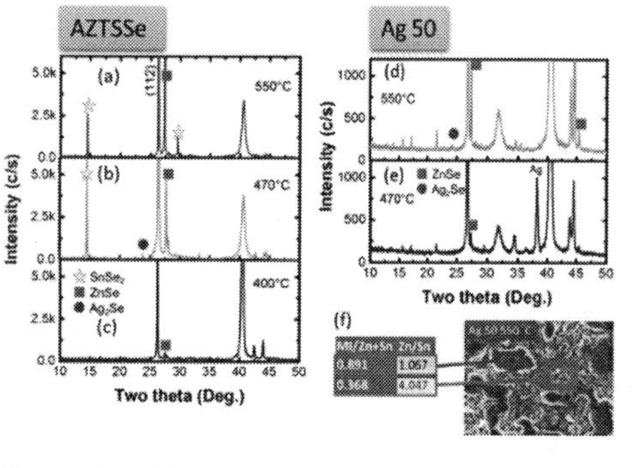

Figure 1: XRD patterns and SEM-EDX results of AZTSSe and Ag 50 with differnet annealing temperature.

3.2 Effect of annealing temperature on ACZTSSe

According to the aforementioned results, the phase stability of the Ag-incorporated kesterite was affected by the annealing temperature. Moreover, annealing temperature also impact the film morphology of the ACZTSSe. For AZTSSe, continuous film can be attained at low annealing temperature, but serious triangular aggregation occurs at high annealing temperature. This is also the case of Ag 50 annealed at 550˚C. The poor morphology is detrimental for the solar cell performance due to the large roughness and the bare Mo, which may induce the shunting path. In order to obtain proper film morphology and reduce the formation of second phases, it is mandatory to choose suitable annealing for ACZTSSe with different Ag content. To reduce second phases, annealing temperature of ACZTSSe should be decreased according to the Ag content. Finally, after the reduction of annealing temperature, continuous ACZTSSe film with limited second phases can be fulfilled. The proper annealing temperature of the AZTSSe, Ag 50, and Ag 20 are 400, 470, and 550˚C, respectively as shown in Fig. 2. Under the same temperature, the grain size became larger as the silver content of kesterite increased. The enhancement of grain size can be explained by either the formation of low melting point Sn-Ag-Cu proposed in the nanoparticle process[7], or the low melting point nature of the silver related chalcogenide.

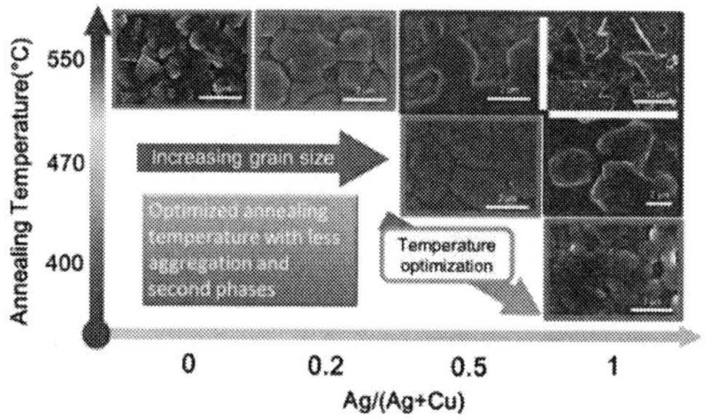

Figure 2: SEM plane view of ACZTSSe thin films with varoous Ag content annealed at differnet temperature.

3.3 Crystal structure

The XRD patterns of the ACZTSSe with different Ag content were shown in Fig. 1(a). Note that the characteristic peak of the Ag is from the screen printed top electrode. As the silver content increases, the strongest (112) peak shift to lower angle, indicating that replacement of Cu by larger Ag atoms expands the lattice spacing. Besides, (220)/(204) split into two peaks. Lattice constants of the tetragonal kesterite structure can be refined using the relation $1/d_{hkl}^2 = (h^2 + k^2)/a^2 + l^2/c^2$, where d_{hkl} is the plane distance between (hkl) planes. The refined lattice constant a, c was shown in the Fig. 2(b). The lattice constant along a-axis demonstrates linear relationship with the Ag content, while non monotonic change of c axis was observed, resulting in the c/a ratio decrease and peak splitting of (220)/(204). This phenomena was also observed in the previous reports of ACZTS/Se[8], indicating that Ag is successfully incorporated into ACZTSSe kesterite lattice. Although lattice constant is slightly smaller than the literature CZTSe because of the residual sulfur in the ACZTSSe, the trend of the lattice change in our sulfide-selenide system is similar and consistent with the selenide case in the powder specimen[8].

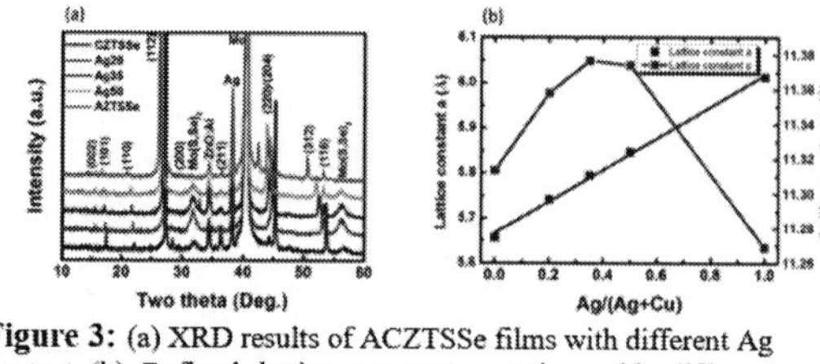

Figure 3: (a) XRD results of ACZTSSe films with different Ag content (b) Refined lattice constant a and c with different Ag/(Ag+Cu)

3.4 Device characteristics

Current density-voltage (J-V) characteristics of the ACZTSSe are summarized in Table 3. As 20% Ag is incorporated, significant improvement of the V_{OC} and J_{SC} are fulfilled, leading to PCE of 6.86%. However, if Ag/IB is over 35 %, significant current reduction and poor fill factor were observed. It was proposed that band alignment or the low carrier concentration may be responsible for the low efficiency [6][7]. To characterize the optical response of the ACZTSSe, external quantum efficiency (EQE) analysis was conducted. The bandgap derived from EQE is shown in Fig. 4(a), demonstrating the increasing trend from 1.06 (for CZTSSe) to 1.20 eV (For Ag 50). To check the effect of Ag on the defect properties, Urbach tail analysis was also conducted and the results are shown in Fig 4(b)[7]. The Urbach tail energy reduces with the increasing Ag content, showing that the ability of Ag to reduce the tailing state, and is possible to further improve the device performance.

Table III: Summary of the device characteristics of ACZTSSe

Device	Voc (mV)	Jsc (mA/cm^2)	FF(%)	η(%)	R_s (Ωcm^2)	R_{sh} (Ωcm^2)	E_g (eV)	E_U (meV)
CZTSSe	388	27.25	50	5.30	1.94	76.86	1.06	23.9
Ag 20	444	29.53	52	6.86	2.75	137.51	1.13	23.1
Ag 35	401	19.70	43	3.39	4.94	73.04	1.16	22.2
Ag 50	390	17.28	31	2.06	12.75	36.68	1.20	21.5

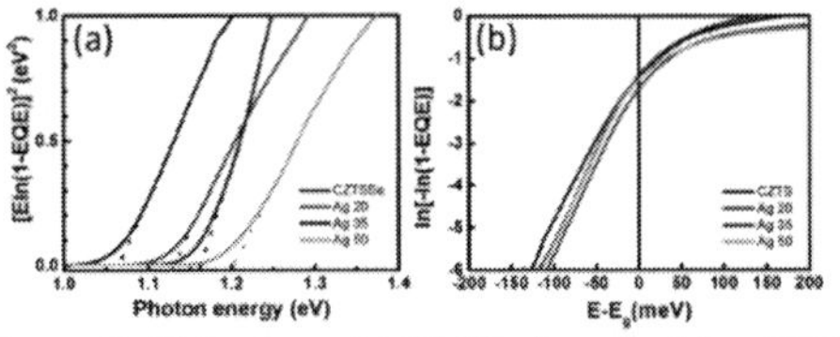

Figure 4: (a) Bandgap of ACZTSSe determined from EQE (b) ln[-ln[1-EQE]] vs. E-Eg for Urbach tail analysis

4 Conclusion

ACZTSSe thin film solar cells with 20%, 35%, 50% and 100% (AZTSSe) Ag incorporation have been successfully fabricated by chemical spray pyrolyis process. The adequate annealing temperature of ACZTSSe that fulfills the criteria "less aggregation" and "reduced second phases" decreases with the increasing Ag content and was designated. Under the same temperature, Ag enlarges the grain size of kesterite. The XRD results reveal the crystal parameters of the ACZTSSe, in agreement with the results of the powder specimen. Finally, significant efficiency improvement of V_{OC} and efficiency can be fulfilled. Device with 6.86% efficiency is achieved with 20% Ag incorporation, but decreased at high Ag content due to the improper band alignment or low carrier concentration. Ag alloying can enlarge the bandgap and reduce the tailing state, and is a promising way to further improve the performance and reduce the process temperature for the kesterite solar cell.

5 References

[1] J. Kim, H. Hiroi, T. K. Todorov, O. Gunawan, M. Kuwahara, T. Gokmen, D. Nair, M. Hopstaken, B. Shin, Y. S. Lee, W. Wang, H. Sugimoto, and D. B. Mitzi, "High Efficiency Cu 2 ZnSn(S,Se) 4 Solar Cells by Applying a Double In 2 S 3 /CdS Emitter," *Adv. Mater.*, p. n/a–n/a, 2014.

[2] T. M. Friedlmeier, P. Jackson, A. Bauer, D. Hariskos, O. Kiowski, R. Wuerz, and M. Powalla, "Improved Photocurrent in Cu(In,Ga)Se<inf>2</inf> Solar Cells: From 20.8% to 21.7% Efficiency with CdS Buffer and 21.0% Cd-Free," *IEEE J. Photovoltaics*, vol. 5, no. 5, pp. 1487–1491, 2015.

[3] S. Chen, A. Walsh, X. G. Gong, and S. H. Wei, "Classification of lattice defects in the kesterite Cu2ZnSnS 4 and Cu2ZnSnSe4 earth-abundant solar cell absorbers," *Adv. Mater.*, vol. 25, no. 11, pp. 1522–1539, 2013.

[4] T. Gokmen, O. Gunawan, T. K. Todorov, and D. B. Mitzi, "Band tailing and efficiency limitation in kesterite solar cells," *Appl. Phys. Lett.*, vol. 103, no. 10, 2013.

[5] Z. K. Yuan, S. Chen, H. Xiang, X. G. Gong, A. Walsh, J. S. Park, I. Repins, and S. H. Wei, "Engineering Solar Cell Absorbers by Exploring the Band Alignment and Defect Disparity: The Case of Cu- and Ag-Based Kesterite Compounds," *Adv. Funct. Mater.*, pp. 1–11, 2015.

[6] T. Gershon, Y. S. Lee, P. Antunez, R. Mankad, S. Singh, D. Bishop, O. Gunawan, M. Hopstaken, and R. Haight, "Photovoltaic Materials and Devices Based on the Alloyed Kesterite Absorber (Ag x Cu 1 – x) 2 ZnSnSe 4," pp. 1–7, 2016.

[7] C. J. Hages, M. J. Koeper, and R. Agrawal, "Optoelectronic and material properties of nanocrystal-based CZTSe absorbers with Ag-alloying," *Sol. Energy Mater. Sol. Cells*, vol. 145, pp. 342–348, 2016.

[8] W. Gong, T. Tabata, K. Takei, M. Morihama, T. Maeda, and T. Wada, "Crystallographic and optical properties of (Cu,Ag)2ZnSnS4 and (Cu,Ag)2ZnSnSe4 solid solutions," *Phys. Status Solidi*, vol. 12, no. 6, pp. 700–703, 2015.

STUDY OF MOO_X BACK CONTACT FOR LOW TEMPERATURE CDTE SOLAR CELLS ON SUPERSTRATE CONFIGURATION

Elisa Artegiani^, Daniele Menossi^, Simone Di Mare^, Andrei Salavei^, Arun Kumar†, Gino Mariotto†, and Alessandro Romeo^.

^LAPS-Laboratory for Applied Physics, Department of Computer Science, University of Verona, Ca' Vignal 1, Strada Le Grazie 15, 37134 Verona, Italy, tel.:+39-045-8027974, e-mail address: alessandro.romeo@univr.it

† Raman Laboratory, Department of Computer Science, University of Verona, Strada Le Grazie 15, 37134 Verona, Italy

ABSTRACT: Copper is a crucial element for the CdTe doping and consequently for high conversion efficiency. To analyse the effect of copper doping it is necessary to remove it from the back contact in order to avoid uncontrolled copper diffusion in the absorber. Recently, MoO_3 has been successfully applied for back contact, because of its work function. In this work we present a study on the effect of MoO_x buffer and copper doping on low substrate temperature vacuum evaporated CdTe.
Starting from our CdTe baseline with copper/gold back contact (with routine efficiencies above 14 %), solar cells have been prepared with MoO_x/Mo and MoO_x/Au back contact deposited by reactive radio frequency sputtering. MoO_x have been applied with different thicknesses and oxygen contents. On the other hand copper has been deposited on the CdTe surface and diffused by annealing in vacuum, prior to the MoO_x/Metal deposition for CdTe doping investigation.
With MoO_x/Metal back contact and a 0,15 nm thick copper layer, efficiency is very similar to our standard cells, by using less than a tenth of copper thickness.

Keywords: CdTe, doping, back contact

1 INTRODUCTION

CdTe thin film solar cells are the most successful thin film photovoltaic devices in terms of production yield, as attested by the remarkable market performance. Nevertheless the device has also shown to be competitive in terms of efficiency as attested by the latest results from First Solar, where a record efficiency of 22.1% has been obtained [1].

One of the main issues for CdTe device stability is the nature of the back contact. Because of the CdTe high electron affinity the insertion of copper has demonstrated to deliver higher efficiency, but on the other side the copper diffuses into the CdTe bulk degrading the cell. Moreover recent works have shown that copper is a crucial element for the CdTe doping and consequently for high conversion efficiency [2].

To analyse the effect of copper doping it is necessary to remove it at the back contact in order to avoid uncontrolled copper diffusion in the absorber. At the same time it is important to apply a metal with a suitable electron affinity to form a good ohmic contact with CdTe. Recently, MoO_3 has been successfully applied for back contact performing good efficiencies on high temperature deposited CdTe by close space sublimation [3]. Due to its high work function of about 6.8 eV, MoO_x is one of the few semiconductors able to form an ohmic contact with p-type CdTe. MoO_x layer has to be optimized in order to produce a performing back contact without application of copper. Then the devices have been prepared with different copper thicknesses in order to optimize its quantity to dope CdTe avoiding shunts.

CdTe solar cells with different copper thicknesses are analysed to understand the copper miscibility in to the CdTe matrix so to find the right compromise between copper doping and copper shunting.

2 EXPERIMENTAL

CdTe thin film solar cells were made in superstrate configuration by low-temperature fabrication process based on vacuum evaporation (VE). On an ITO/ZnO stack as front contact, 300 nm thick CdS thin film is deposited at 100 °C and then annealed at 450 °C in vacuum. Subsequently CdTe layer, with a thickness of 5 to 6 microns, is deposited in the same vacuum chamber and subsequently activated by $CdCl_2$ treatment applied by wet deposition and annealed in air at 380 °C. In our standard cells this treatment is followed by Br-MeOH etching in order to have a Tellurium rich region, but in the case of these devices we only clean the surface of CdTe with distilled water.

We usually deposit 2 nm of copper, which has the task of doping CdTe and making a good contact at the same time. In these cells, in order to analyse the doping effect, copper has been inserted prior the back contact deposition, with thicknesses in the range of 0,15 to 1 nm. The process is applied in a dedicated vacuum chamber with a quartz crystal thickness monitor (Intellemetrics IL150).

After deposition the stacks are annealed at 200°C for 30 minutes in vacuum to favour the diffusion of copper in CdTe.

Back contact is made by reactive sputtering of molybdenum in argon/oxygen atmosphere for MoO_x deposition followed by sputtering of pure molybdenum or by vacuum evaporation of gold. MoO_x deposition is carried on by Radio Frequency (RF) sputtering technique after reaching a vacuum of 10^{-6} mbar, with a power of 100 W and a chamber pressure of $5 * 10^{-3}$ mbar, and Mo with 220 W and 10^{-3} mbar. Substrate temperature is kept for both materials at room temperature (in order to avoid excessive diffusion of copper). Finally the samples are annealed in the furnace at 200°C for 30 minutes in air.

The transmission measurements have been obtained by a Unicam spectrometer Type UV2-100 in UV-visible range. Raman spectra were collected with a Horiba Jobin-Yvon LabRam HR800 microprobe setup (in backscattering geometry) by using an exciting radiation at 632.8 nm (He-Ne laser). The current–voltage characteristics (J-V) were measured at room temperature by a Keithley 2420 SourceMeter, using a calibrated halogen lamp with an irradiation of 1000 W/m^2.

3 RESULTS AND DISCUSSION

3.1 MoO_x preparation

We deposit MoO_x by reactive radio frequency sputtering of molybdenum in argon/oxygen atmosphere at room temperature. Comparing MoO_x and Mo thin films? with similar thicknesses it results that the first one is transparent and shows high resistivity, while the second one is reflective in the visible region showing a good conductivity, roughly showing the positive results of reactive sputtering.

Different MoO_x thin film layers have been deposited on glass and their transparency has been measured with a UV/visible/IR spectrophotometer.

As shown in figure 1, the transparency is very similar for both 20 and 50% oxygen/(argon+oxygen) content in the sputtering system, suggesting a similar band gap. In fact for both films we have calculated a bandgap of about 3.5 eV, in the same range identified by Simchi et al. for MoO_3 [4].

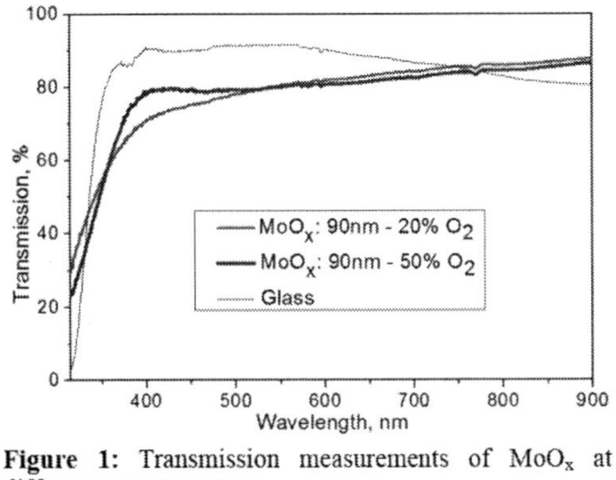

Figure 1: Transmission measurements of MoO_x at different oxygen content.

However in the Raman analysis, different peaks have been detected, showing a slightly different structure of the layers. In the case of 20% oxygen/(argon+oxygen) ratio, the Raman bands assigned to the terminal oxygen stretching at 995 cm^{-1}, triply connected bridge oxygen stretching mode at 666 (cm^{-1}) and the doubly connected bridge oxygen stretching at 821 cm^{-1} (see figure 2, top) are typical of MoO_3 layer. On the other hand, for the 50% oxygen/(argon+oxygen) ratio, the observed peaks (at 777 cm^{-1}, at 852 cm^{-1} and at 950 cm^{-1}) are attributed to a layer of MoO_{x-3} [5]. During micro raman analysis with an incident laser power higher than 1.5 mW, the above mentioned peaks increase with time, showing a laser induced transformation from amorphous to crystalline structure.

Also finished devices have been fabricated with 20% and 50% oxygen/(argon+oxygen) ratio. With excessive amount of oxygen in the sputtering chamber, the current and therefore also the efficiency is limited probably by a high back contact barrier. For this reason the CdTe cell fabrication process has been optimized by applying 20% $O_2/(Ar+O_2)$ deposited MoO_x.

Moreover, solar cells have been grown with different thickness of MoO_x. With a 9 nm thick layer the current-voltage characteristic begins to show a marked roll-over, for this reason our devices have been optimized with a 4-5 nm thick MoO_x layer. The explanation can be attributed to the high resistivity of MoO_x.

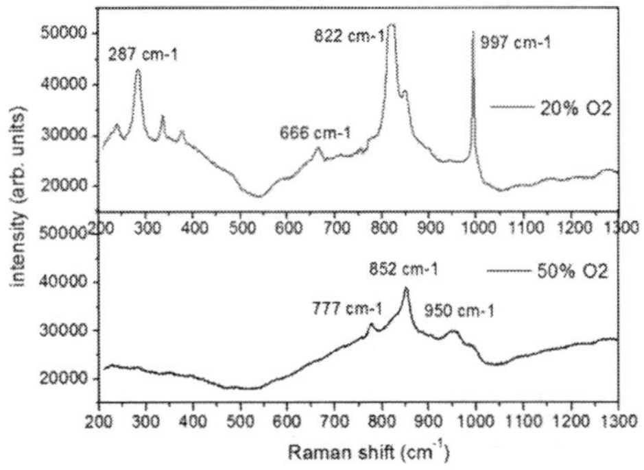

Figure 2: Raman plots of MoO_x deposited by reactive sputtering, top with 20% O_2 over ($Ar+O_2$) ratio, bottom with 50% O_2 over ($Ar+O_2$) ratio.

3.2 Copper thickness

In figure 3 J-V characteristics of solar cells with different copper thicknesses are shown. The cells were prepared with our standard process and prior to the MoO_x/Mo back contact, a very thin layer of copper was deposited by vacuum evaporation, and the stacks are annealed in air at 200 °C for 30 minutes for enhancing Cu diffusion into CdTe and doping it.

First of all, different J-V behaviour in the first quadrant of figure 3 is observed: roll over is inversely proportional to the Cu amount. This shows that copper is still acting as back contact when thickness is above 0,5 nm.

Moreover, the main difference in performance is due to the current density, whereas open circuit voltage and fill factor exhibit similar values. This would suggest a different doping of the CdTe according to the copper thickness, however the samples have also different back contacts so definitive conclusions cannot be made.

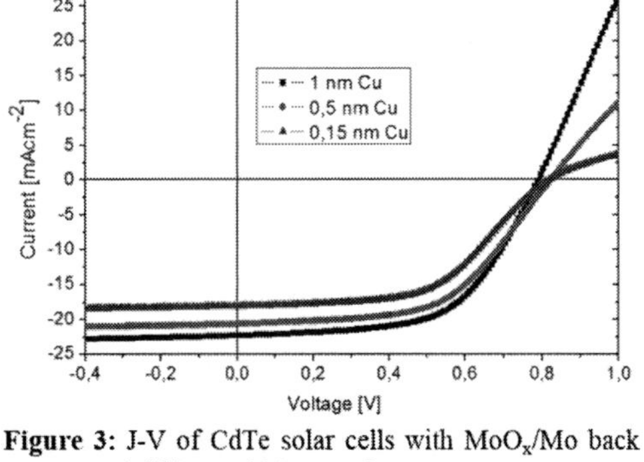

Figure 3: J-V of CdTe solar cells with MoO_x/Mo back contact and different thickness of copper.

Table I: Electrical parameters for cells with different copper content.

Copper content	Eff. (%)	F.F. (%)	J_{SC} (mA/cm^2)	V_{OC} (mV)
0,15 nm	7,9	53,6	18,0	817
0,5 nm	9,4	55,3	20,6	824
1 nm	10,3	58,0	22,3	796

3.3 Cells with MoO_x & Mo contact.

To avoid that copper making contact in place of MoO_x, the cells have been optimized by applying 0.15 nm thick Cu and subsequent annealing in vacuum at 200 °C for 30 minutes. Moreover in order to enhance Cu diffusion into CdTe we do not perform the Br-MeOH etching on the surface of CdTe: Rimmaudo et al. show how the copper reacts with Te forming CuTe and keeping Cu at the back contact [6]. Instead, the sample has been rinsed in water to remove surface residues from the $CdCl_2$ activation treatment. In order to reach good efficiencies, after the deposition of the back contact, a second annealing at 200 °C in air is needed. Comparing cells contacted with and without MoO_x followed by Mo, there are no significant differences in efficiency: cells with MoO_x show an improved current, but also with larger roll over. This could be explained supposing that MoO_x is blocking the carriers, since the electron affinity of the MoO_x made by sputtering could be too low [7]. It could also be that the Cu diffusion during the second annealing in air destroy MoO_x layer. Moreover without copper the samples do not perform sufficient conversion efficiency.

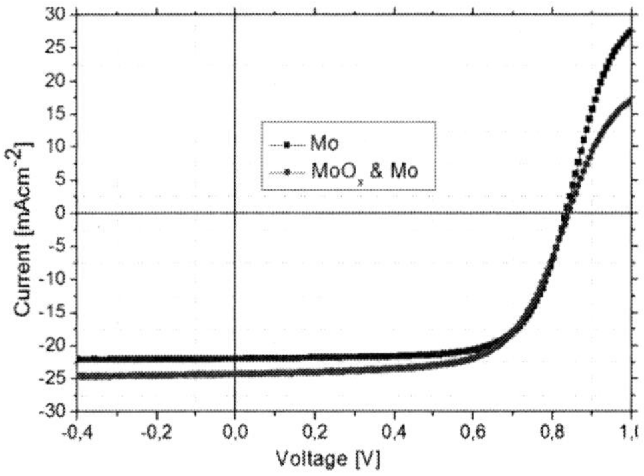

Figure 4: J-V of CdTe solar cells with MoO_x/Mo and Mo back contact.

Table II: Electrical parameters for cells with MoO_x/Mo and Mo back contact.

	Eff. (%)	F.F. (%)	Jsc (mA/cm^2)	Voc (mV)
MoO_x & Mo	13,4	65,2	24,3	845
Mo	12,9	70,2	22,0	838

3.4 Cells with Mo & Gold contact.

Back contacts with Au have been also made, the typical back contact in our laboratories are with 2 nm thick copper and 50 nm thick gold layer, instead in this work copper was reduced to 0.15 nm thick. Cells with and without MoO_x between copper and gold have been compared. A first interesting point is that after the second annealing at 200 °C in air, performed on the completed device, the cells show low efficiencies, very different from our standard devices that after this annealing improve their efficiency up to 15 %. Probably with this annealing still the copper does not diffuse from gold into CdTe, suggesting that Cu is still at the contact working better without MoO_x layer. After a third annealing at 300 °C in air the cells reach fairly good efficiency, but they show a very strong roll over meaning that Cu is diffused into CdTe, doping the absorbent layer, but also reducing performance of the back contact. We can see from the J-V curve, that in both cases after the second and the third annealing, the best contact is without MoO_x: after the second copper reacts forming back contact and after the third probably because Cu diffuses in the bulk and also into the MoO_x layer reducing the energy barrier.

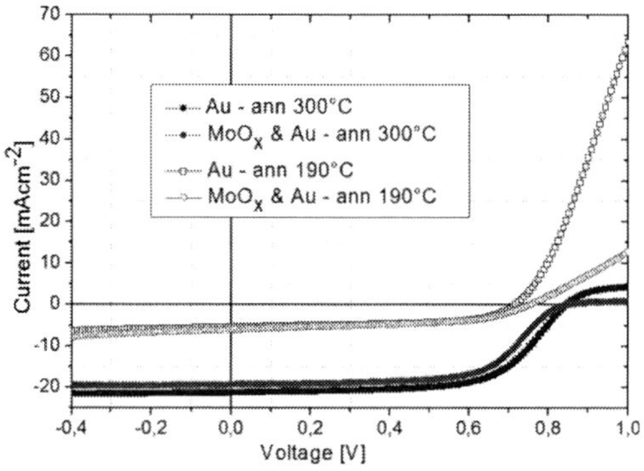

Figure 5: J-V of CdTe solar cells with MoO_x/Au and Au back contact.

Table III: Electrical parameters for cells with MoO_x/Au and Au back contact.

	Eff. (%)	F.F. (%)	Jsc (mA/cm^2)	Voc (mV)
MoO_x/Au 190°C	2,3	49,9	6,0	761
Au 190°C	2,2	58,0	5,4	719
MoO_x/Au 300°C	10,2	61,6	19,4	852
Au 300°C	11,4	63,5	21,3	845

3.4 Aging of the cells.

In order to study the aging, cells contacted by MoO_x/Mo and by Mo have been put for two weeks in accelerated stability test (AST) box under 1 sun illumination and at a controlled temperature of 80 °C. A set of cells has been aged in open circuit conditions (OCC) and a set in short circuit conditions (SCC) by applying a conductive wire between the front and the back contact [8].

We can see that the cells with MoO_x layer aged in OCC suffer from a higher degradation; instead the performance reduction of cells without MoO_x is almost independent on the applied bias. Usually our standard cells have a very stronger degradation in OCC than in SCC, as in case of MoO_x & Mo contact cells.

The fill factor decrease seems to have a greater impact on the degradation of the performance; this could be related to the back contact degradation. Probably the copper diffuses into CdTe, which now directly faces the MoO_x or Mo layers. These devices deteriorate faster than our standard cells, because the copper is not blocked by the Br-MeOH etching (no CuTe compound is formed). Current is not significantly affected by aging, but despite copper diffusion current density does not improve. In any case the temperature of 80 °C in the AST box is not enough to favour the solubility of copper in the CdTe bulk material.

Table IV: Parameters normalized to initial ones in % for aged cells with MoO_x/Mo and Mo back contact after 335 h.

Contact		Eff.	F.F.	J_{SC}	V_{oc}
MoO_x/Mo	SCC	73	80	99	92
MoO_x/Mo	OCC	55	65	97	88
Mo	SCC	75	80	100	93
Mo	OCC	74	81	99	92

4 CONCLUSIONS

We have deposited MoO_x by reactive sputtering of molybdenum in argon/oxygen atmosphere with 20 % and 50 % $O_2/(Ar+O_2)$ ratio: they show the same bandgap of about 3.5 eV, but with 50 % $O_2/(Ar+O_2)$ ratio the efficiency is limited by high back contact barrier. 4-5 nm MoO_x thick layer have been deposited because with thicker layers the J-V characteristics show a marked rollover.

Comparing cells with MoO_x/Mo and Mo contact we can see a small improvement in the current density and consequently an improvement of efficiency, but also an increased roll over which reduces the fill factor. However with this back contact the efficiency is very similar to the standard cells made in our labs but by using less than a tenth of copper thickness, so a greater amount of copper is not needed for doping CdTe. On the other hand these cells degrade faster probably due to a greater instability of the back contact since the copper is not blocked by the Br-MeOH etching procedure. So degradation is not directly linked to the amount of copper content.

Moreover we compared MoO_x/Au with only gold back contact. These samples do not reach the efficiency of those mentioned above. MoO_x deposited by reactive sputtering does not make a perfectly ohmic contact with CdTe, previous studies have shown that electron affinity is too low for sputtered MoO_x [7]: vacuum evaporated MoO_x will be compared in future works.

References

[1] First Solar, First Solar Achieves yet Another Cell Conversion Efficiency World. Record., (2016). http://investor.firstsolar.com/releasedetail.cfm?ReleaseID=956479.

[2] L. Kranz, C. Gretener, J. Perrenoud, R. Schmitt, F. Pianezzi, F. La Mattina, et al., Doping of polycrystalline CdTe for high-efficiency solar cells on flexible metal foil, Nat Commun. 4 (2013). http://dx.doi.org/10.1038/ncomms3306.

[3] I. Irfan, H. Lin, W. Xia, H.N. Wu, C.W. Tang, Y. Gao, The effect of MoO x inter-layer on thin film CdTe/CdS solar cell, Sol. Energy Mater. Sol. Cells. 105 (2012) 86–89. doi:10.1016/j.solmat.2012.04.006.

[4] H. Simchi, B.E. Mccandless, M. T., W.N. Shafarman, Structure and interface chemistry of MoO3 back contacts in Cu(In,Ga)Se2 thin film solar cells, J. Appl. Phys. 115 (2014) 033514.

[5] M.A. Camacho-Lopez, L. Escobar-Alarcon, M. Picquart, R. Arroyo, G. Cordoba, E. Haro-Poniatowski, Micro-Raman study of the m-MoO2 to a-MoO3 transformation induced by cw-laser irradiation, Opt. Mater. (Amst). 33 (2011) 480–484. doi:10.1016/j.optmat.2010.10.028.

[6] I. Rimmaudo, A. Salavei, F. Rossi, C. Ferrari, A. Romeo, Etching effect of CdTe absorber on the stability of thin film solar cell devices, 2013 IEEE 39th Photovolt. Spec. Conf. (2013) 2029–2033. doi:10.1109/PVSC.2013.6744871.

[7] H. Lin, I. Irfan, W. Xia, H.N. Wu, Y. Gao, C.W. Tang, MoOx back contact for CdS/CdTe thin film solar cells: Preparation, device characteristics, and stability, Sol. Energy Mater. Sol. Cells. 99 (2012) 349–355. doi:10.1016/j.solmat.2012.01.001.

[8] I. Rimmaudo, a. Salavei, a. Romeo, Effects of activation treatment on the electrical properties of low temperature grown CdTe devices, Thin Solid Films. 535 (2013) 253–256. doi:10.1016/j.tsf.2012.11.113.

THE INFLUENCE OF COMPOUND TARGET PREPARATION, SPUTTERING POWER AND SUBSTRATE TEMPERATURE ON THE ACHIEVEMENT OF CU(IN,GA)SE$_2$ PRECURSORS SUITABLE TO GET HIGH EFFICIENCY SOLAR CELLS.

A. Bosio[1], G. Rosa[1*], S. Mazzamuto[2], A. Romeo[3], F. Piccinelli[4] and N.Romeo[1]
[1] Thin Film Laboratory, Department of Physics and Earth Sciences, University of Parma, Parco area delle Scienze 7/A, 43124 - Parma, Italy, Tel.: +39-0521-905257; e-mail: greta.rosa@difest.unipr.it
[2] Northumbria Photovoltaic Applications Group, Department of Physics and Electrical Engineering, Northumbria University, Newcastle NE1 8ST, UK; e-mail: samantha.mazzamuto@northumbria.ac.uk
[3] Laboratory for Applied Physics, Department of Computer Science, University of Verona, Ca' Vignal 1, Strada Le Grazie 15, 37134 - Verona, Italy; e-mail: alessandro.romeo@univr.it
[4] Solid State Chemistry Laboratory, Department of Biotechnology, University of Verona, Strada Le Grazie 15, 37134 - Verona, Italy; e-mail: fabio.piccinelli@univr.it
[*] Corresponding author: e-mail: greta.rosa@difest.unipr.it; Tel.: +39-0521-905257

ABSTRACT: In the past, we reported Cu(In,Ga)Se$_2$ solar cells exhibiting an efficiency of about 16% obtained by sputtering and selenization, starting from targets of InSe, GaSe and Cu.
Here, we report on the requirements, which have to be respected in order to get good precursors suitable to give high efficiency by simple selenization in a Se-vapour. First of all, the targets of InSe and GaSe should exhibit a density higher than 97% to avoid their damage during the sputtering deposition. Secondly, especially for InSe, the sputtering power has to be regulated. If it is too low then films with excess of Se are obtained and this can cause the peeling of the film during selenization. If the sputtering power is too high, segregation of In on the surface of the target is observed. We found out that the right sputtering power is about 3.8 W/cm^2 corresponding to a deposition rate of ~ 20 Å/sec. Another important parameter is the substrate temperature. During the InSe deposition, if the temperature substrate is higher than 670 K then the film will contain In-excess. If the temperature is lower than 470 K the film can contain Se-excess. The substrate temperature is less critical when GaSe or Cu are deposited. In this case a higher substrate temperature, in the range 670 ÷ 720 K, is used in order to get a good mixing of InSe, GaSe and Cu.
During the selenization, an important parameter to be considered is the deposition rate of Se. It has been found out that the grain size of the finished Cu(In,Ga)Se$_2$ film is larger when the evaporation rate of Se is not too high and it is around 1 μm/sec.
By controlling the above parameters, high quality Cu(In,Ga)Se$_2$ films with a grain size larger than 1 μm are reproducibly obtained. These films, used to prepare solar cells, gave us the possibility to obtain an efficiency larger than 16%.
Keywords: Thin Film Solar Cell, Cu(In,Ga)Se$_2$ and Sputtering

1 INTRODUCTION

Cu(In,Ga)Se$_2$-based solar cells have reached an efficiency of 22.3% at a laboratory scale [1], which is the highest efficiency ever reported for any thin film solar cell. This is due to the fact that Cu(In,Ga)Se$_2$ exhibits a very high absorption coefficient and its forbidden gap is modulated through its thickness from a high gap at the interface, which generates a high open circuit voltage, to a lower gap in the middle and again to a high gap close to the back contact which acts as a mirror for the minority carriers and thus increasing the photocurrent.

However, in order to get in a large module showing the same efficiency obtained at a laboratory scale, a scalable and reproducible technique has to be used.

For this reason we chose to prepare the Cu(In,Ga)Se$_2$ precursors by sputtering using as targets: InSe, GaSe and Cu [2, 3]. These precursors are then exposed to a pure Se-atmosphere in a vacuum chamber at high temperature. Using this process, efficiencies higher than 16% are obtained.

However, in order to get a high reproducibility, some precautions concerning the method used to prepare the targets, the substrate temperature during the deposition, and the sputtering power on the targets, have to be taken into consideration.

2 EXPERIMENTAL RISULTS

Cu(In,Ga)Se$_2$/CdS solar cells with efficiencies higher than 16% are obtained by using targets of InSe, GaSe and Cu and the deposition sequence InSe, GaSe, Cu and GaSe, as shown in Figure 1 [2]. As substrate, 1 inch square soda-lime glass is used.

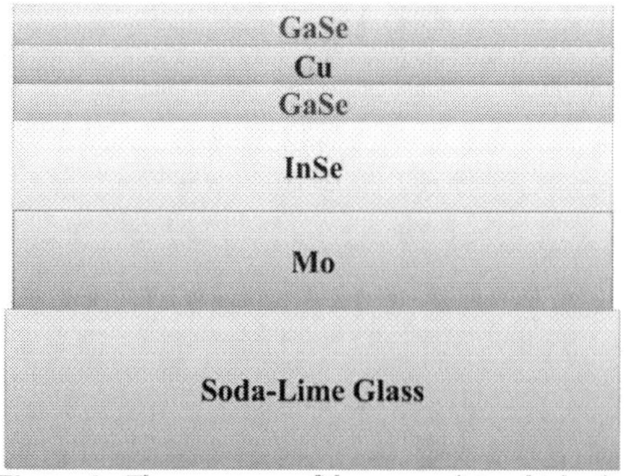

Figure 1: The sequence of layers used to obtain the absorber material.

The first requirement is that the targets have to be very compact, with high density (higher than 97%) and this is achieved by making the targets by high pressure hot pressing.

After several experiments, in which the substrate temperature, the sputtering power for each target and the thickness of the single layers were changed, we found out the following best conditions:

1) Substrate temperature ~ 670 K;
2) Radio Frequency (R.F.) sputtering power density: 3.8 W/cm^2 for InSe, and 3.4 W/cm^2 for GaSe and a deposition rate of 1.9 nm/sec for InSe and 0.6 nm/sec for GaSe;
3) Thickness of the layers: 900 nm for InSe, 150 nm for GaSe, 230 nm for Cu and 250 nm for the GaSe top layer.

Since the most critical layer is InSe, we deposited 1 µm-thick InSe alone on Mo-covered soda-lime glass at three different temperatures, namely 470 K, 670 K and 730 K, with different sputtering power and analyzed the films by an electron microscope, a microprobe and X-Rays. Deposition parameters and the corresponding atomic compositions are shown in Table I.

When the RF sputtering power density is less than 3.2 W/cm^2, then the InSe target surface becomes dark indicating that some Se is sputtered back onto the target surface. On the contrary, when the sputtering power density is higher than 3.8 W/cm^2, some In segregates on the top of the target. At the correct sputtering power, the target surface does not change its color and does not segregate any In-drops. The most uniform and stoichiometric InSe films are obtained by using 3.8 W/cm^2 power density.

When InSe films are deposited at low temperature, namely 470 K, they were uniform and smooth but they contain some Argon englobed and a Se-excess as it has been seen by microprobe analysis. Argon and a Se-excess incorporated into the InSe film can cause sticking problems when the films react with GaSe and Cu. An X-Rays analysis indicates that the low temperature deposited InSe films are amorphous.

InSe films deposited at 670 K are uniform and smooth but at the electron microscope they exhibit some nano-rods, as it can be seen in Figure 2.

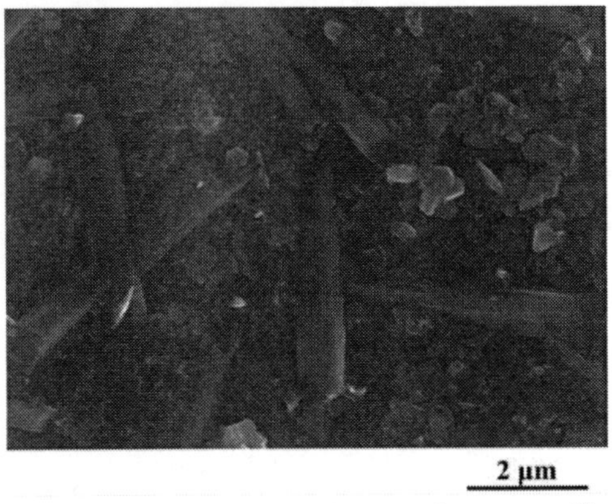

Figure 2: Plan-view SEM image of an InSe film deposited at a substrate temperature of 670 K (sample #2). InSe nanorods are clearly revealed.

The X-Ray analysis indicates that these films are well crystallized but they contain several In-Se phases (Figure 3). In fact, by using a powder diffractometer operating with $CuK_{\alpha 1}$ (λ = 1.5418 Å) in the theta-2theta configuration, the InSe films were characterized in order to determine their crystalline properties. Films deposited at a substrate temperature of 670 K exhibited diffraction peaks corresponding to both InSe and In_2Se_3 polycrystalline materials. Peaks at 10.63° and 21.34°, are related to the (003) and (006) orientations of the InSe phase (R-3m space group). Instead, the 27.62°, 29.4°, 31.37°, 44.13°, 67.3° peaks are associated to the γ-In_2Se_3 polycrystalline phase. The substrate generated the 40.21°, 73.1°, and 86.85° reflections, which correspond to the Mo cubic phase.

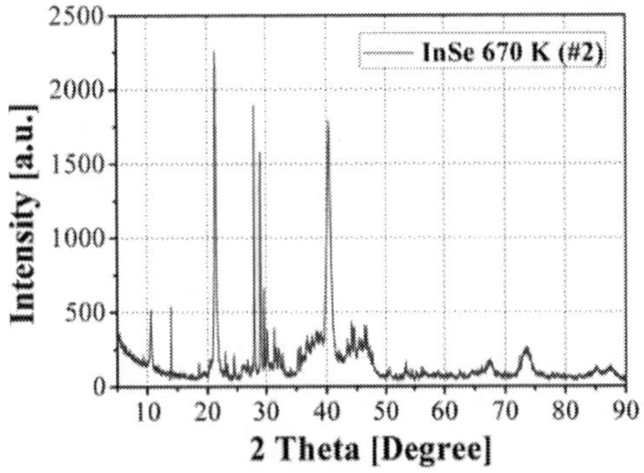

Figure 3: X-Ray Diffraction pattern of an InSe film deposited at a substrate temperature of 670 K (sample #2) on Mo-covered soda-lime glass.

If the InSe deposition is made or carried out at a higher substrate temperature, namely 730 K, even though they contain almost the only InSe phase (Figure 4) they present some unevenness, like voids or ridges, as it is shown in Figure 5.

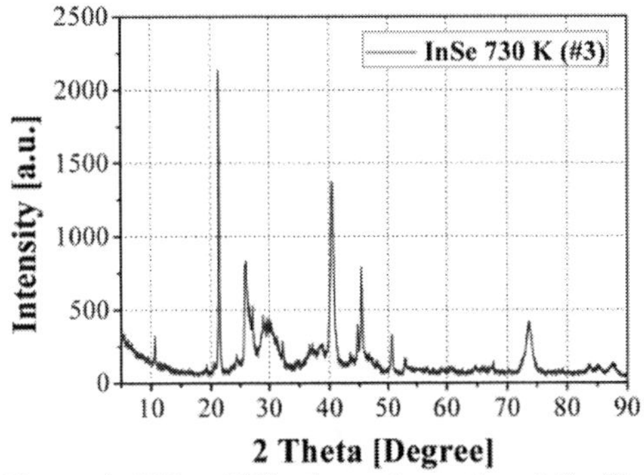

Figure 4: X-Ray Diffraction pattern of an InSe film deposited at a substrate temperature of 730 K (sample #3) on Mo-covered soda-lime glass.

From the XRD pattern showed in Figure 4, we can conclude that together with the peaks located at around 10.6° and 21.3°, also the 44.7°, 45.3° and 50.6° reflections are clearly visible showing a better oriented InSe material. The peaks at 26.7° and 27,3° are related to the In_2Se_3 hexagonal phase.

Uniformity and smoothness are very important in order to have a good mixing of the InSe, GaSe and Cu layers.

Moreover, the chemical and physical stability of the InSe layer is of primary importance when the reactivity with Cu is taken into consideration. In fact, the grain size of the InSe crystallites controls the speed at which In, Cu and Se react. If the grain size is too small (less than 1 µm) then the elements react quickly producing local not uniformities such as spatially separated binary phases (i.e

In_2Se_3 and Cu_2Se). For this reason, the InSe film, shown in Figure 3, is the best suited as In supplier since it presents a lot of crystallites 1 ÷ 2 μm large and 6 ÷ 8 μm long which are bigger if compared with the average dimension of the crystalline grains showed in Figure 5. Despite this, good results are obtained only if the InSe layer is separated from Cu by a GaSe film.

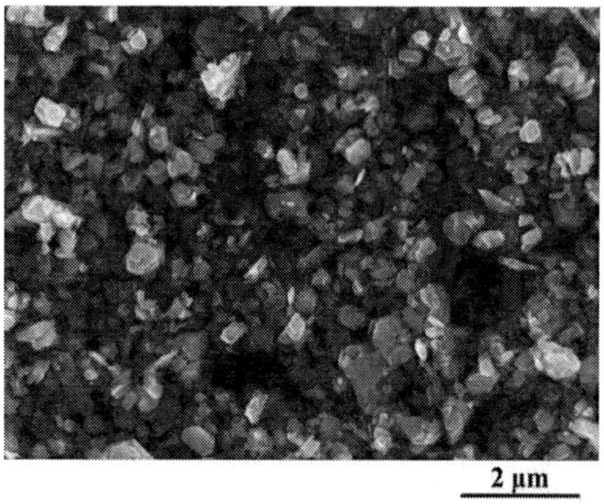

Figure 5: SEM image of an InSe film deposited at a substrate temperature of 730 K (sample #3). The black points in the picture correspond to holes probably due to a separation of crystalline phases.

Concerning the GaSe deposition, we did not find any dependence on the R.F. sputtering power, but we chose to deposit it at a low deposition rate in order to not heat too much the surface of the target, as it is shown in Table I.

Table I: Deposition parameters and the corresponding atomic compositions of the InSe and GaSe films used for this work. All the layers were characterized by Scanning Electron Microscopy (SEM) on a Tescan combined with an Oxford Instruments Energy Dispersive X-ray (EDX) Spectroscopy System in order to measure the elemental compositions of the samples. EDS was carried out using an accelerating voltage of 20kV.

	InSe				GaSe	
Sample	#1	#2	#3	#4	#5	#6
(1)SP [W/cm²]	3.8	3.8	3.8	3.2	3.4	2.2
(2)T_S [K]	470	670	730	670	670	670
In [at. %]	51.9	50.9	50.8	47.8	--	--
Se [at. %]	45.1	49.1	49.2	52.2	50.0	50.1
Ga [at. %]	--	--	--	--	50.0	49.9
Ar [at. %]	3.0	--	--	--	--	--

(1) SP = Sputtering Power Density
(2) T_S = Substrate Temperature

When all the layers are deposited at high temperature, namely 670 K, a complete mixing is obtained and then the precursor is ready to be selenized.

The selenization is carried out evaporating pure Se from a graphite crucible placed in a vacuum chamber. The substrate is heated up by Halogen lamps able to reach 770 K in less than 4 min. It has been found out that a suitable evaporation rate for Selenium is on the order of 1 μm/s, since higher evaporation rates produce $Cu(In,Ga)Se_2$ thin films with bigger grains interspersed with a lot of voids, as it is shown in Figure 6(a) and 6(b). This is probably due to the reaction that is too fast when a high evaporation rate of Se is used. As a result, a chaotic growth is obtained.

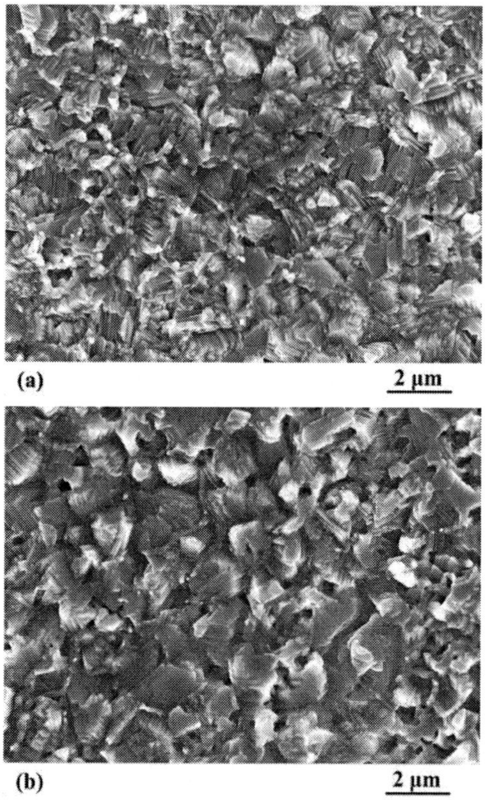

Figure 6: SEM images of the surface morphology of a $Cu(In,Ga)Se_2$ film selenized in a Se vapor.
- **(a):** Se evaporation rate on the order of 1μm/s;
- **(b):** Se evaporation rate higher than 1 μm/s. The black points in the picture correspond to holes, which are not present in the **(a)** case.

By using the above indicated precautions, we are able to obtain high quality $Cu(In,Ga)Se_2$ films, which have given an efficiency higher than 16% [2], when they are used as absorber in a $Cu(In,Ga)Se_2$/CdS solar cells.

3 CONCLUSION

A detailed process suitable to prepare high efficiency $Cu(In,Ga)Se_2$-based solar cells was described, using scalable technique such as sputtering and selenization.

Critical parameters concerning targets, R.F. sputtering power, substrate temperature and Se evaporation rate have been put in evidence. By controlling the above parameters, high quality $Cu(In,Ga)Se_2$ thin films can be obtained with a high reproducibility. These films, used as absorbers in

$Cu(In,Ga)Se_2/CdS$ solar cells, can give efficiencies higher than 16%.

4 REFERENCES

[1] Solar Frontier Achieves World Record Thin-Film Solar Cell Efficiency: 22.3%. Available online: http://www.solar-frontier.com/eng/news/2015/C051171.html (accessed on 06 06 2016).

[2] G. Rosa, A. Bosio, D. Menossi, N. Romeo, "How the Starting Precursor Influences the Properties of Polycrystalline $CuInGaSe_2$ Thin Films Prepared by Sputtering and Selenization", *Energies* (2016), 9, 354.

[3] A. Bosio, D. Menossi, A. Romeo, and N. Romeo, in: Solar Cells – Research and Application Perspectives, edited by: Arturo Morales Acevedo, (INTECH, Rijeka, Croatia, 2013), 79 – 104.

EVOLUTIONARY OPTIMIZATION OF TCO/MESH ELECTRICAL CONTACTS IN CIGS SOLAR CELLS

Paolo A. Losio[1*], Thomas Feurer[2], Stephan Buecheler[2], Beat Ruhstaller[1]
[1] Zurich University of Applied Sciences (ZHAW), School of Engineering, Institute of Computational Physics, Technikumstr. 9, 8401 Winterthur, Switzerland.
[2] Laboratory for Thin films and Photovoltaics, Empa, Swiss Federal Laboratories for Materials Science and Technology, Ueberlandstrasse 129, 8600 Dübendorf, Switzerland
* Corresponding author, paolo.losio@zhaw.ch

ABSTRACT:
State of the art solar cells often require a combination of TCO and metallic grid to efficiently transport the generated current to an external circuit. Optimization of these complex contacts based on several materials with different conductivities and geometries is often still based on a traditional approach consisting of simple analytical formulas and empirical knowledge in spite of the complexity. An evolutionary algorithm combined with an electrical 2D+1D FEM model is used to optimize the shape of the metallic contact considering the material conductivities, the geometry and shadowing.
The performance of two automatically designed contacts is compared with experimental results of CIGS solar cells prepared with two different TCO thicknesses.
Keywords: Thin Film, Modelling, Contact, Metal Grid

1 INTRODUCTION

$Cu(In, Ga)Se_2$ (CIGS) thin film solar cells are promising for large scale efficient PV applications.[1] Collecting and transporting the current generated in CIGS solar cells is a key factor for performance optimization in record devices as well as a critical issue in scaling to large area. Especially CIGS modules, characterized by a comparatively low cell V_{oc} and comparatively high J_{sc} are affected by strong ohmic losses. Therefore minimizing ohmic losses is of paramount importance. Usually the current is transported through a window TCO layer and in recent times a metallic mesh is added for improving the overall conductivity and reducing ohmic losses.

Modelling a solar cell based only on single TCO layers as contacts can be easily achieved by using equivalent circuits in SPICE [2] however adding a highly conductive metallic mesh on top of the TCO makes a simulation by equivalent circuits more complex because a two dimensional spatial discretization is required and the direction of current flow cannot be accurately described [3]. As an alternative it is possible to use the finite element method (FEM) to describe the potential distribution and the current flow in an extended device comprising both TCO and a metallic grid. [4–6]

A common approach to optimize contacting grids and module interconnection layouts is based on analytical models [7,8] relying on several approximations that may hold for simple geometries only. More advanced approaches based on FEM have been used only since recently.[5,9]

2 SIMULATION AND OPTIMIZATION

Simulations using the finite elements method are performed using the Multiphysics FEM package nm-seses developed at ZHAW.[4,10] In the present study the solar cell is decomposed into three 2D domains: a bottom Mo electrode, a top TCO electrode and a metallic grid as shown in Fig. 1. The solar cell is modelled using a one diode model implemented as a one dimensional coupling between Mo and the top TCO, e.g. each mesh point node between top TCO and Mo is connected through a one diode model. The metallic contact grid is modelled as a square mesh where each element can either be metal or void, and the top metal electrode is in perfect contact with the top TCO layer. The presence of a metal mesh element induces a shadowing into the solar cell model, e.g. the photocurrent is locally set to zero. Dirichlet boundary conditions (BC) with voltage = 0V are applied to the bottom Mo electrode, and similarly Dirichlet BC with a voltage corresponding to the external voltage are applied to the edges of a contact pad on the metallic mesh. All other boundaries are defined as insulating. Existing (e.g. experimental) metallic contact grids can be imported as bitmapped images to allow calculating the cell performance.

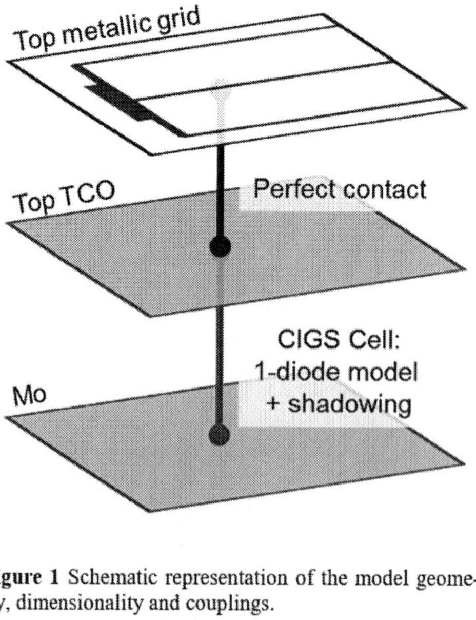

Figure 1 Schematic representation of the model geometry, dimensionality and couplings.

The parameters of the one diode model used for the simulations are obtained by fitting an experimentally measured IV curve. In a first pass the parameters are fitted using a simple one diode model, and then these initial values are used for accurate fitting based on a full FEM model to accurately take into account the potential distribution across a device. The measured resistivity of Al doped Zinc oxide (AZO, $1.1\cdot10^{-3}$ Ωcm) and Mo ($1.2\cdot10^{-5}$ Ωcm) together with tabulated data for Al ($2.7\cdot10^{-6}$ Ωcm) are used for all calculations. Parasitic losses in the TCO layer were estimated by extracting the absorption coefficient of AZO from measured transmission and reflection spectra.

In principle it is possible to automatically optimize the topology and the shape of metallic contacting grids by using topology optimization algorithms used e.g. in civil engineering but adapted to the problem of efficiently collecting current from a surface minimizing shadowing. An intuitive optimization method is the "Bidirectional Evolutionary Structural Optimization" (BESO) method [11] which can be easily adapted to optimize electrical contacts. The optimization algorithm iteratively adds metal mesh elements to the top electrode at points where the ohmic losses in the top TCO layer are highest. This approach is repeated until gains in current collection are compensated by losses due to shadowing and the efficiency of the cell cannot be increased any more. As a successive step, the metallic contacts are iteratively thinned to minimize shadowing losses taking care of not impairing current collection. As this optimization method finds only local minima, it is necessary to repeat the optimization several times and slightly modify the number of metal mesh elements added at each iteration.

The contacts presented in this study were obtained starting from a 0.5x1 cm^2 simulation domain using a resolution of 40 µm defined by the equipment used to produce the shadow masks. Minor modifications allow to apply the optimization method to large area modules.

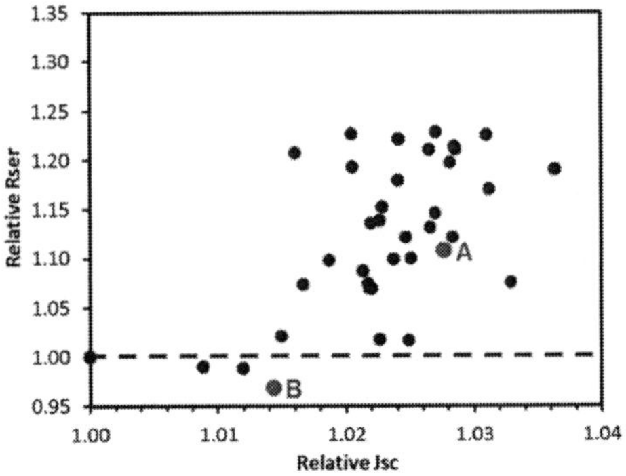

Figure 2: Summary of the improvements in the apparent serial resistance (R_{ser}) and short circuit current (J_{sc}) obtained in about 50 optimization runs for solar cells of 0.5 cm^2 surface. The two contacts (A, B) marked by red dots were realized experimentally.

3 EXPERIMENTS AND RESULTS

CIGS solar cells were prepared at Empa using the procedure described elsewhere.[1,12] Metallic contacts were prepared by evaporation of 4 µm Al through a shadow mask. Shadow masks were prepared by laser-cutting a 100 µm Mo foil according to two grid designs computed by the optimization tool, the nominal dimensions of the experimentally realized cells were 0.54x1.07 cm^2 and are slightly larger than the computed designs. The computed designs were stretched to match the target dimensions and a future optimization series will take into account the new dimensions.

About 50 optimization runs were done using slightly different thresholds for metal addition and different locations of the contact pad to explore the possibilities offered by this method. A standard metallic contact grid was used as benchmark: the results are summarized in Fig. 2 indicating the relative changes in short circuit current (J_{sc}) and in apparent series resistance (R_{ser}) which could be achieved in each run. Ideally, an optimized contact geometry allows reducing the serial resistance and increasing the short circuit current. However, most of the contact shapes lead to an improvement of J_{sc} at cost of a slight increase of R_{ser}.

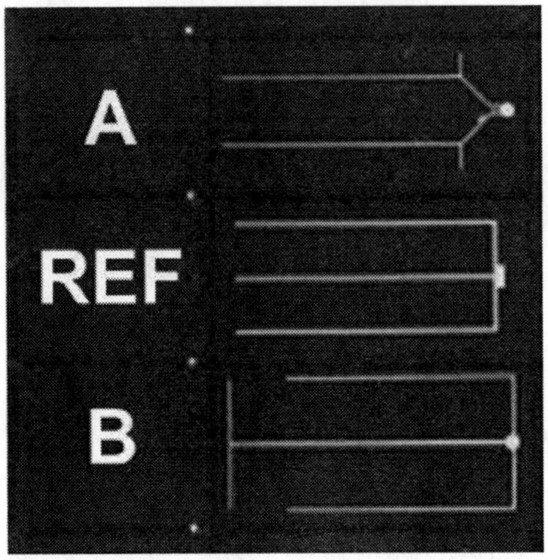

Figure 3: Picture of three solar cells metallic contacts realized by shadow masks. REF: Reference contact, A, B contacts obtained by automatic contact optimization.

Two contacts were selected for being realized experimentally, these are marked with red dots and letters in Fig. 2. The shape of these two contacts together with the reference contact is shown in Fig. 3. Each contact geometry was prepared 6 times on each substrate totaling to 18 cells on each substrate. Two substrates were fabricated: one with a TCO thickness of 200 nm and the second with a TCO thickness of 100 nm thus leading to four different combinations of contact shape and TCO.

Table I: Summary of the absolute efficiency losses calculated for the contact types being studied.

	Ohmic losses	Shadow losses	Total losses
REF	0.2 %	1.0 %	1.2 %
A	0.4 %	0.5 %	0.9 %
B	0.2 %	0.7 %	0.9 %

The absolute loss of efficiency due to ohmic losses and shadowing losses expected for each type of contact can be easily calculated by the FEM tool and a summary is shown in Table I. Additionally, the distribution of the electric potential on the cell surface when the cell is operating at the maximum power point (MPP) is an indication of ohmic losses. In the ideal case of a perfectly conducting TCO and metal grid, the potential on

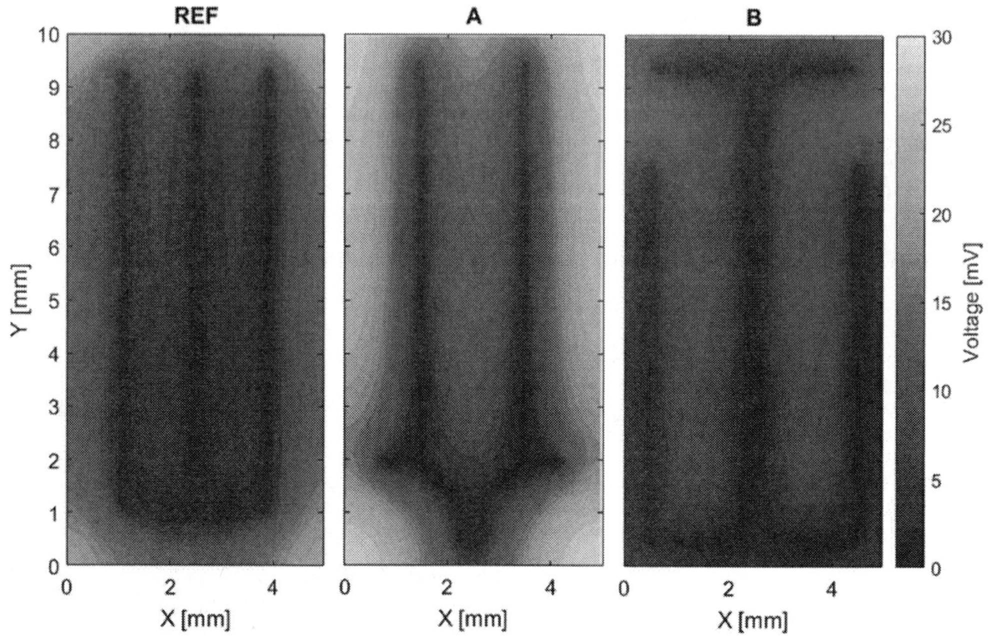

Figure 4: Potential difference distributions on the top TCO for the three contact types being studied. The maximum potential difference from V_{MPP} is 24 mV for the reference contact, 29 mV for the contact A and 15 mV for the contact B.

the whole cell surface corresponds to the voltage at MPP (V_{MPP}). In the presence of ohmic losses the cell surface will have an higher potential than the external contact at V_{MPP}. As an example, the potential differences distributions above V_{MPP} in the three contact types being studied are shown in Fig. 4.

The comparison of predicted and measured changes in R_{ser} and J_{sc} are shown in Fig. 5. The prediction of R_{ser} agrees well with the experimental results thus confirming the validity of this approach to optimize contact geometries.

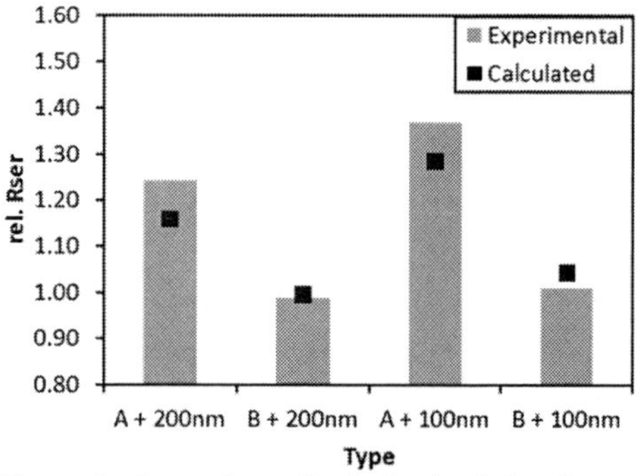

Figure 5: Comparison of calculated relative R_{ser} and measured relative R_{ser} for the different contact types being studied.

Similarly, the relative J_{sc} predicted by simulations and the experimentally measured relative J_{sc} are shown in Fig. 6. The agreement is again good except the first case of contact A combined with 200 nm TCO. The disagreement in the first case is not yet well understood, but it seems to be independent of the contact geometry as the second set of cells produced with the same shadow mask does not show such a large deviation.

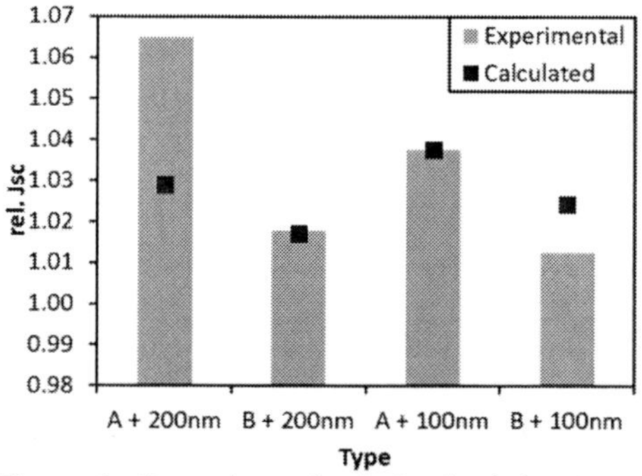

Figure 6: Comparison of calculated relative Jsc and measured relative Jsc for the different contact types being studied.

Finally the absolute gain in efficiency which can be achieved using the optimized contacts is shown in Fig. 7. The cells with 200 nm TCO show a good agreement between simulated gain and experimental gain. However, all cells with thin TCO including the references are affected by a drop of performance which is not fully understood; therefore the agreement is poor in this case.

The simulated improvements indicate an advantage of contacts achieving an improvement of R_{ser}: such contacts can be combined with a thinner TCO to reduce parasitic losses but still obtain a net increase of efficiency. Combining detailed FEM simulations with contact optimization and optical simulations allows to quickly find an optimum combination of TCO thickness and contact shape.

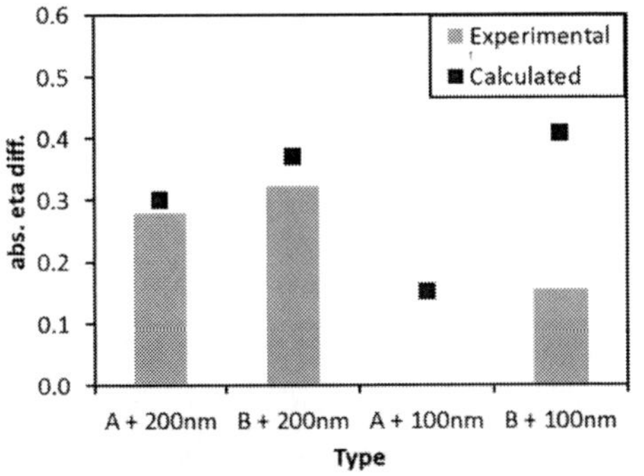

Figure 7: Comparison of calculated absolute increase in efficiency with measured absolute increase of efficiency for the different contact types being studied.

4 CONCLUSIONS

Automatic optimization of metallic contacts has been shown to be able to improve the cell performance compared to reference contacts and the predictions by 2D + 1D FEM simulations can be reliably used to design contacts and optimize solar cells. The method can be used to optimize larger area solar cells and solar modules. Additionally, it can be applied to other PV technologies.

5 ACKNOWLEDGEMENTS

The research leading to these results has received funding from the Swiss Competence Center for Energy and Mobility and from the Swiss Federal Office of Energy (project CONNECT-PV). The authors would like to thank G. Sartoris for useful discussions and valuable advice.

6 REFERENCES

[1] A. Chirilă, S. Buecheler, F. Pianezzi, P. Bloesch, C. Gretener, A.R. Uhl, C. Fella, L. Kranz, J. Perrenoud, S. Seyrling, R. Verma, S. Nishiwaki, Y.E. Romanyuk, G. Bilger, A.N. Tiwari, Highly efficient Cu(In,Ga)Se2 solar cells grown on flexible polymer films, Nat. Mater. 10 (2011) 857–861. doi:10.1038/nmat3122.

[2] K. Brecl, M. Topič, F. Smole, A detailed study of monolithic contacts and electrical losses in a large-area thin-film module, Prog. Photovolt. Res. Appl. 13 (2005) 297–310. doi:10.1002/pip.589.

[3] K. Brecl, M. Topič, Simulation of losses in thin-film silicon modules for different configurations and front contacts, Prog. Photovolt. Res. Appl. 16 (2008) 479–488. doi:10.1002/pip.831.

[4] T. Lanz, M. Bonmarin, M. Stuckelberger, C. Schlumpf, C. Ballif, B. Ruhstaller, Electrothermal Finite-Element Modeling for Defect Characterization in Thin-Film Silicon Solar Modules, IEEE J. Sel. Top. Quantum Electron. 19 (2013) 1–8. doi:10.1109/JSTQE.2013.2250259.

[5] J. van Deelen, L. Klerk, M. Barink, Optimized grid design for thin film solar panels, Sol. Energy. 107 (2014) 135–144. doi:10.1016/j.solener.2014.05.028.

[6] F.W. Fecher, A. Pérez Romero, C.J. Brabec, C. Buerhop-Lutz, Influence of a shunt on the electrical behavior in thin film photovoltaic modules – A 2D finite element simulation study, Sol. Energy. 105 (2014) 494–504. doi:10.1016/j.solener.2014.04.011.

[7] Y. Gupta, H. Liers, S. Woods, S. Young, R. DeBlasio, L. Mrig, Optimization of a-si solar cell current collection, in: Proc. 16th IEEE Photovolt. Spec. Conf., IEEE, San Diego, USA, 1982: pp. 1092–1101.

[8] H. B. Serreze, Optimizing Solar Cell Performance by Simultaneous Consideration of Grid Pattern Design and Interconnect Configurations, in: 1978: pp. 1–8.

[9] Y. Galagan, E.W.C. Coenen, S. Sabik, H.H. Gorter, M. Barink, S.C. Veenstra, J.M. Kroon, R. Andriessen, P.W.M. Blom, Evaluation of ink-jet printed current collecting grids and busbars for ITO-free organic solar cells, Sol. Energy Mater. Sol. Cells. 104 (2012) 32–38. doi:10.1016/j.solmat.2012.04.039.

[10] G. Sartoris, et al., NM-SESES, Numerical Modelling, 2016. http://nmtec.ch/nm-seses/documentation.

[11] G.P. Steven, Q. Li, Y.M. Xie, Evolutionary topology and shape design for general physical field problems, Comput. Mech. 26 (2000) 129–139. doi:10.1007/s004660000160.

[12] A. Chirilă, P. Reinhard, F. Pianezzi, P. Bloesch, A.R. Uhl, C. Fella, L. Kranz, D. Keller, C. Gretener, H. Hagendorfer, D. Jaeger, R. Erni, S. Nishiwaki, S. Buecheler, A.N. Tiwari, Potassium-induced surface modification of Cu(In,Ga)Se2 thin films for high-efficiency solar cells, Nat. Mater. 12 (2013) 1107–1111. doi:10.1038/nmat3789.

ANALYSIS OF BUILD-IN ELECTROSTATIC FIELD IN CDTE THIN FILM SOLAR CELLS BY QE MEASUREMENTS AT BIAS VOLTAGES

Feng Lianghuan, Wu Lili, Li Xinxin, Xu Hang, Cao Sheng， Shu Qing,
Li Wei, Zeng Guanggen, Zhang Jingquan, Li Bing
Collage of Materials Science and Engineering, Sichuan University
29 Wangjiang Road, Chengdu, China

ABSTRACT: The purpose of the work is to understand the relationship between quantum efficiency (QE) at some typical bias voltage (BV) and build-in electrostatic field, and to develop a new method for analyzing the distribution of build-in electrostatic field in solar cells. In general, the external QE is an important characteristic about the performance of solar cells, however, in practice, the good QE curves not always belong to the good devices. It is considerable that QE is measured in short circuit condition under very weak incident light, which may not make any voltage in the both terminals of the cell. In contrast, when measuring the photo I-V curves a photogenerated voltage may apply the both terminals of the cell in a condition that a load resistance is larger than zero. The photogenerated voltage may drive a reverses current to photogenerated current, and also weaken the build-in electrostatic field or change its distribution, which would lead to alter QE curves. Therefore, it is necessary to measure QE of some typical solar cells at the different BVs and to understand the changes in QE and they effects on the performance of the cells.
Keywords: quantum efficiency, build-in electrostatic field, bias voltage, photogenerated current, CdTe solar cells

1 INTRODUCTION

In general, I-V characteristics expression of the solar cells under light can be simplified:

$$I = I_L - I_d(V) \qquad (1)$$

where, I_L is the photogenerated current, I_d is the diode dark current that is a function of V. In the diffusion current model, I_L can be thought a constant. However, in the space charge limited current model, I_L could be a function of the voltage applied the both terminals of a solar cell because that the voltage may decrease the width of the depleted regions, hence, change the build-in electrostatic field, also affect the collection of photogenerated carriers in the cell. It is considerable that the space charge limited current model is appropriate for many kinds of solar cells made of the semiconductors, in which the carrier lifetimes are much less than the dielectric relaxation time[1]. Therefore, the phenomena about I_L changing with the applied voltage is general for CdTe, CIGS, CZTS and amorphous silicon solar cells.

This paper proposes to measure QE curves of CdTe thin film solar cell at different bias voltages to observe the effects of BV on QE curves. The changes in QE curves under BVs reflect the changes in collection efficiency of photogenerated carriers in some wavelength, which results from the changes of build-in electrostatic field in a solar cell. The measured results and analysis may be much helpful to examine the improvements in structure and fabrication conditions.

2 RESULTS AND DISCUSSION

Firstly, QE curves of a current FTO/CdS/CdTe/Au solar cell have been measured at the BVs from -0.5V to 0.6V . The structure of the cell is: CdS (sputtering) 30nm, CdTe (CSS) 4μ, Au 100nm; its performance parameters are: Eff=6.55%, Voc=497mV, Jsc=24.48mA/cm^2, FF=53.82 %.

Figure 1 shows the QE curves. We can see that when the BV is 0.4V the QE in all wavelength range decreases obviously. When the BV is 0.5V the QE in a wavelength range of 800--865nm is minus. According to the photo absorption coefficient of CdTe, it can be thought that the visible light is nearly completely absorbed in the front 2μ of CdTe and the light of 750--865nm is absorbed mainly in the back of CdTe solar cell. The curve means that there is a reverses photogenerated current at the back region of the cell. Obviously, the build-in electrostatic field at back is weaker than that in the front region. When the BV is 0.6V the QE in all wavelength range is minus, which implies that the build-in potential could be lower than 0.6V.

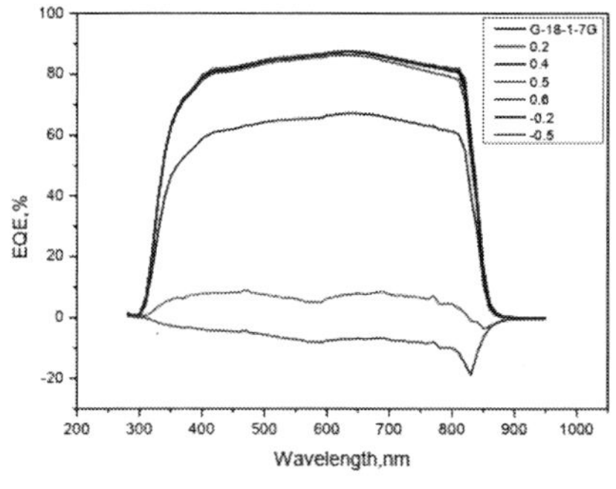

Figure 1: QE curves of a current CdTe solar cell

Figure 2 gives a comparison between QE short circuit current density at different BVs and a light I-V curve of the same cell. It shows that the currents of the light I-V curve are limited by QE short circuit current density.

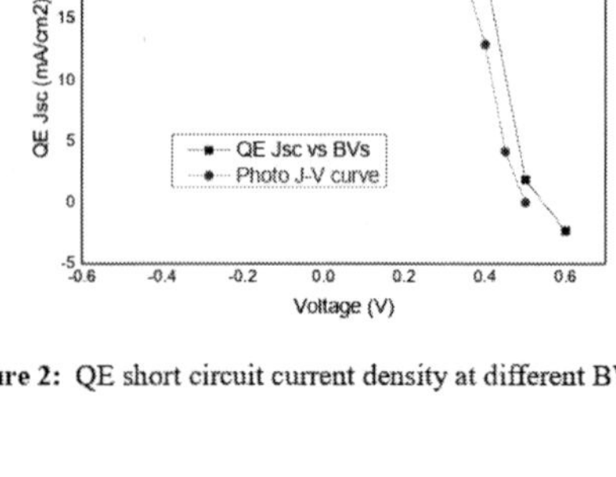

Figure 2: QE short circuit current density at different BVs and

the light J-V curve of a current CdTe solar cell

In the condition of a low Voc, the distribution and intensity of build-in electrostatic field at BVs dominate the performance of the cell. In fact, a low Voc results from a weak build-in electrostatic field.

Figure 3 shows QE curves of two CdTe solar cells 15-10 and 15-10-1. Their CdS/CdTe structures are the same: CdS (chemical bath) 52nm and CdTe (CSS) 3.6μ; but there is a back contact layer ZnTe:Cu in sample 15-10. Their performance parameters are in Table 1.

Table 1 The performance parameters of samples 15-10 and 15-10-1

Sample	Eff, %	Voc, mV	Jsc, mA/cm^2	FF, %
15-10	8.91	723	24.51	50.29
15-10-1	6.84	640	21.73	49.21

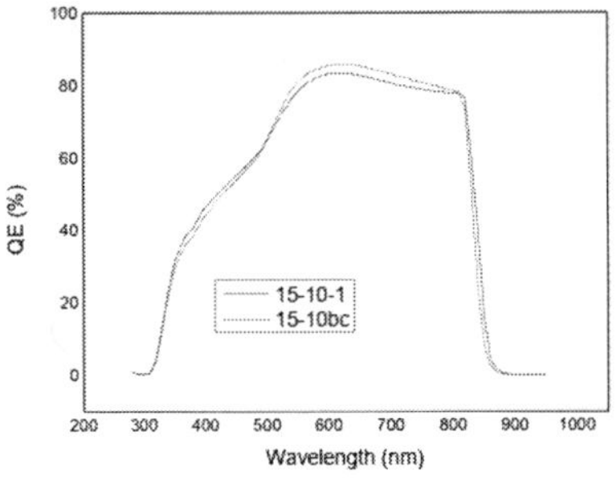

(a)

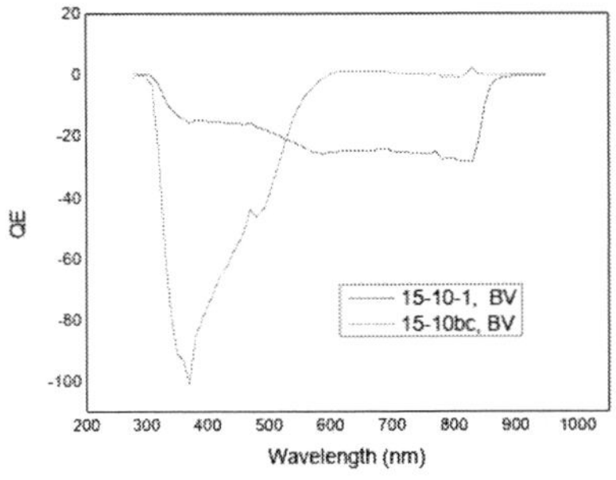

(b)

Figure 3: QE curves of CdTe solar cells 15-10 and 15-10-1. (a): QE curves at 0 BV; (b): OE curves at their open voltages in Table 1.

The experiment results show that it is nearly impossible to infer the changes of QE curves at BV from its QE curves at 0 BV. When the applied BV on both the samples, the most obvious difference between the two QE curves is that QE in long waves is higher than that in short waves for the cell with back contact, and QE in long waves is lower than that in short waves for the cell without back contact, which implies that ZnTe:Cu back contact is helpful to increase the intensity of build-in electrostatic field at back of solar cells.

Figure 4 shows QE curves of two CdTe solar cells. Their structures are the same: CdS (chemical bath) 40nm, CdTe (VTD) 4μ, Au 100nm; but CdS of sample Ta10-3 has been annealed in O_2+N_2 at 380℃ for 30 min and CdS of sample Tb8-1 no annealed. Their performance parameters are in Table 2.

Table 2 The performance parameters of samples Ta10-3 and Tb8-1

Sample	Eff, %	Voc, mV	Jsc, mA/cm^2	FF, %
Ta10-3	10.53	708	25.16	59.13
Tb8-1	8.71	624	24.91	56.03

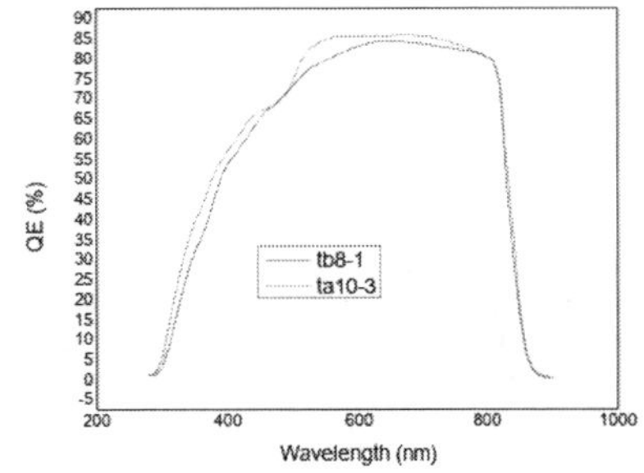

(a)

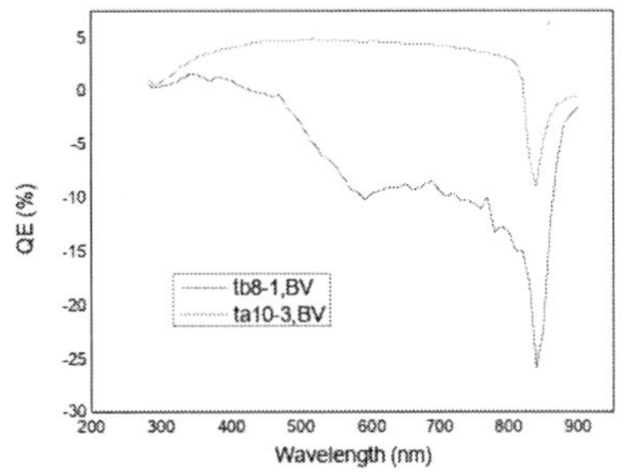

(b)

Figure 4: QE curves of CdTe solar cells Ta10-3 and Tb8-1. (a): QE curves at 0 BV; (b): OE curves at their open voltages in Table 2.

We can see that QE of the cell, in which CdS layer has been annealed, is positive in a wave band from 300nm to 820nm when BV is 0.708V. However, QE for CdS layer has not been annealed is minus in all wave band when BV is lower, 0.624V. The above results imply that a suitable post-treatment of CdS makes CdS/CdTe junction stranger and the build-in electrostatic field distribute nearly over all CdTe layer.

Figure 5 shows QE curves of a relatively high efficiency CdTe solar cell Tq3-1. Its structure and fabrication methods are : sputtering CdS:In/ VTD CdTe/ co-evaporation ZnTe:Cu/ evaporation Au. Its performance parameters are: Eff=15.67%, V_{oc}=0.831V, J_{sc}=25.77mA/cm^2, FF=73.19%. Although BV is high to 0.83V, QE values are positive in all wave band and QE J_{sc} is 10.48mA/cm^2, that equals and offsets the dark diode current density of the cell under 0.83V.

It is worthy to compare the figure with fig.1. When no applied BV, QE in short waves of Tq3-1 is lower than that of G-18-1-7G. This is because of that there is a thicker and doped CdS:In window layer in the sample. This window layer absorbs too much the photons of short waves, however, they generate carriers in CdS:In layer can not be collected to contribute to the photogenerated current. Fortunately, this layer makes the CdS/CdTe junction much stranger, hence, leads to a build-in electrostatic field that can not be easy to offset by higher BV. Therefore, there is a high V_{oc} and FF of the cell.

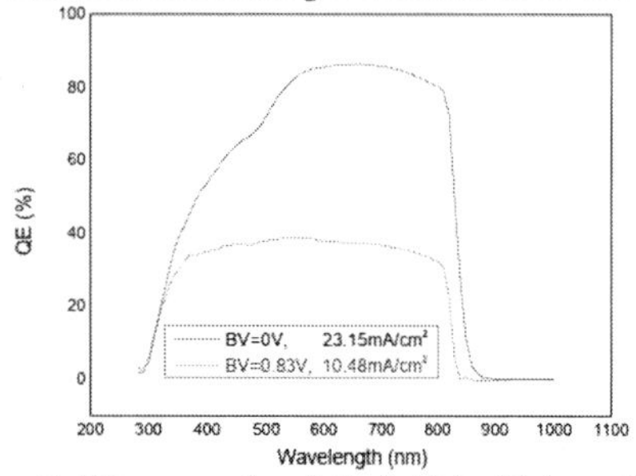

Figure 5: QE curves of a relatively high efficiency CdTe solar cells.

In our studies, all the typical comparative experiments about structure, fabrication and post-treatment method have included the QE measurements at BVs. Some valuable knowledge about QE measurements is obtained from the experiment results.

The first of all, it is tempting to choose the BVs. Exactly, V_{oc} and V_{max}, voltage at the max power point, are specially important. When BV is V_{oc}, the measured QE curves would response much completely the distribution of build-in electrostatic field. In an idea condition that the light resource, which is used to measure light IV curve, is the same with sunlight, the QE J_{sc} at BV=V_{oc} is positive and should equal $I_d(V_{oc})$. Obviously, the larger QE J_{sc} at BV=V_{oc} the higher V_{oc}. In the other hand, the importance of BV=V_{max} is due to that this QE might be the much practical QE, and not be the same one in short circuit condition. Thus build-in electrostatic field at BV=V_{max} would determine the performance and maximum output power of the solar cell immediately. Therefore, these experiment QE curves would be more helpful to analyze and understand if the structure, fabrication and post-treatment of the solar cells are perfect.

The second one is that the two sensitive wavelength ranges for QE curves at BVs are from 350nm to 500nm and from 800nm to 860nm. The changes in QE curves at BVs in the both wave bands are useful to analyze the build-in electrostatic field at two interfaces of window layer/CdTe and CdTe/back contact layer. If the window layer and back contact layer are composite structures that are consist of two or more layers, respectively, the changes in QE curves at these interfaces could be much complex.

3 CONCLUSIONS

In this paper the QE curves of CdTe thin film solar cells with different structures and fabricated in different fabrication conditions have been measured when applying different bias voltage, and the effects of BVs on QE curves have been observed and analyzed. Some important knowledge about determining the typical voltages for applying BVs has been obtained. Our results show that the sensitive wavelength ranges for the changes of QE under BV are from 350nm to 500nm and from 800nm to 860nm, which means that the build-in electrostatic field at window layer/CdTe and CdTe/back contact layer interfaces is easy to be altered by applying BVs. The results also show that the QE curve of the high efficiency CdTe solar cells could not be easily altered by BVs. The facts imply that the studies of QE under BV may be much helpful to examine the improvements in the structure, fabrication and post-treatment conditions. It can be believed that the method would be also useful for the studies of other solar cells made of CIGS, CZTS and amorphous silicon .

ACKNOWLEDGEMENT

This work was supported by the National High Technology Research and Development Program of China (Grant no.2015AA050610).

REFERENCES

[1] Stephen J. Fonash, Solar Cell Device Physics, Academic press, New York, 1981, pp153.

ELECTRICAL PROPERTIES OF THE Al/Cu(InGa)Se_2 JUNCTIONS: PAVING THE WAY TOWARDS SCHOTTKY BARRIER CIGS SOLAR CELLS?

Bertrand Theys*[1, 3, 4], Torben Klinkert[3, 4], Fabien Mollica[2, 3, 4], Enrique Leite[2, 3, 4], Frédérique Donsanti[2, 3, 4], Marie Jubault2[, 3, 4] and Daniel Lincot*[1, 3, 4]
[1] Centre National de la Recherche Scientifique (CNRS), 3 rue Michel-Ange, 75016 Paris (France)
[2] Electricité de France (EDF), 6 quai Watier, 78400 Chatou (France)
[3] Institut de Recherche et Développement sur l'Energie photovoltaïque (IRDEP), UMR 7174 CNRS-EDF-Chimie ParisTech (PSL), 6 quai Watier, 78400 Chatou (France)
[4]Institut photovoltaïque d'Île-de-France (IPVF), 8 rue de la Renaissance, 92160 Antony (France)

*Corresponding authors: Bertrand-externe.Theys@edf.fr; Daniel.Lincot@chimie-paristech.fr

ABSTRACT: Devices based on a metal-semiconductor contact are potential candidates for the fabrication of photovoltaic cells as long as the corresponding junctions exhibit good rectifying properties and a low contact resistance.
This communication presents a detailed electrical analysis of the Al/Cu(InGa)Se_2 junctions. The values of the representative parameters of the quality of the diode such as the barrier height, the ideality factor and the series resistance are deduced from intensity-voltage curves. Barrier height values are comprised between 0.64 V and 0.85 V with n values mostly around 2, depending on the surface treatment of CIGS.
It is concluded that this system may be interesting for being used further in CIGS technologies.

Keywords: CIGS, Schottky diodes, metal/CIGS contacts

1 PURPOSE OF THE WORK

Because of its favorable optical and electronic transport properties $Cu(In_xGa_{1-x})Se_2$ (CIGS) can be used as an absorber in thin layer solar cells exhibiting high performances. A 22.3 % efficiency has been very recently obtained with such devices under standard AM1.5 test conditions [1].

The standard cell structure consists in four layers with the structure [substrate/Mo/CIGS/BL[1](CdS, ZnS, InS)/TCO[2](ZnO)], with a substrate configuration, where light is coming from the top side, through the TCO/BL windows. This work aims to replace these window layers by a metallic contact, with light coming through a transparent contact in a superstrate configuration with the [glass substrate/TCO/CIGS/metal] structure. This is made possible by using thinner CIGS, which is a main stream tendency for the CIGS technology today [2]. More over the structure may become simpler with such a Schottky like configuration.

In this work we present a first step which consists in preparing Mo/CIGS/Al Schottky diodes in order to evaluate the feasability of this approach.

2 SCIENTIFIC INNOVATION AND RELEVANCE

Consisting of a metal layer deposited on semiconductor material, Schottky diodes are certainly, at least theoretically, the simplest rectifying devices. In practice, basic care must be taken when fabricating such devices because of the necessity to obtain a direct contact between the two parts, without any intermediate layer which could impede charge carrier flow.

Additionally, many works are in progress in order to obtain photovoltaic cells made with very thin CIGS absorber layers but still exhibiting a high conversion efficiency. For both resource-preserving and economic reasons, the goal is to reduce the consumption of scarce materials (In, Ga) which are contained in the absorber. However, CIGS-based solar cells with a thin (<500 nm) absorber layer have a rather low conversion efficiency because of a poor quantum efficiency in the infrared part of the solar spectrum due to a lower absorption of the material in this spectral range. By increasing the light path in the absorber this pitfall can be bypassed. This effect can be obtained by using a good rectifying and reflecting contact on the back side of the cell. Aluminum is a good candidate to play this role.

First, from an electrical point of view, the aluminum-CIGS contact is known to be of Schottky type. Secondly, from an optical point of view, the reflection coefficient at the Al/CIGS interface is high in the IR and visible parts of the solar spectrum. Furthermore, the presence of a naturally rough CIGS surface could enhance light scattering after reflections at the CIGS/Al interface. Consequently, the energy conversion efficiency of the device should rise. From a chemical point of view, the Al layer deposited on the back surface of the CIGS may be an efficient barrier against oxygen and/or humidity diffusion into the device and may therefore contribute to increase the lifetime of the cells.

Therefore, it is worth studying in details the electrical behavior of Al/CIGS contacts as they can be obtained using both state of the art semiconductor and metal layers. Such experimental results will be presented and discussed in this communication together with values of some key-parameters of these junctions.

3 EXPERIMENTAL

Schottky diodes fabricated from several different CIGS samples have been measured. All the layers have been grown on Mo coated soda-lime glass by co-evaporation but with different protocols ("one stage" [3] or "three stage" [4] process) at temperatures between 425 and 500°C. The [Cu]/([In]+[Ga]) ratio has been found to be between 0.76 and 0.84 and the [Ga]/([In]+[Ga]) one

[1] BL: buffer layer
[2] TCO: transparent conductive oxide

between 0.28 and 0.34.

Prior the e-beam Al evaporation, some of the CIGS layers have been etched in a 1M KCN solution.

The Al contact was kept at the zero potential and the voltage applied to the Mo contact was varied between -0.4 and +0.7 V.

Measurements were performed at room temperature. All the Al dot areas were 7mm².

4 RESULTS

Current density-voltage curves of several diodes are in figure 1. They have good rectifying properties.

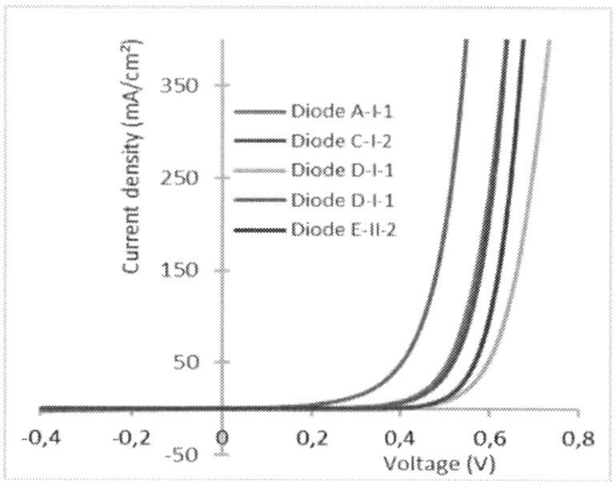

Figure 1: Current density-voltage characteristics of some Al/CIGS Schottky diodes

As a function of the voltage V measured between the 2 terminals of a diode, the current I can be expressed according to the following transcendental equation:

$$I = I_0\left[exp\left(\frac{q(V-R_sI)}{nkT}\right) - 1\right] + \left(\frac{V-R_sI}{R_{sh}}\right) \quad (1)$$

n being the "ideality factor", I_0 the saturation current and R_s and R_{sh} the series and shunt resistances respectively.

When R_s is small, R_{sh} large and qV/nkT>3, the above equation leads to:

$$Log\ I = \frac{qV}{nkT} + Log\ I_0$$

According to the the thermionic emission theory, the saturation current density J_0 (=I_0/A, A being the diode surface) of a metal/semiconductor barrier can be expressed as:

$$I_0 = A^{**}T^2 exp\left(\frac{-q\Phi_b}{kT}\right)$$

where A** is the Richardson constant (84 A/cm²K² for CIGS considering a relative hole effective mass of 0.7). $q\Phi_b$ is the "barrier height" which is the energy barrier that holes have to overcome for travelling from the Al contact to the CIGS bulk.

Therefore, n and Φ_b can be deduced from a plot of Log I as a function of the applied voltage V (from the slope and from the x-axis intercept respectively).

To check the validity of the results found according to this first method, a second determination of the diode parameters has been carried out following a protocol described originally in reference 5. When $R_{sh} \rightarrow \infty$, rearrangement and differentiation of equation (1) lead to the 2 following equations (J is the current density):

$$\frac{dV}{d(LogJ)} = R_sAJ + \frac{nkT}{q}$$

$$V - \frac{nkT}{q}Log\left(\frac{J}{A^{**}T^2}\right) = R_sAJ + n\Phi_b$$

Plotting the left hand side of these linear equations as a function of J allows to quantify R_s, n and Φ_b and therefore to check the consistency of the two different methods of determination of the Al-CIGS junction parameters starting from the intensity-voltage characteristics. Results are shown in Table 1

	Φ_{b1}	Φ_{b2}	n_1	n_2	R_{s1}	R_{s2}
A-I-1	0.69	0.75	2.5	1.9	0.94	0.90
A-I-2	0.73	0.80	2.3	1.7	1.33	1.41
A-II-1	0.78	0.85	2.0	4.2	0.45	3.1
B-I-1	0.72	0.78	2.7	2.0	0.70	0.70
B-I-2	0.75	0.75	5.2	5.2	169	130
C-I-1	0.76	0.79	2.6	2.4	1.14	1.17
C-I-2	0.72	0.75	2.3	2.0	0.27	0.31
D-I-1	0.79	0.79	2.0	1.9	1.31	1.53
D-I-2	0.77	0.80	2.1	1.9	0.94	0.75
E-I-1	0.64	0.66	~2.8	2.4	0.05	0.14
E-I-2	0.64	0.68	2.9	2.3	0.30	0.41
E-II-1	0.79	0.82	1.7	1.7	0.50	0.56
E-II-2	0.81	0.84	1.8	1.6	0.70	0.81
E-II-3	0.84	0.84	1.8	1.6	0.69	0.81

Table 1: Experimental values of the built-in potential (Φ_b, in volts), ideality factor (n) and series resistance (R_s, in ohms) of samples used in this study. The different values of a given parameter have been obtained following the different methods described in the text.

5 DISCUSSION

The experimental barrier-height of the different diodes which were tested is high, varying between 0.65 and 0.85 V. Given the composition of the CIGS samples used in the present study, the theoretical value of this barrier height is ~1.30 eV for the samples of the A series and ~1.35 eV for the samples of the other series. It is less than twice the mean experimental value derived from the current-voltage characteristics measurements.

The discrepancy between experimental and theoretical values of Schottky barrier height is very common whatever the metal and the semiconductor and it has been largely commented in the literature. It is behind the scope of the present paper to discuss in detail the values of the barrier height but, nevertheless, from table 1 it is clear that this parameter is sensitive to the etching of samples in a KCN 1M solution. The value of sample A-II-1 (which underwent this treatment) is 50 to 100 mV higher than those of samples A-I-1 and A-I-2 (which were not etched). The same conclusion can be drawn for the E-family when comparing samples E-I (not etched) and samples E-II (etched). The increase in this barrier height after etching could result from the removal of Se clusters and secondary Cu-Se phases by KCN. Indeed, these Se and Cu compounds are located at the CIGS surface, forming an interfacial layer which may play a role on the distribution of electric potentials at the metal/semiconductor contact. Another consequence of the removal of Cu-Se secondary phases is that the CIGS material becomes Cu-poor and consequently a valence band maximum lowering occurs increasing the bandgap and, therefore, the barrier height.

Except for sample B-I-2, all of the ideality factors n shown in table 1 are close to 2. This value indicates that the diode current is influenced by Shockley-Read-Hall-type recombination in the space-charge region. In the E family samples, the ideality factor decreases after KCN etching indicating reduced space-charge recombination with respect to untreated samples (E-II vs E-I samples). This reduction could result from a decrease of the density of recombination centers in the space-charge region.

It has to be noted that an apparent contradiction may appear between the thermionic emission theory we have used to calculate the barrier height (saturation current dominated by carrier diffusion) and the value ~2 of the ideality factor (saturation current dominated by space charge recombination). This discrepancy has already been discussed in the literature and it has been concluded that the method developed in the present study can nevertheless be applied [6]. This method is very simple and accounts well for our experimental results. It has therefore been chosen preferably to more sophisticated models such that the so-called "two-diode model" which has been elaborated for p-n junction solar cells.

Series resistances are relatively low. In the structures which are reported in the present study, given the CIGS layer resistivity and dimensions the bulk resistance of the diodes ($2x10^{-4}$cm thickness) is ~0.2 Ω. This value is compatible with the total series resistance measured on our devices. For the best "conventional" cells (substrate configuration with the [ZnO/CdS/CIGS/Mo/soda-lime glass] structure) fabricated with the same absorbers as the B, C, D and E samples of the present work, series resistances between 0.04 and 0.08 $\Omega.cm^2$ have been measured. Globally, the R_s values reported in table 1 show, together with the barrier heights, that the quality of the Al/CIGS Schottky diodes whose properties are described in the present papers is encouraging for further use of this approach in solar cells.

6 CONCLUSION

It is shown that Al/CIGS Schottky devices exhibiting good rectifying diode-like current-voltage characteristics can be fabricated following a rather simple process. Barrier heights of 0.7 to 0.8 eV have been obtained which is a good value when compared to 1.35 eV, the theoretical maximum. Series resistances are weak for most of the diodes which have been studied, allowing potentially to use these devices for high intensity current purposes. Regarding the ideality factor, it is of the order of 2, reflecting the important role played by the space-charge recombination current.

In summary, Al/CIGS diodes are promising structures which must be taken into account in view of simplification of the structure of the CIGS-based solar cells.

References:

[1]http://www.pv-agazine.com/news/details/beitrag/solar-frontier-hits-223-on-cigs-cell_100022342/

[2] Z. Jehl Li-Kao, N. Naghavi, F. Erfurth, J.-F. Guillemoles, I. Gérard, A. Etcheberry, J.-L. Pelouard, S. Collin, G. Voorwinden D. Lincot, Prog. Photovolt. Res. Appl. **20**, 582 (2012)

[3] P. Salomé, V Fjällström, P. Szaniawski, J. Leitao, U. Zimmermann, A. F. da Cunha, and M. Edoff, Prog. Photovolt: Res. Appl., **23**, 470 (2015)

[4] T. Klinkert, M. Jubault, F. Donsanti, D. Lincot, J.-F. Guillemoles, J. Renewable Sustainable Energy, **6**, 011403 (2014)

[5] S.K. Cheung, N.W. Cheung, Appl. Phys. Lett., **49**, 85 (1986).

[6] V. Aubry, F. Meyer, J. Appl. Phys., **76**, 7973 (1994)

STUDY OF PROMOTION OF ANTIMONY DOPING TO THE CRYSTALLIZATION OF Cu_2ZnSnS_4 (CZTS) FILMS DURING THE ANNEALING PROCESS

XF. Zhang [1], Y. Umejima [2], and M. Kobayashi [2, 3]
1. International Center for Science and Engineering Program, Waseda University,
2. Department of Electrical Engineering and Bioscience, Waseda University,
3. Kagami Memorial Research Institute for Materials Science
Waseda University, 3-4-1 Ookubou, Shinjyuku-ku, Tokyo 169-8555, JAPAN
Corresponding E-mail: zhangxf@aoni.waseda.jp

ABSTRACT: Cu_2ZnSnS_4 (CZTS) films were fabricated by annealing to nanocrystal ink deposited CZTS precursors in the sulfur atmosphere. A simple approach for overcoming the main challenge of nanocrystal particle processed CZTS film was developed by adding very small amount of antimony to the annealing atmosphere. The grain size and crystallinity of CZTS film was significantly improved by doping antimony to CZTS film. The influence of Sb:S atomic ratio in the annealing atmosphere on grain growth and crystallinity improvement was studied. Even very small amount of antimony plays an important role in promoting the grain growth and recrystallization of CZTS films. Formation of MoS_2 between Mo and CZTS layer was enhanced by antimony as well.
Keywords: CZTS, nanocrystal ink, annealing, Sb-doping

1 INREODUCTION

The quaternary semiconductor CZTS has attracted more and more interest during the past several years, because of its tunable bandgap, high absorption efficiency, nontoxicity and most importantly, utilization of earth-abundant materials. 1.5 eV bandgap of the pure CZTS is nearly ideal for single junction solar cell, and the bandgap can be tuned to less than 1.0 eV by substituting the sulfur atoms by selenium atoms. Another important merit for CZTS solar cell is the applicability of nonvacuum process for fabricating CZTS absorber layer, which is viewed as the most critical and complicated layer in the solar cell, and the a record efficiency of 12.6% has been achieved by a solution based synthesis method recently [1]. The large obstacle on the pathway to high efficiency CZTS solar cells fabricated by nonvacuum process is the small grain size, which induces large amount of grain boundaries, and poor crystallinity of CZTS films.

It has been reported that antimony can significantly improve grain size and device performance of CIGS solar cells [2, 3]. Similar result has been reported to the solution- processed CZTSe films by doping antimony to the nanoparticle inks [4]. This project reports our study on the influence of antimony atmosphere on the grain size and crystallinity of CZTS films.

2 EXPERIMENTAL METHODS

CZTS nanocrystal inks were prepared from CZTS powder by a wet ball-milling method, followed by centrifugal treatment to remove large particles and retrieve just small particles less than 100nm. The CZTS precursor was then deposited on the Mo-coated glass substrate using the as-fabricated inks by a spin-coating approach. The precursor and elemental sulfur powder (99.999%) mixed with elemental antimony (99.999%) were placed in a glass tube, which was evacuated to around 2.0×10^{-2} Pa, and N_2 gas with a pressure of around 0.2 atm was used to fill the tube. Then the glass tube was sealed up to carry out annealing, which was conducted in a cubic annealing furnace with a temperature deviation from the setting value of less than 1%. The annealing temperature was selected as 575°C and the temperature profile as a function of time is revealed in Figure 1. The annealing furnace was set at a heating rate from room temperature to the aim temperature of 15°Cmin^{-1}. The tube was maintained at 575°C for 1.5 hours. After the annealing was finished, the tube was cooled with the furnace naturally to around 400°C to provide the substrate with sulfur atmosphere protection. Then the tube was withdrawn from the annealing furnace to the air, cooling down to less than 200°C within 5 min to get rid of sulfur condensation on the surface of CZTS film. The Sb:S atomic ratio is selected between 0 to 0.3% to check its influence on the annealing process.

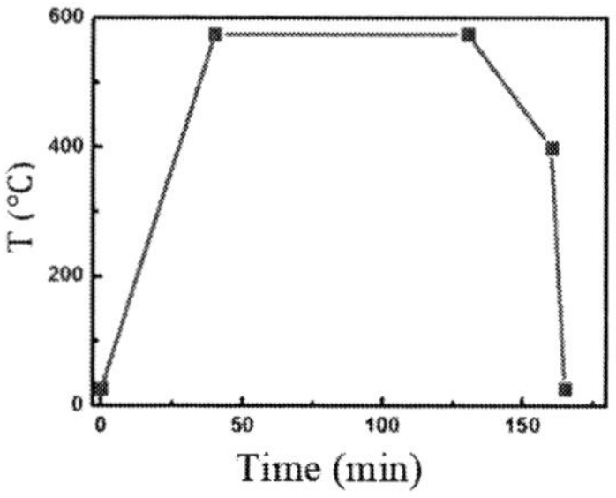

Figure 1 Annealing temperature profile

3 RESULTS AND DISCUSSIONS

3.1 Surface Morphology of CZTS films

Figure 2 shows the surface scanning electron microscope (SEM) images of the CZTS precursor annealed with different Sb:S ratios ranging from 0 to 0.3%. All these samples are annealed with exactly the same conditions. Judging from the figures, one can see significant improvement to grain size. When no antimony is used, the grains of CZTS film show strip shape and the size is very small- less than 200 nm. The density of the film is low as well since many pinholes can be observed. When the sulfur is slightly mixed with antimony, that is when the Sb:S atom ratio is 0.05%, the grain growth is significant. The largest grain goes up to several μm, but surrounded by small grains less than 0.5 μm. When the

Sb:S atomic ratio is 0.1%, the number of small grains decreases and the film grows more compact. With the antimony atom number continues to increase to 0.2%, the grain size also increases and the grain boundary is very clear. When the Sb:S atomic ratio is increased to 0.3%, the grain boundary becomes vague and it seems that the film is covered by a thin layer, which is unclear for us until now.

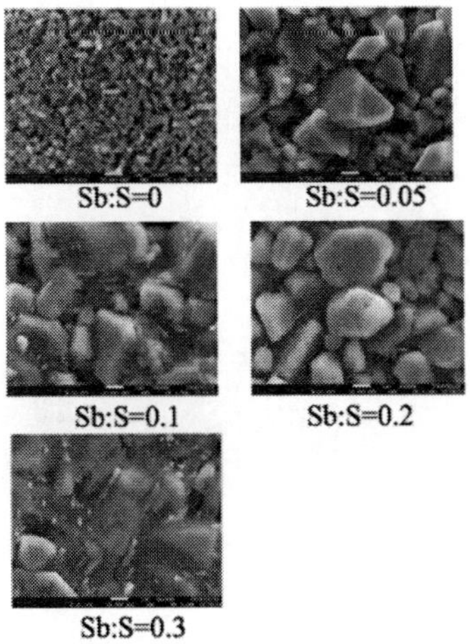

Figure 2 Surface SEM images of the CZTS films with different Sb:S atomic ratios

3.2 Cross-sectional morphology of CZTS films

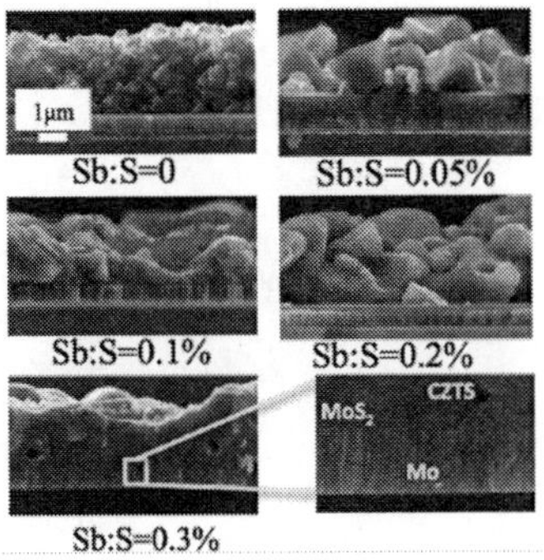

Figure 3 Cross-sectional SEM images of the CZTS films with different Sb:S atomic ratios

Figure 3 show the cross-sectional image of the same CZTS films in Fig. 1. The grain growth is observed obviously. When the atomic ratio is over 0.1%, the film shows significantly large grains covering from the substrate to the surface. When the ratio is 0.3%, one single layer of CZTS is obtained. One possible explanation is the melting down of grain boundaries because of the decrease of melting point caused by the intermediate phase containing antimony [4]. Last figure in Figure 3 is the magnification of the yellow square, which illustrates the interface area between CZTS and Mo. In the Mo layer, there is a bilayer structure, demonstrating the formation of MoS_2 layer. When no antimony is doped, such structure is not observed; meaning antimony also promotes the formation of MoS_2 layer, mechanism of which is still not covered up by any report.

3.3 Composition of the CZTS

Table I Composition of CZTS films

Sb:S Ratio (%)	Cu (%)	Zn (%)	Sn (%)	S (%)	Sb (%)
Precursor	25.6	9.7	15.1	49.6	0.0
0	24.1	9.6	15.0	51.3	0.0
0.05	24.3	9.3	14.9	51.4	0.1
0.1	24.7	9.6	14.0	51.7	0.0
0.2	23.9	9.4	15.1	51.5	0.1
0.3	24.4	9.8	14.5	51.0	0.3

The composition of the precursor and annealed films was determined using energy dispersive spectroscopy (EDS), as shown in Table I. The film precursor showed a sulfur poor composition. After the film precursor was annealed, its composition became sulfur rich. Judging from the Sb atomic ratio, although the value is not zero when the source is Sb doped, it is within the measuring error of EDS. As a result it is difficult to tell whether Sb embodied in the CZTS film or not. In order to judge Sb, further study is carried out.

3.4 Crystallinity of CZTS films

X-ray diffraction (XRD) patterns of CZTS films annealed under different Sb:S atomic ratios are shown in Figure 4. According to the figure, the crystallinity of CZTS is significantly improved after antimony doped. Intensity of the preferred orientation of (112) peak gradually increased, indicating a better crystallization. In the figure, all the main peaks has be assigned to kesterite structural CZTS [5], indication pure CZTS phases. In our other study, a temperature of 625°C is necessary to get comparable intensity to 575°C annealed CZTS film doped with a Sb:S ratio of 0.2%, which shows a clear potential for low temperature annealing. Besides the CZTS peaks, a weak peak at around $2\theta=33°$ is observed, which is assigned to the (100) peak of MoS_2 [6]. The XRD result is in consist with the cross-sectional result.

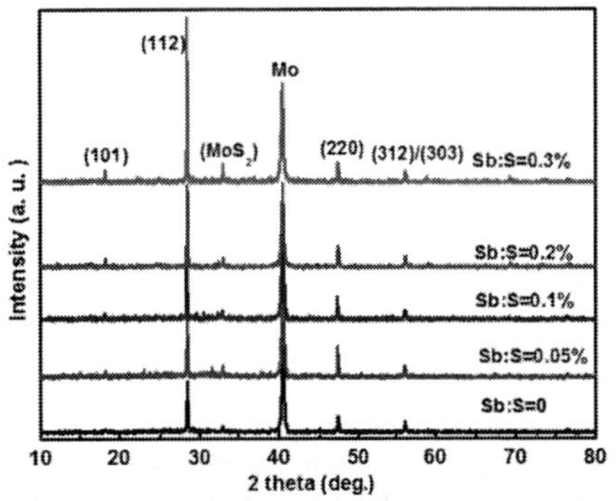

Figure 4 XRD patterns of CZTS films annealed with different Sb:S atomic ratios

3.5 Raman spectrum of CZTS films

The Raman spectra of the precursor and the annealed CZTS thin films are shown in Figure 5. It can be seen from the figure that the film exhibit typical Raman spectra of kesterite CZTS films with three peaks at 285, 330 and 369 cm−1, which correspond to the two A symmetry modes and the B symmetry mode of the CZTS kesterite structure, respectively [7]. However, besides the peak of CZTS, no peaks of Sb and Sb compounds are observed from the spectrum, indicating that Sb does not exist in the film. The result is consist with the EDS and XRD result.

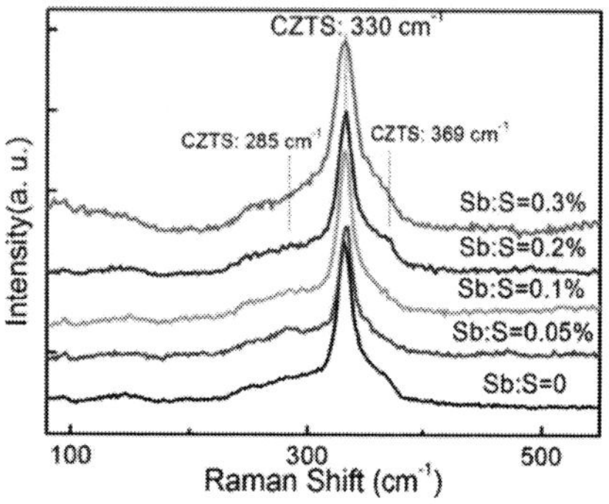

Figure 5 Raman spectrum of CZTS films annealed with different Sb:S atomic ratios.

4 CONCLUSION

Simple antimony doping to the CZTS film during the annealing process approach is developed to achieve significant improvements in the grain structure and crystallization of CZTS films. The grain size of CZTS films grows to μm level from several-ten nm by slight antimony doping. The crystallization is improved as well and pure CZTS peeks are observed, although no Sb or Sb compounds are detected according the the XRD and Raman measurement. Our other research has revealed that the annealing temperature must be over 600°C to get large grain size and better crystallinity, which lead to decomposition of CZTS films. It provides an applicable way for us to deposit high-quality CZTS films at relative low temperature through antimony doping. The growth mechanism of CZTS annealing with antimony remains new topic.

ACKNOWLEDGEMENT
This work is partly supported by Waseda University Initiatives and Waseda University Grand for special Research Project.

REFERENCE

[1] W. Wang, M. T. Winkler, O. Gunawan, T. Gokmen, T. K. Todorov, Y. Zhu, D. B. Mitzi, Adv. Energy Mater. 4 (2014)1301465.
[2] M. Yuan, D. B. Mitzi, W. Liu, D. J. Kellock, S. J. Chey, and V. R. Deline, Chem. Mater. 22 (2010) 285.
[3] M. Yuan, D. B. Mitzi, O. Gunawan, A. J. Kellock, S. Jay Chey, V. R. Deline, Thin Solid Films 519 (2010) 852.
[4] A. Carrete, A. Shavel, X. Fontané, J. Montserrat, JD. Fan, M. Ibáñez, E. Saucedo, A. P. Rodríguez, and A. Cabot, J. Am. Chem. Soc, 135 (15982) 2013.
[5] T. K. Todorov, K. B. Reuter, and D. b. Mitzi, Adv. Matter. 22 (2010) E156.
[6] Y.C. Liu, YP. Zhao, L Jiao, and J. Chen, Mater. Chem. A 2 (2014) 13109.
[7] A. Khare, B. Himmetoglu, M. Johnson, D.J. Norris, M. Cococcioni, E.S. Aydil, J. Appl. Phys. 111 (2012) 083707.

STUDY OF SHUNT DISTRIBUTIONS IN THIN FILM MULTIJUNCTION SOLAR CELLS

Author(s)
Company / Institute(s)
Address(es)

Tomáš Finsterle[1], Jan Klusáček[2], Federico Ventosinos[3], J-W. Schüttauf[4], Pavel Hrzina[1], Ladislava Černá[1], Vítězslav Benda[1] and Jakub Holovský[1,2,*]
[1]Faculty of Electrical Engineering, Czech Technical University in Prague, Technická 2, 166 27 Prague, Czech Republic.
[2]Institute of Physics of the Academy of Sciences of the Czech Republic, Cukrovarnická 10, 162 00 Praha, Czech Republic
[3]LPICM, CNRS, École Polytechnique, Université Paris-Saclay, 91123 Palaiseau, France
[4]École Polytechnique Fédérale de Lausanne (EPFL), Institute of Microengineering (IMT), Photovoltaics and Thin-Film Electronics Laboratory, Maladière 71, CH-2000 Neuchâtel, Switzerland
*Phone: +420 224 352 167; E-mail: jakub.holovsky@fel.cvut.cz

ABSTRACT: The thin film technology of photovoltaic modules is perspective from the low cost point of view and with the respect of special, for example flexible modules. However, within a range of allowed temperature it is hard to prepare a high quality semiconductor material. On the other hand the technology can easily adopt a complex multi-layer structure, particularly a tandem structure, in which the cell with the narrower band gap is covered by the one with the wider band gap. For the characteristics and diagnostics of tandem cells it is necessary to distinguish the contributions of the component sub-cells to the total performance. One of the tasks is the determination of the component contributions to the total voltage, which is strongly influenced by the amount and spatial distribution of shunts in each sub-cell. We show that the total voltage is not a simple sum of component voltages. The experimental results are presented in this paper.

Keywords: a-Si/μ-Si, Multijunction Solar Cell, Tandem

1 INTRODUCTIONLIVERY

The determination of the I-V curves of individual sub-cells in tandem cell is useful but rather complicated. We have published recently an advanced method allowing accurate determination of I-V curves, provided that the component open-circuit voltages are determined [1]. The latter task is especially complicated due to the illumination. This is especially true for the tandem cell composed of the amorphous and microcrystal silicon. This nonexistence of selective illumination can in principle be solved by measuring voltages at more intensity levels and using various mutual mixtures of different illuminations and by additional mathematical procedure [2]. This procedure is based on a simple fact, that the total measured voltage is in every point of characteristics a sum of individual voltages on every cell. This idea does not always apply within the real cells [3]. It has been shown that it is necessary, in the technology of thin film solar cells, to represent a sub-cell not only by one (or two) diode, but by a whole net of parallel diodes interconnected by planar electrode of limited conduction [4].

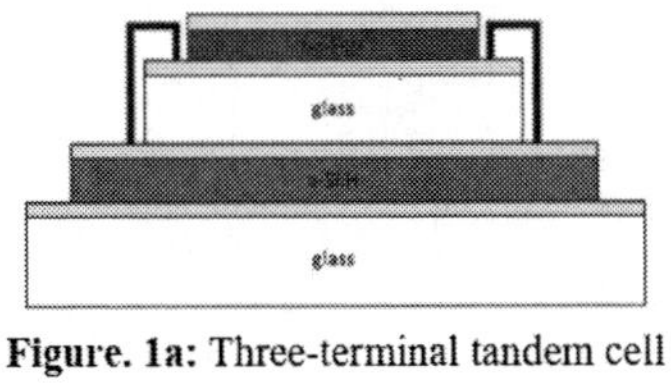

Figure. 1a: Three-terminal tandem cell

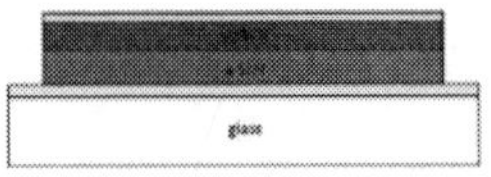

Figure. 1b: Two-terminal tandem cell

2 EXPERIMENTAL DETAILS

2.1 Test structures

To test our hypothesis the measurement were performed on standard two-terminal and three-terminal cells. The three-terminal tandem cells were made by bonding single amorphous silicon (a-Si:H) cell and single microcrystalline (uc-Si:H) cell using carbon conductive paste (Fig. 1a). There is not expected to observe any unusual phenomena on the junction between the amorphous and microcrystalline material, as they are spatially separated. The two-terminal tandem cells (Fig. 1b) were composed from the same amorphous and microcrystalline silicon layers, however the microcrystalline sub-cell was deposited directly on the amorphous one. The two respective cells are connected via tunnel junction extended over whole are of the structure. In such case some unexpected phenomena are observed.

2.2 Measurement procedure

The LED diode illuminating setup was built for the illumination of the cells. The LED colours (wavelength of the emitted light) were selected according to the EQE (external quantum efficiency) of the cells (Fig 2a) in order to maximize the selectivity. The infrared LEDs (wavelength 850nm) were used for the microcrystalline sub-cell and the blue LEDs (wavelength 470nm) were used for amorphous silicon sub-cell. The sets of LEDs were fed by computer-controlled voltage source and the total voltage on the outer terminal was measured simultaneously. For each sample the measurements were taken for blue LEDs and for infrared LEDs separately (= "selective" illumination). The voltages were summed and compared to third measurement in which both LEDs were operated together (= "full" illumination).

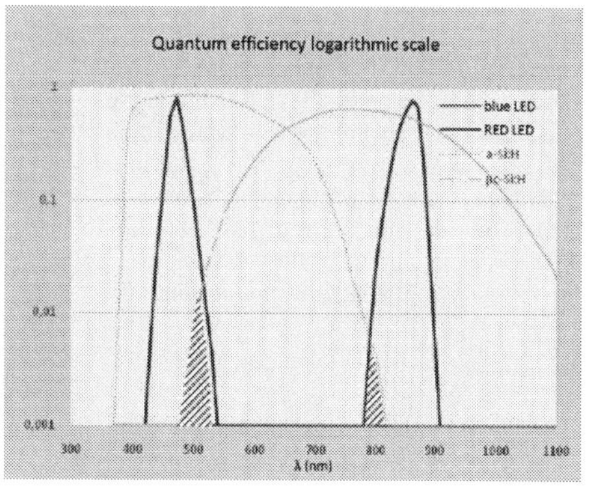

Figure 2a: Quantum efficiency and LED spectra

The measurement of the three-terminal cell (Fig. 2b) confirms that: i) the LEDs are selective enough, and ii) the total voltage taking by full illumination corresponds very well to the sum of the individual voltages taken by selective illumination.

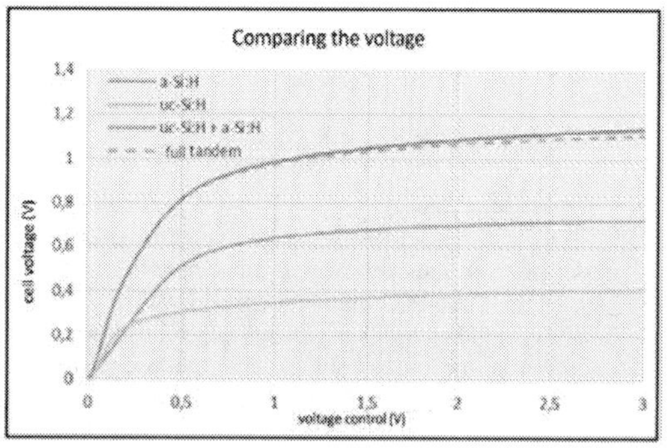

Figure.2b: Measurement of the three-terminal cell

The measurements on the two-terminal cell reveal two different effects that can be observed on selected samples from a batch of identically prepared samples. When measuring with full illumination, almost the same regular voltage evolution was measured on all the tested samples, corresponding to expected behaviour. However, when measuring with a selective illumination separately, making the sum and comparing to the full illumination, the results strongly differ. Both cases are observed: the summed voltage is either much higher than the voltage at full illumination (Fig. 3a) or much lower (Fig. 3b). The fact, that in the case of the three-terminal cell those differences are not present, indicates that the reason of the abnormalities is the direct local junction between the two sub-cells. Such local junction can work efficiently in the vertical direction, however there is no means of lateral conductivity as in the three-terminal cell. Therefore the local shunt of one sub-cell cannot have significant effect unless there is another shunt of the second sub-cell in the vicinity. Due to the fact that the only effective lateral electrodes are at the top and bottom surface, the only effective shunts can be those shunting the whole tandem. This significantly changes the electrical representation of the tandem cell and consequent behaviour.

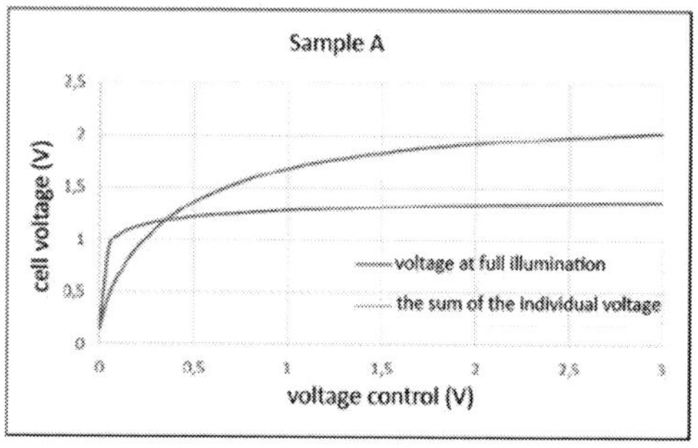

Figure.3a: Measurement of the two-terminal cell

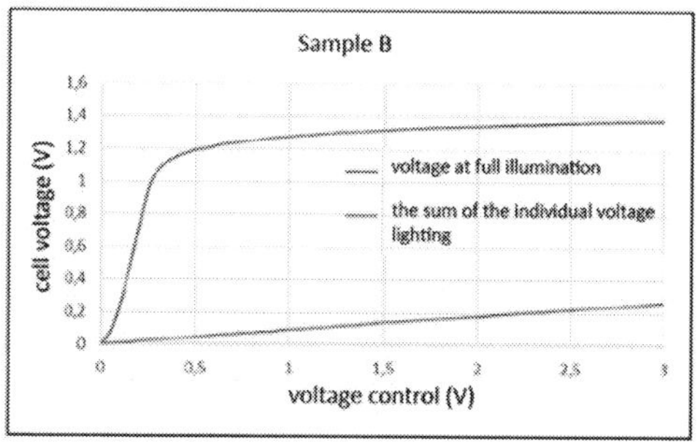

Figure.3b: Measurement of the two-terminal cell

3 CONCLUSION

The quest for method of I-V separation in multi-junction solar cells has led us to revealing interesting effects in illumination dependence of the open-circuit voltage. The measurements with selective illumination showed that the total voltage is not a simple sum of individual voltages excited by selective illumination. The mutual influence of distribution of local shunts is considered. Redefinition of the equivalent circuit of a tandem cell will be necessary to describe this phenomena.

4 ACKNOWLEDGMENT

This work has been supported by CTU Student Grant Competion SGS15/074/OHK3/1T/13, the project LD14011 (HINT COST Action MP1202), the Czech Science Foundation project 16-10429J and the CAS project KONNECT-007.

4.3 References

[1] J. Holovský, M. Bonnet-Eymard, M. Boccard, M. Despeisse, C. Ballif, *Variable light biasing method to measure component I–V characteristics of multi-junction solar cells*, Sol. Energy Mater. Sol. Cells. 103 (2012) 128–133. doi:10.1016/j.solmat.2012.04.014.

[2] J. Holovsky, M. Bonnet-Eymard, G. Bugnon, P. Cuony, M. Despeisse, C. Ballif, *Measurement of the Open-Circuit Voltage of Individual Subcells in a Dual-Junction Solar Cell*, IEEE J. Photovolt. 2 (2012) 164–168. doi:10.1109/JPHOTOV.2011.2178232.

[3] F. Ventosinos, B. Fakes, and E. V. Johnson, *Challenges in advanced characterization techniques for high Voc thin-film tandem solar cells*, 26th International Conference on Amorphous and Nanocrystalline Semiconductors (ICANS 26), Aachen, Germany, 13-18 September 2015, (to be published in PSSA)

[4] Despeisse, M., Boccard, M., Bugnon, G., Cuony, P., Söderström, T., Parascandolo, G., et al., *Low-Conductivity Doped Layers for Improved Performance of Thin Film Silicon Solar Cells on Highly Textured Substrates*, in: Proc. 25th EUPVSEC, Valencia, Spain, 2010. doi:10.4229/25thEUPVSEC2010-3CO.13.4.

CHANGES IN TEMPERATURE-COEFFICIENT OF THE DIODE MODEL CAUSED BY LIGHT-INDUCED DEGRADATION OF a-Si/μc-Si SOLAR CELLS

Johannes A. Weicht, Frank U. Hamelmann, Grit Behrens
University of Applied Sciences Bielefeld
Artilleriestraße 9a, 32427 Minden, Germany

ABSTRACT: In our work we observe the temperature coefficients before and after light-induced degradation of amorphous/micro-crystalline (a-Si/μc-Si) tandem silicon-based solar cells. We show that during light-induced degradation the temperature parameters of the serial and parallel resistance, photo current and the saturation current in the diode model change: the temperature has a stronger effect after the light-induced degradation, the temperature-coefficient of silicon-based thin film solar cells varies during the light-induced degradation.
Keywords: a-Si/μ-Si, Simulation, Degradation

1 INTRODUCTION

The Staebler-Wronski effect describes the degradation of amorphous silicon (a-Si) solar thin film cells. The a-Si solar cell degradation depends strongly on operating temperature [1]. Light-induced degradation (LID), based on the work of Staebler and Wronski, is also observed in a-Si/μc-Si solar cells, mainly because of their amorphous layer [2].

The temperature behavior of solar cells is important for monitoring and planning of PV-systems. This voltage is determined at standard test conditions (STC), with an irradiation of 1000W/m², a cell temperature of 25°C and an air mass of AM1.5 [3]. This is not practicable to characterize the performance of photovoltaic devices under weak irradiation [4], [5], [6].

The performance of a-Si/μc-Si thin film solar cells in different climatic regions depends on the normal operating temperature condition and actual module temperature [2]. During the light-induced degradation, the operating temperature determines the amount of created dangling bonds. The light-induced degradation in a-Si/μc-Si stabilizes after the first 400h at a degradation temperature of 50°C [7]. This degradation is visible at the STC. In our work we measure the temperature-coefficient of the saturation current Is, the photo current I_{ph}, the serial resistance R_s and the parallel resistance R_p before and after the LID at different temperatures.

2 METHODS

The initial temperature-coefficient and the I-V-curves for different illuminations of a-Si/μc-Si mini-modules (active area 38cm², 180nm a-Si layer, 40nm TCO interreflection layer, 850nm μc-Si layer) are measured. Afterwards they are degraded under controlled conditions with constant temperature and constant irradiation. At fixed times the modules are measured at standard test conditions for controlling the status of LID. The key parameters (I_s, I_{ph}, R_s, R_p, the diode factor n and their temperature coefficient) of the IV-curves are extracted.

After stabilizing the solar cells, the temperature-coefficients of the degraded solar cells were measured: The sun simulator has an irradiation of 1000W/m² and the cells are measured at 25°C. We varied the cell temperature stepwise from 20°C up to 50°C. In this approximation we observe the tandem structure as separated cells: Although the observed solar modules have got 11 cell areas connected in series, each area consists of two separated solar cells (one a-Si and one μc-Si cell). Due to the different band gaps, we have to approximate an "effective energy gap".

The diode model for a-Si/μc-Si was described in [7] and the temperature dependent diode model in [8]:

The current can be calculated by the Newton method by the following equation:

$$I = I_{ph} - I_{rec} - I_s\left(\exp\left(e\frac{V-IR_s}{nk_bT}\right) - 1\right) - \frac{V-IR_s}{R_p} \quad (1)$$

with V the applied voltage T the temperature, k_b the Boltzman constant and e the elementary charge. The recombination current I_{rec} for thin film silicon solar cells is described at [9].

The temperature coefficient of the photo current T_{Iph} can be used to calculate the photo current at different temperatures:

$$I_{ph}(T) = I_{ph}(298.15K)\left(1 + T_{Iph}(T - 298.15K)\right) \quad (2)$$

The temperature coefficient of the saturation current T_{Is} can be used to calculate the saturation current at different temperatures $I_s(T)$:

$$I_s(T) = I_s(298.15K)\left(\frac{T}{298.15K}\right)^{T_{Is}/N} \quad (3)$$

The temperature coefficient of the serial resistance T_{Rs} can be used to calculate $R_s(T)$:

$$R_s(T) = R_s(298.15K)\left(\frac{T}{298.15K}\right)^{T_{Rs}} \quad (4)$$

In the same way the temperature coefficient of the parallel resistance T_{Rp} can be used to calculate:

$$R_p(T) = R_p(298.15K)\left(\frac{T}{298.15K}\right)^{T_{Rp}} \quad (5)$$

Analog to the I_{ph} we calculated the built-in voltage V_{bi} and the Lifetime of the charge carrier Lt

$$V_{bi}(T) = V_{bi}(298.15K)\left(1 + T_{Vbi}(T - 298.15K)\right)$$
(6)

$$Lt(T) = Lt(298.15K)\left(\frac{T}{298.15K}\right)^{T_{Lt}} \quad (7)$$

The error between the simulated I_{sim} and the measured I_{meas} current is defined by the amount of measured points #P by the metric:

$$Err = \frac{(I_{meas} - I_{sim})^2}{\#P} \quad (6)$$

3 RESULTS

As shown in figure 1 the difference between the simulated and the measured data is within the tolerance. This is valid for the simulation of the initial and for the degraded a-Si/μc-Si solar module.

In table 1 the different parameters are shown: As expected I_s, n and R_s are increasing and I_{ph} and R_p are decreasing. The temperature parameters T_{Iph} and T_{Rp} are decreasing too. The other parameters T_{Is} and T_{Rs} are increasing. Also

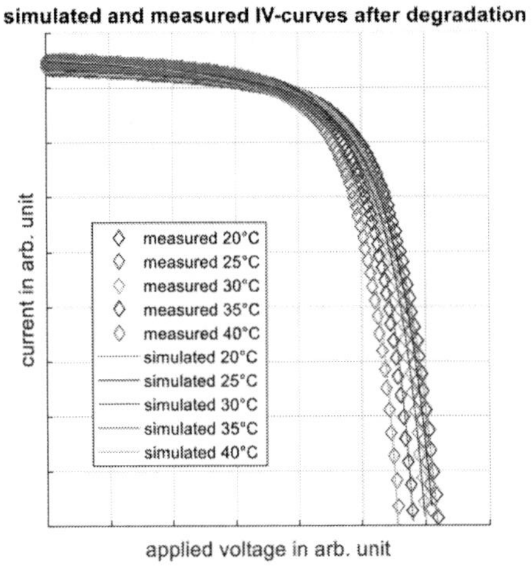

Figure 1: measured and simulated cell current for a-Si/μc-Si solar module at different temperatures after degradation

it should be noticed, that the error is increasing, which can be eliminated by further approximation.

4 CONCLUSION

The light-induced degradation causes changes in the temperature coefficients for several parameters of a-Si/μc-Si solar cell: T_{Iph} and T_{Rp} are decreasing, T_{Is} and T_{Rs} are increasing. An increasing energy gap can also be observed in [10].

This different temperature coefficient will help to improve the simulation and allow a systematic modeling for a-Si/μc-Si at different climatic conditions and their light-induced degradation.

	Changing of the temperature coefficients during the LID normed to their initial values
T_{Iph}	1.04
T_{Is}	2.06
T_{RS}	1.32
T_{RP}	1.47
T_{Vbi}	0.53
T_{Lt}	1.69

Table 1: Changing of the temperature coefficients during LID

ACKNOWLEDGMENTS

We acknowledge support from BMBF through Contact No. 03FH008I2 and the support from Ministry for Innovation, Science and Research of North Rhine-Westphalia

REFERENCES

[1] D. Staebler, C. R. Wronski, "Reversible conductivity changes in discharge produced amorphous Si," Applied Physics Letters, vol. 31, no. 4, pp. 292 - 294, 1977.

[2] J. A. Weicht, F.U. Hamelmann, G. Behrens „Light-induced degradation of a-Si/μ-Si thin film solar cells under different climate conditions and its influence on performance", Proceedings 28th EU PVSEC, 2013

[3] Photovoltaic devices - Part 3: "Measurement principles for terrestrial photovoltaic (PV) solar devices with reference spectral irradiance data", IEC 60904-3:2008; 2009

[4] C. Gemmer, Analytische und numerische Untersuchungen von Solarzellen unter wechselnden Beleuchtungsbedingungen, PhD Thesis, Der Andere Verlag, 2003

[5] J. F. Randall, J. Jacot, "Is AM1.5 applicable in practice? Modelling eight photovoltaic materials with respect to light intensity and two spectra", Renewable Energy, vol. 28 no. 12, pp.1851-1864, 2003

[6] Muller, M., et al. "Characterization of indoor photovoltaic devices and light." Proceedings 34th IEEE Photovoltaic Specialists Conference, 2009.

[7] J. A. Weicht, F. U. Hamelmann, G. Behrens, "Parameter variation of the one-diode model of a-Si and a-Si/μc-Si solar cells for modeling light-induced degradation.", Journal of Physics: Conference Series, IOP Publishing, 559 (2014), 12017

[8] Gow, J.A. and C.D. Manning. "Development of a Photovoltaic Array Model for Use in Power-Electronics Simulation Studies." IEEE Proceedings of Electric Power Applications, vol. 146, no. 2, pp. 193–200, 1999

[9] Merten, Jens, et al. "Improved equivalent circuit and analytical model for amorphous silicon solar cells and modules." Electron Devices, IEEE Transactions vol. 45.no. 2, pp. 423-429, 1998

[10] Kołodziej, A. "Staebler-Wronski effect in amorphous silicon and its alloys." Opto-electronics review, vol. 12, no. 1, pp. 21-32., 2004

ELECTRON BEAM CRYSTALLIZATION OF AMORPHOUS SILICON THIN FILMS IN THE SOLID PHASE REGIME AND ASSISTED SIMULATIONS BY FINITE ELEMENT METHOD

Stefan Saager
Fraunhofer Institute for Organic Electronics, Electron Beam and Plasma Technology (FEP)
Winterbergstr. 28, 01277 Dresden, Germany, phone: +49 351 2586 316, e-mail: stefan.saager@fep.fraunhofer.de

ABSTRACT: The impact of an electron beam (EB) treatment on a-Si coated Si-substrates was investigated by SEM and EBSD analysis. EB processing of a-Si layers by line scanning led to the formation of a narrow grained polycrystalline structure, which is characteristically for an explosive solid phase crystallization. By increasing the EB energy density an undesired delamination of the Si thin film was observed. However, an epitaxial layer regrowth with the similar crystallographic orientation of the (001) Si substrate could be reached by applying a lower EB energy density and by a sequential raise to a tenth of the EB power in comparison to line scanning experiments. By finite element simulations, the reason for layer delamination could be revealed by calculating the temperature and stress field during the EB treatment of the thin film–substrate–system: Due to thermal expansion during the EB line scans the elastic strain energy density accumulates and reaches a critical level. If the value exceeds the a-Si/c-Si interface energy, layer delamination during the EB process is caused.
Keywords: electron beam crystallization, silicon thin films, explosive solid phase crystallization, crucible-free electron beam evaporation

1 PURPOSE

A promising method for low cost production of efficient silicon thin film solar cells is the electron beam physical vapor deposition (EB-PVD) of high purity Si absorber layers with high deposition rates up to 300 nm/s [1, 2] followed by a crystallization step [3]. For liquid phase crystallization the deposited Si layer usually is encapsulated by a SiN_x/SiO_x layer stack for avoiding agglomeration, followed by heating-up in a narrow zone to silicon's melting temperature (1685 K) by laser [4], by EB [5] or by lamps [6]. With this procedure, solar cell efficiencies up to 11.5 % could be reached up to now [7]. But crystallization from liquid phase offers some drawbacks:

- There is a potential risk for taking up contaminations of Si melt from substrate and encapsulation
- SiN_x/SiO_x encapsulation has to be removed after crystallization for final solar cell fabrication
- Formation of elongated grains in lateral pull direction results in recombination defects at grain boundaries
- The maximum crystallization speed in lateral direction is limed to some cm/s [3]
- Due to the high temperature and small process speed the substrate has to withstand large thermal loads [8]

An alternative approach to encapsulation could be kerfless wafering [9] (Figure 1): First cleaning single crystalline substrate (c-Si) *in situ* [10], second depositing an amorphous Si (a-Si) layer on it and third crystallizing the a-Si layer in the solid phase regime by EB. Due to the ability to adjust electron penetration range R_e and corresponding the power-depth relation f_A by tuning the EB energy ($E_{e^-} = eU_B$) an epitaxial regrowth could be reached, starting from the cleaned single crystalline substrate interface (Figure 2). In a fourth step the crystallized layer can be split off by a subsequent EB process step with re-adjusted EB parameters [11]. This paper presents first EB crystallization experiments in the solid phase regime and justifies observed results by finite element (FEM) simulations.

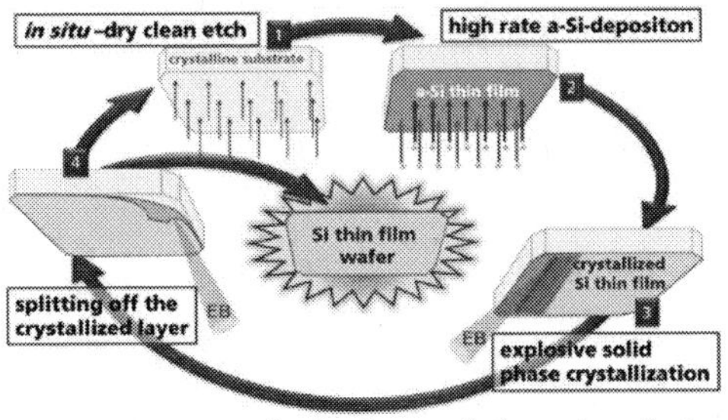

Figure 1: Concept for kerfless wafering using electron beam (EB) technology in a cycle process. The crystalline substrate, e.g. silicon or sapphire wafer, will be reused after splitting of crystallized silicon thin film.

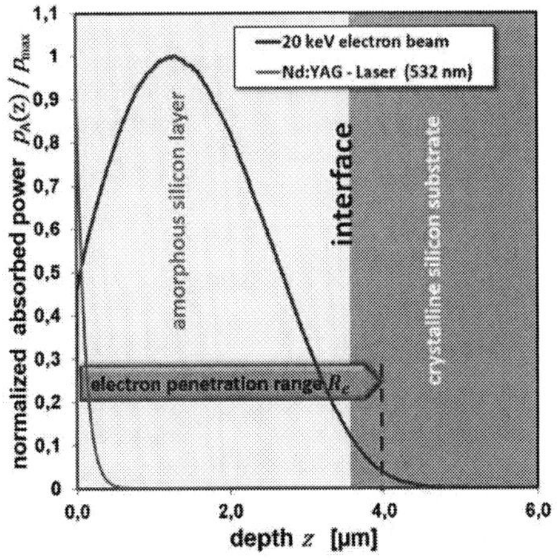

Figure 2: The depth relation of the absorbed power in a thin-film system can be easily tuned for EB (red) by acceleration voltage in contrast to laser (blue).

2 APPROACH

2.1 Experimental

For coating experiments boron doped (001) Si substrates were cleaned by RCA-method *ex situ* and by EB dry clean etch *in situ* [10]. The deposition of $d = 1$ µm and $d = 3.6$ µm thick pure a-Si layers, respectively, were carried out at a substrate temperature of $\approx 200\,°C$ by crucible-free EB-PVD, described in detail earlier [12]. Crystallization was accomplished by EB processing after coating and cooling down the substrates to room temperature (without vacuum interruption for experiments with EB line scans). The focused EB was scanned over the coated substrate at an electron energy of $E_{e^-} = 20\,\text{keV}$. The GAUSSIAN shape of the beam current density profile $j_B(x,y)$ according to Eq. 1, controlled by FARADAY-cup measurements, reach a diameter of $d_F \approx 600\,\mu\text{m}$.

$$j_B(x,y,t) \quad = \quad \frac{I_B}{\pi r_F^2} e^{\frac{x^2+(y-v_y t)^2}{r_F^2}} \qquad \text{Eq. 1}$$

$$p_A(x,y,z,t) = \quad \eta_{th} \cdot U_B \cdot j_B(x,y) \cdot \frac{f_A(z)}{R_e} \qquad \text{Eq. 2}$$

I_B ... electron beam current
v_y ... scanning speed of EB in y-direction
U_B... electron beam acceleration voltage
r_F ... electron beam radius $(1/2 \cdot d_F)$
η_{th}... portion of the EB power transferred into heat
$f_A(z)$... absorbed power-depth relation
(proportional to red normalized curve in Figure 2)

2.2 Theoretical consideration

With these data, the transient temperature field and thermal induced stress during EB treatment were calculated by FEM simulations using *COMSOL Multiphysics* software. Temperature field was calculated by numerically solving the three-dimensional heat equation with temperature dependent material properties. As an inhomogeneous term in the heat equation, a spatial distributed heat generation density $p_A(x,y,z)$ was assumed, according to the absorbed EB power density (Eq. 2). Due to thermal expansion, the stress tensor $\hat{\sigma}$ was deducted by computing the strain tensor $\hat{\varepsilon}^{\sigma} \propto \hat{\alpha} \cdot \Delta T(x,y,z)$ from temperature field. Initial layer tensile stress of $\sigma_{ini} \approx +200\,\text{MPa}$, a common value for the applied deposition conditions [13], was also taken into account.

3 RESULTS AND DISCUSSION

Figure 3-a shows an overview optical micrograph of an a-Si coated substrate (a-Si thickness: 3.6 µm), which was EB processed by scanning seven parallel lines with different scanning speed $v_y = 10 \ldots 58\,\text{m/s}$. As shown in the magnified micrograph Figure 3-b, line scanning with small v_y, which corresponds to large energy density $e_A = \int p_A dt$, leads to delamination of the layer and the formation of detached flakes with a spoke-like microstructure. At $v_y = 26\,\text{m/s}$ delamination occurs only at certain areas and disappears completely for $v_y \geq 50\,\text{m/s}$ and $e_A \leq 7\,\text{kJ/cm}^3$, respectively. The scanning electron microscope (SEM) micrograph in Figure 3-c with overlaid electron backscatter diffraction (EBSD) mappings illustrates, that delamination and crystallization is correlated.

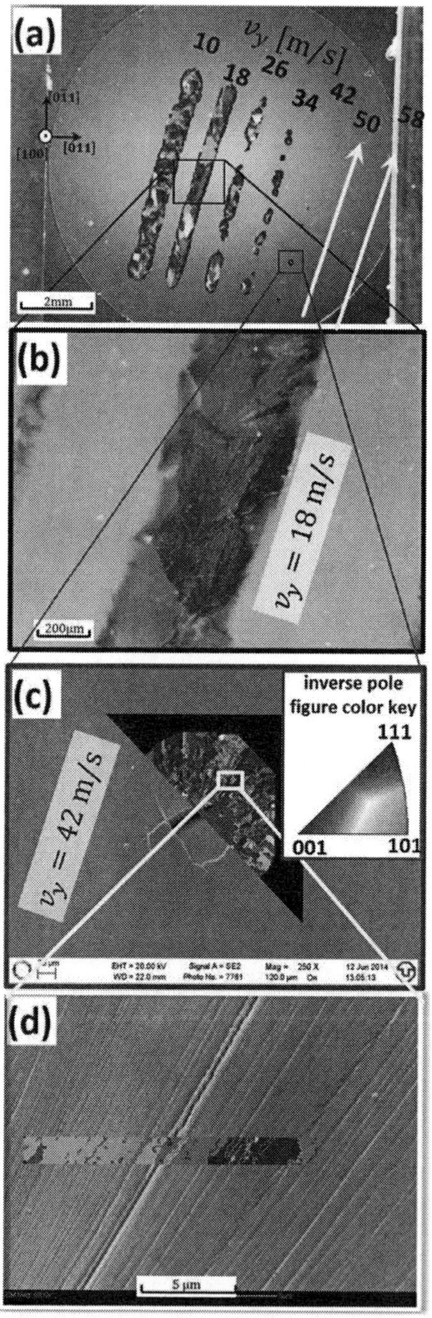

Figure 3: Optical (a, b) and SEM (c,d) micrographs of 3.6 µm thick a-Si layer processed by EB line scans. The overlaid EBSD mappings represent crystal orientation (colored) and amorphousness (black), respectively.

Layer regions, which are still attached to the substrate, are still amorphous as the black color indicates. In the crystallized region, a fine grained structure with elongated crystallites is build up, showing random orientation (Figure 3-d). This structure without agglomeration is typically for explosive solid phase crystallization.

In contrast to line scanning experiments with local power input (Figure 3-a...d), additional crystallization tests with extended scanning pattern were carried out on a 1 μm thick a-Si. The EB was scanned frequently for 180 s in the shape of a spiral line pattern with a line pitch $\ll d_F$. The beam power was raised sequentially to maximum, which was 10x lower than for line scanning experiments. This scanning procedure led to slow heating up of almost the whole sample up to $T_{max} = 1500$ K, which was measured by pyrometer. Figure 4 shows an overview SEM micrograph of the sample after EB processing at extended scanning area.

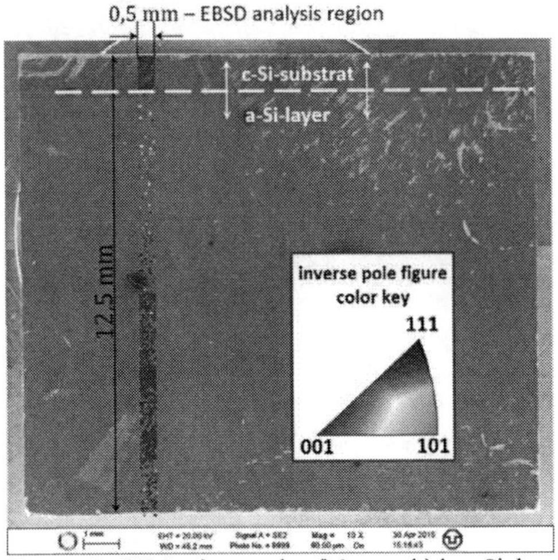

Figure 4: SEM micrograph of 1 μm thick a-Si layer processed by EB scanning pattern over an extended area. The overlaid EBSD mapping represents crystal orientation (colored) and amorphousness (black), respectively.

Because of deposition shadowing, the sample's upper part has an uncoated region with naked (001) wafer substrate, which is also illustrated in Figure 4 by the red color of the overlaid EBSD mapping. Due to the decentering of the spiral pattern, the maximum temperature was reached in the lower part of the sample. In mentioned region, the processed layer is still attached to the substrate and epitaxial regrowth to (001) crystal orientation occurred, which is illustrated in Figure 4 by the red color of the overlaid EBSD mapping.

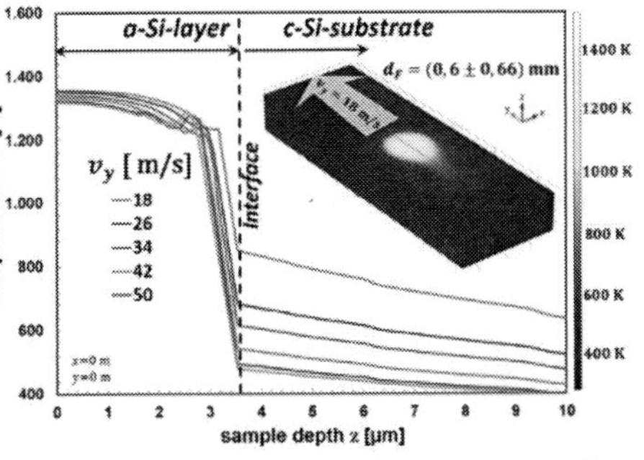

Figure 5: FEM-calculated temperature-depth profile in a-Si/c-Si layer system during EB spot passing for various line scanning speeds. Inset: computed temperature field at $v_y = 18$ m/s.

To understand the mechanism leading to layer delamination during EB line processing, the knowledge of temperature and thermal induced stress field are significant. Figure 5 shows the FEM-calculated temperature-depth profile in the a-Si/c-Si thin film system for various line scanning speeds v_y at the moment of EB spot passing the sample's midpoint ($y = 0$ m).

For mentioned moment the inset in Figure 5 illustrates the three-dimensionally calculated temperature field for $v_y = 18$ m/s by color key. Accordingly, the maximum temperature is below the melting point of a-Si ($T_{a-l} = 1420$ K [14]). Therefore, the initiation of explosive solid phase crystallization would not be expected in contrast to experimental results. The inset in Figure 6 illustrates the computed σ_{yy}-component stress field in color key of sample's half-section. Furthermore sample deformation is displayed 100x expanded. The corresponding σ_{yy}- stress component vs. sample depth in Figure 6 shows only a little variation for the maximum stress value in the a-Si layer for investigated v_y.

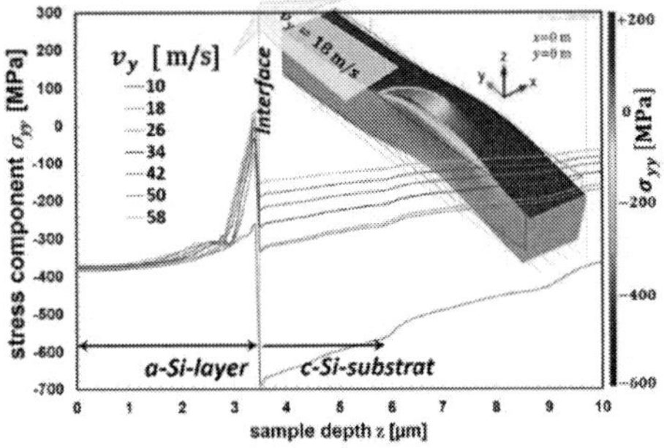

Figure 6: Correlated σ_{yy}-stress component vs. sample depth for various line scanning speeds. Inset: σ_{yy}-stress field with resulting deformation in exaggerated schema.

Nevertheless, in order to evaluate the effects of stress accumulation on layer delamination one has to considere the elastic strain energy density w_σ (Eq. 3).

$$w_\sigma = \frac{1}{2} \cdot \int_0^d \hat{\sigma} : \hat{\epsilon}^\sigma \, dz \qquad \text{Eq. 3}$$

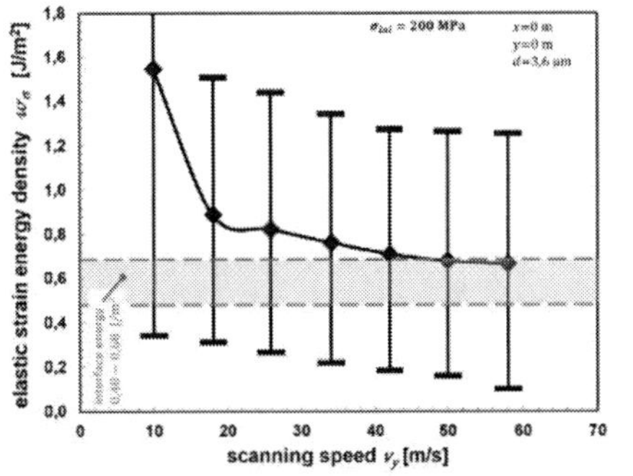

Figure 7: Calculated elastic strain energy density w_σ vs. applied scanning speed. The value range of the a-Si/c-Si interface energy [15, 16] are highlighted for reference.

Layer delamination will be expected if the stored mechanical energy exceeds the interface energy. Figure 7 shows the elastic strain energy density w_σ vs. applied scanning speed. For $v_y < 50\,\mathrm{m/s}$, calculated values of $w_\sigma > 0.68\,\mathrm{J/m^2}$ exceed the a-Si/c-Si interface energy [15, 16].

That means during the experimental EB line processing the a-Si layers spall off due to high elastic energy. Furthermore, delamination will be encouraged by potential oxide film of the non-ideal clean interface. Layer crystallization took place after spalling off because of lower thermal coupling to the c-Si substrate and stronger layer heating. With EB processing by extended scanning pattern at lower EB power, in the EB interaction region the layer grows originated at the substrate interface to a nearly single crystalline structure without layer delamination. Due to the thinner layer thickness, crystallization process may be influenced positively, because:

(i) Corresponding to Figure 2 the maximum of EB power absorption coincide with the substrate interface in this case

(ii) Corresponding to the integral limits in Eq. 3 the accumulation of the strain energy density is more reduced

Epitaxial regrowth occurred in the solid phase regime assuredly, because agglomeration effects were not observed and the measured maximum temperature was 180 K below the melting point of c-Si.

4 CONCLUSIONS AND OUTLOOK

The shown introductory investigations are a part of a research project for kerfless wafering. Associating crystallization tests demonstrate that an epitaxial regrowth of 1 μm EB-PVD deposited a-Si layers in the solid phase regime was reached by EB processing with extended scanning pattern. Enhancing throughput's ability by increasing the a-Si thickness and intensifying EB power density leads to undesired layer delamination. To determine process limits, e.g. maximum layer thickness and minimum EB exposure time, further experiments with raised substrate temperatures by preheating, various a-Si film thickness and adapted electron penetration ranges and EB power have to be carried out systematically. By the investigations fundamentals for an innovative fabrication of Si thin film has been established. With additional investigations the encouraging results could be used for an industrial implementation.

5 REFERENCES

[1] J.-P. Heinß *et al.*, *Proceedings of EU PVSEC 2015*, Hamburg, 3BO.5.4, 1022-1025, **2015**.

[2] S. Saager, PhD thesis, Dresden, **2015**.

[3] C. Becker *et al.*,*Solar Energy Materials and Solar Cells* , vol. 119, no. 0, pp. 112-123, **2013**.

[4] I. Höger *et al.*, *Proceedings of EU PVSEC 2014*, Amsterdam, 3CO.6.2, 1498-1501, **2014**.

[5] D. Amkreutz *et al.*, *Solar Energy Materials and Solar Cells* , vol. 123, no. 0, pp. 13-16, **2014**.

[6] T. Kieliba, PhD thesis, Konstanz, 2006.

[7] J. Haschke *et al.*, *Presentation at EU PVSEC 2014*, Amsterdam, 3CO.6.1, **2014**.

[8] T. Pliewischkies *et al.*, *physica status solidi (a)*, vol. 212, no. 2, pp. 317-322, **2015**.

[9] Chr. Metzner *et al.*,, *Novel Method for the Production of Thin Silicon Wafers*, Fraunhofer FEP Annual Report 2014/15, pp.16-19, **2015**.

[10] E. Bodenstein & D. Temmler, *Proceedings. of EUPVSEC*, Hamburg, 3DV.2.8, 1287-1289, 2015.

[11] D. Temmler *et al.*, Patent WO2014/117890A1, **2014**.

[12] S. Saager *et al.*, *Proceedings of EU PVSEC 2015*, Hamburg, 3BO.6.3, 1036-1041 ,**2015**.

[13] J. Kim *et al.*, *Thin Solid Films*, vol. 518, no. 17, pp. 4908-4910, **2010**.

[14] E. P. Donovan *et al.*, *Applied Physics Letters*, vol. 42, no. 8, pp. 698-700, **1983**.

[15] K. Tu, *Applied Physics A*, vol. 53, no. 1, pp. 32-34, **1991**.

[16] Z. Jian *et al.*, *Acta Materialia*, vol. 54, no. 12, pp. 3227-3232, **2006**.

6 ACKNOWLEDGEMENTS

The author gratefully acknowledges
Prof. Dr. Chr. Metzner, Prof. Dr. J. Weber, Dr. J.-P. Heinß and Dr. D. Temmler for supervising and supporting as well as Prof. Dr. E. Hieckmann for supporting EBSD-analysis. The project was funded by the European Union and the Free State of Saxony (funding reference 100102018).

Europa fördert Sachsen.
EFRE
Europäischer Fonds für regionale Entwicklung

INDUSTRIAL SCALE OPTIMIZATION OF SIOX BOTTOM N-LAYER IN TANDEM SOLAR CELL

Condorelli G.(1); Battaglia A. (1); Canino A. (1); Rapisarda D. (1); Foti M. (2) and Gerardi C.(3)

1. 3SUN, Contrada Blocco Torrazze Zona Industriale 95121, Catania, Italy
2. STMicroelectronics, Contrada Primosole 50, 95121, Catania, Italy
3. Enel Green Power, Contrada Blocco Torrazze Zona Industriale 95121, Catania, Italy

ABSTRACT: SiOx layer n-doped has been developed for industrial application with the aim to increase power conversion efficiency in a-Si/μc-Si thin film tandem solar cells. We firstly optimized n-SiOx layer on industrial scale PECVD reactors depositing on 1.4m2 area glass substrates and then integrate the material into the tandem based modules comparing the performance versus standard μc-Si n layer. CO2/SiH4 gas flow ratio and process pressure have been optimized to achieve n-SiOx films with a wide range of optical and electrical properties. The application of n-SiOx:H improves the Jsc of cells resulting in overall 2% efficiency improvement.

Keywords: a-Si/μ-Si, PECVD, Thin Film Solar Cell

I. INTRODUCTION

Thin-film Si solar cells have received attention due to their potential for large areas and low-cost manufacturing. Despite the many advantages of thin-film Si solar cells their low efficiencies still present a challenge that must be overcome.

For high-efficiency thin film tandem solar cells, maximized light trapping is mandatory to obtain high current density, as the absorption coefficient of μc-Si:H is not sufficient in the long-wavelength near-infrared region[1].

Improved light management is generally obtained using textured Transparent Conductive Oxide (TCO) on the sunny side of the cell and metallic reflector on the backside of the cell.

The texturing of the TCO allows to divert a portion of the light out of the normal direction allowing light longer path in the absorbing material. Generally the ratio between the diffused transmittance and the total transmittance measured at wavelength of 600 nm (also known as haze) is in the range of 12-18% for commercial TCO. For a solar cell with two junctions Shah et al. [8] have theoretically demonstrated that an almost ideal combination is given by the tandem a-Si:H/c-Si:H where Eg top=1.75eV and Eg bottom=1.1 eV. 3Sun tandem cell well match this consideration. The top cell made with amorphous silicon convert mostly the UV and visible part of the solar spectrum (300 nm to 700 nm) while the bottom cell based of microcrystalline silicon with a bandgap of 1.1 eV allow to convert infrared light.

N-layer of the bottom cell in tandem structure has the fundamental role to collect the electrons generated into the intrinsic absorber layer and deliver them to the metal back contact of the cell generally made with silver; it should have enough conductivity to reduce resistive losses and low absorbance in the infrared wavelength region to maximize light absorption in the i layer.

In this paper, we have optimized the properties of n-SiOx nano-crystalline thin film on large area commercial photovoltaic modules by plasma enhanced chemical vapor deposition through tuning deposition parameters and compared the performance versus the standard μc-Si n layer.

The μc-SiOx material is a phase mixture of μc-Si:H and amorphous silicon oxide (a-SiOx:H), where phosphorus doping mainly affects the μc-Si:H phase [2,3]. The n-type μc-Si:H phase ensures sufficient electrical conductivity in the direction of the current flow in the cell device, and the a-SiOx:H phase with different oxygen contents supplies the lower refractive index and optical transparency [2].

The SiOx layer optimization has been firstly optimized on small area commercial corning glass and <100> silicon substrate to ease the characterization of the materials. Then most promising layers in terms of optical and electrical properties have been applied in a large area (1.4m2) tandem a-Si:H/uc-Si:H photovoltaic module.

II. EXPERIMENTAL DETAIL

Doped and undoped silicon-based films have been prepared using an industrial 13.56 MHz multi-chamber plasma-enhanced chemical vapor deposition (PECVD) system.

N-type μc-Si thin films have been deposited from a gas mixture of silane (SiH_4), hydrogen (H_2) and phosphine (PH_3). Carbon dioxide (CO_2) has been added to the gas mixture as oxidant precursor for the μc-SiO_x layer preparation.

After initial investigations we found process conditions to have SiO_x film with good quality and crystallinity ratio higher than 50% are:

- H_2/SiH_4 ratio → 300 – 400
- Substrate temperature → 120° – 180° C
- RF power density → 100 – 200 mW/cm2

To optimize the layer properties in terms of conductivity and refractive index Single Variable Trend (SVT) has been used on the following parameters.

- CO_2/SiH_4 ratio
- Process pressure

Conductivity measurements have been performed using standard 4-probes method. The crystalline phase percentage within the films (not shown here) has been evaluated by using micro-Raman spectroscopy. Spectroscopic ellipsometry (SE) measurements have been carried out over the spectral range of 370–1688 nm. The fit has been done using optical functions from Tauc–Lorentz (TL) equation to acquire the refractive index of the n-SiO_x:H films and to asses samples thickness[4]. Goodness of the fit has been verified by measuring samples thickness using a mechanical profilometer. The tandem device spectral response has been evaluated by measuring the external quantum efficiency (EQE) with a Bunko-keiki tool.

After the material optimization the most promising layers have been tested in the tandem cell replacing the bottom cell n layer as shown in the following picture.

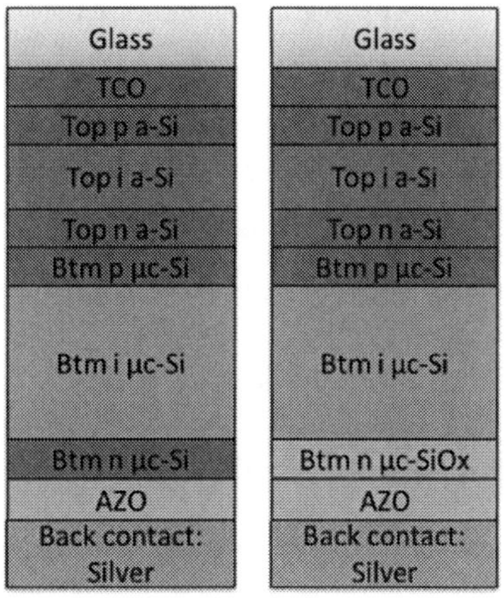

Figure 1: Tandem cell structure standard (left) and with SiO_x layer (right).

The impact of SiO_x layer implementation on PV module electrical parameters: Power, Jsc, Voc and FF has been studied depositing and characterizing large area modules (1.4x1.0) m². The solar modules have been characterized using a Wacom AAA solar simulator at standard test conditions, AM1.5G at 25 °C.

III. Results

For the material optimization SVT method has been applied on the two main parameter: CO_2/SiH_4 ratio, which regulates the chemistry of the reaction, and process pressure, a key factor to control the plasma strength and to regulate also the balance between ion bombardment and deposition rate. For these experiments silane concentration, power density and PH_3/SiH_4 ratio were fixed.

Fig.2 (a and b) shows the refractive index measured at 800 nm and conductivity trends as a function of the pressure nad CO_2/SiH_4 ratio .

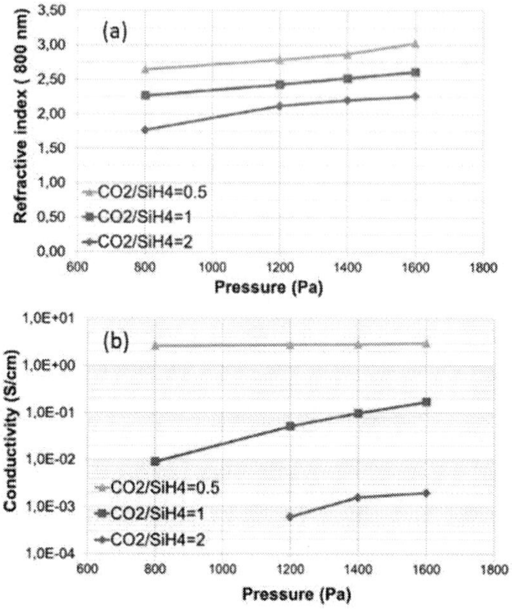

Figure 2: (a) Trend of refractive index at 800nm and (b) Conductivity versus pressure at different CO_2/SiH_4 ratio.

The introduction of CO_2 leads to a significant reduction of the refractive index compared with the standard µc-Si n layer of 3SUN tandem cell, which is 3.5. Such results perfectly in agreement with literature [7] states that the higher CO_2/SiH_4 ratio, the lower is the refractive index of the deposited layer as result of the increased oxidation of the doped layer. In accordance with the increased oxidation, the conductivity of the layer decreases as the CO_2/SiH_4 ratio increases.

Pressure acts as the key parameter to regulate the balance between oxidation chemical process and the ion induced re-arrangements of the oxidized material towards a mixed phase material, which allows to lower the refractive index maintaining reasonable vertical conduction in the material.

For the up scaling of the process, evaluation of film uniformity along the whole surface of deposition is fundamental as this can have huge impact on large area module performances. In Fig.3 we report the resulted SiO_x single layer uniformity map obtained after the process parameter optimization.

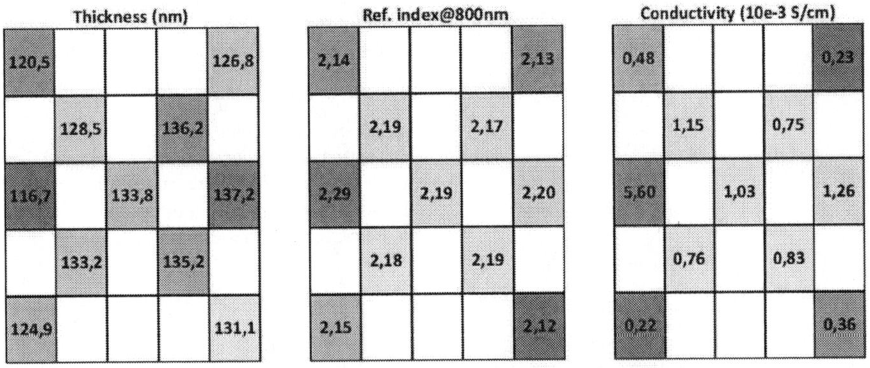

Thickness (nm)	AVG	129,46
	(Max-min)/(Max+min)	8,07%
Refractive index @800nm	AVG	2,178
	(Max-min)/(Max+min)	3,85%
Conductivity (10-3 S/cm)	AVG	1,15

Figure 3:. Typical uniformity maps for a SiO_x layer on 1.0 x 1.4 m^2 substrates

After the material optimization the most promising layers have been tested in the tandem device replacing the standard μc-Si:H n layer.

The electrical results of the modules: maximum power (Pmax), short circuit current (Isc), open circuit voltage (Voc) and fill factor (FF) are reported in fig.4. Each bar represents the average of eight modules (1 x1.4 mt size) measured with Wacom solar simulator and normalized to standard tandem cell (with μc-Si:H based n layer).

An Isc increase has been observed for almost all set of module samples, but the impact of bottom n-layer replacement on Voc and FF for some test is negative because the reduction of conductivity produce an increase of series resistance.

The balancing of positive and negative effect of the introduction of the n-SiO_x layer leads only four tested condition upon six resulted in an effective Pmax improvement which can be estimated in the range of 2%.

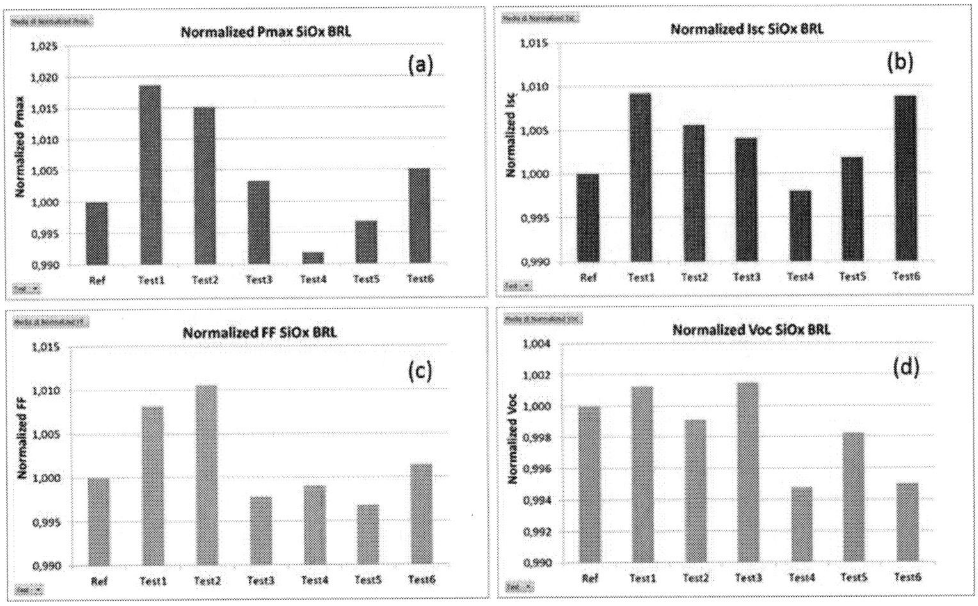

Figure 4a, 4b, 4c and 4d:. Normalized electrical measured on full modules at STC conditions.

The increase in Isc is directly correlated with the index refractive reduction. On Fig. 5 the dependency of Isc of all the test made versus the refractive index, measured on the equivalent single layer, is reported. The trend clearly show a linear tendency as expected.

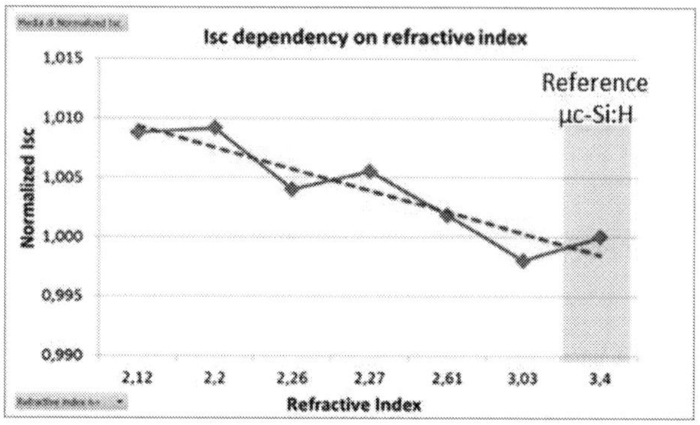

Figure 5: Isc vs refractive index trend.

The increase of Isc has then analyzed through quantum efficiency tool. In Fig.6 we report the EQE measurements performed on small area cells cut from the large area modules.

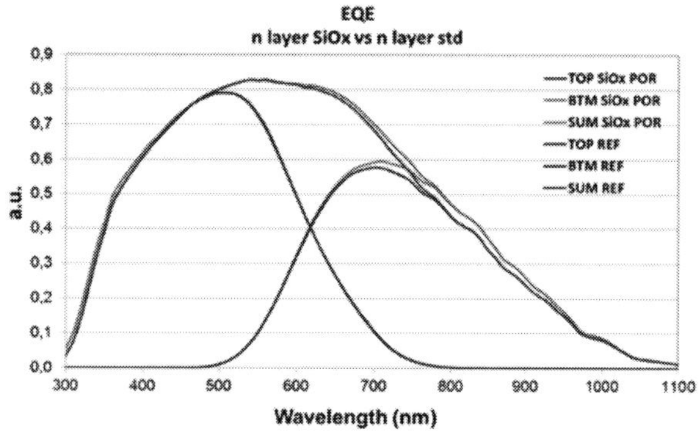

	SiOx POR	REF 3SUN	Delta (%)
TOP Jsc [mA/cm2]	11.27	11.21	0.5
BTM Jsc [mA/cm2]	11.57	11.04	4.8
Ratio Jsc	0.974	1.015	0.041 (net **0.026**)

Figure 6EQE results on tandem cells from large area modules. Results are compared to reference.

The replacement of standard µc-Si n layer with SiO_x based n layer provides an increase up to 5% of the bottom cell current. This is the intrinsic gain of the internal response of the microcrystalline cell. Considering that in any case the limiting current of the device is the lower one between the two junctions, an increase of 5% of the bottom cell only Jsc leads to an unbalanced ratio. This makes the limiting cell is the top one and Top/Btm current ratio became lower than 1. In this configuration, the increase that can be measured in the cell is only the one that allows to reach a top to bottom ratio of 1 so all the current generated by the bottom cell that overcome the current generated by the top cell is excessive. Studies of bottom cell Isc current variation with respect to thickness of the intrinsic absorbing layer with and without the presence of SiO_x back reflectors are going to be performed in the future.

IV. Conclusion

The n layer based on SiO_x material have been studied and obtained by RF-PECVD industrial reactor with a wide range of optical and electrical properties. CO_2/SiH_4 gas flow ratio and process pressure, revealed as key parameter for the material control.

Most promising layers have been tested as bottom cell n layer in a tandem module and four of them lead to an effective Pmax improvement. It has been demonstrated that the application of n-SiO_x:H improves Jsc of the bottom cells, which can be attributed to the enhanced EQE in the wavelength range of 700–1000 nm.

About 2% of efficiency relative improvement has been achieved on tandem module on commercial TCO glass substrate. The result is significant if considered that it leads to an increase of Fab output of more than 4MW/year.

Further optimizations are in progress to take full advantage of the intrinsic improvement observed by the quantum efficiency measurements. In particular the optimization of the bottom cell layers thickness is required to achieve also an improvement of reactor throughput. We can roughly estimate by the data presented in this work a further throughput increase between the 1% and 2% (2÷4MW/y) could be also obtained.

References

[1] J. Müller, B. Rech, J. Springer, M. Vanecek, Sol. Energy 77 (2004) 917.
[2] A. Lambertz, T. Grundler, and F. Finger, *J. Appl. Phys.* 109, (2011) 113109.
[3] V. Smirnov, O. Astakhov, R. Carius, Y. Petrusenko, V. Borysenko, and F. Finger, *Jpn. J. Appl. Phys.* 51, (2012) 022301.
[4] G.E. Jellison Jr., F.A. Modine, *Appl. Phys. Lett.* 69 (1996) 371.
[5] D. Das, M. Jana, A.K. Barua, *Sol. Energy Mater. Sol. Cells* 63 (2000) 285
[6] S Kim, Jin-Won Chung, *Sol. Energy Mater. Sol. Cells* 119 (2013) 26.
[7] X.D. Zhang , Q. Yue, *Thin Solid Films* 520 (2011) 684–688
[8] A.Shah et al., *J. of Non-Crystalline Solids* 639 (2004) 338-340

DEVELOPMENT OF WELL DISPERSED TAPERED ITO NANORODS AS A POTENTIAL LIGHT TRAPPING STRUCTURE FOR AMORPHOUS SILICON BASED SOLAR CELLS

Sukanta Dhar[a]*, Chandan Banerjee[b], and A. K. Barua[a]
[a]Centre of Excellence for Green Energy and Sensor Systems, Indian Institute of Engineering Science & Technology
Shibpur, Howrah 711103, India, *E-mail: sukanto001@gmail.com
[b]National Institute of Solar Energy, Gwalpahari, Haryana, 122003, India

ABSTRACT: In this work, we report the development of well-defined ITO nanorods by pulse DC magnetron sputtering and fabrication of a-Si based p-i-n solar cell by PECVD cluster tool on ITO nanorods. We are able to control the shape (tapered) and dispersion of ITO nanorods by controlling process parameters. Here we focus on few things, like catalyst or mask free low temperature ITO nanorods formation, tapered edge, well dispersion. The solar cells are made on ITO nanorods coated glass substrates. For comparison, the solar cells are also made of regular FTO coated glass substrates with similar thicknesses. Amorphous silicon solar cell fabricated on ITO nanorods by depositing 15 nm p-a-Si:H/150 nm i-a-Si:H/ 25 nm n-a-Si:H layer by PECVD cluster tool CT-150. The solar cells on ITO nanorods coated glass substrates showed an obvious improvement in the light trapping as well as improvement in short circuit current density from 10.05 to 12.50 mA/cm2. Initial efficiency of 4.00% (Voc = 750mV, Jsc = 12.50 mA/cm2, FF = 42.70%) has been achieved on ITO nanorods. Further optimization is needed.
Keywords: Light Trapping, Thin Film Solar Cell, ITO, a-Si:H.

1 INTRODUCTION

Areas of current research attention include arrays of radial junction nanowires to enhance photon absorption and provide efficient carrier collection [1]. In addition to the potential advantages of providing enhanced photon absorption, carrier generation, and carrier collection, nanoscale materials can be configured into more complex architectures, and synthesized directly on or transferred to lowcost substrates. The potential for improved performance and cost reductions of NW arrays over their bulk photovoltaic counterparts is primarily due to (1) increased absorption due to diffuse light scattering in NW arrays, (2) short collection lengths of minority carriers that are radially separated and collected normal to the light absorption direction, and (3) flexibility of cell integration on a variety of low-cost carrier substrates. The radial geometry of the p–n junction allows for the orthogonalization of the light absorption and carrier collection directions. The orthogonalization through 1D nanostructuring is a key factor in the NW approach to solar cells where the optical absorption and charge collection constraints are decoupled. As with the radial junction nanowire solar cell, this allows efficient carrier collection and hence high short circuit currents for low diffusion lengths. However, a-Si:H in thin-film solar cells suffers from a high defect density which leads to a high recombination rate. Bulk recombination is mitigated by using thinner devices, which has also added advantages of increasing fabrication throughput and reducing light induced degradation. On the other hand, cells have to be thick enough in order to efficiently absorb the near-bandgap part of the solar spectrum. Therefore, light-trapping schemes that increase optical absorption are crucial. Light scattering in thin-film solar cells is traditionally achieved by the usage of a textured transparent conductive oxide (TCO) layer, such as commercially available SnO2:F with a randomly textured surface [2], or textured doped ZnO fabricated either by sputtering followed by a post-deposition hydrochloric acid etching [3-5] or by low pressure chemical vapor deposition [6]. The textured TCO layer scatters incident light into off-normal angles to increase the light propagation path in the absorber layer. Recently, radial junction solar cells based on elongated nanostructures, such as nanowires [7-9], nanorods (NRs) [10-12], nano-pillars [13-16], nano-domes [17], and nano-pyramid [18], have attracted strong attention. One main challenge is the fabrication of large-area nanostructure arrays with controllable morphology in a cost-effective and high-throughput way. Another main challenge is that the efficiency and the yield of these innovative solar cells based on elongated nanostructures so far are significantly lower than that for their conventional counterparts. To reach high efficiency, the very rough nanostructures must be conformally coated with device-quality absorber material, which requires smooth features. In contrast, a substrate with steep features is favorable for light scattering. So far, it has been difficult to reconcile these opposing requirements.

2 EXPERIMENTAL

Out of different grown techniques we chose Dc magnetron sputtering methods because sputtering shows the best performance in terms of surface morphology, high deposition rates, low growth temperature, scaling-up ability, and excellent electro-optical performances.

To study the structural, optical and electrical properties, the material was deposited on different substrates like scott glass, FTO coated glass, TEM grid, polish and texture wafers. Before depositing the material all the substrate were properly cleaned. An ITO sintered target with In2O3 and SnO2 in a weight proportion of 9:1 was used. The base pressure of the sputtering system was about 10-6 Torr . The process pressure of the chamber maintained by flowing the Argon gas into the chamber through mass flow controller. We varied different process parameters like Argon flow rate, process pressure, substrate temperature during deposition of the material. We found that the surface morphologies of these types of materials dramatically depend on the deposition temperature and process pressure. The best material was obtained when the chamber pressure was maintained of the order of 10-3 mbar pressure and the substrate temperature was about 320OC. The power density during deposition was 1.5 Watt/cm2.

The a-Silicon based p-i-n structure were fabricated on ITO nanorods coated glass in a conventional capacitive coupled RF-PECVD reactor developed indigenously employed with13.56 MHz RF source for plasma excitation.

3 RESULTS AND DISCUSSIONS

In order to get desired ITO nanorods we vary sputtering process parameters and study the morphology by FESEM.

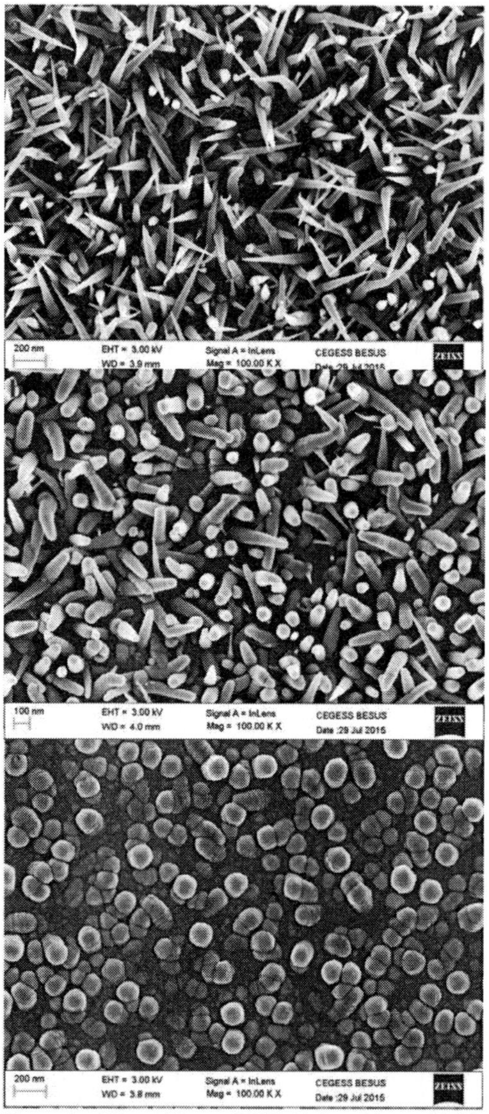

Figure 1: a) Deposition at 320°C, b) deposition at 280°C, c) deposition at 250°C.

Figure 1 shows FESEM images of ITO nano rod developed at different temperature by making other parameter constant. From the above images we can conclude that substrate temperature has an immense effect on ITO nano rod formation. We can see that at 320oC the formation of ITONR is very prominent and have sharp edges, at 280oC the formation is less and at 250oC the formation is negligible. So for synthesis of ITO nano rod by DC sputtering required a minimum temperature. In our sputtering system we are able to create stabilized plasma at process pressure 0.005 mbar, and get well dispersed ITO nanorods with tapered head suitable for conformal p-i-n a-Si layer deposition.

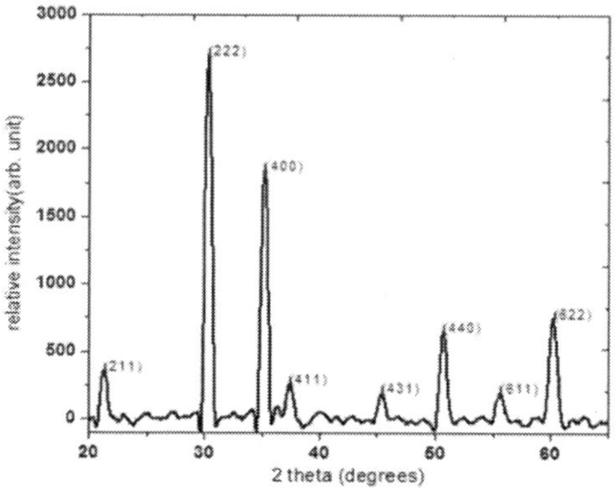

Figure 2: XRD pattern on as grown ITO nano rods on scott glass.

Figure 2 shows an XRD pattern of ITO nano rods grown on scott glass. All the ITO nano rod samples have crystalline structure, which exhibit diffraction peaks. The peak positions were indexed according to JCPDS card No. 76-0152, and they are in good agreement with other reports on ITO nanostructures. There are four major diffraction peaks corresponding to (222) at 30.45o, (400) at 35.28o, (440) at 50.80o and (622) at 60.40o orientations. An obvious trend shown by the XRD diffraction pattern was that the (400) diffraction peak became relatively strong with respect to the planer ITO layer.

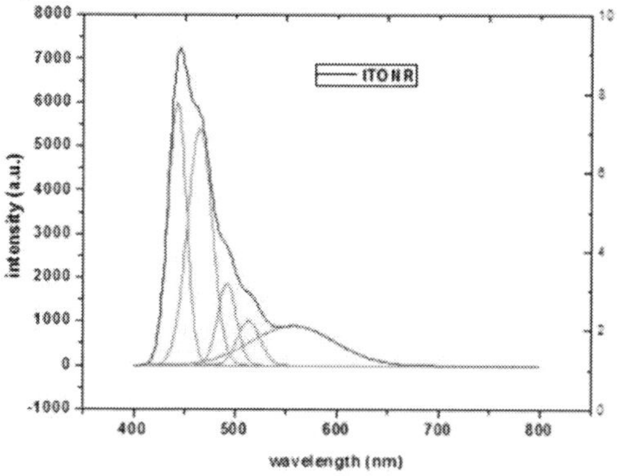

Figure 3: Room temperature PL spectra taken on ITO nano rod sample.

Two strong peaks were observed at 442 nm and 446 nm. Usually, the visible emission of metal oxide 1D nano structure , such as In_2O_3 and SnO_2, is attributed to the surface defect states and oxygen deficiencies. PL property was not observed from ITO thin films although they contain oxygen deficiencies. The PL properties of ITO nano rods is probably related to the surface defect states. And probably the emission due to PL property can contribute to solar cell efficiency.

From the optical properties studied by UV-VIS spectrophotometers shown in figure 4 it is clear that ITO nanorods have better transmission in the blue spectra region compare to planer ITO layer. As we know that in this region a-Si solar cell gives highest Quantum efficiency, so we can say that ITO nanorods can be used as better front TCO for a-Si based solar cell than planner structure.

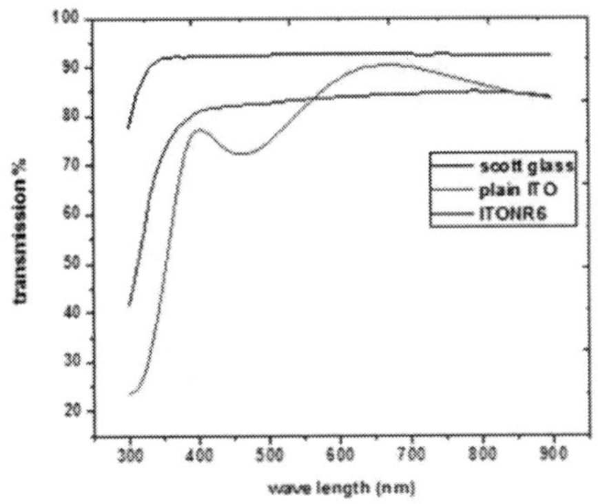

Figure 4: UV-VIS transmittance spectra of different sample on glass.

For comparison, the solar cells are also made of regular FTO coated glass substrates with similar thicknesses.

Table I: Characteristics of a-Si:H thin-film solar cells grown on different substrates.

Substrate	Voc (V)	Jsc (mA/cm^2)	FF (%)	Efficiency (%)
FTO coated glass	0.847	10.05	71	6.04
ITO nanorods coated glass	0.75	12.5	42.7	4

Initial efficiency of 4.00% (Voc = 750mV, Jsc= 12.50 mA/cm2, FF = 42.70%) has been achieved on ITO nanorods. The short-circuit current density Jsc of the nanorod cells is 24 % (10.05 to 12.50 mA/cm2) higher than that of the reference cells.

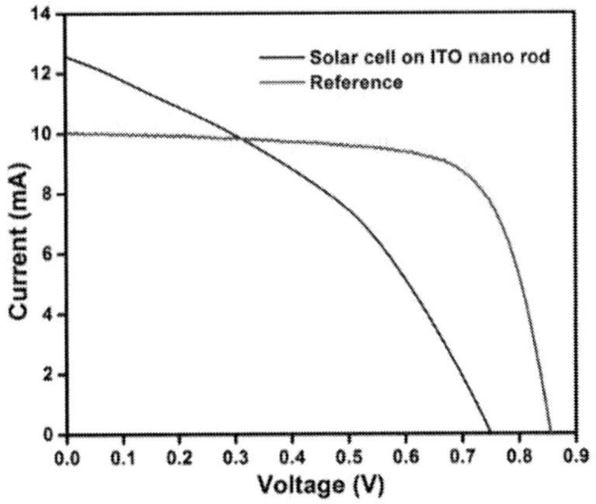

Figure 5: I-V curve of solar cells on ITONR and FTO coated glass substrate.

At the same time the cell is showing substantial decrease in Voc, which may be explained by two factors: an increased interface area between the nanorod front electrode and a-Si:H increases the dark current; and the ITO nanorods have poor interface properties with a-Si:H, especially in the areas with no ITO nanorods. Thus, the conversion efficiency of the nanorod cell is lower than that of the reference cell. The trade-off between optical and electrical properties is important for 3D solar cells. Generally, the Voc of solar cells is not only determined by the bandgap of the intrinsic layer, but also limited by bulk and interface recombination. For the reference cell, the bulk recombination could be a dominating factor for the Voc, whereas for the nanorod cell, the interface recombination could influence the Voc due to an increased internal surface area and poorer interface properties.

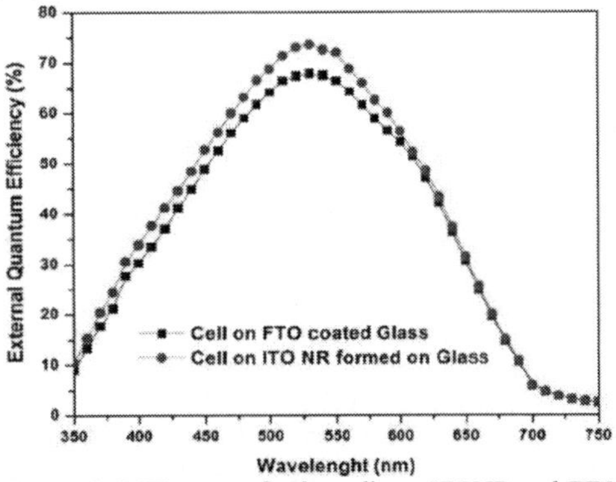

Figure 6: EQE curve of solar cells on ITONR and FTO coated glass substrate.

Table I shows the comparison of the cell fabricated on ITO NR with the reference cell. EQE measurements on the NR cells showed a broad enhancement with respect to the reference cell over almost the entire 350–600 nm spectral range. The EQE results demonstrate that more photo generated charge-carriers are effectively collected in the nanorod devices, due to the unique nano-3D geometry. In the nanorod solar cells, however, the unique nano-3D geometry not only allows the absorber in the axial direction to capture deeply penetrating photons, but there is also considerable enhancement in the blue response, as can be seen in Fig. 6.

4 CONCLUSIONS

In summary, we present an efficient nanostructured thin film a-Si:H solar cell based on a ITO nanorods. ITO nanorods is obtained by a pulse DC magnetron sputtering at comparative lower temperature (320 oC) through self-catalytic growth without any metal catalyst. EQE measurements on the NR cells show a broad enhancement with respect to the reference cell over almost the entire 350–600 nm spectral range. A preliminary conversion efficiency of 4.00% has been obtained for nanorod solar cells with a thin (150 nm) a-Si:H absorber layer. Further optimization is going on for efficiency enhancement. The effective carrier generation in the nano-3D geometry results in a higher efficiency in the nanorods devices even though the absorber layer is two times thinner than that in the planar and the textured solar cells studied. The simple and scalable fabrication process makes our nanorod/a-Si:H opens up a new platform for novel efficient cell design that can be made at low cost.

5 REFERENCFES

[1] E.C. Garnett, M.L. Brongersma, Y. Cui, M. McGehee, Annu. Rev. Mater. Res. DOI: 10.1146/ annurev-matsci-062910—100434.

[2] K. Sato, Y. Gotoh, Y. Wakayama, Y. Hayasahi, K. Adachi, and H. Nishimura, "Highly textured SnO2:F TCO films for a-Si solar cells," Rep. Res. Lab. Asahi Glass Co. Ltd. 42, 129–137 (1992).

[3] O. Kluth, B. Rech, L. Houben, S. Wieder, G. Sch€ope, C. Beneking, H. Wagner, A. L€offl, and H. W.

Schock, Thin Solid Films 351, 247 (1999).
[4] J. Muller, B. Rech, J. Springer, and M. Vanecek, Sol. Energy 77, 917 (2004).
[5] M. Berginski, J. H€upkes, M. Schulte, G. Sch€ope, H. Stiebig, B. Rech, and M. Wuttig, J. Appl. Phys. 101, 074903 (2007).
[6] S. Fa€y, J. Steinhauser, N. Oliveira, E. Vallat-Sauvain, and C. Ballif, Thin Solid Films 515, 8558 (2007).
[7] B. Tian, X. Zheng, T. J. Kempa, Y. Fang, N. Yu, G. Yu, J. Huang, and C. M. Lieber, Nature 449, 885 (2007).
[8] E. C. Garnett and P. Yang, J. Am. Chem. Soc. 130, 9224 (2008).
[9] M. D. Kelzenberg, S. W. Boettcher, J. A. Petykiewicz, D. B. Turner- Evans, M. C. Putnam, E. L. Warren, J. M. Spurgeon, R. M. Briggs, N. S. Lewis, and H. A. Atwater, Nature Mater. 9, 239 (2010).
[10] B. M. Kayes, H. A. Atwater, and N. S. Lewis, J. Appl. Phys. 97, 114302 (2005).
[11] M. Krunks, A. Katerski, T. Dedova, I. Oja Acik, and A. Mere, Sol. Energy Mater. Sol. Cells 92, 1016 (2008).
[12] Y. Kuang, K. H. M. van der Werf, Z. S. Houweling, and R. E. I. Schropp, Appl. Phys. Lett. 98, 113111 (2011).
[13] Z. Fan, H. Razavi, J. W. Do, A. Moriwaki, O. Ergen, Y. L. Chueh, P. W. Leu, J. C. Ho, T. Takahashi, L. A. Reichertz, S. Neale, K. Yu, M. Wu, J. W. Ager, and A. Javey, Nature Mater. 8, 648 (2009).
[14] M. J. Naughton, K. Kempa, Z. F. Ren, Y. Gao, J. Rybczynski, N. Argenti, W. Gao, Y. Wang, Y. Peng, J. R. Naughton, G. McMahon, T. Paudel, Y. C. Lan, M. J. Burns, A. Shepard, M. Clary, C. Ballif, F. Haug, T. S€oderstr€om, O. Cubero, and C. Eminian, Phys. Status Solidi RRL 4, 181 (2010).
[15] M. Vanecek, O. Babchenko, A. Purkrt, J. Holovsky, N. Neykova, A. Poruba, Z. Remes, J. Meier, and U. Kroll, Appl. Phys. Lett. 98, 163503 (2011).
[16] J. Kim, A. J. Hong, J. W. Nah, B. Shin, F. M. Ross, and D. K. Sadana, ACS Nano 6, 265 (2012).
[17] J. Zhu, C. M. Hsu, Z. Yu, S. Fan, and Y. Cui, Nano Lett. 10, 1979 (2010).
[18] A. Mavrokefalos, S. E. Han, S. Yerci, M. S. Branham, and G. Chen, Nano Lett. 12, 2792 (2012).

SOLAR CELLS AND MINI-MODULES BASED ON 40 μm-THICK EPITAXIAL SI FOILS: TOWARDS CONDUCTIVE BONDING ONTO LOW-COST SI POWDER SINTERED CONDUCTIVE CARRIERS FROM THE RECYCLING OF SILICON WASTE

H. Sivaramakrishnan Radhakrishnan[1], K. Van Nieuwenhuysen[1], J. Govaerts[1], V. Depauw[1], T. Bearda[1], M. Debucquoy[1], R. Roozeman[2], J. Heikkinen[2], M. Schumann[3], R. Buchwald[3], H.J. Möller[3], A. Ciftja[4], G. Stokkan[4], E. Øvrelid[4], A. Stonkus[5], P. Dubravskij[5], J. Ulbikas[5], I. Gordon[1], J. Szlufcik[1], J. Poortmans[1], A. Ulyashin[4]

[1]IMEC, Kapeldreef 75, 3001 Heverlee, Belgium.
[2]INKRON, Finland
[3]Fraunhofer Technology Center for Semiconductor Materials THM, Am St.-Niclas-Schacht 13, 09599 Freiberg, Germany.
[3]SINTEF, Material and Chemistry, Trondheim, Norway.
[4]Applied Research Institute for Prospective Technologies, Vismaliukų g. 34, Vilnius LT 10243, Lithuania.
[5]SINTEF, Material and Chemistry, Oslo, Norway.

ABSTRACT: Epitaxial silicon foils (<50 um) lifted-off using the porous silicon-based layer transfer process are used to fabricate heterojunction (HJ) solar cells. Mini-modules with efficiencies above 15% have been fabricated using standard lamination and encapsulation techniques, which have been applied to freestanding epi-Si based HJ solar cells. Analysis of this module highlights the need for bonding these cells onto supporting substrates before module fabrication. We suggest using very low-cost conductive Si carriers as the supporting substrates. Sintering of low-cost Si powder-based ingots, assisted by in-situ doping, has been done by hot pressing or casting of ingots using different types of Si powders and was followed by multi-wire sawing to fabricate thin (~170 μm) supporting highly conductive substrates. The Si material used for these substrates can, in principle, be obtained from Si waste streams, e.g. broken wafers or end-of-line cells in production lines, Si kerf or end-of-life PV modules, making this even more attractive proposition. One important aspect is the development of conductive bonding of epitaxial foils onto low-cost sintered silicon substrates. We evaluated different types of conductive bonding agents. Good mechanical stability was achieved for the bonding processes. These efforts represent the first step towards the wafer-equivalent approach of high efficiency solar cells on thin Si (<50 μm) conductively bonded to low-cost Si substrates. This approach is considered as a promising route for the low-cost Si based PV and a solution module integration of thin Si. This work is done in framework of H2020 EU project CABRISS.

1 INTRODUCTION

Currently, alternatives to Si wafer-based approaches are under consideration towards low-cost c-Si PV. Some potentially promising options are based on the implementation of thinner silicon wafers or foils bonded on low-cost supporting substrates (Si wafer equivalents). The drive to reduce cost per watt-peak has been ever-present in the PV industry. This reduction is being pursued by reducing the costs as well as increasing the efficiency of solar cells. Currently, silicon material costs represent a significant 33% of the final module costs in 2015 [1]. Therefore, efficient usage of silicon is of paramount importance, leading to a keen interest in kerfless technologies which enable the production of very thin silicon (< 50 μm) with minimal kerf loss, using disruptive methods such as the porous silicon-based lift-off of epitaxial silicon [2]–[6]. In addition, there is also an increased interest in the recycling of critical raw materials such as silicon from silicon waste sources such as kerf, broken wafers in the cell manufacturing plants as well as end-of-life modules. This forms the basis of the H2020 EU project CABRISS, which aims to develop a circular economy mainly for the photovoltaic industry, but also the electronics and glass industries.

One of the ideas pursued in CABRISS is that some of the recovered waste silicon can eventually be transformed into low-cost Si substrates by sintering of low-cost Si powders. Such substrates can be used as supporting substrates for the high quality epitaxial foils produced by the porous silicon-based lift-off process, in a wafer-equivalent concept. One of the main challenges for the processing of thin silicon with thickness below 50 μm is that they become highly flexible, fragile and prone to breakage [7]. As a result, it is not possible to handle and process such thin silicon foils as regular wafers. In addition, thinner silicon cells are also more susceptible to breakage during cell interconnection due to stresses arising from the soldering of ribbons in the tabbers-stringers, and to failure during operation through propagation of micro-cracks formed during cell interconnection, eventually leading to reduced reliability and operational lifetime of solar modules [8], [9]. By introducing the wafer-equivalent approach and bonding such high quality epitaxial foils onto low-cost native silicon substrates, a mechanical yield similar to regular wafers can be achieved.

In this work, we report on the first efforts to achieve a highly-efficient wafer-equivalent epitaxial silicon cells-on-sintered low-cost substrates. In the eventual embodiment, the epitaxial foil may be partially processed while it is attached to the parent substrate and then transferred to a low-cost sintered substrate, where the remainder of the cell processing will be completed, before cell interconnection and module lamination. However, at this stage, the cell process and the module fabrication have been performed sequentially.
The heterojunction cell process is completed on a freestanding epitaxial foil of 40 μm thickness. The finished cell is then laminated and interconnected into a mini-module to highlight the challenges in module fabrication for such thin foils. As such, conductive bonding methods are being evaluated and developed in order to bond such epitaxial foil cells onto low-cost sintered Si substrates.

2 POROUS SILICON-BASED LIFT-OFF OF THIN (< 50 μm) EPITAXIAL FOILS

The process leading to the creation of ultra-thin silicon foils with thicknesses (< 50 μm) that are an order of magnitude less than the current wafer thickness (180 μm), is described schematically in **Fig. 1**. Starting with highly-doped p^+ wafers, a double layer of porous silicon is etched electrochemically, and subsequently reorganised at a high temperature, before epitaxial growth of high quality silicon. This epitaxial silicon layer is then exfoliated from the substrate at the porous layer and then subsequently processed into freestanding heterojunction solar cells. The parent substrate is re-used in the next cycle of layer transfer.

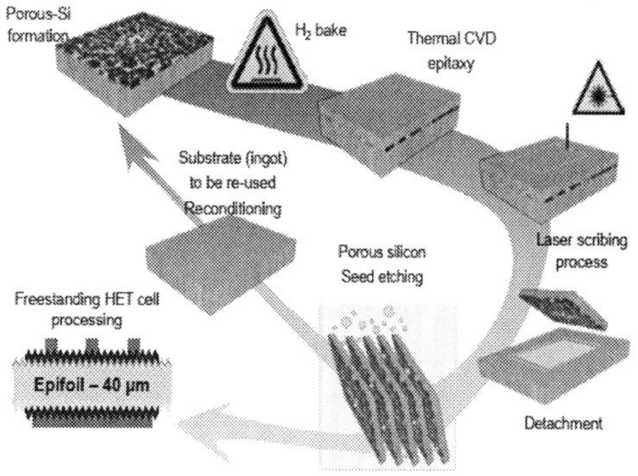

Figure 1: Production of thin epitaxial foils (< 50 μm) using the porous-silicon based lift-off process.

3. MINI-MODULE FABRICATION SEQUENCE WITHOUT BONDING TO SINTERED SUBSTRATES

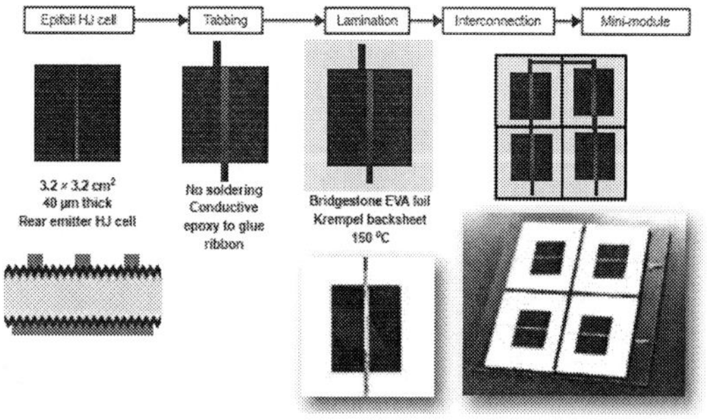

Figure 2: Module fabrication sequence using epitaxial foil-based heterojunction cells without bonding to supporting substrate.

A mini-module was fabricated using 4 epitaxial foil-based heterojunction cells, without bonding to a supporting substrate, in order to highlight the challenges of fabricating modules from freestanding epitaxial foils. Each epifoil cell was first tabbed by gluing ribbon using a conductive epoxy instead of soldering, which can cause high mechanical stresses. The tabbed cells were laminated in a standard way at a temperature below 150°C. The 1-cell laminates were then interconnected together to form a mini-module. This sequence was employed because stringing thin epifoils together before lamination would lead to breakages.

The photoluminescence (PL) and electroluminescence (EL) images of the bonded epitaxial foil cells of the mini-module are shown in **Fig. 3**. The luminescence intensity in the PL is proportional to the lifetime of the Si foils while that of the EL map indicates besides lifetime, the resistance losses as well. Thus, it is clear that two of the cells have cracks which prevent current collection in large parts of one of the cells, despite taking precautions to handle such fragile cells with utmost care. This clearly highlights the main challenge for processing ultra-thin cells, thus motivating the reasons to bond such cells onto low-cost substrates before interconnection and module fabrication and if possible, even at an earlier stage during cell fabrication.

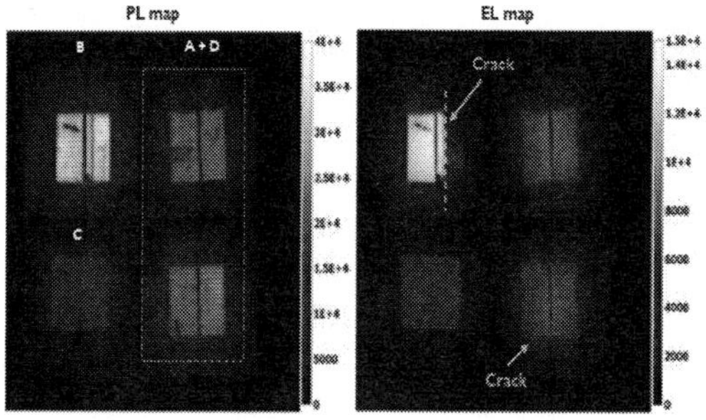

Figure 3: The photoluminescence (PL) and electroluminescence (EL) images of the bonded epitaxial foil mini-module.

The light IV measurements of the individual cells before module fabrication and that of the mini-module is given in **Table I**, where it is clear that J_{sc} is reduced mainly due to through-cracks which cut-off parts of cells from current collection. V_{oc} is largely unaffected, despite the cracks. The main improvement is in the FF which increases by >30% because cell bus bars which have high resistance due to poor design and small metal grid thickness are by-passed by the tabs of the module. Thus, a mini-module efficiency of 15.3% was achieved, which is ~5% absolute increase in efficiency compared to the cells. The potential efficiencies of cells with good FF and mini-module without cracks are also indicated in the table as "pseudo-η".

Table I: Light IV measurements of the individual cells before module fabrication and that of the mini-module itself.

Cell #	J_{sc} $[mA/cm^2]$	*Voc [mV]*	*FF [%]*	*η [%]*	*Pseudo-η [%]*
A	34.5	680	47.9	11.2	19.0
B	34.7	715	40.4	10.0	20.1
C	34.2	657	47.2	10.6	18.2
D	33.4	680	47.2	10.7	18.4
Mini-module	**27.5**	**2745**	**81.2**	**15.3**	**18.6**

3 SINTERING OF LOW-COST SI POWDER BASED INGOTS AND WAFERING

Sintering of low-cost Si powder based ingots, assisted by in-situ doping, has been done by hot pressing or casting of ingots using different types of Si powders and was followed by multi-wire sawing to fabricate thin (~170 µm) supporting highly conductive substrates [10], [11]. Different Si powder feedstocks, including recycled Si powders of 10 and 80 µm particle size distributions have been used. Si powders have been mixed with boron powder to provide in-situ doping upon hot pressing or casting processes of ingots. Hot Pressing involves subjecting the silicon powder to a high pressure at an elevated temperature. The powder was placed in a cylindrical graphite die with graphite punches entering the cylinder from the top and bottom and transferring the pressure from the rams. In order to avoid reaction between the graphite and the silicon and secure release of the ingot from the die after processing, first the inner part of the die and the side of punches facing the silicon material were coated with boron nitride. Then, a molybdenum foil was used to encapsulate the powder. Since the vertical pressure results in horizontal expansion forces, a tight fit between the ingot and the dies results. This assures protection against any chemical reaction and provides a destructible release layer. Before the test, the furnace chamber was evacuated down to 3 mbar, and then filled to atmospheric pressure with Harmix gas, grade 5.0 (95% Ar + 5% H_2; purity 99.999) and flushed continuously during the entire experiment. The thermal cycle for the hot pressing was as follows: (i) heat from room temperature to 1350°C at a ramping rate of 10°C/min, (ii) apply 30 MPa pressure, while holding at 1350°C for 1 h; (iii) after one hour release the pressure; (iv) ramp down to room temperature at 20°C/min.

Casting of Si ingots from the same Si powder feedstock was used as an alternative approach. Multi-wire sawing was used to fabricate thin (~170 µm) supporting highly conductive substrates from the sintered as well as from the casted Si ingots. Resistivity of processed Si powder based wafers in all cases were in the mOhm range, which is sufficient to use them as a supporting part and at the same time a back-side electrode.

4. CONDUCTIVE BONDING OF EPITAXIAL FOILS ONTO LOW-COST SINTERED SILICON POWDER SUBSTRATES

We have started evaluation of conductive bonding of epitaxial foils onto low-cost silicon substrates, manufactured by sintering, described above. For conductive bonding, we are currently evaluating 3 types of bonding agents, differing by their viscosity: (1) 3M conductive adhesive tape, (2) INKRON low-temperature Ag paste, (3) SPI Supplies Ag paint.

Figure 4 shows examples of bonding two types of bonding agents. Important attributes sought after in a conductive bonding agent are the bonding strength, the electrical conductivity and contact resistance through the bonded structure, as well as cost. In addition, a high viscosity bonding agent would present a low risk for shunt. The morphology of the bonding agent after application should be as smooth as possible to avoid voids which can cause cracks.

The properties and evaluation results are qualitatively summarised in **Table II**.

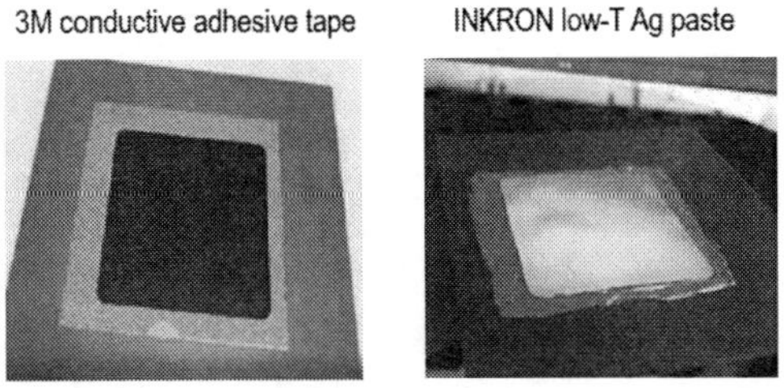

Figure 4: Conductive bonding of epitaxial foils onto low-cost Si substrates using 3M conductive adhesive tape (left) and INKRON's low-T Ag paste (right).

Table II: Summary of the evaluation of 3 types of conductive bonding agents

Glue	*Viscosity*	*Risk of shunt*	*Bond strength*
3M Conductive adhesive tape 9709	Solid	None	Excellent
INKRON IPC-114 Ag paste	High	Low	Excellent
SPI Supplies Ag paint	Low	High	Excellent

Some selected examples of solar cell efficiency changes upon bonding to Al-coated low-cost Si substrates using Ag paint are shown in **Table III**.

Table III: Evolution of epi-Si based solar cell effiencies upon bonding

	J_{SC} $[mA/cm^2]$	V_{OC} $[mV]$	FF $[\%]$	η $[\%]$
Epifoil 1	31.5	615	75	14.5
Epifoil 1 on low-cost multi-c Si	31.6	614	73	14.2
Epifoil 2	30.0	618	76	14.1
Epifoil 2 on sintered substrate	30.3	617	75	14.0

It can be concluded, that bonding of epi-Si based solar cells on top of highly conductive low-cost substrates provides preserving efficiencies of such cells.

5 CONCLUSIONS

Based on the experiments done so far, the following conclusions can be drawn:

- Heterojunction cells fabricated on freestanding epitaxial foils of thickness of ~40 µm were processed into a mini-module with efficiencies > 15%. However, it showed the drawbacks of processing such modules without being supported on a rigid substrate/superstrate.

- Hot pressing of Si powder can be used to process Si ingots and provides in situ doping . Wire sawing of such ingots provides fabrication of highly conductive ~180 µm thick Si supporting substrates with resistivity < 1 mOhm.
- Best efficiency achieved so far for Si epifoil based solar cells bonded on Si powder based substrate is ~14%, proving the feasibility of the wafer-equivalent approach dramatically.

6 ACKNOWLEDGEMENTS

Part of this work is funded by the H2020 EU project CABRISS (grant agreement N° 641972) as well as the IIAP program of IMEC.

REFERENCES

[1] SEMI, "International Technology Roadmap for Photovoltaic (ITRPV) 2014 Results," 2015.

[2] H. Tayanaka, K. Yamauchi, and T. Matsushita, "Thin-film crystalline silicon solar cells obtained by separation of a porous silicon sacrificial layer," in *2nd World Conference and Exhibition on Photovoltaic Solar Energy Conversion*, 1998, pp. 1272–1277.

[3] K. Van Nieuwenhuysen, V. Depauw, R. Martini, J. Govaerts, M. Debucquoy, H. Sivaramakrishnan Radhakrishnan, I. Gordon, T. Bearda, K. Baert, and J. Poortmans, "High Quality Epitaxial Foils, Obtained by a Porous Silicon Based Layer Transfer Process, for Integration in Back Contacted Solar Cells," in *27th European Photovoltaic Solar Energy Conference and Exhibition*, 2012, pp. 2471–2474.

[4] H. Sivaramakrishnan Radhakrishnan, R. Martini, V. Depauw, K. Van Nieuwenhuysen, M. Debucquoy, J. Govaerts, I. Gordon, R. Mertens, and J. Poortmans, "Improving the Quality of Epitaxial Foils Produced Using a Porous Silicon-based Layer Transfer Process for High-Efficiency Thin-Film Crystalline Silicon Solar Cells," *IEEE J. Photovoltaics*, vol. 4, no. 1, pp. 70–77, Jan. 2014.

[5] J. H. Petermann, D. Zielke, J. Schmidt, F. Haase, E. G. Rojas, and R. Brendel, "19%-efficient and 43 µm-thick crystalline Si solar cell from layer transfer using porous silicon," *Prog. Photovoltaics Res. Appl.*, vol. 20, no. 1, pp. 1–5, Jan. 2012.

[6] H. Sivaramakrishnan Radhakrishnan, R. Martini, V. Depauw, K. Van Nieuwenhuysen, T. Bearda, I. Gordon, J. Szlufcik, and J. Poortmans, "Kerfless layer-transfer of thin epitaxial silicon foils using novel multiple layer porous silicon stacks with near 100% detachment yield and large minority carrier diffusion lengths," *Sol. Energy Mater. Sol. Cells*, vol. 135, pp. 113–123, Apr. 2015.

[7] D. Kray, H. Kampwerth, E. Schneiderlöchner, A. Grohe, F. Kamerewerd, A. Leimenstoll, D. Osswald, E. Schäffer, S. Seitz, S. Wassie, S. Glunz, and G. Willeke, "Comprehensive experimental study on the performance of very thin laser-fired high-efficiency solar cells," in *Proceedings of the 19th European Photovoltaic Solar Energy Conference*, 2004, pp. 608–611.

[8] S. Pingel, Y. Zemen, and O. Frank, "Mechanical stability of solar cells within solar panels," in *Proceedings of the 24th European Photovoltaic Solar Energy Conference*, 2009, pp. 3459–3463.

[9] A. Gabor, M. Ralli, S. Montminy, L. Alegria, C. Bordonaro, J. Woods, and L. Felton, "Soldering induced damage to thin Si solar cells and detection of cracked cells in modules," in *Proceedings of the 21st European Photovoltaic Solar Energy Conference and Exhibition*, 2006, pp. 2042–2047.

[10] R. Buchwald, S. Würzner, H. J. Möller, A. Ciftja, G. Stokkan, E. Øvrelid, and A. Ulyashin, "Microstructural characterization of Si wafers processed by multi-wire sawing of hot pressed silicon powder based ingots," *Phys. status solidi*, vol. 212, no. 1, pp. 25–29, Jan. 2015.

[11] G. Stokkan, A. Johanson, E. Øvrelid, A. Ciftja, J. Hennicke, and A. Ulyashin, "SINTERED LOW RESISTIVITY SUBSTRATES FOR HYBRID SOLAR CELLS," in *Proceedings of 28th European Photovoltaic Solar Energy Conference*, 2013, pp. 427–430.

DEPOSITION OF AMORPHOUS AND MICROCRYSTALLINE SILICON IN A VERY HIGH FREQUENCY RANGE UP TO 140 MHZ

B. Leszczynska*[1], C. Strobel[1], S. Leszczynski[1], D.D. Fischer[1], M. Albert[1], J. W. Bartha[1], U. Stephan[2], J. Kuske[2]
[1] Dresden University of Technology, Semiconductor and Microsystems Technology Laboratory, 01062 Dresden, Germany, *Phone: +49 351 463 33151, e-mail: barbara.leszczynska@tu-dresden.de
[2] Forschungs- und Applikationslabor Plasmatechnik GmbH, Gostritzer Str. 67, 01217 Dresden, Germany

ABSTRACT: The very high frequency plasma enhanced chemical vapor deposition technique (VHF-PECVD) enables the highly productive fabrication process of amorphous (a-Si:H) and microcrystalline (μc-Si:H) silicon. The frequency increase leads to an enhancement of electron density in the bulk plasma, while the ion bombardment energy on the growing layer decreases. However, the wavelength of the very high frequency (VHF) signal corresponds to electrode dimensions, which leads to non-uniform deposition. In this work the effect of driving frequency up to 140 MHz on the a-Si:H and μc-Si:H process is studied. For this propose a VHF-PEVCD technique with a linear plasma source has been introduced. The influence of the deposition rate on the a-Si:H and μc-Si:H layer properties for different deposition regimes is analyzed. The role of intrinsic layer quality on the solar cell performance is investigated through the experiments and computer modeling.

Keywords: PECVD, amorphous silicon, microcrystalline silicon, solar cell, very high frequency

1 INTRODUCTION

The growing demand for renewable energy sources has led many research institutes and companies to look for new ways to obtain electrical energy. Although, nowadays a drastic drop in solar panels prices and an increasing competition on the solar energy market puts high pressure on solar cell manufactures. Thus, the cost reduction of silicon based thin film solar cells and the development of highly productive deposition process is more important than ever before.

The plasma enhanced chemical vapour deposition (PECVD) with a frequency of 13.56 MHz is the commonly used silicon thin film deposition technique. In order to enhance the deposition rate and process productivity a higher process power or an excitation frequency in a VHF range has to be used. The higher power level leads to increase in deposition rate, although, at the same time, the plasma potential increases leading to higher ion energies. On the other hand, in the VHF range, the sheath potential is reduced, leading to a reduced peak ion energy and to less energetic ion bombardment [1-3].

2 EXPERIMENTAL

2.1 Deposition system

In this work a-Si:H and μc-si:H layers were deposited in capacitevly coupled inline- deposition system with three interconnected modules, each with separated gas flow and a linear plasma source . The substrates are placed on the carrier and moved horizontally inside a tunnel. Each layer (intrinsic layer and p-, n- doped layers) is deposited inside the separated module. The vacuum system based on two backing pumps and two turbomolecular pumps generates a base pressure of about 10^{-4} Pa. For the process gasses the additional pumping system consisting of dry pumps combined with a roots pump is used. The VHF input signal is generated using the signal generator and amplifier with the frequency range up to 250 MHz. The signal matching to the plasma source is provided by a L-type matching network. The linear plasma source has been specially designed by FAP GmbH to provide a suitable power coupling for plasma ignition in the frequency range up to 140 MHz [4].

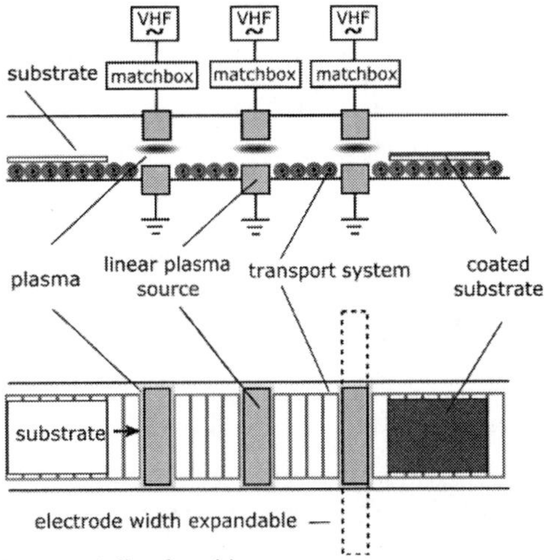

Figure 1: Inline deposition system

2.2 Experimental details

The a-Si:H and μc-Si:H intrinsic single layers were deposited on 1.1 mm thick Corning Eagle XG glass substrates to analyze the semiconductor properties. Additionally the intrinsic a-Si:H layers were deposited on double- side polished <100> Si wafers for structural characterization. The a-Si:H solar cells were fabricated on glass substrates covered with tin oxide (Asahi, U-type) and the microcrystalline solar cells on ZnO:Al covered glass. Evaporated silver served as back contact.

The substrate temperatures during the a-Si:H and μc-Si:H deposition process were 160 and 180 °C, respectively. The main process gases used for the deposition were silane (SiH_4) and hydrogen. Trimethylborane (TMB) and phosphine were used as dopant gases for the p- and n-layer fabrication. The a-Si:H layers were deposited at 45 Pa process pressure, 1200 sccm total gas flow, 33 % silane concentration and varied power densities. The μc-Si:H thin films were fabricated at the power density of 0.14 W/cm^2 and 100 Pa process pressure.

2.3 Measurements

To characterize the intrinsic thin film silicon layers and solar cells several measurement procedures were applied.

The a-Si:H and µc-Si:H layer thickness was measured by a surface profilometer from Veeco. To examine the electrical properties the single layers were covered with the aluminium contacts. Both photo- and dark conductivity were measured. The dark conductivity measurements were performed under vacuum conditions at pressure of 10^{-2} Pa. The photoconductivity values were measured using a sun simulator consisting of a xenon arc lamp combined with an infrared-filter. The photosensitivity of measured samples was estimated as photo- to dark conductivity ratio.

The Fourier transform infrared spectroscopy (FTIR) was applied to examine the structural properties of the amorphous material. The samples were measured from 380 to 3000 cm^{-1} using a Bruker Vertex 80v spectrometer. The microstructure factor was defined by the ratio of integrated absorption bands of the higher stretching mode (HSM) at 2090 cm^{-1} and the lower stretching modes (LSM) at 2000 cm^{-1}. The HSM mode corresponds to hydrogen bonded to internal void surfaces, whereby the LSM mode mainly corresponds to the mono-hydrides in the amorphous network.

The crystallinty of the µc-Si:H layers was estimated by the Raman spectroscopy using 514 nm excitation laser. The crystalline volume fraction was established as the ratio of integrated peaks corresponding to the crystalline phase (500 cm^{-1}, 510 cm^{-1}, 520 cm^{-1}) to the sum of all peaks including the amorphous Raman peak at 480 cm^{-1}.

The solar cell characteristics were measured using the sun simulator. Degradation measurements were performed under AM1.5 illumination conditions with a constant substrate temperature of 50 °C.

3 RESULTS

First, the influence of the excitation frequency on the a-Si:H and µc-Si:H layer properties for different deposition regimes was analyzed. Then, the role of intrinsic layer quality on the solar cell performance was investigated.

3.1 Hydrogenated amorphous silicon

To get further insight into the influence of the excitation frequency and power density on the film properties, the deposition rate and the semiconductor properties of a-Si:H films prepared at the 81.36, 120 and 140 MHz frequencies were analyzed (Figure 2).

The intrinsic layers were deposited at the 81.36, 120 and 140 MHz frequencies at the varied power density while keeping the process pressure and silane concentration constant. The enhancement of the deposition rate up to 70 % was achieved by increasing excitation frequency from 81.36 to 140 MHz at the constant plasma power. The 140 MHz frequency process has no negative influence on the semiconductor properties. At the same plasma power conditions, the good quality intrinsic layers with the photoconductivity of 10^6 and hydrogen concentration under 15 % have been fabricated for all analyzed frequencies. On the other hand, for the same enhancement of the deposition rate up to 70 % at constant excitation frequency an increase of power density up to 0.17 W/cm^2 was necessary. The higher process power leads to deterioration of the semiconductor properties.

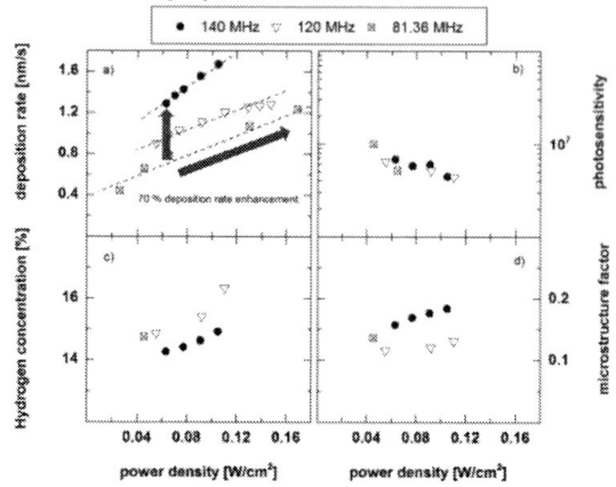

Figure 2: intrinsic a-Si:H: a) deposition rate, b) photosensitivity, c) atomic hydrogen concentration, d) microstructure factor depending on power density

As the next step, the role of intrinsic layer quality on the solar cell performance was investigated. For each plasma excitation frequency, the deposition rate was varied by increasing the RF power (Figure 3). The power density has a huge influence on the solar cells performance. A 70 % increase of the deposition rate by increasing the power density to 0.17 W/cm^2 leads to a decrease of the initial efficiency and to an enhancement of solar cells degradation up to 30 %. At high RF power levels higher values of plasma potential have been reported [3]. An increase in plasma potential leads to ion bombardment damages of the deposited layers and deterioration of the solar cells performance. On the other hand, with an enhancement of the excitation frequency from 81.36 MHz up to 140 MHz, while keeping the other process parameters constant, a 70 % increase in a static deposition rate was achieved without deterioration of the solar cell properties.

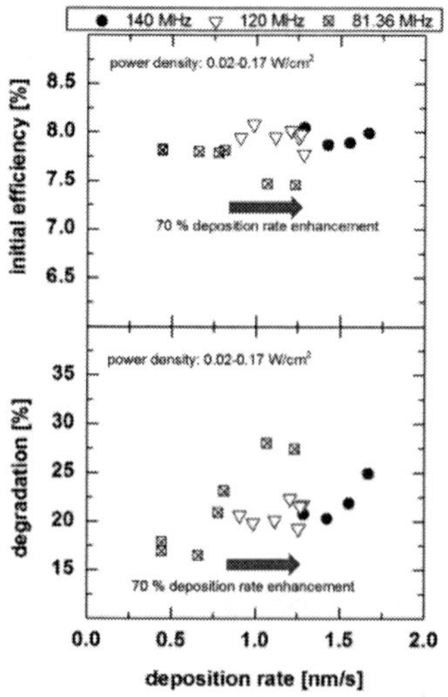

Figure 3: a-Si:H solar cells: initial efficiency and degradation depending on deposition rate.

3.2 Hydrogenated microcrystalline silicon

Next, an influence of the excitation frequency on the properties of intrinsic μc-Si:H material was analyzed. The microcrystalline silicon thin films were fabricated at 120 and 140 MHz excitation frequency and the silane concentration was varied in order to increase the layer crystallinity (Figure 4). A 30 % increase in deposition rate by a frequency increase from 120 up to 140 MHz has been achieved with the good quality i- layers (photosensitivity of 10^2-10^3 in the crystalline volume fraction range of 60-70%).

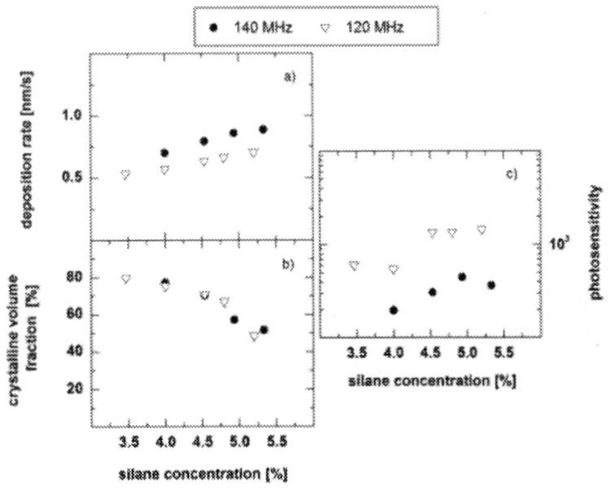

Figure 4: intrinsic μc-Si:H: a) deposition rate, b) crystalline volume fraction, c) photosensitivity depending on silane concentration

As the next step, the μc-Si:H solar cells were produced at 81.36, 120 and 140 MHz frequency. The "low power low pressure" deposition regime was used due to vacuum system specification of the inline deposition system designed for low process pressures.

In Figure 5 the current-voltage characteristics of the illuminated μc-Si:H solar cells are shown. An increase of an excitation frequency from 81.36 up to 140 MHz leads to a 80 % increase in deposition rate without deterioration of solar cells properties. The 140 MHz deposition process has been further developed to ensure very high deposition rate. It has been reported that through combination of VHF regime with high power high pressure process a deposition rates of up to 2.5 nm/s with 8.5 % solar cells efficiency can be achieved [5].

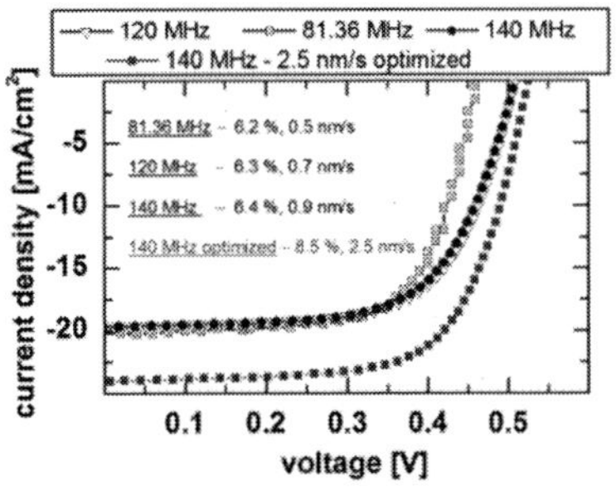

Figure 5: μc-Si:H solar cells: current-voltage characteristics with solar cell efficiency and deposition rate

3.3 Ion bombardment simulations

In order to study the influence of the ions on the deposition process the calculation of projected ranges of low-energy particles in a a-Si:H layer was performed using Monte-Carlo simulation program TRIM (Transport of Ions in Matter) developed by Ziegler et al. [6]. So far, the influence of ion bombardment on defect density [7], electronic properties [8] and intrinsic compressive film stress [9] has been reported. The ion bombardment energy decreases for higher excitation frequency and increases when higher process powers are used [8-10].

Figure 6 shows the results of simulations of the hydrogen and silicon ions incident on an a-Si:H (hydrogen concentration in material 10 %) surface with the energy rage of 5- 45 eV. The projected ranges of silicon (0.4 nm at 5 eV and 0.8 nm at 45 eV) are much smaller than the ones of hydrogen (0.7 nm at 5 eV and 2.5 at 45 eV). The simulation results clearly shows an ion impact on the surface and subsurface layers of a-Si:H films with larger penetration depths for higher ion energies.

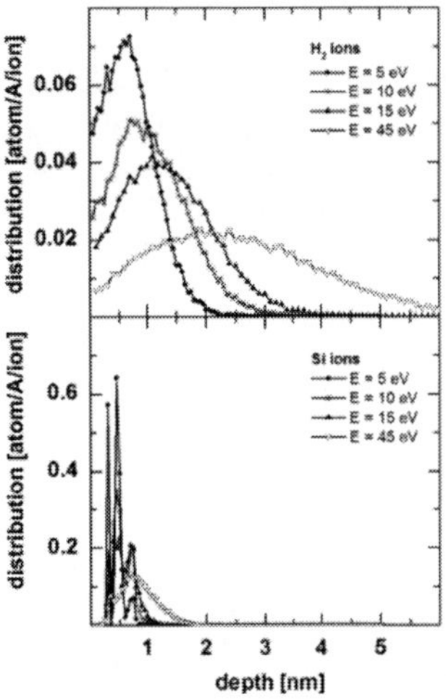

Figure 6: Simulated depth distributions of hydrogen (above) and silicon (below) ions incident on a-Si:H with varied ion energy

3.4 Solar cell simulations

The next study investigated the amorphous silicon solar cell structure by simulations. The aim was to examine the influence of intrinsic layer properties on the solar cell quality.

The solar cell structure was analyzed by performing a parameter sensitivity study with a computer program Advanced Semiconductor Analysis (ASA) [11]. The program solves the set of semiconductor equations for a-Si:H and μc-Si:H materials (Poisson equation and the current continuity equations for electrons and holes) and uses free electron concentration, free hole concentration and electrostatic potential as variables. The model parameters were obtained from measurements and literature [11-14].

Figure 7 shows the influence of the defect density of the intrinsic a-Si:H on the solar cell characteristics. It has been observed that an increase of defect density by a factor of 10 decreases the fill factor and initial efficiency by 20 % and 30 % respectively. The comparable deterioration of fill factor and efficiency has been established by varying the RF power from 10 to 90 W during solar cell deposition at 81.36 MHz.

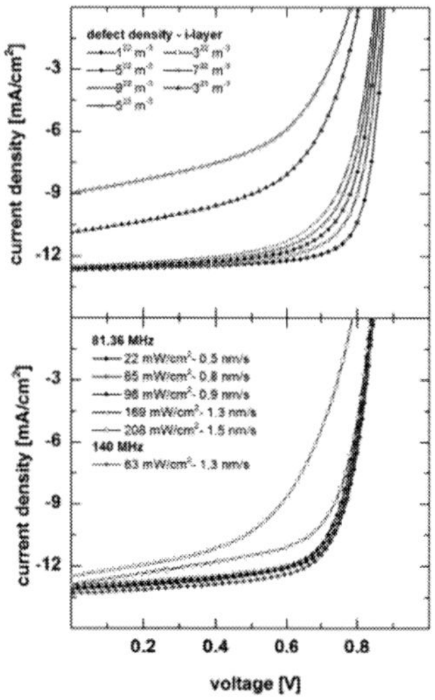

Figure 7: current-voltage characteristics: simulated for varied i - layer defect density (above) and measured for varied power density at 81.36 and 140 MHz (below)

4 CONCLUSION

The deposition rate increase resulting from an enhancement of the excitation frequency shows no negative influence on the layer properties and the solar cell quality for both a-Si:H and µc-Si:H materials. On the other hand higher plasma power leads to a deterioration of semiconductor properties and solar cell parameters. The simulation study shows smaller ion penetration depth for lower ion energy (higher plasma frequencies or smaller plasma power). Furthermore, the solar cells simulations show the role of intrinsic layer quality on solar cell performance.

5 ACKNOWLEDGEMENT

This work is supported by the Bundesministerium für Umwelt (BMU), Germany, under contract number 0325442.

6 REFERENCES

[1] M.J. Colgant, M. Meyyappan and D.E.Murnickt Plasma Sources Sd. Technol. 3 (1994) 181-189

[2] J. Dutta, U. Kroll, P. Chabloz, A. Shah, A. A. Howling, J.-L. Dorier, Ch. Hollenstein, J. Appl. Phys. 72 (7) (1992) 3220

[3] J. Meier, U. Kroll, E. Vallat-Sauvain, J. Spitznagel, U. Graf, A. Shah, Solar Energy 77 (2004) 983–993

[4] B. Leszczynska, C. Strobel, S. Leszczynski, M. Albert, J.W. Bartha, U. Stephan, J. Kuske, Proceedings of 28th European Photovoltaic Solar Energy Conference and Exhibition, Paris, (2013), S. 2568–2572

[5] C. Strobel, B. Leszczynska, U. Merkel, J. Kuske, D.D Fischer, M. Albert, J. Holovský, S. Michard, J.W. Bartha, Sol. Energy Mater. Sol. Cells 143 (2015) 347–353

[6] J. F. Ziegler et al. The Stopping and Range of Ions in Solids (Pergamon, 1985)

[7] W. B. Jackson, N. M. Amer, Phys. Rev. B 25, (1982) 5559–5562

[8] A. S. Abramov, A. Ya. Vinogradov, A. I. Kosarev, and M. V. Shutov, A. S. Smirnov and K. E. Orlov, Tech. Phys. 43 (1998) 180

[9] J. Dutta, U. Kroll, P. Chabloz, A. Shah, A.A. Howling, J.-L. Dorier, Ch. Hollenstein, J. Appl. Phys. 72 (7) (1992) 3220

[10] T. Novikova, B. Kalache, P. Bulkin, K. Hassouni, W. Morscheidt, P. Roca i Cabarrocas, J. Appl. Phys. 93 (2003) 3198

[11] R.E.I. Schropp, M. Zeman, Amorphous and Microcrystalline Silicon Solar Cells, Modelling, Materials and Device Technology, Kluwer Academic Publishers (1998)

[12]G. Munyeme, Experimental and Computer Modelling Studies of Metastability of Amorphous Silicon Based Solar Cells, Proefschrift Universiteit Utrecht (2003)

[13] J.A. Willemen, Modelling of amorphous silicon single- and multi-junction solar cells, Proefschrift Technische Universiteit Delft (1998)

[14] A.M.K. Dagamseh, B.Vet, P. Sutta, M.Zeman, Solar Energy Materials & Solar Cells 94 (2010) 2119-2123

ABSTRACT FOR EUPVSEC 2016
TOPIC: 3.3 (PEROVSKITE, ORGANIC AND HYBRID DEVICES)

DESIGN OF PEROVSKITE/CRYSTALLINE-SILICON TANDEM SOLAR CELLS

S. Altazin[1], L. Stepanova[1], K. Lapagna[2], P. Losio[2], J. Werner[3], B. Niesen[3], A. Dabirian[3], M. Morales-Masis[3], S. de Wolf[3], C. Ballif[3], B. Ruhstaller[1,2]*

[1]Fluxim AG, Technoparkstrasse 2, 8406 Winterthur, SWITZERLAND,
[2]Zurich University of Applied Sciences, Institute of Computational Physics, Technikumstrasse 9, CH-8401 Winterthur, SWITZERLAND,
[3]Ecole Polytechnique Fédérale de Lausanne (EPFL), Institute of Microengineering (IMT), Photovoltaics and Thin Film Electronics Lab, Rue de la Maladière 71, 2002 Neuchatel, SWITZERLAND

1) Summary
While the efficiency of silicon heterojunction solar cells has surpassed 25%, a novel route to high-efficiency wafer-based solar cells is being pursued with perovskite/silicon tandems, with monolithic devices having reached 21.2% [3] and 4- terminal measurements approaching 25% [9] with potential to come close to 30% efficiency. In this context we introduce a simulation tool that combines a 3D ray-tracing algorithm with thin film optics to study light interaction with conformally coated layers on planar and textured silicon wafers. After validating the model with experimental data we employ it for demonstrating further efficiency improvements that can be achieved with a rear-side random pyramid texture and an anti-reflection foil in the front. The simulations show that the light scattering introduced by the rear-side texture and the anti-reflection foil mainly enhances the photocurrent of the silicon heterojunction cell by 1.4 mA/cm^2 and 2.2 mA/cm^2, respectively. With additional adjustment of some layer thicknesses we thus conclude that the overall current enhancement can be as high as 3.9 mA/cm^2, corresponding to a relative boost of about 30% in current density and thus expected power conversion efficiency. Eventually such approaches should provide guidelines on how to further improve the efficiency of perovskite/silicon tandem cells.

2) Purpose of the work
With this contribution we introduce a simulation tool [1] suitable to design and optimize crystalline silicon and silicon heterojunction (SHJ) solar cells [2, 4] as well as monolithic perovskite/silicon tandem solar cells [3] where the perovskite top cell absorbs the visible part and the silicon bottom cell the near-infrared part of the sunlight spectrum. Both single-junction SHJ and tandem cells are based on silicon wafers that are commonly etched to form random pyramid textures (see Fig. 1) on the front and/or rear side in order to enhance light-scattering and absorption at long wavelengths.

Figure 1. Illustration of the random pyramid texture that is obtained by etching the silicon wafer. The blue and green lines visualize the propagation and scattering of light at this interface texture, as calculated by ray-tracing [1].

3) Approach
Light propagation and absorption in these cells is governed by coherence in the conformally coated amorphous silicon thin films of the SHJ cells as well as the thin films of the perovskite top cell. The underlying silicon wafer itself propagates light incoherently and is thus modeled by ray-tracing in 3D. Our two-scale simulation framework [1] combines these regimes and can thus cope with interference in the thin films as well as with light-scattering at the microscopic wafer texture. It thus becomes suitable to simulate and optimize SHJ cells as well as monolithic perovskite/silicon tandems. In order to validate the simulation tool with relevant solar cell structures we present the simulation of absorption spectra of the silicon wafer of a SHJ cell and compare it to measured external quantum efficiency (EQE) spectra of such SHJ cells [4]. For this comparison we consider both planar and double-sided random pyramid wafer textures, see Fig. 2. The comparison of measured EQE spectra with simulated layer absorption spectra is given in Fig. 3.

* corresponding author email: beat.ruhstaller@fluxim.com

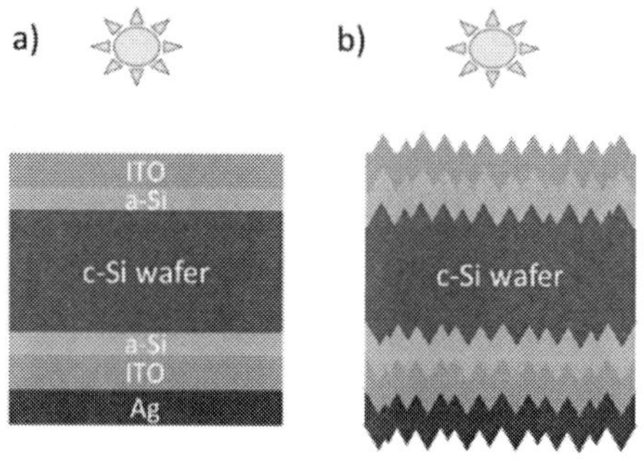

Figure 2. Sketch of SHJ cell structures for model validation: planar case and case of double-sided texture.

4) Scientific innovation and relevance

While planar thin film solar cells can be designed with the help of traditional thin film optics codes or generalized algorithms for multilayer structures including scattering interfaces [5], the simulation of wafer based silicon cells with microscopic textures is only feasible with a ray-tracing algorithm in 3D that accounts for the conformal thin film coatings.

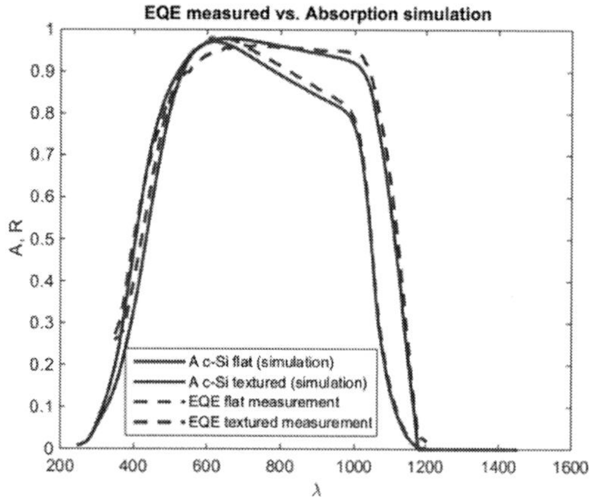

Figure 3. Comparison of simulated absorption (full lines) and measured EQE spectrum (dashed lines) of the SHJ cell structure shown in Fig. 1 both for the planar (blue) and textured (pink) wafer case.

It has recently been calculated by Filipič et al. [6] that (front-) textured perovskite/silicon tandem cells have the potential to exceed 30% power conversion efficiency. Moreover, Asadpour et al. [7] have predicted that bifacial (planar) perovskite/silicon tandems could reach 33% power conversion efficiency. Such reports attract attention to the model-based design and optimization of complex multilayer tandem structures. With the recent achievement of fabricating monolithic perovskite/silicon tandem cells with 1 cm^2 area [3], further improvements by suitable light-absorption enhancement structures that have the potential for industrial manufacturing are sought for.

5) Results

In this section we aim at simulating the absorption spectra of the two active layers in the tandem perovskite/silicon cell and wish to illustrate the room for photocurrent improvement by optical means. Assuming that the considered light-management structures do not affect the charge transport and photovoltage, these improvements translate into power conversion efficiency enhancements.

In Fig. 4 the three tandem structures considered here are illustrated. While structures shown in panels a) and b) were fabricated and reported in Ref. [3], structure of panel c) has not been reported yet in literature. The anti-reflection foil is realized with imprinting a random pyramid structure into a polymer film that is externally laminated to the solar cell [3]. In structure c) the perovskite top sub-cell is processed on the planar front face of a rear-textured silicon wafer for ease of processing and for exploiting scattering of long-wavelength light at the rear-texture. The combination of a rear-texture tandem and lamination of an ARF in the front is not considered yet here.

The refractive index dispersion data of the perovskite absorber layer and the Spiro-OMeTAD charge transport layer were obtained from published ellipsometry measurements [8]. Refractive index data from the other materials were measured by ellipsometry, too, or obtained from literature. The simulation results for these three structures are presented in Figs. 5 to 7.

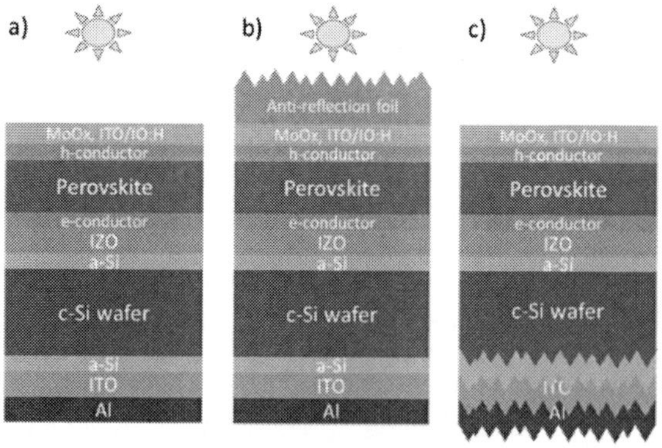

Figure 4. Three monolithic perovskite/silicon tandem structures considered in this manuscript: a) planar reference structure as reported in [3], b) planar reference with external anti-reflection foil (ARF), c) planar perovskite solar cell on rear-textured silicon wafer.

In Fig. 4 the effect of the ARF is clearly visible, leading to both an enhanced top-cell as well as bottom-cell current. With +2.1 mA/cm^2 the

* corresponding author email: beat.ruhstaller@fluxim.com

enhancement is stronger for the bottom crystalline silicon cell, however.
The simulation in the bottom of Fig. 4 was carried out with the Setfos software [1]. In the simulation the anti-reflection foil was modeled with 3D ray-tracing. The photon-to-charge conversion factor (i.e. internal quantum efficiency, IQE) for the perovskite sub-cell was set to 95% for the photocurrent calculation of Fig. 5 (bottom). The agreement between measurement and simulation is very good both in terms of spectral shape and photocurrent.
Having a reliable simulation tool at hand we can further optimize the thicknesses of the active and transport layers to achieve larger and matched sub-cell currents in the presence of the anti-reflection foil. For example, for the given SHJ cell structure a thickness sweep for the perovskite layer and the Spiro-OMeTAD layer is shown in Fig. 6. We achieve an increased and matched photocurrent of 16.1 mA/cm2 for a perovskite layer thickness of 280 nm and a Spiro-OMeTAD thickness of 150 nm, using the ARF. This is an improvement of 0.4 mA/cm2 with respect to the reference tandem stack with 15.7 mA/cm^2 sub-cell current of the silicon wafer.

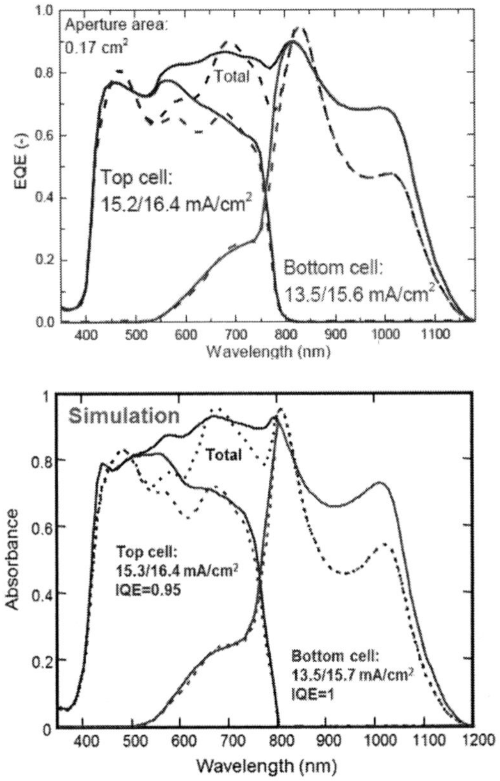

Figure 5. Experiment (top, from Ref. [3]) and simulation by Setfos 4.3 [1] (bottom) of tandem absorption spectra for the planar perovskite/silicon tandem with an anti-reflection foil (ARF) (structure b)) and without ARF (structure a) of Fig. 3).

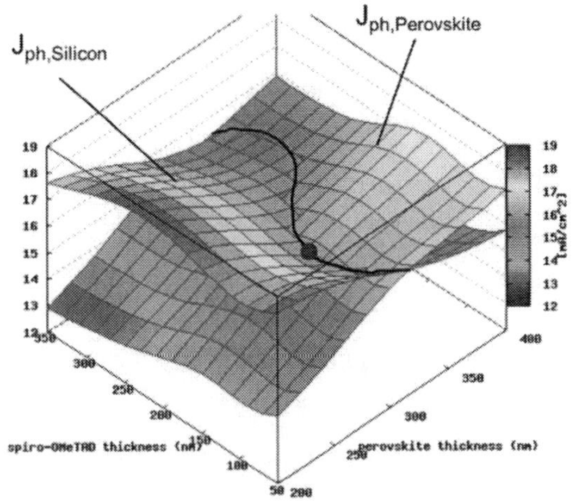

Figure 6. Simulation of the sub-cell currents vs. the thicknesses of the Spiro-OMeTAD layer and the perovskite absorber layer, for the case of using the anti-reflection foil (structure b) of Fig. 3). Black line indicates current matching and the red dot is the optimum with 16.1 mA/cm^2.

Finally, in Fig. 7 we present a prediction of the absorption enhancement that can be achieved when using a random pyramid texture on the rear side of the silicon wafer of the tandem cell, corresponding to structure shown in Fig. 4 (c). We observe that the optical absorption in the SHJ cell is enhanced significantly for $\lambda > 1000$ nm leading to an increase of the SHJ bottom cell photocurrent by 1.4 mA/cm^2. Since this enhancement owing to the rear-side wafer texture occurs in a different spectral range than the enhancement obtained for the anti-reflection foil (see Fig. 5), we conclude that both methods can be combined to achieve maximum boost in photocurrent. A straight-forward combination of the two enhancement mechanisms plus the re-adjustment of layer thicknesses (Fig. 6) would thus lead to a photocurrent increase of about 3.9 mA/cm^2, corresponding to a relative boost of almost 30% in device short-circuit current density.

* corresponding author email: beat.ruhstaller@fluxim.com

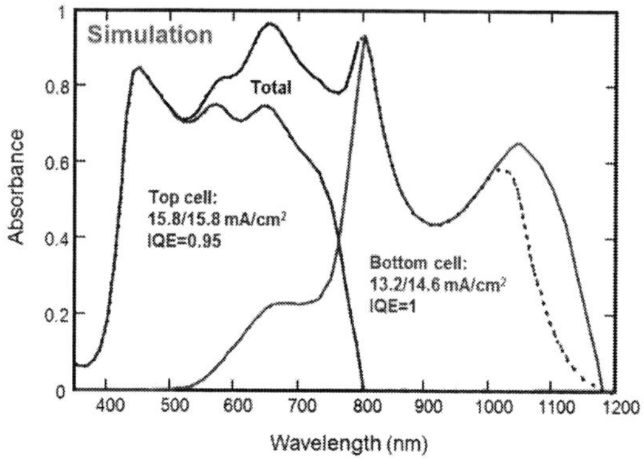

Figure 7. Simulation of tandem absorption spectra for the tandem structure with a planar perovskite cell on rear-textured silicon (structure c) of Fig. 3) compared with the planar tandem (structure a) of Fig. 3).

5) Conclusions

We presented the simulation of planar and textured SHJ single cells and perovskite/silicon tandem cells as well as the optical optimization thereof by means of microscopic textures and anti-reflection foils. We quantitatively assess the improvement of achievable photocurrent due the one- or two-sided wafer texture (in SHJ and perovskite/silicon tandems) as well as due to an anti-reflection foil. With knowledge of refractive index dispersion data of the involved materials, the predicted photocurrent and absorption (i.e. EQE) spectra agree well with the measured results. As the production processes for perovskite/silicon tandems are becoming more reliable and reproducible, the tuning of both top-cell layer thicknesses and light-scattering structures is of major importance. In this context, simulations will guide experimental work. Further microtexture wafer structures, ARF structures, modified refractive index and thickness of the rear TCO or bulk scattering films are conceivable to provide further improvement in the future if they are carefully tuned. Eventually, this shall allow breaking the efficiency record of single crystalline silicon cells by tandem devices.

Acknowledgments

Financial support by Swiss National Foundation for the PV2050 research project as well as by the Swiss Federal Office of Energy for the Connect-PV project is gratefully acknowledged.

References

[1] SETFOS 4.3 software with Absorption and Light-scattering modules, *Fluxim AG*, Switzerland
[2] A. Dabirian, S. Martin de Nicolas, B. Niesen, A. Hessler-Wyser, S. de Wolf, M. Morales-Masis, C. Ballif, *Adv. Mater. Interfaces* 1500462, (2015)
[3] J. Werner, C.-H. Weng, A. Walter, L. Fesquet, J.P. Seif, S. de Wolf, B. Niesen, C. Ballif, *J. Phys. Chem. Lett.* 7, 161 (2016)
[4] S. de Wolf, A. Descoeudres, Z.C. Holman, C. Ballif, Green 2, pp. 7-24 (2012)
[5] T. Lanz, B. Ruhstaller, C. Battaglia, C. Ballif, *J. Appl. Phys.* 110, 3, (2011)
[6] M. Filipič, P. Löper, B. Niesen, S. de Wolf, J. Krč, C. Ballif, M. Topič, *Optics Express* Vol. 23, No. 7, (2015)
[7] R. Asadpour, R.V.K. Chavali, M.R. Khan, M.A. Alam, *Appl. Phys. Lett.* 106, 243902, (2015)
[8] P. Löper, M. Stückelberger, B. Niesen, J. Werner, M. Filipič, S.-J. Moon, J.-H. Yum, M. Topič, S. de Wolf, C. Ballif, *J. Phys. Chem. Lett.* 6, 66 (2015)
[9] J. Werner, S.-J. Moon, P. Löper et al. 31st European PV Solar Conference and Exhibition. Hamburg, Germany, (2015) and private communication

* corresponding author email: beat.ruhstaller@fluxim.com

DYE-SENSITIZED/c-Si AND PEROVSKITE/c-Si TANDEM SOLAR CELLS

M.F. Vildanova, A.B. Nikolskaia, S.S. Kozlov, O.I. Shevaleevskiy
Solar Photovoltaic Laboratory, Institute of Biochemical Physics, Russian Academy of Sciences,
Kosygin St. 4, 119334 Moscow, Russia
e-mail: mvildanova@sky.chph.ras.ru

ABSTRACT: In this paper we present the results for development and a comparative investigation of the photovoltaic characteristics for the two types of tandem solar cells: (1) dye-sensitized/crystalline silicon (DSC/c-Si) tandem solar cell and (2) Perovskite/crystalline (Perovskite/c-Si) silicon tandem solar cell. We demonstrate the advantages of the fabricated tandem solar cells for using in the outdoor conditions. It was shown that under high light intensities both cells of the tandem structure nearly equally contribute to the performance of the tandem device. From the other hand, under low illumination intensities and diffused illumination the main contribution comes from the top DSC (Perovskite) solar cell. The results of long-term outdoor monitoring of the developed tandem structures revealed the advantages of the proposed tandem systems over the conventional solar cells to generate more electricity under real outdoor working conditions.
Keywords: Tandem, Dye-Sensitized, Perovskite, Spectral Response

1 INTRODUCTION

Tandem solar photovoltaic cells attract a considerable attention, mostly in view of the possibility of increasing the energy conversion efficiency up to theoretical limit by combining solar cells with different light absorption characteristics [1].

Crystalline silicon is known to be close to the best possible choice for the bottom cell of a tandem structure. Due to high efficiencies and stability silicon-based solar cells are dominating the PV market for several decades. However, to reach 30% efficiency, Si cells need to be combined with other types of wider bandgap solar cells.

Recent results have demonstrated a remarkable achievement of dye-sensitized solar cells (DSCs) efficiency with a confirmed record of 14.1%, reported by Michael Graetzel and his team from École Polytechnique Fédérale de Lausanne [2]. DSCs show a number of advantages over other PV technologies, such as variety of colors, semi-transparency and low cost fabrication processes, making them of interest as top cells for tandem applications. In our previous publications we have demonstrated the results of a long-term monitoring of DSCs containing cells and modules, those basically show the possibility of utilizing both direct sun irradiation and diffuse light under indoor and outdoor conditions [3].

Tandem cells are commonly divided into either two-, or four-terminal structures. In two-terminal device sub-cells are connected in series. Thus, monolithical integration requires electric current compatibility between top and bottom cells. Incident light spectrum also affects tandem cell performance. Spectral changes throughout the day can lead to mismatch between cell currents. On the other hand, four-terminal structure is flexible in the design and the choice of the sub-cells and their arrangement. Thus, a variety of top cells with different bandgaps may be chosen to enable high efficiencies of a tandem structure [4].

Recently, hybrid organic-inorganic halide perovskites have been introduced in solar cell application due to their absorption characteristics, high built-in voltage and simplicity of the cell fabrication [5]. They show great potential in tandem solar cells, combined with silicon and thin-film CIGS cells, to show efficiencies above 40% [6].

The main goal of this work is the development of the alternative tandem photovoltaic devices with a perfect performance for using at both high light illumination intensities and under cloudy weather conditions.

2 EXPERIMENTAL

2.1 Crystalline and multicrystalline silicon solar cells

Crystalline silicon solar cells used in our studies were provided by the research group of G.G. Untila from the Skobeltsyn Institute of Nuclear Physics, Lomonosov Moscow State University. Multicrystalline silicon solar cells were purchased from Huizhou Ysd Technology Company.

2.2 DSCs and PSCs fabrication

Following the conventional technology [7], TiO_2 based pastes were fabricated from commercially available P25 powder, mixed with a solution of terpineol and ethyl cellulose in ethanol. Solar cells were fabricated on a fluorine-doped tin oxide (FTO) coated glass substrates. The electrodes were prepared from the pastes obtained using screen printing method and sintered at 500°C for 30 min. The electrodes fabricated were dipped in a 0.3 mM of N719 (Sigma Aldrich) dye solution in a mixture of acetonitrile and tert-butyl alcohol (1:1) for 24 h. Platinum-coated FTO glass substrates were used as counter electrodes. The counter and the working electrodes were assembled into the sandwich-type cell and sealed with an ionomer film (Surlyn 1702). Commercially available electrolyte AN-50 Iodolyte (Solaronix) was introduced into the construction. Thus DSCs with the field size of 0.7 cm^2 were fabricated.

For Perovskite-based solar cells (PSCs) we used the fabrication technology described in [8]. Additionally the TiO_2 blocking layer was deposited on the surface of FTO glass substrate by spin-coating and than sintered at 400°C. The TiO_2 nanoparticles based paste diluted at 1:4 mass ratio of paste/ethanol was deposited by spin-coating (2000 rpm, for 60 s) to obtain the thickness of 300 nm and than sintered at 500°C for 1 hour. Perovkite was deposited on the TiO_2 surface by spin-coating (2000 rpm, for 60 s). The compounds of the solution: PbI_2, CH_3NH_3I and DMSO (in molar ratio 1:1:1) were dissolved in DMF and mixed for 1 hour at a room temperature. The TiO_2 layer covered with perovskite was heated at 65°C for 1 min and than at 100°C for 2 min.

2.3 Four-terminal tandem

In the four-terminal tandem cell the top and the bottom compartment were fabricated independently, than mechanically stacked together and electrically connected either in parallel, or in series. The cell performance characterization was carried out independently for each sub-sell and the total efficiency was calculated for the tandem structure. Commercial multicrystalline silicon solar cell was used as a bottom, combined either with DSC, or PSC as a top.

2.4 Cells performance characterization

Prior to the placement in the tandem system, each sub-cell was measured at 1 sun illumination under AM1.5 conditions (1000 W/m^2) using Abet Technologies Solar Simulator and Semiconductor Characterization System 4200-SCS (Keithley, USA). Light filters were used to provide the varied illumination intensities. The measurements of the external quantum efficiency (EQE) were provided using QEX10 Solar Cell Quantum Efficiency Measurement System (PV Measurements, USA) in the wavelength range 300 – 1100 nm. Optical absorption of the materials used was measured with a double-beam Shimadzu 3600 spectrophotometer with integrating sphere (ISR-3100) in the wavelength range 350–900nm.

2.5 Outdoor test facility

We have provided an outdoor monitoring of the developed tandem solar cells performance under varied weather conditions. The homemade test facility included series of DSCs and c-Si solar cells, and was located on the roof of the Institute of Biochemical Physics in Moscow. The PV performances of the cells were measured using Semiconductor Characterization System 4200-SCS (Keithley, USA).

3 RESULTS AND DISCUSSION

3.1 Individual DSC and c-Si cells characterization

First we have fabricated and investigated small laboratory DSC samples with the field size of 0.7 × 0.7 cm^2. The photovoltaic performance of DSCs was measured by recording the current–voltage (I–V) characteristics and energy conversion efficiencies under AM1.5 incident light illumination intensity (1000 W/m^2), the appropriate characteristics are listed in figure 1. Light intensity was adjusted to 1 sun using the reference cell. DSC with the field size of 0.7 × 0.7 cm^2 have demonstrated relatively high efficiency of 8,6%. Further the solar cells under investigation were used for fabricating the tandem DSC/c-Si structures. Current-voltage characteristics for the masked area of 0.7 × 0.7 cm^2 for c-Si cell can be seen in figure 1.

3.2 Individual PSC and c-Si cells characterization

As perovskite has been proved to be a promising candidate material for using in tandem applications, we have fabricated and investigated the laboratory samples to be combined with c-Si in a tandem scheme.

Strong optical absorption is a key component of the outstanding performance of the cells. Spectrum of the absorption coefficient for $CH_3NH_3PbI_3$ layer deposited on the mesoporous TiO_2 surface is given in figure 2.

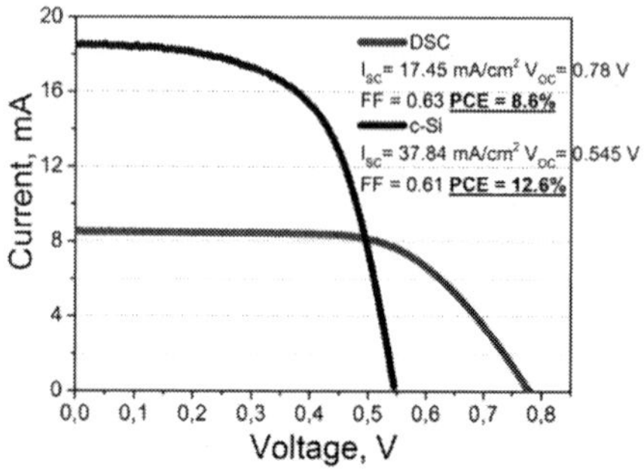

Figure 1: The I-V curves of the individual c-Si (black) and dye-sensitized solar cell (red).

The obtained result shows that due to high optical absorption characteristic the thickness of a perovskite layer used for the fabrication of Perovskite PV cell can be significantly reduced.

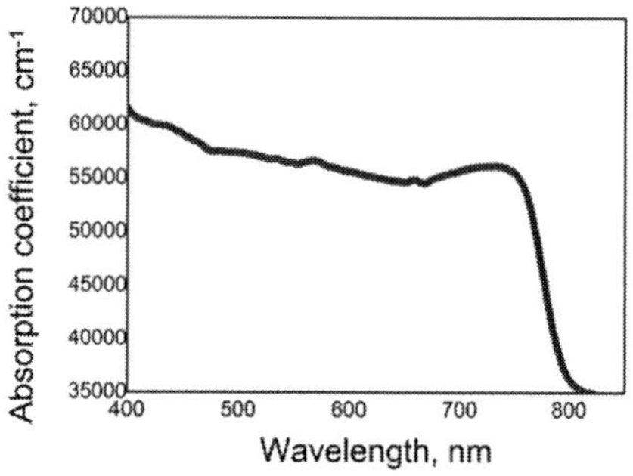

Figure 2: Optical absorption coefficient of $CH_3NH_3PbI_3$ deposited on the surface of TiO_2 mesoporous layer.

With the next step we have fabricated and investigated PSC samples with the field size of 0.5 × 0.5 cm^2. The PV performance and other characteristics of PSC cell are listed in figure 3.

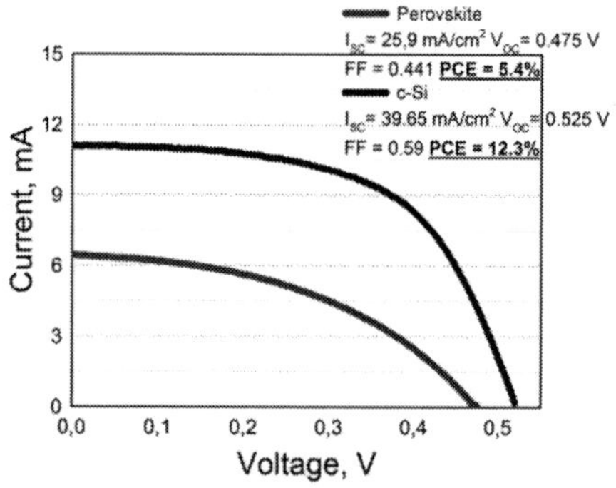

Figure 3: The I-V curves of the individual c-Si and Perovskite solar cells.

IPCE spectrum for a PSC sample, given in figure 4, shows that the spectral area of light energy conversion for this type of PV cell ranges from 400 to 800 nm.

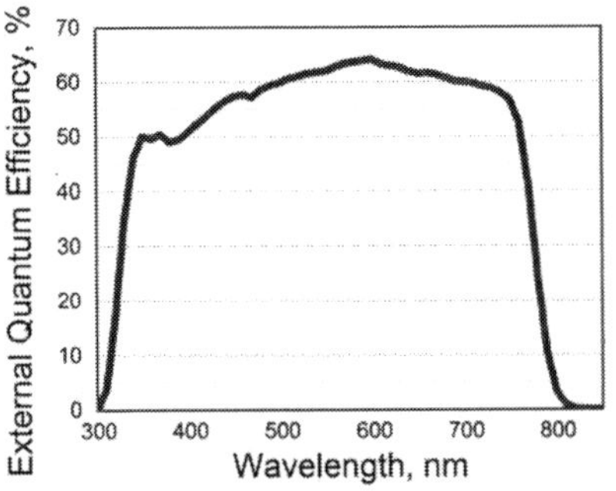

Figure 4: External quantum efficiency of the Perovskite solar cell.

3.3 Tandem DSC/c-Si solar cell characterization

We have fabricated and characterized series of the four-terminal tandem DSC/c-Si and PSC/c-Si solar cells.

The obtained experimental results have shown the advantages of the parallel connection of the individual cells in a tandem construction, that is in agreement with the data, proposed so far [9]. For the combination of a single DSC cell with conversion efficiency of 8.6 % and c-Si cell with 12.6%, the tandem cell have shown the performance of 15.3 % with a short circuit current value of 42,86 mA/cm^2. The resulting efficiency of the tandem system was found to be relatively high and have exceeded each of the sub-cells performances.

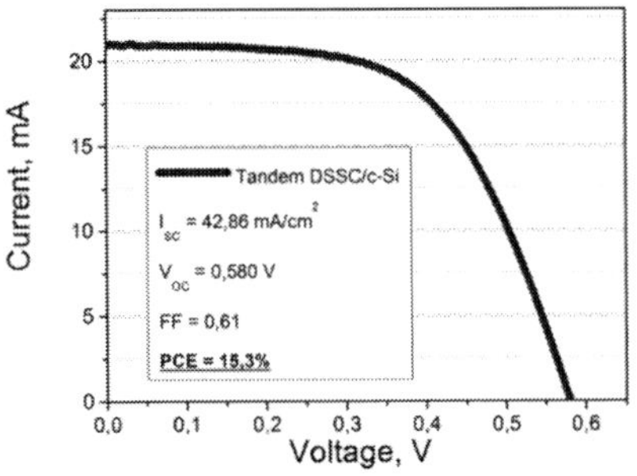

Figure 5: The I-V curve of the four-terminal tandem c-Si/dye-sensitized solar cell.

3.3 DSCs and c-Si solar cells characterization at varied illumination intensities.

Neutral-density light filters, reducing the intensity of all wavelengths equally, were used to perform the indoor characterization of the cells at varied illumination intensities.

We have demonstrated that PV performance of a solar cell shows a strong dependence on the light illumination intensity. The appropriate experimental results that confirm the given above statement are presented in figure 6 for DSC and c-Si solar cells.

It was shown that under high light intensities each of the cell used in the tandem structure nearly equally contribute to the performance of the tandem device. However, with the decrease of the illumination intensity the main contribution should be expected to come from the top DSC solar cell.

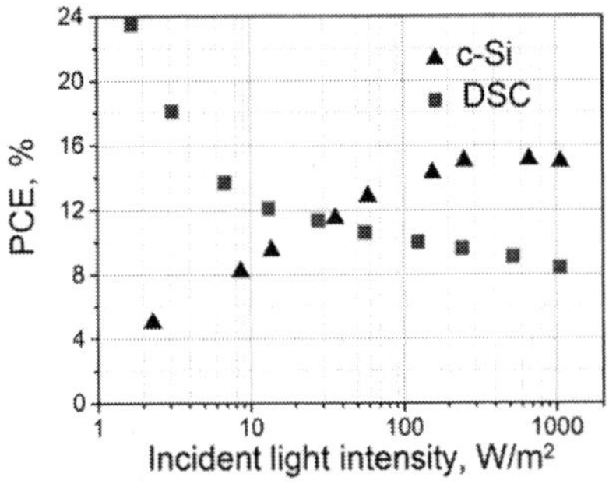

Figure 6: Indoor measurements of the power conversion efficiency values for c-Si and dye-sensitized solar cells carried out at varied illumination intensities.

3.4 Outdoor monitoring of the DSCs and c-Si solar cells performance

The outdoor monitoring was carried out for the demonstration of the improved performance of the developed tandem DSC/c-Si system at real weather conditions. The results of the monitoring are in coincidence with those, obtained within indoor measurements.

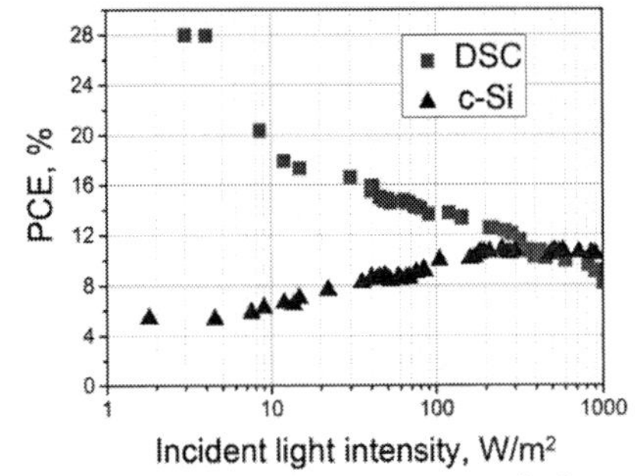

Figure 7: Outdoor measurements of the power conversion efficiency values for c-Si and dye-sensitized solar cells carried out at varied illumination intensities.

The result of the long-term outdoor monitoring of the developed tandem structures have confirmed the given above experimental measurements and revealed the advantages of the proposed tandem systems over the conventional solar cells to generate more electricity under real outdoor working conditions.

4 CONCLUSIONS

We have developed, fabricated and characterized four-terminal tandem DSC/c-Si and PSC/c-Si solar cells with the improved performance characteristics. We have carried out the indoor and outdoor performance measurements of the cells at varied illumination intensities. The results obtained both within outdoor and indoor monitoring are in good coincidence. Thus, we have demonstrated, that the improved performance of the developed tandem structure can be fully realized due to its possibility to utilize both the direct sun irradiation and the diffuse light. We have shown the advantages of the proposed tandem systems over the conventional solar cells to generate more electricity especially under low-light outdoor working conditions. With the results obtained for PSC/c-Si tandem cells, perovskite was proved to be a promising candidate material for using as the top cell in a tandem scheme with crystalline silicon under real outdoor conditions.

5 ACKNOWLEDGEMENTS

The work was partially supported by the Russian Foundation for Basic Research (RFBR) with the grant No. 16-08-00620.

The authors would like to thank the colleagues of G.G. Untila research group from the Skobeltsyn Institute of Nuclear Physics, Lomonosov Moscow State University, for their help in constructing c-Si solar cells for tandems.

6 REFERENCES

[1] O. Shevaleevskiy, Pure and Applied Chemistry 80 (2008) 2079.
[2] S. Mathew, A. Yella, P. Gao, R. Humphry-Baker, B.F.E. Curchod, N. Ashari-Astani, I. Tavernelli, U. Rothlisberger, M.K. Nazeeruddin, M. Gratzel, Nature Chemistry 6 (2014) 242.
[3] M.F. Vildanova, A.B. Nikolskaia, S.S. Kozlov, O.I. Shevaleevskiy, Journal of Physics: Conference Series 643 (2015) 012106.
[4] J.H. Noh, S.H. Im, J.H. Heo, T.N. Mandal, S.I. Seok, Nano Letters 13 (2013) 1764.
[4] W.S. Yang, J.H. Noh, N.J. Jeon, Y.C. Kim, S. Ryu, J. Seo, S.I. Seok, Science 348 (2015) 1234.
[6] M. Green, Proeedings 29th EU PVSEC Vol. 1 (2014) 1489.
[7] N. Tsvetkov, L. Larina, O. Shevaleevskiy, B.T. Ahn, Journal of the Electrochemical Society 158 (2011) B1281.
[8] M.M. Lee, J. Teuscher, T. Miyasaka, T.N. Murakami, H.J. Snaith, Science 338 (2012) 643.
[9] M. Durr, A. Bamedi, A. Yasuda, and G. Nelles, Applied Physics Letters 84 (2004) 3397.

TRAP AND RECOMBINATION CENTERS STUDY IN ORGANOLEAD HALIDE PEROVSKITES

G. Gordillo[1], C. A. Otálora[2], F. E. Guzmán[1], A. A. Ramirez[3],

[1] Departamento de Física, Universidad Nacional de Colombia, Bogotá, Colombia
[2] Departamento de Química, Universidad Nacional de Colombia, Bogotá, Colombia
[3] Departamento de Ingeniería Química, Universidad Nacional, Bogotá, Colombia

The mechanisms affecting the trapping and recombination processes in thin films of methyl ammonium lead iodide ($CH_3NH_3PbI_3$) were studied by means of transient photoconductivity measurements and theoretical simulation of the relaxation curves resulting from the photocurrent measurements. In particular there was studied the influence of pressure inside the system of measurement, temperature and intensity of illumination on the photoconductivity response. By using the real part of Fourier transform, the experimental curves of photocurrent rise were analyzed. The analysis revealed that the photocurrent resulting from measurements performed at room temperature under vacuum includes contributions of both band to band photo generation and generation controlled by delayed trap-assisted processes, being the contribution that the first process makes to the intensity of the photocurrent much greater than that of the second one. However if the measurements are done under atmospheric pressure conditions, the photocurrent arise from contributions originated from the same two processes that contributed to the photocurrent obtained under vacuum and a further contribution related with a non delayed trap-assisted process. Studies of the intensity dependence of steady state photocurrent revealed that the processes of generation and recombination in CH3NH3PbI3 films are presented through a combination of monomolecular and bimolecular traps.

Keywords: Thin films, organolead halide perovskites, photoconductivity, trapping and recombination processes.

1 INTRODUCTION

Recently, methylammonium lead halide ($CH_3NH_3PbX_3$, X = halogen; CH_3NH_3: MA) and its mixed-halide crystals, have been used as light harvesters for solar cells[1-3]. These devices are cost-effective due to their simple fabrication process and high PCE. As a result, these cells have attracted a great deal of attention among researchers in the field.

Perovskite solar cells have shown a rapid efficiency improvement (from 2.2% in 2006 to 20.1% in 2014 [4]) and in December 2015 – Dyesol Limited (ASX: DYE) announced that a research team headed by Professors Michael Grätzel and Anders Hagfeldt at the Ecole Polytechnique Fédérale de Lausanne (EPFL) established a new world record efficiency for its Perovskite Solar Cells, with a certified conversion efficiency (at the laboratories of Newport Corporation) of 21.02% [5] .

In this work, the electrical properties of thin films of methylammonium lead iodide ($MAPbI_3$) deposited in one step by spin-coating using a solution containing methylammonium iodide (MAI) and lead iodide (PbI_2) dissolved in dimethylamine sulfoxide (DMSO) were studied by means of photoconductivity measurements. Special emphasis was made on understanding the mechanisms affecting the trapping and recombination processes of carriers, through theoretical simulation of the relaxation curves resulting from the photocurrent measurements. In particular there was studied the influence of the pressure inside the system of measurement, temperature and intensity of illumination on the photoconductivity response.

2 EXPERIMENTAL

We studied $CH_3NH_3PbI_3$ films deposited on glass slides by spin-coating at 3000 rpm of a mixed halide precursor solution composed of CH_3NH_3I and PbI_2 0.4 M in DMSO. The solution was prepared using commercial PbI_2 (Sigma Aldrich) and MAI synthesized in our laboratory as precursors. The MAI precursor was synthesized from a mixture at 0° C for 2 hours of 5 mL of hydroiodic acid to 67%, 15 mL of solution 2.0 M of methylamine in methanol and 20 mL of water. The obtained solid was purified by washing it with diethylether up to obtaining the pure and dry compound.

Electrical characterization of thin films of $MAPbI_3$ was done through photoconductivity measurements carried out using Ag electrodes and a 250–W Halogen lamp. In particular there was studied the influence of oxygen partial pressure inside the system of measurement, temperature and intensity of illumination (changed by varying the sample-lamp distance) on the photoconductivity response. The type of conductivity was also evaluated, performing photocurrent measurements with lighting focused on one end and changing the direction of the bias voltage.

3 RESULTS

3.1 Theoretical Considerations

The amplitude, time constants, and delayed time (escape time from the traps) have been determined simulating the rise and decay curves of photocurrent, using the method proposed by N.V. Joshi et al., [6], that assumes that if recombination process is other than linear or the time constant varies during the process, then the form of the relaxation curves can be expressed by multiexponential terms. In this case the rise and decay curves can be expressed by [7] :

$$I_{ph} = \begin{cases} \sum_{i=1}^{M} I_i\left(1 - e^{-t/\tau_i}\right) + u(\lambda)\sum_{j=1}^{K} I_j\left(1 - e^{-(t-\tau_{oj})/\tau_j}\right) & (1a) \\ \sum_{i=1}^{M} I_i e^{-t/\tau_i} + u(\lambda)\sum_{j=1}^{K} I_j e^{-(t-\tau_{oj})/\tau_j}, & (1b) \end{cases}$$

Where $u(\lambda)$ is a step function defined as:

$$u(\lambda) = 0 \ \ for \ \ t < \tau_{oj} ,$$
$$u(\lambda) = 1 \ \ for \ \ t > \tau_{oj} .$$

Here, the summation in the first term of Ec. 1 correspond to recombination processes and goes up to M, where M is the number of contributing components and the summation for delayed terms goes from 1 to K, where K is the number of types of traps which hold charge carriers for a certain time, thus, contributing in a delayed signal.

Typical rise and decay curves of photocurrent obtained from measurements carried out in vacuum (2mBar) at temperatures less than 35°C, could be simulated quite well with two non delayed and one or two delayed time constants. However the curves of photocurrent measured under vacuum at temperatures greater than 35°C show an anomalous behavior, characterized by the appearance of a photocurrent additional in the first seconds after the lighting, which decreases rapidly (see Fig. 4). This type of photocurrent curves were simulated by adding a decreasing exponential function of the form $I_{Ad}e^{-t/\tau_{Ad}}$ to the equation (1).

Equations corresponding to the real and imaginary parts of the Fourier transform of Eqs. (1) (see details in Ref. 1) provides sufficient information in order to obtain the values of I_i, I_j, τ_i, τ_j and τ_{oj} uniquely and precisely without ambiguity, in contrast to numerical or curve fitting approaches. Using values of I_i, I_j, τ_i, τ_j and τ_{oj} obtained from the solution of the equations resulting for the real part of the fast Fourier transform with the help of a conventional Cooley-Tukey algorithm, the photocurrent curves were constructed and compared with the experimental one.

3.2 Photocurrent measurements

Initially an experiment was carried out to identify the character p, n or compensated of the $MAPbI_3$ films, that consisted of measuring the photocurrent focusing the illumination close to one of the electrodes and applying direct or reverse bias respectively. it is clear from fig. 1 that the photocurrent under reverse bias is larger than the obtained under direct bias, indicating that the density of photogenerated holes is greater than that of the electrons; these results allowed us to conclude that the $MAPbI_3$ films are slightly p type and that its high resistance is due to a strong compensation.

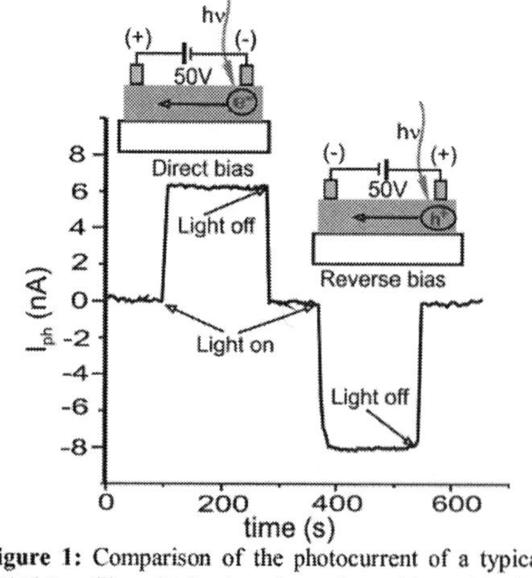

Figure 1: Comparison of the photocurrent of a typical $MAPbI_3$ film obtained under direct bias with that obtained under reverse bias. Inset shows a scheme of the structure used to perform the experiment.

In Fig. 2 is compared a curve of photocurrent rise of a typical $MAPbI_3$ film obtained from measurements performed at normal conditions of temperature and pressure under constant bias of 50 V and excitation power density of 1500 Lux, with a curve simulated assuming two non delayed time constants (τ_i) associated to direct-band-gap and trap assisted processes and one delayed time constant (τ_j) associated to a trap-assisted contribution. It was found that the major contribution to the total photocurrent comes from the direct-band-gap process and the contribution from the traps whose activation process is delayed is significantly less.

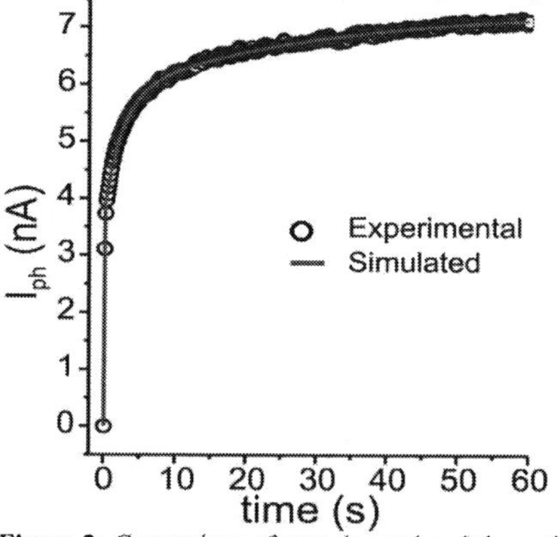

Figure 2: Comparison of experimental and theoretically simulated curves of the rise of photocurrent of a typical $MAPbI_3$ thin film.

Fig. 3 shows rise curves of photocurrent obtained from measurements carried out inside a vacuum chamber under different pressures, using a bias of 50 V and lighting of 1500 Lux. From these results can be observed that the form of the photocurrent curves is significantly affected by the pressure at which the photocurrent is measured. Analysis of these results with the help of theoretical simulation of the experimental photocurrent curves, indicated that the photocurrent resulting from measurements performed under low pressures (<200 mBar) include the contributions of both band to band photo generation and generation controlled by delayed trap-assisted processes, being the contribution of the first process much greater than that of the second one.

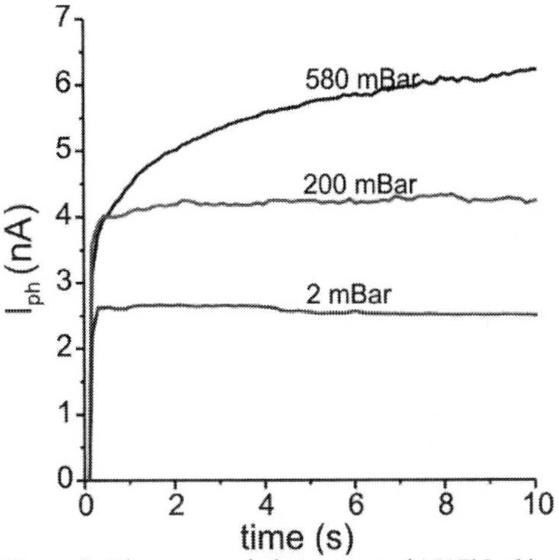

Figure 3: Rise curves of photocurrent of $MAPbI_3$ thin films, measured under different pressures.

When the pressure inside the measuring chamber is increased, the steady-state photocurrent increases as well as the time required to reach its saturation. Theoretical simulation of the photocurrent obtained from measurements done under atmospheric pressure conditions (680 mBar) indicated that in this case, the photocurrent arise from contributions originated from band to band photo generation, delayed trap-assisted processes and of a further contribution related with a non delayed trap-assisted process. It was also found that the time constants of the delayed and non delayed processes increase when the pressure inside the measuring chamber increases. This behavior could be explained assuming that the photoconductivity in air of the $MAPbI_3$ thin films is mainly governed by surface related process induced by adsorption and desorption of the chemisorbed oxygen at the surface of the $MAPbI_3$ ($O_{2(g)} + e^- \rightarrow O^{-2}$ (adsorbed species)), whereas the photoconductivity in vacuum is governed by bulk related process. The initial fast rise in the photocurrent may be attributed to the bulk related process taken onto account that the bulk related process is considered to be fast in comparison to the surface related process [8]. The slow photoresponse of the $MAPbI_3$ films, observed when the photocurrent is measured in air, seems to be caused by shallow traps generated by chemisorbed oxygen.

As can be seen from Fig.3, the steady state value of photocurrent in air is significantly higher than the photocurrent in vacuum. This may be attributed to increased dark current in air which is higher than the dark current in vacuum, because upon illumination, the photogenerated holes lead to a release of the captured adsorbed species ($O^{-2} + h^+ \rightarrow O_{2(g)}$).

Besides the effect of the pressure, we also studied the effect of temperature on the photoconductivity of the $MAPbI_3$ films. Fig. 4 shows the rise of photocurrent obtained from measurements in vacuum at different temperatures under illumination of 1500 Lux, by applying a bias of 50 V.

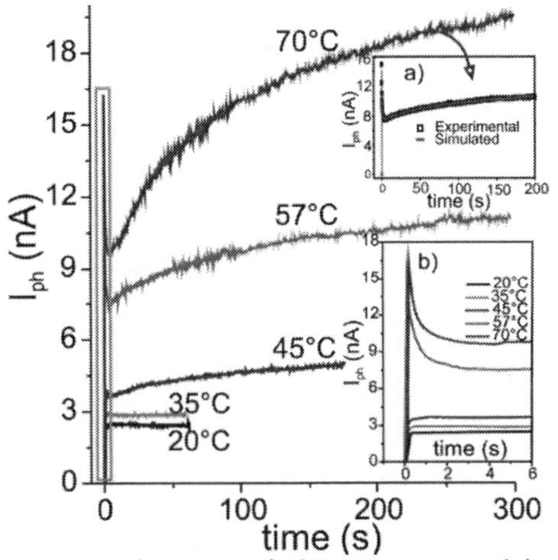

Figure 4: Rise curves of photocurrent measured in vacuum at temperatures ranging from 20 to 70°C. The inset a) shows the photocurrent curve measured at 70°C where the red line represents the fit with eq 1. modified by adding a decreasing exponential function of the form $I_{Ad}e^{-t/\tau_{Ad}}$ and inset b) shows an extended plot of the area enclosed in the orange rectangle.

As can be seen from Fig. 4, the steady-state photocurrent and the time required to reach its saturation increase when the measuring temperature is increased. It is also observed that the photocurrent resulting from measurements done at temperatures greater than 35°C, exhibit an anomalous behavior consisting in a very fast increases of the photocurrent when the light is switched on, due to the quick generation of electron and hole pairs as a result of absorption of photons. However, after attaining a peak, the photocurrent started to decrease very fast even during illumination. This anomalous behavior may be attributed to thermally induced traps characterized by having a very large capture cross section and a very small decay time τ_{Ad} *(~ 1 s)*. The increase of the steady-state photocurrent when the temperature is increased can be attributed to a reduction in the height of the potential barrier in the grain boundaries caused by thermal excitation of electrons trapped in states of the grain boundaries.

Intensity dependence of steady state photocurrent has also been studied to see the nature of recombination processes. It is clear from Fig. 5 that the form of the rise curves of photocurrent and the time required to reach the saturation are not affected by the increase in the intensity of the lighting, indicating that there is no generation of traps by photoinduction mechanisms. The curve of lnI_{ph} *vs.* $ln\Phi$ shown in inset of Fig. 5 is a straight line, which indicate that the photoconductivity follows a power law with intensity :

$$\Phi.I_{Ph\sim\Phi^{\gamma}} \quad (2)$$

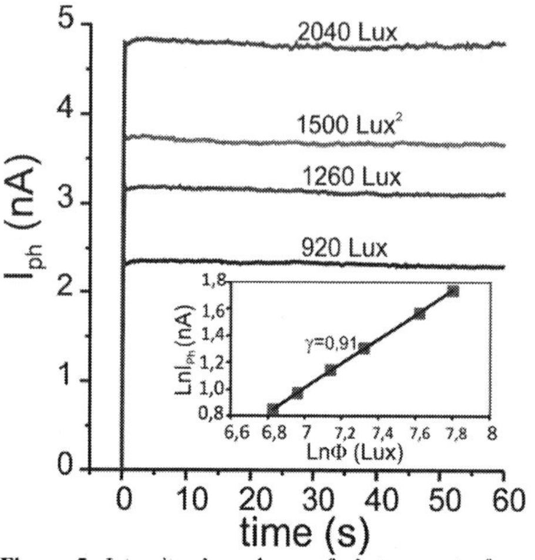

Figure 5: Intensity dependence of photocurrent for a typical $MAPbI_3$ film. Measurements done in vacuum at 20°C.

For $\gamma = 0.5$, the recombination is predominantly bimolecular, for $\gamma = 1.0$, the recombination is monomolecular in nature and for $0.5 \leq \gamma \leq 1.0$, the value of γ can be interpreted by assuming the existence of continuous distribution of trap levels in the band gap [9] and the resulting recombination mechanism will be bimolecular [10]. The value of γ obtained for the studied $MAPbI_3$ films is 0.91, indicating that the generation and recombination processes could be attributed to a combination of monomolecular and bimolecular traps, predominating the processes through the fist one.

4 CONCLUSIONS

By using the real parts of Fourier transform, the rise curves of photocurrent of $MAPbI_3$ thin films have been analyzed. In particular there was studied the influence of the pressure inside the system of measurement, temperature and intensity of illumination on the photoconductivity response. The analysis revealed that the photocurrent resulting from measurements performed under vacuum (<200 mBar) is due to contributions of both band to band photo generation and generation controlled by delayed trap-assisted processes, being the contribution that the first process makes to the intensity of the photocurrent much greater than that of the second one. Theoretical simulation of the photocurrent curve obtained from measurements done under atmospheric pressure conditions indicated that in this case, the photocurrent results from contributions originated from the same two processes that contributed to the photocurrent obtained under vacuum and a further contribution related with a non delayed trap-assisted process.

The photocurrent measured at temperatures greater than 35°C, exhibit an anomalous behavior consisting in a very fast increases of the photocurrent when the light is switched on, followed by a very fast decrease of its maximum, even during illumination. This anomalous behavior may be attributed to thermally induced traps characterized by having a very large capture cross section and a very small decay time.

Studies of the intensity dependence of steady state photocurrent revealed that the processes of generation and recombination in $MAPbI_3$ films are presented through a combination of monomolecular and bimolecular traps, predominating the processes through the latter.

ACKNOWLEDGEMENTS: This work was supported by Colciencias (Contract #038/2013) and Universidad Nacional de Colombia, Sede Bogotá , Facultad de Ciencias, Grupo GMS&ES, Bogotá DC, Colombia (Proy. 20287 supported by DIB).

5 REFERENCES

[1] M. M. Lee, J. Teuscher, T. Miyasaka, T. N. Murakami, H. J. Snaith, Science 2012, 338, 643– 647.
[2] H.S. Kim, C.R. Lee, J.-H. Im, K.-B. Lee, T. Moehl, A. Marchioro, S.J. Moon, R. Humphry-Baker, J.H. Yum, J. E. Moser, M. Gratzel, , N.G. Park, "Sci. Rep. 2012, 2, 591:1–7.
[3] J.H. Noh, S.H. Im, J.H. Heo, T.H. Mandal, S.I. Seok, Chemical Management for Colorful, Efficient, and Stable Inorganic–Organic Hybrid Nanostructured Solar Cells. Nano Letters 2013: 13; 1764–1769.
[4] National Center for Photovoltaics at the National Renewable Energy Laboratory, Research cell efficiency records, www.nrel.gov/ncpv/.
[5]http://www.dyesol.com/media/wysiwyg/Documents/2015-asx-announcements/2015-12-08-DYE0397_ _EPFL_achieves_21_efficiency.pdf
[6] A. Serfaty and N. V. Joshi, Fourier-transform analysis of multiexponential curves observed in photoconductivity studies, Physical Review B Vol. 41, No 12 ,1990, 8545-8548
[7] J. P. McKelvey, Solid State Semiconductors Physics (Harper and Row, New York, (1966)
[8] Zhang D H 1995 *J. Phys. D: Appl. Phys.* **28** 1273
[9] A. Rose, Concepts in Photoconductivity and Allied Problems, Interscience, New York, 1963.
[10] K. Shimkawa , *J. Non-Cryst. Solids* **77** (1985) 1253

PEROVSKITE SOLAR CELL BASED ON $CH_3NH_3PBI_{3-2}CL_2$ /$PC_{61}BM$

Jiří Vaněk, Dávid Strachala, Josef Hylský, Michal Kadlec, Marcela Sionová, Martin Weiter
Brno University of Technology, Faculty of Electrical Engineering and Communication Technologies,
Department of Electrotechnology
Technicka 10, 616 00 Brno, Czech Republic

ABSTRACT: The main purpose of the work is to find out an optimal way for the preparation of the perovskite solar cells based on organic-inorganic structure of $CH_3NH_3PbI_{3-2}Cl_2$ /$PC_{61}BM$. During the manufacture process was created a four solar cell samples with with emphasis on high efficiency and stable structure. Conversion efficiency of the best sample reached approximately η_{MAX} = 7, 22 %. Samples were further analyzed and based on that were implemented other techniques for their improvement.
Keywords: Perovskite solar cell, conversion efficiency, organic-inorganic solar cells.

1 PEROVSKITE SOLAR CELL

Predicted reduction of the fossil energy sources and increasing environmental needs, leads to an intensive research of the new renewable energy sources - especially the methods of conversion the solar energy via alternative photovoltaic cells. The attention was focused on the development of solar cells based on perovskite in the recent years. Well-known scientist Michael Grätzel called perovskite as "unusually cheap" material that could operate successfully in solar cells and combine the seemingly incompatible - low acquisition cost and high efficiency of energy conversion of sunlight into electricity [1]. Advantages of perovskite solar cells lies in their potential use in a broad application such as windowpanes, facade systems and much lower production costs (also availability of materials) in comparison to a traditional silicon solar cells with a relatively easier methods of its preparation. Perovskite solar cells can be considered as promising power source worthy attention.

Perovskite in general is a name for a group of substances with the same crystal structure as $CaTiO_3$. In photovoltaic cells, modified compounds of $CaTiO_3$ are used for absorbing of sunlight. A semiconductor is created as an organic-inorganic hybrid composition as $CH_3NH_3PbX_3$ (X = Cl, I, Br). Solar cells based on perovskite have been continuously developed in the last five years. The first time of publication of this material in the solar cells was in 2009 - energy conversion efficiency of photovoltaic solar cells with liquid electrolyte was 3.13% and 3.81% ($CH_3NH_3PbBr_3$ and $CH_3NH_3PbI_3$) [1]. In the following years, due to optimizations of various parameters of the photovoltaic cells there has been a rapid increase in energy conversion efficiency in a relatively short time [2].

Presented work deals with a hybrid peroskvit solar cell structure of $CH_3NH_3PbI_{3-2}Cl_2$/$PC_{61}BM$. Fig. 1 shows the general device structure of the perovskite solar cells.

2 MANUFACTURING OF PEROVSKITE SOLAR CELL BASED ON $CH_3NH_3PBI_{3-2}Cl_2$ /$PC_{61}BM$ STRUCTURE

An important step of work was to choose an appropriate process of manufacturing of examined peroskvite solar cell. Solar cells were fabricated on PEDOT coated indium tin oxide glass (ITO) with emphasis on the optimal way of heat treatment. The ITO substrates were before this process cleaned in an ultrasonic bath. $CH_3NH_3PbI_{3-2}Cl_2$ film was formed through a spincoating of a mixture of CH_3NH_3PbI, PbI_2 and $PbCl_2$. After verifying that the precursor solutions were successfully synthesized, the organic semiconductor $PC_{61}BM$ was deposited. To prevent the creation of interdiffusion between CH_3NH_3I and $PC_{61}BM$, high temperature wasn´t applied. Finally, aluminum cathode was deposited through a mask on the active layer. A detailed description of the preparation is given in the following text.

a)

b)

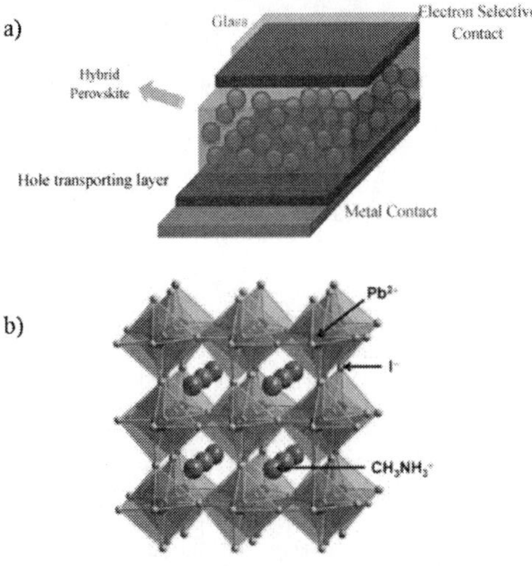

Figure 1: a) General structure of perovskite solar cells [2]; b) Structure of active perovskite layer

Preparation of the carrier substrates with an ITO layer was proceeded according to the following steps: the substrates were firstly put into the beaker with a 5-10% of NaOH solution. Subsequently, after a period of 3-5 minutes, were left in the ultrasonic bath, then rinsed with MilliQ water and finally cleaned with an isopropylalcohol.

PEDOT: PSS was before applying filtered through a 45 μm PVDF filter. A 55 ml of solution was used. Spin-coating was adjusted to a rotational speed of 3000 rpm (with the acceleration speed of 500 rpm) for 30 seconds. The cathode strip was after this process purified by MilliQ water. The final layer thickness after annealing (120 ° C for 30 sec.) was 30 - 40nm.

Perovskite layer was prepared from molar solution of the mass fraction of 1:1:4 with a total concentration of 9 percentage by weight. In numbers it means: of CH_3NH_3PbI (0.19561g), PbI_2 (0.141822g) and $PbCl_2$

(0.08555g). The solution was prepared within the laboratory conditions for 12 hours at the temperature of 60 °C with the stirring speed of 650 rpm. The solution was then filtered with a 45μm filter and applied to a preheated glass slides (125 °C for a 30 sec) by spin-coating in an amount of 35μl with the rotational speed of 3000 rpm with a speed acceleration of 500 rpm. After deposition was on the slides cleaned anode strip. After this step were slides transferred to a hot plate with an 100 °C temperature for a period of 120 minutes. This process wasn't unfortunately successful - created samples has shown a low efficiency in the light spectrum of 280 - 350 nm. It could be due to burn-through of layers which was indicated by the higher amount of carbon contained in the samples. The procedure was therefore adjusted to heating for only 30 seconds and preheating of the glass was reduced from 100 ° C to 70 ° C – figure 2. After this steps did not occur burn-through of layers anymore.

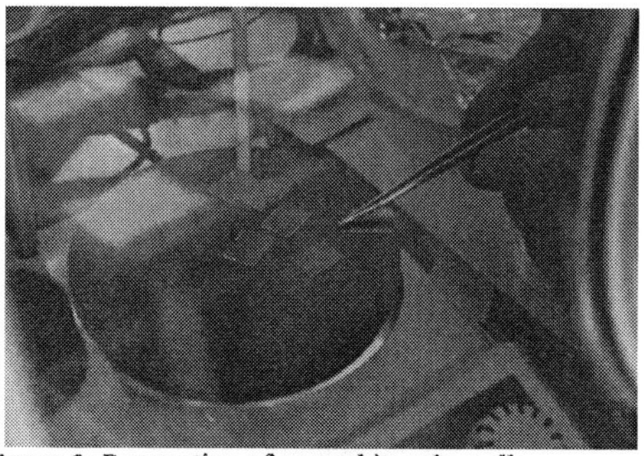

Figure 2: Preparation of perovskite solar cells

The layer of $PC_{61}BM$ was applied similarly. The solution was prepared from 10 mg / ml of $PC_{61}BM$ in a mixture of anhydrous solvents of chloroform-chlorobenzene (1: 1). Deposition was performed through a spin-coating at 1200 rpm for 60 sec. in an amount of 50 ml, but this time without a subsequent annealing. Cathode strip was refined with a chloroform.

In the end an aluminum electrodes were vapor-deposited with a thickness of 100 nm followed by encapsulation of the samples with the epoxy resin. Cover glass with contacts were also attached.

The result of the manufacturing process were 4 created samples. Sample 1 is shown in Figure 3.

Figure 3: Perovskite photovoltaic cell after manufacturing process – sample 1

3 MEASURED DATA

In the structure of the peroskvite solar cells was necessary to monitor various parameters that influenced the resulting device functionality. One of the parameters investigated in the work was the preparation of optimized layer $CH_3NH_3PbI_{3-2}Cl_2$ / $PC_{61}BM$, which was used to form functional and stable reference solar cell with the best electrochemical and physical properties. The structure of these films were scanned by several methods using environmental scanning electron microscope and an atomic force microscope. The chemical composition of the layers was then examined via X-ray diffraction spectroscopy for a suitable setup of deposition processes.

Initial measurement has shown efficiency of the created perovskite solar cells around 5 percent. The highest conversion efficiency was achieved in sample no. 1 with η_{MAX} = 7.22 %, FF = 71,35 %, J_{SC} = 11.02 mA/cm², U_{OC} = 0.918 V . For more parameters see Table I.

Table I: Parameters of manufactured perovskite solar cells – sample 1. and sample 3.

Device	J_{sc} (mA/cm2)	I_{sc} (mA)	V_{oc} (mV)	P_{mpp} (mW)	FF (%)	η (%)
1_P1	10.48	0.49	958.68	0.33	70.93	7.13
1_P2	**11.02**	**0.52**	**918.46**	**0.34**	**71.35**	**7.22**
1_P3	10.70	0.50	646.29	0.20	61.01	4.22
1_P4	8.75	0.41	634.02	0.16	61.39	3.41
1_P5	9.73	0.46	890.33	0.28	68.21	5.91
1_P6	9.48	0.44	936.07	0.30	71.57	6.35
3_P1	5.15	0.24	928.53	0.12	54.92	2.63
3_P2	10.27	0.48	883.63	0.28	65.62	5.96
3_P3	10.88	0.51	475.35	0.13	53.75	2.78
3_P4	10.05	0.47	514.53	0.13	53.88	2.79
3_P5	9.52	0.45	838.21	0.22	59.22	4.72

Current voltage characteristic of the best sample is shown below in Figure 4. Measurements were done through a solar simulator with source of light energy of 1000 W/m².

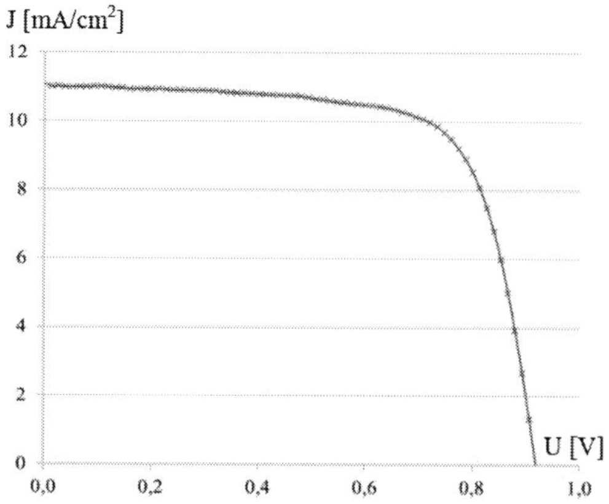

Figure 4: Current voltage characteristic of perovskite solar cell – sample 1.

3.1 Degradation of perovskite solar cell

With a measurement during illumination the efficiency has increased to a certain value, then again has dropped. The reason is probably in the temperature of the sample during measurements (samples were rapidly heating). Samples were also measured consecutively in short intervals. The gradual degradation of the material, well-known phenomenon of perovskite solar cells, has occurred later. The conversion efficiency of the cells has decreased significantly after 20 hours of its function from η_{MAX} = 7.22 % to η_{MAX} = 5.48 %. Degradation of the perovskite solar cells is currently one of the biggest problem of these materials [3], [4].

Degradation and changes in the parameters of the perovskite samples is most influenced by moisture, which despite encapsulation may partially penetrate into the layers. An interesting phenomenon has ocurred related to the gradual increasing and lowering of conversion efficiency. The grains have unified gradually leading to the enlargement of its volume. This may involve into the reducing of the number of grain boundaries which forms a barrier to the flow of current, which could explain the increasing and decreasing of the efficiency. AFM microscope analysis was done because of observation of ongoing processes – Figure 5. This phenomenon will be given additional measurements on perovskite photovoltaic cells.

a)

b)

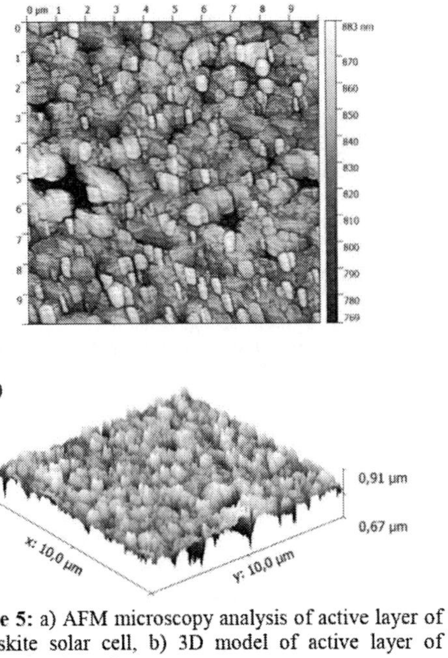

Figure 5: a) AFM microscopy analysis of active layer of perovskite solar cell, b) 3D model of active layer of perovskite solar cell.

Subsequent research will focus mainly on preventing of the degradation by applying the appropriate interlayers between perovskite and its metal contacts. The Cr_2O_3 interlayers will be studied in the context of resistance of the perovskite material against environmental influences [5]. The Cr_2O_3 layer will be applied through magnetron sputtering equipment in the Department of Electrical and Electronic Technology of Brno University of Technology.

4 SUMMARY/CONCLUSION

This work is aimed at optimizing the perovskite solar cells with synthehis of $CH_3NH_3PbI_{3-2}Cl_2$ through a mixture of CH_3NH_3PbI. PbI_2 and $PbCl_2$. In the perovskite solar cell was examined the influence of reactant ratio, time and annealing temperature during its preparation with an emphasis on the absorption and stability of the layers.

The first measurements of realized samples according to chosen procedure has shown an efficiency of peroskvite solar cells around η_{MAX} = 5% percent. The highest conversion efficiency was achieved in sample no. 1 (η_{MAX} = 7.22%)

The main idea of the following work is to achieve the highest quality and most stable structures of synthesized materials and create the long-term stable solar cells with high efficiency of photovoltaic conversion

4.1 References

[1] Kim, H.-S.; Lee, C.-R.; Im, J.-H.; Lee, K.-B.; Moehl, T.; Marchioro, A.; Moon, S.-J.; Humphry-Baker, R.; Yum, J.-H.; Moser, J. E.; Grätzel, M.; Park, N.-G. Nat. Sci. Rep. 2012, 2,591

[2] GRÄTZEL, Michael. *The light and shade of perovskite solar cells*. DOI: 10.1038/nmat4065. ISBN 10.1038/nmat4065. http://www.nature.com/doifinder/10.1038/nmat4065

[3] STRACHALA, D.; HYLSKÝ, J.; KADLEC, M. Organic- Inorganic Hybrid Solar Cells. In *Proceedings of the 22nd Conference STUDENT EEICT 2016*. Brno: Vysoké učení technické v Brně, Fakulta elektrotechniky a komunikačních, 2016. s. 630-634. ISBN: 978-80-214-5350- 0

[4] SONG, Dandan, Jun JI, Yaoyao LI, et al. 2016. Degradation of organometallic perovskite solar cells induced by trap states: degradation and stability. Applied Physics Letters. 108(9), 093901-.DOI: 10.1063/1.4943019. ISBN10.1021/acsami.5b04490. ISSN 0003-6951. http://scitation.aip.org/content/aip/journal/apl/108/9/10.1063/1.4943019

[5] KALTENBRUNNER, Martin, Getachew ADAM, Eric Daniel GŁOWACKI, et al. *Flexible high power-per-weight perovskite solar cells with chromium oxide–metal contacts for improved stability in air*. DOI: 10.1038/NMAT4388. ISBN 10.1038/nmat4388. Dostupné také z: http://www.nature.com/doifinder/10.1038/nmat4388

4.2 Acknowledgement

This research work has been carried out in the Centre for Research and Utilization of Renewable Energy (CVVOZE). Authors gratefully acknowledge financial support from the Ministry of Education, Youth and Sports of the Czech Republic under NPU I programme (project No. LO1210).

PROCESING AND OPTIMIZATION OF THE PEROVSKITE SOLAR CELL BASED ON PEDOT:PSS/$CH_3NH_3PBI_{3\text{-}X}CL_X$

Michal Kadlec, Jiří Vaněk, Dávid Strachala, Josef Hylský, Marcela Sionová, Martin Weiter
Brno University of Technology, Faculty of Electrical Engineering and Communication Technologies,
Department of Electrotechnology
Technicka 10, 616 00 Brno, Czech Republic

ABSTRACT: Purpose of the work is the preparation of perovskite photovoltaic cells and optimization of the structure based on PEDOT:PSS/ $CH_3NH_3PbI_{3\text{-}X}Cl_X$. The main idea is the study of changes in their photovoltaic properties depending on the different composition. As a result were created an eight samples which were further analyzed to improve the manufacturing process.
Keywords: Perovskite photovoltaic cells; PEDOT:PSS; manufacturing process of perovskite solar cell

1 INTRODUCTION

The predicted depletion of fossil energy resources, increasing pressure from environmental and climate changes leads to intensifying research in field of renewable energy sources, especially in photovoltaic solar energy conversion. It is expected that a new generation of solar cells will be able to generate power even in places where it has previously been difficult to install solar cells (e.g. window glass).

Although materials for these solar cells are generally less inexpensive (because of the very small quantity of material used in a relatively large area), increasing the efficiency of photoelectric energy conversion and its durability are technically difficult. In the last two years there has been a rapid development in the field of perovskite solar cells. Currently, it is assumed that the 20% efficiency should be exceeded [1].

Another option is a replacement of silicon solar panels, because the costs of perovskite solar cell manufacturing are much lower. In the structure of these solar cells it is possible to monitor a large number of parameters that have influence on the resulting device functionality. One of these parameters is optimization of $CH_3NH_3PbI3_{\text{-}x}Cl_x/PC_{61}BM$ layer preparation. The purpose of this optimization is to create a functional and stable solar cell with the best electrochemical and physical properties.

2 PEROVSKITE SOLAR CELL

The presented work deals with the perovskite solar cells known as $CH_3NH_3PbI_{3\text{-}x}Cl_x$ / PEDOT:PSS. Its structure is indicated in Fig. 1 [1].

a)

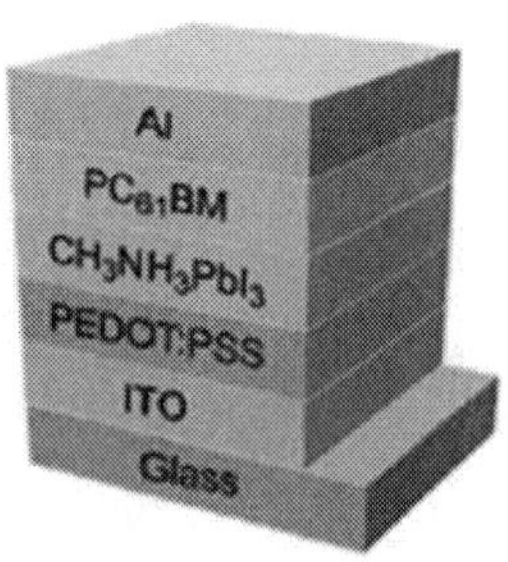

b)

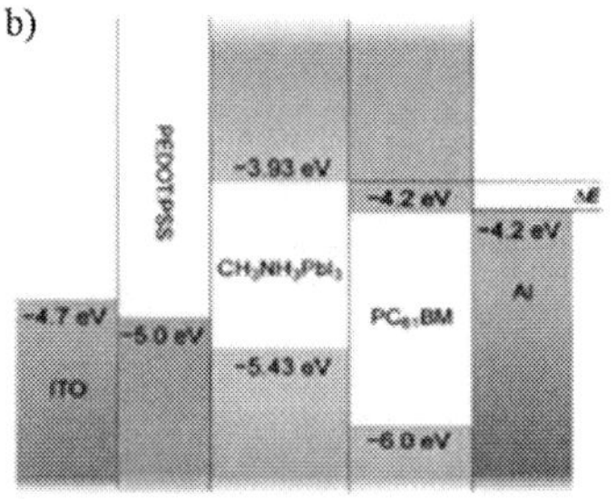

Figure 1: a) General composition of perovskite solar cells; b) energy level diagram of peroskvite solar cell

3 MANUFACTURING OF PEROVSKITE SOLAR CELL

Manufacturing of perovskite solar cells, according to the modified structure shown in Fig. 1 a) was as follows:

Preparation of supporting substrates with ITO layer: at first the substrates were placed into a beaker with 5-10% NaOH solution. Subsequently, there were left in an ultrasound bath for 3-5 minutes and rinsed with MilliQ water and finally cleaned with isopropyl alcohol.

PEDOT: PSS was at first filtered through a 0.45 μm PVDF filter on preheated glass at 140 ° C, then it was applied by spin-coating method. The resulting layer thickness was about 30 – 40 nm.

Perovskite layer was at first prepared from perovskite CH_3NH_3PbI, then from PbI_2 and then form $PbCl_2$. Solutions were prepared in laboratory conditions for 12 hours at 60 ° C and stir-ring speed of 500. After stirring it was filtered through a 45μm filter and applied by spin-coating on prepared microscope slides which has been preheated to 140 ° C in an amount of 35 μl, rotational speed was 3000 rev / min for 30 s with a speed acceleration 500 rev / s. After application it was cleaned cathode strip on slides and the slides were transferred to a hot plate at 100 ° C and heated for 120 minutes.

Similarly, it was applied $PC_{61}BM$ layer. The solution was prepared from 10 mg / ml $PC_{61}BM$ in a mixture of anhydrous solvents Chloroform: Chlorobenzene (1:1). Then is was applied by spin-coating at 1200 rev / min. for 60 s in an amount of 50 μl, this time without a subsequent annealing. After applying it was refined cathode strip by chloroform.

The approximate manufacturing process is shown at Fig. 2.

Figure 2: Shortened process of $CH_3NH_3PbI_{3-x}Cl_x$ / PEDOT:PSS samples manufacturing

3.1 Microscopy analysis of cross section

Microscopic scanning of the cross section of sample number 3 has shown that the sample did not have required structure – see Fig 3. This may have occurred by burn-through of layers during the manufacturing process which could be indicated by a very high amount of carbon contained in the sample. This assumption was also confirmed by images from the SEM microscope Vega3 Tescan and X ray spectroscopy analysis measured at Brno University of Technology. Based on this observation were the manufacturing process of other samples changed - a temperature of preheating during application of $CH_3NH_3PbI_{3-x}Cl_x$ was reduced from 140 °C to 60 °C

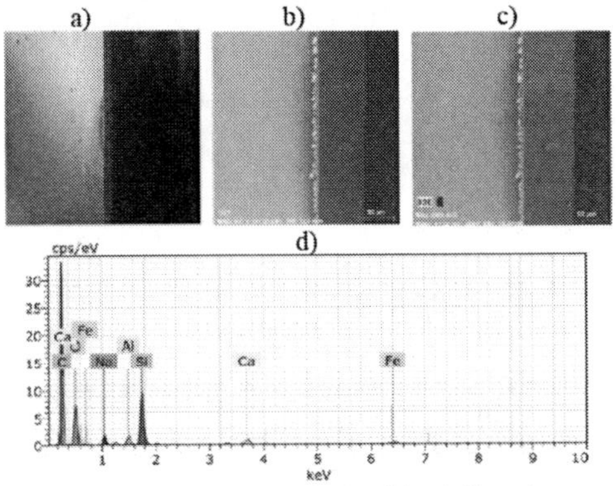

Figure 3: a) Microskopic analysis of the deffected sample 1 by SEM Vega3 Tescan -magnitude 150x, HV 20kV; b) X ray spectroscopy analysis - magnitude 617x, HV 20kV, c) X ray spectroscopy analysis- BSE C magnitude 617x, HV 20kV d) increased incidence of carbon caeused probably by burn-through of applied layers

4 RESULTS

Conversion efficiency of perovskite solar cells was during the initial measurement equal to approximately 6 %. The highest conversion efficiency was reached by cells No. 1 (ηMAX = 7.09%); No.2 (ηMAX = 5.57%) and No.4 (ηMAX = 5.21%). Measurements were performed on the solar simulator with light source of 1000 W / m2 value. The average final values are shown in Table I.

Table I: The average of the measured values perovskite solar cells

Parameter	I_{SC} [mA]	U_{OC} [mV]	I_{MAX} [mA	FF [%]	η [%]
Average value	0,4653	940,3767	0,4095	71,1144	6,6488

Efficiency of solar cells during the measurement fluctuated. Fluctuation was probably caused by heating of the cells during the measurement. The intervals between individual measure-ments were not sufficient.

After some time, perovskite solar cells began to degrade, which is its huge disadvantage. The conversion efficiency of measured structures declined significantly within a few days. Degra-dation is currently one of the largest problems associated with the development of perovskite solar cells. Subsequent research will especially deal with its elimination (for example by sputtering Cr_2O_3 layer between the aluminum electrode and material $PC_{61}BM$ [2])

5 CONCLUSION

Originally, there were perovskite solar cells prepared in combination with a liquid electrolyte, but there occurred rapid degradation due to dissolution of crystals of perovskite. Presented work aims to find the optimal method for the synthesis and preparation of perovskite solar cell based on PEDOT:PSS/ $CH_3NH_3PbI_{3-X}Cl_X$ with promising conversion efficiency of 7 %.

5.1 References

[1] KALTENBRUNNER, Martin, Getachew ADAM, Eric Daniel GŁOWACKI, et al. 2015-8-24. Flexible high power-per-weight perovskite solar cells with chromium ox-ide–metal

[2] Kim, H.-S.; Lee, C.-R.; Im, J.-H.; Lee, K.-B.; Moehl, T.; Marchioro, A.; Moon, S.-J.; Humphry-Baker, R.; Yum, J.-H.; Moser, J. E.; Grätzel, M.; Park, N.-G. Nat. Sci. Rep. 2012, 2,591

4.2 Acknowledgement

This research work has been carried out in the Centre for Research and Utilization of Renewable Energy (CVVOZE). Authors gratefully acknowledge financial support from the Ministry of Education, Youth and Sports of the Czech Republic under NPU I programme (project No. LO1210).

THRESHOLD TRAP DENSITY FOR VALID MOTT-SCHOTTKY ANALYSIS IN CARRIER SELECTIVE OPTOELECTRONIC DEVICES

Vikas Nandal* and Pradeep R. Nair
Department of Electrical Engineering, Indian Institute of Technology Bombay
Mumbai - 400076, India
*phone: +91 9769548217, email: nknandal@iitb.ac.in

ABSTRACT: Solution processed carrier selective devices for optoelectronic applications have gained a lot of recent research interest. Performance optimization of such devices requires detailed understanding of device and material characterization techniques. Capacitance voltage technique using Mott-Schottky (MS) analysis is one such powerful characterization scheme. However, applicability of such analysis for carrier selective PIN devices is not well explained. Here, we report detailed numerical simulations to understand the effect of defect states on CV characteristics. We find that MS analysis is reliable only if trap density is more than a certain threshold trap density, which decreases with thickness of active layer. Apparent linearity in $1/C^2 - V$ plot is often not sufficient, and hence MS analysis should use devices with different active layer thickness to obtain a reliable estimate of trap densities.
Keywords: Mott-Schottky, trap density, HTL, ETL

1 INTRODUCTION

Organic and hybrid materials have attracted a lot of research interest among photovoltaic community due to light weight, low cost, flexibility and ability to get solution processed at low temperature [1]. However, device performance is limited by low mobility and high defect density in such materials [2]. Of the many device configurations, p-i-n based carrier selective device architecture where the active material is sandwiched between doped contact layers (see in fig. 1 a) is now widely explored as solar cells [3] and light emitting diodes [4]. Device and material characterization is essential step towards optimization process to achieve better performance for such device architectures.

Capacitance voltage characteristics are often used in the literature to extract the built in voltage and doping density using Mott-Schottky (MS) analysis – not just for classical PN junction devices, but also for carrier selective p-i-n devices as well [5]. Electrostatics of depletion region plays significant role on the applicability of above said analysis. It provides accurate results only when depletion/space charge region consists of uniform charged density or bands profile are parabolic in nature [6]. In solution processed materials, presence of defect states (reported on the order of $10^{16} - 10^{17}\ cm^{-3}$ [5, 7]) could significantly impact electrostatics of intrinsic layer and thereby influencing the applicability of MS analysis. Therefore, it is important to understand the nature of space charge in the active layer which contributes towards capacitance. To address this issue, we performed detailed numerical simulation [8] to study the CV characteristics. Below, we first describe the model system used for numerical simulations and then explore the validity of MS analysis for carrier selective p-i-n devices.

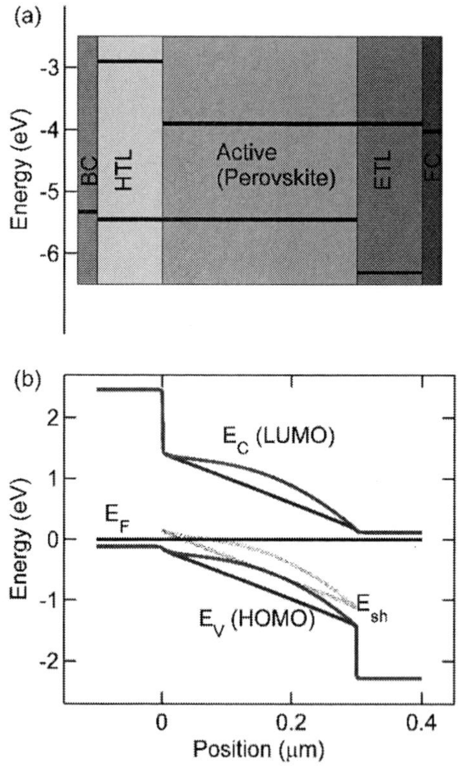

Figure 1 a) Schematic of the modeled device along with the band alignment of different layers, (b) shows the comparison of equilibrium band diagram for trap density, $N_t = 10^{15} cm^{-3} eV^{-1}$ (blue showing linear band profile) and $3 \times 10^{16} cm^{-3} eV^{-1}$ (red curve) at trap energy $E_{tv} = 0.3\ eV$ from valence band.

2 MODEL SYSTEM

Carrier selective device is modeled as 3 layer planar hetero-structure, where the active material (organic-inorganic Perovskite $CH_3NH_3PbI_{3-x}Cl_x$, with $E_g = 1.55$ eV) is sandwiched between hole transporting layer (HTL) and electron transport layer (ETL). Active material is assumed to be intrinsic in nature. Band alignment of ETL and HTL ensure the transport of one type of charge carrier from active layer to the metal contacts or vice versa. Current is mainly dominated by recombination of charge carriers inside active layer because band offset in conduction/valence band at the interface of active layer and HTL/ETL, respectively blocks over the barrier flux of charge carrier. It is assumed that the contact between metal and ETL/HTL is ohmic in nature. Positional disorder in active material is modeled as Gaussian distribution of defect/trap density of states in energy, $N(E)$ and is given by

$$N(E) = N_t \exp\left(-\frac{(E - E_{tv})^2}{2\sigma^2}\right)$$

Where, $N_t\,(cm^{-3}eV^{-1})$ is the maximum trap density of states at trap energy $E = E_{tv}\,(eV)$ from valence band, $\sigma\,(eV)$ is the broadening factor in energy. It has been assumed that trap distribution $N(E)$ was uniform throughout the thickness of active material (L_{ac}). N_t and L_{ac} was varied to study electrostatic effect of trapped charges on band profile whereas broadening factor, σ is fixed to $0.11\,eV$. Modeled device was discretized in one dimension and steady state solution was achieved by solving Poisson's and continuity equation for electrons and holes self-consistently at each bias step. Fig. 1 (b) shows the equilibrium band diagram where non-linearity in band profile was introduced by trapped charges.

4 RESULTS AND DISCUSSION

Capacitance voltage characteristics of the modeled system were obtained by superimposing AC modulating signal at each dc bias step. Frequency of the modulating signal is adjusted in such a way that all trapped charges respond to the modulating signal. In this section, we report the effect of different conditions (N_t, L_{ac}) of active material on the applicability of MS analysis. Capacitance is a measure of response of space charge with the modulating signal. Space charge inside active layer can be mobile and trapped charges implying that capacitance can be thought of as independent additive response of mobile and trapped charges. Trap density was varied from $10^{15} - 10^{18} cm^{-3}eV^{-1}$ at trap energy 0.3 eV in order to understand the effect of trap density on CV characteristics. Fig. 2 (a) shows the effect of trap density on capacitance voltage characteristics ($L_{ac} = 300\,nm$). At low trap density, capacitance in diffusive regime consists of response of injected mobile and trapped charges which was also evident from MS plot (not shown here). Non-linear plot of MS becomes linear with the increase in trap density,

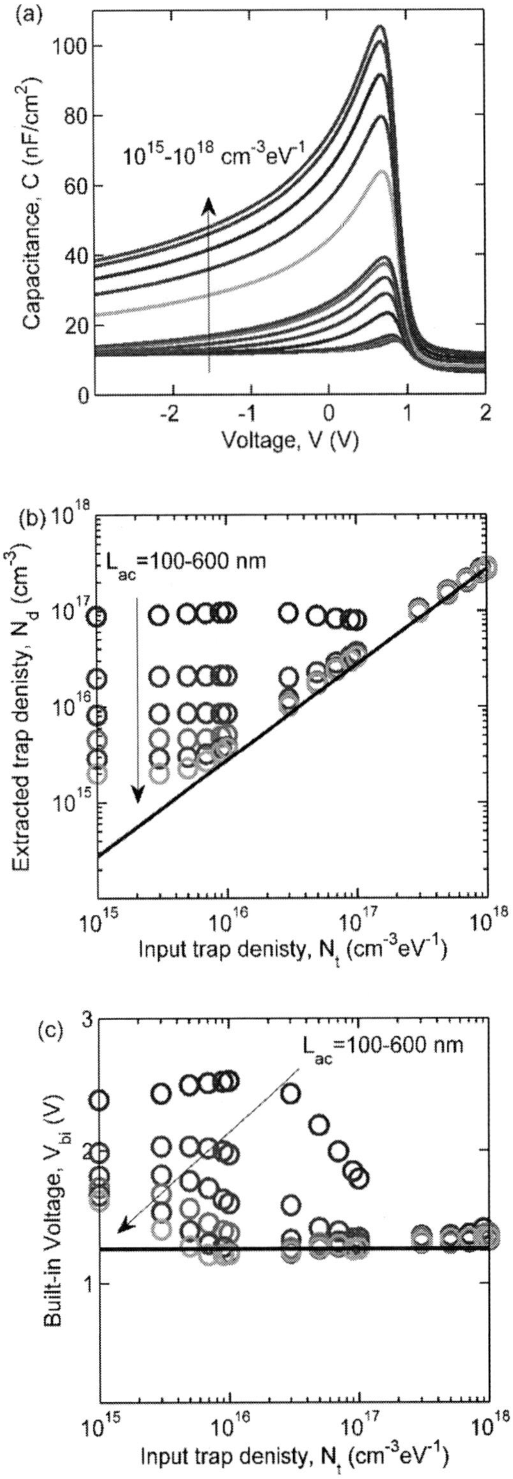

Figure 2: (a) Capacitance voltage characteristics with different trap density ($N_t = 10^{15} - 10^{18} cm^{-3}eV^{-1}$) at trap energy, $E_{tv} = 0.3\,eV$. Comparison of ideal values (solid line) of (b) trap density N_t and (c) built-in voltage, V_{bi}

against the extracted values (circles) from MS analysis at different trap density for different active layer thickness (100-600 nm) with dielectric constant of 4.

which implies that response of trapped charges is solely responsible for capacitance.

Mott-Schottky analysis was performed to extract space charge density and built-in voltage at different trap densities. Fig. 2 (b, c) shows the comparison of extracted values (circles) with ideal values (solid line) and indicate that the extracted trap densities and built-in voltage significantly differ from the ideal values. Curiously, this trend depends on the active layer thickness. Accordingly, we can define threshold trap density as the trap density above which good estimates of built-in voltage and trap density can be determined. Fig. 2 (b, c) shows that this threshold trap density increases with a decrease in the active layer thickness. Further, these results also indicate the need for MS analysis on devices with different active layer thickness to obtain a reliable estimate for trap density. For example, the threshold trap density is $3 \times 10^{16} cm^{-3} eV^{-1}$ for a device with active layer thickness of around 300 nm. This implies that any experimental observation of a trap density lower than the above mentioned for a sample of similar thickness is unreliable, unless supported with additional experimental data using devices with larger active layer thickness.

4 CONCLUSION

Detailed numerical simulations of carrier selective p-i-n devices were performed to explore capacitance voltage characteristics. The limitations of MS analysis to back extract the trap density were explored and a threshold trap density has been identified above which good estimates for trap density and built-in voltage for shallow states can be obtained. This threshold trap density was found to be inversely proportional to active layer thickness.

5. ACKNOWELDGEMENT

This paper is based upon work supported in part by the Solar Energy Research Institute for India and the United States (SERIIUS), funded jointly by the U.S. Department of Energy (under Subcontract DE-AC36-08GO28308) and the Govt. of India's Department of Science and Technology (under Subcontract IUSSTF/JCERDC-SERIIUS/2012).

6. REFERENCES

[1] M. Wright and A. Uddin, "Organic—inorganic hybrid solar cells: A comparative review, *Solar Energy Materials and Solar Cells*", vol. 107, pp 87-111, Dec. 2012.
[2] M. M. Mandoc, F. B. Kooistra, J. C. Hummelen, B. de Boer, and P. W. M. Blom, "Effect of traps on the performance of bulk heterojunction organic solar cells", *Appl. Phys. Lett.*, 91, 263505 (2007).
[3] M. A. Green, A. Ho-Baillie, and H. J. Snaith, "The emergence of perovskite solar cells", *Nat. Photonics*, vol. 8, no. 7, pp. 506–514, Jun. 2014.
[4] J. Huang, M. Pfeiffer, A. Werner, J. Blochwitz, K. Leo, and S. Liu,"Low-voltage organic electroluminescent devices using pin structures", *Appl. Phys. Lett.*, 80, 139-141 (2002).
[5] M. Samiee, S. Konduri, B. Ganapathy, R. Kottokkaran, H. a. Abbas, A. Kitahara, P. Joshi, L. Zhang, M. Noack, and V. Dalal, "Defect density and dielectric constant in perovskite solar cells," *Appl. Phys. Lett.*, vol. 105, no. 15, p. 153502, Oct. 2014.
[6] R. F. Pierret, Semiconductor Device Fundamentals, 5th ed. (Pearson education, 2011).
[7] J. Bhattacharya, R. W. Mayer, M. Samiee, and V. L. Dalal, "Photo-induced changes in fundamental properties of organic solar cells," *Appl. Phys. Lett.*, vol. 100, no. 19, p. 193501, 2012.
[8] Synopsys. Sentaurus Device Simulation Tool, 2011.

ORGANOLEAD HALIDE PEROVSKITE SOLAR CELLS

Aqel Mashot Jafar[1], Mahdi Hasan Suhail[2] ,Falah Mustafa Al-Attar[1*]& Mohammd K. Kalaf[1]
[1]Solar Energy Research Center, Renewable Energy Directorate, Ministry of Science and Technology , Baghdad, Iraq
[2]Physics Department, College of Science, Baghdad University, Higher Education and Scientific Research Ministry, Baghdad, Iraq
aqeel.mashon@yahoo.com; mhsuhail@yahoo.com; falah_im@yahoo.com

ABSTRACT: Organolead halide Perovskite Solar Cells (OPSC) which employed CuI as Hole Transport Layer (HTL) are fabricated .Optical and structure properties of sensitized absorption layer $CH_3NH_3X_3$ in Photovoltaic devices are studied . XRD patterns of $CH_3NH_3X_3$ compounds are indicated to the tetragonal structure at X = I or Br , cubic structure at X is mixed of I and Cl and orthorhombic structure at X is mixed of I and Br . The surface morphology of the films have been observed using Atomic Force Microscope (AFM) and Scanning Electron Microscope (SEM) .Light I-V measurement test reports of (OPSCs) are recorded by Integrated cell tester included I-V Photovoltaic measurements system and Solar simulator. The best Power Conversion Efficiency (PCE) of OPSC is (2.15) to the device have $CH_3NH_3I_3$ as sensitized absorption layer. Measurements are tested at AM1.5 global sunlight (100 mW cm^{-2}).

Keywords: perovskite solar cells, methyl-ammonium Iodide, organo-lead Iodide , Hole Transport Layer

1 INTRODUCTION

During the last two decades ,Solution-processed organic and hybrid organic–inorganic solar cells have been intensely developed [1,2]. Very recently, perovskite solar cells (PSCs) have appeared as the most promising new generation of hybrid photovoltaic devices [3]. The power conversion efficiency of (PSCs) has increased fast and very recently a power conversion efficiency (PCE) higher than 20% has been reported[4].Simon et al [5]are reported a phase of $CH_3NH_3PbI_3$has a large unit cell of the tetragonal structure and direct band gap. The bromine equivalent of $CH_3NH_3PbI_3$, $CH_3NH_3PbBr_3$, has a cubic structure (space groupPm3m) at room temperature. When increasing the fraction of bromine, the $CH_3NH_3PbI_{3-x}Br_x$perovskite undergoes a phase transition from tetragonal to cubic .When chlorine is incorporated in $CH_3NH_3PbI_3$, the band gap remains direct and increases slightly to 1.55 eV with a conduction band minimum of −3.75 eV and a valence band maximum of −5.3 eV [5].In the present work ,we are employed Organic Perovskite Materials (OPM)have different structure as sensitized absorption layer in the Organolead halide Perovskite Solar Cells(OPSCs) addition to study parameters of prepared devices.

Basic theory:

Generic formula of Perovskite materials is ABX_3whereX is an anion and A and B are cations of different sizes (A being larger than B). The degree of distortion of a perovskite from ideal cubic structure is measured by the constant, t , is known as the tolerance factor , t is defined as the ratio of the distance A-X to the distance B-X in an idealized solid-sphere. Therefore, the closer the value of the tolerance factor to cubic structure is to unity and given with eq (1) [6].where (R_A, R_B and R_X are the ionic radii of the corresponding ions).

$$t = \frac{(R_A + R_X)}{\sqrt{2}(R_B + R_X)} \dots (1)$$

For halide perovskites (X = F, Cl, Br, I), generally 0.81 < t < 1.11.If t lies in the narrower range t = 0.89–1.0, the cubic structure of Fig. 1a is likely, with lower t = 0.8 – 0.89 values giving less symmetric tetragonal and orthorhombic structures with lower than t =0.75– 0.8 values. The higher t values than unity (t>1) is giving phase hexagonal. where R_I=2.2A, R_{Br}=1.96A ,R_{Cl}=1.81A ,R_{Pb}=1.19A , methyl-ammonium($CH_3NH_3^+$) with R_A = 0.18 nm [6,7].

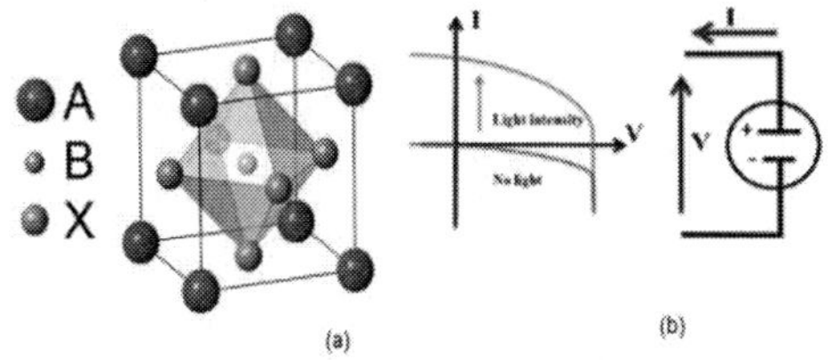

Figure 1:(a): inset depicts schematic to the perovskite structure [6], (b):inset depicts I-V characteristic of PV cell[8].

In general Photovoltaic (PV) cells, can be modeled as a current source in parallel with a diode. As the intensity of light increases, current is generated by the PV cell. Where there is no light, the PV cell behaves like a diode (Fig. 1b)[8].

The total current I in an ideal cell is equal to the current I_l generated by the photoelectric effect minus the diode current I_D in dark , according to the equation 2[8]:

$$I = I_l - I_0\left(e^{\frac{q(V+IR_S)}{nkT}} - 1\right) - \frac{V+IR_s}{R_{SH}} \dots (2)$$

where I_0 is the saturation cell current ,I_l is the photocurrent ,q is the elementary charge 1.6 × 10^{-19} Coulombs, k is a constant of value 1.38 × 10^{-23} J/K, T is the cell temperature in Kelvin and V is the measured cell voltage, R_s and R_{sh} are series and shunt resistances which obtained from Light I-V Measurement Test Reports of solar cells system or calculated from reverse of slope of light I-V curve at I= 0 and V= 0, respectively[9] .

The PCE is estimated by the equation 3[8]:

$$PCE = \frac{FF * V_{oc} * I_{sc}}{P_{in} * A_{ac}} \dots (3)$$

Where FF, V_{oc} , I_{sh} , P_{in} and A_{ac} are full factor , open voltage circuit ,short current circuit ,input power to the cell estimated to one solar illumination which equate to (100 mW/cm^2) and active area of solar cell , respectively.

The optical band gap of the thin film have been investigated for the allowed direct transition in accordance with the theory of Bardeen et al from the equation (4)[10].

$$\alpha h\nu = B * \left[h\nu - E_g\right]^{1/2} \dots (4)$$

Where, h is the Planck constant, (α) is the absorption coefficient ,v is the light frequency , E_g is the optical energy gap and B is empirical constant .

Characterization of samples:

Composition and crystal structure studied by X- Ray Diffraction (XRD Shimadzu 6000, Cu-K_α) for USP coating . The surface morphology of sample is studied by investigation of Atomic Force Microscopy (AFM) (AA3000 Scanning probe microscope, Angstron Advanced Inc.).The difference in the shape of the perovskite crystals studied by images of Scanning Electron Microscope(SEM) (Bruker Nano GmbH, Germany) analysis is performed by using a magnification 1kx and 5kx, high voltage 5kVand SEM in secondary electron mode .

Light I-V Measurement Test Reports are recorded by Photovoltaic measurements system, composed of Oriel I–V test station using an Oriel Solar simulator. The solar simulator is class AAA for spectral performance, uniformity of irradiance ,and temporal stability. The solar simulator is equipped with a450 W xenon lamp. The output power is adjusted to matchAM1.5 global sunlight (100 mW cm^{-2}).

I–V curves are obtained by applying an external bias to the cell and measuring the generated photocurrent with a Keithley model 2400 digital source meter .The transmittance of film coated is measured in the wavelength range of (190 - 1200) nm using a (SPECTRO UV/Vis Double Beam (UVD-3500) Labomed ,Inc). A blank sample of substrate is used as a reference in the measurement of optical transmittance.

Experimental:

Synthesis of Organic Perovskite Materials (OPM) reported in Ref [11]. Methylamine Iodide (CH_3NH_3I) is prepared by reacting Methylamine, 33 wt % in Ethanol (BDH-LTD), with Hydro-Iodic acid (HI) 57 wt % in water (BDH-LTD) under ice bath stirring for 2 h . Typical quantities employed are24 ml of Methylamine, 10 ml of HI, and 100 mL of Ethanol .Upon drying at 100 °C, a white powder is formed, which is placed overnight in a vacuum oven before use. Methylamine Bromide is prepared at the same method and previous quantities .To obtain the perovskite solution precursor , we dissolved both the CH_3NH_3I and the PbI_2 (BDH-LTD) inanhydrous N,N-DiMethylFormamide (DMF) (Sigma Aldrich) at a 3:1 molar ratio , such as in the table (1),with final concentrations of 40 wt %.10 ml of solution preparation added to 0.15 g of Nano particle> 30 nm of Al_2O_3powder (China of origin) to obtain scaffold peroveskite precursor solution of $CH_3NH_3PbI_3$. The reacting quantities of other samples of organo-lead halide pervskite solution precursors of $CH_3NH_3PbX_3$ are inserted in the table (1) .

Table (1) explains the 3/1 Molar ratio of mixing compounds amounts of CH_3NH_3X/PbX_2

Perovskite Compounds	CH_3NH_3I (g)	CH_3NH_3Br (g)	CH_3NH_3Cl (g)	PbI_2 (g)	$PbBr_2$ (g)	$PbCl_2$ (g)
$CH_3NH_3PbI_3$	0.15			0.15		
$CH_3NH_3PbBr_3$		0.11			0.12	
$CH_3NH_3PbBrI_2$		0.11		0.15		
$CH_3NH_3PbICl_2$	0.15					0.092

TiO_2compact layer preparation isa first fabricated applying Aerosol Assisted Chemical Vapor Deposition (AACVD) technique[10] by using Ultrasonic Atomizer(402AI) with ultrasonic frequency (1.5MHz).The precursor solution , which preparation of Nano particle> 20 nm of TiO_2 powder (China of origin) dispersed in ethanol solvent ,is sprayed on per-heated Transparent Conductive Oxide (TCO) glass substrates of Fluorine-doped Tin Oxide (FTO) (Coated Sodaline float glass of Visiontek , sheet Resistance 8 Ω/□) at 450° C, deposited time is 1h and the substrates are left to cold at room temperature , followed by , depositing scaffold peroveskite precursor solution of $CH_3NH_3PbX_3$by spin coated at speed 2000 rpm and annealing at 150^{0}C to obtain 4 samples of (FTO/Compact TiO_2/Scaffold$CH_3NH_3PbX_3$) as shown in fig (2) .The temperature of the (FTO/Compact TiO_2) coated substrates is monitored by an infrared temperature indicator through the all experimental runs.

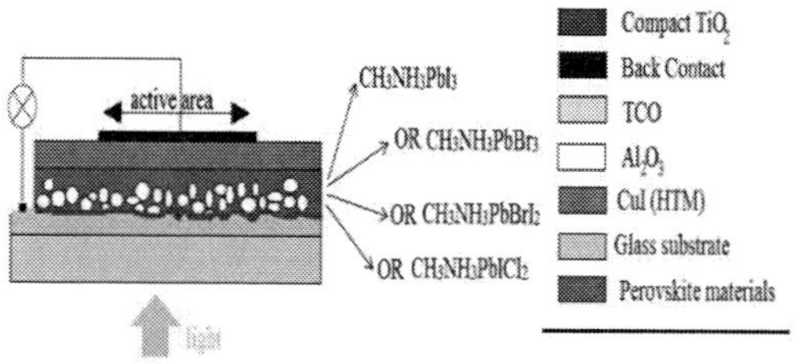

Figure 2; inset depicts schematic 4 samples of OPSCs have different perovskite materials.

CuI thin films are deposited onto (FTO/Compact TiO_2/Scaffold$CH_3NH_3X_3$) as a Hole Transport Layer (HTL) at room temperature by sputtering of pure CuI target in argon gas using DC sputtering technique. CuI target (99.99 % pure)with 50 mm diameter and 3 mm thickness is used for sputtering. The sputter chamber is evacuated by employing diffusion pump and rotary pump combination to achieve base pressure of 3.5×10^{-5} mbar . Argon of 99.99 % purity is used as reactive and sputtering working pressure of gases of 3.5×10^{-1} mbar for deposition of the films . The argon gas flow average is 240 Sccm , the 1.8 kV voltage is supplied to the sputter target using DC power supply , the discharge current of sputtering is 18 mA , the time of sputtering target is 1 h and active electrode spacing is 4cm . For the counter electrode , 100nm-thin film of Al deposited on top HTL by a thermal evaporation ,where Al evaporated under 10^{-5} mbar vacuum condition .

Results and Discussions

Fig 3(a) depicts the XRD pattern of CH3NH3PbI3 layer (red line) and XRD pattern of CH3NH3IPbCl2 layer (blue line), which achieved by deposited perovskite precursor solution using spin coated at speed 2000 rpm and annealing at 150 C on preheated glass substrates .

All reflections is indicated to the tetragonal CH3NH3PbI3 . Reflections positions (black lines) of (002) , (110) , (200) , (202) , (004) , (220) , and (224) corresponding with tetragonal structure Space group I4/mcm (Z=4), a=8.800 Å,c=12.685 Å are indicated by ref [12,13] for comparative . XRD patterns of the mixed halide perovskite structure of CH3NH3PbICl2 sample annealed at 150C was exhibited the tetragonal CH3NH3PbI3 phase and can be confirmed by reflections positions of (110) and (220) . Fig 3(b) depicts the XRD patterns of $CH_3NH_3PbBr_3$ layer (red line) which achieved by deposited perovskite precursor solution using spin coated at speed 2000 rpm and annealing at 150^0Con preheated glass substrates .Reflections positions (black lines) of (100) , (110) , (111) , (200) , (211) and (300) corresponding with cubic structure are indicated by ref [14] for comparative . Our results, XRD patterns of the halide perovskite structure of CH3NH3PbBr3 sample (red line) was exhibited the cubic CH3NH3PbBr3 phase at the reflections positions (black lines) of (100) , (110) , (111) , (200) , (211) and (300) [14] . However, the sample of CH3NH3BrPbI2 film is appeared maximum peak of XRD at reflection position (220) corresponding with tetragonal CH3NH3PbI3 structure which indicated by ref [12,13] as shown in fig 3(c) .

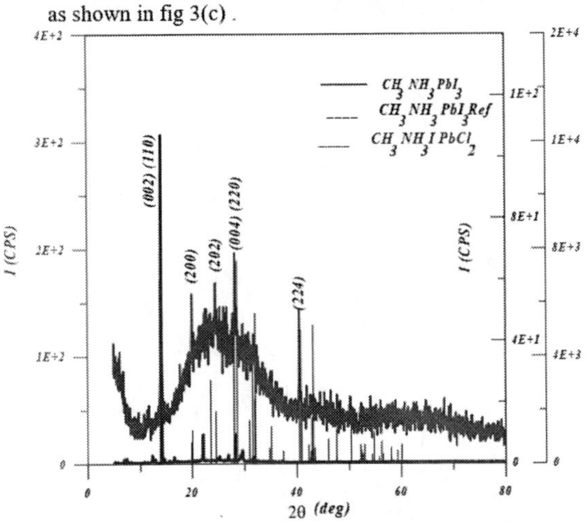

(a)

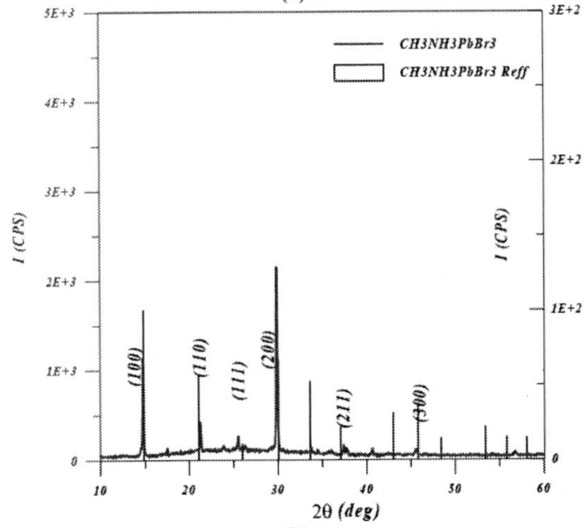

(b)

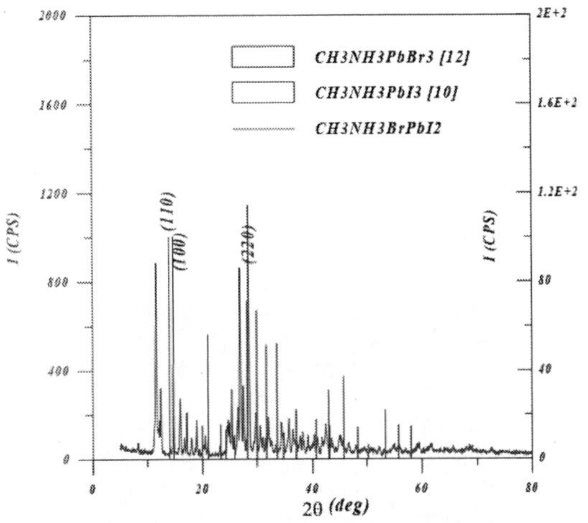

(C)

Figure 3; (a) XRD patterns of $CH_3NH_3PbI_3$ and $CH_3NH_3IPbCl_2$ layer on sheet glass substrates comparative with$CH_3NH_3PbI_3$ Ref [12,13] . (b)XRD patterns of $CH_3NH_3PbBr_3$ and $CH_3NH_3BrPbI_2$layer on sheet glass substrates samples comparative with $CH_3NH_3PbBr_3$ Ref [14]. (c)XRD patterns of CH3NH3BrPbI2 layer on sheet glass substrate sample comparative with CH3NH3PbBr3 Ref [14] and CH3NH3PbI3 Ref [12,13].

We calculated the tolerance factor (t) of equation 1 to identify the distorted perovskite structure and inserted data in the table 1 , where R_I=2.2A, R_{Br}=1.96A ,R_{Cl}=1.81A ,R_{Pb}=1.19A , methyl-ammonium($CH_3NH_3^+$) with R_A = 0.18 nm. On the basis of tolerance factor values , structure of $CH_3NH_3PbI_3$and $CH_3NH_3PbBr_3$ materials in the table2 is tetragonal due to ($0.8 \leq t \leq 0.89$) dentical structure to theresults of XRD pattern. $CH_3NH_3IPbCl_2$ compound have cubic Structure due to ($0.9 \leq t \leq 1$) and $CH_3NH_3BrPbI_2$ have Orthorhombic due to ($t \leq 0.8$) as shown in the table (2).

We calculated the tolerance factor (t) of equation 1 to identify the distorted perovskite structure and inserted data in the table 2 , where RI=2.2A, RBr=1.96A ,RCl=1.81A ,RPb=1.19A , methyl-ammonium (CH3NH3+) with RA = 1.8A . On the basis of tolerance factor values , structure of CH3NH3PbI3 materials in the table2 is tetragonal due to ($0.8 \leq t \leq 0.89$) and identical structure to the results of XRD pattern . The perovskite material of CH3NH3PbBr3 have cubic stricture in the table 2 as consequence of XRD results . The perovskite material of CH3NH3PbICl2 have tetragonal stricture in the table 2 as consequence of XRD results . The perovskite material of CH3NH3BrPbI2 in the table 2 have tetragonal structure as consequence of XRD results .

Table (2) explains the structure and optical parametersof proveskite materials.

Perovskite Compounds	t	Structure	E_g (eV)
$CH_3NH_3PbI_3$	0.834	Tetragonal	1.56
$CH_3NH_3IPbCl_2$	0.942	Tetragonal	1.9
$CH_3NH_3PbBr_3$	0.844	Cubic	2.1
$CH_3NH_3BrPbI_2$	0.784	Tetragonal	1.7

Fig. 4 (a and b) is illustrated top SEM images of $CH_3NH_3PbI_3$ film , (c and d) Top SEM images of $CH_3NH_3PbIBr_2$ film, (e and f) Top SEM images of $CH_3NH_3PbICl_2$ film and (g and h) Top SEM images of $CH_3NH_3PbBr_3$film , scale bars of left images are 10μm and right images are 5μm ,scanning with high voltage 5kV and magnification (5 and 10) kx , respectively, samples are prepared by spray manually deposition of perovskite precursor solution on preheated sheet glass substrate at 100^0 C .

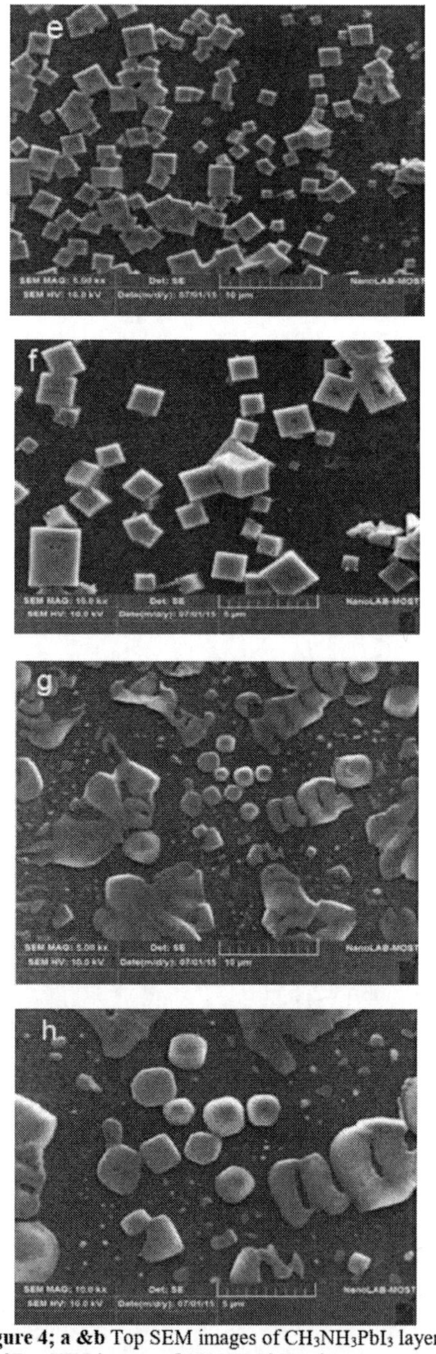

Figure 4; a &b Top SEM images of $CH_3NH_3PbI_3$ layer.
c&dTop SEM images of $CH_3NH_3PbIBr_2$ layer.
e&f Top SEM images of $CH_3NH_3PbICl_2$ layer.
g&h Top SEM images of $CH_3NH_3PbBr_3$.
Scale bars of left perovskite images are 10μm and right images are 5μm .

SEM images of $CH_3NH_3PbICl_2$ film are appeared cubic and tetragonal crystal structures and SEM images of $CH_3NH_3BrPbI_2$ film are appeared tetragonal crystal structure and sphere structure appeared due to precursor solution may be rich PbI_2 . As consequence of results of

SEM images are confirmed to the results of the tolerance factors in the table 2 .

Fig. 5(a) depicts AFM 3D-images of compact layer of TiO_2on FTO electrode deposited by (AACVD) technique on preheated sheet glass substrate at 450 ^{0}C temperature and deposited at 1h time . 5(b) explains chart plot of grain size percentage distribution of the particles of the surface topography of TiO_2 layer. SurfacecompactTiO_2 layer have Average Diameter (D = 89.11 nm) , Peak–Peak of the surface (PP = 66.9 nm) , Root Mean Square (RMS = 18.2nm) and Roughness Average (RA = 15.9 nm) as shown in the fig 5(c).

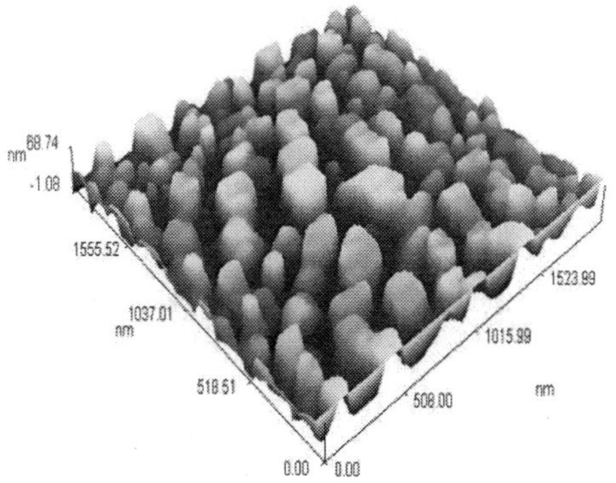

(a)

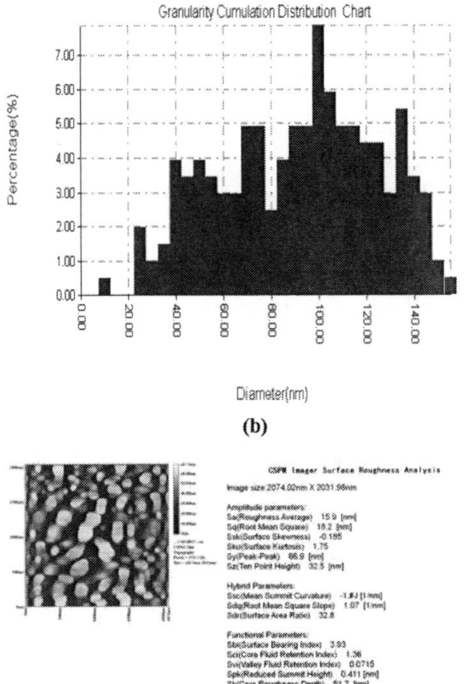

(b)

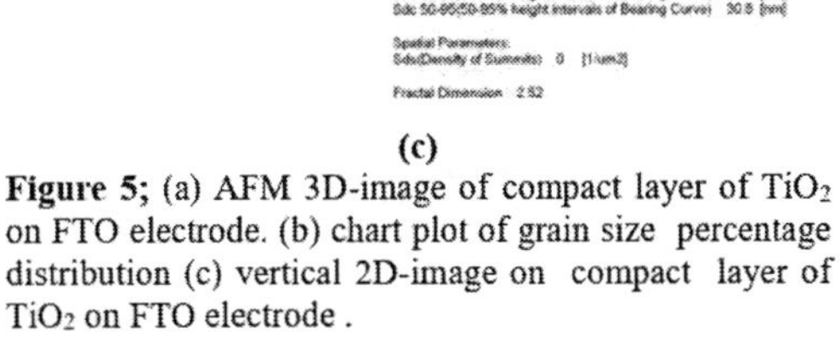

(c)

Figure 5; (a) AFM 3D-image of compact layer of TiO_2 on FTO electrode. (b) chart plot of grain size percentage distribution (c) vertical 2D-image on compact layer of TiO_2 on FTO electrode .

The data of AFM images are confirmed by SEM images of TiO_2 compact layer on FTO substrate which scanned using a magnification 30kx and high voltage 20kV as shown in Fig.6, then (x=46.88nm, y=53.33nm and D=71nm).

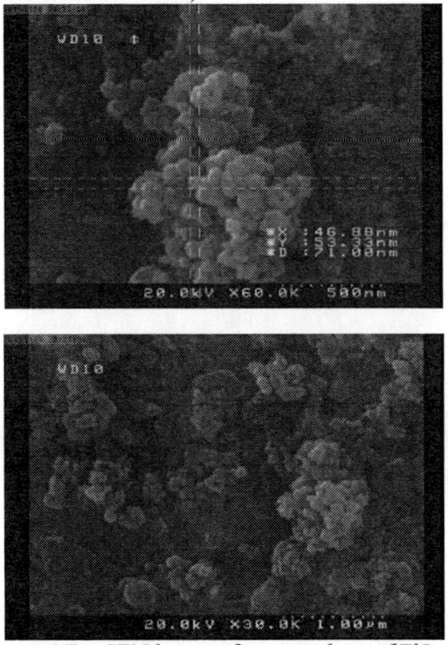

Figure 6;Top SEM images of compact layer of TiO_2 on FTO electrode. Scale bar of left image is 500nm and right image is 1μm.

Fig.7a is explained the absorption of halide perovskite films and appears more absorption to $CH_3NH_3PbI_3$ than other samples that which are measured in the wavelength range (400-800) nm using a (SPECTRO UV/VIS Double Beam (UVD-3500) Labomed, Inc.) and glass substrates as reference samples .

Plots inset of $(\alpha h\nu)^2$ versus photon energy $(h\nu)$ for peroveskite layers on glass substrates are illustrated in Fig .7b. A direct optical band gap energy (E_g) for halide perovskites materials $CH_3NH_3X_3$, (X = Cl, Br, I)is reported by Simon *et al* [5]. (E_g) of samples is determined by fitting the absorption data to the direct transition equation (4). The optical band gap value is obtained by extrapolating the linear part of the curve $(\alpha h\nu)^2$ as a function of photon energy, $h\nu$, intercept the $(h\nu)$ axis at $\alpha = 0$. The inserted E_g values are shown in table 2.

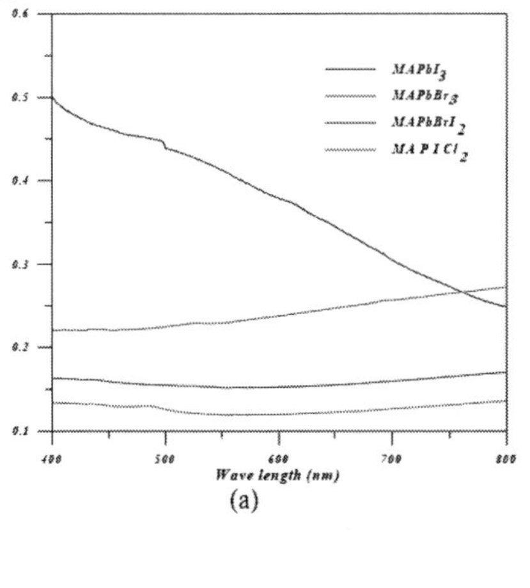

(a)

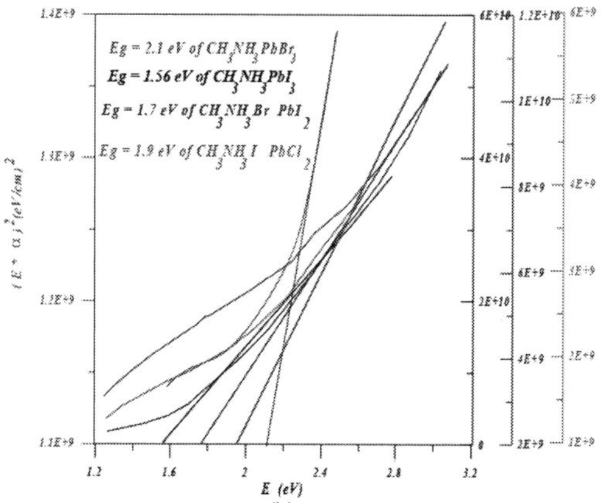

(b)

Figure 7; (a) absorption of Visible light of the $CH_3NH_3PbX_3$ films.
(b) energy gap of the$CH_3NH_3PbX_3$films.

I-V curve of OPSC have best PCE of $CH_3NH_3PbI_3$ as absorption layer of device is depicted in fig.8(a). Measurement is carried out under 1 sun illumination (AM1.5G,100 mW/cm^{-2}) , active area of solar cell is 0.1 cm^{2}and sweeping voltages in the scan-direction with a scan rate of s = 50 mV/s . Fig.8(b)is depicted I-V curves of four devices of OPSC-based on absorption layer materials which inserted in the table (3). Measurements are tested at the same conditions as the fig.8(a) .

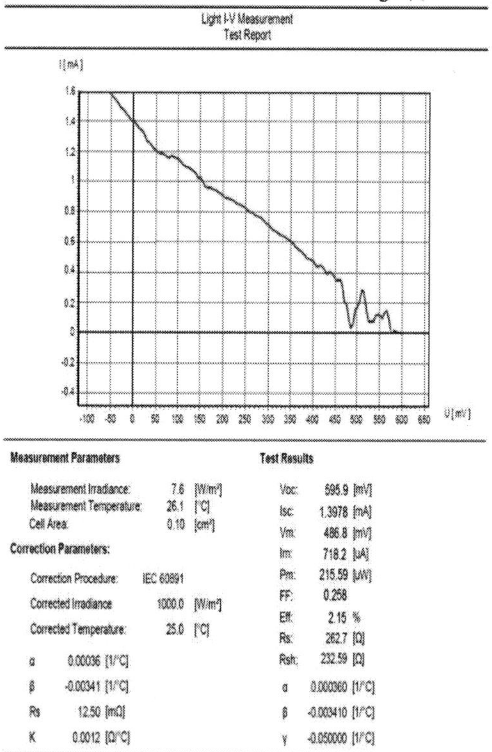

(a)

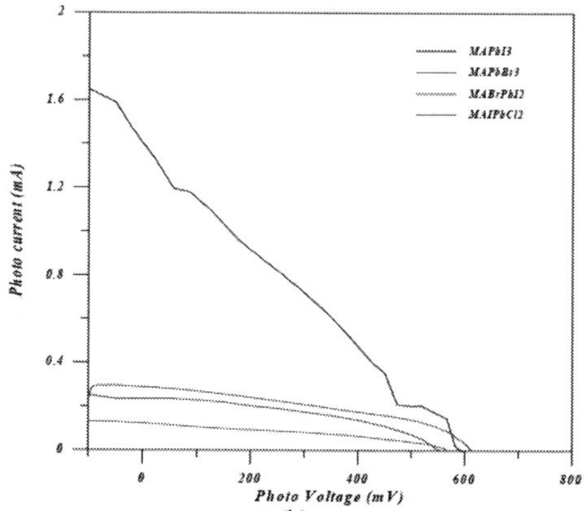

(b)

Figure 8; (a). I-V curve of OPSC has $CH_3NH_3PbI_3$ as absorption layer of device.
(b)I-V curves of four devices of OPSC-based on absorption layer materials.

All parameters of OPSCs are inserted in the table (3). The best(PCE = 2.15) is achieved to the OPSC employing $CH_3NH_3PbI_3$as sensitizer due to the best optical properties ,highest absorption and lowest optical energy gap , of absorption layer $CH_3NH_3PbI_3$, as shown in fig 8(a,b) and the table (3) .To shed light on fig3 and fig8,weare observed increased PCE of OPSC when increased absorption of perovskite layer of device .

Table 3 explains parameters of J–V curve characteristics of OPSCs.

Perovskite Compounds	PCE %	Voc (mV)	Isc (mA)	FF	Rs (Ω)	Rsh (Ω)
$CH_3NH_3PbI_3$	2.15	595	1.39	0.25	262	232
$CH_3NH_3PbBr_3$	0.74	614	0.29	0.41	214	6470
$CH_3NH_3BrPbI_2$	0.56	553	0.24	0.43	505	10010
$CH_3NH_3IPbCl_2$	0.27	569	0.12	0.36	833	14910

Conclusions:

In summary, the OPSCs with construction ($FTO/TiO_2/CH_3NH_3PbX_3$ perovskitewith Al_2O_3 scaffold /CuI/Al electrode) have been successfully fabricated. Our experimental results demonstrated that an optimum PCE = 2.15 % can be achieved by using $CH_3NH_3PbI_3$ as sensitized absorption layer in OPSC . Experimentally , PCE is increased when increased the absorption and decreased band gap energy of pervskite layer of OPSCs . 4. We calculated the tolerance factor (t) of equation 1 to identify the distorted perovskite structure . Results of tolerance are confirmed by SEM images and XRD results to the samples $CH_3NH_3PbI_3$, $CH_3NH_3PbBr_3$, $CH_3NH_3BrPbI_2$ and $CH_3NH_3IPbCl_2$ that images are demonstrated cubic and tetragonal crystalline structures .

References :

[1] Jiandong Fan, BaohuaJia, and Min Gu," Perovskite-based low-cost and high-efficiency hybrid halide solar cells", Photon . Res,Vol. 2(5) (2014) 111-120.

[2] LiozEtgar," Semiconductor Nanocrystals as Light Harvesters in Solar Cells", Materials,Vol. 6, (2013) 445-459.

[3] Askari Mohammad Bagher, Mirzaei Mahmoud AbadiVahid, Mirhabibi Mohsen," Types of Solar Cells and Application", American Journal of Optics and Photonics,Vol.5(3), (2015) 94-113.

[4] Nam-Gyu Park," OrganometalPerovskite Light Absorbers Toward a 20% Efficiency Low-Cost Solid-State Mesoscopic Solar Cell", American Chemical Society, J. Phys. Chem. Lett. Vol. 4, (2013) 2423–2429.

[5] Simon A. Bretschneider , Jonas Weickert , James A. Dorman, and Lukas Schmidt-Mende, " Research Update: Physical and electrical characteristics of lead halide perovskites for solar cell applications ",Applied Materials,Vol.**2**, (2014) 040701-9.

[6] L. Vasylechko, A. Senyshyn, and U. Bismayer, "Perovskite-Type Aluminates and Gallates. In K.A. Gschneidner, Jr., J.-C.G. Bünzli and V.K. Pecharsky, editors": Handbook on the Physics and Chemistry of Rare Earths, Vol. 39, Netherlands: North-Holland, 2009, pp. 113-295. ISBN: 978-0-444-53221-3 © Copyright 2009 Elsevier B.V. North-Holland.

[7] Martin A. Green1, Anita Ho-Baillie and Henry J. Snaith ," The emergence of perovskite solar cells", Nature Photonics ,Vol 8 , (2014) 506-614.

[8] LiozEtgar," Semiconductor Nanocrystals as Light Harvesters in Solar Cells", Materials Vol. 6, (2013) 445-459.

[9] GrégoireSissoko and SenghaneMbodji," A Method to Determine the Solar Cell Resistances from Single I-V Characteristic Curve Considering the Junction Recombination Velocity (Sf)" Int. J. Pure Appl. Sci. Technol., 6(2) , (2011) 103-114.

[10] AqelMashotJafar, Kiffah Al-Amara, FarhanLafta Rashid and Ibrahim KaittanFayyadh," Fabrication and Characterization of Fluorine-Doped Tin Oxide Transparent Conductive Nano-Films", Elixir Nanotechnology 65A , (2013) 20107-20111, www.elixirpublishers.com (Elixir International Journal).

[11] Miguel Anaya,GabrielLozano,Mauricio E. Calvo,Wei Zhang, Michael B. Johnston , Henry J. Snaith and HernánMíguez," Optical Description of Mesostructured Organic–Inorganic Halide Perovskite Solar Cells", J. Phys. Chem.Lett.,Vol.6, (2015) 48–53.

[12] Takeo Oku,"Crystal Structures of $CH_3NH_3PbI_3$ and Related Perovskite Compounds Used for Solar Cells ",Solar Cells - New Approaches and Reviews,Ch.3,83-101,book edited by Leonid A. Kosyachenko, ISBN 978-953-51-2184-8, Published: October 22, 2015 under CC BY 3.0 license.

[13] Baikie, T., Fang, Y,Kadro, J. M. Schreyer, M.; Wei, F.Mhaisalkar, S. G.Grätzel, M.&Whitec, T. J., "Synthesis and crystal chemistry of the hybrid perovskite $(CH_3NH_3)PbI_3$ for solid–state sensitized solar cell applications", Journal of Materials Chemistry A, Vol. 1, (2013) 5628–5641.

[14] D. Priante, I. Dursun, M. S. Alias, D. Shi, V. A. Melnikov, T. K. Ng, O. F. Mohammed, O. M. Bakr, and B. S. Ooi," The recombination mechanisms leading to amplified spontaneous emission at the truegreen wavelength in CH3NH3PbBr3 perovskites", Applied Physics Letters 106, (2015) 081902.

CONDUCTIVE INKS WITH EPOXY RESIN BASED VEHICLES FOR PEROVSKITE SCREEN PRINTING METALLIZATION

C. Montes[1], A. Linares[2], E. Llarena[1], O. González[1], D. Molina[1], A. Pío[1], L. Ocaña[1], C. Quinto[1], M. Friend[1] & M. Cendagorta[1].

[1]Instituto Tecnológico y de Energías Renovables, S. A. (ITER)
[2]Agencia Insular de Energía de Tenerife (AIET)
Pol. Industrial de Granadilla, s/n
E 38600 Granadilla de Abona, España
Tlf. +34 922 747 700
Fax +34 922 747 701
E-mail iter@iter.es, agenergia@agenergia.org

ABSTRACT: The main objective of the present paper is to summarize the work done in the Laboratory of Solar Cells (SiCelLab) form the Instituto Tecnológico y de Energías Renovables (ITER) in order to enable the possibility of using screen printing techniques as an alternative to the actual methods, applied by the scientific community for producing the metal contacts in perovskite based devices. Thus, the deposition of contacts made by screen printing with inks such as the ones based on graphite powders, epoxy resins and anhydrous solvents, due to their inherent low cost and processing temperatures, could lead to reducing the manufacturing costs and, therefore, help with the future commercialization of this kind of solar cell technologies.
Bearing this in mind, optimal proportions of these three components have been combined, in order to obtain a basic ink recipe, suitable for printing purposes, which also provided the possibility to study the effect of using different solvents. The improvement of the derived ink recipes was achieved by producing samples, on the one hand, via screen printing so as to study their electric resistance and, on the other, via doctor blade so as to study their resistivity, in both cases by using a semiconductor characterization system, configured to operate as an ohmmeter. Also, a digital microscope was used to study in detail the homogeneity and adhesion characteristics of the deposited samples.
Finally, it was concluded that the N, N- dimethyl formamide (DMF) and N-methyl-2-pyrrolidinone (NMP) are the solvents that allow to add greater amounts of graphite to the final mixture and, thus, to improve its conductivity, without impairing its printing capabilities.
Keywords: Contact, Metallization, Screen Printing, Perovskite solar cells

1 INTRODUCTION

In the race for achieving higher Power Conversion Efficiencies (PCE) and regarding the need to provide the metal contacts required for completing the diverse fabrication architectures, most of the R&D efforts devoted to Perovskite solar cells have been focused on the use of thermal evaporation techniques [1]-[6]. However, if Perovskite solar cells are to reach the commercial stage, it is necessary to evaluate the introduction of alternative metallization techniques which could address larger active areas, while reducing the manufacturing costs. On the metallization side, one of these techniques is screen printing.

The Laboratory of Solar Cells (SiCelLab) [7][8] form the Instituto Tecnológico y de Energías Renovables (ITER) has been working on the screen printing approach for Perovskite solar cells by, originally, performing trials with different commercial frit based pastes and inks, such as aluminum and silver pastes, commonly used in crystalline silicon solar cell technologies, as well as silver inks, also commonly used in polyester and polymeric film depositions [9]. These trials were conducted applying low temperature and fast curing processes after the deposition stages, in order to preserve the Perovskite substrates qualities. However, it was found that, due to the chemical nature and physical morphology of such substrates, either the kind of vehicles or the size of the powders (or both) used in the composition of these pastes and inks failed to adhere to the substrates (after all, frit based inks require high curing temperatures in order to get tempered onto the substrates) or had water among its constituent elements, rendering them as too aggressive for being of any real use (as soon as they were deposited and cured over the perovskite substrates, they rapidly deteriorated the finished device) [9].

In order to overcome this problem, the Solar Cells Laboratory have been working on developing new conductive ink formulas that use, on the one hand, epoxy resin glues and anhydrous solvents as vehicles and, on the other, conductive powders and flakes of different purities and sizes. The reason for using resin glues lies in that it allows to produce inks with high adhesion capabilities which can be cured at low temperatures. The use of anhydrous solvents enables the possibility of tweaking the ink in order to incorporate higher amounts of conductive materials, while ensuring the absence of water based components in the final mix.

The present paper summarizes the methodology and first results achieved in this ongoing project.

2 METHODOLOGY

The inks were prepared by using a kitchen hand mixer [10], provided with stainless steel beaters, rotating at 200 RPMs, in order to combine the conductive powders with the epoxy resin together with the anhydrous solvent of choice. Finally, in order to perform characterization studies on the resulting contacts, samples were extracted from the blend, with the same mixing time, in order to make two kinds of contacts: via screen printing and via Doctor blade. Finally, the contacts were dried and cured in a drying oven [11] at low temperatures.

2.1 Screen printing study

It was performed over glass substrates with sizes 25mm x 25mm x 2mm by using screens [12] capable of handling inks with up to 67 microns particle sizes together with a screen printer [13], provided with a 70 durometer squeegee blade and a dial indicator for fine adjustment of the screen and the squeegee rod to the tool plate, which allows to control the thickness of the inks and pastes deposited with an accuracy of micrometers.

Figure 1 shows the appearance of the deposited contacts that, on average, were found to have a thickness of 10 microns.

Figure 1: Example of a screen printed contact.

The characterization of the printed samples was performed by studying their visual appearance and electric resistance. The visual appearance assessment was performed by looking at the printed contacts with a microscope [14] in order to evaluate their homogeneity and degree of adhesion. The measurements were performed by using a semiconductor characterization system [15], configured to operate as an ohmmeter in order to obtain the resistance of the printed contacts.

2.2 Doctor blade study

In order to obtain resistivity values of the contacts derived from the different ink recipes, the standard F 1896 Test Method for Determining the Electrical Resistivity of a Printed Conductive Material [16] was considered. In this method it is required to print a determined kind of pattern, consistent of a straight trip, with known thickness. However, since no screen was available for that purpose, the alternative Doctor blade printing technique was performed. Thus, substrates with sizes 25mm x 75mm were prepared by covering them with cellulose tape [17], exposing only the pattern intended to be printed by using a glass rod in order to manually extend the ink over the substrate. The length (L) and width (W) of the printed pattern for the contacts was consistent with the minimum 50:1 size ratio, as it is specified by the above mentioned standard. As per the thickness of the resulting contact, it was estimated by considering the reported thickness of the tape (63.5 µm). As in the previous section, the measurements were performed by using semiconductor characterization system [15], configured to operate as an ohmmeter in order to obtain the resistance of the printed contacts and derive the corresponding resistivity values as per the standard method.

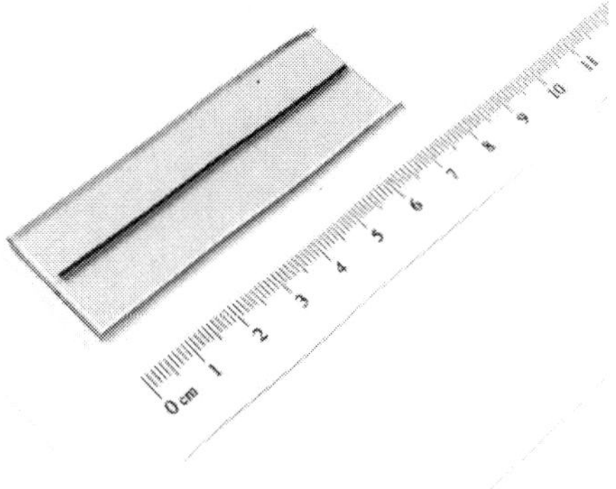

Figure 2: Example of a doctor blade printed contact.

3 EXPERIMENTAL

To get started, a basic ink recipe was obtained by trial and error, via mixing graphite powders [18] with epoxy resin [19] and toluene [20], almost to the point of saturation, before adding the hardener [19]. The reason for using toluene was due to its relatively low boiling point [21], in the assumption that it could speed up the curing process and help reducing the resistance of the resulting contacts.

After several trials, a procedure for elaborating the inks was established by dividing the actual mixing in four stages: At the first one, the resin and a part of the graphite was mixed by steadily stirring with a kitchen hand mixer for 10 minutes. In this phase is where most of the graphite is added. In the second stage, more graphite is added to the mix, followed by another 10 minutes of stirring. At this phase is when it was desired to reach the highest possible degree of saturation of graphite in the mixture. In the third stage, the remaining amount of graphite was added to the mix, together with the solvent of choice, continuing with the stirring but this time for 15 minutes. Finally the hardener was incorporated to the mix.

Notice that, in order to avoid spillage of graphite, every time that it is added to the mix, the first stirring stages were carried out by using a laboratory spatula.

Due to the nature of the chosen ink vehicle, which has a hardening time beyond which it renders quite unusable for printing purposes, a study of the print times for the mixtures was carried out by extracting samples at 5, 10, 20 and 30 minutes after adding the hardener and by trying to interrupt the mixing process as little as possible. For each sample, the two kind of printings explained in the previous section were performed.

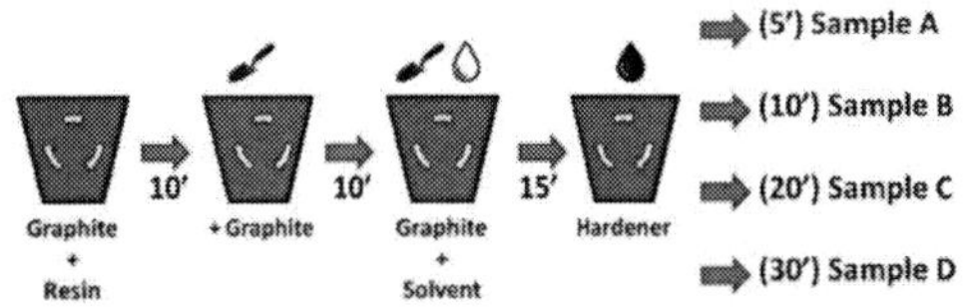

Figure 3: Mixing procedure.

The production of the contacts was concluded by a 10 minute curing process, performed in a drying oven at 100 °C, following the recommendations of the manufacturer of the epoxy resin [19].

Once a basic ink recipe was developed, modifications

were performed on it by keeping the weight concentration of its components, but changing the type of solvent. Thus, trials were performed with solvent with higher boiling points, such as chlorobenzene (ClBz) [22], N,N-dimethyl formamide (DMF) [23], and N-methyl-2-pyrrolidinone (NMP) [24]. To all mixes, the two printing processes were performed, together with their corresponding characterization and print time studies.

Finally, since the inks done with DMF and NMP had the appearance of being able to accommodate more graphite, the basic recipe was modified in order to do so and trials were carried out to conduct such study.

4 RESULTS AND DISCUSSION

4.1 Development of a basic ink recipe

The target was to obtain a basic recipe that could meet the two main requirements, that is, that it was suitable for screen printing and also that it was conductive. In order to do so, different mixtures were prepared, following the procedure described in the previous section, trying to determine an optimal ratio of graphite, resin and toluene.

Of all attempts, Table I summarizes the relevant information from those who gave as a result inks that were useable for screen printing purposes.

Table I: Basic ink recipe trials. "N/C" stands for non conductive and "N/A" for non available.

Mix code	Graphite (wt%)	Resin (wt%)	Solvent (wt%)	Resistance (kΩ)	Resistivity (Ω·cm)
M19	25.97%	74.03%	0.00%	N/C	N/C
M20	35.03%	64.97%	0.00%	N/A	821.82
M21	31.30%	58.13%	10.57%	127.65	50.47
M22	28.76%	53.37%	17.87%	217.92	19.31
M23	33.33%	49.95%	16.72%	300.04	6.40

As it can be seen in the data collected in Table I, different results were found, depending upon the kind of printing technique applied. So, for the screen printed contacts, the one obtained with the ink (M21) showed a better resistance than trials performed by either increasing the amount of vehicle, leading to a more fluid ink (M22) or by increasing the amount of graphite, producing a thicker ink (M23). However, for the doctor blade printed contacts, the key for reducing the resistivity in the printed contacts resided in decreasing the amount of resin in the corresponding ink vehicle (see M23 versus either M22 or M21).

Also, from the inks that were conductive, M21 presented a better behavior during the screen printing itself, showing superior material passage though the sieve of the screen, resulting on much more homogeneous contacts, as it can be seen in Figure 4

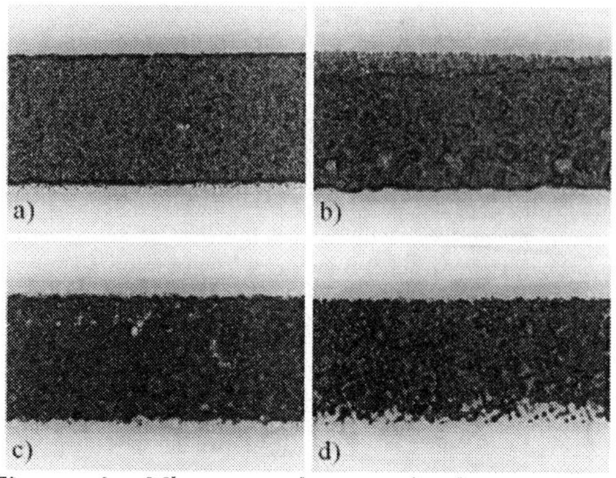

Figure 4: Microscope images showing contacts deposited by screen printing with inks M19 (a), M21 (b), M22 (c) and M23 (d).

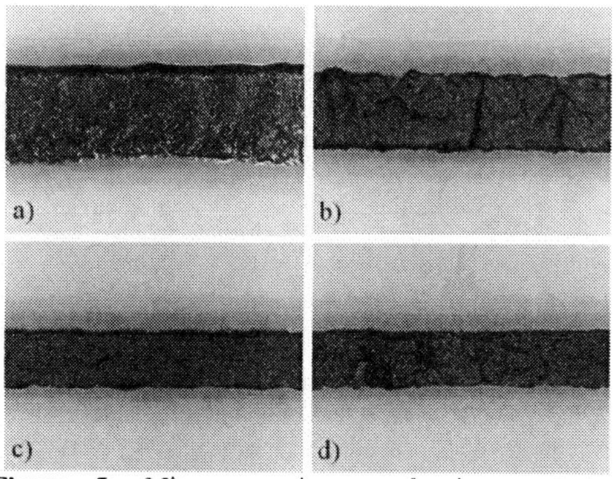

Figure 5: Microscope images showing contacts deposited by doctor blade with inks M19 (a), M21 (b), M22 (c) and M23 (d).

Since the purpose of the project was to develop inks for screen printing purposes, the mix M21, that is, the one with a weight concentration ratio of approximately 32%/58%/10% for graphite, epoxy and toluene, respectively, was considered as the best candidate to carry on with the next stage. Also, despite the differences observed between the contacts printed by screen printing and doctor blade, the possibility of studying the variations in the contact resistivity due to the use different ink recipes, was considered relevant enough to keep producing samples by doctor blade printing in the next stage of the project.

4.2 Ink trials with different solvents and printing times

Throughout the development of the basic ink recipe, it was clear that the mixtures had been affected by evaporating phenomena. Therefore, the addition of solvents with higher boiling point was studied in order to verify whether our original assumption for choosing toluene, so that it could speed up the curing process and help reducing the electric resistance of the resulting contacts, was correct.

Table II: Relevant properties [21][25] of the solvents used.

Properties	Toluene	ClBz	DMF	NMP
Boiling point (°C)	110.6	132	153	202-204
Density (g/mL)	0.867	1.106	0.944	1.028
Vapor pressure @ 20°C (hPa)	29	12	3.5	0.39
Surface tension @ 20 °C (mN/m)	28.4	33.6	37.1	40.79

Also, due to the use of an epoxy resin as part of the chosen vehicle for the inks, it was also required to examine for how long the mixed inks were still suitable for screen printing purposes, and how the differences on the bonding process could affect to the resistance, as well as the resistivity, for the resulting contacts.

Tables III and IV summarize the results obtained with the two printing processes:

Table III: Resistance measured in the contacts deposited via screen printing.

Time (min)	Toluene (kΩ)	ClBz (kΩ)	DMF (kΩ)	NMP (kΩ)
5	142.79	201.01	120.21	153.10
10	395.73	459.83	166.93	165.75
20	447.89	380.88	298.32	366.27
30	1632.20	947.15	273.85	364.18

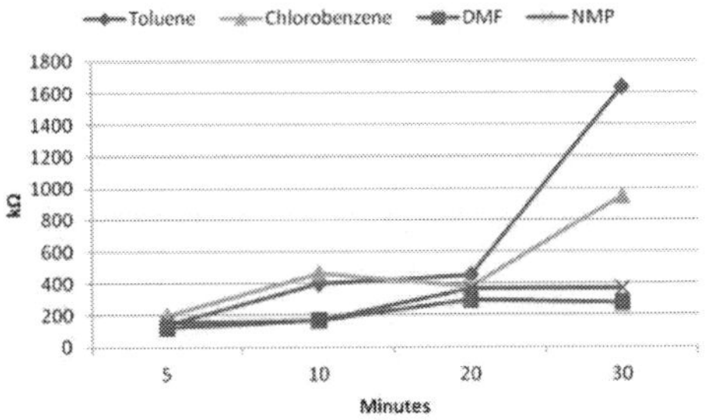

Figure 6: Resistance measured in the contacts deposited via screen printing.

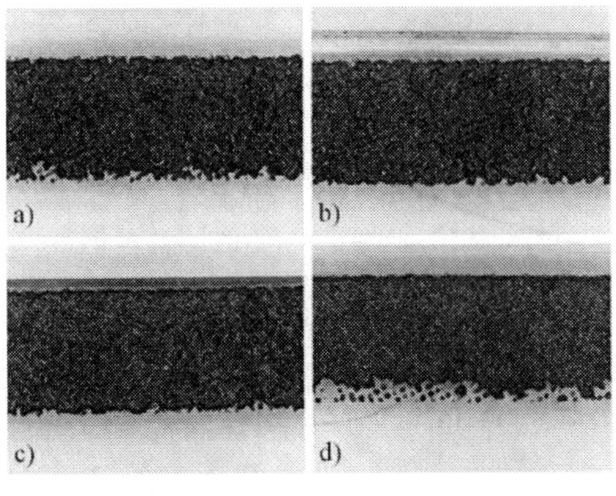

Figure 7: Microscope images showing contacts deposited by screen printing with Toluene (a), Chlorobencene (b), DMF (c) and NMP (d). All the contacts were deposited 20 minutes after adding the hardener.

By looking at the gathered data, it can be deduced that, for the screen printed contacts, which have considerably large surfaces but relatively short thicknesses, the differences in the solvent boiling temperature used are more relevant than the differences in their density, so that the inks which are more exposed to evaporation phenomena, both during the mixing and the contact curing processes, experience an acceleration in the resin bonding process which, due to the geometric nature of the printed contacts, became significant in the samples deposited at longer print times (from the addition of the hardener to the ink mixture). Therefore, contrary to what it was expected, an acceleration on the resin bonding process leads to an increase in the contact resistance, probably because more epoxide groups get in the way between the graphite stacks.

Table IV: Resistivity measured in the contacts deposited via Doctor blade.

Time (min)	Toluene (Ωcm)	ClBz (Ωcm)	DMF (Ωcm)	NMP (Ωcm)
5	27.43	33.79	18.02	15.22
10	40.38	64.58	23.47	23.09
20	59.62	88.08	45.52	46.26
30	57.62	164.51	119.49	59.16

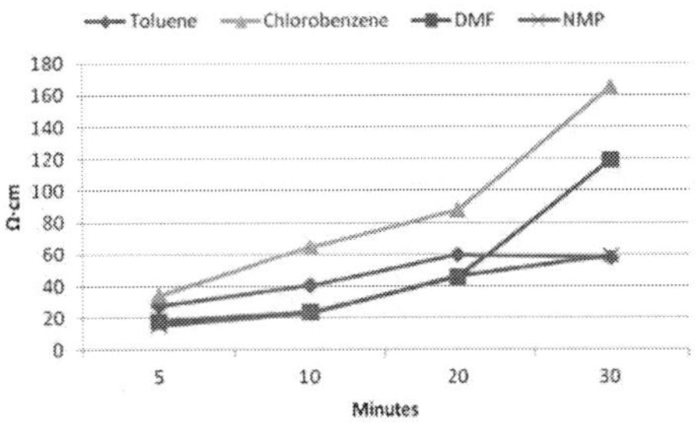

Figure 8: Resistivity measured in the contacts deposited via Doctor blade.

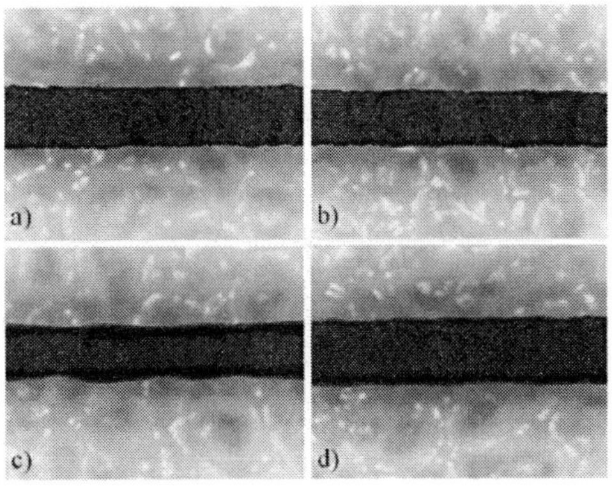

Figure 9: Microscope images showing contacts deposited by doctor blade with Toluene (a), Chlorobencene (b), DMF (c) and NMP (d). All the contacts were deposited 20 minutes after adding the hardener

By contrast, for the contacts printed with doctor blade, which have a much thicker nature than the screen printed ones (63.5 versus 10 µm, respectively), the gathered data suggest that, despite the differences in the boiling temperatures for the solvents, it was their density what helped to reduce the resin bonding speed during the curing process and, hence, also the resistivity of the contacts.

4.3 Ink trials with more graphite

The trials conducted at the previous section resulted in inks that, in the case of toluene and chlorobenzene, had an amount of graphite equal or near to the saturation level. However, this was not the case for the mixtures made with DMF and NMP (see Figure 7). Thus, since the purpose of the entire project was not only to obtain ink recipes suitable for screen printing purposes, but also that they were capable for producing contacts with the lowest resistance possible, a new set of trials were performed with these two last solvents in order to get closer to the desired level of ink saturation with graphite.

Table IV summarize the results obtained with contacts deposited via the two printing processes and by applying a recipe with a weight concentration ratio of approximately 35%/58%/7% for graphite, epoxy and solvent, respectively.

Table V: Resistance and resistivity measured in the contacts deposited via screen printing and doctor blade, respectively.

Time (min)	DMF Resistance (kΩ)	NMP Resistance (kΩ)	DMF Resistivity (Ωcm)	NMP Resistivity (Ωcm)
5	116.28	84.59	10.88	12.24
10	154.75	123.42	28.57	22.98
20	231.94	188.04	22.98	30.12
30	257.37	267.09	22.53	30.04

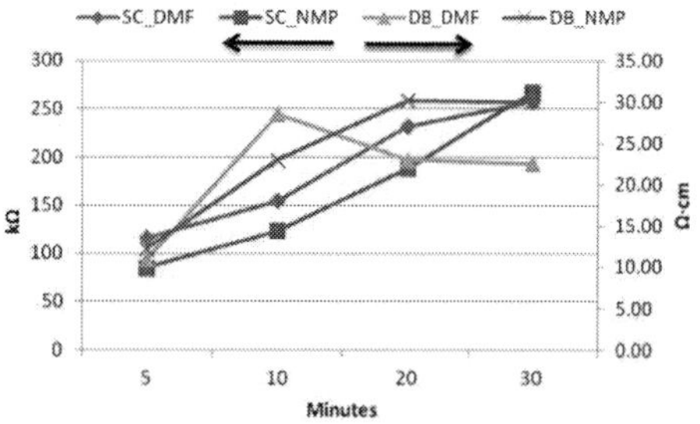

Figure 10: Resistance and resistivity measured in the contacts deposited via screen printing and doctor blade, respectively.

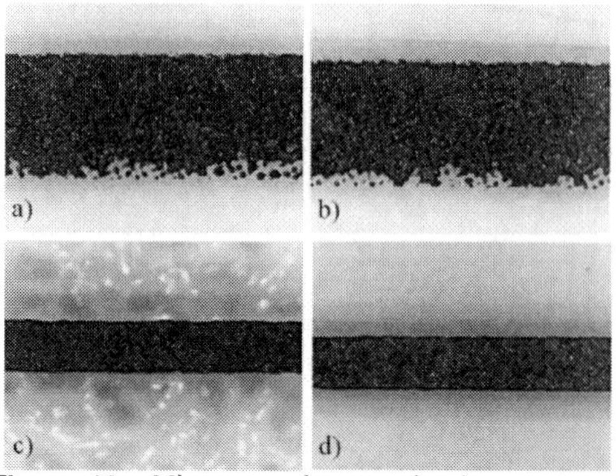

Figure 11: Microscope images showing contacts deposited by screen printing with DMF (a) and NMP (b) as well as doctor blade with DMF (c) and NMP (d). All the contacts were deposited 20 minutes after adding the hardener

As it can be seen in Table V, the obtained results do improve the ones referred in Table III and IV but not significantly. Therefore it can be assumed that the recipe proposed nearly matches the limits of the amount of graphite that these inks are capable of accommodating.

In order to further reduce the resistance of the printed contacts, the following options will be studied in this ongoing project:

- To use graphite powders with flakes made of shorter stacks, halfway between standard graphite and graphene, in order to improve the interconnections throughout the resulting contacts. To this end trials with this kind of graphite [26] have already been scheduled.
- To mix anhydrous solvents of different kinds in order to obtain better density to boiling point ratios so as to help accommodate more graphite during the mix while still facilitating the contact curing.
- To dissolve in the anhydrous solvent a quantity of conductivity improving species, such as ferrocene or graphene, prior to its addition to the ink mixtures. In this way it would be possible to reduce the electric resistance of the deposited contacts, without the need to add large amounts of graphite.
- To directly use powders and flakes of higher intrinsic conductivity, such as aluminium or silver.

On the other hand, and as it was explained in the Experimental section, the ink mixing throughout the project was done by means of a kitchen hand mixer. Although this technique provides fairly homogeneous blends, the mechanical action of the stirring beaters amalgamates the mix with small air bubbles. This phenomenon ultimately may be leading to a deterioration of the resulting inks, once cured, due to the existence of small air pockets within the contact. This was particularly noticeable for Doctor blade printed contacts. A way to overcome this phenomenon could be the use of a three roll mill in the final stages of the mixing process.

Finally, since the results obtained by screen printing seem to differ so much from the ones obtained by doctor blade, particularly at higher application times (that is, more than 20 minutes after adding the hardener to the mix), it would be very useful to obtain the resistivity values by screen printing. To that end, trials for using screens equipped with the necessary type of patterns have been scheduled.

5 CONCLUSIONS

A method for producing inks based upon the use of vehicles made with epoxy resins with anhydrous solvents, suitable for being used with perovskite substrates has been developed.

The entire study was carried out with graphite powders, as the conducting material of choice, which was due to their high purity and screen-compatible particle sizes.

In order to improve the derived recipes, samples were produced, on the one hand, via screen printing so as to study their electric resistance and, on the other, via doctor blade so as to study their resistivity.

With the methodology and the materials chosen, ink recipes were developed which were capable of producing contacts via screen printing with the lowest electric resistance possible. This was achieved by using DMF and NMP as the solvents of choice.

Finally, improvements on the methodology, as well as changes in the materials chosen, were proposed as ways for improving the accomplished results.

6 REFERENCES

[1] Giles E. Eperon, Victor M. Burlakov, Pablo Docampo, Alain Goriely,Henry J. Snaith. Morphological Control for High Performance, Solution-Processed Planar Heterojunction Perovskite Solar Cells. First published: 9 September 2013. doi: 10.1002/adfm.201302090 .

[2] Julian Burschka, Norman Pellet, Soo-Jin Moon, Robin Humphry-Baker, Peng Gao,Mohammad K. Nazeeruddin & Michael Grätzel. Sequential deposition as a route to high-performance perovskite-sensitized solar cells. Nature 499, 316–319 (18 July 2013), doi:10.1038/nature12340.

[3] Qi Chen, Huanping Zhou, Ziruo Hong, Song Luo, Hsin-Sheng Duan, Hsin-Hua Wang, Yongsheng Liu, Gang Li, and Yang Yang. Planar Heterojunction Perovskite Solar Cells via Vapor Assisted Solution Process. J. Am. Chem. Soc., 2014, 136 (2), pp 622–625. doi:10.1021/ja411509g.

[4] Mingzhen Liu, Michael B. Johnston & Henry J. Snaith. Efficient planar heterojunction perovskite solar cells by vapour deposition. *Nature* 501, 395–398 (19 September 2013), doi:10.1038/nature12509.

[5] Low-Temperature Solution-Processed Perovskite Solar Cells with High Efficiency and Flexibility. Jingbi You, Ziruo Hong, Yang (Michael) Yang, Qi Chen, Min Cai, Tze-Bin Song, Chun-Chao Chen, Shirong Lu, Yongsheng Liu, Huanping Zho, and Yang Yang. *ACS Nano*, 2014, *8* (2), pp 1674–1680. doi: 10.1021/nn406020d.

[6] Xiao, M., Huang, F., Huang, W., Dkhissi, Y., Zhu, Y., Etheridge, J., Gray-Weale, A., Bach, U., Cheng, Y.-B. and Spiccia, L. (2014), A Fast Deposition-Crystallization Procedure for Highly Efficient Lead Iodide Perovskite Thin-Film Solar Cells. Angew. Chem. Int. Ed., 53: 9898–9903. doi:10.1002/anie.201405334.

[7] C. Quinto, L. Ocaña, C. Montes, A. Linares, E. Llarena, B. González-Diaz, S. González-Pérez, R. Guerrero-Lemus, M. Friend, M. Cendagorta. Design, start up and society benefits of a PV solar cell laboratory, 30th European Photovoltaic Solar Energy Conference and Exhibition 2014, Amsterdam, Netherlands, doi: 10.4229/EUPVSEC20142014-7AV.6.56

[8] L. Ocaña, A. Linares, E. Llarena, C. Montes, O. González, D. Molina, A. Pío, C. Quinto, M. Friend, M. Cendagorta. Adaptation of a Crystalline Silicon Solar Cell Laboratory to Produce Perovskite Solar Devices, 31th European Photovoltaic Solar Energy Conference and Exhibition 2015, Hamburg, Germany, doi: 10.4229/EUPVSEC20152015-3BV.5.40

[9] C. Quinto, A. Linares, E. Llarena, C. Montes, O. González, D. Molina, A. Pío, L. Ocaña, M. Friend, M. Cendagorta. Screen printing for perovskite solar cells metallization, 31th European Photovoltaic Solar Energy Conference and Exhibition 2015, Hamburg, Germany, doi: 10.4229/EUPVSEC20152015-3BV.5.41

[10] Robot amasadora, LH 12 99, 220-240 V - 50 HZ 300 W. Larry House, by KEC.

[11] Drying oven (19 L capacity). J.P. Selecta, model Conterm.

[12] Stainless steel fabric, 200 mesh, 0.040 D-Ø x 22.5 high tension screens made by SEFAR MAISSA, S.A.U. C/ Alcorcon, 13, 28850 Torrejon Ardoz (Madrid).

[13] MSP-485, Precision Screen Printer. Hary Manufacturing, Inc. Precision Screen Printers and automation equipment (HMI), 24 Cokesbury Road, Lebanon, NJ 08833 USA. www.hmiprinters.com

[14] Digital Microscope U500X, focus range 15mm - 40mm, range 50X-500X. Unimake. Access: Jun 2016: http://www.unimakes.com.

[15] Keithley SCS-4200. http://www.tek.com/keithley-4200-scs-parameter-analyzer. Access: Jun 2016

[16] F 1896 – 98 (Preaproved 2004). Test Method for Determining the Electrical Resistivity of a Printed Conductive Material. Test method © by the American Society for Testing Materials (ASTM) International, 100 Barr Harbor Drive, PO Box C700, West Conshohocken, PA 19428-2959, United States.

[17] Scotch® No. 810 Magic™ Tape. http://docs-europe.electrocomponents.com/webdocs/108d/0900766b8108d212.pdf. Access: Jun 2016

[18] Graphite poder SGA 20 99,9 M. Graphite Kropfmühl GmbH Langheinrichstraße 1, 94051 Hauzenberg. www.gk-graphite.com. Acquired via JLQ ISMAF, polígono torrelarragoiti Pabillion 6 F 48170 - Zamudio - Vizcaya. www.ismaf.com

[19] Fast and clear epoxy system for hot processes. Epoxy resin SR 8500 / hardener SZ 8525. Sicomin Epoxy System, 31 avenue de la Lardière, 13220 Chateauneuf les Martigues, FRANCE.

[20] Toluene puriss (32249). p.a., ACS reagent, reag. ISO, reag. Ph. Eur., ≥99.7% (GC), from SIGMA-ALDRICH

[21] Properties of organic solvents. Access: Jun 2016. www.murov.info/orgsolvents.htm#TABLE_1

[22] Chlorobenzene anhydrous (284513), 99.8%, from SIGMA-ALDRICH.

[23] N,N-Dimethylformamide anhydrous (227056), 99.8%. Synonym: DMF, NSC 5356 from SIGMA-ALDRICH
[24] 1-Methyl-2-pyrrolidinone anhydrous (328634), 99.5%. Synonym: 1-Methyl-2-pyrrolidone, N-Methyl-2-pyrrolidone, NMP, from SIGMA-ALDRICH.
[25] Surface tension values of some common test liquids for surface energy analysis from www.surface-tension.de
[26] Graphite powder SGA 20 OS. Graphite Kropfmühl GmbH Langheinrichstraße 1, 94051 Hauzenberg. www.gk-graphite.com. Acquired via JLQ ISMAF, polígono torrelarragoiti Pabillion 6 F 48170 - Zamudio - Vizcaya. www.ismaf.com

Influence of Titanium Dioxide and Molybdenum Oxide Blocking Layers on Perovskite Solar Cells

Karim Habashy[1], Volker Steenhoff[1], Martin Vehse[1], Markus Becker[2], Manuela Schiek[3], Michael Wark[2], Carsten Agert[1]

1. NEXT ENERGY · EWE Research Centre for Energy Technology at the University of Oldenburg♦
2. University of Oldenburg, Institute of Chemistry, Chemical Technology♣
3. University of Oldenburg, Institute of Physics, Energy and Semiconductor Research Laboratory♣

♦Carl-von-Ossietzky-Straße 15, 26129 Oldenburg, Germany
♣Carl-von-Ossietzky-Straße 9-11, 26129 Oldenburg, Germany

Presented here is a study of titanium dioxide thin films for their use as an electron transport layer (ETL) in perovskite solar cells (PSCs). These studies were performed on both the amorphous and polycrystalline (anatase) phases of titanium dioxide (TiO_2). Amorphous films were deposited by reactive e-beam thermal evaporation of titanium in oxygen atmosphere. To obtain the anatase phase, these films were annealed in air at 500°C for 30 minutes. The effect of the oxygen flow rate on the transmissivity of both the amorphous and anatase films was investigated. It is shown that, for amorphous films, the transmissivity increases with increasing the oxygen flow rate, while the transmissivity is almost constant for the anatase case. Also, the dependence of the resistivity on the oxygen flow rate for the amorphous films was investigated. Both the amorphous and anatase films were integrated into PSCs. It is shown that, for amorphous films, the V_{oc} increases with increasing the oxygen flow rate, while it shows no clear trend for the anatase case. Finally, the effect of a molybdenum oxide (MoO_x) layer between the spiro-MEOTAD and gold back-contact was investigated. This addition leads to an improvement in the cell performance, achieving an efficiency of 8.1%.

Keywords: Blocking Layers, Charge Selective Contacts, Hybrid, Optical Properities, Perovskite, Titanium Dioxide

1 INTRODUCTION

Organo-Metal Halide Hybrid solar cells, commonly known as perovskites, have attracted much attention in the past decade and mainly in the past 5 years due to several factors. The most noticeable one is their rapid development progress reaching a certified efficiency of 22.1% [1] from 2.2% in 2006 [2].

Much effort has been done to characterize the perovskite absorber in order to understand its primary excitation mechanisms and its inner workings in terms of carrier dynamics. In contrast, little attention is given to the electrodes and the carrier selective contacts/blocking Layers, which, together with the absorber, form an integral part of the solar cell. The carrier selective contacts refer to the ETL (electron transport layer), like titanium dioxide, and the HTM (hole transport medium), like the spiro-MEOTAD.

Titanium (IV) dioxide is one of the most extensively studied transition metal oxides. It shows outstanding resistance to corrosion and photocorrosion in water. These properties are invaluably needed in hydrogen generation by water splitting and photocatalytic water purification [3]. It is also employed in DSSC (dye-sensitized solar cells), has a potential for use as a TCO [4] and is used as a pigment, sunscreen and UV-blocking pigment.

One of the recent attracting functions of titanium dioxide is its utilization as an electron transport layer (ETL) in perovskite solar cells. The reason for this is a collection of attributes, the most significant ones being its hole blocking functionality due to its large bandgap and efficient electron collection due to the advantageous location of its conduction band relative to that of the perovskites.

Besides these characteristic properties for titanium dioxide, other general properties are required for a working blocking layer. These properties include high transparency, minimal electrical resistivity, compactness (pin-hole free layers), and the ability to passivate defects at the absorber interface.

In this study, the utilization of both the amorphous and anatase phases of titanium dioxide as an ETL for perovskite solar cells was investigated. The influence of the structure, stoichiometry and thickness of these films on their optical and electrical properties was explored. Additionally, the influence of these hole blocking layers on the performance of the corresponding PSCs in terms of the V_{oc} and J_{sc} was investigated. Finally, employing the optimized ETL, the effect of introducing a 12 nm thick molybdenum oxide (MoO_x) layer between the spiro-MEOTAD and the gold back-contact was investigated.

2 METHODOLOGY

2.1 Fabrication of titanium dioxide layers

Pure titanium was evaporated on top of glass and commercial FTO (fluorine-doped tin oxide) substrates in an e-beam thermal evaporation chamber. This was accompanied by an oxygen flow in order to oxidize the titanium and form amorphous titanium dioxide thin films. The TiO_2 deposition rate was 3 Å/s at a chamber pressure of $1.8 * 10^{-4}$ mbar. The stoichiometry of these as-deposited films changes with changing the oxygen flow rate. This leads to changes in the optical and electrical properities of these films. Some of these as-deposited layers were annealed at 500 °C for 30 minutes to obtain the anatase phase of titanium dioxide. In the case of the anatase films, the final film thickness was also varied.

2.2 Characterizing the fabricated ETL layers

Raman spectroscopy was employed to investigate the morphology of the annealed as-deposited films on FTO.

This was done using "Bruker Senterra Laser Class 1" at an excitation wavelength of 488 nm.

The as-deposited and annealed layers with different stoichiometry and thicknesses were optically characterized by transmission measurements in the 300 nm to 2000 nm range. This was achieved using "Agilent Technologies Cary 5000 UV-VIS-NIR".

The bulk resistivities, at 25 °C, of the amorphous TiO_2 layers with different stoichiometry (oxygen flow rate) were investigated. This was performed by temperature dependent conductivity measurements (TDC). Where, the conductivity as a function of temperature under vacuum and without illumination was measured. During the 2-3 hours of the experiment, the sample is heated up from room temperature up to 100 °C, stabilized at 100 °C for 20 minutes, and then cooled down to 45 °C. the conductivity, and hence the resistivity, at 25 °C were obtained by extrapolating the measurement curves down to this value.

2.3 Fabrication of the perovskite solar cells

Both the amorphous and anatase films were integrated into PSCs. These PSCs were fabricated as shown in Fig. 1. The commercially bought FTO-coated glass was coated with a TiO_2 thin film by reactive e-beam thermal evaporation of titanium (for the anatase layers, the annealing step is introduced here). The perovskite is then deposited by sequential deposition [5]. The perovskite absorber used in the experiments is of the composition $CH_3NH_3PbI_3$ (Methyl Ammonium lead Tri-Iodide).

Figure 1: Perovskite solar cell layout and layer stack

The perovskite deposition process was as follows: An 80µl of lead iodide (PbI_2) solution (0.75 molar PbI_2 in dimethylformamide) was statically spin-coated on TiO_2 at a rotation speed of 2000 rpm for 15 seconds. Both the solution and substrate were at a temperature of 60° C. Then, the film is dried to remove any residual solvent. After the spin coating of PbI_2, The PbI_2-coated substrate (at a temperature of 50 °C) was dipped in a methylammonium iodide (MAI) solution for one minute. The solution was prepared by dissolving MAI in Isopropanol and regulated to achieve a 10 mg/ml concentration at temperature of 50 °C. Then the film is dried to remove residual solvent

The HTM, which is Spiro-MEOTAD, was deposited by spin coating. This was done by dynamically spin coating a 35µl of a spiro-MEOTAD solution (0.034 molar spiro-EMOTAD in chlorobenzene) on the perovskite at a rotation speed of 3000 rpm for 20 seconds. The solution was doped with Lithium Ions. The final PSC is again dried to remove any residual solvent

Finally, a gold layer was deposited by e-beam thermal evaporation. Additionally, a variation of the back-contact was introduced by swapping the Au electrode with a MoOx/Au electrode. The MoO_x/Au back-contact electrode combination was deposited sequentially by resistive thermal evaporation.

2.4 Characterizing the perovskite solar cells

JV measurements under standard test condition illumination were employed to characterize the solar cell performance and to extract the device characteristics (V_{oc}, J_{sc}). From these *JV* measurements, the V_{oc} is extracted and compared for amorphous and anatase TiO_2 layers with varying oxygen flow rates and thicknesses. In the same manner, the J_{sc} is extracted and compared for anatase TiO_2 layers with varying thicknesses. Finally, *JV* curves are shown for the best performing ETL. This ETL was integrated into a PSC with and without the inclusion of a 12 nm MoO_x layer between the spiro-MEOTAD and the gold back-contact.

3 RESULTS AND DISCUSSION

3.1 Raman spectroscopy on titanium dioxide layers

Raman spectroscopy was employed to investigate the effect of annealing on the film morphology. The results of these measurements prior to annealing as well as for annealed films at different thickness are shown in Fig. 2. Clear anatase peaks [6] are observed for the annealed films both in the 40 nm and 80 nm case, while no such peaks are present prior to annealing, meaning that the as-deposited films grow in the amorphous structure and are transformed to anatase during the annealing step.

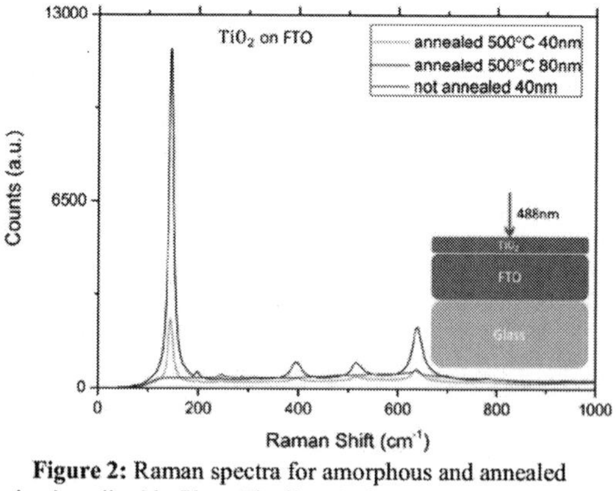

Figure 2: Raman spectra for amorphous and annealed titanium dioxide films. The inset is the measurement structure

The attainment of the anatase peaks signifies a morphological change from the amorphous to the polycrystalline phase, where the charge carrier mobility is enhanced. Indeed, it was found that the mobility, and hence the conductivity, of TiO_2 layers fabricated by different techniques increase with increasing the annealing (sintering) temperature [7, 8]. This increase in conductivity was attributed to an increase in the grain sizes and hence a reduction in the contributions from the grain boundaries [9].

3.2 Transmission measurements on amorphous TiO_2 layers

A major requirement on a blocking layer is high optical transparency in the relevant spectral range. Fig. 3 shows the measured transmission spectra for as-deposited TiO_2 thin films on FTO for different oxygen flow rates at a fixed thickness of 40 nm. The graph shows clearly that a decrease in the oxygen flow rate leads to a decrease in the overall transmission.

Briefly spoken, the decrease in transmission with decreasing oxygen flow rate can be attributed to the formation of oxygen defects, which leads to the formation of unsaturated metallic titanium atoms (or an aggregation of them). By increasing the oxygen flow rate the ratio of saturated-Ti atoms to unsaturated-Ti increases. The saturated-Ti atoms (or an aggregation of them) form regions which are transparent in the visible range, as a stoichiometric titanium dioxide (whether amorphous or polycrystalline) has a bandgap in the range (3.1 eV to 3.5 eV) [10, 11].

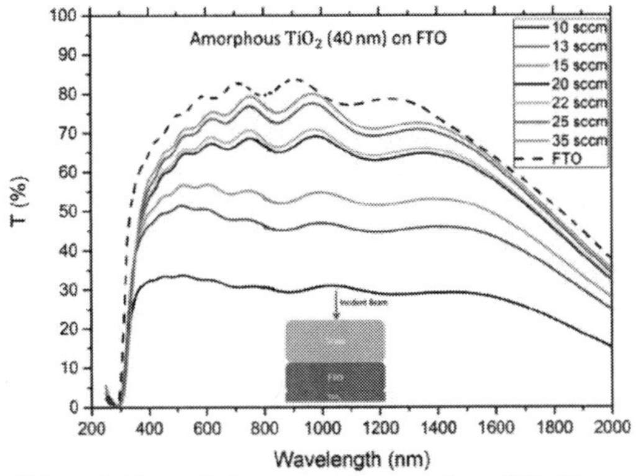

Figure 3: Transmission spectra for amorphous TiO_2 films on FTO different oxygen flow rates at a fixed thickness of 40 nm. The inset is the measurement structure

Under these conditions, the majority of the decrease in the transparency in the visible spectrum can be related to the metallic Ti regions, in which the free-carriers interactions are the main reason for this decrease in transparency.

From this observation, it becomes clear that high oxygen flow rates are required when utilizing amorphous TiO_2 to achieve high charge carrier generation rates inside the absorber material and thus high short circuit current densities.

3.3 Transmission measurements on anatase TiO_2 layers

Fig. 4a shows the transmission measurements performed on anatase TiO_2 layers on FTO for varying oxygen flow rates at a fixed thickness of 40 nm. It is evident that, compared to the amorphous case, the transmissivity depends by far less on the oxygen flow rate during deposition. This can be attributed to a post-oxidation of the layers during the annealing step in air, thus forming layers of nearly identical stoichiometry in all cases. In addition, Fig. 4b depicts the dependence of transmissivity on film thicknesses at a fixed oxygen flow rate of 35sccm. As expected, the graph shows a decrease of transmissivity with increasing thickness, which is particularly pronounced in the visible wavelength range for thickness values above 80 nm.

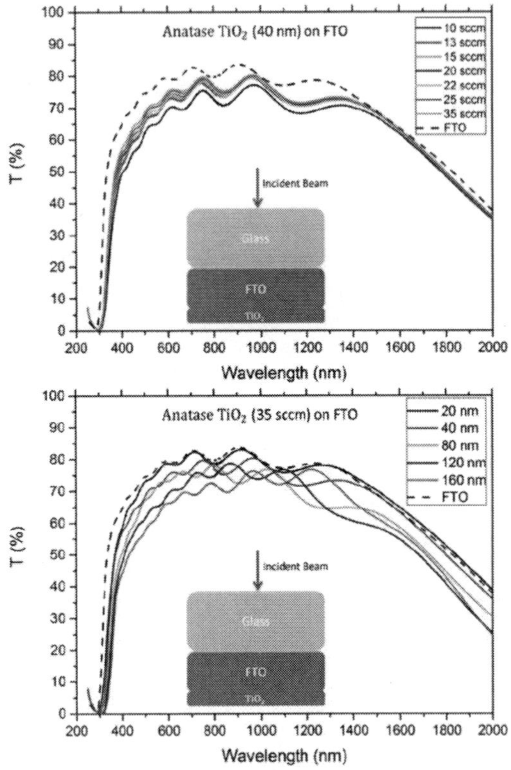

Figure 4: Transmission spectra for anatase films on FTO a) for varying oxygen flow rates at a thickness of 40 nm and b) for varying thicknesses at a fixed oxygen flow rate of 35 sccm. The insets are the measurement structures

3.4 Resistivity vs. transmission for amorphous TiO_2 layers

It is essential for an ETL to have minimal electrical resistivity so as to minimize the ohmic losses. Fig. 5 shows the bulk resistivity as a function of transmission (at 633 nm) for different oxygen flow rates.

It is evident that, the bulk resistivity and the transmissivity (in the visible spectrum) increase with increasing the oxygen flow rate. Hence, there is a tradeoff between the advantages gained from the increase in transmissivity with the disadvantages gained from the increase in resistivity.

3.5 V_{oc} and J_{sc} measurements

The influence of the different blocking layers on the performance of the PSC is investigated through measuring the V_{oc}. Fig. 6 shows the V_{oc} measurements for PSCs incorporating blocking layers with different phases, stoichiometry and thicknesses.

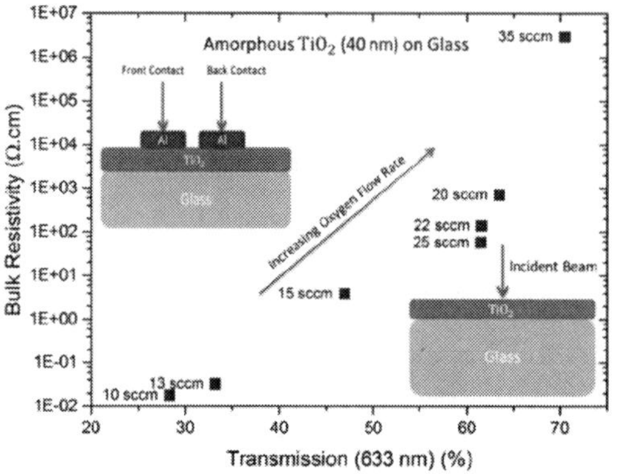

Figure 5: Bulk resistivity vs. transmission at 633 nm for amorphous layers on glass. The insets are the measurement structures

It can be seen from the graph that for amorphous titanium dioxide, the V_{oc} increases with increasing the oxygen flow rate, while in a similarity to the transmission measurements there is no clear trend for the anatase case. Comparing both cases at an equal layer thickness, it can be clearly seen that amorphous films with high oxygen content lead to higher V_{oc}.

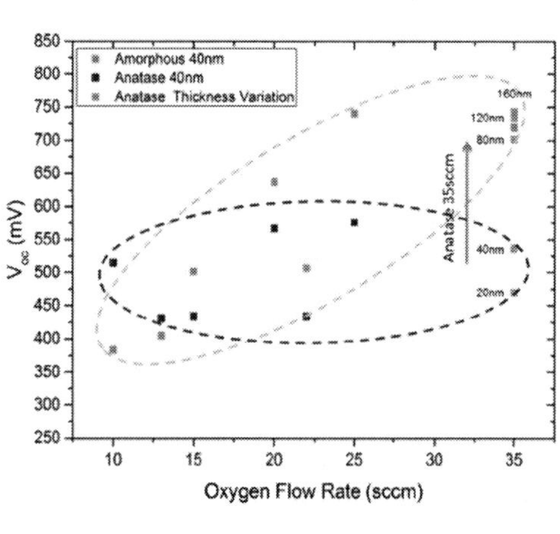

Figure 6: V_{oc} measurements on PSCs incorporating titanium dioxide blocking layers with different phases, stoichiometry and thicknesses

However, for anatase films with a fixed oxygen flow rate of 35 sccm an improvement can be obtained when increasing the layer thickness, so that V_{oc} values similar to the amorphous case are ultimately reached.

As it turned out, devices that were fabricated with amorphous titanium dioxide layers showed less J_{sc} than the anatase ones. Hence, the focus was mainly shifted to the anatase layers. Fig. 7 shows the dependence of J_{sc} on layer thickness for the same anatase films as presented in Fig. 6.

It is evident that an optimum J_{sc} is reached at a TiO_2 layer thickness of 80 nm. The initial increase is related to a thickness-dependent effect, while the latter decrease at higher thicknesses can be attributed to an increase in resistance at a given bulk resistivity. This thickness-dependent effect can be most likely attributed to a better hole blocking functionality and better passivation of defects at the perovskite interface.

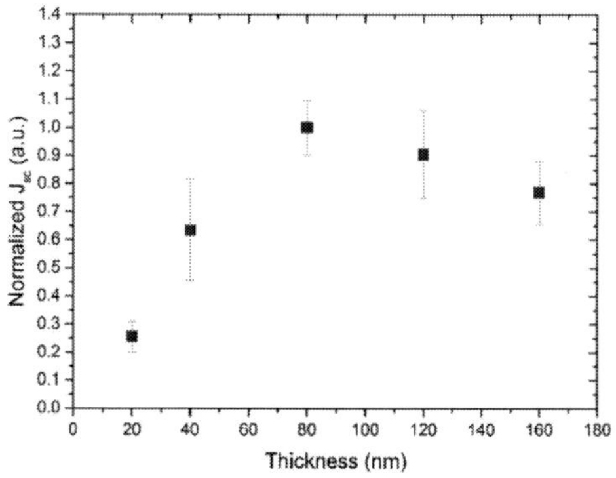

Figure 7: Normalized J_{sc} measurements of PSCs incorporating anatase titanium dioxide blocking layers with different thicknesses at fixed oxygen flow rate of 35 sccm

3.5 Influence of a MoO_x layer

Having identified the optimum ETL layer, the influence of an additional 12nm thick Molybdenum Oxide (MoO_x) electron blocking layer placed between the spiro-MEOTAD and the gold contacts was investigated. *JV*-curves of the corresponding devices are depicted in Fig. 8, and their related electrical parameters are shown Table I.

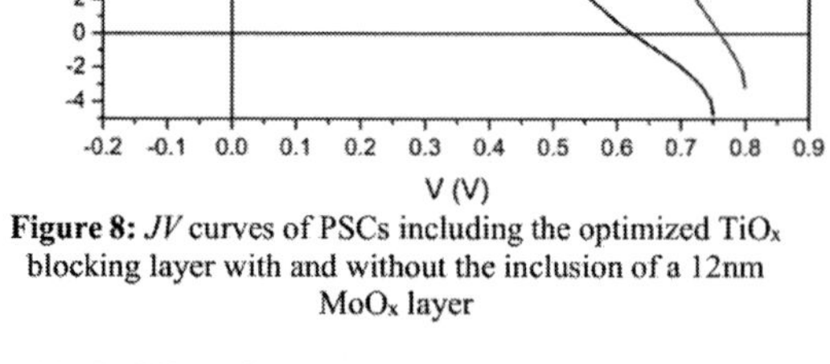

Figure 8: *JV* curves of PSCs including the optimized TiO_x blocking layer with and without the inclusion of a 12nm MoO_x layer

Table I: Effect of a 12 nm MoO_x layer on the electrical parameters of PSCs

	Jsc (mA/cm²)	Voc (mV)	FF (%)	η (%)
Without MoO_x	17	627	32	3.3
With MoO_x	23	760	46.1	8.1

It is evident from the graph that the inclusion of a MoO_x layer significantly increases the efficiency by improving the J_{sc}, V_{oc} and the FF.

It was previously shown by other groups that a thin (10 nm range) MoO_x layer between the spiro-MEOTAD and the metal back-contact works as an effective interlayer for hole extraction [12, 13], allowing a replacement of gold by other metals. According to these results, the inclusion of a thin MoO_x layer does not significantly alter the power conversion efficiency (PCE), but enhances the stability of the PSC and offers a cheap alternative to the expensive Au back-contact via the MoO_x/Al combination. In contrast, a bare Al back-contact without any electron-blocking interlayer lead to poor efficiencies [13], which was attributed to the formation of a back-diode at the spiro-MEOTAD/Al interface [13]. This back-diode effect was not observed in higher work function metals like Ag and Au. However, the inclusion of a thin MoO_x layer between the spiro-MEOTAD and the Al back-contact can greatly suppress this barrier effect. This can be related to the fact that the inclusion of a MoOx interlayer can increase the work function of TCOs like ITO and AZO [14]. Hence, in a rough analogy, it can be said that the MoO_x increases the work function at the spiro-MEOTAD/Al interface, thus avoiding the barrier formation.

Thus, the addition of this MoO_x layer can have several effects: The modification of the anode work function, enhancement of the hole transport process across the interfaces, increasing the stability and in the case of a thin spiro-MEOTAD, minimization of the influence of the Au anode on the perovskite. This list is not conclusive, but further studies need to be implemented in order to pinpoint the exact origin of this effect.

4 CONCLUSIONS

Amorphous and anatase TiO_2 layers were fabricated by e-beam thermal evaporation on FTO and glass. Raman spectroscopy was employed to identify the phase of the annealed substrates. It was demonstrated that annealing at 500 °C for 30 minutes produces the anatase phase as seen from the Raman spectra.

It was found that for the amorphous layers the transmissivity increases with increasing the oxygen flow rate, while for the anatase layers, the transmissivity is nearly independent of the oxygen flow rate.

It was shown that for the amorphous layers, there is a tradeoff between the transparency and the resistivity, as both of them increase with increasing the oxygen flow rate.

Then, these amorphous and anatase TiO_2 films were integrated into PSCs, and it was found out that for the amorphous case, the V_{oc} increases with the oxygen flow rate, while for the anatase case, no clear trend is observed. At fixed thickness and high oxygen flow rates, the amorphous layers show higher V_{oc} than the anatase layers. However, V_{OC} comparable to the amorphous films can be reached by increasing the thickness of the anatase films.

Although the amorphous layers show high V_{oc}, they lead to a lower J_{sc} compared to the anatase films. So, to narrow down the search for an optimum blocking layer, the J_{sc} was measured for anatase layers with varying thicknesses. It was shown that, the J_{sc} initially increases with thickness and then decreases again, after reaching a maximum. The final decrease is related to an increase in the resistance, while the initial increase (which is also observed in the V_{oc}) can be attributed to a better hole blocking functionality and enhanced surface passivation at the perovskite interface.

Finally, the optimum blocking layer was integrated into a PSC to investigate the effect of inclusion a 12 nm MoO_x layer in-between the spiro-MEOTAD and the gold back contact. It was found that, this inclusion enhances the PCE from 3.3% to 8.1%. This observation can possibly be attributed to several factors, such as a modification of the anode work function, an enhancement of the hole transport process and a minimization of the influence of the anode on the perovskite.

5 REFERENCES

[1] Saliba, Michael, Taisuke Matsui, Ji-Youn Seo, Konrad Domanski, Juan-Pablo Correa-Baena, Mohammad Khaja Nazeeruddin, Shaik M. Zakeeruddin, Wolfgang Tress, Antonio Abate, Anders Hagfeldt, Michael Grätzel, Cesium-containing triple cation perovskite solar cells: improved stability, reproducibility and high efficiency, Energy & Environmental Science 9, no. 6 (2016) 1989.

[2] Kojima, Akihiro, Kenjiro Teshima, Tsutomu Miyasaka, Yasuo Shirai, Novel photoelectrochemical cell with mesoscopic electrodes sensitized by lead-halide compounds (2), The Electrochemical Society Meeting Abstracts no.7 (2006) 397.

[3] Janusz Nowotny, Oxide Semiconductors for Solar Energy Conversion, Chapter 3: Titanium Oxide.

[4] Elen, Ken, Boris Capon, Christopher De Dobbelaere, Daan Dewulf, Nick Peys, Christophe Detavernier, An Hardy, Marlies Van Bael, Aqueous chemical solution deposition of doped titanium dioxide films as Transparent Conducting Oxide, Belgian Ceramic Society (2013).

[5] Zheng, Lingling, Danfei Zhang, Yingzhuang Ma, Zelin Lu, Zhijian Chen, Shufeng Wang, Lixin Xiao, Qihuang Gong, Morphology control of the perovskite films for efficient solar cells, Dalton transactions 44 no. 23 (2015) 10582.

[6] Tian, Fang, Yupeng Zhang, Jun Zhang, Chunxu Pan, Raman spectroscopy: a new approach to measure the percentage of anatase TiO2 exposed (001) facets, The Journal of Physical Chemistry C 116 no. 13 (2012) 7515.

[7] Jeong, B-S., D. P. Norton, J. D. Budai, Conductivity in transparent anatase TiO2 films epitaxially grown by reactive sputtering deposition, Solid-State Electronics 47, no. 12 (2003) 2275.

[8] Schattauer, Sylvia, Beate Reinhold, Steve Albrecht, Christoph Fahrenson, Marcel Schubert, Silvia Janietz, Dieter Neher, Influence of sintering on the structural and electronic properties of TiO_2 nanoporous layers prepared via a non-sol–gel approach, Colloid and Polymer Science 290, no. 18 (2012) 1843.

[9] Hanini, F., A. Bouabellou, Y. Bouachiba, F. Kermiche, A. Taabouche, M. Hemissi, D. Lakhdari, Structural, optical and electrical properties of TiO2 thin films synthesized by sol-gel technique, IOSR Journal of Engineering 3, no. 6 (2013) 21.

[10] David O. Scanlon, Charles W. Dunnill, John Buckeridge, Stephen A. Shevlin, Andrew J. Logsdail, Scott M. Woodley, C. Richard A. Catlow, Michael. J. Powell, Robert G. Palgrave, Ivan P. Parkin, Graeme W. Watson, Thomas W. Keal, Paul Sherwood, Aron Walsh, Alexey A. Sokol, Band alignment of rutile and anatase TiO2, Nature materials 12, no. 9 (2013): 798.
[11] Manoj, P. K., P. Koshy, V. K. Vaidyan, Transparent anatase titania films: A critical study on optical properties, Progress in Natural Science: Materials International 22, no. 2 (2012) 79.
[12] Sanehira, Erin M., Bertrand J. Tremolet de Villers, Philip Schulz, Matthew O. Reese, Suzanne Ferrere, Kai Zhu, Lih Y. Lin, Joseph J. Berry, Joseph M. Luther, Influence of Electrode Interfaces on the Stability of Perovskite Solar Cells: Reduced Degradation Using MoO_x/Al for Hole Collection, ACS Energy Letters 1 (2016) 38.
[13] Zhao, Yixin, Alexandre M. Nardes, Kai Zhu, Effective hole extraction using MoO_x-Al contact in perovskite CH3NH3PbI3 solar cells, Applied Physics Letters 104, no. 21 (2014) 213906.
[14] Meyer, Jens, Sami Hamwi, Michael Kröger, Wolfgang Kowalsky, Thomas Riedl, Antoine Kahn, Transition metal oxides for organic electronics: energetics, device physics and applications, Advanced Materials 24, no. 40 (2012) 5408.

OPTIMIZING THE DEPOSITION OF THIN LAYERS OF ORGANIC-INORGANIC HYBRID PEROVSKITE METHYLAMMONIUM LEAD IODIDE (CH3NH3PBI3) ON LARGE SURFACES THROUGH THEIR OPTICAL PROPERTIES

L. Ocaña[1], C. Quinto[1], C. Montes[1], A. Linares[2], E. Llarena[1], O. González[1], D. Molina[1], A. Pío[1], C. Hernández-Rodríguez[3], S. González-Pérez[3] & R. Guerrero-Lemus[3], M. Friend[1] & M. Cendagorta[1].
[1]Instituto Tecnológico y de Energías Renovables, S. A. (ITER), Pol. Industrial de Granadilla, s/n, E 38600 Granadilla de Abona, España
Tlf. +34 922 747 700, Fax +34 922 747 701, E-mail. iter@iter.es
[2]Agencia Insular de Energía de Tenerife (AIET), Pol. Industrial de Granadilla, s/n, E 38600 Granadilla de Abona, España
Tlf. +34 922 747 700, Fax +34 922 747 701, E-mail. agenergia@agenergia.org
[3]Departamento de Física, Universidad de La Laguna, Avda. Astrofísico Fco. Sanchez s/n, 38200 La Laguna Tenerife, Canarias, España.
Tfn: +34 922318243, Fax: +34 922318228, E-mail, chdezr@ull.edu.es

ABSTRACT: Nowadays, most of the Research and Development (R&D) efforts devoted to perovskite solar cells have been focused on achieving higher power conversion efficiencies, by using small sized substrates with even smaller active areas. The present paper proposes a method for optimizing the deposition of thin layers of organic-inorganic hybrid perovskite methylammonium lead iodide ($CH_3NH_3PbI_3$) on large surfaces (up to 75 mm x 75 mm) via spin coating procedures, performed in a clean room environment at room temperature and by considering their optical properties to assess their quality. Thus, in order to map the deposited area and realize studies about their optical properties, the absorption coefficient, the refractive index and layers thicknesses were determined by using a high accuracy spectroscopic ellipsometer system based on rotating compensator ellipsometer (RCE) technology from the J.A. Woolam Co. as functions of the wavelength, within a spectral range from 250 to 800 nm. Also, the samples were studied with a fluorescence spectroscopic system, in order to evaluate their photoluminescence properties and, in order to validate the chosen methodology; an Atomic Force Microscopy was used to verify the layers uniformity and thicknesses.
Keywords: Absorption, deposition, optical properties, photoluminescence.

1 INTRODUCTION

Since its foundation in 2009, the solar cell laboratory from Instituto Tecnológico y de Energías Renovables, S.A. (ITER) has been devoted to R & D on ways to optimize standard manufacturing processes of crystalline silicon solar cells [1]. In addition, a great deal of effort has been realized to extend its know-how to encompass other hot and emerging topics within the photovoltaic technologies. In this sense, Perovskite Solar Cells (PSC) has represented one of its main work activities during the last two years [2]. Throughout that time, the Department of Physics and the Research Support General Service (SEGAI) from the University of La Laguna have been key resources of technical support.

PSC emerged as a promising technology due to the rapidly increase in Power Conversion Efficiency (PCE) registered during the present decade, reaching a PCE record of 22.1% in 2016 [3]. Since most of the R&D efforts devoted to PSC have been focused on achieving higher PCE, by using small sized substrates with even smaller active areas [4], and several authors suggest the use of PSC as a top device in silicon-based tandem solar cells, due to the perovskite (PK) higher bandgap [5][6][7] and the appropriate refractive index [8][9][10], it is clearly necessary the study and optimization of deposition procedures that allow the implementation of PK films on larger substrates, aiming to converge this kind of technologies with the standard fabrication of crystalline solar cells.

The present paper evaluates the deposition of thin films of organic-inorganic hybrid perovskite composed by methylammonium lead iodide (MAPI) on large surfaces via spin coating techniques by considering their optical properties in order to assess their characteristics.

2 METHODOLOGY

All the fabrication procedures were performed in a clean room (ISO 7) environment at room temperature (~21 °C).

Initially the deposition of the PK thin films was undertaken on small size substrates (up to 25 mm x 25 mm) by spin coating from several PK solutions obtained by modifying the MAPI weight concentration. These films were evaluated in order to relate their optical properties with the literature [8][9]. The measurements were performed by mapping the substrate in order to acquire spatial resolution over the surface. In this sense, the refractive index, the extinction coefficient as well as the reflectance and transmittance rates were obtained by spectroscopic ellipsometry techniques as a function of the wavelength and for the film–substrate system. From these measurements, the absorption coefficient (α) and the refractive index (n) of the deposited films were calculated. Also, in order to better characterize the quality of the deposited layers, a FluoroSENS M-11 fluorometer [11] was used so as to determine its fluorescence properties.

Finally, the thickness of the deposited films was estimated from theoretical fit models, implemented by the ellipsometer system software, and evaluated by comparing the obtained results with measurements performed in parallel with an Atomic Force Microscope (AFM) [12].

From these trials, conclusions on the quality of the deposited layers were drawn and, since the objective for this paper was to evaluate the deposition of PK thin layers on larger substrates, the corresponding fabrication parameters were applied to produce and characterize substrates of bigger size.

3 EXPERIMENTAL

3.1 Study over small substrates

For the purpose of studying the PK depositions through their optical properties [8][9], the variables chosen were the weight concentration of the MAPI solutions and the spin coater acceleration. Thus, 25 wt% and 45 wt% MAPI solutions were deposited on 25 mm × 25 mm, 2 mm thick glass substrates, see Figure 1, setting the spin speed to 5000 rpm [7] while varying the acceleration parameter in 2000 rpm/s and 5000 rpm/s.

The relative humidity (RH) inside the chamber of the spin coating equipment was continuously monitored in order to perform the depositions when the Clean Dry Air (CDA) used by the device provided 10% RH within the chamber.

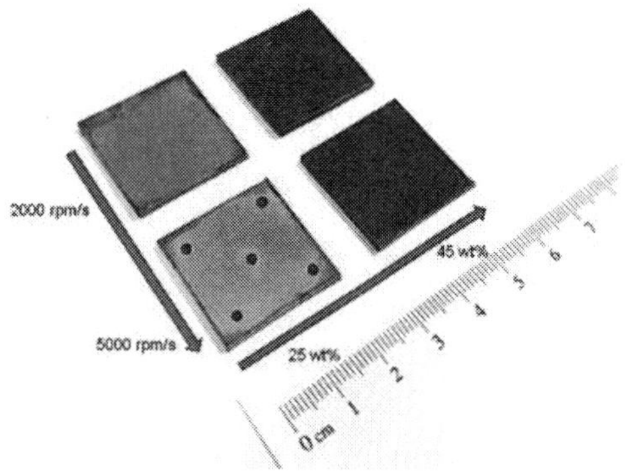

Figure 1: PK thin films deposited on 25 mm × 25 mm substrates, highlighting the measurement points (red dots).

Finally, in order to obtain the necessary spatial resolution, a matrix of measurement points was chosen as a system of reference for all the deposited samples, as it can be seen in Figure 1, in which the red dots represent the considered measurement points.

The samples were characterized, in ambient conditions, under a spectroscopic ellipsometer system composed by a 75 W Xe arc lamp, the beam collimation optics, a fixed polarizer and a compensator located in a continuously rotating motor at the entry arm. The receiver unit at the exit arm consists of a stepper motor driven rotational stage which houses an analyzer. In general, experimental measurements in ellipsometry define the ellipsometric parameters Ψ and Δ. These parameters were then used to calculate, from the theoretical fit models implemented by the data processing software CompleteEASE from J.A. Woolam Co, the refractive index (n), the extinction coefficient (k), the reflectance (R) and the transmittance (T) as functions of the wavelength (λ), as well as the average thickness of the deposited films [13].

3.2 Study over larger substrates.

Glass substrates with sizes 75 mm × 75 mm x 2 mm were chosen and spin coated with the 45 wt% MAPI solution by setting the speed to 5000 rpm, while trying two types of accelerations: 2000 rpm/s and 5000 rpm/s, as it was done in the previous section.

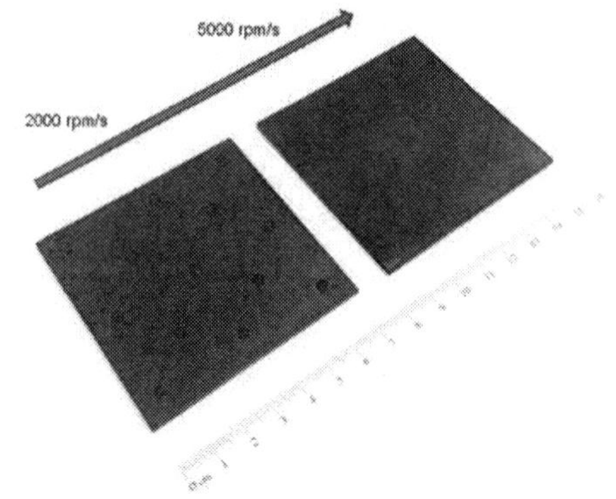

Figure 2: PK thin films deposited on 75 mm × 75 mm substrates, highlighting the measurement points (red dots).

Although visually the deposited samples showed a less uniform appearance, the ellipsometry study was performed by taking measurements in a matrix of equally spaced dots, following the same kind of pattern chosen for the smaller sized ones, as it can be seen in Figure 2 (red dots).

4. RESULTS AND DISCUSSION

4.1 Small substrates

For the purpose of studying the PK depositions through their optical properties [8][9], the variables chosen were the weight concentration of the MAPI solutions and the spin coater acceleration. Thus, 25 wt% and 45 wt% MAPI solutions were deposited on 25 mm × 25 mm, 2 mm thick glass substrates, see Figure 1, setting the spin speed to 5000 rpm [7] while varying the acceleration parameter in 2000 rpm/s and 5000 rpm/s.

The theoretical fit model was derived from the parameterization of the complex dielectric function by using the B-spline function, commonly used for the analysis of the optical constants for absorbing thin layers [13] in the visible range of the spectrum. Optical constants reported in the literature [8][9] were used as control points to match throughout the calculus. Thus, the obtained data was evaluated in a spectral range from 250 to 800 nm to define the material behavior over the visible range with a high degree of precision.

The absorption coefficient (α) for each processed sample was calculated from k, according to the Lambert's Law defined in eq. (1)

$$\alpha = \frac{4 \cdot \pi \cdot k}{\lambda} \quad \text{eq.(1)}$$

Figure 3 shows a graph depicting the values obtained for the absorption coefficient (α) from the different samples, together with values published in the literature. Thus, as it can be seen, the behavior of the 45 wt% solution processed samples coincides with the values reported by Löper et al [8], presenting the same kind of trend reported by Leguy et al [9], in which there is an increase in the values of the absorption coefficient from 300 nm on, peaking by 350 nm, followed by a pronounced decrease which attenuates between 450 and 500 nm. Alike, a 1.59 eV band gap value is observed, confirming results from literature [5]-[10].

It can also be observed the difference in shape of the curves from the 25 wt% solution processed samples, particularly in the range between the 500 to 800 nm, which suggest a better performance for these samples for the purposes of operating within such range, rather than as component of a tandem device.

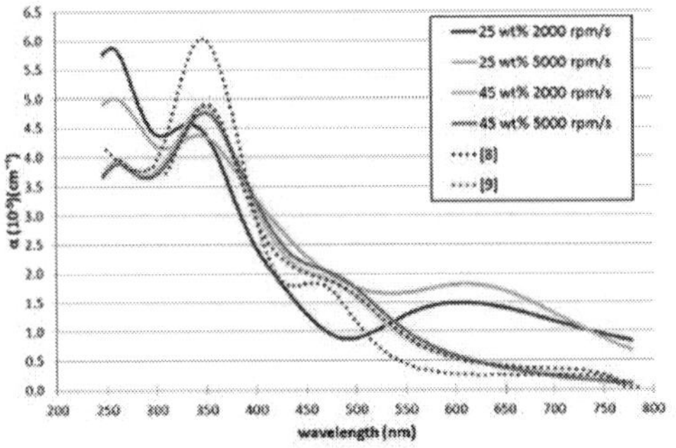

Figure 3: Absorption coefficient (α) of PK thin films deposited on 25 mm × 25 mm substrates.

Similarly, the values obtained for the refractive index (n) from the different samples were compared with the values published in the literature [8][9] and displayed in Figure 4. As it can be seen, the shape of the curves from the 45 wt% solution processed samples, follow a similar patron to the ones available, presenting two peaks clearly visible: one at 400 nm and the other between 500 and 550 nm. As to the curves of the 25 wt% processed samples, it can be seen how the first peak presents a softer shape, possibly due to a merging with the second one, and a rather diverging behavior from 450 nm onwards.

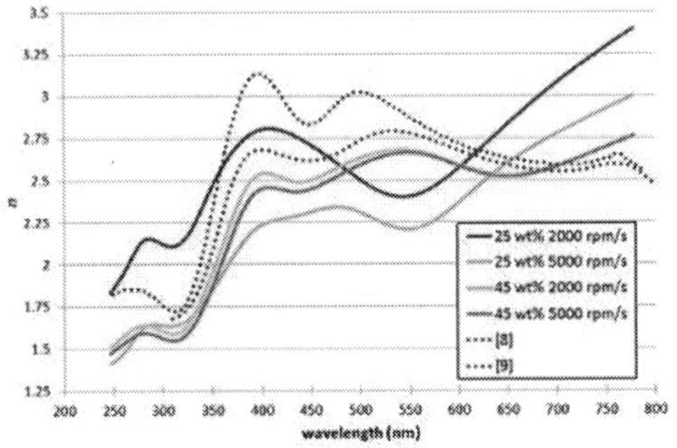

Figure 4: Refractive index (n) of PK thin films deposited on 25 mm × 25 mm substrates.

Regarding the value of the refractive index obtained for the 45 wt% solution processed samples at the wavelength of 633 nm, it is important to note that not only it was quite similar to the one published by Löper et al. [8] and Leguy et al [9] (that is, 2.53 versus 2.61 and 2.65, respectively), but also that it results a more suitable option for an Anti Reflective Coating (ARC) than standard photovoltaic manufacturing ones which, according to eq.(2) [14], addresses a value of 2.41.

$$n_1 = \sqrt{n_0 \cdot n_2} \qquad \text{eq.(2)}$$

Where n_0 is the refractive index of the surrounding material, 1.5 for encapsulant Ethyl Vinyl Acetate (EVA) [15], and n_2 is the refractive index of the semiconductor, 3.8763 for silicon [16].

From the values obtained by the spectroscopic ellipsometer, the reflectance and transmittance were also studied, this time, due to the sample positioning limitations given by the equipment, only at a determined point. As it can be seen in Figure 5, for both kinds of solutions, the reflectance remains stable between 0.20 and 0.30 points over one, from 330 nm to 800 nm, as it has been published [8][17][18]. As for the transmittance, near cero values were found from 330 nm to 360 nm for the 25 wt% solution processed samples, while for the 45 wt% ones the range extended form 330 nm to 500 nm, from where it begins to climb to 800 nm, reaching 0.80 points over one, a result that not only meet the values found in the literature [8][17][18], but also suggest the suitability of using this kind of deposited solutions as a better candidate for photovoltaic purposes.

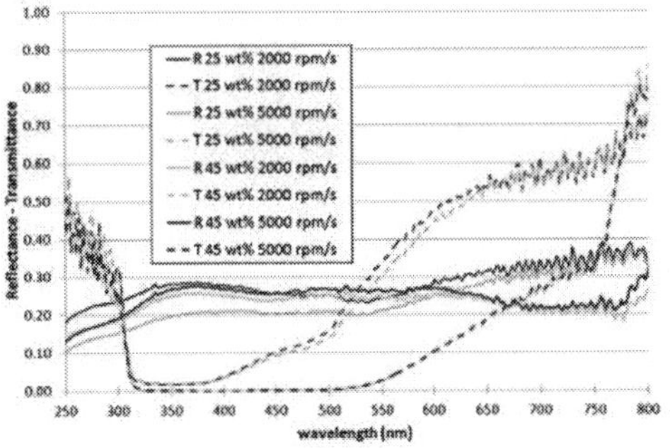

Figure 5: Reflectance and Transmittance versus wavelength of PK thin films deposited on 25 mm × 25 mm substrates.

In addition, as a way to evaluate the behavior of these thin film layers of perovskite when exposed to solar spectral irradiance, a study was performed following the approach described by Xie Ziang et al [10], by analyzing the conversion efficiency of these layers, without considering carrier recombination. This efficiency, known as *ideal efficiency*, is defined by eq. (3):

$$\eta = \frac{\int_0^{\lambda_g} I(\lambda) \cdot A(\lambda) \cdot \frac{\lambda}{\lambda_g} d\lambda}{\int_0^{\infty} I(\lambda) d\lambda} \qquad \text{eq.(3)}$$

Where $I(\lambda)$ is the solar intensity for a given range of wavelength (AM 1.5G) [19], $A(\lambda)$ is the absorptance ratio resulting from eq.(4):

$$A(\lambda) = 1 - R(\lambda) - T(\lambda) \qquad \text{eq.(4)}$$

λ is the wavelength and λ_g is the bandgap wavelength.

As it can be seen in Table I and Figure 6, the ideal efficiency obtained for the 45 wt% solution processed samples have a better response in the visible part of the spectrum than the 25 wt% ones, a result which also support their use as a better option for tandem devices together with crystalline silicon solar cells.

Table I: Ideal efficiency obtained for the PK thin films deposited on 25 mm × 25 mm substrates.

Solution (wt%)	Acceleration (rpm/s)	Substrate size (mm)	Ideal efficiency (η)
25	2000	25 × 25	23.78 %
25	5000	25 × 25	27.36 %
45	2000	25 × 25	45.81 %
45	5000	25 × 25	44.60 %

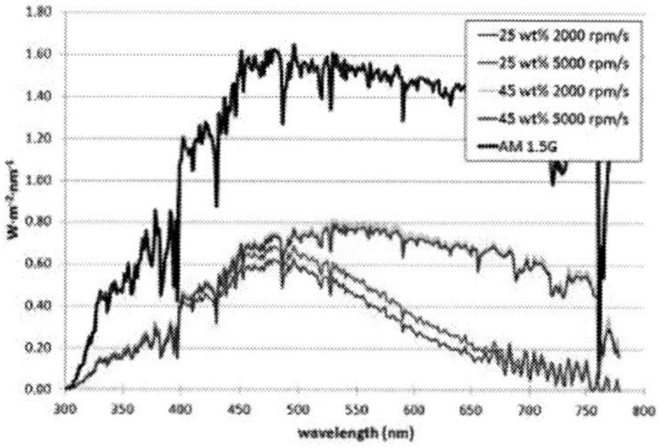

Figure 6: Ideal efficiency study of PK thin films deposited on 25 mm × 25 mm substrates.

On the other hand, in order to determine their fluorescence properties, the samples were studied with a fluorescence spectrometer, composed of two monochromators, a 75W Xenon lamp (THL) and a photo-multiplier tube (PMT) detector. The measurements were carried out by mapping the emission intensity for the samples over an excitation range. The excitation parameters were set from 300 to 800 nm, with 10 nm steps, and the emission ones were scanned from 300 to 800 nm, with an integrating time of 100 ms. In this way, it was possible to detect the wavelengths in which the excitation created conversion phenomena.

Figure 7 shows the results obtained for a 45 wt% solution processed sample, deposited at 5000 rpm/s. These results were quite similar for the rest of the samples, regardless of their weight concentration or the acceleration chosen during their deposition processes. As it can be seen, the higher emission intensity focuses on the visible range. However, some evidence of a conversion phenomenon also appears when the sample is excited between 350 and 400 nm, emitting between 700 and 800 nm.

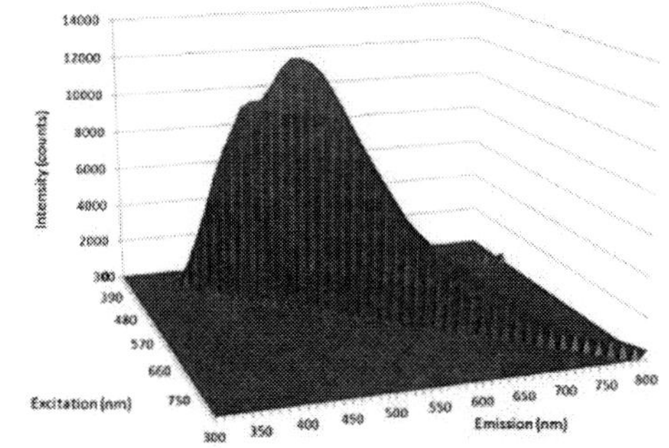

Figure 7: Excitation - Emission Mapping Scan of the sample with a 45wt% and an acceleration of 5000 rpm/s.

From these results, a further study of this phenomenon was carried out, this time by evaluating only the emission produced at different wavelengths from the excitation. In Figure 8 it can be observed, for the four cases studied, that a conversion took place when exciting the sample between 350 and 400 nm, resulting on an emission between 700 and 800 nm. This phenomenon has been extensity reported for organolead trihalide perovskites [20]-[27], although in these cases the excitation was carried out with a laser emitting at a wavelength of visible, and detecting an emission at around 760 nm.

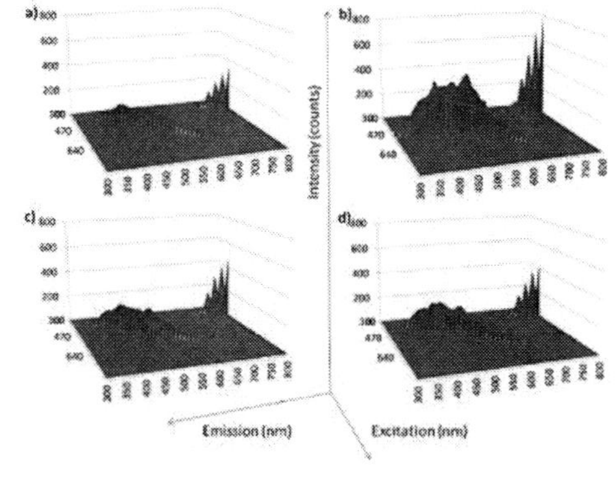

Figure 8: Excitation - Emission Mapping Scan of the sample: (a) Sample 25 wt% 2000 rpm/s, (b) Sample 25 wt% 5000 rpm/s, (a) Sample 45 wt% 2000 rpm/s and (a) Sample 45wt% 5000 rpm/s. The Z axis represents intensity in photon counting, the X axis excitation wavelength and the Y axis emission wavelength.

Table II summarizes the results obtained for the measured samples. By looking at these results, it can be concluded that the samples with 45 wt% have a higher conversion phenomenon than the 25 wt% ones and that, among the 45 wt% samples, the ones with higher acceleration have a slight better conversion as well. These results also support the suitability of using this kind MAPI layers in tandem with crystalline silicon solar cells.

Table II: Percentage of total conversion

Solution (wt%)	Acceleration (rpm/s)	Substrate size (mm)	Total Conversion (%)
25	2000	25 × 25	2.33
25	5000	25 × 25	2.70
45	2000	25 × 25	2.87
45	5000	25 × 25	3.39

Finally, the average thickness was obtained by the theoretical fit model with the ellipsometer, as well as the resulting standard deviation for the 25 mm × 25 mm samples, as it can be seen summarized in Table III. There, it is shown that the approximate thickness for 25 wt% and 45 wt% processed samples was 255 nm and 435 nm respectively. These results confirm the dependence that exists between the solution weight concentration and the resulting deposited thickness [7]. Also, by considering the differences regarding the values obtained for the thickness standard deviation over the studied surfaces, it was possible that the 45 wt% were distributed around the mean value more closely than the 25 wt% ones, which suggest a better uniformity in the deposition processes.

Table III: Deposited thickness estimated by the ellipsometer fit model.

Solution (wt%)	Acceleration (rpm/s)	Substrate size (mm)	Average thickness (nm)	Standard deviation (nm)
25	2000	25 × 25	250.82	43.97
25	5000	25 × 25	262.81	63.16
45	2000	25 × 25	439.07	3.25
45	5000	25 × 25	432.15	4.37

As a way to confirm the validity of the results displayed in Table III, a set of samples were measured by both, the spectroscopic ellipsometer by way of the applied fit model, and a Atomic Force Microscope (AFM) [12]. Given the special requirements of the AFM measurement techniques, these samples were prepared by covering determined areas of the substrate with a polyimide tape before depositing the MAPI solution (45 wt%, with 5000 rpm and 5000 rpm/s as the setups for the wafer spinner). In this way, after peeling off the tape, the samples were ready for measuring the interface between the MAPI and the MAPI-free parts over the substrates.

Figure 9 is an image from the AFM, depicting the interface between the MAPI and the MAPI-free areas. As it can be seen, at the MAPI side, although there is a sort of overgrowth or elevation around 1 µm thick, due to the impact of the MAPI solution against the border of the polyimide tape during the deposition, such effect fades further into the MAPI side and the thickness stabilizes at around 450 nm, a result that concurs with the values obtained by the ellipsometer.

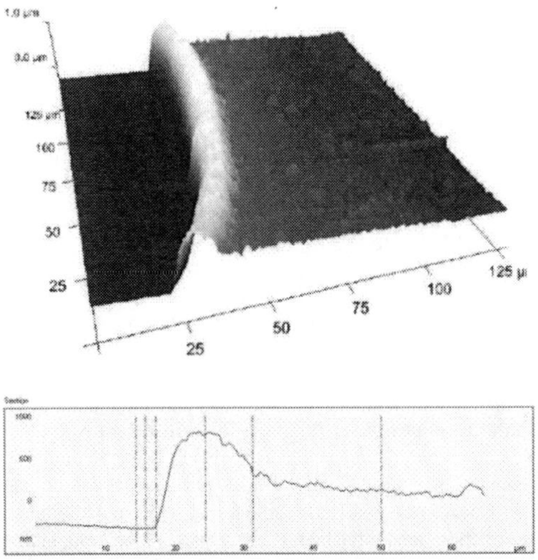

Figure 9: AFM image and cross section, showing the interface between the MAPI-free and MAPI areas in one sample.

Taking into consideration all the above, the 45 wt% MAPI solution was chosen as the most suitable candidate to be evaluated over larger substrates.

4.2 Larger substrates.

As it can be seen in Figure 10, the values derived for the absorption coefficient (α) related to the samples deposited with an acceleration of 5000 rpm/s followed a pattern closest to both, the values obtained for smaller substrates, and the ones reported in the literature [8][9]. This was also true for the refractive index (n), as it can be seen in Figure 11, which shows a slight deviation from the reported behavior that increases considerably at higher wavelengths. Thus, the value obtained for 633 nm was 2.31, which evidences a value not as good as the one obtained for smaller substrates.

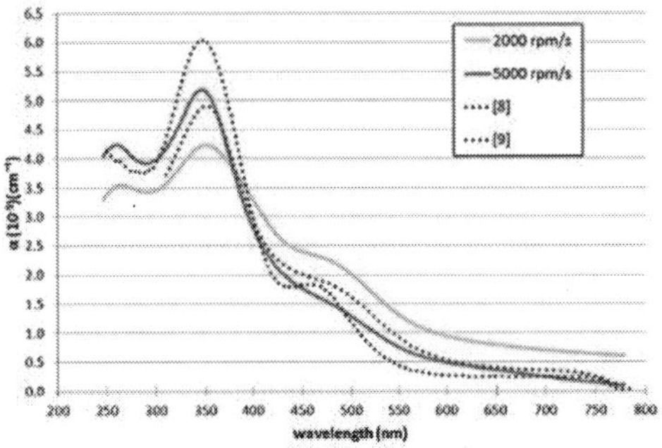

Figure 10: Absorption coefficient (α) of 45 wt% MAPI thin films deposited on 75 mm × 75 mm substrates.

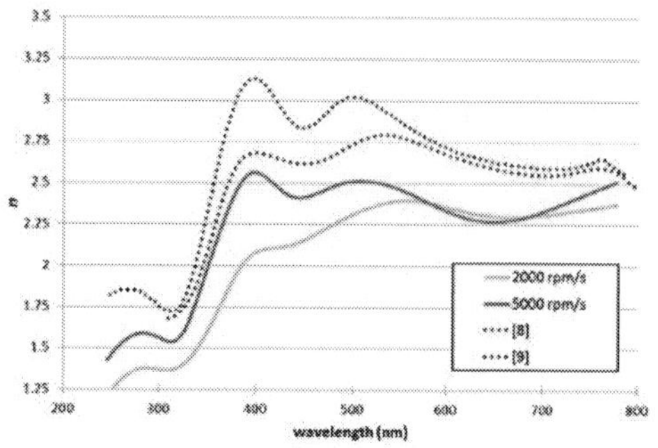

Figure 11: Refractive index (n) for 45 wt% MAPI solution deposited on 75 mm × 75 mm substrates.

Figure 12 shows that both, the transmittance and reflectance obtained for the 75 mm x 75 mm deposited substrates are worse than their respective values obtained for the smaller ones (and reported in the literature [8][17][18]). This can also be addressed as to a worsening in the uniformity of the deposited layers, which suggest the need for a better adjustment in the setting parameters for the wafer spinner, possibly in the line of increasing the acceleration.

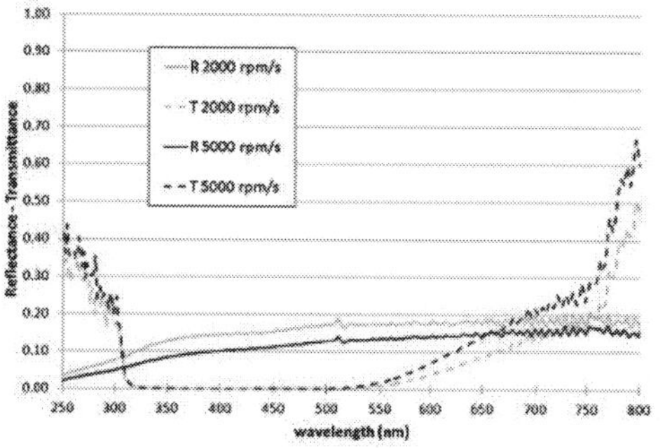

Figure 12: Reflectance and Transmittance versus wavelength of PK thin films deposited on 75 mm × 75 mm substrates.

Table IV and Figure 13 shows that the *ideal efficiency* values obtained for the 75 mm x 75 mm deposited substrates were quite similar to those obtained for the smaller ones with the same weight concentration. If any, they appear to have an even greater efficiency in the visible range, which decrease by the bandgap wavelength. Although this improvement seems to contradict the obtained values for the absorption coefficient, which correlated with worse deposition uniformity, it could be explained by the inherent simplicity of the theoretical approach, as per eq. (3).

Table IV: Ideal efficiency obtenied for the PK thin films deposited on 75 mm × 75 mm substrates.

Solution (wt%)	Acceleration (rpm/s)	Substrate size (mm)	Ideal efficiency (η)
45	2000	75 × 75	55.02 %
45	5000	75 × 75	55.07 %

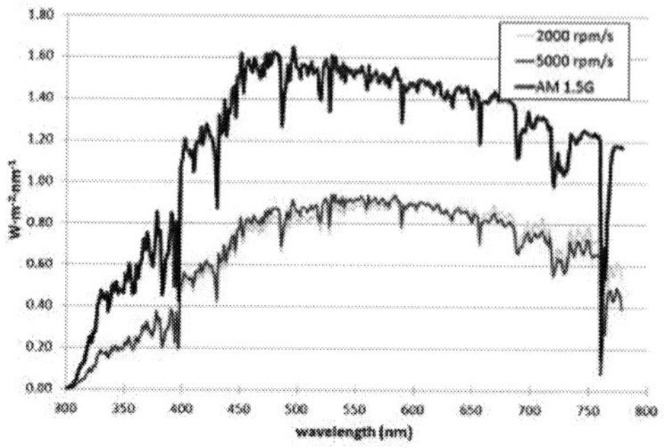

Figure 13 Ideal efficiency study of PK thin films deposited on 75 mm × 75 mm substrates.

Finally, Table V summarizes the results obtained by the ellipsometer fit model for the average thickness, together with their corresponding standard deviation.

Table V: Deposited thickness estimated by the ellipsometer fit model.

Solution (wt%)	Acceleration (rpm/s)	Substrate size (mm)	Average thickness (nm)	Standard deviation (nm)
45	2000	75 × 75	397.21	141.12
45	5000	75 × 75	478.55	117.05

By looking at the resulting data, although for the two accelerations the average thickness roughly corresponds to the values obtained for smaller substrates, their standard deviation was significantly worse, indicating a decrease on the uniformity for the deposited layers, a result which concurs with the visual assessment as it was stated previously. Still and all, these results seem to indicate a possible way for improving the uniformity of the deposited layers by applying higher accelerations.

5. CONCLUSIONS

The use of a spectral ellipsometer as a characterization method for evaluating the optical properties as well as the thickness of thin films of perovskite, deposited over substrates of different sizes, has been demonstrated. This has been achieved by applying a theoretical fit model derived from the parameterization of the complex dielectric function via the use of the B-spline function and by adjusting the calculus in order to comply with the optical constants reported in the literature.

Form the two kinds of MAPI solutions studied, the samples produced with 45 wt% ones were found as a better option for being used not only as an alternative Anti Reflecting Coating material, but also as an active member of a tandem solar cell, together with crystal silicon substrates. This conclusion was sustained by studying their optical properties, such as the absorption coefficient, refraction index, transmittance, reflectance, ideal efficiency and fluorescence properties.

The thickness and uniformity of the resulting deposited samples via spin wafer has been studied over substrates of 25 mm × 25 mm and 75 mm x 75mm and conclusions were obtained for improvement as sizes increase.

6. REFERENCES

[1] Quinto C.; Ocaña L.; Montes C.; Linares A.; Llarena E.; González-Díaz B.; González-Pérez S.; Guerrero-Lemus R., Friend M.; Cendagorta M. Design, start up and society benefits of a PV solar cell laboratory, 30th European Photovoltaic Solar Energy Conference and Exhibition 2014, Amsterdam, Netherlands, doi: 10.4229/EUPVSEC20142014-7AV.6.56

[2] Ocaña L.; Linares A.; Llarena E.; Montes C.; Gozález O.; Molina D.; Pío A.; Quinto C.; Friend M.; Cendagorta M. Adaptation of a Crystalline Silicon Solar Cell Laboratory to Produce Perovskite Solar Devices, 31st European Photovoltaic Solar Energy Conference and Exhibition 2015, Hamburg, Germany, doi: 10.4229/EUPVSEC20152015-3BV.5.40

[3] National Renewable Energy Laboratory (NREL) Best Research-Cell Efficiencies Chart. Access: May 2016 http://www.nrel.gov/ncpv/images/efficiency_chart.jpg

[4] Fan J.; Jia B.; Gu M. Perovskite-based low-cost and high-efficiency hybrid halide solar cells, Photon Res. 2014 2 111-120, doi: 10.1364/PRJ.2.000111

[5] Löper, P.; Niesen, B.; Moon, S.; Martin de Nicolas, S.; Holovsky, J.; Remes, Z.; Ledinsky, M.; Haug, F.; Yum, J.; De Wolf, S.; Ballif, C. Organic-Inorganic Halide Perovskites: Perspectives for Silicon-Based Tandem Solar Cells. IEEE J. Photovoltaics 2014, 4, 1545−1551, doi: 10.1039/c4cp03788j

[6] Mailoa J.; Bailie C.; Johlin E.; Hoke E.; Akey A.; Nguyen W.; McGehee M.; Buonassisi T. A 2-terminal perovskite/silicon multijunction solar cell enabled by a silicon tunnel junction, Appl. Phys. Lett. 2015, 106, 121105, doi:10.1063/1.4914179

[7] Xiao, M., Huang, F., Huang, W., Dkhissi, Y., Zhu, Y., Etheridge, J., Gray-Weale, A., Bach, U., Cheng, Y.-B. and Spiccia, L. A Fast Deposition-Crystallization Procedure for Highly Efficient Lead Iodide Perovskite Thin-Film Solar Cells, Angew. Chem. Int. Ed. 2014 53: 9898–9903. doi:10.1002/anie.201405334

[8] Löper, P.; Stuckelberger M.; Niesen, B.; Werner J.; Filipič W.; Moon, S.; Yum, J.; Topič M.; De Wolf, S.; Ballif, C. Complex Refractive Index Spectra of CH3NH3PbI3 Perovskite Thin Films Determined by Spectroscopic Ellipsometry and Spectrophotometry, The Journal of Physical Chemistry Letters 2015 6 (1), 66-71, doi: 10.1021/jz502471h

[9] Leguy A.; Hu Y.; Campoy-Quiles M.; Alonso M., Weber O.; Azarhoosh P.; Schilfgaarde M.; Weller M.; Bein T.; Nelson J.; Docampo P.; Barnes P., Reversible Hydration of CH3NH3PbI3 in Films, Single Crystals, and Solar Cells, Chemistry of Materials 2015 27 (9), 3397-3407 doi: 10.1021/acs.chemmater.5b00660

[1] Ziang X.; Shifeng L.; Laixiang Q.; Shuping P.; Wei W.; Yu Y.; Li Y.; Zhijian C.; Shufeng W.; Honglin D.; Minghui Y.; Qin G. Refractive index and extinction coefficient of CH3NH3PbI3 studied by spectroscopic ellipsometry, Opt. Mater. Express 2015 5, 29-43, doi: 10.1364/OME.5.000029

[11] FluoroSENS M-11, modular fluorimeter. Laser Technology S.L., Barcelona, Spain.

[12]Atomic Force Microscope (AFM), model Nanoscope III Multimode by Digital Instrument (Santa Barbara, California) available at the Atomic Force Microscope Service, within the Research Support General Service, integrated in the University of La Laguna.

[13] J.A. Woolam Co., Ellipsometry FAQ. Access: May 2016. https://www.jawoollam.com/resources/ellipsometry-faq

[14] Zhao J.; Green M. Optimized antireflection coatings for high-efficiency silicon solar cells, IEEE Transactions on Electron Devices 1991 vol. 38, no. 8, pp. 1925-1934, doi: 10.1109/16.119035

[15] McIntosh K.; Cotsell J.; Cumpston J.; Norris A.; Powell N.; Ketola B. An optical comparison of silicone and EVA encapsulants for conventional silicon PV modules: A ray-tracing study, 34th IEEE Photovoltaic Specialists Conference (PVSC) 2009 Philadelphia, PA, pp. 000544-000549. doi: 10.1109/PVSC.2009.5411624

[162] Green M.; Keevers M. Optical properties of intrinsic silicon at 300 K, Progress in photovoltaics 1995 3, 189-192, doi: 10.1002/pip.4670030303

[17] The Duong, Niraj Lal, Dale Grant, Daniel Jacobs, Peiting Zheng, Shakir Rahman, Heping Shen, Matthew Stocks, Andrew Blakers, Klaus Weber, Thomas P. White, and Kylie R. Catchpole, "Semitransparent Perovskite Solar Cell With Sputtered Front and Rear Electrodes for a Four-Terminal Tandem". IEEE JOURNAL OF PHOTOVOLTAICS

[18] Yajie Jiang , Martin A. Green, Rui Sheng, Anita Ho-Baillie, "Room temperature optical properties of organic–inorganic lead halide perovskites". Solar Energy Materials & Solar Cells, Volume 137, June 2015, Pages 253–257

[19]Air Mass 1.5 Direct+Circumsolar spectrum, American Society for Testing and Materials: http://rredc.nrel.gov/solar/spectra/am1.5/)

[20]Haiping He, Qianqian Yu, Hui Li, Jing Li, Junjie Si, Yizheng Jin, Nana Wang, Jianpu Wang, Jingwen He, Xinke Wang, Yan Zhang & Zhizhen Ye. "Exciton localization in solution-processed organolead trihalide perovskites". Nature communications. Published 21 Mar 2016. DOI: 10.1038/ncomms10896

[21]Michele Saba, Michele Cadelano, Daniela Marongiu, Feipeng Chen, Valerio Sarritzu, Nicola Sestu, Cristiana Figus, Mauro Aresti, Roberto Piras, Alessandra Geddo Lehmann, Carla Cannas, Anna Musinu, Francesco Quochi, Andrea Mura & Giovanni Bongiovanni. "Correlated electron–hole plasma in organometal perovskites". Nature communications. Published 30 Sep 2014. DOI: 10.1038/ncomms6049

[22] Son-Tung Ha, Chao Shen, Jun Zhang and Qihua Xiong. "Laser cooling of organic–inorganic lead halide perovskites". Nature Photonics. PUBLISHED ONLINE: 21 DECEMBER 2015. DOI: 10.1038/NPHOTON.2015.243

[23] Christian Wehrenfennig, Mingzhen Liu, Henry J. Snaith, Michael B. Johnston, and Laura M. Herz. " Homogeneous Emission Line Broadening in the Organo Lead Halide Perovskite CH3NH3PbI3−xClx". Physical Chemistry Letters. Published 24 Mar 2014. dx.doi.org/10.1021/jz500434p | J. Phys. Chem. Lett. 2014, 5, 1300−1306

[24] Zhi-Kuang Tan, Reza Saberi Moghaddam, May Ling Lai, Pablo Docampo, Ruben Higler, Felix Deschler, Michael Price, Aditya Sadhanala, Luis M. Pazos, Dan

Credgington, Fabian Hanusch, Thomas Bein, Henry J. Snaith and Richard H. Friend. "Bright Light-Emitting Diodes based on Organometal Halide Perovskite". Nature Publishing Group, 1748-3387, dx.doi.org/10.1038/nnano.2014.149.

[25]Aditya Sadhanala, Felix Deschler, Tudor H. Thomas, Sian E. Dutton, Karl C. Goedel, Fabian C. Hanusch, May L. Lai, Ullrich Steiner, Thomas Bein, Pablo Docampo, David Cahen,§ and Richard H. Friend. "Preparation of Single-Phase Films of CH3NH3Pb(I1−xBrx)3 with Sharp Optical Band Edges". dx.doi.org/10.1021/jz501332v | J. Phys. Chem. Lett. 2014, 5, 2501−2505

[26]Meng Zhang, Hua Yu, Miaoqiang Lyu, Qiong Wang, Jung-Ho Yun and Lianzhou Wang. "Composition-dependent photoluminescence intensity and prolonged recombination lifetime of perovskite CH3NH3PbBr3−xClx films". Chem. Commun., 2014,50, 11727-11730. DOI: 10.1039/C4CC04973J

[27]Michele Cadelano, Michele Saba, Nicola Sestu, Valerio Sarritzu, Daniela Marongiu, Feipeng Chen, Roberto Piras, Francesco Quochi, Andrea Mura and Giovanni Bongiovanni. "Photoexcitations and Emission Processes in Organometal Trihalide Perovskites". http://dx.doi.org/10.5772/61282

HIGHLY SENSITIVE ORGANIC PHOTODETECTOR BASED ON SI/NIPCTS/PEDOT : PSS BULK HETROJUNCTION BLEND

Morooj A. Abood[1], Falah I. Mustafa Alattar[2], Izzat M. Al-Essa[3]
[1,2] Solar Energy Research Center, Renewable Energy Directorate, Ministry of Science and Technology , Baghdad, Iraq
[3]Physics Department, College of Science, Baghdad University, Higher Education and Scientific Research Ministry, Baghdad, Iraq
falah_im@yahoo.com; morooj_a@yahoo.com; izzat_mohuy@yahoo.com

ABSTRACT: This research reports on the fabrication and characterization of an Si/NiPcTs/PEDOT:PSS bulk hetrojunction blend with different (PEDOT:PSS) concentration. A thin film was deposited by spin coater deposition method on silicon wavers (Si) N-type direction (111) of thickness 100nm. The optical properties were studied by using UV-Vis absorption spectra and photoluminescence spectra (PL) at annealing temperature of wafer 100°C. The decreasing of the (PEDOT:PSS) concentration leads to increase absorbance of films with optical peak shifts position to low energy side of spectra. The values of optical energy gap of (NiPcTs/PEDOT:PSS) thin film between (1.6-1.8)eV for Q band and (2.75-3.2)eV for B band for direct energy band. The optical energy gap obtained from photoluminescence spectra for NiPcTs/PEDOT:PSS bulk heterojunctions blend thin films with excitation wavelength (338- 352) nm is Very approach to energy gap obtained by UV-VIS spectrum. The current–voltage characteristics of the diode were measured at room temperature under dark and light condition . The detectivity and responsivity is very low with conc. 0.5 at 350nm and 550nm but disappear at 800 nm. The times values of Rise time were 0.75μsec, Transit time (400-500)μsec and Life time (160-180)usec.

Keywords: Absorption, Photoluminescence, c-Si, Optical Properties, Polymer Film.

1 INTRODUCTION

Organic semiconductor based electronic devices have already attracted a great deal of attention for various type of electronic devices duo to low cost, light weight, flexibility, feasible for large area applications and many other advantages [1-2]. Among these organic electronic devices; photo-detectors, photo-sensors, photovoltaic and other optoelectronic devices are being heavily investigated by many research groups all over the world [3-4]. Organic electronics and optoelectronics are fast developing branches of modern science and technology that are aiming to replace conventional inorganic materials with light, inexpensive, flexible organic materials. One of the main advantages of organic electronic materials is low-temperature high throughput device formation utilizing solution process able polymers and small molecules. The most widely used organic photo-conducting materials are generally the derivatives of pentacene, thiophene oligomers, regioregular polythiophene and some other related organic semiconducting materials. These materials showed good performance as photo sensor materials. Charge transport in small molecules is via hopping, i.e. the charges have to jump from one molecule to the neighboring one to be transported. In conjugated polymers the charges are partially transported via delocalization states along the HOMO and LUMO levels. Transport properties are usually determined by defects in the 1D-chains (intra molecular) or by hopping from chain to chain (inter molecular) Figure 1 [5].

Organic semiconductor is Containing carbon in composition , electrons strongly localized between the nuclei of C participating to σ bonds , Electrons in HOMO(Highest Occupied Molecular Orbital) (π electron) have higher energy, Absorb visible light but Excited states are localized at shown in the Figure 2, Conduct charge but with low mobility, Easy to deposit but Sensitive to deposition Conditions, Low temperature processes → compatible with plastic substrates . in blend heterojunctions Low leakage current Only photo generated current is desired (no current injected from the contacts), Large area, curved surfaces, mechanical flexibility, deposition from liquid, room temperature processing, Good efficiency (EQE > 10%). Good efficiency (EQE > 10%), Fast < 500 ns, Air stable > 1 year and Matched to scintillators .

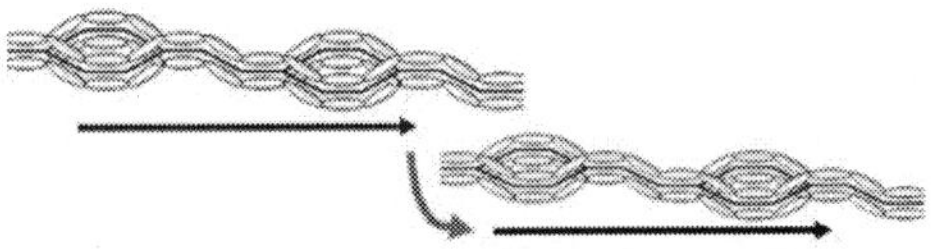

Figure 1: Show chargestransport in conjugated polymers

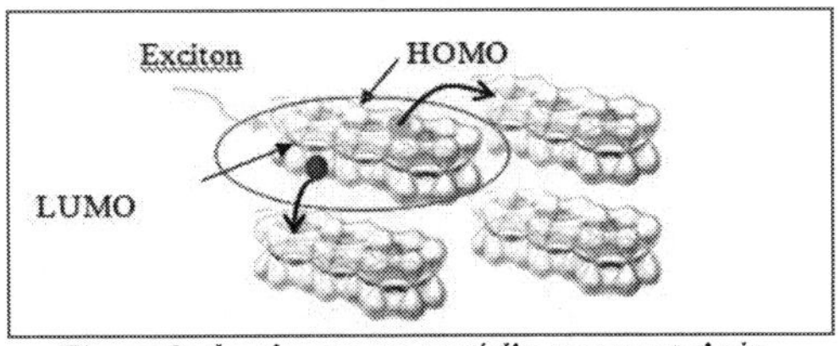

Figure 2: the charges are partially transported via delocalisation along the HOMO and LUMO levels

Experimental

The NiPcTs and PEDOT:PCC was obtained from Sigma-Aldrich. Figure 3, shows the molecular structure of the NiPcTs and PEDOT:PCC molecule used as a p-type organic semiconductor.NiPcTS was dissolved in dyonoyiz water and mix wellusing magnetic stirrer To be homogeneous then was prepered the three concentrations of PEDOT:PSS (0 , 0.5 , 1 ,2) .Siwas etching then was cleaning by acetone and alcohol.the films of blend of thickness of 100 nm, were deposited by spin coatting over Si wafer with 1500 rpm for time 1.5 min. annealing the samples Si/NiPcTS/PEDOT : PSS to an area of 0.5cm to a temperature T_a =100°C for 1 hour at then electrode (Al) deposit, figure 4, shows samples. The optical measurement include UV-Visible and Photoluminescence (PL) at exciting wavelength (332,352)nm. The characteristics of current - voltage ander dark and light conditionwas measured andresponsivity (R_λ), noise equivalent power (NEP) detectivity (D*) measured by the following equations [6]. Rise time was measured for samples by laser pointers (40 mW) and wavelengths (405,532,810) nm, figure 5, at shown in the system.

$R_\lambda = I_{pht} / P_{input}$, $D^* = A\, \Delta f / NEP$, $NEP = I_n / R_\lambda$, $In = (2qI_d\Delta f)^{1/2}$,

P: power of incident light, I_{ph}:light current, I_d:dark current

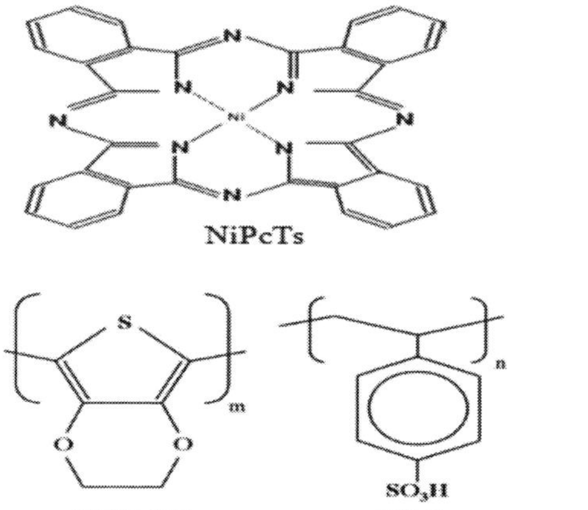

Figure 3: Molecular structureof the NiPc and PEDOT:PSS molecule used as a p-type organic semiconductor

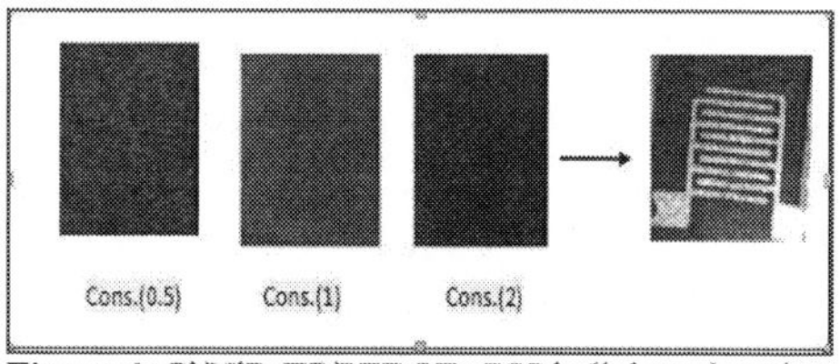

Figure 4: Si/NiPcTS/PEDOT : PSS bulk hetrojunction blend with different (PEDOT:PSS)

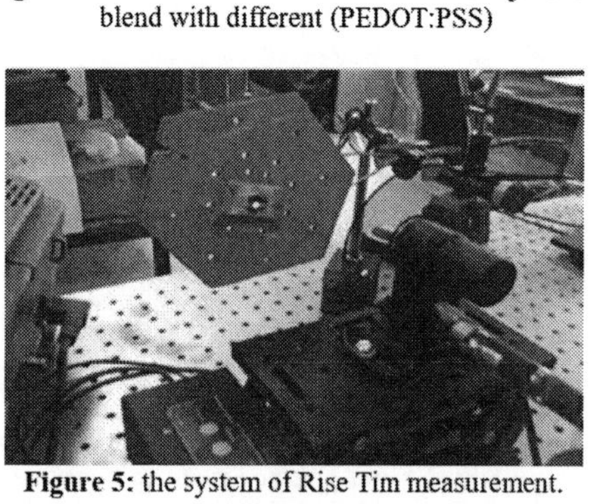

Figure 5: the system of Rise Tim measurement.

Results and Discussion:

The absorption spectra for NiPcTS/ PEDOT:PSS bulk heterojunctions blend thin films of thickness 100nm deposited at (0.5 , 1 ,2) PEDOT:PSS concentrations are given in the figure 6, show that Q and B bands shift towards longer waveleng thin concentrations (1) and maxima absorption intensity. The energy gap value was found by Tauce equation shown in theTable (1) and figure 7, photoluminescence spectrum is more accurate than that calculated by Tauce equation. The photoluminescence spectra for NiPcTs:PEDOT:PSS bulk heterojunctions blend thin films of different concentration are shown in figure 8, and the optical energy gap values are tabled in table (1) with excitation wavelength (332,352)nm , compared with Eg found by Tauce equation. We can see that the lowest value for Eg is in Concentration (1) also shows the intensity higher in Concentration (1).

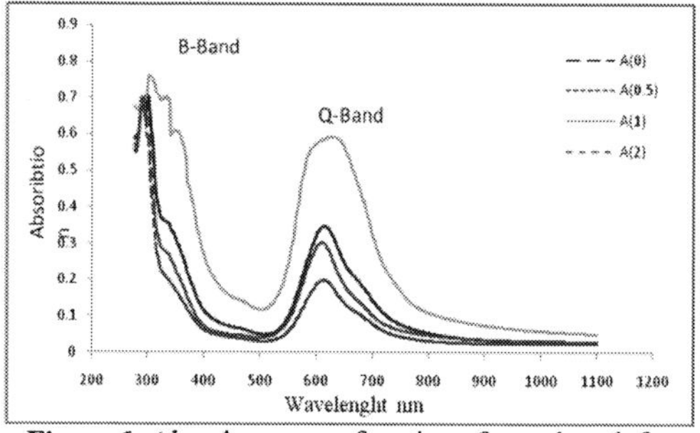

Figure 6: Absorbance as a function of wavelength for NiPcTs/PEDOT:PSSbulk heterojunctions blend thin films of different concentrations

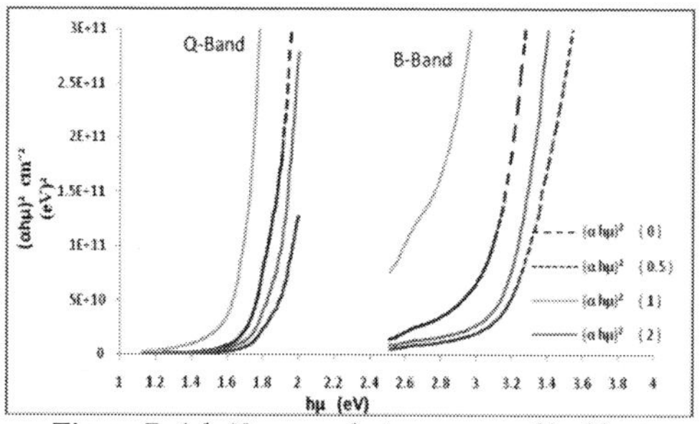

Figure 7: (αhυ)²versus photon energy of incident radiation fordiffernt concentration

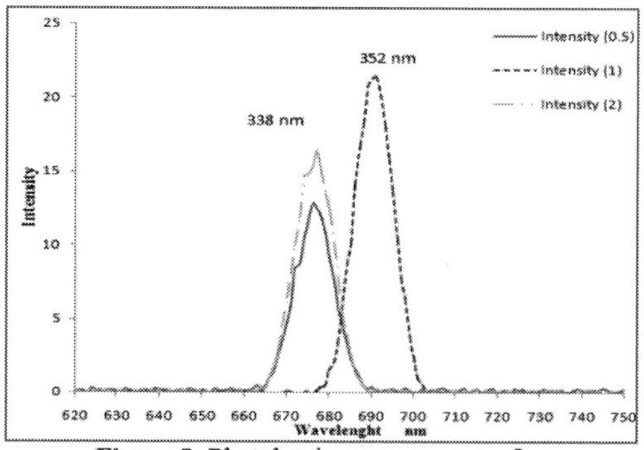

Figure 8: Photoluminescence spectra for NiPcTs:PEDOT:PSS bulk heterojunctions blend thin films at different concentrationat 100°C

Table 1: The optical band gap energies of NiPcTS : PEDOT:PSS bulk heterojunctions blend thin films of thickness100nm deposited at different concentrations at 100°C.

Direct Energy gap (Tauce) (eV)		Energy gap(PL) (eV)	Concentrations of PEDOT:PSS
Q-Band	B-Band		
1.75	2.95		0
1.80	3.2	1.835	0.5
1.60	2.57	1.79	1
1.78	3.12	1.834	2

Figure 9, shows Current –Voltage measurements of Si/NiPcTs/PEDOT:PSS bulk heterojunctions blend thin films at different concentrations of PEDOT:PSS (0.5, 1, 2) in Dark and Light conditions using Tungestn lamp Halogin. Figure 10, shows The spectral response characteristics of the detector, The responsivity parameter R_λ (applicablet o all detectors) reflects the gain of the detector and is defined as the output signal (typically voltage or current) of the detector produced in response to a given incident radiant power falling on the detector [4,5]. Figure 11, shows detectivity D* and therefore the detector output signal-to-noiseratio at I Watt of input UV-VIS radiation normalized to a detector with a unit active area and a unit bandwidth. Figures 10 and 11, shows the higher valu in conc. (1) at 350 nm when the peak in 350 nm disappear in conc. (2) but higher than conc.(1) in 800 nm .The detectivity and responsivity is very low in conc. (0.5) in 350nm, 550nm but disapper in 800 nm shows the figures.12 and 13. The table 2 showing the responsivity and detectivity of the three concentrations .

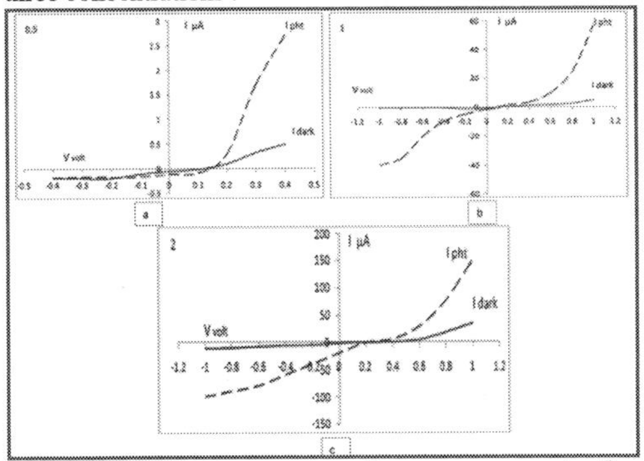

Figure 9: Shows current - voltage measurements of Si/NiPcTs/PEDOT:PSS bulk heterojunctions Blend (a) concentration 0.5, (b) concentration 1 , (c) concentration 2

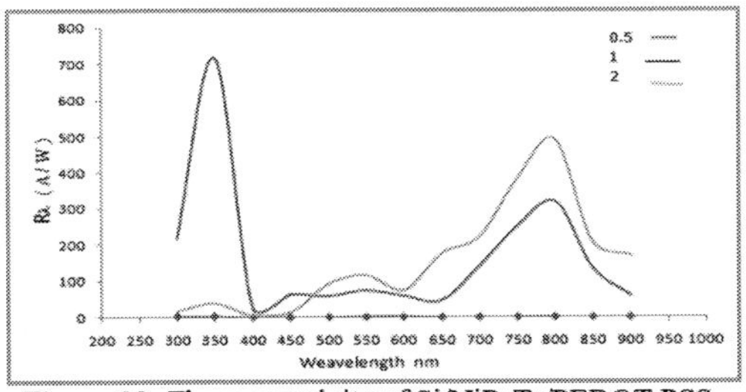

Figure 10: The responsivity of Si/NiPcTs/PEDOT:PSS detector blend

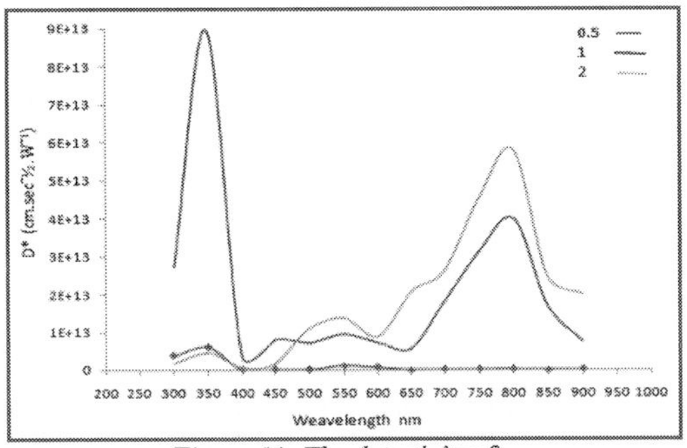

Figure 11: The detectivityof Si/NiPcTs/PEDOT:PSSdetector blend

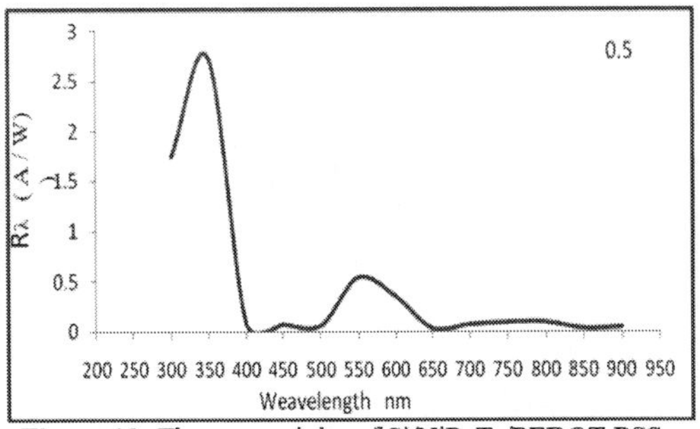

Figure 12: The responsivity of Si/NiPcTs/PEDOT:PSS detector blend in conc. 0.5

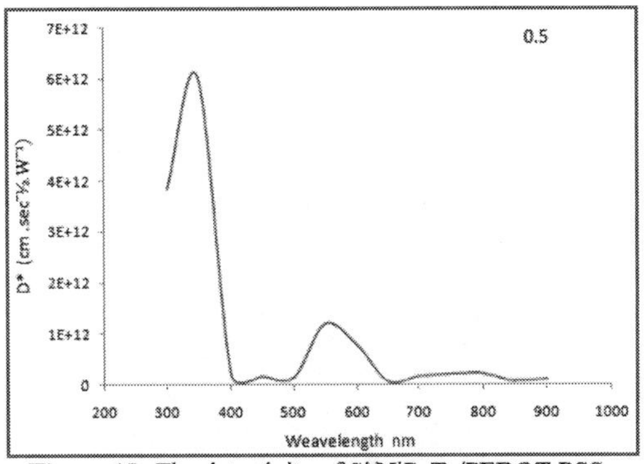

Figure 13: The detectivity of Si/NiPcTs/PEDOT:PSS detector blend in conc.0.5

Detector response speed measurements included Rise time (tr)(Value from 10% to 90% of output signal) , Fall time (tf) (Value from 90% to 10% of output signal) , Transit time (Ttr) (Time required to move all carriers from one electrode to another by applied electric field and incident laser pulse) and Life time (τ) (Average time for a free carrier to be captured (shallow or deep traps) by a localized states) as shown in figure 14. The values of Rise time (0.75μs) , Transit time (400-500)μs and Life time (160-180). Figure 15, shows Signals received by manufactured detector.

Table 2: showing the responsivity and detectivity of Si/NiPcTs/PEDOT:PSS detector blend of the three concentrations .

Conc- of PEDOT: PSS	R_λ A / W		NEP W^{-1}		D* cm $sec^{-½}W^{-1}$	
0.5	2.72 at 350 nm	0.54 at 550 nm	1.17 E-13 at 350 nm	5.94 E-13 at 550 nm	4.25 E+12 at 350 nm	8.43 E+11 at 550 nm
1	715 at 350 nm	318 at 800 nm	7.90 E-15 at 350 nm	1.78 E-14 at 800 nm	6.33 E+13 at 350nm	2.82 E+13 at 800nm
2	488 at 800nm		1.23E-14 at 800nm		4.07 E+13 at 800 nm	

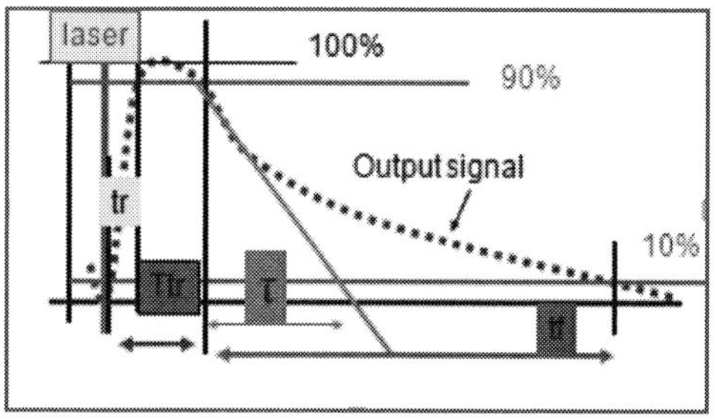

Figure 14: Detector response speed

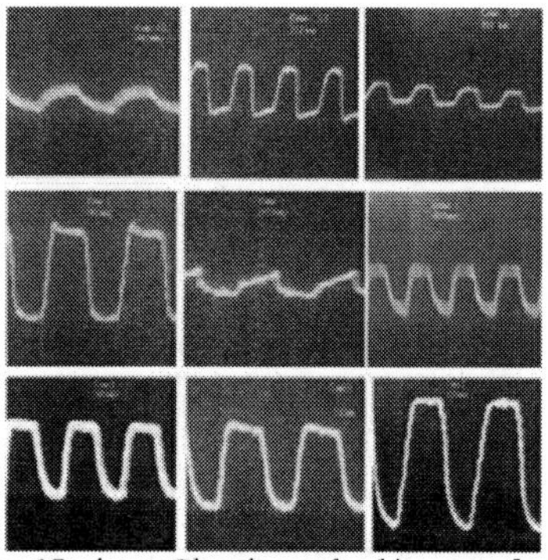

Figure 15: shows Signals received by manufactured detector.

References:

[1] M. El-Gemayel, M. Treier, C Musumeci, C. Li, K. Müllen, P. Samorì, J. Am. Chem. Soc. 134, (2012) 2429.

[2] Kruse, P.W. Uncooled Thermal Imaging, Arrays Systemsand Applications; SPIE Press: Bellingham, WA, 2001

[3] S. A. Moiz, A. M. Nahhas, H. D. Um, S. W. Jee, H. K. Cho, S. W. Kim, J. H. Lee Nanotechnology 44, (2012) 145401.

[4] S. F. Nelson, Appl. Phys. Lett. 72, 1854 (1998).

[5] T. Morimune, H. Kajii, Y. Ohmori, Jpn. J. Appl.Phys., 45(1B) (2006) 546.

[6] E. L. Dereniak and D. G. Crowe, *Optical Radiation Detectors*, Wiley, New York, 1984.

HOMOGENEOUS AND EFFICIENT CO-EVAPORATED MoO_3:CuI ANODE BUFFER LAYER FOR ORGANIC SOLAR CELLS

Mehdi Hssein[1,3], Lamia Barkat[2], Linda Cattin[3], Guy Louarn[3], Mohammed Addou[1], Abdelbacet Khelil[2], Jean Christian Bernède[4].

1- Laboratoire Optoélectronique et Physico-chimie des Matériaux, Université Ibn Tofail, Faculté des Sciences BP 133 Kenitra 14000, Morocco.
2- Université de Oran 1, LPCM²E, BP 1524 ELM Naouer 31000 Oran, Algérie.
3- Université de Nantes, Institut des Matériaux Jean Rouxel (IMN), CNRS, UMR 6502, 2 rue de la Houssinière, BP 32229, 44322 Nantes cedex 3, France.
4- Université de Nantes, MOLTECH-Anjou, CNRS, UMR 6200, 2 rue de la Houssinière, BP 92208, Nantes, F-44000 France.

ABSTRACT: One possibility to increase the power conversion efficiency (PCE) of the organic photovoltaic cells (OPVCs) is to introduce a thin anode buffer layer (ABL) between the anode and the organic electron donor (ED). Among the possible ABLs, it was shown that CuI is very efficient. However, it was shown that the morphology of the CuI layers depends strongly on the film deposition conditions, mainly on the deposition rate. The surface roughness of the CuI layer is often very high if we do not control precisely the conditions of deposition of the layer. This high surface roughness introduces leakage current paths in the devices and therefore poor performances reproducibility. Therefore, in the present work, we proceed to co-evaporation of an hybrid ABL MoO_3/CuI. The co-evaporation process avoids CuI crystallization which prevents rough ABL formation, even for a deposition rate of 0.01 nm/s for each compound. These MoO_3/CuI ABL were probed in planar heterojunction OPVCs based on the couple copper phthalocyanine/fullerene(CuPc/C_{60}). By comparison with OPVCs using reference ABL (MoO_3, CuI or MoO_3/CuI) there is a systematic improvement of their PCE and this for a deposition rate 2 times higher than in the case where CuI is deposited alone.

Keywords: buffer layer, organic solar cell, MoO_3, CuI, CuPc.

1 INTRODUCTION

Given the problems due to greenhouse gas emission in the use of fossil fuels, renewable energies have received a great attention during these last years. Photovoltaic energy is one of the most promising. In that field, organic photovoltaic cells (OPVCs) are, nowadays, extensively studied because of their specific properties such as lightness, flexibility, large area processing low cost. The hearth of these cells is based on an heterojunction electron donor/electron acceptor (ED/EA). Two organic solar cell families are encountered, that using on a polymer as electron donor and that using small molecules. In the present work we used the classical couple of small molecules, CuPc/C_{60}, deposited by sublimation under vacuum. Small molecules OPVCs are classically based on an organic bilayer sandwiched between two electrodes. One of these electrodes must be transparent and the other must be highly reflexive. About the electrodes, while aluminum is often used as reflexive cathode, the transparent conductive anode is usually the indium tin oxide (ITO). One possibility to increase the power conversion efficiency (PCE) of the OPVCs is to introduce thin anode buffer layers between the electrode and the organic material. In the case of planar heterojunctions such as those used in the present work, it was shown that bbetween the Al cathode and the fullerene layer an exciton blocking layer (EBL) [1], here Alq_3, allows improving significantly the OPVC efficiency.

About the interface anode/electron donor, which is the object of the present work, we introduce a thin anode buffer layer (ABL) between the anode and the organic electron donor (ED) to improve the hole extraction. This ABL allows to improve the band matching between the highest occupied molecular orbital (HOMO) of ED and the work function (WF) of the anode. It must also be selective, i.e. it must facilitate the passage of hole and block the electrons. Moreover, its surface must be smooth in order to avoid leakage current path formation [2].

It was demonstrated that when we use CuI as efficient ABL in organic solar cells[3], it permits to control the molecular orientation of CuPc, because the CuI have a high work function(WF_{CuI} = 5.3eV [4]) and it allows a good band matching at the interface anode/ED. However, It was proved that the film deposition conditions [5] and the deposition rate [6] affects highly the growth and the morphology of the CuI. Indeed, when the CuI layer was deposited under vacuum, the surface roughness was very high. On the other hand, if the conditions of deposition of the layers was not controlled precisely, we obtain a rough and inhomogeneous surfaces, which introduces leakage current paths in the devices and therefore low performances reproducibility.

In the present work, we are going to co-evaporate an hybrid ABL MoO_3:CuI, we have chosen the co-evaporation method because it permits to prevent the crystallization of CuI which avoid rough ABL formation, even for a depostion rate of 0.01 nm/s for each compound.

2 EXPERIMENTAL DETAILS

The realization and characterization of the OPVCs has been already described [7] and will not be detailed here. The films are deposited under vacuum, their deposition rate and thickness were measured by a quartz monitor, the final OPVC being as follow: ITO/ABL/CuPc (35 nm)/C_{60} (40 nm)/Alq_3 (9 nm)/ Al (120 nm). In the case of double (MoO_3/CuI) or hybrid (MoO_3:CuI) ABL the ratio CuI/MoO_3 was either 0.5 or 1. The deposition rate of the MoO_3 and the CuI was 0.01 nm/s. For comparison, OPVCs with MoO_3, CuI and MoO_3/CuI ABL were realized during the same deposition processes.

Electrical characterizations were performed with an automated I-V tester, in the dark and under sun global AM 1.5 simulated solar illumination. Measurements were performed in ambient atmosphere. All devices were illuminated through TCO electrodes. The morphology of the different structures used as anode was observed through scanning electron microscopy (SEM) with a JEOL 7600F. The thin films structures were analyzed by X-ray diffraction (XRD) by a Siemens D5000 diffractometer using Kα radiation from Cu ($\lambda_{K\alpha}$= 0.15406 nm). Optical transmission spectra were recorded on a Carry spectrophotometer.

3 RESULTS AND DISCUSSION

It can be seen in table I and Fig. 1 that the best result is obtained with the hybrid MoO_3:CuI layer, while the worst results are obtained with CuI, the ABL being deposited at 0.01 nm/s.

Table I: Performances of the OPVCs, with different ABL configurations, deposition rate of the ABL: 0.01 nm/s.

ABL	Thickness (nm)	Voc (V)	Jsc (mA/cm^2)	FF (%)	η (%)
CuI	1.5	0.39	4.31	47	0.79
MoO_3	3	0.47	5.47	59	1.52
MoO_3 / CuI	3 / 1.5	0.45	6.65	50	1.46
MoO_3 : CuI	3 : 1.5	0.48	8.48	52	2.11

In order to understand these different results we proceeded to some characterizations of the different ABLs. The Fig. 2 presents the surface morphology of the different ABLs. At first glance, it can be seen that the MoO_3:CuI ABL is the most homogenous ABL, while that with MoO_3/CuI and CuI alone are far less homogeneous.

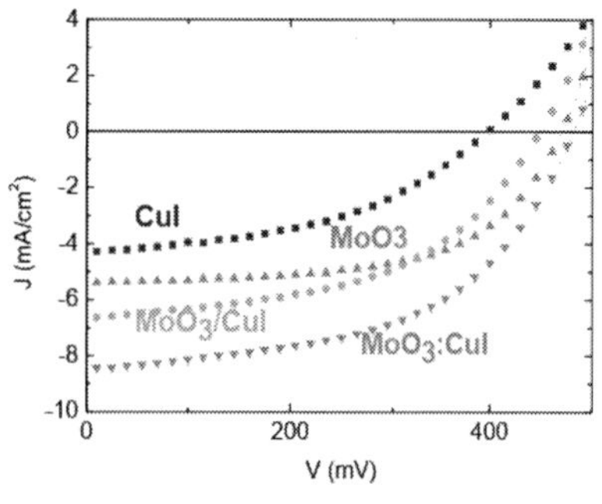

Figure 1: J-V characteristics of OPVCs with different ABL under AM1.5 irradiation

(a)
(b)
(c)
(d)

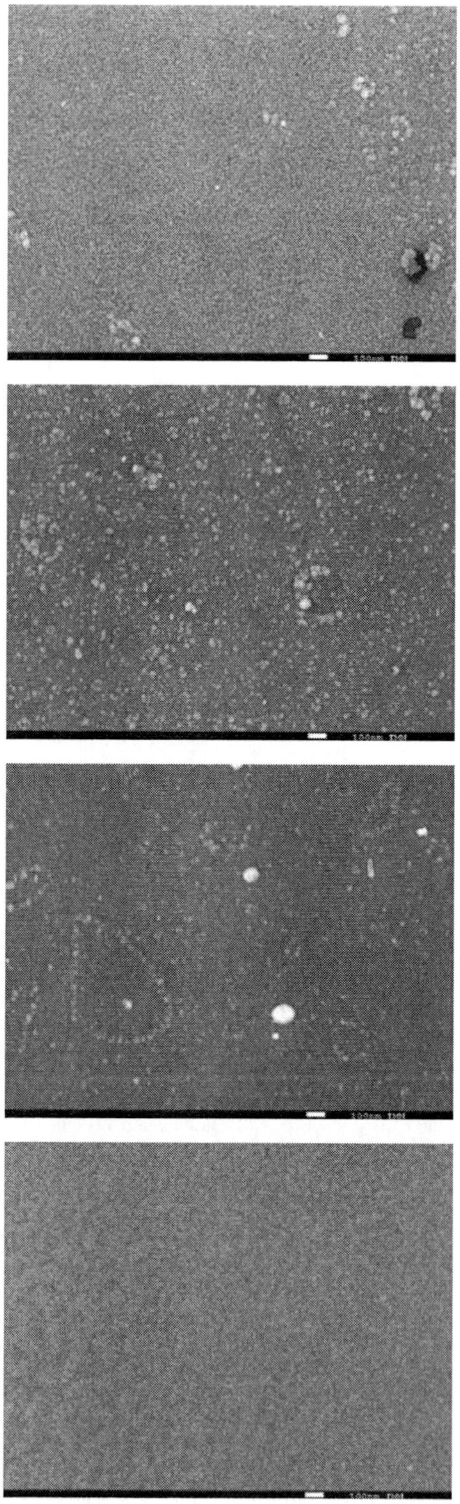

Figure 2: SEM visualization of the surface of MoO_3 (a), CuI (b), MoO_3/CuI (c) and MoO_3:CuI (d)

After showing that the MoO_3:CuI ABL is smoother than the MoO_3/CuI ABL deposited at the same rate, we check the structural properties of CuPc films deposited onto the different ABLs.

In Fig. 3a we can see that the CuPc films deposited onto MoO_3 are more homogeneous than those deposited onto CuI. It also can be seen that the CuPc film deposited onto MoO_3/CuI is similar to that deposited onto CuI.
The grains are well faceted, but some pinholes are randomly distributed in the films. In the case of MoO_3:CuI ABL, we can notice that the CuPc film shows a similar appearance to that elaborated with CuI, and has the smaller grain size, by against we can see that the films are more homogeneous, without holes similar to those present in the layers deposited on CuI alone.

(a)
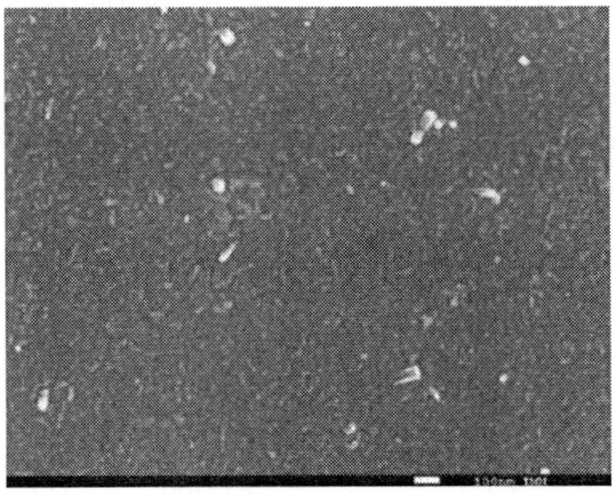

(b)
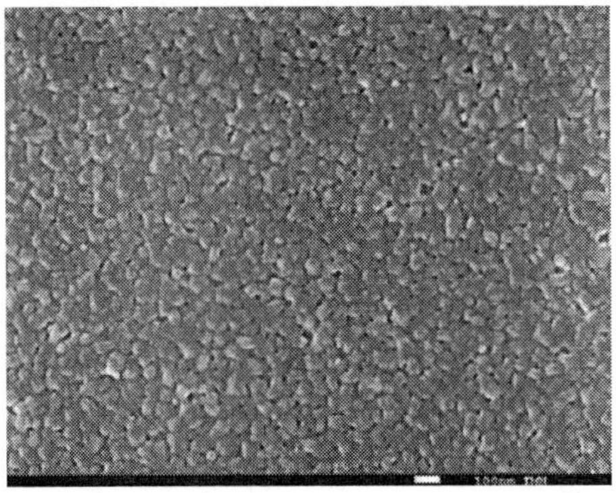

(c)
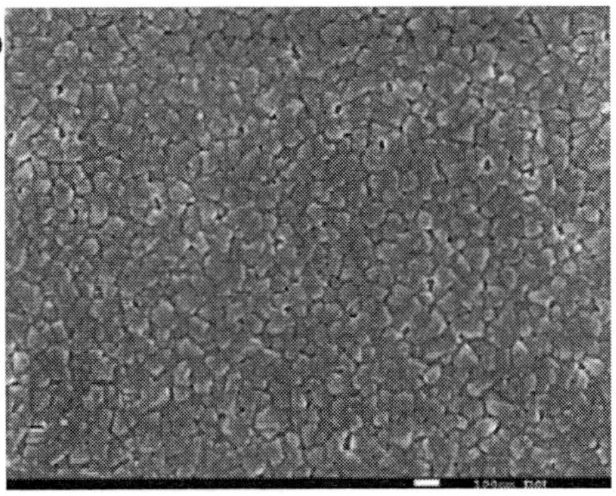

(d)
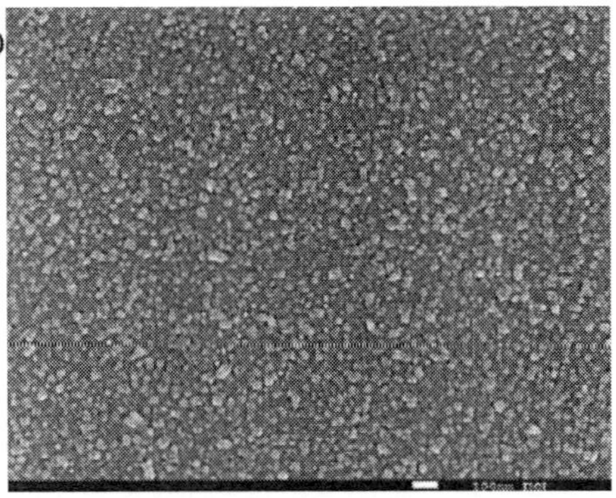

Figure 3: SEM visualization of the surface of MoO_3/CuPc (a), CuI/CuPc (b), MoO_3/CuI/CuPc (c) and MoO_3:CuI/CuPc (d)

As expected, the XRD spectra of CuPc thin films deposited onto MoO_3 (Fig. 4a) or CuI (Fig. 4b and 4c) are quite different. A peak centred at $2\theta = 6.9°$ is visible in the case of ITO/MoO_3 substrates. When the films are deposited onto CuI, a peak centred at $2\theta = 27.8°$ is visible while that situated at 6.9° has completely disappeared. It is known that the peak situated at $2\theta = 6.9°$ corresponds to CuPc molecules perpendicular to the substrate, while that situated at 27.8° corresponds to CuPc molecules which lie down parallel to the plane of the substrate [3]. We can notice that we have the same diffraction peak for CuPc deposited onto MoO3:CuI (Fig. 4d). Therefore, we can conclude that when we make dposition on CuI, MoO3/CuI or MoO3:CuI, the stacking orientation of the MPc molecules is changed, i.e. the CuPc molecules are parallel to the plane of the substrate.

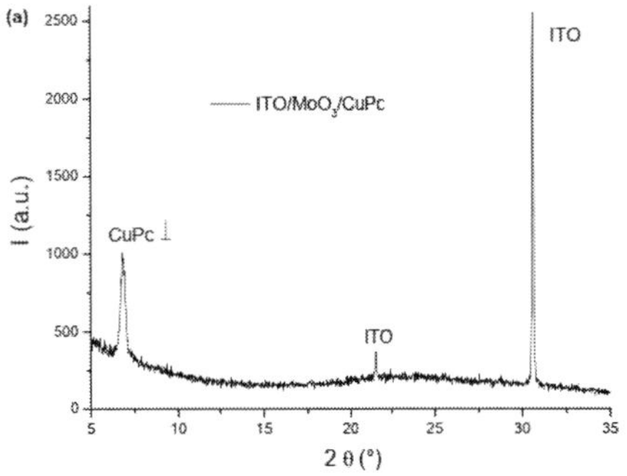

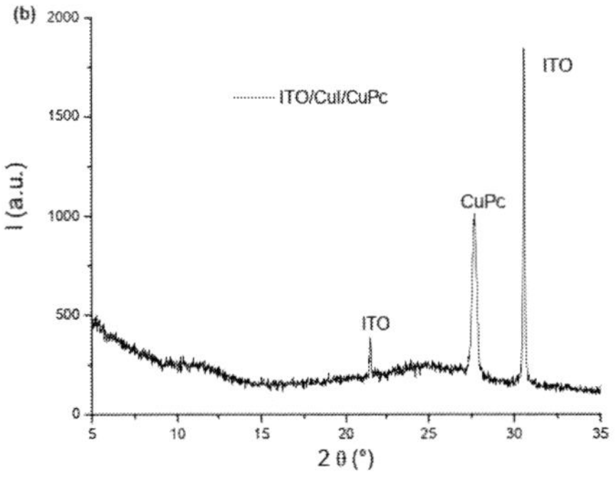

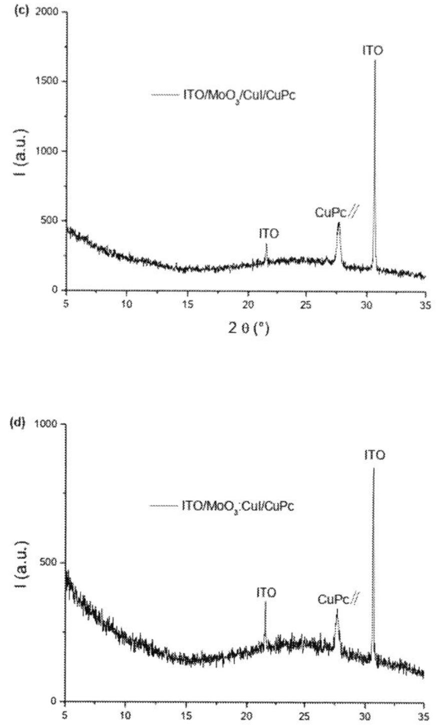

Figure 4: X-Ray diffractograms of the CuPc layer deposited on different anode structures: MoO_3/CuPc (a), CuI/CuPc (b), MoO_3/CuI/CuPc (c) and MoO_3:CuI/CuPc (d)

As expected, this modification of the CuPc molecules orientation influences significantly their optical absorption. As shown in Fig. 5, the optical absorption of the CuPc films deposited onto MoO_3:CuI is similar to that of films deposited onto CuI or MoO_3/CuI. In the visible region, when deposited onto CuI, i.e. when the CuPc molecules are parallel to the plane of the substrate, the absorption and therefore the probability of free-carrier formation in OPVCs increase.

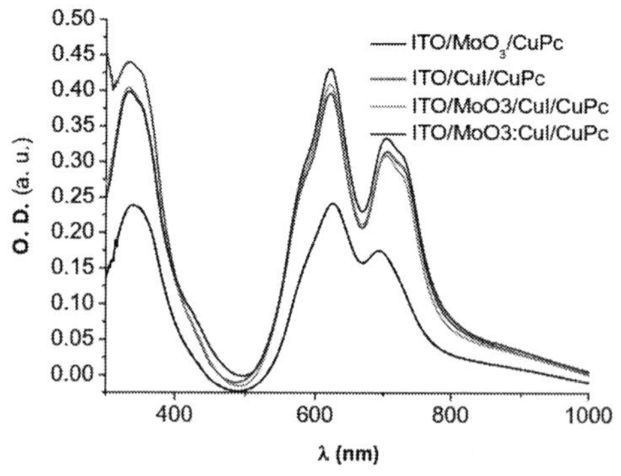

Figure 5: UV-Visible absorption spectra of CuPc layers deposed on the different anode structures

4 CONCLUSION

In this Work, we show that it is easy to achieve a smooth anode buffer layer with codeposition process of MoO_3:CuI and it's more difficult to obtain a smooth ABL with CuI alone. we get an homogeneous layers when we deposit MoO_3:CuI with 0.01nm/s as deposition rate, this number is twice that necessary to obtain acceptable homogenous CuI films, when it is deposited alone.
We demonstrate that the MoO_3:CuI HTL possesses the same influence as CuI or MoO_3/CuI ABL on the structural and optical properties of the CuPc thin films, *i.e.* the CuPc molecules stand parallel to the substrate which increases the absorption. When we inject MoO_3: CuI in OPVCs, we find that this ABL is more homogeneous than the other ABL with CuI, because MoO_3:CuI avoid leakage currents. The efficiency obtained by the OPVCs with the coevaporated MoO_3:CuI HTL is higher than that of the OPVCs using different ABLs.

REFERENCES

[1] P. Peumans, V. Bulovic, S.R. Forrest., Appl. Phys. Lett. 76 (2000) 2650.
[2] F.Z. Dahou, L. Cattin, J. Garnier, J. Ouerfelli, M. Morsli, G. Louarn, A. Bouteville, A. Khelil, J.C. Bernede, Thin Solid Films 518 (2010) 6117.
[3] C. H. Cheng, J. Wang, G. T. Du, S. H. Shi, Z. J. Du, Z. Q. Fan, J. M. Bian, and M. S. Wang, Appl. Phys. Lett. 97 (2010) 083305.
[4] J.-H. Lee, D.-S. Leem, H.-J. Kim, J.-J. Kim, Appl. Phys. Lett. 94 (2009) 123306.
[5] J-H. Lee, D-S. Leem, J-J. Kim, Organic Electronics 9 (2008) 805.
[6] A.S. Yapi, L. Toumi, Y. Lare, G.M. Soto, L. Cattin, K. Toubal, A. Djafri, M. Morsli, A. Khelil, M.A. Del Valle, and J.-C. Bernède, Eur. Phys. J. Appl. Phys. 50 (2010) 30403.
[7] I. Horcas, R. Fernandez, J.M. Gomez-Rodriguez, J.Colchero, J. Gomez-Herrero and A. M. Baro, Rev. Sci.Instrum. 78 (2007) 013705.

P3HT:PCBM-based organic solar cells: the effects of different PCBM derivatives

Burak Kadem, Aseel Hassan and Wayne Cranton
Material and Engineering research institute, Sheffield Hallam University, UK
Tel. +441142253500
Fax. +441142253501
a.hassan@shu.ac.uk

ABSTRACT: P3HT blended with different fullerene derivatives ($PC_{60}BM$, $PC_{61}BM$, $PC_{70}BM$ and $PC_{71}BM$) have been prepared and studied as active layers in organic solar cells (OSCs). These PCBM have different LUMO and HOMO positions which subsequently influence the PV parameters of the OSCs such as open circuit voltage (V_{oc}) and the built-in potential (V_{bi}). Charge carrier mobility is found to vary with the type of fullerene used with maximum value of 9.3 $x10^{-6}$ $cm^2V^{-1}s^{-1}$ for the P3HT:$PC_{61}BM$ based device. The latter has exhibited the highest device performance with power conversion efficiency (PCE) reaching 4.2% corresponding to a fill factor (FF) of 62%, with open circuit voltage (V_{oc}) of 0.65V and short circuit current density (J_{sc}) of 10.4 $mA.cm^{-2}$. The improved fill factor has been attributed to the lower series resistance (R_s) in the studied devices.

Keywords: Organic solar cells; Different PCBM derivatives; Charge carrier mobility; SCLC; P3HT:PCBM.

1 INTRODUCTION

Organic solar cells (OSCs) have been receiving significant attention due to their manufacturing low cost, light weight, as well as possible simple device fabrication on a flexible substrate [1]. Poly (3-hexylthiophene) (P3HT) and fullerene derivatives blends are the most investigated OSCs acceptor-donor active layers; their blends require a nano-scale interpenetrated network, which facilitates charge carrier transport. Among several other parameters, it has frequently been shown that OSCs performance is determined by the electrical characteristics of the organic semiconductor/metal interface [2]. The study of these characteristics could provide valuable information to help improving the shape of the solar cell output. Ideality factor (n), Schottky barrier height (Φ_B) and series resistance (R_s) are the main parameters that can be altered by the nature of this interface. Yakuphanoglu and co-workers [3] have studied the inhomogeneity of Schottky barrier in organic solar cells where higher ideality factor was ascribed to series resistance and interface state properties. The interfacial states at the organic/metal interface affects the barrier height, charge extraction and the carrier transport inside the acceptor/donor blends, which therefore determine the overall performance of OSCs [4]. The organic/metal barrier height varies with the BHJ blend arrangement, depending on which fullerene is used, which leads to the formation of different cathode/PCBM interface properties [5]. In general, the surface disorder of the active layer induces extra energy levels into the band gap, where this is thought to control the charge transport properties in the organic blend [6]. However, further consideration of the physical behaviour of such interface properties such as the understanding of their effect on charge transport in organic layers by space-charge limited current (SCLC) is desired [7]. In the current study different fullerene derivatives have been used to prepare P3HT:PCBM-based OSCs; these are $PC_{60}BM$, $PC_{61}BM$, $PC_{70}BM$ and $PC_{71}BM$. These material have different LUMO and HOMO positions which subsequently have direct impact on the PV parameters such as open circuit voltage (V_{oc}) and the built-in potential (V_{bi}) [8] and therefore the OSCs performance. The interface properties of these devices were studied using Schottky thermionic emission, Mott Schottky analysis and space charge limited conduction theory.

2 EXPERIMENTAL METHODS

P3HT:PCBM active layers in the ratios of 1:1 were prepared using chlorobenzene:chloroform (CB:CF) in the ration 1:1. P3HT, $PC_{60}BM$, $PC_{70}BM$ were purchased from Ossila and $PC_{61}BM$, $PC_{71}BM$ were purchased from Sigma Aldrich and used without any further purification. The solutions were stirred overnight at 45°C followed by filtering through a PTFE filter with pore size of 0.45 µm. ITO coated glass slides with sheet resistance of 8-12 Ω/□ (Sigma Aldrich) were cleaned using DI water, acetone and 2-propanol for 10 min each in ultrasonic bath, respectively, and then blown dry in N_2 gas. 1.3 wt% PEDOT:PSS dispersion in H_2O (conductive grade) were purchased from Sigma Aldrich; the PEDOT:PSS layers were spin coated on the ITO coated glass substrates at 2000 rpm for 30 sec, followed by heating at 150°C on a hot plate in ambient air. PEDOT:PSS coated ITO substrates were then moved to a nitrogen-filled glove box where the P3HT:PCBM active layers were spin coated to obtain approximately 100 nm thickness and then heat-treated inside the glove box for 10 min at 120°C. Aluminium (Al) with 100 nm thickness was thermally evaporated as a top contact using Edward vacuum deposition system. The film thickness was monitored by a quartz crystal thickness monitor at the deposition rate of 0.1-0.2 $nm.sec^{-1}$ under vacuum of about10^{-6} mbar. All the devices were subjected to post deposition heat treatment inside the glove box at 120°C for 10 min and left to cool down for 30 min before any measurements carried out.

3 RESULTS

3. 1. UV-Vis absorption spectra:

The absorption spectra of the P3HT-based layers blended with different fullerene derivatives were investigated using UV-visible absorption spectra as shown in Fig.1. Typical spectra are demonstrated with four different bands for each different blend; bands labelled as A, B and C are ascribed to the absorption of P3HT main polymer [9], whereas band D and/or E below 400nm are related to the absorption of the fullerene derivatives [10]. Band A is attributed to π-π* transition within the P3HT main polymer [11], whereas band B and C are attributed to the extended conjugated polymer and the absorption of the inter-chain stacking of P3HT, respectively [9].

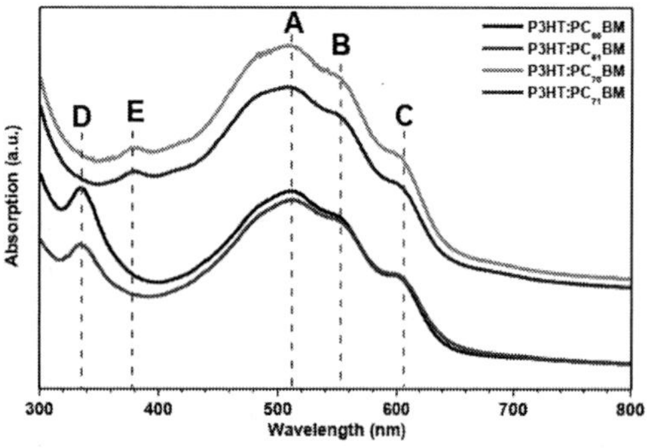

Figure 1: UV-visible absorption spectra of the studied P3HT-based layers

On the other hand, PCBM absorption spectra have shown two different bands (E and D), $PC_{60}BM$ and $PC_{61}BM$ have exhibited an absorption peak (band D) around 340nm while $PC_{70}BM$ and $PC_{71}BM$ have demonstrated an absorption peak (band E) at around 380nm [12]. Nicolaidis and co-workers have stated that an increase of about 13% in generated photo-current could be attributed to light absorption by the fullerene molecules within the P3HT:PCBM blend [13].

3. 2. Structural properties

Figure 2 shows XRD patterns of the P3HT-based films; a single diffraction peak at 2θ of 5.5±0.1 is observed which is correlated with the (100) orientation phase of P3HT main polymer crystallites [14]. In fact, in all measured XRD patterns, the observed reflections peaks could be ascribed to the lamella structure of P3HT which indicates higher degree of crystallinity and edge-on orientation. The latter orientation was ascribed to P3HT inter-chain spacing associated with the interdigitated alkyl chains [15,16]. Generally speaking, PCBM has an amorphous environment which could interrupt the crystallinity of P3HT polymer [17]. All films produced with different fullerenes have exhibited similar peak intensities; therefore, no significant change was induced in the films' crystallinity as a result of changing the fullerene type.

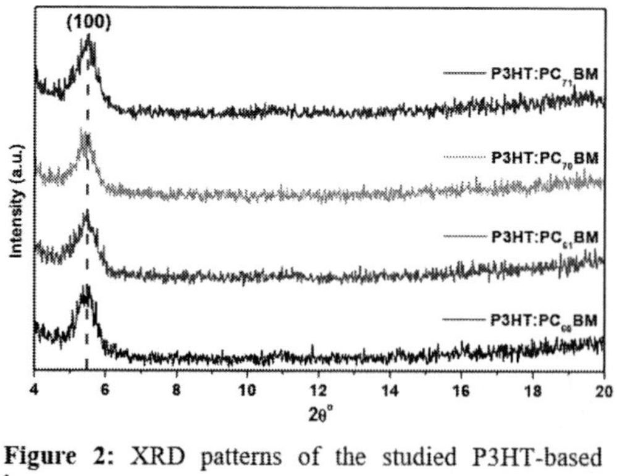

Figure 2: XRD patterns of the studied P3HT-based layers

3. 3. Dark electrical properties

Device electrical characteristics measured in the dark were evaluated using current-voltage (I-V) and capacitance-voltage (C-V) measurements. I-V dark characteristics were evaluated using Richardson-Schottky model and space charge limited conduction theory whereas the C-V measurements were studied using Mott-Schottky analysis. Fig.3(a) shows the energy level alignments of P3HT with different PCBM derivatives. It is expected that different PCBM types with different HOMO and LUMO level positions will have a significant impact on the organic solar cells properties; this includes the device open circuit voltage (V_{oc}) following from relation (1) [8]:

$$V_{oc} = [\,LUMO_{acceptor} - HOMO_{donor}]\text{-}0.3 \qquad (1)$$

The quantity of 0.3V appearing in equ. (1) is the estimated loss of voltage which might be attributed to the blend disorder that induces tail states or the energy loss induced by carriers' recombination [17].

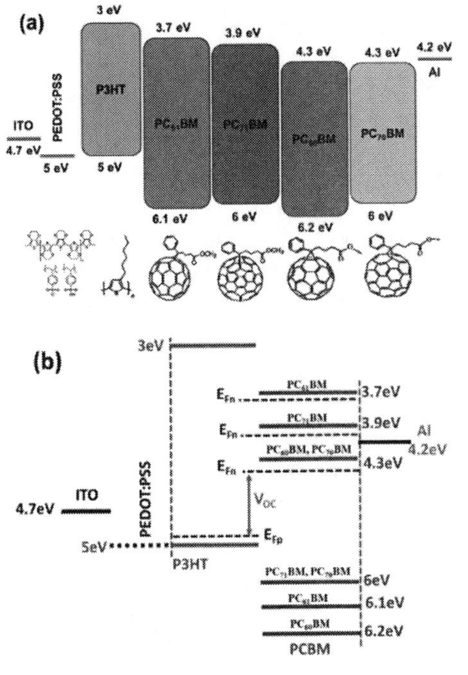

Figure 3: (a) the energy level alignment of the materials under study; The values of ITO, PEDOT:PSS, P3HT, $PC_{71}BM$ and $PC_{61}BM$ were cited from Sigma Aldrich. $PC_{60}BM$ [18], and $PC_{70}BM$ [19]; and (b) is the determination of Voc using Fermi level position.

On the other hand, V_{oc} can be determined as a function of Quasi-Fermi level (EF) using the following equation [8]:

$$V_{oc} = \frac{E_{Fn}-E_{FP}}{q} \qquad (2)$$

where E_{Fn} and E_{Fp} are the Quasi-Fermi levels of electrons and holes under open circuit condition, respectively, as demonstrated in Fig.3 (b). Fig. 4(a) shows a plot of the lnJ versus V in the range -1V to 2V for all studied devices. By assuming the absence of image lowering effect, the thermionic emission model could be used to derive the diode ideality factor (n) as well the barrier height (ϕ_B) as described elsewhere [20,21].

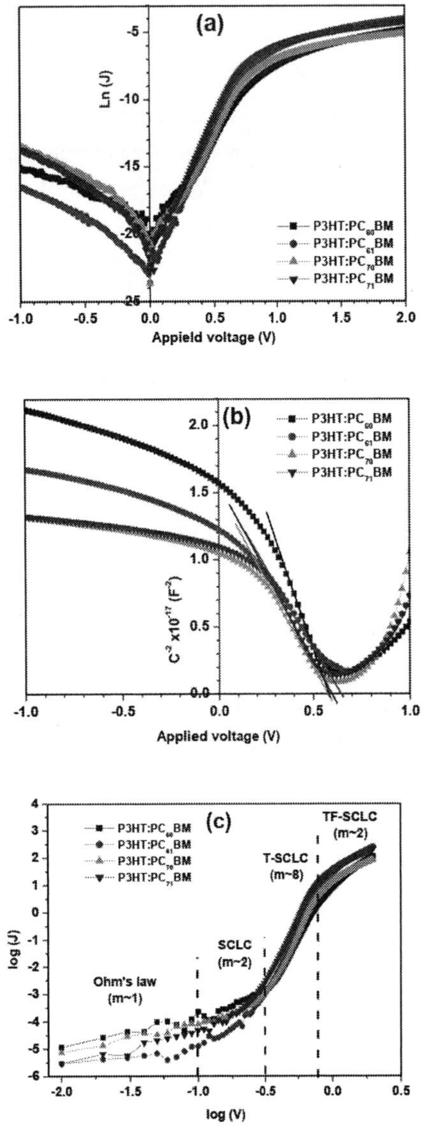

Figure 4: (a) Thermionic emission model in the form of lnJ-V relation; (b) Mott–Schottky analysis; and (c) logJ-logV relation.

$PC_{60}BM$ based device has exhibited the lowest ϕ_B value of 0.84eV whereas $PC_{61}BM$ device has exhibited the highest ϕ_B value of 0.9eV. This increase in ϕ_B is correlated with a decrease in the ideality factor from 2.39 in $PC_{60}BM$-based device to 1.72 in $PC_{61}BM$-based device. Such a decrease in the ideality factor may indicate a lower recombination rate, which could lead to higher charge carrier mobility as demonstrated in Table1. Additionally, ϕ_B was estimated using Mott-Schottky analysis of C-V measurements as shown in Fig.4 (b); the extrapolation of the linear part intersects with the voltage axis giving an estimate value of the built-in potential (V_{bi}) within the blends [22].

The barrier height values estimated from the C-V measurements have shown good agreement with ϕ_B values extracted from the I-V measurements. The small variation in the two estimated values could be attributed to the excess capacitance and inhomogeneity of barrier heights [23]. On the other hand, V_{bi} has demonstrated different values due to different fullerene types. The higher V_{bi} was exhibited for the $PC_{61}BM$ based device with a value of 0.64V whereas $PC_{60}BM$-based device has demonstrated 0.57V as revealed in table (1). Generally in organic solar cells, V_{oc} depends on the solar cell design. i.e. the energy level alignments as described in equ.1, therefore, it could be determined by V_{bi} [24].

It has been reported that the fullerene energetics play great roles in this process; due to the constant P3HT amount in the four blends and different fullerene energetics, different cathode interfaces will exhibit; this may be the main reason for having different V_{bi}, ϕ_B as well as V_{oc} [6]. Another important parameter is the Fermi level position (E_F) within the band gap; this parameter was estimated from the C-V measurements and presented in table (1). It is clear that E_F position has demonstrated different values depending on the fullerene type; this is also an evidence for the variation in V_{oc} between the four prepared devices according to equ.2.

The charge carrier transport mechanism of the prepared OSC devices was studied using space charge limited conduction theory (SCLC) as described elsewhere [25,26]. Generally, all the curves have followed the power law $J\alpha V^m$ where m is the gradient of the double log J-V curves. Four different regions were observed with different gradients over the studied applied voltage range as demonstrated in Fig.4 (c); the first region follows ohms' law with a slope of around 1 whereas a slope of around 2 in the second region can be ascribed to SCLC with charge carrier transport controlled by a single trapping level within the forbidden gap [26]. This region occurs as a result of injected carriers density exceeding the thermally generated carriers within the space charge region.

A significant increase in the gradient values of double log plot of the J-V characteristics within the third region has been observed with slopes in the range 7-9. This region is mainly related to the traps filling and is termed as trap-filled SCLC (T-SCLC) mechanism. The gradient of the curves has exhibited another square-dependence of the J-V plots, at which point where the traps in the space charge region are filled and the conduction mechanism is described as trap-free SCLC conduction (TFSCLC). The charge carrier mobility is found to vary with the type of fullerene used with maximum value of 9.3 $x10^{-6} cm^2V^{-1}s^{-1}$ for the $P3HT:PC_{61}BM$ based device, as shown in table (1).

Table I: Dark electrical characteristic parameters for P3HT-based devices with different fullerene derivatives

DEVICES	V_{bi} (V)	E_F (eV)	ϕ_B (eV)		n	μ x10^{-6} ($cm^2V^{-1}s^{-1}$)	R_s (Ω)
			C-V	I-V			
$P3HT:PC_{60}BM$	0.57	0.28	0.85	0.84	2.39	4.97	93
$P3HT:PC_{61}BM$	0.64	0.269	0.91	0.90	1.68	9.3	25
$P3HT:PC_{70}BM$	0.57	0.274	0.84	0.83	2.1	2.81	67
$P3HT:PC_{71}BM$	0.59	0.274	0.87	0.87	1.72	6.9	42

3. 4. Solar cell performance

OSC power conversion efficiency (PCE) and fill factor (FF) are the most important parameters which determine the solar cell performance as described by the following equations [27]:

$$PCE\ (\%)=\frac{J_{max}\ V_{max}}{P_{in}} \quad (3)$$

$$FF=\frac{J_{max}\ V_{max}}{J_{sc}\ V_{oc}} \quad (4)$$

where J_{sc} is the short circuit current density ($mA.cm^{-2}$), V_{oc} is the open-circuit voltage (V), P_{in} is the incident light power and J_{max} ($mA.cm^{-2}$) and V_{max} (V) are the current density and voltage at the point of maximum power output in the J-V curves, respectively.

Table II: solar cell parameters for P3HT-based devices with different PCBM derivatives

DEVICES	V_{oc} (V)	J_{sc} (mA.cm^{-2})	FF %	PCE %
P3HT:PC_{60}BM	0.61	8.2	52	2.6
P3HT:PC_{61}BM	0.65	10.4	62	4.2
P3HT:PC_{70}BM	0.62	9.3	52	3
P3HT:PC_{71}BM	0.64	10.4	58	3.9

Fig.5 (a) shows the J-V characteristics of the devices under illumination; P3HT:PC_{61}BM-based device has exhibited the highest device performance with power conversion efficiency (PCE) reaching 4.2% corresponding to fill factor (FF) of 62%, open circuit voltage (V_{oc}) of 0.65V and short circuit current density (J_{sc}) of 10.4 mA.cm^{-2}. On the other hand P3HT:PC_{60}BM-based device has shown lower performance with PCE of 2.6%, FF of 52%, V_{oc} of 0.61V and J_{sc} of 8.2 mA.cm^{-2}.

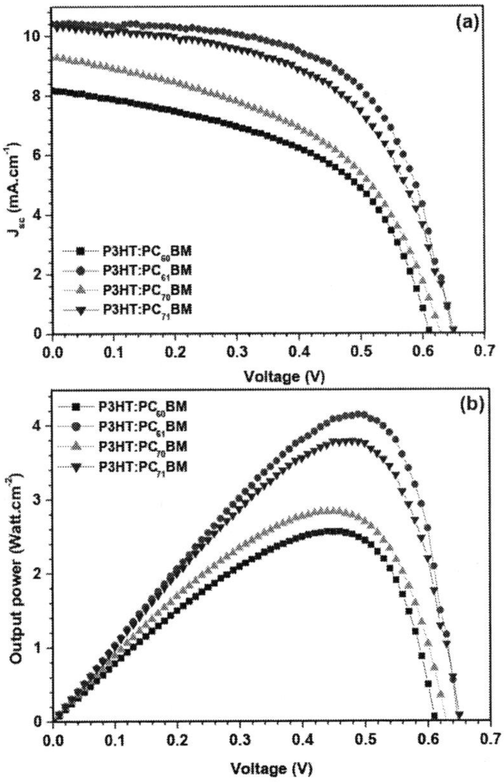

Figure 5: (a) J-V characteristics for P3HT-based devices with different fullerene derivatives under AM 1.5 solar simulator source of 100mW.cm^{-2}. (b) The output power for the devices under study.

Moreover, P3HT:PC_{70}BM-based device has shown PCE of 3%, FF of 52%, V_{oc} of 0.62V and J_{sc} of 9.3 mA.cm^{-2} and P3HT:PC_{71}BM based device has shown PCE of 3.9%, FF of 58%, V_{oc} of 0.64V and J_{sc} of 10.4 mA.cm^{-2}. It could be pointed out that V_{oc} has exhibited different values depending on the used fullerene type, all of which are of sufficiently high value. It has been reported that high V_{oc} could be ascribed to sufficiently low leakage current density in the solar cell devices [28,29].

Fill factor of the studied devices has also exhibited higher values in P3HT:PC_{61}BM compared to other devices, which could be attributed to the lower series resistance (R_s) in this device of 25Ω as shown in table (1) [30]. In this work R_s values have been estimated following our earlier work [31]. The increase in J_{sc} could be related to the higher charge carrier mobility, which might be the main reason as well as the high FF for improving the studied devices PCE. Furthermore, P3HT:PC_{61}BM-based device has exhibited the maximum output power as shown in Fig.5 (b).

CONCLUSION

OSCs based on P3HT blended with different fullerene have been investigated. Optimum device performance was associated with PC_{61}BM as the acceptor material with higher LUMO energy level which has resulted in improving open circuit voltage. Power conversion efficiency of 4.2% and fill factor of 62% have been obtained for such blends with short circuit current density and open circuit voltage of 10.4 (mA.cm^{-2}) and 0.65V respectively. Higher charge carrier mobility of 9.3 x10^{-6} cm^{2}V^{-1}s^{-1} exhibited by the P3HT:PC_{61}BM blend was found to be responsible for the increase in short circuit current (J_{sc}).

ACKNOWLEDGMENT

The authors would like to acknowledge Iraqi Ministry of higher Education and Scientific Research, Babylon University/ Faculty of Science/ Physics department for the PhD sponsorship.

REFERENCES

1. Baek, W. H., Yang, H., Yoon, T. S., Kang, C. J., Lee, H. H., and Kim, Y. S., Solar Energy Materials and Solar Cells 93(2009): 1263-1267.
2. Pfattner, R., Rovira, C. and Mas-Torrent, M., Physical Chemistry Chemical Physics 17 (2015): 26545-26552.
3. Yakuphanoglu, F., and R. S. Anand, Synthetic Metals 160 (2010): 2250-2254.

4. Lou, Y., Wang, Z., Naka, S. and Okada, H., Chemical Physics Letters 529 (2012): 64-68.
5. Guerrero, Antonio, Luís F. Marchesi, Pablo P. Boix, Sonia Ruiz-Raga, Teresa Ripolles-Sanchis, Germà Garcia-Belmonte, and Juan Bisquert, ACS Nano 6 (2012): 3453-3460.
6. Boyuan Qi, Qing Zhou and Jizheng Wang, Scientific reports 5 (2015): 11363.
7. Montero, J.M., Bisquert, J., Garcia-Belmonte, G., Barea, E.M. and Bolink, H.J., Organic Electronics 10 (2009): 305-312.
8. Hui Li, Zhi-Guo Zhang, yongfang Li, and Jizheng Wang, Applied Physics Letters 101 (2012): 163302.
9. Kadem, Burak Y., Mohammed K. Al-Hashimi, and A. K. Hassan, Energy Procedia 50 (2014): 237-245.
10. Kadem, Burak, and Aseel Hassan, Energy Procedia 74 (2015): 439-445.
11. T. Erb, U. Zhokhavets, H. Hoppe, G. Gobsch, M. Al-Ibrahim and O. Ambacher, Thin Solid Films 511 (2006): 483-485.
12. Zhang F, Zhuo Z, Zhang J, Wang X, Xu X, Wang Z, Xin Y, Wang J, Wang J, Tang W and Xu Z., Solar Energy Materials and Solar Cells 97 (2012): 71-77.
13. N.C. Nicolaidis, B.S. Routley, J.L. Holdsworth, W.J. Belcher, X.J. Zhou and P.C. Dastoor, Journal of Physical Chemistry C 115 (2011) 7801-7805.
14. Uladzimir Zhokhavets, Tobias Erb, Harald Hoppe, Gerhard Gobsch and N. Serdar Sariciftci, Thin Solid Films 496 (2006): 679-682.
15. Jong Hwan Park, Jong Soo Kim, Ji Hwang Lee, Wi Hyoung Lee, and Kilwon Cho, J. Phys. Chem. C 113 (2009): 17579-17584.
16. F. Reisdorffer, O. Haas, P. Le Rendu, T.P. Nguyen, Synthetic Metals 161 (2012): 2544- 2548.
17. Qi, B. and Wang, J. Open-circuit voltage in organic solar cells. J. Mater. Chem. 22 (2012): 24315-24325.
18. He, Youjun, and Yongfang Li, Physical chemistry chemical physics 13 (2011): 1970-1983.
19. WooáKim, Gyeong, Young JaeáKim, and Jang HyukáKwon, Journal of Materials Chemistry A 1 (2013): 6895-6900.
20. Güllü, Ö., Ş. Aydoğan, and A. Türüt, Solid State Communications 152 (2012): 381-385.
21. Liu, Chuan, Yong Xu, and Yong-Young Noh, Materials Today 18 (2015): 79-96.
22. N. Simsir, H. Safak, Ö.F. Yüksel , M. Kus, Current Applied Physics 12 (2012) 1510-1514.
23. Zagarzusem Khurelbaatar, Kyu-Hwan Shim, Jaehee Cho, Hyobong Hong, V. Rajagopal Reddy and Chel-Jong Choi, Materials Transactions 56 (2015): 10-16.
24. Uhrich C, Wynands D, Olthof S, Riede MK, Leo K, Sonntag S, Maennig B, Pfeiffer M., Journal of Applied Physics 104 (2008): 3107.
25. Burak Kadem, Aseel Hassan and Wayne Cranton, 31st European Photovoltaic Solar Energy Conference and Exhibition-Hamburg/Germany (2015): 1090.
26. O. Oklobia and T.S.Shafai, Solar Energy Materials and Solar Cells 122 (2014): 158-163.
27. YUE Gen Tian, WU Ji Huai, XIAO Yao Ming, YE Hai Feng, LIN Jian Ming and HUANG Miao Liang, Chinese Science Bulletin 56 (2011): 325-330.
28. Thakur AK, Wantz G, Garcia-Belmonte G, Bisquert J, Hirsch L., Solar Energy Materials and Solar Cells 95 (2011): 2131-2135.
29. Teddy Salim , Hang-Woo Lee , Lydia Helena Wong , Joon Hak Oh , Zhenan Bao and Yeng Ming Lam, Adv. Funct. Mater. 26 (2016): 51-65.
30. Boyuan Qi and Jizheng Wang, Physical Chemistry Chemical Physics 15 (2013): 8972-8982.
31. Kadem, Burak, Wayne Cranton, and Aseel Hassan, Organic Electronics 24 (2015): 73-79.

LIQUID PHASE EXFOLIATED GRAPHENE NANOPLATELETS AS A LOW COST COUNTER ELECTRODE FOR DYE-SENSITIZED SOLAR CELLS

Siva Sankar Nemala[a], Sateesh Prathapani[a], Parag Bhargava[a], Siva Bohm[a b]* and Sudhanshu Mallick[a]*
a Department of Metallurgical Engineering & Materials Science, Indian Institute of Technology Bombay, Mumbai, India
b Tata Steel R&D Global

ABSTRACT: For the development of counter electrodes (CEs) for dye-sensitized solar cells (DSSCs), low cost earth abundant materials like carbon or its allotropes like graphene, carbon nano tubes, carbon nanofibers etc. can be employed as an alternative material to currently using high cost Pt. For the purpose of CE preparation we have chosen graphene nanoplatelets (GNPs) since it can provide good conductivity along with the capability of bulk production. For the bulk production of GNPs we have followed the liquid phase exfoliation of bulk graphite into GNPs and as obtained GNPs in the liquid medium is coated on to the fluorine doped tin oxide (FTO) substrate which serves as CE for DSSCs. The power conversion efficiency achieved with this kind of GNP based CE is about 5 %. The exfoliated GNPs based CEs could be the potential approach towards the low cost module level production of DSSCs.
Keywords: Graphene, DSSC, Counter electrode, Low cost.

1 INTRODUCTION

One of the current problems in the development of DSSCs is the need of cost effective alternative CEs [1]. The highest efficiency reported so far is 13 % using Pt as CE [2]. This Pt has tendency to corrode with electrolyte solution, containing redox couple (I^-/I_3^-) and is costly which is hindering the development of large scale production of DSSCs [3]. From the literature it is understood that various kinds of materials like carbon materials, sulphides, chalcogenides etc. are used as alternative cost effective CEs for DSSCs [4-10]. But, none of them serve the purpose of large scale production due to various reasons like material availability, poor performance and poor stability. After through literature review, we understood that the graphene or GNPs could be a good choice to consider for the CE preparation because of easy material availability, low cost. Only factor we need to look at is its bulk production. To address this here we approached a bulk production method of liquid phase exfoliation (LPE) of graphite into GNPs [11-12] in NMP (N-methyl 2-pyrrolidone). The LPE GNPs can be produced in gram scale, and hence serve the purpose of bulk material availability for cost effective CE preparation.

2 EXPERIMENTAL

2.1. Materials used

Natural graphite AR grade (Timrex), N-Methyl-2-pyrrolidone (NMP) (Merck) and patented surfactants were used for the preparation of graphene nanoplatelets (GNPs). All the chemicals were analytical regent and used without further purification

2.2. Preparation of GNPs

In a typical exfoliation procedure 25 gm of graphite mixed with 150 gm of NMP with the addition of small fraction of surfactants and subjected to high shear mixing by using a Ross mixer. Due to the shear forces during liquid phase exfoliation bulk graphite exfoliates into GNPs. The quality of the GNPs depends on the extent of time duration for which graphite suspension is subjected to shear forces and the rpm at which the shear mixing is carried out. At 10,000 rpm for 8 hrs of mixing we were able to produce GNPs in large quantity.

Figure 1: Photographic image of Ross mixture equipment and schematic illustration of mechanism of exfoliation.

2.3. Preparation of counter electrodes

The FTO glass was cleaned sequentially with acetone; ethanol and deionized water in ultrasonic water bath to remove organic impurities, and then dried in air. The as-prepared GNPs was spread over the spacer between the scotch tape and conducting glass substrate using the doctor blade technique, and then dried in air at 30 °C for 10 min and sintered at 300 °C for 30 min in an inert atmosphere.

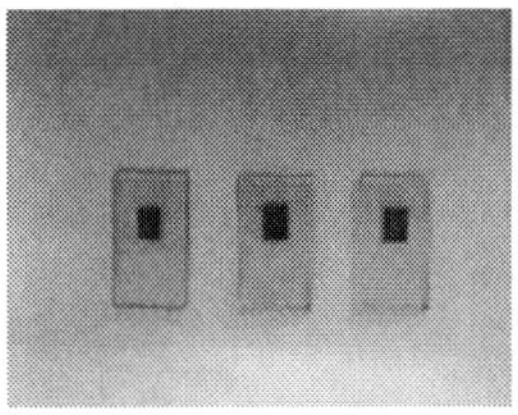

Figure 2: Photograph of prepared counter electrodes.

2.4. Fabrication of DSSC

DSSCs were fabricated with TiO_2 photoanode and Pt and prepared GNPs as counter electrodes by maintaining spacing of approximately 25 μm using a surlyn spacer, which was sandwiched between the electrodes. A drop of electrolyte solution consisted of 0.5 M of LiI, 0.05 M of I_2, 0.5 M of TBP and 0.5 M of 1-butyl-3-methylimidazolium iodide (ionic liquid) in acetonitrile was injected into the cells.

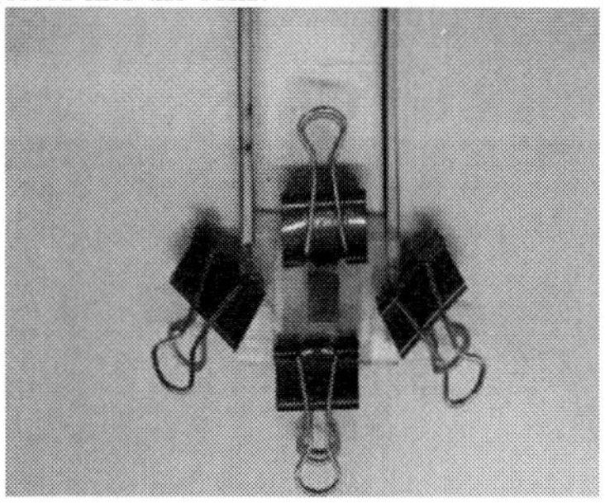

Figure 3: Photographic image of prepared DSSC.

3 RESULTS AND DISCUSSIONS

Raman spectroscopy is a powerful tool to find out the degree of ordered and disordered carbon structures in graphene. The Raman spectra of the GNPs and natural graphite are shown in figure 4. Raman spectra of GNPs show three bands D band, G band and 2D band [13]. A shift is observed with respective to their corresponding bands of the natural graphite. Raman shift and ratio of relative peak intensities are tabulated in the table 1. The Raman shift of D, G and 2D bands from graphite and their relative intensities clearly represents the existence of GNPs [14] obtained from the bulk graphite by high shear exfoliation method.

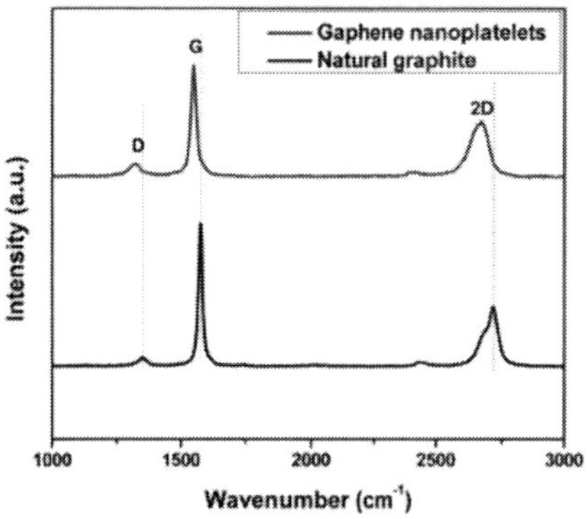

Figure 4: Raman spectra of natural graphite and GNPs.

Table 1: Raman shift and ratio of relative peak intensities of natural graphite and GNPs

Sample	D band (cm^{-1})	G band (cm^{-1})	2D band (cm^{-1})	I_{2D}/I_G
NG	1348.4	1577.5	2717.2	0.41
GNPs	1320.3	1549.5	2672.9	0.52

TEM results: Transmission electron microscopy images are shown in figure 5. Figure 5 represents GNPs of few microns in size bound to one another. Selective area diffraction pattern of the GNPs is shown in figure 5b. The bright spots corresponding to highly order hexagonal lattice confirms the existence of GNPs.

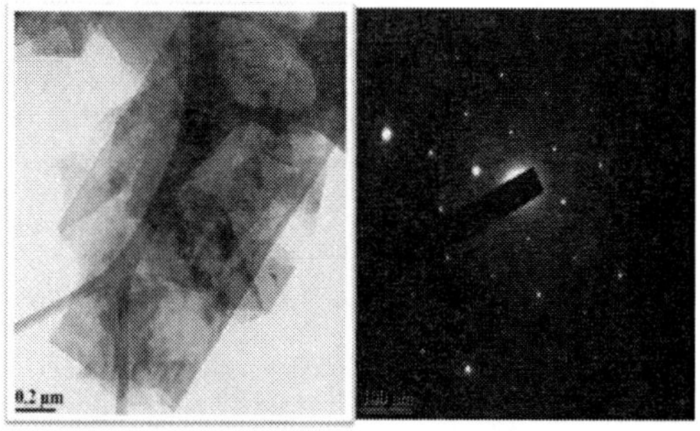

Figure 5: TEM and SAED patterns of the GNPs.

I-V results: The results of current density-voltage curves for the assembled cells using standard Pt and GNPs and are shown in Figure 6. Their corresponding photovoltaic parameters are given in Table 2.

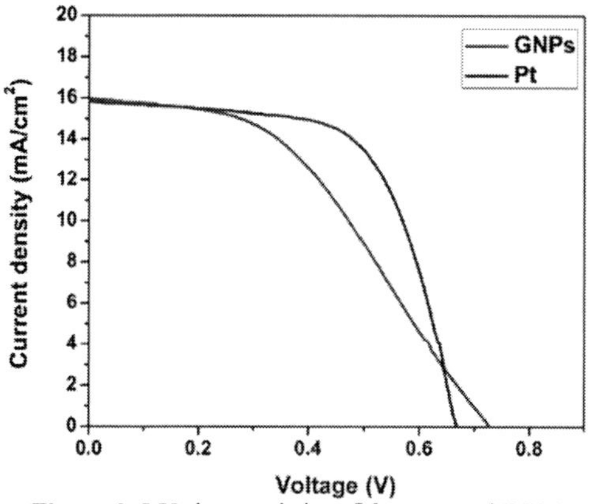

Figure 6: J-V characteristics of the prepared DSSCs

On comparison of performance of GNPs based DSSCs with Pt based DSSCs, it is clearly understood that GNPs could be a potential replacement material for Pt. The open circuit voltage and short circuit current of both GNPs and Pt based DSSCs are comparable but there is a reduction in the fill factor (FF) values in case of GNPs based DSSCs. The optimization of GNPs CE thickness and surface roughness can be expected to improve the FF.

Table 2: Photo-voltaic parameters of the DSSCs

CE	J_{sc} (mA/cm^2)	V_{oc} (mV)	FF	η (%)
Pt	15.8	661	64.5	6.74
GNPs	16.0	721	43.8	5.04

CONCLUSIONS

GNPs have been prepared fromexfoliation of bulk graphite by liquid phase exfoliation method by high shear mixing. GNPs based counter electrodes have been prepared by blade coating technique and DSSCs have been prepared with GNPs based counter electrode and compared with the standard Pt based DSSC for reference. The photo conversion efficiency of GNPs based DSSC is comparable, suggesting its potential as CE material for DSSC.

References

[1] K. Saranya,N. Sivasankar and A. Subramania RSC Adv., 4 (2014) 36226.
[2] S. Mathew, A. Yella, et.al. Nat. Chem., 6 (2014) 242.
[3] Shyama Prasad Mohanty, Venumadhav More and Parag Bhargava RSC Adv., 5 (2015) 18647.
[4] Y. Tang, X. Pan, S. Dai, C. Zhang and H. Tian, Key Eng. Mater., 451 (2011) 63.
[5] T. N. Murakami, et. al. J. Electrochem. Soc., 153 (2006) 2255.
[6] Z. Huang, X. Liu, K. Li, D. Li, Y. Luo, H. Li, W. Song, L. Q. Chen and Q. Meng, Electrochem. Commun., 9 (2007) 596.
[7] B. Ahmmad, Y. Kusumoto, M. Abdulla-Al-Mamun, A. Mihata and H. Yang, J. Sci. Res., 1 (2009) 430.
[8] D. W. Zhang, X. D. Li, S. Chen, F. Tao, Z. Sun, X. J. Yin and S. M. Huang, J. Solid State Electrochem., 14 (2009) 1541.
[9] K. Suzuki, M. Yamaguchi, M. Kumagai and S. Yanagida, Chem. Lett., (2003) 32, 28.
[10] K. S. Novoselov, et. al. Science, 306 (2004) 666.
[11] Andrea C. Ferrari, et.al., NatureNanotechnology, 3 (2008) 563.
[12] Keith R. Paton et al.,Nature Nanotechnology, 13 (2014) 624.
[13] S. Guo, S. Dong and E. Wang, ACS Nano, 4 (2010), 547.
[14] Y. X. Xu, H. Bai, G. W. Lu, C. Li and G. Q. Shi, J. Am. Chem. Soc., 130 (2008) 5856.

TITANIUM OXIDE FILMS DEPOSITED BY E-BEAM EVAPORATION

Rosa Chierchia, ENEA, Casaccia Research Center, via Anguillarese 301, 00123, Roma, ITALY
phone:+3940284208 email: rosa.chierchia@enea.it fax: +3942086405
R. Chierchia[1], *Francesca Menchini* [1], *Luca Serenelli* [1], *Pietro Mangiapane* [1], *Tiziana Polichetti* [2], *Massimo Izzi*[1], *Mario Tucci* [1] rosa.chierchia@enea.it, francesca.menchini@enea.it, luca.serenelli@enea.it, pietro.mangiapane@enea.it, tiziana.polichetti@enea.it, massimo.izzi@enea.it, mario.tucci@enea.it
[1]ENEA, Casaccia Research Center, via Anguillarese 301, 00123, Roma, ITALY
[2]ENEA, Portici Research Center, P.le Enrico Fermi 1, 80055, Portici (Napoli), ITALY

ABSTRACT: In this work structural and electrical properties of titanium oxide thin film obtained by e-beam evaporation and annealed at different temperatures were studied using different characterization techniques, in order to evaluate its use as alternative to ITO as back contact in organic/inorganic solar cells. According to XRD analyses the obtained TiO_2 films showed mainly the presence of anatase crystal structure once annealed at T> 400 °C. Electrical and optical characterization of the annealed samples shows the best results for the films annealed at T> 400 °C with a resistivity of ca.10^{-2} ohm cm, a mobility of 3 cm^2/Vs and carrier concentration of $10^{19}/cm^3$. Therefore, the observed results indicate formation of good quality titanium oxide films by using e-beam evaporation deposition technique followed by thermal annealing, with physical and electrical properties desirable for our applications.
Keywords: see enclosed list of keywords

1 INTRODUCTION

TiO_2 is a n-type oxide semiconductor widely used in many fields of chemical and materials engineering including either traditional catalysis, or photocatalysis, dye-sensitized solar cells, lithium insertion-based devices, integrated circuits, gas sensors and in the paint industry. A number of methods have been employed to prepare TiO_2 films, including e-beam evaporation, sputtering, chemical vapor deposition and sol-gel process.

TiO_2 films microstructures, from anatase to rutile structural transformation were studied by x-ray (XRD) diffraction and Raman spectroscopy [1–3]. The deposition techniques and the processing conditions has been shown to have a strong influence on the microstructures and electronic and vibrational properties of the materials.

2 EXPERIMENTAL

The TiO_2 films have been grown by e-beam evaporation on soda lime glass (SLG) in a Balzer evaporation system followed by an annealing process performed at a temperature of 200° , 400 ° and 500°C in a oven in a N_2 flux . The substrate was not intentionally heated during the deposition. This method is more attractive than others in industrialization applications because it is more suitable for large-scale production.

An heat treatment of these films is necessary to induce crystal growth. Therefore after their deposition, the samples were placed into an oven. The morphology and the crystalline structure of the thin films were characterized with XRD, Raman, transmittance and reflectance measurement. XRD data were collected using a Rigaku MiniFlex diffractometer, with CuKα radiation and Bragg Brentano geometry. Raman spectra have been collected by a Renishaw InVia Reflex Raman spectrometer with an Ar source at 514.5 nm.

3 RESULTS

A number of methods have been employed to prepare TiO_2 films, including e-beam evaporation, sputtering, chemical vapor deposition and sol-gel process. In this work structural and electrical properties of titanium oxide thin film obtained by e-beam evaporation and annealed at different temperatures were studied using different characterization techniques.

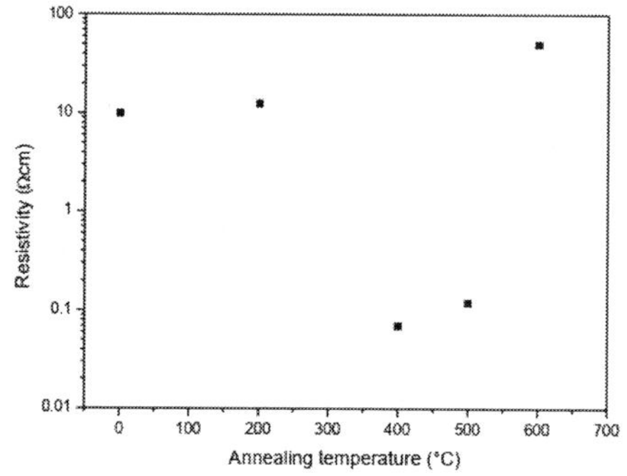

Figure 1 Variation of the resistivity of the TiO_2 *film with temperature*

By annealing at 400 °C of the evaporated samples the resistivity decreases as can be seen in Figure 1, at higher annealing temperature the resistivity increases again. In Figure 2 the transmittance curves are shown, the films become more transparent with increasing temperature.
In the near-IR region, the absorption is due to plasma oscillations of the "free" conduction band electrons (according to the Drude model), and lower transmittance should be related to higher free electron density values.

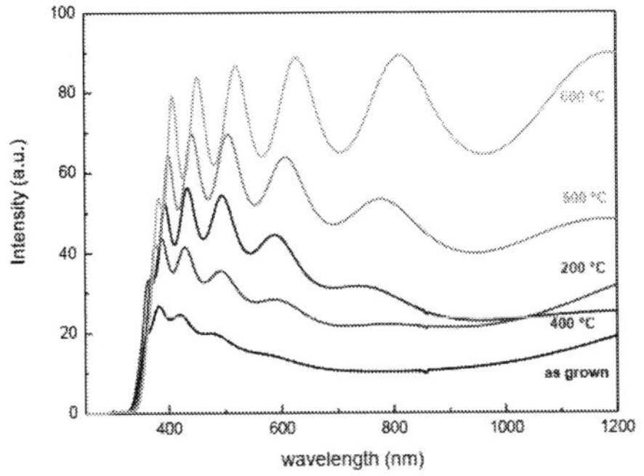

Figure 2: Transmittance of the TiO_2 films vs temperature

The variation of resistivity with annealing temperature could be explained by a variation of the crystalline structure of TiO_2 from amorphous to anatase phase [1-7]. In Fig.2 XRD measurements obtained at different temperatures are reported. XRD curves show the transition from amorphous to anatase TiO_2 at T > 400°C.

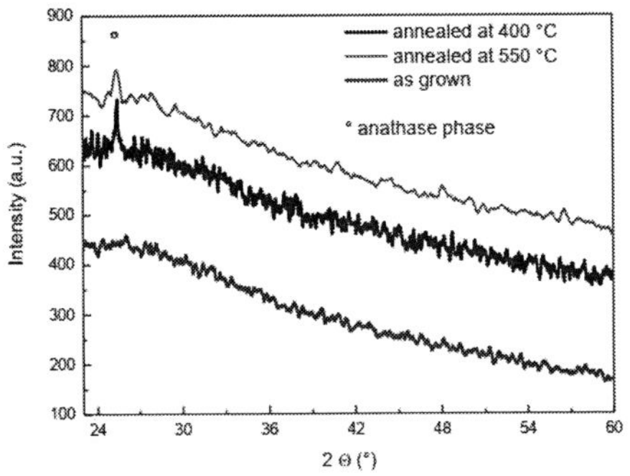

Figure 3: XRD measurements of TiO_2 layer at different temperatures

Raman spectra have been investigated at various annealing temperatures in order to get a better understanding of the lattice vibration characters [3–6]. Figure 4 shows the Raman spectrum of the as grown and annealed films.

The Raman lines at 151, 409, 515 and 633 cm^{-1} can be assigned as the Eg, B_{1g}, A_{1g} or B_{1g}, and E_g modes of the anatase phase, respectively. The strongest E_g mode at 151 cm^{-1} arising from the external vibration of the anatase structure is well resolved, which indicates that an anatase phase was formed at T>400 °C. Such a conclusion is consistent with the XRD result shown in figure 3. Raman lines become weak and broad when the samples have local lattice imperfections.

Figure 4 also exhibits Raman spectra of films annealed at 400 and 600 °C, respectively. As the annealing temperature increases, the intensity of the lowest frequency Eg mode dramatically increases. It can be seen that the high-frequency Raman peaks become well resolved after annealing at 400 °C, and the intrinsically weak Eg mode at 196 cm^{-1} in the anatase phase also looks distinct. The presence of the only anatase phase corresponds to a reduction of the resistivity of the TiO_2 layer as this is the most conductive phase.

These observations demonstrate the annealing-dependent evolution of the anatase phase. It is also found that a weak peak at 450 cm^{-1} and a small shoulder around 610 cm^{-1} clearly appear after annealing at 600 °C, these peaks correspond to the E_g and A_{1g} modes of the rutile phase, respectively. The apparition of the rutile phase corresponds to the increasing of resistivity.

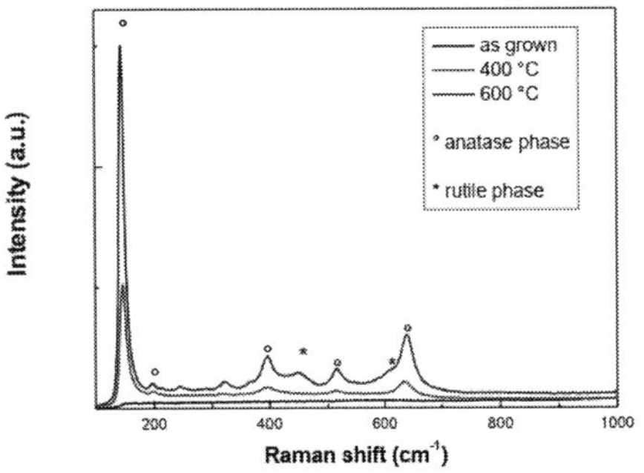

Figure 4: Raman curves of TiO_2 sample at different temperatures

4 CONCLUSIONS

According to XRD and Raman analyses the obtained TiO_2 films showed mainly the presence of anatase crystal structure once annealed at T > 400 °C whereas the rutile phase appears at higher temperatures. Electrical and optical characterization of the annealed samples shows the best results for the films annealed at T > 400 °C with a resistivity of ca.10^{-2} ohm cm, a mobility of 3 cm^2/Vs and carrier concentration of $10^{19}/cm^3$.

5 REFERENCES

[1] Melendres C A, Narayanasamy A, Maroni V A and Siegel R W 1989 *J. Mater. Res.* **4** 1246
[3] Ma W H, Lu Z and Zhang M S 1998 *Appl. Phys.* A **66** 621
[4] Parker J C and Siegel R W 1990 *J. Mater. Res.* **5** 1246
[5] Parker J C and Siegel R W 1990 *Appl. Phys. Lett.* **57** 943
[6] Bersani D, Lottici P P and Ding X Z 1998 *Appl. Phys. Lett.***72** 73
[7] R. Ren, Z. Yang, and L. L. Shaw, Journal of Materials Science, vol. 35, no. 23, pp. 6015–6026, 2000

ALL-SOLUTION PROCESSES FOR MANUFACTURING PHOTOELECTRODES AND DYE-SENSITIZED SOLAR CELLS USING INKJET PRINTING TECHNOLOGY

Chin-Tai Chen*, Bin-Chuen Hu
Department of Mechanical Engineering, National Kaohsiung University of Applied Sciences
No. 415, Chien-Kung Road, Kaohsiung 80778, TAIWAN, ROC
*Correspondence E-mail:chintai@kuas.edu.tw

ABSTRACT: This work explores the emerging dye-sensitized solar cells (DSSCs) by inkjet printing (IJP) technology. Through nanometer materials in a direct manufacturing way, the IJP-based DSSC featuring low-cost, light-weight and flexible advantages is one of promising solar cell types in field of green-power technologies. Pursuing novel non-silicon-based materials and DSSCs still remains. In this study, a diluted titanium dioxide mixtures (TiO_2, 10 wt%) was first filled in an inkjet printhead (Dimatix DMP 2831), directly forming a thin film (area = 4 mm × 4 mm) on a conductive glass (fluorine-doped tin oxide Glass). After thermally sintering at 450 °C for 30 min to form a nano-porous film (thickness ~10 μm) and dyeing (N719, Sigma-Aldrich), a DSSC photo-electrode (PE) was generated. Then, using doctor-blade coating (DBC), a platinum (Pt) film was formed on another FTO glass to generate a DSSC counter electrode (CE), followed by thermally adhesive bonding (Surlyn 1702, DuPont) with the previous PE. Finally, after filling an electrolyte (potassium iodide) into the small gap (60 μm) between PE and CE, a DSSC was eventually achieved. Using a solar simulator (Newport 91160A), the experimental results show that this DSSC exhibited the short-circuit current density (J_{sc}) of 9.17 mA/cm^2, open-circuit voltage (V_{oc}) of 0.72 V, fill Factor (FF) of 0.67, and power conversion efficiency (PCE) of 4.41%.
Keywords: dye-sensitized solar cells (DSSCs), inkjet printing (IJP), photo-electrode (PE), titanium dioxide (TiO_2)

1 INTRODUCTION

Since 1991, dye-sensitized solar cells (DSSCs) chemically composed of TiO_2, dye, and electrolyte have attracted much attention, in which maximum conversion efficiency reached over 12% [1-5]. Incoporating nanotechnology for patterning, many progresses of the solar cells remain with a lot of further efforts to be made [6-9]. For instance, Yang and co-workers made many efforts on Si nanowire P-N junction [10, 11], light trapping [12], core-shell nanowires [13] and silver nanocrystal [14,15]. Menawhile, many research works have been directed to new materials of DSSCs for improving conversion efficiency, which includes D-π-A dye C220 and Z907 [5], ruthenium Complex SJW-E1 and TiO_2 ekectrode [16], triphenylamine (TPA)-based dye and TiO_2 electrode [17], and nano-carbon black [18].

Other the other hand, many different methods for manufacturing DSSCs have been proposed. For example, the typical 10 μm-thick TiO_2 films can be fabricated by spin coating (SIC) [19, 20], screen printing (SCP) [21], inkjet printing (IJP) [22] and spray coating (SRC) [23] to form photo-electrodes of DSSCs. Among them, inkjet printing method (IJP) shows a unique capability of drop-on-demand (DOD) for large-area formation of patterning [22], thereby achieving advantages such as flexibility of fabrication an low cost[24]. Moreover, as adding other materials such as conductive polymers [25] and metal nano-particles [26], this IJP process can be expected to apply for DSSC patterning, microsystem integration and material blending [27] in future. Hence, in the study, it is aimed to explore fabrication of DSSCs using IJP method.

2 DESIGN AND MANUFACTURING

2.1 Inkjet printing process

Fig. 1 shows an inkjet printing process in the study. Using a material printer (Dimatix DMP 2831), digital inkjet printing (IJP) is applied for forming electrodes. Here ink materials for printing were prepared with diluted titanium dioxide mixtures (TiO_2 of ~10wt%, dilute acetic acid of ~89wt%, and surfactant TritonX-100 of ~0.1wt%). Printing of TiO_2 ink was performed at 38 V, 1 kHz , droplet spacing of 15 μm, and gap of 2500 μm thereby yielding a droplet diameter of 31 μm and speed of 17.14 m/s. Wetting angle of TiO_2 on fluorine-doped tin oxide (FTO) glass was obtained with ~38°. Thus nano-porous photo-electrode of TiO_2 was formed after sintering thermally for further absorption of dye (N719).

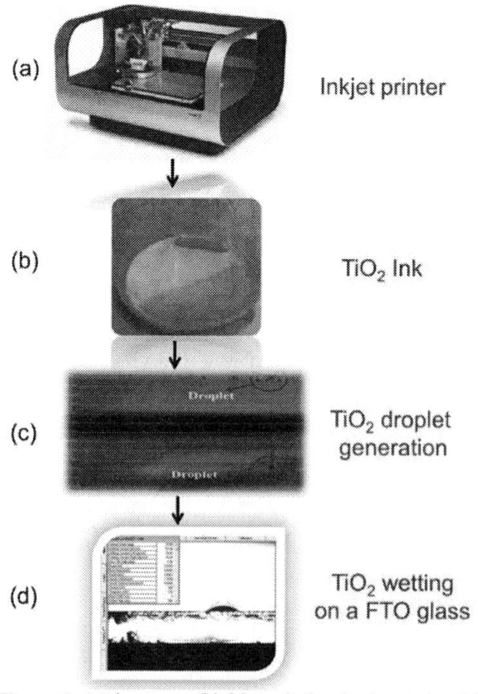

Figure 1: Main steps of inkjet printing in the study: (a) use of a material printer of DMP-2831 (b) preparation of dilute ink material TiO_2 (c) droplet generation of TiO_2 from the printer and (d) accurate droplet deposition of TiO_2 on a FTO conductive glass surface

2.2 Fabrication of dye-sensitized solar cells

Fig. 2 shows a schematic of whole manufacturing processes of an inkjet-printed DSSC in the study. In particular, as given an active area (4 mm × 4 mm) of 0.16 cm^2, the TiO_2 was thermally sintered at 450°C for 30 min. for forming a photo-electrode (PE). By conventional doctor-blade coating (DBC) method, a counter-electrode (CE) of platinum (Pt) was formed on another FTO glass. After bonding of the PE/CE glasses under a pressure of 2 kgf and temperature of 150°C for 5 min., it was injected with electrolyte (potassium iodide) and sealed with a UV-curable adhesive (NOA 89). In this way, a DSSC was eventually achieved.

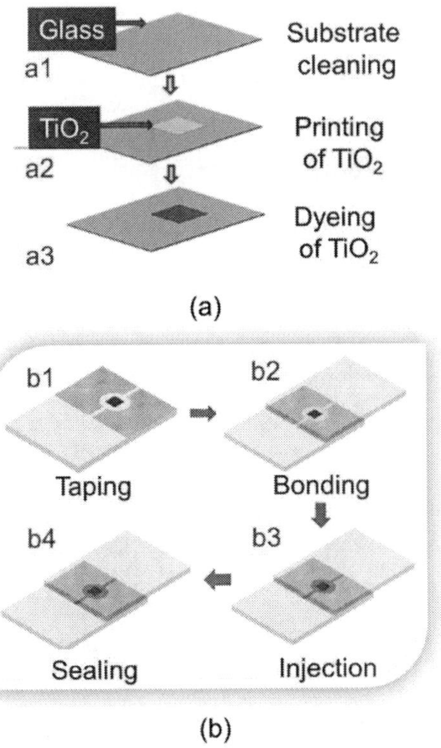

Figure 2: (a) schematic for substrate cleaning, printing of TiO_2, and dyeing (N719) of TiO_2 (b) final assembly steps of a DSSC incorporated with taping (Surlyn 1702, DuPont), bonding (two glasses), injection (potassium iodide), and sealing (NOA 89)

3 RESULTS AND DISCUSSION

3.1 Droplet morphology of microfluidic films

To explore the microfluidic behavior, a Newton-ring pattern nearby the droplet edge was found during droplet evaporation as shown in Fig. 3(a), which indicates the non-uniform film morphology of droplet liquid wetting on the glass surface. This non-uniformity might influence power conversion efficiency of electrodes.

Moreover, Fig. 3(b) shows the further observation of droplet morphology, where (b1) was the dried state at right edge, (b2) and (b3) were the partially dried state in middle regions, and (b4) was the wet liquid state. The effect on electrodes could be explored in experiments.

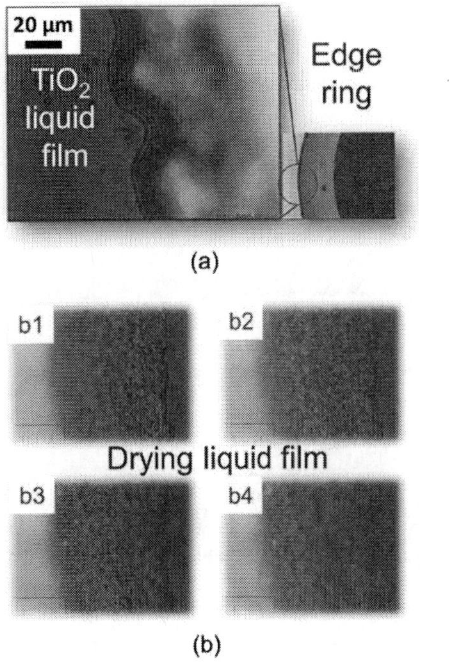

Figure 3: (a) a Newton-ring pattern nearby the droplet edge appeared during evaporation (b) observation of droplet morphology

3.2 Structure morphology of printed nano-porous films

Fig. 4 shows that total six layers of TiO_2 were printed: compared to the 1-printed layer, the nano-porous structure was found for 2 printed layers with particle sizes of 50~100 nm; the densest structures were created for 5-6 printed layers with maximum thickness of 10 µm.

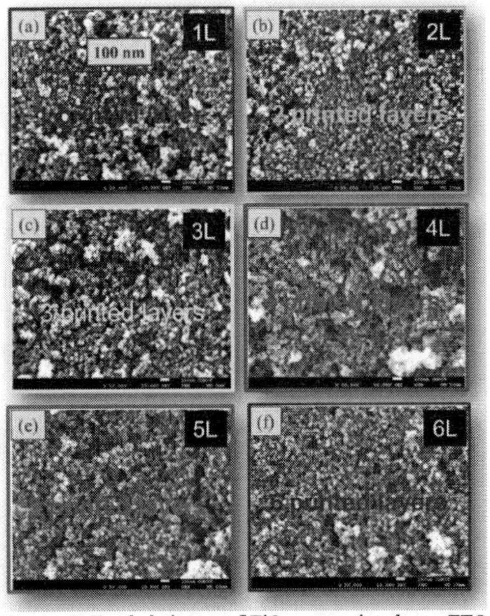

Figure 4: Total six layers of TiO_2 were printed on a FTO glass surface: (a) one layer, 1L (b) two layers, 2L (c) three layers, 3L (d) four layers, 4L (e) five layers, 5L, and (f) six layers, 6L

3.3 Measurement of solar power performances

Using a solar simulator (Newport-91160A), the DSSCs were measured with high short-circuit current densities J_{sc} of 9.17-10.52 mA/cm^2 and open-circuit voltages V_{oc} of 0.72-0.74 V as shown in Fig. 5 and Table I. The fill factor (FF) reached 0.67 and the power conversion Efficiency (PCE) was achieved up to 4.41%. The efficiency of 0.03% for specimen S#1 was due to poor coating, while the failure (PCE=0%) for specimen S#4 was attributed to mishandling during fabrication processes.

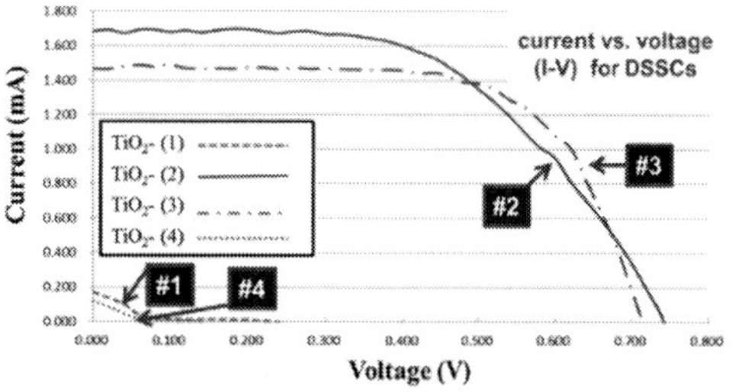

Figure 5: Optical measurement of four DSSCs with electric voltages and currents using a solar simulator

Table I: Optical performances of the inkjet-printed DSSCs in the study

	S#1	S#2	S#3	S#4
J_{sc} (mA/cm^2)	1.25	10.52	9.17	0.05
V_{oc} (V)	0.22	0.74	0.72	0.04
J_{max} (mA/cm^2)	0.68	8.88	8.48	0.01
V_{max} (V)	0.04	0.48	0.52	0.04
FF	0.30	0.55	0.67	0.30
PCE (%)	0.03	4.26	4.41	0.00

4 CONCLUSION

This work explores the emerging dye-sensitized solar cells (DSSCs) by inkjet printing (IJP) technology. Through nanometer materials in a direct manufacturing way, the IJP-based DSSC featuring low-cost, light-weight and flexible advantages is one of promising solar cell types in field of green-power technologies. Through exploration of new manufacturing process on fundamental material properties and conversion efficiency, low-cost and large-area flexible DSSCs can be produced by inkjet printing of the materials of liquid mixtures.

In the study, the experimental results show that this DSSC exhibited the short-circuit current density (J_{sc}) of 9.17 mA/cm^2, open-circuit voltage (V_{oc}) of 0.72 V, fill Factor (FF) of 0.67, and power conversion efficiency (PCE) of 4.41%, respectively. However, many manufacturing parameters of PE and CE were not optimized in the present IJP and DBC process. In very recent future, with more hybrid ways incorporated with versatile composite layered materials, it is expected to report significant increase of J_{sc}, V_{oc}, FF, and PCE for the new development of inkjet-printed DSSCs.

5 ACKNOWLEDGEMENTS

The authors thank research grant for this work partially by the Ministry of Science and Technology (MOST) under MOST 104-2221-E-151-036-MY2, Taiwan, ROC. The Research Centre for MEMS and Precision Machines in National Kaohsiung University of Applied Sciences (KUAS) and Centre for Micro/Nano Science and Technology in National Cheng Kung University (NCKU) are also appreciated for the access of the inkjet-printing and surface analysis equipment.

6 ABBREVIATIONS

V: electric voltage (mV)
J: electric current density (mA/cm^2)
PCE: power conversion efficiency
FF: fill factor
IJP: inkjet printing
DSSC: dye-sensitized solar cell
FTO: fluorine-doped tin oxide
Subscripts
m: maximum value
sc: short circuit
oc: open circuit

7 REFERENCES

[1] B. O'Regan, M. Gratzel, A low-cost, high-efficiency solar cell based on dye-sensitized colloidal TiO_2 films, Nature 353 (1991) 737.
[2] A. Hagfeldt, M. Gratzel, Molecular photovoltaics, Accounts of Chemical Research 33 (2000) 269.
[3] M. Gratzel, Photovoltaic and photoelectrochemical conversion of solar energy, Philosophical Transaction of The Royal Society A 365 (2007) 993.
[4] S. Ito et. al., Fabrication of thin film dye sensitized solar cells with solar to electric power conversion efficiency over 10%, Thin Solid Films 516 (2008) 4613.
[5] N. Cai et al., An organic D-π-A dye for record efficiency solid-state sensitized heterojunction solar cells, Nano Letters 11 (2011) 1452.
[6] N. McElroy et. al., Comparison of solar cells sensitised by CdTe/CdSe and Cd/Se/CdTe core/shell colloidal quantum dots with and without a CdS outer layer, Thin Solid Films 560 (2014), 65.
[7] J. Griffin et. al., Organic photovoltaic devices incorporating a molybdenum oxide hole-extraction layer deposited by spray-coating from an ammonium molybdate tetrahydrate precursor, Organic Electronics 15 (2014) 692.
[8] A. T. Barrows et. al., Efficient planar heterojunction mix-halide perovskite solar cells deposited via spray-deposition, Energy & Environmental Science 7 (2014) 2944.
[9] M. C. Beard et. al., The promise and challenge of nanostructured solar cells, Nature Nanotechnology, 9 (2014) 951.
[10] E. C. Garnett, P. Yang, Silicon nanowire radial p-n junction solar cells, Journal of American Chemical Society 130 (2008) 9224.
[11] E. C. Garnett, P. Yang, Light trapping in silicon nanowire solar cells, Nano Letters 10 (2010) 1082.
[12] M. C. Beard et. al., The promise and challenge of nanostructured solar cells, Nature Nanotechnology, 9 (2014) 951.
[13] J. Tang et. al., Solution-processed core-shell nanowires for efficient photovoltaic cells, Nature Nanotechnology 6 (2011) 568.
[14] S. Brittman et. al., Absorption of light in a single-

nanowire silicon solar cell decorated with an octahedral nanocrystal, Nano Letters 11 (2011) 5189.
[15] P. Yang, J.-M. Tarascon, Towards systems materials engineering, Nature Materials, 11 (2012) 560.
[16] K.-M. Lee et. al., Efficient and stable plastic dye-sensitized solar cells based on a high light-harvesting ruthenium sensitizer, Journal of Materials Chemistry, 19 (2009) 5009.
[17] C.-H. Yang et. al., Characteristics of triphenylamine-based dyes with multiple acceptors in application of dye-sensitized solar cells, Journal of Power Sources 188 (2009) 627.
[18] C.-C. Ting, W.-S. Chao, Efficiency improvement of the DSSCs by building the carbon black as bridge in photoelectrode, Applied Energy 87 (2010) 2500.
[19] Y. Natsume, H. Sakata, Zinc oxide films prepared by sol-gel spin-coating, Thin Solid Films, 372 (2000) 30.
[20] C.-Y. Lin et. al., Design and characterization of novel porphyrins with oligo(phenylethylnyl) links of varied length for dye-sensitized solar cells: synthesis and optical, electrochemical, and photovoltaic investigation, The Journal of Physical Chemistry C, 113 (2009) 755.
[21] S. E. Shaheen et. al., Fabrication of bulk heterojunction plastic solar cells by screen printing, Applied Physics Letters 79 (2001) 2996.
[22] P. Dzik, M. Vesely, J. Chomoucka, Thin layers of photocatalytic TiO_2 prepared by inkjet printing of a sol-gel precursor, Journal of Advanced Oxidation Technologies 13 (2010) 172.
[23] C.-C. Chen, C.-C. Ting, Photoelectrode fabrication of dye-sensitized nanosolar cells using multiple spray coating technique, International Journal of Photoenergy 629059 (2013) 8pp.
[24] Y. Kunugi, K. Shimoyama, S. Umezu, Fabrication of dye-sensitized solar cells using electrostatic inkjet printing, Journal of Photopolymer Science and Technology 26 (2013) 383.
[25] S. H. Eom et. al., Polymer solar cells based on inkjet-printed PEDOT:PSS layer, Organic Electronics 10 (2009) 536.
[26] S. Merilampi et. al., Analysis of electrically conductive silver ink on strentchable substrates under tensile load, Microelectronics Reliability 50 (2010) 2001.
[27] R. A. Reigister, Painting with block copolymers, Nature Nanotechnology 8 (2013) 618.

PREVENTING UV DEGRADATION IN DYE SENSITIZED SOLAR CELLS

Gabriela Gava Sonai[a], Armi Tiihonen[b], Kati Miettunen[b], Peter Lund[b] and Ana Flávia Nogueira[a]
[a] University of Campinas, Institute of Chemistry
P.O. Box 6154, 13083-970 Campinas, Brazil
[b]Aalto University, Department of Applied Physics, New Energy Technologies
P.O. Box 15100, 00076 Aalto, Finland

ABSTRACT: The effects of a UV cut-off filter and a UV additive on the stability of dye-sensitized solar cells were studied. Both approaches to prevent UV, which is a major degradation factor, have been presented in literature, but the effectiveness of these measures have not been compared nor their possible combined effect to reach even higher UV protection tested. Here, the studied cells had conventional iodine based electrolyte or electrolyte with cobalt complex redox couple to which the UV absorber (benzophenone based) was mixed. The cells were subjected to an accelerated aging test using a UV light and the cells with no protection were compared to cells with those having the UV absorber and/or a UV cut-off filter. The initial and final performance of cells were characterized by photovoltaic measurements, electrochemical impedance spectroscopy, and other measurement techniques. The cobalt based cells showed faster degradation compared to iodine based cells, even with the UV filter and their degradation could be related to the degradation of the organic dye. The performance of the iodine based cells remained stable over 111h when a UV cut-off filter was applied. Our study also shows that the UV filter has a superior effect in blocking UV light, whilst the UV additive had only a marginal effect at most in stabilizing the cells.
Keywords: Dye-sensitized solar cells, stability, UV light, cobalt electrolyte, iodine electrolyte

1 INTRODUCTION

Degradation of dye solar cells (DSC) is a major issue in their development towards a commercially viable technology. There are many reasons for their aging, but a degradation related to changes in the electrolyte and in the interface between TiO_2/electrolyte have been identified as major loss factors [1]. One specific issue is the bleaching of the iodine electrolyte which increases by exposure of high temperatures (>50°C) under 1 Sun illumination and due to UV light [2].

TiO_2 is known as a good photocatalyst and when irradiated with UV light, to generate electron-hole pairs which can degrade organic species such as dye molecules and electrolyte [3]. The main crystalline form of TiO_2 used for DSCs is anatase and its band gap is 3.2 eV, being photoactive up to 387 nm [4].

In order to prevent direct band gap excitation of TiO_2 in DSCs, it is common to use a UV cut-off filter or UV free light sources in aging tests. As no UV filter is perfect and as UV light is known to cause serious degradation [2], the effect of UV light should be considered in aging. Therefore, using completely UV free illumination in the aging test does not correlate with real outdoor condition. Instead, the UV illumination combined with UV filters, which would be commercially viable in large scale use, are a proper approach.

Another, less applied type of UV protection is a UV absorber. There are organic or inorganic compounds which absorb UV light (<400 nm). Typically, the UV absorbers are transparent for all visible light and their operation principle is to convert the excitation energy into heat. This kind of compounds are most frequently used in cosmetics to prevent skin cancer, but they could be used to protect material properties of polymers or textiles as well [5]. The UV absorber was firstly applied in DSCs by Cheng et. al (2014) [6]. However, no report about the comparation between these two kind of UV protection in DSCs was made until now.

In the present work, it was compared two approaches for DSCs UV protection: filters and additives. Iodine and cobalt based cells were subjected to an accelerated aging test using UV light under absence or presence of 10% UV absorber (benzophenone based) or UV cut-off filter. Furthermore, a combination of both types of UV protection was used to test if even more superior UV resistance could be achieved.

2 EXPERIMENTAL METHODS

2.1 Materials and methods

All chemicals were used as purchased except 3-methoxy propionitrile (MPN, Sigma Aldrich) that was purified. Additionally, the chemicals and solvents used were purchased from Sigma-Aldrich unless otherwise specified. The cobalt complexes were prepared based on a method described in literature [7].

2.2 Device fabrication

In summary, two test cell series were prepared, one containing Z907 dye (Dyesol) and iodine based electrolyte and the other one containing MK-2 dye and cobalt based electrolyte. The UV additive, 2-hydroxy-4-methoxybenzophenone (BZP), was used in 10 wt.% concentration and the UV cut-off filter was 400 nm (Asmetetc GmbH). The filter was used only when the cells were in the accelerated aging test. Each test series consisted of five groups with 3-6 cell each: two groups with or without UV additive in the electrolyte, two groups with or without UV additive in electrolyte plus UV cut-off filter, and one group of cells prepared with quasi-solid electrolyte with UV additive and UV cut-off filter.

Photoelectrodes for iodine cells consisted of two layers of TiO_2 paste DSL 18NR-T (Dyesol) and one light scattering layer of TiO_2 paste WER2-O (Dyesol). For cobalt electrolyte, the photoelectrodes were prepared with only one layer of TiO_2 paste DSL 18NR-T. The active areas of cells were 0.4 cm^2 and the thickness of TiO_2 layer was approximately 12 and 4 µm for the cells containing iodine and cobalt electrolyte, respectively.

Before the TiO_2 deposition a $TiCl_4$ treatment was applied on top of glass substrate (2.5 mm TEC-15 FTO) by immersing the films in a 40 mM aqueous $TiCl_4$ solution at 70° C for 30 minutes. Then the layers of TiO_2 were screen printed and sintered at 450°C for 30 minutes. A second $TiCl_4$ treatment was applied after the TiO_2 following by another sintering process. Prior to the device assembly, the films were cooled to approximately 80 °C and immersed in the desired dye solution for around 18 hours. The dye solutions consisted of 0.3 mM Z907 dye in acetonitrile and tert-butanol 1:1 or 0.2 mM MK2 dye in toluene and acetonitrile 1:1.

Counter electrodes were prepared by spreading 4 µL of 5 mM chloroplatinic acid hexahydrate in 2-propanol onto the pre-drilled glass substrate and sintered for 20 min at 385 °C. The cells were sealed with 25 µm Surlyn foils and a thin layer of epoxy glue was used on the joints for increasing mechanical strength.

The iodine electrolyte consisted of 0.1 M iodine (I_2), 0.5 M 1-methyl-benzimidazole (NMBI) and 0.6 M 1-propyl-3-methylimidazolium iodide (PMII, IoLiTec), dissolved in 3-methoxypropionitrile (MPN), while the cobalt electrolyte consisted of 0.2 M $Co(bpy)_3(TFSI)_2$, 0.07 M $Co(bpy)_3(TFSI)_3$, 0.05 M LiTFSI and 1.0 M tert-butylpyridine (tBP) in acetonitrile (ACN). For quasi-solid electrolytes 5 wt.% of poly(ethylene oxide-co-2-2-methoxyethoxy) ethyl glycidyl ether-co-allyl glycidyl ether (PEO-EM-AGE, Daiso) with rate of PEO-EM-AGE (78:20:2) and molecular weight of 2.15×10^6 $g.mol^{-1}$ was used.

2.3 Measurements

All the cells were aged at the same time for 111 hours under UV illumination in UV chamber Arctest 150. The lamps used were Osram - Ultra-Vitalux 300W and the amount of the UV light was adjusted equivalent to 1000 Wm^{-2} AM 1.5G (1 Sun).

Before and after the aging test, the cells were characterized by current-voltage curve (IV), impedance spectroscopy (EIS), incident photon to collected electrons (IPCE) and transmittance measurements. The IV curves were also regularly measured during the aging test. The solar simulator used halogen lamps (Philips projection lamp, type 13117) and was calibrated with an official calibrated solar cell with a KG5 filter (PV Measurements) to 1000 Wm^{-2} AM 1.5G (1 Sun).

The IV curves were measured with a Keithley 2420 SourceMeter with black tape masks on top of the cells. EIS at open-circuit voltage V_{OC} under 1 Sun were performed with a Zahner Zennium potentiostat. In the impedance measurements, the frequency range 10^{-1}-10^6 Hz was swept back and forth with 10 mV amplitude. The EIS data were fitted using the ZView2 program, using a suitable equivalent circuit described in literature [8]. The IPCE data were measured by a PV Measurements (QEX7 Solar Cell Spectral Response Measurement System) in DC mode with no bias light, in the wavelength range 300–900 nm. Transmittance spectrum of electrolyte were measured using an Ocean Optics Spectrometer in the wavelength range 300-900 nm.

3 RESULTS

3.1 Aging experiment

Figs. 1 and 2. show the evolution of performance parameters for DSCs prepared with iodine or cobalt based electrolytes containing or not 10% absorber or UV cut-off filter under UV light at 1 Sun equivalent over a period of 111 h.

Initially, the efficiencies of cobalt and iodine cells differ from each other as to be expected. Interestingly, having the absorber in the electrolyte did not affect the initial characteristics of the cells in either of the two series (iodine or cobalt) (Fig. 1-a). After the aging test, for the DSCs without a UV cut-off filter, the efficiency decreased by 20% for iodine cells and more than 90% for cobalt cells. Moreover, there is no clear statistically significant difference in the post-aging efficiencies for DSCs when no UV cut-off filter used, regardless of presence of UV additive. Additionally, V_{OC} (Fig. 1-b) and short circuit current density J_{SC} (Fig. 2-a) followed the same behavior for DSCs without UV filter, these two parameters showed a significant decay. This holds for both series (iodine or cobalt).

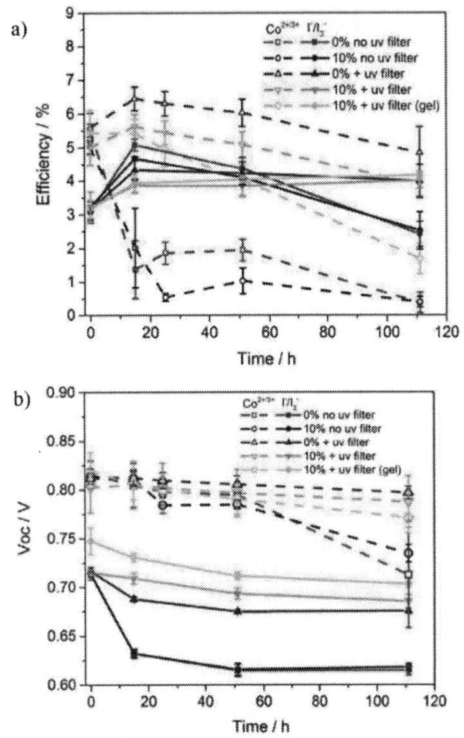

Figure 1: Variation of a) efficiency and b) V_{OC} for DSCs based on iodine and cobalt electrolytes, with Z907 or MK-2 dye, with or without 10% UV absorber and with or without UV cut-off filter under 1 sun (UV light). The error bars indicate the standard deviation between the same type of cells.

On the other hand, when a UV cut-off filter was applied on top of the cells, the presence of the UV absorber BZP showed some minor differences. All the cobalt groups had a decrease in efficiency over the 111 h aging period. The cobalt cells without BZP exhibit decay of 17±14 %, compared to 23±1% for the cells with the additive. In the presence of gel electrolyte and BZP, the performance decrease was more pronounced, with a drop of 70±7%. However, iodine cells showed an increase in efficiency in the range of 26±4% (without BZP), 17±8% (with BZP) and 32±9% for gel electrolyte (with BZP).

Considering only the DSCs prepared with gel electrolyte, a substantial difference on the stability was observed for cells containing cobalt or iodine based electrolyte: whilst iodine gel cells retain stable performance at least up to 111 h similar to those with liquid iodine electrolyte, the cobalt gel cells show a faster decline of efficiency over the time compared to their liquid comparison. One explanation for that is the polymerization of the PEO-EM-AGE, which results in a highly viscous electrolyte. Consequently, this would hinder the mobility of the cobalt complex that is larger than iodine redox couple, and decrease the photocurrent due to the mass transport limitations more quickly [9].

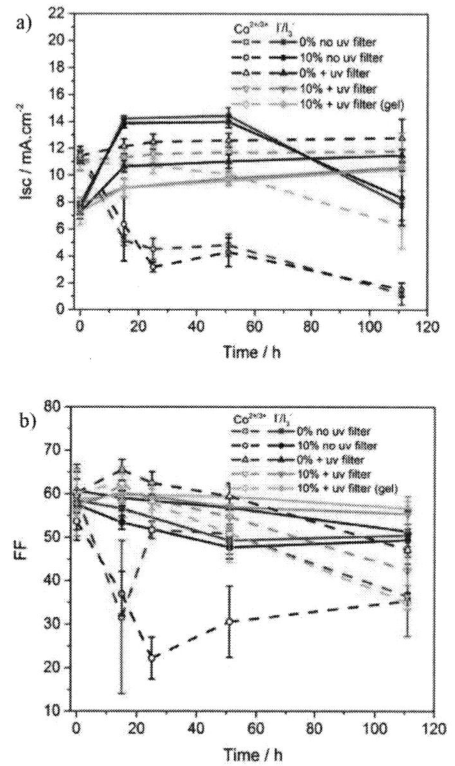

Figure 2: Variation of a) I_{SC} and b) fill factor *FF* for DSCs based on iodine and cobalt electrolytes, with Z907 or MK-2 dye, with or without 10% UV absorber and with or without UV cut-off filter under 1 Sun (UV light). The error bars indicate the standard deviation between the same type of cells.

This aging test has indicated that the differences between the groups without UV cut-off filter for each series are small compared to the standard deviations inside their groups. Thereby, it not possible to affirm that 10% of BZP as UV absorber in electrolyte would provide any clear protection from UV light degradation. The same behavior was observed between the groups with UV filter (except for cobalt gel electrolyte). Additionally, the direct comparison between the results for iodine and cobalt cells cannot be made as these cells were prepared with different dyes and cells configurations.

One reason for why UV absorbers do not appear to have a significant effect could be due to the typical illumination direction: In this case, the UV light reaches the phonoanode first instead of the electrolyte. Even though, photoanode and electrolyte are intermixed, it can be that the absorbers in the electrolyte cannot provide sufficient protection and thereby the TiO_2 could be activated before the UV light gets absorbed by the protection compound in the electrolyte. With reverse illumination, such as used with metal based photoanodes, the additive might have more significant effect.

For cobalt cells, a change in the color of the photoanodes sensitized with MK2 dye was observed only when in the absence of the UV cut-off filter. This behavior indicates dye degradation under UV light, even with BZP. However, even using the UV cut-off filter the cobalt cells resulted in a decrease of the photovoltaic parameters. These results suggest that others kind of degradation processes in addition to the dye degradation could be occurring inside the cobalt cells.

In contrast to the cobalt cell, the iodine cells under the UV cut-off filter do not only retain stable performance but actually improve during the aging test. There was not any changes in dye color, with or without the UV cut-off filter. It is important to remember that the dye is ruthenium based in this case, instead of organic dye for cobalt cells. The main cause of decrease of efficiency for iodine cells without UV cut-off filter is related to the electrolyte bleaching that was detected even by the eye and is typical also in literature [1].

3.2 Transmittance spectroscopy

Figs. 3 and 4 show the transmittance spectrum for cobalt and iodine based cells, respectively.

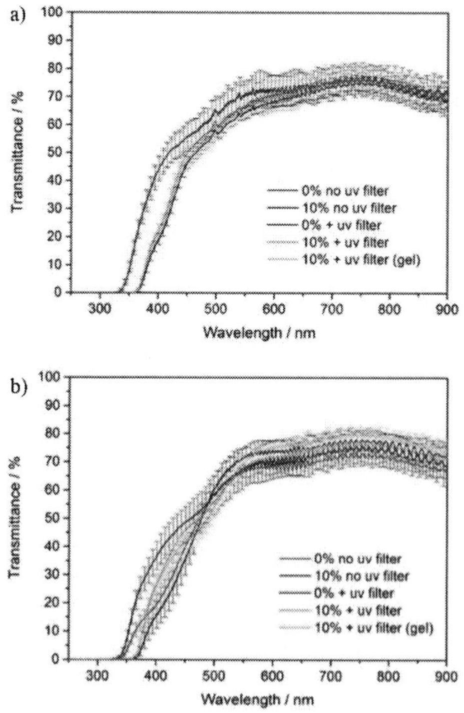

Figure 3: Transmittance spectra for cobalt based electrolyte into DSC, with or without UV absorber and with or without UV cut-off filter. The measurements were made a) before and b) after aging test. The error bars indicate the standard deviation between the same type of cells.

When a compound absorbs at the certain wavelength, the result in transmittance spectrum is a decrease in the observed transmission. For dye solar cells, the transmittance is related to the electrolyte components at shorter wavelengths.

Before the aging, the cobalt electrolyte exhibited almost clear color compared to the yellowish iodine electrolyte as could see in Figure 3-a and Figure 4-a. The iodine electrolyte absorbs most at less than 400 nm while the cobalt electrolyte absorbance is less than 340 nm. However, when the 10% BZP is present the cobalt electrolyte absorb in higher wavelength, approximately until 360 nm, indicating that BZP is absorbing the UV illumination in that range.

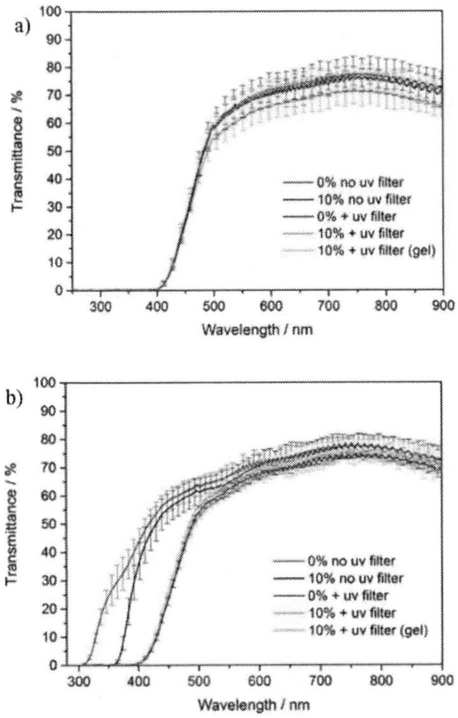

Figure 4: Transmittance spectra for iodine based electrolyte into DSC, with or without UV absorber and with or without UV cut-off filter. The measurements were made a) before and b) after aging test. The error bars indicate the standard deviation between the same type of cells.

After the aging test, some changes could be seen in both electrolytes: The only ones that did not show any change were the cells with iodine electrolyte under UV cut-off filter. Instead, for iodine electrolyte with no BZP and no UV cut-off filter is electrolyte bleaching observed. The bleaching is related with the loss of iodide mediator under UV light irradiation [10]. On the other hand, the electrolyte containing BZP shows absorption in higher wavelength (around 360 nm) which could be related to the BZP absorption.

The differences after aging for cobalt electrolytes are not as clear as for iodine electrolyte but are changes in the spectra for the cells that were aged without UV cut-off filter as well as for the gel electrolyte. These spectrum differences could be related with some electrolyte degradation, generation of a new compound with other wavelength absorption or even dye desorption.

3.3 Incident photon to collected electron efficiency

Figs. 5 and 6 show the IPCE curve for cobalt and iodine based cells, respectively.

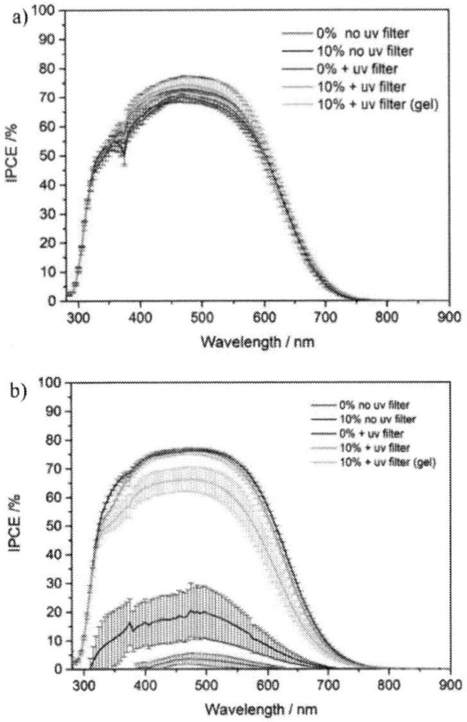

Figure 5: IPCE a) before and b) after the aging test for DSCs based on cobalt electrolytes, with MK-2 dye, with or without 10% UV absorber and with or without UV cut-off filter. The error bars indicate the standard deviation between the same type of cells.

Initially, the groups not differ from each other. For liquid cobalt electrolyte based cells and after the aging test, the original peaks seem to have become wider at range from 300-400 nm for cells under the UV cut-off filter. The effect is more clear without 10% BZP but visible also with BZP. The peak heights stayed quite constant during the aging test. The cobalt cells with gel electrolyte did not show widening of the curve and had also a decrease in IPCE. On other hand, for the cobalt cells without UV cut-off filter the degradation is much more severe: the dye degradation decreases the IPCE values clearly In these groups, there is not enough working dye to produce the same photocurrent as in the initial analysis even with the low light intensity that was used in IPCE without bias light measurement.

A similar behavior was observed in IPCE for iodine based cells except that for these cells there is no dye degradation. Before the aging test, the IPCE were similar but after the aging test is possible to notice an increase in IPCE values. Most increase is observed in the cells that are not protected by the UV cut-off filter. The increased IPCE values suggest that the dye has not suffered in the aging.

At the first sight, the improvement of IPCE for iodine based cells without UV cut-off filter protection was expected to give an increase of I_{SC} of cells, similarly to the cells protected by the UV cut-off filter. However, we found that these cells had lower photocurrent. The increased IPCE behavior is possibly caused by the reduction of iodide mediator in cells, which was observed earlier in transmittance analysis of iodine electrolyte, decreasing the recombination at the photoanode/ electrolyte interface. For the cells without UV cut-off filter, the loss of redox couples (bleaching) has probably progressed so much that the cell currents are limited by that at 1 Sun (high ligh intensity) but not yet at IPCE without bias light measurement (low light intensity).

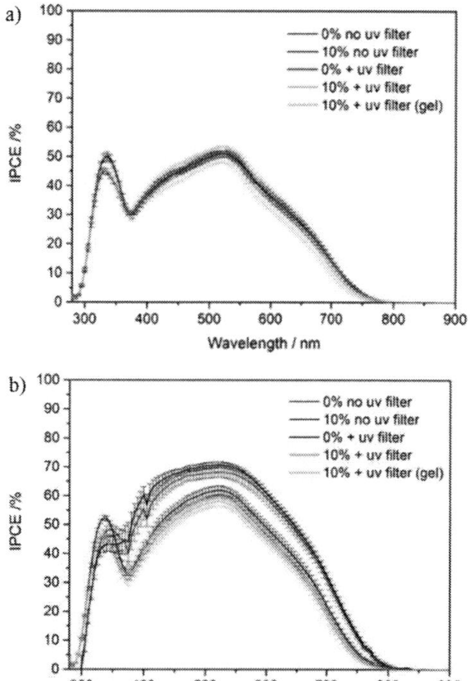

Figure 6: IPCE (a) before and (b) after the aging test for DSCs based on iodine electrolytes, with Z907 dye, with or without 10% UV absorber and with or without UV cut-off filter. The error bars indicate the standard deviation between the same type of cells.

3.4 Electrochemical impedance spectroscopy

Figs. 7 and 8 show the results of fitting the equivalent circuit to the EIS spectra. The EIS analysis was performed at V_{OC} under 1 Sun illumination before and after the aging test for iodine and cobalt based cells, respectively. R_S is the series resistance, R_{CE} is the resistance at counter electrode/electrolyte interface and R_D is the diffusion resistance in the electrolyte at the counter electrode.

R_S has remained stable during the aging test for most cells in both series, indicating a good stability of cell contacts. On the other hand, R_D have increased in both series, mainly for DSCs without UV filter, indicating electrolyte degradation [11]. Initially, the cobalt complexes suffer from higher diffusion resistance when compared to iodine electrolyte mainly due to the size of complexes that are relatively bulky [12]. It should be noticed that the increase of R_{CE} is most pronounced when no UV filter is used.

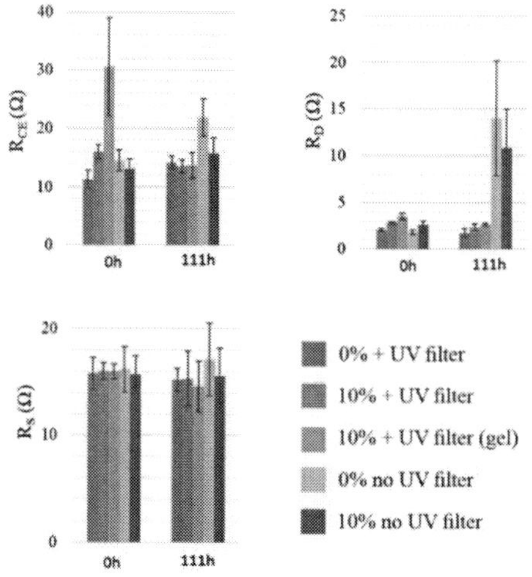

Figure 7: Results of EIS performed at V_{OC} under 1 Sun illumination for iodine based cells. The error bars indicate the standard deviation between the same type of cells.

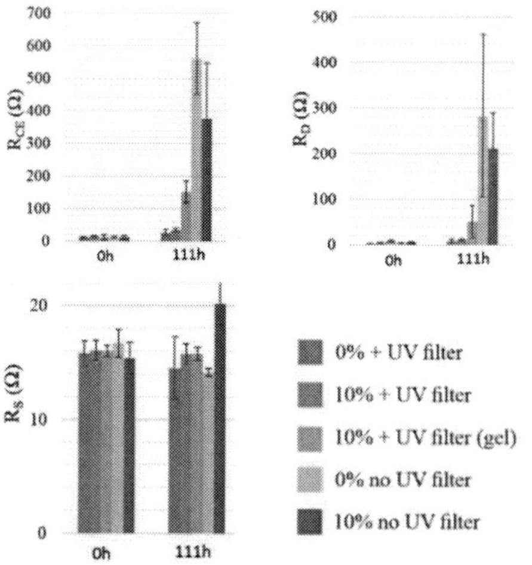

Figure 8: Results of EIS performed at V_{OC} under 1 Sun illumination for cobalt based cells. The error bars indicate the standard deviation between the same type of cells.

R_{CE} has increased more substantially for cobalt based cells than iodine cells, indicating that the interface between the electrolyte and the counter electrode has degraded [9]. While changes at this interface are typically related to changes at the counter electrode catalyst layer, it could be that changes in the electrolyte, namely loss of charge carriers are causing also the increase in R_{CE}.

4 CONCLUSIONS

Two approaches to prevent a major degradation factor, UV light, were tested: UV filter on top of the cell and UV absorber mixed in the electrolyte. Both ways have been suggested in the literature, but here the superiority of the UV filters in comparison to UV absorber was shown. UV cut-off filters are shown to slow down significant the DSCs degradation process both for iodine and cobalt cells. For iodine cells, the filter resulted in stable performance under a short-term aging test while the cobalt cells still degraded. However, the UV absorber used did not give the same protection for DSCs without UV cut-off filter or provide clear extra protection for DSCs with UV cut-off filter as was hypothesized. The situation might be different for back-illuminated cells in which the incoming light would pass through the electrolyte before reaching the photoelectrode.

Another important issue is related to the organic dye degradation under UV light. For the cobalt cells without UV filter and sensitized with MK2 dye, dye discoloration was observed even with the UV additive, indicating that the additive or the filter cannot prevent dye/electrolyte degradation in this case. The ruthenium based dye used in the iodine cells remained very likely stable during all aging test, according to transmittance and IPCE analysis.

5 ACKNOWLEDGMENT

The authors acknowledge for financial support from the Project CNPq/AKA (490241/2012-3), Academy of Finland (SOLID 271081), FAPESP, Aalto University and UNICAMP.

6 REFERENCES

[1] S. Mastroianni, I. Asghar, K. Miettunen, J. Halme, A. Lanuti, T. M. Brown, P. Lund, Phys. Chem. Chem. Phys., 16 (2014) 6092.
[2] M. I. Asghar, K. Miettunen, J. Halme, P. Vahermaa, M. Toivola, K. Aitola, P. Lund. Energy Environ. Sci., 3 (2010) 418.
[3] J. Dewalque, N. D. Nguyen, P. Colson, N. Krins, R. Cloots, and C. Henrist, Electrochim. Acta, 115 (2014) 478.
[4] M. Carnie, T. Watson, D. Worsley, Int. J. Photoenergy, (2012).
[5] B. Mahltig, H. Böttcher, K. Rauch, U. Dieckmann, R. Nitsche, and T. Fritz, Thin Solid Films, 485 (2005) 108.
[6] K.-F. Chen , C.-H. Liu, C.-K. Hsieh, C.-L. Lin, H.-K. Huang, C.-H. Tsai, F.-R. Chen, J. Power Sources, 247 (2014) 939.
[7] W. Xiang, W. Huang, U. Bach, and L. Spiccia, Chem. Commun., 49 (2013) 8997.
[8] J. Halme, P. Vahermaa, K. Miettunen P. Lund, Adv. Mater., 22 (2010) E210.
[9] S. M. Feldt, E. A. Gibson, E. Gabrielsson, and L. Sun, J. Am. Chem. Soc., 13 (2010) 16714.
[10] S. Nakajima, R. Katoh, Chemical Physics Letters, 619 (2015) 36.
[11] A. Tiihonen, K. Miettunen, S. Rendon, D. Mavrynsky, J. Halme, R. Leino, and P. Lund, J. Electrochem. Soc., 162 (2015) H661.
[12] K. Miettunen, S. Jouttijärvi, R. Jiang, T. Saukkonen, J. Romu, J. Halme, and P. Lund, J. Electrochem. Soc., 161 (2014) H138.

HIGHLY TRANSPARENT CELLULOSE SUBSTRATE FOR DYE SENSITIZED SOLAR CELLS

Merve Ozkan [a]*, Syed Ghufran Hashmi [b], Maryam Borghei [a], Armi Tiihonen [b], Killian Lobato [c], António F. da Cunha [d], Orlando J. Rojas [a], Peter. D. Lund [b], Jouni Paltakari [a]

[a] Department of Forest Products Technology, School of Chemical Technology, Aalto University, Finland
[b] New Energy Technologies Group, Department of Applied Physics, Aalto University, Finland
[c] Instituto Dom Luiz, Faculdade de Ciências, Universidade de Lisboa, Portugal
[d] I3N and Department of Physics, University of Aveiro, Portugal
* merve.ozkan@aalto.fi, merve.ozkan85@gmail.com

ABSTRACT: Current electronic devices are fabricated mostly on glass or plastic films which are often hard to recycle. In this study, 2,2,6,6-tetramethylpiperidine-1-oxyl radical (TEMPO) oxidized nano-cellulose (TOCN) film was explored as a green substrate candidate for next-generation electronic applications and its optical, surface and mechanical properties are presented in this report. Its prospective integration in the future dye-sensitized solar cells (DSSCs) was investigated by testing as one of the electrodes in a counter electrode-counter electrode configuration (CE-CE). The TOCN film was able to undergo the device sealing process at 220°C and 140 °C without any change in its color or dimension. The results highlight the potential of environmentally friendly and room temperature processed nano-fiber based films as an alternative substrate for bifacial devices.

Keywords: Substrates, optical properties, TEMPO oxidized cellulose, dye-sensitized solar cells

1 INTRODUCTION

Cellulosic materials are abundant, strong and light-weight. These have recently been received much attention as a substrate in electronic applications [1-2]. Conventional cellulose based products, mostly consisting of 20–50 µm (diameter) size fibers, are opaque, permeable to solvents and have a rough surface contour with low thermal stability. Therefore, their integration in electronic and electrochemical devices has been a challenge particularly in solvent based devices manufactured at high temperatures [3].

The various cellulose processing techniques provide an opportunity to improve its optical, surface and mechanical properties and pave the way to more sustainable and biodegradable products. For example, the nano-fibrillated cellulose (NFC) production [4] and its surface modification with selective 2,2,6,6-tetramethylpiperidine-1-oxyl radical (TEMPO) treatment exchanging the hydroxyl group of the cellulose to carboxyl based group [5] were two breakthroughs in forest products technology. Due to the better dispersion achieved by nano-fibril production and functional group alteration, the film formation was improved and high transparency values, up to 90% transmittance of the incident light for 20 µm films, with acceptable strength values (Young's Modulus value of close to 7 GPa) were reported [6]. Recently numerous electronic devices were successfully integrated on the transparent nano-fiber based films such as, transistors [7], antennas [8] and organic solar cells [9] however, the solvent permeation has remained as the key challenge in the application of TEMPO-oxidized nano-cellulose (TOCN) films to the devices which include liquid components.

We present here cellulose substrates (TOCN films) with high bendability, light transmittance and strength along with low roughness. Slow drying process with two level humidity conditions improved the properties of the TOCN films owing to the better film formation. The physical and chemical properties of this substrate are first investigated through different characterization techniques, discussed in the later sections of this report. The integration in dye-sensitized solar cells (DSSCs) [10] as a counter electrode (CE) was tested by fabricating CE-CE electrochemical cells which exhibited favorable properties such as low charge transfer resistance (R_{CT}) essentially required to produce high performance DSSCs.

2 EXPERIMENTAL SECTION

2.1 Materials

All reagents were purchased from Sigma-Aldrich and used without further purification. EL-HSE high stability electrolyte was obtained from Dyesol.

2.2 Preparation of cellulose films

Hydroxyl groups of nanofibrillated cellulose from birch kraft fibers were selectively exchanged with carboxylate groups by TEMPO catalyst at room temperature and pH 10 [11]. After the mechanical disintegration in a micro fluidizer (Microfluidics M-110EH-30, Microfluidics Int., USA), 0.4 wt % aqueous dispersion was obtained after 24 h of magnetic stirring followed by a mild centrifuge at 4000g for 10 min in order to obtain homogenous dispersion without any aggregates. Casting solution was dispensed onto poly (styrene) Petri dishes and the water was evaporated in 2 steps, i.e. first at ordinary room condition (23°C-35% RH) until the cast solution lost half of its water content and then in a climate room (23 °C-66% RH). A post-heating treatment was carried out at 35 °C for 5 min in order to detach the TOCN film without any damage. The TOCN films were initially sputtered with 300 nm ITO layer (ITOCN film) and then 10 nm platinum layer (PITOCN film).

2.3 Cell Preparation

Conventional CEs were prepared by sputtering 10 nm platinum on FTO-coated glass. The platinized FTO-coated glass-platinized FTO-coated glass (G_{Pt}-G_{Pt}), and the platinized FTO-coated glass-PITOCN (G_{Pt}-PITOCN) CEs were sealed with 25 µm thick thermoplastic foil (Surlyn) on a hot plate at 140°C. The electrolyte was then injected via drilled holes. 25 μm thick cover foil (Surlyn) and thin glass cover were used to seal both type of devices by using hot press at 220°C.

2.4 Measurements

An optical goniometer (CAM 200 from KSV instruments), which was equipped with a camera and dispenser system, was used to capture a sessile drop behavior on TOCN and ITOCN surfaces and the records were analyzed by using a built-in image analysis software to determine the contact angle of the drop based on Young and Laplace equation. Perkin Elmer lambda 950 UV-Vis spectrophotometer was utilized to measure the transmittance of FTO-coated commercial glass, TOCN and ITOCN films. Electrochemical impedance spectroscopy (EIS) data was recorded at open circuit voltage conditions using Zahner-Elektrik IM6 potentiostat and the data was fitted by Zview2 software. The measurement frequency range was from 100 mHz to 100 KHz. The surface properties of flexible films were observed with a scanning electron microscope (Zeiss Sigma VP scanning electron microscope) at 1 kV acceleration voltage, and with a Multimode 8 atomic force microscope (AFM) equipped with a NanoScope V controller (Bruker Corporation, Billerica, MA), operating in tapping mode. The obtained images from AFM were processed by utilizing NanoScope 8.15 software (Bruker). No image correction (except flattening) was employed. The mechanical tests were carried out using a vertical tensile tester, MTS 400/M (MTS Systems, USA), in accordance with ISO 527-3 (Specimen type 5) loading 50 N at 1 mm/min cross-head speed.

3 RESULTS AND DISCUSSION

3.1 Surface Properties

A well-dispersed nano-fiber network along with a smooth surface promote the applicability of the coatings [6]. Investigated with AFM and SEM, TOCN (Figure 1 a) and ITOCN substrates exhibited homogenous topology with the surface RMS roughness values of 8.8±1 nm and 20±1 nm, respectively (Figure 1 b-d and Figure 2). The high water contact angle of 118° (Figure 1 a inset) was measured on the TOCN film which has even surface with low roughness facilitating the even distribution of ITO conductive coating on TOCN surface.

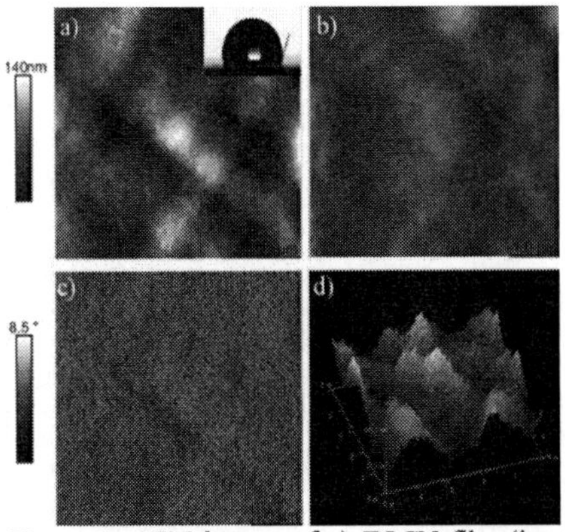

Figure 1: AFM images of a) TOCN film (inset: water contact angle on TOCN), b-d) ITOCN film (height, phase and 3D modes, respectively).

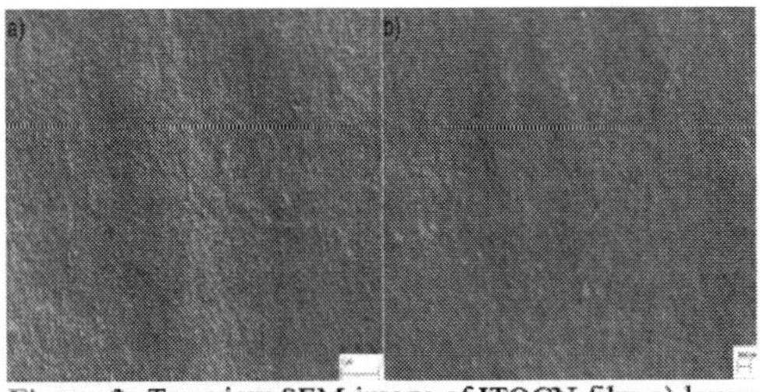

Figure 2: Top view SEM image of ITOCN film a) lower magnification (Scale bar: 1μm), b) higher magnification (Scale bar: 200nm).

The electrolyte absorbance resistance of the substrates was tested with a modified Cobb test in accordance with the ISO 535 and EN 20535 standards as explained in our previous study [13]. Five identical TOCN substrates were exposed to the 3-methoxypropionitrile solvent-based electrolyte for 20 minutes. No alteration in the weight of the TOCN films was detected. The 4 μL electrolyte sessile drop on TOCN surface was monitored in order to further elucidate the interaction of the TOCN surface and the electrolyte. The drop exhibited a stable contact angle (30°) without any difference in its height and base line during the goniometric measurements.

3.2 Transparency

The key requirement for the optical and/or bifacial electronic devices is enhancing the transmittance of the incident light to the photoactive layers, therefore, transparent FTO-coated commercial glass has been widely used in research labs [12]. The TOCN film and ITOCN film with 300 nm ITO coating (Figure 3 a-c) exhibited transmittance values of 86% and 76%, respectively, while FTO-coated glass transmitted 82% of the incident light at 550 nm (Figure 3 d). The transmittance of TOCN film was superior to FTO-coated glass between 570 and 660 nm wavelength values in the optical spectra reaching 85% transmittance at 600 nm whereas the FTO-coated glass had a transmittance value of 80-81% in this region.

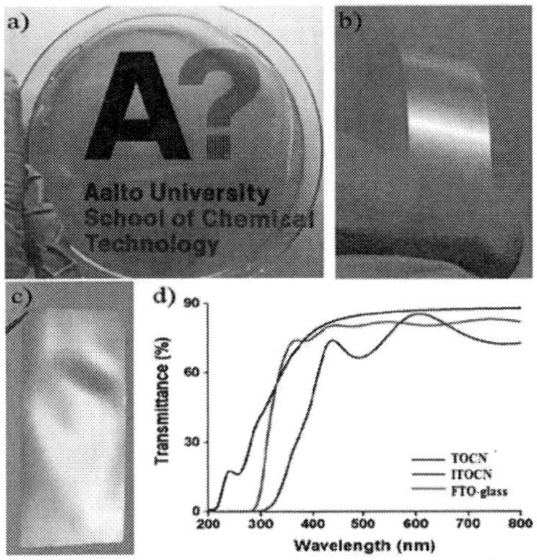

Figure 3: a) TOCN film on Petri dish, b) TOCN stripe, c) ITOCN stripe, d) transmittance of FTO-coated glass and TOCN and ITOCN films.

3.3 Mechanical Properties

The film-forming properties and the durability of TOCN films were characterized by the mechanical parameters recorded during the tensile tests and the bending tests. The average Young's modulus of 8 dog-bone-shaped specimens consisting of 70 g/m^2 grammage was recorded as 9.5±1 GPa and after several bending cycles (2 cm bending radius), no specimen deformation was observed. The physical properties of the TOCN films prepared in this study and the previously reported TEMPO oxidized nano-cellulose films are presented in Table I.

Table I: Physical properties of the TOCN films prepared for this study and reported TOCN films in literature

	TOCN	Reported TOCN
Young's modulus (GPa)	9.5±1	6-15 [14, 15]
Density (g/cm^3)	1.6	0.5-1.4 [16, 14]
Thickness (μm)	44±1	20-50 [6, 17]

3.4 Electrical Impedance Spectra (EIS)

The characteristics of the cellulose substrate presented in earlier sections motivated us to test its integration in the future dye-sensitized solar cells. Before that we fabricated CE-CE cells and investigated the electrochemical impedance characteristics of both types of CE-CE cells (Figure 4). Figure 4 a represents the impedance spectrum of a CE-CE symmetrical cell that was constructed by platinizing two identical FTO-glass substrates whereas in case of Figure 4 b, FTO Glass CE was replaced with the ITO and Pt coated cellulose substrate (PITOCN) and then its electrochemical impedance spectrum was measured. The case for FTO platinized CEs was straight-forward as expected, the CE-CE cells exhibited two semicircles in the Nyquist plot: one appeared at high frequency (1-10 kHz) which corresponds to charge transfer resistance (R_{CT} = 1.9 Ω) whereas the second semicircle appeared at low frequency (1-10 Hz) and relates to the diffusion resistance (R_D = 0.7 Ω) (Table II) [18].

Replacing one platinized FTO with PITOCN substrate did not change any peak position due to the same Pt catalyst layer which was an expected result. The higher series resistance (R_S = 21.8 Ω) was also expected due to higher sheet resistance of PITOCN (R_{SH} = 60 Ω/Sq) which was effectively suppressed after the platinization.

Although the R_{CT} was deduced from the impedance spectra of each type of CE-CE cells, a simpler method can also be applied to calculate the more accurate value. It is to simply subtract the R_{CT} value of the platinized FTO-glass substrate based CE-CE cell since the same platinized FTO-glass substrate was used in platinized FTO-glass and PITOCN based CE-CE cells. Hence with this assumption, we calculated R_{CT} = 3.5 Ω for PITOCN substrate which is comparable to FTO-Pt CE.

These results suggest that the cellulose substrate presented in this report can be potentially integrated as a counter electrode in DSSC. Hence we aim to fabricate DSSCs with this interesting candidate substrate in our future work.

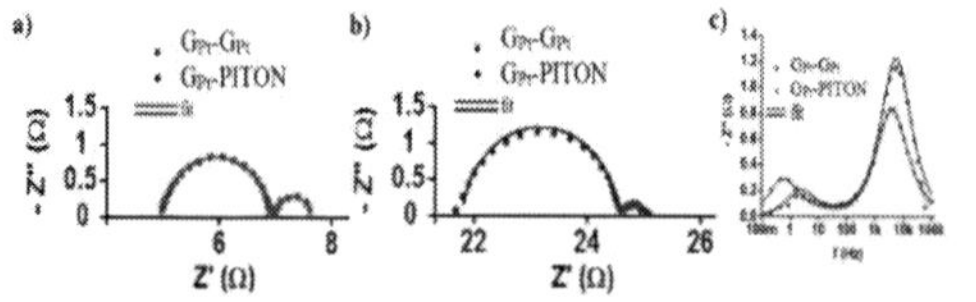

Figure 4: EIS platinized glass- platinized glass (G_{Pt}-G_{Pt}) and platinized glass-PITOCN (G_{Pt}-PITOCN) cells. a) Nyquist plot, b) imaginary impedance (Z") of both cells as a function of frequency.

Table II: EIS parameter of CE-CE cells. The values are calculated for single electrode.

Cell	R_S (Ω)	R_{CT} (Ω)	R_D (Ω)
G_{Pt}-PITOCN	21.8	2.7	0.5
G_{Pt}-G_{Pt}	4.9	1.9	0.7

4 CONCLUSIONS

Flexible, transparent, strong and thermally stable cellulose film with homogenous topology was presented as a potential substrate for electronic applications. Its future integration in the DSSCs was tested by fabricating CE-CE cells which showed comparable R_{CT} to traditional platinized FTO-glass CE. These results highlight the potential application of environmentally friendly and room temperature processed nano-fiber based films as an alternative substrate for DSSC devices.

5 ACKNOWLEDGEMENTS

Merve Ozkan thanks Environomical Pathways for Sustainable Energy Services (SELECT+) doctoral programme for PhD scholarship. Ghufran Hashmi thanks Academy of Finland for Post-Doctoral Research Fellowship (Grant Number: 287641). Maryam Borghei acknowledges Academy of Finland Center of Excellence and the project funding "HYBER". António F. da Cunha acknowledges the financial support of the Portuguese Science and Technology Foundation (FCT) through grant UID/CTM/50025/2013. Armi Tiihonen thanks Academy of Finland for the funding (SOLID project, 271081). This work made use of the Aalto University Nanomicroscopy Center (Aalto-NMC) premises.

6 REFERENCES

[1] H. Olsson, D. O. Carlsson, and M. Strømme, J. Mater Sci., 47, (2012), 5317.
[2] P. Cataldi, I. S. Bayer, F. Bonaccorso, V. Pellegrini, A. Athanassiou, and R. Cingolani, Adv. Electron. Mater., 1, (2015), 1500224.
[3] H. Zhu, Z. Fang, C. Preston, Y. Li, and L. Hu, Energy Environ. Sci., 7, (2015), 269.
[4] B. H. Yano, J. Sugiyama, A. N. Nakagaito, M. Nogi, T. Matsuura, M. Hikita, and K. Handa, Adv. Materials, 17, (2015), 2, 153.
[5] T. Saito, S. Kimura, Y. Nishiyama, and A. Isogai, Biomacromolecules, 8, (2007), 2485.
[6] H. Fukuzumi, T. Saito, T. Iwata, Y. Kumamoto, and A. Isogai, Biomacromolecules, 10, (2009), 162.
[7] Z. Fang, H. Zhu, W. Bao, C. Preston, Z. Liu, and J. Dai, Energy Environ. Sci., 7, (2014), 3313.

[8] M. Nogi, N. Komoda, K. Otsuka, and K. Suganuma, Nanoscale, 5, (2013), 10, 4395.
[9] L. Hu, G. Zheng, J. Yao, N. Liu, B. Weil, M. Eskilsson, E. Karabulut, Z. Ruan, S. Fan, J. T. Bloking, M. D. McGehee, L. Wågberg, and Y. Cui, Energy Environ. Sci., 6, (2013), 2, 513.
[10] S. G. Hashmi, M. Ozkan, J. Halme, K. D. Misic, S. M. Zakeeruddin, J. Paltakari, M. Grätzel, and P. D. Lund, Nano Energy, 17, (2015), 206.
[11] K. D. Misic, A. Puisto, P. Gane, K. Nieminen, M. Alava, J. Paltakari and T. Maloney, Cellulose, 20, (2013), 6, 2847.
[12] S. G. Hashmi, K. Miettunen, T. Peltola, J. Halme, I. Asghar, K. Aitola, M. Toivola, and P. Lund, Renew. Sustain. Energy Rev., 15, (2011), 8, 3717.
[13] S. G. Hashmi, M. Ozkan, J. Halme, J. Paltakari, and P. D. Lund, Nano Energy, 9, (2014), 212.
[14] A. Isogai, T. Saito, and H. Fukuzumi, Nanoscale, 3, (2011), 71.
[15] Y. Qing, R. Sabo, Y. Wu, J. Y. Zhu, and Z. Cai, Cellulose, 22, (2015) 1091.
[16] X. Song and M. A. Hubbe, Carbohydr. Polym., 99, (2014), 617.
[17] H. Sehaqui, N. Ezekiel Mushi, S. Morimune, M. Salajkova, T. Nishino, and L. A. Berglund, ACS Appl. Mater. Interfaces, 4, (2012), 1043.
[18] S. G. Hashmi, J. Halme, T. Saukkonen, E.-L. Rautama, and P. Lund, Phys. Chem. Chem. Phys., 15, (2013), 17689.

REQUIREMENT OF DURABILITY TEST FOR ORGANIC PHOTOVOLTAIC AND DYE-SENSITIZED SOLAR CELL

Shu-Tsung Hsu*, Yean-San Long, Teng-Chun Wu, Hsin-Hsin Hsieh
Center for Measurement Standards, Industrial Technology Research Institute
Rm. 213, Bldg. 3, 321, Sec. 2, Kuang Fu rd., Hsin-Chu 30011, Taiwan, R. O. C.
*Corresponding author, andersonhsu@itri.org.tw, Phone: + (886) 3 5743864, Fax: + (886) 3 5725540

ABSTRACT: Characterizations of organic photovoltaic (OPV) and dye-sensitized solar cell (DSSC) are important in order to determine the reliable measurements of cell efficiencies and inspect the product's defects in early stage. This paper aims to improve the existing durability tests towards a real operating condition for OPV/DSSC. This work developed a durability test according to parallel tests such as thermal cycle (TC), damp heat (DH), light soaking (LS) and combined test (TC50+DH). Each test condition has the stress specification, which differs with IEC 61646 designed for thin-film device. Furthermore, results were applied for SEMI Doc. 5598 [1] to develop an evaluation principle of OPV/DSSC for PV industries to follow up, and harmonized the test guidelines of current-voltage (SEMI PV57 [2]) and spectrum response (SEMI PV69 [3]). Guidance on the choice of the suitable test conditions of durability test in the detailed specification is given. Consequently, this standardized test method for OPV/DSSC can help industry to get reliable performances of emerging PV, and influence the relevant reliability aspect in a long-term perspective.
Keywords: OPV, DSSC, Durability

1 INTRODUCTION

Emerging PV includes organic photovoltaic (OPV) and dye-sensitized solar cell (DSSC). The operation principle of OPV/DSSC is using layers of organic molecules subject to lighting after excitation electronic then pass to the inorganic/organic layer of the wide energy gap nano-layer and voltage. The major differences between OPV/DSSC and p-n junction solar cells are spectrum, absorption range, photoelectric conversion response time and standard test condition (AM1.5G, 25 °C, 1000 Wm^{-2}, STC), etc.

OPV/DSSC is a high-potential product used for energy harvesting, especially in the context of indoor illumination applications. It is important to enhance the quality and reliability of such products, and to overcome the measurement error caused by capacity effect like the hysteresis problem. Industry needs guidelines of test specifications in order to get the reliable measurement. Therefore, it is necessary to establish I-V (current – voltage) and SR (spectrum or spectral response) standardized methods of measurement for OPV/DSSC, and based on this basis, to develop the durability test method is applicable.

SEMI OPV and DSSC TF in Taiwan have developed the detailed testing documents and international standards for OPV/DSSC since 2013. The key members include ITRI, EVERLIGHT, King Design, TDP, INER, NDHU, NCKU, MUST and NCU, etc. ITRI coordinated this TF to set up a series of experiments [4, 5, 12] and discussions, which focused on the evaluation methods of I-V, SR and durability. I-V evaluation was applied to SEMI Doc. 5598 and released as SEMI PV57 [1] by voting in 2014. To be differential with STC used for c-Si product outdoors, SEMI PV57 provides two test reference conditions for low lighting indoors. One is very low irradiance condition (AM1.5G, 25 °C, 60 Wm^{-2}, VLIC), and the other is ultra-low irradiance condition (AM1.5G, 25 °C, 1 Wm^{-2}, ULIC). SR evaluation was applied to SEMI Doc. 5647 and released as SEMI PV69 [2] by voting in 2015.

Therefore, based on these two released standards, this work aims to develop a durability test method to evaluate OPV/DSSC device characteristics, especially in subtropical warm and humidity environment. Durability (or qualification) test can inspect product's defects (e.g., dye leakage, electrode peeling and pin hole, etc.) or weakness in early life stage (see Fig. 1), and provide other benefits such as quickly discover design & process flaw, evaluate & improve design margins, reduce development time & cost and eliminate design problems before release to manufacturing, etc. Results also applied to SEMI Doc. 5598 for SEMI voting globally in May 2016, and aimed to develop an evaluation principle and guideline for industries to follow up. Test procedure include light soaking, thermal cycle, damp heat and combined test, with some specific stress condition. All performance evaluations in this durability test need to follow up the guidelines of I-V [1] and SR [2]. Consequently, OPV/DSSC device makers and buyers, or any other party interested, can thus have the common testing standards to refer to when desire.

Figure 1: Degradation photos (e.g., leakage) of DSSC module after (a) thermal cycle test (b) damp heat test

2 TESTING SETUP SYSTEM

The follows (1)-(6) are the main apparatus for test system for this work.

(1) Visual inspection: each DUT to determine the presence or absence of anomalies or defects. Optical magnification is not required. Apparatus for carefully inspect each test sample under an illumination of not less than 1,000 lx, for the visual check in accordance with IEC 61646 C10.1 [6].
(2) I-V measurement: to determine the maximum power of the DUT before and after the various aging tests.

Repeatability of the test is the most important factor. This apparatus is for measuring I-V curves in accordance with SEMI PV57.

(3) Light soaking (LS): to stabilize the electrical characteristics of OPV/DSSC by simulated solar irradiation such as a class CCC solar simulator, in accordance with the IEC 60904-9 [9].

(4) Thermal cycle (TC): to determine the ability of the DUT to withstand thermal mismatch, fatigue and other stresses caused by repeated changes of temperature.

(5) Damp heat (DH): to determine the ability of the DUT to withstand the effects of long-term penetration of humidity.

(6) Reference device (RD): RD is calibrated indoors using simulated sunlight or outdoors in natural sunlight by reference to the same desired reference spectral irradiance distribution. Report needs to commit traceability including measurement of I-V, SR and IPCE, which are determined in accordance with IEC 60904-2 [10].

3 TEST SEQUENCE AND TEST ITEMS

By referring IEC 61646 to guarantee thin-film module or PV system installed outdoor over 20 years at least, OPV/DSSC needs the different aging test condition due to consumer life time only be few years and application scenario even happened indoor with low lighting. Therefore, three aging tests (TC, LS and DH) and one combined test (TC50+DH) according to specific stress condition and test time (see Table I) is designed for durability test as following.

3.1 Light soaking test (LS)

A LS test can stabilize the electrical characteristics of OPV/DSSC by natural sunlight or simulated solar irradiation or artificial lighting source. In order to simulate the life time of OPV/DSSC under T5 lighting with irradiance 1 Wm^{-2} and room temperature 60 ˚C, stress condition (1000 Wm^{-2}, 60 ˚C) of LS refer to lighting exposure model [7, 8]. e.g., for T5 device, 60 ˚C, 1 Wm^{-2}, intensity ratio = 1/1000 *100 % = 0.1 %, lamp power 80 % (assumption), exposure ratio = 100 – 74.8 = 25.2. Then the life time is estimated as 3/10 years while setting minimum stress time is 1043/3476 hours (see Table I). Therefore, if PV device merely tests with irrandance less than 1000 Wm^{-2} due to some specific materials or device only used for indoor application, then LS needs to enhance the above stress time in order to meet the same life year simulation.

3.2 Thermal cycle test (TC)

TC test can determine the ability of OPV/DSSC to withstand thermal mismatch, fatigue and other stresses caused by repeated changes of temperature. By using Arrhenius model [11] with single-degradation factor (temperature), can simulate the life time of OPV/DSSC during temperature (-10 ˚C ~ 60 ˚C) is 3/10 years according to stress time TC50/TC150 (see Table I), which differ with the stress condition of IEC 61646 C 10.11 (-40 ˚C ~ 90 ˚C and TC200).

3.3 Damp heat test (DH)

DH test can determine the ability of OPV/DSSC to withstand the effects of long-term penetration of humidity. In order to simulate the life time such as keeping OPV/DSSC indoor with 65 ˚C/ 65 %RH, which differ with stress condition (85 ˚C/ 85 %RH/ 1000 hr) of IEC 61646 C10.13. The simulation life time, estimated by inverse power model [11] with multi- degradation factors (temperature, humidity), is 3/10 years while setting minimum stress time is 600/1800 hours (see Table I) with 95% level of confidence (LOC).

Table I: Stress condition and test time (with 1 hr pre- LS)

test	TC		DH		LS	
severity	II	I	II	I	II	I
estimated life year	10	3	10	3	10	3
model	Arrhenius Eq.		inverse power		lighting exposure	
temp	-10~60 ℃		65 ± 2 ℃		60 ± 1 ℃	
RH	NA		65 % ± 10 %		NA	
irradiance	NA		NA		1000 Wm^{-2}	
test time	TC 150	TC 50	1800 hr	600 hr	3476 hr	1043 hr
criteria	Δ P_{max} < 5 %					

3.4 Combined test: thermal cycle and damp heat

To determine the product final packaging ability of OPV/DSSC to withstand thermal mismatch, fatigue and other stresses caused by repeated changes of temperature, and the effects of long-term penetration of humidity. Before installing the product at room temperature in the chamber of damp heat, shall be introduced into the chamber with thermal cycle preconditioning repeat change temperature by 50 cycle times (TC50). The following severities are applied. (a) test temperature: 85 ˚C ± 2 ˚C (b) relative humidity: 85 %RH ± 5 %RH (c) test sequence refers to Fig. 2. Final measurements shall operate both visual inspection and I-V measurement before a minimum recovery time of 1 hr.

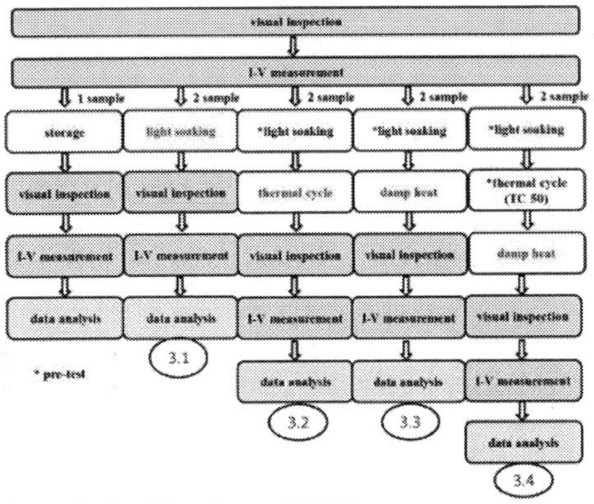

Figure 2: Test flow in parallel [3]

3.5 Test Sequence

In Fig. 2, nine test samples and five-group carried out in the order laid down to the qualification test sequences. Each box refers to the corresponding sub-clause in this test method. Test procedures and severities, including initial and final measurements where necessary. However, with regard to the tests of initial and final measurements, it should be noted that the procedures laid down in SEMI PV57, these tests need be carried out within ± 5 % of the specified irradiance and within ± 2 °C of the specified temperature. Any single test, executed independently of a test sequence, proceed by the initial tests of visual inspection and I-V measurement.

4 DISCUSSIONS

Reliable measurements of I-V [2, 4] and SR [3, 12] are important to evaluate the durability of OPV/DSSC properly. All I-V tests need to follow the same test procedure defined in SEMI PV57, in order to reduce the capacity effect during measurement.

- Final measurements of each aging test before a minimum recovery time of 1 hr, repeat tests include visual inspection and I-V measurement at STC (or standard reporting condition, SRC). Criteria need meet relative performance loss (e.g., P_{max}) less than 5 %. I-V curve in Fig. 3 shows a DSSC sample having leakage phenomenon after the thermal cycle test.
- All estimated stress time in Table I need to follow the demand of LOC for the specific design of product. In general, LOC can keep 95 % for short years, and 80 % ~ 90 % for long years.
- Base on the low-light application indoors, the efficiency of OPV/DSSC module shall be defined as (P_{max} /module area)/input power. The value of input power depends on the choice of irradiation type or lighting source, and keeps 1000 Wm^{-2} in using natural sunlight or simulated solar irradiation.
- However, this work still is relying on the additional field data is available, and the results of accelerated stress tests to predict whether the OPV/DSSC is acceptable for surviving.

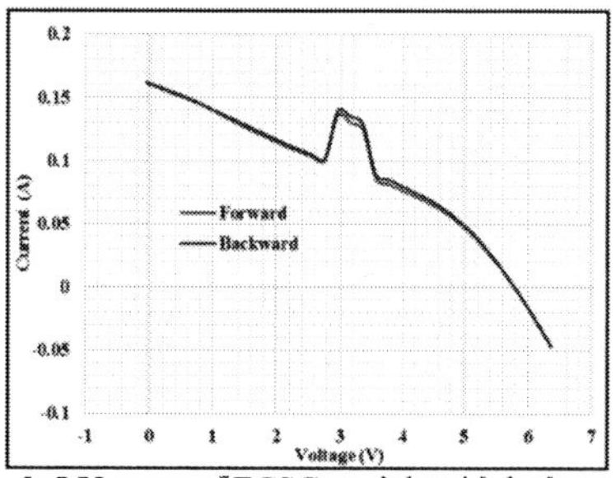

Figure 3: I-V curve of DSSC module with leakage after thermal cycle test

5 CONCLUSIONS

This paper aims to improve the existing durability tests towards a real operating condition for OPV/ DSSC devices, and to broaden the portfolio of very fast durability test beyond the existing standards. This work developed a durability test method to evaluate characteristics for OPV/DSSC device. Test procedures include light soaking test, thermal cycle test, damp heat test and combined test, with some specific stress condition, which differs with IEC 61646. In the meantime, by harmonizing the guidelines of I-V [2] and SR [3], results applied for SEMI Doc. 5598 [1] to develop an evaluation principle of OPV/DSSC for PV industries to follow up. Guidance on the choice of suitable test severity for durability test in the detail specification is also given. Through these standardizations, results can help industry to get reliable performances of emerging PV, and influence the relevant reliability aspect in a long-term perspective.

6 ACKNOWLEDGEMENTS

Authors appreciate SEMI OPV and DSSC TF in Taiwan.

7 REFERENCES

[1] SEMI Doc. 5598, Test Method for Durability of Low Light Intensity Organic Photovoltaic (OPV) and Dye Sensitized Solar Cell (DSSC).

[2] SEMI PV57, Test Method for Current-Voltage (I-V) Performance Measurement of Organic Photovoltaic (OPV) and Dye-sensitized Solar Cell (DSSC).

[3] SEMI PV69, Test Method for Spectrum Response (SR) Measurement of Organic Photovoltaic (OPV) and Dye-sensitized Solar Cell (DSSC).

[4] T. C. Wu, S. T. Hsu, Y. S. Long, New Set-up Procedures and Integrated Measurement System for Organic Photovoltaic (OPV) Module, *International Photovoltaic Science and Engineering Conference (PVSEC23)* (2013).

[5] Y. S. Long, S. T. Hsu, T. C. Wu, Induction of Internal Capacitance Effect in Performance Measurement of OPV (Organic Photovoltaic) Device by RTOSM (Real-Time One-Sweep Method), *Journal of Energy and Power Engineering. 8 (2014) 1059-1066.*

[6] IEC 61646, Thin-Film Terrestrial Photovoltaic (PV) Modules – Design Qualification and Type Approval.

[7] IEC 60068-1, Environmental Testing - Part 1: General and Guidance.

[8] Keith Emery, Measurement and Characterization of Solar Cells and Modules, Chap 16 of Handbook of Photovoltaic Science and Engineering.

[9] IEC 60904-9, Photovoltaic Devices – Part 9: Solar Simulator Performance Requirements.

[10] IEC 60904-2, Photovoltaic Devices – Part 2: Requirements for Reference Solar Devices.

[11] Luis A. Escobar and William Q. Meeker, A Review of Accelerated Test Models, Statistical Science 2006, Vol. 21, No. 4, 552–577.

[12] S. T. Hsu, Y. S. Long, T. C. Wu, Jay Lin, Standardization of Spectrum Response Measurement for OPV and DSSC, *SNEC 10th*, 2016.

A CASE STUDY OF DEVELOPING SEMI STANDARDS FOR ORGANIC PHOTOVOLTAIC AND DYE-SENSITIZED SOLAR CELL IN TAIWAN

Shu-Tsung Hsu*, Yean-San Long, Teng-Chun Wu, Hsin-Hsin Hsieh
Center for Measurement Standards, Industrial Technology Research Institute
Rm. 213, Bldg. 3, 321, Sec. 2, Kuang Fu rd., Hsin-Chu 30011, Taiwan, R. O. C.
*Corresponding author, andersonhsu@itri.org.tw, Phone: + (886) 3 5743864, Fax: + (886) 3 5725540

ABSTRACT: This paper aims to summary the research of pivot SEMI standards about organic photovoltaic (OPV) and dye-sensitized solar cell (DSSC), which are developed by SEMI OPV and DSSC task force (TF) in Taiwan during 2013 to 2016. This work proposes three test methods in detailed specifications for OPV/DSSC product, including current-voltage (I-V), spectrum response (SR) and durability. I-V measurement (SEMI PV57 [1]) is to evaluate product's characteristics without capacity effect for OPV/DSSC. SR evaluation (SEMI PV69 [2]) is to provide a suitable statistic method after solving the slow response time problem for OPV/DSSC, and proposes a guideline to correct performance data measured on an isothermal platform under STC. Based on these two released standards, furthermore, durability test (SEMI Doc. 5598 [3]) for OPV/DSSC can improve the existing aging tests towards a real operating condition, in order to determine its performance accurately and inspect the product's defect or weakness in early stage. Through these standardizations, results can help industry to get reliable performances of OPV/DSSC, and influence the relevant reliability aspect in a long-term perspective.
Keywords: OPV, DSSC, Current – Voltage, Spectrum Response, Durability

1 INTRODUCTION

Emerging PV includes organic photovoltaic (OPV) and dye-sensitized solar cell (DSSC). The operation principle of OPV/DSSC is using layers of organic molecules subject to lighting after excitation electronic then pass to the inorganic/organic layer of the wide energy gap nano-layer and voltage. The major differences between OPV/DSSC and p-n junction solar cells are spectrum, absorption range, photoelectric conversion response time and standard test condition (STC), etc.

OPV/DSSC is a high-potential product used for energy harvesting, especially in the context of indoor illumination applications. It is important to enhance the quality and reliability of such products, and to overcome the measurement error caused by capacity effect like the hysteresis problem. Industry needs test specifications to make the reliable measurement. Therefore, it is necessary to standardize I-V (current – voltage) and SR (spectrum response) test methods for OPV/DSSC.

SEMI OPV and DSSC TF in Taiwan has developed the detailed documents and industry standards for OPV/DSSC since 2013. The key TF members include ITRI, EVERLIGHT, King Design, TDP, INER, NDHU, NCKU, MUST and NCU, etc. ITRI coordinated this TF to set up a series of experiments [4, 5, 9, 10, 11] and discussions, which focused on the performance evaluation method of I-V (current – voltage), SR (spectrum response) and durability for OPV/DSSC. I-V evaluation was applied to SEMI Doc. 5597 and released as SEMI PV57 [1] by voting in 2014. SR evaluation applied to SEMI Doc. 5647 and released as SEMI PV 69 [2] by voting in 2015. Based on these two released standards, SEMI Doc. 5598 [3] aimed to propose the durability test for OPV/DSSC. Results applied for global voting in May 2016 already. Consequently, OPV/DSSC device makers and buyers, or any other party interested, can thus have the common testing standards to refer to when desired.

2 SEMI PV57 [1, 9]

This work presents a new test method, which differs with IEC 60904-1 [8] addressed on single and multi-junction samples, for determining I-V performance of OPV/DSSC, including the indoor scenario in low-level lighting. Results applied for SEMI Doc. 5597 and released as SEMI PV57 by voting in 2014.

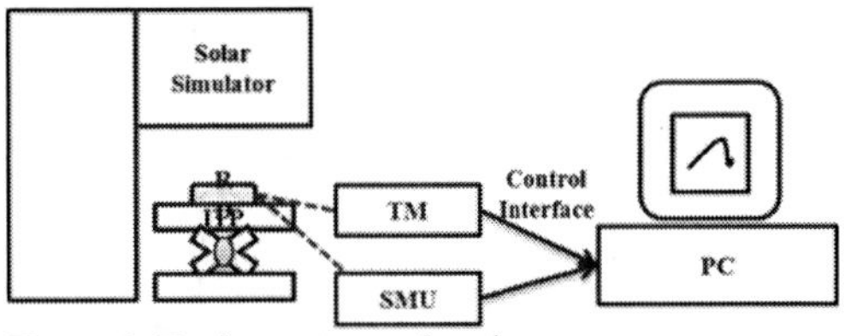

Figure 1: Testing system and equipment

2.1 Main apparatus for test system

Upon I-V measurement, the major apparatus (see Fig. 1) includes a reference device (RD), temperature monitor (TM), source measurement unit (SMU), isothermal test plane (ITP), and solar simulator (SS). A 5-in-1 I-V tester [4] can carry out I-V curves, which also leverage multi-function of I-V, IPCE, and light soaking together. Tester specification summary as follows: spectrums of light radiation AM1.5G with intensity of 1000 Wm^{-2}, variation of uniformity in the plane target (16 cm x 16 cm) is better than 1.5 %, and temporal stability is better than 2 %. Test condition is under STC (i.e., AM1.5 reference spectrum prescribed in IEC 60904-3 [6], 1000 Wm^{-2}, 25 °C) unless otherwise specified. The above tester includes one homemade software [5, 7] to remove the capacitance effect and I-V measurement errors [1, 5].

2.2 Test sequence and test items

To be differential with standard test condition (STC, 1000 Wm^{-2}) used for c-Si devices, there are two reference conditions were defined in SEMI PV57 for low lighting. One is very low irradiance condition (AM1.5G, 25 °C, 60 Wm^{-2}, VLIC), and the other is ultra-low irradiance condition (AM1.5G, 25 °C, 1 Wm^{-2}, ULIC). The testing procedure is written as follows: (1) visual inspection (2) measured active area (3) check I-V/IPCE (incident photon-electron conversion efficiency) measurement system (4) check wire connection (5) light soaking (6) forward/backward I-V at STC or VLIC or ULIC (7) I-V at STC or VLIC or ULIC (8) visual inspection (9) data analysis.

Therefore, the advantage of standardization the testing procedure for OPV/DSSC can reduce the measurement error due to capacitive effect, and evaluate the relationship between short-circuit current (I_{sc}) and spectral correction factor (spectral mismatch factor, MMF). OPV/DSSC needs define the specific basis of reference to make correction for spectral issue and reserve extra time for I-V test as well.

2.3 Experiment and discussion

Performance characterization of OPV/DSSC are important to clarify how to determine their performance accurately. Such characterization requires considering different level lighting on very slow temporal response in the current-voltage (I-V) curves, which are clearly dependent on the voltage sweep direction, even when the sweep time is the order of seconds [4]. Real-time one-sweep method (RTOSM [5] or SEMI PV57 C9.7.3) is a good way in measuring cell performance more accurately and rapidly when evaluates performance of OPV/DSSC.

RTOS method highlights the temporal stability of the output current is measured at a fixed bias voltage as shown in Fig. 2, in order to get the real P_{max} of DSSC, especially for its data influenced in different sweep direction without capacity effect (hysteresis phenomenon). This mean backward sweep has saturation in HUMO (highest occupied molecular orbital), and forward sweep has saturation in LUMO (lowest unoccupied molecular orbital), and its response time is shown to increase while decreasing level lighting in the forward and to keep constant in the backward. This phenomenon also demonstrated the influence from capacity during the measurement of I-V curves by using a steady-state simulator.

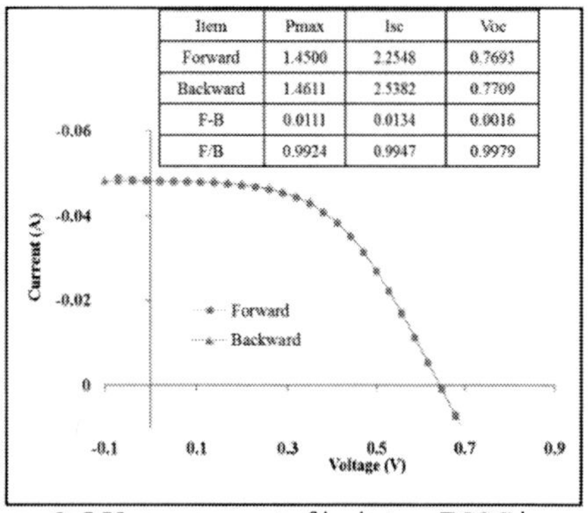

Item	Pmax	Isc	Voc
Forward	1.4500	2.2548	0.7693
Backward	1.4611	2.5382	0.7709
F-B	0.0111	0.0134	0.0016
F/B	0.9924	0.9947	0.9979

Figure 2: I-V measurement of ionic-type DSSC in forward and backward sweep

3 SEMI PV69 [2, 10]

3.1 Main apparatus for test system

The relative spectral responsivity of a test sample is measured by irradiating it with a narrow-bandwidth light source in a series of different wavelengths covering its response range, and measuring the short-circuit current density and irradiance at each of these wavelengths. The light source irradiates the device uniformly, and the temperature of the device to be controlled. The current densities divided by the irradiances or a proportional parameter can plot as a function of wavelength. Alternatively, the irradiance keeps constant, in which case the relative spectral response obtained directly from the current density readings. The irradiance monitor can be a vacuum thermocouple, a pyroelectric radiometer or other suitable detector. Another alternative is a previously calibrated reference photovoltaic device whose relative spectral response covers the required range. Testing equipment (see Fig. 3) include LS (light source), MC (monochromatic), shutter, chopper, LIA (lock-in amplifier), SMU (source measurement unit) and ITP (isothermal test plane).

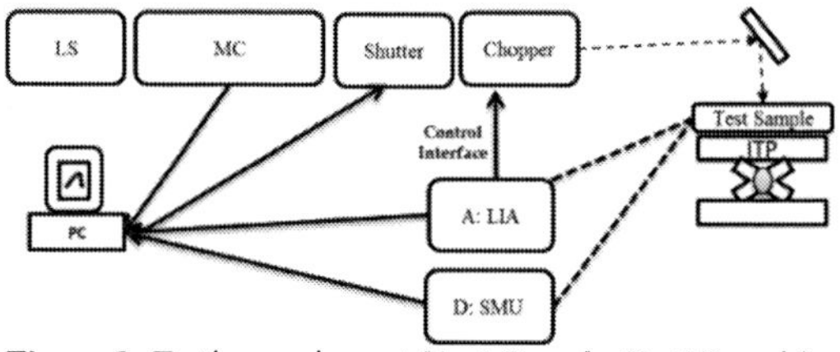

Figure 3: Testing equipment (A: AC mode, D: DC mode)

3.2 Test sequence and test items

Five test samples, which made by same raw materials and process, are required at least. General test sample assembled should be including seal, package and four well-attached wires (or terminal wires). Using terminal wires to multiply set designed for the safety of the rated current, and length is 5 cm.

The performance data of test sample is highly responsive to the external measurement conditions. Reliable evaluation results require the measurement performed under proper conditions. The details of the measurement need clearly described in the report. The testing procedure is written as follows: (1) visual inspection (2) measured active area (3) light soaking (4) check SR/IPCE measurement system (5) check wire connection (6) SR checks AC/DC mode at STC or NTC (normal test conditions) (7) SR at STC or NTC (8) visual inspection (9) data analysis.

3.3 Experiment and discussion

Following the test procedures in 3.2, three SR curves (see Fig. 4) were defined by both DC mode and RTOS method. Due to low response time characteristics of OPV/DSSC, RTOS method can remove the capacitive effect instantly, and amend the spectral response of OPV/DSSC.

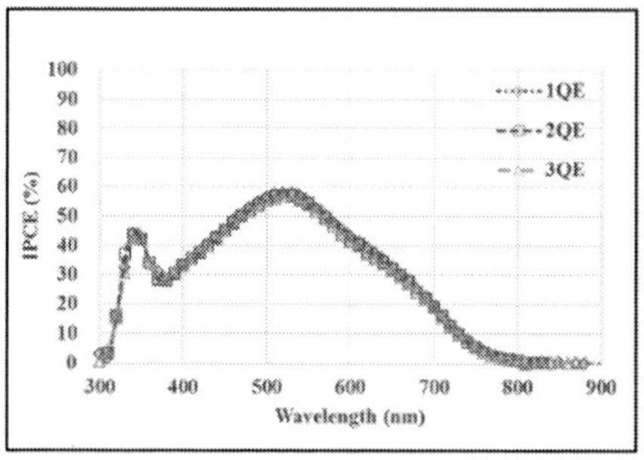

Figure 4: SR curve with difference less than 1.5 %. (Ionic DSSC 1cm x 5 cm, DC mode)

4 SEMI DOC. 5598 [3, 11]

This draft aims to develop a durability test method to evaluate characteristics of OPV/DSSC, especially in subtropical warm and humidity environment. SEMI Doc. 5598 developed an evaluation principle and guideline for industries to follow up. Test procedures (see Fig. 5) include light soaking, thermal cycle, damp heat and one combined test, with some specific stress condition. The performance evaluations of durability test need to follow the guidelines of I-V [1] and SR [2]. Results applied for SEMI global voting in May 2016.

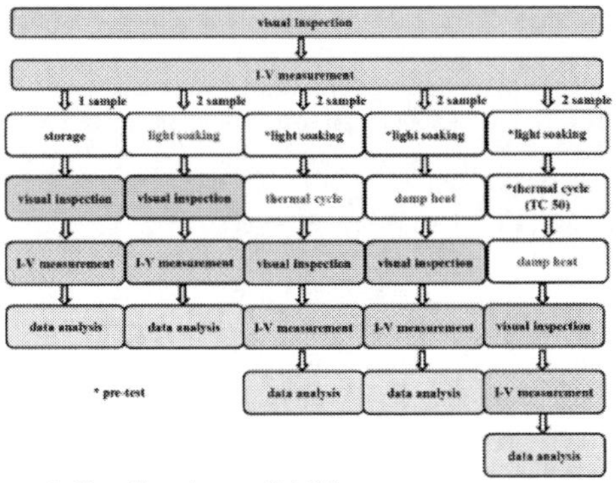

Figure 5: Test flow in parallel [3]

In Fig. 5, nine test samples and five-group carried out in the order laid down to the qualification test sequences. Each box refers to the corresponding sub-clause in this test method. Test procedures and severities, including initial and final measurements where necessary. However, with regard to the tests of initial and final measurements, it should be noted that the procedures laid down in SEMI PV57 [1], these tests shall be carried out within ± 5 % of the specified irradiance and within ± 2 °C of the specified temperature. Any single test, executed independently of a test sequence, proceed by the initial tests of visual inspection and I-V measurement. In carrying out the tests, the tester shall strictly observe the manufacturer's handling, mounting and connection instructions. Final measurements of each aging test before a minimum recovery time of 1 hr, repeat tests visual inspection and I-V measurement at STC. Results need meet criteria of relative performance lose (e.g., P_{max}) less than 5%. All I-V tests need to follow the same rules defined in SEMI PV 57, in order to reduce the capacity effect during measurement for OPV/DSSC.

5 CONCLUSIONS

This paper aims to summary the standardization research of pivot SEMI standards [1, 2, 3] developed by SEMI OPV and DSSC task force in Taiwan during 2013 to 2016. This work proposes three test methods to evaluate the I-V, SR and durability for OPV/DSSC product. Through these standardizations, results can help PV industry to get reliable performances of emerging PV, and influence the relevant reliability aspect in a long-term perspective.

6 ACKNOWLEDGEMENTS

Authors appreciate sincerely for all the discussions and contribution from SEMI OPV and DSSC TF in Taiwan.

7 REFERENCES

[1] SEMI PV57, Test Method for Current-Voltage (I-V) Performance Measurement of Organic Photovoltaic (OPV) and Dye-Sensitized Solar Cell (DSSC).

[2] SEMI PV69, Test Method for Spectrum Response (SR) Measurement of Organic Photovoltaic (OPV) and Dye-sensitized Solar Cell (DSSC).

[3] SEMI Doc. 5598, Test Method for Durability of Low Light Intensity Organic Photovoltaic (OPV) and Dye Sensitized Solar Cell (DSSC).

[4] T. C. Wu, S. T. Hsu, Y. S. Long, New Set-up Procedures and Integrated Measurement System for Organic Photovoltaic (OPV) Module, International Photovoltaic Science and Engineering Conference (PVSEC23), 2013.

[5] Y. S. Long, S. T. Hsu, T. C. Wu, Induction of Internal Capacitance Effect in Performance Measurement of OPV (Organic Photovoltaic) Device by RTOSM (Real-Time One-Sweep Method), *Journal of Energy and Power Engineering, 8 (2014) 1059-1066.*

[6] IEC 60904-3, Photovoltaic Devices – Part 3: Measurement Principles for Terrestrial Photovoltaic Solar Devices with Reference Spectral Irradiance Data.

[7] T. C. Wu, B. N. Chuang, J. S. Song, H. S. Wu, Y. S. Long, U.S. Patent 8,224,598, 2012.

[8] IEC 60904-1, Photovoltaic Devices – Part 1: Measurements of Photovoltaic Current-Voltage Characteristics.

[9] S. T. Hsu, Y. S. Long, T. C. Wu, Standardization of Current - Voltage Test Method for DSSC Products, *12th ISMTII,* 2015.

[10] S. T. Hsu, Y. S. Long, T. C. Wu, Jay Lin, Standardization of Spectrum Response Measurement for OPV and DSSC, *SNEC 10th,* 2016.

[11] S. T. Hsu, Y. S. Long, T. C. Wu, H. H. Hsieh, Requirement of Durability Test for OPV and DSSC, *32nd EU PVSEC,* 3DV.2.58, 2016.

CHARACTERISTICS OF EMERGING PV UNDER LEVELS LIGHTING INDOOR

Yean-San Long[1]*, Shu-Tsung Hsu[1], Teng-Chun Wu[1], Hsin-Hsin Hsieh[1]
[1]Center for Measurement Standards, Industrial Technology Research Institute, Taiwan
*Corresponding author (mickeylong88@itri.org.tw, Phone: +(886) 3 5743844, Fax: +(886) 3 5725540)

ABSTRACT: Nowadays, wireless communication networks focus on indoor applications, Internet of Things (IoT), sensor networks, etc., use batteries as their source of energy. However, batteries have a limited lifetime and to be replaced regularly. The lifetime often is one key limiting factor for the device. Replacing the battery always outweighs the most cost of the device itself. However, if the device itself could be able to harvest energy from renewable resources in the environment, the lifetime of the device will be many times over. The feasibility of PV for indoor low power applications has already been demonstrated, for phone chargers, computer mice and indoor positioning systems. The PV cells for different applications on earth to be rated by their power output under standard test conditions (STC). However, the STC are not relevant for indoor applications. Typically, the light intensity under artificial lighting conditions found in offices and factories is less than 5 W/m^2 as compared to 100–1000 W/m^2 under outdoor conditions, depending on the type of light source and its distance. Characterization of emerging PV requires variable indoor light source and very slow temporal response. In this study, we also made a comparison for different emerging PV in light levels indoor environments. The characterization after test showed it is not necessary to measure performance under all typical artificial light sources. Therefore, we propose the reasonable comment about each solar cell measured by indoor light sources for its characterization and application.

Keywords: Organic Photovoltaic, Real-Time Removing Capacitance Effect, One-Sweep Method.

1 INTRODUCTION

Nowadays, wireless communication networks focus on indoor applications, Internet of Things (IoT), sensor networks, etc., use batteries as their source of energy. However, batteries have a limited lifetime and to be replaced regularly. The lifetime often is one key limiting factor for the device. Replacing the battery always outweighs the most cost of the device itself. Furthermore, from an environmental perspective, to minimize the waste of battery is whenever possible. Moreover, during the last decade battery technology has not improved significantly in terms of energy density and size, especially for low power applications. However, if the device itself could be able to harvest energy from renewable resources in the environment, the lifetime of the device will be many times over.

Solar energy can be an efficient natural energy source. Indoor photovoltaic (PV) applications have been prominent on the market since the first half of the 1970s. The electrical energy spent for lighting purposes can be partly recycled by powering devices with photovoltaic solar cells. Indeed, the radiation (visible and non-visible) emitted from artificial light sources can be used for the production of electricity. Mainly amorphous silicon is used for this purpose in watches and calculators. The latter has been the most successful indoor PV application to date, reaching annual sales of 10 million units by 1990, and thus avoiding the use of millions of alkaline batteries. This volume of solar cell sales represented at the time approximately 10 % of the global photovoltaic shipments. Since then, the solar calculator market has saturated while the outdoor PV market has boomed. Therefore, in order for the indoor PV market to grow, it is necessary for it to find applications in other devices such as IoT and sensor networks. The feasibility of PV for indoor low power applications has already been demonstrated, for phone chargers, computer mice and indoor positioning systems. The PV cells for different applications on earth to be rated by their power output under standard test conditions (STC, AM1.5G/ 25 °C/ 1000 W/m^2). Although these conditions are seldom achieved in practice (except in the lab), this characterization provides a reasonable guideline for comparing different solar cell types under outdoor conditions. However, the STC are not relevant for indoor applications. Typically, the light intensity under artificial lighting conditions found in offices and factories is less than 5 W/m^2 as compared to 100–1000 W/m^2 under outdoor conditions, depending on the type of light source and its distance. Moreover, the spectrum can be totally different from the outdoor solar reference spectrum (e.g. AM1.5G). The spectrum depends not only on the type of light source, but also on the presence of reflected and diffused light. Unfortunately, there are no international norms which determine a way of characterizing solar cells for indoor applications. There has been a lot of attention on emerging PV, such as dye sensitized (DSC), organic (OPV) and inorganic (ex. perovskite solar cell, PSC), given their low-cost and improving efficiency. Accurate characterization of emerging PV requires level lighting consideration on each very slow temporal response in the I-V curves of the emerging PV are clearly dependent on the voltage sweep direction, even when the sweep time is the order of seconds [1]. Furthermore, the temporal response is dependent on different level lighting consideration. In our study, RTOS (real-time one-sweep) method can real-time remove the capacity effect from high to low level AM1.5G lighting for I-V successfully and proposes the forward/backward schematic reaction mechanism for DSC [2, 3, 4]. In this study, an Indoor Multi-Light Source System for indoor PV Measurement (see Fig. 1) homemade by CMS/ITRI, which has leveraged by home-design. Equipment summary as follows: Indoor multi-Light source included D65 (6500K, Average North USA Sky Daylight)/TL84(4100K, European shop Fluorescent)/CWF(4150K, Cool White Fluorescent, shop lighting)/U30(3000K, shop lighting)/A (2856K, Typical Home Lighting), meet CIE Standard illuminants. Additionally, we will be useful to measuring emerging PV performance more accurately and rapidly by RTOS method in different levels indoor lighting condition.

2 EXPERIMENTS

The samples include DSC, PSC, a-Si and Copper Indium Gallium Diselenide (CIGS). Sample size is from 0.1 cm^2 to 1 cm x 5 cm. These I-V curves were carried out Keithley Source Meter (Model 2460), and Indoor Multi-Light Source System for Solar Cell Measurement (see Fig. 1), that equipment summary as follows: Indoor multi-Light source included D65/TL84/CWF/U30/A (meet CIE Standard illuminants), Adjustable light intensity (range from 0 to 2,500 lux), supplied with UV Light Source for aging test, non-uniformity less than 2 % (at 20 cm x 20 cm), and temporal instability less than 2 %.

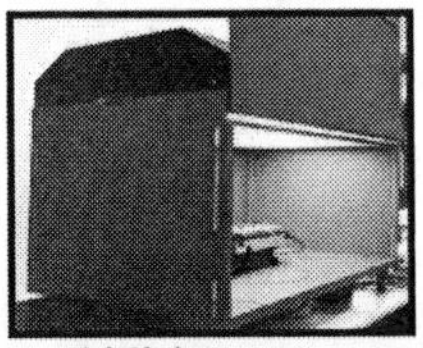

Figure 1: Indoor multi-light source system homemade by CMS/ITRI for solar cell measurement.

In the work, we follow the testing flow [3, 4] in the I-V measure and compared two test methods (delay time and RTOS) to consider their capacitive behavior (see Fig.2): (a) setting delay time, showed no repeatedly and (b) real-time removing capacity effect, is shown best repeatedly for RTOS method. As discussed in reference [1, 2], method B can read simultaneously multi-point forming step after taking the optimization stabilizing test the area at a point on the I-V curve.

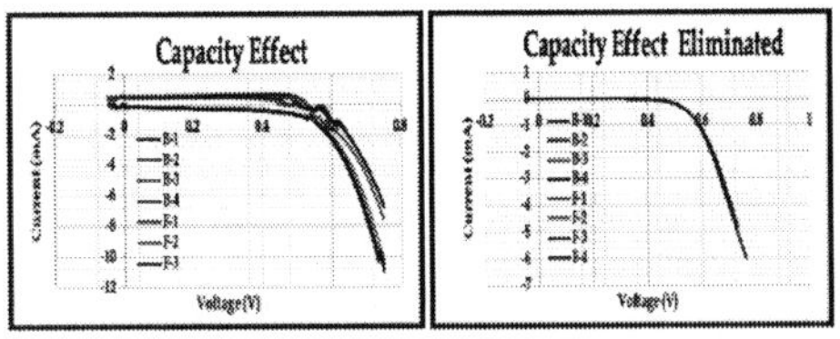

(a) Delay time method (b) RTOS method

Figure 2: I-V curves measured at dark condition.

The measured performance of sample is highly responsive to the external measurement conditions. A reliable evaluation report requires the measurement to be performed under proper conditions and the details of the measurement should be clearly described in the report. A list of proposals [2, 5] for the measurement of sample is given below:

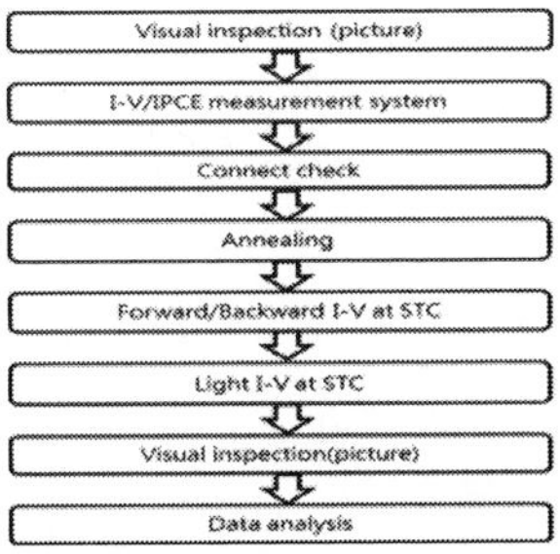

Figure 3: Testing flow-hart[2, 5].

In this study, we design an indoor Multi-Light Source System (Fig. 1). In this system, we check system temporal instability (Table 1), which start check from lamp on, shown temporal instability less than 2 % after 25 mins. Of course, many light sources exist and the question arises which light sources are "typical". The International Commission on Illumination (CIE) is responsible for publishing binding standards for different types of light sources, called illuminants. A standard illuminant represents a mathematical table of relative energy versus wavelength, used for colorimetric calculations. It is a (theoretical) source of visible light with a set spectrum, determined by convention and therefore provides a worldwide basis for comparing images under different lighting. Which illuminants are useful standards for indoor illumination? We consider some typical, widely used, light sources for indoor environments (residential and commercial) and relate them to an appropriate illuminant (standard). The definition of illuminants is relative to each other. If we want to compare the same lighting conditions, we have to scale all light sources to an equal illumination for the human eye to obtain a correct comparison.

Table 1: System of temporal instability (TL84).

Time (min)	Average (lx)	St'd (lx)	RSD (%)
17	2922.08	26.84	0.92
25	2899.11	15.52	0.54
33	2890.50	14.89	0.52
42	2885.06	18.19	0.63
50	2882.33	17.50	0.61
58	2872.76	15.30	0.53

3 ANALYSIS AND RESULTS

We choose a value of 200 lx because family home demand recommended. Where the main task is less demanding, e.g., a corridor, a lower level (e.g., 100 lx) is sufficient. The required illumination can also be higher (e.g., 1,000 lx) in e.g., production rooms in industry where detailed work is necessary and in operation theatres in hospitals. We can test a sample in different intensity (Fig. 4), which shows the linear property in indoor lighting for emerging PV.

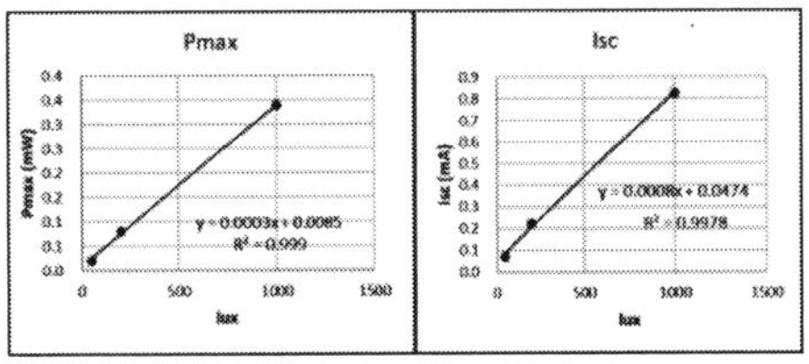

Figure 4: Light linearity of DSC measured by indoor multi-light source system according to TL84 and RTOS method.

We now compare the performance of the different solar cells in light levels indoor environments to the performance of the monocrystalline silicon solar cell. In Table 2, it is Obvious to see the influence from the different indoor light environments (light source at 2,000 lx.) and the value of open-circuit voltage (Voc) is dependent on the light intensity, we approximate the open-circuit voltage to the bandgap of the absorber. Of course, this is an idealized, non-realistic situation, but it allows us to make abstraction of the other influences and thus—by relative comparison—study the influence of one parameter only: the spectrum of the artificial light source. Indeed, performance of five samples (ex. DSC, PSC, Si, a-Si, and CIGS) are different according with each light source showed more power density value at A lighting than other light sources; DSC kept the similar power density value at each light source.

Table 2: I-V characteristics measured by indoor multi-light source system for emerging PV (light intensity at 2000 lx).

Item	A	D65	TL84	U30	CWF
CIGS(Unpackaged)(100 cm²)					
I_{sc} (A/cm²)	0.00007	0.00001	0.00001	0.00001	0.00001
V_{oc} (V)	5.97	2.81	2.77	2.95	2.99
Power Density (W/cm²)	0.00016	0.00001	0.00001	0.00001	0.00002
FF (%)	35.57	35.55	34.83	36.28	37.05
a-Si (6 cm²)					
I_{sc} (A/cm²)	0.00070	0.00039	0.00035	0.00048	0.00042
V_{oc} (V)	1.03	0.92	0.91	0.95	0.94
Power Density (W/cm²)	0.00051	0.00021	0.00018	0.00028	0.00023
FF (%)	70.42	58.34	56.87	61.33	57.15
Si Reference Cell (4 cm²)					
I_{sc} (A/cm²)	0.00153	0.00023	0.00018	0.00024	0.00026
V_{oc} (V)	0.51	0.38	0.38	0.38	0.39
Power Density (W/cm²)	0.00049	0.00005	0.00004	0.00005	0.00006
FF (%)	62.71	55.47	58.30	58.17	57.37
DSC (1) (5 cm²)					
I_{sc} (A/cm²)	0.00022	0.00013	0.00012	0.00016	0.00016
V_{oc} (V)	0.56	0.56	0.56	0.55	0.55
Power Density (W/cm²)	0.00008	0.00005	0.00004	0.00006	0.00006
FF (%)	66.63	64.74	64.91	66.91	66.90
DSC (2) (2.5 cm²)					
I_{sc} (A/cm²)	0.00055	0.00033	0.00033	0.00031	0.00029
V_{oc} (V)	0.67	0.65	0.66	0.64	0.65
Power Density (W/cm²)	0.00027	0.00016	0.00017	0.00015	0.00014
FF (%)	73.48	74.53	76.46	74.58	74.68
PSC(perovskites) (0.11 cm²)					
I_{sc} (A/cm²)	0.00176	0.00072	0.00071	0.00085	0.00088
V_{oc} (V)	0.73	0.75	0.75	0.75	0.78
Power Density (W/cm²)	0.00061	0.00026	0.00025	0.00028	0.00034
FF (%)	47.73	47.54	46.44	48.70	49.57

In this work, RTOS method can real-time remove the capacity effect in an indoor lighting system for I-V measurement successfully (Fig. 2) Results demonstrated the influence from capacity during the measurement of I-V curves. The work also made a comparison for different emerging PV in light levels indoor environments (Table 2). The characterization after test showed it is necessary to measure performance under all typical artificial light sources. Therefore, we propose the reasonable comment about each solar cell measured by indoor light sources for its characterization and application (Fig. 5). Furthermore, our goal is to determine which light source is relevant for a new international standard test method. Upon the indoor characterization of solar cells, it is the spectrum range of the light source that decides the lighting performance, not the light intensity, which is present and affects the cell equally through different light sources.

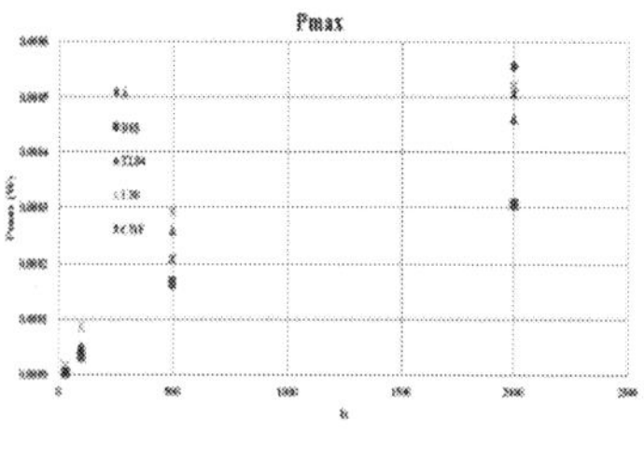

Figure 5: Light linearity of DSC measured by indoor multi-light source system in different lighting source and RTOS method.

4 CONCLUSIONS

RTOS method can real-time remove the capacity effect for I-V measurement. Measurement of the open circuit voltage decay showed that this effect was related to an increased diffusion capacity at higher voltages. However, an appropriate simultaneous voltage and current tracing as well as the measurement are detecting capacity effect. Simultaneous removing capacity effect in sweep I-V, access to more accurate measurement results.

Indoor multi-light measurement system including RTOS method has good performance due to constant of St'd (< 20 lx) and spectrum under varied light intensity, i.e., I-V characteristics is visible to see the influence due to different indoor lighting environment (light source vs power setting), and the value of open-circuit voltage (Voc) is dependent on the light intensity. Performance of test samples (ex. CIGS, Si and PSC) are different according with each light source. Result showed more power density value due to "A" lighting, DSSC kept the similar power density value at each light source.

Furthermore, our goal is to determine which light source is relevant for a new international standard test method. Upon the indoor characterization of solar cells, it is the spectrum range of the light source that decides the lighting performance, not the light intensity, which is present and affects the cell equally through different light sources.

REFERENCES

[1] T. C. Wu, S. T. Hsu and Y. S. Long, "New Set-up Procedures and Integrated Measurement System for Organic Photovoltaic (OPV) Module", International Photovoltaic Science and Engineering Conference (PVSEC23), 2013.

[2] Y. S. Long, S. T. Hsu and T. C. Wu, "Induction of Internal Capacitance Effect in Performance Measurement of OPV (Organic Photovoltaic) Device by RTOSM (Real-Time One-Sweep Method)", Journal of Energy and Power Engineering, Vol 8, pp1059-1066, 2014.

[3] Y. S. Long, S. T. Hsu and T. C. Wu, "Capacitance Effect from High to Low Level Lighting by Real-Time One-Sweep Method for OPV device", IEEE PVSC42, 2015.

[4] Y. S. Long, S. T. Hsu and T. C. Wu, "A Study of Capacitance Effect in DSC from High to Low Level Lighting by Real-Time One-Sweep Method", OPTIC2014.

[5] SEMI PV57-1214, "Test Method for Current-Voltage (I-V) Performance Measurement of Organic Photovoltaic (OPV) and Dye-Sensitized Solar Cell (DSSC)".

[6] US PTO. 8224598, Method for forming optimal characteristic curves of solar cell and system thereof.T. C. Wu, S. T. Hsu and Y. S. Long, "New Set-up Procedures and Integrated Measurement

System for Organic Photovoltaic (OPV) Module", International Photovoltaic Science and Engineering Conference, 2013.

[7] M. A. Green, K. Emery, Y. Hishikawa, W. Warta & E. D. Dunlop, Solar cell efficiency tables (version 39), Prog. Photovolt: Res. Appl. (2012)

[8] V. Shrotriya, G. Li, Y. Yao, T. Moriarty, K. Emery and Y. Yang, Adv. Funct. Mater., 2006, 16, 2016–2023 (2006).

[9] S. Mau, Influence of Solar Cell Capacitance on the Measurement of I-V Curves of PV-Modules, Proc. 20th EUPVSEC (2005).

[10] Cuevas, A., Recart, F., Capacitive effects in quasi-steady-state voltage and lifetime measurements of silicon devices, Journal of Applied Physics, 98(7), pp. 1-7 (2005).

[11] D. Willett and S. Kuriyagawa, The effects of sweep rate, voltage bias and light soaking on the measurement of CIS-based solar cell characteristics, Proc. 23rd IEEE PVSEC, 495-499 (1993).

This is a pre-peer review version of the paper submitted to the journal Progress in Photovoltaics.
It has been selected by the Executive Committee of the EU PVSEC 2016 for submission to Progress in Photovoltaics.

III-V MULTI-JUNCTION METAL-WRAP-THROUGH (MWT) CONCENTRATOR SOLAR CELLS

E. Oliva*, T. Salvetat#, C. Jany#, R. Thibon#, H. Helmers*, M. Steiner*, M. Schachtner*, P. Beutel*, V. Klinger*, J.S. Moulet# and F. Dimroth*
*Fraunhofer Institute for Solar Energy Systems ISE, Heidenhofstraße 2, 79110 Freiburg, Germany
#CEA, LETI, Minatec Campus, F-38054 Grenoble, France

ABSTRACT: III-V multi-junction solar cells are promising devices for photovoltaic applications under very high concentration levels of sunlight. Shadowing losses of the front side metallization and ohmic resistance losses in the metal grid limit the practical cell size typically to around 1 cm^2 at 800 suns. The use of a full back-contact architecture, similar to the Metal Wrap-Trough (MWT) technology known in silicon photovoltaics, can help to overcome this limitation. Furthermore, positioning both the positive and negative contact pads on the rear side of concentrator solar cells opens the possibility for efficient packaging solutions and the realization of dense array receivers with low metal shadowing. The MWT technology addresses conventional concentrating photovoltaics as well as combined PV-thermal applications and offers specific advantages for large area devices at high intensities. This work presents first experimental results for MWT architectures applied to III-V tandem solar cells and discusses specific challenges. An efficiency of 28.3 % at 176 suns and 27.2% at 800 suns has been measured for the best MWT $Ga_{0.51}In_{0.49}P/Al_{0.03}Ga_{0.97}As$ tandem solar cells.

Keywords: III-V Semiconductors, Back Contact, High-Efficiency, Multi-junction Solar Cell, Concentrator Cells

1 INTRODUCTION

The highest light to electricity conversion efficiencies are achieved with photovoltaic devices made of III-V semiconductor materials. For this reason, they are the candidate of choice for many high performance applications such as in space for powering satellites or on earth in high concentration photovoltaics (HCPV) systems or in optical power transmission [1, 2]. Four-junction solar cells with record efficiency up to 46.1% (at 312xAM1.5d) have been demonstrated [3] using wafer bonding technology but several other device architectures had been suggested and demonstrated [4-6]. This includes inverted four-junction solar cells which reached efficiencies up to 45.6 % (at 690xAM1.5d) [7] Typical concentrator solar cell devices in HCPV modules have dimensions of 1 mm^2 up to 1 cm^2 and operate at concentration levels between 200 and 1500 suns. Larger area devices are used in parabolic dish mirror systems within dense array receivers [8, 9]. Such receivers may cover an area of 100 cm^2 up to 1 m^2 and operate up to an intensity of 1 MW/m^2 equivalent to 1000 suns. Ohmic resistance losses scale with the square of the current and therefore become a major concern for large area, high concentration devices. The ohmic resistance must be balanced carefully with the amount of metal on the front side of the solar cell which may lead to significant shading. Monolithic series connection of small solar cell segments has been proposed to minimize shading and resistance losses for large area devices at high concentration levels [10, 11]. Such approaches limited the current through series connection and higher voltage but require trenches for the interconnection which lead to additional inactive area.

The challenge of high currents and metal shading in silicon photovoltaics has led to the development of metal-wrap-through (MWT) technologies in which the emitter and base contacts are both realized on the rear side of the device. Insulated holes are filled with metal to transport the electric current from the front to the back side where both contacts are isolated and collected in an interdigitated metal grid [12, 13]. This technology has led to Si solar cells with efficiency up to 21.0% (AM1.5g) [14]. But this technology has never been applied to III-V multi-junction solar cells. A conceptual study was presented by Zhao et al. [15] in 2012 but the realization of such devices is complex. In this paper we discuss how we adapt the MWT technology to realize first functional devices in an inverted $Ga_{0.51}In_{0.49}P$/AlGaAs tandem solar cell. We further show the influence of design variables like via pitch and pattern. And we discuss the influence of local shunts on the characteristics of concentrator solar cells.

2 BENEFITS OF CONCENTRATOR MWT CELLS

The task of optimizing the front metal grid for conventional concentrator photovoltaic devices includes finding an optimum between resistance and shading losses. Minimum width and maximum height of metal fingers as well as appropriate finger shape for sidewall reflection of photons into the solar cell are typical design variables. Such a front contact grid optimization has been performed at Fraunhofer ISE for GaInP/AlGaAs dual-junction solar cell devices using SPICE network simulation. Detailed descriptions of this modelling method can be found in [16]. Shading losses by the contact grid metallization with different geometries optimized for a concentration ratio of 800 suns are presented in Figure 1. The optimal finger geometry depends on the cell size and shading reaches up to 14 % for a 1x1 cm^2 device with a 10 µm x 2.7 µm finger profile. The shading may be reduced to below 10 % by lowering the metal finger width and increasing the height. But processing finger geometries with aspect ratios >1 or finger width <4 µm is not practical with photolithography and lift-off.

The metal-wrap-through-technology allows a significant reduction in front metal grid shading by conducting the current of the emitter through insulated via-holes to the rear side of the solar cell where large metal areas do not cause additional shading. The possible gains of the MWT concept in III-V multi-junction solar cells are studied in [17]. The absolute efficiency improvement for a 0.49 cm^2 triple-junction GaInP/GaInAs/Ge solar cell at high concentration levels of 500-2000 suns is estimated to be in the range of 1.9-3.0% compared to a conventional design [17]. Higher gains are possible for larger cells and higher concentration levels. An efficiency improvement of 10% (4% absolute) at 500 suns was calculated in [15] for GaInP/GaAs/GaInAs inverted metamorphic (IMM) triple-junction solar cells. Furthermore, a dry etching process for the formation of via-holes in IMM structures has been successfully demonstrated in [15].

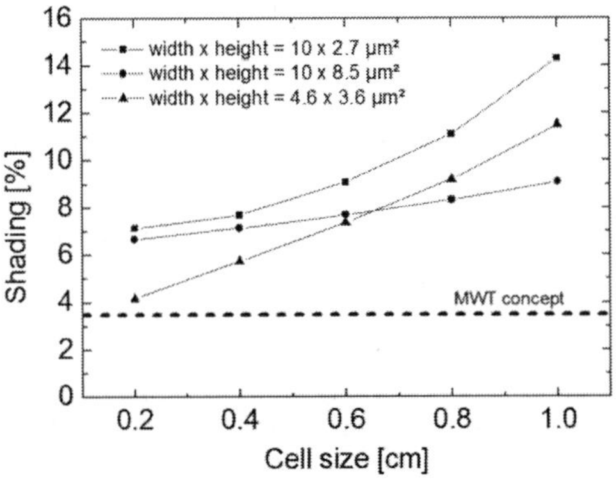

Figure 1: Shading by the front contact grid calculated for a GaInP/AlGaAs tandem solar cell optimized for 800 sun concentration of the AM1.5d spectrum and different metal finger geometries.

We conducted a simulation based on the 2-diode-model and experimental data for GaInP/AlGaAs dual-junction solar cells, assuming a back-contact cell design with a via-hole diameter of 26 µm and a hexagonal via pattern. The shadowing losses obtained for 800-fold concentration is 3.5 % as presented in

Figure 1 by the dotted line. One can see that the MWT technology offers significant advantages in overall shading of large area devices.

3 EXPERIMENTAL APPROACH

Inverted dual-junction $Ga_{0.51}In_{0.49}P/Al_{0.03}Ga_{0.97}As$ solar cell structures were grown on GaAs substrates using an AIX2800G4-TM metal-organic vapour phase epitaxy (MOVPE) reactor with an 8x4" configuration. The hydride sources arsine and phosphine were used for the group-V growth. TMGa, TMIn and TMAl were the group-III precursors. Doping sources were DMZn, DETe, DTSe, SiH_4 and CBr_4. All structures were grown on 4-inch, 450 µm thick GaAs wafers with an orientation of (100) 6° off towards <111>B. Growth temperatures varied between 530-680 °C and the reactor pressure was between 50-100 mbar. V/III-ratios of 30-50 for AlGaInAs and 40-150 for AlGaInP have been used. The total thickness of all compound semiconductor layers was 5.1 µm divided into a 0.9 µm thick $Ga_{0.51}In_{0.49}P$ top cell (E_g = 1.88 eV) and a 2.8 µm $Al_{0.03}Ga_{0.97}As$ (E_g = 1.45 eV) bottom cell. Both subcells have an n-type emitter layer and a p-type base layer surrounded by higher bandgap barriers made of Al-containing alloys. The two subcells were interconnected in series by an AlGaAs/GaInP tunnel diode and the full layer structure was grown on a GaInP release layer which allowed selective removal of the GaAs substrate during processing.

The GaInP top cell was passivated by an n-AlInP window layer followed by a 400 nm n-GaAs cap layer which was highly doped to > $5x10^{18}$ cm^{-3}. The same structure is used for tandem solar cells with a conventional front metal grid and allows reaching low contact resistance in the range of $1\text{-}6x10^{-6}$ Ohm cm^2. For the MWT solar cells, an ohmic metal was applied to the GaAs cap layer and connected to the rear of the device by insulating vias. The density of these vias was selected for a concentration ratio of 800 suns. Simulation results were used as guiding values for via pitch variations in the range of 50-700 µm on cells with three different via layouts as shown in Figure 2. The metal vias were arranged in a squared or hexagonal pattern and in one configuration additional metal fingers were added between the vias to facilitate current collection. This configuration is beneficial in cases where the overall resistance is limited by the emitter sheet and consequently lateral transport of electrons to the vias.

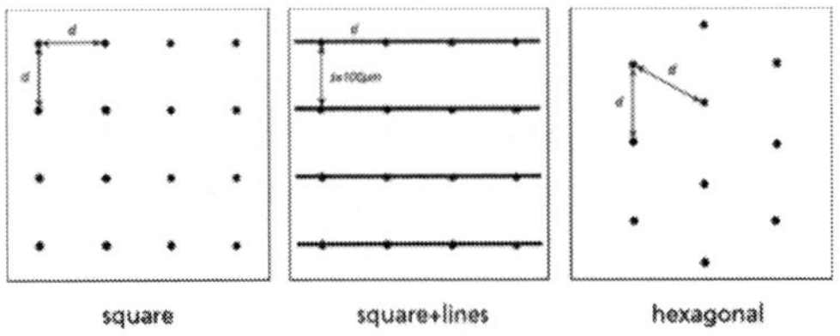

Figure 2: Three different via layouts have been realized to compare the performance under concentrated illumination. Line pitch for the 'square+lines' design was kept constant at 100 µm.

The GaInP/AlGaAs solar cell structure was grown in an inverted direction which gives direct access to the back side of the device structure. During processing, first an ohmic planar back contact was evaporated, followed by dry etching of vias with a diameter of 10 µm. The surface of the metal and the vias were conformal coated with insulating Si_3N_4/SiO_2 and TiN as diffusion barrier before filling the vias with electroplated copper. This resulted in a layer stack of back metal, insulation and front metal. This stack was bonded to a conductive silicon carrier and the GaAs substrate (used for the epitaxial growth of the tandem cell and for mechanical stability) was removed to give access to the GaAs cap. After opening the vias from the top, the GaAs cap layer war etched around the vias and a 2-layer Ta_2O_5/MgF_2 anti-reflection coating was evaporated and structured. As next an ohmic metal contacts were applied with the geometry shown in Figure 2. Finally the cell area was defined by the etched mesa area.

The overall design of the back-contact MWT solar cells is presented in Figure 3. Access to the solar cell back contact was made after mesa etching from the top, whereas the front contact was accessible through the silicon carrier substrate. The different layers and contacts are illustrated in Figure 3 and further details of the processing are published in [18]. It has to be mentioned that the current cell design has not been realized as a full back contact device which is subject to future developments. But the current design offers all the benefits of the MWT concept to minimize shading on large area concentrator solar cells. A substantial number of 4.2x4.2 mm² and 9.5x9.5 mm² MWT cells have been processed for testing.

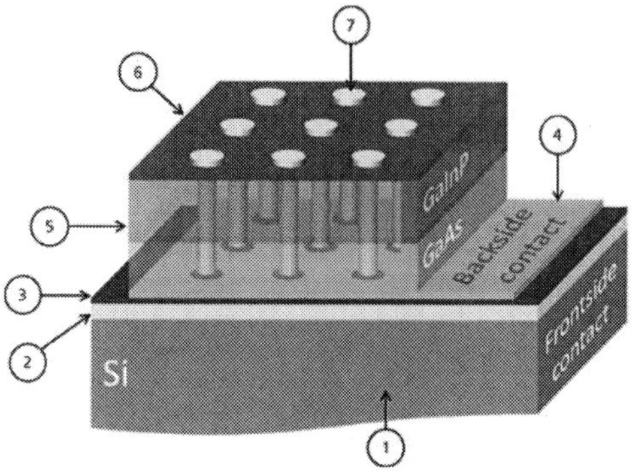

Figure 3: Schematic illustration of the MWT cell architecture with: 1 – silicon carrier, 2 – electroplated copper, 3 – Si_3N_4/SiO_2 insulation (black), 4 – back side contact, 5 – GaInP/AlGaAs tandem solar cell structure, 6 – Ta_2O_5/MgF_2 anti-reflective coating (ARC), 7 – front side contact connected to copper filled insulated vias.

Figure 4 shows a cross-section of a via-hole as the most critical element of the MWT device structure. The via-hole is filled with electroplated copper and insulated carefully from the tandem cell structure and as well as the rear side contact metallization. The diameter of the front ohmic metallization is relatively large (~23 µm) and could be reduced in the future to about 15 µm. The width of grid lines has been 5 µm for the "square+lines" design.

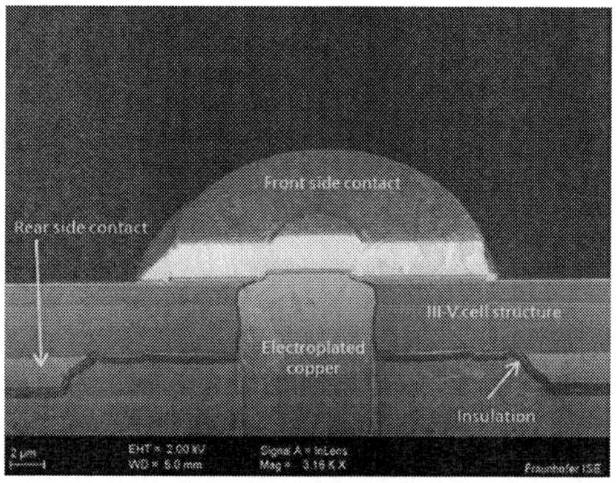

Figure 4: SEM cross-section image of a front contact and via-hole filled with copper and insulated from the GaInP/AlGaAs tandem cell structure.

4 RESULTS AND DISCUSSION

Electrical characterization of processed devices was performed at the Fraunhofer ISE Calibration Laboratory (ISE CalLab PV Cells). External quantum efficiency (EQE) data for the dual-junction cell structure is given in Figure 5 and it shows good carrier collection in the spectral range of both sub-cells. The measurement has been performed on a MWT solar cell. Calculated currents under the AM1.5d standard solar spectrum are 13.3 mA/cm^2 and 12.3 mA/cm^2 for the top and bottom subcell, respectively. The current mismatch of 7.7% exhibits a potential for further optimization.

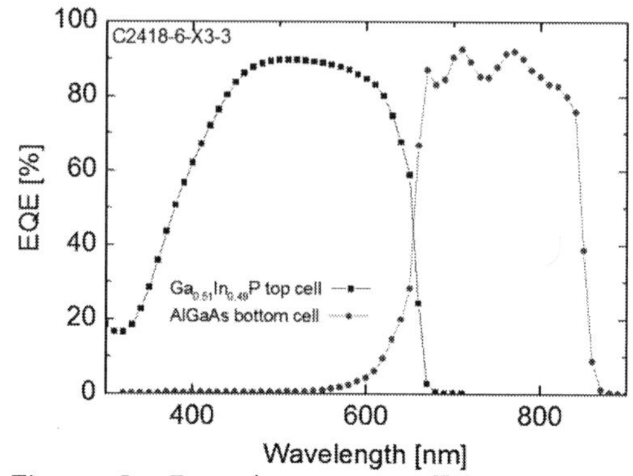

Figure 5: External quantum efficiency for a $Ga_{0.51}In_{0.49}P/Al_{0.03}Ga_{0.97}As$ dual-junction 4.2x4.2 mm^2 MWT solar cell.

One-sun IV-characteristics under the ASTM G173-03 AM1.5d spectrum (normalized to 1000 W/m^2) for MWT tandem solar cells with different via pitch and layout have been measured at T=25 °C using a spectrally matched three-source sun simulator [19]. Measurement results for three cells with hexagonal via arrangement and via pitch of 50, 100 and 200 µm are shown in Figure 6.

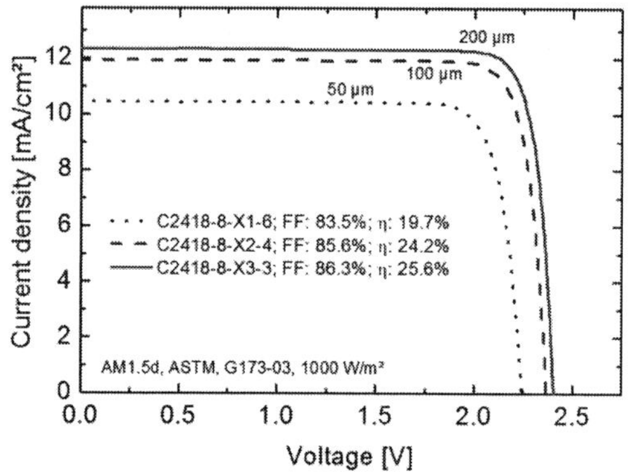

Figure 6: One-sun IV-characteristics of three MWT cells (0.183 cm^2 total area without busbar) with hexagonal via array and different via pitch. All cells were processed on the same wafer.

Excellent IV-characteristics are found with fill factors between 83-86 % indicating a high shunt resistance. The highest performance with an efficiency of 25.6 % is obtained for a device with 200 µm pitch (total area without backside contact pad which is outside the cell mesa area). Open circuit voltage V_{oc}, short circuit current I_{sc} and fill factor FF are reduced for a pitch of 100 and 50 µm. The reduction in current is attributed to additional metal shading, whereas the drop in V_{oc} and FF can be explained by an increasing perimeter with increasing number of metal vias. The additional surface surrounding the metal vias leads to an increase of the dark saturation current and consequently influences the measured V_{oc}.

Some MWT solar cells show a high loss in V_{oc} and fill factor as illustrated by the I-V curves in Figure 7 (left). This behavior indicates a low shunt resistance for one of the sub-cells. Electroluminescence (EL) characterization was carried out and a bright luminescence appears at one of the vias. Subsequent

SEM-FIB analysis confirmed that a damaged via-hole filled with front side metallization has caused this local short-circuit.

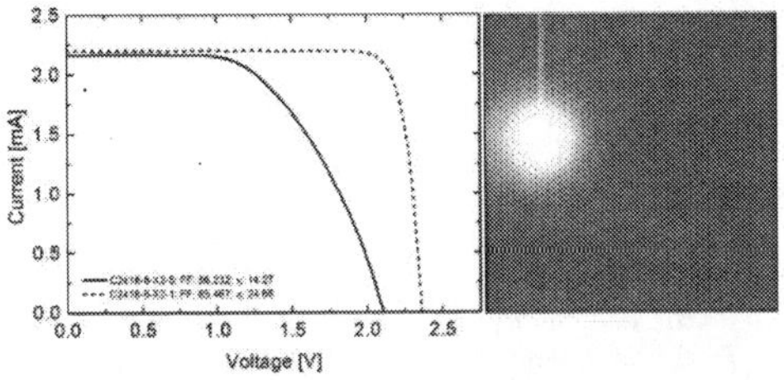

Figure 7: Comparison of two MWT cells with identical layout but significant difference in the IV-characteristics (left). Electroluminescence image of a MWT cell (full line in left figure) showing strong luminescence from a shunted via (right).

The fabricated MWT solar cells were further characterized under concentration using a single-flash simulator. The flash spectrum leads to an excess current generation in the GaInP top cell of approximately 10 % compared to AM1.5d which may results in a slight overestimation in fill factor as described in [20].

Figure 8 compares two of the devices with hexagonal via array (50 and 200 µm pitch from Figure 6) with two additional cells with "square" or "square+lines" pattern. The V_{oc} of all devices increases under concentration. Also it can be seen that the cell with the lowest V_{oc} and highest density of metal vias (50 µm pitch, hexagonal, black triangles) experiences a steeper increase in voltage, approaching nearly the values of cells with 100 µm pitch at 1000 suns. This indicates that the negative voltage effect at one-sun, which was attributed to the high surface area of cells with dense vias, is saturated under concentration. In fact, the power of the hexagonal MWT cell with 50 µm pitch outperforms the cell with 200 µm pitch for C > 400 due to a lower series resistance loss which is also indicated by the higher fill factor. Unfortunately, the hexagonal MWT cell with 100 µm pitch did not survive the measurement and therefore we have included an alternative cell with squared via pattern and 100 µm pitch. The difference between hexagonal and squared via arrays is expected to be negligible and in fact this device shows characteristics in fill factor and voltage which are well between the hexagonal cells with 50 and 200 µm pitch. One can see that 100 µm via pitch outperforms the former devices over a wide concentration range of 50-1000 suns and reaches a maximum efficiency of 28.3 % at 176 suns. This is an excellent achievement for the first MWT solar cells.

Figure 8 also shows the result of a MWT cell with additional metal fingers connecting the vias on the front surface. These metal grid lines are expected to reduce losses due to lateral current conduction in the GaInP top cell emitter. At the same time, the distance between vias could be reduced to 1400 µm which leads to significantly less surface area. In fact, the device shows the highest V_{oc} and fill factor at low concentration but the current is slightly lower due to the additional shading from the grid fingers. At very high concentration factors of 700 suns and above, this device outperforms all other MWT cells and reaches an efficiency of 27.2 % at 800-fold concentration.

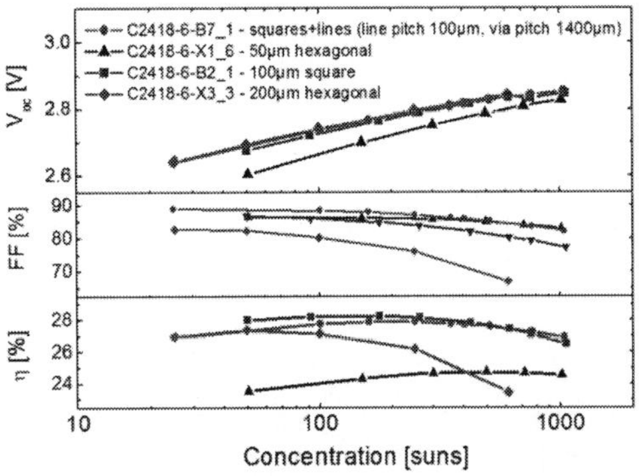

Figure 8: Efficiency, fill factor and open-circuit voltage versus concentration for selected MWT solar cells. The cell size is 18.3 mm², via pitch and layout: X1-6 – 50 µm hexagonal, B2-1 - 100 µm square, X3-3 – 200 µm hexagonal, B7-1 – "square+lines" design (line pitch 100 µm, via pitch 1400 µm).

5 SUMMARY AND CONCLUSIONS

A Metal Wrap Through technology was developed for III-V multi-junction solar cells to demonstrate large area devices operating at high sunlight concentration levels. Shadowing losses for concentrator cells at 800-fold concentration were calculated to be below 4 % and independent of the device area. MWT cells based on an inverted GaInP/AlGaAs dual-junction cell structure were realized. The best MWT devices with an area of 18.3 mm² reach an efficiency of 28.3 % at 176 suns and 27.2% at 800 suns concentration without visible shunts in electroluminescence images. Further improvement of the device performance is possible by adjusting the current matching and by adding additional junctions [21, 22]. Therefore, the MWT technology offers an attractive high performance pathway for solar concentrator applications at high illumination intensities. Candidates are dense array receivers in parabolic dish or solar tower applications.

6 ACKNOWLEDGEMENTS

The authors gratefully acknowledge and thank Rita Freitas, Ranka Koch and Anika Schuette for device processing and Gerald Siefer, Elvira Fehrenbacher, Alexander Wekkeli and Yannick Schreiber for electrical characterization.

7 REFERENCES

[1] Philipps, S.P. and A.W. Bett, *III-V multi-junction solar cells*, in: *Advanced Concepts in Photovoltaics*, A.J. Nozik, G. Conibeer, and M.C. Beard (eds.): The Royal Society of Chemistry: pp. 87-117 (2014).
[2] Philipps, S.P., F. Dimroth and A.W. Bett, *High Efficiency III–V Multijunction Solar Cells*, in: *Practical Handbook of Photovoltaics (Second Edition)*, Vol. Second Edition: Fundamentals and Applications, Boston: Academic Press: pp. 417-448 (2012).
[3] Dimroth, F., T.N.D. Tibbits, M. Niemeyer, F. Predan, P. Beutel, C. Karcher, E. Oliva, G. Siefer, D. Lackner, P. Fuß-Kailuweit, A.W. Bett, R. Krause, C. Drazek, E. Guiot, J. Wasselin, A. Tauzin and T. Signamarcheix,

Four-Junction Wafer-Bonded Concentrator Solar Cells, IEEE Journal of Photovoltaics 6(1): pp. 343-349 (2016).

[4] Chiu, P.T., D.C. Law, R.L. Woo, S.B. Singer, D. Bhusari, W.D. Hong, A. Zakaria, J. Boisvert, S. Mesropian, R.R. King and N.H. Karam, *Direct semiconductor bonded 5J cell for space and terrestrial applications,* IEEE Journal of Photovoltaics 4(1): pp. 493-497 (2014).

[5] Dimroth, F., T.N.D. Tibbits, P. Beutel, C. Karcher, E. Oliva, G. Siefer, M. Schachtner, A. Wekkeli, M. Steiner, M. Wiesenfahrt, A.W. Bett, R. Krause, E. Gerster, M. Piccin, N. Blanc, M. Muñoz, C. Drazek, E. Guiot, J. Wasselin, C. Arena, T. Salvetat, A. Tauzin, T. Signamarcheix and T. Hannappel, *Development of high efficiency wafer bonded 4-junction solar cells for concentrator photovoltaic applications,* 40th IEEE Photovoltaic Specialists Conference: pp. 0006-10 (2014).

[6] King, R.R., D. Bhusari, D. Larrabee, X.Q. Liu, E. Rehder, K. Edmondson, H. Cotal, R.K. Jones, J.H. Ermer, C.M. Fetzer, D.C. Law and N.H. Karam, *Solar cell generations over 40% efficiency,* Progress in Photovoltaics: Research and Applications 20(6): pp. 801-815 (2012).

[7] France, R.M., J.F. Geisz, I. Garcia, M.A. Steiner, W.E. McMahon, D.J. Friedman, T.E. Moriarty, C. Osterwald, J.S. Ward, A. Duda, M. Young and W.J. Olavarria, *Design Flexibility of Ultrahigh Efficiency Four-Junction Inverted Metamorphic Solar Cells,* IEEE Journal of Photovoltaics pp. (99): pp. 1-6 (2015).

[8] Verlinden, P.J., A. Lewandowski, C. Bingham, G.S. Kinsey, R.A. Sherif and J.B. Lasich. *Performance and reliability of multijunction III-V modules for concentrator dish and central receiver applications*, 4th World Conference on Photovoltaic Energy Conversion, Waikoloa, Hawaii, USA: IEEE, pp. 592-7 (2006).

[9] Helmers, H., A. Boos, F. Jetter, A. Heimsath, M. Wiesenfarth and A.W. Bett. *Outdoor test setup for concentrating photovoltaic and thermal (CPVT) systems*, 7th International Conference on Concentrating Photovoltaic Systems, Vol. 1407, Las Vegas, Nevada, USA: AIP, pp. 175-9 (2011).

[10] Helmers, H., E. Oliva, W. Bronner, F. Dimroth and A.W. Bett, *Processing techniques for monolithic interconnection of solar cells at wafer level,* IEEE Transactions on Electron Devices 57(12): pp. 3355-60 (2010).

[11] Borden, P.G. *A monolithic series-connected $Al_{0.93}$ $Ga_{0.07}As$/GaAs Solar Cell Array*, Proceedings of the 14th IEEE Photovoltaic Specialist Conference, San Diego, California, USA, pp. 554-62 (1980).

[12] Van Kerschaver, E. and G. Beaucarne, *Back-contact solar cells: a review,* Progress in Photovoltaics: Research and Applications 14(2): pp. 107-23 (2006).

[13] Fellmeth, T., M. Ebert, R. Efinger, I. Hadrich, F. Clement, D. Biro, P. Noriega, S. Caparros and F.J. Castano, *Industrially feasible all-purpose metal-wrap-through concentrator solar cells,* in series: 2014 IEEE 40th Photovoltaic Specialists Conference: pp. 2106-10 (2014).

[14] Newman, B.K., E.E. Bende, B.B.V. Aken, M. Koppes, L.A.G. Okel, J. Löffler, I.G. Romijn, A. Gutjahr, C.J.J. Tool, N. Guillevin, L.J. Geerligs, J. Wang, J. Zhai, Z. Wang, Y. Chen, Y. Wang and D. Liu. *Manipulating Reverse Current in 21% n-MWT Cells*, 31st European Photovoltaic Solar Energy Conference and Exhibition, Hamburg, Germany, pp. 829 - 834 (2015).

[15] Zhao, Y., P. Fay, A. Wibowo, J. Liu and C. Youtsey, *Via-hole fabrication for III-V triple-junction solar cells,* Journal of Vacuum Science & Technology B 30(6): p. 06F401 (2012).

[16] Steiner, M., S.P. Philipps, M. Hermle, A.W. Bett and F. Dimroth, *Validated front contact grid simulation for GaAs solar cells under concentrated sunlight,* Progress in Photovoltaics: Research and Applications 19(1): pp. 73-83 (2010).

[17] Richard, O., A. Jaouad, B. Bouzazi, R. Arès, S. Fafard and V. Aimez, *Simulation of a through cell via contacts architecture for HCPV multi-junction solar cells,* Solar Energy Materials and Solar Cells 144: pp. 173-180 (2016).

[18] Salvetat, T., E. Oliva, A. Tauzin, V. Klinger, P. Beutel, C. Jany, R. Thibon, P.-H. Haumesser, A. Hassaine, T. Mourier, G. Rodriguez, C. Lecouvey, B. Imbert, F. Fournel, J.-M. Fabbri, J.-S. Moulet, F. Dimroth and T. Signamarcheix. *Metal Wrap Through Like III-V Multi-Junction Solar Cell,* 12th International Conference on Concentrator Photovoltaic Systems, Freiburg, Germany: to be published (2016).

[19] Meusel, M., R. Adelhelm, F. Dimroth, A.W. Bett and W. Warta, *Spectral Mismatch Correction and Spectrometric Characterization of Monolithic III-V Multi-junction Solar Cells,* Progress in Photovoltaics: Research and Applications 10(4): pp. 243-55 (2002).

[20] Siefer, G., C. Baur, M. Meusel, F. Dimroth, A.W. Bett and W. Warta. *Influence of the simulator spectrum on the calibration of multi-junction solar cells under concentration,* 29th IEEE Photovoltaic Specialists Conference, New Orleans, Louisiana, USA, pp. 836-9 (2002).

[21] Miller, N., P. Patel, C. Struempel, C. Kerestes, D. Aiken and P. Sharps. *Terrestrial concentrator four-junction inverted metamorphic solar cells with efficiency > 45%,* 2014 IEEE 40th Photovoltaic Specialist Conference (PVSC), pp. 0014-0016 (2014).

[22] Takamoto, T., H. Washio and H. Juso. *Application of InGaP/GaAs/InGaAs triple junction solar cells to space use and concentrator photovoltaic*, 2014 IEEE 40th Photovoltaic Specialist Conference (PVSC), pp. 0001-0005 (2014).

IMPROVED PERFORMANCE OF III-V MULTI-JUNCTION SOLAR CELLS FABRICATED WITH INDIUM-TIN-OXIDE ELECTRODES

Ray-Hua Horng[1*3], Yu-Cheng Kao[1], Sin-Liang Ou[2], Fan-Lei Wu[1] and Shan-Hui Shi[3]
[1]Graduate Institute of Precision Engineering, National Chung Hsing University, Taichung 402, Taiwan
[2]Department of Materials Science and Engineering, Da-Yeh University, Taiwan
[3]Department of Electronics Engineering, National Chiao Tung University , Taiwan
Tel: +886-3-5712121 ext. 54138 Fax:+886-4-22858362 Email: rhh@nctu.edu.tw

ABSTRACT: In this study, three front-side electrodes consisting of AuGe/Au metal-finger with anti-reflective layer (ARL), ITO-finger with ARL, and ITO-overcoated types without ARL were fabricated for InGaP/GaAs dual-junction solar cells, and samples A, B, and C were used to denote these three devices, respectively. For the sample C with ITO-overcoated electrode, the 200-nm-thick ITO film with a high transmittance of 91% in the visible wavelength range was deposited. The results obtained from the current density-voltage measurements indicated the short-circuit current densities (J_{sc}) of samples A, B, and C are 9.61, 11.16, and 11.65 mA/cm^2, respectively. Meanwhile, the conversion efficiencies of these three solar cells are measured to be 17.50%, 20.41%, and 20.92%, respectively. After replacing the AuGe/Au metal-finger electrode with the ITO-overcoated electrode, there is 19.5% improvement in the conversion efficiency occurred in the device. Moreover, because the ITO front-side electrode can cover overall surface of the device, all regions on the surface of sample C would extract the electrons, resulting in the highest J_{sc}. Not only, the ITO was also used to be ARL and enhanced the device performance.

Keywords: anti-reflective layer, solar cells, multi-junction, InGaP/GaAs, ITO-overcoated electrode.

1 INTRDUCTION

It is well-known that the tandem-type III-V compound multi-junction solar cells have the highest conversion efficiency than that prepared with other absorbing materials. Recently, a high conversion efficiency above 30% under one-sun AM1.5G spectra condition has been achieved in the InGaP/GaAs/Ge triple-junction (TJ) solar cell [1-3]. In the traditional solar cell structure, the metal electrode is usually employed for the device process, which will confine the device performance due to the generation of shading effect. Except for the shading effect, the reflection of incident light would be also occurred in the solar cell fabricated with the metal electrode, causing the degradation of optical characteristics. Fortunately, via the continuous progresses on the transparent conductive oxides (TCOs), the TCOs can serve as the roles of metal electrodes on the solar cells [4-8]. More importantly, the optical performances of solar cells can be improved by using the TCO electrodes. Up to now, various TCO materials have been used as the electrodes in III-V compound and TJ solar cells. However, the problem of shadow loss is still not solved because the metal contact structures exist in the solar cells (i.e., a metal layer should be formed as the ohmic contact between TCO and n-type window layer). In this study, we have presented the ITO film directly contacted to the InGaP/GaAs dual-junction solar cells to replace conventional TCO/metal contact, and the shadow loss arising from the metal contact layer can be overcome. The fabrication processes and the optoelectronic performance of the solar cells were discussed in detail.

2 EXPERIMENTS

In this research, to efficiently decrease the contact resistance between the front-side electrode and the solar cell, the n^+-GaAs layer should be deposited on the device before preparing the electrode. The optimum thickness of the n^+-GaAs layer is determined to be 20 nm. In the epitaxial structure, individual sub-cells were interconnected by using highly-doped tunneling diodes, which can allow the band-to-band tunneling for the series output. Then the device fabrication processes were performed. Firstly, the GaAs ohmic layer was lithographically defined via the selective etching process between the cap GaAs and $Al_{0.5}In_{0.5}P$ window layer. Next, three types of front-side electrodes were individually fabricated on the n^+-GaAs layers of the solar cells, as shown in Figs. 1(a), 1(b), and 1(c), respectively. Fig. 1(a) presents the AuGe/Au metal-finger front-side electrode formed on the cell structure, and this device was denoted as the sample A. Moreover, the ITO front-side electrodes developed in our work consisted of two structures, i.e., the ITO-finger (Fig. 1(b)) and ITO-overcoated (Fig. 1(c)) front-side electrodes. These two devices prepared with the ITO-finger and the ITO-overcoated front-side electrodes were denoted as samples B and C, respectively. The ITO films were deposited by E-beam evaporation at 270 °C, and their thicknesses were 200 nm. For the back-side contact metal deposited on the GaAs substrate for these three samples, the AuBe/Au material was selected. The chip size of these solar cells was 1 mm × 2 mm. Additionally, the block ratio of the bus-bar metal electrode was approximately 8%. Here, it should be mentioned that the conventional SiO_2 anti-reflection layer (ARL; thickness: 75 nm) is deposited both on the AuGe/Au metal-finger front-side electrode and the ITO-finger front-side electrode of samples A and B, respectively. Moreover, because the ITO-overcoated front-side electrode also possesses the ARL function (as discussed later), the SiO_2 layer is not required for the sample C.

The optical transmittance of the ITO film was characterized by the N&K analyzer (model: 1280, N and K Tech.). The associated specific contact resistance between the ITO film and the n^+-GaAs layer was evaluated by the circular transmission line model (CTLM) measurement. Current density-voltage (J-V) characteristics of the solar cells were analyzed by using the standard solar simulator (SAN-EI ELECTRIC XES-40S1) equipped with one-sun air mass 1.5 (AM1.5G) light source (100 mW/cm^2) at room temperature. The responses of external quantum efficiency (EQE) for all cell devices were characterized by a quantum efficiency measurement system (Oriel IQE-200).

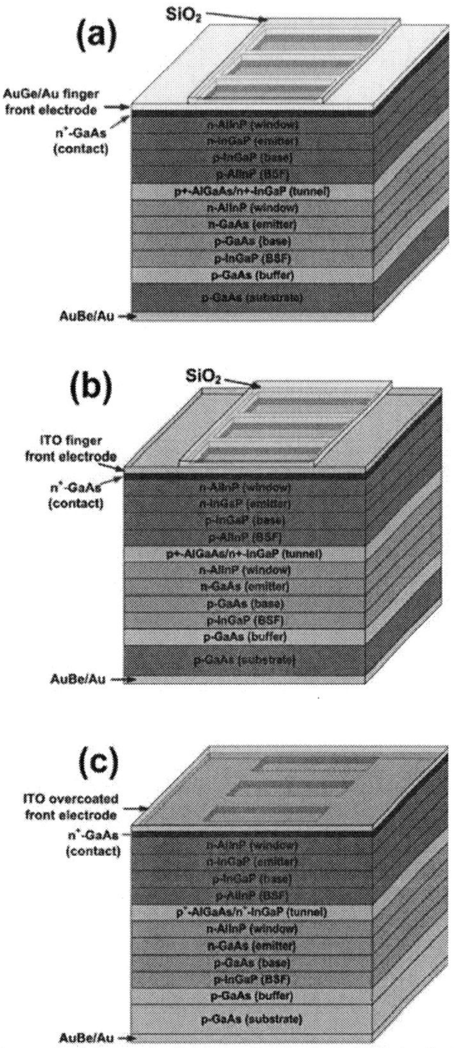

Figure 1: Schematic diagrams for the InGaP/GaAs dual-junction solar cells prepared with (a) AuGe/Au metal-finger, (b) ITO-finger, and (c) ITO-overcoated front-side electrodes.

3 RESULTS AND DISCUSSION

3.1 Transmittance of the ITO film

The transmittance spectrum of the ITO film grown on glass is shown in Fig. 2. It can be found that the transmittance of 77%–86% was obtained when the wavelength was increased to 400-600 nm. Further increasing the wavelength to 600-1000 nm, the ITO film possessed a higher transmittance of 86%–91%. This suggests the ITO film is highly potential for applying the front-side electrode of solar cell, and the shadow loss occurred in the device can be efficiently decreased because of the good transparency of this film.

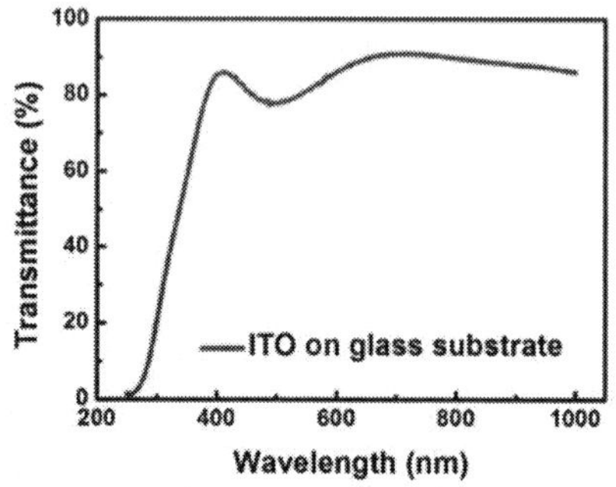

Figure 2: Transmittance spectrum of the 200-nm-thick ITO film prepared on glass substrate in the wavelength ranging from 250 to 1000 nm.

3.2 Contact resistance

The contact characteristic between the ITO and the n$^+$-GaAs layer is the other important factor to estimate the practicability for the front-side electrode of III-V compound solar cells. Fig. 3 shows the result of contact resistance between the ITO film and the n$^+$-GaAs layer. In this figure, the measured resistance values are dependent on the gap spacings (25-350 μm) of contact pads. Based on the current-voltage (I-V) characteristics of this sample, we can observe that there is a good ohmic contact between the ITO film and the n$^+$-GaAs layer without performing the thermal annealing process. The I-V performance of the sample measured over the pad with a spacing of 200 μm is shown in the inset of Fig. 3. After fitting these resistance values, the specific contact resistance between ITO and n$^+$-GaAs can be estimated to 2.89×10^{-4} Ω-cm^2. This reveals that the contact resistance between ITO and n$^+$-GaAs is low enough for III-V compound solar cells. In addition, via the deposition of the ITO film on the n$^+$-GaAs layer, the shadow loss can be indeed reduced.

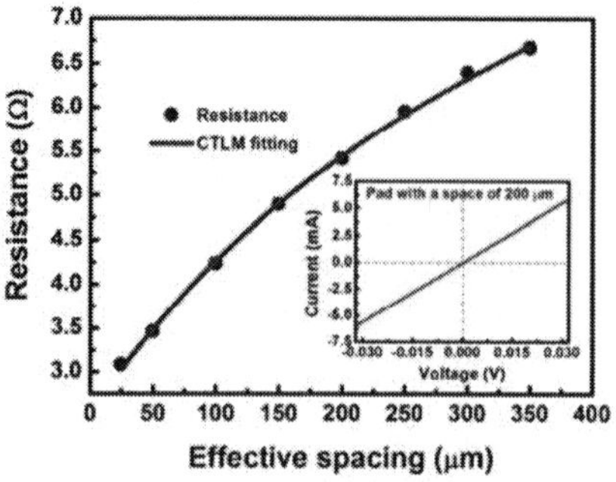

Figure 3: Contact resistance between the ITO film and the n$^+$-GaAs layer. The I–V characteristic of the sample measured over the pad with a spacing of 200 μm is shown in the inset.

3.3 J-V characteristics

Fig. 4 shows the measured J-V characteristics of samples A, B, and C. The open-circuit voltages (V_{oc}) of samples A, B, and C are 2.21, 2.22, and 2.14 V, while their short-circuit current densities are 9.61, 11.16, and 11.65 mA/cm^2, respectively. Obviously, the sample C possesses a lowest V_{oc} and a highest J_{sc} than those of other samples. Since the overall coverage on the device surface was achieved in the ITO-overcoated front-side electrode, all regions on the surface of sample C can extract the electrons, leading to the highest J_{sc}. Additionally, the conversion efficiencies of samples A, B, and C are 17.50%, 20.41%, and 20.92%, respectively. As compared with the sample A, there are 16.6% and 19.5% enhancements in the conversion efficiency for samples B and C, respectively. It indicates the ITO-finger and the ITO-overcoated front-side electrodes are both useful to improve the conversion efficiency of III-V compound solar cells, in particular for the ITO-overcoated electrode.

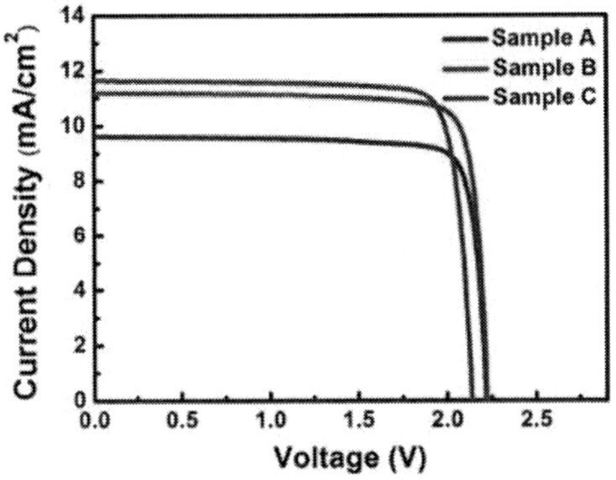

Figure 4: J–V characteristics of samples A, B, and C measured with one-sun AM1.5G light source at room temperature.

3.4 EQE measurement

Fig. 5 displays the EQE spectra of samples A-C with a measured wavelength range of 300-1000 nm. As the ITO-overcoated front-side electrode was used in the sample C, the relatively higher EQE values of 70% and 75% were achieved for top and bottom junctions, respectively, as compared with those of samples A and B. Based on the above result, it can be confirmed that the ITO-overcoated front-side electrode plays both the roles of the TCL and an anti-reflection layer, which would increase the absorption efficiency of the visible light. Thus, by preparing the ITO-overcoated front-side electrode in the sample C, there is an obvious increment in the EQE.

4 CONCLISIONS

In summary, the InGaP/GaAs dual-junction solar cells were prepared with three types of front-side electrode, i.e., AuGe/Au metal-finger, ITO-finger, and ITO-overcoated electrodes. Via the fabrication of the ITO-overcoated electrode, not only the optical characteristic of the solar cell was improved, but also the shadow loss created from the traditional metal electrode can be reduced. According to the J–V measurements, the conversion efficiencies of these three devices were 17.50%, 20.41%, and 20.92%, respectively. Compared to the solar cell with the AuGe/Au electrode, the device prepared with the ITO-overcoated electrode possesses 19.5% improvement in the conversion efficiency. This indicates the ITO-overcoated electrode is highly beneficial for the InGaP/GaAs dual-junction solar cells.

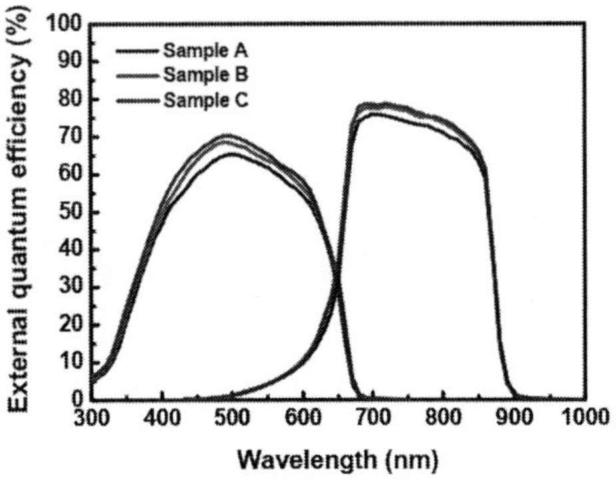

Figure 5: External quantum efficiency (EQE) curves of all samples.

5 ACKNOWLEDGEMENTS

This work was supported by the Ministry of Science and Technology (Taiwan, R.O.C.) under the Contract No. MOST 105-3113-E-009-005.

6 REFERENCES

[1] M.A. Green, K. Emery, Y. Hishikawa, W. Warta, Prog. Photovoltaics 22 (2014) 701.
[2] R.R. King, D.C. Law, K.M. Edmondson, C.M. Fetzer, G.S. Kinsey, H. Yoon, R.A. Sherif, N.H. Karam, Appl. Phys. Lett. 90 (2007) 183516.
[3] H. Cotal, C. Fetzer, J. Boisvert, G. Kinsey, R. King, P. Hebert, H. Yoon, N. Karam, Energy Environ. Sci. 2 (2009) 174.
[4] E. Havard, T. Camps, V. Bardinal, L. Salvagnac, C. Armand, C. Fontaine, S. Pinaud, Semicond. Sci. Technol. 23 (2008) 035001.
[5] S.J. Chang, J.L. Hou, T.J. Hsueh, K.T. Lam, S. Li, C.H. Liu, S.P. Chang, IEEE J. Photovolt. 3 (2013) 991.
[6] C.Y. Tseng, C.T. Lee, Sol. Energy 89 (2013) 17.
[7] C.Y. Tseng, C.K. Lee, C.T. Lee, Prog. Photovoltaics 19 (2011) 436.
[8] B. Yoo, K. Kim, S.H. Lee, W.M. Kim, N.G. Park, Sol. Energy Mater. Sol. Cells 92 (2008) 873.

LUMINESCENT SOLAR NOISE BARRIER – LARGE SCALE TESTING AND MODELING

L.H. Slooff[1], S. Verkuilen[2], M.M. de Jong[3], M.N. van den Donker[3], M. Kanellis[4], M.G. Debije[5]
[1]ECN Solar Energy, P.O. Box 1, 1755 ZG Petten, The Netherlands
[2]Heijmans Wegen B.V., Graafsebaan 67, P.O. Box 5248 JT Rosmalen, The Netherlands
[3]SEAC, High Tech Campus 21, 5656AE Eindhoven, The Netherlands
[4]Smart Energy Buildings & Cities, Eindhoven University of Technology, 5600 MB Eindhoven, The Netherlands
[5]Chemical Engineering & Chemistry, Eindhoven University of Technology, P.O. Box 513, 5600 MB Eindhoven, The Netherlands
Phone: +31 88 515 4314, Fax: +31 88 515 8214, E-mail: slooff@ecn.nl

ABSTRACT: Two modular noise barrier test setups containing photovoltaic modules have been installed in the Netherlands, one in a North/South orientation, and the other in East/West orientation. Both setups contain one set of mono-facial c-Si panels, one set of bi-facial c-Si panels and two luminescent solar concentrator (LSC) panels, one red and one orange/yellow. The performance of all panels is monitored along with the irradiation and temperature data. Here we will focus on the performance of the two LSC panels and present the results of a ray-tracing model that mimics the situation at the demo site.
Keywords: Luminescent Solar Concentrator, Modelling, Field Test, Noise Barrier

1 INTRODUCTION

Noise barriers offer the enormous potential of using the otherwise unproductive area along highways and railways for energy production. There are a few demonstration installations using traditional silicon-based solar panels [1,2,3,4]. Luminescent solar concentrators (LSCs) could play a role in initiating the large scale implementation of energy producing noise barriers. The different colors potentially offer more variety in the design of the barriers compared to traditional PV panels. Furthermore, LSCs are (semi)-transparent, which is generally more appreciated by the community and in some cases even a requirement for noise barriers. Although there have been some larger scale applications of LSCs, not much is known about their annual performance.[5,6,7] This study provides experimental and theoretical data on a large scale test setup which enables performance prediction of an LSC noise barrier with different orientations, sizes and colors. The results form the basis of a study to determine if a solid business case can be made for noise barriers based on LSCs.

2 LUMINESCENT SOLAR CONCENTRATORS

A luminescent solar concentrator normally consists of a polymer plate that contains luminescent species like. organic dyes or inorganic quantum dots. Solar light is incident on the plate and is absorbed by the luminescent particle and subsequently emitted. See Fig. 1

The light is emitted such that a fraction of the light will be trapped inside the plate by total internal reflection. This trapped light will travel through the light guide until it reaches one of the side edges. PV cells are connected to the sides to convert the emission light into electricity. This concept was first introduced in the mid 70's. [8,9,10,11].

The LSC has an important advantage over geometric concentrators in that both direct and diffuse sunlight is collected, making solar tracking unnecessary [12]. This makes them especially interesting for building integration applications.

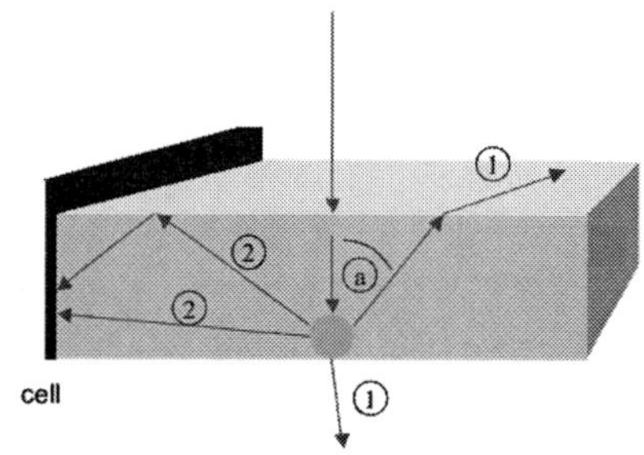

Figure 1: Schematic 3D view of a luminescent concentrator. Light is incident from the top. The light is absorbed by a luminescent particle. The luminescence from the particle is emitted. A fraction of the emission falls within the escape cone (determined by the angle (a)) and is lost from the LSC through the surfaces (1). The remaining fraction of the luminescence is guided to the Si cell by total internal reflection (2)

3 TEST INSTALLATION

A commercial modular noise barrier forms the basis of the test installation. Instead of the standard noise barrier glass or plastic, the setup contains 4 different types of panels: one set of glass/glass mono-facial c-Si panels, one set of glass/glass bi-facial c-Si panels and two luminescent solar concentrator (LSC) panels, one red and one orange/yellow (see Fig. 2). The site consists of two modular noise barriers, one facing North/South and one facing East/West. The barriers are tilted 15° towards the North and East, respectively. The individual panels are 1 m high and 5 meters long. A photograph of the experimental site is shown in Fig. 3. In this paper we will focus on the LSC part of the noise barrier.

Figure 2: Artist impression of the modular noise barrier.

Figure 3: Photograph depicting the test setup of the two modular noise barriers

4 APPROACH

4.1 The ray tracing model

A ray-tracing model was used to simulate the performance of the test setup. In the model a ray represents light of a specific wavelength travelling in a given direction. A ray that is incident on an air- LSC interface can proceed in two ways; it can either be reflected, or transmitted. In ray-tracing, random numbers are generated and used to determine the fate of the ray in such a situation. Consider for example a reflection at an interface with a reflection coefficient of 30%. A random number r is drawn from a uniform distribution in the interval [0,1]. If $r < 0.3$ the ray is reflected, if $r \geq 0.3$ the ray is transmitted. This ray-tracing model has been used previously for modelling lab-scale LSCs and demonstrated accurate predictions when compared to experimental results for LSCs of various sizes, dye content and photovoltaic cell attachments.[13,14,15] Input for the model are the dye absorption and emission characteristics, concentration, luminescence quantum yield and the polymer optical constants. Emission of the dyes is assumed to be in random direction. Only direct illumination of the plates is taken into account. The test set-up contains series connected c-Si cells at various positions along the top and bottom side of the LSC plates. For the PV cells, typical internal quantum efficiency data are used and other parameters were taken from fits to measured I-V data. The basics of the model were first compared to reflection and transmission measurements of the LSC plate material. When transmission and reflection are measured using an integrating sphere, light emitted by the dye will also be detected. An estimate of the dye the luminescent quantum yield (LQY) was obtained by fitting the LQY in the model to match the dye emission part in the experimental reflection and transmission measurements. Subsequently the model was compared with the actual demonstration panel results using the derived LQY values for the dyes. Finally, improvements in LSC noise barrier design are considered to optimize the performance.

4.2 Calculation results

Comparison of the model calculation with data from the test site for a sunny day on May 24, 2015, shows an excellent correlation. Figure 4 shows the normalized power and calculated results for one of the c-Si cell strips at the top of the south-facing red plate. Figure 5 depicts similar results for c-Si cells at the bottom and Fig. 6 the results for the orange plate.

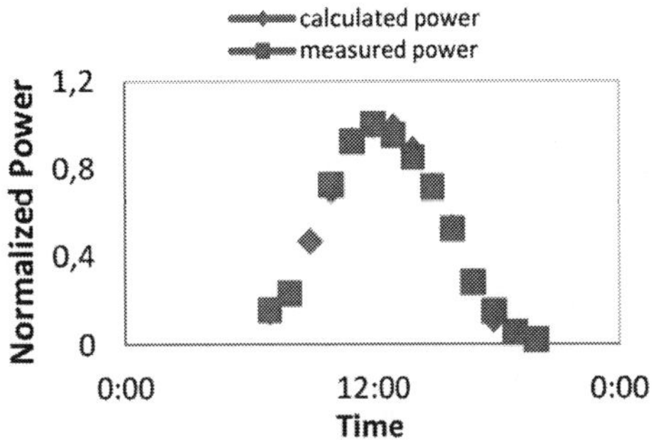

Figure 4: Calculated (blue, diamond) and measured (red, square) power from c-Si cells at the top of the red South-facing LSC panel

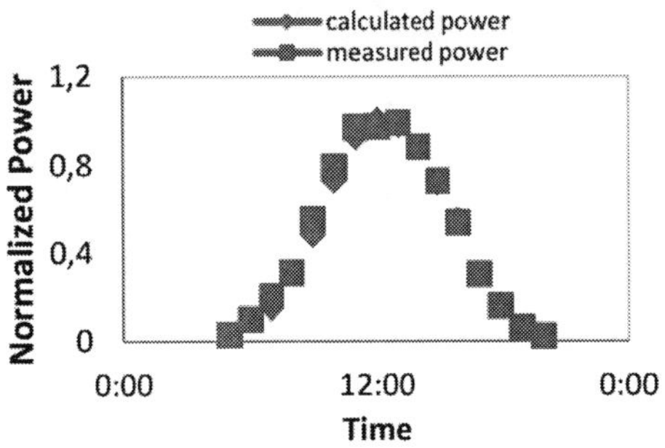

Figure 5: Calculated (blue, diamond) and measured (red, square) power from c-Si cells at the bottom of the red South-facing LSC panel

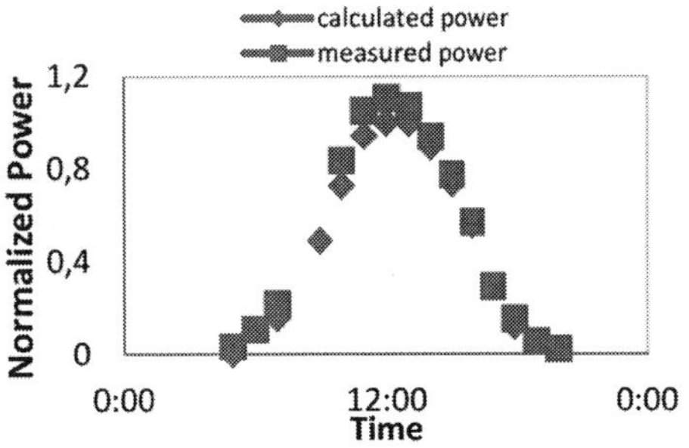

Figure 6: Calculated (blue, diamond) and measured (red, square) power from c-Si cells at the bottom of the orange South-facing LSC panel

In the model there were two fit parameters, the temperature coefficient of the V_{oc} and the dark saturation current I_0. For the c-Si cells a temperature coefficient of -3.8 mV/K was found and an I_0 of 1.1×10^{-11}-1.3×10^{-11} A. As for the overall power, the V_{oc} and I_{sc} matched nicely with the measured data, as can be seen for the V_{oc} in Fig. 7 and 8.

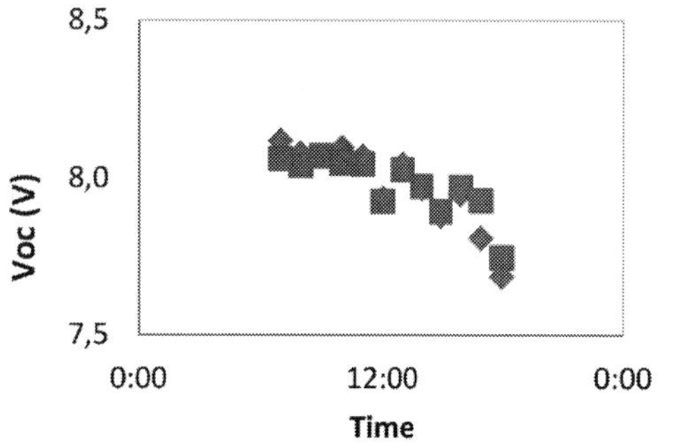

Figure 7: Measured (red, square) and calculated (blue, diamond) V_{oc} of the top c-Si cells on the South-facing red panel

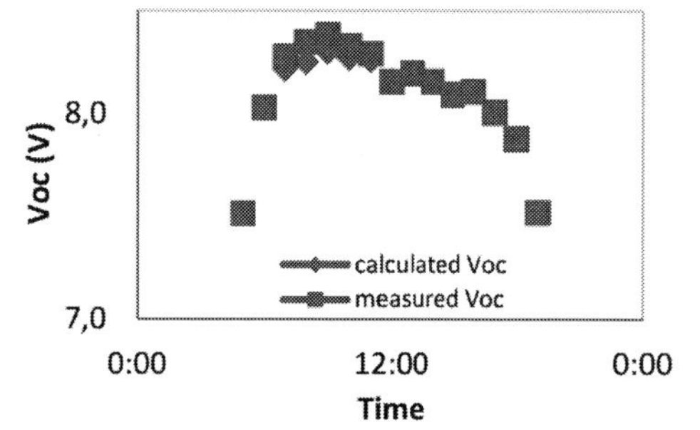

Figure 8: Measured (red, square) and calculated (blue, diamond) V_{oc} of the top c-Si cells on the South-facing orange panel

The observed changes in V_{oc} over the day are due to changes in both the temperature and irradiation conditions.

The results for the East/West oriented panels show some interesting results. Of course, the trend over the day is different as the sun rises in the East, thereby first illuminating the panel from the back side. Near noon it is illuminating the panel from overhead and in the afternoon from the front side. As the supporting aluminum structure at the top extends at the back, this results in shadowing of the rear side as the sun moves to the front. This only affects the cell strips at the top side of the East/West oriented panels, see Fig. 9.

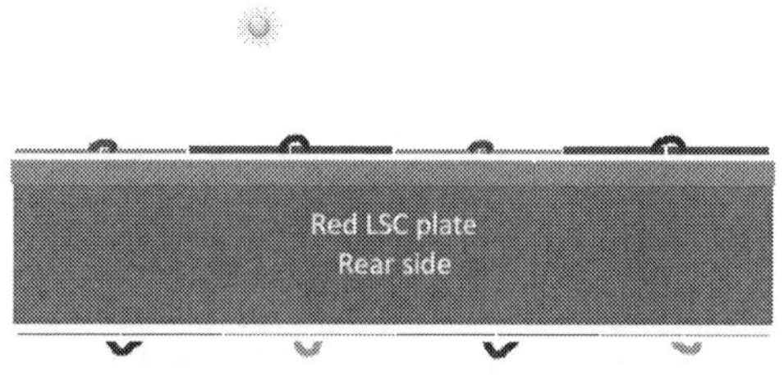

Figure 9: Schematic of the East/West oriented LSC in the morning looking from the East. The supporting construction induces a shadow (grey bar) along the top of the rear side, thereby affecting the output of the top cells

The effect of this shadowing can be seen in Fig. 10-11. In the morning, the modelled data predict a higher output power than was measured, whereas in the afternoon the model matches the measured results. The difference between the model and the measurement in the morning give an indication of the power that is lost due to self-shading by the supporting construction.

These observations makes it clear that for the East/West orientation it is especially important to minimize the self-shadowing of the supporting construction, although complete elimination of shadowing will not be possible. Calculations can be done to estimate the effect of the shadowing on the output of the cells to determine an optimum interconnection scheme for the cells and thereby minimizing the effect of the shadowing.

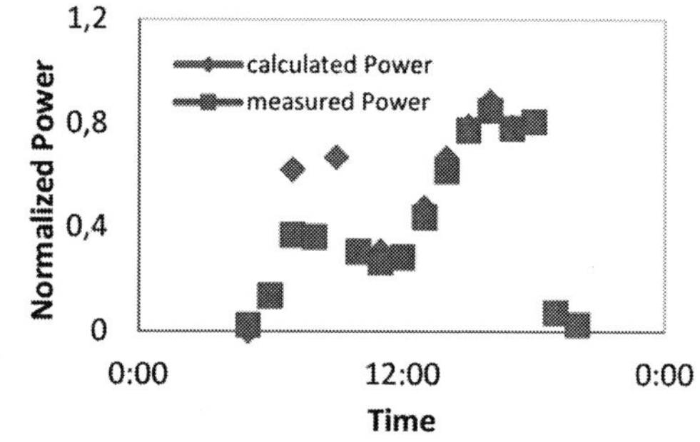

Figure 10: Calculated and measured power from c-Si cells attached to the top left of the West-facing red LSC panel

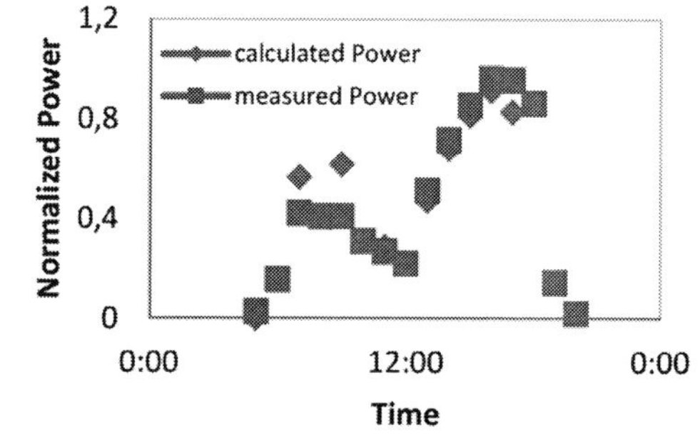

Figure 11: Calculated and measured power from c-Si cells at the top right of the red West-facing LSC panel. The effect of the shadow from the supporting frame on the output is clearly seen around 10:30

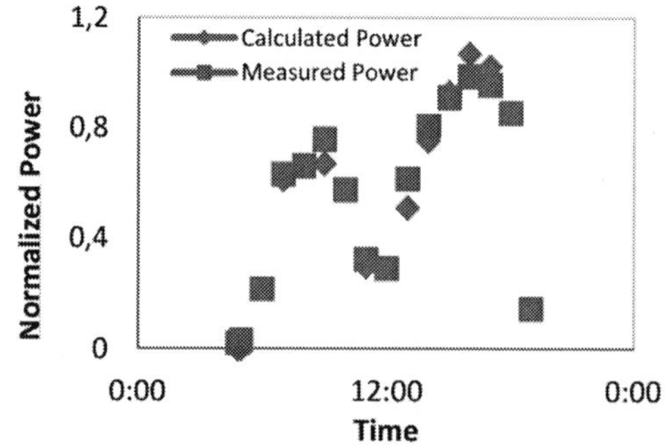

Figure 12: Calculated and measured power from c-Si cells at the bottom of the red West-facing LSC panel

Figure 12 shows the results for one of the bottom mounted c-Si cells. Clearly these cells do not suffer from self-shading by the supporting frame.

4.3 Optimization

Now that the model has been verified with the experimental data, it can be used to perform a parameter study to see if the performance can be optimized. As a start, the effect of varying the panel height and panel thickness has been tested. The result is shown in Fig. 13 where the panel height was varied from 0.3 to 1.0 m and the panel thickness from 12 to 20 mm. Reducing the panel height increases the response. as the shorter travel distance of the emission light results in less light being lost by scattering and (re-) absorption on its way to the cells at the side. An increase in thickness results in an increase in the performance of the panel. Due to the greater thickness, the light encounters less reflections at the front and rear side. As each reflection gives a chance for the light to escape the plate, less reflections will result in a higher performance. In addition, on average the dye molecules are more distant from one another, increasing the path length of emission light between encounters, reducing re-absorption and the associated losses.

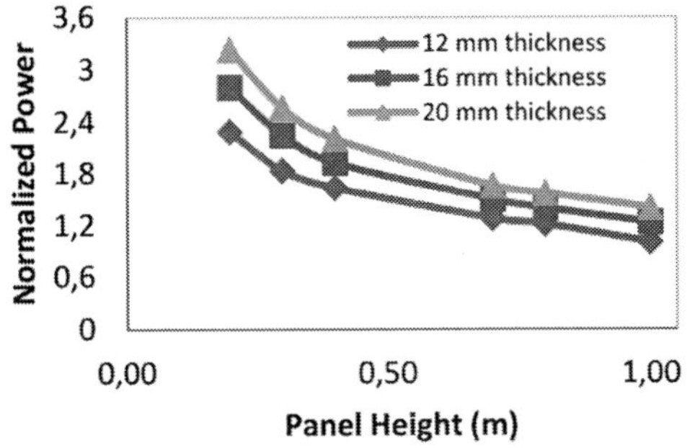

Figure 13: Normalized power output from c-Si cells at the bottom of the red West-facing LSC panel as a function of increased panel thickness and reduced panel height

5 CONCLUSIONS

Experimental results from a large scale luminescent solar concentrator (LSC) noise barrier test setup have been compared with the results from a ray-tracing model for a sunny day. The results show a good comparison between the model and the actual data for both the North/South and East/West barrier orientations. The data from the actual test site show the importance of reducing the shadowing losses caused by the supporting frame. The model can also be used to estimate the performance distribution along the sides of the LSC, thereby proving input for optimization of the electrical interconnection of the cells. Calculations show that the performance of the LSC noise barrier can be improved by increasing the thickness of the panels and by reducing the height. The latter will be at the expense of an increased number of cells. Cost calculations must be performed to determine if the improved performance outweighs the additional cell costs. Further work will focus on the use of the model for annual yield prediction of LSC based noise barriers.

6 ACKNOWLEDGMENTS

This work has been supported by the Dutch Government through the TKI ZEGO program. The authors would also like to acknowledge the other project partners: van Campen Industries BV, and Airbus Defence and Space for their support and the city of Den Bosch for making the test location available.

[1] Noise reduction B31 Germany http://pvdatabase.org/projects_view_details.php?ID=376

[2] Arras Highway21 Noise Barrier France http://pvdatabase.org/projects_view_details.php?ID=91

[3] PV noise barrier A9 The Netherlands http://pvdatabase.org/projects_view_details.php?ID=390

[4] PV noise barrier A27 The Netherlands, http://pvdatabase.org/projects_view_details.php?ID=74

[5] P. Bernardoni, R. Zarcone, M. Tonezzer, D. Vincenzi, M. Brocato, G. Calabrese, V. Guidi, S. Baricordi, Proceedings 31st EU-PVSEC (2015) 29034

[6] P. Bernardoni, M. Tonezzer, D. Vincenzi, V. Guidi, S. Baricordi, G. Calabrese, Proceedings 31st EU-PVSEC (2015) 2639

[7] P. Bernardoni, M. Tonezzer, D. Vincenzi, S. Baricordi, G. Calabrese, V. Guidi, Proceedings 31st EU-PVSEC (2015) 1429

[8] W.H.Weber, J.Lambe, Appl. Opt. 15 (1976) 2299

[9] C. F.Rapp, Norman L.Boling, Proceedings 13th IEEE Photovoltaic Specialists Conference (1978) 690

[10] A. Goetzberger and W. Greubel, Appl. Phys. 14 (1977) 123

[11] M.G. Debije, P.P.C. Verbunt, Adv. En. Mat. 2 (2012) 12

[12] M.G. Debije, V. A. Rajkumar, Solar Energy, Volume 122 (2015) 334

[13] A.R. Burgers, L.H. Slooff, R. Kinderman, J.A.M. van Roosmalen, Proceedings of the 20th EUPVSEC (2005) 394

[14] A., R. Burgers, L.H. Slooff, A. Büchtemann, J.A.M. van Roosmalen, Proceedings of the IEEE 4th WCPEC (2006) 198

[15] L.H. Slooff, R. Kinderman, A.R. Burgers, A. Büchtemann, R. Danz, T.B. Meyer, A.J. Chatten, D. Farrell, K.W.J. Barnham, J.A.M. van Roosmalen, SPIE Proceedings Vol 6197 (2006) 61970k1

DEVELOPING A LOW CONCENTRATION MODULE USING PV ASSEMBLY PROCESSES AND SUITABLE FOR BOTH TERRESTRIAL AND SPACE APPLICATIONS

Clément Weick[1], Pablo Garcia-Linares[1], Philippe Voarino[1] and Mathieu Baudrit[1]
[1]CEA-LITEN, LCPV, clement.weick@cea.fr, INES, Le Bourget du Lac, France

ABSTRACT: Concentrator photovoltaic (CPV) modules are composed of many components and interfaces, which require complex assembling processes, resulting in fabrication complexity and often lack of reliability. The present work addresses these issues, by proposing an innovative low concentration photovoltaic (LCPV) concept. In particular, the purpose here is to develop a module with a high level of integration by lowering the number of components and interfaces. The mirror used as the concentrator optic is multifunctional, as it combines thermal, structural and optical function. Moreover, the proposed design claims to demonstrate the applicability of reliable flat PV processes (such as lamination and cells interconnections), for the manufacturing of this LCPV module. The paper describes both thermal and optical modelling, developed to validate the concentrator design. Then, the main processes qualifications, as well as results on the characterization of the preliminary developed prototypes, are given. The potential of the LCPV concept, for space applications, is also discussed.
Keywords: Concentrators; Cost reduction; Module Integration; Space applications; Tracking.

1 INTRODUCTION

CPV technology takes advantage of low cost concentrator optics, in order to increase the irradiance on reduced-area high efficiency multijunction solar cells (MJSC). In comparison with flat PV, CPV modules are composed of many components and interfaces. Especially for high concentrated photovoltaic (HCPV) mainly represented on CPV market, the optical and thermal constraints require a high assembly accuracy and complex fabrication processes which have a significant impact on the module fabrication cost.

Today, new low cost MJSC are appearing on the market such as III-V on Silicon cells either thin film MJSC with substrate reusing [1]. These cells technologies are expected to cut down the price per watt peak (Wp), with good efficiencies over 30%. Although their targeted cost could not be competitive with Silicon cells within 1-sun applications, they would become affordable for low concentration application [2,3]. Opening the door of LCPV (2X-50X) to such high efficiency cells brings up new opportunities of development and applications, getting rid of the HCPV constraint of process and assembly accuracy. Low concentration modules have larger acceptance angle and, thus, require less tracking accuracy, most of the time even by single axis trackers. Moreover, having the ability to capture a part of the diffused light, LCPV can be deployed under lower DNI (direct normal insolation) level locations, compared to HCPV. Finally, the optical and thermal constraints are simplified, due to the lower flux incident to the cell.

Moreover, the PV market shows a constant decreasing of the manufacturing costs thanks to silicon production improvement but also to the high throughput fabrication methods such as lamination process. This process allows fast and reliable encapsulation of the cells.

The purpose of the present concept is based on lowering the number of involved components and interfaces, thus, simplifying the assembly process. The proposed LCPV design (Fig. 1) includes a parabolic trough reflective concentrator [4] that concentrates the solar flux on a focus line located on the rear of the following mirror. Although such mirrors arrangement, generically known as parabolic trough, has been already studied in past design studies [5], the cell array was placed on a receiver including aluminum fins as heat sink; and the mirrors were usually made of coated glass. Here, the innovation remains in the multi-functionality of the concentrator mirror acting as primary optical element (POE), which, in addition to the optical function, becomes the mechanical support of the cell arrays and the heat sink in charge of the cell cooling. By doing so, we get rid of both the CPV receiver element, and the heat sink, which represent a major part of the CPV module cost [4]. Similar design efforts were developed and demonstrated by Whitfield et al. [6]; however, here, the goal is to optimize a MJSC-based concentrator and its associated fabrication processes. Moreover, this concept relies on well-known, high-throughput, reliable and cost-effective flat-plate PV manufacturing techniques, such as lamination and interconnection, for the cell array assembly into the rear of the POE.

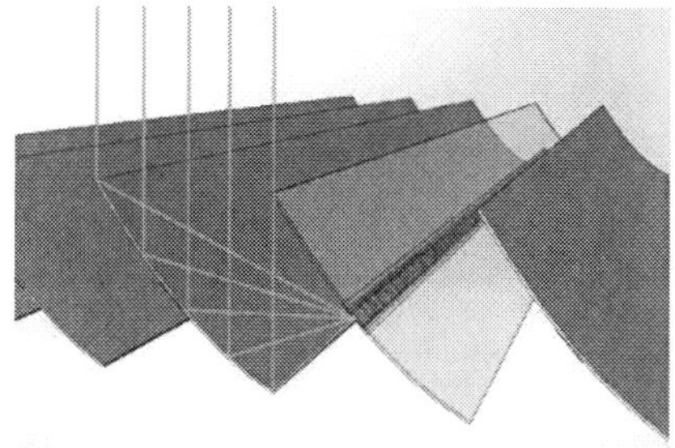

Figure 1: Sketch of the LCPV highly integrated receiverless concept.

In this work, the main actions leading to the design and fabrication of the first highly integrated LCPV module prototypes will be presented, e.g. dimensioning, optical and thermal simulation, prototypes manufacturing, lamination processing, etc. These prototypes have been electrically evaluated indoors by means of I-V curves tracing using collimated light from a Helios 3198 CPV module characterization system.

This module concept represents a good opportunity of lowering the fabrication costs for terrestrial applications. Besides, the lightweight aluminum (Al) mirrors and the reduced number of components could potentially give this concept access to space applications [7].

2 DIMENSIONING

In a first approach, the thermal ability of the mirror to dissipate the heat has been chosen as the dimensioning driver that maximizes cell efficiency and minimizes performance losses and subsequent end-of-life failures due to heating. A first analytical thermal model has allowed calculating the cell temperature as a function of the mirror surface available for convection. It shows that the temperature distribution has to be homogenous through the mirror surface, in order to feature maximum thermal efficiency, in terms of convection. On this basis, the cell array has been placed in the middle of the mirror's back side, equidistantly from the top and bottom edges. A complementary model calculates the relative temperature gradient between the mirror edge and the cell, versus mirror thickness and width (Fig. 2). This model allows defining dimensions that ensure a good thermal spreading throughout the mirror by means of conduction. Given the results, a mirror thickness of 2mm and a width of 20cm were chosen, considering the cell array across the mirror median line. As Fig. 2 shows, these dimensions guarantee a maximal temperature gradient of 90% relative to the maximum value, corresponding to the cell temperature.

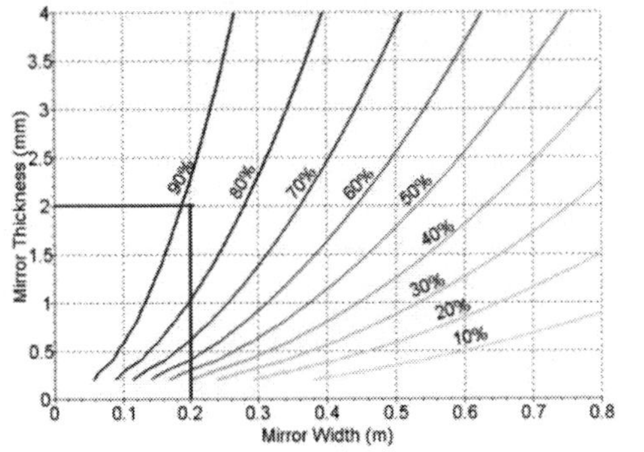

Figure 2: Thermal modeling results: Maximum temperature gradient within the mirror (in % relative to the maximum value) versus width and thickness.

Considering a 1cm² commercial MJSC, the 20cm width mirror matches typical concentrations (10X-50X), based on prior parabolic trough LCPV experiences. The thermal model has been upgraded with finite element simulation, implementing thermal interfaces under worst case convection conditions (no wind effects assumed) and ambient temperature T_{amb}=30°C. The temperature increases by a ΔT_{cell}= 57°C for a simple plain aluminum (Al) sheet, used as mirror, and matches the temperature increase calculated with the simpler previous model.

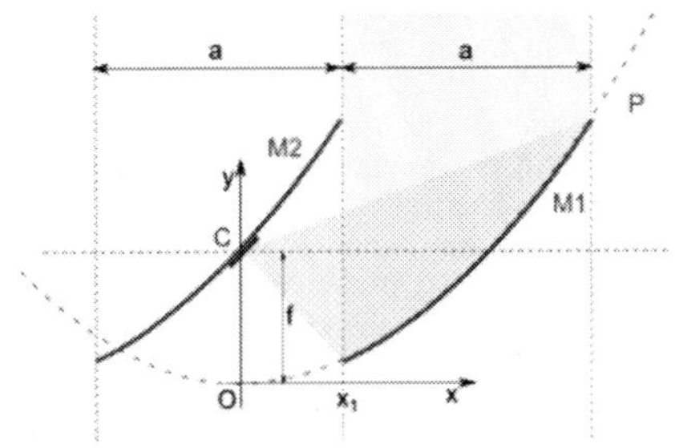

Figure 3: Sketch of the concept. *M1* and *M2* are two adjacent mirrors; *C* is the cell; *f* the focal of the parabola *P*; x_1 is the abscissa of the bottom mirror edge; x_1 defines the limits of the aperture *a* and, thus, the relative position of the cell on the mirror backside.

Such thermal analysis of the described system allowed defining the dimensions of the overall design, which will be taken into account for further steps of prototyping and characterization.

Fig. 3 shows a sketch of the module geometry. The focal line on the back of the adjacent mirror imposes that the focal distance *f* is the half of the aperture *a*, i.e. $f=a/2$. Looking at this geometrical study, an aperture a =150mm matches the desired 20.5cm long mirror profile. The mirror will be then a portion of a 75mm focal parabola *(P)* with an appropriate limit *x1* for which the cell position is equidistant from the edges of the mirror.

3 OPTICAL AND THERMAL STUDY

The obtained design has been simulated by ray-tracing, implementing a source of sunrays within a 0.27° cone angular solar distribution and different deviation angles (reproducing misalignment), with respect to the plane perpendicular to the sunrays. The results for the aforementioned design are presented in Fig. 4. It was found that 95% of the flux is comprised within a 3mm wide line and the peak-to-average ratio is 3.

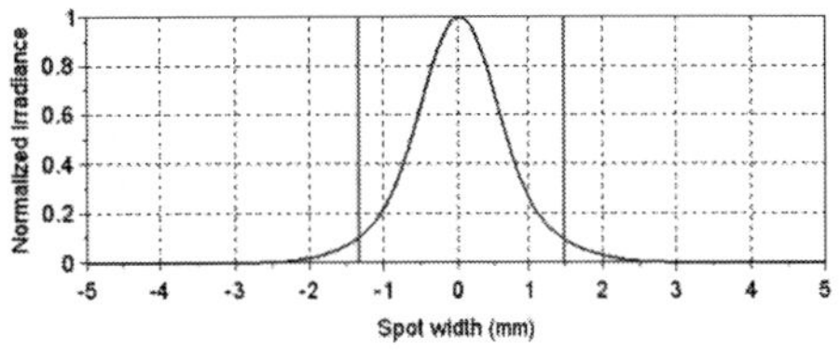

Figure 4: Irradiance distribution on the cross section of a 10mm wide cell, normalized to the maximum value.

For this design, the calculated theoretical acceptance angle (AA) is +/- 0.7°, which is compatible with one axis CPV trackers but low for such concentration ratio.

Fig. 5 shows a geometrical sketch of the system where the parameter x_1 has been lowered. Therefore, the aperture limits are shifted to the left and the cell is set closer to the top edge. It seems that such configuration improves the optical performance of the system, as the cell faces better the mirror. The angles of the rays coming from the mirror are well distributed around the cell front surface normal.

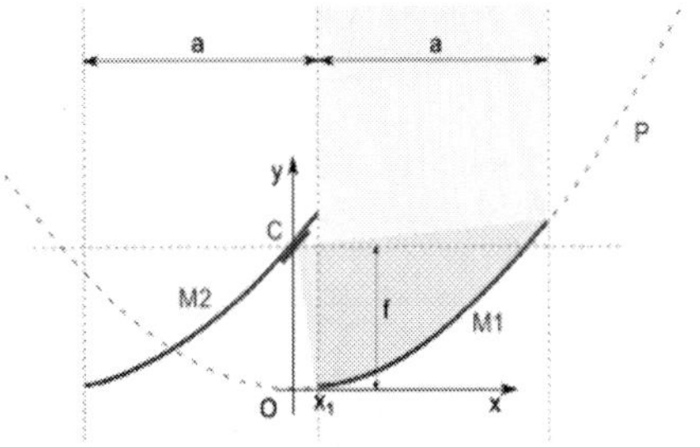

Figure 5: Sketch of the concept in an alternative configuration. x_1 is lower than in Fig. 3 and, thus, the cell moves up on the mirror back.

The AA have been calculated by means of ray tracing simulation for several positions of the cell. The results (Fig. 6a) show that the AA can reach up to +/-1.35° if the center of the cell string is 25mm far from the top edge. In other words, moving the relative position of the cell within the mirror's back surface can double the acceptance angle of the concentrator.

As for the dimensioning, the cell was centered in the aim to maximize the thermal homogeneity within the Al sheet and, thus, to optimize the heat transfer by convection. Moving the cell to the top will decrease the spreading through the mirror; as a result, the cell temperature should be higher. Fig. 6b shows how ΔT_{cell} is affected by the cell position on the mirror backside. This simulation has been done for different thickness values and realized for the worst case scenario (i.e. no wind, where convection coefficient $h=5$ and T_{amb}=30°C).

For the considered thickness, the ΔT_{cell} above ambient slightly rises from 57°C to 64°C. In conclusion, an increase of only 7°C under worst case scenario, allows to almost doubling the acceptance angle.

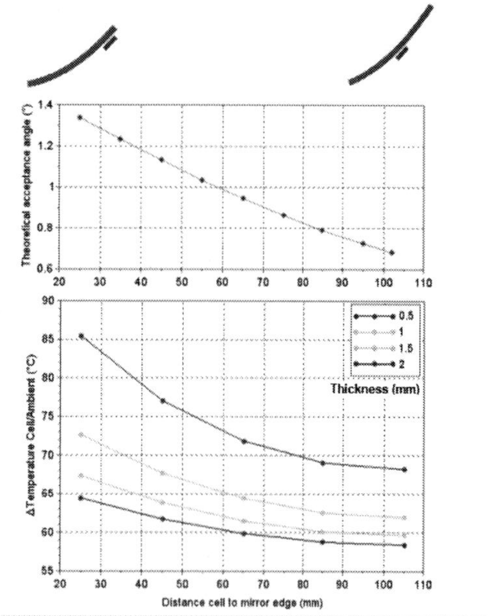

Figure 6: a) (Top) Acceptance angle as a function of the relative cell position on the mirror. This position is measured as the distance between the mirror top edge and the center of the cell. b) (Bottom) Evolution of the cell temperature above ambient versus cell position and thickness.

With the cell string centered, the module is compatible with typical +/- 0.5° accuracy trackers optimized for LCPV systems. However, moving the cell and increasing the AA to +/- 1.35° makes the system able to be tracked with a simple one-axis PV tracker (+/-1° accurate). Evidently, this is a step forward to the (desired) PV-like approach and our ultimate intention to get closer to flat PV cost distribution, in terms of fabrication processes and installation.

As it was previously presented, cell position and tilt cannot be optimized as the string is attached in the mirror backside. The only optimization possible for optical improvement is the one described above. In a conventional parabolic trough concentrator, the cell position is not dependent from any other part and its XY-position and orientation can be optimized. In 1980, A. Luque proposed an AA optimization for reflective parabolic trough design [8]. Therefore we have designed and simulated in ray tracing a conventional parabolic trough system with the same typical dimensions of the module studied, namely f=75mm, a=150mm, with a concentration ratio of 15X. Fig. 7 presents the line profile of the total irradiance incident to the cell string, as function of the misalignment between the sun beam direction and the perpendicular to the aperture area. The dashed line shows the irradiance evolution for the optimized concentration (15X) parabolic trough. The corresponding acceptance angle is +/-1.9°.

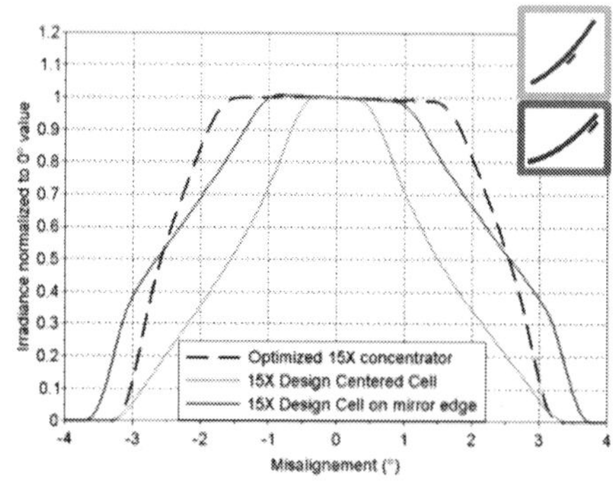

Figure 7: Normalized irradiance incident to the cell area as a function of the misalignment between the perpendicular of the cell aperture and the sun beam direction. Comparison for an optimized 15X concentrator and the two designs studied.

In brief, the simplification of the architecture deteriorates drastically the optical properties of the module, when the cell string is located in the center of the mirror's backside (for optimal heat spreading). On the other hand, shifting up the cell array can double the acceptance angle, getting closer to the optimum AA expected with such parabolic trough system at this concentration ratio.

4 PROTOTYPING AND CHARACTERIZATION

The first prototypes have been manufactured with the first design studied, i.e. with a centered cell. Only the centered design has been studied so far. More prototypes, with better expected AA, will be realized in future work.

Given that the cells are positioned directly on the back surface of the Al mirror, also acting as a heat sink, the cell string will be laminated using a conventional flat-plate PV mass production laminator. We have demonstrated the compatibility of the lamination process on Al mirrors with commercial 1cm² MJSC. The lamination stack is composed of a front sheet, an encapsulant underneath, then the cell array and a thermal conductive and electrical insulating pad as interface with the mirror (Fig. 8a)).

Samples have been laminated with adapted materials and recipe. The bending of the sample did not cause any damage to either the cells or the encapsulant layer. Characterization has been made by means of visual inspection, dark I-V (DIV) and electroluminescence (EL) measurements. The MJSC string encapsulated by lamination shows a similar DIV and EL before and after the process. Moreover, the samples passed typical fatigue testing, i.e. thermal cycling (-40°C to 110°C in climatic chamber). Indeed, no performance degradation was witnessed by EL or DIV; and no delamination of the encapsulant materials is visible.

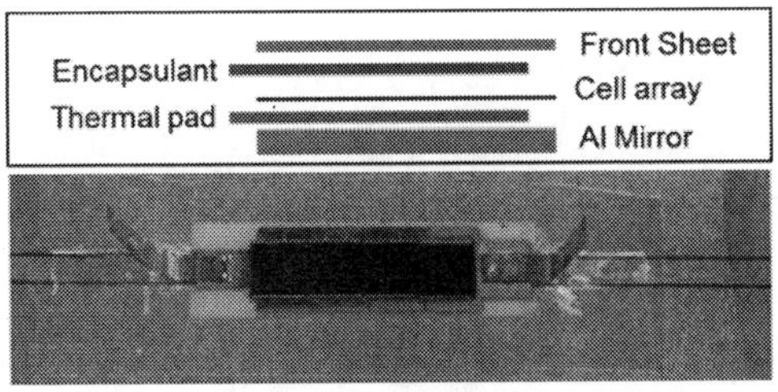

Figure 8: a) (Top) Lamination stack b) (Bottom) The laminated 4-cell string on aluminum mirror.

Two prototypes have been studied. The first one has been made from thin Al sheet, 0.5mm thick. This first prototype version (V1) has allowed developing the lamination process and testing a first functional prototype, prior to the rigid 2mm thick mirror manufacturing. With such thickness, this sheet is flexible and can adopt the curvature of a numerically controlled (NC) machined profile fabricated in purpose. The cell string is laminated on a flat aluminum mirror. Then, the resulting element is mounted on the structure composed of ribs, machined in a parabolic form (Fig. 9 (top)). This method is widely used for large PV/thermal parabolic trough concentrators with Al mirrors. The mirror edges are held by rigid screwed clamps.

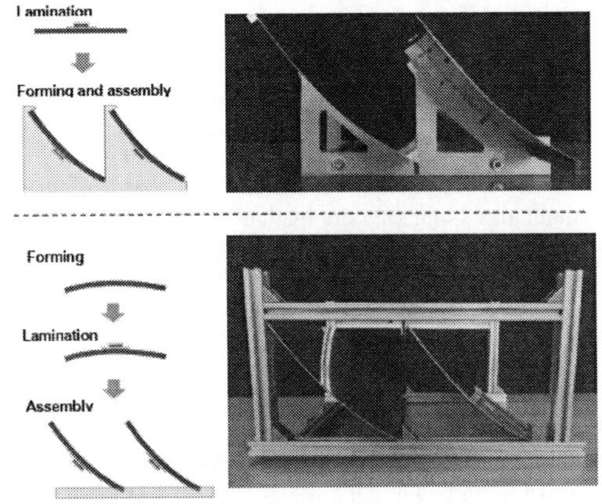

Figure 9: Prototypes simplified process flow sketch and picture for V1 (top) and V2 (bottom).

Different fabrication processes have been considered, for bringing a rigid 2mm thick mirror to the expected dimensions. The rigidity of the Al sheet requires to be formed before performing the lamination (Fig 9(bottom)). NC rolling has been identified as ideal, but no supplier has guaranteed enough precision in the parabolic profile. Cold stamping appears the best alternative solution but such process can only be developed for mass production series, due to tool development and adjustment costs. Indeed, stamping, consisting of pressing the Al sheet between two dies and the shape can be reached in one operation. However the tools adjustment is time-consuming and requires heavy machining processes. Therefore, another process has been explored, called pre-bending. In particular, the parabolic longitudinal profile is obtained by bending the Al sheet step by step, with wide angles. The fabrication is made with a traditional folding press, with a bending step of 5mm, resulting in a linear formed mirror with "faceted" aspect.

Thus, a second prototype (V2) has been manufactured using such mirrors. First, the mirror is shaped. Then the cell array is laminated on the shaped mirror. For this prototype, the mirror reflective coating has been realized by laminating an Al coated reflective polymer sheet. For reliability reasons, the polymer film will not be kept as a long-term choice, but it will be used for proof check of the concept. Other techniques such as physical vapor deposition or polishing will be explored in future works.

More than the good heat spreading offered by a 2mm thick mirror, this optical element is completely rigid once it is formed compared to a 0.5mm thin sheet. This is a good advantage for the shape stability and robustness. Moreover, the mirrors can be self-supported on long distances minimizing the number of holding supports.

Both prototypes presented above were characterized indoors, by means of a solar simulator, type Helios 3198. This CPV module sun simulator fulfils the concentrator standard test conditions (CSTC) requirements (1000W/m^2, 25°C and AM 1.5d spectrum), generating a homogeneous irradiance, with a collimated ray bundle. Fig. 10 shows the I-V characteristics of the two prototypes, with the 4-cell string integrated in the modules, which were also characterized under concentrated light at 15X, using a Xenon multi-flash cell simulator which also provides CSTC conditions. Efficiencies have been calculated for the 4-cell string on both concentrator prototypes and under the flash cell tester, for an aperture area of: i) 60.9cm^2 in the case of the prototypes, at 1X (1 sun) exposure, and ii) 4.06cm^2 for the case of the cell string at 15X. The results are shown in Table I.

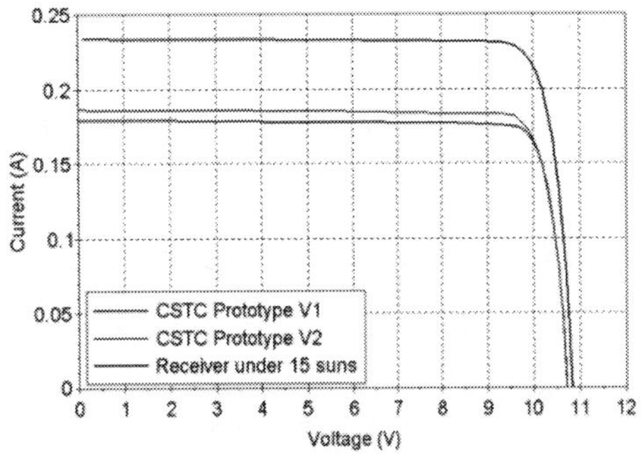

Figure 10: I-V measurements of prototypes V1 and V2 and associated cell string under 15X illumination for an aperture area of 60.9cm^2.

Table I: I-V parameters extracted from I-V curves above.

	CSTC Prototype V1	CSTC Prototype V2	Receiver under 15X
V_{oc} (V)	10.73	10.72	10.83
I_{sc} (A)	0.179	0.186	0.233
P_{pmp} (W)	1.69	1.75	2.19
V_{pmp} (V)	9.70	9.60	9.78
I_{pmp} (A)	0.174	0.182	0.225
FF	0.878	0.874	0.868
Efficiency	**27.7 %**	**28.7 %**	**36.1 %**

Prototype V2 reaches an encouraging 28.7% efficiency, 1% above the first prototype. Comparison of the I-V measurements of the cell array shows a cell-to-module absolute loss of 7.4% providing a global optical efficiency around 80% for the POE. Moreover, the acceptance angle for each prototype was measured and the results are shown in Fig. 11.

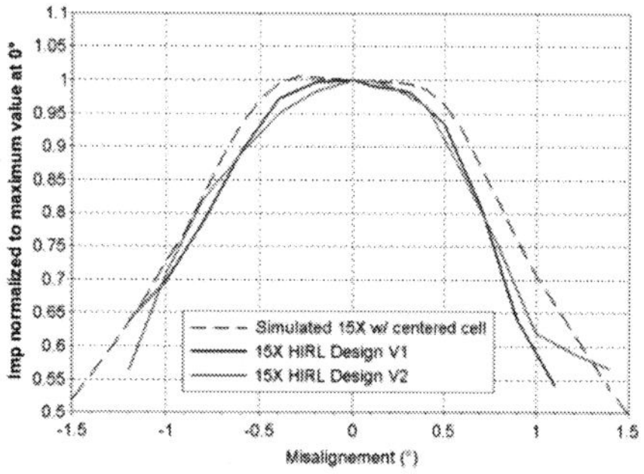

Figure 11: Acceptance angle measurement for prototype V1 and V2 compared with simulation.

The acceptance angle has been measured in the same Helios 3198 CPV sun simulator. The I_{mp} is plotted normalized to the one with perfect alignment (0° misalignment). This half AA is 0.55° for the two designs, compared to the 0.7° theoretical acceptance angle for the selected design. The optical accuracy of the POE needs to be optimized, in terms of fabrication process.

5 SPACE APPLICABILITY

Small satellite trends are shifting away from single mission use, moving towards more regular use in a constellation setting. Over the last years, there is an increasing interest and need for mass production of low-weight satellites. This pivotal moment in the evolution of the commercial satellite industry largely relies on the potential of PV industry to face such new needs and challenges, in terms of production capacity and versatility, to address different requirements imposed to the solar generators. In order to meet the needs of both their individual direct subscribers and those clients leasing large blocks of capacity, satellite operators are insisting on improved technologies, lower prices and faster manufacturing cycles from the satellite and photovoltaic arrays manufacturers. And this is where research and development (R&D) in space PV has a lot to learn from terrestrial photovoltaics.

The potential of the presented concept, towards its space applicability, has been investigated, aiming to its design suitability, in terms of efficiency, weight and cost for use in satellite powering and the involved challenges.

Several publications have shown in the past an interest for concentrator PV in space applications, with mostly low concentration ratios [7,9]. The first advantage is that the III-V MJSC cells efficiency can gain few points of efficiency, at concentrations from 1 to 10 or 15X. Another technical advantage that is often highlighted is that such architecture can protect cells from electromagnetism or particle impacts.

In a context where satellites manufacturing volumes could rise up, and manufacturers are in the pursuit of lower costs, concentration offers a clear economic advantage, reducing the amount of III-V materials in the modules.

On this basis, at first, we investigated the overall weight that such a concentrator module could reach. In particular, Fig. 12 shows the weight per square meter for several explored mirror thicknesses. A typical thickness of 0.3 mm results in a weight below the 1kg/m², which falls within the limits usually aimed by the manufacturers.

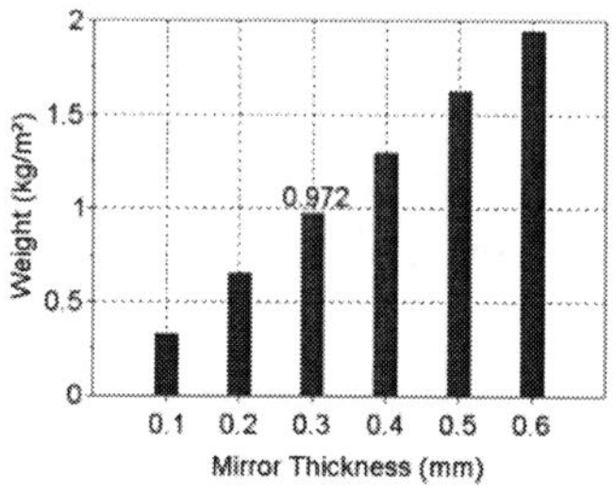

Figure 12: Weight per m² versus mirror thickness (mm), for the case of the concentrator module.

Moreover, using the integrated design presented in this work, the involved mirrors can be easily piled up. This feature offers a significant advantage, in terms of onboard storage for satellites. In practice, given a 5mm gap between each mirror, ensuring that the cell arrays are not damaged, a 2m long module can be folded in just 10cm (Fig. 13).

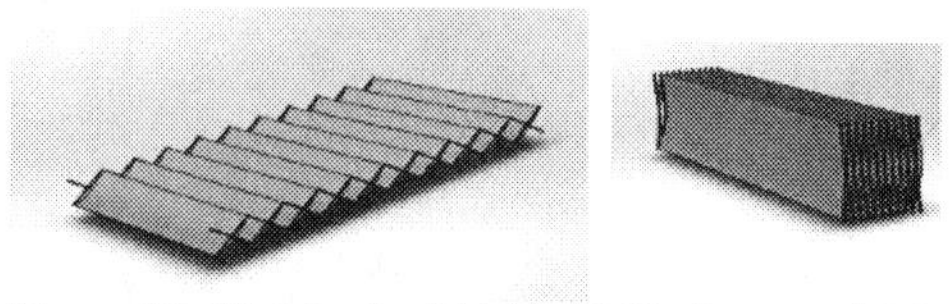

Figure 13: Sketch of a foldable highly integrated LCPV module for space application.

Given the mirror thickness and the emission of aluminum, thermal dissipation by radiation can be different than for traditional satellites flat modules. Indeed, performed thermal simulation showed the significant impact of the surface finishing on the radiative heat transfer in space environment. In principle, raw aluminum features a really low emissivity around 0.1 whereas an anodized finishing can improve it to 0.8.

The simulated cell temperature, for a 0.3mm thickness and assuming optimized emissivity, is around 100°C, which is acceptable for such III-V cells. As a comparison, reported literature shows that the cell temperature can reach up to 70°C, for a traditional flat plate PV module with triple-junction cells [10].

6 CONCLUSIONS

The innovative concept presented in this work intends to cut down the fabrication costs of CPV modules, by limiting the number of involved components, while simplifying the overall assembly. Besides, in such way, the fabrication/processing flow nearly emulates those adopted for flat tracked PV. The mono-element design, combining the receiver and concentrator function, needs mirror forming, stringing and lamination. In addition, storage and, thus, logistics issues are simplified, given the foldability of the designed mirrors.

In further technical terms, the large AA identified through the optical study, makes this module compatible, with commercial one axis PV trackers. Such a feature, comprises an important asset, considering that, today, trackers represent a major part of the CPV operational costs, even for (lower cost) single axis tracking [4].

Larger prototypes will be fabricated, in the aim to study the optical efficiency of the proposed design, for longer distances and for different mirror forming processes. This concept represents a great opportunity for new high efficiency cell designs which could be industrialized at lower cost, in the coming years. Eventually, the simplicity of the assembly could improve the bankability and reliability of LCPV modules.

REFERENCES

[1] J. Adams, V. Elarde, A. Hains, C. Stender, F. Tuminello, C. Youtsey, A. Wibowo, and M. Osowski, "Demonstration of multiple substrate reuses for inverted metamorphic solar cells," in Photovoltaic Specialists Conference (PVSC), Volume 2, 2012 IEEE 38th, 2012, pp. 1–6.

[2] M. Woodhouse and A. Goodrich, "Manufacturing Cost Analysis Relevant to Single-and Dual-Junction Photovoltaic Cells Fabricated with Iii-Vs and Iii-Vs Grown on Czochralski Silicon (presentation)," National Renewable Energy Laboratory (NREL), Golden, CO., NREL/PR-6A20-60126, May 2014.

[3] K. Lee, J. Lee, B. A. Mazor, and S. R. Forrest, "Epitaxial lift-off processed GaAs thin-film solar cells integrated with low-cost plastic mini-compound parabolic concentrators," in Photovoltaic Specialist Conference (PVSC), 2014 IEEE 40th, 2014, pp. 2127–2129.

[4] M. Vivar, I. Antón, D. Pachón, and G. Sala, "Third-generation EUCLIDES concentrator results," Prog. Photovolt: Res. Appl., vol. 20, no. 3, pp. 356–371, 2012.

[5] J. E. Granata, N. R. Sorensen, D. Riley, Z. Judkins, and S. Olsen, "Design for reliability: A low concentration PV case study," in 2012 38th IEEE Photovoltaic Specialists Conference (PVSC), 2012, pp. 001739–001743.

[6] G. R. Whitfield, R. W. Bentley, C. K. Weatherby, A. C. Hunt, H.-D. Mohring, F. H. Klotz, P. Keuber, J. C. Miñano, and E. Alarte-Garvi, "The development and testing of small concentrating PV systems," Solar Energy, vol. 67, no. 1–3, pp. 23–34, Jul. 1999.

[7] T. G. Stern, M. Cornwall, J. F. Lawrence, and D. M.Peterson, "United States Patent: 5131955 - Depolyable retractable photovoltaic concentrator solar array assembly for space applications," 5131955, 21-Jul-1992.

[8] A. Luque, "Quasi-optimum pseudo-Lambertian reflecting concentrators: an analysis," Appl. Opt., vol. 19, no. 14, pp. 2398–2402, Jul. 1980.

[9] M. Cornwall, Bruno NunesVaz, and T. G. Stern, "Solar concentrator array," US5180441 A, 19-Jan-1993.

[10] Célio Costa Vaz, Bruno NunesVaz, and Jaqueline NunesVaz, "Solar concentrator array," 20th International Congress of Mechanical Engineering 2009 by ABCM, 2009, Gramado, RS, Brazil.

PERFORMANCE ANALYSIS OF ECOSOLE HCPV SYSTEM

C. Cancro[1], G.Graditi[1], G.Ciniglio[1], G. Leanza[1], A. Borriello[1], A. Merola[1], S. Ferlito[1], F.Pascarella[1]
M. Carpanelli[2], G. Borelli[2], D. Verdilio[2], D. De Nardis[2], V.Gilioli[2], F. Migali[2]

[1] **ENEA** - Italian National Agency for New Technologies, Energy and Sustainable Economic Development
Research Center of Portici - 80055 Portici (Naples) - Italy; tel: +390817723100; fax: +390817723344
mail to: carmine.cancro@enea.it
[2]**BECAR srl** – Beghelli group's R&D company Viale della Pace 1 – 40050 – Monteveglio (Bologna) – Italy

ABSTRACT: The aim of this work is to report the results of the main characterization activities carried out on the ECOSOLE HCPV prototypal plant realized by BECAR s.r.l. and installed at ENEA Research Center located at Portici in the south of Italy. After a detailed description of the system main components, the experimental results regarding the module characterizations, the tracker accuracy, the system optical acceptance, and the module temperature survey are detailed.
The second part of the paper is focused on the explanation of the performance analysis carried out on the basis of the data collected during the first three months of operation. Particular attention has been paid to the performance comparison of the two sub-arrays of the plant: the first one is realized with module using each one its-own inverter, and the second one is realized with a classic centralized inverter solution. At the end, a procedure for a detailed losses analysis is reported and the results obtained for a whole sunny day are explained.

Keywords: Concentrators, System Performances, Inverter

1 INTRODUCTION

ECOSOLE is an European collaborative Project started in August 2012 and ended in January 2016 that involved several European industrial and institutional partners. It has been leaded by BECAR s.rl. that is a Beghelli group company specialized in industrial researches devoted to products innovation. Main targets of the project were design, development and testing of all CPV system components, and the realization of an high throughput, fully automatized construction line for the photovoltaic C-modules.

These objectives have been reached: the HCPV system main components as optics, receivers, modules, tracker and inverter [1,2] have been developed, the high automation CPV modules assembly line has been realized and it is now operative in a BECAR factory located near to Bologna and three experimental plants have been realized and installed in different sides. In particular a 27kWp prototype has been installed in Italy at BECAR headquarter in Monteveglio (BO) and two smaller 6kWp plants have been mounted and at Ben Gurion University in Israel and at ENEA Research Center in Portici (Italy).

This latter plant has been completed at December 2015, then it has been fully instrumented in order to collect all the meteorological and energetic data needed to analyze its performance parameters.

The aim of this paper is to report an accurate performance analysis of the system carried out during the first working months.

2 DESCRIPTION OF THE SYSTEM

The Ecosole Plant installed at Portici is composed by 32 HCPV modules having an overall active area about of 24sqm and a nominal power about of 6kW.

They are factory pre-assembled on four maxi-modules consisting on the metallic frames used to fix and to parallel align a sub-group of 8 modules.

During this process, precision mechanical connections between maxi-modules and tracker structure are on-factory realized in order to make all the modules parallel.

The advantages of this approach is that the installation becomes very easy since it will be not necessary to make any mechanical alignments on the field.

The whole panel is moved by two perpendicular, series connected slewing drives that are driven by a couple of electric brushless motors controlled by an integrated electronic control system. The tracking method is based on a Hybrid tracking algorithm using a sun camera and the sun ephemeris. Thanks to an embedded camera, the tracker automatically sets it up after the installation process. During the operation of the Plant, data gathered by the camera periodically corrects the algorithm's settings in order to maximize the tracker accuracy.

Figure 1: ECOSOLE CPV Plant installed at Portici ENEA Research Center

2.1 ECOSOLE modules technology

The modules nominal concentration is about of 940X, realized with a SoG Fresnel lens coupled with a prismatic quartz elements (SOE) glued on the cells top, acting as homogenizer and secondary concentrator.

The module housing is composed by an U-shaped back sheet and two lateral sides made of aluminum sheets. The lens parquet, organized as a 6x12 lenses matrix, is glued on the housing upper perimeter, while cell receivers are glued on its inner base surface.

The receivers are realized with commercial III-V PV devices bought by three different manufacturers, namely Azur Space, Emcore and Solar Junction. Each cell, together with its-own by-pass diode, is mounted on an alumina plate using DBC technology.

All the receivers are electrically series connected and consequently the expected module open circuit voltage is about of 230V.

Figure 2: ECOSOLE receiver assembled on an alumina base

Each Ecosole module is connected to its own single module inverter (fig. 3): a DC/AC converter with MPPT tracker able to inject the current into the 230V AC public network. It is mounted in a water-proof box, located on the tracker, beside the HPCV module: it is able to operate outdoors and also in the desert climate. The inverter has 94% conversion efficiency and it outputs AC current synchronously with the AC network voltage.

Figure 3: Single module inverter

This choice theoretically allows to collect the maximum possible energy yield of the HCPV generator, since every single module is independent from the others and there is no problem of module shadowing, mismatching or misalignment.

The prototypal system installed at Portici is split in two parts: the "sub-array A", that is realized with 16 modules, located in the tracker upper part, equipped with local inverters, and the "sub-array B" consisting on the other modules located in the tracker lower area, that are connected with a classic series-parallel arrangement, to an external, centralized inverter. This choice has been done to verify on the field the above mentioned advantages of the use of a distributed inverter approach.

2.2 Monitoring system

Several instrumentation are used for the detection of weather and energetic data:

- a meteo station is used for the detection of meteorological and in particular of the solar radiation data (direct, global, and diffuse components);
- a 3 isotypes spectra-radiometer is used for the measurement of the 3 equivalent DNIs view from each sub-cell: DNI_{top}, DNI_{mid}, DNI_{bot}. These latter are the solar irradiation values with standard spectrum that produce on each of three sub-cells, the same photocurrent generated by the actual irradiation.
- a pyrheliometer installed directly on the tracking plane measures the DNI radiation that really hits the lens parquet;
- an opto-electronic device enables to continuously measure the tracking errors both in azimuth and elevation;
- operative modules temperature are measured by six thermo-resistances attached to their back sheets.

With regards to the system production data, for the sub-array A, ac parameters are directly measured using a grid analyzer coupled with an amperometric transformer. The sub-array B, instead, is monitored also in dc side with current and voltage converters on each of two strings. All the data coming from the plant are measured each minute and collected in a file by a CR6 data logger from Campbell Scientific coupled with an ADAM module to convert thermo-resistance signals. DNI and spectral data are separately registered in two separated files produced by the respective data acquisition systems.

3 EXPERIMENTAL RESULTS

In this paragraph, main results of the experimental campaign will be detailed.

3.1 Modules characterization

In order to calculate the plant nominal power, it has been necessary to characterize all the modules of the plant. Since the modules of the sub-array A produce current directly in ac mode, their efficiency has been evaluated on the basis of the data coming from the local inverters. The modules of the sub-array B have been tested singly with an external electronic load.

The data have been normalized at CSTC conditions implementing the following steps:

- using voltage and current thermal coefficients, the I-V curves have been normalized at 25°C considering as operative temperature the one calculated on the basis of the modules back side temperature ;
- current values have been normalized at DNI of 1000W/m^2, considering as incident irradiance the lowest between the top and middle cell equivalent irradiances measured by the 3-isotype spectra-heliometer [3].

Results shown average efficiencies values of 29,7% and 31,6% considering respectively the measured and the CSTC normalized parameters, and the corresponding average modules powers are of 181W and 192W. Despite the best efficiency value is very high (34,5% in CSTC condition), a not good uniformity of the modules features within the batch is observed. This is due to the still not perfect calibration of the productive line in the very early stage, which has also determined the degradation of some modules due to the water infiltration during raining days.

From the average value of modules power, corresponding to 192W, it is possible to calculate the actual Plant nominal power that is of 6144W.

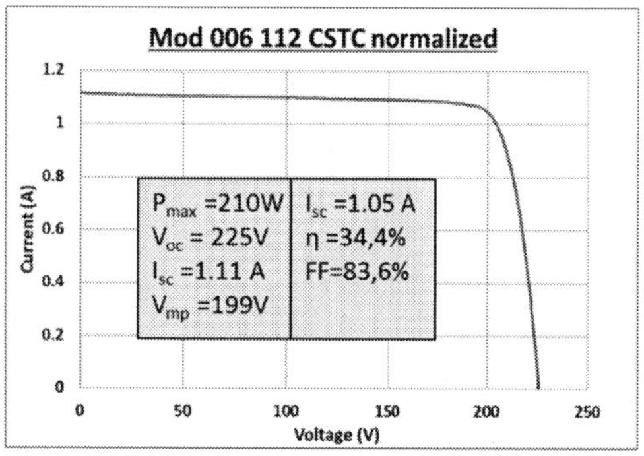

Figure 4: I-V curve and operative features of the best ECOSOLE module CSTC normalized

3.2 Tracking accuracy measurement

As aforesaid, an opto-electronic device ad hoc designed, developed and patented [4] by ENEA has been used to measure the tracker accuracy. It has been mounted (see Fig.5) parallel to the modules parquet and it collects during the day the angular misalignments between the solar beam direction and the parquet perpendicular. The data are elaborated and the corresponding error referred to the azimuth and elevation axis are calculated.

Figure 5: Opto-electronic device used for tracker accuracy measurements

Results, reported on the next figure for a typical sunny day, show both for azimuth and elevation axis, quite constant errors corresponding respectively to 0,35° and 0,3°. These latter are generated from the tracker control system that make the pointing corrections with the aim to maximize the energetic production.

In fact it has been experimental verified that, for problems connected to the production line, the modules real optical axis is not corresponding to the perpendicular of the parquet. In detail, the maximum power is not achieved when the parquet is perpendicular to the sun direction, but when there is a small horizontal misalignment angle between these two directions and this can explicates the revealed azimuth displacement, which has been intentionally applied to maximize performance.

Regarding the vertical error, we verified that most of modules have been mounted on the heliostat with an inclination ranging from 0,25° to 0,40° respect to the ideal plane. Then, the control system optimizing the performances of generators, introduces an angular offset quite constant also in elevation axis.

In correspondence of noon, it is possible to notate a small step in the vertical errors curve (red circle) about of 0,04°. Considering that at noon there is the direction motion inversion of the elevation actuator, this step can be due to a worm gear mechanical backslash.

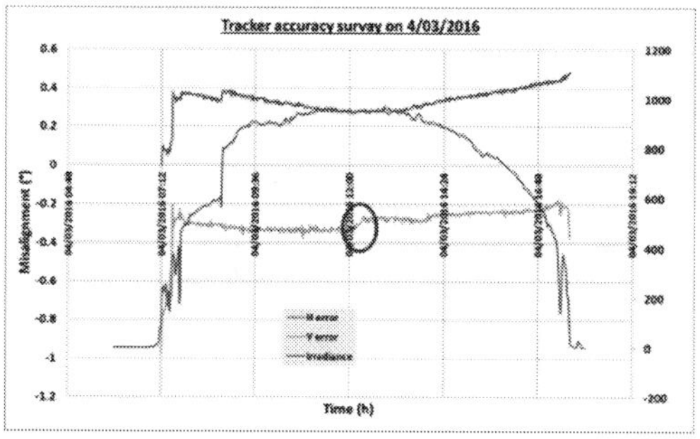

Figure 6: Tracking accuracy measurement

3.3 Module Thermal behavior

The temperature on the modules back sheet (T_{bs}) has been continuously monitored in different points of the tracker. From these data, and tacking into account the amount of dissipated power from each cell and the thermal resistance from the cell and the back sheet, it has been possible to calculate the cell temperature (T_{cell}).

Next figure reports the behavior of these parameters during a clear sky day with an ambient temperature about of 24°C. The maximum operative device temperature doesn't exceed 65°C, that is absolutely compatible with a proper cell operation.

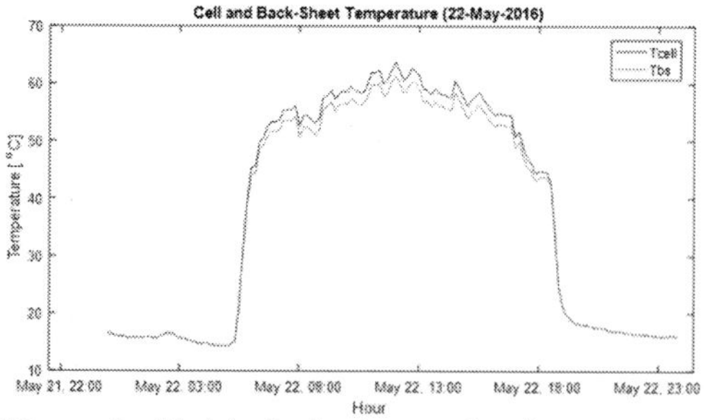

Figure 7: Module back sheet and cell temperature behavior during a sunny day

3.4 Plant acceptance angle measurement

In order to define an acceptance angle value for the system, the plant has been stopped where the tracker was in perfect alignment condition. From this moment, the solar path on the ecliptic determines a misalignment of the solar beam direction vs. the lens parquet perpendicular and at the same time, I-V tests on a complete string of module have been repeated every minute. The measured maximum powers have been correlated with the corresponding misalignment angles and normalized at the starting value, so that the following acceptance curve has been drawn. The misalignment angle corresponding to the 10% of system performance decrease respect to the perfect alignment condition is about of 0,6°.

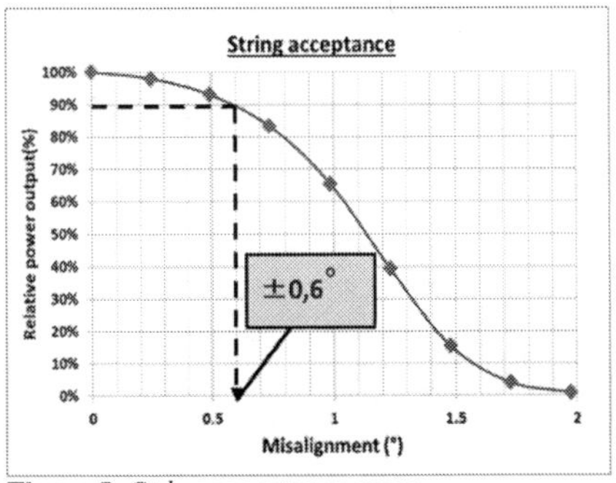

Figure 8: String acceptance curve

4 PERFORMANCE ANALYSIS

Weather and production data coming from the plant, have been collected every minute from the data logger and have been analyzed in order to calculate the main performance indices as prescribed by the current regulation [5]. The analysis has been done separately from the two sub-arrays A and B in order to compare the performances of the distributed and centralized inverter approach.

An ad hoc developed software calculates the daily values of the main indices as Y_r (reference Yield), Y_a (array Yield), Y_f (final PV system Yield), η_{inv}, (inverter efficiency) PR, (Performance Ratio).

Next chart reports the evolution of the daily value of Performance Ratio during first three working months. It is possible to notate that the sub-array A is always more performing than the sub-array B: this is due to the fact that, using one inverter per modules, mismatch losses due to the different features among the modules are avoided. The strong performance difference observable from the day 50 and in particular from day 62 is due to the water infiltration inside three modules of sub-array B. In fact, because the sub-array B is constituted by a parallel of 8 series of two modules, and because the modules with moisture are allocated in three different series, this problem affected the energy production of 6 modules.

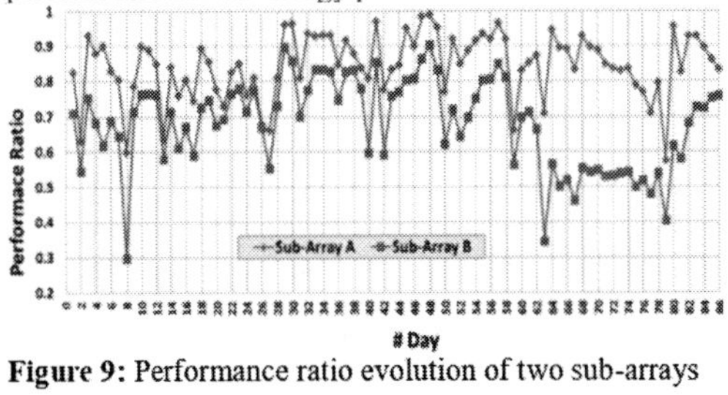

Figure 9: Performance ratio evolution of two sub-arrays

The average values of performance parameters are reported in the following table. Because of the modules of the sub-array A produce directly in ac mode, it has been not possible to calculate Y_a and η_{inv} that are referred to the performances of the plant dc side and of the centralized inverter.

	Sub-array A	*Sub-array B*
Y_r	4,78	
Y_a	NA	3,64
Y_f	4,14	3,33
η_{inv}	NA	90%
PR	84%	68%

Table I: Average values of the performance parameters rat

An ad hoc software tool has been developed [6] to perform the losses analysis of the Plant in dc side (then, it is available only for the modules of the sub-array B).

In particular, it is organized like a sort of logical path having the following calculation steps:

- **tracking losses** that are evaluated on the basis of the difference between the DNI measured by the meteo-station and the one measured by the pyrheliometer parallel to the tracking plane;
- **spectral losses**, calculated on the basis of the difference between the DNI value and the AM1.5 equivalent irradiance that is assumed as the lowest among the equivalent irradiances measured using the 3 isotypes spectra-radiometer;
- **conversion losses,** evaluated as the difference between the energy hitting the modules (considering as irradiance the equivalent AM 1.5 DNI) ant that one theoretically produced by the plant using the CSTC module conversion efficiency;
- **thermal losses**, calculated as difference between the energy produced in CSTC condition and the one calculated taking into account the cell heating over 25°C;
- **mis-match and other losses**, that is the difference between the energy calculated considering the actual cell temperature and the real dc energy produced by the plant.

At the begin of the procedure, the user has to choose the period of the analysis and has to set the plant general information including the STC efficiency of the CPV modules. For each examined period, the tool calculates the incidence percentage of each loss cause as previously listed: this analysis is very usable to understand the causes of a low energy production and to plan the right correction actions, then it could be used by the companies specialized in the monitoring of the photovoltaic plant.

As example we below report the analysis for march 25th that was a complete sunny day.

It shows that, excluding the conversion losses, that depend directly from the nominal conversion efficiency of the modules, the most important losses cause is due to the spectral mismatch of the incoming light respect to the standard spectrum conditions. This is due to the fact that during spring, for most of the hours of the day, the air mass coefficient is considerable higher than 1.5.

Good tracking accuracy revealed by means the opto-electronic device (see paragraph 3.2) is confirmed by the fact that the tracking losses are negligible. Thermal losses are about of 1,6% and this proves the good thermal behavior of the systems. Other losses, due mainly to the performances difference among the modules and to ohmic dissipations, amount to 2,2%. These latter, that can be avoided by using the distributed inverters solution, determine the difference, in terms of performance ratio, between the two sub-arrays of the Plant.

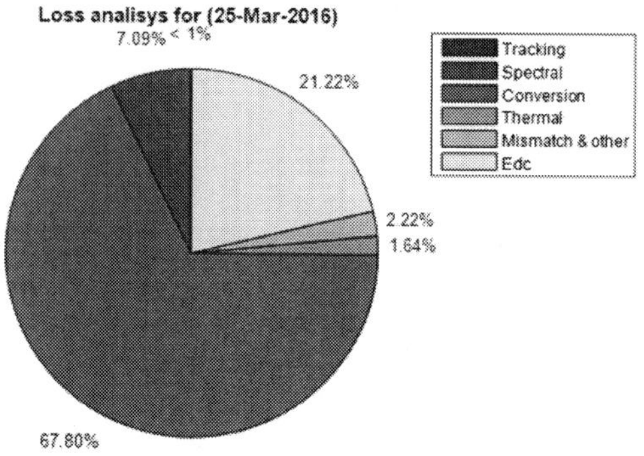

Figure 10: Losses analysis for march 25th

5 CONCLUSIONS

The ECOSOLE prototypal plant installed at ENEA Research Center has been fully instrumented and characterized.

Modules testing shown an average CSTS efficiency about of 31,5%, but, due to the not well initial calibration of the production line at the moment of the prototypes construction, it was found that the features within the modules batch are enough variable and this generates some important mismatch losses.

Tracking accuracy is very high: each motion group has a maximum angular backlash lower than 0,05°. Moreover the feedback algorithm periodically makes adjustments to the theoretical trajectory in order to maximize the plant energy output and in this way the effect of module imperfections and of mounting errors are minimized.

String optical acceptance is about of ±0,6°, then it is quite larger than the expected tracker errors.

The module thermal management is effective: the cells temperature in operative conditions is very far from their working limit.

Plant performance analysis, carried out along of period of three months, revealed that the distributed inverter solution (sub-array A) is more effective than the classic centralized inverter approach. This is due to the minimization of the mismatch losses that in a prototypal plant are relevant due to the not standardization of the production process. Moreover, it has been observed that when a module failure occurs (like the water infiltration inside 3 modules of sub-array B), the centralized solution is affected by a drastic performance reduction, due to the reduction of the energy produced, not only by the fault module, but by all the modules series connected with it.

The Performance Ratio of the sub-array A is of 84%: this is a very high value comparable with a standard flat panel system. Low PR value of the sub-array B, as aforesaid, is strongly affected by the failure occurred to its modules.

Losses analysis confirmed the conclusions of the previous experimentation: tracking losses are negligible, thermal losses are acceptable, while mismatch effects are relevant. The most important losses contribution is due to the difference between the effective and the standard sunlight spectrum.

The Plant monitoring will continue during next month in order to verify the behavior in different seasons and to prove the module reliability.

5 ACKNOWLEDGEMENTS

This project has received funding from the European Union's Seventh Framework Program for research, technological development and demonstration under grant agreement no 295985 (Project acronym: ECOSOLE).

This publication reflects the views only of the author and the European Commission cannot be held responsible for any use which may be made of the information contained therein.

5 REFERENCES

[1] M. Carpanelli et al., "Characterization of the Ecosole HCPV tracker and single module inverter" AIP Conference Proceedings 1679, 120001 (2015)

[2] C. Cancro et al., "Characterization and validation tests on Ecosole C-modules first prototypes" AIP Conf. Proc. 1679, 050005 (2015)

[3] G. Sala et al, "Spectral Solar Radiation Measurements and Models for CPV Module Production Estimation", AIP Proceeding (CPV-6 Conference held in Freiburg on April 2010) 1277, 329 (2010); doi: 10.1063/1.3509223

[4] F. Pascarella et al. "Sistema di Controllo a Correzione Automatica del Puntamento Solare, in Particolare per Eliostati" Italian Patent N° RM2003A000297

[5] IEC 61724 International standard "Photovoltaic system performance monitoring; guidelines for measurement, data exchange and analysis" - April 1998

[6] C. Cancro et al. "Energetic performances of the 50kW CPV plant realized within NGCPV project" Technical digest of 6th WCPEC Conference, Kyoto November 2014

LOW CONCENTRATION GAAS/CUINGASE AND GAAS/SI MULTI-JUNCTION SOLAR CELLS WITH SMART STACK TECHNOLOGY

Kikuo Makita[1], Hidenori Mizuno[1], Ryuji Oshima[1], Takeshi Tayagaki[1], Jiro Nishinaga[1], Hajime Shibata[1], Masaaki Baba[2], Noboru Yamada[2], Hidetaka Takato[1], and Takeyoshi Sugaya[1]
[1]National Institute of Advanced Industrial Science and Technology (AIST) , Japan
[2]Nagaoka University of Technology, Japan
AIST Tsukuba, Central 2, Umezono 1-1-1, Tsukuba, Ibaraki, 305-8568, JAPAN
Phone: +81-29-861-3263, Fax: +81-29-861-5615, E-mail: kikuo-makita@aist.go.jp

ABSTRACT: Low concentration multi-junction (MJ) solar cells have a practical solution to consist with high efficiency and low cost. This paper shows the demonstrations of low concentration GaAs/CuInGaSe(CIGSe) and GaAs/Si MJ solar cells with mechanical stacking method. Our key technology is the direct bonding using conductive nanoparticle alignment, which is named "Smart Stack" technology. The conductive nanoparticles, such as Pd, were aligned through the use of a self-assembled block copolymer template at the bonding interface. Therefore, sufficient reduction of bonding resistance and optical absorption loss can be simultaneously attained. Using this technology, we fabricated a GaAs/CIGSe 3-junction solar cell and observed the maximum efficiency of 25.3% at 5.7 suns (AM1.5g), which is concerned in proper increase of open circuit voltage (Voc) under the concentrated sun. This is the first demonstration as concentration examination for a two-terminal GaAs/CIGSe MJ solar cell. On the other hand, for a GaAs/Si 3-junction solar cell, maximum efficiency was 21.2% at 9 suns. According to our theoretical prediction, these efficiencies can be improved over 30% under the optimized structure design. The obtain results show the possibility of GaAs/CIGSe and GaAs/Si MJ solar cells as next generation solar cell.
Keywords: $Cu(InGa)Se_2$, Silicon, III-V Semiconductors, Nanoparticles, Multijunction Solar Cell

1 INTRODUCTION

Multi-junction (MJ) solar cells have enabled very high efficiencies over 30% due to the effective utilization of the sun's energy. These MJ solar cells are usually composed of complicated multi-layers on high cost GaAs or Ge wafers. Therefore the present cells are very expensive and lacking of material flexibility originated in the involved growth process.

On the other hand, the MJ solar cells using mechanical stacking technology have been interested as next generation solar cell [1, 2]. The mechanical stacking enables high efficiency and low-cost because of flexible and most appropriate combination of different cells. Particularly, GaAs/CuInGaSe (CIGSe) or GaAs/Si MJ solar cell is a useful combination. GaAs-based cell essentially has high efficiency. CIGSe and Si cell have high sensitivity for long wavelength region and hold the promise of low cost bottom cell. According to our theoretical prediction, the current-matched InGaP/GaAs/CIGSe and InGaP/GaAs/Si 3-junction solar cells have a potential to reach more than 30% efficiency. In addition, to improve total performance, low concentration system is a practical solution. Concentrated solar add to the theoretical limit of conversion efficiency. The decrease of high expensive GaAs cell area contributes to cost reduction. Low concentration system does not require a tracking system. According to our cost model, low concentration system below 10 suns realize a competitive cost level (module cost with lens_ <0.4$/W).

Against this background, we have developed a unique mechanical staking method using a Pd nanoparticle alignment, which is named "Smart Stack" technology [3-6]. We successfully reported an InGaP/GaAs/InGaAsP/InGaAs 4-junction solar cell with 31.6% efficiency [7] and an InGaP/GaAs/CIGSe 3-junction solar cell with 24.2% [5]. In this paper, we examined low concentration characteristics of InGaP/GaAs/CIGSe and InGaP/GaAs/Si 3-junction solar cells with Smart Stack technology.

2 DEVICE APPROACH

Figure 1 shows a conceptual diagram of low concentration solar cells, which are based on the combination of GaAs-based top cell of one or more than one junctions, which is typically InGaP/GaAs 2-junction solar cells, and Si or CIGSe bottom cell. In here, our key technology is the semiconductor bonding with Smart Stack technology, which is the integration of conductive nanoparticle (Pd) alignments at the bonding interface. Conductive nanoparticles were aligned on a bottom cell through the use of self-assembled block copolymer (polystylene-block-poly-2-vinylpyridine) templates [3,6]. A top GaAs-based cell was separated from the growth substrate by Epitaxial Lift-off (ELO) technique. Finally, two cells were stacked under the light load (5 N/cm^2). Using this technology, low bonding resistance (< 1 Ω cm^2) as well as low interfacial optical loss (<2%) was reported.

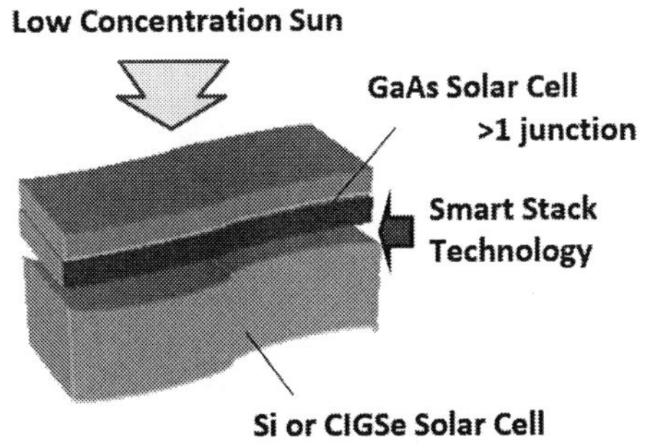

Figure 1: Conceptual diagram of low concentration solar cell.

Figure 2 shows the theoretical efficiency of 3-junction solar cell, which consists of InGaP/GaAs 2-J top cell and 1-J bottom cell (such as CIGSe or Si), for the band-gap energy (Eg) of bottom cell. These values are calculated under 1 sun and current matching conditions. In Fig. 2, the reported best efficiencies of CIGSe solar cell (21.2%) and Si solar cell (25.6%) are plotted together. In here, it is difficult to reach 30% level only for a CIGSe or a Si solar cell. On the other hand, InGaP/GaAs based 3-junction solar cell is predicted to overcome 30% efficiency. In particular, the combination of InGaP/GaAs and CIS (Ga composition=0, Eg=1eV) is possible to realize the best efficiency because of maximum current matching.

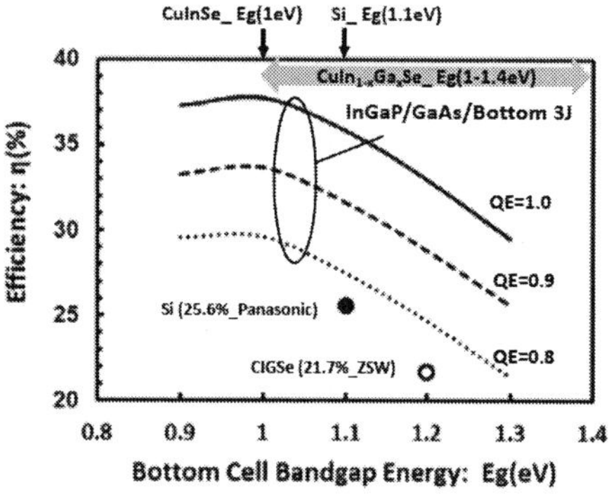

Figure 2: Predicted efficiency of InGaP/GaAs/ bottom (such as CIGSe or Si) 3-junction solar cells under 1 sun (AM1.5g).

3. EXPERIMENT AND DISCUSSION

To verify the feasibility of our concept, we fabricated an InGaP/GaAs/CIGSe 3-junction solar cell, which is shown in Fig. 3. As the top cell, an ELO-released two-junction cell consisting of p-InGaP (Eg-1.89 eV) and p-GaAs (Eg-1.42 eV) absorption layers connected with a tunneling layer was used. As the bottom cell, a CIGSe cell formed on a soda-lime glass (SLG) substrate consisting of a Mo electrode, a CIGSe (Eg-1.2eV) absorption layer, a CdS emitter layer, and a ZnO transparent conductive oxide layer, was used [8]. The surface of the ZnO top layer of the CIGSe cell was flattened by chemical mechanical polishing (CMP) process, and then a Pd nanoparticle alignment was formed. Finally, top and bottom cells were bonded through the load.

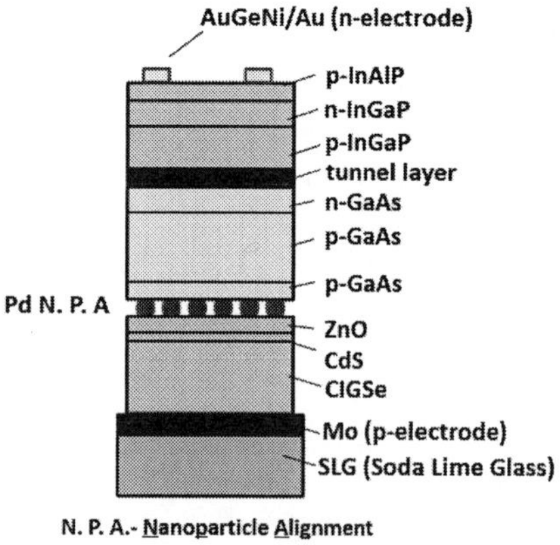

Figure 3: Schematic diagram of the fabricated InGaP/GaAs/CIGSe 3-junction solar cells.

The current-voltage characteristics were measured under the concentrated AM1.5g spectrum. Figure 4 shows the efficiency in dependence of the concentration factor. Under 1 sun, the total efficiency (η) was 22.8% with open-circuit voltage (Voc), short-circuit current density (Jsc) and fill factor (FF) of 2.61V, 11.51mA/cm^2 and 0.76, respectively. The efficiency increases for increasing concentration factor. Maximum efficiency was 25.3% at 5.7 suns. Figure 5 shows the open circuit voltage and the fill factor for concentration factor. The improvement is concerned in proper increase of open circuit voltage. The efficiency decreases along with the deterioration of fill factor in concentration factor 6.0 or more, which is due to the series resistance, such as electrode resistance or bonding resistance. In this structure, the total thickness of top n-electrode (AuGeNi/Au) is as thin as 400nm, which is not enough for high current density of over 100mA/cm^2. In the same experiment for InGaP/GaAs 2-junction solar cells without bonding structure, similar degradation was also observed at 10 suns. Therefore, the main contribution of these losses is attributed to the n-electrode of top cell. As a result, this performance is acceptable level for low concentration system below 10 suns. To our knowledge, this is the first demonstration as concentration examination for a two-terminal GaAs/CIGSe MJ solar cell, and the obtained efficiency of 25.3% is the record value compared to our previous result (24.2%) [5]. Furthermore, we examined an InGaP/GaAs/Si 3-junction solar cell. In this case, the efficiency increases from 19.4% at 1 sun to 21.2% at 9 suns.

The obtained results suggested that these device configurations with Smart Stack technology are useful for low concentration system. The efficiency can be improved by optimizing the device structures.

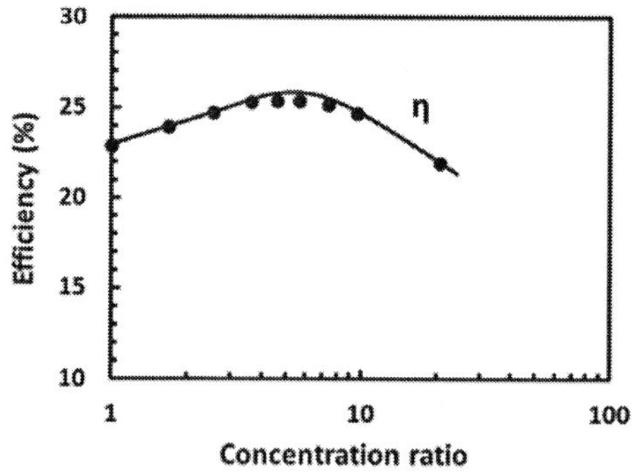

Figure 4: Efficiency of the InGaP/GaAs/CIGSe 3-junction solar cells in dependence of concentration factor.

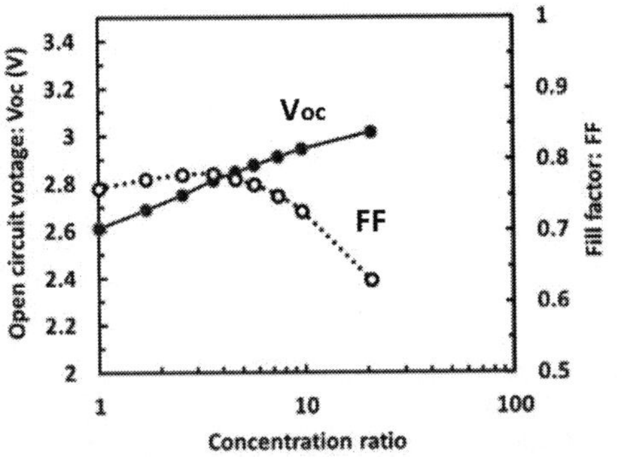

Figure 5: Open circuit voltage and fill factor of the InGaP/GaAs/CIGSe 3-junction solar cells in dependence of the concentration factor.

4 CONCLUSIONS

In summary, we demonstrated low concentration multi-junction solar cells with Smart Stack technology. The maximum efficiency of GaAs/CIGSe and GaAs/Si MJ solar cells were 25.3 % at 5.7 suns and 21.2% at 9 suns, respectively. According to our theoretical prediction, the efficiency can be improved over 30%. The obtain results shows the possibility of heterogeneous combination solar cells, such as GaAs/CIGSe and GaAs/Si MJ solar cells, for next generation solar cell.

This work was supported by the New Energy and Industrial Technology Development Organization.

References

[1] P. T. Chiu, D. C. Law, R. L. Woo, S. B. Stinger, D. Bhusari, W. D. Hong, A. Zakaria, J. Boisvert, S. Mesropian, R. R. King, and N. H. Karam, IEEE Journal of Photovoltaics, Vol. 4 (2014) 493.

[2] A. W. Bett, S. P. Philipps, S. Essig, S. Heckelmann, R. Kellenbenz, V. Klinger, M. Niemeyer, D. Lackner, and F. Dimroth, 28th European Photovoltaic Solar Energy Conference, (2013) 1AP.1.1.

[3] H. Mizuno, K. Makita, and K. Matsubara, Appl. Phys. Lett., Vol. 101 (2012) 191111.

[4] K. Makita, H. Mizuno, H. Komaki, T. Sugaya, R. Oshima, H. Shibata, K. Matsubara, and S. Niki, 2013 MRS Spring Meeting, (2013) FF2.08.

[5] K. Makita, H. Komaki, H. Mizuno, H. Sai, T. Sugaya, R. Oshima, H. Shibata, K. Matsubara, and S. Niki, 29th European Photovoltaic Solar Energy Conference, (2014)3O.4.1.

[6] H. Mizuno, K. Makita, T. Sugaya, R. Oshima, Y. Hozumi, H. Takato, and K. Matsubara, Jap. J. Appl. Phys., Vol. 55(2016)025001

[7] K. Makita, H. Mizuno, R. Oshima, S. Nonaka, A. Furukawa, H. Tkatao, and T. Sugaya, 21st International Photovoltaic Science and Engineering Conference, (2015)GAS. O. 13.

[8] S. Niki, S. Ishizuka, H. Komaki, Y. Kamikawa-Shimizu, S. Furue, S. W. Choi, K. Matsubara, H. Shibata, A. Yamada, H. Nakanishi, N. Terada, T. Sakurai, and K. Akimoto, Proeedings of the 23rd European Photovoltaic Solar Energy Conference, (2008)2846.

CPVMATCH - CONCENTRATING PHOTOVOLTAIC MODULES USING ADVANCED TECHNOLOGIES AND CELLS FOR HIGHEST EFFICIENCIES

Simon P. Philipps[1, a)], Mathieu Baudrit[2], Karla Hillerich[3], Valentine Moreau[4], Rolando Parmesani[5], Eduardo Román[6], Gabriel Sala[7], Bernd Schineller[8], Gianluca Timò[9] and Andreas W. Bett[1, b)]

[1]*Fraunhofer Institute for Solar Energy Systems ISE, Heidenhofstraße 2, 79110 Freiburg, Germany*
[2]*CEA-LITEN, LCPV, INES, 50 avenue du Lac Léman, F-73375 Le Bourget-du-Lac, France*
[3]*AZUR SPACE Solar Power, Theresienstraße 2, 74072 Heilbronn, Germany*
[4]*CYCLECO, 18 Avenue Roger Salengro, 01500 Ambérieu-en-Bugey, France*
[5]*ASSE, Via Milano 5, 34132 Trieste, Italy*
[6]*TECNALIA Research & Innovation, Parque Tecnológico de Bizkaia c/ Geldo, Edificio 700, 48160 Derio, Spain*
[7]*Instituto de Energia Solar-Universidad Politécnica de Madrid, Av. Complutense 30, 28040 Madrid, Spain*
[8]*AIXTRON AG, Dornkaulstrasse 2, 52134 Herzogenrath, Germany*
[9]*RSE SPA, Strada torre della razza, 29122 Piacenza, Italy*

[a)]*Corresponding author: simon.philipps@ise.fraunhofer.de*
[b)]*andreas.bett@ise.fraunhofer.de*

ABSTRACT: This paper presents the project *Concentrating Photovoltaic modules using advanced technologies and cells for highest efficiencies* (CPVMatch), which has received funding from the European Union's Horizon 2020 research and innovation program. The aim of CPVMatch is to bring practical performance of HCPV closer to its theoretical limits with research and development in the field of III-V multi-junction solar cells and CPV modules. Concerning cells, novel wafer bonded four-junction solar cells made of GaInP/GaAs//GaInAs/Ge are optimized with the target of reaching 48% efficiency under concentration. Moreover, multi-junction solar cell technologies with advanced materials - like ternary IV element mixtures (i.e. SiGeSn) and nanostructured anti-reflective coatings - are investigated. Concerning CPV modules the project focuses on both Fresnel-based and mirror-based technologies with a target efficiency of 40% under high concentrations beyond 800x. Achromatic Fresnel lenses for improved light management without secondary optics are investigated. In addition, smart, mirror-based HCPV modules are developed, which include a new mirror-based design, the integration of DC/DC converters and an intelligent tracking sensor (PSD sensor) at module level. A profound life-cycle and environmental assessment and the development of adapted characterization methods of new multi-junction solar cells and HCPV modules complete the work plan of CPVMatch.

1 INTRODUCTION

The EU-funded project *Concentrating Photovoltaic modules using advanced technologies and cells for highest efficiencies* (CPVMatch) is presented here. The collaborative project started in May 2015 with a duration of three and a half years and an EC contribution of 4.95 M€. The consortium consists of four research institutions (Fraunhofer ISE, RSE, CEA, Tecnalia), one University (UPM), two industry partners (AZUR Space Solar Power, AIXTRON) and two SMEs (ASSE, Cycleco) and is coordinated by Fraunhofer ISE. The consortium addresses in their research all topics required to manufacture highly-efficient CPV modules. This includes material issues, manufacturing and equipment aspect and production challenges. University and research institutes are working in close co-operation with industry partners in order to ensure fast industrial exploitation of all results within the whole value chain (see Fig. 1).

The overall aim of this project is to increase the practical performance of High Concentrating Photovoltaics (HCPV) modules and to close the gap to the theoretical limits. This should be achieved through:

- novel multi-junction solar cell architectures using advanced materials and processes for better spectral matching
- innovative HCPV module concepts with improved optical and interconnection designs, thus including novel light management approaches.

The central objective of CPVMatch is to realize HCPV solar cells and modules working at a concentration level above 800x with an efficiency of 48% and 40%, respectively, with a low environmental impact.

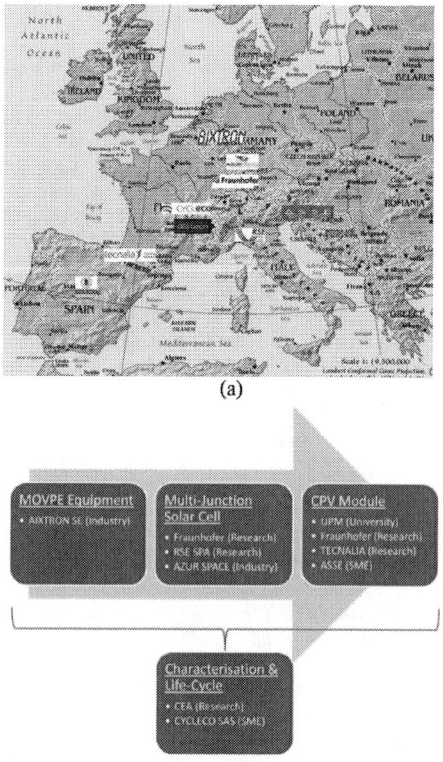

(a)

(b)

Figure 1: Partners of CPVMatch (a). Alignment of the CPVMatch partners along the value chain (b)

2 CONCEPT AND OBJECTIVES OF THE PROJECT

The aim of CPVMatch is to bring practical performance of HCPV closer to its theoretical limits. In order to develop such high performing CPV solar cells and modules, two strategies are adhered to (both for the multi-junction solar cell and module technology). The following gives a short overview on the approach of the scientific work of the project. Note that the description is not necessarily comprehensive and does not present all details.

2.1 Solar cell development

The current landmark for III-V multi-junction solar cells is the $Ga_{0.50}In_{0.50}P/Ga_{0.99}In_{0.01}As/Ge$ triple-junction cell, which only contains lattice-matched layers. Hence all materials have the same lattice constant. This "standard" triple-junction on a Ge substrate suffers from poor current matching under the solar spectrum, leading to a limited efficiency potential. A record efficiency of 41.6% (364xAM1.5d) was achieved in 2009 [1] and currently this device is commercially available with efficiencies around 40% under concentrated sunlight. In order to increase the efficiencies further, hence to bring practical performance of photovoltaics closer to theoretical limits, the number of junctions has to be increased. Recently, a four-junction solar cell with an efficiency of 46.0% at a concentration of 508 suns has been achieved [2]. In CPVMatch two different bandgap engineering approaches for four-junction solar cells are followed:

- the first one concerns cutting edge multi-junction solar cell technologies, namely a combination of the metamorphic growth and wafer bonding to create multi-junction solar cells, which can be considered the latest, most advanced stage of development related to multi-junction solar cells
- the second one concerns frontier multi-junction solar cell technologies, in which new advanced materials, like new alloys of IV elements and nanostructured coatings, are employed in combination with the adoption of novel advanced MOVPE deposition techniques.

2.1.1 Cutting edge multi-junction technologies: metamorphic growth combined with wafer bonding

Lattice-matching strongly restricts the range of materials which can be used and thus limits possible bandgap combinations. Therefore, the concept of metamorphic growth has been developed. Here, materials with different lattice constants are grown on top of each other, widening the range of available bandgaps. However, metamorphic growth is challenging since the change in the lattice constant introduces defects and strain. Therefore, very specially designed buffer structures are necessary. Fraunhofer demonstrated the successful application of the metamorphic approach in 2009 by achieving a world record efficiency of 40.7% (454xAM1.5d) [3]. A similar structure is now available on the market, provided by AZUR SPACE.

In CPVMatch, the metamorphic approach is further implemented and used for the first time to realize the optimal four-junction solar cell with the combination GaInP/GaAs//GaInAs/Ge (Fig. 2 (a)). This structure requires metamorphic growth technology as a buffer between the Ge and the GaInAs subcell for increasing the lattice constant is required. A major challenge of this structure is that the lattice-constant of GaInAs is significantly larger compared to GaInP and GaAs. Realizing this structure in a monolithic metamorphic approach is extremely challenging and would require a second buffer between the GaInAs and the GaAs subcell to decrease the lattice constant.

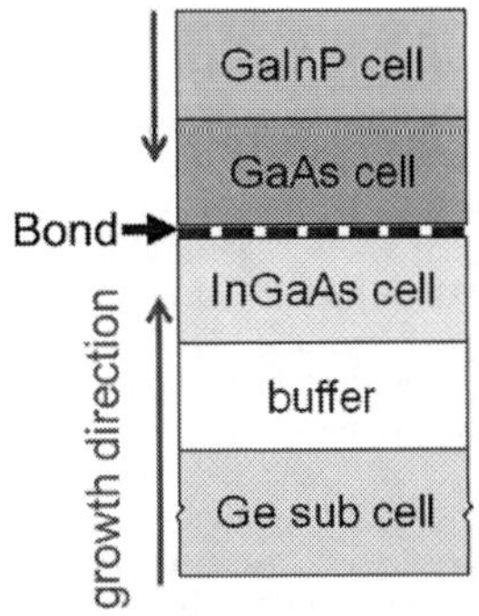

(a)

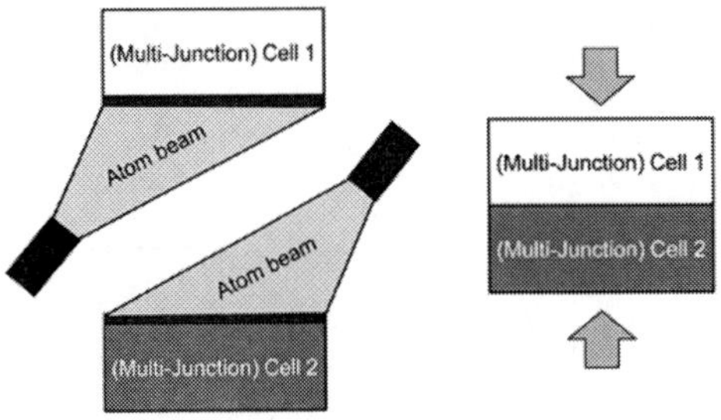

(b)

Figure 2: (a) Concept of a four-junction GaInP/GaAs//GaInAs/Ge solar cell using wafer bonding to combine two independently grown dual-junction solar cells. The upper two subcells are grown in inverse direction. (b) Schematic illustration of Fast Atom Beam Activated Wafer Bonding. Before the actual bonding of the two (dual-junction) solar cells their surfaces are cleaned with atom beams in order to allow for a bond interface with low resistance and high mechanical stability.

Therefore another option, which is investigated in CPVMatch, is to use the advanced post-growth technology, wafer bonding (Fig. 2 (b)), to combine an inversely grown, lattice-matched GaInP/GaAs dual-junction solar cell with an upright, metamorphic GaInAs/Ge dual-junction solar cell. During the bonding process the cell stacks are pressed together without the requirement of monolithical growth of the complete structure. Using this technology, a four-junction solar cell with a record efficiency of 46.0% under 508xAM1.5d has recently been achieved by a partner of the consortium (Fraunhofer) [2]. With respect to the cited record device, the proposed multi-junction structure to be developed in CPVMatch has the obvious advantage of using cheaper substrates and materials, which are more abundantly available in nature. This new four-junction solar cell can be industrialised much faster since it is based on material compositions which are more equal to the current industrial standard. One of the key issues will be to develop a reliable wafer bonding process.

2.1.2 Frontier multi-junction solar cell technologies with lattice-matched approach

Conversely to the presented cutting edge multi-junction solar cells technologies, in which metamorphic materials are exploited to increase the solar cell efficiency value, the frontier multi-junction solar cell technologies proposed in CPVMatch are based on the lattice-matched approach by adopting new materials in order to enlarge the band gap engineering possibilities and then increase the solar device performances. Theoretical calculations show that a 1.0 eV subcell placed in between the $Ga_{0.99}In_{0.01}As$ middle cell and the Ge bottom cell of the standard lattice-matched triple-junction solar cell would lead to a nearly optimal four-junction device. Yet, the realization of such a 1.0 eV material in a lattice-matched configuration is challenging. The promising and thoroughly investigated candidate GaInNAs, e.g. with 7% In and 2% N, suffers from a low minority carrier diffusion length if the material is grown in MOVPE reactors [4]. A promising and novel material alternative to dilute nitrides is the pseudo binary $Ge_{1-x}(Si_{0.8}Sn_{0.2})_x$. A unique feature of these ternary alloys is the possibility of changing compositions while keeping the lattice parameter at a constant value. In particular by increasing the share of Si and Sn – with constant 4:1 ratio – in the Ge matrix, the bandgap of the alloy can be tuned between 0.8 eV and 1.4 eV [5], while maintaining lattice-match to Ge (Fig. 3).

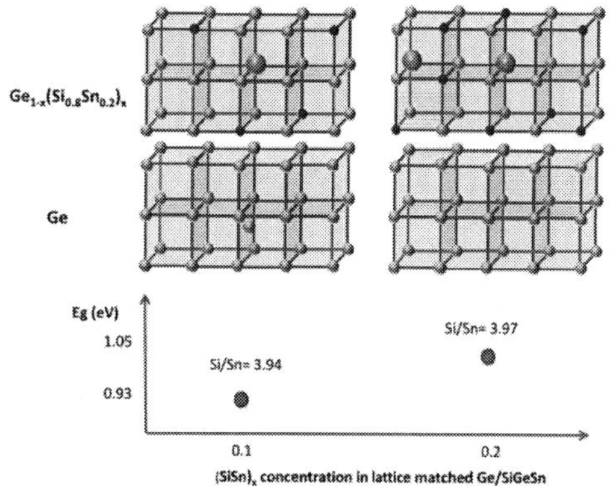

Figure 3: Simplified representation of the crystalline structure of Ge and SiGeSn and of the compositional tuning concept for Ge-rich SiGeSn alloys latticed matched to Ge (yellow circles), in which a ratio of Si (red spheres) to Sn (large grey spheres) of 4 is almost maintained. The plots below illustrate the corresponding increase in bandgap associated with the addition of Si and Sn, while almost preserving the 4:1 ratio.

The central technical challenge for realizing multi-junction solar cells at low cost with this advanced material lies in the growth of III-V with IV elements in the same MOVPE growth chamber. The proof-of-concept has already been demonstrated in the MOVPE growth chamber installed at RSE SPA by growing SiGe and III-V elements [6]. This result brought the partners of the consortium in an outstanding position in the roadmap toward the realization of four-junction GaInP/Ga(In)As/SiGeSn/Ge solar cells, with an ideal energy gap combination of 1.9 eV/1.4 eV/1 eV/0.7 eV. Therefore, starting from the proof of concept demonstrated on SiGe, a novel design of the MOVPE growth chamber is being developed in CPVMatch in order to widen the utilization of IV elements (including Sn) and to reduce the cross over effects between group IV and III-V elements. The target structure for high efficiencies is a GaInP/GaInAs/SiGeSn/Ge four-junction solar cell. After the successful investigation of the advanced material SiGeSn, single and dual-junction SiGeSn/Ge and GaAs/SiGeSn solar cells will be developed. These devices will be overgrown by GaInP/(GaInAs) to form triple and four-junction solar cells with high efficiency potential.

2.1.3 Nanostructured coatings

Incomplete light trapping is another challenge to be faced in order to increase solar cell efficiency and bring practical performance close to the theoretical limit. In CPVMatch, new nanostructured coatings are being developed in order to reduce the light reflection from the surface of the solar cell and then increase the light trapping performance in the multi-junction device. Nanostructured coatings have been already demonstrated for different applications [7], and also on III-V-based solar cells [8, 9]. In these last cases, however, the proposed solutions were not applicable for concentrating solar cells. Based on lessons learned thus far, RSE SPA will apply proper deposition techniques, select different materials and planarization processes, in order to develop broadband and omnidirectional nanostructured coating able to work over a wide spectral range and highly performing at different light incident angles, as needed for high concentration multi-junction solar cells. The more challenging solution will be to realize a nanostructure coating composed of a suitable material (stable and transparent), with a refractive index value near to the value presented by the window layer of the multi-junction solar cells, deposited, after a planarization process, on the multi-junction solar cell. In this case, in fact, the nanostructured coating will allow a gradual change of the refractive index from the value of the air to the value of the semiconductor, strongly minimizing the light reflection. If this first challenging solution does not give rise to the expected improvement, a hybrid nanostructured coating-multilayer film solution will be investigated, in which the nanostructured coating will be deposited over a proper planarization layer and a subsequent multilayer film.

2.2 HCPV module development

The efficiency of III-V multi-junction solar cells is about twice as high as conventional Silicon or thin-film solar cells. However, multi-junction solar cells are currently too expensive for the use in flat-plate modules on Earth. Hence, the development of high concentrating photovoltaic (HCPV) systems with concentration factors above 300 was essential to introduce III-V multi-junction solar cells to the terrestrial market. The use of cost-efficient concentrating optics reduces the need for expensive cell area, which allows competitive levelised cost of energy (LCOE). However, it is a big challenge to realize low cost optics with high optical efficiency and at concentration levels above 800x. Based on the long-standing experience provided by the partners of this project, two promising approaches will be followed: one is based on Fresnel optics and the other one on mirror optics.

2.2.1 The Fresnel optics concept

The CPV industry mainly uses Fresnel lens-based optics, primarily because they greatly simplify the internal architecture of CPV modules, and additionally because they can be manufactured in "parquets" composed of many lenses. However, the standard Fresnel lenses exhibit chromatic aberration due to the variation of the refractive index with the wavelength of light (Fig. 4 (a)). This aberration spreads out the rays, leading to an extended "spot" at the focus, which limits the maximum attainable concentration factor of any given lens. The result is that these lenses can only achieve concentrations of between 100x (with absolutely no chromatic losses) and 350x (with limited losses).

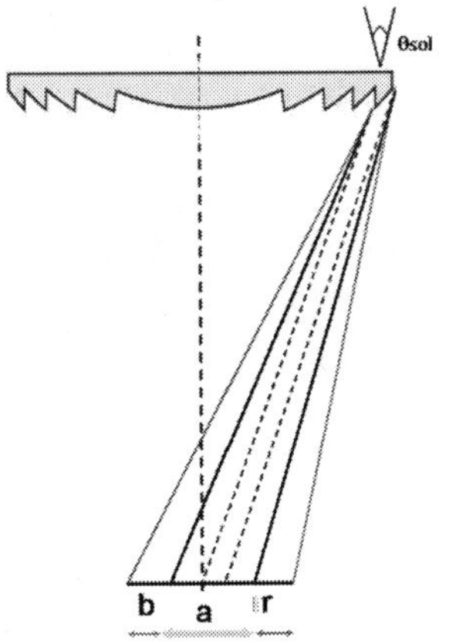

(a)

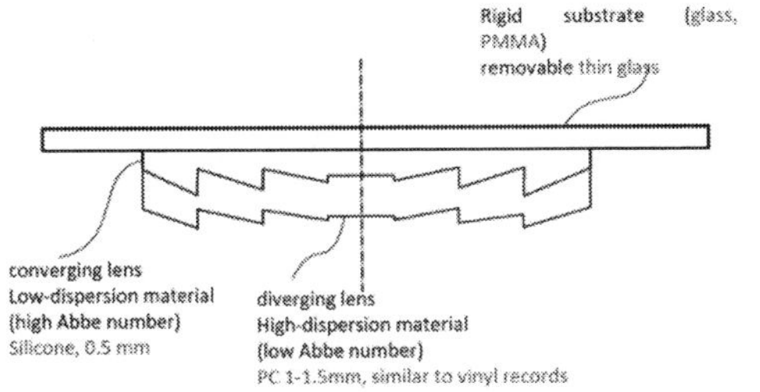

(b)

Figure 4: (a) The sketch illustrates how chromatic aberration (CA) requests larger receivers (b+a+r) for capturing all rays. Avoiding CA the cell receiver could be significantly smaller ("a"). Thus the product "Concentration X Acceptance Angle" increases by a factor between 3 and 4. (b) Proposed achromatic Fresnel lens structure and processing, which could directly replace conventional Fresnel lenses in CPV modules.

In this project a novel low-cost solution for an achromatic Fresnel lens working at high concentration (800x) (patent pending) will be developed. These achromatic lenses (Fig. 4 (b)), which could directly replace conventional Fresnel lenses, have a hybrid configuration that corrects, in large part, the ray deviation caused by chromatic aberration and thereby increases the achievable concentration to around 800x. Achromatic lenses have been used in photographic cameras and other "expensive" optics for decades. The significant novel aspect in this project is the development of a viable method to produce achromatic Fresnel lenses that meets the requirements of solar energy: large area, small f-number, high concentration and most importantly, low cost. In order to reduce the reflection losses caused by the difference in refractive indexes between air and lens materials, Anti-reflective (AR) coatings are being developed and applied to the Fresnel lenses. Anti-reflective (AR) coatings provide not only a way to reduce losses but also can ensure long-term durability against atmospheric conditions and cleaning processes.

The specific objective of this development within the project is to perform detailed designs and analysis, as well as to identify the best materials and techniques for obtaining and validating various prototypes of achromatic Fresnel lenses. The new lenses will be integrated into FLATCON®-type modules, a module technology developed at Fraunhofer [10]. It must be emphasized that the absence of a secondary optical elements (SOE) in the module not only reduces costs but promises increased reliability, less degradation and failure rate. The achromatic improvement is a step in the way of these developments that tries to improve the most proven technologies beyond the already demonstrated performance.

2.2.2 The mirror-based optics concept

Another solution to circumvent the issue of chromatic aberration is to use mirror-based optics. Fig. 5 shows one possible concept for such optics as well as a prototype, which was developed in the EU-project APOLLON (involving four partners of CPVMatch: RSE SPA, AIXTRON SE, TECNALIA and ASSE) [11]. The experimental proof-of-concept of the off-axis mirror-based concentrator has been demonstrated. Such a solution allows working at higher concentration levels around 850x. The major benefits of the proposed mirror-based configuration are the following:

- no central obstruction is present at the input aperture of the concentrator, so better optical and collection efficiency can be achieved, in comparison with other mirror based design;
- the light impinges on the mirrors with high incident angles thus improving the reflection (optical) efficiency;
- the PV cell is located in the cooler position of the concentrator which is not illuminated by the sun; this position also allows a possible replacement of the photovoltaic cell, which will make O&M tasks easier in case receiver replacement is needed, and could also be used for a product update, e.g. in case the PV cell efficiency is improving significantly along the whole lifetime.

Within CPVMatch the lessons learned from the first proof-of-concept must be materialized and validated in the lab and in the relevant environment. In particular the mirror-based smart module design will be innovated with the main goal to improve performances in standard and operative conditions at a competitive cost. This includes new solutions for the thermal management, mirror design and manufacturing. The thermal path is being optimized by using new receiver technologies. In particular the feasibility of the Chip-on-Flex and the Chip-on-Metal technology is being investigated. Moreover, electrical contacts with low resistance and high maintainability are being developed. As the optical efficiency of the mirror is strictly related to the mirror form error, adequate material and appropriate mirror dimension are being assessed.

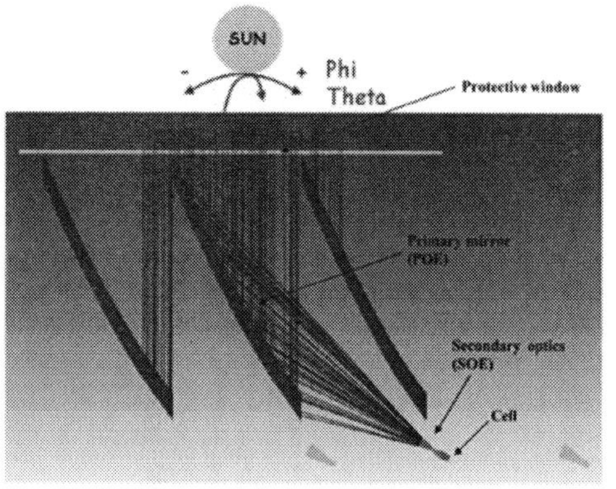

(a)

(b)

Figure 5: (a) Concept of a mirror-based concentrator using off-axis primary optical elements (POE) and secondary optical elements (SOE). (b) Prototype of such an off-axis module developed in the EU-funded APOLLON project.

From the optical point of view, the downscaling of the optic can be performed because the optical components work far from their diffraction limits. Possible limitations related to the manufacturing and mounting precision of the optical components are being defined and considered. The low-cost aim is being pursued throughout several research activities comprising: the selection of cheaper and reliable material, the assessment of the downscaling of the module and the implementation of the smart module design with low cost module integrated point sensitive detectors (PSD) and DC/DC converters. The two main targets for the latter are a cost level of 0.2 €/W_p and a 98% peak efficiency. The smart module mirror-based design also supplies an innovative answer for operating at higher conversion efficiency, since better and reliable tracking accuracies are reached (by the innovative module integrated PSD unit) and power mismatches losses are reduced (by the module integrated DC/DC devices). The module low cost objective is intended not only for the prototype stage, but also towards a future industrial stage, meaning that the laboratory prototype will be designed in such a way that future industrialization aspects are already analyzed and included (advanced manufacturing).

2.3 Characterisation and Testing

The development of novel solar cells and CPV modules needs to be supported by in depth testing and characterization. The main objective is to assess the developments realized throughout the project by means of adapted characterization methods with a special focus on: the analysis of advanced materials developed in the project, the development of adapted tools for the characterization of the multi-junction solar cells and subcells, the development of adapted tools for the characterization of the different optics, the indoor and outdoor testing of the CPV modules.

The different partners are working to adapt their equipment and then compare the results and methods in order to give a relevant feedback on how these future generations of solar cells need to be characterized. In particular the use of four-junction solar cells requires adoption of the measurement tools with respect to the particularities of each subcell material and the higher number of junctions in the device.

Part of this project is focused on the development of HCPV optics and modules. New characterization equipment is being developed for them. In particular, an imaging system is being set-up which is able to quantify the uniformity of the spot light at the cell plane for different ranges of wavelength. This information will help to quantify the losses at the module level implied by the optics and then to give some hint for the redesign of it in order to achieve the maximum module efficiency. Second, equipment is being developed to measure the exact spectra at both the cell and optical plane. The spectral transmission of the lens must be quantified to validate the lens design. This will give information about the matching of the optical transmission spectra to the cell absorption spectra, providing information to better adapt them to each other (i.e. tune design to maximize the overall performance at the module level).

Finally, CEA adopts a prototyping tool able to measure samples of single optics and cells at difference irradiances and geometrical configurations. This tool is being improved in order to measure the performances of single lens / cell coupling at different temperatures, irradiances and spectra in order to optimize the geometrical configuration and to understand what module optical designs will provide the best results at concentrator standard testing conditions (CSTC) conditions. This tool will also give some hint on how to improve the lens/cell coupling and will be used recursively during the project. Finally, the developed module will be measured indoors and outdoors at different partner's sites in order to share the results and ensure a more reliable performance analysis.

2.4 Life-Cycle Assessment

Life Cycle Assessment (LCA) has proven to be a useful tool for environmental impacts assessment in the energy sector [12]. LCA is more and more largely used due to the availability of consensual standards and guidelines such as the ISO 14040-44 (ISO, 2006) and now ILCD handbook (ILCD, 2010). The method is divided into two steps: the environmental screening and a detailed life cycle assessment. The environmental screening is used to identify hotspots and prioritize ecodesign routes. The detailed LCA is used to validate the overall findings and share the results with all stakeholders.

Within CPVMatch a profound LCA study is being carried out to evaluate the environmental impact of the developed multi-junction solar cells and CPV modules. The Life Cycle Assessment is being used to optimize the entirety of the product environmental characteristics and servers as a decision-making tool at the product (re)conception step. The study takes into account raw materials extraction and transformation, production processes, transport, distribution, end product use and reuse, recycling and eventually disposal steps. Currently, inputs and outputs linked to the life cycle are being identified and quantified and their potential environmental impacts are being evaluated. The LCA is an iterative process and the hypothesis will be constantly rethought during CPVMatch depending on new developments and information.

3 CONCLUSION

CPVMatch is an ambitious project, which brings together leading European partners in the field of CPV. All project partners have a high motivation to collaborate and to achieve high-level scientific results, in particular against the background of the current industrial crisis of CPV, which makes technical innovation even more important. Following two parallel, but strongly interlinked routes in the development of multi-junction solar cells and CPV modules it is expected that the project will bring practical efficiencies closer to theoretical limits by reaching 48% at the cell level and 40% at the module level under high concentrations beyond 800x.

4 ACKNOWLEDGMENTS

This work has received funding from the European Union's Horizon 2020 research and innovation programme within the project CPVMatch under grant agreement No 640873. The authors are solely responsible for the content of this work and it only reflects the author's view. The European Commission or any of their agencies are not responsible for any use that may be made of the information it contains.

5 REFERENCES

[1] R. King, A. Boca, W. Hong, D. Larrabee, K.M. Edmondson, D.C. Law, C. Fetzer, S. Mesropian, and N.H. Karam, in *24th European Photovoltaic Solar Energy Conference* (2009).

[2] M.A. Green, K. Emery, Y. Hishikawa, W. Warta, and E.D. Dunlop, Progress in Photovoltaics: Research and Applications. 24 (2016) 3.

[3] W. Guter, J. Schöne, S.P. Philipps, M. Steiner, G. Siefer, A. Wekkeli, E. Welser, E. Oliva, A.W. Bett, and F. Dimroth, Applied Physics Letters 94 (2009) 223504.

[4] K. Volz, W. Stolz, J. Teubert, P.J. Klar, W. Heimbrodt, F. Dimroth, C. Baur, and A.W. Bett, *Doping, Electrical Properties and Solar Cell Application of GaInNAs*, in *Dilute III-V Nitride Semiconductors and Material Systems*, E. Ayse, Editor. (2008), Springer Berlin Heidelberg: Heidelberg. p. 369.

[5] V.R. D'Costa, Y.Y. Fang, J. Tolle, J. Kouvetakis, and J. Menendez, Physical Review Letters. 102 (2009) 107403.

[6] G. Timo, G. Abagnale, N. Armani, E. Malvisi, G. Carbi, F. Farina, and B. Schineller, Crystal Research and Technology. 49 (2014) 606.

[7] F. Trespidi, G. Timo, F. Galeotti, and M. Pasini, Microelectronic Engineering. 126 (2014) 13.

[8] J. Tommila, V. Polojarvi, A. Aho, A. Tukiainen, J. Viheriala, J. Salmi, A. Schramm, J.M. Kontio, A. Turtiainen, T. Niemi, and M. Guina, Solar Energy Materials and Solar Cells. 94 (2010) 1845.

[9] R. Homier, A. Jaouad, A. Turala, C.E. Valdivia, D. Masson, S.G. Wallace, S. Fafard, R. Ares, and V. Aimez, Ieee Journal of Photovoltaics. 2 (2012) 393.

[10] A.W. Bett, C. Baur, F. Dimroth, G. Lange, M. Meusel, S. van Riesen, G. Siefer, V.M. Andreev, V.D. Rumyantsev, and N.A. Sadchikov, in *Proceedings of the 3rd World Conference on Photovoltaic Energy Conversion*. (2003). Osaka, Japan.

[11] APOLLON project. *Multi-APprOach for high efficiency integrated and inteLLigent cONcentrating PV modules (Systems)*. (2013); Available from: http://cordis.europa.eu/result/rcn/55852_en.html.

[12] B. Sørensen, *Life-Cycle Analysis of Energy Systems: From Methodology to Applications*. (2011), Cambridge, UK: Royal Society of Chemistry. 336.

ON THE EFFECT OF OPTICAL CONFIGURATION AND SPECTRAL VARIATION ON THE PERFORMANCE OF III-V TRIPLE-JUNCTION CELLS USED IN H-CPV SYSTEMS.

R.D. Schultz, E.E van Dyk, F.J. Vorster

Nelson Mandela Bay Metropolitan University
R.D. Schultz, Nelson Mandela Metropolitan University, Physics Department
Building 13, room 064, Summerstrand South Campus, NMMU, University Way, Summerstrand, Port Elizabeth, South Africa, 6031, s206029578@live.nmmu.ac.za

ABSTRACT: The power generation and efficiency of High Concentrator Photovoltaic (H-CPV) modules are unsurpassed when compared to other Photovoltaic (PV) technologies. This is achieved by coupling highly efficient III-V multi-junction photovoltaic devices to precise high quality optics. However, depending on their configuration with respect to each other, the system's optimization with regards to electrical performance and efficiency may vary. This is attributed to the partial or full photon absorption by the optical materials in different wavelength regions that are specific to the photo-active regions of the InGaP, InGaAs and Ge subcells. The decrease in photons incident upon the series connected subcells causes a varying degree of (1) current mismatch between the subcells and (2) efficiency of the device during operation. In this paper, the electrical performance of an InGaP/InGaAs/Ge III-V triple-junction device as a function of optical configuration is investigated. Utilizing appropriate algorithms, the current density and Current-Voltage (I-V) characteristic curves of the subcells as a function of the optical system influence and spectrum variation, were simulated. This enabled the resulting device I-V curve to be determined and used to identify the most efficient and highest power generating optics-to-device system.

Keywords: Multi-junction Solar Cell, Concentrators, Simulation, Solar Cell Efficiencies

1 INTRODUCTION

To increase the performance and power output from a semiconductor device, a first generation III-V triple-junction may be used to achieve higher energy generation from the incident spectrum. This is achieved by creating a device which utilizes multiple subcells monolithically connected in series. The subcells are made of different III-V semiconductor materials which absorption photons within different regions of the solar spectrum. Figure 1.1 shows spectral response profiles of a single-junction (Si) and a triple-junction InGaP/InGaAs/Ge device (CTJ) against the background of the solar spectrum (AM 1.5D) [2]. Table 1.1 compares the performance parameters of a typical single-junction Si and first generation CTJ cell.

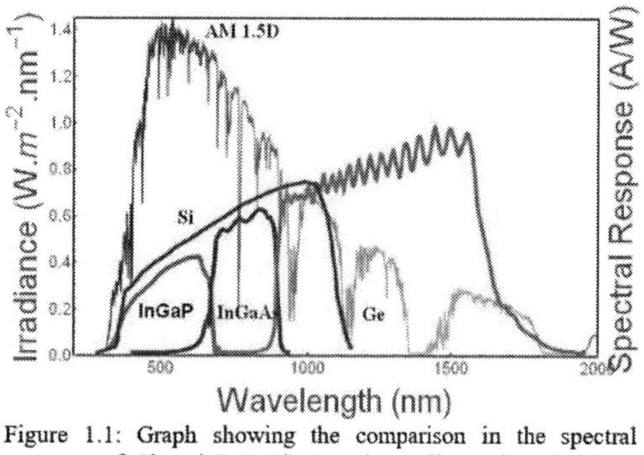

Figure 1.1: Graph showing the comparison in the spectral response of Si and InGaP/InGaAs/Ge cells against the solar spectrum (AM 1.5D).

Table 1.1: Summary of 1-sun performance parameters between a typical silicon and concentrator triple-junction cell.

Parameters	Si	CTJ (1-sun)
Current Density (mA/cm^2)	20.89	13.10
Voltage (V)	0.61	2.61
Power (mW)	10.10	29.40
Efficiency (%)	15.21	31.42

From Figure 1.1 and Table 1.1, the increased performance of the CTJ cell can be seen. This is observed in spectral response over a wider wavelength range which leads to an increased open circuit voltage and power.

In order to achieve the highest possible power and efficiency from a CTJ cell, non-imaging optical elements are used to concentrate and distribute the solar flux across the cell surface as uniformly as possible. However, the intensity distribution and spectral content that is incident on the cell surface is altered by the interaction with the various concentrating optical materials.

This paper investigates only the change in spectral content introduced by the various optical configurations. This study does not report on the intensity distribution impinging on the cell and is reported elsewhere [1].

Optical configurations will include setups that use primary and secondary optics. The primary optics consists of refracting Fresnel lenses made from Poly(methyl methacrylate) (PMMA) or Silicone on Glass (SoG). Secondary optics consist of reflective or refractive lenses. Combining the optical elements together to form the various optical configurations, the most optimal configuration can be identified. This will ensure the highest power generation and long-term system reliable.

2 THEORY

To determine the influence of the optical configuration on the electrical performance of the CTJ, the determination of the current density for each subcell needs to be conducted. The determination of the short circuit current density (J_{sc}) produced by a CTJ cell is more complex than that of a single-junction cell, as the subcells are series connected. The method for the determination of the CTJ current density is described as follows.

For a single cell, the current density is defined as:

$$J_{sc} = e\int QE(\lambda)\Phi_{inc}(\lambda)d\lambda \quad (1)$$

where e is the elementary charge, QE is the quantum efficiency of the material, Φ_{inc} is the incident solar spectrum intensity on the CTJ cell, and λ is the wavelength of a photon [3]. Since the QE and spectral response are related, it is more convenient to use the spectral response (S) and Equation 1 can thus be rewritten:

$$J_{sc} = \int S(\lambda)\Phi_{inc}(\lambda)d\lambda \quad (2)$$

In a CTJ cell, the subcells are connected in series. As a result of this, the J_{sc} of the whole device is equal to the lowest photo-generated current produced by the ith subcell.

$$J_{sc} = Min\{Ji\} \quad (3)$$

By the combining of Equations 2 and 3, and utilising the concentrated spectrum ($C\cdot\Phi_{inc}$), the J_{sc} of the CTJ cell is:

$$J_{sc} = Min(\int S(\lambda)C\cdot\Phi_{inc}(\lambda)d\lambda) \quad (4)$$

Thus, the I_{sc} for the CTJ device is defined as:

$$I_{sc} = J_{sc}.A = Min(\int S(\lambda)C\cdot\Phi_{inc}(\lambda)d\lambda.A) \quad (5)$$

Where J_{sc} is the generated current density and A is the receiver area of the CTJ cell. To maximise the current produced by a CTJ cell and to prevent any of the subcells becoming reverse biased which will have a negative effect on the power production and lifetime of the CTJ cell, all three series connected subcells must be current matched as much as possible.

3 EXPERIMENTAL

To evaluate the electrical performance of the InGaP/InGaAs/Ge III-V triple-junction device, the electrical performance of each subcell as a function of the optical configuration and spectral variation was simulated at 25°C.

This was achieved by combining the transmission spectra of the optical elements, i.e. the primary Fresnel lenses (PMMA and SoG) and secondaries (reflective and refractive) in different configurations, with the measured spectrum (AM1.5D) throughout a typical day. Figure 3.1 shows the schematic diagram of the optical configurations considered in the investigation.

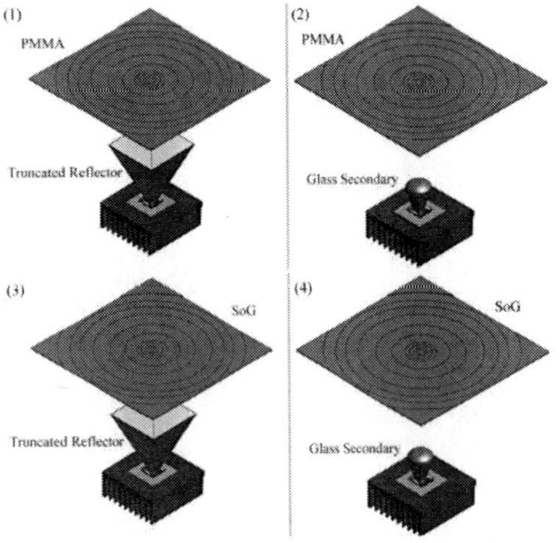

Figure 3.1: Schematic drawing showing the optical configurations considered in the investigation.

Figure 3.2 shows the experimental setup used to measure the spectral transmission of the optical elements and incident DNI solar spectrum throughout an operational day.

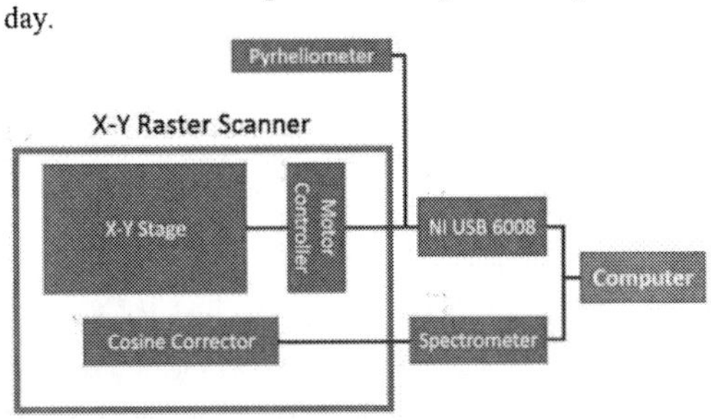

Figure 3.2: Schematic diagram of the experimental set-up used to obtain transmission and spectral measurements.

The above setup comprises of a silicon charged coupled device (CCD) fitted with a collimator to measure the Direct Normal Irradiance (DNI) spectrum. A LabView program that comprised of data acquisition and automation was used to obtain the average transmission spectrum of the optical elements over a sample area throughout the day.

Utilizing appropriate algorithms to combine the transmission spectra of the various optical configurations with the available daily spectra, the resulting transmission spectra were obtained for each of the four optical configurations.

Further combining the transmission spectra with the spectral response of each subcell, the current for each subcell was calculated at 1-sun. Using the semiconductor properties of each subcell, Current-Voltage (I-V) curves for each of the four optical configurations were simulated, and the resulting device I-V characteristic curve was determined [4,5].

Evaluating the I-V curves as a function of spectral variation (time of day), the best performing optical configuration can be determined.

4 RESULTS

The results that follow show the different steps within the simulation process. These steps include the transmission spectra from the optical materials, measured incident solar spectra and simulated electrical performance of the optical configuration.

4.1 Transmission spectra of optical elements

As optical elements are used to concentrate the incident solar spectrum in CPV systems, the current generation potential of each subcell will be influenced by the transmission of photons interacting with the optical materials. Figure 4.1.1 shows a graph of the transmission spectra for each optical element within the respective wavelength regions of the subcells [6].

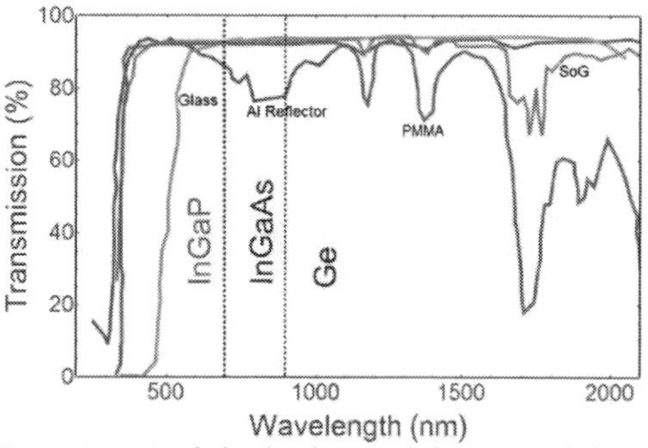

Figure 4.1.1: Graph showing the transmission spectra within the active regions of the subcells for the different optical elements.

Figure 4.1.1 compares the transmission spectra of the four optical element materials in the absorption wavelength regions of the CTJ subcells. The InGaAs absorption wavelength range shows the least amount of variation in the transmission spectra while the InGaP and Ge ranges show the most variation in transmission spectra. Combining the optical elements together in different optical configurations will result in different electrical performances with varying degrees of current mismatch between subcells. This may result in an optical configuration that causes a subcell to become current limiting.

4.2 Incident solar spectrum

Due to the change in air mass, the available solar spectral content will change throughout the day. As the spectral content changes, there will be a difference in the current generation from each subcell. Utilizing Equation 2, the changes in current for the InGaP, InGaAs and Ge subcells were calculated and are summarized below in Table 4.2.1.

Table 4.2.1: Summary of the change in current for the 1cm² subcells within an operational day based on spectral measurements.

Time (h)	InGaP (mA)	InGaAs (mA)	Ge (mA)	Mismatch Ratio
08:00	8.22	12.40	18.34	2.23
09:00	10.23	12.91	18.39	1.79
10:00	11.77	13.10	18.45	1.56
11:00	12.23	12.87	18.47	1.51
12:00	13.05	13.07	18.53	1.41
13:00	12.71	13.19	18.55	1.46
14:00	11.79	13.00	18.50	1.56
15:00	11.14	12.73	18.42	1.65
16:00	10.06	12.42	18.38	1.83
17:00	9.63	12.11	18.21	1.89
18:00	7.38	11.72	17.90	2.42

Table 4.2.1 shows a significant fluctuation in current generation from the InGaP subcell throughout the day. The table also shows that the InGaAs and Ge subcells remained relatively constant throughout the day.

As expected from the changing incident solar spectrum, the degree of current mismatch between the subcells as demonstrated by the mismatch ratio (the ratio of the limiting subcell's current to the highest current generating subcell), is not constant. The highest degree of variation was observed for early morning and late afternoon (2.23 and 2.42). This direct dependence of current on the InGaP subcell and its ability to absorb energies in the visible wavelength region, will result in a varying degree of current mismatch with respect to the other two subcells throughout the day [6, 7].

4.3 Device performance.

Combining the optical transmission spectrum with the spectral results used to determine the current in Table 4.2.1, the operational current generation throughout the day may be deduced. In this section, the results for the PMMA Fresnel and glass secondary configuration will be shown. Figure 4.3.1 shows the reduction in potential current generation of each subcell due to optical absorption. A similar reduction in the current density spectrum is observed throughout the day.

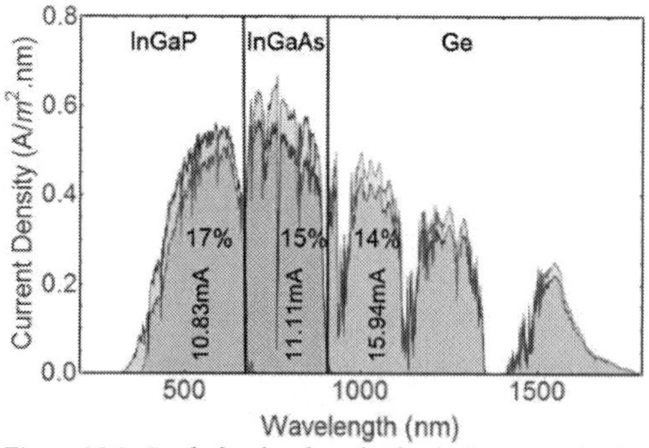

Figure 4.3.1: Graph showing the reduction in the current density spectrum for the subcell due to the absorption by the Primary PMMA lens and glass secondary at solar noon. The integrated current for each the 1cm² subcell is also indicated.

Figure 4.3.1 shows an average reduction in current density of 15.2% across the wavelength range of the CTJ cell. Although the InGaP and InGaAs subcells seem to be relatively current-matched, the Ge subcell remains mismatched by a factor of 1.46 with respect to the limiting InGaP subcell [7]. Based on the result in Figure 4.3.1, the current flowing to the external load circuit is limited by the InGaP subcell. The excess current generated by the InGaAs and Ge subcells will flow through their respective diodes. This will result in a reduction in their subcell voltage. The resulting CTJ cell's V_{max} will shift towards I_{sc}, reducing the maximum power point and yielding a lower power output.

In addition, the power generation of the CTJ cell will be decreased because of:

- The reduction in I_{max} due to the absorption of spectral content by the optical system.
- The reduction in V_{oc} and hence V_{max} due to the reduction in I_{sc} as a function of spectral concentration.
- A further decrease in V_{oc} and hence V_{max} due to the shift of the bandgaps to higher wavelengths because of heating in the cell [8].
- An increase in series resistance with an increase in temperature [9].

Figure 4.3.2 shows the simulated I-V curves highlighting the reduction in electrical performance of the subcells and CTJ cell due to optical absorption.

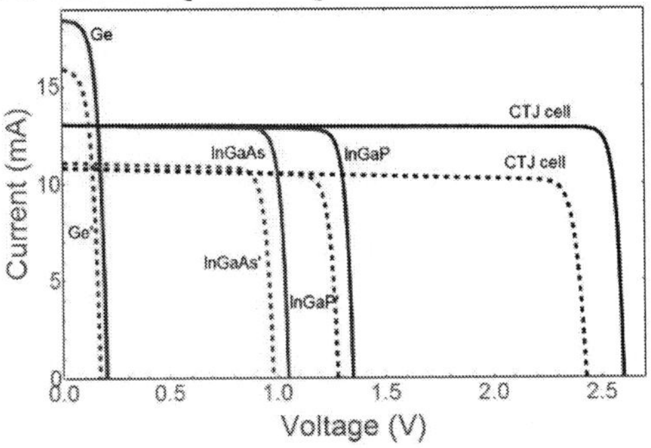

Figure 4.3.2: Graph showing the comparison in I-V curves of the subcells and CTJ due to optical absorption.

As observed in Figure 4.3.2, all the subcells (dashed curves) show a reduction in their current generation and open circuit voltage. The reduction in I_{sc} is attributed to the photon absorption caused by the optical configurations and the V_{oc} is due to the logarithmic dependence of V_{oc} on I_{sc} [3, 4]. Furthermore, the resulting subcell I-V curves (dashed lines) profile was charged with regards to the original I-V curves (solid) and the resulting CTJ I-V curve (black dashed line) will have a different I-V curve to the original. Although the resulting CTJ I-V curve only showed a 2.13% decrease in fill factor, the reduction in power is 24%.

Applying the optical transmission spectrum of the PMMA-glass configuration (2) to the day's spectra, the reduction in current for each subcell can be determined. Table 4.3.1 shows the reduced current for each subcell and the degree of current mismatch created.

Table 4.3.1: Summary of the reduction in current caused by optical configuration 2 for the 1cm² subcells within a typical day based on spectral measurements.

Time (h)	InGaP (mA)	InGaAs (mA)	Ge (mA)	Mismatch Ratio
08:00	6.82	10.54	15.77	2.31
09:00	8.49	10.97	15.82	1.86
10:00	9.77	11.14	15.87	1.62
11:00	10.15	10.94	15.88	1.56
12:00	10.83	11.11	15,94	1.47
13:00	10.55	11.21	15.95	1.51
14:00	9.79	11.05	15.91	1.63
15:00	9.25	10.82	15.84	1.71
16:00	8.35	10.56	15.81	1.89
17:00	7.99	10.29	15.66	1.96
18:00	6.13	9.96	15.39	2.51

Compared to Table 4.2.1, Table 4.3.1 shows a substantial variation in the performance and current generation from a CTJ cell when the influence of the optical system on the incident solar spectra is factored in. The generated current is greatly reduced when compared to that stated by the manufacturer [2]. Around solar noon (12:00) where the highest potential current generation should occur, the 13.05 mA stated for a 1 cm^2 cell was reduced by the optical system to 10.83 mA. This led to a great discrepancy in the performance of the CTJ cell under operational conditions.

One could argue that the decrease in current production due to spectral absorption and fluctuation would be offset by the increase in concentration. However, this is not the case, as current-matching between the subcells is not dependent on the amount of concentration, but rather on the influence of the optical materials, intensity profiles, alignment and reduction in spectral content [10]. The correct conditions and optimal optical configuration will ensure better power generation and current matching between the subcells, leading to a more efficient and reliable concentrator cell and H-CPV system.

4.4 Determination of best optical configuration

To determine the best optical configuration, the optical configurations of (1) PMMA-reflector, (2) PMMA-glass, (3) SoG-reflector and (4) SoG-glass were compared. Table 3.4.1 shows the effect of the different optical systems configurations on the current generation of the subcells and overall performance of the CTJ cell.

Table 4.4.1: Summary of the degree in current reduction due to the different optical configurations at AM1.5D.

Optical Config.	*1*	*2*	*3*	*4*
InGaP (mA)	11.53	10.83	11.56	10.86
InGaAs (mA)	10.38	11.11	10.41	11.16
Ge (mA)	15.68	15.95	17.25	17.32
CTJ cell (mA)	10.38	10.83	10,41	10.86
Mismatch Ratio	1.51	1.47	1.66	1.59
Power (mW)	24.31	22.83	24.37	22.90
Efficiency (%)	26.00	24.42	26.07	24.49

As observed in Table 4.4.1, there is a great degree of current mismatch in each optical configuration. The PMMA-glass and SoG-glass configuration showed that the InGaP subcell was current limiting while the PMMA-reflector and SoG-reflector showed that the InGaAs was current limiting. Each configuration showed an increase in current mismatch with a reduction in power generation. The best performing configuration was determined to be the PMMA-reflector with a 1.51 mismatch ratio, efficiency of 26% and power generation of 24.31mW at solar noon (1-sun).

Hence, it is shown that it would be beneficial to determine the optical transmission of the whole optical configuration. In doing so, the device architecture can be designed in such a way as to minimize high current mismatch ratios and allow for an efficient and higher power generating device. Such a device is the 4-junction cell (InGaP/InGaAs/InGaNAs/Ge) foe which simulations show a power generation of 29mW, a current mismatch ratio of 1.10 with an efficiency of 31% for the PMMA-glass optical configuration [6].

5 CONCLUSION

In conclusion, it is shown that the optical configuration does indeed affect the electrical performance of the CTJ cell over a typical day. Four optical configurations were investigated. Each configuration shows an optical transmission spectrum that varies for each subcell's spectral response region. The InGaP was identified as the most current limiting subcell due to spectral variation throughout the day. The best optical configuration was identified to be the PMMA main- and reflector secondary-lens with a 1.51 mismatch ratio at AM1.5D, efficiency of 26% and power generation of 24.31mW.

6 ACKNOWLEDGEMENTS

The authors would like to the Nelson Mandela Metropolitan University, Centre for Energy Research, Department of Science and Technology and the National Research Foundation for their financial support.

7 REFERENCES

1. R.D Schultz, E.E van Dyk, F.J Vorster. The effect of the optical system on the electrical performance of III–V concentrator triple junction solar cells. Physica B: Condensed Matter, 480:80-83, 2015.

2. Emcore Solar Cell Receiver for Terrestrial Concentrator Photovoltaics (CPV) Datasheet, Accessed: 11 August 2011,http://www.datasheetarchive.com/Emcore/20solar/20cell-datasheet.html.

3. G. Peharz, G. Siefer, and A.W. Bett. A simple method for quantifying spectral impacts on multi-junction solar cells. Solar Energy, 83:1588–1598, 2009.

4. A. Luque and S. Hegedus. Handbook of Photovoltaic Science and Engineering. Wiley, 2002.

5. A.L. Luque and V.M. Andreev. Concentrator Photovoltaics. Springer, 2007.

6. RD Schultz. "On the Characterisation of Diffused Light and Optical Elements in High Concentrator Photovoltaic Modules", PhD Dissertation, Nelson Mandela Metropolitan University, 2014.

7. V.M. Andreev, V.A. Grilikhes, and V.D. Rumyantsev. Photovoltaics Conversion of Concentrated Sunlight. Wiley, 1997.

8. The Basic Physics and Design of III-V Multijunction Solar Cells from NREL, Accessed: 30 May 2012, http://www.uotechnology.edu.iq/eretc/books/NRELokok.pdf.

9. C. Domnguez, I. Anton, and G. Sala. Multijunction solar cell model for translating I-V characteristics as a function of irradiance, spectrum, and cell temperature. Progress in Photovoltaics, 18:272–284, 2010.

10. N. Yeh. Analysis of spectrum distribution and optical losses under Fresnel lenses. Renewable and Sustainable Energy Reviews, 14:2926–2935, 2010.

DESIGN AND PREPARATION OF ANTIREFLECTION COATING FOR INVERTED METAMORPHIC 4 JUNCTION (IMM 4J) SOLAR CELL

Xipeng Sun, Yongchao Du, Zhibin Xiao
Tianjin Hengdian Space Power Co., Ltd.
No.6, Haitaihuake 7 Road, Huayuan Industrial District, Xiqing District, 300384 Tianjin, CHINA
tankerxp@163.com

ABSTRACT: Two designs of antireflection coating (ARC) for high efficiency gallium arsenide (GaAs) solar cell (lattice-mismatch inverted metamorphic 4 junction (IMM-4J) solar cell) are proposed. Design 1 (D1) ARC is composed of titanium oxide (TiO_x), yttrium oxide (Y_2O_3) and cryolite (Na_3AlF_6). Design 2 (D2) ARC is composed of titanium oxide (TiO_x) and cryolite ($Na3AlF_6$). According to the differences of reflectance and external quantum efficiency (EQE) before and after evaporation, the short circuit current gains of each junction are calculated. All the results show that D1 ARC is more suitable for common IMM 4J solar cell.
Keywords: III-V Semiconductors, Solar Cell, Spectral Response, Antireflection Coating

1 INTRODUCTION

Solar energy is an attractive energy because of its cleanness and renewable. As photovoltaic (PV) device plays an important role in energy conversion process, related research is also never stopped. Solar cell is the most representative among PV devices and widely used in various ways. From initial silicon solar cell to present gallium arsenide solar cell, solar cell efficiency improves rapidly as the technology advance. Multi-junction GaAs solar cell shows higher open circuit voltage and better radiation hardened feature than Si solar cell. In order to get the greater efficiency, 4 junctions, 5 junctions even 6 junctions solar cells are also researched. Spectral response ranges of multi-junction GaAs solar cell covers from 300nm to 1800nm or more, so solar spectral energy can be sufficiently used.

It is well known that the surface reflectance could reduce about 30 percent light absorption for solar cell. So the antireflection coating (ARC) is important for solar cell. ARC is usually composed of optical thin films. Optical thin films are easy to prepare and have advantage of high stability. In common use is double layer ARC, this kind of ARC follows 1/4 λ law, the optical thickness of layer is equal to 1/4 reference wavelength and the refractive index of each layer must be matched.

Our study object is lattice mismatched inverted metamorphic 4 junction (IMM 4J) solar cell, which structure is InGaP (1.9eV) / GaAs (1.4eV) / In_xGaAs (1.0eV) / In_yGaAs (0.7eV). Range of spectral response is from 300nm to 1800nm. Short circuit current density (J_{sc}) of each junction is much closed. This reason requires ARC for IMM 4J solar cell should reduce surface reflectance from 300nm to 1800nm equally. Apply double layer ARC (usually composed of 58nm TiO_x and 90nm Al_2O_3) to IMM 4J solar cell, can only get low reflectance from 400nm to 900nm. Reflectance from 900nm to 1800nm is much higher than short wave rang. The short circuit current density of the 4th junction is much lower than other junctions. The J_{sc} of whole solar cell is limited by the 4th junction and the efficiency can't achieve the expected value. It is necessary to find a new type ARC for IMM 4J solar cell to meet requirement of broadband antireflection.

2 DESIGN AND PEPARATION

The new ARC must have a wide antireflective area. One useful method is to increase the ARC layer number. In theory, dozens of layers can perform a good effect of antireflection. But its prepared process is too complex to control. Absorption of these layers is obvious and can't be ignored. So we tend to use fewer layers to get the low average reflectance. According to the formula proposed [1, 2], we know that the surface layer should be low refractive index film and the inner layer should be high refractive index film.

Now we have designed two type of ARC. Both ARC designs don't follow 1/4 λ law any more, each layer thickness is optimized by Essential Macleod. As the optimization results of two designs are much closed, we want to check the actual effects of these two ARC for IMM 4J solar cell by experiment and then chose the better and suitable one.

D1 ARC follows HML structure. We add a low refractive index layer on the surface of initial double layer structure. D1 ARC is composed of titanium oxide (TiO_x), yttrium oxide (Y_2O_3) and cryolite (Na_3AlF_6) (see Table I). All the layers are evaporated by electron beam. In deposition process, we use high temperature baking (240 degrees Celsius) to make sure the adhesive force and refractive index of coatings.

Table I: Parameters of each layer for D1 ARC

Layer	Material	Measured n (630nm) / d (nm)	Optimized n (630nm) / d (nm)
1	Na_3AlF_6	1.32/130	1.33/128
2	Y_2O_3	1.72/88	1.74/85
3	TiO_x	2.26/50	2.29/49

D2 ARC follows HLHL structure. D2 ARC is composed of titanium oxide (TiO_x) and cryolite (Na_3AlF_6) (see Table II). All the layers are evaporated by electron beam too. As the thickness of layer 2 is not thick enough, in order to ensure the refractive index of TiO_2 (layer 2), we use ion-beam aid deposition method to fabricate this layer. The power of ion-beam is less than 100eV. It may ensure that the P-N junctions of solar cell could not be damaged forever. The open circuit voltage of IMM 4J solar cell after ion-beam aid deposition does not reduce obviously. High temperature baking is also need for D2.

Table II: Parameters of each layer for D2 ARC

Layer	Material	Measured n (630nm) / d (nm)	Optimized n (630nm) / d (nm)
1	Na_3AlF_6	1.32/128	1.33/125
2	TiO_x	2.18/21	2.29/20
3	Na_3AlF_6	1.31/36	1.33/35
4	TiO_x	2.26/69	2.29/70

3 RESUIT AND DISCUSSION

We deposited D1, D2 and double layer ARC on IMM 4J solar cell. The IMM 4J solar cell samples we used come from the same epitaxial wafer. So they have much similar optical and electrical characteristics. It's helpful to find the difference between D1 and D2 ARC.

Reflectance is measured when incidence angle is 90° (see Fig. 1). The test equipment is spectrophotometer produced by Varian, Cary 5000 type.

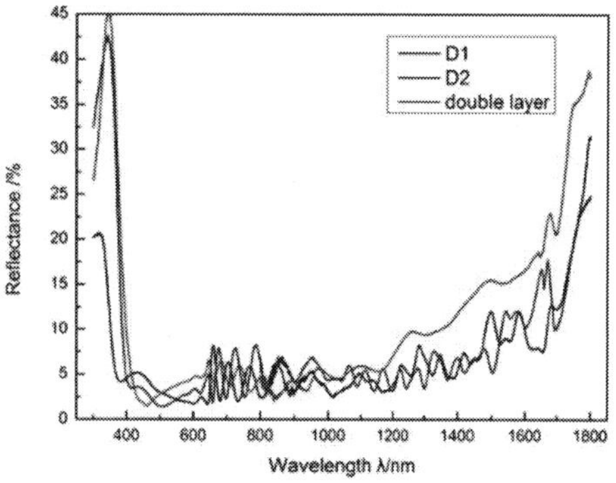

Figure 1: Measured reflective curve of D1, D2 and double layer ARC

It is easy to see that from 400nm to 900 nm double layer ARC shows a good antireflective capacity. But its high reflectance in long wave rang can't be accepted. The short circuit current density of 4th junction will hardly match with the other junctions. So J_{sc} of the whole solar cell is limited by the 4th junction.

In order to get equal and low reflectance among the spectral response range, D1 and D2 ARC all increase the layer number. Although there is no reflectance extreme value in the low reflectance range, the range has been expanded greatly. Reflectance values of D1 and D2 ARC maintain about 5 percent from 400nm-1500nm, increase after 1600nm. D1 ARC has lower reflectance than D2 before 400nm. It is because that D1 has thinner TiO_x layer than D2. Absorption especially for TiO_x can't be regarded as 0 before 400nm.

As reflectance of D1 and D2 ARC are very approximate in other wave band, it's necessary to combine external quantum efficiency (EQE) of these samples to analysis which one is better. EQE (under AM0) of IMM 4J solar cells have been measured before and after ARC deposition. Gain of short circuit current density for each junction is shown in table III and table IV.

Table III: Gain of J_{sc} for D1 ARC

Junction	Range (nm)	IMM-4J$_1$ No ARC/ D1 ARC J_{sc}(mA/cm^2)	Gain
1	300-750	11.47/15.36	33.9%
2	500-900	12.68/16.36	29.0%
3	800-1300	12.23/16.16	32.1%
4	1150-1800	10.59/13.90	31.3%

Table IV: Gain of J_{sc} for D2 ARC

Junction	Range (nm)	IMM-4J$_2$ No ARC/ D2 ARC J_{sc}(mA/cm^2)	Gain
1	300-750	11.51/15.28	32.8%
2	500-900	12.65/16.21	28.1%
3	800-1300	12.10/15.98	32.1%
4	1150-1800	10.54/13.87	31.6%

Among all the junctions, gain of J_{sc} for 2nd junction is the lowest. This result just fits the reflectance curve. The strong fluctuation appears from 650nm to 800nm due to the epitaxial structure, the entire two ARC don't show a very good antireflective capacity.

Compared D1 with D2 ARC, they have the same effect for 3rd junction. D1 ARC has greater gain of J_{sc} for 1st and 2nd junction. Especially for the 1st junction, influence of TiO_x thickness is obvious. But D2 ARC performs a little better for 4th junction.

As our IMM 4J solar cell sample isn't perfect enough, J_{sc} is limited by 4th junction. So D2 ARC is more suitable for these kind samples. But for the perfect IMM 4J solar cell, J_{sc} for each junction matches well, so D1 ARC may be more suitable. On the other hand, preparation process of D1 ARC is easier than D2 ARC. D1 ARC has fewer layer and little absorption, so D1 ARC could perform better for most of IMM 4J solar cell.

We also tried to use silicon dioxide (SiO_2) instead of Na_3AlF_6. Increasing the deposited angle of SiO_2 could reduce its refractive index [3]. In this way refractive index of SiO_2 could be 1.12 or lower. But it's very hard to control the thickness and refractive index of this kind of film in the same time. Questions of uniformity and adhesive force for coating are also not worked out. And this way doesn't fit a large number solar cell product. So we don't choose this design in the end.

4 CONCLUSION

We proposed two designs of ARC for IMM 4J solar cell and measured reflectance and EQE before and after ARC deposition. The result shows that the two designs can all expand the low reflectance range availably. D2 ARC is more suitable for our IMM 4J solar cell samples. But D1 ARC is more suitable for common IMM 4J solar cell. Neither D1 nor D2 ARC can reduce reflectance from 650nm to 800nm effectively. And Reflectance after 1600nm shows evident increasing trend. These problems will be left to solve in the future.

5 REFERENCE

[1] R. R. Willey, "Predicting achievable design performance of broadband antireflection coatings", J. Applied Optics, 32 (28), (1993) 5447-5451.
[2] Baumeiste P, "Starting designs for the computer optimization of optical coatings", J. Applied Optics, 34 (22), (1995) 4835-4843.
[3] Kennedy S R, Brett M J, "Porous broadband antireflection coating by glancing angle deposition", J. Applied Optics, 42(22), (2003) 4573-4579.

INDOOR CHARACTERIZATION OF WIND INFLUENCE ON CPV MODULES THROUGH CELL-TO-AMBIENT THERMAL RESISTANCE MEASUREMENTS

A.V. Chekalin[1], A.N. Panchak[1], N.Yu. Davidyuk[2], V.D. Rumyantsev[1], N.A. Sadchikov[1]
[1]Ioffe Institute, 26 Polytechnicheskaya str., St.-Petersburg 194021, Russia
[2]St Petersburg Academic University, 8/3 Khlopina str, St Petersburg, 194021, Russia
Phone: +7(812)292 7394, e-mail: chekalin@mail.ioffe.ru

ABSTRACT: In the work, an "overall" cell-to-ambient thermal resistance r_{th} is measured indoors in solar concentrator photovoltaic (CPV) modules under "windless" and simulated "windy" conditions. At measurement procedure, forward current through the cells causes moderate heating of the tested module, simulating heat transfer and heat dissipation processes, as it occurring under natural sun illumination. Instantly measured voltage drop value at current flow is used as an indicator of average temperature of the cells. Using the temperature coefficients, comparison of the voltage values under "cold" and "hot" module conditions allows estimating increase in temperature of the cells with respect to ambience. The elaborated method of indoor r_{th} evaluation has been applied at investigations of solar CPV modules based on silicone-on-glass Fresnel lenses and triple-junction GaInP/GaInAs/Ge cells.
Keywords: Multijunction Solar Cell, Module, Thermal Performance

1 INTRODUCTION

In order to further reduce the cost of energy produced by photovoltaic (PV) technology, factors that negatively affect the performance of these systems should be eliminated. One such factor that significantly reduces solar power conversion efficiency of solar cells is increasing temperature under operational conditions. Thermal management plays an important role in PV systems, especially in solar concentrator PV (CPV) modules, where concentration ratios can be as high as 1000x suns. Design optimization of CPV modules becomes possible by obtaining data on thermal behavior of modules in various climatic conditions. During outdoor experiments, meteorological conditions may cause considerable variations of the results at cell temperature evaluation in concentrator modules. The reasons are instabilities in solar irradiation, ambient temperature and wind speed during continuous outdoor experiments. Due to the above reasons, long exposure times as well as outdoor data filtering are required [1]. Also, non-homogeneous temperature distribution on the receiver panel of the concentrator module reduces accuracy of obtained data.
The purpose of the work is to investigate thermal properties of the solar concentrator modules using recently developed indoor cell-to-ambient thermal resistance (r_{th}) measurement method [2]. As it is shown, this method allows comparing the modules differing in structure under "standard" wind conditions simulated indoors.

2 EXPERIMENTAL SPECIMENTS AND MEASUREMENT CONDITIONS

Concentrator modules of the SMALFOC (**S**mall lenses, **M**ultijunction cells, **A**ll from glass, **L**amination, **F**resnel, **O**ptics, **C**oncentration) [3] design have been elaborated in the PV Lab of the Ioffe Institute. A full-size module consists of a Fresnel lens panel with 60x60 mm^2 in size silicon-on-glass lenses (8x16 configuration), and a receiver panel with corresponding number of solar cells. Each group of eight cells is soldered on a common metallic heat sink being connected electrically in parallel. The groups of cells on 16 heat sinks are connected in series. Metallic heat sinks are made of either 0.5 mm thick copper or 1 mm thick steel plates, profiled by a mechanical die. The cells on heat sinks together with interconnecting buses are hermetically sealed with a laminating film on the outer side of the rear glass base.
Measurement conditions were chosen to be close to Concentrator Standard Operation Conditions (CSOC) described by IEC 62670-1 standard [4]. Ambient room temperature during the measurement process was fluctuating insignificantly around 22 °C. Corrections due to these changes were made when calculating overheating temperature of the cells. Range of wind speed 0 ÷ 2 m/s at temperature evaluation was selected based on requirements of CSOC.

3 INDOOR CELL-TO-AMBIENT THERMAL RESISTANCE EVALUATION PROCEDURE

At measurement procedure, forward current through the cells is a source of incoming thermal power. It causes moderate heating of the tested module, simulating heat transfer and heat dissipation processes, occurring under natural sun illumination. In general, by varying the forward current magnitude, one can simulate different illumination conditions, i.e. different DNI. However, there are some restrictions, which will be considered below. To evaluate incoming thermal power, DC current and voltage across the module terminals are measured after a period of time about 30 minutes corresponding to achieving the thermal balance between module parts and environment.
The method of r_{th} evaluation involves two steps. At the first step, the forward voltage data are calibrated in pulse regime relating to actual temperature of the cells using a thermostatic chamber. At the second step, the voltage data for a definitely heated module are compared with those at ambient temperature, so that overheating temperature is evaluated. As a result, r_{th} is calculated using the value of DC incoming power and the corresponding temperature increase.
For pulsed I-V measurements, forward current is used to induce time-dependent voltage drop across the module terminals. This allows recording the I-V curves during 1-3 ms. Hence, obtaining voltage data correspond to temperature of the p-n junctions heated by definite input

electric power. During such a fast measurement process, the temperature of the module does not change considerably owing to finite heat capacity of the solar cell. Implementation of this procedure at the initial moment, before heating of the module, allows obtaining the voltage data related to temperature in the thermostatic chamber [2].

To be able to convert voltage difference into temperature difference, first, one needs to calculate temperature coefficients for the tested modules. It has been done by comparing the forward voltage data to the actual temperature of the module being equal to the temperature inside the thermostatic chamber (Fig.1). Calculated thermal coefficients for investigated modules at different forward current values are presented in Table I.

The comparison of voltage data for a definitely heated module and those at ambient temperature in laboratory environment, with the use of thermal coefficients, allows the evaluation of overheating temperature of the cells. As a result, r_{th} is calculated by formula (1) using the values of incoming thermal power and corresponding temperature increase [2]:

$$r_{th} = \Delta T_{p-n} S_{lens} / P_{EI} [°C \cdot m^2 / W] \qquad (1)$$

where S_{lens} is the lens panel area of the module, ΔT_{p-n} – average overheating temperature of the cells, P_{EI} - incoming thermal DC power.

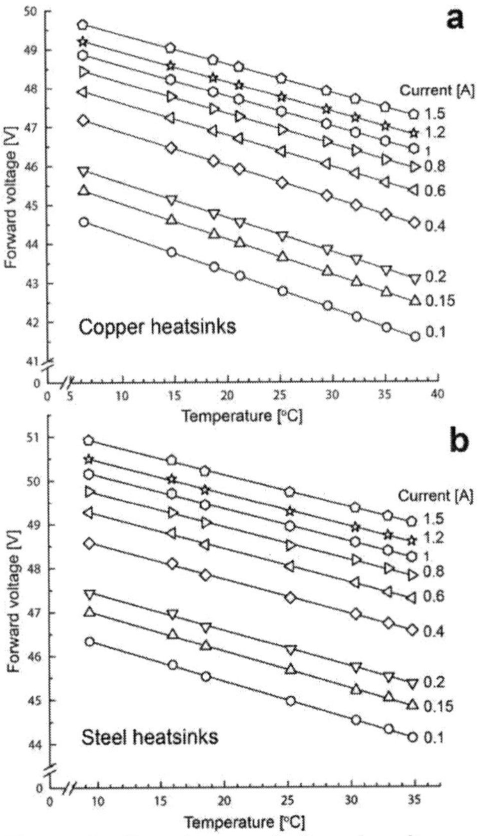

Figure 1: Dependences of forward voltage on temperature for modules with copper (a) and steel (b) heat sinks. Data for different current values were obtained at fast I-V measurements in a thermostatic chamber.

Table I: Forward voltage temperature coefficients β_{cell} for investigated modules with copper and steel heat sinks.

Current [A]	β_{cell} [-mV/°C]	
	Copper Heatsinks	Steel Heatsinks
0.1	5.96	5.43
0.15	5.74	5.33
0.2	5.59	5.15
0.4	5.31	4.95
0.6	5.11	4.88
0.8	4.99	4.77
1	4.89	4.75
1.2	4.81	4.71
1.5	4.71	4.66

The experimental setup used in the work is shown in Figure 2. Wind generator based on two 40 cm fans with a speed regulators and homogenizer allowed simulation of wind speeds in the range of 0 ÷ 3 m/s with homogeneity of about 15%. Airflow direction was horizontal. The position of the modules was varied relating to the airflow direction.

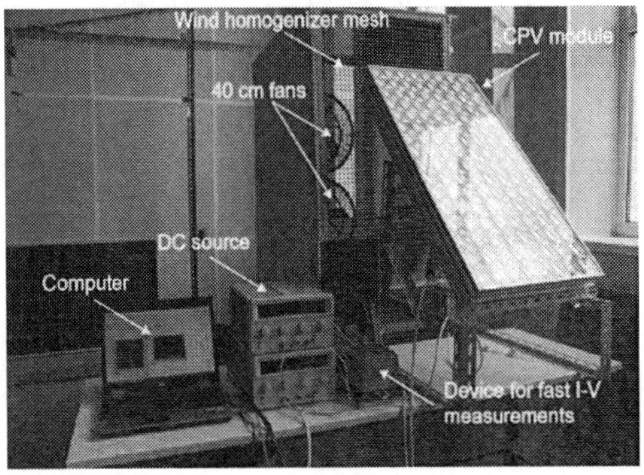

Figure 2: Laboratory setup for indoor wind simulation and I-V measurements: wind generator with 2 x 40 cm fans; homogenizer; DC source for thermal loading of the modules; equipment for fast I-V measurements with specialized software.

4 RESULTS

The initial measurements were carried out for the modules at the 45° tilt angle and windless conditions. In this case, natural air convection plays a major role in heat dissipation process. Non-uniform heating of the receiver panel occurs due to two main reasons revealed by infrared images of the receiver panel. The first reason is a possibility of non-uniform current distribution over the cells of the module. This situation is in contrast with a normal operation conditions under sun illumination, when compulsory equalization of incoming solar energy and of generated photocurrent over the cells is realized. To avoid current non-uniformities, the use of low enough forward current levels is needed, limiting incoming thermal power by about 70 W. The second reason is uprising airflow due to convection inside and outside of the module body [5]. The effect is revealed in Figure 3a.

At wind simulation, temperature distribution becomes more homogenous and the average temperature of the module drops significantly (Fig.3b,c,d). The wind direction almost doesn't affect module temperature, except the case, when the airflow is directed on the lens panel.

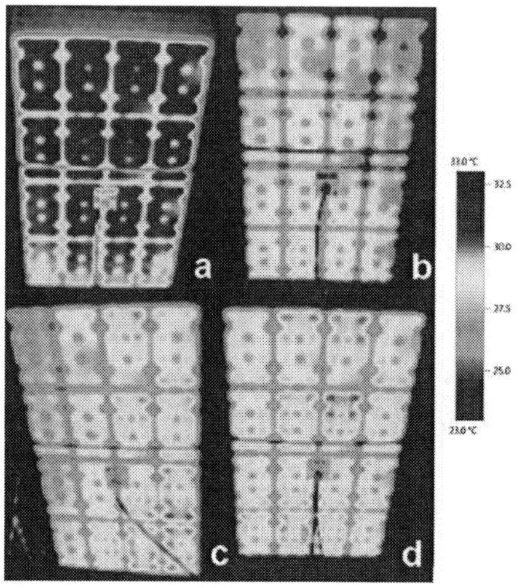

Figure 3: Infrared thermal images of the SMALFOC module receiver panel during thermal loading by electric power of 50 W: (a)- the absence of wind; (b, c)- airflow is directed on receiver panel or sidewall; (d)- airflow is directed on lens panel.

In figure 4, graphs of overheating temperature of the p-n junctions during wind simulation for modules with steel and copper heat sinks are presented. Wind speed and input electric power were varied. One can conclude that overheating temperature of the p-n junctions depends linearly on incoming electric power. The noticeable difference in overheating temperature under wind and windless conditions is revealed. However, this difference becomes insignificant at wind speeds >1 m/s. The reason for this might be that thermal conductivity of the laminating film as a top layer of the receiver panel is quite low so that it limits the heat transfer from the heat sinks to the outer surface of the modules.

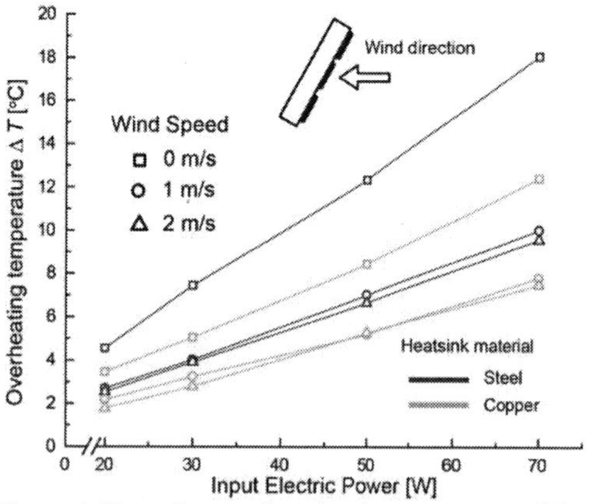

Figure 4: Dependences of overheating temperature of the p-n junctions on input electric power during wind simulation for modules with steel and copper heat sinks. Wind speed was varied being directed on receiver panel.

In Figure 5, comparison of thermal resistance r_{th} values for modules with copper and steel heat sinks during simulation of wind conditions are presented. Direction of wind and wind speed were varied: plot (a) – corresponds to 0.6 A forward current conditions, (b) – 1 A.

At identical environmental conditions and the same amount of incoming thermal power, one can notice a slight difference between obtained r_{th} parameters at different current values (Figure 5). The reason for this is, probably, a slight heating of the solar cells at higher currents during fast I-V measurement process.

Generally, the temperature of the module with copper heat sinks is about 60% the temperature of the one with steel ones, regardless of the wind speed and direction. Despite heat dissipation properties of steel are worse than those of copper, it still can be considered as a heat sink material for CPV modules due to its low cost and better matching of thermal expansion coefficient with that of cell material (germanium) and glass.

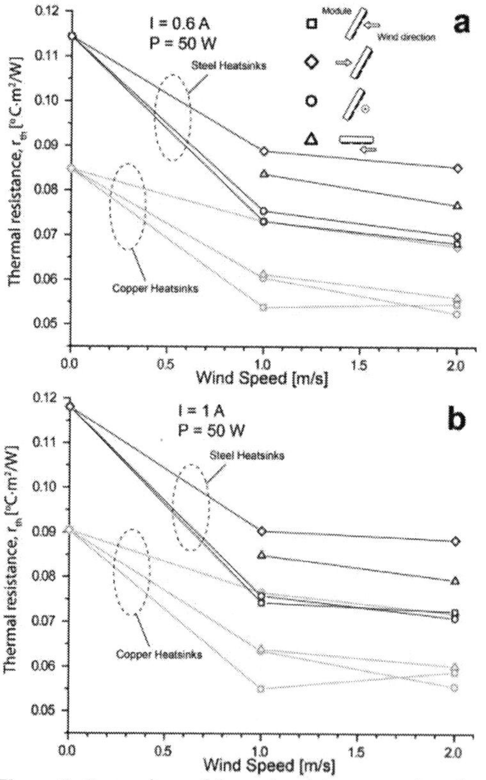

Figure 5: Comparison of thermal resistance r_{th} values for modules with copper and steel heat sinks at varied wind conditions. Direction of wind is indicated in insert of part a. Part a corresponds to 0.6 A forward current conditions, part b – 1 A.

5 CONCLUSION

Wind influence on thermal properties of CPV modules was studied using simple indoor method of cell-to-ambient thermal resistance measurements. Parameter r_{th} was calculated for different module operational positions, different wind speeds and directions. Variation of wind directions does not affect module temperature significantly, except when wind is directed on the lens panel of the CPV module. Increase in wind speed above 1 m/s does not decrease noticeably the module temperature. It seems that thermal conductivity of the laminating film as a top layer of the receiver panel in the investigated modules is quite low, so that it limits the heat transfer from the heat sinks to the outer surface of the modules. The difference in heat dissipation efficiency between copper and steel heat sinks is maintained even in windy conditions.

6 ACKNOWLEDGEMENTS

The work has been supported by the Russian Science Foundation, Agreement No. 14-29-00178.

7 REFERENCES

[1] R. Núñez1, I. Antón, S. Askins, G. Sala, K. Araki, AIP (2014) 144.
[2] V.D. Rumyantsev, A.V. Andreeva, A.V. Chekalin, N.Yu. Davidyuk, D.A.Malevskiy, P.V.Pokrovskiy, N.A. Sadchikov, Proceedings of the 31st European PV Solar Energy Conference (2015) 1403-1406.
[3] V.D. Rumyantsev, V.M. Andreev, A.V Chekalin, N.Yu. Davidyuk, O. Im, E. Khazova, N.A. Sadchikov, AIP Conf. Proc. 1556 (2013) 185.
[4] IEC, Concentrator Photovoltaic (CPV) – Performance testing – Part 1: Standard conditions, in 62670-1, IEC, Editor. (2013) International Electrotechnical Commission.
[5] M. Wiesenfarth, S. Gamisch, H.Kraus, A.W.Bett., AIP 1556 (2013) 189–193.

RADIATION EFFECTS ON ADVANCED MULTI JUNCTION SOLAR CELLS FOR SPACE MISSIONS

R. Campesato[1], G.Gori[1], M. Casale[1], G. Gabetta[1]
M. Sankaran[2], E.P. Suresh[2] , B.R.Uma[2]
[1] CESI SpA , Milan, via Rubattino 54, Italy: campesato@cesi.it
[2] ISRO Satellite Centre, Bengaluru, India

ABSTRACT: InGaP/InGaAs/Ge triple junction solar cells have reached efficiencies above 30% in A.M. 0 conditions. To further improve their performances and their appeal to space missions, several strategies are under development in particular concerning lightweight and low cost approaches. This paper compares the BOL and the EOL performances of the triple junction solar cells named CTJ30, with the 80μm thin (CTJ30-80) and the low cost structures (CTJ-LC) under qualification. The solar cells were electron and proton irradiated at different energies and fluencies and then exposed to photon annealing. The solar cells characteristics were recorded before and after photon annealing: a clear recovery of the damage was observed on all of them.
Keywords: Multijunction Solar Cell, Degradation, Space Cells

1 INTRODUCTION

InGaP/InGaAs/Ge triple junction solar cells have reached efficiencies above 30% in A.M. 0 conditions. To further improve their performances and their appeal to space missions, several strategies are under development:

- Thinning of the substrate to have lighter and flexible solar cells (CTJ30-80) [1]
- Implement low cost processes to reduce their cost (CTJ-LC)
- increase the solar cell area working with 6-inch Ge-substrates
- Develop more than three junctions structures to reach the 32% efficiency target [2]
- Substitute the Ge bottom junction with InGaNAs structures[3]

The knowledge of how the solar cells behave when they get irradiated by charged particles at different energies is a key point to bring the new structures into the consideration of the solar arrays manufacturers.

This paper compares the BOL and the EOL performances of the triple junction solar cells named CTJ30, with the thin (CTJ30-80) and the low cost structures (CTJ-LC) in development.

The solar cells were electron and proton irradiated at the following energies:

- Electrons: 600 keV, 1 MeV, 12 MeV
- Protons: 50 keV, 100 keV, 400 keV, 1 MeV, 4 MeV, 10 MeV.

After irradiation, a combined photon and thermal annealing of 8 hours in open-circuit conditions at 1 AM0 was carried out on all irradiated samples following the Indian Space Research Organisation (ISRO) protocol.

The solar cells characteristics were recorded before and after photon annealing: a clear recovery of the damage was observed on all of them.

2 DESCRIPTION OF SOLAR CELLS

2.1 Standard Triple Junction Solar cells

The triple junction (TJ) solar cells are composed by three sub cells: one InGaP- top junction, one InGaAs-middle junction and oneGe – bottom junction.

The energy gaps of these materials are respectively 1.84, 1.40 and 0.7 eV. Figure 1 shows a cross-sectional scheme of the solar cell structure. The three junctions are series connected by two tunnel diodes. In this configuration, the voltage of the TJ is the sum of the voltage of each sub cell while the current is the minimum current generated by one of the subcells (the "limiting" subcell). Generally, to improve the radiation resistance, the limiting solar cell in a TJ stack is the highest energy gap subcell (the TOP subcell).

The improvement of the TJ efficiency, in both BOL and EOL conditions, was achieved through two main improvements:

- The use of a Bragg reflector under the middle junction;
- The introduction of an electrical field in the base layer of the top and of the middle junctions [1].

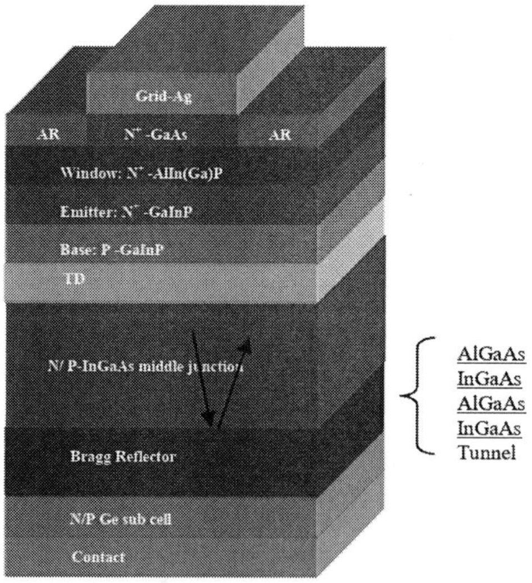

Figure 1: Triple junction solar cell structure with Bragg reflector

This standard Triple Junction solar cell is manufactured at CESI and named CTJ30 because it exhibit a typical efficiency of about 29.5% AM0, 25 °C.

The total thickness of the CTJ30 solar cell is 150 μm and its weight is about 88 mg/cm^2.

The CTJ30 solar cells were fully qualified for LEO and GEO missions and have flight heritage on board of several satellites.

2.2 Thin Triple Junction Solar cells

The requirements for the future generations of space satellites call for solar generators with more power and less mass than in the past. Very high efficiency solar cells, based on III-V compounds are still the best approach to increase the specific power and the solar arrays lifetime . CESI investigated the CTJ30 lattice matched structure on thin Germanium substrates (80 μm) to reduce the mass of the solar cells and to obtain bendable devices [the name of this cell is CTJ30-80]. A 29% efficiency A.M. 0, 25°C and a bending radius of 30 mm have been demonstrated on these devices that reach a specific power of about 1W/g, almost doubled with respect to the standard CTJ30 structure (0.41W/g).

Figure 2 is a picture of a CTJ30-80 wrapped around a cylinder whose radius is 30 mm.

Figure 2: Picture of a CTJ30-80 cell in abent configuration (Radius 30 mm)

These thin solar cells are currently under qualification according to the ECSS E ST20-08C rev.1 standard. One of the main objectives of this campaign is to verify the impact of the wafer thinning to the solar cell radiation resistance against omnidirectional radiation, by comparing their end of life behavior to that of the standard 150μm thick CTJ30 cells.

2.2 Low Cost Triple Junction Solar cells

The new constellations of minisatellites and the widespread use of cubesats call for the development of a highly efficient and reliable III-V solar cell characterized by a cost that is significantly lower than that of the standard CTJ30 devices.

To this respect, CESI developed a new triple junction solar cell, whose efficiency is slightly lower than that of a standard CTJ30 device, but its cost is between 70% and 50% of the CTJ30. This was possible thanks to:

- a strong reduction of the consumption of the most expensive precursors
- a significant reduction of manufacturing time (higher growth rates at MOCVD level and simplified post growth process)
- Higher manufacturing yields
- Higher yearly manufacturing capacity.

This new device is called low cost triple junction solar cell (CTJ-LC) and has a BOL efficiency of about 27.5-28%, high radiation resistance, temperature coefficients in line with the standard CTJ30 triple junction solar cell. The simplification of the epitaxial structure requires a new qualification campaign to assess the solar cell reliability at bare and assembly levels.

3 EXPERIMENTAL RESULTS

All the solar cells were characterized using a dual source solar simulator and spectral responses at BOL and EOL conditions. The test campaign on the CTJ30 standard bare solar cells was concluded and all results are available. The CTJ30-80 and CTJ-LC solar cells are still under qualification and the reported data represent the current state of the characterization. Although still partial, these results are interesting because it is already possible to envisage the effect of the thinning of the substrate on the EOL behavior. The electron and proton irradiation was performed in Europe and in India in cooperation with ISRO.

Solar cells with different sizes were electron irradiated, the achieved Remaining Factors show a slight dependence upon the edge/area ratio. Generally, larger area solar cells appear to be more radiation resistant

3.1 Electrical performances BOL

The electrical performances have been measured using simulated AM0 light at CESI and at ISRO on several samples of CTJ30, CTJ30-80 and CTJ-LC. Solar cell sizes were 4x8 cm^2 and 2x2 cm^2.

During the measurements, the solar cell temperature was kept at 28±1°C by a thermostated chuck.

The solar simulators have been calibrated using the 30 cm^2 reference standards of the top and middle junctions calibrated at INTA-Spasolab (Spain).

The results of the electrical characterization are reported in table I.

Table I: BOL average electrical performances measured at CESI and ISRO

Large sizes (CESI)	Isc(A)	Voc(V)	FF	Eff%
CTJ30 (30 cm^2)	0.54	2.6	0.85	29
CTJ30-80 (26 cm^2)	0.46	2.6	0.85	29
CTJ-LC (26 cm^2)	0.45.....	2.6	0.85	28

Small sizes (CESI)	Isc(mA)	Voc(V)	FF	Eff%
CTJ30 (4 cm^2)	72	2.60	0.84	28.5
CTJ30-80 (4 cm^2)	72	2.58	0.84	28.5
CTJ-LC (4 cm^2)	69	2.61	0.83	27.5

Small sizes (ISRO)	Isc(mA)	Voc(V)	FF	Eff%
CTJ30-80 (4 cm^2)	76	2.58	0.83	29.7

Generally, the small solar cells efficiency is lower because the fill factor is lower. This reduction in FF is to be ascribed to the increase of the edge leakage current. Furthermore the calibration of the solar simulator is carried out with large area solar cells, which introduces a larger uncertainty in the measurement of the small 4 cm^2 area devices.

3.2 Spectral response BOL

The spectral responses have been measured by Spasolab and by University Milano Bicocca on a few samples for each kind of solar cell to identify the limiting subcell.

The typical spectral responses for the CTJ30, the CTJ 30-80 and the CTJ-LC solar cell structures are reported in figures 3, 4 and 5 respectively. With respect to the standard structure in Fig. 3, a clear decrease in the quantum efficiency of the germanium subcell is noticeable for the CTJ30-80. The low cost structure (CTJ-LC), shows a lower quantum efficiency in both the top and the middle junctions whereas the germanium junction seems to improve its behavior.

In any case all the devices (CTJ30, CTJ30-80 and CTJ-LC) are top limited, as requested for space applications.

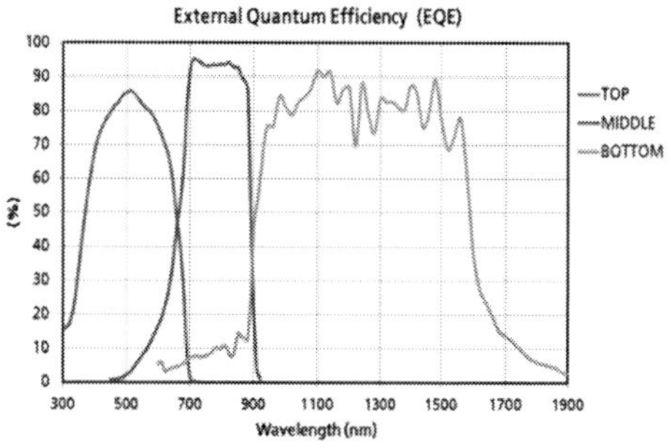

Figure 3: Spectral response of CTJ30 solar cells

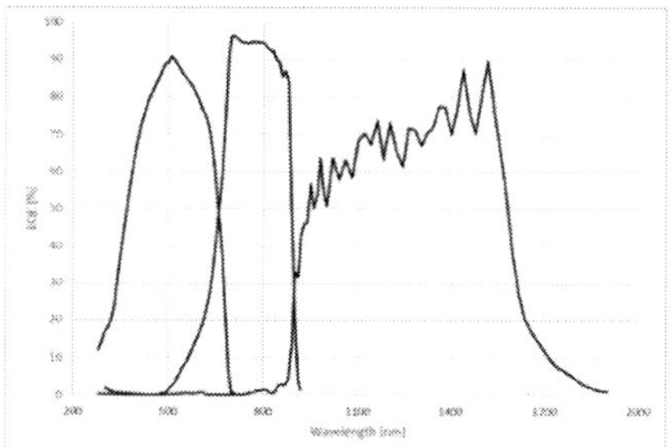

Figure 4: Spectral response of CTJ30-80 solar cells

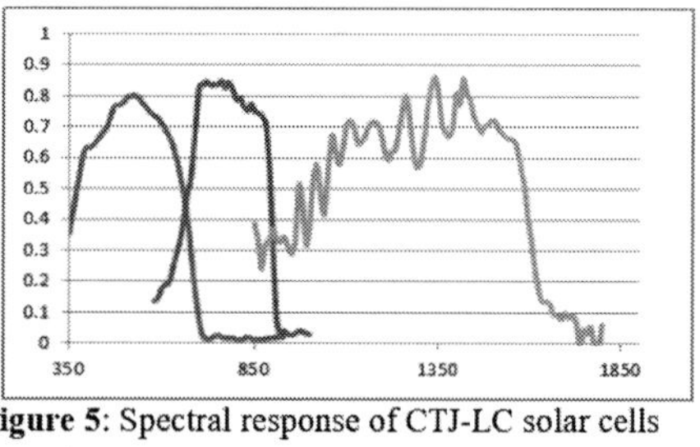

Figure 5: Spectral response of CTJ-LC solar cells

3.3 Irradiation and photon annealing

The electron and protons irradiation was performed at the following facilities:

Electrons:

- 1MeV: Technical University of Delft, ISRO
- 0.6 MeV e 12 MeV: Fraunhofer Institute of Organiche Elektronik, Elektronenstrahl un Plasmatechnik (Dresden)
- 10 MeV: ISRO

Protons:

- 50KeV, 100keV, 400keV: Centre des Science Nucleaires et de Sciences de la Matière, Orsay (F)
- 4MeV: Centro Nacional Aceleradores (Sevilla, Spain)
- 10MeV: National Institute of Nuclear Physics, Legnaro Laboratories (Padua, Italy)

The solar cells electrical performances as recorded immediately after the irradiation, highlighted a strong decrease in efficiency.

The solar cells were subsequently annealed applying ISRO's protocol, consisting in keeping them under sunlight (1 AM0) for 8 hours at a temperature of 60 °C.

After the photon annealing, all the solar cells recovered part of the damages thus improving their performances.

In particular it was noticed that:

- After electron irradiation, Isc and Voc recovered of about 2% and the power recovered of about 4%
- After proton irradiation, the highest recovery is obtained for the 50 keV proton irradiation at the highest fluence where the most damaged subcell is the top: in this case, the recovery in Isc is about 8%, the recovery in Voc is about 2% and the recovery in Pm is about 11%.
- After higher energy proton irradiation, the recovery after photon annealing is lower and at 4MeV proton energy, the same recovery of 1MeV electrons was observed.

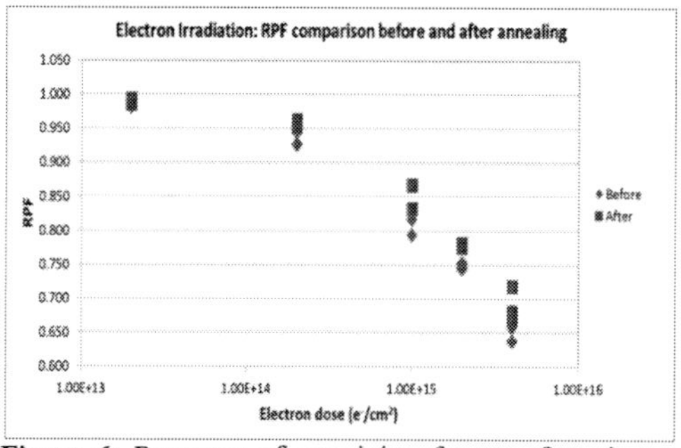

Figure 6: Recovery of remaining factors after photon annealing in solar cells irradiated with electrons

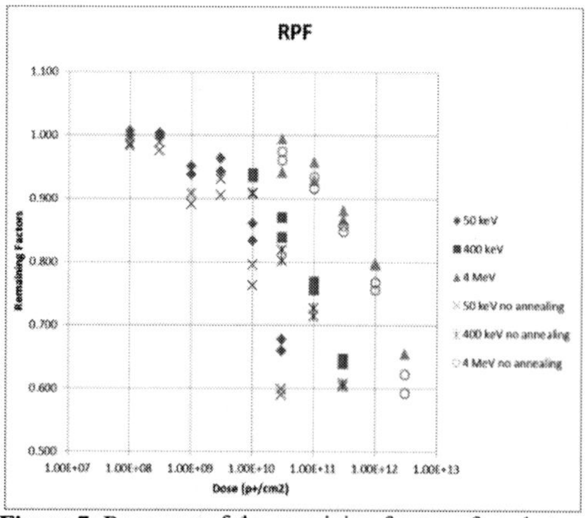

Figure 7: Recovery of the remaining factors after photon annealing in solar cells irradiated with protons

The strongest annealing effect at the low energy protons is related to the better recovery observed in top sub cell that was the most degraded junction after the 50 keV proton irradiation.

3.4 Electrical performances EOL

The Remaining Power Factors for CTJ30 solar cells, 30 cm^2 area, are reported in figure 8. The electron and proton irradiations were performed on a large energy spectrum to determine damage coefficients for these devices.

The CTJ30 solar cells show a very high radiation resistance against high energy protons and a Voc that degrades slowly as a function of the dose under all conditions

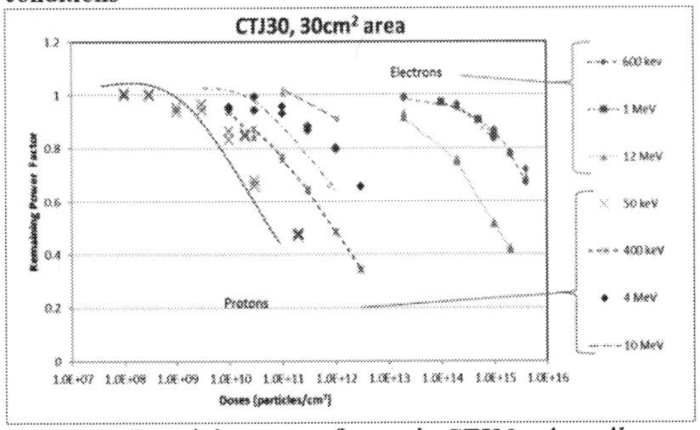

Figure 8: Remaining power factors in CTJ30 solar cells

The thin solar cells were irradiated by 1 MeV and 10 MeV electrons at different doses, while the low cost solar cells were only submitted to 1 MeV electron irradiation at one dose. All cells were then annealed by photon irradiation (8 hours in AM0 conditions open circuited at 60 °C). The results show that:

- CTJ30-80 are slightly less resistant than CTJ30
- CTJ-LC seem slightly more resistant than CTJ30
- RPF are lower as the cell size decreases, mainly because of a reduction in FF after irradiation.

Table II shows the remaining power factors obtained after the 1 MeV electron irradiation on standard CTJ30, CTJ30-80 with two different sizes (4x8 cm^2 and 2x2 cm^2) and CTJ-LC with 26.5cm^2 area.

Table II: RPF for CTJ solar cells after 1 MeV electrons

RPF @1MeV	CTJ30	CTJ30-80	CTJ30-80	CTJ LC
Dose(e/cm^2)	30cm^2	26.5cm^2	4 cm^2	26.5cm^2
1.00E+14	0.97	0.97	0.97	
5.00E+14	0.91	0.88	0.87	0.91
1.00E+15	0.85	0.82	0.81	

Table III reports the results obtained on thin solar cells after irradiation with 10 MeV electrons and after photon annealing.

Table III: RPF for CTJ30-80 solar cells after 10 MeV electrons

MeV	DOSE	RF Isc	RF Voc	RF Pm
10	1.0E14	0.920	0.924	0.832
10	5.0E14	0.806	0.879	0.665
10	1.0E15	0.715	0.838	0.536
10	3.0E15	0.616	0.786	0.369

4 CONCLUSIONS

The CTJ30 solar cells have been extensively investigated as far as radiation resistance is concerned. Remaining factors for Isc, Voc and Pm have been derived after irradiation with protons and electrons at different doses.

The new thin III-V solar cells and low cost solar cells, still based on the Lattice Matched InGaP/GaAs/Ge structure, are now under qualification, the first results are compared with the CTJ30 standard cells.

It was observed that the thinner germanium substrate leads to a solar cell that is more sensitive to the electron irradiation.

Finally, the low cost structure, characterized by a lower BOL efficiency, seems to have a very promising radiation resistance with respect to the standard CTJ30 structure.

4.1 References

[1] R. Campesato et al: "Electron and Proton radiation on high efficiency space solar cells" Paris EUPVSEC 2013 .

[2] R.Campesato et al., "High efficiency InGaP/ InGaAS/Ge triple junction solar cells on semirigid frame for the russian spacecraft" Proc. 25th European PV Solar Energy Conference, Valencia, Sept., 2010

[3] CESI Patent nr. PCT/EP2010/060477

[4] Anspaugh "GaAs Solar Cell Radiation Handbook" NASA 1996

[5] C. Baur et al:"NIEL dose dependence for solar cells irradiated with electrons and protons" Astroparticles, Particle, Space Physics and Detectors for Physics Application 27/11/2014

THEORETICAL CONTRIBUTION TO THE UNDERSTANDING OF PROTON BACK IRRADIATION EFFECTS ON SEMICONDUCTOR DEVICES IN SOLAR PANELS. THE CASE OF SILICON BYPASS DIODES

S. Rodríguez*, J. Plá*,§, J. Duran*, M. Alurralde*,§§
* Dpto. Energía Solar, CAC, CNEA, Avda. Gral. Paz 1499 (1650), Prov. de Buenos Aires, Argentina
§ CONICET, Argentina
§§ Instituto. J. Sábato, UNSAM, Argentina

ABSTRACT: Despite of electrons are the more abundant particles in LEO (low Earth orbit) orbits, the main responsible of solar panels degradation are protons, due to proton mass is similar to target masses and then presents a higher transfer energy efficiency. The object of this work is to study the contribution of proton irradiation on back side of the solar panels to their degradation. Usually, the effect of back irradiation is considered using an equivalent thickness of Al that does not take into account the anisotropy of the panel structure. Here we perform a comparison between the front side irradiation, where usually the method of an equivalent thickness is applied, and a simplified Monte Carlo simulation approximation of the honeycomb to assess the back side irradiation. Considering the results obtained, a simplified method to predict the backside irradiation contribution to damage is proposed.
Keywords: Space, Radiation Damage, Simulation

1 Introduction

Protons in space presents a wide energy range, especially at LEO orbits. Their contribution to degradation of electronic components of the solar panels is hard to assess. On the other hand, most of the satellites have deployable solar panels, with a structure based on a sandwich of carbon fiber with an Al honeycomb (20 μm, thickness) inside to allow rigidity with low weight; over one side the array of solar cells is bonded. This structural support is intrinsically two dimensional. The usual approximation is to estimate an equivalent thickness (in g/cm^2) and consider a single Al layer as shielding to calculate the backside degradation (see for example ref. [1]). In the specialized literature there is not practically any reference about how to estimate the shielding thickness, particularly the honeycomb contribution (see for instance ref. [2]-[4]). In this work, the influence of the panel structure on the back side irradiation effect is explored using the well-known Monte Carlo code TRIM and a simplified model of the honeycomb based on square unit cells.

2 Methodology

The Monte Carlo SRIM [5] code was used to simulate the interaction between protons and solar panel components. The proton spectrum was estimated using the SPENVIS online computational suite [6] for a LEO orbit and a Monte Carlo code developed at group is used to produce the TRIM input file (trim.dat). The structure of the solar panel considered for simulation is the following:

Front face: (100 μm borosilicate glass+6% Ce) + (80 μm silicone) + [(10 μm Ag) + (140 μm Si)]. In square brackets is the bypass diode.

Back face: (600 μm Carbon fiber) + (X μm Al or Al honeycomb) + (600 μm Carbon fiber) + (80 μm silicone) + (100 μm Kapton) + [(15 μm Ag) + (140 μm Si)], in square brackets is the bypass diode.

As far as possible, it was used the chemical composition suggested by SRIM with density corrections from the manufacturers. The X value is 0 for honeycomb simulation or 0, 20, 185 or 370 μm for Al layer simulations. Below, in section xx, the reason to select these thicknesses is explained.

As a first approximation to test the influence of the bidimensional characteristic of the honeycomb, it was consider as formed by squares of two hexagon apothem sizes (4 mm) of side, in order to obtain roughly similar grazing incident angles on the Al walls. This approximation will produce "channels" fifteen percent greater than in a hexagonal geometry. For particles impinging at the honeycomb walls, it is calculated the total thickness (38.2 mm) traversed and this thickness is used to estimate the energy loss in the usual way through using the total stopping power. Thus, for particles with energy greater than zero, a new TRIM.DAT input file for SRIM is written and then the rest of panel structure is simulated.

All simulations were done with 4×10^5 protons, and. taking into account that only particles over 3 MeV can arrive to the diode. This corresponds, over all space spectrum and a honeycomb cell, to a dose of 8.37×10^6 protons by square centimeter.

3 Results and discussion

3.1 Analysis of transmission spectra

The TRIM code was used to simulate the degraded spectra after particles go through different panel layers according to the previous description.

The results are completely general and could be applied to any device placed on a solar panel with a similar structure. In figure 1 the space spectra (grey line; due to solar cell structure only energies over three MeV are considered) and the rest of studied spectra (different symbols, see the inset) are showed. As expected, the front side irradiation is the most important. The honeycomb curve compares well with an Al layer of 185 μm, because most part of the particles goes through one Al wall in an angle close to 6° (representative angle of particles that go through the honeycomb without crossing any other wall).

In figure 2 the angular spectra are showed; as expected, the space one is completely flat. The rest are modified as particles travel through layers of different thicknesses and through the honeycomb. The last one compares well with "without honeycomb" at low angles and with 185 μm at higher ones.

Results presented in figures 1 and 2 suggest that it is possible simulate the honeycomb with a layer of 185 μm. However, there is some missing information because figure 2 does not show the energy distribution at different angles. A way to overcome this situation for some particular cases is described in the next section.

3.2 Analysis of PKA spectra

This part of the analysis could be applied only to silicon devices. The so called CAC method [7] suggests the use of PKA spectra as a meaningful physical measurement of device damage.

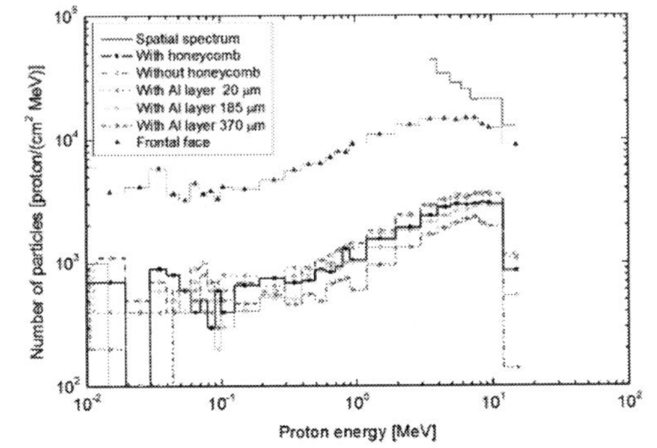

Figure 1: Space and degraded energy spectrum for different simulated cases, specified in the inset.

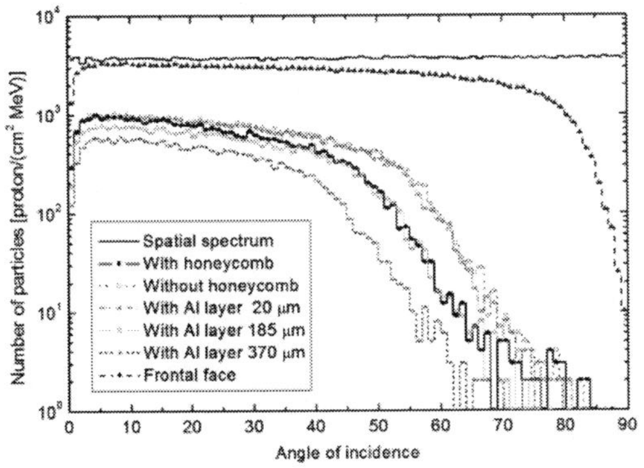

Figure 2: Transmission spectra for different analyzed cases.

It is noticeable that the trend with depth in the case of honeycomb is very similar to the effect of single Al layer, suggesting that it is possible to define an equivalent thickness. To test this possibility, several runs with different thickness of Al layers were done, 0 μm (red empty circles), 20 μm (honeycomb wall thickness, green left peaked triangles) and 370 μm (Al layer thickness of same areal density than of the honeycomb, dark green right peaked triangles); see figures 3 and 4.

Figure 3 shows the PKA spectra for different cases, with and without honeycomb and with different Al layers. Despite of all of them show similar slopes, the number of events changes the trend with depth between frontal face irradiation and the rest of them, as figure 4 shows. The peaks and valleys at begin and end of the layer are artifacts intrinsic to the TRIM code.

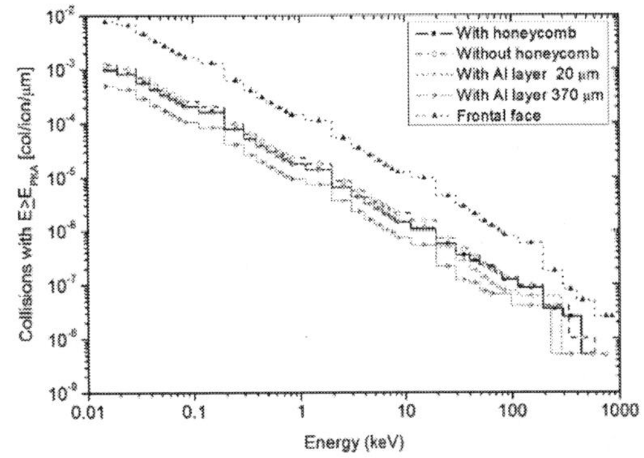

Figure 3: PKA spectra for different cases, with and without honeycomb and with Al layers of different thickness, specified in the inset.

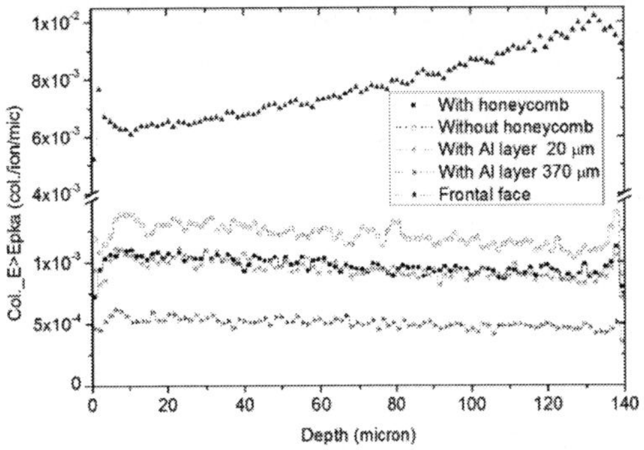

Figure 4: Number of collisions, versus depth, that produce PKA atoms with energies bigger than 15 eV for the different cases studied (see the inset).

4 Conclusions

For a better understanding, in the case of a silicon device, the curves of figures 3 and 4 were redraw normalized to honeycomb curve, as shown in figure 5. From all these results it is possible to conclude:

Most part of the particles affecting devices on a space panel come from the front size (89%, see fig. 5, upper curve). The spectrum shape (figure 1) in all cases is similar, either with honeycomb or Al layer. However, the angular dependence change with energy then is difficult to assess using a layer in place of the honeycomb.

- The front side irradiation is responsible of roughly ninety percent of damage
- The backside irradiation is almost constant in depth, but the front side increases with depth (see fig. 5 lower curve).
- The Al wall thickness of the honeycomb is a reasonable layer thickness for their simulation. It accounts for approximately ninety six percent of damage (compare in fig. 5, black line and the green one).

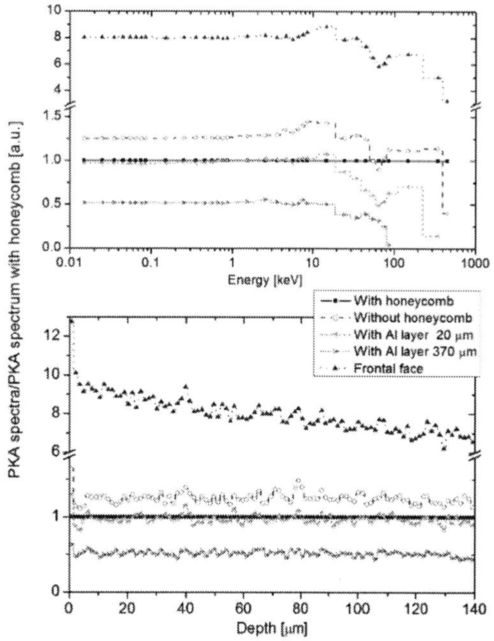

Figure 5: Quotient between PKA spectra and PKA spectra with honeycomb versus PKA energy (upper figure) and depth (lower figure).

5 References

[1] H.Y.Tada, J.R. Carter, B.E. Anspaugh and R.G. Downing, "Solar Cell radiation Handbook", 3rd ed. Page 6-11, NASA, 1982.

[2] Murphy, D. M., Eskenazi, M. I., White, S. F. and Spence, B. R., "Thin-Film and Crystalline SolarCell Array System Performance Comparisons," 29th IEEE PhotoVoltaic Specialists Conference, New Orleans, (2002) 22.

[3] S. R. Messenger, E. A. Burke, J. R. Lorentzen, R. J. Walters, J. H. Warner, G. P. Summers, and S. J. Taylor, "Quantifying low energy proton damage in multijunction solar cells," in Proc. 7th Eur. Space Power Conf., Italy, Stresa, (2005).

[4] R.J. Walters, G.P. Summers, J.H. Warner, S.R. Messenger, J.R. Lorentzen, T.L. Morton, S.J. Taylor, H. Evans, D. Heynderickx., B. Quaghebeur, and F. Lei, "SPENVIS Implementation of End-of-Life Solar Cell Calculations Using the Displacement Damage Dose Methodology", Proc. 19th Space Research and Photovoltaic Technology Conference, Cleveland, OH, USA, (2005).

[5] http://www.srim.org. A description of the code in J.F. Ziegler, J. B. Biersack and U. Littmark, The Stopping and Range of Ions in Solids, Pergamonn, New York, (1985) Vol I.

[6] SPENVIS-Space Environment Information System, is funded by the European Space Agency (ESA) and is registered as contract number 11711-WO and is available on the World Wide Web at http://www.spenvis.ome.be/spenvis/.

[7] M. Alurralde, A method using the PKA spectrum to characterize the electrical degradation of monocrystalline silicon solar cells by space protons, J. Appl. Phys., Vol. 95, 7, (2004) 3391.

RAY TRACING MODELLING OF REFLECTOR FOR VERTICAL BIFACIAL PANEL

Michael Linde Jakobsen[a], Sune Thorsteinsson[a,*], Peter Behrensdorff Poulsen[a], Peter Melchior Rødder[b], Kristin Rødder[b]
[a]Department of Photonics Engineering, Technical University of Denmark, Frederiksborgvej 399, Building 130, 4000 Roskilde, Denmark,
[b]SolarLab, Gunnar Clausens Vej 9, DK-8260 Viby J, Denmark
*Phone: +45 46774521, E-mail: sunth@fotonik.dtu.dk

ABSTRACT: Bifacial solar panels have recently become a new attractive building block for PV systems. In this work we propose a reflector system for a vertical bifacial panel, and use ray tracing modelling to model the performance. Particularly, we investigate the impact of the reflector volume being filled with a refractive medium, and shows the refractive medium improves the reflector performance since it directs almost all the light incident on the incoming plane into the PV panel.
Keywords: Solar cells, Ray tracing, PV applications, Reflector, Energy systems

1 INTRODUCTION

Bifacial solar modules offers in the right orientation a significantly improved energy harvest, while only providing an additional cost of 30 % [1], and these are now an attractive commercially available building block that allows for innovative concepts where both side of the PV-panel can be utilized. Several works have proven that adding reflectors to bifacial panels mounted at a certain optimized tilt, increases the energy harvest even further [2]-[3], and bifacial panels has offered new solutions as e.g. PV based fences [4]. For PV-systems mounted at a tilt, the reflectors will in many cases be placed underneath the PV panel to reflect incident light from surroundings e.g. the albedo. Since these reflectors are not exposed to rain they do not have the same self-cleaning ability as the PV panels itself and the resulting energy harvest is decreased due to soiling [2], [4]. It has also been modeled that the use of bifacial PV panels in solar farms can reduce the needed storage for the power system, since the peak power times can be shifted closer to the power demand [5].

In this work we propose a system using a vertical mounted bifacial panel equipped vertical reflectors at an angle to both reduce the soiling problem, but also to optimize the energy harvest. Here we particularly consider the impact of a refractive medium inside the reflector volume. The reflector is optimized using raytracing modelling of the optical rays, where the details of the reflections at the optical interfaces are taken into consideration.

2 THEORY AND MODEL

2.1 Theory

Two types of optical losses are modelled within this work:

- Fresnel losses, which are optical losses at interfaces between dielectric media, having different refractive indices.
- Material absorption losses, which are losses owing to the absorption in the bulk material and at metallic surface.

The Fresnel losses are determined by using the Fresnel equations, and here the reflection losses at an optical interface are calculated, and within this work averaging the transverse electric (TE) and transverse magnetic (TM) polarization reflection coefficients. Material absorption losses occur either due to absorption, when light propagates inside a transparent dielectric material, or due to absorption, when light is reflected at a metallic surface.

2.2 Model

Initially, a design according to Figure 1 is used for the raytracing model, and in the first design the mirrors are plane surfaces and thus provide no concentration of the sunlight. The system is comprised of a vertical bifacial panel which for simulation purposes is oriented east-west, and two reflector plates extending in angles (α = ±45 deg.) relative to the bifacial panel (a picture of a prototype is shown in Figure 2).

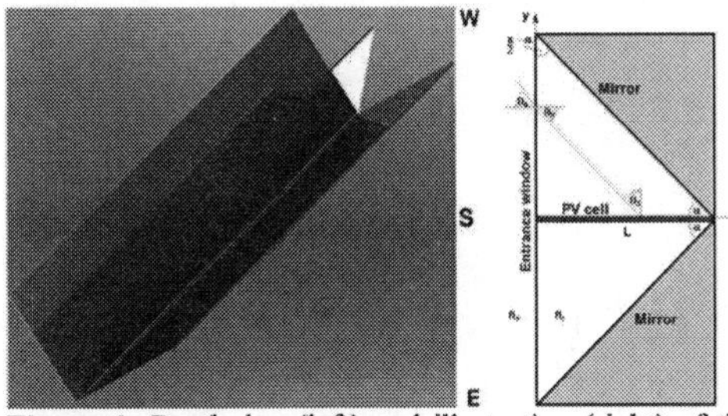

Figure 1: Rendering (left) and illustration (right) of the reflector system.

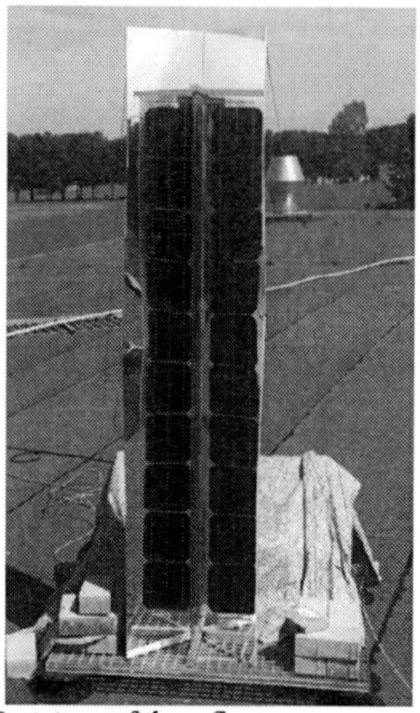

Figure 2: Prototype of the reflector.

Ray tracing, where many optical rays are modelled and finally the weighted output is integrated, are performed on this system. The volume inside the reflector can be bulk and thus be filled with a medium with refractive index *nr*. The medium can be index matched to that of glass, where the Fresnel losses at the surface of the PV-module will be eliminated, or it can be air, removing the optical interface at the entrance window or any other transparent material.

We assume that all light incident into the PV cell and its laminate of glass is absorbed. This allows us to consider only three paths of the incoming light, and these are illustrated in Figure 3. This allows us to omit surface reflections at the PV-module and to omit the optical paths derived from these reflections, which in fact are insignificant.

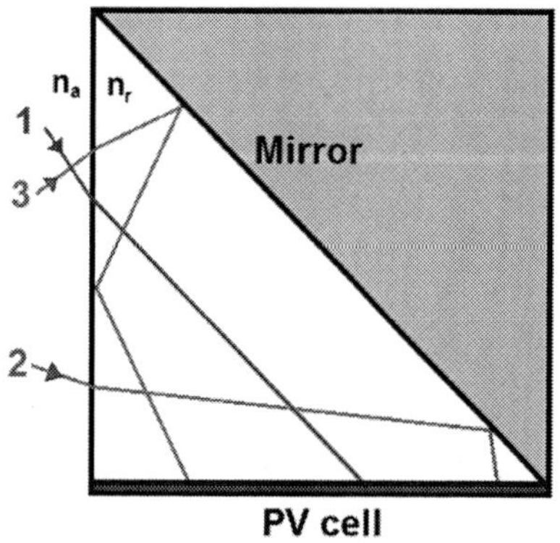

Figure 3: Relevant paths in one half of the retroreflector, when assuming that the PV cell and its glass laminate absorb all light.

The paths are:

1. Refraction at the entrance window and then propagation to the PV-module where refraction occurs again - *the direct path.*
2. Refraction at the entrance window, then propagation to the mirror, where reflection occurs and then propagation to the PV-cell, where refraction occurs again - *the reflected path.*
3. Refraction at the entrance window, then propagation to the mirror, where reflection occurs, then propagation to the entrance window, where refraction and reflection occurs and the reflected part propagates to the PV-cell, where refraction occurs again - *the back multiple reflected path.*

When the ray intersects with a mirror interface we use a reflection coefficient of 95%, as an average value for the reflection coefficient of alumina throughout the spectral range of c-Si. The solar elevation (ψ) is determined from the local latitude (φ), the declination of the sun (δ) and the azimuth angle (θ) of the sun relative to the local true south:

$$\psi = \sin^{-1}[\sin\varphi \sin\delta + \cos\varphi \cos\delta \cos\theta] \qquad (1)$$

The schematic of Figure 1 illustrates the model of the vertical retroreflector, and its orientation relative to earth. At the moment, the model does not include the diffuse solar contribution. The preliminary model does not include variations in sun intensity but considers only the projection of the AM 1.5 spectrum and resulting scattering due to the Fresnel equations.

3 RESULTS

Four cities placed at different latitudes and very similar longitudes are chosen for the simulations: Trondheim (Norway latitude 63.5°N), Copenhagen (Denmark lat. 55.7 °N), Munich (Germany lat. 48.1 °N) and Rome (Italy lat. 41.9 °N). The shortest and the longest days of the year are simulated and the relative irradiances as a function of azimuthal angle are used to normalize for the energy harvest.

As reference for the simulations, a monofacial panel facing south with the same area as a single side of a bifacial panel, is used. A reference scenario is calculated for each position. An example of a reference simulation is shown in Figure 4.

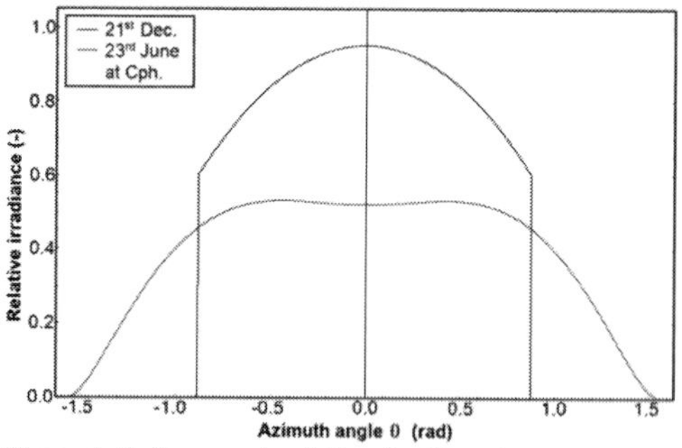

Figure 4: Reference simulation for Copenhagen.

In Figure 5 and Figure 6, two reflectors located in Copenhagen have been simulated. Figure 5 plots the simulation of a reflector filled with air (n_r = 1.0), Figure 6 plots the simulation for a reflector filled with acrylic (n_r = 1.49). The contributions from the three different ray paths are illustrated in Figure 5 and Figure 6 with three different colors. The bulk optical absorption losses are omitted. As can be seen, the 3. ray path contributes significantly to the total light flux. The index matched reflector improves the energy harvest by 32%, when bulk losses are neglected.

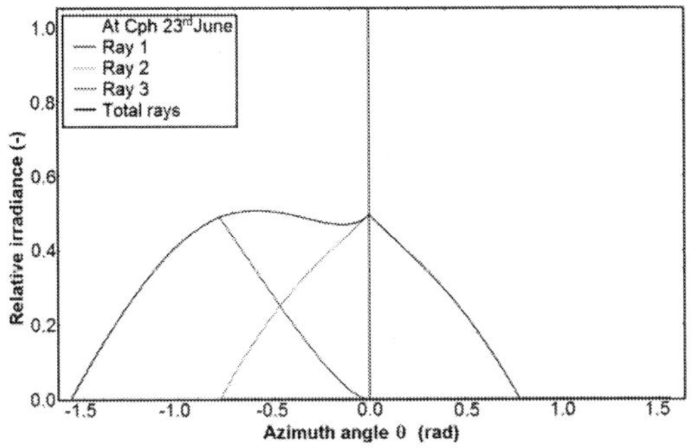

Figure 5: Ray contributions for a reflector filled with air.

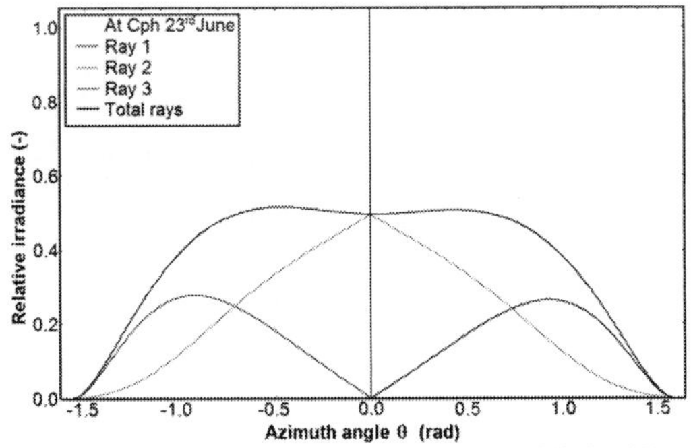

Figure 6: Ray contributions for a reflector filled with acryllic.

The impact of optical absorption and non-ideal mirrors are considered in the following. As it can be seen from table I, going from ideal glass to an acrylic filled reflector and with mirrors with a reflection coefficient of 95 % the power output is reduced by approximately 5 %, and hence fairly optimal materials give acceptable performance.

Table I: Influence of non-idealities

Material	Reflection	Absorption	Loss
Ideal	100 %	0	0 %
BK 7	95 %	0.3151	3.9 %
Acryllic	95 %	0.115	4.9 %

In table II simulations comparing the energy harvest on the longest day and the shortest day of the year at four different latitudes are shown. Bulk optical absorption in the acrylic material is taken into account and the mirrors are assumed to reflect 95 % of the light.

Table II. Simulation for 4 different latitudes, comparing energy harvest with an air filled reflector with an acrylic filled reflector.

Location	Latt.	Improvement			
		December		*June*	
		Air	*Acrylic*	**Air**	*Acrylic*
Trondheim	63.5	1.61	1.91	1.36	1.92
Copenhagen	55.7	1.45	1.91	1.35	1.92
Münich	48.1	1.39	1.91	1.33	1.92
Rome	41.9	1.38	1.92	1.35	1.92

It can be seen that the air filled reflector is performing significantly worse compared to the acrylic filled reflector and the performance of the air based is strongly dependent on the location and the time of year – i.e. the solar angles.

4 DISCUSSION

The results from table II (visualized in Figure 5 and Figure 6), shows that for this initial reflector design (filled with air) that approximately 30 % of the light entering the reflector is lost when reflected back out through the entrance window. Table I shows, that that low optical absorption and good reflectance gives the best performance, however good performance can be achieved with less than the best performing materials.

5 CONCLUSION AND FUTURE WORK

In this work we have proposed an energy system using a vertical bifacial PV-panel with a reflector system that captures almost all the light incident on the entrance window. The reflector volume has to be filled up with a medium with a refractive index matched to that of glass and with low optical bulk absorption. Future work includes testing the fabricated prototype to validate the model, upgrading the model to take diffuse light into account, and use simulations based on real weather data. Further developments include a new reflector design which will direct the reflected light propagating back out through the entrance window into the PV-panel and thereby omitting the need for the air gap being filled with an index matched material.

6 REFERENCES

[1] G. João, "Testing bifacial PV cells in symmetric and asymmetric concentrating CPC collectors," Engineering, vol. 05, no. 01, pp. 185–190, 2013.
[2] A. Moehlecke, F. S. Febras, and I. Zanesco, "Electrical performance analysis of PV modules with bifacial silicon solar cells and white diffuse reflector," Sol. Energy, vol. 96, pp. 253–262, 2013.
[3] A. Luque, E. Lorenzo, G. Sala, and S. López-Romero, "Diffusing reflectors for bifacial photovoltaic panels," Sol. Cells, vol. 13, no. 3, pp. 277–292, 1985.
[4] R. J. Magasrevy, "Bifacial efficiency at monofacial cost Building Integrated photovoltaic Energy Solutions for the World," p. 25, 2007.
[5] S. Obara, D. Konno, Y. Utsugi, and J. Morel, "Analysis of output power and capacity reduction in electrical storage facilities by peak shift control of PV system with bifacial modules," Appl. Energy, vol. 128, pp. 35–48, 2014.

INTEGRATION OF SPECTRAL SPLITTING IN A CPV-T RECEIVER CONCEPT

Robert Reinbrech[1,2], Robert Höller[1,*], and Hartmut Schneider[3]
[1]University of Applied Science Upper Austria, School of Engineering, Stelzhamerstrasse 23, 4600 Wels, Austria
[2]now at: Fronius International GmbH, Wels, Austria
[3]Fresnex GmbH, Ricoweg 1, 2351 Wiener Neudorf, Austria
[*]Corresponding author: Email: robert.hoeller@fh-wels.at; Phone: +43 5 0804 44250

ABSTRACT: The combined use of thermal and electrical power makes it possible to optimize the total energy output in a CPV-T receiver. By implementing a spectral filter in front of the PV-cell, the part of the solar irradiation that cannot be converted by the solar cell can be directly transformed into thermal energy and thereby reducing the solar cell temperature. This paper presents simulation results of a spectral splitting CPV-T receiver concept compared to a conventional CPV-T receiver, and the influence of different physical parameters on the electrical and thermal output. Simulations were performed for a Fresnel-type collector for the two locations Wels (Austria) and Lleida (Spain). For the spectral splitting receiver concept the simulation showed an annual average electrical efficiency of 10.1%, and a thermal annual average efficiency of 86%. On the other hand, the conventional CPV-T receiver reached a higher electrical efficiency with 14.7%, but a lower yearly thermal efficiency with 76.1% had to be accepted.
Keywords: concentrators, receiver, spectral splitting

1 INTRODUCTION

Concentrated photovoltaics (CPV) has a great potential to improve the economic efficiency of a PV-system by reducing the expensive PV-cell area. This concentration can be realised by lenses or reflective systems. Above a certain concentration ratio, an active cooling is needed to reduce the thermal load and further the PV-Cell temperature to facilitate a high electrical efficiency. By using the thermal energy of such a CPV System (then called CPV-T) the overall efficiency and the land utilization can be increased. This can be useful in situations where the area is limited, such as in rooftop situations. Due to the temperature coefficient of a PV-cell, the meaningful temperature range of such a CPV-T is restricted. Here the spectral splitting concept is applied to allow a separate optimisation of the thermal and the electrical efficiency by utilizing the unusable part of the solar spectrum for heat generation and thereby minimize the thermal load on the PV-Cell [1]. By replacing the backside mounted passive cooling fins with a back flow channel, the thermal output of the CPV-T receiver increases. The purpose of this work is to show the influence of different physical parameter on a spectral splitting CPV-T receiver by using an hourly based MATLAB simulation over a period of one year. These results are used for a comparison with a conventional CPV-T receiver. This paper also provides results for different spectral filters, fluids, and locations.

2 METHODS

An hourly MATLAB simulation for a period of one year was performed to investigate the annual performance of the considered spectral splitting CPV-T receiver. Also, a conventional active back cooled CPV-T receiver was simulated to compare and verify the impact of the spectral splitting concept. This receiver concept was described in a previous work by Hangweirer et al. [2].
Based on this concept, the thermal behaviour of the two receivers is modelled and compared. The irradiance input is calculated for two different locations by assuming a Fresnel-type concentrator. The location Wels (Austria) was selected as a central European site with a low fraction of direct normal irradiance (DNI), Lleida (Spain) on the other hand stands for a region with higher potential for concentrating PV due to its high share of direct irradiation.
Fig. 1 shows the solar AM1.5 irradiation spectrum, the spectral response of a Si PV-Cell and the transmission from a Schott spectral splitting filter, as used in the simulations.

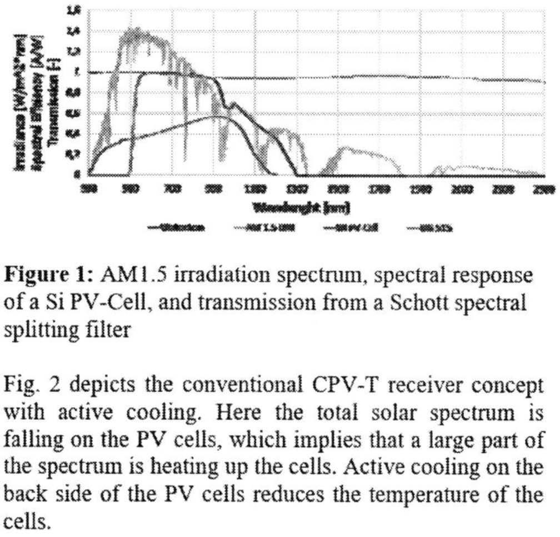

Figure 1: AM1.5 irradiation spectrum, spectral response of a Si PV-Cell, and transmission from a Schott spectral splitting filter

Fig. 2 depicts the conventional CPV-T receiver concept with active cooling. Here the total solar spectrum is falling on the PV cells, which implies that a large part of the spectrum is heating up the cells. Active cooling on the back side of the PV cells reduces the temperature of the cells.

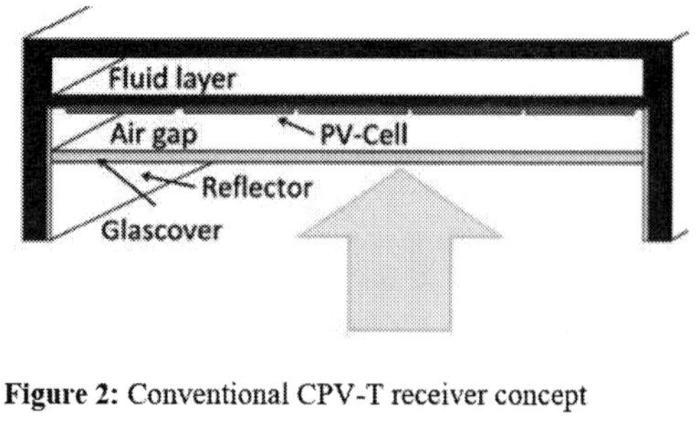

Figure 2: Conventional CPV-T receiver concept

Fig. 3 shows the CPV-T receiver with integrated spectral splitting. Here, before hitting the PV cells the solar irradiation is going through a spectral filter and two fluid channels. Fig. 4 shows transmission functions of different filter types that have been investigated, as well as the AM1.5 solar DNI spectrum and the spectral response (SR) of the PV cells.

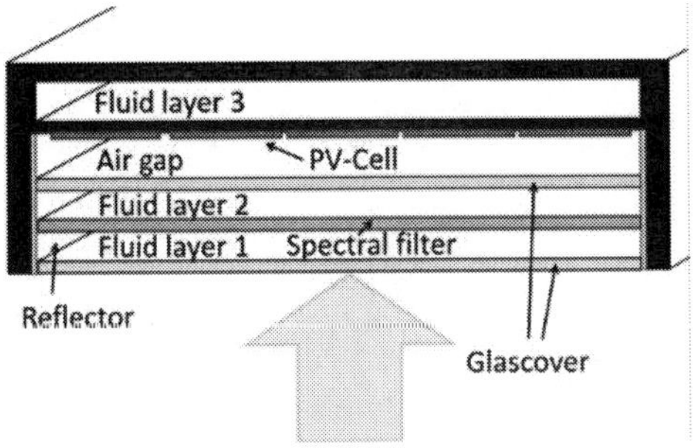

Figure 3: Spectral splitting CPV-T receiver concept

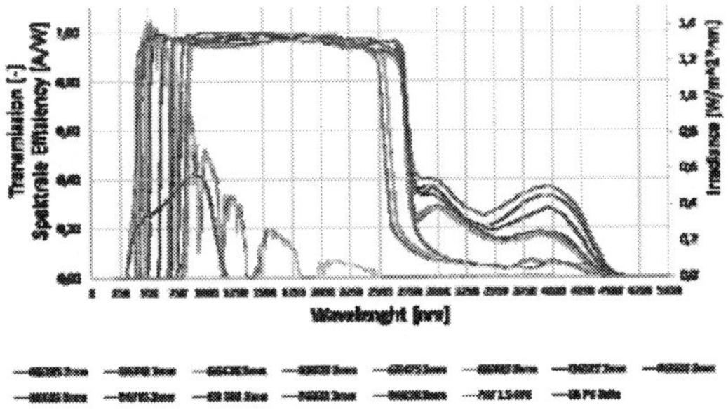

Figure 4: Transmission of different spectral splitting filters

In the MATLAB simulation several sub-programs were created to calculate the energy consumption and temperature in each layer for every hour of a year for the locations Wels (Austria) and Lleida (Spain). The iteration loop runs until the difference of the results are under a certain value (Fig. 5).

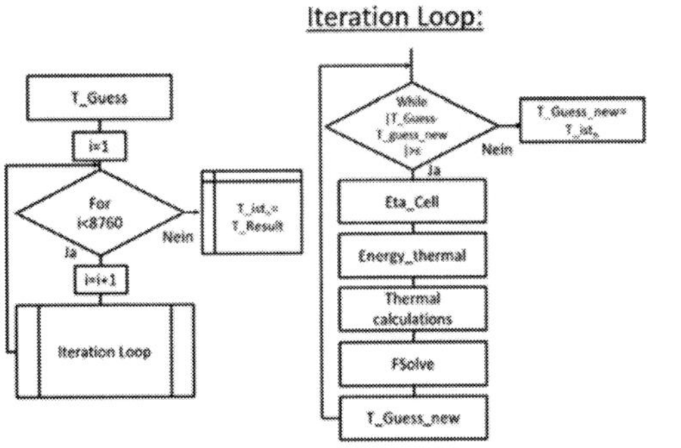

Figure 5: Flowchart of the MATLAB simulation

Energy thermal:
The absorbed energy in each layer is calculated. This is done by a wavelength based calculation of the transmission behaviour of each layer.
Eta cell:
The PV-Cell efficiency is calculated w.r.t. to the spectral irradiation and the PV-Cell temperature.
Thermal calculations:
The temperatures of the receiver layers and the fluids are calculated.

3 RESULTS AND DISCUSSION

The advantage of the applied spectral splitting concept is shown in situations with high fluid temperatures. In this situation the PV-Cell in the conventional CPV-T receiver concept generates more heat and therefore reduces the electrical efficiency. By implementing a spectral filter in front of the receiver, the generated heat in the PV-Cell could be reduced which resulted in an increase of the electrical efficiency. Fig. 6 shows the simulated yearly energy output for the two observed receiver designs in Wels. It can be seen that the spectral splitting receiver concept has a higher thermal output and a lower electrical output compared to the conventional receiver concept. Fig. 7 shows the yearly efficiency of both investigated receiver concepts for the location Wels Austria. As shown, an increase of the input fluid temperature of 60 K caused a drop of 3.2% of the electrical efficiency with the spectral splitting CPV-T concept.
It can be seen that the electrical efficiency of the spectral splitting receiver is less affected by the input fluid temperature than the conventional receiver. This is due to the fact that the spectral splitting receiver can reduce the thermal load in the PV-Cell layer and therefore increase the electrical efficiency.
In comparison a 4.6% electrical efficiency drop by the conventional receiver could be shown. A variation of the pass frequency of the filter resulted in a reduction of the yearly electrical efficiency from 11.7% at 395 nm to 2.4% at 850 nm. However, the thermal yearly efficiency increased from 84.4% at 395 nm to 93.7% at 850 nm.

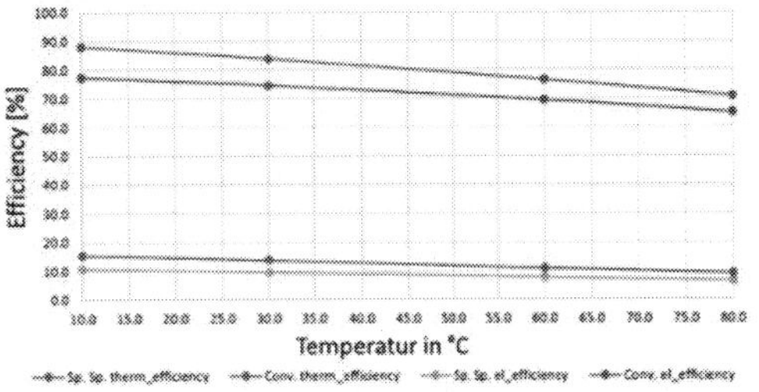

Figure 6: Comparison of yearly efficiency of the spectral splitting and conventional CPV-T receiver concepts at different input fluid temperatures.

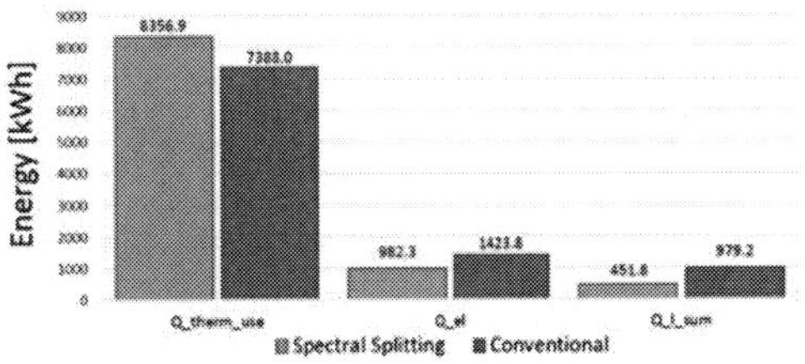

Figure 7: Annual energy output of the spectral splitting and conventional receiver concepts for Wels (Austria).

Fig. 8 shows the yearly energy output for the spectral splitting receiver concept for the two investigated locations. More details are described in Reinbrech [3].

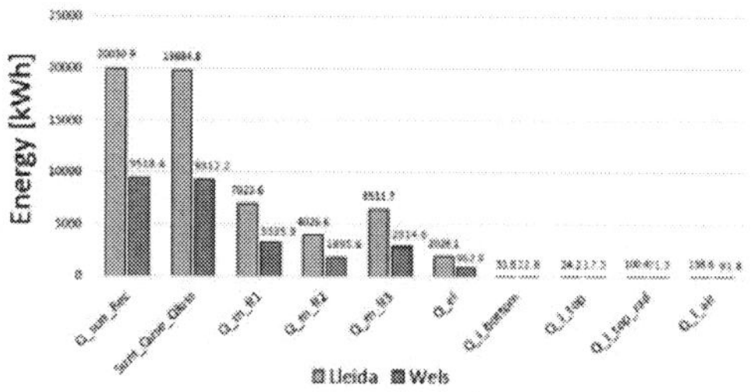

Figure 8: Annual energy output for the spectral splitting receiver concept in Lleida and Wels

A summary of the electrical and thermal output of the spectral splitting receiver concept over one year in Wels Austria is shown in Table I. It can be seen that with an increase in the pass frequency the thermal efficiency also increases and the electrical efficiency decreases. This is due to the fact that with an increase of the pass frequency the overall energy amount at the PV-layer decreases. For an optimization of the receiver, the temperatures of the different layers and fluids have to be considered, because the thermal losses are directly influenced by them.

Table I: Thermal and electrical energy output of different spectral splitting filters

Filter	Q_{th} [kWh/a]	Q_{el} [kWh/a]	Q_{loss} [kWh/a]	$\eta_{opt,a}$ [%]	$\eta_{therm,Kolt,a}$ [%]	$\eta_{therm,Recv,a}$ [%]	$\eta_{el,Kolt,a}$ [%]	$\eta_{el,Recv,a}$ [%]	$\eta_{el,PV,a}$ [%]
GG395	8199.1	1140.1	182.6	36.3	30.7	84.4	4.3	11.7	20.3
GG400	8193.9	1145.4	182.6	36.3	30.6	84.3	4.3	11.8	20.4
GG420	8217.6	1121.7	180.9	36.3	30.7	84.6	4.2	11.5	20.7
GG435	8232.4	1106.9	179.9	36.3	30.8	84.7	4.1	11.4	20.9
GG455	8258.6	1080.7	178.2	36.3	30.9	85.0	4.0	11.1	21.2
GG475	8284.8	1054.6	176.6	36.3	31.0	85.3	3.9	10.9	21.5
GG495	8320.8	1018.5	174.6	36.3	31.1	85.7	3.8	10.5	21.8
OG515	8356.9	982.3	172.7	36.3	31.3	86.0	3.7	10.1	22.2
RG610	8561.4	777.5	163.8	36.3	32.0	88.1	2.9	8.0	23.4
RG665	8768.3	570.0	156.4	36.3	32.8	90.3	2.1	5.9	23.9
RG715	8813.0	525.2	155.0	36.3	33.0	90.7	2.0	5.4	23.9
RG780	8951.9	385.9	150.7	36.3	33.5	92.2	1.4	4.0	23.5
RG830	9059.3	278.2	147.7	36.3	33.9	93.3	1.0	2.9	22.5
RG850	9102.7	234.7	146.4	36.3	34.0	93.7	0.9	2.4	21.8

4 CONCLUSION AND OUTLOOK

This work shows the potential of spectral splitting in CPV-T receivers. Because of the described behaviour, the pass frequency of the spectral filter has to be optimized respectively to its application. Due to the combined spectral absorption behaviour of the fluid and the filter, a reasonable combination of both components has to be found to minimize the costs of such a system. Furthermore, other fluids with different spectral absorption coefficients have to be analysed. Also different PV-Cell technologies have to be investigated to verify the influence of the external quantum efficiency. Due to some simplifications in the simulation model of the receiver, parameters such as the ambient temperature and the wind speed will have to be considered for each time step in the simulation.

4 REFERENCES

[1] A. Mojiri, R. Taylor, E. Thomsen, und G. Rosengarten, Spectral beam splitting for efficient conversion of solar energy review; Renewable and Sustainable Energy Reviews, 28 (2013) 654.

[2] M. Hangweirer, R. Höller, und H. Schneider, Design and Analysis of a Novel CPV-T Receiver Concept, Japanese Journal of Applied Physics, 54 (2015) 08KE01

[3] R. Reinbrech, Integration von Spectral Splitting zur Ertragsoptimierung eines solaren CPV-T Hybridreceivers, Master Thesis, FH OÖ-Campus Wels (2015).

HIGH PERFORMANCE GaAs SOLAR CELL USING HETEROJUNCTION EMITTER AND ITS FURTHER IMPROVEMENT BY ELO TECHNIQUE

Soohyun Kim*, Sun-Tae Hwang, Wonki Yoon, Heon-Min Lee
LG Electronics, Materials & Devices Advanced Research Institute,
38 Baumoe-ro, Seocho-gu, Seoul 06763, Korea
*soohyun94.kim@lge.com

ABSTRACT: We report on two types of single-junction GaAs solar cells that have achieved a record efficiency. We achieved an NREL-confirmed conversion efficiency of 27.5%@1cm^2 in substrate-type GaAs solar cell under AM1.5G illumination at 1 sun concentration, which exceeds the previously published records for devices on single-crystalline substrates. In addition, we also achieved 28.7%@1cm^2 of conversion efficiency in ELO-type GaAs solar cell under AM1.5G illumination, which is very close to the world record 28.8% of Alta Devices. In this study, the p-type $Al_{0.3}Ga_{0.7}As$ heterojunction emitter with 1.82 eV bandgap was applied to increase V_{oc} and conversion efficiency. In order to relieve the valence band offset between p-$Al_{0.3}Ga_{0.7}As$ (emitter) and n-GaAs (base) in the p-n junction, the E_g-graded buffer layer was applied to the p-n junction region. In addition, the thickness and Al composition of the E_g-graded layers were optimized considering band bending in the carrier depletion region. A simulation program 'AFORS-HET', designed especially for the solar cells with heterojunction structure, was used to understand the subtle differences in the band bending of the band diagrams and the performance of the solar cells.
Keywords: III-V semiconductors, Gallium Arsenide Based cells, Heterojunction, AlGaAs/GaAs, ELO

1 INTRODUCTION

Despite many advantages of GaAs solar cell such as optimal bandgap for high conversion efficiency and superior absorption coefficient, the considerably expensive GaAs wafer cost still remains a challenge to overcome and its applications are usually limited to CPV or space applications. To overcome these limitations, many researchers have tried to recycle the GaAs wafer substrate to reduce the manufacturing cost by using epitaxial-lift off (ELO) process [1-3].

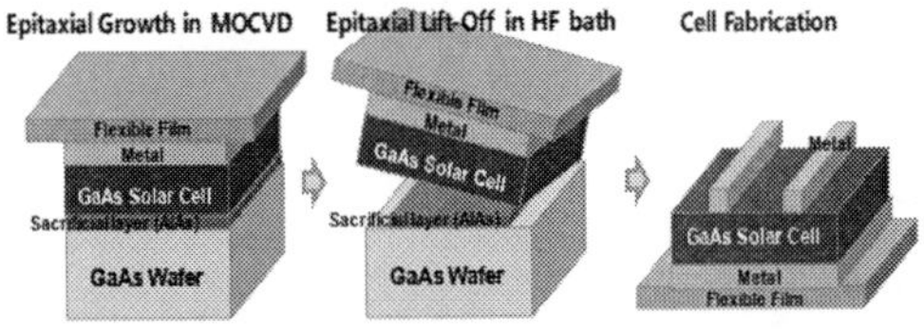

Figure 1: Schematic illustration of epitaxial lift-off (ELO) and device fabrication process in this study.

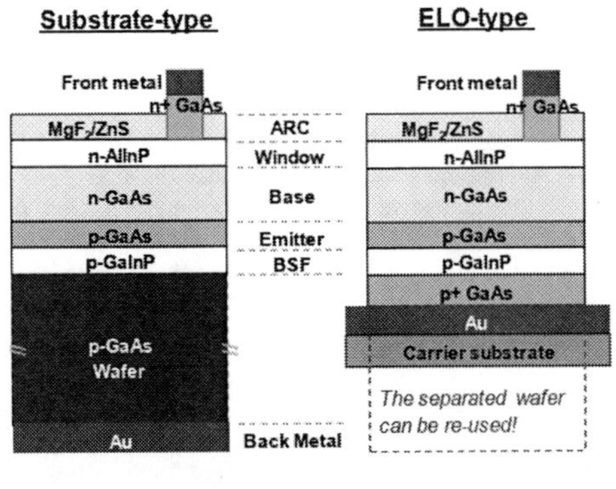

Figure 2: Comparison of device structures between substrate-type and ELO-type GaAs solar cells. The GaAs wafer of 'ELO-type' is separated by ELO process, and then it can be reused by special surface treatment

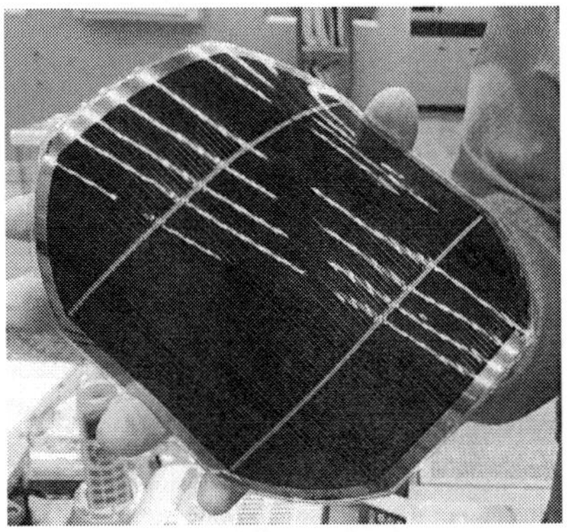

Figure 3: The world-largest 125x125mm^2 thin-film GaAs solar cell fabricated by ELO process in LG Electronics.

The ELO cell structure can be separated from the substrate by selective wet etching of a thin sacrificial layer (generally AlAs layer), and therefore this technique is offering the possibility of substrate re-use for the repeated growth of thin-film solar cell structures.

Figure 1~3 show the schematics of ELO process and its advantageous characteristics for solar cell application. The solar cells by the ELO process have additional advantage to improve V_{oc} and J_{sc} by photon recycling effect, because there is no current loss of the light absorption by the GaAs substrate.[4]

2 EXPERIMENTALS

Each layer of the GaAs solar cells was deposited on the GaAs substrate by the metal organic chemical vapor deposition (MOCVD) system by reacting some of following precursors: trimethylgallium (TMGa), arsine (AsH_3), trimethylaluminium (TMAl), phosphine (PH_3), and trimethylindium (TMIn). Silicon, from disilane (Si_2H_6) gas, was used as n-type dopants. Zinc, from diethylzinc (DEZn) gas was mixed for the p-type layers. The bandgap (E_g)-graded layers in heterojunction structure were deposited automatically controlling the amount of the Al source by the time during deposition, in the MOCVD system.

After deposition, the GaAs substrate was removed from the deposited layer after bonding with the supporting substrate. All cells were defined as size of 1 cm^2 by mesa etching. The conversion efficiency was measured by WACOM WXS-200S-20 solar simulator under AM1.5G illumination at 1 sun (100 mW/cm^2). Anti-reflective coating layers were applied on the champion cell, where the zinc sulfide (ZnS) and magnesium flouride (MgF_2) were deposited by thermal evaporation.

For the solar cell simulation, AFORS-HET, developed by Helmholtz-Zentrum Berlin (HZB), was used [5-6], which can simulate the heterojunction easily. We tried to design the similar device structures as the experiment, which includes the thicknesses, stacking structure, and doping conditions of each layer

3 RESULTS AND DISCUSSION

3.1 Optimization of heterojunction structure

In this study, the p-type $Al_{0.3}Ga_{0.7}As$ heterojunction emitter with bandgap of 1.82 eV was applied to increase V_{oc} and conversion efficiency. In order to relieve the valence band offset between p-$Al_{0.3}Ga_{0.7}As$ (emitter) and n-GaAs (base) in the p-n junction, the E_g-graded buffer layer was applied to the p-n junction region. In addition, the thickness and Al composition of the E_g-graded layers were optimized considering band bending in the carrier depletion region.

Heterojunction on the p-n junction is usually applied to control the built-in potential in the solar cells. When the heterojunction emitter which has higher bandgap (E_g) than the GaAs base is applied in the solar cells, the open-circuit voltage (V_{oc}) normally increases due to higher built-in potential. However, the band offset in the hetero p-n junction may hinder the drift of the carriers, leading to higher carrier recombination rate and deteriorate the performance of the solar cells. Therefore, the E_g-graded layer, which can minimize or remove the band offset at the p-n junction, is applicable with the p-$Al_{0.3}Ga_{0.7}As$ emitter.

Furthermore, considering the band gap and lattice constant of alloyed semiconductors, application of Al-graded $Al_xGa1_{-x}As$ layer allows to control the band offset of p-n junction by E_g-graded layers. Figure 4 shows the device structure applying AlGaAs heterojunction emitter together with E_g-graded layer.

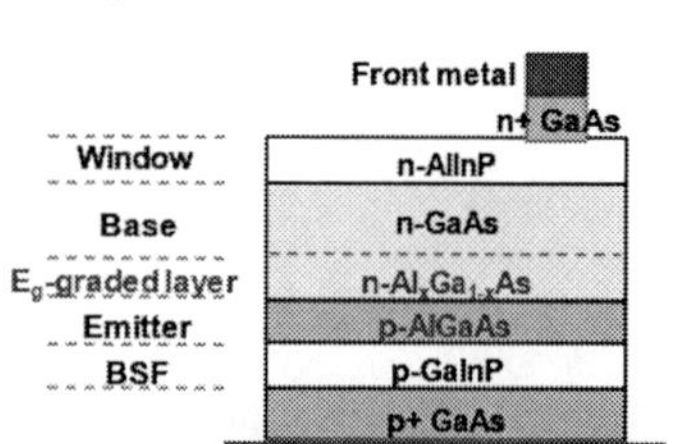

Figure 4: The device structure applying AlGaAs heterojunction emitter together with E_g-graded layer.

A simulation program 'AFORS-HET', designed especially for the solar cells with heterojunction, is used to easily understand the subtle differences in the band bending of the band diagrams, the carrier recombination rate, and the performance of the solar cells.[5] AFORS-HET, developed by Helmholtz-Zentrum Berlin (HZB), is mainly used for the simulation for the solar cell with a-Si/c-Si heterojunction. However, it can be also applied for the GaAs-based solar cells.[6] This simulation program draws the band diagrams, calculates the carrier recombination rates, and estimates the I-V characteristics, such as open circuit voltage, fill factor, and conversion efficiency, of the solar cells. Then, these simulation data are matched to the experimental data.

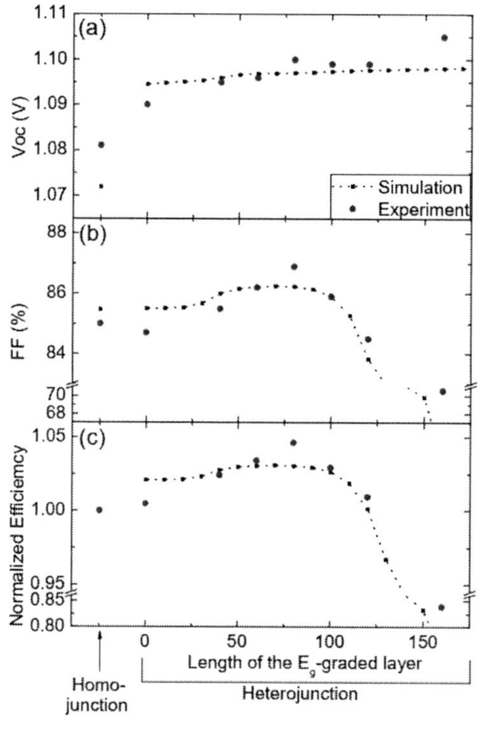

Figure 5: The simulation and the experimental cell data of (a) open-circuit voltage, (b) fill factor and (c) normalized efficiency as a function of the length of E_g-graded layer.

Through our preliminary experiments and AFORS-HET simulations, we could find that the heterojunction with E_g-graded layer in the 'n-type' base exhibited the better solar cell performance than the other heterojunction devices with E_g-graded layer in the 'p-type' base and without any E_g-graded layer. As expected, the V_{oc} of the solar cells with heterojunction was higher than that with homojunction due to higher built-in potential by higher E_g of heterojunction emitter, and in addition, the FF of heterojunction device without any E_g-graded layer was lower compared with other cells because the band offset in the valence band was observed near the interface of the p-n junction.

The thickness of the E_g-graded layer in the base was then optimized by changing the band bending properties on the band diagram. Figure 5 shows the performance of the solar cells as a function of the thickness of the E_g-graded layers. The J_{sc} is not shown because the difference of J_{sc} among samples (for the experiment and for the simulation, respectively) is less than 0.1 mA/cm^2.

The V_{oc} slightly got higher as the E_g-graded layer got thicker, both for the experiment and the simulation. The tendency for the V_{oc} increment in the experiment was more obvious than in the simulation. For FF, it got lower if the thickness of the E_g-graded layer got too thick (thicker than 100 nm) or too thin (thinner than 40 nm), and best at 60~80 nm in the simulation. In the experiment, the tendency of FF looked similar, and as a result, the FF was the best when the E_g-graded layer is 80 nm, more obviously than in the simulation. The more detailed discussion will be described elsewhere by another publication. [9]

3.2 Performance of champion cell

The champion cell was fabricated by applying the $Al_{0.3}Ga_{0.7}As$ heterojunction emitter and 80 nm thick E_g-graded layer in n-type base layer. The anti-reflective coating (ARC) was applied to minimize the reflection loss on the front surface. Here we report on two types of single-junction GaAs solar cells that have achieved a record efficiency. We achieved an NREL-confirmed conversion efficiency of 27.5%@1cm^2 in substrate-type GaAs solar cell under AM1.5G illumination at 1 sun concentration, which exceeds the previously published records for devices on single-crystalline substrates. In addition, we also achieved 28.7%@1cm^2 of conversion efficiency in ELO-type GaAs solar cell under AM1.5G illumination, which is very close to the world record 28.8% of Alta Devices.[7-8] The J-V curves are shown in Figure 6.

3.3 Further Improvement for GaInP solar cell

As known, the GaInP solar cell is widely used as a top cell within multi-junction III-V compound solar cells. Our novel results can be also applied to this another III-V solar cell. The p-type $Al_xGa_{1-x}As$ emitter in GaAs solar cell can be simply replaced with p-type $Al_xGa_{1-x}InP$ emitter in GaInP solar cell, and n-type Al-graded $Al_xGa_{1-x}As$ layer can be replaced with n-type Al-graded $Al_xGa_{1-x}InP$ layer in GaInP solar cell. This is one of our undergoing research and Table I shows our recent achievement in GaInP solar cell performance. The V_{oc} of GaInP solar cell with heterojunction is also much higher than that with homojunction.

Table I: Solar cell performances of GaInP solar cells with $Al_xGa_{1-x}InP$ heterojunction emitter. ARC layers were not applied for all samples.

Base	E_g-graded buffer layer	Emitter	V_{oc}	FF
n-type GaInP	-	p-type GaInP	1.399 V	87.1%
n-type GaInP	No	p-type $Al_xGa_{1-x}InP$	1.417 V	85.9%
n-type GaInP	Al-graded $Al_xGa_{1-x}InP$	p-type $Al_xGa_{1-x}InP$	1.498 V	87.8%

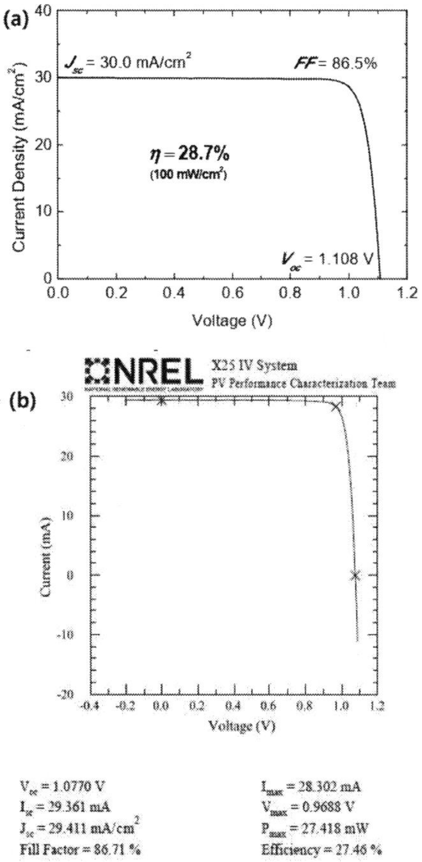

Figure 6: The I-V curves of (a) ELO-type GaAs cell and (b) substrate-type GaAs cell with the best efficiency applying the heterojunction with the E_g-graded layer. The (b) 27.5% efficiency of substrate-type GaAs cell is the new world record efficiency confirmed by NREL.

4 CONCLUSION

In this study, an $Al_{0.3}Ga_{0.7}As$ heterojunction emitter and the E_g-graded layer were designed and optimized by the computer simulation (AFORS-HET) and fabricating the GaAs solar cell by the metal-organic chemical vapor deposition (MOCVD) system. The heterojunction emitter exhibited the best performance when the E_g-graded layer was applied in the n-type base with optimized thickness (~80 nm). In addition, we believe that this results can be applied in many kinds of III-V compound solar cells, not only for the GaAs single junction solar cell, but also the multi-junction solar cell regardless of whether the GaAs substrate exists or not.

5 REFERENCES

[1] E. Yablonovitch, T. Gmitter, J. P. Harbison, R. Bhat, Appl. Phys. Lett., 51 (1987) 2222.
[2] J. Yoon, S. Jo, I. S. Chun, I. Jung, H. –S. Kim, M. Meitl, E. Menard, X. Li, J. J. Coleman, U. Paik, J. A. Rogers, Nature, 465 (2010) 329.

[3] G. J. Bauhius, P. Mulder, E. J. Haverkamp, J. C. C. M. Huijben, J. J. Schermer, Sol. Energy Mater. Sol. Cells 93 (2009) 1488.
[4] G. J. Bauhuis, J. J. Schermer, P. Mulder, M. M. A. J. Voncken, P.K. Larsen, Sol. Energy Mater. Sol. Cells 83 (2004) 81.
[5] R. Varache, C. Leendertz, M. E. Gueunier-Farret, J. Haschke, D. Muñoz, L. Korte, Sol. Energy Mater. Sol. Cells 141 (2015) 14.
[6] A. S. Gudovskikh, K. S. Zelentsov, N. A. Kalyuzhnyy, V. M. Lantratov, S. A. Mintairov, Proceedings of the 25th European Photovoltaic Solar Energy Conference, Valencia, Spain (2010) 472.
[7] M. A. Green, K. Emery, Y. Hishikawa, W. Warta, E. D. Dunlop, Prog. Photovolt.: Res. Appl. 24 (2016) 3.
[8] Hui Nie, T. Twist, S. G. Spruytte, F. Reinhardt, I. C. Kizilyalli, G. S. Higashi, Photovoltaic Specialists Conference (PVSC), 37th IEEE, Seattle, WA (2011) 4.
[9] S.-T. Hwang, S. Kim, H. Cheun, H. Lee, B. Lee, T. Hwang, S. Lee, W. Yoon, H.-M. Lee, B. Park, *accepted in* Sol. Energy Mater. Sol. Cells (2016)

This is a pre-peer review version of the paper submitted to the journal Progress in Photovoltaics.
It has been selected by the Executive Committee of the EU PVSEC 2016 for submission to Progress in Photovoltaics.

IDENTIFICATION OF TECHNICAL RISKS IN THE PV VALUE CHAIN AND QUANTIFICATION OF THE ECONOMIC IMPACT

David Moser*[1], Matteo Del Buono[1], Ulrike Jahn[2], Magnus Herz[2], Mauricio Richter[3], Karel De Brabandere[3]
[1]EURAC research, Institute for Renewable Energy (Italy), david.moser@eurac.edu,
[2]TÜV Rheinland (Germany),
[3]3E sa, Quai à la Chaux 6, 1000 Brussels, Belgium

ABSTRACT: In PV projects it is important to establish a common practice for professional risk assessment, which serves to reduce the risks associated with related investments. The objective of this paper is to present a methodology in how to improve the current understanding of several key aspects of technical risk management during the PV project lifecycle, with the identification of technical risks and their economic impact. To achieve this, available statistical data of failures during a PV project have been collected with the aim to i) suggest a guideline for the categorisation of failure, and ii) develop a methodology for the assessment of the economic impact of failures occurring during operation but which might have originated in previous phases. The risk analysis has the aim to assess the economic impact of technical risks and how this can influence various business models and the LCOE. This paper presents the first attempt to implement cost-based Failure Modes and Effects Analysis (FMEA) to the PV sector and to define a methodology for the estimation of economic losses due to planning failures, system downtime and substitution/repair of components. The methodology is based on statistical analysis and can be applied to a single PV plant or to a large portfolio of PV plants in the same market segment. The quality of the analysis depends on the amount of failure data available and on the assumptions taken for the calculation of a Cost Priority Number (CPN). The overall results can be linked to the cost of periodic and corrective maintenance and form the basis to estimate the impact of various risk and mitigation scenarios in PV business models.

1 INTRODUCTION

Historical performance data for PV systems on which to base technical risks assessments and investment decisions are not easily accessible by all market players, such as investors, PV plant owners, EPC contractors, insurance companies, etc. The reasons for this difficulty are to be found in the short time that most PV systems have been operational (GWs cumulative installations in many countries were only reached after 2010) and a tendency among system operators and component manufacturers to keep available performance data as confidential. In addition, detailed performance data are in most cases not available for PV plants of low capacity (e.g. residential-commercial market segments up to 250 kWp) as the cost of monitoring is still perceived as an added cost. Finally, although the description of failure and corrective measures is common practice in the field of operation and maintenance (at least in paper form), this is not often carried out with the sufficient level of details to derive meaningful statistical analysis due to missing cost information and lack of a common approach in the assignment of failures to a specific category. For the PV industry to reach mature market level, a better understanding of technical risks, risk management practices and the related economic impact is thus essential to ensure investor's confidence.

Description of typical failures at PV module level was subject to extensive studies within the first phase of the IEA PVPS Task 13 "Performance and Reliability" and the results were presented in the deliverable "Review of Failures of PV Modules" [1]. In the document, the most common failures of PV modules are described together with the measurement methods to assess the impact on the performance and safety with a particular relevance given to visual inspection. Hasselbrink recently summarized data for returns from a fleet of >3 million module-years [2]. The study found that 0.44% of front contact modules were returned after an average deployment of 5 years, with the majority (~66%) of these returned because of problems with laminate cell/ribbon/solder failures (primarily cell interconnections). The second most common reason (~20%) for returns was because of problems with the backsheet or encapsulant (e.g. delamination). Thus, the vast majority of the returns were associated with failures that can usually be identified visually. More analyses are needed to understand if the lower rate of return associated with other types of failures are due to the low detectability by visual inspections (e.g. hotspots, cracked cells, PID, etc.) thus leading to a biased conclusion.

Modules that have failed and been returned to the manufacturers are not the only factor to be considered; modules are usually observed to degrade slowly in the field. Jordan et al have summarized ~400 reports in the literature on the subject of the degradation rates for crystalline silicon modules [3] where the degradation is dominated by a loss of short-circuit current. In most cases, they observed that this decrease in short-circuit current is associated with discolouration and/or delamination of the encapsulant material. Thus, both statistics on returns of modules and statistics on slow degradation appear to be correlated to mechanisms that can be observed visually. The systematic use of visual inspection would enable the collection of a large dataset of failures.

Regarding the determination of the reliability at inverter level, it involves taking a look at the failure rate (including the bathtub curve of failure), the infant mortality rate, the useful life of a solar inverter and the meantime between failures (MTBF). The vast majority of PV system failures are believed to be inverter related [4, 5]. Interestingly, a 1994-1997 study on 126 PV systems found that 75% of the failures were due to inverters with an MTBF of 1.65 years. Module MTBF was 552 years for residential and 6666 years for utility scale system, i.e. one would expect one module of every 552 or 6666 to fail every year, respectively. The MTBF of inverters are thus not comparable with the values for modules and inverters must be replaced one or more times during the course of the PV system service life.

The failure modes that mostly affect PV inverters are related to units exposed to high thermal and electrical stress as well as the thermal management system itself [4],

[6]–[8]. Electronic components such as bus capacitors, electronic switches (e.g. IGBTs) and printed circuit boards (PCBs) are found to be responsible for the majority of PV inverter failures reported in the literature. Furthermore, maximum power point tracking (MPPT) schemes are also identified as an important factor impacting the overall reliability of PV inverters [9].

A fan failure could cause the inverter to overheat affecting its overall lifetime and reliability. However, it is shown in literatures that even under extreme operating conditions, state-of-the-art fans used in PV inverters may work without failing over a period of at least twelve years [10]. Typical estimated life expectancy of integrated circuits (ICs) and optical components is around ten years [10]. However, this will strongly depend on the quality of the materials used and on the design topology. For example, new developments with high quality materials used for special applications like, for e.g. micro-inverters, are designed to work under extreme conditions and are claimed to have longer lifetimes.

In terms of efforts in failure collection, extensive work was carried out in the USA in the framework of the PVROM project [11] where a rigorous data collection, analysis and feedback mechanism is developed and considered a best practice for PV plant owners and operators looking to go beyond simple data collection and immediate incident response. The database allows for detailed analysis of component failures and indicators such as the average active repair time, mean downtime and maintenance actions. The database builds on the commercially available software tool XFRACAS for failure reporting and corrective actions [12]. Collins et al described [13]–[15] the minimum data necessary for reliability and availability analyses of PV systems.

Ideally, during the process of failure detection and correction, an automatic ticketing system should be in place. This is not always common practice and the collection of a high number of failure data for statistical analysis should also consider existing failure reports, which might come in paper form and might not include all the necessary information.

In this paper, we will present a methodology developed within the Solar Bankability project to assess the economic impact of failures on the calculation of the LCOE and business models. The initial results of the assessment are presented and critically discussed. We believe that this first attempt to apply a cost based FMEA represents an important step towards an increased confidence in the operation of PV systems based on large-scale failure analysis. Extended results of this work are presented in the public project report *Technical Risks in PV Projects* [16], which is available on www.solarbankability.eu. This work has been developed within the Solar Bankability project, financed by the European Commission under its Horizon 2020 research and innovation programme. The Solar Bankability project aims at contributing to the reduction of the risks associated with investments in sustainable energy projects.

2 METHODOLOGY

The typical approach in risk analysis for technical projects is to apply a classic Failure Modes and Effects Analysis (FMEA) where the various risks, belonging to a certain phase and component, can be prioritized through their Risk Priority Number (RPN). In the FMEA, each identified risk is evaluated for its severity (S), occurrence (O) and detectability (D); numbers are used to score each of these evaluation parameters. The RPN is then obtained by multiplying these three factors with the following formula:

RPN = S x O x D

The classic FMEA with RPNs, although important, is incomplete and needs to be enhanced to include a method to assess the cost impact of each risk. A classic FMEA is thus deemed as not adequate for this specific objective when the technical risk analysis needs to provide a framework for the calculation of the economic impact. Many articles are available for cost priority FMEA applied to other fields with a special coefficient called CPN (cost priority number) corresponding to RPN for the classic FMEA. To the best of our knowledge, no analysis is present in the literature for photovoltaic plants. The literature is mainly related to automotive or wind turbine market [17].

In this paper, we present the implementation of a cost-based FMEA by introducing a Cost Priority Number (CPN) which includes cost consideration directly in the risk assessment. A CPNs ranking prioritizes risks, which have a higher economic impact. However, it might not be applicable to each type of risk. To this extend, technical risks were first listed in a so-called Risk Matrix divided by project phase (product testing / development, planning, transportation / installation, operation & maintenance, decommissioning) and component (module, inverter, mounting structure, cabling, etc.).

2.1 Technical risks before operation

The technical risks identified before the operational phase (Testing / Planning / Transportation & Installation) can be defined as Year 0 risks in the LCOE calculation – and consequently to a business model - if they have an impact since the beginning of a plant operation. Other technical risks can be defined as root cause of failures occurring during operational phase, i.e. precursors of failures. In this section, only Year 0 risks are considered. For each of these risks it is important to understand how the variability and associated uncertainty are calculated and how the values are distributed in terms of probability. These aspects are essential for the calculation of the exceedance probability of the energy yield and how this is influenced by the overall uncertainty (see Figure 1 where the exceedance probability is calculated based on a normal distribution).

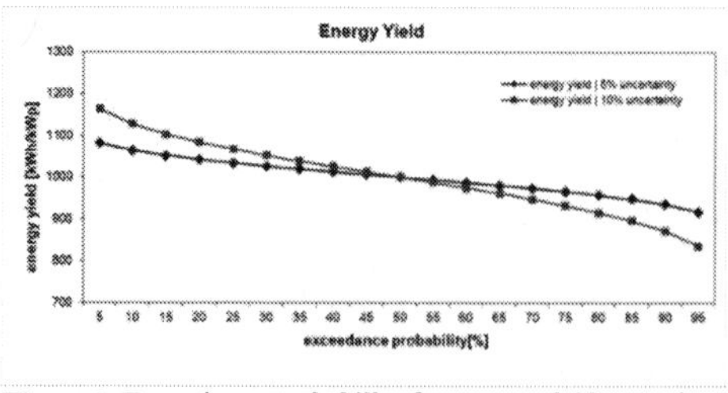

Figure 1: Exceedance probability for energy yield assuming different uncertainties calculated with a normal probability distribution function

A reduction in the uncertainties can lead to a higher value of energy yield for a given exceedance probability and hence a stronger business case. Reich et al [18] estimated the combined overall uncertainty of the energy yield to fall in a range between 5 and 11%; in the study, the uncertainty on various effects such as irradiation, shading, soiling, inverter losses, etc. were taken into account. In another study, [19] have calculated the variation of the overall uncertainty of the energy yield over the lifetime of a PV plant. These efforts show the importance of having a common framework that can assess the impact of technical risks on the economic performance of a PV project. In the coming years, as the availability of measured data will exponentially increase, it will be important to build large databases to increase the confidence level of the statistical analysis and thus reduce the perceived risk from investors related to the initial yield assessment. The uncertainty contribution of each identified effect can be found as given by Moser et al [16] where the overall uncertainty is based on values available in the literature. [20] reported an analysis based on 26 systems located in Germany and Spain where the measured energy yield was 4% higher than initially estimated. This was mainly due to an overestimation of the annual insolation for the selected location with a +4.9% and an underestimation of the PR of -0.9%. This is of particular importance for systems where a guaranteed yield is requested together with a PR. While in [18] an overall uncertainty on the energy yield due to various effects is given in the range of ±5-11%, [21] give a value of ±6-8% for the energy yield and ±2-6% in PR.

In Figure 1 the impact of the energy yield uncertainty on the exceedance probability was shown with a difference of around 60 kWh/kWp at P90 assuming a P50 value of 1000 kWh/kWp (>5% difference at P90). The curve was created based on a normal distribution of the energy yield where the median is equal to the mean value. Positively or negatively skewed distribution (due for e.g. to not normally distributed irradiance or degradation) will also have an impact on the exceedance probability.

2.2 Technical risks during operation

For the calculation of the economic impact of risks, which are likely to occur during the implementation phase, namely during operation and maintenance, the Occurrence and Severity were calculated following a well-defined procedure. The methodology is designed to allow for generalization and flexibility in order to maximise the use of the methodology. The severity, S, is calculated as the total plant(s) production over one year in absence of failures, the occurrence, O, is calculated based on the downtime of a specific failure normalised over the number of components and the total hours, and the calculation of downtime costs as missing production / savings:

$C_{down}=L\times(FIT+PPA+RCE)$, where

L	Production losses due to down time	PPA	Power Purchasing Agreement
FIT	Feed-In Tariff	RCE	Retail Cost of Electricity

For the calculation of the costs due to downtime, it is in fact important to consider the missing income of feed in tariffs, the missing income from PPA, and/or the missing savings generated by PV plants installed on roofs/facades.

Specifically, the downtime costs are calculated considering the time to detection, t_{td}, the time to repair/substitution, t_{tr} or t_{ts}, and the time to fix the problem, t_{fix}. The steps that lead to the downtime costs are the following:

- Calculation of downtime caused by a specific failure

$$t_{down,fail} = (t_{td} + t_{tr,ts}) \cdot PL \cdot M + t_{fix} \cdot M$$

where PL is the performance loss, expressed in fraction (therefore PL=1 for failures causing total downtime). M is a multiplier to consider failures that cause problems at higher component level (e.g. 1 module takes down the whole array).

- calculation of total downtime for the n number of failures (hours)

$$t_{down} = t_{down,fail} \cdot n_{fail}$$

- calculation of total downtime normalised by components (hours/component)

$$t_{down,comp} = t_{down}/n_{comp}$$

- calculation of occurrence over a time t_{ref} (%)

$$O = t_{down,comp}/t_{ref}$$

t_{ref} could be either the equivalent hours (specific yield), the total number of hours per year or the number of sun hours.

- calculation of production losses, L, due to downtime (kWh)

$$L = O \times S$$

The costs related to fixing the failure results from the sum of the costs of repair/substitution, costs of detection, costs of staff, costs of transport and cost of labour.

$$C_{fix}=(C_{det}+C_{rep/sub}+C_{transp})\cdot n_{fail}+C_{lab}\cdot t_{fix}\cdot n_{fail}$$

C_{down}	Downtime costs as missing production	C_{transp}	Costs of transport
C_{fix}	Costs related to fixing the failure	C_{lab}	Cost of labour
C_{det}	Costs of detection	t_{fix}	Time to fix the failure
$C_{rep/sub}$	Costs of repair/substitution	n_{fail}	Number of failures

The sum of C_{fix} for various components is then equal to the cost of monitoring/detection and corrective maintenance. Preventive maintenance can be included as a detection cost, and its impact can be assessed using the methodology as it reduces the detection time.

As a final step, the calculation of the Cost Priority Number is then given by $CPN=C_{down}+C_{fix}$.

The division into the various categories allows for the calculation of CPNs for very generic cases or to plant specific figures depending on the type of input data available (specific plant related figures or statistical analysis of failures). The parameters used for the calculation of the CPN can also be given as country dependent by applying country based coefficients to take into account different FITs schemes, retail cost of electricity, annual insolation, cost of labour, etc.

This methodology can only be applied to the failures with a direct economic impact to the business plan either in terms of the reduced income due to downtime or the costs for repair or substitution. The technical risks included in the risk matrix which cannot be described with an exceedance probability or with a CPN are very important and have to be considered as they might have an impact on the CPN value of other component failures. For example, the technical risks related to monitoring system, spare parts, normative and documentation, insurance reaction time, operation and maintenance contract, video surveillance, detailed field inspection (IR, EL, etc.), just to name a few, can reduce or increase the time to detection or the time to repair/substitution and might have an impact on the detection costs. The determination of the impact of mitigation measures is the focus of ongoing activities and will be covered by the future deliverables of the project Solar Bankability which will be available in August 2016.

3 RESULTS AND DISCUSSION

The analysis is based on the described CPN method, the collected failure data and the defined downtime costs and fixing costs for each failure. Thus, all developed results are strongly depending on the database and the defined conditions. An overview of the total amount of analysed plants, components and detected failures is given in Table 1.

Table 1: Overview of the main features of the database of failures used for the analysis

	Total number of plants	Total Power [kWp]	Average number of years
TOTAL	772	441678	2.7
Components	**No. tickets**	**No. Cases**	**No. Components**
Modules	473	678801	2058721
Inverters	476	2548	11967
Mounting structures	420	15809	43057
Connection & Distribution boxes	221	12343	20372
Cabling	614	367724	238548
Transformer station & MV/HV	53	220	558
Total	2257	1077445	2373223

In total 1,077,445 failure cases were included in the analysis of 772 PV plants including 2,373,222 components. The failure mode database contains data of around 442 MW_p of PV plants nominal power corresponding to around 0.5% of the installed peak power in Europe.

In terms of occurrence the failures were distributed as shown in Table 2 where all failures were summed up for the components modules and inverters.

Table 2: Occurrence statistics considering all failures for modules and inverters

	no. cases	no. components	Years	Occurrence	Occurrence/ year
Modules	678,640	2,058,721	2.68	33%	12%
Inverters	2,474	11,967	2.68	21%	8%

For the calculation of the CPN, for the first scenario, a value of 8760 hours was chosen for the parameter "time to detect" (t_{td}) corresponding to an entire year (it is the maximum number of hours per year). This scenario is defined as the "never detected" scenario and the corresponding CPN is thus equal to C_{down} as the economic impact of the failure is only related to the downtime (i.e. performance loss/missing production) caused by it.

$$CPN_{never\_detected} = \frac{(t_{td} \cdot PL \cdot M) \cdot n_{fail}}{n_{comp} \cdot t_{ref}} \cdot S \cdot (\text{FIT} + \text{PPA} + \text{RCE})$$

For the analysis of the base scenario, to give also an indication of the overall costs of repair or substitution, we have also calculated the corresponding CPN for a "failure fix" scenario equal to C_{fix} (cost of fixing the failures) with in addition a downtime term C_{down} (downtime after detection) due to the time to repair (t_{tr}) and repair time (t_{fix}), 744 hours (1 month) and up to 2 hours for PV modules, respectively. The value of 1 month was selected assuming that no spare parts are available.

$$CPN_{failure\_fix} = \frac{(t_{tr/ts} \cdot PL \cdot M + t_{fix} \cdot M) \cdot n_{fail}}{n_{comp} \cdot t_{ref}} \cdot S \cdot (\text{FIT} + \text{PPA} + \text{RCE}) + (C_{\text{det}} + C_{\text{rep/sub}} + C_{\text{transp}}) \cdot n_{fail} + C_{\text{lab}} \cdot t_{fix} \cdot n_{fail}$$

This allows for a comparison of the two scenarios "never detected" and "failure fix" with the parameters C_{down} and C_{fix} for different failures belonging to the same component. All parameters used for the calculation of the CPN value are summarised in the tables at the end of the manuscript.

3.1 Technical risks for PV modules

The first look is into the total quantity of PV module failures for all market segments (see Figure 2).

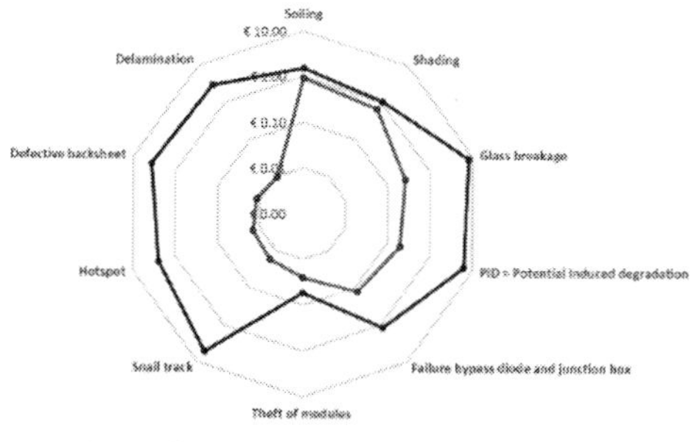

Figure 2: Top 10 technical risks for PV modules and all market segments. CPN is given in Euros/kWp/year for both scenarios.

The ranking of the risks is based on the never detected scenario, which describes the frequency and the economic impact of the specific failure. For PV modules, the impact of production loss (blue/continue line, $CPN_{never\_detected}$) appears to be minimal for the module dominant failures except of soiling, shading, glass breakage, and PID. The dominant factor here is the cost of substitution (red/dashed line, $CPN_{failure\_fix}$), since repairing of a defective module is only possible in particular cases, e.g. cleaning of the surface or changing of defective bypass diodes. The second reason is that repairing a defective module might actually violate the module manufacturer's warranty restriction (i.e. unauthorised modification or tempering of module) resulting in warranty claim exclusion and thus substitution of the defective module is the preferred procedure. Overall, the common failures like glass

breakage, PID, snail track, hotspots, delamination and defective backsheets bear a higher level of economic risk. The same analysis carried out only for the residential segment shows that the predominant failures are those which can be visually detected.
It is important to highlight that a lower CPN value for the "never detected" scenario does not mean that this strategy is more cost-effective than fixing the problem. Power losses will increase over the years and the existing or impending failure could also pose safety risks. For this reason, we have also calculated the impact for the worst case scenario which shows how the impact of the "never detected" scenario can vary depending on the chosen Performance Loss, PL, parameter (PL) (see Table 4). In Figure 3, the increased impact of downtime to the CPN becomes apparent with an order of magnitude comparable with the $CPN_{failure\_fix}$.
Overall, by summing the CPN off all module related failures for both scenarios, we obtain a value of around 60 Euros/kWp/y.

Table 3: Power losses range for various module failures

Failures	Power loss [%]	Max power loss [%]
Hotspot	2,00%	20,00%
Delamination	1,00%	30,00%
Glass breakage	10,00%	50,00%
Soiling	10,00%	30,00%
Shading	10,00%	40,00%
Snail track	1,00%	8,00%
Cell cracks	1,00%	15,00%
Defective backsheet	1,00%	20,00%
Overheating junction box	1,00%	33,00%
PID = Potential induced degradation	10,00%	70,00%
Failure bypass diode and junction box	33,00%	33,00%
Corrosion in the junction box	1,00%	33,00%
EVA discoloration	0,0%	10,0%
Theft of modules	100,00%	100,00%
Broken module	100,00%	100,00%
Damage by snow	100,00%	100,00%
Corrosion of cell connectors	1,00%	15,00%
Unsufficient theft protection	0,00%	100,00%
Improperly installed	5,00%	20,00%
Module damaged due to fire	100,00%	100,00%
Missing modules	100,00%	100,00%

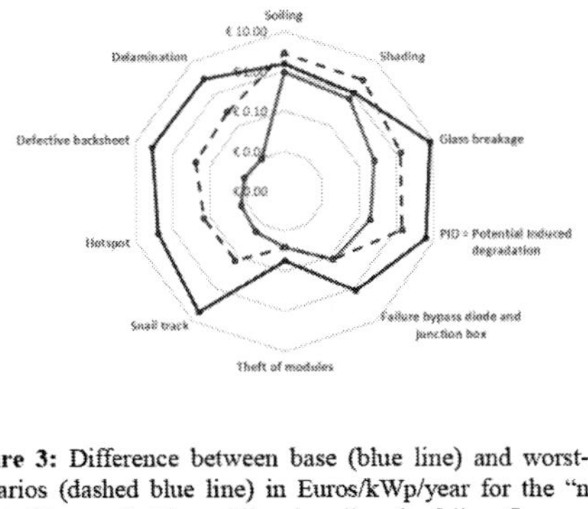

Figure 3: Difference between base (blue line) and worst-case scenarios (dashed blue line) in Euros/kWp/year for the "never detected" scenario. The red line describes the failure fix scenario.

The results presented so far were normalised considering the total number of PV plants and thus over the total number of components. The database allows us to analyse also how a specific failure is distributed, e.g. if it is present in most of the PV plants or if it had an impact on just a few PV plants. For this analysis we have calculated the ratio between the CPN calculated for the affected plants with the CPN calculated for the total number of plants. Looking at the CPN ratio, it can be stated that for high CPN ratio, if one of the failures is detected for a specific plant, the risk to detect the same failure type for other components increases by a significant multiplier. This can be observed for product failures, e.g. potential induced degradation, delamination and failures of the junction box, or external hazards on the overall PV plant, e.g. destruction by fire or theft of modules as shown in Figure 4.

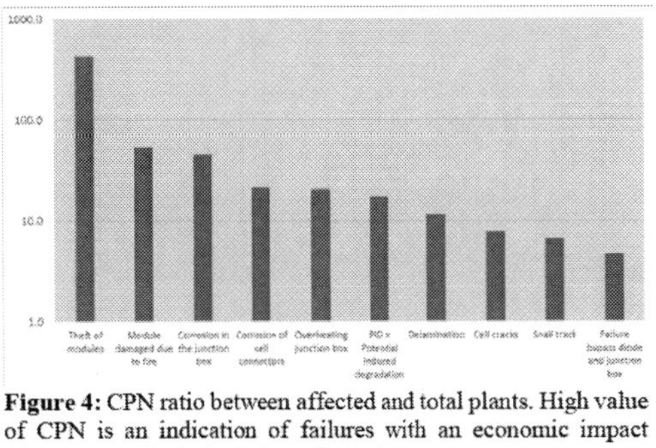

Figure 4: CPN ratio between affected and total plants. High value of CPN is an indication of failures with an economic impact localized to a limited number of PV plants

For example, for the risk "theft of modules", the overall CPN would increase from 0.08 to 34 Euros/kWp/y, while for PID the value would move from 6 Euros/kWp/y to 114 Euros/kWp/y if we only consider affected plants.

3.2 Technical risks for inverters

Unlike the PV module risks, the inverter related risks appear to have a significant impact on the production due to downtime (red/right bars in Figure 5). The production losses caused by the missing production are higher than the overall repair costs (blue/left bars). In this case, it is clear that repair or substitution of the component should be addressed as early as possible once detected. The same conclusions apply for almost all the market segments separately since the risks are following the same pattern as it is shown in the Figure 5.

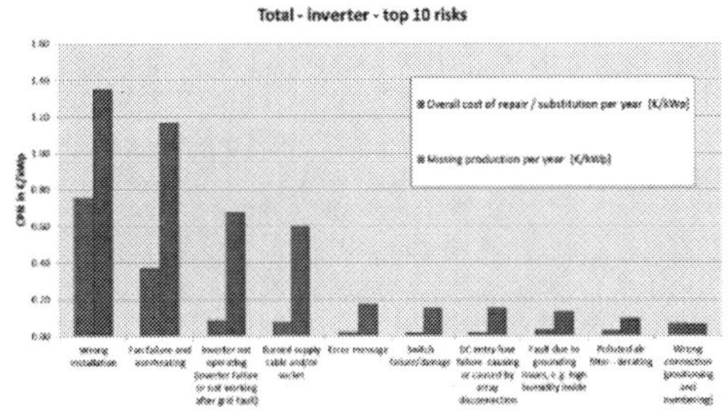

Figure 5: Top 10 risks of inverters; CPN is given in Euros/kWp/year

3.3 Overall technical risk analysis

The analysis can be expanded for all components to have a broader overview for the whole PV Project during the Operation & Maintenance phase. Figure 6 shows the Top 10 technical risks when all components are considered for the "never detected" scenario. The sum of the annual value for the CPN varies from 17 Euros/kWp for the "never detected" scenario up to 100 Euros/kWp for the "failure fix" scenario. A real case scenario will include a combination of the detection of a failure and the fixing of the failure with a sum well above 100 Euros/kWp/y. The impact of mitigation measures can then be assessed to evaluate the decrease of the annual overall CPN value thanks to an effective implementation and will be subject of further studies. In addition, considerations will be added in terms of risk ownerships.

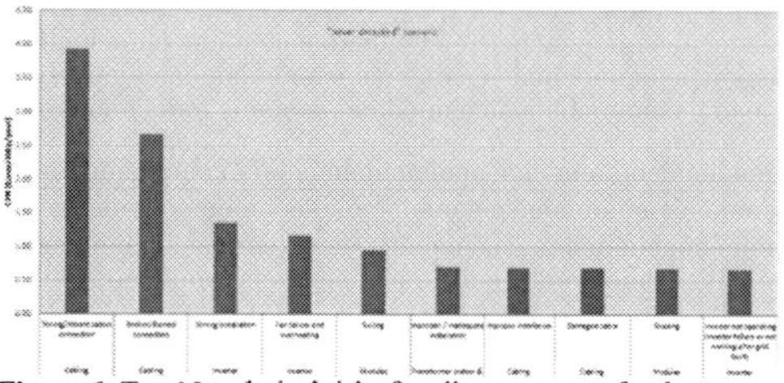

Figure 6: Top 10 technical risks for all components for the "never detected" scenario

4 CONCLUSIONS

In this paper the most important technical risks related to PV projects were identified and included in a risk matrix organised by components and divided into 5 categories to cover the whole PV value chain: Product testing / development, PV plants planning / development, Transportation / installation, PV plant operation and maintenance, Decommissioning. The inclusion of the risks into a risk matrix is considered a fundamental step to enable the possibility to share failure data based on an agreed nomenclature and definition by all different stakeholders. The prioritisation of the risks was not estimated by following a classical FMEA approach by assigning a RPN value, but by developing a methodology that was never previously applied to PV systems, a Cost based FMEA with Cost Priority Numbers. CPNs are given in Euros/kWp or in Euros/kWp/year and can thus directly give an estimation of the economic impact of a technical risk.

The CPN methodology was defined in order to assess two main economic impacts of a specific failure: impact due to downtime and impact due to repair / substitution cost. For the calculation of C_{down}, parameters such as time to detection, time to repair, repair time were considered, while for C_{fix}, cost for detection, labour cost, cost of repair / substitution, cost of transportation were included. As a result, the overall CPN value for various components and failures would correspond to the cost of O&M for various scenarios. The methodology also considers the year of installation, the year of failure and the nominal power in order to be able to run analysis for different market segments and to evaluate the distribution of failure probability once the available data in the database reaches statistical relevance to this type of granularity. The methodology also considers other statistical parameters such as the number of affected plants and the number of components in affected plants; in this way it is possible to understand if a specific failure is PV plant dependent or if it is equally present over the whole PV plant portfolio.

The CPN methodology can only be applied to failures with a direct economic impact to the business plan either in terms of reduced income due to downtime or costs for repair or substitution. However, the technical risks included in the risk matrix which cannot be described with a CPN are also very important and have to be considered as they might have an impact on the CPN value of other component failures. For example, technical risks related to monitoring system, spare parts, normative and documentation, insurance reaction time, operation and maintenance contract, video surveillance, detailed field inspection (IR, EL, etc.), can reduce or increase the time to detection or the time to repair/substitution and might have an impact on the detection costs. A thorough analysis will be carried out in relation to mitigation measures and will be covered by the future deliverables of the project Solar Bankability which will be available in August 2016. Other risks can have an impact on the overall uncertainty during the initial energy yield calculation and assessment. A reduced uncertainty can in fact correspond to a higher level of energy yield for a given exceedance probability value (given as Px) and thus directly impact the business model. In this publication, the typical uncertainties related to the main parameters affecting the energy yield were given together with an analysis on how the distribution function of a specific parameter might influence the exceedance probability. Works in the literature have shown the importance of providing the industry operating in the field with a common framework that can assess the impact of technical risks on the economic performance of a PV project.

In the next years, as the availability of measured data will exponentially increase, it will be important to build large databases with potentially a uniform method to increase the confidence level of the statistical analysis and thus reduce the perceived risks from investors. With the availability of these large databases the necessary information (minimum requirement) can be filtered out to perform tailored analysis in a uniform way i.e. same granularity, same data, same formulas. Our methodology and CPN attempts to provide such benchmark.

5 ACKNOWLEDGEMENTS

The research leading to these results has received funding from the European Union's Horizon 2020 research and innovation programme under the grant agreement No 649997 (Solar Bankability).

6 REFERENCES

[1] M. Köntges, S. Kurtz, C. Packard, U. Jahn, K. A. Berger, K. Kato, T. Friesen, H. Liu, and M. Van Iseghem, "Review on Failures of Photovoltaic Modules," IEA PVPS, IEA-PVPS T13-01:2013, Mar. 2014.

[2] E. Hasselbrink, M. Anderson, Z. Defreitas, M. Mikofski, Y.-C. Shen, S. Caldwell, A. Terao, D. Kavulak, Z. Campeau, and D. DeGraaff, "Validation of the PVLife model using 3 million module-years of live site data," in *Photovoltaic Specialists Conference (PVSC), 2013 IEEE 39th*, 2013, pp. 0007–0012.

[3] S. R. Kurtz, J. H. Wohlgemuth, and D. C. Jordan, "Technology and Climate Trends in PV Module Degradation," *27th European Photovoltaic Solar Energy Conference and Exhibition*, pp. 3118–3124, Oct. 2012.

[4] A. Ristow, M. Begovic, A. Pregelj, and A. Rohatgi, "Development of a Methodology for Improving Photovoltaic Inverter Reliability," *IEEE Transactions on Industrial Electronics*, vol. 55, no. 7, pp. 2581–2592, Jul. 2008.

[5] A. Maish, "Defining requirements for improved photovoltaic system reliability," *Prog. Photovolt: Res. Appl.*, vol. 7, no. 3, pp. 165–173, May 1999.

[6] J. Flicker, "PV Inverter Performance and Component-

Level Reliability," presented at the 2014 Photovoltaic Module Reliability Workshop, Aug-2014.

[7] J. Flicker, R. Kaplar, M. Marinella, and J. Granata, "PV inverter performance and reliability: What is the role of the bus capacitor?," in *Photovoltaic Specialists Conference (PVSC), Volume 2, 2012 IEEE 38th*, 2012, pp. 1–3.

[8] R. Kaplar, R. Brock, S. DasGupta, M. Marinella, A. Starbuck, A. Fresquez, S. Gonzalez, J. Granata, M. Quintana, M. Smith, and others, "PV inverter performance and reliability: What is the role of the IGBT?," in *Photovoltaic Specialists Conference (PVSC), 2011 37th IEEE*, 2011, pp. 1842–1847.

[9] G. Petrone, G. Spagnuolo, R. Teodorescu, M. Veerachary, and M. Vitelli, "Reliability issues in photovoltaic power processing systems," *Industrial Electronics, IEEE Transactions on*, vol. 55, no. 7, pp. 2569–2580, 2008.

[10] Z. J. Ma and S. Thomas, "Reliability and maintainability in photovoltaic inverter design," in *Reliability and Maintainability Symposium (RAMS), 2011 Proceedings-Annual*, 2011, pp. 1–5.

[11] G. T. Klise, R. R. Hill, C. J. Hamman, P. H. Kobos, V. Gupta, and B. B. Yang, "PV Reliability Operations and Maintenance (PVROM) Database Initiative: 2014 Progress Report," Sandia National Laboratories, Albuquerque, New Mexico, Unlimited Release SAND2014-20612, Dec. 2014.

[12] C. J. Hamman, "Using Xfracas as a Pvrom Database Tool.," Sandia National Laboratories (SNL-NM), Albuquerque, NM (United States), SAND2014-2914, Apr. 2014.

[13] E. Collins, J. Mahn, M. Mundt, J. Granata, and M. Quintana, "Field data collection for quantification of reliability and availability for photovoltaic systems," in *2010 35th IEEE Photovoltaic Specialists Conference (PVSC)*, 2010, pp. 002386–002391.

[14] E. Collins, M. Dvorack, J. Mahn, M. Mundt, and M. Quintana, "Reliability and availability analysis of a fielded photovoltaic system," in *2009 34th IEEE Photovoltaic Specialists Conference (PVSC)*, 2009, pp. 002316–002321.

[15] J. S. Stein, J. E. Granata, M. J. Mundt, S. P. Miller, M. A. Quintana, E. W. Collins, and N. R. Sorensen, "A Reliability and Availability Sensitivity Study of a Large Photovoltaic System.," Sandia National Laboratories, SAND2010-5782C, Aug. 2010.

[16] D. Moser, M. Del Buono, W. Bresciani, E. Veronese, U. Jahn, M. Herz, E. Janknecht, E. Ndrio, K. de Brabandere, and M. Richter, "Technical Risks in PV Projects." 15-Mar-2016.

[17] M. Shafiee and F. Dinmohammadi, "An FMEA-based risk assessment approach for wind turbine systems: a comparative study of onshore and offshore," *Energies*, vol. 7, no. 2, pp. 619–642, 2014.

[18] N. Reich, J. Zenke, B. Muller, K. Kiefer, and B. Farnung, "On-site performance verification to reduce yield prediction uncertainties," in *Photovoltaic Specialist Conference (PVSC), 2015 IEEE 42nd*, 2015, pp. 1–6.

[19] B. Müller, L. Hardt, A. Armbruster, K. Kiefer, and C. Reise, "Yield predictions for photovoltaic power plants: empirical validation, recent advances and remaining uncertainties: Yield predictions for photovoltaic power plants," *Progress in Photovoltaics: Research and Applications*, vol. 24, no. 4, pp. 570–583, Apr. 2016.

[20] B. Muller, W. Heydenreich, N. Reich, C. Reise, and B. Farnung, "Investment risks of utility-scale PV: Opportunities and limitations of risk mitigation strategies to reduce uncertainties of energy yield predictions," in *Photovoltaic Specialist Conference (PVSC), 2015 IEEE 42nd*, 2015, pp. 1–5.

[21] E. Richter, M, T. Schmidt, J. Kalisch, A. Woyte, K. de Brabandere, and M. Lorenz, E, "Uncertainties in PV Modelling and Monitoring," *31st European Photovoltaic Solar Energy Conference and Exhibition*, pp. 1683–1691, Nov. 2015.

[22] G. Belluardo, P. Ingenhoven, W. Sparber, J. Wagner, P. Weihs, and D. Moser, "Novel method for the improvement in the evaluation of outdoor performance loss rate in different PV technologies and comparison with two other methods," *Solar Energy*, vol. 117, pp. 139–152, Jul. 2015.

[23] G. Belluardo, G. Barchi, D. Baumgartner, M. Rennhofer, P. Weihs, and D. Moser, "Uncertainty analysis of a radiative transfer model using Monte Carlo method within 280–2500 nm region," *Solar Energy*, vol. 132, pp. 558–569, Jul. 2016.

	[illegible]	[illegible]	[illegible]	[illegible]	[illegible]	[illegible]
Modules	Hotspot	8760	744	1	[illegible]	1
	Delamination	8760	744	3	1.00%	1
	Glass breakage	8760	744	1	[illegible]	1
	Soiling	8760	744	0.01	[illegible]	1
	Shading	8760	744	0.01	[illegible]	1
	Snail track	8760	744	2	1.00%	1
	Defective backsheet	8760	744	3	1.00%	1
	PID = Potential induced degradation	8760	744	3	[illegible]	1
	Failure bypass diode and junction box	8760	744	2	[illegible]	1
	Corrosion in the junction box	8760	744	1	1.00%	1
	EVA discoloration	8760	744	3	6.0%	1
	Theft of modules	8760	744	[illegible]	100.00%	1
	Broken module	8760	744	1	100.00%	1
	Corrosion of cell connectors	8760	744	1	1.00%	1
	Improperly installed	8760	744	1	[illegible]	1
	Module damaged due to fire	8760	744	[illegible]	100.00%	1
	Missing modules	8760	744	3	100.00%	1
Inverter	Fan failure and overheating	8760	744	4	20.00%	1
	Switch failure/damage	8760	744	4	100.00%	1
	Fault due to grounding issues, e.g. high humidity inside	8760	744	4	[illegible]	1
	DC entry fuse failure causing or caused by array disconnection	8760	744	4	100.00%	1
	Inverter not operating (inverter failure or not working after grid fault)	8760	744	4	100.00%	1
	Polluted air filter - derating	8760	744	4	20.00%	1
	Error message	8760	744	4	100.00%	1
	Wrong connection (positioning and numbering)	8760	744	4	[illegible]	1
	Burned supply cable and/or socket	8760	744	4	100.00%	1
	Wrong installation	8760	744	4	[illegible]	1

	Failures	[illegible]	[illegible]	[illegible]	[illegible]
Modules	Hotspot	[illegible]	[illegible]	[illegible]	[illegible]
	Delamination	[illegible]	[illegible]	[illegible]	[illegible]
	Glass breakage	[illegible]	[illegible]	[illegible]	[illegible]
	Soiling	[illegible]	[illegible]	[illegible]	[illegible]
	Shading	[illegible]	[illegible]	[illegible]	[illegible]
	Snail track	[illegible]	[illegible]	[illegible]	[illegible]
	Defective backsheet	[illegible]	[illegible]	[illegible]	[illegible]
	PID = Potential induced degradation	[illegible]	[illegible]	[illegible]	[illegible]
	Failure bypass diode and junction box	[illegible]	[illegible]	[illegible]	[illegible]
	Corrosion in the junction box	[illegible]	[illegible]	[illegible]	[illegible]
	EVA discoloration	[illegible]	[illegible]	[illegible]	[illegible]
	Theft of modules	[illegible]	[illegible]	[illegible]	[illegible]
	Broken module	[illegible]	[illegible]	[illegible]	[illegible]
	Corrosion of cell connectors	[illegible]	[illegible]	[illegible]	[illegible]
	Improperly installed	[illegible]	[illegible]	[illegible]	[illegible]
	Module damaged due to fire	[illegible]	[illegible]	[illegible]	[illegible]
	Missing modules	[illegible]	[illegible]	[illegible]	[illegible]
Inverter	Fan failure and overheating	[illegible]	[illegible]	277.00 €	[illegible]
	Switch failure/damage	[illegible]	[illegible]	277.00 €	[illegible]
	Fault due to grounding issues, e.g. high humidity inside	[illegible]	[illegible]	[illegible]	[illegible]
	DC entry fuse failure causing or caused by array disconnection	[illegible]	[illegible]	277.00 €	[illegible]
	Inverter not operating (inverter failure or not working after grid fault)	[illegible]	[illegible]	[illegible]	[illegible]
	Polluted air filter - derating	[illegible]	[illegible]	277.00 €	[illegible]
	Error message	[illegible]	[illegible]	277.00 €	[illegible]
	Wrong connection (positioning and numbering)	[illegible]	[illegible]	[illegible]	[illegible]
	Burned supply cable and/or socket	[illegible]	[illegible]	277.00 €	[illegible]
	Wrong installation	[illegible]	[illegible]	277.00 €	[illegible]

MEAN DEGRADATION RATES IN PV SYSTEMS FOR VARIOUS KINDS OF PV MODULE FAILURES

Marc Köntges[1], Sascha Altmann[1], Tobias Heimberg[1], Ulrike Jahn[2], Karl A. Berger[3]
[1]Institut für Solarenergieforschung Hameln (ISFH), Am Ohrberg 1, D-31860 Emmerthal, Germany
[2]TÜV Rheinland Energy GmbH, D-51105 Köln, Germany
[3]Austrian Institute of Technology GmbH, Energy Department, 1210 Vienna, Austria

ABSTRACT: We analyse the impact of various photovoltaic (PV) module failure modes in PV systems in the field to rank the impact of the failure types on the power generation. Therefore we designed a PV system failure survey based on a spread sheet programme. The survey is distributed to international PV experts who fill in the data. The failure data is collected since Oct. 2015 and the data collection is still in progress. The module failure types "cell cracks", "potential induced degradation by shunting (PIDs)", "defect bypass diodes" and "discolouring of pottant" have the highest fraction in the database. The PIDs effect has a mean annual degradation rate of about 16%/a and affects about 3/5 of a system in moderate climate. This degradation rate is calculated for the PV modules PV systems which show degradation. Cell cracks show a mean degradation rate of around 5%/a for systems in the moderate climate zone and a higher degradation rate of about 8%/a in cold and snow climate. Cell cracks also often affect only some modules in a string showing a relevant power loss. Defect bypass diodes may show in some cases a very high impact on the module and system power degradation. They occur in the first 10 years of system operation. The wear out failure discolouring of pottant occurs during the whole system life but accumulates for very old systems. Its degradation rate is below 1%/a for all climatic zones.

Keywords: degradation rate, statistic, PV module failure

1 INTRODUCTION

Plenty of data on photovoltaic (PV) system and PV module reliability has been collected in the past [1][2][3][4][5][6][7][8][9][10]. Some studies focus on the PV module failure in great detail [10], other focus on the power losses of PV systems, but loose the origin of the power loss [11][12]. Jordan showed mean degradation rates for x-Si in the 0.8–0.9%/year range, ~1% for HIT and microcrystalline silicon and for the sum of thin film technologies 1.38%/year. There are strong variations by thin film technologies indicated. For example there are some studies on Cu(In,Ga)Se with low degradation rates like x-Si technology.

To collect and evaluate data on a specific failure one has to define the most important failures in advance. As basis for the failure definitions in the survey we used the failure description given in the TASK13 report "Review of Failures of Photovoltaic Modules" [13]. Besides the PV module failures also methods like visual inspection, IV curve analysis, electroluminescence, thermography, UV fluorescence and other techniques are described to detect and identify the failures. A clear definition is a prerequisite to make the failure classification more reliable.

2 EXPERIMENTAL

We designed a survey format to collect failure data of PV systems for various climate zones with the focus on both, the origin of failure and the power loss. A full list of failure types used in the survey is given in Tab.1 in the Appendix. The long-term goal of the survey is to evaluate the possible different impacts of the failures for various climate zones. The survey is implemented in a Microsoft Excel table worksheet. The file and the corresponding help file can be accessed via internet at the homepage of the Photovoltaic Power Systems Programme of the International Energy Agency [14],[15]. The format of the survey is shown in Fig. 1.

2.1. Target group of the survey

The survey is designed to be distributed to international PV experts who fill in the data. However, the primary origin of the data may be from other sources such as system owners, installers, other experts, manufacturers, scientific publications and internet surveys. Each of the provided datasets is checked by a second expert for plausibility at the ISFH. The survey input parameters are reduced to a minimum so that it is still possible to evaluate the data for power loss and degradation rates for a specific failure. We developed the survey to ensure a maximum of anonymity of the PV system. In this work we only evaluate PV module data. **Tab. 1** shows publications used in survey data as data source.

Tab. 1: List of survey data extracted of publications. Some authors submit additional information (marked with *).

Power loss dominating failure	Ref.
Discolouring of pottant, Delamination, Disconnected cell or string interconnect ribbon	[1]
Delamination	[16]
Defect backsheet, Discolouring of pottant*	[17]
Cell cracks*	[18]
Dust soiling	[19]

2.2 Design of the survey

For the database the climatic zone, system data, the component types, and start and analysis date are collected. However, the categorization of each item is as rough as possible in order to be as anonymous as possible.

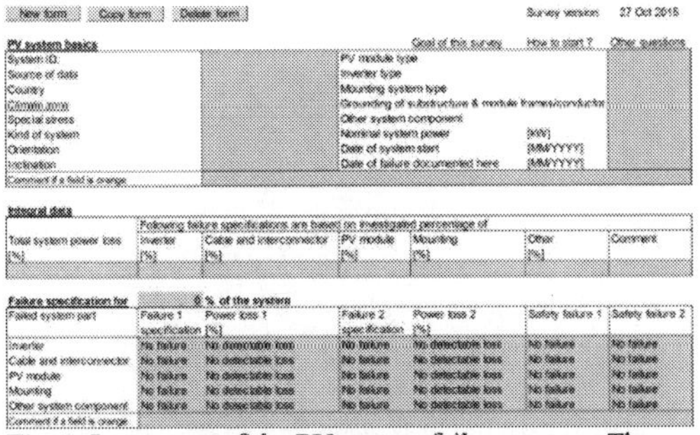

Fig. 1: Input page of the PV system failure survey. The section "Failure specification for" is available five times to specify system parts with different failures. Underlined blue text leads to linked help files. E.g. the climate zone is linked to the Wikipedia classification of the Köppen-Geiger climate system.

The characterization of the PV failures allows specifying only two failures for one system part. This is for the reason that if there are more than two failures in one part of the system it is very difficult to specify the power loss for each type of failure. Therefore, it is only allowed to specify two failures which are most important for the power loss of the system. If there are various failures for different parts of the system it is possible to specify the various failures for at maximum five system parts. Sometimes it is not possible to find out all reasons for a power loss. In this case one should only state the power loss of the well analysed failures in the system.

In the appendix the selectable failure types in the survey are visualized. The light induced degradation is not added to the choice list of possible failures. This early power degradation in the modules life must be incorporated in the nominal power of the name plate. Even if the manufacturer does not incorporate this degradation in the nominal power or it is higher than assumed it is very difficult to classify this failure. There is no easy way to detect this failure type. Therefore, we do not account for light induced degradations.

In addition to the climate zones it is possible to specify special stress loads at the location of the system. Besides the standard item here are two possible items to be chosen: "Island, coastal region (10 km)" and "Agricultural environment".

3 EVALUATION OF THE SURVEY

Due to the fact that it is very difficult to obtain power loss data for specific failure types, the power loss choices in the survey are roughly subdivided into the following items and intervals "Unknown", „No detectable loss",]0%-3%],]3%-10%],]10%-20%],]20%-30%] and so on until]90%-100%]. The intervals are defined in the bracket notation of the International standard ISO 31-11. Because of the rough interval definition we do not differentiate between measured initial power of the PV system or modules and nominal power of the PV system or modules. Fig. 2 illustrates the definition of various partitions of a PV system being important for the definition of degradation rates of power losses. For each survey dataset i the total installed power Pi is collected in the database. For the fault analysis of a failure x very often only a part yi of the total system is investigated. Furthermore, the failure x effects only a part $z_{i,x}$ of the system. The partitions y_i and $z_{i,x}$ are given in percentage of the installed power. These values are the basis of the following degradation rate calculations.

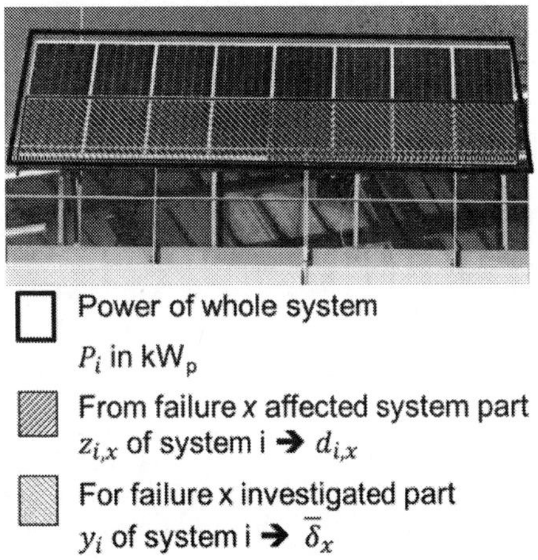

Fig. 2: Definition of various parts of a PV system. The nominal system power P_i is given in kW_p. All other parts are given in % of P_i.

3.1 Degradation rate

The power loss $\Delta P_{i,x}$ of a specific module failure x is documented in percent of the affected nominal module power sum. The equation of the degradation rate di,x of a specific module failure type x of dataset i is given by:

$$d_{i,x} = \frac{\Delta P_{i,x}}{\tau_{b,i} - \tau_{a,i}}. \quad (1)$$

The parameter $\tau_{b,i}$ is the date of failure documentation of dataset i and $\tau_{a,i}$ is the date of system start of dataset i.

The equation for the degradation rate of the whole system is given by:

$$\delta_{i,x} = d_{i,x} \frac{z_{i,x}}{y_i}, \quad (2)$$

where z_i is the percentage of the system being affected and y_i is the investigated system part in percent from the total nominal system power P_i. For the further evaluations it is assumed that the investigated system part is chosen large enough that the investigation result is representative for the whole system. In this case $\delta_{i,x}$ shows how much the total system is affected by the failure type x.

We define two mean degradation rates for the PV modules:

1. Unweighted mean degradation rate $\overline{d}_x$ of PV module failure x of all datasets n_x with the module failure x:

$$\overline{d}_x = \frac{\sum d_{i,x}}{n_x}. \quad (3)$$

2. Degradation rate $\overline{d}_{xw}$ of failure x weighted by the part $z_{i,x}$ of the nominal power P_i of the PV system affected by the failure x:

$$\overline{d}_{xw} = \frac{\sum_{i=1}^{n} d_{i,x} \cdot P_i \cdot z_{i,x}}{\sum_{i=1}^{n} P_i \cdot z_{i,x}}. \quad (4)$$

The first degradation rate $\overline{d}_x$ is unbiased of the size of the systems in the dataset. In this case small systems have the same impact on the mean module degradation rate for a failure type than large systems. The second degradation rate $\overline{d}_{xw}$ accounts for the system size of the failure affected part.

Furthermore, we define two mean degradation rates for the PV system affected by a failure x:

1. Unweighted mean degradation rate $\overline{\delta}_x$ of the investigated PV system part:

$$\overline{\delta}_x = \frac{\sum \delta_{i,x}}{n_x}. \quad (5)$$

2. Degradation rate $\overline{\delta}_{xw}$ of the investigated PV system part weighted by the investigated part $y_{i,x}$ of the nominal power $P_{i,x}$ of the PV system:

$$\overline{\delta}_{xw} = \frac{\sum_{i=1}^{n} \delta_{i,x} \cdot P_i \cdot y_i}{\sum_{i=1}^{n} P_i \cdot y_i}. \quad (6)$$

The first system degradation rate $\overline{\delta}_x$ is again unbiased of the size of the systems in the dataset. The second system degradation rate $\overline{\delta}_{xw}$ accounts for the system size of the failure affected part. These degradation rates allows to assess how a failure x affects the whole system power.

Our survey data does not allow checking if the degradation rates of the found failures are constant over time. However the degradation rates calculated with Eq. 1-6 allow comparing and averaging slowly developing power loss effects.

3.2 Mean power loss of sudden events

For sudden power losses caused by a storm, a hail storm or a lightning stroke the calculation of a degradation rate makes no sense. Therefore we evaluate how much percent of the investigated system power is affected by any power loss after the sudden event:

$$\overline{p}_{xa} = \frac{\sum_{i=1}^{n} P_i \cdot z_{i,x}}{\sum_{i=1}^{n} P_i \cdot y_i}, \quad (7)$$

and how much is the mean power loss relative to the investigated system power:

$$\overline{p}_{xd} = \frac{\sum_{i=1}^{n} \Delta P_{i,x} \cdot P_i \cdot z_{i,x}}{\sum_{i=1}^{n} P_i \cdot y_i}. \quad (8)$$

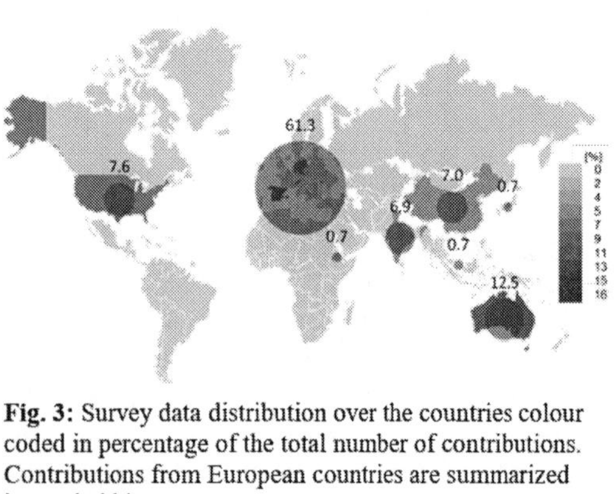

Fig. 3: Survey data distribution over the countries colour coded in percentage of the total number of contributions. Contributions from European countries are summarized in one bubble.

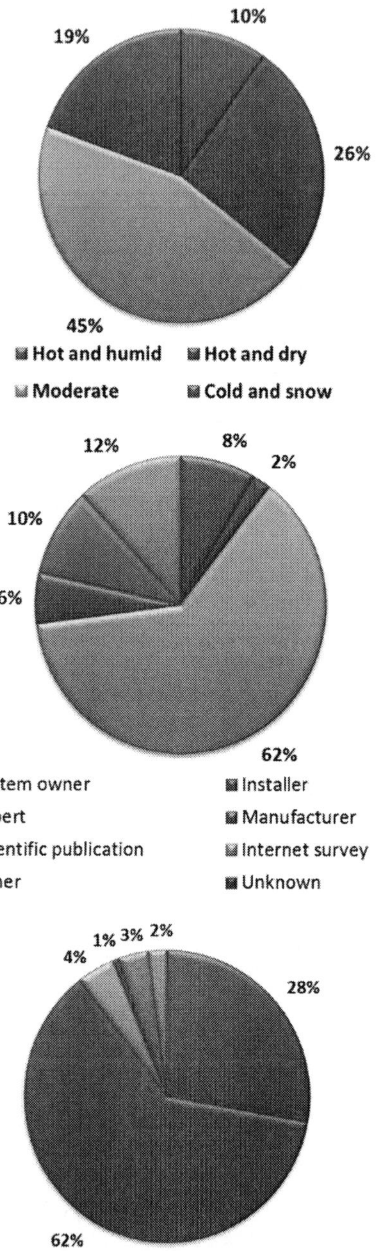

Fig. 4: Composition of the input data sorted by climate zones (top), by contributor profile (middle) and by PV module technology (bottom).

4 COMPOSITION OF SURVEY DATA

Fig. 3 shows the survey data distribution over the countries. In total 144 failure-survey-data sets from 18 countries have been evaluated. **Fig. 4** shows the composition of the survey data. The moderate climate zone dominates the database with 45 % of data sets and unfortunately we have only 10% data from hot and humid climate. Most of the data is generated by experts from plant investigations or taken from scientific reports (72%). The distribution into mono- with 28%, multi- with 62% and thin film technology with 8% approximately equals the capacity share of these technologies in years 2011-2015 with ~24% for mono, ~66% for multi and ~10% for thin film [20].

5 RESULTS

Before we evaluate the degradation rates we show the occurrence of failures over the years of operation in Fig. 5. The upper diagram shows the occurrence of all reported failures and the lower graph shows only failures causing a power loss. Each diagram is split into the occurrence of degrading failures and sudden occurring failures. The occurrence of both failure types accumulates at the first 7 years.

When we focus on special types of failures we see that cell crack failures are mostly reported in the very early stage of PV system operation from year 1 to year 2. Systems with PIDs failure are mainly reported during year 3 and year 4. Disconnected cells or strings in the module are reported after year 4 spread over the whole operation time. Discolouring of pottant is spread over the years but power relevant discolouring starts after year 3 with a high accumulation after 18 years of system operation. Defect bypass diodes are spread over the first 10 years of operation. The total occurrence of the other failures is too seldom for further discussion.

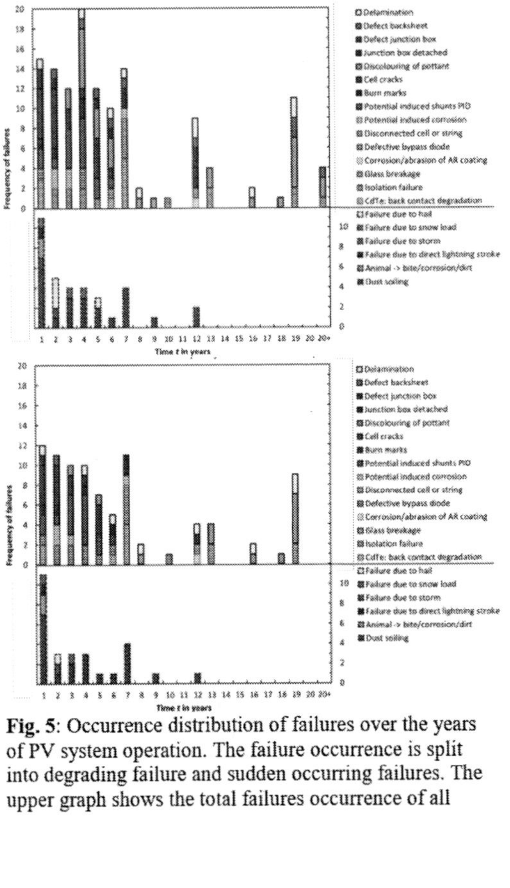

Fig. 5: Occurrence distribution of failures over the years of PV system operation. The failure occurrence is split into degrading failure and sudden occurring failures. The upper graph shows the total failures occurrence of all detected failure. The lower graph shows the occurrence of detected failure which causes a measurable power loss.

For the current status of collection of failure reports only some mean degradation rates can be analysed. Results based on less than four reports are not discussed. Therefore in all figures presenting statistical data the basic population of PV systems is shown as number above the mean value. The weighted degradation rates $\bar{d}_{xw}$ and $\bar{\delta}_{xw}$ are currently too much influenced by large single PV systems. Therefore, we do not show the results of the weighted degradations here.

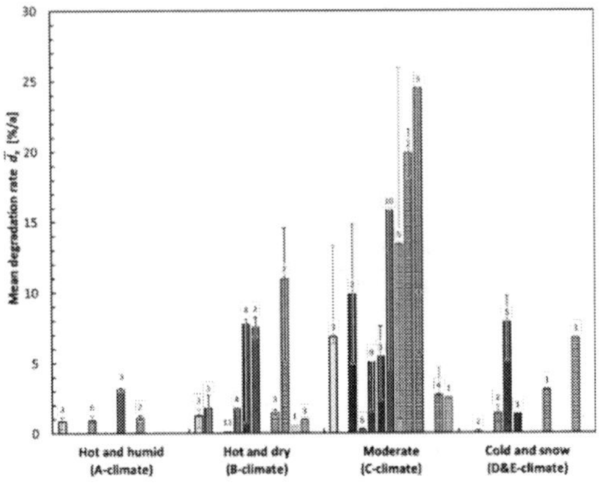

Fig. 6: Mean unweighted degradation rates $\overline{d}_x$ of PV module failures sorted by climatic zones. The numbers show the quantity of data per failure in the database. The whiskers indicate the upper and lower quartile of the degradation rates. Failures with no data or 0%/a degradation rate are not shown. Right: legend of figure.

Delamination
Defect backsheet
Defect junction box
Discolouring of pottant
Cell cracks
Burn marks
Potential induced shunts PID
Potential induced corrosion
Disconnected cell or string
Defective bypass diode
Corrosion/abrasion of AR coating
Glass breakage
CdTe: back contact degradation

Fig. 6 shows the mean unweighted degradation rates for the affected part of the PV systems. The highest impact on the performance of PV modules have defective bypass diodes in the hot and dry climate with 11%/a, and in moderate climate with 25%/a. The degradation rate $\overline{d}_x$ caused by cell cracks in the cold and snow climate is about 3%/a higher than in the moderate climate and 6%/a higher than in hot and dry climate. The PIDs effect shows a mean degradation rate of about 15% per year. In the moderate climate it is the most often found failure together with a high degradation rate. Unfortunately there are not enough PIDs events documented from other climate zones. The discolouring of pottant failure is found in the hot and humid, hot and dry and in the moderate climate as well. In the three climate zones this degradation mechanism is in the mean below 1%/a. Therefore, this effect is most often not the cause for warranty claims.

How much a failure dominates the total system power can be seen by the unweighted degradation rate of the investigated system part shown in **Fig. 7**. For the PIDs effect in the moderate climate the degradation of the investigated system part is reduced by 3/5 compared to the degradation of the effected system part. Therefore, PIDs affects in average about 3/5 of a system in the moderate climate zone. For cell cracks the same evaluation shows that cell cracks affect about 3/5 of the system in the moderate zone if they cause a power loss.

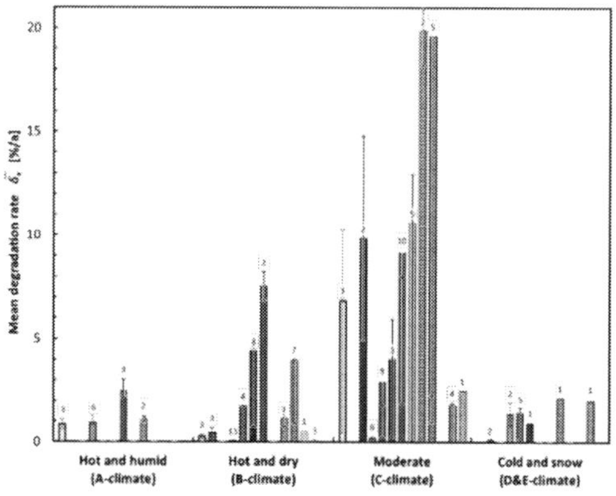

Fig. 7: Mean unweighted degradation rates $\bar{\delta}_x$ of the investigated PV system part of PV module failure sorted by climatic zones. The numbers show the quantity of data per failure in the data-base. The whiskers indicate the upper and lower quartile of the degradation rates. Failures with no data or 0%/a degradation rate are not shown. Right: legend of figure.

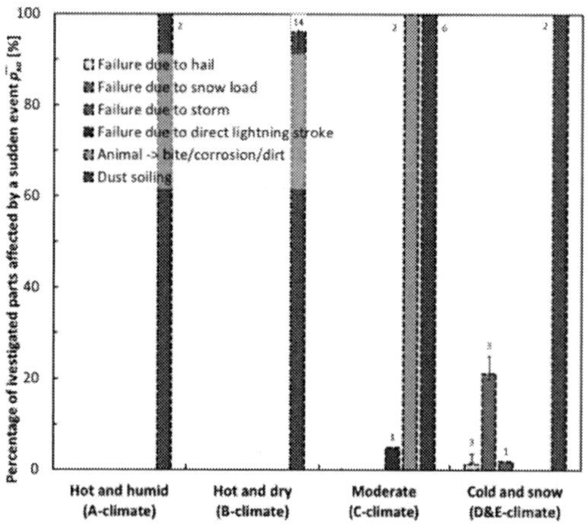

Fig. 8: Mean system part $\overline{p}_{xa}$ affected after a sudden event. The numbers show the quantity of data per failure in the database. The whiskers indicate the upper and lower quartile of the affected system part.

Fig. 8 shows how much of a system is affected by a sudden event calculated by Eq. (7). As expected, soiling affects almost the whole systems in nearly all cases. There is one event of "Animal -> bite/corrosion/dirt" where animals also soil all of the modules. The events "lighting stroke", "storm" and "hail" only cause a power loss on less than 10% of the modules in the system. The failure "snow load" affects about 20% of the modules in the system.

The power loss of a system caused by a sudden event is calculated by Eq. (8) and is shown in **Fig. 9**. Besides soiling the snow load event has the highest impact on the system power loss. The other sudden events only affect less than 1% of the system power. Except for soiling the calculations in **Fig. 8** and **Fig. 9** are mostly based on single to three entries in the database and are therewith only show cases.

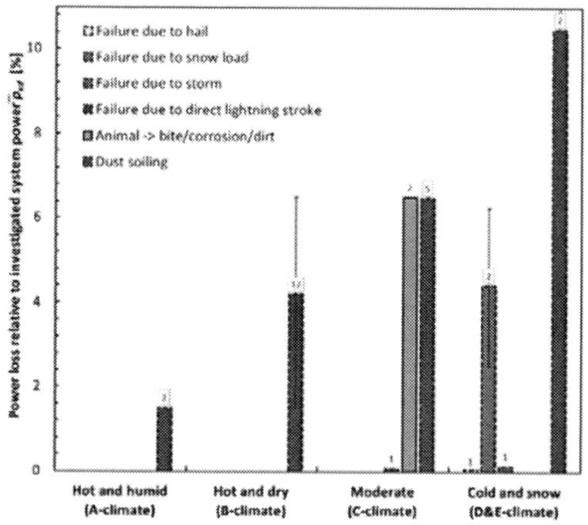

Fig. 9: Mean system power $\overline{p}_{xd}$ lost from the investigated system part after a sudden event. The numbers show the quantity of data per failure in the database. The whiskers indicate the upper and lower quartile of the lost system power.

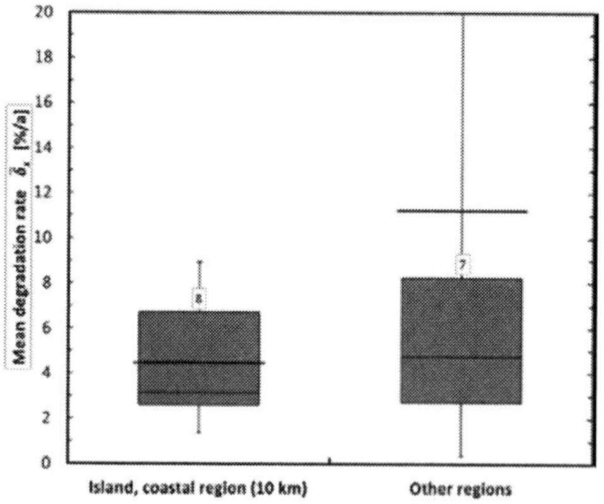

Fig. 10: Box-Whisker-Plot of degradation rate of the investigated part of the system for the PIDs effect sorted by the special stress coastal region and other. The numbers show the quantity of data per failure in the database. The mean value is indicated by a thick line. The whiskers indicate the upper and lower quartile of the degradation rates.

72% of all reported failures with the special stress loads "Island, coastal region (10 km)" show a PIDs failure. From the rest data only 4.6% shows PIDs. This indicates much higher PIDs risk for island and coastal regions as also found by Berghold [21]. Therefore, one would expect a much higher PIDs degradation rate for "coastal regions/island" systems. But this is not the case. The total degradation rate of the coastal regions/island systems is shown in **Fig. 10** in comparison to all other systems. The mean PIDs degradation rate is much higher for other regions than for "coastal regions/island" systems. However this is caused by some systems in other regions with an extreme high degradation rate.

5 DISCUSSION

The occurrence distribution of failures over the operational years deviates from a typical bath tube curve. A reason for the difference may be that the number of installed systems changes exponentially with the year of installation. We have much more young systems in the world than old ones as shown in **Fig. 11**. Therefore the failure occurrence is highly biased by not equally distributed annual system installations. This may be the reason for the high failure level at the first seven years in **Fig. 5**. Especially the sudden occurring failures may be highly biased by this effect.

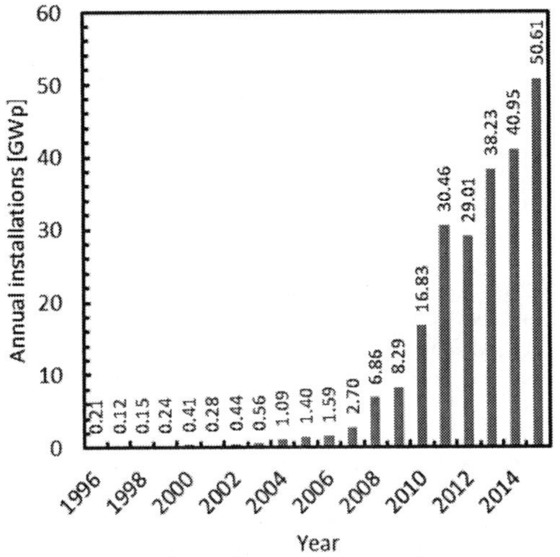

Fig. 11: Evolution of global annual PV system name plate installations, composed from numbers in [22].

The failure type soiling does not fit into degrading and sudden failures category, because the power of soiled module degrades over time but can fully be recovered. We see a dependence of dust soiling on the climate zones in **Fig. 9**. However the high mean power loss due to dust in the moderate zone (6%) does not coincident with the expectation to find higher dust soiling losses the hot and dry climate (4%). Probably dust soiling is strongly influenced by local conditions. Herrmann developed a model on dust soiling based on weather data and other influencing factors [23]. He showed a correlation between desert and no desert regions. To prove this correlation we would need a much more detailed climate zone resolution as performed in this study.

The anti-correlation between mean degradation rate and occurrence of PIDs in coastal regions/island systems is unexpected. Berghold presented a classification of PIDs modules into "Frame PID" where only the cells close to the frame are affected and "Surface PID" where almost all cells are affected [21]. Therewith the different affected area partitions influence the degradation rate. Berghold suggests that the "Frame PID" with the low degradation rate is found in hot and dry locations. But in our database only two island/coast region system are from hot and dry locations. Therefore, this "Frame PID" effect cannot explain the relative low degradation rate for island/coast region systems compared to other regions. But there may be some systems among the other region systems that show "Surface PID" leading to very high degradation rates that influence the statistic.

Maybe in coastal region already systems are affected by PIDs if the PIDs effect would not occur for the same systems in the other regions. And in the other regions we find almost heavily degrading PIDs affected systems.

The degradation rates $\overline{d}_x$ of failure x affected modules are important for module manufacturers. It shows how fast modules degrades caused y failure x. The degradation rate $\bar{\delta}_x$ of the investigated PV system part is important for the bankability of PV systems. Under the assumption that the people who analysed the failures in the system choose a representative part of the PV system for the failure analysis the degradation rates $\bar{\delta}_x$ are representative for such systems.

7 SUMMARY AND CONCLUSIONS

The current status of the data collection of PV system failures is preliminary. For some cases results can be drawn from the data. The degradation rate due to cell cracks is highest in cold and snow climates. Cell cracks dominate the early failures in year one and two after installation. However the system degradation rate stays below 3%/a for all climatic zones, while the degradation rates of the affected part is highest with about 8%/a in the cold and snow climate. The next dominant failure is the potential induced shunts effect in year three and four with mean degradation rates of the system of 9%/a and 16%/a for the affected parts in the moderate zone. The PIDs effect occurs 15 times more often in coastal/island regions than in other regions, but is less severe than in other region. Defect bypass diodes heavily affect the module and the system power. The failure occurrence is spread over the first 10 years of system operation. The discolouring of the pottant is a quite often seen wear out failure. But for all climate zones this is effect is in the mean well below 1%/a for the system and the affected part. In our future work we will expand the analysis also for PV system components.

8 ACKNOWLEDGMENTS

Funding was provided by the Federal Ministry for Economic Affairs and Energy (BMWi) under contract no. FKZ 0325786C/A Task13 Projekt, the State of Lower Saxony, and by the Austrian Climate and Energy Fund (KliEn) within the 1st e!MISSION energy research call 2014, project Infinity, under contract no. 850.414. We like to thank Iris Kunze and Arnaud Morlier for supporting the data collection. Furthermore we thank all TASK13 members and experts who contributed survey data for the database.

8 REFERENCES

[1] R. Dubey, S. Chattopadhyay, V. Kuthanazhi, J. J. John, B. M. Arora, A. Kottantharayil, K. L. Narasimhan, C. S. Solanki, V. Kuber, V. Vasi, A. Kumar, and O. S. Sastry, "All-India India Survey of Photovoltaic Module Degradation : 2013," 2013. [Online]. Available: www. ncpre. iitb. ac. in/uploads/All_India_Survey_of_Photovoltaic_Module_ Degradation_2013. pdf.

[2] A. Jacobson, D. Kammen, R. Duke, and M. Hankins, "Field performance measurements of amorphous silicon photovoltaic modules in Kenya," in The Solar Conference, 2000, pp. 95–100.

[3] A. B. Maish, C. Atcitty, S. Hester, D. Greenberg, D. Osborn, D. Collier, and M. Brine, "Photovoltaic system reliability," in Conference Record of the Twenty Sixth IEEE Photovoltaic Specialists Conference - 1997, 1997, pp. 1049–1054.

[4] K. Stokes and J. Bigger, "Reliability, cost, and performance of PV-powered water pumping systems: a survey for electric utilities," IEEE Trans. Energy Convers., vol. 8, no. 3, pp. 506–512, 1993.

[5] K. Kato, "PVRessQ!: a research activity on reliability of PV systems from an user's viewpoint in Japan," in SPIE 8112, Reliability of Photovoltaic Cells, Modules, Components, and Systems IV, 81120K (21 September 2011), 2011, p. 81120K–81120K–9.

[6] A. Skoczek, T. Sample, and E. D. Dunlop, "The results of performance measurements of field-aged crystalline silicon photovoltaic modules," Prog. Photovoltaics Res. Appl., vol. 17, no. 4, pp. 227–240, Jun. 2009.

[7] R. J. van der Plas and M. Hankins, "Solar electricity in Africa: a reality," Energy Policy, vol. 26, no. 4, pp. 295–305, Mar. 1998.

[8] S. Kumar, S. C. Bhattacharya, and M. A. Leon, "A survey on PV systems and accessories in Asia," in In: Sayigh AAM, editor. World Renewable Energy Congress VI. Oxford: Pergamon, 2000, pp. 860–863.

[9] S. Djordjevic, D. Parlevliet, and P. Jennings, "Detectable faults on recently installed solar modules in Western Australia," Renew. Energy, vol. 67, pp. 215–221, Jul. 2014.

[10] E. D. Dunlop, D. Halton, and H. A. Ossenbrink, "20 years of life and more: where is the end of life of a PV module?," in Conference Record of the Thirty-first IEEE Photovoltaic Specialists Conference, 2005., pp. 1593–1596.

[11] D. C. Jordan and S. R. Kurtz, "Photovoltaic Degradation Rates-an Analytical Review," Prog. Photovoltaics Res. Appl., vol. 21, no. 1, pp. 12–29, Jan. 2013.

[12] D. C. Jordan, S. R. Kurtz, K. VanSant, and J. Newmiller, "Compendium of photovoltaic degradation rates," Prog. Photovoltaics Res. Appl., vol. 24, no. 7, pp. 978–989, Jul. 2016.

[13] M. Köntges, S. Kurtz, C. Packard, U. Jahn, K. A. . Berger, K. Kato, T. Friesen, H. Liu, M. Van Iseghem, J. Wohlgemuth, D. Miller, M. Kempe, P. Hacke, F. Reil, N. Bogdanski, W. Herrmann, C. Buerhop-Lutz, G. Razongles, and G. Friesen, Review of Failures of Photovoltaic Modules. IEA, 2014.

[14] M. Köntges, "PV-failure_survey, 27th Oct. 2015," 151027__PV-failure_survey.xlsm, 2015. [Online]. Available: http://iea-pvps.org/index.php?id=344&eID=dam_frontend_push&docID=2918. [Accessed: 29-Jan-2016].

[15] M. Köntges and S. Altmann, "Explanation of the PV System Survey sheet," 151027_Explanation_PV-System_Survey.pdf. [Online]. Available: http://iea-pvps.org/index.php?id=344&eID=dam_frontend_push&docID=2919. [Accessed: 26-Jan-2016].

[16] P. Sánchez-Friera, M. Piliougine, J. Peláez, J. Carretero, and M. Sidrach de Cardona, "Analysis of degradation mechanisms of crystalline silicon PV modules after 12 years of operation in Southern Europe," Prog. Photovoltaics Res. Appl., vol. 19, no. 6, pp. 658–666, Sep. 2011.

[17] T. Friesen, "PV Modules Failures Modes Observed in Real Use and Long Term Exposure," in PV-Module Reliability, Berlin, April 5-6, 2011.

[18] F. Spertino, A. Ciocia, P. Di Leo, R. Tommasini, I. Berardone, M. Corrado, A. Infuso, and M. Paggi, "A power and energy procedure in operating photovoltaic systems to quantify the losses according to the causes," Sol. Energy, vol. 118, pp. 313–326, Aug. 2015.

[19] H. Häberlin and P. Schärf, "Langzeitverhalten von PV-Anlagen über mehr als 15 Jahre Prof.," in 25. Symposyum Photovoltaische Solarenergie, 2010, p. 0043_8050_3.

[20] B. Burger, K. Kiefer, C. Kost, S. Nold, S. Philipps, R. Preu, J. Rentsch, T. Schlegl, G. Stryi-Hipp, G. Willeke, H. Wirth, I. Brucker, A. Häberle, and W. Warmuth, "Photovoltaics Report," Fraunhofer Institute for Solar Energy Systems, ISE, Freiburg, 2015.

[21] J. Berghold, S. Koch, S. Pingel, S. Janke, A. Ukar, P. Grunow, and T. Shioda, "PID: from material properties to outdoor performance and quality control counter measures," in SOPHIA Workshop, p. 95630A.

[22] BP, "BP Statistical Review of World Energy June 2016," Excel workbook, 2016. [Online]. Available: http://www.bp.com/content/dam/bp/excel/energy-economics/statistical-review-2016/bp-statistical-review-of-world-energy-2016-workbook.xlsx.

[23] J. Herrmann, K. Slamova, R. Glaser, and M. Köhl, "Modeling the Soiling of Glazing Materials in Arid Regions with Geographic Information Systems (GIS)," Energy Procedia, vol. 48, pp. 715–720, 2014.

APPENDIX

Table 1: List of PV module failures being selectable in the survey.

Type	Image	Type	Image
Delamination Photo: c-Si solar cells delaminte from the cell interconnect ribbon.		**Burn marks** Photo: Burn mark along cell edge on module front side	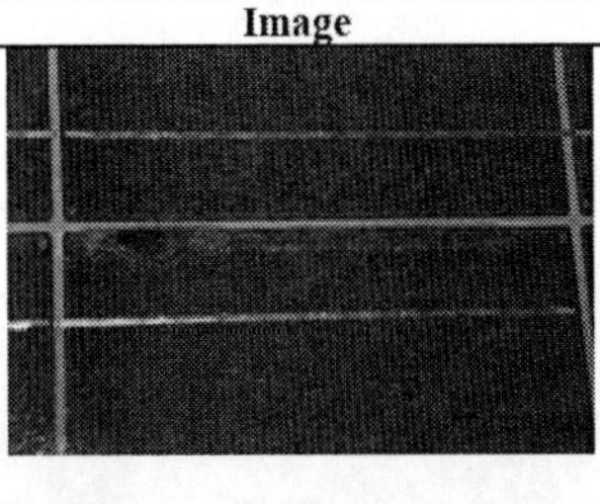
Defect junction box Photo: lid of junction box fell of and pottant discolor		**Potential induced shunts** (often named PID) Photo: some spots indicate the shunt positions	
Junction box detached Photo: junction box is not fixed to the module any more	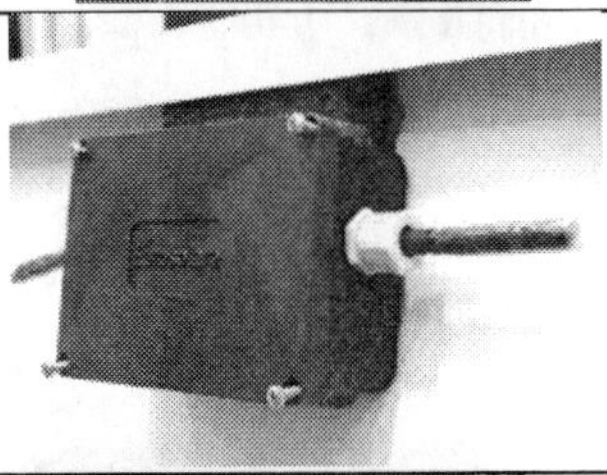	**Potential induced corrosion** (often with thin film modules) Photo: thin film module, corrosion near edge below clamp	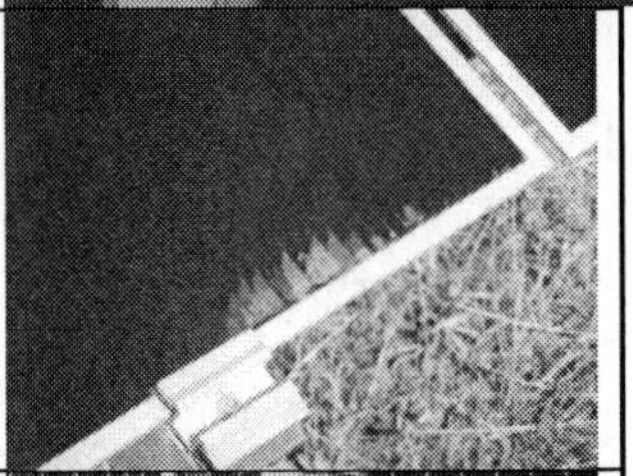
Discolouring of pottant Photo: back sheet of module discolored	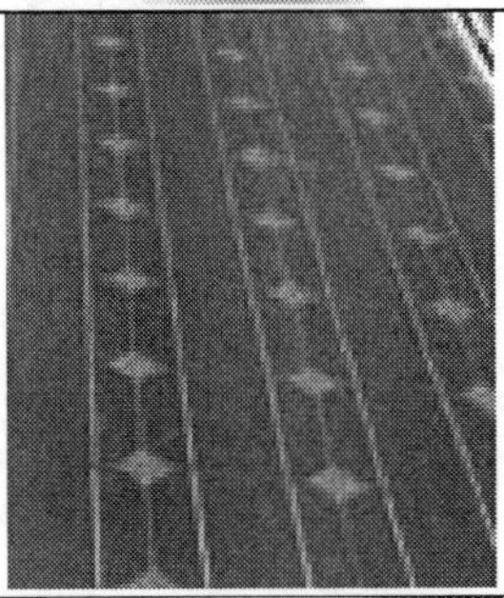	**Disconnected cell or string interconnect ribbon** Photo: disconnected cell interconnect ribbon	
Cell cracks Photo: diagonal cell cracks visible in browned EVA on top of a cell	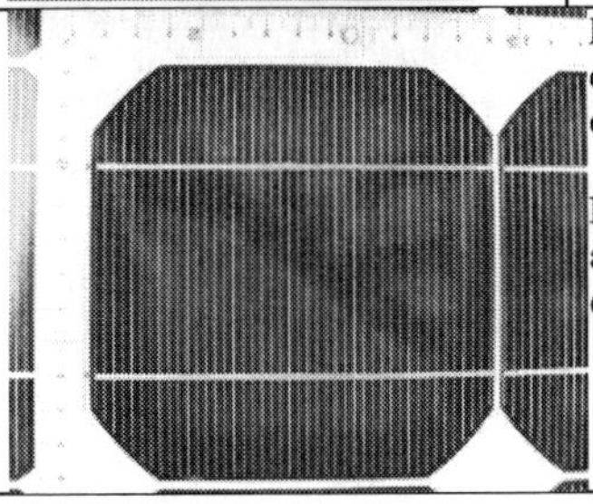	**Defective bypass diode/wrong dimensioned** Photo: molton plastic around the bypass diodes	

Type	Image	Type	Image
Defect backsheet	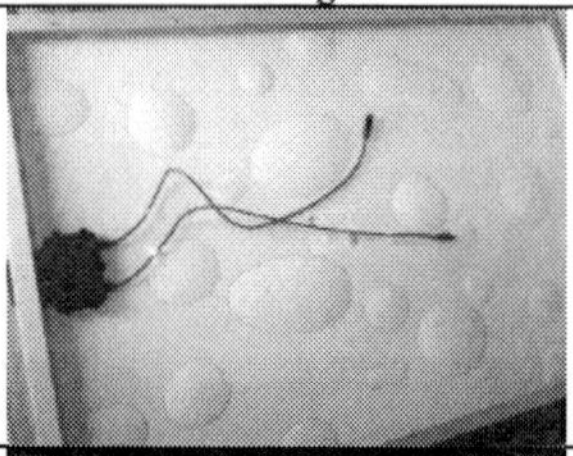	**Hail -> glass breakage/cell breakage**	
Potential induced shunts (often named PID) Electroluminescence image.	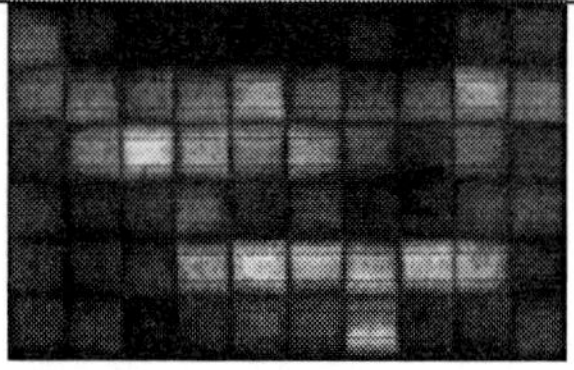	**Snow load -> deformed frame/glass- /cell-breakage**	
Corrosion/abrasion of AR coating		**Direct lightning stroke -> defect glass/frame and defect bypass diodes**	
Glass breakage Excluding other glass breakage reasons also listed here.		**Biofilm soiling** Source: Sulaiman bin Shaari	
Isolation failure	Any part of the module shows an isolation failure incl. the cables and connectors of the module.	**Dust soiling** Source: Uwe Hupach	
CdTe: back contact degradation	Temperature and voltage induce degradation of the CdTe back contact. Can be identified by fill factor loss and "roll over" of the *IV* curve.	**Animal -> bite/corrosion/dirt**	Any damage, corrosion or dirt caused the system to loos power.
Frame breakage/bent/defect	Any damage on the frame, exept caused by special evens also listed here.	**Storm -> deformed frame/glass-/cell-breakage**	Any damage caused by a storm.

FORECASTING AND OBSERVABILITY: CRITICAL TECHNOLOGIES FOR SYSTEM OPERATIONS WITH HIGH PV PENETRATION

Pierre-Jean ALET[1,*], Venizelos EFTHYMIOU[2], Giorgio GRADITI[3], Norbert HENZE[4], Mari Juel[5], David MOSER[6], Franko NEMAC[7], Marco PIERRO[6], Evangelos RIKOS[8], Stathis TSELEPIS[8], Guangya YANG[9]
[1]CSEM, PV-center, Jaquet-Droz 1, CH-2000 Neuchâtel, Switzerland, Pierre-Jean.ALET@csem.ch, +41 32 720 5251
[2]University of Cyprus, Panepistimiou 1 Avenue, P.O. Box 20537, 1678 Nicosia, Cyprus
[3]ENEA Portici Research Center, P.E. Fermi, 80055 Portici (Naples), Italy
[4]Fraunhofer IWES, Institut für Windenergie und Energiesystemtechnik, Königstor 59, 34119 Kassel, Germany
[5]SINTEF Materials and Chemistry, Alfred Getz' vei 2, Trondheim, Norway
[6]EURAC Research, Institute for Renewable Energy, Viale Druso, 1, 39100 Bolzano, Italy
[7]ApE - Agencija za prestrukturiranje energetike d.o.o., Litijska cesta 45, 1000 Ljubljana, Slovenia
[8]Centre for Renewable Energy Sources and Saving, 19th klm Marathonos Ave.,19009 Pikermi, Greece
[9]DTU Electrical Engineering, Building 325, DK-2800 Kgs. Lyngby, Denmark

ABSTRACT: Forecasting and monitoring technologies for photovoltaics are required on different spatial and temporal scales by multiple actors, from the owners of PV systems to transmission system operators. In this paper the Grid integration working group of the European Technology and Innovation Platform – Photovoltaics (ETIP PV) reviews the different use cases for these technologies, their current status, and the need for future developments.
Power system operations require a real-time view of PV production for managing power reserves and for feeding short-term forecasts. They also require forecasts on all timescales from the short (for dispatching purposes), where statistical models work best, to the very long (for infrastructure planning), where physics-based models are more accurate. Power system regulations are driving the development of these techniques. This application also provides a good basis for a cost/benefit analysis since the forecasting error can be linked to the prices charged for energy imbalance.
Keywords: Forecasting, PV System, Funding and Incentives, Grid Integration, System Performance

1 INTRODUCTION

In its review of the challenges and opportunities associated with massive deployment of solar PV generation [1], the Grid integration working group of the European PV Technology Platform (now ETIP PV) identified forecasting and observability as critical technologies for the planning and operations of the power system with large PV penetration. In this paper we set out to spell out in more details what features are needed from these technologies and, after an assessment of their current status, how they need to be developed.
Some very good reviews of forecasting techniques have been published in recent years [2,3]. We have built on these by taking a step back and analysing the different use cases for forecasting in relation to PV.
To estimate the economic value of further improvements in forecasting, we linked the effect of forecast errors with the current imbalance settlement prices charged by balancing authorities in Europe.

2 RATIONALE FOR PV POWER FORECASTING

2.1 Dynamics of power system

At any times in any power system, consumption (including charging of storage systems, and losses) and production (including discharging of storage systems) need to be equal. In a conventional power system operating in alternating current (AC) the frequency is a real-time indicator of that balance. However grid assets have a finite dynamics so to ensure the balance any fluctuation of production or consumption needs to be anticipated as much as possible before it translates into frequency deviations.
Indeed, the characteristic time constants of power system components range from less than a second to ten years or more, as summarized on Table 1:

Table 1: Characteristic time constant of power system components

< 10 s	Inertia response Protection system operations Switching of power electronics Battery switching between charge and discharge
1 min	Fast start of pumped hydropower plant [4] Fast start of some combustion engines [5]
15 min	Gas power plant from 1/3 to full power [6]
1 h	Start-up and shutdown of most power plants
24 h	Commitment of generation units
1 year	Maintenance planning
10 years	Expanding transmission infrastructure
20+ years	Economic lifetime of PV systems Economic lifetime of grid assets

Prior to the introduction of variable renewable sources (wind and PV), power consumption was the only stochastically variable component in the power system balance. Forecasting its variations was already introduced in the 1940s. It has since been refined to take into account "seasonal" variations (day of the year, day of the week, hour of the day) and the specific characteristics of different electricity uses (heating and cooling, cooking, industrial equipment, lighting, etc.) [7]. The focus has always been on regional or national aggregates.
The deployment of variable renewable generation is introducing new requirements on forecasting techniques. First of all PV and wind generators are much more sensitive to weather conditions. The main weather parameter with an influence on electricity demand – where heating or cooling is powered with electricity – is temperature. This parameter varies slowly in time and space. PV and wind on the other hand strongly depend on

rapidly changing variables: in first approximation, PV power is in first approximation proportional to $G \cdot (1 + k \cdot \log G)$ [8] where G is the global irradiance in the plane of array, and wind power varies as V^3 where V is the wind speed. As a result, the geographic distribution of the generators matter more for the aggregate variations than that of the loads.

In addition, PV generation is highly distributed in terms of locations and ownership. It is therefore often necessary to forecast generation with a higher spatial resolution than demand. Indeed single MW-scale plants may be exposed to market trades, and microgrid operations with self-consumed PV electricity require forecasts at the buildings or district levels. Such granularity increases the forecasting difficulty: the standard deviation of PV power production is reduced as $1/\sqrt{S}$ and $1/\sqrt{N}$, where S is the surface area of a PV power plant and N is the number of aggregated plants [9,10].

2.2 Drivers for PV forecasting

An important concept when dealing with forecasting in the power system is the balance group. Balance groups can include generation units, consumption units, or be "virtual" when operated by financial actors who only trade. Forming a balance group is a requirement to participate in wholesale electricity markets. All balance group reports to a balancing authority, which in Europe is generally the transmission system operator (TSO). This authority ensures that trades on the electricity market are balanced i.e., that contracted generation matches forecast consumption. Balancing group managers are responsible to ensure that at each time step of market operations their contracted production and/or consumption matches the realised values. In case of mismatch between prediction and realisation, balance group managers are penalised based on intraday market price; if the imbalance is in the same direction as the whole system (e.g., a producer under-delivering when there is a shortage in production), the penalty will be above the intraday market price and if the imbalance is in the opposite direction the penalty will be below.

PV generators were until recently shielded from this balancing responsibility. In Germany for example, transmission system operators carry the responsibility and operate a balance group for PV systems which are connected under the Renewable Energy Sources Act (EEG) in their area [11]. Regulators are now pushing to increase exposure of PV generators to market conditions and increase their responsibility in the balancing mechanisms. A 2014 ruling by the Italian regulator introduced imbalance charges for renewable power generators of more than 1 MW in capacity; the mechanism is similar to that applied to conventional balance groups but the fees are modulated to take into account the inherent volatility of the different sources [12]. The resulting cost for PV generators is estimated around 5 €/MWh, which is still significantly lower than imbalance prices applied to regular balance groups in Europe [13,14].

In addition, support mechanisms for large PV generators are evolving from feed-in tariffs to market premiums in France, Germany and the UK [15] under which these generators receive a regulated payment on top of market prices. As illustrated on Figure 1, these premiums can be fixed, or floating i.e., cover the difference between the average market price over a certain period of time – generally one month – and a reference price set by the regulators. In both cases generators have a direct interest in maximising the value on the market of the electricity they produce and the volumes they can effectively sell. Since a generator can only commit on the market power it is confident it can produce, accurate forecasts are essential to maximising these sold volumes.

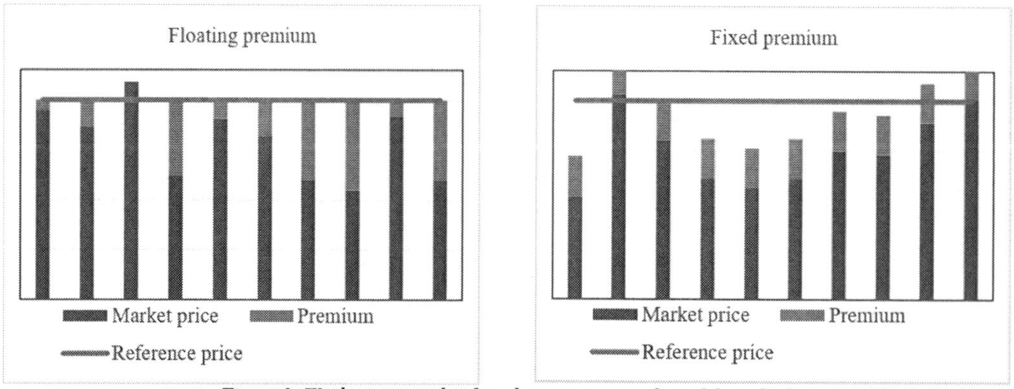

Figure 1: Working principle of market premiums; adapted from [15]

Finally, the development of micro-grids and of combined PV+storage systems require local energy management which, for optimal operations, relies on predictive control. Single-system or neighbourhood-level power forecasts on timescales from a few minutes to 24 h are therefore necessary.

These drivers and the dynamics of power system components described earlier together create a range of use cases for forecasts on time horizon ranging from 15 min or less to decades, and on geographical scales ranging from the single site to an entire region or country. These use cases are summarised on Table 2.

Table 2: Summary of use cases for PV power forecasting

Time horizon	Single site (10 m – 100 m) **PV plant owners** **PV plant operators**	MV distribution grid (1 km – 10 km) **DSOs** **Microgrid operators**	Transmission system (100 km – 1000 km) **TSOs** **Market operators**
15 min	Management of storage system	Management of active/reactive power	Activation of reserves
1 h	Management of storage system Intra-day trades	Storage and load management	Intra-day trades
24 h	Management of storage system Compliance with regulations Day-ahead trades	Storage and load planning	Booking of reserves Transmission scheduling Day-ahead trades
1 year	O&M contract	Planning of maintenance operations	Long-term trades
20+ years	Investment case	Infrastructure planning	Infrastructure planning

3 FORECASTING APPROACHES AND PERFORMANCE

3.1 Performance criteria

Because the use cases are so diverse, there is not a single metric which could characterise an absolutely "good" forecast. Instead, any of the three most commonly used metrics, which are listed in Table 3, can be preferred depending on the target application. These metrics are generally reported in a normalised way; particular attention must be paid to the normalisation factor and to the integration period. It is in particular good practice to integrate the error only over day hours, since PV production is sure to be zero in the night. And while errors in irradiance forecasts are generally normalised by the average measured irradiance, those on power forecasts are often normalised by the nominal peak power of the system. This difference mechanically results in errors for power generation which are about three times lower than for irradiance.

Table 3: Main performance metrics used to assess forecasting methods

Metric	**Formula**	**Application**
Mean bias error	$MBE = \frac{1}{N}\sum_{i=1}^{N}(Y_{forecast} - Y_{realised})$	Investment decision
Mean absolute error	$MAE = \frac{1}{N}\sum_{i=1}^{N}\lvert Y_{forecast} - Y_{realised}\rvert$	Balance group management
Root-mean-square error	$RMSE = \frac{1}{N}\sqrt{\sum_{i=1}^{N}(Y_{forecast} - Y_{realised}}$	Optimisation of generation reserves

3.2 Current forecasting techniques

The first approach in PV power forecasting relies first on the **prediction of relevant weather parameters** (at least temperature and irradiance), followed by a calculation of the corresponding power output. This approach can build on existing weather forecasting tools. The most appropriate tool to predict irradiance depends on the desired time horizon.

For **resource assessment** i.e. to predict patterns of energy generation over the lifetime of the system, statistically representative time series of weather parameters are generated based on interpolation of ground-level measurements (weather stations) or satellite images to produce "typical meteorological years"

For time horizons **between six hours and three days**, numerical weather prediction (NWP) is preferred. NWP data are generated by global or mesoscale simulation models which provide the numerical integration of the coupled differential equations describing the dynamics of the atmosphere and radiation transport mechanisms [16]. The initial conditions are given by satellite, radar, radiosonde and ground station measurements. NWP data are often corrected by post-processing algorithms called Model Output Statistics (MOS) which use historical ground measurements to partially remove systematic errors [17].

For time horizons **between two hours and six hours**, visible and/or infrared images are acquired by satellite-based sensors. A cloud index is computed based on reflectance measurements and is typically used to derive ground-level global and direct irradiances [18]. As compared to NWP, only a few relatively simple modelling assumptions have to be applied to derive the solar resource. Persistence of cloud speed and direction (as derived from the two last images) is generally assumed. The dynamic nature of clouds challenges cloud-motion vector approaches as cloud distribution can change substantially within the 30 min horizon which is the typical rate of image refresh. It is indeed challenging to account for cloud convection, formation, dissipation, and deformation. However, since large-scale cloud systems (such as those associated with a cold front) are more persistent, satellite-based forecasts typically perform more accurately than NWP-based forecasting models up to 6 h ahead, mostly because of ingestion, data assimilation, and latency of calculations required to "spin up" NWP-based forecasts. As classical satellite methods use only the visible channels (i.e., they work only in daytime), morning forecasts are less accurate than daytime ones because of a lack of time history; to overcome this issue, images from infrared channels (which work day and night) have to be taken into consideration [19].

For time horizons **below 30 minutes**, total sky imaging is the preferred method. It consists in four steps:

1. Acquisition of the sky image from a ground-based, wide-angle camera
2. Analysis of the sky image to identify clouds
3. Estimation of cloud motion vectors
4. Prediction of future cloud cover and ground irradiance.

The maximum accuracy with this method is generally obtained between 5 min and 20 min; with low and fast-moving clouds it can be reduced to 3 min and for high and slow-moving clouds it can be extended to 30 min.
The state-of-the-art accuracy for these methods is summarised on Figure 2.

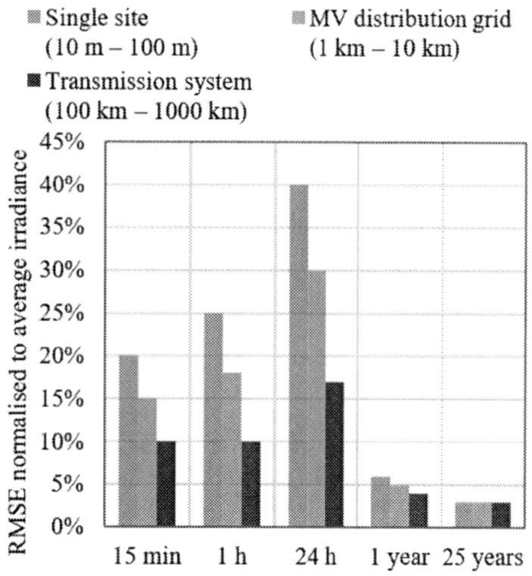

Figure 2: Error obtained with state-of-the-art physical forecasting methods for irradiance.

The uncertainty on PV power modelling from irradiance and weather data comes then on top. In a review of major modelling tools, the hourly RMSE on AC power output was found to be below 7% in all situations [20].
To avoid this addition of errors and to deal with time horizons between 30 min and 2 h where there is no satisfactory physical forecasting technique for irradiance, **stochastic learning techniques** are used. These methods can be separated between:

- **Univariate methods** i.e., methods where only time series of the target variable (here, PV power) are fed into the model. These include:
  - Persistence: $P(t+1) = P(t)$
  - STL: seasonal decomposition of time series by Loess
  - Holt-Winters seasonal method
  - TSLM: linear model fit with time series components
  - ARIMA: autoregressive integrated moving average
  - BATS: exponential smoothing state-space model with Box-Cox transformation, ARMA errors, Trend and Seasonal components
  - Nnetar: Feed-forward neural networks with a single hidden layer and lagged inputs for forecasting univariate time series
- **Multivariate methods** i.e., methods where exogenous variables such as measurements of ground irradiance, temperature or humidity levels are fed into the model in addition to the target variable. These include:
  - MLR: Multi-Linear Regression Model
  - SVM: Support Vector Machine
  - ANN: Artificial Neural Network
  - Regression Tree

3.3 Value of forecasting

To estimate the value of forecasting, and of improvement in forecasting techniques, the best analogy is the operations of balance groups, since for them forecasting errors have a well-defined cost. Indeed, European TSOs currently charge a typical imbalance price of 20 €/MWh. If a 1 MW_p plant in the North of Italy were a balance group on its own it would then be charged this price. The mean absolute error over four years for such a plant is 11.6% of nominal power with clear-sky persistence, and 7.1% with an advanced forecasting technique (numerical weather forecast plus support vector machine) [21]. Since only daytime is taken into account (12average), these errors translate into an annual imbalance of 0.50 MWh/kW_p and 0.31 MWh/kW_p, respectively. So the annual imbalance cost would be 10'000 € and 6'200 €, respectively. As a comparison, with power-purchase agreements at 80 €/MWh as are now contracted in Germany, annual income for this plant would be 80'000 €. So two conclusions can be drawn:

- Forecasting errors can reduce the value of PV electricity by more than 12%
- Advanced forecasting techniques can generate a value of almost 4'000 € per year for a 1 MW_p plant.

4 CONCLUSION

Accurate forecasting of PV power production has many use cases in both current power system operations and foreseen evolutions towards a more PV-centric system. Many of these cases require day-ahead forecasting, which is also the time horizon among those considered for which forecast errors are the largest. Research and development efforts should therefore focus on this horizon.
Other promising developments for using forecasts in power system operations include the communication of confidence intervals in addition to forecast values [22], and regional clustering to improve the accuracy of estimates of current power production and of forecasts.
Both physical and stochastic learning techniques are available to forecast PV power. Their choice mainly depends on the target time horizon and on the availability of sensors.
In a simple case, the lost value of PV electricity due to forecast errors can be estimated at more than 12% of annual revenues. Using advanced forecasting techniques can significantly reduce this loss and generate a value of almost 4'000 € per year for a 1 MW_p plant based on power system balancing only. In smaller, weaker power systems than those considered here this value would be even higher.

REFERENCES

[1] P.-J. Alet *et al.*, in *31st European PV Solar Energy Conference and Exhibition*, 2015, 2937.

[2] S. Pelland *et al.*, Photovoltaic and Solar Forecasting: State of the Art. International Energy Agency, 2013.

[3] E. Lorenz *et al.*, in *31st European Photovoltaic Solar Energy Conference and Exhibition*, 2015, 1524.

[4] Black & Veatch, Cost and performance data for power generation technologies. National Renewable Energy Laboratory, 2012.

[5] J. López, *Wärtsilä*,

[6] M. Fuchs, W. Timpf, *VGB Congress Power Plants 2011*, 2011.

[7] A. Grandjean, J. Adnot, G. Binet, *Renewable and Sustainable Energy Reviews*, 2012, **16**, 6539.

[8] T. Huld *et al.*, *Solar Energy*, 2010, **84**, 324.

[9] J. Marcos *et al.*, in *31st European Photovoltaic Solar Energy Conference and Exhibition*, 1549.

[10] J. Munkhammar *et al.*, in *31st European Photovoltaic Solar Energy Conference and Exhibition*, 1560.

[11] TenneT TSO GmbH, *TenneT TSO*, 2016.

[12] Autorità per l'energia elettrica il gas e il sistema idrico, Disposizioni in materia di dispacciamento delle fonti rinnovabili non programmabili a seguito della sentenza del Consiglio di Stato - Sezione Sesta - 9 giugno 2014, n. 2936. Autorità per l'energia elettrica il gas e il sistema idrico, 2014.

[13] Elexon, *BM Reports*,

[14] RTE, Historique des tendances et des prix de règlements des écarts validés. RTE Réseau de transport d'électricité, 2016.

[15] SDE TF, Key support elements of RES in Europe: moving towards market integration. Council of European Energy Regulators (CEER), 2016.

[16] R. Perez *et al.*, *Solar Energy*, 2013, **94**, 305.

[17] M. Pierro *et al.*, *Solar Energy*, 2015, **117**, 99.

[18] T. Cebecauer, M. Šúri, R. Perez, in 2010.

[19] R. Perez *et al.*, in *ASES Annual Conference*, 2010.

[20] J. Freeman *et al.*, in *2014 IEEE 40th Photovoltaic Specialist Conference (PVSC)*, 2014, 1932.

[21] M. Pierro *et al.*, *Solar Energy*, 2016, **134**, 132.

[22] X. Le Pivert, *Renewable Distributed Generation Forum*, 2016.

PV BIFACIAL YIELD SIMULATION WITH A VARIABLE ALBEDO MODEL

Matthieu Chiodetti[1], Amy Lindsay[1], Patrick Dupeyrat[1], Didier Binesti[1], Eric Lutun[2], Khalid Radouane[2], Sophie Mousel[2]
[1]EDF R&D – ENERBAT, France, [2]EDF EN, France
[1]Ecuelles - 77818 Moret sur Loing - France
[2]100 Esplanade du Général de Gaulle - 92932 Paris La Défense Cedex - France

ABSTRACT: Bifacial solar cells convert photons into electricity on both their front and rear sides, which makes the assessment of solar resource trickier than for classical solar cells that are active on their front side only. Among other parameters, **the albedo around a bifacial installation has a major impact on the irradiation** received on the rear side, and thus on the yield. For this reason, it is crucial to have an accurate characterization of the ground albedo for a bifacial system. While a constant and default albedo value is generally a reasonable approximation for the simulation of conventional PV, **a more accurate albedo model is required in the case of bifacial PV**.
In this paper, experiments and simulations on albedo that have been carried out at several of EDF R&D's research facilities are presented. It confirmed that **the ground albedo is not constant** as described in the majority of PV simulation systems, but varies throughout the days and the seasons.
A simplified **variable albedo model** is proposed and has been tested with measurements on several surface types. It can be calibrated with only a small set of data and is easy to use. It has been integrated to our bifacial PV model developed in a Dymola simulation environment and used to compute rear side irradiations and yield estimations.
Keywords: Bifacial, Modelling, Energy Performance

1 INTRODUCTION

Albedo can be defined as the reflective power of a surface. It is the ratio of reflected radiation from the surface to incident radiation received upon it. A great majority of photovoltaic simulation environments only provides the possibility of having a constant albedo – or sometimes depending on the season to take into account the snow-covered time periods. This is generally a reasonable approximation for the simulation of conventional low-tilt PV. On the other hand, measurements carried out at EDF R&D's test facilities have shown that the albedo of several surfaces varies throughout the days and seasons. This is confirmed by a great number of publications in the domain of climatology, recognizing that albedo is not an intrinsic property of a surface. It is actually quite complex to precisely describe the albedo as it depends on the spectral and angular distributions of the incident light, which are linked to the atmospheric composition and the position of the sun. [1-4]

While the approach considering a constant albedo seems to be sufficient in the majority of PV applications, there are some specific cases where a more accurate description of the albedo is of great importance. The simulation of the yield of tilted bifacial PV modules is one example as the energy generated by the rear side will be mainly due to the reflected irradiance. Bifacial simulations of a large scale plant have showed the great influence of albedo as the bifacial gain could vary from 5 to 15% for an albedo range of 0.2 to 0.4. [5]

For this reason, a simplified model for the albedo has been adapted from the literature to describe the variability of the reflective properties of the surface with a small set of data for calibration. It has been tested and calibrated with long-term measurements on several sites representing several types of ground. The albedo model was integrated into EDF R&D's internal bifacial model. The resulting irradiances calculated with the variable albedo model have been compared to the ones resulting from the constant albedo model.

2 SIGNIFICANCE OF THE ALBEDO FOR BIFACIAL PV

For a standard PV power plant, the ground-reflected irradiance linked to its albedo value has only a limited impact. However, albedo is of great importance for a bifacial PV plant. The relative contribution of the different components of the irradiances received on the front side and on the back side of a module is shown in Figure 1 and Figure 2.

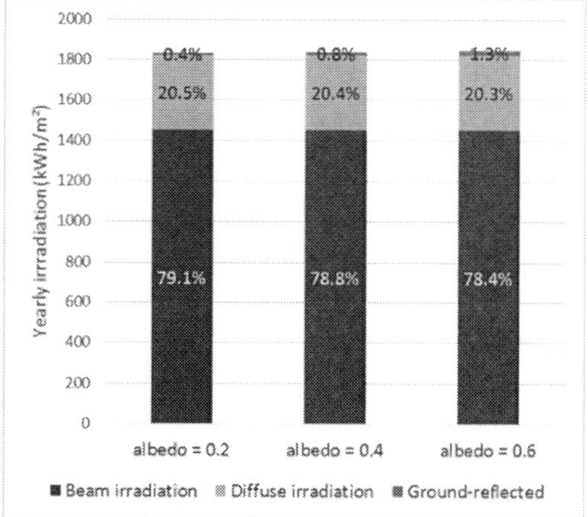

Figure 1: Relative contribution of the different components of light to the irradiance received on the front side of a PV module for different albedo values

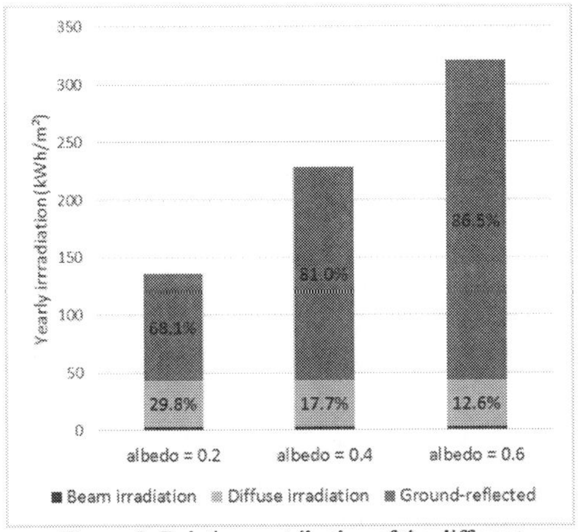

Figure 2: Relative contribution of the different components of light to the irradiance received on the rear side of a PV module for different albedo values

The reflected irradiation coming from the ground is the major contributor to the irradiation received by the back side of the bifacial module, whereas in a conventional photovoltaic plant the direct and diffuse components totally overrule the reflected irradiation.

The relative importance of the albedo on the rear side irradiances simulated above increases with the reflective power of the surface. This is consistent with the bifacial gain measured at EDF R&D's test facility, compared to monofacial modules of the same type (cf. Figure 3). [10]

The gain was measured on a daily basis for four ground surfaces (grey membrane, gravel, concrete and white membrane) over several weeks of measurements. The average gain for an isolated module varies between 12% and 49% (white membrane) while it reaches a maximum of 21% for a string of 8 bifacial modules (white membrane).

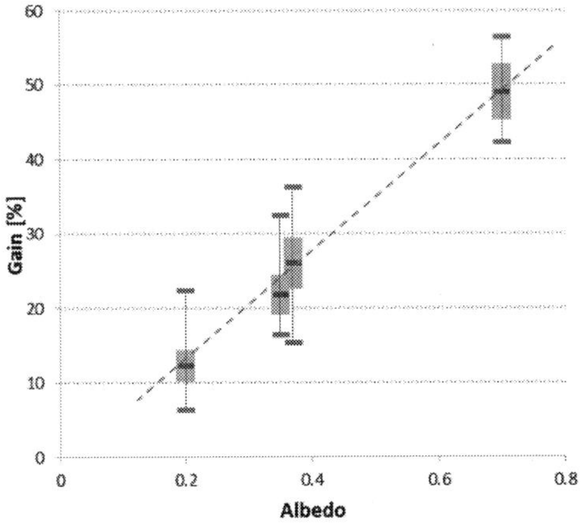

Figure 3: Measured gain on an isolated bifacial module for different surfaces at EDF R&D's test facility

Even at a larger scale, the albedo will be a key factor for the gain expected from a bifacial installation, even though the shadow of the whole installation cast on the ground impacts the irradiation received by the rear of the module. [5] For a bifacial PV array of 10x3 modules in the South of France with a 30° tilt, an elevation of 1m above the ground and a 7m spacing between the array before and behind, the bifacial model gave **a difference in yearly production of 4.9% between a "standard" albedo of 0.2 and an albedo of 0.4** (which can correspond to clear sand).

Other researchers have showed the importance of the albedo and have looked into its impact on bifacial PV [11-16] or on conventionnal PV [17-19].

For these reasons, an in-depth understanding of the albedo behaviour is required.

3 EXPERIMENTAL ALBEDO MEASUREMENTS

In several of EDF R&D's test facilities, albedo measurements have been conducted on various surfaces and for time periods up to one year.

3.1 Instrumentation

The albedo of the ground was monitored with an albedometer (cf. Figure 4) which is composed of two horizontal pyranometers, one facing the ground and the other one facing the sky. Long period measurements were made on concrete and on grass at two different latitudes.

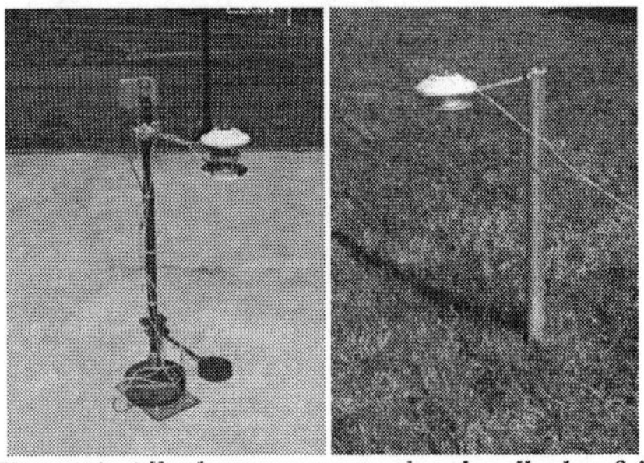

Figure 4: Albedometers measuring the albedo of the concrete and the grass at EDF test facilities

Literature gives typical constant albedo values of 0.2 to 0.5 for concrete and 0.15 to 0.25 for grass [20].

With an instantaneous measurement, a first approximation of the albedo was 0.37 for the concrete and 0.25 for the grass. However, it appears that the measured albedo differs significantly from a constant value over time, even if it stays around the values given above. Different types of variations are presented in the next parts.

3.2 Long term variations of the albedo

Over long time periods, albedo is subject to significant changes and it seems difficult to define a constant value for a one year-period. Over several months, the albedo of the grass differs depending on the height of the grass or its color (yellowing of the grass because of drought results in a lower albedo). This difference can be seen on the following picture for two identical days in terms of irradiance and season of the year, and results in an albedo value lower by 22% for the year that is most affected by the drought (cf. Figure 5 and Figure 6).

24/10/2014 21/10/2015

Figure 5: Two photos at a one year interval showing the change in appearance of the grass due to the drought

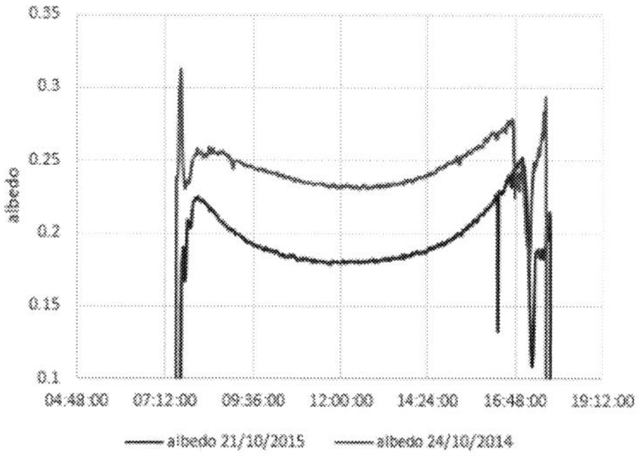

Figure 6: Albedo variation over a year corresponding to the two photos of Figure 5

Not taking into account such a behaviour for a grass covered ground could result in overestimation of the albedo during the summer and thus in the overestimation of the rear irradiances. EDF's internal bifacial model has been used to simulate the production of a bifacial plant (10x3 stand, 30° tilt, 7m spacing between rows, 0.4m elevation) in these two different albedo conditions. Simulations for September 2014 and September 2015 have shown a production drop of 2.5% while the albedo of grass varies from 0.25 to 0.17.

As expected, the rain can also strongly affect the albedo: 0.37 for dry concrete but only 0.2 for wet concrete (cf. Figure 7). A higher amount of energy is absorbed by the wet ground resulting in a lower reflective power. The same effect has been observed on grass with a drop from an averaged 0.25 albedo on a clear day and 0.20 the next day after rainfall. This means that bifacial simulations for climate subject to rainfall will tend to overestimate the production on rainy days (although the relative contribution of these days to the yearly yield will be low as it corresponds to low irradiance days). On the day given as an example, the overestimation would reach 9% in terms of production.

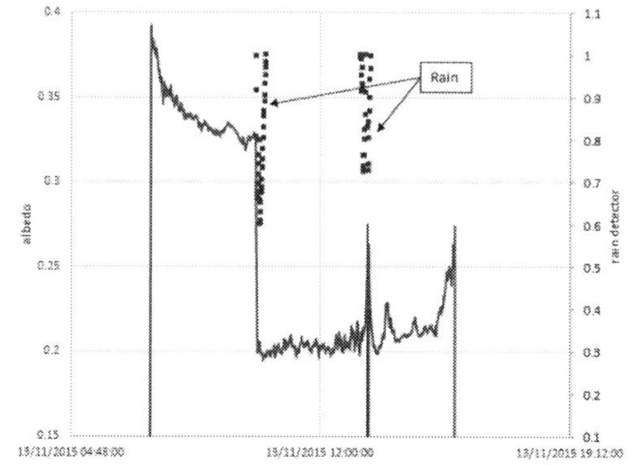

Figure 7: Measured impact of rain on the albedo of concrete. The rain detector is a device displaying a voltage – when rain occurs, the voltage drops from 1.2V to a value depending on the rain intensity

3.3 Diurnal variation of the albedo

On almost every day of measurement, it can be noted that the albedo presents a U-shape, both for concrete and grass. This is particularly noticeable on clear days, for which albedo is much higher in the morning and evening than at noon. This can be seen for example in Figure 8.

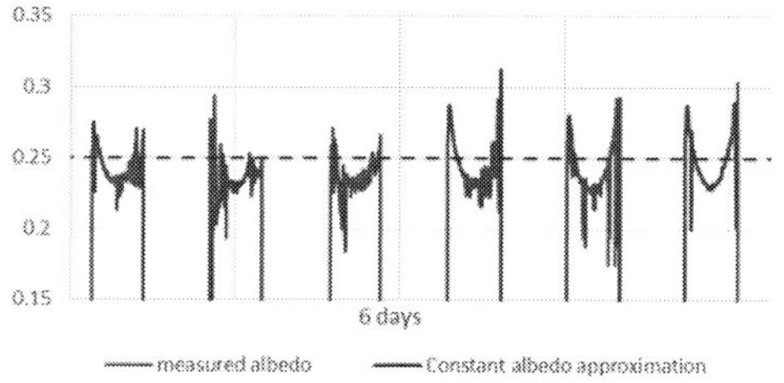

Figure 8: Diurnal variation of the albedo of grass compared to a constant value

On clear days, daily albedo variations can reach 25% for concrete and up to 60% for grass (between morning and noon). It has been noted that the U-shape of the albedo is deeper for the grass. This means that the incident angle of the light has a greater influence on the reflection of grass than that of concrete.

Over several days of measurements, it appears that these daily variations are also dependent on the sky conditions: for cloudy sky conditions, albedo show variations of a lower magnitude (the U-shape is less present for cloudy conditions) and is globally lower (cf. Figure 9 and Figure 10). Explanations of this phenomena can be found in part 4.

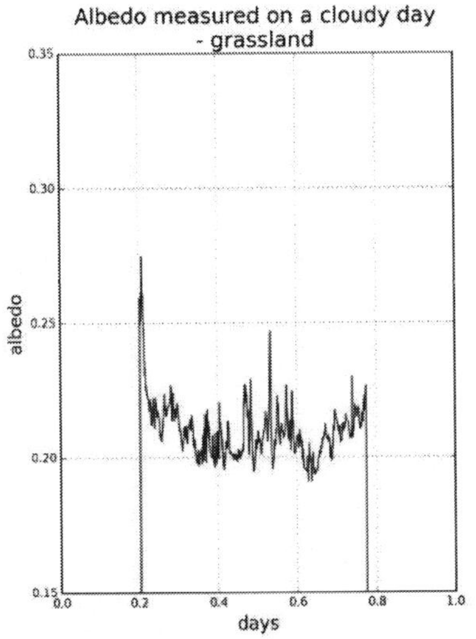

Figure 9: Albedo measurements of grass under diffuse illumination over a day.

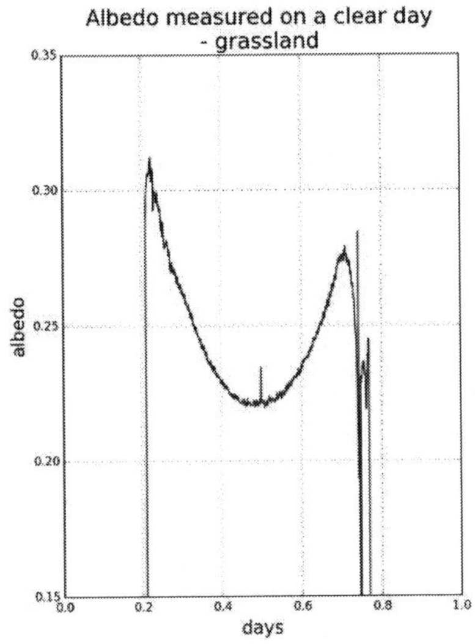

Figure 10: Albedo measurements of grass under direct illumination over a day.

Once more, these results show that a constant albedo model is an important simplification for bifacial PV, as the variations in albedo can be significant and can cause consequent production over or underestimations. These experimental observations call for a model describing the behaviour of the albedo.

4 PROPOSED MODEL FOR A VARIABLE ALBEDO

A literature study has shown that the albedo depends on – among other parameters – the solar incidence angle, the diffuse ratio of irradiation, and spectral response of the surface [4] [1]. These dependences are complex and are still the object of many researches. Different approaches for modelling the albedo are presented in the following parts.

4.1 Simplified approach

Obviously, the albedo is mainly dependent on the type of land-cover. The surface type can change considerably according to the location and throughout the season. This makes it difficult to develop a model which covers all possible cases. Dickinson described a scheme which he used for a semi-infinite forest-covered land: [5]

$$\alpha(\theta) = \alpha_0 \frac{1+C}{1+2\cdot C\cdot\cos(\theta)} \qquad \{1\}$$

Where Θ is the solar zenith angle, α_0 is the albedo at $\Theta=60°$ and C is a constant depending on the type of surface.

It has been improved by several authors including Hou *et al.* [8] or Wang *et al.* [1] who added a spectral dependence among others:

$$\alpha_{dir}^{m,n}(\theta,\lambda) = \alpha_{dir}^{m,n}(60°,\lambda)\frac{1+c^n}{1+2\cdot c^n\cdot\cos(\theta)} \qquad \{2\}$$

Where λ designates a spectral wavelength band and albedo is defined differently for each vegetation class (index n) and for each month (index m), c^n is a constant depending on the type of surface (0.4 is proposed by Briegleb for grassland [3]).

These equations only describes the albedo under a direct illumination. For diffuse illumination, the albedo is usually considered independent of the solar zenith angle.

4.2 BRDF approach

The approximation made in the approaches above is that the surface reflects isotropically, which means that the reflected intensity is independent of the view angle. No surface is truly isotropic although snow comes close to this hypothesis due to the multiple scattering of light in the elements constituting the snow. For typical surfaces, as the elements which produce the largest amount of scattered light are usually larger than the wavelength of light, the scattering is often anisotropic.

To precisely assess the reflectance, a bidirectional reflectance distribution function (BRDF) is necessary. The BRDF defines the irradiance reaching a point as a function of both the angle of incidence and the angle of viewing. It can be defined according to Rahman et al. by: [9]

$$\rho(\theta',\theta,\phi) = \rho_0 \frac{\cos^{k-1}\theta'\cos^{k-1}\theta}{(\cos\theta'+\cos\theta)^{1-k}} P(g,\Omega)\cdot(1+R(G,\Delta)) \qquad \{3\}$$

Where Θ' is the view zenith angle, Θ is the solar zenith angle and Φ is the relative azimuth angle, P is the Henyey-Greenstein phase function. ρ_0, k, g, Δ are parameters adjusted to fit the observed reflectances.

For more details about this function, please refer to references [4] [7].

4.3 Chosen variable albedo model

The choice was made to keep a simple approach which would not require an extensive set of data to calibrate the parameters of the albedo model. The proposed model is directly adapted from {1} and takes into account the ratio of diffuse light:

$$\alpha(\theta) = (1-\frac{DHI}{GHI})\cdot\alpha_0\frac{1+C}{1+2\cdot C\cdot\cos(\theta)} + \frac{DHI}{GHI}\cdot\alpha_{diff} \qquad \{4\}$$

Where Θ is the solar zenith angle, α_0 is the albedo under direct illumination at Θ=60°, C is a constant depending on the type of surface – varying between 0.1 (surface with a weak solar angle dependency) and 0.4 (surface with a strong solar angle dependency), α_{diff} is the albedo under a 100% diffuse illumination (assumed independent of the solar angle), DHI is the diffuse irradiance and GHI the global horizontal irradiance.

The three unknown parameters of the model (α_0, α_{diff}, C) can be found by fitting the model to on–site measurements. Only a few days or weeks of measurements are theoretically needed provided they include various irradiation conditions (clear sky, clouds…).

A parameter has also been included for the wetness of the ground which can lower considerably the albedo as seen in part 3.2.

5 EXPERIMENTAL VALIDATION

The albedo can be divided into two parts, a direct component which mainly depends on the solar position and a diffuse component which is supposed approximately constant (cf. Figure 11). It appears from measurements under 100% diffuse illumination that there is actually a dependency to soalr zenith angle, lower than for direct illumination, which has been taken into account.

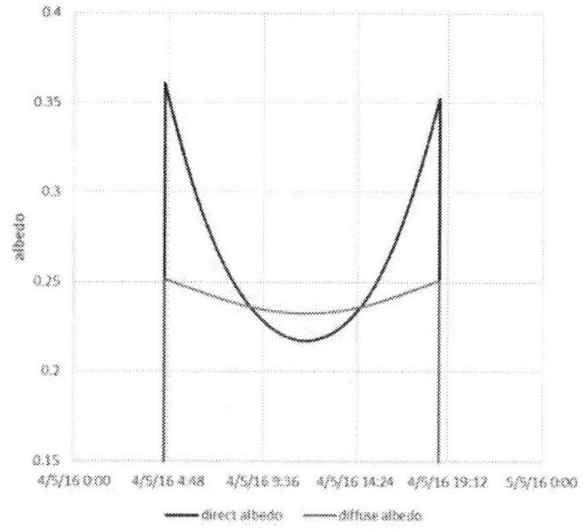

Figure 11: Separation of the albedo into two components for light concrete

The predictability of the model has been tested by comparing it to the measured albedo over 45 days and it resulted in a global Root Mean Square Error (RMSE) of 6.6% for the concrete, and preliminary results for grassland give a RMSE of 5.5%.

On Figure 12, results of the model on a characteristic week with clear days, cloudy days and rain are shown.

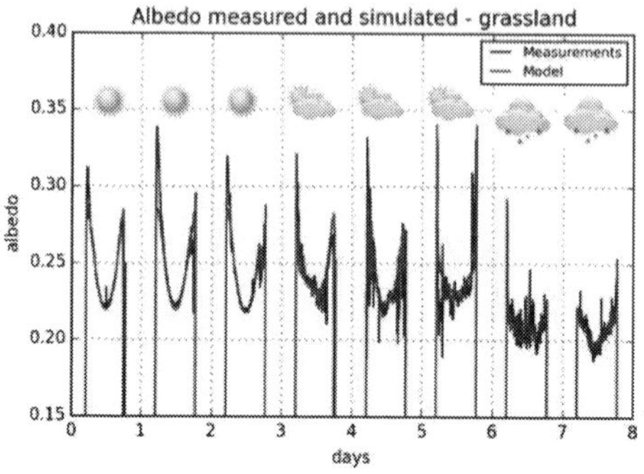

Figure 12: Comparison between model and albedo measurements for a characteristic set of days. The three first days are clear days, the last two are rainy days.

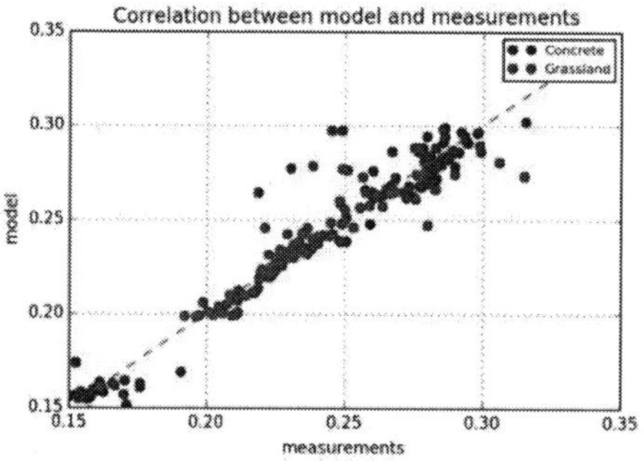

Figure 13: Correlation between model and measurements for concrete and grass.

6 ACCURACY OF THE REAR SIDE IRRADIATION CALCULATION FOR A BIFACIAL ARRAY WITH A VARIABLE ALBEDO

6.1 Experimental setup

The performance of the variable albedo model has been compared to that of the constant albedo model for the calculation of the irradiation received by the rear side of bifacial PV modules. The resulting irradiances have then been compared with the data acquired by the following experimental setup: six pyranometers have been positioned at the back of a standard photovoltaic stand thus mapping the irradiation received by the rear side of PV modules (cf. Figure 14 and Figure 15) [5].

Figure 14: Experimental setup: 6 pyranometers placed on the back side of a PV array to map the irradiance (3 on

the Eastern edge, 3 in the center).

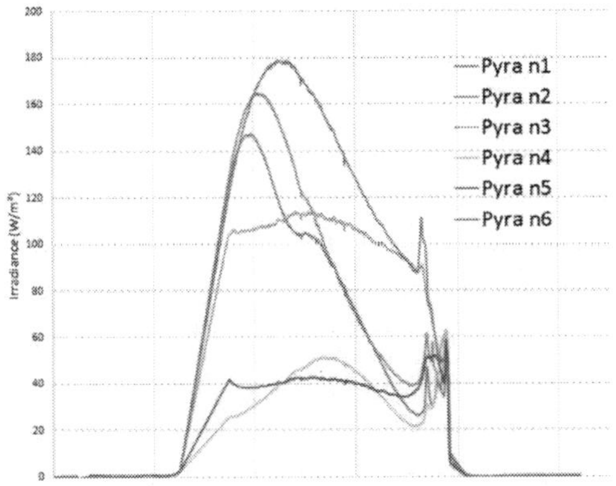

Figure 15: Experimental setup: irradiances measured by the 6 pyranometers from Figure 14 on a clear day.

As expected, the experiments show that the irradiances at the center are lower than at the edge of the array, and symmetrical during the day. On the other hand, the irradiance received by the modules at the edge drops dramatically during the day, as the shadow cast on the ground moves closer to these modules. Globally, the energy received is higher for the modules at the edge and at the top of the array [5].

6.2 Results

The variable albedo model described above has been used with EDF's bifacial model to compute the rear side irradiance of the monitored array.

From these experiments, it appears that the variable albedo model reduces the absolute RMSE of the computed irradiances, when compared to a constant albedo value. This is mainly due to the fact that a large part of the ground is actually shadowed, as mentioned above, and the "diffuse" albedo is usually lower than the average albedo used in a constant model (as observed for concrete and grass).

Over several months of data, the mean bias error is slightly positive – usually between 0 and +10 W/m² – and the RMSE reaches 11.2 W/m². These values are close to the uncertainties of our pyranometers (about 10 W/m²).The relative MBE (MBE*100/Avg) drops from +11.2% to +5.6% (cf. Figure 16 and Figure 17).

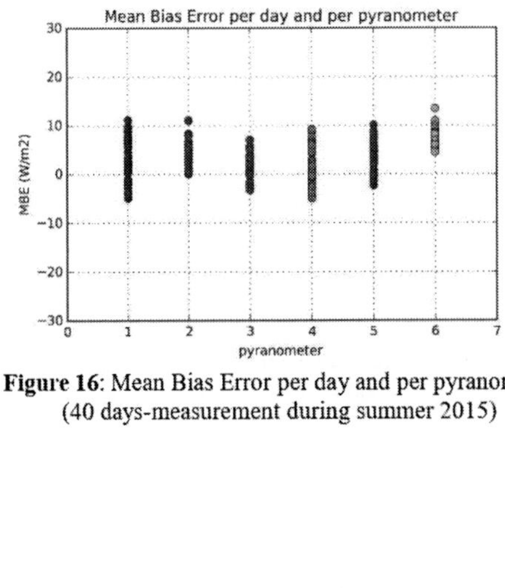

Figure 16: Mean Bias Error per day and per pyranometer (40 days-measurement during summer 2015)

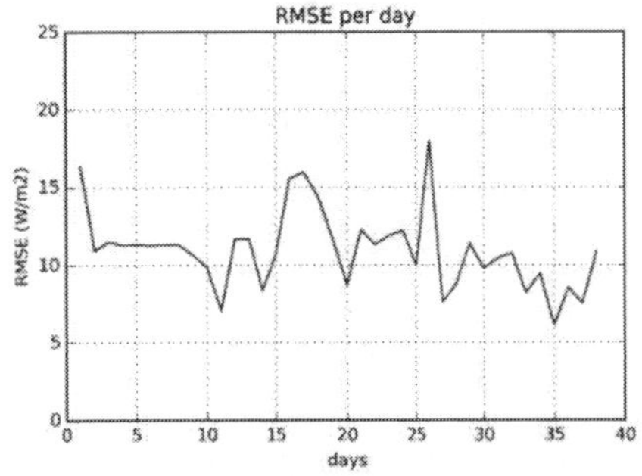

Figure 17: Root Mean Square Error per day for all the pyranometers (40 day-measurement during summer 2015)

6.3 Impact of a variable albedo on the production of a bifacial array

Using EDF's bifacial model, the production of a bifacial array has been simulated over a year, with a constant albedo and a variable albedo as defined in the previous parts.

The albedo chosen was that of a light-colored concrete, with a mean albedo under direct illumination of 0.37 which will be considered as the value of the constant albedo. The parameters of the model for the variable albedo were: α_0=0.37, α_{diff}=0.3, C=0.4. Impact of the rain was considered, by reducing the albedo by 45% (observed at EDF's test facilities, cf. Figure 7).

The simulations were located in:

- Marseille (South of France), annual irradiation ~1500 kWh/m², 81 days of rain
- Paris (North of France), annual irradiation ~1000 kWh/m², 170 days of rain

The array was considered between two other rows, with a 7m spacing, 30° tilt, 0.4m elevation, and constituted of 10x3 modules (3 strings of 10 modules, 7.8 kWp in total).

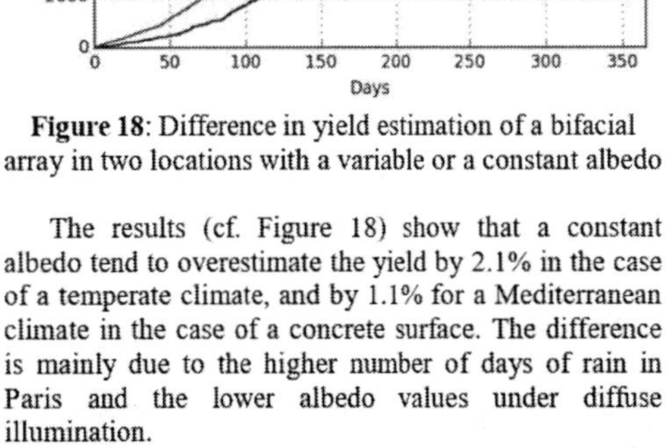

Figure 18: Difference in yield estimation of a bifacial array in two locations with a variable or a constant albedo

The results (cf. Figure 18) show that a constant albedo tend to overestimate the yield by 2.1% in the case of a temperate climate, and by 1.1% for a Mediterranean climate in the case of a concrete surface. The difference is mainly due to the higher number of days of rain in Paris and the lower albedo values under diffuse illumination.

By taking into account the dependency of the albedo to the solar zenith angle, the diffuse of ratio and the influence of rain, the variable albedo model allows a more accurate estimation of the yield of a bifacial PV plant. It is easy to implement and calibrate such a model, as its inputs (diffuse ratio, solar zenith angle) are already used in PV simulation models.

7 SUMMARY

For conventional PV yield estimations, ground-reflected light is of secondary importance which explains why albedo is usually taken constant.

However, albedo of a surface is not constant and can vary substantially during a given time period. These variations are due to the spectral and angular dependencies of the reflective properties of the surface, the change in appearance of a surface through a season, or the impact of rain – among other explanatory variables. These albedo variations have been observed and can reach in some cases 45% for a wet ground compared to a dry ground for instance.

Modelling the albedo is complex and is still the subject of on-going research, especially in climatology. The most accurate way of defining the scattering properties of a surface is by using a bidirectional reflectance distribution function. However, a simplified model of a variable albedo – with only a few calibration parameters and explanatory variables – achieves sufficient accuracy for irradiance and yield estimations for PV applications with high impact of the ground-reflected irradiation, such as bifacial PV or vertically mounted PV. These parameters can easily be determined by monitoring the albedo of a given surface during a few characteristic days.

In the particular case of bifacial PV, a simplified variable albedo has enabled to obtain a more accurate estimation of the rear side irradiations of a bifacial PV array with EDF's internal bifacial model [5]. It shows that modelling the albedo with a short time-step is pertinent for bifacial PV. Consequently, the accuracy of the yield estimation in the case of bifacial PV has been improved.

The impact on the yield estimation can be important: in a case with a high albedo strongly dependent to the effect of rain, a constant albedo can overestimate the yield by 2.1% when compared to a variable albedo.

8 REFERENCES

[1] Z. Wang, X. Zeng, M. Barlage, R. Dicksinson, F. Gao and C. Schaaf, «Using MODIS BRDF/albedo data to evaluate global model land surface albedo,» *Hydrometeor,* 2004.

[2] F. Yang, K. Mitchell, Y.-T. Hou, Y. Dai, X. Zeng, Z. Wang and X.-Z. Liang, «Dependence of Land Surface Albedo on Solar Zenith Angle: Observations and Model Parameterization,» *American Meteorological Society,* 2008.

[3] Briegleb, «Delta-Eddington approximation for solar radiation in the NCAR community Climate Model,» *Geophys. Res.,* 1992.

[4] J. Coakley, «Reflectance and albedo, surface,» 2003.

[5] A. Lindsay, M. Chiodetti, P. Dupeyrat, D. Binesti, E. Lutun and K. Radouane, «Key elements in the design of bifacial PV power plants,» at *EU PVSEC,* 2015.

[6] R. Andrews, A. Pollard and J. Pearce, «Photovoltaic system performance enhancement with non-tracking planar concentrators: experimental results and bi-directionnal reflectance function (BDRF) based modelling,» 2013.

[7] Y. T. Hou, S.Moorthi and K. Campana, «Parametrization of solar radiation transfer in the NCEP models,» *NCEP Office Note 441,* 2002.

[8] H. Rahman, B. Pinty and M. Verstraete, «Coupled surface-atmosphere reflectance (CSAR) model 2. Semi-empirical surface model usable with NOAA Advanced Very High Resolution Radiometer dat,» *Journal of Geophysical Research 98,* 1993.

[9] R. Dicksinson, «Land surface processes and climate-surface albedos and energy balance,» *Advance in Geophysics,* 1983.

[10] P. Dupeyrat, C. Lucas, A. Lindsay, A. Plotton and K. Radouane, «Investigations on Albedo Dependency of Bifacial PV Yield,» at *EU PVSEC,* 2014.

[11] U. A. Yusufoglu, T. Pletzer, L. J. Koduvelikulathu and C. Comparotto, «Analysis of the Annual Performance of Bifacial Modules and Optimization Methods,» *IEEE Journal of Photovoltaics,* vol. 5, pp. 320 - 328 , 2015.

[12] U. A. Yusufoglu, T. H. Lee, T. M. Pletzer, L. J. K. Andreas Halm b, C. Comparotto, R. Kopecek and H. Kurz, «Simulation of Energy Production by Bifacial Modules with Revision of Ground Reflection,» at *4th International Conference on Crystalline Silicon Photovoltaics,* 2014.

[13] C. Reise and A. Schmid, «Realistic Yield Expectations for bifacial PV Systems - an Assessment of Announced, Predicted and Observed Benefits,» at *PV Performance Modelling and Monitoring Workshop,* Cologne, 2015.

[14] L. Kreinin, N. Bordin, A. Karsenty and A. Drori, «Outdoor evaluation of power output improvement of the bifacial module,» at *Photovoltaic Specialists Conference (PVSC),* Seattle, 2011.

[15] B. B. V. Aken, M. J. Jansen, A. J. Carr, G. Jansen and A. A. Mewe, «Relation between indoor flash testing and outdoor performance of bifacial modules,» at *29th European Photovoltaic Solar Energy Conference and Exhibition,* Amsterdam, 2014.

[16] A. Luque and E. Pigueiras, «Diffusing Reflectors for bifacial Photovoltaic Panels,» *Solar Cells,* 1985.

[17] A. L. A. M. P. Brennan, R. Andrews and J. M. Pearce, «Effects of spectral albedo on solar photovoltaic devices,» *Solar Energy Maetrials and Solar Cells,* vol. 124, pp. 111-116, 2014.

[18] R. W. Andrews and J. M. Pearce, «The effect of spectral albedo on amorphous silicon and crystalline silicon solar photovoltaic device performance,» *Solar Energy,* vol. 91, pp. 233-241, 2013.

[19] P. Ineichen and R. Perez, «Ground-reflected radiation and albedo,» *Solar Energy,* 1990.

[20] T. Markvart and L. Castañer, Practical Handbook of Photovoltaïcs : Fundemantals and Applications., 2003.

HIGH RESOLUTION SOLAR RADIATION DATABASE. SOLAR ATLAS FOR SOUTH AFRICA

Ana Gracia Amillo[1], Lucky Ntsangwane[2], Thomas Huld[1], Jörg Trentmann[3]
[1] European Commission, Joint Research Centre, Institute for Energy and Transport. Renewables and Energy Efficiency, TP 450, 21027 Ispra (VA) Italy
[1] Ana Gracia Amillo, Phone +39 0332 783989, Fax +39 0332 789268, ana.gracia-amillo@ec.europa.eu
[1] Thomas Huld, thomas.huld@jrc.ec.europa.eu
[2] South African Weather Service, 442 Rigel Ave S, Erasmusrand, 0181 Pretoria, South Africa
[2] Lucky Ntsangwane, lucky.ntsangwane@weathersa.co.za
[3] Deutscher Wetterdienst, 135 Frankfurter Strase, 63067, Offenbach, Germany
[3] Jörg Trentmann, joerg.trentmann@dwd.de

ABSTRACT: Solar radiation data can be estimated at high spatial resolution over large geographical areas using satellite images. In this paper, two different satellite based methodologies have been applied to calculate global horizontal and beam normal irradiance values for South Africa. The accuracy of the estimates has been analyzed by comparing the modelled values with ground measurements recorded at thirteen stations distributed over the country. The best method results in an overall relative mean bias of 1.64% for the global horizontal irradiance and 2.93% for the beam normal irradiance estimation. This method has been implemented into an open GIS software that allows users to obtain for free, estimates of the available solar resource over large regions of their choice. In addition to this, the implemented methodology outputs the photovoltaic production of various PV technologies and different settings from fixed modules to sun tracking systems. The said estimation model takes into account the effect of the module's temperature, the cooling effect of wind, the angle of incidence to the module's surface and the spectral distribution of the received solar radiation. Furthermore, the GIS module contains high resolution terrain data that allows considering the effect of the surrounding topography on the above mentioned calculations.
Keywords: modelling, photovoltaic, software, pyranometer, solar radiation

1 INTRODUCTION

Information about the solar resource available at any location can be obtained from various means. The most accurate description is obtained from on-site measurements, as long as the quality of the data and maintenance of the instruments used in the measuring campaign is guaranteed. However, this option is not always feasible, due to the time and dedicated resources required to obtain representative and high quality datasets. If data from nearby stations is available, a spatial interpolation could be used. Notwithstanding, this method is limited by the distance between the stations and the location under analysis which should be less than 35 km when working with hourly values, and 50 km when daily values are considered [1]. When distance exceeds these thresholds, other techniques report better estimates of solar radiation. For instance, there are several empirical models that estimate solar radiation using information of other meteorological variables like temperature or sunshine duration [2]. These models' estimates could be improved using geostatistical interpolation and soft-computing techniques [3]. When ground measurements at nearby locations are not accurate enough, solar radiation data could be obtained from reanalysis and retrospective weather prediction models which are based on long time series of climatic parameters data. The ERA-Interim reanalysis [4] is a good example and contains solar radiation data since 1979 to near real time. In recent years, satellite based models have become a very common and powerful tool for estimating the solar radiation data thanks to the satellites' high spatial and temporal resolution. The EUMETSAT's Climate Monitoring Satellite Application Facility (CM-SAF) provides continuous and validated solar radiation data from first and second generation geostationary satellites since 1983 to present. Satellite retrieved solar radiation data have been extensively validated in the scientific literature, see for instance [5, 6]. Some of the satellite based methods perform better than some of the previously mentioned techniques, like interpolation or reanalysis, especially in locations where there are no nearby measurements available.

In this paper, two of these satellite based models have been applied to estimate the solar resource available in South Africa. In this location, due to its large extension and low number of stations with ground measurements, the satellite retrieved estimates become the best option to obtain solar irradiance data.

2 SATELLITE BASED RADIATION DATA

Two different satellite based methodologies have been considered, which will be named hereafter Operational and SARAH. Both were developed by the CM SAF Collaboration (www.cmsaf.eu), a collaboration between a number of European meteorological services led by Deutscher Wetterdienst (DWD).

2.1 Operational dataset

The Operational dataset is derived from measurements of the SEVIRI instruments onboard the Second Generation of Meteosat satellites (MSG) using a look-up table (LUT) retrieval algorithm as described in [7]. The Operational dataset contains estimates of global horizontal (GHI) and beam normal (DNI) irradiance. The spatial resolution of the data set is the original satellite resolution but for this study the data have been reprojected to a latitude-longitude grid with a resolution of 1'30''.

2.2 SARAH dataset

The SARAH data is obtained by applying the SPECMAGIC algorithm [8], which in addition to the broadband irradiance, also provides the global and beam irradiance distributed in 30 spectral bands in the range from 330 nm to 4600 nm. The spatial resolution is 3', lower than the Operational data record.

For the present study satellite images from the MSG have been used. Even though, there are various images per hour, only one image has been treated every hour to obtain one hourly value of broadband global horizontal and beam normal irradiance from both datasets, Operational and SARAH.

3 GROUND MEASURED DATA

In order to validate the accuracy of the solar radiation data retrieved from the two satellite based methods described in Section 2, ground measurements from thirteen stations located in South Africa have been used. The stations are distributed evenly at the varying climatic zones of the country over the region bounded between latitudes 23° to 34° south and longitude from 18° to 31° east. Their elevation ranges from 190 m to almost 1700 m. They are located in six different climatic regions as described in Table I.

Table I: Climatic regions of the South Africa stations

Climatic region	Stations
Cold interior	Bethlehem, De Aar
Temperate interior	Mahikeng, Irene
Hot interior	Thohoyandou, Polokwane, Nelspruit
Arid interior	Upington, Prieska
Sub-tropical coastal	Umthatha, Durban
Temperate coastal	Cape Point, Prieska

Most stations were installed during 2014. However, some, like Durban or Polokwane, started registering data only in 2015. Therefore, the length of the measured time series depends on the station, ranging from 23 months available in Upington to 11 months with recorded data in Durban.

The solar radiation network stations are equipped with Kipp & Zonen instrumentation. Common at all stations is one SOLYS 2 sun tracker with shading ball assembly and one first class ISO - 9060 classified CHP1 pyrheliometer that measures beam normal irradiance. Except for De Aar BSRN station, all other 12 stations are equipped with two high performance research grade pyranometers (CMP11) for measurements of the global (unshaded) and the diffuse horizontal irradiance (shaded). The CMP11 are fully compliant with all ISO 9060:1990 secondary standard instrument performance criteria. As a BSRN station, De Aar employs two ventilated CMP21 which are considered to be in the highest possible ISO pyranometer performance category. The ventilation units help keeping the pyranometer's domes clean from dust and water. Some of the stations, in addition, measure ultraviolet radiation-A (UVA) and ultraviolet radiation-B (UVB). Moreover, all stations are equipped with an Automatic Weather Station (AWS) that registers basic weather parameters.

The measured database contains 1-minute values of global and diffuse irradiance on the horizontal plane and beam normal irradiance. In contrast, the satellite retrieved datasets contain one value per hour of global horizontal and beam normal irradiance, corresponding to the single image analyzed every hour.

3.1 Measured irradiance values used in the validation

Due to the different time stamp between measured and satellite datasets, before comparing estimated and measured irradiance values, the measured database has to be slightly modified.

In order to compare the hourly satellite values to the closest in time measured data, for every satellite derived data, a ten minute window centered at the time of the image has been defined. The measured values inside this interval will be averaged to obtain the final "measured" value to be used in the validation process. Since Operational and SARAH datasets derive from different images, the measured data used in the validation of each of the satellite data records is different. As a result, the number of data points considered in the validation may be different. Also, the number of data points used for the validation of the global horizontal and beam normal irradiance may be different. There may be moments when the sun tracking system where pyranometers and pyrheliometers are located did not work properly causing wrong beam irradiance measurements, even though the global irradiance measurements are still valid.

In addition to the quality control process applied to the data at the measuring stations, before doing the present analysis, an additional simple filtering procedure has been applied as described below.

- Negative values at night time are replaced by 0 and kept for the validation. However, negative values during day time are removed.
- Irradiance values higher than 1300 Wm^{-2} are removed.
- The coherence between the three irradiance values is checked in order to remove measurement suspicious of coming from moments when the sun tracking system was not working properly.

If less than 5 valid minute data points remain after applying these filters, the ten minute window is removed entirely and the corresponding satellite data for that hour is dismissed. If there are enough valid minute data, the average value is calculated and kept as the "measured" value to be used to validate the satellite derived values.

4 RESULTS

Estimated and measured irradiance values of GHI and DNI have been compared in every station for every year and month independently. The Mean Bias Difference (MBD) and Mean Absolute Difference (MAD) in absolute (Wm^{-2}) and relative (%) values have been calculated. However, due to the large amount of resulting values, only a summary and the most significant results will be presented in this paper. The results as a whole can be obtained from the authors.

Considering all the stations, the SARAH data record performs better for both the GHI and the DNI estimation, with an overall relative MBD for all the stations (grMBD) of 1.64% for the GHI and 2.93% for DNI. The Operational data record results in rgMBD values of 5.18% and 13.86% for GHI and DNI respectively. Both satellite datasets present higher relative deviations in the estimation of the beam component even though the average measured DNI values are higher than the GHI values.

If the absolute difference is considered, the

differences between SARAH and Operational are not as significant as with the rMBD values. For the GHI validation, the grMAD for SARAH is 11.47%, while the Operational dataset results in 11.49%. Differences are slightly higher for the DNI, with 21.20% for SARAH and 23.52% for the Operational data record. In fact, for most stations the rMAD for the GHI validation is lower in the Operational than in the SARAH dataset. This shows that the errors of the GHI estimates compensate (overestimation and underestimation) better in the SARAH dataset, even though the absolute value of the differences is higher than those from the Operational data record.

For those stations with registered data during both 2014 and 2015, the errors barely change although a slight increase in the errors is observed in 2015. The worst case occurs in the Operational DNI estimates where overall relative MBD value for all the stations (grMBD) goes from 9.50% in 2014 to 14.19% in 2015, and the grMAD increases from 18.56% to 23.99%. For the SARAH validation and the GHI analysis, the difference between errors in 2014 and 2015 are smaller.

When monthly results are considered, a tendency in all the stations can be observed when the Operational dataset present an increase in the rMBD values for both GHI and DNI estimates between August and October 2015, which could indicate some kind of problem with the satellite data. As an example, we present two figures in which the monthly relative MBD values for the GHI and DNI validation are shown. Figure 1 shows the results for Upington, while Figure 2 contains the results for Irene. Both stations have data during 2014 and 2015. Each figure contains two figures, the upper one with the results of the GHI validation and the lower figure with the DNI validation results. Every figure shows four lines, two for the Operational data record (OP) and other two for the results using the SARAH data record (SA). Each of these two sets of lines corresponds to every year available (2014 and 2015) as described in the legend. The values that accompany the legend are the yearly rMBD values, in %.

Deviations tend to be lower during autumn and winter months (April - July) as can be observed in the two figures shown as example.

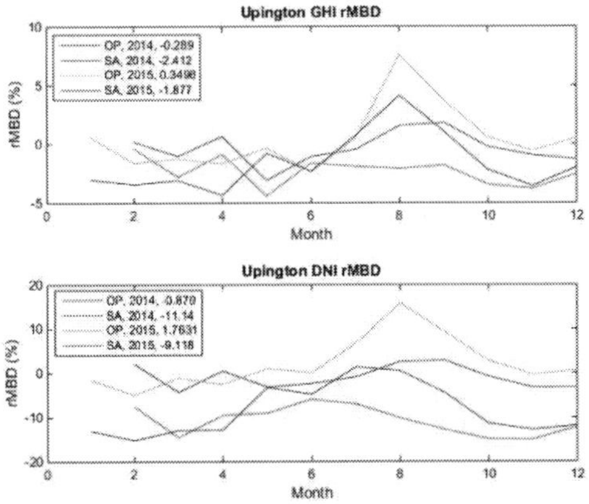

Figure 1: Monthly rMBD values for the validation of the GHI and DNI estimates for Upington

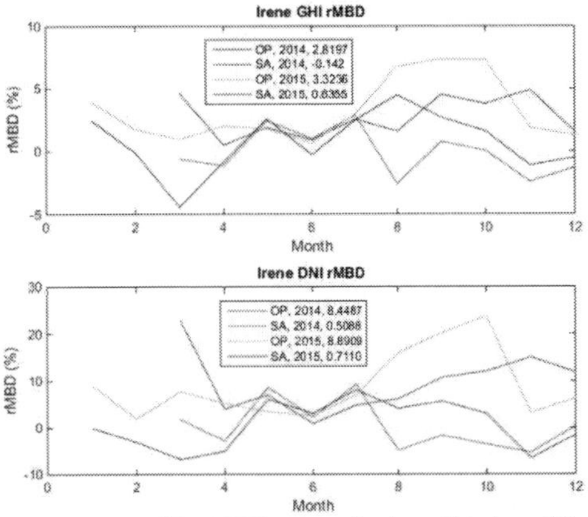

Figure 2: Monthly rMBD values for the validation of the GHI and DNI estimates for Irene

The deviations observed in Irene (Figure 2) for the months between May and July, are very similar for both years and satellite datasets. This is not the case for the rest of locations, where in general the Operational monthly rMBD are higher than the SARAH ones.

4.1 Global horizontal irradiance validation

Table II contains the relative MBD and MAD values for every station, as well as the average measured irradiance and number of data points considered in the validation of the GHI values retrieved from the SARAH dataset.

Table III contains the corresponding values for the validation of the GHI estimates from the Operational dataset.

GHI is overestimated in all stations by the Operational records, while SARAH data underestimates the global irradiance measured in Prieska, Upington and De Aar. rMBD values from the Operational data tend to be twice or even higher than those from SARAH dataset. This tendency is not observed in the rMAD values. In fact, the absolute deviations obtained from the Operational dataset are in general lower.

Table II: Results from the validation of the GHI values retrieved from the SARAH dataset

GHI SA Station	rMBD (%)	rMAD (%)	Av Meas (Wm^{-2})	N
Prieska	-0.06	8.13	237.02	14061
Upington	-2.12	7.79	256.63	15595
Irene	0.30	11.07	231.19	15675
De Aar	-0.29	9.46	248.06	13778
Nelspruit	3.71	13.17	204.21	14091
Umthatha	1.71	13.22	196.91	12531
Mahikeng	0.67	10.58	255.98	12336
Cape Point	1.33	11.66	215.89	8889
George	3.52	14.69	192.88	8635
Bethlehem	0.50	10.43	240.56	8580
Thohoyandou	4.99	12.32	208.59	7260
Polokwane	0.68	11.10	233.94	7018
Durban	6.37	15.49	177.89	7317

For both satellite datasets, both MBD and MAD, tend to be higher in the stations with lower elevation such as

Durban (105m), George (197m) and Cape Point (228m). It is also true that the length of the time series available in these locations is shorter than in other stations, so it would be interesting to analyse more data from these locations. Although the higher deviations observed in these stations could indicate a possible underestimation of the attenuation suffered by solar radiation in the atmosphere. This is more significant in the Operational dataset.

Table III: Results from the validation of the GHI values retrieved from the Operational dataset

GHI OP Station	rMBD (%)	rMAD (%)	Av Meas (Wm^{-2})	N
Prieska	2.50	7.39	232.79	14012
Upington	0.05	6.17	251.08	15502
Irene	3.10	10.34	228.15	15553
De Aar	0.92	7.71	244.05	13653
Nelspruit	5.95	12.73	202.25	14005
Umthatha	7.22	13.95	193.09	12534
Mahikeng	2.93	10.23	250.59	12237
Cape Point	5.80	10.50	210.17	8731
George	8.65	16.93	188.15	8481
Bethlehem	3.43	10.94	234.86	8479
Thohoyandou	6.71	12.02	207.21	7071
Polokwane	1.87	9.54	230.53	6808
Durban	18.24	20.89	173.60	7250

Although not shown in Tables II and III, the absolute MBD values are below the threshold of 15 Wm^{-2} for the SARAH dataset. In fact, in eight of the thirteen stations analysed, the MBD value is lower than 5 Wm^{-2}. As for the Operational values, only in Durban and George, the 15 Wm^{-2} threshold is not respected.

4.2 Beam normal irradiance validation

Tables IV and V contain respectively the rMBD and rMAD values observed from the validation of the DNI values retrieved from the SARAH and Operational datasets. The tables also contain the average beam irradiance value registered at every station and the number of data points considered at every location.

Table IV: Results from the validation of the DNI values retrieved from the SARAH dataset

DNI SA Station	rMBD (%)	rMAD (%)	Av Meas (Wm^{-2})	N
Prieska	-5.82	17.34	315.30	13794
Upington	-10.07	19.63	343.81	15484
Irene	0.62	19.37	258.73	15566
De Aar	-6.29	20.36	328.07	13667
Nelspruit	10.64	22.52	202.28	13580
Umthatha	2.06	22.28	202.28	12353
Mahikeng	0.54	19.78	275.66	11400
Cape Point	10.91	25.23	217.05	8745
George	8.06	23.09	190.17	8000
Bethlehem	2.88	20.90	232.55	7651
Thohoyandou	8.21	20.56	219.00	7178
Polokwane	3.70	19.36	256.94	6865
Durban	12.67	25.22	103.06	5242

For the beam component estimation, both rMBD and rMAD values are lower for the SARAH data than for the Operational, with the exception of the estimates for Upington and De Aar, where the Operational data performs better.

The differences between the rMBD values derived from the SARAH validation are lower than the Operational ones to a higher extent than what was observed in the validation of the global irradiance values.

Both satellite data records tend to overestimate the beam component, except for the SARAH estimates for Prieska and De Aar, where this dataset also underestimated the global irradiance values.

Table V: Results from the validation of the DNI values retrieved from the Operational dataset

DNI OP Station	rMBD (%)	rMAD (%)	Av Meas (Wm^{-2})	N
Prieska	5.87	14.74	310.88	13736
Upington	0.49	12.85	338.25	15373
Irene	8.69	20.36	257.02	15434
De Aar	1.42	15.40	325.04	13534
Nelspruit	15.85	24.44	203.13	13477
Umthatha	16.99	27.17	200.65	12359
Mahikeng	9.22	20.25	271.52	11308
Cape Point	27.02	31.85	211.18	8585
George	29.24	35.29	186.41	7852
Bethlehem	10.06	24.16	230.28	7592
Thohoyandou	14.38	22.32	220.88	6976
Polokwane	7.52	19.50	256.93	6652
Durban	33.43	37.44	100.84	5155

The largest deviations, both as MBD and MAD values, are also observed in the stations mentioned in section 4.1 located at low altitude. These stations also have the lowest average DNI values of the set of stations considered. It should be pointed out here that in Durban there are only data available from February 2015, meaning almost no data from summer months are included.

If we consider the absolute MBD values, for the SARAH dataset, only in six stations the deviations are lower than 15 Wm^{-2}. For the Operational estimates, only in Upington and De Aar, the absolute deviation is below that value. In fact in these two stations, the Operational data record performs better than the SARAH dataset in the estimation of the beam component. Stations where SARAH estimates underestimate the solar resource, in opposition to the overestimation observed in the other stations. The change in the trend in these two stations could indicate that either both satellite methods retrieve significantly lower DNI values than in the other stations, or that the measurements for whatever reason are higher than in the other locations. Since the overestimation shown by the Operational dataset is much lower in these two stations. And the SARAH tendency to overestimate DNI recordings is not followed in these two stations.

5 PVMAPS: software and data for the estimation of solar radiation and PV performance

PVMAPS is a free software recently developed in the Joint Research Centre of the European Commission [9]. In addition to the set of calculation tools, it also provides datasets of climatic variables and topographic information.

Users can obtain long-term monthly average values of global horizontal and beam horizontal irradiation data estimated by the two satellite based methodologies discussed in the present paper. The solar radiation data have been calculated for Europe, Africa, most parts of Asia and the eastern part of South America. Temperature data is also provided.

The software also provides estimates of the PV performance of various technologies applying the methods described in [10]. The user can define different settings, from fixed systems to sun tracking options, even though the software can define the module's best inclination and orientation for every location. The models used in the estimation of the PV output can take into account the cooling effect of the wind on the modules' temperature and the effect of the angle of incidence of solar radiation with regard to the modules' surface. Since the satellite models provide information about the spectral distribution of solar irradiance, the spectral response of the different PV technologies is also considered when estimating the PV output. The real spectral distribution of irradiance, in opposition to the assumed standard, can affect the final PV production of the modules [11, 12]. Besides, a high resolution terrain data, allows considering the surrounding topography in the calculations. The terrain data contain not just the elevation but also the horizon height in 48 directions around each point in the database. The spatial resolution of the terrain data is 3 arc-seconds (about 90m) which makes it possible to produce very high resolution maps.

All the estimation models and datasets are integrated into GRASS GIS modules, so users can easily perform the calculations of solar radiation and PV production for any location.

5.2 PVMAPS applied to South Africa

The estimation models integrated in PVMAPS have been applied to South Africa. Based on the validation results we decided to use the SARAH solar radiation data set for the map calculations within PVMAPS. As an example, we present here the results for the area shown in Figure 3, bounded by 27° 30' S, 30° S, 27° 30' E, 30° E.

Figure 3: Topographical map of the area bounded by 27° 30' S, 30° S, 27° 30' E, 30° E.

One of the possible outputs of PVMAPS is shown in Figure 4, which presents a comparison of the results from the said region of South Africa without (a) and with (b) the terrain effects. The effect of shadows is clearly visible in mountainous areas.

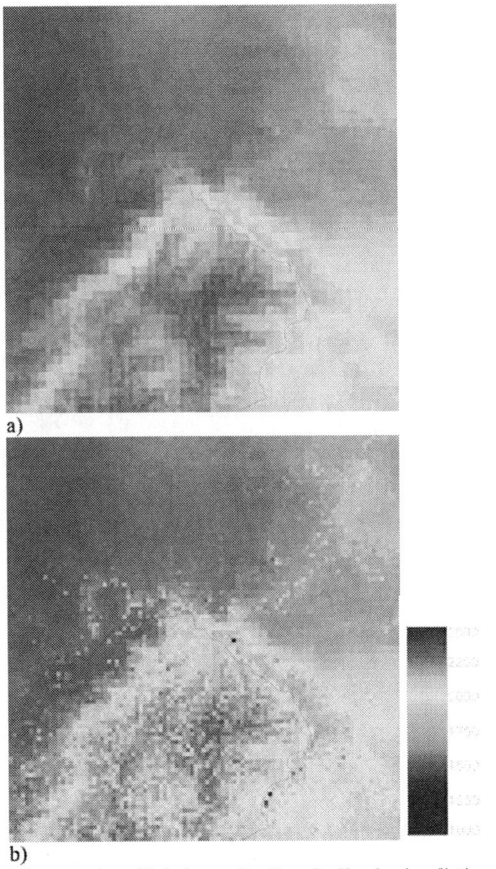

Figure 4: Global optimally inclined irradiation (kWh/m^2/year) for the area bounded by 27° 30' S, 30° S, 27° 30' E, 30° E, without (a) and with (b) terrain effects.

6 CONCLUSION

We have presented here the performance of two different satellite based estimation models to retrieve solar irradiance data over large regions, in this case South Africa. The estimates contain data of the global horizontal and beam normal irradiance, providing a complete description of the solar resource available at any location. One of the methodologies can also provide the irradiance values distributed in spectral bands.

The best estimation model for South Africa provides estimates of the global irradiance with an average rMBD of 1.64%, less than 4 Wm^{-2} in absolute terms. As for the estimation of the beam component, the average deviation is 2.93%, around 7 Wm^{-2} deviation with regard to the average measured value. Data from thirteen stations have been considered, although the length of the time series varies considerably. As more data become available, we plan to extend this analysis.

We have also presented here a GIS based tool that provides free estimates, not only of the solar resource over large regions but also of the PV production of different technologies. This estimation model of the PV production takes into consideration the effects of wind, angle of incidence and spectral irradiance distribution and topography, which increase the accuracy of the estimates.

The software and data package is open source and is available at this link:

http://re.jrc.ec.europa.eu/pvmaps/index.html

The said tool has been used to create a solar atlas for South Africa. It is envisaged that the solar energy atlas will provide much needed information primarily to policy and decision makers on the potential of solar energy and suitable areas. Solar energy project developers and the general public will likewise have access to the information.

7 ACKNOWLEDGMENTS

Part of the work presented here was carried out within the European Metrological Research Programme (EMRP) ENG55 project "Towards an energy-based parameter for photovoltaic classification". The EMRP is jointly funded by the EMRP participating countries within EURAMET and the European Union.

8 REFERENCES

[1] R. Perez, S. Seals, Z. Zelenka. Comparing satellite remote sensing and ground network measurements for the production of site/time specific irradiance data. *Solar Energy* 60, 2, 89-96, (1997).

[2] F. Antoñanzas-Torres, A. Sanz-Garcia, FJ. Martinez-de-Pison-Ascacibar, O. Perpiñan-Lamiguiero. Evaluation and improvement of empirical models of global solar irradiation: case study northern Spain. *Renewables Energies* 60, 604-614, (2013).

[3] J. Antoñanzas, R. Urraca, FJ. Martinez-de-Pison-Ascacibar, F. Antoñanzas-Torres. Solar irradiation mapping with exogenous data from support vector regression machines estimations. *Energy Conversion and Management* 100, 380-390, (2015).

[4] DP. Dee et al. The ERA-Interim reanalysis: configuration and performance of the data assimilation system. *Quartely Journal of the Royal Meteorological Society* 137, 553-597, (2011).

[5] P. Ineichen, CS. Barroso, B, Geiger. R, Hollmann, A. Marsouin, R, Mueller. Satellite application facilities irradiance products: hourly time step comparison and validation over Europe. *International Journal of Remote Sensing* 30, 5549-5571, (2009).

[6] A. Gracia-Amillo, T. Huld, R, Mueller. Hourly solar radiation database for Eastern Europe and Asia. Validation and Applications to PV performance estimates. Proceedings 29th European Photovoltaic Solar Energy Conference and Exhibition, 2584-2588.

[7] R. Mueller, C. Matcoukas, A. Gratzki, HD. Behr, R. Hollmann. The CM-SAF operational scheme for the satellite based retrieval of solar Surface irradiance – A LUT based eigenvector hybrid approach. *Remote Sensing and Environment* 113, 1012-1024, (2009).

[8] R. Mueller, T. Behrendt, A. Hammer, A. Kemper. A new algorithm for the satellite-based retrieval of solar surface irradiance in spectral bands. *Remote Sensing* 4, 622-647, (2012).

[9] T. Huld. PVMAPS: Software tools and data for the estimation of solar radiation and photovoltaic module performance over large geographical areas. Submitted to *Solar Energy*. Preprint available at: http://re.jrc.ec.europa.eu/pvmaps/pvmaps.pdf

[10] M. Suri, T. Huld, E. Dunlop, H. Ossenbrink. Potential of solar electricity generation in the European Union member states and candidate countries. *Solar Energy* 81, 1295-1305, (2007).

[11] A. Gracia Amillo, T. Huld, P. Vourlioti, R. Mueller, M. Norton. Application of satellite-based spectrally resolved solar radiation data to PV performance studies. *Energies* 8, 3455-3488, (2015).

[12] T. Huld, A. Gracia Amillo. Estimating PV module performance over large geographical regions: the role of irradiance, air temperature, wind speed and solar spectrum. *Energies* 8, 5159-5181, (2015).

Fast All-sky Radiation Model for Solar applications (FARMS): A Brief Overview of Mechanisms, Performance, and Applications

Yu Xie and Manajit Sengupta
National Renewable Energy Laboratory
15013 Denver West Parkway, Golden, CO 80401, USA
Yu.Xie@nrel.gov
Manajit.Sengupta@nrel.gov

ABSTRACT: Solar radiation can be computed using radiative transfer models, such as the Rapid Radiation Transfer Model (RRTM) and its general circulation model applications (RRTMG), and used for various energy applications. Due to the complexity of computing radiation fields in aerosol and cloudy atmospheres, simulating solar radiation can be extremely time-consuming, though many approximations, e.g., two-stream approach and delta-M truncation scheme, can be utilized. To provide a new fast option for computing solar radiation, we developed the Fast All-sky Radiation Model for Solar applications (FARMS) by parameterizing the simulated diffuse horizontal irradiance and direct normal irradiance for cloudy conditions from RRTM runs using sixteen-stream discrete ordinates radiative transfer method. The solar irradiance at the surface was simulated by combining the cloud irradiance parameterizations with a fast clear-sky model, REST2. To understand the accuracy and efficiency of the newly developed fast model, we analyzed FARMS runs using cloud optical and microphysical properties retrieved using GOES data from 2009–2012. The global horizontal irradiance for cloudy conditions was simulated using FARMS and RRTMG with a two-stream approximation and compared to measurements taken from the U.S. Department of Energy's Atmospheric Radiation Measurement Climate Research Facility Southern Great Plains site. Our results indicate that the accuracy of FARMS is comparable to or better than the two-stream approach; however, FARMS is approximately 400 times more efficient because it does not explicitly solve the radiative transfer equation for each individual cloud condition. Radiative transfer model runs are computationally expensive, but this model is promising for broad applications in solar resource assessment and forecasting. It is currently being used in the National Solar Radiation Data Base, which is publicly available from the National Renewable Energy Laboratory at http://nsrdb.nrel.gov.
Keywords: Radiative transfer, solar energy

1. Purpose and Approach

An atmospheric radiative transfer (RT) model simulates light scattering and absorption within air molecules, aerosols, and clouds by numerically solving for the transmission of electromagnetic radiation in the Earth's atmosphere; consequently, it has broad applications in many fields of science and industry. The radiation computed by broadband RT models in the solar region provides quantitative information on the solar resource at specific locations and is required by people endeavoring to make effective use of solar energy. RT computations for cloudy skies are time-consuming because of the complexity in solving the RT equation. Although the two-stream approximation has been applied to solar resource assessment and forecasting, the increasing spatial and temporal resolutions of satellite retrievals as well as general circulation model and numerical weather prediction experiments demand RT models with significantly higher efficiency. Rapid development of satellite remote sensing has led to the corresponding development of fast cloudy-sky RT models aimed at simulating upwelling radiances for specific satellite channels (forward RT models); however, there is still a need to develop fast models that efficiently simulate all-sky solar irradiance at the land surface.

The Fast All-sky Radiation Model for Solar Applications (FARMS) solves for solar radiation under all-sky conditions—i.e., clear, cloudy overcast, and partially cloudy skies. The clear-sky solar radiation in the direct and diffuse directions are computed by REST2 [1] because it has been proven to be computationally efficient yet reasonably accurate. The commonly used single-layer cloud model [2, 3] is assumed as a first-order approximation to derive solar radiation under a cloudy sky. The light scattering and absorption by clouds are assumed as occurring at the top of the atmosphere to simplify the formulation and may result in minor uncertainties in simulating solar radiation at visible wavelengths. The uncertainties may not be ignorable for extreme cloud conditions—e.g., opaque clouds—but these are less interesting for solar energy applications.

The Rapid Radiative Transfer Model (RRTM) [4, 5] with sixteen-stream Discrete Ordinates Radiative Transfer (DISORT) [6] has been shown to produce radiative transfer calculations of high accuracy when compared to observations. Therefore, the sixteen-stream RRTM is used to compute a lookup table (LUT) of cloud transmittance for diffuse flux under all possible cloud conditions. The LUT is parameterized using an exponential function of solar zenith angle, cloud optical thickness, and particle size. Similar to transmittance, cloud reflectance for diffuse flux can also be simulated using RRTM with sixteen-stream DISORT. The cloud transmittance and reflectance are combined with those for clear-sky conditions to simulate the all-sky solar radiation at the land surface.

2. Method

Figure 1 shows the flowchart of FARMS. The LUTs of cloud transmittance and reflectance of irradiance are computed for both water and ice clouds. To simplify the application and reduce the computing time, the LUTs are parameterized by functions of solar zenith angles and cloud microphysical and optical properties. A clear-sky radiative transfer model, REST2, and atmospheric properties are used to compute the light scattering and absorption by aerosols and trace gases in the atmosphere. By utilizing REST2 and LUTs of cloud transmittances and reflectances precomputed by RRTM, FARMS is developed to simulate surface radiation for solar applications. The all-sky surface radiation is calculated by accounting for the influences of the clear-sky atmosphere, clouds, and the multiple reflections between cloud and land surface. Compared to previous models, this model is much more efficient yet provides very high accuracy.

3. Results

RRTM with sixteen-stream DISORT is used as a benchmark to simulate cloud total transmittances for all possible cloud conditions. FARMS and the RRTM for global circulation modeling (RRTMG) with a two-stream approximation are used to perform the same computations and evaluated using cloud transmittances from the sixteen-stream approximation. As shown in Table 1, it is found that both the two-stream approximation and FARMS show reasonable performance in computing cloud transmittance under clear and cloudy conditions. The percent errors (PEs) of FARMS are 0.0294% and 2.828% for typical water and ice clouds, respectively. FARMS significantly reduces PE and mean bias error (MBE) compared to the two-stream approximation, whereas its absolute percent error (APE) and mean absolute error (MAE) are slightly smaller. The comparison between satellite-based retrievals of atmospheric properties and surface measurements of solar radiation can also evaluate the accuracy of FARMS. As shown in Table 1, the PE and MBE for the FARMS computation of global horizontal irradiance (GHI) (-1.8936% and -5.4068 W/m^2, respectively) are much better than those of the two-stream approximation (-7.109% and -20.2983 W/m^2, respectively). The two-stream approximation and FARMS perform similarly when evaluated using APE, MAE, and root mean square error (RMSE).

We compare the computing time of FARMS to the two-stream approximation and RRTM with four- to 16-stream approximations for the surface radiation over the Atmospheric Radiation Measurement (ARM) Southern Great Plains (SGP) site. The computing time associated with FARMS and RRTMG is 2 seconds and 12 minutes and 32 seconds, respectively. The computing time for RRTM with sixteen-stream approximation is 2 hours and 2 minutes. Thus, for computations that result in similar accuracy levels, FARMS can dramatically reduce the computing time.

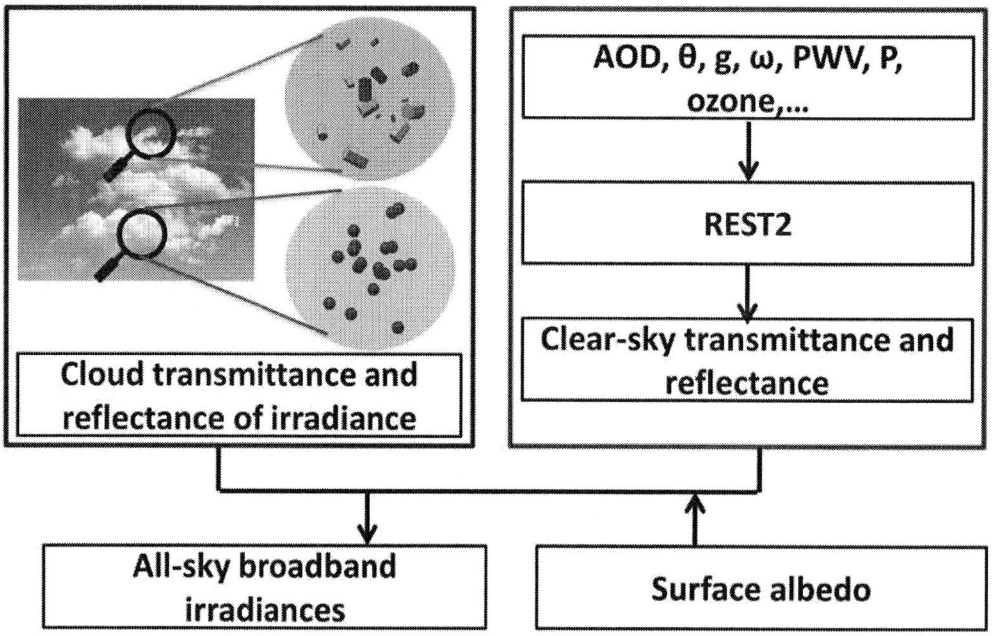

Figure 1. Flowchart of FARMS.

Table 1. Statistics of PE, APE, MBE, MAE, and RMSE of simulated cloud transmittance and GHI using two-stream approximation and FARMS.

RT Model	PE (%)	APE (%)	MBE	MAE	RMSE
Transmittance for Water Cloud					
Two-Stream	3.7195	4.6311	0.0156	0.0195	0.0319
FARMS	0.0294	2.9558	0.0012	0.0124	0.0165
Transmittance for Ice Cloud					
Two-Stream	7.0986	7.0986	0.0289	0.0289	0.0319
FARMS	2.8280	5.3852	0.0115	0.0220	0.0268
RT Model	PE (%)	APE (%)	MBE (W/m^2)	MAE (W/m^2)	RMSE (W/m^2)
GHI over ARM SGP Site					
Two-Stream	-7.109	31.7935	-20.2983	90.7798	132.749
FARMS	-1.8936	31.886	-5.4068	91.044	130.286

4. Conclusions

We have developed a fast all-sky radiative transfer model, FARMS, that has accuracy comparable to a two-stream radiative transfer model but is nearly 400 times computationally faster. This model is expected to be widely used in various applications that require rapid computations of solar radiation at the earth's surface under all atmospheric conditions. The model is currently being used to compute solar radiation for the National Solar Radiation Database (http://nsrdb.nrel.gov), which provides 4-km half-hourly resolution solar data for a 17-year period covering 1998–2014. The FARMS model is also expected to be included as an option in the version of the Weather Research and Forecasting model that is specially designed to produce accurate solar forecasting.

5. Acknowledgements and Disclaimer

This work was supported by the U.S. Department of Energy under Contract No. DE-AC36-08GO28308 with the National Renewable Energy Laboratory. The U.S. Government retains and the publisher, by accepting the article for publication, acknowledges that the U.S. Government retains a nonexclusive, paid-up, irrevocable, worldwide license to publish or reproduce the published form of this work, or allow others to do so, for U.S. Government purposes.

6. References

[1] C. Gueymard, "REST2: High-performance solar radiation model for cloudless-sky irradiance, illuminance, and photosynthetically active radiation - Validation with a benchmark dataset," *Sol. Energy,* vol. 82, pp. 272-285, 2008.

[2] J. A. Coakley, M. A. Friedman, and W. R. Tahnk, "Retrieval of cloud properties for partly cloudy imager pixels," *J. Atmos. Oceanic Technol.,* vol. 22, pp. 3-17, 2005.

[3] V. Ramanathan, "The role of earth radiation budget studies in climate and general circulation research," *J. Geophys. Res.,* vol. 92, pp. 4075-4095, Apr 20 1987.

[4] E. J. Mlawer, S. J. Taubman, P. D. Brown, M. J. Iacono, and S. A. Clough, "RRTM, a validated correlated-k model for the longwave," *J. Geophys. Res.,* vol. 102, pp. 16663-16682, 1997.

[5] L. Oreopoulos and H. W. Barker, "Accounting for subgrid-scale cloud variability in a multi-layer 1-D solar radiative transfer algorithm," *Q. J. R. Meteorol. Soc.,* vol. 125, pp. 301-330, 1999.

[6] K. Stamnes, S. C. Tsay, W. Wiscombe, and K. Jayaweera, "Numerically stable algorithm for discrete-ordinate-method radiative transfer in multiple scattering and emitting layered media," *Appl. Opt.,* vol. 27, pp. 2502-2509, Jun 15 1988.

PRELIMINARY RESULTS OF THE FIFTH INTERNATIONAL SPECTRORADIOMETER COMPARISON FOR IMPROVED SOLAR SPECTRAL IRRADIANCE MEASUREMENTS.

R. Galleano[1], W. Zaaiman[1], D. Alonso Alvarez[2], A. Minuto[3], N. Ferretti[4], R. Fucci[5], M. Marzoli[6], L. Manni[6], M. Halwachs[7], M. Friederichs[8], F. Plag[9], D. Friedrich[9], E. Haverkamp[10]
[1] EC, DG JRC, IET, Renewables and energy efficiency Unit, ESTI, roberto.galleano@ec.europa.eu, Ispra, (I), [2] EXSS group, Department of Physics, Imperial College, London, (UK), [3] RSE spa, Milan, (I), [4] Photovoltaik-Institut Berlin, (D), [5] ENEA Research Center of Portici, (I), [6] SUPSI, Campus Trevano, CH-6952, Canobbio, (CH), [7] AIT, Austrian Institute of Technology, 1210 Vienna, (AT), [8] PV Lab Germany, 14482 Potsdam, (D), [9] Physikalisch-Technische Bundesanstalt, Arbeitsgruppe "Solarzellen", 38116 Braunschweig, (D), [10] Radboud University Nijmegen, AMS, IMM, Toernooiveld 1, 6525 ED Nijmegen, (NL)

ABSTRACT: The wider portfolio of today's available photovoltaic (PV) technologies on the market makes the measurement of the spectral content of the incoming sun or artificial light a key parameter for the characterization, calibration, and energy yield estimation of these devices. At present there is a growing request for harmonization of good measurement practices and for knowledge transfer in the field of spectrally resolved solar radiation for solar energy applications (e.g. photovoltaics) in order to make these measurements comparable and directly traceable to SI units. A group of European research institutes active in the PV field for research, characterization and engineering, set up the fifth comparison of spectroradiometers for solar spectrum measurements. In this paper the preliminary results of such exercise are reported.
Keywords: PV calibration, spectroradiometers, intercomparison

1 INTRODUCTION

The wider portfolio of today's available photovoltaic (PV) technologies on the market makes the measurement of the spectral content of the incoming sun or artificial light a key parameter for the characterization, calibration, and energy yield estimation of these devices. Nowadays, spectroradiometers with different principles of operation (e.g. single-, double- stage rotating grating monochromator or fixed single grating polychromator with photodiode (PD) array or CCD detectors) are routinely used for solar spectrum measurements. At present there is a growing request for harmonization of good measurement practices and for knowledge transfer in the field of spectrally resolved solar radiation for solar energy applications (e.g. photovoltaics) in order to make these measurements comparable and directly traceable to SI units. Moreover, periodical intercomparisons are part of performance-based quality-control checks for laboratories working according to ISO-IEC 17025 or 9000 standards. Within this context, a group of European research institutes active in the PV field for research, characterization and engineering, set up the fifth comparison of spectroradiometers for solar spectrum measurements. Aims of the intercomparison were: exchange and compare instrument calibration procedures, assess laboratory measurement capabilities, establishing equivalence figures for solar spectra measurement and put in practice lessons learnt from previous editions. This paper describes the intercomparison campaign and reports on the preliminary results.

2 PURPOSE OF THE WORK, EXPERIMENTAL APPROACH

There is a growing request for harmonized, traceable and low uncertainty solar spectrum measurements for calibration and energy yield estimation in photovoltaics. This intercomparison, whose preliminary results are summarized in this work, is a good opportunity to raise the awareness and exchange of good practices on reliable, traceable and low uncertainty measurements of solar spectrum. Moreover, for the participating Institutions applying a quality system or having an accreditation according to the ISO/IEC17025 standard [1], the comparison is an implementation, together with round robin, of the required checks for establishing a performance-based quality-control system. The intercomparison took place for the second time at the 'Instituto Nacional de Técnica Aeroespacial' laboratory in Madrid (Spain) from 18th to 22nd of May 2015. Twelve spectroradiometers systems from six different manufacturers and covering two different technologies (single-stage rotating-grating and fast fixed-grating polychromator with single or CCD array detectors) were set to simultaneously measure global normal incidence (GNI) spectral irradiance from 300 to 1700 nm or 300 to 1100 nm, depending on the available instruments. The large variety of manufacturers and principles of operation represents a good cross section of today's most used spectroradiometers in the PV community. A selected number of instruments, equipped with suitable collimation tubes to reduce their angle of view [2], were set to measure also the direct normal incidence (DNI) spectral irradiance in the range from 300 to 1700 nm. In this paper only the GNI measurements results from ten laboratories will be reported. Due to the differences among various instruments in the measurement timing, bandwidth and spectral resolution, specific procedures for data acquisition, synchronization and analysis were developed in order to make the spectroradiometers' output data comparable to each other. Prior to the intercomparison each participating laboratory calibrated their own spectroradiometer(s) following their usual procedures, thus allowing evaluating the instrument performance together with its traceability chain and calibration procedure. Some spectroradiometers were calibrated by an external accredited calibration laboratory, while others were calibrated in-house using a calibrated radiometric standard lamp, or at the manufacturer. Table 1 summarizes the main characteristics of the instrument involved: commercial name, instrument type (poly- or mono- chromator and detector configuration) the wavelength range covered, and the calibration source. All participating instruments were mounted on high-accuracy solar trackers in order to reduce errors due to instruments pointing (e.g. different instruments entrance optics cosine response). In parallel to the intercomparison a set of cavity radiometers were also in use as reference detectors for total irradiance data. These last assure the direct link of solar irradiance measures to the SI units as these cavity radiometers take part to the international pyrheliometer comparison under the aegis of the world radiometric reference (IPC - WRR), held every 5 year at PMOD-Davos (CH) [3].

Table I: Summary of the characteristics of the spectroradiometers involved in the intercomparison. Commercial name, principle of operation, measuring wavelength bands and calibration method are reported.

Laboratory	Instrument name	Instrument type	Wavelength band nm	Calibration
AIT	Ocean Optics	Polychromator, CCD array detector	300-1600	In house, standard lamp
ENEA	Stellarnet	Polychromator, two CCD array detectors	300-1700	External accredited cal. lab.
Imperial College	Ocean Optics	Polychromator, CCD array detector	250-1100	In house calibration
JRC	OL750	monochromator double PD detector	250-2500	In house, standard lamp
PI Berlin	Tec5	Polychromator, CCD array detector	300-1700	In house, standard lamp
PTB	Instrument System CAS 140CT	Polychromator, three CCD array detectors	250-2200	In house, standard lamp
RSE	Stellarnet EPP2000	Polychromator, two CCD array detectors	300-1700	Outdoor, AM1.5
Radboud University	EKO MS711	Polychromator, CCD array detector	300-1100	Manufacturer
PV Lab	Instrument System CAS 140CT	Polychromator, CCD array detector	300-1100	Manufacturer
SUPSI	Eko wiser system	Polychromator, two CCD array detectors	300-1700	In house, standard lamp

3 PRELIMINARY RESULTS

In order to compare solar spectra acquired by 'fast' and 'slow' measuring instruments, several sets of average spectra, measured during 7-minute acquisition time series, were analyzed. The duration of the time series were determined by the acquisition time of the slowest spectroradiometer. During the time series, the irradiance must remain stable to 1 %, or better, to consider them 'stable' and flagged for analysis. The stability constraint avoids adding errors arising from fast changing weather conditions affecting the output of spectroradiometers in different ways. For instance, a cloud rapidly passing close to the sun disk may affect and invalidate only few spectra in the series acquired by a fast spectroradiometer while, for a slow spectroradiometer, it may results in a spectrum shape distortion and invalidate the whole measurement series. This constraint limited the useful sky conditions to clear or almost clear. Moreover, the acquired spectra were also convoluted in order to reduce and harmonize the spectral bandwidth to 4 nm full width half maximum (FWHM); this is done to reduce artefacts in comparing spectra in the atmospheric absorption bands [4]. Several analyses were performed on acquired data both in terms of absolute spectral irradiance and of spectral shape deviation. Diversified data analyses can better separate errors or uncertainty components arising from calibration systematic effects or from instrument non linearity, internal stray light or drifts. Figure 1 shows a typical example of wavelength-by-wavelength spectral irradiance deviation analysis performed on a set of simultaneous GNI spectra measured by the 10 partners' instruments. The graph (a) shows a group of ten simultaneously acquired spectra during a 7-minute acquisition series. Graph (b) in the same figure shows the wavelength-by-wavelength percent deviation of each spectrum with respect to Lab A spectrum and normalized to its peak irradiance. The average difference values for the reported spectra lie in a band of ± 2.3 % with associated standard deviations up to 5.1 %.

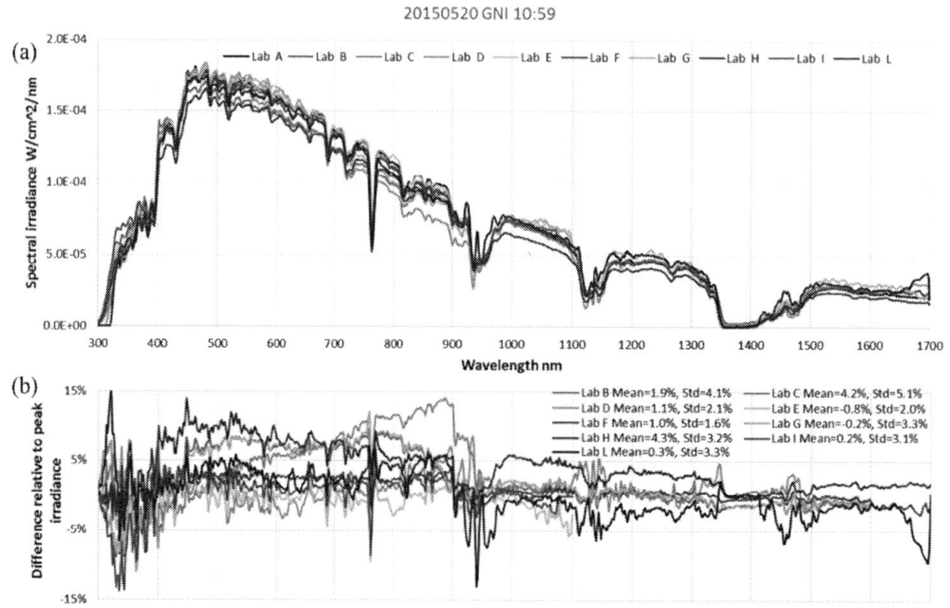

Figure 1: (a) Ten GNI solar spectra simultaneously measured by participating spectroradiometer systems. (b) wavelength-by-wavelength difference with respect to the Lab A spectrum and normalized to its peak irradiance, calculated average differences and standard deviations are also reported.

Table II reports the average wavelength-by-wavelength differences and average standard deviation for 21 stable GNI spectra each measured during a 7-minute acquisition time series and distributed along a full day measurement session. For this day, the average difference values lie within an interval of ± 3.1 %. A 1-day average standard deviation value much larger than the corresponding 1-day average difference denotes a possible instrument drift and/or instability. For another day of measurements the 1-day average differences values lie within an interval of ± 2.7 %. From an accurate analysis of these information collected all along the comparison campaign, participants can gain knowledge of their instrument behavior such as temperature stability, repeatability and reproducibility.

Table II: Average wavelength-by-wavelength differences and average standard deviation values calculated with respect to Lab A spectra and expressed in percentage of its peak irradiance. Reported values refer to a group of 21 stable spectra acquired during 7-minute time series and distributed along a full day session of measurements.

Laboratory	1-day Average difference	1-day average std. dev.
Lab B	1.8%	4.0%
Lab C	3.9%	5.0%
Lab D	0.8%	2.1%
Lab E	-1.2%	2.0%
Lab F	0.8%	1.4%
Lab G	-0.4%	3.4%
Lab H	5.1%	3.7%
Lab I	-0.3%	3.1%
Lab L	-0.2%	3.3%

Previous data analysis focusses on the absolute spectral irradiance differences among participating instruments; a different approach can be used to separate systematic effects (e.g. arising from instrument calibration or from instrument time-drift), from non-linearity or spectra distortion due to intrinsic instrument limitation. This is important in solar spectrum measurement applied to PV field, where a correct measurement of the shape of incoming solar light is fundamental, whereas the absolute irradiance value is usually measured by other means, often with lower uncertainty (e.g. cavity radiometers, reference solar cells, pyrheliometers, pyranometers etc.). The calibration of a generic PV device at standard test conditions (STC) entails, among others, the correction to the standard spectrum AM1.5 [5]. This correction is performed by applying a mismatch correction factor (MM) accounting for difference in spectral responsivities between reference and under calibration device(s), and for the relative spectral difference between actual- and standard-spectrum conditions. An easy comparison of the relative spectral differences among acquired spectra can be done slightly modifying the guidelines described in [6]. The spectral irradiance data of a specific measurement from each participating spectroradiometer were integrated into five 100-nm bins from 400 to 900 nm plus one 200-nm bin from 900 to 1100 nm and expressed as percent ratio to the total irradiance of the same spectrum as integrated in the 400 to 1100 nm band. The afore mentioned standard compares, then, the irradiance percentage in each bin with the same bin percentage, of the AM1.5 standard spectrum to assess the spectrum quality class of a generic solar simulator. In this exercise we substitute the AM1.5 spectrum with the Lab A spectrum taken as reference, and applied the E_n performance statistics analysis method [7] to each bin comparison result. The E_n number is a performance statistics tool and is defined in Equation (1) as:

$$E_n = \frac{MLab_i - Mref}{\sqrt{(ULab_i * MLab_i)^2 + (Uref * Mref)^2}} \quad (1)$$

Where En is the normalized error for the M_{th} bin (unitless), $ULab_i$ and $Uref$ are the reported percent measurement expanded uncertainty for the i_{th} spectroradiometer and the reference, respectively. $MLab_i$ and $Mref$ represent the ratio of the M_{th} bin to the total irradiance for the i_{th} spectroradiometer and the reference, respectively. The calculated E_n number in eq. 1 involves the establishment of acceptance limits of ± 1; E_n values within acceptance limits are considered satisfactory because consistent with declared/assigned uncertainty. E_n values outside acceptance limits highlight inconsistency with estimated measurement uncertainty and/or severe instrument drift from expected performance. In this exercise the measurement expanded uncertainty expressed in percent, $ULab_i$ and $Uref$, were estimated to be 5 % and 3 %, respectively. Figure 2 shows performance statistics results as applied to the spectra reported in figure 1. Six out of nine compared instruments show all E_n values within the ± 1 acceptance limits (Labs B, D, E, F, H and I), two instruments (Labs G and L) show one E_n value outside acceptance limits, and one instrument (Lab C) has severe deviations from the acceptance limits. It is worth notice that the proposed combined data analysis allowed to highlight that Lab H despite exhibiting the worst wavelength-by-wavelength average difference in Figure 1(b), showed E_n values well within consistency threshold, suggesting a systematic effect due, probably, to a scaling factor in the calibration process; considering data from Labs G and L an inconsistency in part of the acquired spectrum was spotted, despite showing in Figure 1(b) low wavelength-by-wavelength average difference. The same approach applied to Lab C data, confirmed highly divergent results due to instrument non-linearity or very poor calibration standard and or procedures.

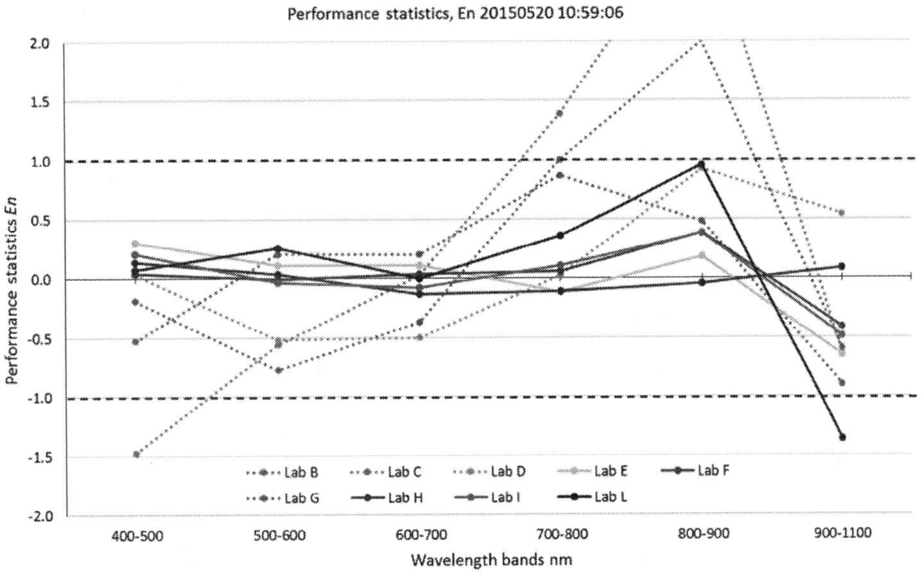

Figure 2: Performance statistics results using Equation 1. Percent spectral irradiance contents, as integrated in 100 nm bins from 400 to 900 nm plus a 200 nm bin from 900 to 1100 nm, computed from spectrum measured by various partner's instruments, are compared with those of the reference instrument and related to declared/assigned measurement uncertainty. Dashed lines at E_n 1, -1 represent acceptance/consistency limits.

4 CONCLUSIONS

A spectroradiometers intercomparison was performed among ten European scientific and industrial partners. The intercomparison aimed at exchanging and comparing instrument calibration procedures, measurement capabilities, at putting in practice lessons learnt from previous editions and at establishing equivalence figures for solar spectra measurement. Different and complementary data analyses were applied to measured spectra so that partners can have comprehensive knowledge about their instrument's behaviour, calibration stability, and measurement procedures output. Preliminary results showed wavelength-by-wavelength average difference values lying within an interval of ± 3.1 % for a whole day of measurements. Similar values of ± 2.7 % were found for another day of GNI measurements with similar sky an weather conditions. Performance statistics E_n number analysis showed that six spectroradiometers had their output results consistent with declared measurement uncertainty in all the considered wavelength bands, two other systems showed inconsistent data in one wavelength band, and one showed larger deviation from the ± 1 acceptance limits. Absolute and relative spectral irradiance comparison methodologies proved to be a good approach to separately assess instrument non linearity and distortion effects, from systematic effects due to calibration or due to drift with time and/or temperature which may not be present in all of the performed analysis.

REFERENCES

[1] *ISO/IEC 17025:2005, General requirements for the competence of testing and calibration laboratories, International Organization for Standardization, ISO Central Secretariat, Chemin de Blandonnet 8, CP 401 1214 Vernier, Geneva, Switzerland.*

[2] *Guide to meteorological instruments and observations, WMO-No-8 Secretariat of the World Meteorological Organisation, 2006, Geneva, Switzerland.*

[3] *Finsterle, W., (2011), WMO International Pyrheliometer Comparison IPC-XI Final Report, WMO IOM Report No., 108.*

[4] *Galleano, R, Zaaiman, W, Strati, C, Bartocci, S, Pravettoni, M, Marzoli, M, Fucci, R, Leanza, G, Timò, G, Minuto, A, Catena, M, Aleo, F, Takagi, S, Akiyama, A, Nuñez, R, and Belluardo, G (2015), Second international spectroradiometer intercomparison: results and impact on PV device calibration. Prog. Photovolt: Res. Appl., 23, 929–938. doi: 10.1002/pip.2511.*

[5] *ISO/IEC 60904-7 Photovoltaic devices - Part 7: Computation of spectral mismatch error introduced in the testing of a photovoltaic device, International Organization for Standardization, ISO Central Secretariat, Chemin de Blandonnet 8, CP 401 1214 Vernier, Geneva, Switzerland.*

[6] *ISO/IEC 60904-9:2007 Photovoltaic devices - Part 9: Solar simulator performance requirements, International Organization for Standardization, ISO Central Secretariat, Chemin de Blandonnet 8, CP 401 1214 Vernier, Geneva, Switzerland.*

[7] *ISO/IEC 17043:2010 Conformity assessment— General requirements for proficiency testing, ISO Central Secretariat, Chemin de Blandonnet 8, CP 401 1214 Vernier, Geneva, Switzerland.*

THE QUALITY OF SATELLITE-BASED IRRADIATION DATA FOR OPERATIONS AND ASSET MANAGEMENT

Achim Woyte, Karel De Brabandere, Babacar Sarr, Mauricio Richter

3E sa, Quai à la chaux 6, B-1000 Brussels, Belgium
Tel. +32 22 17 58 68, Fax +32 22 19 79 89, E-mail: Achim.Woyte@3E.eu

ABSTRACT: This paper evaluates solar irradiation data from different satellite-based irradiance data services. For seven satellite irradiance models, we have evaluated global horizontal irradiation by comparison with data from meteorological stations for the years 2011 to 2015. The reference data cover measurements from 203 stations in the Netherlands, Belgium and France. We separately evaluate the systematic error and the random error. Our results show that the systematic error of most models ranges between 3 and 5% of the measurement. For monthly irradiation, the random errors can be as low as 2%. The overall normalized root mean square errors of monthly global horizontal irradiation for the best performing models are of 3 to 6%. For monthly plane-of-array irradiation they range between 5.4 and 8.1%. For daily and hourly irradiation the errors are much higher. We conclude that satellite-based irradiation can be a reliable and valuable alternative to on-site measurements for monthly and quarterly performance reporting for small to medium-size PV plants. For fault detection with hourly or daily resolution, on-site sensors are the first choice. Satellite data are less precise and may be considered as back-up when the sensors fail or appear to be badly maintained.
Keywords: Calibration, Performance, Photovoltaic, Pyranometer, Reference Cell, Solar Radiation

1 INTRODUCTION

Irradiation data derived from satellite images are increasingly used as input for long-term yield assessment and as reference yield for monitoring and business reporting. The reference yield is usually used as a basis for verifying the fulfilment of contractual key performance indicators (KPIs), e.g., performance ratio (PR) or energy-weighted availability. The under- or overachievement of these KPIs often also triggers penalties or bonuses, respectively. Therefore, before relying on a reference yield that is derived from satellite data, asset managers and O&M contractors expect that the fidelity of these data is confirmed independently and with scientific rigour.

Several authors have already evaluated the quality of satellite-based irradiance data in the past and some comprehensive overviews for further reading can be found in [1]–[3]. Typical normalized root mean square errors are situated between 4 to 8% for monthly and 2 to 6% for yearly irradiation values. However, up to now these publicly available measures of precision could not fully satisfy the expectations of many asset managers and O&M contractors.

Recently, several new or improved satellite-based irradiance services have become available. In several of these services, the underlying cloud models increasingly take into account the physical properties of the clouds. Particularly for the more recent of these models no large-scale evaluation has been available up to now. Therefore, a rigorous and methodologically well-described evaluation based on a reliable and representative set of data was missing from a scientific point of view. With this paper, we wish to make such evaluations available to the market as a reference and to the scientific community as a stepping stone for further improvement.

In short, the purpose of this paper is to quantify the precision of the different models and services available for the practical use as reference yield.

2 METHODOLOGY

2.1 Satellite Irradiance Models under Evaluation

All different satellite irradiance services for Europe are derived from images taken by the European Meteosat Second Generation (MSG) series. Currently Meteosat 7, 8, 9 and 10 are operational. Since 2012, irradiation data for Europe and Africa as used for PV applications usually originate from Meteosat 10. This satellite essentially covers Europe, Africa and the eastern part of Brazil. Images are taken with a time resolution of 15 minutes and a nadir spatial resolution of 3 x 3 km^2. The images are taken by a so-called SEVIRI (Spinning Enhanced Visible and Infrared Imager) instrument [4].

Data from seven different satellite irradiance models have been evaluated as listed in Table I.

Table I: Models under evaluation

Model (*abbreviation*)	Available Through	Ref.
MACC-RAD (*maccrad*)	CAMS Radiation service, through SoDa	[5],[6]
HelioClim-3 v3 (*hc3_v3*) HelioClim-3 v4 (*hc3_v4*) HelioClim-3 v5 (*hc3_v5*)	SoDa	[7]
MSG-CPP (*cpp*)	KNMI	[8]
GSIP (*gsip*)	NOAA (since 3/2014)	[9],[10]
EnMetSOL (*enmetsol*)	Uni Oldenburg, turbidity model according to [11]	[12]

2.2 Reference Data for the Evaluation

The satellite-based irradiation data have been compared to pyranometer measurements from 203 meteorological stations maintained by the national public weather services of France, Belgium and the Netherlands. The available reference data sets are listed in Table II with the spatial distribution of the stations illustrated in Figure 1.

Table II: Reference data set of global horizontal irradiation and minimum aggregation period available for the evaluation

Provider	Sites	Aggregation	Coverage
MétéoFrance (FR)	160	daily	2012-2013
RMI (BE)	12	daily	2012-2013
KNMI (NL)	31	hourly	2011-2015

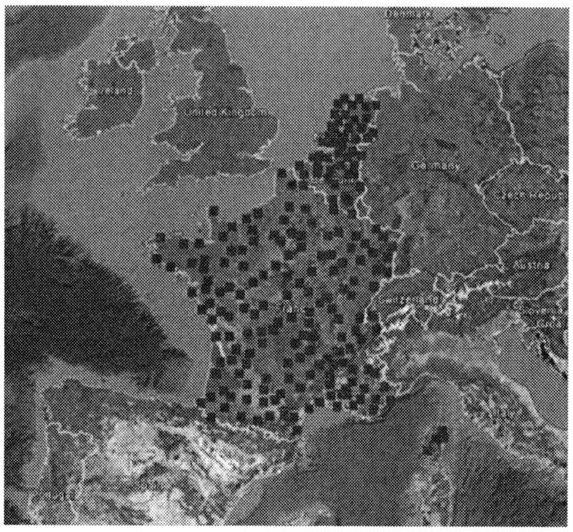

Figure 1: Meteo stations for the evaluation

Based on the available data sets, daily and monthly irradiation for all 203 sites has been evaluated for 2012 and 2013. Hourly, daily and monthly irradiation for the 31 stations in the Netherlands has have been evaluated for 2011 to February 2015. The focus on the Netherlands is for several reasons, namely, the consistently high quality of the available measurements, the absence of horizon shading due to the generally flat landscape, the relatively rare occurrence of snow and the free availability of hourly values over a long period.

2.3 Metrics for the Evaluation

The satellite irradiation data have been evaluated against the reference data by their root mean square error (*RMSE*), the standard deviation of error (*SDE*) and the *Bias* with

$$RMSE^2 = SDE^2 + Bias^2. \quad (1)$$

Equation (1) allows to split the *RMSE* into a random error (*SDE*) and a systematic error (*Bias*). It describes a circle. When setting out *SDE* against *Bias* in Cartesian coordinates, the *RMSE* is the distance from the origin. In practice, when computing irradiation by integrating irradiance over long times, random errors are averaged out while the bias will remain the same.

For each of the irradiance data services under test we have computed the three error measures separately for each station and per year. We have evaluated the three error measures for monthly, daily, and in the Netherlands also hourly irradiation values. Finally, we have computed the expected error values $E(x)$ for each service, year and time aggregation from the geometric mean of error x over all stations. These values can be interpreted as mean error values for each service over the region. The principle is illustrated in Figure 2.

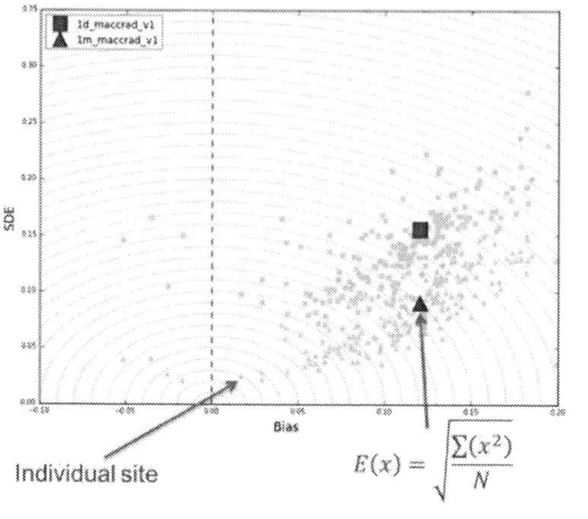

Figure 2: Exemplary distribution of the error measures *RMSE*, *SDE* and *Bias* for data with daily (square) and monthly (triangle) aggregation

3 RESULTS

3.1 Results for Daily and Monthly Data

The results for daily and monthly data are shown in Figure 3 for 2012 and Figure 4 for 2013. The results are consistent for both years. We can observe a relatively high *Bias* for MACC-RAD. The HelioClim model improves with each version. The best performing models for monthly irradiation are HelioClim-3 v4 and v5, EnMetSOL and MSG-CPP. For hourly data, EnMetSOL and MSG-CPP perform a little bit better than the HelioClim models.

3.2 Results for Hourly Data

The results for hourly, daily and monthly data for the Netherlands are shown in Figure 5 to Figure 8 for four different years. Notably, in Figure 8 we have chosen to show the results for data running for 11 months running from April 2014 to February 2015 rather than from January to December 2014. This allowed us to include the results from the GSIP model. Data from GSIP for Western Europe have only been fully available since April 2014.

3.3 Overall Results

The *Bias* of most data services lies consistently between 3 and 5%. However, for individual sites, the *Bias* rather ranges between -5 and 10%.

The *SDE* can be as low as 2% for monthly irradiation values from the MSG-CPP method. For daily and hourly irradiation, the *SDE* and, hence, also the *RMSE* are much higher. The average *SDE* for daily values lies above 10% for all services and for hourly value it lies even above 20%.

Accordingly, the *RMSE* values for the best performing models (EnMetSol, HelioClim-3 v5, MSG-CPP and GSIP) are of 3 to 6% for monthly and 9 to 11% for daily irradiation. The *RMSE* for hourly irradiation for the better performing models is much worse, in the range of 19 to 23%.

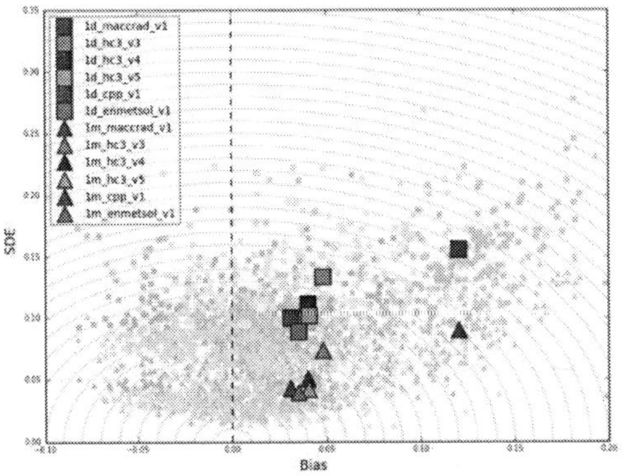

Figure 3: Error measures for the different models for daily (squares) and monthly (triangles) irradiation in 2012

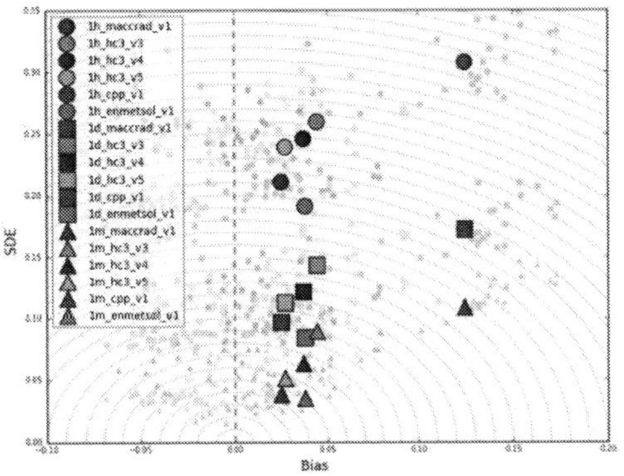

Figure 6: Error measures for the different models for hourly (circles), daily (squares) and monthly (triangles) irradiation in 2012, Netherlands only

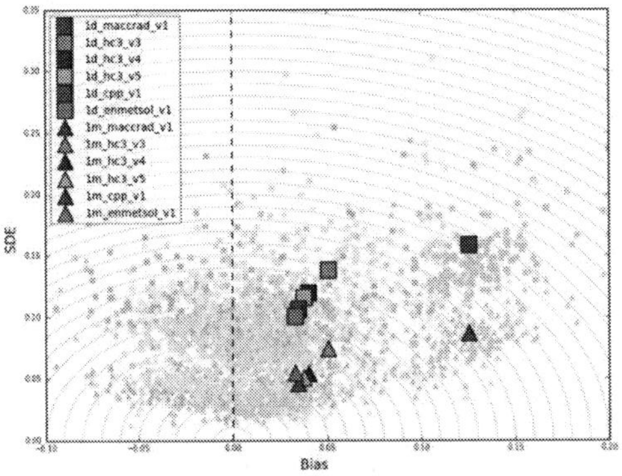

Figure 4: Error measures for the different models for daily (squares) and monthly (triangles) irradiation in 2013

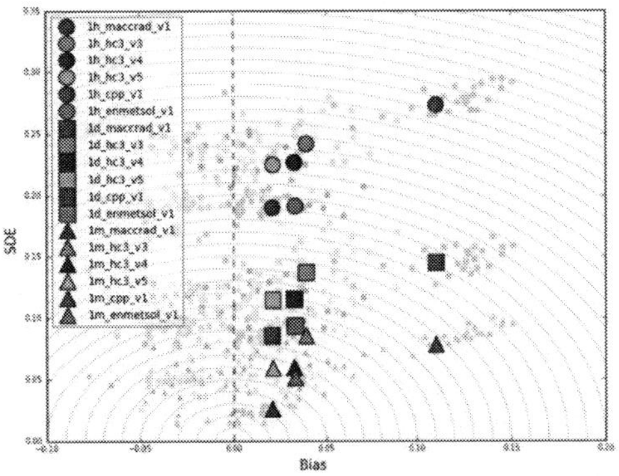

Figure 7: Error measures for the different models for hourly (circles), daily (squares) and monthly (triangles) irradiation in 2013, Netherlands only

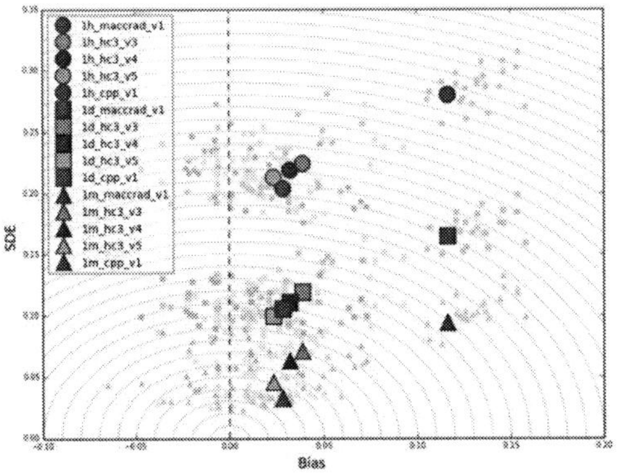

Figure 5: Error measures for the different models for hourly (circles), daily (squares) and monthly (triangles) irradiation in 2011, Netherlands only

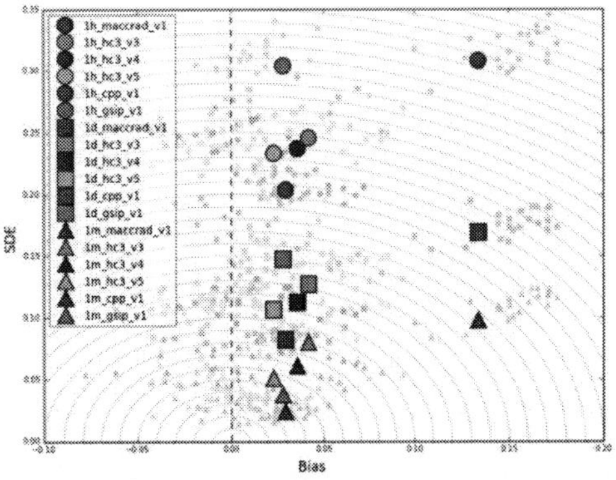

Figure 8: Error measures for the different models for hourly (circles), daily (squares) and monthly (triangles) irradiation; 11 months from April 2014 to February 2015, Netherlands only

The average bias values of the different models are compared in Figure 9. For MACC-RAD the *Bias* is around 10% and much higher than for all other models. Notably, no default bias correction is applied to the results of the MACC-RAD model by Copernicus or SoDa. Being the direct outcome of an R&D project, the idea is that users of the model results would correct for the bias themselves and depending on their particular situation.

The results of the other models show much lower biases, around 2.5% for HelioClim-3 v3 and below 2% for the others.

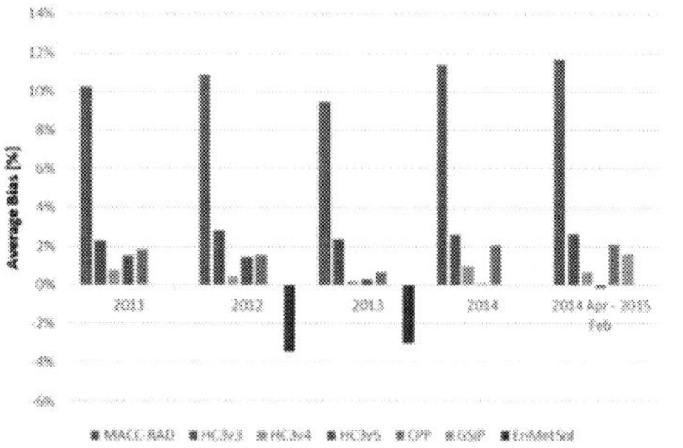

Figure 9: Arithmetic average *Bias* over all stations in the Netherlands for the different models and years

4 DISCUSSION

4.1 Possibilities for further Improvement

The resulting *RMSE* and *Bias* for the best performing sensors are reasonably low. The *Bias* may still be further reduced through so-called model output statistics. The data are then adjusted through a stochastic filter, taking into account time of the day, or other parameters. If available, the data may also be calibrated specifically for each site by means of measurements from nearby stations [3], [13].

4.2 Conversion to the Plane of Array

The error measures determined here apply to global horizontal irradiation. For irradiation in the plane of a PV array the irradiation needs to be converted. The *RMSE* introduced by this conversion has been found to be between 4.5% and 5.4% [14].

When evaluating satellite-based irradiation on the plane of array, this uncertainty has to be combined with the one of horizontal global irradiation. This is done by adding the squared *RMSE* values for each step and computing the square root of this sum.

Accordingly, the *RMSE* for monthly plane-of-array irradiation for the best-performing models as listed in Section 3.3 ranges between 5.4 and 8.1%.

4.3 Comparison to On-site Sensors

When comparing the results to on-site measurements in the plane of array, it is clear that for hourly and daily irradiation, on-site measurements with calibrated and well-maintained instruments will be much more precise than satellite-based data. Also for monthly irradiation, the precision of a well-maintained secondary standard pyranometer will be higher than the one from the satellite. However, for first and second class pyranometers as well as for silicon irradiance sensors as we often see them in small to medium size PV plants, the precision of the satellite service is generally comparable and sometimes even better than the one of the on-site measurement.

Generally speaking, the advantages of on-site sensors are their precise localization of the measurement in time and space and the well-defined precision of the instrument. A disadvantage is the need for regular cleaning, maintenance and calibration, which are not always done correctly, especially for small to medium-size PV plants. Satellite-based irradiation is independent from local maintenance activities.

5 CONCLUSIONS AND RECOMMENDATIONS

We have evaluated solar irradiation data from different satellite-based irradiance data services. The evaluation covers seven different satellite irradiance models in total. The evaluation relies on a comparison of hourly, daily and monthly irradiation with measurements of global horizontal irradiation from public meteo stations. In total, we have used data from 203 stations in France, Belgium and the Netherlands for several years of data.

For global horizontal irradiation, the systematic error of most models ranges between 3 and 5%. For monthly irradiation, the random errors can be as low as 2%. The overall *RMSE* for the best performing models are then situated between 3 to 6%. For daily and hourly irradiation the errors are still much higher.

For irradiation in the plane of the PV array, the overall RMSE also includes the uncertainty introduced by the receiver plane conversion. Accordingly, the *RMSE* for monthly plane-of-array irradiation for the best-performing models ranges between 5.4 and 8.1%.

When comparing the precision to that one of ISO 9060 Secondary Standard thermopile pyranometers, we can formulate the following conclusions.

- Fault detection by O&M contractor requires good data with hourly or daily resolution. On-site sensors are the first choice but require appropriate maintenance. Satellite data may be used as back-up when the sensors fail or appear to be badly maintained.
- For monthly to annual reporting on the overall plant performance by O&M contractors and asset managers, satellite data are a valid and reliable reference. In case of doubt, the satellite should be evaluated with one or two years of data from a well-maintained meteo station in the neighbourhood.
- For long term yield estimates as computed by investors, installers and consultants, satellite data are a valid and reliable reference. In case of doubt, the satellite should be evaluated with one or two years of data from a well-maintained meteo station in the neighbourhood.
- Finally, satellite data may also serve to validate the proper calibration and configuration of irradiance sensors in case of doubt. Particularly for large deviations, cleaning needs or shadowed sensors, the satellite may spare the operator a site visit and can already indicate what is wrong.

6 ACKNOWLEDGEMENTS

The research leading to these results has received funding from the European Union's Horizon 2020 research and innovation programme under the grant agreement No 649997 (Solar Bankability).

The authors are grateful to the different suppliers of satellite-based irradiation data as listed in Table I. Special thanks are due to Annette Hammer from University of Oldenburg, Germany for the preparation of the EnMetSOL data set for the coordinates of our reference sites.

The authors also thank MétéoFrance, the Belgian RMI and the KNMI for making available the measurements from their stations.

REFERENCES

[1] P. Ineichen, 'Five satellite products deriving beam and global irradiance validation on data from 23 ground stations', University of Geneva, Geneva, Switzerland, Feb. 2011.

[2] P. Ineichen, 'Long term satellite hourly, daily and monthly global, beam and diffuse irradiance validation. Interannual variability analysis', University of Geneva, Geneva, Switzerland, Dec. 2013.

[3] K. De Brabandere, M. Richter, F. Assiandi, and B. Sarr, 'Engineering Models for PV System Operations', Performance Plus WP2 Deliverable D2.3, Jul. 2014.

[4] 'Meteosat — EUMETSAT'. [Online]. Available: http://www.eumetsat.int/. [Accessed: 16-Jun-2016].

[5] M. Schroedter-Homscheidt, 'The Copernicus Atmosphere Monitoring Service (CAMS) Radiation Service in a nutshell', DLR, Service Contract No 2015/CAMS_72/SC1, Feb. 2016.

[6] B. Espinar, C. Hoyer-Klick, M. Lefèvre, M. Schroedter-Homscheidt, and L. Wald, 'User's Guide to the MACC-RAD Services on Solar Energy Radiation Resources v3 - July 2014', Public report MAC-II Deliverable D_122.6, Jul. 2014.

[7] 'SoDa - HelioClim3'. [Online]. Available: http://www.soda-is.com/eng/helioclim/helioclim3_versions.html. [Accessed: 31-Dec-2013].

[8] H. M. Deneke, A. J. Feijt, and R. A. Roebeling, 'Estimating surface solar irradiance from METEOSAT SEVIRI-derived cloud properties', *Remote Sensing of Environment*, vol. 112, no. 6, pp. 3131–3141, 2008.

[9] 'GSIP version 3, Office of Satellite and Product Operations'. [Online]. Available: http://www.ospo.noaa.gov/Products/land/gsip/index_v3.html. [Accessed: 16-Jun-2016].

[10] A. Habte, M. Sengupta, and S. Wilcox, 'Validation of GOES-Derived Surface Radiation Using NOAA's Physical Retrieval Method', *Contract*, vol. 303, pp. 275–3000, 2013.

[11] J. Remund, 'Aerosol optical depth and Linke turbidity climatology, Description for final report of IEA SHC Task 36', Meteotest, Bern, Switzerland, 2009.

[12] A. Hammer, 'Documentation - EnMetSOL', Carl von Ossietzky University of Oldenburg, Mar. 2016.

[13] J. Betcke and H. G. Beyer, 'Accuracy Improvement of Irradiation Data by Combining Ground and Satellite Measurements', presented at the Eurosun, 2004.

[14] M. Richter, K. De Brabandere, A. Woyte, J. Kalisch, T. Schmidt, and E. Lorenz, 'Uncertainties in PV Modelling and Monitoring', in *EU PVSEC 2015*, Hamburg, Germany, 2015.

Multi-Model Ensemble for day ahead PV power forecasting improvement

Marco Pierro[a,e], Francesco Bucci[a], Matteo De Felice[d], Enrico Maggioni[c], David Moser[e], Alessandro Perotto[c], Francesco Spada[c], Cristina Cornaro[a,b]

[a]Department of Enterprise Engineering, University of Rome Tor Vergata, Via del Politecnico 1, 00133 Rome, Italy, e-mail: cornaro@uniroma2.it, marco.pierro@gmail.com, frabucci@gmail.com
[b]CHOSE, University of Rome Tor Vergata, Via del Politecnico 1, 00133 Rome, Italy 7
[c]Ideam Srl, via Frova 34 Cinisello Balsamo, Italy, e-mail: alessandro.perotto, enrico.maggioni, francesco.spada@ideamweb.com
[d]Casaccia R.C., ENEA Climate Modelling Laboratory, Rome, Italy e-mail: matteo.defelice@enea.it
[e]EURAC Research, Viale Druso, 1, 39100 Bolzano, Italy e-mail: david.moser@eurac.edu

ABSTRACT: This work aims to compare several data-driven models using different Numerical Weather Prediction (NWP) input data and then to build up an outperforming Multi-Model Ensemble (MME) and its prediction intervals. Statistic, stochastic and hybrid machine-learning algorithms were developed and the NWP data from IFS and WRF models were used as input. It was found that the same machine learning algorithm differs in performance even if it make use of NWP data with comparable accuracy. This apparent inconsistency depends on the capability of the machine learning model to correct the bias error of the input data. The stochastic and the hybrid model using the same WRF input, as well as the stochastic and the non-linear statistic models using the same IFS input, produce very similar results. The MME resulting from the averaging of the best data-driven forecasts, improves the accuracy of the outperforming member of the ensemble, bringing the skill score from 42% to 46%. To reach this performance, the ensemble should include forecasts with similar accuracy but generated with the higher variety of different data-driven technology and NWP input.
Keywords: TBD

1 INTRODUCTION

The accuracy in forecasting PV power is of great relevance when high PV generation requests greater secondary reserves and ready supply to overcome the unpredictability and variability of the residual load. Moreover this large amount of energy needed to ensure electrical balancing has to be purchased at a higher price on the Real-Time Energy Market. Thus improvement of accuracy, reducing unpredictability, reduces also energy costs.

In the last five years the data-driven approach has been extensively tested for PV power generation forecast for the 24/72 hours horizon. This approach involves a wide range of machine learning methods that can be built using historical data. These statistical or stochastic models try to reconstruct the relationships between input and output data sets. They do not require knowledge of the physical laws describing the phenomenon and none or very little plant information. Once trained, these algorithms can use a variety of NWP products to directly provide the PV generation forecast. Hybrid models could be obtained using different models in series. If different predictions are combined together, a Multi-Model Ensemble can be obtained (see Figure 1).

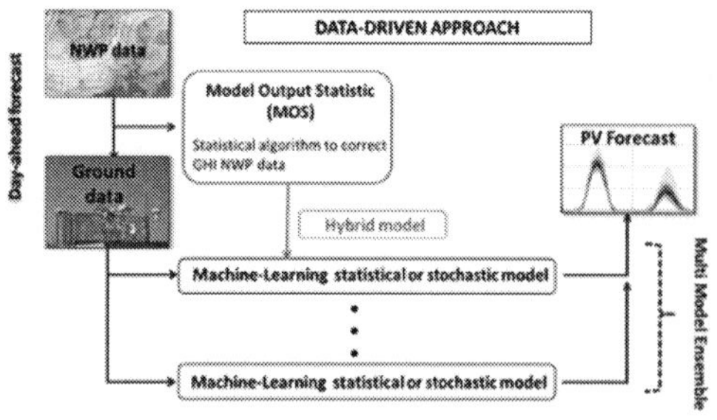

Figure 1: Schematics of data-driven approaches for day ahead PV power forecast

A variety of machine-learning models using different techniques were developed by various authors. For site day-ahead power forecasts, Yona et al. [1] implemented models based on a multilayer perceptron (MLPNN), radial basis function (RBFNN) and recurrent neural networks (RNN). Chen et al. [2] and Tao et Al. [3] used RBFNN and Non linear autoregressive exogenous neural network (NARX) while Wang et al. [4] coupled MLPNN with the gray model (GM). Mellit et al. [5] introduced adaptive feed-forward back-propagation network (AFFNN) for short-term forecasting.

For site and regional forecast, Zamo et al. [6] compared the accuracy of many different techniques: bagging, random forest, boosting, support vector machine (SVM) and generalized additive model (GAM), concluding that the random forest was the outperforming model. For regional forecast, Gari da Silva Fonseca Junior et al. [7] explored an interesting technique based on support vector regression (SVR) coupled with a principal component analysis (PCA) pre-processing.

Although literature and methodology are established for single models, a Multi-Model Ensemble approach has not been fully explored yet. Our results could contribute to improve the knowledge of solar and PV forecasting, deepening the concept of multi-model ensembles. In particular, some relevant questions related to data driven approaches have been investigated: using different irradiance predictions (NWP) as input for the same model, does the outperforming NWP produce also the best power forecast? How much could the accuracy change, if, on the contrary, the same NWP input is used for different kind of models? How can an outperforming Multi-Model Ensemble be constructed and what is the improvement with respect to the best forecast model of the ensemble? How can prediction intervals be estimated?

Thus, two different NWP data coming from the global model IFS of the ECMWF and from the mesoscale model WRF, were used. Moreover, the WRF-RRTM irradiance forecast was refined with a Model Output Statistic post-processing algorithm (MOSRH), to improve prediction accuracy. A SVM model was developed (GTSVM). Two stochastic models based on an ensemble of Artificial Neural Network (ANN) were also built (RHNN and GTNN). Then a hybrid model based on the model chains of MOSRH and GTNN was used to study the impact of the different irradiance predictions on the accuracy of the power forecast. Finally an outperforming Multi-Model Ensemble (MME) was generated together with its prediction intervals. Four years of monitored weather and production data from a 662 kWp Cadmium Telluride PV plant, located in Bolzano (Italy), were employed to train and test the models.

Moreover, in the paper a new parameter called clear sky performance index, was defined. It was used both to develop the stochastic model and to set up an outperforming persistence model. This parameter could be considered as the equivalent of the clear sky index for PV power generation.

2 DATA DESCRIPTION

2.1 Local Experimental Data

Four years of experimental data were used to train and test the forecast models: from 01/01/2011 to 31/12/2014. Measurements were performed in the outdoor test facility of Airport Bolzano Dolomiti (ABD) (position ca. 46.46N, 11.33E, elevation: 262 m) located in South Tyrol, Italy [8]. In particular, the power generation measurements are the hourly average production of a 662 kWp commercial PV plant. This plant uses Cadmium Telluride thin film technology. The modules are 8.5° South West oriented and tilted by 30°. The DC-side electrical parameters are measured every 15 minutes by commercial inverters that assure a good level of accuracy in current (Impp) and voltage (Vmpp), with an average difference from a dedicated system of less than 5% and less than 2%, respectively, further decreasing at higher irradiance [9]. Moreover, an independent meteorological station provides measurements of horizontal global and diffuse irradiance, global plane-of-the-array (GPOAI) irradiance, wind speed, air temperature and back of the module temperature with a frequency of 1 minute, which is then averaged on a 15-min time interval. The sensors are systematically cleaned and periodically calibrated in order to comply with the standard IEC61724:1998 (1998).

2.2 NWP data

2.2.1 Weather Research and Forecasting (WRF–ARW)

The NWP model used is Weather Research and Forecasting (WRF–ARW) version 3.6.1 developed by NCAR (National Center of Atmospheric Research).
The shortwave (SW) radiation scheme chosen is the Rapid Radiative Transfer Model (RRTM) because of its skill to describe sub grid cloud variability through a Monte-Carlo independent column approximation and due to its ability to distinguish near- infrared, visible and UV wavelengths. Radiation absorption and scattering by other gases and aerosols are calculated through various parameterizations. Concentration and chemical composition come from parametric values taken from pre-built tables.
Daily hindcasts were performed for the period in exam (2011-2014). The model was initialized at 12 UTC, analyzing the 24 hour forecasts starting from the following 00 UTC, which is typical procedure for the NWP solar day-ahead forecast. The model domain is centered over Italy with a horizontal resolution of 12 km, a more resolute inner domain is nested centered on the region of interest. The horizontal resolution is approximately 3 km due to the complex orography of the region.
The global horizontal irradiance (GHI) provided by WRF was post processed with an original model output statistic (MOS) called MOSRH. This MOS is described in detail in Pierro et al. [10], nevertheless for this work an outperforming version was developed (equation 1). MOSRH use GHI ground measurement and relative humidity prediction of 20 vertical levels (RH_{level}) to correct the clear sky irradiance provided by WRF-RRTM:

$$GHI(WRF)=d\ GHI_{cs}*(1-a*PCC^{b})+c \qquad (1)$$

Where PCC is the "Pseudo Cloud Cover" obtained by a weighted average of RH_{level} and the parameters (a,b,c,d) are regression coefficients obtained by fitting observational data (see Table 1).

Table 1: MOSRH regression coefficients

	a	b	c	d
PCC<0.05	0	1	22	0.96
0.05<PCC<0.7	0.36	0.88	0	1
PCC>0.7	0.68	2.33	0	1

2.2.2 Integrated Forecasting System (IFS) model of ECMWF

The Integrated Forecasting System (IFS) is the operational global weather forecasting model from European Centre for Medium-Range Weather Forecasts (ECMWF). The model runs twice per day providing the deterministic forecasts up to ten days. In the latest upgrade the spatial horizontal resolution reached 9 km although in this study data from the previous version of the model was used, with a spatial resolution of 16 km.
The radiation scheme is based on the rapid radiation transfer model (RRTM) and the cloud scheme is an enhanced version of the original scheme by Tiedtke described by Forber and Tompkins [11].
For an analysis on the predictive skills of ECMWF IFS for solar radiation we refer to the scientific works by Troccoli and Morcrette [12]. In this paper we use the hourly solar irradiance (surface downwelling shortwave).

3 CLEAR SKY PERFORMANCE INDEX

The clear sky performance index can be considered as the equivalent of the clear sky index for PV power generation and it was used to set up an outperforming persistence model, to develop the stochastic model, and to calculate the prediction intervals:

$$PK_{cs} = \frac{P}{P_n} / \frac{GPOAI_{cs}}{1000} \qquad (1)$$

where P is the PV power generated, Pn is the plant nominal power and GPOAIcs is the clear sky global tilted irradiance. The variability of this index is in between the performance ratio (PR) and the clear sky index (Kcs) depending on the sky conditions. In a perfect modelled clear sky condition (GPOAI = GPOAIcs) the PKcs is equal to the performance ratio of the PV plant: PR = (P/Pn)/(GPOAI/1000). On the contrary, in overcast conditions PKcs behaves like the clear sky index projected on the plane of array: Kcs = GPOAI/GPOAIcs.

4. FORECAST MODELS DESCRIPTION

In this section all the developed models together with the persistence model (used as benchmark) are described in details. Four data driven models were constructed:

- A statistical model based on the support vector machine algorithm (GTSVM);
- Two stochastic models based on artificial neural network ensembles (RHNN and GTNN).
- A hybrid Model obtained coupling the GTNN with the statistical post-processing MOSRH.

4.1 Support Vector Machine model

SVMs were developed by Cortes & Vapnik [13] for binary classification and then extended to regression problems (Support Vector Regression).

In our case, we use a Support Vector Regression method called ε-SVR.

After an extensive grid search on more than 400 combinations, we selected the following parameters: $C = 1$, $\gamma = 0.25$ and $\varepsilon = 0.03$. The inputs of the SVM are the same of the GTNN model that will be described in the next subsection.

4.2 Artificial Neural Network models

Artificial Neural Networks are computational modelling tools that mimic the basic structure and operation of biological neurons. These tools are often employed in solving forecasting problems; an extensive review can be found in Zhang et al. [14]. For what concerns solar radiation forecast using ANN, a review can be found in Mellit [15]. In this paper the MLPNN was adopted to develop two different forecast models: RHNN and GTNN.

The optimization procedure used to generate the MLPNN is described in detail in Cornaro et al. 2014 [16]

4.3.1. RHNN

The model RHNN provides the day-ahead forecast of the clear sky performance index (PKcs defined in section 3) with an hourly granularity. The MLPNN architecture used has seven input features, one hidden layer and PKcs output. The inputs are:

- cosine of the angle of incidence to account for the sun position with respect to the plant and the modules irradiance reflection;
- air temperature forecast at the ground (2 meters above) to consider the effect of temperature on the plant performance;
- average values and the standard deviation of the relative humidity predicted by WRF for the vertical levels below 775 hPa and between 775 hPa and 400 hPa.
- relative humidity predicted by WRF at a higher level corresponding to 300 hPa.

The last five parameters take into account the cloud formation at different atmosphere levels: low, mid and high clouds.

The PV power forecast can be derived from the PKcs prediction as following:

$$P^{for} = P_n \, PK_{cs}^{for} (GPOAI_{cs}/1000) \qquad (2)$$

Where P^{for} is the hourly power generation, P_n is the plant nominal power, $GPOAI_{cs}$ is the clear sky irradiance predicted by the WRF model and PK_{cs}^{for} is the RHNN output.

4.3.2. GTNN

The model GTNN has the same architecture of RHNN but provides directly the day-ahead forecast of the PV power generation (P^{for}) with an hourly granularity. It has five input variables:

- sun elevation and sun azimuth;
- air temperature NWP forecast at the ground (2 meters above)
- clear sky irradiance and NWP irradiance

The GTNN model was forced both with the NWP data from ECMWF and with the WRF predictions post-processed by the MOSRH algorithm.

4.4 Persistence Model

This is a trivial model that assumes that the weather conditions on the present day and on the day of the forecast are identical. There are several persistence models used in day ahead solar irradiance predictions that lead to different RMSE [17]. The simple persistence model (PM) assumes the persistence of global horizontal irradiance while other more accurate models presume the persistence of the clear sky index (KPM). In literature, regarding PV generation forecast, only the simple persistence is used.

In this paper, a better suited reference model based on the persistence of the daily PKcs index was defined:

$$PK_{cs}^{for}(day + 1) = PK_{cs}(day) \qquad (3)$$

Thus the power forecast can be derived from equation (1).

5 METRICS

According to solar forecast literature, the main metrics used to evaluate the models accuracy are reported in Table 2:.

Table 2: accuracy metrics

Acronym and formulae
$e_h = (X_h^{for} - X_h^{obs})$
$RMSE = \sqrt{\frac{\sum_{h=1}^{n} {e_h}^2}{n}}$
$MAE = \frac{\sum_{h=1}^{n} \lvert e_h \rvert}{n}$
$MBE = \frac{\sum_{h=1}^{n} (e_h)}{n}$
$Skill\,Score = 100 \left(\frac{RMSE(PM) - RMSE(for)}{RMSE(PM)}\right) \%$

Where

$X_h^{obs} = Observed\ hourly\ PV\ power\ yield\ [kW/kWp]\ or\ Irradiance\ [W/m^2]$

$X_h^{for} = Forecast\ hourly\ PV\ power\ yield\ [kW/kWp]\ or\ Irradiance\ [W/m^2]$

$Pn = Plant\ capacity\ [kWp]$

$n = number\ of\ sun\ hours$

6 RESULTS AND DISCUSSION

6.1 Forecast models accuracy analysis

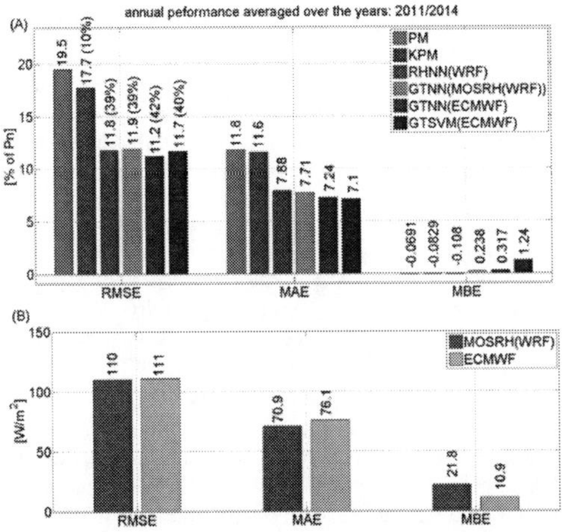

Figure 2: (A) Accuracy comparison of PV power generation forecast (yearly average), in brackets the skill score of the model; (B) Accuracy comparison of GHI numerical weather prediction (yearly average)

Figure 2A shows the yearly average of accuracy values of all the models together with the PM skill score (in the brackets). In irradiance forecast community the persistence model used as reference is usually the one obtained by the clear sky index since it achieves a lower RMSE with respect to the simple persistence [17]. Similarly, for the power forecast, the persistence calculated using the clear sky performance index (KPM) shows a 10% skill score with respect to the simple persistence model (PM), so that it could be used as a better benchmark model. Nevertheless, in literature on PV generation forecast, the simple persistence is adopted as reference model, so the skill scores reported in this section are calculated with respect to the RMSE of the PM.

The RHNN(WRF) model that uses the WRF prediction of relative humidity to provide the PV power forecast achieves the same accuracy of the hybrid model GTNN(MOSRH(WRF)) that uses the GHI prediction of WRF-RRTM post-processed by the MOSRH algorithm. This means that MOSRH and RHNN provide a similar correction to the WRF output. As a whole the RHNN(WRF) and GTNN(MOSRH(WRF)) forecast achieve a skill score of 39%.

The model based on the support vector machine GTSVM(ECMWF) uses the same five inputs and the same NWP data of GTNN(ECMWF). It obtains a skill score of 40%, with an RMSE and a MBE higher than the ANN based model but a slightly lower MAE. Nevertheless, it should be pointed out that the GTSVM(ECMWF) performs poorly mainly during the year 2011 while for the other years the RMSE is very similar to the GTNN model. Thus the statistic model based on the support vector machine could be considered essentially equivalent to the stochastic one based on the ANN ensemble.

The same stochastic model GTNN was used with different NWP inputs (ECMWF and MOSRH(WRF)). From Figure 2B it can be observed that ECMWF prediction of GHI exhibits a Mean Absolute Error (MAE) of 76 W/m^2 while MOSRH(WRF) forecast achieves a MAE of 70 W/m^2. Nevertheless the low performing NWP input (ECMWF) produces better power performance. Indeed the skill score of GTNN(ECMWF) is 42% while the skill score of GTNN(MOSRH(WRF)) is 39%. This apparent inconsistency depends on the capability of the machine learning model to correct the errors of the input data.

Figure 3 shows the GHI prediction of the different NWP (ECMWF and MOSRH(WRF)) and the power generation forecast of the two stochastic approaches, GTNN and RHNN, using respectively ECMWF and WRF data. It visually summarizes the main concepts of the present analysis. For the day 9/08/2013 the WRF model provides a higher value of the medium level relative humidity (775 – 400 hPa) and the MOSRH post processing as well as the RHNN stochastic model cannot correct this erroneous meteorological prediction. This leads to the lower accuracy of RHNN(WRF) model with respect to GTNN(ECMWF). On the other hand, for the day 10/08/2013, the time shift of the GHI predicted by ECMWF is well visible, while this hourly error disappears in the power forecast achieved by the GTNN(ECMWF) model. Thus the stochastic approach is able to correct this kind of bias providing an outperforming power forecast with respect to the NWP irradiance forecast.

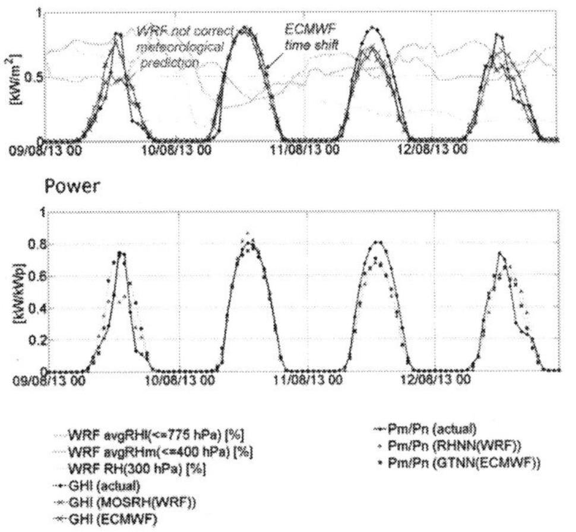

Figure 3: hourly trend of GHI and power generation forecast

As a whole, the GTNN using the GHI prediction of ECMWF is the outperforming model, reaching a 42.5% of skill score.

6.2 Multi-Model Ensemble construction end evaluation

To construct a Multi-Model Ensemble a preliminary analysis of the models behavior in a different typology of days was developed. Figure 4 shows the errors of the models versus the daily clear sky predicted by WRF. Since all the forecasts obtain similar RMSE and MBE trends, the Multi-Model ensemble was built simply averaging the different prediction trajectories.

Figure 5(A) reports the accuracy of four different Multi-Model Ensemble obtained averaging different forecast models output as follows:

1. forecasts obtained using same NWP input and same data-driven technique but different models: MME(RHNN(WRF),GTNN(WRF));

2. forecasts obtained using same NWP input but different data-driven algorithms: MME(GTSVM(ECMWF),GTNN(ECMWF));
3. forecasts obtained using same data-driven model but different NWP input: MME(GTNN(WRF),GTNN(ECMWF));
4. forecasts obtained using all the different NWP and algorithms: MME.

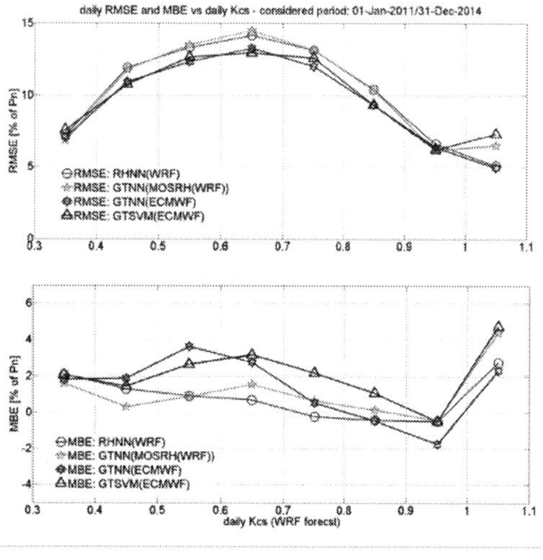

Figure 4: Errors of the models selected for the ensemble versus the daily clear sky predicted by WRF

First of all, all the multi-models achieve a higher accuracy than the outperforming member of the ensemble (RMSE improvement is reported in bracket in Figure 5 (A)). Thus, every multi-model instead of reaching an accuracy between the worst and the best member of the ensemble (as expected from the average operation), improves the forecast performance. This smoothing effect is possible since all the forecast of the ensemble exhibit similar yearly accuracy values (around 1% difference in RMSE) so that the average could effectively reduce the noise of each single predictor.

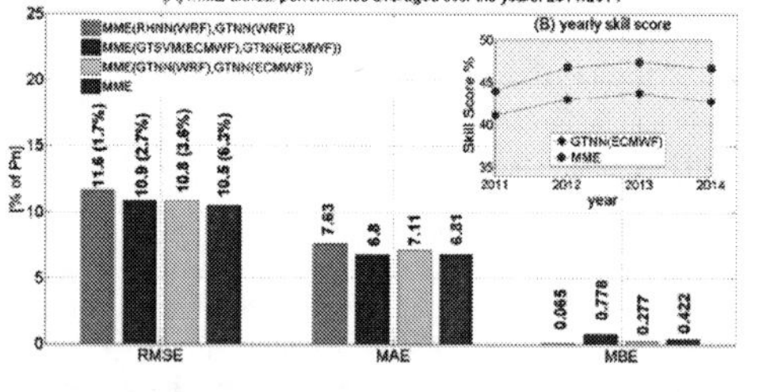

Figure 5: (A) yearly average of the accuracy of different Multi-Model Ensemble (in bracket the improvement with respect to the RMSE of the outperforming member of the ensemble); (B) yearly skill score of the MME and of the best forecast of the ensemble

The MME(RHNN(WRF),GTNN(WRF)) shows the lowest improvement (1.7%) with respect to the RMSE of RHNN(WRF). Indeed, this multi-model averages forecasts obtained with the same NWP input (WRF) and based on the same machine learning technique (ANN). The MME(GTSVM(ECMWF),GTNN(ECMWF)), built averaging the forecast based on different data-driven algorithms (SVM and ANN) but using the same NWP input (ECMWF), achieves an improvement of 2.7% with respect to the RMSE of GTNN(ECMWF).

The MME(GTNN(WRF),GTNN(ECMWF)) based on the same algorithm (ANN) but using input data generated by different NWP models (WRF and IFS) obtains a greater improvement of 3.6% with respect to the RMSE of GTNN(ECMWF).

The outperforming Multi-Model is the one built using all the four forecast trajectories (MME), which doubles the improvement reaching the value of 6.3% with respect to the RMSE of GTNN(ECMWF). Figure 5(B)) shows the yearly skill score of the MME and of the best forecast of the ensemble GTNN(ECMWF).

Thus, the best performance of the multi-model approach could be achieved averaging the higher variety of different algorithms and different NWP models (to obtain different hourly errors time series) with the only condition that all the ensemble members should have similar RMSE.

As a whole the MME reaches an yearly average RMSE of 10.5% of Pn and a skill score of 46% with respect to the RMSE of PM.

Unfortunately in literature there are no European accuracy benchmarks on PV power generation forecasts, like [18] for the GHI prediction, so that it is not easy to understand the quality of the reported forecast methods. Nevertheless, Lorenz et al. [19] reports a German average value of RMSE for single site one day ahead forecast of 12.8%, achieved using a deterministic approach. Always using a deterministic approach with GEM post-processed NWP, Pelland et al. [20] found a RMSE in the range of 6.38%-9.17% of Pn corresponding to a Skill Score between 36.7% and 64% with respect the RMSE of PM, for three Canadian PV plants. Da Silva Fonseca Junior et al. [7], report some accuracy values for regional PV generation forecast in Japan. Nevertheless, from the study, the single site RMSE obtained by Support Vector Regression model could be deduced: 9% of 14% of Pn. For a single plant in a French counties, Zamo et al. [6], experiment different forecast approaches. They obtain a skill score between 29.4% and 47%. Thus, the performance of MME is perfectly inside the range of accuracy reported in literature.

6.4. Prediction Intervals estimation and evaluation

The prediction intervals can be calculated forecasting the standard deviation of the residuals (σ^{for}) under the hypothesis that the residuals are normally distributed with zero expected value. In the case of solar irradiance forecast, Lorenz et al. [21], show that the error standard deviation could be predicted with a good accuracy by a four order polynomial function of the zenith angle cosine and the predicted clear sky index.

Following this idea, the predicted standard deviation of the PV generation error was here supposed as a function of the cosine of the angle of incidence (*cos(AOI)*) and the predicted clear sky performance index (PK_{cs}^{for}). Then σ^{for} was predicted using the same machine learning technology used to provide the forecast. Thus a ANN ensemble was trained using *cos(AOI)* and PK_{cs}^{for} as input and the square forecasted error (e^2) as a target. As for the power forecast the model was trained on 2012 to forecast the years 2011, 2013 and 2014 while it was trained on 2011 to predict the error of 2012.

Similar approach was used for solar irradiance forecast by Marquez and Coimbra [22].
Figure 6(A) shows the shape of the predicted standard deviation. As in [21], the maximum MME error appears at low incidence angles in overcast and variable sky conditions (predicted PK_{cs} in between 0.4 and 0.7).

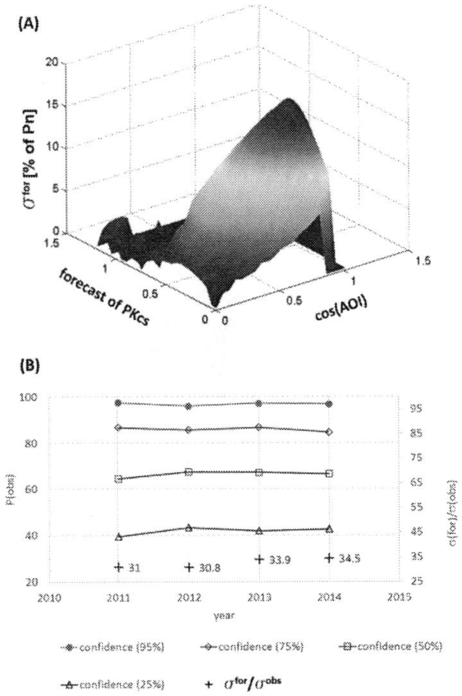

Figure 6: (A) Shape of the predicted standard deviation; (B) Yearly trend of observed frequency and the ratio between the predicted and observed standard deviation

Figure 6 (B) reports the yearly trend of the observed frequency for different confidence levels and the ratio between the predicted and observed standard deviation. It can be noted that the frequency of the observations falling inside the prediction interval is always greater than the corresponding confidence level, so that a reliable estimation of prediction intervals was provided.
The yearly average of the predicted standard deviation is between 30% and 35% of the standard deviation of the observations.
The prediction interval for solar irradiance estimated by Lorenz et al. [21] achieves a $< \sigma^{for} > / < GHI^{obs} >$ of 28.4% while in this case the value of 30.8% was found. Considering that the $< GHI^{obs} >$ for Germany is lower than Bolzano but also that the error on optimal tilted surface is greater than the one obtained on the horizontal plane, this could be considered a comparable result.
Figure 7 shows the trend of the MME forecast and its prediction intervals for five different typologies of days in 2011.

7 CONCLUSIONS

In this paper several data-driven models were compared using different NWP input data and then an outperforming Multi-Model Ensemble was built.
Two different Numerical Weather Prediction data, coming from the global model IFS of the ECMWF and from the mesoscale model WRF, were used. Moreover, the WRF-RRTM irradiance forecast was refined with a Model Output Statistic post-processing algorithm (MOSRH), to improve the accuracy of the prediction.

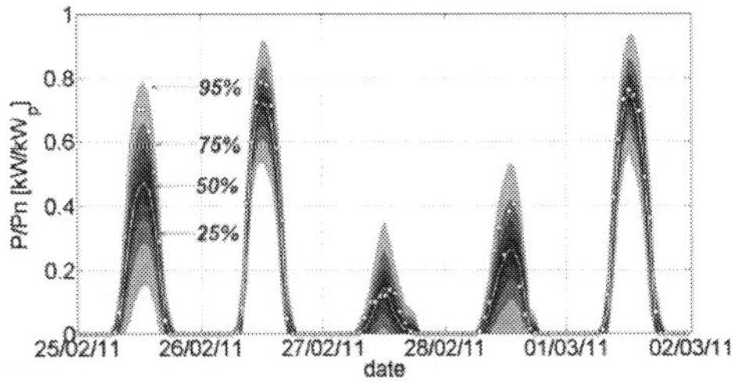

Figure 7: Observation (dots), MME forecast (white line) and prediction intervals (grey lines) for five days of 2011

One statistical model based on SVM algorithm was developed: GTSVM(ECMWF). Two stochastic models based on ensemble of ANN were built: RHNN(WRF) and GTNN(ECMWF). Then a hybrid model based on the model chains of MOSRH and GTNN: GTNN(MOSRH(WRF)) was used to study the impact of the different irradiance prediction on the accuracy of the power forecast. Finally an outperforming Multi-Model Ensemble (MME) was generated together with its prediction intervals.
It was found that the performance of a stochastic model are not always related with the accuracy of the NWP information used. For example, the accuracy of the GTNN(ECMWF) was higher than the accuracy of GTNN(MOSRH(WRF)) even if the GHI predicted by the MOSRH(WRF) post-processing outperforms the ECMWF GHI forecast. This apparent inconsistency depends on the capability of the machine learning model to correct the bias errors of the input data.
On the contrary, models based on different no linear machine learning algorithms (stochastic or statistic) but making use of the same NWP data provided forecast with similar accuracy. Indeed the models GTNN(ECMWF) and GTSVM(ECMWF) as well as the model RHNN(WRF) and hybrid GTNN(MOSRH(WRF)) showed similar performance.
As a whole the GTNN(ECMWF) provided the best forecast with a RMSE of 11.2% of Pn and a skill score of 42%. Finally the MME obtained averaging the four forecasts achieved a considerable improvement in performance. It reached a RMSE of 10.5% of Pn and a skill score of 46%. The MME outperformed the best model of the ensemble, since the average reduced the noise of each single predictor. In particular, to reach this improvement, the ensemble should include all forecast trajectories with similar accuracy (around 1% of difference in RMSE) but generated by the highest variety of different algorithms and NWP input. It should be remarked that these results cannot be considered a general conclusion on multi-model approach. This is only a first case study and further research should be done using different NWP and data-driven techniques in other locations.

8 ACKNOWLEDGEMENTS
DM and MP are indebted to the funding received from the EURAC internal

7 REFERENCES

[1] A. Yona, T. Senjyu, A. Y. Saber , T. Funabashi and H. Sekin, «Application of neural network to 24hours ahead generating power forecasting for PV system», in *Proceedings of IEEE power and energy society 2008 general meeting: Conversion and delivery of electrical energy in the 21st century*, 2008.

[2] C. Chen, S. Duan, T. Cai and B. Liu, «24-h solar power forecasting based on weather type classification using artificial neural network», *Solar Energy*, n. 85, pp. 2856-2870, 2011.

[3] C. Tao, D. Shanxu and C. J. Changsong, «Forecasting power output for grid-connected photovoltaic power system without using solar radiation measurement», in *The 2nd International Symposium on Power Electronics for Distributed Generation Systems. Institute of Electrical & Electronics Engineers (IEE)*, 2010.

[4] S. Wang, N. Zhang, Y. Zhao and J. Zhan, «Photovoltaic system power forecasting based on combined grey model and BP neural network», in *ICECE proceedings, 4623-4626*, 2011.

[5] A. Mellit, A. M. Pavan and V. Lughi, «. Short-term forecasting of power production in a large scale photovoltaic plant.», *Solar Energy*, n. 105, pp. 401-413, 2014.

[6] M. Zamo, O. Mestre, P. Arbogast and O. Pannekoucke, «A benchmark of statistical regression methods for short-term forecasting of photovoltaic electricity production part I: Deterministic forecast of hourly production», *Solar Energy*, n. 105, pp. 792-803, 2014.

[7] J. G. Da Silva Fonseca Junior, T. Oozeki, H. Ohtake, K. Ichi Shimose, T. Takashima and K. Ogimoto, «.Regional forecasts and smoothing effect of photovoltaic power generation in Japan: An approach with principal component analysis», *Renewable Energy*, vol. 68, n. 403-413, 2014.

[8] G. Belluardo, P. Ingenhoven, W. Sparber, J. Wagner , P. Weihs and D. Moser, «Novel method for the improvement in the evaluation of outdoor performance loss rate in different PV technologies and comparison with two other methods», *Solar Energy Journal*, vol. Volume 117.

[9] D. Bertani, S. Guastella, G. Belluardo and D. Moser, «Long term measurement accuracy analysis of a commercial monitoring system for photovoltaic plants», in *In: IEEE Workshop on Environmental, Energy and Structural Monitoring Systems (EESMS)*, 2015.

[10] M. Pierro , F. Bucci , C. Cornaro , E. Maggioni, A. Perotto, M. Pravettoni and F. Spada, «Model output statistics cascade to improve day ahead solar irradiance forecast», *Solar Energy Journal*, vol. 117, pp. 99-113, July 2015.

[11] R. Forbes and A. Tompkins, «An improved representation of cloud and precipitation», ECMWF Newsl. 129, 1318, 2011.

[12] A. Troccoli e J. J. Morcrette, «Skill of direct solar radiation predicted by the ECMWF global atmospheric model over Australia», *J. Appl. Meteorol. Climatol*, vol. 11, n. 53, p. 2571–2588, 2014.

[13] C. Cortes and V. Vapnik, «Support-vector networks», *Mach. Learn.* , vol. 3, n. 20, p. 273–297, 1995.

[14] G. Q. Zhang, B. E. Patuwo and M. Y. Hu, «Forecasting with artificial neural networks: the state of the art», *International Journal of Forecasting* , n. 14, p. 35–62.

[15] A. Mellit, «Artificial Intelligence technique for modelling and forecasting of solar radiation data: a review», *Int. J. Artif. Intell. Soft Comput.* , vol. 1, n. 1, p. 52–76, 2008.

[16] C. Cornaro, M. Pierro and F. Bucci, «Master Optimization Process based on Neural Networks Ensemble for 24-h solar irradiance forecast,» *Solar Energy*, n. 111, pp. 297-312, 2014.

[17] H. G. Beyer, J. Polo Martinez, M. Suri, J. L. Torres, E. Lorenz , S. C. Müller, C. Hoyer-Klick and P. D. Ineichen, «Report on Benchmarking of Radiation Products,» Report under contract no. 038665 of MESoR, January 12, 2012.

[18] E. Lorenz, J. Remund, S. Müller, W. Traunmüller, G. Steinmaurer, D. Pozo, J. A. Ruiz-Arias, V. Lara Fanego, L. Ramirez, M. G. Romeo, C. Kurz, L. M. Pomares and C. G. Guerrero, «Benchmarking of Different Approaches to Forecast Solar Irradiance,» in *Proceedings of the 24th European Photovoltaic Solar 35 Energy Conference*, Hamburg, Germany, 2009.

[19] E. Lorenz, «PV production forecast of balance zones in Germany,» EA PVPS Task 14 & SHC Task 46, 2015.

[20] S. Pelland, G. Galanis and G. Kallos, «Solar and photovoltaic forecasting through post-processing of the Global Environmental Multiscale numerical weather prediction model.,» *Prog. Photovolt: Res. Appl.*, vol. 3, n. 21, p. 284–296, 2011.

[21] E. Lorenz, J. Hurka , D. Heinemann and H. G. Beyer, «Irradiance forecasting for the power prediction of grid connected photovoltaic systems», *IEEE Journal of Selected Topics in Applied Earth Observations and Remote Sensing*, vol. 1, n. 2, pp. 2-10, 2009.

[22] R. Marquez and C. F. Coimbra, «Forecasting of Global and Direct Solar Irradiance Using Stochastic Learning Methods, Ground Experiments and the NWS Database», *Solar Energy*, n. 85, pp. 746-756., 2011.

OPTIMAL SELECTION OF TRAINING DATASETS FOR SOLAR NOWCASTING MODELS

Antonio Sanfilippo, Luis Pomares, Daniel Perez-Astudillo, Nassma Mohandes, Dunia Bachour
{asanfilippo, lpomares, dastudillo, nsalim, dbachour}@qf.org.qa
Qatar Environment and Energy Research Institute, Hamad Bin Khalifa University, Qatar Foundation
P.O. Box 5825, Doha, Qatar

ABSTRACT: Most solar nowcasting approaches use regression analysis to fit a model to historical solar irradiance data which predicts solar irradiance in near-real time (e.g. intra-hourly). A priori, any amount of historical data can be used as training material, e.g. months to years. However, evidence suggests the hypothesis that fitting a solar nowcasting model to a data partition that presents some form of solar irradiance cohesion (e.g. clear vs. cloudy sky) can help select the appropriate forecasting technique (e.g. linear vs. non-linear) to achieve increased accuracy. In this paper, we present a systematic study to test this hypothesis by (a) developing different methods to select solar data partitions to train nowcasting models, (b) assessing the relative performance of each method by evaluating the same forecasting algorithms with the different data partitions, (c) developing an optimal technique to identify and match partitions in solar time series data with forecasting algorithm.

Keywords: solar forecasting; solar irradiance data; data de-trending; data de-seasoning.

1 PURPOSE OF THE WORK

The ability to forecast solar irradiance in near-real time (now-casting) is crucial in managing the integration of solar energy in power grids. Stochastic techniques (e.g. autoregressive and neural network models) have been shown to be more effective than physics-based approaches (e.g. numerical weather prediction models) for intra-hour solar forecasts [1, 2]. Such stochastic techniques typically rely on regression methods to learn coefficients which provide the basis for prediction by measuring the relationship between an observation at time t and the observations at previous times.

For example, within an autoregressive modeling approach, AR(N), the current term x_t of a time series $x_i, \dots, x_N$, for $N \geq t \geq i$, can be estimated as the linear weighted sum of previous terms $x_{t-1}, \dots, x_{t-N}$ in the series, as shown in (1) were c is a constant, ε_t is white noise, and $a_i, \dots, a_N$ are the auto-regression coefficients that function as the weights in the sum. AR coefficients are usually calculated using Yule–Walker equations or regression techniques such as least squares estimation or maximum entropy. For additional details see [3].

$$x_t = c + \sum_{i=1}^{N} a_i \, x_{t-i} + \varepsilon_t \qquad (1)$$

The coefficients of a forecasting model such as AR(N) are typically learned from a collection of historical time series data. A priori, any amount of historical time series data can be used as training material, e.g. several months, a whole year, or several years. However, evidence suggests the hypothesis that fitting a solar nowcasting model to a data partition that presents some form of solar irradiance cohesion (e.g. seasons, clear vs. cloudy sky) can improve model performance. For example, Diagne et al. [1] observe that when using statistical analysis for forecasting purposes, it is desirable to de-trend and de-season the training datasets. This is because most statistical time series analysis assume that the input time series data are stationary or at least weakly stationary. In this paper, we present a systematic study to test this hypothesis by

- developing different methods to de-trend solar data for training nowcasting models
- assessing the relative performance of each method by evaluating the same forecasting algorithms with the different data partitions
- defining a technique to identify the de-trended class of solar time series data (e.g. in terms of their variability) so that each time series can be processed using the most appropriate forecasting algorithm.

2 SCIENTIFIC INNOVATION AND RELEVANCE

Several approaches have been developed to de-trend hourly solar irradiance. Jain [4] and Baig et al. [5] propose a Gaussian function to fit the experimental ratio of hourly to daily solar radiation data, and an alternative method is presented by Kaplanis [6]. Akarslan and Hocaoglu [7] use two partitioning techniques, one based on seasons and the other on the clearness index, to establish whether a linear or an empirical model should be used to predict hourly solar radiation. Ji and Chee [8] measure the stationarity of a de-trended series using the Augmented Dickey Fuller test to establish when to switch from a linear to a non-linear solar prediction model.

These studies have provided useful insights on how de-trending and de-seasoning can help select the appropriate prediction model. However, a systematic study of various de-trending techniques and their relative merits with reference to diverse solar prediction algorithm is still missing. The purpose of this paper is to address this gap.

3 APPROACH

In addition to the techniques used in the literature hitherto, we investigate the use of data mining methods for de-trending solar irradiance time series data. One such technique is the use of clustering algorithms to detect latent classes of varying stationarity in solar irradiance time series data, and then using classification to verify the goodness of the clusters identified using precision and recall measures.

Using dataset of 5-minute averages of normalized clearness index measurements [9] (K^*_T) collected in Education City, Doha, Qatar, in 2014 [10], we develop models that provide partitions of the data in several sets of clusters (e.g. 12 to 2 clusters) using the expectation–maximization algorithm. Each record in the input dataset to clustering consists of a date field (e.g. month, day, hour, and minute) and 12 consecutive 5-minute K^*_T measurements, as shown in Table I.

Table I: Data record sample

Month	*Day*	*Hour*	*Min*	KTp_1	...	KTp_{12}
...	...	...	...	...	...	...
1	1	6	45	0.86	...	0.74
1	1	6	50	0.84	...	0.72
...	...	...	...	...	...	...

We then train a Bayesian net classifier with each clustered dataset. In a preliminary study, we have used 10-fold cross-validation to evaluate the classification effort, and obtained average precision-recall results from 93.4% for a 12-clusters dataset up to 96.9% for a 2-clusters dataset. These results indicate that any of these cluster datasets provide a reliable partitioning of the solar irradiation time series data.

Next, we train and evaluate several forecasting models that provide 12 steps-ahead predictions at 5 minutes intervals, using training datasets obtained through de-seasoning and the data mining de-trending approach described above. We use persistence as baseline for model comparison. We also train and evaluate the models with full year dataset to provide a term of comparison for the models developed with the other data partitions. We use the *relativized Root Mean Square Error* (rRMSE) shown in (2) as one of the main measures to evaluate model accuracy.

$$rRMSE = \frac{\sqrt{\frac{\sum_{t=1}^{n}(x_{predicted\ value_t} - x_{observed\ value_t})^2}{n}}}{mean(observed\ values_{1,\dots,n})} \quad (2)$$

4 RESULTS

Table II shows the evaluation of forecasting results for the whole year and by season, relative to a 12 steps-ahead horizon, for an autoregressive model of order 11 – AR(11). These results indicate that when trained on data partitions that are more stationary (e.g. season 2 and 3), the error rate for AR(11), which is a linear model, is lower than when the model is trained on the whole year dataset. When the model is trained on data partitions that are less stationary, i.e. seasons 1 and 4 covering the months (Nov-Mar) which display the greatest K^*_T variability in Qatar [11], the opposite results obtain.

Table III shows the evaluation of forecasting results for an Artificial Neural Network model (ANN) [12] using data partitions for the whole year, by season, and by cluster (see description above), averaged across a 12 steps-ahead horizon. As in the evaluation results shown in Table I, the error rate for seasons 1 and 4 is higher than those for other seasons and the whole year due to the higher variance in the data. Interestingly, the ANN results for seasons 1 and 4 (Table III) are better than the AR(11) results (Table II), probably due to the capability of ANN to approximate non-linear functions of their inputs.

Table II: Evaluation results for the autoregressive model of order 11 expressed as rRMSE percentages.

Horizon	*Whole year*	*Season 1*	*Season 2*	*Season 3*	*Season 4*
1	8.68	11.58	8.87	5.31	10.92
2	11.27	14.47	11.35	7.46	14.08
3	12.76	15.83	12.66	8.97	15.92
4	13.83	16.70	13.57	10.24	17.12
5	14.68	17.32	14.23	11.35	18.12
6	15.43	17.86	14.75	12.32	19.04
7	16.09	18.29	15.22	13.24	19.84
8	16.70	18.74	15.65	14.08	20.52
9	17.25	19.17	16.03	14.85	21.06
10	17.75	19.55	16.39	15.55	21.56
11	18.23	19.94	16.75	16.18	22.04
12	18.69	20.36	17.04	16.79	22.45
Average	15.11	17.48	14.38	12.20	18.56

Table III: Evaluation results for the Artificial Neural Network model by season and cluster across (average rRMSE percentages across 12 steps ahead).

Whole year	*Season 1*	*Season 2*	*Season 3*	*Season 4*
12.89	16.81	13.06	7.66	16.33

Whole year	*Cluster 1*	*Cluster 2*	*Cluster 3*	*Cluster 4*
12.89	2.04	5.42	27.57	1.92

The cluster results in Table III corroborate the observation that error rate in the ANN forecasting model is highly correlated with increasing data nonlinearity. Cluster 3 is both the cluster with the highest error rate for the ANN model (Table III) and the cluster with the highest variance, as show in the standard deviation measurements of the data by horizon and by time series (Table IV).

Most importantly, a comparison of forecasting results in Table III shows that clustering is a more effective de-trending method than a season-based approach in separating time series data in terms of ease of modeling due to their variability/non-linearity. This improved de-trending functionality has a significant beneficial effect on overall performance: the average rRMSE across cluster results is 4.23% lower than the average rRMSE across season results (9.24% vs. 13.47%). Moreover, a de-trending approach that is more effective in concentrating data variability offers the opportunity to develop algorithmic solutions that are specifically designed to address the ensuing data complexities, with potential additional gain in performance through error reduction (see section 5).

The results of persistence model (Table V), according to which the value for each step-ahead is by definition the same as the present value, upholds the trend of the forecasting results obtained with the AR(11) and ANN models. More specifically, the error rate of forecasting results for seasons 2 and 3 and clusters 1, 2 and 4, which present less variability, are lower than the error rate for the whole year. Conversely, error rate of forecasting results for seasons 1 and 4 and cluster 3, which present greater variability is higher than the error rate for the whole year. Overall, the AR and ANN models perform significantly better than Persistence.

Table IV: Average standard deviation by cluster (%).

	Cluster 1	*Cluster 2*	*Cluster 3*	*Cluster 4*
By horizon	2.04	3.62	18.8	1.63
By time series	0.77	2.10	10.1	0.79

Table V: Evaluation results for the Persistence model by season and cluster across (average rRMSE percentages across 12 steps ahead).

Whole year	*Season 1*	*Season 2*	*Season 3*	*Season 4*
16.23	19.61	15.82	12.78	20.17

Whole year	*Cluster 1*	*Cluster 2*	*Cluster 3*	*Cluster 4*
16.23	2.62	8.55	35.87	2.67

5 FINAL REMARKS

As the variance in the input time series data increases, the solar forecasting task becomes more complex and simple linear approaches such as autoregressive modeling tend to underperform, as shown by comparing AR and ANN results by season in Tables II and III. We expect that with solar irradiance data that present higher variability, forecasting algorithms capable of approximating non-linear functions of their inputs or specifically designed to model nonlinear data will yield a lower error rate. These include the ANN approach explored in this paper and Support Vector Regression [13], as well as algorithms specifically designed to model nonlinear data, such as those based on nonlinear regression [14]. Another possibility is to use stochastic models such as ANN that integrate ground and satellite solar irradiance measurements with outputs of numerical weather prediction model [15]. Multi-model approaches that combine different stochastic models [11, 16] also offer a promising solution.

Once the appropriate algorithmic solution for each level of data complexity is found, its application is straightforward. For example, according to the cluster-based de-trending approach proposed in this paper, each cluster would be matched with a forecasting algorithm appropriate to address the data complexity of the cluster. New input time series data would be processed by the classifier trained on the results of the clustering model developed with historical solar irradiance data (see section 3). This classification step will assign a cluster to each input time series record. Thereafter, the classified time series data will be processed by the forecasting algorithm appropriate to the cluster class of the time series to provide solar irradiation predictions at the appointed horizons.

6 REFERENCES

[1] Diagne, M., David, M., Lauret, P., Boland, J., & Schmutz, N. (2013). Review of solar irradiance forecasting methods and a proposition for small-scale insular grids. Renewable and Sustainable Energy Reviews, 27, 65-76.

[2] Inman, R. H., Pedro, H. T., & Coimbra, C. F., 2013. Solar forecasting methods for renewable energy integration. Progress in energy and combustion science, 39(6), 535-576.

[3] Brockwell, P. J., & Davis, R. A., 2006. Introduction to time series and forecasting. Springer Science & Business Media.

[4] Jain, P. C. (1984), Comparison of techniques for the estimation of daily global irradiation and a new technique for the estimation of global irradiation, Solar Wind Technol. 1: 123-134.

[5] Baig, A., P. Akhter and A, Mufti (1991), A novel approach to estimate the clear day global radiation, Renewable Energy, l:(1) 119-123.

[6] Kaplanis, S. N. (2006). New methodologies to estimate the hourly global solar radiation; comparisons with existing models. Renewable Energy, 31(6), 781-790.

[7] Akarslan, E., & Hocaoglu, F. O. (2016). A novel adaptive approach for hourly solar radiation forecasting. Renewable Energy, 87, 628-633.

[8] Ji, W., & Chee, K. C. (2011). Prediction of hourly solar radiation using a novel hybrid model of ARMA and TDNN. Solar Energy, 85(5), 808-817.

[9] Perez, R., Ineichen, P., Seals, R., & Zelenka, A. 1990. Making Full use of the Clearness Index for Parameterizing Hourly Insolation Conditions. Solar Energy, 45, 111-114.

[10] Perez-Astudillo, D., Bachour, D., 2014. DNI, GHI and DHI Ground Measurements in Doha, Qatar. Energy Procedia, 49(0), 2398-2404.

[11] Sanfilippo, A., Martin-Pomares, L., Mohandes, N., Perez-Astudillo, D., & Bachour, D. (2016). An adaptive multi-modeling approach to solar nowcasting. Solar Energy, 125, 77-85.

[12] ANN. http://www.mathworks.com/products/neural-network/ . Visited on 20 June 12016.

[13] Müller, K. R., Smola, A. J., Rätsch, G., Schölkopf, B., Kohlmorgen, J., & Vapnik, V. (1997). Predicting time series with support vector machines. In Artificial Neural Networks—ICANN'97 (pp. 999-1004). Springer Berlin Heidelberg.

[14] Seber, George AF. (2015). Nonlinear Regression Models. In *The Linear Model and Hypothesis*. Springer International Publishing, 2015. 117-128.

[15] Aguiar, L. M., Pereira, B., Lauret, P., Díaz, F., & David, M. (2016). Combining solar irradiance measurements, satellite-derived data and a numerical weather prediction model to improve intra-day solar forecasting. *Renewable Energy*, *97*, 599-610.

[16] Lauret, P., Rodler, A., Muselli, M., David, M., Diagne, H., & Voyant, C. (2012). A Bayesian model committee approach to forecasting global solar radiation. *arXiv preprint arXiv:1203.5446*.

MATHEMATICAL PARAMETRISATION OF IRRADIANCE TRANSITIONS CAUSED BY MOVING CLOUDS FOR PV SYSTEM ANALYSIS

Kari Lappalainen and Seppo Valkealahti
Tampere University of Technology, Department of Electrical Engineering
P.O. Box 692, FI-33101 Tampere, Finland
E-mail: kari.lappalainen@tut.fi, seppo.valkealahti@tut.fi

ABSTRACT: Irradiance transitions caused by moving clouds can have considerable negative effects on the operation of photovoltaic (PV) systems. They may lead to failures in maximum power point tracking causing extra losses and to mismatch power losses due to partial shading. Further, they can cause significant fluctuations in the output power of PV systems.

This paper presents a method to parametrise irradiance transitions caused by moving clouds based on a mathematical model of the transitions. Irradiance transitions were parametrised by four variables: shading strength, shape related parameter *b*, speed and direction of movement which have no correlation with each other. The applicability of the parametrisation method in PV system analysis was demonstrated by simulations. The simulation results show that the developed parametrisation method is suitable for long-term PV system analysis.
Keywords: Shading, Solar Radiation, PV System

1 INTRODUCTION

Fast variability of irradiance is the main cause of fluctuating photovoltaic (PV) power production and overpassing cloud shadows are the main reason of such variability. Irradiance transitions caused by edges of cloud shadows can be very steep and large and might lead to failures in maximum power point tracking causing extra losses and to mismatch power losses due to partial shading of the PV system. With fast growth of PV power production, there is a growing potential of PV output power variability having a negative effect on the power quality and reliability in the grid. Although the geographic dispersion of PV production dampens the effects of irradiance fluctuations [1]–[3], this issue is of special importance locally and in small grids with high PV penetration.

The effects of overpassing cloud shadows on the operation of PV power systems have been studied e.g. in [4]–[7]. Variability of solar radiation due to moving clouds has been studied e.g. in [8]–[11]. The characteristics and identification of irradiance transitions caused by edges of cloud shadows have been studied in [12]–[14]. In [12] a comprehensive study of identification, analysis and mathematical modelling of irradiance transitions caused by moving clouds has been presented. It has been found that irradiance can change over 300 W/m^2 in 0.1 s during an irradiance transition. Determination of shadow velocity has been studied e.g. in [15]–[17]. In [18] a comprehensive analysis of shading periods caused by moving clouds has been presented and the Linear Cloud Edge (LCE) method presented in [16] has been used to determine the velocity of shadows. It has been found that the speed of shadows varies considerably with an average value of around 13 m/s.

Irradiance transitions caused by moving clouds can have considerable negative effects on the operation of PV systems and there is an apparent need to understand them better. Irradiance transitions can be extremely fast and their characteristics vary greatly and have no correlation with each other. Therefore, the examination of these effects considering local weather conditions is not straightforward. For example, calculation of mismatch losses during one year for a selected PV generator layout by using measured irradiance transitions is computationally demanding. Thus, especially for long-term PV system analysis, more effective methods are needed for analysing the operation of PV systems under changing irradiance conditions based on measured data to ensure reasonable computing time and appropriate accuracy. Mathematical parametrisation of irradiance transitions offers a solution for this.

In this paper, a method is presented to parametrise irradiance transitions caused by moving clouds based on a mathematical model of the transitions. Irradiance transitions can be parametrised by four variables: shading strength SS, shape related parameter *b*, speed and direction of movement which have no correlation with each other. All the other variables needed to define irradiance transitions mathematically can be calculated from SS, *b* and initial irradiance. The symmetry of practical PV generator layouts simplifies the parametrisation process.

In Section 2, the mathematical model of irradiance transitions and the used measurement dataset are introduced. Further, the distributions of the shading strength, *b*, speed and direction of movement of all the identified irradiance transitions are presented. The used simulation model is presented in Section 3. In Section 4, the developed parametrisation method is presented and its applicability in PV system analysis is demonstrated. Finally, conclusions are presented in Section 5.

2 IRRADIANCE TRANSITONS CAUSED BY MOVING CLOUDS

2.1 Mathematical model of irradiance transitions

The developed method for parametrisation of irradiance transitions is based on the mathematical model of transitions presented in [12]

$$G(t) = \frac{a}{1+e^{(t-t_0)/b}} + c, \tag{1}$$

where G is the irradiance, t the time, a the change of irradiance and c the minimum irradiance during the transition. Parameter b is related to the shape and duration of the transition and the sign of it defines whether the transition is a fall, i.e., a decreasing irradiance transition, or a rise, i.e., an increasing transition. Parameter t_0 adjusts the transition time defining the midpoint of the transition. In [12] the operation of the model has been validated with around 40 000 measured irradiance transitions.

Shading strength SS of an irradiance transition, can be written as

$$SS = \frac{G_{us} - G_s}{G_{us}}, \quad (2)$$

where G_{us} is the irradiance of an unshaded situation and G_s the irradiance under shading.

2.2 Measurement data

The chosen dataset consists of 15 months (457 days) of data measured by irradiance sensors S2, S5 and S6 of the TUT solar PV power station research plant (see [19]) with a sampling frequency of 10 Hz. The dataset contains all the months of the years 2011–2014 with good summer time insolation conditions for PV power production from which comprehensive measurements (over 95% of time) are available. The same dataset have been used in [18]. The used irradiance sensors are photodiode-based SP Lite2 pyranometers (Kipp&Zonen) oriented nearly towards due south and mounted with a tilt angle of 45° from the horizontal plane which provides the maximum annual output power for fixed PV power plant installations at the Tampere region.

Irradiance transitions were identified in the data by using the method presented in [12]. The method uses a sampling frequency of 1 Hz during identification to ensure reasonable computing time and a sampling frequency of 10 Hz when analysing the identified irradiance transitions. The irradiance before and after a transition was checked to be steady, to ensure that the transition is not occurring during another irradiance fluctuation or is not only a part of a larger transition. In order to identify only irradiance transitions meaningful to the operation of PV generators, the shading strength was checked to be more than 40%. It has been shown in [12] that moving shadows with lower shading strength have no significant effect on the operation of series-connected PV strings.

2.3 Characteristics of irradiance transitions

In total, 27 210 irradiance transitions, i.e., shadow edges, (13 241 falls and 13 969 rises) were identified by all the three sensors S2, S5 and S6. The distributions of the shading strength and absolute value of parameter b of all the identified transitions are presented in Fig. 1. The velocities of the identified shadow edges were calculated from the time lags between shading of the three irradiance sensors. The curve fits of (1) were used to identify the same irradiance transition out of the transitions identified by all the three sensors and to define the time lags between shading of the sensors. A clearly erroneously high speed of over 150 m/s was obtained for one shadow edge and it was thus rejected. The distributions of the speed and direction of movement of all the identified shadow edges are presented in Fig. 2. Angle 0° denotes the movement of the shadow edge towards the north and the angle increases clockwise. The median values of the SS, absolute value of b and speed of the identified shadow edges are 57.8%, 1.48 s and 7.86 m/s, respectively. Light shadows are more common than dark shadows and the maximum shading strength is around 90%. Parameter b has a positive skew distribution and the speed has also to some extent. Dominant shadow edge movement appears to be towards the directions from north-east to east.

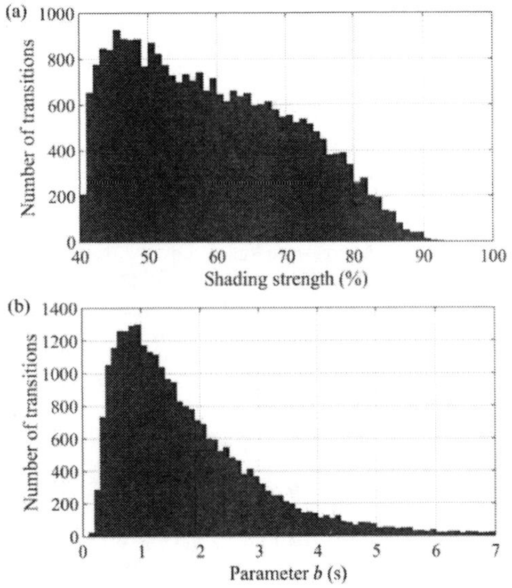

Figure 1: The distributions of the shading strength (a) and absolute value of parameter b (b) of all the identified irradiance transitions.

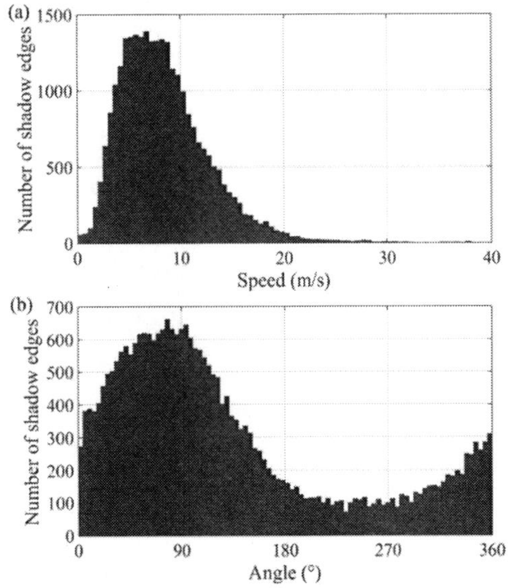

Figure 2: The distributions of the speed (a) and direction of movement (b) of all the identified shadow edges.

3 SIMULATION MODEL

Electrical operation of PV modules were modelled by an experimentally verified MATLAB Simulink model a PV cell based on the model presented by Villalva et al. [20]. The model is based on the well-known one-diode model providing the following relation between the current and voltage of a PV cell:

$$I = I_{ph} - I_0\left(e^{\frac{U+R_sI}{AU_T}} - 1\right) - \frac{U + R_sI}{R_{sh}}, \quad (3)$$

where I is the current, I_{ph} the light-generated current, I_0 the dark saturation current, U the voltage, R_s the series resistance, A the ideality factor, U_T the thermal voltage, and R_{sh} the shunt resistance of the PV cell [21]. Thermal voltage of a PV cell can be written as $U_T = kT/q$, where k is the Boltzmann constant, T the temperature of the cell, and q the elementary charge. The model for a PV module was obtained by scaling the parameter values used in the model of a PV cell by the number of cells in the module.

The characteristic of the simulation model was fitted to the electrical characteristics of the NAPS NP190GKg PV module. Bypass diodes were modelled using (3) by assuming that the light-generated current I_{ph} is zero and the shunt resistance R_{sh} is infinite. The dark saturation current $I_{0,\ bypass}$, the series resistance $R_{s,\ bypass}$, and the ideality factor A_{bypass} of the bypass diodes were determined by means of curve fitting to a measured *I-U* curve of a Schottky diode. The temperature of bypass diodes was assumed to be constant and the same as the module temperature. The experimental verification of the simulation model has been presented in [22].

The studied PV array consists of six series-connected strings of 28 modules connected in parallel with each other. The nominal power of the generator under standard test conditions (STC) is 31.92 kWp. The PV modules are mounted with a tilt angle of 45° from the horizontal plane. There are no gaps between the series-connected PV modules and the distance between the PV module strings is 2.0 m.

The simulations were conducted by time steps of 0.1 seconds. The irradiance at the centre of each module with an accuracy of 0.1 W/m^2 was used as the irradiance of that module during a time step. A simulation period starts when a shadow edge moves over the first modules of the PV array and ends when the shadow edge has moved across the array, i.e., when all the modules are uniformly shaded. The PV array was chosen to be under STC (irradiance of 1000 W/m^2 and temperature of 25 °C) before each irradiance fall and the temperature of the modules is assumed to be constant during the irradiance falls. During fast irradiance transitions, changes of module temperatures are small having only a negligible effect on the operation of the modules.

4 MATHEMATICAL PARAMETRISATION OF IRRADIANCE TRANSITIONS

4.1 Parametrisation method

The symmetry of PV array configurations simplifies the parametrisation process. In this study, the parametrisation is implemented for analysis of PV systems with physical and electrical array layouts which have two orthogonal lines of symmetry. An example of this kind of layout is aligned series-connected strings of PV modules of equal length connected in parallel forming a rectangle. For that kind of layouts, a situation of an overpassing irradiance fall is symmetrical to a situation of a similar overpassing irradiance rise, i.e., the sign of parameter b has no effect. It has been found in [12] that the characteristics of irradiance falls and rises are in practise similar. Thus, absolute value of b was used in this study. Moreover, the symmetry reduces the amount of studied movement directions of shadow edges. If the layout has one line of symmetry the amount of movement directions will reduce to 180° and if it has two orthogonal lines of symmetry, like in this study, the amount of movement directions will reduce to 90°.

Irradiance transitions can be parametrised by four variables: SS, b, speed and direction of movement. All the other variables needed to define irradiance transitions can be calculated from SS, b and initial irradiance. Because absolute value of b was used in this study, parameters a and c in (1) could be calculated from SS and irradiance before the transitions. The correlation between the absolute value of b and the experimentally obtained duration of all the identified transitions is presented in Fig. 3. The Pearson product-moment correlation coefficient (PCC) is 0.806 showing that parameter b and the duration of the transition are closely related. Linear least squares regression for b and the experimentally obtained duration leads to regression coefficient of 7.67. The duration of a transition is calculated by multiplying b with the regression coefficient.

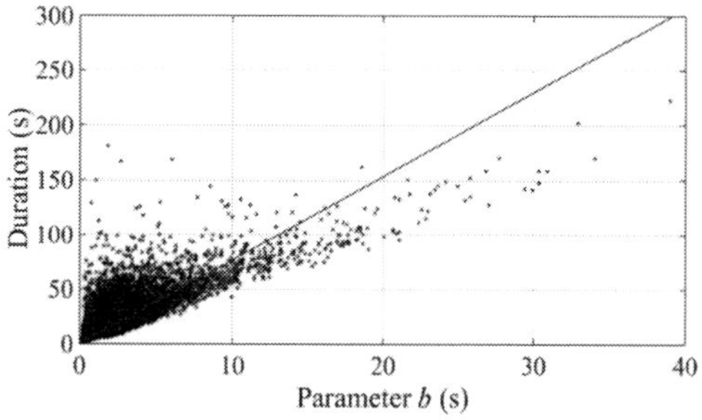

Figure 3: The scatter plot and regression line between the absolute value of b and the experimentally obtained duration of all the identified irradiance transitions.

No correlation was noticed between the shading strength, absolute value of b, speed and direction of movement of the identified transitions. The PCCs for all the combinations of these four variables are presented in Table I.

Table I: Pearson product-moment correlation coefficients for variables used in the parametrisation.

Variables	PCC
SS, b	-0.045
SS, speed	-0.074
SS, direction of movement	-0.052
b, speed	-0.194
b, direction of movement	0.086
Speed, direction of movement	-0.145

In the parametrisation, the direction of movement was divided into ten groups at regular intervals. Other variables were divided into groups by the following procedure. First, two quantiles of a variable were picked at a range of certain margin from the end points of the cumulative distribution function. For example, in case of 2% margin second and 98th percentiles were selected. Then, the space between the selected percentiles was divided to regular intervals and the median value of each group was used to represent that group.

The margins and the numbers of groups used for SS, b and speed were chosen based on the error caused by the parametrisation with respect to the initial values. The mean absolute errors of the initially obtained values with respect to the median values of the corresponding groups of SS, *b* and speed are presented in Tables II–IV, respectively, with different margins and numbers of groups. It seems that the error caused by the parametrisation decreases with the increasing number of the groups and that there is an optimal margin for each number of the groups which decreases with the increasing number of the groups. Errors for *b* and speed are higher than for SS, because the shapes of their distributions are less flat (see Figs. 1 and 2). Therefore, their parameter interval had to be divided into more groups than of SS to achieve small enough errors close to the error of the used combination for SS.

Table II: The mean absolute errors (%) of the initially obtained values of SS with respect to the median values of the corresponding groups with different margins and numbers of groups. The smallest error with each number of the groups is bolded and the error of the used combination is underlined.

		Margin (%)					
		1.5	2.0	2.5	3.0	3.5	4.0
Number of groups	5	3.92	3.89	3.87	3.86	**3.86**	3.86
	6	3.27	3.24	3.23	3.23	**3.23**	3.23
	7	2.82	2.80	2.79	**2.79**	2.79	2.80
	8	2.50	2.49	2.48	**2.47**	2.47	2.48
	9	2.22	2.21	**2.21**	2.21	2.22	2.23
	10	1.99	**1.98**	1.98	1.98	1.99	2.00

Table III: The mean absolute errors (%) of the initially obtained values of *b* with respect to the median values of the corresponding groups with different margins and numbers of groups. The smallest error with each number of the groups is bolded and the error of the used combination is underlined.

		Margin (%)					
		5.0	5.5	6.0	6.5	7.0	7.5
Number of groups	5	21.70	21.54	21.29	21.11	21.03	21.01
	6	18.99	18.85	18.68	18.58	18.50	**18.48**
	7	16.94	16.88	16.74	16.69	16.67	**16.66**
	8	15.41	15.39	15.31	**15.29**	15.32	15.31
	9	14.11	14.14	**14.07**	14.09	14.13	14.20
	10	**13.07**	13.13	13.14	13.14	13.20	13.29

Table IV: The mean absolute errors (%) of the initially obtained values of speed with respect to the median values of the corresponding groups with different margins and numbers of groups. The smallest error with each number of the groups is bolded and the error of the used combination is underlined.

		Margin (%)					
		3.5	4.0	4.5	5.0	5.5	6.0
Number of groups	5	13.31	13.22	13.15	**13.11**	13.11	13.12
	6	11.53	11.46	11.43	**11.40**	11.42	11.46
	7	10.26	10.22	**10.21**	10.21	10.24	10.30
	8	9.27	9.23	**9.21**	9.23	9.29	9.35
	9	8.42	**8.41**	8.43	8.48	8.53	8.62
	10	7.81	**7.80**	7.82	7.88	7.95	8.04

Based on the errors presented in Tables II–IV, SS, *b* and speed were divided into 5, 10 and 8 groups, respectively. Thus, the total number of possible combinations of the variables is 4000, which is a feasible number of cases for simulating PV system operation compared to the initial number of the identified transitions of 27 210.

The median parameter values and the shares (weights) of the identified irradiance transitions falling into chosen groups of the variables are presented in Tables V and VI. The identified irradiance transitions fell more evenly into the groups of SS than of the other variables as could be expected based on the shapes of their distributions. The highest differences between the shares of different groups exist in the case of *b*. This is expected since it has the most skew distribution.

Table V: The median values of SS and speed groups used in the parametrisation and the shares of the identified irradiance transitions falling into the groups.

SS (%)	Share (%)	Speed (m/s)	Share (%)
45.59	28.38	3.51	17.38
53.78	22.03	5.45	17.56
61.87	18.88	7.19	17.52
69.66	16.26	8.92	16.07
78.45	14.45	10.64	11.20
		12.44	7.66
		14.15	5.17
		17.69	7.43

Table VI: The median values of parameter *b* and direction of movement groups used in the parametrisation and the shares of the identified irradiance transitions falling into the groups.

Parameter *b* (s)	Share (%)	Angle (°)	Share (%)
0.61	24.90	0	4.23
1.09	19.22	10	9.02
1.53	14.98	20	9.56
1.97	11.51	30	10.23
2.42	8.55	40	10.84
2.86	6.18	50	11.25
3.31	3.92	60	12.33
3.76	2.55	70	12.59
4.22	1.85	80	13.28
5.95	6.34	90	6.67

4.2 Applicability of the parametrisation method in PV system analysis

In order to illustrate the significance of the errors presented in Tables II–IV, the effects of a change of 10% of SS, *b* and speed on the produced energy of the PV array were studied. The effects were studied during 10 irradiance transitions with directions of movement from perpendicular to parallel with respect to series connections of the PV array at intervals of 10°. The value of the studied variable was chanced ±10% from the median value of the identified transitions while the values of the other variables were kept fixed at the median values. The corresponding relative changes of the produced energy of the PV array are presented in Table VII.

Table VII: The effects of a change of ±10% of SS, *b* and speed on the produced energy of the PV array (%).

Variable	-10%	10%
SS	4.98	-5.05
b	-7.66	7.67
Speed	2.54	-2.01

As can be seen from Table VII, the change of the produced energy is the lowest when the speed of shadow edges is changed, i.e., the energy production does not vary strongly with changing shadow edge speed. It varies much more with changing shading strength and duration of the transition (parameter *b*).

Furthermore, the error caused by the parametrisation was studied during all the identified irradiance falls. The differences in the produced energy, average power and mismatch losses of the PV array were calculated between the use of the parametrisation method and the direct use of the mathematical model with respect to the latter for all the identified irradiance falls. The average differences in the produced energy and average power of the array are 10.5% and 2.29%, respectively. The differences in the produced total energy and mismatch losses during all the identified irradiance falls are 9.28% and 21.9%, respectively. The biggest differences of the produced energy and the mismatch losses occur during irradiance transitions with very low or high values of speed or *b* which are either smaller than the lowest or higher than the highest value used in the parametrisation. With these exceptionally low or high values there is a significant difference in the duration of the shading of the array, i.e., the simulation period, between the use of the parametrisation method and the direct use of the mathematical model. A difference in the length of the simulation period, naturally, leads to a difference in the produced energy without considerable difference in the average power.

The differences show that the developed parametrisation method is accurate enough for long-term PV system analysis. The developed method is not regionally bounded. The numbers and representative values of the variable groups can be adjusted to the local weather conditions.

Naturally, the error caused by the parametrisation can be reduced by increasing the number of the groups of the variables. Further, the accuracy can be increased by adding the duration of irradiance transitions as the fifth parametrisation variable. However, these actions make the parametrisation computationally more demanding.

5 CONCLUSIONS

In this paper, a method was presented to parametrise irradiance transitions caused by moving clouds based on a mathematical model of the transitions. Irradiance transitions were parametrised by four variables: shading strength, duration related parameter *b*, speed and direction of movement which have no correlation with each other. All the other variables needed to define irradiance transitions mathematically were calculated from SS, *b* and initial irradiance.

In the parametrisation, SS, *b*, speed and direction of movement were divided into 5, 10, 8 and 10 groups, respectively, based on experimentally obtained distributions of the variables. The total number of possible combinations of the variables was 4000, which is a feasible number of cases for simulating PV system operation.

The applicability of the parametrisation method in PV system analysis was demonstrated by simulations. The simulations were conducted by using an experimentally verified MATLAB Simulink model of a PV module based on the well-known one-diode model of a PV cell. The simulation results showed that the developed parametrisation method is accurate enough for long-term PV system analysis. The developed method is not regionally bounded and can be easily applied to different weather conditions.

6 REFERENCES

[1] J. Marcos, L. Marroyo, E. Lorenzo, M. García, Progress in Photovoltaics: Research and Applications 20 (2012) 226.

[2] M. Lave, J. Kleissl, E. Arias-Castro, Solar Energy 86 (2012) 2190.

[3] O. Perpiñán, J. Marcos, E. Lorenzo, Solar Energy 88 (2013) 227.

[4] K. Lappalainen, A. Mäki, S. Valkealahti, Proceedings of 28th European Photovoltaic Solar Energy Conference, Vol. I (2013) 4071.

[5] K. Lappalainen, A. Mäki, S. Valkealahti, Proceedings of 28th European Photovoltaic Solar Energy Conference, Vol. I (2013) 4081.

[6] A. Mäki, S. Valkealahti, Renewable Energy 71 (2014) 89.

[7] C.R. Sánchez Reinoso, D.H. Milone, R.H. Buitgaro, Applied Energy 103 (2013) 278.

[8] J. Marcos, L. Marroyo, E. Lorenzo, D. Alvira, E Izco, Progress in Photovoltaics: Research and Applications 19 (2011) 218.

[9] T. Tomson, M. Hansen, Theoretical and Applied Climatology 106 (2011) 171.

[10] R. Perez, S. Kivalov, J. Schlemmer, K. Hemker, T. Hoff, Solar. Energy 85 (2011) 1343.

[11] M. Lave, M.J. Reno, R.J. Broderick, Solar Energy 118 (2015) 327.

[12] K. Lappalainen, S. Valkealahti, Solar Energy 112 (2015) 55.

[13] T. Tomson, Solar Energy 84 (2010) 318.

[14] T. Tomson, Theoretical and Applied Climatology 112 (2013) 403.

[15] J.L. Bosch, J. Kleissl, Solar Energy 95 (2013) 13.

[16] J.L. Bosch, Y. Zheng, J. Kleissl, Solar Energy 87 (2013) 196.

[17] V. Fung, J.L. Bosch, S.W. Roberts, J. Kleissl, Atmos. Meas. Tech. 7 (2014) 1693.

[18] K. Lappalainen, S. Valkealahti, Solar Energy 135 (2016) 188.

[19] D. Torres Lobera, A. Mäki, J. Huusari, K. Lappalainen, T. Suntio, and S. Valkealahti, International Journal of Photoenergy 2013 (2013).

[20] M.G. Villalva, J.R. Gazoli, E.R. Filho, IEEE Transactions on Power electronics, 24 (2009) 1198.

[21] S.R. Wenham, M.A. Green, M.E. Watt, R. Corkish, Applied Photovoltaics 2nd ed., Earthscan, London 2007.

[22] A. Mäki, S. Valkealahti, J. Leppäaho, Progress in Photovoltaics: Research and Applications, 20 (2012) 298.

SHORTEST TERM FORECASTING OF DNI FOR CONCENTRATED SOLAR TECHNOLOGIES

Stefan C. Müller[1], Jan Remund[1]
[1]Meteotest
Fabrikstrasse 14, CH-3012 Bern, Switzerland

ABSTRACT: Accurate knowledge of the future state of direct normal irradiance (DNI) is needed for the successful operation of concentrated solar technologies like concentrating solar power (CSP) or concentrating photovoltaics (CPV). For different applications within a power plant forecast horizons of 0 to 4 hours with spatial resolutions from meters to kilometers are required. The model presented here uses satellite images from MSG satellite as a basis for a cloud motion nowcasting to derive the DNI. Forecasts are evaluated at four locations in Southern Spain and Northern Africa for two three monthly time periods in 2010 and 2013. Uncertainties were found for the North African locations of 120 to 250 W/m^2 (22 - 55%) root mean square error and larger deviations in Southern Spain mostly due to complexity of topography.

Keywords: solar radiation, forecasting

1 INTRODUCTION

Direct normal irradiance (DNI) is the most important meteorological parameter for the electrical output of concentrated solar technologies (CST) like concentrating solar power (CSP) or concentrating photovoltaics (CPV). Different operational phases have different requirements for the knowledge of the DNI at a power plant site in space and time in terms of the forecast horizon and forecast update. E.g. for the decision to start a CSP plant a longer forecast is needed, than for the control of the stability of single loops when the plant already runs or for the load management of a CPV plant with battery storage.

Thus for the successful and optimized operation of CST power plants a good knowledge of the future DNI for the next hours is needed. A nice and promising method is to use satellite images and trajectories for the cloud movement, which was already tested for shortest term forecasts of global radiation [1].

Within the EU project DNICast (www.dnicast-project.org) different nowcasting methods to forecast DNI are developed. In this paper we present a forecasting model for DNI for the next four hours based on satellite images. The paper is devided into a requirements section, model description, result section and a conclusion.

2 NOWCAST SYTEM REQUIREMENTS

For the operation support of a CSP or CPV plant different specifications for the forecast system are given. There are different requirements in the spatial and temporal resolution as well as in the forecast horizon. Some operation examples are given in Table I. The nowcasting model described in this paper can serve some but not all of them.

Table I: Some specifications for different operation aspects [2]. The nowcast horizon and the spatial resolution are shown here.

Application	horizon	spatial res
CPV with batteries load management	15 min to 4 hours	1-3 km
CPV with buffers for ramp mitigation	1 min to 4 hours	10 m – 1 km
CPV energy output estimation	sec to 4 hours	10 m – 1 km
CSP operation support Startup, shutdown	up to 4 hours	1-3 km
CSP, Operational stability (e.g. loop controls)	75 min	1-3 km
High accuracy DNI Distribution (R&D)	0 min	5-20 m

3 MODEL DESCRIPTION

3.1 Overview

DNI is the part of the sunlight that transfers the atmosphere straight-lined and arrives at the surface on a plane normal to the beam. DNI, as it is used here includes the circumsolar radiation which is the diffuse radiation coming from the vicinity of the sun. Main sources of reducing the amount of DNI is scattering by clouds and aerosols. High amounts of DNI are found in sunny places. The focus of DNICast is on the area of Southern Spain and Northern Africa, where several CSP plants are located. Model test locations are given in Table II.

Table II: Test locations used for the model development in Spain, Algeria and Tunisia.

Abr.	Location name	Lat [°]	Lon [°]
PSA	Plataforma Solar de Almeria (SP)	37.096	-2.356
GHA	Ghardaia (ALG)	32.386	3.780
ADR	Adrar (ALG)	27.878	-0.270
TN	Tataouine (TUN)	32.974	10.485

3.2 Model description

As satellite images have a spatial coverage on a continental scale, the actual cloud state is detected in a wide area around a specific location. This allows an estimation of the future cloud states for a cloud forecast. Here the 15 minute satellite images from MSG/SEVIRI are used. During daytime we use images from the high resolution visible (HRV) channel which has a resolution of approximately 3 km in the considered area. For late night and the time around sunrise the lower specially resolved infrared channel at 10.8 nm is used to detect clouds. In the area of the PSA in addition the near-infrared channel at 1.6 nm is needed for snow detection on the ground in the mountains. With a visible channel snow and clouds cannot be differentiated. If snow is detected clear sky condition is implied.

The optical thickness of the clouds is determined by the Heliosat method [3] and turned into a clearness index (CI). CI describes the amount of the clear sky radiation (Gc) arriving at ground level, see (1) where Gh is the global radiation horizontal.

$$CI = Gh/Gc \qquad (1)$$

With the knowledge of these radiation components we can calculate DNI as done with the BRL model [4]. In the following referred as the indirect method. We also tested a direct conversion from CI to DNI following the formulation of [5] which we name as the direct method.

Once the cloud state is derived from the images they need to be propagated into the future. We use the wind vector components U and V from the mesoscale numerical weather model WRF (www.wrf-model.org) for the calculation of trajectories for each satellite pixel. Our setup of WRF is driven by boundary conditions from the Global Forecasting Model (GFS). The wind information is taken at a fixed height of 4000 m above sea level. The trajectories are calculated in temporal resolutions of 5 minutes.

A clear sky model is needed to calculate the irradiation under clear sky condition to finally calculate DNI. We use the Simple Solis [6] model for clear sky as well as the model of the European Solar Radiation Atlas ESRA [7]. The water vapor column as an input is taken from GFS. For the second and more important input parameter, Aerosol Optical Depth (AOD), two different sources were tested, an aerosol climatology based on satellite and ground measurements from MODIS, MISR and AERONET [8] and aerosol forecasts from the MACC project (www.gmes-atmosphere.eu/).

With the different options of input sources and clear sky and DNI calculation choices we defined three approaches and numbered them M1, M2 and M3 which is shown in Table III.

Table III: Input data sets and models.

Nr	DNI method	aerosol	clear sky model
M1	indirect	climatology	ESRA
M2	direct	climatology	ESRA
M3	direct	MACC	Simple Solis

3.3 Post processing

To reduce forecast uncertainty onsite measurements of DNI from the previous hour are compared to all of the DNI values derived from the satellite image of the previous hour, which are the forecast starting points at horizon zero minutes to account for biases due to wrong inputs of e.g. the aerosol load.

3.4 Technical boundaries

The calculation of the forecasts from satellite image scan to the delivery of the forecasts need approximately 12 minutes. This means the first available forecasted value which is not yet in the past is for a horizon of 15 minutes. Every 15 minutes a new satellite image is available and the forecast is updated. The maximum spatial resolution is the size of a satellite image pixel, which is at the sub satellite point 1 km. We interpolated on a regular latitude-longitude grid of 1/40 degree.

4 RESULTS

Forecasts for two three monthly test periods were calculated. The first period was January to March 2010 and the second March to May 2013. Ground measurements of DNI served as reference values. The root mean square error was used as the statistical parameter for calculating the uncertainty. It weights large deviations stronger, which makes sense as these uncertainties will produce more loss in the sense of economic aspects. Only daytime values as five minute averages were used for the analysis

Figure 1 and Figure 2 show for each model approach the overall uncertainties with the forecast horizon. In Figure 1 the forecasts without post processing were evaluated. The RMSE is range from 200 to 400 W/m^2. There is a higher uncertainty and a strong increase after 30 minutes for the forecasts for PSA. For the other locations the increase in uncertainty is constant. Figure 2 in comparison show the evaluation of the same forecasts with the post processing applied. There is a significant improvement for the locations in Northern Africa, which can be explained by aerosol uncertainty. With the post processing the RMSE is lowered by the amount of approximately 40% and is 120 W/m^2 for horizon 0 and goes up to 250 W/m^2 at 4 hours. There is no significant effect for the PSA area.

Figure 3 shows a comparison of the uncertainty of the three approaches. All of them are in the same range. Best performer was model number M1 with the indirect method and the aerosol climatology, but the differences are rather small between the approaches.

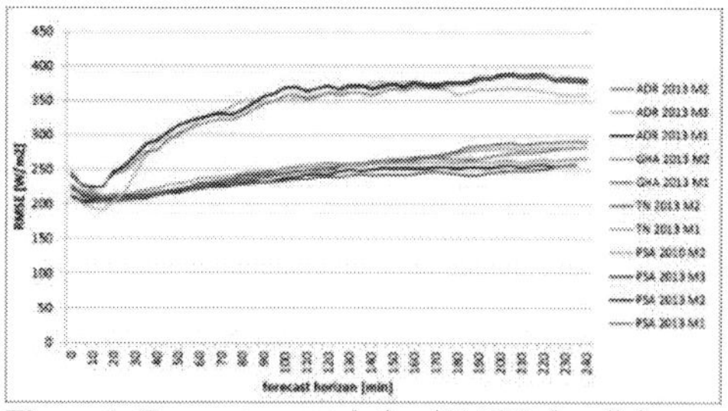

Figure 1: Forecast uncertainties (RMSE) for all forecast horizons for the forecasts <u>without</u> post processing.

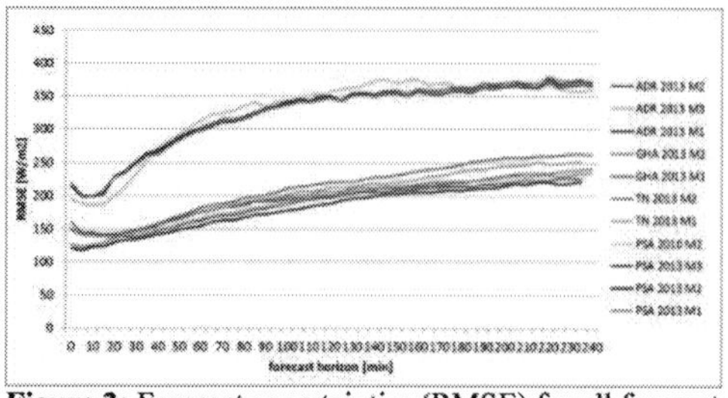

Figure 2: Forecast uncertainties (RMSE) for all forecast horizons for the forecasts <u>with</u> post processing.

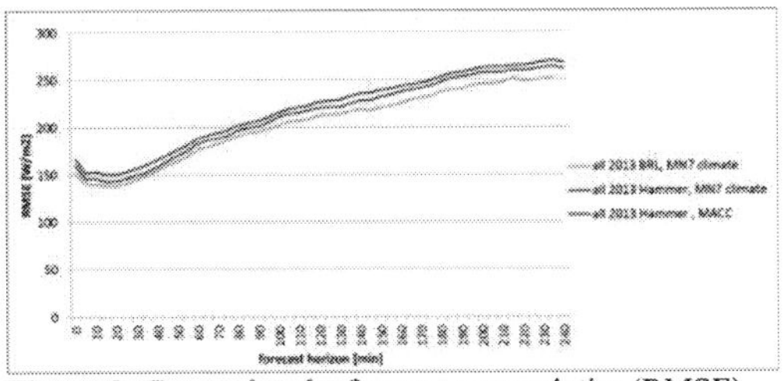

Figure 3: Comparing the forecast uncertaintics (RMSE) for all forecast horizons for the three approaches.

5 DISCUSSION

We presented a DNI nowcasting method with a forecast horizon of 4 hours for the operation of CSP and CPV plants. The model is based on satellite images to detect the optical thickness of clouds in terms of the clearness index CI. Wind vectors from a numerical weather model are used to calculate trajectories which are applied to the CI for the future cloud pattern. Three different approaches are tested with different input sources for the DNI calculation, which had similar performance. A huge effect up to 40% was found with a post processing method for North African locations. Uncertainties are from 120 to 250 W/m^2 (22 - 55%) for 0 to 4 hours forecast horizon there. There is a significantly poorer performance for the PSA area in Southern Spain. The main reason for that is the complexity of the topography. There is much more cloud formation and dissolving, which is not modelled. Standing clouds due the mountain lee waves or fog are transported forward by the trajectories. Thus for complex terrain locations this method would need adaptions mainly concerning the choice of the altitude level of the wind vectors for improving the performance.

BIBLIOGRAPHY

[1] Müller S., Remund J., 2013: Shortest term regional solar energy forecast for the next 1 to 6 hours, PVSEC 2013, Paris.

[2] Hirsch et al., DNICast 2014, Direct Normal Irradiance Nowcasting methods for optimized operation of concentrating solar technologies, Deliverable Nr. 2.1

[3] Rigollier C. et al., 2004: The method Heliosat-2 for deriving shortwave solar radiation data from satellite images. Solar Energy, 77(2), 159-169

[4] Ridley B. t al.., 2010: Modelling of diffuse solar fraction with multiple predictors. Renewable Energy, 35(2), 478-783.

[5] Hammer A. et al.., 2009: Direct Normal Irradiance for CSP based on satellite images of Meteosat Second Generation, SolarPACES 2009, 15-18.9, Berlin

[6] Ineichen P., 2008: A broadband simplified version of the Solis clear sky model. Solar Energy, 82, 756-762

[7] Rigollier, C., Bauer, O., & Wald, L. (2000). ON THE CLEAR SKY MODEL OF THE ESRA — EUROPEAN SOLAR RADIATION ATLAS — WITH RESPECT TO THE HELIOSAT METHOD. Solar Energy, 68(1), 33–48.

[8] Remund et al., 2015: meteonorm – Global Meteorological Database Version 7, Handbook part II: Theory.Aeronet Holben

ACKNOWLEDGMENT

This project has received funding from the European Union's Seventh Programme for research, technological development and demonstration under grant agreement No [608623].

APPLICATION OF WHOLE SKY IMAGERS FOR DATA SELECTION FOR RADIOMETER CALIBRATION

Stefan Wilbert[1], Bijan Nouri[2], Christoph Prahl[3], Ginés García[4], Lourdes Ramírez[5], Luis Zarzalejo[6], Rita Valenzuela[7], Francisco Ferrera[8], Nora Kozonek[9], José Liria[10]

[1] Deutsches Zentrum für Luft- und Raumfahrt (DLR), Institute of Solar Research, Plataforma Solar de Almería (PSA), Ctra. de Senés s/n km 4, Apartado 39, 04200 Tabernas, Spain. Tel.: +34 950 277619. Fax: +34 950 260315, stefan.wilbert@dlr.de

[2] DLR, Institute of Solar Research, PSA, Ctra. de Senés s/n km 4, Apartado 39, 04200 Tabernas, Spain. bijan.nouri@dlr.de

[3] DLR, Institute of Solar Research, PSA, Ctra. de Senés s/n km 4, Apartado 39, 04200 Tabernas, Spain. christoph.prahl@dlr.de

[4] CIEMAT, PSA – Data and Instrumentation, PSA, Ctra. de Senés s/n km 4, Apartado 39, 04200 Tabernas, Spain. ggarcia@psa.es

[5] CIEMAT, Department of Energy – Renewable Energy Division, Av. Complutense, 40, 28040 Madrid – Spain. Lourdes.Ramirez@ciemat.es

[6] CIEMAT, Department of Energy – Renewable Energy Division, Av. Complutense, 40, 28040 Madrid – Spain. lf.zarzalejo@ciemat.es

[7] CIEMAT, Department of Energy – Renewable Energy Division, Av. Complutense, 40, 28040 Madrid – Spain. r.valenzuela@ciemat.es

[8] CIEMAT, Department of Energy – Renewable Energy Division, Av. Complutense, 40, 28040 Madrid – Spain. Francisco.Ferrera@externos.ciemat.es

[9] DLR, Institute of Solar Research, Paseo de Almería 73, 04001 Almería, Spain. nora.kozonek@dlr.de

[10] CIEMAT, PSA – Data and Instrumentation, PSA, Ctra. de Senés s/n km 4, Apartado 39, 04200 Tabernas, Spain. jose.liria@psa.es

ABSTRACT: The accurate measurement of shortwave solar irradiance is essential for the planning and operation of photovoltaic plants. The measurement uncertainty caused by the calibration is the most relevant contribution to the overall uncertainty of radiometers. During the outdoor calibration of radiometers the meteorological conditions have to be observed carefully and data points collected under unfavorable conditions must be rejected. The rejection criteria concern the level of the involved irradiances, wind speed and –direction, and also the cloud conditions. The latter are quantified in terms of the minimum angular distance of clouds from the sun and are usually determined visually by an observer. A purely manual visual determination of the cloud's angular distance from the sun is time consuming and often inaccurate. We developed an algorithm for whole sky imagers (WSIs) that allows the determination of the minimum angular distance of clouds from the sun and the automatic filtering of the calibration data. The high importance of the cloud filtering for pyranometer calibration is shown with an exemplary calibration result for nine pyranometers. The effect of the cloud filtering for pyrheliometer calibration has shown to be negligible for this example. The filtering method increases the calibration quality and noticeably reduces the personnel effort for pyranometer calibration.

Keywords: whole sky imager, radiometer, pyranometer, calibration, cloud detection

1 PURPOSE OF THE WORK AND APPROACH

Accurate measurements of shortwave solar irradiance are essential for the planning and operation of photovoltaic plants. High quality measurements can be obtained with well-maintained and thoroughly calibrated field pyrheliometers and pyranometers with thermopile sensors. The uncertainty caused by the calibration is the most relevant contribution to the overall uncertainty of well-maintained sensors. By means of outdoor calibration with World Radiometric Reference traceable absolute cavity radiometers (ACR) the uncertainties can be reduced considerably compared to indoor calibrations. Outdoor calibrations use the sun as the light source and can hence compare measurements of the reference and the test instrument that are obtained at the same time. Indoor calibrations use lamps as radiation source and compare measurements that are not obtained at the same time. Hence, small instabilities of the lamp cause additional calibration errors.

During the outdoor calibration the meteorological conditions have to be carefully observed and data points collected under unfavorable conditions must be rejected. The rejection criteria concern the level of the involved irradiances, wind speed and –direction, and also the cloud conditions. The cloud conditions are indirectly controlled via the DNI (Direct Normal Irradiance) and DHI (Diffuse Horizontal Irradiance) and must be controlled directly in terms of the minimum angular distance of clouds to the sun. This is usually done manually by a human observer. Depending on the calibration method the minimum angular distance between clouds and the sun must be greater than 15°, 30° or 45°. A purely manual visual determination of the cloud's angular distance from the sun is time consuming and often inaccurate. This effectively restricts calibration to conditions with nearly no clouds. Nowadays, whole sky imagers (WSIs) can be used to automatically detect clouds ([1]–[3]). Hence, we developed an algorithm for WSIs that allows the determination of the minimum angular distance of clouds from the sun and provides this input to the calibration software.

The WSI system has been included in a calibration bench for pyrheliometers and pyranometers at the Plataforma Solar de Almería (PSA). Besides the WSI, the calibration bench consists of a heliostat which is used as tracker for the reference and field pyrheliometers, a Kipp&Zonen 2AP tracker with shading structures for the DHI reference pyranometer, and wind speed, wind direction, pressure, ambient temperature and humidity measurements. In this work we present the calibration bench, the WSI system and an example for the calibration

with the WSI information compared to a calibration without it. We also manually correct the automatic WSI filter results by visual inspection of the WSI images and compare the calibration with the automatic WSI filter to this manually filtered reference case.
Section 2 of this paper describes the calibration test bench and the calibration procedure itself. After that the WSI system used in this context is presented. In a third section we show a comparison of a calibration with the WSI information compared to the calibration without this system. This latter example proves the importance of the cloud filtering and the applicability of the developed WSI system for this purpose.

Figure 1: Solar tracker with reference pyranometer.

2 DESCRIPTION OF THE CALIBRATION TEST BENCH AND THE CALIBRATION PROCEDURE

At PSA pyrheliometers are calibrated following the ISO 9059 [4] by comparing the voltage signal of the tracked test pyrheliometer to the DNI from one or a group of reference pyrheliometers. Pyranometers can be calibrated with three different methods. One option from ISO 9846 [5] compares the voltage signal of the test pyranometer obtained in the GHI (Global Horizontal Irradiance) measurement position to the GHI calculated from the DNI and DHI measurements of a reference pyrheliometer and a shaded reference pyranometer. We basically discuss these two types of calibration in this work. Another option from ISO 9846 is to compare the DNI from a reference pyrheliometer to the DNI derived from the test pyranometer which is subsequently used to measure voltages proportional to GHI (unshaded) and DHI (shaded). The third option for pyranometer calibration from ISO 9847 [6] compares a test pyranometer to a reference pyranometer while both sensors are in GHI measurement position.

As reference instruments PMOD ACRs and a Kipp&Zonen CMP22 pyranometer are used at PSA. There are two ACRs devoted to the calibration facility. The PMO6-CC 0301 (CIEMAT) took part in the International Pyrheliometer Comparisons IPC-XI and IPC-XII. The PMO6-CC 0807 (DLR) took part in the American National Pyrheliometer Comparison (NPC-2014). For the pyranometer facility, Kipp&Zonen CMP22 pyranometers with SN 140048 (CIEMAT) and SN 110288 (DLR) are available as additional reference instruments (Fig. 1). The sensors are traceable to the World Radiometric Reference. The heliostat based tracking system of the calibration set-up has been designed in-house by the Instrumentation and Data Group of CIEMAT (Fig. 2). The reference and field pyrheliometers are mounted on a plate with a surface of 1.02 m² attached to the tracking system.

For the reference pyranometer calibration Black Photon Instruments' sunscanner tracking system is used. The tracker has two separate elevation axes, so that the ACR can be tracked to the sun permanently while the pyranometer can be periodically shaded and unshaded as described in ISO 9846. The shading cycles are programmed such, that the cycles of the chopped ACR DNI measurements are synchronous to the automatic shading cycle.

Figure 2: Ciemat's solar tracker with reference and test pyrheliometers. Reference pyrheliometers are mounted in a weatherproof case with windowless automatically closeable apertures.

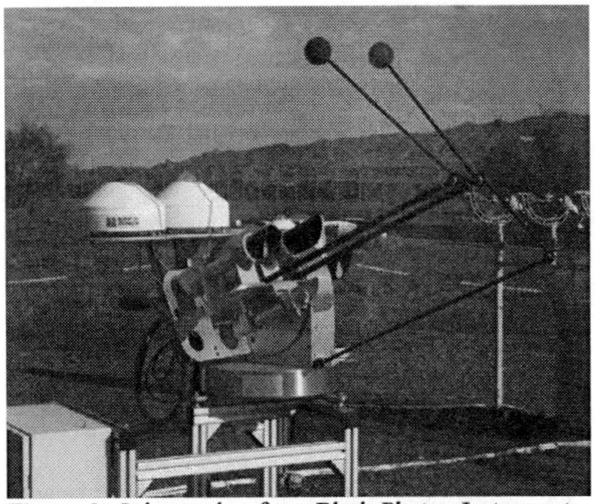

Figure 3: Solar tracker from Black Photon Instruments used for the calibration of the reference pyranometer with separate elevation axes for the shading structure and the absolute cavity radiometer allowing automatic shading and unshading.

2.1 Data filtering according to ISO standards

The data filtering is performed according to the used ISO standards. The sort out criteria are listed in Table 1 and described in the following.

In the case of the pyrheliometer calibration following ISO 9059, all data with a DNI below 700 W/m² are excluded. Values above 300 W/m² could be used, but the text advises that "irradiance values exceeding 700 W/m² are preferred". For the pyranometer calibration no DNI limit is specified, but we use the DNI lower limit of 300 W/m² in this work.

Table I: Sort out criteria for radiometer calibration

	Pyrheliometer calibration (ISO 9059)	Pyranometer calibration with ref. DNI and DHI (ISO 9846, continuous method)
DNI	DNI < 300 W/m² recommendation: DNI < 700 W/m²	None defined (used: DNI < 300W/m²)
Wind	Too high wind speed particularly when blowing from sun azimuth ±30° (wind speed larger 5 m/s, ±30°)	
Linke Turbidity	> 6	-
DHI variation	-	DHI varies more than "1 % in about 10 s"
Clouds	Angular distance from sun ≤ 15° Recommendation: cloud cover ≥ 12.5 %	angular distance from sun ≤ 15°
Outliers	Results for preliminary calibration factors F_{ij} which deviate by more than ±2 % from average calibration factors F_j of series j	Results for F_{ij} which deviate by more than 5 % from the average F_j in the series j
Series length	Series with less than 10 valid measurement values	
	-	series if more than 50 % of its values have been eliminated

Wind is problematic if the wind direction coincides with the solar azimuth angle within ±30°. No concrete limits are specified in the ISO 9059 or 9846. As for NREL national pyrheliometer comparisons an upper limit for the wind speed of 5 m/s is used if the wind direction matches the solar azimuth within ± 30° [7].

The Linke turbidity has an effect on DNI and DHI measurements if the angular acceptance functions of the test and reference instruments are not identical. For high turbidity higher circumsolar contributions to the measured DNI are found and hence, deviations occur if the acceptance functions are different [8]. Therefore, the Linke turbidity is recommended to be less than 6 for pyrheliometer calibration. The acceptance function of the reference pyrheliometer and the acceptance function corresponding to the sun blocking geometry of the shading ball and the DHI pyranometer deviate from each other for most cases and all solar zenith angles above zero. Hence it can be useful to avoid high turbidities also for pyranometer calibrations following ISO 9846. However, no limit is defined in this case.

For pyranometer calibration with the continuous shade method it is furthermore required that the variation of the DHI caused by cloud movement is less than 1 % in 10 seconds. This statement from the standard can be understood in different ways. We evaluate it with three different criteria in order to exclude as much inadequate data as possible. Only if all three tests are passed the data is used for the calibration. The first test is to evaluate the 2-second average DHI measurements contributing to the 10-second averages that are used in the calibration. If the difference between the maximum and the minimum of the 2-second DHI measurements is greater than 1 % of the 10-second average the 10-second average is excluded. This is not sufficient in our opinion as the response time of the reference pyranometers is 5 seconds (95 % signal level). Hence, we also investigate the change of the 10-second DHI averages. If the change is greater than 1 % for two consecutive DHI 10-second averages both values are excluded. The third test is to investigate the temporal gradient of the DHI between two reference DHI measurements that are obtained in parallel to the ACR readings. ACR readings are obtained only every 2 minutes. If the gradient is greater than 1 % per 10 seconds the data is excluded, too.

As mentioned in the first section of this work also the smallest distance of clouds from the sun has to be monitored. It is mentioned in ISO 9846 that "clear sky conditions reduce the variance" of the calibration results, but measurements are acceptable if a minimum distance of 15° is not underrun. ISO 9059 for pyrheliometers also uses 15° as this limit. Interestingly, for pyrheliometer calibration 12.5 % maximum cloud cover is recommended, although clouds outside the field of view of the pyrheliometers should not affect the measurements. The method for the determination of the cloud's distance from the sun is explained in the next section.

After taking into account all these filters, outliers are rejected. Calibrations consist of several series j which contain several parallel measurements i of test and reference instruments. Filters are applied to reject outliers by comparing the calibration constants for one series of measurements F_j to the calibration constants derived for each single measurement point F_{ij} within the series. If less than 10 or less than half of the full series remains after applying all filters the complete series is excluded.

3 WHOLE SKY IMAGER SYSTEM FOR CLOUD FILTERING

For the cloud filtering an automated evaluation of WSI images is applied. The WSI is a Mobotix Q24Sec fisheye camera that is also used to derive real time and nowcasted DNI and GHI maps of PSA [9]. The image acquisition and processing is carried out with standalone software compiled from MATLAB code. The camera's internal orientation is calibrated using images of checkerboards. The external orientation is calibrated based on an image series acquired during a clear sky day by detecting the sun path and optimizing the camera orientation such that the calculated sun path in the image series fits the detected sun path.

3.1 Cloud segmentation

The image processing of the WSI aims to detect cloud positions via image segmentation methods. A clear sky library containing red to blue ratios is created and applied. The images' red to blue ratios (RBR) are compared pixel-wise to the clear sky RBR. If the RBR in the pixel under examination is greater than the RBR of the clear sky library plus a given threshold the pixel is assumed to be cloudy. After this comparison to the RBR the segmented image is cleaned using erode and dilate filters. Saturated areas around the sun and artefacts as lens flares are removed and these areas are filled based on information obtained from surrounding pixels.

A four dimensional clear sky library is used .The first two dimensions of the clear sky library are the distance of the pixel to the zenith (zenith-pixel-angle) and the distance of the pixel to the sun (sun-pixel-angle). While these two dimensions are also used in other publications as, e.g. [3], the third dimension is the pressure corrected air mass and not the solar zenith angle. We also enhanced the clear sky library method by adding the Linke turbidity as a fourth dimension. This is useful as the RBR of clear sky pixels increases with the Linke turbidity because of atmospheric scattering. Linke turbidity is derived from DNI following [10].

Three exemplary layers of the clear sky library are shown in Fig. 4. The variation of the RBR with the air mass and the Linke turbidity is clearly visible.

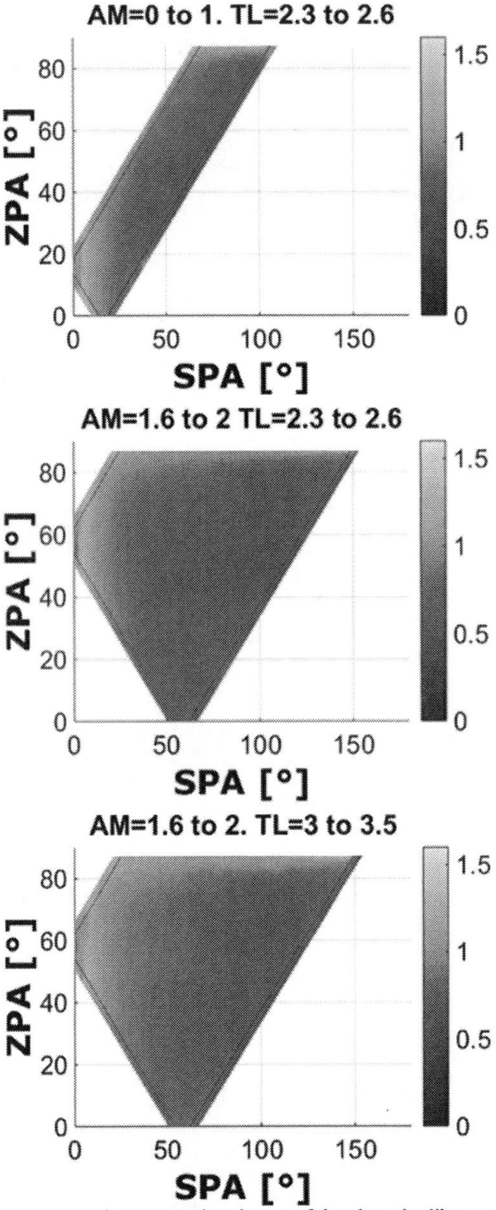

Figure 4: Three exemplary layers of the clear sky library for different bins of the air mass (AM) and the Linke turbidity (TL). The color bars show the RBR. Lines mark the region of angles occurring in the AM bin.

3.2 Determination and visualization of the distance of clouds from the sun

The above mentioned calibration of the camera orientation allows to associate each image pixel with a certain elevation and azimuth angle. With this information, a region around the sun can be projected to the image to check and visualize the required minimum distance of the cloud to the sun. The images from the calibration campaign are processed and the minimum angular distance of clouds to the sun and the cloud cover are saved for the further filtering of the irradiance and voltage signals.

For the calibration of pyrheliometers and the calibration of pyranometers with the continuous sun-and-shade method there should be a minimum angular distance from clouds to the sun of 15° according to the standard. For the alternating sun-and-shade method a higher angular distance of > 45° is required. When the limits are exceeded, timestamps are saved for the subsequent data exclusion during the evaluation. The upper image of Fig. 5 shows the acceptance region of 15° around the sun (blue). The distance of the closest detected cloud is represented by the red line. The lower image of Fig. 5 shows an invalid timestamp which was filtered out automatically during the calibration.

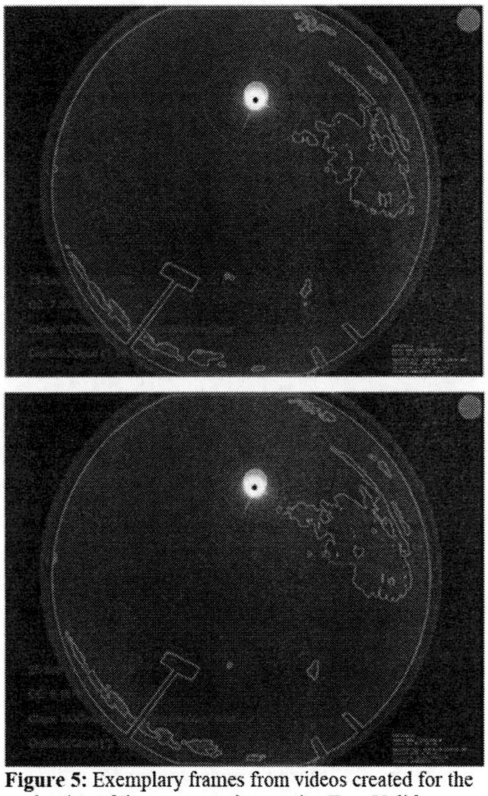

Figure 5: Exemplary frames from videos created for the evaluation of the segmentation results. Top: Valid conditions. The blue ring around the sun shows the region within 15° from the sun and the red ring indicates the distance of the closest clouds. Bottom: Invalid conditions.

Videos of such images have been used for manual verification and correction of the automatically calculated angular distance of the clouds to the sun. The manually corrected data set will be used in the following as a reference data set to evaluate the performance of the automatic segmentation.

4 APPLICATION OF THE WSI FOR EXEMPLARY RADIOMETER CALIBRATIONS

The WSI system was tested during a calibration campaign carried out between 2015-09-22 and 2015-10-09 at PSA. A pyrheliometer calibration following ISO 9059 and a continuous sun and shade calibration following ISO 9846 have been performed once with and once without the WSI based filtering. A third calibration with the same data and the manually corrected WSI based filters is used as a reference result. In order to show the importance of the cloud filtering, the case without the WSI based filter uses the complete data set and all other automatic filters, but no manual filtering for the angular cloud distance to the sun. This situation corresponds to a worst case scenario for the quality of the subjective cloud observations of the human observer.

The results of the pyrheliometer calibration remain unchanged for all of the 15 calibrated pyrheliometers. The sort out criteria using DNI and outlier rejection based on the deviation of the single point calibration factors F_{ij} from the factor for the series F_j are apparently sufficient for the sky conditions faced during the exemplary campaign.

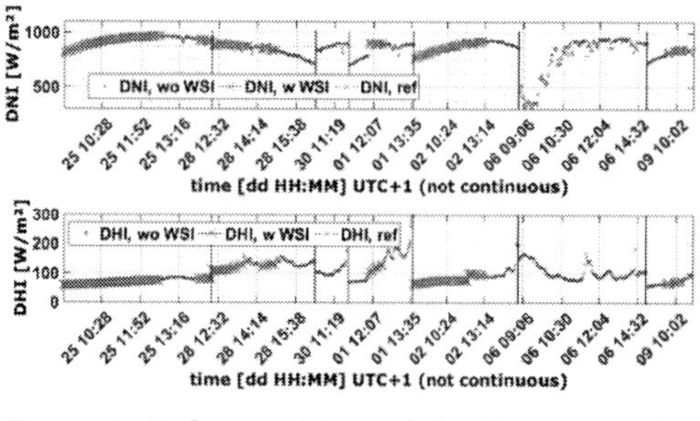

Figure 6: Reference data used for the pyranometer calibration in the three cases (without cloud distance filter "wo WSI"; with automatic WSI based cloud distance filter "w WSI"; manually corrected WSI filter as reference "ref"). Top: DNI. Bottom: DHI.

The results for the pyranometer calibration on the other hand change noticeably due to the cloud filtering. The used reference data after applying the three different data filtering options are shown in Fig. 6. The reference DNI and DHI are plotted vs the time of the measurement point number of the used calibration data. Note that the length of the time axis is not proportional to the time as only the used data points are shown. The used data after the WSI filtering are shown in magenta, the results without the WSI filtering are shown in red and the data after applying the manually corrected WSI filters are shown in green.

It can be seen that the number of used data points is reduced strongly when the WSI based filter is applied. The reference and the automatic WSI filters lead to nearly the same results. Only for nine of 656 timestamps the decision whether or not a cloud is more than 15° away from the sun was not correct. The automatic sort out was correct in 98.6 % of the cases.

The other automatic filters applied for the continuous sun-and-shade method do not exclude data points even if a high DHI or low DNI was measured. The outlier rejection for F_{ij} that deviate from the preliminary average within one series j is less strict than in the case of the pyrheliometer calibration–instead of 2 % maximum deviation for the pyrheliometer 5 % are defined for the continuous sun-and-shade method.

The effect of the filtering is also clearly visible in the results of the calibration of the nine field pyranometers that were calibrated in this example. This holds for the resulting calibration constants and also for the standard deviations of the F_j from different series for one instrument. Figure 7 shows the responsivities R of the test pyranometers at a solar zenith angle of 45° obtained with the three different cloud filters. Also the relative deviation $\Delta_{rel,wo2ref,45°}$ between the evaluation without the WSI data and the reference data set is shown. Without cloud filtering high relative deviations of R to the reference of up to 0.42 % occur. The relative deviation between the result for the automatic WSI based filter and the reference is small ($\Delta_{rel,w2ref,45°} < 0.01$ %).

These results show that the cloud filtering is crucial for the calibration quality. The manual corrections of the automatic WSI based filtering only have a small effect. Of course, the effect of the cloud filtering depends on the calibration conditions. For a calibration campaign in 2014 only a negligible effect of cloud sort out was found due to the much better sky conditions.

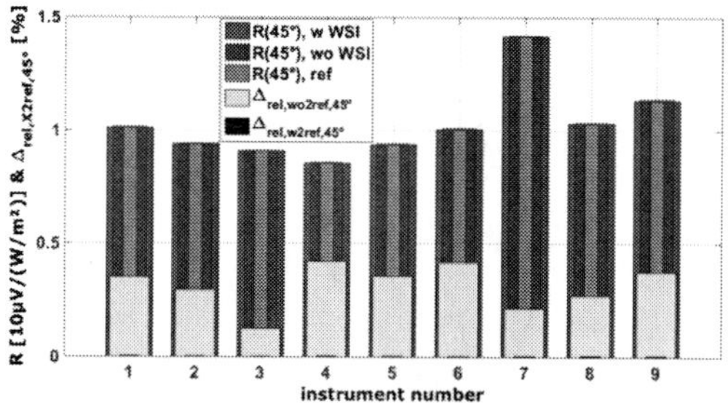

Figure 7: Comparison of pyranometer responsivities derived for a solar zenith angle of 45° with the reference filter (green), with WSI based filtering (magenta) and without WSI based filtering (red). The relative deviations of the results without WSI filtering to the reference are shown in yellow. The relative deviations of the results with automatic WSI filtering to the reference are shown in black.

4 CONCLUSION AND OUTLOOK

A new WSI based data filtering method for radiometer calibration has been developed and demonstrated. The effect of the cloud distance filtering for pyrheliometer calibration has shown to be negligible. The importance of the cloud distance filtering has been applied for pyranometer calibration using the continuous sun-and-shade method. For an exemplary calibration campaign with complicated conditions deviations of the determined pyranometer responsivities of up to 0.42 % were found if no cloud distance filter was used. In another calibration campaign the effect was negligible. It is expected that the filtering will also affect other pyranometer calibration methods. The filtering method increases the calibration quality and noticeably reduces the personnel effort for pyranometer calibration.

Future versions of the ISO calibration standards should further stress the importance of the cloud filtering. Also stricter outlier rejection might be considered for pyranometer calibration as the 5 % limit was found to be too tolerant.

The new WSI based filtering will be applied at the calibration facility at Plataforma Solar de Almería. Although the automatic WSI filtering was correct in 98.6 % of the cases we will continue to use the WSI videos to control and correct the WSI filters. Similar WSI data should be collected during calibration campaigns if they include complex sky conditions.

We thank the EU for partially funding the activity within the EU FP 7 project SFERA II.

REFERENCES

[1] T. L. Shields, J.E., Johnson, R.W., Koehler, "Automated whole sky imaging systems for cloud field assessment," in *Fourth Symposium on Global Change Studies*, 1993, pp. 228–231.

[2] G. Seiz, E. P. Baltsavias, and A. Gruen, "Cloud mapping from the ground: Use of photogrammetric methods," *PE RS-Photogramm. Eng. Remote Sens.*, vol. 68, no. 9, pp. 941–951, 2002.

[3] C. W. Chow, B. Urquhart, M. Lave, A. Dominguez, J. Kleissl, J. Shields, and B. Washom, "Intra-hour forecasting with a total sky imager at the UC San Diego solar energy testbed," *Sol. Energy*, 2011.

[4] "ISO 9059: 1990 (E)," *Solar energy - Calibration of field pyrheliometers by comparison to a reference pyrheliometer.* ISO (International Organization for Standardization), p. 10, 1990.

[5] "ISO 9846: 1993(E) ," *Solar energy - Calibration of a pyranometer using a pyrheliometer.* International Standard Organization, International Standard, p. 24, 1993.

[6] "ISO 9847: 1992 (E)," *Solar Energy - Calibration of field pyranometers by comparison to a reference pyranometer.* International Standard Organization, International Standard, p. 20, 1992.

[7] I. Reda, M. Dooraghi, and A. M. Habte, *NREL Pyrheliometer Comparison: September 16 to 27, 2013 (NPC-2013).* National Renewable Energy Laboratory, 2013.

[8] S. Wilbert, R. Pitz-Paal, and J. Jaus, "Comparison of measurement techniques for the determination of circumsolar irradiance," *AIP Conf. Proc.*, vol. 1556, no. 1, pp. 162–167, 2013.

[9] D. Oberländer, C. Prahl, S. Wilbert, S. Müller, B. Stanicki, and N. Hanrieder, "Cloud shadow maps from whole sky imagers and voxel carving," in *ICEM*, 2015.

[10] P. Ineichen and R. Perez, "A new airmass independent formulation for the Linke turbidity coefficient," *Sol. Energy*, vol. 73, no. 3, pp. 151–157, 2002.

SAFE PV PLANTS WITH PANEL LEVEL ELECTRONICS?

C. Merz, G. Bettenwort, M. Hopf, H. Knopf, J. Laschinski
SMA Solar Technology AG
Sonnenallee 1, 34266 Niestetal, Germany
Phone: +49 561 9522 1169, Fax: +49 561 9522 421003, E-Mail: Christopher.Merz@SMA.de

ABSTRACT: The safety philosophy of PV plants has changed in the last 30 years. With technological progress, preventive safety measures have been established to increase plant safety as well as personal safety, i.e. double electrical insulation, residual current monitoring or the arc fault detection. In the last years, requirements for a rapid shutdown of the PV generator at string level during an emergency came up. In order to comply with the new requirements, a variety of vendors entered the market with technical solutions. Contrary to intentions, an increase of serious emergency situations, even fires [3], had to be observed. This shows that the integration of single safety measures would not increase the plants safety necessarily. The recent requirement for a rapid shutdown on module level even tightens the situation because it leads to even more components in the PV system using Panel Level Electronics (PLE). Hence, this paper discusses the harsh operating conditions and system failures that can occur within a PV plant and which directly affect the design of PLE.
Keywords: Durability, Panel Level Electronics, PV System, Reliability, Safety

1 SAFETY OF PRESENT PV PLANTS

The requirement for Rapid Shutdown of PV generators on module level is attended by the recent publication of the NEC 2017. Figure 1 shows a drawing of the most important requirements regarding voltage limits and boundaries.

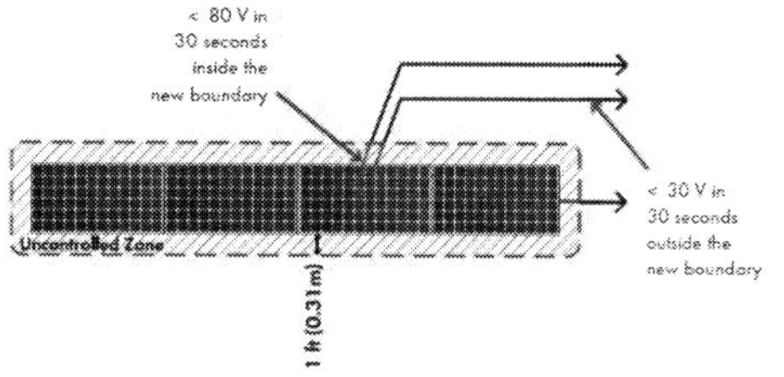

Figure 1: Voltage limits and boundaries of the NEC 2017 according to [6]

In summary the voltage within a PV generator is limited to module level (80 V) in case of an emergency shutdown. This requirement automatically leads to the need for electronic circuits and switches in every junction box: Panel Level Electronics (PLE). Motivation and main argument for this guidance is the protection of firefighters from being exposed to high string voltages during rescue operation.

Doubts about those arguments come up if relevant statistics on safety issues of PV plants are considered. A Fraunhofer ISE study [1] provided statistics relating to PV-related fires and the number of fatalities in Germany over the last 20 years. According to this study only 0.006 % of all PV plants have caused a fire resulting in serious damage. Furthermore, no firefighter has been injured by PV power so far while putting out a fire. These numbers are even more considerable due to the fact, that in Germany not even a string level rapid shutdown is common practice.

Following best practice while extinguishing a fire in an electrical system seems to be very efficient to protect first responders. According to a study from TÜV Rheinland and Fraunhofer ISE [2] there are minimum distances between the fire hose and a 1500 Vdc electrical system to ensure a safe work of the firefighters (see Table I).

Table I: Minimum distances between fire hose and fire in a 1500 Vdc electrical system according to [2]

	Minimum distance
Spray water jet	1 m
Full water jet	5 m

The statistics and examples of best practice mentioned above indicate that conventional PV systems with string technology are safe and can be managed in an emergency.

Nevertheless, in the last years a large number of vendors entered the market with a variety of DC disconnectors that were meant to comply with the NEC 2014 requirement for rapid shutdown on string level. But not all of these products were well designed and some had to be recalled because they were causing a fire instead of making the PV plant safer [3]. Compared to the conventional string technology, a rapid shutdown at module level using PLE will automatically lead to a higher number of components in the PV system. Hence, an accurate analysis of the operating conditions and system failures is mandatory to assess the possible risks of PLE concerning the PV plant.

2 OPERATING CONDITIONS & SYSTEM FAILURES

Those operating conditions and system failures have been defined and analyzed in the public funded project “PV-Firebreaker” [4] under the guidance of TÜV Rheinland with collaboration of SMA et al.. The operating conditions and system failures are shown in Table II.

Table II: operating conditions & system failures

Operating conditions
• extended environmental conditions
• reverse and bypass currents due to shading
• capacitive switching currents
• asynchronous switching
• ageing

System failures
• ground faults within the PV generator
• short circuits within the PV generator
• installation faults (i.e. reverse connection)
• lightning discharge

A system failure could eventually damage the PLE itself or could cause a lack of its power supply. Therefore it is important, that the PLE is able to turn into a safe operation mode. In such an operation mode the output voltage of the PLE could still be limited to the 80 V limit required by the NEC 2017.

However, the potential risks that arise from the operating conditions are probably underrated. For instance, reverse currents in PV plants can last for a long time during the day and can reach values of twice the short-circuit current of the PV modules. The components inside the PLE of course have to withstand this reverse current.

Below, two of the operating conditions are analyzed in detail, the *bypass currents due to shading* and the *capacitive switching currents*. Figure 2 shows a PV module which is part of a string. A module level switch is attached to the module. At the output of the switch there is an additional bypass diode which takes over the string current if the PV module is shaded. The power dissipation in this diode could amount to several Watts which leads to a high risk of overheating. The PLE must either withstand these dissipation losses by an appropriate cooling concept or the bypass path could be designed as an active solution to reduce these power losses.

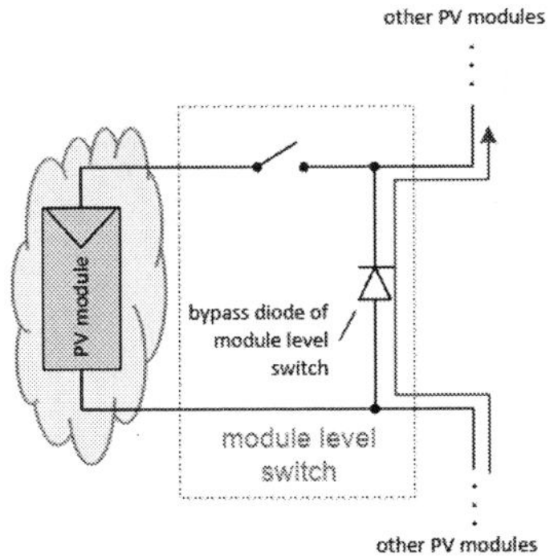

Figure 2: Bypass currents due to shading according to [4]

Figure 3 illustrates, how capacitive switching currents occur in a PV system comprising a PV module, a switching device and an inverter. Initially, the input capacitance of the inverter is uncharged if the switch is still open. The PV module then is in open-circuit mode. The PV module itself also has a parasitic capacitance which could reach values of more than 1 mF in open-circuit mode (in-house measurements at SMA). If the switch is closed, the capacitance of the PV module is connected to the discharged input capacitance of the inverter. This can result in a switching current of more than 1 kA.

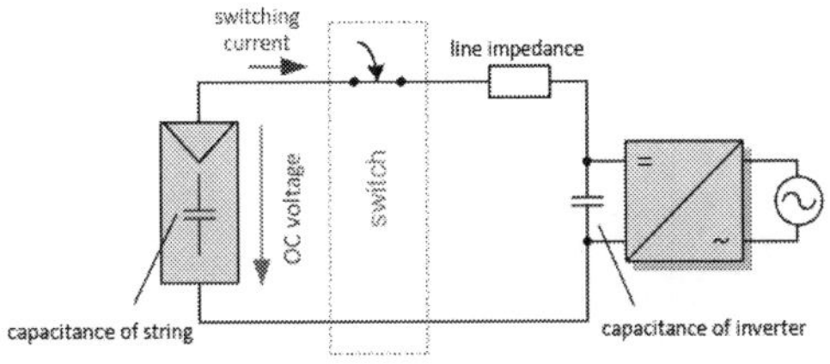

Figure 3: Capacitive switching currents according to [4]

The examples above show, that PLE has to withstand a variety of operating conditions and system failures which are specific for PV.

3 TEST CRITERIA

The current national and international standards concerning switching components are lacking some of the PV specific challenges that have been shown in the previous chapter. Hence, new test criteria have to be developed by the public funded project "PV-Firebreaker" [4] taking into account those operating conditions and system failures.

The next step would be to improve the existing standards with these experience and new test criteria. This task is now up to the relevant standardization committees.

4 RELIABLITY OF PANEL LEVEL ELECTRONICS

Renewable Energies already plays an important role at least in some countries. According to a study of GTM Research [5] there will be an installed worldwide PV capacity of 321 GW by the end of 2016. It is obvious, that PV will become a fundamental part of the future energy supply.

To ensure the availability of electricity, a high reliability of all components is essential comparable to the quality of conventional power stations. With respect to the PV plant this means that every part of the PV system has to fulfill the reliability targets and the usual service life.

Concerning PLE, there are some key factors in the design and application to achieve those reliability and durability targets.

The PLE has to be well designed and technically mature. This results from the harsh operating conditions and system failures that have been mentioned above. Moreover adequate trigger algorithms should be implemented in order to avoid nuisance tripping. It is absolutely necessary, that those PV specific requirements are considered during the design phase to ensure a reliable operation of the PLE.

A low component count could help to reduce the failure rate of the device. This can be achieved by a modular design which is tailored to the specific requirements of the PLE. For instance, a pure switching device doesn't necessarily need an optimizing function.

Another advantage is a selective application of the PLE. An example is shown in Figure 4. The PLE (here: power optimizer) is only applied to those modules that are definitely shaded during the day. This situation often appears on rooftops with dormers or satellite dishes which shade the same area of the PV generator every day.

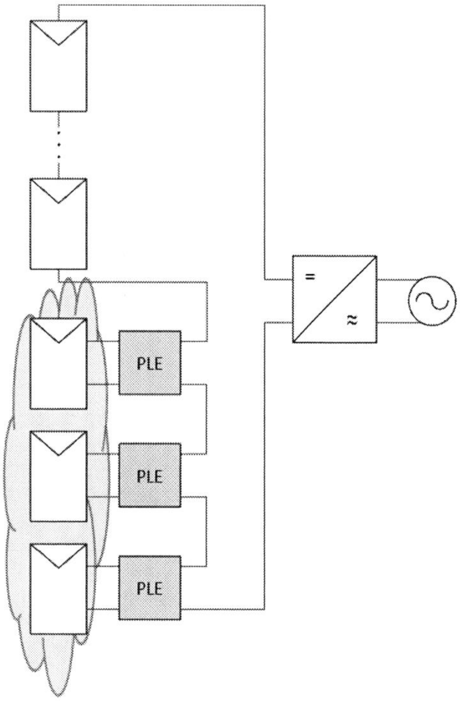

Figure 4: Selective application of PLE in shading situations

Finally, high durability is achieved if active parts of the PLE stay inactive most of the time. Power optimizers with buck converter topology only have to perform buck converting if the corresponding PV modules are shaded, while other parts of the PV generator see full insolation. At virtually homogeneously insolation, all buck converters switch off. Thus, the power loss is reduced to a minimum.

5 CONCLUSION

The experience of 20 years [1] shows that conventional PV systems with string technology are safe and can be managed even in an emergency. The consequent use of best practice allows a safe work of first responders while extinguishing a fire.

Current standards often don't consider the operating conditions and system failures that are specific to PV installations. Panel Level Electronics have to withstand and manage those conditions and failures to allow a safe and reliable operation.

On the basis of the mentioned operating conditions and system failures test criteria have been defined [4]. Those test criteria have to be fixed in future test standards to help the designers of PLE to develop safe and reliable devices.

PLE has to be well designed and technically mature with the lowest component count and low power loss to reach the reliability and durability targets.

6 REFERENCES

[1] Recent Facts about Photovoltaics in Germany, Fraunhofer ISE (2015)
[2] Assessment of the Fire Risk in Photovoltaic Systems and Elaboration of Safety Concepts for Minimization of Risks,TÜV Rheinland, Fraunhofer ISE (2015)
[3] http://www.recalls.gov.au (Australian Competition and Consumer Commission), (06/2016)
[4] "PV-Firebreaker" final report (FKZ 0325596), TÜV Rheinland et. al. (2016)
[5] http://www.greentechmedia.com, GTM Research: Global Solar PV Installations grew 34% in 2015 (22/01/2016)
[6] Evaluating the Case for Module-Level Shutdown, SMA Solar Technology AG (2015)

7 ACKNOWLEDGMENTS

Parts of this work were obtained within the framework of the project "PV-Firebreaker" (FKZ: 0325596) promoted by the German "Bundesministerium für Wirtschaft und Energie" (Federal Ministry for Economic Affairs and Energy). Only the authors are responsible for the content of this paper.

A MPPT ALGORITHM FOR PARTIAL SHADING CONDITIONS EMPLOYING CURVE FITTING

Efstratios BATZELIS[(1)] Georgios KAMPITSIS[(2)] Stavros PAPATHANASSIOU[(3)]

National Technical University of Athens, Electric Power Division, 9 Iroon Politechneiou st., 15780, Athens, Greece
[(1)]batzelis@mail.ntua.gr,[(2)]gkampit@central.ntua.gr,[(3)]st@power.ece.ntua.gr

ABSTRACT: Standard maximum power point tracking (MPPT) algorithms often fail to locate the global maximum of a photovoltaic (PV) system under partial shading conditions, while other more sophisticated approaches usually involve extra perturbation of the operating point, which entails undesired output power fluctuation. In this paper, a new MPPT method is introduced, which continuously detects the shading parameters and estimates all power peaks (MPPs) on the *P-V* curve, guaranteeing continuous operation at the global maximum. The algorithm applies least squares (LSQ) curve fitting (CF) to measurements at the current MPP, utilizing the inherent ripple, without the need for additional perturbation on the operating point. The calculations performed are entirely mathematical and no extra measurement equipment is required, such as irradiance or temperature sensors. The method is designed for PV strings illuminated at two irradiance levels.
Keywords: Curve fitting, global maximum, least squares (LSQ), Levenberg-Marquardt, maximum power point (MPP), maximum power point tracking (MPPT), partial shading, photovoltaic (PV), power peaks.

1 INTRODUCTION

Several local maximum power points (MPPs) arise on the *P-V* curve of a photovoltaic (PV) string, when operating under partial shading conditions. Standard MPP tracking (MPPT) methods, like Perturb & Observe (P&O) or Incremental Conductance (INC), often fail to locate the global maximum and lock to a suboptimal MPP, leading to significant power losses.

There are some more sophisticated approaches in literature especially designed for partial shading, which are based on empirical formulae or evolutionary optimization algorithms [1]–[5]. Yet, all these methods involve periodic perturbation of the operating point, leading to undesirable fluctuation in the output power and extra power losses. In order to balance the pros and cons, the execution frequency of these algorithms is adjusted accordingly in these papers, still compromising between losses due to partial shading and the aforementioned implications.

A relevant global MPPT (GMPPT) strategy that avoids this kind of perturbation is presented in [6], which monitors the power peaks on the *P-V* curve and switches from one local MPP to another whenever needed in order to ensure operation at the global MPP. It utilizes measurements of the terminal voltage and current of the PV string at the actual operating MPP in order to mathematically estimate the shading conditions and evaluate the voltage and current of all MPPs. However, this method is designed for nominal temperature and assumes that the intensity of a particular shading event remains constant, which limits the universal applicability.

These restrictions arise from the use of only the operating voltage and current as input data, from which only two shading parameters can be determined by the equations in [6]. In this paper, this concept is enhanced employing least squares (LSQ) curve fitting (CF) on a larger set of voltage and current samples. The inherent switching ripple and perturbation introduced by a constantly running P&O algorithm are utilized, and the shading parameters and properties of all MPPs are estimated. This is a purely mathematical procedure that introduces no additional variation of the operating point, thus permitting frequent execution in order to keep track of very fast shading variations (duration of a few seconds). The proposed technique does not require an irradiance sensor or other additional components and is designed for PV strings illuminated at two irradiance levels. Its effectiveness is validated and compared to a typical P&O approach via simulations in MATLAB/Simulink.

2 PV STRING MODELING UNDER PARTIAL SHADING CONDITIONS

A PV string consists of several PV modules connected in series, while a PV module comprises series connected PV cells. A group of series connected cells within a module, with a bypass diode connected in parallel at its terminals, is denoted a *cell string*. Under partial shading conditions, assuming two irradiance levels (i.e. one level of shade), the shaded cell strings are illuminated at irradiance G_{sh}, while the unshaded ones at irradiance G_{un} ($G_{un}>G_{sh}$). The shaded and unshaded parts of the string (denoted as *cell string groups*) comprise N_{sh} and N_{un} cell strings respectively ($N_{tot}=N_{sh}+N_{un}$).

As explained in [7], [8], in such conditions up to two MPPs may appear on the *P-V* curve of the string, denoted MPP1 and MPP2. An indicative case is illustrated in Figure 1(a)-(b). When operating at MPP1, the shaded cell strings are bypassed and only the unshaded ones generate power (red square marker in Figure 1(a)-(b)). On the other hand, at MPP2 all cell strings contribute to the power generation at the reduced current imposed by the shaded cell strings (green circle markers in Figure 1(a)-(b)).

Using the single-diode model, the equivalent circuit of the PV string is depicted in Figure 1(c) in black line color, comprising the unshaded (white background) and shaded (grey background) components. Each subcircuit employs a set of five parameters I_{ph}, I_s, a, R_s and R_{sh} for the single-diode model and the bypass diode *BP* coefficients I_{sbp}, a_{bp} (the subscript "un" or "sh" refers to the unshaded or shaded group respectively).

It is worth noting that, while operating at MPP1, the bypass diode of the unshaded cell strings BP_{un} is not conducting, while the p-n junction of the shaded cell strings D_{sh} is reverse biased. Therefore, the equivalent circuit for operation at MPP1 may be simplified as shown in Figure 1(d) in red color. Similarly, at MPP2, none of the bypass diodes conducts, leading to the simplified

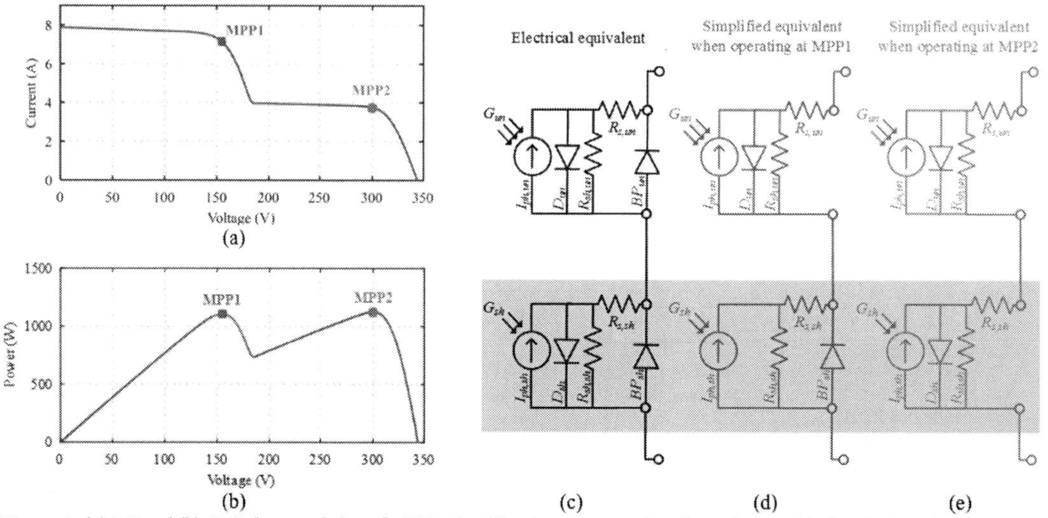

Figure 1. (a) *I-V* and (b) *P-V* characteristics of a PV string illuminated at two irradiance levels, (c) electrical equivalent circuit and (d)-(e) simplified equivalents for operation at MPP1 and MPP2 respectively.

equivalent depicted in Figure 1(e) in green color.

In order to describe the operation of the aforementioned circuits, the single-diode model equation is used [7], [9]:

$$V_{cs} = R_{sh} I_{ph} - \left(R_s + R_{sh}\right) I - aW\left\{ \frac{R_{sh} I_s}{a} e^{R_{sh}\frac{I_{ph}-I}{a}} \right\} \quad (1)$$

where V_{cs} and the five parameters I_{ph}, I_s, a, R_s and R_{sh} refer to a *single* cell string. Eq. (1) is expressed in explicit form by means of the Lambert W function $W\{.\}$.

According to the above analysis, when operating at MPP1 the unshaded part of the string may be modeled through (1) (Figure 1(d)). On the other hand, the shaded component is bypassed, thus it is described by the bypass diode equation [7]:

$$V_{bp} = -a_{bp} \ln\left(\frac{I - I_{SC,sh}}{I_{sbp}} \right) \quad (2)$$

where $I_{SC,sh}$ is the short circuit current of the shaded cell strings, approximated as:

$$I_{SC,sh} = \frac{R_{sh,sh}}{R_{s,sh} + R_{sh,sh}} I_{ph,sh} \quad (3)$$

For MPP2 operation (Figure 1(e)), eq. (1) applies to both the shaded and unshaded part, albeit using different sets of five parameters. Therefore, the entire PV string equation is given by:

$$\begin{cases} \text{MPP1}: \; V_{str} = N_{un} V_{cs,\text{un}} - N_{sh} V_{bp} \\ \text{MPP2}: \; V_{str} = N_{un} V_{cs,un} + N_{sh} V_{cs,sh} \end{cases} \quad (4)$$

where V_{str} denotes the total string voltage, $V_{cs,un}/V_{cs,sh}$ is the voltage of one unshaded/shaded cell string using (1) and the respective five parameters, and V_{bp} is the voltage drop on the bypass diode via (2).

Eq. (4) describes the entire PV string when operating under two irradiance levels, depending on the actual operating region (MPP1 or MPP2). The parameters involved are: five parameters for the unshaded cell strings ($I_{ph,un}$, $I_{s,un}$, a_{un}, $R_{s,un}$ and $R_{sh,un}$), five parameters for the shaded ones ($I_{ph,sh}$, $I_{s,sh}$, a_{sh}, $R_{s,sh}$ and $R_{sh,sh}$), the coefficients of the two bypass diode (I_{sbp} and a_{bp}) and the number of shaded and unshaded cell strings (N_{sh} and N_{un}).

3 CURVE FITTING MPPT (CFMPPT) ALGORITHM

The main concept of the proposed algorithm is to estimate the aforementioned parameters by applying curve fitting on a set of past measurements. According to the previous analysis, the unknowns to be determined are 14 in total, which are too many to be determined via a curve fitting procedure. Given that the unshaded cell strings number is the complement of the shaded ones ($N_{tot} = N_{sh}+N_{un}$), and considering the bypass diodes' coefficients known and unaffected by the operating conditions, still there are 11 unknown parameters.

In order to further reduce the number of unknowns, the five parameters of each group are correlated with their STC values and the operating conditions (irradiance and temperature). For this, the extrapolation equations proposed in [10], as reworked in [11], are employed:

$$I_{ph0} = G\left(I_{ph0} + a_{Isc}\left(T - T_0\right)\right) \quad (5)$$

$$I_s = I_{s0}\left(\frac{T}{T_0}\right)^3 e^{47.1\left(1-\frac{T_0}{T}\right)} \quad (6)$$

$$a = a_0 \frac{T}{T_0} \quad (7)$$

$$R_s = R_{s0} \quad (8)$$

$$R_{sh} = \frac{R_{sh0}}{G} \quad (9)$$

where I_{ph0}, I_{s0}, a_0, R_{s0}, R_{sh0} denote the reference parameters (STC values), a_{Isc} is the temperature coefficient of the short circuit current (A/K) and T_0 = 298.15 K is the nominal temperature at STC. The incident irradiance is denoted as G (p.u.) and the cell temperature as T (K).

The five reference parameters at STC are known beforehand, calculated from datasheet information via a method such as the one proposed in [11]. Therefore, only the two irradiance levels and the temperature have to be determined in order to obtain all ten circuit parameters (a common cell temperature across the PV string is assumed). Consequently, the final *shading parameters* are limited to four: [G_{un}, G_{sh}, T, N_{sh}].

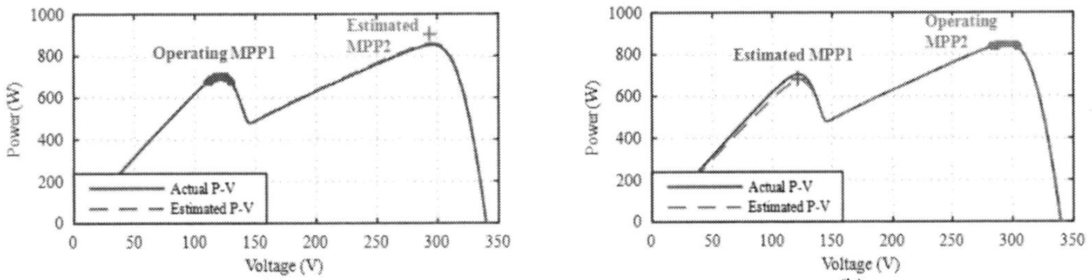

(a) (b)

Figure 2. Operating point variation around (a) MPP1 and (b) MPP2, exploited in order to estimate the shading parameters and the other MPP.

These 4 parameters are determined by applying least squares curve fitting (LSQ CF) on a set of previous measurements [12], included in a *measurement window*. To facilitate understanding, an indicative scenario is illustrated in Figure 2(a). A standard P&O algorithm is constantly running, continuously perturbing the operating point of the PV string, thus creating a measurement window around MPP1 indicated by red circle markers in Figure 2(a). The LSQ CF method fits (4) (upper branch in this case) to the measurements in order to extract the shading parameters.

The resulting $[G_{un}, G_{sh}, T, N_{sh}]$ are then used to calculate the other MPP's voltage and power (green cross marker in Figure 2(a)), using the simple expressions given in [7] and enhanced in [8] for arbitrary temperature:

$$\text{MPP1:}\begin{cases} V_{mp1} = N_{un}V_{mp}^{T} - N_{sh}\Delta V_D \\ I_{mp1} = G_{un}I_{mp}^{T} \end{cases} \quad (10)$$

$$\text{MPP2:}\begin{cases} V_{mp2} = N_{un}\left(\dfrac{G_{sh}}{G_{un}}V_{mp}^{T} + \left(1-\dfrac{G_{sh}}{G_{un}}\right)V_{oc}^{T}\right) + N_{sh}V_{mp}^{T} \\ I_{mp2} = G_{sh}I_{mp}^{T}\left(1+\lambda\dfrac{N_{un}}{N_{tot}}\right) \end{cases} \quad (11)$$

where V^T_{mp}, I^T_{mp} and V^T_{oc} are the MPP voltage, MPP current and open circuit voltage at *nominal irradiance but arbitrary temperature T*, ΔV_D denotes the voltage drop on one bypass diode (typically 1 V) and λ is an empirical coefficient equal to 0.06 [7]. The parameters V^T_{mp}, I^T_{mp} and V^T_{oc} are related to their nominal values via [8]:

$$\begin{aligned} I_{mp}^{T} &= I_{mp0} + \alpha_{Imp}\left(T - T_0\right) \\ V_{mp}^{T} &= V_{mp0} + \beta_{Vmp}\left(T - T_0\right) \\ V_{mp}^{T} &= V_{oc0} + \beta_{Voc}\left(T - T_0\right) \end{aligned} \quad (12)$$

where α_{Imp}, β_{Vmp} and β_{Voc} are the respective *absolute* temperature coefficients. If, α_{Imp} is not provided in the datasheet, it may be set equal to α_{Isc} or simply zero, while in the absence of β_{Vmp}, β_{Voc} may be used instead.

Using the shading parameters $[G_{un}, G_{sh}, T, N_{sh}]$ and the aforementioned equations, the other MPP's voltage and current are determined. Thereafter, the operating point is shifted to MPP2, if the latter provides more power than the current operating point MPP1 (Figure 2(a)). A similar procedure is performed when operating at MPP2 (Figure 2(b)).

It is worth noting that the variation of the operating point around the current MPP is due to the inherent ripple caused by the switching of the DC/DC converter and the perturbation introduced by the P&O algorithm. The duration that corresponds to the samples of the measurement window is denoted window period T_{win}. It can be set from several tens to a few hundreds of milliseconds, to ensure curve fitting robustness and enhanced tracking capability of fast shade variations.

Regarding the calculation algorithm employed at LSQ CF, the Levenberg-Marquardt method is adopted which is an iterative procedure [13]. Nevertheless, in order to reduce computational cost and permit implementation in a simple microcontroller, *only one iteration* is executed in each step of the proposed CFMPPT algorithm, i.e. every T_{win}. A few iterations per second (one every T_{win}) are sufficient to continuously adapt to the varying environmental conditions, even the most rapidly changing ones.

The flowchart of the proposed CFMPPT technique is illustrated in Figure 3. It is worth noting that a standard P&O is constantly in operation, in order to continuously fine tune the operating point and create the measurement window required by the CFMPPT algorithm. The latter is executed at a lower frequency, to track the MPPs and ensure operation at the global maximum at all times.

At each cycle of the CFMPPT execution (Figure 3), the ten circuit parameters (five for the shaded and five for the unshaded group) are first extrapolated to the previously estimated irradiance and temperature values, applying the equations (5)-(9). Thereafter, a *single iteration* of the LSQ CF algorithm is performed (Levenberg-Marquardt) to update the shading parameters using the upper or lower branch of (4), depending on the current MPP.

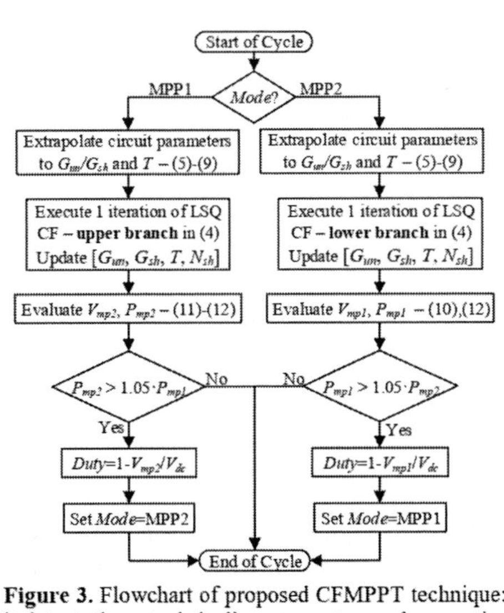

Figure 3. Flowchart of proposed CFMPPT technique: it detects the actual shading parameters and properties of all MPPs and manages transitions to global MPP.

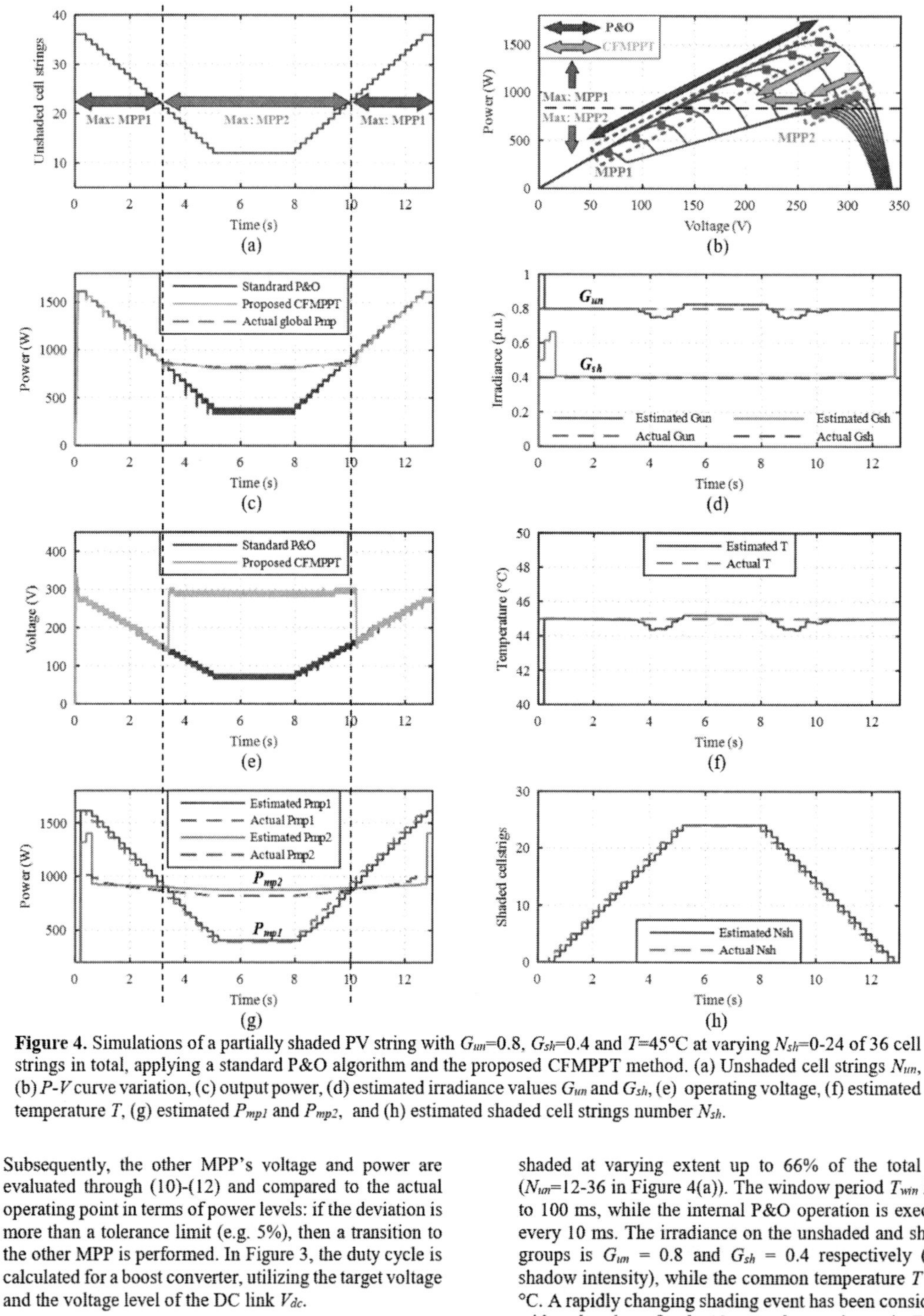

Figure 4. Simulations of a partially shaded PV string with G_{un}=0.8, G_{sh}=0.4 and T=45°C at varying N_{sh}=0-24 of 36 cell strings in total, applying a standard P&O algorithm and the proposed CFMPPT method. (a) Unshaded cell strings N_{un}, (b) *P-V* curve variation, (c) output power, (d) estimated irradiance values G_{un} and G_{sh}, (e) operating voltage, (f) estimated temperature T, (g) estimated P_{mp1} and P_{mp2}, and (h) estimated shaded cell strings number N_{sh}.

Subsequently, the other MPP's voltage and power are evaluated through (10)-(12) and compared to the actual operating point in terms of power levels: if the deviation is more than a tolerance limit (e.g. 5%), then a transition to the other MPP is performed. In Figure 3, the duty cycle is calculated for a boost converter, utilizing the target voltage and the voltage level of the DC link V_{dc}.

4 SIMULATIONS

In order to study the proposed CFMPPT algorithm and compare its effectiveness to a standard P&O method, simulations are performed in MATLAB/Simulink. A PV string composed by 12 PV modules is considered, each having 3 bypass diodes (36 cell strings in total), partially shaded at varying extent up to 66% of the total area (N_{un}=12-36 in Figure 4(a)). The window period T_{win} is set to 100 ms, while the internal P&O operation is executed every 10 ms. The irradiance on the unshaded and shaded groups is G_{un} = 0.8 and G_{sh} = 0.4 respectively (50% shadow intensity), while the common temperature T = 45 °C. A rapidly changing shading event has been considered with a duration of only 13 seconds. As shown in Figure 4(a), MPP1 is the global maximum at the time intervals 0-3.2 s and 10-13 s (red arrows), whereas MPP2 provides more power at the rest of the duration (green arrow).

The *P-V* curve variation over time is illustrated in Figure 4(b) in blue colored curves, indicating MPP1 and MPP2 with red square and green circle markers respectively. While the system is unshaded, the standard P&O method operates at the sole MPP of the *P-V* curve (upper red square marker MPP1). As the shadow appears and extends, the P&O continuously fine tunes the operating point around MPP1 while it shifts to lower voltages. Essentially, the algorithm is locked at this MPP for the entire duration of the event (blue arrow in Figure 4(b)), even when it is not the global one. On the contrary, the developed CFMPPT initially operates at MPP1, but always keeps track of MPP2, switching to the latter when MPP1 is no longer the global maximum, as the orange arrows indicate in Figure 4(b).

This is clearly illustrated in Figure 4(c), where the output power over time is depicted. While the typical P&O (blue color) stays locked on MPP1 for the entire duration, the CFMPPT method (orange color) performs a transition to MPP2 at t = 3.4 s and back to MPP1 at t = 10.2 s, guaranteeing optimal operation throughout the simulation. This way, it provides output power almost identical to the maximum possible (red dashed line). Consequently, the MPPT efficiency of the standard P&O algorithm is measured 83.5% for this particular scenario, as opposed to 98.2% of the proposed CFMPPT method (the small deviation from 100% is due to the unavoidable ripple of the operating point). Notably, the CFMPPT algorithm performs reliably even for fast illumination transients, lasting only a few seconds, as would occur during fast moving cloud conditions.

The operating voltage when applying the two methods is depicted in Figure 4(e). The standard P&O algorithm (blue line) follows the variation of MPP1 voltage, which is proportional to the unshaded area (Figure 4(a)). On the other hand, the operating voltage at the CFMPPT technique presents two distinct levels that correspond to the two different MPPs. The short delay in the transitions is associated to the duration of the window period T_{win}.

Regarding the accuracy in the calculation of the shading parameters, the estimated and actual irradiance levels G_{un} and G_{sh} are depicted in Figure 4(d). After t = 0.2 s, the estimated G_{un} (continuous blue line) coincides with its actual value (red dashed line) as long as the operating point is MPP1, while a small deviation is observed during operation at MPP2. In the latter case, the effect of the unshaded group is quite limited and thus the parameters of this group (G_{un}) are more difficult to be extracted by curve fitting. Similarly, the G_{sh} estimation in the same figure (green continuous line) is perfectly aligned with the actual irradiance (purple dashed line), as long as at least one cell string is shaded. When there is no shadow at all (0-0.6 s and 12.8-13 s), G_{sh} takes arbitrary/meaningless values, yet always lower than G_{un}.

In Figure 4(f), the estimated temperature T (blue continuous line) quickly converges to its actual value (red dashed line) within the first 0.2 s. During the rest of the simulation, a slight deviation is observed from 3.4 to 10.2 s, for the same reasons as G_{un}. Regarding the number of shaded cell strings N_{sh} in Figure 4(h), perfect coincidence between actual and estimated values is observed, albeit with a small time delay associated with the window period T_{win}.

As a result, the estimation of the two MPPs power is quite satisfactory in Figure 4(g). The estimated P_{mp1} (blue line) presents a small delay compared to its actual value (red dashed line), due to N_{sh} calculation (Figure 4(h)). Similarly, P_{mp2} (green line) takes meaningless values when the system is unshaded. Overall, the proposed CFMPPT algorithm correctly determines the global maximum power over the entire duration, performing the appropriate transitions of the operating point when needed.

5 CONCLUSIONS

In this paper, a new MPPT strategy for a PV string under partial shading is introduced, which applies curve fitting on electrical measurements recorded by the dc/dc converter, to continuously monitor the global maximum. The method is capable of tracking fast shading variations, while it introduces minimum additional perturbation to the operation of the PV string. The proposed algorithm supports any temperature and irradiance (two different irradiance levels) and is readily implementable in a microprocessor. The simulations presented validate its superiority over a standard P&O approach.

6 ACKNOWLEDGMENTS

Mr. E. Batzelis and Mr. G. Kampitsis are supported in their PhD studies by "*IKY Fellowships of Excellence for Postgraduate Studies in Greece - Siemens Program*".

7 REFERENCES

[1] H. Patel and V. Agarwal, "Maximum power point tracking scheme for PV systems operating under partially shaded conditions," *IEEE Trans. Ind. Electron.*, vol. 55, no. 4, pp. 1689–1698, Apr. 2008.

[2] Y.-H. Ji, D.-Y. Jung, J.-G. Kim, J.-H. Kim, T.-W. Lee, and C.-Y. Won, "A real maximum power point tracking method for mismatching compensation in PV Array under partially shaded conditions," *IEEE Trans. Power Electron.*, vol. 26, no. 4, pp. 1001–1009, Apr. 2011.

[3] K. Ishaque, Z. Salam, M. Amjad, and S. Mekhilef, "An improved particle swarm optimization (PSO)–based MPPT for PV with reduced steady-state oscillation," *IEEE Trans. Power Electron.*, vol. 27, no. 8, pp. 3627–3638, Aug. 2012.

[4] E. Koutroulis and F. Blaabjerg, "A new technique for tracking the global maximum power point of PV arrays operating under partial-shading conditions," *IEEE J. Photovoltaics*, vol. 2, no. 2, pp. 184–190, Apr. 2012.

[5] M. Miyatake, M. Veerachary, F. Toriumi, N. Fujii, and H. Ko, "Maximum power point tracking of multiple photovoltaic arrays: A PSO approach," *IEEE Trans. Aerosp. Electron. Syst.*, vol. 47, no. 1, pp. 367–380, Jan. 2011.

[6] E. Batzelis and S. Papathanassiou, "An efficient MPPT algorithm for partially shaded PV strings," in *Proc. 31st Eur. Photovolt. Sol. Energy Conf. Exhib. (EU PVSEC 2015)*, Hamburg, Germany, Sep. 2015, pp. 1615–1619.

[7] E. I. Batzelis, I. A. Routsolias, and S. A. Papathanassiou, "An explicit PV string model based on the Lambert W function and simplified MPP expressions for operation under partial shading," *IEEE Trans. Sustain. Energy*, vol. 5, no. 1, pp. 301–312, Jan. 2014.

[8] G. N. Psarros, E. I. Batzelis, and S. A. Papathanassiou, "Partial shading analysis of multistring PV arrays and derivation of simplified MPP expressions," *IEEE Trans. Sustain. Energy*, vol. 6, no. 2, pp. 499–508, Apr. 2015.

[9] E. I. Batzelis, G. E. Kampitsis, S. A. Papathanassiou, and S. N. Manias, "Direct MPP calculation in terms

of the single-diode PV model parameters," *IEEE Trans. Energy Convers.*, vol. 30, no. 1, pp. 226–236, Mar. 2015.

[10] W. De Soto, S. A. Klein, and W. A. Beckman, "Improvement and validation of a model for photovoltaic array performance," *Sol. Energy*, vol. 80, no. 1, pp. 78–88, Jan. 2006.

[11] E. I. Batzelis and S. A. Papathanassiou, "A method for the analytical extraction of the single-diode PV model parameters," *IEEE Trans. Sustain. Energy*, vol. 7, no. 2, pp. 504–512, Apr. 2016.

[12] "Least squares." [Online]. Available: http://en.wikipedia.org/wiki/Least_squares.

[13] "Levenberg-Marquardt algorithm." [Online]. Available: https://en.wikipedia.org/wiki/Levenberg-Marquardt_algorithm.

DEVIATIONS OF RESULTS FOR ENERGY YIELD FROM EFFICIENCY RANKINGS OF MICRO-INVERTERS

Stefan Krauter and Jörg Bendfeld
University of Paderborn, Electrical Energy Technology – Sustainable Energy Concepts
Pohlweg 55, D-33098 Paderborn, Germany

ABSTRACT: To determine and rank the performance of micro-inverters, DC-AC conversion efficiency over the full range of load conditions (including inherent MPPT accuracy) is usually used as the principal indicator. However, for the end-user the AC energy yield fed into the grid is the most important value for benchmarking. To compare efficiency and yield of most micro-inverters available on the world market in 2014, an in- and outdoor test field at the University of Paderborn have been set up. The inverters have been fed by identical and calibrated crystalline silicon PV modules of 215 W at STC. To monitor accurately AC power output and yield, each of the micro-inverters has been equipped with a calibrated precision electricity meter. For micro-inverters that require control units for grid-feeding this has been purchased also. The test field is equipped with pyranometers in the module plane. The comparison includes energy yield over almost one year, efficiency-load characteristics, recovery times after low irradiance levels, cost consideration: While the market is quite new, the range of purchase costs varies considerably between the models in comparison, sometimes inverter costs are higher than module costs, particularly if an additional grid-connection or interface device is necessary for operation. The weighted conversion efficiency according to EU and CEC standards has been measured and compiled. While some inverters have been optimized for high irradiance levels, they ranked better at the CEC-efficiency, others performed very well also for low irradiance levels, thus ranking higher at in the EU-efficiency tables. These results deviated from the actual energy yield measurements, which showed a slightly different ranking. One device showed an extreme deviation of the efficiency rating from the yield results. An accurate but very slow MPPT algorithm that hardly could follow quickly changing irradiance conditions has possibly caused this effect. Apparently, some inverters are been optimized to show excellent EU and CEC efficiency ratings Further investigations are on the way, including temperature effects.

Keywords: Inverter, Small Grid-connected PV Systems, Power Conditioning

1 INTRODUCTION

Micro-inverters are inverters that are connected principally to a single PV module (sometimes to two modules), so each module-inverter combination acts as an independent power plant. The micro-inverter consists of a maximum power point tracker (MPPT), the DC-AC inverter, and an islanding protection unit (see e.g., [1]). To achieve higher power outputs, several module-inverter combinations are interconnected in parallel on the AC output side. This configuration offers various advantages: Easier planning and easier installation, easy up- and downscaling of the power plant, including extensions or repair that could be carried out during power plant operation. Effect of shadowing is very limited, and due to low system voltages, potential induced degradation (PID) does not occur. Logistics is simplified. However, costs of power plants based on micro inverters are about 20% higher. Some of the inverters cannot be operated by themselves and require a control unit (often combined with a remote shutdown option and a monitoring system), thus adding extra costs. Also, conversion efficiency may not be as high as for central inverters. Due to smart master-slave concepts centralized solutions with multiple but relatively large inverters may offer higher yields under weak light conditions.

2 INDOOR TESTS

Due to the reproducible test conditions in the indoor lab, the inverters have been examined individually with predefined and controlled input data. Input has been a PV module simulator with data set according to the modules used in the outdoor test. The principal output data recorded has been the delivered AC power of the inverters. Besides input power, output is also a function of input voltage. If input voltage is getting too low, the inverters even stop operating. The following examinations are based on the possible range of input data (including voltage) given the specific PV module also used for the outdoor investigation.

2.1 Peak and Weighted Efficiency

Peak efficiency is often reached close to the maximum load of the inverter. Peak efficiency in data sheets is not a helpful value since most of the time the inverters operate in the range of 20% to 30% of their rated power. Consequently, an adequately weighted efficiency is a more adequate measure. One type of weighted efficiency is the so-called "European Efficiency"; the other is the "CEC efficiency" by the California Energy Commission (CEC). CEC efficiency is computed as an average value of DC-AC conversion efficiencies at six pre-defined outputs between 10% and 100% of the rated power. For the European Efficiency the weighting factors are lower for the high relative power values.

2.2 Test Setup & Results

Ten different output power values from zero up to the maximum operational power of the PV modules have been considered (they are roughly identical for all seven micro inverters). The values used (adjusted by controlling the current) are: 10 W, 25 W, 50 W, 75 W, 100 W, 125 W, 150 W, 175 W, 200 W, and 225 W. The averaging time for the power meter used (Zimmer LMG670) was set to 500 ms. One micro inverter showed an unstable AC power output for values below 100 W, for this device the averaging time had to be extended to 2 s.

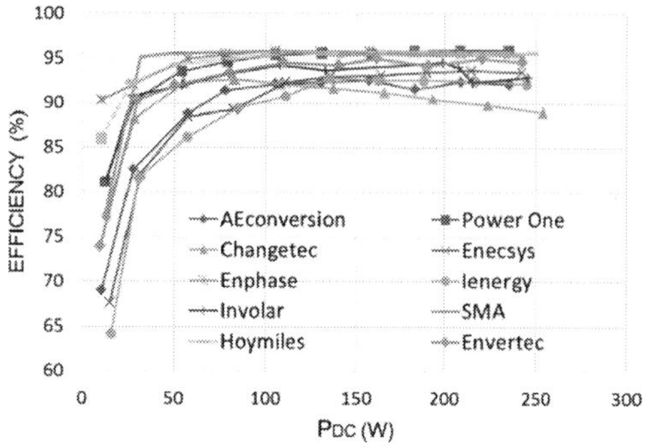

Figure 1: Measured DC-AC conversion efficiencies as a function of DC input power for ten micro-inverters.

The European conversion efficiency η_{Euro} is calculated according to:

$$\eta_{Euro} = 0.03 \cdot \eta_{5\%} + 0.06 \cdot \eta_{10\%} + 0.13 \cdot \eta_{20\%} + 0.1 \cdot \eta_{30\%} + 0.48 \cdot \eta_{50\%} + 0.2 \cdot \eta_{100\%} \quad (1)$$

Table I: Ranking of all micro-inverters by European conversion efficiency (same types as Table III)

Rank	Manufacturer	European efficiency	Relative eff. vs #1
1	SMA	95.4 %	100.0 %
2	Enphase	95.2 %	99.8 %
3	Hoymiles	95.0%	99.5 %
4	Power One	94.6 %	99.2 %
5	Envertech	93.2 %	97.7%
6	Involar	92.7 %	97.2 %
7	Changetec	90.9 %	95.3 %
8	AEconversion	90.3 %	94.7 %
9	Enecsys	90.3 %	94.7 %
10	Ienergy	89.9 %	94.3 %

The weighting for η_{CEC}, as proposed by the California Energy Commission (CEC) is determined by:

$$\eta_{CEC} = 0.04 \cdot \eta_{10\%} + 0.05 \cdot \eta_{20\%} + 0.12 \cdot \eta_{30\%} + 0.21 \cdot \eta_{50\%} + 0.53 \cdot \eta_{75\%} + 0.05 \cdot \eta_{100\%} \quad (2)$$

Table II: Ranking of all micro-inverters by CEC conversion efficiency (same types as Table III)

Rank	Manufacturer	CEC efficiency	Relative eff. vs #1
1	Enphase	95.6 %	100.0 %
2	Power One	95.5 %	99.9 %
3	Hoymiles	95.4 %	99.8 %
4	SMA	95.1 %	99.5 %
5	Envertech	94.1%	98.4 %
6	Involar	93.9 %	98.2 %
7	Enecsys	92.0 %	96.3 %
8	Ienergy	91.5 %	95.7 %
9	AEconversion	91.2 %	95.5 %
10	Changetec	90.9 %	95.1 %

For each power-step 25 values have been recorded. The measured DC-AC conversion efficiencies of all inverters are shown in Fig. 1. Based on those measurements, the European efficiencies and the CEC efficiencies for the micro inverters have been calculated according to (1) and (2). The ranking considering the European conversion efficiency is shown in Table I, the ranking for the CEC conversion efficiency is given in Table II.

3 OUTDOOR TESTS

3.1 Set-up for testing

For the PV outdoor test lab (see Fig. 1) installed on the roof of the University of Paderborn (51.707°N, 8.771°E), a specific test system comprising of eight photovoltaic modules with eight micro-inverters has been employed. To examine ten inverters, two of them have been interchanged.

Figure 2: PV outdoor laboratory at the University of Paderborn for micro-inverter comparison using eight equal, calibrated PV modules.

Each PV module (QC-CO2 by Hanwa Q-cells) consists of 60 solar cells. In stated plant, these equal and calibrated modules are the input for each micro inverter. The goal of this investigation is to analyze the performance of the inverters in the lab and under real operating conditions comparing their energy yield simultaneously at the same climatic conditions and solar irradiance. The climatic conditions have been monitored during the whole test period. The meteorological monitoring equipment consists of two calibrated pyranometers in the plane of the module (CMP 21 and SP2 lite by Kipp&Zonen), a 3-D ultrasonic anemometer (by Thies), and a thermo-hygro sensor (by Thies).

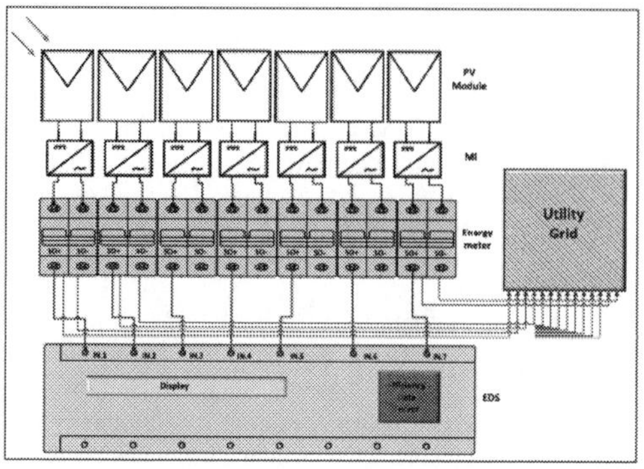

Figure 3: Scheme of the outdoor test setup for the micro-inverters (MI) investigated.

Fig. 3 shows the electrical measurement setup. Each micro inverter has been directly connected to a calibrated electrical energy meter with a S_0-interface. All S_0-interfaces have been connected to an efficiency recording data server. Fig. 5 shows an example of the collected data for single day (data recording every 5 min). Due to the high time resolution, some specific characteristics could be observed: E.g., starting behaviors, MPP tracking (accuracy and speed), dropouts and performance in high resolution can be perceived.

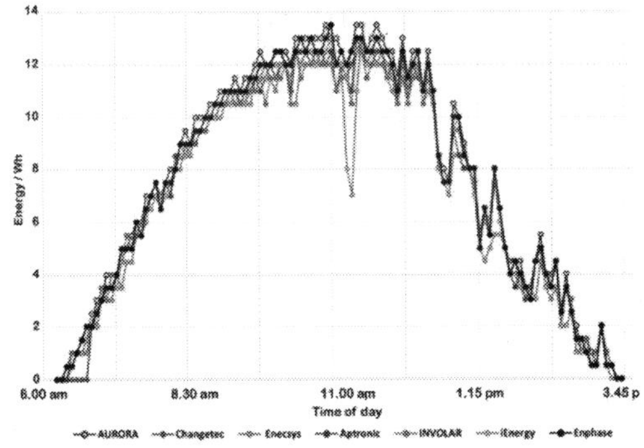

Figure 4: Actual recorded data of AC energy generation (over an integral of 5 min) for seven micro-inverters during the day of 10/31/2013.

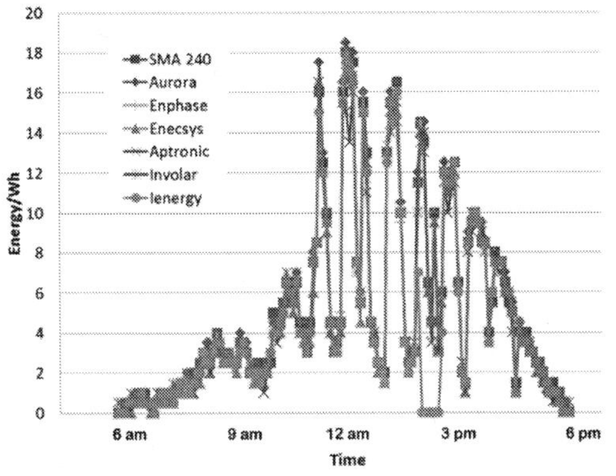

Figure 5: Actual recorded data of AC energy generation of seven micro-inverters (over an integral of 5 min) during the day of 4/6/2015.

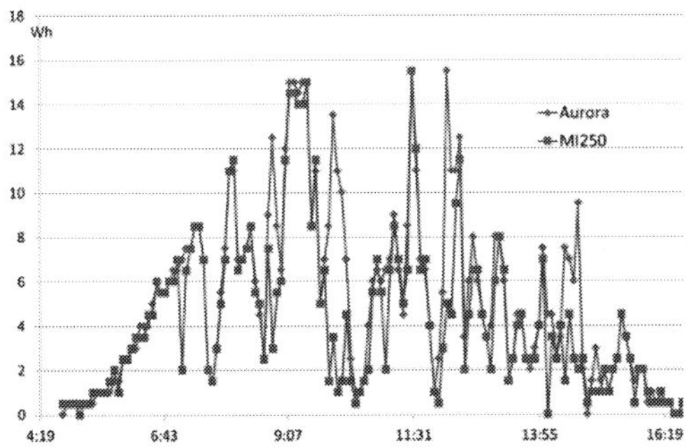

Figure 6: Actual recorded AC-energy generated within a 5 min interval during a day with changing irradiance levels. One inverter adapts quickly (blue line; ABB/Aurora), the other not (red; Hoymiles MI 250).

Fig. 5 shows an example of the collected data for single day during spring of the year 2015 (including the new SMA micro-inverter). It is visible that some inverters have difficulties to follow the quick change of irradiance levels during that day (e.g. Ienergy, and Hoymiles – see also Fig. 6)

A ranking list according to the total AC energy yield of the micro inverters during the common operation period is shown in Table III (July 2015 until June 2016, (Envertech 6/1/2016 until 6/15/2016 only) relative energy yield vs. #1.

Table III: Ranking of all micro-inverters by measured AC energy yield (in Paderborn 2015/16, 50 weeks, *2 weeks only)

Rank	Manufacturer	Type, Model	rel. yield vs #1
1	Power One / Aurora / ABB	Micro-0.25-i	100.0 %
2	SMA	Sunny Boy 240	95.2 %
3	Involar	MAC 250	94.4 %
4	Enphase	M 215	95.0 %
5	AEconversion / Aptronic	INV 250-45	92.4 %
6	IEnergy	GT 260	91.8 %
7	Enecsys	SMI-S-240W	88.8 %
8	Envertech	EVT248	88.0 %*
9	Hoymiles	MI 250	78.0 %
10	Changetec	ELV 300-25	75.6 %

4 DISCUSSION AND CONCLUSION

4.1 Results

Some of the results could be expected by taking a close look at the efficiency curves of Fig. 2: For example: If highest performance occurs at relatively low input power (and the corresponding low irradiance levels), the ranking for the European Efficiency (which puts a higher weight on lower irradiance levels) is higher, while the CEC ranking is worse due to the higher weighting of high irradiance levels (this is the case for the Changetec inverter). The best in the CEC ranking has been the Enphase inverter, which was second in European Efficiency ranking. For the SMA inverters the ranking positions have been the other way round: SMA is optimized for European irradiance conditions, Enphase for Californian conditions.

4.2 Correlation of energy yield with efficiency

Some of the rankings by yield (as shown in Table III) cannot be explained by the curves of Fig. 2 and the European or the CEC efficiency. We observed significant differences of the MPPT devices in their capabilities to adapt to fast changes of irradiance. Best in this discipline has been the Power One / Aurora inverter, which has not been the best in the static conversion efficiency measurements, but in energy yield. The device MI 250 showed an extreme deviation of its efficiency rating from the yield results. An accurate but slow MPPT algorithm that had difficulties to follow changing irradiance conditions has possibly caused this effect.

Fig. 7a shows the relative energy yield as a function of European efficiency (which is often an important criterion for the purchase decision of a potential customers). As can be observed, there is just a quite vague correlation between those two values with high deviations. For the example indicated at Fig. 7b it can be observed that there may not be just a significant quantitative deviation, but also a qualitative one. For an increase of 0.4% in Euro-efficiency, a decrease of -22% in yield may occur. The outcome for yield vs. CEC efficiency is similar.

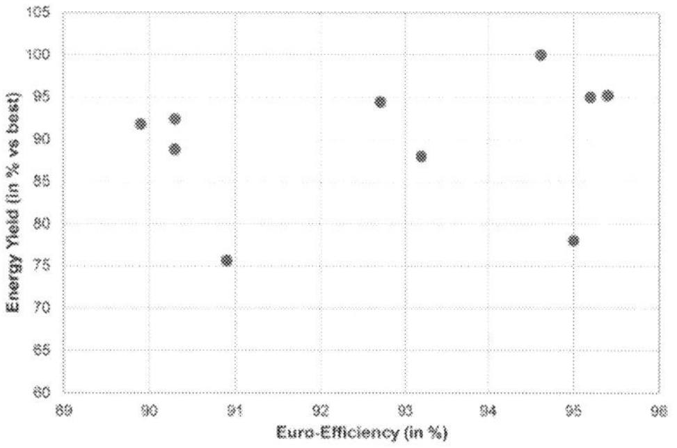

Figure 7a: Electrical energy yield as a function of European efficiency for all tested microinverters.

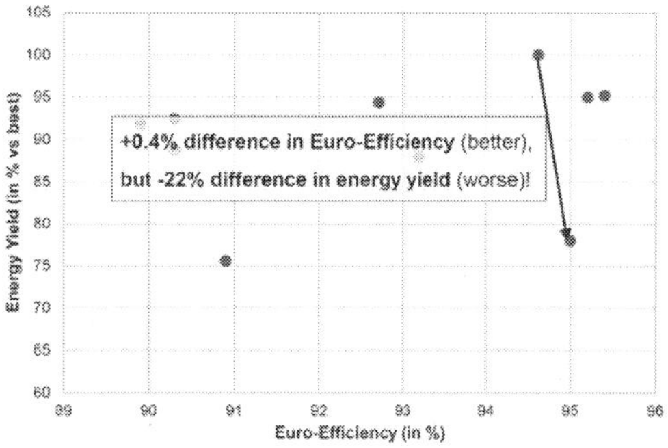

Figure 7b: Electrical energy yield as a function of European efficiency: The user cannot expect a higher yield if he purchased an inverter with a higher European efficiency – the yield can be significantly lower.

4.3 Outlook

To overcome the considerable deviation between yield and efficiency measurement results, a dynamic test will be deployed to find a better indicator for the relative yield to be expected by the different inverters.

While the lab tests have been made at a laboratory temperature conditions (20°-22°C) and the yield measurements at actual operating temperature, additional efficiency tests at elevated temperatures (e.g. 40°C and 60°C) will be carried out.

5 REFERENCES

[1] W-F. Lai, S-M. Chen, T-J. Liang, K-W. Lee and A. Ioinovici, "Design and Implementation of Grid Connection Photovoltaic Inverter," in *IEEE Energy Conversion Congress and Exposition (ECCE)*, 2012, pp. 2426-2432.

[2] S. Krauter; J. Bendfeld, "Cost, performance, and yield comparison of eight different micro-inverters", in *42nd IEEE Photovoltaic Specialist Conference (PVSC)*, 2015.

5.3 BALANCE OF SYSTEM COMPONENTS

PERFORMANCE OF RECENT INVERTER SYSTEMS UNDER PARTIAL SHADING CONDITIONS

RESULTS FROM SIDE-BY-SIDE COMPARISON IN A FIELD TEST

Ralph Lingel • Thomas Nordmann • Thomas Vontobel
lingel@tnc.ch • nordmann@tnc.ch • vontobel@tnc.ch
TNC Consulting AG
General Wille-Strasse 59 • 8706 Feldmeilen • CH – Switzerland
www.tnc.ch • P: +41 44 991 5577

ABSTRACT: Three common but fundamentally different inverter concepts are investigated under two types of partial shading conditions. One string inverter stystem, one microinverter system and one inverter system with power optimizers are installed in a side-by-side configuration at an existing PV plant. During summer one third of the modules experiences a similar shading situation due to self-shading. During winter season the whole plant is shaded heterogeneously by several objects in the vicinity. Due to the string design and the homogeneous shading in summer no increase in yield could be measured for the two decentralized inverter systems compared to the string inverter system. The system with power optimizers achieves equal performance values compared to the string inverter. The microinverter system with a lower inverter efficiency achieves lower yield values of -10 % on month base compared to the string inverter. Under heterogeneous shading conditions in winter both decentralized inverter systems can achieve significantly higher yields compared to string inverters. On a month base + 92 % for the system with power optimizers and +57 % for the microinverter system were measured.

Keywords: PV systems, partial shading, power optimizers, microinverter

1 PURPOSE OF THE WORK

Partial shading on PV systems with string inverters can cause significant reduction in yield. Manufacturers of decentralized inverter systems, such as inverters with power optimizers and microinverters hold out the prospect of an increase in yield of more than 25 %. There is little operational experience and data from PV plants published comparing these fundamentally different inverter systems under the same shading conditions. In this work three different inverter concepts are investigated under two different types of partial shading at a "real-life" PV-plant.

2 INVERTER CONCEPTS TESTED

Three fundamentally different inverter concepts are investigated. The string inverter system characterizes a central DC/AC conversion together with MPP-tracking on string level. Partial shading within one string affects the performance of all other modules in this string. With microinverters DC/AC conversion and MPP-tracking takes place per PV module. Inverters with power optimizers have a MPP-tracking per module (DC/DC) and a central DC/AC conversion.

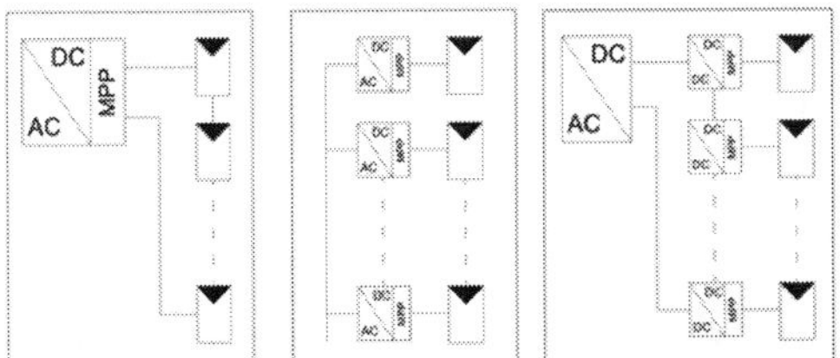

Fig. 1: Three different inverter Systems

3 PV-PLANT AND SHADING SITUATION

The tests are carried out at an existing PV testing plant near Zurich, which was built in 1998. The plant consists of 45 modules (210 W_p) arranged in three rows and has a nominal power 9.5 kW_p (Fig. 2).

Fig. 2: Testing plant: partial shading in January

During winter season the plant is affected by shading caused by houses and trees in the vicinity (Fig. 2 & 3). Using simulation tools (Meteonorm; PV*SOL) the calculated reduction of irradiation compared to an open horizon is 48% in December for the whole plant with a maximum of 78 % for the module at the most shaded position.

Fig. 3: houses and trees in the southern vicinity of the plant

Due to the characteristic zig-zag shape of the PV plant the undermost row experiences shading caused by the rows above during summer months (Fig. 4). The calculated reduction of irradiation onto this row is 20 % in June compared to open horizon. The shading losses for the whole plant are 15% on annual base.

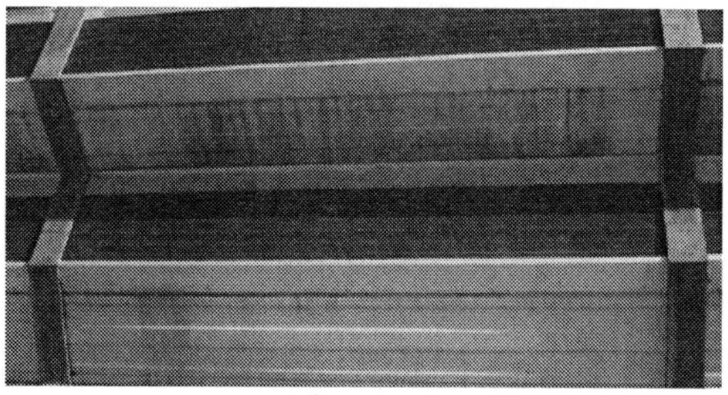

Fig. 4: Shadow on undermost module during summer

4 IMPLEMENTATION AND MONITORING

The inverter with power optimizers (15x) and the string inverter (2 MPP trackers) were installed in February 2014 together with the measurement equipment.
To ensure that all three inverter systems experience comparable shading conditions the modules are assigned in a "checkered" manner to the inverters (Fig. 4 & 5).
The modules assigned to the string inverter (blue) are distributed to two strings. The string inverter has a global maximum power point tracking which is useful especially under shading conditions. The modules of the upper rows are connected to the first MPP-tracker and the modules of the undermost row to the second MPP-tracker.
The 15 microinverters were installed in March 2015.

Fig. 4: Allocation of modules to the inverter systems

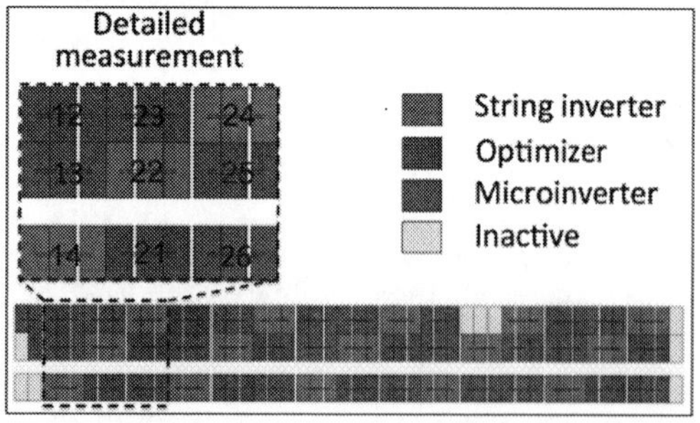

Fig. 5: Plant overview and modules with detailed measurement

The plant is equipped with precise measurement equipment, recording DC and AC values of the plant. Beside inverter and string values the most shaded modules "12" to "26" are monitored in detail. Two pyranometers measure irradiation in horizontal and module plane. Module and ambient temperature is also collected. The values are recorded on a minute base.

5 RESULTS

Preliminary results comparing operational behavior of the string inverter and the optimizer system were presented in [1]. With the installation of microinverters in Q1/2015 three fundamentally different inverter concepts can be compared under partial shading conditions. In June 2016 one year of operation and measurement has been completed. In the following chapters operational data of all three systems are presented and analyzed.

5.1 OBSERVING SINGLE DAYS

Figure 6 shows the DC-power output of three modules with string inverter on a sunny day in July (10/07/15). The effect of shading onto the module of the undermost row "14" between 8:00 h and 11:00 h is clearly visible. Figure 7 illustrates the normalized AC-power output of the three inverter systems for the same day. No difference in performance can be seen between the three systems for the time period when the undermost row is shaded. The string design leads to similar shading/irradiation conditions within the two strings and therefore there is no advantage for the two decentralized inverter systems. Around midday it can be seen that the string inverter and the optimizer system show a similar performance whereas the microinverters show a slightly lower power output (-5 %). This correlates well with the nominal inverter efficiencies: string inverter: 96 %; optimizer system: 96 %, microinverter: 92 %.

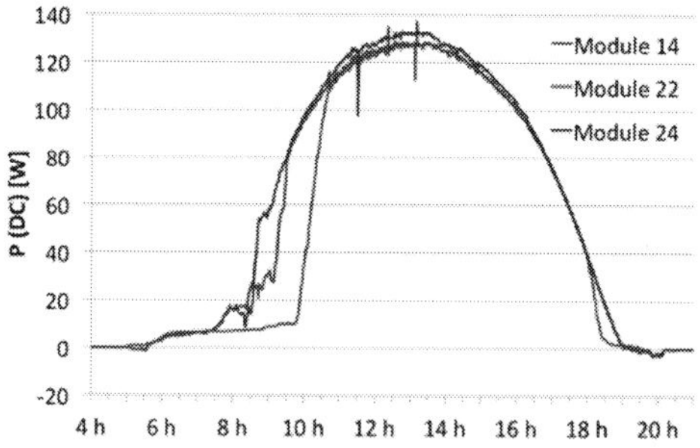

Fig. 6: DC-power of three modules with string inverter on a sunny day in July

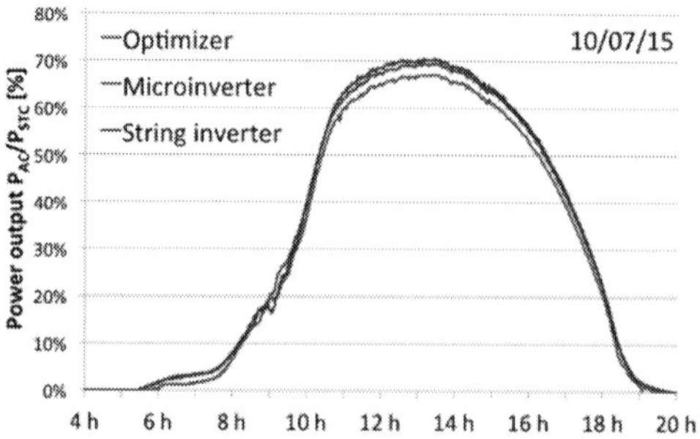

Fig. 7: Normalized AC-power output of the three inverter systems on a sunny day in July

Figure 8 shows the DC-power output of the three modules with string inverter on a sunny day in October. It can be seen that the power output of the modules breaks down according to their position in the PV-plant and the course of the shadows caused by the objects in the vicinity. The power outputs even reach negative values. This can be explained by the voltage drop at the bypass diodes of the modules.

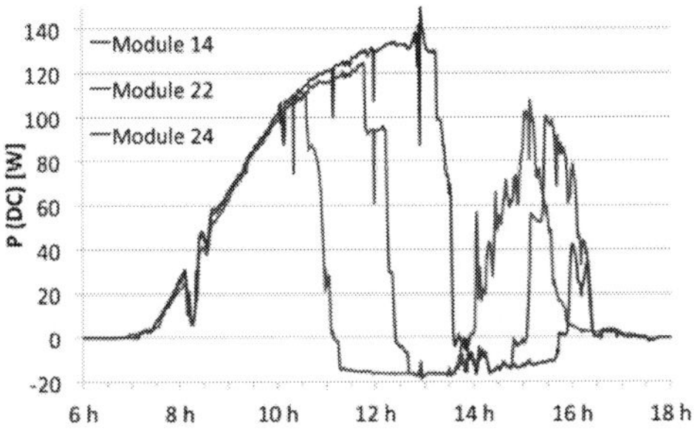

Fig. 8: DC-power of three modules with string inverter on a sunny day in October

The PV-plant experiences most severe shadow during December and January when sun is low on the horizon. Figure 9 illustrates the normalized AC-power output of the three inverter systems on 24/12/2015. Due to the heterogeneous shading situation, both, the optimizer system and the microinverters show a significantly higher power output compared to the string inverter. The difference in power indicates that not only the voltage drop of the bypass diodes causes this lower performance. It can be assumed that due to the great differences of irradiation conditions within the strings, most modules operate outside their optimum operating point. For this day the yield of the optimizer system is 133 % higher compared to the string inverter. The microinverter system achieves a 110 % higher yield.

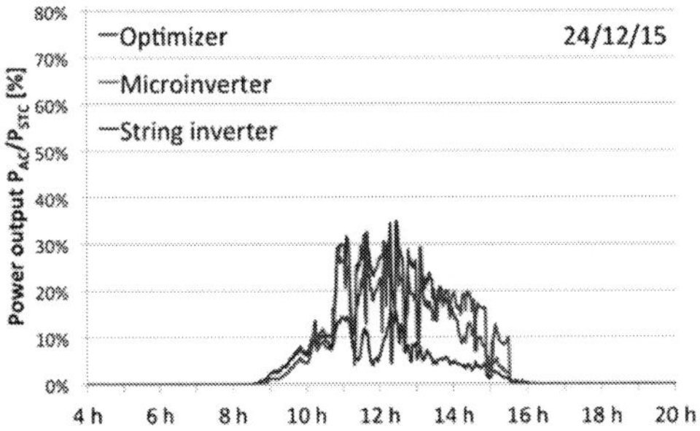

Fig. 9: Normalized AC-power output of the three inverter systems on a sunny day in December

5.2 SEASONAL CHANGE OF SHADING CONDITION AND ITS IMPACT ON PERFORMANCE RATIO

Figure 10 shows the final yield plotted versus the reference yield (irradiation) for the three systems in July 2015. It can be seen that irrespective of irradiation the string inverter and the optimizer system perform in a very similar way. The microinverters show a slightly lower final yield also irrespectively of daily irradiation values. This indicates the lower inverter efficiency of this system.

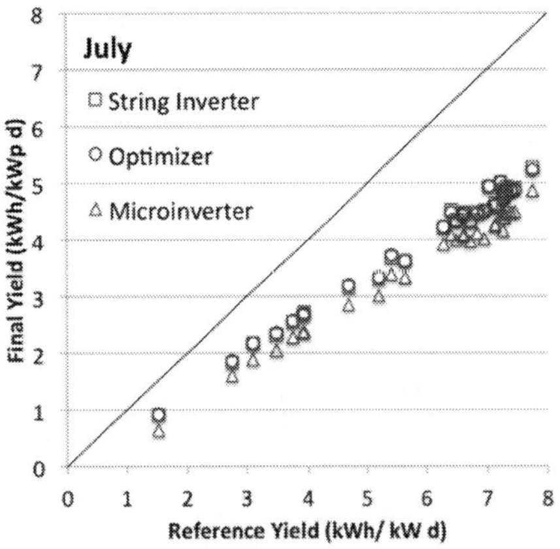

Fig. 10: Final yield vs. reference yield for the three inverter systems in July 2015

Figure 11 shows the final yield plotted versus the reference yield for the three systems in December 2015. The optimizer system shows the highest final yield values both for high and low irradiation values. On sunny days now the microinverter system shows a significantly higher performance compared to the string inverter. Under the very heterogeneous irradiation situation at this time of the year the advantage of decentralized MPP-tracking is overcompensating the lower inverter efficiency. On cloudy days with low irradiation this advantage disappears and the microinverter system shows lower final yield values compared to the two other systems.

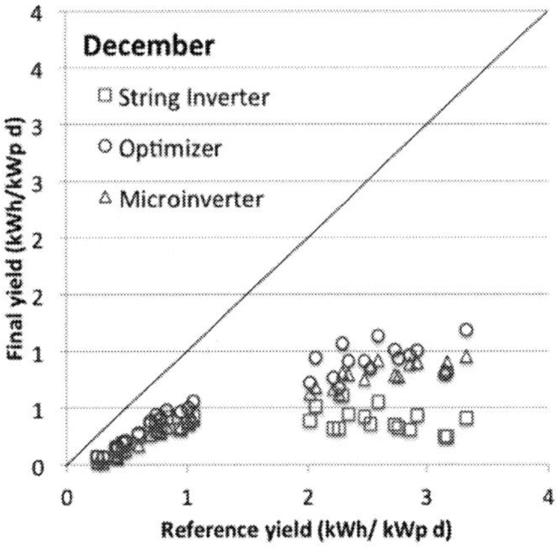

Fig. 11: Final yield vs. reference yield for the three inverter systems in December 2015

Figure 12 shows the average daily yield values per month of the three systems for one year of operation (4/6/2015 – 3/6/2016). The green and the red line indicate performance of the optimizer system and the microinverter system relative to the performance of the string inverter as a benchmark. During summer under homogeneous shading conditions the optimizer system shows similar values compared to the string inverter. With the beginning of partial shading in September the optimizer system shows higher average daily yield values. In December the yield of this system is 92 %

higher compared to the string inverter. For the whole year the optimizer system achieves a 8 % higher yield.
During summer the microinverter system shows a lower performance of 9 % - 10% compared to the string inverter. In October the advantage of the decentralized MPP-tracking under increasing heterogeneous shading compensates the lower inverter efficiency compared to the string inverter. In December the microinverter system achieves a 57 % higher yield than the benchmark system. For the whole year the microinverters have a slightly lower yield of -5 % compared to the string inverter.

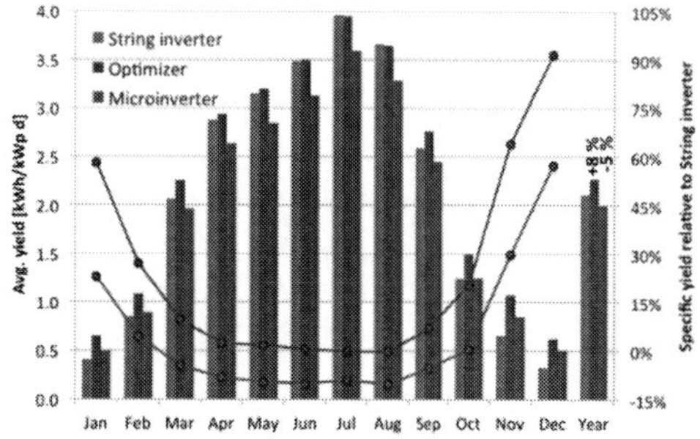

Fig. 12: Average daily yield per month of the three systems (4/6/15 – 3/6/16)

4 CONCLUSIONS

- Reduction in yield depends on the type of shading and the string design.
- Under heterogeneous shading conditions within strings decentralized inverter systems can achieve significantly higher yields compared to string inverters. On a month base + 92 % for the system with power optimizers and +57 % for the microinverter system were measured.
- Under homogeneous shading conditions and adjusted string design no increase in yield could be measured for decentralized inverter system. In this case the lower inverter efficiency of the microinverter system leads to lower yield values compared to the string inverter.

5 OUTLOOK/WORK IN PROGRESS

- Continuing measuring campaign to obtain long-term operational data
- Comparison of measured values with results obtained from simulation tools
- Comparison with historical data of the plant and evaluation of repowering

6 REFERENCES/PUBLICATIONS OF MAIN AUTHOR

[1] R. Lingel, T. Vontobel, T. Nordmann: *Wechselrichterkonzepte bei Teilverschattung in einem unabhängigen Feldtest*, 30. Symposium Photovoltaische Solarenergie, Bad Staffelstein, 2015

[2] Nordmann, T.; Frölich, A.; Clavadetscher, L.: *Drei integrierte PV-Schallschutz Versuchsfelder*, Schlussbericht PV P+D, DIS 17225 / 59391, 2002

PERFORMANCE EVALUATION OF HOUSEHOLD LI-ION BATTERY STORAGE SYSTEMS

Nina Munzke, James Barry, Bernhard Schwarz
Karlsruhe Institute of Technology (KIT), INT-PCE
Hermann-von-Helmholtz-Platz 1, Gebäude 276
76344 Eggenstein-Leopoldshafen, Germany

ABSTRACT: The increase in electricity prices along with a decrease in the price of storage systems has led to a rapid expansion of the PV-battery storage system market in German households. In order to be economically viable, PV-storage systems must fulfil certain performance criteria, and the system control strategies as well as the control speed have a large impact on the overall system performance. While the system control software should be regulated in such a way as to maximize the self-supply ratio as well as the battery lifetime, the control speed and strategy has a direct influence on the self-supply ratio and as a consequence on the overall self-sufficiency. An intelligent control strategy also has added benefits for the grid operator. Several commercially available PV-battery systems have been analyzed with respect to these specific performance criteria. The test results presented here provide a representative picture of the current status of the storage system market and offer useful pointers for future system development.
Keywords: Batteries, Battery Storage and Control, Lifetime, Storage, System Performance

1 INTRODUCTION

Efficient and economic energy storage technologies are key elements for a sustainable future energy supply. Attractive and economical applications can be established through the successful implementation of long-lasting and cost-efficient storage systems for electrical energy.

Since electricity prices in Germany have increased in the last few years and storage prices are starting to decline, small German household applications will soon become economically feasible [1], [2]. There are several critically important aspects affecting commercial viability, such as the system design and dimensioning, the level of development of the system control software and the calendar and cycle life of the battery and electronic components. In this work the effect of system control software is focused on, using the results of field tests performed on commercially-available storage systems.

2 METHODOLOGY

The system control strategies as well as the control speed have a large impact on the overall system performance. For this reason, research into the development of control software that is able to fulfil the prerequisites for the successful commercial operation of stationary storage applications is pushed forward at KIT. The software is being tested at three different PV energy storage systems at KIT and the Helmholtz Institute in Ulm. The first two systems started their operation at the beginning of 2013. The development of autonomous KIT software also serves as a model to understand the difficulties involved and the interdependencies between the quality of the software and the performance of a PV-storage system.

On top of that 18 different commercially available PV-battery systems are analyzed with respect to specific performance criteria.

2.1 Intelligence of the system control software

The control software of the systems developed at KIT is composed of two software modules: one that includes forecast algorithms in order to determine when the batteries should be charged and discharged during the day and another one that controls the effective power flows according to real time measurements and the forecast.

To find out whether the commercial home storage systems possess any sort of intelligent and/or predictive charging/discharging algorithm, they are tested within a hardware-in-the-loop environment. The measurements are done for typical reference days with measured solar PV data from KIT with a time resolution of one second and load curves for single-family households which can be generated based on the VDI 4655 reference profiles [3]. The initial tests were performed with an annual electricity demand of 4000 kWh and two inhabitants per household. The resulting household load profiles have a time resolution of one minute. The PV data comes from the 1 MW PV plant at KIT north campus, located at 49.1° N, 8.44° E, which corresponds to climate zone TRY12 in the VDI 4655 classification system.

The main objectives are to evaluate the level of development of the system control software and overall performance of commercially available systems, in order to ascertain the required future developments. To find out whether the systems possess an intelligent charging algorithm several similar days are tested in a row.

2.2 Response time and the ability to smooth load and PV-fluctuations

Storage systems react to changes in load and/or PV-production by adjusting the charge or discharge power of the battery, and they always have a finite response time due to delays in measurement and/or communication. Tests can be carried out to determine the response time as well as the resulting energy flows to and from the grid. These response tests were carried out with steps of different magnitude, ranging from 0.2 kW to 3.2 kW for both PV and load.

Figure 1 shows an example of such a measurement and illustrates the defined time intervals. Error bands around the primary signals (PV power and load) are defined, so that five different time points can be defined:

- t_1: the source signal leaves the error band (PV or load) at the beginning of the step
- t_2: battery power leaves the error band
- t_3: the step response of the battery power reaches 10 % of the required change in power

- t_4: the step response of the battery power reaches 90 % of the required change in power
- t_5: the step response of the battery power reaches and remains within the error band

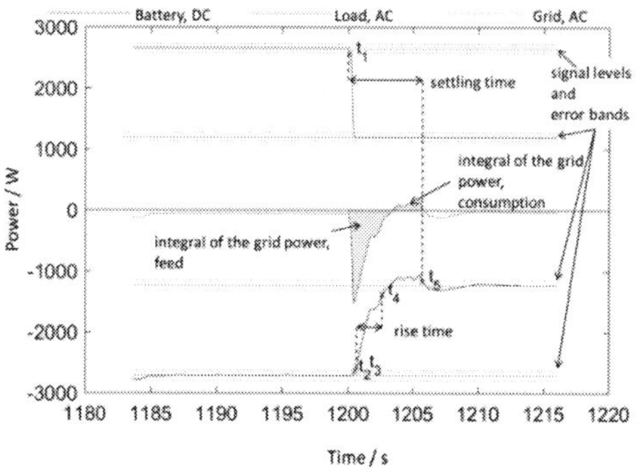

Figure 1: Time definitions and integrals used to determine energy flows to and from the grid

In order to compare storage systems with each other the following values can be calculated:

- Rise time (fall time) from 10% to 90% (90% to 10%) of the signal level ($t_4 - t_3$)
- Dead time (t2 – t1)
- Settling time (t5 – t1), i.e. time span until the step is completely compensated

Based on these definitions the energy flows to and from the grid can be obtained by integrating the power curves, resulting in (*i*) the total energy fed into the grid from the battery or PV (instead of being consumed) and (*ii*) the total energy unnecessarily drawn from the grid, even though the battery could have provided it. Similar tests are suggested by [4].

3 INTELLIGENCE OF THE SYSTEM CONTROL SOFTWARE

In this section the intelligence of the system control software of commercially available stationary storage is discussed. The first subsection gives a short introduction to the basic principles of such software, and in the second and third subsection the results of empirical tests on several such systems are shown and discussed.

3.1 Intelligent control of PV-battery storage systems

Firstly, the system should be operated in such a way so as to maximize the self-supply ratio. The underlying algorithms must also ensure the longest possible operational lifespan of the battery; in particular the various electrochemical processes that take place during the normal operation of a lithium-ion battery must be taken into account. A vital component of the software is therefore its ability to predict both the energy needs of the customer as well as the energy supplied by the PV array throughout the day. In essence one needs to be able to predict the load as well as the available power, to deduce how the system should best be regulated in order to increase profitability. In the system at KIT, predictions of power and energy are made at regular intervals throughout the day at 15 minutes intervals. The prediction is autonomous in that it does not need any external weather forecast data. Figure 1 and 2 show different PV power and energy predictions from the 18th of May 2014. The predicted curves match the real values quite well: the comparison of the power and energy predictions at 08:00 am and 01:00 pm shows that the algorithm adapts to the effect of clouds on radiation level. This example demonstrates an accurate PV prediction, which is an important component of high-level storage system control software.

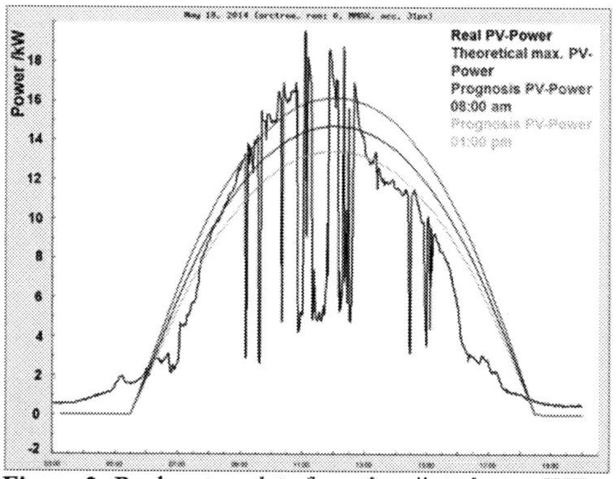

Figure 2: Real system data from the pilot plant at KIT on 18 May 2014, with measured and predicted PV power

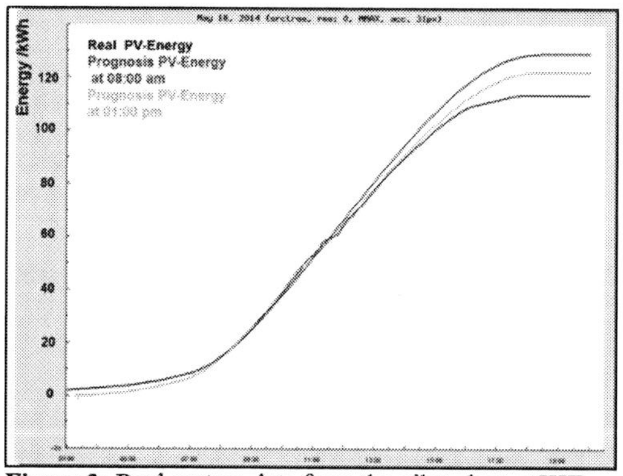

Figure 3: Real system data from the pilot plant at KIT on 18 May 2014, with both measured and predicted PV energy

In additional to the PV prediction shown above, a storage system should possess some sort of load prediction to be able to determine how the storage system should be controlled throughout the day. A description of the load prediction as well as its integration into the overall system control is beyond the scope of this work; several examples of autonomous control programming are described in [5] and [6].

3.2 Results

Three out of the eight PV-battery systems tested showed signs of the existence of intelligent charging algorithms. In Figure 5 an example of this is shown, whereas Figure 4 shows an example of a system without such intelligence. Note that the time of year, day of the week and weather conditions are denoted as "transition", "workday" and "fine" respectively, which are part of the VDI load profile convention.

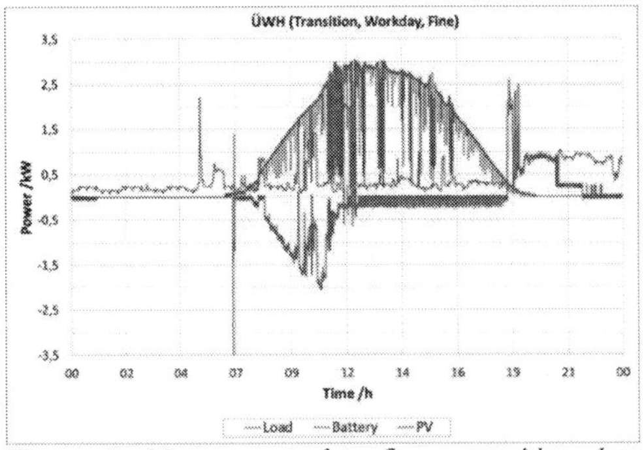

Figure 4: Measurement data for a transition day: example of a system that does not seem to have an intelligent charging algorithm

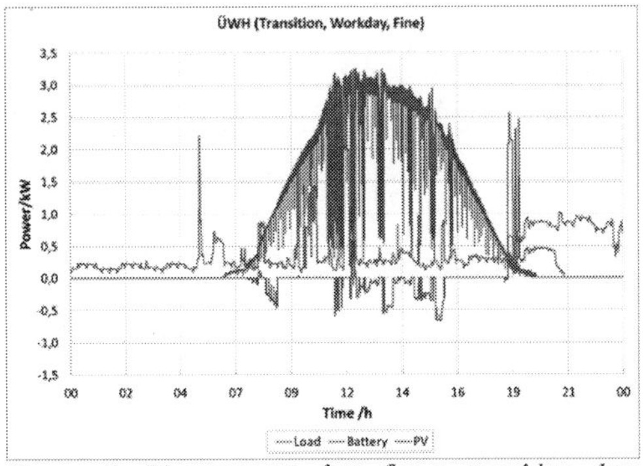

Figure 5: Measurement data for a transition day: example of a system that displays intelligent behavior regarding its battery charging algorithm

So far only 8 of 18 systems have been evaluated regarding the intelligence of their control system. By comparing the data sheets and additional available information of all systems only three more seem to possess any kind of intelligent system control algorithm (see for example Figure 5), which controls the system in such a way that the battery is charged in order to maximize lifetime (see the discussion in the subsection below). Thus only a third of systems tested and under evaluation possess this capability.

Further functions such as the ability to incorporate additional components like heat pumps within the system or a winter mode to limit to depth of discharge during winter can be found in four different systems. These functions require additional intelligence and could result in a higher system performance as well as contributing to grid stability.

3.3 Discussion

Figures 6 and 7 show different control strategies for a stationary energy storage system; in both cases the battery is not big enough to store all of the surplus solar energy. In the case shown in Figure 6 the batteries are charged much faster than in the one shown in Figure 7. As a consequence the batteries are fully charged for a longer period of time without being used. If this is done regularly their cycle lifetime will decline tremendously, resulting in a loss of profitability of the whole storage system.

This means that excess energy should not be stored in the battery right at the start of the day, but rather only after midday, since lithium-ion batteries age faster when their state of charge (SOC) is high over a long period of time [7]. All of these factors require not only extensive expertise but also serious brainpower in the control system, since one has to decide in advance when and with what power the battery should be charged or discharged in order to simultaneously maximize self-supply and ensure a long battery life.

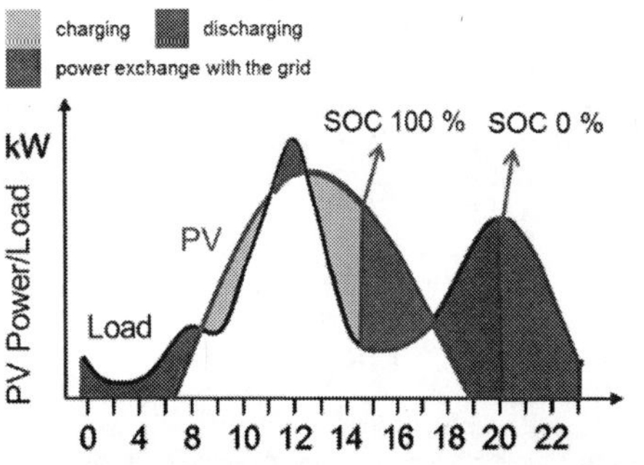

Figure 6: PV production and load curve during the course of the day – system without an intelligent charging algorithm

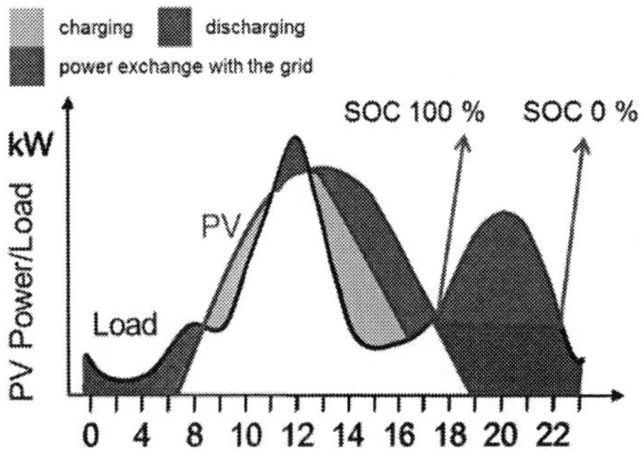

Figure 7: PV production and load curve during the course of the day – system with an intelligent charging algorithm

The tests revealed that there are only a few systems on the market that possess a fully developed control strategy similar to that shown in Figure 7. The strategy of many systems is just to increase the amount of self-consumption, which results in the strategy represented by Figure 6 and the measurements shown in Figure 5. The losses in storage capacity caused by repeatedly keeping the battery at high SOC levels have not yet been quantified, but will be evaluated during the course of the project.

An intelligent charge/discharge regime also benefits the grid operator: due to delayed charging the battery is not completely full around midday, enabling the system to perform peak-shaving of excess PV production that usually occurs around this time. A reduction in such peaks of up to 40 % of the installed PV power has been shown to be possible, but in each case this obviously depends on the system's specific dimensions [8]. Of the storage systems tested, only a small proportion were shown to possess such intelligent control algorithms, and the expected future increase in the amount of installations in the field makes the integration of such software all the more important.

Peak-shaving can also increase the overall system

profitability in cases where the maximum power that can be fed into the grid is limited by law: during times of high PV production the predictive control system ensures that the excess energy is stored without having to throttle the PV generators [9].

4 CONTROL SPEED AND THE ABILITY TO EQUALISE LOAD AND PV-FLUCTUATIONS

This section deals with the control response time, or in other words the ability to react fast enough to fluctuations. After a brief introduction, the results of measurements of commercially available storage systems are presented.

4.1 Introduction to the control speed requirements and load and PV power fluctuations

The chosen power electronics and sensors as well as the controllers used in such a system have an influence on control speed and the resulting performance of the overall system. Real systems always have a finite response time due to delays in measurement and/or communication.

4.2 Results

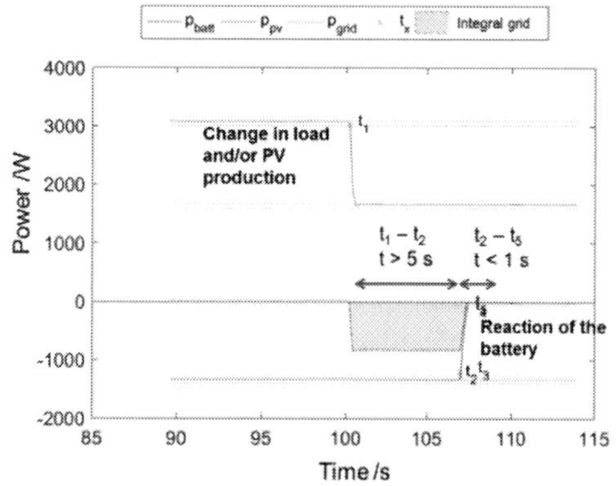

Figure 8: Measurement of the response time of storage system A.

Figure 8 shows a measurement of the response time of one of the storage systems tested. At time t_1 the PV production drops down from 3200 W to 1700 W and the system responds only 6,8 seconds later. In contrast, the rise time between t_2, where the battery leaves the error band, and t_5, where the battery power finally reaches the new target error band is less than 0,4 second. This long dead time could be due to slow communication between the grid sensor and the overall system control or slow measurements of the grid sensor. On the other hand Figure 9 shows an example where the dead time is 0,1 s and the settling time between t_2 and t_5 is 10,0 s. As described in Section 2, the settling time determines the amount of energy unnecessarily drawn from the grid, even though the battery could have provided it, as well as the energy from the battery or PV fed into the grid instead of being used. Nevertheless in the present tests a slow respond time of the system resulted in a larger amount of energy exchanged with the grid than a long time between t_2 and t_5.

By taking into account the frequency of load variations as well as the frequency of PV fluctuations the amount of energy unnecessarily exchanged with the grid during the year can be calculated (see Table 1). As the annual electricity demand of the example household was determined to be 4000 kWh, the frequency of load variations was selected from measured household load data with an annual electricity demand between 3500 kWh and 4500 kWh. The data stem from two weeks of measurements, one in winter and one in summer, from the project „ADRES-CONCEPT“ [10].

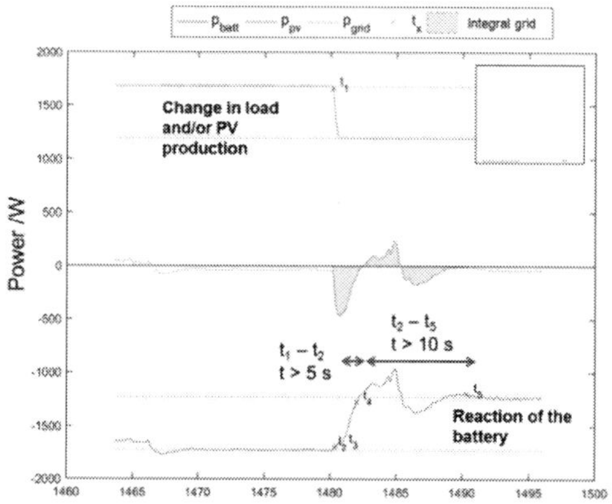

Figure 9: Measurement of the response time of storage system B.

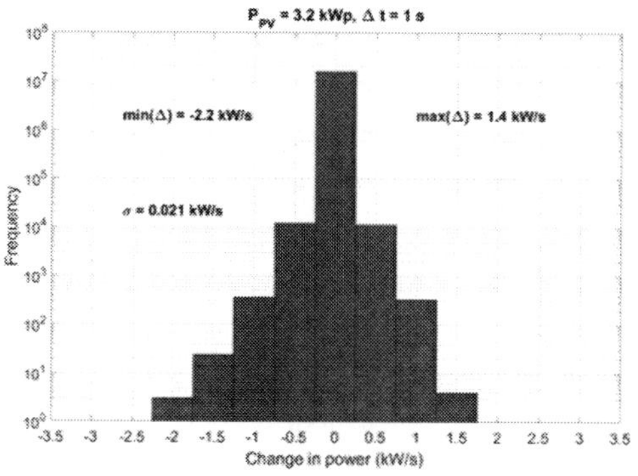

Figure 10: Variations in PV power production for a PV plant with an installed capacity of 3.2 kWp.

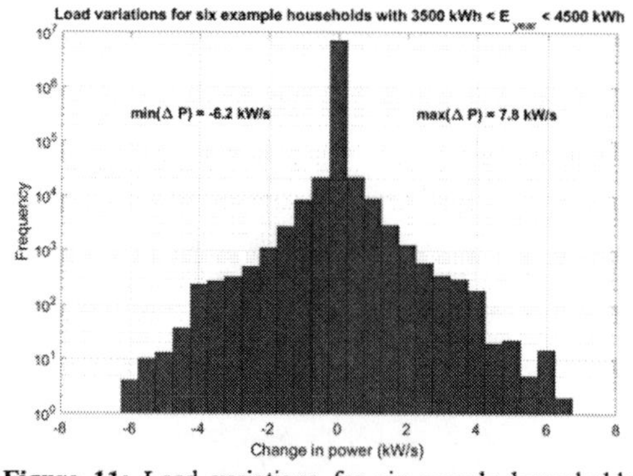

Figure 11: Load variations for six sample households with a yearly electricity demand of between 3500 kWh and 4500 kWh

Table 1: Energy flows in Wh to or from the grid due to a finite response time, for four different systems

	System no.	1	2	3	4
Changes in PV power	Energy unnecessarily drawn from the grid	2,3	1,3	2,3	6,6
	Energy from battery or PV is fed into the grid	0,7	1,5	3,5	4,4
Changes in Load	Energy unnecessarily drawn from the grid	27,8	36,9	47,6	122,0
	Energy from battery or PV is fed into the grid	24,3	37,3	63,8	100,5
Total	Energy unnecessarily drawn from the grid	30,1	38,2	49,9	128,6
	Energy from battery or PV is fed into the grid	25,0	38,8	67,3	104,9

4.3 Discussion

There are systems on the market that react to changes in load and/or PV-production within 100 ms, and others that only react within a few seconds. Indeed, by comparing the performance of the commercial storage systems with each other it can be shown (see Table 1) that the energy exchanged with the grid is four times higher in system 4 than in system 1. The frequency distributions in Figure 10 and 11 show that large fluctuations in load are more frequent than those in PV power, i.e., the standard deviation is higher so that the distribution is broader. This is primarily due to the fact that cloud-induced ramp rates are not as large as those that come from changes in electrical load [11]. On top of that changes in PV only occur during daytime, whereas changes in load happen during the whole day. As a consequence the amount of energy exchanged with the grid caused by changes in load is much higher than that due to PV fluctuations.

If the system as a whole cannot smooth fluctuations fast enough it can happen that energy is unnecessarily exchanged with the grid. As self-supply is often economically more favorable than feeding into or sourcing from the external grid this effect can reduce the overall profitability. By taking into account energy prices of 29 Cent/kWh and a feed-in tariff of 12 Cent/kWh, this makes a difference of 17 Cent/kWh for the energy that is fed into the grid instead of being consumed. For the tested systems this would result in maximum losses of between 10 € and 40 € per year.

It should be taken into account that the measured values are maximum values for the tested load curves. This is due to the fact that as soon as the battery is fully charged or completely empty it cannot compensate changes in load and PV power anymore. As a consequence the actual power exchanged with the grid would be smaller than the values shown in Table 1.

The results shown here agree well with those found via simulation by Weniger et al [12]. In that work the annual amount of energy exchanged with the grid is calculated as a function of the dead time and the time constant. While the energy fed in to the grid for systems with a settling time of 5 s is between 60 kWh and 90 kWh per year the energy unnecessarily drawn from the grid is between 60 kWh and 100 kWh.

5 CONCLUSION

The work described in this paper quantifies the extent to which commercially available home storage systems implement intelligent system control software as well as their ability to equalize PV and load fluctuations.
Of the 8 systems evaluated only three possess an intelligent algorithm control. As a consequence this means that further development of the system control software is needed to achieve better performance and enhance battery lifetime. As soon as more and more systems are installed this will also be needed to contribute to grid stability. Regarding the control speed some systems can still be optimized to increase system performance. However, for the systems under study the financial loss due to unwanted energy flows are not higher than 40 € per year.

6 OUTLOOK

The same tests which were carried out with the first 8 systems regarding intelligence of the system control software and the 4 systems regarding control speed will be repeated with the rest of the 18 systems. In addition to that further performance tests as well as long-term tests will be done to determine the status quo of the storage market and the required future developments in the field. It is expected that the long-term measurements will also reveal how well the software (with and without prognoses) works to enhance battery lifetime and at the same time maximize self-consumption.

ACKNOWLEDGEMENTS

The authors thank the German Federal Ministry for Economic Affairs and Energy (BMWi) and the Projektträger Jülich (PtJ) for support of the project "SaftyFirst" (grant no. 03ET6055A). Furthermore, the authors thank the Vienna University of Technology for the provision of measurements. The authors are solely responsible for the content of this publication. The load data was generated based on measurements from the research project "ADRES-CONCEPT" (EZ-IF: Development of concepts for ADRES –Autonomous Decentralized Regenerative Energy Systems, project no.815674). This project was funded by the Austrian Climate and Energy Fund and performed under the program "ENERGIE DER ZUKUNFT".

REFERENCES

[1] European Commission, "Database - Eurostat," *eurostat*, 12-Jan-2016. [Online]. Available: http://ec.europa.eu/eurostat/data/database. [Accessed: 14-Jan-2016].

[2] O. Wollersheim and A. Gutsch, "Transparenz ist gefragt," *Pv-Mag.*, Jun. 2015.

[3] VDI, "VDI-Richtlinie: VDI 4655 Referenzlastprofile von Ein- und Mehrfamilienhäusern für den Einsatz von KWK-Anlagen." Beuth Verlag, 2008.

[4] C. Messner, J. Kathan, and J. Mayr, "Effizienz und Effektivitaet von netzgekoppelten PV-Heimspeicher Systemen," presented at the 30 . Symposium Photovoltaische Solarenergie, Kloster Banz, Bad Staffelstein, 2016.

[5] J. Bergner, J. Weniger, T. Tjaden, and V. Quaschning, "Verbesserte Netzintegration von PV - Speichersystemen durch Einbindung lokal erstellter PV - und Lastprognosen," presented at the 30 . Symposium Photovoltaische Solarenergie, Kloster Banz, Bad Staffelste, 2015.

[6] J. Bergner, J. Weniger, T. Tjaden, and V. Quaschning, "Feed-in Power Limitation of Grid-Connected PV Battery Systems with Autonomous Forecast-Based Operation Strategies," presented at the 29th European Photovoltaic Solar Energy Conference and Exhibition, Amsterdam, 2014.

[7] M. R. Palacín and A. de Guibert, "Why do batteries fail?," *Science*, vol. 351, no. 6273, p. 1253292, Feb. 2016.

[8] R. Hollinger, B. Wille-Haussmann, T. Erge, J. Sönnichsen, T. Stillahn, and N. Kreifels, "SPEICHERSTUDIE 2013 Kurzgutachten zur Abschätzung und Einordnung energiewirtschaftlicher, ökonomischer und anderer Effekte bei Förderung von objektgebunden elektrochemischen Speichern." Jan-2013.

[9] J. Weniger, J. Bergner, T. Tjaden, and V. Quaschning, *Solarspeicherstudie: Dezentrale Solarstromspeicher für die Energiewende*. Berlin: BWV Berliner Wissenschafts-Verlag, 2015.

[10] Vienna University of Technology, "EA : ADRES-Concept," 17-Sep-2015. [Online]. Available: http://www.ea.tuwien.ac.at/projects/adres_concept/EN/. [Accessed: 24-Jun-2016].

[11] J. Barry, N. Munzke, and J. Thomas, "Short-term power fluctuations in densely clustered PV-Battery Systems," presented at the 32nd European Photovoltaic Solar Energy Conference and Exhibition, München, 2016.

[12] Weniger, J., Tjaden, T., Bergner, J., and Quaschning, V., "Auswirkungen von Regelträgheiten auf die Energieflüsse in Wohngebäuden mit netzgekoppelten PV-Batteriesystemen," presented at the 30 . Symposium Photovoltaische Solarenergie, Kloster Banz, Bad Staffelste, 2016.

Photovoltaic Emulator for High-performance Multi-Subtring Simulations

Tuan Dat Mai, Sven De Breucker, Kris Baert, Paul Van Tichelen, and Johan Driesen
EnergyVille (KU Leuven - VITO - Imec)
Thor Park 8300, 3600 Genk, Belgium

***Abstract*—This paper presents the design and implementation of a photovoltaic (PV) emulator which can act as a multiple sub-string PV module under various irradiation levels, shading conditions and cell temperatures. Based on one-cell diode model approach, the modeling of a PV module takes into account the avalanche breakdown characteristics at the level of shaded cells as well as at substring level. Operation tests of the developed emulator are performed with electronic loads and a DC-DC converter.**

***Index Terms*—PV module, emulator, partial shading, real-time systems.**

I. Introduction

The operation of PV systems under partial shading conditions which reduces the total energy yields becomes one of highlighted-study topics in recent years. With the potential growing trend of urban-area Building Integrated PV systems (BIPV) as well as the decentralized modular PV converters, laboratory facilities for testing PV DC-DC power converters are highly required to study the operation of PV systems under non-uniform irradiation conditions [1]. Testing a PV module with different bypass diode configurations in a solar simulator chamber is not a productive works and spend a lot of time on setting up. Nevertheless, the effectiveness of new designed Maximum Power Point Tracking (MPPT) algorithm and various PV power control strategies under different irradiation levels over a year should validated for PV systems installed at specific geographical locations [2], [3]. Therefore, simulating the PV characteristics is necessary to study PV systems in laboratory.

PV emulator is one of devices in the test facilities which reproduces electrical output characteristics of either a PV panel or PV arrays. Conventional emulators have been introduced in [4], [5], [6], [7], [8]. It has been found that such emulators contain some limitations such as low-response, no shaded PV cell behaviors and multiple series-substring connection. Those limitations can violate the accurate evaluation of PV interfacing converters when being tested under different partial shading conditions, such as soft-shading and hard shading conditions which correspond respectively with low-irradiation and fully illuminated dark cell modeling[9], [10].

The developed work in this paper aims to emulate PV characteristics under multiple shading conditions where few PV cells are accidentally shaded by buildings or plants. The modeling IV characteristics of an identical PV module is developed from PV cell-level approach and taken into account the avalanche breakdown. The performance assessment of DC/DC converters can be done by using this developed emulator instead of installing a PV module in solar simulator's chamber. Not only the non-uniform irradiation simulation substring of PV module but also the fast-response, low-error at the emulator's output are accomplished in this paper's work.

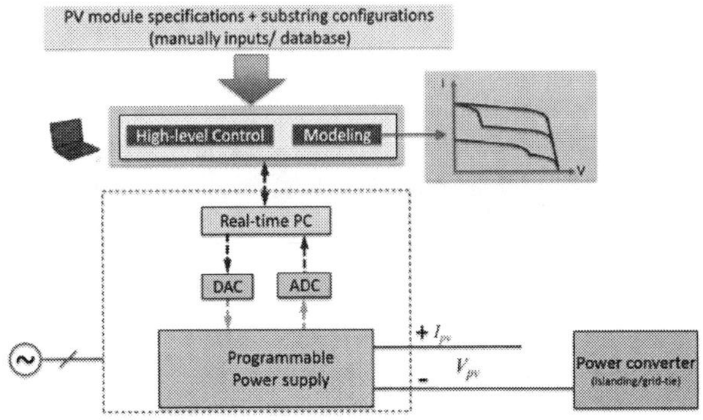

Fig. 1. Hardware implementation of PV emulator

Extensively, it is also able to emulates output of a PV module under irradiation variation according EN50530 standard with substring level[11].

II. Hardware structure, Modeling and Design Approach

The developed emulator is a modular device which consists of hardware components as introduced in Fig. 1. The power stage which is actually a commercial power supply extracts controlled DC power at the output. A signal processing module conditions measurements and controlled reference setpoints between the power stage and the real-time target computer (RTT). The RTT is a main element which controls and monitors the power stage's output during PV emulations. At the supervisory level, a computer running under Matlab/Simulink environment performs the modeling of IV characteristics in accordance with input parameters, and then compiles the executing tasks operated by the RTT [12]. The interface of firmware and hardware is designed at $16kHz$ sampling, which is much faster than power controllers of commercialized power converter interfaces. Real-time measurements and active-filter are applied to eliminate the measurement latency and errors respectively.

In this paper 's work, modeling a PV module is based on one-diode model at cell level [13], [14]. Parameters of a solar cell are extrapolated according to the PV module's datasheet which characterize the physical behavior within the PV module under Standard Test Conditions (STC). The IV characteristics of each cell then adjusted by photovoltaic current depending on irradiation conditions, and the modified cell's parameters influenced by cell temperature [15]. Avalanche breakdown causes negative voltage breakdown at shaded cells is also included in this modeling approach [16].

PV cells are connected in series within a PV module. However,an equal number of PV cells are grouped which are called sub-strings. A bypass diode is placed in parallel with each substring to eliminate power mismatch when irradiated power of substring is different. Not only that, non-uniform irradiation also can happen within a substring. Therefore modeling PV substring characteristics with and without bypass diode also developed in this paper work.

Since all cells in a substring carry the same current at shading conditions, a shade cell produces less power and may cause hot spot depending on the mismatch between the highest short-circuit current I_{sc} in this substring and the one of whole module passed by the diode. Shading conditions in a substring are classified into soft and hard shading [17]. In the soft shading condition, it is considered that PV cells in a substring receive the same irradiation level as an example of a 60-cell, 3-substring PV module presented in Fig. 2 (a). The three sub-strings receive the irradiation of $200, 500, 1000W/m^2$ respectively. As a result, the IV characteristics of each 20-cell substring without and with bypass diode are respectively illustrated in Fig. 2 (b, c).

Contradictory, the hard shading condition is the result of different irradiation levels to PV cells within one substring [17] in which several PV cells can operate as fully shaded cells when operating in avalanche breakdown region where negative voltage appears. It brings about IV characteristics which are completely different from those of a uniform irradiation PV substring. An identical PV module might have different bypass diode configurations. An example of hard-shading conditions which will be presented in later chapters in which the modeling of IV characteristics demonstrates the bypass diode behaviors under partial and fully-shaded PV cells.

The operation of the emulator is extensively developed in two emulation modes. For the static PV emulations, the IV curve of a simulated PV module doesn't change as solar levels irradiated to each cell remain constant values over time. The number of sub-strings in an emulated PV module is configurable as many as cell number. For the dynamic emulation mode, the solar irradiation have to be specified in separate solar irradiation profiles for each substring.

III. Results and Discussion

The validation is conducted by wiring the developed emulator to an electronic load and a PV DC-DC converter, respectively.

A. *PV emulation under constant irradiation conditions*

The interconnection between the emulator and the electronic load is conducted to benchmark the transient responses and errors. Datasheet of the 60-cell PV module Kyocera KC200GT is modeled in Matlab, and then extracted into the RTT [18]. The IV characteristics corresponding to single panel, two sub-strings, and three sub-strings are modeled as in Fig. 3 where soft shading condition is assumed under different uniform-irradiation levels. The output error is calculated as the following

Fig. 6. Testbed of the developed PV emulator interfacing to a STEVAL board

$$\varepsilon = \frac{I_{mea} \cdot V_{mea}}{I_{mod} \cdot V_{mod}} [\%] \quad (1)$$

where the subscript *'mea'* and *'mod'* denote measured and modeled quantities, respectively.

The comparison of IV points obtained by the developed emulator and the modeling of IV characteristics are reported in Fig. 4 in order to evaluate the developed emulator's accuracy which is less than 4% for all simulation cases.

Five hard shading scenarios (SSN) are applied for the same 60-cell, 3 substring Kyocera GC200GT PV module which is introduced in the upper side of Fig. 5. As a result, electric current in the SSN1 significantly increases near by $0V$ area due to the partially shaded cells at three sub-strings. The MPPs in SSN 2, 4 and 5 are almost equal. In SSN 2 and 4, the same number of shades cells at all sub-strings causes two very similar IV curves. The SSN3 is actually contains one fully-shaded substring and two fully-irradiated ones, that results two local peak power as the case of two-substring under two uniform irradiation levels.

An extent modeling case with the same shading pattern scenarios is also carried out the 10 substring case which is less in favor in the current PV market. The experimental validation tests with a DC/DC converter connected to the emulator's terminals are illustrated in Fig. 6. The configuration of sub-strings, bypass diodes causing IV characteristics are presented in Fig. 7. The solid lines represent the IV characteristics while cross-points show working points of the DC converter from starting-up to steady-MPP operation.

Each cell receives sun level conreesponding to either $1000W/m^2$ (high irradiation) or $500W/m^2$ (low irradiation). The non-uniform irradiation to PV cells of identical substrings is reflected by smooth $I-V$ curves in the first and fifth

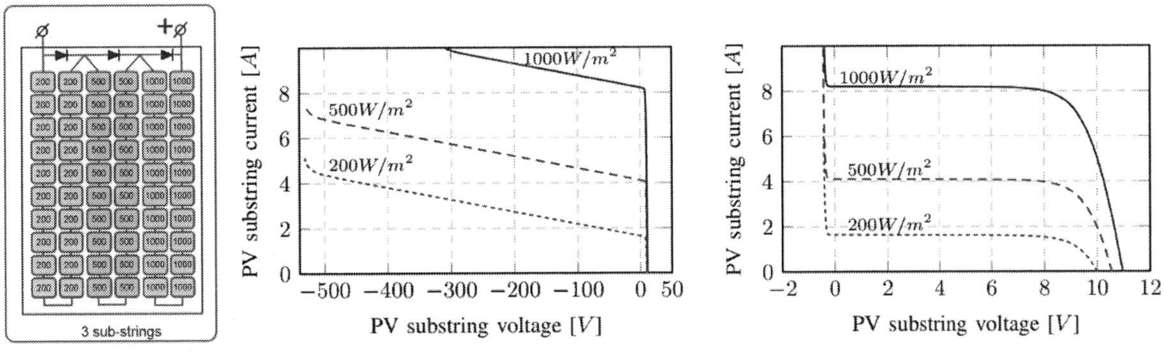

(a) Diagram of a 60-cell PV module with 3 substrings

(b) w/o bypass diode

(c) with substring bypass diode bridge.

Fig. 2. Modeling IV characteristics of a 20-cell substring at 200, 500, and $1000W/m^2$ without (b) and with (c) bypass diode for PV module KC200GT

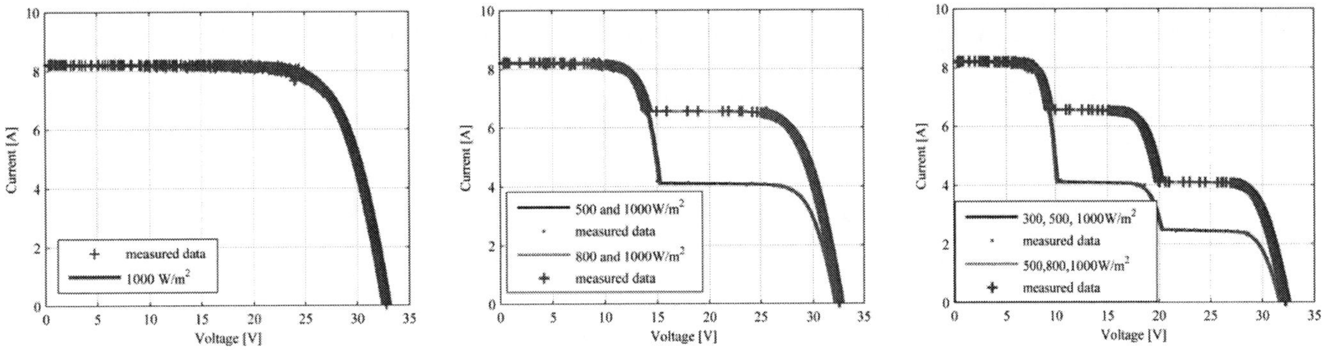

(a) single PV simulation illuminated at $1000W/m^2$.

(b) containing two substrings with 2 cases of partial shading.

(c) containing three substrings with 2 cases of partial shading.

Fig. 3. Experimental IV characteristics of 60 cell PV module KC200GT under partial shading conditions (25^oC) and done by the developed PV emulator

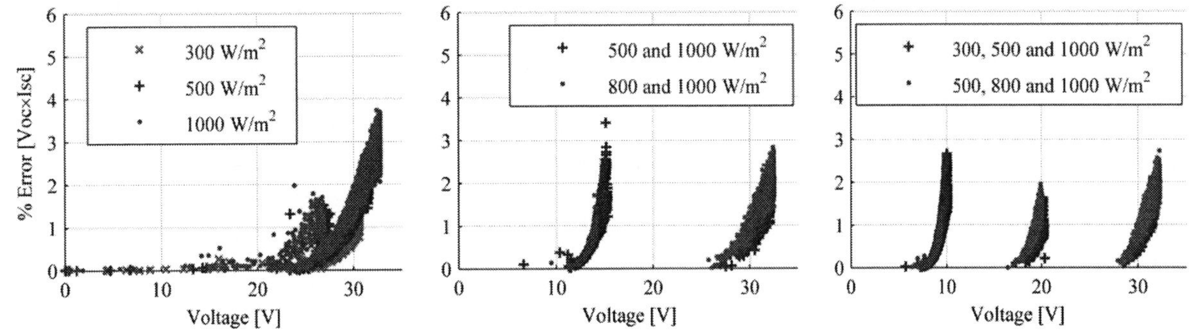

(a) under different uniform irradiation in Fig. 3(a)

(b) two substrings under different irradiations in Fig. 3(b)

(c) three substrings under different irradiations in Fig. 3(c)

Fig. 4. Errors of experimental IV points compared to the modeled characteristics in (a) uniform irradiation , (b) two-substring, and (c) three-substring PV simulation, respectively.

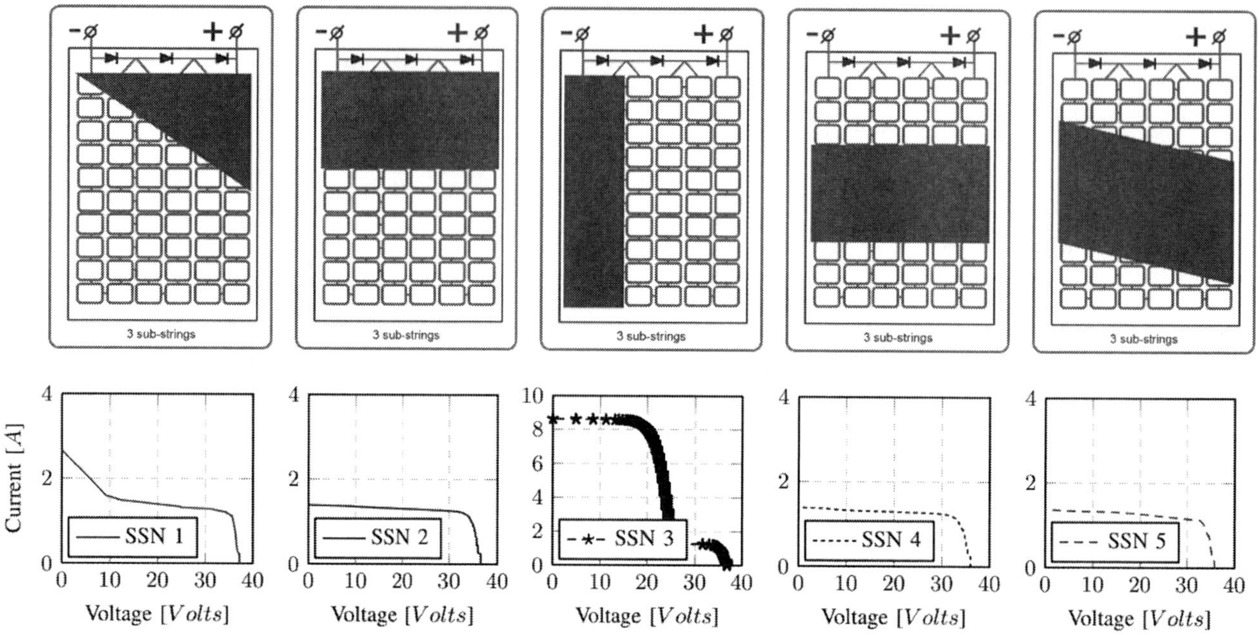

Fig. 5. Typical characteristics of a 3-substrings,60-cell PV panel tested in five shading scenarios (SSN)

scenarios. The obtained $I - V$ curves are also shown to verify the effectiveness of the proposed modeling method. The obtained global peak power of these five shading scenarios are approximately $133.53W$, $145.04W$, $133.93W$, $134.32W$, and $127.80W$ respectively. The highest global peak power corresponds to the scenarios 2. Nevertheless, P&O algorithm controlled by the STEVAL board are trapped at local power peak points closely to the V_{oc} of the PV module which is significantly less than the global ones. It can be explained by the Maximum Power Point Tracking algorithm Perturb and Observe (P&O) implemented in the ST DC-DC converter demonstration board (STEVAL) [19]. Measurements converge around local peak power points which are close to the open-circuit voltages of different PV simulations.

B. PV emulation under dynamic irradiation profiles

The PV emulator is able to emulate the dynamic characteristic of a PV module in case solar irradiation level varies over time. An assessment to evaluate the PV emulator's performance is illustrated in Fig. 8 according the same testbed shown in Fig. 6. When the irradiation level continuously increases from $0s$ to $132s$, the STEVAL starts-up its operation at $36s$ and executes MPPT from $48s$. However, the MPPT convergence fails so that STEVAL only restarts as soon as the solar irradiation remains constant at $1000W/m^2$. The operating points fluctuates around the maximum power points which can be seen by the voltage variation between the instants $160s$ and $340s$.

The P&O converges after $180s$ and continues to track perfectly the peak power of the PV emulator until the instant reaches $580s$. It can be explained that the irradiation value at this instant is too low to be applied to a suitable power production, therefore the STEVAL shuts down since PV production supplies too low power for its operation.

IV. Conclusion

The paper presents the development of a PV emulator which reproduces IV characteristics of an identical PV module under partial-shading conditions. The implementation of a suitable control method is also vital to tackle the IV curves with discontinuous breakpoints in sub-string PV modeling, and the output's error is 4% less than the modeling.

Experimental results illustrate that the $I - V$ measured points when the emulator interconnects to the STEVAL apparently lay on the performance of the PV emulator in this work. In addition, the PV emulator precisely simulates $I - V$ characteristics as a real PV module in which power output is determined by the power converter interface such as STEVAL.

The emulator is not only used to test the interconnected operations of PVs with converters under static irradiation conditions, but also applied to energy yield assessment according solar irradiation variations. The configurations of sub-strings, bypass diodes, cell temperatures and cell irradiation levels can be rapidly modified to execute a test, so that provides the versatility and cost-effectiveness in studies involving PVs and PV converter interfaces.

V. Acknowledgments

We would like to thank all co-workers at EnergyVille and especially G. van den Broeck for support in solar cell measurements. This work has been supported by an internal

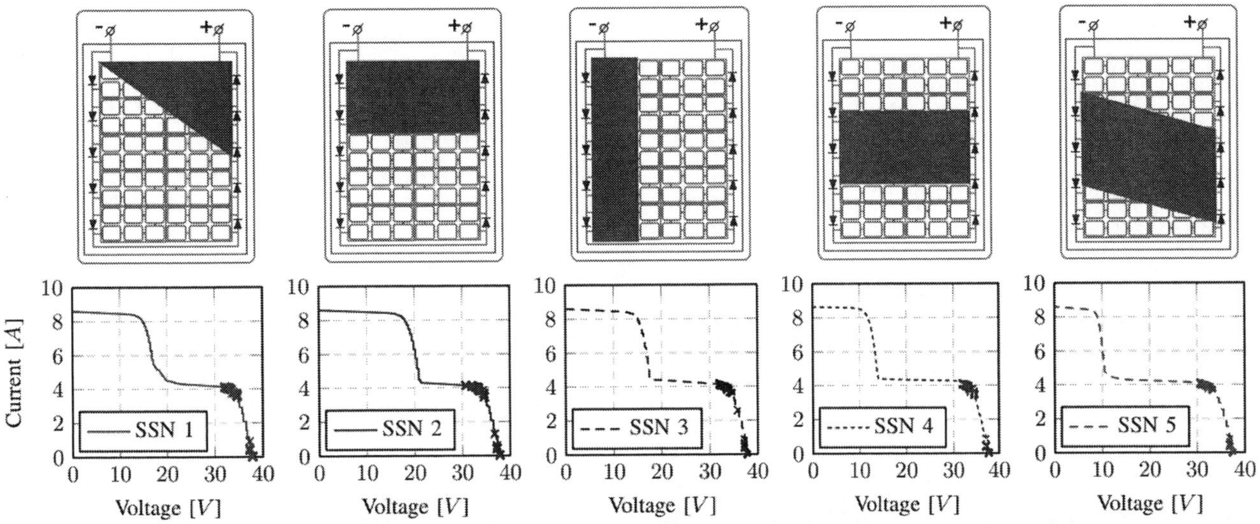

Fig. 7. Typical characteristics of a 10-substrings, 60-cell PV panel tested in five shading scenarios (SSN) and operating measurements when the emulator interconnects to the STEVAL shown in Fig. 6.

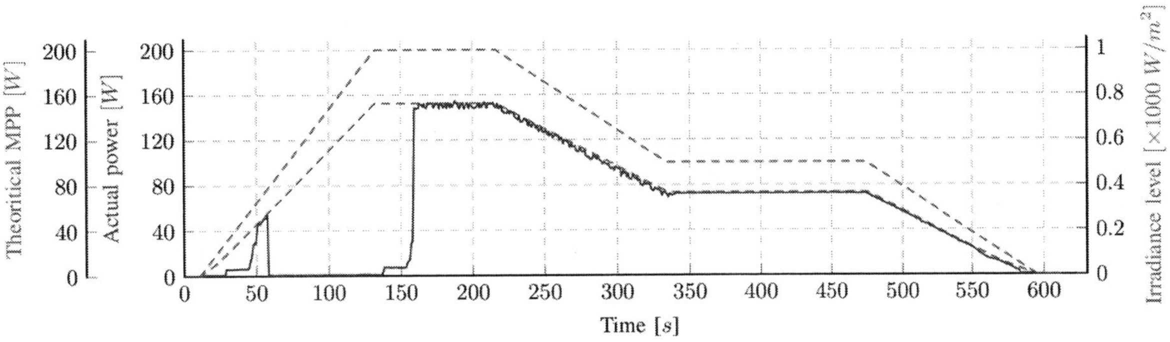

(a)

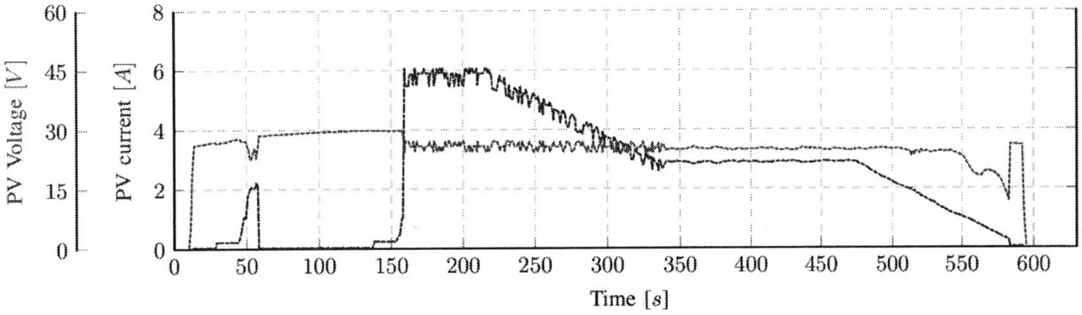

(b)

Fig. 8. PV emulator interconnected to a STEVAL DC/DC converter under the simulation of single PV module's characteristics according variable solar irradiation levels.

part of the SBO project "SmartPV" funded by the Flemish Agency for Innovation by Science and Technology (IWT).

REFERENCES

[1] J. Parnell, "EU PVSEC: Building sector buy-in required to mainstream BIPV," 2015. [Online]. Available: http://www.pv-tech.org

[2] E. Koutroulis and F. Blaabjerg, "A New Technique for Tracking the Global Maximum Power Point of PV Arrays Operating Under Partial-Shading Conditions," *IEEE Journal of Photovoltaics*, vol. 2, no. 2, pp. 184–190, 2012.

[3] M. Uno and A. Kukita, "Single-switch Voltage Equalizer Using Multi-stacked Buck-boost Converters for Partially Shaded Photovoltaic Modules," *IEEE Transactions on Power Electronics*, vol. 30, no. 6, pp. 3091–3105, 2015.

[4] R. Kadri, J.-P. Gaubert, G. Champenois, and M. Mostefaï, "Real-time Emulator of Photovoltaic Array in Partial Shadow Conditions Based on Closed-Loop Reference Model," in *2010 World Energy System Conference (WESC)*, vol. 7004, Targoviste, Romania, 2010, pp. 71–77.

[5] M. C. Di Piazza and G. Vitale, "Photovoltaic Field Emulation Including Dynamic and Partial Shadow Conditions," *Applied Energy*, vol. 87, no. 3, pp. 814–823, 2010.

[6] J. Agrawal and M. Aware, "Photovoltaic System Emulator," in *2012 IEEE International Conference on Power Electronics, Drives and Energy Systems (PEDES)*, 2012, pp. 1–6.

[7] Chroma, "Programmable DC Power Supply Solar Array Simulation, model 62000H-S series," p. 8, 2012.

[8] D. S. L. Dolan, J. Durago, and Taufik, "Development of a Photovoltaic Panel Emulator using Labview," in *Conference Record of the 37th IEEE Photovoltaic Specialists Conference*, Seattle, WA , US, 2011, pp. 001 795–001 800.

[9] S. Guo, T. M. Walsh, A. G. Aberle, and M. Peters, "Analysing Partial Shading of PV Modules by Circuit Modelling," in *Conference Record of the IEEE Photovoltaic Specialists Conference*, 2012, pp. 2957–2960.

[10] K. Ishaque, Z. Salam, and Syafaruddin, "A Comprehensive Matlab Simulink PV System Simulator with Partial Shading Capability based on Two-Diode Model," *Solar Energy*, vol. 85, no. 9, pp. 2217–2227, 2011.

[11] CEN/CENELEC, "EN50530-Overall efficiency of grid connected PV inverters," 2010.

[12] N. Femia, G. Petrone, G. Spagnuolo, and M. Vitelli, "Optimization of Perturb and Observe Maximum Power Point Tracking Method," *IEEE Transactions on Power Electronics*, vol. 20, no. 4, pp. 963–973, jul 2005.

[13] C. Carrero, D. Ramírez, J. Rodríguez, and C. a. Platero, "Accurate and Fast Convergence Method for Parameter Estimation of PV Generators Based on Three Main Points of the I-V Curve," *Renewable Energy*, vol. 36, no. 11, pp. 2972–2977, 2011.

[14] A. Jain and A. Kapoor, "Exact analytical solutions of the parameters of real solar cells using Lambert W-function," *Solar Energy Materials and Solar Cells*, vol. 81, no. 2, pp. 269–277, 2004.

[15] M. G. Villalva, J. R. Gazoli, and E. R. Filho, "Comprehensive Approach to Modeling and Simulation of Photovoltaic Arrays," *Power Electronics, IEEE Transactions on*, vol. 24, no. 5, pp. 1198–1208, 2009.

[16] H. Kawamura, K. Naka, N. Yonekura, S. Yamanaka, H. Kawamura, H. Ohno, and K. Naito, "Simulation of I-V Characteristics of a PV Module with Shaded PV Cells," *Solar Energy Materials and Solar Cells*, vol. 75, no. 3-4, pp. 613–621, 2003.

[17] L. A. Lamont and L. E. Chaar, "Enhancement of a stand-alone photovoltaic systems performance: Reduction of soft and hard shading," *Renewable Energy*, vol. 36, no. 4, pp. 1306–1310, 2011.

[18] Kyocera, "High-efficiency multi-crystal photovoltaic module KC200GT," 2009.

[19] STMicroelectronics, "Appl. Note STEVAL-ISV013V1 250 W DC-DC solar module demonstration board for distributed photovoltaic architecture," p. 5.

YELLOWING OF PV BACKSHEETS IN ACCELERATED TESTS CAN BE USED AS A REALISTIC INDICATION OF POSSIBLE FIELD FAILURES – FACT OR FICTION? EU PVSEC 2016

Emily Parnham, Amy Seaman, Anthony Whitehead, William Brennan, Emma Ashford
DuPont Teijin Films UK
The Wilton Centre, Redcar, TS10 4RF, UK

ABSTRACT: Photovoltaic (PV) modules currently have lifetime guarantees of 25 years or more. Ideally all components of the modules would be tested in field conditions for at least 25 years, however this is unrealistic, so accelerated, laboratory based testing is required. When using accelerated testing it is important to ensure the test does not result in false negatives, leading to the ruling out of good materials, and more importantly, does not give rise to false positives, which could lead to material failure in the field. A key question to answer is: "Can yellowing be used as a reliable method to rank materials after accelerated testing and to predict failures in the field?" This paper compares eight accelerated test conditions combining different temperatures, irradiance and condensation and their impact on a range of PV backsheet materials with regards to yellowing and loss of mechanical properties. The final part of this paper will compare unstabilised white polyethylene terephthalate (PET), UV stabilised PET and polyvinyl fluoride (PVF) films in Xenon weatherometer aging with Florida weathering and backsheet failures reported in the field. Using these data, looking at colour change and mechanical properties, we are able to increase the confidence in the predictions being made from correctly conceived accelerated tests.
Keywords: Backsheet, Degradation, Accelerated Weathering

1 INTRODUCTION

Photovoltaic (PV) modules are exposed to a wide range of environmental processes over the course of their lifetime. These processes degrade the various components of the module and may eventually lead to failure of the PV cell. The module backsheet is one component upon which the overall failure of the module is dependent as it protects the inner electrical components from being damaged by environmental factors (humidity, dust etc.) and provides safety-critical electrical insulation. There are numerous ways in which these backsheets can be degraded; hydrolytically, thermally, mechanically and via UV exposure. It is the last of these which is the main focus of this paper.

Throughout the literature there has been extensive work investigating the effect of weathering conditions on the materials typically used as PV backsheets, in particular under UV light exposure [1-6]. Today PV modules have guarantees of 25 years or more, so it is important that the materials used in the backsheet maintain their functionality for at least 25 years. In a research setting, accelerated testing is used to simulate real-life conditions. Tests use cycles of high temperatures and irradiances as well as moisture and following exposure the backsheet materials are then tested to assess any degradation using a variety of methods.

The purpose of this work is to show the initial results where different conditions varying intensity of irradiance (UVA), temperature and condensation have been investigated, and show the importance of choice of both test method and pass/failure criteria when assessing PV backsheet materials. These results will mainly focus on PET, but samples of polyvinylidene fluoride (PVDF) and polyvinyl fluoride (PVF) have been included for comparison.

PET absorbs UV light in the region of 300-350 nm leading to two main reaction pathways (Figure 1). One results in strongly coloured conjugated molecules [7] and the other leads to chain scissioning [8] which ultimately results in mechanical failure via embrittlement and cracking.

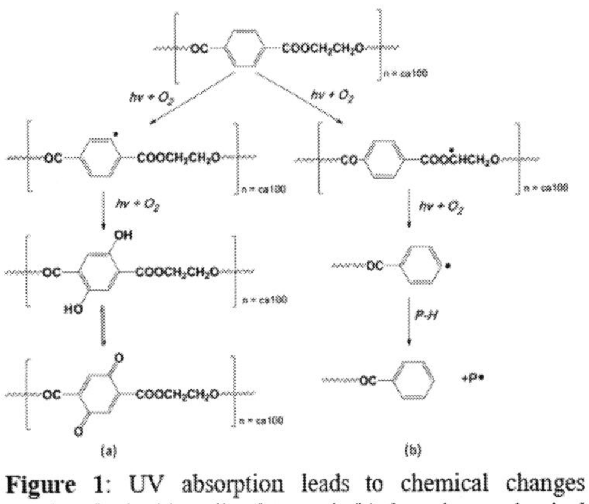

Figure 1: UV absorption leads to chemical changes causing both (a) yellowing and (b) loss in mechanical performance

Numerous data have been published [3-6] recording the yellowing of PET backsheets in the field and laboratory. This yellowing has been postulated to be indicative of simultaneous mechanical degradation of the PET studied. However, these previous studies appear to be trying to correlate a diverse range of coated, pigmented, UV stabilised and non-UV stabilised PET films. As a result the correlations shown are distinctly multi-modal with no particular relationship between yellowing and mechanical degradation.

There are also numerous examples within the literature concluding that it is not possible to correlate the two processes. Fechine et al. [1] studied samples which contained a variety of additives and UV stabilisers and measured the colour and mechanical properties of the films. It can be seen from their data that although both processes occurred over the 1000 hour period of testing, the majority of yellowing took place in the first 200 hours of exposure, whereas the decline in mechanical integrity occurred after the initial 200 hours. Further to this, a 2014 study by Gong et al. [2] investigated various PET backsheets under damp heat testing and UV exposure. The authors conclude that there was no correlation between colour change and extent of cracking and speculated that certain additives in backsheets can reduce the extent of yellowing even if the materials have been degraded.

This paper illustrates the importance of understanding the mechanisms that are being accelerated when carrying out laboratory weathering such that tests can be selected that are suitable for all material types. We need to ensure there are no false negatives (materials fit for purpose ruled out), and no false positives (bad materials ruled in).

2 EXPERIMENTAL

10 polymer films, specifically engineered for PV backsheet applications were tested for 30 kWhrs using a QUV accelerated weathering tester with fluorescent UVA-340 lamps. 8 different sets of conditions, varying temperature, irradiance and condensation were used as detailed in Table I.

Table I: Weathering program conditions

Test	Temp (°C)	Condensation	Irradiance @ 340nm
1	60	N	1.2
2	80	N	1.2
3	80	Y	1.2
4	60	Y	1.2
5	60	N	0.76
6	80	N	0.76
7	80	Y	0.76
8	60	Y	0.76

Xenon arc light source accelerated weathering testing was carried out in an Atlas Ci-5000 with optical filters conforming to the requirements of ASTM D7869. The irradiance level at 340 nm was 0.80 W/m^2. Black Panel Temperature (BPT) and Chamber Air Temperature (CAT) were set to 95 and 70 °C respectively. This gives a nominal sample temperature of about 75 °C. Relative humidity was set to 20%.

Outdoor weathering was performed in Miami, Florida (26°N) at a tilt angle of 26° from the horizontal facing south. The specimens were mounted on special black-coated aluminum sheets. A thermocouple was fixed to the back of a control sample in order to monitor sample temperature.

Colour was measured using a Konica Minolta Spectrophotometer CM-3600A with a 25.4 mm aperture and the parameters *CIE* (D65/10°) L* a* b* reported using specular components included and 100 % UV.

Mechanical properties of the film were measured on a 10 mm wide strip cut using a JDC precision cutter. Tensile strength and elongation to break were measured on an Instron (models 4464, 4301 and 3365) according to the ASTM D882-83 method, with the jaws set at 50 mm separation, and the sample pulled at a rate of 25 mm/min.

3 RESULTS

Figure 2 shows a plot of yellowing (Δb*) against % loss of mechanical properties in terms of % loss in the material's elongation to break (ETB). It can be seen from this graph that there is no general correlation between yellowing and loss of mechanical properties.

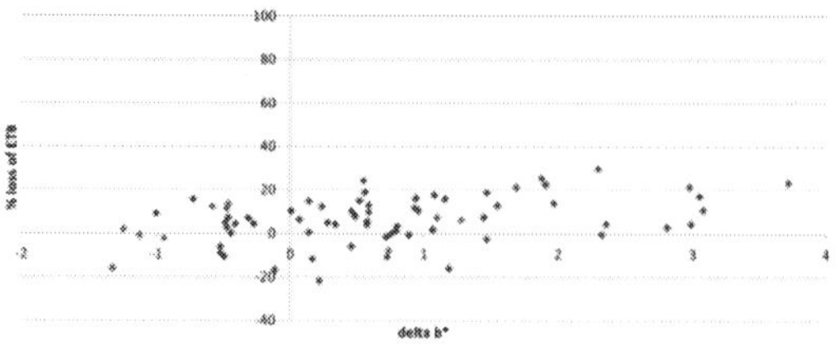

Figure 2: Graph of delta b* (yellowing) against % loss of mechanical properties (ETB)

Looking at the data in greater detail it is possible to identify trends in conditions used and materials. Figure 3 shows a graph of five different UV stabilised PET materials, two PVDF and one PVF, exposed to 30 kWhrs UVA at 60 or 80 °C and an irradiance of 0.76 or 1.2 W/m^2 at 340 nm. Increasing irradiance and temperature had no effect on PVDF and PVF with regards to delta b*. Increasing temperature or irradiance alone did not affect PET. However, it was found that simultaneously increasing irradiance and temperature resulted in increased yellowing across all 5 PETs tested. Mechanical properties were also tested on these films and it was found that there was no significant loss in mechanical properties of the PET samples exposed to high temperature and high irradiance. This indicates when high temperature and high irradiance are used in accelerating weathering; the colour-forming degradation mechanism in PET is over-accelerated when compared to the chain scissioning pathway. It should also be noted that it only takes small quantities of colour forming species in white film to give a significant delta b*.

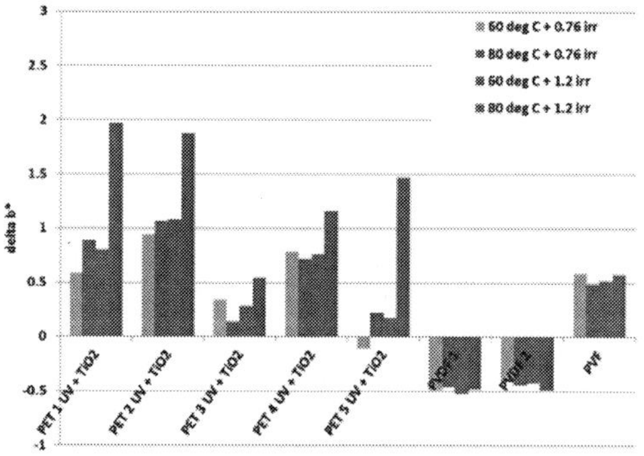

Figure 3: Delta b* of 8 materials used for PV backsheets after accelerated testing under 4 different test conditions

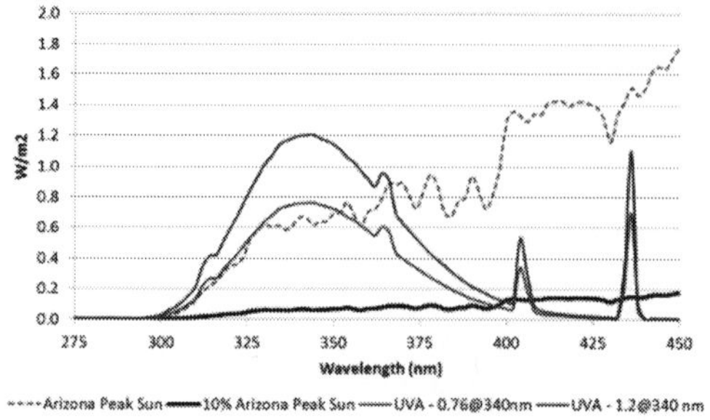

Figure 4: Graph to compare irradiance of Arizona peak sun, 10% (backsheet exposure) and the two irradiance levels used in this paper for accelerated testing.

When designing accelerated testing, consideration needs to be given to whether the correct mechanisms are being accelerated and the appropriate properties measured. In the field PV backsheets are exposed to high temperatures, but they are never directly exposed to high irradiance. Figure 4 clearly shows that the irradiance actually experienced, taking into account a 10 % "albedo" factor, is very low for modules in the field. It is important to ensure that accelerated test methods do not result in false negatives and the data presented in this paper show that if high temperatures and high irradiance are to be used in accelerated testing, colour change should not be used for PET as an indication of failure, but rather mechanical properties should be tested. This can be even more evident from experiments using coated PET film structures, where the delta b* remained less than 2, yet on testing mechanical properties of the backsheet, the coating cracked, flaked off and the adhesive between the backsheet layers failed. If colour change alone was tested in this example, the result could lead to a false positive and a possible failure in the field.

4 VALIDATION OF ACCELERATED TESTING

Ideally we would wait 25 years and test samples recovered from the field following real application exposure. From a product development perspective this is a highly impractical timescale hence accelerated testing is used. It is, however, important to validate any accelerated testing that may be selected. There are several ways to do this, one of which is to test against a known reference material with "good" performance, for example PVF for PV backsheets, as this has been in the field for more than 30 years. [9]

Another way to validate accelerated weathering test chambers and conditions for PV backsheet applications is to compare with samples from outdoor weathering in direct sun which when accounting for the accepted industry-standard "albedo" factor for backsheets, has an acceleration factor of 10 times with regards to UV dosage. It should be noted that the mechanism of polymer degradation by hydrolysis of the outdoor exposed samples does not have an acceleration factor.

Figure 7 shows a comparison of delta b* and relative ETB % for a UV stabilized, hydrolysis stabilized PET in an accelerated Xenon test vs. Florida direct sun weathering. The data show that this Xenon test accelerates initial colour formation at about double the rate of the Florida outdoor test. The data also show that Florida samples have a greater retention of mechanical properties for the same UV dosage. It is important to note that for the Florida samples the temperature fluctuates (0-55 °C), but for the Xenon test it is constant at 75 °C. This higher and constant temperature is most likely the cause of the increased acceleration of both the yellowing and the mechanical degradation mechanisms as it is known that temperature has a significant effect in increasing damage resulting from UV exposure.

Higher temperatures are often used to accelerate weathering. The use of temperature to accelerate UV degradation has been found to reduce the correlation between laboratory and real outdoor field weathering.[10] The temperature that a PV backsheet experiences in the field varies depending on many factors including location, module mounting geometry and so on. A module backsheet will not see a constant temperature, but rather it will fluctuate between day and night and even during the day depending on cloud cover variations.

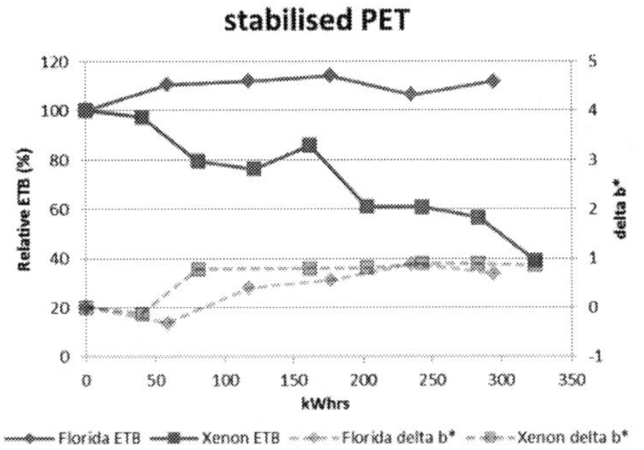

Figure 7: Florida outdoor weathering and Xenon arc accelerated weathering, comparing mechanical properties (ETB %) and colour change (delta b*).

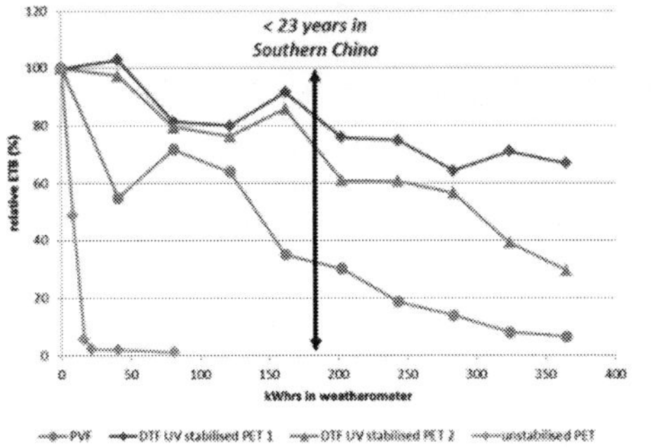

Figure 8: Mechanical properties for non-UV stabilised PET against UV stabilised PET and PVF in the Xenon arc weathering test.

PVF is the only polymeric material that has been widely used for PV backsheets in the field for more than 30 years. This makes it an ideal material for benchmarking other novel or improved materials for backsheet applications. Some environments are particularly harsh regarding UV and humidity levels. One such example reported at the 31st PVSEC 2015 considered 100 µm single layer PVF backsheet modules exposed for just under 23 years in South China (subtropical climate).[11] Whilst the performance of the modules was still good, almost all of the backsheets exhibited cracks, leading to concerns regarding possible safety issues from lack of electrical insulation. We can conclude from this that for the toughest environments, there is a need for innovation and the development of new materials with increased performance.

Figure 8 compares non-UV stabilised PET against PVF and two commercially available UV stabilised PET films used for PV backsheet applications. It can be seen that in this Xenon arc weathering test both of the UV stabilised PET films outperform the PVF material with regards to mechanical properties, which is important to prevent cracking and maintain electrical insulation of a PV module. Taking the example of Sanya (Southern China) [12] having a total UV per year of 288 MJm^2 and an "albedo" of 10 % for backsheet, we can calculate that these PVF modules have been exposed to less than 184 kWhrs of total UV exposure. Figure 8 shows that at 184 kWhrs in the Xenon weathering test, there is a clear difference with regards to mechanical strength after UV exposure between the PVF and UV stabilised PET materials. These test results indicate that for harsh, tough environments, UV stabilised PET film could well outperform materials such as PVF that have been in the field for 30 years.

5 CONCLUSION

This paper has demonstrated that when using accelerated weathering, an observed colour change does not necessarily correlate with loss of mechanical properties. The rate of colour change and loss of mechanical properties for a backsheet depend upon material make up and accelerated test conditions chosen. Whilst measuring colour change after UV aging is often favored due to the non-destructive test method, this could lead to both false negatives and the ruling out of good materials, or false positives and possible early failure in the field. To increase confidence in material reliability after UV aging, it is important to test the mechanical properties of the material directly.

As with any accelerated testing it is also important to correlate this data to field examples. For new materials for PV backsheets, an ideal comparative material is PVF as this is the only polymer that has been used in the field for this application for over 30 years. This paper highlights that even PVF has suffered mechanical degradation when exposed to the harshest environments. Accelerated weathering tests can be used to help develop, engineer and identify materials that may withstand these environments. This paper has shown that when testing mechanical properties of UV-stabilized PET, the UV properties of the material can be tuned to give an optimum performance and by testing against PVF as a control, we are able to increase the confidence in the predictions being made from the accelerated testing.

6 REFERENCES

[1] G. J. M. Fechine, M.S. Rabello and R. M. Souto-Maior, Poly Degrad. Stab., 2002, 75, 153-159

[2] H. Gong, G. Wang, M. Goa and L. Zheng, Proceedings 31st EU PVSEC, Hamburg, Germany, September, 2015, 2066-2072

[3] W. Gambogi, Y. Heta, K. Hashimoto, J. Kopchick, T. Felder, S. MacMaster, A. Bradley, B. Hamzavytehrany, L. Garreau-Iles, T. Aoki, K. Stika, T. Trout and T. Sample, IEEE Journal of Photovoltaics, 2014, 4, 935-941

[4] W. Gambogi, J. Kopchick, T. Felder, S. MacMaster, A. Bradley, B. Hamzavytehrany, B-L. Yu, K. Stika, Y. Heta, L. Garreau-Iles,C-F. Wang, H. Hu and T. Trout, Proceedings 31st EU PVSEC, Hamburg, Germany, September, 2015, 1930-1934

[5] W. Gambogi, J. Kopchick, T. Felder, Y. Heta, S. MacMaster, A. Bradley, B. Hamzavy, B-L. Yu, K. Stika, L. Garreau-Iles, H. Hu and T. Trout, Proceedings 30th EU PVSEC, Amsterdam, The Netherlands, September, 2014, 2471-2476

[6] W. Gambogi, J. Kopchick, T. Felder, S. MacMaster, A. Bradley, B. Hamzavy, B-L. Yu, K. Stika, L. Garreau-Iles, C-F. Wang, H. Hu, Y. Heta and T. Trout, IEEE, 2015 (New Orleans)

[7] M. Edge, N.S. Allen, R. Wiles, W. MacDonald, S.V. Mortlock, Polymer, 1995, 36, 227-234

[8] M. Day, D.M. Wiles, Journal of Applied Polymer Science, 16, 175-215.

[9] http://www.dupont.com/products-and-services/solar-photovoltaic-materials/photovoltaic-backsheet-films.html

[10] IEC 62788-7-2 (draft) Measurement procedures for materials used in photovoltaic modules – Environmental exposures
Part 2: Accelerated weathering tests of polymeric materials

[11] X. Dong, H. Wang, Y. Jin, J. Huang, H. Shen, Proceedings 31st EU PVSEC, Hamburg, Germany, September, 2015, 1842-1846

[12] B. Hu, Y. Wang, G Liu, Atmospheric Environment, 2007, 41, 5707-5718

METHOD TO MEASURE LIGHT RECOVERY PROBABILITY OF PV MODULE BACKSHEETS ENABLING 20.2% MODULE EFFICIENCY WITH PASSIVATED EMITTER AND REAR SOLAR CELLS

Marc Köntges[1], Henning Schulte-Huxel[1], Susanne Blankemeyer[1], Malte R. Vogt[1], Hendrik Holst[1], Rolf Reineke-Koch[1], Robert Witteck[1], Karsten Bothe[1], and Rolf Brendel[1,2]
[1]Institut für Solarenergieforschung Hameln (ISFH), Am Ohrberg 1, D-31860 Emmerthal, Germany
[2]Institute of Solid-State Physics, Leibniz Universität Hannover, Appelstr. 2, D-30167 Hanover, Germany

ABSTRACT: We measure the short circuit current improvement of various commercial backsheets for highly efficient passivated emitter and rear solar cells (PERC) in photovoltaic (PV) modules. In our test modules an electrically contacted centre cell is surrounded by 8 cell parts of the same type to simulate the optical neighbourhood of a middle cell in the centre strings of a standard sized module. In order to reduce the impact of variations in the solar cell performance, we use the same glass/EVA/cell/EVA stack for all tests. The various backsheets are successively attached to the stack by a thin film of liquid paraffin. A measurement without backsheet and a "light trap" behind the module stack allows the calibration of the recovery probability of the cell gaps with nearly zero light recycling. This procedure allows a fast and precise measurement of the light recovery probability for the cell gaps covered by various backsheets. The light recovery probability k quantifies how much of the light reflected by the backsheet around the cell is converted into electrical current by the solar cells in the module. A black backsheet has a recovery probability close to 0%. Typical white scattering backsheets show a light recovery probability between 45% and 55% for 2 mm cell gaps. Higher recovery probabilities of about 75% are achieved by using highly reflective structured materials in the cell gaps. Using such structures in combination with half cells we process a world record standard size (1.5 m x 1 m) PERC PV module with an energy conversion efficiency of 20.2%.

Keywords: Backsheet, light harvesting, high efficiency module, PERC

1 INTRODUCTION

Passivated Emitter und Rear Cells (PERC) from p-type crystalline Si wafers are probably the cell technology with the largest gain in market share in the next years [1]. The performance of industrial-type screen-printed 6" PERC solar cells significantly increased during the past years. Efficiencies of up to 22% have been published [2], [3]. In the field of large area PERC modules the highest reported module efficiency is 19.5% for a module with 294 W [4]. Since this publication new results on higher module power have been reported. The present power record is 335.2 W at an efficiency of 19.1% [5]. An increase in module power often comes along with a larger sized module area in order to benefit from additional light scattering onto the cells by multiple reflections in the module. However, this results in an increase in consumption of materials during module fabrication and in increased installation costs since both scale with the module area [6]. To obtain high module efficiencies the whole PV module area must use the light for energy conversion. The optical recovery of photons being reflected by the backsheet between the cells has an approximate potential of 5.6 W_p to 11.3 W_p for a standard 120 half-cell module with 20% cell efficiency using 2 mm or 4 mm spacing, respectively. There are many methods to measure the light recovery of inactive module areas. Some methods use light beam induced current [7] or local external quantum efficiency [8] measurements to determine the position dependent recovery probability of cell metallization, cell interconnect ribbon or wire as well as various backsheets. Another method uses the reciprocity theorem of the (electro-)luminescence to measure optical recovery of inactive module parts [8]. Last but not least it is possible to measure the recovery probability by simply producing multiple laminates and evaluate the relative current gain difference for various backsheets [9]. All these techniques need to have a finalized PV module for each test. In order to overcome this limitation we present a simple method using a glass/EVA/cell/EVA stack to which various backsheets are attached successively by liquid paraffin. Variations in the module performance due to the different backsheets are determined by measuring the short circuit current. On the bases of the result of this study we manufacture a 20.2% efficient world record PERC module with a power of 303 W.

2 METHOD TO MEASURE THE RECYCLING PROBABILITY OF BACK SHEETS

2.1 Experimental setup

In a standard laminate without backsheet a center 156 mm x 156 mm sized three busbar cell is surrounded by 8 cell parts to simulate the optical neighborhood of a middle cell as in the center strings of a standard sized module. The rear metallization of the cells does not cover the complete cell rear side. There is a distance between rear metallization and wafer edge of 1 mm. To avoid light coupling into the module from surrounding glass or the glass edges a black aperture is laminated onto the front of the laminate. **Fig. 1** shows a photograph of such a test laminate with aperture and 4 mm cell gaps. Such laminates are made for 2 mm, 4 mm and 6 mm cell gaps. We measure the cell width, the cell distance and the distance between cell edge and rear side metallization by a vernier caliper with an uncertainty of 0.1 mm.

Various backsheets are successively attached to the stacks by a thin (20 µm-50 µm) film of liquid paraffin. The liquid paraffin film is prepared without bubbles between the laminates and the backsheet. **Fig. 2** shows a schematic cross sectional view of the stacks. Before applying the backsheet to a test module or any optical measurement the backsheet materials are pre-treated by a standard lamination process to catch optical changes due to the lamination procedure. Possible rough adhesive layers on the backsheet become mirror like flat after the procedure.

This procedure avoids typical errors caused when comparing various backsheets by producing a test laminate for each backsheet and comparing their current output to each other (e.g. variation of cell distance from laminate to laminate, different cells, different cell shading due to variation in cell interconnect ribbons, cell breakage during lamination).

To compare the effect of the optical coupling between EVA, liquid paraffin and the backsheet we use one extra sample SB4 with 4 mm cell gaps. We measured the light recovery probabilities with one and the same stack first with a paraffin coupled backsheet and later on with the laminated backsheet. We determine the error of the method by comparing these two measurements. **Table 1** shows all samples prepared for this paper and **Table 2** provides a list of the used backsheets. Here we introduce a new measurement method and the used backsheets only serve as examples for the technique.

Table 1: Overview of prepared samples.

Sample	Description	Paraffin and laminated
SA2	2 mm cell gaps	no
SA4	4 mm cell gaps	no
SA6	6 mm cell gaps	no
SB4	4 mm cell gaps	yes
WR	120 half cell module	no

Table 2: Overview of used backsheet materials.

Sample name	Explanation
LRF	2.5 mm wide strips of Light Redirecting Film (LRF) from 3M. for 4 mm and 6 mm cell gaps the strips are assembled
White BSn	Seven commercial standard white backsheets
Al-foil	Al-foil (50 μm)
Black BS	Black backsheet

The *IV* curve parameters of the test modules are measured by a flasher system from h.a.l.m. The mean value of three measurements is taken as result. The used measurement range has an uncertainty of the reproducibility of 0.01 A. A measurement without backsheet as well as a measurement with a light trap behind the module stack allows the calibration of the recovery probability of the cell gaps with nearly zero light recovery. An important factor of the measurement is that no stray light from the flash or other light sources reach the rear of the module. Additional light from the rear distort the measurement, because module backsheets may be partly transparent. The rear EVA foil is prepared optically flat to avoid stray light form the EVA/air interface during the light trap (LT) reference measurement.

We measure the angle dependence from -80° to 80° in steps of 10° of the current gain of the samples according to the indoor method described in standard IEC 61853-2 [10]. For the angle adjustment mechanical predefined angles are used. The angle uncertainty is 1°.

To compare the new test method with the current standard IEC 62788-2 Ed.1 we measure the hemispheric $\rho_h(\lambda)$ and direct reflection $\rho(\lambda)$ of all backsheet samples except for the LRF sample in the range from 300 nm to 1250 nm in intervals of 5 nm. The diffuse reflection $\rho_d(\lambda)$ is calculated by subtracting the direct reflection from the hemispheric reflection.

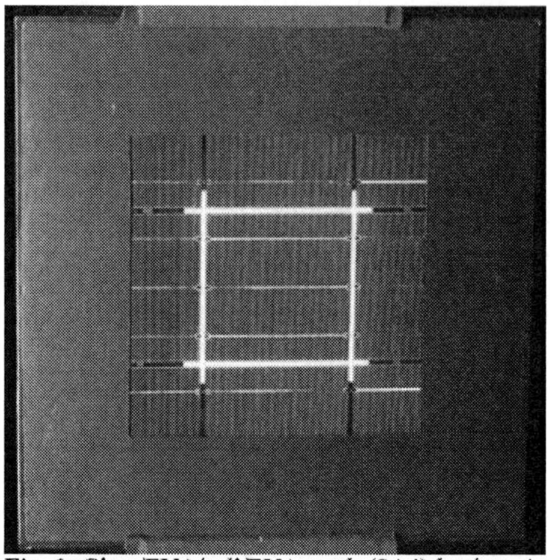

Fig. 1: Glass/EVA/cell/EVA stack (SA4) having 4 mm cell gaps with aperture and partly covered backsheet around the cell. For storage the rear side is protected by a adhesive foil.

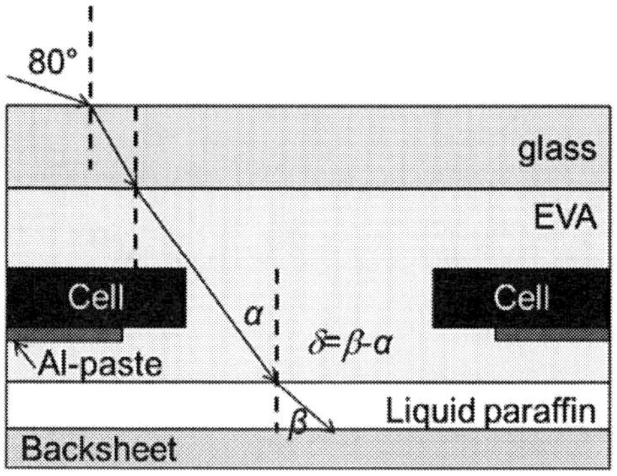

Fig. 2: Cross sectional view of a test sample in the area of the cell gap. The arrows visualise the definition of the maximum angle error δ for angle dependent measurements using liquid paraffin for optical backsheet coupling.

2.2 Calculation of recovery probability

The measured nominal current of the test module with backsheet $I_{sc}(BS)$ subtracted by the nominal current $I_{SC}(LT)$ measured without backsheet and light trap (LT) behind the module results in the current gain ΔI_{SC} for the backsheet,

$$\Delta I_{SC} = I_{sc}(BS) - I_{SC}(LT). \quad (1)$$

We define the light recovery probability k as:

$$k = \frac{\Delta I_{sc}}{I_{sc}(LT)} \frac{A_{cell}}{A_{BS}} \cong \frac{j_{BS}}{j_{cell}}, \quad (2)$$

with the measured cell area A_{cell} and the unit area A_{BS} of the backsheet around the cell. The areas A_{cell} and A_{BS} are defined in **Fig. 3**. The light recovery probability k provides information how much of the light reflected by the backsheet around the cell is used by the solar cells in the module for current generation. It is a measure of the current density j_{BS} generated indirect from the backsheet relative to the current density of the cell j_{cell}.

The angle dependence of k can be calculated with Eq. (2) with angle dependent nominal currents $I_{SC}(BS,\alpha)$ and $I_{SC}(LT,\alpha)$. The angle dependence of the areas A_{cell} and A_{BS} compensates each other in the equation.

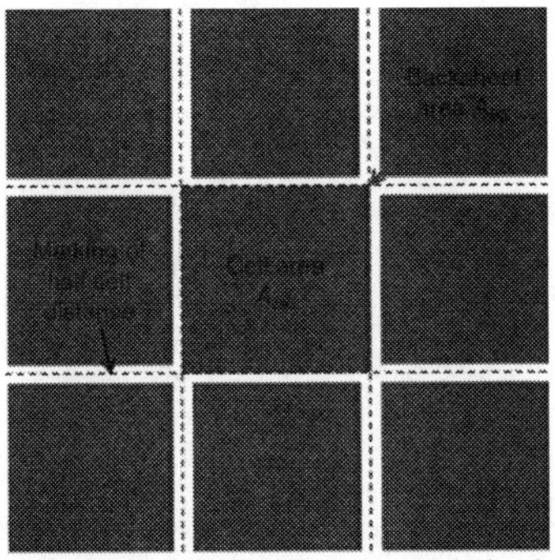

Fig. 3: Definition of cell area A_{cell} and the backsheet area A_{BS} belonging to the cell.

2.3 Calculation of weighted reflections

We define the weighted diffuse reflection ρ_{wd} and the weighted hemispheric reflection ρ_{wh} similar to the standard IEC 62788-2 Ed.1 as

$$\rho_{wd/wh} = \sum_{300\ nm}^{1250\ nm} \frac{\rho_{d/h}(\lambda)\phi_P(\lambda)SR(\lambda)\Delta\lambda(\lambda)}{\phi_P(\lambda)SR(\lambda)\Delta\lambda(\lambda)}. \quad (3)$$

$\rho_{d/h}(\lambda)$ is the spectral diffuse (index d) or the hemispheric (index h) reflectance of the film at wavelength λ(%), $\phi_p(\lambda)$ is the reference spectral photon irradiance at wavelength λ as given in IEC 60904-3 for the AM1.5 global terrestrial spectrum in $m^{-2}\cdot s^{-1}\cdot nm^{-1}$, λ is the wavelength in nm and $\Delta\lambda$ is the wavelength interval in nm. Additional to the solar spectrum we weight the reflectance by the spectral response $SR(\lambda)$ of a 3 busbar PERC cell [11].

2.4 Simulation of recovery probability

We simulate the recovery probability of the backsheet with the ray tracing program DIADALOS [7]. **Table 3** shows the used simulation parameters. Within the simulation we focus on backsheet BS3 because all optical parameters are measured for this backsheet type.

Table 3: Measured sample parameter of sample SA2, SA4 and SA6 with white BS3. Assumed parameter are marked with *. These parameters are used for the simulation of the recovery probability with DIADALOS.

Parameter	Value in Simulation	Value in Sample
EVA thickness below cell	450 µm	~400 µm
EVA thickness above cell	450 µm	~400 µm
Wafer thickness	170 µm	180 µm
Cell gaps SA2, SA4, SA6	2 mm, 4 mm, 6 mm	2 mm, 4 mm, 6 mm
Glass thickness	3.2	4.0 mm
Glass parameter n and k	Low iron float glass [12]	Low iron float glass
EVA parameter n and k	EVA_{UV-T} in [13],[14]	EVA_{UV-T} in [13],[14]
Parameter of the cell optics	3 Busbar cell [15]	3 Busbar cell
Distance between cell edge and rear metallization of cell	1.0 mm	1.0 mm
Total reflectivity R of white BS1	Backsheet of BS3 [13]	BS3
Assumed lambertian factor for backsheet*	1	
Backsheet coupling	direct	Liquid paraffin oil

3 RESULTS

3.1 Measured recovery probability

The recovery probability for various backsheet materials are shown in **Fig. 4**. A black backsheet shows a recovery probability of close to 0%. A typical household aluminum foil shows a low light recovery probability because most of the light is reflected out of the module. Typical diffuse reflecting backsheet materials show a recovery probability between 45% and 55% for 2 mm cell gaps. In this case the diffuse reflected light with a reflection angle higher than 42° to the backsheet normal is total reflected on the glass/air interface and might be collected by a solar cell. The recovery probability decreases with increasing cell gap width for all measured materials with a polynomic function of an order of 2. The highest light recovery probability of about 75% is measured for the LRF.

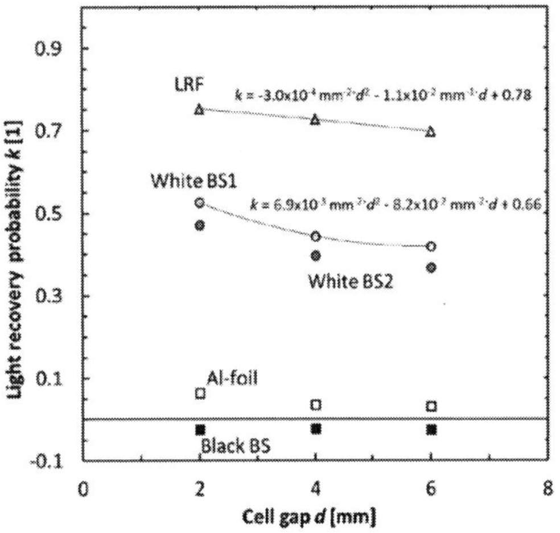

Fig. 4: Light recovery probability k of various backsheet materials as function of the cell gap. For clarity the results of backsheets BS3 and BSn are not shown, however BS1 and BS2 give the upper and lower limits of k for all investigated backsheets of this work. The backsheet BS3 is shown separately in Fig. 5.

3.1 Simulated recovery probability

The measurement and simulation of BS3 is compared in Fig. 5. The uncertainty bars are calculated by error propagation calculation of Eq. 2. The measurement uncertainties given in chapter 2.1 are used for the measurement errors and two times the standard deviation of the numerical uncertainties of the Monte-Carlo algorithm of the ray-tracing simulations. The horizontal uncertainty bar of the distance measurement (0.1 mm) is equal to the measurement symbols size. The simulation has no uncertainty in cell distance. The results match within the range of the uncertainty bars. Even the polynomial fit of the measurement can be used as prediction for smaller cell distance down to 0.5 mm. For the very small cell distance of 0.2 mm the light recovery probability is dominated by other effects than for greater cell distances. We assume that for this small gap most of the light is trapped between both cells and is utilized by the 1 mm wide metallization free cell rear edge.

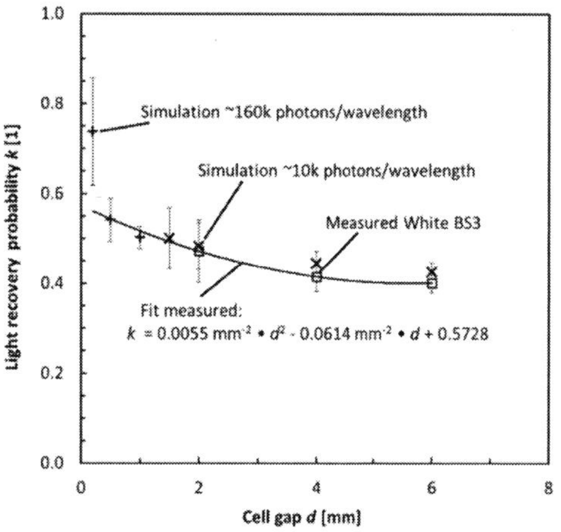

Fig. 5: Comparison of light recovery probability of backsheet BS3 measured (SB4) and simulated with the parameter set of Table 3.

3.2 Recovery probability and reflection

To test the correlation of the light recovery probability to the weighted hemispheric reflection and the weighted diffuse reflection we compare them in **Fig. 6** and **Fig. 7**, respectively.

The weighted hemispheric reflection does not differentiate between direct reflected and diffuse reflected light, see insets in Fig. 6. Therefor the weighted hemispheric reflection for an Al-foil is similar to white backsheets. However the direct reflected light cannot be recovered in a PV module. Therefore the weighted hemispheric reflection and the recovery probability do not correlate.

The weighted diffuse reflection and the recovery probability correlate as shown in **Fig. 7**. Therefore we prefer using the weighted diffuse reflection for benchmarking various backsheet materials. The correlation is not linear because the diffuse reflection does not account for the angle dependence of the diffuse reflection. However the angular dependents cannot be characterized by this method but angular dependents influences the recovery probability in the PV module.

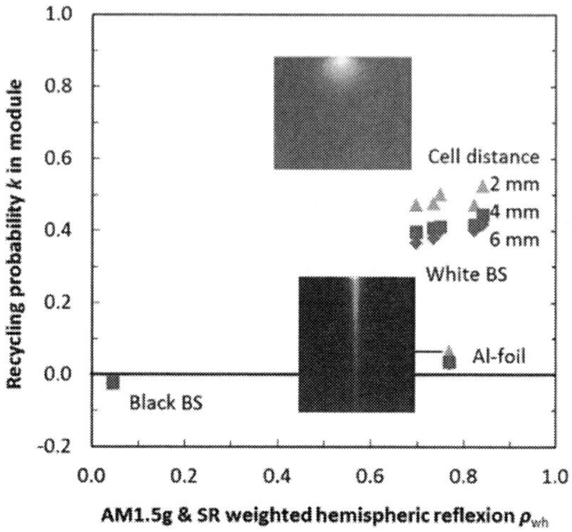

Fig. 6: Comparison of the recovery probabilities and the weighted hemispheric reflection ρ_{ht} of the backsheets listed in **Table 2**. The picture insets visualize the reflection type of white backsheets (upper image) and the direct reflection of the Al-foil (lower image).

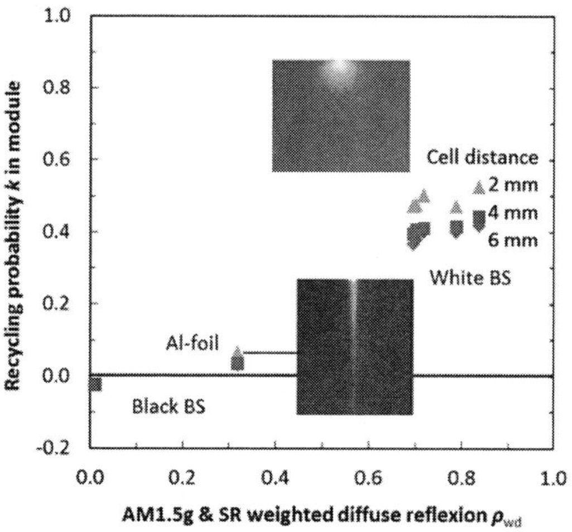

Fig. 7: Comparison of the recovery probabilities and the weighted diffuse reflection ρ_{ht} of the backsheets listed in Table 2. The picture insets visualize the reflection type of white backsheets (upper image) and the direct reflection of the Al-foil (lower image).

3.3 Angle dependence

To reduce the influence of a slightly inhomogeneous light field during the angle dependent measurement and the influence of a systematic angle error we average the corresponding measurements with positive and negative angle. The resulting recovery probability for sample SB4 with backsheet "White BS3" is shown in **Fig. 8**. The backsheet is coupled once with liquid paraffin to the EVA rear side and finally laminated with the same backsheet. As previous the uncertainty bars are calculated by error propagation calculation of Eq. 2. The uncertainty of the cell gap measurement is set to 0 cm to highlight the errors of the changes made with and without liquid paraffin. For perpendicular incident light the relative difference in k is 1.7%. For increasing angles the uncertainty of k increases because the total current amplitudes of $I_{sc}(BS)$ and $I_{SC}(LT)$ decreases while the current measurement uncertainty stays constant. For angles up to 60° the measured recovery probabilities with and without liquid paraffin are equal in the range of the error bars.

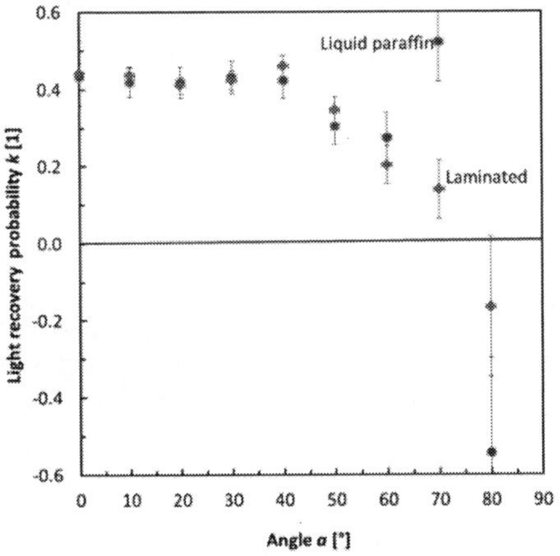

Fig. 8: Comparison of angle dependence of the recovery probability k for 4 mm cell gaps measured with liquid paraffin for optical coupling of the backsheet BS3 and afterwards directly laminated back sheet.

4 DISCUSSION

For angles higher than 60° this measurement method seems not to be applicable. For these angles most of the light enters the light trap behind the sample by traveling besides the sample and back to the sample. We assume that during the reference measurement with no backsheet a small light intensity is reflected from the light trap back to the module boosting the module current more than the backsheet. Therefor we get a negative recovery probability for the 80° angel. Furthermore the high measurement uncertainty fully hides the k value for 80°. For perpendicular incident light (0°) some light from a reflection at the EVA/air interface is reflected. The total reflection at the EVA/air interface is about 4%. This might be the reason why the black backsheet has a slightly negative light recovery probability (-2.5%). Therefor it is reasonable to use a standardised light trap foil directly coupled with liquid paraffin to the modules rear for future work. However currently there is no standard for that purpose.

Fig. 9 and **Fig. 10** show a comparison of the optical properties of EVA [13] and liquid paraffin [16]. The liquid paraffin has a similar refraction index as EVA and low absorption in the region of interest. The extinction coefficient of liquid paraffin is lower or equal to the one of EVA in the wavelength range from 300 nm to 900 nm. The thickness of the liquid paraffin (20 µm – 50 µm) is equal to the thickness variation of 0.45 mm thick EVA sheets. Therefore, the additional absorption in the liquid paraffin is negligible and the method is well applicable for direct incident light.

However for high incidence angle as shown in Fig. 2 the difference in refraction index changes the angle of impact on the backsheet. For the highest measurable incident angle of 80° the maximum angle error δ is calculated as defined in Fig. 2. The refraction index of solar grade glass is taken from Vogt [12], the refraction index from UV transparent EVA is measured [13] and from liquid paraffin is taken from Li [16] shown in **Fig. 9**.The resulting maximum angle error δ is also shown in **Fig. 9**. For wavelength up to 670 nm the maximum angle error is below 2°. For higher wavelength the maximum angle error reaches up to 5° for 900 nm. Therefor this error has to be taken into account when using this method for measuring an angle dependent recovery probability.

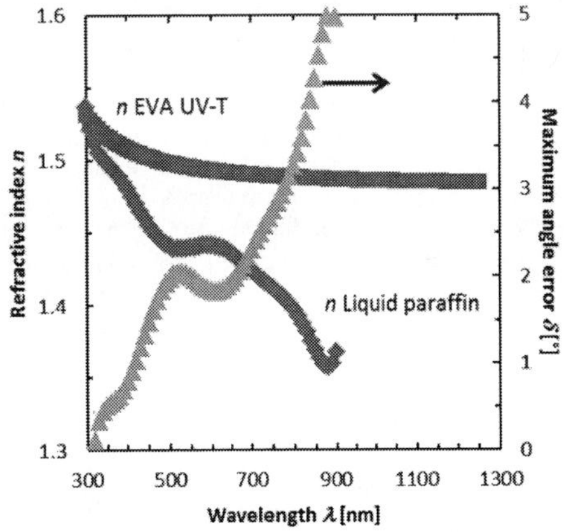

Fig. 9: Comparison of the refractive index of UV transparent EVA and liquid paraffin. The optical data for the liquid paraffin is taken and interpolated from Li [16]. The calculated maximum angle deviation δ caused by the liquid paraffin at the EVA/paraffin interface for 80° incident light on the glass plate of the module is shown in green.

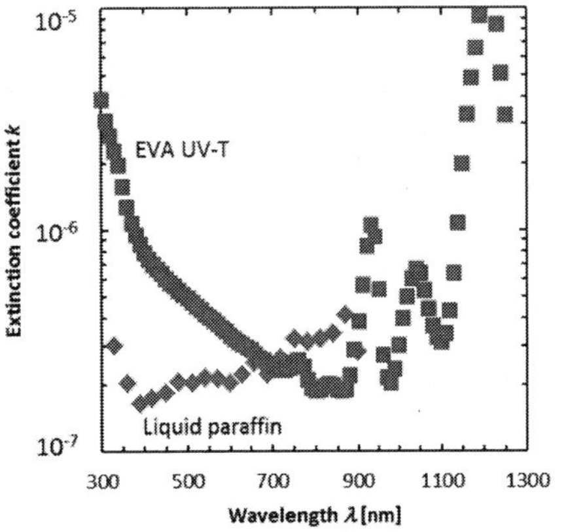

Fig. 10: Extinction coefficients of liquid paraffin [16] and UV transparent EVA.

Furthermore due to the lower refraction index of liquid paraffin compared to EVA there is some additional reflection at the paraffin/EVA interface. For the highest measurable incident angle of 80° the mean reflection is 3.4% at the paraffin/EVA interface.

For Lambertian reflecting backsheets the coupling with liquid paraffin should result in no error, because they scatter independent from the incident angle. Even most "realistic" Lambertian reflecting backsheets should be measureable with small error with this method, because due to the glass cover the first incident ray hits the modules back for a flat front glass in any case within an angle of less than 45°. Realistic Lambertian reflector behaves Lambertian for incident angle in the range between 0° and 22.5° and start to deviate from the Lambertian characteristic at about 45°[17].

5 APPLICATION OF LIGHT RECOVERY MATERIAL

With the optimum light reflecting material, LRF, we build a large-area PERC module optimized for module efficiency by decreasing optical losses. The cell gaps between interconnected cells are 1.5 mm and between the strings 1 mm. Additionally, we use cell interconnect ribbon with a structured front side (Light Harvesting String, Schlenk, 1.5 × 0.2 mm²). Light reflected at the ribbons' front side is totally reflected at the air-glass interface and therefore can be collected by the solar cell. The cell strings are connected in series with string interconnect ribbons with a cross-section of 0.2×6 mm². These are located behind the solar cells to maximize the module area covered by solar cells.

We use PERC solar cells fabricated in our SolarTeC technology center using industrial production tools, details are given in [18]. The front and the rear side metallization is realized using screen printing technology. Instead of the typical three busbars, the cells feature four 1 mm wide busbars on the front side in order to reduce the series resistance losses in the metal fingers. We apply four parallel rows each consisting of six Ag solder pads for the cell interconnection on the rear side. After firing the solar cells are laser diced in half cells and receive a permanent deactivation of the B-O-defects.

Using 120 half 6" PERC solar cells with an average efficiency of 20.8%, we reach a module power of 303.2 W on a module aperture area of 1.50 m², see **Table 4**. The aperture area only excludes the Al frame. This power corresponds to a module efficiency of 20.2%. To our knowledge this is the highest module efficiency reported so far for large-area modules with industrially-processed PERC solar cells on p-type monocrystalline Si.

Table 4: Parameters measured under standard test conditions (AM1.5G, 1000 W/m², 25°C).

	P_{mpp} [W]	η [%]	FF [%]	V_{oc} [V]	I_{sc} [A]
120 half PERC cells	303$^{\Sigma}$	20.8^{Ø}	79.7^{Ø}	79.3$^{\Sigma}$	4.80^{Ø}
Module$^{\#}$	303*	20.2	79.1*	78.9*	4.86*

$^{\Sigma}$ sum over all cells, Ø average of all cells, $^{\#}$ module measurements on aperture area (ap.) (excluding frame)
* independently confirmed by TÜV Rheinland, Cologne, Germany

A detailed analysis of the electrical and optical cell-to-module gains and losses is given in [19]. Related the effect of the light recovery form the cell spacings the analysis based on the above mentioned results shows a boost in power by 6.2 W or 2.1 W for a LRF material in the cell gaps compared to no backsheet or white backsheet, respectively.

6 SUMMARY AND CONCLUSION

We presented a new method for precise, cost effective and fast measurement of the light recovery probability of backsheets in PV modules. We use liquid paraffin to couple the backsheet on special test modules. This method avoids typical uncertainties by comparing backsheets with separate laminated test modules like variation of cell distance from laminate to laminate, different cells, different cell shading due to variation in cell interconnect ribbons, and cell breakage. The method is also applicable for angle dependent measurements of the recovery probability of backsheets up to an angle of 60°. The method is used to identify the optimum light recovery material to build high efficient PERC solar modules. The light recovery probability correlates with solar weighted reflectance of backsheet as defined in IEC 62788-2 Ed.1 when using diffuse reflectance instead of hemispheric reflectance as required in the standard. With the Light Redirecting Film (LRF) we build a world record PERC module with a module power of 303.2 W on a module aperture area of 1.50 m² with 20.2% efficiency. The LRF with k=0.75 boosts the module power by 6.2 W compared to no backsheet and 2.1 W compared to a white backsheet with k=0.5 recovery probability. The aperture area only excludes the Al frame.

7 ACKNOWLEDGEMENT

We gratefully acknowledge the support by the German Federal Ministry for Economic Affairs and Energy under FKZ 0325641 (PERC-2-Module) and by the state of Lower Saxony.

8 REFERENCES

[1] SEMI, "International Technology Roadmap for Photovoltaic 2014 Results," 2015.

[2] "Trina press release," 2015. [Online]. Available: http://ir.trinasolar.com/. [Accessed: 04-Feb-2016].

[3] "Rekordzelle mit 22 Prozent Wirkungsgrad," *SolarWorld press release*, 2016. [Online]. Available: http://www.solarworld.de/konzern/presse/aktuelles/pressemitteilungen/single-pressemitteilung/article/solarworld-praesentiert-solartechnik-der-naechsten-generation/. [Accessed: 13-Jun-2016].

[4] A. Metz, D. Adler, S. Bagus, H. Blanke, M. Bothar, E. Brouwer, S. Dauwe, K. Dressler, R. Droessler, T. Droste, M. Fiedler, Y. Gassenbauer, T. Grahl, N. Hermert, W. Kuzminski, A. Lachowicz, T. Lauinger, N. Lenck, M. Manole, M. Martini, R. Messmer, C. Meyer, J. Moschner, K. Ramspeck, P. Roth, R. Schönfelder, B. Schum, J. Sticksel, K. Vaas, M. Volk, and K. Wangemann, "Industrial high performance crystalline silicon solar cells and modules based on rear surface passivation technology," *Sol. Energy Mater. Sol. Cells*, vol. 120, pp. 417–425, Jan. 2014.

[5] S. Zhang, X. Pan, H. Jiao, W. Deng, J. Xu, Y. Chen, P. P. Altermatt, Z. Feng, and P. J. Verlinden, "335-W World-Record p-Type Monocrystalline Module With

20.6% Efficient PERC Solar Cells," *IEEE J. Photovoltaics*, vol. 6, no. 1, pp. 145–152, Jan. 2016.

[6] P. Verlinden, "Cost analysis of current production and strategy for future silicon PV modules," in *28th European Photovoltaic Solar Energy Conference*, 2013.

[7] H. Holst, M. Winter, M. R. Vogt, K. Bothe, M. Köntges, R. Brendel, and P. P. Altermatt, "Application of a New Ray Tracing Framework to the Analysis of Extended Regions in Si Solar Cell Modules," *Energy Procedia*, vol. 38, pp. 86–93, 2013.

[8] Z. Liu, M. Peters, V. Shanmugam, Y. S. Khoo, S. Guo, R. Stangl, A. G. Aberle, and J. Wong, "Luminescence imaging analysis of light harvesting from inactive areas in crystalline silicon PV modules," *Sol. Energy Mater. Sol. Cells*, vol. 144, no. JANUARY 2016, pp. 523–531, Jan. 2016.

[9] I. Haedrich, U. Eitner, M. Wiese, and H. Wirth, "Unified methodology for determining CTM ratios: Systematic prediction of module power," *Sol. Energy Mater. Sol. Cells*, vol. 131, pp. 14–23, Dec. 2014.

[10] "IEC 61853-2 Ed. 1.0 Photovoltaic (PV) modules performance testing and energy rating - Part 2: Spectral response, incidence angle and module operating temperature measurements." .

[11] H. Hannebauer, T. Dullweber, U. Baumann, T. Falcon, and R. Brendel, "21.2%-efficient fineline-printed PERC solar cell with 5 busbar front grid," *Phys. status solidi - Rapid Res. Lett.*, vol. 8, no. 8, pp. 675–679, Aug. 2014.

[12] M. R. Vogt, H. Hahn, H. Holst, M. Winter, C. Schinke, M. Köntges, R. Brendel, and P. P. Altermatt, "Measurement of the Optical Constants of Soda-Lime Glasses in Dependence of Iron Content and Modeling of Iron-Related Power Losses in Crystalline Si Solar Cell Modules," *IEEE J. Photovoltaics*, no. 99, pp. 1–8, 2015.

[13] M. R. Vogt, "Development of Physical Models for the Simulation of Optical Properties of Solar Cell Modules," Phd thesis Leibniz University Hannover, 2015.

[14] M. R. Vogt, H. Holst, H. Schulte-Huxel, S. Blankemeyer, D. Hinken, R. Witteck, M. Winter, B. Min, C. Schinke, I. Ahrens, M. Köntges, K. Bothe, and R. Brendel, "Optical constants of UV transparent EVA and the impact on the PV module output power under realistic irradiation," in *accepted SiliconPV*, 2016.

[15] H. Holst, M. Winter, H. Schulte-Huxel, S. Blankemeyer, R. Witteck, T. Booz, F. Distelrath, M. Köntges, K. Bothe, and R. Brendel, "Increased light harvesting by structured cell interconnectors: an optical ray tracing study under realistic daylight conditions," *to be Publ. Energy Procedia*, vol. 00, 2016.

[16] D. Li, Y. Zheng, Z. Li, and H. Qi, "Optical properties of a liquid paraffin-filled double glazing unit," *Energy Build.*, vol. 108, no. 2015, pp. 381–386, Dec. 2015.

[17] A. Höpe and K.-O. Hauer, "Three-dimensional appearance characterization of diffuse standard reflection materials," *Metrologia*, vol. 47, no. 3, pp. 295–304, Jun. 2010.

[18] C. Kranz, B. Lim, U. Baumann, and T. Dullweber, "Determination of the Contact Resistivity of Screen-printed Al Contacts Formed by Laser Contact Opening," *Energy Procedia*, vol. 67, no. 0, pp. 64–69, Apr. 2015.

[19] H. Schulte-Huxel, R. Witteck, H. Holst, M. R. Vogt, S. Blankemeyer, D. Hinken, T. Brendemühl, T. Dullweber, K. Bothe, M. Köntges, and R. Brendel, "High efficiency modules with passivated emitter and rear solar cells – an analysis of electrical and optical losses," *IEEE J. Photovoltaics, Submitt. 2016.*

DEVELOPMENT OF ADHESIVE AND COHESIVE FAILURES IN EVA-BACKSHEET STRUCTURES IN ENVIRONMENTAL TESTING

Jiang Zhu*, Daniel Montiel-Chicharro, Tom Betts and Ralph Gottschalg
Centre for Renewable Energy Systems Technology (CREST), Wolfson School of Mechanical, Electrical and Manufacturing Engineering, Loughborough University, Leicestershire, UK, LE11 3TU
*Tel: +44(0)1509 635313, Fax: +44(0)1509 635341, E-mail: J.Zhu@lboro.ac.uk

ABSTRACT: The development of adhesive and cohesive failures at the EVA-backsheet interface under different damp-heat (DH) testing condition is investigated. Failures are classified as adhesive and cohesive by the surface roughness of the peeled off backsheet strips. Different DH testing condition leads to different dominating failure modes. The adhesive failure is the main failure mode at lower testing temperature, which are also the predominant realistic operating conditions (by time). These failures can be masked by mixed failures which start to dominate at higher testing temperatures. This is due to different thermal acceleration factors of the two processes. Development of accelerated environmental testing protocol requires the failure mode analysis to ensure the target failure mode or degradation mechanism is accelerated and not masked by any other processes. The results indicate that adhesion testing is difficult to accelerate thermally as this might accelerate one mode which is not the predominant one being seen in realistic operation.

Keywords: adhesive, cohesive, EVA, backsheet, acceleration

1 INTRODUCTION

PV module packaging consists of front cover, encapsulant, backsheet and edge seal. This forms a multilayer system which results in a number of interfaces. Different interfaces as well as the bulk materials degrade during outdoor operation. Good adhesion between these interfaces is necessary to minimise contaminants ingress (and generation) that will further degrade modules and may cause safety issues [1]. Integrity of the multi-layer system is critical to ensure a long service lifetime.

Backsheet delamination, as a result of loss of adhesion, is a commonly observed failure mode in PV modules. Loss of adhesion strength depends on operating conditions. Currently there are some minimum requirements being set for adhesion strength in new as produced module. There is no clear scientific foundation for these values, i.e. no link of having lower values resulting failure. Different failure modes have also been observed and so far there is no distinction between adhesive and cohesive failures at any of the interfaces investigated.

Ethylene Vinyl Acetate (EVA) is still the most commonly used encapsulant in the PV industry. A complex cocktail of additives is used to enhance the performance of pure EVA resin. This may include adhesion promoter, normally in the form of silane-coupling agents that are used to enhance adhesion between EVA and glass by forming silicon-oxygen covalent bonds [2]. For the adhesion between EVA and the backsheet, coupling agents or primers are often added at the inner side of the backsheet to enhance adhesion by either forming chemical bonds or increasing physical absorption.

In order to test the performance and durability of the entire module as well as encapsulation materials, manufacturers rely on standardised environmental tests [3], e.g damp-heat (DH), thermal-cycling (TC), humidity-freeze (HF) and etc. The environmental testing is designed as a screening test for specific ageing mechanisms. The tests have been adopted from other industries and there is the possibility of them giving misleading results. This is particular true if multiple degradation mechanisms exist with different thermal acceleration, i.e. different activation energy (E_a). Given that environmental tests are conducted at elevated temperature to gain some thermal acceleration, this may result in skewed assessment in failure modes. This paper investigates this phenomenon for two failure modes that are the adhesive and cohesive failures within EVA-backsheet structures observed during DH testing. At this interface, moisture ingress may cause bond decomposition. This depends on the type of the bond, but will also result in reduced adhesion strength. Such a de-bonding reaction is normally accelerated by temperature. Backsheet material will also suffer from hydrolysis and become brittle upon moisture absorption which produces extra stresses within the interface. Those processes occurred at the same time. Either one of them might become a dominating process that leads to a specific failure mode depending on the environmental testing condition and the testing time.

2 EXPERIMENTAL SETUP

2.1 Testing samples

In order to test the adhesion failure at the EVA and backsheet interface, PV laminates with a backsheet/EVA/EVA/backsheet structure were laminated at Centre for Renewable Energy Systems Technology (CREST) using a 2BG laminator L176A. The size of the samples is 20cm by 12.5cm.

The EVA is 460µm fast cure EVA with 34% of vinyl acetate. The backsheet is a two-layer insulating polymer consisting PET/primer layer. More typical tri-layer backsheet was not used in order to reduce the number of interfaces. A silicone coated release liner was inserted between backsheet and EVA during lamination to enable consistent separation of layers when initiating the peel test.

2.2 T-peel test

Adhesion at EVA/backsheet interface was measured by conducting the T-Peel test with a Chatillon motorised force tester at CREST. Each testing sample was cut into

Table I: DH conditions and measurement intervals

Test name	Temperature (°C)	Relative humidity (%)	Test time intervals (hours)
60/85	60	85	0, 50, 100, 250, 500, 1000
85/85	85	85	0, 50, 100, 250, 500, 1000
95/85	95	85	0, 50, 100, 250, 500, 1000
120/100	120	Maximum (85-100)	0, 12, 24, 50, 100

strips of 10mm width for the peel test. The test was carried out at a constant crosshead speed of 50mm/min in room condition. A total of 15-20 stripes from two to three laminate samples were measured for each testing intervals from which the average adhesion strength and frequency of failure mode were calculated. Effects of the environment, peel speed, peel angle and etc. on the peel strength values are not the scope of this paper as the focus is on the comparison study of the failure mode in the EVA/backsheet structures during the peel test.

2.3 Damp-heat test and measurement intervals

The damp-heat tests were carried out in environmental chambers at CREST. Four damp-heat conditions with temperature ranging from 60°C to 120°C implemented in this work are summarised in Table I. The test duration is 1000 hours except that it is 100 hours for the testing at 120°C/100%RH as the samples degraded quickly and became too brittle to peel at this stress condition. The measurement intervals are defined in Table I. At the beginning of the DH tests, 10-12 laminate samples were loaded into the environmental chamber. At each measurement interval, 2-3 laminates were taken out for T-peel test in order to track the degradation process over ageing time.

3 DEVELOPMENT OF ADHESIVE AND COHESIVE FAILURES

3.1 Peel strength and failure types

Peel strength values are likely dependent on a number of factors including the peel angle, backing material, peel speed and etc. Typical peel strength measurements plotted as measured load (in Newton) against time are shown in Fig. 1. They are the measurements for all strips after 50 hours ageing at 85°C/85%RH. The solid red line is the average of these measurements. Three spikes can be seen around 120s and 270s. These variations are difficult to avoid in peel strength measurements and they might be related to different failure types. In order to confirm whether a different failure mode was developed, surface analysis was carried out for the area A and B of the peeled off backsheet strips, respectively.

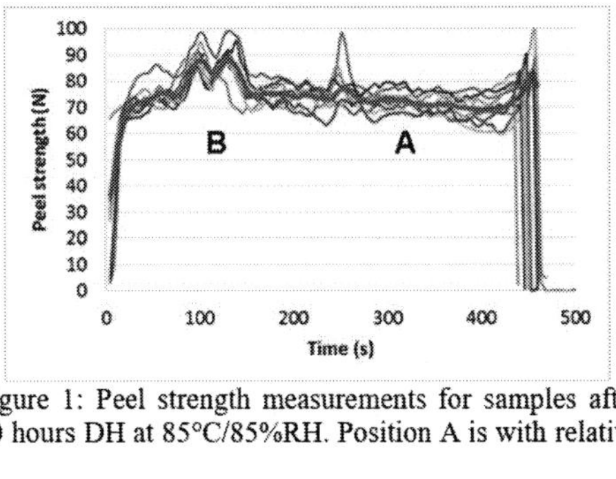

Figure 1: Peel strength measurements for samples after 50 hours DH at 85°C/85%RH. Position A is with relative flat peel strength values, whereas position B is with spikes in peel strength.

3.2 Surface analysis of peeled backsheet

As shown in Fig. 1, position A is with relative flat peel strength values, whereas position B is where the spikes occurred. The inner side surface of the peeled off backsheet strips for the two positions was checked and clear differences were found.

From microscopic images, the surface of peeled off backsheet strips at position A is smooth, while the surface at position B is coarse. Furthermore, the surface roughness was also measured for the two positions by using the Coherence Scanning Interferometry (CSI). Multiple measurements were taken. The surfaces at position B have relatively bigger peaks and valleys, whereas the surfaces at position A have relatively smaller peaks and valleys. One parameter used to describe the surface roughness is the averaged deviation of the surface height. The averaged deviation calculated for position A is 0.48-0.49µm, which is much lower than that at position B that varies between 0.8µm and 1.12µm.

With the surface analysis, the adhesive and cohesive failures are defined and able to be distinguished. The failure at the position A and B are classified as adhesive and thin cohesive failures, respectively. It is likely that the different failure modes are due to the factors such as the variations in manufacturing process of the material, imperfect lamination quality due to uneven temperature and pressure distribution, and etc.

4 FAILURE TYPES OVER AGEING TIMES

4.1 Observed failure types under different DH conditions

Applying the failure mode analysis for all the peeled off backsheet strips after ageing under different DH conditions, the frequencies of failure modes over ageing time are plotted in Fig. 2 and investigated. Three failure modes are defined in this investigation, including the adhesive failure mode, the mixed mode of adhesive and thin cohesive failures and the failure mode due to backsheet strips being snapped during the peel test.

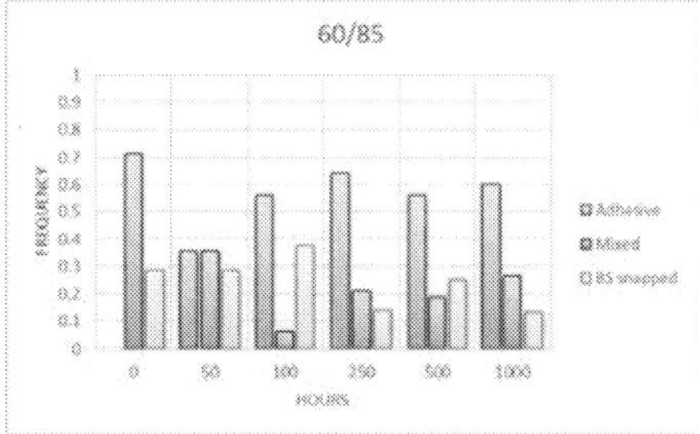

(a)

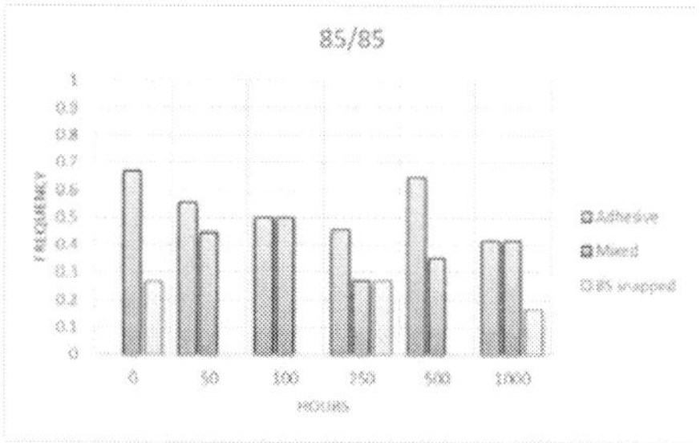

(b)

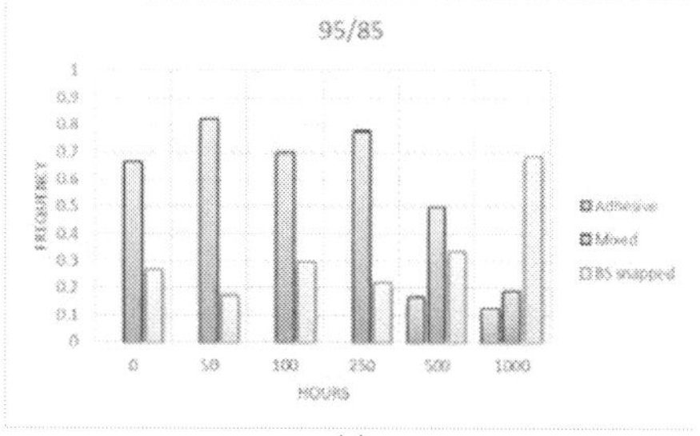

(c)

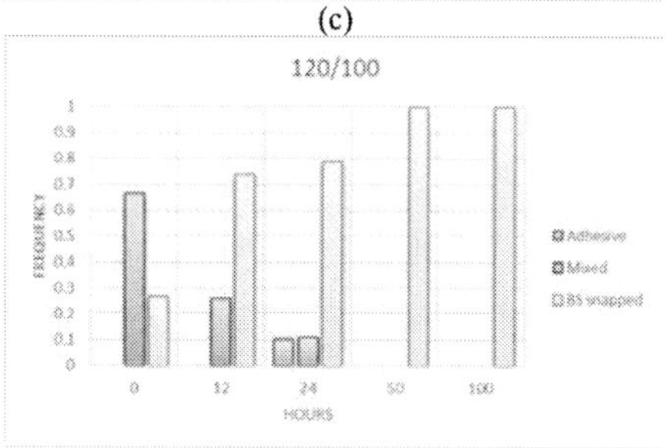

(d)

Figure 2: Frequencies of observed failure types over ageing time under different DH conditions at a) 60°C/85%RH, b) 85°C/85%RH, c) 95°C/85%RH and d) 120°C/100%RH.

As clearly seen from Fig. 2, more adhesive failures were observed at the DH testing condition of 60°C/85%RH. The frequency of mixed failure mode increased when testing at 85°C/85%RH. Under the testing condition of 95°C/85%RH., the mixed failure mode became dominating and about 2 out of 3 backsheet strips were snapped during peel test after 1000 hours testing. At the highest testing condition of 120°C/100%RH, the samples degraded fast and became brittle due to the hydrolysis of backsheet. Thus, the embrittlement led to the failure of snapped backsheet strips as the main failure mode after 12 hours.

4.2 Probability analysis for failure mode analysis

A statistical approach of probability analysis for the adhesive and mixed failure modes was carried out in attempt to show the effect of temperature acceleration on the failure modes.

For simplicity, this approach assumes the Gaussian distribution to represent the probability of occurrences of a specific failure mode. In reality, the probability could be in different distributions depending on the properties of material, manufacturing process and etc. As depicted in Fig. 3, the distribution of adhesive failure exhibits bigger probabilities than the distribution of mixed failure mode. This means at this testing condition, the samples are more likely to suffer the adhesive failure than the mixed failure. Both distributions might move towards higher probabilities if higher stresses or longer testing time is imposed upon the testing samples as heavily degraded samples are more likely to fail.

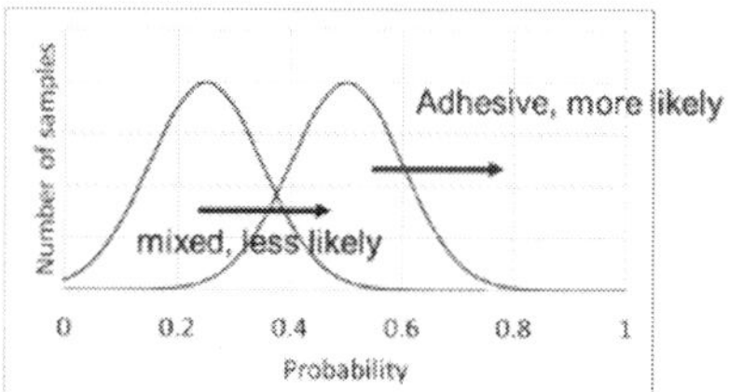

Figure 3: Probability analysis used for adhesive and mixed types of failure.

The probability analysis is applied to all the tested

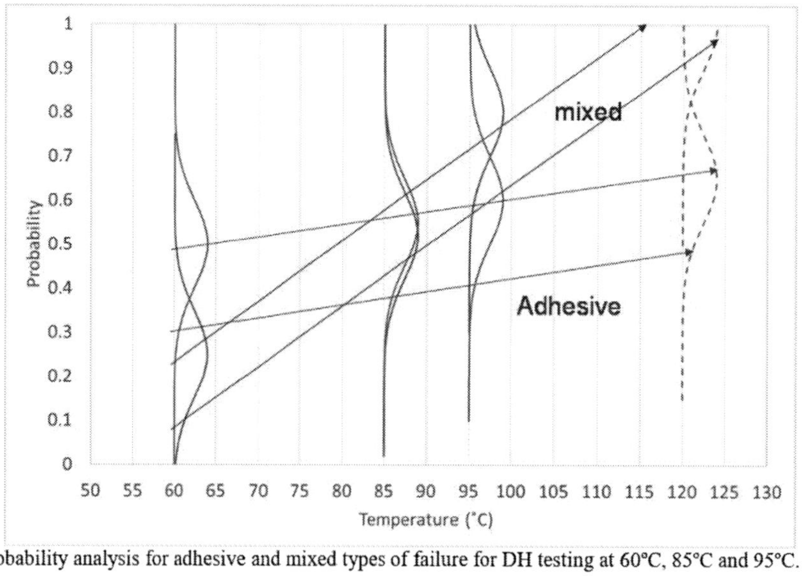

Figure 4: Probability analysis for adhesive and mixed types of failure for DH testing at 60°C, 85°C and 95°C.

samples. Based on the observed failure modes as reported in Fig. 2, the distributions of the adhesive and mixed failure modes in dependency of testing temperature are shown in Fig. 4. The failure probabilities increase with the increasing testing temperature. This is due to the fact that temperature acts as an accelerator in both failure modes. However, the acceleration factors due to temperature are not at the same level for the two failure modes. At 60°C, the samples are more likely to suffer adhesive failure, which is the main failure mode. At 85°C, the probabilities for the adhesive and the mixed failure are at a similar level which means there is no dominating failure mode. At 95°C, on the contrary, the mixed failure is more likely to happen and became the main failure mode. The potential reason behind this can be explained that the thermal acceleration for the mixed failure mode is with a relatively larger E_a, while that for the adhesive failure mode is with a relative smaller E_a. Therefore, the mixed failure process with a large E_a is accelerated significantly at higher temperatures and masks the process of adhesive failure with a smaller E_a. There is no measurement available to show the probabilities of adhesive and mixed failure mode at 120°C as all samples saw snapping backsheet strips, i.e. a third failure mode masks those two failures at 120°C.

Considering the outdoor PV module temperatures under different climates [4] are normally up to 80°C, the tests carried out at temperature above this, e.g. the pressure cooker test (PCT), might lead to a different dominating failure mode and masks the realistic failure mode that is seen by fielded modules.

5 CONCLUSIONS

Failure mode analysis at the EVA/backsheet interface for PV laminates underwent DH testing at different temperature and relative humidity conditions are carried out. Different failure modes including the adhesive and thin cohesive failure are identified and the development of the adhesive and cohesive failure mode is found to be directly related to the DH testing condition. Lower testing temperature at 60°C leads to adhesive failure at EVA/backsheet interface, whereas the failure mode changes to the mixed failure at testing temperature above 85°C. Therefore, the development of accelerated testing protocol for PV modules or components requires the comprehensive failure mode analysis for the tested samples in order to ensure the target failure mode or degradation mechanism is accelerated and not masked by other processes. Generally, testing at reasonable lower temperatures is more realistic though the acceleration factor might be small and the tests duration is relatively longer.

REFERENCES

[1] M.D. Kempe, Modeling Rates of Moisture Ingress into Photovoltaic Modules, Solar Energy Materials and Solar Cells, 90, 2006, 2720-2738.
[2] F.J. Pern, S.H. Click, Adhesion Strength Study of EVA Encapsulants on Glass Substrates, in Proc. of NCPV and Solar Program Review Meeting, March 24-26, 2003, Denver, CO.
[3] IEC 61215-2: 2016, Terrestrial photovoltaic (PV) modules - Design qualification and type approval - Part 2: Test procedures.
[4] M. Owen-Bellini, J. Zhu, T.R. Betts, R. Gottschalg, Environmental Stress Potentials of Different Climatic Regions, in Proc. of 31st European Photovoltaic Solar Energy Conference and Exhibition, 1854 – 1857, 2015, Germany.

AUTHOR INDEX

AUTHOR INDEX

AUTHOR INDEX

AUTHOR INDEX

AUTHOR INDEX

AUTHOR INDEX

AUTHOR INDEX

AUTHOR INDEX

AUTHOR INDEX

AUTHOR INDEX

AUTHOR INDEX

AUTHOR INDEX

AUTHOR INDEX

AUTHOR INDEX

AUTHOR INDEX

AUTHOR INDEX

AUTHOR INDEX

AUTHOR INDEX

AUTHOR INDEX

AUTHOR INDEX

AUTHOR INDEX

AUTHOR INDEX

AUTHOR INDEX

AUTHOR INDEX

AUTHOR INDEX

32nd European Photovoltaic Solar Energy Conference and Exhibition (EU PVSEC 2016)

Munich, Germany
20-24 June 2016

Volume 3 of 4

Editors:

M. Topic
P. Helm
N. Taylor

ISBN: 978-1-5108-3651-8

Printed from e-media with permission by:

Curran Associates, Inc.
57 Morehouse Lane
Red Hook, NY 12571

Some format issues inherent in the e-media version may also appear in this print version.

Printed by Curran Associates, Inc. (2017)

WIP – Renewable Energies
Sylvensteinstr. 2
81369 Munchen
Germany

Phone: +49 89 72012735
Fax: +49 89 72012791

wip@wip-munich.de

Additional copies of this publication are available from:

Curran Associates, Inc.
57 Morehouse Lane
Red Hook, NY 12571 USA
Phone: 845-758-0400
Fax: 845-758-2633
Email: curran@proceedings.com
Web: www.proceedings.com

32nd European Photovoltaic Solar Energy Conference and Exhibition

Proceedings of the International Conference held in Munich, Germany

20 June - 24 June 2016

Edited by:

M. TOPIČ
University of Ljubljana
Chairman of European Technology & Innovation Platform Photovoltaics
Ljubljana, Slovenia

N. TAYLOR
European Commission - DG Joint Research Centre
Ispra, Italy

P. HELM
WIP Renewable Energies
Munich, Germany

Coordination of the Technical Programme:

European Commission - DG Joint Research Centre
Via E. Fermi 1
21020 Ispra (VA)
Italy

Institutional Support:

European Commission

UNESCO – United Nations Educational, Scientific and Cultural Organization, Natural Sciences Sector

WCRE – World Council for Renewable Energy

Supporting Organisations:

ARE – Alliance for Rural Electrification

EREF – European Renewable Energies Federation

ESA – European Space Agency

EUFORES –The European Forum for Renewable Energy Sources

EUREC – The Association of European Renewable Energy Research Centres

European Photovoltaic Technology Platform

IEA PVPS – International Energy Agency Photovoltaic Power Systems Programme

REN21 – Renewable Energy Policy Network for the 21st Century

Institutional PV Industry Cooperation:

SolarPower Europe

SOLARUNITED

Supporting Associations:

AIE – European Association of Electrical Contractors

AREA – African Renewable Energy Alliance

CABA – Continental Automated Buildings Association

EPIC – European Photonics Industry Consortium

SASIA - Saudi Arabia Solar Industry Association

EU PVSEC realised by:

WIP
Sylvensteinstr. 2, 81369 München, Germany
Tel: +49 89 720 12 735, Fax: +49 89 720 12 791
Email: pv.conference@wip-munich.de
www.photovoltaic-conference.com
www.wip-munich.de

Proceedings produced and published by:

WIP
Sylvensteinstr. 2, 81369 München, Germany
Tel: +49 89 720 12 735, Fax: +49 89 720 12 791
Email: pv.conference@wip-munich.de
www.photovoltaic-conference.com
www.wip-munich.de

ISSN 2196-0992

32nd EUROPEAN PHOTOVOLTAIC SOLAR ENERGY CONFERENCE AND EXHIBITION

20 JUNE – 24 JUNE 2016

MUNICH, GERMANY

EU PVSEC COMMITTEES

INTERNATIONAL SCIENTIFIC ADVISORY COMMITTEE (ISAC)

Chair
G.F. De Santi, European Commission, DG JRC, Director of the Institute for Energy and Transport, Ispra, Italy

Secretary
N. Taylor, European Commission, DG JRC, Institute for Energy and Transport, Ispra, Italy

Committee Members
P. Frankl, Head of the Renewable Energy Division, International Energy Agency, France
P. Hofer-Noser, CTO, Meyer Burger Technology, Switzerland
P. Malbranche, CEA, Le Bourget du Lac, France
S. Nowak, Managing Director of NET Nowak Energy & Technology, Switzerland
W. Sinke, Program Development Manager, ECN Solar Energy, The Netherlands
P. Verlinden, Chief Scientific and Vice Chair, State Key Lab. TRINA Solar, China
E. Weber, Director, Fraunhofer ISE, Germany
P. Helm, WIP, Munich, Germany
A. Grassi, ETA-Florence Renewable Energies, Florence, Italy

CONFERENCE EXECUTIVE COMMITTEE

Conference General Chairman
M. Topič, University of Ljubljana, Chairman of European Technology & Innovation Platform Photovoltaics, Ljubljana, Slovenia

Technical Programme Chairman
N. Taylor, European Commission, DG JRC, Institute for Energy and Transport, Ispra, Italy

Committee Members
H. Ossenbrink, European Commission, DG JRC, Institute for Energy and Transport, Ispra, Italy
S. Nowak, Managing Director of NET Nowak Energy & Technology, St. Ursen, Switzerland
W. Sinke, Program Development Manager, ECN Solar Energy, The Netherlands
P. Helm, WIP, Munich, Germany
A. Grassi, ETA-Florence Renewable Energies, Florence, Italy

2016 SCIENTIFIC COMMITTEE
Topic Organisers and Paper Review Experts

Topic 1 – New Materials and Concepts for Solar Cells and Modules

G. Beaucarne, Dow Corning, Seneffe, Belgium
N.J. Ekins-Daukes, Imperial College London, United Kingdom
A. Martí Vega, UPM, Madrid, Spain
J. Poortmans, imec, Leuven, Belgium
M. Rusu, HZB, Berlin, Germany
J.A.M. Van Roosmalen, ECN, Petten, Netherlands

V. Bukauskas, Center for Physical Sciences and Technology, Vilnius, Lithuania
F. Falk, IPHT, Jena, Germany
J. Govaerts, imec, Leuven, Belgium
J.F. Guillemoles, CNRS, Chatou Cedex, France
M.A. Hernandez-Fenollosa, UPV, Valencia, Spain
J.C. Hummelen, University of Groningen, Netherlands
D. König, UNSW, Sydney, Australia
I. Konovalov, Ernst Abbe University of Applied Science Jena, Jena, Germany
C. Lévy-Clément, CNRS, Thiais, France
Y. Okada, University of Tokyo, Japan
B. Rech, HZB, Berlin, Germany
F. Roca, ENEA, Portici, Italy
V. Saly, Slovak University of Technology, Bratislava, Slovakia
H. Shirai, Saitama University, Japan
W. Warta, Fraunhofer ISE, Freiburg, Germany

Topic 2 – Wafer-Based Silicon Solar Cells and Materials Technology

O. Anspach, PV Crystalox Solar, Erfurt, Germany
D.L. Bätzner, Meyer Burger Research, Hauterive, Switzerland
R. Brendel, ISFH, Emmerthal, Germany
F. Ferrazza, Eni, San Donato Milanese, Italy
S.W. Glunz, Fraunhofer ISE, Freiburg, Germany
G. Hahn, University of Konstanz, Germany
D. Muñoz, CEA, Le Bourget du Lac, France
S. Riepe, Fraunhofer ISE, Freiburg, Germany
A.W. Weeber, ECN, Petten, Netherlands
P. Wohlfart, Singulus Technologies, Kahl am Main, Germany

A.G. Aberle, SERIS, Singapore
N. Azkona, UPV/EHU, Zamudio, Spain
J. Bagdahn, Fraunhofer CSP, Halle (Saale), Germany
R. Barinka, Fill Factory, Rožnov pod Radhoštm, Czech Republic
P.A. Basore, PV Specialist Services, Solana Beach, USA

D. Biro, Fraunhofer ISE, Freiburg, Germany
S. Bowden, Arizona State University, Tempe, USA
T.M. Bruton, TMB Consulting, Woking, United Kingdom
B. Dauksher, Arizona State University, Tempe, USA
M. Debucquoy, imec, Leuven, Belgium
T. Desrues, CEA, Le Bourget du Lac, France
D. Dimova-Malinovska, Bulgarian Academy of Sciences, Sofia, Bulgaria
S. Dubois, CEA, Le Bourget du Lac, France
T. Dullweber, ISFH, Emmerthal, Germany
P. Fath, RCT-Solutions, Konstanz, Germany
A.J. Galdikas, Modernios E-Technologijos, Vilnius, Lithuania
M. Hermle, Fraunhofer ISE, Freiburg, Germany
K. Hesse, Wacker Chemie, Burghausen, Germany
J. Horzel, CSEM, Neuchâtel, Switzerland
J. John, imec, Leuven, Belgium
S.D. Kim, SunPower, San Jose, USA
J. Libal, ISC Konstanz, Germany
A. Metz, h.a.l.m. elektronik, Frankfurt am Main, Germany
J.W. Müller, Hanwha Q CELLS, Bitterfeld-Wolfen, Germany
Y. Ohshita,TTI, Nagoya, Japan
S. Peters, Hanwha Q CELLS, Bitterfeld-Wolfen, Germany
K. Petter, Hanwha Q CELLS, Bitterfeld-Wolfen, Germany
R. Preu, Fraunhofer ISE, Freiburg, Germany
J. Rentsch, Fraunhofer ISE, Freiburg, Germany
E. Sauar, Brighterlite, Oslo, Norway
H. Schlemm, Meyer Burger, Hohenstein-Ernstthal, Germany
M.C. Schubert, Fraunhofer ISE, Freiburg, Germany
B. Terheiden, University of Konstanz, Germany
Y. Veschetti, CEA, Le Bourget du Lac, France
G.P. Willeke, Fraunhofer ISE, Freiburg, Germany
D. Wood, Hemlock Semiconductor, Hemlock, USA
A. Yamada, Tokyo Institute of Technology, Japan
D.L. Young, NREL, Golden, USA

Topic 3 – Thin Film Solar Cells and Modules

C.J. Brabec, University of Erlangen-Nuremberg, Germany
J. Cárabe, CIEMAT, Madrid, Spain
A. Meeder, Flisom, Dübendorf, Switzerland
J. Meier, PV Consultant, Berlingen, Switzerland
S. Siebentritt, University of Luxembourg, Belvaux, Luxembourg
M. Topic, University of Ljubljana, Slovenia
S.C. Veenstra, ECN, Eindhoven, Netherlands
W. Wischmann, ZSW, Stuttgart, Germany

C. Ballif, EPFL&CSEM, Neuchâtel, Switzerland
R. Birkmire, University of Delaware, Newark, USA
D. Cheyns, imec, Leuven, Belgium
S. Cros, CEA, Le Bourget du Lac, France
T. Dalibor, AVANCIS, Munich, Germany
P. Delli Veneri, ENEA, Portici, Italy
B. Dimmler, Manz CIGS Technology, Schwäbisch Hall, Germany
K. Durose, Stephenson Institute for Renewable Energy, Liverpool, United Kingdom
S. Gall, HZB, Berlin, Germany
I. Gordon, imec, Leuven, Belgium
S. Huang, UNSW, Sydney, Australia
S.J.C. Irvine, Swansea University, St. Asaph, United Kingdom
R. Klenk, HZB, Berlin, Germany
C.-F. Lin, NTU, Taipei, Taiwan
M.C. Lux-Steiner, HZB, Berlin, Germany
E. Mellikov, Tallinn University of Technology, Tallinn, Estonia
M. Meuris, imec/IMOMEC - ATFPV, Leuven, Belgium
C. Molpeceres, UPM, Madrid, Spain
J. Parisi, University of Oldenburg, Oldenburg, Germany
R. Po, Eni, Novara, Italy
M. Powalla, ZSW, Stuttgart, Germany
A. Romeo, University of Verona, Verona, Italy
V. Sittinger, Fraunhofer IST, Braunschweig, Germany
A. Slaoui, InESS, Strasbourg, France
A.N. Tiwari, EMPA, Dübendorf, Switzerland
A. Virtuani, SUPSI, Canobbio, Switzerland
M. Zeman, Delft University of Technology, Netherlands

Topic 4 – Concentrator and Space Applications

R. Campesato, CESI, Milan, Italy
G. Flamand, imec, Leuven, Belgium
F. Rubio Berenguel, Freiburg, Germany
C. Signorini, ESA-ESTEC, Noordwijk, Netherlands

I. Anton Hernandez, UPM, Madrid, Spain
A.W. Bett, Fraunhofer ISE, Freiburg, Germany
V. Khorenko, AZUR SPACE, Heilbronn, Germany
R. King, Arizona State University, Tempe, USA
G. Siefer, Fraunhofer ISE, Freiburg, Germany
T. Takamoto, SHARP, Nara, Japan
G. Timò, RSE, Piacenza, Italy

Topic 5 – Operations, Performance, Reliability and Sustainability of PV

J. Binder, ZSW, Stuttgart, Germany
M. de Wild-Scholten, SmartGreenScans, Groet, Netherlands
R. Gottschalg, Loughborough University, United Kingdom
U. Jahn, TÜV Rheinland Energy, Köln, Germany
N.M. Pearsall, Northumbria University, Newcastle upon Tyne, United Kingdom
M. Perrin, CEA, Le Bourget du Lac, France
C. Protogeropoulos, Phoenix Solar, Athens, Greece
K. Radouane, EDF EN, Paris La Defense, France
T. Sample, European Commission DG JRC, Ispra, Italy
W.G.J.H.M. van Sark, Utrecht University, Netherlands
K. Wambach, Wambach-Consulting, Aindling, Germany

C. Alonso-Tristán UBU, Burgos, Spain
M. Bächler, PerVorm, Ulm, Germany
T.R. Betts, Loughborough University, United Kingdom
P. Blanc, MINES ParisTech, Paris, France
R. Einhaus, Apollon Solar, Lyon, France
G. Friesen, SUPSI, Canobbio, Switzerland
V. Fthenakis, Columbia University, New York, USA
W.J. Gambogi, DuPont, Wilmington, USA
L. Gomes de Freitas, Federal University of Uberlândia, Brazil
G. Graditi, ENEA, Portici, Italy
M. Grottke, WIP - Renewable Energies, Munich, Germany
Y. Hishikawa, AIST, Tsukuba, Japan
T. Huld, European Commission DG JRC, Ispra, Italy
A.R. Lagunas, CENER, Sarriguren, Spain
O. Mayer, GE Global Research, Garching, Germany
M.A. Muñoz, UPM, Madrid, Spain
H. Nagel, Fraunhofer ISE, Freiburg, Germany
H. Nussbaumer, Zurich University of Applied Sciences, Winterthur, Switzerland
C. Nyman, Soleco, Porvoo, Finland
S. Oktik, Sisecam, Istanbul, Turkey
W. Palitzsch, Loser Chemie, Zwickau, Germany
A.H.M.E. Reinders, University of Twente, Enschede, Netherlands

J. Remund, Meteotest, Bern, Switzerland
J.N. Roy, IIT Kharagpur, Kharagpur, India
M. Schroedter-Homscheidt, DLR, Wessling, Germany
M. Sengupta, NREL, Golden, USA
R. Sinton, Sinton Instruments, Boulder, USA
J. Sutterlueti, Gantner Instruments, Schruns, Austria
S. Tselepis, CRES, Pikermi, Athens, Greece
M. Van Iseghem, EDF R&D, Moret-sur-Loing, France
E. Voroshazi, imec, Leuven, Belgium
A. Wade, First Solar, Berlin, Germany
J. Wohlgemuth, NREL, Golden, USA
M. Zehner, Rosenheim University of Applied Sciences, Germany

Topic 6 – PV Applications and Integration

F.P. Baumgartner, Zurich University of Applied Sciences, Winterthur, Switzerland
F. Bonemazzi, ENEL, Rome, Italy
F. Frontini, SUPSI, Canobbio, Switzerland
P. Malbranche, CEA, Le Bourget du Lac, France
A. Scognamiglio, ENEA, Portici, Italy
X. Vallvé, Trama TecnoAmbiental, Barcelona, Spain
I. Weiss, WIP - Renewable Energies, München, Germany

R. Burkhardt, Energy Depot, Kreuzlingen, Switzerland
S. Caneva, WIP - Renewable Energies, Munich, Germany
I. Da Silva, Strathmore University, Nairobi, Kenya
J.P. de Barros Duarte Fonseca, University of Cape Verde, Mindelo, Cape Verde
B. de Boer, TNO, Eindhoven, Netherlands
H. Fechner, University of Applied Sciences Technikum Wien, Breitenfurt, Austria
I.B. Hagemann, Architekturbüro Hagemann, Aachen, Germany
C. Helmke, Management Consultancy Research, Erfurt, Germany
E. Kaijuka-Okwenje, Makerere University, Kampala, Uganda
G. Mütter, Alternative Energy Solutions, Vienna, Austria
K. Peter, ISC Konstanz, Germany
T. Reijenga, BEAR-iD Sustainable Urban Planners + Architects, Gouda, Netherlands
S. Wittkopf, HSLU, Horw, Switzerland

Topic 7 – PV Economics, Markets and Policies

C. Breyer, Lappeenranta University of Technology, Finland
M. Getsiou, European Commission DG RTD, Brussels, Belgium
G. Masson, Becquerel Institute, Brussels, Belgium
T. Nordmann, TNC Consulting, Feldmeilen, Switzerland
S. Nowak, NET Nowak Energy & Technology, St. Ursen, Switzerland
E. Perezagua, Consultores de Energía Fotovoltaica, Madrid, Spain

I. Kaizuka, RTS, Tokyo, Japan
D. Mayer, MINES ParisTech,Paris, France
P. Mazzucchelli, EUREC,Brussels, Belgium
D. Stickelberger, Swissolar, Zürich, Switzerland
S. Szabó, European Commission DG JRC, Ispra, Italy

SUBJECT INDEX

New Materials and Concepts for Solar Cells and Modules
Sessions 1AP.1, 1AO.1, 1AO.2, 1AO.3, 1BO12, 1CO.9, 1CO.10, 1CO.11, 1BV.5, 1BV.6

Wafer-Based Silicon Solar Cells and Materials Technology
Sessions 2BP.1, 2BO.1, 2BO.2, 2BO.3, 2BO.4, 2BO.9, 2CO.1, 2CO.2, 2CO.3, 2CO.4, 2DO.1, 2DO.2, 2DO.3, 2DO.4, 2DO.16, 2AV.1, 2AV.2, 2AV.3, 2BV.7, 2DV.3

Thin Film Solar Cells and Modules
Sessions 3CP.1, 3AO.4, 3AO.5, 3AO.6, 3BO.5, 3BO.6, 3BO.7, 3BO.8, 3CO.5, 3CO.7, 3CO.8, 3DO.9, 3CV.2, 3CV.4, 3DV.1, 3DV.2

Concentrator and Space Applications
Sessions 4CP.2, 4CO.6, 4CV.1

Operations, Performance, Reliability and Sustainability of PV
Sessions 5DP.1, 5AO.7, 5AO.8, 5AO.9, 5BO.10, 5BO.11, 5CO.13, 5CO.14, 5CO.15, 5CO.16, 5DO.10, 5DO.11, 5DO.12, 5DO.15, 5EO.3, 5BV.1, 5BV.2, 5BV.3, 5BV.4, 5CV.3

PV Applications and Integration
Sessions 6DP.2, 6CO.12, 6DO.5, 6DO.6, 6DO.7, 6DO.8, 6EO.1, 6AV.4, 6AV.5, 6AV.6,

PV Economics, Markets and Policies
Sessions 7EP.1, 7DO.13, 7DO.14, 7EO.2, 7DV.4

Topic Code

1 New Materials and Concepts for Solar Cells and Modules
2 Wafer-Based Silicon Solar Cells and Materials Technology
3 Thin Film Solar Cells and Modules
4 Concentrator and Space Applications
5 Operations, Performance, Reliability and Sustainability of PV
6 PV Applications and Integration
7 PV Economics, Markets and Policies

Session Type

P = Plenary Session
O = Oral Session
V = Visual Session

Day Codes

A = Monday, 20 June 2016
B = Tuesday, 21 June 2016
C = Wednesday, 22 June 2016
D = Thursday, 23 June 2016
E = Friday, 24 June 2016

e.g. 3CP.1 ⇒ 3=Thin Film Solar Cells and Modules, C=Wednesday, P=Plenary Session, 1=Session 1

TABLE OF CONTENT
of EU PVSEC 2016 Proceedings Papers

New Materials and Concepts for Solar Cells and Modules

Visual PRESENTATIONS 1BV.6 New Materials and Concepts for Cells

Wafer-Based Silicon Solar Cells and Materials Technology

Oral PRESENTATIONS 2DO.3 Silicon Solar Cell Characterisation and Modelling (I)

Oral PRESENTATIONS 2DO.4 Silicon Solar Cell Characterisation and Modelling (II)

Oral PRESENTATIONS 2DO.16 Advanced c-Si Solar Cell Architectures

Visual PRESENTATIONS 2AV.2 Silicon Solar Cell Improvements and Innovation (II)

Visual PRESENTATIONS 2AV.3 Silicon Solar Cell Improvements and Innovation (III)

Concentrator and Space Applications

Oral PRESENTATIONS 4CO.6 III-V-based Devices for Terrestrial and Space Applications / Concentrator and Space Systems

Operations, Performance, Reliability and Sustainability of PV

Visual PRESENTATIONS 5BV.2 Operation of PV Systems

Visual PRESENTATIONS 5CV.3 Solar Resource and Forecasting / Sustainability and Recycling

PV Applications and Integration

Plenary SESSION 6DP.2 PV as Part of Our Built Environment: Solutions for Integration into Building Envelops and Energy Systems

Oral PRESENTATIONS 6CO.12 Grid and Energy System Integration (III) - Technology Solutions

Visual PRESENTATIONS 6AV.6 Utility-Scale PV / PV Applications without a Centralised Grid

PV Economics, Markets and Policies

Visual PRESENTATIONS 7DV.4 PV Economics and Markets / PV Global Issues, Policies and Strategies

FOREWORD

The EU PVSEC 2016 in Munich concluded with the strong message that photovoltaic solar energy is here to stay. Research and innovation across the sector, together with its deployment and endless application option, are paving the way to putting PV at the center-stage of sustainable energy systems. 1,742 conference participants from 73 countries participated in this most inspiring platform. It provided a vital forum for information exchange in the field of PV solar energy research, demonstration and applications.

The EU PVSEC 2016 Proceedings give a comprehensive overview of the photovoltaic solar sector, its current status and future prospects in science, research, innovation, development and deployment on more than 5,000 pages. Selection for inclusion in the conference was made by paper review experts and topic organisers (see the listing on pages III-V), to whom we express our sincere gratitude for their comprehensive review work and overall contribution to the success of the conference. In addition to the 658 submitted papers, the proceedings includes 82 presentations (slides) shown during the plenary and oral presentations as well as 269 poster files of the visual presentations. In total this amounts to 1,009 publications!

The Proceedings constitute a powerful tool for targeted and quick information search and retrieval, enabling you to search by keyword, paper title and authors. We are sure that these features will help to simplify the use and exploit the full potential of this extensive source of information.

The Conference Proceedings are also published online. **A DOI code (Digital Object Identifier) has been assigned to each paper. This ensures unequivocal and permanent identification and full citability**. The papers can be viewed and downloaded from the EU PVSEC Proceedings website www.eupvsec-proceedings.com.

We are confident that these Proceedings will play an important role in providing a comprehensive overview of the current actors and activities in the global PV sector and that they will disseminate information on the state-of-the-art of technologies and applications. This can generate further research, add momentum to innovation and promote interest in PV worldwide.

We would like to cordially thank all authors and participants of the EU PVSEC 2016 for their contributions and look forward to welcoming you in 2017 at the next EU PVSEC.

The Editors

Joint Research Centre

the European Commission's in-house science service

Serving society
Stimulating innovation
Supporting legislation

EU PVSEC 2016 Highlights

Nigel Taylor
Technical Programme Chair

European Commission

First some stats ...

Participants: 1742

Presentations: 1053 (334 oral, 719 posters)

Parallel Events: 6

European Commission

European Commission

Silicon: strength in depth

European Commission

Q.ANTUM
Champion R&D module results

Champion module **features**

- Use current production tools
- Standard soldering technique
- Same BOM as in standard modules (high quality ARC glass, back sheet)
- Standard 60 cells module size
- No additional light-trapping schemes

- Usage of champion cell technology with module optimized ARC coating
- Half cell technology and splitted junction box
- Minimized intercell distance

- η = 19.5 % (ap)*
- V_{oc} = 41.5 V
- J_{sc} = 9.30 A
- FF = 77.4 %
- P = 301 W (ta)*

Classification	Efficiency (%)	Area (cm²)	V_{oc} (V)	I_{sc} (A)	FF (%)	Test centre (date)	Description
Si (crystalline)	23.8±0.5	11562 (ap)	53.4	6.32*	81.6	AIST (1/16)	Panasonic (72 cells) [11]
Si (multicrystalline)	19.5±0.4	15349 (ap)	41.53	9.299*	77.4	FhG-ISE (12/15)	Hanwha Q Cells (120 cells) [12]

M. Green et al, Prog. Photovolt: Res. Appl. 2016; 24:905–913

Current Status of High-Efficiency Q.Antum Technology with New World Record Module Efficiency of 19.5%

M. Scherff, P. Kowalzik, C. Gerber, K. Duncker, M. Junghänel, C. Fahrland, S. Kunath, S. Hörnlein, M. Schütze, L. Niebergall, B. Klöter & J.W. Müller
Hanwha Q CELLS, Bitterfeld-Wolfen, Germany

European Commission

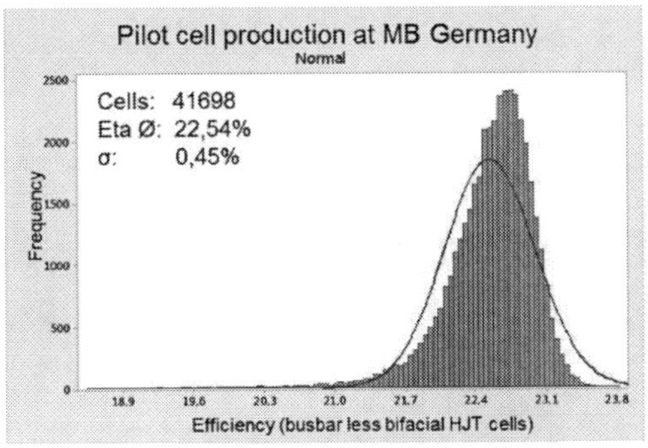

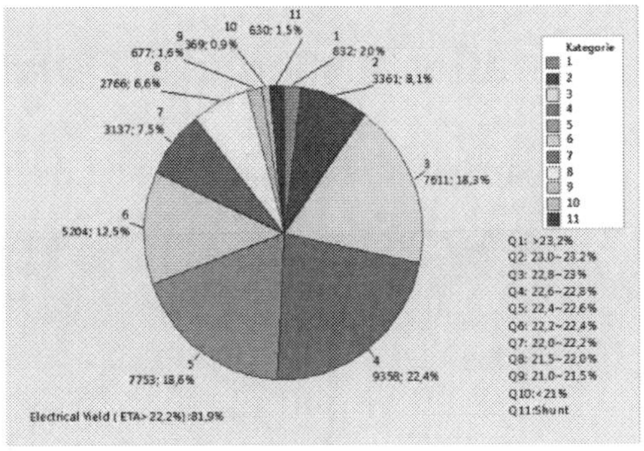

Golden Cell

Status	Cell	Area (cm²)	Eff. (%)	V_{oc} (mV)	FF (%)	J_{sc} (mA/cm²)
09/2015	Cz	239,5	23,3	742,5 ⇑	81,3 ⇑	38,64

- In this pilot production, total 41698 SWCT(busbar less) bifacial cells were processed
- Average ETA reached 22,54%
- 34617 cells (yield of 82%) are over 22,2%
- Record cell reached 23,3% cell efficiency, which is from its high V_{oc} and FF.

Silicon Heterojunction Solar Cells in Meyer Burger's Demo line: Results of Pilot Production on Mass Production Tools

J. Zhao et al

European Commission

Tipping point:

2BO.3 Heterojunction Solar Cell Concepts

Speaker gender balance for the first time!

European Commission

Vibrationally Induced Fluidization

1. CapEx, Full Cost, and Cash Cost will be lower than legacy processes

China: $/Kg on a Polysilicon Plant Basis[1]			
	Genesis™	H-FBR	HC-TCS/Siemens
CapEx	43	65	68
Full Cost	10.25	14.00	18.45
Cash Cost[2]	7.35	9.65	13.90

2. Advantages include
   - Only 1% yield loss to dust
   - Only 1 KwHr/Kg for V-FBR process unit – only 8 KwHr/Kg for entire plant
   - 10X higher reactor productivity
   - Self-seeding and in-situ particle size classification
   - On/off fluidization and "Ok" to temperature cycle
   - Continuous self-regulating product withdrawal

2BP.1.2 New Monosilane Fluid Bed Decomposition Technology for the Production of Solar Quality Silicon Feedstock, M. Dassel, SiTec, Augsburg, Germany

European Commission

Material Limits of Si from State-of-the-Art PLI Techniques

WAFER ELBA

Exact specification of loss mechanisms by ELBA

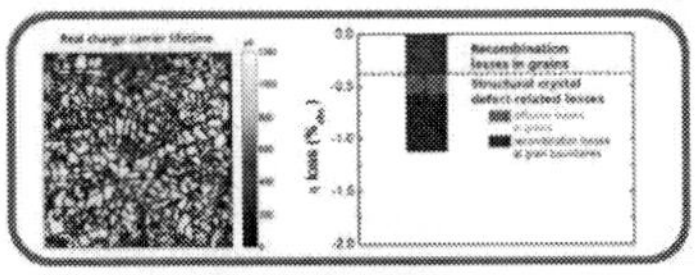

INGOT

Precise PL imaging of τ and [Fe_i] on ingot level

2CO.1.2 Material Limits of Silicon from State-of-the-Art Photoluminescence Imaging Techniques, F. Schindler, J. Giesecke, B. Michl, W. Warta & M.C. Schubert Fraunhofer ISE, Freiburg, Germany

European Commission

Thin film competes on efficiency

European Commission

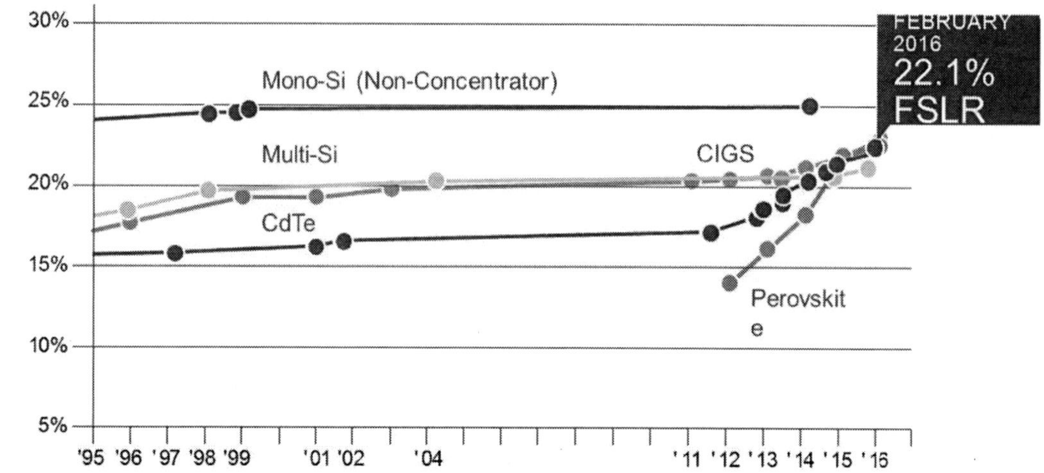

First Solar: current production module average 16.2%, top 18.2% and set to increase

European Commission

AVANCIS' Track Record of Certified Champion Efficiencies

LATEST 2016 EXTERNALLY CERTIFIED CHAMPION MODULE

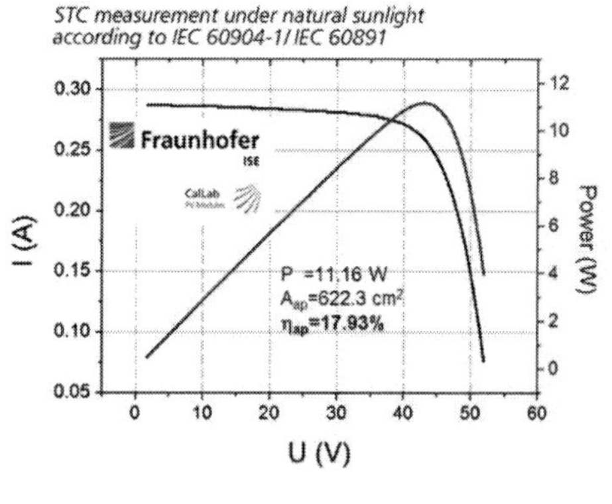

Comparison of champion modules:

	η_ap (%)	Jsc (mA/cm2)	Voc (mV)	FF (%)
2014 Champ (NREL)	16.6	38.5	621	69.4
2016 Champ (ISE)	**17.9**	39.7	627	72.2
Improvement	+7.8%	+3.1%	+1%	+4%

▶ Key improvements 2016 at a glance:

- Optimized cell width and dead area
- Improved TCO layer

▶ Major improvements on FF and Jsc

3BO.8.2 Efficiency Improvement of CIGSSe/Cd-Free Solar Module by Optimized Cell and Interconnect Design

P. Eraerds, C. Schubbert, T. Kwast, M. Grave, F. Braun, M. Algasinger, R. Lechner, T. Dalibor & J. Palm
AVANCIS, Munich, Germany

European Commission

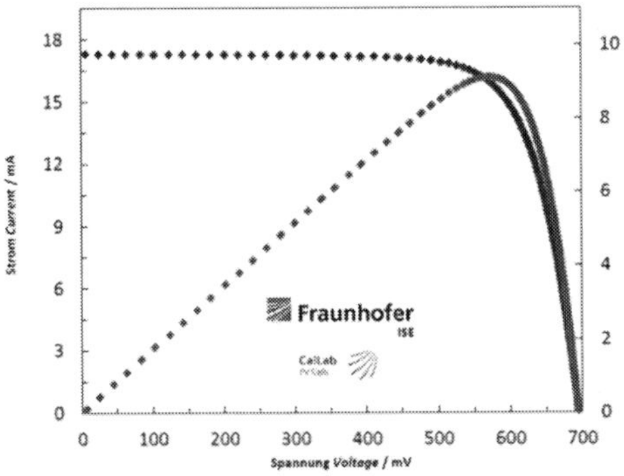

V_{OC}	= (	694.5	±	2.4	) mV
I_{SC} (Ed.2 - 2008)/3/	= (	17.28	±	0.33	) mA
J_{SC}	= (	33.58	±	0.64	) mA/cm²
I_{MPP}	=	15.94			mA
V_{MPP}	=	572.0			mV
P_{MPP}	= (	9.12	±	0.18	) mW
FF	= (	75.98	±	0.49	) %
η	= (	17.72	±	0.37	) %

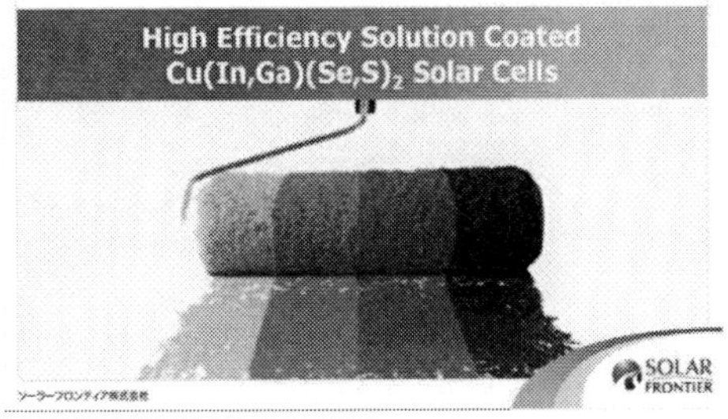

- **Eff 17.72% has been approved by Fraunhofer Institute**
- **Best result for the solution coated CIGSSe cell**
  (Nano solar : Eff 17.1%)

Photo conversion efficiency of 18.7% was achieved with solution coated $Cu(In,Ga)(S,Se)_2$.

3BO.8.3 High Efficiency Solution Coated Cu(In,Ga)(Se,S)2 Thin Film Solar Cells

T. Aramoto & Y. Kawaguchi
Solar Frontier, Atsugi, Japan
Y.-C. Liao, Y. Kikuchi, T. Ohashi, H. Iida & A. Nakamura
Tokyo Ohka Kogyo, Koza-Gun, Japan

European Commission

Lab scale PSC baseline development: mini-modules

- **Perovskite module development: lab scale**
  - Glass/sputtered ITO/P1 laser scribe /e-beam TiO_2/spin coat/Perovskite/spin coat OMeTAD/P2 mechanical scribe/evaporate Au/P3 mechanical scribe
- **Status Solliance**: **16 cm²** mini-module performance of **14,2%**

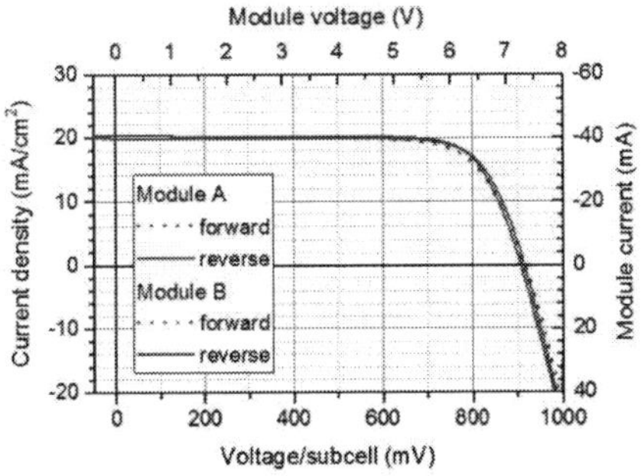

For further information:
3CO.7.5
Loss Analysis and Optimization for High Efficiency Perovskite Photovoltaic Modules
R. Gehlhaar, T. Merckx, C. Masse de la Huerta, L. Rakocevic, W. Qiu & M. Jaysankar
imec, Leuven, Belgium

Towards Roll-to-Roll Manufacturing of Perovskite Based PV Modules

R. Andriessen & P. Poodt, TNO, Eindhoven, Netherlands, T. Aernouts, imec, Leuven, Belgium. S. Veenstra, ECN, Eindhoven, Netherlands, R. Janssen & A. Createore, TU/e, Eindhoven, Netherlands, D. Vanderzande, University of Hasselt, Diepenbeek, Belgium T. Kirchartz Forschungszentrum Jülich, Germany

European Commission

Computational Screening of A_2MX_6 Compounds

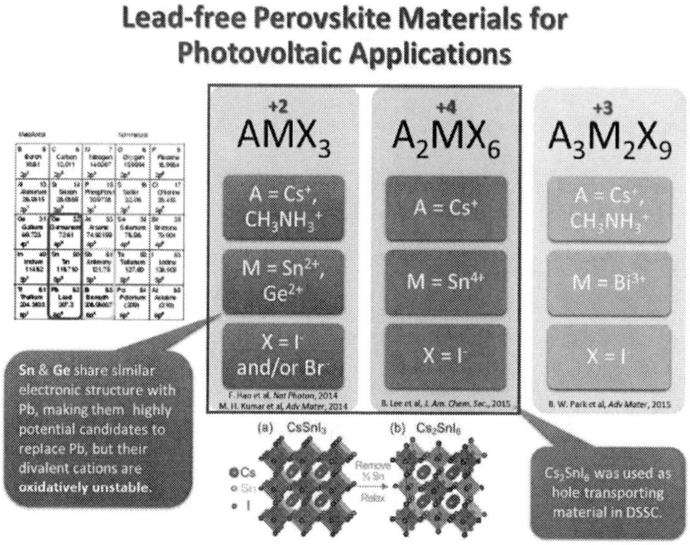

Tin(IV)-based Iodide Perovskites A2SnI6 (A = Rb+, Cs+, MA+, FA+)
Y. Chen, T. Krishnamoorthy, C. Yao, T. Baikie, S. Dharani, J.H. Fu, S.T. Chien, M. Asta, N. Mathews, L.H. Wong & S.G. Mhaisalkar
Nanyang Technological University, Singapore, Singapore & University of California, Berkeley, USA

European Commission

Concentrators:

new paradigms

European Commission

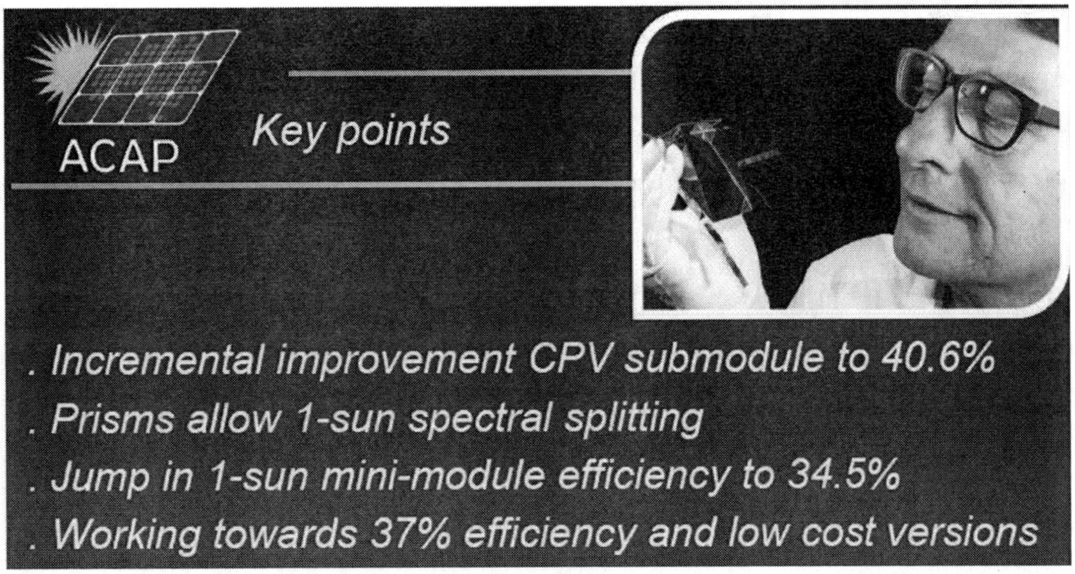

1AP.1.1 37% Efficient One-Sun Minimodule and over 40% Efficient Concentrator Submodules, M.A. Green, M.J. Keevers, B. Concha-Ramon & J. Jiang, UNSW, Sydney, Australia, P.J. Verlinden, Y. Yang & X. Zhang, Trina Solar, Changzhou, China

European Commission

4CO.6.4 Luminescent Solar Noise Barrier – Large Scale Testing and Modeling L.H. Slooff ECN, Petten, Netherlands S. Verkuilen Heijmans Wegen, Rosmalen, Netherlands M.M. de Jong & M.N. van den Donker SEAC, Eindhoven, Netherlands M. Kanellis & M.G. Debije Eindhoven University of Technology, Netherlands

European Commission

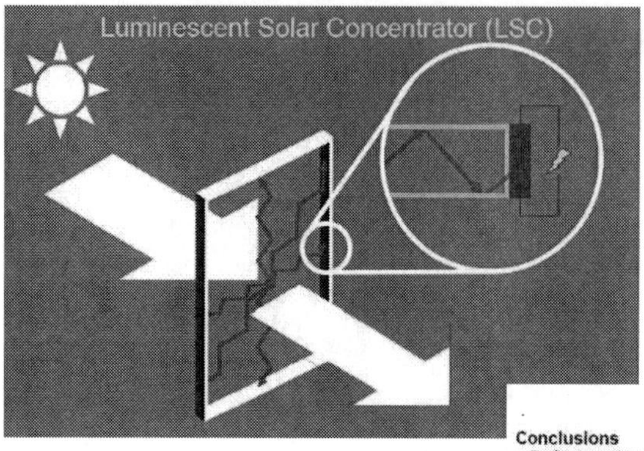

Conclusions
- Tm^{2+} doped halides discovered
  - Colorless
  - No self-absorption
  - >60% sunlight absorption
  - Expected ~15% power efficiency
- Made as both powders and thin-films
- Scattering still a challenge for BIPV

Outlook
- Not a lot known/understood (only ~4 major studies)*
  - Fundamental luminescence properties
  - Optimizing thin film growth
- Realize a lab-scale Tm^{2+} LSC prototype

1AO.2.4 Efficient Luminescent Solar Concentrators Based on Self-Absorption Free, Tm2+ Doped Halides

O.M. ten Kate, M. ten Kate, M. De Jong, O.M. ten Kate & E. van der Kolk
Delft University of Technology, Netherlands
K.W. Krämer
University of Berne, Switzerland

European Commission

Operation

European Commission

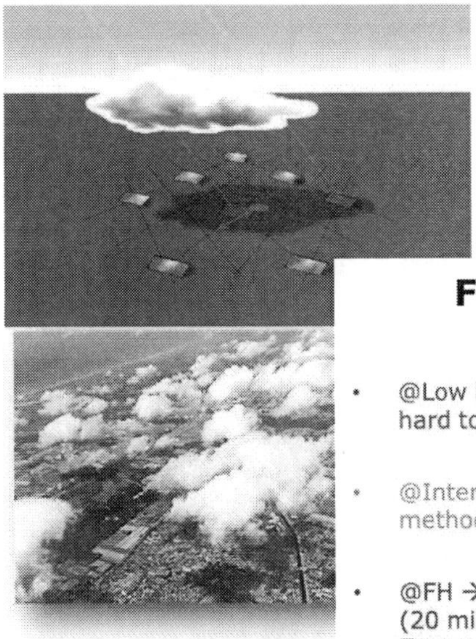

Peer-to-Peer (P2P) Forecasting:

Forecast Skill vs. Forecast Horizon

- @Low FH: **persistence** hard to beat
- @Intermediate FH: P2P method shows its **value**
- @FH → T_{Obs} (20 min. in this example.): **Fewer** to **no** available periph. $k_j^*(t-\tau_{ij})$ timeseries for P2P FC

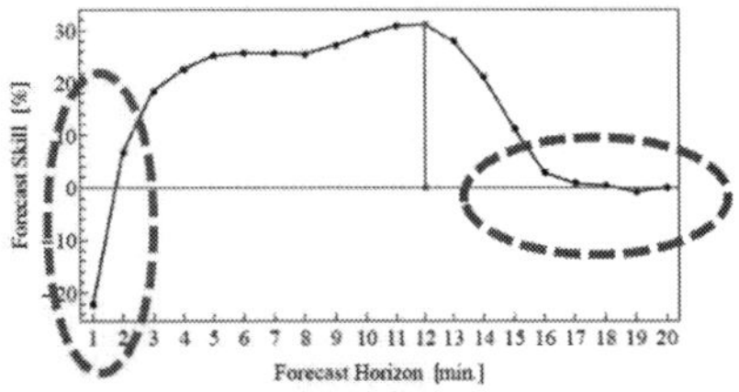

Dependence of Peer-to-Peer Solar Forecast Skill on Irradiance Variability

B. Elsinga & W.G.J.H.M. van Sark
Utrecht University, Netherlands

European Commission

Analysis of Failure Occurrence

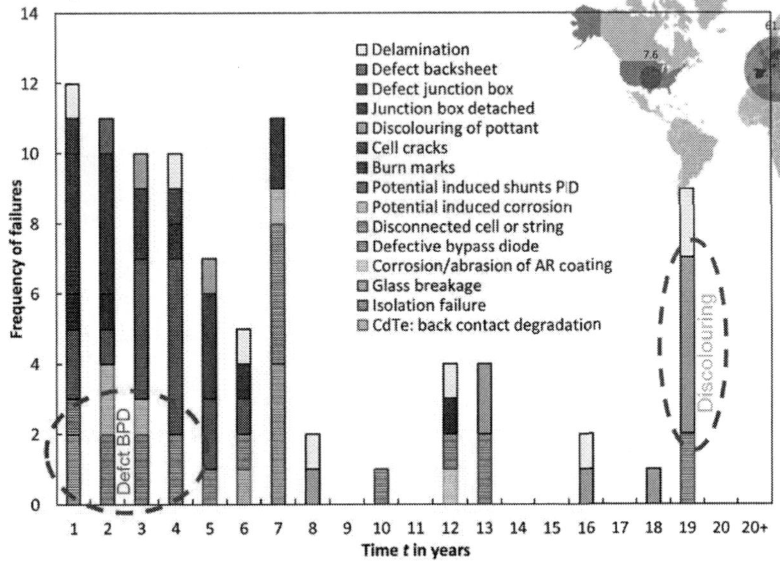

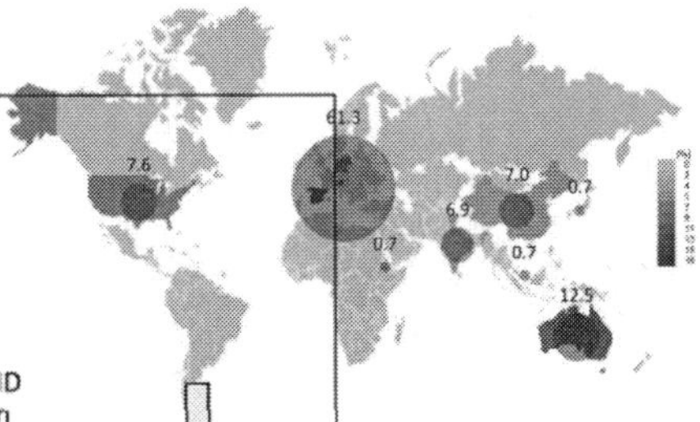

5DP.1.2 Mean Degradation Rates in PV Systems for Various Kinds of PV Module Failures

M. Köntges & S. Altmann
ISFH, Emmerthal, Germany
U. Jahn
TÜV Rheinland, Cologne, Germany

European Commission

Implementation and Monitoring

Detailed measurement

2015/16

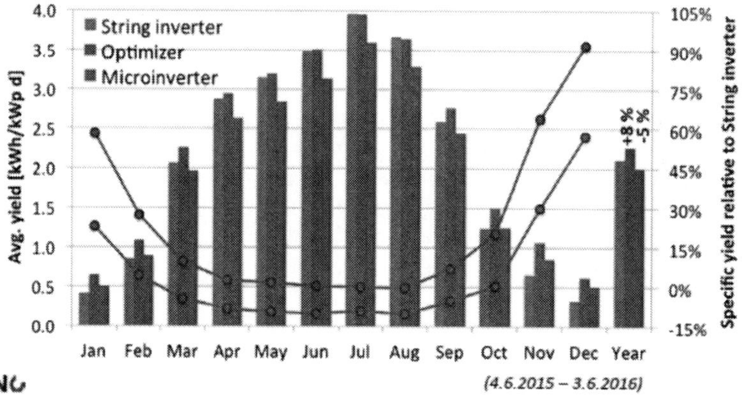

TNC

12

5AO.9.4 Performance of Recent Inverter Systems under Partial Shading Conditions

R. Lingel, T. Nordmann & T. Vontobel
TNC Consulting, Feldmeilen, Switzerland

European Commission

Energy Solutions

European Commission

IS "BIPV" ALREADY "VINTAGE"...?

SO... BIPV "MAINSTREAM POTENTIAL"

.TECHNOLOGICAL INTEGRATION | PV AS A PART OF A WHOLE ENERGY SOLUTION SET
.POLICY DRIVERS ARE CRUCIAL

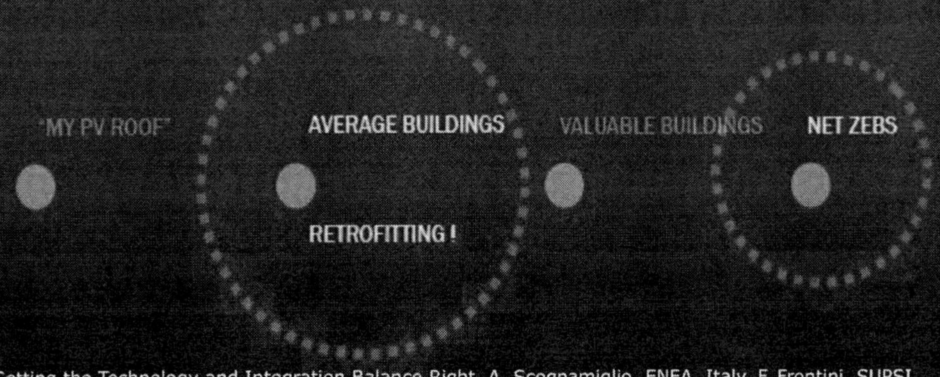

Kinetic envelope for heating cooling and light comfort + PV generation

Case Study

- Single Zone Office
- Zurich Weather Data
- 50 Panels in a Single Cluster
- Heating COP: 4
- Cooling COP: 3
- Lighting Load: 11.74 W/m^2

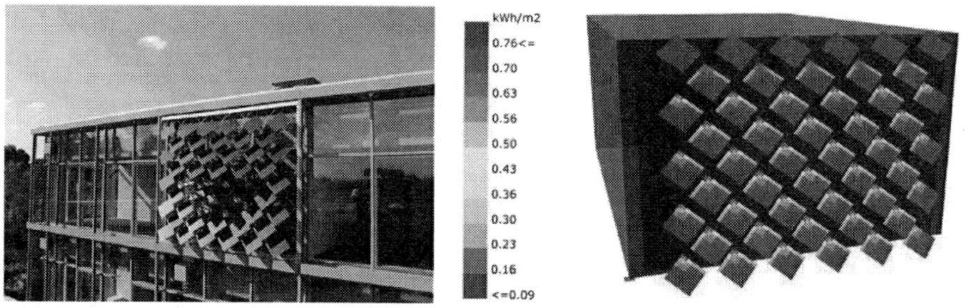

6DO.8.5 Energy Performance of PV Modules as Adaptive Building Shading Systems

J. Jayathissa, J. Schmidli, J. Hofer & A. Schlueter
ETH Zurich, Switzerland

European Commission

New metrics for PV + storage performance!

Definition of the Storage Performance Index (SPI)

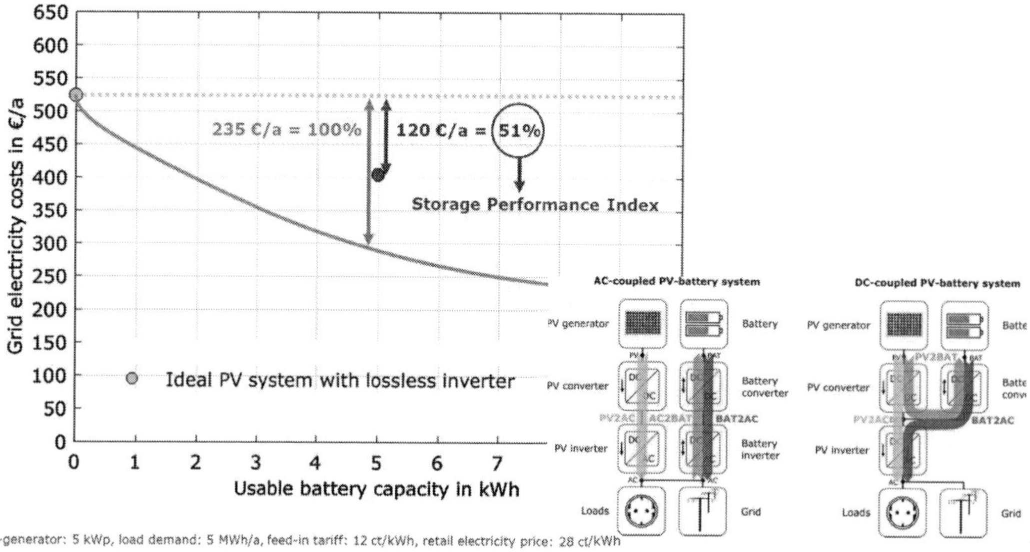

PV-generator: 5 kWp, load demand: 5 MWh/a, feed-in tariff: 12 ct/kWh, retail electricity price: 28 ct/kWh

Emerging Performance Issues of Photovoltaic Battery Systems, Johannes Weniger, Tjarko Tjaden, Joseph Bergner, Volker Quaschning, HTW Berlin - University of Applied Sciences, Germany

European Commission

Thanks!

European Commission

Start working on excellent abstracts for EU PVSEC 2017!

European Commission

We thank the EU PVSEC 2016 Sponsors

INVESTIGATION OF THE EVA DEGRADATION AND PREDICTION OF RELIABILITY BY THE RAMAN SPECTROSCOPY

Mohammad Aminul Islam[1], Kazuki Noguchi[1], Hidenari Nakahama[2], Yasuaki Ishikawa[1]
[1]Nara Inst. Sci. & Tech., 8916-5, Takayama-cho,Ikoma, Nara 630-0192, Japan
[2]Nisshinbo Mechatronics Inc., Miai-cho, Okazaki, Aichi, 444-8560, Japan.

ABSTRACT: Reliability study of photovoltaic modules is a most important term in determining the economical efficiency of the energy generation business. The investigation of the degradation on a molecular level could give information about the degradation process even before the electrical properties of the module are changed, is important for the reliability of PV-modules. In this study, the more dedicated analytical technique of Raman Spectroscopy was applied to qualitatively determine the performance degradation of the PV modules through the investigation of de-acetylation of ethylene-vinyl acetate (EVA). The IV characteristics, Electroluminescence (EL) and Dark Lock-in Thermography (DLIT) imaging, and IR spectroscopy have also been investigated in the relevance of Raman Spectroscopy. The local distribution of desorbed acetic acid correlated with the Raman fluorescence has been determined from the pressure cooker test and from the dipping of the module in water-acetic acid solution. The correlation of Raman fluorescence with the acetic acid desorption, years in service and loss of performance have also been investigated for the several numbers of field aged modules. It was found that the module reliability could be predicted from the correlation of Raman spectroscopy analysis even though before exposed in the field.
Keywords: Raman Spectroscopy, IR Spectroscopy, PCT, Field Aged Modules, Acetic Acid, EVA Degradation

1 INTRODUCTION

Understand the degradation mechanism of PV module involved in the encapsulant is definitely very important, especially de-acetylation of EVA for predicting the lifetime of PV module because the central cause of PV module power degradation is optical and chemical change of EVA [1]. Various degradation mechanisms related to EVA have been investigated over the last few years, however, the aging process under the influence of heat, humidity and UV radiation and its effect on module performance and reliability is not fully understood yet [2]. Therefore, appropriate sensitive methodologies for the correlation of performance deviations at outdoor exposure with the results of accelerated life testing finds great demand.

A long-term stability of PV modules with a maximum total loss in module performance of 20% over a 20 year period is greatly demanded in energy generation business [3]. However, the present results are showing the opposite situation as the module efficiency is observed to deteriorate unexpectedly in the field. The stability of a module is deteriorated by the different kinds of the effect such as encapsulant degradation [4], PID [5] and other natural causes [6]. Among these affect, the de-acetylation of EVA considered as the leading cause of PV module performance degradation, which also influenced the other degradation causes [4]. Therefore understanding the degradation of EVA can help greatly to predict the reliability of the PV modules. The aging behavior and degradation mechanism of EVA as PV encapsulation material are well described in the literature [3, 7-9].

Recently, ion chromatograph (IC), Hot Water Extraction Method (HWEM), Infra-Red (IR), and Thermal Desorption GC-MS were used to evaluate the amount of free acetic acid desorbed in EVA [9-11]. The above processes are destructive, not suitable for an in-line investigation of PV modules and cannot measure early degradation before becoming visible or electrically shutdown. In this study, Raman Spectroscopy as the dedicated analytical technique was applied to qualitatively determine the performance degradation of the PV modules with field exposed and lab accelerated tested. Raman Probe Spectroscopy study is found as a potential technique for analyzing the degradation behavior of encapsulates materials in PV module. It is a non-destructive and quick testing method which is desirable to follow material changes during the aging of PV modules. The correlation in between acetic acid, Raman fluorescence intensity, module aging time, and module performance losses have been made to predict the reliability of the module. Finally, a model test of Raman spectroscopy has been done for two set of the module and their reliability has been compared. We have also attempted to predict the distribution of acetic acid forms on the EVA encapsulant by pressure cooker test (PCT) and module immersed in hot water and acetic acid solution.

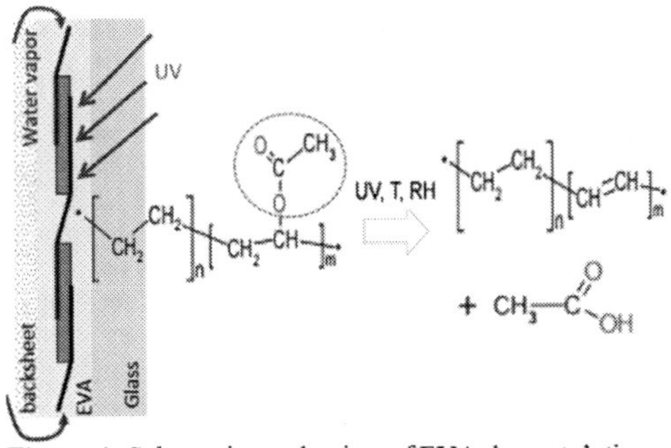

Figure 1: Schematic mechanism of EVA de-acetylation

2 RAMAN STUDY

When a photon of an incident light beam collides with molecules within a material, most of the light is scattered elastically (Rayleigh scattering). A small fraction of the scattered photons (approximately 1 in 10 million) transfer part of their energy to the material, resulting in a molecular vibration and/or excitation. The energy of the in elastically scattered photon, E_R (Stokes Raman scattering) is thereby lowered by exactly the energy of the molecule vibration, ΔE_{vib} compared to the initial energy of light:

$$E_R = h\nu_o - \Delta E_{vib} \quad (1)$$

The electromagnetic wave $\vec{E}$ associated with radiation of frequency ν_o and time t can be written as:

$$\vec{E} = \vec{E}_0 \cos 2\pi\nu_o t \quad (2)$$

If, $\vec{E}$ interact with the electron cloud of a molecule and induce a dipole moment $\vec{\mu}_{ind}$, then:

$$\vec{\mu}_{ind} = \alpha\vec{E} = \alpha\, \vec{E}_0 \cos 2\pi\nu_o t \quad (3)$$

The intensity of the Raman scattering is proportional to the change in polarization during the vibration [12]:

$$I_{Raman} = (\delta\alpha/\delta q)^2 \quad (4)$$

$$I_{Raman} \propto (1/\lambda^4) \quad (5)$$

Since Raman intensities are inversely proportional to the 4th power of the incident wavelength as shown in eqn. 5, the usage of short excitation wavelengths is reasonable to obtain intense Raman peaks. The Table 1 shows the Raman frequency in non-aged EVA and their assignment [12].

Table I: Raman frequencies (cm^{-1}) of EVA and their assignment

Raman shift (cm^{-1})	Group	Intensity
629	Acetates	Weak
1068	Ethylene unit	Medium
1086	Acetates	Medium
1102	Acetates	Medium
1126	Ethylene unit	Medium
1295	Ethylene unit	Medium
1340	Acetates and Ethylene unit	Weak
1362	Ethylene unit	Weak
1438	Acetates and Ethylene unit	Strong
1461	Ethylene unit	Medium
1635	C=C stretching vibration	Very weak
1724	Acetates	Weak
2832	Acetates and Ethylene unit	Very strong
2861	Acetates	Very strong
2880	Acetates and Ethylene unit	Strong
2895	Acetates and Ethylene unit	Strong
2913	Acetates and Ethylene unit	Very strong

3 METHODOLOGY

The analyzed modules used in this study were crystalline Si-PV commercialized by different Japanese manufacturer. The investigated modules have a similar design, consisting of solar-glass glazing, EVA encapsulation, monocrystalline (mc) Si-cells and a TPT back sheet. Also, measurements were carried out on a non-aged reference module and the lab accelerated tested module using of similar solar cells. The lab accelerated test has been done in two ways, (i) pressure cooker test (PCT) (100% RH) using temperature of 130 °C and 2 atm. pressures for 168 hours, and (ii) dipping tabbed cell in water and acetic acid solution (4%) for 72 hours, 168 hours and 720 hours of duration, respectively. Raman measurements as well as electroluminescence (EL) and lock in thermography image have been carried by in line system developed by Nisshinbo Mechatronic Inc. as shown in Fig. 2. In Raman study, the 532 nm line of a beam doubled Nd:YAG laser with an excitation power of 16.6 mW has been used as the excitation source. The EL image has been taken in forward bias with a current of 30mA/cm^2 by using the Si-CCD camera at frame/512 pixels.

However, in support of Raman measurement, mid-IR analysis has also been carried out at the same point of interest. Power loss has been investigated through Current–voltage (I–V) measurement using standard system under AM 1.5.

Figure 2: EL/Thermo/Raman Spectroscopy 3-in-1 System Developed by NISSHINBO Mechatronics Inc.

4 RESULTS AND DISCUSSION

The formation of acetic acid has been observed from the combo pressure cooker test (PCT) (100%RH) using temperature of 130 °C and 2 atm. pressures for 168 hours. Fig. 3(a) and 3(b) show the results from Raman spectroscopy and IR spectroscopy of the sample corresponding to the different positions indicated at the inset of Fig. 3(a). Esters pyrolysis mechanism is observed in the Fig. 3(b) and the esters in EVA decreases while carbonyl increase towards the edge indicating that the higher EVA degradation occurred at the edge and/or EVA degradation is started from the edge of the module. As Raman fluorescence intensities are varied followed by Ester pyrolysis, it could be noticed that the fluorescence intensity can indicate the degradation of EVA. The detail on the variation Raman fluorescence by the quantitative values of acetic acid is presented in the next sections.

To investigate the distribution and effect of acetic acid, six of four cell mini-modules have been intentionally immersed in warm water and 4% of acetic acid solution for 72 hours, 168 hours and 720 hours of duration, respectively. The photographic image of warm water immersed module (Fig. 4(a), immersed for 168 hrs.) shows that the color is changed at the edge of the module which moves to the center with the time. These phenomena indicated that the edge is affected first by the humidity, which supports the results shown in Fig. 3. Likewise, Fig. 4(b) and 4(c) shows the clear effect of acetic acid on the cell surface. From the photographic and

EL images, there is no significant change was observed in the module immersed for 68 hrs. (not shown in here), however, it is clearly observed that the Ag finger is partially dissolved in the module immersed for 172 hrs. and completely dissolved in the module immersed for 720 hrs. as an effect of acetic acid.

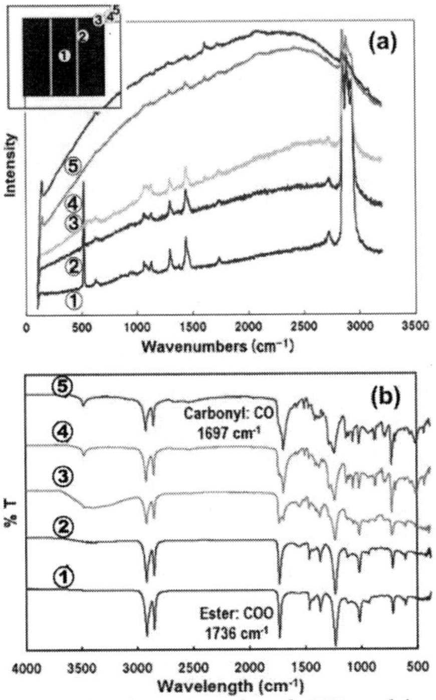

Figure 3: Experimental results of PCT module, (a) Raman spectra and (b) IR-spectra (inset: point of interest for examining the IR and Raman spectroscopy

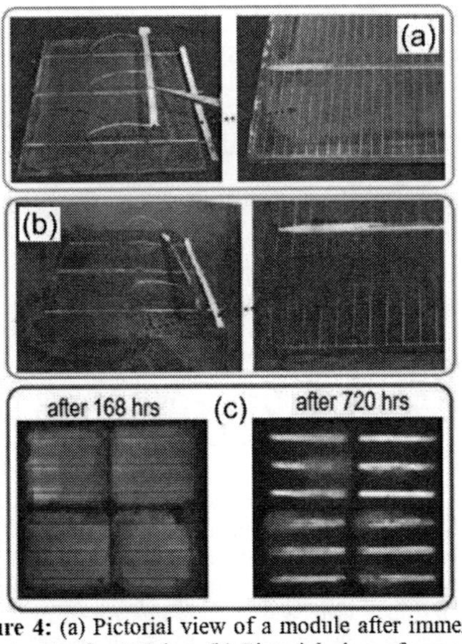

Figure 4: (a) Pictorial view of a module after immersed in hot water for 168 hrs. (b) Pictorial view of a module after immersed in water-acetic acid solution for 168 hrs. (c) EL image of the module of the module immersed in 4% acetic acid (effect of acetic acid is clearly observed in the figures)

The results shown in Fig. 5 correlates the aging time, performance drop and Raman fluorescence intensity. To compare the aging effect, fluorescence intensity at 1724 cm^{-1} is evaluated for all three modules and has found to increase with the increase of field exposed time and/or performance drop ratio. The module that unused for around 3 years shows a typical Raman spectrum of EVA where the peaks are well matched with the assigned peaks in Table 1. However, the Raman fluorescence background becomes very intense for the field aged modules. Comparing with Fig. 3, it could easily realize the amount of acetic acid formation on the modules. The performance drop and effect of acetic acid has also seen in EL (not shown in here) and thermography image (Fig. 6) of the modules.

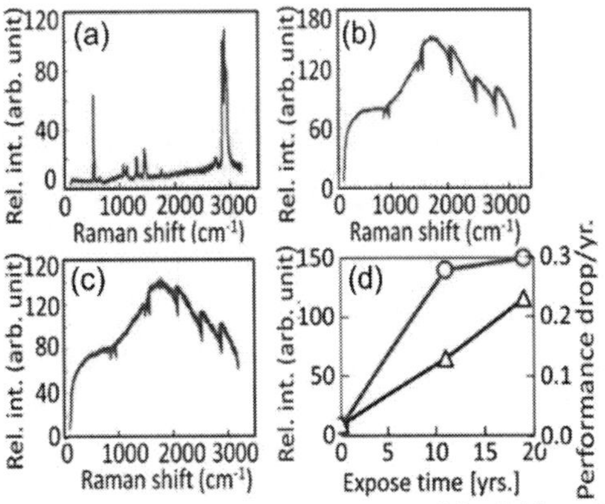

Figure 5: Raman fluorescence spectra of the field aged module, (a) not used but remaining in room for around 3 years, (b) used 11 years in the field, (c) used 19 years in the field, and (d) exposed time vs. fluorescence intensity and module's performance drop

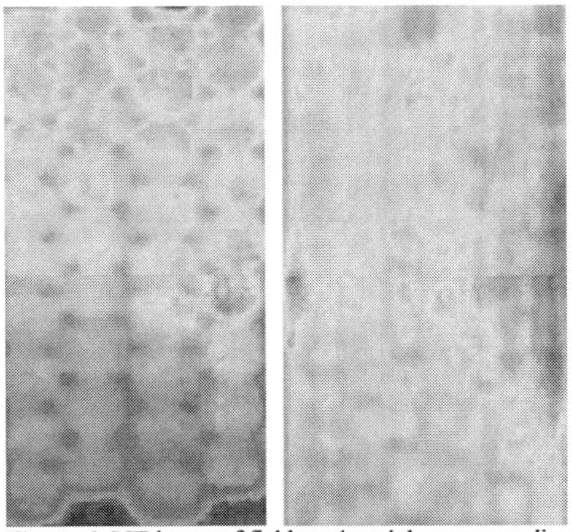

Figure 6: LIT image of field aged module corresponding to the Figure 5, (left) used for 11 years in the field, and (right) used for 19 years in the field

Several no. of field aged modules have been investigated again in contrast to Raman fluorescence intensity ratio, IR absorbance, acetic acid formation, years in service and loss of performance. Raman fluorescence intensity ratio has been calculated as dividing the Raman fluorescence intensity at wavenumber 1724 cm^{-1} by intensity at wavenumber 2861

cm^{-1}. Fig. 7(a) shows the amount of desorbed acetic acid versus fluorescent intensity ratio in three different types and different aged module manufactured in 1993 (Mfg. A), 1989 (Mfg. B) and 2011 (2011), respectively. From the trend line, it is seen that the Raman fluorescence increased followed by the amount of acetic acid. It is also found that the higher amount of acetic acid is desorbed at the edge which depicts the result found in PCT as shown in Fig. 2.

Fig. 7(b) shows the IR absorbance vs. Raman fluorescence intensity ratio found in the PCT test. It is clearly seen that both are following a linear relation. The amount of acetic acid has been determined by the IR spectroscopy analysis. The detail process will be explained in the extended version of the manuscript. Power losses of several groups of PV Modules (made in different years) with the different field exposed time correlated with the Raman fluorescence intensity ratio has been presented in Fig. 8(a). Comparing the result of a module with its group members, it is seen that higher Raman fluorescence intensity noticed the higher power loss in the module. However, every group of module shows different amount of power loss in same fluorescence intensity and vice versa. To get the characteristics of a module with Raman fluorescence intensity, six modules of same characteristics with different service lifetime has been investigated as shown in Fig. 8(b). The result shows that fluorescence intensity ratio increases along with years of service except for one module.

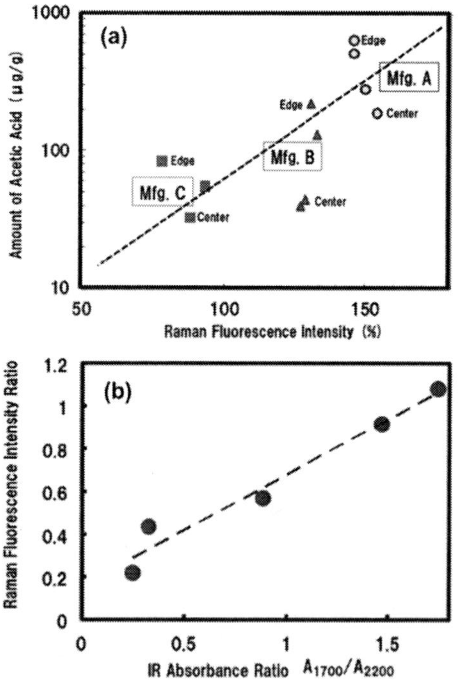

Figure 7: (a) Amount of desorbed acetic acid vs. fluorescent intensity ratio in three different types and aged of module manufactured in 1993 (Mfg. A), 1989 (Mfg. B) and 2011 (2011), respectively, (b) relation between IR intensity and Raman intensity found in the accelerated tested (PCT) module

The results shown in Fig. 7 and Fig. 8 indicating that any type of module has a direct relation with the Raman fluorescence intensity which increases with the increase of power loss as well as service lifetime. Accordingly, it could be concluded that Raman fluorescence test could be the indispensable technique to predict the performance degradation as well as the reliability of the primary module. From the above prediction, a model acceleration test has been employed for two sets of modules as shown in Fig. 8(c). These two different types of module with similar aging time has been investigated as a function of Raman Fl. Intensity vs. performance degradation. The predicted degradation for set A has found around 10%/yr., where for B is around 17%/yr. From the result, it is expected that the modules of set A are more reliable than modules of set B.

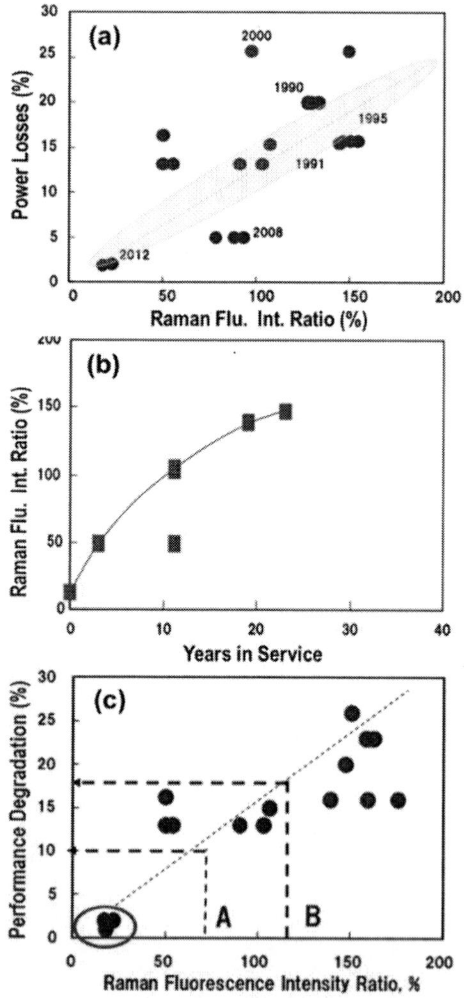

Figure 8: (a) Relation between power loss and Raman fluorescence intensity ratio for group of different aged module, (b) relation between service lifetime and Fluorescence intensity ratio for a similar type of module, (c) relation between Raman fluorescence intensity ratio and performance degradation in lab accelerated tested modules of two different types (the results in circle for the fresh module)

5 CONCLUSION

The aims of this work were to analysis and understand the EVA degradation mechanism linking with the field expose and lab tested modules. Therefore, a detail investigation considering the number of points of view, such as power loss, years in service, manufacturing quality; acetic acid formation and distribution through the EVA degradation mechanism has been conducted in this study. Pressure cooker test, water and acetic acid dipping methods show that the EVA degradation is started from the edge of a module; and higher amount of acetic acid is existed at the module edge. However, the sub-contact fingers are dissolved as an effect of acetic acid, as an effect, the series resistance increase and current collection from the module is reduced. The field aged module shows intense IR and Raman spectra where both are increased linearly with the exposed time. The lab tested and field aged module shows intense Raman florescence spectra where the intensity is varied proportionally with the test or exposer time, amount of acetic acid and module power loss. The relation of Raman fluorescence intensity ratio with the IR absorbance, acetic acid formation, years in service and loss of performance indicates that Raman fluorescence study could be a non-destructive potential technique to predict the reliability and analyze the degradation behavior of encapsulates materials of the PV module.

Acknowledgements
This work was supported by The New Energy and Industrial Technology Development Organization (NEDO), JAPAN.

6 REFERENCES

[1] S. R. Kurtz, John Wohlgemuth, Michael Kempe, Nick Bosco, Peter Hacke, Dirk Jordan, D. Miller, National Renewable Energy Laboratory, 2014.
[2] C. Peike, T. Kaltenbach, K.-A. Weiß, M. Koehl, Solar Energy Mat. & Sol. Cells 95 (2011) 1686.
[3] A. W. Czanderna, F. J. Pern, Solar Energy Materials & Solar Cells 43 (1996).
[4] S. Jonai, K. Hara, Y. Tsutsui, H. Nakahama, & A. Masuda, Japanese Journal of Applied Physics 54(8S1) (2015) 08KG01.
[5] V. Naumanna, C. Hagendorfa, S. Grossera, M. Wernera, J. Bagdahn, Energy Procedia 27 (2007) 1.
[6] G. Oreski, G. M. Wallner, Solar Energy 79 (2005) 612.
[7] B. A. Sultan, E. So¨rvik, Journal of Applied Polymer Science 43 (1991) 1737.
[8] P. Klemchuk, M. Ezrin, G. Lavigne, W. Hollex, J. Gallica, S. Agro, Polymer Degradation and Stability 55 (1997) 347.
[9] M. D. Kempe, G. Jorgensen, K. M. Terwilliger, T. McMahon, C. E. Kennedy, T. T. Borek, Solar Energy Materials and Solar Cells 91 (2007) 315.
[10]Zhao Wenwei, Zhong Xiaoguang, Yu Li, Zhang Yuefang, Sun Jiazhen, Polymer 35 (1995) 3348.
[11]T. Shioda, NREL PV Module Reliability Workshop, 2012.
[12]C. Peike, T. Kaltenbach, K.-A. Weiß, M. Koehl, Solar Energy Materials & Solar Cells 95 (2011) 1686.

DIRECT EVIDENCE FOR HOT-CELL-INDUCED MODIFICATIONS IN PV MODULE ENCAPSULANTS

Christian Camus*[1], Claudia Buerhop-Lutz[1], Simon Wrana[1], Jens Adams[1], Tobias Pickel[1], Hans Scheuerpflug[1], Cornelia Zetzmann[2], Enno Malguth[3], Jens Hauch[1] and Christoph J. Brabec[1,4]
[1] Bavarian Center for Applied Energy Research, Haberstraße 2a, 91058 Erlangen, Germany
[2] Rauschert GmbH, Bahnhofstraße 1, 96332 Pressig, GmbH
[3]LayTec AG, Seesener Str. 10-13, 10709 Berlin, Germany
[4] Materials for Electronics and Energy Technology (i-MEET), Friedrich-Alexander-Universität Erlangen-Nürnberg, Martensstraße 7, 91058 Erlangen, Germany
e-mail: christian.camus@zae-bayern.de

ABSTRACT: In this paper, the influence of the heat dissipated by a broken cell in a photovoltaic module on the encapsulation properties is investigated. Therefore, a combined study based on drone-based infrared thermography, electroluminescence and indentation tests of the mechanical properties of the encapsulant is presented. For the first time, direct evidence is given that the viscoelastic properties of the encapsulant are shifted towards increased stiffness and brittleness by the long-term presence of hot cells. Therefore, the dissipated heat may lead to a local damage of the encapsulant on the long-term, which in turn may trigger the degradation of the module due to insufficient encapsulation.
Keywords: Encapsulation, Reliability, Degradation

1 INTRODUCTION

It is well known that broken cells in crystalline silicon photovoltaic (PV) modules can lead to local heating (hot cells) in the modules. Besides the reduction of the output power caused by the broken cell it is also known that local heat dissipation can yield detrimental burnings leading to a complete module failure. However, so far little is known about the influence of less severe hot cells on the long-term stability of the module.

Whilst defect-free crystalline silicon PV modules usually exhibit module temperatures in the range of 40-50°C for a sunny day in Bavaria defects such as broken or cracked wafers can easily lead to significantly higher temperatures of 70°C or even more [1]. Note that for regions with higher solar irradiance also higher temperatures will be reached.

As most PV modules are encapsulated using a thermal process, in which the solar cells are laminated by two sheets of Ethylene-Vinyl Acetate Copolymer (EVA) foil between the cover glass and a polymeric backsheet at temperatures of $T \approx 120°C$ or above [2] it can be expected that the elevated temperatures in operating PV modules will also affect the encapsulation properties to some extent when approaching this temperature regime.

The purpose of the work presented in this paper is to assess the long-term influence of the presence of hot cells in crystalline silicon PV modules beyond their direct electrical influence on the module output.

2 TECHNICAL DETAILS

This work aims at understanding the effect of the presence of hot cells on the encapsulant properties. Therefore, several experimental methods are combined in order to investigate the local encapsulant properties in the vicinity of hot cells. Firstly, drone-based infrared (IR) thermography (Fig. 1) was used to investigate a large PV plant and to identify a suitable module with a single hot cell for further investigations. The drone as well as the respective IR setup is described elsewhere [3]. This approach ensured that the module was known to be operated for a long time under conditions in which the hot cell indeed exhibited an elevated temperature. This aspect is of special importance as the actual temperature of hot cells depends on the configuration of the particular string in which they are operated [4]. Therefore, lab IR thermography measurements of isolated modules may be deceiving.

a)
b)

Figure 1: a) Drone-based IR-thermography setup used in this work [1, 3]; b) IR thermography image of the PV plant investigated. The module selected for this study is marked by the white box and shown on larger scale.

Once a module was identified it was taken out of the plant and characterized by electroluminescence (EL) using a silicon detector. Here, a broken cell could be identified as the root-cause for the hot cell. In a next step, the distribution of the viscoelastic properties of the EVA/backsheet stack was investigated by means of an indentation measurement (LayTec X Link, [5]). In this measurement, a quantity called LXM is derived, which is directly correlated to the dynamic modulus $E^* = E' + iE''$ of the material probed by indentation, whilst being heated locally to beyond the melting point of the EVA. LXM scales with increasing storage modulus E' (elastic properties) and with decreasing loss modulus E'' (viscous properties). It could be shown that the hot cell led to a local change of the viscoelastic properties of the encapsulant shifting its properties towards higher LXM values, i.e. to higher stiffness and brittleness. In general, LXM values can be converted to gel content values based on an empirical correlation for a given EVA / backsheet stack (e.g. in module production lines). However, as the module of this study was taken from the field and has already been operated no systematic correlation was feasible within this study. Therefore, only LXM values are stated which still demonstrate the relative material changes within the module area.

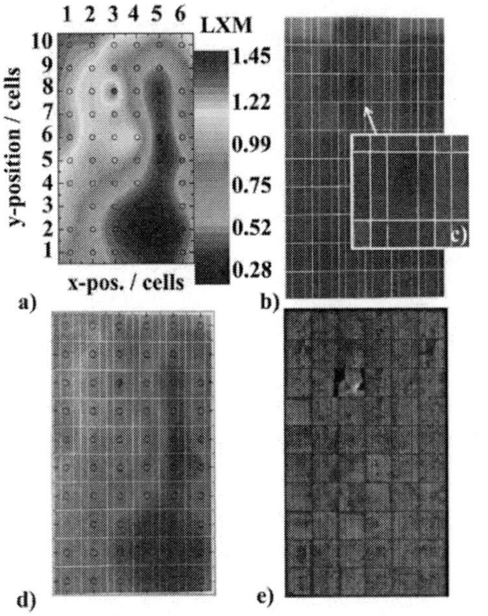

Figure 2: a) LXM map derived from the X Link measurements; b) photograph of the module; c) close-up of the hot cell with visible browning; d) superposition of image a) and image b); e) EL image of the same module revealing a broken cell.

3 RESULTS

In this work, an industrial PV-Plant was analyzed by drone-based IR-thermography. The PV plant consisted of commercial polycrystalline silicon 60-cell-modules. The modules had a standard back sheet configuration. The drone (Fig. 1a) as well as the optical setup is described elsewhere [1]. From the IR data acquired from this plant a module with a very high temperature gradient was identified (Fig. 1b). The module selected for this study exhibited a temperature of T = 100.1 °C at the location of the hot cell, which was the highest temperature observed in this measurement. This module was taken out of the plant for further characterization. By visual inspection a visible crack as well as local browning at the position of the hot cell was observed (Fig. 2b and c). Lab EL measurements clearly revealed a broken cell as the root-cause of the hot cell (Fig. 2 e)).

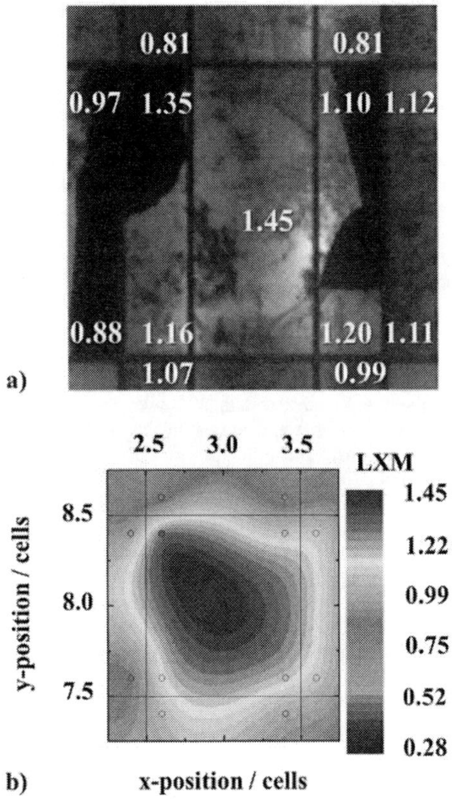

Figure 3: a) EL image of the fractured cell and its direct vicinity. The LXM values are stated at the positions of the respective indentation measurement. b) LXM map derived from the X Link measurements on the hot cell and its vicinity.

In order to investigate the heat effect of the hot cell on the encapsulant, the module was mapped with respect to the viscoelastic properties of the EVA / back sheet stack using LayTec's X Link indentation tool. This tool is described in [5]. In a first step, an overview map was acquired with one measurement in the center of each cell. In Fig. 2 a) the resulting map of the derived LXM values is shown. Here, the circles mark the actual measurement whereas the area in between was interpolated based on the measurements. As mentioned above, higher LXM values indicate a higher degree of stiffness and brittleness of the encapsulant stack. In particular, they correlate to a higher degree of cross linking of the EVA. However, since the indentation method always probed the entire stack, changes of the back sheet could also influence the LXM similarly [5, 6]. From Fig. 2a-d it can be clearly seen that the global maximum of the LXM values coincides with the position of the hot cell. Superimposed is an LXM gradient that is typical for 2x2 laminators, which exhibit an inhomogeneous heat profile with higher temperature in the center (resulting in higher LXM values near one corner of the module) and lower temperature along the edges (resulting in lower LXM values along two edges of the module).

For closer analysis of the influence of the local heating on the encapsulant properties an LXM map of higher resolution was obtained from the region of the hot cell and its direct vicinity. Fig. 3a shows the LXM values measured at the various locations on and around the hot cell, while Fig. 3b illustrates the LXM distribution as an interpolated LXM map. These measurements show that the high LXM values are strongly localized to the area of the fractured cell and that all surrounding cells exhibit values well below the maximum value of 1.453 measured in the center of the hot cell. This localization strongly indicates that the modification of the viscoelastic properties of the EVA / back sheet encapsulant stack is a direct consequence of the local heat dissipation caused by the fractured cell.

4 CONCLUSIONS

The results presented in the previous sections give direct evidence that a local heating caused by a hot cell can lead to permanent changes in the properties of the EVA / back sheet encapsulant of a PV module by shifting its viscoelastic properties towards increased stiffness and brittleness. As the encapsulant is the most important protection of the module, modifications of its properties impose a direct risk on the module lifetime. In particular, the fact that the encapsulant becomes more brittle indicates its durability with respect to humidity protection might be compromised. Also, the observation of local browning also indicates accelerated aging at the position of the hot cell.

As the cross linking of EVA and similarly heat-induced changes in the backsheet are temperature-dependent processes it is expected that heat dissipation by hot cells does not necessarily always lead to changes or damages of the EVA / backsheet stack. Furthermore, it can be expected that a certain threshold temperature exists for each configuration of EVA and backsheet, at which changes of the viscoelastic properties occur.

Interestingly, the hot cell investigated in this study exhibited a significant degree of browning, which is usually attributed to UV irradiance [7]. However, the fact that here this browning only occurred locally, already indicates that the properties of the encapsulant are modified locally and that ageing and degradation might occur accelerated at this location. This finding agrees well with earlier studies of the yellowing / browning behavior of polymers in general [8] and EVA as a PV encapsulation material in particular [9]. Generally, yellowing induced by UV irradiation increases linearly in time [8]. However, the process is also strongly temperature-dependent, as it was observed that the yellowing / browning rate follows an Arrhenius relationship with a reaction rate doubling for approximately every 10°C increase in temperature for temperatures >85°C. Thus, a locally increased temperature should lead to an increased yellowing/ browning rate at the given location as it is observed in this work.

From material testing of EVA encapsulants is known that the simultaneous exposure of EVA to UV and heat leads to discoloration and loss of tensile properties as found in the present paper. Furthermore, it has been shown that this behavior is observed for temperatures above 85°C but not below 50°C [10 and references therein]. This also indicates the existence of a certain threshold temperature for encapsulant degradation. Thus, only hot cells which lead to local heating beyond this threshold temperature would have to be considered critical with respect to module encapsulation stability. This also agrees to further investigations of the present study, in which hot cells detected by IR thermography did only result in local browning if high temperatures >85°C were detected.

To our knowledge, direct evidence for the permanent modification of the heat dissipated from hot cells (originating from cell breakage) on the viscoelastic properties of the EVA / back sheet encapsulant stack is given for the first time. Based on this study further experiments will aim on the identification of a threshold temperature at which this modification starts. If this temperature can be determined, data acquired by drone-based IR thermography could be directly used to identify modules whose long-term stability is questionable due to local heatings above this threshold temperature. Thus, quantitative outdoor assessment of PV plants could become even more powerful.

SUMMARY

In this work, a combined study of indoor (EL, indentation) and outdoor (IR) measurements of a PV module exhibiting a hot cell was carried out in order to investigate the changes in the encapsulant induced by the heat dissipated by the hot cell. This approach has the advantage that the real operating temperature of the module is known.

ACKNOWLEDGEMENT

Parts of this work have been funded by the German Federal Ministry for Economic Affairs (AQUAM, Förderkennzeichen 0325807A).

Dr. M. Heindl of SKZ (Süddeutsches Kunststoffzentrum, Würzburg, Germany) is gratefully acknowledged for helpful discussions and advise.

REFERENCES

[1] C. Buerhop, H. Scheuerpflug, Proceedings 29th European Photovoltaic Solar Energy Conference (2014) 3260.

[2] F. J. Pern, S. H. Glick, Proceedings 25th Photovoltaic Specialist Conference (1996) 1251.

[3] C. Buerhop, H. Scheuerpflug, Proceedings 29th European Photovoltaic Solar Energy Conference (2014) 2975.

[4] C. Buerhop, T. Pickel, M. Dalsass, H. Scheuerpflug, C. Camus, and C. J. Brabec, Proceedings 43rd Photovoltaic Specialists Conference (2016) *accepted.*

[5] E. Malguth, B. Buhl, T. Schenk, Proceedings 28th European Photovoltaic Solar Energy Conference (2013) 472.

[6] C. Lux, U. Blieske, E. Malguth, N. Bogdanski, Proceedings 29th European Photovoltaic Solar Energy Conference (2014) 2462

[7] W. H. Holley, S. C. Agro, J. P. Galcia, L. A. Thoma, R. S. Yorgensen, M. Ezrin, P. Klemchuk, G. Lavigne, Proceedings 1st World Conference of Photovoltaic

Energy Conversion (1994) 893
[8] B. R. Hinderliter, S.G. Croll, JCT Research, Vol. 3, (2006) 203
[9] R. G. Ross, Jr., Proceedings 39th Photovoltaic Specialists Conference (2013) 1
[10]G. Wypych, Handbook of Material Weathering, 4th edt., ChemTec Publishing (2008) 360f.

REGENERATION OF POTENTIAL INDUCED DEGRADATION AFFECTED MODULES

Cyril Hinz, Simon Koch, Thomas Weber and Juliane Berghold
PI Photovoltaik-Institut Berlin AG, Wrangelstraße 100, D-10997 Berlin, Germany
Phone: +49 (30) 814 52 64 106, Fax: +49 (30) 814 52 64 101,
E-mail: hinz@pi-berlin.com, koch@pi-berlin.com, weber@pi-berlin.com, berghold@pi-berlin.com

ABSTRACT: There are certain regeneration methods to recover photovoltaic modules which are affected with potential-induced degradation (PID) in a PV-plant. The aim of these investigations is to compare the effectiveness of the solutions of two well-established providers with the concept of grounding as well as a high voltage power supply, which will serve as a Reference. The focus will be on the degree of regeneration of the module power output at STC (Standard Test Conditions) and weak light conditions as well as the influence of the modules degradation history on the recovery-potential. The experiment was conducted with 40 PID-affected modules. As expected, the applied anti-PID-devices induced a regeneration of the analyzed modules within the 70 days of the test period. Despite different functional principles, the application of the two anti-PID-devices as well as the Reference design led unexpectedly to very similar test-results. The worst recovery results were achieved with the grounding configuration. Within the observation period of 70 days between 20% and 33% of the modules achieve the power warranty threshold for PID recovery with the investigated devices and the reference. In case of solid grounding 0% could reach the power warranty threshold of the module manufacturer (at STC) within the observation period. In general, very severely degraded modules achieved a low degree of regeneration. Additionally, the weak light performance (at < 400 W/m²) of the modules after recovery was still significantly reduced for all four methods tested.
Keywords: PID, Recovery, Module reliability, Performance, P-type crystalline

1 INTRODUCTION

The topic of PID has become an important issue in the photovoltaic (PV) industry for the last years. This degradation effect can lead to massive yield losses in a PV-power plant within a short period after grid connection. This problem is not restricted to certain locations, modules all over the world are affected and the financial damage can be accordingly high.

Influencing factors of potential-induced degradation were extensively examined and the most important driving factors identified. These are for instance a high negative potential of the cells relative to ground as well as high temperatures and humidity (disregarding several existing material and manufacturing influencing factors).[1-4, 12]

It is also known that PID is (partly) reversible at certain conditions once the negative voltage exposure has stopped [1, 5]. The recovery process can also be accelerated by e.g. high temperature, irradiation and particularly with the application of positive voltage [1, 2, 4, 5, 6, 12]. The latter is utilized by anti-PID device providers which proclaim a quick and nearly complete recovery of affected modules. The functional principle of the systems varies as well as their prices.

Furthermore, it is known that PID regeneration also depends on the 'degradation history' of the affected modules (e.g. number of cycles or depth of degradation) and that the weak light performance of the modules is generally even more affected [4-9].

Therefore, the effectiveness of four different recovery methods will be investigated at real outdoor conditions with 40 modules degraded originally in a PV plant. Besides the recovery-trend, the focus will be also on the weak light behavior as well as the regeneration from deep degradation.

2 EXPERIMENTAL

2.1 Examined recovery methods

In the following, the 4 examined recovery methods were investigated. Following basic information about the connection method and applied voltage for the recovery process are given:

Device A is connected in series to the module string, inactive during the day and applies a positive voltage during the night (only automatic mode possible). Sequential measurements have revealed an alternating course from +542 V to +815 V in a period of 32 seconds for a time span of 6 h.

Device B is connected parallel to the module string, inactive during the day and applies a positive voltage on the modules between sunset and sunrise. The level can be set manually gradational and was set to a value at about half of the maximum system voltage (+590 V).

The **Reference** was realized via a high voltage power supply and controlled by a light sensor. During the day, a voltage of -200 V was applied and +550 V (in order to simulate a formerly available anti PID-device) in the night. The modules were short circuited with each other.

Solid grounding (improved) was also simulated with a high voltage power supply and controlled by a light sensor. A positive voltage of +200 V was applied to the modules during the day. In the nights the modules were free from voltage. The test specimens were also short circuited with each other within this setup.

The applied voltages have been chosen (not possible for Device A) in order to be comparable among the systems as good as possible and without stressing the modules with high positive voltage near maximum system voltage.

2.2 Test modules

The used test specimens are p-type polycrystalline modules with 60 cells. They originate from a PV plant located in the Mediterranean area. The 40 modules were four years operating in field. After one year grid connection, they showed PID and solid grounding was selected as a countermeasure. After three years of grounding these modules still showed significant PID and were therefore selected for testing. Results of the initial measurement and degree of PID are presented in section 3.1.

For the investigation of the recovery process after a severely degradation, 8 modules were selected for a further degradation in the climate chamber at 85 °C, 85% rel. humidity and -1000 V for 24 h. Results are presented in section 3.2.

2.3 Experimental set-up and measurement period

The test specimens were mounted in southern direction with an inclination angle of 30° on top of the PI-Berlin laboratory.

For Devices A and B, two strings with 15 modules each were set up in series and connected to an inverter of the exact same type (floating potential, transformerless). For the Reference as well as the improved solid grounding, 5 modules were connected to a high voltage power supply.

The severely degraded modules were distributed equally to all four methods (two each). Test specimens with comparable PID-levels were installed on the same string-position to keep the comparability as high as possible. This enables a comparative analysis of the recovery progress along string positions among the methods A and B.

The test duration was from September 2015 until November 2015, for altogether 70 days.

2.4 Measuring methods

For quantitative monitoring of the recovery process the power output (at maximum power point) of the test specimens was measured frequently with a Flashlight Sun Simulator AAA from PASAN SA under STC as well as 700, 400, 200 and 100 W/m² (2,87% relative measurement uncertainty for power measurement). Besides an initial and final measurement of all modules, 16 test specimens were measured every two weeks (alternating 8 modules each week) for tracking the recovery-course.

Electroluminescence pictures (EL) were taken as a qualitative method to visualize the course of the recovery process.

To examine the PID status, equivalent to irradiances at STC conditions, following settings were chosen:

- high current of 7,96 A (I_{mp}) and shutter speed of 16 s

In addition, to examine the PID status equivalent to weak light conditions of 100 W/m², following settings were also chosen:

- low current of 0,796 A (10% of I_{mp}) and shutter speed of 200 s

3 RESULTS

The presentation of the results is divided into six subtopics. First, the results of the initial measurements of the test specimen are presented. Subsequently the results of the further degraded modules are shown in the same way.

In the third section, the results of the regeneration examination are shown. The focus will be on the regeneration level depending on the string position, followed by the examination of the regeneration trend with time and the weak light performance in the 5th section. In the final section a comparison of the module recovery rate of all devices is given.

3.1 Initial measurements field modules

Figure 1 presents the initial power output results for the field modules at all measured irradiances. The bar chart shows the mean values of all field modules already allocated to their testing method. The average power output was normalized on the nominal values from the module data sheet and adjusted for other degradation effects over 4 years of operation in the field.

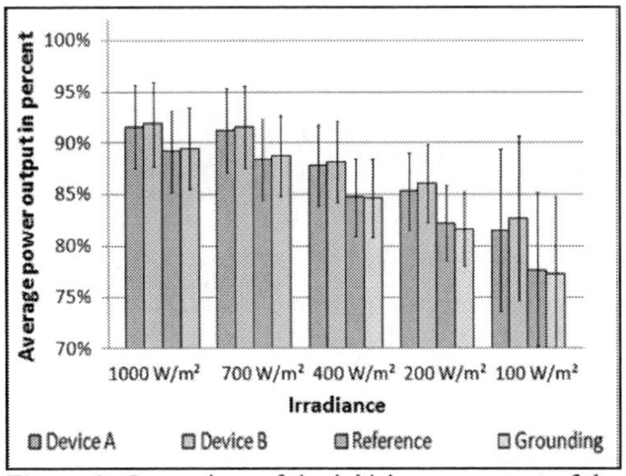

Figure 1: Comparison of the initial power output of the moderately degraded field modules sorted by the applied recovery method

The mean values of the initial measurement reveal a moderate state of power output degradation at STC due to PID, visualized and confirmed with EL-pictures (both left images from Module 1 and Module 2 in Figure 2). The bar chart also reveals the decreasing power output with a declining irradiation, especially <700 W/m². Compared to the results under STC-conditions, the biggest influence of PID can be recognized for measurements at 100 W/m² with a 10% higher degradation rate. The low current EL-pictures (both right pictures Figure 2) correspond with the findings in the measurements with an increased number of dark cells compared to the high current pictures.

Module 1) Module 2)

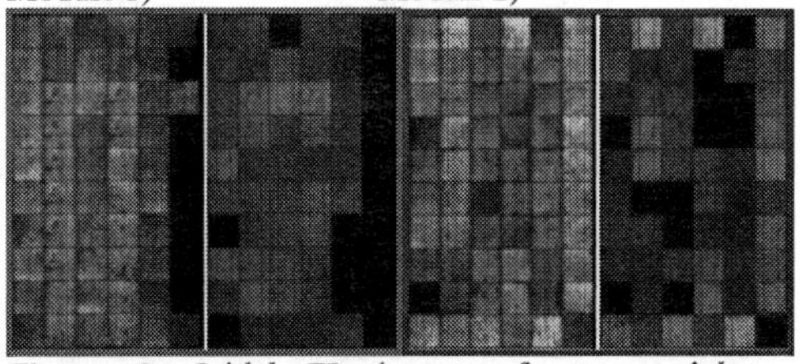

Figure 2: Initial EL-pictures of two modules - exemplarily for all test specimen (Module 1, green frame and Module 2, red frame) taken with high- (left) and low current (right), are shown to illustrate the two prevailing PID pattern

3.2 Further degradation in climate chamber

Figure 3 presents the initial power output results for the severely degraded modules after the treatment in the climate chamber at all measured irradiances. The bar chart shows the mean values of each two modules already allocated to their testing method. Also for this comparison the average power output was normalized on the nominal values from the module data sheet and adjusted for other degradation effects over 4 years of operation in the field.

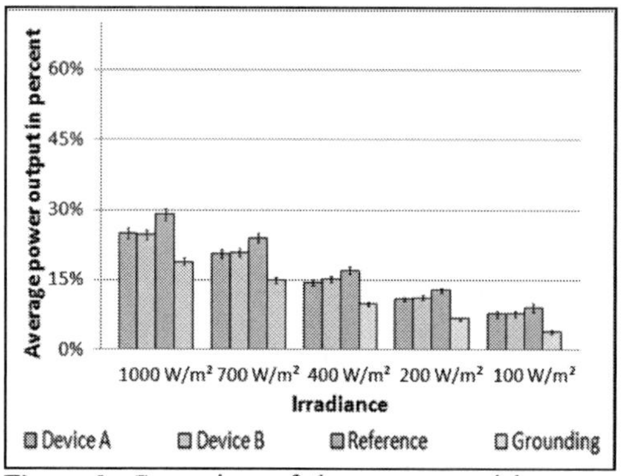

Figure 3: Comparison of the average module power output of the severely degraded modules sorted by recovery method

The degradation was very inhomogeneous despite of homogenous conditions in the climate chamber. The degradation depths of the 8 test specimens differ from 10% to 40% of the nominal value at STC. The influence of degradation increases with declining irradiations. The EL-pictures for the severely degraded modules correspond with the measurement results. The stronger the degradation, the darker are the cells, especially in the direction from the frame to the center due to the grounding of the frame (see Figure 4).

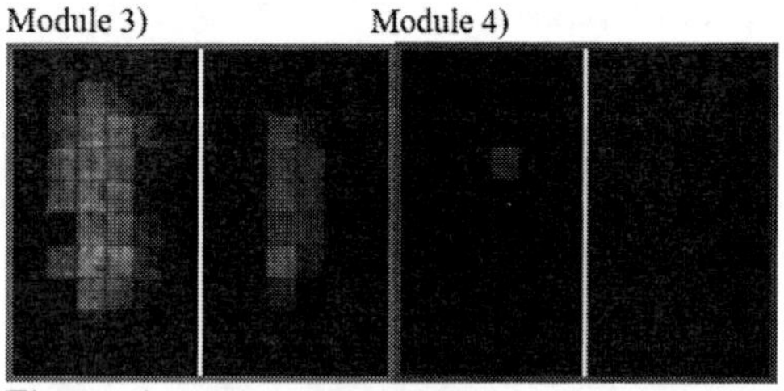

Figure 4: An exemplary selection of two severely degraded modules after the climate chamber cycle; Module 3 (green) reveals about 40% power output at STC and Module 4 (red) about 10%; each picture was taken with the same settings as in Figure 2

3.3 Regeneration depending on string position

The results of each single module connected to Device A show that none of the test specimen reveals a further degradation after the test period. Taking into account the measurement uncertainties, four modules exceeded the threshold of the warranty. Figure 5 shows that all modules roughly reach the same level after 70 days of testing period. It is also recognizable that slightly more degraded modules in the negative part of the string (position -1 to roughly -9) reveal also the highest extent of regeneration. (The skipped module positions belong mainly to severely degraded modules and they are skipped due to a better solution in Figure 5 and Figure 6).

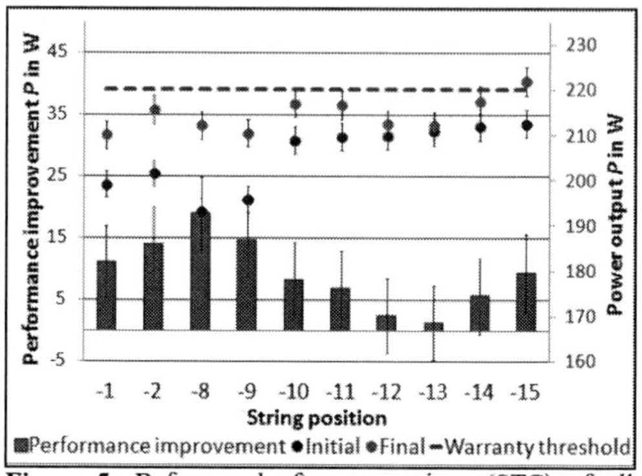

Figure 5: Before and after comparison (STC) of all moderate degraded field-modules from the string connected to Device A

Comparable results could also be determined for the modules connected with Device B. No further degradation could be measured. The highest extent of regeneration could be reached mainly in the negative string part (position -1 to roughly -9) and an approximate performance level at STC could be measured. Taking into account the measurement uncertainties, five modules exceeded the threshold of the warranty.

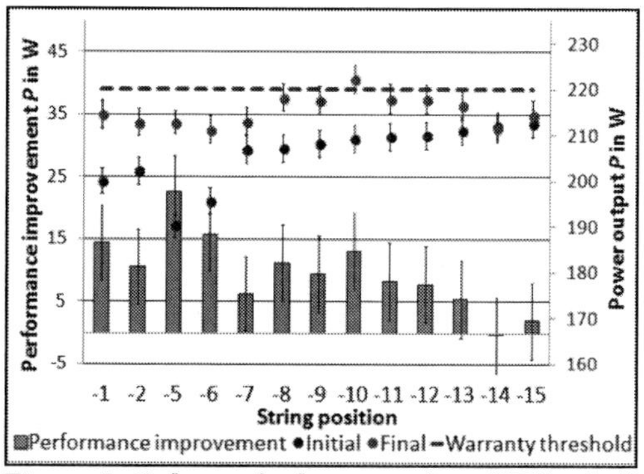

Figure 6: Before and after comparison (STC) of all moderately degraded field-modules from the string connected to Device B

3.4 Regeneration trend

First, the results of the regeneration trend of the moderately degraded field-modules are shown in Figure 7 (STC-measurements).

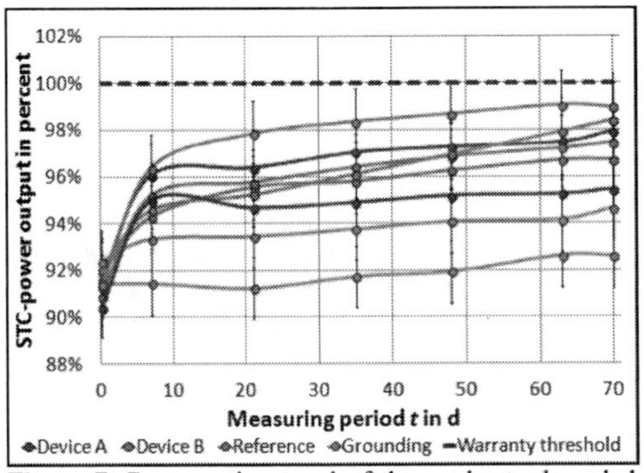

Figure 7: Regeneration trend of the moderate degraded field modules, two modules each method; warranty threshold is set as guiding value for 100% recovery

Except of the improved grounding, all recovery-methods revealed a significant increase of power output already after one week. During the remaining measuring period a slight but ongoing regeneration of the modules was recognizable, also for the improved grounding configuration.

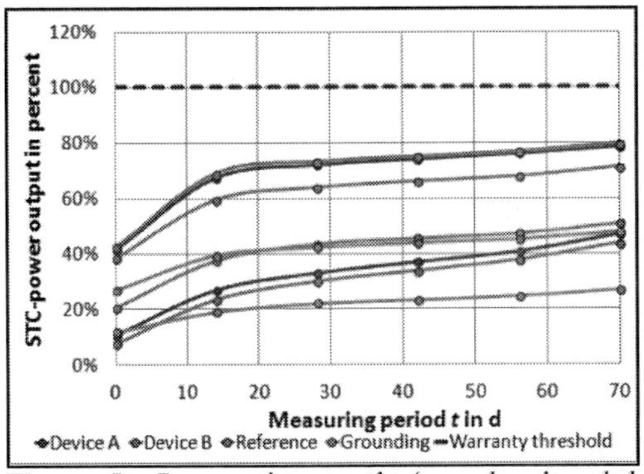

Figure 8: Regeneration trend (severely degraded modules), two modules for each method; warranty threshold is set as reference value for 100% recovery

The regeneration trends of the severely degraded modules in Figure 8 are very similar. However, for the solid grounding the extent of recovery was smaller within the observation period. The first measurement after two weeks has also shown a steep increase of the power output and a clearly recognizable further regeneration over the rest of the time. Especially striking are the nearly equal power trends from the modules connected to the anti PID-Devices A and B over the whole period. By far, none of the deeply degraded modules could exceed the warranty value given by the manufacturer after 70 days of recovery.

3.5 Weak light performance

For the comparison of the weak light performance evolution, mean values were calculated for all modules from each method (as in 3.1 and 3.2). The slight differences of the initial values between the Devices A as well as B compared to the grounding simulation and Reference are due to a higher number of test specimens with slightly higher initial values.

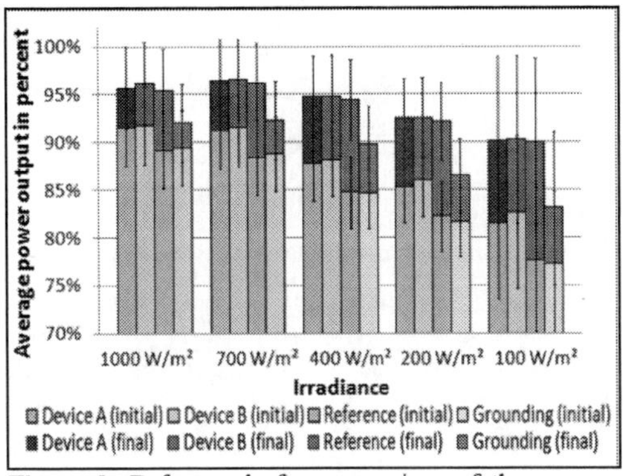

Figure 9: Before and after comparison of the average module power output of the moderate degraded field modules at all measured irradiations (further development Figure 1)

Figure 9 shows that the modules connected to Device A, Device B and the Reference gained most performance improvement at low irradiances at 200 and 100 W/m² (between around 8% and 12%). The smallest improvement was measured at STC with only about 5% to 7%. In spite of the higher performance increase at the weak light conditions, the modules still reveal the strongest impact of degradation at weak light conditions. The bar chart also shows clearly the similar regeneration behavior for all three methods tested and at all irradiance levels. The grounding configuration shows consistently the smallest magnitude of regeneration (at about half of the amount compared to the other three methods).

Module 1) Module 2)

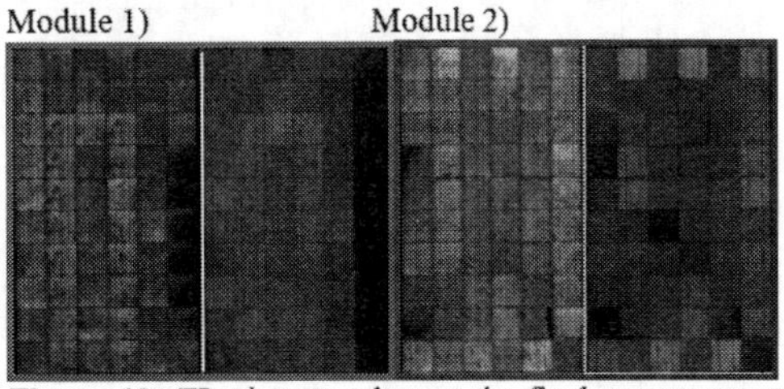

Figure 10: EL-pictures taken at the final measurement from the same modules and settings as in Figure 2.

The EL-pictures in Figure 10 clearly show a reduction of dark cells, especially with the use of 10% of the I_{MP}. Nevertheless, the images with the reduced current still reveal a higher amount of PID affected cells which confirm the results of the bar chart in Figure 9.

The same comparison was also conducted for the severely degraded modules as shown in the following diagram (Figure 11).

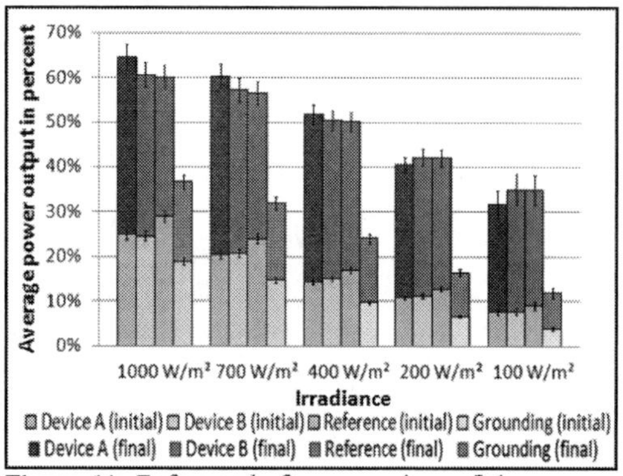

Figure 11: Before and after comparison of the average module power output of the severely degraded modules at all measured irradiations (further development Figure 3)

In comparison to Figure 9, the biggest performance increase of the severely degraded modules could be mainly detected at high irradiation intensities. An increase of performance at about 30% to 40% could be measured for all devices, except of the grounding simulation, at STC and about 20% to 25% at an irradiance of 100 W/m². The modules from the grounding configuration show in principle the same behavior like the other modules, but in a much smaller magnitude (less than half of the values compared to the others).

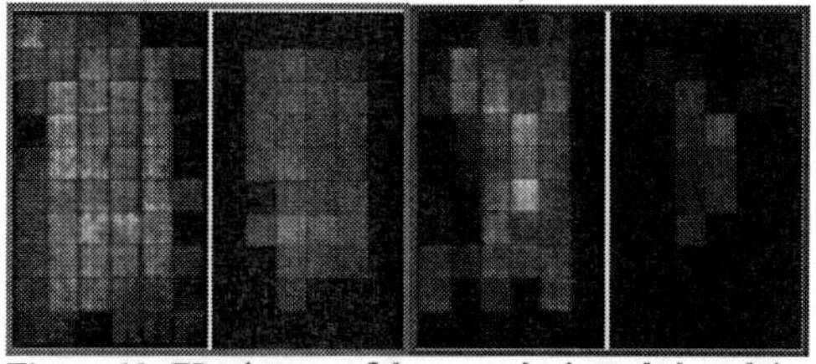

Figure 12: EL-pictures of the severely degraded modules taken at the final measurement, from the same modules and settings as in Figure 4

The EL-pictures of the severely degraded modules reveal a recovery process mainly in the center of the modules. Corresponding to the highest power output increase at high irradiations (Figure 11), most of the cells reappeared at high current. According to the lower amount of regeneration at weak light conditions, also fewer cells reappeared at 10% I_{MP} current as with the use of high current.

3.6 Comparison of the recovery rate

A comparison of the recovery rate (STC-power-measurement) of all modules connected to the four recovery methods was made. The following Table I shows the percentage of modules which exceed the warranty threshold after a measuring period of 70 days.

Table I: Comparison of recovery rate

Device A	Device B	Reference	Grounding
25%	33%	20%	0%

None of the severely degraded modules has exceeded the threshold.

4 SUMMARY & CONCLUSION

The results have shown that all investigated recovery methods have lead to a regeneration of PID-affected modules. The nightly positive bias voltage has a prevailing effect on the modules, therefore no further degradation could be recognized due to daytime degradation. The recovery effect was found to be higher for higher initial degradation levels of the modules. The corresponding power increase was in accordance to the higher number of 'reappearing' cells in EL imaging.

Despite different functional concepts, different voltage trends over night and very different pricing for the devices, no significant differences in the recovery results could be found for the different investigated Devices A and B and the reference method (simulating a high voltage application).

In objection to device supplier announcements, the total number of completely recovered modules after 70 days was found to be comparatively low. Only a small number of test specimens could exceed the warranty threshold and the results for the severely degraded modules are generally not sufficient. None of them could reach the power warranty threshold or even get close to it. This confirms the bad regeneration ability of severely degraded modules, even with the use of anti PID devices.

Referring to the weak light performance, the anti PID-solutions could improve the performance but not recover it on the same level as at STC. Moderately degraded field modules show the biggest improvement at very low irradiance levels below 400 W/m². Nevertheless, their power output level is still significant lower than at STC. The severely degraded modules showed the biggest improvement at irradiations above 400 W/m², but the results for this group are overall very poor especially at irradiations of 100 and 200 W/m².

The selected grounding configuration with a voltage shift of about half of the string-voltage has shown the worse recovery results within the observation period – compared to the other three methods. This is mainly due to the lower applied voltage compared to the other methods and the disadvantageous climate conditions for recovery with grounding in autumn (short daytime and colder temperatures).

The results of this experiment show a clear trend but the recovery process is not completed after 70 days. This outdoor experiment will be further continued.

Another conclusion of the partly poor recovery results -despite of the use of anti-PID devices- is the need of early detection of PID and therefore the high importance of an accurate monitoring of PV-plants and the immediate countermeasures when PID is identified. In this way long-lasting recovery periods and module replacement procedures could be avoided.

5 OUTLOOK

As the regeneration process is not finished yet, this experiment will be continued further also in order to investigate the influences of the seasons on the different recovery methods and therefore to increase the comparability. For example, it is expected that the grounding configuration has a higher efficiency in the summer month during long days, high irradiation and temperatures [11].

The concept of most of these recovery-devices is the inactivity during the day (modules are exposed to a negative voltage) and a positive voltage for regeneration at night. Therefore the impact of repeating degradation and recovery cycles will be further investigated and the test set up will be partly adapted in order to investigate this issue at real outdoor conditions.

As the results have shown, the biggest improvement of performance output could be observed at weak light conditions at the moderately degraded field modules. Therefore it needs to be observed whether this trend will be ongoing and if (and when) they can achieve the same recovery level as under STC.

6 REFERENCES

[1] S. Pingel, O. Frank, M. Winkler, S. Daryan, T. Geipel, H. Hoehne and J. Berghyold, "Potential Induced Degradation of solar cells and panels", 35th IEEE Conference, 2010

[2] M. Schütze, M. Junghänel, M.B. Koentopp, S. Cwikla, S. Friedrich, J.W. Müller, and P. Wawer, "Laboratory study of Potential Induced Degradation of silicon photovoltaic modules", 37th IEEE Conference, 2011

[3] J. Hattendorf, R. Loew, W.-M. Gnehr, L. Wulff, M.C. Koekten, D. Koshnicharov, A. Blauaermel, J.A. Esquivel, "Potential Induced Degradation in mono-crystalline silicon based modules an acceleration model", 27th EUPVSEC, 2012

[4] P. Hacke, K. Terwilliger, R. Smith, S. Glick, J. Pankow, M. Kempe, S. Kurtz, I. Bennett, M. Kloos, "System voltage Potential-Induced Degradation mechanisms in PV modules and methods for test", 37th PVSC , 2011

[5] G. Mathiak, M. Schweiger, W. Herrmann, "Potential-Induced Degradation– comparison of different test methods and low irradiance performance measurements", 27th EUPVSEC, 2012

[6] Simon Koch, Juliane Berghold, Cyril Hinz, Stefan Krauter and Paul Grunow, "Improvement of a prediction modle for Potential Induced Degradation by better understanding the regeneration mechanism", 31th EUPVSEC, 2015

[7] Juliane Berghold, Simon Koch, Anja Böttcher, Asier Ukar, Paul Grunow, "PID and correlation with field experiences", Präsentation at 3rd International PV module QA forum, Tokyo, 2012

[8] Simon Koch, Daniel Nieschalk, Juliane Berghold, Stefan Wendlandt, Stefan Krauter and Paul Grunow, "Potential Induced Degradation effects on crystalline silicon cells with various antireflective coatings", 27th EUPVSEC, 2012

[9] A. Raykov, H. Hahn, K.-H. Stegemann, M. Kutzer, O. Storbeck, H. Neuhaus, W. Bergholz, "Towards a root cause model for Potential-Induced Degradation in crystalline silicon photovoltaic cells and modules", 28th EUPVSEC, 2013

[10] M. Martin, R. Krause, H. Eckert, M. Pfeifer, D. Kohake, „Investigation of Potential Induced Degradation for various module manufacturers and technologies“, 27th EUPVSEC, 2012

[11] S. Pingel, S. Janke and O. Frank, „Recovery methods for modules affected by Potential Induced Degradation (PID)“, 27th EUPVSEC, 2012

[12] J. Berghold, S. Pingel, O. Frank, M. Winkler "Potential induced degradation of solar cells and panels," Proc. 25th European Photovoltaic Solar Energy Conference, pp. 3753-3759 (2010).

MODELING POTENTIAL-INDUCED DEGRADATION (PID) IN CRYSTALLINE SILICON SOLAR CELLS: FROM ACCELERATED-AGING LABORATORY TESTING TO OUTDOOR PREDICTION

Eleonora Annigoni[1], Marko Jankovec[2], Federico Galliano[1,3], Heng Yu Li[1,3], Laure-Emmanuelle Perret-Aebi[3], Marko Topič[2], Fanny Sculati-Meillaud[1], Alessandro Virtuani[1], Christophe Ballif[1,3]
[1] École Polytechnique Fédérale de Lausanne (EPFL), Institute of Microengineering (IMT), Photovoltaics and Thin Film Electronics Laboratory (PV-lab), Rue de la Maladière 71b, 2000 Neuchâtel, Switzerland
[2] University of Ljubljana, Faculty of Electrical Engineering, Laboratory of photovoltaics and optoelectronics, Trzaska cesta 25, SI-1000 Ljubljana, Slovenia
[3] CSEM, PV-center, Jaquet-Droz 1, 2000 Neuchâtel, Switzerland

eleonora.annigoni@epfl.ch

ABSTRACT: We present a mathematical model to predict the effect of potential-induced degradation (PID) on the power output of c-Si modules in different climates. For the experimental part, we manufacture mini-modules made of two c-Si p-type cells, and use accelerated ageing laboratory testing performed at different combinations of stress factors (temperature, relative humidity, and voltage). By modeling the effect of each stress factor in a step-wise approach, we obtain a model for the PID at constant stress conditions, which agrees well with models that can be found in the literature for full-size modules. Our model is obtained complementing existing models by introducing a term that describes a linear dependence of module's power degradation on the magnitude of the applied voltage. Since in field installations PV modules are connected in strings and exposed to different potential – and, therefore, stress – levels, this latter term is needed to approach real field conditions. Finally, we present the first attempts to model PID outdoor degradation in different climate conditions based on the proposed model and on the indoor-determined coefficients for the devices tested. The outdoor prediction model makes use of Typical Meteorological Year (TMY) data for a specific location.
Keywords: Potential induced degradation, PID, Modeling, Lifetime, Crystalline silicon

1 INTRODUCTION

Predicting lifetime and power delivery over time of photovoltaic (PV) modules in real outdoor conditions is essential to all stakeholders. During their operational life, modules are exposed to a variety of stresses that can degrade their performance, the exact type of degradation depending on the module technology and the climate where the modules are operating. One particular degradation mode is potential-induced degradation (PID), which is increasingly evident not only in tropical climates but also in temperate ones (as e.g. in Germany, [1]). Results published in the last few years about accelerated tests on commercial modules gave evidence that commercial modules presently on the market can be prone to large levels of degradation. This was confirmed by other authors showing that even modules sold as "PID-free" fail the proposed standard test with power losses of up to 50% (see [2]). For all these reasons, a more extensive investigation on PID is required to assess its impact on modules lifetime.

In the last few years different models were proposed to simulate field power degradation due to PID. Some authors (e.g. [3], [4], and [5]) based their approach on the modeling of the shunt resistance, derived from laboratory testing in climatic chamber first, and then simulating its evolution in outdoor conditions. In other studies (such as [6] and [7]), empirical equations for the power loss were deduced from accelerated testing in climatic chambers and then used in combination with climatic data to simulate outdoor degradation. Here, we performed accelerated PID tests on 2-cells mini-modules at different stress levels (temperature, relative humidity, and voltage), from which we developed an empirical mathematical model that describes the power degradation. Such degradation model is mainly based on the one introduced by Hacke et al in [8], where in particular the power loss is described with a parabolic behavior over time. We further complemented this model with the introduction of a term describing the linear dependence of the degradation mechanism on voltage. The impact of different encapsulants is assessed by comparing results of accelerated tests on EVA (ethylene vinyl-acetate) and TPO (thermoplastic polyolefin)-based samples. In a second step, we propose an implementation of such model in outdoor conditions using typical meteorological year (TMY) data as input parameters, giving a first approximation of PID-related power degradationin different climates.

2 PID

Different types of PID occur depending on the module technology. A polarization-induced degradation was described in [9] for back-contacted (Sunpower) modules. They discovered the so-called "surface polarization effect": when n-type cells operate at high positive voltage towards ground, negative charges accumulate in the SiN_x antireflective layer increasing the front surface recombination, and consequently enhancing the recombination of the photo-generated carriers. This mechanism was proved to be reversible by, for instance, making the cells operating at negative voltage towards ground. For conventional c-Si cells, a similar phenomenon involving an ionic transport between the solar cells and the module frame through the front glass and encapsulant leading to modules' power degradation is in some cases observed as well (see [10], [11], [12] among others). In the case of conventional p-type cells, a negative voltage towards ground drives positive ions (predominantly sodium Na^+) from the front glass towards the solar cells. Here they can affect cell performance not only by increasing surface recombination [13], but also by migrating to stacking faults in bulk silicon that may potentially intersect the junction and lead to shunt creation [14].

PID depends on the polarity (i.e. conventional p-type cells are affected by negative voltages only) and on the magnitude of the potential applied between cells and ground [11]. It is also clearly linked to environmental factors as relative humidity [12] and temperature [15]. Moreover, it depends on the cell or module design and material properties, such as the conductivity (or Si/N ratio) of the anti-reflective coating [16], the resistivity of the encapsulant [17], or the presence of a module frame.

3 EXPERIMENTAL WORK

Two cells mini-modules (with a size of20 × 40 cm) were laminated at EPFL PVLAB. All samples feature the same cell type, commercial mono crystalline silicon p-type cells with a single p-n junction. We employed two encapsulation schemes, glass/glass (G/G) and glass/backsheet (G/B),with the same soda-lime glass for all samples and the same polyethylene-based backsheet for G/B samples. Two different encapsulants were used in the encapsulation process: an EVA and a TPO. The metallic frame was simulated by means of an electrically conductive aluminum tape, (see [18]). Accelerated aging tests were then performed in a climatic chamber according to the procedure described in the proposed IEC Technical Specification for PID in c-Si modules (IEC TS 62804-1:2015). In order to model the contribution of the various stress factors, we varied the stress levels in each test, as illustrated in Table 1. Moreover, we extended the tests duration to 192h, twice as much the prescribed time of 96 h. The voltage levels refer to the potential difference of the module's cells with respect to the grounded frame.

Table 1 Accelerated-aging testing conditions to investigate PID dependence on temperature T, relative humidity RH, and voltage V. Tests at 85°C/85% RH were repeated with additional voltage levels of -600 V, -400 V, and -200V. Duration of all tests was 192h.

	V=-1000 V			
T / RH	**65%**	**75%**	**85%**	**95%**
45°C				
60°C				
75°C				
85°C			V=-200V V=-400V V=-600V	

The electrical characterization of the mini-modules was performed at standard test conditions (STC: AM1.5G, 25°C, 1000W/m²) by means of current–voltage (IV) characteristics using a LED-halogen based sun simulator. Mini-moduels were tested (i) prior to climatic chamber testing, (ii) at an intermediate stage of 96h, and (iii) after 192h.

3.1 Leakage current

The leakage current measured between the frame and the cells was proposed by some authors as a possible measure of PID ([18], [5]). However, some studies ([19], [20]) proved that a leakage current increase cannot always be directly correlatd to modules' power loss. Results of our tests also proved this lack of correlation, as shown in Figure 1. Therefore, even though leakage current is representative of the accumulated positive charge on the cell, it is not a good indicator for PID, and we consequently chose the mini-module maximum power as more suitable parameter to model PID.

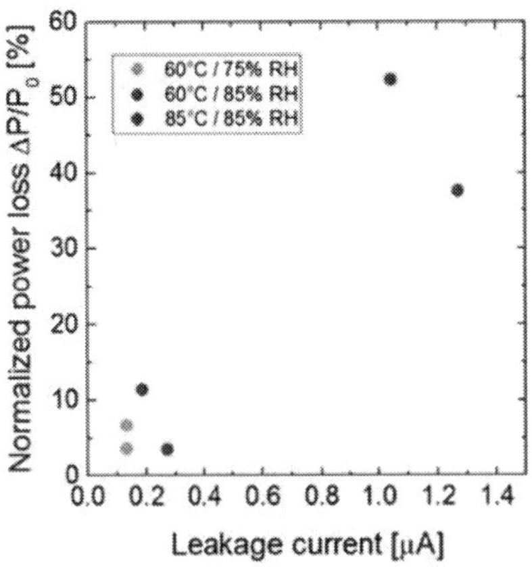

Figure 1 Power loss after 192h of PID test, for G/B samples with EVA, versus leakage current (obtained as the average of the leakage current measured all over the test duration): no correlation is observed.

In agreement with what previously found by other authors such as [10], [12], and [18], we observed that leakage current follows an Arrhenius behavior with respect to temperature, with an activation energy of 0.69 eV. This value is similar to values previously published in the literature (see [7]). Instead, the absolute values of leakage current are lower to theirs considering that they were testing 1-cell mini-modules. Values up to 10 times higher were reported by [18] for their 4-cells mini-modules with aluminum frame: such a difference can be attributed not only to the different number of cells, but to the use presumably of a different type of encapsulant too.

3.2 Encapsulant influence

It is known that PID can be minimized at module level by using an encapsulant material that reduces the transport of sodium ions ([11], [21]). In Figure 2 the performance of G/G samples with the two different encapsulants is shown.

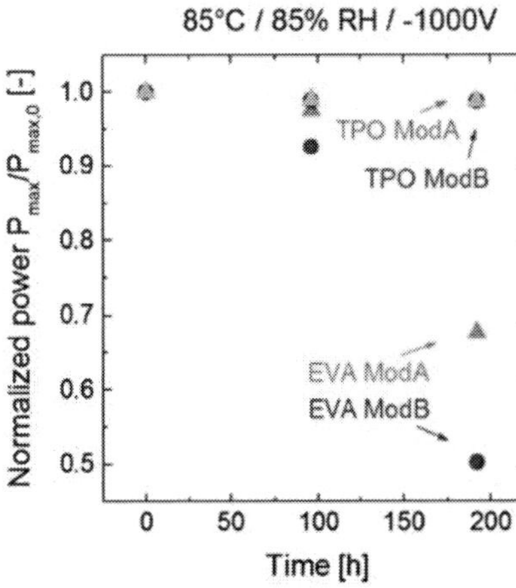

Figure 2 Power time-evolution of four G/G mini-modules tested at 85°C / 85% RH / -1000V Samples with EVA degraded more than those with TPO. The points for the two TPO samples are overlapped.

A significant degradation is observed for EVA-based mini-modules, where a standard commercial EVA was used. On the other hand, samples with TPO showed very good resistance to PID also under harsh test conditions such as 192h at 85°C/85% RH. The better performance of TPO is attributed to its higher electrical resistivity and better resistance to water ingress, as for instance given by the water vapor transmission rate (WVTR). These properties are listed in Table 2.

Table 2 Properties of the EVA and TPO employed in PID tests on mini-modules. WVTR stays for Water Vapor Transmission Rate.

	WVTR [g/(m^2·day)]	Water absorption [%]	Volume resistivity [Ohm·cm]
EVA	15-25	~0.04	>1·10^{15}
TPO	<5	<0.01	>2·10^{16}

To investigate TPO's long-term performance, we plan to extend the duration of the testing for samples laminateed with this material. Here, with the purpose of developing a model for PID, we restricted our attention to EVA that is still the mostly used encapsulant in PV. Moreover we focused our attention on mini-modules with G/B construction as more representative of the PV industry for c-Si modules.

4 PID MODEL FROM LABORATORY TESTING

In this Section we use a step-wise approach and present the mathematical description of each stress factor's contribution to the modules's power loss. First, this brings to the validation of the degradation equation given by Hacke in [8]. Then, the contribution of voltage is added.

Each point in the following plots represents the average of measurements performed on a number of samples, which varies from two to four samples, depending on the test conditions.

4.1 Temperature

As documented in several studies, [20] among others, temperature is one of the degradation factors for PID, mostly under humid conditions. Indeed, high temperature promotes ion migration from the glass to the cell as a result of an increased electrical conductivity of both the glass surface and the encapsulant material. The experimental tests performed in this work show an Arrhenius dependence of the power loss on temperature (see Figure 3). The activation energy we derived for the power loss is 0.86 eV.

4.2 Relative humidity

A complete description of PID requires modeling the effect of relative humidity on the power loss. Monitoring of modules in outdoor conditions [18], [22] showed a strong degradation during periods of high relative humidity, when the surface's glass conductivity increases making the glass surface nearly equipotential to the frame. As an example, a relevant reduction in shunt resistance together with high values of leakage currents were in fact obeserved in [22] during the mornings, when water condensates on the module surface, and during rainy days. In this work, tests performed in climatic chamber at different relative humidity levels suggest a polynomial behavior of the power loss with respect to RH, as depicted in Figure 4.

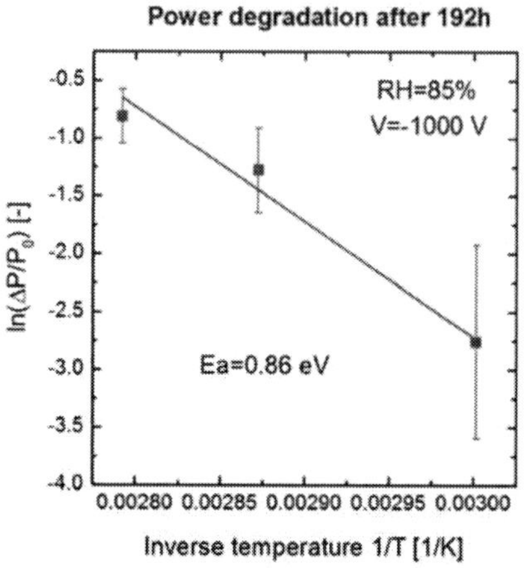

Figure 3 Normalized power loss for G/B mini-modules with EVA after 192h of PID test at 85% of RH and -1000V applied: an Arrhenius behavior with respect to temperature is deduced, with an activation energy of 0.86 eV.

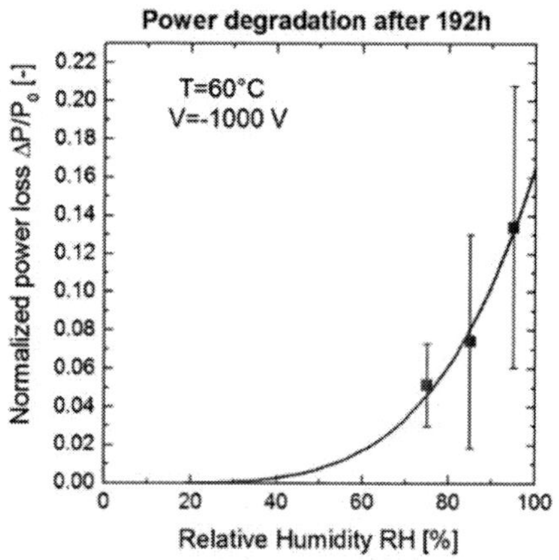

Figure 4 Normalized power loss for G/B mini-modules with EVA after 192h at 60°C and -1000 V exhibits a polynomial dependence on relative humidity.

4.3 Time

As mentioned, mini-modules IV characterization was performed after 96h and 192h in climatic chamber. In agreement with what found in [8] for commercial modules, the evolution of the maximum power of our mini-modules is quadratic in time, in our case during the first 192h, as shown in Figure 5. Tests at longer duration are planned to verify if such parabolic behavior still gives a good description of power evolution on longer time scales.

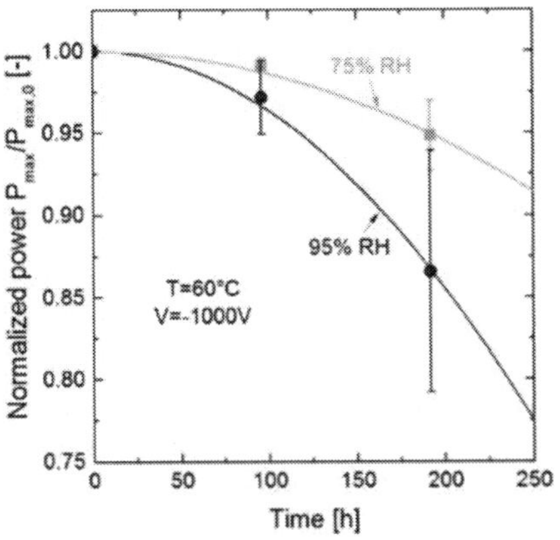

Figure 5 Evolution in time of maximum power for some of the test conditions performed: the behavior in the first 192h can be fitted with a quadratic curve.

The models for the contributions of the stress factors discussed above (temperature, relative humidity, and time) were introduced by Hacke in [8], who proposed equation (3) in the reference to describe power degradation during accelerated tests at constant T and RH conditions with an applied constant voltage of -1000V. Sections 4.1, 4.2, and 4.3 provided further experimental validations of this model as a good agreement was found with results of accelerated tests on our 2-cells mini-modules. In the next section, we introduce a term assessing the impact of voltage on the PID effect.

4.4 Voltage

Several inspections on PV plants (e.g. [11]) clearly indicate that, for modules made with conventional p-type cells, the degradation of modules within a string is stronger when moving towards higher negative potentials with respect to ground. This latter term is, therefore, needed to approach real field conditions, as field PID degradation will depend on the module's position within the string.

Here, we performed tests at several voltage levels and observed that the power losses at a fixed time (e.g. after 96h) vary linearly as a function of the applied voltage, as Figure 6 shows.

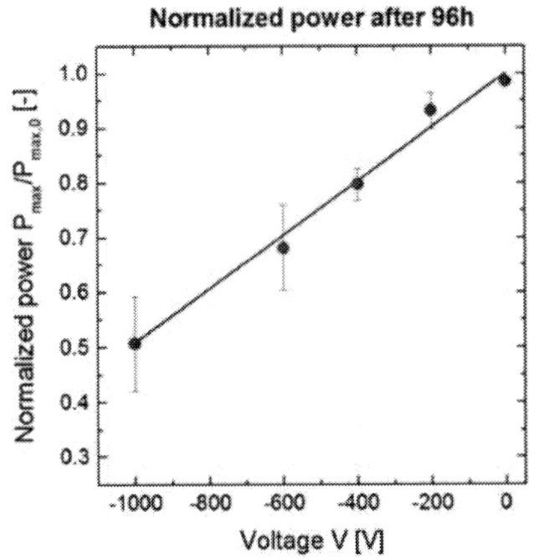

Figure 6 A linear behavior is observed for maximum power after 96h of PID tests as a function of the voltage.

We finally bring together the effect of all distinct contributions and propose a simplified relation (Eq. (1)) describing the power degradation as function of PID at constant stress test conditions:

$$\frac{P_{max}(t)}{P_{max}(0)} = 1 - C_V \cdot V \tag{1}$$

where $P(t)$ is the modules' measured power after t hours of exposure to constant stress levels, $P(0)$ is the initial power, and C_V is a coefficient, which includes the effect of temperature, relative humidity, and time t. The coefficient C_V represents the equation previously proposed by Hacke et al. in [8], recalled hereunder in the form of Eq. (2), where κ is the Boltzmann constant.

$$C_V(T, RH, t) := A \cdot e^{-\frac{E_a}{\kappa T}} \cdot RH^B \cdot t^2 \tag{2}$$

By fitting Eq. (1) for our mini-modules, we obtained the following values for the parameters: $A=6.0188 \cdot 10^{-5}$ h^{-2}, B=4.43 [-], and E_a=0.86 eV as pointed out in Section 4.1. One should notice that the parameter E_a is interpreted here as the activation energy of the power loss. Other authors concentrate, instead, on the activation energy of the time to reach a certain level of power loss (let us call it the "time-to-failure"). One can check that Eq. (1) and (2) imply in particular that the activation energy of the power loss is twice the activation energy of the "time-to-failure". Values reported by Hacke in [8] and Raykov in [7] are 0.85 eV and 0.75 eV respectively: this would therefore mean an activation energy of 1.7 eV and 1.5 eV for the power loss. We conclude that the value of 0.86 eV obtained by us is lower compared to the mentioned references. Reasons for this might lie for instance in the particular type of encapsulant employed or in the samples configuration. This will be subject of further investigations.

We are currently subjecting samples to experimental work (and different combinations of temperature and irradiance) to model a regeneration term for the PID phenomenon. Regeneration is in fact generally experienced by degraded modules when stored under controlled laboratory conditions or can be experienced by modules in the field. This part is not presented here and will be subject of a future contribution.

5 SIMULATING OUTDOOR PID DEGRADATION

The model presented in Eq. (1) describes the power degradation due to PID at constant stress laboratory conditions.

In this section, we attempt to describe what the power degradation of mini-modules would be, if they were installed at a constant negative voltage (-1000 V) in sites representative of different climatic conditions (Riyadh = Hot and Dry, Sigonella (Sicily) = Mediterranean, Miami = Hot and Humid). To do this, we use the coefficients of the distinct ageing contributions (temperature, relative humidity and time) as determined in Section 4 by means of indoor laboratory testing.

Using Eq: (1) and (2) and the indoor determined parameters to model outdoor PID degradation is, however, not straighforward. Firstly, because stress conditions outdoors are not constant. Secondly because, whereas indoors module's power undergoes a continuos degradation described by Eq. (1) and by a fixed initial power value $P_{max}(0)$, outdoors the module power $P_{max}(t)$ at time t fluctuates continuosly depending on irradiance, module's temperature and other environmental parameters.

For the different locations, to model the istantaneous power $P_{max}(t)$ of the devices, we use Typical Meteorological Year (TMY) data from Meteonorm with time-resolution of 1 hour as input weather parameters and PV_LIB Toolbox for Matlab (from Sandia National Laboratories). A c-Si glass/backsheet module was considered, with open-rack configuration. The degradation Eq. (1) was then applied in an iterative way, taking as stress factors at each time-interval $[t_i, t_{i+1}]$ the instantaneous temperature T_i and relative humidity RH_i values and supposing a constant voltage of -1000 V over daylight. No degradation was applied during the night and no regeneration was considered. First results are shown in Figure 7.

In order to make the implementation consistent, per each time interval the starting point on the degradation curve was calculated as the equivalent time needed at temperature T_i and relative humidity RH_i to reach the level of degradation of the previous time-step $[t_{i-1}, t_i]$. A similar approach has previously been applied by [5]. For the first time-interval of each day, the starting point was defined as the equivalent time to reach the average degradation level of the previous day (this choice is responsible of the non monotonously decreasing behavior of the normalized power that can be observed in Figure 7). This methodology for implementation differs from that by Hattendorf [6] in particular in the power loss equation, in the method used to estimate $P_{max}(0, i)$ and because it makes no use of threshold values for RH to trigger a degradation.

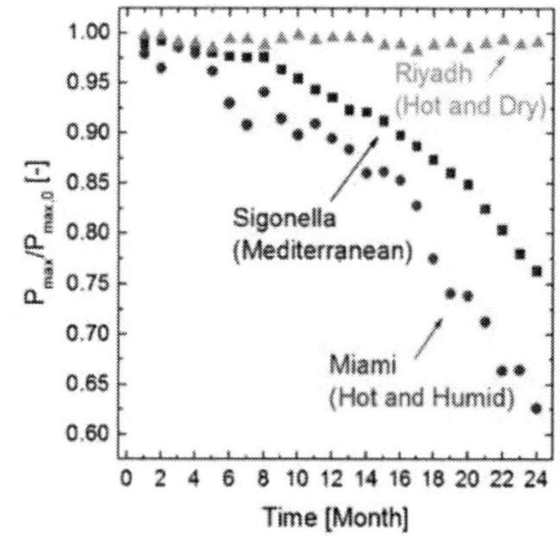

Figure 7 Simulated evolution of maximum power of a module in three different climates as consequence of the degradation mechanism of PID.

As mentioned, what we present here are the first attempts to model outdoor degradation of PID-affected modules. In fact a regeneration term is still missing from the model. As shown in other studies, ([15], [23]), in real outdoor conditions the effect of PID is a continuous trade-off between a degradation and a regeneration mechanism, which may occur for example during hot dry periods. Some models have been proposed for the regeneration mechanism ([6], [3], [7]). At the same time, we are collecting data from different PV plants in order to validate this model.

The possibility to manufacture and work with mini-modules carries considerable advantages: the manufacturer process is flexible, relatively fast and cost-effective, and we can vary materials (encapsulant, baksheets, glass) and cells properties. On the downside, extending the results obtained on mini-modules to large-area modules is not always straightforward and may require some adjustment to the model. This has been the object of recent analyses, where for example in [24] it was shown that in desert climates the degradation typically occurs only in the edge cells, while a more homogeneous degradation pattern can be observed in humid climates due to increased surface's glass conductivity.

6 CONCLUSIONS

Laboratory PID testing was performed on 2-cells mini-modules at several stress conditions. On the one hand, tests results suggest that mini-modules behavior can be described by the model that Hacke et al developed in [8] for commercial modules. On the other hand, we added to their power loss equation a term that describes the dependency of degradation on voltage, obtaining Eq. (1). Such equation, which describes the power loss at constant stress conditions, was then used to model outdoor PID degradation for different climates. To do this we use the indoor determined coefficients of the distinct ageing contributions (temperature, relative humidity and time) and use TMY data to model the power in operation of outdoor exposed modules in different locations. A strong influence of the climate is evident in simulations results where, as expected, a more severe degradation is predicted for hot and humid climates. Such simulations give a preliminary approximation of the time-evolution of power performance and are not complete yet as they do not consider the contribution of power recovery. Experimental work to model the regeneration as function of the climate factors is ongoing, as well as accelerated testing on commercial modules of different manufacturers. Future work will include investigation of any discrepancy between mini-modules and full-size ones, and of how to take this into account in the model.

In parallel to this experimental work, field data are collected in order to provide a validation of our simulations.

ACKNOWLEDGMENTS

We would like to gratefully acknowledge EOS Holding for their financial support to this work. We are grateful particularly to Xavier Niquille and Lionel Domon for support in the experimental work and to the all PV-lab team. We thank Peter Hacke and Jens Hattendorf for the fruitful discussions.

REFERENCES

[1] J. Berghold, S. Koch, A. Ukar and P. Grunow, "PID testing standard IEC62804 Outdoor correlation & prediction," 2014.

[2] N. Riedel, L. Pratt, E. Moss and M. Yamasaki, "600 Hour Potential Induced Degradation (PID) Testing on Silicon, CIGS and HIT Modules (Poster)," NREL, Golden (Colorado, US), 2015.

[3] S. Koch, Berghold, Juliane, C. Hinz, S. Krauter and P. Grunow, "Improvement of a Prediction Model for Potential Induced Degradation by Better Understanding the Regeneration Mechanism," in *31st European Photovoltaic Solar Energy Conference and Exhibition*, Hamburg, 2015.

[4] C. Taubitz, M. Kröber, D. Buß and M. B. Koentopp, "Potential Induced Degradation: Measurement and Modeling of Progression Under Field Conditions," in *31st European Photovoltaic Solar Energy Conference and Exhibition*, Hamburg, 2015.

[5] B. Braisaz, C. Duchayne, M. Van Iseghem and K. Radouane, "PV ageing model applied to several meteorological conditions," Amsterdam, 2014.

[6] J. Hattendorf, W.-M. Gnehr, R. Loew, T. Roth , D. Koshnicharov and M. Zentgraf, "Potential-induced degradation and temperature-driven regeneration: a realistic simulation," Paris, 2013.

[7] A. Raykov, H. Nagel, D.-J. Amankwah and W. Bergholz, "Climate Model for Potential-Induced Degradation of Crystalline Silicon Photovoltaic Modules," in *27th European Photovoltaic Solar Energy Conference and Exhibition*, Frankfurt (Germany), 2012.

[8] P. Hacke, S. Spataru, K. Terwilliger, G. Perrin, S. Glick, S. Kurtz and J. Wohlgemuth, "Accelerated Testing and Modeling of Potential-Induced Degradation as a Function of Temperature and Relative Humidity," *IEEE Journal of Photovoltaics*, vol. 5, no. 6, pp. 1549-1553, 2015.

[9] R. Swanson, M. Cudzinovic, D. DeCeuster, v. Desai, J. Jürgens, N. Kaminar, W. Mulligan and S. Corporation, "The Surface Polarization Effect in High Efficiency Silicon Solar Cells," 2005.

[10] J. A. del Cueto and T. J. Mc Mahon, "Analysis of leakage currents in photovoltaic modules under high-voltage bias in the field," *Progress in Photovoltaics: Research and Applications*, vol. 10, pp. 15-28, 2002.

[11] S. Pingel, O. Frank, M. Winkler, S. Daryan, T. Geipel and H. B. J. Hoehne, "Potential Induced Degradation of Solar Cells and Panels," 2010.

[12] P. Hacke, M. Kempe, K. Terwilliger, S. Glick, N. Call, S. Johnston, S. Kurtz, I. Bennett and M. Kloos, "Characterization of Multicrystalline Silicon Modules with System Bias Voltage Applied in Damp Heat," in *25th European Photovoltaic Solar Energy Conference and Exhibition*, Valencia, 2010.

[13] J. Bauer, V. Naumann, S. Großer, C. Hagendorf, M. Schütze and O. Breitenstein, "On the mechanism of potential-induced degradation in crystalline silicon solar cells," *Phys. Status Solidi,* vol. 6, no. 8, pp. 331-333, 2012.

[14] V. Naumann, D. Lausch, A. Graff, M. Werner, S. Swatek, J. Bauer, A. Hähnel, O. Breitenstein, S. Großer, J. Bagdahn and C. Hagendorf, "The role of stacking faults for the formation of shunts during potential-induced degradation of crystalline Si solar cells," *Phys. Status Solidi* , vol. 7, no. 5, pp. 315-318, 2013.

[15] P. Hacke, R. Smith, K. Terwillinger, S. Glick, D. Jordan, S. Johnston, M. Kempe and S. Kurtz, "Testing and Analysis for Lifetime Prediction of Crystalline Silicon PV Modules Undergoing Degradation by System Voltage Stress," *IEEE Journal of Photovoltaics,* vol. 3, no. 1, pp. 246-253, 2013.

[16] V. Naumann, D. Lausch , A. Hähnel, J. Bauer, O. Breitenstein , A. Graff, M. Wer, S. Swatek, S. Großer , J. Bagdahn and C. Hagend, "Explanation of potential-induced degradation of the shunting type by Na decoration of stacking faults in Si solar cells," *Solar Energy Materials & Solar Cells,* vol. 120, p. 383–389, 2014.

[17] Y. Lee, M. Tsai, K. Lu, W. Yang, S. Chen, M. Lin, H. Sung and H. Wu, "Potential Induced Degradation Experiments on Different Encapsulated Silicon Mini-Solar Modules," Amsterdam (Netherlands), 2014.

[18] S. Hoffmann and M. Koehl, "Effect of Humidity and Temperature on the Potential-Induced Degradation," *Progress in Photovoltaics: Research and Applications* , vol. 22, no. 2, pp. 775-783, 2014.

[19] J. Hattendorf, R. Loew, W.-M. Gnehr, L. Wulff, M. C. Koekten, D. Koshnicharov, A. Blauaermel and . J. A. Esquivel, "Potential induced degradation in mono-crystalline silicon based modules: an acceleration model," Frankfurt (Germany), 2012.

[20] P. Hacke, R. Smith, K. Terwilliger, G. Perrin, B. Sekulic and S. Kurtz, "Development of an IEC test for crystalline silicon modules to qualify their resistance to system voltage stress," *Progress in Photovoltaics: Research and Applications,* vol. 22, pp. 775-783, 2014.

[21] J. Berghold, S. Koch, A. Böttcher , M. Leers , A. Ukar and P. Grunow, "PID and correlation to field experiences," in *Sophia Workshop PV Module Reliability*, Chambery, 2013.

[22] S. Koch, J. Berghold, B. Abdullah, C. Hinz and P. Grunow, "Prediction model for potential induced degradation effects on crystalline silicon cells," in *Proceedings of the 29th European Photovoltaic Solar Energy Conference and Exhibition*, Amsterdam, 2014.

[23] P. Lechner, S. Hummel and J. Schnepf, "Evaluation of Recovery Methods after Potential Induced Degradation of PV Modules," in *31st European Photovoltaic Solar Energy Conference and Exhibition*, Hamburg, 2015.

[24] J. Berghold, S. Koch, C. Hinz, P. Grunow and T. Shioda, "PID: From Material Properties to Outdoor Performance & Quality Control Counter Measures," in *SOPHIA PV-Module Reliability Workshop*, Vienna, 2016.

POTENTIAL-INDUCED DEGRADATION: AN IMPROVED UNDERSTANDING OF MECHANISM AND INFLUENCE FACTORS

Christian Taubitz*, Achim Schulze**, Max B. Koentopp
Hanwha Q CELLS, Sonnenallee 17-21, 06766 Bitterfeld-Wolfen
**Fachhochschule Rosenheim, Hochschulstraße 1, 83024 Rosenheim
*corresponding autor: c.taubitz@q-cells.com, +49 (0)3494-6699-52162

ABSTRACT: Potential-induced degradation (PID) is a critical failure mode in today's photovoltaic (PV) system architecture with very large to catastrophic impact on PV-module performance and energy yield. Potential-induced shunting (PID-s) is one of the most severe types of PID leading to shunting of p-type crystalline silicon solar cells. In previous works a preliminary phenomenological model was developed describing the progression of shunt resistance (R_{sh}) under realistic field conditions. Comparison between measured and predicted R_{sh} progressions revealed good qualitative agreement. Nevertheless quantitatively some differences between measured and simulated R_{sh} progressions were found.
In order to better understand these differences various aspects of R_{sh} - measurement and - modelling are investigated in this paper. Firstly a cell-type and temperature dependence of the temperature coefficient of R_{sh} was found revealing a possible source of error in the correction of the temperature influence. In addition for a temperature of 70°C the grounded module-frame is found to influence the R_{sh} progression of the investigated one-cell mini-module even under dry conditions, which is not yet part of the preliminary model. Finally a physical model based on a description of the shunting and recovery process as a combination of drift and diffusion of sodium ions and atoms is presented improving upon the preliminary model.
Keywords: Degradation, Module, Reliability, PID

1 INTRODUCTION

Solar cells incorporated in photovoltaic (PV) modules of modern PV systems are usually exposed to bias voltages of several hundred volts with respect to the grounded module frames/mounting. This bias voltage can cause various kinds of different module damage, which are commonly called potential-induced degradation (PID). One of the most severe types of PID is potential-induced shunting (PID-s) leading to shunting of p-type silicon solar cells [1] [2]. This failure mode can cause a catastrophic performance loss of affected modules. In order to prevent modules from PID/PID-s related power loss recently a test method (IEC TS 62804-1) was established. Nevertheless it remains difficult to correlate laboratory PID test results with actual PID-s behavior of fielded modules. This is due to the fact that fielded modules are exposed to both degradation causing conditions and recovery from PID-s depending on current weather conditions (e.g. [3]-[8]). To understand better the correlation between laboratory and field behavior, a preliminary model was developed. Recently this model was employed to simulate the shunt resistance (R_{sh}) progression of one cell mini-modules and 60cell modules with different levels of PID-s resistance exposed to field conditions and a negative bias voltage of -1 kV [9]. A comparison between simulation and measurement results revealed good qualitative agreement. Nevertheless also differences between model predictions and measured R_{sh} behavior were found. In this work reasons for these discrepancies are investigated in order to improve the model results.

2 EXPERIMENTAL

2.1 Laboratory test setup

In order to investigate the temperature dependence of R_{sh} and the frame influence on the R_{sh} progression during PID-s, a laboratory test setup as shown in Fig. 1 was used. Mini modules consisting of single p-type solar cells were produced. These modules were PI shunted in an environmental chamber at different temperatures at high negative voltage (-1 kV). During the experiment the front glass was temporary covered with a grounded aluminum foil inducing an elevated and constant PID-s stress. The mini modules were equipped with an aluminum frame, which was grounded during the experiment. To monitor the R_{sh} progression the solar cell was biased with a voltage in reverse direction V_{rd} (< 0.3 V) at regular time intervals and the current $I(V_{rd})$ was measured. A detailed description of the experimental setup was published previously [2].

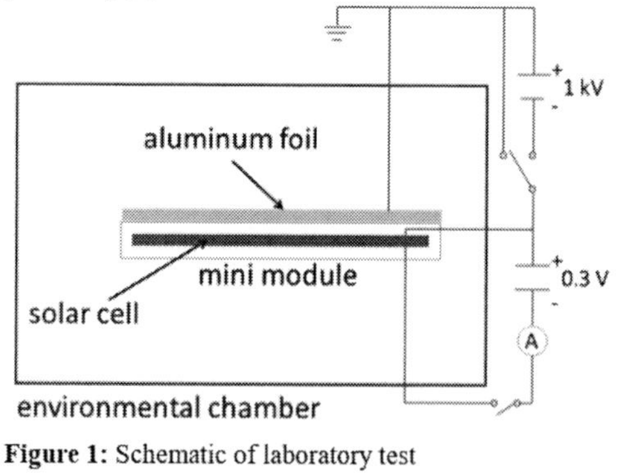

Figure 1: Schematic of laboratory test

3 RESULTS AND CONCLUSIONS

In order to reveal reasons for the differences found between R_{sh} progressions predicted by the preliminary model and actual field measurements three aspects are investigated in this paper: 1. sources of error; 2. additional influence factors for PID-s, which are not yet part of the PID-s model; 3. improvements to the model functions based on description of physical processes.

3.1 Sources of error

The shunt resistance (R_{sh}) of a solar cell is

temperature dependent. Therefore R_{sh} measurements of fielded modules show fluctuations due to temperature variations. In order to reveal the PID-s influence on R_{sh} only, the measurements have to be corrected for temperature [2,10]. In the laboratory two cell-types incorporated in one cell mini-modules were exposed to various temperatures without application of a bias voltage. During the treatment R_{sh} of the modules and the module temperature was measured. In Fig. 2 and Fig. 3 results are shown. Data presented here stem from samples exhibiting different degrees of degradation due to previous stress events, i.e. they show different starting values of R_{sh}. In Fig. 2 it is shown that between 22°C and 49°C the R_{sh} of cell-type A shows a negative temperature coefficient (TC) (Fig. 2a) while the R_{sh} of cell-type B shows a positive TC (Fig. 2b).

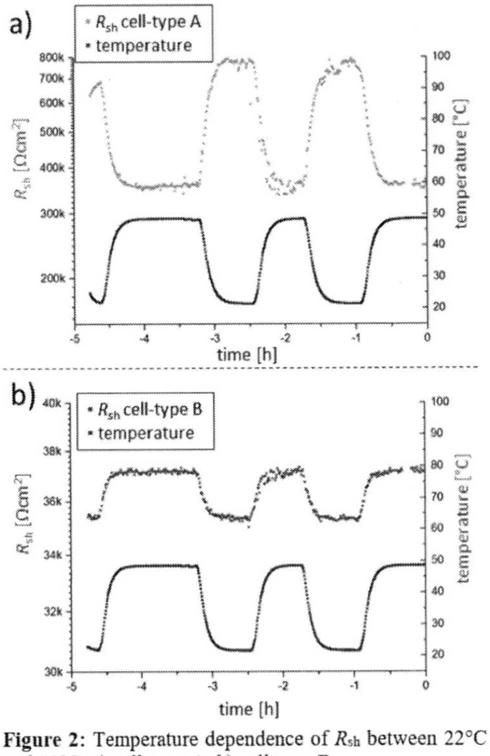

Figure 2: Temperature dependence of R_{sh} between 22°C and 49°C. a) cell-type A; b) cell-type B.

In the temperature range between 22°C and 90°C for cell type A the TC of R_{sh} remains negative (Fig. 3a). Surprisingly for cell type B between 22°C and 90°C the sign of the TC of R_{sh} changes (Fig. 3b). While between 22°C and 49°C TC is positive, at some temperature above 49°C the sign of TC changes to negative and remains like this up to 90°C. Cooling from 90°C down to 22°C again shows the change of the sign of TC. Note that a previous degradation of cell type B explains the recovery of R_{sh} during periods of constant elevated temperatures.

The measurements in Fig. 2 and Fig. 3 reveal that the temperature influence on the R_{sh} value can differ significantly depending on the cell-type and temperature range. This has to be taken into account when correcting R_{sh} for temperature. If a detailed analysis of the temperature dependence of the investigated samples is missing temperature correction of R_{sh} measurements might be inaccurate leading to errors in the description of the progression of R_{sh}. A possible explanation of the differences in the TC are changes in the dominant shunting mechanism at different temperatures [11].

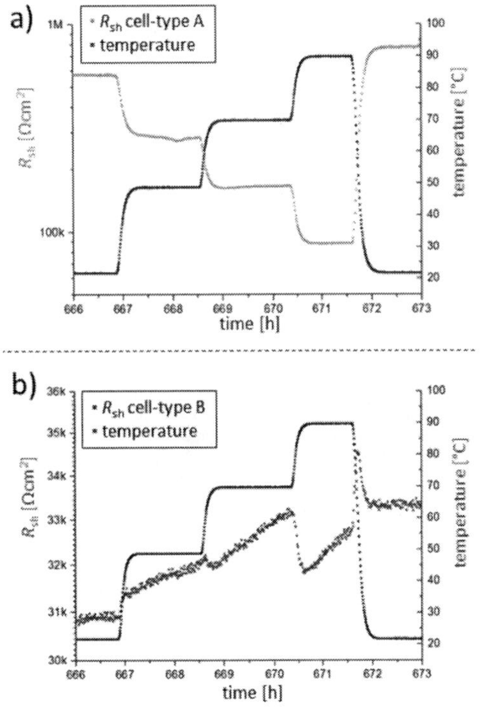

Figure 3: Temperature dependence of R_{sh} between 22°C and 90°C. a) cell-type A; b) cell-type B.

3.2 Additional influence factors for PID-s

Regarding the second aspect the influence of a grounded module-frame on PID-s even under dry conditions reported recently [6] is investigated in more detail. A mini-module with a grounded frame is exposed to a bias voltage of -1 kV, shunted via a grounded aluminum foil covering the front glass and then recovered by removing the foil at a temperature of 49°C and 70°C, respectively. In Fig. 4 the results are shown. At 49°C after removing the aluminum foil the R_{sh} recovers without visible influence of the grounded frame (Fig. 4a). In Fig. 4b bias voltage is applied again at 49°C without grounding the module surface by Aluminum foil, which does not influence the R_{sh} value. With increase of the sample temperature from 49°C up to 70°C R_{sh} starts to decrease slowly even without aluminum foil on the sample glass, whereas one normally would expect no shunting or even recovery to occur. Then the module glass is covered with a grounded aluminum foil leading to fast shunting. Shortly after that the foil is removed, instantly recovering the module. But in contrast to the behavior at 49°C at 70°C the recovery stops at a lower R_{sh} value followed again by a slow decrease of R_{sh}.

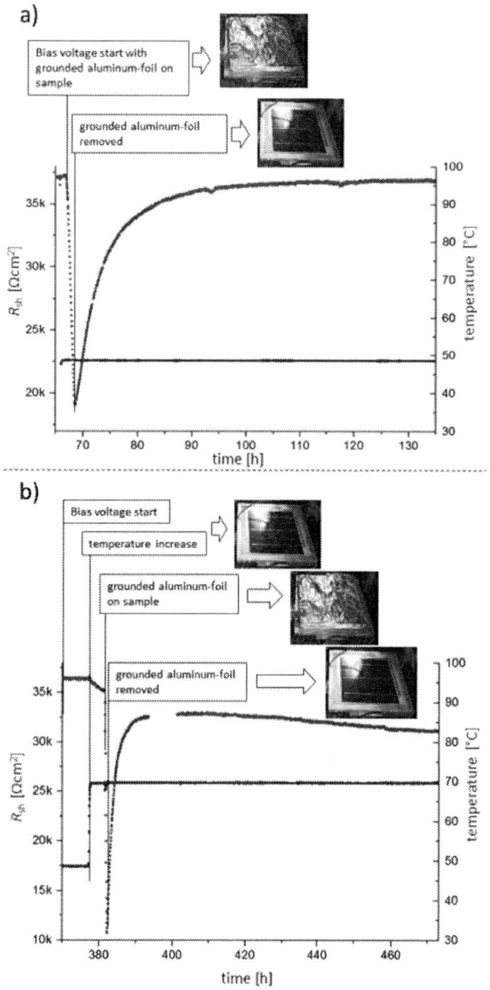

Figure 4: R_{sh} measurement during shunting and recovery of a framed mini module at a) 49°C and b) 70°C.

These measurements reveal that for the investigated mini-module there is a temperature dependent influence of the framing, i.e. the grounding geometry on PID-s. The effect can be explained by an increase of the glass conductivity at elevated temperatures. This leads to an increased area of the glass surface being at a potential closer to the grounded module frame causing a potential difference between glass and cell and due to this PID-s progression. In Fig. 5 this is illustrated. Previously it was published, that the potential distribution between cell and glass surface is dependent on the glass and encapsulation resistance [12]. Therefore the temperature dependent distribution of the potential found here may change for different glass or encapsulation materials.

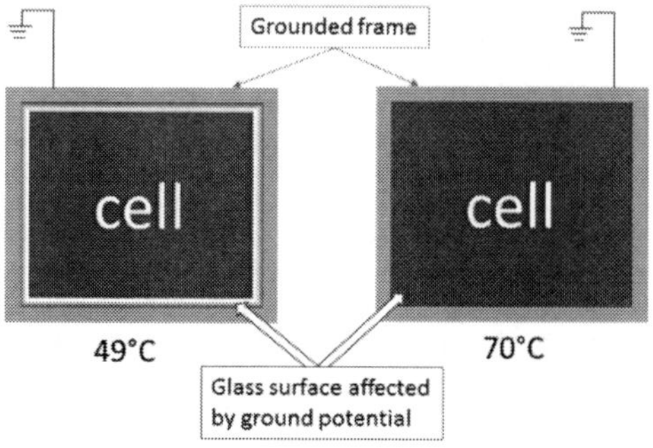

Figure 5: Illustration of the temperature dependent influence of the grounded frame on the potential of the glass surface of a mini module.

3.3 Improvements to the model functions

The last aspect treated in this paper is the mathematical description of the PID-s kinetics. The previously published preliminary model uses empirically chosen functions to describe the R_{sh} progression [9]. Here data using a physical model is presented improving the preliminary model by describing the shunting and recovery processes as a combination of drift and diffusion of sodium ions and atoms during shunting and recovery. This allows for a uniform description of shunting AND recovery using one set of common parameters and gives insight into the kinetics of PID-s. In Fig. 6 the basic principle is illustrated. Due to a bias voltage mobile sodium ions drift in the electric field through the SiNx layer and Na atoms can accumulate in stacking faults. The accumulation process inside the stacking fault is assumed to be fast compared to the drift process through the SiN-Layer. The shunting is described as a combination of drift and diffusion process, whereas recovery is described with the same equations with no drift process present. A detailed description of the model is currently in preparation and will be published elsewhere.

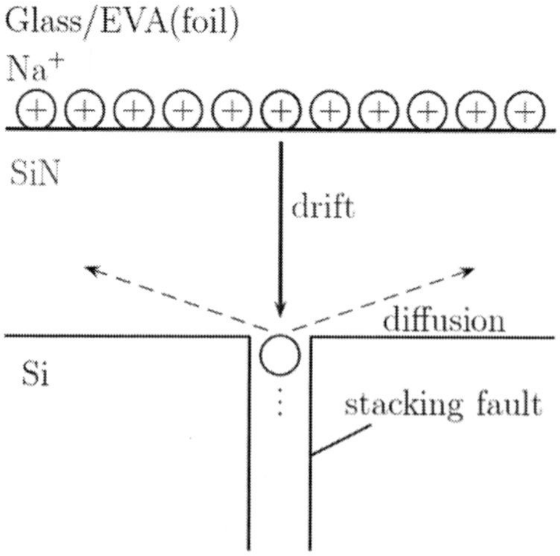

Figure 6: Illustration of cross section of the front side of a solar cell in a solar module with drift and diffusion paths of sodium.

In Fig. 7 laboratory measurement of the R_{sh} progression during shunting and recovery is compared to calculation of the physical model. It is revealed that the model is able to describe shunting (S)-, transition (T)- and recovery (R)-phase and gives a detailed description of the measurement results.

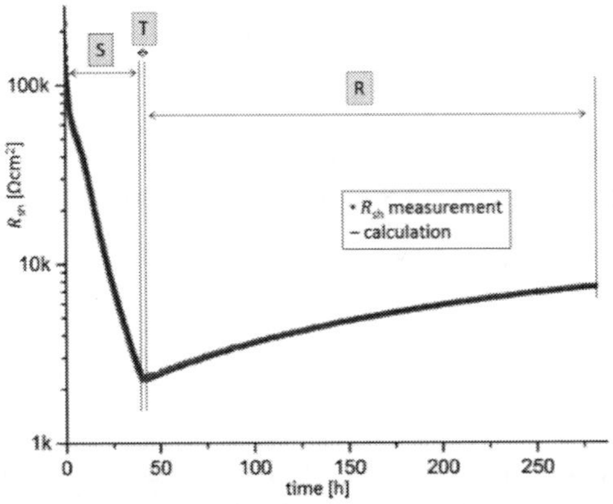

Figure 7: Comparison of R_{sh} measurement and calculation of the physical model.

4 SUMMARY AND OUTLOOK

Previously a preliminary phenomenological model describing the PID-s progression under field conditions was developed. In this work the improvement potential of this model is investigated. Three aspects were investigated: 1. sources of error; 2. additional influence factors for PID-s, which are not yet part of the PID-s model; 3. improvements to the model functions.

Regarding the first aspect the temperature influence on the R_{sh} measurement was investigated. The sign of the temperature coefficient (TC) was found to be both cell-type and temperature dependent. This reveals that a detailed analysis of the TC of the investigated samples is necessary in order to correct the temperature influence on the R_{sh} measurements without error.

Concerning the second aspect the influence of a grounded module-frame on R_{sh} was investigated in more detail. It was found that the test sample showed PID-s due to the frame at elevated temperatures, which is not yet included in the preliminary model.

The last aspect treated in this paper is the improvement of the mathematical description of the PID-s kinetics. A physical model was developed describing the shunting and recovery processes as a combination of drift and diffusion of sodium ions and atoms during shunting and recovery. First comparison of measurement and calculations show that with this model a uniform and detailed description of both shunting and recovery is possible. Details of the physical model will be published in an upcoming paper.

In a next step the frame influence and the physical model will be included in an improved version of the preliminary PID-s model.

4.1 References

[1] M. Schütze, M. Junghänel, M. B. Koentopp, S. Cwikla, S. Friedrich, J.W. Müller, P. Wawer, Proceedings 37th IEEE Photovoltaic Specialists Conference, Seattle, Washington, USA, 2011, pp. 821-826.

[2] C. Taubitz, M. Schütze, M. B. Koentopp, Proceedings 27th European PV Solar Energy Conference, Frankfurt, Germany, 2012, pp. 3172-3176.

[3] P. Hacke, K. Terwilliger, R. Smith, S. Glick, J. Pankow, M. Kempe, S. Kurtz, I. Bennett, M. Kloos, Proceedings 37th IEEE Photovoltaic Specialists Conference, Seattle, Washington, USA, 2011.

[4] P. Hacke, R. Smith, K. Terwilliger, S. Glick, D. Jordan, S. Johnston, M. Kempe, S. Kurtz, Proceedings 38th IEEE Photovoltaic Specialists Conference, Austin, TX, USA, 2012.

[5] M. Schütze, M. Junghänel, O. Friedrichs, R. Wichtendahl, M. Scherff, J. Müller, P. Wawer, Proceedings 26th European PV Solar Energy Conference, Hamburg, Germany, 2011.

[6] S. Koch, J. Berghold, C. Hinz, S. Krauter, P. Grunow, Proceedings 31th European PV Solar Energy Conference, Hamburg, Germany, 2015.

[7] P. Lechner, S. Hummel, D. Geyer, H.-D. Mohring, Proceedings 28th European PV Solar Energy Conference, Paris, France, 2013.

[8] P. Hacke, R. Smith, K. Terwilliger, G. Perrin, B. Sekulic, S. Kurtz, Proceedings 28th European PV Solar Energy Conference, Paris, France, 2013.

[9] C. Taubitz, M. Kröber, D. Buß, M. B. Koentopp, Proceedings 31th European PV Solar Energy Conference, Hamburg, Germany, 2015.

[10] S. Koch, J. Berghold, B. Abdullah, C. Hinz, P. Grunow, Proceedings 29th European PV Solar Energy Conference, Amsterdam, Netherlands, 2014.

[11] K. Bothe, K. Ramspeck, D. Hinken, C. Schinke, J. Schmidt, S. Herlufsen, R. Brendel, J. Bauer, J.-M. Wagner, N. Zakharov, O. Breitenstein, Journal of applied physics, 106, 104510, 2009.

[12] S. Pingel, S. Janke, J. Seydewitz, R. Alam, S. Koch, J. Kupke, J. Berghold, Proceedings 29th European PV Solar Energy Conference, Amsterdam, Netherlands, 2014.

PV MODULE TEST FOR ARID CLIMATES INCLUDING SAND STORM AND DUST TESTING

G. Mathiak, M. Hansen, M. Schweiger, L. Rimmelspacher, W. Herrmann, J. Althaus, F. Reil
TÜV Rheinland Energy GmbH
Am Grauen Stein, 51105 Cologne, Germany
Email: Gerhard.Mathiak@de.tuv.com

ABSTRACT: More and more PV systems are installed in deserts with arid climate. Test procedures for PV modules for hot and dry climates are under development. In a comprehensive standard for arid climate conditions, it is necessary to include sand abrasion and dry cleaning resistivity test for PV modules. Two sand abrasion test methods (MIL STD 810G: Blowing sand, DIN 52348: Sand trickling) are compared concerning effects on transmittance and reflectance of PV glasses and PV modules. Additionally, the evaluation of power plants should not only address the modules, but also a proper cleaning concept and anti- sand storm measures.

Keywords: Qualification and testing, Arid climate, Sand abrasion, Soiling

1 INTRODUCTION

More and more PV systems are installed in deserts with arid (hot-dry) climate. Higher module temperatures, ultraviolet radiation, sand storms and soiling by dust are the main challenges for the special module design and the module test.

From real weather data and module damages in the field, appropriate test parameters for standard test procedures have to be deduced: A combination of sand and dust tests with UV exposure pre-conditioned test samples is recommended to determine the resistance of PV components against abrasion effects from wind-blown sand.

2 ARID CLIMATE – TEST STANDARDS

The main type approval standards for PV modules (IEC 61215[1] / IEC 61730[2]) were developed for a moderate climate. The module and environmental temperatures of Cologne with moderate climate (Fig. 1) correspond better to the standards than the module and environmental temperatures of Tempe with arid climate (Fig. 2). The extrapolation of module temperatures to higher irradiances and environmental temperatures can lead to temperatures significantly above 85°C in particular the temperatures inside the laminate and junction box housings may considerably exceed material ratings.

Apart from higher module temperatures and higher daily temperature changes, higher daily and annual irradiance with an increased portion of UV radiation affect the modules more severely. More sun does not only mean more energy yield, but also more stress to the components.

A test scenario with adapted test parameters for the resistivity of PV modules to desert conditions [3] is presented in Fig. 3. Higher UV load is applied. The maximum temperature of the thermal cycling is increased. A sand abrasion test is included. The irradiance for the hot spot endurance test and the temperature test is raised to 1100 W/m². The bypass diode test is more severe.

Performance testing for arid climates is under development in new standards. A new approach is made in IEC 62892 (draft) [4] where testing of PV modules is described to differentiate performance in multiple climates, including arid climates (Fig. 4 and Tab. 1). The PVQAT High Temperature Group started a systematic analysis of IEC 61215 / IEC 61730 where the effect of higher temperatures to each test is estimated.

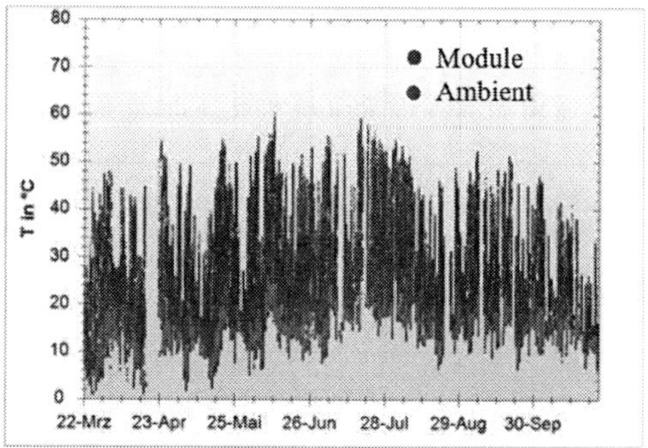

Fig 1: Module (backsheet) and ambient temperature [°C] at the test site in Cologne (Germany) during the summer half year [3].

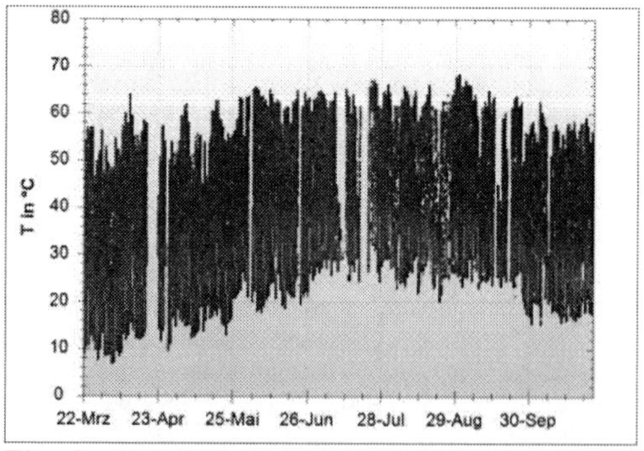

Fig. 2: Module (backsheet) and ambient temperature [°C] at the test site in Tempe (Arizona) during the summer half year [3].

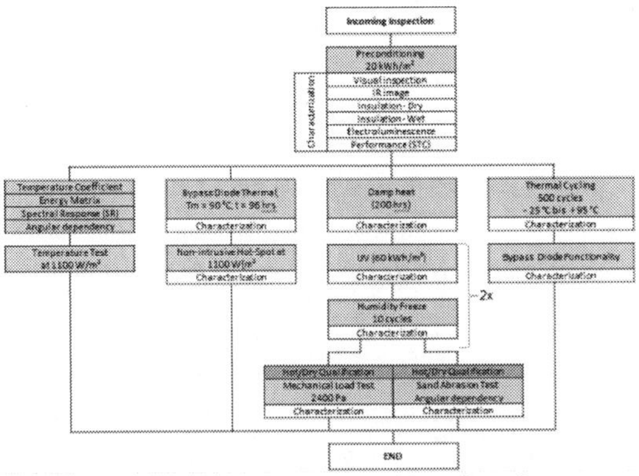

Fig. 3: Possible test scenario with adapted test parameters for the resistivity of PV modules to desert conditions with higher UV load, adapted TC testing and sand abrasion test [3].

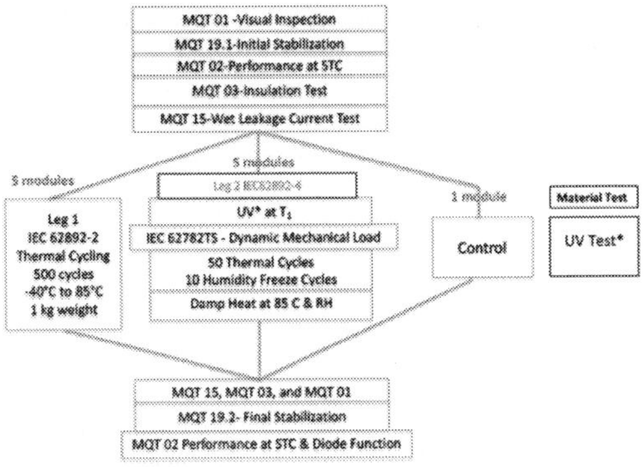

* UV Exposures TBD

Fig. 4: IEC 62892 (draft) [4]: Flow Chart for Testing with two legs testing five modules. UV exposures are to be defined. MQT Tests are described in IEC 61215-2.

IEC 60721-2-1 Climate Zones	Mounting Classes	
	Rack Mount	Roof Mount
Extremely Warm Dry	Leg 1: 500 thermal cycles Leg 2: UV exposure at 80 °C, DML, 50 TC, 10 HF, and 1000 hours of DH (85/40)	Leg 1: 500 thermal cycles Leg 2: UV exposure at 100 °C, DML, 50 TC, 10 HF, and 1000 hours of DH (85/40)

Tab. 1: IEC 62892 (draft) [4]: Test levels for different climates and mounting configurations. Extremely warm dry climate corresponds to arid desert climate.

3 SAND ABRASION TESTS

3.1 Different sand abrasion test methods

Each location of a PV power plant in a desert should be analyzed for its sand type (hardness, grain size distribution), its wind conditions (speed, frequency) and its history of sand and dust storms [5].

In this study two different sand abrasion tests were applied and analyzed. A key parameter for the adaptation of a sand abrasion test to a specific desert region is the momentum per area which means the product of the sand mass per area and the wind speed during the whole operation time [6].

The standard MIL STD 810G was originally developed for environmental tests of military material. Its method 510.5 Procedure II: Blowing Sand [7] uses as abrading sand silica sand with at least 95 % SiO_2 particles with diameters from 150 µm to 850 µm. The test is performed inside a closed wind tunnel chamber. The test air temperature can be chosen between ambient and high operating temperature (up to 70°C). Wind speeds of 18 m/s capable of blowing large particle sand are common, whilst gusts reach up to 29 m/s. Sand concentration for natural conditions is 0.18 g/m³. Sand concentration of 2.2 g/m³ is applied for material likely to be used close to helicopters operating over unpaved surfaces. Typical test duration is 1.5 hours per vulnerable area i.e. on the front glass and the backsheet with the higher sand concentration (see Tab. 2).

MIL STD 810G – Method 510.5, Proc. II: Blowing Sand	
Duration	1.5 h
Concentration	2.2 g/m³ (most severe)
Wind speed	18 m/s
Angle	0…25°
Mass per area	213 kg/m²
Momentum per area	3849 kg /m/s
Exposition area, diameter	200 mm
Temperature	60°C

Tab. 2: Test parameters for a typical blowing sand test over 1.5 hours with a momentum per area of 3849 kg/m/s.

The sand trickling tests according to DIN 52348 [8] and ASTM D968-05 [9] are usually used in the industry to investigate the abrasion resistance of glass and surface coatings. The test setup consists of a funnel for the sand and a tube with two sieves to guide the free fall of the sand. The tube is 1.65 m high and has a diameter of 120 mm. The maximum speed of the free falling sand particles is 5.7 m/s. The sieves homogenize the particle concentration over the exposition area. Standardized natural silica sand (ASTM 20/30 sand) with 95 % SiO_2 and grain diameter of 0.5 mm to 0.7 mm is mainly used for this test. Silica sand produces a higher rate of abrasion compared to the most natural desert sands. The test is adapted to tempered glass sheets and modules in original size and is without continuous sample rotation (Tab. 3).

DIN 52348: Sand Trickling	
Mass	3 kg
Duration	approx. 7 min
Exposition area, diameter	120 mm
Particle speed	5.7 m/s
Angle of impact	45°
Mass per area	265 kg/m²
Momentum per area	1512 kg/m/s
Temperature	25°C (room temperature)

Tab. 3: Test parameters for a typical sand trickling test with sand mass of 3 kg and a momentum per area of 1512 kg/m/s without consideration of the angle of impact. The duration of such an experiment is approx. 7 minutes. The main difference of ASTM D968-05 is the smaller tube diameter of 25 mm.

The sand trickling test is simpler and faster than the more realistic sand blowing test. In the applied configuration the exposition area of the sand trickling test is smaller than the solar cell size.

3.2 Optical characterization

After a sand abrasion experiment PV modules and PV glasses with and without anti-soiling or anti-reflection layers are characterized optically.

The aim of the measurements is to determine the optical transmission through the solar glass to a laminated solar cell behind the glass. Therefore, three solar glass measurements with a 10 mm diameter direct light beam and an integrating sphere were performed: 1) Transmission through the solar glass τ_tot 2) Reflection of the solar glass with reflections of both sides ρ_tot and 3) Reflection of the front side with light trap behind the glass r_1. From these measurements, it is possible to calculate the reflection at the backside r_2, the internal transmission t and the interface-corrected transmission $\tau\_Kor,G2$ which is used for the evaluation of the abrasion test [10]. For solar glasses, a specular reflectance analysis with haze and gloss measurements in comparison with surface roughness measurements is under discussion [11].

The transmission of a solar glass of a full-size PV module is characterized by the short-circuit current measured with a pulsed solar simulator. Due to the size of the exposition area of the sand abrasion test, single cells are measured: directly contacted through the backsheet foil or nondestructively with partly shading the cell under test [12]. The angular dependency of the transmission according to IEC 61853-2 [13] is determined by Isc measurement in dependence of the angle of incidence in the range from -85° to +85°, subsequent normalization to vertical incidence and division by the cosine of the angle of incidence.

3.3 Results of sand abrasion tests with modules

3.3.1 Sand blowing of PV modules

After the sand abrasion test (MIL STD 810G) of complete cells, the short-circuit current in dependence of the test duration and the angle of incidence was measured (Fig. 5). The relative lower losses at angles above 60° can be explained by roughening of the front glass.

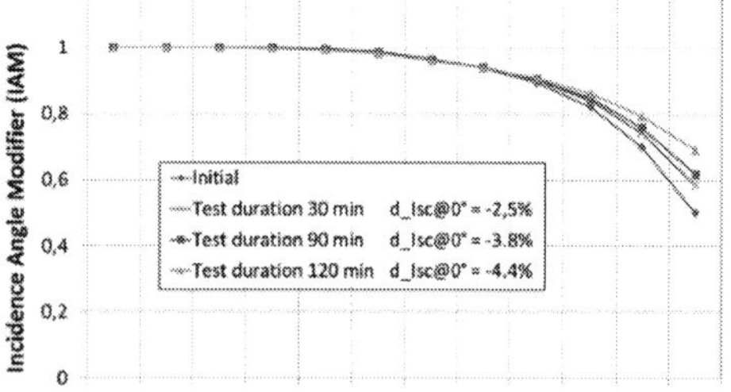

Fig. 5: Incidence angle modifier (IAM) over the angle of incidence. For the calculation of IAM, the Isc is divided by the Isc at normal incidence and by the cosine of the angle of incidence. The loss of Isc at normal incidence as a function of test duration is given in the legend.

3.3.2 Sand trickling of modules

Three cells of a PV module with AR-coated front glass were contacted through the backsheet foil without interrupting the circuitry of the module. The three cells were partly sand trickled with different sand masses.

Both methods: partly shaded cell (Tab. 4) versus directly contacted cell (Tab. 5) showed low transmission degradation (<1 %). One reason is that the cells are partly (approx. 60%) sand trickled.

Sand mass	Isc, shaded before	Isc, shaded after	Difference
9 kg	6,118 A	6,079 A	-0,64 %
6 kg	6,093 A	6,048 A	-0,74 %
3 kg	6,077 A	6,050 A	-0,44 %

Tab. 4: Measurement of the short-circuit current of an approx. 1/3 shaded cell with the nondestructive test method before and after sand trickling.

Sand mass	Isc before	Isc after	Difference
9 kg	8,965 A	8,938 A	-0,30 %
6 kg	8,916 A	8,896 A	-0,22 %
3 kg	8,902 A	8,894 A	-0,09 %

Tab. 5: Measurement of the Isc directly at the contacts before and after sand trickling.

3.3.3 Polishing of modules with eccentric sander

Three cells of a module with and a module without antireflective coating were polished with an eccentric sander. The grain sizes were K80, K120 and K240. After this process, the glass in front of the cells appeared frosted. The transmission change was measured with the partly shaded cell method.

The module without AR-coating (Tab. 6) showed a transmission gain of maximum 0.65% due to roughening of the glass surface. The module with AR-coating (Tab. 7) showed a transmission loss of maximum 1.72% due to the abrasion of the AR-coating.

Without AR-Coating			
Grain size	Isc, shaded before	Isc, shaded after	Difference
K80	5.881 A	5.919 A	0.65 %
K120	5.877 A	5,908 A	0.53 %
K240	5.866 A	5.889 A	0.39 %

Tab. 6: Measurement of the short-circuit current of an approx. 1/3 shaded cell with the nondestructive test method before and after polishing with an eccentric sander of a module without AR-coating.

With AR-Coating			
Grain size	Isc, shaded before	Isc, shaded after	Difference
K80	6.157 A	6.051 A	-1.72 %
K120	6.142 A	6.037 A	-1.70 %
K240	6.198 A	6.115 A	-1.34 %

Tab. 7: Measurement of the short-circuit current of an approx. 1/3 shaded cell with the nondestructive test method before and after polishing with an eccentric sander of a module with AR-coating.

3.4 Results of sand abrasion tests with solar glass

3.4.1 Sand trickling of solar glass

The transmission and reflection of a solar glass with antireflective coating was measured and the interface-corrected transmission $\tau\_Kor,G2$ was calculated (Fig. 6).

The degradation of the transmission $\tau\_Kor,G2$ was determined after sand trickling test with 3 kg, 6 kg and 9 kg (Fig. 7). The transmission loss is 0.5 % per test mass of 3 kg corresponding to 1% per 3000 kg/m/s.

Measurements according to ASTM D968-05 with a smaller tube diameter showed a transmission loss of 6 % [14, Fig. 8] for glasses with antireflective coating at a total momentum of 16 kg m/s which corresponds to a test mass of 3 kg and -4.5 % [15] with the same test mass. In these cases the transmission losses are 1% per 5700 kg/m/s resp. 1% per 7700 kg/m/s.

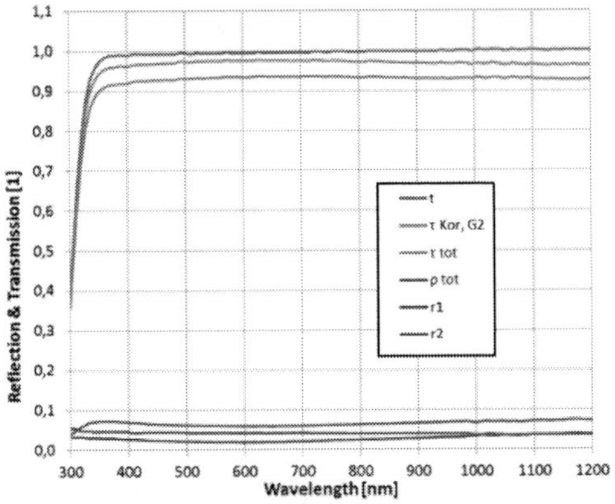

Fig. 6: Reflection and transmission of a one-side AR-coated PV glass. τ_tot is corrected by the internal transmission t (calculated) and the reflection losses of both glass surfaces r_1 (measured) and r_2 (calculated). The most relevant result for PV applications is the interface-corrected transmission $\tau\_Kor,G2$.

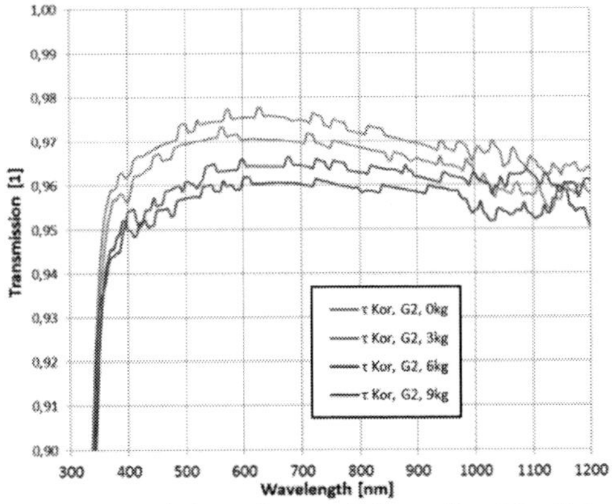

Fig. 7: Degradation of transmission $\tau\_Kor,G2$ of the solar glass as in Fig. 6 after sand trickling test with 3 kg (0.5 % transmission loss), 6 kg (1.0 % loss) and 9 kg (1.5 % loss).

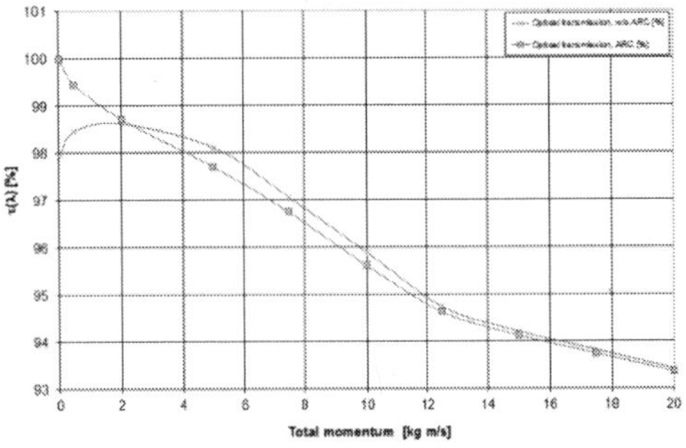

Fig. 8: Degradation of transmission of two solar glasses a) with and b) without antireflective coating according to ASTM D968-05. A total momentum up to 2 kg m/s leads to better transmission of the solar glass without ARC, before the transmission worsens. Both glasses converge to the same transmission after 12 kg m/s [14].

4 OUTDOOR PERFORMANCE AND SOILING

TÜV Rheinland operates two outdoor test facilities under arid climate conditions: Tempe in Arizona and Thuwal in the Kingdom of Saudi Arabia. The performances of several modules of the same types were continuously monitored and the modules were regularly inspected and cleaned.

In Thuwal, considerable dust settlement and missing rainfall lead to substantial daily performance losses in average of 0.5 % per day, in the worst case more than 7 % per day during a dust storm (Fig. 9). At lower angle of incidence the relative soiling power loss is higher. The severe soiling effect requires periodical cleaning for economic operation of a power plant.

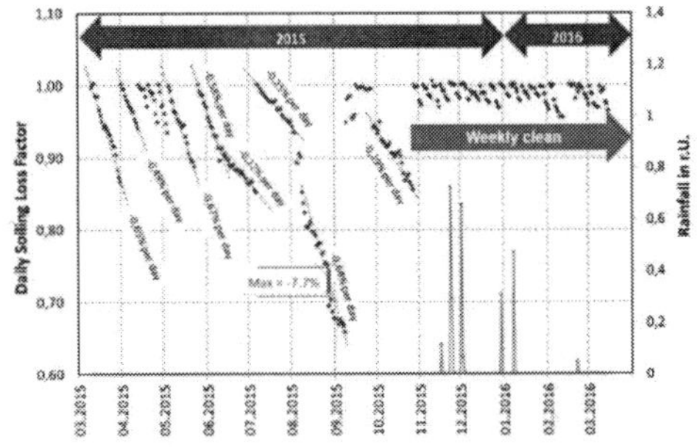

Fig. 9: Soiling pattern for Thuwal (Saudi Arabia) during the period March 2015 to March 2016 with occasional and weekly (from 28 OCT 2015) wet cleaning. High ambient dust concentration leads to an average daily decrease of -0.22 % and -0.67 % per day. During a dust storm in August 2015 a maximum daily decrease of 7.7 % per day was measured [16].

In Tempe, moderate dust settlement between periodic rainfall events leads to daily performance losses of approx. 0.1 % per day (Fig. 10). Effective cleaning by rain with complete recovery after rain and no permanent soiling is observed.

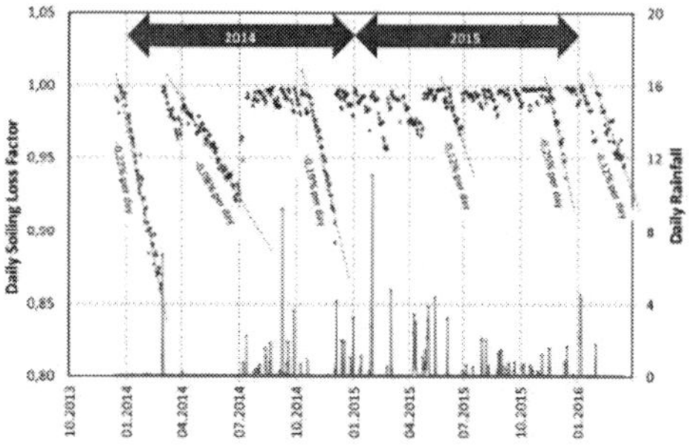

Fig. 10: Soiling pattern for Tempe (Arizona) during the period December 2013 to April 2016. Average daily soiling loss in dry periods is -0.08 % to -0.25 % per day [16].

Permanent soiling can occur if water condensates on the modules and dust sticks to the surface. Due to the lack of water in arid climates, wet cleaning can be performed on test facilities but dry cleaning (e.g. by brushes) has to be performed on commercial sites. Abrasion due to cleaning brushes has to be considered.

Both test sites showed complete recovery after rain or wet cleaning during the whole exposure duration. This agrees with experiences of [17] with no degradation of the glass in the Negev desert.

5 CONCLUSIONS

- Test procedures for PV modules for hot and dry climates are under development. In a comprehensive standard for arid climate conditions, it is necessary to include sand abrasion and dry cleaning resistivity tests.
- Two sand abrasion test methods (MIL STD 810G Blowing Sand, DIN 52348 Sand trickling) are compared concerning effects on transmission of PV glasses and PV modules. Glass with antireflective coating is more sensitive to abrasion than glass without coating, which can gain transmission by minor roughening of the surface.
- Additionally, the evaluation of power plants should not only address the PV modules, but also a proper cleaning concept and anti-sand storm measures (e.g. wind shield nets, module alignment considering wind directions, module tilt angle, module tracking with safety position in the case of sand storm).
- At the test sites Thuwal and Tempe, complete performance recovery was observed after rain or wet cleaning in the first year of operation. No permanent soiling was observed during the exposure duration.

6 REFERENCES

[1] IEC 61215-2: Terrestrial photovoltaic (PV) modules - Design qualification and type approval - Part 2: Test procedures, 2016.
[2] IEC 61730-2: Photovoltaic (PV) module safety qualification - Part 2: Requirements for testing, 2004.
[3] J. Althaus: "PV-Module unter Wüstenbedingungen", 11. Workshop Photovoltaik-Modultechnik, Cologne 2014.
[4] IEC 62892 (draft): Testing of PV modules to differentiate performance in multiple climates and applications - Part 1: Requirements for testing. 82/929/CD Date of circulation: 2015-01-23.
[5] F. Reil, I. Baumann, J. Althaus, S. Gebhard: "Evaluation of Current Standards and Practices for the Simulation of Wind-Blown Sands and their Applicability as Accelerated Ageing Tests for PV Modules", IEEE 39th PVSC, 2013, 1537-1541.
[6] C. Holze, A. Brucks: „Accelerated lifetime modeling on the basis of wind tunnel analysis and sand storm aging", Energy Procedia 49 (2014) 1692 – 1699.
[7] MIL STD 810G: Department of Defense Test Method Standard: Environmental Engineering Considerations and Laboratory Tests, 2008. – Method 510.5, Proc. II: Blowing Sand.
[8] DIN 52348: 1985:02: Testing of glass and plastics, abrasion test, sand trickling method.
[9] ASTM D 968-05:2010: Standard Test Methods for Abrasion Resistance of Organic Coatings by Falling Abrasive.
[10] F. Ruesch, L. Omlin, S. Brunold: "Solarglas PV. Zertifizierungsverfahren für Gläser im Einsatz in c-Si Fotovoltaikmodulen", Schlussbericht BFE-Projekt SI/500614, 2012.
[11] R. Alexander-Katz, R. G. Barrera: "Surface Correlation Effects on Gloss", Journal of Polymer Science: Part B: Polymer Physics, Vol. 36, 1321–1334 (1998).
[12] W. Herrmann, M. Schweiger, L. Rimmelspacher: "Solar simulator Measurement Procedures for Determination of the Angular Characteristic of PV Modules" Proc. 29th EU PVSEC 2014, 5CO.16.2, 2403-2406.
[13] IEC 61853-2 (draft): Photovoltaic (PV) module performance testing and energy rating - Part 2: Spectral responsivity, incidence angle and module operating temperature measurements.
[14] C. Holze: www.toughtrough.com, 2016.
[15] C. Völker, D. Philipp, M. Masche, T. Kaltenbach "Development of a Test Method for the Investigation of the Abrasive Effect of Sand Particles on Components of Solar Energy Systems", Proc. 29th EU PVSEC 2014, 5BV.4.12, 3142-3146.
[16] W. Herrmann: "Time evolution of PV soiling loss at test locations in different climates", Workshop on "Soiling Effect on PV Modules, 5-7 April 2016, Dubai.
[17] E. Klimm, T. Kaltenbach, D. Philipp, M. Masche, K.-A. Weiß, M. Köhl: "Soiling and Abrasion Testing of Surfaces for Solar Energy Systems Adapted to Extreme Climatic Conditions", Proc. 31st EU PVSEC 2015, 5CV.2.22, 2521-2523.

7 ACKNOWLEDGMENT

The work presented has been elaborated in the frame of the project 'Fidelitas' (FKZ: 0325735A) supported by the German Federal Ministry for Economic Affairs and Energy.

ADVANCEMENTS IN THE DEVELOPMENT OF "ATAMO": A SOLAR MODULE ADAPTED FOR THE CLIMATE CONDITIONS OF THE ATACAMA DESERT IN CHILE- THE IMPACT OF SOILING AND ABRASION

Enrique Cabrera[1], Andreas Schneider[1], Eckard Wefringhaus[1], Jorge Rabanal[1], Pablo Ferrada[2], Diego Thaller[1], Francisco Araya[2], Aitor Marzo[2], Mauricio Trigo[2], Douglas Olivares[2], Jannik Haas[3], Edward Fuentealba[2], Radovan Kopecek[1]

[1]International Solar Energy Research Center -ISC- Konstanz, Rudolf-Diesel-Str. 15, 78467 Konstanz, Germany
[2]University of Antofagasta, Av. Angamos 601, 1270300 Antofagasta, Chile
[3]University of Chile, Av. Tupper 2007, 8370451 Santiago, Chile
Corresponding author: Phone: +49 7531 3618356; Fax: +49 7531 3618311; Email: enrique.cabrera@isc-konstanz.de

ABSTRACT: Dust is currently one of the major problems of PV plants in Chile. In the present study we investigate the influence of abrasion and soiling on solar modules using different combinations of solar module materials and solar cells under simulated test conditions. Indoor and outdoor tests were performed at ISC Konstanz, Germany, and at the Solar Platform of the Atacama Desert (PSDA) in Yungay, Chile, respectively. We fabricated half-size and full-size cell modules employing different cell types, glass thickness, encapsulants, backsheets or glass-glass configurations. Modules were measured in the field for eight months. During exposure time, cleaning was performed every 60 days. Preliminary outdoor results indicate that ARC glass type exhibits higher Isc losses than standard (STD) glasses due to soiling, regardless of the combination of other materials used for the modules. Transparent backsheets (TBS) or glass on the rear side help to compensate the soiling losses due to increased internal reflection of light coming from the rear side of the module. In dirty condition, higher performance ratio (PR) was observed on half-size than on full-size cell modules under monofacial operation condition (the rear side covered: 97% compared to 94%). Under bifacial operation much higher but similar PRs for half- and full-size modules (both 107%) were reached. For the abrasion outdoor test, no damage was observed until now, even on vertical installed glasses which were installed perpendicular to the wind speed and close to the ground. For the indoor abrasion and soiling test, we adapted the standard testing conditions for sand (IEC and MIL-STD) using representative parameters such as temperature and humidity registered by 12 years of measurements and dust from eight PV plants located in the Atacama Desert. For the abrasion indoor test, the ARC layer on the glass was removed by all dust types after one minute of exposure and for 200-500 μm dust grain size. Increasing the exposure time, higher transmissivity (T), reflectivity (R) and Isc losses were detected until the abrasion effect almost reached saturation. When changing the tilt angle from 90° to 30° and reducing the grain size to 63-112 μm, lower losses were observed. Regarding the indoor test, a model was performed to predict the T and R losses according to the geographic place, glass type and sand density on the glass surface. The model predicts 3.5% higher T losses and 2.4% R losses for ARC compared to STD glasses with dust of less than 63μm grain size, regardless of the location.

Keywords: PV Desert Module, Abrasion Test, Soiling Test, ARC Glasses, Bifaciality, Silicon Solar Cell

1 INTRODUCTION

With regard to the quality of solar modules and performance, standard testing conditions as IEC were not developed for solar module applications in climates such as the Atacama Desert which is characterized by extremely high irradiation levels (in particular in the UV-B range), large amount of sun hours per day, low temperature, corrosive environment, partial high humidity known as "Camanchaca" (sea fog) and a really fine dust called "Chusca".

In February 2016 the PV installed capacity in Chile reached 1 GW, positioning PV as the first source of renewable energy in the country [1]. Currently one of the major problems of PV plants in Chile is the fine "Chusca" dust which - in combination with the hard environment condition in the Atacama Desert - results in huge losses in system performance. A study reported up to 40% soiling losses in the PV plants in Chile [2].

This fine Chusca dust in the presence of Camanchaca creates a hard dust layer on the surface of the PV modules which has a strong effect on the module lifetime, degradation and power achieved by the modules. Furthermore, it strongly increases the cleaning frequency required. As reported by GIZ [3], the maximum clean frequency of 14 PV farms located in the Atacama Desert, which are currently operating and injecting energy to the Chilean grid, is 1.2 times per month. Water consumption is in the order of a couple of liters of water per m^2 for each cleaning interval. On the other hand, according to a study from SISS-Chile (supervision of health services) [4] a common person in Chile consumes an average of 121 liters per day. Thus, with the amount of water to clean a 1 MW PV plant it is possible to supply water for a typical village in the Atacama Desert, where the scarce of water is a problem.

Recently, studies regarding the performance degradation due to dust accumulation on PV systems in the Atacama Desert have been carried out [5-7]. In order to quantify the performance of PV systems, the IEC defines in indicator called performance ratio (PR) [8]. The PR is a quality factor that describes the relationship between the actual and theoretical energy outputs of the PV plant considering the availability of the solar resource and size (capacity) of the installation. In [5], the PR, , decreased at rates of 4.0%/month (5.1%/month) for CdTe and 4.4%/month (6.0%/month) for mc-Si PV plants installed at 1700 m above sea level (the coast zone of Atacama), respectively. More studies of PV technologies operating at the coastal zone of the Atacama Desert found that a-Si/μc-Si and mono c-Si technologies degraded up to 0.18%/day [6]. Further studies regarding soiling in the

Capital of Chile (Santiago) reported degradation rates of 0.13%/day and 0.56 %/day [7].

The present study corresponds to advancements in the development of "AtaMo" [9]. In this case, we focus our investigation on the performance of different setups of half-size and full-size cell modules under indoor and outdoor abrasion and soiling test in order to find the best material combination against the impact of abrasion and soiling.

2 EXPERIMENT

2.1 Outdoor tests

Outdoor measurements were performed at the Solar Platform of the Atacama Desert (PSDA) [10] in Yungay, Chile, see Figure 1.

Figure 1: Testing under outdoor conditions at PSDA

Half-size and full-size single cell modules based on 6" p-type mc-Si and bifacial 6" n-type pseudo-square Cz-Si solar cells with standard 3 busbar metallization patterns, different glasses, encapsulants, and backsheets were designed. The ribbon soldering process on the front and rear side was performed manually or semi-automatically by a stringer machine using the same type of ribbon and flux. The glass thickness was varied using 1.5, 2.0 and 3.2 mm glass. For the 3.2 mm thick solar glass an antireflection coating layer (ARC) was also incorporated. For the encapsulation material ethylene vinyl acetate (EVA), a thermoplastic material (TM) and low UV light cut-off EVA (U) were used. For the backsheet, standard, transparent, and desert type sheets (which are optimized in terms of abrasion and thickness for desert regions), as well as glass, were employed, see Tables I and II.

For characterization, Electroluminescence (EL) measurements were carried out to control the soldering and lamination performance. Peel tests were carried out to monitor the soldering process. IV characteristics were measured by a h.a.l.m. cell- and module-flasher at STC conditions before and after lamination to determine the cell-to-module losses, and to quantify the losses in the environmental tests. For optical investigations, spectral photometer measurements of reflectivity (R) and transmissivity (T) versus wavelength from 300 to 1200 nm were carried out after and before the tests. In addition, for the abrasion test, the depth and the roughness of the glass surface generated by the dust was measured by Laser scanning microscopy (LSM).

Table I: Experimental setups of p-type monofacial modules for environmental testing

Glass thickness [mm]	Backsheet (BS)	Encapsulant
1.5	Desert (D)	EVA
1.5	Standard (S)	EVA
2.0	Desert (D)	EVA
2.0	Standard (S)	EVA
3.2	Desert (D)	EVA
3.2	Standard (S)	EVA
3.2	Desert (D)	TM
3.2	Standard (S)	TM
2.0	Transparent (T)	EVA
3.2	Transparent (T)	EVA
3.2	Standard (S)	U
3.2ARC	Standard (S)	EVA

Table II: Experimental setups of n-type bifacial modules for environmental testing (Enc = Encapsulant)

Front side glass [mm]	Rear side glass [mm]	Bifacial cell	Enc
2.0	2.0	FULL	EVA
2.0	2.0	HALF	EVA
1.5	1.5	FULL	EVA
1.5	1.5	HALF	EVA
1.0	1.0	FULL	EVA
1.0	1.0	HALF	EVA
3.2ARC	Transparent (T)	FULL	EVA
3.2ARC	Transparent (T)	FULL	TM
3.2ARC	Transparent (T)	FULL	U
3.2ARC	Transparent (T)	FULL 4 mc-Si	EVA
3.2ARC	Transparent (T)	FULL 4 Cz-Si	EVA
3.2	Transparent (T)	FULL 4 Cz-Si	EVA

The modules were measured in the field for a period of 8 months. During exposure time the modules were cleaned every 60 days (see Figure 3).

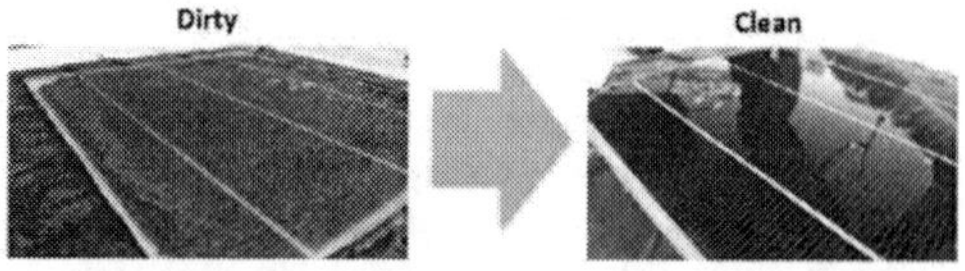

Figure 3: Before and after cleaning every 60 day

In addition to the modules, 3.2 mm glasses with and without ARC were installed at different inclinations. In particular vertical glasses were faced perpendicular to the wind speed and at 15 cm above from ground level. The maximum average wind speed measured at the PSDA was 14 m/s. The glass after abrasion exposure time of 100 days was returned to ISC Konstanz for T, R, and LSM analysis.

2.2 Indoor tests

Dust was collected from the ground at eight PV plants located in the 1st Region Tarapacá, 2nd Region Antofagasta and the 3rd Region Atacama y Coquimbo in Chile: 100 MW Amanecer, 50.3 MW San Andrés, 24 MW Diego de Almagro, 2.8 MW Esperanza, 1.4 MW Huayca I, 7 kW PSDA Yungay, 5.8 kW San Pedro and 5.8 kW Antofagasta (see Figure 3). The dust was separated with a dry sand sieve shaker in order to use grain sizes in the range of 200 to 500 μm and 63 to 112 μm for the abrasion indoor test and less than 63 μm for the soiling indoor test. The dust characterization was carried out by EDX (Energy Dispersive X-ray Spectroscopy) and XDR (X-Ray Diffraction) using the same set of parameters to measure each sample. Note that for grain sizes less than 500 μm, the EDX and XDR analysis of dust from module surface and from the ground provided the same soil composition. This result validates the procedure used to collect the dust and to perform the indoor tests.

The design of the abrasion test adapted to the climatic conditions in Chile was made after reviewing the blow sand standardizations MIL-STD-810G, IEC60068-2-68 and IEC60529 [11]. In the present study, the wind speed was set to 17 m/s according to the maximum average wind speed as registered between 2000 and 2012 by Meteorological Direction of Chile in the central valley of the Atacama Desert [12]. In this region most of the PV farms are installed. The tilt angle was adjusted to 30° and 90° for 3.2 mm glass with and without ARC using the different types of dust collected. The exposure time was 1, 2 and 3 min for all different dust types. For the Yungay samples, additional bifacial 5x5 cm^2 modules were made for 3, 6 and 9 min exposure time and 90° tilt angle.

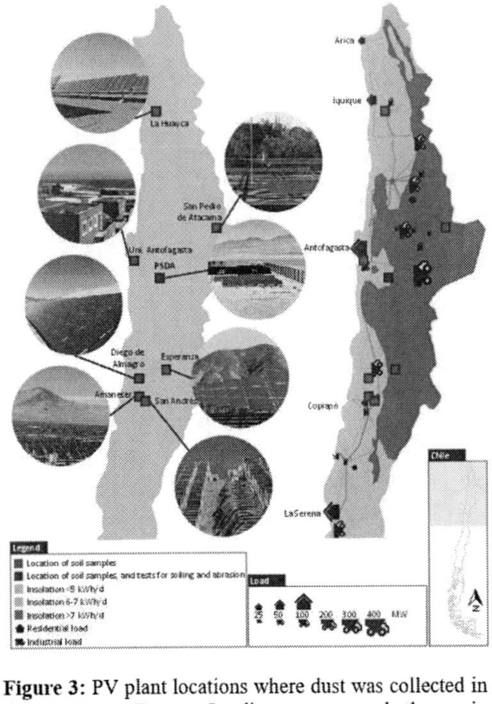

Figure 3: PV plant locations where dust was collected in the Atacama Desert. Irradiance map and the main industrial and residential consumers are shown. The red square indicates the PSDA in Yungay, where the outdoor tests were performed

To simulate the hard dust layer formed by "Chusca" dust in combination with "Camanchaca" fog, the glass samples with and without ARC were placed in the climatic chamber. This chamber was operated with a humidity (RH) variation according to the metrological data of 12 years as reported in [12]. This data ranges between 61 and 83% and temperatures between 55 and 65 °C corresponding to the operating temperature for c-Si modules registered by the University of Antofagasta. The grain size used was less than 63 μm according to the sand grain found on the modules surface in Yungay. After the exposure in the climatic chamber, the sand deposition was performed by free fall from 1 m height using a cylindrical tube of 8 cm diameter based according to [13]. After sand deposition on the glass, the sample was dried and the dust on the surface weighted with a balance of 0.1 mg sensitivity. Finally, each glass sample was encapsulated with 1.5 mm glass in order to avoid the loss of sand from the glass surface and for characterization.

3 RESULTS

3.1 Outdoor tests

The following results correspond to the short circuit current (Isc) losses and performance ratio (PR) before and after cleaning with an average of a clean step every 60 days in a period of 8 months of exposure time.

Regardless of the module configuration the modules with ARC glasses led to higher Isc losses compared to STD modules, figure 4.

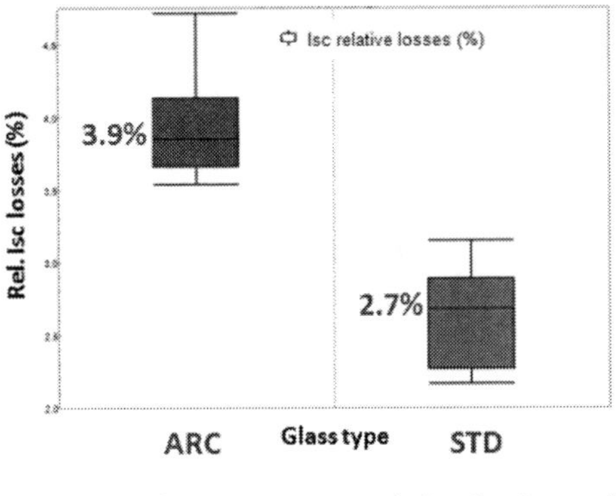

Figure 4: Glass Type versus relative Isc losses (%) showing higher losses for ARC than STD glasses

The soiling effect on the Isc losses was larger for standard and desert backsheets than for transparent backsheets possibly due to increased albedo light through the rear side that reflects from the accumulated dust layer on the glass and penetrates into the front side of the solar cell, see Figure 5. Additionally, reduced Isc losses for thermoplastic material and thinner glasses were measured. Previous studies have shown that the module performance is mainly dependent on the encapsulation materials, regardless of the backsheet used [9] under high humidity and temperatures after 6000 hours exposure time.

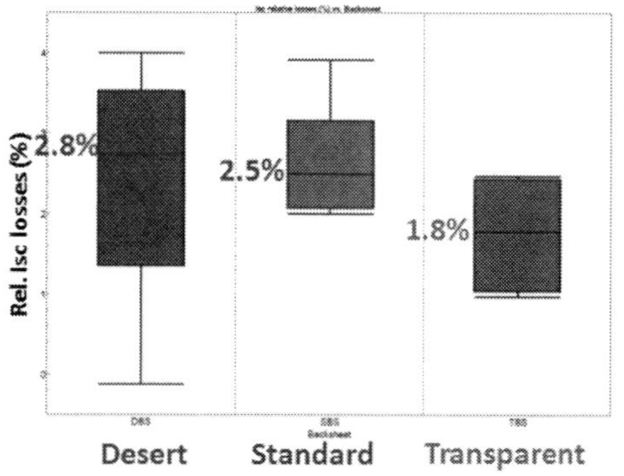

Figure 5: Relative Isc losses (%) for 3 different encapsulants. Lower losses for TBS than for SBS and DBS can be observed

Regarding the PR of different solar cell designs, in dirty condition and covering the rear side, better PR values were measured with half- compared to full-bifacial cell modules with 97.2% and 94.2%, respectively. If the rear side is not covered the PR of bifacial modules increased even more, achieving same values in dirty condition, 107.3% for half-size configurations and 107.0% for full cell design.

No damage in terms of abrasion were found for any glass group installed at different tilt angles and heights distributed in several places at the PSDA in Yungay, see Figure 6. Even for the most challenging case, vertical glasses perpendicular to the wind speed and 15 cm height from the ground, no abrasion effect was observed. The reason could be that the frequency of sandstorm in the Atacama Desert in comparison with others deserts is low.

Figure 6: Glass configurations at the PSDA in Yungay

3.2 Indoor tests

Soiling and abrasion indoor experiments were carried out at ISC Konstanz, Germany. As the sand deposition on the glass surface increased, higher losses in T and R were measured. Losses were larger with the use of ARC compared to STD glass type, see Figure 7.

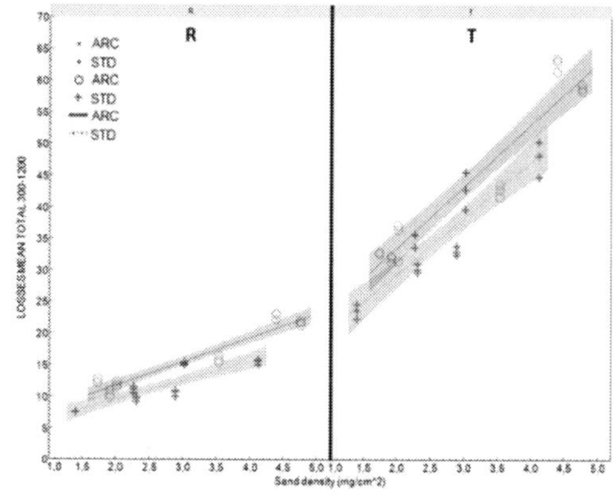

Figure 7: Absolute R & T mean losses (%) versus sand density (mg/cm^2) using ARC and STD glasses

Based on the experimental data generated by the use of different dust, a model was developed in order to simulate the losses in T and R for every place. A relationship between glass type and sand deposition was found, see Figure 8. Setting the model to 2 mg/cm^2, T losses ranged from 34 to 43% and R losses from 30 to 39%. Changing the ARC glass type to STD, a loss reduction of 3.5% in T and in 2.4% R were obtained, regardless the place.

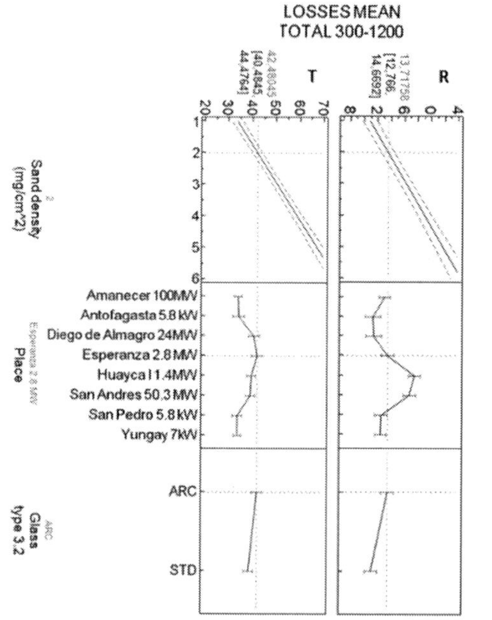

Figure 8: Model to predict the absolute R & T mean losses (%) versus sand density (mg/cm^2) using ARC and STD glasses

Although no damage in terms of abrasion was found at the PSDA in Yungay, abrasion indoor tests were performed using parameters from meteorological and measured data in the Atacama Desert. The goal was to predict the damage that could occur. The wind speed was set to 17 m/s according to the maximum average wind speed reported after 12 years of measurements in the central valley of the Atacama Desert, where most PV plants are located.

Figure 9 illustrates the results for tests with glass in a vertical position with respect to the wind direction. As the test time increased, the transmissivity curve went down. LSM pictures before and after three minutes exposure time show that the abrasion test was very aggressive, not only removing the ARC layer, but also producing deep holes in the glass surface..

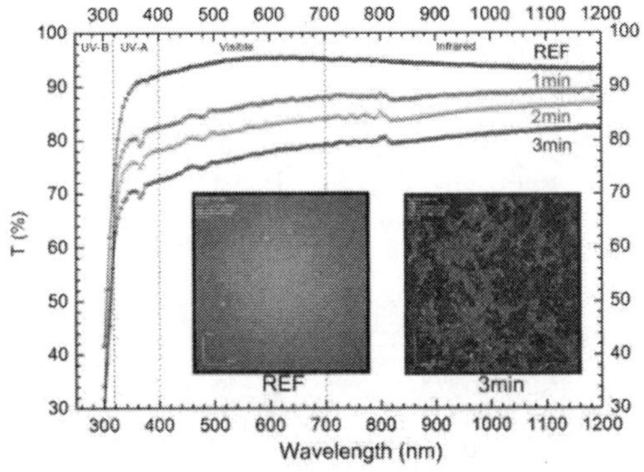

Figure 9: Transmissivity versus wavelength and increasing exposure time. Shown are LSM images before and after a test time of three minutes on ARC glass

The test for tilt angles of 30° was less aggressive than for 90°, changing the slope of the losses in T and R, and reducing the depth of the holes on the glass surface. In terms of Isc losses, 3.2 mm front glass bifacial modules with and without ARC reveal Isc losses of 3.5% for standard and 6% for ARC modules after three minutes of exposure time. After six and nine minutes, Isc losses slightly increased and the difference between ARC and STD glass type remained almost the same (Figure 10), i.e. the abrasion effect reaches almost saturation after three minutes.

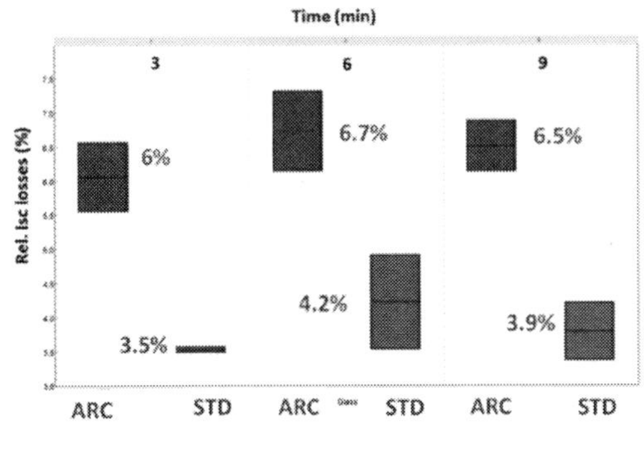

Figure 10: Short circuit current (Isc) losses in (%) versus exposure time for bifacial glass modules (red) and bifacial glass modules with ARC (blue)

In order to characterize the impact of dust abrasion in more detail, 72 LSM measurements were carried out, inside and outside of the abrasion area (Figure 11). If the ARC layer thickness is assumed to be 0.15 µm, it becomes obvious that the ARC layer disappeared. Reducing the grain size of the dust from 200-500 µm to 63-112 µm, which is more realistic, the dust impact still remove the ARC layer.

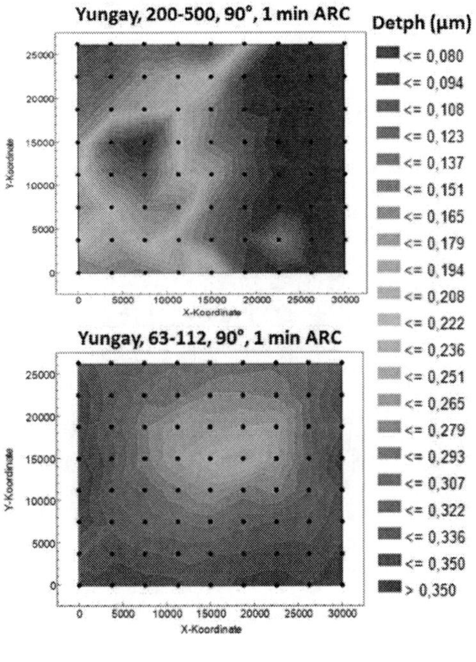

Figure 11: Depth map (µm) plotted over the position of LSM measurement coordinates in µm for two different grain size ranges, 90° tilt angle, after one minute of exposure for ARC glass

4 CONCLUSION

Soiling and abrasion indoor and outdoor tests were carried out in order to determine the best combinations of materials preventing soiling and abrasion losses.

For outdoor and indoor soiling tests, higher losses in T, R and Isc for ARC compared to STD glass type were measured. An Isc loss reduction was determined for the transparent backsheet or a glass sheet as a rear side.

For abrasion no damage was found at the PSDA in Yungay, possibly due to the low frequency of dust storm on the Atacama Desert (in comparison to other deserts such as the Sahara). For the adapted abrasion indoor test, using parameters that have been reported and measured in the Atacama Desert, good analysis and characterization have been shown. Again ARC glass type shows larger T, R and Isc losses compared to STD glass type. The ARC layer on the glass surface was partially removed for all studied dust samples (places). Nevertheless, long term outdoor characterization is required to validate the indoor tests.

5 ACKNOWLEDGEMENTS

This work was supported by the German Ministry of Education and Research (BMBF) under contract no. 01DN14005 (SolarChilD). This investigation is supported by the University of Antofagasta, Universidad of Chile and the Chilean Solar Energy Research Center (SERC Chile) under the framework of the BMBF project SolarChilD. The authors would like to thank Brigitte Lutz and Dr. Renate Zapf-Gottwick from ipv of the University of Stuttgart for kindly providing the EDX analysis and to Dr. Hilmar Hofmann from the Limnology Institute Konstanz of the University of Konstanz for proving the sand sieve shaker.

6 REFERENCES

[1] Report CIFES, "Energías Renovables en el Mercado Eléctrico Chileno", February 2016. [http://cifes.gob.cl]

[2] P. Darez et al., "Soiling Losses in the Atacama Desert", Intersolar Europe 2014, München, http://www.gter.cl/wp-content/uploads/2015/10/20140513-Soiling-losses-in-the-Atacama-intersolar.pdf

[3] GIZ report 2014, "Experiencias de Plantas Solares en Chile en Operación y Conectadas a la Red", http://www.fch.cl/recurso/sustentabilidad/documentos-plan-estrategico-solar/

[4] SISS Chile, Super Intendencia de Servicios Sanitarios, ''SISS da a conocer nivel de consumo de agua potable en el país'' 2009, pp. 11-13, http://www.siss.gob.cl/577/articles-7663_recurso_5.pdf

[5] F. Araya et al., ''Performance Analysis of Photovoltaic Systems Installed at different Sites in the Atacama Desert''. 32nd EU PVSEC Munich, Germany, 2016

[6] P. Ferrada et al., "Performance analysis of photovoltaic systems of two different technologies in a coastal desert climate zone of Chile", Solar Energy 114 (2015) 356–363, DOI:10.1016/j.solener.2015.02.009

[7] E. Urrejola et al., ''Effect of Soiling and Sunlight Exposure on the Performance Ratio of Photovoltaic Technologies in Santiago'', Chile. Energy Conversion and Management 114 (2016), 338-247

[8] IEC 61724. 1998. Photovoltaic system performance monitoring – guidelines for measurement, data exchange and analysis, IEC.

[9] E. Cabrera et al., "Advancements in the Development of ''AtaMo'': a Solar Module Adapted for the Climate Conditions of the Atacama Desert in Chile", Proceeding of 31st EUPVSEC, Hamburg, Germany, 2015, pp. 1935-1941, DOI: 10.4229/EUPVSEC20152015-5EO.1.6

[10] P. Ferrada et al., "AtaMo: PV meets the high potential of the Atacama Desert", PV Tech Power Volume 05, 2015, pp. 55-60.

[11] F. Reil, ''Evaluation of Current Standards and Practices for the Simulation of Wind-Blown Sands and their Applicability as Accelerated Ageing Tests for PV Modules'', 39th IEEE PVSC, pp. 1537 - 1541, 2013, Tampa, FL

[12] DM Chile, Dirección General de Aeronáutica Civil, Dirección Meteorológica de Chile, ''Anuario Climatológico 2000- 2013'', Chile, http://164.77.222.61/climatologia/

[13] H. Qasem et al., ''Dust-induced shading on photovoltaic modules''. Prog. Photovolt: Res. Appl. 2014; 22:218–226. DOI: 10.1002/pip.22

INVESTIGATION OF DAMP HEAT AGING ON SODA - LIME GLASS FOR PHOTOVOLTAIC APPLICATIONS

V. Guiheneuf, F. Delaleux, O. Riou, P-O. Logerais, J-F. Durastanti

University of Paris-Est, CERTES Laboratory (EA 3481)
IUT de Sénart-Fontainebleau, 36 rue Georges Charpak, 77567 Lieusaint, France
Tel: +33 1 64 13 46 86; E-mail: vincent.guiheneuf@u-pec.fr;

ABSTRACT: Damp heat test was performed on commercial soda-lime glass to characterize functional properties of glass in photovoltaic applications and to define aging mechanisms. In addition to the measurements of optical properties, FTIR, SEM-EDX and XPS analyses were carried out. The damp heat exposed samples presented an increased transmittance in UV, visible and near-IR ranges. The results were explained by hydration of glass network with heat. Molecular water adsorption involves a leaching process of network modifier cations and especially sodium ions. Hydrolysis also occurs in silica network with free molecular water reaction. This phenomenon enhances concentration of hydroxyl groups in glass highlighted by FTIR measurements and promotes formation of silanol groups. The tin side dependence on sodium leaching which acts as a passivating barrier for ionic transport is emphasized which would be worth taking into consideration for photovoltaic module design.
Keywords: soda-lime glass; photovoltaic; damp heat; aging; leaching

1 INTRODUCTION

A photovoltaic (PV) module consists of a stack of different kinds of materials: glass, polymer (e.g. ethylene vinyl acetate for encapsulation and polyvinyl fluoride for backsheet) and photovoltaic cell. Glass plays a major role in most of the photovoltaic systems such as substrate and cover. The glass front side of photovoltaic panels is exposed to many environmental stresses (UV radiation, heat, humidity, shading, abrasion by dust, snow load, etc.) depending on the localization and the weathering. Jordan et al. reported degradation rates in their publications. The averages of reported rates for Si and thin-film technologies are of 0.7%/year and of 1.5%/year respectively [1]. It is found that the main issue is caused by failure of cell interconnects, solder bonds or bypass diodes [2]. Other challenges concern the transmittance changes because of encapsulant discoloration, delamination at glass/encapsulant interface and anti-reflection coating degradation [3-4]. The aim of the present work is to investigate the influence of damp heat test on the properties of commercial float soda-lime glass. Efforts were made to characterize the functional properties of glass used in PV applications (i.e. transmittance) and to understand the degradation mechanisms of glass by physicochemical analysis.

2 EXPERIMENTAL

A commercial soda-lime glass was considered for this study (see table I) with a 2 mm thickness without texture nor coating. Aging accelerated tests were performed in an environmental chamber (MKF 240 model from Binder) under damp heat conditions, i.e. 85°C ± 1°C with 85% r.h. ± 5% for up to 1000 h. The glass samples were installed in a vertical position to similarly expose both their sides.

Table I: Composition of glass samples (at.%)

Al_2O_3	CaO	MgO	Na_2O	SiO_2	Others
0.7	9.0	3.5	11.9	74.4	0.5

The aging behavior was characterized by several methods which can be divided into optical and chemical techniques. Transmittance and reflectance measurements were performed by using a spectrophotometer (Perkin Elmer Lambda 750 UV/Vis/NIR) equipped with an integrated sphere of 60 mm diameter. The absolute accuracy was assessed at 0.002 transmittance or reflectance units. Infrared spectra were obtained by averaging 32 scans taken with a Fourier transform infrared (FT-IR) spectrophotometer (Perkin Elmer Frontier) in Attenuated Total Reflectance (ATR) mode. The chemical measurements were carried out by using analytical scanning electron microscopy (ASEM, Jeol JSM 6301-F) with energy dispersive X-ray fluorescence (EDX). Due to the electrical insulating behavior of the glass, charge neutralization was made by platinum sputtering. SEM analysis provides chemical information within 1 μm depth. The accuracy is less than 0.5%. Moreover, two glass samples (unexposed and after 1000 h under DH) were analyzed by X-ray Photoelectron Spectroscopy (XPS) using VersaProbe (PHI 5000) Scanning XPS microprobe with a monochromated Al K_α anode (hν=1486.6 eV, 200 μm diameter). Some survey scans were performed on each sample in order to check the homogeneity of the surface composition. Next, one detailed scan was realized on each sample to obtain accurate quantitative analysis. The probed depth is less than 10 nm which allows to analyze the near-surface layer of glass.

3 RESULTS

Fig. 1 exhibits the transmittance and reflectance spectra of glass samples before and after 1000 h of damp heat treatment. A significant increase in the optical transmission after 1000 h of damp heat exposition in the 300–2300 nm wavelength range can be observed. This fact is interesting for photovoltaic conversion of Si cell, typically in the 400-1100 nm range. Nevertheless, the transmittance increase in the UV range could accelerate the polymer degradation. Hence, the efficiency of the PV device would be diminished. Concerning reflectance measurements, the behavior is opposed but stays in the same proportion. So, the absorption of glass samples remains unchanged after this accelerated aging.

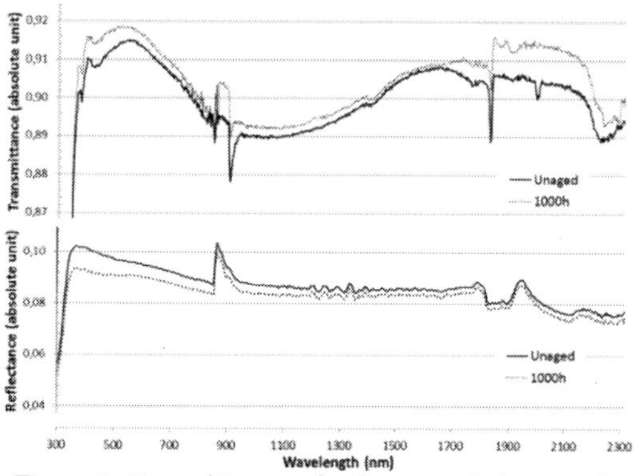

Figure 1: Transmittance and reflectance of glass sample before and after 1000 hours of damp heat treatment.

Fig. 2 shows the superposed FTIR spectra of each sample to visualize shifts of the characteristic peaks. By studying infrared reflectance spectra, structural changes can be noted during hydration of the three glass samples at different instants. The bending vibration of molecular water occurs at 1611 cm^{-1} but it can be modified by hydrogen bond stretch [5]. The band at 1646 cm^{-1} which is expected for molecular water (v_B (H_2O)), shows raised peak heights for samples after 500 and 1000 h under DH respectively. The peak shift towards lower wavenumbers can be explained by the weakening of the hydrogen bonds. The broad O-H absorption of silanol at 3600 cm^{-1} is due to the specificity of hydroxyl groups that can be freed or hydrogen-bonded.

The chemical analyses of the glass bulk approximately within 1 µm depth were performed by SEM-EDX to compare with those based on XPS analysis (see Table 2). Two kinds of results are displayed according to the Sn element concentration in near-surface glass composition for three samples: unaged, after 500 h and also after 1000 h under DH. On the tin side, SEM-EDX measurements reveal a leaching phenomenon of Na element during hydration. The Na_2O content decreases from 11.4 at.% to 9.9 at.% after 1000 h. The Na leaching does not appear linear because of the more pronounced drop in the Na content after the first 500 h contrary to the period between 500 h and 1000 h. The leaching did not occur for other intermediates (Al_2O_3) or network modifiers (CaO, MgO) which can be explained by lower migration kinetics in glass. The mechanism is similar for the no-tin side but in a more important proportion. The Na content decreases from 11.9 at.% for the unexposed sample to 9.2 at.% after 1000 h and here too, no significant change is observed for the other element. The tin side has a passivating barrier ability which reduces the sodium leaching and the glass network hydrolysis.

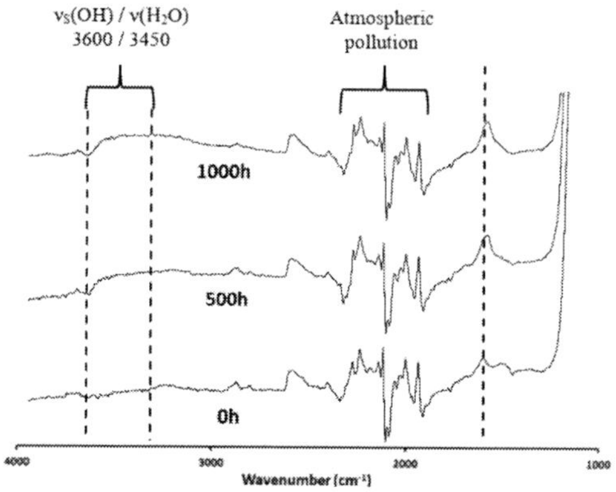

Figure 2: Infrared reflectance spectra of 3 glasses samples under damp heat conditions as a function of exposure time: 0, 500 or 1000 hours.

Table 3 shows the variation of intensity ratios of different chemical elements as a function of Si element for the non-aged or aged glass samples. The surface composition determined by XPS is above all very different from the bulk composition provided by SEM-EDX characterization. The Mg element is missing in significant proportions and so is the K element. It may be due to a high thickness dependent gradient of glass chemical composition. The strong presence of potassium could be explained by an assumed external and accidental pollution for example with KCl (see Fig.7). The Mg element is not detected although it was the third component in glass composition in the bulk analysis. One might expect its lack due to its good solubility in glass during float or annealing process. After this aging, there was a slight decrease in the surface Na concentration, of approximately 10 percent of the initial value. Concerning the I Ca2p3 / I Si2p and I K2p / I Si2p ratios, both are decreased in the same proportion. The I Sn3d5 / I Si2p ratio dropped off with regard to the initial ratio. The migration kinetics vary greatly according to the chemical

Table II: Variation of relative concentration (at.%) of components as a function of damp heat (DH) exposition duration obtained by SEM-EDX measurements

Side of float glass	DH exposition	Al_2O_3	CaO	MgO	Na_2O	SiO_2	SnO	Others: K_2O, Fe_2O_3
Tin side	Unaged	0.6	9.1	3.4	11.4	73.3	1.9	0.3
	500h	0.6	9.0	3.5	10.5	74.3	1.8	0.3
	1000h	0.7	9.0	3.6	9.9	74.7	1.9	0.3
No-tin side	Unaged	0.7	9.0	3.5	11.9	74.4	0.2	0.3
	500h	0.7	8.9	3.6	10.6	75.8	0.2	0.3
	1000h	0.6	9.2	3.7	9.2	76.8	0.2	0.3

elements because the oxides have a variable chemical reactivity with molecular water. In comparison with SEM-EDX analysis, one can suppose that Na element has stronger mobility than Ca or Mg elements.

Table III: Variation of intensity ratios as a function of damp heat exposition by XPS measurements

Intensity ratio	Unexposed	After 1000h	Variation (relative unit)
I Al2p / I Si2p	0	0.01	-
I Ca2p3 / I Si2p	0.08	0.01	0.81
I K2p / I Si2p	0.10	0.02	0.81
I Na1s / I Si2p	0.20	0.18	0.11
I Sn3d5 / I Si2p	0.22	0.02	0.91

4 DISCUSSION

The increase of the glass transmittance can be explained by the fact that molecular water penetrates into the glass. One might suggest that molecular water reacts with glass network as explained below and which is confirmed by FTIR, SEM-EDX and XPS analyses.

The study of glass aging during damp heat test also shows changes of surface and bulk chemistry representative of leaching process accelerated under a hot and humid atmosphere. The following model proposed by the authors and depicted in Fig. 3 is based on Schaut-Pantano's model [6]. The glass degradation mechanism can be described in three stages: 1) water gel film coating, 2) molecular water adsorption with reaction in glass and 3) corrosion and deposits of carbonates. Fig. 3 illustrates the following descriptions for each stage of glass degradation mechanism.

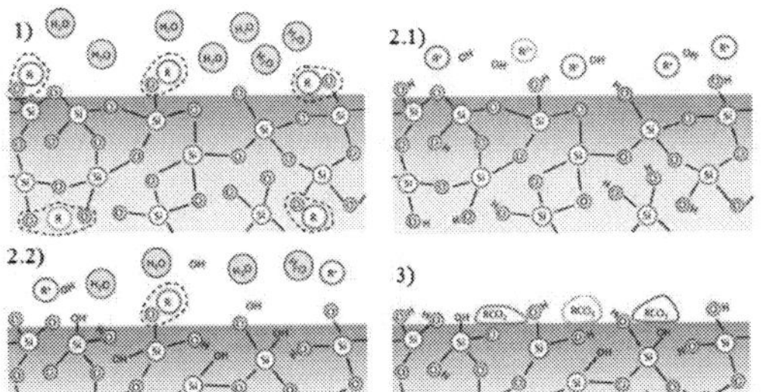

Figure 3: Aging mechanism of soda-lime glass under damp heat as described in Ref. [6]: 1) water gel film coating, 2.1) leaching of cations, 2.2) hydrolysis of glass network and 3) glass corrosion and deposits of carbonates.

A water film coating is formed on glass during damp heat aging by condensation and adsorption of water vapor. In this water film, we can distinguish the physisorption for which water molecules bind on glass by means of Van der Waals forces and the chemisorption at the near water/glass interface for which chemical reactions take place with cations and silica to generate new specific chemical bonds. Then free molecular water diffuses through silica network. At this moment, two types of chemical reactions can take place. The first one based on the reaction of molecular water with monovalent cations, R^+, like Na^+ or K^+ and divalent cations, R^{2+} , like Ca^{2+} or Mg^{2+} on non-bridging oxygen (NBO) sites. The cationic atoms are free and released apart from glass network to water film where they accumulate as soluble precipitates on a mixture of molecular water and hydroxide ions, which is the well-known mechanism of leaching:

$$\equiv SiO^-R^+ + H_2O \rightarrow \equiv SiOH + R^+ + OH^-$$
$$\equiv SiO^-R^{2+}O^-Si\equiv + 2H_2O \rightarrow 2\equiv SiOH + R^{2+} + 2OH^-$$

The ionic migration aims to reduce the electrochemical potential between bulk and surface, so the pH will be raised in the liquid film during glass aging process.

But at the same time, molecular water also reacts with silica network on BO sites with three major chemical reactions to allow hydrolysis of silica network. They imply Si-O-Si or SiO^- groups with molecular water and hydroxide ions. The latter react within silica network providing a linkage of hydrogen bonds to form silanol groups:

$$\equiv SiOSi\equiv + H_2O \rightarrow \equiv SiOH + HOSi\equiv$$
$$\equiv SiOSi\equiv + OH^- \rightarrow \equiv SiOH + \equiv SiO^-$$
$$\equiv SiO^- + H_2O \rightarrow \equiv SiOH + OH^-$$

When liquid film reaches pH 9, degradation of glass surface becomes more important with formation of precipitates of insoluble carbonates like Na_2CO_3 and $CaCO_3$. The formation of carbonates involves a permanent staining which reduces the optical performance of top glass. However, transmittance measurements and visual inspection do not reveal permanent staining. The cationic leaching and hydrolysis of silica network provide the corrosion layer formation with lacunas left by leaching. The increasing porosity involves a most effective removal of soluble constituents, especially of network modifiers, which probably improves the transmittance properties. This fourth step has not been highlighted in the present study and it is certainly due to a restricted aging time.

5 CONCLUSIONS

A set of technical characterizations was used to investigate the aging mechanisms of soda-lime glass under damp heat. The results suggest that the glass aging is induced by hydration of glass network during damp heat treatment. The network modifiers (Na, Ca, K) are more or less leached at the glass surface and corrosion involved by hydrolysis of Si-O-Si network occurs. A transmittance increase for optical properties is highlighted. This can have a positive impact on photovoltaic conversion. Nevertheless, the increase of transmittance in the UV range may enhance degradation of the screened materials (encapsulant, cell, back-sheet) and should be investigated in a future work.

It has also been shown that tin side of glass plate modifies the damp heat impact on functional properties. The tin side has a passivating barrier ability which reduces the sodium leaching and the glass network hydrolysis. It would be interesting to design a photovoltaic module with tin side at glass/polymer interface. It induces a decrease of the cationic migration which could permit to avoid the devastating PID phenomenon [7].

6 ACKNOWLEDGEMENTS

Financial funding of the project was provided by the French government through the Ministry of Higher Education and Research (MESR). The authors would like to thank Mr. David Alamarguy and Mrs. Anne Migan-Dubois (GeePs Laboratory, France), for having provided us with the XPS analysis. This analysis was supported by the grants from Region Ile-de-France, CNRS and Agence Nationale de la Recherche. Their thanks also go to Mr. Patrick Ausset (LISA Laboratory, France) for the SEM-EDX analysis and the interesting discussions on natural glass aging and to Mr. Laurent Ibos (CERTES EA 3481, France) for his help in FTIR measurements.

7 REFERENCES

[1] D. Jordan, S. Kurtz, "Photovoltaic degradation rates – an analytical review", in NREL Journal Article 5200-51664, 2012.

[2] C. Peike, S. Hoffmann, P. Hülsmann, B. Thaidigsmann, K.A. Weib, M. Koehl, P. Bentz, "Origin of damp-heat induced cell degradation", in Solar Energy Materials & Solar Cells 116, 2013, pp. 49–54.

[3] S. Kurtz, J. Wohlgemuth, M. Yamamichi, T. Sample, D. Miller, D. Meakin, C. Monokroussos, M. TamizhMani, M. Kempe, D. Jordan, N. Bosco, P. Hacke, V. Bermudez, M. Kondo, "A Framework for a Comparative Accelerated Testing Standard for PV Modules", in NREL Conference Paper 5200-57838, 2013.

[4] C. Osterwald, T. McMahon, "History of Accelerated and Qualification Testing of Terrestrial Photovoltaic Modules: A Literature Review", in Progress in Photovoltaïcs: Research and Applications. vol. 17, 2009, pp. 11–33.

[5] K. Davis, M. Tomozawa, "Water diffusion into silica glass: structural changes in silica glass and their effect on water solubility and diffusivity", in Journal of Non-Crystalline Solids 185, 1995, pp. 203-220.

[6] R. Schaut, C. Pantano, Acid interleave coatings inhibit float-glass weathering, corrosion, The Glass Researcher 14 (2005) 44-49.

[7] P. Hacke, K. Terwilliger, S. Kurtz, "In-Situ Measurement of Crystalline Silicon Modules Undergoing Potential-Induced Degradation in Damp Heat Stress Testing for Estimation of Low-Light Power Performance", in NREL Technical Paper 5200-60044, 2013.

IMPACT OF RIBBON SPECIFICATION AND HANDLING DURING PV MODULE MANUFACTURING TO MODULE RELIABILITY

*Andreas Schneider[1], Rüdiger Fernada[1], Thomas Link[2], José Ignacio Fidalgo Martínez[3], Jörg Schmauder[1], Rudolf Harney[1]
[1]International Solar Energy Research Center (ISC Konstanz), Rudolf-Diesel-Str. 15, 78467 Konstanz, Germany
[2]SI Module GmbH, Bötzinger Straße 21c, 79111 Freiburg im Breisgau, Germany
[3]Specialized Technology Resources Espana S.A., Parque Tecnológico de Asturias, Parcela 36
33428 – Llanera, Asturias, Spain
*Corresponding author: Phone: +49 7531 36183302, Fax: +49 7531 361831, Email: andreas.schneider@isc-konstanz.de

ABSTRACT: The current BOM of crystalline silicon solar cell modules appear to be very small if compared to other products for example in consumer electronics or car industry. Nonetheless almost each item listed in the BOM is a key material for the lifetime of +20 years for PV panels. Currently available characterization equipment being used by module makers only partly allows to give a clear interpretation and statement of the capability of the module to withstand climatic impact such as thermo-mechanical stress for many years. In most cases defects on cell level such as cracks or metallization defects grow in time and degrade the fill factor hence the power on a longer time-scale. This paper investigates the effect of ribbon thickness to the for EL invisible defect rate and evaluates how this effect increases by applying thermo-mechanical stress to the module. Clear measures are presented on how such stress hence power degradation can be reduced. The results indicate that depending on the encapsulation layer thickness a critical ribbon thickness exists which shall not be exceeded. In this situation the ribbon yield strength has only a negligible influence on the defect reduction. Beside a ribbon thickness reduction the encapsulation foil thickness can be increased or the lamination membrane pressure accordingly adjusted. The amount of defects appearing only after thermo-mechanical stress are a very good indication for the expected power losses and can be used to a certain degree as a prediction for such.
Keywords: Manufacturing and Processing, Module, PV Materials, Degradation, Encapsulation, Ribbons

1 INTRODUCTION

The worst defects on solar module level are those who exist and degrade module output power in the field inevitably but are not detectable by standard characterization equipment being currently used by industry. One category of these defects is certain soldering and lamination induced damages which cannot always be detected by means of Electroluminescence (EL) after module production. Some techniques exist on lab scale which combine mechanical load and EL allowing to detect such defects [1]. Another means is the application of thermo-mechanical stress such as thermo-cycling to the modules. Typically already after very few cycles the crack growth progresses and the defects are visible by EL but this procedure is not feasible even for smaller module makers. The gravity and impact of soldering and soldering material induced cracks are mainly influenced by the physical ribbon parameters such as ribbon thickness and yield strength Rp0,2 (corresponding to the hardness of the material) and the soldering process parameters. A non-matching set of material or process parameters may jeopardize the longevity of the product and lead to significant power losses, enhanced by thermo-cycling [2, 3, 4]. Power losses during operation and dangerous hot spots which reduce product lifetime are the main outcomes of these damages. In order to minimize these damages the ribbon thickness and yield strength beside the thickness of the encapsulation foil are key parameters for the material selection. The ribbon thickness strongly impacts the lamination defect rate and has to stay below an upper limit to prevent any larger "invisible" defects even if this means that the module maker cannot optimize the module power by reducing the electrical cell to module (CTM) losses any further. Product safety, lifetime and quality are important factors a manufacturer has to thrive for.

This paper presents a study on 1-cell mini-modules comprising cells soldered with ribbons of two different thicknesses: 0.20 mm and 0.24 mm in combination with varying yield strengths: soft (50 MPa) and hard (90 MPa) to investigate its effect on damages to the metallization pattern on the front side and generation of cracks after soldering. Beside this the encapsulation foil thickness was increased and the membrane pressure lowered to see if the defect rate can be lowered by optimizing other material and process parameters. To make such defects visible thermo-cycling was applied to all samples and EL as well as IV measured prior, after 100 and after 200 cycles. Furthermore EVA and a POE (polyolefin elastomer) encapsulation foil of 450 µm thickness were used for the investigation.

Results and observations after TC100 and TC200 are presented which show the challenges the module maker has to overcome to guarantee the customers solar modules with a least amount of soldering induced cracks specifically under consideration that such cracks are not visible after production.

2 DESIGN OF EXPERIMENT

The input parameters for the experiments of section 3.1 were solar ribbon of two different yield strength (hardness) and thickness, Cz and mc solar cells as well as two different encapsulation materials (an EVA and POE material) resulting in an experimental matrix of 16 input parameters. The ribbon yield strength was 90 MPa and 50 MPa with thicknesses of 0.20 mm and 0.24 mm respectively and therefore covered typical values for ribbon used in industry. The ribbon details are summarized in Table I. For the encapsulation material a standard EVA and a POE material was used as being currently seen in many industrial production lines worldwide. The experiments of section 3.2 were performed with Cz and mc cells, both encapsulation materials of standard and increased thickness (+ 200 µm only on front side) and the 0.24 mm thick ribbon of 50 MPa yield strength. The incoming parameters for section 3.3 were mc cells, both encapsulation materials of standard thickness, the thickest and hardest ribbon and varying lamination membrane pressures. The industrial study of section 3.4 was conducted by varying the cell and encapsulation material and applying various stretch factors to the same ribbon prior ribbon soldering. The output parameters of the experiment are initial power and power loss after thermo-cycling including the EL pictures.

It has to be mentioned that for the sake of simplicity the lamination cycles for the EVA and the POE encapsulation materials were the same. It is known that the non-EVA based material usually requires a longer evacuation time due to its higher melting point. Therefore the obtained results given below in terms of additional crack generation can be caused by the higher melting point of the POE material. This would lead to a larger viscosity of the material when the pressure is applied during lamination. The difference in crack generation rate might therefore be explained and minimized if the evacuation time is slightly extended (glass transition temperature of the POE is lower).

We processed standard 6" multi-crystalline (mc) and mono-cyrstalline (Cz) solar cells using a metallization pattern with three busbars on front and pads on rear side for the power loss study. To allow for good statistics five mini-modules for each input parameter were processed. Furthermore the flux and soldering process (manual or in one case automatic) were the same for all samples.

Table I: Ribbon details

Ribbon Type	Thickness [µm]	Yield Strength [MPa]
1	0.20	50
2	0.24	50
3	0.20	90
4	0.24	90

3 EXPERIMENTAL RESULTS

The focus of the following experiments was set on determining the fill factor losses after thermo-cycling as well as the change in defect pattern and number of defects visible in EL. Beside this we performed a cell to module loss analysis of 1-cell mini modules exhibiting standard p-type solar cells. For all experiments electrical IV and Electroluminescence measurements were taken before and after the test sequences.

3.1 Variation of ribbon thickness and encapsulant

Cell to module losses were determined for all samples and are displayed for mc wafer material in Table II. The lower CTM fill factor losses for the ribbon thickness d=0.24 mm compared to d=0.20 mm are caused by the different ribbon cross sections. CTM fill factor and power losses are close and standard deviation small for all samples indicating that if any cracks initially already exist on module level these do not damage the cells mechanical and electrical integrity. This was also confirmed by initial EL pictures which show only minor defects such as small soldering induced defects.

Table II: CTM fill factor and power loss for mc cell mini-modules (averaged over five samples)

Ribbon Type	CTM_{FF} EVA [%]	CTM_{FF} POE [%]	CTM_{Pmpp} EVA [%]	CTM_{Pmpp} POE [%]
1	-3.8±0.1	-3.7±0.1	-3.9±0.2	-3.7±0.3
2	-2.7±0.3	-2.8±0.2	-3.4±0.5	-3.7±0.2
3	-3.8±0.3	-3.8±0.1	-3.6±0.2	-3.6±0.1
4	-2.3±0.4	-2.8±0.1	-3.3±0.8	-3.4±0.3

Figure 1 and 2 show EL pictures of the same Cz and mc cell module samples each before thermo-cycling, after 100 cycles and after 200 cycles. The pictures clearly demonstrate that no or only very few defects initially exist on module level. After 100 cycles the situation has changed completely: a large amount of finger to busbar interruptions occur which is further increased after another 100 cycles.

Figure 1: EL of Cz sample (left: initial, middle: after TC100, right: after TC200)

Figure 2: EL of mc sample (left: initial, middle: after TC100, right: after TC200)

The POE material has superior material properties compared to the EVA based material in terms of light transmissivity in visible and UV range (specifically the UV cut-off is significantly reduced to lower wavelength). Table III shows the cell to module losses related to the short circuit current for both encapsulation materials as well as for Cz and mc cells. The data given confirms the improved encapsulation material data on mini-module level for all materials with a larger gain for the Cz cells due to a cells quantum efficiency significantly higher in the lower wavelength compared to mc cells.

Table III: CTM short circuit current (Isc) loss for mc and Cz cell mini-modules (averaged over all samples)

CTM_{Isc} EVA [%] Cz	CTM_{Isc} POE [%] Cz	CTM_{Isc} EVA [%] mc	CTM_{Isc} POE [%] mc
-3.99±0.34	-3.67±0.39	-0.46±0.55	-0.37±0.45

Figure 3 shows the fill factor degradation after TC200 for all samples. For almost all samples we see a slightly increased FF loss for the POE material compared to the EVA material of approximately 1% for Cz material and 2% for mc material. Most of all the data shows that for the chosen thickness range the ribbon yield strength has only a small impact on the FF losses: the softer ribbon reduces the losses significantly for the mc material by about 2%. Mini-modules with Cz cells and 0.24 mm thick ribbon show already after only 100 cycles a degradation of 3-4% which only slightly increases after TC200. The thinner ribbon samples (d=0.20 mm) pass all cycles with approx. 1% FF losses. For the mc material the losses between TC100 and TC200 significantly increase which is also visible in EL showing growing and proceeding cracks and finger to busbar interruptions on cell level. The losses for mc material are 2-3 times higher compared to Cz material. After TC200 only samples with 0.2 mm thick ribbon pass the TC200 test with losses of 2%. The defects visible in EL after TC testing are mainly finger to busbar interruptions and soldering induced cracks at the very end of the busbar (discontinuity of busbar). It has to be mentioned that even though 1-cell mini-modules have been used for this study no physical reason exist why these results may not be directly transferred to large modules (e.g. a 60 cell module).

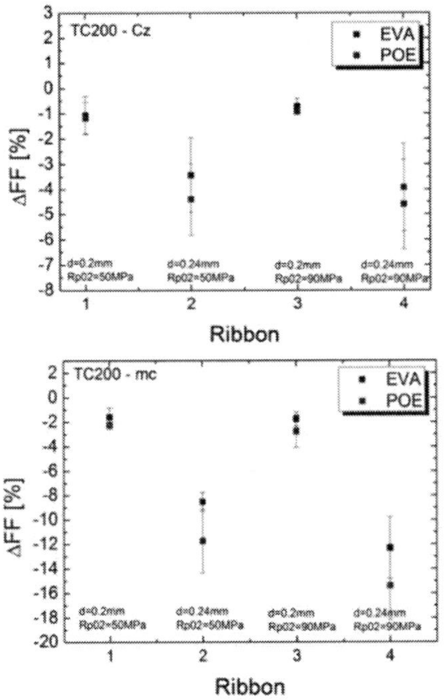

Figure 3: Fill Factor loss of samples after 200 cycles of thermo-cycling

3.2 Variation of the encapsulation foil thickness

The results of the previous section clearly indicated that a critical ribbon thickness exist. Exceeding such will lead to severe defects on cell level inside modules which are not necessarily detectable by EL prior any temperature cycling or mechanical load testing. The usage of super soft ribbon (meaning ribbon with very small yield strength) cannot prevent crack generation during soldering or lamination since the limiting factor in this situation is the thickness of the encapsulation foil. STRE not only produces encapsulation materials in industry-wide used standard thickness of 400 - 450 µm but also in thickness of 200 µm which might be used for rear contact cell applications to reduce the absorption losses on module level. By neglecting the inevitable absorption losses and using the 450 µm thick material in combination with a foil of thickness 200 µm on the front side it was possible to process mini-modules with an encapsulation material thickness increased by 50% (front side) compared to the samples of section 3.1.

The experimental design of this study was similar to the previous investigation with the only difference that the focus was set on 0.24 mm thick ribbon thereby omitting the 0.2 mm thick ribbon. Again 5 samples per group for averaging were prepared and measured after 100 and 200 cycles of thermo-cycling. Figure 4 shows the results for the fill factor loss for all groups related to the data obtained in Figure 3.

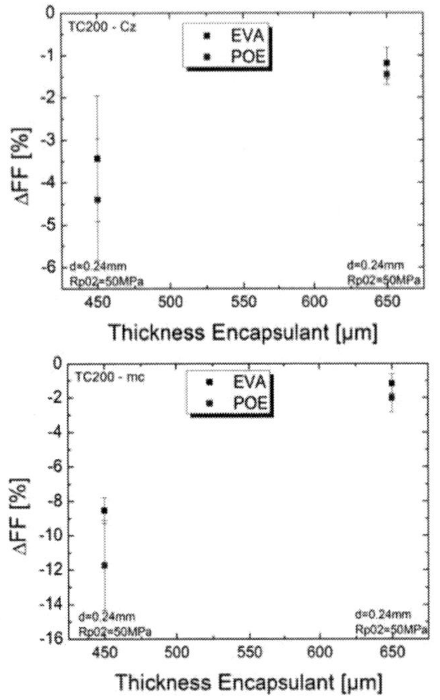

Figure 4: Fill Factor loss of samples after 200 cycles of thermo-cycling

A slightly reduced fill factor loss for the EVA based encapsulation material compared to the POE samples are found which is in line with the results obtained earlier. The losses from TC100 to TC200 slightly increase only due to the contribution of the fill factor which is also visible in EL by an increase of crack generation or crack growth. The results clearly indicate that the use of thicker encapsulation material immediately leads to an improvement in fill factor losses after climatic testing for the mc material from values as large as 12% up to only 2%. Samples with Cz cells show fill factor losses of around 1.5% for the thick encapsulation layer compared to losses of around 4% for the standard foil thickness. From the previous results it can be concluded that in case of 0.2 mm thick ribbon only a small or no gain is expected for samples with thicker encapsulation material but results also show that two layers of encapsulation material (or one in case of 650 µm material is available) will reduce the losses to an extend to prevent any excessive crack formation caused even by thick ribbon material. In terms of potential applications this means solutions exist in case the module maker has to use ribbons of large thickness, e.g. for reducing the electrical losses for solar cells with narrow busbars.

3.3 Variation of the lamination membrane pressure

In terms of physical material data which influences the crack generation rate during lamination the ribbon thickness and the encapsulation layer thickness were both varied with results showing that either the ribbon thickness or the encapsulation layer thickness can be used to achieve solar modules with a strongly reduced defect rate. On the other hand it is known that the defect generation happens during the soldering and lamination process hence the specific lamination recipe seems to provide a number of parameters such as membrane pressure, temperature and time to optimize the process. From these three available parameters it is assumed that the membrane pressure is the parameter which will have the largest effect on any mechanical defects generated on cell level and was therefore used for another study.

6" mc wafer material was processed in an industrial standard cell process as has been used for cell processing in the previous sections. MC material was used since it has shown a stronger tendency to crack generation during the soldering process if compared to the mono-crystalline wafer material. All cells were laminated into mini-modules using the known EVA and POE encapsulation materials and leaving the backsheet and glass the same. During lamination the membrane pressure was varied from 700 mBar to 1000 mBar in 100 mBar steps. For each group at least 5 mini-modules were used for statistics. The emphases was set on crack generation. For this reason a hard ribbon of 0.24 mm thickness was used to electrically connect the cells. Results of fill factor change after 200 cycles of TC are displayed in Figure 5.

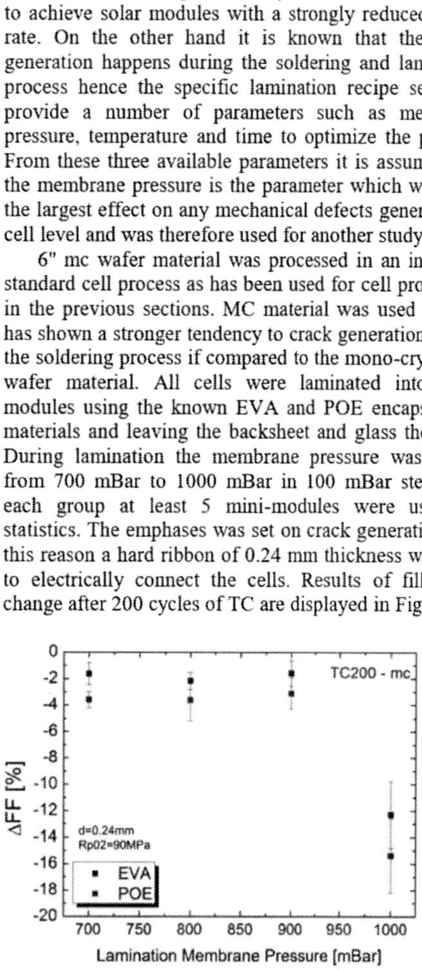

Figure 5: Fill Factor loss of samples after 200 cycles of thermo-cycling

The results clearly show how strongly the lamination membrane pressure influences any fill factor degradation after thermo-cycling. Based on these results we believe that most of the defects must have existed prior TC testing but not before ribbon soldering. As has been seen in sections 3.1 and 3.2 the defects were neither visible in IV nor in EL after lamination and only thermo-mechanical stress was able to proceed crack growth hence degrading the electrical cell performance on module level. As in all experiments seen before the samples with EVA performed slightly better than samples with the POE encapsulation material probably due to the fact that the same lamination cycle was used for materials with very different viscosity.

3.4 Stretch factor variation

All industrial type stringers have to pre-stretch the ribbon prior ribbon layup and soldering process to ensure a certain ribbon flatness hence a decent quality for the soldering process. The ribbon usually is taken from a spool with a curvature far too large to allow down holding pins to press the ribbon homogenously without gaps to the busbars during soldering. Without this stretching the ribbon might only be partly soldered to the cells busbar. This pre-stretching negatively affects the hardness of a ribbon: the larger the stretch factor is the flatter the ribbon gets but the higher the ribbon yield strength becomes hence the hardness increases. This means even if the module maker starts with a super soft (e.g. $Rp_{0.2}$=50 MPa) ribbon after pre-stretching, depending on the stretching factor, the ribbon may not be soft anymore. To investigate the influence of pre-stretching (meaning the stretch factor) to the ribbon yield strength ribbon material was pre-processed in an industrial Somont stringer and $Rp_{0.2}$ determined by using a Zwick mechanical tester (the ribbon geometry was experimentally determined after stretching for each ribbon sample since any stretching will affect the ribbon height and width). The initial ribbon specification as given by the supplier was: thickness: 0.18 mm, width: 1.8 mm and $Rp_{0.2}$=60.48 MPa. The typical stretch factor for this specific stringer (depending on the ribbon type and supplier) ranges in between 0.6% and 1%. Worst case scenarios to the lower and higher range are 0.3% and 2%. For this reason all variations were investigated including the un-stretched ribbon.

For averaging 10-20 ribbon samples per group were used. The results in Figure 6 clearly demonstrate how strongly the yield strength is affected by the stretch factor. A soft ribbon easily becomes a hard ribbon if the stringer equipment operator will use larger stretch factors. In terms of yield and throughput increasing the stringer stretch factor may make sense for the module maker since the flat ribbon will lead to less quality issues after soldering (as non-properly soldered ribbons or ribbon to busbar mis-alignments). This statement is valid as long as the effect of crack generation visible only after thermo-mechanical stress application is not considered.

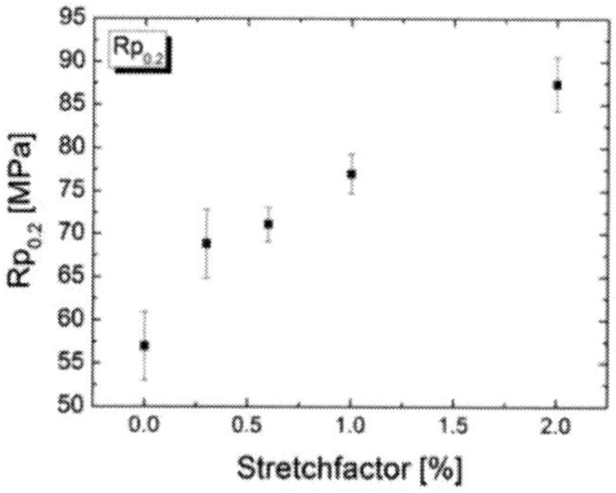

Figure 6: Yield strength in dependence of the stretch factor set on a Somont industrial stringer

In a next step Cz and mc cells were soldered with the

same Somont stringer by varying the stretch factor from 0.3% to 2%. The one-cell strings were laminated using POE and EVA as well as the backsheet and solar glass from the previous experiments. For averaging several mini-modules per group were processed. IV and EL measurements were taken prior and after the application of thermo-mechanical stress inside a climatic chamber.

Figure 7 shows the results for the changes in fill factor after climatic testing for the Cz samples. A tendency for the fill factor losses can be stated from the data, nonetheless specifically due to the smaller ribbon thickness of 0.18 mm the losses are far less pronounced as before. For the EVA material the losses increase from 1.74% (stretch factor of 0.3%) to 2.38% (stretch factor of 2%). The POE material shows losses increasing from 1.42% (stretch factor of 0.3%) to 1.79% (stretch factor of 2%) with an exception for stress factor 0.6%. For the mc samples no significant changes were found therefore this data was omitted. In general the losses for POE are significantly lower than for EVA which confirms that if a ribbon of thickness below the threshold for the specific material combination (here 0.2 mm) is used the POE material will outperform.

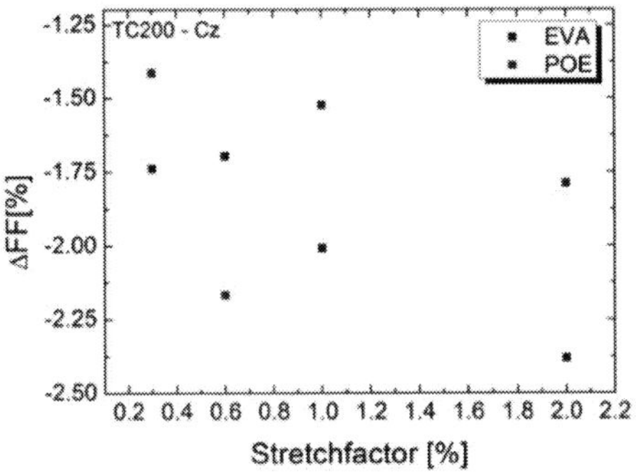

Figure 7: Fill factor losses after climatic testing for mini-modules comprising Cz cells

3.5 In depth EL investigation on defects

In all conducted experiments of this work a fill factor degradation of the mini-modules was found after thermo-cycling. This fill factor degradation can be caused by two kind of defects, categorized into two groups: metallization defects (e.g. finger to busbar interruptions) and mechanical defects (e.g. cracks). To further evaluate if one or both defect kinds contribute stronger to the electrical degradation all EL pictures were evaluated. For this the defects visible in EL prior climatic cycling were evaluated, counted and compared to the defects visible in EL pictures after cycling. Any change in either defect size or number or in case defects appeared during cycling was reported and quantified. Figure 8 shows the delta of only metallization defects meaning the difference between defects before and after thermo-cycling of the experimental study shown in section 3.1. The delta for the amount of cracks correlate with the results of Figure 3 and are not displayed here.

From the achieved results the fill factor losses (Figure 3) can be fully explained by the defects found in EL. Furthermore a good correlation exists between fill factor losses and the amount of metallization defects generated or pronounced during thermo-cycling.

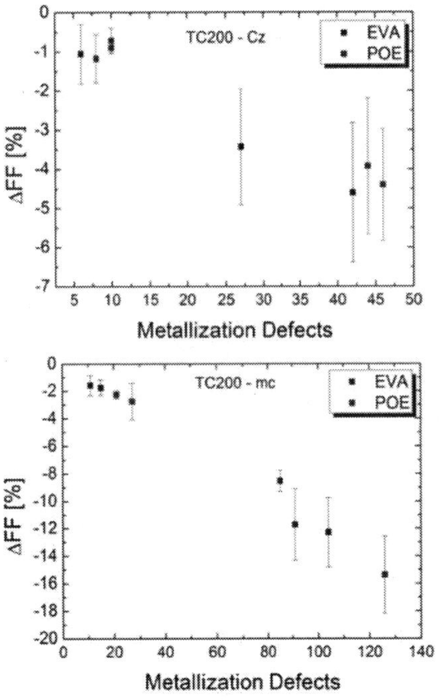

Figure 8: FF loss after TC200 in dependence of the amount of metallization defects appearing during TC

The same evaluation was performed for the experimental study of section 3.2 (variation of the encapsulation foil thickness) and are displayed in Figure 9. The amount of metallization defects directly compares to the fill factor degradation seen in Figure 4.

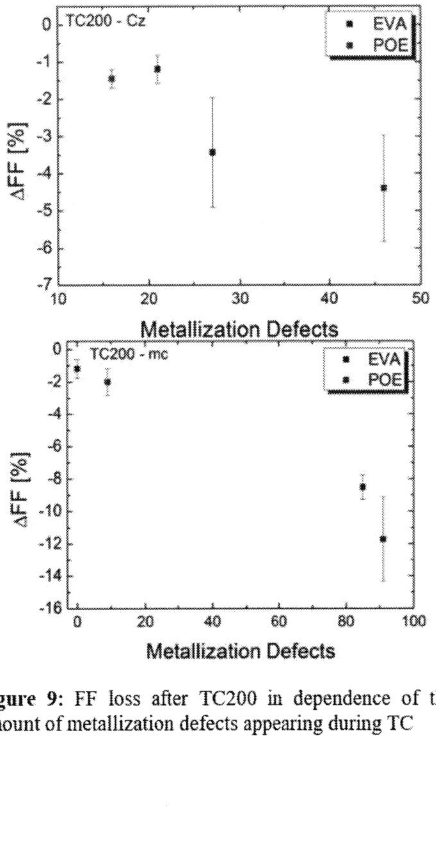

Figure 9: FF loss after TC200 in dependence of the amount of metallization defects appearing during TC

4 SUMMARY

In this work we presented results showing how detrimental the ribbon thickness may contribute to fill factor losses on mini-module level once exceeding a certain ribbon thickness threshold. Obtained results indicate that the generated defects are mainly caused by the lamination process nonetheless a certain contribution from the soldering process cannot completely be excluded. The ribbon yield strength (meaning the hardness factor) has only a small positive effect once this threshold is exceeded where a softer ribbon would mainly improve the soldering quality but not the defects generated during the lamination step. The results further indicate and proof that module makers are not able to detect such defects after lamination by means of electroluminescence. Only after thermo-cycling crack growth and front side metallization defects such as finger to busbar interruptions proceed hence being detectable. Moreover the results show that various parameters exist to reduce the number of such defects: first of all the thickness of the encapsulation material can be used to reduce the amount of cracks. Only by increasing the encapsulation layer thickness by 200 µm will lead even for the thick ribbon type to very moderate fill factor losses after thermo-cycling. Beside the foil layer thickness the membrane pressure during lamination is a critical factor according to our results. By reducing the pressure to a lower value the defects can be almost completely prevented. The other factor is the choice of encapsulation material itself, but we have already mentioned here that most probably comparable results for encapsulation materials of different chemical composition can be achieved if the lamination recipe is optimized.

Special attention has to be put on the cells wafer material: results shown in this study indicate that defect growth significantly change if mono- or multi-crystalline material is used. For this individual optimization studies shall be performed by module makers to make sure PV panels will not show defects after thermo-mechanical stress is applied.

Between the defects appearing only after thermo-cycling and the fill factor losses a correlation exist which further can be used to predict power losses in advance if the defect pattern is controlled and evaluated on a regular interval.

5 ACKOWLEDGEMENTS

The authors gratefully acknowledge the financial support by the German Federal Ministry for Economic Affairs and Energy for the sponsorship of the research project "PVScan" (FKZ 0325588C) which made this work possible.

6 REFERENCES

[1] M. Sander, S. Dietrich, M. Pander, M. Ebert, J. Bagdahn, Solar Energy Materials and Solar Cells, 01/2012; 111:82–89

[2] B. Lalaguna, P. Sánchez-Friera, H. Mäckel, D. Sánchez, J. Alonso, Proceedings of the 23rd PVSEC, Valencia, Spain, 2008

[3] J.Wendt, M. Träger, M. Mette, A. Pfennig, B. Jäckel, Proceedings of the 24th PVSEC, Hamburg, Germany, 2009

[4] A. Schneider, M. Pander, T. Korvenkangas, S. Aulehla, R. Harney, T. Horttana, Proceedings of the 31st PVSEC, Hamburg, Germany, 2014

FATIGUE ANALYSIS OF SOLAR CELL INTERCONNECTORS DUE TO CYCLIC MECHANICAL LOADING

Matthias Pander[a], Sascha Dietrich[a], Rico Meier[a,b], Matthias Ebert[a]
[a]Fraunhofer Center for Silicon Photovoltaics, Otto-Eissfeldt-Str. 12, 06120 Halle (Saale), Germany
[b]Anhalt University of Applied Sciences, Bernburger Strasse 57, 06366 Koethen, Germany

phone: +49-345-5589-5215, fax: +49-345-5589-5999, matthias.pander@csp.fraunhofer.de

ABSTRACT: Fatigue of solar cell interconnectors is one of the key failure modes for PV modules. Interconnectors are loaded when cell displacement takes place due to temperature changes and mechanical loading of the PV module. Currently, time consuming thermal cycling tests are used to identify early failure and design issues of PV modules related to ribbon fatigue. Furthermore, a cyclic mechanical load test is proposed as technical standard for accelerated fatigue testing on module level.

This publication has the purpose to investigate ribbon fatigue due to mechanical loading with help of finite element simulations. For a specific module design (concerning materials, geometry and mounting) ribbon loading is investigated with respect to amplitude and temperature. Ribbons are loaded inhomogeneous inside the module with specific positions of high failure probability. With detailed analyses investigations are carried out in order to evaluate cyclic plastic deformation in the ribbon, used as a failure criterion. The results can be examined to reduce the number of wind loading events that need to be considered for fatigue analysis, in order to perform lifetime predictions for specific climate conditions.

Keywords: Ribbons, Cyclic Mechanical Loading, Lifetime Modeling, Fatigue, Finite Element Analysis

1 INTRODUCTION

Fatigue of interconnectors for solar cells is identified as one of the key failure modes for PV modules [1]. Interconnectors are loaded mechanically when cell displacement takes place due to temperature changes and mechanical loading of PV modules [3]. In order to identify early failure and design issues of PV modules related to ribbon fatigue time consuming thermal cycling tests are carried out. Furthermore, a cyclic mechanical load test is newly introduced as technical standard for accelerated fatigue testing on module level [4].

Yet there are few systematic investigations concerning the load levels contributing to ribbon fatigue. Under field conditions loads from both - wind and temperature - are present. Daily temperature changes are superimposed by wind loads.

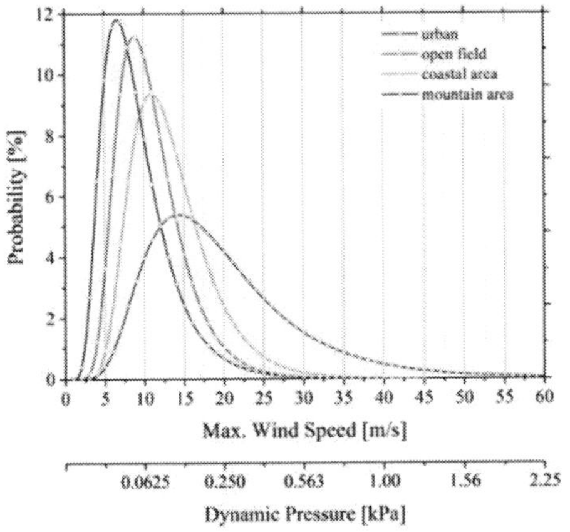

Figure 1: Maximum wind speed distribution for different regions in Germany [5]

There are specific location-dependent wind speed distributions (Figure 1) regarding amplitude and frequency of occurrence. High wind speeds are equivalent to high (static) surface loads on the module, assuming frontal inflow. Resonance effects from dynamic wind gusts or turbulent wind flow are neglected in this first order approach. The resulting deformation further depends on module temperature due to temperature-dependent stiffness of the encapsulant.

Therefore, the following questions can be formulated:

1. What's the dependency between wind speed (load amplitude) and ribbon stress?
2. Which load amplitude leads to significant damage and contributes to ribbon fatigue and thus has to be considered for lifetime estimation?
3. How many cycles to failure can be expected for specific geographical locations?

2 METHODS

2.1 Wind load data

Wind loading is a climate related variable impact on solar modules. Usually wind acts as surface load and can lead to pressure and suction effects. In accordance with Bernoulli's principle frontal inflow on a module leads to velocity reduction of the flow and results in excess (positive) pressure. Reverse flow leads to stall and results in negative pressure (suction). Turbulence additionally induces negative pressure areas.

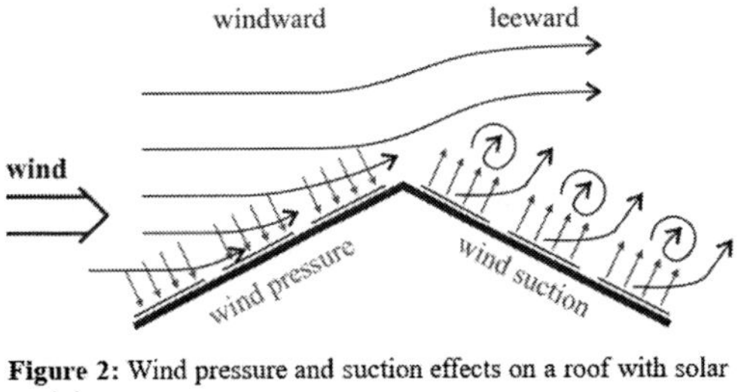

Figure 2: Wind pressure and suction effects on a roof with solar modules

Wind loading is strongly seasonal and location dependent. Factors are:

- regional wind climatology
- roughness of the surrounding terrain
- nearby sheltering obstacles
- local orography

Climatology is related to historical wind data of an interesting site. The surface roughness can be divided into 4 classes ranging from open areas to urban districts and farm land with windbreaks [6].

For Europe five landscape types are defined:

- Plains, water areas and lowland regions far from mountains
- gently undulating and hilly regions far from mountains
- strongly undulating and highland regions
- foothill regions
- high mountain massifs cut by deep valleys

Due to turbulence in the atmospheric boundary layer wind speed may change within seconds. In order to define a measured wind speed, it must be referred to the averaging period [6]. In practice 10 minutes to 1 hour averaged data is normally available. It is important to note that this way only static wind loads are considered; sudden wind gusts are averaged out. Further the wind direction is given as one of 12 sectors (see Figure 3).

In order to describe the frequency of occurrence, a two-parameter Weibull distribution is used in [6]:

$$f(w) := \frac{k}{A}\left(\frac{w}{A}\right)^{k-1} \exp\left(-\left(\frac{w}{A}\right)^{k}\right) \tag{1}$$

where $f(w)$ is the frequency of occurrence of wind speed w. The two Weibull parameters are the shape parameter k and the scale parameter A.

In a wind rose measured wind speeds are arranged according to probability of direction and wind speed distribution parameters.

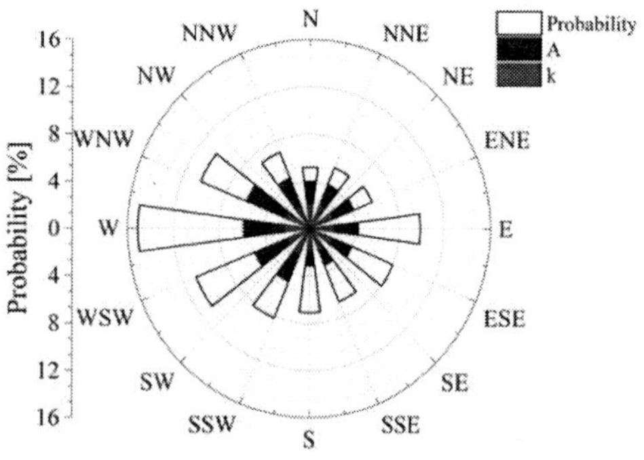

Figure 3: Wind rose for Berlin (Germany), data from [6]

In order to calculate the dynamic pressure p_{dyn} from the wind speed data Bernoulli's equation is used:

$$p_{dyn} = c_g \cdot \frac{\rho}{2} \cdot w^2, \tag{2}$$

with air density ρ and wind speed w. In standards air density is 1.25 kg/m^3, which corresponds to 1013 hPa atmospheric pressure and 10°C temperature at sea level [7]. Furthermore the gust coefficient c_g is introduced to account for turbulence related overload effects that are usually not captured in averaged 10 minute mean data. During this averaging period it is possible to have gusts of factors 1.5 to 2.5 within the wind speed value [8]. This is illustrated in Figure 4 where the gust factor (GF) is defined as the maximum peak wind speed U_{1sec} (~1 s) divided by the maximum time-varying mean wind speed U_t at a specific averaging time t [8].

$$GF = \frac{U_{1sec}}{U_t} \tag{3}$$

Due to the quadratic relationship to pressure the gust coefficient c_g can take on values from 2 to 6 for 10-minute mean data. For type approval of PV-modules a c_g of 3 is assumed for mechanical load testing [9].

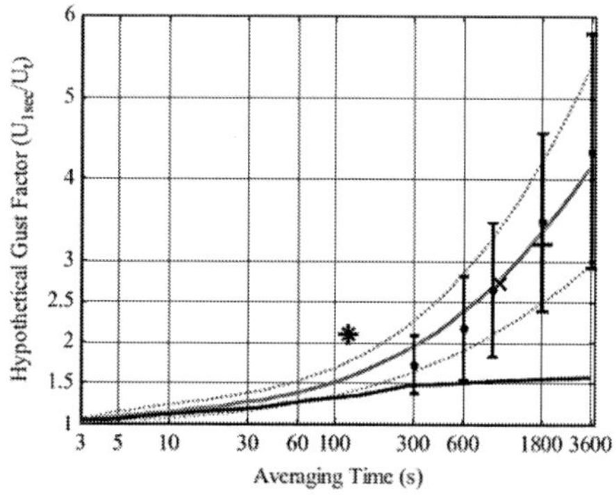

Figure 4: Mean ramp-up GF curve (gray, solid line) Dotted gray lines indicate ±1% standard deviation of the ramp-up GFs, [8]

2.2 Ribbon fatigue investigations

Some material fatigue basics are summarized for introduction. Engineers usually utilize Woehler diagrams to represent lifetime of ductile materials under oscillating load. According to the number of cycles to failure N_B three regions are defined:

- Low cycle fatigue (LCF, $N_B < 5 \cdot 10^4$)
- High cycle fatigue (HCF, $5 \cdot 10^4 < N_B < 2 \cdot 10^6$)
- Fatigue limit or very high cycle fatigue (VHCF, $N_B > 2 \cdot 10^6$)

In contrast to steel, copper and aluminum do not have a fatigue limit [10]. HCF was first investigated by O.H. Basquin [11], who formulated a correlation between stress amplitude $\frac{\Delta\sigma}{2}$ and cycles to failure:

$$\frac{\Delta\sigma}{2} = \sigma'_f \cdot (N_B)^b, \tag{4}$$

where b is a material-specific constant. For LCF Manson and Coffin have found the correlation between plastic strain amplitude and cycles to failure:

$$\frac{\Delta\varepsilon_{pl}}{2} = \varepsilon'_f \cdot (N_B)^c, \tag{5}$$

where c is a material-specific constant and determines the slope of the Coffin-Manson-graph. Hence, fatigue life under a given total strain ε_t can be expressed by [10]

$$\frac{\Delta\varepsilon_t}{2} = \frac{\Delta\varepsilon_{el}}{2} + \frac{\Delta\varepsilon_{pl}}{2} = \frac{\sigma'_f}{E}(N_B)^b + \varepsilon'_f(N_B)^c \tag{6}$$

These considerations are shown in Figure 5. Based on these fatigue laws extensions and enhancements were introduced, e.g. mean stress effects.

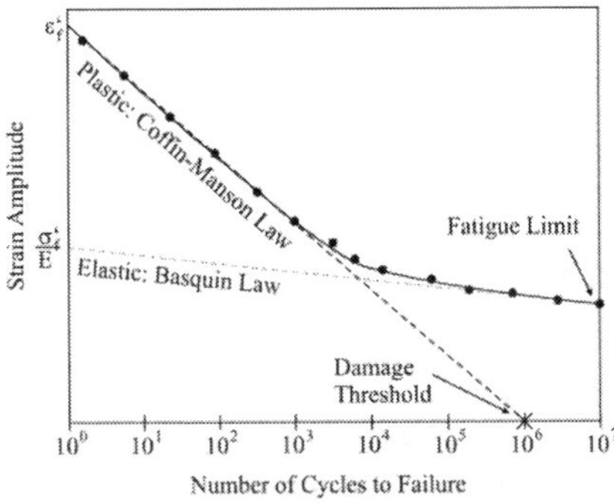

Figure 5: Schematic Strain-Life diagram adapted from [10]

From preliminary experiments fatigue data and two approaches to describe low cycle fatigue (LCF) behavior are available (Figure 7, [13]). The first one uses the change of cell gap and is assumed to be more conservative by generally overestimating the risk of ribbon fatigue in the laminate by neglecting the encapsulant. The formulation is as follows:

$$\Delta gap_{rel} := \frac{\Delta gap}{initial\ gap} = a \cdot (N_B)^c, \quad (7)$$

where Δgap is the peak-to-peak amplitude, N_B are the cycles to failures and a and c fit parameters. The cycle definitions used are depicted in Figure 6. For asymmetric loading it is more convenient to substitute full cycles by two half cycles.

The second model uses the increment of plastic strain energy density during a load cycle:

$$\Delta W_{pl} = W_f \cdot (N_B)^d \quad (8)$$

It requires more detailed information and effort to calculate the strain energy, but is yet more accurate [19].

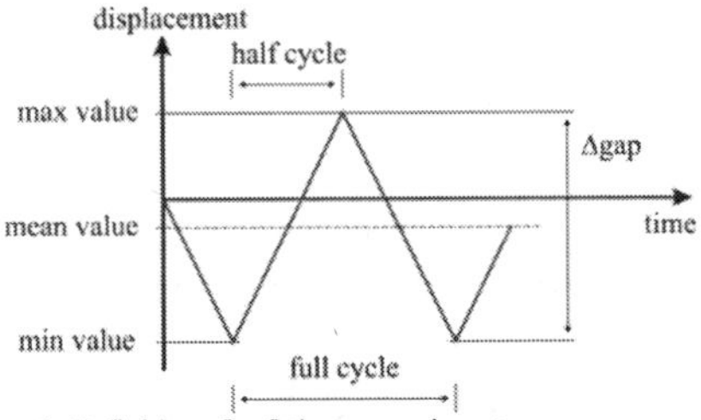

Figure 6: Definitions for fatigue experiment

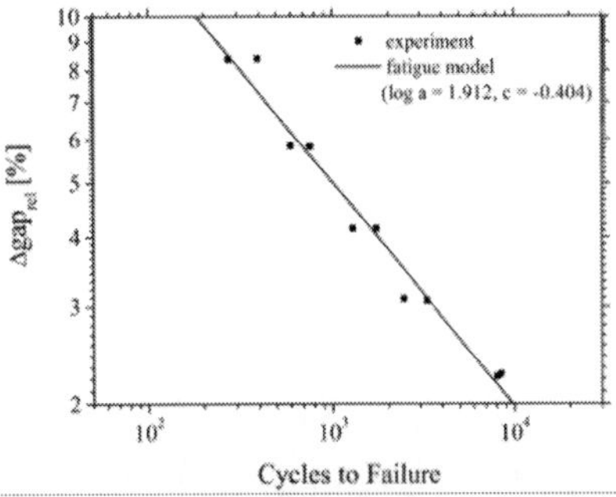

Figure 7: Results of fatigue experiments of solar cell interconnectors, adapted from [13]

Since in PV module operation the module and therefore the ribbons are loaded by different stress levels a damage accumulation law is necessary to evaluate the corresponding cycles to failure. Miner's rule is used for a qualitative analysis:

$$D = \sum_{i=1}^{k} \frac{n_i}{N_i} \quad (9)$$

D – cumulative damage, failure occurs when $D = 1$
k – number of load levels
n_i – number of cycles at load level i
N_i – mean cycles to failure at load level i

However, it is important to mention that linear damage accumulation law has some principle limitations: Linear damage accumulation is not always present and the probabilistic nature of fatigue is ignored.

"Significant damage" is meant in a sense of the loading amplitude that will result in ribbon failure within the first 10^6 cycles. As a conservative assumption the low cycle fatigue data can be extrapolated to 10^6 cycles to failure, corresponding to a damage threshold (see Figure 5). Using these considerations, a limit of 0.3 % for the relative change of cell gap was found for experimental data [13].

2.3 Simulation model

Finite element simulations are used to analyze the strain in the ribbon inside PV modules under cyclic mechanical loading. The finite element analysis consists of two parts. First a simplified 60 cell standard solar module is simulated. Geometry and material models are summarized in Table I. The frame is mounted by clamps along the long edge. Quarter symmetry is used in order to save

Table I: Overview of simulation material parameters

Component	thickness	Material model	Dimensions	
Glass	3.85 mm	linear-elastic [21]	Module size	1650 mm x 985 mm
Silicon	0.18 mm	linear-elastic [22]	Initial cell gap	3 mm
TPT back sheet	0.38 mm	visco-elastic	Busbar distance	52 mm
EVA encapsulant	0.4 mm	visco-elastic [12]	Busbar width	1.7 mm
Copper ribbon	150 µm	Chaboche nonlinear kinematic hardening [13]	Clamp width	40 mm
Sn62Pb36Ag2	25 µm	Creep, Generalized Garofalo [23]	Bearing distance	980 mm
Aluminium frame		linear-elastic	Frame moment of inertia I_{yy}	70663 mm^4
Al-/Ag paste	30 µm	bilinear-elastic-plastic [24]	Frame moment of inertia I_{zz}	11839 mm^4

resources (Figure 8). Wind loads are approximated by equally distributed surface loads on the front glass. Loads up to 1100 Pa (including 100 Pa tolerance of IEC TS 62782 load) are applied in pressure (positive) and suction (negative). From this model the change of cell gap inside the module can be evaluated. The temperature is set to 20°C.

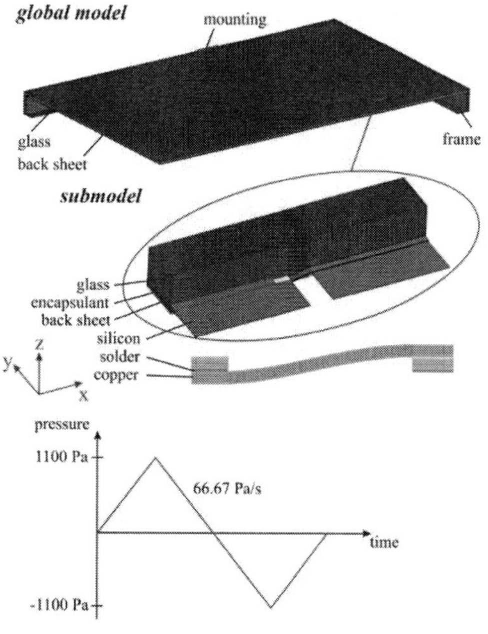

Figure 8: Finite element simulation models and pressure load profile

In a second step a detailed sub-model is implemented to evaluate plastic deformation in the copper ribbons at selected positions. This model uses the calculated displacement from the global model as boundary conditions and includes a finely meshed interconnector (Figure 8).

3 RESULTS

3.1 Evaluation of simulation results

At first, it is focused on the cell displacement that will occur due to external load. Therefore, the displacements on opposing cell edges are used. The x-axis represents deformation in the cell plane while the z-axis describes out-of-plane deformations. The y-axis considers in-plane shear deformation. The current cell gap is calculated by Pythagorean Theorem:

$$gap = \sqrt{(gap_{ini} + \Delta u_x)^2 + \Delta u_y^2 + \Delta u_z^2} \qquad (10)$$

$$\Delta u_i = (u_{2i} - u_{1i}), i = x, y, z$$

The change of cell gap Δgap is the difference between the current cell gap and its initial value gap_{ini}. Relative change of cell gap is calculated by normalization:

$$\Delta gap_{rel} = \frac{gap}{gap_{ini}} - 1 \qquad (11)$$

For several positions the plastic strain energy density introduced during the load cycle is analyzed. The positions of the cell gaps in the module are labeled in parenthesis with the first element giving the cell string number i and the second giving the gap number j. Within this investigation cells with three busbars and interconnectors are used, indicated as left, mid and right (Figure 9).

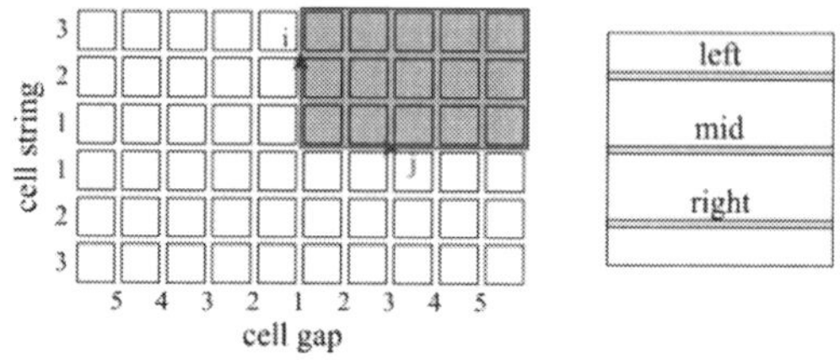

Figure 9: String and busbar labeling

3.2 Results for global model

Figure 10 shows the deflection at the middle of the module. There is a 2.5 mm offset due to modules dead weight that was previously subtracted. The graph is S-shaped implying structural nonlinearity. Further the deformation under pressure is lower compared to suction. This effect is attributed to the asymmetric stiffness of the frame.

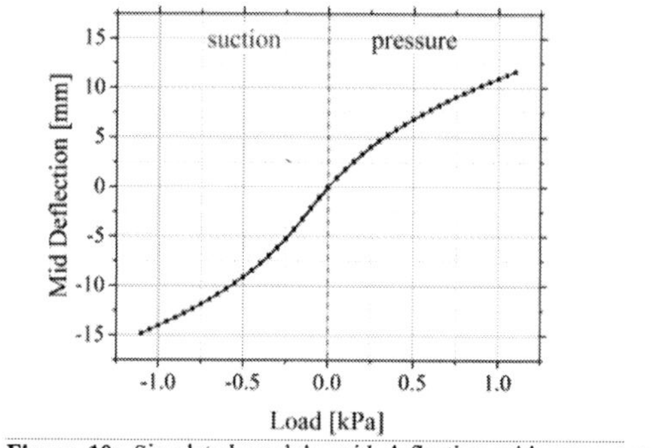

Figure 10: Simulated module mid deflection with respect to surface wind load

Figure 11 shows the changes of cell gap at different positions in the module for 1 kPa uniform pressure load. There are significant differences between the displacements. The highest ribbon strains are expected in the inner strings with a maximum change of cell gap at the end (position (1,5)). Further the cells move closer together at the outer strings under pressure load for positions (3,1-4), which presumes the ribbons are compressed. In suction all gap changes are negative, but the highest change is present at position (2,5). The ribbons in string 3 are also compressed (Figure 12).

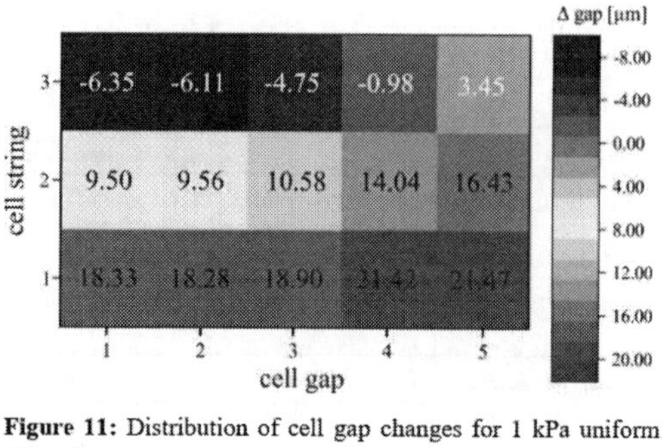

Figure 11: Distribution of cell gap changes for 1 kPa uniform pressure

In Figure 13 to Figure 15 the relative changes of cell gap are shown with respect to the applied load. The horizontally dashed orange lines indicate a 0.3 % gap change from initial state (dead weight). The corresponding vertical lines indicate the load levels when this particular gap change is reached at one position in the string. Results for these threshold load levels and corresponding wind speeds are summarized in Table II and Table **III**.

The nonlinear trend with respect to suction and pressure is primarily due to the asymmetric module layer structure. Highest cell displacement and ribbon stress is found under pressure. Under suction the load characteristic of the ribbons changes. At position (1,1) and (1,2) there is no significant cell displacement (in suction).

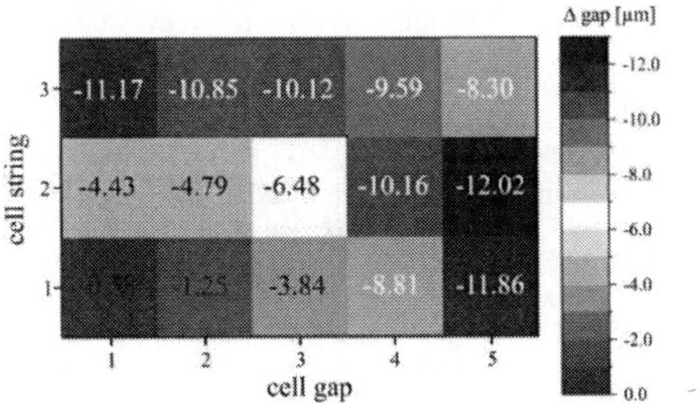

Figure 12: Distribution of changes in cell gap for -1 kPa pressure load (suction)

It is also observed that the most critical position in pressure and suction are identical in most cases. Though there are also some crossovers at position (1,4) and (1,5) as well (2,4) and (2,5), while the absolute values are close to each other. In string 3 the cell gaps are mainly compressed in both loading cases (suction as well as pressure) and the cell displacement is the lowest within the module.

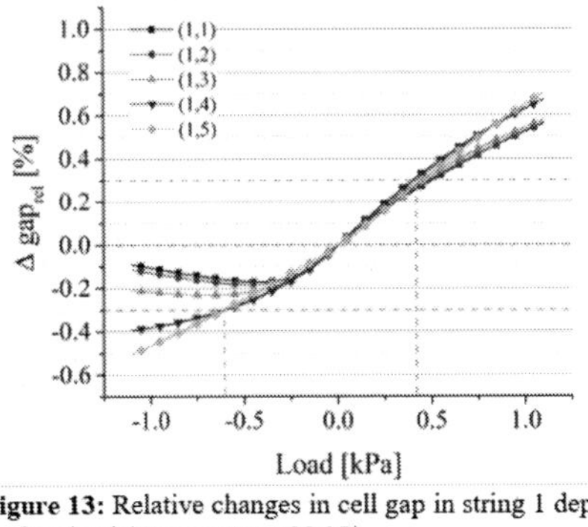

Figure 13: Relative changes in cell gap in string 1 depending on surface load (Temperature: 20 °C)

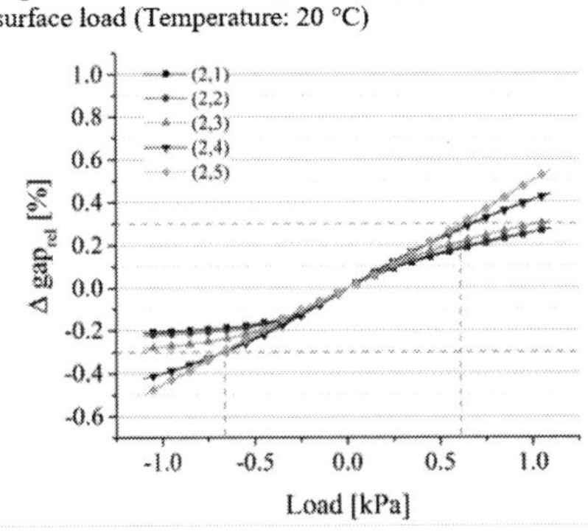

Figure 14: Relative change of cell gap in string 2 depending on surface load, (Temperature: 20 °C)

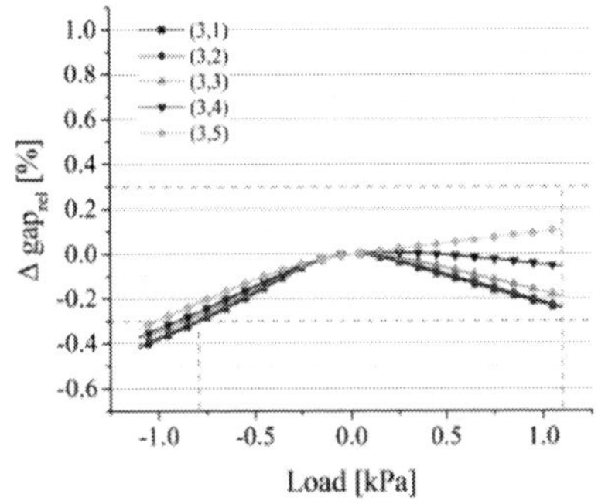

Figure 15: Relative change of cell gap in string 3 depending on surface load, (Temperature: 20 °C)

Table II: Threshold load levels and corresponding wind speeds for Δgap_{rel} = 0.3 %, (20 °C, pressure load)

String	Surface Load	Pos.	Wind speed ($c_g = 1$)	Wind speed ($c_g = 3$)
1	420 Pa	(1,4)	25.9 m/s	15 m/s
2	610 Pa	(2,5)	31.2 m/s	18 m/s
3	>1100 Pa	(3,1)	42 m/s	24 m/s

Table III: Threshold load levels and corresponding wind speeds for Δgap_{rel} = 0.3 %, (20 °C, suction load)

String	Surface Load	Pos.	Wind speed ($c_g = 1$)	Wind speed ($c_g = 3$)
1	-600 Pa	(1,4)	31.1 m/s	18 m/s
2	-660 Pa	(2,4)	32.5 m/s	18.8 m/s
3	-790 Pa	(3,1)	35.6 m/s	20.5 m/s

Influence of temperature

For studying the effect of temperature, 0 °C and 30 °C have been chosen in order to remain within a realistic temperature range. At high wind speeds the difference between module temperature and ambient temperature is below 10 K [14] and high irradiation is unlikely. The gradient of the change of cell gap is the same for all strings; therefore only results from string 1 are discussed here. Due to the lower stiffness of the encapsulant at 30 °C (Young's modulus is reduced from 5.4 MPa to 3.8 MPa) the shear bond between glass and cells is lowered and therefore cell shift is increased (Figure 16). This effect is small with a maximum increase of 0.042 % abs. at position (1,5). As a result, the loading amplitude on the cell interconnectors is increased, leading to earlier fatigue. This behavior is inverted for 0 °C, where the increased stiffness of the encapsulant (Young's modulus is increased to 10 MPa) reduces cell shifting (Figure 17). The cell shift reduction is 0.1 % abs. at position (1,5).

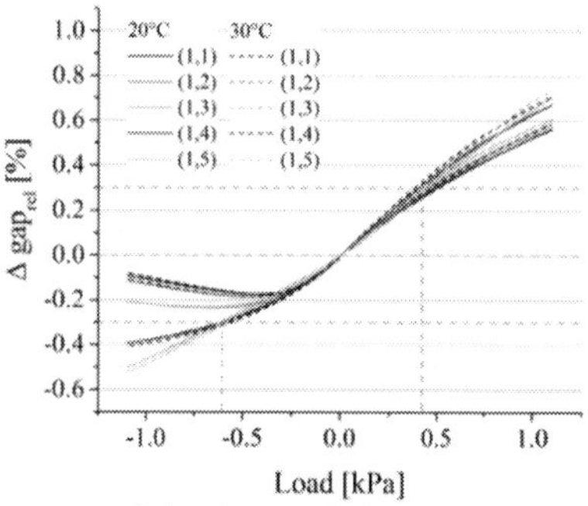

Figure 16: Relative changes of cell gap in string 1 depending on surface load; Comparison of 30 °C and 20 °C

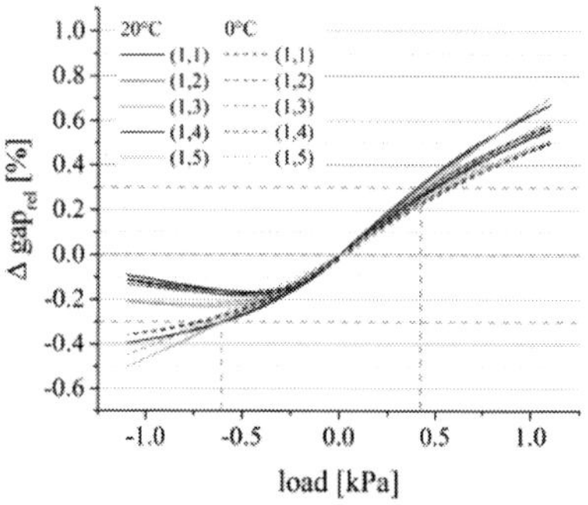

Figure 17: Relative changes of cell gap in string 1 depending on surface load; Comparison of 0 °C and 20 °C

Influence of glass thickness

Another study considers the glass thickness, which has been gradually reduced for standard modules during the last years to a nominal thickness of 3.2 mm. Due to manufacturing tolerance the lower bound of 3.05 mm is used for simulations. As result the whole module is more slender hence flexible and the cell shift is increased by a distinct amount. This is shown for string 1 in Figure 18. For this configuration an increase of 0.22 % abs. is found at position (1,5) at maximum pressure load. The threshold load for a 0.3 % change of gap is reduced to 300 Pa (~12.6 m/s).

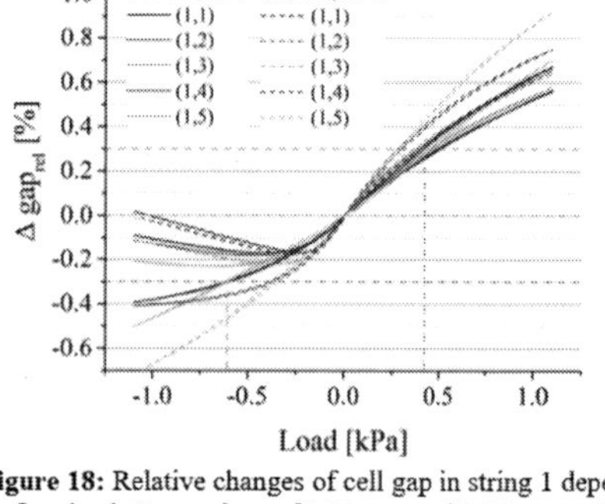

Figure 18: Relative changes of cell gap in string 1 depending on surface load, Comparison of 3.05 mm and 3.85 mm at 20°C

3.3 Estimation of cycles to failure

In the next step the results of the change of cell gap shall be used to estimate the cycles to failure of the interconnector loaded by wind for a specific location. One major problem is that when using Miner's rule the number of load cycles at a load level n_i, as well as the correct load level itself is unknown, when only daily or hourly data is available. The hourly data is averaged over time without any further information of maximum wind speed, but with information about its orientation. This data yields only poor information about gust events. From daily data a maximum wind speed for each day generated from 10-minute average data was available. This data contains no information about wind direction. The daily data with the maximum wind provides an indicator of the maximum loads of a day, but lag the number of cycles (only 1 cycle a day). In order to make a more general study about relative differences of locations this data is used to calculate the damage over 20 years for 4 regions specified in Figure 1. For evaluation, the most critical position (1,4) with the highest change of cell gap has been chosen.

The procedure is as follows:

- extract daily maximum wind speeds for a period of 20 years (1995 – 2015, except Kassel: 1992 – 2012)
- calculate the pressure load of wind speed with a gust factor of three
- calculate Δgap_{rel} from load data by interpolation of the simulated data
- calculate the damage parameter according to Miner's Rule (eq. (9)) and neglecting all values of $\Delta gap_{rel} \geq 0.3$ %

The results are shown in Table IV. As expected the introduced damage is quite low (<<1) and no failure is expected from these values. According to Miner's rule, damage scales by increasing the number of cycles at each load level. If the maximum wind speed has occurred ten times at one day, the additional damage (at this load) would be increased by a factor of ten. The comparison of four locations shows a significant increase of damage for windy regions. This is obvious regarding the cumulative distribution function of daily maximum wind speed in Figure 19. There is a distinct higher probability of wind speeds above 15 m/s for coastal and mountain area.

Table IV: Estimated damage for four locations in Germany for 20 years

Climate	Weather station	Damage after 20 years (daily data)	Normalized to Urban
Urban	Kassel	0.00123	1
Open field	Leipzig/Halle Airport	0.00274	2.23
Coastal area	Cuxhaven	0.0062	5.04
Mountain area	Zugspitze	0.0451	36.7

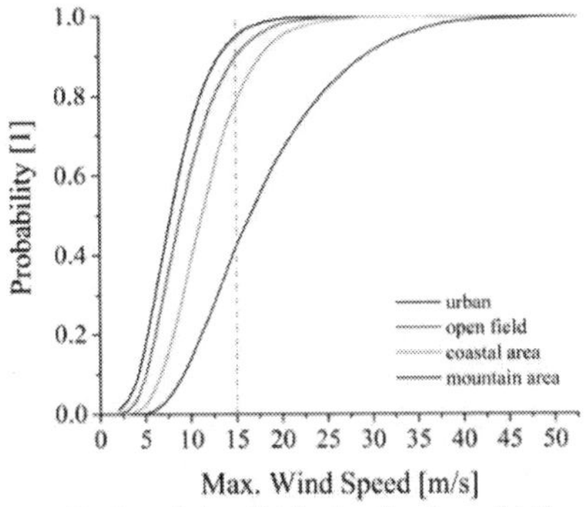

Figure 19: Cumulative distribution function of daily maximum wind speed for a period of 20 years

3.4 Results of submodeling

After solving the global finite element model in order to derive the displacement of the cells a second simulation with a submodel has been performed in order to get the plastic strain energy density that is introduced in the interconnectors. The results are given in Figure 20. On the second ordinate the amplitude of change of cell gap during the cycle is given (maximum minus minimum value) to see if these values correlate.

In contrast to the highest amplitude of relative change of cell gap at positions (1,4) and (1,5), the highest plastic deformations were found at positions (2,4), (2,5) and (3,5) (Figure 20). Especially position (3,5) has an unexpected high amount of plastic deformation compared to the observed cell shift. At these positions an additional bending of the ribbon increases the local plastic deformation. This bending occurs due to the relative displacement of the cells perpendicular to the longitudinal axis of the string (shear displacement). Plots of the plastic strain energy density after one cycle are shown Figure 21. The shear displacement is indicated by the red arrows. In Figure 22 the deformation of the ribbons in y-direction is depicted at maximum pressure load to illustrated the shear displacements. For basic understanding of the introduced load, the bending strain due to the relative displacement can be approximated by a simple analytical model (double-sided fixed beam):

$$\varepsilon = \frac{3h}{l^2}\delta \tag{12}$$

ε – strain
h – ribbon width
l – cell gap
δ – shear displacement

For the values in Figure 22 this estimate delivers 0.12 % for (1,5), 0.52 % for (2,5) and 1.02 % strain for (3,5). The latter cases have a distinct contribution to calculated von-Mises equivalent plastic strain and indicate the multiaxial stress state.

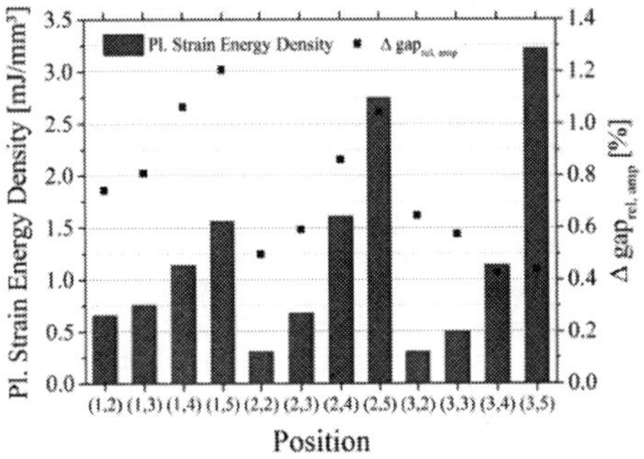

Figure 20: Plastic strain energy density in the ribbon after one cycle and the corresponding amplitude of change of cell gap

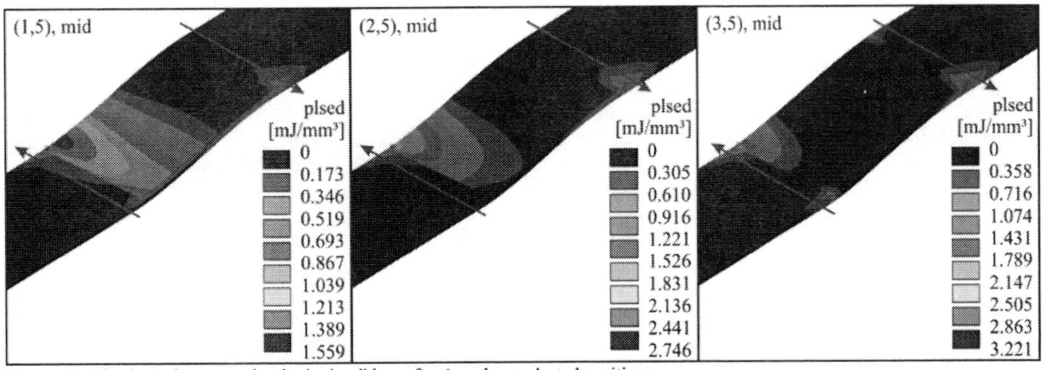

Figure 21: Plastic strain energy density in the ribbon after 1 cycle at selected positions

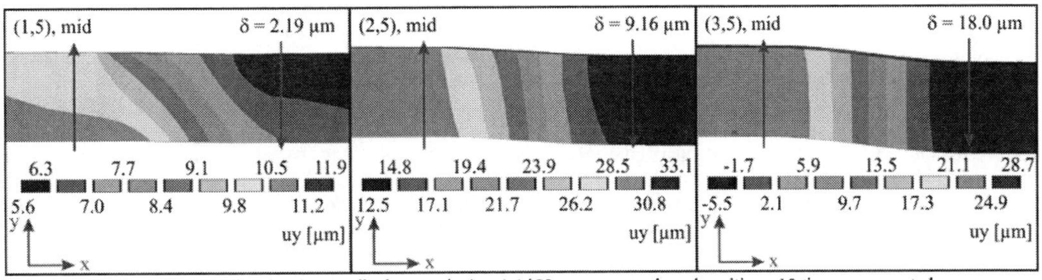

Figure 22: Deformation in y-direction (perpendicular to string) at 1.1 kN pressure at selected positions, 10 times exaggerated

Furthermore, it was found that the plastic strain energy density differs between the three ribbons within a cell gap. As example, at position (2,5) the ribbon in mid position has the highest values compared to left and right (Figure 23). This is due to the deformation shape of the whole module. The superposition of the highest shear displacement and second largest change of cell gap results in the highest plastic strain energy density.

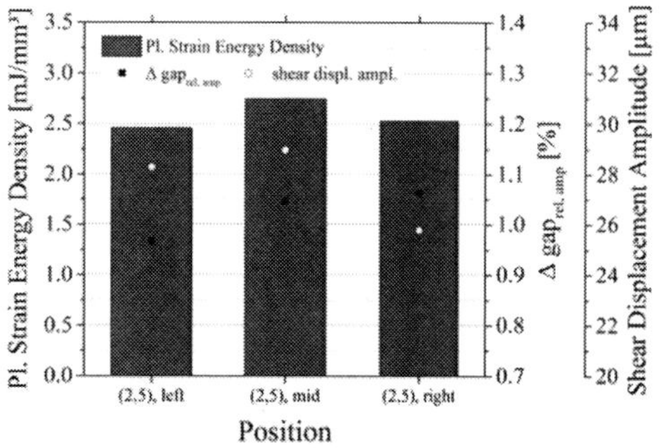

Figure 23: Comparison of plastic strain energy density in the ribbon after 1 cycle for different positions of the same cell gap

4 DISCUSSION

The threshold load levels to achieve 0.3% cell displacement give an inside of the wind events that contribute to ribbon fatigue in the field. Without consideration of wind gusts the minimum wind speed seems to be quite high with 25.9 m/s (93 km/h, whole gale in Beaufort scale) and therefore a rare event in normal as well as extreme climates. However, with increasing wind speed the stream includes turbulences that increases resulting pressure loads. For simple quantification a factor of 3 is used to take wind gusts into account. As result a nominal wind speed of 15 m/s (54 km/h, moderate gale) already may contribute to ribbon fatigue, which is in the range of spring and autumn stormy days in Germany. These events occur frequently and may play a role in overall fatigue of ribbons. Additional factors are temperature and glass thickness. Elevated module temperatures further increase the stress on the ribbons at lower wind speed due to the lower stiffness of the encapsulant, therefore even lower wind speeds than 54 km/h have to be taken into account at higher temperatures. Another important factor is the glass thickness, which influences the bending stiffness of module due to cubic dependency of the area moment of inertia. Using a thinner nominal glass thickness of 3.2 mm, which becomes more common on standard PV-modules, reduces the threshold value by 16 % to 12.6 m/s (45 km/h, strong breeze) compared to a standard 4 mm glass layout. This shows that further measures have to be taken in order to maintain long-term reliability. Such measures may be using an adjusted module design, using other material for the laminate or changing the structural design such as back rails.

For accurate damage estimation the available weather data and therefore the load information was too inaccurate. In the current approach only daily maximum wind data was used. Wind direction and load reversals (suction) are neglected and only one cycle per day is present. The estimated damage is rather low and is only usable for qualitative comparison of the locations. There is no doubt that gust events occur more often so the conclusion is that the damage is underrated, because there are a lot more load events happening. From data in [16] it can be seen how hourly averages are smoothing the wind speed curve. In [8] wind stations that record a wind speed value every 3 s and record the maximum wind speed over a 5 min period are used to analyze thunder storms. If only a factor of ten is assumed for the number of loading events for the mountain region 42 % of the ribbon damage is done by wind load. This damage is further increased by the daily temperature changes. The importance of wind loads contribution to ribbon fatigue can be proofed only by more extensive data analysis for a better estimation of the load cycles.

Detailed models show that there is no straight forward correlation between the change of cell gap evaluated in the string direction and the plastic strain energy density introduced into the ribbons during loading for all gaps. An additional bending due to a shear displacement is present that leads to more severe plastic deformation. Therefore, the shear displacement should be considered to evaluate ribbon stress on module level without detailed local simulation models. This also leads to the question: what happens when a module is twisted or inhomogeneous loaded? It is also concluded from the approximation of bending strain (eq. (12)) that for smaller ribbons this local bending might be smaller.

5 SUMMARY AND OUTLOOK

This paper focuses on stress on the interconnector ribbons inside a PV module during cyclic mechanical loading by means of finite element analysis. The simulations allow a detailed description of the interconnector loading conditions within the PV module laminate, with respect to surrounding materials and arbitrary mechanical load. The focus was on creating an understanding of the loading conditions.

Considering the constitutive and fatigue properties of the solar cell interconnectors the minimum cyclic loading amplitude for continuous progress of damage can be simulated. Regarding the change of cell gap a threshold of 0.3 % was found for a common interconnector material to generate failure within 10^6 cycles.

An evenly distributed surface load leads to different cell shifts inside the module. The result is an uneven distributed loading for the individual ribbons, with a maximum at the end of the string. This differs compared to cell shifting during thermal loading. Here, cell gaps change almost equally within the module when temperature changes [15]. Only the cell at the end of each string shows increased displacement. These results in conjunction with the mechanical results suggest that ribbon failures in the field most likely are localized between the last two to three cells of each string, since daily temperature changes are superimposed by wind loads. This correlates with observations from failed modules in the field shown in Figure 24. At the end of the strings there is an accumulation of breakage events and the inner strings have already failed.

For the investigated module structure the threshold criterion is equivalent to a wind speed of 15 m/s. This result is used to reduce the number of wind loading events that need to be considered for fatigue analysis and detailed simulations. In urban regions the high number of moderate wind loads seems to be insignificant for ribbon fatigue. On the other hand, wind loads may play an important role in coastal areas or windy mountain regions. In addition, it is shown that module temperature and glass thickness have distinct influence on ribbon loading and can further reduce the minimum wind speed.

Figure 24: Electroluminescence image of PV-module extracted from a PV installation after approximately 5 years

Detailed models of the cell gap region show that a shear displacement between the cells introduces a bending load in the ribbon leading to an increased localized plastic deformation. Therefore, there is no straight forward correlation between change of cell gap and plastic strain energy density. This is important when only change of cell gap measurements or global model results are present. Even if there is no general function to transfer the global measurements to estimate local damage parameters, it is necessary to find one for each cell gap for efficient damage calculation and fatigue estimation.

In future work it is necessary to acquire wind load data with higher resolution in order to access the number of load cycles and to understand its quantitative contribution to ribbon fatigue in the field. A cycle counting analysis method, e.g. rainflow-counting needs to be performed with that data. Further module temperature, wind direction and specific module mounting needs to be considered.

6 ACKNOWLEDGEMENTS

The financial support by the German Federal Ministry for Economic Affairs and Energy within project "LAURA" (contract no. 0325716B) is gratefully acknowledged.

7 REFERENCES

[1] W. Herrmann et al, "PV Module Degradation Caused by Thermo-Mechanical Stress: Real Impacts of Outdoor Weathering versus Accelerated Testing in the Laboratory", Proc. of SPIE Vol. 7773, 2010

[2] R. Meier et al, "Thermo-mechanical Behaviour of Copper Ribbon Materials", Proc. 24th EU PVSEC 2009, Sept. 21-25, 2009, Hamburg, Germany, pp. 3413-3419

[3] R. Meier et al., "Reliability of copper-ribbons in photovoltaic modules under thermo-mechanical loading", 35th IEEE Photovoltaic Specialists Conference (PVSC), 2010

[4] IEC TS 62782:2016, Photovoltaic (PV) modules - Cyclic (dynamic) mechanical load testing, March 2016

[5] Data from Deutscher Wetterdienst, ftp://ftp-cdc.dwd.de/pub/CDC/, Feb. 2016

[6] Troen, I., & Lundtang Petersen, E., European Wind Atlas, Roskilde: Risø National Laboratory, 1989

[7] DIN 1055-4:2005-03, Action on structures, Part 4: Wind loads, 2005

[8] F. T. Lombardo, "Thunder storm characteristics of importance to wind engineering", J. Wind Eng. Ind. Aerodyn. 125, 2014, pp. 121–132

[9] IEC 61215-2:2016, Terrestrial photovoltaic (PV) modules - Design qualification and type approval - Part 2: Test procedures

[10] H. Mugrabi, Werkstoffverhalten und Bauteilbemessung, DGM-Informationsgesellschaft, 1987

[11] O. H. Basquin, "The Exponential Law of Endurance Tests," American Society for Testing and Materials Proceedings, Vol. 10, 1910, pp. 625-630.

[12] M. Pander et al, "Thermo-mechanical assessment of solar cell displacement with respect to the viscoelastic behaviour of the encapsulant", 12th. Int. Conf. on Thermal, Mechanical and Multiphysics Simulation and Experiments in Microelectronics and Microsystems, EuroSimE, 2011

[13] M. Pander et al, "Lifetime Estimation for Solar Cell Interconnectors", 28th EU PVSEC, 2013

[14] U. Zeller et al, "The Influence of Air Flow on the Temperature Distribution within Crystalline Solar Modules," 28th EU PVSEC, 2013

[15] U. Eitner, M. Pander, "Thermomechanics of PV modules including the viscoelasticity of EVA", 26th EUPVSEC, 2011

[16] R. Chiod, F. Ricciardelli, "Three issues concerning the statistics of mean and extreme wind speeds", J. Wind Eng. Ind. Aerodyn. 125, 2014, pp. 156–167

[17] J.L. Chaboche, "Time-independent constitutive theories for cyclic plasticity", International Journal of Plasticity, Volume 2, Issue 2, 1986, Pages 149-188

[18] J. Morrow, Cyclic plastic strain energy and fatigue of metals

[19] D. Kujawski, "Fatigue Failure Criterion based on strain energy density", Theoretical and Applied Mechanics, Vol. 7, 1, 1989, pp.15-22

[20] M. Sander, S. Dietrich, M. Pander, M. Ebert, J. Bagdahn, "Systematic investigation of cracks in encapsulated solar cells after mechanical loading", Solar Energy Materials and Solar Cells, 111, (2013), 82-89

[21] DIN EN 572-1, Glas im Bauwesen – Basiserzeugnisse aus Kalk-Natronglas – Teil 1: Definitionen und allgemeine physikalische und mechanische Eigenschaften, 2004-09

[22] R. Hull, Properties of crystalline silicon, INSPEC, Institution of Electrical Engineers, 1999

[23] R. Darveaux, K. Banerji, "Constitutive relations for tin-based solder joints," *IEEE Transactions on Components, Hybrids, and Manufacturing Technology*, Dec 1992, doi: 10.1109/33.206925

[24] Kohn et al.; "Analyses of warpage effects induced by passivation and electrode coatings in solar cells", 22nd EUPVSEC, Milano, 2007

RELIABILITY OF LOW TEMPERATURE CONDUCTIVE FILM INTERCONNECTION PROCESS FOR PV MODULES

Shu Zhang, Yan Xie, Haijun Jiao, Jianmei Xu, Zhiqiang Feng and Pierre J. Verlinden
State Key Laboratory of PV Science and Technology, Trina Solar
No.2 Trina Road, Trina PV Park, New District, Changzhou, Jiangsu, China, 213031, shu.zhang@trinasolar.com

ABSTRACT: A detailed investigation of the reliability of PV modules using a low-temperature conductive film interconnection process is presented. An advanced low-temperature conductive film bonding technology is compared to the conventional soldering process. The reliability of PV modules with different interconnection processes is analyzed by up to two times the standard module reliability testing sequences. At the material level, the stability of the contact resistance is analyzed under different temperatures. The results show that, compared to the conventional soldering process, the low-temperature conductive film bonding technology exhibits a comparable reliability, benefiting from low stress thanks to a low-temperature process. This analysis demonstrates the potential for long durability of the cross-linked conductive film. Besides, low temperature interconnection technology may be combined with other advanced technologies to achieve higher module power.
Keywords: low temperature, conductive film, contact resistance, interconnection

1 INTRODUCTION

There is a strong trend within the PV industry to minimize silver consumption to reduce manufacturing cost and LCOE. The silver consumption for the front side of conventional 6-inch silicon solar cells is still as high as 90~120 mg/piece. Typical amounts of silver required for busbars is approximately 25~35 mg/piece. There are several ways to reduce silver, including plating technology [1-3], Cu electrode technology [4] and interconnecting with multiple wires rather than busbars. Another research focus of many institutions and companies is the development of high-efficiency Heterojunction with Intrinsic Thin-layer (HIT) solar cells [5-9]. HIT cells are sensitive to high temperature, so they should be interconnected at low temperature below 200℃ .

On the module side, about 3% of the light is reflected out by conventional flat ribbons. Thanks to the total internal reflection effect, light capture ribbons (LCR) with a grooved front surface can recapture 60%-80% of the reflected light that would otherwise be lost [10]. LCR should be bonded to the cell at a low temperature, lower than the melting point of the solder which is typically 183 ℃ for the conventional Sn-Pb alloy solder.

The conventional interconnection process is soldering. Soldering can interconnect ribbons only with conventional silver busbar electrodes. Other metal electrodes are difficult to interconnect and require the soldering temperature to be higher than the melting point of solder. For typical Sn-Pb alloy soldering, the temperature should be 210~230℃, while the temperature for typical Pb free soldering is 30~50℃ higher. The high-temperature soldering process creates aditional stress in a solar cell due to differences in thermal expansion coefficients of silicon and the soldering metal. Due to the above requirements of the conventional soldering process, the application of new technologies cannot be fully utilized and calls for a need to develop advanced low-temperature interconnection technologies.

There are solutions for interconnections that could potentially overcome issues the industry is facing today. Conductive glue and conductive films CF) have been studied for a long time. For example in Ref. [11], the cells were interconnected by ribbons with a different interconnection technique, so mini modules built with busbar-free cells and only the conductive film are capable to withstand moisture, thermally induced mechanical stress, and corrosion during the TC200, DH1000 and HF10 tests [11].

In this paper, we verify the reliability of conductive film (CF) interconnection process for standard industrial 60-cell PV modules made inhouse. The study includes module reliability results and material analysis.

2 EXPERIMENTAL

For module reliability testing, experiments were performed on modules containing 60 p-type 156×156 mm^2 industrial multicrystalline (mc-Si) cells , encapsulated in EVA and a TPE backsheet. Cell thickness is 190 μm ± 20 μm. Standard solar cells were processed using two metallization patterns, one with continuous busbars for the reference soldering process and one without busbars on the front side for the low temperature CF process. The pattern on the front and rear sides of the two are shown in Tab.I. The ribbons applied for the reference soldering interconnection and the low temperature CF interconnection were Sn63Pb37 coated ribbons made with the hot-dip process or the plating process, respectively. The plated Sn63Pb37 ribbon is a Light Capture Ribbon (LCR) with regular grooves on the front surface with a thickness of 5±1 μm. The thickness of the conventional soldering ribbon is 20±2 μm. The above mentioned ribbons have the same cross sectional area. The dimension of the conventional ribbon is 0.25 mm x 1.50 mm, while that of the plated LCR is 0.26 mm x 1.60 mm. Before and after the reliability testing, the module performance at STC was measured with a Pasan pulsed solar simulator of Class AAA, according to IEC60904-9:2007. The reference module used here is a mc-Si PV module with the same spectral response as modules in the experiment and has been calibrated by Frauhofer ISE. The test conditions of all reliability tests (TC200, DH1000, HF10) were set according to IEC 61215. To further evaluate the long-term reliability of CF interconnection, modules were subject to twice the number of cycles or twice the time usually required by the IEC standard (TC400, DH2000, HF20).

Table I: Front and rear side patterns for two interconnection processes.

Group	Front side pattern	Rear side pattern	Note
1			Conventional soldering
2			CF+LCR

For the CF material analysis, the contact resistance for both front and rear sides was measured. The environmental testing chamber Testsky was employed to vary the ambient temperature. During testing, samples were put in the chamber and biased with a typical I_{mmp} of 8.5 A. We varied the temperature to evaluate the stability of the contact resistance. The instrument and the test method are shown in Fig. 1. Samples were fabricated according to the real-world condition: front and rear electrodes with a width of 8 mm were printed and fired onto the wafer. To measure the contact resistance more comparatively, the same pattern as the cell with fingers was printed on busbar-free cells. The interconnection between the electrodes and the plated LCR with an area of 0.26×1.6 mm^2 was realized by a low-temperature CF bonding process. For comparison, the hot-dipped ribbon with an area of 0.25×1.5 mm^2 was manually soldered. The sample size for each group is 5. The calculated resistance included three parts - the resistance of both the ribbon and the busbar, as well as the contact resistance. Since the combined resistance of the ribbon and the busbar for the interconnecting part is less than 1% of the contact resistance, we can ignore these two parts.

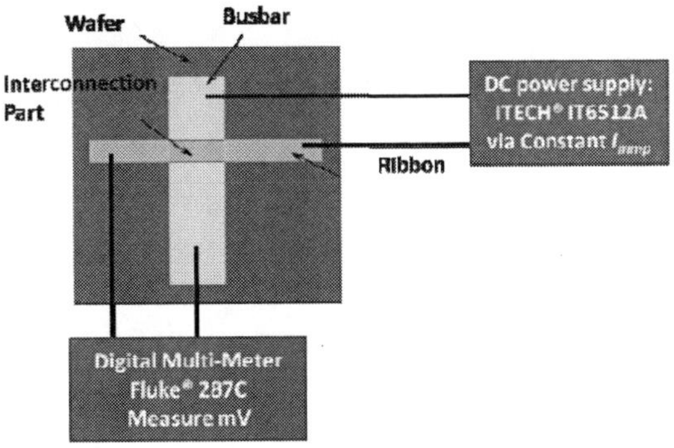

Fig. 1: Set-up and geometry of the contact resistance testing.

Reference soldering was done using an automatic soldering process with a IR heater. The soldering temperature is 210-220℃. For CF, the specified curing time is 10 seconds and the curing has been achieved at 170-180℃ but 2 MPa pressure. The cross-linking ratio for the resin within the CF post curing was higher than 80%. Such a high cross-linking ratio ensured that the connection was stable during the lamination process. After lamination, the cross-linking rose to 100%.

On the front side, after final bonding, fingers were pressed into the solder as shown in Fig. 2(a). Interconnection was mainly achieved by the contact between fingers and the solder, where elastic conductive particles strengthened the interconnection. The contact depth is around half the height of finger, which is higher than 5 μm. For the rear side, the contact geometry is shown in Fig. 2(b). The solder at the peak of the LCR is pressed into the rear electrode. In most situations, the contact depth is higher than 2 μm.

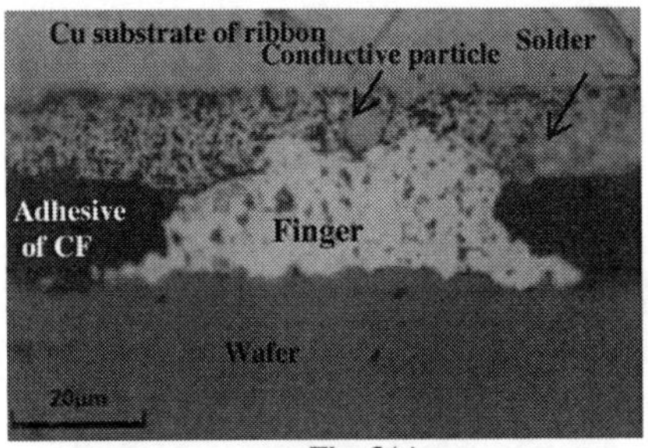

Fig. 2(a)

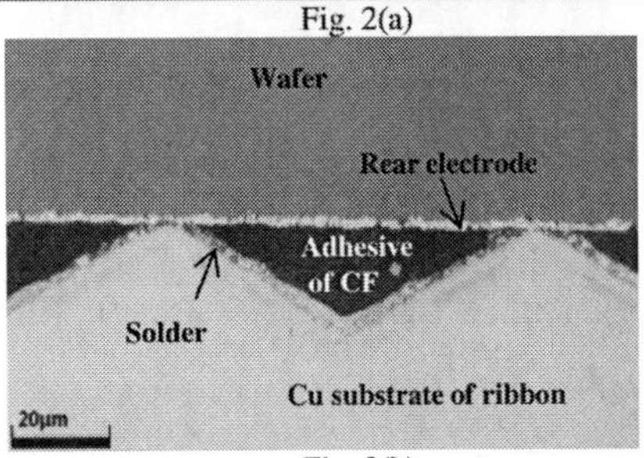

Fig. 2(b)

Fig. 2 Contact geometries of the CF interconnections at the front side (a) and the rear side (b).

The storage modulus curve, which was provided by the CF supplier, is shown as Fig. 3. The storage modulus was measured in tensile mode with a thermal analysis (TA) instrument (ARES-G2 environmental systems). The curve illustrates that the operating temperature of the CF within a PV module is lower than the glass transition temperature of the 100% cross-linked CF. In the high range of the operating temperature, the storage modulus of the cross-linked CF is stable.

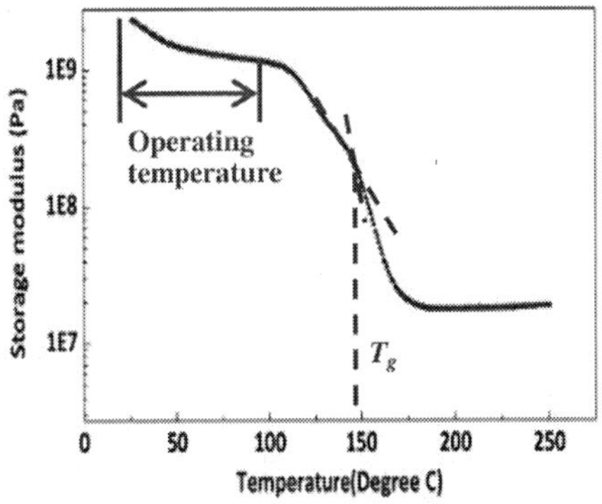

Fig. 3 The storage modulus curve of the 100% cross linked CF.

3 RESULTS AND DISUSSION

3.1 Performance characteristics of cells

2000 solar cells for each group were fabricated in-house. The median electrical parameters of both cell groups are shown in Table II. It is apparent that V_{oc} of the solar cells with no-busbar pattern on the front side is slightly higher than that of the reference group. It is mainly due to a smaller recombination area and to a better passivation for solar cells with no front busbars. Both Isc and Eff of no-busbar cells were lower than those of the reference group, however, resulting from a higher shadow ratio for our busbar-free measurement frame.

Table II: Median *I-V* parameters of the two groups of solar cells, containing 2000 cells each.

Processes	*Eff.* (%)	V_{oc} (mV)	I_{sc} (A)	*FF* (%)	R_s (mohm)
No front BB	18.28	634.4	8.835	79.36	2.0
Reference	18.31	633.7	8.865	79.30	2.1

3.2 Initial *I-V* characteristics of modules

To accurately compare the I-V parameters of the two interconnection solutions, the data at the module level was measured. PV modules with completed solar cells for each group were fabricated. To avoid mismatch, cells within a module were sorted according to their *I-V* parameters. Mainly I_{sc}, but to some extent also R_s change due to the LCR and the applied new cell bonding process, as listed in Table III. The relative I_{sc} gain for the LCR module with the CF technology compared with no-LCR modules with standard soldering process is 1.50%. As the total shading area ratio of the LCR is 3.07%, the optical yield of the LCR is 55.0%. The loss is mainly caused by the limited reflectance of the tin coating. Although the cross sectional area of the LCR is the same as that of the ordinary soldering ribbons, the line resistance of the LCR is 10% lower than that of ordinary ribbons. Hence, the LCR results in a $4.2\%_{relative}$ reduction in series resistance for the entire module. Also, the low processing temperature induces minimal stress on the cell and a slightly higher V_{oc} in the module. These factors result in a 1.83% relative gain in P_{mmp}.

Table III: The median electrical parameters of the two groups of PV modules.

Groups	P_{mmp} (W)	V_{oc} (V)	I_{sc} (A)	*FF* (%)	R_s (ohm)
2	272.2	38.06	9.208	77.67	0.434
1	267.3	37.99	9.072	77.55	0.453

3.3 Thermal cycling Results

Thermal cycling testing is critical to evaluate the reliability of interconnections. During the testing, the temperature cycle ranged between -40 and 85℃. The mismatch of deformation between ribbons and wafers causes heat stress and the relative motion of ribbons along the longitudinal direction. The electrical parameter degradation of the PV modules for two groups after TC200 and TC400 tests are shown in Table IV. The result shows that the plated LCR combined with the CF suffers almost the same power degradation ratio as the soldering process, even after 2 times the TC200 standard testing.

Table IV: The median relative change in electrical parameters for two groups of PV modules after TC200 and TC400 tests (relative to initial parameters).

Groups		ΔP_{mmp} (%)	ΔI_{sc} (%)	ΔR_s (%)	ΔV_{oc} (%)	ΔFF (%)
2	TC200	-1.09	-0.05	4.01	-0.11	-0.93
	TC400	-3.27	-0.65	10.48	-0.67	-1.97
1	TC200	-0.94	-0.37	2.84	-0.15	-0.43
	TC400	-2.87	-0.25	8.33	-0.71	-1.93

3.4 Contact resistance of the interconnections

To measure the reliability of the contact resistance of the interconnecting part, the contact resistance was measured at a variable temperature ranging from 0 to 85℃. The nominal contact area for each sample was 8 × 1.50 or 1.60 mm^2 (12.0 mm^2 or 12.8 mm^2) . For cells with the CF combined with the LCR, the contact area is determined by the finger and the peak of the LCR, which is quite different to conventional soldering. Therefore, only the resistance (μΩ) could be evaluated and the contact resistivity (μΩmm^2) could not be determined sufficiently precisely. Parameters for each sample are shown in Figs. 4 and 5. The result indicates that the contact resistance of soldered samples has slight variances due to the different contact area. However, for samples with the CF interconnection method, the coherence of the contact resistance was much better. At various temperatures, the contact resistance of the CF samples was comparable to that of soldered samples. This result indicates that the module with the CF interconnections has a reliable power at various ambient temperatures. In addition, the contact resistance of both interconnection processes was extremely low compared to the total series resistance of the module.

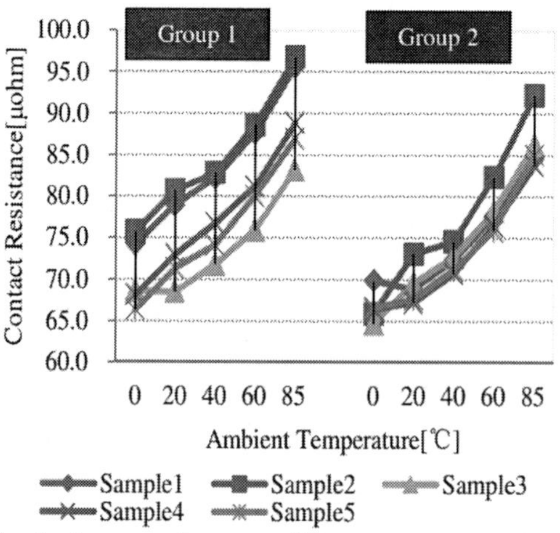

Fig. 4: Contact resistance at different temperatures for the front side contact.

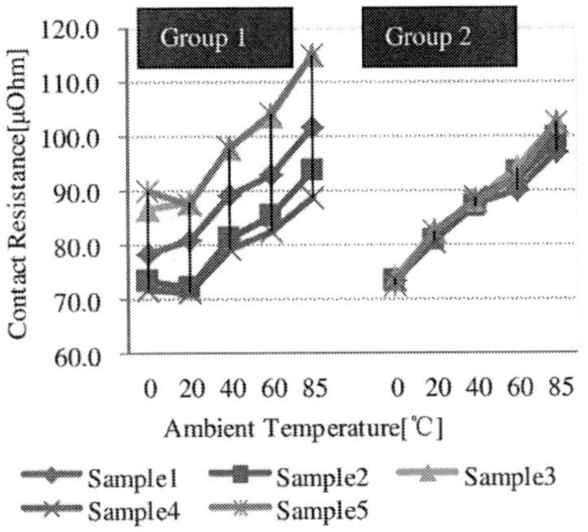

Fig. 5: Contact resistance at different temperatures for the rear side.

3.5 Humidity freeze test results

Humidity freeze testing is important to verify the reliability of interconnections. During each HF cycle, the relative humidity is maintained at 85% ± 5% R.H. and 85℃. The moisture can easily diffuse into the module, including the interconnection parts. When the temperature goes down to -40℃, the moisture freezes and expands, and induces stress in interconnections perpendicular to the interconnecting surface. The change in electrical parameters of the PV modules after HF10 and HF20 tests are shown in Table V. The result shows that the plated LCR combined with the CF is as reliable as the soldering process, even after 2 times the HF standard testing.

Table V: Median relative change in electrical parameters of two groups of PV modules after HF10 and HF20 tests (relative to initial) parameters).

Groups		ΔP_{mmp} (%)	ΔI_{sc} (%)	ΔR_s (%)	ΔV_{oc} (%)	ΔFF (%)
2	HF10	-0.98	-0.42	3.68	-0.03	-0.55
	HF20	-2.26	-0.73	8.46	-0.04	-1.51
1	HF10	-1.06	-0.51	4.08	-0.05	-0.50
	HF20	-2.52	-1.06	8.05	-0.09	-1.38

3.6 Damp heat results

The presence of moisture in metal contacts for a long time can cause metal corrosion and contact failure. Both the standard DH1000 and DH2000 tests were conducted. The relative change in electrical parameters of the PV modules after DH1000 and DH2000 tests is illustrated in Table VI. The result shows that the plated LCR ribbon combined with the CF was a reliable solution, even after DH2000 testing.

Table VI: Median relative change in electrical parameters of two groups of PV modules after DH1000 and DH2000 tests (relative to initial parameters).

Groups		ΔP_{mmp} (%)	ΔI_{sc} (%)	ΔR_s (%)	ΔV_{oc} (%)	ΔFF (%)
2	DH1000	-0.67	-0.42	0.91	-0.10	-0.15
	DH2000	-2.76	-2.25	12.87	-0.21	-0.32
1	DH1000	-0.58	-0.38	1.52	-0.14	-0.06
	DH2000	-2.48	-1.78	10.78	-0.18	-0.53

4 CONCLUSION

The low-temperature CF interconnection technology combined with plated LCR ribbons increased the Pmmp of the PV modules by 1.83%. The result of a double sequence of reliability testing shows that the power degradation of modules with a low-temperature conductive film interconnection is comparable to that of conventionally soldered modules. The analysis of the contact resistance measured under different temperatures also shows comparable resistance values and similar relative changes in resistance for both interconnection processes. Hence, the low-temperature CF interconnection technology, combined with plated LCR ribbons, can be used in mass production of PV modules.

ACKNOWLEDGEMENT

This work is supported by the National "863" Project under the Project Number of 2015AA050302 and the Changzhou Sci&Tech Program with the Grant Number of CC20150005.

REFERENCES

[1] D. Pysch, et al. Comprehensive Analysis of Advanced Solar Cell Contacts Consisting of Printed Fine-line Seed Layers Thickened by Silver Plating, Progress in Photovoltaics, 17(2), 101-114 (2009)
[2] A. Mette, et al. Increasing the Efficiency of Screen-Printed Silicon Solar Cells by Light-Induced Silver Plating, Photovoltaic Energy Conlversion,1, 1056 - 1059 (2006)
[3] L. Alison, et al. Evolution of Metal Plating for Silicon Solar Cell Metallisation, Progress in Photovoltaics, 21(7), 1454–1468 (2013)
[4] R. Russell, et al. A Simple Copper Metallization Process for High Cell Efficiencies and Reliable Modules, 27th EU PVSEC, 538-543 (2012)
[5] T. Mishima, et al. Development status of high-efficiency HIT solar cells, Solar Energy Materials and Solar Cells, 95(1), 18-21(2011)
[6] M. Taguchi, et al. 24.7% Record Efficiency HIT Solar Cell on Thin Silicon Wafer, IEEE Journal of Photovoltaics, 4(1), 96-99(2014)
[7] E. Fortunato, et al. Transparent Conducting Oxides for Photovoltaics, MRS Bulletin, 32(3), 242-247(2007)
[8] F. Hiroyuki, et al. Impact of epitaxial growth at the heterointerface of a - Si : H c - Si solar cells, Applied Physics Letter, 90(1), 013503 (2007)

[9] W. Guo, et al. Microstructure of Hydrogenated Amorphous Silicon Layers Studied by Spectroscopic Ellipsometry for the Surface Passivation in Heterojunction Solar Cells, 42nd IEEE PVSC, (2015)
[10] S. Zhang, et al. 335-W World-Record p-Type Mono-crystalline Module with 20.6% Efficient PERC Solar Cells, Journal of Photovoltaics, 6(1), 145-152(2016)
[11] R. Markus, et al. Qualification Testing of Monolithically Assembled Modules Using MWT PV Cells and Contactfoil®, A Large Area, Laser Patterned Foil System, 27th EU PVSEC, 3456 – 3460 (2012)
[12] A. Schneider, et al. Conductive Gluing as Interconnection Technique towards Solar Cells without Front Busbars and Rear Pads, 27th EU PVSEC, 335-229 (2012)

GEOGRAPHICAL MAPPING OF THE PERFORMANCE OF VERTICALLY INSTALLED BIFACIAL MODULES

Masakazu Ito** and Eric Gerritsen*
* Univ. Grenoble Alpes, INES, F-73375 Le Bourget du Lac, France
CEA, LITEN, Department of Solar technologies, F-73375 Le Bourget du Lac, France
eric.gerritsen@cea.fr

** JSPS Fellow for Research Abroad assigned at CEA-INES
currently at: ACROSS, Waseda University, 3-4-1, Okubo, Shinjuku-ku, Tokyo, 169-8588, Japan

ABSTRACT: Bifacial modules offer the unique potential of vertical installation with benefits in reducing soiling or snow coverage. Moreover, mixing East/West and North/South oriented vertical modules will level out the energy production over the day, aligning energy production to demand and so reducing the required storage capacity. We apply modelling based on global irradiance (direct / diffuse / land reflection) and albedo data maps to quantify energy performance of vertically installed modules as a function of their geographical location.

We compare vertically installed bifacial modules against tilted monofacial and bifacial equator-oriented (EO) tilted modules. In regions at higher northern latitudes (>40°) and the polar regions, vertical bifacial installations are found to be energy efficient compared to monofacial tilted modules and equally useful to levelling of energy production over the day by mixing E/W and S/N orientations. Also large (semi-) arid regions below the 20° latitude line are found to be favorable for vertical bifacial installations due to their high albedo.
Keywords: Bifacial, Energy Performance, Vertical Installation, Global Analysis

1 BACKGROUND AND OBJECTIVE

The amount of installed PV systems is rapidly growing [1], since they are environmental friendly compared to conventional fossil fuel power plants [2]. In addition, there is a large potential in arid regions where abundant sunshine and land are available [3] and also in cold regions with strong albedo of snow [4]. However, in these regions there are challenges to tackle such as soiling in arid regions [5] and snow coverage in cold regions [6]. For these challenges, bifacial modules offer the unique potential of vertical installation with benefits in reducing soiling or snow coverage. Moreover, mixing East/West and North/South oriented vertical modules will level out the energy production over the day, aligning energy production to demand and so reducing the required storage capacity [7]. Guo et al. [8] have investigated global analysis of vertically installed bifacial PV modules. They have assumed equal efficiencies for front and rear side of the bifacial module, which will strongly affect the results. We apply modelling based on global irradiance (direct / diffuse / land reflection) and albedo data maps which can consider different efficiencies for front and rear side to quantify energy performance of vertically installed modules as a function of their geographical location.

2 METHODOLOGY

To compare the potential of a bifacial PV module and a monofacial PV module, the energy yield at all locations in the world needs to be calculated. This would be possible if in-plane irradiation data of all locations would be available, which is not the case. Therefore, surface irradiation data measured by satellite were used in this study. Their accuracy is lower than ground measured data, but they do cover the complete surface of the earth. Then, to convert the surface irradiation data into in-plane irradiation data for calculation of the energy yield, different models were applied. The Liu-Jordan model [9] is used to distinguish between direct and diffuse irradiation. Then the two types of daily irradiation are divided into hourly irradiation. For the in-plane irradiation of tilted surfaces, three additional models are used. These are the Direct method [10] which is known as a method to calculate direct irradiation theoretically, the Hay model [11] for diffuse irradiation and the isotropic reflection model [10] for land reflection. Next, assuming equator equator-oriented (EO) PV modules, facing south in the northern hemisphere and facing north in the southern hemisphere, the in-plane irradiation of each location by 1 degree latitude and longitude is calculated by the methods above. The methodology of these calculations is schematically shown in figure 1. Details will be explained in sections 2.1 and 2.2.

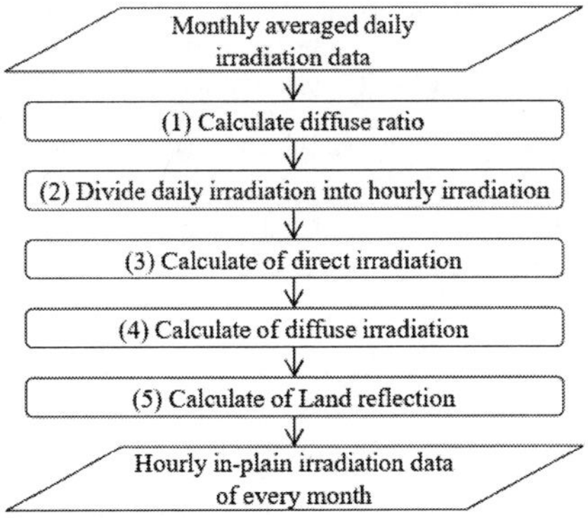

Figure 1: A scheme to calculate hourly in-plane irradiation data from monthly averaged daily irradiation data

2.1 Irradiation data

Global surface irradiation data are produced by several databases like the HelioClim Project by MINES Paris/CNRS [12], SeaWiFS Surface Solar Irradiance by the NASA Goddard Institute for Space Studies [13] or Surface meteorology and Solar Energy (SSE) by NASA/POWER [14]. We selected the SSE database for this study because of its easy access and because it contains a detailed accuracy evaluation. An example of irradiation data is shown in figure 2.

For land reflection, albedo data are required. The NASA also provides an albedo dataset called ISLSCP II MODIS (Collection 4) Albedo [4] which are supplied as a 1 degree mesh at 16 day time intervals. An example of albedo data is shown in figure 3. These data are used as input for the models indicated below.

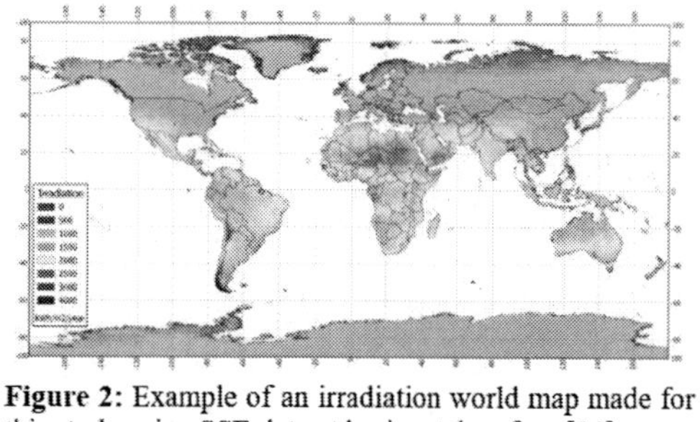

Figure 2: Example of an irradiation world map made for this study using SSE data at horizontal surface [14]

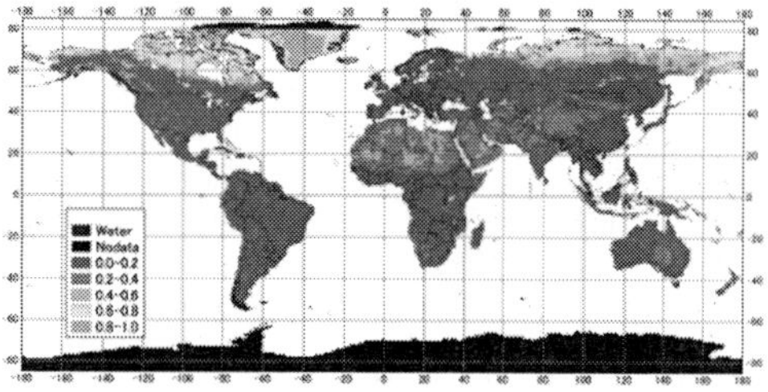

Figure 3: Example of an albedo world map for the month of May made for this study using albedo data of ISLSCP II MODIS (Collection 4) Albedo [4]

2.2 Models to calculate in-plane irradiation for monofacial and bifacial PV modules

The different irradiation models available are explained in [10, 15]. The models for this study are selected to calculate in-plane irradiation on rear and front surfaces of monofacial and bifacial PV modules. Equation 1 expresses the basic idea to calculate in-plane irradiation $I_{\beta\gamma}$ from direct irradiation $I_{b\beta\gamma}$, diffuse irradiation $I_{s\beta\gamma}$ and land reflection $I_{r\beta\gamma}$. β is the tilt angle in degrees, and γ is the azimuth direction in degrees starting at south and increasing clockwise.

$$I_{\beta\gamma} = I_{b\beta\gamma} + I_{s\beta\gamma} + I_{r\beta\gamma} \quad \text{(eq.1)}$$

2.2.1 Calculation of the diffuse ratio

The ratio of the diffuse irradiation to the global irradiation is used to divide global irradiation data into direct and diffuse irradiation. With global irradiation given by equation 2 as

$$H = H_b + H_d \quad \text{(eq.2)}$$

where,
H is daily global irradiation,
H_b is daily direct irradiation,
H_d is daily diffuse irradiation.

The diffuse ratio is then expressed by

$$K = H_d/H \quad \text{(eq.3)}$$

where,
K is the diffuse ratio,
$\overline{K}$ is the diffuse ratio averaged over the month.

$\overline{K}$ can be calculated by the Liu-Jordan model [9] as shown by equation 4.

$$\overline{K} = 1.390 - 4.027\overline{K_T} + 5.531\overline{K_T}^2 - 3.108\overline{K_T}^3 \quad \text{(eq.4)}$$
$$(0.3 < \overline{K_T} < 0.7)$$

where $\overline{K_T}$ is the ratio of global irradiation to extraterrestrial irradiation averaged over the month, as in equation 5.

$$K_T = H/H_0 \quad \text{(eq.5)}$$

where H_0 is the extraterrestrial global daily irradiation.

2.2.2 Dividing daily irradiation into hourly irradiation

Monthly averaged hourly global irradiation $\overline{I}$ and diffuse irradiation $\overline{I_d}$ can be calculated from the calculated monthly averaged daily irradiation data above.

$$\overline{I} = \overline{H} \cdot rt \quad \text{(eq.6)}$$

$$\overline{I_d} = \overline{H_d} \cdot rd \quad \text{(eq.7)}$$

where rt is the hourly allocation ratio of global irradiation, which divides daily irradiation into each hour, and rd is the hourly allocation ratio of diffuse irradiation. The values of rt and rd can be calculated by the following equations using hour angle ω and hour angle at sunset time ω_s.

$$rt = (\pi/24)(a + b\cos\omega)(\cos\omega - \cos\omega_s)/(\sin\omega_s - \omega_s\cos\omega_s)$$
$$a = 0.4090 + 0.5016\sin(\omega_s - \pi/3)$$
$$b = 0.6609 + 0.4767\sin(\omega_s - \pi/3) \quad \text{(eq.8) [16]}$$
$$rd = (\pi/24)(\cos\omega - \cos\omega_s)/(\sin\omega_s - \omega_s\cos\omega_s) \quad \text{(eq.9) [9]}$$

2.2.3 Direct irradiation

When the direct irradiation on a perpendicular plane is given by I_n and direct irradiation on a horizontal surface by I_b then the direct irradiation part of in-plane irradiation $I_{b\beta\gamma}$ can be calculated by equations 10 and 11, as shown in figure 4.

$$I_b = I_n\cos\theta_z \quad \text{(eq.10)}$$
$$I_{b\beta\gamma} = I_n\cos\theta \quad \text{(eq.11)}$$

The ratio rb , of direct irradiation on a horizontal surface relative to a tilted surface, is calculated by eq.12.

$$rb = I_{b\beta\gamma}/b = \cos\theta/\cos\theta_z \quad \text{(eq.12)}$$

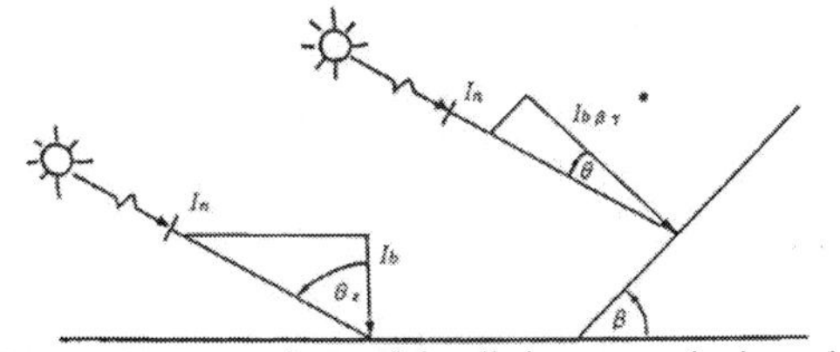

Figure 4: Comparison of irradiation on a horizontal surface and a tilted surface. Figure by JSES [10].

The direct irradiation can be calculated subtracting the diffuse radiation from the global irradiation, thereby converting equation 12 into equation 13.

$$I_{b\beta\gamma} = (I - I_d)\cos\theta/\cos\theta_z \quad \text{(eq.13)}$$

$\cos\theta$ and $\cos\theta_z$ are obtained from equations 14 and 15 where δ is the declination angle. The diagram in figure 6 visualizes the incident angle θ.

$$\cos\theta = (\sin\phi\cos\beta - \cos\phi\sin\beta\cos\gamma)\sin\delta + (\cos\phi\cos\beta + \sin\phi\sin\beta\cos\gamma)\cos\delta\cos\omega + \cos\delta\sin\beta\sin\gamma\sin\omega \quad \text{(eq.14)}$$

$$\cos\theta_z = \sin\phi\sin\delta + \cos\phi\cos\delta\cos\omega \quad \text{(eq.15)}$$

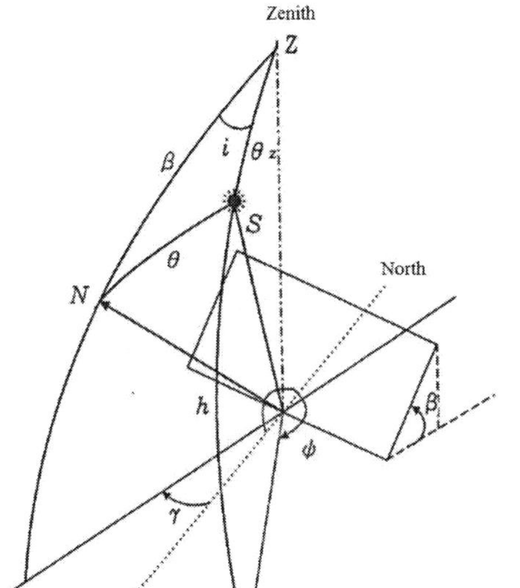

Figure 5: Diagram to calculate the incident angle θ. Figure by JSES [10].

2.2.4 Land reflection

The isotropic reflection model assumes that the diffuse irradiation is of uniform distribution [10]. Therefore, the land reflection is calculated by equation 16, where ρ is the albedo, expressing the ratio of reflection of irradiation [4].

$$I_{r\beta\gamma} = I\rho(1 - \cos\beta)/2 \quad \text{(eq.16)}$$

2.2.5 Diffuse irradiation

For the estimation of diffuse irradiation, the Hay model [11] is used. In this model, to consider non-uniformity, the ratio calculated by $(I - I_d)/I_0$ of the diffuse part is assumed to be irradiation from around the sun. Therefore calculation of this part is similar to the direct irradiation and the other part $1 - (I - I_d)/I_0$ is assumed to be uniformly distributed.

$$I_{s\beta\gamma} = I_d \begin{bmatrix} \{I - I_d/I_0\}\cos\theta/\cos\theta_z + \\ \{(1 - I - I_d/I_0\}(1 + \cos\beta)/2 \end{bmatrix} \quad \text{(eq.17)}$$

3 STUDIES TO COMPARE BI-FACIAL AND MONO-FACIAL PV MODULES

To evaluate the potential of vertical and tilted bifacial PV modules, seven different comparative studies were performed. The comparison is done on the basis of the annual irradiation (kWh/m²/year) received on front and rear side of the modules.

- Study 1 (figure 6): Comparison of E/W and equator equator-oriented (EO, facing south in north hemisphere and north in south hemisphere) vertically installed bifacial modules.
- Study 2 (figure 7): The E/W oriented vertically installed modules are now compared to monofacial equator-oriented modules, tilted at latitude angles.
- Study 3 (figure 8): Effect of the rear side performance of the bifacial modules, as expressed by the back-to-front ratio BTFR (BTFR is 100% for a perfectly symmetrical bifacial module and 0% for a monofacial module). The E/W oriented vertically installed bifacial modules with different BTFR are compared to monofacial equator equator-oriented modules tilted at latitude angle
- Study 4 (figure 9): Comparison of seasonal effects on bifacial vertical E/W and tilted EO monofacial modules.
- Study 5 (figure 10): Output performance of S/N oriented tilted bifacial modules as function of tilt angle and ground albedo
- Study 6 (figure 11): Comparison of E/W oriented vertically installed bifacial modules to EO monofacial modules tilted at 30°.
- Study 7 (figure 12): Comparison of bifacial and monofacial modules, both EO and tilted 30°.

The different configurations of these studies are summarized in table I.

Table I: Overview of the 7 configurations studied to evaluate the global potential of bifacial PV systems.

study figure	type	orien-tation	tilt angle	BTFR	albedo	period
1	bi	E/W	90°	100%	dataset	year
figure 6	bi	S/N	90°	100%	dataset	year
2	bi	E/W	90°	100%	dataset	year
figure 7	mono	S/N	latitude	0%	dataset	year
3 figure 8	bi	E/W	90°	100, 90, 80, 55%	dataset	year
	mono	S/N	latitude	0%	dataset	year
4	bi	S/N	90°	100%	dataset	seasons
figure 9	mono	S/N	latitude	0%	dataset	seasons
5 figure 10	bi	S/N	0, 30, 60, 90°	90%	0, 0.2, 0.4, 0.6	year
6	bi	E/W	90°	100%	dataset	year
figure 11	bi	S/N	30°	100%	dataset	year
7	bi	S/N	30°	100%	dataset	year
figure 12	mono	S/N	30°	0%	dataset	year

4 RESULTS AND DISCUSSION

The comparison in figure 6 between vertically oriented bifacial modules in both E/W N/S orientation shows that E/W orientation is seen to give a far better (green) or at least equivalent (yellow) performance compared to N/S orientation (except for polar region above 70° latitude). The dark green regions where the E/W orientation is performing up to 40% better than N/S are located below the 40° latitude line. The yellow regions at higher latitudes (between 40 and 70°) show equivalent performance for N/S and E/W vertical bifacial orientation. These latter regions could thus ideally benefit from a mix of N/S and E/W oriented modules to adapt hourly PV production over the day to requirements on electricity demand as well as from vertical applications with varying orientations like in sound barriers or fences.

Because of the results in figure 6 the bifacial vertical (BV) modules studied in the following will be oriented E/W. The comparison of BV-E/W versus tilted monofacial modules in figure 7 shows their better performance in regions above 40 degree latitude as well large (semi-)arid desert like regions at lower latitude with high natural ground albedo. The first regions are specifically interesting for BV to reduce the effects of snow coverage whereas the latter arid regions can benefit from reduced soiling effects for BV modules. Soiling effects in these arid regions, where cleaning is difficult because of the lack of water, can lead to a performance decrease of 1% per day or much larger after severe dust storms [17], leading to accumulated losses of typically 10-20% and up to 50%.

Figure 8 shows how the beneficial effect of BV modules rapidly declines with reduced back-to-front ratio. As typical BTFR ratios for bifacial module types entering the market are typically well above 85% this benefit should be feasible.

Modelling of the seasonal performance of BV modules in figure 9 shows a clear peak of their beneficial effect in the northern hemisphere for the month of June, a lowest effect in January and moderate beneficial effects for both March and September.

A geographical mapping of the effect of tilted equator equator-oriented bifacial modules is shown in figure 10. The effect of ground albedo appears about equally strong for all tilt angles and geographical locations.

Tilted equator equator-oriented bifacial modules are seen to perform on a global average 7.5% better than BV-EW (figure 11).

Finally bifacial modules are found to perform from 15 up to 35% better than their monofacial counterparts, when both equator equator-oriented at 30° tilt (figure 12).

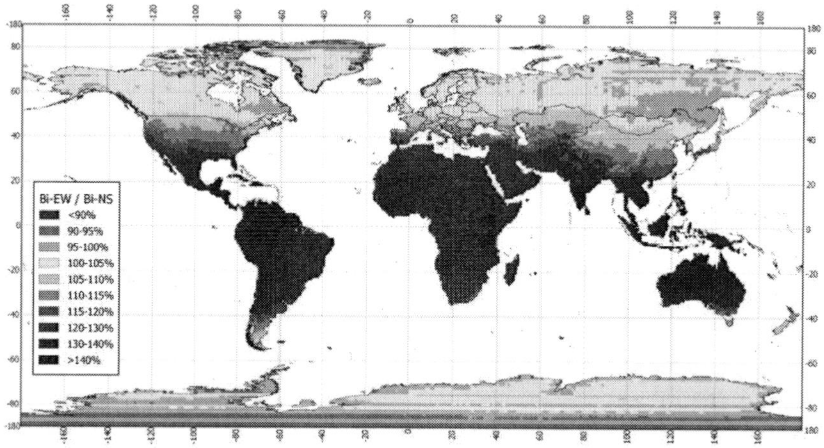

Figure 6: Annual kWh performance of bifacial vertical: E/W compared to S/N (equator) oriented GREEN: up to 40% higher performance for E/W orientation. YELLOW: similar performance

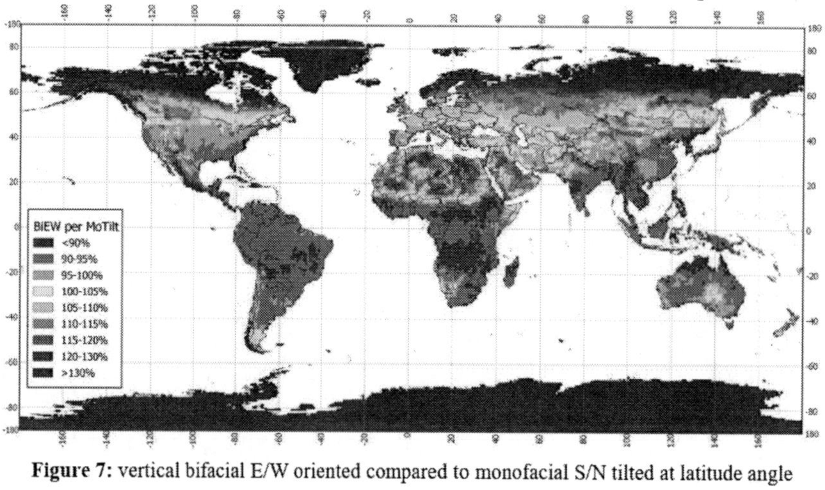

Figure 7: vertical bifacial E/W oriented compared to monofacial S/N tilted at latitude angle

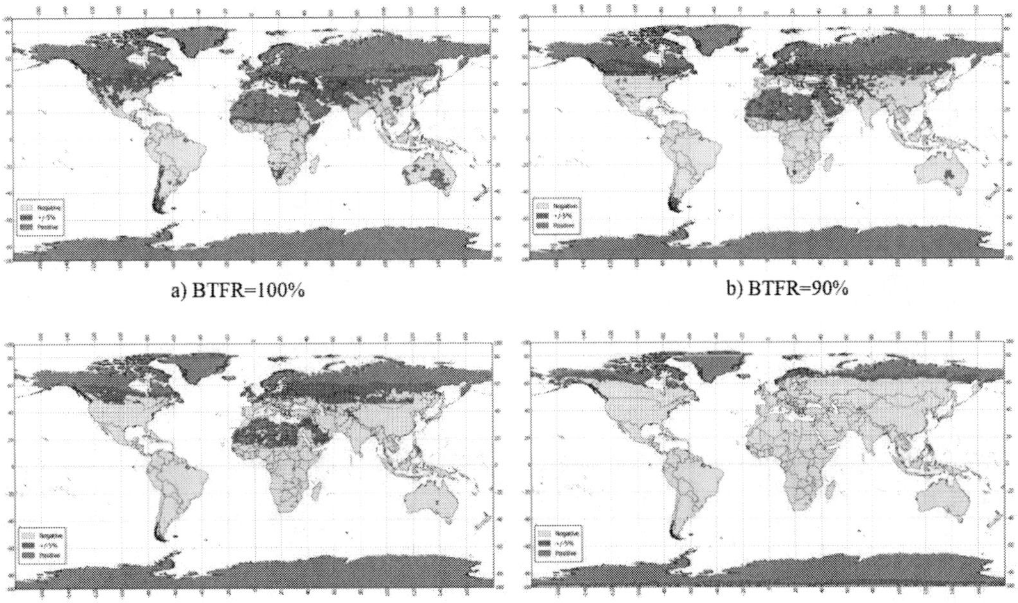

a) BTFR=100% b) BTFR=90%

c) BTFR=80% d) BTFR=55%

Figure 8: Effect of Back-to-Front Ratio (BTFR) on the annual energy gain of vertical bifacial E/W oriented modules compared to monofacial tilted modules. GREEN > +5% gain, 5% loss < GREY < +5% gain, YELLOW > 5% loss

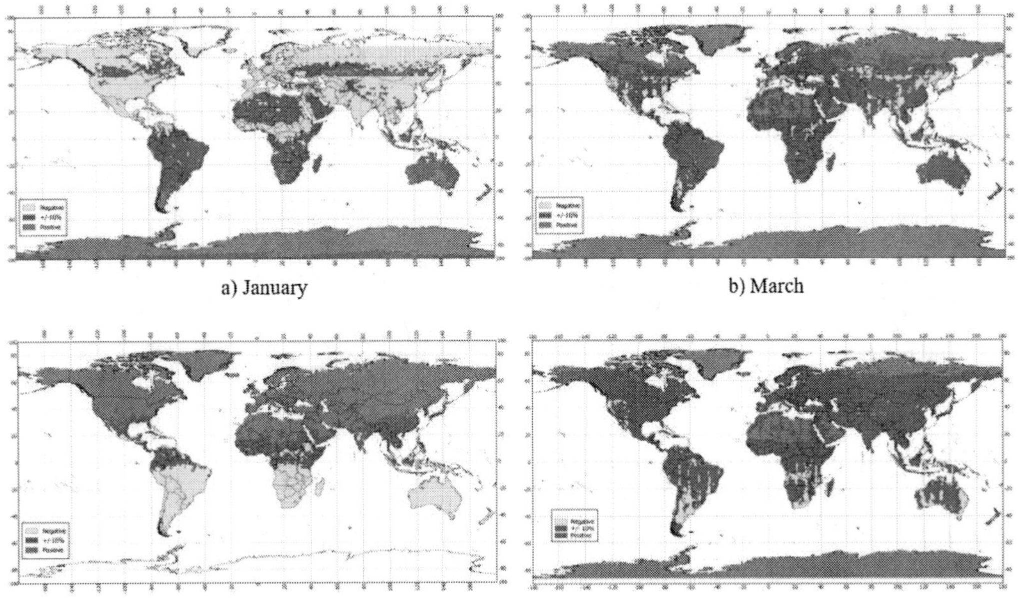

a) January b) March

c) June d) September

Figure 9: Seasonal effects (January, March, June and September) on vertical bifacial E/W oriented modules compared to tilted monofacial modules.
GREEN > +10% gain, 10% loss < GREY < 10% gain, YELLOW > 10% loss

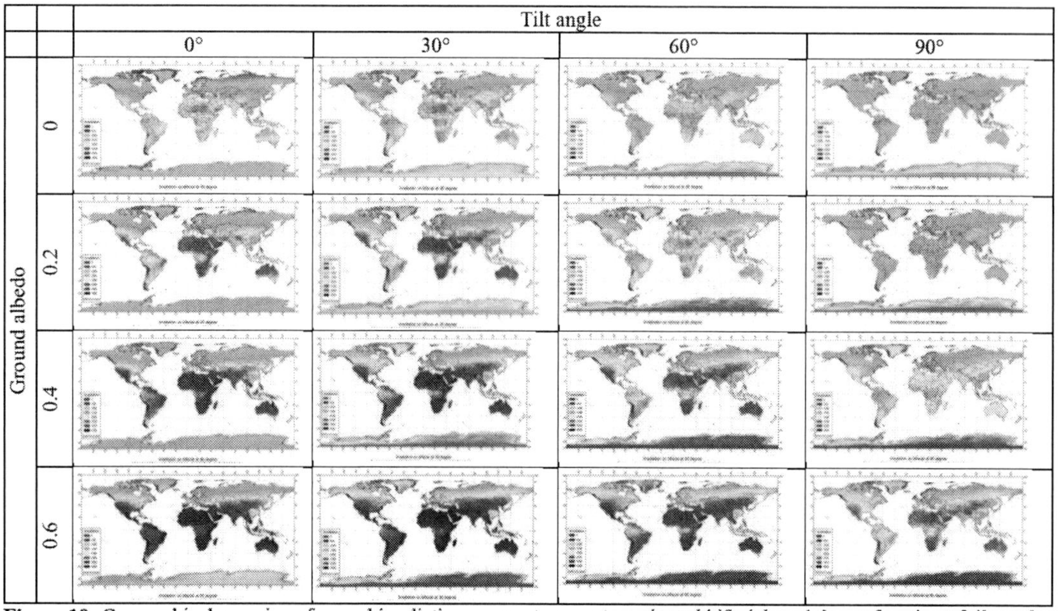

Figure 10: Geographical mapping of annual irradiation on equator equator-oriented bifacial modules as function of tilt angle and ground albedo. Total irradiation = front + rear ; BTFR=90%; light blue = 1000 and dark brown= 4000 kWh/m^2/year

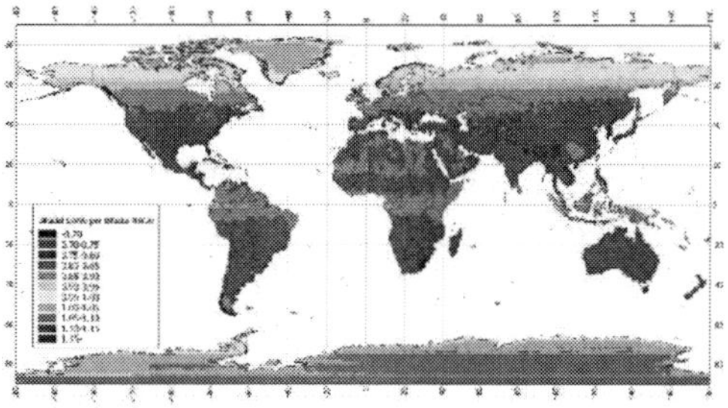

Figure 11: Vertical bifacial modules in E/W orientation compared to 30° tilted equator oriented bifacial modules

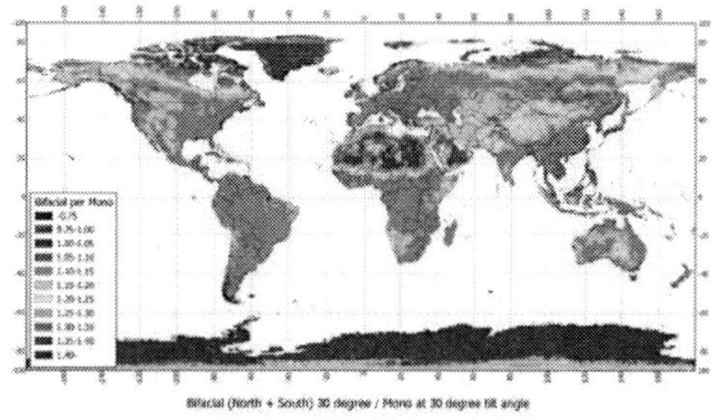

Figure 12: Bifacial and monofacial modules, both tilted at 30° and oriented equator-oriented

Table II: Comparison of estimated irradiation in kWh/m^2/year for monofacial modules at various tilt angles and vertical bifacial modules.
Values in bold are calculated in this study.

		0° mono	**30° mono**	**90° mono**	**90° bi (E + W)** (error %)
Paris **France** 48,82N 2,33E	Meteonorm ground and satellite data	1013	1141	786	637(E) + 622(W) = 1259
	PVGIS ground and satellite data	1194	1373	953	---
	our SSE based model result	1145	**1319**	**909**	**1247** (-1%)
Tokyo **Japan** 35,68N 139,77E	Meteonorm	1174	1304	872	677(E) + 677(W) =1354
	Monsola-11 ground data	1213	1362	920	705(E) x 2 = 1410
	our SSE model	1371	**1570**	**1004**	**1346** (-5 to -1%)
Taman-rasset **Algeria** 22,78N 5,52E	Meteonorm	2366	2537	1353	1311(E)+1309(W) = 2620
	PVGIS	2327	2465	1235	---
	our SSE model	2129	**2222**	**1216**	**2336** (-11%)
Tromso **Norway** 69,65N 18,95E	Meteonorm	634	769	652	502(E)+524(W) = 1026
	PVGIS	709	866	705	---
	our SSE model	744	**950**	**801**	**1067** (+4%)

5 ACCURACY

The accuracy of the SSE dataset is discussed in [18] reporting an accuracy of 10.62% RMSE for the monthly mean all sky insolation at 60° equator-ward, 32.20% RMSE for 60° poleward, whereas the accuracy of the global insolation is 13.94%.

Accuracy for the models applied to calculate in-plane irradiation with the Hay model is discussed in [10], reporting an accuracy of monthly averaged daily irradiation from 0.015 to 0.044 kWh/m^2/h in RMSE.

Although the various accuracies are not straightforward to combine, a comparison can however be made of in-plane irradiation measured and calculated by different databases and models, as shown in table II. An accuracy report of Meteonorm 7.1, a climatological database for solar energy applications gives an variable uncertainty varying from 2 to 10% [19]. PVGIS, the Photovoltaic Geographical Information System developed by Joint Research Center (JRC), reports that the standard deviation of individual station MBE values is 5.0% [20] .

Monsola-11 is a ground measured database for Japanese irradiation data [21]. The database which gives maximum differences compared to our SSE based results is the Meteonorm of Tamanrasset, located in the Algerian Sahara. It is 11 % larger than our SSE results. In summary our results compare reasonably well to ground based date with an accuracy of on average 3% for vertical bifacial module configurations.

6 CONCLUSION & OUTLOOK

In regions at higher northern latitudes (>40°) and the polar regions vertical bifacial installations are found to be energy efficient (+10 to +30%) compared to monofacial tilted modules. Also large (semi-)arid regions below the 20° latitude line are identified to be favorable (+10 to +30%) for vertical bifacial E/W installations due to the high albedo in these regions.

Reduction of soiling in the latter areas by vertical operation is expected to bring an additional gain of 10-20 % and would likely outperform the effect of existing anti-soiling coatings [17, 22].

For the northern latitudes >40°, vertically oriented modules perform equally well in E/W and S/N orientation, thus allowing a mixture of these orientations to level out energy production over the day, aligning energy production to demand and thereby reducing the required storage capacity.

Finally, it should be noted that an E/W vertical installation may require larger space compared to a S/N tilted installation. The authors are developing a method to calculate the required land space and will report an area based comparison for vertically installed bifacial installations.

7 ACKNOWLEDGEMENT

The authors thank the Japan Society for the Promotion of Science (JSPS) for supporting the assignment of one of the authors (M.I) at CEA-INES in France.

8 REFERENCES

[1] Trends 2015 in Photovoltaic Applications, International Energy Agency/Photovoltaic Power System Programme (2016).

[2] IPCC, IPCC Special Report on Renewable Energy Sources and Climate Change Mitigation: Summary for Policymakers, (2011).

[3] Masakazu Ito, Sylvain Lespinats, Jens Merten, Philippe Malbranche, and Kosuke Kurokawa, Life cycle assessment and cost analysis of very large-scale PV systems and suitable locations in the world, Progress in Photovoltaics: Research and Applications, Vol.24, No.2, Feb (2016) pp.159-174.

[4] A.H. Strahler, C.L.B. Schaaf, and F. Gao., ISLSCP II MODIS (Collection 4) Albedo, (2002).

[5] Travis Sarver, Ali Al-Qaraghuli, and Lawrence L. Kazmerski, A comprehensive review of the impact of dust on the use of solar energy: History, investigations, results, literature, and mitigation approaches, Renewable and Sustainable Energy Reviews, Vol.22, (2013) pp.698-733.

[6] Rob W. Andrews, Andrew Pollard, and Joshua M. Pearce, The effects of snowfall on solar photovoltaic performance, Solar Energy, Vol.92, (2013) pp.84-97.

[7] S. Obara, D. Konno, Y. Utsugi, and J. Morel, Analysis of output power and capacity reduction in electrical storage facilities by peak shift control of PV system with bifacial modules, Applied Energy, Vol.128, Sep 1 (2014) pp.35-48.

[8] S. Y. Guo, T. M. Walsh, and M. Peters, Vertically mounted bifacial photovoltaic modules: A global analysis, Energy, Vol.61, Nov 1 (2013) pp.447-454.

[9] Benjamin Y. H. Liu and Richard C. Jordan, The interrelationship and characteristic distribution of direct, diffuse and total solar radiation, Solar Energy, Vol.4, No.3, (1960) pp.1-19.

[10] Japan Solar Energy Society, New Solar Energy Utilization Handbook (2006).

[11] J.E.Hay, Study of Shortwave Radiation on Non-horizontal Surfaces, Atmospheric Environment Service (1979).

[12] Philippe Blanc, Benoît Gschwind, Mireille Lefèvre, and Lucien Wald, The HelioClim Project: Surface Solar Irradiance Data for Climate Applications, Remote Sensing, Vol.3, No.12, (2011) pp.343-361.

[13] NASA Goddard Institute for Space Studies, SeaWiFS Surface Solar Irradiance, Version 2 ed. 30 October 1994 (1994).

[14] POWER and NASA, Surface meteorology and Solar Energy (release 6.0), POWER and NASA (2013).

[15] John A. Duffie and William A. Beckman, Solar Engineering of Thermal Processes (Fourth Edition): Wiley (2013).

[16] Manuel Collares-Pereira and Ari Rabl, The average distribution of solar radiation-correlations between diffuse and hemispherical and between daily and hourly insolation values, Solar Energy, Vol.22, No.2, (1979) pp.155-164.

[17] Arash Sayyah, Mark N. Horenstein, and Malay K. Mazumder, Energy yield loss caused by dust deposition on photovoltaic panels, Solar Energy, Vol.107, (2014) pp.576-604.

[18] POWER and NASA, Surface meteorology and Solar Energy (SSE) Release 6.0 Methodology Version 3.1.1, (2013).

[19] Jan Remund, Accuracy of Meteonorm (7.1.6.14035), Meteotest (2015).

[20] Thomas Huld, Richard Muller, and Attilio Gambardella, A new solar radiation database for estimating PV performance in Europe and Africa, Solar Energy, Vol.86, No.6, Jun (2012) pp.1803-1815.

[21] NEDO, MONSOLA-11, (2012).

[22] M. Piliougine, C. Cañete, R. Moreno, J. Carretero, J. Hirose, S. Ogawa, and M. Sidrach-de-Cardona, Comparative analysis of energy produced by photovoltaic modules with anti-soiling coated surface in arid climates, Applied Energy, Vol.112, (2013) pp.626-634.

MODELLING OF SINGLE-AXIS TRACKING GAIN FOR BIFACIAL PV SYSTEMS

Amy Lindsay[1], Matthieu Chiodetti[1], Didier Binesti[1],
Sophie Mousel[2], Eric Lutun[2], Khalid Radouane[2],
James Christopherson[3]
[1]EDF R&D – ENERBAT, France,
[2]EDF EN, France,
[3]EDF RE, USA

ABSTRACT:
Bifacial gains that can be reasonably achieved for **utility-scale PV plants on fixed structures** are estimated to be **between 5% and 15% for natural ground albedos** (0.2 for grassland to 0.5 for light-coloured sand). [1-2]
In areas of the globe with **high Direct Normal Irradiance**, **Horizontal Single-Axis Trackers** are now sufficiently **cost-effective** to be installed for standard PV modules, unlocking increased irradiances on the front side of the PV modules. Bifacial modules can obviously also benefit from higher front irradiances. However, the **tracking motion will affect the irradiance collected by the rear of the bifacial modules** and this impact needs to be quantified.
A **model** for bifacial modules on a single-axis tracker has been developed with the Modelica language in a Dymola simulation environment in order to calculate the back-side solar irradiations received by the different modules in a PV installation and the power that would be delivered at each time step.
This paper aims at assessing **the additional gain that can be expected for bifacial modules on horizontal single-axis trackers**, depending on season, ground-coverage ratio, diffuse/global horizontal irradiance ratio and albedo of the ground.
Keywords: Bifacial, Design, Energy Performance, Modelling, Tracking

1 INTRODUCTION

Bifacial modules will be part of the future of photovoltaics, with an **increasing share of bifacial cells being de facto bifacial.** [3] However, the implementation of bifacial modules at a large scale depends on the cost-benefit analysis. For each location, the **bifacial gain** – defined as the rear-side generated producible over front-side generated producible – needs to be assessed for different configurations, fixed-tilt structures and Horizontal Single-Axis Trackers being some solutions among others.

Should it be for fixed-tilt or tracking bifacial surfaces, **the estimation of the irradiances collected by the rear side** of the installation is tricky as it depends to a large extent on the **ground-reflected and diffuse** irradiances. Thus, precise modelling of the ground and sky view factors is required, along with the shadow cast on the ground.

Few PV simulation tools calculate these view factors with a high level of accuracy and most studies for bifacial gain have been performed on fixed-tilt **single modules** or **vertical installations**. [4-7] More recently, some publications have arisen for the assessment of **bifacial gain in large-scale fixed-tilt** PV installations. [1-2, 8-9] To our knowledge, there are however **no recent papers investigating the feasibility of implementing bifacial PV on sun-trackers** apart from experimental data collected from low concentration PV with reflectors. [10-12]

In areas of the globe with **high Direct Normal Irradiance, Horizontal Single-Axis Trackers** are now sufficiently cost-effective to be installed for standard PV modules, unlocking increased irradiances on the front side of the PV modules. In these zones, the bifacial gains for fixed tilt structures – between 5 and 15% - are not high enough compared to tracking gain – between 20 and 30% - to switch from a sun- tracking system back to a fixed-tilt structure with bifacial modules. For these reasons, **an in-depth analysis of the technical pertinence of bifacial PV on horizontal single axis trackers needs to be carried out.**

The **model** that has been developed for fixed-tilt bifacial installations has been **extended** in order to take into account the **tracking of the sun and the impact on the rear irradiances.** Through simulations, an estimation of the additional gain that can be expected for bifacial modules on horizontal single-axis trackers is given, depending on the diffuse/global horizontal irradiance ratio and the albedo of the ground.

This study focuses **on side-by-side comparison of utility-scale monofacial and bifacial installations**, for **fixed-tilt and sun-tracking structures**, depending on season, ground-coverage ratio, diffuse/global horizontal irradiance ratio and albedo of the ground.

2 BIFACIAL PV HSAT MODEL

A numerical model for bifacial PV power plants is under development at EDF R&D since 2014. [1] It has been extended to take into account Horizontal Single Axis Trackers (HSAT) in addition to fixed-tilt structures.

2.1 Sun-tracking and backtracking

By default, the horizontal single-axis trackers are supposed to have their rotation axis running along the North-South direction, thus tracking the sun from East to West. Nonetheless, the model is valid for any orientation of the single-axis trackers.

At each time step, for a North-South tracker, the position of the tracker is given by:

$$azim = -90° \text{ if } azim_{sun} \leq 0°$$
$$azim = 90° \text{ if } azim_{sun} \geq 0°$$
$$\tan(tilt) = \tan(\theta_z) * |\cos(azim - azim_{sun})|$$

Where $azim_{sun}$ is the azimuth angle of the sun (S=0°, E=-90°, O=90°, N=180°), θ_z is the zenith angle and *(azim, tilt)* describe the azimuth and tilt of the tracker.

In order to avoid mutual partial shadings between tracker rows (see Figure 1), most tracker developers implement **backtracking strategies,** which reduce the tilt angle compared to the equation expressed here above in the morning or in the evening when the solar elevation angle is too low. In practice, **the energy output is slightly reduced compared to an isolated module on an ideal tracker**, as the front irradiance is no longer maximal. All in all, **avoiding partial shadings improves significantly the annual yield** while maintaining a high ground coverage ratio.

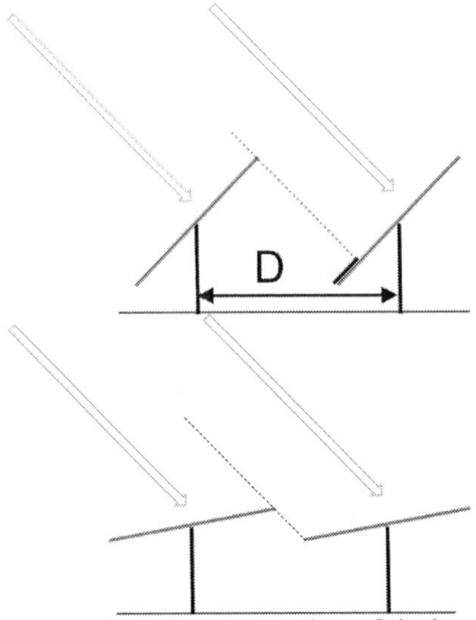

Figure 1: Schematic representation of backtracking implementation: mutual shadings without backtracking (above) and no mutual shadings with backtracking (below)

The **impact of inter-row distance D** on the activation of the backtracking strategy can be seen in Figure 2. Reducing inter-row distance unleashes higher ground coverage ratios but solicits the backtracking for longer periods.

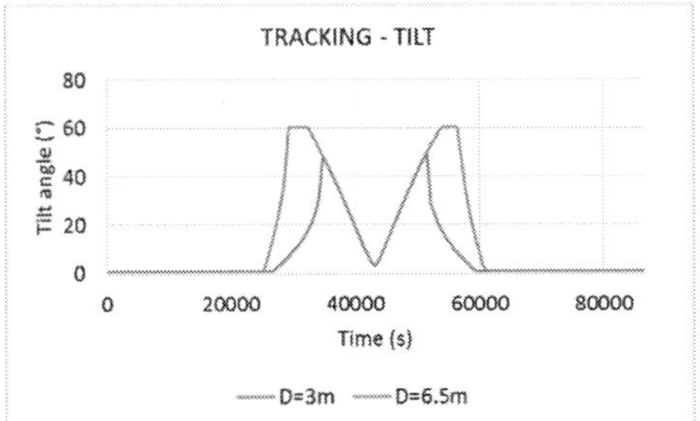

Figure 2: Illustration of East-West tracking algorithm for different inter-row distances D

2.2 Irradiance model

The irradiation received by a photovoltaic surface is made up of three elements: the beam component, the diffuse component and the ground-reflected irradiation.

The irradiance model is roughly the same one as presented in [1], with one noticeable difference concerning view factor calculation: since the tracking surface is per definition mobile, **all view factors must be calculated throughout the day for each module**, whereas for fixed surfaces, only the view factors for the shadowed ground need to be computed dynamically. View factors represent the proportion of the ground (or the sky) which is "seen" by the rear of the modules, i.e. the proportion of the radiation which leaves the ground and strikes the given module. For these reasons, the modelling is complex.

As a reminder, the irradiation received by the back of a bifacial module depends on its position within the PV installation (1), the shadow cast on the ground (2), the albedo of the ground (3) and the surrounding strings (4) as illustrated on Figure 3.

Figure 3: Explanatory figure for the model developed at EDF R&D

In order to calculate the rear irradiance, the **view factors from the module to the shadowed and non-shadowed ground, the stand behind and the sky need to be computed.**

The different components of the back irradiance are given by the following equations:

$$I_{ground-reflected} = \alpha * GHI * VF_{module \to non-shadowed\ ground} + \alpha * DHI * VF_{module \to shadowed\ ground}$$

$$I_{diffuse} = DHI * VF_{module \to sky}$$

$$I_{direct} = \max(BNI * \cos(i), 0)$$

View factors are purely geometrical but there are no analytical expressions for some configurations, for instance in the case of two finite non parallel surfaces such as a module and the ground. A meshing of one of the surfaces must be performed to be able to calculate the view factors [13-14].

The **albedo is separated into two components: a direct component and a diffuse component**. The proposed model is directly adapted from the literature:

$$\alpha(\theta) = \left(1 - \frac{DHI}{GHI}\right) . \alpha_0 . \frac{1 + C}{1 + 2.C.\cos(\theta)} + \frac{DHI}{GHI} . \alpha_{diff}$$

Where θ is the solar zenith angle, α_0 is the albedo under direct illumination, C is a constant depending on the type of surface, α_{diff} is the albedo under a 100% diffuse illumination, DHI is the diffuse irradiance and GHI the global horizontal irradiance. The three unknown parameters of the model $(\alpha_0, \alpha_{diff}, C)$ can be found by fitting the model to on-site measurements. [15]

Good predictability of modelled irradiance by comparison with experimental data has been tested **for fixed-tilt installations** and has been thoroughly documented in publications [1] and [15].

2.3 Electrical model

Each bifacial cell is represented by **a two-diode equivalent circuit**. The electrical diagram of the model is presented in Figure 4.

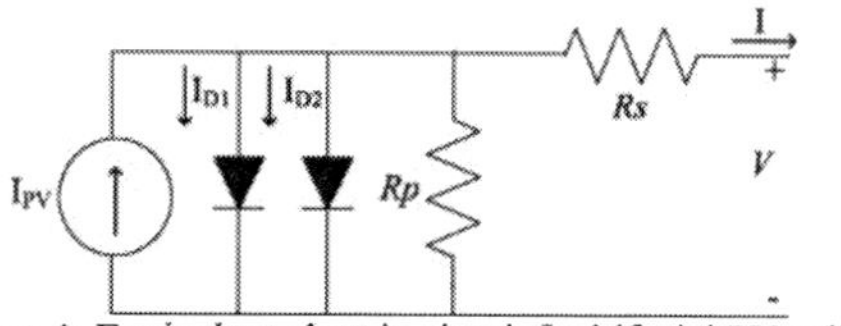

Figure 4: Equivalent electric circuit for bifacial PV cells

Each side of the bifacial module is flashed independently at STC, the opposing side covered by a black sheet. The two-diode parameters are extracted from the I-V curves by the least squares method for the front side of the module.

The **bifaciality** factor is defined as rear-to-front efficiency ratio and is calculated **for different irradiance levels** as follows :

$$BFF(G) = \frac{\eta_{front,\,25°C,\,AM1.5}(G)}{\eta_{rear,\,25°C,\,AM1.5}(G)}$$

An example of irradiance-dependent bifaciality factor for a fake bifacial module is given on the graph in Figure 5:

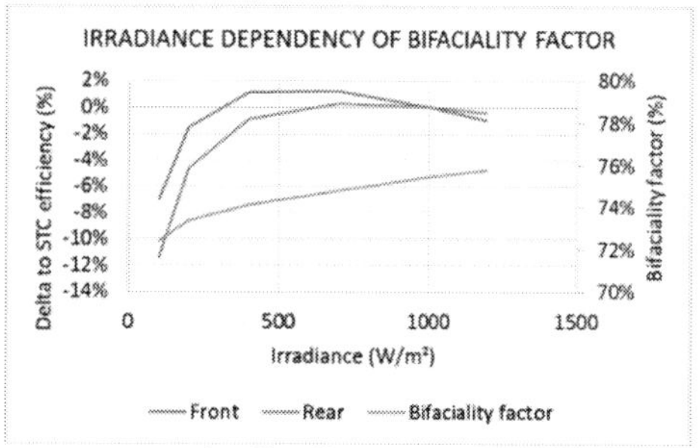

Figure 5: Example of irradiance dependency of bifaciality factor

Partial shadings are also accounted for in the case of fixed-tilt installations with high ground coverage ratios that lead to direct shadows cast from one row onto the next.

3 INTERMEDIATE SIMULATION RESULTS

The following results have been obtained for PV installations in **Cairo (DNI=1760 kWh/m²/year) with a ground albedo of 0.2, typical of grassland.** In such a location with high Direct Normal Irradiance (DNI), horizontal single-axis trackers are already cost-effective in comparison to fixed tilt installations.

For fixed tilt installations, the **tilt angle is 30°**, optimum for monofacial annual yield in Cairo when not considering ground coverage ratio or partial shading issues (see Figure 6).

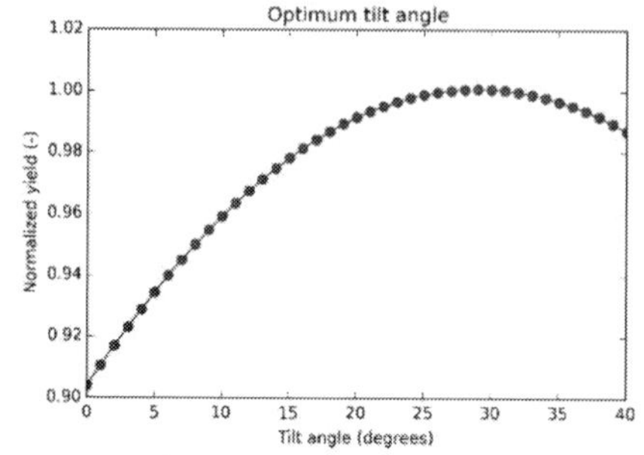

Figure 6: Optimum tilt angle for monofacial PV yield in Cairo

In-depth investigations on impact of elevation, inter-stand distance, ground albedo and tilt angle for fixed-tilt small and large-scale bifacial PV installations is documented in publications [1, 4-5, 16].

Four rows of ten 72-cell bifacial modules are installed in landscape at **0.6 meters from the ground** and are cabled in U-shapes, as illustrated in Figure 7 below:

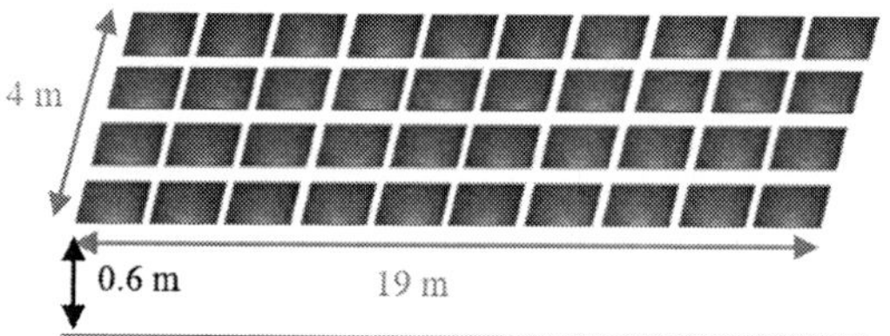

Figure 7: Fixed-tilt installation considered in Cairo

For **Horizontal Single Axis Trackers**, the **maximum tracking angle is 60°** with **vertical posts of 1 meter**, and twenty 72-cell bifacial modules in portrait that track from East to West (see Figure 8). The distance from the ground is 0.2 meters in maximum tracking position, and backtracking is considered.

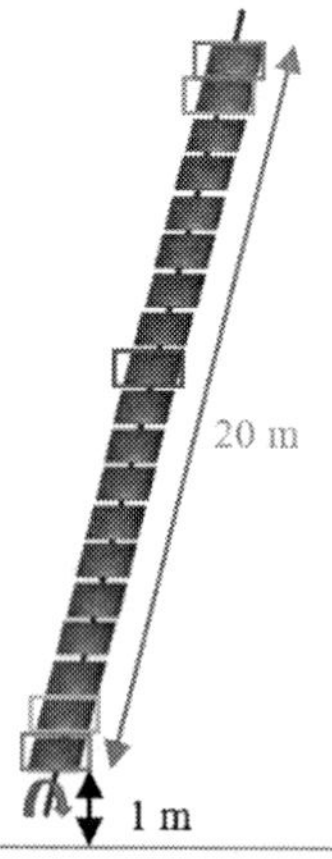

Figure 8: Tracker installation considered in Cairo – contoured panels will be used for illustration purposes, orange is furthest South, light blue furthest North

In order to avoid all irradiance edge effects and present results that are those of large-scale PV installations, the yield estimations are performed for the least-irradiated modules (module in the middle for tracker, and 4 modules in height in the middle for fixed-tilt).

3.1 View factors

The following figure (Figure 9) shows that the shadow view factors for the rear side of different modules of the installation evolve during the day. The modules that are the furthest South have relatively low shadow view factors compared to those in the middle or the furthest North in the installation. These results are consistent with the way the shadow moves during the day. This illustration is performed for an inter-row distance of 6.5 meters (GCR=29%).

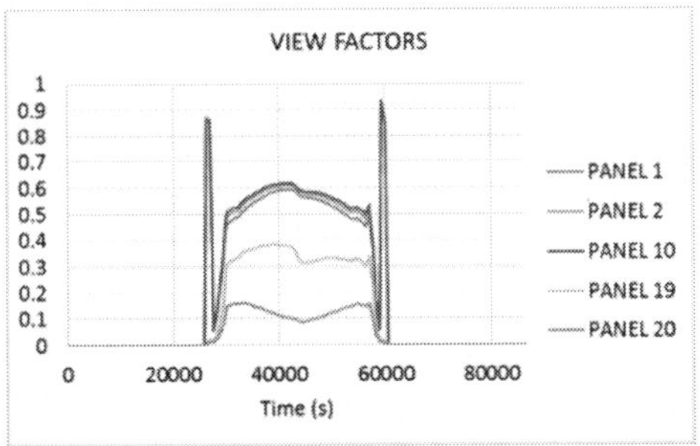

Figure 9: Rear-side shadow view factors for HSAT system in Cairo – orange furthest South, light blue furthest North

Contrary to fixed tilt installations, all other view factors that do not depend on shadow cast on the ground are also variable, since they evolve with tilt angle of the tracker at any given moment, as can be seen in Figure 10 for the panel in the middle of the tracker row.

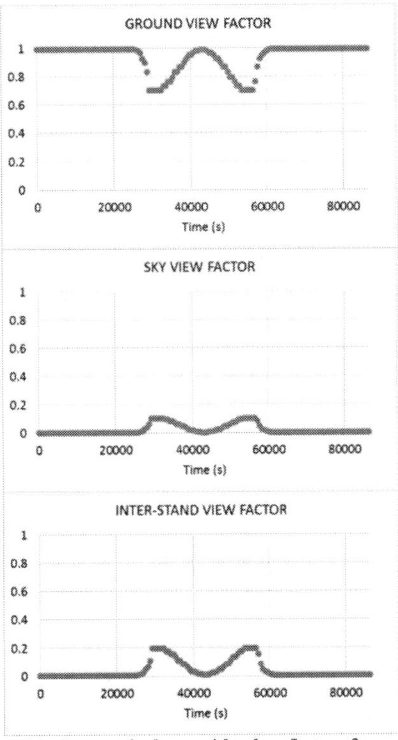

Figure 10: Geometrical rear-side view factors for panel in middle of tracker that depend on time of day: ground view factor, sky view factor and view factor of row behind

The view factors are calculated in a similar way for the front side of the HSAT installation.

3.2 Irradiances

With the help of the view factors previously calculated and the irradiance model explained in part 2.2, it is possible to calculate the different components of irradiance received for each module, on the front and rear sides: beam, diffuse and ground-reflected.

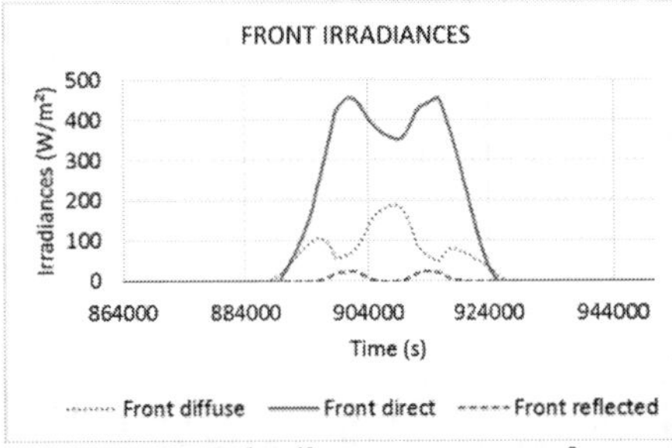

Figure 11: Front irradiance components for panel in middle of tracker in Cairo

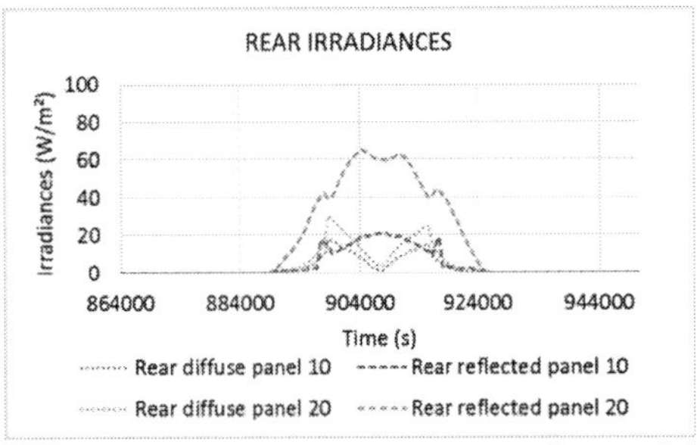

Figure 12: Rear irradiance components for panel in middle and furthest South of tracker in Cairo

As can be seen on Figures 11 and 12, **the rear side irradiances** are significantly lower than front side irradiances: 95 kWh/m²/year vs. 2035 kWh/m²/year for an inter-row distance of 3 meters (GCR=63%). There is an **important contribution of the reflected irradiance** especially for the panels at the extremity of the row, that benefit from lower shadow view factors as discussed in part 3.1. Even for the least-exposed module, the **reflected irradiance mounts up to 55% of total rear side irradiance** on average over a year, insisting once again on **the importance of a good knowledge of ground albedo.**

When summing the irradiances received on the rear side of a Horizontal Single Axis Tracker installation and of a fixed-tilt installation over a year, a mapping can be performed (see Figures 13 and 14), pointing out the **heterogeneity of rear irradiances especially in the case of fixed-tilt systems.**

239	165	131	120	116	116	120	131	165	239
214	138	109	101	98	98	101	109	138	213
188	111	90	84	82	82	84	90	111	187
185	116	106	103	101	101	103	106	116	184

Figure 13: Mapping of the rear irradiances received on the fixed-tilt installation, GCR=63% [kWh/m²/year]

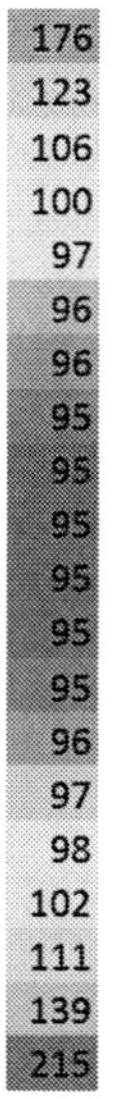

Figure 14 (left): Mapping of the rear irradiances received on the HSAT installation, GCR=63% [$kWh/m^2/year$]

A few elements that are intuitive but are worth noticing when comparing these two maps are the following:

- The lowest rear-side irradiance is achieved by the fixed-tilt system, more precisely for the 2nd row in height on the stand (for this location, due to the elevation off the ground by 0.6 meters and the general position of the sun, the shadow cast on the ground impacts more severely row n°2 than row n°1)
- The irradiance map for fixed-tilt is perfectly symmetrical whereas the irradiance map for HSAT presents a positive bias for the South-side of the tracker

An interesting point to be highlighted is the **seasonality of this heterogeneity map**. Whereas both types of installations will obviously present a seasonality of bifacial gain as discussed later in part 4.1, the distribution of rear-side irradiances will remain roughly the same in aspect for HSAT and vary for fixed-tilt structures. This is illustrated in Figure 15, the ranking of least to most rear-side-irradiated module according to elevation of module from ground varying through time.

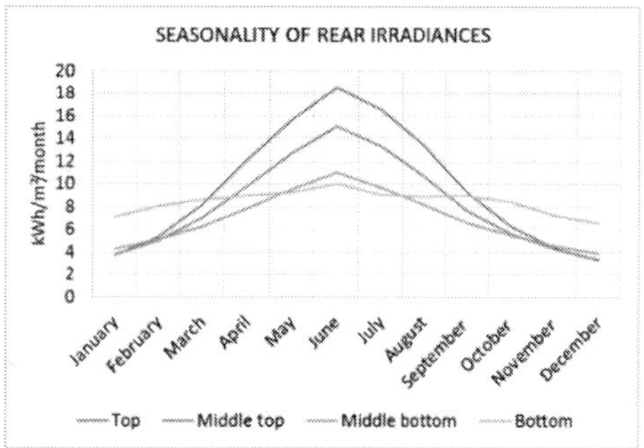

Figure 15: Seasonality of the rear irradiances received on the fixed-tilt installation, GCR=63% [$kWh/m^2/month$]

For instance, the bottom module receives roughly the same amount of irradiation throughout the year as the shadow cast on the ground is of little impact.

3.3 Power output

Based on the irradiances, the electrical model is applied and allows us to compare bifacial and monofacial options on both fixed-tilt structures and Horizontal Single Axis Trackers.

On a clear-sky winter day, the HSAT systematically performs less well than fixed-tilt, due to low solar elevation angles and high solar incidence angles on the front of the tracker, thus triggering significant reflection losses (Figure 16). On a clear-sky summer day, the sun-tracker increases to a large extent the yield by adding minutes of PV production in the morning and in the evening (Figure 17).

The seasonality of bifacial and tracking gains will be detailed in part 4.1.

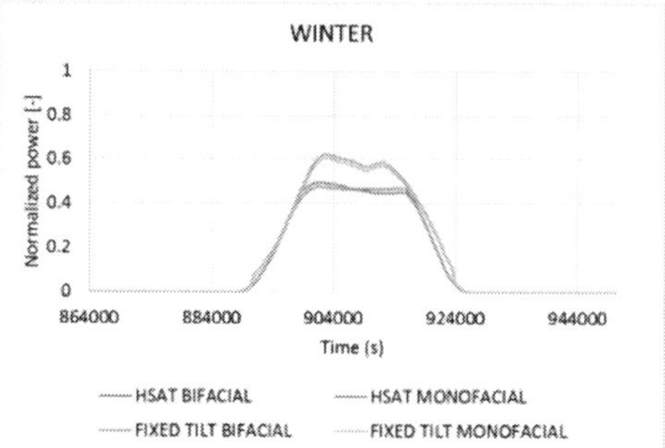

Figure 16: Normalized power for bifacial and monofacial modules on HSAT and fixed-tilt on a sunny winter day in Cairo

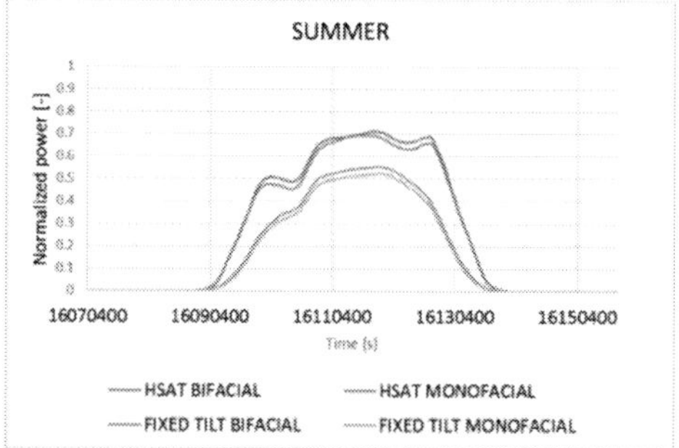

Figure 17: Normalized power for bifacial and monofacial modules on HSAT and fixed-tilt on a sunny summer day in Cairo

4 ENERGETIC AND ECONOMIC GAINS

All the models used to estimate yield of bifacial and monofacial modules on fixed-tilt and horizontal single axis tracking structures have now been presented in parts 2, in theory, and 3, in practice.

For decisions on investments in bifacial tracking systems, a good **estimation of bifacial gain** and, consequently, of **Levelized Cost of Electricity (LCOE)** is necessary. Indeed, if the bifacial gain is sufficient, it can justify a switch from bifacial to monofacial modules.

In this part, the impact of two variables on bifacial gain and LCOE is looked into: **ground coverage ratio** (defined as PV surface over total ground surface for PV installation) and **ground albedo**.

4.1 Bifacial and tracking gain

Bifacial gain is the relative additional yield generated from the rear side compared to the yield generated by a monofacial module with the same front-side STC power. For instance, if a monofacial module generates 1 kWh while a bifacial module generates 1.05 kWh over a year, both having a front-side STC power of 1 kW, the bifacial gain is 5%.

Bifacial gain varies throughout the seasons, with an amplitude that is **more significant for fixed-tilt structures** since the rear side irradiances depend quite strongly on sun elevation angle (see Figure 15) whereas the sun tracking system is always going to maximize front side production in priority.

Tracking gain (or loss) is the relative additional yield generated for the same type of module (bifacial or monofacial), when switching from a fixed-tilt structure to a horizontal single axis tracker. The **tracking gain is higher for monofacial modules** as bifacial modules already benefit from higher reference yields thanks to the additional irradiances received on their rear side. The **tracking gain is higher in summer than in winter**, as expected.

Bifacial and tracking gains are illustrated on Figure 18, for an equivalent GCR of 63% for both systems (HSAT and fixed-tilt).

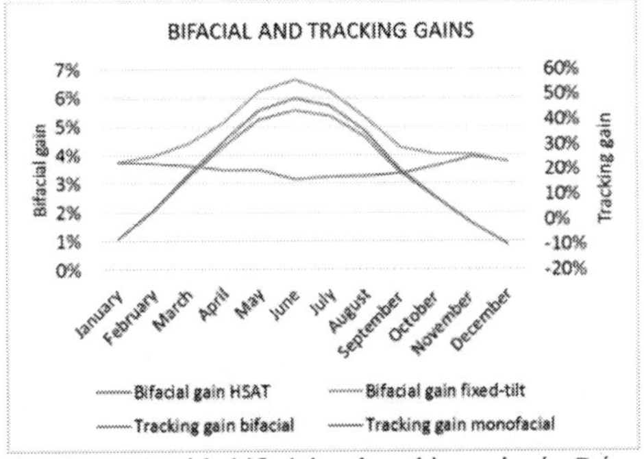

Figure 18: Monthly bifacial and tracking gains in Cairo

An interesting correlation to look into is **dependency of bifacial gain to diffuse content of global horizontal irradiance** received over a day. Figure 19 casts a light on the possibility of deriving a simple linear correlation as a first approach of **bifacial gain as a function of daily diffuse content in global horizontal irradiation for horizontal single axis trackers**. This is explained by the low seasonality of the irradiance map on the rear side of a HSAT and the fact that the mobile plane is always going to favor the front side of the panel normal to the direction of the sun rays.

A trend can also be detected for fixed tilt systems with a clear exponential curve, especially for high diffuse contents for which the irradiance heterogeneity on the rear side of the PV plant will be less significant. The scattering of points at low diffuse ratios is due to strong heterogeneity of the rear irradiances for clear-sky days, with a seasonality looked into in part 3.2.

For both HSAT and fixed-tilt systems, the bifacial gain increases with diffuse content in the global horizontal irradiance, the front side of the module benefiting from relatively less light on cloudy days.

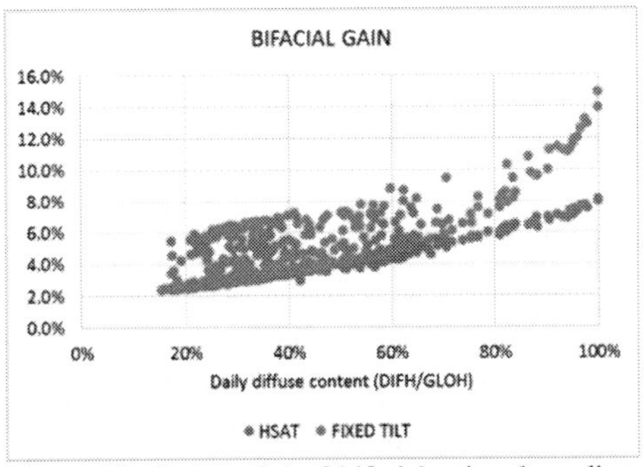

Figure 19: Scatter plot of bifacial gains depending on daily diffuse content in Cairo

For a first approach of bifacial gain, a simple formulation of bifacial gain as a function of albedo, diffuse ratio and ground coverage ratio can be imagined. For a finer understanding of the **intraday variations of bifacial gain**, the precise modeling of irradiances as exposed in part 2.2 is necessary. Indeed, the bifacial gain can change quite significantly over the course of a day, as illustrated in Figure 20 below.

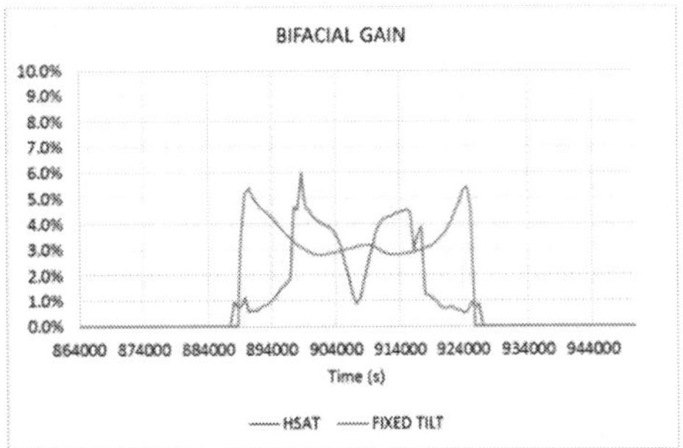

Figure 20: Intraday variations of bifacial gain in Cairo

Finally, a variation on ground coverage ratio and on ground albedo was performed to assess the impact on specific production (kWh/kWp) of the bifacial PV installation, and on bifacial gain relative to an equivalent monofacial installation.

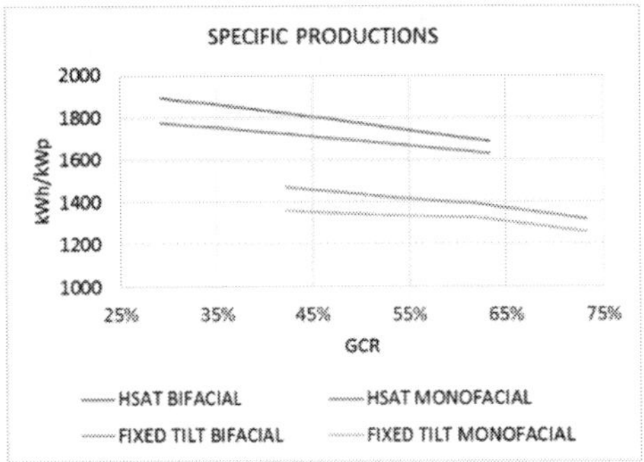

Figure 21: Specific productions for the different technical combinations tested in Cairo

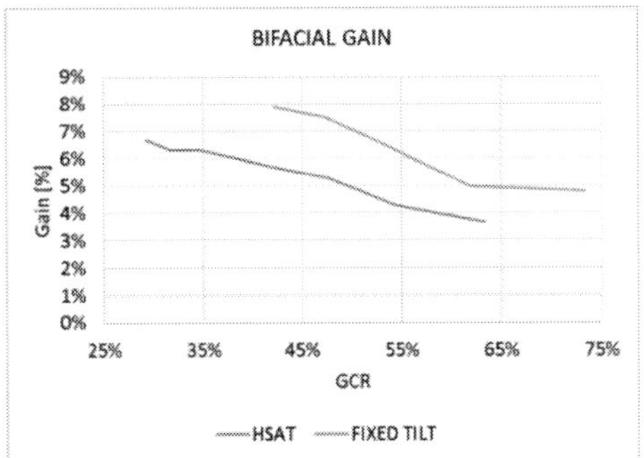

Figure 22: Bifacial gain for HSAT and fixed-tilt structures with varying ground coverage ratios in Cairo

The **tracking gain is between 21% and 27%** depending on ground coverage ratio, with a higher gain for monofacial modules than for bifacial modules.

The **bifacial gain varies between 5% and 8% for fixed-tilt structures on a ground albedo of 0.2**, as can be seen on Figure 22. The **bifacial gain for HSAT is lower than that for fixed-tilt for equivalent ground coverage**

ratios, and varies **between 3% and 7%** on the tested interval. This is due to the fact that horizontal single axis trackers already produce more relatively to fixed-tilt installations, thus reducing the relative bifacial gain. Globally, **bifacial gain is higher for low ground coverage ratios**.

4.2 LCOE

For the market to adopt the use of bifacial modules at a utility scale plant level, the criteria for investment decisions will be the CAPEX and **the Levelized Cost of Electricity** over the lifetime of the plant. Indeed, specific production and bifacial gain are not sufficient information to decide whether bifacial is a cost-effective solution in comparison to monofacial modules.

This part uses public sources of costs in France and in the USA that are summarized in Table 1. [17-18] The idea is not to discuss the economic hypothesis but rather to determine whether in a given economic context, bifacial could be a viable option.

Generally, the costs for a PV power plant are described in €/Wp, which is a problem when varying the Ground Coverage Ratio in order to find best GCR for a minimum LCOE. Indeed, for a given installed power, this would suggest that the land costs will be the same no matter how low the ground coverage ratio is, leading to selecting low GCR for maximum yield. For this reason, we have introduced a hypothesis on **inverse variation of land costs and labor costs** in €/Wp as a **function of Ground Coverage Ratio** (a highly packed PV installation will cost relatively less).

For the costs of bifacial PV, two scenarios have been looked into – a low cost hypothesis (0.6 €/Wp) and a high cost hypothesis (0.8 €/Wp, offers available online).

Table I: Economic hypothesis

€/Wp	ADEME - FR	SUNSHOT – US
€/$ conversion		0.89
-PV modules	0.53	0.58
-Inverter	0.12	0.13
-eBOS	0.03	-
-BOS	-	0.31
-Fixed-tilt structure	0.09	-
-Additional cost of HSAT	0.14	0.13
-Soft costs	0.1	-
-Grid interconnection	0.15	0.04
-Installation	0.11	0.18
-Land	-	0.03
TOTAL	1.36	1.41
-WACC	5.2%	4.4%
-Inflation	1.3%	2.5%
-O&M as % CAPEX	2.4%	1.0%
-Years	20	30
-Degradation	0%	-0.75%/y

The distribution of Levelized Cost of Electricity for different scenarios and ground coverage ratios is summarized in Figure 23. For the **low cost scenario** of bifacial PV modules, **the LCOE of bifacial is very close to that of the monofacial equivalent** (on HSAT or on fixed-tilt structures), and can even be **lower for high albedos or for low ground coverage ratios**. This means that the **bifacial gains** estimated in part 4.1 are **sufficient to justify the additional initial capital expenditure for bifacial PV**, for relatively low bifacial module costs and no matter the other economic hypothesis selected. Furthermore, there is a **higher benefit in switching from monofacial modules to bifacial modules for high ground albedos / low land costs** as low land costs are suitable for low ground coverage ratios.

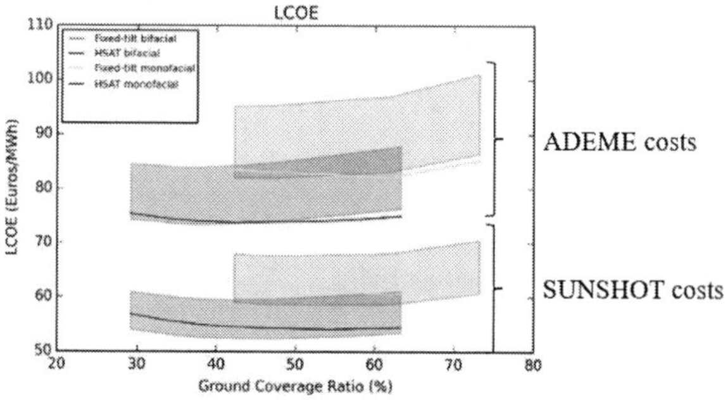

Figure 23: Levelized Cost of Electricity for the different technical combinations tested in Cairo

In Cairo, the **LCOE is lowest for tracking systems due to high DNI,** and this LCOE can be **further reduced if the bifacial module prices are close to that of monofacial modules or if the Ground Coverage Ratio is below 45%** for an albedo of 0.2.

5 CONCLUSIONS

This paper has shown the **feasibility of modelling bifacial PV power plants** in order to assess the gains that bifacial modules could bring in the next few years, for both fixed-tilt and horizontal singles axis tracking structures. A **simulation tool** has been developed at EDF R&D for precise estimation **of large-scale bifacial PV installation yield**, based on – among others – **precise modelling of rear irradiances**, irradiance dependency of bifaciality factor and a variable albedo model.

In comparison to standard PV plants, **the gains in terms of specific production (kWh/kWp) can be estimated between 5% and 15%** for "normal" ground albedos (0.2-0.5) for fixed-tilt structures, and **between 3% and 11% for HSAT.** The bifacial gain is higher for fixed tilt than for HSAT for reasons explained in detail in part 4.1.

The last part proves that **the economic reality of bifacial PV power plants needs to be investigated closely** as the viability of switching from monofacial to bifacial could **arise in the near future, depending on relative additional cost of bifacial modules, ground albedo and land costs.** An accurate estimation of bifacial yield is obviously key to good and reliable prediction of LCOE of the power plant. Finally, the case study has pointed out that **in locations with high DNI for which HSAT is already economically viable, bifacial PV on tracking systems can be the most cost-effective option in terms of LCOE.**

7 REFERENCES

[1] A. Lindsay et al., Key Elements in the Design of Bifacial PV Power Plants, Proceedings of EU PVSEC (2015).

[2] C. Reise et al., Realistic Yield Expectations for Bifacial PV Systems – An Assessment of Announced, Predicted and Observed Benefits, Proceedings of EU PVSEC (2015).

[3] International Technology Roadmap for Photovoltaic (ITRPV) 2015 Results, 7th edition (2016).

[4] U. Yusufoglu et al., Analysis of the Annual Performance of Bifacial Modules and Optimization Methods, IEEE Journal of Photovoltaics Vol.5 (2015) 320-328.

[5] H. Nussbaumer et al., PV Installations Based on Vertically Mounted Bifacial Modules Evaluation of Energy Yield and Shading Effects, Proceedings of EU PVSEC (2015).

[6] B. Soria et al., Vertical Facade Integration of Bifacial PV Modules: Outdoor Testing & Optical Modelling, Proceedings of EU PVSEC (2013).

[7] G. J. M. Janssen et al., Outdoor performance of bifacial modules by measurements and modelling, Energy Procedia Vol.77 (2015) 364 – 373.

[8] J. Appelbaum, Bifacial photovoltaic panels field, Renewable Energy Vol.85 (2016), 338–343.

[9] J. E. Castillo-Aguilella et al., Multi-Variable Bifacial Photovoltaic Module Test Results and Best-Fit Annual Bifacial Energy Yield Model, IEEE Access Vol.4 (2016) 498-506.

[10] M. A. Egido et al., Bifacial Photovoltaic Panels with Sun Tracking, International Journal of Solar Energy Vol.4 (1986) 97-107.

[11] J. Perezoria et al., A good combination: Tracking of the sun in polar axis and bifacial photovoltaic modules, Solar & Wind Technology Vol.5 (1988) 629-636.

[12] V. Poulek et al., Innovative low concentration PV systems with bifacial solar panels, Solar Energy Vol.120 (2015) 113–116.

[13] H. Bouia et al., Application de la triangulation de Delaunay au calcul des facteurs de forme entre deux surfaces planes quelconques, Proceedings of SFT (2002).

[14] G. Ritoux et al., Evaluation numérique des facteurs de forme, Revue de Physique Appliquée Vol.17 (1982) 503-515.

[15] M. Chiodetti et al., PV Bifacial Yield Simulation with a Variable Albedo Model, Proceedings of EU PVSEC (2016).

[16] P. Dupeyrat et al., Investigations on albedo dependency of bifacial PV yield, Proceedings of EU PVSEC (2014).

[17] ADEME, Etude technico-économique filière photovoltaïque française : bilan, perspectives et stratégie (2015).

[18] R. Jones-Albertus et al., Technology advances needed for photovoltaics to achieve widespread grid price parity, Progress in Photovoltaics Research and Applications (2016).

PERFORMANCE ANALYSIS OF PV GREEN ROOF SYSTEMS

Thomas Baumann[1]*, Daniel Schär[1], Fabian Carigiet[1], Andreas Dreisiebner[2],
Franz Baumgartner[1]
[1] ZHAW, Zurich University of Applied Sciences, SoE, Institute of Energy Systems and Fluid Engineering,
Technikumstrasse 9, CH-8401 Winterthur, Switzerland, www.zhaw.ch/~bauf
*phone: +41 (0) 58 934 71 87; e-mail: thomas.baumann@zhaw.ch
[2] Solarspar, Bahnhofstrasse 29, CH-4450 Sissach, Switzerland, www.solarspar.ch

ABSTRACT: The common assumption is that plants underneath PV modules will cause a cooling effect resulting in a higher energy yield due to the negative temperature coefficient of PV power. The PV module temperature is measured to verify this thesis on a flat roof PV plant in Winterthur, Switzerland. Thirteen different mounting test field types with two different mechanical PV mounting systems are used, 20° or 15° tilted angles. Each type of mechanical mounting system is combined with planting beneath and with and without irrigation in the different test fields. The difference of the module temperatures, weighted according to the energy production, of all 13 analysed test fields is between ± 1.8 °C. This temperature difference results in a calculated energy yield difference of about ± 0.7 % for the used crystalline silicon modules. This value is in the range of the measurement uncertainty of the power measurement (± 1.2 %). Therefore the green roof has only negligible influence on the temperature reduction of the PV modules on the base of the used system components. However, a combination of PV and green roof is absolutely doable and recommendable provided that the mounting system is optimized for green roofs. Other more powerful benefits of such combination of green roofs and PV are the improvement of water retention in the city.
Keywords: PV System, System Performance, Thermal Performance, Green Roof

1 INTRODUCTION

According to the energy strategy 2050 of the Federal Council and Parliament of Switzerland electricity originated from nuclear power has to be completely covered by electricity from new renewable energies and 21 % attributed to PV power until 2050 [1]. Thereof, the share PV corresponds to twelve square meters PV module area per capita, which is available on the present Swiss roofs.

Cities all over the world are promoting green roofs to reduce further inner city heat generation. Chicago forced to build over half a million square meters green roofs stipulated by law in recent years [2]. The Swiss cities Zurich, Basel, Bern, Winterthur, Lucerne and St. Gallen have some requirements about goals for green roofs. In some cities, the laws are set into force recently and they are not same stringent everywhere [3]. Today, there are nearly no PV installations on the green roofs, yet.

Because of the space requirement conflict between PV on the roof and green roofs, it's essential to combine these two systems. The goal of this project is to analyse the potential and benefits of PV panels with plants underneath. Outdoor measurements of PV module temperature and power are needed to estimate the potential energy surplus if there is a significant cooling effect observable.

2 APPROACH

2.1 Project goal

A project was set up to verify the PV performance on a flat green building roof in Switzerland. This was achieved with project partners consisting of green roof experts and photovoltaic specialists [4]. One sub goal was to check if the PV panels are cooled caused by the plants on the green roof.

2.2 Setup

A PV plant of 78 kWp was installed on a flat roof (84 m x 16 m, -3° azimuth) located in Winterthur Switzerland and the situation is shown in Figure 1. The roof is divided in 13 test fields. Eight test fields are equipped with 20° tilted commercial ZinCo mounting systems and five are equipped with 15° tilted commercial Hilti mounting systems. The two mounting types are shown in Figure 2.

Figure 1: Aerial photo of the test facility with the installed PV capacity of 78 kWp located in Winterthur, Switzerland. All modules have an azimuth orientation of -3°.

Figure 2: The two used mounting systems on the PV plant. On the left side the 20° tilted ZinCo mounting system for green roofs and on the right side the 15° tilted Hilti mounting system mostly used for flat roofs without greening.

Each test field consists of different soil conditions on the ground including gravel, different substrates and different substrate layer thickness, respectively. The whole green roof part was seeded with the herbs seed mixture "UFA-Kräuter Solardach CH" [5] from the Swiss company "UFA Samen".

2.3 Measurements

Several module temperature sensors are installed in every test field, total 48 PT100 sensors over the whole PV plant site. The measurement uncertainty of the temperature sensors depends on the actual temperature resulting in ± 0.55 °C (k=1) at 50 °C. The used measurement electronic in the data logger contributes another 0.3 °C to the measurement uncertainty (k=1). The ambient temperature, the wind speed and direction and the power of three PV modules are measured each 10 seconds as well. The measurement accuracy of these three power measurements are ± 1.2 % (k=1). The measurements campaign was started in autumn 2014. The power of the PV modules is also measured and controlled by the SolarEdge module optimizers in approximately 15 minute intervals with a measurement accuracy of ± 5 % (k=1). Additionally three modules were selected and their power was measured with a resolution of 10 second as a reference to the SolarEdge measurement.

3 INNOVATION AND RELEVANCE

Sealed floor surfaces in cities, waste heat of vehicles and houses which build heat islands enhance the warming of the air during the day and decrease the cooling of the air during night. Thus, summer nights are four to five degrees warmer in cities compared to rural regions [3]. The economic and ecological advantage of green roofs is undisputed. Today, the decision between PV and green roof is often made on flat roofs. If it's decided to install PV, about 16 % of heating power is converted into electricity and then bitumen or gravel is used regularly under the PV panels. The green roof has several advantages like water retention, reduction of peak water runoff, protection of the roof seal, additional insulation, cooling and air humidification of the ambient atmosphere, habitat for animals especially insects, filtering air pollutant etc [6]. With PV on green roofs, the symbiosis of these two components can be combined. In this project, the advantages and disadvantages of this symbiosis is analysed with focus on the cooling effect on the PV panels caused by the plants.

4 ANALYSES AND MEASUREMENT RESULTS

The energy weighted module temperatures were analysed at the different test fields from 01.10.2014 to 31.05.2016. The days were classified into seven classes according to the operating hours. In each of these classes, the differences in the energy weighted module temperatures between the test fields were analysed.

4.1 Data availability

Some days of data were missing due to a hardware failure of the module temperature measurement system in summer 2015. Furthermore, the days taken into account are those that have every 10 second one measurement between 08:30 in the morning and 16:00 in the afternoon available. Thus, 527 days of measurement data are analysed in the period from 01.10.2014 to 31.05.2016 resulting in an electricity production of 99 MWh for the whole PV plant according to the SolarEdge web portal data for these 527 days. Thus a specific AC yield of 974 kWh/kWp results for the annual period starting in June 2015.

4.2 Calculation of the energy weighted module temperatures

The temperature measurements are available in a resolution of 10 seconds as well as three PV module power measurements. The SolarEdge module power measurements are available for all installed modules. Unfortunately, the SolarEdge power measurements, exported from the online monitoring portal, have no constant time interval. The calculation of the energy weighted module temperatures according to Formula 1 needs a synchronised data set between the SolarEdge data and the temperature data.

$$T_{m,w} = \frac{\sum(T_m \cdot P_{m,DC})}{\sum P_{m,DC}} \quad [°C] \tag{1}$$

Three methods are analysed by calculating the energy weighted module temperature using the SolarEdge power data. The results is compared to the the energy weighted module temperature of one module that has power and temperature measurements with 10 seconds resolution available. The methods are applied on to the SolarEdge data set using different time periods.

- Method A "Linear interpolation": The interpolated value at a query point is based on linear interpolation of the values at neighbouring grid points in each respective dimension. The time resolution of the received values is 10 seconds.
- Method B "Nearest neighbour interpolation": The interpolated value at a query point is the value at the nearest sample grid point. The time resolution of the received values is 10 seconds.
- Method C "closest own measurement point": For each SolarEdge point the closest own measurement point is searched in a range of ± 10 seconds around the SolarEdge point. The time resolution of the received values is the same at the SolarEdge values.

Table 1 Energy weighted module temperature for one module with 10 second power measurement (Ref) compared with the three methods linear interpolation (A), nearest interpolation (B) and closest own measurement (C) calculated over different time periods.

Calculation Method / Time period	Energy weighted module temperature for 1 Module			
	Ref [°C]	A [°C]	B [°C]	C [°C]
2 days (08-09.11.2014)	21.08	21.20	21.20	21.22
1 month (September 2015)	32.32	32.39	32.41	32.48
Whole analyse period 01.10.2014 to 31.05.2016	31.65	31.67	31.68	31.72

Table 1 shows the results for these three methods compared to the values calculated with the 10 second data (Ref). The linear interpolation method A provides the most accurate result with a deviation of + 0.07 % compared to the Ref value.

Figure 3 illustrates the result for the 4th August 2015.

The green graph corresponds to the 10 second power measurement, the SolarEdge power measurements are indicated by black x-markers and the blue graph shows the linear interpolated SolarEdge power measurements.

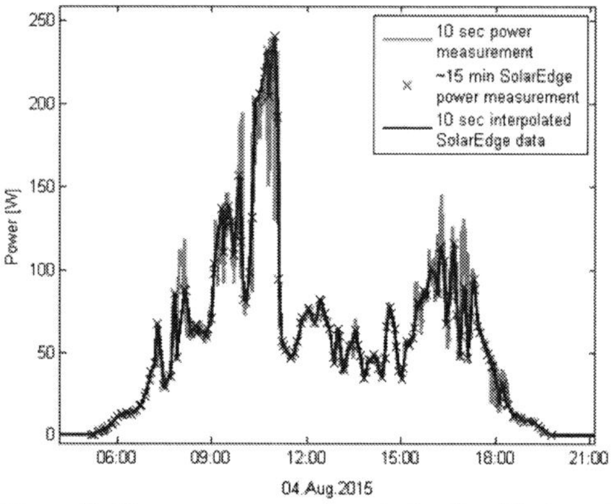

Figure 3: The power measurements for the 4th August 2015 are shown for the 10 second power measurement (green), the SolarEdge power measurement (x) and the linear interpolated SolarEdge measurement (blue).

4.3 Results

In Table 2 and Figure 4, the resulting energy weighted module temperatures calculated according to Formula 1 are shown for the whole analysed period and for each test field. The average energy weighted module temperature over all test fields is 31.7 °C. Table 2 shows also the configuration of all test fields.

Table 2 Overview over the test fields data with name, characteristics and energy weighted module temperature analysed from the 1st October 2014 to the 31th Mai 2016. Two different mounting systems are used, 8 test fields from ZinCo and 5 test fields from Hilti, respectively. At each type of mounting system, there is one field without planting and there are fields with and without irrigation (* = raincover).

Test field	Energy weighted module temperature [°C]	Deviation from the average [°C]	Mounting system	Irrigation	Planting	Vegetation layers [mm]	Vegetation material
Z50K	31.6	-0.1	ZinCo	No	No	50	Gravel
Z70R	30.4	-1.3	ZinCo	No	Yes	70 (60+10)	RieFa/ZinCo
Z70	30.3	-1.4	ZinCo	No	Yes	70	ZinCo
Z100VA	31.8	0.2	ZinCo	No	Yes*	70/100/130	ZinCo
Z100V	30.5	-1.2	ZinCo	No	Yes	70/100/130	ZinCo
Z70B	33.3	1.6	ZinCo	Yes	Yes	70	ZinCo
Z70BR	31.3	-0.4	ZinCo	Yes	Yes	70 (60+10)	RieFa/ZinCo
Z100VB	30.6	-1.1	ZinCo	Yes	Yes	70/100/130	ZinCo
H70	33.5	1.8	Hilti	No	Yes	70	ZinCo
H100V	32.1	0.5	Hilti	No	Yes	70/100/130	ZinCo
H70R	32.7	1.0	Hilti	Yes	Yes	70 (60+10)	RieFa/ZinCo
H70Ro	30.8	-0.9	Hilti	No	Yes	70 (60+10)	RieFa/ZinCo
H50K	32.9	1.2	Hilti	No	No	50	Gravel
average	31.7						

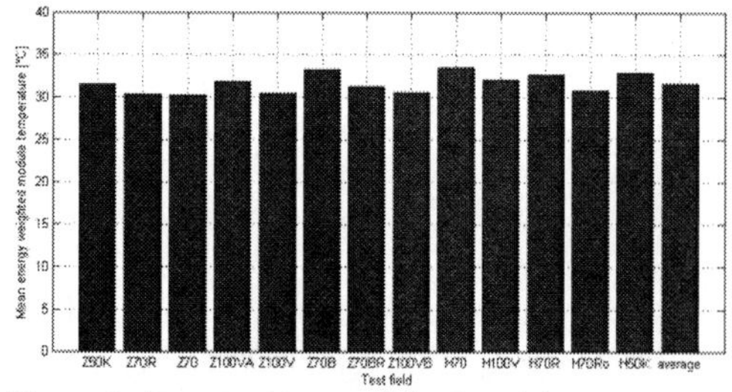

Figure 4: Average energy weighted module temperatures in °C of each test field during the measurement period from the 1st October 2014 to the 31th May 2016. The overall averaged energy weighted module temperature is shown in the last bar and corresponds to 31.7 °C.

The temperature differences result in a lower or higher energy yield because of the negative temperature coefficient (-0.39 %/K according to the module datasheet). Figure 5 shows these energy yield differences relative to the overall average.

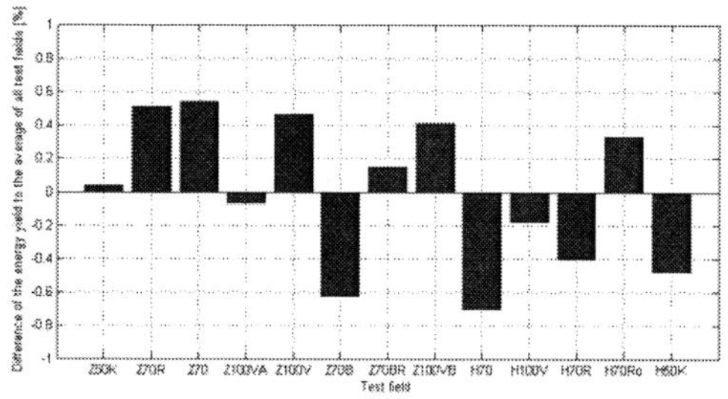

Figure 5: Differences of the energy yield relative to the average of all test fields in % calculated with the power temperature coefficient (-0.39 %/K) and the energy weighted module temperatures during the measurement period from the 1st October 2014 to the 31th May 2016.

4.4 Distribution of the resulting energy weighted module temperatures according to the operating hours of all test fields

For each day, the nominal operating hours are calculated resulting in 1356 hours over the 527 days. The distribution of the number of days according to the nominal operating hours is build and illustrated in Figure 6.

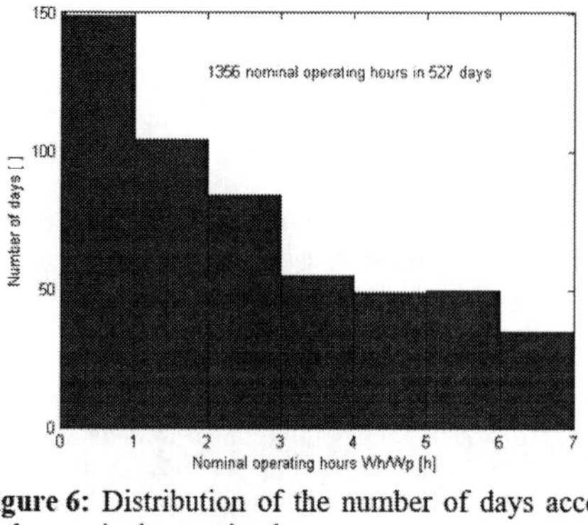

Figure 6: Distribution of the number of days according to the nominal operating hours.

Next, the daily average energy weighted temperature is calculated for each group of daily measurements that have the same nominal operating hour. This is shown in Figure 7.

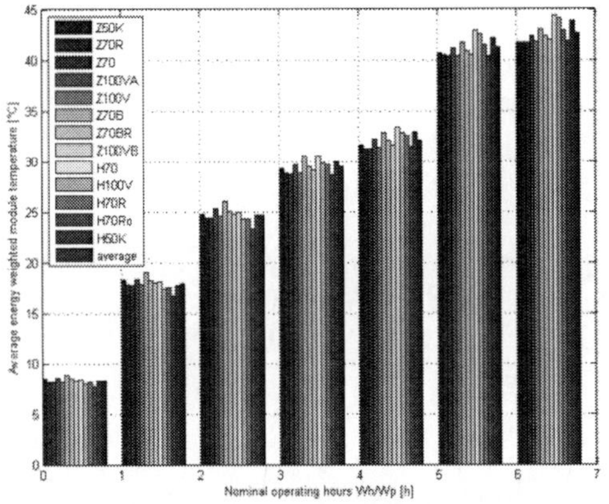

Figure 7: Distribution of average energy weighted module temperatures according to the nominal operating hours of all test fields during the measurement period from the 1st October 2014 to the 31th Mai 2016.

Figure 8 illustrates the distribution of the energy yield difference compared to the average of each range with the same nominal operating hour. This is calculated by building the differences between the energy weighted module temperatures and its corresponding average values in the same nominal operating hour range and by multiplying with the power temperature coefficient of - 0.39 %/K.

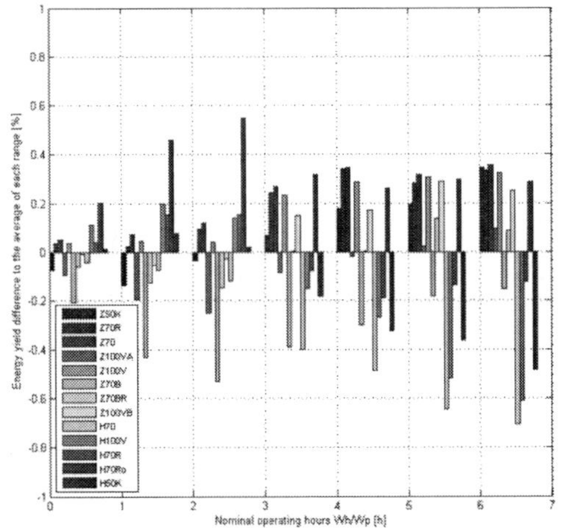

Figure 8: The distribution of the energy yield difference in % is calculated using the power temperature coefficient and the energy weighted module temperatures and it is compared to the average yield of each range with the same nominal operating hours from the 1st October 2014 to the 31th May 2016.

5 CONCLUSION

The main result of the analyses during the first one and a half year is that the difference of the energy weighted module temperatures between all the thirteen test fields is below ± 1.8 °C. Therefore, the energy yield difference is below ± 0.7 % calculated with the standard temperature coefficient for crystalline silicon modules. This difference is in the range of the measurement uncertainty of a power measurement of ± 1.2 %. It's relevant that the mounting system is optimised for green roofs in order that the plants do not cause shades on the PV modules and thus leading to expensive maintenance [7].

Because of the space requirement conflict between PV on the roof and green roofs, it's essential to combine these two systems. Due to the fact that the green roof has a negligible influence on the temperature reduction of the PV modules on the base of the used system components a new follow-up project with Solarspar was startet. The vertical East-West facing bifacial modules shown in Figure 9 combine PV and green roof and allow a cost-effective maintenance of the green roof.

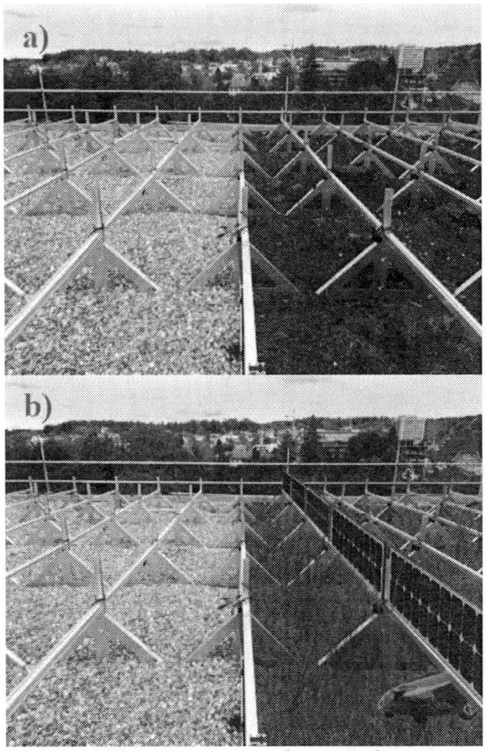

Figure 9: The PV power plant "Seniorenzentrum Wiesengrund" a) with vertical East-West facing bifacial modules before module mounting b) schematic of mounted bifacial PV modules and picture of the PV powered lawnmower working autonomously beneath the PV modules, located in Winterthur, Switzerland is a follow-up project of ZHAW with Solarspar and will be start operation on July 11th, 2016.

ACKNOWLEDGMENT

Thanks to the founding agency climate funds Winterthur and project partners Solarspar, ZinCo, intelli solar GmbH, RieFa – BAWES GmbH, Fenaco, A777 garden design, Fritz Wassmann, public utility Winterthur, PlantCare, ZHAW Life Sciences and Facility Management and Markus Klenk, IEFE contributing to the design of the new bifacial vertical module for project "Wiesengrund".

REFERENCES
[1] Swiss Federal Office of Energy SFOE; "Botschaft zum ersten Massnahmenpaket der Energiestrategie 2050"; http://www.admin.ch/opc/de/federal-gazette/2013/7561.pdf
[2] City of Chicago Planning and Development; "Chicago Green Roofs"; http://www.cityofchicago.org/city/en/depts/dcd/supp_info/chicago_green_roofs.html
[3] Michael Soukup und Stefan Häne; "Mit grünen Dächern gegen die Hitze"; Tagesanzeiger vom 07.09.2015; http://www.tagesanzeiger.ch/leben/gesellschaft/mit-gruenen-daechern-gegen-die-hitze/story/24108679
[4] Markus Chretien, Ralf Walker, Marcel Okle, Matthias Delker, Tobias Probst, Walter Schmidt, Fritz Wassmann, Stephan Brenneisen PV green roof project funded by the Klimafonds Stadtwerk Winterthur, Switzerland, 2015
[5] UFA Samen; "UFA-Kräuter Solardach CH"; http://www.ufasamen.ch/de/dachkraeuter-mischungen/product/dachkraeutermischungen/ufa-kraeuter-solardach-ch
[6] Berlin Senate for Urban Development Communications; "Rainwater Management Concepts, Greening buildings, cooling buildings"; http://www.gebaeudekuehlung.de/SenStadt_Rainwater_en.pdf
[7] VESE Tagung April 2015; Presentation Tecton "Pflegeintensive PV-Dächer"; http://www.vese.ch/wp-content/uploads/Tecton20150418.pdf

PERFORMANCE MONITORING OF DIFFERENT MODULE TECHNOLOGIES AND DESIGN CONFIGURATIONS OF PV SYSTEMS IN SOUTH AFRICA

Tshepo Serameng[1,2], Kittessa T. Roro[3,*], Manjunath Basappa Ayanna[3], Kumaresan Cunden[1], Senzo Myeni[1], Tobias Bischof-Niemz[3], E.E van Dyk[2], Stephen Koopman[3]
[1]ESKOM, Research Testing and Development, Private Bag X40175, Cleveland, 2022, South Africa
[2] Physics Department, Nelson Mandela Metropolitan University, P.O. Box 77000, Port Elizabeth, 6031, South Africa
[3]CSIR Energy Centre, Council for Scientific and Industrial Research, P O Box 395, Pretoria 0001, South Africa
[*]Author for correspondence: Phone: +27128414927, E-mail: KRoro@csir.co.za

ABSTRACT: Performance monitoring of different module technologies and system configurations of Photovoltaic (PV) systems in South Africa is rare, resulting in-few reports based on field results of PV systems installed and operated in South Africa. The goal of this work is to analyse the performance of PV systems, evaluate the energy output of various PV system configurations and module technologies in the South African climatic conditions. To achieve this, a 400 kWp PV Solar plant has been installed and monitored since January 2015 at the Research and Development Centre of ESKOM in Rosherville, Gauteng. The plant consists of polycrystalline silicon (c-Si) and copper indium gallium selenide (CIGS) thin film technologies. The plant comprises of two 25° fixed north facing inclination, north-south single axis tracking and two 10° fixed shed orientation each for both module technologies. The analysis was conducted for duration of 1 year. The evaluation is achieved by studying the operation of PV plant under different weather conditions and the impact of different configurations and technologies on systems overall output, and performance ratio was calculated. The CIGS module technology performed better than c-Si technology in both fixed north facing and tracking configurations when considering specific or final yield. The study provides an insight to identify the optimal configuration and suitable PV technology for deployment of solar photovoltaic system in South Africa. The monitored data and operating experience of PV system can be applied for future projects.
Keywords: PV system, Performance, Analysis

1 INTRODUCTION

Amongst the various renewable energy sources, Photovoltaic (PV) technologies that convert sunlight directly to electricity are becoming a major force reshaping the future energy mix, especially in countries with high solar radiation such as South Africa. South Africa has a solar irradiation ranging from 1,700 kWh/m^2 to 2,300 kWh/m^2, ranking among the top countries in the world in terms of expected solar energy production. With the South African government's new drive to promote and implement renewable energy projects, the solar photovoltaic market is one of the industries with highest growth potential in the country.

Solar PV is starting to play a substantial role in electricity generation in many countries as rapidly falling costs have made unsubsidised solar PV-generated electricity cost-competitive with fossil fuels in an increasing number of locations around the world. The cost of PV systems declined by about 75% in just less than 10 years making solar PV a cost competitive, reliable and sustainable electricity source in many countries [1]. It is estimated that at least 200 TWh, or 200 billion kWh of electricity will be produced in 2015 by PV systems installed and commissioned until January 2015 globally. This represents about 1% of the electricity demand of the planet, though some countries have reached rapidly significant percentages. From 2000 to 2014, global solar PV deployment has increased from 0.26GW to 178 GW [2] with an annual growth rate of more than 45%, due to both technological innovations that have reduced manufacturing costs by 100 times and various government incentives for consumers and producers.

Moreover, a growing number of countries including South Africa (through its Renewable Energy Independent Power Procurement Programme (REIPPPP)) has established renewable energy targets for clean energy for a number of reasons including: to diversify energy mix, archive low carbon emissions, and to reduce environmental issues associated with fossil energy resources such as greenhouse gas emissions and climate change. Many sources of renewable energy such as wind power, hydropower, solar energy, biomass, biofuels and geothermal energy, which come either directly or indirectly from the sun, are considered to be environmentally friendly and are abundantly available throughout the world.

Due to the international drive for clean energy which has seen many countries participating in the renewables space and also due to the PV market growth globally and especially in South Africa over the past few years, it has become very important to compare, analyse and evaluate the performance of PV modules especially in actual South African climatic conditions. Performance monitoring of different module technologies and system configurations of Photovoltaic (PV) systems in South Africa are rare and as a there are a limited number of reports based on field results of PV systems installed and operated in South Africa. The goal of this work is to analyse the performance of PV systems, evaluate the energy output of various PV system configurations and module technologies in the South African climatic conditions and hence levelised cost of electricity (LCOE). To achieve this, a 400 kWp PV Solar plant has been installed and monitored since January 2015 at the Research and Development Centre of ESKOM in ERIC, Gauteng. The plant consists of polycrystalline silicon (c-Si) and copper indium gallium selenide (CIGS) thin film technologies. The plant comprises of two 25° fixed north facing inclination, north-south single axis tracking, east-west single axis tracking and two 10° fixed shed orientation each for both module technologies. This analysis was conducted for a 1-year monitoring period. The evaluation is achieved by studying the operation of PV plant under different weather conditions and the impact of different configurations and technologies on systems overall output, and performance ratio and other performance metrics were calculated to facilitate future LCOE determination.

2 ERIC SOLAR PV SYSTEM FACILITY

2.1 FACILITY DESCRIPTION

The ERIC Solar PV System facility is a 400kWp PV plant localised at Eskom Research and Innovation Centre (ERIC), south of Johannesburg, South Africa. The facility is geographically located at 26°5'5" South, 28°58'1" East and approximately 1625 m above sea level. The site generally has a flat terrain and the facility occupies a total surface area of approximately 1 hectare.

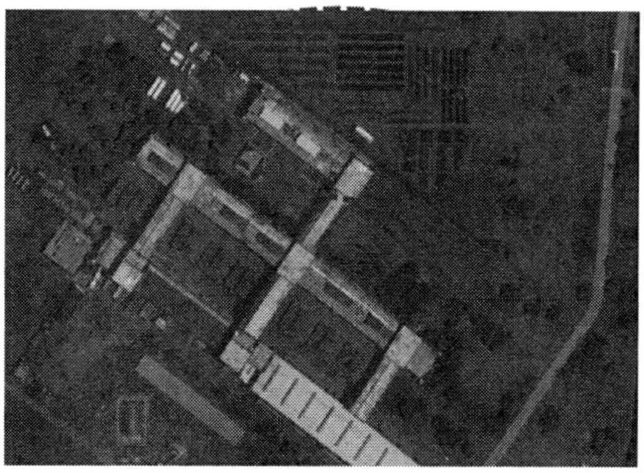

Figure 1: Ariel View of ERIC Solar PV System Facility

The PV plant utilizes ground mounted, multi-configuration PV systems comprising of crystalline and thin film technologies. It comprises of four sections of 100kWp of installed capacity each. Figure 1 shows an aerial photograph of the PV facility. Figure 2 is a close zoom at the various orientations available at the ERIC Solar PV plant.

The various PV configurations are subject to the same environmental conditions in terms of temperature, radiation, humidity and wind speed. South Africa in general has excellent solar resource providing a favourable condition for the development of solar projects. The facility has annual average global horizontal solar irradiation (GHI) of 2,400 kWh/m^2. The minimum and maximum monthly average temperature is estimated at 10.4°C and 21.7°C respectively. The long term monthly average wind speed is approximately 4 m/s.

Figure 2: Aerial Photograph of ERIC Solar PV plant

The various PV configurations at the facility are illustrated in Table 1.

Table 1: ERIC PV System set-up

PV System Number	Orientation	Sub-orientation	Total kWp of modules at STC (kWp)	Tilt Angle (°)
1	North facing fixed Crystalline		100.80	25
2	North facing fixed Thin film		102.66	25
3	Fixed E/W Crystalline			
		West Tilted Crystalline	33.60	10
		East Tilted Crystalline	33.60	10
4	Fixed E/W Thin film			
		West Tilted Thin film	17.11	10
		East Tilted Thin film	17.77	10
5	Tracking E/W Crystalline		33.60	
6	Tracking E/W Thin film		33.60	
7	Tracking N/S Crystalline		34.22	

2.2 SYSTEM DESCRIPTION

The PV systems 1 and 2, north facing fixed multi crystalline and north facing fixed thin-film, are adjacent to each other within the facility as illustrated in Figure 1. The PV modules are ground mounted with a tilt of 25° and azimuth angle of 0°. The modules are mounted on a portrait layout along the width of the structure. The row to row distance for the multi crystalline configuration is 3.39 meters with a pitch of 6.4 meters. The pitch for thin film configuration is 4.5 meters and row to row distance is 2.3 meters. Both, tilt angle and row to row distance are selected in order to optimize the performance (minimize near shading effect) and installed capacity.

PV System 1, north facing fixed multi crystalline PV system layout consists of 4 strings of 14 multi crystalline silicon modules in series. 4 strings are connected to one

15kW inverter. A basic section of mounting structure consists of 14 modules of two rows in portrait layout, this arrangement makes 28 modules (for c-Si) mounted per section of structure. 14 modules are connected in series to make one string for c-Si modules. Hence, a basic section of mounting structure consists of two complete strings. The multi-crystalline PV module configuration is illustrated in Table 2.

Table 2: Fixed Tilted Crystalline - North Facing configuration

Inverter No.	No. of strings	No. of modules/string	DC (kWp)	Number of modules
1 (15kW)	4	14	16.8	56
2 (15kW)	4	14	16.8	56
3 (15kW)	4	14	16.8	56
4 (15kW)	4	14	16.8	56
5 (15kW)	4	14	16.8	56
6 (15kW)	4	14	16.8	56
Total AC = 90kW		Total DC = 100.8kW		
Array to inverter ratio (AIR) = 112% (16.8/15)				336

PV System 2, north facing fixed thin-film PV system is 102.66 kWp and each thin film PV module can generate 140Wp with 7.71 A of IMP and 29.25 VMP. The thin film PV system layout consists of 2 different string configurations which are both connected to the same inverter. Configuration 1 constitutes of 2 strings of 9 thin film modules in series and configuration 2 constitutes of 10 strings of 10 thin film modules in series. A basic section of mounting structure for configuration 1 consists of 9 modules of two rows in portrait layout, this arrangement makes 18 modules (for thin film) mounted per section of structure. 9 modules are connected in series to make one string for thin film modules. Hence, a basic section of mounting structure consists of two complete strings. A basic section of mounting structure for configuration 2 consists of 10 modules of two rows in portrait layout, this arrangement makes 20 modules (for thin film) mounted per section of structure. 10 modules are connected in series to make one string for thin film modules. Hence, a basic section of mounting structure consists of two complete strings. The thin film PV module configuration is illustrated in Table 3.

The PV systems 3 to 4 are crystalline silicon modules and thin film modules mounted on a fixed structures with half of the modules at an azimuth angle of 90° (Due West) and the other half of the modules at an azimuth angle of -90° (due East). The inclination angle of these modules is 10° in order to maximise annual energy yield. The modules are mounted on a portrait layout along the width of the structure.

The PV systems 6 to 7 crystalline silicon and thin film modules mounted on single axis tracking structures. The tracking structures have the following orientations:

a) 1-axis trackers installed in an E-W direction with a N-S tracking function.

b) The rest of this sub-section consist of 1-axis trackers installed in a N-S direction with an E-W tracking function.

Table 3: Fixed Tilted Thin Film - North Facing configuration

Inverter No.	No. of strings	No. of modules/string	DC (kWp)	Number of modules
1 (15kW)	2	9	2.61	18
	10	10	14.5	100
2 (15kW)	2	9	2.61	18
	10	10	14.5	100
3 (15kW)	2	9	2.61	18
	10	10	14.5	100
4 (15kW)	2	9	2.61	18
	10	10	14.5	100
5 (15kW)	2	9	2.61	18
	10	10	14.5	100
6 (15kW)	2	9	2.61	18
	10	10	14.5	100
Total AC = 90kW		Total DC = 102.66kW		
Array to inverter ratio (AIR) = 114%				708

PV Systems 1 & 2 which are fixed north facing orientations are most preferred in South Africa as seen from most large scale REIPPPP plants.

The PV systems 6 to 7 are tracking systems and these are expected to indicate the maximum possible energy yield. The PV System number 7 (tracking N/S Crystalline) has been included for comparison reasons. These systems are generally horizontal during the day and their tracking movement is significant only in the morning and evening. These systems are normally not installed, as they tend to lose their performance very fast due to dirt, which cannot be washed away by rain.

The PV module properties for the multi crystalline and thin film modules installed at ERIC Solar PV facility are shown in Table 4.

2.2.1 DC SYSTEM

The DC system comprises of the following; PV modules, DC cabling, combiner boxes, protective devices and disconnect switches. PV modules are first connected in series to make a string. A group of strings (arrays) is then connected to inverter through combiner boxes. For the ERIC PV Solar system multi-crystalline modules are not connected to the combiner boxes, strings are connected directly to the DC side of the inverter. The combiner boxes are utilized by the thin film modules only.

Table 4: ERIC Solar PV module properties

Specifications	Technology	
	Poly-crystalline Silicon	Thin film (CIGS)
Manufacturer	Astronergy	tsmc solar
Model	CHSM6612P-300	TS-145C2
Maximum Power	300Wp	145Wp
Voltage at Pmax	8.91V	63.6V
Current at Pmax	34.74A	2.28A

The combiner box collects individual strings to form an array. It houses the fuses for overcurrent protection of individual strings, on-load disconnect switch for isolation and surge arrestors for over voltage protection. The combiner box is located between the solar field and the inverter. Combiner box is utilized thin-film technology only because strings need to be combined in order for the minimum inverter voltages and currents to be met. Poly crystalline technology is connected directly the inverter as it satisfies the minimum requirements for voltages and currents.

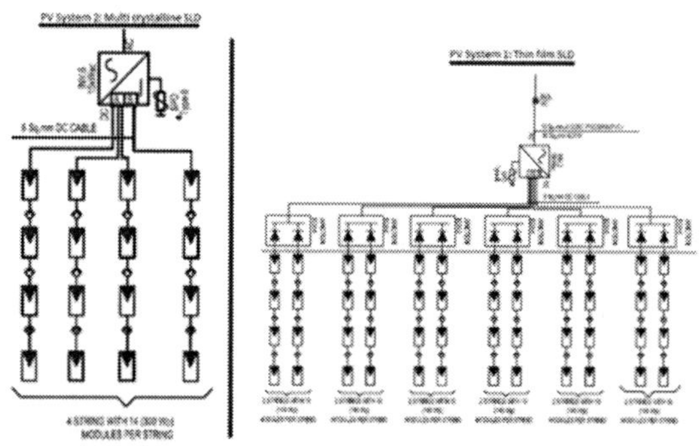

Figure 3: Typical SLD for PV System 1 and PV System 2

2.1.2 INVERTER DESCRIPTION

As an example, the dc design includes 14 modules per string for multi-crystalline (PV System 1) and 9 and 10 modules per string for thin film technology (PV System 2). The number of modules per string is selected based on the module characteristics and inverter mppt input voltage range. The configuration is made in such a way that the string voltage will always be within inverter's mppt operating voltage range.

Same type of inverter is used at ERIC Solar PV facility, which is a three phase SMA Sunny tripower 15000TL. The Sunny Tripower 15000TL has enhanced features. This includes the integration of grid management functions such as , Integrated Plant Control that allows the inverter to regulate reactive power at the grid-connection point. This means that upstream regulator units are no longer needed. Inverter maximum efficiency is 98.2 %. It also provides enormous design flexibility with the two MPP inputs in connection with a broad input voltage range. DC input voltage of up 1000V is achievable [3].

All Inverters communicate via RS485 to the Webbox which is installed inside the Server Panel [3]. The SCADA then communicates to the Webbox via Ethernet to read all information from the Inverters.

Figure 4: Typical Thin Film Configuration for the Inverter at ERIC

Figure 5: Typical Multi Crystalline Configuration for the Inverter at ERIC

2.1.3 METEOROLOGICAL MONITORING EQUIPMENT

Measurements of the meteorological parameters (global horizontal irradiance (GHI), ambient and module temperature is done at the facility weather station. All meteorological parameters are measured for each PV system orientation and transmitted to the server via a RS485 to the MOXA serial to Ethernet converter cards. The SCADA then reads the information via Ethernet from the MOXA cards.

All pyranometers are secondary standard (i.e. Compliant with ISO 9060) with an accuracy of at least +/- 3%. The pyranometer measurements are sampled at 60 second intervals and averaged and recorded at 15 minute intervals [4].

The temperature sensors are PT 100 sensors which measure the ambient temperature as well as the PV module temperature. The temperature sensors have a minimum accuracy of +/- 1°C [4]. The temperature sensor measurements are recorded at 15 minute intervals.

The orientation and inclination (tilt) of PV modules determines the average daily irradiation on the system. The irradiation is directly linked with the power generation of the PV system through size and efficiency of the modules.

3 PERFORMANCE METRICS

The performance metrics used for the comparison are: final yield, reference yield and performance ratio.

3.1. Final yield

The final yield (Y_f) is defined as:

$$Y_f = \frac{E}{P_{STC}} \quad (1)$$

where:

E is the total energy output of the PV system measured in kWh AC;

P_{STC} is rated power (kWp) at standard test conditions STC being irradiation of 1 kW/m², cell temperature of 25°C, and air mass of 1.5..

The reference yield (Y_r) is given by:

$$Y_r = \frac{H_m}{G_{STC}} \quad (2)$$

where:

is the equivalent hours of STC experienced by the Site;

H_m is the total in-plane solar radiation in kWh/m²;

G_{STC} is the reference solar radiation at STC of 1 kW/m².

3.2 Performance ratio

The PR is defined in accordance with IEC 61724 [5] as the ratio of energy fed into the grid to the energy that the system could have produced had it operated at STC of 1 kW/m².

The PR is given by:

$$PR = \frac{Y_f}{Y_r} \quad (3)$$

3.3 Yield gain

The yield gain Y_{Gain} of a tracking system versus a fixed tilt is defined as:

$$Y_{Gain} = \frac{Y_{f,tracking} - Y_{f,fixed}}{Y_{f,fixed}} = \frac{Y_{f,tracking}}{Y_{f,fixed}} - 1$$

where:

$Y_{f,tracking}$ is the monthly specific yield of a tracking system;

$Y_{f,fixed}$ is the monthly specific yield of a fixed system.

4 APPROACH FOR PERFORMANCE COMPARISON

The approach used in comparing the system performance was to keep the system design parameters constant as far as possible and to only consider variables which are pertinent to the specific system. Table 6 shows the module technology and design configuration used for comparison of performance in this study.
(2)

Table 6: Module technology and design configuration used in this study

Module technology
• c-Si vs CIGS: Fixed tilt
• c-Si vs CIGS: E-W tracking
Design configuration
• Fixed vs E-W tracking
• Fixed vs E-W fixed

5 PERFORMANCE RESULTS AND DISCUSSIONS

5.1 Comparison of module technologies

5.1.1 Fixed tilt orientation

In order to compare the outputs from different PV module technologies, the monthly specific yield (in kWh/kWp) is calculated and the results are shown in Figure 6. It can be noted that through out the year, the thin film produced more energy when compared to the c-Si modules. It is worth mentioning that the specific yield during winter time seems to be lower compared to other seasons. However, one has to take into consideration when comparing the month to month yield data since the values might be affected by the data cleaning process and the duration of operation that has been considered for performance comparison. Nonetheless, the physical response of the different technologies will still holds true. For example, the December data was available for only half the duration of December due to power outage. May to June 2015 also experienced heavy power outages from time which affected hours of operation for the PV plant.

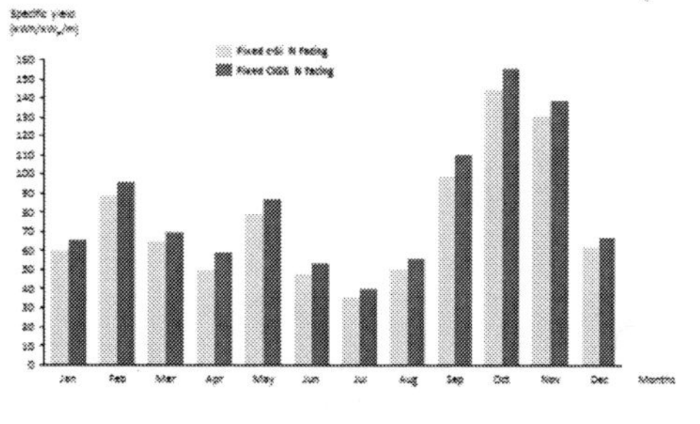

Figure 6: Monthly specific yield of c-Si module performance compared with CIGS modules on fixed tilt configuration

The comparison between the performance ratios of c-Si and CIGS modules in fixed tilt configuration are given in Figure 7. It can be seen form the figure that the PR values vary between 72 to 93 %. The highest value of PR is found to be around 93 % in the month of June for CIGS technology. It is observed that the performance ratio of CIGS thin film is higher for all of the months throughout the year.

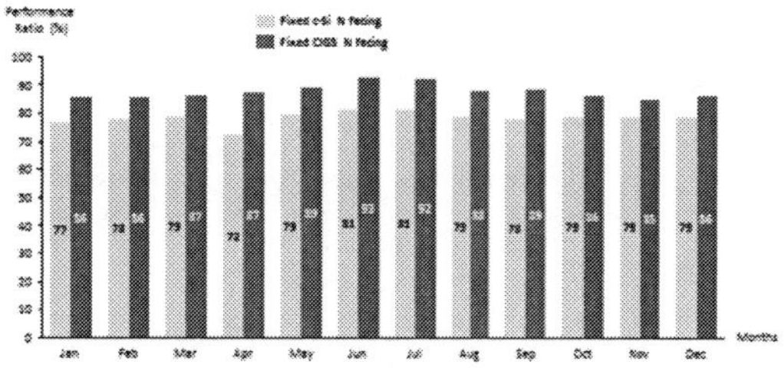

Figure 7: Comparison of monthly performance ratio of c-Si and CIGS modules

5.1.1 Tracking

Figure 8 shows the comparison between the monthly specific yield of c-Si and CIGS modules in east-west tracking configuration. From the graph one can see that the tracking yields more energy every month throughout the year except August where the tracking system malfunctioned.

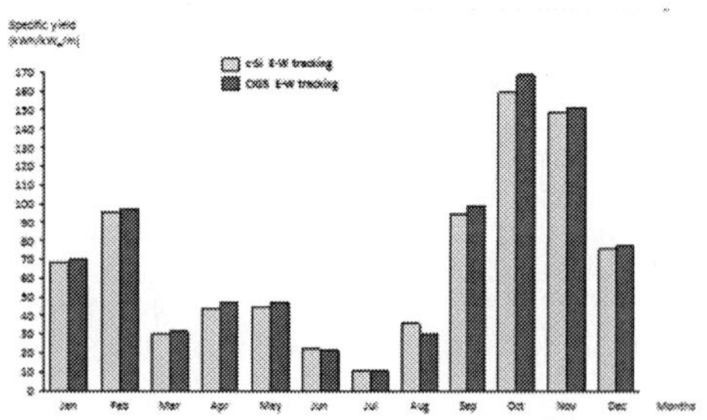

Figure 8: Monthly specific yield of c-Si module performance compared with CIGS modules on fixed tilt configuration

Figure 9 depicts the comparison of the monthly PR values for c-Si and CIGS thin films in tracking configuration. It can be seen from the figure that the PR values of the CIGS technology is higher for the whole year except August.

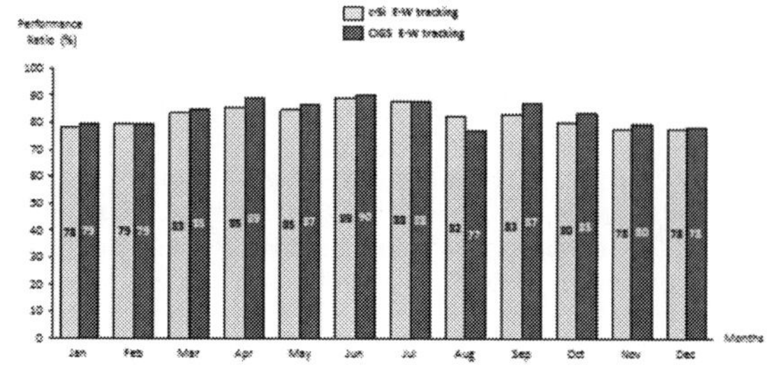

Figure 9: Comparison of monthly performance ratio of c-Si and CIGS modules

As observed in Figures 6 – 9 the CIGS module technology has a higher specific yield and performance ration than the c-Si technology in both fixed north facing and east-west tracking configurations.

5.2 Comparison of different design configuration

5.2.1 Fixed vs tracking

The monthly specific yield of optimally aligned fixed tilt compared with east-west tracking for the same module technologies is shown in Figure 10. It can be seen, that the tracking configuration outperforms the fixed tilt installation in summer months, when the sun stands high in the sky and the days are long whereas in winter months the fixed tilt installations perform better.

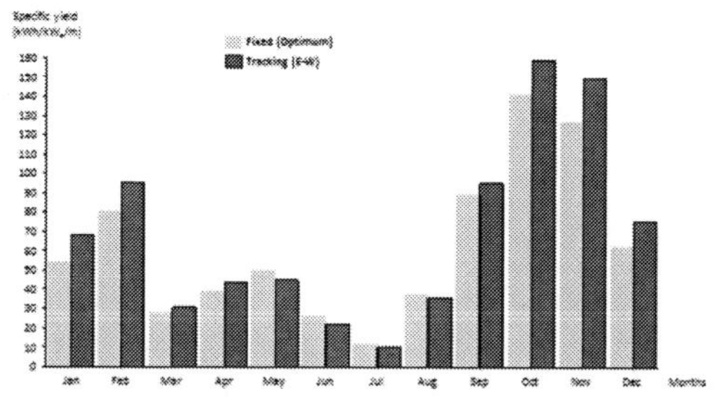

Figure 10: Comparison of fixed tilt yields with east-west tracking yields for each month of a year

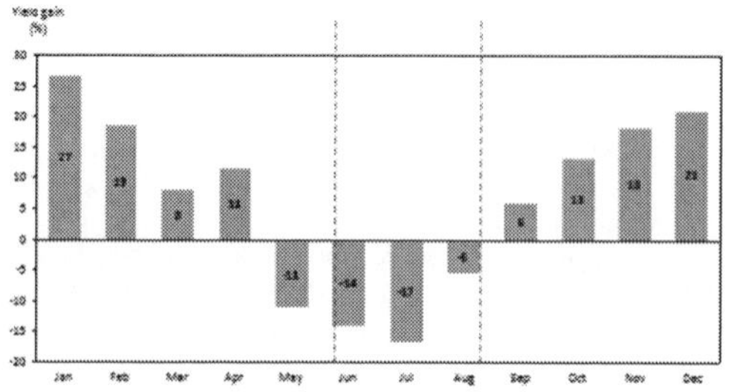

Figure 11: Monthly yield gain of tracking system with respect to the fixed tilt system

As expected, it is observed that the monthly yield gain of the tracking versus fixed installations during winter months is negative (Figure 11).

5.2.2 Fixed Vs E-W

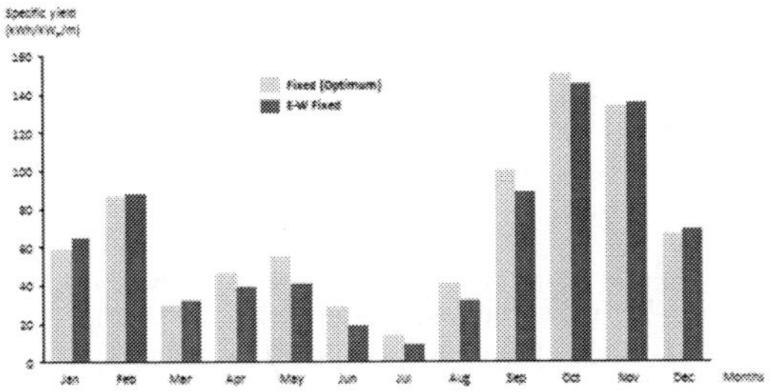

Figure 12: Comparison between fixed tilt installations and east/west tilt installations Monthly specific yield.

Figure 12 depicts the monthly specific yield comparison between fixed installation and east-west oriented PV system. Due to the different orientations in an east-west PV system, the solar modules are exposed to various irradiation levels which results in lower yield in other seasons. This is clearly evident in the yield gain of east/west configuration with respect to the fixed tilt (Figure 13).

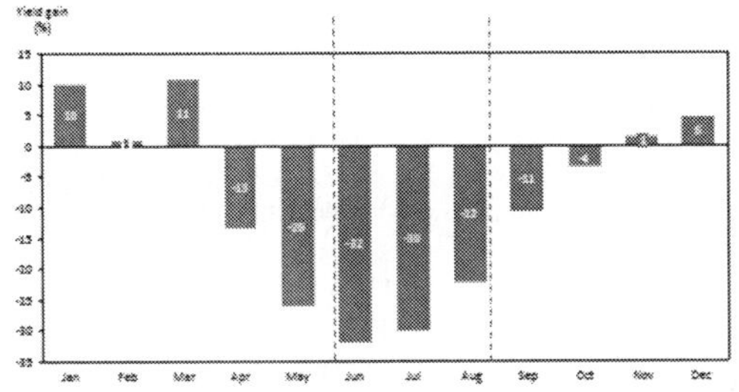

Figure 13: Monthly yield gain of east-west tilted installations with respect to the fixed tilt installations

6 CONCLUSIONS

A performance monitoring of 400 kWp solar park installed at Rosherville was evaluated for one year from January to December, 2015. The final yield and performance ratio were calculated. It is found that the CIGS module technology has a higher specific yield than c-Si technology in both fixed north facing and east-west tracking configurations. The study provides an insight to identify the optimal configuration and will also give an indication of suitability of PV technology for deployment in South Africa once all design parameters are considered. This information is useful in evaluating the operational benefits of the plant based on the net energy output. The monitored data and operating experience of PV system can be applied for future projects.

5 REFERENCES

[1] Department of Energy, (2010). Integrated resource plan for electricity (IRP) 2010-2030: Update report. Government of South Africa.

[2] Department of Energy, (2013). Integrated resource plan for electricity (IRP) 2010-2030 – update report 2013

[3] SMA product datasheet, 2015

[4] ESKOM PV Scada report, 2015

[5] International Electrotechnical Commission: Photovoltaic system performance monitoring - Guidelines for measurement, data exchange and Analysis International Standard IEC 61724, Geneva, Switzerland, first edition, April 1998

THE NEED OF FRAMELESS MOUNTING STRUCTURES FOR VERTICAL MOUNTING OF BIFACIAL PV MODULES

Jorge Rabanal-Arabach[1*], Andreas Schneider[1], Milica Mrcarica[2], Radovan Kopececk[1], Martin Heckmann[3]
[1] International Solar Energy Research Center (ISC), Rudolf-Diesel-Str. 15, 78467 Konstanz, Germany
[2] DSM Innovation Center, Poststraat 1, 6135 KR Sittard, The Netherlands
[3] Technische Universität Berlin - Campus El Gouna, El Gouna, Egypt
*Corresponding author: Phone: +49 7531 36183341, Fax: +49 7531 3618311, Email: jorge.rabanal@isc-konstanz.de

ABSTRACT: The vertically mounted bifacial modules (VMBM) allow attractive applications, for example as a complement for smoothing the power peak at noon time (getting 2 peaks of power: morning and afternoon). Studies [1-3] indicate that in specific places, VMBM can obtain better performance than conventionally mounted mono-facial modules (CMMM). In the present work, the relevance for choosing an adequate mounting structure for VMBM facing east-west in desert climate conditions is investigated to reduce the self-shading impact, on module power, to a minimum. Five 60 cell modules were installed in El Gouna, Egypt, and their IV curves measured every minute. Results indicate that the mounting frame shading effect can easily turn on the bypass diode, diminish the power output up to approx. $10\%_{rel}$. The DSM anti-reflective coated glass improves the mean energy yield for the VMBM by at least (4.2±0.2)%. The soiling test demonstrates that VMBM are practically unaffected by dust accumulation for this specific desert location. Extrapolating the results to a 1 MW PV plant, the tilted mounted bifacial modules (TMBM) generates more energy than CMMM and VMBM but this last one has the chance to produce more net energy than CMMM if the soiling effect is considered.
Keywords: Antireflection Coating, Bifacial, Performance, PV Module, Shading.

1 INTRODUCTION.

An increasing number of photovoltaic (PV) cell and module makers are currently introducing bifacial products to the market. The bifacial solar module offers yield gains of up to (and exceeding) 30% when compared to monofacial module installations [1]. This gain allows for the bringing down of the levelized cost of energy (LCOE) significantly [2]. We are now facing a time where more and more bifacial installations on Megawatt scales are being built at different locations world-wide. One of the key items now is to clarify how bifacial module materials, installations and mounting can be optimized in a way to achieve the highest possible performance of the bifacial setup. According to a recent study [3], a minimum albedo value of 0.2 is required for vertically mounted bifacial modules (VMBM) to outperform conventionally mounted mono-facial modules (CMMM), which according to [4] is valid only at higher latitudes. However both studies indicate that specific places have a chance to obtain better performance in VMBM than CMMM especially due to the expected low soiling effect for the vertical mounting. [5] A 1 kW_P vertically mounted (strings in vertical position) and grid connected system was analyzed, finding that "the frame of the panels causes significant shading on the east-side as the sun altitude increases".

Even if bifacial PV becomes more and more attractive these days with low prices in terms of kWh per installed kW_P and extremely large power benefits, field data for desert VMBM applications exist mainly as theoretical work done in simulations. In this work, we investigate the effect of the mounting frame on the module power output for VMBM in landscape (strings horizontally positioned) and different module glass coating technologies. The self shading effect is studied in detail for VMBM and its effect on the daily power curve evaluated and for yield calculations taken into consideration. Furthermore is evaluated on how VMBM, in comparison to CMMM, reduces the soiling effect for desert places such as El Gouna, Egypt.

2 DESIGN OF EXPERIMENT

Figure 1 shows part of the experimental setup, which is composed in its totality of five modules: three tilted and two vertical installed. The installation is located close to El Gouna, in the Egyptian desert. For the data acquisition, a Papendorf maximum power point (MPP) meter system [6] is used and IV curves of all modules taken every minute. Four ISET sensors, cleaned every two days, are employed to measure the in-plane global irradiance (G): two for the tilted and two for the vertical mounting. Global net irradiance (G_{net}) is obtained by adding the values of each pair of sensors respectively.

The VMBM group consists of two bifacial PV modules, each one manufactured with full-size bifacial solar cells (6" semi-square, n-type, 3 busbars). The modules are made out of 3 strings of 20 cells in a series connected and each string protected by a bypass diode. Thus, both modules are made out of 60 cells, 3 bypass diodes and have been manufactured with the same kind of encapsulation material (EVA) and same kind of ribbon interconnections (1.5mm width and 0.2 mm thickness).

The tilted modules group has same characteristics as those described before, except for one module with two different components: PERC mono-Si cells and white foil as the backsheet instead of bifacial cells and rear glass respectively, meaning a "standard" mono-facial module.

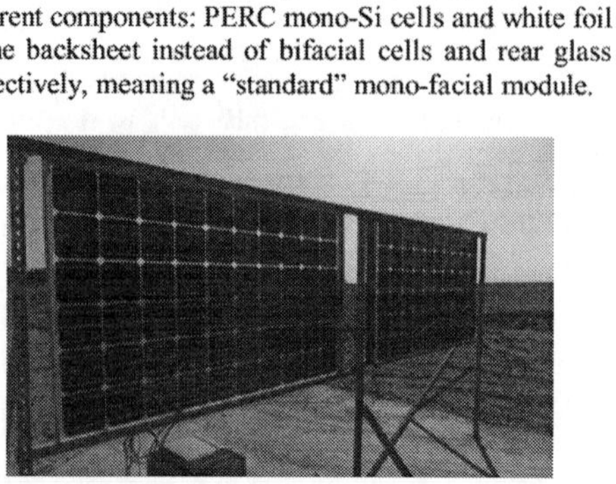

Figure 1: View of the two VMBM for the experimental setup installed in Egypt.

The modules under test varied by the coating applied to the glass and are summarized in Table I. In this case SM stands for structured matt glass (one side of the glass is matt and smooth while the other side has a light scattering structure) and SM-AR means SM glass with DSM anti-reflective coating (ARC).

Table I: Summary for the experimental module setup.

Module	Module type	Front glass	Rear glass	Tilt
AR_AR	Bifacial	SM-AR	SM-AR	90°
SM_SM	Bifacial	SM	SM	90°
AR_AR	Bifacial	SM-AR	SM-AR	20°
SM_SM	Bifacial	SM	SM	20°
AR_WB	Mono facial	SM-AR	-	20°

The investigation is divided into three parts, one where the VMBM are mounted in a structure with an upper beam as support, which introduces additional shading to the rear side (Figure 2). In the second part the mounting structure is modified in order to diminish the rear side shading to a minimum, although the module's frame still creates shading close to solar noon (Figure 3). The third part corresponds to the soiling test, meaning only the ISET sensors are cleaned, letting the modules (tilted and vertical) accumulate dust.

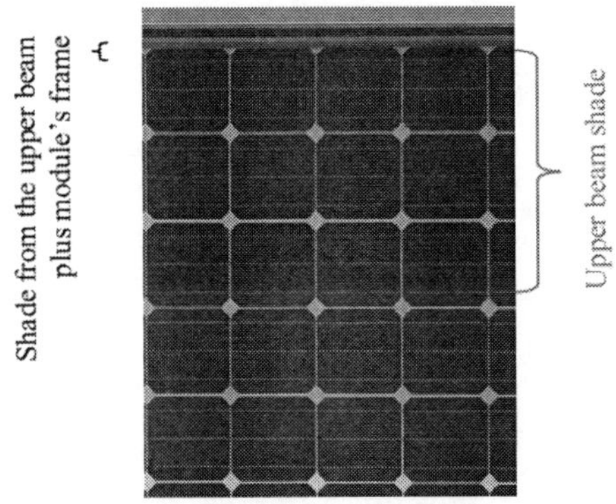

Figure 2: Example for VMBM mounted with upper beam. Picture taken at 11:15 local time.

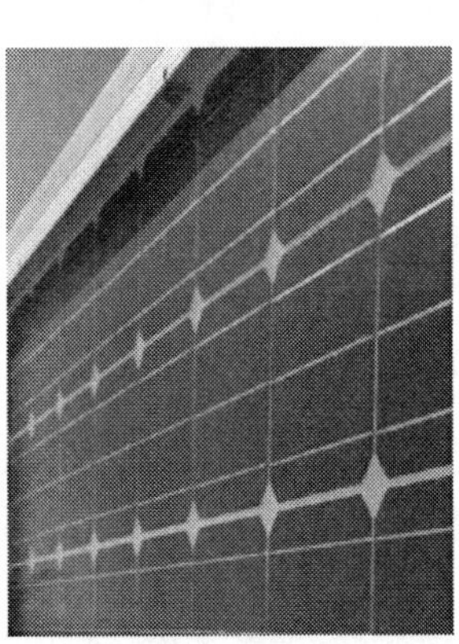

Figure 3: Example of VMBM mounted with minimized shading but self-shading effect. Picture taken at 10:00 local time.

2.1 Parameters comparison

In this work the Performance ratio (*PR*) is calculated as the IEC standard 61724 suggests, meaning the ratio between the module yield (Y_A) and the reference yield (Y_R) for each module.

$$Y_A = \frac{E}{P_{STC}} \quad (1)$$

$$Y_R = \frac{H}{G_{STC}} \quad (2)$$

$$PR = \frac{Y_A}{Y_R} \quad (3)$$

Where E is the generated electrical energy (in Wh), H is the in-plane irradiation (in Wh/m²), and P_{STC} is the measured power (in W) at standard test conditions (STC), meaning an irradiance G_{STC} equal to 1000 W/m² and at 25°C module temperature.

If the measured power output (P_{MPP}) and in-plane irradiance (G) are used instead of electrical energy and in-plane irradiation, in (1) and (2) respectively, the results refer to instantaneous yields (y_A, y_R) and consequently instantaneous performance ratio (pr).

The "cell level instantaneous performance ratio" (pr_C) is defined as:

$$pr_C = \frac{P_{MPP}}{N_{cell}\, P_{MPP,cell}} \frac{G_{STC}}{G} \quad (4)$$

Where N_{cell} is the number of solar cells embedded into the module (60 for all modules in this work) and $P_{MPP,cell}$ is the mean STC power for the used batch of cells.

In order to quantify the soiling effect, the I_{SC}:G_{net} ratio is calculated, which is expected to decline with increasing amount of dust deposited on the module.

3 RESULTS AND DISCUSSIONS

3.1 Frame shading effect in VMBM

The shadow created by the upper mounting structure (specifically in the morning for the mechanical setup exposed in Figure 2) results in a stepped IV curve shown in Figure 4. The stepped IV curve makes it almost impossible for a controller unit (MPP tracker) to allocate and work at P_{MPP}. In our case the micro-controller logic of the P_{MPP} tracker is only controlling one module, for larger installations the tracker is watching and controlling dozens of in series/parallel connected solar modules instantaneously, meaning the shading effect will have a serious impact on the produced kWh of a Megawatt size solar plant.

The stepped IV curve reveals three local maxima which causes the MPP tracker to jump in between these local maxima. Under these conditions, MPP is normally achieved at the highest voltage and hence a lower current. Close to solar noon one of the three bypass diodes is activated, reflected in the shape of I_{MPP} and V_{MPP} in Figure 5: once this diode is turned on, the current jumps and the voltage drops, causing a further drop in power.

During an average day for the setup without the mounting frame ("module_frame" section in Figure 6), the I_{MPP}:G_{net} ratio declines in the morning while the voltage does not, meaning the bypass diode does not turn on with only the self shading. This means the effect of excessive shading caused by the mounting structure (see Figure 3) is the main driver for the bypass diode to be switched on. Translated to numbers, the further shade by the upper mounting beam causes the output power to diminish by up to approx. $8\%_{rel}$ for the SM_SM and up to $10\%_{rel}$ for AR_AR, while the self shading effect of the module does it by approx. $1\%_{rel}$ and $5\%_{rel}$ respectively.

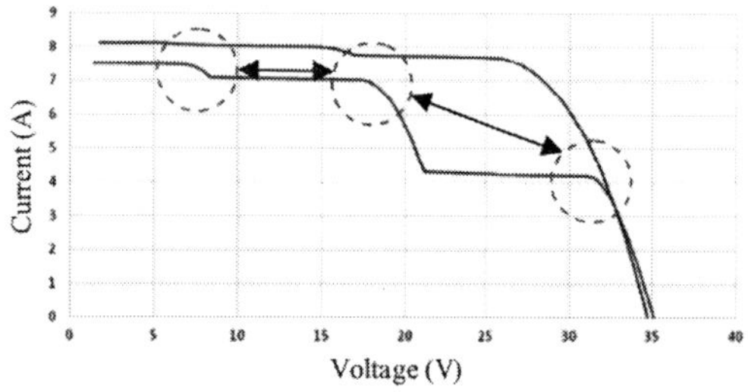

Figure 4: Shading effect caused by the upper mounting structure demonstrated by a stepped IV curve.

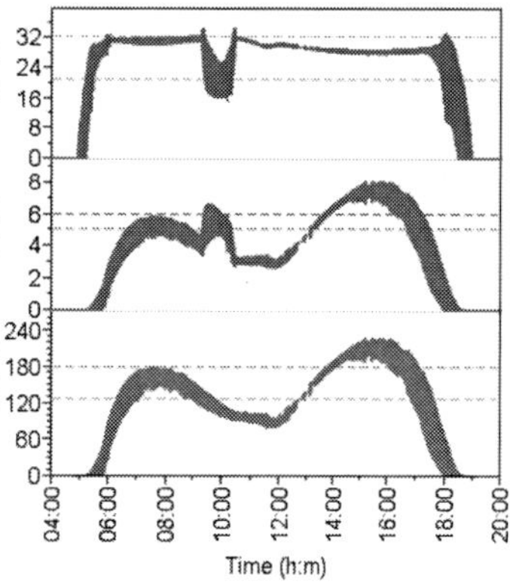

Figure 5: Mean of V_{MPP}, I_{MPP} and P_{MPP} for the SM_SM module mounted with a non optimum structure triggering power losses due to shading.

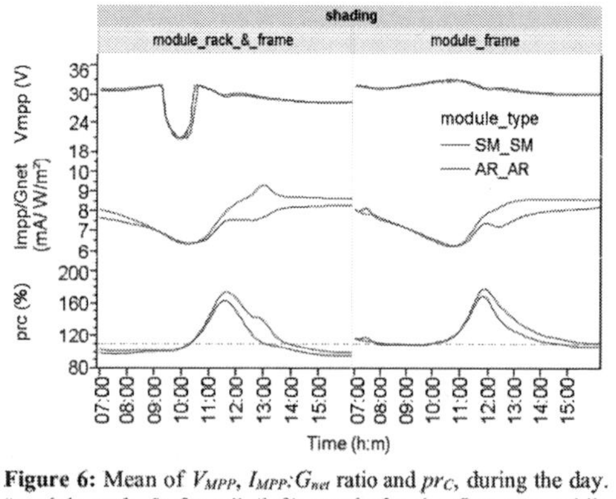

Figure 6: Mean of V_{MPP}, I_{MPP}:G_{net} ratio and prc, during the day. "module_rack_&_frame" (left) stands for the first part, while "module_frame" (right) stands for the second part.

3.2 Anti-reflective coating for VMBM

A large yield gain for the module with DSM ARC glass compared to the other modules is found. Figure 7 compares the data of the AR_AR module with the SM_SM module, analyzing the afternoon energy production (front side power output and no shading effect). On average a gain of $(4.2 \pm 0.2)\%$ in module yield for the months of July, 23rd to September 24th 2015 is seen.

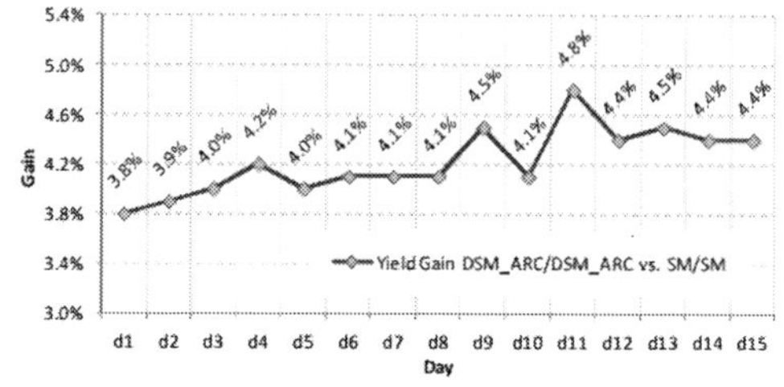

Figure 7: Example of the afternoon-yield gain for AR_AR module compared to the SM_SM module.

The mean I_{MPP}:G_{net} ratio for the VMBM group without the additional shade, exposed in the "module_frame" section of Figure 6, is summarized in Table II for an average typical day. Here, the AR_AR module shows an I_{MPP}:G_{net} ratio approx. $7.7\%_{rel}$ higher than the SM_SM, basically due to the current gain for light on wider incident angles, next to the solar noon within 10:00 to 14:00 hours (local time). On that timeframe the AR_AR module has a mean I_{MPP}:G_{net} ratio higher than the SM_SM by approx. $10.2\%_{rel}$. Likewise the mean P_{MPP} is higher for the AR_AR than for the SM_SM even when the FF for AR_AR is lower by up to $(1.0 \pm 0.1)\%$. From the graphs plotted in Figures 6 and 8 it is noticeable how the power output is affected by the series electrical losses and by the module temperature, where the last is the main driver for the FF to decrease.

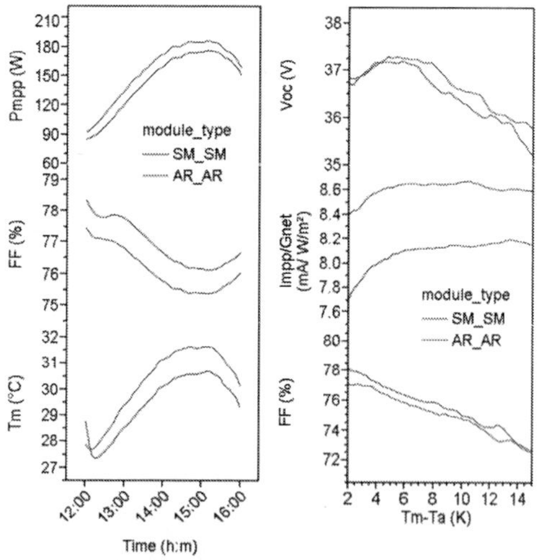

Figure 8: Mean values for the VMBM group during the timeframe within 12:00 to 16:00 hrs in an average day for the second part of the investigation. Left: mean of P_{MPP}, FF and module temperature (T_m) over time; Right: Mean of V_{OC}, I_{MPP}:G_{net} ratio and FF plotted over the temperature difference between the module and environment ($T_m - T_a$).

Notice that Voc is still large for the AR_AR due to the higher net current (the mean temperature difference between modules is only 1.5 K).

Table II: Comparison between AR_AR and SM_SM modules in term of I_{SC}:G_{net} and I_{MPP}:G_{net} ratios (in mA W^{-1} m^2) for a typical averaged day under no shading conditions.

Module	$I_{SC} : G_{net}$	$I_{MPP} : G_{net}$
SM_SM	8.56 ± 0.36	7.96 ± 0.38
AR_AR	9.25 ± 0.22	8.57 ± 0.28
AR gain	$(8.1 \pm 0.4)\%_{rel}$	$(7.7 \pm 0.4)\%_{rel}$

3.3 Soiling

As mentioned before, global net irradiance (G_{net}) is obtained adding the values of each pair of sensors: the two tilted or the two vertical respectively. It is noteworthy that the modules are not cleaned during the period of soiling investigation, but the ISET sensors are still regularly cleaned. All exposed modules are cleaned on day 0 and then later on day 58, these two days are excluded from the data analysis, but days 59 and 60 are incorporated.

Previous studies conducted by University of Antofagasta, in northern Chile [7, 8] show a decrease in *PR* due to dust accumulation between 0.07% to 0.10% per day for mono-Si CMMM. Nevertheless, the authors found the module temperature as the main *PR* loss driver in the coastal zone of the Atacama Desert. Therefore in this work the ratio between short circuit current and net irradiance is compared instead of *PR*, in order to reduce the influence of the module temperature. Even if I_{SC} varies proportionally with temperature, the temperature effect on V_{OC} and *FF* is significantly higher [9].

The trend for the daily mean of the I_{SC}:G_{net} ratio is plotted in Figure 9 and summarized in the Table III, suggesting that after 57 days the VMBM have no significant soiling degradation with a difference between the dusty and the clean module of $-0.1\%_{rel}$ for AR_AR module and $0.0\%_{rel}$ for the SM_SM. In the case of tilted modules the history is a little adverse, with differences of $-17.3\%_{rel}$ for the AR_WB, followed by the AR_AR with $-12.6\%_{rel}$ and later by the SM_SM with $-12.5\%_{rel}$. The foregoing can be translated to a reduction in energy generation due to dust accumulation in the same order of magnitude.

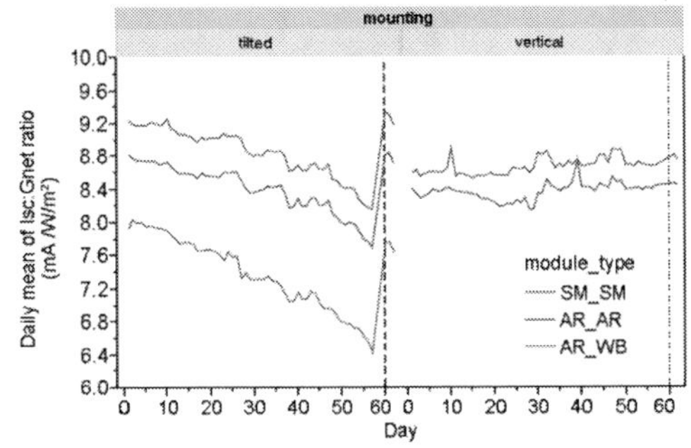

Figure 9: Daily mean of the I_{SC} : G_{net} ratio plotted over 60 days.

If a lineal fit is done to the trend in Figure 9, a reduction in the I_{SC}:G_{net} ratio of 0.026, 0.017 and 0.016 $mAW^{-1}m^2d^{-1}$ for AR_WB, AR_AR and SM_SM modules respectively is seen, meaning a daily I_{SC} drop of (24 ± 1), (19 ± 0) and (21 ± 1) mA per day.

Table III: Summary of results after 57 days of soiling. I_{SC}:G_{net} ratio (in mA W^{-1} m^2) and relative difference (Rel. Diff.)

Module	Mounting	I_{SC}:G_{net} clean	dusty	Rel. Diff. ($\%_{rel}$)
AR_AR	Tilted	9.34	8.16	-12.6
AR_AR	Vertical	8.78	8.77	-0.1
SM_SM	Tilted	8.81	7.71	-12.5
SM_SM	Vertical	8.45	8.45	0.0
AR_WB	Tilted	7.77	6.43	-17.3

By translating these results to a 1 MW PV plant installed in this region, based on frameless AR_AR modules (without module or rack frame shading), it is possible to generate a mean electrical energy of up to 2.76 GWh per year for tilted mounted bifacial modules (TMBM) and up to 2.19 GWh for VMBM. Both previous estimations are done with no losses introduced by the power conversion system or by dust accumulation. If a bi-monthly cleaning interval is considered, an average loss up to $10\%_{rel}$ of the total annual production for a TMBM is calculated, which means 260 MWh for the AR_AR and 250 MWh for the SM_SM. Meanwhile for the case of CMMM employing the AR_WB module an average annual energy generation of up to 2.16 GWh (no soiling) is expected, losing up to approx. $13\%_{rel}$ because of the soiling (280 MWh). Figure 10 shows that the VMBM can generate more energy than CMMM if the dust accumulation is included in the annual energy estimation. In the simulated scenarios, a sensitivity analysis shows that the VMBM can generate more annual energy than the homologous in TMBM only if the last one has a relative loss caused by the dust accumulation higher than $19\%_{rel}$.

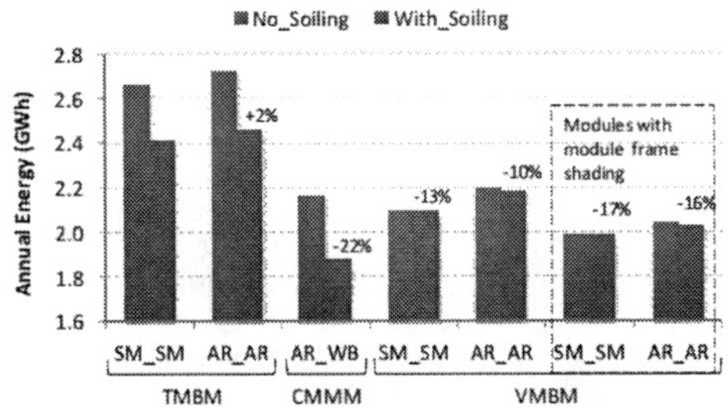

Figure 10: Estimated annual energy generation for the tested modules in a quasi-ideal 1 MW PV plant with bi-monthly cleaning program. The "No_Soiling" state stands for the ideal case of no dust accumulation, while the "With_Soiling" is based on the results of the soiling test performed.

4 CONCLUSIONS

The results show the importance to implement frameless modules on vertical mounting in order to avoid self shading during the time lapse when the sun is facing the rear side (shaded side). It is demonstrated that this effect decreases the output power by up to approx. $10\%_{rel}$, revealing the importance for having frameless modules especially for places with high direct irradiance. The most appropriate, fast and flexible solution seems to be frameless glass/glass module, mechanically supported by a horizontal beam at the bottom and holding the module at the edges with vertical pillars using rubber clamps, with a module in portrait orientation to the ground (strings vertically oriented). Under these conditions it is expected to obtain at least a 7% gain in the daily yield introduced by the DSM anti-reflective coated glass.

The soiling test proved that VMBM are not heavily affected by soiling after 57 days of exposition for this specific desert location. Likewise bifaciality helps to diminish the influence of dust accumulation in tilted mounting due to the rear side power contribution, giving up to 5% less power drop compared to a mono-facial module with the same coating on front glass, making bifaciality more attractive for applications in places with expected high soiling impact.

Comparing a 1 MW bifacial PV plant for CMMM, TMBM and VMBM, in terms of annual energy generation and soiling accumulation, preliminary results indicate that VMBM can generate more energy than the TMBM if the dust accumulation provokes energy losses higher than $19\%_{rel}$. While for a bi-monthly cleaning interval maintenance the VMBM can generate more energy than CMMM, it still cannot compete with the TMBM. In any simulated scenario the PV plant with bifacial modules outperform mono-facial installations but only if the soiling effect is included. It has to be noted that the LCOE calculation including the cost for the cleaning, which can be very high for desert locations will be in favour of the VMBM.

5 ACKOWLEDGEMENTS

The authors want to thank Engelbert Lemp (from ISC-Konstanz) for the valuable support provided.

6 REFERENCES

[1] J.E. Castillo-Aguilella, P.S. Hause. Multi-Variable Bifacial Photovoltaic Module Test Results and Best-Fit Annual Bifacial Energy Yield Model. IEEE Access 2016, 4:498-506. doi: 10.1109/ACCESS.2016.2518399

[2] R. Kopecek et al. Bifaciality: One Small Step for Technology, One Giant Leap for kWh Cost Reduction. Photovoltaics International, 26th ed., 2015, pp. 32-45.

[3] S. Guo, T.M. Walsh, M. Peters. Vertically mounted bifacial photovoltaic modules: A global analysis. Energy 2013, 61:447-454.

[4] I. Shoukry. Bifacial Modules - Simulation and Experiment. MSc Thesis, University of Stuttgart, 2015, pp. 58-61.

[5] J. Johnson et al. Modeling and Analysis of a Bifacial Grid-Connected Photovoltaic System. Power and Energy Society General Meeting, 2012 IEEE, 2012, pp. 1-6.

[6] T. Glotzbach, J. Kirchhof. Novel Measuring System for Long Term Evaluation of Photovoltaic Modules. 24th European Photovoltaic Solar Energy Conference and Exhibition, 2009.

[7] E. Fuentealba et al. Photovoltaic performance and LCoE comparison at the coastal zone of the Atacama Desert, Chile. Energy Conversion and Management 2015, 95:181-186. doi: 10.1016/j.enconman.2015.02.036

[8] P. Ferrada et al. Performance analysis of photovoltaic systems of two different technologies in a coastal desert climate zone of Chile. Solar Energy 2015, 114:356-363. doi: 10.1016/j.solener.2015.02.009

[9] O. Dupré, R.Vaillon, M.A. Green. Physics of the temperature coefficientes of solar cells. Solar Energy Materials & Solar Cells 2015, 140:92-100. doi: 10.1016/j.solmat.2015.03.025

PERFORMANCE ANALYSIS OF PHOTOVOLTAICS SYSTEMS INSTALLED AT DIFFERENT SITES IN THE ATACAMA DESERT

Araya Francisco[1], Ferrada Pablo[1,2], Rabanal Jorge [1,3], Marzo Aitor [1,2], Fuentealba Edward [1,2]

[1]Centro de Desarrollo Energético Antofagasta (CDEA), Av. Universidad de Antofagasta 02400, Antofagasta, Chile
[2]Solar Energy Research Center (SERC Chile), Av. Tupper 2007 4th floor, Santiago, Chile
[3]International Solar Energy Research Center – ISC Konstanz, Rudolf Diesel-Str. 15 D-78467 Konstanz, Germany

ABSTRACT: This work reports on the performance of photovoltaic systems installed at a coastal and inner zones of the Atacama Desert (Chile), at 1700 and 2400 m.a.s.l. Seven photovoltaic systems were studied for 19 months. The technologies installed at each place were a cadmium telluride thin film and a multicrystalline silicon PV plant at fixed angle of 10° for the coastal zone, 25° for the 1700 m high and 10° for the 2400 m high location. A Single Axis Tracking multicrystalline PV plant is also installed at 1700 m.a.s.l. The global horizontal irradiation reached 2488 kWh/m^2·year. The solar irradiation at the plane of array for both plants at fixed angle was 2367 and 2863 kWh/m^2·year for the coastal and inner zone respectively. The plant with single axis sun tracking received 3510 kWh/m^2·year. The maximum rate by which the performance ratio decreased was –4.4 %/month for the plant at 1700 m due to the soiling over 11.5 months. The energy yield for location 1, 2 and 3 resulted in 1774-1590-1902 kWh/kWp·year for CdTe, 1440-1790-1665 kWh/kWp·year for mc-Si at fixed angle, respectively, and 2320 kWh/kWp·year for mc-Si with one axis tracking. The levelized cost of electricity was reduced up to 30% after cleaning.

Keywords: System performance, PV system, environmental effects, CdTe.

1 INTRODUCTION

Solar energy technologies, e.g photovoltaics (PV), implemented in desert or arid regions in the world can be attractive for large-scale plants [1]. However, dust existing in the soil and air can accumulate in the presence of humidity causing a decrease in the transmittance of PV glass. As a result, power and energy yield losses can be a consequence. A previous study for a maritime desert climate found that PV modules experimented a degradation rate of 7 %/month and that the optical efficiency of thermal panels decreased at a rate between 2.8 %/month and 7 %/month [2]. The dust accumulation on PV modules installed in Riyadh (Saudi Arabia), exhibiting a hot desert climate, resulted in a 21.05% output reduction [3]. The application of an anti-soiling coating for PV modules was investigated in [4] showing that in dry periods without rainfall, daily energy soiling losses are greater for uncoated modules. It was found that the masses of adhered sand were 0.005 g for samples with coating and greater than 0.010 g for samples without coating. Recommendations regarding the cleaning frequency required for PV modules were summarized in [5]. It was pointed out that, regions with dusty desert environments, where often occur dust storms, needed to be clean once a week. As seen in those recommendations, the study of the soiling effects on performance degradation in correspondence with the type of dust associated to a region are relevant to determine the cleaning frequency, the cleaning approach, the energy to produce, and evaluate the project. Such as study has not been conducted in different places at the Atacama Desert (Chile), which is experiencing a strong development with new installations and accounts on extreme different environmental conditions within the same region [6]. The particular conditions of Atacama Desert are presented in the next section.

The performance of a PV system does not only depend on the selected technology but also on the environment. External factors such as ultra violet solar radiation, corrosive surroundings, temperature, humidity, ventilation and dust can strongly affect the energy, which PV modules can produce. The Atacama Desert in Chile, is one of the most suitable places in the world for the use of solar energy due to the high solar radiation levels. It exhibits a large number of days with clear skies and a low absorption ozone or/and water vapor columns. Thus, the global horizontal irradiation (GHI) can exceed 8 kWh/m^2 per day, which results in more than 2500 kWh/m^2 per year. Solar energy technologies, e.g photovoltaics (PV), implemented in desert or arid regions in the world can be attractive for large scale plants. The unique environmental conditions in Atacama motivate basic and applied research for this location. The PV capacity installed by April 2016 grows up to 1113 MWp and 2011 MWp are under construction in Chile [7]. However, the mentioned external factors can vary even within the same region. Consequently, performance evaluation and so energy predictions for PV systems, which consider the impact of the environment, must be accompanied by studies of the specific place.

1.1 Location Chile

The Andean mountains and the Humboldt Current have a dominant effect on environmental variables such as temperature and precipitation. The Antofagasta Region [8] exhibits four different climates, determined by the altitude respect the sea level. (1) Coastal or Cloudy Desert at altitudes below 1000 m, (2) Normal Desert at altitudes between 1000 m and 2000 m, (3) High Desert at altitude between 2000 m and 3000 m and (4) High Steppe at altitude above 3000 m. They determine the cloud frequency and the solar resource at ground level.

The Humboldt Current is a cold, low-salinity ocean current that flows along the west coast of South America from southern Chile to northern Peru. The density changes cause deep water to emerge cooling the water surface. The low temperature of the water hinders the evaporation resulting in low rainfall and the occurrence of dense fog at ground level along the coast due to the condensation of mist. This phenomenon is the reason of the abundance of cloudy days along the coastline.

The climates (3) and (4) are influenced by Monsoon in summer by the South American Monsoon which produces the 60% of the annual precipitation of the Andes. The phenomenon is called "Bolivian Winter" [9].

1.2 Evaluating PV plants

The performance ratio (PR) is a well-known indicator to evaluate PV systems. It is defined as the ratio of the final energy yield or final yield (Y_f) to the reference yield Y_r [10]. The final yield is the net energy output of the system (kWh at the AC side) divided by the capacity or size of the PV plant (kWp) at standard testing conditions (STC). The Y_r is defined as the total in-plane solar irradiance H (kWh/m2) divided by irradiance at standard testing conditions or STC (GSTC), which means, 1 kW/m2. Y_r corresponds to the average number of peak sun hours representing the availability of the solar resource for the PV plant. Thus, the PR, in Eq. x, is independent of local radiation conditions and module orientation, quantifying the overall effect of losses and external environmental factors [11].

$$PR = \frac{Y_f}{Y_r} = \frac{E_x}{P_{STC}} \Big/ \frac{H_{POA}}{G_{STC}} \quad (1)$$

One of the most important parameter in the purchase of electricity is the cost. In order to demonstrate reliable, clean and low costs solutions for customers and industry, the Levelized Cost of Electricity (LCOE) is one of the preferred indicators to investigate the feasibility of an energy generation system. It considers the effect of lifetime, performance and financial policies, allowing for a comparison of the cost of electricity produced from different technologies [12]. The LCOE can be used as an indicator for evaluating the competitiveness of different sources of energy. It is equal to the sum of all the cost incurring during the lifetime of the project divided by the units of energy produced during its lifetime [13] as Eq. 2 shows:

$$LCOE = \frac{Total\ Life\ Cycle\ Cost}{Total\ Lifetime\ Energy\ Production} \quad (2)$$

The LCOE can be determined as the average cost of energy over the lifetime of the project such that the net present value (NPV) becomes zero in the discounted cash flow analysis [14]. The LCOE can be expressed as follows:

$$LCOE = \frac{\sum_{i=0}^{n} \frac{C_i}{(1+r)^i}}{\sum_{i=0}^{n} \frac{E_i}{(1+r)^i}} \quad (3)$$

In Eq. 3, the net annual cost of the project, C_i, includes the investment, fixed operation and maintenance (O&M) costs, which are paid along of the project [15, 16]. In C_i, the interest expenditures (Fi) may also be included (not in this work). The energy generated in a given year is represented as E_i. As the energy output of a PV plant can degrade year after year, the energy at year i was calculated as $E_i=E_0(1-d)^i$, where d is the degradation rate in %/year, obtainable from datasheets. The discount rate is represented as r and n is the life of the project (in years). In detail, the LCOE calculation can include PV modules, inverters, balance of system, installation labor, installation, land, installation margin and overhead, permitting and environmental studies, grid interconnection, and O&M costs. The LCOE can also consider incentives [16].

2 EXPERIMENTAL SETUP

2.1 PV System

This study was performed on grid connected PV systems, based on cadmium telluride (CdTe) thin films and multicrystalline silicon (mc-Si) modules, installed at three different locations (L1, L2 and L3) of the Antofagasta Region. Each type of technology is installed next to each other and their characteristics are detailed in Table I.

Location 1 (L1) corresponds to two plants installed on the roof of a building at the University of Antofagasta, at the coastal zone (with coordinates 23°42'02.37''S, 70°25'13.28''W), with a 10° fixed tilt angle (FT10) and 0° azimuth, pointing north and at 50 m above sea level (m.a.s.l.). As described, PV plants consist of a CdTe and a mc-Si PV plant.

The second place (L2 with coordinates 23°25'52.5''S, 69°30'05.58''W) refers to an industrial environment, where CdTe and mc-Si PV plants with a fixed angle of 25° (FT25), azimuth 0° pointing north and at 1700 m.a.s.l. are installed. Additionally, a mc-Si plant with a single axis tracking (SAT) is also installed at this place.

The third location is San Pedro de Atacama (L3 with coordinates 22°54'46.67''S, 68°11'47.98''W) at 2400 m.a.s.l. where PV systems are installed with a 10° fixed tilt angle (FT10) and 0° azimuth pointing the Equator, replicating the PV technologies installed at L1.

2.2 Instrumentation

For PV system at L1 and L3, a photocell with an accuracy of ±5%, and a pyranometer installed next to the PV plants were used to measure the Global Tilted Irradiance (GTI) at the plane of array (POA). The use of both reference cell and a pyranometer in a single system brings along the benefits of measuring the relative influence of spectral effects and angular losses combined. For the PV system at L2, a photocell installed next to the PV plants with an accuracy of ±8% was used to measure the GTI at the POA of the PV plants at fixed angle, and the 1-Axis Sun-Tracking Global Irradiation (TGI). The value of the irradiation H, H_{POA} and H_{track} were obtained from GHI, GTI and TGI, respectively. PT100 sensors with an accuracy of ±0.5% measured the ambient temperature (Tamb). The data from the inverters (current and voltage) and from the sensors were stored in a datalogger, which collected them averaging the values of GHI, GTI, TGI and temperature every 5 min.

Table I: Parameters of the PV plants.

Parameter	Technology			
	CdTe (L1, L3)	mc-Si (L1, L3)	CdTe L2	mc-Si L2
Power kWp	3.0	3.23	5.6	5.52
Efficiency at STC (%)	10.4	13.9	10.4	14.14
K_V (%/°C) for voltage	-0.2	-0.34	-0.25	-0.33

Parameter	Technology			
	CdTe (L1, L3)	mc-Si (L1, L3)	CdTe L2	mc-Si L2
Power kWp	3.0	3.23	5.6	5.52
Efficiency at STC (%)	10.4	13.9	10.4	14.14
K_I (%/°C) for current	+0.04	+0.074	+0.04	0.06
K_P (%/°C) for power	-0.25	-0.45	-0.25	-0.45

3 RESULTS AND DISCUSSION

This section shows the records of 19 months global irradiation data at the POA as well as the energy yield of the PV plants. Some particular working conditions are presented focusing on the PR degradation on each site of study. Then, the monthly PR is depicted in Section 3.5 for technologies at L1 and L2 throughout the period studied. The PR of PV plants at L3 is not shown since the photocell and pyranometer were not under a strict maintenance schedule. Therefore the solar resource for the PR calculation is not accurate enough.

3.1 Solar resource

Figure 1 depicts the monthly-cumulated values of irradiation (H) when evaluating the measured values of GTI, for each place. Similar to most PV modules, however, reference cells are encapsulated by polymers, thus, they are supposedly as sensitive to angular losses as the PV modules monitored. If the spectral response of the reference cell and the PV technology monitored is equal, also the spectral mismatch cancels out. On the other hand, due to the geometry of the dome of pyranometers, relative little directional errors result and fewer errors due to sensor soiling are produced. Due to the tilt angle of the PV modules and photocells at each place, and due to the sun position close to the zenith in summer, the cosine error produced a reduction in the energy captured compared to the horizontal or tracked measurements. For the coastal zone (L1), after a year of evaluation, the global irradiation measured at the POA using a reference cell (henceforth denoted as POA_{si}) resulted in 2200 kWh/m^2·year and 2367 kWh/m^2 measured using a pyranometer. For summer months a maximum of 8 kWh/m^2·day, a mean of 7.3 kWh/m^2·day and a minimum of 3.5 kWh/m^2·day was determined. For winter these values were 4.7 kWh/m^2·day, 6.9 kWh/m^2·day and 1.4 kWh/m^2·day, respectively. For L2, Figure 1 shows values for the global irradiation at POA_{si} above those found for L1 and L3 throughout the year. It can be explained by the difference on the tilt angle set for each system at each location, as described in section 2.1. A maximum of 8.8 kWh/m^2·day, a mean of 8.3 kWh/m^2·day and a minimum of 1.6 kWh/m^2·day were found for summer season at the POA_{si} in this place. It is noteworthy that the mean values of H_{track} reached at this place were 11.6 kWh/m^2 for summer and 7.7 kWh/m^2 for winter, accumulating 3500 kWh/m^2·year. Finally, for L3 the cumulated irradiation reached 2340 kWh/m^2·year at POA_{si} and 2660 kWh/m^2·year using a pyranometer, considering the lack of maintenance.

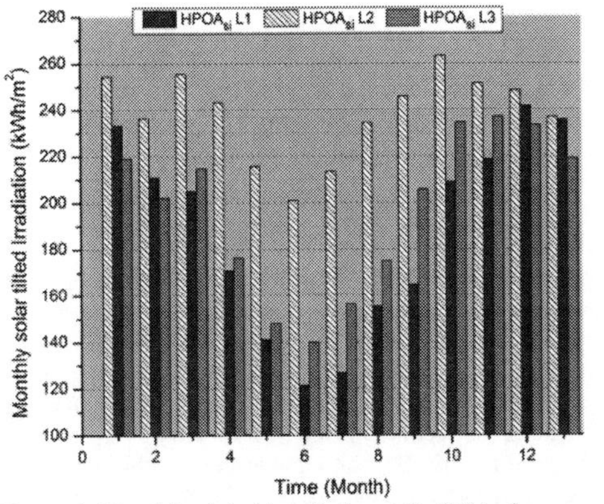

Figure 1: Monthly global irradiation at the POA along one year for each place at the corresponding tilt angle.

3.2 Ambient temperature

The annual maximum, mean and minimum ambient temperature (Tamb,max, Tamb,mean, Tamb,min) were determined for each place. Temperatures averaged 26.5 ± 3.6, 22.5 ± 3.2 °C and 15.6 ± 2.8 °C, respectively for L1. A difference of -5.9 ± 1.1 °C, 3.3 ±0.8 °C and 6.1 ±0.5 °C were found for the L2 (Tamb at L1 – Tamb at L2) and 5.5 ± 1.1 °C, -2.2 ± 1.2 °C, 9.7 ± 0.9 °C for L3 with respect to those found at L1 (Tamb at L1 – Tamb at L3). The Tamb,max and Tamb,min reached values of 35 and 9 °C for L1, 40 °C and -2.8 °C for the L2 and 38.6 and 0.1°C for the L3, respectively, within a year of analysis. Figure 2 depicts the ambient temperature for each place. The above mentioned Tamb,max and Tamb,min clearly show the extreme Tamb values found within the same day for L2.

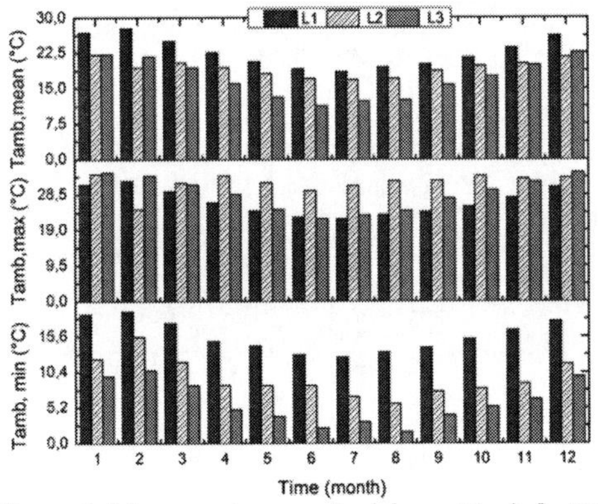

Figure 2: Mean, maximum and minimum Tamb for L1, L2 and L3.

3.3 Energy Yield

According to the International Standard (IS) IEC 61724, the Energy yield on field (Y_f) represents the normalized energy to the size of a PV system, i.e $Y_f = E/P_{STC}$ (kWh/kWp), where E is the total AC energy produced by the PV system in kWh and PSTC is the DC power capacity of the PV system at STC. Thus, Y_f can be used to compare energy produced by PV systems of different sizes. Energy yield was calculated for each technology within the 19 month of evaluation. Figure 3 shows the monthly difference regarding Y_f for each technology and location. The difference was calculated between the Y_f of the PV plants installed at the same place, considering the mc-Si technologies as the reference. For instance at L2 ($Y_{fmc\text{-}Si}$ SAT – $Y_{fmc\text{-}Si}$ FT25)/ $Y_{fmc\text{-}Si}$ SAT. A remarkable fact is that the CdTe at L2 (FT25) exhibited a similar behavior to mc-Si FT25 but their production substantially differed from that of the mc-Si SAT plant. A maximum difference of 48.2 and 40.2 % was found between mc-Si SAT – CdTe at FT25 and mc-Si SAT – mc-Si at FT25 on Dec-13 respectively. When comparing Yf in winter, the mc-Si SAT plant did not show any advantage with regard to the energy production. This result suggests that plants with a 2-axis tracking could significantly increase their energy production. It was pointed out that the gain in electrical energy produced by a tracked PV system compared with fixed modules depend on the used tracker, sky conditions and seasonal evolution of the sun horizontal coordinates [17]. In general, the mc-Si SAT plant exhibited substantially a higher energy production in the period comprehended from the late winter (Aug - 13) to the beginning of autumn (May -14). Particularly, Y_f values of the mc-Si plant with tracking in May, June, and July 2013 and 2014 were not higher than those of the plants at fixed angle. This fact can be explained by the lower solar irradiation (H_{track}) of the tracked mc-Si plant compared to HPOA of the fixed CdTe and mc-Si PV plants (up to 8% less). On the other hand, when evaluating the Y_f values for the coastal zone, the CdTe FT10 kept above that of mc-Si FT10 along the period of study. A similar result is found when comparing (a-Si/μc-Si) tandem and mc-Si at the same place as reported in [18]. A seasonal variation was observed for each technology as Tamb changes. The difference between both technologies decreased during winter season. The latter can be explained by the temperature coefficient factors for power, less negative for CdTe than that of mc-Si. Particularly in Apr-14 the difference is higher than that in Apr-13 due to the more accumulated dust (see dashed lines on Figure 4)

The final values for the cumulated Y_f were 1440.5 kWh/kWp·year for mc-Si at FT10 and 1774.64 kWh/kWp·year for CdTe at FT10. Meanwhile 2532.5 kWh/kWp·year for mc-Si SAT, 1969.5 kWh/kWp·year for mc-Si at FT25 and 1813.9 kWh/kWp·year for CdTe thin films at FT25, were obtained at L2. Finally cumulated Y_f of 1665.6 kWh/kWp·year for mc-Si at FT10 and 1902.8 kWh/kWp·year for CdTe at FT10 were found at L3.

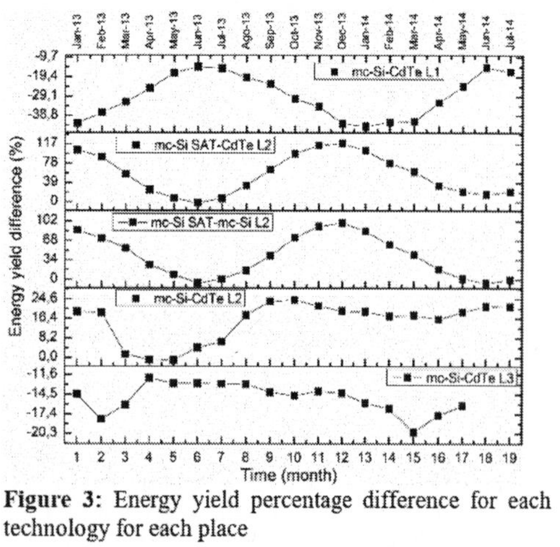

Figure 3: Energy yield percentage difference for each technology for each place

3.5 Performance Ratio

Figure 4 depicts the PR calculated for 19 months of analysis. The standard deviation for PR of each technology is displayed by the error bar. Panels were cleaned randomly, as indicated by vertical lines. Throughout the period of study no precipitation was registered at each place.

A first look on the graphs shows that in the coastal zone (L1), CdTe arrays resulted in PR values above those obtained with mc-Si technologies during the entire period of study. Conversely at L2, mc-Si technologies reached higher PR values than the neighbor technology, CdTe (i.e. during summer, where both mc-Si at FT25 and SAT averaged above 80%) in the same period.

Once obtained monthly performance ratio (Figure 4) it is possible to calculate the PR slope representing the rate by which PR decreases due to local conditions for each technology. For the period between Feb-13 and Jun-13 at L1, from a clean condition PR slope degraded at a rate of -4.1 ± 0.2 %/month for CdTe and -2.8 ± 0.1 %/month for mc-Si both at FT10, whereas the temperature gradient was -2.2 ± 0.1 °C/month. Thus, after five months dust accumulated, PR dropped 16.1 % and 11.0 % for CdTe and mc-Si, respectively. For the period where temperature gradient was +2.2 ± 0.1 °C/month (Oct-13 and Dec -13), PR decreased at a rate of -3.6 ± 0.2 %/month for CdTe, and -5.1 ± 0.1 %/month for mc-Si at FT10. After six month without cleaning (Oct13-Apr14) PR reduced 15.5 and 20.1 % for CdTe and mc-Si respectively. During this period mc_Si FT10 plant showed a higher PR degradation compared to that found in the first period (Feb-13 and Jun-13).

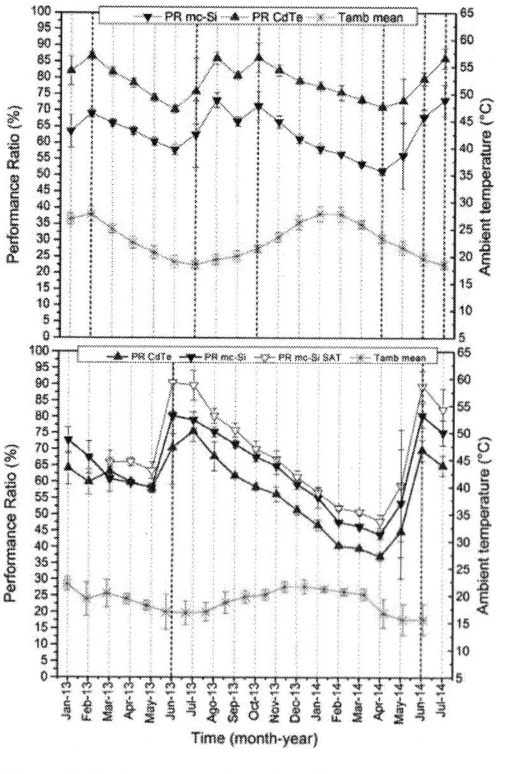

Figure 4: Performance ratio and ambient temperature at L1 (Top) and L2 (Bottom).

On the other hand, for the PV plants installed at the interior of the region (L2), these values resulted in PRslope equal to -4.4 ± 0.2 %/month, -4.1 ± 0.1 %/month and -3.9 ± 0.2 %/month for CdTe FT25, mc-Si FT25 and mc-Si SAT, respectively. Multiplying these values by 12 months, it comes out that the theoretical performance reduction was close to 53%, 49% and 50%, respectively, without cleaning an entire year. The performance ratio and PR degradation through the PR slope were also determined for a-Si/μc-Si plant, mono-Si and the mc-Si PV plants installed at the coastal zone of the Atacama Desert [19]. Using the PR values of that work, it can be established that in 12 months the PR could reduce up to 64%, 62% and 70%, respectively, considering worse situation when temperature increases (temperature gradient) and without cleaning. Thus, the coastal zone produced at least 10% higher performance degradation compared to the mining environment in the Desert.

3.7 LCoE analysis

The LCOE was calculated for 2 situations: (1) the real case, considering the maintenance schedule (displayed in Figure 4). (2) A hypothetical case in which the PR kept near the maximum possible. The second case means that a "stronger" cleaning schedule was required. i.e. more frequent. Table 3 summarizes the values used to estimate the LCOE and compares situation (1) and (2). The goal of the calculation was to determine the difference obtained when modules are uncleaned and clean. Absolute cost values regarding the Chilean market were not available.

All values in Table 3 are according to [15]. Modules and inverters are international average prices. Balance of system considers solar irradiance sensors, anemometer, wiring and mounting system. Installation labor is required for any system as well as a fixed O&M value. Variable O&M costs due to fuels do not apply. Further costs such as grid interconnection and environmental studies are not required for the size of the PV plants in this work (installed capacity of PV plants at L1, L2 and L3 are below 10 kWp).

For (1) the actual total energy produced E_t is known. For (2), the Et needs to be estimated using the PR. The values close to the clean condition were used to obtain an annual PR value at clean condition for each PV plant. As the irradiation and size of the plants are known, Et can be calculated according Eq. 1. The LCOE calculation for case (1) and (2) is then performed following the methodology in section 2.

For L1, under unclean condition, the LCOE of CdTe, mc-Si both at FT10 were 16.8 cUS$/kWh, 18.8 cUS$/kWh, respectively. Whereas the clean condition led to 15.7 cUS$/kWh and 16.9 cUS$/kWh. The LCOE reduced 6% for CdTe and 10% for mc-Si when comparing clean condition with respect to the unclean condition.

For L2, with the modules unclean, the LCOE of CdTe FT25, mc-Si FT25 and mc-Si SAT were 18.2 cUS$/kWh, 15.7 cUS$/kWh and 14.2 cUS$/kWh for the real dirty condition. In case of cleaned modules, the LCOE values were 14.8 cUS$/kWh, 12.5 cUS$/kWh and 9.5 cUS$/kWh, respectively. Likewise, the LCOE reduced 18% for CdTe FT25 and 20% for mc-Si FT25 and 33% for mc-Si SAT when comparing clean condition with respect to the unclean condition

The calculation of the LCOE for clean and unclean condition revealed that its reduction was more significant at the location with higher tilt angle (25° instead of 10°). A LCOE reduction up to an 18% with the CdTe and 20% with the mc-Si plant were obtained.

4 CONCLUSIONS

In this work the performance of PV installed at three different location (L1, L2 and L3) in northern Chile, Atacama Desert, was studied. The technologies consisted of cadmium telluride (CdTe) and multicrystalline silicon (mc-Si).

Regarding the solar resource at 23% more solar irradiation at the plane of array (POA) was found al L2 compared to L1, considering a 15° tilt angle difference (higher at L2).

CdTe showed higher energy yield (Yf) at the coastal zone (L1) and above 2400 m.a.s.l. (L3), contrasting to the values found at 1700 m.a.s.l (L2), higher for mc-Si technologies. A maximum difference of almost 50 % was found between Yf of mc-Si with Single Axis Trracking (SAT) and plants at fixed angle. When comparing Yf in winter, the mc-Si SAT plant did not show any advantage with regard to the energy production.

CdTe showed a 16% higher performance ratio than mc-Si at L1 throughout period studied. Conversely, it was mc-Si the technology achieving a 2% higher PR than CdTe at fixed tilt angle. However, as dust accumulated this difference increased up to a maximum of 7 %.

The calculation of the LCOE for clean and unclean condition revealed a more significant reduction at the location with higher tilt angle (25° instead 10). An LCOE reduction up an 18% with CdTe and 20% with the mc-Si plant were obtained. On the other hand, the one axis tracking showed the highest LCOE reduction from unclean to a clean condition.

5 REFERENCES

[1] M. Ito, K. Kato, H Sugihara, T Kichimi, J Song, K Kurokawa. Sol. Energy Mater. Sol. Cells (2003) 75: 507–517.
[2] SAM Said. Applied Energ (1990) 37: 73-84.
[3] S Alajlan. Appl Energy (1999) 63(2): 91–9.
[4] M. Piliougine, C. Cañete, R. Moreno, J. Carretero, J. Hirose, S. Ogawa, M. Sidrach-de-Cardona. Applied Energy (2013) 112: 626–634.
[5] M. Mani, R. Pillai. Renewable Sust Energ Rev (2010); 14: 3124–3131.
[6] E. Urrejola, J. Antonanzas, P. Ayala, M. Salgado et Al. Energy Conversión and Management (2016) 114: 338-347.
[7] CIFES report. Renewable Energy in the Chilean Electricity Market. May 2016.
[8] J. Rech, B. Currie, E. Shullenberger, S. Dunagan, et all. Earth and Planetary Science Letters 292 (2010) 371–382
[9] R. Garreaud, A. Molina, M. Farias. Earth and Planetary Science Letters 292 (2010) 39–50.
[10] Photovoltaic System Performance Monitoring - Guidelines for measurement. data exchange and analysis. IEC 61724. IEC. 1998.
[11] NH Reich, B Mueller, A. Armbruster, K. Kiefer, C. Reise. Progress in Photovoltaic Res. Appl. 2012: 717-726.
[12] K. Kost, JN Mayer, J.Thomsen, N. Hartmann C. Senkpiel Philipps, et all. Fraunhofer Institute for Solar Energy Systems. ISE. Levelized cost of electricity renewable energy technologies. study; 2013. Last access on July 30, 2015.
[13] K. Branker, MJM Pathak, JM. Pearce. Renewable & Sustainable Energy Reviews 15, pp.4470-4482 (2011)
[14] M. Said, M. EL-Shimy, M. Abdelraheem. Sustainable Energy Technologies and Assessments (2015) 9: 37–48.

[15] M. Said, M. EL-Shimy, MA. Abdelraheem. Sustainable Energy Technologies and Assessments (2015) 9: 37–48.
[16] M. El-Shimy. Renewable Energy (2009); 34(10): 2187–96.
[17] M. Koussa, A. Cheknane, S. Hadji, M. Haddadi, S. Noureddine. Applied Energy (2011); 88: 1756–1771.
[18] E. Fuentealba, P. Ferrada, F. Araya et all. Energy Conversion and Management 95 (2015) 181-186.
[19] P. Ferrada, F. Araya, A. Marzo, E. Fuentealba. Solar Energy 2(015) 114: 356–363.

LOW COST MAXIMUM POWER POINT TRACKER REPLACES BYPASS DIODE

T. Czarnecki, A. Schneck, R. Merz*
University of Applied Science Karlsruhe
Moltkestraße 30, 76133 Karlsruhe, Germany
*corresponding author: rainer.merz@hs-karlsruhe.de

ABSTRACT: Photovoltaic strings are sensitive to partly shades. In order to prevent power losses due to limitation of string current, string inverters activate bypass diodes and short circuit shaded modules and the power of partly shaded module becomes lost. Using power optimizers or module inverters rises number of devices and increases costs and failure probability. The paper presents the Low Cost Module Maximum Power Point Tracker (LCMPPT). The low cost maximum power point tracker only operates in case of the module is shaded, without shade the device is inactive and prevents losses. Therefore, only modules which potentially became shaded need the LCMPPT, modules without potential shade do not need the LCMPPT. The LCMPPT does not need any communication, and its activated by global maximum power point search of string inverters. When the shade disappears the LCMPPT stops operation and the module directly provides its power to the string without losses due to limited efficiency of the LCMPPT.

Keywords: module maximum power point tracker, partly shade, photovoltaic string, bypass diode, string loss

1 INTRODUCTION

Globally, the installed photovoltaic (PV) capacity rises by about 40 - 50 GWp/a the last years [1]. Usually a PV-system consists of a number N_M of modules connected in series, forming a PV-string. Additionally, a number N_{ST} of strings connected in parallel rises the output power of a PV-system. A partial shade by objects in the near environment lowers the current I_{MY} of a number N_Y of shaded modules M_Y ($1 \leq Y \leq N_M$). Due to the series connection of PV-modules, shaded modules limit the current of all N_M modules of the string. The module with the lowest current $I_{MY,MIN}$ limits the string current $I_{St} = I_{MY,MIN}$. The limitation of the string current prohibits to operate the number N_Z ($Z = 1 \ldots N, Z \neq Y$) of non-shaded modules at their maximum power point and results in significant power losses of the whole string.

In order to minimize power losses due to partial shades, every module uses up to three bypass diodes (BP) connected in parallel to a typical number $N_{CBP} = 20$ of solar cells. In case of partial shades, the diodes short circuit the number N_{CBP} of solar cells in order to rise the string current $I_{St} > I_{MX,MIN}$ [2].

Modern technologies like maximum power point (MPP) tracker (MPPT) or module inverter at each module are able to operate each module M_N at its individual MPP_{MN}. Unfortunately, module MPPTs or inverters at each module rise system costs and failure probability due to the huge number N_M of devices [3]. Additionally state of the art devices need some communication in order to control the PV-String.

This contribution presents the Low Cost Maximum Power Point Tracker (LCMPPT), which individually operates the shaded module at the MPP only if the module limits the string current and rises the current of the string to the MPP current $I_{MZ,MPP}$ of non-shaded modules. In order to rise lifetime and to reduce power losses, the LCMPPT is inactive without any shade. Additionally, only for modules which potentially become shaded, the LCMPPT replaces the bypass diode. Modules, known from shadow simulation of the near environment, without the possibility to become shaded, do not need a LCMPPT device and the number of devices is as low as the number of potentially shaded modules N_Y of the string. Therefore, the use of LCMPPT will rise the annual energy yield of PV-systems with partial shades compared to bypass diodes. Compared to the use of a number N_M of state of the art module MPPTs or module inverters, using a number $N_Y < N_M$ of LCMPPT saves costs.

First chapter 2 analyses shading effects to output power of PV-strings and chapter 3 introduces the LCMPPT and proves that modules without probability to become shaded do not need an additionally LCMPPT, and are protected only by bypass diodes. Chapter 4 presents an experimental proof of concept and compares the gain of output power P_{ST} using LCMPPT of a string of two modules without shade and one module equipped with a LCMPPT device.

2 PV-STRING WITH SHADED MODULES

2.1 Shaded and unshaded modules

The current I_{MX} of PV-modules M_X significantly depends on irradiation compared to voltage V_M. Objects in near environment result partly shades, and limit I_{MX}.

Figure 1 compares the output currents of ideally modeled modules with and without bypass diodes in respect to their output voltage V_{MX} for two different illuminations. The current I_{M50BP} represents a module M_{50BP} with commonly used bypass diodes in parallel to all the number $N_{CBP} = 60$ cells and its shaded by 50% compared to the current I_{M100BP} of a module M_{100BP} without shade. The current I_{M100} represents a 100 % illuminated module and I_{M50} a module, which is shaded by 50 % without bypass diodes. The reduced illumination results in current $I_{M50BP} = 50\ \%\ I_{M100BP}$ and $I_{M50} = 50\ \%\ I_{M100}$ compared to the 100 % illuminated module. Due to the ideal model neglecting series and parallel resistance and breakdown voltage, the short circuit currents $I_{M50,SC}$ and $I_{M100,SC}$ limit the current $I_{M50}(V_M < 0\ V) = I_{M50,SC}$ and $I_{M100}(V_M < 0\ V) = I_{M100,SC}$ also for reverse voltage $V_M < 0$ V. Adding bypass diodes to the modules results in the currents I_{M50BP} for a 50 % shaded module and I_{M100BP} for a module without shade. At module voltages $V_M > 0$ V the bypass diodes are inactive. Operating the module at voltages $V_M < 0$ V the bypass diode becomes active and limits the reverse voltage $V_M > -1$ V. Commonly PV-modules are equipped with bypass diodes in order to prevent hot spot effects due to break down voltage $-30\ V < V_B < -5\ V$ of solar cells [4].

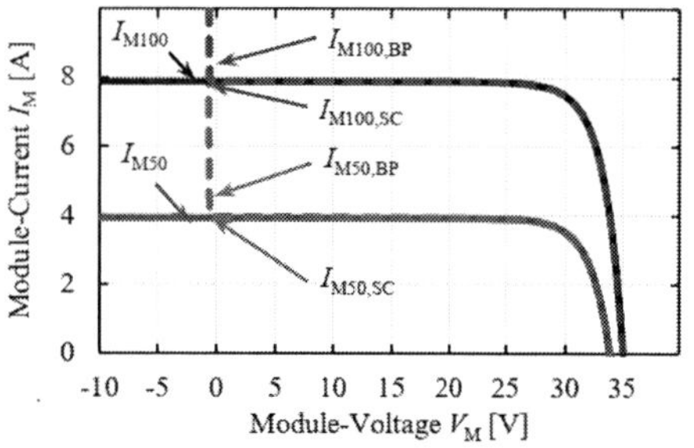

Figure 1: Module current I_{M50} of 50 % illuminated module without and I_{M50} with bypass diode. I_{M100} 100 % illuminated module without and I_{M100BP} with bypass diode. Without bypass diode short circuit currents $I_{SC,50}$ and $I_{SC,100}$ limit $I_{M50} \leq I_{SC50}$ and $I_{M100} \leq I_{SC100}$. With bypass diodes there is no limit.

2.2 Module string without bypass diode

Figure 2a and 2b draw the schematic of PV-strings with a number $N = 3$ of modules M_N in series. The series connection of modules forces the module current $I_{MX} = I_{ST}$ of the module X ($1 \leq X \leq N$) to match the string current I_{ST} and the string voltage $V_{ST} = V_{M1} + \ldots + V_N$ increases with number N of modules. In order to maximize the string power, commonly a string MPPT will adjust $I_{ST} = I_{X,MPP}$ to operate the modules M_X at their common MPP-current $I_{ST} = I_{X,MPP} = I_{ST,MPP}$.

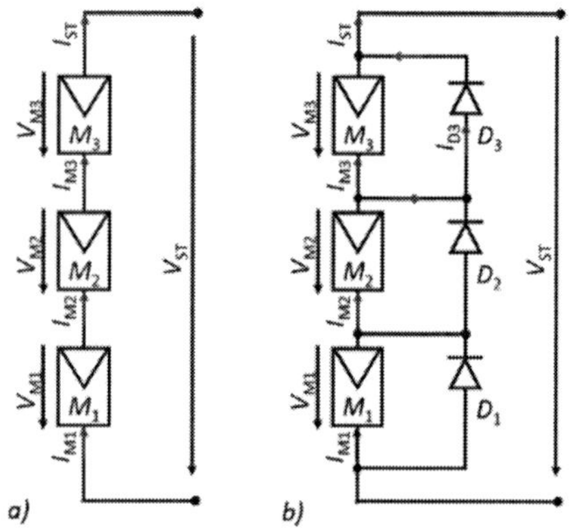

Figure 2: a) Three PV-modules M_X (X = 1..3) connected in series forming a string with voltage $V_{ST} = V_{M1} + V_{M2} + V_{M3}$. Currents $I_{MX} = I_{ST}$ match string current I_{ST} b) Three PV-modules with antiparallel bypass diodes D_X in series connection. Bypass diodes D_X prevent limit of string current $I_{ST} = I_{M3} + I_{D3}$ in case M_3 becomes partly shaded.

Figure 2a) does not use bypass diodes. If one module M_Y ($1 \leq Y \leq N = 3$) becomes shaded and limits the current I_{ST}, the modules M_Z (Z = 1 … N, Z ≠ Y) without shade are not able to operate at their maximum power point current $I_{Z,MPP}$ or voltage $V_{MPP,Z}$. Figure 2b) uses bypass diodes D_N connected parallel to each module in reverse direction. As long as the module voltages $V_{MX} > 0$ V, the bypass diodes D_X of the modules operate in reverse direction. Neglecting the saturation currents of the bypass diodes, the current I_{MX} matches the string current I_{ST}. If one module e.g. M_3 becomes shaded and the voltage $V_{M3} > 0$ V prevents a current $I_{BP3} \approx 0$ A, I_{M3} limits the current $I_{ST} = I_{M3}$. If the module voltage $V_{M3} < 0$ V drops, the diode D_3 operates in forward direction and the current $I_{D3} + I_{M3} = I_{ST}$ allows $I_{ST} = I_{MZ}$ to match the maximum power point current $I_{MZ} = I_{Z,MPP}$ of the non-shaded modules M_Z.

2.3 String power

Figure 3 compares the output power P_{M50} of modules M_Y shaded by 50% to the output power P_{M100} of modules M_Z without shade. In matters of fig 2a a string of three non-shaded modules result in a string power P_{ST100} providing a maximum output power $P_{ST100,MPP}$. If one module is shaded by 50% the string power P_{ST50} significantly drops and the possible maximum string power $P_{ST50,MPP} \approx 50$ % $P_{ST100,MPP}$ is as low as the whole string would be shaded by 50%.

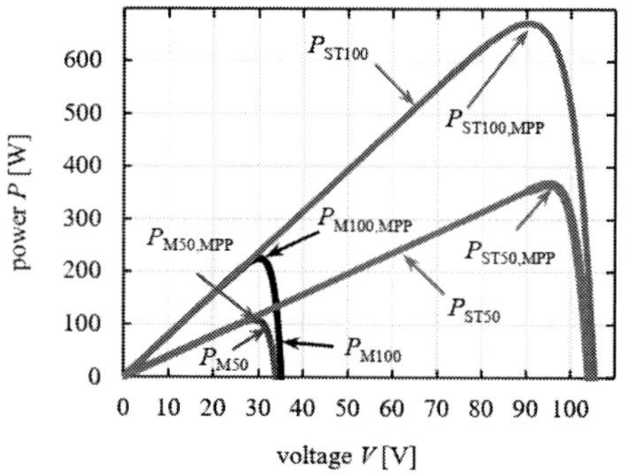

Figure 3: Power P_{M100}, with maximum $P_{M100,MPP}$, of 100 % and P_{M50}, with maximum $P_{M50,MPP}$, 50 % illuminated module. Three modules in series, P_{ST100}, with maximum $P_{ST100,MPP}$, all modules 100 % illuminated and $P_{ST,50}$, with maximum $P_{ST100,MPP}$ two modules 100 % and one module 50 % illumination.

Figure 4 calculates the module power P_{M50BP} of shaded modules $M_{Y,BP}$, equipped with bypass diodes D_Y and P_{M100BP} of non-shaded modules $M_{Z,BP}$ with bypass diodes D_Z. Referring to fig 1) in case of module voltages $V_{MN} > 0$ V, the modules $M_{Y,BP}$ and $M_{Z,BP}$ with bypass diodes do not affect the output power P_{M50BP} and P_{M100BP} of the modules. In matters of fig 2b connecting three modules in series results in an output power $P_{ST100BP}$. When no module is shaded, the bypass diodes do not limit the maximum power $P_{100BP,MPP} = P_{M100MPP}$. In case one module M_Y (Y = 3) becomes shaded, while $V_{MY} > 0$ V, the current $I_{MY} = I_{ST}$ limits the string current and limits the string power to $P_{LOCAL,MPP}$. Reducing the string voltage V_{ST} reduces the module voltage $V_{MY} < 0$ V and the current I_{DY} of the bypass diode D_Y rises the string current $I_{ST} = I_{MY} + I_{DY}$ and the power of the string increases to $P_{ST50BP,MPP} > P_{LOCAL,MPP}$.

To reduce power losses due to the shaded modules, state of the art modules use bypass diodes connected in parallel to the modules as described in fig 2b. The string inverter activates the bypass diodes of shaded modules, like module M_3, by doing a sweep measurement of the power voltage characteristic of the string [5,6,7]. Unfortunately, active bypass diodes operate modules at short circuit conditions and they provide no power to the string.

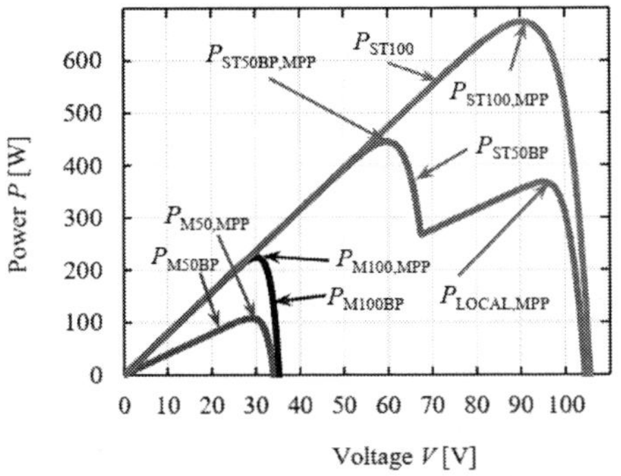

Figure 4: Power P_{M100BP}, with maximum $P_{M100BP,MPP}$, of 100 % illuminated module with bypass diode D, and P_{M50BP}, with maximum $P_{M100,MPP}$, 50 % illuminated with bypass diode D. Three modules in series, $P_{ST100BP}$, with maximum $P_{ST100,MPP}$, all modules 100 % illuminated and $P_{LOCAL,MPP}$ local maximum $P_{ST,MPP50}$ and global maximum $P_{ST50BP,MPP}$ with bypass diode active for two modules 100 % and one module 50 % illumination.

3 OPERATION OF THE LCMPPT

In order to prevent short circuit conditions of partly shaded modules the local maximum power point tracker operates partly shaded modules at their individual MPP. Without shade the LCMPPT is in standby mode and the module current $I_{MY} = I_{ST}$ matches the string current I_{ST}. In case of a shade, the string inverter activates the LCMPPT during Global MPP Tracking Sweep Measurement (GSM). The LCMPPT automatically will turn OFF when the shade disappears. Due to the LCMPPT only operates during partly shading conditions, without the need of communication, it is possible to add the LCMPPT only to modules M_Y which are potentially become shaded. In order to save costs modules M_Z without possibility to become partly shaded by the environment, are protected by common used bypass diodes D_Z.

3.1 Local maximum power point tracker

Figure 5 draws the schematic of the LCMPPT, which bases on the topology of a buck converter. The switch S_1, the diode D and the inductor L_1 form the buck converter. A buck converter adjusts an output voltage $V_{OUT} = dV_{MY}$ in respect to its input voltage V_{MY} varying the duty cycle $d = T_{ON}/(T_{ON}+T_{OFF})$ of a pulse width modulated (PWM) signal. Due to the PWM signal the switch S_1 turns ON for a period of time T_{ON} or OFF for the period of time T_{OFF}. Capacitor C_1 buffers the input voltage V_{MY} and C_2 the output voltage V_{OUT}. Without any shade the currents I_{MY} of modules equipped with a LCMPPT match the direct current (DC) $I_{OUT} = I_{ST}$. In this case the duty cycle $d = 100$ % turns switch S_1 continuously ON and the diode D_1 operates in same way as bypass diodes as described in fig 2b. The closed switch S_1 applies the voltage V_{MY} of the PV-module to the diode D and prevents a current I_D. Due to direct current (DC) conditions $I_{MY} = I_{ST}$, and neglecting the series resistance R_L of L_1, the inductor L_1 does not affect the module current I_{MY} or voltage V_{MY}. In case the module becomes shaded and the string inverter activates the diode D, the voltage $V_{OUT} < 0$ V $< V_{TH}$ drops below a threshold voltage V_{TH} and activates the MPPT of the LCMPPT.

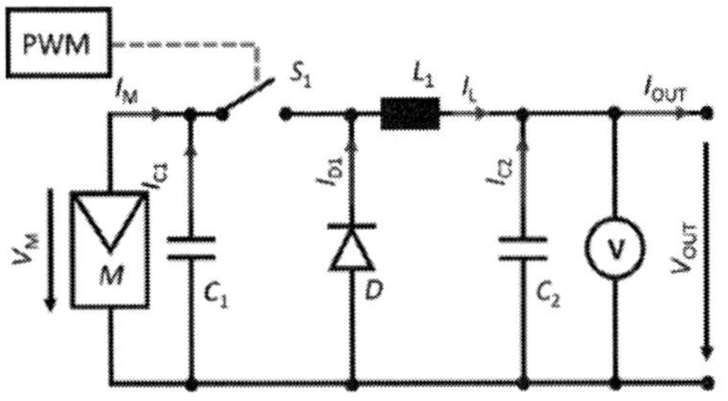

Figure 5: Schematic of LCMPPT. C_1 and C_2 stabilize V_M and V_{OUT}. S_1, L_1 and D forming a buck converter. A PWM signal with duty cycle d boosts I_M to I_{OUT}. Voltage measurement for MPPT

Figure 6 explains the state machine of the LCMPPT. For $V_{OUT} < V_{TH}$ the LCMPPT first switches S_1 OFF and sets the duty cycle $d = d_{MIN}$ to an initial minimum $d_{MIN} << 1$ and the current $I_M = -I_{C1}$ charges C_1. Charging C_1 increases the voltage V_{MY} and buffers the output power $P_{MY} = V_{MY}I_{MY}$ of the module M_Y. During S_1 is switched OFF, the current $I_{D1} = I_L = I_{ST}$ of diode D matches the current I_L of the inductor L_1 and the string current I_{ST}. Controlled by the duty cycle $d = T_{ON}/(T_{ON}+T_{OFF})$ of PWM signal, after a period of time T_{OFF} the switch S_1 turns ON again. During the period of time T_{ON}, the voltage $V_{MY} = V_{D1}$ turns OFF the diode D and the current $I_L = I_{C1} + I_{MY}$ discharges C_1.

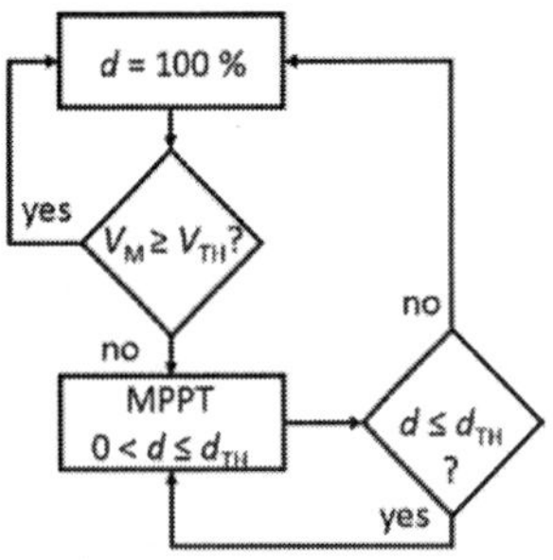

Figure 6: Activation algorithm of MPP tracking. If $V_{MY} < V_{TH}$, MPPT activates. When the shade disappears $d > d_{TH}$ disables the MPPT and $d = 100$ connects MY in series to the modules M_Z.

The transfer function $V_{OUT}(V_{MY})^{-1} = I_{MY}(I_{ST})^{-1} = d$ of the buck converter offers the possibility, that a maximum power point tracker algorithm adjusts the duty cycle d to $d = d_{MPP} = V_{OUT}/V_{MY,MPP}$ to the optimum duty cycle d_{MPP} in order to operate the module M_Y at its individual maximum power point current $I_{MY,MPP} = dI_{ST}$. Neglecting the efficiency $\eta_{LCMPPT} < 1$ of the LCMPP, the modules M_Y add their MPP power $P_{MY,MPP}$ to the string power P_{ST}. In order to maximize the string power, $P_{ST} = \Sigma P_{MY,MPP} + \Sigma P_{MZ,MPP}$ the MPPT of the string inverter adjusts the string current $I_{ST} = I_{MZ,MPP}$ to the maximum power point current $I_{MZ,MPP}$ of the non-shaded modules. Once the MPPT of the LCMPPT operates partly shaded modules at their individual MPP, the LCMPPT continuously proves, if the shadows disappeared. In case the shade disappears the current $I_{MY} \cong I_{ST}$ of the partly shaded module increases and matches the string current I_{ST} provided by the non-shaded modules M_Z. For $I_{MY,MPP} = I_{ST,MPP}$ the MPPT increases the duty cycle

$d = I_{MY}(I_{ST})^{-1} \cong 1$. Detecting a duty cycle $d > d_{TH}$ close to 100 %, the MPPT of the LCMPPT stops operation and sets d = 100 % in order to permanently connect the module M_Y in series to the modules M_Z of the string.

Figure 7 depicts the series connection of potentially be shaded modules M_Y (Y=3), equipped with a LCMPPT introduced by fig. 5, in series with two non-shaded modules M_Z (Z = 1, 2) protected by commonly used bypass diodes D_1 and D_2.

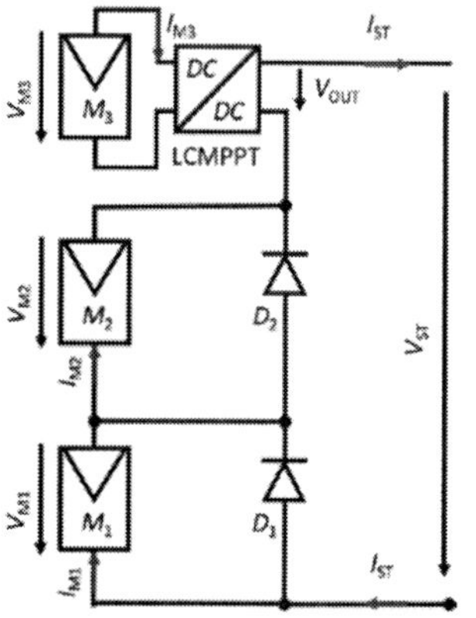

Figure 7: Module M_3 equipped with LCMPPT connected in series to M_1 and M_2 with bypass diodes D_1 and D_2 result in String voltage V_{ST}. In case of shade LCMPPT bucks V_{M3} to V_{OUT} and boosts I_{M3} to match $I_{M1} = I_{M2} = I_{ST}$ of modules M1 and M2, protected by diodes D_1 and D_2.

Due to the MPPT of the LCMPPT stopped operation and the series connection of modules M_Y in series to the modules M_Z, the MPPT of the string inverter adjusts the voltage $V_{ST} = V_{M1} + V_{M2} + V_{OUT} = V_{ST,MPPT}$ and operates the string and also the modules M_Y at MPP. If the party shade of the modules M_Y recurs, the GSM of the string inverter again activates the diode D and the LCMPPT again starts individual maximum power point tracking of module M_Y.

4 EXPERIMENTAL RESULTS

With respect to fig. 7 an experiment proves the use of the LCMPPT technology. The experimental setup uses three modules consisting of 20 solar cells connected in series. Two modules M_1 and M_2 are equipped with bypass diodes D_1 and D_2 and one uses a LCMPPT instead of bypass diodes. Due to laboratory conditions, power LEDs homogeneously illuminate the three modules. A first experiment varies the string voltage V_{ST} without any shade and measures the power $P_{ST,100}$ of the string. A second experiment partly shades module M_3 by 50% In order to reduce the short circuit current $I_{M3,SC} \approx 50$ % $I_{M1,SC} = 50$ % $I_{M1,SC}$ of module M_3 by 50 %.

Figure 8 compares the measured power $P_{ST50,BP}$ of a string using three modules M_1, M_2 and M_3 with bypass diodes to the power P_{LCMPPT} of a string with one module M_3 equipped with the LCMPPT and two modules M_1 and M_2 protected by bypass diodes D_1 and D_2 introduced by fig. 7. Emulation the GSM-measurement of a string inverter the experiment starts measurement at open circuit voltage $V_{ST} = V_{OC} = 32.76$ V. During the sweep measurement to find the MPP of the partly shaded string, the string power P_{LCMPPT} rises to the local-MPP power $P_{LOCAL,MPP} = 10.9$ W. By tracking string voltage V_{ST} in downwards direction, V_{ST} continuously decreases, and the

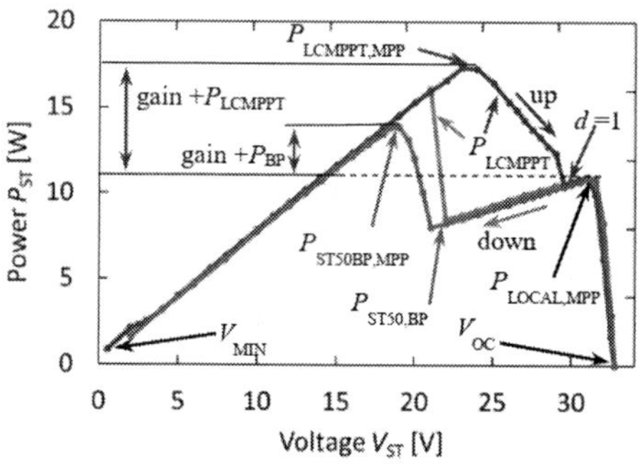

Figure 8: String power $P_{ST50,BP}$ of three modules with bypass diodes, one partly shaded. Bypass diodes rises MPP-power from $P_{MPP,local}$ to $P_{MPP,BP}$ and results power gain $+P_{BP}$. P_{LCMPPT} with LCMPPT during sweep measurement starting at V_{OC}. LCMPPT inactive results $P_{LOCAL,MPP}$, activated string power rises. Starting measurement at V_{MIN} with activated LCMPPT results a continuous power function with maximum power $P_{LCMPPT,MPP}$. LCMPPT rises string performance from $P_{local,MPP}$ to $P_{LCMPPT,MPP}$ and results in power gain $+P_{LCMPP}$ compared to $+P_{BP}$ using bypass diodes to increase power to $P_{ST50BP,MPP}$. $d = 1$ indicates switching OFF of LCMPPT at low string current.

diode D of the LCMPPT, explained by fig. 5, activates at a voltage $V_{ST} = 22$ V and operates M_3 at short circuit voltage $V_{M3} \leq 0$ V. Triggered by $V_{M3} < V_{TH}$ the LCMPPT starts MPP-tracking and operates the shaded module M_3 at MPP and the string power immediately rises. The GSM of the string inverter decreases V_{ST} to minimum voltage $V_{MIN} = V_{ST} = 0.6$ V. After $V_{ST} = V_{MIN}$ the GSM increases the string voltage V_{ST} in order to emulate the MPPT of a string inverter. Usually now the string inverter adjusts the string voltage $V_{ST} = V_{ST,MPP}$ to the detected maximum power point voltage $V_{ST,MPP}$ and starts a MPPT algorithm. Due to the active LCMPPT increasing V_{ST} from $V_{ST} = V_{MIN}$ towards V_{OC} the power P_{LCMPPT} of the string continuously increases to $P_{LCMPPT} = 17.4$ W > $P_{ST50,BP,MPP}$. Increasing the string voltage to $V_{ST} > V_{ST,MPP}$ reduces the string current $I_{ST} = I_{M3,SC}$ to match the short circuit current of module M_3. For $I_{ST} = I_{M3,SC}$ the duty cycle $d > d_{TH}$ of the MPPT of the LCMPPT turns of the LCMPPT and sets d = 100 %.

Comparing $P_{LOCAL,MPP}$ to $P_{LCMPPT,MPP}$ results in power gain $+P_{LCMPPT} = 6.5$ W compared to the power gain $+P_{BP} = 3$ W of $P_{ST50,BP}$ using bypass diodes. The additional power $+P = P_{LCMPPT,MPP} - P_{ST50BP,MPP}$ proves the for surplus of power using LCMPPTs instead of bypass diodes of the LCMPP. Due to the continuous function of the string power P_{LCMPPT} with activated LCMPPT state of the art MPPT are able to track the string to maximum output power.

5 CONCLUSION

The contribution introduced topology, the operation and the system integration of the Low Cost Maximum Power Point Tracker (LCMPPT). The LCMPPT prevents losses of unshaded modules in a string due to shaded modules. The LCMPPT only operates in case the module equipped with the LCMPPT is shaded, without shade the LCMPPT is inactive. Due to the LCMPPT only operates in case a shadow limits the current of a module, only modules which potentially became shaded need the LCMPPT. Modules without potential shade do not need to use a LCMPPT. In order to activate the LCMPPT in case of a shade at the module, the global MPPT search of the string inverter results a voltage drop and the LCMPPT tracks the shaded module to its individual maximum power point. In case the shade disappears, the LCMPPT detects that the MPP current of the shaded module matches the string current and stops operation. The LCMPPT does not need any communication and operates in combination with state of the art string inverters. First experiments result that the LCMPPT prevents power losses due to bypassing shaded modules and provides a maximum string power close to the theoretical power.

6 REFERENCES

[1] Adrian Whiteman et. al., Renewable Capacity Statistics 2016, (IRENA, 3.2016) p. 29

[2] E. Karatepe, M. Boztepe, M. Colak, Development of a suitable model for characterizing photovoltaic arrays with shaded solar cells, Solar Energy 81, (2007) p. 977–992

[3] S. Harb and R. S. Balog, Reliability of Candidate Photovoltaic Module-Integrated-Inverter (PV-MII) Topologies—A Usage Model Approach, *IEEE Trans. Power Electronics*, 28, 6, (2013) p. 3019 doi:10.1109/TPEL.2012.2222447

[4] M.C. Alonso-García, J.M. Ruíz, Analysis and modelling the reverse characteristic of photovoltaic cells, Solar Energy Materials and Solar Cells 90, (2006) p. 1105-1120 DOI: 10.1016/j.solmat.2005.06.006

[5] COOLCEPT, http://www.steca.com/index.php?coolcept-StecaGrid-1500-4200-de

[6] Datasheed, Fronius Primo3.0-1, www.fronius.com (20.6.2016)

[7] SMA, Keine macht den Schatten, http://www.sma.de/partner/expertenwissen/keine-macht-den-schatten.html

POWER BALANCE CONTROL FOR A TWO-STAGE SOLAR INVERTER WITH LOW VOLTAGE RIDE THROUGH CAPABILITY

Georgios KAMPITSIS(1), Efstratios BATZELIS(2) and Stavros PAPATHANASSIOU(3)

National Technical University of Athens, Electric Power Division, 9 Iroon Polytechneiou st., 15780, Athens, Greece
(1)gkampit@central.ntua.gr, (2)batzelis@mail.ntua.gr, (3)st@power.ece.ntua.gr

ABSTRACT: The latest grid codes require the renewable energy sources (RES) to provide ancillary services during fault and post fault conditions. More specifically, in case of a short-duration voltage dip, the grid-tied photovoltaic (PV) system should stay connected and support the grid by injecting reactive power. However, meeting these requirements during voltage sags is a challenge for two-stage systems, due to the power imbalance between the dc/dc converter and the inverter, resulting in dc-link voltage excursions and output current overshoots. In this paper, a power balance control scheme is proposed, by which, a successful low voltage ride through (LVRT) and smooth dc-link voltage variation are achieved, while the output current is kept within the predefined limits. Two reactive power injection strategies are investigated that exhibit different dynamic response during voltage sags. The effectiveness of the proposed LVRT control is verified though simulations of a 2 kVA solar system.
Keywords: Grid code, grid connected inverter, low voltage ride through (LVRT), photovoltaic (PV), power control, voltage support

1 INTRODUCTION

The high penetration level of grid-connected solar inverters raises concern regarding their effect on grid stability and response under disturbances [1]. Although interconnection standards of the recent past (e.g. IEEE 1547 and UL1741) required disconnection of PV power plants in case of voltage sags (e.g. when the voltage at the point of common coupling (PCC) drops below a certain threshold [2]), the high PV penetration levels reached have imposed a fundamental differentiation of requirements in new grid codes, according to which:

- Distributed generation (DG) should not disconnect in case of grid faults, i.e. it should exhibit Fault Ride-Through (FRT) capability
- It should support the grid voltage by injecting reactive current during voltage depressions.

LVRT requirements are quantified via voltage-time characteristics, such as the ones shown in Figure 1 for certain grid codes, [3], [4]. The required reactive current injection as a function of the magnitude of the voltage dip is presented in Figure 2.

PV generators are interfaced to the network through either a single-stage inverter or a two-stage converter with an intermediate dc-link [5]. The later offers the possibility of transformerless interconnection to the low voltage (LV) grid, leading to higher system efficiency. In addition, two independent/decoupled controllers can be adopted [6], with the dc/dc converter tracking the maximum power point (MPP) of the PV source and the inverter being responsible for dc voltage and output current control.

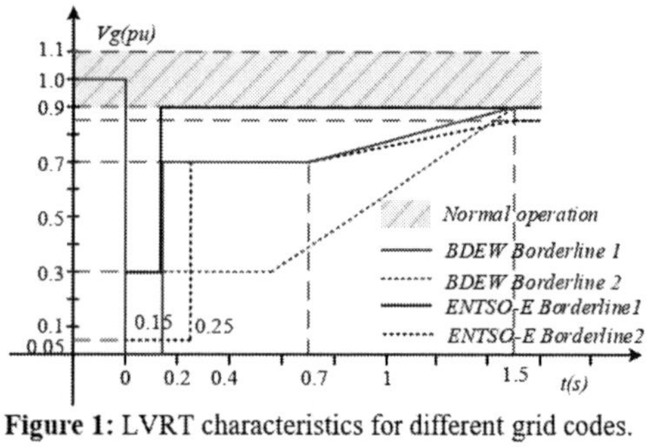

Figure 1: LVRT characteristics for different grid codes.

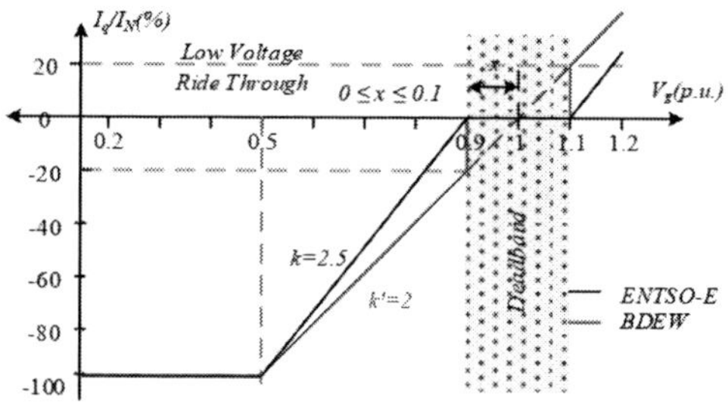

Figure 2: Reactive current injection against voltage.

A common approach for meeting the LVRT requirenents in single-stage systems is to disable the MPPT algorithm and shift the operating point of the PV generator according to a constant output current command, [7]–[9]. However, these techniques cannot be applied in a two-stage topology, due to the imbalance between the power provided by the PV generator and that fed to the grid, creating dc overvoltages. One solution would be to utilize energy storage systems (ESS) across the dc-link, such as batteries [10] or super capacitors [6], in order to absorb excess power during voltage sags. This, however, inevitably impacts cost and complexity. In [11] and [12] it is proposed that the dc/dc converter pauses MPP tracking and shifts the operating point of the PV system to a suboptimal power level. The PV current I_{PV} is reduced by the ratio of the voltage sag to the nominal voltage. However, these techniques rely on a linear approximation of the relation between PV current and power and are susceptible to the power losses of the two-stage converter.

A more sophisticated control scheme has been presented in [13], where a combination of a proportional-integral (PI) and a proportional-derivative (PD) controller is used to determine the operating point of the PV generator during faults. This is an effective, yet complex approach that requires real time gain readjustment (gain scheduling strategy) to meet the response times imposed by the grid codes.

The purpose of this work is to propose a simple and effective control scheme that meets LVRT requirements with minimum dc voltage variations, limited output current overshoot, short response time, without employing additional ESS, switching devices or sensors. Two reactive current injection strategies are investigated, namely *active current priority*, in order to effectively limit dc voltage overshoot, and *reactive current priority*, in order to minimize the response time of reactive power injection. The performance of the proposed controller is validated through simulations using Matlab/Simulink.

2 POWER BALANCE CONTROLLER

The overall diagram of the grid-tied solar inverter is shown in Figure 3, in which the power circuit is indicated in black color and measurement signals in blue. The front-end stage is a boost dc/dc converter, responsible for regulating the operating point of the PV generator. The standard perturb and observe (P&O) MPPT algorithm is applied during normal operation for tracking the MPP, while a PI controller becomes active during fault mode to shift the operating point of the PV generator in order to keep the inverter output current within limits. In all cases, the transformerless inverter maintains the dc-link voltage to its reference.

2.1 Phase locked loop

The phase locked loop (PLL) algorithm is a key feature for a successful LVRT response. A double second order generalized integrator (DSOGI) PLL is used, which cancels out harmonics at twice the line frequency, introduced by the negative sequence components, [14], [15]. The block diagram of the PLL is presented in Figure 4. This technique is based on the quadrature signal generator (QSG) equations (1) - (2) with an appropriate sequence extraction relation, as in (3), where ξ denotes phase shift of -90 deg and ω is the line frequency.

$$H_d = \frac{V_{out}}{V_{in}} = \frac{k \cdot \omega \cdot s}{s^2 + k \cdot \omega \cdot s + \omega^2} \quad (1)$$

$$H_q = \frac{\xi V_{out}}{V_{in}} = \frac{k \cdot \omega^2}{s^2 + k \cdot \omega \cdot s + \omega^2} \quad (2)$$

$$\begin{bmatrix} V_\alpha^+ \\ V_\beta^+ \end{bmatrix} = \frac{1}{2} \cdot \begin{bmatrix} 1 & -\xi \\ \xi & 1 \end{bmatrix} \cdot \begin{bmatrix} V_\alpha \\ V_\beta \end{bmatrix} \quad (3)$$

2.2 Inverter current controller

A proportional resonant (PR) regulator is selected for the inverter inner current control loop [13], which has significant gain around the line frequency and sufficient attenuation at higher frequencies. The block diagram of the current control is illustrated in Figure 5. The PR controller is described by the transfer function (4).

$$H_{PR}(s) = K_p + K_i \frac{s}{s^2 + \omega_0^2} \quad (4)$$

Alternativelly, the integral term of the PR controller can be expressed by (1), in which case, an adjustable gain at the resonant frequency is obtained, [16]. The grid voltage in the stationary reference frame is added to the output of the PR controller to perform a feed-forward loop that enhances the controller response, [17] (Figure 5).

A common PI controller regulates the power to be transferred to the grid, P_g, in order to keep V_{dc} at its reference, as depicted in Figure 6. Given that the q component of the grid voltage is zero via the PLL [18], the active current is given by (5).

$$i_{d\_PI} = \frac{2}{3} \cdot \frac{P_g}{V_d} \quad (5)$$

In Figure 6, the output of the PI controller is saturated to a maximum power level determined by the actual grid voltage and the maximum permissible current I_{max}, according to (6). Thus, i_{d_PI} cannot surpass I_{max}.

$$P_{max} = \frac{3}{2} \cdot V_d \cdot I_{max} \quad (6)$$

2.3 Dc/dc converter power controller

Figure 7 illustrates the proposed power balance controller of the front end converter. The voltage sag detection mechanism is based on continuous monitoring of the d component of the grid voltage. While V_d is greater than 90 % of the nominal value (normal operation), the P&O algorithm is executed, otherwise the MPP tracking is paused and the PI controller is activated to ensure operation at a reference power, P_{PV}^*.

The reference power is calculated as the minimum of *the pre-fault power at MPP*, P_{mp}, and a limit to the permissible PV power, P_{lim}, given in (7), where i_{q_LVRT} is the reactive current determined in Figure 2.

$$P_{lim} = \frac{3}{2} \cdot V_d \cdot \sqrt{I_{max}^2 - i_{q\_LVRT}^2} \quad (7)$$

According to this reactive power injection (RPI) strategy, the operating point of the PV generator is shifted in order to ensure that the output current magnitude, i_s, never exceeds I_{max}. The PI controller remains active after fault clearance, until P_{PV} is restored to its pre-fault level.

It is worth noting that the new suboptimal power

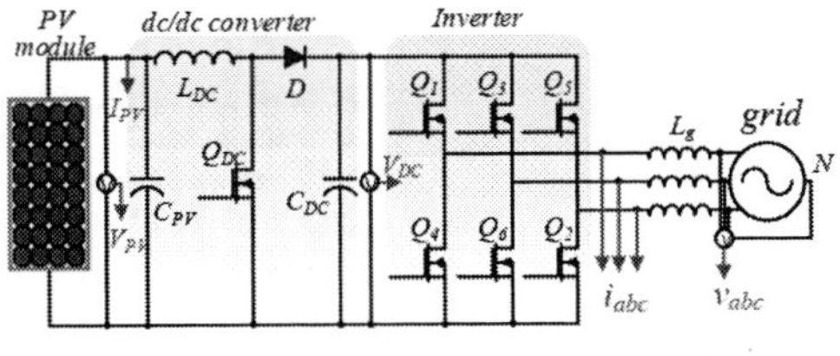

Figure 3: Schematic diagram of a two-stage PV inverter.

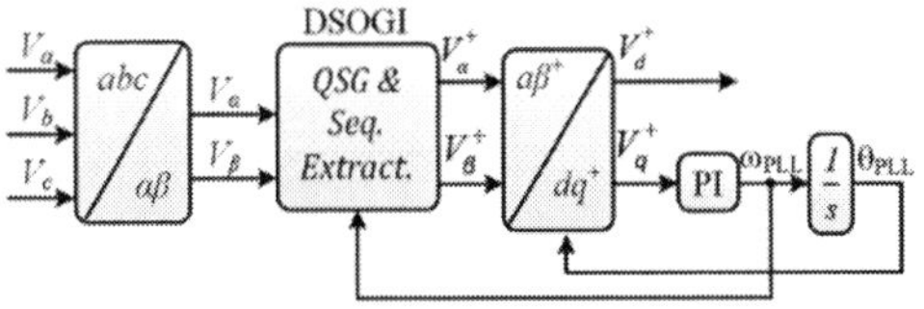

Figure 4: Block diagram of the PLL.

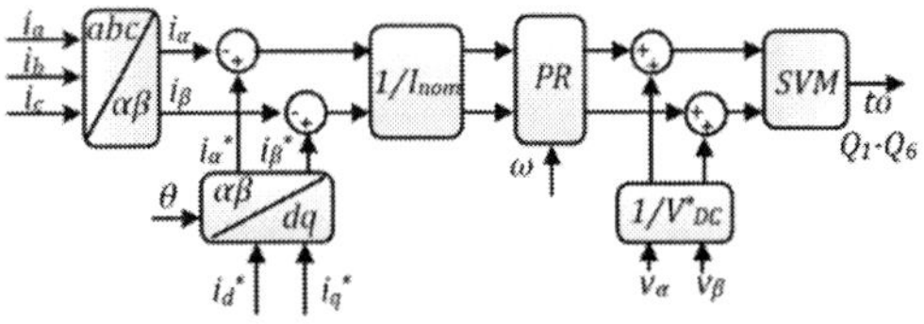

Figure 5: Block diagram of the inverter current control.

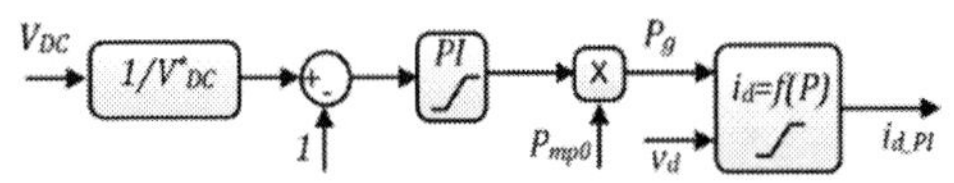

Figure 6: Control scheme of the dc-link voltage controller.

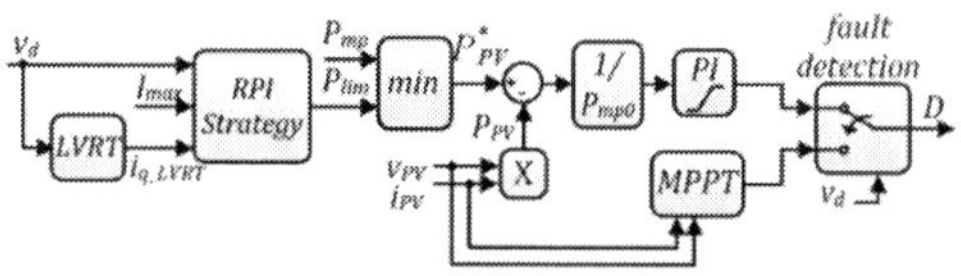

Figure 7: Power balance controller of the dc/dc converter.

level corresponds to a PV voltage greater than that at the MPP (right hand side of the *P-V* curve). This operation is preferable for several researchers compared to the left part of the characterisitc [19], [20]. The transition is controllable by adjusting the maximum allowable duty-cycle step and the gains of the PI controller, thus avoiding surge currents in the power device, [12]. Throughout this process, the inverter control remains unaffected. However, the actual power fed to the grid is slightly lower than P_{lim}, due to the system power losses.

Concerning the calculation of the reference currents, i_d^* and i_q^*, two alternatives are proposed and evaluated, exhibiting different response during sag transients.

- *Case A: active current priority*

In this case, the active and reactive current references are given in (8). During the voltage sag transient, the inverter provides the active current needed to contain V_{dc} variations, setting the active reference current i_d^* (Figure 5) equal to the output of the dc-link controller i_{d_PI} (Figure 6). On the contrary, the reactive reference current i_q^* is saturated in order to keep the total current within limits, essentially giving a lower priority to the reactive power injection.

$$\begin{cases} i_d^* = i_{d\_PI} \\ i_q^* = min\left(i_{q\_LVRT}, \sqrt{I_{max}^2 - i_d^{*2}}\right) \end{cases} \quad (8)$$

- *Case B: reactive current priority*

The (i_d^*, i_q^*) references are now determined by (9). The provision of the required i_{q_LVRT} by the inverter is prioritized, setting the reactive reference current i_q^* (Figure 5) directly equal to i_{q_LVRT} as determind via Figure 2. On the other hand, the active current reference i_d^* is now saturated to keep i_s within limits. This approach may lead to increased V_{dc} overshoot.

$$\begin{cases} i_d^* = min\left(i_{d\_PI}, \sqrt{I_{max}^2 - i_q^{*2}}\right) \\ i_q^* = i_{q\_LVRT} \end{cases} \quad (9)$$

The main advantage of the proposed control technique, compared to previous works, is the robustness and ease of implementation of the dc/dc converter controller. The reference power during voltage sags is directly calculated, in contrast to earlier works [13], where the power set point is indirectly applied, based on the dc voltage variation, through additional controllers that increase the complexity of the system. Further, with the developed strategy the inverter control remains unaffected by grid voltage sags, resulting in smoother fault and post-fault transients. No gain scheduling is required, rendering the proposed scheme suitable for implementation in a digital signal processor (DSP).

3 SIMULATION RESULTS

The effectiveness of the proposed controller is evaluated through simulation of a 2 kVA PV inverter, connected to the LV grid. The values of the circuit parameters and the controller gains are listed in Table I. As shown, the maximum allowable current is 1.5 times the nominal one.

Table I: Simulation parameters.

C_{PV}	220 μF	$k_{P\text{-}Pow}$	0.04
C_{DC}	560 μF	$k_{I\text{-}Pow}$	40
L_{DC}	600 μH	$k_{P\text{-}DC}$	25
L_g	10 mH	$k_{I\text{-}DC}$	2500
I_{nom}	3.93 A	k_{PR}	0.05
I_{max}	$1.5 \cdot I_{nom}$	$k_{P\text{-}PR}$	0.6
P_{mp0}	1920 W_p	$k_{I\text{-}PR}$	1.8
Irrad	800 W/m²	k_{PLL}	3
Temp	25 °C	$k_{P\text{-}PLL}$	0.05
F_{sw}	20 kHz	$k_{I\text{-}PLL}$	0.2

A symmetrical 60 % voltage sag (0.4 *pu* remaining voltage) is considered, as shown in Figure 8. The responce of the DSOGI PLL technique during the voltage dip is quite satisfactory, as illustrated in Figure 9. Both active/reactive current prioritization approaches (*Case A* and *B*) are examined.

Prioritizing active current injection (*Case A*) leads to a more effective control of V_{dc}, as observed in Figure 10 (blue line), associated with the immediate increase of the active current reference, i_d^*, up to I_{max}, shown in Figure 11(a). On the other hand, a delay in reactive current injection, i_q^*, of ~20 ms is noted in Figure 11(b). With the alternative *Case B*, the required i_q^* is provided with no delay, leaving a reduced margin for i_d^* increase, which in turn, has a notable impact on V_{dc} regulation (red line in Figure 10). In both scenarios, the output current magnitude never exceeds I_{max}.

Figure 12(a) illustrates the active power produced by the PV source and injected to the grid for both cases. The PV power (yellow line) is reduced at the same rate in both examined scenarios, while the dc/dc converter effectively reaches the new set point, P^*_{PV}, (purple line) within 20 ms. An increased amount of active power is initially transferred to the grid in *Case A* (blue line) compared to *Case B* (red line), that discharges the dc-link capacitors and thus mitigating the dc voltage overshoot (blue line in Figure 10). The reactive power injected is

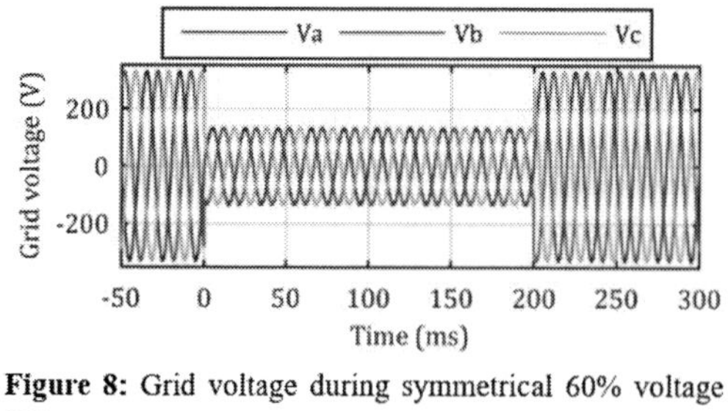

Figure 8: Grid voltage during symmetrical 60% voltage sag.

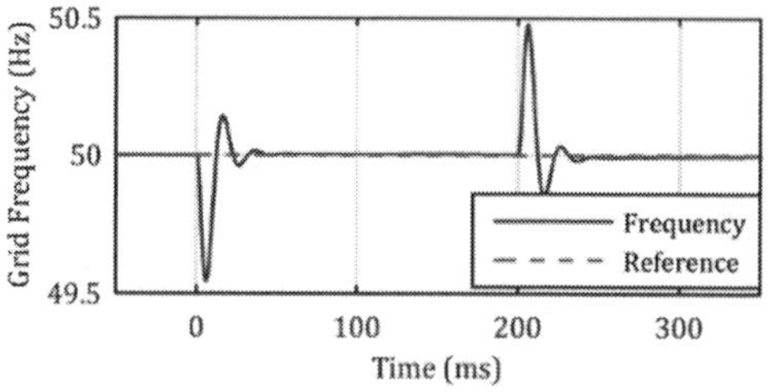

Figure 9: Responce of the DSOGI PLL algorithm.

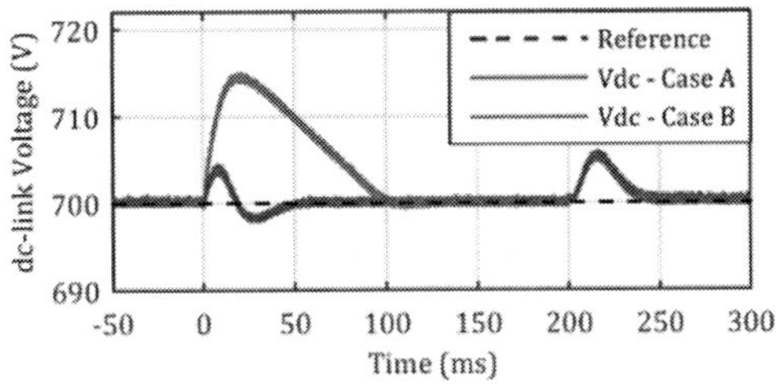

Figure 10: Dc-link voltage variation during voltage sag.

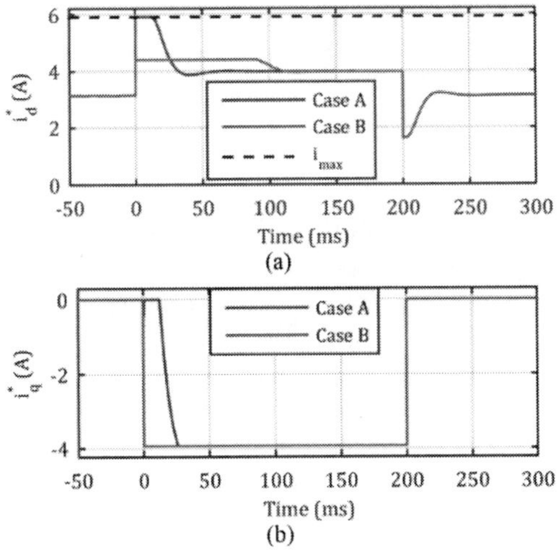

Figure 11: (a) Active and (b) reactive output current references.

shown in Figure 12(b), exhibiting a delay time of 20 ms in *Case A* as expected.

The ac-side currents are illustrated in Figure 13. It is evident that the output currents are in good agreement with the reference values of Figure 11, verifying the effectiveness of the inverter controller.

The voltage and current of the PV generator, along

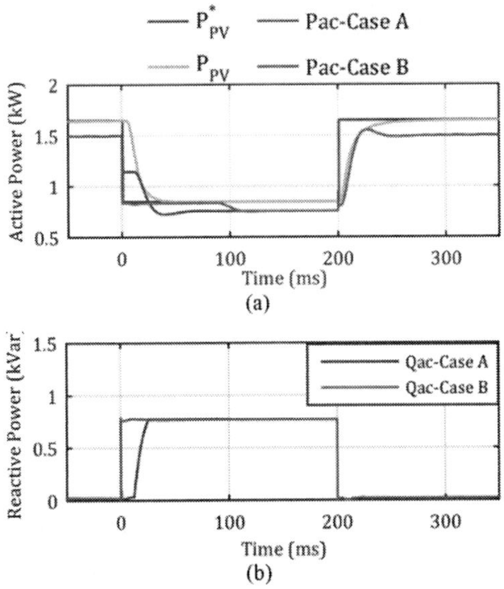

Figure 12: (a) Active and (b) reactive power at the dc and ac side.

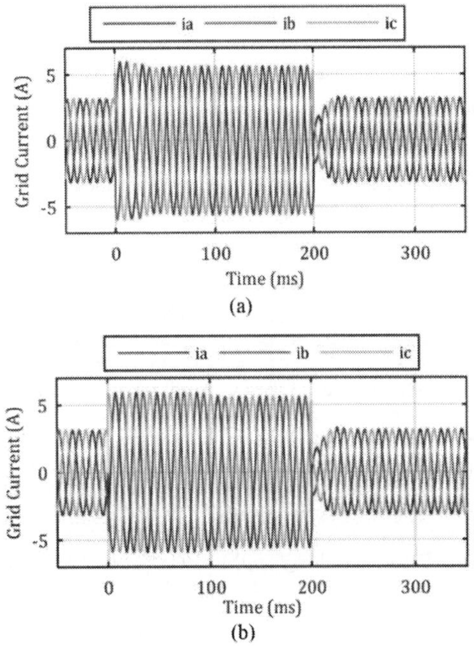

Figure 13: Output currents in (a) *Case A* and (b) *Case B*.

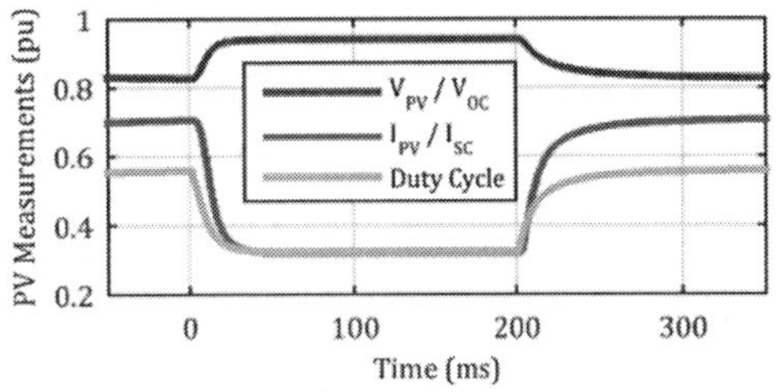

Figure 14: PV voltage and current variation along with the respective duty cycle.

with the duty cycle command of the dc/dc converter, are depicted Figure 14.

4 CONCLUSIONS

In this paper, a power balance control strategy for two-stage PV systems with LVRT capability is introduced. Low dc-link voltage variation is achieved by giving priority to the active current reference, while a faster response of reactive power is accomplished by giving priority to the reactive current reference. The developed technique does not rely on any approximations or other limiting assumptions and does not require any additional sensors. The robustness and ease of implementation renders the developed technique suitable for practical applications. Simulation results have confirmed that the inverter meets LVRT requirements, with very good overall response characteristics.

5 ACKNOWLEDGMENTS

Mr. G. Kampitsis and Mr. E. Batzelis are supported in their PhD studies by "*IKY Fellowships of Excellence for Postgraduate Studies in Greece - Siemens Program*".

7 REFERENCES

[1] A. Marinopoulos, F. Papandrea, M. Reza, S. Norrga, F. Spertino, and R. Napoli, "Grid integration aspects of large solar PV installations: LVRT capability and reactive power/voltage support requirements," in *IEEE Trondheim PowerTech*, 2011, pp. 1–8.

[2] IEEE, "Standard for Interconnecting Distributed Resources with Electric Power Systems," *IEEE Std 1547-2003*. pp. 1–27, 2003.

[3] Technical Guidline, "Generating Plants Connected to the Medium-Voltage Network." Bundesverband der Energie und Wasserwirtschaft (BDEW), 2008.

[4] Network Code, "Network Code for Requirements for Grid Connection Applicable to all Generators." European Network of Transmission System Operators for Electricity (ENTSO-E), 2012.

[5] J. L. Kirtley, M. S. El Moursi, and W. Xiao, "Fault ride through capability for grid interfacing large scale PV power plants," *IET Gener. Transm. Distrib.*, vol. 7, no. 9, pp. 1027–1036, 2013.

[6] Hao Tian, Feng Gao, and Cong Ma, "Advanced performance control of two-stage grid-tied photovoltaic inverter with fast energy storage component," in *3rd IEEE International Symposium on Power Electronics for Distributed Generation Systems (PEDG)*, 2012, pp. 403–409.

[7] X. Bao, P. Tan, F. Zhuo, and X. Yue, "Low voltage ride through control strategy for high-power grid-connected photovoltaic inverter," in *28th Annual IEEE Applied Power Electronics Conference and Exposition (APEC)*, 2013, pp. 97–100.

[8] Chen Yaai, Liu Jingdong, Zhou Jinghua, and Li Jin, "Research on the control strategy of PV grid-connected inverter upon grid fault," in *International Conference on Electrical Machines and Systems (ICEMS)*, 2013, pp. 2163–2167.

[9] P. Zhang, G. Zhang, and H. Wang, "Control strategy of low voltage ride-through for grid-connected photovoltaic inverter," in *6th International Symposium on Power Electronics for Distributed Generation Systems (PEDG)*, 2015, pp. 1–6.

[10] X. Wang, M. Yue, and E. Muljadi, "Modeling and control system design for an integrated solar generation and energy storage system with a ride-through capability," in *IEEE Energy Conversion Congress and Exposition (ECCE)*, 2012, pp. 3727–3734.

[11] Y.-S. Wu, C.-H. Chang, Y.-M. Chen, C.-W. Liu, and Y.-R. Chang, "A Current Control Strategy for Three-Phase PV Power System with Low-Voltage Ride-Through," in *9th IET International Conference on Advances in Power System Control, Operation and Management*, 2012, pp. 104–104.

[12] C. Tang, Y.-T. Chen, and Y. Chen, "PV Power System With Multi-Mode Operation and Low-Voltage Ride-Through Capability," *IEEE Trans. Ind. Electron.*, vol. 62, no. 12, pp. 7524–7533, Dec. 2015.

[13] S. I. Nanou and S. a. Papathanassiou, "Modeling of a PV system with grid code compatibility," *Electr. Power Syst. Res.*, vol. 116, pp. 301–310, 2014.

[14] F. J. Rodriguez, E. Bueno, M. Aredes, L. G. B. Rolim, F. a S. Neves, and M. C. Cavalcanti, "Discrete-time implementation of second order generalized integrators for grid converters," in *34th Annual Conference of the IEEE Industrial Electronics Society (IECON)*, 2008, no. 1, pp. 176–181.

[15] G. E. Kampitsis, A. P. Tsoumanis, K. C. Gallos, and S. A. Papathanassiou, "Experimental Investigation of the Response of Different PLL Algorithms," *Mater. Sci. Forum*, vol. 856, pp. 291–296, 2016.

[16] D. N. Zmood and D. G. Holmes, "Stationary frame current regulation of PWM inverters with zero steady-state error," *IEEE Trans. Power Electron.*, vol. 18, no. 3, pp. 814–822, May 2003.

[17] G. M. S. Azevedo, G. Vazquez, A. Luna, D. Aguilar, and A. Rolan, "Photovoltaic inverters with fault ride-through Capability," in *IEEE International Symposium on Industrial Electronics*, 2009, pp. 549–553.

[18] Y. Bae, T.-K. Vu, and R.-Y. Kim, "Implemental Control Strategy for Grid Stabilization of Grid-Connected PV System Based on German Grid Code in Symmetrical Low-to-Medium Voltage Network," *IEEE Trans. Energy Convers.*, vol. 28, no. 3, pp. 619–631, Sep. 2013.

[19] S. I. Nanou, A. G. Papakonstantinou, and S. a. Papathanassiou, "A generic model of two-stage grid-connected PV systems with primary frequency response and inertia emulation," *Electr. Power Syst. Res.*, vol. 127, pp. 186–196, Oct. 2015.

[20] H. Xin, Y. Liu, Z. Wang, D. Gan, and T. Yang, "A new frequency regulation strategy for photovoltaic systems without energy storage," *IEEE Trans. Sustain. Energy*, vol. 4, no. 4, pp. 985–993, Oct. 2013.

MODULE-LEVEL POWER ELECTRONICS: THE BUSINESS CASE FROM AN END-USER PERSPECTIVE

M.N. van den Donker, G. Verberne, K. Sinapis and W. Folkerts
Solar Energy Application Centre, High Tech Campus 21, Eindhoven, The Netherlands

ABSTRACT: Over the past decade power optimizers and micro inverters have gained a significant market share over traditional string inverter approaches. In this paper we will present a study on the business case for these Module Level Power Electronics from the perspective of a small residential end-user. A techno-financial model was constructed that calculates the net present value based on technical input data such as cost, lifetime and energy yield. Input for the model was gathered by conducting an end-user price study on PV systems and components in The Netherlands, whereas energy yield numbers were taken from a field test study and simulation model. Key variables we considered are the system architecture (string inverter or MLPE) and shading scenario (roof with no shading or heavy shading). A Monte Carlo simulation was set up in which 10.000 simulations were run within the ranges of the variables. We conclude that the traditional string inverter with a 'global mpp tracking' software gives the best economic value for the end user. However, the differences are very small and additional value divers such as monitoring, safety, reliability, O&M, degradation can easily tip the scale in the direction of MLPE. Future research should concentrate on the impact of these value drivers.
Keywords: Economic Analysis, Inverter, System Performance

1 INTRODUCTION

Over the past decade power optimizers and micro inverters have gained a significant market share over traditional string inverter approaches. As an example, Figure 1 shows the micro-inverter. Although significant research has been conducted on technical aspects, the economic aspects regarding these Module Level Power Electronics are still largely unknown. In this paper we will present a study on the business case from and end-user perspective for Module Level Power Electronics. The end-user perspective we chose is that of a residential home owner. Besides shedding light on the economic viability as such, we will also investigate what the key parameters are that influence the outcome of the business case in a sensitivity study.

The study is a result of the Dutch publically financed MLPM research project. Earlier work resulting from this project included a full year field test of three system architectures [1], accelerated lifetime study [2], and a simulation model development [3]. The knowledge on yield of partially shaded PV systems presented in these preceding papers will be used in the economic assessment of this paper.

2 TECHNO-FINANCIAL MODEL

2.1 Model design

The techno-financial model consists of three main parts. The first part holds the technical, financial and scenario input. The second part holds calculations and equations. The last part is the output, captured in key financial parameters. Figure 2 provides an overview of the main parameters that are used in the model.

The simulation was executed using the Monte Carlo method. Some input parameters were kept fixed. Other input parameters were randomly varied between a minimum and maximum expected value in order to gain insight in the expected uncertainty levels and sensitive parameters.

2.2 Definition of the Input Parameters

Installed power refers to a certain PV system size in kWp.

Annual yield refers to the expected kilowatt-hours per kilowatt-peak installed per year.

Module Degradation refers to the degradation of the PV system and is expressed in %/a of yield loss.

CAPEX (capital expenses) refers to the investment costs for the PV system.

OPEX (operational expenses) refers to the costs for

Figure 1: The Micro-Inverter, an example of Module Level Power Electronics

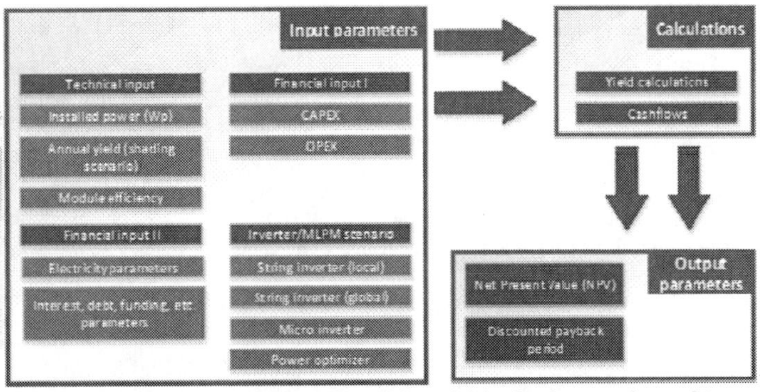

Figure 2: Design of the Techno-Financial Model in the three sections of Input, Calculation and Output.

keeping the PV system in operation.

Weighted Average Value of Electricity (WAVE) refers to the value that the generated electricity has. It is a weighted average of the feed-in tariff and self-consumed electricity.

Discount rate refers to the percentage by which future cash flows are annually lowered in the cash flow analysis. It is made up of the cost of the investment capital plus a highly situational and sector-specific 'risk percentage'. For consumers investing in a residential PV system the discount rate can be understood as follows. The cost of capital for a consumer equals the interest rate on the bank savings account. However, the PV system should preferably return a bit more than that as incentive for all the organizational burden and financial risks involved in having a PV system on the roof.

2.3 Definition of the Output Parameters

We used one main output parameter in this study: *Net Present Value (NPV)*. It is a measure of the PV system's profitability. All future cash flows are calculated into present value using the discount rate r. If $NPV > 0$ it is profitable to invest in the PV system, if $NPV < 0$ it is not profitable to invest in the PV system.

$$NPV = \sum_{t=0}^{N} \frac{income - costs}{(1+r)^t}$$

3 SIMULATION SETTINGS

3.1 System definition

We chose the small scale residential end-user perspective. The system is defined as a 3.2 kWp Dutch rooftop system, facing south under 35° inclination angle.

3.1 System architecture and shading scenarios

In the simulations we consider three scenarios for the system topology:

(A1) String inverter with conventional mpp tracking algorithm. This is the historical scenario used in PV systems built before 2010. These string inverters use the conventional 'perturb and observe' algorithm also known as the 'Fraunhofer algorithm'. This method has the danger of getting stuck in a local maximum of the IV curve.

(A2) String inverter with global mpp tracking. Most high quality inverter brands have introduced a global mpp tracking algorithm in their firmware in recent years. When this algorithm is switched on, the string inverter still employs the perturb and observe algorithm to keep track of the local maximum, but in addition to that performs a broader voltage scan every 15 minutes or so. The advantage of this method is that it enables bypassing of shaded substrings, limiting the effect of partial shading on the annual yield. The danger of this method is that some solar energy might get lost while performing the voltage scans. We estimate this additional loss to be less than 1 kWh per year.

(A3) Module Level Power Electronics (MLPE). In this scenario, the PV system is built using mpp tracking per module. This can be done by using either a power optimizer or a string inverter. Compared to the scenario (2), shaded modules do not need to be bypassed and still produce energy. On the negative side these approaches have an energy conversion efficiency about 1% less than string inverters.

In addition to the three system architecture scenarios, we defined four shading scenarios described below.

(S1) "No shading". Here we assume that the modules are free of shade for the whole year.

(S2) "Pole shading". Here we assume that the system is shaded by a pole-like object, such as a lamp post, chimney, exhaust pipe or TV antenna. The pole only blocks direct light and its effect is more severe in winter than in summer months.

(S3) "Horizon shading". Here we assume that the modules are shaded horizontally in the winter months, corresponding to the shading from a nearby building, wall or row-to-row shading in flat roof mounted PV systems.

(S4) "Soil shading". Here we assume that the modules are to some extent polluted at the edges. We assume algae or dirt covers the first 2 centimeters of the cells around all edges of the module where it blocks all the irradiance, both direct and diffuse, all year round.

The electricity yield of the PV system in these four shading scenarios was taken from our full year field test and simulation model developed in earlier work [1,3].

3.3 Input values

Table I lists all the input values we collected for the simulations. They were derived from earlier work and literature values [1-7] as well as a newly conducted market survey. As we chose the end user perspective, the found price levels resemble the end-user prices including the margins for wholesalers and installers.

Table I: Input values used in the simulation.

Quantity	Value
System size	3.2 kWp
Module degradation	0.5%/a
CAPEX excl. inverter	€ 3900
CAPEX conventional inverter	€ 825 - € 1560
CAPEX MLPE inverter	€ 1545 - € 2318
O&M costs excl. inverter replacement	0.0-0.5% of the CAPEX
Price inflation	0.8-2.8%
Conventional inverter lifetime	10 - 12 years
MLPE lifetime	15 - 25 years
Inverter replacement costs	Price deflation of 2%/a
Weighted Average Value of Electricity	Net metering: 0.23 €/kWh at annual increase of 1.8-3.8%/a
Energy yield conventional inverter in various shading conditions	845-1015 kWh/kWp
Energy yield inverter with global mppt in various shading conditions	950-1014 kWh/kWp
Energy yield MLPE in various shading conditions	965-1005 kWh/kWp
Economic lifetime N	25 years
Discount rate r	2.5 - 5.0%

4 RESULTS

Figure 3 shows the result of the Net Present Value analysis.

Figure 3 – Net Present Value analysis results. Scatter plot on left y-axis shows the Net Present Value in worst (triangle), median (square) and best (diamond) case. Bar graph on right y-axis shows the contribution to the NPV variance of the most influencial parameters.

Figure 3 is a holistic graph from which many results can be derived:

All NPVs are highly positive, whether best case, worst case, string inverter or MLPE. Apparently under the current net metering scheme in The Netherlands, all PV systems are profitable even in heavy shading scenarios, expensive financing and low yield.

The system architecture with the overall highest NPV is the string inverter with global mpp tracking. On average, this is the most profitable PV system and therefore seems to be the rightful market leader.

The parameter that has the highest influence on the outcome of the business case is the discount rate, with a contribution of ~50%. PV systems of low risks and low cost financing are the most profitable. This is more important than the yield or shading condition.

The parameter of the second highest importance is the electricity price development with a contribution of ~30%. This is quite a risk for residential investors, as electricity retail price prediction is very difficult and on top of that regulatory changes might occur in the future.

The shading scenario is the third most important parameter. In particular in the case of a string inverter with conventional mpp tracking (no global peak scanning) the influence is very high and about 29%. This means that in this system architecture the presence of an exhaust pipe or tree can have a big impact on the profitability of the system. In the string inverter with global mpp tracking or the MLPE system the influence is strongly reduced to 2-5% and these type of shades no longer influence the output significantly.

The fourth parameter of influence is the purchase price of the inverter. In particular in MLPE systems, prices still vary wildly and large margins are accumulated over the distribution chain, leading to 16% impact of this parameter on the NPV variance. The price development of MLPE should be critically followed in years to come as it is a volatile parameter that can show strong variations from year to year.

Surprisingly, the least influential parameter is the inverter replacement costs. This parameter is a combination of the initial inverter costs, inverter lifetime, inverter price deflation over that lifetime and discount rate. In particular the last two lead to the small influence: Whether an inverter should be replaced in 10, 12, 15 or 25 years does not have a big impact because inverters have become a lot cheaper *and* the cash flows are already heavily discounted.

5 CONCLUSIONS

We executed an economic analysis on the profitability of a PV system from the perspective of a small residential end-user. Scenarios were made for the system architecture (string inverter or Module Level Power Electronics) and shading conditions (pole, horizon, soiling or no shading).

We conclude that the traditional string inverter with a 'global mpp tracking' software gives the best economic value for the end user. However, the differences are very small. Future research should focus on additional value divers such as monitoring, safety, reliability, O&M, degradation which could tip the scale in the direction of MLPE.

ACKNOWLEDGEMENT

This work is supported by RVO and the Dutch Top Team Energy via the project Module Level Power Management (MLPM) with grant number TKIZ01005.

REFERENCES

[1] K. Sinapis, G.B.M.A. Litjens, M.N. van den Donker, W. Folkerts, *Outdoor Characterization of Three PV Architectures under Clear and Shaded Conditions*, PVSEC-29, Amsterdam, The Netherlands, 2014

[2] N.J.J. Dekker, M.J. Jansen, W. Eerenstein, R. Schaacke, R. Kockelkoren, R. Jonkman, R. van der Sanden, *DH Test of a Femtogrid Power Optimizer and Microinverter under Operation*, PVSEC-29, Amsterdam, The Netherlands, 2014

[3] K. Sinapis, C. Tzikas, G.B.M.A. Litjens, M.N. van den Donker, W. Folkerts, W.G.J.H.M. van Sark, A. Smets, *Yield Modelling for Micro Inverter, Power Optimizer and String Inverter under Clear and Partially Shaded Conditions*, PVSEC-30, Hamburg, Germany, 2015

[4] D. C. Jordan and S. R. Kurtz, *Photovoltaic Degradation Rates — An Analytical Review*, Prog. Photovolt: Res. Appl. 21 (2013) 12–29

[5] G. Verberne, *BIPV price benchmark 2014*, online available at www.seac.cc/downloads

[6] R. Kannan, K. C. Leong, R. Osman, H. K. Ho, and C. P. Tso, *Life cycle assessment study of solar PV systems: An example of a 2.7kWp distributed solar PV system in Singapore*, Sol. Energy, vol. 80, no. 5, pp. 555–563, May 2006.

[7] D. Perez, V. Cervantes, M. J. Baez, and F. T. Dominguez, *PV Grid Parity Monitor - Residential Sector 2nd issue*, Berlin, 2013.

FIELD AND LABORATORY PERFORMANCE CHARACTERISATION OF MICROINVERTER AND POWER OPTIMIZER SYSTEMS

Dirk Stellbogen, Peter Lechner, and Martin Senger
Zentrum fuer Sonnenenergie- und Wasserstoff-Forschung Baden-Wuerttemberg (ZSW)
Industriestrasse 6, D-70565 Stuttgart, Germany
dirk.stellbogen@zsw-bw.de

ABSTRACT: Module level power conditioning devices such as micro inverters or power optimizers have recently attracted a lot of attention due to claimed advantages such as energy gains in cases of shadowing as well as the possibility of module level status monitoring and security switching. This study addresses the assessment of conversion efficiency as well as operational behaviour at elevated temperatures for such devices and compares their energy performance to that of a reference string inverter in a field test, which is conducted in a side-by-side evaluation in unshaded and shaded configurations. The tested module level inverters show to be comparable to typical string inverters in terms of power conversion efficiency and exhibit convincing operating temperature behaviour. They conduct power-derating at much higher temperatures than the typical string inverter. The AC energy yield of the module level inverter systems equals the yield of the reference string inverter in unshaded conditions. In a typical shading condition with a pole, the module level inverters and power optimizers show 5%-10% gain. Significant surplus is only achieved in extreme shading with a gable wall set-up.

Keywords: Module level power electronics, Laboratory measurements, Field testing, Performance, Durability

1 INTRODUCTION

Module level power conditioning devices such as micro inverters or power optimizers have recently attracted a lot of attention and reached a relevant market share especially in the U.S. domestic PV sector. Claimed advantages of these types of devices over PV string or array inverters include energy gains in cases of shadowing or highly non-uniform illumination as well as the possibility of module level security switching.

However, the assessment of individual power conversion efficiency as well as reliability and durability of the devices in elevated temperature conditions which are experienced when mounted at the back surface of the modules should also be considered for informed choices on particular applications.

Some work has already been published on the outdoor performance of various power device options [1] as efficiency and yield measurements of micro-inverters in laboratory and field tests [2].

This study contributes further knowledge on module level power electronics by collecting and evaluating measurement data from laboratory tests at different operating conditions and temperatures as well as from field evaluations.

The laboratory investigations include climate chamber tests at and above the specified operation temperature limits of 65°C to assess the operational behaviour and reliability of these devices.

A side-by-side setup for the field tests allows for meaningful evaluations of the relative performance of the module level oriented power conditioner systems to a reference string inverter under the same operating conditions.

2 APPROACH

2.1 Systems in Test

In the study, two module level inverter types have been investigated in detail:

- Enphase M250, and
- SMA SB240,

which are displayed in Figure 1.

Additionally, one type of module power optimizers has been assessed in the field evaluation:

- SolarEdge P300 (together with its dedicated system inverter).

As a reference system, a typical string inverter of 3kW AC rating with global MPP tracking option was applied:

- SMA SB3000TL-21 with 'OptiTrac Global Peak' function.

These devices are shown in Figure 2.

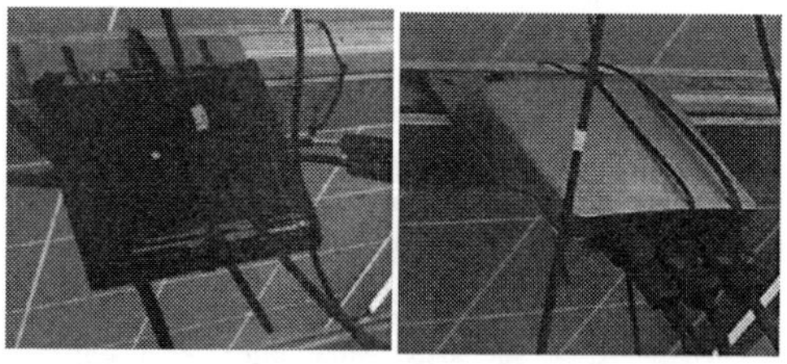

Figure 1: Photographs of tested devices: Enphase M250 module inverter (left) and SMA SB240 module inverter (right).

Figure 2: Photographs of the evaluated SolarEdge P300 power optimizer (left) and the string inverter SMA SB3000TL-21 (right), which served as the reference system.

Some characteristics of the module level power electronic devices under test are summarised in Table 1. The two module inverters are rated at 250W AC and 230W AC, respectively and have similar values of rated EUR efficiency. Notable are the input voltage ranges stating relatively high thresholds of 23V and 27V, respectively. In contrast, the power optimizer quotes a minimum input voltage level of only 9V.

Table 1: Characteristics of the tested MLPE devices

model	Enphase M250	SMA SB240	SolarEdge P300
type	inverter	inverter	power optimizer
power rating	250W AC	230W AC	300W DC
input range	23V – 39V	27V – 39V	9V – 48V
η EUR	95.7%	95.3%	-

2.2 Laboratory measurements

In the laboratory, the static DC/AC conversion efficiency of the module level inverters has been determined following the testing protocols of standard EN 50530. This requires a setup with an electronic PV simulator acting as the power source to the equipment under test (Figure 3).

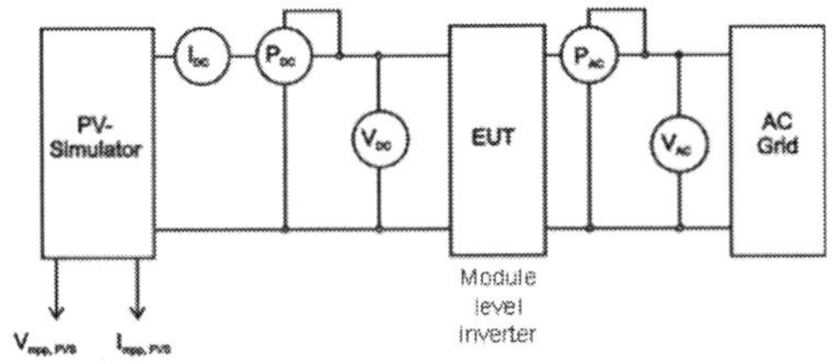

Figure 3: Measurement set-up acc. to EN 50530 for the determination of the static DC/AC conversion efficiency of (module level) inverters.

The efficiency measurements have been conducted both at room temperature (25°C) as well as at elevated temperatures in a climate chamber.

2.3 Field evaluation

A PV systems test stand has been established at the solar test field Widderstall which has been operated by ZSW since more than 25 years in Southern Germany (9,72°E, 48,53°N, 750 m altitude) [3]. This has been adopted for a field experiment to evaluate the performance of module level power electronic systems in a side-by-side comparison to a reference system with a typical string inverter.

This includes two sets of 10 PV modules (type: Canadian Solar CS6P 245P) with 60 c-Si in series comprised in 3 substrings, each of which carries a bypass diode. The modules have a nominal power of 245Wp and have been individually power measured with the ZSW class AAA indoor flash simulator.

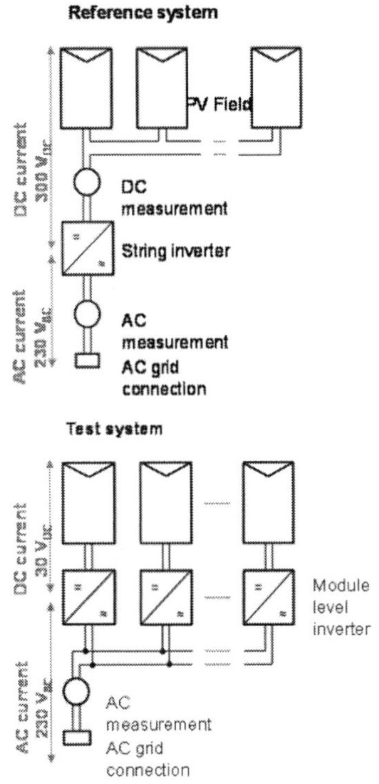

Figure 4: Electrical layout of the reference system (top) and the test system (bottom) of the field experiment.

Figure 5: Experimental set-up for field evaluations at the ZSW test site Widderstall; artificial shading configuration with wooden poles. The markings show the two sets of 10 PV modules which are connected to the evaluated power electronic systems.

Figure 6: Experimental set-up for field evaluations at the ZSW test site Widderstall; severe shading configuration with two identical gable wall mock-ups.

All modules have been placed onto a test rack in one long row. One set of 10 modules was wired in series and then connected to the string inverter. The combined DC input to the inverter and its AC output to the grid have been measured. This reference system configuration is sketched in Figure 4 (top).

From the second module set, each module was individually connected to a module level power device. In the case of the module inverters, their AC outputs have been combined and the total AC yield has been recorded as displayed in Figure 4 (bottom). When testing the module power optimizers with DC output, these were wired in series and the DC was connected to the input of the respective system inverter. In this case, the electric power was measured both on the DC and the AC side of that inverter.

In order to eliminate potential bias in the power performance of the two module sets applied, the power electronics systems in examination are swapped between each measurement period of about one week.

A view of the experimental set-up is displayed in Figure 5. It shows the two sets of PV modules side-by-side in one row in a shading configuration. Two wooden poles are erected in front of the module sets in the exactly equal geometric configuration.

The field evaluations have been conducted in unshaded conditions, with the shading poles (Figure 5) representing a typical obstruction, and in a severely shaded situation with two identical gable wall mock-ups (Figure 6).

3 RESULTS

3.1 Laboratory assessments

The static DC/AC conversion efficiencies of the two module inverters in test as measured in the laboratory at 25°C are shown in Figure 7 together with the efficiency data of the reference string inverter declared in its data sheet.

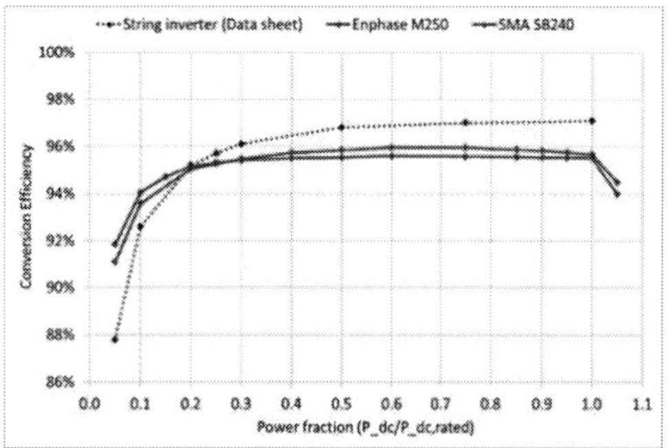

Figure 7: Static DC/AC conversion efficiencies measured for the Enphase M250 (red) and SMA SB240 module inverters together with the efficiency values of the reference string inverter quoted in the data sheet (dotted line).

The efficiencies of the two module inverters are quite similar with the Enphase M250 being somewhat advantageous at medium to high power levels whereas the SMA SB240 has slightly higher efficiencies at low power levels. The module inverters are about 1% less efficient than the string inverter in the upper power range whereas the module level devices clearly outperform the string inverter at low power.

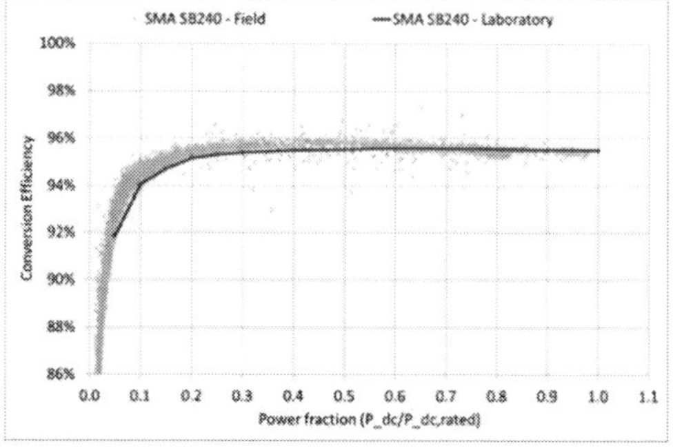

Figure 8: Conversion efficiency data of the SMA SB240 module inverter from the field test together with the efficiency curve measured in the laboratory at 25°C.

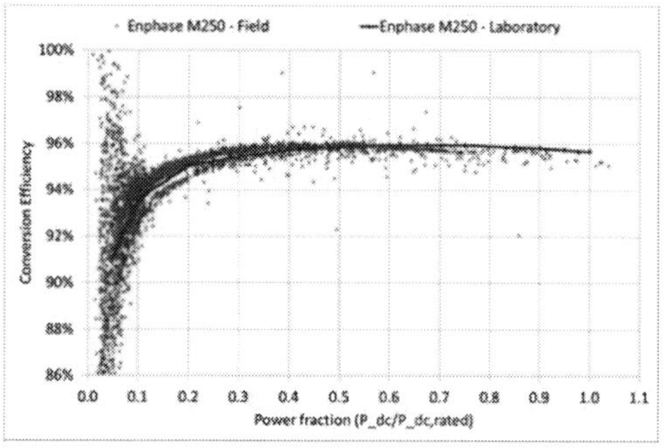

Figure 9: Conversion efficiency data of the Enphase M250 module inverter from the field test together with the efficiency curve measured in the laboratory at 25°C.

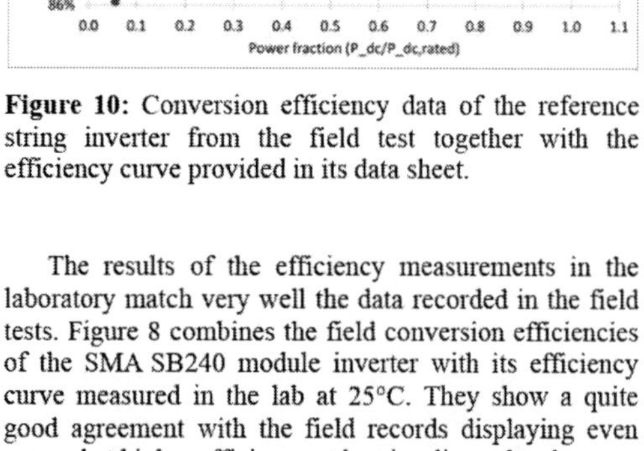

Figure 10: Conversion efficiency data of the reference string inverter from the field test together with the efficiency curve provided in its data sheet.

The results of the efficiency measurements in the laboratory match very well the data recorded in the field tests. Figure 8 combines the field conversion efficiencies of the SMA SB240 module inverter with its efficiency curve measured in the lab at 25°C. They show a quite good agreement with the field records displaying even somewhat higher efficiency at low irradiance levels.

The combined field and laboratory efficiency values of the Enphase M250 module inverter as presented in Figure 9 also agree very well. The notable scatter of the data at low power levels is due to oscillations at the DC input caused by the Enphase device control.

The comparison plot for the reference string inverter (Figure 10) proves that the field measured efficiencies are well represented with the curve specified in the data sheet with the limitation that the field data show somewhat lower efficiency at low power levels.

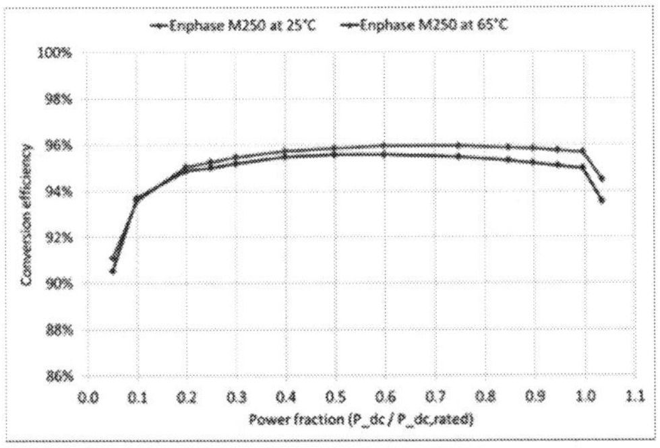

Figure 11: Conversion efficiencies of the Enphase M250 module inverter measured in the laboratory at 25°C and 65°C, respectively.

Backed by the good match between indoor and outdoor measurement data, the laboratory investigations of the module inverters have been extended to address their operational behaviour in elevated temperature conditions.

The operating ambient temperature range for both tested module inverter types extends to 65°C. Hence, the measurements of the static DC/AC conversion efficiency have additionally been conducted at this temperature level while placing the tested devices in a climatic chamber. The results displayed in Figure 11 for the Enphase M250 module inverter demonstrate a small decrease of the efficiencies at 65°C ambient temperature of about 0.3% at medium to high power levels.

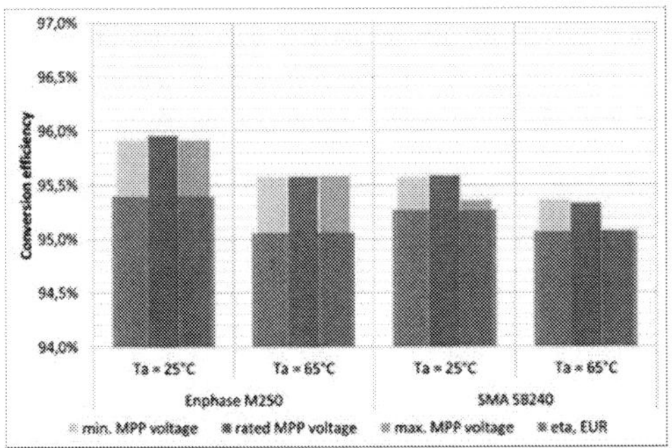

Figure 12: Maximum static conversion efficiencies measured at minimum, rated and maximum MPP voltage and temperature levels of 25°C and 65°C for the module inverters Enphase M250 (left) and SMA SB240 (right); additionally, the combined EUR weighted efficiencies are indicated (grey bars).

The results of the conversion efficiency measurements performed at different voltage levels and temperatures are summarized in the chart in Figure 12. The small bars represent the measured maximum conversion efficiencies at minimum, rated and maximum MPP voltage. The graph shows that for the Enphase M250 the peak efficiency is independent of voltage and temperature and for the SMA SB240 only at maximum voltage somewhat lower peak efficiencies are observed.

The grey bars in the chart represent the combined EUR weighted efficiencies at rated voltage. They are quite similar for the two device types and at 65°C only a moderate reduction of 0.2% to 0.3% with respect to the values at 25°C is experienced.

For further investigations into the operational temperature limitations of the module inverter devices, we evaluated their power derating characteristics.

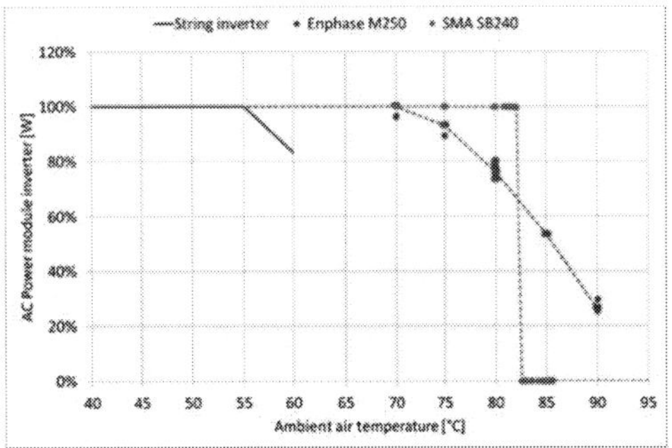

Figure 13: Power derating characteristic of the reference string inverter according to its data sheet (blue) and measured power derating of the module inverters Enphase M250 (red) and SMA SB240 (green).

The reference string inverter has a declared ambient temperature limit at 60°C, which is typical for such devices. Its data sheet states that there is a power derating starting at 55°C ambient temperature which reduces the convertible power by about 17% at 60°C (Figure 13).

The data sheets of the module inverters in test do not give any information on their temperature related power derating. Therefore, the temperature behaviour was determined a climate chamber test where the chamber air temperature was increased in a stepwise manner.

The power transmission of the Enphase M250 was unchanged at about 70°C. From 75°C ambient temperature, the maximum converted power was gradually decreased with the device still operating at 90°C (Figure 13) with reduced power. In contrast, the SMA SB240 transformed the full power up to a threshold of about 82°C ambient air temperature. Above that limit, the module inverter was switched off (Figure 13).

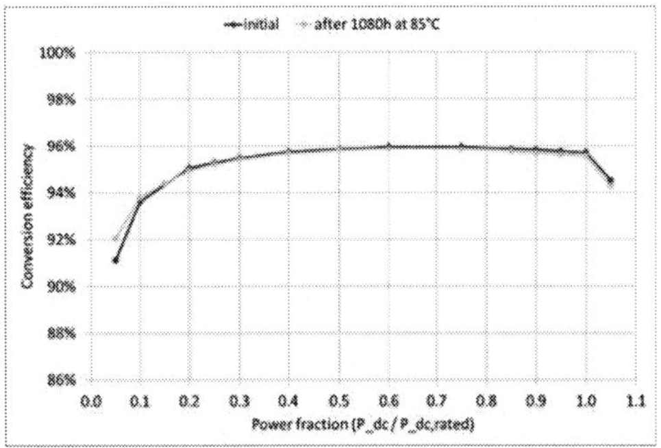

Figure 14: Static DC/AC conversion efficiencies of an Enphase M250 module inverter measured before (red) and after (yellow) a climate chamber stress test of about 1000h at 85°C.

The issue of durability of the devices is difficult to address in laboratory tests. As a first approach, we conducted a climate chamber test with one of the Enphase M250 module inverters. The device was operated for about 1000 hours at 85°C chamber temperature with the maximum converted power. The efficiency curves measured before and after the climate chamber stress did hardly show any deviations (Figure 14), suggesting that the device did not suffer a degradation in this test.

3.2 Field evaluations

In the field assessment, the two types of module inverters, Enphase M250 and SMA SB240, respectively, have been evaluated in side-by-side tests with the string inverter reference system in unshaded conditions as well as in the artificial pole and gable wall shading configurations. Additionally, the SolarEdge P300 module power optimizers have been compared to the reference in the pole shading set-up.

In shading situations, the IV curves of a series connected module string can be very complex with a number of "steps" exhibiting several local maximums of the extractable power. Figure 15 shows IV curves of the test string with 10 modules measured over the course of a half day in the gable wall shading configuration. The curve of the MPP points shows a large variation of the optimal operating voltages which in some cases are beyond the MPP tracking range of the inverter. Noticeable is the large change of the MPP voltage in the afternoon which is hard to follow for the inverters MPP control even with global searching strategies.

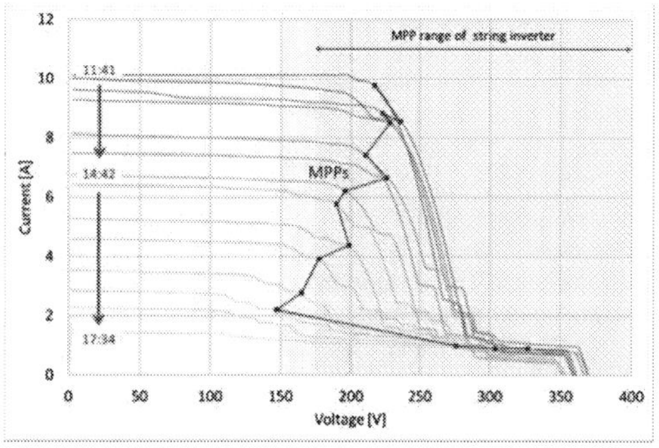

Figure 15: Measured IV curves of one string of 10 modules in the gable wall shading configuration; the red curve indicates the trajectory of the MPP points.

It is expected, that the sets of module inverters should be advantageous over the string inverter in any shading instances. However, a detailed analysis of the measured data showed that this is not always the case. An example is presented in Figure 16, which shows the development of the AC performance ratios of the test system of Enphase M250 module inverters together with the string reference system on a clear day.

When the shadow begins to fall onto the module row, the performance of the two systems drops strongly due to the long shadow affecting a high number of modules. Afterwards, as the sun moves, the performance of the string system increases in a stepwise manner. Each of the steps corresponds to one module substring being no longer affected by the shadow.

As expected, the module inverters perform better than the string inverter in the morning. But at some point, the situation changes and the combined output of the Enphase M250 devices is lower than that of the string inverter. This is due to the high MPP voltage threshold of 27V which prevents the module inverters from making use of one single unshaded substring within the module.

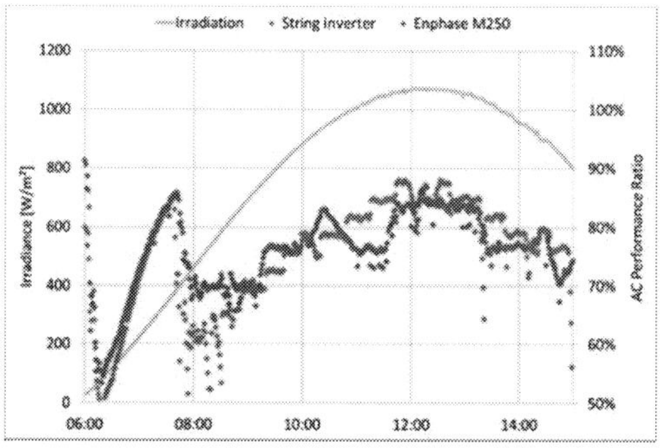

Figure 16: Course of the AC performance ratio of the string inverter reference system (blue) and the test system with Enphase M250 module inverters (red) in the experimental pole shading configuration on a clear day.

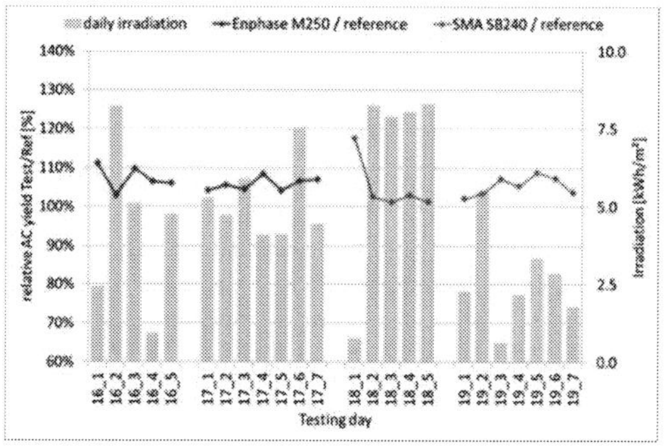

Figure 17: Daily irradiation and relative AC yield of the tested module inverter systems with respect to the reference string inverter system for the test campaign in unshaded conditions.

The graphs in Figure 18 to Figure 20 display the relative daily energy gain of the tested module inverter and module power optimizer systems with respect to the reference string inverter system for unshaded conditions and for the pole and gable wall shading configurations. In general, there is a substantial variation and no significant distinction among the different device types.

For unshaded situations (Figure 18), the module inverters are about equal to the reference string inverter system for bright days. At days with low irradiation, the module level devices are clearly advantageous (at low total outputs) which originates in their higher conversion efficiencies at low power fractions.

In the shading pole set-up (Figure 19) representing a typical shadowing configuration, the module level power devices, both micro inverters and the optimizer, exhibit an advantage of about 3% at clear days and relative gains of 5% to 10% at medium irradiation conditions.

More substantial gains are only observed in the severe gable wall shading configuration (Figure 20). Again, the advantage increases from clear days to less bright irradiance situations, reaching up to about 20% for some of the testing days.

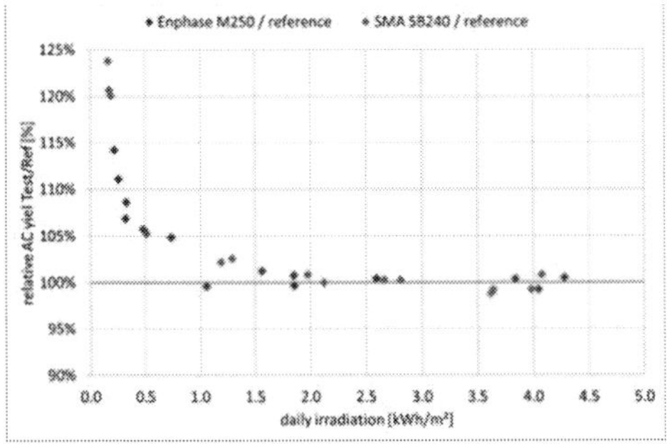

Figure 18: Unshaded configuration: Daily relative AC yield of the test systems with module inverters Enphase M250 (red) and SMA SB240 (green) with respect to the string inverter reference system in dependence of the daily in-plane irradiation.

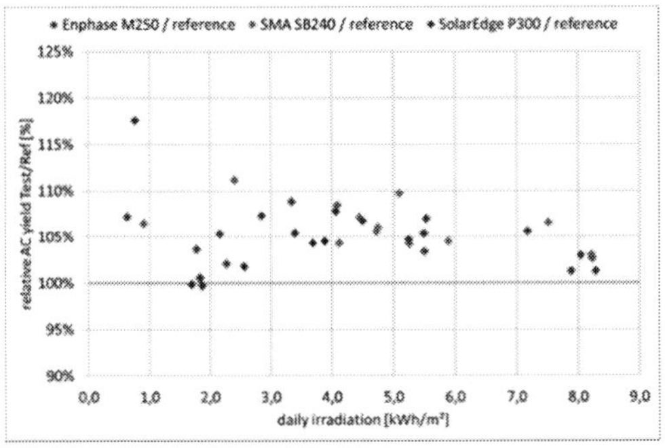

Figure 19: Pole shading configuration: Daily relative AC yield of the test systems with module inverters Enphase M250 (red) SMA SB240 (green) and Solaredge P300 (violet) with respect to the string inverter reference system in dependence of the daily in-plane irradiation.

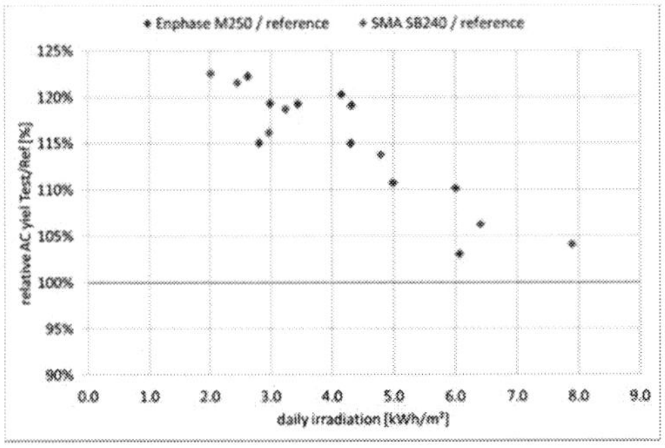

Figure 20: Gable wall shading configuration: Daily relative AC yield of the test systems with module inverters Enphase M250 (red) and SMA SB240 (green) with respect to the string inverter reference system in dependence of the daily in-plane irradiation.

4 SUMMARY

Two types of module level inverters and one type of module power optimizers have been assessed in laboratory measurements and field evaluations with artificial shading configurations.

The tested module level inverters are comparable to typical string inverters in terms of power conversion efficiency and show convincing operating temperature behaviour. They conduct power-derating at much higher temperatures than the typical string inverter.

The AC energy yield of the module level inverter systems equals the yield of the reference string inverter in unshaded conditions. In a typical shading condition with a pole, the module level inverters and power optimizers show 5%-10% gain. Significant surplus is only achieved in extreme shading with a gable wall set-up.

REFERENCES

[1] K. Sinapis, G.B.M.A. Litjens, M.N. van den Donker, W. Folkerts, Outdoor Characterization of Three PV Architectures under Clear and Shaded Conditions, 29th European Photovoltaic Solar Energy Conference and Exhibition Proceedings, 2014

[2] S. Krauter, J. Bendfeld, Micro-Inverters: Comparison of Costs, Performance and Yield, European Photovoltaic Solar Energy Conference and Exhibition Proceeings, 2015

[3] D. Stellbogen, H.-D. Mohring, A. Jagomägi, E. Mõttus, G. Friesen, D. Domine, F. Fabero, T. Betts, R. Gottschalg, T. Zdanowicz, M. Prorok, W. Herrmann, J.L. Martin, A.G. de Montgareuil, J. Merten, D. Faiman, Comparative outdoor characterisation of PV Modules across Europe, 25th European Photovoltaic Solar Energy Conference and Exhibition Proceedings, 2010

COMBINING SOLAR IRRADIANCE DATABASES AND PV PERFORMANCE MODEL FOR PV SYSTEM PERFORMANCE ANALYSIS

Blaž Kirn and Marko Topič
Faculty of Electrical Engineering, University of Ljubljana
Tržaška 25, SI-1000 Ljubljana, Slovenia

ABSTRACT: To analyze the PV system potential for different locations and orientations, PV performance models coupled with solar irradiance databases are used. We have recently developed direct-diffuse-power-rating (DDPR) model that separately addresses the diffuse and direct irradiance components supplemented by angular loss compensation. In this study, the DDPR PV module performance model is extended to PV systems. The 17 kW PV system located in Ljubljana, Slovenia is used in this experiment to test the accuracy of the model which rendered 3.6% yearly root mean square error (RMSE) compared to 4.3% RMSE of basic model used as a reference. The major improvement occurs in the early mornings and late evenings when the diffuse component dominates. Also, the effect of longer averaging time intervals (which are usually utilized in many solar irradiance databases) on the accuracy of the PV performance models was researched. The averaging interval analysis showed, that up to 0.28% additional positive bias is induced for 1 hour averaging intervals compared to 5 minute one. With the use of the DDPR model's feature of separating diffuse and direct irradiance, the model was used to estimate the loss or gain of PV system performance at different locations as well as orientations, different from optimal. The results for Ljubljana, Slovenia show that yearly temperature compensated performance ratio (PR_T) can be 1.8% lower for certain orientations (e.g. vertical orientation facing south) compared to optimal orientation (south, 30° inclination) due to higher share of direct irradiance with higher solar incident angles.
Keywords: PV System, Modelling, System Performance, Solar Radiation

1 INTRODUCTION

Accurate prediction of PV system performance requires detailed models with many parameters [1–5]. We have recently developed direct-diffuse-power-rating (DDPR) model [6] that separately addresses the diffuse and direct irradiance components supplemented by angular loss compensation. The model shows good accuracy in field, however it requires two pyranometers measuring global and diffuse irradiances. The second application where the model is easy to implement is evaluating global or local PV potential at different orientations, using many available solar irradiance databases (PVGIS provided by Joint Research Centre of European Commission, Solar Radiation Data (SoDa), National Solar Radiation Database of NREL and others). These databases usually provide global as well as diffuse irradiance in horizontal plane at certain intervals (e.g. half hour). Using solar irradiance models [7] those values can be translated to arbitrary tilt and azimuth and coupled with the DDPR PV performance model to evaluate the PV potential and/or performance at different orientations and locations.

To evaluate the performance of PV systems, the PV module performance models have to be extended to PV system usage. This can be achieved by considering additional system losses (wires, mismatch, partial shading, etc.) as well as adding a PV inverter model [8]. The simpler approach is to use the DC power measurements of the PV system array(s) that are available within the PV inverter and thus taking the PV inverter out of consideration. In this study, the DDPR model was extended form single module to PV system use according to second approach and one year measurements of the 17 kW PV system were used to evaluate the accuracy of the model. Next, the DDPR model was used to calculate the PV system performance for different locations (Ljubljana - Slovenia, Berlin - Germany and Denver – Colorado, USA). For the case of Ljubljana, the PV system performance was evaluated also at various orientations in order to analyze the decrease of performance for orientations with higher solar incident angles.

2 DDPR MODEL

The DDPR model has been recently developed in the Laboratory of Photovoltaics and Optoelectronics at the Faculty of Electrical Engineering within University of Ljubljana. The model utilizes widely used linear correction for operating temperature with power temperature coefficient γ (%/°C). However, compared to other models, the DDPR model separately addresses diffuse irradiance and direct irradiance together with solar incident angle. For silicon PV modules, the diffuse irradiance ($G_{poa\text{-}dif}$) power model is simple linear equation (Eq. 1), while direct irradiance ($G_{poa\text{-}dir}$) power model is represented by polynomial equation for direct irradiance (Eq. 2) and exponential equation for solar incident angle (Θ_s) loss factor (Eq. 3) using only one parameter a_r [9].

$$P_{model-dif}(G_{poa-dif}) = k_{dif1} \cdot G_{poa-dif} \tag{1}$$

$$\begin{aligned} &P_{model-dir}(G_{poa-dir}, \theta_s) = \\ &= \left[k_{dir1} \cdot G_{poa-dir} + k_{dir2} \cdot G_{poa-dir}^2\right] \cdot F(\theta_s) \end{aligned} \tag{2}$$

$$F(\theta_s) = \left[1 - \frac{e^{\left(-\frac{\cos\theta_s}{a_r}\right)} - e^{\left(-\frac{1}{a_r}\right)}}{1 - e^{\left(-\frac{1}{a_r}\right)}}\right] \tag{3}$$

To calculate the PV module (or PV system) electrical power, both diffuse and direct light model are summed up and compensated for operating cell temperature. All parameters of the model can be obtained from outdoor measurements. In this research, the 17 kW PV system located in Ljubljana, Slovenia was used to extract the parameters on monthly basis for the period from October 2014 to September 2015. Two pyranometers were used, measuring irradiance in the plane of array (G_{poa}) and horizontal diffuse irradiance ($G_{H\text{-}dif}$) which was later recalculated for the tilt of PV system using Tian's approach [10]. First, the diffuse light power model is obtained by filtering the measurements with direct irradiance component being equal to 0. Then, the diffuse light induced power is subtracted from measured points to obtain direct light power model (Figures 1 and 2).

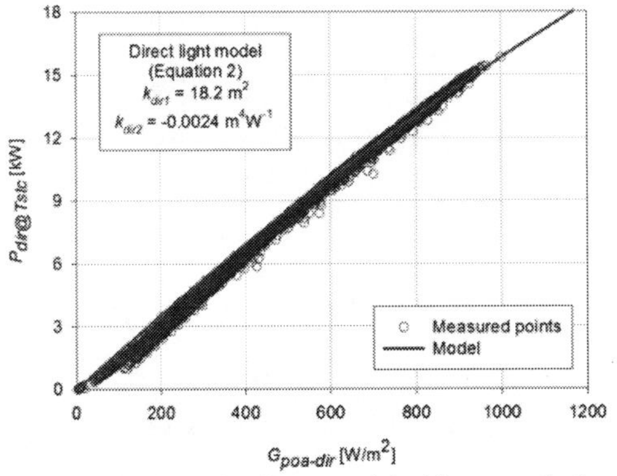

Figure 1: The direct irradiance model without angular loss compensation for the month of August 2015. Note that the model is further multiplied with angular loss factor (Eq. 3) to compensate for angular losses (which are seen at lower irradiances).

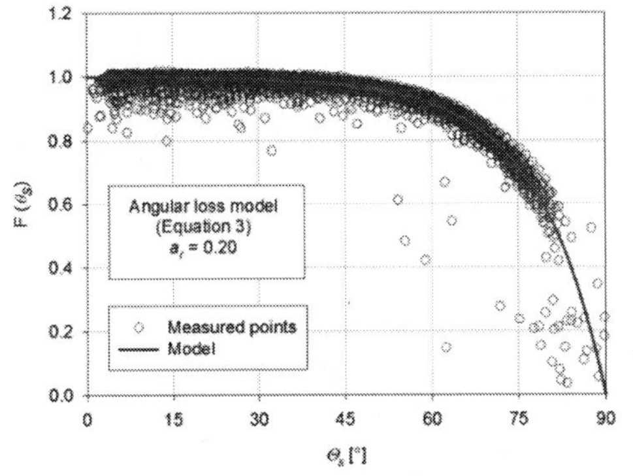

Figure 2: The angular loss model obtained from the PV system measurements of the month of August 2015

2 THE VALIDATION OF DDPR MODEL FOR PV SYSTEMS

The original DDPR model was intended to be used with instantaneous irradiance measurements (and power measurements obtained through *I-V* curve sweeps at outdoor PV test site) due to the fact that sun changes its position through time and thus changes the solar incident angle affecting the angular loss of the module. However, PV systems usually gather weather and power data in certain averaging interval. The evaluated PV system has an averaging interval of 5 minutes, but other (usually longer) periods are also widely used. This can induce errors due to change of sun's position and solar incident angle during that period. On the other side, since the data is averaged, fast dynamics are leveled out and the possibility of outlier appearances are reduced. To evaluate the performance of DDPR model for PV systems with 5 minute averaging interval, the model was compared to nonlinear basic model [2] which includes logarithmic term for better accuracy at lower irradiances (Eq. 4).

$$P_{model}(G_{poa}) = k_{nl1} \cdot G_{poa} + k_{nl2} \cdot G_{poa}^2 + k_{nll} \cdot G_{poa} \cdot \ln(\frac{G_{poa}}{G_{STC}}) \quad (4)$$

The models utilizes G_{poa} as an input and in our case irradiance was measured both using pyranometer as well as with Equivalent Technology Solar Cell - ETSC (mc-Si) solar sensor, both orientated in the plane of the array. The accuracy validation of the evaluated PV performance models was performed on the one-year measurement period (October 2014 to September 2015) of 17 kW PV system. The accuracy of the models was calculated in form of root-mean-square error (RMSE) at different irradiances (Eq. 5). Systematical error or bias was also considered.

$$RMSE(G_{poa}, \pm 5\ \mathrm{W/m^2}) = \frac{\sqrt{\frac{1}{n} \cdot \sum_{i=1}^{n} (P_{model,i} - P_{measure,i})^2}}{\frac{1}{n} \cdot \sum_{i=1}^{n} P_{measure,i}} \quad (5)$$

The overall RMSE in 8 months without snow for tested models was 3.6% for DDPR, 4.3% for basic model (pyranometer) and 2.5% for basic model with ETSC solar sensor. Lowest RMSE was expected for the model with ETSC sensor, due to a matching angular and spectral response of the sensor and the modules. In Figure 3 RMSE of all three tested models is shown together with final yield (Y_F) histogram for the month of August 2015. All models perform satisfactory at higher irradiances, while at lower irradiances the basic model with ETSC sensor performs best, followed by the DDPR and the basic model with pyranometer. The histogram of relative systematical error of modelled Y_F is shown in Figure 4, revealing higher systematical errors at lower as well as higher irradiances. Higher systematical errors at higher irradiances are a consequence of dynamic conditions starting with cloudy weather and relatively cool modules followed by clouds moving away from position of the sun coupled with additional over-irradiance events [11].

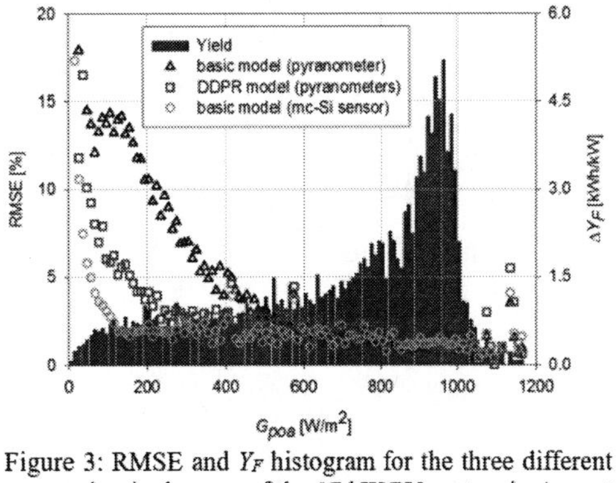

Figure 3: RMSE and Y_F histogram for the three different approaches in the case of the 17 kW PV system in August 2015

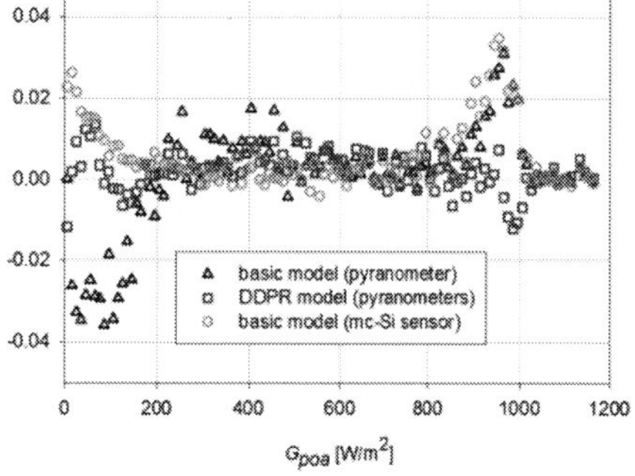

Figure 4: The histogram of systematical error in yield calculation using three different approaches in August 2015

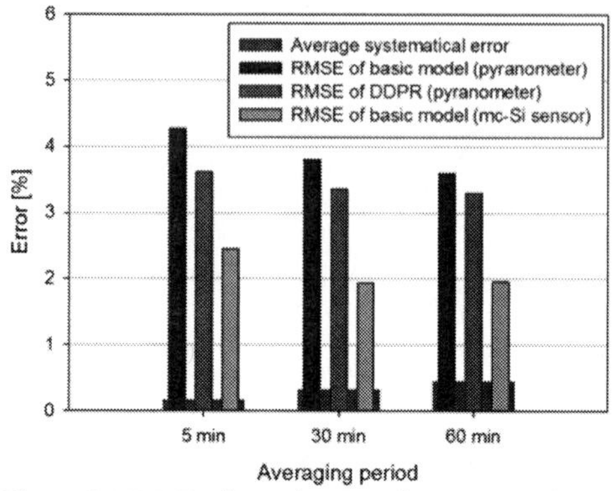

Figure 5: RMSE of tested PV performance models and average systematical error for yearly analysis with different averaging intervals

3 SOLAR DATABASES AND THE DDPR MODEL

Lately, more and more weather resource services are available which can provide regional or global solar irradiation data based on ground stations, satellite observation data as well as various solar and weather numerical models, for example PVGIS from Joint Research Centre of European Commission, NSRDB from NREL, USA or SoDa project. By merging those data with the PV performance model one can obtain a reliable source of global PV potential. In this study we used DDPR model which is very suitable for such application since it considers the diffuse and direct irradiance separately what is in agreement with the available data. However, the data is usually provided with intervals of 15 minutes or longer, what could present a problem for the PV performance model(s) due to change of solar incident angle and thus angular response of the PV modules during that period. In this aspect, the tested PV performance models were evaluated in yearly yield simulation of 17 kW PV system for three different averaging time periods: 5, 30 and 60 minutes. The results showed lower yearly (without months with snow) RMSE error at longer averaging periods for all tested models which can be explained with the averaging effectively leveling out the outliers. However, the systematical error of all three approaches increased in average from 0.16% for 5 minutes to 0.44% for 60 minutes interval. The reason for this behavior is the averaging effect on the irradiance which lowers the maximal irradiances and thus lowers the effect of negative second order polynomial coefficient (which represents the PV module serial resistance) of the performance models. It should be noted, that for all tested averaging periods the used parameters were extracted from 5 minute interval measurements. This indicates that the PV performance models should be corrected for larger time periods as a countermeasure for this biasing effect.

4 PV SYSTEMS PERFORMANCE SIMULATIONS

The combination of available solar irradiance data and the DDPR model can be used to research and evaluate the performance of PV system at given locations as well as different orientations. The results of temperature compensated performance ratio (*PR_T*) simulations for three different locations (Berlin, Germany; Denver, Colorado, USA; and Ljubljana, Slovenia) are shown in Figure 6 together with irradiation bars. The performance of PV systems is presented with *PR_T* to exclude the effect of operating temperature of PV modules which can vary between mounting orientations as well as mounting methods (open rack, building integrated, with or without insulation, etc.) [12].

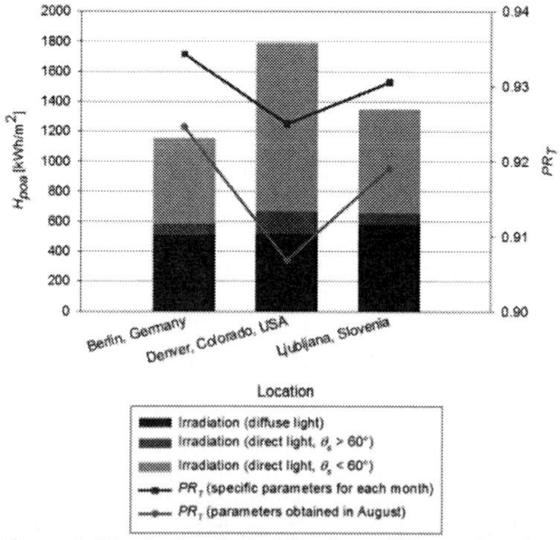

Figure 6: Yearly irradiation for three different locations and temperature compensated performance ratio (*PR_T*) simulated with DDPR model using parameters from August and with seasonal approach using parameters for each month separately

The solar irradiation data was obtained for Berlin from Meteonorm 7 software [13], for Denver from NSRDB [14] and for Ljubljana from measured values. Additionally the solar irradiation for Ljubljana was simulated based on measured values in horizontal plane in 2014 in 30 minutes intervals for 5 different orientations and further used as an input to DDPR model to simulate the yearly yields. The results of PR_T simulations are shown in Figure 7 together with irradiation bars for all evaluated orientations. It can be seen, that PR_T is lower in orientations with larger share of direct irradiation at solar incident angles higher than 60°, where the angular response starts to decrease sharply (Figure 2). Two scenarios were evaluated – in first approach DDPR model parameters extracted in month of August 2015 were used for entire year and in the second approach the parameters were extracted for each month separately from measurement of 17 kW PV system and applied for each month separately in the yearly yield simulations. The second approach include the sessional effect into consideration, and the resulted yearly PR_T is generally higher for all orientations, 0.9% in average. This can be explained with spectral response of the module, which can render up to 3% spectral losses in summer clear sky conditions (for example in August) [15] and thus the extracted parameters in such conditions result in under-prediction of calculated power for other (spring and autumn) months of the year. The effect that enables DDPR model to include the spectral influence is the air mass which is directly dependent to solar zenith angle (θ_z) which in turn is closely related to solar incident angle [16]. Thus, the solar incident angle loss factor can already include some part of air mass correction, what can be observed by difference in obtained DDPR model parameters for different months of the year. We are however continuing the research by also considering separate spectral correction to both direct and diffuse irradiance spectra to further consider seasonal behavior of PV modules.

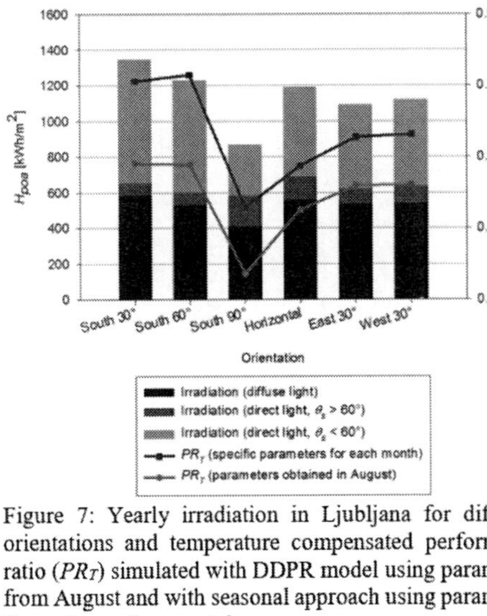

Figure 7: Yearly irradiation in Ljubljana for different orientations and temperature compensated performance ratio (PR_T) simulated with DDPR model using parameters from August and with seasonal approach using parameters for each month separately

The results of orientation analysis were further analyzed in the view of relationship between the share of direct irradiance conditions at higher incident angles and the loss of PR_T. The results are shown in Figure 8 and the relationship is modelled with exponential (for monthly model parameters) and linear functions (for parameters form August). In average, approximately 1.8% loss in PR_T is detected for orientations with 20% share of higher incident angle direct light irradiation compared to optimal orientation with only 5% share.

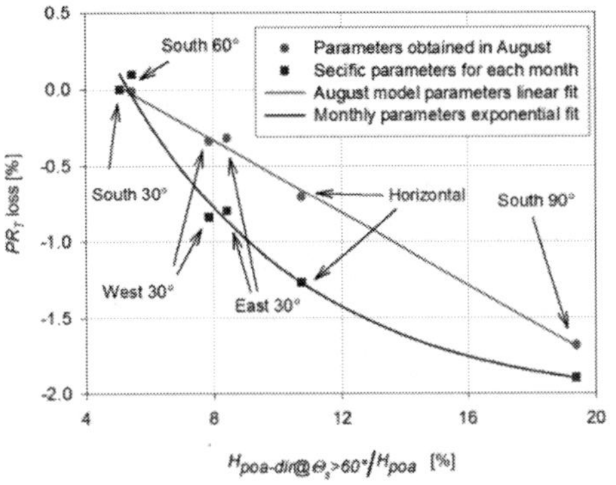

Figure 8: The relationship between share of direct irradiation at higher solar angles (>60°) and PR_T decrease for different orientations in Ljubljana, Slovenia

5 DISCUSSION AND CONCLUSION

By separating the irradiance into diffuse and direct components and adding the angular response of the module, the DDPR model shows better performance (yearly RMSE of 3.6%) compared to the basic model which only considers G_{poa} (yearly RMSE 4.3%). Both models however are surpassed by a simple model if ETSC solar sensor is used (yearly RMSE 2.5%), what is largely the consequence of matched spectral and angular response of the sensor and evaluated PV modules. This makes the ETSC sensor a good option for real-time performance monitoring and fault detection of PV systems. However, when studying the long term PV potential based on solar irradiation data covering almost full solar spectrum the DDPR models shows good potential, since diffuse and direct irradiance components are usually easily available within the databases. In this respect, longer averaging intervals of available solar data may present unwanted systematical error, for example, if averaging interval of 1 hour is used instead of 5 minutes, the systematical error without correcting the (DDPR) model parameters is higher for 0.28%, absolutely. However, in the case of longer interval the RMSE of PV performance models decreases, in average for 0.5%.

The DDPR model was successfully coupled with solar irradiation database to simulate the PV system yield for different locations (Berlin, Denver and Ljubljana) as well as orientations (Ljubljana). The results showed that location or orientations with higher share of direct irradiance at higher solar incident angles render lower PR_T. In the case of orientation analysis for Ljubljana, the difference in PR_T (between horizontal and south orientation with 30° inclination) can reach 1.8%. There was also approximately 0.9% difference in calculated PR_T if seasonal approach was used with parameters used for each month separately instead of using unified parameters. This indicates the importance of spectral correction in PV performance modelling and in this respect we are already working in upgrading the DDPR model to further increase accuracy.

6 ACKNOWLEDGEMENTS

The authors acknowledge the financial support from the Slovenian Research Agency (Research Programme P2-0197 and Project J2-7105). B. Kirn personally acknowledges the Slovenian Research Agency for providing PhD funding.

7 REFERENCES

[1] De Soto W, Klein S a., Beckman W a. Improvement and validation of a model for photovoltaic array performance. Sol Energy 2006;80:78–88. doi:10.1016/j.solener.2005.06.010.
[2] Ding K, Ye Z, Reindl T. Comparison of Parameterisation Models for the Estimation of the Maximum Power Output of PV Modules. Energy Procedia 2012;25:101–7.
[3] Huld T, Gottschalg R, Beyer HG, Topič M. Mapping the performance of PV modules, effects of module type and data averaging. Sol Energy 2010;84:324–38.
[4] King DL, Boyson WE, Kratochvill JA. Photovoltaic Array Performance Model. Sandia National Laboratories: 2004.
[5] Marion B. A method for modeling the current-voltage curve of a PV module for outdoor conditions. Prog Photovoltaics Res Appl 2002;10:205–14. doi:10.1002/pip.403.
[6] Kirn B, Brecl K, Topic M. A new PV module performance model based on separation of diffuse and direct light. Sol Energy 2015;113:212–20. doi:10.1016/j.solener.2014.12.029.
[7] Perez R, Ineichen P, Seals R, Michalsky J, Stewart R. Modeling daylight availability and irradiance components from direct and global irradiance. Sol Energy 1990;44:271–89. doi:10.1016/0038-092X(90)90055-H.
[8] King DL, Gonzalez S, Galbraith GM, Boyson WE. Performance Model for Grid-Connected Photovoltaic Inverters, SAND2007-5036. Contract 2007;38:655–60.
[9] Martin N, Ruiz JM. Calculation of the PV modules angular losses under field conditions by means of an analytical model. Sol Energy Mater Sol Cells 2001;70:25–38.
[10] Tian YQ, Davies-Colley RJ, Gong P, Thorrold BW. Estimating solar radiation on slopes of arbitrary aspect. Agric For Meteorol 2001;109:67–74. doi:10.1016/S0168-1923(01)00245-3.
[11] Yordanov GH, Saetre TO, Midtgard O-M. 100-millisecond Resolution for Accurate Overirradiance Measurements. IEEE J Photovoltaics 2013;3:1354–60. doi:10.1109/JPHOTOV.2013.2264621.
[12] Luque A, Hegedûs S, editors. Handbook of Photovoltaic Science and Engineering. 2nd ed. John Wiley & Sons, Ltd; 2011.
[13] Meteotest. Meteonorm 7 2015. http://www.meteonorm.com/ (accessed March 14, 2015).
[14] NREL. National Solar Radiation Data Base 2016. https://maps.nrel.gov/nsrdb-viewer/ (accessed January 13, 2016).
[15] Martín N, Ruiz JM. A new method for the spectral characterisation of PV modules. Prog Photovoltaics Res Appl 1999;7:299–310. doi:10.1002/(SICI)1099-159X(199907/08)7:4<299::AID-PIP260>3.0.CO;2-0.
[16] Eicker U. Solar technologies for buildings. Wiley; 2003.

IMPACT OF WIND ON INTRA-MODULE ENERGY YIELD VARIATIONS

H. Goverde[1,2,3,*], G. Van Den Broeck[2,3], D. Anagnostos[1,4], B. Herteleer[2,3], J. Govaerts[1], D. Goossens[2,3], E. Voroshazi[1], K. Baert[2,3], F. Catthoor[1,2], J. Driesen[2,3] and J. Poortmans[1,2,3]
[1]Imec, Kapeldreef 75, 3001 Leuven, Belgium
[2]KULeuven, Oude Markt 13 bus 5005, 3000 Leuven, Belgium
[3]EnergyVille, Dennenstraat 7, B-3600 Genk, Belgium
[4] National Technical University of Athens (NTUA), Herron Polytechneiou 9, Zographou Campus 157 80 Athens
*corresponding author: hans.goverde@imec.be, Tel: +32-16-28 85 23, Fax: +32-16-28 10 97

ABSTRACT: The energy yield performance of a novel PV module with 10 strings of 6 cells was investigated. The IV characteristics of each individual 6-cell string were measured and used to determine the monthly energy yield of each sub-string. Differences up to +3.7 % compared to the average value were observed. However, analysis of the daily yield showed variations up to +17.6 %, depending on the wind conditions. These daily yield variations are critical, as the variations may be masked when using longer-term values, such as monthly or yearly yield. As a result, models may give a yearly yield value that is close to the measured value but the daily variation in yield may easily exceed 15%. Cumulatively, this will result in large model errors. Forced convection introduces considerable and often neglected intra-module temperature variations, leading to differences in conversion efficiencies at sub-string level. These temperature variations need to be taken into account for accurate energy yield evaluations. In addition, this insight provides guidelines for novel smart PV module technologies that aim to improve energy yield harvesting.

1 INTRODUCTION

With a worldwide annual growth of installed photovoltaic (PV) capacity of more than 50 GW in 2015 and an expected growth of at least 50 GW each year, accurate (short-term) energy yield predictions become more and more important for a balanced and stable grid [1]. Next to that, energy yield estimations become more critical for technology and installation selection for maximal energy yield. Current state-of-the-art models reported in the literature predict or assess an average module temperature and energy yield for the coming hour or 15 min [2]. These models are able to predict short-term energy yield with an average accuracy of 3 - 5 % absolute percentage error (APE) but they assume a uniform module operational temperature. However, wind tunnel measurements and FEM modelling have shown that strong intra-module temperature differences can occur due to the effect of wind [3]. These temperature variations influence the conversion efficiency and hereby the energy yield of individual cells in the module. Numerous studies investigated the effect of wind on the solar cell's operation temperature [4,5,6]. However, either the heat transfer was assumed uniform over the PV panel or it was assumed that most of the heat was extracted via an insulated air layer, hereby neglecting the convection of heat from the sunny side of the panels

A 10 string of 6 cell (10x6) PV module was installed at a state-of-the-art measurement setup in Ghent, Belgium [7]. Current-voltage (IV) characteristics of each sub-string as well as the environmental conditions were measured. In this paper, sub-string level energy yield variations within the PV module will be investigated based on outdoor measurements, focusing mainly on differences caused by wind effects.

2 EXPERIMENTAL SETUP

A 1.60 m x 1.00 m PV module was constructed using standard 3.2 mm tempered and textured glass, EVA as encapsulating material and a standard black back-sheet. Full Al-BSF solar cells [8] were used to build the PV module. The module consisted of 10 substrings with 6 cells connected in series. This configuration was chosen to investigate the energy yield of novel snake-like module configurations, as discussed in [9]. Fig. 1 shows a schematic overview of the connection scheme and the numbering of the strings. Each string was connected to its own connection box enabling sub-string level IV measurements. The IV characteristics of each individual sub-string were measured every 15 s. A detailed description of the measurement setup can be found in [10].

The PV module was installed at a test site in Ghent, Belgium [7], at a tilt angle of 18° due south. Fig. 2 shows a picture of the outdoor measurement setup. At thes test site, global-horizontal and in-plane irradiance, ambient temperature, wind speed and wind direction are measured at 1 Hz.

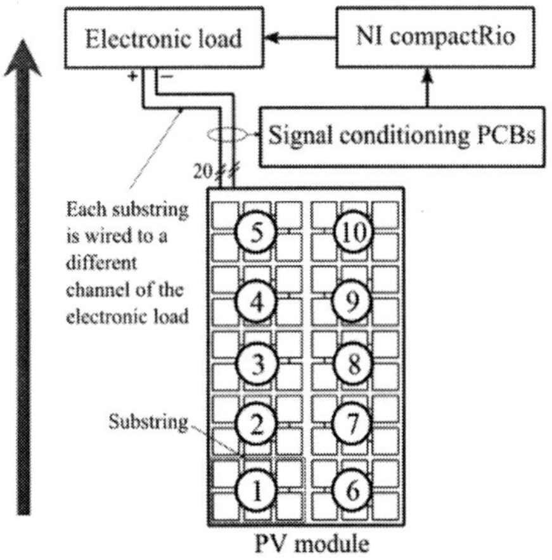

Fig. 1: Schematic overview of the PV module. Numbers indicate the position of the different sub-strings analyzed in this paper.

3 SUB-STRING LEVEL ENERGY YIELD

The individual 15-s IV characteristics of July 2015 were used to calculate the *monthly* energy yield. Fig. 3 shows the energy yield of July 2015 for each sub-string. The average sub-string energy yield is shown by the dashed line. The deviation from to the average energy yield is given by the number on top of the bars. Small deviations compared to the average energy yield were observed. The maximum deviation measured was +3.7 % compared to the average energy yield.

Fig. 2: PV module test site in Ghent. The module was installed at a tilt angle of 18° due south. The green arrow indicates the module analyzed in this paper and the red arrow corresponds to the arrow indicated in Fig. 1.

Fig. 4 shows the *daily* energy yield for each individual sub-string for two different days (1st and 29th July 2015). The average sub-string energy yield is shown by the dashed line. Fig. 4 (a) shows that strong and highly varying differences in energy yield up to +17.6 % compared to the average energy yield were observed on July 1st, while the greatest deviation on 29th July was -2.5 %.

Looking at the weather characteristics of these two days (Table 1), one can see that the average illumination intensity was slightly higher on 1st July. This resulted in a higher average daily energy yield. However, important differences in wind speed and wind direction occurred between the two days. Table 1 shows an *average* daily wind direction and wind speed of 131° (SE) and 1.7 m/s respectively for July 1st, while an average wind direction and wind speed of 269° (W) and 3.0 m/s were observed on 29th July. Since the module was not affected by (partial) shading, we hypothesize that the energy yield deviations were affectedby wind effects.

Table 1: Average weather conditions for 1st and 29th July

Day	*Illumination [W/m²]*	*Wind direction [°]*	*Wind speed [m/s]*
01-07	544	131 (SE)	1.7
29-07	430	269 (W)	3.0

4 EFFECT OF WIND ON PV MODULE CHARACTERISTICS

As mentioned in the previous section, wind conditions differed significantly between the two reported days. In a previous study [3] indoor (wind tunnel) experiments were performed to get more insight into the effect of wind on PV module temperature. These measurements showed important intra-module temperature differences on an illuminated PV module subjected to wind. Fig. 5 shows the spatial temperature distribution of such an illuminated (300 W/m²) PV module, mounted at an angle of 10° and with a gap of 5.5 cm between the module and its supporting frame.

As expected, the cooling of the module depends significantly on the wind speed; lower module temperatures were observed at higher wind speeds. Note that, within the module, greater temperature differences were observed at *lower* wind speeds. This is in line with the outdoor measurements shown in Fig. 4. The average wind speed was lower on 1st July compared to 29th July. This led to a larger temperature gradient, and therefore, to higher energy yield differences at sub-string level. Temperature differences were most likely also affected by the angle of incidence of the wind.

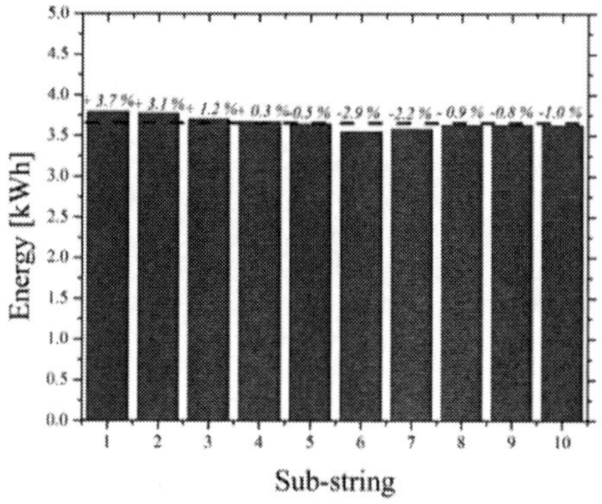

Fig. 3: Monthly energy yield of the different sub-strings in the module. The dashed line indicates the average monthly yield (3.66 kWh) and the numbers above the bars show the deviation (in relative percentage) from the average.

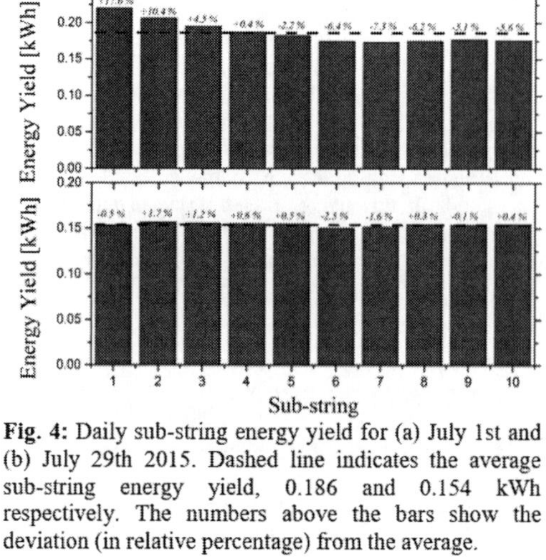

Fig. 4: Daily sub-string energy yield for (a) July 1st and (b) July 29th 2015. Dashed line indicates the average sub-string energy yield, 0.186 and 0.154 kWh respectively. The numbers above the bars show the deviation (in relative percentage) from the average.

The effect of the temperature distribution on the energy yield of the different substrings in the Ghent setup is also visible in the IV characteristics shown in Fig. 6. For the two experimental days, the IV curves were recorded at 15:00. Fig. 6a shows that, on July 1st, the short-circuit current was equal for both strings, indicating that both strings were equally illuminated. However, an important difference in open-circuit voltage (Voc) was observed. A difference in voltage of 0.71 V was measured. Since both strings were equally illuminated, one can therefore conclude that, most likely, the difference in Voc was caused by an intra-module temperature gradient, induced by wind effects.

The IV characteristics of sub-string 1 and 10 that were measured on July 29th are shown in Fig 6b. Here too the short-circuit current of both curves indicates that both strings were equally illuminated. The open-circuit voltage indicates that both string operated at the same solar cell temperature.

The differences in sub-string IV characteristics caused by the intra-module temperature differences resulted in different energy yields on sub-string level. On July 1st, large difference in energy yield were measured, while on July 29th, small energy yield differences were recorded. These intra-module energy yield variations are important for accurate short-term simulations [3,11-14], especially for those that focus on novel module topologies with multiple DC/DC converters and smart components as suggested in [9].

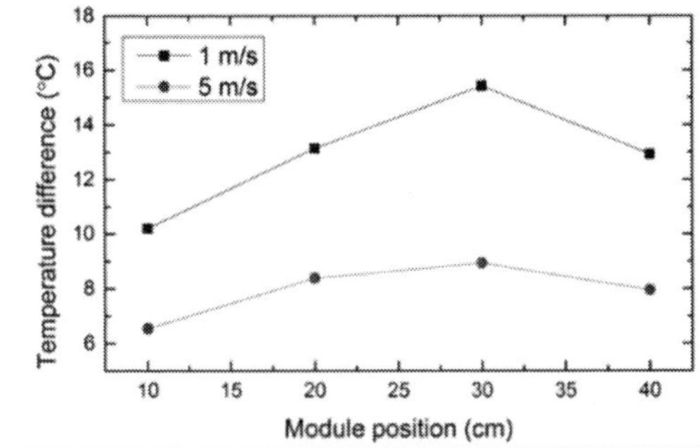

Fig. 5: PV module (illuminated at 300 W/m²) temperatures for two wind speeds. The x-axis shows the distance from the leading edge of the module, the y-axis the temperature difference compared to ambient temperature.

5 RELATIVE STDV COMPARISON

The daily amounts of energy produced by each sub-string and the environmental conditions were further analyzed to gain more insight into the effect of wind speed and direction on intra-module temperature differences. The daily amount of energy produced by each sub-string was used to calculate the average sub-string energy production as well as its standard deviation (STDV) for the sub-strings. This standard deviation is considered to be a measure of the intra-module difference since a higher value indicates higher sub-string energy variations.

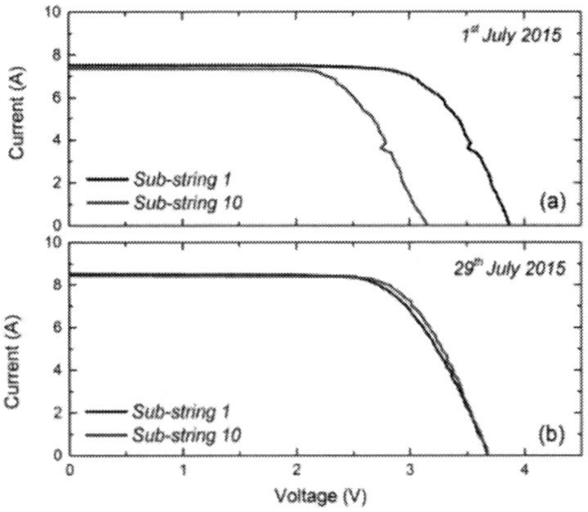

Fig. 6: IV characteristics of sub-string 1 and 10 on the 1st and 29th of July 2015.

The average daily sub-string energy production varied due to the varying weather conditions. Therefore the relative STDV (the so-called coefficient of variation) was calculated by dividing the STDV by the mean energy sub-string energy production. This enables us to make inter-day comparisons. For example, the relative STDV was 8.4 % for July 1st and 1.2 % for July 29th. The relative STDV value was then calculated for all 31 days of July 2015 and plotted against the average wind direction for each day (Fig. 7).

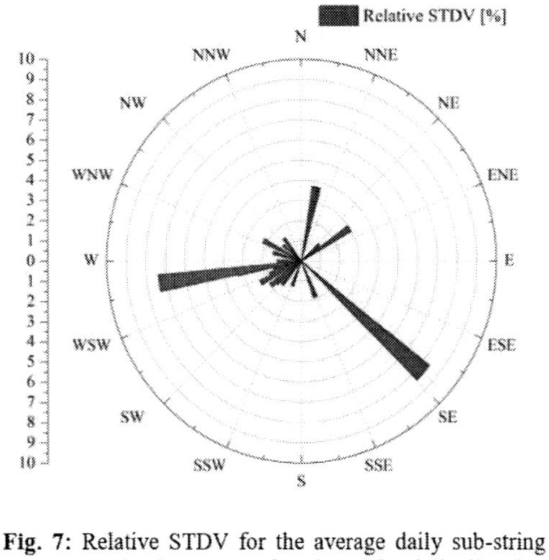

Fig. 7: Relative STDV for the average daily sub-string energy production as a function of wind direction (period: July 2015).

Two days (July 1st and 4th) had a relative STDV of more than 7 % The average wind directions for these days were respectively 131 and 261°. No strong relation between wind direction and the relative STDV for the average daily energy production was observed.

In a second step the relative STDV was plotted as a function of the wind speed (Fig. 8). The figure shows that the wind speed clearly influences the intra-module differences. It is not a proportional correlation because also other factors play a role (like irradiance and ambient temperature) but the trend is clear: the intra-module temperature differences, and therefore the energy yield differences, mainly occur at low wind speeds. Both the wind tunnel measurements and the outdoor measurements thus show that the highest sub-string, and thus intra-module differences occur at low wind velocities. We suggest that at low wind speed, natural convection starts to dominate the convection of heat. As a result , hot air is transported from the bottom to the top of the module hereby hampering the convection of heat from the top sub-strings. It follows that the strings that are located at the bottom of the module (e.g. sub-strings no. 1 and 6) will perform better compared to the strings at the top of the PV module (strings 5 and 10). For accurate energy yield evaluations these intra-module temperature differences, and the resulting differences in energy yield, difference, need to be taken into account.

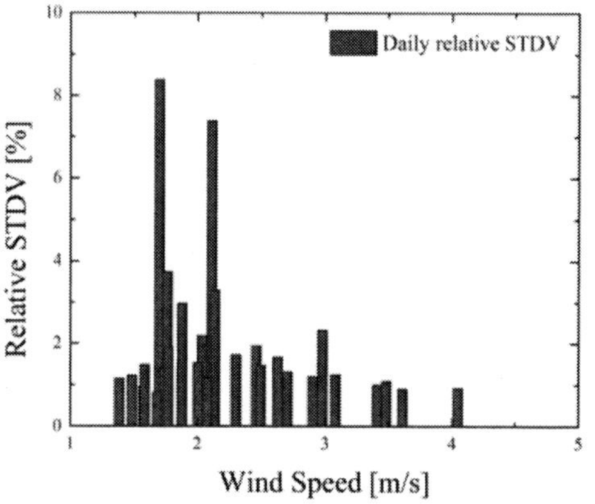

Fig. 8: Relative STDV for the average daily sub-string energy production as a function of wind speed (period: July 2015).

6 CONCLUSION

The effect of wind on sub-string level energy yield was discussed. Measurements showed a small *monthly* energy yield variation on sub-string level, however, significant variations were observed for *daily* sub-string energy yield measurements. These variations were interpreted as caused by wind effects. It has been suggested that especially the wind speed induces intra-module temperature differences, leading to different sub-string conversion efficiencies followed by sub-string level energy yield variations. Depending on the environmental conditions, an open-circuit difference of up to 0.71 V was observed, leading to daily energy yield differences of up to +17.6 % compared to the average energy yield. Therefore, intra-module temperature gradients need to be taken into account for accurate energy yield modelling.

7 ACKNOWLEDGEMENTS

The authors gratefully acknowledge the financial support of imec's industrial affiliation program for Si-PV. This work receives the support of the European Union, the European Regional Development Fund ERDF, Flanders Innovation & Entrepreneurship and the Province of Limburg.

8 REFERENCES

[1] SolarPower Europe, Solar Market Report & Membership directory, 2016 editions, available at www.solarpowereurope.org

[2] G. Makrides, B. Zinsser, M. Schubert, G. Georghiou, *Energy Yield Prediction Errors and Uncertainties of Different Photovoltaic Models*, Progress in photovoltaics: research and applications, 21, 500-516

[3] H. Goverde, D. Goossens, J. Govaerts, V. Dubey, F. Catthoor, K. Baert, J. Poortmans and J. Driesen, *"Spatial and temporal analysis of wind effects on PV module temperature and performance"*, Sustainable Energy Technologies and Assessments, 11, 36-41

[4] S. Armstrong, W. Hurley, *A thermal model for photovoltaic panels under varying atmospheric conditions*, Applied Thermal Engineering 2010; 30: 1488-1495.

[5] A. Jones, C. Underwood, *A thermal model for photovoltaic systems*, Solar Energy 2001; 70: 349-359.

[6] G. Tina, *A coupled electrical and thermal model for photovoltaic modules*, Journal of Solar Energy Engineering 2010; 132: 024501.

[7] B. Herteleer, B. Morel, B. Huyck, J. Cappelle, R. Appels, B. Lefevre, F. Catthoor, J. Driesen, *High Frequency Outdoor Measurements of Photovoltaic Modules Using an Innovative Measurement Set-Up*, 29th European Photovoltaics Science and Engineering Conference, 2700-2706

[8] TSEC semi-square 6" Mono *c*-Si solar cell, http://www.rits-solar.com/images/products/cell/E-Cell/TSS62TN_ECELL.pdf

[9] M. Baka, F. Catthoor, D. Soudris, *"Smart PV Module Topology with a Snake-Like Configuration"*, in European Photovoltaic Solar Energy Conference and Exhibition, 2015, 104-107

[10] G. Van den Broeck, B. Herteleer, H. Goverde, F. Catthoor, K. Baert, and J. Driesen, *"Substring-level Energy-yield Assessment of Photovoltaic Modules subject to Partial Shading Conditions"*, in European Photovoltaic Solar Energy Conference and Exhibition, 2015, 1–5.

[11] H. Goverde, J. Govaerts, J. Baert, F. Catthoor, J. Driesen, J. Poortmans, *Optical-Thermal-Electrical Model for a Single Cell PV Module in Non-Steady-State and Non-Uniform Conditions Built in SPICE*, 28th European Photovoltaic Solar Energy Conference and Exhibition, 3291–3295

[12] H. Goverde, B. Herteleer, D. Anagnostos, G. Köse, D. Goossens, B. Aldalali, J. Govaerts, K. Baert, F. Catthoor, J. Driesen, J. Poortmans, *Energy Yield Prediction Model for PV Modules Including Spatial and Temporal Effects*, 29th European Photovoltaics Science and Engineering Conference, 3292-3296

[13] D. Anagnostos, H. Goverde, B. Herteleer, F. Catthoor, D. Soudris, J. Driesen, J. Poortmans, *Demonstration and validation of an energy yield prediction model suitable for non-steady state non-uniform conditions*", 6th World Conference on Photovoltaic Energy Conversion

[14] H. Goverde, D. Anagnostos, B. Herteleer, J. Govaerts, K. Baert, B. Aldalali, F. Catthoor, J. Driesen and J. Poortmans, *"Model requirements for accurate shot term energy yield predictions during fast-varying weather conditions"*, 31st European Photovoltaic Solar Energy Conference and Exhibition, Hamburg, Germany, 2015

QUANTIFICATION OF LOSSES CAUSED BY DYNAMICALLY CHANGING SHADOWS IN MULTI-MW PV PLANTS BASED ON COMPUTATIONS ON TOP OF MONITORING

Gerhard Mütter[1], Benjamin Eizinger[1] , Michael Edelbacher[2]
[1] AES, Alternative Energy Solutions GmbH, A-1010 Vienna, Austria
[2] greentec services GmbH, CH-9444 Diepoldsau, Switzerland

Alternative Energy Solutions GmbH
1010 Vienna, Austria.
T: +43-1-253 10 10
F: +43-1-253 10 10 20
muetter@alteso.at
http://www.alteso.at

ABSTRACT:
This paper is a report out of practical life with a huge theoretical background and the experience of more than 5 years of operating and monitoring of more than 850 MWp large scale PV plants, located all over Europe, O&M Supervision done by AES, and more than 70 MWp small and medium sized PV plants located in central Europe, O&M coordinated by Greentec Services.

All parks have overall performance already beyond the predicted energy yields.

Fig 1: Basic situation with trees on border of park

The goal was the development of a methodology to quantify the effective losses on dynamically changing shadows like

- Growing plants in or near outside the plant
- Leaf stripping and leaf growing effects on broad-leafed trees shadowing the modules
- Missing grass cutting

in large PV-Plants with size of more than 5 MWp.

The challenge was to get a tool applicable in all our parks that delivers a precise as possible result to decide based on cost/benefit analysis about corrective activities like cutting vegetation at the economically most feasible time.

We will present results and impressive special effects of large PV plants verified in all our parks including:

- Reasons why simulation is not economical, especially in large plants
- Reasons for underestimation of the real amount of the losses in usual situations
- Basics of the methodology, including gimmicks to avoid side effects
- Impressive details of some local effects
- Details of sample parks including pictures and annual effect distribution
- Summary of all our parks

Purpose of the work: Development of a tool to decide about need and best time for cutting activities based on best possible evaluation of effective losses caused by shadows from unsteady growth of vegetation.

Approach: Using historical data of string monitoring to build for every string a unique model and use this for the evaluation of the losses caused by the dynamical changing shadowing effect.

Scientific innovation and relevance: Unique huge amount of monitored values, handled with new techniques for generating local models to quantifying losses in Multi MW PV Plants.

Results and Conclusions: Significant improvement of accuracy for deciding cost effective activities to keep and improve performance of Multi-MW PV Plants.

Keywords: large Grid-connected PV systems, Energy performance, System performance, Monitoring, Defects, Shading, Plant Control

1 INTRODUCTION

The long term yield stability and operational safety of PV plants are primarily depended on the longevity and reliability of the modules and system technology components used. Quality control of single components is usually included visual checks, performance measurement, thermal imaging and electroluminescence testing, ect. However, in Multi-MW PV Plants this can only safeguard that all components meet the necessary requirements at construction and operational phase. Recognizing of faults and further risk minimizing of technical failures during the 20-25 years operation time even permanent monitoring is not enough to guarantee optimum performance. One of the most underestimated failure is the impact of shadows, which did not exist at construction time or during the usual Due Diligence at start of operation. Vegetation in and near outside the park is growing away from the scope of the technicians operating the park and taking care of the electrical perfection.

AES Alternative Energy Solutions GmbH acts as an Owner Representative Advisor and coordinates in this function asset management as well as operation and maintenance of more than 850 MWp photovoltaic. Based on the needs to handle these huge assets a special software toolset, the Performance Improvement Tools "PIT", a special analysis toolset with computations on top of monitoring, was developed and helps other O&M supervisor controlling and improving their assets.

Greentec services gmbh is one of the technical most competent AES´s clients that uses PIT to get a more precise knowledge about details in the PV-parks that they support with operation & maintenance services. Combining the different size and location of the serviced plants guarantees that the single tools suit to a wider range of different PV plants.

The main focus of our work was the development of a tool to decide about need and best time for cutting activities. Therefore the best possible evaluation of effective losses caused by shadows from unsteady growth of vegetation was used to get clear decisions about economically feasible activities of the corrective activities.

We showed that the major problem in evaluation of the complete loss in large plants are many local differences on the effects. They do not allow to pick up single situations and extrapolate to the whole PV-plant. This is one of the main reasons why a solution by taking simulation software is economically not feasible, especially in large plants with more than 5MWp per unit.

2 CHALLENGES TO GAIN ACCEPTABLE RESULTS

On a more detailed view to the mission some pitfalls got obvious.

Fig 2: Unsteady border of shadow

Caused by the size of the parks we had to consider a lot of discontinuous side effects like

- Different grow rates of different kinds of vegetation a specific locations
- Different distribution of leaves and branches on different trees
- Different module cooling behavior near the border of the shadows
- Local wind caused by local temperature differences between sunny and shadow regions

Fig 3: impact on little change of sun angle

Away from the challenges of the effects of discontinuous growing vegetation we had to consider the irregularities of local weather situation like

- Local weather effects like sunny afternoon hours on cloudy weather rest of the day
- Unexpected long time of shadow impact during sunny winter days
- Foggy days with only ambient irradiation for a longer time

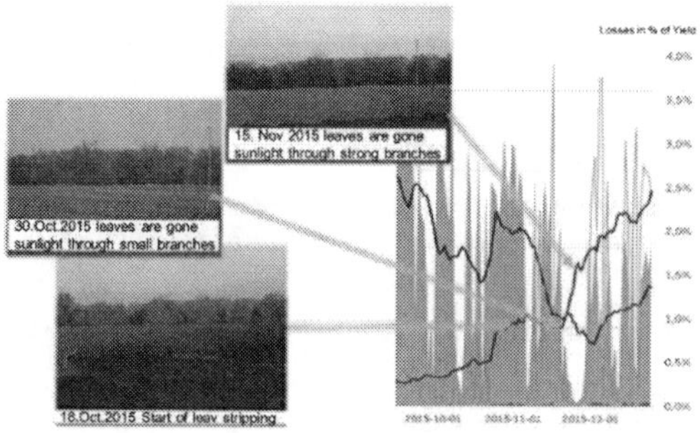

Fig 4: Reduction of shadow effect when leaves are falling

3 METHODOLOGY

Taking all these effects in consideration a dynamic loss profile of the park has to be evaluated as a major step to get for every string his unique profile. The following chart shows the number of infected strings at defined dates where clear weather situation allowed a precise evaluation of the sun height where shadow starts to have significant impact.

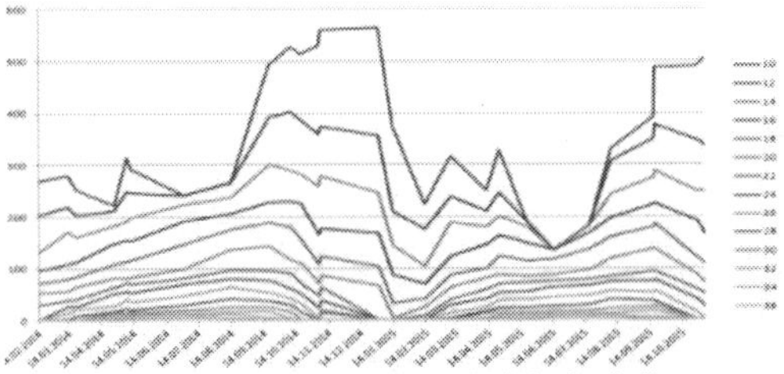

Fig.5: Number of strings with sun height for influence at defined level

Based on this global impact indication for every string the critical sun height for every string is calculated by generating a function for start of impact with respect to sun height.

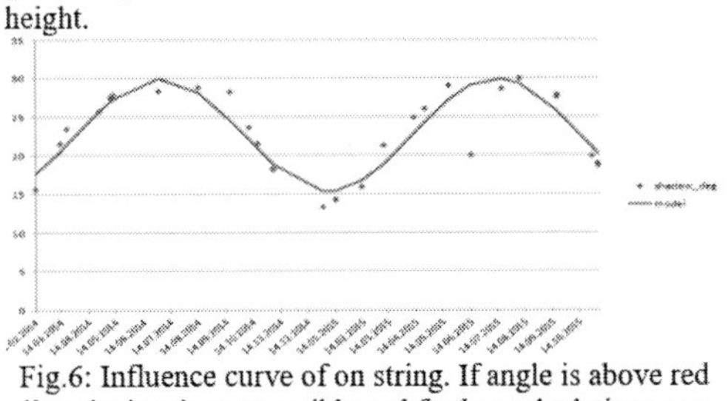

Fig.6: Influence curve of on string. If angle is above red line shadow is not possible and further calculations can be skipped.

This unique function for every string is taken for comparison between normal illuminated strings and shadow infected strings during all timeframes where effect is relevant for specific string. The example in the following chart shows the impact of trees over a segment of 600 m near outside the border of a 34 MW park. The annual losses in 2015 only by the effect of the growing trees have been 165.000 kWh, which was quite unexpected high. The yellow arrows point to the effect of leaf growing and leaf stripping which is compensated as soon as the sun height is at a level that the shadow is released more by branches than be leaves (orange arrows)

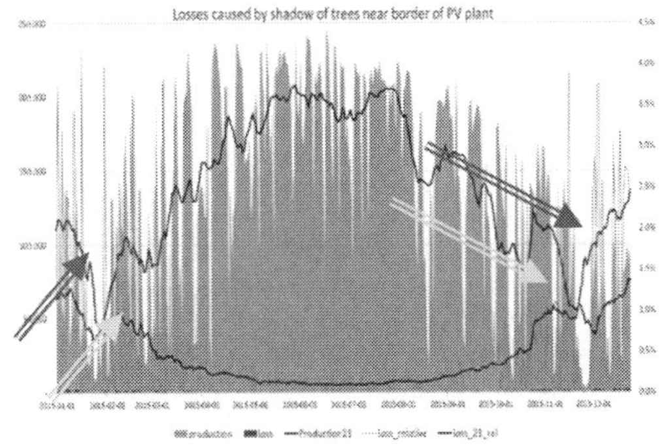

Fig 7: Park production and losses for 2015

For getting a better impression of the different effects we used also methods to smoothen the results without losing accuracy on the loss summary.

To evaluate the losses we used the monitoring values of every string in 1 minute timeframes and filtered all values where the effect is not valid for each string.

We showed the overview of the results of all our parks and highlight some specific details.

There are parks, where the costs for cutting will not exceed the losses, as well as unexpected high losses in underestimated situations.

4 EXAMPLES OF APPLICATION

We analyzed all parks with significant vegetation at the border and presented the results of 2 Parks

PARK 1

- Size: 34,14 MW,
- Location: Odessa Region Ukraine
- 141.812 Crystalline modules (230Wp-245Wp)
- 52 Inverter a 630kVA
- Start of Production: 2013

Fig 8: aerial view to PARK 1 including locations of trees and infected strings

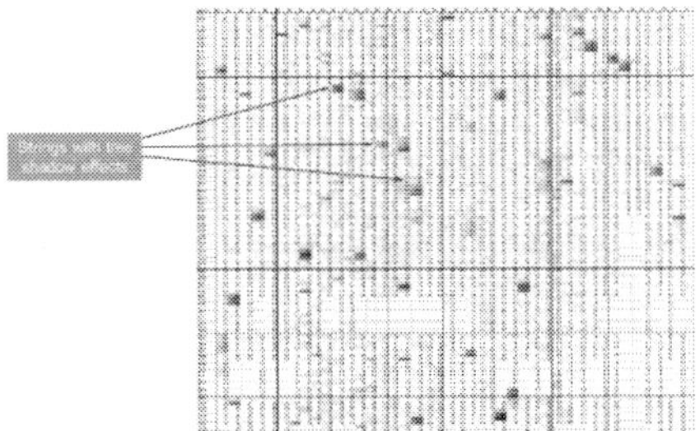

Fig 9: String Performance map for Park 1 logical view to infected strings

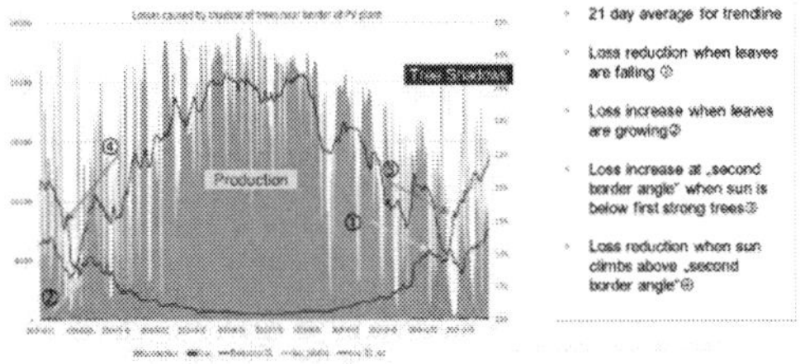

Fig 10: Daily production and level of loss

It is clear visible, that on days with low production the percentage of impact is growing dramatically

PARK 2

- Size: 6,1 MW,
- Location: Germany
- 24.680 Crystalline modules (230Wp)
- 44 Inverter a 125kVA
- Start of Production: 2011

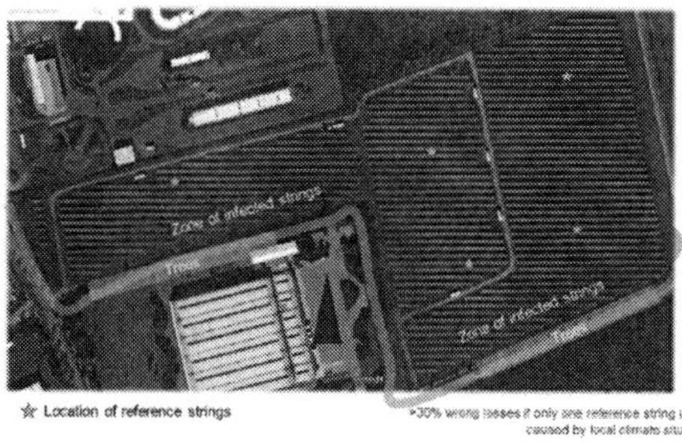

Fig 11: aerial view to PARK 2 including locations of trees and infected strings

An unexpected result of our investigation was caused by different local climate situation the need for more reference strings to compare local losses.

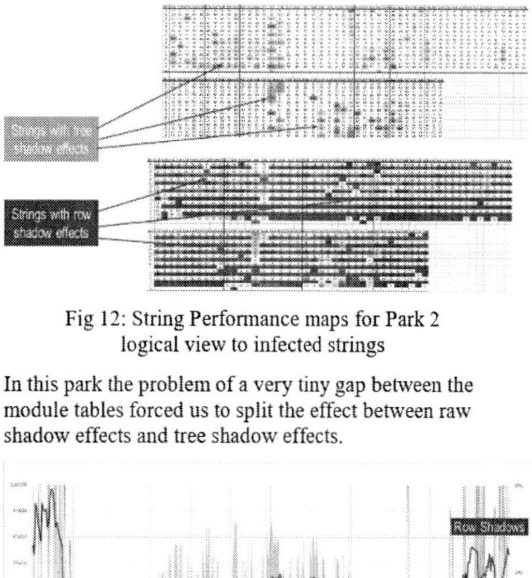

Fig 12: String Performance maps for Park 2 logical view to infected strings

In this park the problem of a very tiny gap between the module tables forced us to split the effect between raw shadow effects and tree shadow effects.

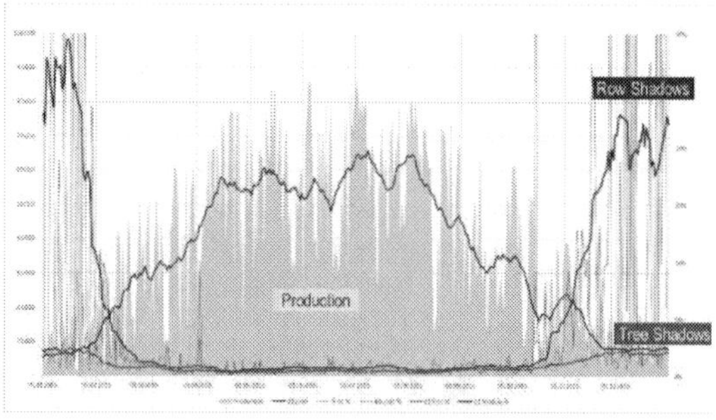

Fig 13: Daily production and level of loss

Here it is clear visible that the impact of raw shadow in this park is significant more than the tree shadow effects.

5 FINALLY QUANTIFIED RESULTS / IMPACT 2015

We got a final results on the losses that we could identify as a tree loss with:

PARK 1:

Tree losses 190.000 kWh (0,4% of annual yield)

PARK 2:

Row shadow losses

257.600 kWh (1,96% of annual yield)

Tree losses 71.600 kWh (0,54% of annual losses)

References

[1] M. Ali, M. Khan, K. S Karimov, "Design of a cost efficient pv parameters monitoring system", Sci., Tech. and Dev., Vol. 31 3 (2012) 244-250.

[2] S. Silvestre, A. Chouder, E. Karatepe, "Automatic fault detection in grid connected PV systems", Solar Energy 94 (2013) 119–127.

[3] N.H. Reich, B. Müller, A. Armbruster, K. Kiefer, W.G.J.H.M. van Sark, Ch. Reise, "Performance Ration Revisited: Are PR>90% Realistic?", 26th European Photovoltaic Solar Energy Conference and Exhibition (2011) 3922.

[4] B. Müller, Th. Reis, A. Driesse, Ch. Reise, "Maximizing the yield of large pv power plants: what can we learn from monitoring and simulation?" 27th European Photovoltaic Solar Energy Conference, (2012) 3775.

[5] F. Vignola, F. Mavromatakis, J. Krumsick, "Performance of PV inverters", Proc. of the 37th ASES Annual Conference, San Diego, CA, (2008).

[6] N. Allet, F. Baumgartner, J. Sutterlueti, S. Sellner, M. Pezzotti, "Inverter performance under field conditions", 27th European Photovoltaic Solar Energy Conference and Exhibition, (2012) 3110.

[7] T. Krametz, B. Kubicek, G. Mütter, "Analyse der Ursachen für Ertragsverlust in der Anlaufphase des Einspeisebetriebs von pv Kraftwerken", 29. Symposium Photovoltaische Solarenergie, (2014).

[8] R. Bruendlinger, B. Bletterie, M. Milde, H. Oldenkamp, "Maximum power point tracking performance under partially shaded PV array conditions", 21st European Photovoltaic Solar Energy Conference, (2006) 2157.

[9] B. Mathur, "Effect of Shading on Series and Parallel Connected Solar PV Modules", Modern Applied Science, Vol. 3 10 (2009) 32.

[10] R.Ramaprabha, Dr.B.L.Mathur, "Impact of Partial Shading on Solar PV Module Containing Series Connected Cells", International Journal of Recent Trends in Engineering, Vol 2 7 (2009) 56.

[11] E. Díaz-Dorado, A. Suárez-García, C. Carrillo, and J. Cidrás, "Influence of the shadows in photovoltaic systems with different configurations of bypass diodes", International Symposium on Power Electronics, (2010).

[12] G. Mütter, Y. Voronko, T. Krametz, B. Kubicek, P. Steirer, "Evaluation of influence factors on park performance in large PV-plants based on monitoring of up to 700 MWp over 2 years" " 29th European Photovoltaic Solar Energy Conference", (2014).

ADVANCED ANALYSES OF LOSS MECHANISMS FOR PV SYSTEMS IN DELHI, INDIA

André M. Nobre[1,#], Daksh Dave[1], Alan Khor[1], Raghav Malhotra[1], Shravan Karthik[1,2], Ian Marius Peters[3] and Thomas Reindl[2]
[1] Cleantech Energy Corporation Limited, 25 Church Street #03-04, Capital Square 3, Singapore 049482
[2] Solar Energy Research Institute of Singapore (SERIS), National University of Singapore, 7 Engineering Drive 1, Block E3A, #06-01, Singapore 117574
[3] Massachusetts Institute of Technology, 77 Massachusetts Avenue, 02139 Cambridge, USA
[#] andre.nobre@cleantechsolar.com

ABSTRACT: Although the vast majority of systems in India are still deployed in multi-MW scale in solar farms, cities like New Delhi have also seen a surge in industrial rooftop PV facilities, especially in the suburbs and the greater surroundings of the city center (within a 100 km radius). Due to its massive population and the combination of millions of vehicles and many heavy-industry zones, it is only natural that air pollution exists. The smog gets further accentuated by agriculture fires at certain portions of the year. This work aims at investigating two loss mechanisms that prevent PV systems in the Delhi area from achieving their maximum power output potential: soiling due to dust accumulation on the modules' surfaces; and air pollution (i.e. blockage of irradiation from reaching the surface of PV systems), a much less investigated parameter in PV system technology research. In this work, we found that soiling accumulation rates for Delhi can lead to losses between 0.3% and 0.6% per day. Air pollution related blockage could represent yield losses of ~6% on an annual basis. Such information is crucial for PV system owners and developers alike when evaluating PV asset deployments in Delhi, as well as maintaining these.

Keywords: PV System, System Performance, Loss Mechanisms, Air Pollution, Soiling

1 INTRODUCTION

Worldwide photovoltaic system deployment has seen a shift over the past years away from traditional markets such as Europe or the United States towards Asia, and here predominantly China, Japan and India. This also means that PV systems are installed in climate zones where less experience has been gathered and where new challenges arise, such as year-round high ambient temperatures and humidity levels.

In India, as an example, the Prime Minister has set a target of 100 GW_p of photovoltaic capacity by 2022, of which 60 GW_p would be utility scale projects and 40 GW_p rooftop installations and other small grid projects [1]. India has very recently crossed the 5 GW_p mark for cumulative installed capacity, securing it a 9th position globally by cumulative volume, and still amongst the top five for annual installations [2]. The Indian government has plans to increase the power generation capacity from renewables to 40% by 2030, translating to approximately 350 GW_p of clean energy sources [3]. Unfortunately, the nation faces some rudimentary issues to reach maximum PV efficiency, as outlined in this paper.

Apart from ground-mounted solar farms, which are normally installed a few hundred kilometers away from major city centers (where land is less expensive), rooftop PV installations have grown considerably, especially in the commercial and industrial sector. Cities like New Delhi have also seen a surge in commercial and industrial rooftop PV facilities within a 100 km radius of the city center, especially in the suburbs.

Due to its massive population and the combination of millions of vehicles and many heavy-industry zones, air pollution in New Delhi has increased in the past years, which can act in two-fold to deter optimal PV system performance – i) the reduction in power output by soiling accumulation from dust particles present in the air which get deposited onto panels (see example of such an accumulation in Fig. 1) and, ii) a less investigated topic, the blockage of sunlight from reaching PV panels via scattering or absorption of pollutant aerosols in the air, thus reducing the expected energy yield (see also comparison images in Fig. 2). The smog in the Delhi region gets further accentuated by agriculture fires at certain portions of the year, namely during the months leading to winter (Sep-Nov) and during the first part of the winter (Dec-Jan).

Figure 1: Advanced soiling observed at a PV system in Delhi, India. On the left, panels after a three-month period without cleaning (Aug-Oct/2015). On the right, a section of the PV system is seen after a cleaning (Nov/2015).

Figure 2: Visual air pollution at a PV system in Delhi, North India (left image, Dec/2015), compared with another site in Chennai, South India, with blue skies present (also taken in Dec/2015).

This work aims at separately investigating the two described loss mechanisms that prevent PV systems in the Delhi area from achieving their maximum power output potential. The investigation further intends to benchmark PV systems with and without these two loss mechanisms to comprehensively evaluate how these factors impact PV systems' performance.

Chapter 2 discusses further motivational points for the creation of this work, with Chapter 3 addressing the method implemented. Chapter 4 presents the results and discussions for the investigations as well as recommendations for further works. Chapter 5 concludes the paper.

2 MOTIVATION

Pollution and soiling are two deterrents to the effective functioning of PV systems, and there is sufficient and critical evidence that they need to be addressed.

Soiling has been a topic more thoroughly studied in the recent past, but early beginnings date back to the 90s, all the way to the major growth period in the past two decades. The thesis in reference [4] present a comprehensive overview of soiling as a loss mechanism and its influence on system power output.

Present literature shows little research work that has been conducted regarding the effects of air pollution on the performance of PV systems. A recent study by some of the authors of this work has shown though that losses in PV power output in Singapore can be in the range of 15-25% during seasonal haze episodes triggered by agriculture fires in nearby countries [5].

India is a typical example of the many other countries that face such challenges triggered by air pollution, especially China, but also Indonesia, The Philippines or Malaysia and, generally, any megacity center in Asia. The common hurdles go far beyond air pollution problems [6]. The so-called "pollution standards index" (PSI) is a metric commonly used in countries to gauge levels of air pollution in city skies. PSI data obtained from readings at the United States Embassy in New Delhi and Chennai show the critical situation at hand which is object of this study (see Fig. 3).

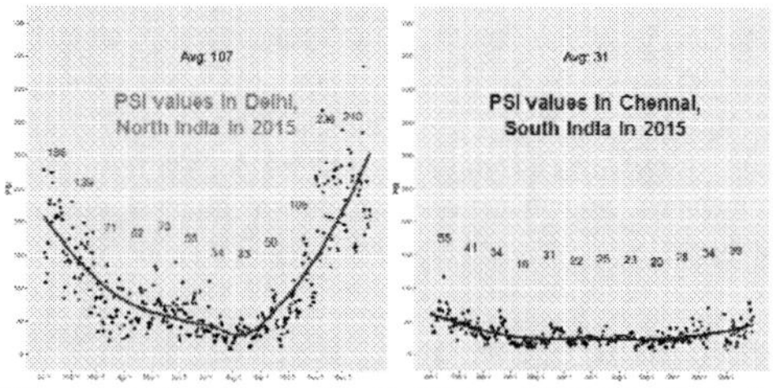

Figure 3: PSI levels for Delhi for 2015 highlight the peaks in air pollution in the late fall and winter months (left). The PSI levels for a relatively cleaner city in the South of India (Chennai) is shown as comparison (right).

New Delhi has been pegged as one of the most polluted city centers in the world [7]. With an ever-increasing fleet of PV installations to be commissioned in emerging economies [8], the loss mechanisms described in this work have to be better understood to early determine targeted and cost-effective mitigation strategies. It is evident that a resolution needs to be put forward, not only to prevent detrimental commercial and economic impacts on the growth of a nascent PV industry, but foremost to protect the population of city centers around the world from health hazards.

3 METHOD

This paper addresses reduction in the power output of PV power systems due to both soiling and air pollution. To identify the impact of these effects, we use analytical procedures to process ground data collected from various PV systems (AC power output) and sensors (in-plane global irradiation) for the period of three months in Delhi, north India.

3.1 Soiling analysis

To gauge the immediate effect of soiling (see Fig. 1), PV systems' performance ratios are evaluated at several sites, with cleaning routines in place for certain sections of the system (see example of a cleaned/uncleaned section in Fig. 4). The cleaning routines allow for an inter-comparison exercise of different sub-systems with identical characteristics (e.g. same tilt and orientation) on the same rooftop. One of such soiling tests is prepared whereby particular system sections were left uncleaned for three months during the months of Aug-Oct/2015 in order to gauge the soiling accumulation impact.

Figure 4: Two PV system sections are shown, with a 3-month uncleaned section at the far end of the image, and a freshly-cleaned portion presented in the foreground.

As soiling accumulates fast on ground-based irradiance detectors, hourly satellite data is used to derive a concept of performance ratio (PR), based not on a tilted irradiance sensor, but rather on tilted values of data coming from the satellite model, adjusted to a tilt/orientation of the system at hand. This step allows for a refinement in the analysis, thus comparing PRs before and after cleaning cycles without irradiation sensor biases. An average PR of five days prior to the cleaning was contrasted with the PR of five days post-cleaning. Satellite data were provided by GeoModel Solar.

In a second investigation step, a meteorological station was then deployed (Mar/2016), among its sensors are silicon sensors for global horizontal irradiation measurement that are cleaned on a daily basis. An extra sensor was deployed to facilitate this study, acting as a soiling baseline. The sensor was not cleaned for the first two months of lifetime of the station which coincided with the dry season of New Delhi. The exercise was carried until a major rain shower took place and effectively cleaned the soiling sample sensor.

3.2 Air pollution analysis

To determine the effect of air pollution (see Fig. 2), readings from irradiance sensors were filtered according to the methodology presented in [5], conducted in Singapore. The findings are matched against a clear sky model generated for Delhi, where the air pollutant standard index is usually above 300 for concentration of PM 2.5 (PM = particulate matter) during heavy periods of air pollution. Comparisons are drawn to a city in South India (Chennai) where air pollution is minimal. A weighted yearly loss value is estimated for the Delhi region, based on previous literature.

Silicon sensors in the station are factory-calibrated (±5.0% of precision, from Mencke & Tegtmeyer). A pyranometer was also deployed for future investigations and has a precision of ±2.0% (SMP10, from Kipp & Zonen). Ambient temperature and relative humidity sensors are also present to facilitate secondary analysis (e.g. identification of rain showers due to humidity spikes).

4 RESULTS & DISCUSSIONS

4.1 Soiling investigations

Four PV systems deployed in the period of Jul-Sep/2015 were evaluated in the Delhi region, all within a 100 km radius from the center of the city. All systems had no cleaning routines implemented since deployment. The period between deployment of the systems and the end of the exercise (early Dec/2015, thus ~3-4 months) did not present many occurrences of rain showers.

The average 5-day performance ratio of the systems, based on satellite-derived data (and not via ground-sensors), was evaluated. Satellite data from the Delhi region was chosen as a baseline for the study especially as some sites did not have irradiance sensors or had sensors that had not been cleaned regularly.

Table I displays the results found for performance ratio gains in absolute terms after the cleaning interventions. Results of PR improvement ranged between 27-39%. As the systems had been left uncleaned for 3-4 months, it could be inferred that potential soiling accumulation rates ranged between 0.3-0.6%/day.

Table I: Sites where cleaning routines were applied after a period of 3-4 months without interventions. An uncleaned system section was used as a baseline at each of the sites. PR gains obtained for a sequence of 5 days pre- and post-cleaning.

Site Location/ Number	Performance ratio, satellite-derived [%, before clean]	Performance ratio, satellite-derived [%, after clean]	Absolute gains in PR_{sat} [%]
Delhi #1	53%	81%	+28%
Delhi #2	54%	81%	+27%
Delhi #3	35%	70%	+35%
Delhi #4	43%	82%	+39%

With the goal of assessing the soiling accumulation rate in more detail, the meteorological station deployment was performed in order to provide further insights. As discussed in 3.1, one of the silicon sensors was left uncleaned since the start of operations of the station (17th March 2016). Fig. 5 (left) shows the front glass surface of the sensor fully cleaned at the date of commencement of the exercise. The image in Fig. 5 (center) shows the soiling accumulated on the surface of the sensor after two months. Irradiation losses in relation to a daily-cleaned baseline sensor stabilized at about 20-23% (see Fig. 6), with the slope indicating soiling accumulation rates of 0.3-0.4%/day.

Figure 5: Uncleaned silicon sensor at the date of the meteorological station deployment (Mar 17th left), two months later (center) and a few days later after consecutive heavy rainy days (right).

Heavy rain showers occurred in New Delhi closer to the end of May/2016. The rain was sufficient to almost entirely clean the sensor (see Fig. 5, right, also shown in Fig. 6 with values of irradiation ratio against the clean sensor nearly back at day 1 levels).

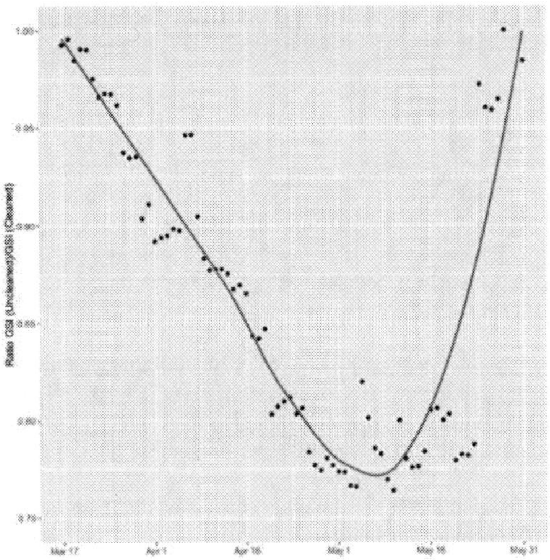

Figure 6: Loss of irradiation at the uncleaned silicon sensor in relation to a daily-cleaned identical sensor. The duration shown covers a period of ~2 months when rain showers occurred, bringing levels nearly back to those for a clean sensor. The curve fit serves as guide to the eye.

Although soiling accumulation rates varied between the two exercises, the investigations serve as a general guide for the quantitative impact of soiling for the Delhi region, which in turn could facilitate cleaning interval timings for system owners and developers alike. Additionally, the visual outcomes of the exercise indicate that PV systems in this area require regular cleaning.

It could be further discussed that the silicon sensor surface does not match the one of a solar PV panel, in terms of its area and even the type of glass cover. However, such a second exercise with two benchmarked silicon sensors is of relatively lesser complexity/cost to stakeholders and could be further replicated elsewhere.

4.2 Air pollution investigations

With the absence of a meteorological station for the hazy months of November through February, one PV site with a silicon sensor was selected whereby the sensor was cleaned at regular intervals at least for the duration of time until a dedicated station was launched. The results here arise from those readings. As an example, Fig. 7. shows the filtered values utilizing the methodology presented in [5], with the generation of typical meteorological days (TMDs). When compared to a clear sky model for Delhi for November, the values in Fig. 7 indicate a drop of ~20% in irradiation for the period.

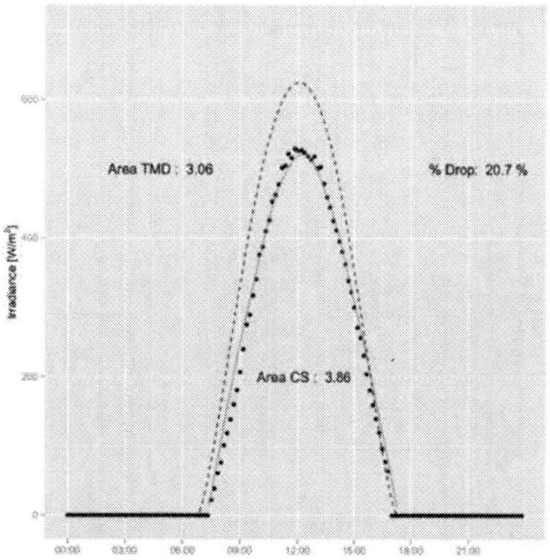

Figure 7: Reduction of ground irradiation for the month of November in Delhi, as per methodology previously implemented in Singapore [5], using ground readings and sequential filters.

For assessing the impact of air pollution, data from the hazy months of November through February allowed some initial gauging of irradiation loss levels, which we estimate to range from 20% to 40% for single days, and up to 30% for the heavy polluted months on average. The estimated losses for the other months of the investigation are presented in Table II.

Table III summarizes the findings of the previous table. When weighing the potential loss levels of irradiation due to haze for the months of the year, a total loss of ~6% is estimated.

Because some sites are rather remote, defined by the availability of operation & maintenance staff, cleaning of irradiance detectors cannot happen on a daily basis when deployed next to the PV systems. This likely caused errors in the readings when taking a whole month into account (in relation to previous investigations conducted for Singapore, which took place under optimal conditions). Such investigations could be further refined with the new data after the deployment of the meteorological station, with daily maintenance routines in place. The station will allow for further investigations, especially with the arriving of the next winter and haze seasons. The young age of the station at the time of this work completion, coupled with the presence of low levels of PSI during the Mar-Jun/2016 months in Delhi, did not allow for a continuous assessment which would include a potential return of high air pollution levels in the city (see Fig. 8).

The newly deployed meteorological sensor would also allow for more in-depth analysis of the influences of air pollution since the station instrumentation followed the one used in the paper in [5] and moreover, in the Delhi context, has its pyranometer and silicon sensor cleaned on a daily basis, which would eliminate any influences to the readings arising from soiling.

Table II: Monthly PSI average values for Delhi in 2015, with values for Jan to May taking 2016 figures into account (denoted by *). Measured values of irradiation loss using figures from a silicon sensor at a PV site are shown for Nov/15 to Feb/16. Values from Mar/16 to May/16 are available from the new meteorological station. Values for the other months of the year are estimated based on the findings in [5].

Air pollution losses	Jan*	Feb*	Mar*	Apr*	May*	Jun
PSI average in Delhi	286	142	26	17	24	55
Measured/Expected average monthly % losses (ground)	**28%**	**10%**	**2%**	**2%**	**2%**	1%
Monthly irradiation, averages [kWh/m^2]	104.2	122.1	172.4	195.9	217.0	190.2
Weighted monthly losses [kWh/m^2]	29.2	12.2	3.5	4.0	4.3	1.9
Net monthly irradiation [kWh/m^2]	75.0	109.9	168.9	191.9	212.7	188.3
Air pollution losses	Jul	Aug	Sep	Oct	Nov	Dec
PSI average in Delhi	34	33	50	105	236	240
Measured/Expected average monthly % losses (ground)	~0%	~0%	1%	5%	**15%**	**20%**
Monthly irradiation, averages [kWh/m^2]	170.8	152.8	155.4	150.7	122.4	102.6
Weighted monthly losses [kWh/m^2]	-	-	1.6	7.5	18.4	20.5
Net monthly irradiation [kWh/m^2]	170.8	152.8	153.8	143.2	104.0	82.1

Table III: Summary of findings from Table II, with average annual losses at ~6%.

Air pollution losses	Average/Total
PSI average in Delhi	104
Monthly irradiation, long-term, averages [kWh/m^2]	1,856
Weighted monthly losses [kWh/m^2]	103.1
Net monthly irradiation [kWh/m^2]	1,753 (-5.6%)

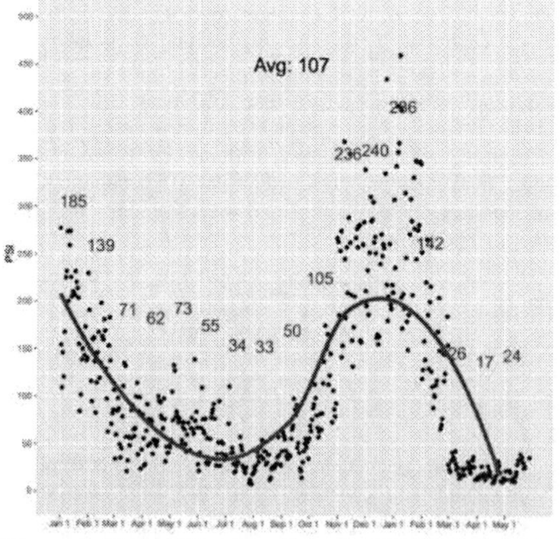

Figure 8: Daily PSI values from Jan/2015 to May/2016 for Delhi (dots) with the numbers as monthly averages.

Delhi has implemented a car rotation system in late 2015 with the aim to reach lower gas emissions and better air quality levels [9]. In the system, it is intended that a major portion of the fleet of nearly 9 million vehicles would be off the roads. Such a move shall be put to test once other sources of air pollution come into play in the late fall and early winter seasons.

6 CONCLUSIONS

With negligible precipitation in Delhi during the months of November 2015 to January 2016, results for the soiling exercises were validated by a single cleaning cycle as well as via findings form a newly deployed meteorological station. Daily soiling loss accumulation rates of 0.3% to 0.6% were assessed. The cleaning cycle revealed gains in performance ratios of up to ~40%, highlighting the necessity of regular cleaning routines to guarantee profitability to investors and system owners alike in Delhi. The rainy season start could provide further insights into the level of self-cleaning potential for the region.

Reductions of irradiation due to air pollution of up to 40% were detected for single days under analysis. More investigations are needed with irradiance sensors that are cleaned on a daily basis and which are now available in a recently deployed meteorological station, also benchmarked against other sensors (such as a research-grade pyranometer). Lower levels of air pollution during the summer months in Delhi suggest that the effect throughout the year, weighted against historical irradiation levels for the city, would equal to total annual yield losses of ~6% in PV systems deployed within a 100 km radius of the city center.

Deployments of further stations elsewhere in India could assist the effort of gauging soiling accumulation at different locations, but also the air pollution aspects at cleaner sites or even at other problematic ones.

ACKNOWLEDGEMENTS

A.M. Nobre would like to acknowledge the field support of Ankur Ghosh, Sahil Kumar, Rohit Jaswal and Rupesh Baker. Additionally, the lead author would like to thank Gavin Adda and Geoffrey Jahnke for their support in this work.

REFERENCES

[1] India Times. "India likely to be among top solar markets in 3 years". http://economictimes.indiatimes.com/industry/energy/power/india-likely-to-be-among-top-solar-markets-in-3-years-research-firm-bridge-to-india/articleshow/47599509.cms

[2] Trends 2014 in photovoltaic applications - Survey report of selected IEA countries between 1992 and 2013. International Energy Agency, 2014.

[3] http://cleantechnica.com/2015/09/23/india-aims-350-gw-renewable-energy-capacity-2030/

[4] H. QASEM, T. R. B., H. MÜLLEJANS, H. ALBUSAIRI, R. GOTTSCHALG. Effect of Dust Shading on Photovoltaic Modules. In: 26th European Photovoltaic Solar Energy Conference and Exhibition, 2011. Conference Proceedings. Hamburg, Germany, 2011.v.p. 3652 - 3656.

[5] NOBRE, A.M., KARTHIK, S., LIU, H., YANG, D., MARTINS, F.R., PEREIRA, E.B., RÜTHER, R, REINDL, T., PETERS, I.M. On the impact of haze on the yield of photovoltaic systems in Singapore. Renewable Energy, Vol. 89 (2016). http://dx.doi.org/10.1016/j.renene.2015.11.079.

[6] VELASCO, E.; ROTH, M. Review of Singapore's air quality and greenhouse gas emissions: Current situation and opportunities. Journal of the Air & Waste Management Association. Issue 6, v.62, p.625-641, 2012.

[7] CNN. "I live in the world's most polluted city". http://money.cnn.com/2016/02/17/news/economy/india-new-delhi-air-pollution-health/index.html?iid=EL

[8] Solar Power Europe. Global Market Outlook for Solar Power - 2015-2019. 2015

[9] CNN. "India's pollution control measures". http://edition.cnn.com/videos/world/2016/01/12/india-new-delhi-pollution-controls-field-pkg.cnn

This is a pre-peer review version of the paper submitted to the journal Progress in Photovoltaics.
It has been selected by the Executive Committee of the EU PVSEC 2016 for submission to Progress in Photovoltaics.
32nd European Photovoltaic Solar Energy Conference and Exhibition

PV DEGRADATION CURVES: NON-LINEARITIES AND FAILURE MODES

D.C. Jordan*, T. J. Silverman, B. Sekulic, S.R. Kurtz
National Renewable Energy Laboratory (NREL), 16253 Denver West Parkway, Golden, CO 80401
*Tel: 1-303-384-6762; Fax: 1-303-384-6790; email: dirk.jordan@nrel.gov

ABSTRACT: PV reliability and durability have seen increased interest in recent years. Historically, and as a preliminarily reasonable approximation, linear degradation rates have been used to quantify long-term module and system performance. The underlying assumption of linearity can be violated at the beginning of the life, as has been well documented, especially for thin-film technology. Additionally, non-linearities in the wear-out phase can have significant economic impact and appear to be linked to different failure modes. In addition, associating specific degradation and failure modes with specific time series behavior will aid in duplicating these degradation modes in accelerated tests and, eventually, in service life prediction. In this paper we discuss different degradation modes and how some of these cause roughly linear degradation (e.g. modules that were mainly affected by encapsulant discoloration) while other degradation modes lead to distinctly non-linear degradation (e.g. hot spots caused by cracked cells or solder bond fatigue and failure). The various behaviors are summarized with the goal of aiding in predictions of what may be seen in other systems.
Keywords: photovoltaic, reliability, durability, failure mode

1 INTRODUCTION

PV durability and reliability questions have attracted increased interest in recent years because of their technological and economic significance. Reliability is the ability to perform a required function for a given time interval and is often measured in terms of failure rate or as a probability for failure. [1]In contrast, durability relates to the time interval a system is performing its desired task and is in PV commonly measured as the degradation rate, the slow gradual loss of performance. Literature degradation rates were summarized and analyzed by some of the authors and were recently updated. [2,3] However, the fallacy lies in the word "rate" because of the intrinsic linearity assumption. While the linearity assumption may be a preliminarily appropriate choice, it is often violated in the infant or wear-out phase of a system's life cycle. This can have significant economic consequences, as illustrated in Fig. 1. Four different degradation curves, each of them observed (approximately) in field performance, are used in Monte Carlo simulation to quantify the effect on levelized cost of energy (LCOE). [4] The spider graphs of Fig. 1 (b) constitute the sensitivity analysis of the simulations. The most significant factors (for the ranges explored) on LCOE are the discount rate and initial cost indicated by the greatest range on the ordinate for each of the graph compartments. The curve characteristic of all the input variables does not change for the different input degradation curves. However, the mean of the graph shifts for the different degradation curves. The impact of the different degradation curves is ca. 1.5 c/kWh making it the third most important factor after the discount rate and the initial cost. Therefore, measuring and including linear degradation rates in models may not be sufficiently accurate.

Besides the illustrated economic motivation, the technically compelling reason to determine degradation non-linearities is the development of lifetime prediction models. One pitfall of accelerated testing is that it may over-accelerate a specific failure mode while another failure mode may be masked that could turn out to be dominating in the field under potentially different use conditions. [5]

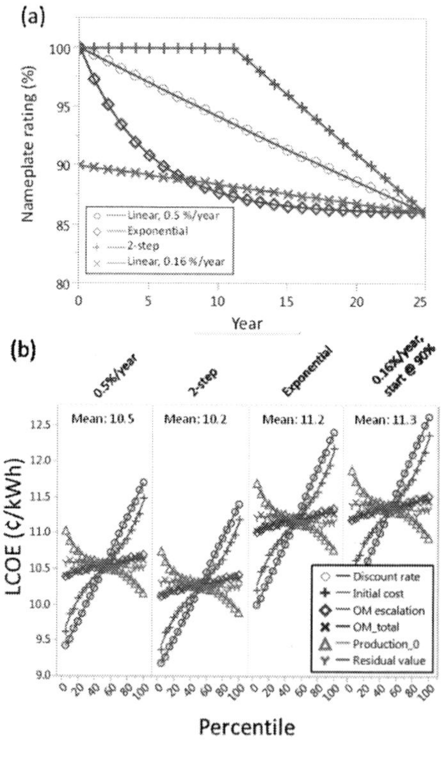

Figure 1: Monte Carlo simulation of various PV degradation curves' impact on LCOE after 25 years. Four different input degradation curves potentially encountered in field measurements (a) and sensitivity analysis impact on LCOE (b). Discount rate is the rate at which future cash flows are discounted to present day values. Operational & maintenance (O&M) escalation takes into account increased O&M expenses with increasing age and residual value is the value of the project after 25 years.

To guard against this apparent danger, the underlying failure mechanism, the physical and or chemical cause, must be understood. Therefore, understanding specific failure or degradation modes and their function of time is a crucial step in synchronizing chamber and field test observations. Non-linearities are typically more easily observed in accelerated tests than in field observations because the faster changes are easier to detect. Perhaps because of the variety of conditions that modules are exposed to in field tests or because of the synergistic nature of some degradation modes, fielded modules often display a variety of degradation modes making it difficult to correlate a specific failure mode with a specific time series behavior.

In this paper we aim to provide some examples of degradation modes and their different behavior as a function of time, though in some cases the exact function remains to be determined. We will start with the infant phase of a PV system's life cycle and then discuss examples in the wear-out phase.

2 METHOD AND OVERVIEW

Data were collected for a variety of modules and small systems deployed at the Outdoor Test Facility at the National Renewable Energy Laboratory in Golden, Colorado. The data collection and analysis methodologies were described in references [6,7]. The soiling observed in this test yard is typically 1% - 2%. In order to reduce the effects of the soiling on the degradation rate data, the irradiance sensors were cleaned only when the PV modules were also cleaned, typically by rainfall or at the time of I-V measurements.

A high-level summary of the results is shown in Table 1. Although only one data set is shown for each line in the table, these observations are consistent with other observations either in the literature or in our own experience. Table 1 differentiates some of the changes seen immediately after deployment (infant phase) from those that may be seen after years in the field (wear-out phase). The one degradation mechanism we have seen to be mostly linear over the years is noted in the "Phase" column as "consistent" because we have observed this degradation to be constant over many years, though there is some evidence that after more than a decade the degradation rate may slow slightly.

2 INFANT PHASE

Many PV technologies, especially thin-film technologies, exhibit non-linearities at the beginning of their useful life. The initial rapid decline for amorphous silicon (a-Si) has been well documented. [8] The initial rapid decline occurs during several months before the onset of the long-term trend, as shown in Fig. 2. Cadmium telluride (CdTe) – inverted blue triangles – may also pass through a transient regime that can range from 1 to 3 years before stabilization and normal performance commences. [9,10] Light-induced degradation (LID) can also affect crystalline silicon (x-Si) systems, particularly those fabricated with Czochralski based wafers. [11,12] Yet, the time scale is much shorter than thin-film systems making it in general difficult to detect in outdoor data. The time series behavior can be modeled exponentially or separated into two distinct phases, the initial decline followed by a more stable phase. The former approach may be more accurate, yet the latter is often more convenient despite the subjectivity to determine the exact start of the more stable period. In contrast, the CIGS system shown here, which does not represent all CIGS technologies, shows a distinctly different behavior; an initial increase during the first several months of light-exposure is followed by the onset of the long-term degradation. It is evident from these examples that the initial trend can be significantly different from the long-term behavior. In addition, if the initial phase were included in the evaluation, the long-term prediction would include a substantial, yet unintended bias. In this case, a sensible choice may be to exclude the initial phase that commences with light-exposure to determine the ultimate long-term behavior. This choice may also be viable for known start-up issues such as incomplete connectivity of all strings in a system, for example.

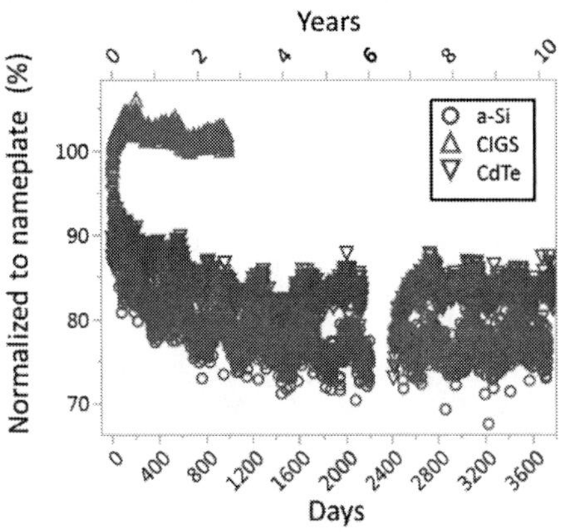

Figure 2: Initial non-linearities for thin-film technologies upon first light-exposure. Nameplate ratings are often adjusted to align with the more stable performance.

3 WEAR-OUT PHASE

In the wear-out phase of a product the failure rate increases as a function of time. The increasing number of failures may display a single mode or several; in addition, the failure modes may not be completely independent of each other. For example, delamination may initially only lead to extra current-dominated power loss because of the formation of an extra optical interface. As the delamination becomes more severe, the probability increases for moisture ingress. The moisture ingress may lead to internal circuitry corrosion and eventually even lead to internal circuitry failure. Each of these failure modes may exhibit a different power decline as a function of time, yet isolating these different functional forms is problematic. During the block buy program of the Jet Propulsion Laboratory (JPL) in California, USA, the interdependency between different durability performance and failure modes was first recognized. [13] More recently the International Energy Agency PV Power Systems Program – Task 13 – published a detailed review of failures of PV modules based on literature and site visits. [14] The treatise discusses at length inspection tools, observed failures and test methods. Different failure modes are classified by safety and their different time series behavior, e.g. linear versus non-linear, although corroborating time series data were not provided. Several studies from different continents have emerged that investigated an ensemble of modules at different times of their life cycle. [15,16,17,18] Of these ensembles the better performing modules and the central tendency of the modules appear to decline fairly linearly, however, the worst performing modules show non-linear behavior. [3]

3 DISCOLORATION

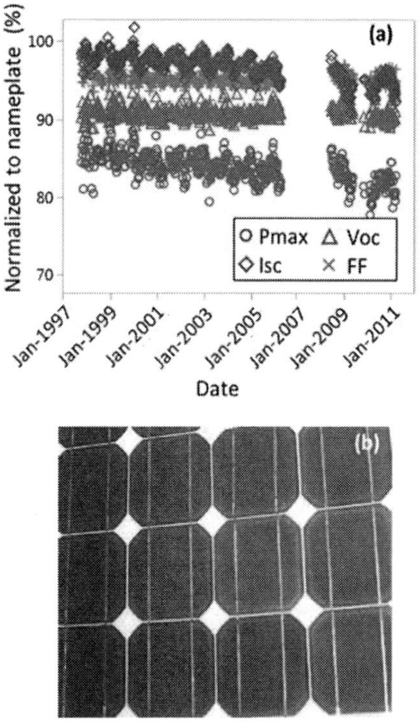

Figure 3: Mono-Si module I-V performance over 20 years of field exposure (a). The vertical axis shows the I-V parameter as a percentage of the nameplate rating. Discoloration in the center of the cells is shown (b) and some bleaching at the cell edges. Photo credit: Dirk Jordan, NREL PIX 36886.

The most commonly reported degradation mode is encapsulant discoloration, which may be aided by the fact that discoloration is also the most noticeable mode by visual inspection. [19] We refer to this as "degradation" rather than "failure", as discoloration leads typically to lower performance but not necessarily to failure, even when considering soft failure limits such as a typical warranty. Figure 3 (a) shows a typical example of a linear degradation curve for a mono-crystalline (mono-Si) module that was fielded for 14 years. The module was installed at NREL's performance and energy rating testbed with current-voltage measurements taken every 15 min. The maximum power (Pmax) decline of (0.33±0.02) %/year is dominated by short-circuit current (Isc) decline (0.31±0.01) %/year. The fill factor (FF) and the open circuit voltage (Voc) display relative stability: (0.01±0.01) %/year and (0.02±0.01) %/year, respectively, within the statistical uncertainty. The Pmax loss, which is Isc dominated, totals to less than 5% over the lifetime of the module; the transmission loss from the discoloration in the center of the cells is shown in Figure 3 (b) with some bleaching at the cell edges.

4 SOLDER BOND FATIGUE

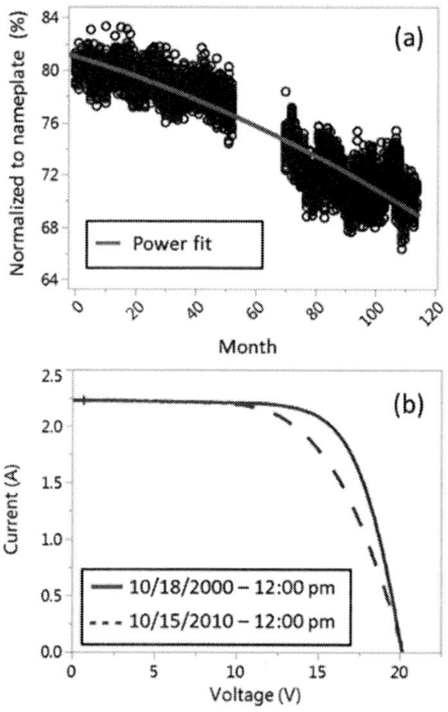

Figure 4: Time series behavior for series resistance impacted x-Si module over 10 years (a). Initial and end-of-life I-V shows clear signs of series resistance increase (b).

In contrast, the x-Si module of Fig. 4 that was exposed next to the module of Fig. 3, shows initially a fairly stable behavior that appears to be followed by a concave decline. Data acquisition problems led to ca. 1 year gap during which the module was exposed outside and continued to degrade. A second interruption starting at month 71 was caused when the module was stored inside while the data collection system was upgraded. During this time no additional degradation was observed; for clarity, this section was removed from Fig. 4. It is interesting to note that a power fit provides a better fit to the data than a linear fit. The I-V characteristics that are taken at the beginning and almost exactly after 10 years display clear signs of series resistance increase. Series resistance increase has been associated with thermal mechanical fatigue of solder joints, solder corrosion or ribbon fatigue. [12,20,21,22]

This module exhibited many cracks in the cell and string interconnect ribbons, an example of which is shown in Fig. 5 (a). The cracks appear to have not caused a complete failure, however they affirm the increased series resistance and the hot-spots in the infrared image that occur along the cell interconnect ribbons, Fig 5 (b). The interconnect connections could be improved by applying pressure to the front side of the module. The electroluminescence images of Fig. 6 demonstrate the change in connection pointed out by the arrows.

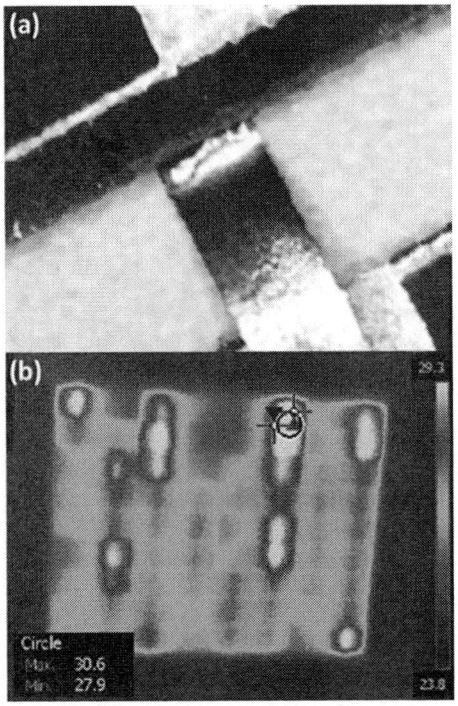

Figure 5: Visual image of one of the cracks in the cell interconnect ribbon (a). Infrared image showing hot-spots along interconnect ribbons (b). Photo credit: Dirk Jordan, NREL PIX 38042 and 38044.

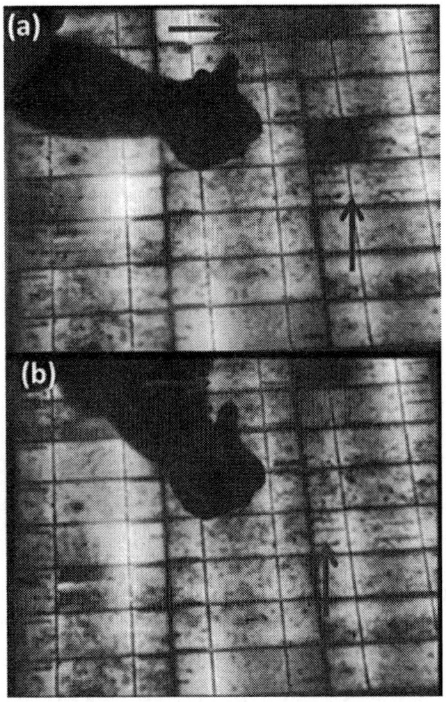

Figure 6: Electroluminescence images before (a) and while applying (b) pressure to the front side. The arrows point out changes in the luminescence because of improved conductivity through bridging of the gaps. Photo credit: Tim Silverman, NREL PIX 38045 and 38046.

5 SOLDER BOND FAILURE

Hot spots are important because of safety considerations. In side-by-side comparisons, modules affected by hot spots displayed greater power loss compared to unaffected modules. [11,23,24] Hot spots as a failure mode can have different causes, such as solder bond failures or cracked cells, as we will discuss in this section.

Thermal cycling can eventually lead to solder bond failures. Figure 7 (a) shows I-V parameters of a single string over nearly 22 years of field exposure that was recently affected by a hot spot. The hot spot was caused by solder-bond failure between the ribbon and the cell busbar. A characteristic burn mark was observable on the front and the backsheet of the module (b). The hot spot was first detected in 2014 but the temperature was measured to be in the 50°C–60°C range. After an additional 1½ years of field exposure the hot-spot had significantly increased in temperature to 110°C–130°C. I-V parameter analysis of the affected string shows a fairly linear decline during 20 years that was correlated equally to FF and Isc losses. However, the more recent evaluation concluded that the power loss is now following a different path that is more significantly correlated to FF loss.

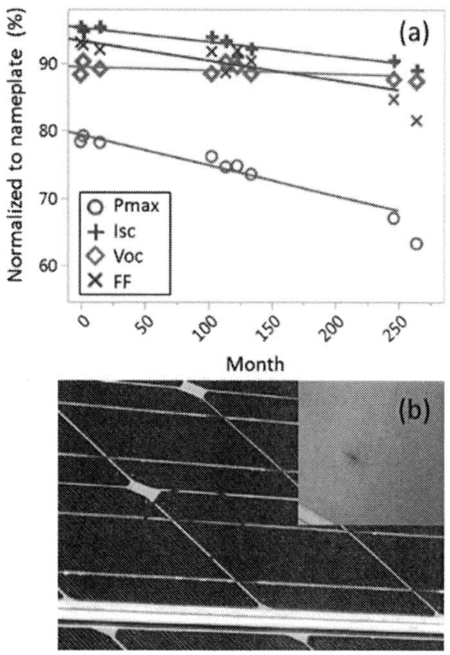

Figure 7: Solder-bond-failure-induced hot spot and its time series effect (a) and visual front image burn mark on the backsheet (b). This particular module was part of a 20-year-old system. The I-V traces may show increased degradation. Photo credit: Dirk Jordan, NREL PIX 36887.

6 CRACKS

Cracked cells can be another mechanism for hot spots, as illustrated in this section by a moderate hot spot (ca. 60°C) in a mono-Si module after more than 10 years of field exposure. The I-V parameters for this particular module are presented in Fig. 8 (a), which were measured by 4 different light sources. Initially, a drop of several percent can be seen that appears to be dominated by current loss and could be associated with light-induced degradation. For the subsequent 5 years very little change in all parameters can be discerned until the appearance of the crack-induced hot spot. The degradation slope changes and the power loss become FF-dominated. More frequent data points are needed to ascertain the exact functional form, as it could follow a 2-step or a more gradual decline, possibly a power decline. The overlay from the large area continuous solar simulator (LACSS) show the increases in series resistance for the module.

Electroluminescence image, not shown here, exhibited cracks in several cells but only the crack in one cell resulted in a hot-spot. The loss in power is related to the loss of current flow and is a complex function of the direction of the cracks as well as the integrity of the metallization (grid lines and busbars). It appears that some cracks have a more deleterious impact on the module power performance than others. [25]

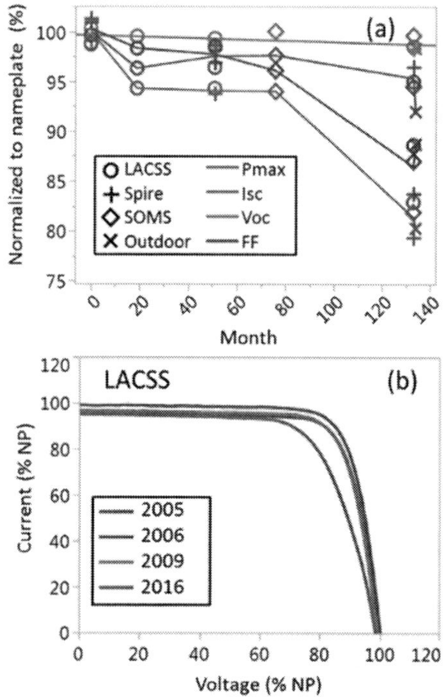

Figure 8: I-V parameters of the hot-spot module for more than 10 years exposure, as measured under four different light sources (a). Large area solar simulator measurements (b) show series resistance increase.

Figure 9: Infrared image of a mono-Si module that has been fielded for more than 10 years with a hot spot caused by a cell crack. Photo credit Dirk Jordan, NREL PIX 36888.

More frequent data are necessary, however best analytical practices are also required to accurately determine the functional form and ascribe them to specific failure modes, as is illustrated in the next example.

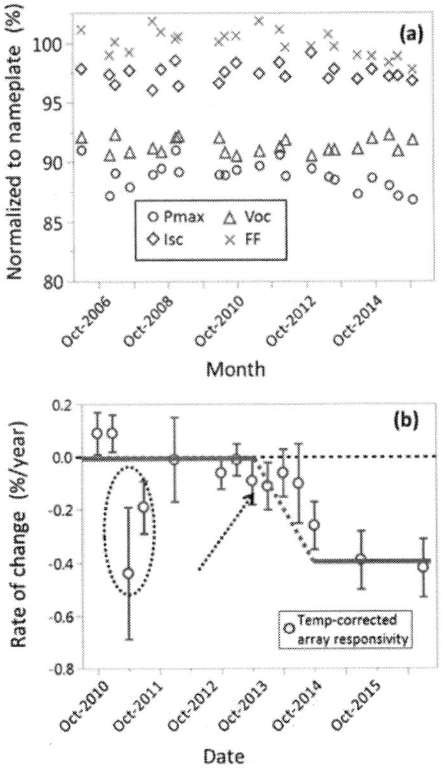

Figure 10: Outdoor measured I-V parameters for a Shell CIGS system for a 10-year period (a). Continuous data for the system analyzed the temperature-corrected array responsivity (blue circles). The dashed oval indicates the time period of a drifting current transducer and the dashed arrow identifies when cracks first were observed.

I-V parameters that were taken quarterly (a) and continuous data (b) for a Shell E80 CIGS system are shown in Fig. 10. [26] The measurements were made without washing the system and therefore display some noise. Despite this noise in the data the curve for Pmax appears to consist of two distinct sections. In the first 6 years the system appears without any significant degradation followed by a quick transition to a degradation near -0.4 %/year that is mostly caused by FF losses. During that period visible cracks in 3 modules were observed from the front, Fig. 11. Isc and Voc appear to be stable over the entire 10 years field exposure.

The temperature-corrected array responsivity (blue circles) shows the two distinct degradation sections clearly; this technique provides reduced uncertainty relative to the more commonly used performance ratio. [7] Furthermore, a third section becomes clear with the temperature-corrected array responsivity that is indicated by a dashed oval. This time preceded our implementation of daily, automated data quality checks and was caused by a drifting current transducer. It was not reflective of the module performance but indicated data collection problems. After replacement of the transducer, the system returned to near zero degradation before the cracking of the modules. At first, the severity of the decline is surprising, as the cracks appear not very severe, yet the cracks are not located in the front glass but the active cell area. Photoluminescence corroborates the significant decline dominated by FF losses, as the cracks extended far more into the module than is visible.

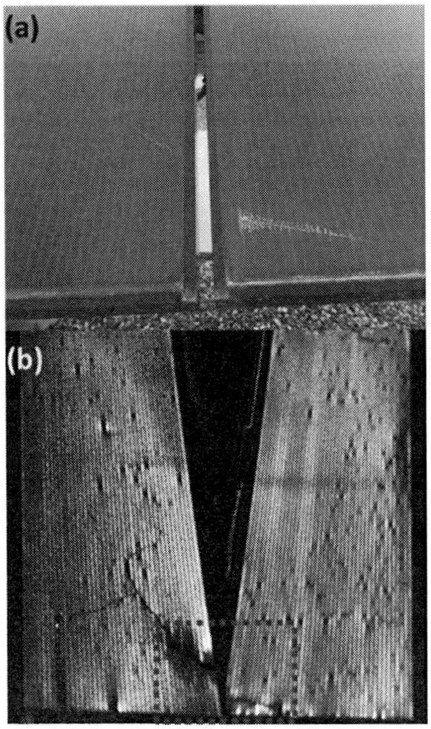

Figure 11: Two modules of a Shell E80 CIGS system after 10 years of field exposure. Visual image (a) is close up of the photoluminescence image (b) indicated by a red dashed rectangle. Photo credit Tim Silverman, NREL PIX 36890.

It is conceivable that more 2-step profile curves will become observable in the future, not as a result of different underlying degradation mechanisms but as a result of system design choices. Systems with high DC-to-AC ratios, as is becoming more common, may exhibit a 2-step profile regardless of the functional form of the underlying degradation mechanism. As module prices have fallen, the balance-of-system (BOS) components have become a larger percentage of the overall project cost. As a result, project costs can be reduced if the inverter is undersized limiting the input to the inverter's capability during the highest producing periods of the year. The effect of this so-called "inverter clipping" is that such a system may not exhibit obvious degradation until many years of field exposure.

7 CONCLUSION

Degradation rates are important for financial projections and may be indicative of different degradation modes. However, as modeling of PV performance becomes more sophisticated, degradation rates based on the assumption of linearity may not be sufficiently accurate. Non-linear trends at the beginning of life or during the wear-out phase are important to detect and understand economically and technically. Monte Carlo simulation of LCOE showed the significant effect of the shapes of degradation curves. Technically, accurately identifying non-linearities in degradation paths is essential for service lifetime predictions. The most commonly reported degradation mode, encapsulant discoloration, is associated with linear decline that is correlated to Isc losses. In contrast, series resistance increases caused by thermal mechanical solder bond fatigue appear to follow a delayed degradation trajectory that may be difficult to detect over common variations in outdoor measurements. Cracked cells or complete solder bond failure show a more significant decline after the failure occurrence. In general, more frequent acquisition of data needs to be coupled with optimized analytical methods to delineate subtle trends that often are masked by measurement noise.

8 ACKNOWLEDGEMENT

We would like to thank Jose Rodriguez for measurements and maintenance of the systems. This work was supported by the U.S. Department of Energy under Contract No. DE-AC36-08-GO28308 with the National Renewable Energy Laboratory.

Table I: Summary of degradation types, time scales and how they tend to appear for some technologies in specific phases..

Phase	Technology	Time scale	Degradation type
Infant (decrease)	a-Si	Months	LID, Staebler-Wronski
Infant (decrease)	CdTe	1 – 3 years	Metastability
Infant (decrease)	x-Si	Hours	LID, O-B complex
Infant (decrease)	CIGS	Days, Months	Metastability
Consistent	all	Years	Discoloration
Wear-out	all	Months, Years	Series resistance
Wear-out	all	Months, Years	Cracked cells, solder bond fail.

[1] International Electrotechnical Commission 60050-191, Dependability and quality of service, 1990.
[2] Jordan DC, Kurtz SR, PV Degradation Rates – an Analytical Review, Progress in Photovoltaics: Research and Application, 2013, 21(1).
[3] Jordan DC, Kurtz SR, VanSant KT, Newmiller J, Compendium of Photovoltaic Degradation Rates, Progress in Photovoltaics: Research and Application, 2016.
[4] Short W, Packey DJ, Holt T, A Manual for the Economic Evaluation of Energy Efficiency and Renewable Energy Technologies, Report NREL/TP-462-5173, March 1995.
[5] Meeker WQ, Escobar LA, Pitfalls of Accelerated Testing, IEEE Transaction on Reliability, 47(2), 1998, 114 – 118.
[6] del Cueto, JA, Review of the Field Performance of One Cadmium Telluride Module, Progress in Photovoltaics: Research and Application, 1998, 6, 433-446.
[7] Jordan DC, Kurtz SR, The dark horse of evaluating long-term field performance—data filtering, IEEE Journal of Photovoltaics, 2014, 4, 317–323.
[8] Staebler DL, Wronski CR, Optically induced conductivity changes in discharge-produced hydrogenated amorphous silicon, Journal of Applied Physics, Vol. 51 (6), 1980, 3262–3268.
[9] Ngan L, Strevel N, Passow K, Panchula AF, Jordan D, Performance Characterization of Cadmium Telluride Modules Validated by Utility-Scale and Test Systems, Proceedings of the 40th IEEE Photovoltaic Specialists Conference, Denver, CO, USA, 2014, 1957 – 1962.
[10] N. Strevel, L. Trippel, M. Gloeckler, Performance Characterization and Superior Energy Yield of First Solar PV Power Plants in High-Temperature Conditions, Photovoltaics International, 2012.
[11] Broek KM, Bennett IJ, Jansen MJ, van der Borg NJC, Eerenstein W, Light and current induced degradation in p-type multi-crystalline cells and development of an inspection method and a stabilisation method, Proceedings of the 27th European Photovoltaic Solar Energy Conference and Exhibition, Frankfurt, Germany, 2012, 3167–3171.
[12] B. Sopori, P. Basnyat, S. Devayajanam, S. Shet, Y. Mehta, J. Binns, and J. Appel, "Understanding light-induced degradation of c-Si solar cells," in Photovoltaic Specialists Conference (PVSC), 2012 38th IEEE, 20 12, pp. 1115-11.
[13] Ross Jr. RG. Reliability and performance experience with flat-plat photovoltaic modules. Proceedings of the 4th European Photovoltaic Solar Energy Conference, Stresa, Italy, 1982; 169–179.
[14] Köntges M, Kurtz S, Packard C, Jahn U, Berger KA, Kato K, Friesen T, Liu H, Van Iseghem M, Performance and Reliability of Photovoltaic Systems, Subtask 3.2: Review of Failures of Photovoltaic Modules, International Energy Agency, Report IEA-PVPS T13-01:2014, 2014.
[15] Jordan DC, Sekulic B, Marion B, Kurtz SR, Performance and Aging of a 20-Year-Old Silicon PV System, Journal of Photovoltaics, Vol. 5 (3), 2015, 744 – 751, DOI: 10.1109/JPHOTOV.2015.2396360.
[16] Chamberlin CE , Rocheleau MA, Marshall MW, Reis AM, Coleman NT, Lehman PA, Comparison of PV Module Performance Before and After 11 and 20 Years of Field Exposure, , Proceedings of the 37th IEEE Photovoltaic Specialists Conference, Seattle, WA USA, 2011, 101 – 105.
[17] L. Abenante, F. De Lia, S. Castello, Long-term Performance Degradation of c-Si Photovoltaic Modules and Strings, Proceedings of the 25th European Photovoltaic Solar Energy Conference, Valencia, Spain, 4023 - 4026.
[18] Friesen T, Chianese D, Realini A, Friesen G, Burà E, Virtuani A, Strepparava D, Meoli R, TISO 10 kW: 30 Years Experience with a PV Plant, Proceedings of the 27th European Photovoltaic

Solar Energy Conference, Frankfurt, Germany, 2012, 3125 – 3131.

[19] Jordan DC, Kurtz SR, Field experience: Degradation Rates, Lifetimes & Failures, PV Module Reliability Workshop, Lakewood, CO , USA, 2016.

[20] Quintana MA., King, D.L., 2002. Commonly observed degradation in field-aged photovoltaic modules, Proceedings of the 29th IEEE Photovoltaic Specialists Conference, New Orleans, USA, 1436–1439.

[21] Bosco N, Silverman TJ, Wohlgemuth J, Kurtz S, Inoue M, Sakurai K, Shioda T, Zenkoh H, Hirota K, Miyashita M, Tadanori T, Suzuki S, Evaluation of Dynamic Mechanical Loading as an Accelerated Test Method for Ribbon Fatigue, Proceedings of 29th European Photovoltaic Solar Energy Conference, Amsterdam, Netherlands, 2014. 2484–2489.

[22] Dubey R, Chattopadhyay S, Kuthanazhi V, John J, Arora BM, Kottantharayil A, Narasimhan KL, Solanki CS, Kuber V, Vasi J, Kumar A, Sastry OS, All-India suvery of PV module degradation: 2103, National Centre for Photovoltaic, Research and Education Indian Institute of Technology Bombay National Insititue of Solar Energy, 2013.

[23] Bradley A, Hamzavy B, Gambogi W, Analysis of the degradation and aging of a commercial photovoltaic installation, Proceedings of the SPIE, 9179, 2014, doi: 10.1117/12.2062046.

[24] Chattopadhyay S, Dubey R, Kuthanazhi V, John JJ, Vasi J, Kottantharayil A, Arora BM, Narsimhan KL, Solanki CS, Bora B, Singh YK, Sastry OS, Effect of Hot Cells on Power Degradation Rate of PV Modules, PV Module Reliability Workshop, Lakewood, CO, USA, 2016.

[25] M. Köntges, S. Kajari-Schröder, I. Kunze, U. Jahn, Crack statistic of crystalline silicon photovoltaic modules, Proceedings of the 26th European Conference on Photovoltaic Solar Energy, Hamburg, Germany, 2011, 3290–3294.

[26] Jordan DC, Kurtz SR, Thin-film Reliability Trends toward Improved Stability, Proceedings of the 37th IEEE Photovoltaic Specialists Conference, Seattle, WA, USA, 2011, 827 – 832.

ACCELERATION FACTORS FOR MOISTURE INDUCED DEGRADATION OF FLEXIBLE PV MODULES AND PREDICTION OF FIELD PERFORMANCE

Kedar Hardikar, Todd Krajewski, Kristopher Toivola
MiaSolé Hi-Tech Corp., Santa Clara, CA, USA

ABSTRACT: The work presented here is aimed at assessing the field performance of flexible PV modules with regard to moisture-induced degradation through the use of Simulation and Mechanism based Accelerated Reliability Testing (SMART) empirical approaches. Semi-empirical acceleration factor models are presented based on an analytical solution and empirical models from literature. Testing at different temperature and humidity conditions is carried out for representative flexible thin film module configurations. Test results highlight the limitations of such accelerated testing and it is shown that for the high-moisture-barrier systems tested, the test times can become prohibitive for establishing humidity dependence. Through the use of empirical models based on the theory of rate processes, a method is proposed to translate accelerated test results to field performance using typical meteorological year (TMY) data.
Keywords: photovoltaic cells, moisture ingress testing, acceleration factor model, lifetime model

1 INTRODUCTION

There have been several attempts in the past to develop techniques and models to predict the life of PV modules with moisture ingress as a proposed mechanism leading to failures or degradation in performance [1-2]. For consumer electronics products a typical approach to characterizing the field performance of encapsulated electronic components based on accelerated testing with exposure to temperature and humidity, is an empirical approach commonly referred to as the Hallberg-Peck model [3]. While it may seem conceivable that a similar approach can be taken to assess field performance of flexible PV modules, this approach is not well established or validated for PV products. In this work, semi-empirical approaches are adapted within a framework that provides a structured approach for such exploration.

In this paper we illustrate a methodology that we call Simulation and Mechanism based Accelerated Reliability Testing (SMART) empirical aproaches to model field performance of flexible PV modules for moisture induced degradation, using results from accelerated testing. Candidate module configurations chosen for accelerated testing are copper indium gallium selenide (CIGS) thin film flexible modules with different moisture barrier performance levels. The method involves the following-

1) Through simulation, analytical modeling, review of experimental data or root cause analysis conjecture a functional form for an acceleration factor model.
2) Develop an accelerated test plan to specifically validate or refine the proposed acceleration factor model.
3) Based on acceleration factor model that has an intermediate validation through accelerated tests predict field perofmance specific to a location.
4) Final validation and refinement of the model from actual field data.

In this approach the first three steps are compatible with product development cycle while the last step is expected to be a long term activity. This is specifically relevant to newly introduced products such as flexible PV modules for which extensive field data is not likely to be available. As the field data becomes available over time it can be used to refine the acceleration factor models, however, in lieu of extensive field data it is necessary to have a framework to assess the field performance from accelerated testing.

2 MODEL DEVELOPMENT

2.1 Analytical model

There are several potential moisture ingress paths in a flexible PV module. A methodology to assess moisture resistance of an edge seal was developed in [4] for a glass-glass module. The method was based on an acceleration factor model developed analytically and validated by experiments. In the present work, learning from [4] is used in conjunction with additional testing to first establish that front barrier is the primary moisture ingress path for the samples chosen.

Diffusion of moisture through a barrier material is a process which involves adsorption of moisture on the exposed surface followed by diffusion of moisture through the barrier. The primary driving force for any diffusion process is the gradient in concentration of the diffusing species (Fick's Law). Moisture ingress through front barrier as primary path can be modeled using a 1-D diffusion equation. For a high barrier system used in PV modules the moisture emerging near the solar cell after passing through the barrier and encapsulant can be assumed to be consumed through a reaction with materials constituting the PV cell (e.g. reaction with TCO or CIGS material). This leads to a candidate model described by the following initial boundary value problem

$$\frac{\partial C}{\partial t} = D\frac{\partial^2 C}{\partial x^2} \tag{1}$$
$$C(x,0) = C_0 = 0; C(x=0,t) = C_1; C(x=l,t) = 0$$

Here x=0 represents the exposed surface of the module while x=l represents the thickness of the encapsulant and front barrier through which the moisture must diffuse to reach the PV cell. Details of this model are available in [5-6]. Using solution to this initial boundary value problem, and assuming that the degradation in performance is proportional to the amount of moisture that reached the cell over time 't' (equivalent to first order reaction) it can be shown that the time to failure (TTF) of the module under temperature T and relative humidity RH must have the functional form –

$$TTF \propto (RH)^{-1} \exp\left(\frac{\varepsilon_a}{kT}\right) \quad (2)$$

The corresponding form for the acceleration factor is

$$AF = \left(\frac{RH_{field}}{RH_{test}}\right)^{-1} \exp\left[\frac{\varepsilon_a}{k}\left(\frac{1}{T_{field}} - \frac{1}{T_{test}}\right)\right] \quad (3)$$

$AF = Acceleration\ factor$

$k = Boltzman\ const.$

$\varepsilon_a = Activation\ energy$

This analytical model provides a basis for understanding the structure of empirical models chosen for further study.

2.2 Candidate empirical models

Based on the insight gained from analytical model following two candidate empirical models are chosen for further exploration.

1. The Hallberg-Peck model

This is a corrosion rate model based on relations proposed in [3]. The functional form for the acceleration factor in this case between the field conditions and test conditions is given below in eq(4). Based on the analytical model it is clear that the humidity exponent in eq(4) below is expected to be negative. The acceleration factor relation is -

$$AF = \left(\frac{RH_{field}}{RH_{test}}\right)^{n} \exp\left[\frac{\varepsilon_a}{k}\left(\frac{1}{T_{field}} - \frac{1}{T_{test}}\right)\right] \quad (4)$$

$AF = Acceleration\ factor$

2. The Eyring model

This model is based on statistical mechanics considerations involving entropy of activation from the theory of rate processes as detailed in [7]. This model enables developing a functional form for the acceleration factor which accounts for thermal (T) as well as non-thermal (e.g. RH) effects on rate processes. We note that the negative RH exponent in Hallberg-Peck model is expected to result in singular behavior in acceleration factor model applied to dry environment. This insight is the motivation to look for another acceleration factor model which can account for rate processes but does not have singular behavior at low RH values. Based on [8] the functional form of acceleration factor between the test conditions and field conditions appropriate for PV module degradation due to temperature and humidity is chosen to be -

$$AF = \exp\left[\frac{\varepsilon_a}{k}\left(\frac{1}{T_{field}} - \frac{1}{T_{test}}\right)\right] \exp\left[-\gamma\left(RH_{field} - RH_{test}\right)\right] \quad (5)$$

$AF = Acceleration\ factor$

$\varepsilon_a, \gamma = Model\ parameters$

3 ACCELERATED TESTING

There are two sets of experiments performed in this study. First, the barrier material permittivity, diffusivity and solubility are measured at different temperatures through tests carried out at MOCON testing services. The second set of experiments is testing of sample PV modules of identical construction at different temperature and humidity conditions – (85°C/85%RH, 75°C/85%RH, 95°C/85%RH, 85°C/75%RH, 85°C/95% RH)

Table I: Barrier types chosen for study

Name	Water Vapor Transmission Rate
Barrier-1	0.0071g/sq.m/day @37°C/100%RH
Barrier-2	<0.0022g/sq.m/day@37°C/100% RH
Barrier-3	<5e-4g/sq.m/day @50°C/100%RH (No MOCON testing)

There are a few things noted in results of accelerated testing. First we note that the MOCON testing of barrier-1 and barrier-2 materials does exhibit an Arrhenius behavior with temperature as expected, however, barrier 3 is not amenable to MOCON testing due to equipment limitations. Test times for barrier-2 at low temperatures are seen to be prohibitive. In testing of mini-modules, it was seen that the degradation in module performance as characterized by %power change (ΔPmax) can be assumed to be approximately linear only up to 20% degradation and beyond that it exhibits a "crash behavior". The scatter in data is higher beyond 20% degradation and hence critical value of ΔPmax=-20% is taken as criterion for "failure" in analysis of time to failure.

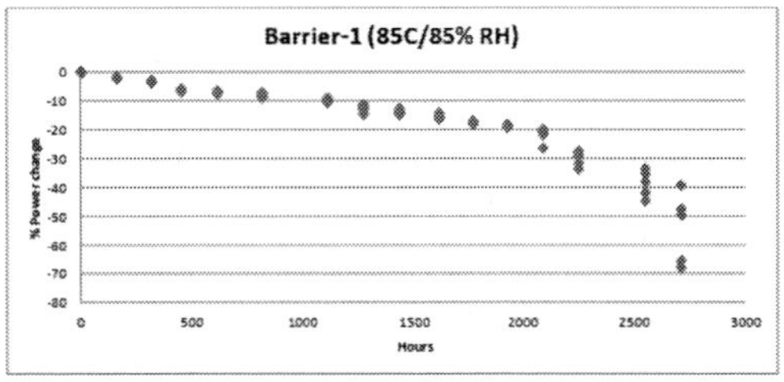

Figure 1: Degradation data for barrier-1

Fig. 1 shows a typical degradation curve while Fig. 2 and Fig. 3 show that the observed time to failure in these experiments complies with acceleration factor models proposed. The results of accelerated testing were in compliance for both the Hallberg-Peck model as well as the Eyring model. Additional information is available in [5-6]. Similar behaviors were observed for barrier-2. Barrier-3 was seen to outperform both barrier-1 and barrier-2, however, the test times for barrier-3 were prohibitive and tests were terminated. The humidity exponents obtained for barrier-1 and barrier-2 were n=-3.41 and n=-1.22 respectively while the corresponding activation energy values were 0.63 eV and 0.73 eV respectively. For the Eyring model fit the humidity constants for barrier-1 and barrier-2 were γ=-0.041 and γ=-0.014 respectively.

Table II: Module lifetime (years) considering moisture induced degradation only (ΔP=-20%)

Location	Barrier-2 (n=-3.41)	Barrier-2 (n=-1)	Barrier-1 (n=-3.41)	Barrier-1 (n=-1)	Barrier-2 (Eyring, γ=-0.0142)	Barrier-1 (Eyring, γ=-0.041)
Denver	45664	487	16489	163	363	455
Arizona	3332	86	1649	36	69	100
Bangkok	27	21	13	10	21	12
San Jose	175	85	72	33	84	48

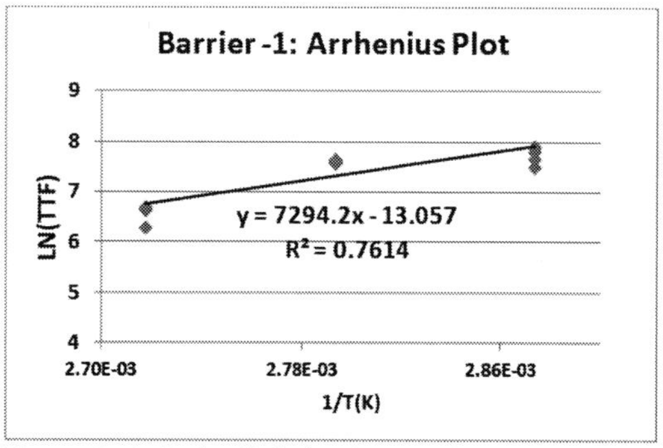

Figure 2: Arrhenius fit for barrier-1 results

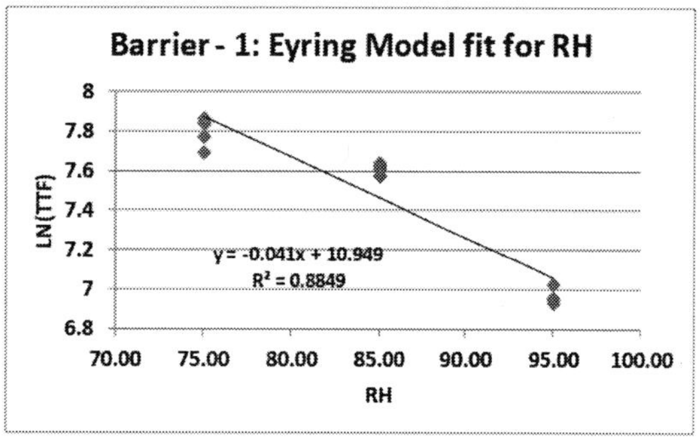

Figure 3: Eyring model fit for barrier-1 results

It should be noted that a negative exponent for relative humidity in Hallberg-Peck model implies that the corresponding times to failure in dry environments with low RH values will be very high due to the singular behavior. This is a limitation of the model. In dry environments, thermal degradation mechanisms are expected to dominate. This limitation is overcome by the use of Eyring model which correctly transitions to Arrhenius behavior at low RH values.

4 ASSESSMENT OF FIELD PERFORMANCE

Once we have acceleration factor models with intermediate validation based on results of accelerated testing the same framework can now be used to carry out a location specific assessment of the field performance. The idea is to use TMY data to predict module temperature using the Sandia model and apply the acceleration factor relations over the hourly intervals to get an average acceleration factor for one TMY. This enables translating results from accelerated tests to prediction of field performance with regard to moisture induced degradation for the modules under consideration. Details can be found in [5-6].

Results of such calculations are summarized in Table II. It is observed that the Hallberg-Peck model, as expected, exhibited a singular behaviour in dry environments leading to unrealistic time to failure. Note that in the framework of SMART empirical approches the refinement to use Eyring model was a result of first conjecturing the Hallberg-Peck model, executing the test plan and subsequently refining the conjecture when discrepancy and model limitation for singular behaviour in RH was seen. Eyring model overcomes the limitation of singular behaviour in RH correctly and is more consistent with the expected behaviour. Since the available experimental data is limited these results should be taken as the best estimates of product performance. The quality of the data and the quality of the fit can be improved by using larger modules, increasing the sample size, maintaining consistent electrical contacts through the tests to name a few improvements possible. However, in spite of these limitations it is seen that the SMART empirical approach chosen enables assessment of field performance of these barriers in a structured manner so as to enable decisions during product development phase. It is concluded based on this study that barrier-1 and barrier-2 are likely not adequate to meet the typical warranty requirements of 25 years in aggressive environments like Bangkok while barrier-3 is judged adequate.

5 CONCLUSIONS

Application of a SMART empirical approach was demonstrated for development of acceleration factor models for thin film flexible modules to assess field performance based on results of accelerated testing for moisture induced degradation. It was shown that Eyring model was a better candidate for modeling moisture induced degradation in flexible modules compared to the Hallberg-Peck model. Hallberg-Peck model, although consistent with results obtained in accelerated testing, has a singular behavior in dry environments with low RH values. This limitation is overcome by use of Eyring model. Based on results obtained it was concluded that barrier-1 and barrier-2 are likely not adequate to meet typical warranty requirements for 25 years in an aggressive environment like Bangkok while barrier-3 is judged adequate. It is also concluded that extensive study of high barrier systems with wide range of conditions and in particular test conditions involving low RH values can make test times prohibitive.

REFERENCES

[1] Coyle, D. J., "Life Prediction of CIGS Solar Modules Part 1: Modeling Moisture Ingress and Degradation" Progress in Photovoltaics: Research and Applications, 2011, DOI: 10.1002/pip.1172.:

[2] Coyle, D. J., "Life Prediction of CIGS Solar Modules Part 2: Degradation Kinetics, Accelerated Testing and Encapsulant Effects" Progress in Photovoltaics: Research and Applications, 2011, DOI: 10.1002/pip.1171

[3] Hallberg O., Peck D. S., "Recent Humidity Accelerations, a Base for Testing Conditions," Quality and Reliability Engineering International, pp. 169-180, Vol. 7, 1991.

[4] Hardikar, K., Vitkavage D., Saproo A., Krajewski T., "A Methodology for Testing, Characterization and Prediction of Edge Seal Performance in PV modules,"

PV International, Vol. 22, Jan. 2014.
[5] Hardikar, K., Krajewski, T., Toivola, K., "Predicting Moisture Induced Degradation of Flexible PV Modules in the Field," PV International, 30th Ed., pp. 72-83, Dec 2015.
[6] Hardikar, K., Krajewski, T., Toivola, K., Assessing Field Performance of Flexible PV Modules for Moisture Induced Degradation from Accelerated Testing, PVSC-43, Portland, June 2016.
[7] Glasstone S., Laider, K. J., Eyring H., "Theory of Rate Processes," McGraw-Hill Book Company, London, 1941.
[8] Luis A. E., Meeker W. Q., "A Review of Accelerated Test Models," Statistical Science, pp. 552-577, Vol. 21, 2006.

BIAS AND IRRADIATION DEPENDENCIES OF CIGS MODULE RELIABILITY DURING HEAT TESTS

Keiichiro Sakurai, Kinichi Ogawa, Hajime Shibata, Atsushi Masuda
National Institute of Advanced Industrial Science and Technology
AIST #2-20273080, 1-1-1 Umezono, Tsukuba, Ibaraki 305-8568, Japan

Hiroshi Tomita, Darshan Schmitz, Shuuji Tokuda
Solar Frontier K.K.
123-1 Shimokawairi, Atsugi, Kanagawa 243-0206, Japan

ABSTRACT: We have found that conventional damp-heat (DH) test on a commercial CIGS (a.k.a. CIS, CIGSS) module causes an irreversible "Test-specific" degradation that is not observed in modules deployed in fields. We have tested full-size CIGS modules with/without forward bias, light irradiation and humidity during heat tests. The results clearly show that adding forward bias, or white light irradiation during DH tests suppresses this irreversible degradation. Based on these results, we recommend to add forward bias and/or light irradiation during DH tests of CIGS modules, to make the test condition closer to real fields and suppress degradations not observed in the field.
Keywords: CIGS, CIS, Degradation

1 BACKGROUNDS

Damp-heat (DH) tests in dark have been widely used for testing the reliability of PV modules, including the IEC 61215, used worldwide for type testing and qualification. Recently, this IEC 61215 is undergoing a thorough revision, including amendments to pass/fail criteria of DH tests. The DH test protocols in the new IEC61215-1 series will include a light-soaking process to stabilize the module, both before and after keeping the module in DH condition for 1000 hours. This process is known to work for market-dominating crystalline silicon modules. However, in the case of CIGS modules, which sometimes show a large change in performance by dark storage or light soaking, it was not known if the conventional DH process in dark was appropriate for accelerating the same type of degradations observed in real fields. The DH tests in the new IEC 61215-1 series should avoid any degradations not observed in fields, while accelerating the degradations observed in fields, both within practical boundaries. However, as far as we know, heating of the module without light illumination is unnormal in the field (Fig. 1). From this aspect, conventional DH tests that heat modules in dark without bias differs with real conditions in the field.

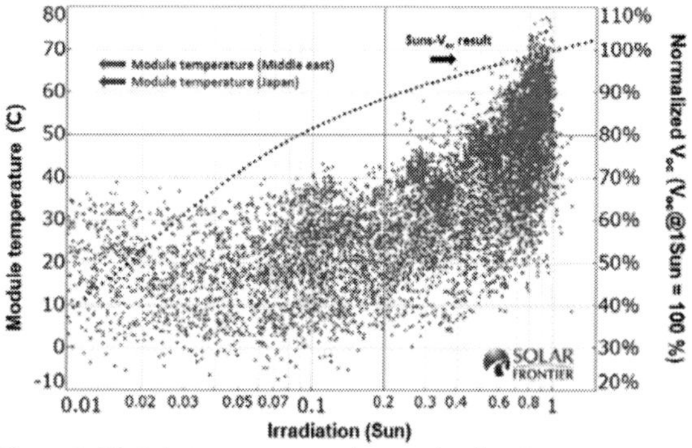

Figure 1. Module temperature and open-circuit voltage (Voc) logged in middle east and Japan. When the module temperature was higher than 50°C, no data point was logged to be darker than 0.2 Sun.

2 EXPERIMENTAL

We have heated over 100 full-size CIGS modules in environmental chambers, with/without forward bias, light irradiation and humidity during heat tests (Fig. 2). Before and after the chamber tests, light soaking (LS) process was applied to all modules for stabilization.

	Test type	Irradiation	Bias
Wet	Normal DH	—	—
	DH with irradiation	0.1~0.65 Sun*	—
	DH with forward bias	—	$V_{pm@NMOT}$** $V_{pm@STC}$*
Dry	Normal dry heat test	—	—
	Dry heat test with forward bias	—	$V_{pm@NMOT}$** $V_{pm@STC}$*

* Performed at AIST. Others were performed at SF.
**NMOT : Nominal Module Operating Temperature.

Test sequence

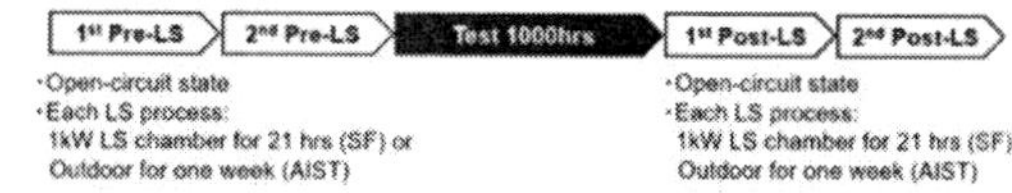

Figure 2. Table of test conditions. 5 types of tests were conducted on 170 W-class commercial CIGS modules from Solar Frontier.

For heating with light illumination (="illuminated") tests, LED (blue+yellow) light with an intensity of 100~650 W/m^2 was used (Figure 3).

Figure 3. Environmental chamber with LED used for illuminated tests.

For heating with forward bias (="bias") tests, forward bias was applied to the modules. The current was approximately 0.2 A for each module. For both illuminated and bias tests, temperature of the chamber was adjusted to keep the module temperature at 85 °C. For illuminated tests, the chamber's relative humidity was kept at 85% r.h. The modules were connected to a load so that the modules operate at a point close to the maximum power point. For bias tests, the chamber's relative humidity was adjusted so that the humidity at module surface would be kept close to 85% r.h.

3 RESULTS AND DISCUSSION

Illuminated tests, which are the closest to real field conditions, have invoked no degradation in maximum power (Pmax) of the modules tested (Fig. 4). The results did not show any clear dependency to light intensity so far. Meanwhile, modules tested in the dark (with no bias voltage) have shown about 5% of degradation in average Pmax, and did not recover even with additional LS for up to 126 kWh. We call this degradation as "test-specific degradation" (TSD). When the module was short-circuited (by connecting the lead wires) during the illuminated test, TSD was observed. However, we note that the sample numbers for illuminated tests are still small, and thus further data are needed for firmer conclusions.

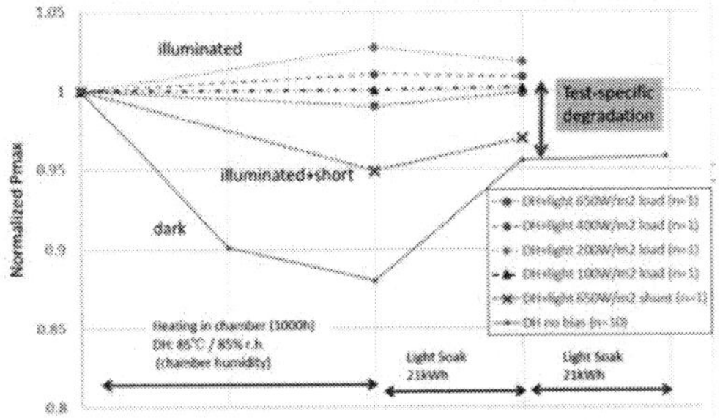

Figure 4. Normalized Pmax of heated modules, with and without light illumination, with one sample short-circuitted.

Bias tests, which is easier to test multiple modules than illuminated tests, gave essentially the same results to illuminated tests (Fig. 5). The humidity invoked no change in Pmax after stabilization by LS, suggesting that the TSD is not caused by humidity ingress.

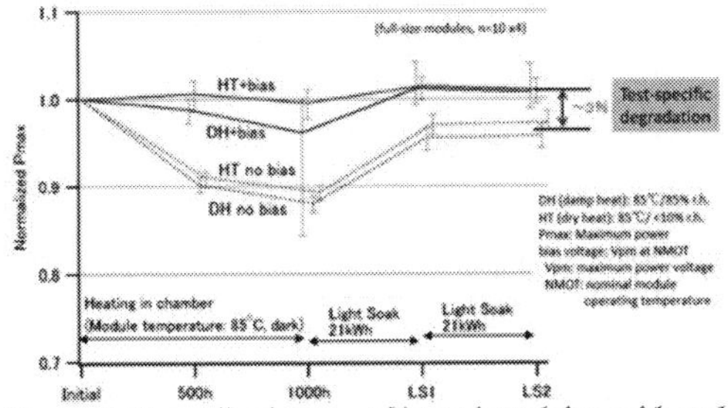

Figure 5. Normalized Pmax of heated modules, with and without forward bias during heating.

As far as we know, degradations similar to TSD have not been observed in real fields. Considering that illuminated or bias tests are closer to real field conditions, at least for the most major CIGS product in the market now, we propose that heat tests should not be conducted under dark and unbiased state, but with light illumination or forward bias conditions. However, we also note that this may be different in other CIGS products, which may differ in cell structure and/or fabrication processes.

4 SUMMARY

We have compared 5 types of heat tests on a major product of commercial CIGS module. Results show that DH tests with forward bias or light illumination are more suitable for this CIGS product, compared to the conventional DH test in dark and unbiased state which caused degradations not observed in real fields. Further investigations on other products and mechanisms of TSD will be required for future works.

PV MODULE DAMAGES CAUSED BY HAIL IMPACT AND NON-UNIFORM SNOW LOAD

G. Mathiak, J. Sommer, W. Herrmann, N. Bogdanski, J. Althaus, F. Reil
TÜV Rheinland Energy GmbH
Am Grauen Stein, 51105 Cologne, Germany
Email: Gerhard.Mathiak@de.tuv.com

ABSTRACT: Extreme weather events as hail impact or heavy snow loads can destroy PV modules by glass breakage. But also hidden damages as cell cracks can lead in the combination with thermo-mechanical stresses caused by wind load and temperature changes to follow-up failures and energy yield losses.
A five-year equivalent sequential stress testing of PV modules from a system, which suffered from hail impacts showed power losses up to 4%. Severely affected modules from these stresses were more sensitive for sub sequential testing under accelerated weathering in the laboratory.
Non-uniform snow load tests according to IEC 62938 (draft) simulated failure types similar to the field. The key parameter for the durability during the non-uniform snow load test is the frame adhesive, which revealed load resistances between 4600 Pa for acrylic foam tape to 8400 Pa for silicone.

Keywords: Qualification, Testing, Hail Impact, Snow Load, Electroluminescence, Microcracks, Accelerated Weathering

1 INTRODUCTION

Strong hail impacts and heavy snow loads are weather events which can cause severe module damages. The outcome of such extreme weather conditions can destroy the modules resulting in glass breakage due to mechanical impacts. But also hidden damages as cell cracks can lead, in the combination with thermomechanical stresses caused by wind and temperature changes, to follow-up failures and energy yield losses.

In central Europe, especially close to the Alps, many PV module damages are caused by hail impacts and snow loads acting on roofs and field-mounted systems. The international standards for the type approval of PV modules (IEC 61215 [1]) require various tests inducing mechanical loads. In Switzerland, hail impact resistance tests are claimed by VKF (Association of Swiss Canton Fire Insurance Companies) which are more severe than the minimum requirements of IEC 61215. In IEC 61215 a snow load and wind load test is demanded with horizontally oriented modules and an application of homogeneously distributed loads of 2400 Pa and 5400 Pa. A current draft for non-uniform snow load testing is now being considered under the registration of IEC 62938 [2].

A first comparison of hail and snow events in the field with a quantitative analysis of hail impacts and non-uniform snow loads in the laboratory has been elaborated.

Glass breakage and star cracks (Fig. 1) within the cells are the typical failure modes of hail events in the field and in the lab. We found that the glass thickness and the frame thickness are key parameters for both failure types due to different stresses and strains of the glass after the hail impact.

Bent frames and resulting glass breakage are typical failures of heavy snow events. We figured out that the durability of the adhesive junction between frame and glass is the key parameter for the determination of more resistant modules.

2 HAIL IMPACT

2.1 Laboratory Tests and Field Experience

A quantitative analysis of hail impact as a function of diameter and standard-compliant velocity of the ice balls has been started and first results were given in [3]. The results indicated the module behavior of six different module types. Glass breakage occurred only by impacts close to the frame with 50 mm diameter ice balls. The glass-glass module withstood 50 mm ice balls impacts without cell cracks. PV-modules with 4 mm glass thickness endured hail impacts considerably better than PV-modules with 3.2 mm thick glass.

a. 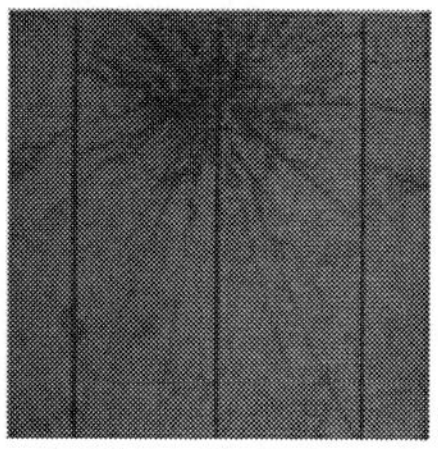 b. 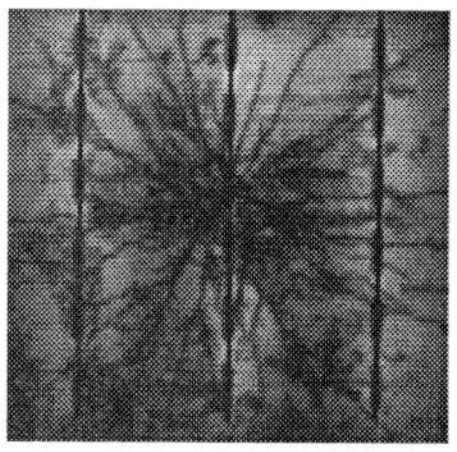

Fig. 1: Electroluminescence image of star cracks induced a. in the field by hail and b. in the lab with a 35 mm diameter ice ball [3].

2.2 Ageing Simulation for Follow-up Failures with Modules from the Field

A plant in middle Europe with several 100 kWp power was affected by a major weather event concluding in a strong hail impact. Four different module types were installed (Manufacturer A: monocrystalline cells, Manufacturer B: 6x8 multicrystalline cells; 6x10 monocrystalline cells; 6x10 multicrystalline cells).

Many modules had glass breakage. Cell cracks are visually non-detectable and can strongly affect performance and may further progress with time.

The task was to develop and to perform a sequential testing for modules from the field, to evaluate current and potential future performance losses. The test sequence is to stress the modules in a realistic way with thermal cycles and mechanical loads based on an analysis of the real environmental conditions and impacting stresses at the specific location resulting from temperature cycles and wind load. Fig 2 shows the temperature gradients from averaged occurring values and frequency distribution at various climates. In this particular case, the moderate climate from Cologne was evaluated and decided for the most frequent gradient to transfer these states later to the laboratory.

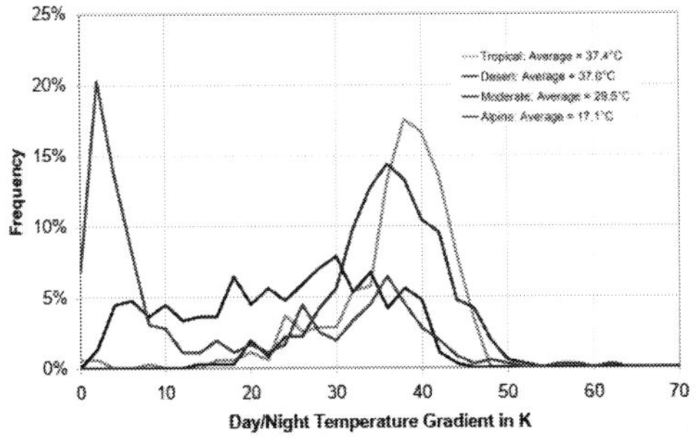

Fig. 2: Module temperatures day/night differences at four different test sites. For the analysis the data of Cologne were used [4].

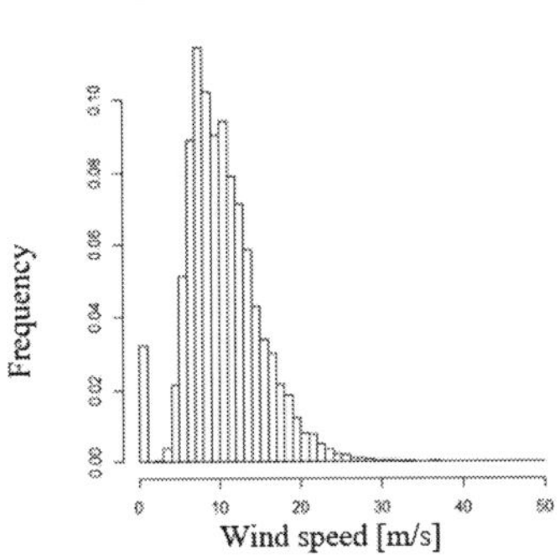

Fig. 3: Wind speed histogram of a town close to the location. A wind speed of 30 m/s induces a load of approx. 600 Pa [5].

Based on the wind speed histogram from Fig. 3 the peak loads (wind gusts) were identified and the repeating occurrence over a year (frequency). Accordingly, the amount and amplitude of mechanical load cycles for lab testing could be identified.

1-year equivalent			
Thermal cycles		Mechanical load cycles	
number	cycles	number	cycles
88x	-10°C...+20°C	1x	600 Pa
88x	0°C...+30°C	5x	500 Pa
189x	10°C...+40°C	7x	400 Pa
		23x	300 Pa
		59x	250 Pa
		80x	200 Pa
		243x	150 Pa

Tab. 1: Thermal Cycle and Mechanical Load Cycle Sequential Testing: the 1-year equivalent test was performed in 1 week.

With the knowledge of the local lows and highs of environmental temperatures at the plant site, the peak temperature gradients were connected to the most frequent occurring changes (see Tab. 1) simulating a one year from the local conditions. The test scenario in the lab was applied to simulate 5 years at real plant conditions.

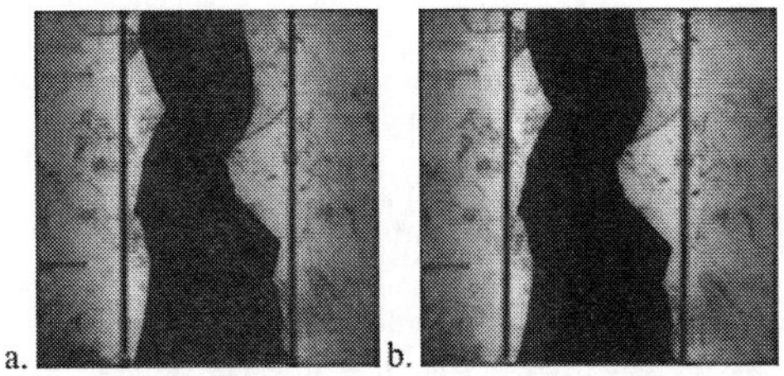

Fig. 4: Electroluminescence images of a cell under test (type 2, 6x8 multi, severely affected) a) directly from the field and b) after 5-year equivalent testing with an additional power loss of -4.0 % of the whole module. Especially dark parts of the cell became darker.

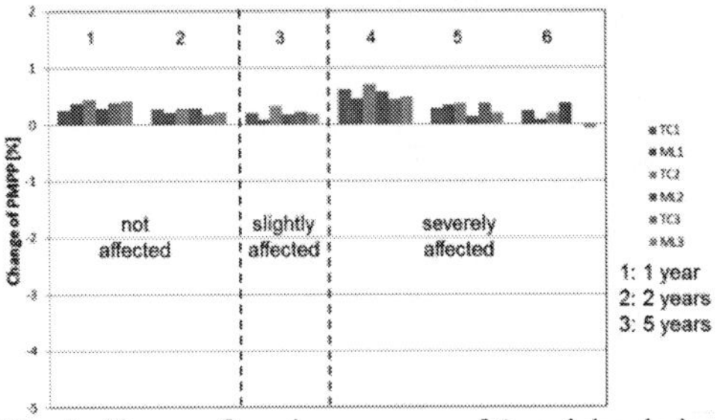

Fig. 5: Change of maximum power of 6 modules during 5-year equivalent testing: Type 1 (manufacturer A, 6x12 mono).

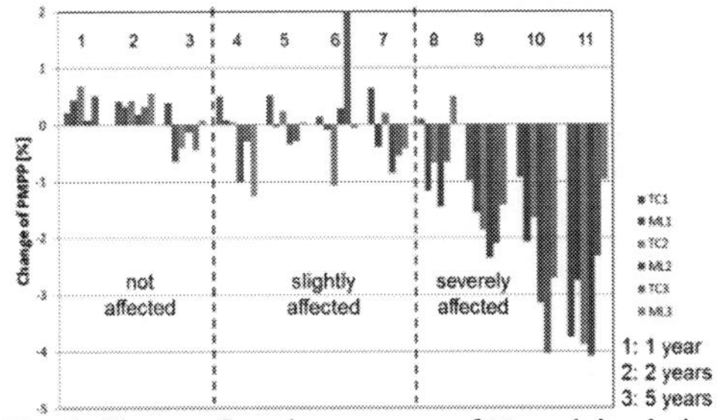

Fig. 6: Change of maximum power of 11 modules during 5-year equivalent testing: Type 2 (manufacturer B, 6x8 multi).

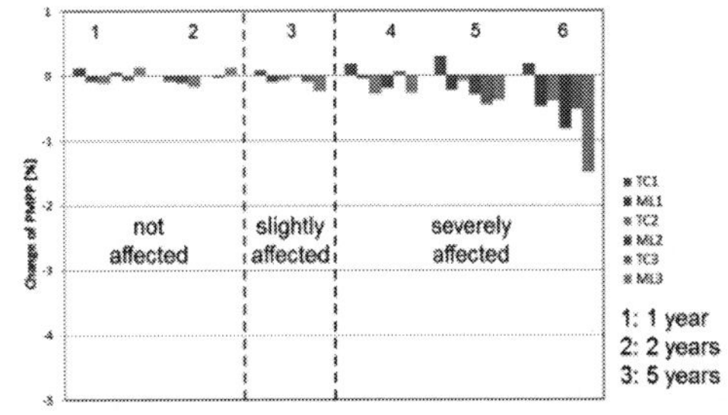

Fig. 7: Change of maximum power of 6 modules during 5-year equivalent testing: Type 4 (manufacturer B, 6x10 multi).

The 5-year equivalent sequential stress testing showed power losses up to 4%. Fig. 4 showed only minimum differences in the EL image. The investigated module types were differently sensitive (Fig. 5 to 7). Severely affected modules were more sensitive for further stress testing (Fig. 4 and 6). Additional hail impact tests showed slight power losses up to 0.6%. Safety tests (e.g. hot spot endurance test) and UV-fluorescence imaging for the cell crack age determination are planned for further analysis.

3 NON-UNIFORM SNOW LOAD

Snow load damaged modules from the field typically have bended frames and broken glass (Fig. 8). The increased load on the lower part of the module affects mostly the modules in the bottom row of roof-top mounted systems. Creeped ice, which has accumulated between the laminate and the frame, can cause additional stress. This chapter enables an overview of a first material intercomparison of such simulated conditions in the laboratory.

Fig. 8: Snow load damaged modules from the field typically have bended frame parts at the bottom and broken glass. The arrow shows the down-hill force component [6].

3.1 Test Set-up

In IEC 61215-2 [1] a snow load and wind load test with horizontal modules and homogeneous loads of 2400 Pa and 5400 Pa is described. A draft for non-uniform snow load testing is now being considered as IEC 62938. The test procedure was developed at TÜV Rheinland [7] as 2Pfg 2310/11.12 and conducted under the inclinations of 37° from EN 1991-1-3 [8] at 25°C and a non-uniform load distribution on the module (Fig. 9).

The purpose of the test after IEC 62938 [2] is to determine the load limit of the module type. The test plan is shown in Fig. 10. For six modules, a preconditioning with 10 cycles humidity freeze is foreseen. Five modules are 'tested to failure' and a sixth module is stressed with approx. 63 % of the mean load of the five modules after the indication of their failure limit. Normally, the modules are mounted in portrait orientation with 4 clamps on the long frame parts as defined by the manufacturer.

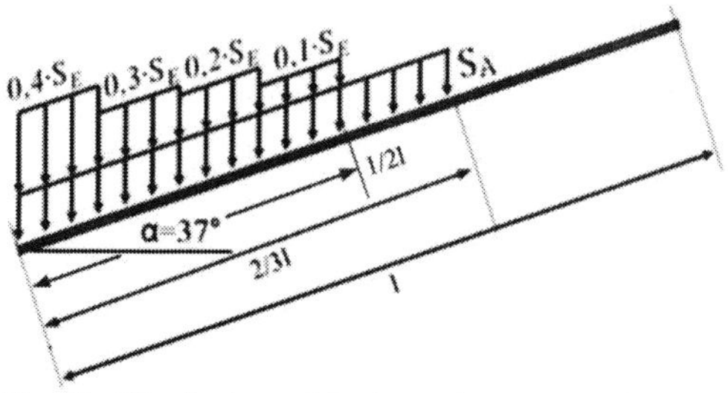

Fig. 9: Distribution of load on the test specimen at inclination of 37° [2].

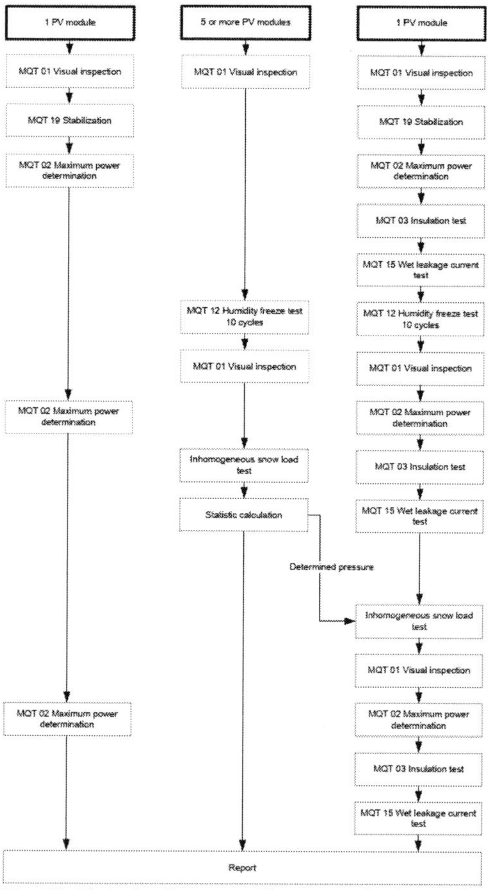

Fig. 10: Test plan for non-uniform snow load test after IEC 62938 draft [2]. The numbers (MQT xx) relate to the clause numbers in IEC 61215-2 [1].

The initial load is 2400 Pa. After 24 hours of holding, further weight elements out of steel on sliders out of Teflon for low static friction, are placed on the module in 200 Pa steps until breakage of the module (Fig. 11 and 12). The process is controlled by distance sensors (Fig. 13 and 14).

Fig. 11: Non-uniform snow load simulation on an inclined module with the application of load elements made of steel and therefore bending of the bottom frame part.

Fig. 12: Irreversibly bent frame and broken glass after a test sequence

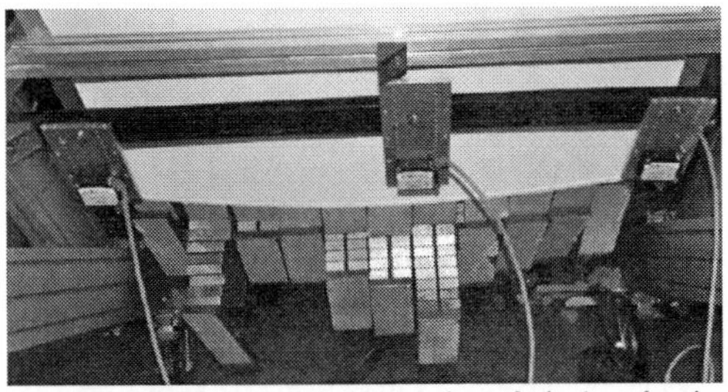

Fig. 13: Distance sensors (Laser triangulation) under the module after the end of the stress test.

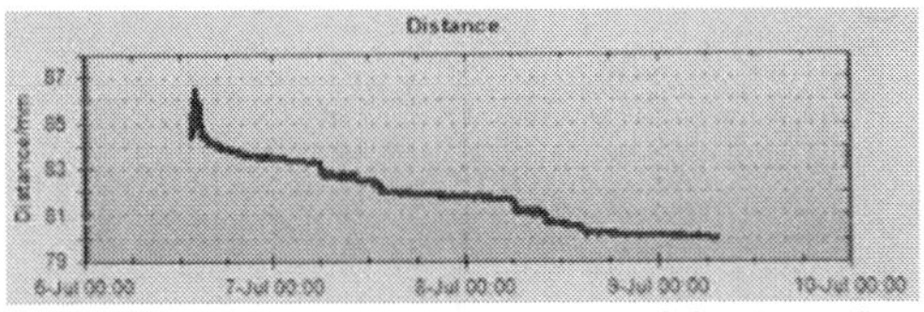

Fig. 14: Distance sensors under the module: Exemplary distance curve over 3 days of testing with several holding periods.

Other approaches have been published by Schletter [6] who developed a test facility with loosely connected movable load holders with an inclination of 30°, and by HSR Rapperswil [9] with a cover by air cushion and parallel forces by straps.

3.2 Results (Maximum Load)

A test series (Tab. 2 and 3) with 14 module types with different frame dimensions, adhesives and module sizes was conducted under the inclination of 37°. At room temperature, modules with silicone based frame adhesives resisted loads of up to 8400 Pa (Fig. 15) without frame bending or any damages, while modules with acrylic foam tape based adhesives allowed frame bending and a glass breakage at lower loads up to 4600 Pa. One module type with acrylic tape showed poor quality. The second order effects are glass and frame stability.

Reasons for deviations up to 1100 Pa (resp. 20%) may result from the tape position between start and end edge of the tape and for silicone voids in silicone and EVA bursts.

Tests at temperatures at -10°C were less severe, during the warming up phase the adhesive weakens and the module breaks.

	Adhesive	**Precond.**	**Load [Pa]**	**Comment**
1	Acrylic tape	DH	1880	>5% loss
2	Tape	HF	2840	
3	Silan-modified	---	3600	
4	Acrylic tape	DH	4280	
5	Acrylic tape	DH	4333	
6	Silicone	HF	5240	
7	Silicone	---	5400	
8	Tape	HF	6080	Landscape
9	Silicone	DH	7000	
10	Silicone	HF	7360	
11	Silicone	HF	7680	
12	Silicone	HF	7800	
13	Silicone	HF	7920	
14	Silicone	HF	8000	

Tab. 2: Test series with 14 module types: the frame adhesive, the preconditioning by damp heat or humidity freeze, the average maximum load and a comment for 2 module types are given. The module types are sorted after the average load. Module type 1 failed the power loss criterion and module type 8 was mounted in landscape orientation.

	Mechanical Load [Pa]					
	1	2	3	4	5	Average
1	0	2800	3200	3400	0	1880
2	3400	3000	2800	2600	2400	2840
3	3600					3600
4	4600	4200	4200	4000	4400	4280
5	4200	4200	4600			4333
6	4200	4200	6200	7000	4600	5240
7	5400					5400
8	6000	6000	6200	6200	6000	6080
9	6400	7200	7000	7200	7200	7000
10	6400	7200	8400	7200	7600	7360
11	7200	7800	7400	7800	8200	7680
12	7800	7800	7400	8000	8000	7800
13	7800	7400	8000	8200	8200	7920
14	8000	8000	8000	8000	8000	8000

Tab. 3: Test series with 14 module types: The maximum of every module and the average for the module type are

given.

Fig. 15: Module loaded with 8000 Pa. 8400 Pa are approx. 1500 kg steel loads on 1,8 m² module area. The weights were kept from falling of the test-deck by safety chains.

4 CONCLUSIONS

- Hail impact tests with more severe test conditions than usually applied for PV module product qualifications simulated similar cell damages in the field rather than hail impacts do according to IEC 61215 (25 mm ice balls).
- The 5-year equivalent sequential stress testing of PV modules from a system which was stressed from a hail event showed power losses up to 4%. Module types were differently sensitive and severely affected modules were more sensitive for further stress testing.
- Non-uniform snow load tests after IEC 62938 simulate failure types similar to the field and reflect closely real snow impacts to the modules.
- Key parameter for the non-uniform snow load test is the frame adhesive with maximum loads of 8400 Pa for silicone and 4600 Pa for acrylic foam tape.

5 REFERENCES

[1] IEC 61215-2: Terrestrial photovoltaic (PV) modules - Design qualification and type approval - Part 2: Test procedures, 2016.

[2] IEC 62938 Non-uniform snow load testing for photovoltaic (PV) modules, draft 2016.

[3] G. Mathiak, L. Pohl, J. Sommer, U. Fritzsche, W. Herrmann, F. Reil, J. Althaus, M. Köntges: „PV Module Damages caused by Hail impact – Field Experience and Lab Tests" Proc. 31st EU PVSEC, Hamburg, 2015, 5DO.12.5, 1915-1919.

[4] W. Herrmann, N. Bogdanski: "Outdoor Weathering of PV modules – Effects of various climates and comparisons with accelerated laboratory testing", IEEE PVSC 37, 2011 2305-2311.

[5] F. Reil: "Dynamic Stress Tests on PV Modules – Derivation of Extended Stress Scenarios", Photovoltaics International, 2012, 136-141.

[6] Schletter GmbH, Extended module testing procedure in accord. with RAL GZ 966, www.solar.schletter.eu, 2008.

[7] F. Reil, G. Mathiak, K. Strohkendl, S. Raubach, C. Schloth, B. v. Wangenheim, K. Fotak: "Experimental Testing of PV Modules under Inhomogeneous Snow Loads", Proc. 27th EU PVSEC, Frankfurt, 2012, 3414-3417.

[8] EN 1991-1-3:2010: Eurocode 1 - Actions on structures - Part 1-3: General actions - Snow loads, 2010.

[9] A. Bohren, T. Friesen: "Realistic Snowload Testing for Extreme Alpine Conditions", Proc. 31st EU PVSEC, Hamburg, 2015, 5BO.9.2, 1635 – 1638.

6 ACKNOWLEDGMENT

The work presented has been elaborated in the frame of the project 'Fidelitas' (FKZ: 0325735A) supported by the Federal Ministry for Economic Affairs and Energy

INVESTIGATION ON THE IMPACT OF MODULE CLEANING ON THE ARNTIREFLECTION COATING

Nicoletta Ferretti*[1], Klemens Ilse+, Aylin Sönmez*, Christian Hagendorf+, Juliane Berghold*
*PI Photovoltaik-Institut Berlin AG, Wrangelstr. 100, 10997 Berlin, Germany
+Fraunhofer-Center für Silizium Photovoltaik CSP, Walter-Hülse-Str. 1, 06120 Halle (Saale), Germany

[1]Phone: +49(30)8145264-119/Fax: +49(30)8145264-101/E-Mail: ferretti@pi-berlin.com

ABSTRACT: Soiling can cause strong losses in the performance of PV modules, especially in regions with high dust loads like deserts. Therefore, the modules are periodically cleaned in many cases, which can damage the surface of the modules, especially the typical anti-reflection coating (ARC) on the glass surface. For evaluating the impact of the cleaning on the surface of the modules, accelerating aging tests are performed at PI Berlin. Within these tests, the glass surface reflectivity is measured initially and after a number of cleaning cycles which corresponds to a defined number of years of operation [1]. Moreover, for comparing the mechanical resistance of different antireflection coatings, reflection measurements are performed before and after abrasion tests. In this work, the effect of the cleaning and abrasion on the antireflection coatings (ARC) is analysed on a microstructural level with optical microscopy and scanning electron microscopy (SEM) at Fraunhofer CSP. The results are correlated to the reflection measurements. Additionally, the effect of sand exposure during the accelerated cleaning test – equivalent to realistic field scenarios – is investigated. A comparison of the reflection results after accelerated aging test performed with sand of different grain size is performed on different types of ARC.
Keywords: Antireflection Coating, Evaluation, Lifetime, Qualification and Testing, Stability

1 INTRODUCTION

In the last years there was an increasing interest in PV investments in desert regions. In these areas, because of potentially high performance losses due to massive soiling [2], a regular cleaning of the modules is applied.

In order to effectively reduce the about 4% reflection loss from the module glass surface [3,4], today most of the module manufacturers utilize solar glass with ARC (Anti Reflection Coating).

However, for applications in desert regions, the coatings need to be mechanically resistant to both the abrasion due to (sand) particles and the applied cleaning procedure.

In the previous work [1], the impact of accelerated cleaning on the glass surface of PV modules was investigated with reflectance measurements. Furthermore, reflection measurements after abrasion tests using a TABER ® linear abraser 364 were performed in order to compare the mechanical resistivity of different ARC and for relating the effect of the abrasion to the cleaning.

In this paper, the reflection measured after cleaning with spreading of sand is compared to the results obtained after cleaning without sand. The impact of different kind of sand is also investigated by spreading during the accelerated cleaning fine and coarse sand on glasses with different coatings.

Moreover, we investigated the behavior of glasses with functional coatings after mechanical abrasion and cleaning on a microscopic level. Therefore, changes in reflectance can be related to microstructural changes.

This is the basis for fundamental understanding of the detailed processes during abrasion and cleaning.

Microstructural investigations permit to determine damage patterns of abrasion and cleaning.

In a first step, light-microscopic measurements on module level were conducted to visualize optical variations due to scratches or removal of the functional coating. After microscopic inspection, separation of small samples of each test setup and target preparation is done for surface analysis with scanning electron microscopy (SEM).

The impact on the module surface of the spreading of sand during the accelerated cleaning tests is also microscopically investigated.

2 EXPERIMENTAL

2.1 Methodology

Several types of coated glasses were exposed to accelerated aging by abrasion and a standard cleaning procedure. The cleaning was performed either without sand or with fine or coarse sand. Before and after the stress tests, reflection measurements were performed and the reflection deviation of final to initial was evaluated.

Successively, one glass type was selected for microscopic analysis (with both light microscope and scanning electron microscope) in order to relate the changes of the reflection after the stress tests to microstructural changes.

A comparability of cleaning and abrasion was also made by both reflection and microscopic analysis.

2.2 Test items

First, flat glasses with two different high durable hydrophobic anti reflection coatings were tested. The coatings are both single layer sol-gel based coatings about 100 nm thick with refractive index between 1.35 to 1.40. These glasses are 30 cm x 30 cm and non-tempered.

Then 8 different structured coated glasses of the same type as in [1] were here further tested by additional sand exposure during the accelerated cleaning. These glasses have module size and are tempered.

For the microstructural investigations, the glass H was chosen, which performed best in durability tests of anti-reflection [1].

2.3 Aging methods

Abrasion - The abrasion test is performed with the

device TABER ® linear abraser 364 basing on the EN 1096-2:2012 norm. For the abrasion the 'elastic soft' abradant CS-10 from TABER with a diameter of 7 mm is used. The speed is of 60 cycles per minute and the force applied of 3.5 N. After each cycle the abrading stick rotates by 30°. The abraded stripe has a length of 10 cm. Glass abrasion was performed with 25, 50, 250, 500 and 1000 cycles (one cycle corresponds to two directions).

Cleaning - With an internal developed cleaning procedure (Figure 1) 100, 500 and 1000 cleaning cycles were performed on the glasses. Here one cycle corresponds to the cleaning of the modules in one direction.

1000 cleaning cycles correspond to 20 years of weakly cleaning.

Figure 1: Internally designed cleaning device.

Three glasses for each coating type were exposed to dry cleaning with addition of either fine sand or coarse sand.

The sand was spread every 10 cleaning cycles, corresponding to 5 dust storms a year. 28 g sand was distributed with a strainer on each position (~25 cm^2) where the reflection was measured.

The sand employed is according to the standard AECTP 300-3, method 313 and it is 99% composed by SiO_2. The fine sand has an average diameter of 0.16 mm while the coarse sand of 0.42 mm.

The fiber of the brushes of the cleaning device are constituted by Polyamide 6. The fiber length is of 4 cm and the diameter of 0.3 mm. During the cleaning three cm of the fiber are above the glass.

2.4 Investigation methods

Reflectance - Reflection measurements are performed in the wavelength range 400-1000 nm with an Avantes spectrometer. The initial and final reflection (after aging tests) are measured at the same positions of the glass. The percentage deviation between final and initial reflection is averaged for each position over the wavelength range 400-1000 nm. The average of the deviations over the measured positions is finally given.

The repeatability error for the averaged reflection deviation is of 3%.

Light Microscopy - Using a special test setup at Fraunhofer CSP, all positions on module sized samples are accessible to take vibration-free microscopic images with magnifications up to 500 fold. For this, a digital microscope *Keyence VHX-2000* with light diffuser was used.

Scanning Electron Microscopy – For detailed inspection of surface properties, the scanning electron microscope *Hitachi SU70* was utilized with an acceleration voltage of 3.0 kV.

3 REFLECTION RESULTS

3.1 Sol-gel based coated glasses

The impact of spreading of dust during the accelerated cleaning and of the dust type was first tested on the two different single layer sol-gel based coatings.

For each coating type three glasses were aged by accelerated cleaning. One without sand, the second one by spreading fine sand and the third one by spreading coarse sand. For each glass the reflection was measured at the same positions initially, after 100 cycles, after 500 cycles and after 1000 cleaning cycles. The sand was applied every 10 cleaning cycles.

The reflection deviations are shown in the Figures 2 and 3 with respect to the initial measurements after 100, 500 and 1000 cleaning cycles relative to the two different coatings..

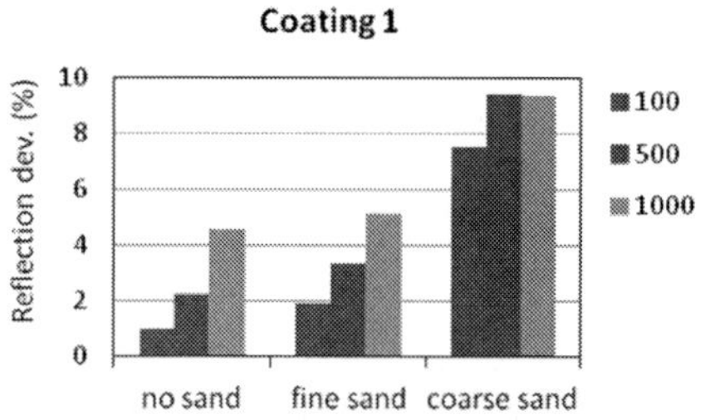

Figure 2: Reflection deviations after accelerated cleaning on three glasses with the coating 1.

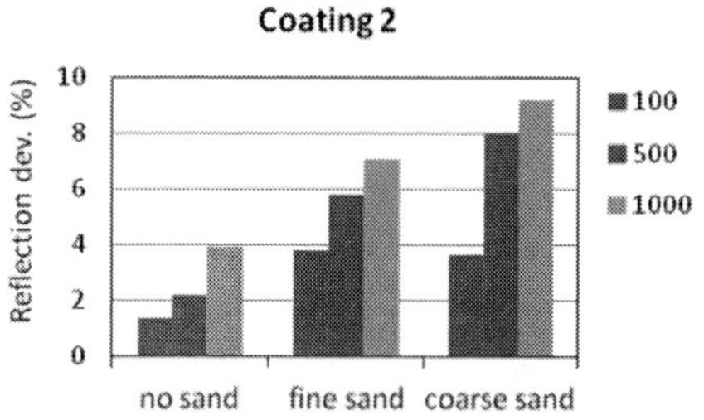

Figure 3: Reflection deviations after accelerated cleaning on three glasses with the coating 2.

It can be observed that for these coatings the reflection deviation increases when applying sand during the accelerated cleaning and is even higher in the case of coarse sand. This implies a stronger removal of the ARC.

3.2 Different structured coated glasses

8 different structured coated glasses of the same type as the ones exposed to dry cleaning in [1]were tested by spreading fine and coarse sand as in the experiment above. In Figure 4, the reflection deviations after 1000 cleaning cycles relative to the different glass types are shown.

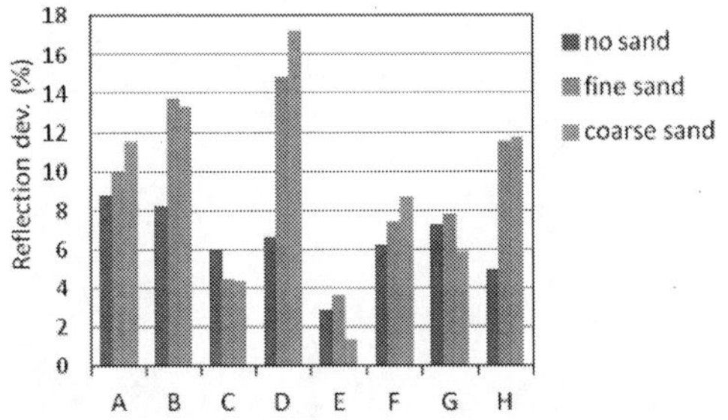

Figure 4: Reflection deviation after 1000 cycles dry cleaning with respect to the initial value.

Here, for some glass types, the cleaning by spreading of sand causes a higher reflection deviation with respect to the pure cleaning. For other glasses, the reflection deviations are similar within the measuring error.

The weak influence of sand type on the removal of the coating can be attributed either to the structure of the glasses or to the fact that these glasses, differently than the previous one, were tempered.

3.3 Structured glass H

Figure 5 shows the reflection deviations after 100, 500 and 1000 cleaning cycles for the glass H, which was the best performing coating in [1].

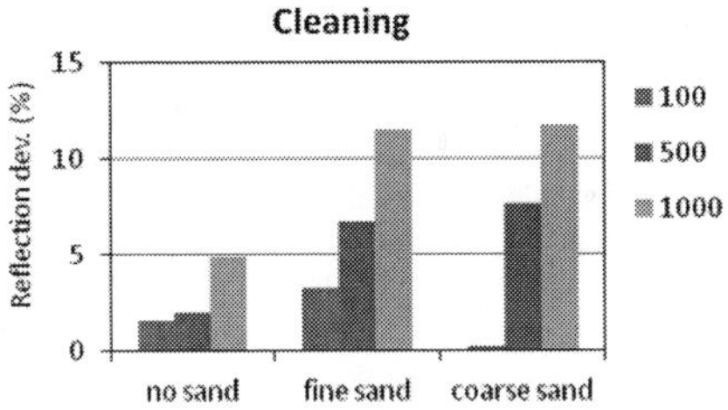

Figure 5: Reflection deviations after 100, 500 and 1000 cleaning cycles for the glass type H.

Here the reflection deviations are similar after 500 and 1000 cleaning cycles for the cleaning with fine and coarse sand.

Figure 6 shows the results after different abrasion cycles for the glass type H from [1].

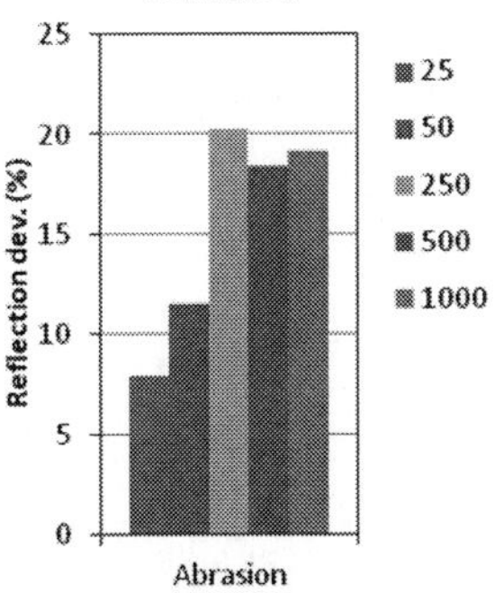

Figure 6: Reflection deviations after 25, 50, 250, 500 and 1000 abrasion cycles for the glass type H.

By comparing the figures 5 and 6, it can be observed that for this glass the reflection deviations after 500 and 1000 cleaning cycles are comparable to the deviations after 25 and 50 abrasion cycles, respectively.

From Figure 6, as the reflection deviations stay constant for abrasion cycle numbers higher than 250, it can be deduced that after 250 cycles the coating is already removed.

4 MICROSTRUCTURAL INVESTIGATIONS ON GLASS H

The glass type H was successively analyzed at a microscopical level at Fraunhofer CSP.

The impact of cleaning and abrasion on the ARC was observed and compared by using light microscopy and a scanning electron microscopy.

4.1 Microstructural analysis after Cleaning

Figure 7 images shows light microscopy images at 200 fold optical magnification. Images were taken on a reference glass, were no stress test was performed, and on glasses exposed to 1000 cycles of pure cleaning without soiling, of cleaning with fine sand and cleaning with coarse sand.

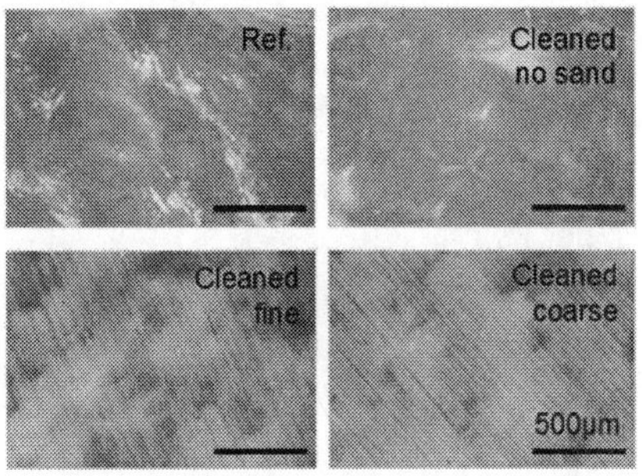

Figure 7: Light microscopy at x200 magnification after 1000 cleaning cycles on a reference glass and on glasses exposed to cleaning without sand and cleaned with fine and coarse sand.

The lighter regions visible on the reference sample can be attributed to the ARC. In the image relative to the glass exposed to pure cleaning, only few scratches can be recognized.

The images taken on the glasses exposed to cleaning with fine and coarse sand show a similar density of scratches. In these images, the scratches are more uniform and visible than for pure cleaning. From this, a stronger removal of the coating can be concluded. Therefore, these images validate the reflection results after 1000 cleaning cycles shown in Figure 5.

Figure 8 shows SEM images of the same samples shown in Figure 7.

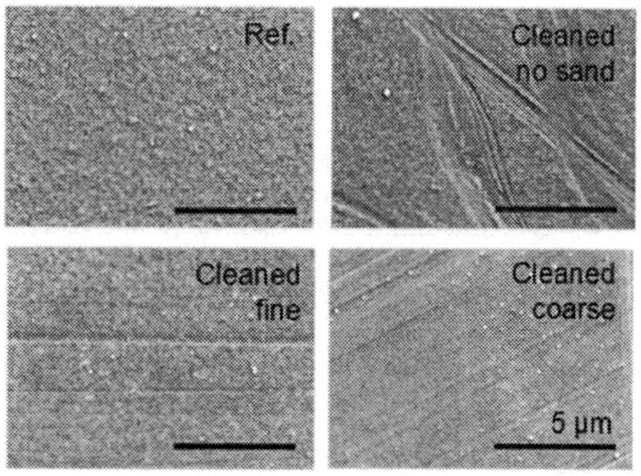

Figure 8: SEM at x10.0k magnification after 1000 cleaning cycles on a reference glass and on glasses exposed to cleaning without sand and cleaned with fine and coarse sand.

From the image taken on the reference, a sponge-like structure of the coating can be observed, see also Figure 10 (25 and 50 abrasion cycles). On the picture of the glass cleaned without sand the scratches of the single fibers of the brush can be recognized. The images of the the glasses cleaned with fine and coarse sand show both more uniform scratches.

4.2 Microstructural analysis after Abrasion

In Figure 9, pictures from areas abraded with 25, 50, 250 and 1000 cycles are shown.

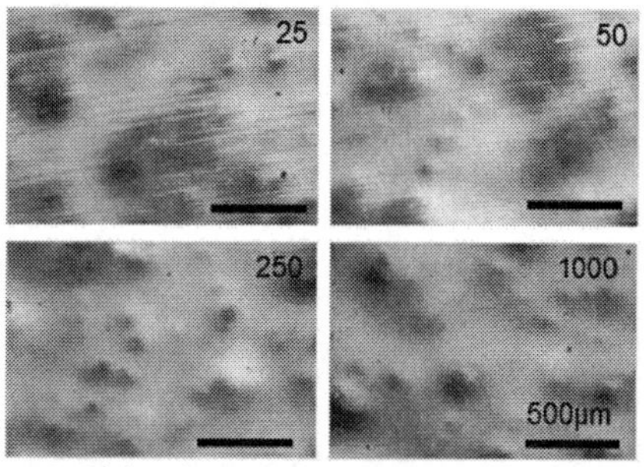

Figure 9: Light microscopy at x200 magnification after 25, 50, 250 and 1000 abrasion cycles.

The pictures of the areas abraded with 25 and 50 cycles show scratches and are comparable to the images after 1000 cleaning cycles with sand of Figure 7, especially regarding the scratch structures. Therfore, these images support the comparability of the reflection deviation after abrasion to the one after cleaning of Figures 5 and 6.

From the images it can be also observed that after 250 abrasion cycles the ARC is almost completely removed, as it was already deduced from the reflection results of Figure 6.

In the SEM pictures after 25 and 50 abrasion cycles of Figure 10 the sponge-like structure of the coating is still visible. This structure can not be recognized anymore after 250 and 1000 abrasion cycles, implying, as already observed by both reflection and optical analysis, that after 250 cycles the coating is removed. The deeper scratches visible can be attributed probably to damages in the glass surface.

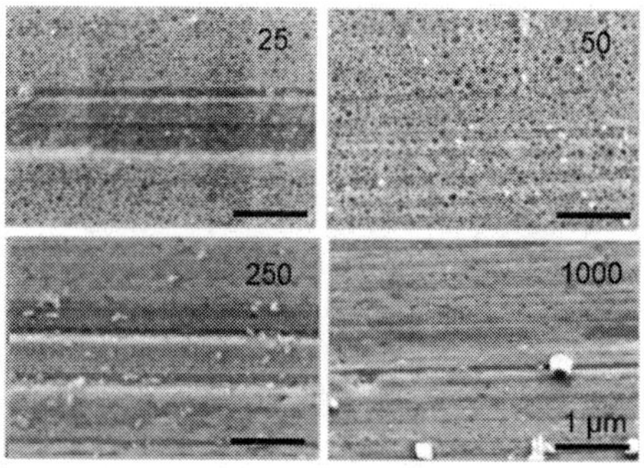

Figure 10: SEM at x30.0k magnification after 25, 50, 250 and 1000 abrasion cycles.

5 CONCLUSIONS

The reflection results after accelerated cleaning performed with sand showed a stronger removal of the coating than for cleaning without sand. The testing of different glass types with accelerated cleaning showed that the size of the used sand can have an impact on the coating, depending on the ARC type.

The optical and scanning electron microscopy analysis validated the reflection results.

The reflection after 1000 cleaning cycles with sand showed similar results as after abrasion with 50 cycles. Comparable damage patterns were also observed by the microscopic analysis. This implies that the TABER abrasion test gives comparable estimation for accelerated cleaning, but with much higher damage impact per cycle than cleaning (~ factor 20).Therefore, we conclude, that abrasion testing can be used for qualification of long-term cleaning behavior of coated glasses.

Reflection measurements, optical microscopy and SEM all indicated a removal of the coating after 250 cycles of abrasion.

6 REFERENCES

[1] N. Ferretti *et. al.*, Reflectance measurements for testing the effect of module clearing on the glass surface, Proceeding of the 31st EU PVSEC pp. 2530-2536, Hamburg, 2015.
[2] T. Sarver *et al.*: A comprehensive review of the impact of dust on the use of solar energy: History, investigations, results, literature, and mitigation approaches, Renewable and Sustainable Energy Reviews 22, 698-733, 2013.
[3] A. Artigao *et al.*: 4% higher energy conversion from BP 7180 modules, 21st EU PVSEC, Dresden, 2006.
[4] W. Hermann *et al.*: Optical characterization of PV modules, 6th WCPEC, Kyoto, 2014.

ELECTRICAL PERFORMANCE CHARACTERISATION INTERCOMPARISON OF HIGH EFFICIENCY C-SI PV MODULES WITHIN EUROPEAN AND ASIAN LABORATORIES

C. Monokroussos[1], D. Etienne[1], J. Ha[1], S. Dittmann[2], K. Morita[3], J. Stang[4], T. Herbrecht[4], V. Fakhfouri[5], N. Rebeaud[5], E. Salis[6], D. Pavanello[6] and H. Müllejans[6]
1 TÜV Rheinland (Shanghai) Co., Ltd., No.177, Lane 777, West Guangzhong, 200072, Shanghai, P.R. China
2 Scuola Universitaria Professionale della Svizzera Italiana (SUPSI-ISAAC), Canobbio, CH-6952, Switzerland
3 TÜV Rheinland Japan Ltd., 4-5-24 Chigasaki-higashi, Tsuzuki-ku, 224-0033, Yokohama, Japan
4 TÜV Rheinland Energy GmbH, Am Grauen Stein, 51105, Cologne, Germany
5 PASAN SA (Meyer Burger Group), Jaquet-Droz 8, Neuchatel, CH-2000, Switzerland
6 Joint Research Centre, European Solar Test Installation (ESTI), v. E. Fermi 2749, I-21027, Ispra, Italy

ABSTRACT: Progress in production and installation of high-efficiency (HE) photovoltaic (PV) modules has made the accuracy of their power rating and hence calibration increasingly important. This work presents the comparison of measurement results for eight HE c-Si PV modules performed at six European and Asian laboratories over a wide range of measurement methods, which include well established procedures that represent good industry practice, as well as recently introduced ones that have not been verified yet. A preliminary analysis revealed that differences in P_{MAX} remained within 1.5% for all participants and methods applied. Furthermore, the results obtained through step-like methods like *DragonBack®* and *dynamic IV* are compared to the standard procedures, which foresee long-pulse solar simulators, multi-flash method or measurement under natural sunlight. The high correlation of deviations between I_{SC} and P_{MAX} shows that differences can be largely attributed to reference irradiance measurement. The SR comparability of measurements is also examined showing an agreement within 4% for all participants in the range of 400-1100nm.

Keywords: round-robin, intercomparison, high-efficiency c-Si photovoltaic module, electrical performance, characterization

1 INTRODUCTION

Reliable measurements of maximum power of high-efficiency (HE) c-Si photovoltaic (PV) devices have been a challenge for test laboratories for over two decades [1]. Specifically, the calibration of HE c-Si PV devices is mainly affected by uncertainty components related to: i) their slower response to junction voltage changes compared to traditional c-Si and ii) spectral mismatch. The first may cause hysteresis effects in current-voltage (IV) measurements on (short-) pulsed solar simulators [2-6]. The spectral mismatch affects the measurement of effective irradiance, as the spectral mismatch of HE PV devices compared to typical reference cells made of standard monocrystalline Si may be substantial. Such deviations have been previously reported for both solar simulator [7] and outdoor measurements [8, 9].

Recent progress in production and installation of HE PV modules has made the accuracy of their power rating and hence of their calibration increasingly important. In parallel, advances in measurement systems [10-13] have opened new possibilities in characterization compared to proven standard procedures involving long-pulse solar simulators, multi-flash method and steady-state conditions such as those provided by natural sunlight. In this context the HE PV module round-robin was initiated.

The aim of this paper is to compare the measurement results of eight HE c-Si PV modules obtained by six European and Asian laboratories over a wide range of measurement methods, which include well established procedures that represent good industry practice, as well as recently introduced ones that have not been verified yet for measuring this specific PV technology. The present round-robin covers also a wide variety of traceability chains to WPVS (World Photovoltaic Scale) (AIST, ESTI, NREL, PTB), commercially available HE c-Si PV modules with efficiencies within 16.5%-19.0%, as well as measurement setups (i.e. short, long-pulse solar simulator and natural sunlight) and procedures (i.e. linear sweep rate, sectional IV with linear sweep rate, *DragonBack®* [11] and *dynamic IV* [10]). A preliminary comparison is presented for the electrical performance parameters at standard test conditions (STC) and the measured spectral responsivity (SR).

2 METHODOLOGY

2.1 Participant laboratories and measurement setups

The participants were five ISO/IEC 17025 [14] accredited laboratories located in Europe and Asia and one industrial laboratory based in Europe. Namely, in alphabetic order, the European Solar Test Installation (ESTI) of the European Commission's Joint Research Centre, PASAN SA of the Meyer Burger Group (PASAN), the University of Applied Sciences and Arts of Southern Switzerland (SUPSI-ISAAC), TÜV Rheinland Energy GmbH (TR-E), TÜV Rheinland Japan Ltd (TR-J) and TÜV Rheinland (Shanghai) Co. Ltd (TR-SHG).

More information on the accreditation scope, setup, specifications and procedure applied by each laboratory in the round-robin is listed in Appendix I. The measurement uncertainties for the electrical parameters at STC are reported in Table I and Table II for cases with and without spectral mismatch corrections respectively.

Table I: Laboratory measurement uncertainty of electrical parameters at STC (k=2) with spectral mismatch correction.

	P_{MAX}	I_{SC}	V_{OC}
ESTI	1.30%	0.84%	0.66%
PASAN	2.12%	1.65%	1.13%
SUPSI, Multiflash	1.70%	1.60%	0.35%
SUPSI, DragonBack®	1.70%	1.60%	0.35%
TR-E	1.90%	1.70%	0.70%
TR-JP	1.80%	1.60%	0.70%
TR-SHG	1.90%	1.70%	0.70%

Table II: Laboratory measurement uncertainty of electrical parameters at STC (k=2) without spectral mismatch correction.

	P_{MAX}	I_{SC}	V_{OC}
ESTI	1.60%	1.30%	0.66%
PASAN	N/A	N/A	N/A
SUPSI, Multiflash	2.6%	2.5%	0.37%
SUPSI, DragonBack®	2.6%	2.5%	0.37%
TR-E	2.50%	2.30%	1.00%
TR-JP	3.60%	3.40%	1.10%
TR-SHG	2.50%	2.30%	0.90%

All participant laboratories measured the SR either directly for the tested PV module or for an encapsulated mini-module of the same bill of materials. This is a fair approximation for as-prepared modules as long as consistent materials and manufacturing procedures are employed [15]. Additionally, electroluminescence (EL) images were acquired for the PV module under test.

2.2 Test Samples

The test specimens were three commercial and one R&D prototype HE c-Si PV modules, with total-area efficiencies varying between 16.5% and 19.0%. The PV modules were either submitted by manufacturers or purchased on the market. More details about the test specimens are listed in Table III. All types of modules consisted of triplets: two samples per type were used for the round-robin measurements, and one sample per type was stored in the dark under controlled conditions of temperature and humidity between 2013 and 2016 at the initiator's laboratory. In addition to these samples, encapsulated solar cells of the same bill of materials were available only for two types of them and were optionally used for determining the SR of the test modules. Prior to any measurements all PV-modules were subjected to $30kWh/m^2$ preconditioning at the initiator laboratory. Other light-induced stabilisation procedures (e.g. IEC 61215-2) were excluded, since all involved PV modules were made of n-type mono-crystalline Si wafers, which are known to be less sensitive to LID effects [16] than p-type material.

2.3 Intercomparison Approach and Project Timeflow

The intercomparison was performed from December 2013 to January 2016. The round-robin approach was employed here, i.e. the test samples were measured in turn by each participant and shipped to the subsequent laboratory. The agreed test protocol required electrical characterization at STC as described in IEC 61215, clause 4.6 [17] and EL imaging for all participants. All measurements were accomplished by all parties. A timeflow of the round-robin is shown in Table IV. The identity of participants is not disclosed to ensure the confidentiality of results. Participant laboratories are rather identified with capital letters A-F assigned by order of testing. Due to shipment issues the initiator laboratory (A) repeated its final measurements before laboratory (F) concluded the round-robin. It is also noted that laboratory D participated in the round-robin submitting multiple results, one for multi-flash measurements and another for *DragonBack®* method. PV-module handling in transportation is usually a concern in round-robins, but no significant damage was detected with the exception of sample #YL2, which will be discussed in more detail later.

Table III: List of PV modules used in the round-robin

Manufacturer / Module Type	Identification	Use	Efficiency(*)	Cell Type / Size	No. of Cells	Module Size	Encapsulated Cells Availability
Jiawei / JW5660/1850SP	JIA#1	round-robin	18.3%	mono c-Si / 5″	60	130cm × 78cm	No
	JIA#2	round-robin					
	JIA#3	control					
Meyer Burger / SWCT HJT	MB#1	round-robin	19.0%	mono c-Si / 6″	60	166cm × 100cm	Yes
	MB#2	round-robin					
	MB#3	control					
Panasonic / VBHN235SE10	PAN#1	round-robin	18.7%	mono c-Si / 5″	72	158cm × 80cm	No
	PAN#2	round-robin					
	PAN#3	control					
Yingli Solar / YL270C-30b	YL#1	round-robin	16.5%	mono c-Si / 6″	60	165cm × 99cm	Yes
	YL#2	round-robin					
	YL#3	control					

(*)Efficiency is calculated based on nominal power assuming the total projected area of the PV-module including frame.

Table IV: Time flow of measurement in the round-robin between participant laboratories (A-F) highlighted with different colour code; inter-laboratory shipment time is indicated with arrows and highlighted in grey.

Year	**2013**	**2014**											
Month	**12**	**1**	**2**	**3**	**4**	**5**	**6**	**7**	**8**	**9**	**10**	**11**	**12**
Lab.	A1	→		B		→		C				→	
Year	**2015**												**2016**
Month	**1**	**2**	**3**	**4**	**5**	**6**	**7**	**8**	**9**	**10**	**11**	**12**	**1**
Lab.		D		→	E	→			A2			→	F

2.4 Methods for statistical analysis

Inter-laboratory measurement deviations of PV module electrical parameters are commonly derived against the arithmetic average of participant laboratories [18-20]. The downside of this approach is that it does not account for the measurement uncertainty of the stated values. Strictly speaking, values with a lower associated uncertainty must carry higher weight.

The approach used for this work differs from that because it considers as reference value for each electrical performance parameter the mean of all measurements weighted by their measurement uncertainty (from now on just referred to as "weighted mean"), as already applied to other intercomparisons [21]. The disadvantage of such an approach is that it depends on each laboratory's stated uncertainties; therefore it relies on these uncertainty estimates to be reliable. A reliable uncertainty estimate requires firstly traceability (i.e. an unbroken chain of prior measurements to international standards) and secondly a rigorous analysis of all possible contributions in the measurement itself. Since all but one participant laboratories are ISO/IEC 17025 accredited we assume that their stated uncertainties are realistic and the approach is justified. An additional assumption made for this work is that all measurements are uncorrelated, although this may not always be the case when for example different laboratories are traceable to the same primary reference, therefore carrying a certain degree of systematic deviation.

Deviations for SR measurements are more complex to be calculated, firstly because differences in measurement setups do not allow direct comparison for each single wavelength and secondly because the measurement uncertainty was not always provided by participants as wavelength-dependent function. Therefore, the arithmetic mean of all measurement (after suitable interpolation) was used as reference for comparing the measured spectral responsivities.

2.4.1 STC performance analysis

The weighted mean for each electrical parameter (either I_{SC}, V_{OC}, or P_{MAX}) was used as its reference value. In addition, corrections were applied to consider multiple results submitted by a single laboratory. In particular, laboratory A measured twice all the modules, and laboratory D submitted multiple results. Therefore, a formula accounting for both reliability weights (i.e. measurement uncertainty) and frequency weights (i.e. number of independent measurements submitted per laboratory) was employed. Specifically, the weighted mean, $\bar{x}^*$,of measurements was derived as follows [22, 23]:

$$\bar{x}^* = \frac{\sum_{i=1}^{n} w_i x_i}{\sum_{i=1}^{n} w_i} \quad (1)$$

with weights w_i defined as:

$$w_i = \frac{1}{N_{lab} \cdot {U_{95_i}}^2} \quad (2)$$

and x_i is a measurement with expanded measurement uncertainty (k=2), U_{95} [24]. In eq. (2) N_{lab} is the number of independent measurement sets submitted by each laboratory. The notation n represents the total number of independent measurements submitted by all participants in the round-robin. The derived values are considered as the best estimate of the physical true value of the electrical parameters.

The uncertainty of the weighted mean can then be expressed as [25]:

$$\bar{U}_{95} = \frac{1}{\sqrt{\sum_{i=1}^{m} \frac{1}{{U_{95_i}}^2}}} \quad (3)$$

The notation m represents the number of participant laboratories, i.e. six in our case.

Inter-laboratory deviations were then expressed in terms of their deviation from the weighted mean as defined above. This is a more conservative approach as compared to the maximum absolute laboratory differences, because it accounts for the dispersion of measurements, and therefore can provide a more comprehensive estimate of laboratory agreement. The weighted deviation was used rather than the standard deviation, which gives statistical weight to differences measured with lower uncertainty. As the sample size was relative small (<30), the unbiased estimator correction was used for the population variance [26]:

$$\bar{s}_w = \sqrt{\frac{\sum_{i=1}^{n} w_i (x_i - \bar{x}^*)}{\sum_{i=1}^{n} w_i - \frac{\sum_{i=1}^{n} w_i^2}{\sum_{i=1}^{n} w_i}}} \quad (4)$$

A validity and consistency of all results was also tested in order to identify and, if necessary, eliminate outliers. It is noted that for this purpose all measurements were compared to ESTI results, which was the only laboratory in this round-robin accredited for PV calibration according to ISO/IEC 17025. A proficiency test was performed employing E_n numbers in accordance with ISO/IEC 17043 [27], as:

$$E_{n_i} = \frac{x_i - x_{ESTI}}{\sqrt{{U_{95_i}}^2 + {U_{95_{ESTI}}}^2}} \quad (5)$$

An agreement between results within stated UCs is obtained when the E_n number is between -1 and 1; the results do not agree with the reference, and therefore the UCs are to be revised, in case the E_n number is above 1.

2.4.2 Spectral responsivity comparison

Since measured SR data points are usually not available at the same wavelength for all the laboratories, interpolation is necessary in order to compare the derived SR function. In this paper the same interpolation method is applied to all measured SR datasets to avoid possible differences due just to the selected method. Firstly, mean SR values for each data point were computed based on laboratory submitted SR measurements for both PV modules and encapsulated cells. Subsequently, piecewise cubic spline interpolation method [28, 29] was employed. Such common and consistent approach is more suitable when the comparison has to be made between SR data points obtained by different measurement procedures, e.g. by a continuous monochromatic light beam on one side and by band-pass filters on the other, or by discrete points at different wavelengths. The resulting interpolated SR functions were then scaled such that their maximum was equal to 1 (i.e. relative SR) and compared to each other at the same wavelengths for which the reference spectrum, AM1.5, is tabulated. This is a fair approach, which would be anyway necessary to compute the spectral mismatch by following a common method. Extrapolation was not applied, therefore the SR functions are compared just over the common wavelength range. Differences were then expressed in terms of the mean of all measurements per each wavelength excluding outliers.

3 PRELIMINARY RESULTS

A preliminary analysis of the results is discussed in the following paragraphs in terms of stability of the reference samples, interlaboratory deviations for the electrical performance parameters, robustness of results and SR deviations. A more comprehensive analysis of the results including the E_n numbers will be presented elsewhere.

3.1 Drift and metastability of test samples

Potential change in modules performance due to shipment was monitored by comparing EL images acquired by each laboratory to those of the initiator. Modules did not show noticeable damage with the exception of sample #YL2, which probably sustained mechanical shock during transportation from laboratory A to laboratory B. A comparison of EL images taken in December 2013, February 2014 and October 2015 by laboratories A and B is shown in Appendix II. Sample #YL2 exhibited performance degradation within 1.0% in PMAX between December 2013 and October 2015 as measured by laboratory A, and remained within 0.4% if compared to the reference sample #YL3. A relative comparison of all modules against their respective reference sample is shown in Figure 1 according to the standard IEC 61215, clause 10.2 [30] for measurements recorded between December 2013 and October 2014. The relative change in performance remained within 0.4% for all modules, suggesting that they remained stable over the considered period. Therefore, on the basis of the considerations made until here, results can be presented in this work assuming that no change occurred during transport or due to intrinsic metastabilities.

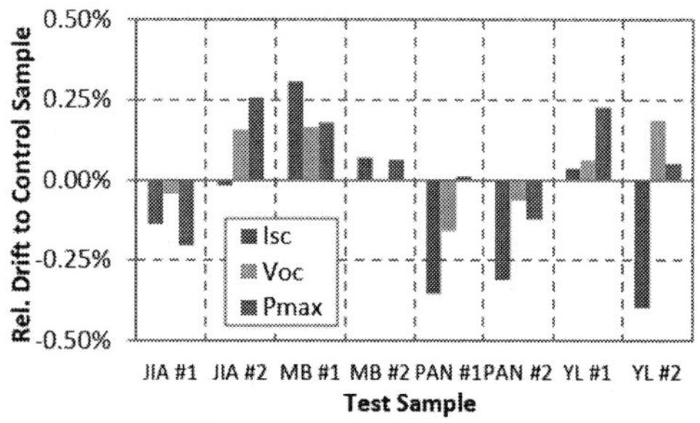

Figure 1: Relative drift of modules tested by laboratory A between October 2013 and December 2015 against the respective reference sample in accordance to IEC 61215; 10.2.

3.2 STC performance deviations

Interlaboratory deviations in the electrical performance parameters I_{SC}, V_{OC} and P_{MAX} are shown in Figure 2, 3 and 4 respectively.

Absolute differences in measured I_{SC} to the weighted mean remain below 1.12% for all laboratories. The highest deviations are obtained when spectral mismatch corrections are not included. Specifically, samples with reference JIA#1 and #2 showed the highest I_{SC} deviations in the round-robin for laboratory A, but these measurements did not include spectral mismatch corrections.

The weighted deviations of each module varied between 0.85% and 1.50% (k=2). The calculated uncertainty of the weighted mean for I_{SC} varies between ±0.55% and ±0.62% (k=2) depending on the module. Considering that traceability to WPVS for this round-robin was broad (AIST, ESTI, NREL, PTB), with reported uncertainties varying within 0.5%-1.0% (k=2) [25], the overall result for I_{SC} indicates a good agreement among the laboratories. This can be also understood by the E_n numbers against ESTI, which did not exceed 0.8 for any sample and any participant.

In general, disagreements in I_{SC} values are mainly due to the calibration of the reference device, to spectral mismatch calculation and to irradiance non-uniformity on the test plane. The latter did not exceed ±1% for any of the participants, which makes the contribution of the reference cell calibration and of the spectral mismatch calculation more relevant. It should be taken into account however, that the majority of reported deviations appear systematic in nature; meaning laboratories inclined to overestimate or underestimate I_{SC} results irrespectively to the PV-module under test. The latter would imply some systematic effects arising either from the reference cell calibration or its integration into the system (eg non-uniformity differences between the test plane and the reference cell, systematic electronic inaccuracies, or systematic errors in spectral mismatch calculation). In actual fact, tracing the origin of systematic errors, which remain below one standard deviation of the reported uncertainty, can be fairly complex.

Lastly, sweep time management is critical for step-like methods and can affect I_{SC} if initial delay is undersized [10]. Nevertheless, both *DragonBack®* and *dynamic IV* show good agreement in I_{SC} with the other methods, thus indicating adequate measurement procedures for this parameter.

Laboratory weighted deviations in V_{OC} remain within 0.62% for all modules, with weighted deviations varying between ±0.23% and ±0.57% (k=2) for all samples. The uncertainty of the weighted mean of V_{OC} varies between ±0.24% and ±0.26% (k=2).

The major sources of uncertainty contribution to V_{OC} measurement are device temperature and device hysteresis. As for the first component, IV measurements performed outdoors under natural sunlight have highest measurement uncertainty for V_{OC}; however, the comparison results do not show significant differences between indoor and outdoor measurements. In fact, the level of agreement in V_{OC} makes relevant other uncertainty components, such as for example electronic load specifications and IV curve translation. When step-like methods are considered, for which V_{OC} is obtained by interpolating distant discrete measured points, the spacing between them near V_{OC} and the interpolation method may influence the result. Laboratory E applied a correction technique based on diode modeling [31], which reduced the error of V_{OC} fitting as compared to simple linear approximation. In general, reported V_{OC} results agree within UCs for all the modules and in the worst case E_n number is 0.84.

Absolute differences in measured P_{MAX} to the weighted mean are within 1.50%, whose uncertainty is estimated to be ±0.7% - ±0.8% (k=2) depending on the module. Weighted deviations though are higher and vary between 0.99% and 2.24% (k=2) depending on the module. P_{MAX} is affected by uncertainties related to cali-

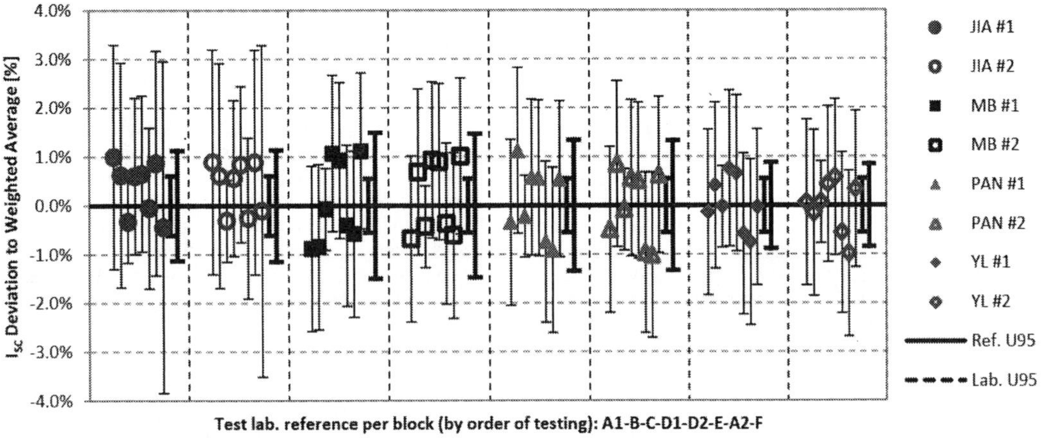

Figure 2: Short-circuit current deviations to the weighted mean for all modules as measured blindly by each laboratory at STC. The estimated values of measurement uncertainty (k=2) are shown as error bars for each laboratory. The uncertainty of the weighted average representing the best estimate of physical truth is shown as thick solid error bars (Ref. U_{95}). The laboratory agreement (Lab. U_{95}) expressed as the weighted deviation of all measurements is shown with dotted error bars.

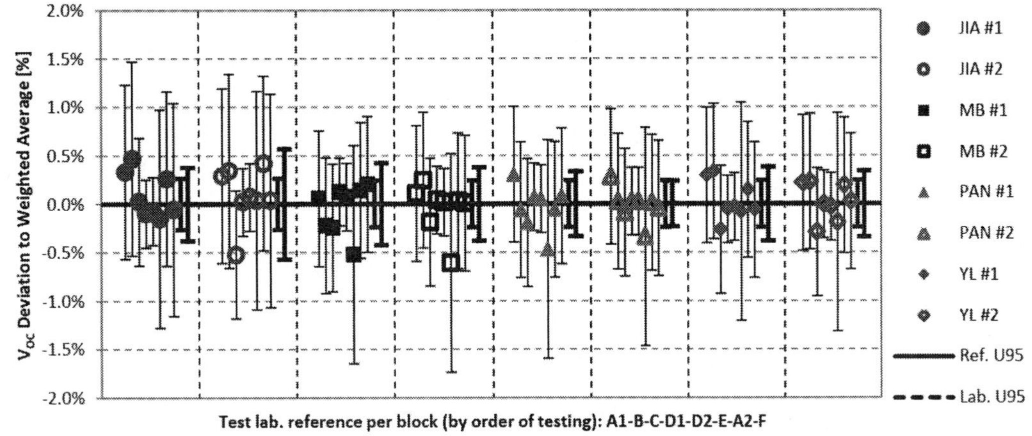

Figure 3: Open-circuit voltage deviations to the weighted mean as measured blindly by each laboratory at STC. The representations used are the same as in Figure 2.

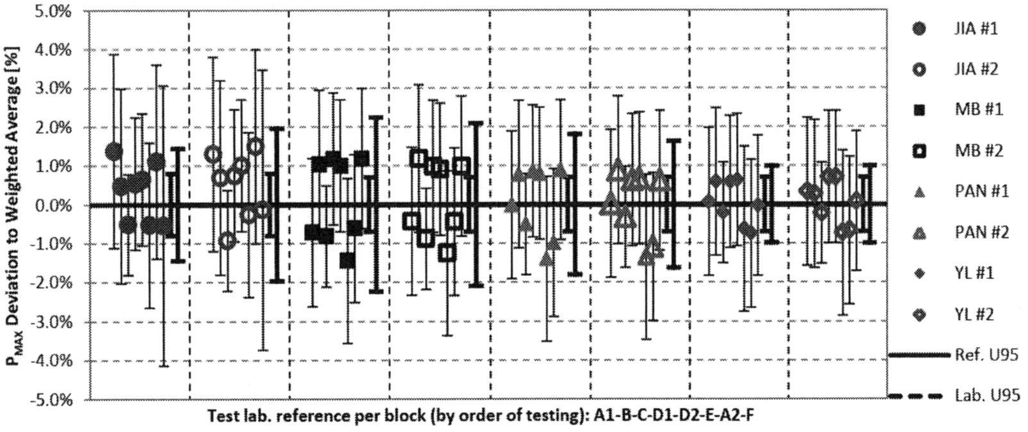

Figure 4: Maximum power deviations to the weighted mean for all modules as measured blindly by each laboratory at STC. The representations used are the same as in Figure 2.

bration of the reference device used for irradiance measurement, to capacitive behavior of HE c-Si, as well as to temperature effects and electrical connections. Isolation of the single contribution from each of these sources would require additional measurements and data analysis, which goes beyond the scope of the current paper. At present it can be observed that I_{SC} and P_{MAX} deviations are strongly correlated for all participants and modules (as shown in Figure 5), which may suggest that reference irradiance calibration had a major influence in the present round-robin. This is crucial for step-like methods, as it indicates that hysteresis artefacts are not evident as they agree with the results of long-pulse, multiflash and outdoor measurements. Overall, all measurements appear consistent with E_n scores below 0.89 in all cases.

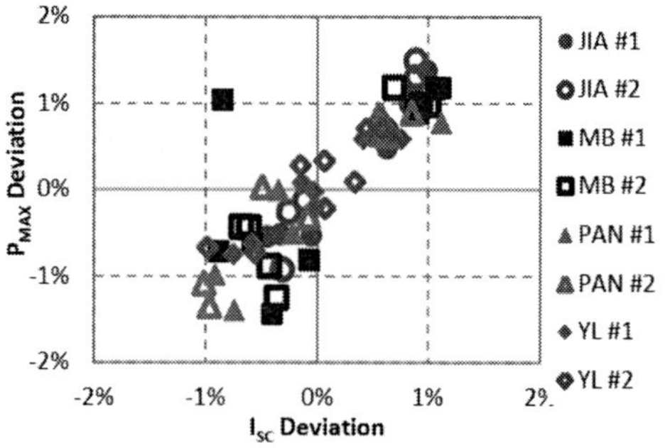

Figure 5: Correlation of I_{SC} and P_{MAX} deviations in the round-robin. The linearity shown demonstrates the dependence of power rating on reference irradiance calibration.

3.3 SR deviations

The SR deviations between laboratories were computed assuming the methodology detailed in 2.4.2. An example of SR measurement overlay for Meyer Burger type is shown in Figure 6. Meyer Burger type was chosen because the sample SR extended from 340nm to 1250nm and showed the highest deviations. The baseline reference curve represents here the mean of all measurements. Measurements of laboratory D above 930nm were identified as outliers and did not contribute. Deviations were computed by interpolating the measured data points and projecting to AM1.5 wavelengths, as shown in Figure 7. The majority of results from 400nm to 1100nm remained in reasonable agreement within ±4%. Importantly, the results were comparable for all setups (based on either monochromators or filter-wheels), methods (under-illumination vs. over-illumination) and samples (encapsulated cells or actual test modules). However, it is noted that systems, which employed monochromators mostly agreed within 2% in the range of 400-1100nm. Deviations were significantly higher in the low signal regions of 300-400nm and 1100-1200nm. A practical difficulty for low signal analysis is that only few systems were able to capture these ranges, which makes comparisons difficult. A comprehensive analysis of SR results including low signals and their impact on spectral mismatch will be presented elsewhere.

4 CONCLUSIONS

A round-robin intercomparison of high efficiency c-Si PV-modules was carried out within five Asian and European ISO/IEC 17025 accredited laboratories and one industrial laboratory based in Europe. The scope of the round-robin was to examine the measurements comparability with respect to ISO/IEC 17025 laboratory conformity assessment. Both standard and novel methods were compared. A preliminary analysis revealed that the maximum measured deviations in I_{SC} and V_{OC} varied for all samples within ±0.76%-1.12% and within ±0.29%-±0.62%, respectively. The maximum measured deviations to mean value in P_{MAX} varied within ±0.73%-±1.5% with sample type. The weighted P_{MAX} deviations per sample type ranged within ±0.99%-±2.24% (k=2), which can be seen as a conservative estimation of interlaboratory agreement for HE c-Si PV. These numbers are encouraging and indicate good agreement for all participants. Furthermore, a proficiency test was employed to examine the consistency of results and their corresponding uncertainties. The E_n numbers of each measurement was calculated with respect to ESTI (the only accredited calibration laboratory for the measurements reported here); all results were below 1 indicating consistency. Essentially this work validated all applied methods and their stated measurement uncertainties. The SR comparability of measurements was also examined showing general agreement within 4% for all participants in the range of 400-1100nm. A more comprehensive analysis of all results will be presented elsewhere.

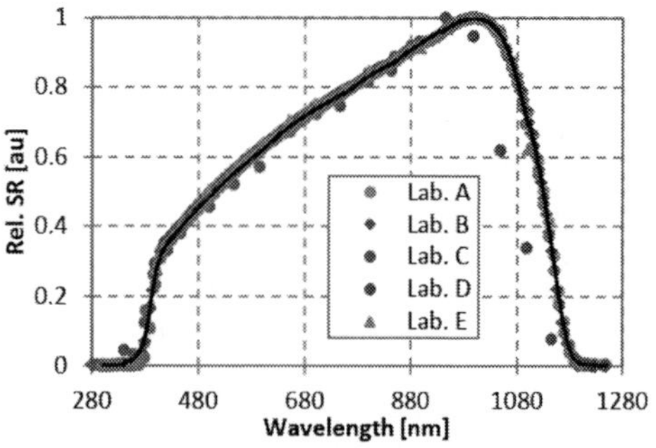

Figure 6: Variation in SR measurements in round-robin for Meyer Burger type for different participants. SR curve of each laboratory represents the mean of test samples and includes both PV-modules and encapsulated cells, when available. The mean reference curve of all measurements excluding outliers is shown.

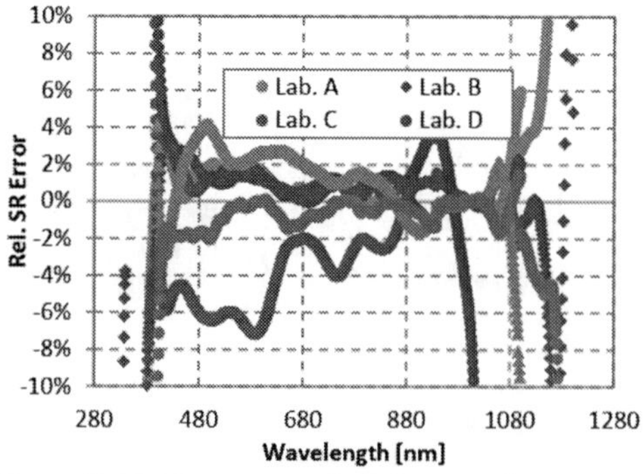

Figure 7: SR Measurement deviations to reference curve for all round-robin participants. Deviations were computed by interpolating and projecting to AM1.5 wavelengths. Most differences from 400nm to 1100nm remained within 4%.

ACKNOWLEDGEMENTS

The authors would like to acknowledge the contribution of Meyer Burger Group, Panasonic Corporation, Shenzhen Jiawei Photovoltaic Lighting Co., Ltd and Yingli Green Energy Holding Co., Ltd.

REFERENCES

[1] H. A. Ossenbrink, W. Zaaiman . J. Bishop, "Do multi-flash solar simulators measure the wrong fill factor?" (1993); pp. 1194-1196.

[2] G. Friesen, H.A. Ossenbrink, "Capacitance effects in high-efficiency cells", Solar Energy Materials & Solar Cells, Vol. 48, Issues 1–4, (1997); pp. 77–8

[3] J. Metzdorf, A. Meier, S. Winter, T. Wittchen, "Analysis and correction of errors in current-voltage characteristics of solar cells due to transient measurements", 12th EUPVSEC (1994); pp. 496-499

[4] Y. Hishikawa, "Precise Performance Measurement of High-Efficiency Crystalline Silicon Solar Cells", 4th IEEE World Conference on Photovoltaic Energy Conversion (2006); pp. 1279-1282

[5] C. Monokroussos, R. Gottschalg, A. N. Tiwari, G. Friesen, D. Chianese, S. Mau, "The Effects of Solar Cell Capacitance on Calibration Accuracy When using a Flash Simulator", 4th IEEE World Conference on Photovoltaic Energy Conversion (2006); Vol. 2 (2006); pp. 2231 – 2234

[6] A. Virtuani, H. Müllejans, F. Ponti, E. Dunlop, "Comparisonof indoor and outdoor performance measurements of recent commercially available solar modules", 23rd EU PVSEC (2008); pp. 2713-2718

[7] C. Monokroussos, J. Lau, D. Etienne, K. Morita, C. Dreier, and U. Therhaag, "IEC 60904-9 spectral classification and impact on industrial rating of c-Si devices", 6th World Conference on Photovoltaic Energy Conversation (2014)

[8] H.Müllejans, W. Zaaiman, E.D. Dunlop, H.A. Ossenbrink, "Calibration of photovoltaic reference cells by the global sunlight method", Metrologia42(2005); pp 360–367 (doi:10.1088/0026-1394/42/5/004)

[9] D. Dirnberger, G. Blackburn, B. Müller, C. Reise, "On the impact of solar spectral irradiance on the yield of different PV technologies", Solar Energy Materials and Solar Cells, DOI: 10.1016/j.solmat.2014.09.034

[10] C. Monokroussos, D. Etienne, K. Morita, C. Dreier, U. Therhaag, W. Herrmann, "Accurate power measurements of high capacitance PV modules with short pulse simulators in a single pulse", 27th EUPVSEC (2012)

[11] A. Virtuani, G. Rigamonti, G. Friesen, D. Chianese, P. Beljean, "Fast and accurate methods for the performance testing of highly-efficient c-Si photovoltaic modules using a 10 ms single-pulse solar simulator and customized voltage profiles", J. Meas. Sci. Technol., 23 (2012); doi:10.1088/0957-0233/23/11/115604

[12] H. Kojima, K. Iwamoto, A. Shimono, J. Abe and Y. Hishikawa, "Accurate and Rapid Measurement of High-Capacitance PV Cells and Modules Using a Single Short Pulse Light", IEEE PVSC (2014)

[13] K. Ramspeck, S. Schenk, L. Komp, A. Metz, M. Meixner, "Accurate efficiency measurements on very high efficiency silicon solar cells using pulsed light sources", Proceedings of the 29th EUPVSEC (2014)

[14] ISO/IEC 17025:2005, "General requirements for the competence of testing and calibration laboratories"

[15] C. Monokroussos, D. Etienne, J. Theophil, J. Lau, L. Xu, S. Kera, K. Morita, C. Dreier, R. Struwe, T. Hasegawa, "Effect of Glass/EVA/Backsheet Encapsulation Package on the Industrial Rating of c-Si PV Modules", Proceedings of 31st EUPVSEC (2015); doi: 10.4229/EUPVSEC20152015-5DO.11.4

[16] B. Lim, K. Bothe, V. Voronkov, R. Falster, J. Schmidt, "Light-induced degradation of the carrier lifetime in n-type Czorchralski-grown silicon doped with boron and phosphorus", 27th EUPVSEC (2012); doi10.4229/26thEUPVSEC2011-2BO.6.6

[17] IEC 61215 – 2, Ed. 1, "Terrestrial photovoltaic (PV) modules - Design qualification and type approval - Part 2: Test procedures", 4.6 Performance at STC and NMOT (MQT 06)

[18] T.R Betts, T. Zdanowicz, M. Prorok, W. Kolodenny, H.d. Moor, N.v.d. Borg, D.Stellbogen, J.Hohl-Ebinger, W.Warta, G. Friesen, D. Chianese, A.Guerin de Montgareuil, W. Herrmann, J.D. Berrade, J. Moracho, A.B. Cueli, A.R. Lagunas, R. Gottschalg, "Photovoltaic performance measurements in Europe: PV-catapult round robin tests", Proc. 4th IEEE World Conference on Photovoltaic Energy Conversion 2238-41 (2006)

[19] W. Herrmann, S. Mau, F. Fabero, T.R. Betts, v.d.N. Borg, K. Kiefer, G. Friesen, W. Zaaiman,"Advanced intercomparison testing of PV modules in European test Laboratories", Proc. 22nd EUPVSEC (2006); pp. 2506-2510

[20] Y. Hishikawa, H. Liu, H.-H. Hsieh, T. Inoue, K.-S. Kim, C. Limsakul, S.-W. Kim, R. Ninae, K. Morita, "Round-robin measurement intercomparison of c-Si PV modules among Asian testing laboratories", Prog. Photovolt. Res. Appl. 21 (2012); doi: 10.1002/pip.2255

[21] D. Dirnberger, U. Kräling, H. Müllejans, E. Salis, K. Emery, Y. Hishikawa, K. Kiefer, "Progress in PV module calibration – results of a world-wide intercomparison between four reference laboratories", Measurement Science and Technology 25 (2014); 105005 (17pp)

[22] M.G. Cox, "The evaluation of key comparison data", Metrologia 39 589 (2002)

[23] M.G. Cox, C. Eiø, G. Mana and F. Pennecchi, "The generalized weighted mean of correlated quantities Metrologia", 43 S268 (2006)

[24] ISO/IEC Guide 98-1:2009 "Uncertainty of measurement -- Part 1: Introduction to the expression of uncertainty in measurement"

[25] H. Müllejans, W. Zaaiman, E. D. Dunlop, "Reduction of uncertainties for photovoltaic reference cells", Metrologia 52 (2015); pp. 646–653, doi:10.1088/0026-1394/52/5/646)

[26] C.F. Dietrich, "Uncertainty, Calibration and Probability", 2nd Ed. (1991); pp. 37-42

[27] ISO/IEC 17043:2010, "Conformity assessment—general requirements for proficiency testing"

[28] MATLAB 2016a, The MathWorks, "Piecewise Cubic Hermite Interpolating Polynomial" (2016)

[29] Fritsch, F. N. and R. E. Carlson, "Monotone Piecewise Cubic Interpolation," SIAM J. Numerical Analysis, Vol. 17, 1980, pp.238-246.

[30] IEC 61215 – 2, Ed. 1, "Terrestrial photovoltaic (PV) modules - Design qualification and type approval - Part 2: Test procedures", 4.2 4.2 Maximum power determination (MQT 02)

[31] B. Paviet-Salomon, J. Levrat, V. Fakhfouri, Y. Pelet, N. Rebeaud, M. Despeisse, C. Ballif, "Accurate Determination of Photovoltaic Cells and Modules Peak-Power from Their Current-Voltage Characteristics", 32nd EU PVSEC (2016)

Appendix I: Summary of scope, apparatus, specifications and methodology applied per laboratory in the round-robin (in alphabetic order)

Laboratory, location	**JRC-ESTI, Ispra, Italy**	**PASAN SA of Meyer Burger Group, Neuchâtel, Switzerland**	**SUPSI-ISAAC, Canobbio, Switzerland**	**TÜV Rheinland Energy GmbH (TR-E), Cologne, Germany**	**TÜV Rheinland Japan Ltd (TR-J), Yokohama, Japan**	**TÜV Rheinland (Shanghai) Co. Ltd (TR-SHG), Shanghai, China**
ISO 17025 accreditation applicable to round-robin	ACCREDIA as Calibration Laboratory	Not accredited	Swiss Accreditation Service SAS, testing laboratory (STS)	DAkkS (testing laboratory)	ASNITE (testing laboratory), DAkkS (testing laboratory)	DAkkS (testing laboratory), CNAS (testing laboratory)
Accredited scope applicable to round-robin	IEC 60904-1:2006 IEC 60904-2:2015 IEC 60904-3:2008 IEC 60904-4:2009 IEC 60904-7:2008 IEC 60904-8:2014 IEC 60891:2009	n/a	IEC 60904-1; IEC 60904-3; IEC 60904-7; IEC 60904-8; IEC 61215, 10.4; IEC 61646, 10.4; IEC 61853-1, Cpt. 8; IEC 61853-1, 7.5; IEC 61215, 4.6	IEC 60904-1:2006 IEC 60904-2:2015 IEC 60904-3:2008 IEC 60904-7:2008 IEC 60904-8:2014 IEC 60891:2009 IEC 61215:2016 (Clause 4.6)	IEC 60904-1:2006 IEC 61215:2016 (Clause 4.6)	IEC 60904-1:2006 IEC 60904-3:2008 IEC 60904-7:2008 IEC 60904-8:2014 IEC 60891:2009 IEC 61215:2016, (Clause 4.6)
Light source	Continuous natural sunlight	Pulsed Solar Simulator Light	Pulsed Solar Simulator Light	Pulsed Solar Simulator Light	Long Pulse Solar Simulator	Pulsed Solar Simulator Light
Pulseduration , ifpulsed:	NA	10.5ms	10.5ms	10ms	800 ms	10.5ms
Spectrum	close to AM1.5G IEC 60904-3:2008	Filtered Xenon, A acc. to IEC 60904-9	Filtered Xenon, A acc. to IEC 60904-9	Filtered Xenon, A acc. to IEC 60904-9	Filtered Xenon, A acc. to IEC 60904-9 (<25%)	Filtered Xenon, A acc. to IEC 60904-9
Non-uniformity of Irradiance	0% (assumed)	0.42%	1% (acc. meas. uncertainty calculation, meas.0.6%)	<0.5%	<2%	0.50%
LTI	typically 0.25%, max 0.4%	0.08%	0.20%	<0.2%	<2%	0.20%
STI	simultaneous acquisition of irradiance, voltage and current signals	multiplexing time between channels lesser than acquisition time (<0.1%)	multiplexing time between channels lesser than acquisition time (<0.1%)	multiplexing time between channels lesser than acquisition time (<0.1%)	<0.1%	multiplexing time between channels lesser than acquisition time (<0.1%)
Angle field of view from lamp housing	direct<1° , diffuse: 90°	95%, <20°	95%, <20°	95%, <20°	<12°	95%, <20°
Irradiance sensor	c-Si 2cm x 2cm reference cell	c-Si 2cm x 2cm reference cell	c-Si 2cm x 2cm reference cell	c-Si 2cm x 2cm reference cell in WPVS design	c-Si 2cm x 2cm reference cell	c-Si 2cm x 2cm reference cell
Irradiance traceability	IEC 60904-4: DSR, DSM, GSM, SSM via ESTI, NREL, AIST, PTB	via ISE-Fraunhofer	via PTB	via PTB	via AIST	via PTB
Temperature sensing of reference	internal PT100, traceable via UKAS	internal PT1000	extern PT1000, traceable via Aptomet	internal PT100	internal PT100	internal PT100 (or PT1000), traceable via NIM
Temperature sensing of test PV module	PT100 centrally backside of module, ISO 17025 external calibration accredited by UKAS	1 infra-redsensors, traceable via Raytek	1 PT1000 centrally backside of module, traceable via Aptomet	4 infrared sensors, traceable to national standards	PT100 centrally backside of module, traceable via AIST	4 infra-redsensors, traceable via NIM

Laboratory, location	JRC-ESTI, Ispra, Italy	PASAN SA of Meyer Burger Group, Neuchâtel, Switzerland	SUPSI-ISAAC, Canobbio, Switzerland	TÜV Rheinland Energy GmbH (TR-E), Cologne, Germany	TÜV Rheinland Japan Ltd (TR-J), Yokohama, Japan	TÜV Rheinland(Shanghai) Co. Ltd (TR-SHG), Shanghai, China
IV elect. load	ESTI setup; internal calibration traceable via UKAS / ACCREDIA	Internalcalib. traceable via METAS	external calibration via Pasan and Elcal	calibration by PASAN, traceable to national standards (Swiss)	Internalcalib. traceable via UKAS	External calibration, internal ISO 17025 assessment traceable via NIM
Data points acquired during measurement	approximately 200 points	Approx. 2000 points	approx. 1800 points	Multiflash & DragonBack®	250 points	Approx. 1000 points
Method of IV acquisition	linear voltage sweep	DragonBack® (step-like with auto-adapted step duration)	Multiflash	Multiflash & DragonBack®	linear voltagesweep	Dynamic IV (non-linear, step-like with progressive step duration)
Sweep direction	from Isc to Voc including both zero crossings	from Isc to Voc	from Isc to Voc	from Isc to Voc	from Isc to Voc	from Isc to Voc
Sweep duration	ca 1 second	10.5ms	10.5ms (approx. 0.8ms delay)	10ms	375ms	1.8ms delay, 8.7ms IV sweep
Number of sweeps / flashes per meas.	1	1 or 2	DragonBack®: 1, Multiflash: 25 single measurments in stationary mode	up to 20	1	1
Spectral irradiance	Spectroradiometer internally calibrated against standard lamp traceable via NPL	external calibration traceable via PTB	Spectroradiometer internally calibrated against standard lamp traceable via METAS	internal calibration, traceable to national standards	internal calibration (spectroradiometer) based on standard lamp traceable via AIST	ISO 17025 external calibration via DAkkS accredited laboratory and NIM
Spectral Responsivity	IEC 60904-8:2014 against c-Si reference cell traceable via PTB	-	IEC 60904-8:2014 against c-Si reference cell traceable via PTB	internal calibration, traceable to national standards	IEC 60904-8:2014	Internalcalibrationtraceable via PTB
IV correction	IEC 60891:2009 correction method 1 for irradiance; no translation for temperature which is accounted as uncertainty	Dynamic IV correction, LTI & effective irradiance corrections acc. to IEC 60891	IEC 60891:2009 correction method 1 for irradiance; no translation for temperature which is accounted as uncertainty (25°C ±0.5°C)	Dynamic IV correction, LTI & effective irradiance corrections acc. to IEC 60891	LTI irradiancecorrections	Dynamic IV correction, LTI & effective irradiance corrections acc. to IEC 60891
Average of measurements	3	3	1	1	3	4

Appendix II: EL images acquired in the round-robin

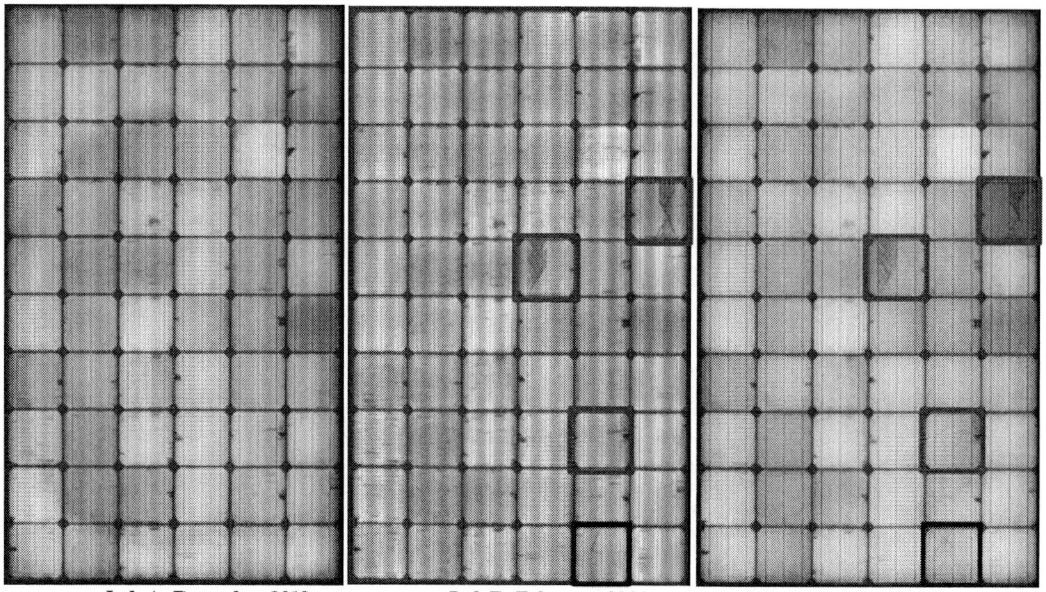

Lab A: December 2013 **Lab B: February 2014** **Lab A: October 2015**

Figure 8: EL-images of sample with identification #YL2, which exhibited the worst case degradation in the round-robin as measured in December 2013 (left) and October 2015 (right) by laboratory A and in February 2014 by laboratory B. Cell cracks (highlighted in red) appear for the first time in the image taken in February 2014. The cracks appear not to have progressed further in the image taken in October 2015. Minor crack progression is observed at the cell highlighted in black between February 2014 and October 2015. It is noted that the sample exhibited degradation in performance, which did not exceed 1% between December 2013 and October 2015 as measured by laboratory A, and remained within 0.5% comparing to the control sample #YL3.

COMPREHENSIVELY CHARACTERIZED SOLAR CELLS: IMPACT OF ANGULAR, SPECTRAL, AND NON-LINEAR EFFECTS

Thomas Fey, Ingo Kröger, Florian Witt, Stefan Winter
Physikalisch-Technische Bundesanstalt (PTB)
Bundesallee 100, 38116 Braunschweig, Germany
Tel. +49 531 592 4185, Fax: +49 531 592 694185, E-mail: Thomas.Fey@ptb.de

ABSTRACT: The prediction of the short-circuit current and hence the efficiency of solar cells is one of the main tasks for the energy rating (IEC 61853). For the complex simulation of the short-circuit current in the course of a day a comprehensive characterization of solar cells and a detailed knowledge about the spectral distribution of the natural sunlight are necessary. PTB recently developed a multi-functional laser-based Differential Spectral Responsivity (DSR)-facility for characterization and primary calibration of solar cells with lowest uncertainties. Furthermore PTB developed an outdoor facility for secondary calibration of solar cells and spectral irradiance measurements. We investigate the spectral responsivity, the linearity, the *I-V*-curves and the angular dependence of a reference solar cell using the multi-functional DSR-facility for several irradiance levels and temperatures. The obtained experimental results are combined with calculated solar spectra for a simulation of the short-circuit current and further the efficiency for a course of a day. These calculations are done for several orientations of the solar cell (roof: east, south, west and flat). The algorithm takes the orientations of the solar cells towards the sun and the spectral distribution (SMARTS, direct and diffuse light, including albedo) into account. For validation of these simulations we measured the short-circuit currents for the same orientations with the outdoor facility over a complete day. The results show a good agreement of the measurements and the simulations. This study demonstrates, that a realistic simulation of short-circuit current and hence the efficiency can be obtained for various kind of solar cells at any outdoor or indoor conditions, if typical properties of the solar cells (spectral responsivity, linearity, the *I-V*-curves, angular dependence) and the spectral distribution are known.
Keywords: Characterisation, Energy Rating, Experimental Methods, Performance, Simulation

1 INTRODUCTION

Indoor and outdoor measurements of solar cells can show significant differences in the short-circuit current and hence in the efficiency. In general, the measurement conditions for indoor measurements are reproducible and well defined. For outdoor measurements, this is not the case. For calibrations, the Standard Test Conditions (STC) have to be complied with: the reference solar spectrum (AM1.5G spectrum according to IEC 60904-3), the irradiance level of 1000 W m^{-2} and the solar cell temperature of 25 °C.

These parameters do not represent the conditions in the field. The temperature is usually higher, the irradiance is lower than 1000 W m^{-2} and the spectra (AMx) have direct and diffuse components. Consequently, calibration values give a good qualitative overview for comparing solar cells, but there may be major differences in the case of other conditions. However, the prediction of the short-circuit current and hence the efficiency is one of the main tasks for the energy rating (IEC 61853).

We performed indoor high-precision primary calibrations and angular characterizations of reference solar cells with a laser-based DSR (Differential Spectral Responsivity) facility [1]. Complementary we performed outdoor measurements of the short-circuit current (I_{SC}) for a selected c-Si reference solar cell in different orientations (roof: east, south, west and flat) with an outdoor facility [2, 3].

In our study, we focused on the comparison of calculated I_{SC} for the chosen orientations (roof: east, south, west and flat) and the measured I_{SC} of a reference solar cell in the course of a day. The calculated I_{SC} consider the spectral responsivity, the linearity, the angular dependence and the spectral distribution (SMARTS [4, 5]). The influences of the correction procedures will be discussed. Also the solar cell efficiency was calculated under consideration of *I-V*-curve measurements.

2 APPROACH

The short-circuit current (photocurrent) can be calculated by the following equation:

$$I_{SC} = \int E_{\lambda}(\lambda) \cdot s(\lambda) \mathrm{d}\lambda \qquad (1)$$

with the spectral irradiance - $E_{\lambda}(\lambda)$ and the spectral responsivity - $s(\lambda)$. If the most important influences on the spectral responsivity (temperature, linearity, angular dependence) and the spectral distribution are known, a realistic simulation of short-circuit current and hence the efficiency can be obtained for various kind of solar cells at any outdoor or indoor conditions.

In a next step we calculated the short-circuit currents for natural sunlight in the course of a day. Subsequently we compare the calculated course of a day with outdoor measurements. Finally, we calculate the efficiency of a solar cell under consideration of *I-V*-curve measurements for a course of a day.

2.1 Indoor-Measurements

The facilities at PTB [1, 6] are based on the successful application of the differential spectral responsivity (DSR) method [7] which allows $s(\lambda)$, the temperature coefficient of the short-circuit current (I_{SC}), and the linearity of the solar cells to be determined with the lowest uncertainties. The short-circuit current under Standard Test Conditions (I_{STC}) of a reference solar cell can be calibrated with an uncertainty of about 0.4 % (k = 2) [1]. Additionally we measured the angular dependence of the short-circuit current with a white light source and a goniometer and *I-V*-curves at well-chosen irradiance levels and temperatures [3].

2.2 Simulated spectra

The results obtained from the indoor measurements were combined with calculated spectra. Natural-sunlight spectra in the course of a day are calculated by SMARTS (Version 2.9.2) [4, 5]. SMARTS was configured in addition to standard values with the tilted surface orientation (roof: east, south, west), the local Albedo (Dry grass, concrete slab), the geographic coordinates and the time. The selected output values were: direct-tilted and diffuse-tilted spectral irradiances.

2.3 Outdoor facility

At PTB a fully automated outdoor facility for reference solar cells was designed and built up [2]. The measurement capabilities are:

- Cosine-weighted solar spectrum ($220\ nm < \lambda < 1600\ nm$)
- Global irradiance
- Temperature of the solar cells
- Short-circuit current.

Furthermore, the setup can biaxially track the Sun or can be set to chosen orientations. The temperature control of the solar cells was realized by a water cooled mounting plate [8]. Excluding extensive temperature effects, the solar cell temperature was kept constant at approx. $T = 25$ °C (± 0.5 °C).

3 RESULTS

With the purpose simplify the calculations and measurements, both were performed with a crystalline-silicon (c-Si) reference solar cell in WPVS design at the device temperature $T = 25$ °C. The active area of the reference solar cell covers 20 x 20 mm².

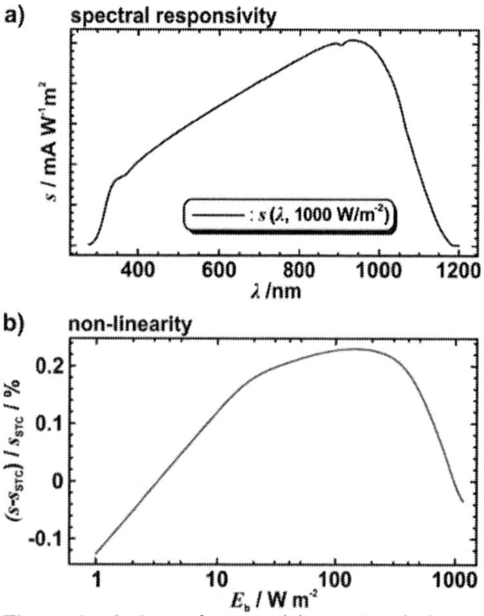

Figure 1: a) Spectral responsivity at Standard Test Conditions (STC) and b) AM1.5G weighted non-linearity of an encapsulated c-Si reference solar cell.

3.1 DSR Method

We performed measurements of the differential spectral responsivity (DSR) at well-chosen irradiance levels. Afterwards the spectral responsivity ($s(\lambda)$, see Figure 1 a)) and the non-linearity of a reference solar cell were performed with the DSR method [7]. As expected of, a low non-linearity of the c-Si reference solar cell was found. The non-linearity relative to the value at Standard Test Conditions (STC, Figure 2 b)) is only between approx. -0.1 und 0.2 %

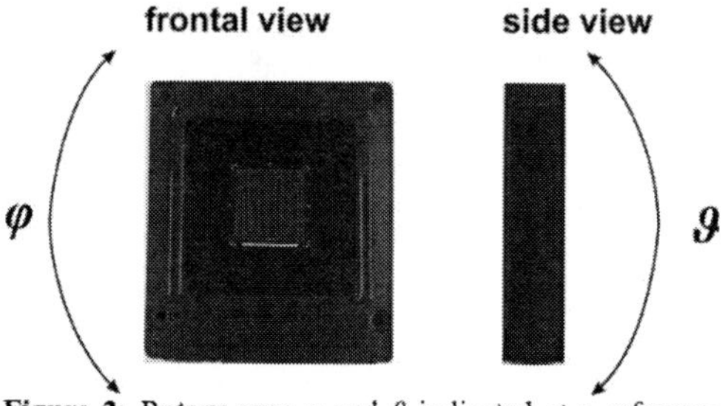

Figure 2: Rotary axes φ and ϑ indicated at a reference solar cell

3.2 Angular dependence

In addition the angular dependence of the short-circuit current is measured with a white light source, a goniometer, an amplifier, and a multimeter. The goniometer rotates the solar cell around the φ- and ϑ -axis. The rotary axes are indicated in Figure 2. We measure the short-circuit currents ($I_{SC}(\varphi, \vartheta)$) for selected combinations of φ and ϑ. The angular dependence of the I_{SC} is nearly symmetrical around φ (Figure 3 a)). Therefore we regard the ϑ-axis angular dependence (Figure 3 b)).

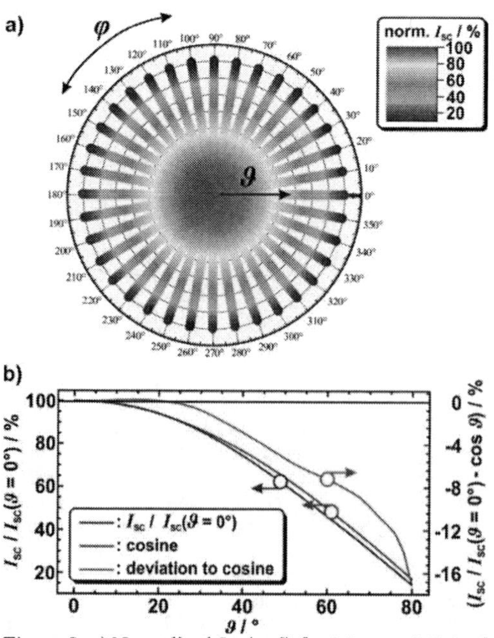

Figure 3: a) Normalized $I_{SC}(\varphi, \vartheta)$ for $0° \leq \varphi < 360$ °and $0° \leq \vartheta < 80$ °. b) Angular dependence of a c-Si reference solar cell. The measured $I_{SC}(\vartheta)$ (blue), the ideal cosine (red) and the deviation between both (green) are shown.

The deviations of the angular dependence to the ideal cosine are negligibly small for angle of incidence below 25 °. The deviation to the cosine increases with increasing incidence angles. At an incidence angle of ϑ = 80 ° the deviation to the cosine is approx. -17 %.

3.3 Comparison of calculated and measured short-circuit currents of a selected c-Si reference solar cell

The determined $s(\lambda)$, non-linearity and angular dependence of the short-circuit current (I_{SC}), as well as the calculated solar spectra (2.2, SMARTS: "direct-tilted spectral irradiances" - $E_{AMx,\,direct}(\lambda,\,t)$ - and "diffuse-tilted spectral irradiances" - $E_{AMx,\,diffuse}(\lambda,\,t)$) were combined to expand equation (1) and calculate the corrected I_{SC} by the following equation:

$$\begin{aligned} I_{SC}(t) = (&\int E_{AMx,direct}(\lambda,t)\cdot s(\lambda)\,d\lambda\cdot a_{angle,direct}(\vartheta) \\ &+\int E_{AMx,diffuse}(\lambda,t)\cdot s(\lambda)\,d\lambda\cdot b_{angle,diffuse}) \\ &\cdot c(\int E_{AMx,direct}(\lambda,t)\,d\lambda+\int E_{AMx,diffuse}(\lambda,t)\,d\lambda) \end{aligned} \quad (2)$$

with the angular corrections in accordance with Figure 3:

$$a_{angle,direct}(\vartheta)=\frac{I_{SC}(\vartheta)}{I_{SC}(\vartheta=0°)\cdot\cos(\vartheta)} \quad (3)$$

(correction due to the incidence angle of the sun) and

$$b_{angle,diffuse}=1+\int_0^{90°}(a_{angle,direct}(\vartheta)-1)\sin 2\vartheta\,d\vartheta \quad (4)$$

for 0° ≤ ϑ < 90° [9] (integral angular correction of the diffuse light) as well as the non-linearity correction due to the irradiance level (E) and AM 1.5 weighted responsivity (s) under consideration of the results of the DSR characterization (Figure 1b)):

$$c(E)=\frac{s(E)}{s(E_{AM1.5})}. \quad (5)$$

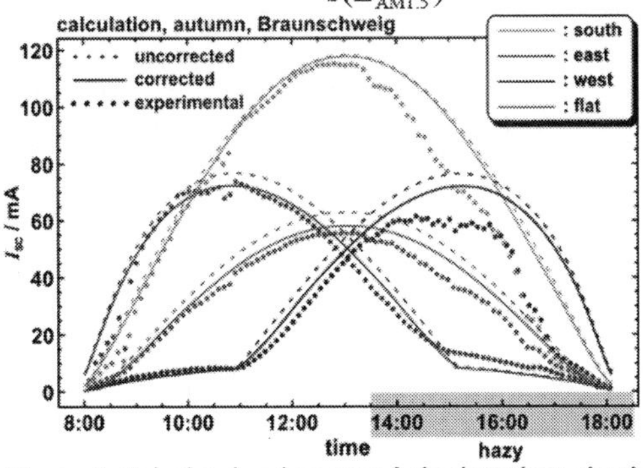

Figure 4: Calculated and measured absolute short-circuit currents of a c-Si reference solar cell on the 19th October 2012 (autumn) at PTB (Braunschweig, Germany).

Equation (2) was used to calculate the corrected I_{SC} (lines, Figure 4) of a c-Si reference solar cell on the 19th October 2012 (autumn) at PTB (Braunschweig, Germany). The roof orientation was 36° for the east (red), the south (green), and the west (blue) and 0° for the flat roof (pink). Furthermore we measured the I_{SC} (dots) at this day. The corrected and the measured I_{SC} are shown in Figure 4. Additionally the uncorrected I_{SC} (dashed line) is shown to emphasize the impact of the angular and non-linear corrections.

The obtained experimental results show a good agreement with the corrected I_{SC} the morning hour to lunchtime. The measured I_{SC} decreased compared with the calculated I_{SC} in the afternoon. The decrease is caused by a hazy sky in the afternoon, indicated by the grey region in Figure 4. Nevertheless the comparison encourages, that a realistic simulation of the I_{SC} can be obtained.

However, the uncorrected I_{SC} is higher than the corrected $I_{SC}(t)$ and would lead to an overestimation of the obtainable $I_{SC}(t)$. To identify the most important influences, the daily influences (spectral, angular and non-linear) were calculated as an example for the east roof orientation.

Because of the good agreement between model and outdoor measurement for clear sky conditions, we use for the following investigations the calculated SMARTS spectra, to have well defined representative spectra and to get rid of effects caused by hazy sky.

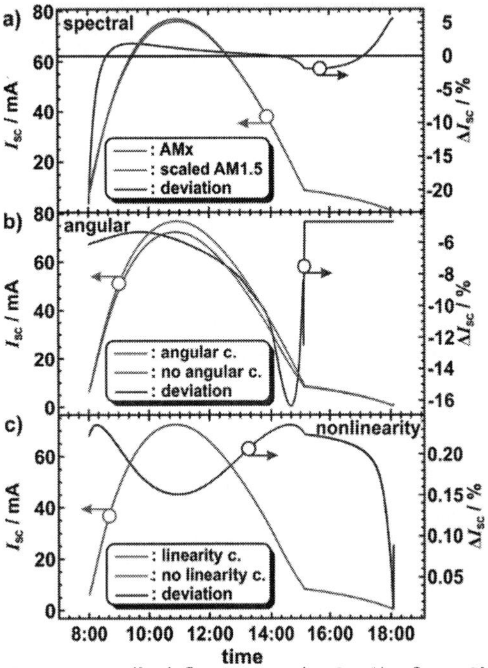

Figure 5: Daily influences on the I_{SC} (t) of a c-Si reference solar cell on the 19th October 2012 (autumn) at PTB (Braunschweig, Germany) for the east roof orientation.

a) Daily influence of the **spectra** according to the equations (6) and (7) on the I_{SC}.

b) Daily influence of the **incidence angle**. The angular corrected I_{SC} was calculated by (8) and the angular uncorrected I_{SC} was calculated by (6).

c) Daily influence of the **non-linearity** of the c-Si reference solar cell. The non-linearity uncorrected I_{SC} was calculated using equation (8) and the non-linearity corrected I_{SC} by equation (2).

The daily spectral influence compare the I_{SC} calculated by the SMARTS solar spectra $I_{SC,AMx}$ (t) (Figure 5 a), green line):

$$I_{\mathrm{SC,AMx}}(t)=\int E_{\mathrm{AMx,direct}}(\lambda,t)\cdot s(\lambda)\,\mathrm{d}\lambda + \int E_{\mathrm{AMx,diffuse}}(\lambda,t)\cdot s(\lambda)\,\mathrm{d}\lambda \quad (6)$$

with an I_{SC} calculated by scaled AM1.5 reference spectra (Figure 5 a), pink line):

$$I_{\mathrm{SC,scaled\,AM1.5}}(t)=\int E_{\mathrm{AM1.5}}(\lambda)\cdot s(\lambda)\,\mathrm{d}\lambda\cdot \frac{\int E_{\mathrm{AMx,direct}}(\lambda,t)\,\mathrm{d}\lambda+\int E_{\mathrm{AMx,diffuse}}(\lambda,t)\,\mathrm{d}\lambda}{\int E_{\mathrm{AM1.5}}(\lambda)\,\mathrm{d}\lambda} \quad (7)$$

The scaling ensures the same irradiation levels for both spectral inputs. The formulas simulates the results, one would obtain with a pyranometer, but without using a spectroradiometer in comparison to a setup with an additional spectroradiometer. The deviation (Figure 5a, blue line) between both currents differ by up to 22 % due to the strong spectral deviation at high AMx at the sunrise. Subsequently it reaches a maximum in the morning and indicate the best spectral match during the time with a direct incidence of the natural sunlight. Afterwards, the deviation decreases until the evening. Influenced by the increasing red diffuse light the $I_{SC,\,AMx}$ (t) increased compared with the $I_{SC,\,scaled\,AM1.5}$ (t) in the evening.

In addition to (6) the angular corrected needed to be taken into account. Therefore the angular corrections (3) and (4) are applied to (6) and lead to the following equation:

$$I_{\mathrm{SC,ang.c.}}(t)=\int E_{\mathrm{AMx,direct}}(\lambda,t)\cdot s(\lambda)\,\mathrm{d}\lambda\cdot a_{\mathrm{angle,direct}}(\vartheta) + \int E_{\mathrm{AMx,diffuse}}(\lambda,t)\cdot s(\lambda)\,\mathrm{d}\lambda\cdot b_{\mathrm{angle,diffuse}} \quad (8)$$

The angular corrected I_{SC} (green line), the angular uncorrected I_{SC} (pink line) and the deviation (blue line) between both are shown in Figure 5 b). The deviation starts at approx. -6 % and decreases slightly. Afterwards it increase to more than -16 % due to the increasing incidence angle. Finally the deviation decrease to approx. -5 % due to the reduced influence of the direct light and the gained constant value from equation (4).

The impact of the non-linear correction was worked out by comparing the non-linear corrected I_{SC} by equation (2) with the non-linear uncorrected I_{SC} by equation (8). In analogy to above the deviation was calculated and the related results are shown in Figure 5 c). The daily non-linearity is between 0 and 0.23 % due to the desirable small non-linearity of the reference solar cell. Under consideration of Figure 1 b) it is obvious that the reason for the non-linearity in the course of the day is the change of the irradiation due to the direct and diffuse irradiance.

Finally the daily-I_{SC} weighted deviations of the spectral, angular and non-linear deviations are calculated summarized in Table 1.

The non-linearity and the spectral influence are approx. one order of magnitude lower than the angular influence. Hence the results indicate the impact of angular measurements in the case of the investigated c-Si reference solar cell at $T = 25$ °C.

Table 1: Daily-I_{SC} weighted influence of the spectral, angular and non-linear correction for the east-roof orientation and a tilt of 36° to the horizontal.

influences	weighted influence / %
spectral	0.7
angular	-7.0
non-linearity	0.2

3.4 Efficiency over the course of a day

Finally the efficiency for the course of a day was calculated. Therefore we performed *I-V*-curve measurements at well-chosen irradiance levels and temperatures. For simplicity, we study the efficiency at the device temperature $T = 25$ °C.

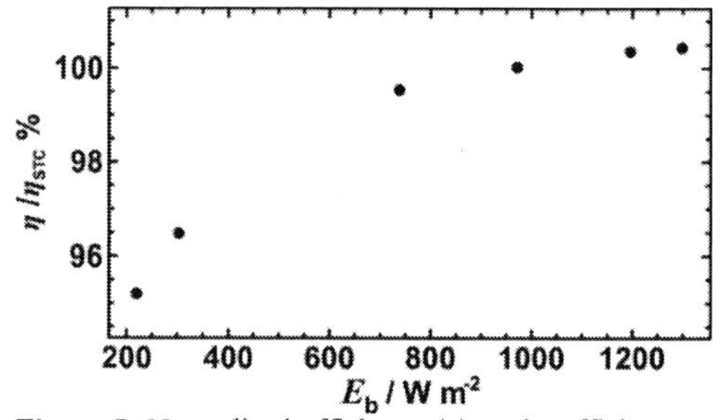

Figure 5: Normalized efficiency (η) to the efficiency at Standard Test Conditions (η_{STC}) for a c-Si reference solar cell at T = 25 °C and well-chosen AM1.5G weighted irradiance levels (200 W m^{-2} < E_λ < 1300 W m^{-2}).

Based on the measured *I-V*-curves the efficiency was calculated and normalized to the value at STC (see Figure 5). Furthermore, every obtained efficiency represent also the related non-linear electrical parameters: $I_{SC}(E)$, open-circuit voltage ($U_{OC}(I_{SC})$) and fill factor ($FF(I_{SC})$).

Subsequently with a calculated I_{SC} (t) in the course of a day and a corresponded irradiance level ($E(t)$)

$$E(t)=\int E_{\mathrm{AMx,direct}}(\lambda,t)\,\mathrm{d}\lambda+\int E_{\mathrm{AMx,diffuse}}(\lambda,t)\,\mathrm{d}\lambda \quad (9)$$

the efficiency ($\eta(t)$) over the course of a day of a reference solar cell (active area - A_{SC}) can be calculated by using the following equation.

$$\eta(t)=\frac{P_{\mathrm{Out}}(t)}{P_{\mathrm{In}}(t)}=\frac{I_{\mathrm{SC}}(t)\cdot U_{\mathrm{OC}}(t)\cdot FF(t)}{E(t)\cdot A_{\mathrm{SC}}} \quad (10)$$

Finally the equations (2), (9) (10) were used to calculate the normalized short-circuit current and the normalized efficiency of a c-Si reference solar cell on the 24th July 2013 (summer) at PTB (Braunschweig, Germany). The roof orientation was 36° for west. The results are shown in Figure 6.

First we regard the calculated I_{SC}. The initial increase of the I_{SC} is due to the rise of the Sun and consequently the rise of the diffuse light. Subsequently with the incidence of the direct sunlight I_{SC} increases faster until it reaches a maximum in the afternoon. Afterwards, the I_{SC} decreases until the sunset.

First the normalized efficiency starts high, i.e. above 97 %, and decreases due to the spectral shift of the diffuse spectrum in the sunrise hours. The angular dependence leads to an additional decrease with the incidence of the direct light. Surprisingly the lower incidence angles (in summer) and a more red sunlight (direct + diffuse) causes the normalized efficiency to increase over 100 % in the afternoon. However in the normalized efficiency decreases due to the higher AMx, higher incident angles, and at lower irradiance levels in the evening.

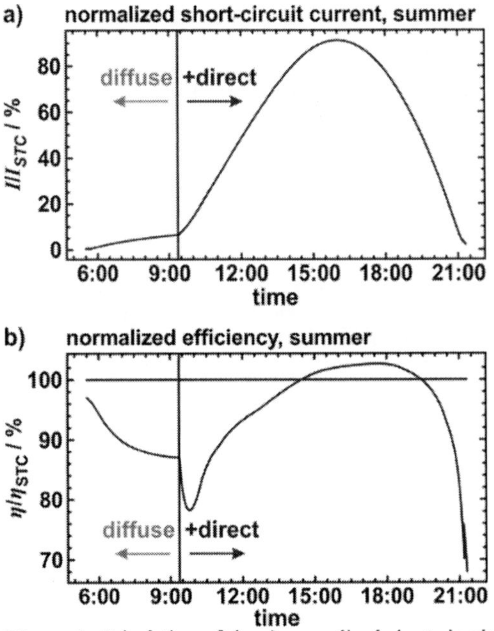

Figure 6: Calculation of the a) normalized short-circuit current and b) the efficiency of a c-Si reference cell on the 24th July 2013 (summer) at PTB (Braunschweig, Germany). The roof orientation was 36° for west.

4 SUMMARY

We investigated the spectral responsivity, the linearity, the *I-V*-curves and the angle dependence of a reference solar cell with the multi-functional DSR-facility for several irradiance levels and temperatures. The obtained results were combined with SMARTS calculated spectral irradiance distributions to calculate the intraday short-circuit current. To validate our calculation, we compared them with outdoor measurements acquired under realistic conditions. We observe a good agreement between both investigated methods.

Furthermore the short-circuit current and the efficiency for a day in summer was calculated to show the impact of diffuse and direct irradiation components.

5 ACKNOWLEDGEMENTS

The research work leading to this article was carried out within the EMRP ENG55 project „Towards an energy-based parameter for photovoltaic classification“. The EMRP is jointly funded by the EMRP participating countries within EURAMET and the European Union.

6 LITERATURE

[1] S. Winter, T. Fey, I. Kröger, D. Friedrich, K. Ladner, B. Ortel, S. Pendsa, F. Witt, “Design, realization and uncertainty analysis of a laser-based primary calibration facility for solar cells at PTB”, Measurement, Vol. 51, (2014) 457-463.

[2] T. Walzak, Master's thesis: “Untersuchung des Einflusses der Einstrahlungsbedingungen auf das Ergebnis von PV-Outdoorkalibrierungen“, PTB, December 2012

[3] T. Fey, PhD thesis: “Charakterisierung von organischen Solarzellen an einem neu aufgebauten Laser-basierten DSR-Messplatz“, PTB, September 2015

[4] C. Gueymard, "SMARTS, A Simple Model of the Atmospheric Radiative Transfer of Sunshine: Algorithms and Performance Assessment". Professional Paper FSEC-PF-270-95. Florida Solar Energy Center, 1679 Clearlake Road, Cocoa, FL 32922, 1995.

[5] C. Gueymard, "Parameterized Transmittance Model for Direct Beam and Circumsolar Spectral Irradiance", Solar Energy, Vol. 71, No. 5, pp. 325-346, 2001.

[6] S. Winter, T. Wittchen, J. Metzdorf: Primary Reference Cell Calibration at the PTB Based on an Improved DSR Facility; in “Proc. 16th European Photovoltaic Solar Energy Conf.”, ed. by H. Scherr, B. McVelis, E. Palz, H. A. Ossenbrink, E. Dunlop, P. Helm; Glasgow, (2000) James & James (Science Publ., London), ISBN 1 902916 19 0

[7] J. Metzdorf, "Calibration of solar cells. 1: The differential spectral responsivity method," Appl. Opt. 26, 1701-1708 (1987)

[8] T. Fey, I. Kröger, D. Friedrich, D. Schlüssel, S. Winter, “Indoor and outdoor measurements on various solar cells at the DSR- and the new outdoor facility of PTB”, Proc. 28th European Photovoltaic Solar Energy Conf., (2013)

[9] S. Winter, D. Friedrich, T. Gerloff: „Effect of the Angle Dependence of Solar Cells on the Results of Indoor and Outdoor Calibrations“, in “Proc. 25th European Photovoltaic Solar Energy Conf.”, (Valencia 2010), pp. 4304-4306, 4AV.3.97.pdf, ISBN 3-936338-26-4

PRECISE DETERMINATION OF THE STC I-V CURVES BY WIDE-RANGE LINEAR EXTRAPOLATION OF OUTDOOR I-V CURVES ON PARTLY SUNNY DAYS

Y. Hishikawa, H. Ohshima, M. Higa, K. Yamagoe, and T. Doi

National Institute of Advanced Industrial Science and Technology (AIST), Research Center for Photovoltaics (RCPV), Central 2, Umezono 1-1-1, Tsukuba, Ibaraki, 305-8568 Japan

ABSTRACT: A practical and precise determination of the I-V curve of PV modules at STC (i.e., under AM 1.5G, 1 kW/m^2, 25 °C), from outdoor measurements by using translation equations, is investigated. It is shown that the linear interpolation formula described in IEC 60891 can be used for extrapolation to precisely translate the data to STC. Only three experimental reference I-V curves are enough for the translation, without prior knowledge of temperature coefficients or fitting parameters. Moreover, experimental results at much higher temperatures than STC, such as 50 °C - 65 °C, can be used as the reference curves. This enables determination of the STC I-V curves from experiments under various climates and seasons throughout the year. The outdoor reference I-V curves were measured by a procedure including a short sweep time of 0.2 s and accurate module temperature measurement etc., which is applicable even on partly sunny days, hence significantly improving the available measurement conditions and precision. Good agreement with indoor results within about ±2% - ±3% is demonstrated, and further improvement is expected.
Keywords: I-V curve, Outdoor, Characterization, Performance, Translation, Irradiance, Temperature, STC

1 INTRODUCTION

Recent outspread of PV has led to increased need of their precise outdoor performance measurements. Their performance at the standard test conditions (STC; AM1.5G, 1 kW/m^2, and device temperature of 25°C) is important for many purposes such as installation test and performance monitoring. However, its outdoor measurement is not straightforward, since the PV modules usually operate at higher module temperature (T_m) than STC. Conventional precise outdoor performance measurements were usually limited to clear sunny days [1][2], which are not frequently available in many locations. The purpose of the present study is to explore precise determination of the STC I-V curves even under unstable irradiance of partly sunny days, thereby increasing the chance of measurement and reducing the required period. Translation of the experimental I-V curves to STC by linear interpolation formulas [3]-[6] is used for the extrapolation. Precision of the measured reference I-V curves are very important for the translation. Therefore, experimental conditions for enabling precise measurements even on partly sunny days are employed, in order to achieve both the frequent measurement opportunity and high precision. The conditions included fast current-voltage (I-V) measurements within 0.1 s - 0.2 s, appropriate attachment of temperature sensors to the backsheet, etc. [7]-[9].

2 EXPERIMENT

The outdoor I-V curve measurements were carried out on the rooftop of a building at AIST Tsukuba Center, Japan [7]-[9]. I-V curves of a commercial crystalline silicon (c-Si) PV module, hereafter denoted as the module under test (MUT), were measured at a sweep time of 0.2 s and an interval of 2 s - 20 s. Its backsheet temperature was monitored by three resistive temperature detectors (RTD (Pt100)) attached by polyimide tapes, whose average was used as the module temperature T_m. The solar irradiance was estimated by the short circuit current I_{sc} of the MUT. Its I_{sc} at STC (I_{sc0}) was determined by indoor precision measurement at AIST, by using a long pulse solar simulator [7]. Thus, the irradiance corresponds to 1 kW/m^2 when the I_{sc} is equal to I_{sc0}, and estimated to be proportional to I_{sc}. The outdoor I-V curves of the MUT were compared with indoor precision measurement results.

3 RESULTS AND DISCUSSION

3.1 Range of irradiance and module temperature in outdoor performance measurements

Typical irradiance and temperature of the MUT are shown in Fig. 1. Their appearance frequency is shown by different colors. Although the irradiance and temperature vary in wide ranges, the T_m is usually higher than 25 °C when the I_{sc}/I_{sc0} is 1, and the I_{sc}/I_{sc0} is usually smaller than 1 when the T_m is 25 °C. Therefore, the STC rarely occur in the present setup and location. Similar situation is expected in many areas of the world. Therefore, translation of the experimental I-V curves to STC from high temperatures such as 30 °C - 65 °C or higher has much practical importance.

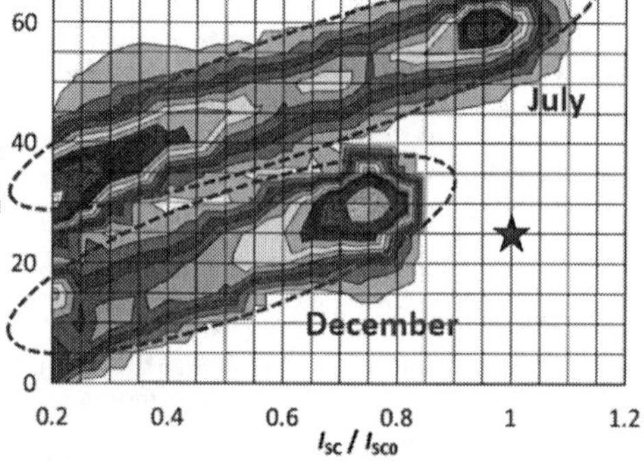

Figure 1 Appearance frequency of the I_{sc} and backsheet temperature T_m of the MUT in July and December 2014 in Tsukuba [7]. STC is shown by a star.

3.2 Translation of the I-V curves to STC by linear interpolation/extrapolation method based on experimental results measured at about 1 kW/m^2 irradiance.

Examples of the I_{sc}/I_{sc0} and T_m of three experimental reference I-V curves used for the translation to STC are shown by open circles in Fig. 2(a). The translated I-V curve by the linear interpolation/extrapolation method [3] [6], is shown by a red line in Fig. 2(b), together with the reference I-V curves and indoor result. Good agreement of 0.09% in P_{max} with the indoor result is obtained. Another example is shown in Fig. 3, where the T_m values of reference I-V curves are higher than those of Fig. 2. The difference in the P_{max} of the translated and indoor results is -1.75%. On clear sunny days, the available range of reference I-V conditions are small, as shown in Fig. 4(a). Therefore, a large factor of extrapolation for temperature is necessary, which is prone to increase translation error. However, reasonable agreement of -1.99% between the translated and indoor P_{max} was obtained, as shown in Fig. 4(b). These are very good agreement, considering that the T_m of the reference I-V curves are much higher than STC, i.e., 40 °C - 65 °C, and large amount of extrapolation is included in the translation, hence demonstrate the flexible applicability of the linear interpolation/extrapolation method.

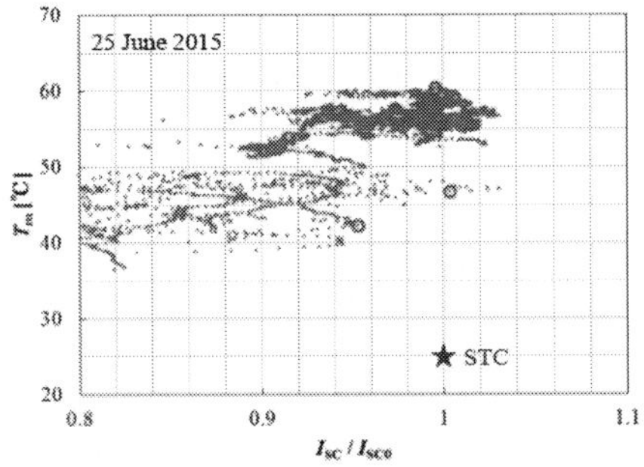

(a)

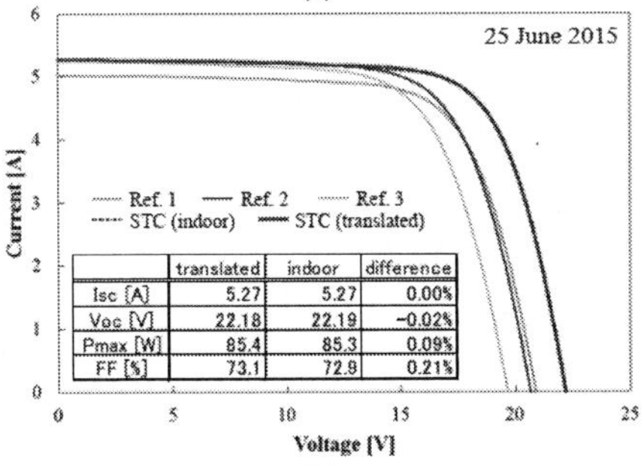

(b)

Figure 2 (a)I_{sc}/I_{sc0} and T_m of the experimental I-V curves measured on a partly sunny day are shown by dots. Those of three reference I-V curves used for the translation to STC are shown by open circles. (b) The translated I-V curve is shown by a red line, together with the reference I-V curves and the indoor result.

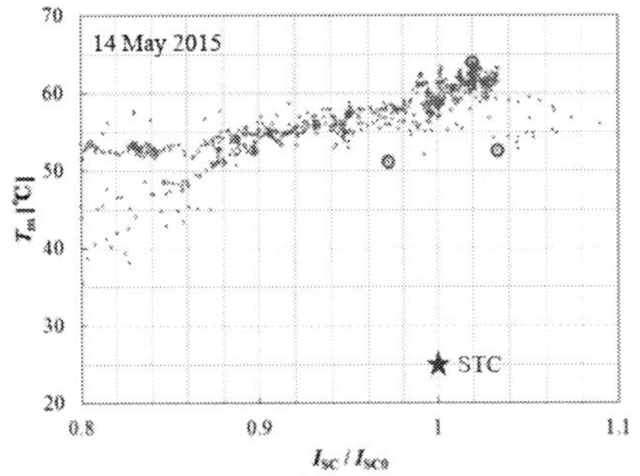

(a)

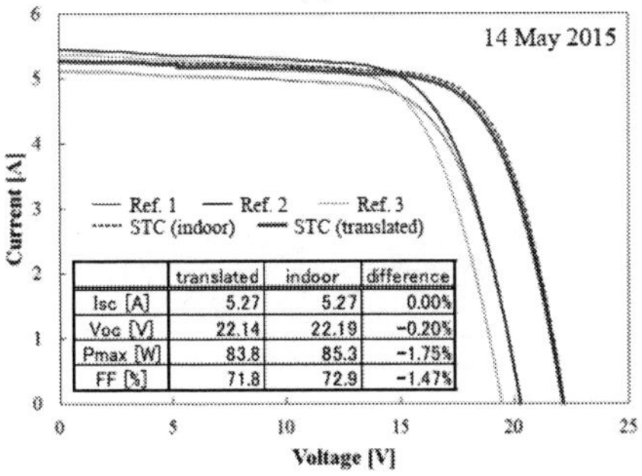

(b)

Figure 3 Similar results as Fig. 2 on another partly sunny day.

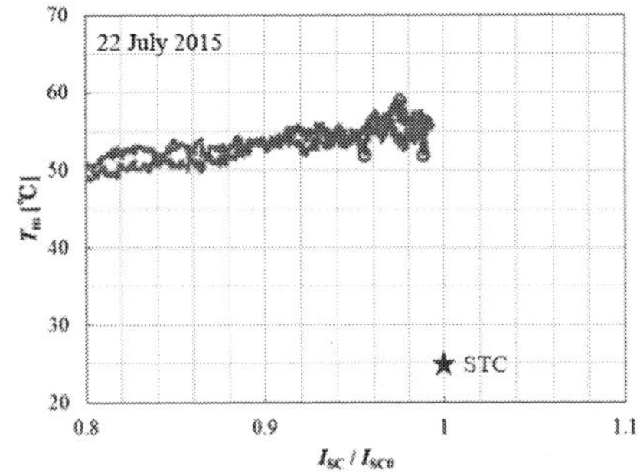

(a)

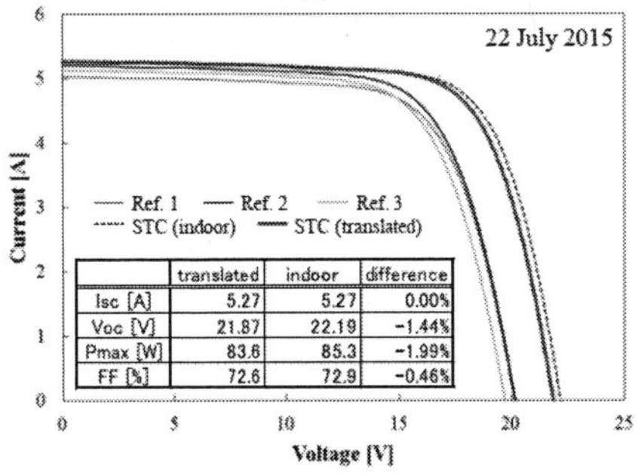

(b)

Figure 4 Similar results as Fig. 2 on a clear sunny day.

In Figs. 2 - 4, the reference I-V curves were chosen so that their difference in T_m is as large as possible within $0.95 < I_{sc}/I_{sc0} < 1.05$, in order to make the extent of extrapolation small. Values of I_{sc}/I_{sc0} which are smaller than 1 and larger than 1 are both chosen, in order to avoid extrapolation for irradiance. Solid circles of Fig. 5 are relative values of the translated P_{max} compared to the indoor results on 19 different days, showing good agreement within about ±2%. Open circles represent other translation method, which will be discussed in the next section. It should be stressed that a major advantage of the linear interpolation/extrapolation method is that it only uses two or three reference I-V curves, and does not require prior knowledge of the temperature coefficient or fitting parameters [5][6]. It is also readily applicable to various kinds of PV technologies. The present results suggest that the I-V curves measured in a short period such as a few hours and even shorter period can be utilized for precisely determining the STC performance within reproducibility of about ±2%.

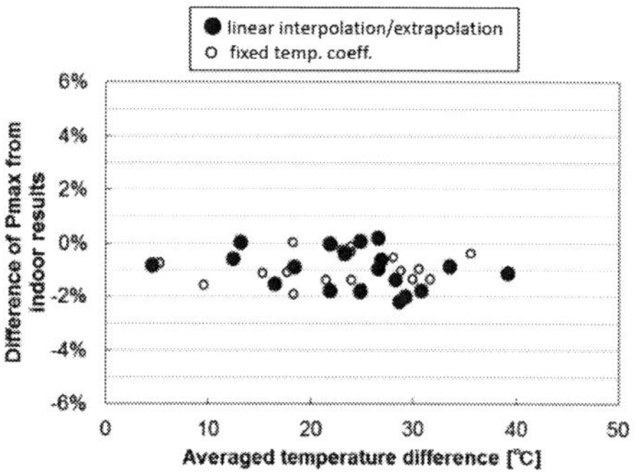

Figure 5 Relative difference in the P_{max} at STC between the I-V curves measured indoors and translated from three outdoor data on 19 different days (including Figs. 2 - 4) is shown by solid circles. The horizontal axis is the average of temperature differences between the reference I-V curves and STC, or 25 °C. The difference between the indoor data and calculated results assuming a temperature coefficient of -0.46 %/°C is shown by open circles.

3.3 Translation to STC by other translation method

Translation methods described in ref. [3] other than the linear interpolation/extrapolation require prior knowledge of the parameters such as the temperature coefficients, series resistance and curve correction factor etc., each of which should be determined from multiple experimental I-V curves. Therefore, they are not appropriate for strictly being applied for determining the I-V curve at STC from a few experimental data. However, simultaneous determination of those parameters and translation to STC by the outdoor data may be possible when many experimental data are available as shown in Figs. 2-4, which remains to be confirmed by further studies. It is noted that estimation of performance such sas P_{max} is possible, if its temperature coefficient is known, as shown in Fig. 6. For example, comparable results as Figs. 2-4, i.e., agreement with the indoor results within ±2%, are obtained, if the temperature coefficient of -0.46 %/°C is assumed, as shown in the open circles in Fig. 5. Although the estimation error can be large if the assumed temperature coefficient is not correct, it may be practical if appropriate value is known for the corresponding type of MUT.

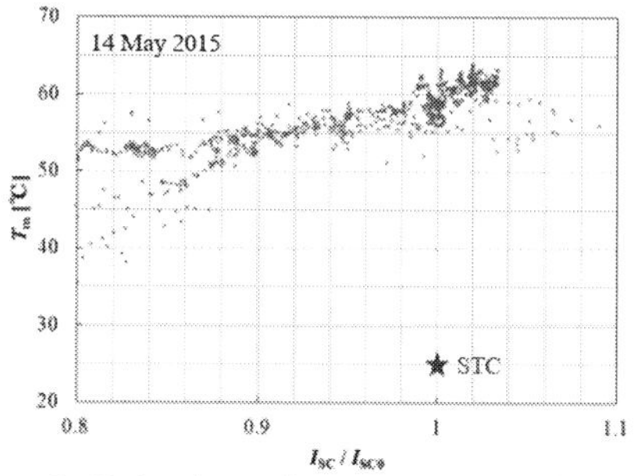

Figure 6 Estimation of P_{max} at STC from one experimental I-V curve is possible, if only the value of P_{max} is necessary, and its temperature coefficient is known.

4 SUMMARY AND PROSPECT FOR FURTHER IMPROVEMENT

The present results have demonstrated that the I-V curve of MUT at STC can be determined by using the linear interpolation/extrapolation method from only three experimental I-V curves. The method is applicable even when the T_m is much higher than 25 ° C, without prior knowledge on the temperature coefficient etc.. Good precision of the reference I-V curves are essential for the translation, especially when large factor of extrapolation is carried out such as in Fig. 4. It is noted that the good agreement (within about ±2%) of the translated I-V curves with the indoor results owes to the good basic precision of the outdoor I-V measurements of the present study, thanks to the fast I-V measurement of 0.2 s and precise T_m measurement using RTDs attached by polymer tapes, etc. [7]-[9].

Translation mainly for temperature was investigated in Figs. 2 - 5, based on the reference I-V curves at near 1 kW/m^2. Not only that, the linear interpolation/extrapolation formulas are basically applicable for any combination of reference and target I-V curves, which allows simultaneous translation for irradiance and temperature [3]-[6]. Figure 7 shows preliminary results of the simultaneous translation based on outdoor I-V data measured within one day. Three out of the eight I-V curves at various irradiance and temperature, as shown in Fig. 7(a), were used as the reference to determine the STC I-V curve by the linear interpolation/extrapolation. Figure 7(b) shows the results of translation where all the combinations are used except for those being located nearly in a line in Fig. 7(a). They show agreement within ±3% of the indoor result, thereby suggests that determination of the STC I-V curve is possible with reproducibility of ±2% - ±3% also by the simultaneous translation for irradiance and temperature, although further study is necessary for confirming the precision. It should be noted that the translation for irradiance requires data in the negative current region of low-irradiance reference I-V curve, if the whole I-V curve including the part around V_{oc} is necessary. The negative current region is not usually necessary, for calculating the curves around P_{max}. It is also noted that the interpolation/extrapolation possibly become unstable, when the reference I-V curves are noisy or have steps, which suggests a room for improvement of the procedure.

Present results such as Figs. 2(a), 3(a), 4(a), 6, and 7(a) are normalized by the I_{sc0}, which is the I_{sc} of MUT under STC. This corresponds to the situation that the I_{sc0}

is known, or that only relative value of I_{sc} is relevant. When the absolute performance of the MUT such as I_{sc} and P_{max} is to be characterized, the irradiance should be independently measured by irradiance sensors such as the PV module irradiance sensors PVMS [5] - [7]. In that case, factors such as the difference in the temperature, spectral response, irradiance, angular response, and linearity between the MUT and PVMS will affect the measurement uncertainty. Recent results suggest that the precision of the I-V curve measurement is very good when the irradiance is simultaneously determined by PVMS [10]. However, the detail remains to be assessed by future study.

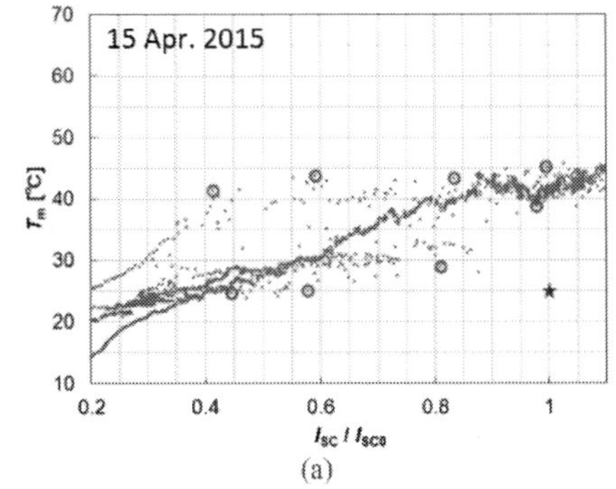

(a)

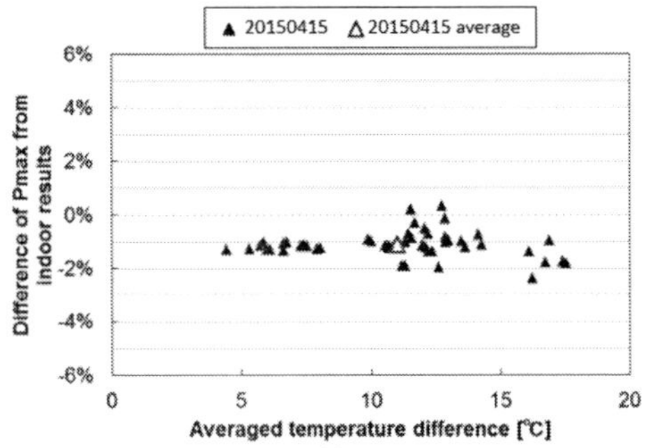

(b)

Figure 7 Simultaneous translation of the outdoor I-V curves for irradiance and temperature to STC by the linear interpolation/extrapolation method. (a) I_{sc}/I_{sc0} and T_m of eight I-V curves, measured on a partly sunny day are shown by open circles. (b) Relative difference of the STC P_{max} between the indoor I-V curves measured indoors and translated from three of the eight data in (a); i.e., 51 combinations except for those being located nearly in a line.

ACKNOWLEDGEMENT

This work was supported by NEDO under METI.

REFERENCES

[1] European Commission Joint Research Center, "Guidelines for PV Power Measurement for Industry" (2010) JRC Scientific and Technical Reports EUR24359 EN

[2] K. Paghasian, G. TamizhMani, "Photovoltaic module power rating per IEC 61853-1: A study under natural sunlight" Solar ABC Study Report (2011) (www.solarabcs.org)

[3] IEC 60891:2009 "Photovoltaic devices - Procedures for temperature and irradiance corrections to measured I-V characteristics"

[4] B. Marion, S. Rummel, A. Anderburg, "Current–voltage curve translation by bilinear interpolation", Prog. Photovol. Res. Appl. 12-8 (2004) 593-607

[5] Y. Tsuno, Y. Hishikawa, K. Kurokawa, " Modeling of the I– V curves of the PV modules using linear interpolation/ extrapolation", Sol. Energ. Mat. Sol. Cells 93-6 (2009) 1070-1073

[6] Y. Hishikawa, Y. Tsuno, " Calculation Formula for Irradiance and Temperature Correction of the I-V Curves of Solar Cells and Modules by Linear Interpolation/Extrapolation", 24th EUPVSEC, Hamburg (2009) 21-25

[7] Y. Hishikawa, A. Fukabori, T. Takenouchi, Y. Tsuno, " Accurate outdoor measurement technology of PV devices", Proceedings of the JSES JWEA Joint Conference, Fukushima (2014) 289-292, in Japanese

[8] Y. Hishikawa, K. Yamagoe, H. Ohshima et al., "New Technology for Precise Outdoor PV Module Performance Measurements", 42nd IEEE PVSC, New Orleans (2015)

[9] Y. Hishikawa, T. Doi, M. Higa, K. Yamagoe, " Precise Outdoor PV Module Performance Characterization Under Unstable Irradiance", IEEE J. Photovolt. (2016) DOI:10.1109/JPHOTOV.2016.2571620

[10] Y. Hishikawa, T. Doi, M. Higa, K. Yamagoe, H. Ohshima, K. Matsuda, H. Wakabayashi, "Precise Outdoor PV Performance Measurements at Various Irradiance Levels ", 43rd IEEE PVSC, Portland (2016).

UNCERTAINTY ANALYSIS IN THE POWER RATING MEASUREMENT OF SOLAR CELL AS PER IEC 61853-1

Rashmi Singh, Birinchi Bora, O.S. Sastry*, Supriya Rai, Manander Bangar, Renu, Rahnuma Siddqui
National Institute of Solar Energy, Gurgaon- Faridabad Road, Gurgaon, Haryana-122003, India
*sastry284@gmail.com, Tel: +91-0124-2579213, Telefax No. : +91-124-2579207

ABSTRACT: An estimation of source of uncertainty in the measurements of power matrix values using indoor test conditions as per IEC 61853-1 is performed. During the experiment it is found that the spectrum of light source is changing with irradiation. In this paper an uncertainty analysis of the measurements have been carried while estimating the power output for 23 test conditions as per the standard. During this study it is observed that the power temperature coefficient, MMF of solar cell does not remain constant and is dependent on irradiation level. Therefore use of one power temperature coefficient (@ STC) for all irradiation levels mentioned in the standard leads to significant error in the energy estimations. It needs to be estimated MMF for each irradiance level for different PV technologies for indoor testing also. The estimation of power by using translation method as per IEC 60891 also leads to increase in uncertainty
Keywords: Uncertainty analysis, IEC 61853, 23 Test condition, Spectral distribution, temperature coefficients

1 INTRODUCTION

Energy generation from photovoltaic (PV) devices such as solar cells and modules are influenced by many environmental factors such as temperature of the device, incident irradiance, and spectral irradiance distribution [1-3] under which the device is operating. The lifetime energy generation and levelized cost of energy (LCOE) of modules strongly depends on these conditions. However PV devices are normally tested and characterized under standard test conditions (STC) (1000 W/m^2 irradiance, 25 ^{0}C cell temperature), low irradiance (200 W/m^2 irradiance, 25 ^{0}C cell temperature), and nominal operating cell temperature (800 W/m2 irradiance, 20 ^{0}C ambient temperature) which are different than the actual field conditions. In estimation of energy using IEC 61853 standards interpolation or extrapolation within the range is carried out to estimate the total energy for a given period which is shown in table 1. It is known that the irradiation influence the short circuit current linearly and open circuit voltage logarithmically. At the same time the module temperature effects significantly the open circuit voltage [4]. These factors influence the output power of the device substantially.

Table 1 23 test condition according to IEC 61853-1

Irradiance	Module Temperature (^{0}C)			
(W/m^2)	15(^{0}C)	25(^{0}C)	50(^{0}C)	75(^{0}C)
1100	NA	1	2	3
1000	4	5	6	7
800	8	9	10	11
600	12	13	14	15
400	16	17	18	NA
200	19	20	21	NA
100	22	23	NA	NA

Therefore accurately predicting the behavior of a representative solar cell of the module under varying solar irradiation and temperature condition is necessary. To characterize the PV cells within the range of the matrix of temperature, irradiance and AM 1.5 spectral distribution to generate power matrix is necessary as per IEC 61853 standard. The importance of spectral effect on solar cells has also been investigated by several authors such as [5] on the basis of indoor measurement. And these investigations reported are based on useful fraction defined as the ratio of the observed spectral irradiation in the useful spectral range of solar cell under investigation to the observed global irradiance [6-7] spectrum. Effect of change in illumination is also found to influence temperature coefficient of voltage, power and current [5,6,7]. Most module data sheets indicate these temperature coefficients under STC only. Along with efficiency and power output, the temperature coefficient also varies with irradiance. While reporting the result of any measurement, it is necessary that quantitative indication of the quality of results in terms of uncertainty should be given to assess the reliability of data. In the uncertainty calculation the following points needs to be considered:

a. Irradiance and spectrum measurements during irradiance variation
b. Data acquisition and calibration
c. Device temp measurement
d. Device mounting and connection.

The paper presents the estimation of source of uncertainty in the measurements of power as per 23 tests conditions of IEC-61853-1 standard for indoor solar cell testing is performed. The data in the case of different PV technologies are presented in the paper.

2 APPROACH

To conduct this objective, the facility has an ORIEL's class AAA "SUN Simulator" which provides a continuous study state source and an electronic load capable of testing 6 inch X 6 inch solar cells of any type. The cell can be tested under standard test conditions (1000 W/m^2, cell temperature 25 ^{0}C, and air mass 1.5). The spectrum of the source qualify IEC 60904-9; 2007 Edition, or ASTM E 927-05 standards. The current - voltage (I-V curves) measurements, both dark and light conditions of solar cells can be studied at different intensities and temperatures. A mono crystalline solar cell of 2 x 2 cm is used in the experiment. The spectral measurements were performed by using a LICOR 1800 spectrophotometer. Each characteristic spectrum measured comprises 651 measurement points at 1 nm steps across the spectral range 300–1100 nm.

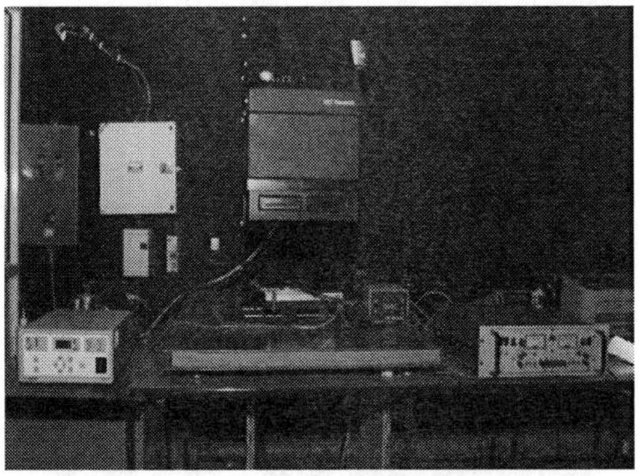

Figure 1: Test setup for performing Measurement

3 SCIENTIFIC INNOVATION & RELEVANCE

At the time of energy modelling and prediction, if any error introduced in the measurement or experiment then the economics & the power prediction is certainly not accurate. So there's an essential need for estimation of the uncertainty in the measurements. This study leads to estimate the uncertainty in the solar cell for various operating range of Temperature and irradiance in indoor condition.

(A) For temperature coefficients of Voltage, current and power.
Select I-V curves at constant irradiance and different temperatures and temperature coefficients are estimated. The temperature coefficients vary due to the variations in the parasitic resistances. These variations in the parasitic resistances are due to the manufacturing process. The parasitic resistances affect the power; fill factor, and efficiency of that module. The types of parasitic resistive losses are the series resistance and shunt resistance. Module should have a low series resistance and high shunt resistance.

(B) For power matrix at 23 test conditions.

The test standard comprises of four parts but here study is conducted only on two parts;

- IEC 61853-1: It describes requirements for evaluating PV module performance in terms of power (watts) rating over a range of irradiances and temperatures.
- IEC 61853-2: It describes test procedures for measuring the effect of varying angle of incidence and sunlight spectra, the estimation of module temperature from irradiance, ambient temperature, and wind speed.

Data from the sun simulator for all the ranges of Irradiance and varying temperature is taken and recorded using lab view software. Some combination is not maintained in the environment so it should translated over the desired points. The four translation procedures are used to translate from one curve to any other target curve the first procedure is used to translate a single measured I-V characteristic to select temperature and irradiance or test conditions by using equations (1) and (2), [8-9]

$$I_2 = I_1 + Isc\left(\left(\frac{G2}{G1}\right) - 1\right) + \alpha(T2 - T1)$$
(1)

$$V_2 = V_1 - R_S(I_2 - I_1) - \kappa * I_2 * (T_2 - T_1) + \beta * (T_2 - T1)$$
(2)

where:

I_1 (A) and V_1 (V) are coordinates of the measured I-V curve
I_2 (A) and V_2 (V) are the coordinates of the translated I-V curve
G_1 (W/m^2) is the irradiance measured with the primary reference cell
G_2 (W/m^2) is the irradiance at desired conditions in the matrix
T_1 (^{0}C) is the module temperature
T_2 (^{0}C) is the desired temperature in the matrix
Isc (A) is the measured short circuit current of the test specimen for measured I-V curve
Rs (Ohm) is the internal resistance of the test module
k is the curve correction factor derived from measured conditions α (A/k) and β (V/k) are temperature coefficients of Isc and Voc respectively.

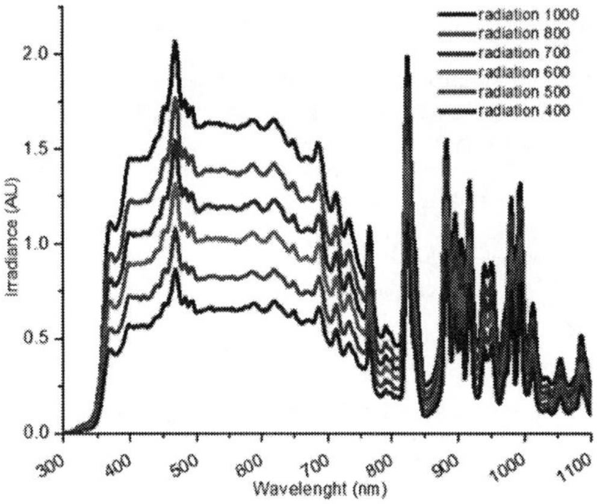

Figure 2: Spectrum at different Irradiance

(C) Uncertainty Measurement of the measured points.

The basic uncertainty analyses principles can be applied to any electrical parameters of solar cell of interest (e.g., short-circuit current, open-circuit voltage and fill factor). For commercial labs, this need may be acute in order to control measurement quality. This paper presents a method for estimating source of uncertainty of solar cell performance using sun simulator. This treatment adopts the analysis and terminology contained in the Guide to the Expression of Uncertainty in Measurement, which is an ISO standard (International Organization for Standardization). Consider that uncertainty in its broadest sense is a characterization of the possible dispersion of measured values around the true value of a variable. Evaluation of uncertainty based on statistical analysis of repeated observations obtained under the same conditions of measurements is considered as Type A, and expressed as Standard Deviation. Alternatively, a single measurement with an instrument implies a possible dispersion that can be quantified if the instrument's accuracy (or tolerance) and calibration are known (Type B). If a few measurements (e.g., 10) are performed with instruments, it is clear that a combination of both Type A and Type B uncertainties have to be considered to fully quantify the possible dispersion of a measurement. Several data sets are required as function irradiance and module temperature to: (1) allow for a regression of data to STC cell temperature, and (2) estimate the variance of the Pmax data about the STC operating point.

ASSUMPTIONS

The primary assumptions needs to be made are:

1. PV device current varies linearly with irradiance (i.e., irradiance is restricted to near STC values, and at constant spectral distribution).
2. Covariance between voltage and current is negligible.
3. Probability density distributions are symmetric and Type B distributions are rectangular.
4. Expanded uncertainty is evaluated assuming the shape of the combined uncertainty probability density function is approximately normal (this assumption is often found to be acceptable for the convolution of three or more rectangular distributions).
5. The reference cell is at or near thermal equilibrium with the testing environment.

Type B analysis of uncertainty follows these basic (abbreviated) procedures:

Step 1: Combine all the uncertainties arising from different component of system and from the Type A uncertainty.

Step 2: Assume an uncertainty component probability distribution based on all available information on the each component.

Step 3: Calculate the combined Uncertainty and later from the effective uncertainty and later the Expanded Uncertainty is presented.

Following all these steps the uncertainty budget for the different electrical parameters can be estimated.

4 RESULTS

The power matrix for 23 test condition is provided as follows in Table 1.

Table 2 Power matrix of 23 test condition

	Temperature			
Radiation	15°C	25 °C	50 °C	75°C.
1100W/m²	NA	33.3	32.4	30.3
1000W/m²	35.8	32.3	30.4	29.3
800 W/m²	29.7	28.3	26.6	26.0
600 W/m²	22.0	20.8	19.5	19.4
400 W/m²	14.9	13.3	13.1	NA
200 W/m²	6.4	6.6	NA	NA
100 W/m²	2.8	2.9	NA	NA

The short circuit current is dependent on the spectral distribution of the incident radiation. If any modification of the spectrum of the incident radiation, subsequent modification of the short circuit current should be noticed. It has been observed that at lower intensity the IR content of the spectrum goes on increasing as compared to higher intensity range.

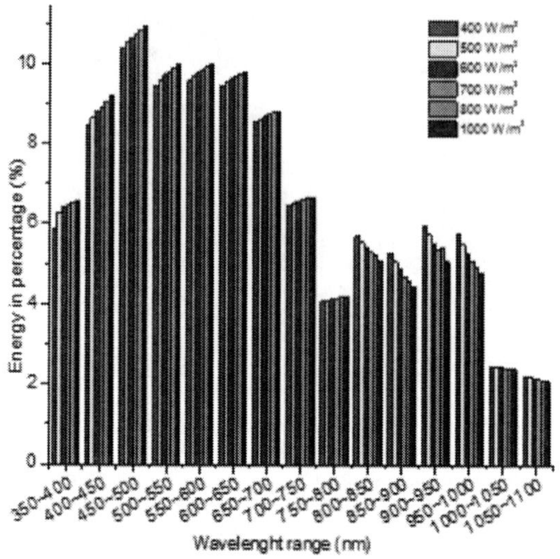

Figure 4 Spectral distribution in percentage at the different Intensity

However some of the technologies like mono and multi crystalline silicon shows good response in the IR range.The variation in spectrum could lead to the higher in the uncertainty of measurement. The MMF of different PV technologies CdTe, HIT, a-Si and m-Si at different irradiance range has been calculated using the spectrum of the solar simulator and produced in the figure 5. It has been found that the MMF is depending on the irradiance of the simulator. For m-Si and HIT solar cells the value of MMF is decreasing with increasing the irradiance and for CdTe and a-Si it is opposite. So there will be a variation of accuracy with change in irradiance in terms of spectrum also.

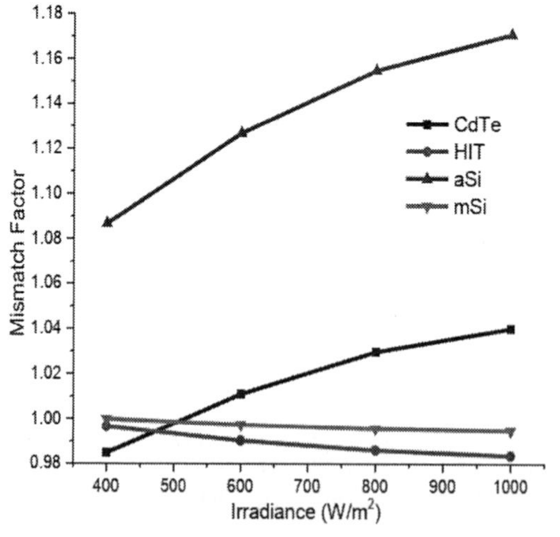

Figure 5: MMF of different PV technologies at different irradiance range of solar simulator

For the estimation of power as per the 23 test conditions people uses the translation procedure as described above. But however it has been reported by Tsuo & Hushikawa [9], and Sowmya Priya et al [10], that there is large variation in the estimation as per the different procedures of IEC 60891. Both the two author was agreed that procedure 3 is the best option for the translation of data with ±1% error. For lower intensity level the error of estimation with the different procedures are high as compared to high intensity level. So there will be a chance of increasing uncertainty. So it is suggestable that use of light source in solar simulator in which controlling the spectrum with reference to the change in irradiance are possible. Also there should be options of testing the solar cell and module over a wide range of irradiance and temperature distribution without increasing the uncertainty.

CONCLUSION

This study concludes that the efficiency of the PV modules is not constant at all irradiance levels and it could decrease at very low irradiance levels depending on the quality of the cells used in the module. The short circuit current is dependent on the spectral distribution of the incident radiation. If any modification of the spectrum of the incident radiation is done, subsequent modification of the short circuit current should be noticed. The power temperature coefficient of the module is not constant. It has a dependence on the irradiance level and hence the use of single power temperature coefficient for all irradiance levels is not accurate. The major conclusions resulting from this project are: Accurate test results require meticulous experimental setup and rigorous test procedures.

ACKNOWLEDGEMENT

This research is based upon work supported in part by the Solar Energy Research Institute for India and the U.S. (SERIIUS) funded jointly by the U.S. Department of Energy subcontract DE AC36-08G028308 (Office of Science, Office of Basic Energy Sciences, and Energy Efficiency and Renewable Energy, Solar Energy Technology Program, with support from the Office of International Affairs) and the Government of India subcontract IUSSTF/JCERDC-SERIIUS/2012 dated 22nd Nov. 2012

REFERENCES

[1] Shaltout, M.A.M., Hassan, A.H., Ghitas, A.E., 1992. The spectrum of the solar global radiation on horizontal solar cells in polluted and pure air zones of great Cairo. In: Proceedings of the 11th European Photovoltaic Conference, Montreux, Switzerland.

[2] Takashi, M., Todaa, M., et al, 2007. Effect of spectral irradiance distribution on the outdoor performance of amorphous Si//thinfilm crystalline Si stacked photovoltaic modules. Sol. Energy Mater. Sol. Cells 91, 120–122.

[3] Nagae S, Toda M, Minemoto T, Takakura H, Hamakawa Y. Evaluation the impact of solar spectrum and temperature variations on output power of Silicon based photovoltaic modules. Solar Energy Materials & Solar Cells. 2006; 90: 3568 – 75.

[4] IEC Standard 61853-1,2.

[5] Gottschalg, R., Infield, D.G., Kearney, M., 2003. Experimental study of variations of the solar spectrum or relevance to thin film solar cells. Sol. Energy Mater. Sol. Cells 79, 527–537.

[6] Durisch, W., Urban, J., Smestad, G., 1996. Characterisation of solar cells and modules under actual operating conditions. In: Proceeding of World Renewable Energy Congress, vol. 1. Pergamon Press, Oxford, Denver, CO, USA, pp. 359–366

[7] Kenny, R.P., Ioannides, A., Ullejans, H.M., Zaaiman, W., Dunlop, E.D., 2006. Performance of thin film PV modules. Thin Solid Films 511–512, 663–672.

[8] IEC Standard 60891 Edi.2.0.

[9] Tsuno, Y., & Hishikawa, Y. (2012). Comparison of curve correction procedures for current-voltage characteristics of photovoltaic devices. Japanese Journal of Applied Physics, 51(10 PART 2), 2–5. http://doi.org/10.1143/JJAP.51.10NF02

[10] Sarah Sowmya Priya, O.S. Sastry, Birinchi Bora, Avinash Kumar, Comparison of curve correction procedure of current and voltage as per IEC 60891for thin film technology, 42 IEEE PVSC, New Orleans, USA.

Characterisation of n-type bifacial silicon PV modules

Juan Lopez-Garcia*, Bereket Haile, Diego Pavanello, Alberto Pozza, Tony Sample
European Commission, DG JRC, Institute for Energy and Transport, Renewables and Energy Efficiency Unit, Via Fermi 2749, 21027 Ispra (VA), Italy
*Tel: +39 0332783720, Fax: +39 0332789268, e-mail: juan.lopez-garcia@ec.europa.eu

ABSTRACT: Bifacial PV modules have shown the potential to increase the performance of traditional photovoltaic module architectures using relatively conventional silicon manufacturing techniques. Bifacial modules can produce additional energy from the rear surface of the module, collecting not only light incident on the front surface but also sunlight scattered or reflected from the ground or environment onto the rear surface. To assess their performance and quality, PV modules are characterized using international standards to rate the modules in terms of output power and efficiency. However, at present, no specific standards exist for bifacial modules. Furthermore, these modules are often characterized by a high efficiency and an increased capacitance and the measurement with standard pulsed solar simulators may be affected by strong measurement artefacts resulting in increased or reduced maximum power. This work presents the indoor characterisation of bifacial c-Si PV modules under varying irradiance levels using different methods such as a single-sweep flash simulator, the so-called multi-flash (MF) method and a steady state sun simulator for the accurate measurement of the module parameters. The effect of different rear surface coverings, from matt black painted wood to gloss white card and the differences among the characterisation method are also investigated.

Keywords: Bifacial; PV Modules; High-efficiency; Indoor testing;

1. INTRODUCTION

To reduce the cost of solar power generation, the main focus areas of the photovoltaic (PV) industry are improving the solar cells and module efficiency, lifetime, cost and PV module's energy yield. Bifacial PV modules can produce additional output energy in comparison to conventional mono-facial modules because both sides of the cell/module, front and rear, can absorb solar radiation, utilizing the scattered light from the ground and surroundings. This capability of converting light from either side of the module into electricity makes these high-efficiency bifacial modules suitable to reduce the cost of PV electricity [1-3]. Replacement of a conventional free standing glass-backsheet module by a glass-glass module with bifacial cells can already result in an up to 20% annual energy gain, depending on the albedo [4]. To assess their performance and quality, PV modules are measured and characterised in a controlled environment under standard test conditions (STC) as defined by the International Electrotechnical Commission (IEC) [5]. Although the IEC is currently working on the development of a draft technical specification (IEC TS 60904-1-2) for the measurement of current-voltage characteristics of bifacial photovoltaic devices [6], there is currently no standard. Therefore, different groups and laboratories have proposed different methods [1,7]. As the PV modules are generally sold according to the nominal power output and efficiency under STC conditions, the lack of bifacial standards leads each manufacturer to claim different amounts for the "added value" of the bifaciality that makes it difficult for customers to directly compare between bifacial manufacturers. In absence of standards, many of the bifacial PV module manufacturers measure using only front-side illumination by a conventional pulsed simulator while covering the rear side with a non-reflecting cover, and vice versa, and tabulate the efficiency/power with a linear addition of the front and rear side values for particular rear side irradiance conditions [1,3,8-10]. However, most of the literature reports are on outdoor measurements where the environment and reflecting surfaces are usually not under control but they enable a "real value" of the bifacial module performance to be determined for a specific set of local environmental conditions [8-11].

Moreover, the indoor testing of high-efficiency and highly-capacitive bifacial PV c-Si modules with standard pulsed solar simulator may be affected by strong measurement artefacts and lead, depending on the pulse duration and sweep speed, to under or over-estimations of the measured power. These modules usually require solar simulators with a pulse duration longer than 100-200 ms in order to make a single sweep measurement of an I-V curve. Alternatively, the I-V curve can be obtained using the so called multi-flash (MF) method or steady-state solar simulators (continuous light) and slow sweep speeds in order to avoid measurement artefacts and obtain accurate electrical parameters [12-14].

This work reports on the indoor performance testing of highly-efficient n-type silicon bifacial modules by means of different characterisation methods and/or solar simulators and rear covers. The objective is to show the differences between the alternative methods and the influence of opaque or reflective rear covers in the electrical parameters of the c-Si bifacial PV modules.

2. EXPERIMENTAL METHOD

2.1. I-V measurement setup.

A selected number of modules (ESTI codes QV61, QV62 and QV66) from a batch of 8 commercially available bifacial n-type c-Si glass/glass modules consisting of 60 cells in three sub-strings, which have been light soaked to eliminate possible light induced degradation (LID), were characterized indoors at the European Solar Test Installation (ESTI). I-V curve measurements were performed with a PASAN IIIB pulsed simulator (with sweep times or flash duration of the order of 10 ms) at different irradiances from 100 W/m^2 to 1200 W/m^2 and with a steady state solar simulator (Apollo) with single I-V sweep times of around 1 second, at a temperature of 25.0 ± 0.1 °C. The measurements using the flash simulator were performed using a standard forward sweep single flash method and a multi-flash (MF) method in which 35-40 individual points are taken to construct the final I-V curve. In this case, each point of the I-V curve was obtained by applying a constant voltage to the module which was then subjected to a flash from the solar simulator at the desired irradiance level. With this method, capacitive effects are avoided by considering only stabilized current and voltage output values for each measurement. The single I-V point is obtained from the stable portion of the measured current and voltage of the module.

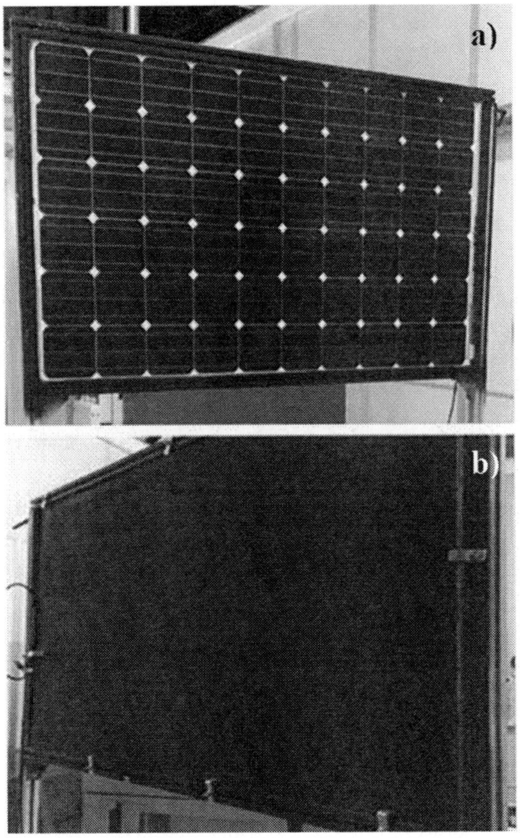

Figure 1 Setup for the measurement of bifacial glass/glass c-Si modules with a) the white rear cover observed through the front side and b) a black painted rear cover in contact to the rear surface of the module.

The resultant module I-V curve was then obtained by performing an interpolation of the measured points [12,13].

In order to investigate capacitive effects in the module performance due to the insufficient sweep time of the standard simulators, forward (I_{sc} to V_{oc}) and reverse (V_{oc} to I_{sc}) sweep single flash measurements were performed.

In order to ensure the reproducibility of the measurements and to avoid uncertainty due to the set-up of the system, a special structure was designed to hold the modules. The set-up consists on a black painted aluminium support structure at a height to align the module centre with the centre of the illuminated area as shown in figure 1. The front and rear side of the module has been defined by the manufacturer, in practice it means that the rear side is that incorporating the junction box or labels. The electrical characterization of the modules was carried out with a black ("non-reflective") painted wooden rear cover and with a gloss white reflective rear cover. The rear cover has the same dimensions as the module and was placed in direct contact to the surface of the module not exposed to the flash, thus only light that is transmitted through the module can be scattered or reflected back to the module. I-V curves have been obtained by illuminating alternatively the front or the rear side while covering the rear with the different covers. This has been done for different irradiances for MF, forward sweep I-V and constant solar simulator.

The spectral responsivity of representative modules (QV61, QV62 and QV66) were measured using a differential spectral responsivity technique with a large area pulsed solar simulator equipped with a number of bandpass filters to obtain illumination of modules with quasi-monochromatic light.

2.2. Electroluminescence and thermal imaging.

Electroluminescence (EL) and infrared thermography (IR) characterization was performed in order to characterise the initial state of the modules. EL images were performed (Sensovation SVSB14-M) in darkness with the module forward biased at I_{sc} with an exposure of 120 seconds. IR images were performed with a Fluke Ti55 camera under two conditions:

- With the module forward biased to current equivalent to I_{sc}.
- Under continuous exposure to light ($\approx$1000 W/m^2) provided by a steady state solar simulator and short circuited with an appropriate resistance.

3. RESULTS AND DISCUSSION

3.1. Electroluminescence and thermography analysis

The Electroluminescence and thermography IR images are shown in figure 2, respectively. This show the EL image of a representative c-Si bifacial module and the corresponding IR thermography images obtained on the module subjected to continuous illumination, with an appropriate resistive load and on the module biased at I_{sc}. EL imaging has shown minor defects such as finger interruptions. However, major defects that have a direct influence on the module performance such as cracked cells, dark areas in cells corresponding to inactive areas or scratches on the cell surface have been not detected. IR thermography also presents a homogeneously distributed temperature without any gradient. This analysis shows that the module characteristics are not influenced by intrinsic defects and the initial state of the modules was defect-free.

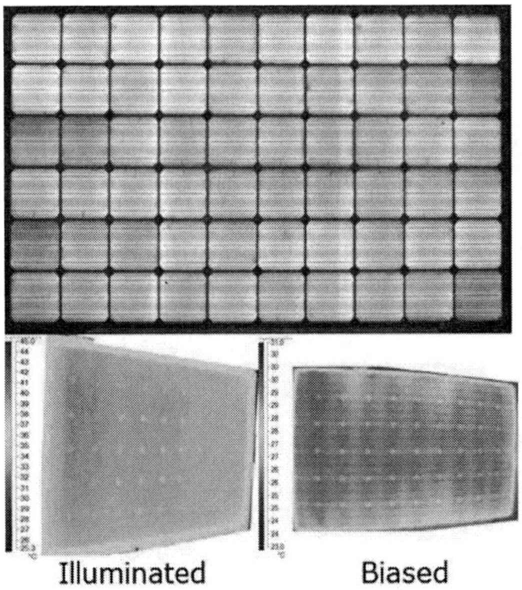

Figure 2 EL image of a bifacial module and the respective IR thermography images using continuous light exposure (left) and forward biased modules (right).

3.2. Different I-V measurement methods.

Conventional flash light simulators are generally used for the measurement of module performance due to the possibility to have very controlled environmental conditions and their ease of use. Moreover, class AAA solar simulators nowadays lead to a spectral mismatch factor close to unity for standard c-Si modules. However, for high-efficiency c-Si modules with increased capacitance, the effects of the highly-capacitive devices can lead to inaccurate measurements of the PV module electrical parameters as shown in Figure 3. A marked difference in the P_{max} and FF between the forward and reverse measurements is evident in figure 3. This means that a longer sweep time (conventional pulsed simulators are in the order of 10 ms) is needed for loading the cell capacitance and correctly evaluating the module maximum power. The suitable sweep time is achieved when both forward and reverse measurements have the same value within the repeatability error.

A representative module (QV66) was selected in order to analyse the data and represent the results. There are slight differences in the absolute values between the measured modules due to the manufacturing process of each cell/module but in general the tendency of the parameters is the same and can be extrapolated. Figure 4 shows the I-V curves obtained with three different methods and corrected to STC: forward sweep in a 10 ms pulsed solar simulator, multi-flash method in a pulsed simulator and by mean of a slow sweep (~1sec) using a continuous or steady state simulator. A similar behaviour of the steady state and the multi-flash method was expected.

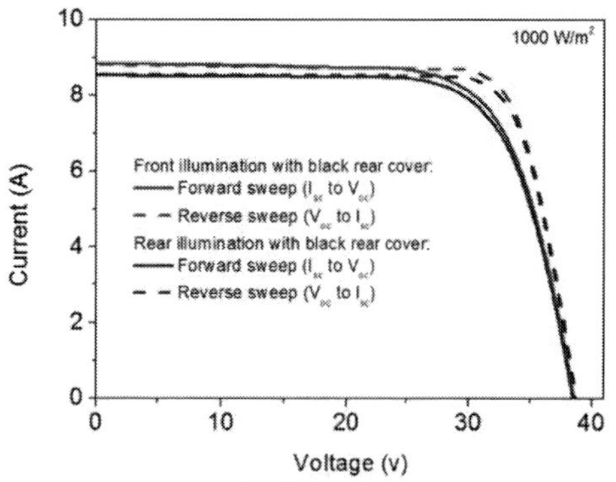

Figure 3 I-V curves of a bifacial c-Si module with a flash simulator in forward (I_{sc} to V_{oc}) and reverse (V_{oc} to I_{sc}) sweep illuminated from the front and rear side, respectively.

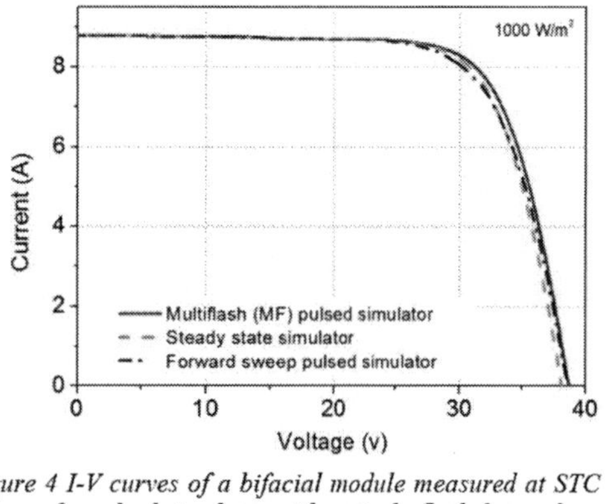

Figure 4 I-V curves of a bifacial module measured at STC by means of a pulsed simulator with a single-flash forward sweep and a multi-flash method and a steady state simulator, using a black rear cover.

Table I Change (%) for a bifacial module (QV66) between the steady state and the multi-flash method.

Change (%)	Front ill. Black rear cover	Front ill. White rear cover	Rear ill. Black rear cover	Rear ill. White rear cover
I_{sc}	0.0	+0.2	+0.2	+0.3
V_{oc}	-1.4	-1.8	-1.0	-1.1
P_{max}	-1.5	-2.0	-0.4	-0.4
FF	-0.1	-0.5	+0.4	+0.4

However, it was observed for all modules measured slight differences in the P_{max} (table I) due to a shift of the V_{oc} towards lower values. This can be explained because the temperature control in the steady state is not trivial and the module can heat up quickly during the measuring time (~1 sec) shifting the V_{oc} to lower values with a minor change in the I_{sc} and FF.

The detailed analysis of the change in the module parameters between the three measurement methods is shown in figure 5. The single-sweep flash method is taken as the baseline and the change (in %), for the different configurations, are given in terms of illumination side and rear cover, for I-V curves obtained at 25.0±0.1°C and corrected to 1000W/m². A positive value means a higher value for the steady state or multi-flash method relative to the forward sweep. A P_{max} average change of around +3.9% was obtained between the forward sweep single flash and the multi-flash method, independently of the rear cover used and the side of the illumination. However, the difference was reduced below +2.2% between the forward sweep and the steady state simulator measurements for the front illumination. The changes in the I_{sc} and V_{oc} are relatively low between forward sweep and multi-flash method (less than +1% or even less than +0.5%) which implies that the P_{max} change is due to the change in FF. As shown in the I-V curve of figure 3, the standard single flash method is less accurate in the region near P_{max}. It should be taken into account this systematic error introduced in the measurement for the rated output power if the modules are measured with a forward sweep using a short time pulsed simulator. This means that the multi-flash method is the most accurate method to characterise the high-capacitance bifacial modules but it is time consuming, requiring multiple flash measurements.

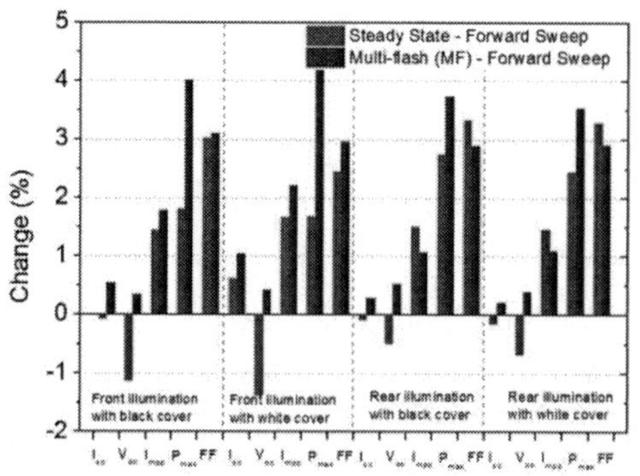

Figure 5 Change (%) in the module parameters between the three measuring methods taking the forward sweep single flash as baseline.

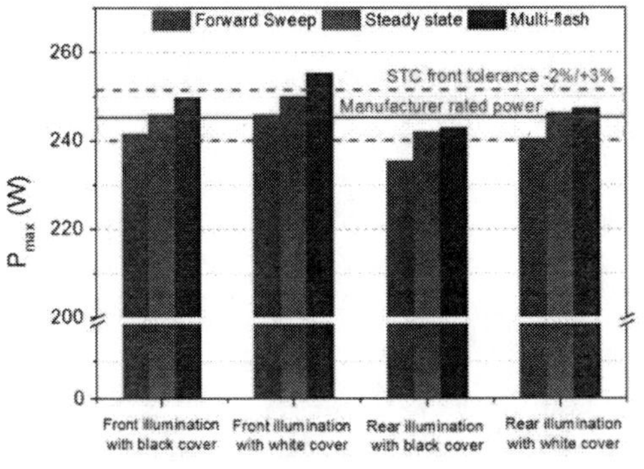

Figure 6 P_{max} values in (W) of a bifacial module for different configurations measured by three different methods. The manufacturer front rated power and the STC front tolerance are also plotted.

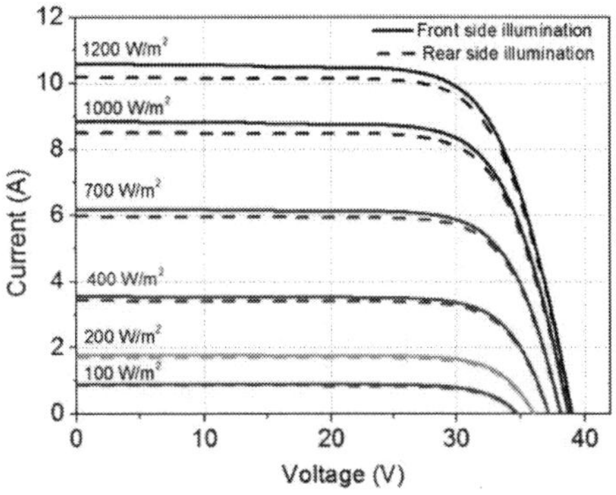

Figure 7 Irradiance dependence of I-V curves of a bifacial module.

Figure 6 presents the absolute value of P_{max} obtained with the three measurement methods together with the manufacturers front rated power and the STC front tolerance. It is shown in figure 6 that for the standard configuration front illumination and black cover, the P_{max} is within the tolerance. However, it is evident in figure 6 that under certain measurement conditions that the module power is out of the STC front tolerance range (-2% / +3%) stated in the datasheet of the manufacturer. This highlights the need, even with a high bifaciality of >97% as will be shown later, to quote the STC_{front} and STC_{rear} values for the module together with the measurement setup, ie type of rear cover (if used).

3.3. Analysis of front and rear side illumination.

In order to analyse the different performance of the front and rear of the module as a function of the incident irradiance, a representative module (QV66) and the more accurate multi-flash method has been selected. Figure 7 shows the I-V curves of a bifacial module as a function of the incident irradiance for front and rear side illumination, respectively.

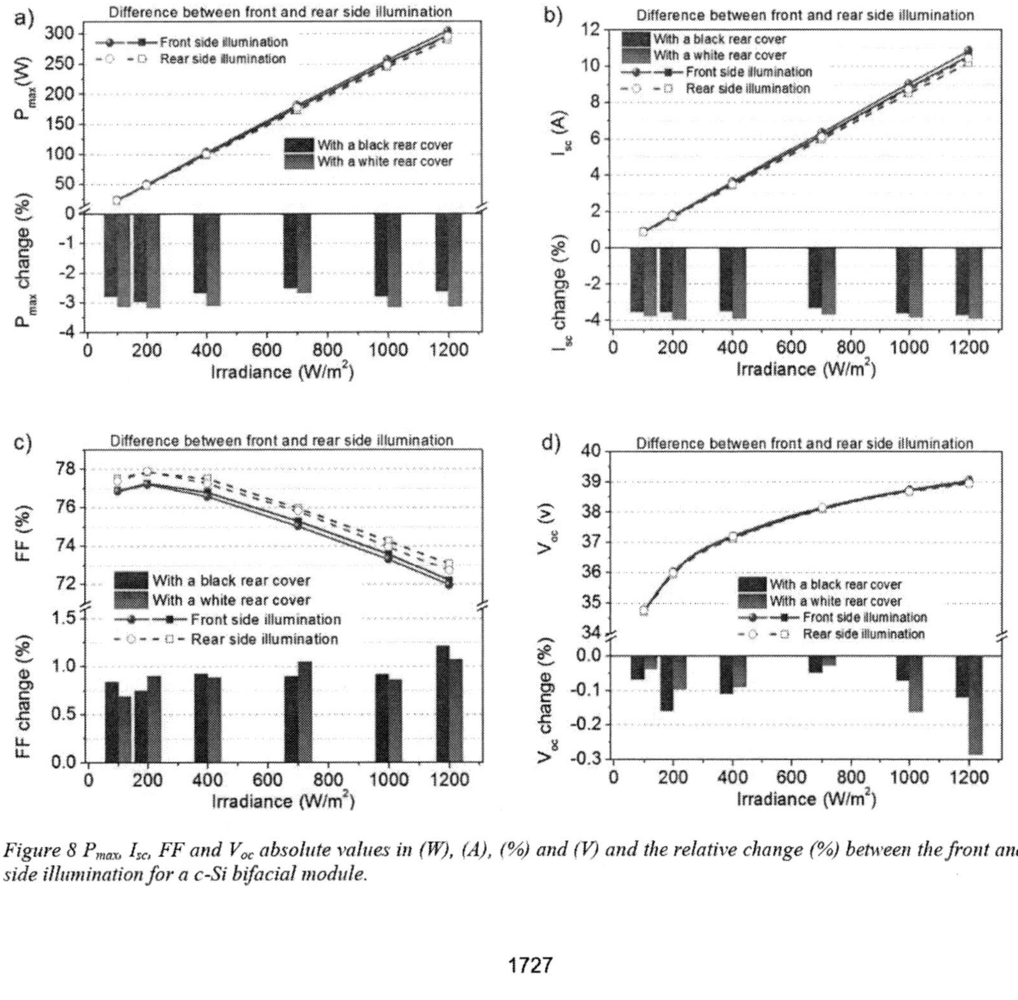

Figure 8 P_{max}, I_{sc}, FF and V_{oc} absolute values in (W), (A), (%) and (V) and the relative change (%) between the front and the rear side illumination for a c-Si bifacial module.

As expected, current from the rear illumination is lower than from the front illumination and follow a linear dependency with the incident irradiance. In some modules a kink in the I-V curve for the rear side illumination, which can be due to a mismatch in the rear cells due to the manufacturing process was observed. This behaviour is under analysis and will be discussed in more detail in the future.

The absolute values of the characteristic parameters of a bifacial module measured by a multi-flash method and the change (in %) in the electrical data when the module is front or rear side illuminated are plotted in Figure 8 a, b, c and d. The front side illumination is taken as the reference for the change. The measurements have been carried out with the black and the white rear cover. A linear dependency of the P_{max}, I_{sc}, and FF, and a non-linear dependency of the V_{oc} with the incident irradiation was observed. A negative P_{max} change implies that the P_{max} measured with the front illumination is higher, as expected. The P_{max} change is around -3% and is greater for the white rear cover. The observed variation in P_{max} was mainly due to a change in the I_{sc} (-3.5 to -3.8%) between the front and the rear side illumination. This can be due to a lower efficiency of the rear surface of the cells as a consequence of the manufacturing process that imply a lower current.

The bifacial ratio or bifaciality can be defined as the ratio between the P_{max} generated by the rear side of the PV module and the one generated by the front side:

$$Bifaciality\ (\%) = \frac{P_{maxREAR}}{P_{maxFRONT}} \times 100 \quad (1)$$

where $P_{max\ REAR}$ and $P_{max\ FRONT}$ are the module output power measured when illuminating the rear and the front side, respectively. Figure 9 shows the bifacial ratio as a function of the incident irradiance for a non-reflective cover and a white reflective rear cover. An average bifaciality above 97% was obtained which was independent of the incident irradiance. It was observed that the bifaciality is slightly higher when a non-reflective rear cover is on the non-illuminated side. This value is higher than other reported values for the bifaciality of Si-based bifacial modules ranging from 90 to 95% [9,10,15]. Bifacial ratio up to 98.5 % has been reported for cells by some manufacturers [16].

3.4. Analysis of the rear side cover.

The effects of a "non-reflective" (black) and a reflective (white) rear sheet covering the rear of the module have also been investigated for a representative module (QV66) and the multi-flash method. Normally, bifacial Si modules are tested indoors with a black rear cover to ensure that no stray light (for

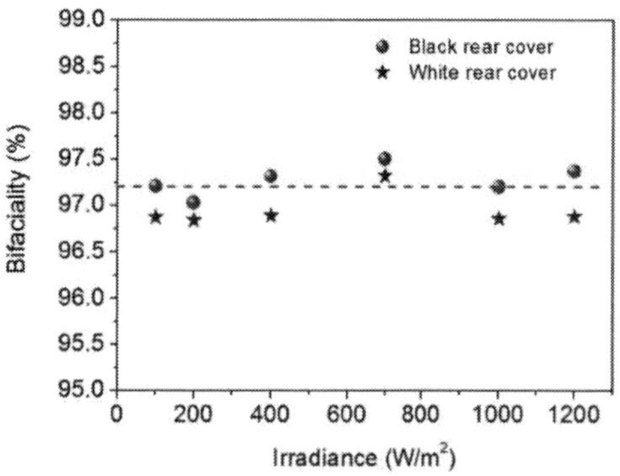

Figure 9 Bifaciality of c-Si bifacial module as a function of the incident irradiance.

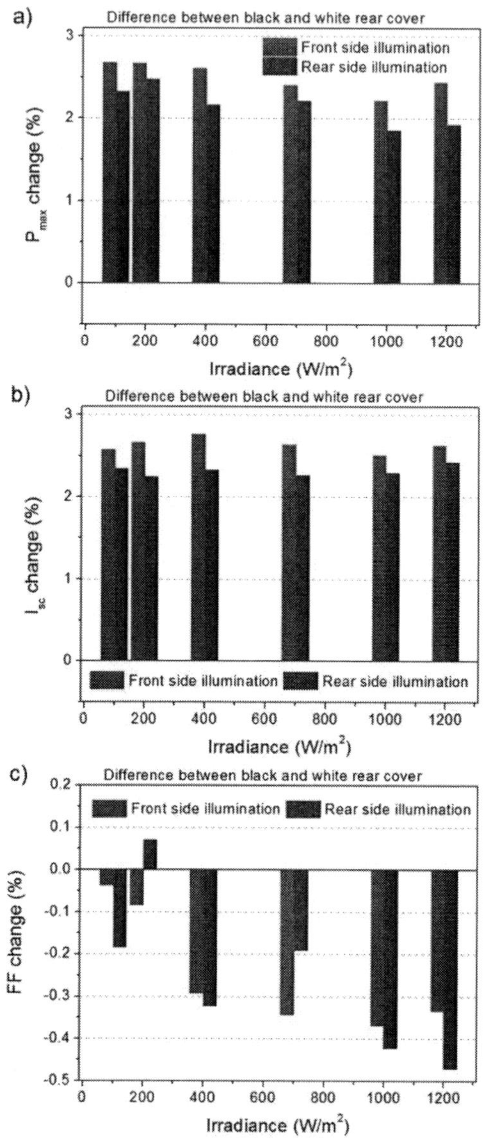

Figure 10 a) P_{max}, b) I_{sc} and c) FF change (%) between a black rear cover and a white rear cover of a bifacial module as a function of the incident irradiance.

example due to reflection from the wall/measurement structure) can enter the module [1]. In this section we present the effects of a white cover at the rear side. Figures 10a, 10b and 10c present the P_{max}, I_{sc} and FF gain between the black and white rear cover, expressed as change in % (equation 2) taking the black rear cover as the reference, as a function of the irradiance.

$$P_{max}change\ (\%) = \frac{P_{max\,white\,cover} - P_{max\,black\,cover}}{P_{max\,black\,cover}} \times 100 \quad (2)$$

where the $P_{max\ white\ cover}$ and $P_{max\ black\ cover}$ are the maximum output power of the bifacial module with a white sheet and a black sheet on the non-illuminated side of the module. An average P_{max} gain between +2.2 and +2.5% (rear side illumination and front side illumination, respectively) was observed caused by the white back sheet and it is due to an increase in the I_{sc} (+2.3 to +2.6%) and independent of the irradiance. The V_{oc} remains almost constant with a slight difference within the uncertainty and the FF change decreases to lower values (more negative) with the irradiance but with negative values below -0.5%. The obtained gain in I_{sc} is in good agreement with reported values for similar experiments ranging from +1.3 to +3% with white sheets at 0cm from the rear module surface [3,8,11,17]. This should be noticed that the difference between the output power and I_{sc} with the black and white cover is higher in the case of front illumination. This shows that the background and the surroundings have a significant influence in the electrical parameters of the bifacial modules even when the distance between the reflective ground and the back surface of the module is small, which in theory reduce the albedo and the gain[11].

4 CONCLUSIONS

Three different test methods for characterising the modules have been used: a standard pulsed simulator with a forward sweep, a multi-flash method and a steady state simulator with continuous light and slow sweep time (~1sec). It has been shown that for these high-efficiency and highly-capacitive bifacial modules, the output power is underestimated by around 3.9% and, in some cases, is out of the tolerance range given by the manufacturer, when they are measured using a forward sweep of 10 ms in a commonly used flash simulator. Using alternative methods such as steady state and multi-flash leads to a more accurate determination of the I-V curves mainly due to a correct assessment of the FF. A slight difference between the steady state and the multi-flash methods was found due to a shift of the V_{oc} to lower values as a consequence of the heating up of the module in the continuous light simulator.

Around a 3% greater P_{max} from the front, compared to the rear, was obtained mainly due to a change in the I_{sc} (that correlates with a measured bifaciality of above 97%). It has been found that the electrical parameter change to be independent of the incident irradiation.

A P_{max} and I_{sc} gain in the range of 2.2 to 2.5% for the white rear cover compared to the black rear cover was found.

According to the results and despite the lack of specific bifacial standards, some general recommendation can be suggested for implementation in the manufacturer datasheets in order to allow the customer to compare the different bifacial manufacturers:

- To mention the set-up and measurement condition in which the rated values at STC has been obtained (i.e. if a rear cover has been used, or the simulator employed)
- To include the bifaciality or the front and rear output power as a function of the irradiance to enable particular customers the possibility of calculating the specific gain from the bifaciality depending on the specific application or site location.

Some manufacturers have started to implement part of this information in the datasheet of the new models together with some information regarding to the measurement conditions [18]. However, it is important the general implementation for the rest of the manufacturers in order to make to the customers compare bifacial manufacturers.

5 REFERENCES

[1] J.P. Singh, A.G. Aberle, T.M. Walsh, Electrical characterization method for bifacial photovoltaic modules, Sol. Energy Mater. Sol. Cells, 127 (2014) 136-142.

[2] J. Appelbaum, Bifacial photovoltaic panels field, Renew. Energy, 85 (2016) 338-343.

[3] J.P. Singh, S. Guo, I.M. Peters, A.G. Aberle, T.M. Walsh, Comparison of Glass/Glass and Glass/Backsheet PV Modules Using Bifacial Silicon Solar Cells, IEEE J. Photovoltaics, 5 (2015) 783-791.

[4] S. Guo, T.M. Walsh, M. Peters, Vertically mounted bifacial photovoltaic modules: A global analysis, Energy, 61 (2013) 447-454.

[5] International Electrotechnical Commission, IEC 61836 ed. 2.0: Solar photovoltaic energy systems – Terms, definitions and symbols, 2007.

[6] International Electrotechnical Commission, IEC TS 60904-1-2: Measurement of current-voltage characteristics of bifacial photovoltaic (PV) devices, on-going, discussed by the TC82.

[7] K. Emery *et al.*, Rating photovoltaics, 39th IEEE Photovoltaic Specialists Conference, Tampa, FL, 2013, pp. 1-6.

[8] C. Erben, J.B. Heng, X. Zhigang, J. Fatehi, B. Yang, High efficiency bifacial modules in outdoor testing, 42nd IEEE Photovoltaic Specialist Conference, New Orleans, LA, 2015, pp. 1-3.

[9] J.E. Castillo-Aguilella, P.S. Hauser, Multi-Variable Bifacial Photovoltaic Module Test Results and Best-Fit Annual Bifacial Energy Yield Model, IEEE Access, 4 (2016) 498-506.

[10] C. Comparotto *et al.*, Bifacial n-Type Solar Modules: Indoor and Outdoor Evaluation, 29th European Photovoltaic Solar Energy Conference and Exhibition, Amsterdam, Netherlands, 2014, pp. 3248-3250.

[11] B.B. Van Aken, M.J. Jansen, N.J.J. Dekker, Reliability and energy output of bifacial modules, 39th IEEE Photovoltaic Specialists Conference, Tampa, FL, 2013, pp. 1610-1614.

[12] N. Ferretti, Y. Pelet, J. Berghold, V. Fakhfouri, P. Grunow, Performance Testing of High-Efficient PV Modules Using Single 10 ms Flash Pulses, 28th European Photovoltaic Solar Energy Conference and Exhibition, Paris, France, 2013, pp. 3184 - 3187.

[13] A. Virtuani, G. Rigamonti, P. Beljean, G. Friesen, M. Pravettoni, D. Chianese, A fast and accurate method for the performance testing of high-efficiency C-Si photovoltaic modules using A 10 Ms single-pulse solar simulator, 38th IEEE Photovoltaic Specialists Conference, Austin, TX, 2012, pp. 496-500.

[14] H.A. Ossenbrink, W. Zaaiman, J. Bishop, Do multi-flash solar simulators measure the wrong fill factor?, 23rd IEEE Photovoltaic Specialists Conference, Lousville, KY, 1993, pp. 1194-1196.

[15] G.J.M. Janssen, B.B. Van Aken, A.J. Carr, A.A. Mewe, Outdoor Performance of Bifacial Modules by Measurements and Modelling, 5th International Conference on Silicon Photovoltaics, 2015, pp. 364-373.

[16] PVG solutions inc., EarthOn Cells, http://www.pvgs.jp/en/earthon.html, 2011.

[17] L. Kreinin, N. Bordin, A. Karsenty, A. Drori, D. Grobgeld, N. Eisenberg, PV module power gain due to bifacial design. Preliminary experimental and simulation data, 35th IEEE Photovoltaic Specialists Conference, Honolulu, HI, 2010, pp. 2171-2175.

[18] http://www.prismsolar.com/pdf/bi60specs.pdf, 2015.

ACKNOWLEDGEMENTS

Authors would like to thank E. Salis for the technical support in the I-V curve measurements and R. Kenny for discussion of the results.

IMPLEMENTATION OF AERIAL THERMOGRAPHY INSPECTION OF PV MODULES IN THE O&M ACTIVITIES IN LARGE PV PLANTS

* Jorge Coello[1], Pablo Gutierrez[2], Alvaro Velasco[1], Alberto Cristobal[2], Vicente Parra[1] and Miguel Rosa[2]
[1] Enertis, Av. Bruselas, 31. Alcobendas (Madrid). 28108, Spain. www.enertis.es
[2]AEROTOOLS-UAV, Av. Bruselas, 31. Alcobendas (Madrid). 28108, Spain. http://www.aerotools-uav.com/
*Phone: +34 91 651 70 21, email: jorge.coello@enertis.es

*Author for correspondence

ABSTRACT: This paper identifies and analyses relevant issues involving the different stages of the aerial thermography inspection in utility-scale plants: flight plan, flight operation, analysis of radiometric information and the final ground level inspection. The document also proposes guidelines and considerations to have into account to perform aerial thermography with high quality results. The proposed procedures allow detection of hot spots of high and medium severity with inspection rates of 18 MW/day in tracker facilities and 27 MW/day in facilities with fixed structure. Furthermore, the paper includes the comparative analysis detecting hot spots by ground level and aerial thermography.

Keywords: Thermography, PV modules, Quality Control

1 INTRODUCTION

Unmanned Aerial Vehicles (UAV) are increasingly used in the PV field because of its high inspection rates, practical data registration and thus, cost-effectiveness, mostly when multi-MW or utility-scale plants are involved.

Both visual and IR thermography-UAV inspections are typically conducted by adapting the suitable high resolution cameras to the aerial vehicle. These systems can be effectively implemented during commissioning, PAC/FAC tests or O&M activities during the common lifespan of a PV solar plant.

However, when performing these inspections within the activities of O & M in large PV plants, the usefulness of the results is based on the realization of a high quality work and in short time. It is not useful to have a large volume of radiometric information that does not allow, for example, the posterior location of defective modules at ground level.

The same applies when results are images that show reflections of solar radiation that hinder the identification of hot spots, or have little sharpness and resolution for subsequent analysis.

Actually, aerial thermography is not simply install a thermal imaging camera on a drone and fly it on a PV system. The reality is that in carrying out the work of aerial thermographic inspection of large photovoltaic plants, the authors have found a number of problematic issues related to the planning own work, with the completion of flights, and with the treatment and analysis of the obtained radiometric and RGB information.

In this document, the aforementioned relevant aspects of this type of thermography inspection are identified and proposes guidelines and considerations to have into account to perform aerial thermography with high quality results.

2 STAGE 1: FLIGHT PLAN

2.1 Relevant aspects

The aspects to be considered at this stage are those related to the Plant's configuration, the aerial vehicle and sensors, and the flight plan aspects.

Firstly, the mounting structure where the modules are installed must be considered, whether a fixed structure or a tracker: This is an aspect which bears an important influence on the flight plan, as one of the key factors for the service is the angle of the thermal camera with respect to the panels (see Figure 1), because of the effect of this variable in the appearance of reflection, phenomenon that does not allow for a correct interpretation of the termal images of the modules. In the case of fixed structures, the angle of the panels remains unchanged, whereas in the case of trackers it varies along the inspection day. Thus, it is important to know the Plant's configuration to be able to plan correctly the optimum camera angles as well as the times for inspection. In the case of trackers, it might be necessary to modify the angle of the camera along the day if the inspection is to be performed without interfering in the normal operation of the trackers, not manually modifying their tracking functioning.

Aiming at minimizing potential sun reflections or drone shading on the panels and avoid mistakes in the analysis of images, the camera's angle will be defined based on the angle of incident irradiance on the plane of array. Such angle will be determined based on the time of the day, the time of the year and the latitude if the Plant.

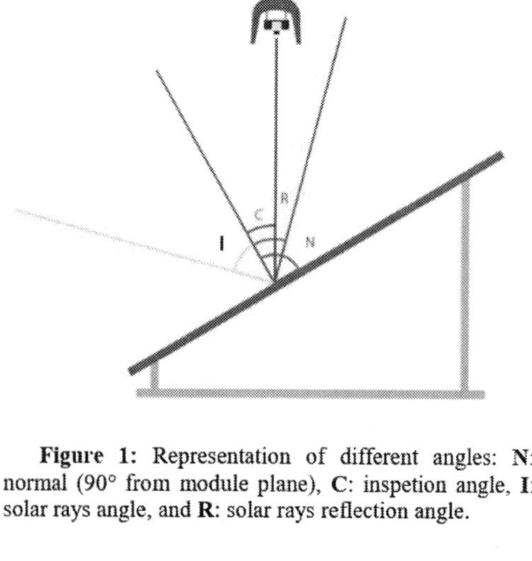

Figure 1: Representation of different angles: **N:** normal (90° from module plane), **C:** inspetion angle, **I:** solar rays angle, and **R:** solar rays reflection angle.

The physical location of the plants is also another factor to be considered to comlpy with applicable RPAS regulations. In the case of Spain, for example, the regulatory framework [1] determines, amongst others, a series of restrictions in regard to the areas over which drone flying is not permitted:

- Building congregations in cities, towns or inhabited places.
- Controlled aerial space.
- Areas located within 8 km. from an Airport or aerodrome for VLOS flights (Visual Line of Sight) and, in the case of BVLOS (Beyond Visual Line of Sight) at less than 15 km. in case that the airport o aerodrome has any type of instrumental flight procedures.

Additionally, limitations exist with respect ti the image taking in the so-called Restricted Areas for Photographic Flights (RVF), which are areas defined by the Army which feature restrictions in this regard for containing of being close to areas with classified objects.

In the case of countries with a regulation in force, one of the issues which might be encountered is the obtaining of the necessary licenses requested per such regulation (documentation, timelines, etc) or even the impossibility or carrying out the work. It will be necessary, therefore, to carry out a study first in terms of regulations for use of drones in public spaces.

As regards the planning of the flight operation itself, and in the case of choosing to do it automatically, it is necessary to introduce into the aerial vehicle the specific flight plan through appropriate planning software. To do that we need to have cartographic information of the plant, in order to know its spatial location based on a given reference system, the distribution of the modules in the system, and its size in terms of area. This is essential in order to plan the flight path to be followed by the drone and the dimensioning of the operation itself (number of flights necessary) to cover the entire surface of the plant, which in turn influences the battery management and development of the inspection in regard to the possible locations of the ground station, takeoff and landing areas. This information can be: 1. Available through a map server; 2. Available in CAD or GIS format type; 3. Not available.

In the first two cases, it is necessary to ensure that such information is updated and corresponds to the reality at the time of the operation. Also, in the case of having information as drawings, it is necessary to check that the information is georeferenced and proceed to its introduction into a map server or into the own planning software to be used. In the event that this information did not exist, it would be necessary to generate it in situ by generating an orthophotography from RGB images obtained in a pre-flight inspection using photogrammetric processing software such as Pix4D.

In the case of the flight vehicle, it is essential to study the kind of drone to be used, its features and benefits, in order to identify the flight range and its ability to carry the cameras that will be used in the operation. The choice of the characteristics of the sensor to be used will also determine the operational work and subsequent analysis of the information collected. In this respect it will be necessary to determine the minimum resolution to obtain in the images to identify possible anomalies in the modules. This resolution will determine the height of the flight based on the resolution of the sensor, focal length and FOV (Field Of View).

It will also be necessary to consider whether or not recording radiometric data to select the type of sensor to use. Also if this recording can be performed autonomously by the camera itself, or is it necessary to have a management and recording team installed on the drone.

2.2 Methodology for flight planning

The methodology begins with the collection of technical information on the plant: installation type, location, surface, drawings/ mapping, etc.

Once the information of the location is obtained, it must be verified if it is located within one of the areas restricted for flight RPAS according to the relevant regulations in the country where the work takes place. In this case it will be necessary to arrange the necessary authorizations to perform the operation.

In the case of not having cartographic information of the plant, or this not being updated, a pre-flight inspection will be conducted in order to obtain the map needed to feed the planning software to be able to perform the flight, both in terms of flight paths and in terms of choice of sites for the ground station and landing and takeoff zones.

Regarding the aerial vehicle, and although there might cases of very large plants, the most suitable for this type of work type is the multirotor despite they generally having less flight range with respect to fixed-wings aircrafts. The main advantages of these vehicles applied to the inspection of photovoltaic plants, is the possibility of carrying different types of sensors either individually or simultaneously and the ability to perform hover for obtaining information detail installation. For example, the results presented in this article were obtained using the AT8-XL (Figure 2) platform whose characteristics are indicated in Table I.

Figure 2: Drone multirotor AT8-XL

Table I: Multirotor specifications

Drone	Mutirotor AT8-XL
Manufacturer	Aerotools-UAV
Configuration	8 arms; 8 rotors
MTOM	11 kg
Autonomy	20 min (total mass 8.5 kg)
Operation mode	Manual / Autonomous by waypoints

The characteristics of the thermographic sensors which are considered more suitable for carrying out the work are:

- Camera Core type (reduced weight and size) with spectral range 7,5-13,5 µm (LW).
- Resolution: 640x480 pixels.
- Accuracy: ± 2 °K.
- FOV: 33°x 45° (horizontal x vertical).

At a flight altitude of 40m, the pixel size for a camera of this type would be between 3.6 and 5.2 cm.

Looking at the subsequent inspection it will be required that the sensor has the ability to store the radiometric data. So a sensor with recording module is used, that works autonomously and does not need any external module. The output format will therefore be radiometric video or image in which the value associated with each pixel represents a temperature value based on a certain scale.

For obtaining visual images it is recommended to use simultaneously to the capture of radiometric information a RGB photo or video camera. This information will allow for having extra information about the status of the installation, which will facilitate further analysis.

3 STAGE 2: FLIGHT OPERATION

3.1 Preliminary factors

There are important factors to take into account into the operational flight phase: weather conditions (wind and relative humidity), irradiance level, location of the control station, areas of landings and takeoffs (in case of several locations) and management and handling of batteries. All these aspects will directly affect working time and therefore the inspection yield itself.

The problems associated with wind speed have to do with the flying of the drone itself, as strong wind conditions may hinder the capture of information, and not even allow operation in the event that the safety thresholds are exceeded of the operation. Also, the presence of wind is a factor that can distort the data collected by the thermographic sensor element assuming a change in the heat flow conditions that occur in the environment of inspection. In the same way, moisture conditions can also alter the sensor readings

The ambient temperature conditions in which the operation is carried out will affect both the personnel in charge of the operation, and the equipment and auxiliary elements themselves. Especially delicate is the handling of batteries, whose exposure to high temperatures can pose a risk to personnel and equipment.

The existence of minimum levels of solar irradiance for conducting the inspection will directly affect the number of hours available for operations and, not surprisingly, yields on aerial inspection. In this sense, the presence of clouds not only affects irradiance levels, but changing conditions can alter the data collected from the modules depending on whether they have been covered or not during data collection by clouds.

Beyond the precautions that need to be taken in the case of handling of batteries, it is necessary to take into account the number of batteries needed for each day of flight and load cycles thereof and providing access to a power source for charging the batteries that have already been consumed.

Initial configuration times for the drone and installed cameras as well as times of displacement and positioning in the successive areas where it is to be placed the ground station, landings and takeoffs (if necessary), will also influence inspection yields in terms of MW/day inspected.

The arrangement of the modules on the ground (forming continuous or separate rows) must be taken into account at the time of designing the flight paths, as these will affect the number of modules that can be captured in each frame by the sensor and therefore the surface area to be able to cover for the same flight height. In relation to these paths, the orientation of the camera must be defined adequately in order to facilitate further processing of the images.

3.2 Procedure in flying stage

First, and prior to the preparation and performance of flights, it is necessary to ensure that environmental conditions are right in terms of primarily wind speed and solar irradiance levels. The maximum wind speed of operation is determined by the own platform to be used and the desired quality of the final images. In the case of the platform and the onboard camera used in this publication, the maximum wind speed for a correct thermographic analysis is 20 km/h. The minimum irradiance level was set at 600 W/m^2.

Once environmental conditions are deemed suitable, the steps to be followed during the flight can be found in Table II.

In regard to flight types, the following options are described:

- Option 1: Autonomous flight + Radiometric video + continuous RGB photos or video.
- Option 2: Autonomous flight + Radiometric video + RGB photos by "Trigger" in GPS-defined waypoints.
- Option 3: Manual flight.

From the options considered above, option 1 is considered the most appropriate, carrying out a pre-planned automatic flight, and by continuously capturing information in the infrared and RGB imaging. The second one is not considered a viable option in the case of working with standard GPS accuracy, due to proximity and a large number of waypoints that would be necessary to be predetermined. Finally, the manual flight, although considered an alternative to consider in the event of need to obtain detailed information of any element, it is not considered appropriate for the systematization of work.

Table II: Flight procedure

Flight Procedure steps	
1. Drone configuration.	6. Landing
2. Ground station set up.	7. Batteries substitution
3. Flight path introduction.	8. Charge of batteries
4. Take off	9. Displacement to the next ground station
5. Flight	

In order to minimize flight times to cover the area of work, the flight path in the case of automatic flight over the modules considered more suitable to be performed is longitudinally in the direction where the separation between these (if any) is smaller. Similarly, the orientation of the camera is set to capture the image match the flight path to facilitate later viewing and analysis.

The viewing angle sensor relative to the plane of the module will be between +/- 30 degrees from the normal, always bearing in mind that it should be taken into consideration the time of day, time of year and the latitude at which the installation and type of structure are found. The objective is to avoid that at all times that the solar reflection angle does not match the observation one to avoid reflections.

Regarding energy needs to carry out the work, it will be necessary to have connection to the electricity grid or have alternative means such as an autonomous generator to provide power to the equipment composing the ground station and batteries charger.

Given the operation described, the number of flights per working day (8 hours) is normally between 8 and 9 flights at a speed of 4 m/s, lasting 15-20 min per flight. With this configuration and considering modules with 72 cells and 310-320 Wp, the approximate performance of inspection is 18 MW/day in tracker facilities and 27 MW/day in facilities with fixed structure if use flight altitude of 40 m.

The battery needs, considering a recharge cycle approximately of 1.5 hours, would be in the range of 3-4 sets of batteries per day. However, it is recommended to have as many sets of batteries as flights are planned as far as possible, in order to avoid any potential difficulties in terms of battery charging.

4 STAGE 3: ANALYSIS OF RADIOMETRIC INFORMATION

4.2 Introduction

The analysis of information obtained in the thermographic inspection will depend on the identification of hot spots, the spatial location of these and the estimation of its temperature.

The problems that we can find to make a correct identification of hot spots that may indicate a deficiency in the modules will be determined by the ease with which we are able to achieve the distinction of contrast levels between modules that work correctly and those which show an abnormality in their thermal behavior.

On the other hand, it is not only essential to know the number of detected anomalies, but also its spatial location. Being able to know their position will allow for drastic reduction of inspection times on the ground, by generating a global database of information for the whole plant about the position of the cells with deficiencies.

The position of these deficiencies can be estimated based on a local reference system or based on a geodetic reference system using GNSS technology.

Finally, the conditions in which the data captured with the thermographic sensor installed on the aerial platform occur makes the temperature data recorded for a particular hot spot within the module do not exactly correspond to the actual temperature it reaches [2]. The fact of taking images at a distance causes the radiation received by the sensor (and which is converted into temperature values) to differ from that emitted by the module (which is what is related to the actual temperature), due to being influenced by the conditions of the atmospheric column located between the sensor and the panels, and the least amount of pixels per cell receiving information.

It is necessary to note also that the analysis of images obtained by aerial thermographic sensor can encounter false hot spots, which can be due, among other things, bird droppings on the module, the presence of strings with reversed polarity or short-circuited, or reflections due to an observation on the angle of reflection of solar radiation. A subsequent manual inspection at ground level is considered recommended depending on the conditions of the inspection and its goal.

4.2 Activities during the image analysis

The procedure for identification of potential hot spots in the modules begins with the downloading of information obtained by the sensor. Once it is done, and in order to improve their visualization, it is appropriate to adjust the temperature range to be displayed on the radiometric image to improve contrast and help to identify hot spots.

It is advisable also to have the GPS data of each of the images obtained (infrared and RGB) in order to spatially locate potential hot spots and thus facilitate subsequent work land inspection position.

Identifying areas with potential deficiencies from the aerial image can done in two ways (see Figure 3): a) Using the graphic display material obtained and the location of hot spots on the floor plan; b) By generating georeferenced mosaics of images to obtain global thermal mapping of the installation. From this mapping , potential individual hot spots will be determined, allowing for identifying their GPS position based on a given reference system.

The accuracy of the GPS coordinates of hot spots will be determined by the accuracy in capturing these positioning data of the image, which is an aspect to take into account when performing the manual inspection on the ground.

As already mentioned above, the temperature data of hot spots detected on the module and that of the collecting sensor installed on the aerial platform may be not coincident [2, 3]. This implies that while the aerial observation of the modules brings us massive and very fast information on the number and location of anomalies therein, determining temperature gradients of the hot points accurately should be complemented with a manual inspection on the ground. However, in the case of large area photovoltaic plants, the use of aerial platforms equipped with thermal imaging sensors saves significant time by allowing to generate initial information on the number and location of potentially defective modules.

4 STAGE 4: GROUND INSPECTION

Once the modules with hot spots have been located, whether by their position on the drawing or by GPS coordinates, the next step would be to go to the module position for a deeper thermographic analysis and a visual inspection to check if the hot spot identified it is not due to bird droppings or shadows caused by vegetation or the like, etc.

This activity can be done by the staff of O & M of the installation itself. However, in case that these results need to be used in a warranty claim process, it is advised that the ground inspection is conducted by a testing laboratory accredited for this task.

In Figure 3 it can be seen that a thermographic inspection at 40 meters flight altitude allows for detection of hot spots of high and medium severity. This means that the aerial thermography technique allows a very good detection of hot spots in a very short time in large photovoltaic plants.

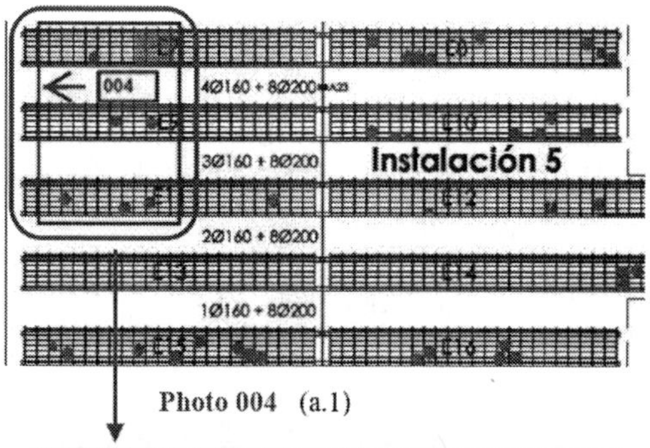

Photo 004 (a.1)

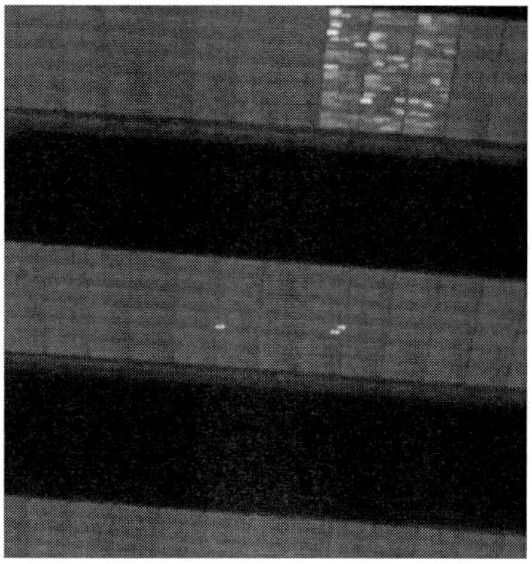

(a.2)

(b)

Figure 3: (a.1) Example up of an inspected facility where the modules that have hot spots are identified; (a.2) Aerial thermography image used for the analysis. The image is taken with a thermal imaging camera sensor size 640x480 pixels and flight altitude of 40 m; (b) Mosaic made with 92 georeferenced images. The mosaic allows to determine the GPS position of each defective module in the plant.

However, for the accurate measurement of the temperature increase ΔT of the hot spots it is recommended a manual thermography inspection because the aerial thermography tends to underestimate the values of ΔT hot spots.

As an example of this, Figure 4 shows two comparative analysis detecting hot spots by ground level and aerial thermography. This comparative analysis was performed in a sunny day without clouds with a stable irradiance of 900 W/m2 in a large PV plant located in the South of Spain. Both thermographic techniques (aerial and ground level thermography) were performed in a short period in order to maintain a stable irradiance to ensure that no change would occur in the photocurrent generated by the module that could affect the temperature of the hot spots.

In both examples, a quantitative analysis of ΔT in 4 different hot spots was performed. Table III shows the results, where ΔT is calculated as the difference between the absolute temperature of each hot spot and the temperature of a same reference point (point 5 in both examples).

The results show that the general trend of aerial thermography is to underestimate the temperature increase of the hot spots. This situation is particularly important if the inspection is to be used in a contractual acceptance and rejection criteria based on hot cells gradients.

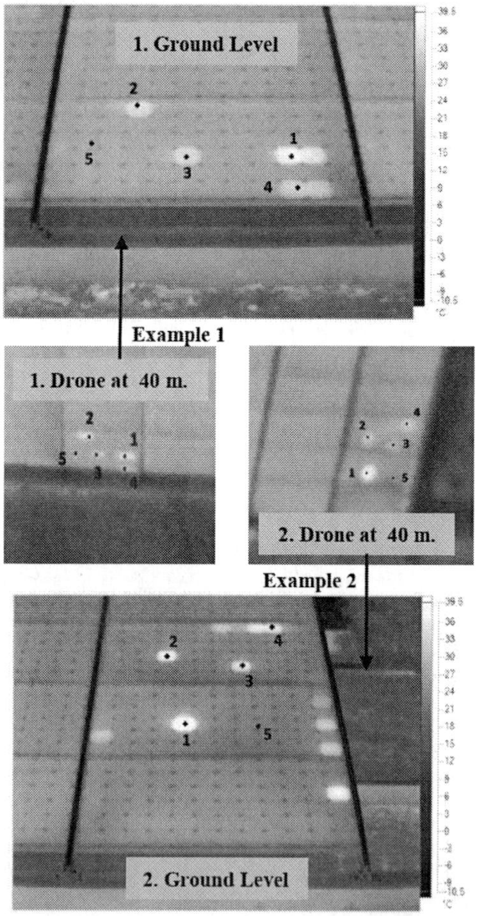

Figure 4: Detection of two areas with hot spots in a photovoltaic facility by manual ground level thermographic inspection and automatic thermographic inspection with drone (40 meters of flight altitude).

In both examples, a quantitative analysis of ΔT in 4 different hot spots was performed. Table III shows the results, where ΔT is calculated as the difference between the absolute temperature of each hot spot and the temperature of a same reference point (point 5 in both examples).

The results show that the general trend of aerial thermography is to underestimate the temperature increase of the hot spots. This situation is particularly important if the inspection is to be used in a contractual acceptance and rejection criteria based on hot cells gradients.

However, from the point of view of preventive maintenance activities in large PV plants, the aerial thermography is a versatile tool for fast and effective monitoring of the thermal behavior of photovoltaic modules.

Table III: Hot spots ΔT comparison obtained by aerial and ground level thermography (from Figure 3 photos).

Example 1			
Hot Spot	ΔT_{Ground} (°C)	ΔT_{Drone} (°C)	Difference (°C)
1	10,6	6,4	-4,2
2	7	6,9	-0,1
3	6,3	4,1	-2,2
4	4,8	2,3	-2,5
5	Reference	Reference	
Example 2			
Hot Spot	ΔT_{Ground} (°C)	ΔT_{Drone} (°C)	Difference (°C)
1	35,2	10,9	-24,3
2	12,4	3,7	-8,7
3	8,1	1,7	-6,4
4	9,3	3,5	-5,8
5	Reference	Reference	

5 CONCLUSIONS

This work identifies a number of relevant issues that should be considered to perform a success aerial thermography inspection. These issues are related to the planning own work, with the completion of flights, and with the treatment and analysis of the obtained radiometric and RGB information.

Furthermore, inspection procedures based on the authors' expertise are proposed. With these procedures, inspection rates of 18 MW/day in tracker facilities and 27 MW/day in facilities with fixed structure have been reached with flight altitude of 40 m.

Results shows that a thermographic inspection at 40 meters flight altitude allows for detection of hot spots of high and medium severity, being very useful in the O&M activities. However, comparative analysis detecting hot spots by ground level and aerial thermography shows that the general trend of aerial thermography is to underestimate the temperature increase of the hot spots. This situation is particularly important if the inspection is to be used in a contractual acceptance and rejection criteria based on hot cells gradients.

6 REFERENCES

[1] Ley 18/2014, de 15 de octubre, de aprobación de medidas urgentes para el crecimiento, la competitividad y la eficiencia. Secc. 6ª, art. 50 y 51.

[2] Jorge Coello, Leonardo Pérez, Vicente Parra and Ruperto Gomez, Proceedings 31st European Photovoltaic Solar Energy Conference, pp. 2327 - 2331.

[3] Claudia Buerhop, Hans Scheuerpflug, Proceedings 29th EUPVSEC, (2014) 2975.

7 ACKNOWLEDGEMENTS

The authors gratefully acknowledge the assistance of Francisco Dominguez and Jose Alberto Florez in the performance of the radiometric analysis of aerial and manual thermograpic inspections.

The activities performed in this paper have been partially financed by the *Centre for the Development of Industrial Technology (CDTI) of the Spanish Ministry of Science and Innovation.* Enertis and the authors acknowledge the support received from this institution.

OUTDOOR ELECTROLUMINESCENCE IMAGING OF CRYSTALLINE PHOTOVOLTAIC MODULES: COMPARATIVE STUDY BETWEEN MANUAL GROUND-LEVEL INSPECTIONS AND DRONE-BASED AERIAL SURVEYS

Simon Koch, Thomas Weber, Christian Sobottka, Andreas Fladung[1], Patrick Clemens[2] und Juliane Berghold
PI Photovoltaik-Institut Berlin AG, Wrangelstr. 100, 10997 Berlin, Germany
Phone: +49 (30) 814 52 64 204, Fax: +49 (30) 814 52 64 101, koch@pi-berlin.com
[1]Fladung Solartechnik GmbH, Im Johannistal 31a, 52064 Aachen, Germany
[2]safeTwork GmbH, Am Hauptbahnhof 4, 66111 Saarbrücken, Germany

ABSTRACT: The electroluminescence (EL) scan is a powerful method to detect solar cells or modules with various defects. The electroluminescence measurement is used by default already in most cell and module production although an international standard determining pass and fail criteria is still missing. Furthermore, so far there was no suitable approach available for large-scale investigation of complete plants over 1 MW. One of the reasons is varying ambient conditions in real power plants that have a significant impact on the EL image quality during data acquisition in field. However, there is an increasing interest and demand in high volume EL-field inspections. Different systems are available, whereas most of them only operate during the night, because of less interfering radiation. In this paper two different methods for high volume EL image capture are compared including the usage of a drone generating EL videos of solar plants. The pros and cons and especially the influence of irradiation, wind and day time are investigated. Exemplarily results are presented obtained in a project where 12500 modules – potentially crack affected – have been investigated. The software-based data analysis revealed that crack-affected cells were mainly located in the center of the modules and PID affected modules were manly located at the edge.

Keywords: Electroluminescence, EL, IR, thermography, crack, PID, grid-connected, large PV-systems

1 PURPOSE OF THIS WORK

The primary purpose of this work is to conduct a comparative study between manual ground-level and drone-based aerial inspection method, in order to investigate all the potential benefits of using Unmanned Aerial Vehicles (UAVs). Therefore, different EL systems were analyzed with regard to high volume measurements. The main focus of such electroluminescence inspection methods is on the quality of crystalline PV modules. Special attention is paid to the ability of acquiring high quality images down to the PV cell level, where small defects must be recorded in order to clearly detect failures like micro cracks.

2 INTRODUCTION

Photovoltaic (PV) modules are engineered to produce electricity for 25-30 years – also for the increasing usage in large-scale PV plants. Quality assurance is essential for assuring the long-term reliability of PV modules in order to maximize financial and energy returns. This implies the need for failure tracking once the modules are installed. For that reason, technical inspection methods, such as I-V curve tracing, electroluminescence (EL) imaging and infrared (IR) thermography are being put into practice in order to assess the quality and performance of PV modules on-site.

However, the implementation of a cost-effective EL-imaging for scanning very large PV-plants represents different challenges including the cost and time effective picture capture, the defect detection and classification and the localization of defect panels in the solar plant. In this work, a study is conducted in order to compare three EL inspection methods.

3 THE NEED OF HIGH VOLUME INSPECTION OF PV MODULES ON-SITE

Four main factors have been found that lead to the necessity of inspecting large amounts of installed PV modules on-site. These factors show the great importance of PV module quality control, which has become vital in the recent years.

3.1 Increasing size of large scale grid connected PV plants

In 2014 a cumulative capacity of 178 GWp was reached worldwide [1]. Large-scale PV plants comprise approximately 45% of the global cumulative installations and can exceed several hundred MWp [2]. They started to appear around the world in the 2000 and have been rising substantially year after year. Now, 500 MWp utility-scale PV-power plants are not the exception anymore.

3.2 The need of tracking down the source of failures

The cost of the PV modules represents approx. 54% of the total cost of a PV installation, so their correct operation assures the efficient operation of PV plants [3]. Köntges et al. presented three categories of PV module failures according to the time of their occurrence: infant-failures, midlife-failures and wear-out-failures. They predict in their field study that approx. 2% of the modules fail after 11-12 years and therefore, do not meet the manufacturer's warranty [4]. Thus, quality inspections are necessary to detect and diagnose infant and midlife-failures. For warranty claim issues, it is essential to identify - with sufficient proof - whether the responsibility falls on the PV module manufacturer, on the delivery company or on the EPC company in charge of the installation.

3.3 Financial support and subsidy reduction

In the case of mature markets, e.g. Germany and Italy, financial support to PV is being decreased and so the revenues per kWh produced reduces. Therefore, the risks concerning the failure of PV modules become increasingly important, i.e. if the PV modules fail, the financing of the project will be in danger.

3.4 Increasing awareness of investors regarding module failures

Investors and banks are becoming more aware concerning PV module failures that can occur and the corresponding financial risks. This is especially relevant in sell/purchase agreements and re-financing.

4 INFRARED THERMOGRAPHY VS. ELECTROLUMINESCENCE IMAGING

Infrared Thermography and Electroluminescence imaging are two well established imaging techniques in the PV industry since years. Some defects can be found by both methods (e.g. PID, inactive cells) other defects are specific for each method (e.g. Hot spots for IR and micro cracks for EL). Based on the experience of several years' in-field analysis, one can say that PID and broken cells are the most requested defects, which can be analyzed with both methods. But comparing the images of both methods, as seen in Figure 1, it becomes clear that EL gives much more detailed images. An evaluation of the shown IR images might even lead to a misinterpretation of the containing errors. The shown comparison of the IR- and EL-images lead one conclude that the higher resolution in EL-imaging enable a more reliable and more accurate data assessment.

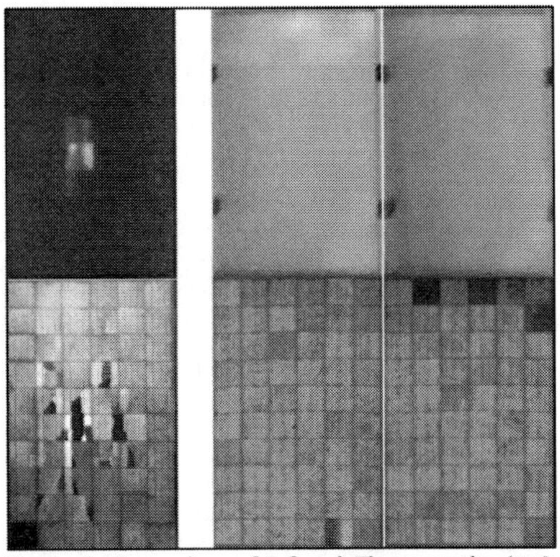

Figure 1: Comparison of Infrared Thermography (top) and Electroluminescence (bottom) images showing that EL reveals more failures then IR due to a higher resolution

Moreover IR and EL are technologies, which work at different operating points of the solar modules. Generally speaking, IR is an analyzing tool, which is working during the day, when the sun is shining and EL is a method, which works during the night. Additionally, different thresholds exist for each technology. These are listed in Table I.

Table I: Comparison of the thresholds for IR and EL

IR (IEC 6090 4-12 TS Ed.1)	**EL** (PI-Berlin experience)
Min. 600 W/m² *(influenced by the sun)*	Max 100 W/m² *(influenced by IR safety cameras and full moon)*
Max 4 Bft *(wind influences the module temperature distribution)*	Max 6 Bft
Max 2 okta (cloudiness)	-
Low soiling	Moderate soiling
No rain	No heavy rain
Angle of view not below 30° and not 90°	Angle of view not below 30°

To visualize the availability of both methods weather data of six different locations were filtered by the boundary conditions given in Table I. Availability is defined here in terms of time prevailing operation conditions enable performing a testing and expressed as possible theoretical value of working hours the specific technology is applicable (see Figure 2). One can see that the availability of EL is significantly higher in each location. Especially, in the winter time, the radiation is low and the nights are long and EL screenings are without any alternative. For the most investigated locations the possibility to perform on-site examinations with EL (possible working hours) is more than doubled or even quadrupled.

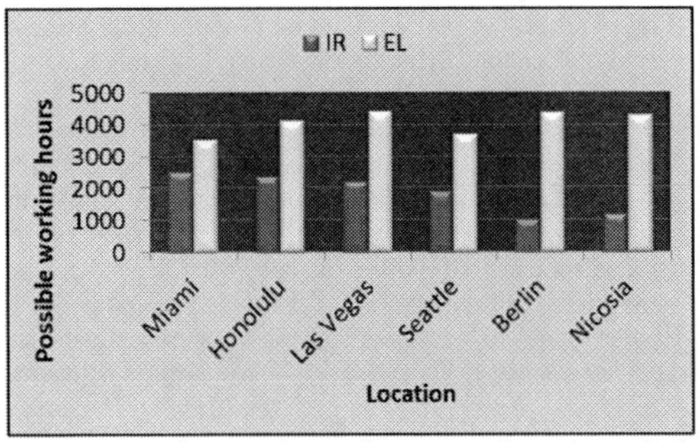

Figure 2: Comparison between the theoretical availability of IR (red) and EL (grey) imaging

5 EL FOR PV MODULES FAILURE DETECTION

Theoretically, different EL recording technologies are possible for large scale EL analysis. There are two main approaches, manual ground-level and aerial inspection systems. Besides the presented setups in this paper, other systems are also available on the marked. For example: Daysy [5], Suncycle [6] or MBJ [7]. These approaches were not investigated in this work, but however are likely not practical for high volume EL measurements because of low resolution or the need to dismount the modules from the supporting structure.

5.1 Manual ground-level inspections

Manual ground-level inspections consist of the scanning of PV modules by means of fixed EL cameras on portable constructions, which are manually moved by field engineers. The investigated ground-level inspection includes two different systems.

5.1.1 One Camera-Tripod-System

A tripod system consists of one tripod and one camera. It features a high flexibility. It can be used as a high-tripod, a "small" camera-tripod combined with a cherry picker (see Figure 3) or just as a conventional tripod.

The system is easy to transport in a plane with the ordinary luggage. Therefore, this system does not need a long preparation. The operation can be assured by one or two engineers, which can adjust the setup configuration easily. This allows detailed images of unclear or unknown defects. The biggest disadvantage of the system is the place of the measurement and its maybe unsuitable ground. Only a few modules fit on one image at once making the system vary slow (max 0.2 MW/night) and hence is only restrictedly applicable for large scale analysis.

Figure 3: EL ground-level system application example based on a cherry picker and one camera approach

5.1.1 Multi Camera-Tripod-System

Multi camera systems are customized for particular power plant layouts. Thus a lead time to adjust the construction is necessary. Contrary to the tripod system the transport is bulky. Also the need of at least 4 persons is required. But the number of modules per image is also higher, even the quality of the pictures is adequate and last but not least the position of the modules within the images is always the same. The last characteristic specially helps during the after processing steps, where cropping and equalizing are conducted. As an example a fast and standardized capturing of EL-images is presented here, for a project a specific set-up was developed with two high-resolution EL-cameras working in parallel (see Figure 4). The upper camera captured 2 x 3 modules and the lower camera captured 2 x 2 modules.

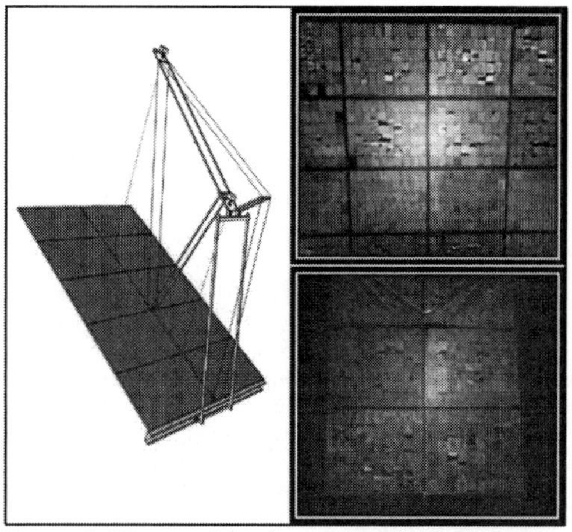

Figure 4: EL ground-level system application example with a custom made tripod and multi camera approach

With the presented two camera system, which was moved on a rail system using the supporting structure, up to 0.8 MW could be recorded per night. This system can also be adjusted for other power plants but needs a redesign to fit the specific needs of every plant.

5.2 Drone-based aerial surveys

The newest EL system, which is applied in cooperation with its developer Fladung[1] aerial pv inspection, is based on an octocopter drone with one high resolution EL camera. The angel of the camera is adjustable per remote control (see Figure 5). Using a drone opens a new variety of applications. Working in difficult terrain and with non-standard designs becomes feasible.

Figure 5: EL camera mounting and associated octocopter drone

It simplifies the EL measurements on roof tops and on steep and rough ground. Depending on the flight height, which is easily adjustable, the numbers of modules per picture is nearly limitless. It reaches from single module images to one picture of the whole plant. Additionally, EL-videos could be made with a drone for the first time, which also opens new opportunities for the power plant analysis.

Along with the technical development new environmental influencing factors were discovered. One main factor are sun winds, which make it difficult for all GPS guided systems to determine their position. This means for the flying drone that the drone is constantly correcting their position during strong sun winds and therefore doesn't manage to stand stable in the air, which creates blurred images.

Due to the constant technical development further improvements in the drone technic are expected. In the next step flights during light rain will be possible and the flight duration, which is currently at around 60 min, will be enhanced significantly.

For EL, it is generally necessary to connect solar modules with a power supply. The ground level inspection systems are using commercially available power supplies, which can be connected with individual modules or with whole strings. During the measurements the individual modules or strings have to be manually sequentially connected to the power supply and disconnected again. This is usually done directly on the modules or on the generator junction boxes.

For Drone based EL analysis this is not practical anymore. The short flight time needs a faster switching between the strings. Therefore an automated wireless remote controlled power supply is necessary. Due to the low flight times (currently around 60 minutes), it is necessary to carry out as many shots as possible. In contrast, an automated energization is in the case of ground level arrays not mandatory but also increases the working time. For this reason string boxes have been developed by Fladung Solartechnik GmbH[1] in master-slave configuration, which enable to connect up to 100 module strings (500 kWp).

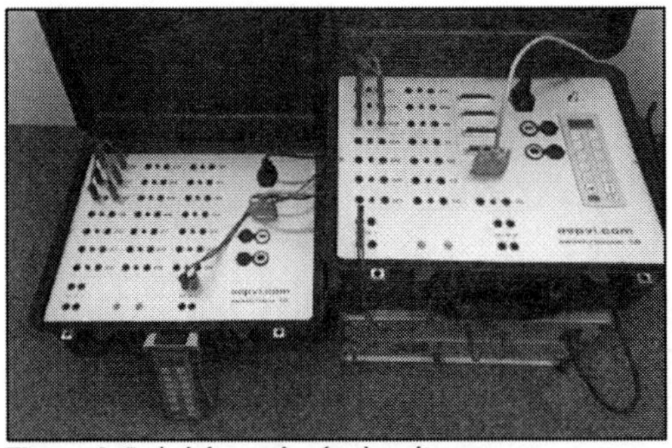

Figure 6: Switch boxes in plastic suitcases

The drone in combination with the string boxes makes it possible to measure up to 1 MW per night and this with just a small amount of engineers. One aspect, which is difficult to estimate, is the statutory frame work for drones in every country. During the last years drones became affordable and are used as toys and recreational fun or planned as transport units in the near future. The governments of each country will react on these developments and will pass laws for regulation. In Germany every federal state has to gives its flight permissions at the moment. How this will develop in the next years is not predictable.

5.3 Comparison

One can say that each power plant has its specific difficulties and every EL system fits different needs (see Table II). But in terms of large scale EL analysis, a normal tripod system is not applicable but irreplaceable for quick analysis on a random basis.

Table II: Comparison of three different EL systems

	Tripod	Multi-camera construction	Drone based photovoltaic inspection
*Throughput in **MW/night***	*Max. 0.2*	*0.3 – 0.8*	*0.5 – 1*
Module/image	1 - 4	10 - 15	1 - 100
Operator needed	2 - 3	4 - 6	2 - 3
Advantage	No lead time	Identical pictures	No plant design limitations / fast

Multi camera and drone based analysis are both applicable for large scale measurements. Both technics have the potential to generate thousands of pictures in one night. On the strength of their flexibility, drone systems will be the choice of the future.

6 DATA ANALYSIS

As a result of the high volume data acquisition with the presented EL systems it becomes necessary to handle, analyze and categorize thousands of EL images or videos equivalents to gigabytes of data. A manual processing is not possible anymore and software assisted steps become necessary. Figure 7 visualizes the typical processing steps from the raw image/video to the result presentation for the customer.

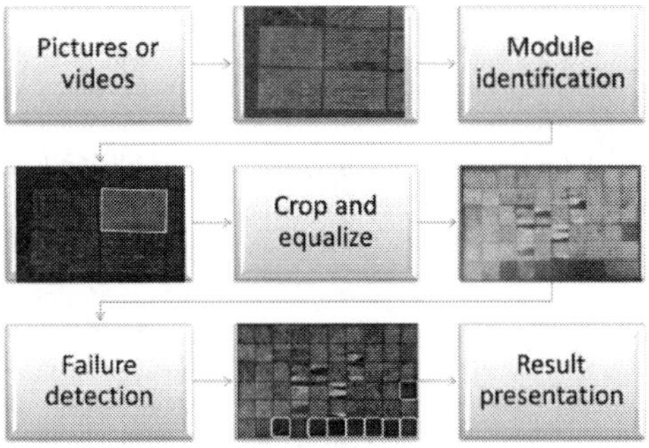

Figure 7: Flow chart showing the software assisting analysis steps for high volume EL investigations

In the PI-Berlin work flow, a software identifies all modules shown in an image. After identification each module gets cropped and equalized in order to depict it fully in one image and additionally has the same fixed dimensions. This is necessary, thus the failure detection can be executed automatically with the help of the software, which was developed in cooperation with the company safeTwork. The actual version of the software can identify isolated cell fragments and PID affected cells. A semi-automatic evaluation can identify up to 22 different defects, according to an internal PI-Berlin failure catalog.

An advantage of the highly standardized image detection and analysis is the possibility to generate failure distribution, of defects of interest. In Figure 8 two examples are shown. The graphic on the left hand side presents the results of a project with an EL-scan of 12500 modules. The graphic shows a module in its mounting orientation and each square represents a cell on this position. Every cell, of each module, at this position with a mechanical defect was counted and summarized. The result shows a typical distribution of mechanical stress caused by transportation and rough handling. Most of the affected cells are in the center of the module were the mechanical stability is low and therefore the bending radius is high. Starting point of this project was a snail-trail/crack issue in the plant.

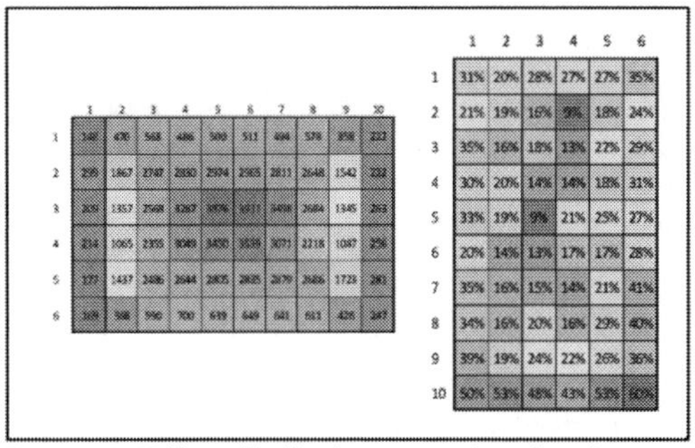

Figure 8: Failure distributions examples of real power plants analyzed with the high-volume EL Technology showing mechanical defects on the left and PID on the right

On the right hand side of Figure 8, a PID distribution pattern of a power plant affected by PID is presented, which is completely different distributed. One can see that the dispersion shows the opposite distribution compared to the mechanical defects. In detail, most unaffected cells are in the module center and most affected cells are on the module edge or more precise on the lower module edge. This result showed the strong influence of humidity in the progress of real PID in an earlier work [8].

With the help of different defect distribution, the origin of particular failure origins can be researched specifically for each power plant.

7 SUMMARY

For huge power plants thermography and electroluminescence are the two main used inspection tools. Due to its high resolution, EL is the preferred tool for power plant inspection. For High Volume inspection different EL systems can be used and adjusted, according to the specific needs in a solar project.

In summary, the drone inspection shows the biggest potential in high volume EL evaluations. The huge amount of pictures and videos require a software-assisted data analysis.

Solving most of the problems in the past, PI-Berlin and their partners can provide its customers an EL analysis for every module, even in huge power plants.

8 AKNOWLAGEMENT

The authors gratefully acknowledge the co-authors and particularly the cooperation partners from aerial pv inspection Fladung[1] and the safeTwork[2] company.

9 REFERENCES

[1] SolarPower Europe (2015): Global Market Outlook. For Solar Power / 2015-2019.
[2] Electric Power Research Institute (EPRI): *Utilizing Unmanned Aircraft Systems as a Solar Photovoltaics Operations and Maintenance Tool,* 2015.
[3] ITRPV (2015): *International Technology Roadmap for Photovoltaic.* 2014 Results.
[4] Köntges, Marc; Kurtz, Sarah; Packard, Corinne; Jahn, Ulrike; Berger, Karl A.; Kato, Kazuhiko et al. (2014): *Report IEA-PVPS T13-01:2014 - Review of Failures of Photovoltaic Modules*
[5] L. Stoicescu, M. Reuter, J.H. Werner; *DaySy: Luminescence Imaging of PV Modules in Daylight*; 29th European Photovoltaic Solar Energy Conference and Exhibition
[6] http://www.pv-magazine.de/nachrichten/details/beitrag/suncycle--el-messung-vor-ort-bei-tageslicht_100020417/
[7] http://www.mbj-services.com/
[8] S. Koch, J. Berghold, C. Hinz, S. Krauter, P. Grunow; *Improvement of a Prediction Model for Potential Induced Degradation by Better Understanding the Regeneration Mechanism*; 31st European Photovoltaic Solar Energy Conference and Exhibition

TOPIC 5.2 OPERATION OF PV SYSTEMS - Oral presentation

OUTDOOR NON-DESTRUCTIVE INFRARED THERMOGRAPHY OF PHOTOVOLTAIC MODULES AND PLANTS FOR INSPECTION: IEC 62446-3

Bengt Jaeckel[1], Bernhard Weinreich[2], Claudia Buerhop-Lutz[3], Ulrike Jahn[4]

1 UL International GmbH, Admiral-Rosendahl-Strasse 9, 63263 Neu-Isenburg (Zeppelinheim), Germany, Bengt.Jaeckel@gmail.com, Tel. +49-162-2621901
2 HaWe Engineering GmbH, Mühlthaler Weg 1, 82131 Gauting-Hausen, Germany, weinreich@hawe-eng.com
3 ZAE Bayern, Haberstraße 2a, 91058 Erlangen, Germany, Claudia.Buerhop-Lutz@zae-bayern.de
4 TÜV Rheinland Energie und Umwelt GmbH, D-51101 Cologne, Germany, ulrike.jahn@de.tuv.com

ABSTRACT:
The use of thermography in Photovoltaic started around 2010 and is almost main stream today. Due to that development quite different results are achieved currently while carrying out thermography in a PV power plant. To increase the level of understanding and more importantly comparability between thermograms standardization is urgently needed.
During the past years standardization projects were ongoing on national and international level to improve the evaluation of thermograms taken of PV components. In this contribution we will first give an overview about the current situation of standardization by reviewing the history of thermography standardization.
Second we will focus on further achievements in thermography and the implementation of them into an international IEC standard IEC 62446-3. That includes some, in principle, basics of thermography, personal qualification and reporting. The purpose of this work and the standardization is to describe the physical and technical parameters that are required for reliable fault detection in the field when conducting infrared (IR) imaging of PV plants.
Although IR imaging of PV modules and plants in operation is used for several years to locate thermal abnormalities of BOS components (PV modules, here in particular cells, diodes and connections, and other components like combiner or fuse boxes), the evaluation and conclusions drawn from such collected thermograms differ significantly. With the help of this document and the upcoming IEC standard the comparability is aimed to increase and help to establish a common understanding. At the end an outlook is given for further work which might result in edition two of IEC 62446-3.

Keywords: Infrared, thermography, imaging, interpretation,

1. INTRODUCTION

When the usage of PV thermography in the field started to be popular around 2010 there have been many discussions between module manufacturers and installers about the validity and meaning of thermograms (often titled infrared (IR) images). This is mainly caused by different level of understanding and knowledge of the physics of thermography, the quality of images in general and their interpretation. Based on that the range of issued reports (quality, details and recommendations) is varying form a child picture book style to highly valuable scientific correct reports that will also be accepted e.g. for claims. Weinreich et al summarized those findings within a project review [2][3].
One major outcome of that study was that a general accepted guideline is needed for:

- Qualification of personnel
- Thermography procedure
- Evaluation of thermograms
- Interpretation of thermograms

Based on those identified gaps an INS-project was initiated. The further history of standard preparation is summarized in the next chapter. Together this nicely shows the background of the new IEC 62446-3 document [4]. This following paper will highlight the most relevant topics and will discuss in the end the next tasks.

2. HISTORY and FUTURE of IEC 62446-3

The concept of thermography is quite old. It is well established in many industries especially in the electrical industry since more than 30 years to e.g. quickly check electrical cabinets. In Photovoltaic it was attracting more attention starting in 2010. The following highlights the most important steps towards the IEC standard:

- **2007:** Publication of first handbook for Photovoltaic thermography was published by Buerhop [1]
- **2010:** First discussions of the necessity of a PV thermography guideline to start in collaboration with DGS-Berlin.
- **Since September 2011:** Several parallel guideline and standardizing activities were started as e.g. by Solarschmiede (today: HaWe Engineering)
- **February 2012:** constitutive meeting for the VDI guideline 2883 "maintenance of PV-Plants" including main chapters to IR-imaging of PV-plants initiated by different companies (module manufacture, insurances, PV-plant designer)
- **March 2012:** Continuous work for the INS project by Solarschmiede and DGS-Berlin and parallel within the German Thermograph Association VATh
- **March 2014:** Publication of the first guideline by the German Thermograph Association VATh [5]
- **July 2014:** Formation of DKE / AK 373.0.30 Committee to work on international IR document
- **February 2015: IEC** New work item Proposal (NWIP) for Outdoor thermography: IEC document number: 82/956/NP
- **September 2015:** Approval of NWIP as IEC 62446-3 TS with the title: Grid connected photovoltaic (PV)

systems – Part 3: Outdoor infrared thermography of photovoltaic modules and plants

- **January 2016:** Draft document that included all comments from the NWIP phase was circulated as IEC-CD document number 82/1073/CD
- **February 2016:** English Translation of the VATh guideline
- **May 2016**: International comments received and during IEC WG 3/6 Meeting in Phoenix a status report including next steps was given.

To finish the history section a short outlook into the future:

- **By September 2016:** Include all international comments into current IEC draft
- **Oct 2016:** Submit next draft to IEC
- **Q1 2017:** Publication as IEC TS
- **2017:** align VATh document(s) [5] with IEC requirements where applicable for Photovoltaic applications

3. DETAILS and BACKGROUND

During the preparation of the VATh guideline and also during the preparation of the international IEC document several topics were discussed controversial. This chapter will give a short summary of the most important ones and give some background and inside into the discussion within the project team and the national/international community.

a. Personal qualification

Qualification of personal conducting the thermographic imaging is described in Annex B. It reads as follows:
"The testing personnel shall have sufficient knowledge in electrical installations, PV plants and also regarding thermography including the analysis of the thermographic findings in electrical systems."
What does that mean? Persons who work in and on PV power plant equipment should have the sufficient knowledge to be able to work in electrical installations. In addition to that fundamental knowledge in electro thermography is mandatory. A certificate according to ISO 9712 can serve as confirmation and will ensure that thermograms are taken, evaluated and interpreted correctly.

b. Inspection equipment

Obviously an infra-red camera is required to conduct a thermographic study of a PV power plant. The camera should be suitable to fulfill at least the given minimum requirements like resolution. But this is only one piece of the equipment necessary to conduct a high quality thermographic inspection. Besides the IR camera, at least the following equipment is required: Irradiance sensor, a mean to monitor the state of operation of the plant (MPP-tracking) e.g. by using a current probe logger, an air temperature sensor, a wind speed and direction sensor and a visual camera to document any findings of the generator but also to record clouds that will impact the thermogram and are needed for image evaluation and interpretation.

c. Environmental conditions

For image evaluation and interpretation it is essential to know the environmental conditions (irradiance, temperatures, operation condition of plant) when the thermogram was recorded.

Besides that also the operator can introduce image errors resulting in not correct temperature readings. Figure 1 shows a typical photo while taking a thermogram. It is a sunny day with almost constant environmental conditions (no clouds and constant winds). Although having perfect conditions the thermogram might bear artefacts. This is illustrated in the following. As indicated with the angles α and β the view angle of the camera between module and camera detector varies. Depending on the angles (α and β) and the size of the object a-b (here a standard c-Si PV module) the effect of recoded temperature in a thermogram can be significant. An example thermogram is shown in Figure 2. A temperature gradient across the module is obviously visible. In this particular image this is an artefact. The measured module temperature is totally homogeneous.

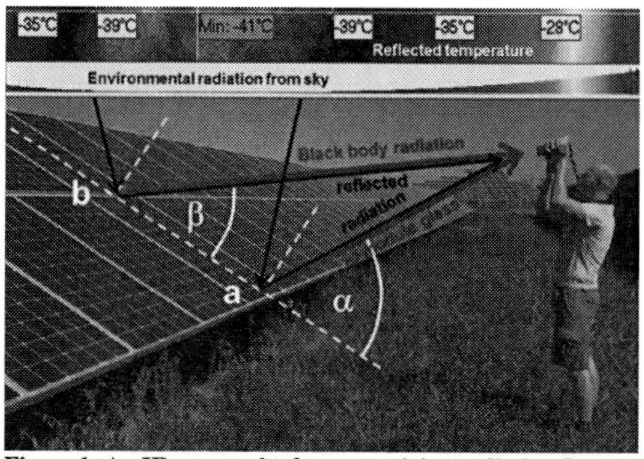

Figure 1: An IR camera is always receiving radiation from at least two sources. Within thermograms of PV modules that are: 1. the black body radiation from the module (red arrows) and 2. the by the object (module) reflected environmental (sky) radiation. To calculate the true glass surface temperature the influence of the environmental radiation has to be measured by a picture of the sky. The black body equivalent temperature of the sky is typically below 0°C. It is an average of all traversed partly transparent parts of the atmosphere and space temperature.

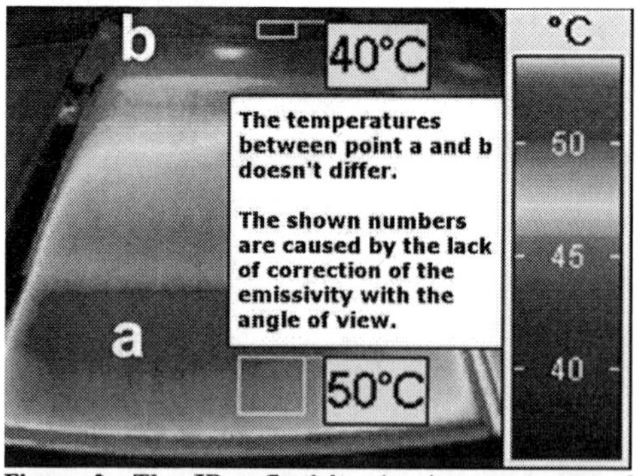

Figure 2: The IR reflectivity (or its counterpart the emissivity) of all materials (glass) is changing with the angle of view. Therefore an uncorrected emissivity is causing misleading temperature gradients caused by the different angle of view at point a and b, which have to be respected.

This behavior is a result of the changing emissivity of objects by absolute temperature and herein more importantly of the view angle. A typical emissivity vs view angle plot is given in Figure 3. As the plot shows the emissivity is reduced significantly towards lower view angles. The result in recorded temperature is even more severe. For better comparability of thermograms it is therefore essential that the angles α and β are as large as possible and that all thermograms are taken under the same set of angles so that all images are affected similarly.

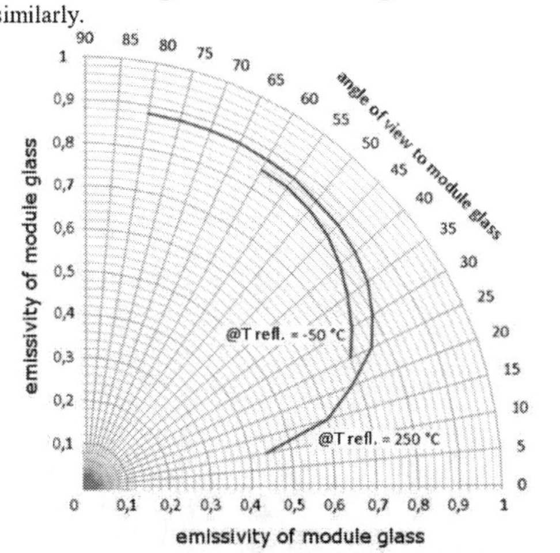

Figure 3: The emissivity (ε) differs not just with the angle of view but also many other influences as soiling on the surface and the reflected temperature (T refl.) itself. In most cases the use of ε = 0,85 @ 90-50°; ε = 0,80 @ 45-35° and ε = 0,7 @ around 25° is a sufficient approximation for module glass.

The necessity to document environmental conditions and to measure precise absolute temperatures is exemplarily shown in the next two figures (Figure 4 and Figure 5). Figure 4 shows a part of a PV generate with thin film modules. The displayed temperature scale is equally in both images. Obviously there is a temperature between both images. During the imaging the environmental conditions were almost the same and can be ruled out as a cause for the observed temperature. It turned out that the result was due to different operating states of the PV generator. The temperature difference of a PV generator in MPP and in open circuit at 1000 W/m² is just around 4K for the shown thin film modules (see Figure 4). Therefor it is important to constantly monitor the status of the generator itself.

The same 4K temperature differences can be caused by changing wind conditions or reflected temperatures e.g. clouds as shown Figure 5. In the one case the crucial instrument was the wind sensor to understand the image in the other just a visual image from the sky in combination of the thermogram.

The presented images are examples but nicely highlight the importance of collecting continuously information about the environmental conditions.

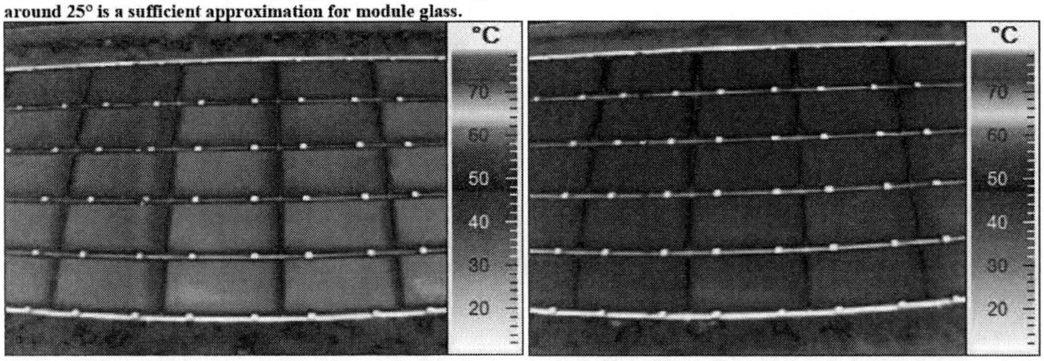

Figure 4: These figures show module tables under identical environmental conditions like irradiance, air temperature and wind speed. The temperature scale of both images is the same. Obviously the right image shows a lower module temperature (about 4K) versus the left image. Root cause: in the left image the PV generator was not in operation whereas in the right the generator was at maximum power point. Important: any change in the environmental conditions would have lead to similar changes in temperature difference and the problem would have not been detected. This underlines also the important of monitoring the state of the generator (see chapter 3b).

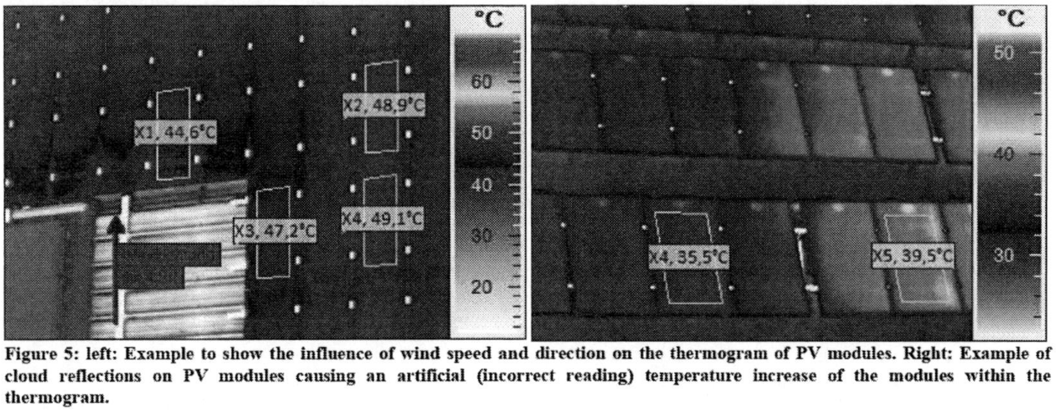

Figure 5: left: Example to show the influence of wind speed and direction on the thermogram of PV modules. Right: Example of cloud reflections on PV modules causing an artificial (incorrect reading) temperature increase of the modules within the thermogram.

e. Imaging and software evaluation procedure

The base for a high quality image evaluation process is a defined and sufficient imaging procedure and a sufficient software tool for imaging evaluation.

Besides the requirements from chapter 3c the geometrical resolution of a thermogram is important. Based on the survey [6] the minimum resolution is about 5x5 pixels per 6"x6" c-Si solar cell (see Figure 6). For thin film the resolution requirements are the same (=> size of pixel 3x3 cm²).

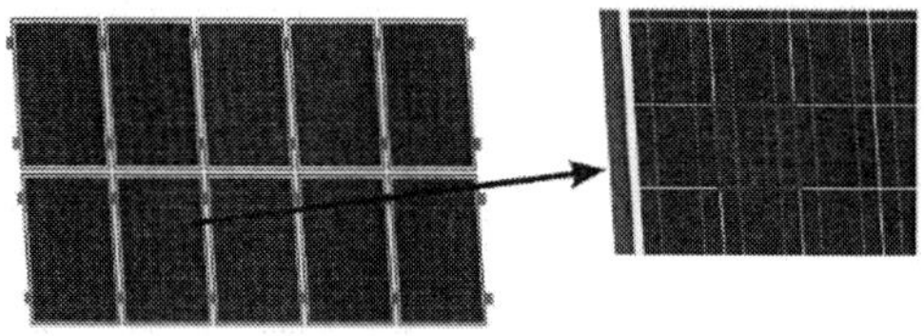

Figure 6: Sketch of a PV array and to illustrate the 5x5 pixel on a 6"x6" c-Si solarcell.

But the calculated resolution not only is important for image quality. Fast camera carriers e.g. aerial drones can cause a negative effect on the image quality as the typically used bolometer detectors have a long time constant and high camera movement velocities above 3m/s [7] may cause smearing as seen in the picture.

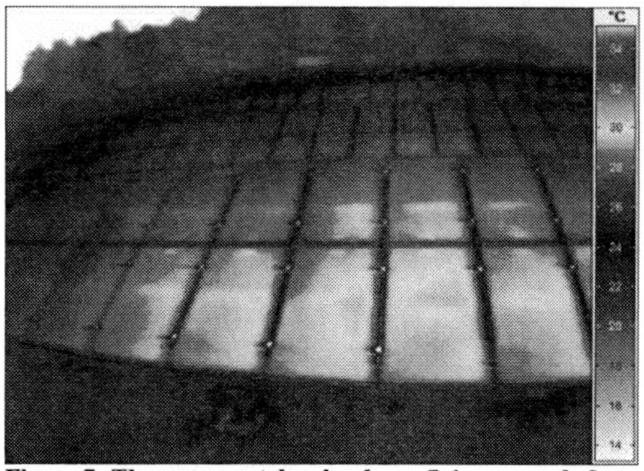

Figure 7: Thermogram taken by drone flying a speed above 3m/s. All modules actually have the same temperature pattern if measured stationary. The visible smearing is caused by the speed of the drone (> 3 m/s) and by the too long time constant of the bolometer detector for the used speed of the drone.

The evaluation process has to be supported by analysis software. The software should be enable the user to change the values of emissivity ε and the reflected temperature (Trefl.; also "ambient temperature"). Furthermore the color scale of every picture called level and span should be easily changeable. For complex assessments of abnormalities the software has to offer different measuring tools as highlighting the maximum temperature spot and also tools to measure the average value within a polygon field as shown in Figure 8.

f. Content of inspection report

Chapters 3a to 3e are essential for taking and evaluating thermograms. The last important topic is the reporting of the findings. A thermogram without the proper given information is just an image. The value and the possibility to compare thermograms between inspections is set by the quality of the report, especially important if different companies are carrying out the inspection in the field. All collected information (e.g. status of plant/strings – MPP, Isc or open circuit; location, date, weather, personal qualification) should be given in conjunction with the thermogram.

Based on the kind of inspection (simplified vs detailed, commissioning vs repeated) more details have to be given of the procedure but also the interpretation of what kind of anomalies are found during the inspection. Based on the findings recommendations, if safety relevant mandatory, should be given to resolve the found issue.

Figure 8: The thermogram shows two modules in the center with a quantitative evaluation of their mean temperatures by using a polygon software tool and a spot measuring tool. These kinds of detailed assessments can be important to do validity checks by controlling if the thermal pattern also shows the same temperature rises as expected in theory.

4. CONCLUSIONS

Significant standardization progress was achieved in the last years with respect to thermography. During the development new insight and new knowledge was developed and included in the standardization documents. Herein we highlighted aspects like personal qualification and focused on difficulties and physics of the imaging procedure itself (see Figure 1 to Figure 3). Recording and monitoring of the other BOS components and environmental parameters is essential to correctly evaluate thermograms as the examples show in Figure 4 and Figure 5. Taking good images is a basic requirement. Evaluation, interpretation and reporting are also crucial. Reports without proper description of the thermogram including all environmental parameters are not comparable. To increase the trust in thermography of PV power plants it is essential that reporting is done in the best possible way that different parties can reliably verify anomalies found once. Without proper reporting e.g. the example in Figure 4would not have been found.

Still there are many open topics that will need continuous scientific work in the future to further improve the standards.

5. OUTLOOK

The main focus in this contribution but also in IEC 62446-3 lays on PV module inspection. PV modules are still a large percentile of the total costs of a PV asset. But for operation and maintenance the other BOS components are of similar importance, since their failure will lead directly to downtime, other fire hazards and performance losses. Work is ongoing to look into other BOS components and there thermographic features, see e.g. Figure 9.

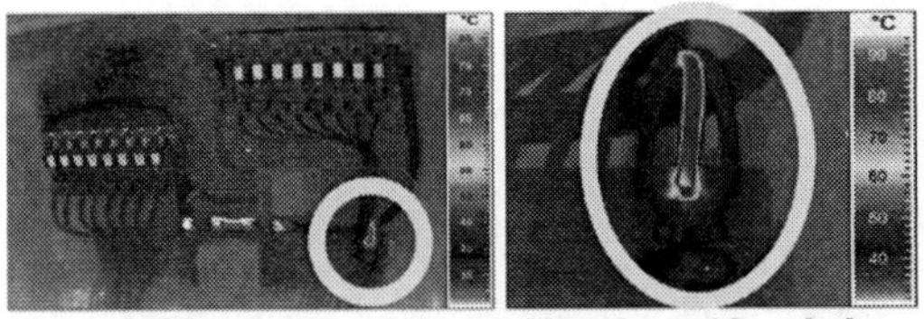

Figure 9: Thermograms of a combiner box with a faulty connection. A close-up is shown on the right side. This kind of heating will not only cause a power loss, more importantly it could cause a fire within the combiner box. This would result in costly repairs and downtime of a large part of the power plant.

Besides ongoing efforts on integrating BOS components the normalization of critical temperatures to e.g. 1000W/m² is a high priority. Only with that normalization more stringent pass/fail criteria can be developed and internationally agreed on.

6. REFERENCES

[1] C. Buerhop, et al., Abschlußbericht der Machbarkeits-studie zur Überprüfung der Qualität von Photovoltaik-Modulen mittels Infrarot-Aufnahmen (2007), www.zae-bayern.de

[2] Bernhard Weinreich, Björn Schauer, Rudolf Haas, Eva Schubert, Udo Siegfriedt, and Ralf Haselhuhn; „Prozess der Standardisierung in der Photovoltaik – Thermografie", PV Symposium Bad Staffelstein, 2015

[3] C. Buerhop, D. Schlegel, M. Nieß, C. Vodermayer and C. J. Brabec; Solar Energy Materials and Solar Cells, 107 (2012) 154

[4] IEC/TS 62446-3, Photovoltaic (PV) systems - Requirements for testing, documentation and maintenance - Part 3: Outdoor infrared thermography of photovoltaic modules and plants

[5] VATh- Guideline: Electrical thermal imaging:
German link: http://vath.de/docs/richtlinien/VATh-Richtlinie_Elektro_NS+PV_engl_web.pdf;
English link: http://vath.de/docs/richtlinien/VATh-Richtlinie_Elektro_NS+PV_web.pdf

[6] Weinreich B., (Solarschmiede) „Geometrische Auflösungen in der PV-Thermografie aus technischer und wirtschaftlicher Sicht" Vortrag 8, DGZfP-Thermografie-Kolloquium 2013, Stuttgart

[7] Reich T., „Vollradiometrische PV-Thermografie aus der Luft" Fachartikel „ep Photovoltaik" 2013

DATA MINING METHODS FOR FAILURE CLASSIFICATION ON PV-MODULES MONITORED UNDER FIELD-CONDITIONS

G. Behrens[1], A. Dercho[1], H. Quakernack[1], T. Wächter[1], S. Hempelmann[2], I. Kruse[2]
[1]Solar Computing Lab - University of Applied Sciences Bielefeld, Artilleriestraße 9, 32427 Minden, Germany
[2]STORM Energy GmbH, Ludwig-Feuerbach-Str. 69, 90489 Nürnberg, Germany
Telephone +49.571.8385-184, Email: gbehrens@fh-bielefeld.de

ABSTRACT: The focus of this work is to demonstrate the advantage of collecting modul based PV-power system monitoring data with modern communication technology and the intelligent data analysis algorithms of computer science. Prediction and recognition of different faults in large PV-arrays are very important for the effective operation of PV-plants. Different PV-Systems are monitored with a large amount of data collected. The monitoring data is combined with weather data from free internet sources. The challenge in this work is using intelligent data analysis methods of data mining technology for a fault detection of different failure types in real PV-plants. The focus is on using neural network classifier for failure types of partial shading, defect bypass diodes, hailstorm damage and mechanical damage.
Keywords: Monitoring, Performance, PV-Modules, Failure Classification, Data Mining

1 PURPOSE OF THE WORK

Since PV became a significant part of world energy production, more effective monitoring is needed to ensure reliability and a fast and precise method of detecting faults in PV fields. Some photovoltaic systems are equipped with sensors for measuring voltage, power, irradiance and partially module temperature, like it is done by SunSniffer®-Sensor which is used and described by Kilper et al in [1]. Sensor data is normally sent to server collecting all information. This data can be stored and analyzed with the aim to predict module performance [2] and to detect malfunctioning systems automatically. Data Mining Technologies are able to analyze huge amount of time series of monitoring data to recognize and predict faults on the base of large data amounts taken by monitoring PV-plants, what was shown from Braun et al [3]. Marion et al [4] provides a mathematical model of PV-modules performance prediction and tried for the first time to apply Machine Learning algorithms on this field. Based on time series of module based data taken under field conditions by SunSniffer®-Sensor a recognition of failures caused by partial shading effects in PV-fields is possible with high accuracy, as we had shown in our previous work [5]. This work is focused on Data Mining Methods using free WEKA-libraries from University of Waikato [6] with the aim to distinguish between different typical failure types on PV-modules in field. We combined module based monitoring data from SunSniffer®-Sensor with time series of weather data from free web-based interface Forecast.io [http://forecast.io]. Based on the extended information in consolidated long term data series we developed a new data mining application, which is able to classify three different failure types caused by: defect bypass diode, by hailstorm and by simulated mechanical destruction.

2 APPROACH

Typical Data Mining Projects consist of for steps: **Data Collection**, **Data Preprocessing**, **Data Analysis** and **Data Classification**.
In the first step of **Data Collection** measurements are done with sensors and data sets are collected from other available sources. In our application we wrote a web based data interface to data server of SunSniffer®-measurements from the firm StormEnergy. We took data from all modules of three PV-plants over a year. In two of the PV-plants failures on PV-modules like shadowing and defect bypass diodes are known. In one special PV-plant for testing (as shown in figure 1) we could provide some experiments for failure simulation.

Figure 1: Test-site for Failure simulation on PV-modules in Isseroda, Germany. PV- modules are from Scheuten on the top (1S-10S) and PV-moduls from BP Solar (1BP-10BP). All modules are equipped with SunSniffer®-Sensors.

We destroyed on two modules the bypass diodes, damaged one module like a hailstorm, the other like a man went for a walk over the PV-modul by accident. The other PV-moduls worked very well.
The SunSniffer®-sensors (figure 2) measure every 10 minutes for every module the module voltage and the module temperature. Also String current values are collected by SunSniffer®- collection box and sent to data server.

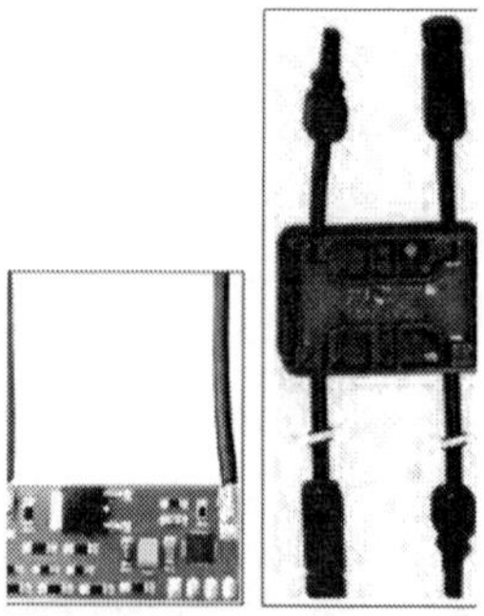

Figure 2: SunSniffer®-Sensor on left side and Sunsniffer®-Retrofit on the right side, as it was mounted on PV modules of experimental test site in Isseroda.

Additionally we imported the values forecast.io like temperature, cloud covering, timestamp of sunrise and sunset and solar efficiency. Value of solar efficiency is calculated as the arithmetic product of the time between sunrise and sunset multiplied with the difference of one minus cloud covering. Forecast.io obtains weather data beginning with the day of 01.01. 1970 and provides up to 1000 answers for free a day. The request URL is shown in figure 3.

Forecast.io

HTTP-GET Request
Request-URL:
https://api.forecast.io/forecast/APIKEY/LATITUDE,LONGITUDE,UNIXTIMESTAMP

- APIKEY = User Verification
- LATITUDE = degree of latitude
- LONGITUE = degree of longitude
- UNIXTISTAMP = time in seconds since 01.01.1970

HTTP Response:
JSON-Object with weather data of one day in the past.

Figure 3: HTTP Request and response for weather data interface from Forecast.io

All Data of long term measurement are stored in Universal Solar database of the Solar Computing Lab as described in [7].
Data collection in step two means also data consolidation from different sources. It was done by database programming an SQL-question for matching the time stamps of measurement series approximately. Generally it is important in Data mining to collect all possible data, which could have some information about the failure events on PV-modules, we want to classify in automatized applications. Data Mining Applications always can give only results as good, as the learning data sets include information about classification aims.
In the second step of **Data Preprocessing** we filtered out data outliers, implemented smoothing algorithms for new calculation of empty and incorrect measuring values, and we normalized all data types into comparable accuracy and value ranges.
For example temperature measurements obtained outliers of -273°C, which were filtered out and replaced by arithmetic middle of the next neighbors with "normal" values (figure 4):

$$x_f^{(i)} = \frac{x_{f-1}^{(i)} + x_{f+1}^{(i)}}{2}$$

Figure 4: Interpolation by using next neighbors

Results of interpolation of module temperatur values are shown in figure 5.

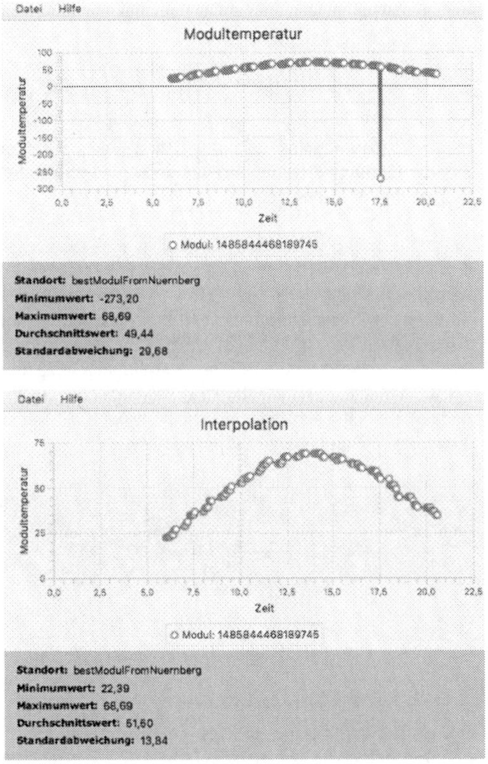

Figure 5: Modul temperature values before (upper chart) and after interpolation operations (lower chart).

Normalization we provided in an easy way by calculation the reciprocal of every value type and received an equal value range of [0.0 , 1.0]. For example modul temperature values after normalization are shown in figure 6.

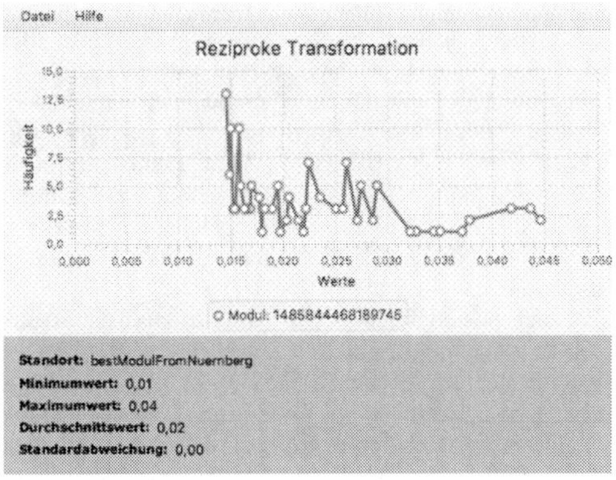

Figure 6: chart of normalized module temperature values

In third step of **Data Analysis** data sets will formed in different feature vector sets, which will be analyzed and optimized.
Feature vectors will be analyzed with focus on statistical correlations between different data types and clusters of data types with statistical correlations between their values with the aim to find out from the measurements the significant data types for failure classification. Values

of correlation matrix between all value types are calculated as shown in figure 7:

$$c_{ij} = \frac{1}{n-1}\sum_{k=1}^{n}\left(x_k^{(i)} - \bar{x}^{(i)}\right)\left(x_k^{(j)} - \bar{x}^{(j)}\right), \quad mit \quad i,j = 1,..,p$$

Figure 7: Correlation matrix calculation function, where n is the number of feature vectors and p the number of components of feature vector.

For example the correlation between modul temperatur and weather temperature was the highest with an value over the full data sets by 82,02. The twice biggest correlation we calculated between modul temperatur and string current with a value of 43,98. The third biggest correlation was found between modul temperatur and module voltage. A chart showing correlation between modul temperatur and string current visualized for an exemplary PV-modul is given in figure 8.

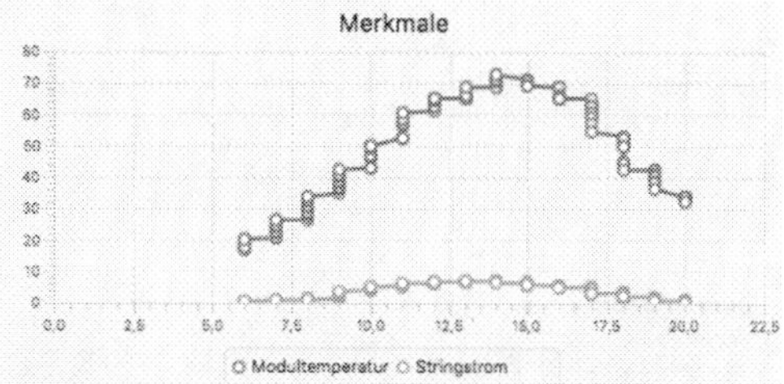

Figure 8: Chart with correlation example of values for module temperature and string currency.

The resulting feature vector of our application after correlation optimization consist of the six following components: modul temperature, modul voltage, string current, weather temperature, solar efficiency and cloud covering.

In the fourth step of **Data Classification** the data set was labeled for different failure event types: no failure, defect bypass diode, hailstorm, simulated mechanical destruction. Different machine learning algorithms from WEKA-Library [6] are applied and parameterized in a huge amount of calculating experiments on the manually labeled data sets. Best results we got on neural network classifier, which we parameterized experimentally in the parameters of the epochs, the momentum, the learning rate and the ratio of the number of feature vectors in training data dividing by the number of feature vectors of the testing data sets.

We investigated mainly three classification experiments. The first one trained into four failure types (top reference module, defect bypass diode, hailstorm, mechanical destroyed by walking over), the second one was trained for the two classes (top module, all defect module types) and the third experiment trained on the four failure types, but feature vectors only for good sunny conditions (string current > 5,9 A) were taken into training data set. All parameters of neural network classifier were permutated in their value ranges and classification results were written in a data file. Afterwards the data files were analyzed for getting the best parameter set. For example the first experiment the best parameter sets are given in table 1, the corresponding classification accuracies are given in figure 9. The classification results for third experiment are exemplarily given in figure 10. A accuracy of classification by 96% could be achieved.

Table 1: Neural Network parametrization values for best classification results of a classification experiment in all four different failure types

Right classified	Epochs	Momen-tum	Learning Rate	Training Data Ratio
75,21%	572	0,6	0,9	87%
75,15%	695	0,4	1,0	85%

```
=== Confusion Matrix ===

  a   b   c   d    <-- classified as
194   4  14   1 |  a = Destroyed
 70 199  13   0 |  b = Topreference
 44   0  18   1 |  c = Hailstormed
  0   0   0  35 |  d = Defect diode
```

```
=== Confusion Matrix ===

  a   b   c   d    <-- classified as
235   4   2   0 |  a = Destroyed
101 230   0   0 |  b = Topreference
 56   6   9   1 |  c = Hailstormed
  0   0   0  40 |  d = Defect diode
```

Figure 9: Classification results for the neural network classifier trained on four different failure types.

The classification results for third experiment are exemplarily given in table 2 and figure 10. An accuracy of classification by 96% could be achieved, if the conditions for training data sets with 66% are realistic to adopt in an real PV-plant monitoring.

Table 2: Neural Network parametrization values for best classification results of a classification in all four different failure types by taking only sunny conditions under account.

Right classified	Epochs	Momen-tum	Learning Rate	Training Data Ratio
100%	584	0,1	0,1	98%
96,88%	69	0,2	0,5	66%

```
=== Confusion Matrix ===

a b c  <-- classified as
2 0 0 | a = Topreference
0 0 0 | b = Destroyed
0 0 2 | c = Defect diode
```

```
=== Confusion Matrix ===

 a  b  c   <-- classified as
27  1  0 | a = Topreference
 1 17  0 | b = Destroyed
 0  0 18 | c = Defect diode
```

Figure 10: Classification results on four different failure types by taking only sunny conditions into calculation.

3 SCIENTIFIC INNOVATION AND RELEVANCE

The scientific innovation of our work is that our application provides Data Mining Algorithms based on a huge amount of big data combined from modul based monitoring data collected with SunSniffer®-Sensor and webbased source of weather data. The software application is easy to adopt to any monitoring system with the possibility of data collection and failure labeling. The recognition rates of our data mining algorithm process for four different tested failure types are high with 96% accuracy. We could show, that it is possible to recognize different failure types in PV-plants in fully automatic application based on different data types.

4 RESULTS AND CONCLUSIONS

Modul based monitoring of PV-modules by SunSiffer®-Sensor gives the possibility to recognize different failure types like shadowing, defect bypass diodes, hailstorm damage and mechanical damage in PV-Plants. The implementation of Data Mining technology can realize it. The results would be better, than more labeled failure types the algorithm could use for training. On very huge training data sets we can apply deep learning algorithms, which can achieve recognition rates of failure types of nearly 100%.

5 REFERENCES

[1] T. Kilper, I. Kruse, C. Feser, U. Kirstein, D. Peters, M. Tapia, K. Maydell, S. Yilmaz, G. Behrens "A new generation of PV Monitoring system with high-grade remote diagnostics based on module level monitoring and integrated yield simulation" Proceedings of 32th European Photovoltaic Solar Energy Conference and Exhibition, Hamburg 2015

[2] Herbort, Brecht, te Heesen. Insolation dependent solar module performance evaluation from PV monitoring data. Proceedings of 27th European Photovoltaic Solar Energy Conference and Exhibition, Frankfurt 2012, pp.4154-4157

[3] H. Braun, S.T. Buddha, V. Krishnan, A. Spanias, C. Tepedelenlioglu, T. Yeider and T.Takehara. "Signal processing for fault detection in photovoltaic arrays." In 37th IEEE International Conference on Acoustics, Speech and Signal Processing, pages 1681-1682, 2012

[4] B. Marion. "Comparison of predictive models for PV module performance." In 33rd IEEE Photovoltaic Specialist Conference, pages 1-6, 2008

[5] G. Behrens, A. Domnik, S. Hempelmann, S. Yilmaz, I. Kruse "Maschine Learning Methods for partial Shading Detection in Monitoring Data on PV-System", Proceedings of 31th European Photovoltaic Solar Energy Conference and Exhibition, Amsterdam 2014

[6] I.H. Witten, E. Frank, M. A. Hall "Data Mining: Practical Machine Learning Tools and Techniques" Elsevier Inc. 2011

[7] G. Behrens, F. Hamelmann, L. Niemann, J. Weicht, S. Yilmaz, I. Kruse „Universal location based data sets for monitoring and scientific data analysis on PV-systems", Proceedings of 30th European Photovoltaic Solar Energy Conference and Exhibition, Paris 2013

ASSESSMENT FOR IR INSPECTION CYCLES AND EFFORTS RELATED TO SYSTEM DESIGN

Adrian Häring [1], Tobias Henne [1], Stefan Dobler [2]
[1] SolarEdge Technologies GmbH, [2] DOSOL GmbH&Co KG
Werner-Eckert-Str. 6, D-81829 Munich, Germany, Adrian.Haering@SolarEdge.com

ABSTRACT: Operation and maintenance of PV systems is getting more and more important. IR thermography is getting a standard maintenance measure. However, there is no difference in procedures depending to system design. But depending to the existence of string-level or module-level monitoring, inspections cycles and efforts in procedure can be reduced. The more detailed the monitoring is, the less important are short cycles for running a detailed IR inspections. An assessment is shown how IR inspections could be done considering system design. Reduced costs for operation and maintenance by saving costs for thermography can be achieved without loss in safety or yield.
Keywords: Monitoring, System Performance, IR, Operation, Thermography

1 MOTIVATION

IR thermography is used in PV widely. There is lot of scientific work to improve methods for detection of failures creating yield loss, also presented at the 31st EU PVSEC in a dedicated Infrared Imaging session (5CO.15, [1]). All these contributions show how to improve measurements and try to get as much information from the pictures to see effects having impact to performance. But there is no discussion yet about taking existing measurement into account and adapting the procedures accordingly. At one presentation an IR image was shown, including the question, if anyone has an idea, what this means for the power output of these PV modules (Fig. 1).

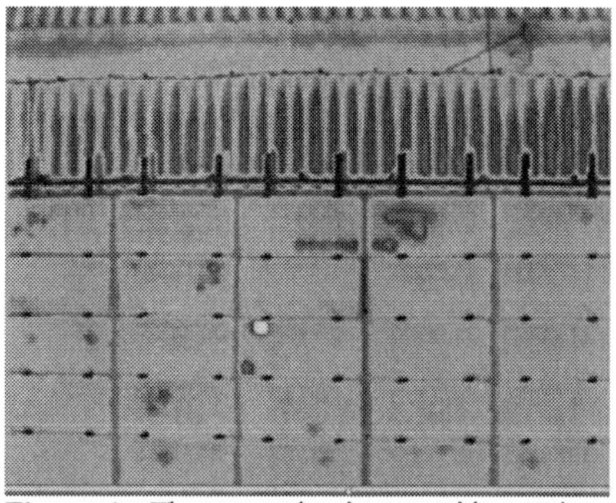

Figure 1: Thermography image with suspicous spots (Source: EU PVSEC 2015; 5CO.15.3, C. Buerhop et al)

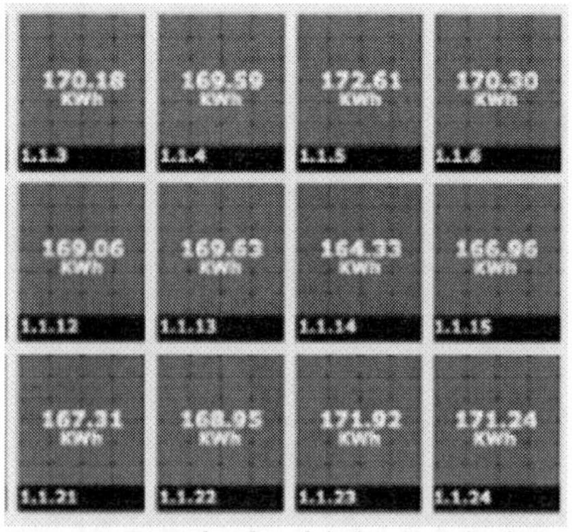

Figure 2: Example of module-level monitoring

If you have module-level monitoring, you would exactly know what this would mean for the power output of each PV module (Fig. 2). You can compare the IR image with the power output of the PV modules. If you have such a monitoring system you can see, that there is not the same need for IR inspections to detect performance problems. The monitoring level has a major impact in defining inspection cycles and inspection approaches. The purpose of this work is to show how IR inspection cycles and efforts can be adapted to the existing monitoring infrastructure and other influences.

2 REASONS FOR THERMOGRAPHY INSPECTION

The first important reason for IR inspections is to find fire risks. The main part (but not all) fire risks show up as a hot spot long time before they turn into a fire. Due to the high temperature difference, they can be detected quite easy. In the upcoming thermography standard IEC 62446-3 TS [2], this kind of inspection is considered as simplified thermographic inspection. The requirements for this inspection are quite low. You need good irradiance so the critical locations get heated, but compared to a performance related inspection, you reduce efforts a lot (correct angle, detail level of camera, clouds and others). In spite of the lower efforts, the benefit regarding risks is high. The fire risks is at the same time a high financial risk. You can lose up to 100% of your PV system and in addition affect the surrounding, e.g. the whole building where the system is mounted. This simplified inspection with lower efforts has a high benefit from a financial point of view.

The second reason for IR inspections is to find performance issues at PV modules. Especially in big PV systems, monitoring efforts are reduced to save costs. To limit the risks of having losses to due low performing PV modules IR inspections are performed to detect such modules. In the draft of IEC 62446-3 TS, this kind of inspection is considered as detailed thermographic inspection and analysis. The get IR images which really help to detect performance issues you have to spend a lot efforts. You need e.g. a good camera (high resolution, showing small temperature differences), good weather conditions with no or almost no clouds and a suitable angle to the PV modules (which often means use of lifters or UAVs). The financial risk is lower, compared to the fire risk. There is a risk of undetected performance issues, but high impacts will be detected by the overall plant performance. So here, it will be a financial tradeoff between the detail level of the monitoring system and the inspection efforts (cycles and costs).

It is obvious that deviation between these two inspection types is necessary to define reasonable inspection cycles. In the current draft of the thermography standard, there is a recommendation for 4 years, but without reference if this applies for the simplified or the detailed inspection. Further, there is written, that applied intervals should be agreed upon with the owner/operator. But how can you fix such an interval. Which influences have to be considered?

3 INFLUENCES TO INSPECTION INTERVAL

3.1 Detail level of monitoring

As already mentioned the detail level of the monitoring system has a major impact to inspection intervals. In PV systems, you can find monitoring levels from module level (e.g. 250W), string level (e.g. 5kW, string inverter) and sub-array level (e.g. 60kW, inverter or combiner). In Table 1, you can see what the loss of a module or a quarter of the module power does mean in power difference depending to the monitoring level.

Table I: Relative power differences

	250W	5kW	60kW
Loss of 1 PV module	100%	5%	0,42%
Loss of 25% of a module	25%	1,25%	0,1%

In practice at field measurements, it has hard to find differences below 5% due to measurement accuracy and the variation in power of the single modules. Depending to the monitoring level there is a higher of lower impact to the risk of undetected issues. The longer a problem remains undetected, the higher is the financial loss. So a high low monitoring level (e.g. sub-array level) will shorten inspection intervals, a high level (as module-level) will extend inspection cycles.

3.2 Efforts/costs for thermography inspections

Costs for the thermography itself have also a big impact. The lower the relative costs are, the more often you can afford. Here the most relevant aspect is the system size. The smaller the size of the PV array is, the higher are the specific costs. That is one of the reasons, that detailed inspections are common at big PV systems. It depends also to the accessibility of the PV system. E.g., ground-mounted systems are easier to access. A further aspect is the kind of inspection. Due to lower efforts simplifies inspection can be done easier and so they are cheaper. Therefore, you can afford to have more often simplified inspections.

Regarding the costs for IR inspections, the inspection interval can be decreased, if the inspection efforts are low. If there are high inspection costs, the interval has to be increased to get a reasonable tradeoff between risks and costs.

3.3 Accepted risk level

Another influence is the accepted risk level. Here one important aspect is the risk assessment for the PV modules. Do you only consider some single faults or do you also consider faults considering the whole system, e.g. as PID. It is also a difference if you use new PV modules, where you have no experience. In case of using PV modules, where you have good experience, your risks are lower and you can think about extending the IR inspection cycle. In addition, the warranty conditions may influence the cycles or a special timing.

Here it depends also to the readiness to assume risks by the investor. A PV system is a financial invest. Some of the investors accept higher risks to get a better return rate. Depending to the overall accepted risk level you can adapt the inspection interval, e.g. if you are willingly to accept higher risks, you can extend the interval and reduce costs for thermography.

3.4 Fire risk

As already mentioned at the reasons for thermography, inspections for reducing fire risks are very important. They needed inspections are cheaper and they cover a high financial risk at the same time. However, PV systems are installed in different surroundings and so there are different needs.

First aspect is what kind of materials are around the PV system. How easy can they catch fire? Ground mounted systems typically have low fire risks at all, in some regions bush fires have to be considered. Also at systems and buildings, there are differences. In case of fire- and arc-resistant materials as e.g. stone tiles, the fire risk for the surrounding is still low. However, in case of bitumen roofs or flat roofs with polystyrene insulation, there is a high fire risk. As soon these materials catch fire, you have a very fast spread of the fire including a high risk to lose the whole building and all facilities inside.

Further aspect is the use of the building. In case of a warehouse or solar carports, you may accept higher fire risks. Having critical buildings such as schools or hospital as well as facilities, where stopping production creates huge costs, you have to keep the fire risk as low as possible. Having short inspection cycles will reduce the fire risk.

In addition, additional preventive measures have an impact here. If you have additional measures, which reduce the risk, you may extend your inspection intervals. One typical example is, if there is an arc-fault detector. This detector can take out many fire risks. Also the module-level monitoring reduce the fire risk by detecting some of the critical cases early by detecting the power loss of a hot spot at the faulty module.

Regarding fire risks a high fire risk will lead to shorter intervals, lower risks e.g. due to additional measures or non-critical surroundings as given for ground mounted systems may allow longer intervals.

Looking at these influences, you can see that the assessment depends to plant-specific conditions (Fig. 3). At the overview, you can see the influences to the determination of a reasonable inspection interval

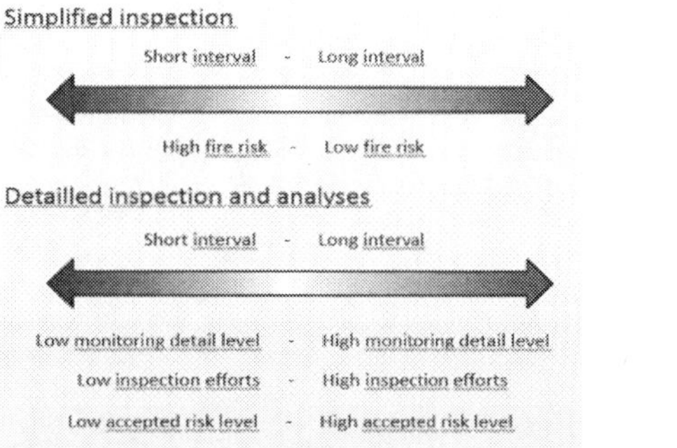

Figure 3: Overview of influences and impact to intervals

4 EXAMPLES FOR ASSESMENT

In the following an example for the assessment of the IR inspection cycle is performed. The intention is not to give an overall value for the time in between, but to show have the determination can be done and what conditions may change this interval.

It is assumed to have a PV system as shown in Fig. 4. The PV array power is 60kWp, connected to two inverters with 3 MPPT-Tacker each. There are two strings with each 20 PV modules. The surrounding is a flat-roof with polystyrene insulation. The building is a commercial building, where the facilities inside are a additional risk regarding stopping operation. The owner is a professional investor. The accessibility is good.

Figure 4: Roof-mounted PV system

The assessment could look like this:

- Regarding the monitoring level, you have 2-string monitoring. This means 2,5% power difference, in case of losing the power of a whole module, or 0,6% power difference, in case of losing 25% of a module. Compared to a 55 gap, which you probably would detect, you can see that there is a risk of having low performing modules.
- The system is good accessible and not to small, so you can maybe afford to have some IR inspections. Moreover, a professional investor may pay for this.
- The overall fire risk is high. The roofing material is critical and stopping the operation of the facilities inside will not only mean costs for losing the building and the PV system, but also high additional costs for losses of the company inside the building.

The outcome of such an assessment could be that a simplified thermographic inspection is done every two years, to keep the fire risk as small as possible. In addition, there is a risk to have undetected low performing PV modules. A possible interval could be 4 years (depending mainly to the costs for the inspection).

However, what would happen, if the system designer has decided, to use another system technology at the same on-site specific conditions? This may change the determination of the intervals. Looking at the total cost of ownership this could be an interesting option. In the following, the assessment is done again using another system technology:

- A module-level monitoring is used. You can see the power differences for each PV module. Each low performing PV module will be detected early by continuous monitoring
- Additional fire prevention measures are part of the system. The use of a serial arc-fault detector takes out the main part of the fire risk. In addition the module-level monitoring will decrease the risk in addition by detection some of the hot spots in an early stage, allowing to repair the system before a fire is caused.

The outcome of the assessment with the changed conditions by system technology could be as follows. Due to decreased fire risks, you can consider a simplified inspection each 5 years. There is no risk for low performing modules. Therefore, there is no need for a detailed inspection. If you want to be sure, not to have very homogeneous faults with a slight power loss of less than 10%, a detailed inspection after 10 years could be worth to consider.

5 CONCLUSION

Many people are making no deviation between simplified and detailed thermographic inspection today. Looking at the needs it is obvious, that a deviation between these both kinds is necessary, including different inspection cycles.

The investigation has shown that the plant-specific conditions have a major influence to IR inspection intervals. There is no reasonable one-fits-all interval for IR inspections. The best solution would be, also to add this assessment to the considerations for the selection of the site of a PV system. However, typically there is no influence or only small influence to the site selection, so most of the site-specific conditions are given. Regarding system technology, the designer has the opportunity to compare different system technologies to find an optimum for each plant. Especially at residential and commercial systems, where detailed inspections are not affordable due to the system size, using system technologies with additional functions can be a better choice. By accepting maybe higher initial costs, the total cost of ownership may be lower due to lower inspection costs.

Making plant-specific assessments for the intervals of IR inspections instead of fixed overall intervals, allows finding the optimum between risks of the plant and costs for IR inspections.

4 REFERENCES

[1] Program EU PVSEC 2015, 5CO.15: Infrared Imaging

5CO.15.1; Eva Schubert et al. "Standardization of infrared imaging of photovoltaic plants in operation"
5CO.15.2; Ioannis A. Tsanakas et al. "Fault diagnosis and classification of large-scale PV plants through arial orthophoto thermal mapping"
5CO.15.3; Claudia Buerhop et al. "Characterization of defects in PV modules by their temperature development using IR-thermography"
5CO.15.4; Manuel Dalsass et al. "Correlation between the monitoring data of a photovoltaic power plant and module defects detected by drone-mounted thermography"
5CO.15.5; Giuseppe Leotta et al. "Post processing technique for thermo-graphic images provided by drone inspections"
5CO.15.6; Manuel Lanz et al. "Infrared (IR) drone for quick and cheap PV inspection"

[2] New IEC standard project: IEC 62446-3 TS Ed.1: Grid connected photovoltaic (PV) systems – Part 3: Outdoor infrared thermography of photovoltaic modules and plants

ESTIMATION OF THE DEGRADATION RATE OF FIELDED PHOTOVOLTAIC ARRAYS IN THE PRESENCE OF MEASUREMENT OUTAGES

Alexander Phinikarides, George Makrides and George E. Georghiou
Photovoltaic Technology Laboratory, FOSS Research Centre for Sustainable Energy,
Department of Electrical and Computer Engineering, University of Cyprus.
1, Panepistimiou Avenue, 2109 Aglantzia, Nicosia, Cyprus.
(Telephone: +357 22 894398, Fax: +357 22 895370, email: phinikarides.alexander@ucy.ac.cy)

ABSTRACT: This paper presents a study of the sensitivity of the estimated energy degradation rates of grid-connected crystalline Silicon (c-Si) PV arrays to measurement outages and missing data. The arrays under study have been operating side-by-side since June 2006 in Nicosia, Cyprus. Fifteen-minute average field measurements over their first ten years of operation were used to create a data set, in which invalid measurements were removed and logged system downtimes were corrected based on past performance. The resulting data set was randomly sampled to select data points which were designated as Missing Completely at Random (MCAR) in order to create unbiased artificial outage periods. The same data set was treated with missing data imputation techniques such as imputation by the mean, linear interpolation and imputation by bootstrap. The resulting data sets were then analysed with linear regression (LR), classical seasonal decomposition (CSD) and regARIMA models to extract the trend and estimate the energy degradation rates. The results have shown that regARIMA was the most robust method for up to 10 % of missing data, the LR and CSD methods were sensitive starting from 2 % of missing data and that imputation by the bootstrap enhanced the accuracy of the estimated degradation rates up to 20 % for regARIMA and 40 % for LR.
Keywords: degradation, photovoltaic, performance, monitoring, outages, imputation.

1 INTRODUCTION

A major requirement for the uptake of PV is assurance of the lifetime energy yield and formulation of strict warranties in order to reduce investment risk. This requires accurate evaluation of the energy degradation rate, R_{DE}, of fielded PV systems [1]. The term energy degradation rate includes shading and soiling effects, module mismatches and module degradation and is a positive quantity. Factors such as length of field exposure, module technology, measurement accuracy and precision and measurement outages all affect the degradation rate estimates [2].

Measurement outages cause missing measurements and incomplete data sets. The outages can be due to data logger faults, measurement noise, data transmission errors, connection errors and data storage faults. Missing data is also caused by discarding invalid measurements and outliers due to inverter clipping, misconfigured sensors, sensor drifts and invalid calibration, array or irradiance sensor shading and soiling and PV system downtimes. Additionally, outliers can be determined from abrupt changes and high-frequency content in the measurements and also from highly volatile measurements (higher than the response time of the sensors.)

As the energy degradation rate is estimated with statistical methods, a multitude of statistical analyses were developed to test the sensitivity and the bias introduced to the resulting R_{DE} estimated from data sets with missing measurements. In the presence of measurement outages, these incomplete data sets can bias the R_{DE} estimate, either by exacerbating the effect of the outage factors when the missing data point is of an instance of optimal operation, or by overestimating PV system performance and therefore underestimating degradation, when the missing data point is of a period of increased system losses.

2 METHODOLOGY

2.1 Experimental setup

At the outdoor PV Technology test site of the University of Cyprus (UCY), grid-connected PV systems of different technologies and approximately 1 kWp capacity each, were installed and commissioned in June 2006. The performance of each PV system and the prevailing meteorological conditions were recorded according to IEC 61724 [3] at a resolution of one second and stored as fifteen-minute averages. The meteorological measurements include the global irradiance on the plane of array (POA), G_I, wind direction, α_W, and wind speed, S_W, as well as ambient, T_{am}, and module temperature, T_m. The electrical measurements include maximum power point (MPP) DC current of the array, I_A, voltage, V_A, and power, P_A, and AC power to the utility grid, P_{TU}.

2.2 Data sets of observed measurements

Measurements of the DC array power at MPP, P_A, of three different c-Si PV systems and global irradiance on the POA, G_I, from a Kipp & Zonen CM21 pyranometer were selected for the first ten years of operation (June 2006 – May 2016). The long testing period ensured that the estimated energy degradation rates represented their long-term operation and were not skewed by reversible degradation factors [4].

Invalid measurements and outliers were removed from the P_A and G_I time series by preset thresholds and maintenance logs were used to correct periods of system and sensor downtimes. The downtimes were corrected using past measurements for short outage periods (less than one day). For longer periods, the outages were not corrected to avoid introducing bias. Calibration factors were also applied to the P_A and G_I measurements. The resulting data sets were defined as the complete data sets of observed values, D^{obs}.

The complete data sets were used to create time series of monthly Performance Ratio, PR [3], as shown in Figure 1. The PR plots show the typical seasonal profile of c-Si PV systems, with high *PR* in the winter and lower in the summer, due to higher module temperatures.

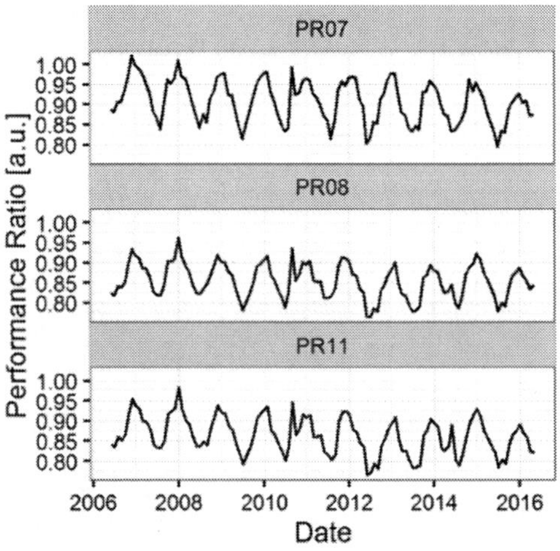

Figure 1: Monthly Performance Ratio (PR) of the three c-Si PV arrays under study from June 2006 to May 2016

2.3 Generation of artificial missing data

A Monte Carlo approach was employed, in order to create data sets with artificial missing data from the complete data set of observed values, D^{obs}, for each PV system. The index of the complete data set was randomly sampled at different levels in order to select artificial outage periods. Random sampling was performed from 1 to 40 % of the total amount of data in the complete data set. The data points, M, were then designated as missing from the complete data sets. The resulting incomplete data sets contained instances where all variates were missing at once (case deletion), to simulate multiple failures in the whole measurement chain.

Since random sampling was utilized, an unbiased set of data points was selected. This method represents the case where data is Missing Completely at Random (MCAR) [5] and the distribution of missing data points did not depend on either the observed values or the missing values (see Eq. 1.)

$$P(M \mid D^{obs}) = P(M) \qquad (1)$$

Lastly, as with the case of the complete data, the incomplete data sets were used to create monthly *PR* time series for each percentage of missing data. In total, forty incomplete data sets were created for each of the PV systems in Figure 1.

2.4 Imputation of missing data

As the estimation of the degradation rate relies on statistical analysis, the statistical properties of the time series must be retained without introducing bias.

When using PV models such as PVUSA [6], single-point efficiency or others [7] to interpolate missing values, all the required explanatory variables may not always be available and may not always be measured (irradiance, module temperature, wind speed, humidity.) Additionally, since data logger, data transmission and storage related errors affect all measured variables, PV models were not considered in this study. Instead, the handling of missing data was based on univariate imputation of missing G_I and P_A data points, where the basic assumptions were that the variables were normally distributed and that the pattern of missing data was independent of the underlying data set.

The missing data points were imputed with a) the unconditional mean of the variable, b) linear interpolation and, c) bootstrapping in order to fill in the gaps in the fifteen-minute data sets.

Firstly, the missing data points were imputed by the unconditional mean of the data set. This ensured that the probability distribution of the measurement variable remained the same, but at the expense of large errors due to the seasonal profile of P_A and G_I.

Similarly, the same missing data points were imputed by linear interpolation with time, in order to better model the effect of the trend present in the complete data set (see Eq. 2.)

$$P_A = \beta_1 + \beta_2 t + \varepsilon \qquad (2)$$

where β_1 and β_2 are regression coefficients, t is the time and ε are the residuals.

2.5 Imputation by the bootstrap

In addition to imputation by the mean and linear interpolation, the bootstrap method was employed in order to use as much information from the probability distribution of each variable and be able to assess the uncertainty of the imputed values [8]. The bootstrap method relies on sampling from the posterior distributions and replacing missing data points with the sampled values. The algorithm then re-evaluates and resamples the posterior distribution, and replaces the missing data points with the newly sampled values. This iterative procedure is performed until the Expectation Maximization algorithm [9] converges. The result is that the missing values are filled in with a distribution of imputations that reflect the uncertainty about the missing data, for each data point.

The energy degradation rate of PV arrays was observed to vary smoothly over continuous exposure time [10], either linearly or with inflection points. In order to incorporate linear degradation in the bootstrap algorithm, the effect of outdoor exposure time was added by building a general model of variables across time by creating a sequence of polynomials of the time index [8]. In this case, observed values close in time to any missing data had the largest contribution to the imputation of the missing data point. The P_A data sets for each PV system were therefore described as a first-order polynomial of time (see Eq. 3):

$$P_A \propto \beta_1 t \qquad (3)$$

The algorithm was thus modified so that the imputed data sets, constructed from each run of the bootstrap algorithm, were regressed on Eq. 3 and the missing data points interpolated based on this linear relationship.

2.6 Estimation of the energy degradation rate

A large number of data sets were created by applying the methodology described in the previous sections. For each PV system, 161 different data sets were analysed, in order to estimate the energy degradation rate: a) one complete data set, b) forty incomplete data sets with 1-40 % of missing data, c) forty data sets imputed by the mean, d) forty data sets imputed by linear interpolation and, e) forty data sets imputed by the bootstrap.

Each data set of G_I and P_A values was used to create monthly *PR* time series which were then analysed with different statistical methods over the first ten years of field exposure [11], [12]. The *PR* time series of each PV array was seasonally adjusted with Regression models with ARIMA errors (regARIMA) [13], [14] and classical seasonal decomposition (CSD) in order to separate it into the trend, the seasonal and the irregular components and was also modelled with linear regression (LR) using ordinary least squares.

CSD was used to fit an additive model on the data and separate the trend, *T*, the seasonal, *S*, and the residual, *R*, components (see Eq. 4):

$$Y_t = T_t + S_t + R_t \quad (4)$$

regARIMA models were applied on the time series through the X-12-ARIMA algorithm. These model the mean of the time series by a linear combination of regressors and the residual component by an ARIMA process (see Eq. 5):

$$Y_t = \sum_i \beta_i x_{it} + z_t \quad (5)$$

where the residual component, z_t, is modelled by the seasonal ARIMA model given in Eq. 6:

$$\Phi(T)\Phi_S(T_S)\nabla^d \nabla_S^D z_t = \theta(T)\theta_S(T_S)e_t \quad (6)$$

Lastly, the completed data sets were used to estimate the linear energy degradation rate, R_{DE}, using linear regression (see Eq. 7):

$$R_{DE} = -\alpha * 12 * 100\% \quad (7)$$

where α is the slope of the linear equation. The monthly R_{DE} was multiplied by 12 to convert it to an annual rate.

3 RESULTS

The results of the analysis for 1-40 % of sampled data points missing completely at random are shown in Fig. 2, Fig. 3 and Fig. 4 for the LR, CSD and regARIMA methods respectively. It can be seen that imputation by the mean and linear interpolation performed very poorly for all PV systems as these two methods underestimated the R_{DE} consistently with increasing amount of missing data. On the other hand, imputation by the bootstrap has been shown to provide a large improvement to all three degradation estimation methods. Therefore, the bootstrap method provided robustness to the estimation, regardless of the degradation estimation method employed. For LR, the bootstrap provided robustness for up to 20 % of missing data for the ucy07 and ucy11 systems, whereas for the ucy08 system, the estimates were stable even at 40 % of missing data.

The results of applying CSD were shown to be very sensitive to the amount of missing data and whether imputation was used or not. Without using imputation, the estimates were underestimated by 0.22 %/year for the ucy07 system and overestimated by 0.05 %/year for the ucy08 and ucy11 systems. By using the bootstrap, the under/overestimation was reduced to 0.05 %/year for all systems.

In the case of regARIMA, the results were shown to be the most robust to missing data for all systems, regardless of whether imputation was employed. For the ucy07 and ucy11 systems, the maximum differences without imputation were 0.05 %/year and for the ucy08 the maximum difference was 0.02 %/year. When imputed by the bootstrap, the differences were slightly reduced.

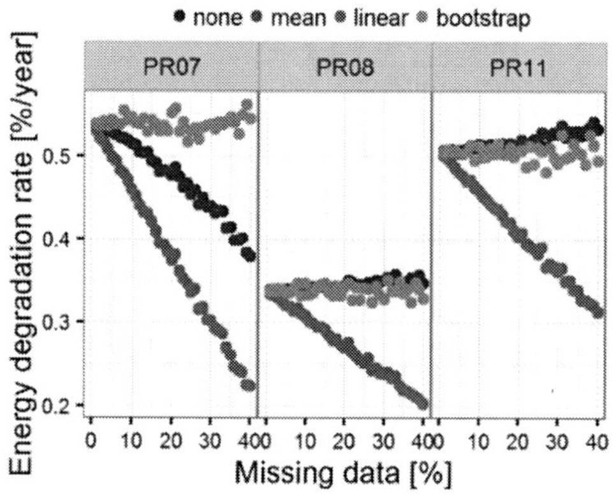

Figure 2: Energy degradation rate estimated with linear regression for 1 – 40 % missing data points and imputation by the mean, linear interpolation and bootstrap.

A comparison between degradation estimation methods with incomplete data sets, where no imputation was applied, has shown that regARIMA was the most robust, for up to 10 % of missing data for the ucy07 system and 20 % of missing data for the ucy08 and ucy11 systems. Linear regression and CSD were shown to be very sensitive to missing data, starting from 2 %.

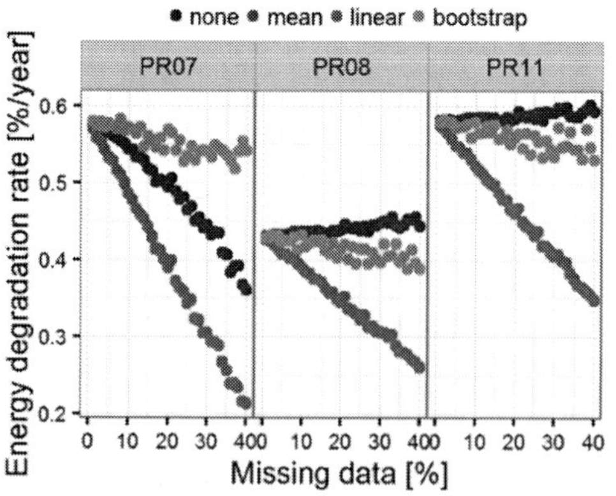

Figure 3: Energy degradation rate estimated with CSD for 1 – 40 % missing data points and imputation by the mean, linear interpolation and bootstrap.

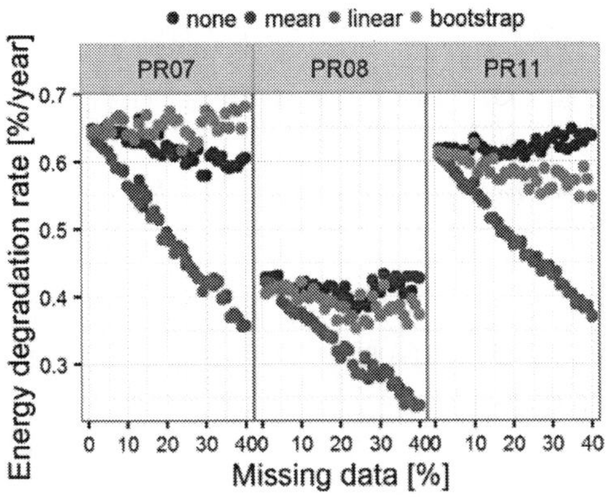

Figure 4: Energy degradation rate estimated with regARIMA for 1 – 40 % missing data points and imputation by the mean, linear interpolation and bootstrap.

4 CONCLUSIONS

The results of the investigation into the energy degradation estimation's sensitivity to missing data have shown that missing data as defined in this work greatly affected the estimation of degradation rates

To mitigate the effects of missing data, imputation by the bootstrap was shown to be the most successful in providing robust energy degradation rate estimates. In contrast, imputation by the mean or linear interpolation were incapable of providing any improvement and resulted in worse estimates than not applying any imputation.

Finally, the application of statistical methods to estimate the degradation rate was also shown to be sensitive to the amount of missing data. regARIMA was the most robust method, whether imputation was applied or not. LR was more robust when missing values were imputed by the bootstrap and CSD was the most sensitive method to missing data.

5 ACKNOWLEDGEMENTS

This work was funded through the IPERMON project (KOINA/SOLAR-ERA.NET/1214/08) which was co-financed by the European Regional Development Fund and the Republic of Cyprus through the Cyprus Research Promotion Foundation (DESMI 2009-2010)

6 REFERENCES

[1] A. Phinikarides, N. Kindyni, G. Makrides, and G. E. Georghiou, "Review of photovoltaic degradation rate methodologies," *Renew. Sustain. Energy Rev.*, vol. 40, pp. 143–152, Dec. 2014.

[2] D. C. Jordan and S. R. Kurtz, "Photovoltaic Degradation Rates-an Analytical Review," *Prog. Photovoltaics Res. Appl.*, vol. 21, no. 1, pp. 12–29, Jan. 2013.

[3] IEC 61724:1998, *Photovoltaic system performance monitoring - Guidelines for measurement, data exchange and analysis*, 1st ed. Geneva, Switzerland: IEC, 1998.

[4] A. Phinikarides, N. Philippou, G. Makrides, and G. E. Georghiou, "Performance loss rates of different photovoltaic technologies after eight years of operation under warm climate conditions," in *29th EU-PVSEC*, 2014, pp. 2664–2668.

[5] J. L. Schafer and J. W. Graham, "Missing data: Our view of the state of the art.," *Psychol. Methods*, vol. 7, no. 2, pp. 147–177, 2002.

[6] A. Kimber, T. Dierauf, L. Mitchell, C. Whitaker, T. U. Townsend, J. Newmiller, D. L. King, J. E. Granata, K. Emery, C. R. Osterwald, D. R. Myers, B. Marion, A. Pligavko, A. F. Panchula, T. Levitsky, J. Forbess, and F. Talmud, "Improved test method to verify the power rating of a photovoltaic (PV) project," in *34th Annual Conference of IEEE Industrial Electronics, IECON*, 2009, pp. 316–321.

[7] G. T. Klise and J. S. Stein, "Models Used to Assess the Performance of Photovoltaic Systems," SAND2009-8258, Sandia National Laboratories, 2009.

[8] J. Honaker and G. King, "What to Do about Missing Values in Time-Series Cross-Section Data," *Am. J. Pol. Sci.*, vol. 54, no. 2, pp. 561–581, Apr. 2010.

[9] A. P. A. Dempster, N. M. N. Laird, and D. D. B. Rubin, "Maximum likelihood from incomplete data via the EM algorithm," *J. R. Stat. Soc. Ser. B Methodol.*, vol. 39, no. 1, pp. 1–38, 1977.

[10] D. C. Jordan, S. R. Kurtz, K. VanSant, and J. Newmiller, "Compendium of photovoltaic degradation rates," *Prog. Photovoltaics Res. Appl.*, 2016.

[11] A. Phinikarides, G. Makrides, B. Zinsser, M. Schubert, and G. E. Georghiou, "Analysis of photovoltaic system performance time series: seasonality and performance loss," *Renew. Energy*, vol. 77, pp. 51–63, 2015.

[12] A. Phinikarides, G. Makrides, N. Kindyni, and G. E. Georghiou, "Comparison of trend extraction methods for calculating performance loss rates of different photovoltaic technologies," in *40th IEEE PVSC*, 2014, pp. 3211–3215.

[13] A. Phinikarides, G. Makrides, and G. E. Georghiou, "Estimation of annual performance loss rates of grid-connected photovoltaic systems using time series analysis and validation through indoor testing at standard test conditions," in *42nd IEEE PVSC*, 2015, pp. 1–5.

[14] B. C. Monsell, "The X-13ARIMA-SEATS Seasonal Adjustment Program," in *Proceedings of the Federal Committee on Statistical Methodology Research Conference*, 2007.

ECO-SOLAR FACTORY: 40% PLUS ECO-EFFICIENCY GAINS IN THE PHOTOVOLTAIC VALUE CHAIN WITH MINIMISED RESOURCE AND ENERGY CONSUMPTION BY CLOSED LOOP SYSTEMS

M.P. Bellmann[1], R. Roligheten[2], G.S. Park[3], J. Denafas[4], F. Buchholz[5], R. Einhaus[6], I. Lombardi[7], B. Ehlen[8], K. Wambach[9], P. Romero[10], A. Bollar[11]

[1]SINTEF Materials and Chemistry, Trondheim, Norway;
e-mail: Martin.Bellmann@sintef.no
[2]Steuler Solar Technology AS, Porsgrunn, Norway
[3]Norsun AS, Oslo, Norway
[4]UAB Soli Tek R&D, Vilnius, Lithuania
[5]International Solar Energy Research Center - ISC Konstanz, Konstanz, Germany
[6]Apollon Solar, Lyon, France
[7]Garbo Srl, Cerano, Italy
[8]Boukje.com Consulting BV, EM Bleiswijk, The Netherlands
[9]bifa Umweltinstitut, Augsburg, Germany
[10]Aimen Laser Application Centre, Porrino, Spain
[11]Ingesea Automation SL, Elgoibar, Spain

ABSTRACT: The European H2020-FOF-13-2015 Eco-Solar project, that started in October 2015 envisions to increase resource and energy efficiency over the entire photovoltaic value chain, while simultaneously maximising recycling and remanufacturing possibilities, by introducing design for recovery, repair and reuse, and collaborating for improvements in waste reduction. Reusing materials and reducing the consumption of raw materials will make solar cell panels both cheaper and greener. When fewer resources, including critical materials are needed, the emissions of greenhouse gases from their production will decrease. Likewise, the energy consumed by these processes will be paid off faster than it is today. All this should improve market penetration for European PV producers. The overarching aim is to strengthen European PV companies who are driven by innovation and who are able to secure Europe's power supply in a sustainable way. The paper presents the methodology followed by Eco-Solar to reach this ambitious goal.

Keywords: crystalline silicon solar cells, remanufacturing, repair, reuse, recycling, eco-efficiency analysis, life cycle assessment, life cycle costing, living lab

1 INTRODUCTION

Eco-Solar is a 3 years project, which involves 11 partners from 7 member states and administers the resources in terms of personnel, knowledge and equipment that are required for success. Details can be found on the project website: http://www.ecosolar.eu.com. It is a medium size focused research project carried out in the HORIZON 2020 EU Framework Programme.

Eco-Solar is an ambitious project. It aims to develop concepts to increase the remanufacturing potential, including repair of PV-modules and cells, while simultaneously improving reuse of resources and components and increasing resource efficiency over the whole value chain. The Eco-Solar project strives towards a PV module that has:

• High yield of resources per PV-module produced and per kWh -electricity generated by these modules thanks to a more eco-friendly Bill of Materials (BOM).

• High potential for repair: if the module is not properly functioning after manufacturing, or declines function during use, repair of the module is the most sustainable and economic viable solution. This includes a strong argument for implementation of module monitoring systems during operation

• High remanufacturing potential, that allows for the extraction of useful components from obsolete, malfunctioning and "end-of-life" modules, to reuse useful parts and/or materials for new PV-modules or to extend their lifespan.

• High potential for reuse of the recovered materials, either within the PV-module value chain or within other markets, including a high material and resource recovery potential during the manufacturing process.

Eco-Solar will focus on further developing remanufacturing, resource efficiency, recovery and reuse aspects over the full value chain. Moreover, a pan-industrial approach to identify by-products, recycled and recovered wastes into the best possible utilisation with respect to the eco-efficiency assessment results and will present solutions for poorly recovered and reused materials. The results for the individual process steps and their impact over the integrated value chain will be assessed with LCA and eco-efficiency analysis methods to provide proof of their success to reduce the environmental footprint.

2 IMPORTANCE OF ECO-MANUFACTURING

As a result of global efforts to reduce CO_2-emissions and dependence on fossil fuels, the photovoltaic industry is one of the fastest growing economic sectors. The global PV module production capacity has grown with approximately 39 GW (28%) in 2013 [1], and with approximately 62.7 GW in 2014 [2]. The IEA reports that

PV systems installed by the end of 2013 are generating 160 TWh/yr of clean electricity and thus avoiding about 140 million tonnes of CO_2 per year ($MtCO_2$/yr). Annual emissions from the power sector would increase from 13 Gt CO_2 in 2011 to about 22 Gt CO_2 in 2050 in the ETP 2014 6DS (IEA, 2014b, see Box 3). Solar PV could be responsible for avoiding 4 Gt CO_2/yr of emissions, or 19% of the total power sector emission reductions, by 2050, and 20% of cumulative emission reductions over the entire scenario period, according to the hi-Ren Scenario [3]. However, all of these scenarios imply (shown in Fig. 1) that the PV sector will continue growing rapidly in the next decade, which means the industry will grow to 860 -1,721 GWp until 2030 according to IEA Technology Roadmap (2DS Scenario and hi-Ren Scenario respectively) with a market share of Si-based PV modules of about 70%. Frost & Sullivan forecasted a cumulated global installation of 668 GW in 2025 [4].

Figure 1: Market outlook for the next decades: cumulative installed capacity worldwide based on different scenarios.

Even though solar power is pollution-free during use, production of PV-modules consumes considerable energy and natural resources. Moreover, recycling is hardly considered during module production, and therefore still cumbersome and inefficient. Considering that, the fast growth of the PV-industry entails similarly fast growth in resource consumption with growing production capacity: currently modest amounts of use can become very high.

When projecting the materials consumption in line with the projected PV-growth from Fig. 1, materials reduction (of the targeted materials in Eco-Solar) in 2030 will be as shown in Fig. 2.

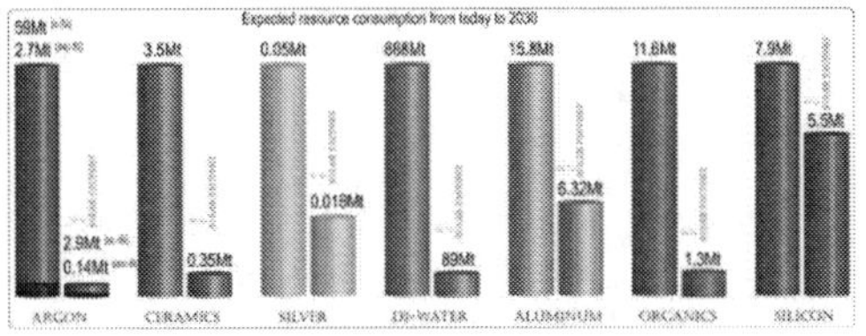

Figure 2: Expected material savings by 2030 based on Eco-Solar innovations (in Megatonnes). (Note: worldwide silver resources in 2014 belong to 0.52 Mt.)

3 CONCEPT AND APPROACH

3.1 Recovery & reuse during Si feedstock crystallisation

3.1.1 Reuse of argon purge gas

Pure argon gas is used to remove contaminants during ingot crystallization and it is currently being vented into the air. Though perhaps argon is abundantly available in the earth's atmosphere (approximately 1% of mass), there are still (ecological and financial) costs involved in the production of pure argon gas. For solar applications, approximately 5% of the wafer costs are attributed to argon gas. Even if in other industries, argon gas recycling has become state of the art, in PV-production this is not yet the case. The recycling processes from the semi-conductor industry are too refined and hence too costly, but they could be modified for PV-requirements. Within Eco-Solar, we will evaluate an existing argon gas recycling system, with respect to its potential for using it within both Cz and DS crystallisation processes aiming at a recycling rate of >95%. SINTEF will determine crucial parameters from the baseline process, and will make modifications to make the system applicable and cost-effective for PV. Once a working prototype has been established at SINTEF, Norsun will install it, and validate the results in an industrial setting.

3.1.2 Reusable crucibles

Crucibles are the main containers for the (molten) silicon feedstock. As silicon at high temperature reacts easily with almost everything, there are only a few crucible materials that will meet purity requirements for manufactures. One of the main drawbacks of using silica, as a current standard practice, is the (ecologic and financial) cost due to single use. Silica crucibles contribute up to ~30% of the conversion cost from Si-feedstock to the as-grown ingot.

STEULER SOLAR has developed a concept for reusable crucibles based on advanced silicon nitride ceramics, applicable for both crystallization processes, DS and Cz. In Eco-Solar the impact of crucible manufacturing recipes and crystallization process parameters to meet the requirements for Cz and DS ingot production will be determined, aiming at 10 times reuse.

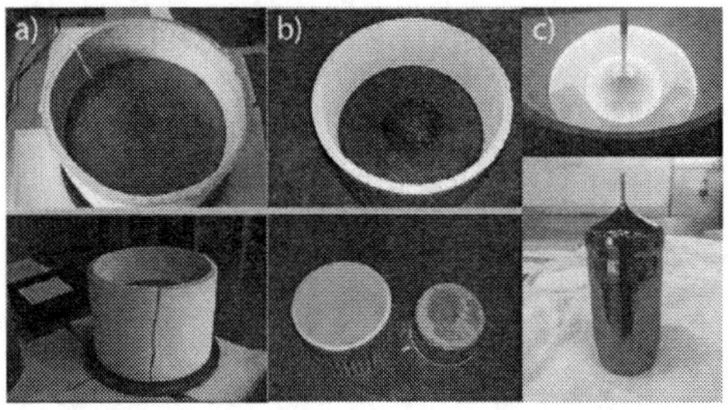

Figure 3: Directional solidification in conventional crucibles (a) and in reusable crucibles (b), Cz-ingot pulled from a reusable crucible (c).

3.2 Recovery & reuse of Si-kerf-loss

After crystallisation, side-, top- and bottom-parts are removed before the ingot is cut into blocks and then wafered. A considerable amount of silicon is lost in form of fine powders (kerf-loss) during the mechanical processing of silicon ingots to obtain wafers: up to 50% of the material is lost in air and water used to rinse the wafers.

The ability to recycle the Si-kerf-loss for solar ingot production will have beneficial effect in terms of savings of precious poly-silicon consumption and of waste reduction. Therefore, Garbo has patented and implemented a silicon recycling process which removes contaminations and brings silicon to "six nines" (or 6N), i.e. 99,9999% purity. Then purified silicon is compacted through another patented technology in order to obtain a mechanically stable, pelletized material, which can be used in the standard production of solar ingots and cells, or as fine powder for manufacturing lithium ion batteries and silicon nitride crucibles. SINTEF will carry out crystallization experiments of the recycled compacted silicon powder in order to assess the melting properties of the compacted silicon powder and final ingot properties. The potential of manufacturing lithium ion batteries and silicon nitride ceramics from fine powders will be investigated as well.

3.3 Remanufacturing, resource efficiency & reuse in solar cell processing

Wafers are being processed into solar cells, causing waste like etching chemicals, broken wafers and broken solar cells. By reducing and avoiding certain chemicals or metals (e.g. HF, Pb, Ag, Al) and using less pure or recycled chemicals, waste can be reduced and the ecologic footprint of solar cells and panels can be improved. Moreover, as it is extremely costly to undo the processing steps and recover the chemicals and metals as separate elements after solar cell processing, solar cells represent a very valuable half-product. Thus, repair of solar cells can provide significant environmental benefit, while reducing the outcast of solar cells can provide a cost benefit simultaneously.

3.3.1 Solar cell repair

In order to capitalise on the value of a solar cell, repair of defect solar cells within the solar cell process can represent a significant financial and ecological impact. AIMEN coordinated the EU-funded project REPTILE, in which ISC and INGESEA participated as beneficiaries. The project resulted in a fully functional prototype system, able to select and cut or isolate non-defective areas automatically in defective cells and wafers, a so-called Cell-Doctor (see Fig. 4).

Within the Eco-Solar project, ISC will evaluate the rejected solar cells with an automated system for defects recognition. AIMEN and INGESEA will further develop the accuracy of the Cell-Doctor prototype, aiming to avoid 50% scrapped cells.

Figure 4: Integrated laser processing, machine vision and cell handling equipment used for cell repair.

3.3.2 Reducing silver consumption

Silver is used in most solar cells (>98%) currently produced, providing the metal contact that "collects" and "drains" the current from the solar cell. It is of great importance to establish good electrical contact between the metal grid and the emitter surface with a low contact resistance, while having narrow, well conducting, fingers to minimise shadowing losses to produce higher efficiency solar cells. Today's silver consumption for PV is 3.5 – 15% of the total silver market and is expected to grow further assuming a market share of crystalline Si PV of approx. 70% in 2030.

ISC will experiment with different solar cell architectures, to minimise the use of Ag. In addition, an innovative interconnection scheme enabled by Apollon Solar's module design, will avoid soldering of the contact tape onto the rear side Ag pads and the front side busbars. Only the small contact fingers on the front side are needed. The full rear contact area is realized by aluminium.

In the Eco-Solar project, experiments will be carried out that will test: different solar cell architectures, the novel interconnection scheme from Apollon Solar and metallisation pastes that contain less Ag than current standard, aiming to save at least 66% of Ag.

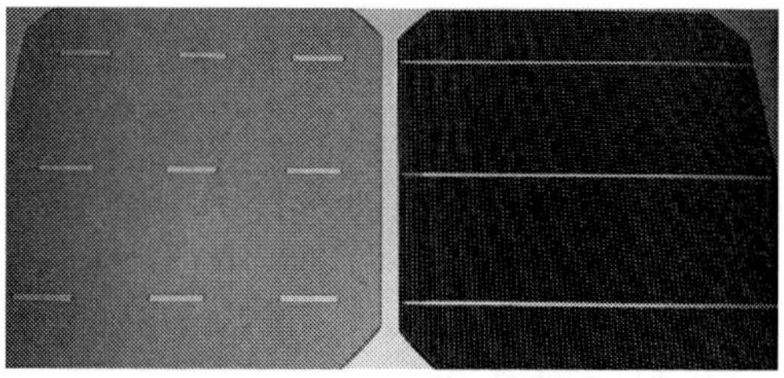

Figure 5: Current solar cell design with the silver rear-side pads, front side busbars and contact fingers indicated.

3.3.3 Reducing chemicals consumption through resource efficiency and chemicals recovery

|In crystalline solar cell manufacturing, several wet chemical etching and cleaning steps are implemented, which have a negative impact on the environmental footprint. ISC, together with SoliTek, will develop an alkaline method for saw damage removal, texturisation and cleaning, that will replace the current state of art concentrated HF:HNO3. Moreover, ISC and SoliTek will investigate advanced processes for emitter formation that will enable higher throughput, resource efficiency and avoid the needs for chemical phosphorous glass layer removal (PSGremoval).

The separation of front and rear of the solar cells is normally carried out with wet chemistry. AIMEN will further develop advanced laser treatment, in order to minimize the cut-off area at the wafer edges to a negligible fraction of the whole wafer area. This will be possible due to the small volume of laser-ablated silicon on the real wafer edges (and not on the front surface), using high speed scanning strategies.

3.3.4 Recycling water

Tap water is being de-ionised by the solar cell producers (and by solar cell research institutes), resulting in very clean water. This de-ionised water is mainly used for several cleaning steps in the solar cell process: after each chemical processing step, the wafers are flushed with this ultraclean water. The water that goes into the drain, is mainly of higher quality than the water contracted from the tap. As an example, solar cell processing at SoliTek, with a production capacity of 80 MW, consumes about 54.000 m^3 de-ionised water/year.

The protection of water resources, of fresh and salt-water ecosystems and of the water we drink and bathe in is one of the cornerstones of environmental targets in Europe. ISC will look into industrial viability of recycling systems for the waste water.

3.4 Module design for remanufacturing

After completion of the cells, they are electrically series connected and assembled into a solar module. Moreover, a junction box is attached that will ensure suitable power output. Within current solar modules, the biggest eco-challenge is that they cannot be disassembled into their components and the only potential for recycling is through destructive processes (see Fig 6.). Current PV modules also contain a vast amount of organic materials (EVA for encapsulation, PVF in backsheets)

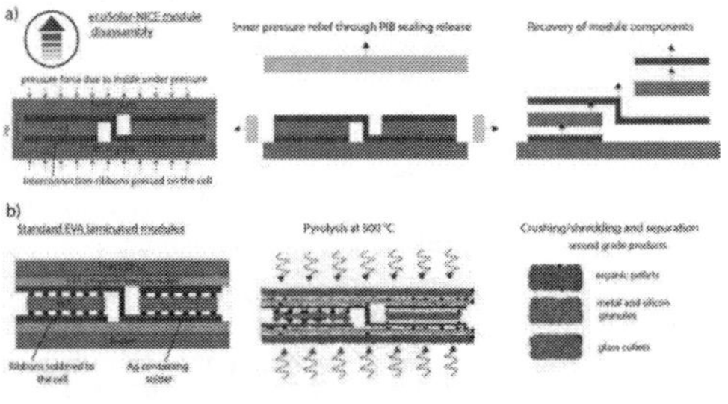

Figure 6: Standard PV module recycling versus the Eco-Solar disassembly into components

As Apollon Solars NICE-modules are already designed for disassembly (see Fig. 6 and 7), in Eco-Solar Apollon Solar will focus on the following:

• Further reduction of the module BOM (Bill of materials) focussing on a frameless version of the PV module

• Specifications for automated module disassembly equipment including diagnosis tools for quality screening of module components.

• Integration of electronics in the junction box, that can monitor the performance of the individual module and provide warning if performance drops drastically.

• Investigating the ageing and degradation aspects of solar modules during installation; and assess the technological viability to capture and recycle components at end of life, while considering environmental and economical aspects as well.

Moreover, Apollon Solar will demonstrate that module assembly of the Eco-Solar cells will result in comparable module output as with current non-eco-friendly state of the art module technology.

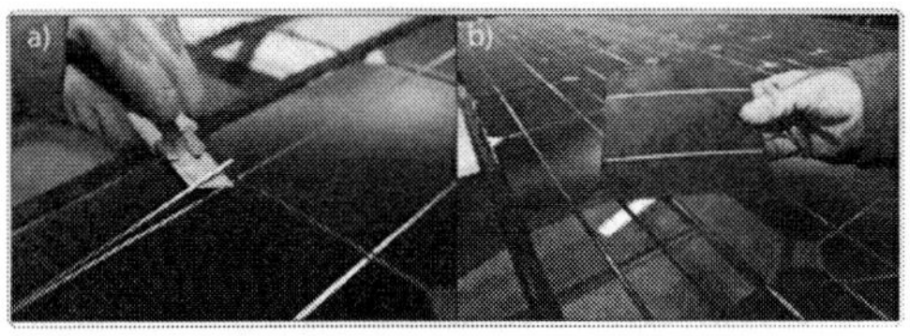

Figure 7: Manual NICE-module disassembly at Apollon Solar.

4 ENVIRONMENTAL IMPACT

In order to be able to determine the actual environmental impact of the Eco-Solar results, a thorough ecological validation is necessary. Moreover, the only chance for implementing eco-friendly production methods is when they are economically viable as well. Therefore, it is of crucial importance, that all processes are assessed for their ecological and economical viability and that the recovered waste products find (new) markets, where they can be purposefully introduced as (new) resources.

Thus bifa carries out an eco-efficiency analysis, based on data from the project partners and from published data. One strength of the eco-efficiency analysis is the simultaneous investigation of economical and ecological aspects that can be used to identify the most promising development routes and products. It is also applied to develop recycling processes that allow the highest possible recycling yield and quality that will be feasible from an economic point of view for the individual materials used in production and products.

The eco-efficiency analysis starts with setting the system boundaries and defining the processes. After that the collection of environmental and economical data starts, the life cycle inventory (LCI) is built up and state of the art ecological assessment (LCA - life cycle assessment) and cost consideration (LCC – life cycle costing) are carried out (see Fig. 7).

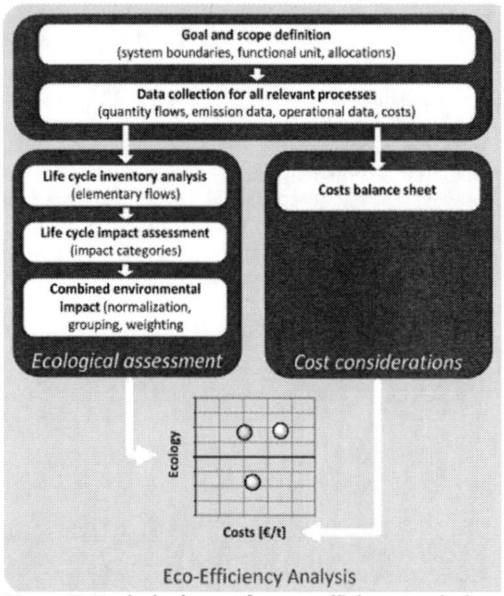

Figure 7: Typical scheme of an eco-efficiency analysis.

4.1 Input data collection for LCI

A standard EVA laminated module (60 6-inch solar cells, about 270Wp) with front glass and polymer back sheet including aluminium framing was defined as baseline for the comparison with the new products developed in this project. The research of recently published LCI data is finished and data gaps were filled by using expert assumptions. All project partners provided input for the LCI. The results were reviewed by the project team and the PV experts in the consortium.

An example section (ingot production) of the preliminary LCI baseline of the module production is presented in Table I.

Table I: Section of initial data set for ingot production.

Parameter	Unit	sc-Si	mc-Si
Reference product			
CZ single crystalline silicon	kg	1.00	-
Casted multi crystalline silicon	kg	-	1.00

Materials			
Argon (liquid)	g	1,000	186
Ceramic tiles	g	167	289
Hydrogen fluoride	g	10	-
Lime (hydrated)	g	22	-
Nitric acid (50 % in H_2O)	g	66.8	-
Helium (liquid)	g	-	0.1
Silicon	g	781	896
Sodium hydroxide (50 % in H_2O)	g	41.5	5
Water			
Deionised water	kg	4.01	
Cooling water	m³	5.09	1.06
Tap water	kg	95.1	
Energy/Fuels			
Electricity	kWh	68.2	7.7
Natural gas (burned)	MJ	68.2	
Waste			
Inorganic waste	g	167	289
Water emissions			
Biological oxygen demand 5	g	130	
Chemical oxygen demand	g	130	
Dissolved Organic carbon	g	40.5	
Total Organic carbon	g	40.5	
Hydroxide	g	367	
Nitrate	g	83.5	
Nitrogen oxides	g	33.9	

Due to the progress made in the PV industry since the writing of the proposal, first improvements were already observed. The initial data sets for the individual life cycle sectors ingot, wafer, cell, module and BOS therefore were updated and the expected material reductions based on Eco-Solar innovations were adjusted (see Fig. 8). The reference process used as a benchmark includes recent developments, the absolute project targets are expected to be achieved.

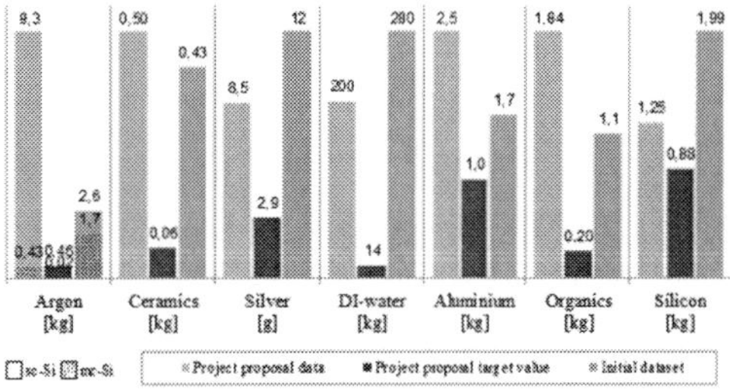

Figure 8: Reduction of waste and resource consumption per PV-module (60 6-inch solar cells) envisaged in Eco-Solar.

4.2 Preparation of midterm LCA

The next steps are the definition of goal and scope (functional unit, scenarios, system boundaries, geographic and time references, dealing with data lacks and credits/benefits, etc.) as well as selection of impact categories (i.e. climate change, acidification, eutrophication, photochemical ozone formation, resource depletion, toxicity) in close consultation with the whole project team.

In the meantime the preparation of the material and energy flow model based on the LCI baseline has already started. The first step – the development of the model structure – is already finished (see Fig. 9).

The model structure is designed to facilitate a sectoral evaluation of the results of the LCA. On the one hand it will make it possible to identify the contributions of the individual life sectors ingot, wafer, cell, module and BOS to the individual impact results. On the other side the model structure enables to use the results to evaluate direct emissions from sources that are owned or controlled by the reporting entity (scope 1 emissions), indirect emissions from consumption of purchased electricity, heat or steam (scope 2 emissions) and all other indirect emissions (scope 3 emissions).

5 REPURPOSING OF WASTE PRODUCTS

5.1 Living Lab – Establishment of Pan Industrial Material Re-use Opportunities

The target is to identify reuse and recovery routes for the main products and the by-products within the solar industry and if this will not be possible in other industries. The major material streams are assessed with respect to reuse, remanufacturing, recyclability and utilization of the output for different applications to optimize the eco-efficiency utilizing the Living Lab Methodology [5]. The use of the output may take place in completely different industrial branches. Potential customers and users will be invited to test the materials and provide feedback on the experience made. This will be done via a living lab, project webpage, trade fairs and conferences.

Living Lab is a kind of research in order to develop and to apply new products, services and systems by involving users to improve the products. A user can be everyone who deals with the object, especially consumers, workers, producers, suppliers, craftsmen or waste treatment companies. The target is to introduce the object successfully in a defined branch in order to save time and money. Living Lab is based on a four-step-cycle-system which includes the steps: co-creation (ranking and filtering of ideas), exploration (preparation of models/prototyping), experimentation (testbeds and observation) and evaluation (discussion of results). In the application for the solar industry it is necessary to determine the waste products/byproducts with the highest environmental impact and to identify possible customers (co-creation). Afterwards, negotiations with identified customers are required (exploration). The next step is to integrate the determined waste products/byproducts in a re-use process (experimentation) followed by an assessment of this process (evaluation). If the implementation of the determined waste products/byproducts in a re-use process is sufficient, the Living Lab process will be finished. If the implementation is not suitable, the four-step-cycle-system will be restarted.

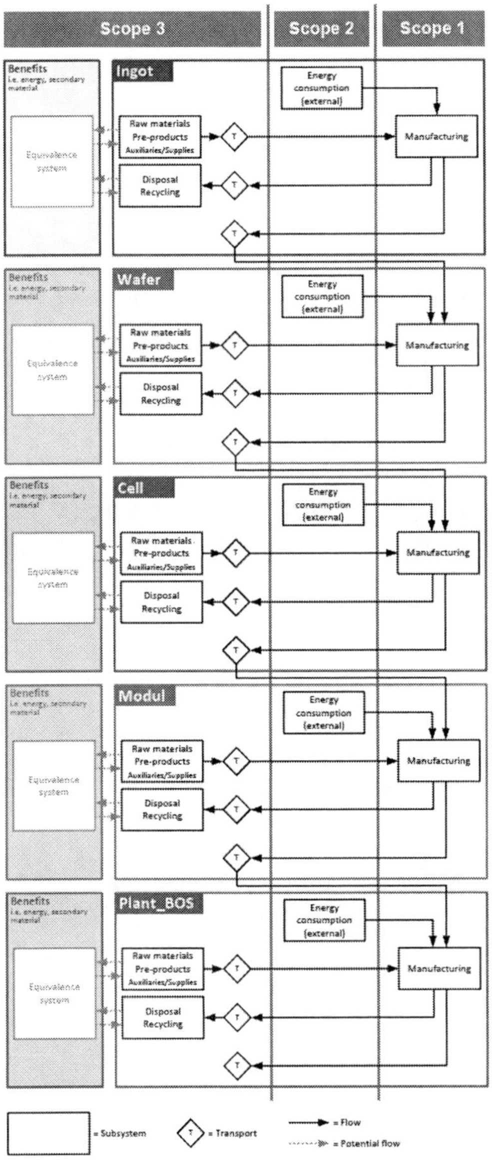

Figure 9: Structure of the flow model.

5 CONCLUSION

Europe wants to reduce its needs for raw materials and raise the level of recycling of resources in the solar power industry. Our target is that after the successful completion of this project the greenhouse gas emissions from solar panel manufacturing will be reduced by 25 to 30 per cent and the waste generated will be decreased by 10% minimum.

6 REFERENCES

[1] IEA PVPS TRENDS 2014 IN PHOTOVOLTAIC APPLICATIONS.
[2] International Technology Roadmap for Photovoltaic (ITRPV) 2013 Results, Revision 1, 24 March 2014.
[3] IEA, Technology Roadmap Solar PV Energy 2014 p.20.
[4]http://ww2.frost.com/news/press-eleases/photovoltaic-wind-and-hydro-star-top-renewables-finds-frost-sullivan/, accessed 2015-29-01.
[5] S. Vicini, S. Bellini, A. Sanna: The City of the Future Living Lab, International Journal of Automation and Smart Technology, Vol. 2, No. 3 (2012).

7 ACKNOWLEDGEMENT

This project has received funding from the European Union's Horizon 2020 research and innovation programme under grant agreement No. 679692.

NEW FINDINGS IN FIRE PREVENTION AND FIRE FIGHTING OF PV INSTALLATIONS

Prof. Urs Muntwyler
Berne University of Applied Sciences (BFH), Engineering and Information Technology
Photovoltaic Laboratory (PV-Lab), Jlcoweg 1, CH-3400 Burgdorf, Switzerland
phone: +41 (0)34 426 68 37, fax: +41 (0)34 426 68 63, urs.muntwyler@bfh.ch

ABSTRACT: Fire prevention of PV installations is a key research field of the PV Laboratory of the Berner Fachhochschule[1]. In a comprehensive research project of TÜV-Rheinland (with participation of the PV laboratory of the Berner Fachhochschule) a certain number of fire incidents with and on PV installations had been reported. In the state of Berne, the state owned insurance company "Berner Gebäudeversicherung GVB" started together with the PV Laboratory a three year research project on the topic. The research should lead in practical advices for the firemen and their instructors in the state of Berne[2].
Investigations about the cross interoperability of PV connectors showed that this creates a serious quality problem in PV installations[3]. Through fires on farm houses with roof integrated PV installations, we got aware that, after a fire, a serious amount of silicon particles will be spread other a wide area in the direction of the wind. We investigated the process and analyzed all the particles and their impact on the farmers land. We investigated several aspects of this process and how he has to be treated to avoid problems for the farmer and his animals.
Keywords: Photovoltaics, fire prevention, PV connectors, cross interoperability, silicon particles

1 RESEARCH TOPIC: PV FIRE HAZARD

The PV laboratory of the BFH collaborated on the TÜV-Projekt „PV-Fire Hazard". The final report can be downloaded since 2015 under: www.pvtest.ch[4].

2 RESEARCH ON CROSS INTEROPERABILITY OF PV CONNECTORS

Components of PV installations as PV modules, connection boxes and PV inverters are connected more and more with connectors. The industry developed special "PV – connectors" for that purposes for high DC-voltages up to 1'000 VDC, DC-currents up to 30 ADC and the ambient conditions of a PV installation as radiation, humidity and changing temperatures. The PV market offer several standards or products. Normally these connectors are not interchangeable. The problem are copies of the „standard" connectors. They are normally cheaper than the „original connector". Connections with the „orginal-" and the „copy-" – connector are a "Cross connection". In a diploma work[5] we tested these connections according the industry norm EN50521. We used the PV Laboratory infrastructure and two climate chambers. We measured the connections before and after the artificial aging. All connectors would be checked with a scanning electron microscopy SEM. With that we could investigate the material and the thickness of the layers. With this information we can calculate the life time of the connector.

3 SURVEY BY PV INSTALLERS IN SWITZERLAND

To have an idea how relevant the problem would be, we asked 190 PV installers from Switzerland. 30 Installers sent the questionnaire back (15,8%)[6].

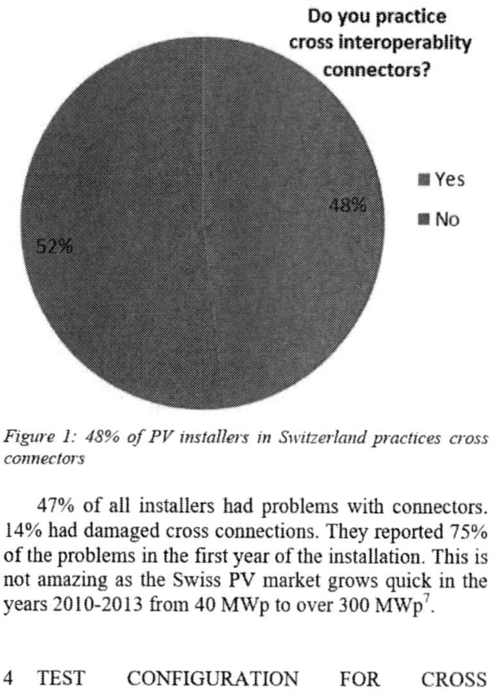

Figure 1: 48% of PV installers in Switzerland practices cross connectors

47% of all installers had problems with connectors. 14% had damaged cross connections. They reported 75% of the problems in the first year of the installation. This is not amazing as the Swiss PV market grows quick in the years 2010-2013 from 40 MWp to over 300 MWp[7].

4 TEST CONFIGURATION FOR CROSS CONNECTION TEST

We choosed connectors with a relevant market share in mid Europe:

- Amphenol
- Lin Yang
- Multicontact
- QC Solar
- SMK
- Tyco
- Weidmüller

Cycle-Test

After the first measurement of the contact resistance, the cycle test will be started according the Norm EN 50521 „Connectors for PV – Systems – safety requirements and tests“. The test cycle includes:

- 30 min in the upper temp. range +85°C +/-2°C;
- 30 min in the lower temp. range of -40°C +/-2°C;
- Transfer time: < 3min;
- cycles: 200.

Humidity-/ heat test

The artificial aging is done by the test conditions:

- Test temperature: +85°C +-/2°C;
- Relative air moisture: +85% +/- 5%;
- Test time: 1000h.

Final measurement

Similar to the initial measurement is the finale measurement. According Norm EN 50521 the resistance should not be higher than 50% of the reference value or not more than 50%. The higher value is allowed.

We used three samples for all the combinations to have enough samples for further tests. The test combinations were:

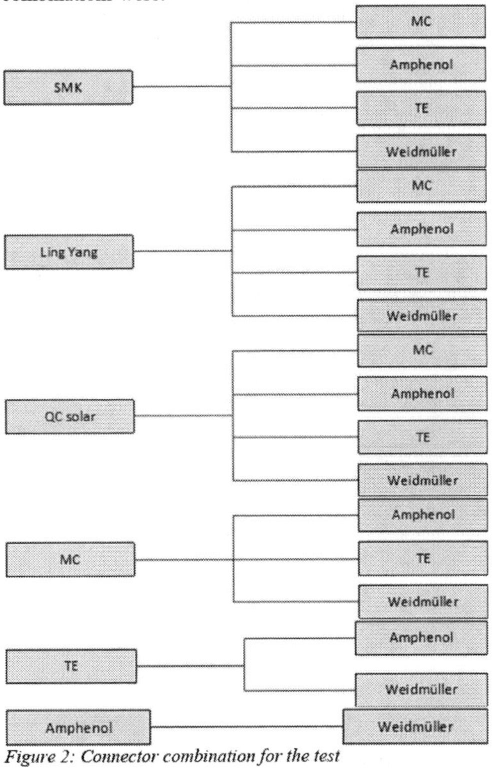

Figure 2: Connector combination for the test

Measurement of the resistance

The resistance test was done according Norm EN 60512-2-2:2013. The current has to be measured in both directions with the formula:

$$\frac{|\Delta U_{Vor} + \Delta U_{Rück}|}{|I_{Vor} + I_{Rück}|} = R_{Brutto}$$

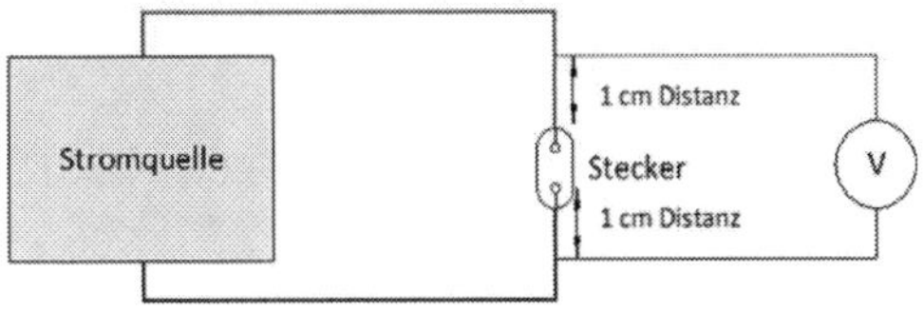

Figure 3: Measurement of the resistance

Results of the measurement before the aging:

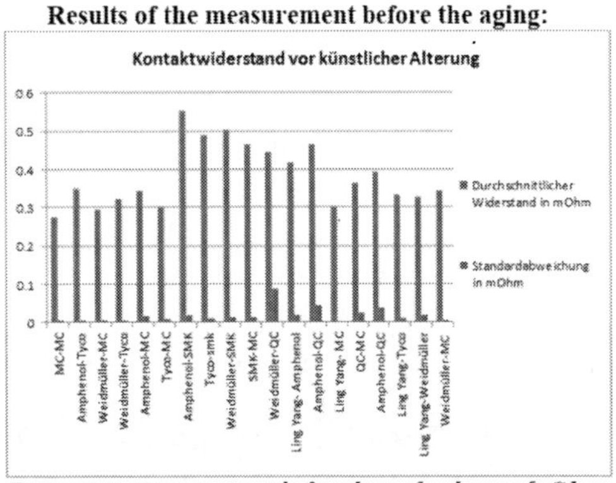

Figure 4: Contact resistance before the artificial aging [mΩ]

The differences in the resistance of the different combinations are relevant, but still in a small band. With the nominal current of 30 A the temperature enhancement is small:

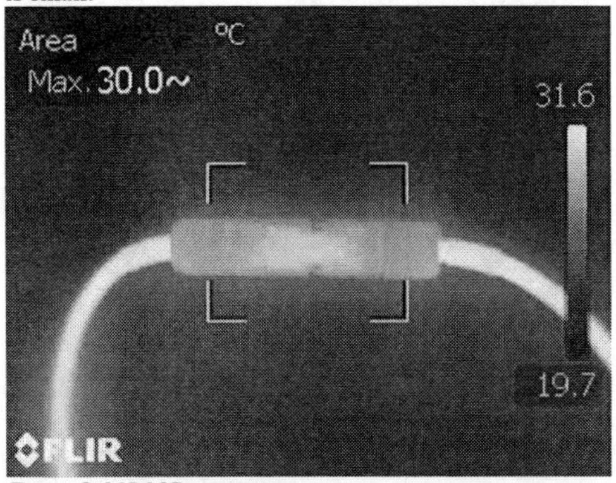

Figure 5: MC-MC connection

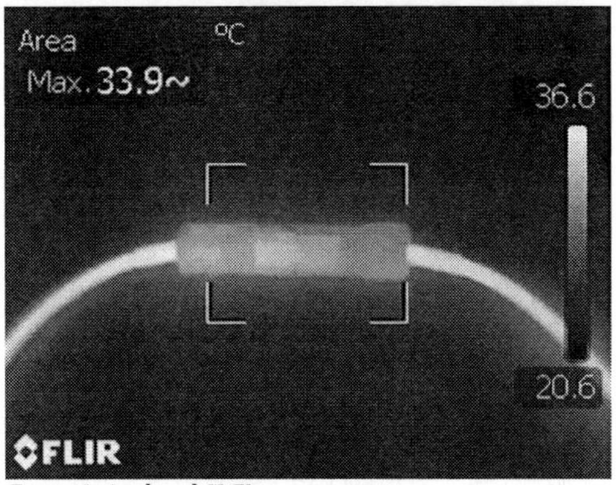

Figure 6: Amphenol-SMK cross connection

Results of the measurement after the aging:

The resistance went up to more than 180%. Many of the cross connections are over the allowed value of 50%. Some of the changes are very significant.

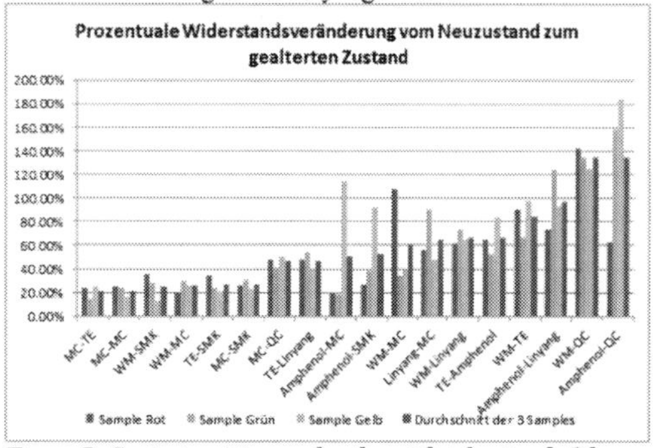

Figure 7: Contact resistance after the artificial aging [mΩ]

An example after the cross connection test after aging shows a drastic enhancement of the temperature which could cause a fire:

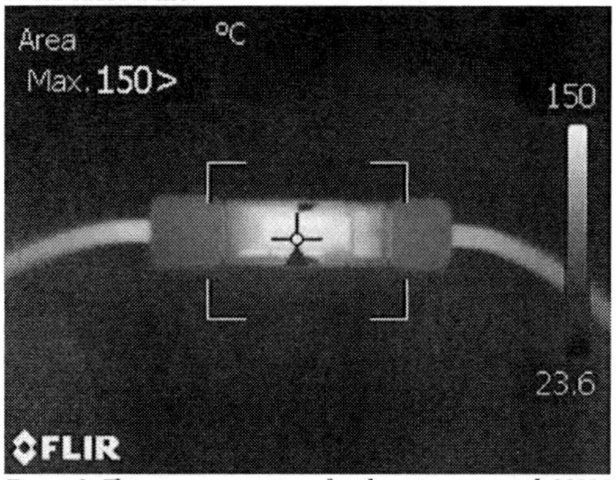

Figure 8: This cross connection after the aging can reach 1500C with nominal current. This could cause a fire!

Practical example: Cross connection failed in the field

In a 131 kWp in-roof installation in Switzerland, in spring 2015 this cross connectors were found. This could cause a fire.

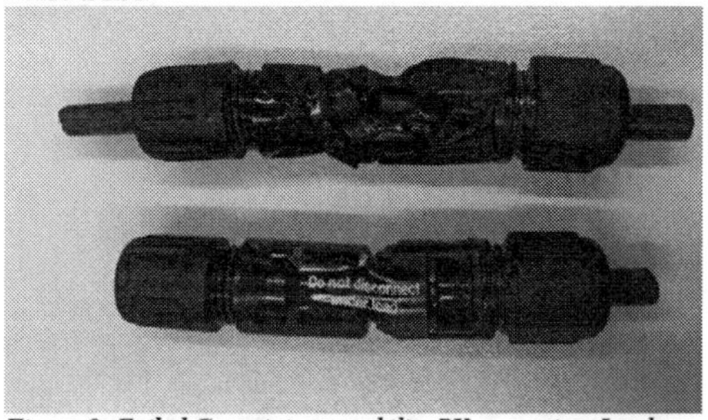

Figure 9: Failed Cross interoperability PV connectors Lumberg-MC of a BIPV-installation

Summary:

Cross connections with different PV connectors must be avoided. It has to be seen as a serious technical deficit and must be replaced. Liability through insurance companies for such incidents has to be rejected. With a strict policy to ban the cross connectors, a weak point of PV installation could be eliminated.

5 RELEASE OF SILICON PARTICLES AFTER A FIRE

Through fires on farm houses we got aware that, after a fire, a serious amount of silicon particles will be spread over a wide area in the direction of the wind. The farmer organizations of the state of Berne are concerned, that this area could be contaminated by the silicon particles.

Fire on a farmer workshop

We took the opportunity after a fire in April 2015 in Lanzenhäusern near Berne. A farmer workshop got into a fire after welding activities. The roof integrated "Solrif" - PV installation of about 60 kWp with 205 Wp (CNPV-205-BLK PV modules was completely destroyed. Through the strong west wind silicon particles would be distributed up to 2 km away from the site.

Figure 10: Workshop after the fire seen from east (position of the photographer first red circle on the map)

Figure 11: Situation in Lanzenhäusern (red surface – the workshop building) – first circle on the street position of the photographer/ second red circle: place of finding of silicon particles (see photo)

Figure 12: Silicon particle in the farmland – no further restriction for the use of the farmland

Figure 13: Finding spot (right circle) with silicon particles

In the debris all components and materials of the PV installation can be found. In the farmland only silicon particles could be found. Investigation of the particles with SEM showed that mainly Al/ Si and O could be detected. This is not a problem for a further use of the land for agricultural purposes.

Compatibility of silicon particles towards animals

To check the compatibility of the silicon particles towards animals, especially cows, we contacted Prof. Dr. med veterinarian Samuel Kohler from the agricultural department of the Berner Fachhochschule[8]. Problems with horses, sheep, goats etc. could be quickly excluded. After further investigations also the "cow" got a "green light". The silicon particles are so fragile that they will further smashed in the stomachs of the cow. Therefore silicon particle after a fire doesn't cause serious problems for the farmers.

Figure 14: Silicon particles for the „cow" – investigation – „green light" even for cows

6 RESULTS AND CONCLUSIONS

Fire prevention in PV installations can be done in a better way. Our investigations under PV installers showed that several influences are not well respected by the PV installers. We expect some more problems in the future. Cross interoperability of PV connectors is a serious threat for PV installations. Technical components as "light arc detectors" and "infrared-detection" of PV installations could help to prevent fires through the PV installations himself.

The training of fireman should be done seriously. The work of the PV laboratory and our new checklist we publish together with the insurance company Berner Gebäudeversicherung GVB leads the way.

Together with material specialist of the Berner Fachhochschule we investigated the mixture of elements in those particles with a scanning electron microscope SEM. We could show that only silicon and aluminum oxide from the back side of the solar cell can be found in the particles. These particles could be eaten by animals as horses, sheep, pigs and cows. All the animals, without the cows, choose their food very carefully and wouldn't eat such particles. The cows however would bring the particles into their stomachs. The silicon particles are so fragile that they will further smashed in the stomachs of the cow. Therefore silicon particles, after a fire, don't cause serious problems for the farmers.

7 ACKNOWLEDGEMENTS

These investigation could be done thank to the support of the Bernische Gebäudeversicherung GVB. We had further support from the Swiss Competence Energy Research programm for further grids „SCCER-FURIES" of the federal commission for technical innovations CTI and the Berner Fachhochschule.

1 Neue Erkenntnisse in Brandprävention und Brandbekämpfung von Photovoltaik-Anlagen
Prof. Urs Muntwyler, Fachhochschule Bern, Burgdorf, PV Konfernz Staffelstein 2016 /
PV and Fire Brigade Safety: No Panic, but Realistic Assessment of Danger and Possible Countermeasures, Heinrich Haeberlin, Luciano Borgna and Philipp Schärf; EUPVSEC 2011

2 Photovoltaikanlagen im Feuerwehreinsatz (Merkblatt), www.gvb.ch, 2016

3 Qualität von PV-Kraftwerken – Technische Risiken, Reduktionsmaßnahmen und Felderfahrungen
Dipl.-Ing. Willi Vaaßen, TÜV Rheinland. PV-Konferenz Staffelstein 2016

4 PV Fire Hazard – Analysis and Assessment of Fire Incidents, Herrmann Laukamp and others, 28th EU PVSEC 2013, Paris

5 Kreuzverbindungen von PV-Steckern, Lukas von Ballmoos, Berner Fachhochschule Juli 2015

6 Kreuzverbindungen von PV-Steckern, Lukas von Ballmoos, Berner Fachhochschule Juli 2015

7 Markterhebung Photovoltaik, Juni 2015, Bundesamt für Energie

8 Correspondence from Prof. Dr. Kohler to Prof. Muntwyler, February 2016

RECYCLING OF BROKEN SI BASED STRUCTURES AND SOLAR CELLS

M. Syvertsen[1], B. Ryningen[1], M. Di Sabatino[1], W. Palitzsch[2], H.J.Moeller[3], C. Audoin[4], M.Sérasset[4], D.Pelletier[4], J.P.Rakotoniaina[4] ,J.Dieguez[5], A.Souto[6], J.Denafas[7], L.Petrėnienė[7], M. Pranaitis[7], V.Čyras[7], R. Zolubas[7], A. Ulyashin[8]

[1]SINTEF, Material and Chemistry, Trondheim, Norway,[2] Loser Chemie GmbH, 08056 Zwickau, Germany,
[3]Fraunhofer Technology Center for Semiconductor Materials THM, 09599 Freiberg, Germany
[4] CEA, LITEN, Department of Solar Technologies – National Institute of Solar Energy, F-73375 Le Bourget du Lac, France,[5]Silicio FerroSolar SL, 15142-Arteixo (A Coruña) Spain,[6]FerroAtlantica I+D, 15142-Arteixo (A Coruña) Spain
[7]JSC "Soli Tek R&D", LT-08412, Vilnius, Lithuania,[8]SINTEF, Material and Chemistry, Oslo, Norway

ABSTRACT: This paper presents works where broken solar cell structures early in the PV production process chain as well as broken finished solar cells have been recycled into new Si feedstock through demetallization step, purification and directional solidification. The paper discuss two different routes for removal of diffusion layers and anti-reflection coating (ARC). It also presents characterizations of ingots produced with the Si-feedstock from the two routes by directional solidification.

1 INTRODUCTION

The CABRISS project is a European collaboration between 16 partners from 9 countries; 6 SMEs, 5 Industry and 5 RTO (Research Organization) [1]. The main vision of the project is to develop a circular economy mainly for the photovoltaic, but also for other industries such as electronic or metallurgy. It will consist in the implementation of recycling technologies to recover In, Ag and Si for the sustainable PV technology and others applications.

At the moment, there is few quantity available of end-of-life modules and processes for dismantling are not fully operate, e.g. recovery of cells from the encapsulates is at lab-scale level. On the other hand, broken cells and wafers are available even at little quantity (1 to 3% of the production) due to a combination of rough handling and mechanical weaknesses in PV-material. Until now, these cells have been considered to be waste and not valuable enough for recycling.

The paper investigates two routes for the removal of ARC and diffusion layers on broken cells either coming from the cell manufacturing or from end of life module. One route use etching of ARC and diffusion layer before melting and directional solidification whereas the other route use melting, vacuum treatment and subsequent directional solidification.

The results of this study can be transferred to cells from end-of-life modules.

2 EXPERIMENTAL SETUP

This section describes the two routes for removal of ARC and diffusion layers. The first route was explored at SINTEF in Trondheim, Norway at Lab scale, and at FEROANTALTICA (FAID) at larger scale, whereas the second route was tested at CEA-INES.

2.1 Input materials

LOSER Chemie, one of the partner of CABRISS, is able to remove and recover the aluminium rear contact and the silver and indium from broken solar cells [2,3]. The solar cell scraps provided by LOSER Chemie are shown in Figure 1.

Figure 1: Broken solar cell pieces (Si-shards) delivered from LOSER Chemie. The front finger and Al back contacts have been removed. The size of pieces are about 2 - 20 mm. (courtesy LOSER Chemie GmbH).

By use of a fresh solution of aluminium chloride and water, LOSER Chemie has developed a procedure to remove the aluminium from the back side of Si solar cell and simultaneous recover a valuable product. During the reaction, [Aln(OH)mCl3n-m] (poly-aluminium-hydroxide-chloride) is formed which is usually produced by pressure digestions of aluminium hydroxide and concentrated hydrochloric acid at high temperatures.

LOSER's hydrometallurgical extraction for the recycling of silver from waste silicon cells operates at room temperature and the used alkane sulfonic acid is readily biodegradable (OECD 301 A). The alkane sulfonic acid is used as a transport system, and can be recovered.

The process runs without formation of nitrous gases, as they normally will, with use of nitric acid in the field of silver recovering. The recovery of the silver using sulfonic acids is a novel and the most effective, economical and ecologically method. A major advantage is particularly the extremely high solubility of silver salts of the corresponding sulfonic acids.

After the aluminium back and silver front contacts has been removed, the ARC layer and a front (P) and back (Al) doping field still remains and has to be removed before the shards can be used as in a conventional melting and directional solidification step.

Figure 2 shows Tof-SIMS analysis of the back layer of a cell piece from LOSER. It is clear that aluminium has a diffused thickness about 10µm into the bulk silicon from the rear contact.

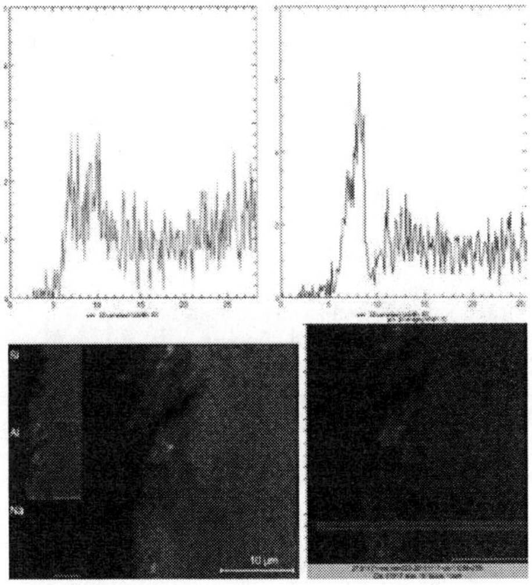

Figure 2: Tof-SIMS analysis of the back side of Si solar cell structure after removal of front and back side contacts. The upper graphs are two line scans for aluminium along the lines in lower right picture.

The other type of material tested was broken wafers from an early stage in the cell processing line at SOLITEK. These cells were taken out after PECVD and contain phosphorus diffusion layer and ARC (Si_3N_4 anti-reflection coating) in front, but no aluminium diffusion layer at the back. The thickness of these front layers are considered to be about 1µm.

It is clear that from the layer thickness measurements, the LOSER-material should have a thickness reduction of about 20µm (we are not able to etch 10µm from the back side only), and the SOLITEK material would require only 2µm to be removed [4, 5].

2.2 Etching procedure

Four different etching techniques have been tested in small scale and a method for larger batches were developed for two of the etching technique tested. The challenge was to choose an etching solution with fast enough reaction, environmental friendly, and yet not too violent (exothermic), to be suitable for etching away the layers in relatively large batches. First small scale experiments (1 liter acid + 65 g Si scrap) were done to verify if a composition of acids was suitable. If suitable, an upscaling to 5 liters acid and 600 g Si scrap was done. The composition of the four etching techniques are shown in Table 1.

Table 1: Composition of the four etches that were tested.

Content	Used polish etch	CP4	"Alexander's etch"	Used and new 9:1 etch
HNO_3 (65%)	3	5	4.6	9
HF (50%)	1	3	1.5	1
CH_3COOH	3	3		
H_3PO_4 (85%)			1.5	
DI-water			2.3	

A schematic of the etching set-up experiments is shown in Figure 3.

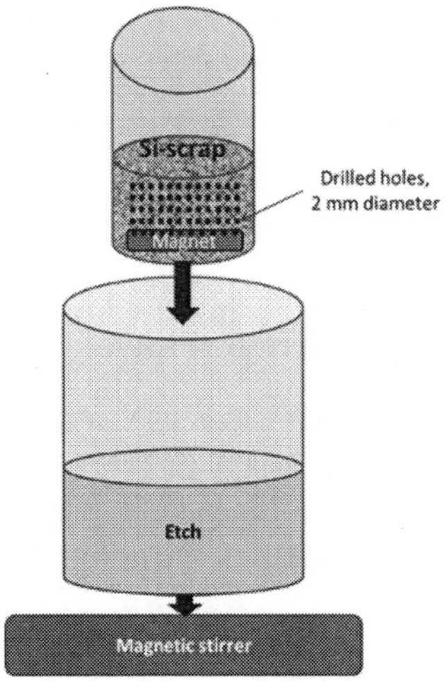

Figure 3: Principle sketch of the setup used for the small scale experiments at SINTEF (not to scale).

All four etch types were tested in small scale. The chemical reaction with the use of "Used polish etch"[1] and the "9:1" etch were reaction was gentle with little heat generation. The other two etches were considered too fast and violent.

After deciding the etch type, the setup was upscaled to 600g batches and in total 12 kg scrap was etched for use in the Crystallox DS250 furnace at SINTEF. According to thickness measurements done during the etching, about 20 – 30µm was removed from both sides of the flakes. This should be good enough for removal of both front and back layers.

2.3 Vacuum treatment process

An alternative route to remove emitter and layers is by vacuum treatment, as proposed by CEA-INES. In the case of p-type cells technology, diffusion layer contain high phosphorous concentration. As it shown on Figure 4, vapour pressure of phosphorous remain higher than silicon vapour pressure beyond melting point of silicon. Phosphorous removing by evaporation from melted silicon bath would be easy under vacuum treatment. This process is also efficient to remove some metallic impurities as such aluminium, or silver which are expected impurities in Si scraps.

CEA has done such vacuum treatment of material coming from LOSER.

[1] The "Used polish etch" has been used to do chemical polishing of other silicon samples previously.

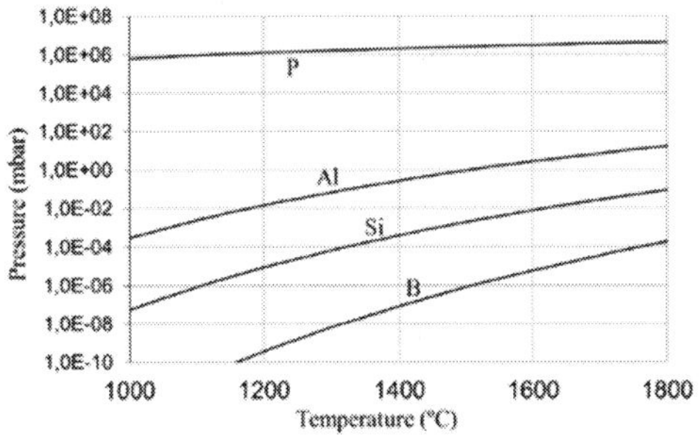

Figure 4: Vapour pressure of pure elements as function of temperature [6,7].

Experimental conditions and results are detailed in next paragraph.

3 ETCHING AND SOLIDIFICATION EXPERIMENTS WITH DISCUSSION

3.1 Directional solidification of lab scale ingots

A schematic of the furnace is shown in Figure 5. A typical charge size with poly-silicon is 12 kg, but since the charge tested here consisted of thin flakes, only 8.3 kg was possible to charge in the crucible.

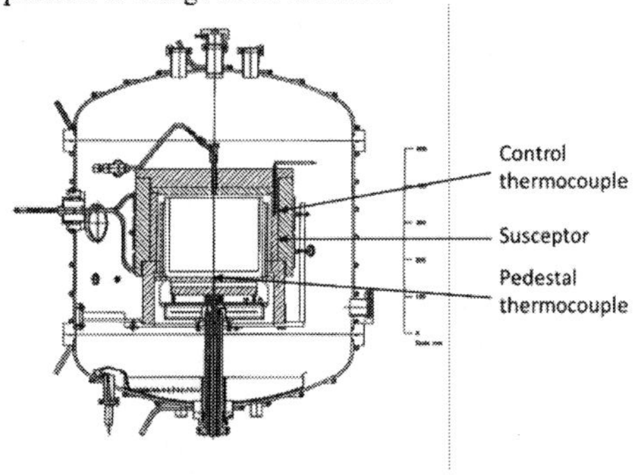

Figure 5: Schematic of the Crystalox DS-250 furnace at SINTEF.

The charge melted and solidified in a normal way except for shorter time for solidification than a standard charge due to lower ingot height.

After the ingot was produced, a central brick of 158mm*158mm was cut out for wafering by THM Fraunhofer in Freiberg, Germany. At the time of writing, the brick has been wafered, cell processing or characterisations will been done soon.

From one of the side cuts of the ingot, three samples were taken out and send for GDMS analysis at NTNU in Trondheim. The resulting measurements of boron, phosphorus and aluminium are shown in Table 2.

Table 2: GDMS measurements of B, Al, and P at three different (relative) heights in the ingot.

Height fraction	B (at/cm^3)	Al (at/cm^3)	P (at/cm^3)
0,28	1,80E+16	1,10E+15	1,59E+16
0,49	1,83E+16	1,13E+15	2,00E+16
0,87	2,45E+16	4,97E+15	5,74E+16

The concentration of P is higher than the sum of B+Al, indicating n-type material. It also shows that not all of the diffusion layers were removed.

The question was then: What went wrong?

The answer was found in the thickness measurements. New thickness measurements was done on the non-etched and large scale etched material from LOSER. 30 randomly picked samples were chosen from both batches. The resulting histograms (with fitted normal distributions) are shown in Figure 6 and Figure 7. These show that only between 5 and 10µm was removed from each side of the silicon flakes. This thickness reduction was too little in order to remove the ARC and both diffusion layers.

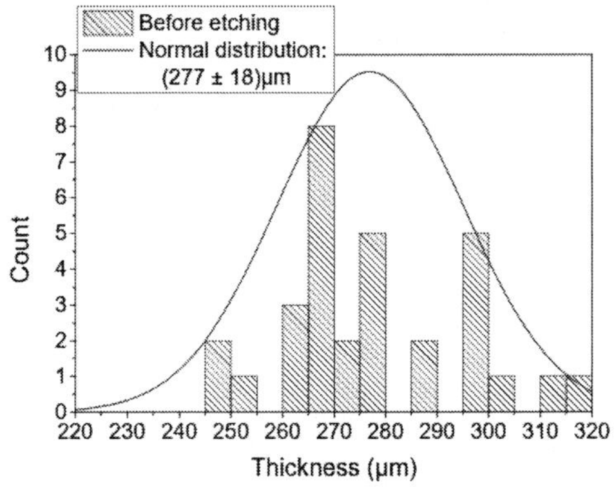

Figure 6: Thickness distribution histogram of 30 samples from the LOSER material before etching at SINTEF.

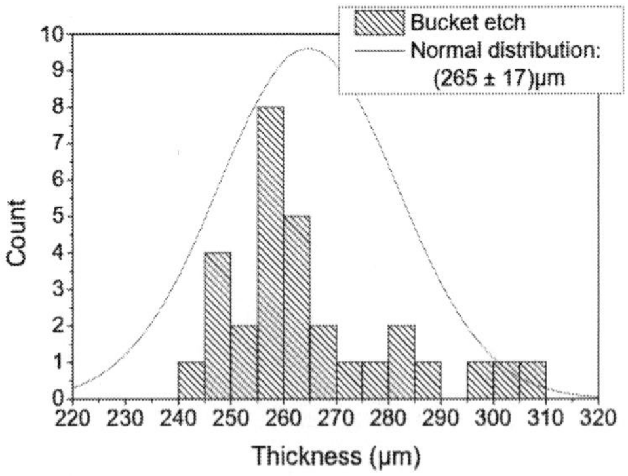

Figure 7: Thickness distribution histogram of 30 samples from the LOSER material after large scale etching at SINTEF.

After this it was decided to do more etching, but this time only small scale, and then rather do small scale directional solidification experiments with subsequent proper thickness and GDMS measurements.

3.2 New lab scale etching of LOSER and SOLITEK material

Three more etching batches were done at SINTEF. Again a "used polish etch" was used. One new batch of LOSER material and two batches of the SOLITEK material (taken after PECVD) was chosen. The thickness reduction is given by Figure 6 and 8 – 11.

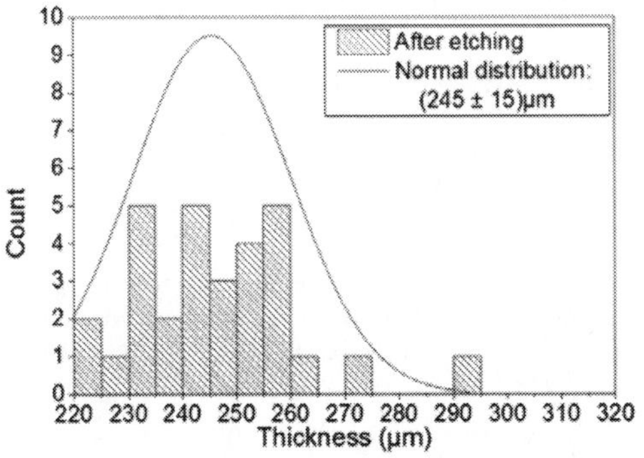

Figure 8: Thickness distribution histogram of 30 samples from the LOSER material after small scale etching at SINTEF. Thickness before etching is given in Figure 6.

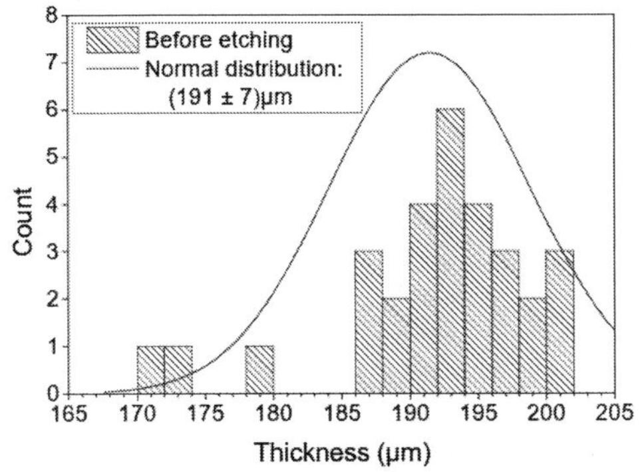

Figure 9: Thickness distribution histogram of 30 samples from the SOLITEK material before etching.

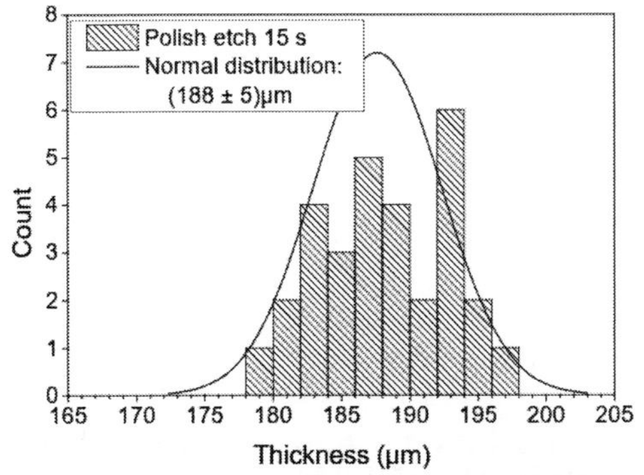

Figure 10: Thickness distribution histogram of 30 samples from the SOLITEK material after small scale etching in 15 seconds. Thickness before etching is given in Figure 9.

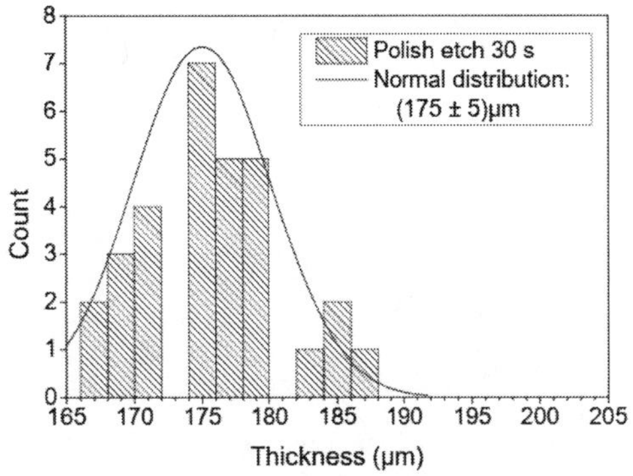

Figure 11: Thickness distribution histogram of 30 samples from the SOLITEK material after small scale etching in 30 seconds. Thickness before etching is given in Figure 9.

From the thickness distributions (Figures 6, 8 – 11), it is seen that the average thickness of the LOSER material has been reduced by about 16μm on both sides, and according to the Tof-SIMS measurements in Figure 2, this is sufficient for removal of the diffusion layer containing aluminium.

The two batches of SOLITEK material showed an average thickness reduction of about 1.5 and 8 μm on both sides. This should be enough to remove the Si_3N_4 ARC and P-diffusion layer.

3.3 Directional solidification test in small tube furnace

Directional solidification test in a tube furnace was performed at NTNU, Trondheim, Norway. A schematic of the tube furnace is shown in Figure 8. The furnace is designed as a traditional Bridgeman furnace, that is, the crucible with charge is slowly pulled down and out of the hot zone of the furnace leading to a controlled directional solidification of the metal. The sample size of the finished ingots are about 50 – 60 g, with a diameter of about 40 mm.

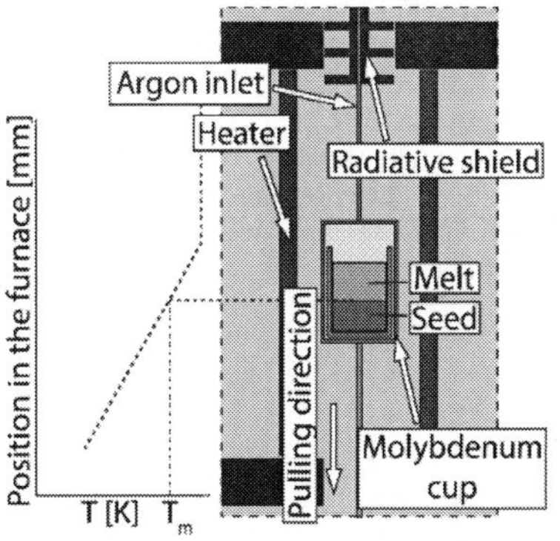

Figure 12: Bridgeman tube furnace at NTNU in Trondheim, Norway [8].

After the melting and crystallisation, two planes were cut from each samples and sent for GDMS analysis.

In total 6 different material samples were run in the small Bridgeman furnace. An overview of the different material is given in Table 3.

Table 3: Overview of the material tested in the small Bridgeman furnace.

Si-shards from LOSER	Wafers from SOLITEK
1. As received (with ARC, P, and Al)	1. As received (with ARC and P)
2. SINTEF etch	2. Etch 15 sec
3. Ferroatlantica etch	3. Etch 30 sec

The GDMS results from melting of the Si-shards from LOSER in the small Bridgeman furnace tests are shown in Table 4.

Table 4: Concentration of boron, aluminium, and phosphorus for the Si-shard samples (from LOSER) and the two heights. All concentrations are given in units of 10^{16} atoms/cm^3.

	Lower part			Upper part		
	B	Al	P	B	Al	P
As received	2.1	0.81	7.5	3.7	1.5	38
Ferroatlantica etched	2.3	0.49	0.16	3.0	1.1	0.38
SINTEF etched	1.6	0.17	2.0	2.9	0.71	6.1

For all the three small casts, there are higher concentration of the dopant elements in the upper region than in the lower one. This proves that the small ingots have been directional solidified.

The etching done at Ferroatlantica was more efficient than the etching done at SINTEF. Calculation of the net dopant level in the three ingots (B+Al-P) show that the ingots made from "As received" and "SINTEF etched" are n-type material whereas the ingot produced from the "Ferroatlantica etched" material is p-type. This concludes that even though about 16µm was removed from each side during etching at SINTEF, this was not enough. There is an uncertainty whether the sample analysed with Tof-SIMS (Figure 2) is representative.

The GDMS results from melting of the wafers from SOLITEK (containing only ARC and P-layer) are shown in Table 5. Since there were no aluminium diffusion layer in the SOLITEK-material, only boron and phosphorus were measured.

Table 5: Concentration of boron and phosphorus for the wafer samples (ARC and P layer containing wafers from SOLITEK) and the two heights. All concentrations are given in units of 10^{16} atoms/cm^3. The GDMS-signal for two of the samples had too much noise to be reliable.

	Lower part		Upper part	
	B	P	B	P
As received	0.66	3.5	0.73	6.6
Etched 15 sec	-	-	0.91	0.24
Etched 30 sec	-	-	1.0	0.10

The concentration of boron and phosphorus in the small ingots from the SOLITEK material shows that removing 1.5µm is enough to remove all the ARC and most of the P-diffused layer. For both ingots of the etched material, the material are p-type, whereas the ingot from the "As received" material is n-type.

A general comment for the small scale experiments is that larger scale testing must be performed in order to conclude regarding electronic properties of the material.

3.4 G2 ingot crystallisation from etched silicon

CEA produced one G2 ingot (55kg) by conventional directional solidification process using refined silicon from FAID etch process.

Picture of the ingot is shown in Figure 13. Applied recipes were standards one and no specific observations was done during the crystallisation.The ingot has been cut into two larges vertical cut of around 160mm thickness and observe by IR transmission imaging.

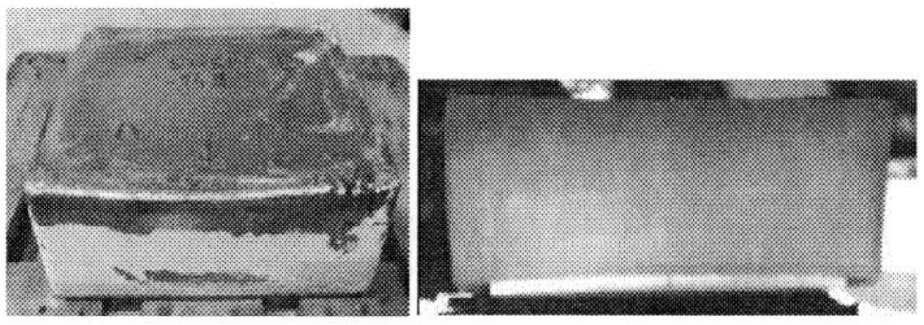

Figure 13: Picture of the ingot casting by directional crystallization done at CEA (at left) and IR transmission imaging.

The IR transparent image in Fig. 13 shows that no precipitates have been located.

The corresponding resistivity along the relative ingot height was measured using SEMILAB WT 2010D device. The resistivity remain low in the range of 0.4 to 0.6 Ω.cm indicating high dopant concentration.

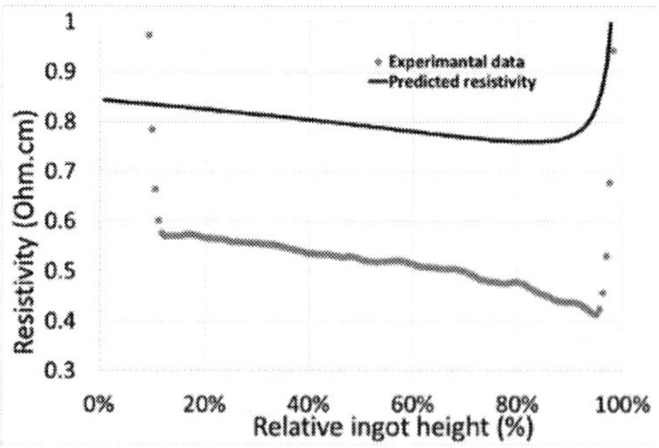

Figure 14: Resistivity profile of G2 ingot from etched silicon along relative ingot height.

4 PYROMETALLURGICAL REFINING AND SOLIDIFICATION EXPERIMENT

4.1 Vacuum treatment experiment

In this work, de-metallized broken silicon cells were

treated under vacuum refining condition. The silicon is introduced in a laboratory vacuum induction furnace. Crucible filling capacity is around 2-3kg depending material shape.

2.8kg of de-metallized broken silicon cells were melt in a graphite crucible. Experiment was done at temperature high above silicon melting temperature, under vacuum conditions and during a sufficient time to ensure high phosphorous content extraction. At the end of the treatment, the purified silicon is directly poured and quickly cooled in graphite mold. A purified silicon ingot of 2.46kg was obtained (Figure 16). Overall silicon mass yield is about 88%.

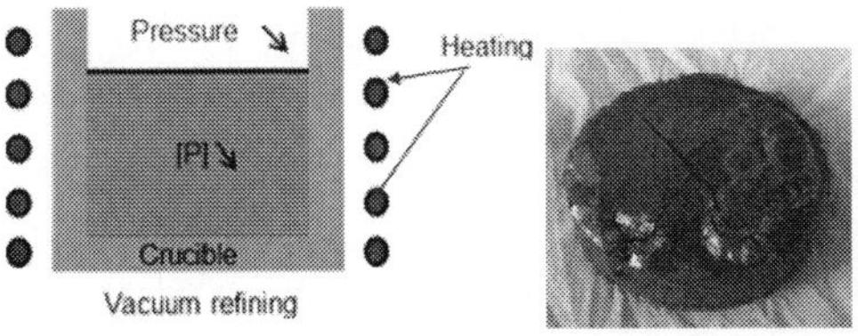

Figure 15: Schematic vacuum induction melting process (at left) and picture of the purified Si-ingot (2.46kg) (at right).

The concentration of the impurities was measured before and after refining respectively by ICP-OES (Inductively Couple plasma-Optical Emission Spectrometry) and GDMS (Glow Discharge Mass Spectrometry).
Samples have been directly taken for the silicon melt, assuring sample homogeneity. Chemical analysis results are given on the following table 5.

Table 6: Chemical analysis of de-metallized silicon broken cells before and after vacuum refining.

	before vacuum refining	after vacuum refining
elements	ICP-OES analysis Content (ppm wt)	GDMS analysis Content (ppm wt)
P	7.24	0.13
B	< 0.06	0.24
Al	29.02	0.16
Ag	125.23	0.14
Cu	0.43	< 0.01
Fe	0.36	2.3
Pb	1.30	< 0.005
Zn	0.28	< 0.01
Ti	0.90	0.17
Ni	< 0.32	0.38

Table 7: Light element concentration measurement before and after vacuum refining.

	Light element analysis by IGA	
elements	Content (ppm wt) Before refining	Content (ppm wt) After refining
C	90	82
N	26	<1
O	29	4.7

The main impurities quantified in the de-metallized broken silicon cells are aluminum, phosphorous and silver. A small amount of lead was detected and few quantities of iron and titanium are present. Other impurities are in lower trace or under detection limit.

After refining, except for boron and iron, all impurities content are reduced. This reduction is particularly significant for P, Al and Ag. This demonstrates the efficiency of the vacuum refining process.

Light element concentration (C, O and N) were measured by IGA (Interstitial Gas Analysis). Low nitrogen content were measured. This proves that no pre-treatment to remove ARC layer is necessary. Oxygen contamination was widely reduced. Only carbon contamination remains in the matrix due to the use of graphite crucible.

4.2 G1 ingot crystallisation from vacuum process

After refining, CEA used the purified silicon to produce one G1 ingot size by conventional DSS crystallization.

Blending of purified silicon with electronic grade silicon was performed to produce an ingot of 10kg. 2.4kg of purified silicon was mixed with 7.6kg of electronic grade silicon and loaded in a silica crucible. The overview of the ingot is shown in the Figure 17.

After lateral side cropping, the ingot was observed by IR transmission imaging to localized defects or precipitate into the silicon ingot as it can be seen on the Figure 17b. Some defects can be observed at the bottom part of the ingot as dark zones.

First optical microscopy observations reveals cavity-like defects. More investigation will be performed to understand the origin and the nature of those cavity-like defects.

Figure 16: Silicon ingot picture (at left), IR transmission imaging of the ingot (at right) and optical microscopy imaging (c) of the dark bottom part (Fig 15.b) revealing cavity like defects.

Predicted resistivity without and with blending versus solidified fraction have been calculated from Arora mobility model regarding dopant concentration profile according to Scheil law.

Resistivity profiles along relative ingot height are plotted on Figure 18. Resistivity mapping was performed by SEMILAB WT 2010D.

The lower curve shows a simulation of the resistivity profile of the purified silicon without blending. This is one of a reason why blending has been suggested to achieve high resistivity compatible for solar cells processing.

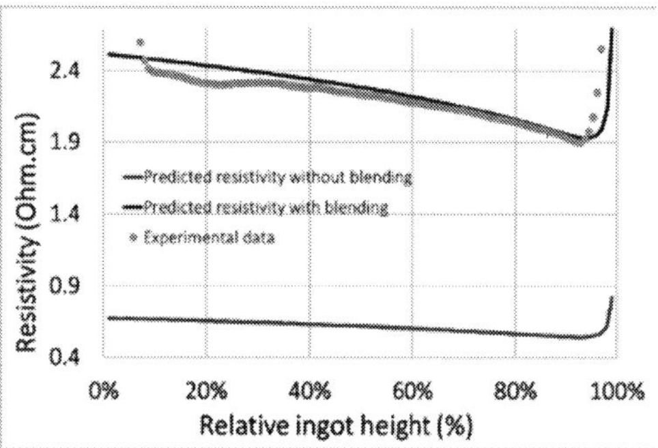

Figure 17 : Resistivity profile along relative ingot height (reporting data of middle of one lateral side). Comparison with predicted resistivity without and with blending with EG-Si.

Measured resistivity fitted well with the resistivity prediction with blending (top curve). Obtained resistivity are in the range 1.9 to 2.4 Ohm.cm and so in-spec with standard solar cells processing.

The next steps of the study will be to make wafers and cells from this ingot to test the material all along the production line. Processes and materials need more investigations to have a better control of final doping levels and limit the blending.

5 CONCLUSIONS

Based on the experiments done so far, the following conclusions can be drawn:

- Regarding the etching, "Used polish etch" or "9:1 etch" are suited for the etching of ARC and diffusion layers. A proper stirring of the mixture "etch – Si-shards". plays also an important role. And strong time vs. ΔT dependency has been observed.
- It has been shown that vacuum process is very efficient to remove ARC and diffusion layers for P-type waste (Phosphorous emitter layer can be easily removed by the vacuum treatment). It is also important to note that control of final doping level works better due to the homogeneous liquid state step.

Recycled silicon from the two routes have been crystallized using DSS process. Electrical and IR characterizations of ingots showed first good results. However more investigations are needed to understand cavity-like defect found on the ingot produced after the vacuum process.

Larger scale is needed for wafer production and testing of electronic properties to provide economical suitability of the different processes and routes.

Finally, the two process routes described in this paper will be transferred to solar cells coming from end-of-life modules.

6 ACKNOWLEDGEMENTS

The project has received funding from the European Union's H2020 research and innovation programme under grant agreement N° 641972 with the consortium: CEA, SINTEF, IMEC, LOSER CHEMIE, SOLITEK, PYROGENESIS, RHP TECHNOLOGY, RESITEC, TU VIENNA, FERROATLANTICA I&D, SUNPLUGGED, THM FRAUNHOFER, PROJECT KOMPETENZ, PV CYCLE, INKRON, and ECM GREENTECH.

7 REFERENCES

[1] http://www.spire2030.eu/cabriss/
[2] Patent number: DE 10 2014 105 143 A1
[3] Patent number: DE 10 2007 034 441 A1
[4] Werner et. Al, **Proceedings from the 25th EU-PVSEC conference,** Valencia Spain, 2010. pp. 1942 – 7
[5] Komatsu et. Al, **Proceedings from the 25th EU-PVSEC conference**, Valencia Spain, 2010. pp. 1924 – 9
[6] Safarian and Tangstad, **Metallurgical and Materials Transactions B**, Vol 43B, Dec. 2012, pp. 1427 – 45.
[7] http://www.iap.tuwien.ac.at/www/surface/vapor_pressure
[8] Ekstrøm, PhD-thesis, NTNU Trondheim 2016:65

FRELP 2 Project – Full Recovery End of Life Photovoltaic

Dr. Pietro Ercole
Consultant
e-mail: percole@atlink.it

ABSTRACT: The abstract has been changed and the presentation also because SASIL believed to enhance the delegate attention on the main critical phases of its FRELP process:
1. Glass detachment,
2. Organic combustion and heat recovery,
3. Original process for pure useful metals (Ag, Cu) recovery by means of HNO3 leaching using a minimal excess.

The reason of this change is to update the Conference delegates on the SASIL jobs done in the last 15 months after the shortage of PV end of life PV modules based on silicon technology.
The presented results need, in the future, some experimental confirmation but the tests already done are promising,

1. Introduction

In this article, the actual state of the art of SASIL studies on FRELP Project are presented.
It makes clear the interesting results achieved in the past two years which are very promising even if need some confirmation and pilot plant continuous and integrated trials.
The new technologies studied and experimented to overcome the problems met in the more critical phases of FRELP project, who are determinant to realize a complete, economical and feasible process for full recovery or reuse of materials present in the end of life PV modules based on crystalline silicon are explained and discussed.
Other reasons for SASIL effort are the following:

- The conviction that PV panels, based on mono and/or polycrystalline metallic silicon, are and, could be, in the future, the more efficient solution for electric energy production from sun light and, also, that its future should be, in some way, linked to end of life panel's materials recovery.
- Last probable future improvement could be achieved recovering metal silicon from its mix with silicon carbide, byproduct of wafers cutting.

This ambitious program could improve the competitiveness of electric energy from this kind of PV panel's against the current lower cost of conventional systems.
In this paper, the essential phases of SASIL new process, which aims to achieve the full recovery of materials present in end of life PV panels are presented.
They are:

- Glass detachment, after aluminum structure and electrical connectors mechanical separation, to obtain 98,5% of pure glass and a residual panel, named sandwich, which contains: the back sheet, EVA partially cured, silicon, metallic conductors and glass residues (1,5% of the initial content).
- Studies and results achievable using available technologies (hot cracking at 500°C or controlled combustion at 850°C) to achieve, on one side, the best heat recovery from the organics complete combustion and metals conductors full oxidation.
- Chemical process, actually in progress, aimed to recovery from the combustion ashes:
  - Metallic silicon by filtration,
  - Useful conductors (Silver and Cupper) by electrolysis of their nitrates solution,
  - Other metals (Aluminum, Tin, Lead etc.) as hydroxides,
  - Neutralization, at the end of chemical process, of diluted HNO_3 solution with $Ca(OH)_2$ obtaining a 5% in weight $Ca(NO_3)_2$ solution, hopefully, sellable as fertilizer.

2. Glass detachment after recovery of aluminum structure and electrical devices

This process phase, must achieve the recovered glass highest purity and the maximum glass extraction yield (98 – 99%) for the following reasons:

- The recovered glass must be completely silicon free because, recycling it in glass batches for both: flat and containers glasses furnaces, creates serious problems. In fact the presence of micro particles of metallic silicon, which have a dilatation coefficient much higher than glass, generates cracks with consequent breakages.
- The glass residues on the sandwich surface will remain with the metallic silicon until the process end creating depreciation of the silicon itself.

This critical situation makes glass detachment one of the most difficult phase of our kind of end of life PV modules recovery process, because the operation is complicated by the old EVA, already

highly cured, and by the consequent difficulty to soften it by heat, avoiding, on the same time, the crosslinking progress, which could achieve irreversible solid state, blocking the operation.
The studies on different technologies for feeding heat to the system, considering, in particular, the infrared ray's absorption behaviors of each material present in the end of life PV module, convinced our team to start a preliminary experimental program which gave comfortable results.
In fact, has been confirmed that the infrared rays cross the glass thickness without heating up the glass itself which is warmed, at the contact surface with EVA, by conduction only.
In this way the main part of energy emitted by infra rays selected lamps is absorbed by EVA first and back sheet later on.
This favorable situation allowed the infrared rays utilization, minimizing the heat absorbed by glass, maximizing the one fed to EVA and, on the same time, reducing the energy losses.
In other words, the sentence was that the heating system with infrared rays can be exploited for enhancing the temperature gradient between EVA and glass, improving, in a surprising way, the glass detachment (quality and yeld).

a. **Study to favor glass detachment exploiting the not complete EVA cross linkage**

The knowledge of crosslinking reactions behaviors of both: fresh EVA and the one present inside the modules end of life, allows calculation of the time available for detaching glass, at various temperatures, avoiding EVA curing progress and back sheet temperature increase over its softening point, which could interfere with the operation.
At the same time, the thermodynamic study allowed to define the back sheet working temperature at which it doesn't start to soften and, so, continues to support the module structure during glass detachment.
The back sheet structure is a thermoplastic polymer as: polyester, fluorinated aliphatic polymers, etc. So, when the temperature overcomes 120°C, it softens creating working continuity problems and frequent intolerable production stop and go.
The following technical articles plaid, an important role for helping us to define the detachment best solution:

b. <u>Hau and others: "Thermoanalytical Characterization of Ethylene Vinyl Acetate (EVA) for Lamination Process Simulation and Gel Content Determination in Photovoltaic Modules, 2012 PV Module Reliability Workshop Golden Colorado (2nd February – 1st March) Denver Colorado Marriott West, NREL USA.</u>

c. <u>Hirschl and others: "Determining the degree of crosslinking of Ethylene Vinyl Acetate photovoltaic modules encapsulates. A comparative study, Solar Cells Energy Materials and Solar Cells (2013; 203 – 218).</u>

Both studies, performed with the scope to understand the EVA curing behavior during PV modules lamination, have been useful to define the cured end of life EVA residual unsaturation, and, consequently, to calculate and optimize both working temperatures and times, to be adopted for glass continuous detachment device.
The graphic inserted below, has been constructed for two types of EVA.
The first one is the original (virgin) polymer, employed for silicon PV modules lamination, which does not contain crosslinking bonds.
The second one is EVA extracted from an end of life PV module in which 65% of the initial double bonds have been spent for crosslinking reactions during the module lamination.

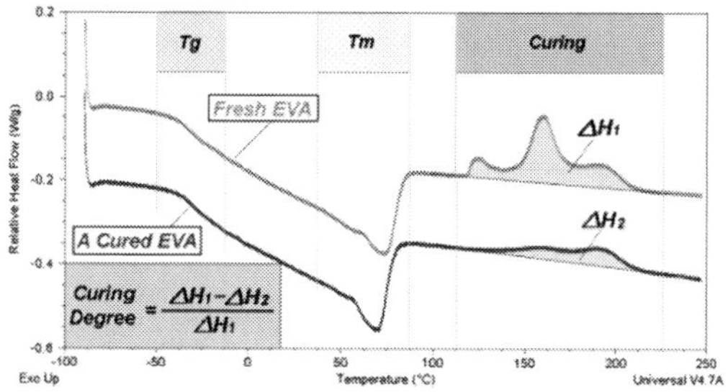

The graphic analysis allows the following important considerations:

- Fresh EVA crosslinking reactions take place at 130°C, blocking the PV module main components (Silicon, Glass and Metallic Filaments supported by the Back Sheet).
- Both polymers behaviors in function of temperature (green curve for fresh EVA and red one, for EVA partially cured) have been defined measuring the heat emitted from the exothermic crosslinking reactions after deduction of the one spent to achieve the curing temperature.
- The behaviors from about 90°C getting along are different because the green one shows three high and well separated picks, while the red one shows the same picks but shorter and flat.

Following the end of life EVA behavior (red curve), the completion of crosslinking reactions starts at 120°C and finish at 220°C before burning.

This study allowed acquiring the following data:

- The behavior of EVA extracted from end of life module, highlights that, when the material achieves 88 – 92 °C, its viscosity starts to slope down, while the crosslinking completion will start again only when the mass temperature achieves 120°C.
- Therefore, the time available to complete glass detachment before the crosslinking reactions will start again, is safe enough. In fact, considering the heating speed adopted for graph construction (10°C/minute), the available time to increase the temperature from 90°C up to 120°C is long enough (3 minutes).
- In the above temperature range, the EVA partially cured softens without curing, helping the glass detachment.
  Obviously, when the glass detachment operation starts at temperature lower than 90°C, more time is, in principle, available.
- Has been proved that, when EVA, partially cured. achieves 130°C, at which the reticulation is not started again or is going very slowly, the detachment operation continuity is strongly compromised because, the back sheet, after few minutes, softens blocking the sandwich band mobility and so the system running.

The drawing, inserted below, shows the pre-prototype used for continuous detachment tests. It represents the view from which the sandwich, free from aluminum structure and electric devices, is fed in the machine. It shows the guides on which it flows to the heating system. The module bands, 25 cm. wide, fed by hand, are then trained on from a roll to the area where the vibrating knife, opportunely located at the junction glass – EVA, provides detachment.
Using this machine, a glass detachment of 1,0 – 1,1 m/minute has been achieved and confirmed. These values guaranty machine productivity high enough for an industrial unit (60 module/hour for each line).

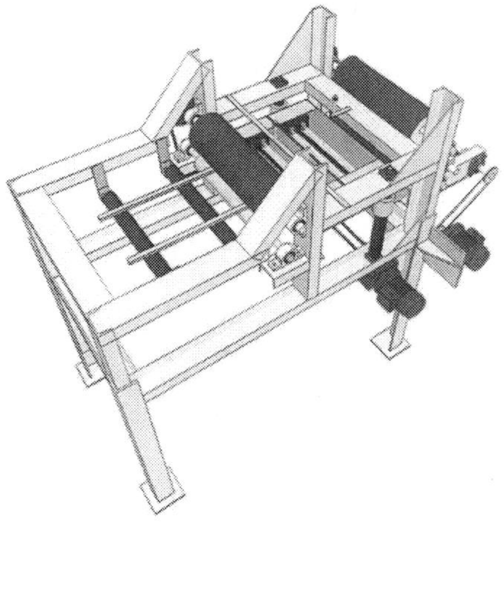

During these experiences, has been tested a glass and EVA temperatures control system based on an algorithm which, taking in account the power absorbed by both materials in function of the distance from the infrared lamps and the PV module, moves a continuous mechanism on which the infrared lamps are linked, up when the temperatures is going higher than the set point, and down when EVA temperature is mowing in the other direction.
The pictures inserted below represent the glass cullet discharged during trials.
The one on the left is the glass cullet as it is at the exit from the machine. The one on the right is the fraction smaller than 5 mm (2 – 3% of the total) after treatment by the optical electronic machine to take out all the dark particles.
The particular, on the middle, refers to the same glass (< 5mm) not treated with the optical electronic system in which the dark particles are present.

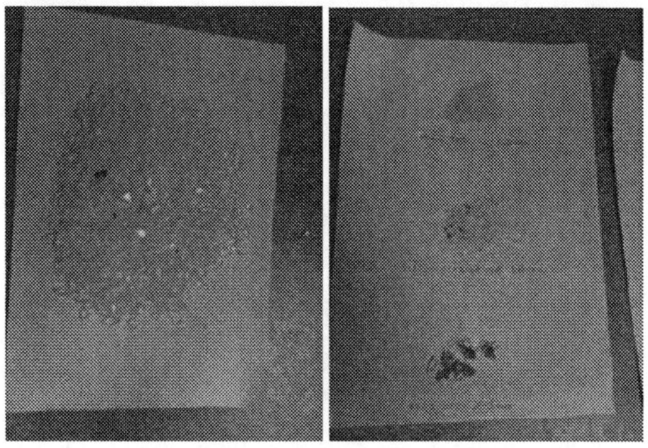

The sandwich, almost free from glass, is, then, fed to the direct controlled continuous combustion device, working at 850 °C, obtaining ashes without any reducing agent inside them.

3. **Energy recovery by controlled combustion at 850°C instead hot cracxking**

The first approach done by Stazione Sperimentale del Vetro, has been the sandwich thermogravimetric analysis, inserted below, which put in evidence, that the weight loss start at 120 °C and finish at 500°C.
The behavior evidences two picks. The first one starts at 120°C and finish at 400°C with a weight loss of 10,5% and the second from 400 and 500°C with a loss of 58,5%. The total weight loss achieves 69,1%.
Respecting the above described thermogravimetric results, the laboratory tests have been organized at a maximum temperature of 500°C for hot cracking and, at the range 600 – 700°C for the direct controlled combustion.

As a matter of facts, EVA releases, first, the acetic acid and, then, aliphatic hydrocarbons unsaturated and poliunsaturated which, at the hot cracking temperature, recombine themselves forming new macromolecules different from the starting ones.

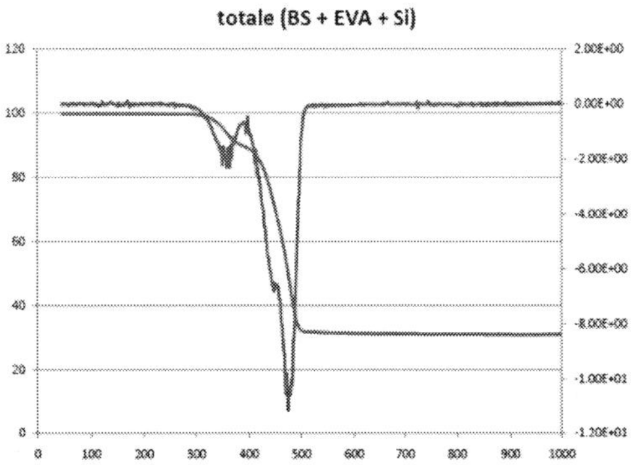

Therefore the thermogravimetric analysis does not take in account the EVA hot cracking ripolymerization phenomenon and, probably for this reason, the defined hot cracking temperature (500°C) and, consequently, the controlled combustion one were to low and wrong.
To clarify the origin of some unsolved problems, define the mechanisms of both solutions and achieve objective conclusions, SASIL started a cooperation with "Stazione Sperimentale per i Combustibili", San Donato Milanese (Italy).

a. Hot Cracking experimental tests and results

The hot cracking tests have been done, using a quartz tubular reactor equipped with a metallic support on which, the sample to be tested, has been propped up.
By means of an electric heating system, the reactor has been heated up at 500°C with a controlled speed of 10°C/minute. Then, the achieved temperature has been maintained for 10 minutes.
The following test conditions have been applied:

- The thermal cracking must be done in discontinue because the risk of air leakages, during ashes feeding and, consequently, of organic materials losses by parasite combustion reactions, are not controllable working in continuous.
- The hot cracking reactor must be maintained under a constant pure nitrogen flow to contrast the organic materials combustion.
- The produced hydrocarbons must be separated by condensation from the nitrogen which must be recycled back to the reactor. This, energy expensive, operation can be done, only, if the cracked intermediary compounds are stable.

The tests, done in cooperation with SS per i Combustibili, put in evidence the following main problems

- At the tubular reactor bottom, where the temperature achieved in the range 500 – 550°C, the unsaturated and poliunsaturated hydrocarbons from EVA, recombine themselves producing new polymers which can't be extracted from the reactor and so are not transferable to the combustion unit.
- At the end of the test, 40,26% of the sandwich sample has been discharged as new polymers cereus materials.
  This product has been utilized for two soxhlet extractions, by isopropyl alcohol and xilol. The materials extracted by the two solvents are morphologically different. The one, from isopropyl alcohol, looks as a light colored rubber containing –COOH groups not present in xilol extraction.
- The hot cracking residue (ashes), 40% by weight of the starting sample, contains small glass fragments, metallic silicon and conductor metals.
- The hot cracking results and the study done, for defining the reactions mechanisms, put in evidence that:
  - The thermal cracking starts at 120°C producing acetic acid from EVA (poor heat power).
  - From 120 – 500°C almost all the unsaturated aliphatic hydrocarbons from EVA, came out.
  - At 500°C, the back sheet cracking (mainly polyester) was not started yet.

As is known, the panels sandwich, after glass detachment, contains useful metals, as Silver, Cupper, etc. which could be separated from hot cracking ashes by means of a proper acid leaching followed by the electrolytic cathodes depositions. Due to the extremely reduced conditions, generated by thermal cracking, the metals, obviously, remain at the metal state, with the following negative consequences:

- The metals are more resistant to the HNO_3 attack, so the reaction, necessarily, needs more HNO_3, more time and higher temperatures.
- The metals oxidation, by means of HNO_3 oxidant activity, increases the acid specific consumption around 25%.
- Due to these problems, the sandwich hot cracking viability is compromised and so, SASIL decided to burn the polymers (EVA and back sheet) by direct controlled combustion at 650 - 700°C, using their heat power to achieve and maintain, the desired combustion temperature.

Results discussion

All the results are collected and explained in the following tables:

Table n°1 reports the hot cracking mass balance.
It clarifies that, at 500°C, the obtained products, collected at the reactor exit, mainly aliphatic unsaturated and poliunsaturated monomers and oligomers, recombine themselves and crosslink, as

the rubber nature of the isopropyl alcohol extract, clearly explain.
The result is a solid cereus material which represents 40% of the initial sandwich weight; 25% of it soluble in isopropyl alcohol and 15 - 16% soluble in xilol.
So the new polymers, soluble in isopropyl alcohol are coming from EVA and the materials soluble in xilol from the back sheet as polyester, intermedia oligomers and aromatic aldehydes .
These data say also that, at 500°C, the back sheet does not crack at an appreciable level.

Table n°1 Hot cracking mass balance		
Frazioni	Ripartizione delle tre frazioni	
	Grammi	% Sul peso iniziale
Residuo solido	12,7785	39,78
Isopropanolo	7,9836	24,85
Estratto xilolo	4,9505	15,41

Table n°2 refers to acetic acid and fluorides emissions in the atmosphere. The data are expressed in grams and in percentages referred to the sandwich weight fed to the hot cracking.

Table n°2 Emissioni di Acido Acetico e Fluoruri				
Intervallo di T °C	Acido Acetico		Fluoruri	
	mg	% Sul peso iniziale	mg	% Sul peso iniziale
20 – 360	417368	1,299	237,336	0,001
360 - 500	390432	1,215	6772.000	0,021

Must be underlined that both compounds concentrations are small because the waste gas were strongly diluted by nitrogen and, particularly, the fluorine concentrations are, obviously, a lot smaller than the acetic acid ones.
The data indicate also that fluorine released at the temperature range 20 to 360°C is 20 times smaller than the ones released from 360 to 500°C. On the contrary, acetic acid concentrations don't change along the temperature increase.

Table n°3 reports the chemical analysis of gas captured in plastic bags at the following temperature ranges 300 – 350°C and 415 – 450°C.
The data put in evidence:

- At 300 – 350 °C only a small content of methane was present in reaction flow but at 415 – 450°C, saturated, unsaturated and poliunsaturated aliphatic hydrocarbons appear.
- The aromatic compounds absence, which could come, only, by polyester from back sheet structure, say that, until 450°C, the back sheet thermal cracking is not yet started.
- The release (around 40% of sandwich sample weight) collected at the temperature range 415 – 450°C shows the presence of unsaturated aliphatic hydrocarbons which, at higher temperatures recombine themselves forming the cereus material.

These results allow concluding that the hot cracking of polymers present in the sandwich coming from PV panel's end of life, can't be adopted because:

- Difficulty to dominate the different polymers (EVA and Back Sheet) hot cracking.
- Poor energy recovery due to nitrogen separation.
- Evident need of doing the hot cracking at temperature over 800°C to overcome the recombination of unsaturated monomers.
- High cost of production.
- Problems at the project successive phases caused by the mass very reducing state.

Table n°3 Analisi prelievi effettuati, nell'intervallo di T°C 300 – 360 e 415 – 450				
Composiz. della miscela	Analisi dei gas in sacca alle temperature indicate			
	Prelievo a 300 – 360°C		Prelievo a 415 – 450°C	
	Rilevato mg/Nm3	A 100 mg/Nm3	Rilevato mg/Nm3	A 100 mg/Nm3
CH4	0,08	0,09	0,12	0,13
CO2	0,15	0,14	0,97	1,03
O2	0,37	0,40	0,33	0,35
N2	91,70	99,36	90,39	96,19
CO	0,01	0,01	0,73	0,78
H2	-	-	0,22	0,23
C2H6	-	-	0,11	0,12,
C2H4	-	-	0,38	0,40
Propano	-	-	0,12	0,13
Esani e superiori	-	-	0,18	0,19
n.Butano	-	-	0,08	0,09
+ Butene 1	-	-	0.30	0.30
Transbut2	-	-	0,08	0,09
Cisbutene2	-	-	0,01	0,01
n. pentano	-	-	0,03	0,03
Altro	-	-	O, 19	0,20

b. Controlled combustion experimental tests and results

As has been already said, the selected temperatures were not high enough to obtain the complete combustion and the full metals oxidation.
Despite the wrong selected temperatures (650 – 700°C) the tests done allowed to achieve the following advantages:

- The direct controlled combustion put in play all the polymers thermal power reducing, at minimum, the heat losses and, of course, minimizing the outside combustible consumption.
- Moreover the following benefits have been foreseen:
  - No problems during the controlled combustion in continuous.
  - No problems to complete 100% of combustion, at the right temperature.
  - For evident reasons, no problems for post polymerization.
  - Easier waste gases analysis control.

- The residual ashes from controlled combustion allowed to foresee the benefits listed below:

  - According with the technical literature, the oxidation of Silicon to SiO2 is unappreciable.
  - According with instrumental chemical analysis, at 650 – 700°C, the metals oxidation, is not complete but, at temperature and time suggested by SS per i Combustibili should be possible.
  - If not, after combustion, the inorganic ashes must be heated up, by means of an induction furnace device, to 1000°C, in presence of oxygen for converting 100% of metals to their oxides, controlling reaction time to avoid consecutive reactions creating compounds insoluble in HNO3, as corundum by Al2O3. .

It must be said that studies and experiments carried out confirmed the expected results and opened the door to overcome the initial FRELP process setting. For the above listed reasons, has been decided to continue the cooperation with "SS per Combustibili" with the scope to make clear:

- Study on the Sandwich direct controlled combustion,
- Mechanisms of the combustion reactions,
- Identify the intermediary compounds coming from different polymeric materials,
- Define combustion temperature and time for quantitative continuous combustion of the intermediary compounds

The direct EVA and back sheet polymers combustion tests have been done in a furnace, designed for wood combustion studies, equipped with 6 KW power unit and fed by hands.
For the reasons reported before, the initial combustion temperature was in the range 650 – 700°C.

- During the tests the combustion temperature was maintained under control all the time. The waste gases sampling has been organized to identify each compound present and its concentration.
- Three tests have been done with three subsequent charges of material 0,5 kg each.

Results discussion
During the furnace heating up, the combustion camera temperature grew up from 330 to 650 – 700°C.
The final temperature (650 – 700°C) was maintained constant for each test time (0,5 kg in 15 minutes).
Has been confirmed that the cut Sandwich burns very easy and spontaneously

Table n°4 reports the concentrations of inorganic compounds present in the combustion waste gases. The values on the left are data obtained at the tests oxygen content and the ones, reported on the right, have been normalized for 13% oxygen imposed by the European law. The presence in the waste gases of fluorides, are in concentrations higher than law limits (average 73 mg/Nm3, a lot over the European law limit).
That means that the eventual future industrial combustion device must be equipped with a proper system for its abatement.

Tabella n°4
Emissioni di HF – HCl - NH3 all'ossigeno di prova ed all'ossigeno di riferimento (13%)

Prova	O2 di prova			O2 di Riferimento (13%)		
	HF mg/Nm3	HCl mg/Nm3	NH3 mg/Nm3	HF mg/Nm3	HCl mg/Nm3	NH3 mg/Nm3
1	54,7	5,5	4,2	73,0	7,3	5,6
2	74,9	13,6	22,0	94,9	17,3	27,8
3	22,4	5,5	4,6	30,3	7,5	6,2

Table n°5 contains the gas analysis results obtained at the O2 content measured during the trials and Table n°6 contains the same data normalized for the O2 content of 13%.
These results, apparently negative, really, are indicating that, at the working temperature (650 – 700°C), the combustion can't be completed.

- The CO emissions overcame largely the 7000 ppm
- The COT was extremely high particularly in test n°1 and 3 in which the real value was over the instrument limit (2000 mg/Nm3),
- The NOx emissions were, on the contrary, very small and this is the consequence of the low combustion temperatures.
- The PTS (dust emission) overcame 1500 mg/Nm3.

Tabella n°5
Analisi dei fumi eseguite nelle 3 prove di combustione controllata del sandwich da fotovoltaico a fine vita. I valori riportati sono riferiti all'O2 rilevato analiticamente

Prova	Fattori determinati nel corso della prova					
	NOx Ppm	CO ppm	CO2 %	O2 %	COT mg/Nm3	PTS mg/Nm3
1	30,5	5387,8	5,3	15,2	3502,0	2101,2
2	26,4	6492,2	5,7	14,7	5435,7	1535,4
3	25,3	6345,8	5,4	15,1	5302,5	1607.1

Tabella n°6
Analisi dei fumi effettuate nelle 3 prove di combustione controllata del sandwich da fotovoltaico a fine vita. Analisi omogeneizzate al 13% di O2

Prova	Fattori determinati nel corso della prova					
	NOx ppm	CO ppm	CO2 %	O2 %	COT mg/Nm3	PTS mg/Nm3
1	42,1	7431,5	7,3	13,0	4830,3	2898,2
2	33,4	8223,7	7,3	13,0	6885,4	1944,9
3	34,3	8596,2	7,4	13,0	7181.2	2176,5

In fact, as is shown in table n°7, inserted below, a series of aldehydes are still present in the waste gases. Some are aliphatic unsaturated or not, some are aromatic.
Most of them come from PET (polyethylentereftalate) which, when heated up, could generate both families of aldehydes which, as the Table 7 data are showing, could be stable at temperature lower than 650 – 700°C, bat not at 850°C.
So, all these compounds will burn at higher temperatures and residence times (850°C and 3 seconds).
The three columns on the left show the values, at the O2 content measured during tests (violet); the three on the right, the ones standardized for the O2 content of 13% in volume (light blue). The aldehydes dangerous for human health have been highlighted in yellow.
It must be underlined that:

- All the aldehydes are burnable and so at higher temperatures they will disappear.
- At the maximum working temperatures (700°C), the aromatic aldehydes which come from tereftalic acid, component of back sheet polyester structure, start to appear.

So the conclusions of this process critical phase are:

- The combustion tests put in evidence that our material is a good combustible. It burns easily, in 15 minutes, 0,5 kg of sandwich.
- The high contents of reducing agents (CO, COT, PTS, and aldehydes) depend, clearly, from the temperatures (650 – 700°C), not enough to complete the combustion.

Tabella n°7
Emissioni di Aldeidi all'O2 di Prova e all'O2 di riferimento

Aldeidi	Ossigeno di Prova			Ossigeno 13%		
	Prov a 1 mg/N m3	Prov a 2 mg/N m3	Prov a 3 mg/N m3	Prov a 1 mg/N m3	Prov a 2 mg/N m3	Prov a 3 mg/N m3
Formalde ide	7,9	1,8	3,8	10,9	2,3	5,2
Acetaldei de	81,1	90,0	106,4	111,9	114,0	144,1
Acetone	0,6	3,0	15,2	0,8	3,8	20,6
Acroleina	12,6	23,4	28,8	17,4	29,7	39,0
Propional deide	7,2	6,8	9,8	9,9	8,6	13,2
Crotonald eide	1,2	9,8	13,4	1,7	12,5	18,2
Butirralde ide	2,8	2,9	5,3	3,8	3,7	7,2
Benzaldei de	1,1	16,9	21,1	1,6	21,4	28,6
Isovaleral deide	0,0	0,1	0,3	0,0	0,1	0,4
Valer aldeide	1,3	1,8	7,6	1,8	2,3	10,3
o con l'aldeide	0,6	1,7	2,0	0,8	2,1	2,6
m.p.tolual deide	5,9	11,7	11,2	8,2	14,8	15,1
Xilolaldei de	0,7	1,4	1,8	1,0	1,8	2,4

- The recent tests and the achieved results allow to conclude that at higher temperatures (850°C) and adopting a waste gases residence time inside the combustion chamber (over 3 seconds) .is possible to complete the combustion and so to exploit all the heat power coming from polymers.
- Has been demonstrated that temperature and residence time could be achieved and maintained constant in a continuous system.
- Besides, it must be underlined that the obtained ashes are easily transportable and contain small glass fragments.
- On the light of the results discussed in this presentation the decision to follow the direct combustion line has been taken.

4. Useful metals (Silver and Cupper) recovery by means of an original HNO3 attack using a minimal acid excess.

The dissolution of metals, deposed on the cathodes by the free HNO3 action caused very bad results of the first electrolysis tests due to the huge HNO3 excess employed for ashes acid leaching.
Moreover, the presence of a maxima azeotrope (65% HNO3 - 35% H2O boiling at 122°C), doesn't give any possibility to recycle back to the reactor the HNO3 excess.
In fact, the HNO3 excess recovery, by concentration, is, economically impossible, because, during concentration, the HNO3 percentage in liquid phase containing the metallic nitrates increases.
On the other hand, does not make any sense to dilute with water the nitric solution to achieve a HNO3 content around 3%, before feeding the solution to the electrolytic devices, for the following main reasons:

- Loss of the HNO3 excess,
- Environment problems,
- Realize an uneconomic electrolysis device due to the very low and insufficient productivity and, consequently, high depreciation and running costs.

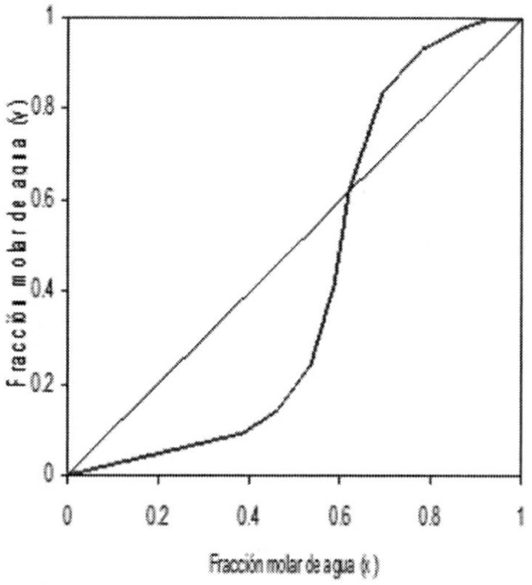

This situation stopped the experiments and opened a period of serious reflection on how to overcome these problems, apparently unsolvable.
So, SASIL had to face a series of difficult studies with the aim to meet the following results:

- Use a very low HNO3 excess to achieve a final HNO3 content of 3% max.
- Realize the ashes acid attack conditions strong enough to dissolve all the metal oxides,
- Increase the nitrates content in solution fed to electrolysis, avoiding solution concentration technologies,
- Improve the electrolysis efficiency.
- Minimize the electrolysis device investment and, consequently, both depreciation and running costs,
- Maximize the productivity of electrolytic metals separation devices.

Below, the adopted solutions are described:
The first two points to be assessed were:

- The HNO3 water solution at 50% by weight, for acid leaching, must contain 1,2 equivalents/1 equivalent of overall metals present in sandwich ashes coming from controlled combustion.
- The small excess (0,2 equivalents), at the reaction end, represents around 3% of HNO3 free in nitrates solution to be fed to the electrolysis devices.

Obviously, the small quantity of HNO3, water solution at 50% by weight, fed to the reactor is not enough to create a mixable silicon suspension and, on the same time, the acid excess itself seems, in principle, too low for guarantying the complete conversion of metal oxides in their nitrates.
So, has been necessary, to find a solution capable to increase both:

- Ratio Liquid/solid and HNO3 excess inside the reactor without changing the HNO3 equivalents imposed for the fresh ashes fed to the system.
- The practicable reaction conditions consist on heaping up, in regime condition, a reacted solution quantity big enough for recycling a significant part of it to the reactor itself. In so doing, is possible to control both ratios to levels good enough for the complete oxides salinization and so achieving the expected results

This recycle creates the following benefits:

- The silicon suspension is diluted thanks to the recycled reacted solution, having exactly, the same composition of the one generated by the ashes and HNO3 imposed quantities.
- Controlling the ratio between the solution recycled back and the one corresponding to the fed raw materials (ashes and HNO3), the reaction conditions can be optimized.
- It must be underlined that each part of regime solution contains the initial excess of HNO3 (0,2 equivalent of HNO3/1 equivalent of metal oxides).

So, increasing the above mentioned ratio, the HNO3 excess will increase more and more and the silicon concentration in the suspension will decrease.

- Adopting a ratio = 5/1, the HNO_3 equivalent in the reactor achieve 2,2 equivalents/1 equivalent of metal oxides. At the reaction end, 1,2 equivalents (1,0 as metals nitrates and 0,2 as free HNO_3) are present inside the reactor and 1 is added, continuously, by means of the 5 volumes of recycled regime solution.

To date, we could say that the actual state of the art is very close to meet the goals indicated at the studies beginning.

In the SASIL Process flow sheet inserted below all the new technical provisions are presented.
It is, anyway, interesting and useful to pay attention to some significant expedients studied to collect and feed to the reactor all the needed HNO_3, which is used, first, to wash the metal silicon and complete recovery and quality metals nitrates and, then, to achieve a quantitative recovery of the small HNO_3 excess (0,2 equivalent) and its utilization.

- The first is that all the necessary HNO_3 is fed in water solution at 50% by weight to the second stage of the belt filter to meet, there, the silicon separated by the reaction solution, wash it for taking away the metals nitrates and other silicon pollutants present. The collected solution is then fed continuously to the reactor.
- The second is that the water necessary to dilute the HNO_3 in the reactor from 50 to 25% is fed to the third stage of the filter where washes the silicon taking away the HNO_3 present in it. The diluted HNO_3 solution is then fed to the reactor.

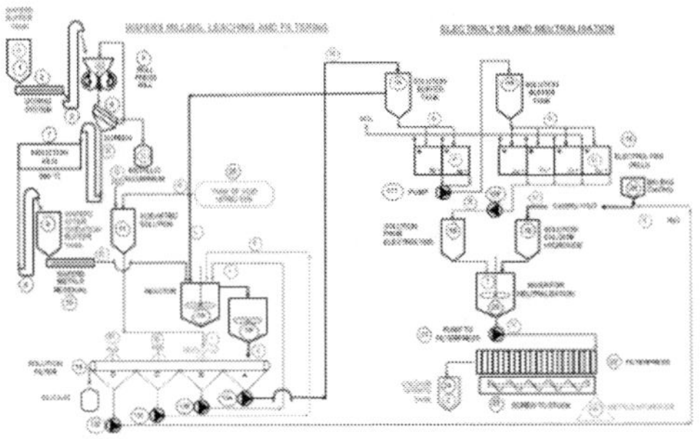

The silicon, with glass not taken away during glass detachment (very small if referred to the total glass (1,5% max) but big if referred to the controlled combustion ashes (20 - 25%) is washed, once again, in the fourth stage of the filter.
This water is then collected with the bottom solution coming from the electrolytic cells in another reactor in which a controlled quantity of $Ca(OH)_2$ suspension for HNO_3 excess neutralization obtaining a 5% $Ca(NO_3)_2$ solution to be sold, hopefully, as a fertilizer.
During this final reaction, the metals as Pb, Sn etc. precipitate as idroxydes, not soluble in water at neutral pH and so separable from the useful solution by means of a press filter.

5. Conclusions

The work done for facing all the emerged problems has been interesting and satisfactory.
The achieved results are good enough to design a pilot unit respecting the inserted flow sheet.
The small unit will test, working in continuous, the described solutions bonded each other in order to harmonize and optimize the overall process.
For sure, other problems will be faced and properly solved.
The most important, because involves several millions of end of life PV modules, will be the large differences between the PV silicon panels produced, time by time, in the last 25 – 30 years.
The changes, in terms of chemical composition, glass types and their thicknesses, conductor metals and ratios between them, back sheet materials etc., done, during the overall period, to reduce initial costs, panels weight and to increase power generation capability, will generate different scenarios.
This situation needs practical, economic and feasible technologies for treating each type of end of life PV panels, as they are, adopting a proper solution.
The feasibility of potential solutions would be faced easier in cooperation with metallic silicon, PV panels, EVA and back sheet producers for discussing materials optimal recycle or reuse.

ANALYSIS OF THE ENERGY AND ECONOMIC INFLUENCE OF THE O&M ANNUAL COST IN THE PROFITABILITY OF PV SYSTEMS

J.C. Lomas[1], E.Muñoz-Cerón *[2], J.Montes-Romero[3], G. Nofuentes[3] and J. de la Casa[3]

[1] Gerión Ingeniería S.L. Cl San Antón 72, 5th Floor. Office 18. 18005 Granada, Spain
www.gerion.es Phone: +34 958 088 090. E-mail: clomas@gerion.es

*[2] Corresponding author. IDEA Research Group (Research and Development in Solar Energy)
Electrical Engineering Department. University of Jaén
Campus las Lagunillas s/n. Building A3, Office 234. 23071 Jaén, Spain
Phone.: +34 953 212 462. E-mail: emunoz@ujaen.es.

[3] IDEA Research Group (Research and Development in Solar Energy)
Electronics and Automation Engineering Department. University of Jaén

ABSTRACT: In this paper, the energy production of a PV plant located in the South of Spain is compared with the annual O&M cost. This comparison has been made since its commissioning in late 2007, where the plant was subscribed under the most favorable legislative framework that has ever existed in Spain for these systems (RD 661/2007). The energetic and economical influence of this O&M cost in the profitability of the PV plant will be addressed since then. In the literature it has been superficially covered that the reduction of the annual expenditures in the O&M tasks has a positive effect in the profitability of the investment, precisely in the Levelised Cost of Electricity. On the other hand, this reduction in the maintenance also means a decrease in the energy generated by the system, which has a negative effect in the electrical production of the plant. Consequently, the incomes received for the energy injected to the grid also decrease. Which of these two effects of the annual O&M expenditure is larger, the increase in the profitability of the investment or the decrease in electricity sales? In the case of Spain, according to the existing legal framework, unfortunately, seems to be more profitable to "abandon" the O&M tasks of the plant rather than keep on investing in improving its performance yield.
Keywords: Economic Analysis, Grid-connected, Operation and Maintenance

1 INTRODUCTION

Boosted by a favorable regulatory framework, in the period 2007-2008 PV plants in Spain were understood more like an investment product rather than a sustainable awareness-raising of the investors [1]–[3]. In this sense, the profits obtained were obviously proportional to the electrical production of the PV system, because under the Feed-in Tariff schemes existing at that time in Spain, the more you generate and inject into the grid, the larger the revenues should be. Therefore, the annual investment in Operational and Maintenance (O&M) tasks is justified by the aim of maximizing the energy generated, thus increasing the incomes resulting from the sale of the electricity.

Figure 1: PV plant under analysis

In this paper, the energy production of a PV plant (see figure 1) located in the South of Spain [4] is compared with the annual O&M cost. The characteristics of the plant are summarized in table I.

This comparison has been made since its commissioning in late 2007, where the plant was subscribed under the most favorable legislative framework that has ever existed in Spain for these systems (RD 661/2007), when an unit of electricity injected to the grid was remunerated at 0.44 €/kWh during the following 25 years [5]. The energetic and economical influence of this O&M annual cost in the profitability of the PV plant will be addressed since then.

Table I: Main characteristic of the PV system

Configuration of the 100kW generator	
Series-connected modules	18
Modules in parallel	32
Short-circuit current (A)	257.6
Open-circuit Voltage (V)	574.2
Current at the maximum power point (A)	236.48
Voltage at the maximum power point (V)	462.6
Maximum Power (W)	109,440
Number of trackers (one-axis, polar based)	48
Modules per tracker	12

In the literature it has been superficially covered that the reduction of the annual expenditures in the O&M tasks has a positive effect in the profitability of the investment, precisely in the Levelised Cost of Electricity (LCOE) [6]–[8]. Nevertheless, this reduction in the maintenance also means a decrease in the energy generated by the system, which has a negative effect in the electrical production of the plant as it can be observed in figure 2, where the differences in the energy data from 2014 and 2015 are large compared to the previous years, matching a dramatic reduction in the O&M expenditures. Consequently, the incomes received for the energy injected to the grid also decrease. Which of these two effects of the O&M cost is larger, the increase in the profitability of the investment or the decrease in electricity sales? In the case of Spain, the answer to this question directly depends on the existing legislative framework.

2 O&M EXPENDITURES VS. PV ELECTRICITY

The first step of this work has been the identification of the annual expenditure in O&M tasks from a realistic point of view. Traditionally, at least in the specialized PV reports, the annual O&M cost of a standard PV plant has been assumed to be around 1-1.5% of the initial investment cost [9], but there is scarce data from real operation costs of PV plants. In the Grid-Connected PV system analyzed in this work, the annual O&M expenditure raised up to around 16,760€, which meant approximately 2.2% of the initial investment In the maintenance tasks different items have been included such as taxes, insurance, alarms and security, stock of small electrical and mechanical elements and the hiring of a technician for the surveillance and reparation of minor issues (see table I).

Although the PV plant studied has been object of a proper maintenance protocol, since 2013, as a consequence of the unstable legislative framework existing in Spain, where several royal decrees with retroactive application were endorsed, the annual budget for the maintenance tasks was dramatically reduced in order to afford the payment of the investment loans.

Table II: O&M annual costs distribution

Concepts	Expenditure
Salary of a technician	12000 €
Stock of small electrical and mechanical elements	1000 €
Alarms and surveillance	420 €
Insurance	1396 €
Taxes	250 €
Auxiliary services	900 €
Renting of machinery and equipment	800 €

The annual energy production of the PV plant is graphed in figure 2, where the effect of the decrease in the O&M annual is noticeable in 2014 and 2015. Nevertheless, although the reduction in 76% of the expense in O&M tasks the drop in energy production did not follow the same descending trend.

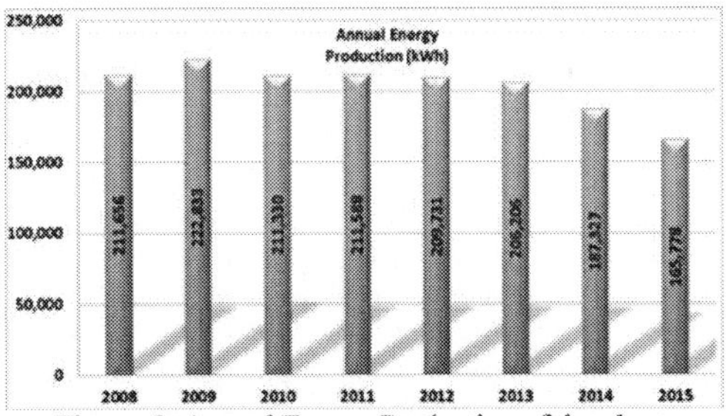

Figure 2: Annual Energy Production of the plant

Table III shows a comparison among the best year of generation, which was recorded in 2009, and the average production in the period when the O&M was not decreased, which correspond to 2008-2012. Therefore, it allows us to analyze the differences in the electricity generated and compare it with the O&M variations. The first conclusion is that the decrease in the energy generated is much lower that the difference in the annual O&M cost, so current owners may wonder whether it is interesting to maintain the level of annual O&M cost at their PV plants

Table III: O&M expenditures vs. PV electricity generated

O&M annual cost		Difference
Annual O&M 2008-2012	16766 €	-
Annual O&M 2013-onwards	3966 €	-76.34%
Electricity		**Difference**
Maximum Electricity 2009	222833 kWh	-
Average Electricity 2008-2012	213427 kWh	-
Electricity 2013	206206 kWh	-3.38 % (max) -7.46 % (aver.)
Electricity 2014	187327 kWh	-12.22 % (max) -15.93 % (aver.)
Electricity 2015	165778 kWh	-22.32 % (max) -25.6 % (aver.)

Although it is remarkable that the difference in the annual O&M cost are much larger than the electricity drop, a proper analysis should be undertaken in future years to analyze whether the energy keeps on falling or it stabilizes, which would be a signal that maybe there is no need to dedicate so much annual budget in these O&M tasks.

3 ECONOMIC INFLUENCE

The last Royal Decree passed in Spain in June 2014 (RD 413/2014) modified dramatically the retribution mechanism with the aggravating circumstances that its application had a retroactive effect in the already operating PV plants. The traditional Feed-in Tariff scheme was cancelled and it was substituted by a fixed term (Investment Return) which depends on the PV power installed and a variable term (Operational Retribution) which depends on the energy injected. This last term (Ro) is limited to a certain number of operational hours each year, beyond which, the owner will received no retribution. Besides, the owner of the plant also receives a variable amount depending on the electricity injected to the grid and sold at market price but in this case, there is no upper limitation [10].

This maximum threshold of operational hours for the variable retribution (R_O) can lead to a situation whether, from an economic point of view, it is more feasible to decrease the O&M annual cost at the expenses of reducing the electricity produced, but maintaining the profitability of the plant according to the limitations of the applicable legislation.

In 2014 and 2015, the energy generated was below the expectations and the operating data from previous years as it was stated in figure 2. The reason for that decrease was justified because the trackers were either broken or manually stopped at a fixed position to avoid mismatching losses. The cost for the reparation was unacceptable for the owner. In other words, the annual O&M proportion was dramatically reduced.

Presumably this drop in electricity production should turn into a reduction in the retribution received, because there is a variable term depending on the electricity injected to the grid (Ro+market). Nevertheless, according to the last Spanish law, there is a maximum threshold for the number of hours of operation, therefore, a comparison between the maximum hours of operation achievable in our plant with the O&M cost reduction and its resulting energy decrease should be undertaken.

In those years, the plant reached 1873 and 1658 hours of operation respectively, so the incomes received for the energy variable terms (Ro+market) achieved 14114 and 12490€, where Ro was remunerated at 23.344€/MWh [11] and the term corresponding to sale of the electricity produced has been averaged to 51 €/MWh [12] . During this period, the O&M outcomes were reduced to 3966€, just to maintain the minimum and unavoidable expenditures. The balance between the energy related incomes and the O&M costs was positive in both cases (10148€ and 8524€).

On the other hand, it is interesting to analyze a simulated scenario where the plant is perfectly maintained, and therefore it could produce all the electricity according to the original design specifications. If the record production of the PV plant is taken, with its corresponding original maintenance cost, and it is translated to the limitation of hours of the last Royal Decree the result is surprising.

The record electricity production of the plant was achieved in 2009. This energy production meant 2228h of operation, which it is above the limit imposed by the RD 413/2014, fixed at 2102 hours. In this scenario, the energy remuneration estimated under the ideal energy scenario translated to current year, but maintaining the original O&M cost, meant a negative balance of -284€, which is below the ones obtained in the presumably worse scenarios from 2014 and 2015. Comparing these results, under the current Spanish legal constrains, it is more profitable to "abandon" the maintenance of the PV plant than keeping on the original O&M expenditures.

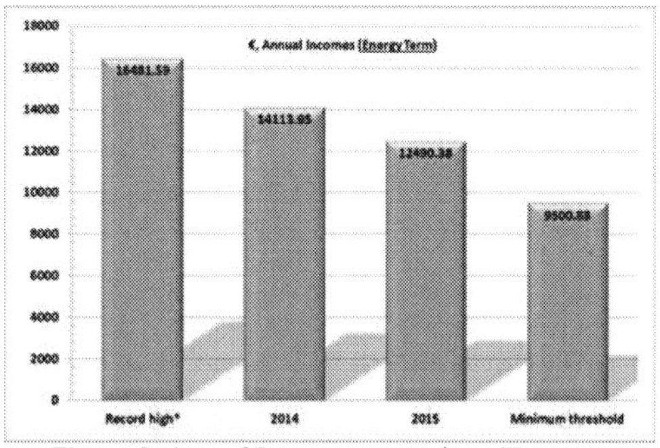

Figure 3: Annual Incomes comparison (Energy term)

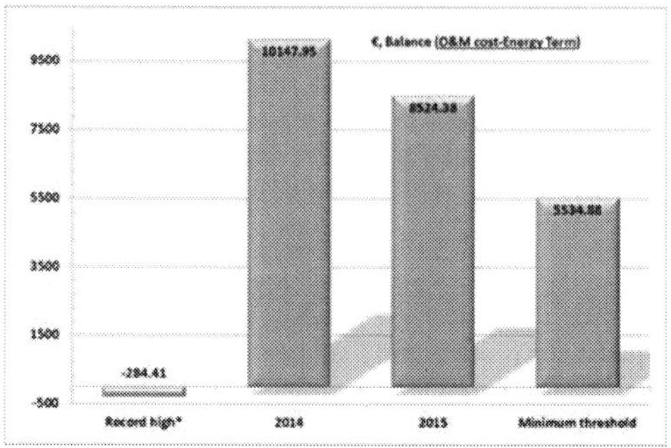

Figure 4: Annual Incomes balance (Energy term-O&M cost)

Moreover, a lowest scenario possible is also included. The current PV laws in Spain define a minimum threshold in order to receive any remuneration at all corresponding to the Ro term. Therefore, assuming the plant only produces this minimum, it would generate 126100kWh, which corresponds to the minimum threshold of 1261hours for receiving the variable retribution. In this case, and with the minimum O&M expenditures possible, it is still more profitable for the annual liquid asset than the results obtained in the record scenario with the current state of the law.

Table III: O&M annual costs distribution

Incomes vs O&M cost	**Record** 222833 kWh	**2014** 187327 kWh	**2015** 165778 kWh	**Threshold** 126100 kWh
Hours operation	*2102*	1873	1658	1261
Ro (€)	5117.1	4560.3	4035.7	3069.8
Pool (€)	11364.5	9553.7	8454.7	6431.1
O&M (€)	16766	3966	3966	3966
Balance (€)	-284.4	10147.9	8524.4	5534.9

4 CONCLUSIONS

The major side effect that the current legislation is causing in Spain is the abandonment of the maintenance protocols of PV plants, resulting in a decrease of the energy generated but not in the profitability. The maximum threshold of operational hours for the variable retribution (Ro) can lead to a situation whether, from an economic point of view, it is more feasible to decrease the O&M annual cost at the expenses of reducing the electricity produced, but maintaining the profitability of the plant according to the limitations of the applicable legislation.

The lesson learned after this analysis is that under the economic situation of the PV investment made by the owner in 2007 and the legal framework existing is Spain, the lack of liquidity is more influencing that improving the energy yield of the plant by investing more annual budget in O&M tasks.

5 REFERENCES

[1] R. Gellings and I. Rutschmann, "Limitless growth," *Phot. Int.*, vol. 4/2008, p. 32, 2008.

[2] P. del Río, "Ten years of renewable electricity policies in Spain: An analysis of successive feed-in tariff reforms," *Energy Policy*, no. 36, pp. 2917–2929, 2008.

[3] J. De La Hoz, O. Boix, H. Martín, B. Martins, and M. Graells, "Promotion of grid-connected photovoltaic systems in Spain: Performance analysis of the period 1998-2008," *Renew. Sustain. Energy Rev.*, vol. 14, no. 9, pp. 2547–2563, 2010.

[4] J. C. Lomas, E. Muñoz-Cerón, J. V. Muñoz Díez, and M. and de la C. Fuentes J., "Are the Spanish PV plants installed under the RD661/2007 still profitable?," 2014.

[5] Ministry of Industry, Energy and Trade. Government of Spain, "Royal Decree 661/2007." Boletín Oficial del Estado, 2007.

[6] E. Muñoz-Cerón, *New trends in Photovoltaic Economy and development of a different technological spectral approach in Concentrator Photovoltaics*. Spain: University of Jaen, 2014.

[7] D. L. Talavera, J. de la Casa, E. Muñoz-Cerón, and G. Almonacid, "Grid parity and self-consumption with photovoltaic systems under the present regulatory framework in Spain: The case of the University of Jaén Campus," *Renew. Sustain. Energy Rev.*, vol. 33, pp. 752–771, May 2014.

[8] D. L. Talavera, E. Muñoz-Cerón, J. de la Casa, M. J. Ortega, and G. Almonacid, "Energy and economic analysis for large-scale integration of small photovoltaic systems in buildings: The case of a public location in Southern Spain," *Renew. Sustain. Energy Rev.*, vol. 15, no. 9, pp. 4310–4319, 2011.

[9] S. Danchev, G. Maniatis, and A. Tsakanikas, "Returns on investment in electricity producing photovoltaic systems under de-escalating feed-in tariffs: The case of Greece," *Renew. Sustain. Energy Rev.*, vol. 14, no. 1, pp. 500–505, 2010.

[10] Ministry of Industry, Energy and Trade. Government of Spain, "Royal Decree 413/2014." Boletín Oficial del Estado, 2014.

[11] MINETUR, "Ministerial Order IET/1045/2014," *Boletín Of. del estado*, vol. 150, pp. 46430–48150, 2014.

[12] OMIE, "Operador del Mercado OMI-POLO ESPAÑOL, S.A. (Spanish Electricity Market Operator)." [Online]. Available: http://www.omie.es/inicio.

RELIABILITY OF PHOTOVOLTAIC SOLAR SYSTEMS THROUGH REAL O&M FOLLOW-UP DATA

Isidoro Lillo[1], Antonio Palomo[1], Jose Guasumba[2] and Manuel Silva[1]
1 Department of Energy Engineering. University of Seville. Spain
2 Renewable Energy Laboratory. University of the Armed Forces. Quito. Ecuador
Camino de los Descubrimientos s/n, 41092 Sevilla, Spain. Phone: (+34) 954487236.
Fax: (+34) 954487233. E-mail: isidorolillo@us.es

ABSTRACT: This paper provides a detailed analysis of failures and incidents that occurred in 218 PV systems, in 24 PV plants and 17 PV parks, for 15 months of performance, located in Spain and Italy. The results show that the photovoltaic technology is the cause of the 20.3% of the incidents while the 44% of failures are caused due to external causes of the PV system installation. The 56.7% of the failures affected the energy production of the PV system. The major cause of failure is given by the monitoring systems, followed by low power inverters (between 5-90 kW). Despite these failures, the efficiency of the facilities is high with a performance ratio between 69% and 83%.
Keywords: Reliability, Photovoltaic, Fault.

1 INTRODUCTION

The evolution of photovoltaic systems has been exponential for the last few years [1]. The detailed knowledge of these facilities operation under real conditions contributes to improving the design of equipment and auxiliary elements, to optimize the operation and maintenance tasks and to obtain more accurate economic evaluations.

For this reason, diagnostic methods to identify inefficiencies, failures and incidents in photovoltaic installations are being optimized. Numerous studies on diagnostic methods have been developed. A first group of studies is mainly based on the analysis of climatological and electrical data obtained through the monitoring system and the neuronal networks [2-6]. Another group proposes a method for the detection of failures based on the absolute performance ratio error (APRE) which represents the difference between measured and the simulated performance ratio [7]. Also specific methods to determine failures in equipment have been developed. Diagnostic methods based on the analysis of the dI/dV - V curve are proposed. This curve detects partial shading [8], others detect the MPP evolution or faults in a inverter [9, 10].

On the other hand, the operation and maintenance costs included in the calculate of the levelized cost of energy (LCOE) or used in the economic feasibility studies of photovoltaic installations are generally based on a fixed percentage of the total PV plant cost, between 0.8% and 1.2% per year [11,12].

The influence of the impact of uncertainties and statistics of failures in the financial and economical models of these facilities are reflected in other documents [13].

However, in contrast to the numerous studies on fault detection facilities, there are few studies of failures and incidents in real photovoltaic systems [14] that can feed back the learning process.

The current study focuses on the analysis of failures and incidents produced in grid-connected photovoltaic systems, without storage system. The reflected incidents are those that require corrective maintenance.

Some incidences that produce inefficiencies or impairments in photovoltaic systems as degradation of PV module, shading phenomena, soiling effect, mismatch effect, inverter power limitation by temperature, maximum power point tracking losses, show losses or module temperature effect are not included in this study.

The failures and incidents of this study have been obtained from real data and provided by the responsible companies for the operation and maintenance of PV systems. These failures have been mainly detected through the analysis of climatological and electrical data obtained from the monitoring system, the alarm system, the monitoring of the inverter and from the visual inspection of installations (discolouration, browning, delamination, glass and cell broken in module).

It is intended to contribute to improving the knowledge of the true reliability of photovoltaic systems through their behavior under real conditions.

2 METHODOLOGY

Operation and maintenance data for 15 months, from January 2014 to March 2015, of 218 PV systems distributed in 17 PV parks in Spain and Italy have been analyzed. Their characteristics and the incidents detected are described in Table I.

The characteristics of the analyzed facilities are summarized in Table II:

Table II. - Characteristics of the analyzed facilities.

	[illegible]	[illegible]		[illegible]		[illegible]
17 PARKS → 218 SYSTEMS	Ground: 11 2010-2011	SILICON				7
		NUMBER OF PANELS	PEAK POWER	NUMBER OF INVERTERS	POWER	
		23220	5340 kWp	10	5000 kW	
		Cadmium Telluride (CdTe)				27
		NUMBER OF PANELS	PEAK POWER	NUMBER OF INVERTERS	POWER	
		346854	26880 kWp	90	24780 kW	
	Covering: 6 2010-2011	SILICON				10
		NUMBER OF PANELS	PEAK POWER	NUMBER OF INVERTERS	POWER	
		25902	6000 kWp	114	5320 kW	
		Cadmium Telluride (CdTe)				1
		NUMBER OF PANELS	PEAK POWER	NUMBER OF INVERTERS	POWER	
		5139	400 kWp	4	400 kW	
		Total	Total	Total	Total	Total
		401115	38620 kWp	218	35500 kW	45

Si	CdTe
12.20%	87.80%

Table I. - General data and system failures.

PV PARKS CHARACTERISTICS

LOCATION			TYPE	SOLAR FIELD				INVERTER		ST	PANT	
PV Park	PV Plant	Province	Ground/Roof	Nº PV Modules	Nº Strings	Material	Peak Power (Wp)	Nº Inverters	Nominal Power (kW)	Nº TSs	Peak Power (kWp)	Nominal Power (kW)
S	S-1	Valencia	Roof	3266		Si PC	230	1	630	1	751,2	630
	S-2		Roof	3266		Si PC	230	1	630	1	751,2	630
C	C1	Zamora	Ground	57888	6432	CdTe	75	4	1000	4	4341,6	4000
V	V-1	Cádiz	Ground	29106	3234	CdTe	75	21	90	3	2183,0	1890
	V-2		Ground	29106	3234	CdTe	75	21	90	3	2183,0	1890
	V-3		Ground	26334	2926	CdTe	75	19	90	3	1975,1	1710
A	A-1	Mallorca	Ground	44928	1664	CdTe	75/77,5	2 4	500 540	3	3425,8	3160
MA	MA-1	Murcia	Ground	25812	2868	CdTe	80	4	500	3	2065,0	2000
MB	MB-1	Murcia	Ground	9540	477	Si PC	235	4	500	4	2241,9	2000
MC	MC-1	Murcia	Ground	13680	684	Si PC	230	6	500	3	3146,4	3000
B	B-1	Castellón	Roof	3465	231	Si MC	238	7	100	1	824,7	700
	B-2		Roof	2590	259	Si MC	305	7	100	1	789,9	700
T	T-1	Tarragona	Roof	5374	260/270	Si PC	215-235	10	100	1	1218,6	1000
TS	TS-1	Lérida	Roof	1587		Si PC	215-235	22	15	1	354,6	330
	TS-2		Roof	1560		Si PC	235	24	15	1	366,1	360
TA	TA-1	Lérida	Roof	520	2	Si PC	220	20	5	1	114,4	100
	TA-2		Roof	520	2	Si PC	220/225	14	5	1	81,3	70
	TA-3		Roof	5139	141/148	CdTe	77,5	4	100	1	398,0	400
F	F-1	Lérida	Roof	3754	22,24,25	Si PC	235	8	100	1	880,1	800
CR	CR-1	Cremona	Ground	34080	12	CdTe	77,5	2 2	500 760	2	2641,2	2520
D	D-1	Cremona	Ground	16200	12	CdTe	77,5	2	630	1	1255,5	1260
PP	PP-1	Palazzo P.	Ground	46200	12	CdTe	77,5	2 3	800 630	3	3580,5	3490
P2	P2-1	Parmense	Ground	20520	12	CdTe	77,5	2	800	1	1590,3	1600
P3	P3-1	Parmense	Ground	16680	12	CdTe	77,5	2	630	1	1292,7	1260
TOTAL 17	24		Ground 13 Roof 11	401115		Si 12,2% CdTe 87,8%		218	MAX 1000 MIN 5	45	38452	35500

FAILURES AND INCIDENTS

PV Park	INVERTER: Operation Failures	INVERTER: Starting and Stop Failures	INVERTER: Monitoring and Communication Failures	INVERTER: Total Incidents	INVERTER: %	MONITORING SYSTEM: Total Incidents	MONITORING SYSTEM: %	SOLAR FIELD: Replacement of PV Modules	SOLAR FIELD: Wiring Failures (Strings)	SOLAR FIELD: String Boxes Failures	SOLAR FIELD: Vandalism Incidents	SOLAR FIELD: Robbery Incidents	SOLAR FIELD: Wind Impacts Incidents	SOLAR FIELD: Total Incidents	SOLAR FIELD: %	ST: Total Incidents	ST: %	ST: Meteorological Incidencies	%	ELECTRICAL GRID: Total Incidents	ELECTRICAL GRID: %	Total Incidents
S	0	0	0	0	0,0%	4	66,7%	0	0	0	0	0	2	2	33,3%	0	0,0%	0	0,0%	0	0,0%	6
C	1	0	2	3	10,7%	4	14,3%	13	0	2	0	0	0	15	53,6%	4	14,3%	0	0,0%	2	7,1%	28
V	31	16	24	71	47,3%	41	27,3%	6	2	0	1	4	0	13	8,7%	2	1,3%	1	0,7%	22	14,7%	150
A	6	2	9	17	32,1%	3	5,7%	26	0	1	1	0	0	28	52,8%	3	5,7%	2	3,8%	0	0,0%	53
MA	0	1	0	1	25,0%	2	50,0%	0	0	0	0	0	0	0	0,0%	0	0,0%	0	0,0%	1	25,0%	4
MB	3	0	0	3	27,3%	4	36,4%	0	0	2	0	0	0	2	18,2%	0	0,0%	0	0,0%	2	18,2%	11
MC	0	2	0	2	28,6%	2	28,6%	0	0	3	0	0	0	3	42,9%	0	0,0%	0	0,0%	0	0,0%	7
B	2	0	0	2	5,6%	34	94,4%	0	0	0	0	0	0	0	0,0%	0	0,0%	0	0,0%	0	0,0%	36
T	2	1	0	3	10,3%	26	89,7%	0	0	0	0	0	0	0	0,0%	0	0,0%	0	0,0%	0	0,0%	29
TS	38	30	2	70	59,8%	47	40,2%	0	0	0	0	0	0	0	0,0%	0	0,0%	0	0,0%	0	0,0%	117
TA	2	17	1	20	45,5%	24	54,5%	0	0	0	0	0	0	0	0,0%	0	0,0%	0	0,0%	0	0,0%	44
F	0	2	0	2	8,3%	20	83,3%	0	0	2	0	0	0	2	8,3%	0	0,0%	0	0,0%	0	0,0%	24
CR	2	0	4	6	14,3%	15	35,7%	16	0	0	1	0	0	17	40,5%	4	9,5%	0	0,0%	0	0,0%	42
D	0	0	0	0	0,0%	11	55,0%	8	0	0	0	0	0	8	40,0%	0	0,0%	0	0,0%	1	5,0%	20
PP	1	2	0	3	5,2%	3	5,2%	45	0	0	1	0	0	46	79,3%	5	8,6%	0	0,0%	1	1,7%	58
P2	1	0	1	2	8,0%	5	20,0%	12	0	1	0	0	0	13	52,0%	2	8,0%	2	8,0%	1	4,0%	25
P3	0	0	0	0	0,0%	4	20,0%	11	0	0	0	0	0	11	55,0%	2	10,0%	3	15,0%	0	0,0%	20
TOTAL	89	73	43	205	30,4%	249	36,9%	137	2	11	4	4	2	160	23,7%	22	3,3%	8	1,2%	30	4,5%	674

Failures and incidents are grouped according to the following criteria for this analysis:

Criterion 1: Depending on the equipment or group that is affected, grouped in solar field, inverter, transformer, drainage network and monitoring system.

Within the solar field incidents have been grouped into failures that have led to the replacement of the photovoltaic module, faulty wiring, faulty junction box, theft or vandalism faults and failures by wind.

Inside the inverter incidents are grouped into operation failures, failures of starting and stopping and monitoring system failures.

Within the transformation center, incidents have been grouped in failures per share of transformer protections and failures resulting from weather.

Within the evacuation network incidents have been grouped in inherent network failures or network failures resulting from extreme weather conditions.

Criterion 2: Internal and external elements of the PV system distinguishing those that affect production or not. Solar field, inverter and transformation center are considered internal elements, drainage network and monitoring system are considered external elements.

Of all failures, derivatives from global monitoring installation system and monitoring inverter system are the only ones that do no affect to the electricity production. Rest of failures are included into the group that affect the electricity production installation.

Criterion 3: This classification is based on the technology linked to the failure, distributed on PV, electrical, electronic and telecommunications technologies. PV technology failures are associated to PV module failures. In the case of electrical technology failures included problems related to network, faulty installation wiring and junction boxes failures. In the group associated with electronic technology failures by operation, start and stop the inverter are included. In telecommunications failures associated with the communication systems and data acquisition are included. In the group of external causes, have been included, other causes such as theft, vandalism and wind actions, of which no technology is responsible.

3 RESULTS AND ANALYSIS

According to Table I, we disaggregate and analyse failures according to the 3 indicated criteria.

Criterion 1: The 674 incidents are distributed according to Figure 1.

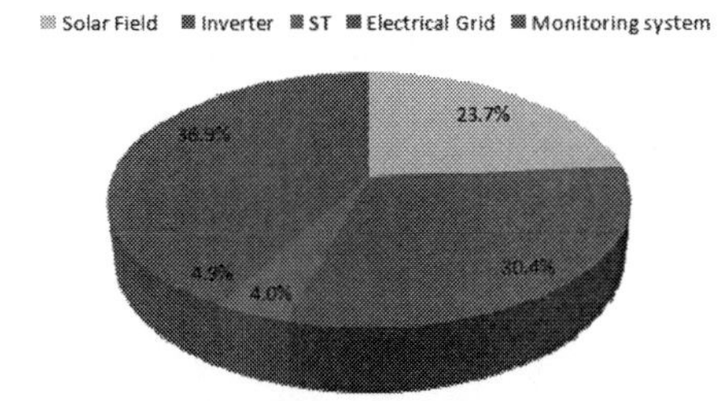

Figure 1. Distribution of equipment failures.

Note how the monitoring system is the responsible of most of the failures, but those failures do not have an impact in the production.

Figure 2 presents the distribution of the 160 incidents of the solar field. Notice how most of them are due to the replacement of modules. This percentage would seem to be high a priori, but should not be considered thay way, understanding that there are 401115 modules in all the facilities. Actually replacing modules represented 0.35 ‰ throughout the entire period. Taking into account the impact on production, incidents by theft and vandalism are much more significant.

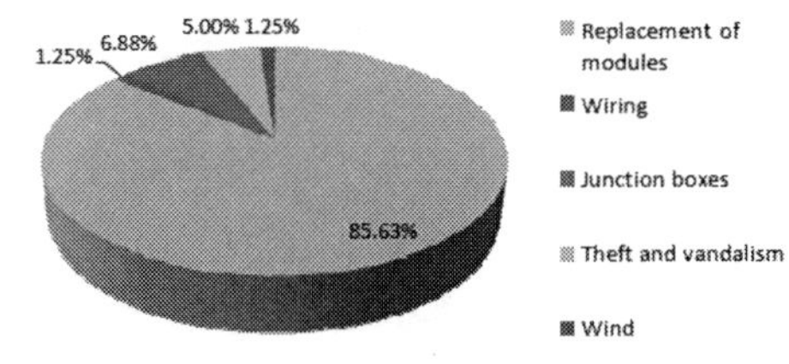

Figure 2.- Distribution of PV array failures.

In the case of the inverter, according to Figure 3, approximately 21% of the incidents are caused due to the monitoring system that does not affect the production. Incidences called "operation failures" are those that have a greater impact on production.

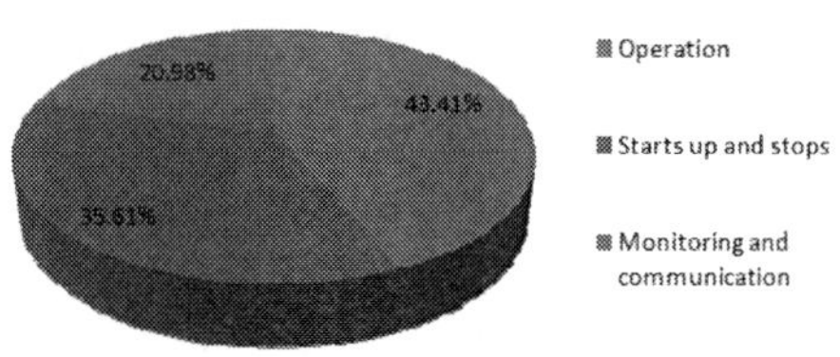

Figure 3.- Distribution of inverter failures

In the case of the inverter it is shown in Table III a breakdown of failures in operation and start-up and shutdown depending on the inverter power. It is noteworthy that start-up and shutdown have a greater impact in inverters with lower power.

Table III. Incidences according to the inverter power.

Power (kW)	Number of inverters	Number of failed inverters	Percentage of failured inverters	Operation failures	Start-up and shutdown failures	Total failures	Total failures/Total inverters
5	34	10	29.4%	1	12	13	0.4
15	46	23	50.0%	38	30	68	1.5
90	61	22	36.1%	31	16	47	0.8
100	36	9	25.0%	5	8	13	0.4
500	19	8	42.1%	6	4	10	0.5
540	4	3	75.0%	4	2	6	1.5
630	8	1	12.5%	1	1	2	0.3
760	2	1	50.0%	1	0	1	0.5
800	4	1	25.0%	1	0	1	0.3
1000	4	1	25.0%	1	0	1	0.3
Total	218	79		89	73	162	

Complementarily presented in Figure 4 the probability density function where it is revealed that approximately 60% of the inverters have not presented any failure and that there are no inverters with more than 8 faults recorded in the period analyzed.

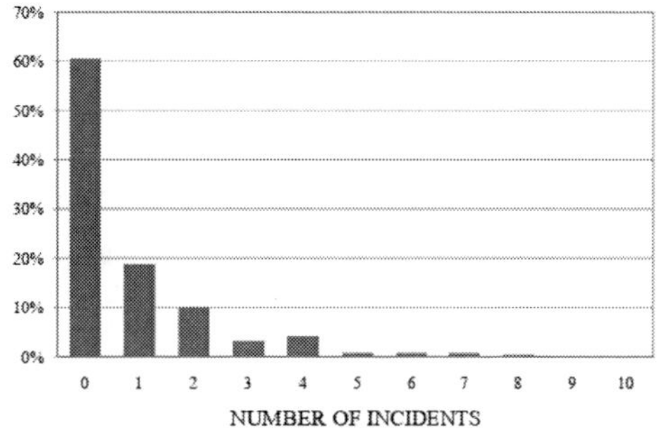

Figure 4. Distribution of the number of inverter incidents.

In the case of the transformer station, according to Figure 5, approximately 82% of the failures are caused due to actions of the protection system. Note that some of these actions have been derived from the own grid.

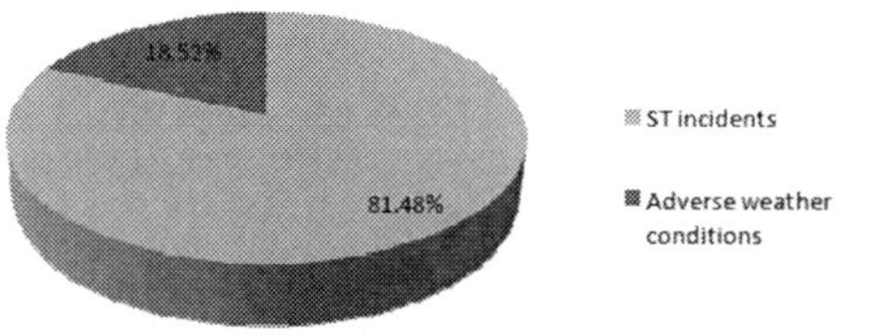

Figure 5. Distribution of transformer failures.

In the case of the grid, according to Figure 6, over the 90% of the failures are caused due to network quality.

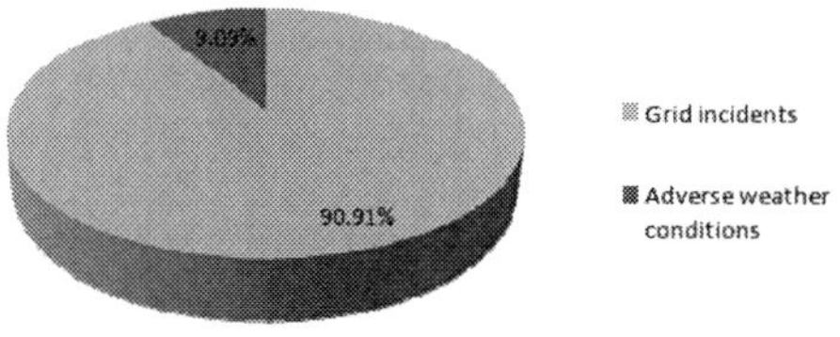

Figure 6. Distribution of failures in the power grid.

Criterion 2: Figure 7 shows the distribution of the failures based on the origin of the incidence (internal and external) and whether has impact on production or not within each of these groups.

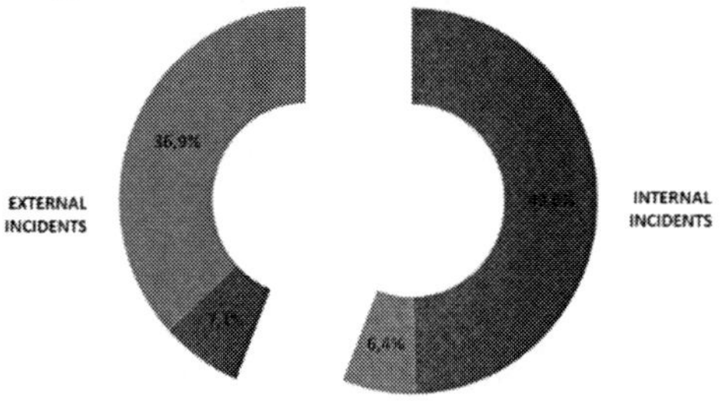

Figure 7. Distribution of failures (internal and external) and whether or not have impact on production.

56.7% of failures affect production while 43.3% do not affect production.

A more detailed breakdown by photovoltaic park is shown in Table IV. Of note, two photovoltaic parks, V and TS, where the number of both internal and external incidents are far superior to the rest.

Table IV. Distribution of failures (internal and external) and whether or not have impact on production in each PV park.

PV PARKS	Internal Incidents: Number of incidents	Internal Incidents: Number of incidents/Month	External Incidents: Number of incidents	External Incidents: Number of incidents/Month	A. P. Incidents: Number of incidents	A. P. Incidents: Number of incidents/Month	NOT A.P. Incidents: Number of incidents	NOT A.P. Incidents: Number of incidents/Month
S	0	0.0	6	0.4	2	0.1	4	0.3
C	22	1.5	6	0.4	22	1.5	6	0.4
V	81	5.4	69	4.6	85	5.7	65	4.3
A	47	3.1	6	0.4	41	2.7	12	0.8
MA	1	0.1	3	0.2	2	0.1	2	0.1
MB	5	0.3	6	0.4	7	0.5	4	0.3
MC	5	0.3	2	0.1	5	0.3	2	0.1
B	2	0.1	34	2.3	2	0.1	34	2.3
T	3	0.2	26	1.7	3	0.2	26	1.7
TS	70	4.7	47	3.1	68	4.5	49	3.3
TA	20	1.3	24	1.6	19	1.3	25	1.7
F	4	0.3	20	1.3	4	0.3	20	1.3
CR	26	1.7	16	1.1	23	1.5	19	1.3
D	8	0.5	12	0.8	9	0.6	11	0.7
PP	53	3.5	5	0.3	55	3.7	3	0.2
P2	17	1.1	8	0.5	19	1.3	6	0.4
P3	13	0.9	7	0.5	16	1.1	4	0.3
Σ	377	1.5	297	1.2	382	1.5	292	1.1

Criterion 3: Figure 8 shows the distribution of failures depending on the technology.

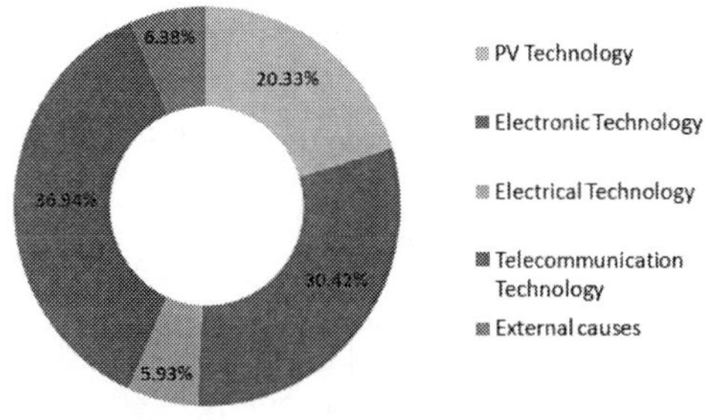

Figure 8.- Distribution of failures depending on the technology.

Notice how communications technology is the one presenting most failures while photovoltaic technology represents 20.33% despite the large number of existing photovoltaic units (401115).

An important aspect to note is that the distribution of the number of failures is not evenly spread across all the photovoltaic parks, as shown in Figure 9.

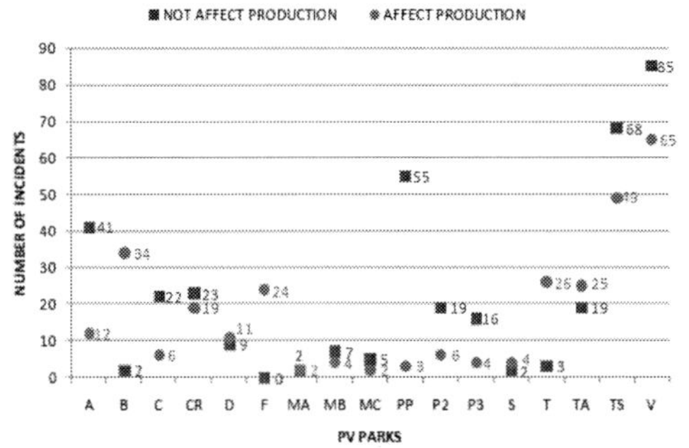

Figure 9.- Number of failures per park affecting or not to the production.

The average of the total failure in all facilities is 19.8 while the standard deviation is 21.5. Where failures concern energy production the average is 17.4 and the standard deviation is 17.6. In the case of failures that do not affect energetic production average is 22.2 and the standard deviation is 24.6.

Note that there are parks with no incidences or only a few and some others with more than 100.

4 INFLUENCE IN ENERGY FIELD

The calculation of the actual influence of these failures in the energy loss of the installation requires a very detailed information on the different faults. For example, to identify energy loss due to a failure in a photovoltaic module would require to know how long it has been previously failing to be identified as a failure, not only counting from the day that the failure was identified and the module was replaced.

In Figure 10 we show the Performance Ratio (PR) and the ratio efficiency (RE) defined according to expressions 1 and 2 respectively. We have determined them for 15 of the 17 parks from which we had quality information available on production and other required data. These parameters include the set of real energy losses of the system.

$$PR(\%) = \frac{E(kWh)}{\frac{H_g(kWh/m^2)}{1000(W/m^2)} P_p(W_p) \left[1-\alpha(\%/^{\circ}C)\cdot(\bar{T}_P - 25)(^{\circ}C) \right]} 100 \quad (1)$$

$$RE(\%) = \frac{E(kWh)}{H_g(kWh/m^2)\cdot A_{cap}(m^2)} 100 \quad (2)$$

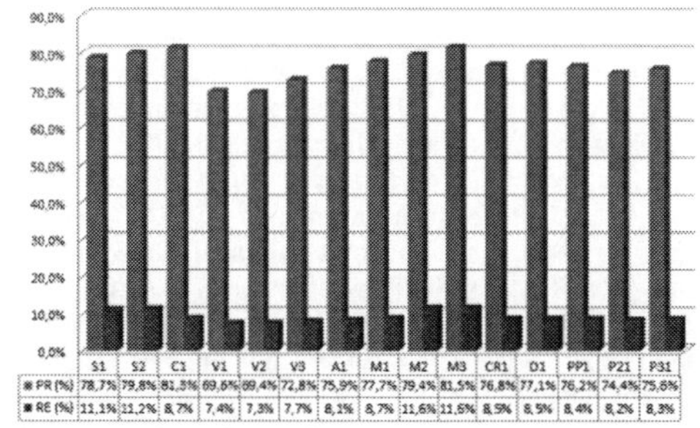

Figure 10.- PR and RE of the 15 parks on the analyzed period.

It is noted that the PR varies between 69% and 83% and according to Figure 11 where the failures that affect the installation are represented, it is observed that there is no direct relationship between the number of failures and the loss of energy as was expected. Thus, the park that has more failures (PP1) is not the park with greater energy losses. This is because most incidents of this park have occurred in the solar field and such losses are the least affecting energy loss compared to the impact of the inverter, transformer or power supply.

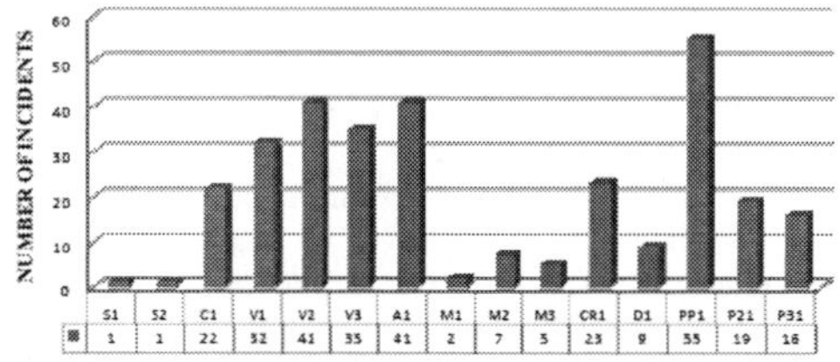

Figure 11. Failures affecting production in the different photovoltaic parks.

5 CONCLUSIONS

In general, the photovoltaic systems are quite reliable. The probability that an inverters fails over one year of operation is 33.2% and the probability of a photovoltaic module replacement during the 15 months was 0.39 ‰.

The highest percentage of incidents, 36.9%, are caused due to the monitoring system. Although these failures have no influence on the energy production of the installation itself, it has important implications for ensuring compliance with the contracts between the owner of the photovoltaic installation and the maintenance company. Besides the failures in the monitoring system avoid to identify, in many cases, other failures and incidents since much of the fault detection methods are based on the data obtained through the monitoring system itself.

There are particular circumstances that lead certain photovoltaic parks to have a significant number of failures. For example, it is remarkable the relatively high failure rate of the grid compared to the 17 grids analyzed and the importance that these failures have on the loss of energy production of the photovoltaic plant. However, almost all network failures have been caused by a single power grid, the photovoltaic park called V. Hence the importance of an optimum quality assessment of the electrical network that will be connected to the photovoltaic plant and its maintenance.

56.7% of failures affect production while 43.3% do not affect production. Of the issues affecting production, 49.6% were internal and 7.1% were external.

The probability of occurrence of at least an annual failure in a photovoltaic installation is 53.6%, meaning that this is the probability of a PV installation to fail over one year. This issue affects production in 49.6% of the cases, therefore the probability of occurring at least one fault that adversely affects the annual output is approximately 27%.

6 REFERENCES

[1] GTM Research/SEIA, Q3 2015: U.S. Market insight. December 2015.

[2] Drews A., de Keizer, A.C., Beyer. H.G.., Lorenz, E. Betccke, J., Van Sark, W.G. et al., 2007. Monitoring and remote failure detection of grid-connectedPV systems base don satellite observations. Solar Energy 81, 548-564

[3] W. Chine, A Mellit, V. lugi, A. Malek, G. Sulligoi, A. Massi Pavan. 2016. A novel fault diagnosis technque for photovoltaic systems based on artificial neural networks. Renewable Energy 90. p. 501-512.

[4] W. Chine, A. Mellit, A. Massi Pavan, S.A. Kalogirou. 2014. Fault detection method for grid-connected photovoltaic plants. Renewable Energy 66, 99-100.

[5] Yihua Hu, Bin Gao, Xueguan Song, Gui Yun Tian, Kongjing Li, Xiangning He. 2013. Photovoltaic fault detection using a parameter based model. Solar Energy 96. Pp 96-102.

[6] Santiago Silvestre, Aissa Chouder, Engin Karatepe. 2013. Automatic fault detection in grid connected PV systems. Solar Energy 94. pp 119-127.

[7] Choulder, S. Silvestre. 2010. Automatic supervisión and fault detection of PV systems based on power losses analysis. Energy conversion and Management. 51. pp. 2929-1937.

[8] M. Wiwa, S. Kawamura, H. Ohno.2006. Diagnosis of a power output lowering of PV array with a (dI/dV)-V characteristic. Proceeding of IEEE 4th World Conference on Photovoltaic Energy Conversion. Waikoloa. 2442-2445.

[9] Khomfoi, S., Tolbert, L.M., 2007. Fault diagnostic system for a multilevel inverter using a neural network. Power Electron., IEEE Trans. 22 (3), 1062–1069.

[10] Kim, T.J., Lee, W.C., Hyun, D.S., 2009. Detection method for opencircuit fault in neutral-point-clamped inverter systems. Ind. Electronics, IEEE Trans. 56 (7), 2754–2763.

[11] Cameron, C.P., Goodrich, A.C., 2010. The levelized cost of energy for distributed PV: a parametric study. 2010 35th IEEE Photovoltaic Specialists Conference (PVSC). Institute of Electrical & Electronics Engineers (IEEE), pp. 529–534. http://dx.doi.org/10.1109/pvsc.2010.5616811.

[12] Darling, S.B., You, F., Veselka, T., Velosa, A., 2011. Assumptions and the levelized cost of energy for photovoltaics. Energy Environ. Sci. 4 (9), 3133. http://dx.doi.org/10.1039/c0ee00698j.

[13] IEA. Photovoltaic power systems programme annual report. 2015. task 13.Performance and reliability of PV systems. Subtaks 3.3. Characterization of PV Module Condition in the Field - Guidelines on IR and EL in the Field. http://www.iea-pvps.org/index.php?id=164.

[14] Lukamp, H., March 2002. Reliability Study of Grid-Connected PV Systems: Field Experience and Recommended Design Practice. Technical report Task 7, IEA PVPS.

WEATHER SENSITIVITY ANALYSES IN LAYOUT PLANNING

Dr. Martin Bischoff[1], Dr. Joachim Bamberger[2], Dr. Tino Fleuren[3], Dr. Kai Plociennik[3]
[1]: Siemens AG, Corporate Technology, Otto-Hahn-Ring 6, 81739 München, Germany
[2]: Siemens AG, Energy Management Division, Freyeslebenstr. 1, 91058 Erlangen, Germany
[3]: Fraunhofer ITWM, Fraunhofer-Platz 1, 67663 Kaiserslautern, Germany
[martin.bischoff, joachim.bamberger]@siemens.com, [tino.fleuren, kai.plociennik]@itwm.fraunhofer.de

ABSTRACT: PVplanet is a planning software for large-scale ground-mounted photovoltaic power plants, developed by Siemens AG in cooperation with Fraunhofer ITWM. The software integrates layout planning, yield calculation, and financial calculations. Several hundred layout variants are automatically generated by the software. For each variant, the energy output and financial objectives are calculated. With this information, the planner obtains an overview on the planning possibilities and avoids trial-and-error planning iterations. With PVplanet, the time for planning layouts can be drastically reduced, whereas the quality of the results increases. One of the most recent developments is a weather sensitivity analysis where, based on multiple different weather data sets for the project site, the impact of weather variations on technical objectives such as energy output and performance ratio (PR) and on financial objectives such as levelized costs of energy (LCoE), net present value (NPV), and internal rate of return (IRR) is evaluated. Thus, EPCs can quantify the risk of attaining their performance ratio guarantee, and investors can evaluate the impact of weather variations on their business case.
Keywords: Economic Analysis, Financing, Large Grid-connected PV systems, Simulation, Weather Data

1 PV plant layout planning software *PVplanet*

The *PVplanet* software development project was initiated in 2011 by the Siemens Energy Sector to significantly reduce the engineering effort for offer preparation and to improve economic efficiency of Siemens' photovoltaic power plants. Currently, PVplanet is applied within the business unit Smart Generation Solutions for planning the photovoltaic power generator in hybrid power plants. Since three years, Siemens offers a technical consulting service for applying the software to optimize plant layouts, or to review existing plant layouts for their customers. Only recently Siemens started evaluating business opportunities to commercialize the software outside of Siemens.

1.1 Planning Approach

In contrast to conventional PV software, which is focused on a single layout that can be modified and improved in a stepwise manner, PVplanet generates hundreds of layout variants automatically, according to the planner's specifications. Subsequently, the energy output is calculated with a physical plant simulation, and economical objectives are determined based on a financial model. All this is done based on actual weather data for the project site. The characteristics of all layout variants are then clearly laid out in an interactive user interface (see Figure 1), which is a multiobjective decision support tool in which the planner can determine the layout with the best compromise between the different planning objectives. In other words, the planner can identify advantages and drawbacks of each layout and understand interdependencies and sensitivities of planning parameters. With respect to the project-specific requirements and customer requests, in the end, the best solution, and optionally a set of alternatives can be selected. For the preferred solutions, a whole set of planning documents such as CAD layouts, cable cross section calculations, energy output report, and finance calculations can be exported.

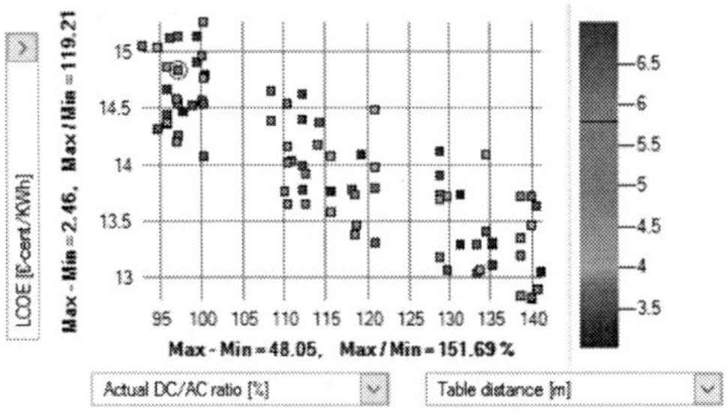

Figure 1: The multiobjective decision support window

In Figure 1, the multiobjective decision support window, every point represents a complete plant layout. The layouts differ in parameters such as tilt angle, DC/AC ratio, table distance, and inverter or module model. With drop-down-menus, the planner can select the objectives for the x-axis, the y-axis, and the color scale. In the figure, DC/AC ratio [%] has been selected for the x-axis, LCOE (levelized costs of energy) [€-Cent/kWh] for the y-axis, and table distance [m] for the color scale. It can be seen that in tendency, the best (lowest) LCOE values of around 13.5 €-Cent/kWh are achieved by the layouts with DC/AC ratio between 130 and 140 %, while table distance (color scale) seems to have no simple relation to the former two parameters. Even though low LCOE values are desirable, it might be that the corresponding layouts perform not so well in other objectives such as performance ratio. Hence, it is the task of the planner to investigate the values of the layouts in the different planning objectives and determine the layout with the best overall compromise between objectives.

By double clicking on one of the points in Figure 1, the corresponding plant layout is shown in a layout preview dialog, as can be seen in Figure 2:

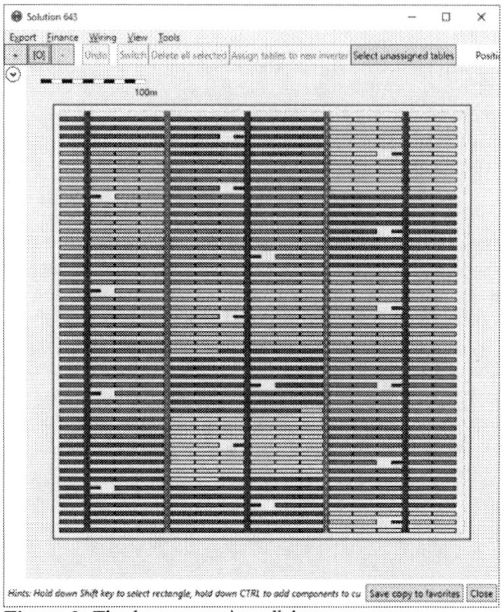

Figure 2: The layout preview dialog

In this window, the cable- and service ways are depicted as red and grey stripes, respectively. The tables (differently colored rectangles) carrying the PV modules are colored according to the inverter areas in the plant, and the inverters are shown as green rectangles. In this window, the planner can examine the plant layout in detail and make modifications like re-locating individual tables or inverters or changing the electrical allocation. After each modification, all objectives are recalculated and compared to the original values. When all adjustments have been made, the set of planning documents can be exported for the layout.

Results from Siemens-internal application as well as external consulting workshops show a reduction of engineering effort for offer preparations of up to 80 % compared to layout planning without the software. This drastic time reduction is achieved by automation routines, e.g. Excel export of cable cross section calculations or exports of whole DC layouts in CAD files. In the conventional planning process, layout generation, energy output and finance calculations takes a significant amount of time such that only one, or a small set of layout alternatives can be compared. With PVplanet, the planner obtains an overview on the planning possibilities and the corresponding technical and economical consequences. This leads to better plant layouts.

2 Weather Sensitivity Analyses in PVplanet

For qualified energy output calculations, multiple weather datasets from different sources and data providers should be taken into account. The datasets can be based on pure measurement data (e.g. from local weather stations or satellite images), or processed data (e.g. via climate models, spatial interpolations, or reference year algorithms). There exists no clear best choice of datasets in general, as their qualities vary with the location.

PVplanet does not rely on any particular weather data provider. Instead, the planner can import any available project-specific dataset. Currently, three data formats can be imported: TMY3, Meteonorm Standard Minute, and PVweather, a proprietary csv-based file format. Datasets in other formats must be converted prior to the import.

PVplanet currently supports energy output calculations in hour- or minute resolution. Minute resolution is of particular importance for an accurate layout optimization, e.g. a correct determination of inverter de-rating losses.

In PVplanet, several weather data sets can be imported for the project site, and the planner can assign a probability of occurrence for each data set. Then, a Monte Carlo simulation can be performed, where a large number (e.g., 1000) of weather data paths over the operating period of the plant (e.g., 25 years) is generated. This is done by drawing the weather data for each year randomly from the set of weather data sets according to the configured probabilities. For each weather data path, the technical and financial behavior of a plant layout is evaluated, and from the obtained values, probability distributions of the corresponding objectives are presented to the user. To be precise, cumulative distribution functions (CDF) of the objectives are plotted, as can be seen in the following figure for energy yield:

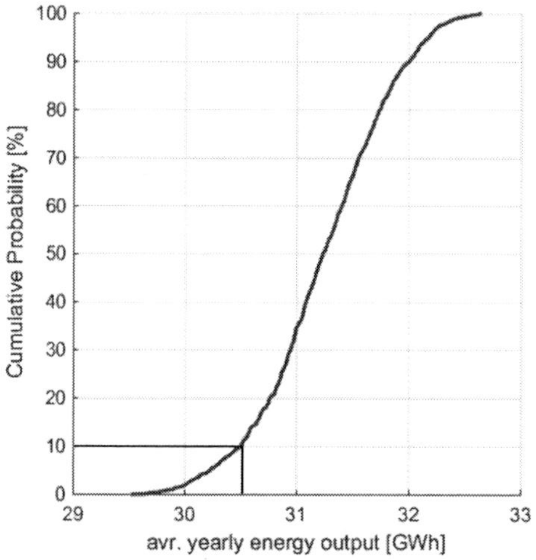

The figure can be interpreted as follows: The plotted value P(x) is the probability that the energy yield is at most x GWh per year. For example, as indicated by the two black lines, the probability that the energy yield is at most 30.5 GWh per year is approximately 10 %. Consider the following three figures:

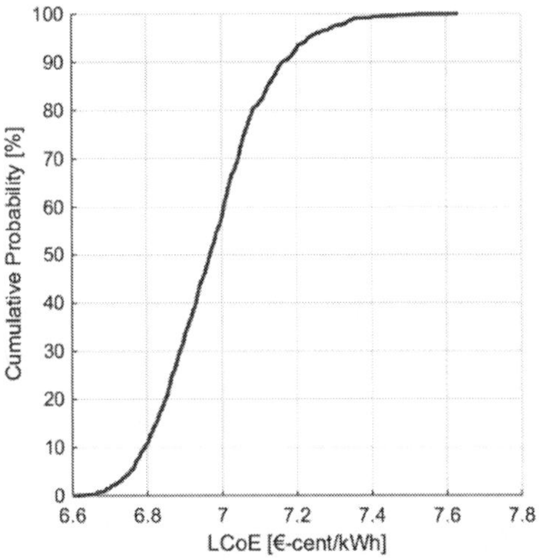

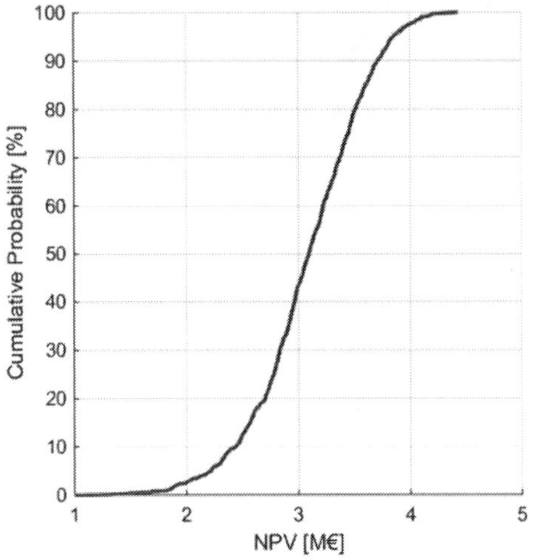

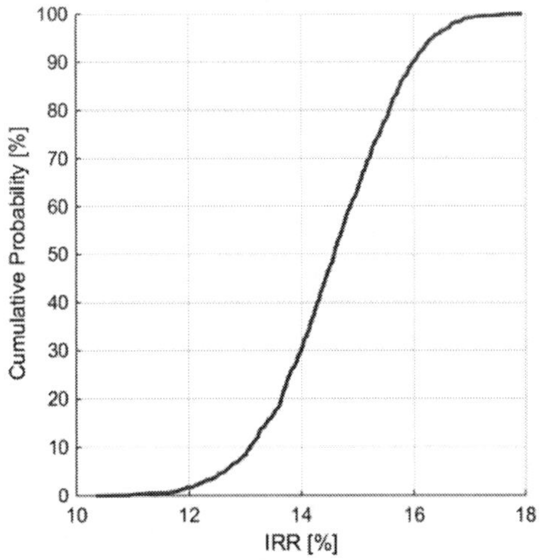

All CDF plots shown here are the results of an example sensitivity analysis performed for a plant layout located close to the equator. For this particular location, 18 different weather datasets where available, each dataset contains (among others) time series of direct irradiation, diffuse irradiation and ambient temperature, for one whole year in minute resolution. Three suppliers provided datasets for the years 2009 to 2013, calculated with their proprietary climate model. The three remaining datasets have been generated with the software Meteonorm, where the weather on that particular site is determined via spatial interpolation of nearby weather stations. With this software we generated one reference dataset which resembles the typical (P50) weather, as well as an optimistic and pessimistic dataset, each with a statistical probability of occurring once per decade.

We performed the weather sensitivity analysis for one layout variant generated with PVplanet, using 1000 Monte Carlo simulations and an operating period of 25 years. The resulting cumulative distribution functions for the average yearly energy output, LCoE, NPV and IRR are illustrated in the above figures. Note that also the well-known P10, P50 and P90 values can be read from these plots. They are the x-values of the points on the CDFs with y-values 10%, 50% and 90%, respectively.

Project-specific weather data analyses as the one presented here can be performed with PVplanet within minutes. They are of great avail for all decision makers to evaluate the impact of weather on the criteria of their business case or for determining, which guarantees can be reasonably be given on the technical and financial performance of a plant.

3 References

Selected Articles:

[1] *A software-based planning approach for photovoltaic power plant layouts.* M. Bischoff, J. Leitner, K. Plociennik, EU PVSEC 2014 Conference Proceedings, 2315-2318 (2014)

[2] *Economic evaluation of two alternative layout planning concepts.* I. Schüle, K. Plociennik, H. Ewe, M. Bischoff, EU PVSEC 2013 Conference Proceedings, 4143-4147 (2013)

[3] *Utilizing full planning potential.* M. Bischoff, M. König, I. Schüle, K. Plociennik; PV Magazine 02/2014, 60-62 (2014)

[4] *Multi-Objective planning of large-scale photovoltaic power plants.* M. Bischoff, K. Plociennik, H. Ewe, I. Schüle; Operations Research Proceedings 2012, Selected Papers of the International Conference on Operations Research 2012, 333-338 (2013)

Published Patent Applications:

[5] *Simplified PV plant construction with a continuously placed system block.* DE 102012106130 A1 06/2013, Karl-Heinz Küfer, Kai Plociennik, Ingmar Schüle, Martin Bischoff, Oliver Hennig, Jens Weber

[6] *PV system design (procedures for the construction and design of a solar system).* DE102012106124 A1 05/2013, Karl-Heinz Küfer, Kai Plociennik, Ingmar Schüle, Martin Bischoff, Oliver Hennig, Frederik Brandes

5EO.3.5

NEW APPROACH TO ANALYZING LONGTERM PERFORMANCE OF LARGE POPULATIONS OF PV SYSTEMS IN FIT MARKETS WITH MINIMAL EFFORTS AND COSTS

CASE STUDY SWITZERLAND: DEVELOPMENT, IMPLEMENTATION AND APPLICATION, ANALYSIS AND FIRST RESULTS

TNC Consulting AG •
(*) Activity leader PVPS T13 2.1
Thomas Vontobel • Thomas Nordmann (*) • Ralph Lingel
vontobel@tnc.ch • nordmann @tnc.ch • lingel@tnc.ch
General Wille-Strasse 59 • 8706 Feldmeilen • CH – Switzerland
www.tnc.ch • P: +41 44 991 5577

ABSTRACT: Longterm and standardized performance monitoring of PV plants is essential for future PV deployment. The value of performance monitoring is widely underestimated and lack of willingness to share operational data make it difficult to obtain reliable project data. IEA PVPS T13 maintains a performance database with predefined and standardized performance analysis. In this work, a new approach to assessing and analyzing performance data of large populations of PV plants is developed, using existing yield data from Feed-in-Tarif (FiT) databases in combination with calculated solar irradiation values based on measured hourly solar irradiation data.
This approach is applied exemplary to the Swiss PV market, producing 1'170 PV projects for performance analysis and confirming the potential to monitor large amounts of PV projects in all countries with FiT. The number of entries in the IEA PVPS performance database over all countries can be doubled with the assessed Swiss PV projects of this work. First findings confirm average PR of 0.75, differences in performance between small and large PV plants in the range of 2…5% and around 6.8% higher average specific yield than assumed in FiT models.
Keywords: Performance, PV systems, Monitoring, Operation of PV plants, IEA PVPS

1 PURPOSE OF THE WORK

Monitoring of performance of operating PV plants is essential to assess effectiveness of FiT schemes, energy strategies and goals, confirming PV plants as reliable energy sources to political and other decision makers.

Nonetheless it is not done on a large scale. This is partly due to the general lack of interest in performance of PV plants as well as the lacking willingness of operators to share operational data, but also because there is a perception that the process of performance surveillance is too complicated. IEA PVPS T13 2.1 operates such a database for performance analysis.

This paper describes a new method to use existing data to verify and confirm essential questions concerning performance of PV plants and shows results such as Performance Ratio, development of performance over time, comparison of performance for different sizes and types of PV plants with little effort and for large numbers of PV plants.

The method is applied to Switzerland, where a sample of 1'170 PV plants is evaluated and first results are presented. The method described can be applied to every country with a FiT scheme or similar, centrally organised systems.

2 APPROACH

While in-depth monitoring of PV plants implies (additional) need of hardware to monitor all of the necessary data as well as a plant operator who is willing to share the operational data, there are other approaches.

Every nation with a Feed-n-Tariff (FiT) or a similar system has a database with the yield of the PV plants using FiT. For Switzerland, the grid operator Swissgrid collects the yield of PV plants in the FiT scheme. These databases are used for accounting for produced solar electricity and paying out the FiT and therefore have to at least account for the overall yield of a PV plant.

Basic information about the PV plant such as location, orientation and inclination of modules and nominal power are mostly available.

An agreement with the Swiss Federal Office of Energy and the national grid operator Swissgrid allowed the use of data of the operational PV plants under the Swiss FiT scheme in an anonymous form. The following questions are of specific interest:

- Do the PV plants perform well? How do they perform compared to the yield in FiT assumptions?
- Can a decrease in performance for a PV plant over time be observed?
- Is there a difference in performance between new and old PV plants?
- Is there a difference in performance between large and small PV plants?

In order to calculate a Performance Ratio PR for PV plants, solar irradiation at the location is needed.

$$PR = Yf \div Yref$$

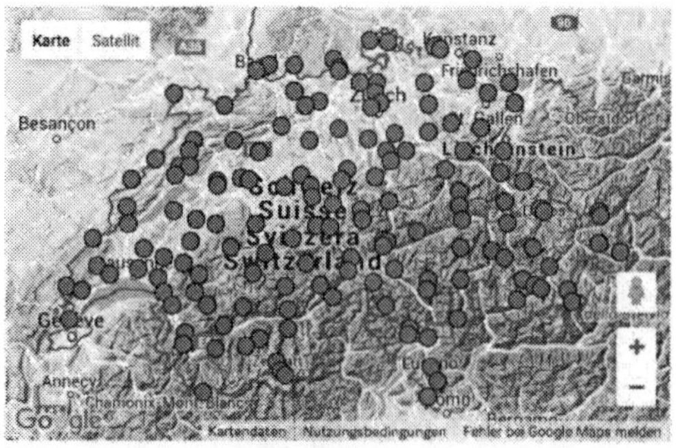

Fig 1: Around 160 meteo-stations in Switzerland are used as baseline for calculating solar irradiation at PV sites.

Values for solar irradiation are usually not available in FiT databases (neither measured nor calculated). For the approach described in this paper, measured data from a network of around 160 meteo-stations is used to calculate the solar irradiation for the locations of the PV plants. This task is performed by MeteoTest, a Swiss company specialized in meteorology. (www.meteotest.ch) [2]

The available 1'170 projects, corresponding to a nominal power of 49.3 MWp, have been split in two groups: One group of 801 PV plants, which consist of a single module array, and a group of 369 PV plants, which consist of several module arrays with different orientation and/or inclination. The selected evaluations shown here base on the first group of 801 projects.

3 RESULTS

The following numbers apply to Switzerland. A total number of more than 8'900 PV plants was operated under FiT when this project started. Not counted are PV projects on the waiting list (cap of FiT in CH). An overall set of 1'170 records was identified for which data on yield was available suitable to answer the above questions (for example having complete data on yield over a full year available). FiT in Switzerland exists since 2009, giving a timeframe of 6 years of operational data for analysis and evaluation.

Fig 2: Geographical distribution of the 1'170 PV plants corresponding to 49.3 MWp used for this evaluation. Every dot represents one PV plant.

3.1 POTENTIAL INFLUENCE ON PERFORMANCE MONITORING –IEA PVPS T2 AND T13

IEA PVPS performance database sets a benchmark on performance monitoring. In T13 and T2 an in-depth analysis of performance is conducted. There are 19 countries providing operational data of PV plants for performance analyses, leading to an overall of 594 projects in the IEA PVPS database.

The described approach using FiT databases for yield data and combining it with calculated solar irradiation delivers a total of 1'170 additional PV plants in this batch processed for Switzerland only. This is an increase of 197% (double) in total number of PV plants available in the IEA PVPS database. FiT PV plants in Switzerland potentially allow for 10'512 PV plants for Switzerland only (current state Q4 2015) to be evaluated.

The database developed for this project is highly flexible and can easily be adapted to different countries and combinations of data available, offering the potential of a very large number of PV plants for integration in performance monitoring such as IEA PVPS. Requirement for availability of good meteorological data is in the view of the authors the only potentially restricting criteria.

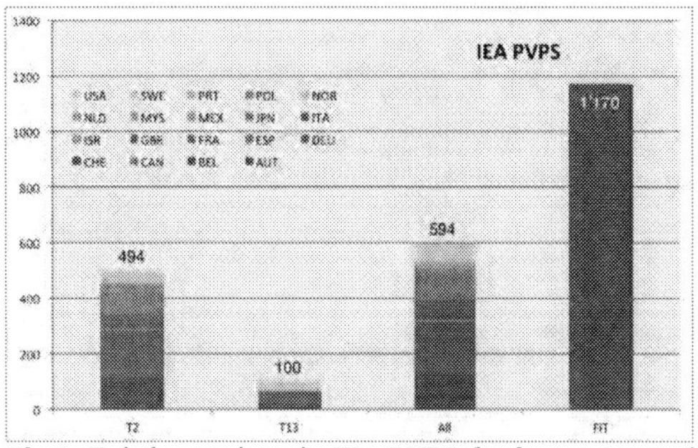

Fig 3: Existing projects in IEA PVPS database compared to projects processed in this batch for Switzerland.

For validation of the method described, the available data of the IEA PVPS database shall in parts be recalculated using the method described and then compared with the obtained results from this work.

3.2 DATA REALIABILITY AND QUALITY

Several calculations and evaluations have been performed in order to verify data quality from FiT database.

In the FiT database nominal installed power and module area are specified. Where nominal power is reliable, since it is verified upon entering FiT by a third party, module area is a value given by the FiT applicant. Module efficiency is not specified in FiT database, but can be calculated using these two available parameters.

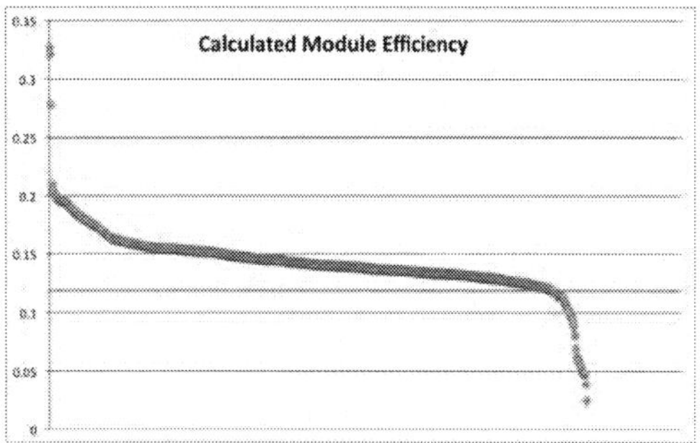

Fig 4: Calculated module efficiency in decreasing order for verified nominal power and user input module area.

49 datasets have a calculated module efficiency of less than 12%, and 5 datasets have a calculated module efficiency of above 25%. These datasets have to be assumed to be of poor quality leading to these unrealistic module efficiencies.

These 54 datasets make up for 6.7% of the 801 overall datasets, supporting the impression of reliability for information provided by FiT applicants and allowing better judgment of overall quality of derived assumptions.

Other inputs are also validated. For example, calculated solar irradiation is compared to measured solar irradiation for exemplary PV plants where measured data is available. Existing values of the IEA PVPS database are also used to verify the obtained results.

3.3 HOW WELL DO THE PV PLANTS PERFORM? OVERALL PERFORMANCE RATIO

Overall performance ratio as defined by IEA PVPS

$$PR = Yf \div Yref$$

is calculated on a yearly base for all PV plants with a full year of yield data available as well as on a monthly base.

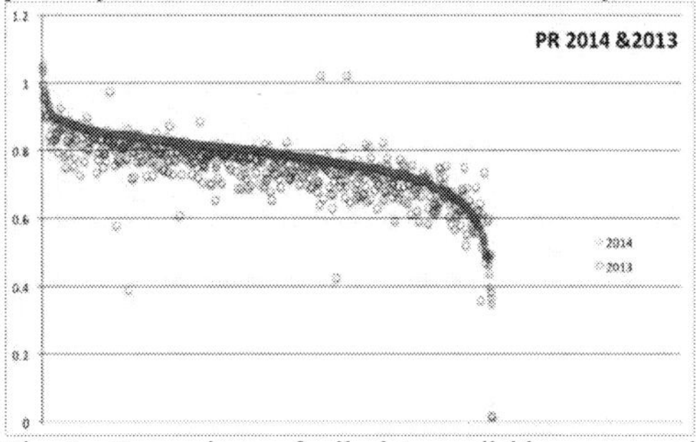

Fig 5: Comparison of all data available on annual performance ratios for 2014 (blue) and 2013 (red). The same PV plant is located at the same position on the x-axis for both years.

In this evaluation, performance data for the year 2014 (blue) is sorted declining. Performance data for the year 2013 (red) are shown in the same order of projects, allowing a comparison of development for each project.

Most PV plants show constant annual performance. Those PV plants with low annual performance in 2014 and average performance in 2013 seem to have had a case-based issue leading to performance decrease in 2014, while those which have low performance as well in 2014 as in 2013, a persistent performance issue has to be assumed. This second case accounts for about 1.9% of the evaluated PV plants.

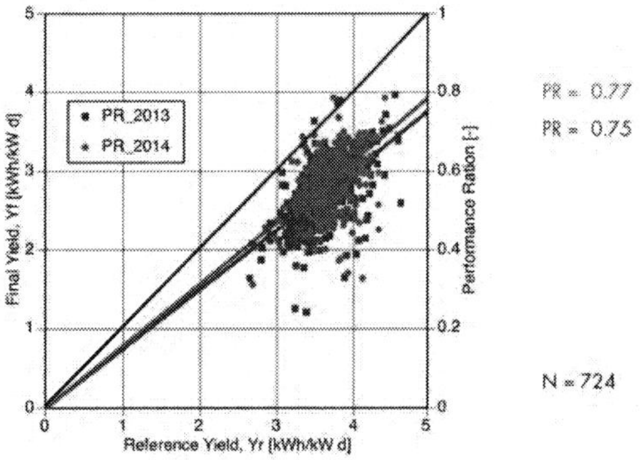

Fig 6: Annual Performance ratio according to IEA PVPS standards with solar irradiation vs normalized yield plotted for 2013 (blue) and 2014 (red).

Average performance ratio over all years results in 0.75. As expected, some values clearly lie above and some below average performance ratios. Possible reasons for such values are: faulty data, poor performance of the PV plants, difference in calculated solar irradiation vs on site solar irradiation.

Annual PR show a rather broad variance in values. To understand this variance better, PR calculations are performed on monthly base for more detailed insight in distribution of PR.

For further improvement of PR calculations, available module temperature data shall be taken into account.

3.4 ARE THE FiT ASSUMPTIONS CORRECT? SPECIFIC YIELD

FiT models are based on assumptions concerning annual specific yield for "good" PV plants. In Switzerland, FiT calculations assume 950 kWh/kWp specific yield. Evaluation of the data sample shows a higher specific yield than assumed.

	Average specific Yield [kWh/kWp]	Comparison to FiT assumptions [%]
2009	1'000	+ 5.2%
2010	945	- 0.5%
2011	1'085	+ 14.2%
2012	1'049	+ 10.5%
2013	964	+ 1.4%
2014	1'038	+ 9.2%
Overall	1'013	+ 6.7%

Table 1: Average specific yield over analyzed PV plants compared to FiT assumptions (950 kWh/kWp).

Although not necessarily representative for a lifetime of a PV plant, for the 5 years evaluated, the average specific yield is 6.7% higher than the assumed specific yield in FiT calculations. This fact implies, that in FiT success predictions, it can be expected that more solar electricity is generated with the same amount of installed nominal capacity of PV plants than is assumed. This has to be taken into account by decision makers.

The findings are compared to the findings with the second group of projects, containing PV plants with several module arrays, such as east/west mounted PV systems.

3.5 IS THERE A DECREASE IN PERFORMANCE OVER TIME?

PV plants with complete yield data as well for 2009 as for 2014 are compared to each other, giving the largest possible time between datasets. This leaves a dataset of 88 PV plants. Overall performance decrease is calculated at 0.79%. This is far less than could be expected for example based on module efficiency warranties (25a, 80%).

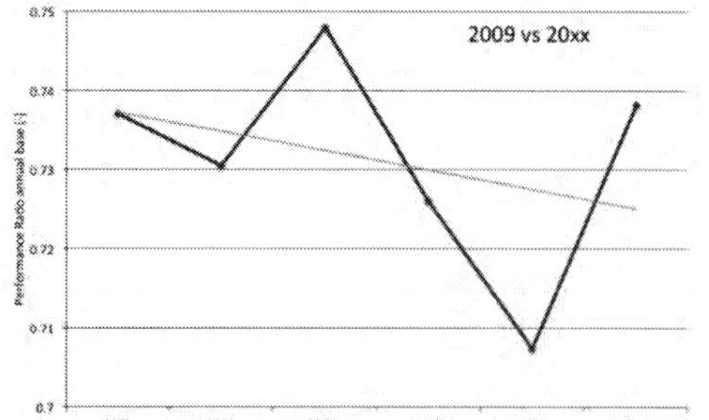

Fig 7: Decrease of performance over time for the same PV plants is lower than could be expected from component warranties.

In order to quantify the finding, other sets of data are compared. Performance 2013 is compared to performance 2009, showing a decrease of performance of 4.23% for the same PV plants. Performance 2012 is compared to performance 2009, showing a decrease of performance of 1.43%. For performance 2011 there is even an increase in performance of 1.25%, while for 2010 there is a decrease of 1.52%

3.6 IS THERE A DIFFERENCE IN PERFORMANCE BETWEEN SMALL AND LARGE PV PLANTS?

The database allows for easy selection of different sets of project criteria. One example is the comparison of efficiency of small to large PV plants.

PV plants with a nominal installed power <15 kWp are considered small, where as PV plants with a nominal installed power >50 kWp are considered large PV plants. The database provides 51…273 small and 19…313 large projects for 2009…2014. The difference in performance varies from year to year, but shows a tendency that smaller PV plants are performing slightly less well as larger PV plants. For 2010, the difference in performance was at 2.4%, for 2011 it was at 5.4%, in 2012 and 2013 it was at 5.0% and in 2014 it was at 6.6%.

A more restricting definition with small PV plants below 10 kWp (46…241 projects) and large PV plants above 100 kWp (5…115) leads to the following results: For 2010, the difference in performance was at 3.0%, for 2011 it was at 9.9%, in 2012 it was at 4.8%, in 2013 it was at 4.3% and in 2014 it was at 6.4%.

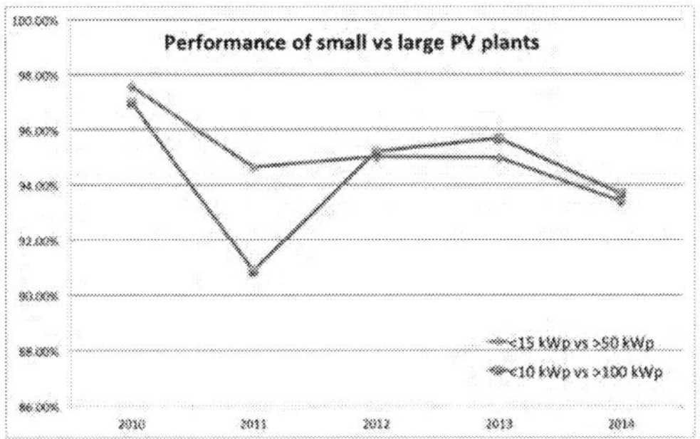

Fig 8: Visualization of two different sets of assumptions for comparing performance of large and small PV plants. Smaller PV plants perform 2-5% worse than medium to large PV plants. 100% on the y-axis is the average PR of large PV plants for the corresponding year.

The results found show that smaller PV plants are around 2-5% less efficient than larger PV plants.

Using the geographical data in the database, integration into GIS can easily be achieved in order to empirically control representativeness in geographical distribution of the PV plants analyzed.

Fig 9: Geographical distribution of PV plants analyzed: Green: PV plants >100 kWp, pink: PV plants <10kWp all with at least one complete year of data on yield.

3.7 HOW FAST IS PERFORMANCE IMPROVING FOR NEW PLANTS?

Large populations of PV plants of different sizes, locations, technologies and operators allow for a monitoring of performances of average populations. These overall average performances take into account that not all PV plants are operated professionally. An average performance of different PV plants in operation also takes into account improvements in nominal performance of components as well as reductions in performance due to reduced yields (lower tilt angles, soiling, snow, etc).

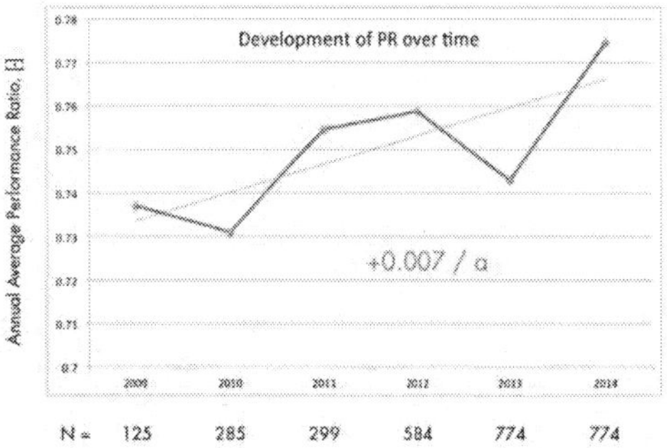

Fig 10: Increase of 0.7% per year over the whole population of PV plants examined (large and small plants, taking into account existing older PV plants)

3.8 PERFORMANCE ANALYSIS ON MONTHLY BASE

Performance analysis as was done on annual base can also be done on monthly base for more detailed insights. Performance Ratio can vary strongly from one year to another depending on conditions such as snow during winter and differences between winter and summer months.

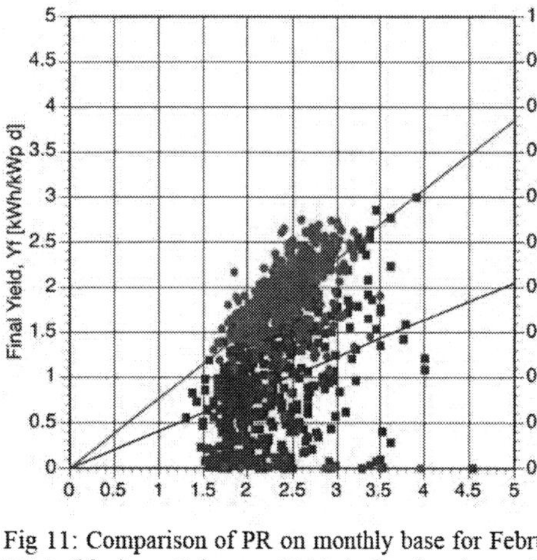

Fig 11: Comparison of PR on monthly base for February 2013 (blue) vs February 2014 (red), showing a large amount of PV plants with almost no yield in 2013 due to snow on modules.

Average PR for February 2013 (blue line) is at 0.41 almost half as low as for February 2014 (red line) with a value of 0.77. This is due to snow on modules, leading to a large amount of PV plants with no yield at all, shown by blue dots on the x axis or very close to it. The graph also shows that not all PV plants are affected in the same amount.

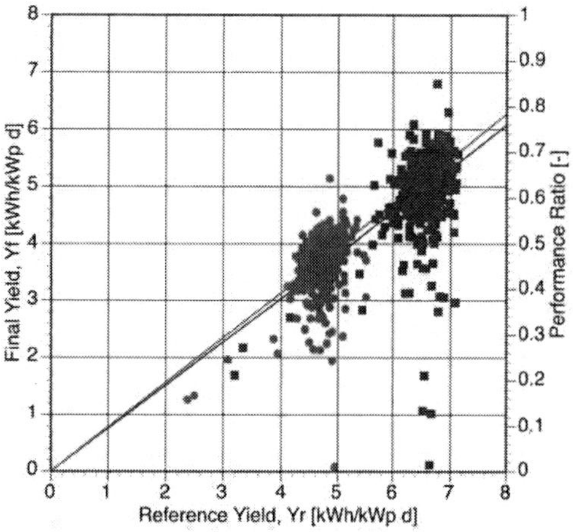

Fig 12: Comparison of PR on monthly base for July 2013 (blue) vs July 2014 (red), showing a very similar average PR but at different absolute yield levels.

The example of PR for the month of July shows very similar average PR of around 0.76…0.78. The shift of the data points toward the point of origin in the graph shows the different meteorological conditions 2013 and 2014. With lower solar irradiation 2014 (red) a lower yield results.

Yield data is available on monthly base, quarterly base or annual base. Regulations in Switzerland oblige PV plants larger 30 kVA to monitor production on a 15 minutes time resolution. For all larger PV plants, monthly data is therefor available. Some of the smaller PV plants still provide monthly yield data, depending on the local grid operator. This has to be taken into account considering the findings on performance of small PV plants compared to larger PV plants.

3.9 ARE THERE PATTERNS RECOGNIZABLE IN PERFORMANCE?

The large number of PV plants available for analysis allows for new approaches in evaluating the data. Using color-coding for different levels of PR, patterns can be detected. Where large shares of the population of PV plants is affected at the same time, this shows in a pattern of similar colors deviating from average PR.

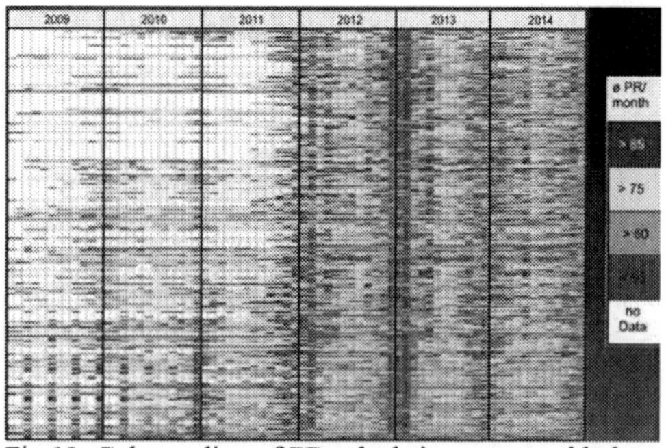

Fig 13: Color-coding of PR calculations on monthly base allow the search for patterns. In this example, December 2012 and January/February 2013 show very low PR over all PV plants.

In Figure 13 every row depicts one PV plant. The PV plants are sorted by nominal power. The columns are months over the period of time where data is available. The color-coding can be adapted, red color in this case shows PV plants with PR below 0.5.

December 2012 together with January and February 2013 show an obvious pattern in red, where almost all PV plants in Switzerland were affected and showed a very low PR. During these months, there was a large amount of snow in Switzerland, leading to covered PV modules and therefore low yields.

Patterns for other reasons may look different. Sorting of PV plants also influences the patterns, nominal power seems to be a promising start for sorting..

4 CONCLUSIONS

- A new and efficient approach to assess and analyze performance data has been developed and implemented, using existing yield data from FiT database in combination with measured and converted solar irradiation for performance analysis delivering a vast mass of projects with small efforts.
- The approach has been successfully applied to Swiss FiT projects producing 1'170 projects with monthly yield and performance available over a period of 6 years. The approach can easily be adapted for any country with a FiT scheme.
- The described approach delivers results, which can be classified somewhere between a global (annual) and analytical monitoring. With monthly performance data available, the approach can be used to complement the data acquisition for IEA PVPS performance database on a large scale.
- This approach allows a confirmation of the assumptions made for performance and yield of PV plants based on large numbers of existing PV plants of all sizes and types, not only large PV plants with professional operation.
- The approach shall be applied to other PV markets with large populations and available yield data.

First findings include:

- Average yield for the evaluated projects with one module array is 6.7% above FiT assumptions. PV is underestimated.
- Small PV plants perform 2…5% worse than large PV plants.
- Average of annual performance for datasets analyzed is at 0.75. About 1.9% of the PV plants seem to have a consistently lower than expected performance.
- Increase in overall performance for PV plants population in Switzerland is at around 0.7%/a
- Global effects such as snow covering PV plants can easily be detected and verified with large smaples of performances of PV plants

5 OUTLOOK, WORK IN PROGRESS

Performance analysis is being further developed: in-depth on more detailed monthly base and specific aspects of performance and in terms of numbers of PV projects with inclusion of multi-array PV projects.

Monthly performance data is integrated into IEA PVPS T13 performance database. Existing data from IEA PVPS database is used to verify this new approach. The approach will be propagated to other IEA members with FiT schemes.

Data and findings are further validated and additional parameters taken into account to obtain more detailed and proficient results.

Visualization and performance indicators from standardized IEA PVPS evaluations are applied to the 1'170 PV projects from Swiss FiT program.

6 REFERENCES

[1] Nordmann, Clavadetscher, van Sark, Green; Analysis of Long-Term Performance of PV Systems; IEA-PVPS T13-05:2014

[2] Zelenka, A., G. Czeplak.,V. D'Agostino, J. Weine., E. Maxwell., R. Perez, M. Noia, C. Ratto and R. Festa (1992): Techniques for supplementing solar radiation network data, Volume 1-3. IEA Report No.IEA-SHCP-9D-1

A FAST AND EFFECTIVE APPROACH TO MODELLING SOLAR ENERGY POTENTIAL IN COMPLEX SHADING ENVIRONMENTS

I.R. Cole*, D. Palmer, T.R. Betts & R. Gottschalg

CREST
Loughborough University
LE11 3TU
UK

*Corresponding Author: I.R.Cole@lboro.ac.uk, +44 (0)1509 635605

ABSTRACT: A fast and effective model for the computation of solar energy potential in complex shading environments is presented. Accurate calculation and identification of solar energy potential profiles is demonstrated over large areas. Calculation time is exceptionally fast, even on an average specification PC (typically under 1 min per 1 km^2). Problems with commonly used low-resolution sky domes that can lead to irradiance calculation errors of ~5% are identified. Ideal placements are easily visually identified from resultant irradiance/irradiation profile images. Image processing techniques for spatially distributed optimization problems are described and an example of energy value optimization is presented by means of individual dwelling demand separation & comparison.

Keywords: see enclosed list of keywords

1 MOTIVATION

The deployment of rooftop photovoltaic (PV) systems has seen rapid expansion in recent years, vastly exceeding predictions. The UK scenario is presented for reference in Figure 1 [1]. Almost all of these systems are grid connected, supplying electricity to the national grid when not consuming it on site. There is no definable pattern to the geographic distribution of these installations and information on the vast majority of these systems is not documented, yet alone accessible.

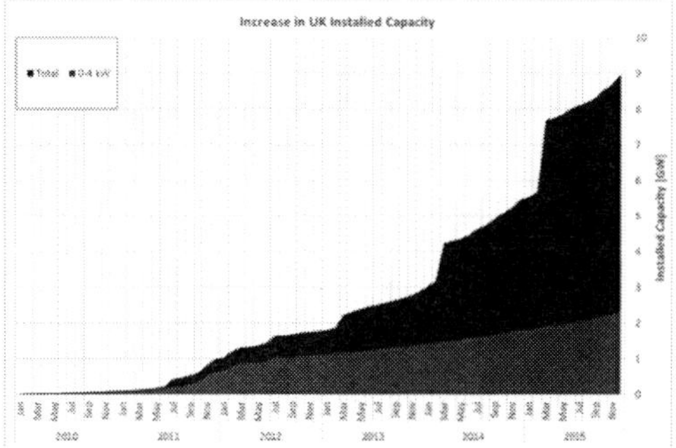

Figure 1: UK PV Deployment from 2010 to 2016 by Capacity [1]

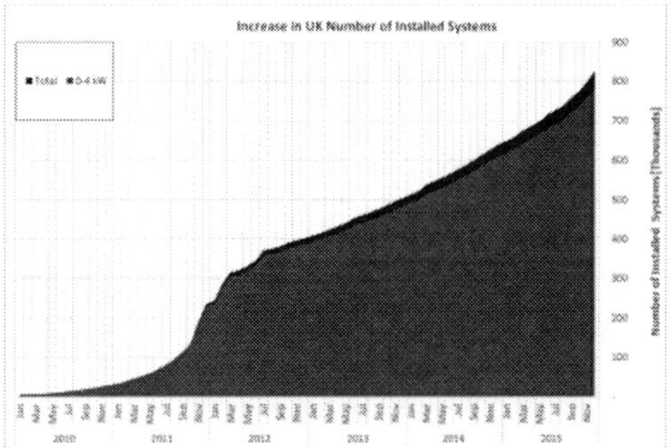

Figure 2: UK PV Deployment from 2010 to 2016 by Count [1]

In built up areas especially, optimisation of the deployment of PV is a non-trivial task. Shading can have a depredatory effect on system performance and, with residential installations in particular, system owners are often ill equipped to identify performance issues. Furthermore, at the system design or fault identification stages, the person time involved in the assessment of installation conditions by surveying is often prohibitively expensive, particularly for large scale applications.

Figure 3 demonstrates a relevant issue of note. A citywide set of approximately 2000 PV system installations in Nottingham, UK, show performances relative to the expected mean that are negatively skewed. Shading and sub-optimal installation conditions are likely significant contributors to this skew.

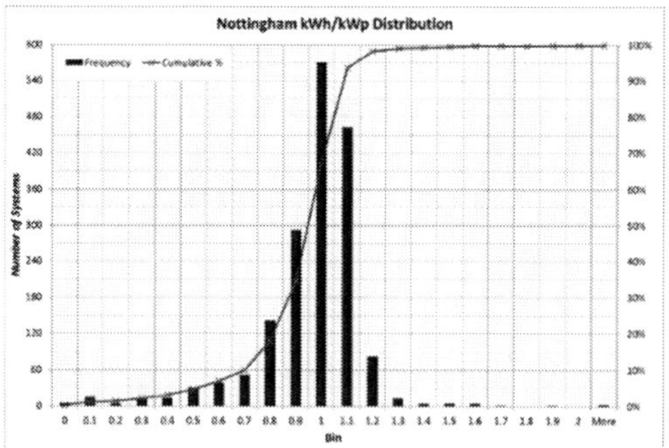

Figure 3: Performance spread of installed systems in Nottingham, UK, as indicated by kWh/kWp values normalised to a 'standard' system

Example problem installations are shown in Figure 4 and Figure 5. Unfortunately, these examples are two of many and represent a significant problem in rooftop system installations.

Figure 4: Building to Building Shading System Example (Credit: Nigel Monk, CREST)

Figure 5: Shading by a Nearby Tree Example (Credit: Brian Goss, CREST)

This work presents and describes a model, solarscene.xyz. The purpose of this model is to automate the identification of optimum and poor installation conditions, mitigating the problems described above.

2 MODEL DESCRIPTION

2.1 Objective

There exist a number of tools to accelerate system design and decision making processes. These tools do not typically include coherent shading analysis, but do offer a commonplace infographic to help designers arrive at the optimum installation angle for a given location. This is a relative energy harvest for varying installation angles, the equivalent as generated by solarscene.xyz is given in Figure 6.

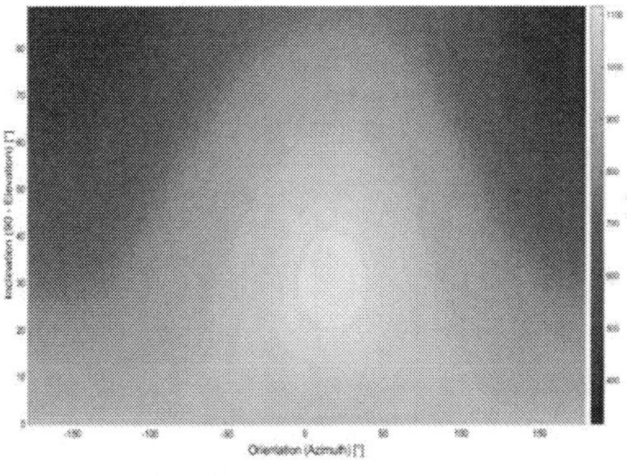

Figure 6: Net annual solar irradiation collection dependency on module orientation and inclination for Sutton Bonnington, UK, in 2014

Such information offers a useful guide for choice of PV farm module orientation or in the identification of suitable roofs in situations where there is next to no light obstruction and data is available on roof orientation. However, it is rarely the case that much is known about roof designs and vectors for wide scale applications and gathering such information can be both time consuming and costly. Furthermore, particularly in built-up areas, light obstruction is commonplace.

The key objective of this model was to generate a map of spatially distributed irradiation collection that can be used to easily identify optimum installation locations as well as potential problem areas. This has been achieved. An example of such a map compared to a Google Maps image of a given location is shown in Figure 7.

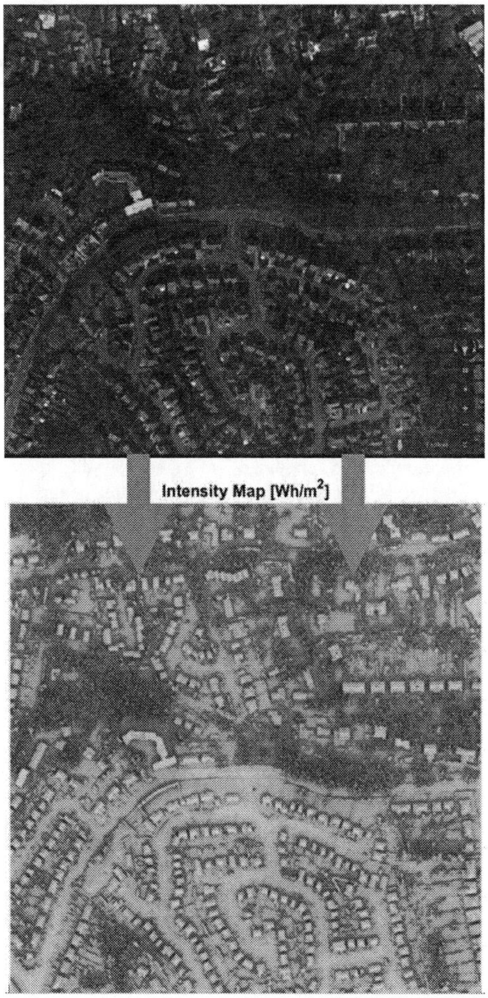

Figure 7: Net annual solar irradiation collection for each pixel in a 0.25 km² LiDAR map of Prestwich, UK, for 2014 using interpolated Met Office irradiation data

A secondary objective of the model was to be computationally efficient and to run in little time on an average specification PC. This has also been achieved, the above map took under 10 s to generate.

2.2 Model Description

The model can be considered in three distinct stages, as shown in Figure 8. Input data describing the environment is split into mapped 3D data describing the terrestrial surface and meteorological data describing the local weather variations. These input data are processed and passed through to a ray trace algorithm which generates mapped solar energy data. Each phase is described separately in this section.

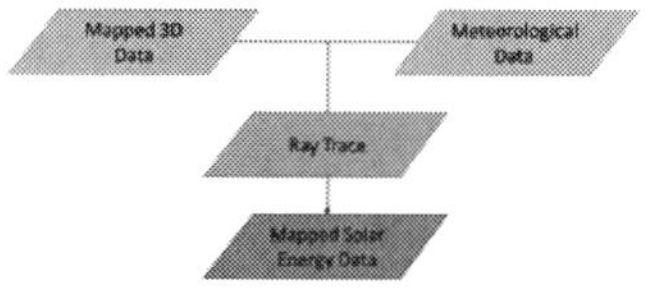

Figure 8: Model Flowchart

3D Environment Data

The input 3D environment primary data is in the form of a map of pixel heights. Metadata map layers are also used for specific analysis, for example, highlighted regions of interest can be used to provide specific focus to output data. Test case data from LiDAR has been used here for illustration. CAD drawings of proposed building plans or .stl conversions are also possible. It is also possible to place a proposed building plan within a set of LiDAR (or other existing) data to construct a contextualised placement scenario.

An interpolated surface is produced from the input LiDAR points mesh. This surface is used to form surface normal vectors from each pixel that are used in intensity correction calculations. Here, a mesh resolution of 1 m^2 was used. The mesh resolution is variable and is dependent on user requirements/available data. A one km^2 example is presented in Figure 9, the region presented is the same as that shown in Figure 7, expanded is the northerly and easterly directions.

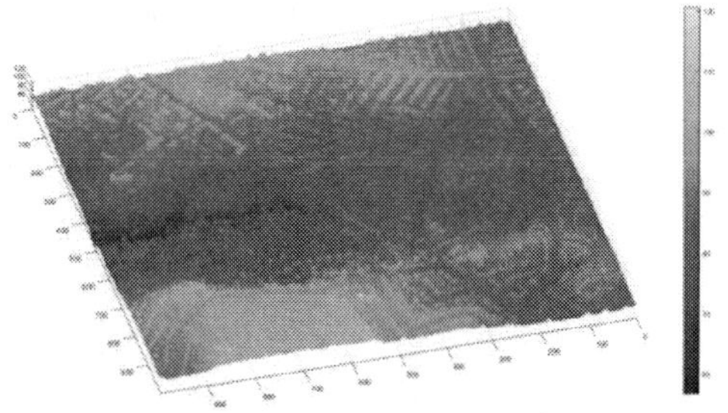

Figure 9: Points to Surface Map from LiDAR – 1 km^2 of Prestwich, UK

A cuboidal columnar representation of the data is also created (Figure 10). It is this surface that is used in the ray trace analysis for shading profile calculation and is key to the computational efficiency of the simulation.

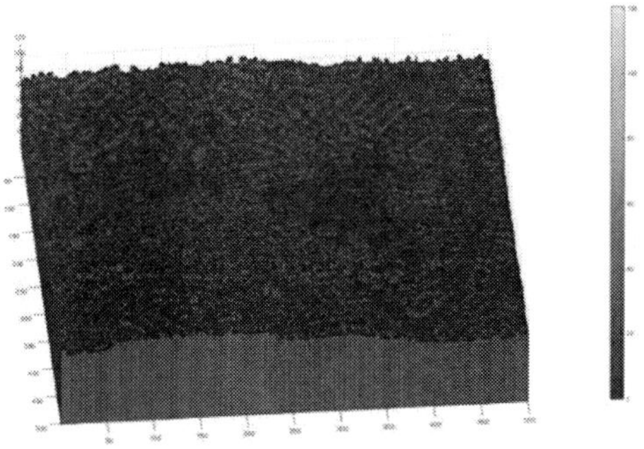

Figure 10: Points to Cuboidal Columnar Map from LiDAR – 1 km^2 of Prestwich, UK

Meteorological Data

Input meteorological data is in the form of horizontally received beam and diffuse irradiation. The temporal resolution of this is variable, depending on user requirements and available data. Currently, the model does not use albedo irradiation as there is little data available for this. However, this is planned as an extension in future development of this model. The irradiance values used here have been derived by weighted interpolation of met office hourly irradiation values for nearby and neighbouring sites.

The irradiance/irradiation (in instantaneous/ time-extended analyses) is distributed within a hemispherical representation of the sky, termed a sky dome (Figure 11). Each segment of the sky dome is individually traced through the 3D environment and the contribution of its power/energy to each surface pixel calculated as a multiplication of its value, shading factor and angle of incidence correction.

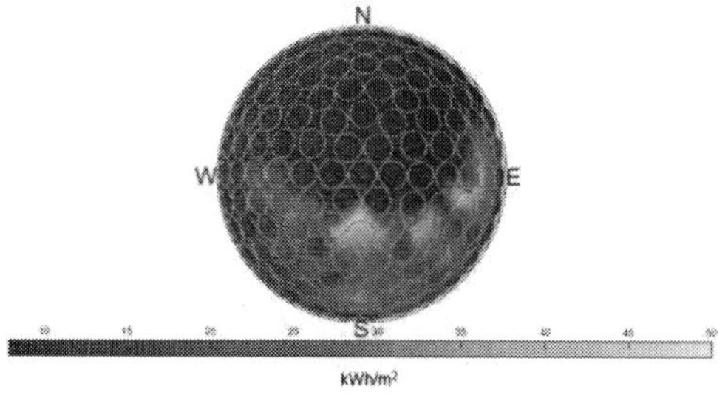

Figure 11: Net Annual Sky Dome Example with Tregenza Overlay

This tool was developed at CREST for the generation of high resolution maps of solar energy potential for existing and/or proposed (building plans) locations. These calculated solar energy potential maps include global and localised shading effects. Time resolved output maps allow for the specific investigation of shading scenarios, rather than limiting the results to averaged values. Diffuse and beam irradiance components are specifically considered and the model is easily extended to include spectral effects.

For ease of presentation, Figure 11 uses a common segmentation of the sky into 151 patches known as the Tregenza dome [2]. In practice, use of the Tregenza dome is not acceptable for energy prediction scenarios as the granularity of the dome, especially regarding elevation, causes unnatural stepped values in in-plane irradiance, as shown in Figure 12. These can lead to errors in calculated irradiance values of ~5%.

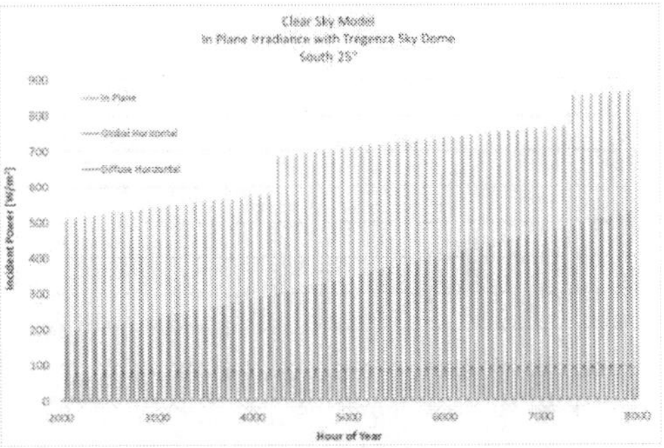

Figure 12: Stepped In-plane Irradiance Artefacts from Low Resolution Sky Dome

Alternative sky domes derived by similar means are presented in [3], although in practice any sky dome of suitable resolution (mitigating the artefacts shown in Figure 12) can be used. The relative contribution of diffuse irradiance can corrected for by calculating the relative projected area contribution of the confined surface by, for example, Equation 1. The selection of an appropraitely defined sky dome can also lead to a significant increase in computational efficiency in the sky patch location of sun position.

$$\int_{\gamma_{s1}}^{\gamma_{s2}} \int_{h_1}^{h_2} \cos h \,.\, dh \,.\, d\gamma_s$$

Equation 1

Where γ_s is the solar azimuth angle and h the solar elevation angle (see [4] for explanation of terms)

Ray Trace Method

The ray trace method employed here is one developed at CREST, building on the work presented in [5-7]. Comparisons with existing methods such as [8 & 9] are presented in [10]. The procedure is optimised for the specific calculations performed here. This is a marked advantage over using off-the-shelf solutions as any data beyond that required is simply not calculated here, rather than computed and not used. Much computational time is thus saved in this approach. Further to the streamlined computation process, the ray trace procedure here is divided in 16 special cases: the four major compass directions, the four minor compass directions and the 8 spaces between Figure 13. The case is first identified and the corresponding optimised calculation procedure triggered in the code accordingly.

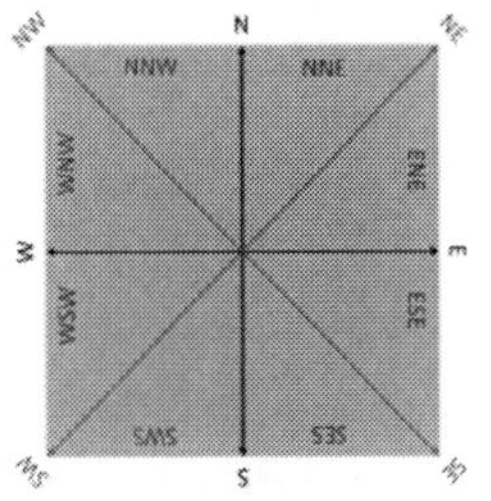

Figure 13: Compass Directions Broken Into 16 Special Cases for Efficient Computation

Using this method, the net annual irradiation values on each 1 m^2 pixel in a 1 km^2 region using hourly meteorological data distributed in a Tregenza dome can be calculated on an average specification PC in under 1 min. A further optimisation advantage to this method is that relative map trace values for a given location and sky dome combination can be stored and later recalled for investigation of a different meteorological dataset – bypassing the ray trace procedure altogether – at which point the computational time reduces to a value that is primarily dependent on data drive read/write speeds.

3 MODEL PERFORMANCE & UTILIZATION

3.1 Validation

The most visually significant validation for the key model objective is a comparison of model shadow identification with those visible from aerial photography. Here, for clarity, the 0.25 km^2 image of Figure 7 is used. Figure 14 shows the shading pattern produced by the model. It can be seen that the resultant shadows are extremely well aligned. The corresponding sun position used in this instantaneous test case is shown in Figure 15.

Figure 14: Intensity Map Produced by solarscene.xyz Showing Matching Shadow Profiles (see Figure 7)

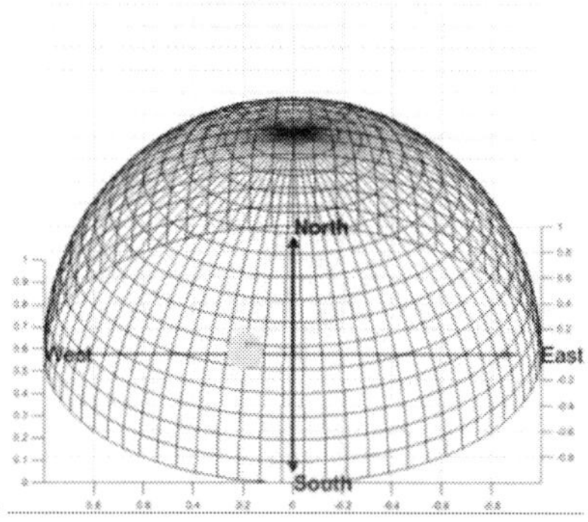

Figure 15: Depiction of the Sun Position used in the Generation of Figure 14

It is these instantaneous profiles that are integrated in order to obtain the primary model objective – irradiation maps, as shown in Figure 7.

3.2 Model Utilization

With further processing of the images generated by the model or the use and analysis of related spatially distributed information, the model can be further developed as a tool for optimization problems. Two example model developments that are currently being explored are shown below.

Automated Identification of Optimum Installation Locations – Preliminary Investigation

It can clearly be seen from Figure 7 that certain areas are significantly brighter than others. If the image is generated in greyscale then the brightness of each pixel is directly related to an annual irradiation value. Thus this lends itself particularly well to an image processing for analysis.

Figure 16 shows the process of taking a given location, generating a greyscale irradiation map, applying a threshold brightness filter to extract pixels above 700 W/m^2 and then applying a morphological filter (2 m*4 m rectangle) to determine a space suitable for solar panel installation. This process will be further developed and published in future work.

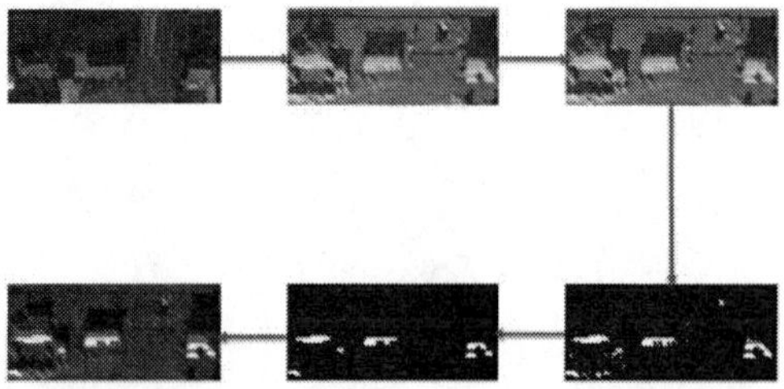

Figure 16: Image Processing for Optimum Installation Location Determination - Preliminary Example

Relative Value by Dynamic Assignment – Preliminary Investigation

The irradiation map presented in Figure 7 is useful in the context of optimization or energy harvest. However, the reality of energy value is dynamic. For example, peak generation time of south facing residential rooftop PV systems at 12:00 is mostly wasteful. A simple investigation of energy value optimization is shown below. Figure 17 shows normalized energy harvest for a given location, with only rooftop pixels shown for a 200 m by 200 m area in Nottingham, UK.

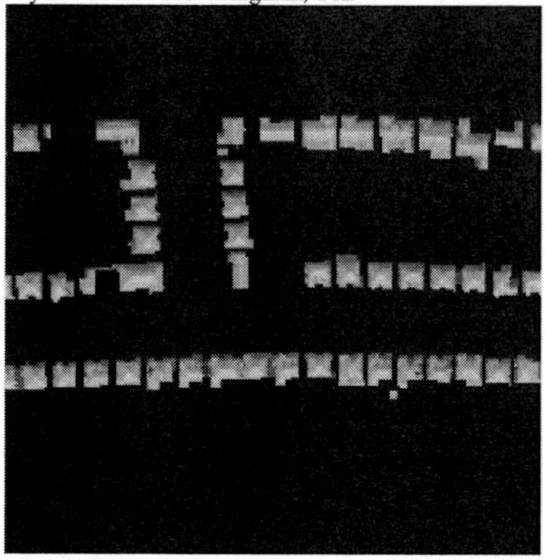

Figure 17: Normalised Irradiation Harvest on Nottingham Rooftops

Data separating each individual dwelling was then obtained and 5 realistic energy usage profiles were generated using [11]. These profiles are shown in Figure 18.

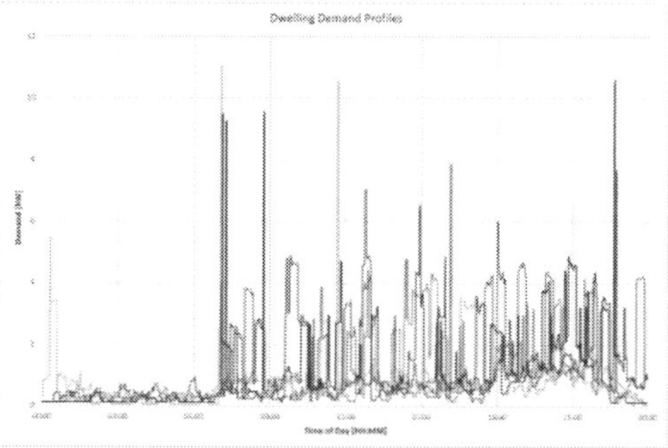

Figure 18: 5 Different Example Energy Demand Profiles

The simulation was then rerun, randomly assigning one of the 5 demand profiles to each of the dwellings above and producing a normalised map of energy value per pixel (see Figure 19). The resultant map is significantly different to Figure 17. This will be investigated further and published in future work.

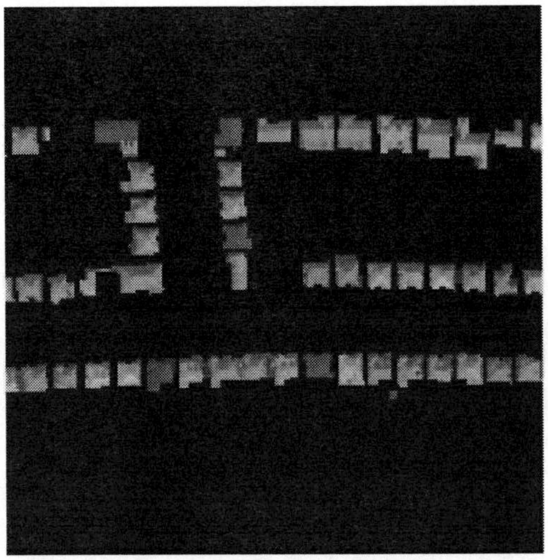

Figure 19: Normalised Energy Value on Nottingham Rooftops

4 CONCLUSIONS

- A Fast and Effective Approach to Modelling PV Potential in Complex Shading Environments has been Developed and Presented.
- Accurate Pixel by Pixel Irradiance and Irradiation Maps Are Calculated by the Model
- Key mistakes in existing models have been Identified (Can Lead to ~5% Irradiance Calculation Error)
- The Automated Identification of Optimum Installation Conditions has been Demonstrated
- The Identification of Potential Problem Areas and System Underperformance due to Shading has been Demonstrated

There will be much further development of this model. Some opportunities for development have been identified here alongside preliminary investigations, though there are many more possibilities for useful model expansion and utilization. solarscene.xyz will be made publically available later in the year. The authors welcome collaborative works and feedback.

5 REFERENCES

[1] National Statistics – Solar photovoltaic deployment, Department of Energy and Climate Change (DECC), UK Government, 25 February 2016

[2] TREGENZA, P. R. 1987. Subdivision of the sky hemisphere for luminance measurements. Lighting Research and Technology, 19 (1), 13-14.

[3] GOSS, B., COLE, I., BETTS, T. & GOTTSCHALG, R. 2014. Irradiance modelling for individual cells of shaded solar photovoltaic arrays. Solar Energy, 110, 410-419.

[4] ISO 9488:1999 – Solar Energy – Vocabulary

[5] COLE, I. R., BETTS, T. R. & GOTTSCHALG, R. 2012. Solar Profiles and Spectral Modeling for CPV Simulations. IEEE Journal of Photovoltaics, 2, 62-67.

[6] COLE, I. R. & GOTTSCHALG, R. 2015. Optical Modelling for CPV Systems: Insolation Transfer Variations with Solar Source Descriptions. IET Renewable Power Generation, 9, 412-419.

[7] COLE, I. R. & GOTTSCHALG, R. 2016. Improved Model for Circumsolar Irradiance Calculation as an Extended Light Source and Spectral Implications for High-Concentration Photovoltaic Devices. IEEE Journal of Potovoltaics, 6(1), 258-265.

[8] BERGAMASCO, L., & ASINARI, P. 2011. Scalable methodology for the photovoltaic solar energy potential assessment based on available roof surface area: Further improvements by ortho-image analysis and application to Turin (Italy), Solar Energy, 85, 2741–2756.

[9] TARINI, M., CIGNONI, P. & MONTANI, C. 2006. Ambient Occlusion and Edge Cueing to Enhance Real Time Molecular Visualization, IEEE Transactions on Visualization and Computer Graphics, 12(5), 1237-1244.

[10] PALMER, D., GOSS, B., COLE, I., BETTS, T. & GOTTSCHALG, R. 2015. Detection of Roof Shading for PV Based on Lidar Data Using a Multi-Modal Approach. Proceedings of the 31st Photovoltaic Solar Energy Conference, 1753 – 1759.

[11] Richardson, I and Thomson, M (2012) Integrated simulation of photovoltaic micro-generation and domestic electricity demand: a one-minute resolution open-source model, Proceedings of the Institution of Mechanical Engineers, Part A: Journal of Power and Energy, -, DOI: 10.1177/0957650912454989.

COMPARATIVE STUDIES IN DEGRADATION BEHAVIOR OF SINGLE-CELL MODULE BY PRESSURE COOKER TEST (PCT) AND EXTENDED DAMP HEAT (DH) TEST

Yu-Tai Li[a], Chao-Ming Tung[b], Behrang H. Hamadani[c],Xiao-Hong Gu[c],Wei-Lun Yang[a], Chin Lien[a]
Hung-Sen Wu[a] , and Pei-chen Yu[b]
[a] Photovoltaic Metrology laboratory, Center for Measurement Standards, Industrial Technology Research Institute, Hsinchu, Taiwan
[b] Department of Photonics and Inst. of Electro-Optical Engineering, National Chiao-Tung University, Taiwan
[c] Engineering Laboratory, National Institute of Standards and Technology, Gaithersburg, Maryland 20899, USA

ABSTRACT: Recently, standard of IEC61215 for PV industry which designed to identify early failures of photovoltaic (PV) modules within the first several years are insufficient to the market. More and more stringent accelerated aging protocols, such as extended Damp Heat (DH) tests or Pressure Cooker Tests (PCT), are proposed in order to ensure longer PV modules lifetime. Compared to DH tests, PCT provided an alternation to evaluate durability of PV encapsulation quickly through higher temperature, humidity and pressure. Nevertheless, it is still unclear about the relationship between these two accelerated tests. Thus, the primary object of this work is to investigate suitable encapsulation of single-cell-laminates PV modules for PCT and extended DH test and compare their degradation behaviors. Besides, we also study the main degradation mechanism for PV modules aged by PCT through electroluminescence (EL), Spectrum Response (SR) and SEM measurement. Our result shows only less than 3 % of power degradation is measured for 4000 hours of extended DH test while around 30% of power degradation for 450 hour of PCT. Similar increase trend for these two tests are measured in series resistance (Rs) and it could be mainly attributed to corrosion of Sn and Ag along ribbon and fingers, which could be observed from EL image and SEM measurement.

Keywords: Pressure Cooker Test, Damp Heat Test, Qualification and Testing, Reliability, Encapsulation

1 INTRODUCTION

Lifetime of photovoltaic (PV) modules is highly related to the cost of generated electricity and PR value, which impacts total generation of power plant. But the certification for PV modules for industry, standard of IEC61215 tests, is only designed to identify early failures within the first several years. In these test protocols, DH test (85°C and 85% relative humidity, 1000 hours) is one of most key protocols to accelerate potential degradation mechanisms. However, recently, it seems that only DH 1000 hours cannot satisfy manufactures to ensure lifetime of PV modules more than 25 years. Therefore, more stringent accelerated aging protocols such as extended DH tests (>3000 hours) are proposed [1]. Besides, degradation mechanism analysis of extended damp heat is also investigated. [2]

However, in order to shorten test time of extended DH test, PCT have also been adopted. Through higher relativity humidity (100%), water vapor pressure and higher temperature (105°C~130°C), PCT can be used to quickly evaluate durability of PV encapsulant, and further investigate the relevant degradation mechanisms of PV module. [3] Theoretically, PCT is more efficient to evaluate the moisture resistance of packaged solid state devices through the moisture condensing or moisture saturated steam environments. [4-5]

Suitable test conditions included encapsulation and test temperature (105°C and 121°C) are also investigated in our previous work, which is helpful to evaluate reasonable requirement or test result. [6] Our pervious results suggests a condition of 121°C of PCT test for 450 hours could induce corrosion interconnects or Si-cell quickly. However, in order to improve reliability of PV modules by PCT tests, we need to understand the mechanism behind these failures under these conditions and compared to extended DH tests. Therefore, in this work, we investigate suitable encapsulation of single-cell-laminates PV modules for PCT (121 °C, 450 hours) and extended DH test (4000 hours); and compare their degradation behaviors. Moreover, the main degradation mechanism are also studied through electroluminescence (EL), Spectrum Response (SR) and SEM measurement.

2 EXPERIMENTAL

The suitable size for PV modules test in pressure cookers should be single-cell-laminates. Raw materials adopted in this work are poly-crystal cell, EVA and backsheet (TPT). Figure 1 shows two types of single-cell-laminates in the test. One shown in figure 1(a) is a single-cell-laminated PV module without sealing around the edge while the other PV modules shown in figure 2 (b) sealing with Al foil. Both types of PV modules are aged by DH 4000 hours and PCT 121 °C. Using different sealing types may cause different degradation of PV cell because different moisture ingression from edge. Degradation behaviors were recorded every 90 hours for PCT while every 500 hours for DH test including electrical performance, change in appearance, EL imaging. Besides, from change of IV curve, Rs change could be measured. In order to understand the reason of increasing Rs, distribution of Rs image and spectrum response for Si cell are also investigated. The local series resistance of aged solar cell are mapped with the help of a high and a low voltage EL-image.[7] Total response of full cell

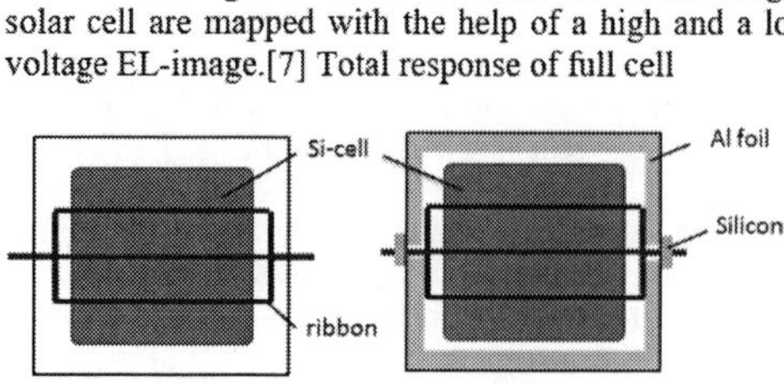

Figure 1: Single-cell-laminates for tests (a) non-sealed (b) sealed with Al foil and Silicone

3 RESULTS AND DISCUSSIONS

3.1 Appearance change and EL images

Figure 2 (a) shows appearance and their EL image for non-sealed sample after DH 4000 hours. No obvious changes are observed in these both images. Figure 2 (b) and (c) show PCT, 121°C, 450 hours and sealed with Al foil and PCT, 121°C, 450 hours separately. Compared EL image of 2 (b) with 2 (c), serious dark region could be found around edge of non-sealed modules due to the accumulated acid generated from ingress water. Such phenomenon could not be observed for modules after 4000 hours DH test even sealing or non-sealed one. Figure 2(d) shows enlarged appearance of DH test (Left) and their PCT test (Right). Some white powder or particles could be observed from top side of PV modules aged by PCT, which could be attributed to damaged PET particles ingress with water from Backsheet and edge of cell. Such side effect of particles ingress is only happened in PCT test, not for DH tests. From the distribution of these white particles and dark region of EL image, we could guess water ingress of PCT test is from edge of cell and alone the ribbons.

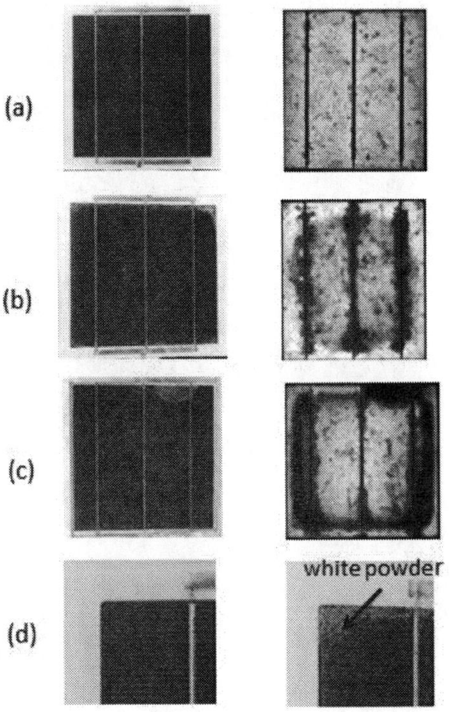

Figure 2: Appearance of modules and their EL image for (a) DH 4000 hours (b) non-sealed, PCT, 121°C, 450 hours (c) sealed with Al foil, PCT, 121°C, 450 hours, and (d) enlarged appearance of DH test (Left) and PCT test (Right)

3.2 Electrical performance change

Power degradation of DH test and PCT tests and their I-V curve are shown in figure 3(a) and 3(b) separately. Power degradation of DH test 4000 hours for non-sealed and Al foil sealing samples are 1.5 % and 2.1 % separately. As for the PCT tests 450 hours, 18.8% and 30.4 % of power decay are measured for non-sealed and Al foil sealing samples. Higher decay could be measured for Al foil sealing modules under both DH and PCT tests. From change of I-V curve, we can get change of Rs for DH and PCT tests, which are shown in figure 4 (a) and (b). From figure 3 (a), trend of power decay for these two test methods is not very consistent. However, from figure 4 (a) and (b), similar trend of Rs change for DH and PCT tests could be found.

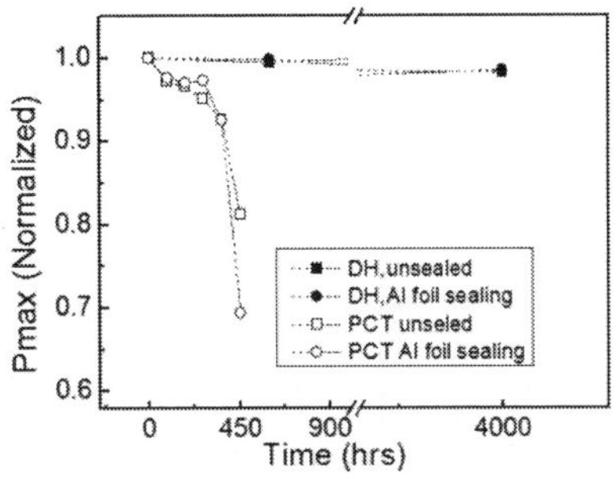

(a)

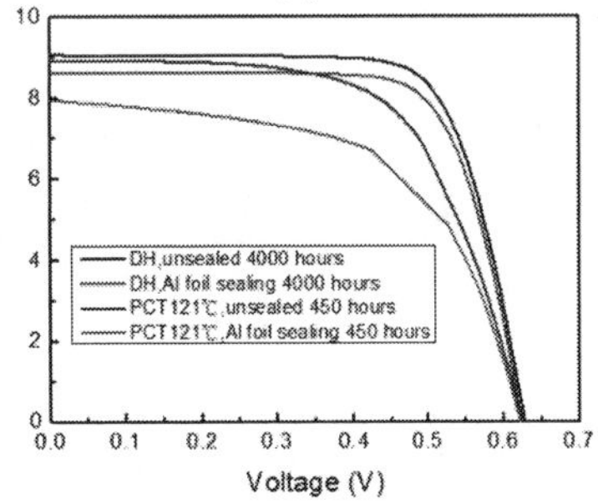

(b)

Figure 3: (a) Power degradation of DH and PCT tests (b) I-V curve changes of DH and PCT tests

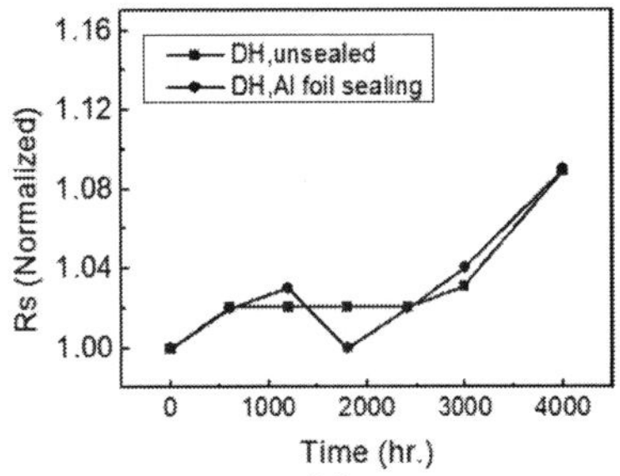

(a)

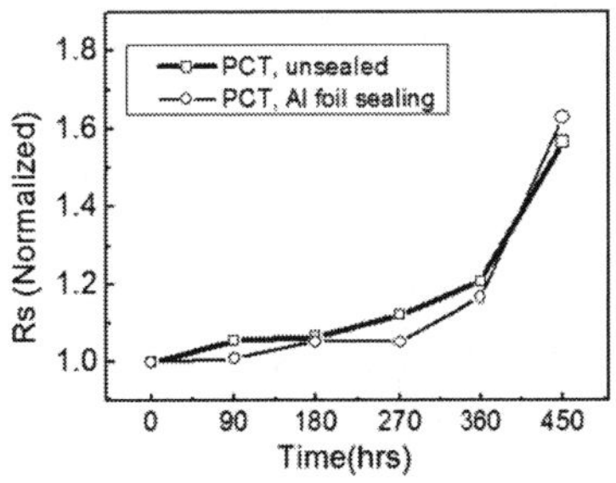

(b)

Figure 4: Changes of series resistance for (a) DH test and (b) PCT

3.3 Spectrum Response (SR) and Rs distribution

Figure 5 shows spectral response for (a) total cell and (b) local area for non-sealed PV modules (PCT, 121°C, 450 hours). Total response measurement for the hall cell is according to an LED-based integrating sphere-coupled spectral response measurement system. [8] Compared to fresh samples, PCT aged modules show little degradation in spectral response. Besides, though monochromatic source, we also measured local spectral response of aged PCT module in several points where shown different brightness in its EL image marked as point A to E in inset of figure 5 (b). This figure suggest that there is no obvious relationship between dark area of EL image and spectrum response, that means dark region of EL image may be attributed to corrosion of contact metal (finger or ribbons) not corrosion of Si cell.

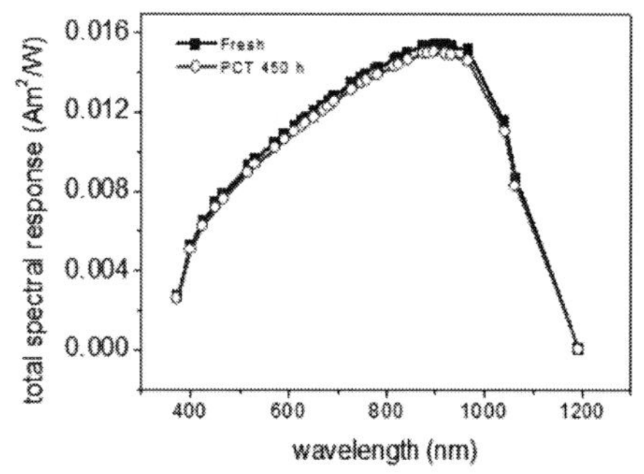

(a)

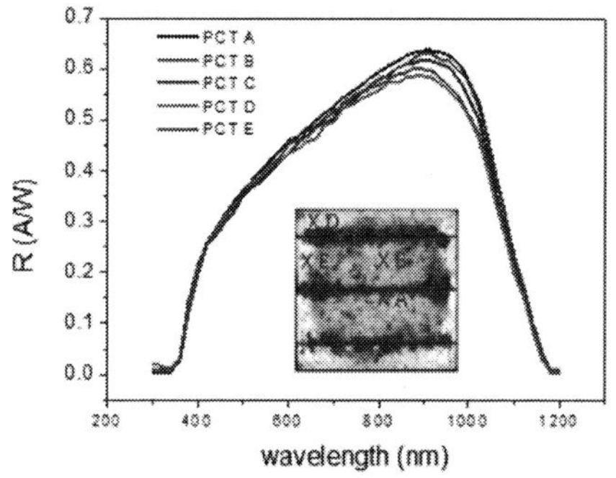

(b)

Figure 5: Spectral response for (a) total cell and (b) local area for non-sealed PV modules (PCT, 121°C, 450 hours)

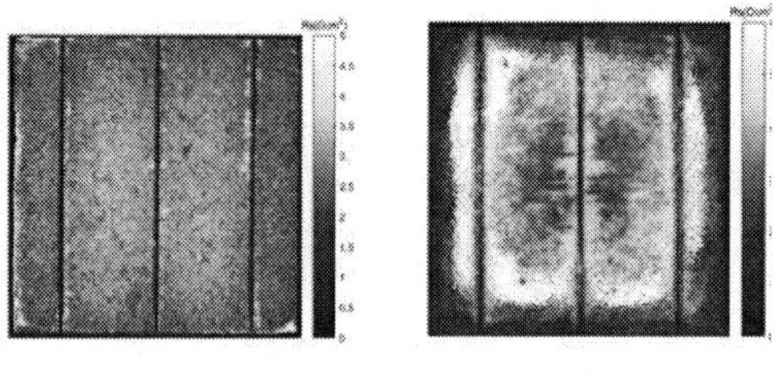

(a) (b)

Figure 6: Distribution of Rs for (a) DH 4000h (b) PCT, 121°C, 450 hours

In order to investigate distribution of increasing Rs, we calculated local Rs with the help of a high and a low voltage EL-image and absulte of Rs fitted from curve I-V in figure (3). Figure 6 (a) and (b) show distribution of Rs of samples for DH test 4000 hours and PCT 121°C, 450 hours separately. It's still very uniform for DH 4000h while high resistance 6 Ω/cm² shown around edge of PCT aged samples. Compared to EL image of figure 2 (c), some corrosion along fingers could be observed around center area of cell which shown in figure 6 (b). Such distribution of Rs increasing could be explained by moist ingress and accumulated of water around cell while aged by PCT test.

3.4 Analysis of SEM measurement

In Figure 7 (a)~(d), we measured SEM image of cross section of aged PCT samples where shown higher Rs distribution in figure 6 (b). Figure 6(a) shows optical image of PV modules and figure 6 (b) shows its SEM image. Sn and Ag distribution could be observed in figure 7 (c) and (d), which show element drifting to other area. Some separated space between ribbon and Si cell could also be found. Such phenomenon could be attributed to corrosion by moisture ingress. More study by SEM alone fingers should be studied in the future.

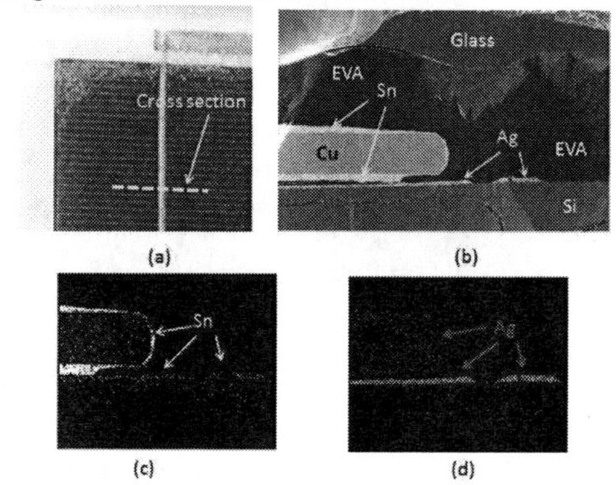

Figure 7: (a) optical image of PV module and sampling area. (b) Cross-sectional SEM micrograph, (c) Dissolved of Sn (d) Distribution of Ag

4 Conclusion

In conclusion, we investigate suitable encapsulation of single-cell-laminates PV modules for PCT and compared the result to extended DH test. The power degradation after 4000 hours of DH tests is less than 3 %, while around 30 % after 450 hours of PCT. Although power degradation is very slow compared to PCT, similar increase trend for these two tests could be measured in series resistance. Through EL image and series resistance mapping, we could find that increase of Rs is main reason for the power degradation of the modules compared to seldom degradation of response for Si-cell. The increase of Rs could be observed alone ribbon and through EL image and it could be mainly attributed to corrosion of Sn and Ag along ribbon and fingers. Such corrosion could be observed from cross section view alone ribbon and fingers of SEM images.

6 Reference

[1] Ketola, Barry, and Ann Norris. "Degradation mechanism investigation of extended damp heat aged PV modules." *26th EUPVSEC* (2011).

[2] Haidan Gong, Guofeng Wang , and Minge gao, "Degradation mechanism analysis of damp heat aged PV modules." *27th EUPVSEC* (2012).

[3] JEDEC Standard JESD22-A102-C, "Accelerated Moisture Resistance - Unbiased Autoclave", Dec. 2000.
[4] Kim, Namsu, et al. "Study on the degradation of different types of backsheets used in PV module under accelerated conditions." *Solar Energy Materials and Solar Cells* 120 (2014): 543-548.
[5] Haidan Gong, Guofeng Wang , and Minge gao, "Degradation mechanism analysis of damp heat aged pv modules." *27th EUPVSEC* (2012).
[6] Y.T. Li, C.-M. Tung, Y.-H. Lee, W.-L. Yang, H.-H. Hsieh, H.-S. Wu, P. Yu, E.Y. Wang ,"Investigation of Standardized Test Condition for PV Encapsulants by Pressure Cooker Test (PCT)", 31th *EUPVSEC* (2015)
[7] Walter, Johann, et al. "A Method to Detect Defective Solder Joints by R S-Electroluminescence Imaging." Energy Procedia 27 (2012): 652-657.
[8] Hamadani, Behrang H., et al. "Large-area irradiance-mode spectral response measurements of solar cells by a light-emitting, diode-based integrating sphere source." Applied optics 53.16 (2014): 3565-3573.

INFLUENCE OF BACKSHEET TYPE ON FORMATION OF ACETIC ACID IN PV MODULES

A. Mihaljevic[1], G.Oreski[1*], G. Pinter[3]
[1] PCCL Polymer Competence Center Leoben, Roseggerstraße 12, 8700 Leoben,
Corresponding author: E: Gernot.oreski@pccl.at, T: +43 3842 42962-51
[3] University of Leoben, Institute of Material Science and Testing of Plastics, Otto Glöckl Straße 2, 8700 Leoben,
Gerald.pinter@unileoben.ac.at

Abstract:

At the moment, the dominating encapsulation material for PV modules is chemically crosslinked ethylene vinyl-acetate copolymer (EVA). Besides the time and energy consuming module lamination process, the major drawback of EVA is the formation of corrosive degradation products like acetic acid during aging. Acetic acid has a great impact on ageing mechanisms like yellowing of EVA, PID effect, corrosion of metal parts or delamination, and, it is known to accelerate the oxidation processes in EVA. However, ingress of permeates like oxygen, water or acetic acid, which are assumed to be reaction partners in degradation processes inside of PV module, is governed by the permeation properties of the polymeric backsheets or encapsulation materials used. In this paper the main purpose was to correlate permeation properties of backsheets with formation of acetic acid in the EVA encapsulant layer. Results from previous works showed significant differences in the permeation behavior of backsheet laminates depending on different layer composition. The lowest permeation rates (up to 100 times lower) were found for laminates containing a polyester (PET) core layer in comparison to co-extruded backsheets based on polyamides and polyolefines. Therefore, test laminates consisting of EVA encapsulant and backsheet laminates with low (AAA) and high (APA) barrier layers were prepared and exposed to UV irradiation and temperature. The concentration of formed and permeated acetic acid was measured via GC/MS method and correlated with different AATR values of the backsheet laminates. The results were assigned to autocatalytic effect of acetic acid which retained at interfaces of test laminate. Degradation of EVA in test laminates was investigated after aging via Raman confocal microscopy.

Keywords: *ethylene vinyl-acetate (EVA), backsheet, acetic acid, permeation, degradation*

1. INTRODUCTION

Polymeric materials are the most common material used in crystalline silicon PV module for protection of brittle solar cell from the weather from front and back in form of encapsulant and backsheet. The most employed encapsulating material is ethylene vinyl acetate copolymer (EVA). Being the most capable to fulfill many of requirements for encapsulating material (low modulus, high damping capacity, optical transparency), ethylene – vinyl acetate (EVA) has become the most popular encapsulation material for PV industry. Polymeric backsheet materials, on the other hand, are commonly used in PV modules for physical protection, enhanced encapsulation, light reflection, electrical insulation and aesthetic purposes. They are multilayer laminates usually containing PET and fluoropolymer layers. Although being lightweight, cheap and easily available, polymers are prone to degradation under extended exposure to environmental conditions (UV, humidity, high T, dust, hail…) and play a critical role in determining the long performance of PV modules and service life time which is predicted to be more than 20 years. [1] The major disadvantage of EVA, however, is the degradation via Norrish I and II reactions which leads to formation of chromophores and acetic acid as by-products. [2] Acetic acid is linked to several PV module failure mechanisms like corrosion of interconnectors or potential induced degradation (PID). Moreover, the acetic acid has an autocatalytic effect and promotes further formation of acetic acid in the EVA, especially under UV exposure. [3] In several studies backsheets with presumably high acetic acid permeation rates were correlated with better PV module performance. [3] [4] Therefore, the main purpose of this work is to correlate permeation properties of backsheets with formation of acetic acid in the EVA layer of test laminates.

2. EXPERIMENTAL

For the preparation of the test laminates two types of EVA encapsulants from different producers and two types of backsheet materials with high and low barrier layer were chosen (see Table 1.).

Table 1. Materials used in preparation of test laminates

Layer	Material composition	Remark
Encapsulant	EVA 1 EVA 2	different producer
Backsheet	AAA (PA/PA/PA)	low barrier
	APA (PA/PET/PA)	high barrier

In Table 2 the composition of each test laminate is shown. Test laminates were cured in vacuum lamination press together with glass (thickness ≈ 4 mm) by standard lamination process (t=9 minutes and 20 seconds; p= 850 mbar and T=150°C). It is necessary to note that also test laminates without backsheet were prepared in order to detect production of acetic acid in EVA itself.

Table 2. Composition of test laminates

Sample	EVA1	EVA2	AAA	APA	UV/T	T
1	+	-	-	-	+	-
2	+	-	+	-	+	-
3	+	-	-	+	+	-
4	-	+	-	-	+	-
5	-	+	+	-	+	-
6	-	+	-	+	+	-

Afterwards, the test laminates were placed in specially designed stainless steel capsules with diameter around 8 cm with septum on the bottom. Special precision O-rings and grease were used in each capsule to accomplish sufficient sealing. Capsules were exposed to artificial ageing in Suntester CPS+ (Atlas Materials Testing Technology GmbH, Germany) in order to examine photo-oxidative ageing processes according to method ISO 4892-2 with an irradiance of 765 Wm^2 and the black standard temperature of 90°C (adjusted temperature due to short testing time) in the range of λ= 250-765 nm up to 200h. In order to support thermal ageing processes, capsules were exposed to temperature of 90°C. Samples were taken out periodically and vapor containing acetic acid from the backside of test laminate was collected via a gas proof syringe (50 µL, SGE Analytical Science, Australia) without opening capsule. Afterwards, the content of syringe was investigated by gas chromatography coupled with mass spectroscopy (GC/MS) (Shimadzu GC/MS-QP 2010 Plus) to examine the concentration of formed and permeated acetic acid. Additionally, the degradation of EVA layers in the test laminate was examined with Raman confocal spectrometer LabRAM HR 800 (Horiba JobinYvon GmbH, Germany) with green laser (λ=514 nm) in the Raman shift range of 650-4500 cm^{-1}. Since the increase in baseline was used as an indicator of EVA degradation (due to fluorescence background originating from degradation by-products of EVA), no baseline or peak normalization of Raman spectra was done.

3. RESULTS AND DISCUSSION

Present measurements of acetic acid transmission rates (AATR) showed how AATR values of backsheets depend on the temperature, thickness and layer composition. [5] As expected, AATR values of EVA are much lower in comparison to AATR value of backsheets (see Figure 2.). At room temperature, AATR values of backsheets are comparatively low, especially for those containing high barrier (PET) layers. Conducting experiments at temperatures similar to the operational temperature of the PV module (85°C) revealed in increase of permeation rates of acetic acid by factor of 100 for AAA laminate (1900 g/m^2day) and by factor of 1000 for APA laminate (200 g/m^2day) in comparison to the values at room temperature.

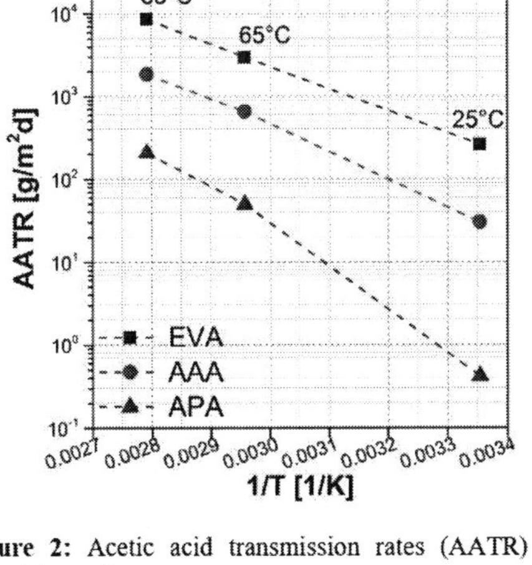

Figure 2: Acetic acid transmission rates (AATR) of materials used.

However, despite lower AATR value at experimental temperature, GC/MS measurements revealed a significantly higher concentration of formed and permeated acetic acid in the test laminates containing the APA backsheet in comparison to test laminate with AAA backsheet (see Figure 3).

This peculiar behavior is a direct consequence of the autocatalytic effect of acetic acid that was retained at interfaces of the test laminate and caused additional production of acetic acid and therefore degradation of EVA. Also a difference between the two types of EVA was observed which is probably result of different composition of EVA i.e. vinyl-acetate content or additives which are vulnerable components in EVA. [6] Furthermore, in Figure 3 it can be seen that the concentration of the produced acetic acid is much higher after UV/T exposure, which confirms tendency of EVA to degrade faster under UV light than temperature. [2] However, degradation of EVA, i.e. increase in baseline in test laminates without backsheet (laminate 1 and laminate 4) as shown in Figure 4. is not significant when compared to degradation of EVA in test laminates with backsheet (Figure 5.-6.) As a result of the higher concentration of the formed acetic acid, the test laminate with the APA backsheet (laminate 3 and 6) showed higher increase of the baseline in Raman spectra due to fluorescence background caused by degradation by-products (see Figure 5.-6., blue curve) indicating degradation of EVA. Moreover, a higher increase in baseline was observed after UV/T exposure confirming higher degradation of EVA under UV/T than T exposure. The highest increase in baseline was observed for the test laminate which showed the highest concentration of formed acetic acid via GC/MS (follow the star).

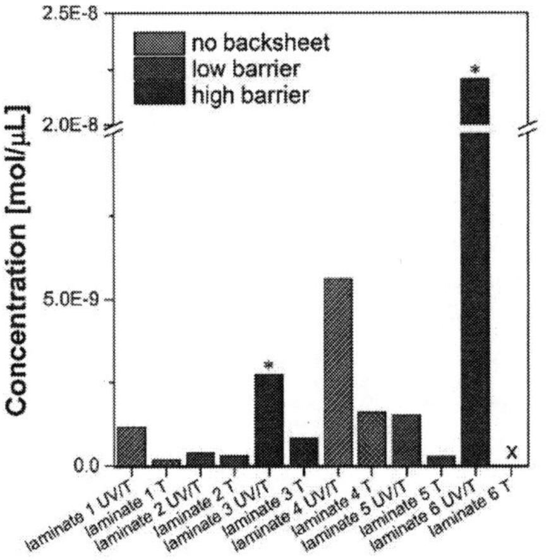

Figure 3: Concentration of formed and permeated acetic acid (GC/MS); star is pointing test laminate with the highest concentration of acetic acid (*Note: no results for laminate 6 T labeled as "x"*).

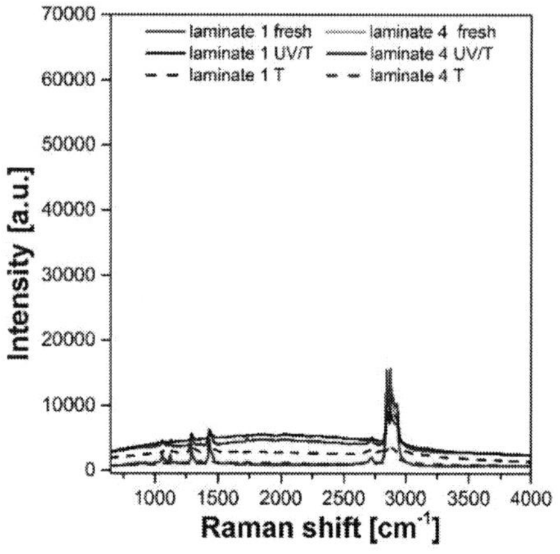

Figure 4: Raman spectra of EVA in test laminate without backsheet

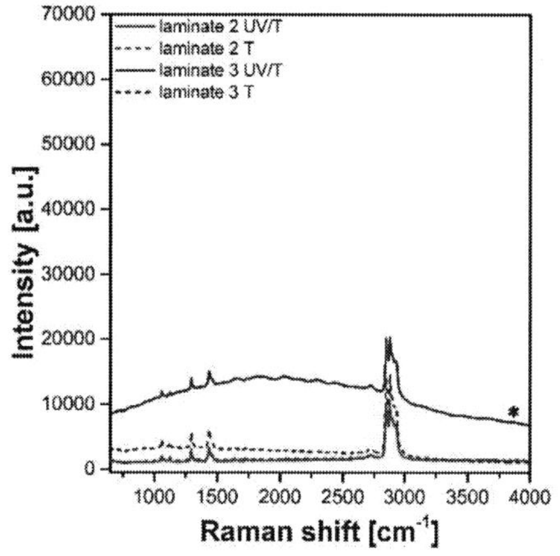

Figure 5: Raman spectra of EVA1 in test laminates with backsheet; star is pointing the highest increase in baseline (compare with Figure 3).

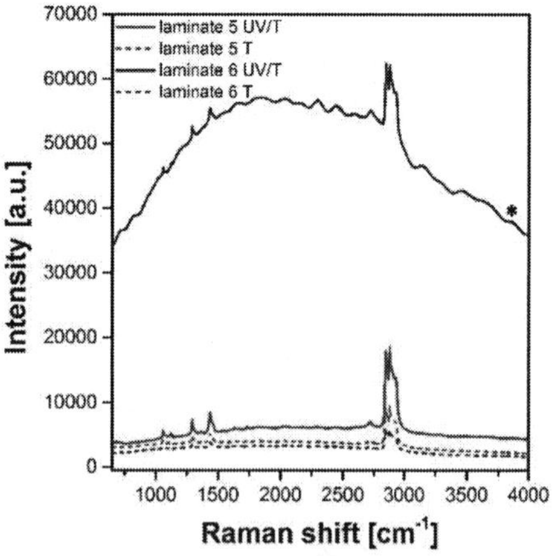

Figure 6: Raman spectra of EVA2 in test laminates with backsheet; star is pointing the highest increase in baseline (compare with Figure 3).

4. CONCLUSION

The main aim of this work was to correlate permeation properties of backsheets with formation of acetic acid in the EVA encapsulant layer of test laminates. An excellent correlation between GC/MS and Raman confocal microscopy was obtained. It was shown how type of exposure, composition of test laminate and permeation properties of backsheet layers have great influence on formation of acetic acid. EVA is more susceptible to degradation when exposed to UV light in comparison to temperature exposure, which results in higher formation of acetic acid. Due to the autocatalytic effect of the retained acetic acid on the interfaces of the test laminate, more acetic acid is formed in test laminates with high barrier backsheet layer. At the end this caused higher degradation of EVA indicated by increase in baseline in Raman spectra. It can be concluded that right combination of "breathable" backsheet with EVA encapsulant influences reliability of PV modules.

5. REFERENCES

[1] **Liu, F.; Jian, L. Y. S.:** UV degradation behaviour of polymeric backsheets for photovoltaic modules, Solar energy, *pp. 88-100*, 2014.

[2] **Czanderna, A.; Pern, F.:** Encapsulation of PV modules using ethylene vinyl acetate copolymer as a pottant: A critical review, *pp. 101-181*, 1995.

[3] **Peike, C.; Kaltenbach, T.; Weiss, K.-A.; Koehl, M.:** Non-destructive degradation analysi of encapsulants in PV modules by Raman spectroscopy, *pp. 1686-1693*, 2011.

[4] **Peike, C.; Kaltenbach, T.; Weiss, K.-A.; Koehl, M.;:** Lateral distribution of the degradation of encapsulants after different damp-heat exposure times investigated by Raman spectroscopy, 2010.

[5] **Oreski, G.; Brasch, M.; Mihaljevic, A.; Hirschl, C.; Pinter, G.:** Acetic acid transmission rates of PV backsheets, 31st European Photovoltaic Solar Energy Conference and Exhibition, Hamburg, 2015.

[6] **Peike, C., Purschke, L.;Weiss, K.-A.; Koehl, M.; Kempe, M.:** Towards the origin of photochemical EVA discoloration, 39th IEEE Photovoltaic Specialists Conference, PVSC, Tampa Bay, USA, 2013.

A NEW APPROACH TO DETERMINE THE CROSSLINKING IN POLYETHYLENE-VINYL ACETATE VIA RAMAN SPECTROSCOPY

Sebastian Jäger*[1], Stephan Wittmann[1], Marcus Heindl[2], Aswin Linsenmeyer[3], Thomas Kunz[1], Christian Camus[1], Jens Hauch[1] and Christoph J. Brabec[1,4]
[1] Bavarian Center for Applied Energy Research, PV-Systems, Haberstraße 2a, 91058 Erlangen, Germany
[2] SKZ – Testing GmbH, Friedrich-Bergius-Ring 22, 97076 Würzburg, Germany
[3] Sunset Energietechnik GmbH, Industriestraße 8-22, 91325 Adelsdorf, Germany
[4] Materials for Electronics and Energy Technology (i-MEET), Friedrich-Alexander-Universität Erlangen-Nürnberg, Martensstraße 7, 91058 Erlangen, Germany
*Corresponding author, e-mail: sebastian.jaeger@zae-bayern.de

ABSTRACT: In the lifetime of photovoltaic (PV) modules, the quality and the stability of the encapsulation are important characteristics as they affect the electrical isolation as well as the weathering protection of the photovoltaic cells. One of the most commonly used encapsulation materials is ethylene vinyl acetate (EVA), which is crosslinked during the lamination process. The degree of crosslinking is decisive regarding the long-term stability of photovoltaic modules. Therefore, manufacturers look for technologies for inline evaluation of the EVA crosslinking degree. In this work, the authors present a new method to evaluate the crosslinking degree of EVA based on Raman spectral analysis. We found a direct correlation between the degree of crosslinking and the Raman peak at 1649 cm^{-1}, which presumably originates from residual C=C double bonds in EVA.
Keywords: Encapsulation, Ethylene-Vinyl Acetate, Degree of Crosslinking, Raman Spectroscopy, Process Analysis

1 INTRODUCTION

The production of photovoltaic (PV) modules has to satisfy a multitude of demanding needs. The production process has to be cost efficient while the final PV-modules have to fulfill certain quality standards, as they are intended to be in service more than 25 years. Concerning the module's lifetime limit, the lamination foil is crucial, as it provides electrical isolation and protection against environmental influences. Still, the most commonly used material for the encapsulation is polyethylene-vinyl acetate (EVA) that is crosslinked by peroxide agents during the lamination [1]. By crosslinking the EVA is converted from a thermoplastic into an elastomeric rubber. For the long term stability of this elastomeric form, and therefore the properties of the embedding material of the PV-module, the degree of crosslinking is most important.

The standard procedure to evaluate the degree of crosslinking is the gel content measurement according to the standards ISO 10147 or ASTM D2647 [2], [3]. Here, a piece of EVA is boiled under reflux in an organic solvent to remove uncrosslinked polymer chains from the elastomeric polymer web. Finally, the degree of crosslinking is measured by determining the percentage of mass of the polymer piece that is lost during this process. This standard procedure has some drawbacks in the context of process analysis: it is slow, the duration of one analysis is at least 12 hours and it is a destructive method. To compensate these drawbacks for a real process analysis of PV modules in the production line, different methods, which are calibrated according to the standardized chemical extraction method, have been developed. One of these methods is the evaluation of the shear modulus by performing a dynamic mechanical analysis (DMA), at a measurement temperature of 60°C, through the back sheet of the module.[4]–[6] This limits the utilization of this method to modules with back sheet. In Addition, the modules have to be heated up locally for the measurement limiting the capability of reducing measurement times. Besides, this method doesn't deliver any chemical information of the system.

To overcome the limitation of this system to modules with back sheet, an optical method is preferred. Here, one method published uses the increase of intensity in luminescence in PV-grade EVA when excited by UV light [7]. Another approach is the utilization of Raman spectroscopy to gain information about the desired crosslinking degree. In this context, the utilization of the Raman peak, ratio between CH_2 and CH_3 peaks of EVA has been published.[8]

We found an alternative Raman peak ratio, utilizing the vanishing of a Raman peak that is likely to be assigned to the residual C=C stretching vibration band during the lamination. This peak is a very promising candidate for a process analysis system, as it is assumed to show a reduced dependence on the exact ratio between ethylene and vinyl acetate in the used EVA material.

2 MATERIAL AND METHODS

2.1 Polyethylene-vinyl acetate

Crosslinked EVA is the most commonly used embedding materials for photovoltaic modules and is available in a broad variety e.g. with different curing behaviors.[1] To enable crosslinking in EVA, organic peroxides, peroxy acids or peroxy esters are mixed into the formulation of the EVA raw material.[9] This peroxidic substances decompose at temperatures around 150°C to radicals, starting the crosslinking reaction.

In this paper the EVA encapsulant "Vistasolar 550.68 – Ultra Fast Cure – UV Blocking" (Eastman) was used. To obtain specimens with different cross linking degree, the EVA foils were laminated in 11 x 9 cm² test modules containing of front glass, EVA encapsulation, a Si solar cell (maximum 7 x 4 cm², electrically not connected) and a back sheet. For a correlating gel content analysis, a 4 x 5 cm² piece of EVA is separated after lamination.

The lamination process consists of three phases:

1. Thermal equilibration: Temperature of hotplate: 150°C, modules resting on spacers above hotplate, duration: 300 s.

2. Heating: Module on hotplate at 150°C, establishing of vacuum, duration: 120 s
3. Compression: Module on hotplate at 150°C with pressure from above, duration 100 s to 550 s in steps of 50 s

In addition to the test lamination procedure, one of the used test modules was laminated according to the lamination procedure used for the production of PV modules. Measurements of this sample are highlighted or marked as "standard recipe".

2.2 Raman Spectroscopy

The Raman measurements were performed with a WITec alpha 500 RA confocal Raman microscope with a 10x microscope lens (NA: 0.25) and a fiber coupled spectrometer. For excitation, a frequency doubled Nd:YAG Laser operated at 532 nm and an excitation power of 30 mW was used. To obtain sufficient signal through the front glass of the test modules, the integration time under these illumination conditions was 40s. After acquisition, the processing of the spectra is done by a custom made, automatic fitting program. Here, the single peaks are fitted by a Loretzian function with a constant or linear background.

2.3 Gel-content Measurement

According to the standard ISO 10147 the gel contents are measured by boiling the EVA samples in Xylene at 140°C for 8 hours. After this extraction, the sample is dried for at least 3 hours at 140°C in a drying cabinet. To derive the gel content G, the masses of the sample before (m_1) and after extraction (m_2) are used:

$$G = \frac{m_2}{m_1} \times 100\%$$

For every test module, the degree of crosslinking is determined several times at different locations. The results of these individual measurements of G are used to calculate the average value G_a as well as the standard deviation of the measurements.[2]

2.4 Differential Scanning Calorimetry (DSC)

Differential Scanning Calorimetry (DSC) measures the heat flow of a sample versus the sample temperature. Therefore the heat flow from sample in a measurement pan towards an empty measurement pan is recorded. This method can be utilized to gain access to reaction enthalpies of e.g. thermally triggered crosslinking reactions. The values of these reaction enthalpies in EVA are correlated to the degree of crosslinking.[10]

3 RESULTS AND DISCUSSION

3.1 Raman spectrum, of EVA

The Raman spectrum of EVA is well known and most of the peaks are described in detail in literature for either EVA, polyethylene or polyvinyl acetate.[11]–[14] When measuring through the front glass of a photovoltaic module, the Raman spectrum changes in the spectral regime up to approximately 1000 cm^{-1}. Here the EVA Raman peaks are reduced in intensity while the Raman peak of Si at 521 cm^{-1} can be found. As shown in Figure 1, a slightly higher background of the entire spectrum was found. In the spectral range around 2850 cm^{-1}, the CH stretching vibrations are representing the most dominant peaks of EVA. This peak area consists of at least 5 different CH vibrations, while the exact intensity ratios in this area may be related to composition and crosslinking.[8], [14]

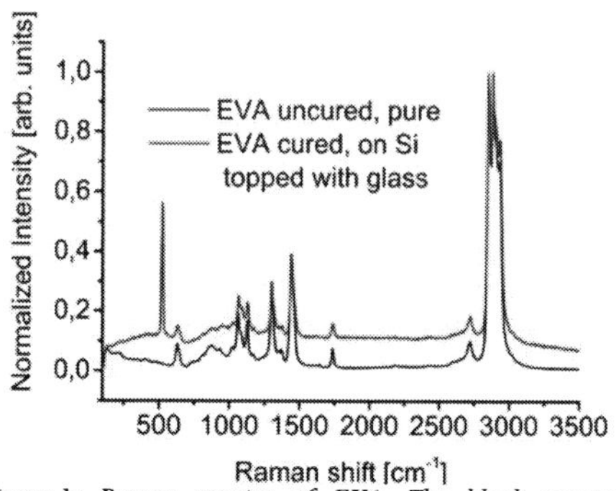

Figure 1: Raman spectra of EVA. The black spectrum is measured on a pure, uncured EVA foil while the red spectrum is measured on a PV module through the front glass. Note that the prominent peak at around 520 cm^{-1} in the red spectrum is the Raman peak originating from the Silicon PV-cell.

Taking a closer look into the Raman shift area around 1700 cm^{-1} (see Figure 2), the dominant peak originating from the C=O stretching vibration can be found at 1738 cm^{-1}.[14] In addition, another peak is visible at 1649 cm^{-1} which can be found in uncured EVA as well as in different kinds of EVA polymers (e.g. ELVAX 40W and ELVAX 250A, DuPont). The most likely origin of this peak are residual C=C double bonds left inside of the polymer, as this peak can also be used for monitoring the polymerization of vinyl acetate[11] and is found for C=C double bonds in poly butadiene [15]. Additionally, a strong indication for this origin of this peak is its disappearance during the crosslinking process. The weaker bonding energies of double bonds are a good starting point for radical reactions like polymerization or crosslinking processes.[16]

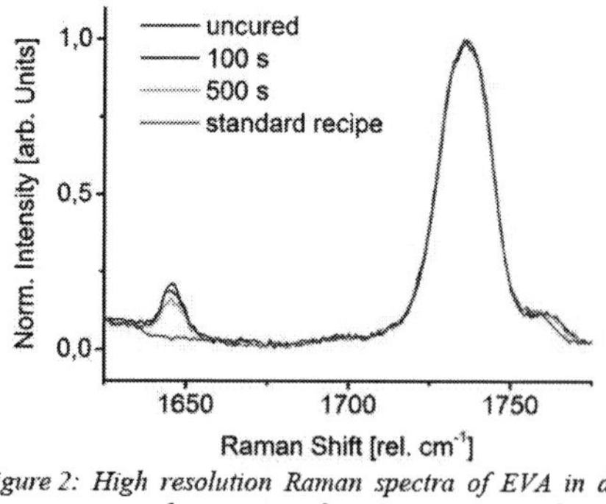

Figure 2: High resolution Raman spectra of EVA in different curing states after various lamination times. The peak at 1649 cm^{-1} is declining with curing of the EVA. Note that the "standard recipe" means the curing conditions according to the lamination procedure used for lamination of PV modules.

In Figure 2 spectra at various curing states of EVA are shown. The spectra contain some noise, which influences the accuracy of the crosslinking measurement via Raman spectroscopy and has a high influence onto the error bars shown in chapter 3.2.

3.2 Crosslinking Evaluation via Raman Spectroscopy

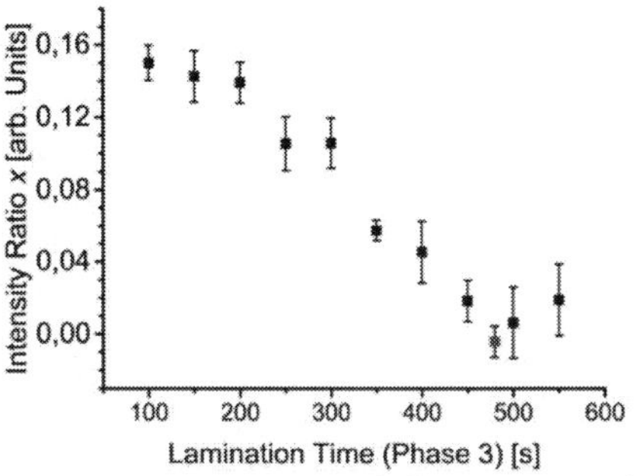

Figure 3: Intensity ratio x versus the lamination time (phase 3) of the test modules. The pink data point represents the measurement acquired from the sample that was laminated according to the "standard recipe".

The absolute intensity of Raman signals depends on several conditions like the intensity of the laser, the polarization of the incoming laser beam with respect to the molecular axis or, in the case of confocal Raman, the exact position of the excitation spot on the sample. Therefore, it is very useful to correlate the intensity of the interesting Raman peak, here the one at 1649 cm^{-1}, to another Raman peak as internal standard. In this case, the internal standard is the Raman band of the C=O vibration at 1738 cm^{-1}. In Figure 3, the intensity ratio $x = \frac{I_{1649\,cm^{-1}}}{I_{1738\,cm^{-1}}}$ is plotted versus the corresponding lamination time (phase 3) of differently cured EVA specimens. In general, the peak ratio x shows a clear trend towards zero for higher lamination times and so for higher crosslinking degrees.

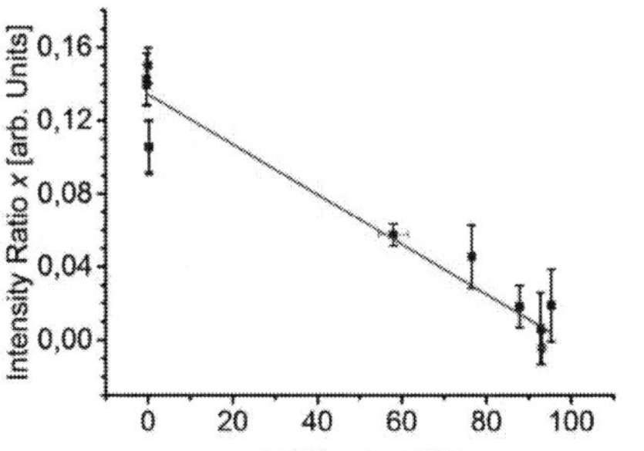

Figure 4: Intensity ratio x versus the gel content of the test modules. The red line represents a linear fit to the data points with a slope of -0.00137. The pink data point represents the measurement acquired from the sample that was laminated according to the "standard receipt".

In order to correlate the intensity ratio x to real crosslinking data, gel-content measurements as well as measurements of the residual crosslinking reaction enthalpy via differential scanning calorimetry (DSC) are performed on EVA samples extracted of the test modules.

In

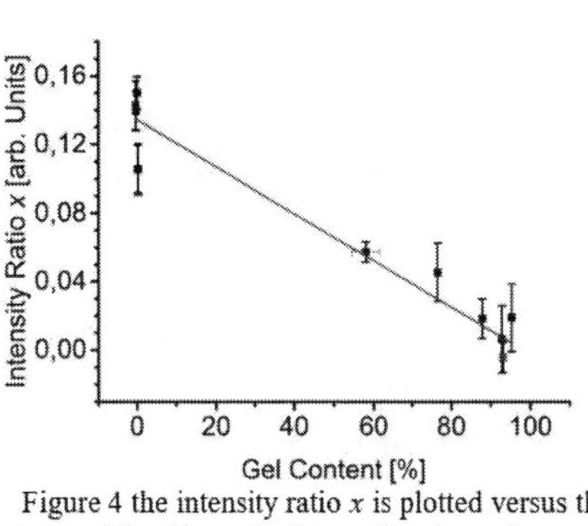

Figure 4 the intensity ratio x is plotted versus the gel-content while Figure 5 shows the dependency of the intensity ratio x versus the DSC residual enthalpy values. The results of the gel-content measurement reveal that in all test modules with a lamination time below 300 seconds the degree of crosslinking is at nearly 0%. On the contrary, while looking closer at the Raman and DSC results, a difference in reaction enthalpy and crosslinking is observable leading to the conclusion that the crosslinking for the gel content measurement have to exceed a certain threshold. A linear correlation between the gel content and the intensity ratio x with a slope of -0.00137 and was found.

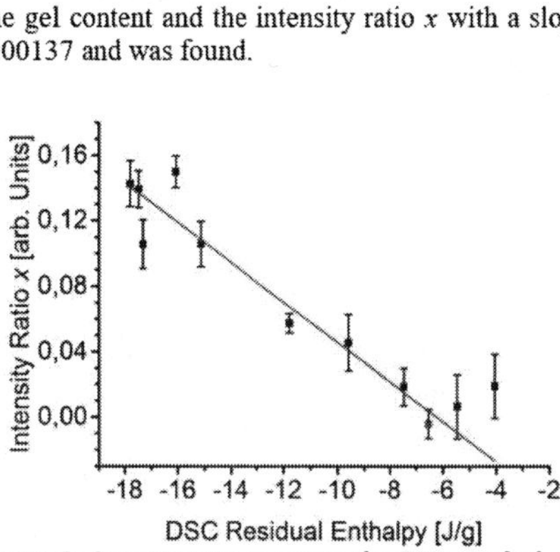

Figure 5: Intensity ratio x versus the DSC residual enthalpy values of the test modules. The red line represents a linear fit to the data points. The pink data point represents the measurement acquired from the sample that was laminated according to the "standard receipt".

4 CONCLUSION AND OUTLOOK

In this paper we have shown that the intensity of the Raman peak at 1649 cm^{-1} recorded of poly ethylene vinyl acetate has a direct correlation towards the degree of crosslinking. These results indicate that Raman spectroscopy could become a valuable tool for EVA lamination process control. Especially the process analysis of the encapsulation material is a challenging task for other methods and in the case of glass-glass module architecture Raman spectroscopy seems to be the only feasible method. Here we have presented one possible solution to this analytical problem, by utilizing the Raman band at 1649 cm^{-1}, that presumably originates from the stretching vibration of residual C=C double bonds in uncured EVA. This Raman peak has a strong correlation to the degree of crosslinking in EVA and can therefore be utilized as a method for non-destructive evaluation of the degree of crosslinking in PV modules.

In order to ensure the usefulness of this method further EVA samples of different origin have to be evaluated. Our next step is the transfer of this technique from small test modules, towards a setup that can measure full size PV modules. Furthermore, possibilities to decrease the measurement time per spectrum will be tested and evaluated.

5 ACKNOLEDGEMENT

The authors acknowledge the funding of this work by the "Federal Ministry of Economic Affairs and Energy" in the framework of the ZIM-Network (Zentrales Innovationsprogramm Mittelstand) PV-ZUM and the project InoVEVA, contract no. 16KN 040 131.

6 REFERENCES:

[1] C. Peike, C. Peike, I. Hädrich, K. Weiß, I. Dürr, and F. Ise, *Photovoltaics International*, vol. 22 (2013) 85.

[2] "DIN EN ISO 10147:2011, Rohre und Fittings aus vernetztem Polyethylen (PE-X) - Berechnung des Grades der Vernetzung durch Bestimmung des Gel-Gehaltes (ISO 10147:2011)."

[3] "ASTM D2647-10, Standard Specification for Crosslinkable Ethylene Plastics, ASTM International, West Conshohocken, PA, 2010, www.astm.org." .

[4] J. Lloyd, T. Christian, D. Doble, and R. A. Mickiewicz, "Non-destructive measurement of the degree of cross-linking of EVA solar module encapsulation," in *2011 37th IEEE Photovoltaic Specialists Conference*, (2011), 2273.

[5] R. A. Mickiewicz, E. Cahill, and P.-I. Wu, "Non-destructive determination of the degree of cross-linking of EVA solar module encapsulation Using DMA shear measurements," in *2012 38th IEEE Photovoltaic Specialists Conference*, (2012), 710.

[6] E. Malguth, B. Buhl, and T. Schenk, *EU PVSEC Proceedings*, (2012) 472.

[7] A. Morlier, M. Köntges, S. Blankemeyer, and I. Kunze, *Energy Procedia*, vol. 55 (2014) 348.

[8] C. Hirschl, L. Neumaier, G. Eder, and M. Kraft, "Raman Spectroscopic Determination of the Degree of Encapsulant Crosslinking in PV Modules," in *EU PVSEC Proceedings*, (2013), 3025.

[9] C. Peike, W. Phondongnok, T. Kaltenbach, and M. Koehl, *Open Journal of Renewable Energy and Sustainable Development*, vol. 1 (2014) 14.

[10] C. Hirschl, M. Biebl–Rydlo, M. DeBiasio, W. Mühleisen, L. Neumaier, W. Scherf, G. Oreski, G. Eder, B. Chernev, W. Schwab, and M. Kraft, *Solar Energy Materials and Solar Cells*, vol. 116 (2013) 203.

[11] T. Özpozan, B. Schrader, and S. Keller, *Spectrochimica Acta Part A: Molecular and Biomolecular Spectroscopy*, vol. 53 (1997) 1.

[12] C. Peike, T. Kaltenbach, K.-A. Weiß, and M. Koehl, *Solar Energy Materials and Solar Cells*, vol. 95 (2011) 1686.

[13] Y. Takahashi, *Macromolecules*, vol. 34 (2001) 7836.

[14] K. B. R. Devi and R. Madivanane, *Engineering Science and Technology: An International Journal*, vol. 2 (2012) 795.

[15] S. W. Cornell and J. L. Koenig, *Macromolecules*, vol. 2 (1969) 540.

[16] W. Walter and W. Francke, *Beyer-Walter: Lehrbuch der Organischen Chemie*, 23rd ed. Stuttgart - Leipzig: S. Hirzel Verlag, (1998).

DSC METHODS FOR A FAST DETERMINATION OF THE CROSSLINKING DEGREE OF EVA.

Stéphane Ogier[a bcd], David Chapron[a b], Isabelle Royaud[c], Marc Ponçot[c], Marion Vite[de], Manuel Hidalgo[e]
[a] LMOPS, University of Lorraine, EA n°4423, 2 rue Edouard Belin, Metz, 57070, France
[b] LMOPS, Centrale-Supélec, 2 rue Edouard Belin, Metz, 57070, France
[c]IJL, University of Lorraine, UMR 7198 CNRS, Parc de Saurupt CS 50840, Nancy, 54011, France
[d]University Grenoble Alpes, INES, F-73375 Le Bourget du Lac, France
[d]CEA, LITEN, Department of Solar Technologies, F-73375 Le Bourget du Lac, France
[e]SOLLIA, Joint laboratory ARKEMA-CEA, LITEN, Department of Solar Tech., F-73375 Le Bourget du Lac, France
Author for correspondence: Stéphane Ogier Email: stephane.ogier@cea.fr, Phone: +33 (0)4.79.79.20.69

ABSTRACT: The determination of the crosslinking degree is a key factor to ensure EVA reliability when used as a PV encapsulant. Several methods exist. Some of them, such as rheological measurements or Soxhlet extraction (gel content) are rather difficult to perform, or time-consuming. To simplify the determination of the crosslinking (XL) degree, calorimetric measurements have been developed. A first method called the Melt/Freeze method (MF) is based upon parameters reflecting the crosslinking influence on the crystallization behavior. Thermal fractionation of EVA crosslinked at several degrees, performed with the Differential Scanning Calorimetry – Successive Self-nucleation and Annealing (DSC-SSA) technique, shows a narrowing of the crystallite size distribution towards smaller sizes. This confirms the link between the crosslinking reaction, and its effect on the crystallization behaviour. An alternative method, the Residual Enthalpy (RE) method, follows the crosslinking agents' depletion as a function of crosslinking. A comparison is made between both DSC methods and rheology, taken as a reference. The MF method fits better with rheological measurements than the RE one. However the latter method is very useful to identify the crosslinking agent-type of mix.
Keywords: EVA, crosslinking degree, DSC, crystallite size

1. INTRODUCTION

Photovoltaic (PV) modules, comprising crystalline silicon cells have become a widespread, and familiar object in most parts of the world. Today's PV modules feature typically a multilayer foil structure of the type {glass / EVA encapsulant / PV cells / EVA encapsulant / Polymeric Backsheet (BS)} (Fig. 1).

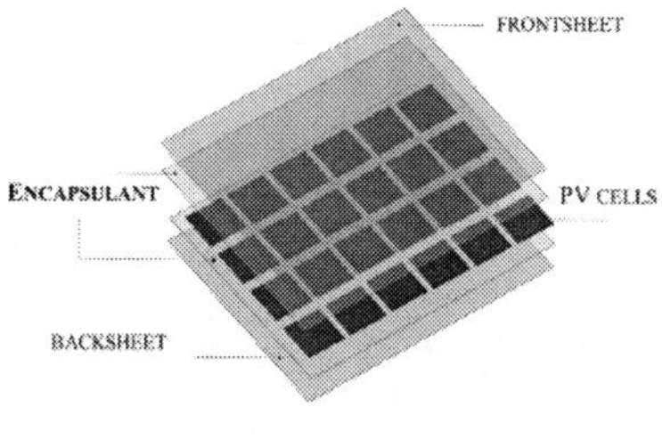

Figure 1: Structure of a standard PV module.

During the so-called lamination process (Fig. 2), the EVA foils are melted to encapsulate the cells, and then crosslinked to form a polymer network. This process is generally composed of three steps: (1) first, vacuum is applied to remove gas, while beginning to heat up the stack (2) then a c. 1 atmosphere pressure is applied to the stack while continuing to heat; curing occurs during this step since the crosslinking agent (peroxide) decomposes. Finally the PV module is cooled down (3) either using a cooling press or by waiting long enough at room temperature.

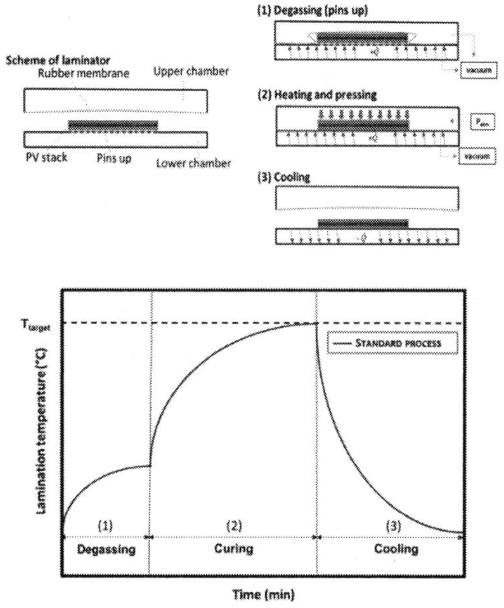

Figure 2: The different steps of the lamination process in a standard laminator (on the top), and the thermal evolution of the PV stack (on the bottom).

Crosslinking is necessary in order to ensure that the EVA encapsulant will not flow with time or when the temperature of the module rises, and it also improves the optical, chemical-resistance, and mechanical properties of the encapsulant (Fig. 3).

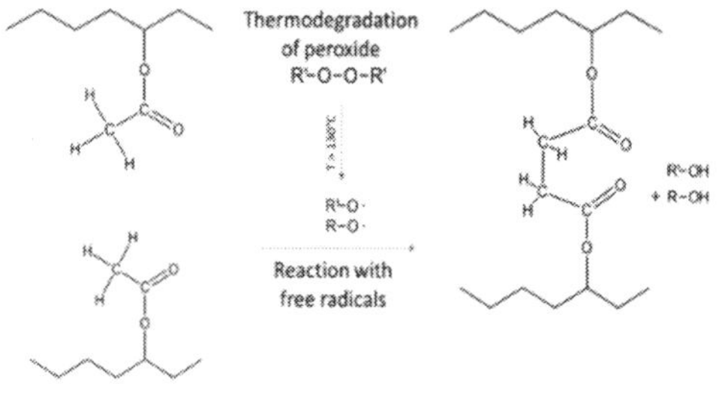

Figure 3: The crosslinking chemical reaction of EVA as suggested by Hirshl *et al.* 2013 [1].

Several analytical methods to determine the degree of crosslinking of ethylene vinyl acetate copolymers (EVA) have been developed over the past decades [1-3]. Some of them, such as solvent extraction or rheological methods, are known to be consistent but tedious, complex, or requiring toxic chemicals, which makes them difficult to apply in the PV industry. The purpose of this work is to demonstrate the accuracy of a new thermal method to determine crosslinking degrees by analyzing the EVA crystallization using Differential Scanning Calorimetry (DSC) thermograms. This technique, named the "Melt-Freeze" Method (MF) [2-3], is based upon the follow-up of parameters representing the crystallite size distribution.

2. MATERIAL AND METHODS

Two EVA commercial PV films, supplied by STR (Spain) were studied: 28%wt. (Photocap® 15580 P), and 32%wt. (Photocap® 15420 P/UF) of vinyl acetate-derived units.

EVA samples were cured in a 3S laminator between two polytetrafluoroethylene (PTFE) sheets and upon a glass sheet. The EVA films underwent different thermal treatments leading to a wide range of crosslinking (XL) degrees. The curing step was either carried out at 140°C, 150°C or 160°C for 3, 4, 5, 6, 8, or 10 min. The degassing step was always finished after 2 min while the cooling step was done at room temperature.

The two DSC methods evaluating the XL degree (described below) were conducted in a TA Q20 instrument and operated at 10°C.min^{-1}.The so-called MF method is run from 100°C to 0°C, while the so-called RE method is performed from 40°C to 210°C.

The rheological measurements were carried out using an Anton Paar MCR 302 rheometer. The EVA samples were heated up to 100°C to be at their rubber-like state. Then, a dynamic shear stress was applied from 100 Hz to 0,1 Hz, in order to determine the loss factor, tan δ at 1 Hz.

3. DIFFERENT INSIGHTS GIVEN BY THE DSC METHODS

EVA copolymers are semi crystalline materials, for which the crystallinity is influenced by their crosslinking degree [4-5]. Differential Scanning Calorimetry is an easy-to-use technique which can be used to detect all thermal events related to the physical, and chemical transformations of a material (melting, crystallization, glass transition, chemical reactions). A typical thermogram of EVA is shown in Fig. 4.

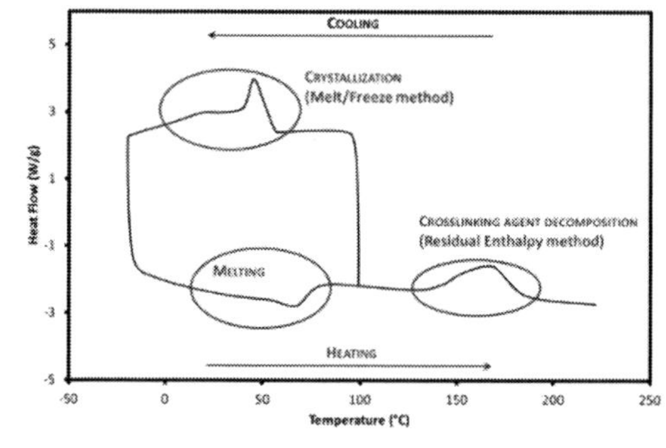

Figure 4: Global EVA thermogram with a 10°C.min^{-1} heating/cooling rate. Three calorimetric events happened: one endothermic peak (melting), and two exothermic peaks (crystallization and crosslinking agents' decomposition).

3.1 The Melt Freeze method (MF)

The MF method takes advantage of the crystallization phase-transition, which reflects the influence of the EVA crosslinking. This method is based on three parameters following the evolution of the crystallization peak: the T_{onset} temperature where crystallization begins, the T_c crystallization peak temperature, and SF, corresponding to a peak shape factor. The latter exists only for copolymers. SF is influenced by the formation of small crystallites. This is due to vinyl acetate-derived units, which are considered as defects and thus, rejected into the amorphous phase.

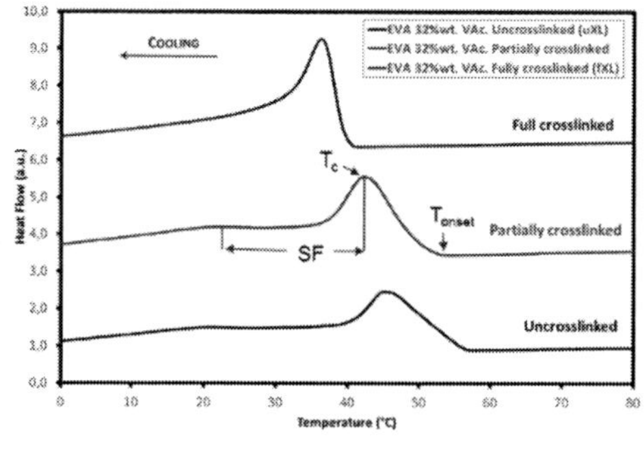

$$\%XL^{Melt/Freeze} = \frac{(\%XL_{T_c} + \%XL_{T_{onset}} + \%XL_{SF})}{3}$$

$$\%XL_Y = \frac{Y_{uXL} - Y_L}{Y_{uXL} - Y_{fXL}} \times 100 \text{ with } Y = T_c, T_{onset} \text{ and Shape Factor(SF)}$$

Figure 5: Crystallization thermograms of EVA at different crosslinking degrees (on the top). The Melt Freeze equation (on the bottom) developed as described in reference [2] is based upon the changes in the crystallization peak. uXL = uncrosslinked, fXL = fully crosslinked, L = laminated test sample (partially crosslinked).

The crystallization peak reflects the crystallite size distribution which is influenced by the polymer crosslinking. In order to highlight this change, a thermal fractionation technique carried out in the DSC equipment, and known as Successive Self-nucleation and Annealing (SSA) method [6], is applied to EVA samples with different XL degrees (Fig. 6). A tightening of the crystallite size distribution towards smaller sizes is observed when the samples are more and more crosslinked.

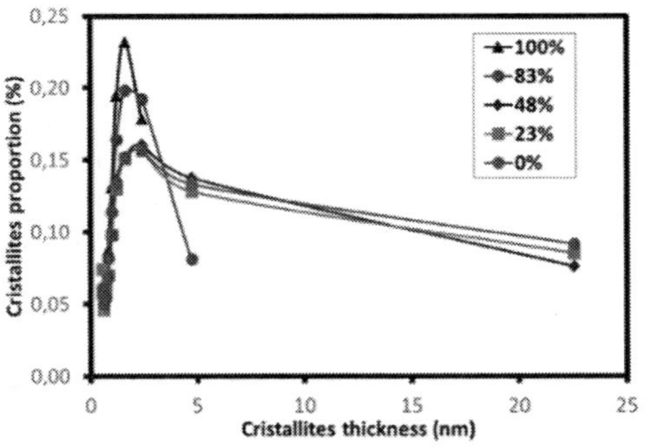

Figure 6: Evolution of the crystallite size distribution of EVA, crosslinked at several degrees as obtained by the DSC SSA method; the XL degrees were determined by the rheological method described before.

3.2 The Residual Enthalpy method (RE)

Alternatively, the so-called Residual Enthalpy DSC method (RE) uses the enthalpy of crosslinking of EVA as the to-be-followed parameter to monitor the conversion rate of this chemical reaction (Fig. 7) [7].

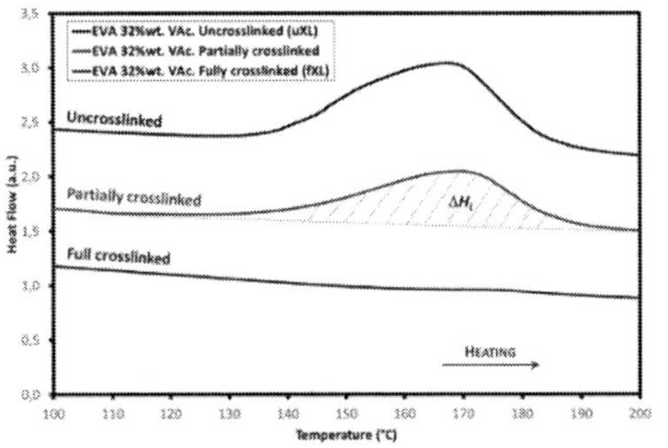

$$\%XL^{RE} = \frac{\Delta H_{uXL} - \Delta H_L}{\Delta H_{uXL} - \Delta H_{fXL}} = \frac{\Delta H_{uXL} - \Delta H_L}{\Delta H_{uXL}}$$

Figure 7: Depletion of the residual enthalpy during the crosslinking reaction (on the top) according to the Residual Enthalpy equation [7], and the equation used to evaluate the crosslinking degree (on the bottom). uXL = uncrosslinked, fXL = full crosslinked, L = laminated test sample (partially crosslinked).

The RE method considers a linear relationship between the XL degree, and the remaining enthalpy. However, in some cases, EVA films feature a complex mix of crosslinking agents (Fig. 8). This implies that the depletion of this complex mixture might follow a non-linear behavior, inducing some error in the estimation of the XL degree. Still, the RE method is very useful to identify the crosslinking agent-type of mix.

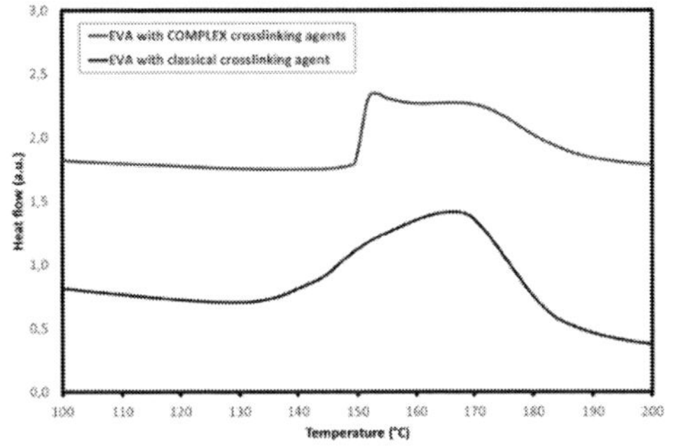

Figure 8: DSC thermograms of two different EVA encapsulants. Composition of crosslinking agents differs and may follow a non-first order thermal decomposition of them.

4. COMPARISON BETWEEN THE MF, AND RE METHODS

In order to compare the two DSC methods, we used a third method as a reference. The latter can be the very well-known gel-content method, which has been used for years by the PV industry to monitor the degree of crosslinking of EVA encapsulants in modules. The reference method can also be the more rigorous, although more complex rheological measurement of the mechanical moduli of the crosslinked materials (Fig. 9). This method gives access to the actual crosslinking density of the material by linking the shear modulus to the crosslinking density [8].

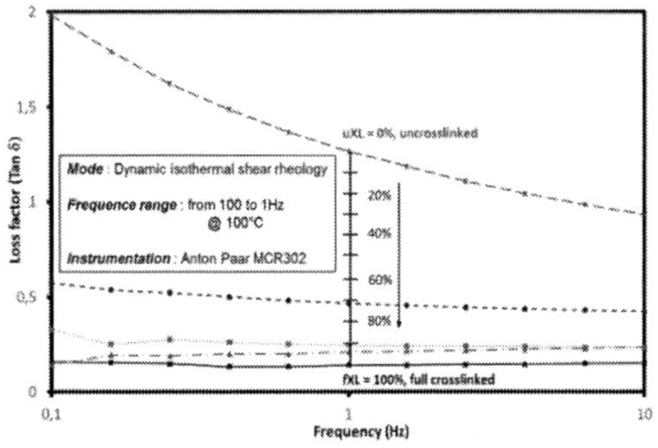

Figure 9: Data output from the rheological method used as reference method to compare with the DSC methods.

We compare both DSC methods to the rheological reference method (Fig.10). A divergence between crosslinking degrees obtained by rheology (used as the reference method) and by the RE DSC method, is observed when EVA samples contain a large excess of crosslinking agent. The same behaviour can be observed for EVA samples with a complex crosslinking agents' mix. This is not the case when using the MF DSC method, based upon the crystallization behaviour of crosslinked EVA's, which fits very well with the results of the reference methods (rheology, and gel-content measurements). We conclude that this is due to the fact that the MF method probes the crystallization behaviour of samples, which is better linked to the physical structure obtained after crosslinking, while the RE method is based on a more indirect measurement, that of the enthalpy of the crosslinking reaction.

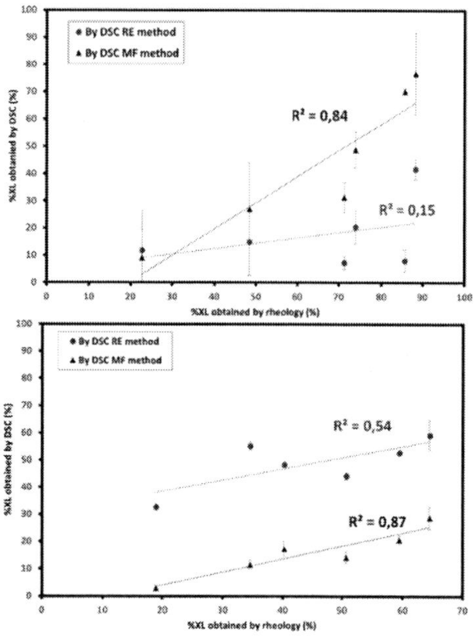

Figure 10: Comparison between several techniques to evaluate the crosslinking degree for two commercial EVA's for PV modules (up: 32%wt. VAc. Photocap® 15420 P/UF, down: 28%wt. VAc. Photocap® 15580 P). It can be observed for both samples that the MF method is in better agreement to the rheological reference method than the RE one.

5. CONCLUSION

Several techniques allow to evaluate the crosslinking degree of EVA. A few of them are directly linked to crosslinking densities but they are rather complex and require experimental expertise. On the other hand, DSC methods are fast and easy to use. The MF method fits very well with the reference measurements from rheology. It is linked to the crystallites size-distribution, which is influenced by crosslinking. The RE method keeps a correlation with the rheological reference measurements, although the fit is less good than that for the MF method. Still, this last method is very useful in order to identify the crosslinking agent-type of mix.

This project has received support from the State Program "Investment for the Future" bearing the reference ANR - 10-ITE-0003

6. REFERENCES

[1] Ch. Hirschl, M. Biebl-Rydlo, M. DeBiasio, W. Mühleisen, L. Neumaier, W. Scherf, G. Oreski, G. Eder, B. Chernev, W. Schwab, M. Kraft, Solar Energy Materials and Solar cells 116 (2013) 203

[2] M. Hidalgo, F. Medlege, M. Vite, C. Corfias-Zuccalli, P. Voarino, J. Gonzãlez-Leon, Photovoltaïcs International 14th (2011) 130

[3] M. Hidalgo, D. Thil, C. Dugué, Photovoltaïcs International 21rst (2013) 98

[4] N. Okui, T. Kawai, Die Makromolekulare Chemie 154 (1972) 161

[5] R.G. Alamo, E.K.M. Chan, L. Mandelkern, I.G. Voigt-Martin, Macromolecules 25, n24 (1992) 6381

[6] A.J. Müller, R.M. Michell, R.A. Pérez, A.T. Lorenzo, European Polymer Journal 65 (2015) 132

[7] Z. Xia, D.W. Cunningham, and J.H. Wohlgemuth, PV International *5th* (2009) 150

[8] A. Morlier, M. Köntges, S. Blankemeyer, I. Kunze, Energy Procedia 55 (2014) 348

EFFECT OF DIFFERENT UV CUT OFF WAVELENGTHS IN EVA ENCAPSULANT ON CR-SI PV MODULE'S PERFORMANCE & RELIABILITY

Ashish Singh & Rajesh Singh
RenewSys India Pvt Ltd., Bommasandra – Jigani Link Road Industrial Area, Bangalore – 560 105, India
Phone: +91-80-33494545, Email: ashish.singh@renewsysindia.com

ABSTRACT: High energy UV radiations are harmful for polymeric materials such as Encapsulant & Backsheet used in the manufacture of PV modules. To protect the PV modules against UV degradation, EVA Encapsulant is added with additive, which cuts off the harmful wavelength present in the sunlight. The PV industry is gradually moving towards Lower UV cut-off wavelengths for the front side of EVA Encapsulant in order to harness the UV radiation up to 300nm, as against 360nm conventionally. The present study was performed to validate (i). The effect of lowering the UV cut off wavelength on the optical & mechanical properties of EVA Encapsulant. (ii). The effect of lowering the UV cut off wavelength on the performance & durability of PV modules.

The cured EVA sheet of 360 nm, 300 nm & 234 nm UV cut-off wavelengths were exposed to UV radiation of 340 nm wavelength up to 1500 hrs. The performance of the EVA sheets were analyzed by measuring optical transmittance, yellowness index & tensile strength. It was found that there is no effect of lower UV cut-off wavelength on optical & mechanical properties of EVA Encapsulant.

The solar PV modules were made using 360 nm, 300 nm & 234 nm UV cut-off wavelengths EVA encapsulants on the front side of the module and 360 nm on the back side. Each modules with these EVA Encapsulant were exposed to UV radiation of 280nm -385 nm wavelengths for 2000 hrs. It was found that up to 1500 hrs of exposer, the power loss in all there type modules were comparable. However, the power loss after 2000 hrs was higher in case of lower UV cut-off wavelengths Encapsulants. The UV cut-off wavelength of 360 nm for the front side of EVA is recommended to exhibit better reliability.

Keywords: PV module, Reliability,

1. INTRODUCTION

The UV cut off wavelength of an Encapsulant is the irradiation up to which the wavelength of the spectrum is absorbed / stopped by the Encapsulant. The irradiation having wavelength more than UV cut off wavelength range is transmitted through the Encapsulant and generate electricity when fall on a PV solar device.

The UV spectrum consists of short wave (100 nm – 285 nm, UV-C), medium wave (285 nm-315 nm, UV-B) and long wave (315 nm-400nm, UV-A). The short waves are almost completely absorbed by ozone layer of atmosphere and do not reach to earth. The medium waves are also partially absorbed by ozone layer of atmosphere and only the small portion reaches on the earth. These irradiations vary with respect to geographical locations and seasons. The long waves (black light) which are not absorbed by the ozone layer, reach to earth almost completely and we need to deal with it always, be it health, fading of colours, degradation of synthetic or natural beings.

The solar irradiations falling on solar PV modules reach to solar cells via Glass & EVA. The solar glass has UV cut off wavelength of 290 nm which is blocking almost 100 % of medium wavelength UV up to 290 nm.

Currently in PV Industry, people started using front EVA Encapsulant with lower UV Cut-off wavelength assuming that it generates some extra power. The back EVA has 360 nm UV cut-off wavelengths. The lower UV cut-off wavelength targeted between 300nm- 200nm.

There were several studies already done on EVA degradation caused by UV exposer in the form of loss in transmittance [1] and degradation of solar cell ARC coating [2].

The present study performed by us is to understand the effect of lowering the UV cut-off wavelength on optical & mechanical properties of EVA Encapsulant and the performance and durability of solar PV modules.

2 EXPERIMENTAL PROCEDURE

2.1 Experiment

A- UV exposer of cured EVA sheet having UV cut-off wavelength of 360 nm, 300 nm & 234 nm was performed up to 1500 hrs in QUV at 340 nm. The Yelloness Index, Optical Transmittance & Tensile Strength of exposed EVA samples were measured at an onterval of 500 hrs.

B- Three coupon modules (one cell module) were made using 360 nm, 300 nm & 234 nm UV cut-off wavelength EVA as a front and 360 nm as a back EVA for each sample. UV exposer of all coupon modules were done at: Intensity-1.36 W/m2, Wavelenghth-340 nm, temperature-60°C, duration- 2500 hrs. Visual inspection of all there samples was done after 500, 1000, 1500, 2000 & 2500 hrs

C- Two modules were made using 360 nm, 300 nm & 234 nm UV cut-off wavelength EVA in front and 360 nm in back EVA for each module. UV exposer on all modules was done at: Wavelength 280nm -385 nm, temperature-60°C, duration- 2000 hrs. Visual inspection and power measurement of all samples were done after 500, 1000, 1500 & 2000 hrs.

3 RESULTS & DISCUSSIONS

3.1 UV exposer test of cured EVA sheet

UV exposer test was done on Eva having regular UV cut-off wavelength and lower UV cut off wavelengths to understand the phenomenon of UV dehradation. The reduction in transparancy, incraese in yellowing & reduction in retention of tesile strength are the sign of degradation of polymer under UV exposer. The all these parameters were measured post exposing 1500 hrs in UV test and results are shown in figure(1) & (2).

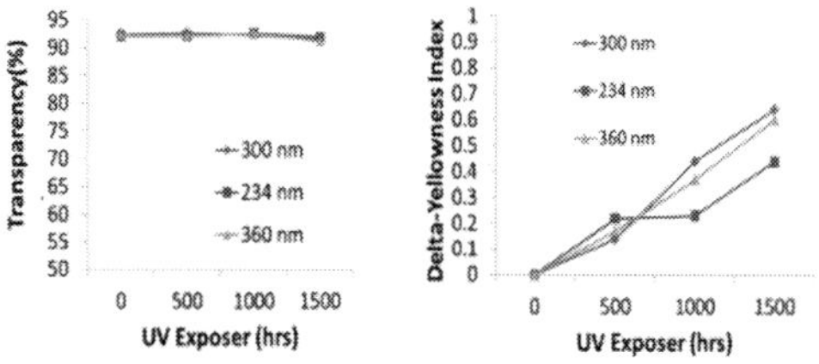

Figure 1: Optical performance of cured EVA sheet

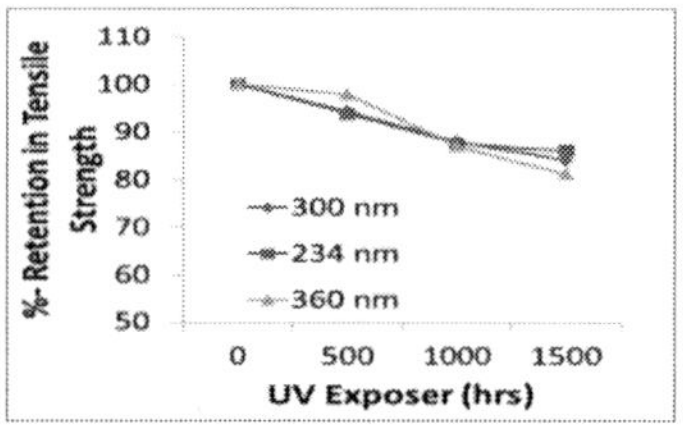

Figure 2: Mechanical performance of cured EVA sheet

The transparency of the EVA films have no significant reduction and it is consistent > 91% after 1500 hrs of exposer. The yellowness index has also identical trend in all EVA and change in yellowness index is <1.

The retention of tensile strength of all EVA is also identical and it has retention of 80 % which seems to be good. From above all data it is clear that the effect of lowering UV cut of has no significant effect on performance parameters of EVA sheet after 1500 hrs UV exposer test.

3.2 UV exposer of coupon modules

To simulate the effect of lower UV cut off in actaul module UV exposer test was conducted on coupan modules. In the actual module, the light isfalling on EVA via glass. The UV cut off wavelength of glass may change the results. The UV cut off wavelength of glass was also measured and it was found to be having 290 nm, this means that glass blocks UV light up to 290 nm.

To see the effect of Glass's UV cut-off on all EVA tested in section 3.1, coupan modules were made with same EVA. The front EVA was 360nm , 300 nm & 234 nm UV cut-off wavelength while back EVA is having 360 nm UV cut off wavelength in all coupons.

The UV exposer was done up to 2500 hrs at same test condition as mentioned in sec 2.1 (B). The visual inspection of cpupan modules were done to see any remarkable colour change due to EVA degradation

Figure (3): UV degraded cpoupon modules

There was no significant colour change found in any coupon modules post 2500 hrs. It seems that all EVA are having similar degradaion rate in UV exposer irrespective of UV cut-off wavelength.

3.3 UV exposer of Solar PV modules

The effect of lowering UV cut-off wavelength was done in section 3.1 & 3.2.

To understand the effect of lower UV cut-off wavelength on module performance & durability, UV exposer test was done on solar PV modules having 360 nm, 300 nm, & 234 nm UV cut-off wavelength in front Encapsulant & 360 nm UV cut-off wavelength in back Encapsulant in all modules. There were two modules made using each type of front EVA. Apart from front EVA, remaining materials of the modules were identical.

The UV exposer test was done at test condition mentioned in section 2.1(C). The module's visual inspection, insulation test and power measurement was done after at 500 hrs interval. The control module was used to offset the module power data to prevent any error in module power due to flasher uncertainty.

There was no visual defect or color change observed in any nodule after 2000 hrs UV. The insulation resistant was found as per specification and having no effect of UV cut-off wavelength.

The power of each module after the interval of 500 hrs is shown in figure (4).

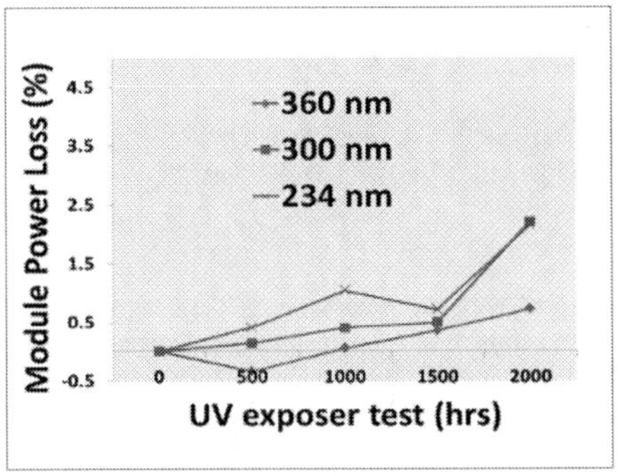

Figure 4: Module power after UV test exposer hrs

From the figure (4) it seems that the module output power loss is below 0.5% up to UV exposer of 1500 hrs in all the three module.

There was a sudden increase in power loss of modules made by lower UV cut off EVA after 2000 hrs exposer. The module made with 360 nm front EVA has also power loss after 2000 hrs but is very minimal as compared to lower UV cut off wavelength EVA modules. The two possible reasons for power loss may be loss in transmittance of EVA [1] and due to solar cell ARC coating [2] The lower UV cut-off wavelength EVA allows more UV to fall on cell as compared to 360 nm UV cut off wavelength.

From the visual inspection, it was found that there was no color change observed in any module and suspect of transmission loss has very little probability.

4 CONCLUSION

The effect of lowering UV cut off wavelength was not found on EVA performance parameter but there was small power loss observed in module made with lower UV cut-off wavelength EVA as compared to conventional 360 nm UV cut-off wavelength. The reason may be due to loss in transmission of EVA or degradation of ARC coating or both. Lowering UV cut off wavelength seems a risky call for module durability and performance. The UV cut-off wavelength of 360 nm looks a better choice to have better EVA and hence module durability against UV light.

5 REFERENCES

1- C. G. Zimmermann, "Time dependent degradation of photovoltaic modules by ultraviolet light," Applied Physics Letters, vol. 92, no. 24, p. 241110, 2008.

2- Performance degradation of c-Si solar cells under UV exposure. Kim H, Choi P, Kim K, Kuh H, Beak D, Lee J, Yi J, Choi B. J Nanosci Nanotechnol. 2014 May;14(5):3561-3.

MODULE INSPECTION USING LINE SCANNING PHOTOLUMINESCENCE IMAGING

Iskra Zafirovska[1], Mattias K. Juhl[1], Jürgen W. Weber[2], Oliver Kunz[1], and Thorsten Trupke[1,2]
[1] School of Photovoltaic and Renewable Energy Engineering, The University of New South Wales, Sydney, Australia
[2] BT Imaging Pty Ltd, Sydney, Australia

ABSTRACT: A custom built laboratory scale tool that enables line scanning photoluminescence (PL) imaging of full size photovoltaic modules was demonstrated for the purpose of manufacturing quality control. The tool allows fast on-the-fly acquisition with exceptional image quality, and can identify series resistance faults without making contact to a sample. This is due to the use of localised line illumination for photoexcitation, which induces lateral current flow between the illuminated and non-illuminated regions in a sample. We show that line scanning PL allows a more reliable differentiation of failure mechanisms than the electroluminescence (EL) imaging systems currently favoured by the industry.

Keywords: PV Module, Photoluminescence, Electroluminescence, Characterisation, Silicon

1 INTRODUCTION

Silicon Photovoltaic (PV) module manufacturing is prone to a number of efficiency limiting defects, manufacturing faults and degradation mechanisms, including material and process induced defects and series resistance (R_s) faults. For example, the high temperature interconnection (soldering) and lamination processes create thermo-mechanical stress in the module, due to the differing thermal expansion coefficients of the materials present [1]. These stresses can form cracks in the cells or metallisation of the module, which have been shown to act as starting points for defect propagation during accelerated stress testing [2].

These faults pose significant reliability risk once a module is exposed to the environmental pressures in the field. Most module faults are highly localised, and cannot reliably be identified by a final IV test or by visual inspection. In response to this, manufacturers have begun to use electroluminescence (EL) imaging [3], due to its simplicity in acquiring spatially resolved quality information. EL measurements require a dark environment and electrical contact to the module. Such contacting increases complexity, risk of mechanical damage to a sample [4] and can have a negative impact on production throughput.

A faster, more robust and more versatile imaging approach that does not significantly affect the production throughput or the time required to complete the production process would be beneficial. Photoluminescence (PL) imaging [5] applied using a line scanning methodology can fulfil these requirements, since it is contactless and allows on-the-fly inspection at inline manufacturing belt speeds. Line scanning PL imaging systems are used in production today for the inspection of as-cut silicon wafers. This paper demonstrates line scanning PL images of silicon modules, and exposes some fundamental differences between PL and EL for module manufacturing quality control.

2 BACKGROUND

PL imaging on modules was first demonstrated by Ebner, *et al.* [6] with a system that utilised a silicon-CCD area camera. The system was designed for and tested on thin film modules (a-Si, CIS and CdTe) in a laboratory setting. Peng, *et al.* [7] developed an experimental PL imaging system that could be mounted within a robotic cleaning device used to maintain modules once installed in the field. The system used an Andor iKon-M series silicon-CCD area camera, particularly because of its low noise and NIR responsivity. However, it images modules on a cell by cell basis, as both illumination and detection is restricted to a 6 inch region.

A line scanning approach requires use of a line scan camera, in which the sensor is reduced to a single line of pixels rather than a 2D matrix. Imaging involves moving a sample through the field of view, whilst the camera continuously acquires line images in sync with the sample motion. The line images are then assembled to form the final image.

The illumination source for photoexcitation can also be reduced to a line. This allows for imaging at higher carrier injection and has the additional benefit of allowing current extraction during image acquisition, despite the cells being under open circuit conditions externally, i.e. zero current between the cell's terminals. Excess carriers are only generated within the illuminated line, forming a voltage difference between illuminated and non-illuminated regions that is the driving force for lateral currents. This effect is demonstrated in Figure 1, which shows a snapshot in time of the luminescence distribution across a uniform silicon solar cell during line scanning PL acquisition, simulated with the solar cell finite-element model program *Griddler* [8]. The line width is strongly exaggerated here for demonstration purposes. As can be seen in the horizontal cross-section plot, the non-illuminated areas outside of the red dashed lines emit a luminescence signal due to lateral current flow induced by the voltage difference. In this case, the luminescence intensity from the non-illuminated regions is almost identical to the intensity from illuminated cell areas, demonstrating the effectiveness of the lateral currents. Note that at each point in time the line scan camera inspects the illuminated area, from which current is extracted. The final PL image is therefore expected to resemble a conventional PL image that is measured with current extraction.

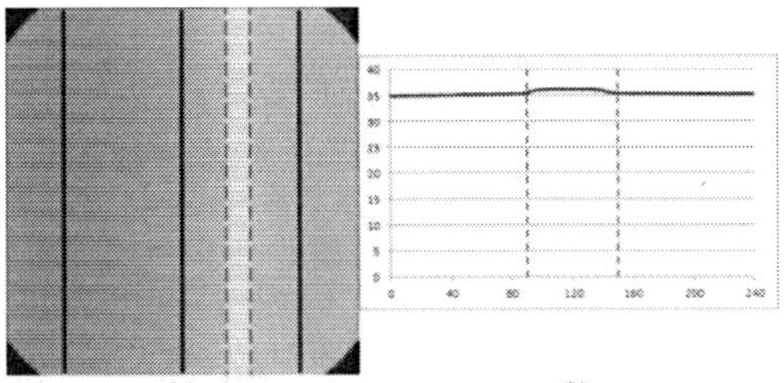

(a) (b)

Figure 1: (a) PL intensity from a monocrystalline cell simulated with *Griddler*, applying 0.05 Suns illumination intensity within the red dashed lines (width exaggerated). (b) Horizontal cross-section of (a).

3 EXPERIMENTAL SETUP

A custom built laboratory scale automated line scanning PL imaging tool was developed, capable of measuring individual silicon cells and mini modules as well as full size industrial silicon modules . A schematic of the tool and its main components is shown in Figure 2. Photoexcitation is achieved using a high-powered illumination source that was focused parallel to the short edge of modules. A line scanning silicon-CCD camera with an enhanced NIR response sensor is used for detection. Specialised optics were added to the camera to prevent both the detection of spurious luminescence (for example from encapsulation material) and detection of excitation illumination.

The sample stage consists of a support frame, which is conveyed underneath the camera to mimic movement of modules on a manufacturing line. All images presented in this paper were both flat-field and background corrected.

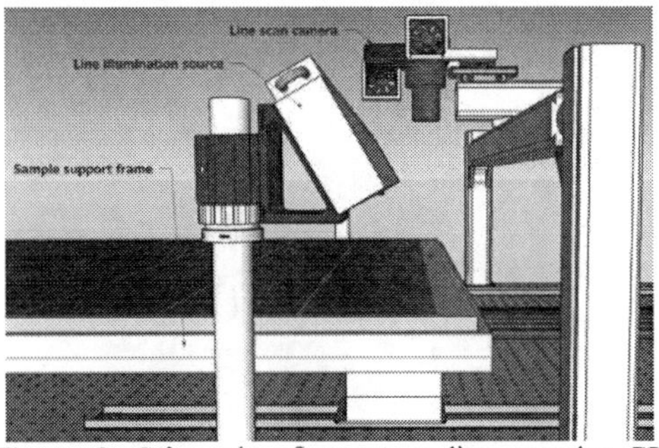

Figure 2: Schematic of prototype line scanning PL imaging tool used in this study.

4 RESULTS AND DISCUSSION

4.1 Line scanning PL module imaging

The performance of the prototype tool was evaluated by imaging a commercial module consisting of 60 multicrystalline silicon 156 mm x 156 mm cells. The module has a rated power output (P_{max}) of 260 W. A PL image of this module acquired with the prototype tool is shown in Figure 3. A number of cells in the central region of the module contain many irregularly shaped dark regions originating from a busbar, which correspond to cell cracks in Chaturvedi, *et al.* [1]. In this specific module these cracks were induced during prior mechanical load testing.

Figure 3: Line scanning PL image of an industrial multicrystalline silicon PV module taken on-the-fly with approximately 4 Suns illumination intensity. Three triangular regions of interest are highlighted by red boxes.

4.2 Comparison to EL imaging

The prototype line scanning PL tool is capable of acquiring EL images, when an external DC power supply is connected for excitation. An EL image of the same module is shown in Figure 4.

In both images areas of high recombination such as grain boundaries and dislocation clusters appear dark, but a reverse in contrast is observed when comparing high R_s regions. Regions of high R_s appear darker than their surroundings in the EL, and brighter than their surroundings in PL with current extraction [9]. As an example we can identify three triangular regions in the EL image (Figure 4), which appear entirely black due to the fact that the current flow into those regions is impeded by local interrupts in the metal fingers caused by large cell cracks. In contrast, the same regions appear brighter than their surroundings in the PL image (Figure 3). This is expected from the previous discussion, since areas with enhanced series resistance remain at a higher local voltage due to the restrictions on local current extraction. Differentiation of failure mechanisms, specifically the distinction of recombination sites from series resistance problems is therefore more reliable when using line scanning PL imaging, since defects attributed to high recombination and those attributed to high R_s appear with inverted contrast, whereas in an EL image all defects appear dark. This is highlighted in Figure 5, which compares PL and EL images of the specific cell highlighted in colour in Figure 4. These images were taken directly from the full module images and highlights the exceptional image quality that our prototype system enables.

Figure 4: EL image of the module shown in Figure 3, taken with forward bias of 39.5 V (equivalent to 1.045 x V_{oc}). Three triangular regions of interest are highlighted by red boxes.

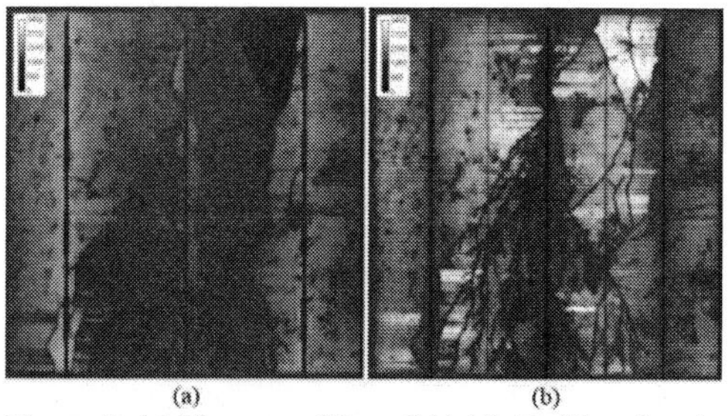

Figure 5: (a) Close up of the cell highlighted in colour in Figure 4. (b) Line scanning PL image of the same cell.

Finger disconnections create areas of increased R_s, which are identified as regularly shaped rectangles extending from the affected busbar [1]. In the PL image these broken finger patterns appear bright and clearly discernible from the neighbouring dark cell cracks, whilst in the EL image they are dark and hence blend in. In addition the PL image provides information on the electrically isolated region in the top right of the cell, whilst in the EL image it emits no signal.

4.3 Local current extraction

To emphasise the effect of local current extraction in an image acquired with an on-the-fly line scanning PL tool, we compare luminescence image data from the PL line scanning system with conventional images measured on a BT Imaging LIS-R1 tool. The LIS-R1 uses a conventional area scanning (2D) camera and full area illumination. Images acquired on a 125 mm x 125 mm CZ silicon p-type cell, that has a finger interruption near the busbar, are shown in Figure 6.

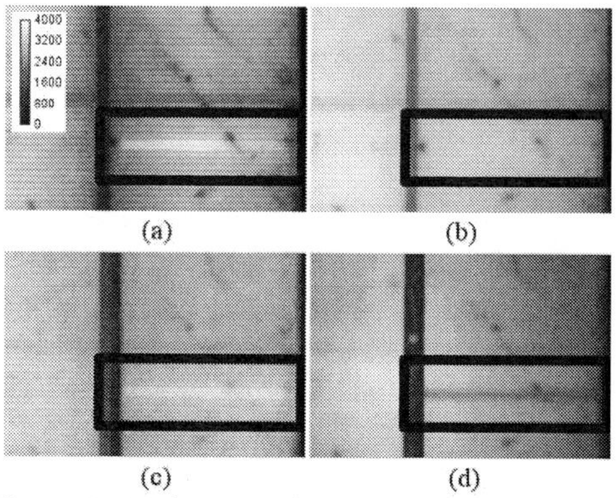

Figure 6: Luminescence images of a monocrystalline silicon solar cell. (a) line scanning PL image taken with approximately 4 Suns illumination intensity at a measurement speed of 135 mm/s. (b) conventional open circuit PL image. (c) PL image with current extraction at 0.55V. (d) EL image with forward bias of 0.625 V. The impact of a finger break is highlighted by the black box.

As expected the finger disconnection highlighted by the black box has similar contrast in the line scanning PL and in the PL image taken with current extraction, whereas it appears as a dark line in the EL image. In the conventional open circuit PL image the defect is not visible, since no current flows across the finger break.

Finally, the impact of the finger disconnection is confirmed by the quantitative R_s image shown in Figure 7 [10]. The affected area has an elevated average R_s value of 1.12 Ωcm^2 whilst the surrounding unaffected cell area has an average R_s value of 0.64 Ωcm^2.

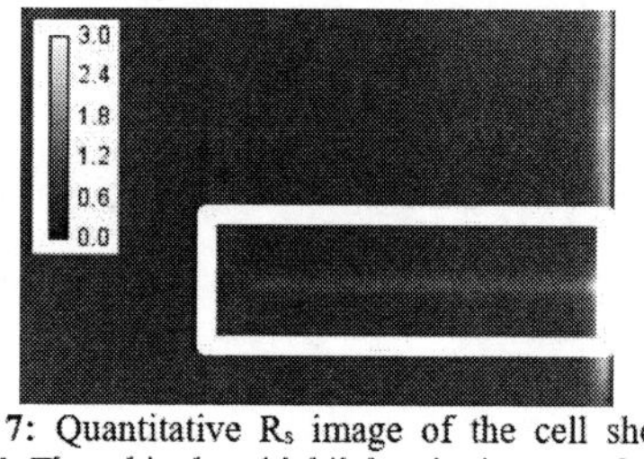

Figure 7: Quantitative R_s image of the cell shown in Figure 6. The white box highlights the impact of a finger disconnect. Scale in Ωcm^2.

5 CONCLUSION

The use of line scanning PL imaging was demonstrated on a full size commercial multicrystalline silicon PV module using a prototype system. Images of exceptional quality can be acquired on-the-fly for quality control in a manufacturing setting. PL images exhibit a local current extraction effect due to localised line illumination. This image acquisition mode and the resulting image features are expected to be more effective than EL imaging in discerning R_s effects from recombination effects, particularly when automated image pattern recognition techniques are applied. This technology has the potential to become a key application in commercial module quality control, which is of increasing relevance in an ever-growing and highly competitive market.

ACKNOWLEDGEMENTS

The authors would like to thank Dr Johnson Wong for support with *Griddler* simulations.

This program has been supported by the Australian Government through the Australian Renewable Energy Agency (ARENA) grants 7-F008 and RND009. Responsibility for the views, information or advice expressed herein is not accepted by the Australian Government.

REFERENCES

[1] P. Chaturvedi, B. Hoex, and T. M. Walsh, "Broken metal fingers in silicon wafer solar cells and PV modules," *Solar Energy Materials & Solar Cells,* vol. 108, pp. 78-81, 2013.

[2] M. Sander, S. Dietrich, M. Pander, M. Ebert, S. Thormann, J. Wendt, *et al.*, "Investigations on Cracks in Embedded Solar Cells After Thermal and Mechanical Loading," in *27th European Photovoltaic Solar Energy Conference and Exhibition (27th EU PVSEC)*, Messe Frankfurt and Congress Center, Frankfurt, Germany, 2012, pp. 3188-3193.

[3] T. Fuyuki, H. Kondo, T. Yamazaki, Y. Takahashi, and Y. Uraoka, "Photographic surveying of minority carrier diffusion length in polycrystalline silicon solar cells by electroluminescence," *Applied Physics Letters,* vol. 86, pp. 1-3, 2005.

[4] T. Trupke, B. Mitchell, J. W. Weber, W. McMillan, R. A. Bardos, and R. Kroeze, "Photoluminescence Imaging for Photovoltaic Applications," *Energy Procedia,* vol. 15, pp. 135-146, 2012.

[5] T. Trupke, R. A. Bardos, M. C. Schubert, and W. Warta, "Photoluminescence imaging of silicon wafers," *Applied Physics Letters,* vol. 89, pp. 1-3, 2006.

[6] R. Ebner, B. Kubicek, and G. Újvári, "Photoluminescence measurement system for thin film modules," in *31st European Photovoltaic Solar Energy Conference and Exhibition (31st EU PVSEC)*, Hamburg, Germany, 2015, pp. 39-41.

[7] C.-Y. Peng, S.-Y. Wen, Z.-J. Ji, and C.-P. Huang, "Non-Destructive Displaying Defects for Luminescence Image of Photovoltaic Panel Arrays," in *2015 International Conference on Advanced Robotics and Intelligent Systems (ARIS)*, National Taipei University of Technology, Taipei, Taiwan, 2015, pp. 1-4.

[8] J. Wong, "Griddler: Intelligent computer aided design of complex solar cell metallization patterns," in *IEEE 39th Photovoltaic Specialists Conference (PVSC)*, Tampa, FL, USA, 2013, pp. 933-938.

[9] T. Trupke, R. A. Bardos, M. D. Abbott, F. W. Chen, J. E. Cotter, and A. Lorenz, "Fast Photoluminescence Imaging of Silicon Wafers," in *4th World Conference on Photovoltaic Energy Conversion (WCPEC)*, Hilton Waikoloa Village, Waikoloa, Hawaii, USA, 2006, pp. 928-931.

[10] H. Kampwerth, T. Trupke, J. W. Weber, and Y. Augarten, "Advanced luminescence based effective series resistance imaging of silicon solar cells," *Applied Physics Letters,* vol. 93, pp. 1-3, 2008.

Optical Simulation for Ribbon with Optical Structure in c-Si PV Module

Chao-Wei Yang, Chun-Ming Yang, Ming-Yu Chen, Chien-li Cheng
AU Optronics Corporation, No. 1, JhongKe Rd., Taichung 40763, Taiwan, R.O.C.
E-mail: lepton.cw.yang@auo.com , Phone: +886 4 2460-8800 #1242, Fax: +886 4 2460-8077

ABSTRACT: The ray-tracing optical simulation model was used to analyze several optical structures, which constructed on ribbon inside photovoltaic module. Two kinds of optical structures, groove-like & diffuse white, were analyzed with indoor and outdoor conditions. For groove-like structure, the outdoor performance was correlative to mounting configurations of module, portrait and landscape. Moreover, such results were strongly depended on different seasons. For diffuse white optical structure constructed on ribbon, significant difference was shown during simulation with outdoor condition when compared with groove-like structure. But diffuse white structure was poor than groove-like structure about the efficiency of light capture. Three kinds of full-size photovoltaic modules (60-cells), groove-like structure mounting with portrait and landscape configuration, and diffuse-white structure mounting with portrait configurations, were produced and examined in practical field during summer. The results of field test matched well with ray-tracing optical simulation.

1 Introduction

Simulation models done based on several kinds of numerical method were already performed and studied[1-2]. Detail physical and electrical mechanisms were realized through the setup of numerical simulation for solar cells and PV modules. However, such simulation models based on the ray tracing in one dimensional, layer-by-layer approach could only represent a simplified module. When implementing the complex geometry of practical PV module, such as metal fingers on solar cell, matrix of cells, different module dimensions, into the simulation model, the resource capacity on the machine running the simulation increased significantly. In the past study[3], it is successful to apply commercial software; Trace-Pro which are widely adapted on LCD and LED industry, to solve CTM loss issue of PV modules about module margins, cell spacing, and effectiveness of ARC-glass...etc. In addition, such study could be applied and achieved on 60-cell c-Si PV module with full complex 3D model. In advanced, it is possible to simulate the outdoor performance of PV modules by applying the sun trajectory.

Recent years, several innovational technology about PV module, aimed on enhanced the ability of light mangnment to increasing absorption effeciency of solar power. A major enhanced technic was applied optical captural strcture on tapping ribbon of cell inside PV module. Through well-defined optical structure, the incident light on tapping ribbon could be re-directed into solar cell. Without such optical structure on tapping ribbon, incident light were directly reflected and escaped outside PV module which have no contribution on the effeciency of PV module. However, such optical structure on tapping ribbon was designed, which based on ideal condition with normal incident light. And such structure would be highly sensetive to the angle of incident of light.

The purpose of this paper is aimed to simulate the real condition of power gain under outdoor condition which incident angle of light vary with time when the optical structure add-on ribbon inside PV module.

2 Simulation Methodology

The ray-tracing model of optical simulation was built upon the commercial software, TracePro[4], with practical geometry and dimension based on conventional design of c-Si PV module. Detail information was showed as figure.1. There were two kinds of optical structure constructed and simulated on tapping ribbon, which showed inside figure.1. The optical structure-1 was built-up with plastics substrate, which grooved with tri-angular shape coating with high reflective material. The 60-degree angle on groove was designed to induce total reflection of normal incident sun light and trapped it inside PV module. The optical structure was designed and built-up with flat polymer stripe, which behaved high and diffused white reflection. Such diffused white reflection was treated as ideal Lambertian condition during optical simulation.

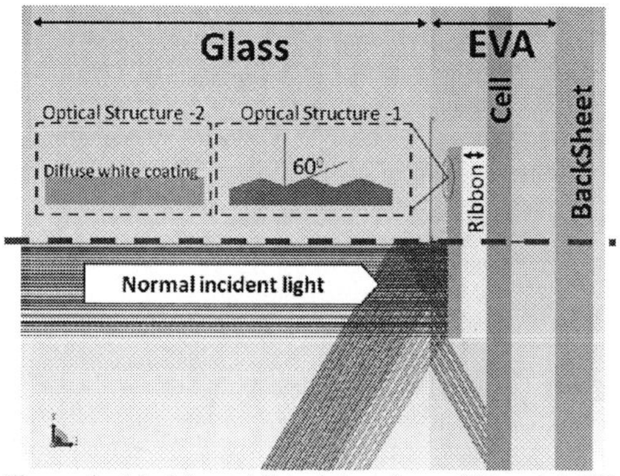

Figure 1. Model established based on 1X1 mini PV module with two kinds of optical structure add-on ribbon.

The benefit of such model was capable to conduct simulation of practical PV module. In addition, the anti-reflective coating including ARC-glass and SiNx could be modified with refractive indices (n), extinction coefficient (k) and film thickness by the built-in function of the software. The properties at interface such as EVA to backsheet, and ribbon on solar cell were measured with UV/Vis/NIR spectrophotometer, Lambda 750 and inputted into the program. The dimension and component information of the module were based on the BENQ PM250M01 model[5]. For outdoor condition, the simulation was setup at Taichung city (120E37'00'' 24N11'00''), Taiwan. The time interval of simulation was setup at May 20, 2015. Moreover, the sun trajectory could be evaluated through such geographical and time condition for PV module mounted with 0^0 tilt angle.

Through applying all the information into the model, the behavior of ray-tracing could be simulated and behaved as practical c-Si PV module.

3 Result and discussion

Based on the model established via commercial software (TracePro) with PV module using ARC glass, UV-transmittance encapsulant, and PET-based backsheet, the ray-tracing optical simulation aimed to study optical structure on tapping ribbon was realized. Figure 2. showed the ray tracing results for PV module with optical structure-1 when incident angle of light varied from 0 to 70 degree. For the normal incident (0 degree) condition, the incident light almost reached Si-surface of solar cell (85.5%) because of total reflection by glass. Only 14.5% light lose due to bulk absorption of glass and EVA, and reflection when light incident from air to glass. Only 0.0176% light, which reached optical structure-1 on tapping ribbon, could escape outside the module. By increasing of incident angle, the escape% of incident light would also increase. When the angle of incident greater than 17 degree, the escape% would increase very significantly. For 10, 20, and 30 degree of incident angles, the escape% would be 0.02%, 60.97%, and 82.77%. At the same angles, the absorption% which light reached Si-surface of solar cell would be 84.228%, 22.147%, and 0.369% which decreasing drastically. For higher incident angle (> 30 degree), the escape% kept in high level (> 80%) and absorption% kept in very low level (<1%).

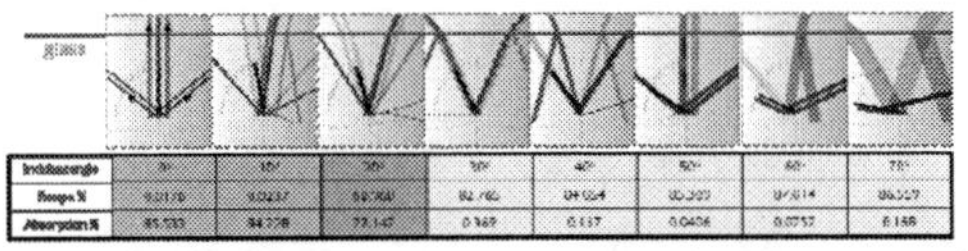

Figure 2. Simulation were performed with 1X1 mini PV module and optical structure-1 add-on ribbon.

From such simulation result, it was obviously that the performance of optical structure-1 on tapping ribbon inside module would be fullly active only at midday (around AM 11:00 to PM 1:00). But it would be highly relative to the mounting configuration of PV module in the field. For two kinds of mounting configuration of PV module, portrait and landscape, the simulation results were showed as figure3. For portrait and landscape mounting configuration, the sun trajectory and optical structure-1 on tapping ribbon were presented as the schematic diagram on the right side of figure3. For portrait mounting configuration, the sun rays would fit the simulation results showed as figure2. For simulation of the timeline showed with figure3, it could be found that the absorption% of portrait mounting configuration would match the profile of the strength of sun light only during am.10:30 to pm.1:00 at May.20th, 2016. In opposited, the absorption% of landscape mounting configuration would match well with the profile of the strength of sun light during all the daytime in the same setting of time. For reference, the simulation of tapping ribbon without any optical structure was performed. It could be found that much less absorption% of Si-surface of solar cell would be contributed from reflection of the reference ribbon.

In advanced, the study of mounting configuration for optical structure-1 on tapping ribbon was performed not only with daytime period, but also with season interval.

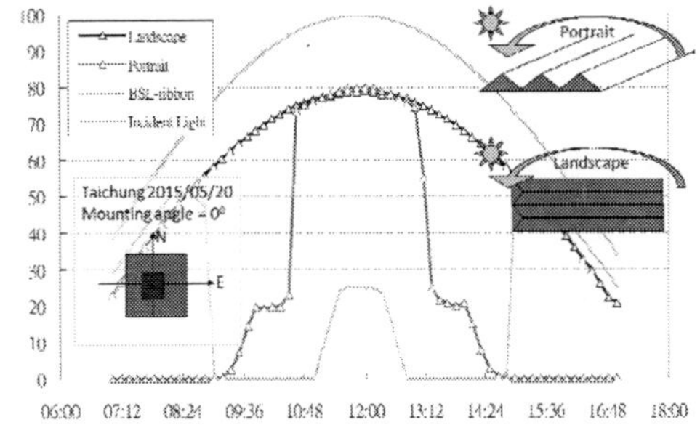

Figure 3. Simulation performed with 1X1 mini PV module and optical structure-1 add-on ribbon for different mounting type within daytime.

From January to July, the simulation result of first day of each month were showed as figure4. For portrait mounting configuration of module with optical structure-1 on tapping ribbon, the behavior of each month follow the provious result showed in figure2 which absorption% of incident sun light kept effective only during the midday. For January, sun trajectory would be shift to lower angle and result longer effective period of high absorption% of incident sun light. However, sun trajectory shift to lower angle also resulted lower strength of incident sun light. Resulting from both of the two reasons, the domain of effective absorption% would vary from lower but wider to higher but narrower ones as time passed from January to July. Comparing with portrait, the lanscape mounting configuration of module with optical structure-1 on tapping ribbon showed different behavior of absorption%. For summer time of May, Jane, and July, the behavior of absorption% would keep in the same level and similar to the previous simulation result showed in figure3. When the time approach to spring and winter like Januray to Apirl, sun trajectory would be shift to lower angle and resulting the absorption% dropped significantly during midday. For lower sun trajectory like January, the dropped region would become seriously wider. It meaned the lanscape mounting configuration of module with optical structure-1 on tapping ribbon would have better performance only during summer time (May to September). For the winter (November to Feburay), the opposited result about absorption% would happen.

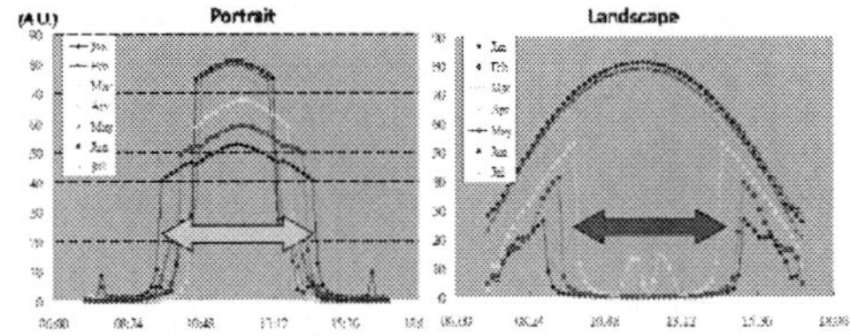

Fig.4 Simulation performed with 1X1 mini PV module and optical structure-1 add-on ribbon for different mounting type within daytime for different month.

For optical structure-2 on tapping ribbon designed and built-up with flat polymer stripe, which behaved high and diffused white reflection, the simulation result was showed with figure5, which following with same condition of optical structure-1 showed in figure3. Comparing with landscape mounting of optical structure-1, the behavior of absorption% was similar but lower for optical structure-2. It was resulted from the diffused white paint property which near Lambertian scattering. Only part of incident light could be always satisfied the total reflection and then absorpted by the Si-surface of solar cell. However, the situation of only part incident light absorpted always happened. Moreover, it was independent of time and mounting configuration of module. No significantly dropped and increased behavior of absorption% was found for optical structure-2.

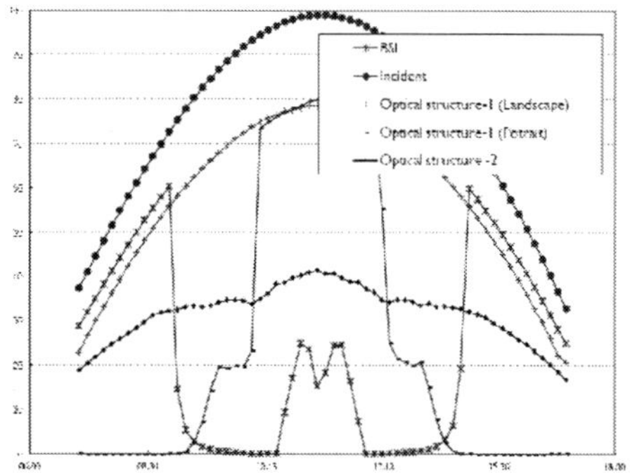

Fig.5 Simulation performed with 1X1 mini PV module and optical structure-2 add-on ribbon within daytime.

4 Conclusion

The ray-tracing simulation model was applied into practical PV module to study the difference of optical structure on tapping ribbon under outdoor condition about different mounting configuration of module. In addition, the outdoor simulation was aimed to analysis the influence between optical structure and the practical sun trajectory during different time at fixed geographical position, Taichung city at Taiwan. For the same time at May 20th, 2016, the grooved optical structure (structure-1) have better light absorption performance only at midday for portrait mounting configuration of module. Moreover, the same optical structure mounting with landscape configuration would have better light absorption performance all daytime. However, such result of light absorption performance about different mounting configuration with optical structure-1 only happened during summer time. For the time nearby winter, the behavior of landscape mounting configuration with optical structure-1 would become worse due to the lower of sun trajectory. For optical structure-2, which designed and built-up with flat polymer stripe behaved high and diffused white reflection, the behavior of absorption% would be similar to landscape mounting configuration of module with optical structure-1 on tapping ribbon. However, lower absorption% was showed, which due to diffused white reflection near Lambertain condition. Moreover, it was independent of time and mounting configuration of module. And no significantly dropped and increased behavior of absorption% was found for optical structure-2.

5 References

1. K. R. McIntosh , J. N. Cotsell , J. S. Cumpston , A. W. Norris , N. E. Powell and B. M. Ketola "An optical comparison of silicone and EVA encapsulants for conventional silicon PV modules: A ray-tracing study", *Proc. 34th IEEE Photovolt. Solar Conf.*, pp.1 -7 2009
2. McIntosh, K.R.; Cotsell, J.N.; Norris, A.W.; Powell, N.E.; Ketola, B.M. "An optical comparison of silicone and EVA encapsulants under various spectra", *Photovoltaic Specialists Conference (PVSC), 2010 35th IEEE,* On page(s): 000269 – 000274
3. Chao-Wei Yang ; Chen-Wei Lin ; Ming-Yuan Huang ; Zhen-Cheng Wu "An optical ray-tracing model for module short circuit current", European Photovoltaic Solar Energy Conference (EUPVSEC), 2014
4. TracePro : Lambda Research Corporation. http://www.lambdares.com/tracepro
5. BenQ PM250M00 : http://www.benqsolar.com/?sn=1264&lang=en-US

INFLUENCE OF PHOTOVOLTAIC MODULE MOUNTING SYSTEMS ON THE THERMO-MECHANICAL STRESSES IN SOLAR CELLS BY FEM MODELLING

Andreas J. Beinert[1], Matthieu Ebert[1], Ulrich Eitner[1] and Jarir Aktaa[2]
[1] Fraunhofer Institute for Solar Energy Systems ISE, Heidenhofstraße 2, 79110 Freiburg, Germany;
[2] Karlsruhe Institute of Technology (KIT), Hermann-von-Helmholtz-Platz, 76344 Eggenstein-Leopoldshafen, Germany.
Corresponding author: Andreas J. Beinert | Phone: +49 (0)761 4588 5630 | E-mail: Andreas.Beinert@ise.fraunhofer.de

ABSTRACT: The mounting system of photovoltaic (PV) modules has a significant impact on the thermo-mechanical stress in PV modules. In this work the clamping of framed PV modules is compared to the clamping of unframed PV laminates by a simulation study using the finite element method (FEM). The FEM modelling allows to calculate the local stress distribution in the solar cells directly. We present results from a model of a standard glass-backsheet PV module with 3 mm glass under homogenous mechanical pressure load of up to 5400 Pa. The thermal stress from the lamination process is considered as a pre-study, similar to [1]. The frameless clamped PV laminate shows a significantly larger displacement of 147 mm than the framed PV module with 54 mm for 2400 Pa. In line with the findings of Kajari-Schroeder [2] we simulate an elliptic deflection distribution for the framed PV module, whereas the clamped PV laminate shows a wave-like shape. Consequently, the area of high tensile stresses in the silicon solar cells, with a maximum value of 142 MPa at 2400 Pa load, is narrowly located around the highest curvature at the center of the framed PV module. In case of the frameless clamped PV laminate we identify four areas of high tensile stresses with a higher maximum value of 218 MPa. The results show that the frame reduces the tensile stresses in the solar cells significantly compared to unframed laminates.
Keywords: PV Module Design, Thermo-Mechanical Stress, FEM Modelling, Reliability

1 INTRODUCTION

PV module manufactures offer a linear performance warranty of up to 30 years. In order to enable a sufficient service life time, the mechanical stresses in the solar cells have to remain below critical values. A measure for the criticality of stresses is the characteristic fracture stress, at which 63.2% of the solar cells fail. Kaule determined a fracture strength of about 200 MPa [3] for standard solar cells, depending on the orientation. Cracks can cause a significant loss in the module power depending on the type of crack [4]. Hence it is essential to minimize the stress in the solar cells.

1.1 Thermo-mechanical stresses

In PV modules mechanical stresses are induced by mechanical loads, like snow or wind [2], or/and thermally. The latter occur whenever there is a change in temperature of the PV module, starting from the interconnection [5, 6] and lamination process [1] and continuing to the daily and yearly temperature cycles. These stresses are due to the mismatch of the coefficients of thermal expansions (CTE) of the different PV module materials. Often the mechanical stresses which occur in a PV module are a combination of mechanically and thermally induced stresses, hence the terminology thermo-mechanical stresses is used. They are influenced by various factors like the used materials, design and interconnection technique and were investigated thoroughly in the past, e.g. [1, 7]. An overlying effect however is the mounting technology, since it defines the boundary conditions for the module's deformation. In this work we investigate two standard technologies, the clamping of framed and unframed laminates by means of finite element method (FEM) analysis. The aim is to assess the thermo-mechanical stresses in encapsulated solar cells resulting from these systems. For this purpose two three dimensional FEM models of full size PV modules containing 60 solar cells are set up, which differ only in the mounting system. The effect of interconnection techniques is neglected.

2 FINITE ELEMENT MODEL

2.1 Sub-modeling and geometry

The large aspect ratio of the solar module edge length to its thickness demands for high computational effort when using classical continuum mechanics on 3D volumes. Usually plate- or shell-type formulations are used for efficient mechanical modeling of thin structures. However, these descriptions are not designed to describe local stresses in discontinuous layers like the solar cell matrix and are thus not applicable for our purposes. In order to minimize the computational effort of our 3D model, the sub-modeling method is applied in this study. With this method the model is separated in different problems: one for the computation of the displacement, the global model. The second model is a minimized geometry in order to use a finer mesh for the computation of the mechanical stress. This is the so called sub-model.

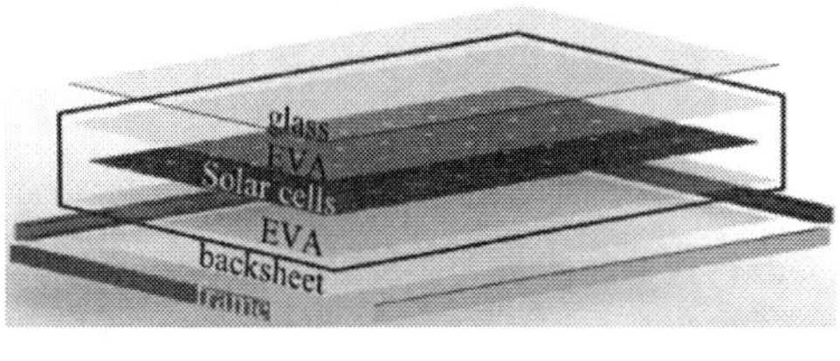

Figure 1: Schematic of a PV laminate with a frame, which is implemented as a sub-model in the FEM model. The red box indicates the sub-model for stress analysis.

The global model consists of the PV laminate, as shown schematically in figure 1, without any mounting system. By exploiting the symmetry of a PV module, a quarter of the PV laminate containing 15 solar cells is modelled. The corresponding materials and layer thicknesses are given in table I. The mounting system is implemented as another fully coupled sub-model to the laminate. Thus the laminate could be meshed identically in both FEM models.

Table I: Materials of the PV laminate used in the FEM model with the applied material model and thickness.

Layer	Material	Material model	Thickness [µm]
Front Glass	solar glass	linear elastic	3000
Encapsulant	EVA	linear elastic	400
Solar Cells	monocrystalline silicon	linear elastic, anisotropic	200
Backsheet	PVF-PET-PVF	linear elastic	350

In the sub-model for the stress computation the laminate is reduced to the solar cell matrix embedded in the encapsulant, as indicated by the red box in figure 1. The deformation of the sub-model is given from the global model on the boundary encapsulant/glass and encapsulant/backsheet.

2.2 Materials and mesh

The solar cells are modelled as full-square monocrystalline silicon cells without any metallization. This allows for the use of rectangular mesh elements. The global model exhibits 7.7x10^6 and the sub-model 18.6x10^6 degrees of freedom. Figure 2 shows the cross section of the aluminum frame and the aluminum laminate clamp. Sliding at the interface is prohibited and the displacement field is continuous. The rubber inlay is meshed with a finer resolution.

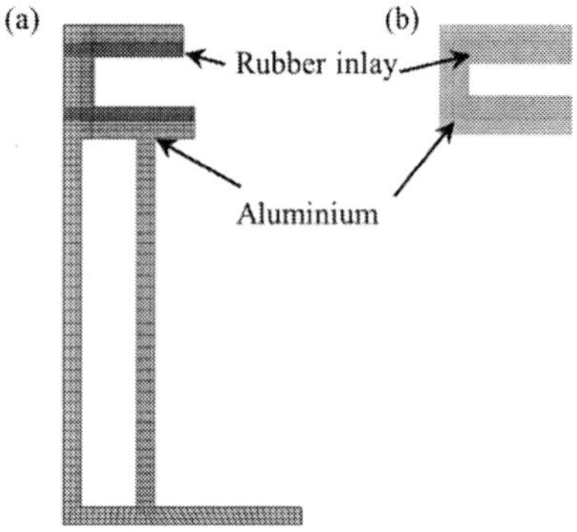

Figure 2: Cross section of the meshed aluminum frame (a) and the aluminum laminate clamp (b) with a finer meshed rubber inlay.

For all materials linear elastic material models are used. The corresponding material parameters are given in table II. Instead of modeling the full viscoelastic behavior of EVA [8] we take into account only the temperature dependency by using a function of T for the Young's modulus. The monocrystallinity of silicon is modelled by a cubic-symmetric elasticity matrix given in [1].

The computation itself is carried out in two successive steps. First the lamination process is modeled by cooling down from a stress-free state at 150°C to 25°C. In this step the mounting structure is not considered. Secondly the mechanical load test at room temperature is applied onto the thermally stressed module. According to the IEC 61215 [9] the glass surface is homogenously loaded with 2400 Pa and 5400 Pa. The mounting of the frame and the clamp is modelled by a fixed constraint, with a distance of 20% of the PV module length from the short edge. This boundary condition represents the screwing on a rack and is comparable to model 8 in the study of Schicker [10].

Table II: Material properties used in the FEM model; *: provided by manufacturer.

Material	Density ρ [kg/m³]	Young's modulus E [GPa]	Poisson's ratio ν [-]	CTE α [10^{-6} K^{-1}]
Glass [11]	2500	70	0.2	9
EVA [7]	960	T-dep.	0.4	270
Silicon [1, 12, 13]	2329	elasticity matrix		T-dep.
Back Sheet [1]	2520	3.5	0.29	50.4
Aluminum [14]	2700	70	0.33	23
Rubber inlay*	67	0.0074	0.3	769

3 RESULTS

3.1 Lamination

The compression due to the cooling down process after lamination is shown in figure 3. A maximum compressive stress of 81 MPa is obtained on the rear surface of the solar cells. The compressive stresses show a superposition of the compression of the laminate and the solar cells themselves. Therefore the compressive stress is slightly higher at the edge of the laminate than in the center. Since compressive stress is less crucial for silicon solar cells than tensile stress, the stress obtained from lamination is not critical. In the encapsulant no significant stresses are obtained, which reflects the buffering character of EVA.

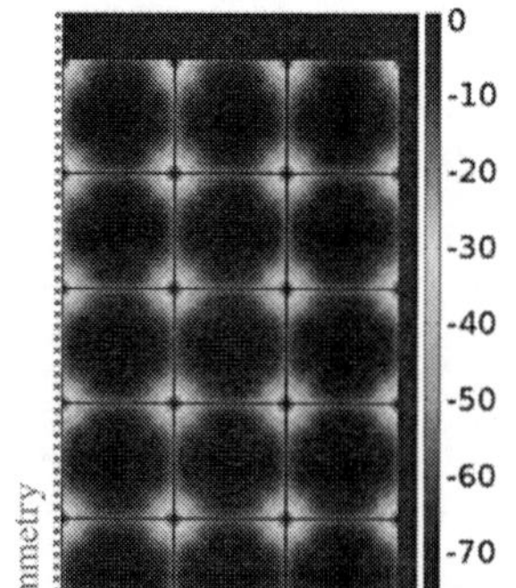

Figure 3: Thermal compression stress from lamination process on the rear surface of the solar cell layer. The color table shows the third principal stress σ_{III}. The dotted line represents the symmetry plane.

3.2 Mechanical load of 2400 Pa

In this section the results of the FEM model for the mechanical pressure load test with 2400 Pa are presented. Since the distribution at 5400 Pa is very similar, only the maximum values are shown in figure 6.

Figure 4 shows the total displacement d at 2400 Pa homogeneous mechanical pressure load on a glass-backsheet PV laminate mounted with four aluminum clamps (b) and the same PV laminate with a frame mounted on a rack (a). The corresponding maximum values are given in figure 6 (left). With a maximum displacement of 54 mm the framed PV module shows a much smaller displacement compared to the maximum displacement of 147 mm of the frameless PV laminate mounted by four clamps. The displacement of the frameless laminate is a superposition of a deflection in x-direction and a deflection in y-direction. It exhibits a wave like character, while the framed PV module shows an elliptic distribution. The different deflection shape arises from the additional stiffness which the frame provides to the laminate. The frame supports the PV laminate along all four edges while the clamps support the module locally at four positions. Therefore the displacement at the edges almost vanishes for the framed PV module, while the frameless PV laminate shows a significant displacement between the clamps.

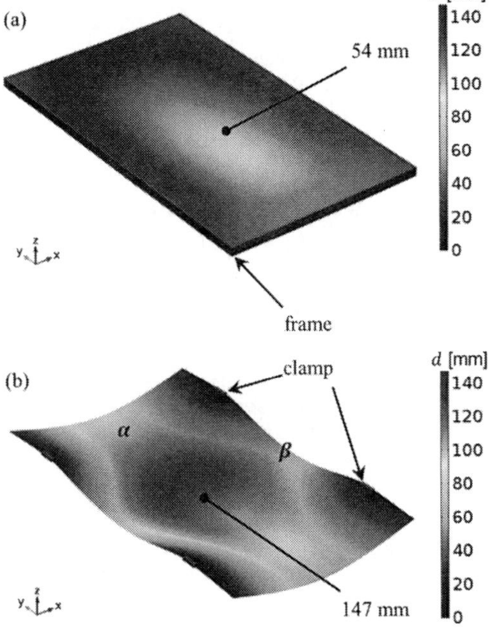

Figure 4: Total displacement d from homogenous mechanical load with 2400 Pa for the full framed PV module (a) and full unframed PV laminate mounted with clamps (b). The color legend shows the displacement values in mm. The displacement is represented by the deformation of the geometry (to scale).

Due to the deflection, the highest tensile stresses in the solar cells occur on their rear side. The distribution of the first principal stress σ_I on the rear side of the solar cells at 2400 Pa is shown in figure 5. The maximum values are depicted in the right part of figure 6. For the framed module the maximum value of 142 MPa is below the fracture strength of 197 MPa [3]. Therefore no significant cell breakage is expected for framed modules. For unframed modules the fracture strength is exceeded in 16 of the 60 cells with a maximum of 218 MPa, as indicated by the dashed ellipse in figure 5. Hence we expect a critical risk of cracks for unframed glass-backsheet modules.

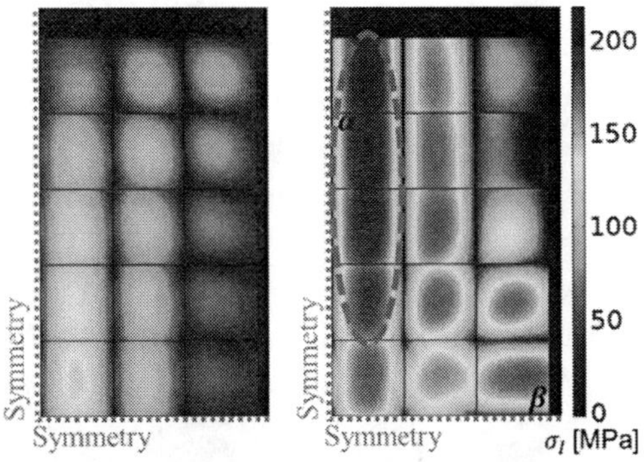

Figure 5: Tensile stress on the rear side of the solar cells resulting from homogenous mechanical load with 2400 Pa for the framed PV module (left) and the unframed laminate mounted with clamps (right). The color legend shows the first principal stress σ_I and the dashed ellipse indicates cells with stress values above the fracture strength reported in [3].

The stress distribution in the solar cells corresponds to the different displacement shapes seen in figure 4. For the homogenous displacement distribution of the framed PV module, the maximum curvature occurs at the maximum deflection in the center of the PV module. Hence the highest tensile stress in the solar cells is found at the symmetry corner of the framed PV module. The stresses in the module corner arise from high diagonal strains, as described in [2]. Contrary in the case of the frameless PV module: due to the sinus-like deflection, the maximum curvature does not occur at the maximum deflection. Additional to the global maximum in the module center, we find two local maxima, which are marked by "α" and "β" respectively in figure 4. These correspond to the deflections in x- and y-direction. At these local maxima, the principal stress directions are dominated by the resulting curvature. Therefore two areas of high tensile stresses occur in the solar cells. Consequently, the overall area of high tensile stresses is larger for unframed laminates than for framed modules.

3.2 Mechanical load of 5400 Pa

With increasing load the deflection increases and hence the difference in the deflection shape becomes more crucial. According to the FEM results of 5400 Pa load, 52 of the 60 solar cells in the unframed, clamped PV laminate are exposed to tensile stresses above the fracture strength with a maximum of 443 MPa. Consequently a severe cell breakage is anticipated at high loads for clamped PV laminates. Also in the framed PV module the fracture strength is exceeded with a maximum tensile stress value of 268 MPa at 5400 Pa load. However only 12 of the 60 cells show tensile stresses above the fracture strength. Hence cell cracking is expected to a less extend than for the frameless laminate. By increasing the glass stiffness, i.e. increasing the thickness and/or the Young's modulus, the stress values will decrease. For framed modules the frame geometry can be optimized additionally. With these measures the tensile stresses in framed PV modules could be decreased below the fracture strength given in [3]. The tensile stresses in frameless glass-backsheet laminates can hardly be decreased below the fracture strength. Therefore glass-glass PV laminates should be investigated for heavy loads.

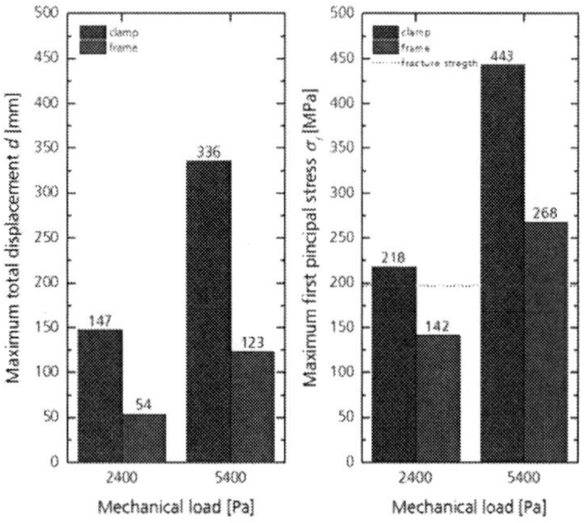

Figure 6: Total displacement d (left) and first principal stress σ_1 (right) depending on applied mechanical load.

4 CONCLUSION AND OUTLOOK

A simulation model for a straightforward thermo-mechanical assessment of different mounting systems is presented. In this study the clamping of framed and unframed glass-backsheet PV modules is compared by FEM modelling in regards to the mechanical stress in the solar cells due to mechanical load from the glass side.

The deflection shape of the two concepts shows a different pattern. Due to supporting at four local positions, the unframed laminate has a wave-like deflection. The support which the frame provides to the laminate along all edges leads to an elliptic deflection. Consequently the area of high tensile stresses is larger in frameless PV laminates compared to framed PV modules.

With 142 MPa compared to 218 MPa at 2400 Pa mechanical load, the results show significantly higher tensile stresses in the cells for the clamping of unframed PV laminates compared to clamped framed PV modules. With 16 of 60 solar cells showing tensile stresses above the fracture strength given in [3], it is most likely that cell cracking will occur for the clamping of unframed laminates. At the same load level, the stresses in a framed module are well below the fracture strength. Therefore the stiffness provided by the frame is sufficient to avoid cell cracking at a load of 2400 Pa.

At a load of 5400 Pa tensile stresses above the fracture strength arise in both concepts. With 52 of the 60 solar cells in the clamped laminate showing critical stress values, severe cell cracking is expected. This is reduced by the frame to 12 out of 60 cells with significantly lower maximum values of 268 MPa compared to 443 MPa. A further stress reduction can be achieved by replacing the backsheet with an additional glass. This will be investigated by the presented model in the future. Also further mounting technologies as well as PV module concepts can easily be investigated.

The presented results prove that, from a mechanical point of view, a frame is essential for glass-backsheet laminates.

5 REFERENCES

[1] U. Eitner, S. Kajari-Schroeder, M. Koentges, and H. Altenbach, "Thermal stress and strain of solar cells in photovoltaic modules," in *Advanced Structured Materials*, vol. 15, *Shell-like Structures: Non-classical Theories and Applications*, H. Altenbach and V. A. Eremeyev, Eds, Berlin/Heidelberg: Springer, 2011.

[2] S. Kajari-Schröder, I. Kunze, U. Eitner, and M. Koentges, "Spatial and orientational distribution of cracks in crystalline photovoltaic modules generated by mechanical load tests," *Solar Energy Materials and Solar Cells*, vol. 95, no. 11, pp. 3054–3059, 2011.

[3] F. Kaule, W. Wang, and S. Schoenfelder, "Modeling and testing the mechanical strength of solar cells," *Solar Energy Materials and Solar Cells*, vol. 120, Part A, pp. 441–447, 2014.

[4] S. Kajari-Schröder, I. Kunze, and M. Köntges, "Criticality of cracks in PV modules," *Energy Procedia*, vol. 27, pp. 658–663, 2012.

[5] J. Wendt, M. Träger, M. Mette, A. Pfennig, and B. Jäckel, "The link between mechanical stress induced by soldering and micro damages in silicon solar cells," in *Proc. of the 24th EUPVSEC*, 2009, pp. 2705–2708.

[6] B. Lalaguna, P. Sánchez-Friera, H. Mäckel, D. Sánchez, and J. Alonso, "Evaluation of stress on cells during different interconnection processes," in *Proc. of the 23rd EUPVSEC*, 2008, pp. 2705–2708.

[7] U. Eitner, "Thermomechanics of photovoltaic modules," Dissertation, Zentrum für Ingenieurwissenschaften, Halle-Wittenberg, Martin-Luther-Universität, Halle, 2011.

[8] U. Eitner, S. Kajari-Schröder, M. Köntges, and R. Brendel, "Non-Linear Mechanical Properties of Ethylene-Vinyl Acetate (EVA) and its Relevance to Thermomechanics of Photovoltaic Modules," in *Proc. of the 25th EUPVSEC/ 5th World Conference on Energy Conversion*, 2010, pp. 4366–4368.

[9] *IEC 61215: 2005-04. Crystalline silicon terrestrial photovoltaic (PV) modules - Design qualification and type approval*, 2005.

[10] J. Schicker, C. Hirschl, and R. Leidl, "Effect of PV Module Frame Boundaries on Stresses in Solar Cells," *Journal of Energy Challenges & Mechanics*, vol. 1, no. 3, pp. 155–162, 2014.

[11] f|solar, *data sheet f | solarfloat*. Sülzetal: f | solar GmbH.

[12] R. B. Roberts, "Thermal expansion reference data: silicon 300-850 K," *Journal of Physics D: Applied Physics*, vol. 14, no. 10, pp. L163, 1981.

[13] R. B. Roberts, "Thermal expansion reference data: silicon 80-280K," *Journal of Physics D: Applied Physics*, vol. 15, no. 9, pp. L119, 1982.

[14] W. M. Haynes, *CRC handbook of chemistry and physics*: CRC press, 2014.

NON-STATIONARY OUTDOOR EL-MEASUREMENTS WITH A FAST AND HIGHLY SENSITIVE INGAAS CAMERA

Jens Adams[1], Bernd Doll[1], Claudia Buerhop[1], Tobias Pickel[1], Janine Teubner[1], Christian Camus[1], Christoph J. Brabec[1,2]
[1] Bavarian Center for Applied Energy Research, Haberstraße 2a, 91058 Erlangen, Germany
[2] Materials for Electronics and Energy Technology (i-MEET), Friedrich-Alexander-Universität Erlangen-Nürnberg, Martensstraße 7, 91058 Erlangen, Germany
E-mail: bernd.doll@zae-bayern.de

ABSTRACT: A fast and non-destructive failure inspection is essential to ensure the reliability of PV plants. This might only be achieved by using highly sensitive imaging techniques which can detect local electrical losses in solar modules. Here, EL imaging used for outdoor PV inspection gained large interest, since it can be used to characterize local cell damages, series resistances, and shunts in solar modules. However, currently available EL-cameras base on Si-CCD-detectors. Besides high spatial resolutions they exhibit long image integration times (IIT) of about 1 s for obtaining evaluable EL-images. To overcome this limitation we present, for the first time, an InGaAs camera which enables a complete EL-characterization of field-installed PV-modules with IITs between 1 ms and 20 ms. The short IITs result in frame rates of about 100 Hz which even offer the opportunity of recording dynamic EL-movies and daylight Electroluminescence Lock-In (ELLI) measurements. To test the outdoor applicability we compared EL-movies of the InGaAs camera with EL-images of a Si-CCD camera and show that all relevant power reducing cell failures (e. g. cracks) can be detected and characterized.

Keywords: Performance and Reliability measurement, Electroluminescence of PV Modules, Energy rating, Degradation, Ageing, Lifetime Characterization of PV Cells and Modules

1. INTRODUCTION

A fast and non-destructive failure inspection is essential for ensuring the reliability of PV plants. This might only be achieved by using highly sensitive imaging techniques which can detect local electrical losses in solar modules. Here, EL imaging used for outdoor PV inspection gained large interest, since it can be used to characterize local cell damages, series resistances, and shunts in solar modules [1–3].

State-of-the-art EL-imaging methods suffer from large image integration times (~1 s) as well as the need for a reference measurement for reducing unwanted noise signals and to obtain evaluable EL-images from PV modules. Furthermore, Si-CCD detectors show only a weak spectral sensitivity in the EL emission range of Si solar cells, since these detectors are optimized for the spectral range of visible light. All this limits a fast and effective characterization of large PV stations comprising thousands of PV modules as well as the integration of Si-CCD cameras into unmanned aerial vehicle based quality control systems. Despite a spatial resolution of less than 1 MPixel of the InGaAs detector we demonstrate evaluable EL-images with image integration times between 1 ms and 20 ms which show all important cell features. The short image integration times allow recoding EL-movies and the integration of InGaAs cameras into non-stationary characterization tools as well as the characterization of non-stationary and transient current distributions in solar cells and modules. With our work we want to demonstrate the first step of an EL-camera which can be easily integrated into a fully automatized unmanned aerial vehicle (UAV) suitable as failure control tool for large area in-flight PV station inspection. Therefore, the purpose of our work is to quantify the physical and optical limits of a fast InGaAs EL-camera which provides evaluable EL images and measurements with IIT of about 1 ms and a spatial resolution of the detector which is less than 1 MPixel. Furthermore, we want to increase the knowledge of fast EL-cameras and measurements for outdoor applications and show that our results can also be used for understanding transient changes of non-stationary current injection into solar cells, modules, and module strings. Furthermore the Electroluminescence Lock-In (ELLI) method is introduced for outdoor daylight applications. Previously, it was shown that this imaging technique is sensitive enough to detect not only several radiative recombination processes in Si solar cells but also moisture induced degradation phenomena in organic solar cells [1,4]

2. TECHNICAL DETAILS

For comparison we performed EL measurements with two different EL camera systems. The first system consisted of an EQUUS 327k N EL-camera (IRCAM GmbH, Erlangen, Germany), equipped with an InGaAs focal plane array (FPA) detector. The InGaAs detector is highly responsive in the spectral range between 0.8 µm and 1.6 µm with a spatial resolution of 640 x 512 pixels. Therefore, the frame rate can be controlled between 100 Hz for 1 ms of IIT and 50 Hz for 20 ms of IIT, respectively. The EL-camera was equipped with a 25 mm focal lens imaging system featuring a spectral transparency >91% (IRCAM GmbH, Erlangen, Germany).

As reference EL-camera we used a standard GE 2048 2048 FI EL-camera (greateyes GmbH, Berlin, Germany), equipped with a magnesium fluoride based and front illuminated charge coupled device (CCD) detector. The CCD and FPA detector is cooled by a peltier element with a spatial resolution of 2048 x 2048 pixels. The sensor is highly responsive in the spectral range between 400 nm and 1000 nm with an absorption maximum at 700 nm. The GE 2048 2048 FI EL-camera was equipped with a planar 1.4/50 ZF-IR lens (Carl Zeiss AG, Oberkochen, Germany). To stimulate EL radiation of a module string we connected 13 modules in series to a module string and used a PS 8000 3U power supply (Elektro-Automatik, Viersen, Germany) for current injection. For current injection of singles modules we used a BOP-1000 power supply (KEPCO Inc., Flushing, NY, USA) which can be controlled by a trigger signal provided from the used EL-cameras.

3. EXPERIMENTAL DETAILS

In order to qualify our imaging systems for outdoor applications we designed a special test string, comprising 13 defective and non-defective silicon solar modules connected in series. The modules consist of 36 mono-crystalline solar cells, connected in series and arranged in a 4 x 9 matrix, providing a rated module power P = 80 W, an open circuit voltage V_{oc} = 22 V, and a short circuit current I_{sc} = 4.7 A at 1000 W/m² irradiation. The defective modules were chosen to guarantee a large variety of different defective cells which may be caused by environmental influences such as thermal cycling or physical stress. To demonstrate that InGaAs cameras can be easily integrated into non-stationary inspection tools we mounted the camera on a mobile carrier and recorded EL-movies by moving the camera continuously and parallel to the length of the module string.

4. RESULTS

In a first step, the optimum measuring parameters for outdoor EL measurements were investigated. Therefore, Fig. 1 shows the EL-signal of a non-defective cell located within a Si-PV solar module comprising 36 cells as a function of increasing image integration time (Fig. 1a) and increasing injection current (Fig. 1b). For all EL-measurements in Fig. 1 the EL-signal of the non-defective cell was averaged over the cell area. Fig. 1a) shows a linear increase (note logarithmical x-axis) of the measured EL-signal with increasing image integration time. Note, that during all EL-measurement the same injection current (4.8 A) was used. It can be seen that no signal change between both indoor (lab, darkness) and outdoor (night) EL-measurements of the InGaAs EL-camera was observed. The highest EL-signal was measured with an image integration time of 20 ms which was also the highest possible image integration time for the InGaAs EL-camera used here. Interestingly, to generate comparable EL-images with the Si CCD-camera image integration times between 1-2 s were needed (see therefore black dashed line in Fig. 1a). In a second experiment, shown in Fig. 1b), we investigated the dependence of the EL-signal of the same cell on the current injection. To guarantee comparable EL-measurements we chose an image integration time of 20 ms for the InGaAs EL-camera and an image integration time of 2000 ms for the Si-CCD camera.

For both EL-cameras a linear increase of the cell's EL-signal was observed with increasing injection current. Again both measurement series of the InGaAs EL-camera show a similar linear increase of the EL-signals. However, the slope of the EL-signals measured with the Si-CCD-camera is 1.3 times higher compared to the slope of the InGaAs EL-measurements. Here, one should note that the higher EL-signals measured with the Si-CCD camera were measured with a 100 times higher image integration time compared to the image integration time of the InGaAs camera.

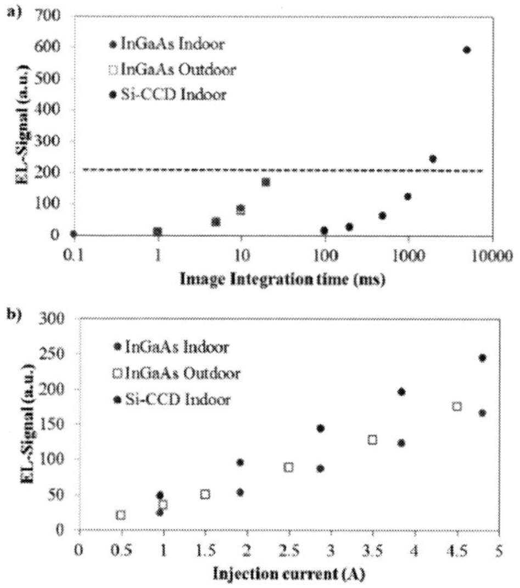

Figure. 1: EL-signal of a non-defective cell within a module as a function of different measurement parameters. a) EL-signal of the same cell measured with both an InGaAs and a Si-CCD EL-camera as a function of the image integration time. For all EL-measurements the injection current was set to 4.8 A (I_{sc}), b) Increase of EL-signal as a function of different injection currents. For the InGaAs EL-measurements the image integration time was set to 20 ms whilst for Si-CCD EL-measurements the image integration time was set to 2000 ms.

In order to compare the imaging quality of the two EL-cameras, Fig. 2a) shows one EL-image taken with the InGaAs camera and three EL images taken with the Si-CCD-camera with different image integration times but same injection current. Comparing the EL-images the InGaAs camera reveals - despite the lower spatial resolution - all relevant cell defects. Furthermore, from the magnification of a defective cell located in the module at position 2 x 6 see Fig. 2b), one can see that even smaller cell defects such as partial cracks can be visualized with the InGaAs camera.

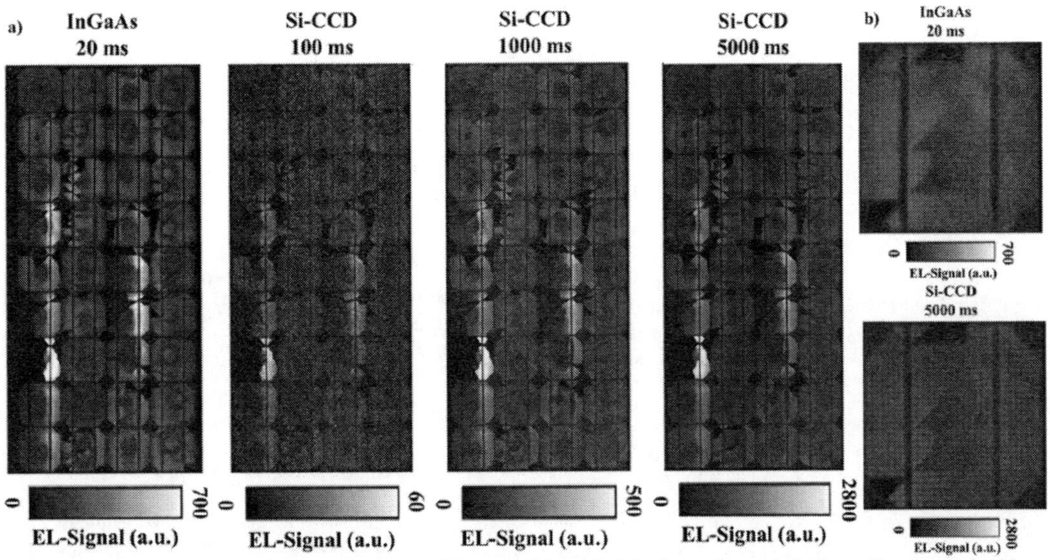

Figure. 2: Different EL-images of a defective module comprising 36 defective and non-defective cells measured with an InGaAs camera and a Si-CCD camera, respectively. a) EL-images of the same PV module measured with same injection current but different EL-cameras and image integration times. b) Magnification of a defective cell containing different defects which are commonly known for reduction of an efficient power extraction.

From the previous experiments discussed above (see Fig. 1) we determined the optimum measuring parameters (20 ms image integration time and 4,8 A injection current) for recording EL-movies of an outdoor module string. For this movie we placed the InGaAs camera on a mobile carrier and moved the camera in parallel to the length of the test string with a speed of approximately 0.4 m/s. Fig. 4 shows an image montage of single EL-images extracted from a recorded EL-movie (InGaAs) as well as a montage of EL-images taken with a Si-CCD camera. Similarly to Fig. 2, all important cell failures, affecting the electrical properties of cells, could be visualized and characterized thereafter. Note, that the InGaAs camera provides signal-to-noise ratios sufficient for neglecting additional reference measurements (see color scale in Fig. 4). Furthermore, the averaged EL-line scans in Fig. 3 prove the generallycomparable results of the contrast for the different camera systems.

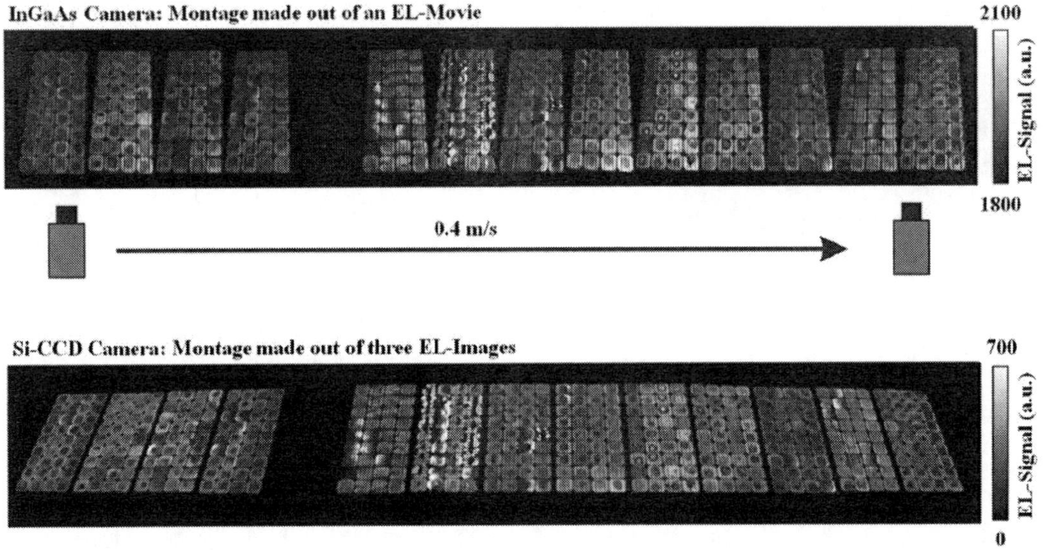

Figure. 3: Image montages reconstructed of an EL-movie, screenshots and conventional EL-images. (top) EL-movie was recorded with an image integration time of 20 ms and an injection current of 4.8 A. For the montage single images were extracted from a recorded EL-movie. (bottom) 3 single EL-images were measured with an image integration time of 2000 ms and an injection current of 4.8 A.

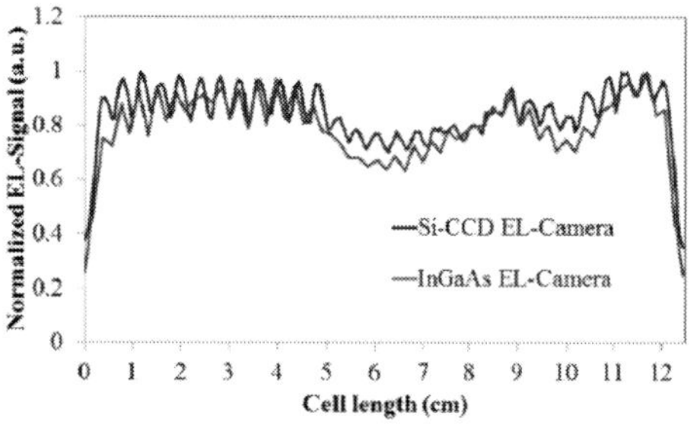

Figure. 4: Averaged EL-line scans over the active area of the same non-defective cell. The line scans were extracted from two EL-measurements of the same solar module which was EL-characterized with both a Si-CCD and an InGaAs EL-camera. For the EL-measurements same current injection conditions (I_{sc}) were used while the image integration time of the Si-CCD EL-camera was 2000 ms and 20 ms for the InGaAs EL-camera.

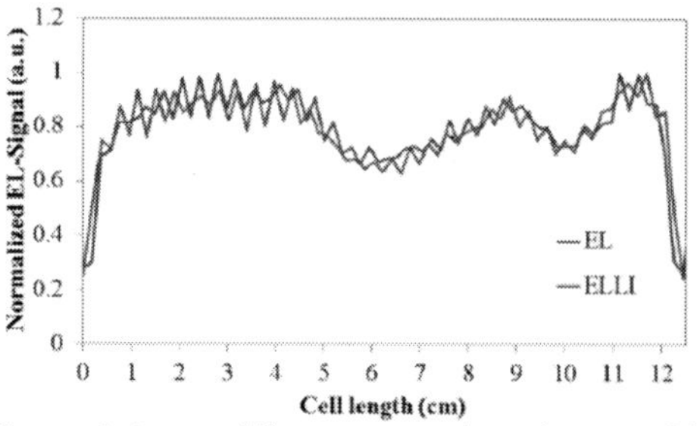

Figure. 6: Averaged line scans over the active area of the same non-defective cell. The line scans were extracted from either one EL or ELLI-measurement of the same solar module which was EL-characterized with the InGaAs EL-camera. For the measurements same current injection conditions (I_{sc}) were used. The image integration time of the camera was 20 ms, for the ELLI-measurement one period was recorded.

5. APPLICATIONS

The EL-movies can be used for daylight ELLI measurements. For this application the power supply can be controlled by a pulsed trigger signal provided from the InGaAs camera. In Fig. 5 the schematic sketch of the ELLI method is shown. Within each period *T* ideally four images with an integration time t_{int} are recorded. Every single image is multiplied with a rectangle correlation function. The first two images are recorded with applied power. These images are multiplied with the rectangle correlation function as "+1". The second two images are recorded without applied power. These images are multiplied with the rectangle correlation function as "-1". For each period the four correlated images are summed up. Then the final ELLI image is the result of calculating the average over all period images taken over the measurement time.

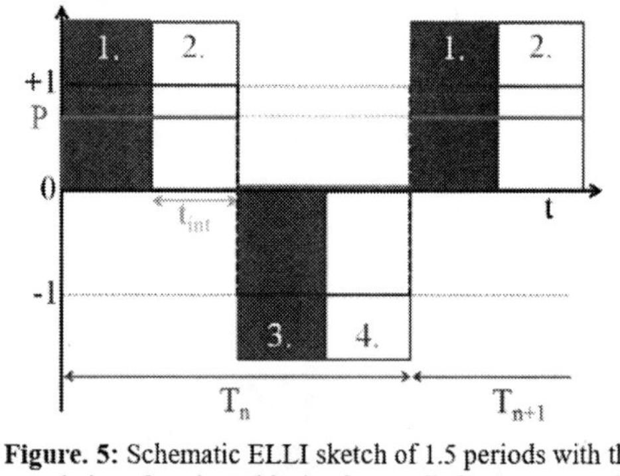

Figure. 5: Schematic ELLI sketch of 1.5 periods with the correlation function (blue), the applied power on the single module or module string (green) and the image integration time area (orange). The images are numbered from 1.-4. for each period (red n. period and dark red n+1. period)

For comparison of the methods indoor EL and ELLI measurements with the InGaAs camera are carried out. The normalized average line scans of the same solar module in Fig. 6 show matching results.

With the lock-in based method strongly noise reduced EL-images of PV-modules and module strings are recordable. Using filtering, disturbing noise signals which are not correlated to the lock-in signal are minimized. To identify the quality of the images typical the signal-to-noise ratio *SNR* is used. The *SNR* is calculated from the mean signal value m_S, the mean noise value m_N and the standard deviation of the noise σ_N [5]. To validate the ELLI method we calculated the *SNR* for the reference EL-image and for ELLI-images measured with one and 100 lock-in periods with the equation:

$$SNR = \frac{m_S - m_N}{\sigma_N} \qquad (1)$$

The calculated values for the *SNR* from EL-images measured indoor equal 33.95 for the reference EL-image, 24.78 for one ELLI period and 221.04 for 100 ELLI periods . Those *SNR*s indicate that ELLI improves the image quality. Thus, with periodical excitation of the sample and a lock-in algorithm, the *SNR* can be increased strongly, i.e. ELLI has the potential to visualize even low radiating structures of solar cells.

6. SUMMARY

The experiments presented in this work demonstrate the benefits of InGaAs EL-cameras for EL-characterization of field-installed Si solar modules. The short image integration times (1 ms - 20 ms) fulfil the requirements for integrating InGaAs cameras into non-stationary inspection tools such as UAVs or robots. This capability enables the fast, reliable, and efficient EL-characterization of large area PV power stations to determine the degradation, ageing and lifetime, even in daylight when used with the ELLI lock-in algorithm. Furthermore, choosing optimum measuring parameters two dimensional and transient EL-measurements are possible which can be used to characterize non stationary current distributions in solar cells and modules.

7. ACKNOWLEDGEMENT

We gratefully thank the German Federal Ministry for

Economic Affairs and Energy (BMWi) for financial funding this project (AQUAM, Förderkennzeichen 0325807A).

8. REFERENCES

[1] M. Köntges, S. Kajari-Schröder, I. Kunze, U. Jahn, in: 26th European Photovoltaic Solar Energy Conference and Exhibition, Hamburg, (2011) pp. 3290 – 3294.
[2] C. Buerhop-Lutz, R. Auer, C.J. Brabec, in: 28th European Photovoltaic Solar Energy Conference and Exhibition, Paris, 2013: pp. 3243 – 3245.
[3] S. Kajari-Schröder, I. Kunze, U. Eitner, M. Köntges, 95 (2011) 3054–3059.
[4] Adams, J., Salvador, M., Lucera, L., Langner, S., Spyropoulos, G.D., Fecher, F.W., Voigt, M.M., Dowland, S.A., Osvet, A., Egelhaaf, H.-J., Brabec, C.J., Advanced Energy Materials 5 (2015), 1501065
[5] R. C. Gonzales, R. E. Woods, Pearson. (2008) ISBN: 978-0131687288, p. 354f

IMPEDANCE SPECTROSCOPY AND ITS POSSIBLE USE FOR DEFFECTS DETECTION

Ladislava Černá[1,2], Tomáš Finsterle[1], Pavel Hrzina[1,2] and Vítězslav Benda[1,]

[1]Department of Electrotechnology, Faculty of Electrical Engineering, Czech Technical University in Prague, Technicka 2, 166 27 Prague, Czech Republic.

[2] University Centre of Energy Efficient Buildings, Czech Technical University in Prague, Trinecka 1024, 273 43 Bustehrad, Czech Republic.

Phone: +420 224 352 090; E-mail: ladislava.cerna@fel.cvut.cz

finsttom@fel.cvut.cz, hrzinap@fel.cvut.cz, benda@fel.cvut.cz

ABSTRACT: A common problem to be solved during the operation of photovoltaic (PV) power plants is early detection of PV modules defects. Usual methods are based on both measurement of electric parameters and visualization of possible defects (thermography). The common electric measurement usually doesn't show the initial damage like less serious cracks inside the module. For these purposes is necessary to use some advanced visualization method like electroluminescence. But this requires either special equipment or the dark chamber. In this text is presented a new approach to revealing the hidden defects using impedance spectroscopy. The results of measurement are presented as well as comparison between the individual methods.
Keywords: Photovoltaic, Electroluminescence, Capacitance, Flash Test, Impedance Spectroscopy

1 INTRODUCTION

There exist many diagnostic methods used for PV cells and modules diagnostics. Most of them are based on the measurement of static parameters (I-V curve measurement, e.g. [1]) or currently also by visualization of defects (thermography, electroluminescence – e.g. [2]). Except these methods exist also methods for dynamic parameters extraction which are currently in the scope of researchers for its using in multijunction cells. When applying these methods they are used for shunt resistance and its propagation determination but there are also other advantageous properties of this characterization method which are shown in the following text.

2 PHOTOVOLTAIC MODULE

The photovoltaic (PV) cells can be modelled by PN junction behaviour, as is obvious from the Volt-Ampere characterization of cell [3]. PV module consists of cells which are serially connected in order to reach a sufficiently high voltage output, so the module can be modelled using same circuit and parameters which values represent the serial connection of the parameters of individual cells. Every cell (module) is characterized by its basic parameters which are obtained by measuring volt-ampere characteristics - short circuit current I_{SC} (A), open circuit voltage V_{OC} (V), shunt resistance R_{SH} (Ω) which represents defects inside the cell, series resistance R_S (Ω), which represents ohmic losses, maximum power P_{MAX}, also *MPP* (W) / maximum power point, fill factor *FF* (-), $FF = P_{MAX}/(I_{SC} \cdot U_{OC})$ and efficiency η (%) given by the ratio of P_{MAX} and cell's irradiation P_{IN}. Moreover, there exist also other parameters based on the AC

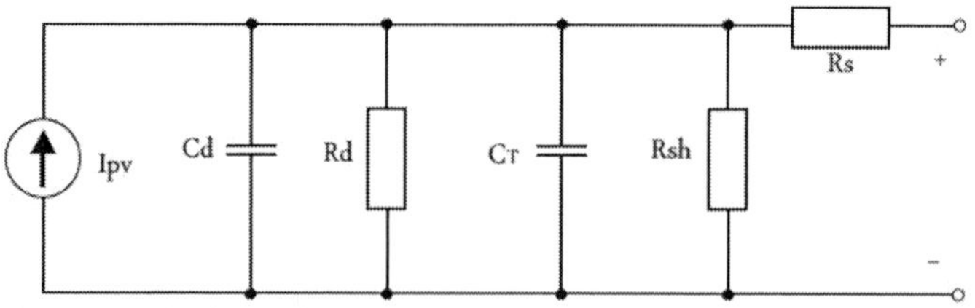

Figure 1: Dynamic equivalent circuit of PV cell (module)

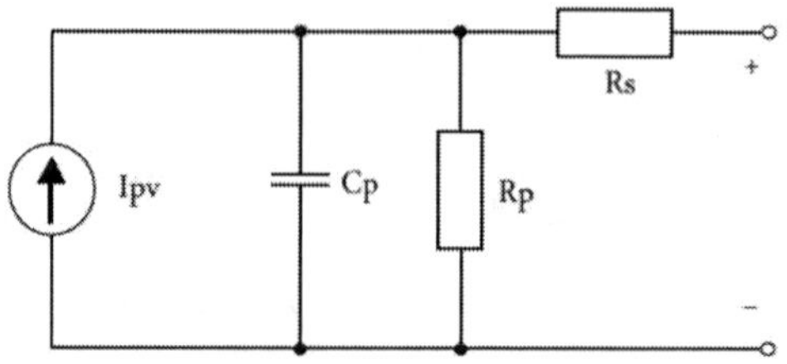

Figure 2: Simplified dynamic equivalent circuit of PV cell (module)

equivalent circuit like transition capacitance (C_T) and diffusion capacitance (C_D). A solar cell (module) can be modeled using the equivalent circuit shown at Fig. 1 [2]:

When combining C_D with C_T and R_D with R_{SH} the simplified dynamic model of PV cell, module respectively, can be obtained as shown on Fig. 2 [4].

The most problematic parameter is the capacitance which is the representation of physical capacitances inside the structure, particularly space charge capacitance of depletion region and capacitance due to minority carrier oscillation in response to the AC signal. Especially the second capacitance represents relatively difficult problem because it shows strong dependence on injection level. Diffusion capacitance is defined as $C_D = dQ/dV$, where Q represents injected positive charge and V the voltage [5]. For lower frequencies can be expressed by [6]:

$$C_D = \frac{e}{kT}\left(\frac{eL_p p_{n0}}{2} + \frac{eL_n n_{p0}}{2}\right)\exp\left(\frac{eV}{kT}\right) \quad (1)$$

where e is electron charge, k Boltzmann constant, T temperature, L_p and L_n diffusion lengths of holes and electrons and p_{n0} and n_{p0} initial minority carriers concentrations.

From the equation above it is apparent that injection level is tightly connected with minority carrier generation, it means, in the case of PV structures, also with irradiance. Because the irradiation of structure is never sufficiently constant, different results can be obtained when evaluating these parameters, so the results than are unrepeatable. For this reasons, it can be recommended to perform the AC measurement in dark conditions.

3 IMPEDANCE SPECTROSCOPY

The AC parameters of a solar cell can be measured either by a frequency domain technique or a time domain (transient response measurement) technique. In the frequency domain technique, a small signal is applied around the operating point and the steady state values of the AC parameters are measured. In the time domain technique, the cell voltage varies from short circuit to open circuit or vice versa.

The Impedance Spectroscopy is characterized by the measurement and analysis of some or all impedance related functions of an electronic device. In impedance spectroscopy, the complex impedance Z(ω)=R(ω) + jX(ω) of a device is measured directly within a large range of frequencies. A purely sinusoidal voltage with varying frequency is applied to the terminals of the device under test, and the phase shift and amplitude of the voltage and current signals are measured. The ratio between the applied voltage and the resultant current is calculated and this gives the impedance Z(ω) of the device under test. The plotting of R(ω) and X(ω) on a complex plane, in function of the varying frequency, gives the impedance spectrum of the device. In the photovoltaic area, this method is used for many years, mainly for AC parameters estimation such a R_P determination (e.g. [7], [8]), but in [9] is presented also the usage for the donor concentration determination. For measurement, the LCR meter is usually used.

Cole-Cole Diagram

The graphical representation of X(ω) vs. R(ω) is called the "Cole-Cole diagram" of the solar cell, and it gives the impedance arcs from which the different AC parameters can be extracted. The construction of Cole-Cole diagram allows simple evaluation of parameters and also the quality of the tested cell, module respectively [4]:

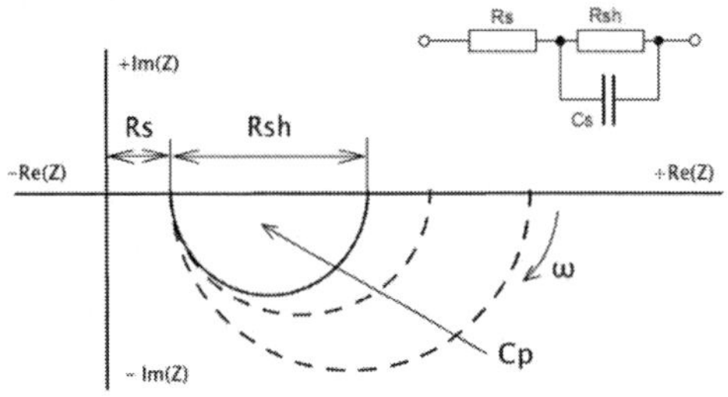

Figure 3: Cole-Cole diagram of PV cell (module)

4 EXPERIMENTS

In this research, impedance spectroscopy was used not for AC parameters extraction, but for defects evaluation. There were examined 9 monocrystalline and 9 multicrystalline silicon PV modules consisting of 4 serially connected PV cells. All modules were the new with the same dimensions, configuration, age and storage conditions. First, flash tests and electroluminescence was performed. Than all the modules were measured using impedance spectroscopy. For measurement, the LCR meter HP 4284A was used.

A B C

Figure 4: Example of electroluminescence images - typical mono c-Si (A) and multi c-Si (B), defective multi c-Si (C)

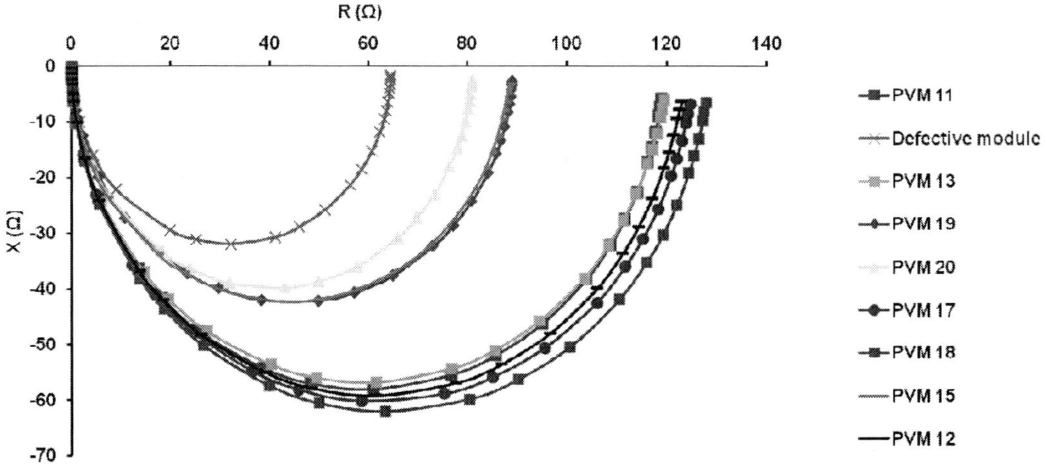

Figure 5: Cole-Cole diagrams of measured multicrystalline PV modules

Electroluminescence showed that all of the modules except one are without visible defects – see Fig. 4. One of multicrystalline PV modules showed cracks inside, but the affected parts are still electrically connected to the rest of module (no "dark places"). This presumption was confirmed also by *I-V* curve where affected module didn't show any significant difference from the rest of modules of the measured group. On the other hand, the impedance spectroscopy showed significant difference which is well observable at Cole-Cole diagrams. For easier evaluation, the average of measured values at "good" modules as well as parameters were calculated and compared with the "wrong" one (Table I).

Table I: Parameters obtained by I-V curve measurement

		sample no. 16	average value of samples 11 – 20
G	(kW/m^2)	1.0	1.0
I_{SC}	(A)	8.8	8.8
V_{OC}	(V)	2.6	2.5
η	(%)	11.2	11.3
FF	(%)	70.2	70.9
P_{MAX}	(W)	15.8	15.9
V_{Pmax}	(V)	2.0	1.9
I_{Pmax}	(A)	8.1	8.2
R_S	(Ω)	0.0	0.0
R_{SH}	(Ω)	66.5	72.8

5 CONCLUSION

The basic description of Impedance spectroscopy method was presented. The measurements using flash test, electroluminescence and Impedance Spectroscopy at two sets of modules (multi and monocrystalline silicon) were performed. The results proved that if there exist some distortion inside the module which does not influence the performance yet, IS can be very useful tool for revealing it, because this diagram takes into account also capacitances that can be strongly influenced by PV module internal structure. Although there still exist many problems connected with application of this method (temperature, configuration of measurement apparatus and injection level dependence), which will be closely investigated in the future, IS seems to be very promising in terms of timely diagnostics. Big advantage of this method is performing in dark and possibility of using for TF technologies which will be investigated in the future research as well.

6 ACKNOWLEDGMENT

This work has been supported by project of University Centre for Energy Efficient Buildigs, SGS SGS15/074/OHK3/1T/13 and Laboratory of Photovoltaic Systems Diagnostics.

7 REFERENCE

[1] IBRAHIM, A., Athapol KITIYANAN, Takashi FUYUKI a Ayumi TANI. 2006. *Analysis of Electrical Characteristics of Photovoltaic Single Crystal Silicon Solar Cells at Outdoor Measurements*. Editor Frank Kreith, D Goswami. New York: Nova Science Publishers, ix, 256 p. DOI 10.4236/sgre.2011.22020.

[2] HOYER, Ulrich; BUERHOP, Claudia; JAHN, Ulrike. Electroluminescence and infrared imaging for quality improvements of PV modules. In: *23rd European Photovoltaic Solar Energy Conference and Exhibition, Valencia*. 2008.

[3] LUQUE, Antonio; HEGEDUS, Steven (ed.). *Handbook of photovoltaic science and engineering*. John Wiley & Sons, 2011.

[4] CHEVNIDHYA, D., K. KIRTIKARA a C. JIVACATE Dynamic Impedance Characterization of Solar Cells and PV Modules Based on Frequency and Time Domain Analyses. *Trends In Solar Energy Research*. Hough, T.P. New York: Nova Science Publishers, 2006, s. 21-45. ISBN 1-59454-866-8; LCCN: 2005034740.

[5] KASAP, Safa O. *Principles of electrical engineering materials and devices*. McGraw-Hill Companies, 2000.

[6] FRANK, H.; ŠNEJDAR, V. *Principy a vlastnosti polovodičových součástek.* SNTL-Nakladatelství technické literatury. Prague 1976.

[7] THONGPRON, J.; KIRTIKARA, K. Voltage and frequency dependent impedances of monocrystalline, polycrystalline and amorphous silicon solar cells. In: *Photovoltaic Energy Conversion, Conference Record of the 2006 IEEE 4th World Conference on*. IEEE, 2006. p. 2116-2119.

[8] JOHNSON, Jay, et al. Creating dynamic equivalent PV circuit models with impedance spectroscopy for arc fault modeling. In: *Photovoltaic Specialists Conference (PVSC), 2011 37th IEEE*. IEEE, 2011. p. 002328-002333.

[9] AL ABDULLAH, K., et al. Low Frequency and Low Temperature Behavior of Si, Solar Cell by AC Impedance Measurements. *Energy Procedia*, 2012, 19: 183-191.

QUANTITATIVE OUTDOOR IMAGING: ANALYSIS OF SOLAR MODULES IN DAYLIGHT

Yael Augarten*, Andrew Wrigley, Andreas Gerber, Bart Pieters, and Uwe Rau
IEK5, Forschungszentrum Juelich, Wilhelm-Johnen Strasse, 52428 Juelich.

ABSTRACT: Photoluminescence imaging is often used to image photovoltaic cells and modules under relevant working conditions. However, it is underrepresented in outdoor imaging methods, owing to the small signal to noise ratio, and the requirement for steady state environmental conditions. Here we make use of the 'lock in' outdoor imaging technique, and present a modulation sequence for better stability against the influence of environmental conditions, allowing quantitative analysis of outdoor photoluminescence images. The differential photocurrent collection efficiency of a silicon heterojunction module is calculated at several working conditions between maximum power and open circuit conditions.
Keywords: Photoluminescence imaging, PV module, outdoor

1 BACKGROUND

Outdoor imaging techniques such as outdoor steady state thermal imaging methods are already well used by industry, providing useful data to improve the operating efficiency of solar plants, for example [1]. Outdoor luminescence imaging is less used, as the luminescence signal is weak, and under daylight, not visible with steady state imaging. Electroluminescence (EL) images can be captured at night time, or by individually measuring each module in a mobile lab, for example [2, 3], but not at realistic (illuminated) working conditions.

Somewhat recently, 'lock in' luminescence imaging methods have been developed to measure the emitted photoluminescence (PL) during the day, under actual working conditions [4]. This technique solves the low signal to noise problem, however measurement over a long time period is still necessary in order to get good image quality.

This is not so much of an issue for qualitative imaging, or relative comparison of individual cells within each module. However in order to obtain quantitative data, comparable measurements at different module working points are required. This generally requires that the environmental conditions (the irradiation and module temperature) are comparable over the whole measurement time.

This is also the case for making relative comparisons between modules; for example comparison of modules with respect to their position in the string.

We present, (1) measurement setup and modulation sequence for better stability against environmental conditions, and (2) Calculation of the differential photocurrent collection efficiency of outdoor modules, for a silicon heterojunction module. This parameter in installed modules can provide quantitative information on the efficiency of the modules and individual cells [7].

2 EXPERIMENTAL SETUP

The experimental setup is shown in figure 1. A Xenics Xeva InGaAs camera, with a resolution of 256 by 320 pixels, cooled to -40°C was used. A daylight filter, central wavelength of 1150 nm, with a FWHM of 30 nm, was mounted in front of the camera lens, to help increase the signal to noise ratio. Photoluminescence images were taken at a frequency of 30 frames per second.

The module voltage was modulated using a resistive load. Absolute irradiation was recorded using a temperature controlled reference cell, and module temperature was recorded using a PT100 sensor mounted on the rear side of the module.

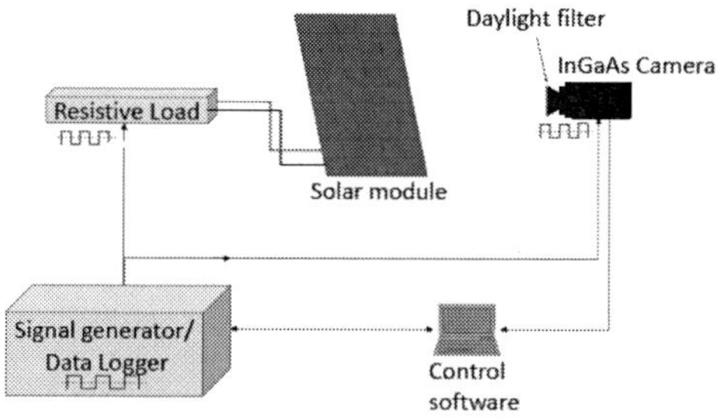

Figure 1: outdoor photoluminescence imaging measurement system

A sample of the voltage modulation cycle is shown in Fig. 2. In every cycle, a V_{oc} image and an I_{sc} image were taken as reference for the module temperature and irradiation, followed by a series of images between $0.7V_{oc}$ and V_{oc}. This modulation sequence proved to be an efficient method for collecting data under comparable environmental conditions.

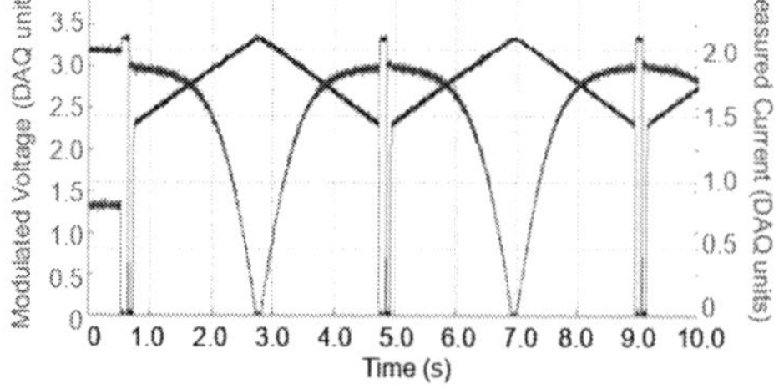

Figure 2: Sample modulation sequence showing modulated voltage (blue) and corresponding measured module current (green). Voltage and current units are given in the scale of the analog to digital converter.

3 THEORY AND METHOD

Photocurrent collection efficiency imaging using photoluminescence was recently developed [5]. In this work, the process by which solar cells convert of sunlight into electricity can be described as a three-stage process:

1. The absorption of light, and the subsequent creation of charge carriers

2. The collection of created carriers at the junction

3. Transport of the collected charge carriers from the junction to the external circuit.

The photocurrent collection efficiency (f_{pc}) is in this case defined as the efficiency of process (3). For a linearised network model, it is defined as:

$$f_{pc}(x,y) = \left.\frac{\delta V_j(x,y)}{\delta V_{Term}}\right|_{\delta I_L=0} \qquad (2)$$

Where (x,y) refers to a local point, δV_j is a small change in the junction voltage, and δV_{Term} is the corresponding change in the terminal voltage, provided the light generated current (I_L), is held constant.

Equation 2 was originally derived by Wong and Green [5], and can be used to calculate the differential photocurrent collection efficiency from PL images. An approximation to the method has also been derived in [6], and calculation of the f_{pc} using both [5] and [6] has been evaluated for single cells and series-connected modules [7]. In this work, only the results using the Wong method are presented.

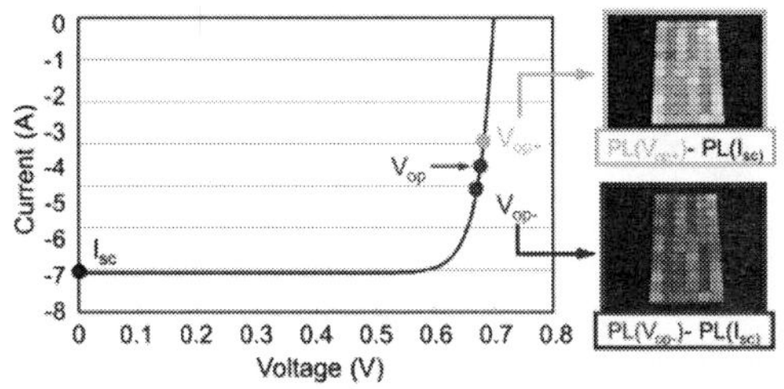

Figure 4: Graphical representation of the three PL images required for calculating f_{pc} of V_{op}, using the Wong [4] method.

The procedure for calculating f_{pc} from PL images using Eq. 1 requires 3 PL images, shown in Figure 4. For an arbitrary operating point, V_{op}, the required images are: $PL(V_{op+})$, and $PL(V_{op-})$, and an image $PL(I_{sc})$. The short circuit image is subtracted from the two $V_{op\pm}$ images, which are then converted to voltage images using the relationship:

$$V_j = V_{th} \cdot \exp\left(\Phi_{PL}/C\right) \qquad (3)$$

Where Φ_{PL} is the average module PL signal, Vj is the local junction voltage, V_{th} is the thermal voltage at the measured module temperature, and C is the proportionality constant, equal to the emission from an ideal blackbody multiplied by the sample EQE [8]. C is assumed to be constant across the module, and is calculated using Eq.3, with the measured module voltage, V_{module} in place of V_j. The module f_{pc} image is then calculated using Eq. 2.

4 RESULTS AND DISCUSSION

4.1 Measured data

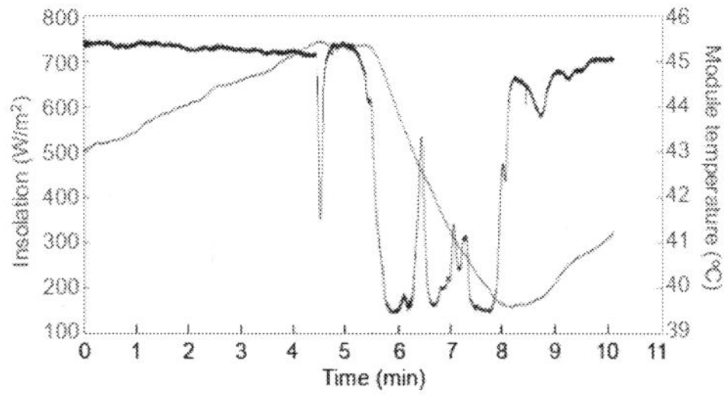

Figure 3: Environmental conditions for ten minutes measuring time, showing the irradiation intensity measured by the reference cell (blue), and the module rear side temperature (green).

Figure 3 shows the measured irradiation (blue) and temperature (green) over ten minutes. Irradiation varied between 150 W/m^2 and 750 W/m^2, and module temperature varied between 39.5°C and 45.5°C over the measurement period. Such large variation in both irradiation and temperature, is clearly non ideal, and makes the acquisition of sufficient data for good quality, quantitative measurements difficult. The data used for the quantitative calculation in section 3.2 was taken from the first 4 minutes (8000 frames) of measured data, as marked in Figure 3. In this region, irradiation was 730 W/m^2 ± 15 W/m^2, and module temperature was 44 °C ± 1 °C. The final PL images used in the calculation of the photocurrent collection efficiency were thus extracted from the measured data by the following:

$$PL(V_{op}) = \frac{\sum_N PL_N(V_{op} \pm V_{err})}{N} \qquad (1)$$

Where N is the number of measured cycles (as shown in Fig. 2), V_{op} is the desired module voltage, and V_{err} is the error range about V_{op}. This error range is necessary as the operating voltage is only indirectly controlled by the electronic load. i.e. The voltage modulation is precisely controlled, but the exact voltage at which the images are taken is not. In this work, the maximum V_{err} was 0.36% of V_{op}.

4.2 Photocurrent collection efficiency calculation

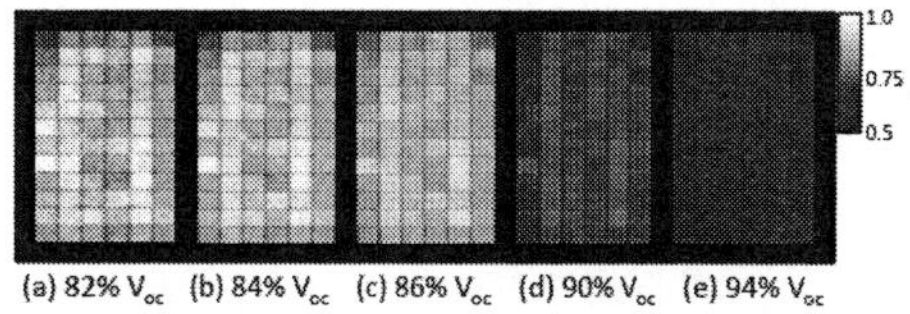

Figure 5: Photocurrent collection efficiency f_{pc} images for operating voltages between 82% and 94% of the module V_{oc}. Images were calculated from outdoor photoluminescence images of a silicon heterojunction module, using the Wong method [2]. Total imaging time was 4 minutes. All images are scaled between 0.5 and 1.

Fig. 5 shows the f_{pc} calculated from outdoor data, for V_{op} between 82% and 94% of the module V_{oc}. Closer to short circuit conditions, $f_{pc} \approx 1$, as most of the generated current is collected. As the module approaches open-circuit conditions, f_{pc} decreases. These results show the expected trend as simulated and measured in [7].

5 CONCLUSION

Quantitative analysis of outdoor PL images is difficult, as outdoor conditions are often unstable. To this end, we developed and demonstrated our modulation cycle. Using this cycle we were able to quantitatively analyse the outdoor operation of a silicon heterojunction module.

Using a total measurement time of 4 minutes, outdoor photocurrent collection efficiency images for operating voltages between 82% - 94% of the module V_{oc}, were calculated. The module behaves as expected.

REFERENCES

[1] Infratec, Thermographic inspection of photovoltaic installations, http://www.infratec.eu/thermography/application-area/predictive-maintenance/photovoltaic-thermography.html (accessed 05.02.2016)

[2] Solar Tester, www.solartester.nl, (accessed 28.04.2016)

[3] J. Coello, L. Pérez, F. Domínguez, M. Navarrete, On-site Quality Control of Photovoltaic Modules with the PV MOBILE LAB, *Energy Procedia*, **57**, pp89-98 (2014)

[4] L. Stoicescu, M. Reuter, J.H. Werner, DaySy: Luminescence Imaging of PV Modules in Daylight, *29th EUPVSEC Amsterdam*, pp 2553-2554 (2014)

[5] J. Wong and M. A. Green, From junction to terminal: Extended reciprocity relations in solar cell operation, *Phys. Rev. B* , **85**, 235205 (2012)

[6] U.Rau, V.Huhn, L.Stoicescu, M. Schneemann, Y.Augarten, A.Gerber, and B.E.Pieters, Photocurrent collection efficiency mapping of a silicon solar cell by a differential luminescence imaging technique, *Applied Physics Letters*, **105**, 163507 (2014)

[7] V. Huhn, A. Gerber, Y. Augarten, B.E. Pieters and U. Rau, Analysis of Cu(In,Ga)Se2 thin-film modules by electro-modulated luminescence, *J. Appl. Phys.*, **119**, 095704 (2016)

[8] U. Rau, Reciprocity relation between photovoltaic quantum efficiency and electroluminescent emission of solar cells, *Phys. Rev. B*, **76**, 085303 (2007)

IMPEDANCE CHARACTERIZATION OF PV MODULES IN OUTDOOR CONDITIONS

Oprea Matei I.[*1], Spataru Sergiu V.[1], Sera Dezso[1], Thorsteinsson Sune[2], Poulsen Peter B.[2], Andersen Anders R.[3], Basu Ronni[3], Frederiksen Kenn H. B.[4]
[*]mio@et.aau.dk , [1]Aalborg University, Department of Energy Technology
Pontoppidanstraede 111, DK-9220 Aalborg, Denmark
[2]Technical University of Denmark, Department of Photonics Engineering
Frederiksborgvej 399, DK-4000 Roskilde, Denmark
[3]EmaZys Technologies ApS
Green Tech Center, Lysholt Alle 6, DK-7100 Vejle, Denmark
[4]Kenergy
Grønningen 43, DK-8700 Horsens, Danmark

ABSTRACT: Impedance spectroscopy (IS) has been used for laboratory characterizations of photovoltaic (PV) technologies under well controlled conditions. This work applies IS for outdoor characterization of PV panels, in order to observe the effect of ambient conditions on the PV impedance spectrum, and further construct an impedance model that can link environmental changes to the model's parameters. To achieve this, a setup has been developed for long-term impedance spectra monitoring synchronized with accurate irradiance and temperature data. Preliminary results show clear correlation between the determined parameters and irradiance conditions.
Keywords: impedance spectroscopy, photovoltaic panel, photovoltaic string, outdoor characterization, AC model

1 INTRODUCTION

It has been shown before that impedance spectroscopy [1, 2] can be used as characterization technique [3-8] for solar cells and PV modules [9]. However, previous works typically focused on materials characterization in laboratory conditions (i.e. in dark or under controlled irradiance and temperature) and not performance or diagnostics. In this study, the focus is on identifying a correlation between environmental variations and equivalent PV module, or string, AC parameters.

Current-voltage (I-V) characterization is perhaps the most common diagnostic tool for both PV strings and modules, but IS methods may have the potential of detecting and locating additional faults and degradation from an early stage of development. The parameters extracted from the impedance model can be used in diagnostic methods, long term monitoring, or to identify arc faults and short-circuits in PV strings.

However, application of IS on operating PV strings, i.e. in outdoor conditions faces the challenge of uncontrolled environmental conditions. The actual irradiance and temperature will affect the overall impedance spectra of the PV string.

A straightforward effect of increased irradiance will be the reduction of the apparent parallel resistance, due to the forward characteristic of the parallel diode (see Fig 1).

On the other hand, the dependency of the parallel capacitance on the ambient conditions is less direct. The capacitance value varies with the operating conditions of the PV cells, meaning irradiance level as well as the operating point on the IV curve [6].

The purpose of this work is to investigate the effect of the ambient conditions on the impedance spectrum of PV modules. These results can be further used to detect system failures when these appear [10].

In a broader perspective, we aim to determine if IS can be an effective field test tool for detecting incipient degradation of the solar cells and modules, which can be difficult to detect outdoor by traditional characterization methods.

IS measurements have been performed outdoor on a crystalline silicon (c-Si) PV module and a string, using a commercial IS analyzer for PV modules and arrays [11].

In order to eliminate the effect of the load (operating point) on the measurement of the parallel capacitance and resistance, all our measurements have been performed in no load (open circuit) conditions.

The measurement procedure is an automated test-setup that is able to continuously monitor the impedance spectra of several standard sized PV modules. The measurement data is processed through fitting for extracting the equivalent circuit parameters (Fig. 1) that are correlated to the information provided by irradiance sensors available on the setup.

The paper is structured as follows: a short description of the AC model used will be discussed in section 2, together with the description of the setup and the parameters used to perform the IS measurements. Further on, in section 3, the measurement results will be presented. Together with the measurement, the parameter fitting results will be discussed. At the end, the study's findings are concluded to offer an overview of this work.

2 METHODOLOGY

2.1 Modelling approach

The model considered for determining the parameters is a single diode one [12], as shown in Fig. 1 A). The corresponding AC model [13][6] is shown in Fig. 1 B), where the AC behavior of the diode in the equivalent circuit is modeled by two parallel capacitances (C_{diff} and C_{dep}) and an equivalent resistance (R_d). The two parallel capacitors and resistors can be represented by a single component of each type, (C_p and R_p) resulting in the simple AC equivalent circuit in Fig. 1 C), based on which the parameters C_p and R_p will be determined [10]. Due to the variability of both C_{diff} and C_{dep}, as well as R_d with the current through the diode D (Fig 1 A), both C_p and R_p will be affected by irradiance conditions as well as the operating point of the PV panel or string.

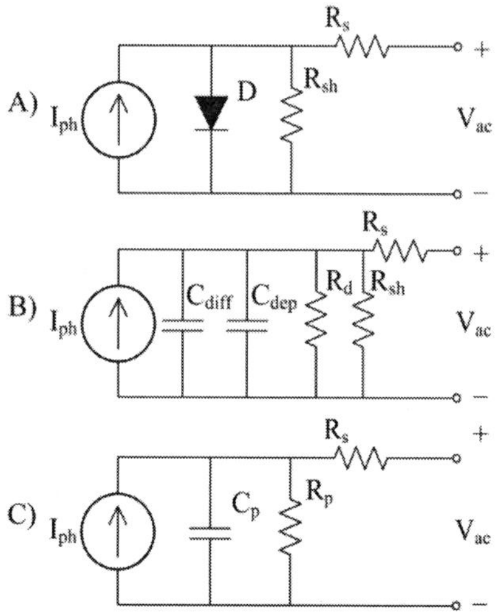

Fig. 1 The single diode model is considered for this study and is presented in A). This DC model can be translated into an AC model presented in B). The two capacitances are considered as a single parallel capacitance (C_p). The same approach is applied for the two resistances in parallel (R_p), resulting in C).

The series resistance (R_s) is considered known and will not be further discussed in this study. On the other hand, the diode's equivalent parameters are dependent on its operating point: R_d is large when the diode is blocked, while its value decreases significantly with increasing forward current. The C_{diff} is dominant when the diode is in conduction state and the C_{dep} can be observed when the diode is blocking state.

The photo-generated current (I_{ph}) is dependent on the irradiance and sets the operating point of the diode.

Eq. (1) describes how the parameters of the equivalent circuit (Fig. 1 C) are fitted form the actual measured data. In this study only the variation of the parameters with regard to the irradiance is considered.

The measurement of impedance is made under absolute (norm) and phase (angle) form. In order to apply the fitting, the results are transformed in real and imaginary values. The relations between the two forms of the impedance are presented in eq. (2) and (3). Fig. 3 is given as an example of such a measurement in two different irradiance conditions.

$$Z_{pv} = \left[R_s + \frac{R_p}{1 + \left(\omega R_p C_p\right)^2}\right] - i\left[\frac{\omega R_p^2 C_p}{1 + \left(\omega R_p C_p\right)^2}\right] \quad (1)$$

In eq. (1) the notations used are: Z_{pv} – PV panel impedance, R_s – series resistance, R_p – parallel resistance, C_p – parallel capacitance , ω – angular frequency ($\omega = 2\pi f$) and i – imaginary operator. The relation between the real and imaginary parts of the impedance and the norm – phase form is given by:

$$\left|Z_{pv}\right| = \sqrt{\left\{Z_{pv}\right\}_{Re}^2 + \left\{Z_{pv}\right\}_{Im}^2} \quad (2)$$

$$\theta = atan2\left(\left\{Z_{pv}\right\}_{Im}, \left\{Z_{pv}\right\}_{Re}\right) \quad (3)$$

2.2 Automatic test setup

Results have been achieved using a single c-Si module mounted on a rooftop installation, which is located in the proximity of an advanced solar irradiance measurement station. Furthermore a string of 12 PV panels composed by 60 c-Si cell has undergone the same routine, but in this case the environmental conditions have been recorded through the string inverter.

The measurements have been performed with a Z100 impedance analyzer that has been automated through software to perform and record impedance spectra at every 9 minutes, meaning around 6 measurements per hour.

The frequency range used for these measurements is from 200 Hz to 100kHz. The lower limit is motivated by hardware limitations. The upper limit is imposed by the increasing effect of the wires inductance as discussed in section 3.

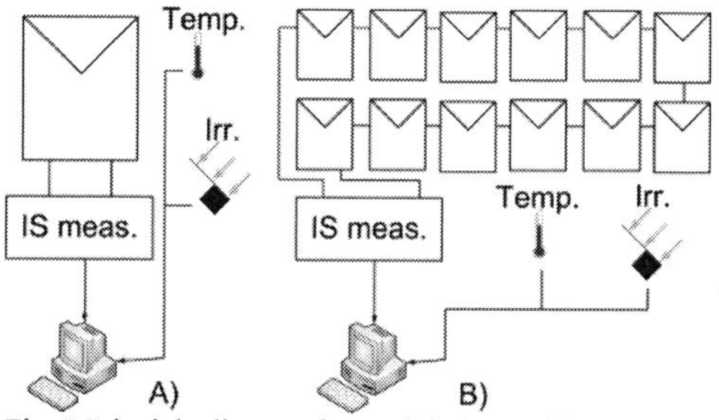

Fig. 2 Principle diagram for module (A) and for string (B) characterization. The measurements are performed over 24 hours using a Z100 impedance analyzer. Both the single panel and string's panels are c-Si modules.

3 RESULTS

Performing the measurements as stated at point 2.2 results in a multitude of curves as the ones presented in Fig. 3. Using the model depicted in Fig. 1 C) and characterized by eq. (1) the equivalent parameters of the PV module, or string, are determined. The focus is on the parallel elements of the circuit, R_p and C_p, thus the R_s will not be presented.

The variation of absolute value is similar in both cases, for the panel and for the string. The results are shown in Fig. 4 in the case of the single panel respectively Fig. 8 for the string. The maximum absolute value is greater in the case of the sting, but this was to be expected as there are more panels connected together. In the case of phase, the measurements also yield similar result. The panel's measurements are shown in Fig. 5, while the string's measurements are presented in Fig. 9. Although the general trend is similar, there is a strong inductive influence that can be observed in the case of the sting. This influence can be accounted by the more interconnecting cables used in the string's case. Electromagnetic interference can be observed in both phase measurements in the region of low impedance (around midday).

The fitting results show a change of the operation point of the PV characteristic that is documented in the literature. The parallel resistance, dominated in some cases by the diode's dynamical resistance, is decreasing with regard to irradiance values. Fig. 6 and Fig. 10

present the results for the panel and string.

The opposite can be said about the capacitance. As the PV starts generating current the capacitance grows in direct relation with the irradiance. This can be interpreted considering the model from Fig. 1 B). The capacitance of the depletion region of the diode diminishes as the diffusion capacitance grows, exceeding the value of the first. The direct relation of the Cp with regard to irradiance is shown in Fig. 7 for the panel and in Fig. 11 in the case of the string.

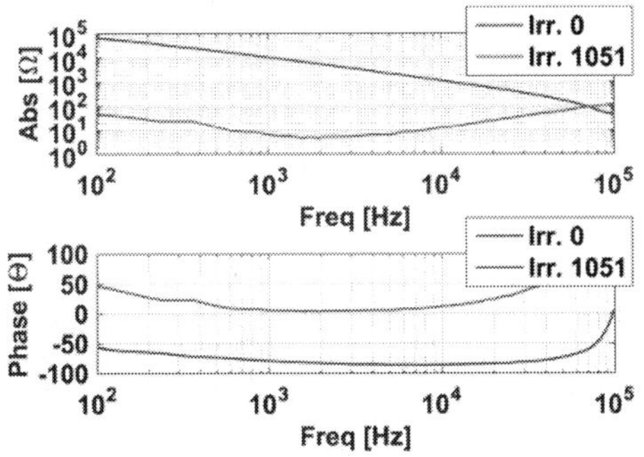

Fig. 3 Impedance spectrum measurement of a PV string, composed by absolute value and phase. Two cases are presented for comparison, and to further ease the understanding of next figures. The first measurement is performed in dark (0 W/m^2), during the night at 23:53 hours. The second case is a measurement made during high irradiance (1051 W/m^2), during mid-day at 11:58 hours.

The difference is visible for both, absolute and phase values. The norm decreases almost four orders of size, while the phase's characteristic changes from a capacitive one to more resistive one. It is worth mentioning that over a frequency of 50kHz the inductance of the circuit becomes dominant in relation with the capacitance of the PV modules, and the characteristic becomes strongly inductive. Thus one can conclude that the characteristic of the PV's changes significantly with the irradiance variation. In Fig. 4, Fig. 5, Fig. 8 and Fig. 9 will be shown measurements of impedance spectra over one day, creating a more complex 3D representation of absolute value and phase.

3.1 Panel characterization

The results of the IS measurement are presented in Fig. 4 and Fig. 5 for absolute value and phase. The figures show the change in the operating characteristic of the PV panel that goes from dark conditions to light conditions in the open voltage point.

The fitted parameters C_p and R_p reflect very well this transition. The resistance is shown in Fig. 6, while the capacitance is shown in Fig. 7. Both the capacitance as well as the resistance experiences a change of a few order of size, even though their changes are not the same. The capacitance is direct proportional with the irradiance, while the resistance is inverse proportional.

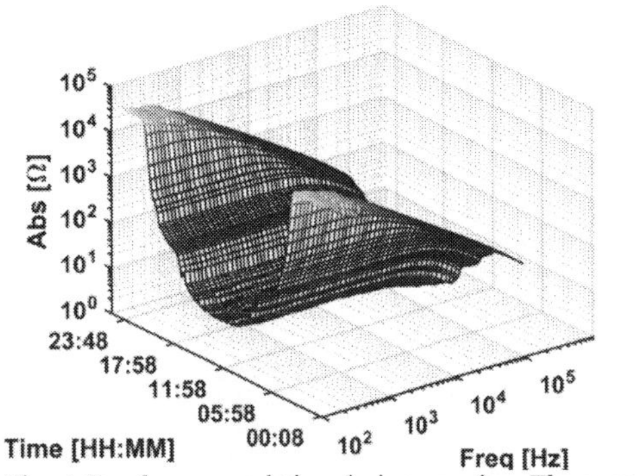

Fig. 4 Panel norm variation during one day. The x axis represents the measurement frequency range. The y axis shows the time point when the measurement has been performed. The z axis represents the absolute value. The shape of the impedance spectra corresponds with the expected variation since the PV panels is entering conduction, thus reducing its impedance. The values can be considered characteristic for a c-Si.

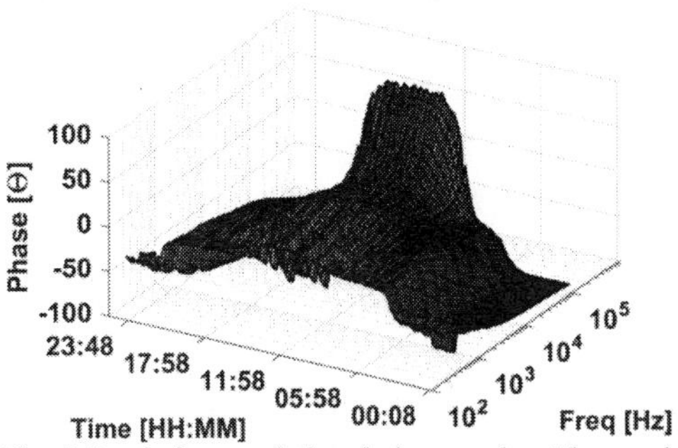

Fig. 5 Panel phase variation during one day. The x axis represents the measurement frequency range. The y axis shows the time point when the measurement has been performed. The z axis represents the phase value. The phase transformation during the day is significant, changing from a capacitive one, in the dark (early hours of the morning and night), to a more resistive-inductive during the day. A strong inductive influence can be observed at measurements during light conditions over the 50kHz threshold. Noise from EMI can be observed in the low impedance region, during midday at low frequency measurements.

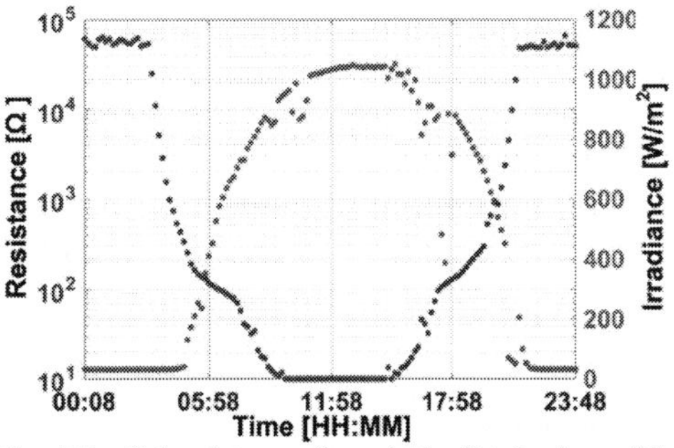

Fig. 6 Parallel resistance of panel. The fitted values of the IS measurements are presented on the left y axis, while the right y axis shows the irradiance variation during the day. The x axis presets the time point when the

measurement used to determine the parameter has been performed. The R_p varies inverse proportional with the irradiance. The lower value of saturation around 10Ω is set by the fitting algorithm used for the determination.

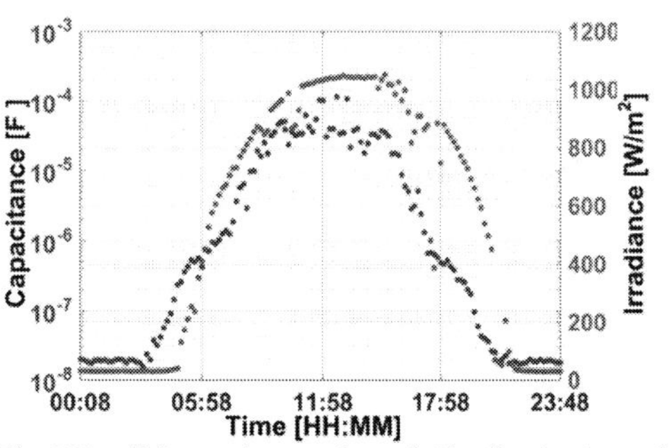

Fig. 7 Parallel capacitance of panel. The fitted values of the IS measurements are presented on the left y axis, while the right y axis shows the irradiance variation during the day. The x axis presets the time point when the measurement used to determine the parameter has been performed. The C_p parameter is direct proportional to the irradiance variation.

3.2 String Characterization

String measurements yield similar results as in the case of the panel. These are show in Fig. 8, for the absolute value, and in Fig. 9 for the phase. For the sting measurements, the absolute value tends to have higher values in the low frequency area. This can be explained as a result of having more panels connected together.

The phase also experiences a much more inductive character due to the interconnecting cables. As in the case of the panel, there is some EMI present in the low frequency area in the region of low impedance.

The fitted results for C_p and R_p show the same general trend as in the case of the panel determination. These are presented in Fig. 10 for the resistance and Fig. 11 for the capacitance. The values differ though from the ones seen in the case of the panel. The parallel capacitance is higher, while the parallel resistance is lower.

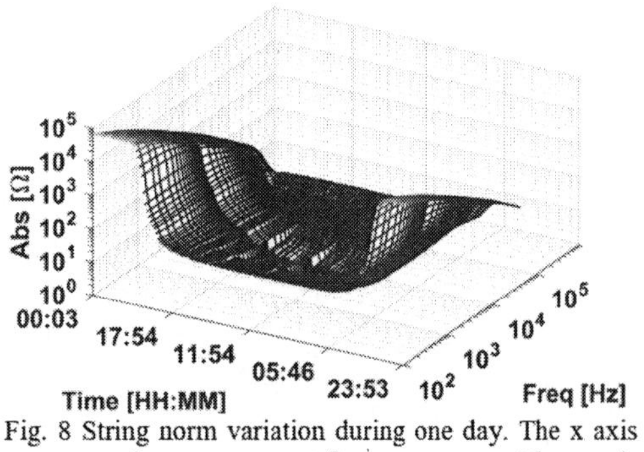

Fig. 8 String norm variation during one day. The x axis represents the measurement frequency range. The y axis shows the time point when the measurement has been performed. The z axis represents the absolute value. The measurement is consistent with the panel measurement presented in Fig. 4.

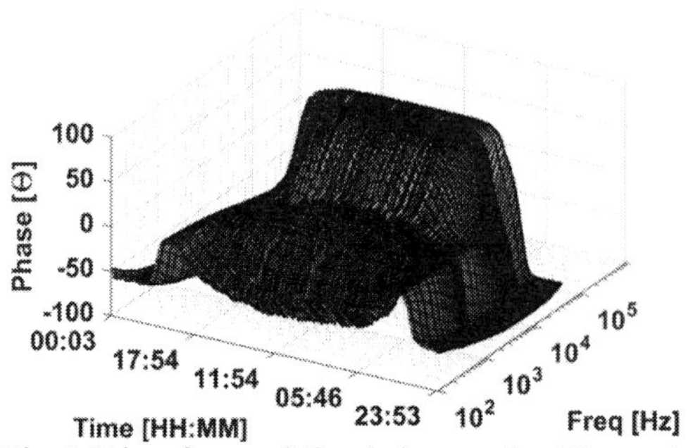

Fig. 9 String phase variation during one day. The x axis represents the measurement frequency range. The y axis shows the time point when the measurement has been performed. The z axis represents the phase value. The variation follows the same trend as the panel's phase in Fig. 5. But the more inductive values of the cables used for connection has a visible effect at a frequency higher than 10kHz in low impedance conditions.

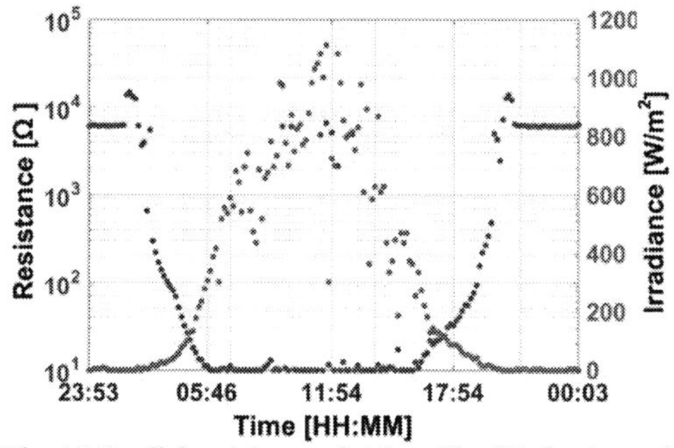

Fig. 10 Parallel resistance of string. The fitted values of the IS measurements are presented on the left y axis, while the right y axis shows the irradiance variation during the day. The x axis presets the time point when the measurement used to determine the parameter has been performed. Once again the string determination is consistent with the panel's determination. The trend is similar, although the lower values of R_p in dark conditions can be caused by the different characteristics of each of the panels in the string.

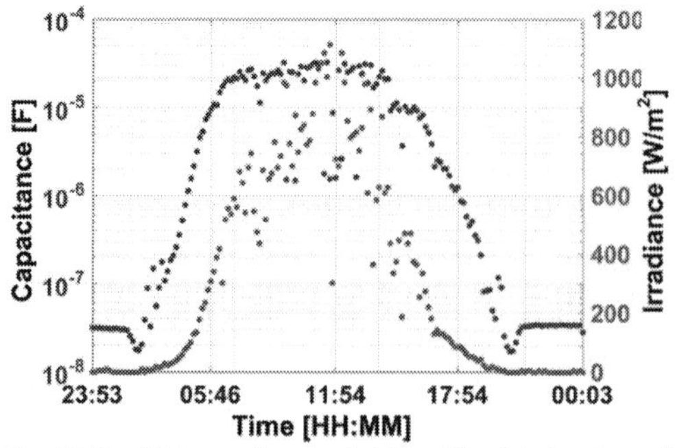

Fig. 11 Parallel capacitance of string. The fitted values of the IS measurements are presented on the left y axis, while the right y axis shows the irradiance variation during the day. The x axis presets the time point when the measurement used to determine the parameter has been

performed. In the case of the sting measurement, a greater value of the capacitance can be observed.

The IS measurements are consistent with the laboratory results simulating different irradiance conditions [4, 9]. Furthermore the parameters variation, both in the case of the parallel resistance as well as in the case of parallel capacitance is as expected from the previous studies [6].

Both the parallel capacitance and the parallel resistance show a very strong variation with irradiance. In both cases the change is around 3-4 orders of magnitude, therefore this significant variation needs to be accounted when AC impedance test are performed in outdoor conditions.

4 CONCLUSION

The results show that IS can be applied to extract equivalent circuit parameters of PV panels and strings in the field, however environmental conditions strongly affect the extracted key parameters, which must be taken into account when drawing conclusion about the state of the panel or string. It is the aim that this analysis can give basis to a dynamical model that can use environmental measurements to account for the PV panel's parameters variation, to be further used for diagnostics.

Acknowledgement
This research was partially funded by the Danish Energy Agency within the EUDP program with the project journal number 64014-0507 “New technology for localization and characterization of faults in solar panels”.

5 REFERENCES

[1] J. R. Macdonald, "Impedance spectroscopy," *Annals of Biomedical Engineering,* vol. 20, pp. 289-305, 1992.

[2] E. Barsoukov and J. R. Macdonald, *Impedance Spectroscopy*: Wiley, 2005.

[3] L. a. Mallette and R. L. Phillips, "Modeling solar cells for use as optical detectors: background illumination effects.," *Applied optics,* vol. 17, pp. 1786-1788, 1978.

[4] R. L. Mueller, M. T. Wallace, and P. Iles, "Scaling nominal solar cell impedances for array design," in *Photovoltaic Energy Conversion, 1994., Conference Record of the Twenty Fourth. IEEE Photovoltaic Specialists Conference - 1994, 1994 IEEE First World Conference on* vol. 2, ed, 1994, pp. 2034-2037 vol.2.

[5] R. Anil Kumar, M. S. Suresh, and J. Nagaraju, "GaAs/Ge solar cell AC parameters under illumination," *Solar Energy,* vol. 76, pp. 417-421, 2004.

[6] I. Mora-Seró, G. Garcia-Belmonte, P. P. Boix, M. a. Vázquez, and J. Bisquert, "Impedance spectroscopy characterisation of highly efficient silicon solar cells under different light illumination intensities," *Energy & Environmental Science,* vol. 2, p. 678, 2009.

[7] Y. Y. Proskuryakov, K. Durose, M. K. Al Turkestani, I. Mora-Seró, G. Garcia-Belmonte, F. Fabregat-Santiago, *et al.*, "Impedance spectroscopy of thin-film CdTe/CdS solar cells under varied illumination," *Journal of Applied Physics,* vol. 106, pp. 1-9, 2009.

[8] D. Chenvidhya, K. Kirtikara, and C. Jivacate, "A new characterization method for solar cell dynamic impedance," *Solar Energy Materials and Solar Cells,* vol. 80, pp. 459-464, 2003.

[9] D. Chenvidhya, K. Kirtikara, and C. Jivacate, "PV module dynamic impedance and its voltage and frequency dependencies," *Solar Energy Materials and Solar Cells,* vol. 86, pp. 243-251, 2005.

[10] J. Johnson, D. Schoenwald, S. Kuszmaul, J. Strauch, and W. Bower, "Creating dynamic equivalent PV circuit models with impedance spectroscopy for arc fault modeling," *Conference Record of the IEEE Photovoltaic Specialists Conference,* pp. 002328-002333, 2011.

[11] (2016). *EmaZys Z100□□ PV Test Instrument.* Available: http://emazys.com/z100-pv-test-instrument/

[12] T. Markvart and L. Castañer, *Practical Handbook of Photovoltaics: Fundamentals and Applications: Fundamentals and Applications*: Elsevier, 2003.

[13] S. M. Sze and K. K. Ng, *Physics of Semiconductor Devices*: Wiley, 2006.

LIGHT INDUCED DEGRADATION OF P-MONO PERC FROM INGOT, CELL, MODULE TO SYSTEM

Maoyi Chang, Haomin Chen, C. H. Hsueh and Chienyu Chen
AU Optronics Corporation
No.1 JhongKe Rd, Taichung, 40763, Taiwan

ABSTRACT: To ensure the massive deployment for PERC technology application, light induced degradation (LID) is one of the most crucial issue. The investigation of P-mono PERC LID is performed from ingot, solar cell, module to system performance. All cells fabricated with four different ingot sources have high possibility of over 3% LID in output power, and the oxygen concentration is not correlated with power drop. Only one of the seven cell types surveyed from the worldwide market has LID less than 3%. The process window for LID is optimized in AUO by inspecting the external quantum efficiency (EQE) for each cell processing step. In module level, indoor light soaking test shows smaller LID, while it owns much faster LID recovery rate than outdoor test. Module LID is larger when it is with resistive load than without mode (at Voc) for both indoor and outdoor test condition. The optimized PERC module owns less than 2% degradation during 19 month long-term operation in outdoor test system, although it is applied without any LID regeneration technique in cell or in module level. The module also demonstrates excellent performance ratio than other types in the field.
Keywords: light induced degradation (LID), PERC, reliability

1 INTRODUCTION

Passivated emitter rear cell (PERC) is one of the most promised technologies to achieve high efficiency with reasonable production cost, while the obstruction of higher degree of light induced degradation (LID) for PERC must to be conquered [1]. Although there are few approaches for LID regeneration in cell or in module level [2, 3, 4], the mechanism and the correlation of PERC LID among each PV sector still need to be investigated.

In this work, we evaluate P-type single crystalline silicon (P-mono c-Si) PERC LID through ingot, cell, module and outdoor system. Various sources of ingots and PERC cells from the worldwide market are tested, and most samples own LID in output power higher than 3%. The influence of oxygen in ingot on LID is also inspected [5, 6, 7]. EQE analysis is applied for optimizing PERC LID by tuning each process step during cell fabrication without any regeneration technique. Both indoor light soaking and outdoor natural sunlight LID are performed in module level. Indoor test shows much smaller LID than outdoor test resut, while indoor test shows higher LID regeneration rate than outdoor test. It is also found that higher module LID occurs with resistive load mode than with Voc mode for both indoor and outdoor test condition. We suggest to perfom LID in outdoor exposule with load to fit the real PV operation scenario. From the outdoor test system, AUO module demonstrates lower degradation after 19 month monitor. In system level, it also owns much higher performance ratio than other types during the previous 15 months.

2 EXPERIMENT

Xenon lamp light soaking chamber with Class BBB according to IEC 60904-9 is applied for evaluating light induced degradation of PERC solar cell as in Figure 1 (left). Effective irradiation dimension is 2.4 m x 2.0 m in the chamber, and the temperature is controlled at 60℃ ±5℃. 24KWHr/m^2 irradiation dosage is applied for indoor cell LID test. For module level LID, indoor test is the same as test condition and equipment as in cell level, and outdoor exposure is performed with nature sunlight. Module LID is tested in both with attached resistive load (4Ohm) and without load (open circuit, at Voc) mode during outdoor exposure. In system level, modules are installed in the outdoor test system to monitor outdoor performance as shown in Figure 1 (right). Samples are periodically inspected by Class AAA flashing type sun simulator at STC condition in the laboratory for output power measurement. In our experiment, pass/fail criteria for LID is 3% of power degradation in both cell and module level.

P-mono c-Si wafers are provided by four different suppliers from Europe and from Asia, while seven types of P-mono PERC solar cells are collected from worldwide market. AU Optronics (AUO) owns its solar cell and module production lines to optimize PERC process. In the experiment, modules are also assembly with the same materials except cell types.

Figure 1: Indoor Xeon-lamp light soaking chamber (left) and outdoor test system (right) for LID experiment.

3 INGOT / WAFER - LEVEL

P-mono PERC cells are fabricated with the identical process from four different Cz wafer supplies. Several test runs are performed for each wafer source in various material incoming batches during a year. The cell LID test result will be judge as failure, if power degradation is over 3% from any one of 20 piece cells in every test run. In Figure 2, all types have high failure rate more than 50% among various wafer supplies, although the oxygen concentration is less than 18ppm for all sample types.

The oxygen concentration is measured by SIMS (secondary ion mass spectroscopy) as the location distribution in the ingot. In the general for P-mono c-Si cell, higher oxygen contain usually accompanies with higher LID, however there is no specific trend between LID and oxygen concentration (10.9ppm ~ 18.2ppm) in Figure 3. Type-C ingot has lower oxygen concentration than type-T ingot, while type-C suffers more LID than type-T does. It infers that boron oxygen complex mechanism is not the only factor for PERC LID [5, 6, 7].

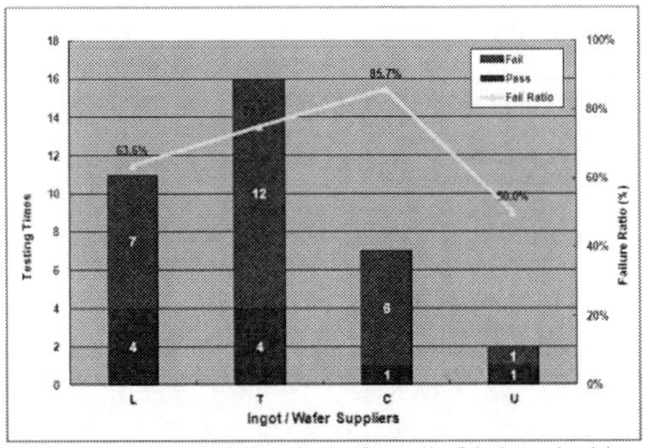

Figure 2: The LID failure ratio for cells fabricated with four different ingot (wafer) sources.

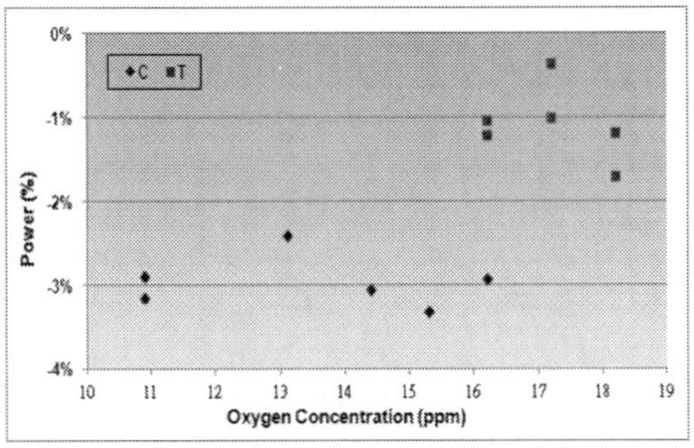

Figure 3: The correlation between LID and the oxygen concentration distribution in the ingots.

4 CELL - LEVEL

Various types of P-mono PERC cells are tested by seven different cell suppliers from worldwide. We test 20 cells of each incoming batch per time. The test result will be judge as failure, if power degradation is over 3% from any one of 20 pieces in each test run. Only one of seven vendors passes the LID test, and the other six types own high failure ratio as in Figure 4. The quality of LID effect is not well controlled among vendors in the market. Type-C is utilized with regeneration technique in cell level to ensure the stability of LID.

We try to investigate the influence of each process step during cell manufacturing on PERC LID. As shown in Figure 5, it is found that LID is quite sensitive with some certain processes. To enlarge the process window and to fine tune process among Anti-reflection coating (Arc) layer, diffusion, co-firing, passivation layer and laser ablation are very crucial.

External quantum efficiency (EQE) is a power tool to evaluate the influence of those processes on LID. After optimize those process, LID is dramatically suppressed without any extra regeneration technique as in Figure 6. Process tuning of Arc layer and diffusion improves short wavelength EQE. Optimizing co-firing and diffusion process increase the EQE at 450nm - 900nm wavelength. Passivation thin-film layer structure (AlOx/SiNx) and laser ablation process play the key characteristics in long wavelength region. Adequate thermal budget in diffusion / co-firing process, hydrogen concentration in Arc SiNx layer and proper passivation film quality might all the factors regarding to PERC LID.

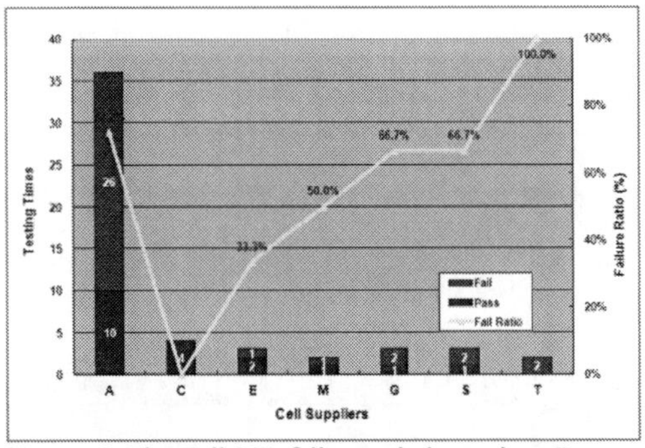

Figure 4: The Cell LID failure ratio by various P-mono PREC cell suppliers.

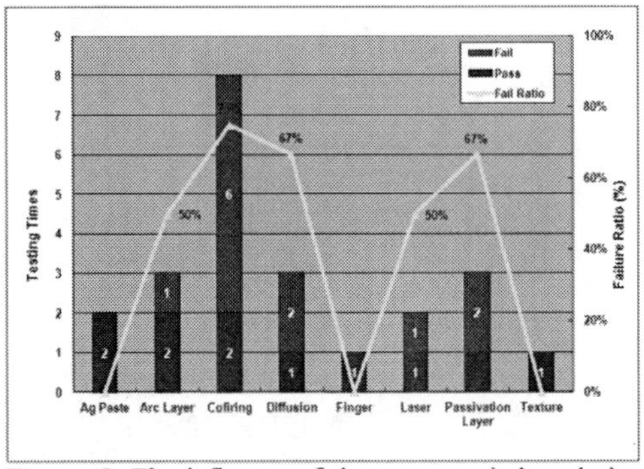

Figure 5: The influence of the process window during cell fabrication on LID effect.

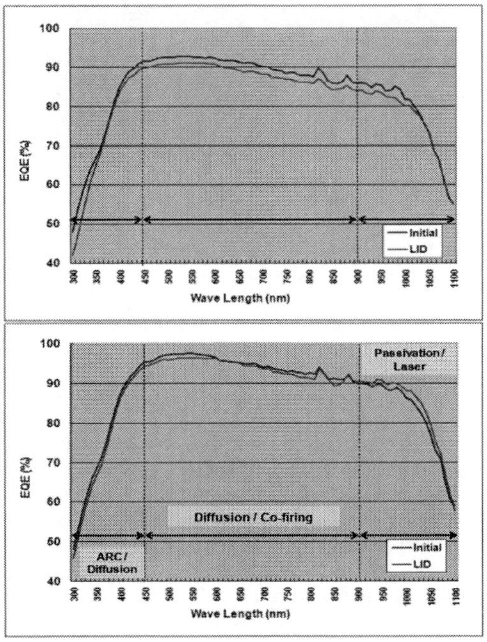

Figure 6: The external quantum efficiency (EQE) before (top) and after (bottom) cell process optimization.

5 MODULE - LEVEL

5.1 Indoor module LID

Module LID is performed with the indoor light soaking chamber with $24KWHr/m^2$ irradiation dosage at open circuit mode. All modules use the same component materials except cell types from four different cell suppliers. It is surprised that module LID is smaller than 3% for most types of modules in Figure 7, while most cell types have higher failure rate at cell LID test in Figure 4. After irradiation some cells degrade more than other cells in the module of type-A, but all cells degrade evenly in the module for type-C as shown in EL image of Figure 8.

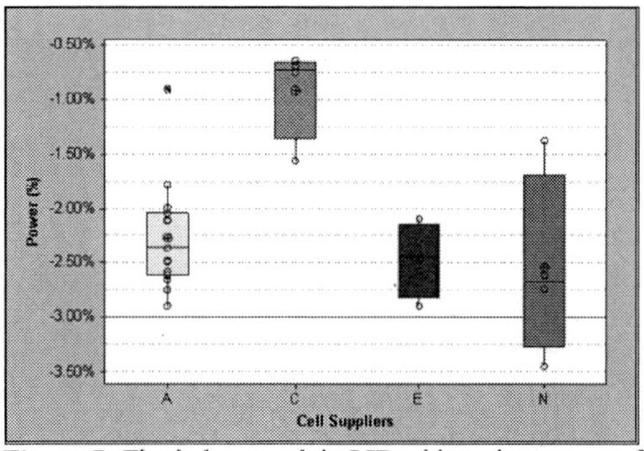

Figure 7: The indoor module LID with various types of P-mono PERC cell.

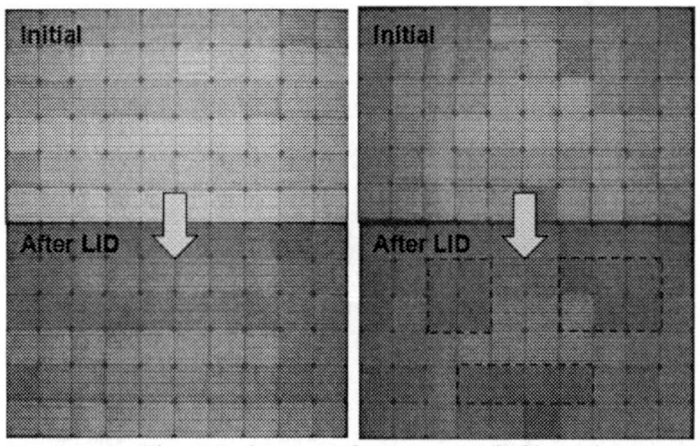

Figure 8: The EL images for type-C (left) and type-A (right) PERC module.

5.2 Outdoor Module LID

In figure 9, module LID recovery is observed after $5KWHr/m^2$ and stabilizes after $50KWHr/m^2$ irradiation for indoor test at Voc mode. We put the same batch sample to perform outdoor LID at Voc mode. Outdoor LID recovery starts from $15WHr/m^2$ and stabilizes after $150KWHr/m^2$ irradiation. Indoor exposure owns around 3 times of accelerated factor faster than outdoor exposure at our test site in the summer time in Taiwan. This would be explained, because nature sunlight is not always on the normal direction to the module surface. $5KWHr/m^2$ irradiation of pre-condition according to IEC 61215 is obviously not enough for PERC module.

As in the real PV application scenario in the field, module is always operated with load, therefore it is more reasonable to perform LID test with attach load. Module shows lower LID recovery rate with load mode than at Voc status for both indoor light soaking and outdoor exposure in Figure 10. Indoor test shows much smaller LID (-2.6%, -2.9%) than outdoor LID (-3.6%, -4.2%). In addition, module owns higher LID recovery rate at Voc is due to higher carrier injection during light irradiation [3]. This is correspondent to why module indoor LID test at Voc mode is usually pass the criteria (-3%) in Figure 7. It is suggested to apply outdoor exposure with load to evaluate LID for PERC module.

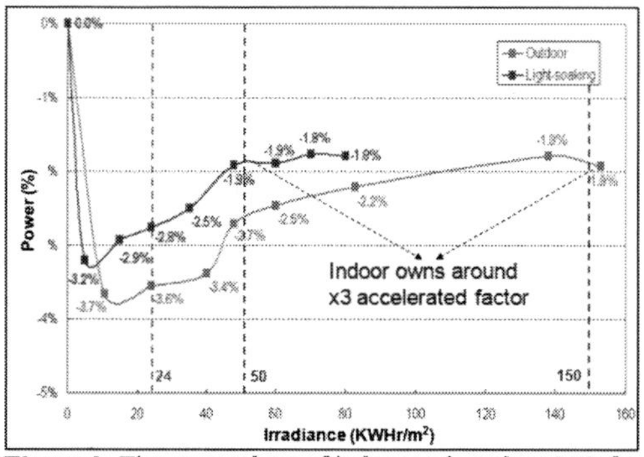

Figure 9: The comparison of indoor and outdoor LID for PERC module at Voc mode.

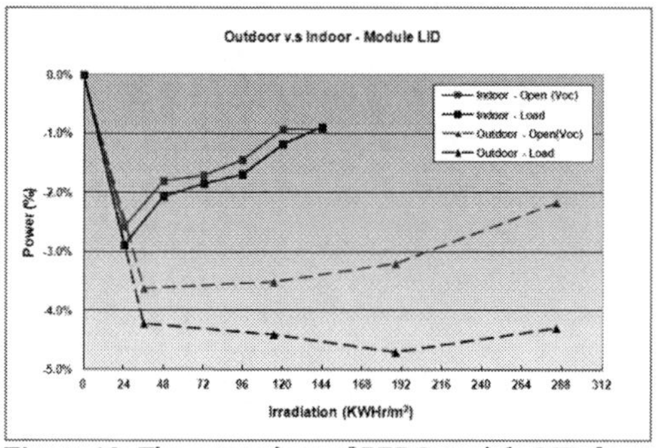

Figure 10: The comparison of PERC module LID for at open mode and with load mode.

6 SYSTEM - OUTDOOR PERFORMANCE

To evaluate the outdoor performance, PERC modules are installed in outdoor test system with load for long-term monitoring. Samples are periodically taken off from the system and, module power is measured at STC condition in the laboratory. The optimized PERC modules (PERC-AUO) show less than 2% LID in 19 month operation in Figure 11. Type-A modules show steep degradation till 3.4% power drop in the first two months and quicker recovery rate than type-S, while type-S modules show slower degradation rate till 4.1% power drop in the first nine months and slower recovery rate after. Form the IV cures in Figure 12, type-A has dramatic decade in both Isc and Voc in the first two months, but type-S has only drop in Isc.

Module DC performance ratio (P.R.) is monitored automatically by the outdoor test system AUO PERC module demonstrates excellent outdoor performance as in Figure 13. AUO PERC module owns only 1.5% to 2% LID degradation, and it gains more 5% P.R. than other types during 15 months in the field. This results from optimizing each PERC solar cell fabrication step corresponding to the EQE improvement of LID in the whole wavelength range as in Figure 6.

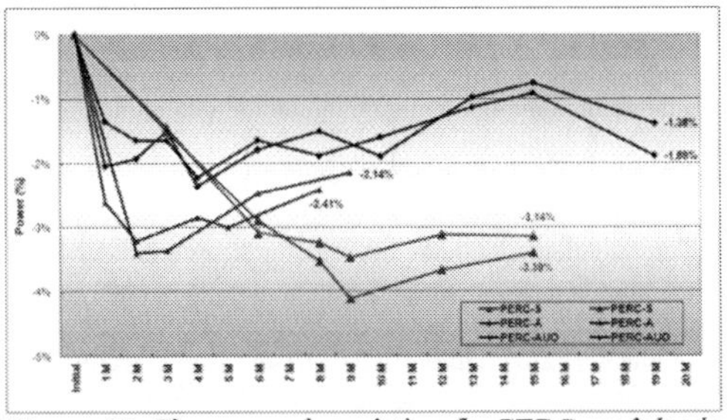

Figure 11: The power degradation for PERC modules in the outdoor test system.

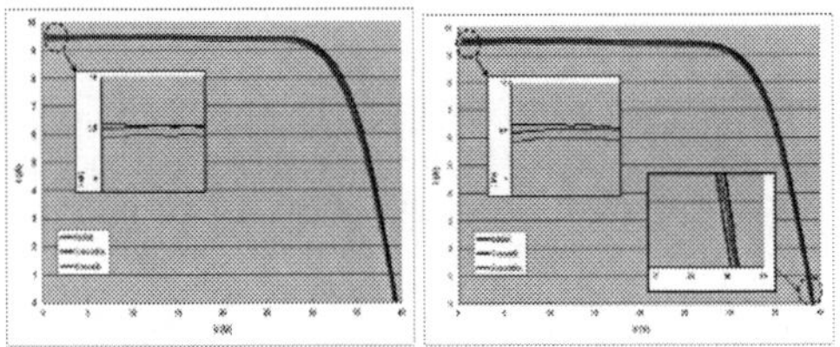

Figure 12: The IV cures for type-S (left) and type-A (right) module during outdoor LID.

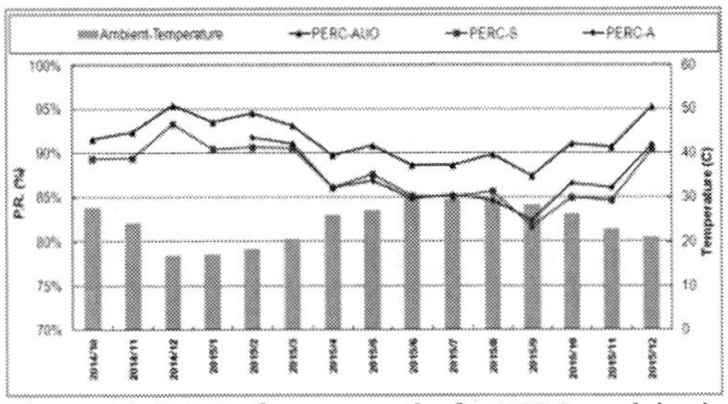

Figure 13: DC Performance ratio for PERC modules in the outdoor test system.

7 CONCLUSION

The influence of P-mono PERC LID on ingot, cell, module and system are evaluated. Various sources of ingots and PERC cells from worldwide market are tested, and most samples show large power degradation higher than 3%. Boron oxygen complex mechanism is not the only factor for PERC LID. EQE analysis is performed for tuning process window of each process step during cell fabrication without any regeneration technique.

In module level, indoor light soaking shows smaller LID power drop, and it owns much faster LID recovery rate than natural sunlight irradiation in outdoor. Module LID is more severe with resistive load than without mode for both indoor and outdoor test condition. We suggest performing LID test in outdoor exposure with load, and pre-condition irradiation should to be extended for PERC. In system level, AUO module owns much smaller LID during long-term operation in the outdoor test system. The module also demonstrates outstanding performance ratio than other types of P-mono PERC module in the field.

REFERENCE

[1] Klaus Ramspeck et al., "Light induced degradation of rear passivated mc-Si solar cells", 27th EU-PVSEC, 2012, p. 861-865

[2] F. Kersten et al., "A New Light Induced Volume Degradation Effect of mc-Si Solar Cells and Modules", 31st EU-PVSEC, 2015, p. 1830-1834

[3] Kyumin Lee et al., "Natural recovery from LID: regeneration under field conditions?", 31st EU-PVSEC, 2015, p. 1827-1829

[4] A. Herguth, G. Schubert, M. Kaes and G. Hahn, "Avoiding Boron-Oxygen Related Degradation In Highly Boron Doped CZ Silicon", 21st EU-PVSEC, 2006.

[5] J. Schmidt and R. Hezel, "Light-Induced Degradation in Cz silicon solar cells: Fundamental Understanding and Strategies for its Avoidance", 12th Workshop on Crystalline Silicon Solar Cell Materials and Processes 2002, Breckenridge, Colorado: 2002.

[6] J. Schmidt, "Light-Induced Degradation in Crystalline Silicon Solar Cells", Solid State Phenomena, vol. 95-96, 2004, p. 187-196.

[7] M. Munoz, F. Chenlo and M. C. Alonso-García, "Influence of initial power stabilization over crystalline-Si photovoltaic modules maximum power", Progress in Photovoltaics: Research and Applications, 2010, p. 417-422.

MEASURING ANTI-REFLECTION COATINGS ON PATTERNED GLASS

Brenor Brophy, Ze'ev R. Abrams, Peter Gonsalves
Enki Technology Inc., 2192 Bering Drive, San Jose, CA 95131, USA
bb@enkitech.com, Phone 1 408 383 9034

ABSTRACT: Most solar panels use patterned glass with a textured back-side. This back-side texture complicates measurements of transmission and reflected light during characterization of Anti-Reflection (AR) coatings. This paper describes a method of distinguishing between the front-side and back-side reflections by using a simple handheld reflectometer and a semi-index-matching stack of water and glass. This water-table method is described in theory and demonstrated in practice. Glass and AR coatings from different suppliers can be compared on an absolute reflectance scale, with an error of no more than 0.2%. This result shows that the water-table method is very effective in isolating the AR performance from confounding factors such as glass-type and hence is an excellent way to accurately compare AR coatings from different glass suppliers. The method is also useful for durability testing of AR coatings by isolating the performance of AR coatings from optical effects of degradation on the uncoated glass surfaces.
Keywords: Optical Properties, Anti-Reflection, Reflectance, Transmission

1 INTRODUCTION

Most solar panels use patterned glass with a textured back-side primarily to improve adhesion to the laminating material [1]. When laminated, this texture becomes transparent due to index-matching with the encapsulant and therefore inconsequential optically (see Figure 1). However, when measuring bare glass, the textured back-side will trap light producing an optical effect similar to an AR coating [2]-[4]. This effect is impossible to isolate from the performance of an AR coating on the front-side. Additionally, the textured back-side varies considerably between different glass manufacturers and even varies to some extent across single pieces of glass. This variation from supplier to supplier and from sample to sample makes it very difficult to perform accurate comparisons of AR coating performance from different suppliers, or to evaluate different AR coating materials applied to glass from the same supplier. In addition, when performing durability testing of AR coatings, such as damp heat (85 °C at 85% relative humidity), degradation of the uncoated back-side of the glass surface can produce optical effects that can't be isolated from the durability performance of the front-side AR coating. Therefore, in order to accurately characterize the performance of AR coatings a method of eliminating all optical effects from the back-side is desirable.

Figure 1: Example of a piece of patterned low-iron solar-glass showing the textured side. Underneath is a solar-module with similar glass where the textured side disappears when the glass is laminated into a module.

One common method of increasing the measurement accuracy for solar-glass is to coat both the front-side and back-side of a sheet of glass and to divide the transmission result in two [5]. However, this method is problematic for three reasons. First, float (non-patterned) glass is used as a substrate, as the textured back-side of patterned glass cannot be uniformly coated. Second, there is an intrinsic difference between the two sides of float glass that will be elaborated upon in section 3, and therefore the two coated sides will not be identical and finally as will be discussed in section 3.5, the transmission through glass into air is not the same as transmission through glass laminated into a module. Another method is to measure the sample using incident light at an angle that prevents internal reflections [6]. However, this method requires a complex setup involving a goniometer and large integrating sphere that would be difficult to implement for measuring patterned solar-glass.

In this paper, we describe a general method for measuring the performance of AR coatings using a water-table and a handheld reflectometer. This method is insensitive to variation in the textured back-side, isolates the front-side reflectivity and is easy to implement. The theory of the method is described and experimental results are presented. Using this method, variations in AR coatings at or below 0.1% absolute transmission or reflection can be reproducibly measured on the same glass type, and variability between glass types can be reduced to below 0.2%. This result shows that the water-table method is very effective in isolating the AR performance from confounding factors such as glass-type and hence is an excellent way to accurately compare AR coatings from different glass suppliers.

2 EXPERIMENTAL

All measurements were made used a handheld Konica-Minolta CM-2500 or CM-2600 Spectrophotometer, with an 8 mm aperture and 52 mm internal integrating sphere. This device measures the reflectance between 360 nm – 740 nm, which is approximately half of the relevant solar spectrum for a silicon solar cell (380 nm – 1100 nm). The device samples every 10 nm. The solar-weighted reflection is calculated by using the global spectral photon irradiance given in IEC 60904-9, $S(\lambda)$. The weighted transmission, reflection, transmission gain and reflection gain all follow the same formula, replacing the term $T(\lambda)$ with $R(\lambda)$, $\Delta T(\lambda)$ and $\Delta R(\lambda)$, respectively:

$$T_{AM} = \frac{\int_{380}^{1100} S(\lambda)T(\lambda)d\lambda}{\int_{380}^{1100} S(\lambda)d\lambda} \cong \frac{\sum S(\lambda)T(\lambda)\Delta\lambda}{\sum S(\lambda)\Delta\lambda} \quad (1)$$

where the transition to a sum is done at intervals ($\Delta\lambda$) of 10 nm. Since the device measures from 360 nm to 740 nm, the denominator can be estimated at $1.42\text{x}10^{21}$ photons$\cdot$m$^{-2}\cdot$s$^{-1}\cdot$nm^{-1} over that range for the 10 nm intervals measured here.

Substrates used in this study include high-transmission Starphire® float glass from PPG Industries, as well as patterned glass from multiple suppliers from around the world. The back-side texture on most patterned glass is hexagonal, however the aspect ratio, size and depth of these textures varies (see Figure 2). The texture is imparted by a metal roller during manufacturing. Rollers can vary over time and roller position, creating variations in texture even on the same sheet of glass. Due to the approximate 1 mm scale of these imprints, the number of these shapes can change within the field of view of the measurement device.

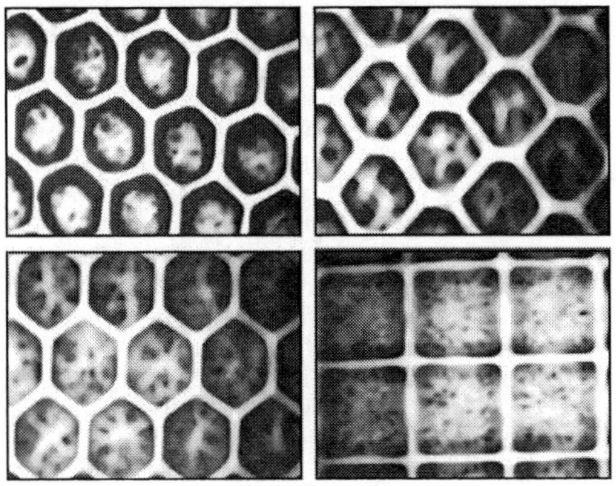

Figure 2: 4X optical images of the back-side texture from 4 different glass suppliers. The shape, size and orientation of these textures differs significantly between suppliers, and varies to some extent within the same glass from the same supplier.

Float glass is made by floating molten glass on a molten tin (Sn) bath. The side of the glass that contacts the tin bath absorbs some tin and is called the "tin-side". It is easily distinguishable from the non-tin (NT) side using a 254 nm UV light source which causes it to fluoresce slightly. The refractive index of the tin-side is higher than the NT side, and is measureable using the technique described here.

The water-table apparatus shown in Figure 3 consists of a piece of Starphire float glass with the tin-side facing up, and surrounded by a plastic frame glued to the surface with a silicone adhesive. To suppress stray-light the water-table apparatus is placed on top of a tall (~150 cm) enclosed box with a small (~5 cm diameter) aperture at the top. The interior and top-exterior of the box is coated with black, light-absorbent material (paint or fabric). Measurement is made over the location of the aperture. De-ionized water (DIW) was used as the partial index matching material.

Figure 3: Sample under measurement using the water-table apparatus described in this paper. The water-table is placed on top of a stray-light suppression box.

3 THEORETICAL BACKGROUND

3.1 Reflectance of a Single Sheet of Glass

For a sheet of glass with flat surfaces on front and back sides (with transmission and reflection terms of t_f, r_f, t_b, r_b, respectively), there can be multiple internal reflections for incoming light, as shown in Figure 4. The reflection for normally incident light off a single surface follows the formula:

$$r = \left[\frac{(n_1 - n_2)}{(n_1 + n_2)}\right]^2 \quad (2)$$

with the refractive index n_1 of air being unity at all relevant wavelengths. The terms here are all wavelength dependent, with the refractive index of solar-glass ranging from approximately 1.55 to 1.49 in the visible wavelength range.

The reflection off the glass/air and air/glass surface is the same, however the two sides of the glass are typically not equivalent. The textured back-side of patterned glass can act as an AR coating due to light-trapping and will typically have an effective refractive index that is lower than that of the bulk glass. For float glass, the tin-side has a higher refractive index and reflection than the non-tin side:

$$n_{tin} > n_{NT}$$
$$r_{tin} > r_{NT} \quad (3)$$

The total reflection off a surface includes the initial reflection, as well as the series of internally reflected rays off the back surface of the glass:

$$R_T = R_1 + R_2 + R_3 + \cdots \quad (4)$$

The transmission through each surface is the complementary term $t = 1 - r$, and the total transmission is likewise a sum, $T_T = \sum(1 - r_i)$.

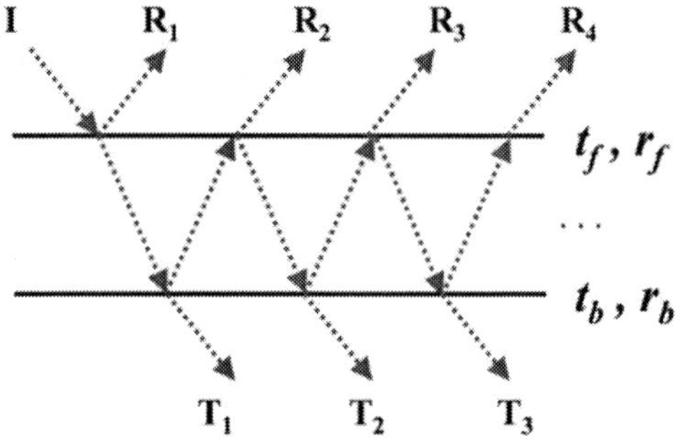

Figure 4: Schematic of the reflection and transmission components in a sheet of glass, here depicted as having flat surfaces. The reflection measured using a reflectometer consists of the R components at the top.

For standard low-iron solar-glass, the refractive index at 630 nm is approximately 1.52, and therefore the interface with air will have a reflection of approximately 4.3%. As a result, measuring the total reflection from the top side of the glass, as is done with a reflectometer will be influenced by the back-side reflection. For the simplified case of float glass appearing in Figure 4, with incoming intensity I, each reflection term measured at the front-side is:

$$R_1 = Ir_f$$

$$R_2 = Ir_b(1 - r_f)^2$$

$$R_3 = Ir_f r_b^2(1 - r_f)^3$$

$$\dots \qquad (5)$$

The total reflection measured from the front is:

$$\frac{R_T}{I} = r_f + r_b(1 - r_f)^2 + r_f r_b^2(1 - r_f)^3 + \cdots \qquad (6)$$

Since each reflection term, r, is of the order of 4%, then all cubed (or greater) terms are negligible (of the order of 10^{-5} or less), and therefore can be simplified:

$$\frac{R_T}{I} = r_f + r_b(1 - r_f)^2 + O(r^3) \qquad (7)$$

As can be seen from Eq. 7, the reflection measured will always be influenced by the back-side reflection, and the two terms cannot be de-coupled.

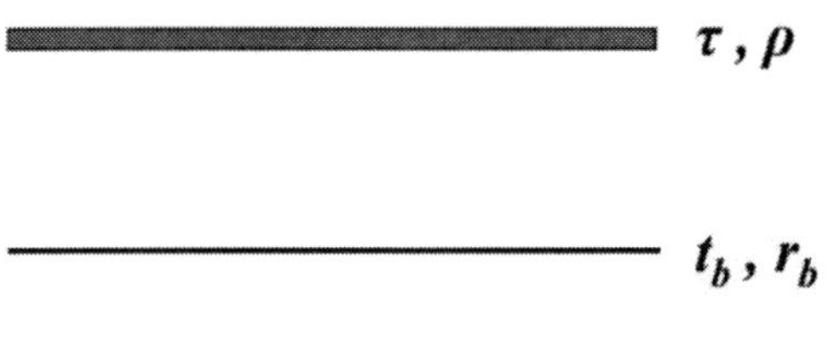

Figure 5: A sheet of glass having an AR coating will have a new effective transmission and reflection component of the front-side, $\tau = 1 - \rho$.

For an AR coating, we wish to measure the difference between the initial reflection of the front surface, r_f, and the new reduced reflection, denoted here as ρ, as shown in Figure 5. The gain in photon flux due to the AR coating is only a function of the front-side reflectance. However, the measured difference between a control sheet and an AR coated sheet (assuming identical back-side reflections) using a reflectometer will be:

$$\Delta R_{AR,meas} = R_{control} - R_{AR}$$

$$\cong r_f + r_b(1 - r_f)^2 - \rho - r_b(1 - \rho)^2$$

$$= (r_f - \rho)[1 - 2r_b + r_b(r_f + \rho)]$$

$$= \Delta R_{AR,real}[1 - 2r_b + O(r^2)] \qquad (8)$$

As can be seen in Eq. 8, the real AR gain, ($\Delta R_{AR,real}$) has a correction term associated with the measured AR performance of the order of $1 - 2r_b$, which is approximately 91%, with terms of the order of r^2 neglected. This coupling of the back-side reflection, r_b, introduces significant variation to the measurement caused by variation in patterned glass back-side textures across suppliers and even across sample sheets.

3.2 Index-Matching the Back-side using Water

The ideal method of removing the back-side texture's effect on the optical path would be to use an index-matching material that is equivalent to glass, for example mineral oil or EVA (Ethylene-Vinyl Acetate), however, these are somewhat inconvenient, in that the EVA cannot be removed after lamination, and oils are difficult to remove. A compromise is to use water, with a refractive index at 630 nm of 1.33. This will reduce the glass/air reflection by an order of magnitude, from approximately 4% to 0.4%. Using water with sugar (glucose/sucrose) can increase its refractive index at the cost of making the samples more difficult to handle and clean. A schematic of the described water-table method is shown in Figure 6. The patterned glass is placed within a shallow bath of DIW atop a piece of high-transmission Starphire glass, such that the textured surface is mostly index-matched and therefore its back-side reflection is reduced.

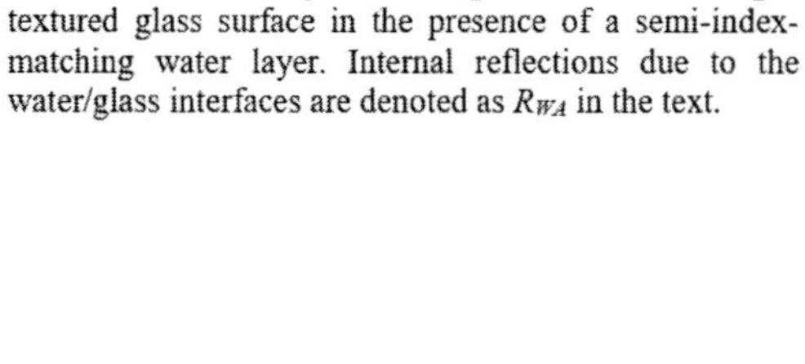

Figure 6: Depiction of the water-table method, using a bottom sheet of Starphire float glass, and measuring a textured glass surface in the presence of a semi-index-matching water layer. Internal reflections due to the water/glass interfaces are denoted as R_{WA} in the text.

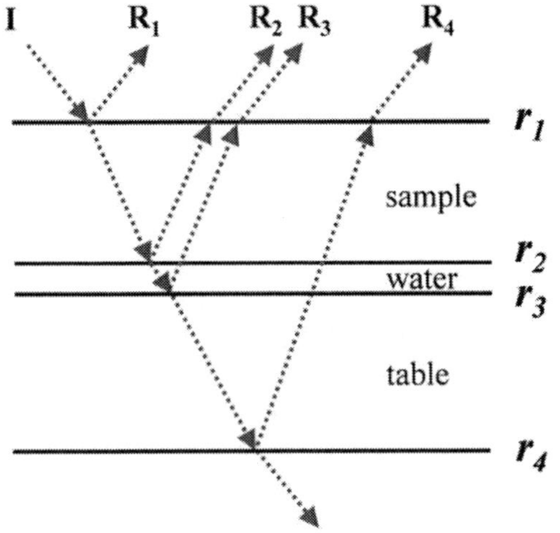

Figure 7: Diagram of the different surfaces using a water layer, including only the first-order reflection terms. The surface reflectivity notations are as used in the text.

Figure 7 displays the first order rays of reflection from each surface. To analyze the effect of the water layer we can ignore the internal reflections which are negligible. Using the notation in Figure 7, the measured reflection, including only the first reflection terms, is:

$$R_{meas} = r_1 + r_2(1-r_1)^2 + r_3(1-r_1)^2(1-r_2)^2 + r_4(1-r_1)^2(1-r_2)^2(1-r_3)^2$$

$$= r_1 + r_2 + r_3 + r_4 - 2r_1r_2 - 2r_1r_3 - 2r_1r_4 - 2r_2r_3 - 2r_2r_4 - 2r_3r_4 + O(r^3) \quad (9)$$

Since the front-side and back-side surfaces have the larger terms of reflection, another approximation would be to neglect all multiples of r_2 and r_3, such that:

$$R_{meas} \cong r_1 + r_2 + r_3 + r_4 - 2r_1r_4$$

$$= r_1 + r_4 - 2r_1r_4 + R_{WA} \quad (10)$$

where $R_{WA} = r_2 + r_3$, and is the contribution of the internal reflections due to the water layer. Note that Eq. 10 reverts to Eq. 7 when $R_{WA} = 0$.

When measuring the AR performance of a coating with respect to a control piece of glass, the measured difference in reflection between an AR coated sample and a control sample is:

$$\Delta R_{AR,WA,meas} = R_{control} - R_{AR}$$

$$\cong r_1 + r_2 + r_3 + r_4 - 2r_1r_4 - (\rho + r_2 + r_3 + r_4 - 2\rho r_4)$$

$$= (r_1 - \rho)[1 - 2r_4 - 2R_{WA}] \quad (11)$$

Note again that Eq. 11 reverts to Eq. 8 when $R_{WA} = 0$. Eq 11 shows that with the water-table method we have replaced the variable back-side reflection from the measured sample (r_b) with reflection terms for the water layer (R_{WA}) and the back-side of the water-table (r_4) which may both be quantified and which do not vary between measurements.

3.3 Quantifying the Contributions of the Water Table

To quantify the individual contributions of the internal reflection layers, an initial set of calibration experiments is performed. For a single glass sample, as depicted in Figure 4, the unknown parameter is the back reflection component, r_b. In the water table system, there are now three more unknowns: r_2, r_3 and r_4. To isolate these terms, a set of measurements is first done on the table, following the schema in Figures 8 – 10.

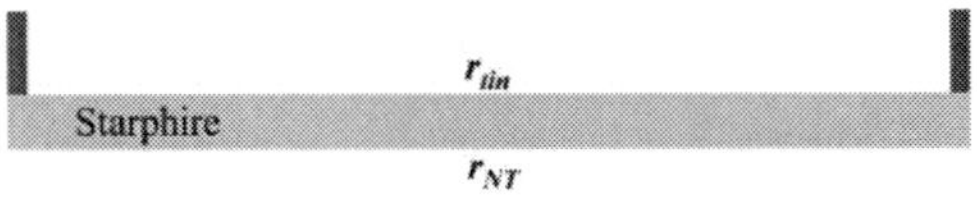

Figure 8: Step 1 of water-table calibration. Measure the reflection from the top surface (tin-side) of the table glass.

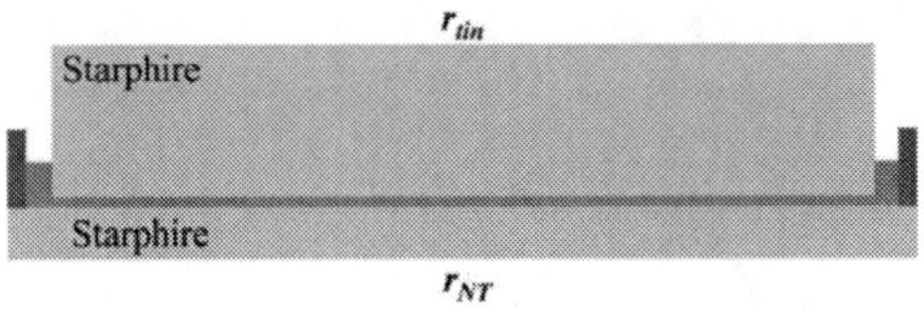

Figure 9: Step 2 of water-table calibration. Measure the reflection from the top surface (tin-side) of a piece of the same type of glass. This supplies the water related component, R_{WA}.

First, the reflection of the table itself is taken (Figure 8). In this step, the contribution of r_{tin} as a front surface, and r_{NT} as a back surface is measured. Next, water is added to the table, and a piece of the same type of glass as that of the table material is used. In this case, a piece of Starphire glass is cut from the same large sheet of float glass. The sheet is placed into the water with the tin-side facing up, and measured again (Figure 9). This measurement captures the additional internal reflection components, $r_2 + r_3 = R_{WA}$. Therefore, subtracting the measurement for the glass in water from the glass in air (tin-side facing up) isolates the first order water related component:

$$R_{M,tin,water} - R_{M,tin,air} = R_{WA} \quad (12)$$

This assumes that the absorption in the glass is negligible, which is true for solar-glass that is 3.2 mm thick or less.

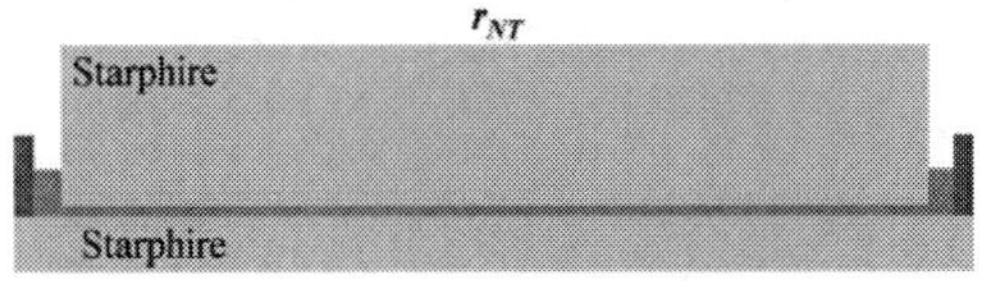

Figure 10: Step 3 of water-table calibration. Measure the reflection of the same glass sample as the previous step with the non-tin (NT) side facing up. This step provides a symmetrical system used to isolate the contribution from the air/NT surface, r_4.

The next component to isolate is the back surface reflection, r_4. For this, the same sheet of glass (Starphire) is dried and placed with the NT side facing up in the water (Figure 10). This measurement in water, $R_{M,NT}$, now has a symmetric set of parameters:

$$r_1 = r_4 = r_{NT,air}$$

$$r_2 = r_3 = r_{tin,water} \quad (13)$$

Using the Eqs. 12 and 13 in Eq. 10, one can then isolate the component r_4:

$$r_4 = \frac{1-\sqrt{1-2(R_{M,NT}-R_{WA})}}{2} \qquad (14)$$

Having isolated the back surface component (Eq. 14), and the water addition component (Eq. 12) using this calibration scheme, a patterned glass substrate can then be measured (Figure 11).

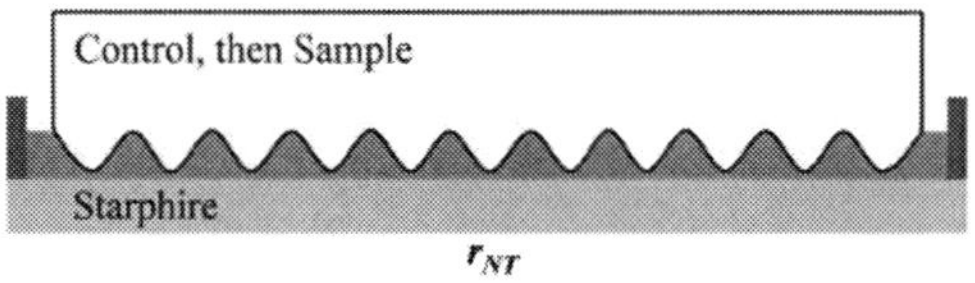

Figure 11: After calibration in the previous steps, samples can be measured and corrected for the contributions of the water-table.

The patterned glass index-matching is not identical to the float glass, both due to the light trapping from the texture as well as the slight difference in index between glass types. However, the corrections are small compared to the variation between glass types when measured in air.

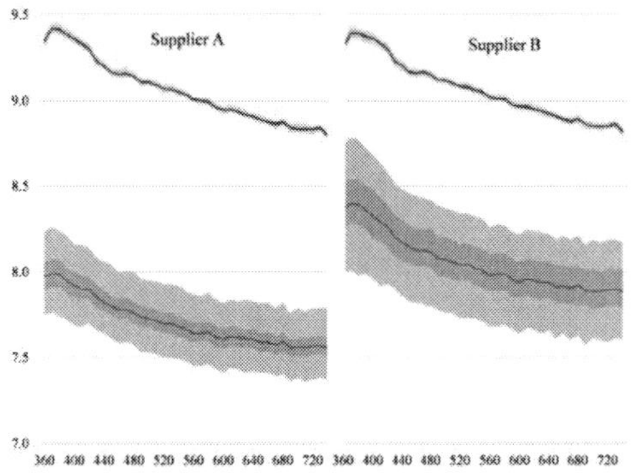

Figure 12: Comparison of reflection spectra of patterned glass from two suppliers measured with (red/top series) and without (blue/bottom series) the water-table method. Light shaded colors indicate data in the 1st and 4th quartiles. Dark shaded colors indicate the 2nd and 3rd quartiles. Solid lines are the mean. Three sample sheets from each supplier were measured at 4 points in two perpendicular orientations of the reflectometer relative to the texture on the glass.

Figure 12 shows the reduction in variation of the reflectance spectra of patterned solar-glass from two different suppliers, measured in air (blue/bottom series) and with the water-table method (red/top series). Three samples sheets from each supplier were measured under both orientations of the hexagonal back-side texture, along the short axis of the hexagon, and perpendicular to it (refer to the top two images of Figure 2). The data shows the large difference in variability of reflection values measured in air versus values measured with the water-table method. This variation is due to 1) differences in the back-side texture at different locations on sheets of glass from the same batch, 2) the specific view of the back-side texture that the instrument sees in its aperture both in position and in 3) the orientation of the instrument relative to the texture. The range of reflectivity measured in air across all wavelengths is 0.43 for supplier A and 0.62 for supplier B. Using the water-table method the measured ranges fall to 0.06 for both suppliers, close to an order of magnitude reduction in variability. This data also shows that glass from supplier A measured in air has approximately 0.35 lower reflectivity compared to that of supplier B. This is due to the better light-trapping performance of the back-side texture used by supplier A. However this difference between glass types drops to just 0.01 when measured in water, which eliminates the effect of the back-side texture. Lastly, there is offset between data measured in air and in water. This offset is due to the addition of R_{WA} and the difference in back-side reflection of each supplier's glass and the back-side reflection (r_4) of the water-table. The measurement of these different glass types with the water-table method therefore demonstrates that the front-side reflectivity is essentially the same, and the differences measured in air are caused by the back-side texture.

3.4 Modeling Anti-Reflection

Using the data from the measurements described in the previous section, the performance of an AR coating can be matched to that of a single-index material. The underlying assumption is made that the AR layer has a single refractive index that is nondispersive, i.e. it remains the same value over all relevant wavelengths. The formula for an AR coating depends on the refractive index of the film, n_{AR}, the thickness, t, and substrate refractive index, n_{sub}. Most calculations of AR coatings typically assume a single-value for the substrate refractive index. However, the method described here allows back-calculation of the control substrate's refractive index. By measuring a control sample of glass in water ($R_{M,control}$) and then using Eqs. 10, 12 and 14, the corrected front-side reflectivity of the control sample can be determined to a first order approximation:

$$R_{sub} = \frac{(R_{M,control}-R_{WA}-r_4)}{(1-2r_4)} \qquad (15)$$

The refractive index can then be determined using Eq. 2:

$$n_{sub} = \frac{(1+\sqrt{R_{sub}})}{(1-\sqrt{R_{sub}})} \qquad (16)$$

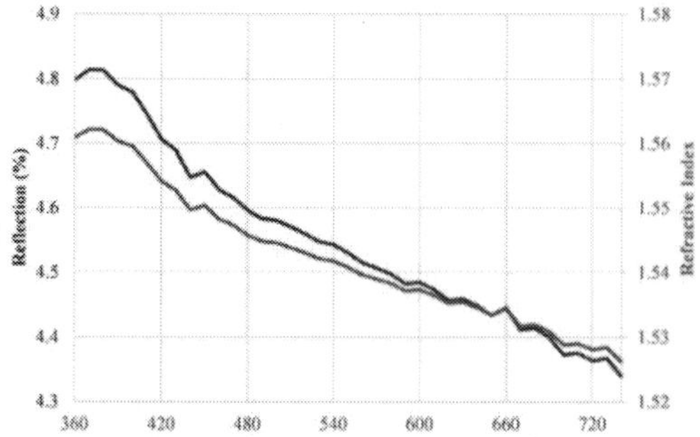

Figure 13: Control reflectance and refractive index of a sheet of patterned glass, as isolated using Eq. 15. The index of the control glass is then calculated using Eq. 16, and follows the same trend.

Using these wavelength dependent values (plotted in Figure 13 for a sample of patterned glass), one can then use the formula for an AR coating:

$$R_{AR,model} = \frac{r_a^2+r_b^2+2r_a r_b \cos\left(\frac{4\pi t n_{AR}}{\lambda}\right)}{1+r_a^2 r_b^2+2r_a r_b \cos\left(\frac{4\pi t n_{AR}}{\lambda}\right)} \quad (17)$$

with r_a being the air-AR reflectivity, and r_b being the AR-substrate reflectivity using Eq. 2. The gain of the AR coating can be determined using Eqs. 15 and 17:

$$\Delta R_{model} = R_{sub} - R_{AR,model} \quad (18)$$

ΔR data generated with Eq. 18 can be compared to the measured performance of an AR sample vs a control that has been corrected for R_{WA} and back-side reflection from the water table (r_4) as shown in Eq. 19:

$$\Delta R_{corrected} = \frac{(R_{M,control}-R_{M,sample})}{(1-2r_4-2R_{WA})} \quad (19)$$

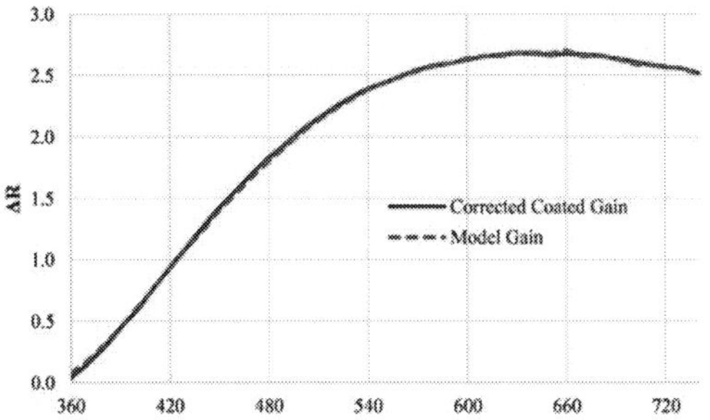

Figure 14: Corrected coated sample measurement data and corrected anti-reflection model for a piece of coated glass. The thickness of this coating is 121nm with a refractive index of 1.41.

The model parameters of thickness (t) and AR coating refractive index (n_{AR}) can be varied to fit the corrected reflection data (Figure 14). For the sample shown in Figure 14, the measured data had an optimal fit to the model with a thickness of 121 nm and a refractive index of 1.41. Once the model parameters are determined the absolute solar-weighted transmission gain (T_{AM}) of the material can be calculated using the model for the full spectral window (380 nm – 1100 nm) using Eq. 1. Within the approximation errors involved in the method, T_{AM} can be used to compare AR coatings on different glass substrates. A comparison was performed for glass from 6 different suppliers. All samples (3 sheets per supplier) were coated in the same batch, with the same coating material and were cured together. Four measurements were made per sheet. The resulting data is shown in Figure 15. As can be seen, the largest difference is 0.2%. This result shows that the water-table method is very effective in isolating the AR performance from confounding factors such as glass-type and hence is an excellent way to accurately compare AR coatings from different glass suppliers.

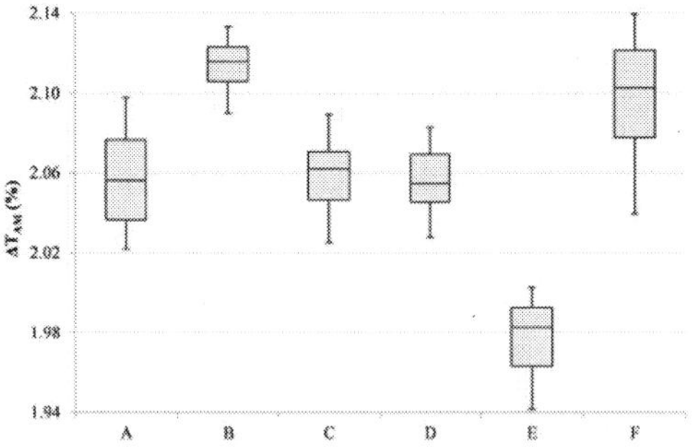

Figure 15: Weighted AR performance for 6 different patterned glass suppliers (A-F), with 3 sheets per supplier and 4 measurement points per sheet, coated at the same time. The largest difference is < 0.2% (absolute).

3.5 Relation to Transmission Data

The method described in the proceeding section relates to reflection measurements, which can be measured using a handheld reflectometer. However, much of the solar glass industry uses spectrophotometers that directly measure transmission and output a weighted value (T_{AM}) using Eq. 1.

For transmitted light, when simplifying for flat glass as in Figure 4, the total transmission of all the terms is:

$$T_{total} = t_1 t_2 + t_1 t_2 (r_1 r_2) + t_1 t_2 (r_1 r_2)^2 + \cdots$$

$$= t_1 t_2 [1 + r_1 r_2 + (r_1 r_2)^2 + \cdots]$$

$$\sum_{k=0}^{n-1} a q^k = a\frac{1-q^n}{1-q} \quad (20)$$

Summing the transmission terms, and using the formula for an infinite geometric series Eq. 20 with $q = r_1 r_2 = (1-t_1)(1-t_2)$, the transmission through a sheet of glass is:

$$T_{total} = \frac{1}{\left(\frac{1}{t_1}+\frac{1}{t_2}-1\right)} \quad (21)$$

The corresponding transmission gain due to an AR coating with front surface transmissivity, τ, is:

$$\Delta T_{AR,trans} = \frac{\tau - t_1}{t_1 \tau \left(\frac{1}{t_1}+\frac{1}{t_2}-1\right)\left(\frac{1}{\tau}+\frac{1}{t_2}-1\right)} \quad (22)$$

Note that for glass laminated in a solar module, the transmission of the back surface approaches unity, $t_2=1$, and then the gain is equivalent to the difference $\tau - t_1$. Likewise, making the identical calculation in terms of front-side reflectivity, such that $t=1-r$, the total transmission is:

$$\Delta R_{AR,trans} = (r_1 - \rho)\frac{(1-r_2)^2}{(1-r_1 r_2)(1-\rho r_2)} \quad (23)$$

Eq. 23 also reverts to the correct gain, $r_1 - \rho$, when the back-side is laminated, and r_2 approaches zero. Using either formula, one can see that the AR performance, $\tau - t_1 = r_1 - \rho$, must have a correction term due to the back-side reflectivity.

This means that measuring AR coating performance in transmission mode does not remove the need to isolate the back-side reflection component. And hence, transmission mode measurements will also be subject to measurement variation caused by the back-side texture. Given that the water-table method would be difficult to implement with a transmission mode spectrophotometer, reflection mode spectrophotometry is a more suitable metrology tool for accurate characterization of AR coating performance.

3.6 Accelerated Aging Use

In addition to measuring the absolute performance from an AR coating, the water-table method described also assists in measuring samples that have undergone accelerated durability testing, such as damp heat according to IEC 61215. During the test, the back-side of the glass may be damaged such that it will act as an AR layer (see Figure 16). This layer will thicken with continuous testing, and will be most effective when it is in the thickness range of 100 nm. At that point, it will be difficult to distinguish the front AR coating from the back-side damage when measuring a sample in air.

Figure 16: Schematic of an induced back-side anti-reflection layer due to accelerated aging.

Using the water-table method reduces the effect of the back-side damage, and allows one to isolate the performance of the front-side alone. Figure 17 shows a patterned glass sample measured in air and in water after 500 hours in a damp heat chamber. When in air, the gain has a double peak, signifying front and back-side AR layers of different thicknesses, and with overall higher gain. However, when placed in water, a single peak emerges, indicative of a single-index material. Measuring differences in AR performance before and after durability testing with the water-table method is more accurate, and removes the artifact of the back-side damage and apparent higher gain.

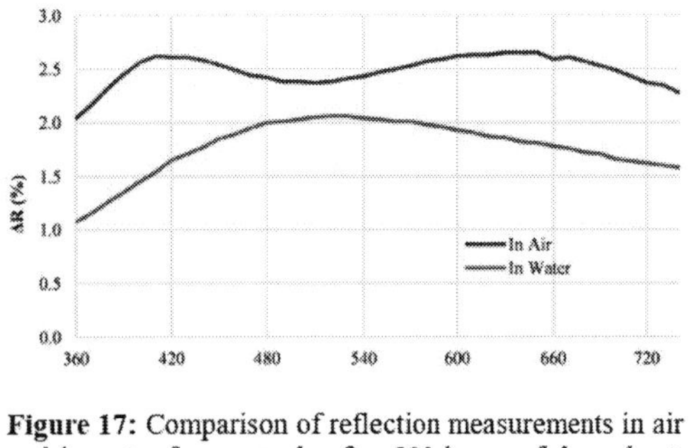

Figure 17: Comparison of reflection measurements in air and in water for a sample after 500 hours of damp heat. The measurement in air shows a double peak, due to back-side damage acting as an AR coating.

4 CONCLUSION

The water-table method described in this paper produces an absolute AR performance measurement that is de-coupled from the back-side reflection, which is shown to produce large variations in transmission spectra. The method is easy to implement in a manufacturing setting using a handheld reflectometer. Using this method on glass samples with similar refractive index enables an accurate comparison between AR coatings from different suppliers and on different substrates. We have compared the same coating material on glass samples from 6 different suppliers and reported a 0.2% (absolute) difference in AR performance across all samples. This residual difference may include actual differences in the glass and instrumentation accuracy.

The ability to distinguish between the front-side and back-side reflectivity is critical in order to distinguish the actual performance of different AR coatings. The ability to compare different types of glass enables a fair comparison between AR coatings from different suppliers. Furthermore, the ability to reduce the effects of back-side damage caused by accelerated durability testing allows a better understanding of the durability performance of AR coatings in isolation from degradation of the glass substrate.

5 REFERENCES

[1] F.J. Pern and S.H. Glick, "Adhesion Strength Study of EVA Encapsulants on Glass Substrates", presented at the National Center for Photovoltaics and Solar Program Review Meeting, Denver, CO, USA, March 24-26, 2003.

[2] E. Yablonovitch and G. D. Cody, "Intensity Enhancement in Textured Optical Sheets for Solar Cells", IEEE Transactions on Electron Devices, ED-29, p. 300, 1982.

[3] A. Deinega, I. Valuev, B. Potapkin, and Y. Lozovik, "Minimizing light reflection from dielectric textured surfaces", J. Opt. Soc. Am. A, Vol. 28, p. 770, 2011.

[4] T. von Versen, "Quality above Quantity: Opportunities for Differentiation – Using the Example of Solar Glass Choice", EUPD Research White Paper, 2011.

[5] A. Nositschka, "Sun-Light Harvesting with Surface Patterned Glass for Photovoltaics", International Workshop on Glass for Harvesting, Storage and Efficient Usage of Solar Energy, Pittsburgh, PA, USA, November 16-18, 2008.

[6] P. Nostell, "Preparation and Optical Characterisation of Antireflection Coatings and Reflector Materials for Solar Energy Systems", Acta Universitatis Upsaliensis, Uppsala 2000.

MEASUREMENT STATION TO EVALUATE ANTI-SOILING PROPERTIES OF PV MODULE COATINGS

Michael Gostein[1], Bill Stueve[1], Brenor Brophy[2], Kichang Jung[3], Alfredo Martinez-Morales[3], Shu Zhang[4], Yeyi Jin[4], Jianmei Xu[4]

[1]Atonometrics, Inc., Austin, TX, 78757, USA
[2]Enki Technology, Inc., San Jose, CA, 95131, USA
[3]University of California Riverside, Riverside, CA 92507, USA
[4]State Key Laboratory of PV Science and Technology, Changzhou Trina Solar Energy Co., Ltd., Changzhou, 213031, China

ABSTRACT: Soiling can be a significant loss factor for PV systems, accounting for >5% annual energy loss in locations without frequent rainfall. Therefore, many manufacturers are investigating the potential benefits of incorporating anti-soiling or self-cleaning properties into PV module coatings. In order to study the potential anti-soiling behavior of various surface treatments, we have deployed a test station with custom-made PV modules having coatings with both hydrophilic and hydrophobic properties that may inhibit soiling and/or induce self-cleaning. The soiling of each sample is determined by comparison to an automatically washed reference cell in the station. We present the motivations for and design of the station, along with preliminary results regarding the impact of the coatings on soiling.
Keywords: photovoltaic systems, coatings, soiling

1. INTRODUCTION

Soiling, the accumulation of dust and dirt on PV modules, can be a significant loss factor in PV power plants, accounting for >5% annual reduction in PV system energy output in locations without frequent rainfall. Therefore, many manufacturers are investigating the potential benefits of incorporating anti-soiling or self-cleaning properties into PV module coatings, particularly when such properties can be directly integrated into anti-reflective coatings. Reducing the average annual energy loss due to soiling by even a small amount could have tremendous value in terms of increased energy production, especially for utility-scale projects.

One approach to achieving anti-soiling behavior is to alter the surface energy and water contact angle of the exposed surface. For example, studies have shown that both hydrophilic and hydrophobic surfaces can exhibit anti-soiling and/or self-cleaning effects on solar modules [1]-[5].

In order to further study the potential anti-soiling behavior of various surface treatments, we have deployed a test station in southern California to compare modules with several commercially available anti-reflective coatings having both hydrophilic and hydrophobic properties. The soiling measurement station is designed to make high-resolution measurements of the soiling-related power loss of each of the side-by-side samples.

2. EXPERIMENTAL

Four 60-cell multi-crystalline photovoltaic modules were custom assembled in the same batch from matching PV cells. The only significant difference between the four modules is the anti-reflective (AR) coating on the front side of the cover glass. As listed in Table I, one module uses uncoated glass, one has a hydrophilic pre-tempered coating typical of coatings used on the majority of PV modules currently produced, and the final two have hydrophobic post-temper coatings with similar surface energy but slightly different optical properties. The cover glass was coated at a commercial PV glass manufacturer and the modules were assembled in a production module assembly facility. Table I lists the water contact angle of each of the samples, illustrating the different surface properties.

Figure 1: Station for testing impact of module coatings on soiling.

The test station, depicted in Fig. 1, is installed outdoors near the University of California Riverside campus, and follows a previously described concept [6][7] in which the electrical outputs of the naturally soiled test modules are compared to the output of a reference device which is routinely cleaned. In this station the clean reference device, visible in the left side of the figure, is a crystalline silicon reference cell. The reference cell and the four modules are installed south-facing at a 35 degree tilt. A measurement and control system (Atonometrics RDE300) measures irradiance, via the reference cell, and I-V curves and back surface

Table I: Module descriptions

Module	Surface Treatment	Water Contact Angle
1	Uncoated	37°
2	Hydrophobic 1	90°
3	Hydrophobic 2	90°
4	Hydrophilic	30°

temperature of each of the four modules. From the I-V curves the control unit calculates the short circuit current I_{sc} and the maximum power P_{max} of each module. In addition the control unit operates an automated washing system which cleans the reference cell daily with a pressurized water spray. Measured data are recorded once per minute and logged at a remote server for later analysis.

3. ANALYSIS METHODOLOGY

As described in [7], a soiling ratio metric SR is calculated for each module for each minute in the data set by dividing the module's measured maximum power by its expected maximum power if it were clean and free of soiling. Thus, at each minute, for each module,

$$SR = \frac{P_{max}}{P_{max,0} \cdot \left(1 + \gamma \cdot (T - T_0)\right) \cdot (G/G_0)} \qquad (1)$$

where P_{max} is the measured maximum power of the module in its soiled state, $P_{max,0}$ is a calibration value for the module's maximum power when clean and at a reference condition (e.g. STC), γ is the temperature coefficient of maximum power, T is the measured module temperature, T_0 is the module temperature at the reference condition (e.g. 25° C at STC), G is the measured plane-of-array irradiance determined from the clean reference cell, and G_0 is the irradiance at the reference condition (e.g. 1000 W/m^2 at STC).

To compute the average impact of soiling for a given day, which we denote $\langle SR \rangle_d$, an irradiance-weighted average of the instantaneously measured values of SR is calculated as discussed in [6] and [7]:

$$\langle SR \rangle_d = \frac{\sum SR \cdot G}{\sum G} \qquad (2)$$

Note that $\langle SR \rangle_d$ is equivalent to the ratio of the actual measured energy to the expected energy over the summation period.

In using (2) we exclude from the summation the minutes during which any of the modules was shaded by nearby structures. This condition typically occurred only early in the morning. We also exclude any minutes during which data were missing for any of the modules, in order to ensure all modules are treated equally.

The $P_{max,0}$ value of each module was determined from a regression of the measured temperature-corrected P_{max} of each module versus the irradiance G from the reference cell, using data from selected clear-sky days following rain events when all the modules were known to be clean. This self-calibration method ensures that the $\langle SR \rangle_d$ values equal 1.0 for the selected clear-sky clean days. The resulting $P_{max,0}$ values also compared favorably with flash test results for each module determined prior to installation at the site.

4. RESULTS

The station was brought online in August 2015. Data collection is ongoing as of June 2016 and is expected to continue throughout 2016 in order to capture results from the dry summer months.

Results to date are shown in Fig. 2. The top panel shows daily average soiling ratio $\langle SR \rangle_d$ for each module versus time, while the bottom panel shows daily rainfall up to 20 mm. Rainfall data were taken from a nearby meteorological station available via the Weather Underground website. In mid-October and early December the soiling measurement station was offline for maintenance.

The data in Fig. 2 show several examples of the progressive accumulation of soiling followed by cleaning due to rainfall. For example, in the period from August to early September the measured soiling ratios fall gradually,

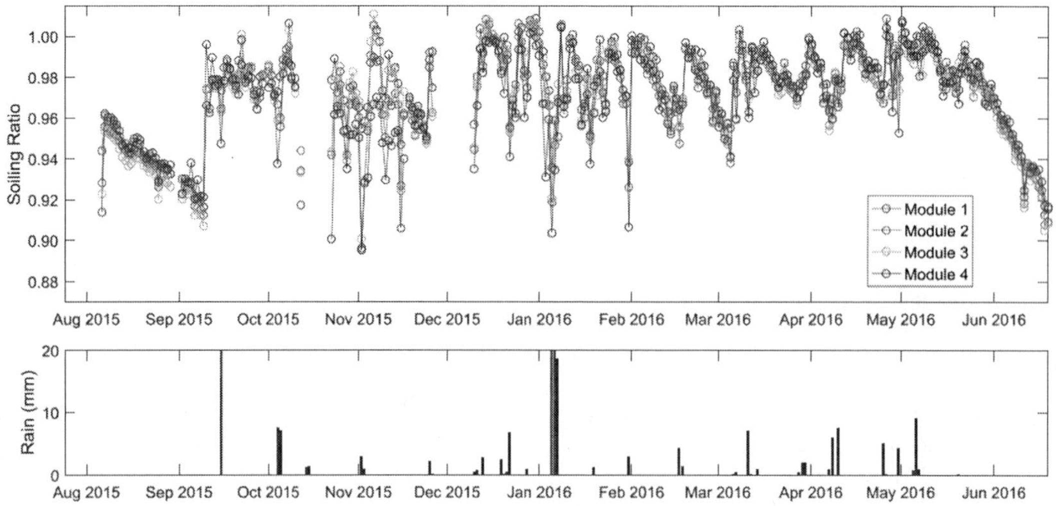

Figure 2: Daily average soiling ratio measured by soiling station (top) and rain at the site (bottom).

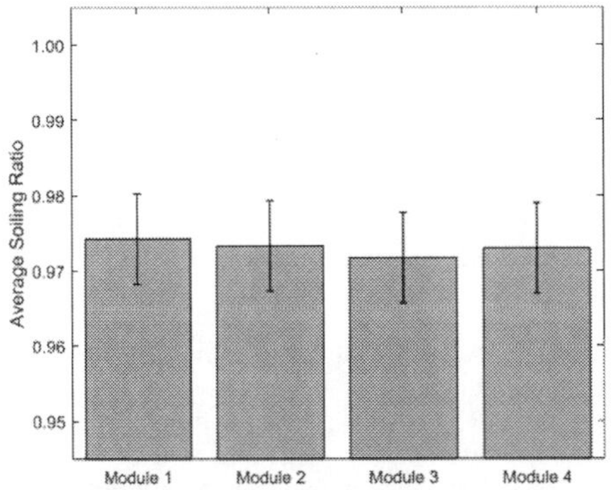

Figure 3: Average soiling ratio for each module over the entire 10-month measurement period to date.

reaching a low point of about 0.92. Following the rain event in mid-September, the ratios recover to values near 1.00 indicating that the modules are clean. This characteristic sawtooth pattern is also visible in the period from February through June. Note that the soiling rates during these periods range from 4-10 % per month, or 0.1-0.3 % per day, which is comparable to published results for the region [8]. Note also that rain events of only a few millimeters, such as in mid-February, were sufficient to clean the modules. In some cases it appears that rain events are not synchronized with the module cleaning; this is likely due to the weather station being located some distance from the soiling station site.

In contrast to the pattern just described, from November through January the upper panel of Fig. 2 is more irregular, which is due to low irradiance and frequent rain conditions.

To assess the overall impact of soiling on the four modules, we take the average of the soiling ratios for the entire period from August 2015 through June 2016. These results are plotted in Fig. 3. All four modules show a soiling ratio of approximately 0.973, indicating approximately 2.7% energy loss over the measurement period. The error bars in Fig. 3 are based on an uncertainty analysis which takes into account the effects of uncertainty in the module calibration constants (the $P_{max,0}$ values used in (2)) and temperature non-uniformities. Within the uncertainties, there was no discernible difference in the average impact of soiling on the performance of the four modules over the current test period.

5. DISCUSSION

The potential anti-soiling behavior of various coatings depends on the characteristics of soiling at the site, because the adhesion of dust particles to hydrophobic or hydrophilic surfaces depends on particle composition as well as meteorological conditions. In particular, hydrophilic versus hydrophobic surface treatments have been found to respond differently to organic particles such as soots versus inorganic particles such as soil minerals. Previous studies [9] have shown different soiling responses to hydrophobic and hydrophilic coatings at different locations.

For the present study the soiling performance of all four modules so far appears equivalent at the test site, within the resolution of the measurement station. These results differ from those recently found in [5], where a hydrophilic surface treatment was found to give a noticeable reduction in soiling. However, the site locations in [5] were characterized primarily by desert conditions, while in the present study the site location is urban and is situated near both a highway and an agricultural area, which together may provide both organic and inorganic soil particles. The soiling behavior at the site is therefore more complex. Furthermore, the data collection so far was concentrated mostly in the rainy season, during which the average soiling loss was small. Data collection is ongoing and we will continue to evaluate the soiling behavior of the modules through a greater range of seasonal conditions, including the dry summer months, to determine how the coatings respond.

6. REFERENCES

[1] C. Cañete, R. Moreno, J. Carretero, M. Piliougine, M. Sidrach-de-Cardona, J. Hirose, S. Ogawa, "Effect Of The Self-Cleaning Coating Surface In The Temperature And Soiling Losses Of Photovoltaic Modules", in the 27th EU PVSEC, 2011.

[2] E. Klimm, T. Lorenz, K.-A. Weiss, "Can anti-soiling coating on solar glass influence the degree of performance loss over time of PV modules drastically?," in the 28th EU PVSEC, 2011.

[3] K. Midtdal, B. P. Jelle, "Self-cleaning glazing products: A state-of-the-art review and future research pathways", *Solar Energy Materials & Solar Cells*, vol. 109, pp. 126–141, 2013.

[4] A. R. Hoffman, C. R. Maag, "Photovoltaic Module Soiling Studies May 1978 - October 1980," JPL Publication 80-87.

[5] M. A. Grammatico and B. W. Littmann, "Anti-Soiling Benefits of Anti-Reflective Coatings on First Solar Cadmium Telluride PV Modules," PVSC, 2016.

[6] M. Gostein, J. R. Caron, B. Littmann, "Measuring Soiling Losses at Utility-scale PV Power Plants," in the 40th PVSC, 2014.

[7] M. Gostein, T. Düster, C. Thumann, "Accurately Measuring PV Soiling Losses With Soiling Station Employing Module Power Measurements," in the 42nd PVSC, 2015.

[8] A. Kimber, L. Mitchell, S. Nogradi, and H. Wenger, "The Effect of Soiling on Large Grid-Connected Photovoltaic Systems in California and the Southwest Region of the United States," in the IEEE 4th World Conference on Photovoltaic Energy Conversion, Waikoloa, HI, 2006.

[9] B. Brophy, Z. R. Abrams, P. Gonsalves, K. Christy, "Field and Lab Verification of Hydrophobic Anti-Reflective and Anti-Soiling Coatings on Photovoltaic Glass" in the 31st EUPVSEC, 2015.

GUIDELINES FOR THE DEVELOPMENT OF ABRASION-RESISTANT AR COATINGS: INPUT FROM MODELLING AND EXPERIMENTAL WORK

Romain Cauchois[1], Marcel Meuwissen[1], Changjian Shen[2], Heidrun Keul[1], Paul Steeman[1], Damien Reardon[1†]
[1] DSM Ahead R&D, Urmonderbaan 22, P.O. Box 18, 6160 MD Geleen, The Netherlands
[2] DSM, Performance Materials Research Center, Zhangjiang Hi-Tech Park, Shanghai, 201203, China
[†]Corresponding author: Dr. Damien Reardon, damien.reardon@dsm.com

ABSTRACT: In this paper, the impact of abrasion on the performance of thin AR coatings over the years is studied by the development of accelerated life tests. The analysis of the surface morphology after abrasion sheds light on the different failure mechanisms that entail a loss of efficiency of the AR properties: ranging from a roughness modification of the surface up to the total removal of the coating. Finite element modelling is used to define guidelines for the development of abrasion-resistant coatings based on different scenarios like a modification of the void fraction or of the pore size. The result of these simulations indicates that the void fraction of the coating has significantly more impact on the hardness, than for example the void dimension. Simulations are confirmed experimentally using accelerated abrasion tests, and emphasize that mechanical reinforcement strategies have to be defined based on the lifetime requirements and the environmental conditions at the end-user location.

Keywords: Antireflection Coating, Modelling, Durability, SiO_2, Optical Losses

1 INTRODUCTION

DSM is a world leading supplier of a nano-porous silica based coating as used in the solar glass industry for anti-reflective coatings for PV module cover glass. The use of sol-gel derived silica layers has been reported extensively in the recent past for multiple applications in relation to their improved low-k dielectrics, anti-reflective or bio-compatible properties. Coatings in such applications are subject to degradation by abrasion and erosion throughout their lifetime due to mechanical actions, such as installation or cleaning operation.

PV module surface is indeed soiled by rain/pollution, dust, pollen, or sand over time. As a consequence, solar light is partially blocked and less energy is produced. Depending on the climate zone and local soiling conditions, cleaning operation is therefore required to maintain energy production.

In this paper, the impact of abrasion on the performance of model AR coatings over the years is studied by the development of accelerated life tests which simulate different cleaning operation conditions. The analysis of the surface morphology after abrasion sheds light on the failure mechanisms. According to the conditions used during the test (mild to harsh conditions), failure mechanisms are observed for various model coating systems.

2 EXPERIMENTAL

2.1 Coating preparation

In this study, the porous AR coatings were prepared from silica sols dip-coated on soda lime glass substrates. The sols are based on pre-oligomerized tetraethyl orthosilicate (TEOS) in an alcoholic solvent at low pH. Variation of porosity is achieved by adding varying amount of latex nanoparticles, as described earlier in [1]. The film is then deposited by dip-coating on both sides of the substrates under normal conditions of temperature and pressure, under a relative humidity of 30%. The coating thickness can be controlled via the dip speed. Typical coating speeds for dip-coating are in the range of 0.5 to 1 $m\cdot min^{-1}$. The final AR properties of the coating are achieved by a final conversion step in an oven at a temperature of 650°C.

The optical performance, the density and the surface morphology of AR coatings are mainly characterized by structural information such as their porosity, the pore size, and the pore geometry. Several model AR coatings for solar application have been reported in the literature [2] and the more widely used are schematized in Fig. 1.

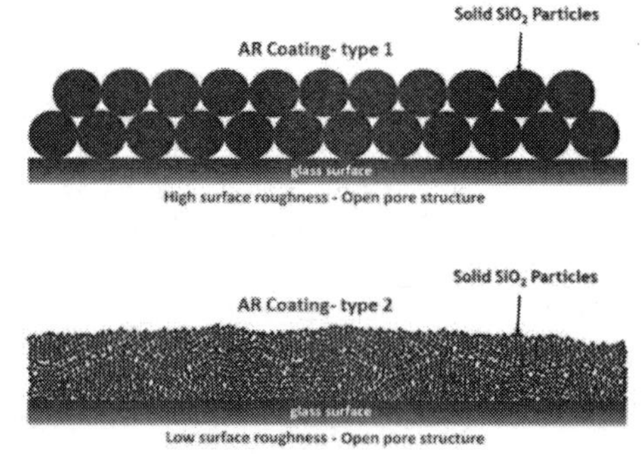

Fig. 1: Some common configurations of model AR coatings reported in literature for solar applications.

2.2 Performance tests

The spectral transmittance of AR coatings considered in this paper is measured using a dual-beam UV-VIS-NIR spectrophotometer (Shimadzu UV-2600) equipped with an integrating sphere. Transmittance is performed under normal incidence irradiation with air as reference, over a wavelength range of 370-1200 nm.

The resistance to abrasion test is performed following the standard EN1096-2 where a pad is rubbed on the surface of the coating under dry conditions. The pad is moved in an alternating forward and backward motion with a frequency of 60 strokes per minute, over a stroke length of 120 mm. Additionally to the linear movement, the pad continuously rotates at 6 rpm. The test is carried with a vertical load of 4N applied to the coating surface via the pad. Two materials are investigated in this paper: a standard felt pad with a density of 0.52g/cm^2 and an abrasive pad where abrading particles are embedded in resilient clay.

3 RESULTS AND DISCUSSION

3.1 Modelling

A 3D finite element modelling (FEM) approach is used to define guidelines for the development of abrasion-resistant coatings, based on different scenarios like for instance the impact of a structural reinforcement by a hard coat, or the impact of the morphological reinforcement by a variation of the porosity or of the pore size. The FEM is based on standard ABAQUS package, using representative volume elements (RVE), and assuming a linear elastic behaviour of SiO_2 matrix with a Young's modulus of 60 GPa and a failure strength of 8 GPa. For the calculations, the pore diameter is varied between 40 and 80 nm, the porosity from 5 to 55% and the top reinforcing layer thickness from 5 to 15 nm. The 3D geometry of the unit cell used for the modelling is shown in together with the axes of symmetry.

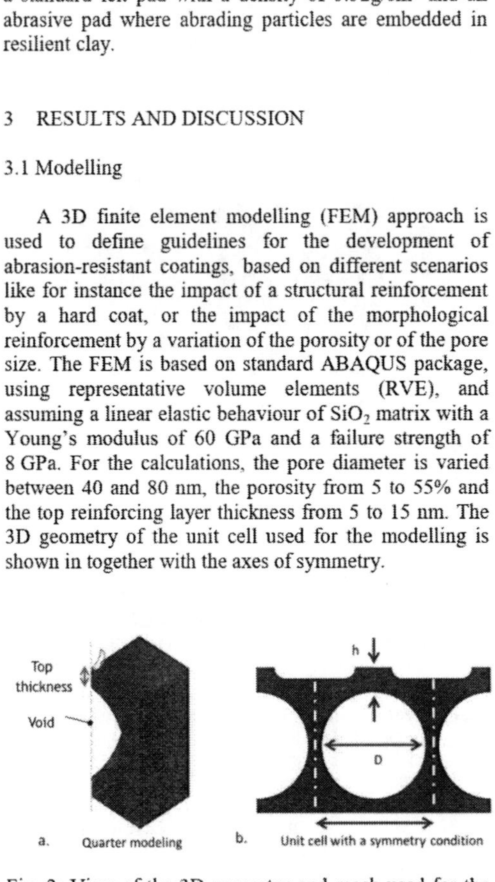

Fig. 2: View of the 3D geometry and mesh used for the FEM (a), as well as a 2D projection (b).

For the development of AR coatings with improved abrasion resistance, three strategies are considered from a modelling perspective: a modification of the coating porosity, a variation of the pore diameter, or the addition of a thin dense layer on top of the coating. Based on the distribution of the principal stress, the hardness of these three approaches is derived and reported in Fig. 3.

It is observed that the porosity has a major impact on the hardness of the coating with higher porosity leading to lower hardness (Fig. 3.a). This correlation levels off at higher values of void fraction. On the contrary, for a constant porosity, when the pore radius is changed the hardness varies only little, with larger pore radius leading to lower hardness (Fig. 3.b). For instance, doubling the size of the pore diameter, from 40 to 80 nm, leads to a reduction of the hardness from 1.9 GPa to 1.6 GPa. With respect to a reinforcement with the addition of dense silica top layer on top of the AR coating, an increase of its thickness has an intermediate impact on the hardness value (Fig. 3.c). The result of these simulations therefore indicates that the porosity (i.e. the void fraction) of the coating has significantly more impact on the hardness, than for example the pore dimension or the thickness of a dense top layer.

3.2 Abrasion experiments

These predictions are confirmed experimentally using the accelerated abrasion test developed for AR coatings. Since a higher void fraction is directly related to a lower refractive index, the development of the optimum AR coating is a trade-off between its optical and mechanical performance.

Accelerated abrasion tests have been developed based on the standard EN1096-2, and the impact of different materials (felt and abrasive pads) on the optical properties has been investigated on a model AR coating. Several mechanisms of degradation are observed according to the type of AR coating: roughness modification, complete or partial stripes, thinning, deformation of the coating, etc. The surface morphology of an AR coating (type 2) after an accelerated abrasion is presented in Fig. 4. The loss of optical properties of the model AR coating upon abrasion is ~1000 times quicker when using abrasive pads than with felt pads (Fig. 5). The acceleration factor enables therefore the development of quicker screening tests.

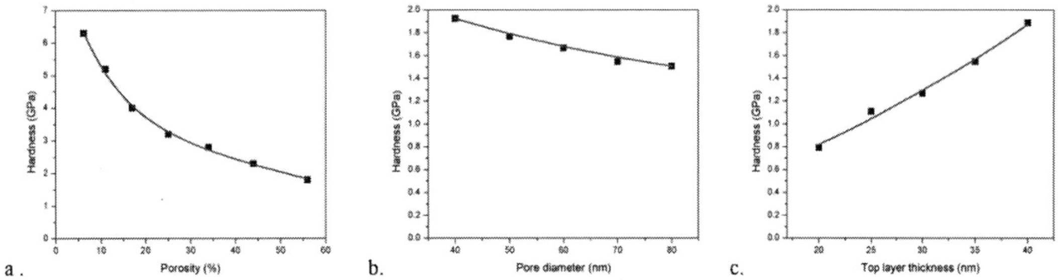

Fig. 3: Evolution of the hardness of a silica-based AR coating as a function of the porosity (a), the pore diameter (b), or the thickness of a dense top layer (c).

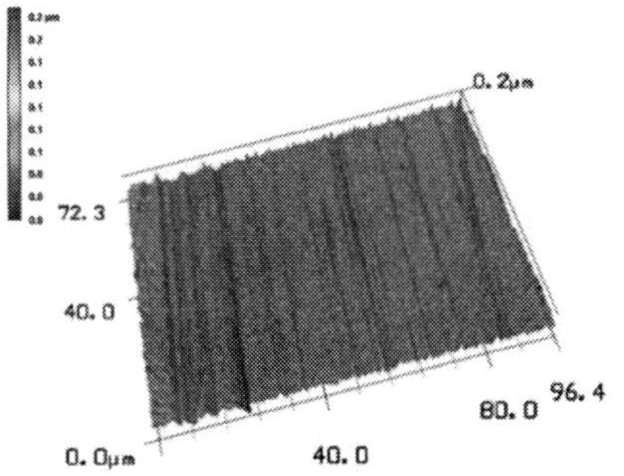

Fig. 4: 3D profile of the model AR coating surface (type 2) after an accelerated abrasion test

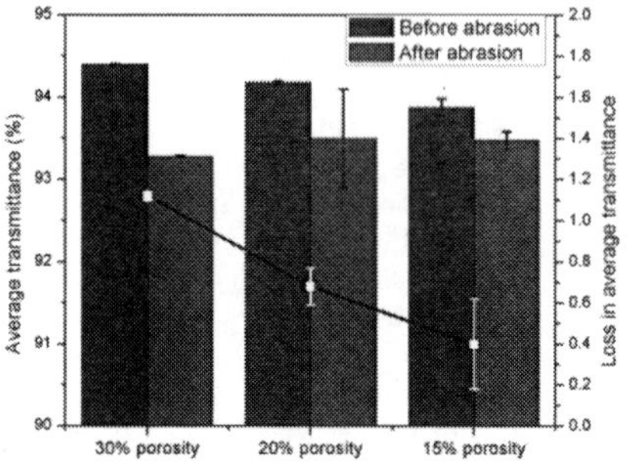

Fig. 6: Impact of the porosity of the layer on the optical transmittance and the loss during an abrasion test.

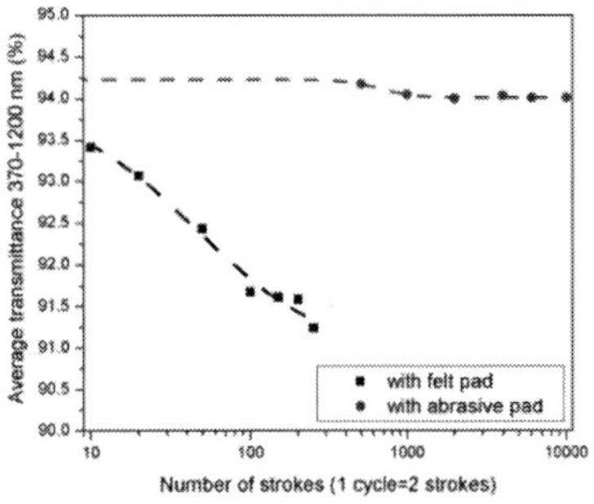

Fig. 5: Evolution of the average transmittance of the model AR coating after abrasion as a function of strokes.

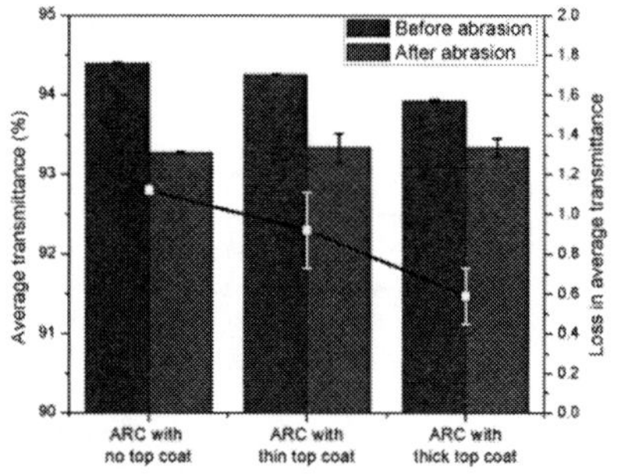

Fig. 7: Impact of the thickness of a dense top coat layer.

The two major contributing factors for a harder coating identified by the FEM approach were experimentally studied via accelerated abrasion tests using an abrasive pad. Experimental results of these two parameters for three different combinations are presented in Fig. 6 and Fig. 7. The measurements show that the optical transmittance is increasing with a higher porosity as a direct effect of the decrease of the effective refractive index of the layer, as predicted by the EMA theory. A lower porosity proves to be beneficial to abrasion resistance as the optical loss during the test is limited and the residual AR property is higher than for coatings with higher initial transmittance (Fig. 6). The same trend is observed when a dense top layer of silica is deposited on the AR layer (Fig. 7).

These experiments validate the trends observed upon modelling. A significant improvement of the mechanical properties is therefore achievable with either a 2-layer system (porous bottom coat / dense top coat), or a system with low porosity. In both cases, the optical performance of the AR coatings is similar.

4 CONCLUSIONS

The impact of abrasion on the performance of thin AR coatings was studied by the development of accelerated abrasion tests. The analysis of the surface morphology after abrasion sheds light on the different failure mechanisms that entail a loss of efficiency of the AR properties: ranging from a roughness modification of the surface up to the total removal of the coating. Finite element modelling is used to define guidelines for the development of abrasion-resistant coatings based on different scenarios like a modification of the pore fraction or of the pore size. The result of these simulations indicates that the porosity of the coating has significantly more impact on the hardness than for example the pore dimension. Simulations are confirmed experimentally using accelerated abrasion tests, and emphasize that mechanical reinforcement strategies have to be defined based on the lifetime requirements and the environmental conditions at the end-user location.

5 REFERENCES

[1] F. Guillemot, A. Brunet-Bruneau, E. Bourgeat-Lami, T. Gacoin, E. Barthel, and J.-P. Boilot, "Latex-Templated Silica Films: Tailoring Porosity to Get a Stable Low-Refractive Index," Chem. Mater., vol. 22, no. 9, pp. 2822–2828, May 2010.

[2] H. K. Raut, V. A. Ganesh, A. S. Nair, and S. Ramakrishna, "Anti-reflective coatings: A critical, in-depth review," Energy Environ. Sci., vol. 4, no. 10, p. 3779, 2011.

EFFECT OF CHANGE IN SPECTRAL TRANSMITTANCE DUE TO DUST ON CDTE AND MONO CRYSTALLINE SILICON MODULES

Supriya Rai[1], Birinchi Bora[1], O.S. Sastry[1], Rashmi Singh[1], Manander Bangar[1], Renu[1], Gopal Jha[1], Tika Ram Khadka[2]
[1]National Institute of Solar Energy, Gurgaon-122003, India,
[2]Pondicherry University, Pondicherry- 605014, India
*sastry284@gmail.com, Tel: +91-0124-2579213, Telefax No. : +91-124-2579207

ABSTRACT: This paper investigates the effect of dust on transmitted spectrum in terms of APE and MMF. The effect on spectral transmittance due to dust accumulated at different tilt angles has also been studied. Maximum effect is found to be for the 13^0 tilt, which is latitude- $15^0(28^0\text{-}15^0=13^0)$. Two SPV modules each representing crystalline (Mono) and thin film (CdTe) are used in the study. It has been found that the spectral mismatch factor increases rapidly with the concentration of dust. The measured data showed a decrease in transmittance at different wavelengths. The MMF gain increase in CdTe as compared to monocrystalline module which enhance the absorbance spectrum more efficiently but the lower tilt is mostly effected due to high concentration of dust and decrease of transmittance.
Keywords: MMF, APE, Spectrum, Dust, Soiling ratio

1 INTRODUCTION

The response of Photovoltaic devices is spectrally selective, and is also influenced by irradiance, temperature, wind speed and humidity. Difference between outdoor and standard solar spectrum plays an arching role for power rating. The solar radiation incident on the Earth's atmosphere is relatively constant but the radiation at the Earth's surface varies widely due to atmospheric effects, including absorption, scattering, reflections due to local variations in the atmosphere, water vapor, clouds, latitude of the location and dust particles.

In large power plant scale the dust accumulation is higher where we contrive maximum sunshine hour. The energy yield decreased due to the fraction of radiation losses received at the particular tilt .William et.al [1] studied the spectrum for different semiconductor devices and made the modeling for specific site condition. The variation was analyzed taking the seasonal effect. Dust affects the performance of PV module because of the change in transmittance, angular effect and change in spectral transmittance. The spectrum can be quantified by using two different factors:

1.1 MMF

Solar radiation available at the earth's surface is a function of time, location (latitude) and also attenuation effects of the earth's atmosphere. The miss match factor are depends on spectral response of the test cell and spectral response of the reference cell. The spectral mismatch correction factor, M, as per IEC 60904-7 [2] can write as:

$$M = \int \frac{\int_{\lambda 1}^{\lambda 2} (E_S(\lambda) * S_t(\lambda) d\lambda)}{\int_{\lambda 3}^{\lambda 4} (E_S(\lambda) * S_r(\lambda) d\lambda)} * \frac{\int_{\lambda 3}^{\lambda 4} (E_{ref}(\lambda) * S_r(\lambda) d\lambda)}{\int_{\lambda 1}^{\lambda 2} (E_{ref}(\lambda) * S_t(\lambda) d\lambda)} \quad (1)$$

Where:

$E_s(\lambda)$ = the absolute spectral irradiance of the Solar simulator

$E_{ref}(\lambda)$ = the Reference spectral irradiance

$S_t(\lambda)$ = Spectral response of the test cell, and

$S_r(\lambda)$ =Spectral response of the reference cell.

Simon et al. [3] reported that the spectral mismatch of crystalline silicon.

1.2 APE

Average photon energy (APE) is defined as the total irradiance contained in the spectrum divided by the total photon flux density. The APE of each spectrum is calculated from the wavelength range of 350–1100 nm of the measured spectrum. The APE value for the AM 1.5 standard solar spectrum calculated for 350-1050 nm wavelength range is 1.878 eV [5,9]. APE can also be defined as the integrated irradiance divided by the integrated flux density, which yields the average energy per photon

$$APE = \int_a^b E(\lambda) d\lambda / q \int_a^b \phi(\lambda) d\lambda \quad (2)$$

Where, q is the electron charge, E is the spectral irradiance, and Φ is the spectral photon flux density. In this study, the wavelength ranges are set to 350 nm and 1100 nm respectively.

2 EXPERIMENTAL APPROACH

NISE has established three test structure installed at 13^0, 28^0& 43^0 as shown in figure (Fig. 1) for study the dust accumulation and its effect on the PV performance. In same mounted structure clean and dust module of same technology has been installed to study the effect in variation of short circuit current, spectrum. The cleaning of one module from each technology is done on daily basis in the early morning. The glasses are taken out every day to study the spectrum effect of dust, calculate MMF and APE. The experiment is firmly established in

Figure 1: Experimental setup for testing effect of dust at different tilt

outdoor test bed exposing 90 glasses for 1 month at different tilt which is subdivided in 3 parts at angle of 13^0, 28^0 and 43^0. The glasses collected regularly experimented under solar stimulator, Specto-radiometer to see the variation in spectrum mismatch MMF, APE. The spectral mismatch should be unity for ideal case, the mismatch show the change of spectrum with respect to reference spectrum. APE decreased the diffuse irradiation which increases light scattering and AM enhances to the blue shift. Dust play a paramount role which act as 1st layer interference between clean and dust module and change in the spectrum occurred in the different tilt. This mismatch occurred due to the accumulation of soil of dust which induce double layer interface between glass and ambient layer. Spectral losses in the module change the absorption range of the spectrum due to different colour profile of the dust. Patrick D. Burton et al [4] studied the variety of coloured soil and measured its UV/VIS /NIR spectroscopy to find its attenuation. In this study the spectrum variation of exposed glasses kept in outdoor condition are used.

The effect on the performance of PV module has been analyzed by using three ways:

2.1 Transmissivity

The drop in I_{sc} due to clean glass and dusty glass was measured using the formula below:

$$\frac{1000}{G} = \frac{I_{SC} of\, Ref}{I_{SC} of Glass} \frac{Sample}{sample}$$

Where G is the extrapolated radiation at which the reference cell will give the same Isc as it is with glass on it.

2.2 Spectrum

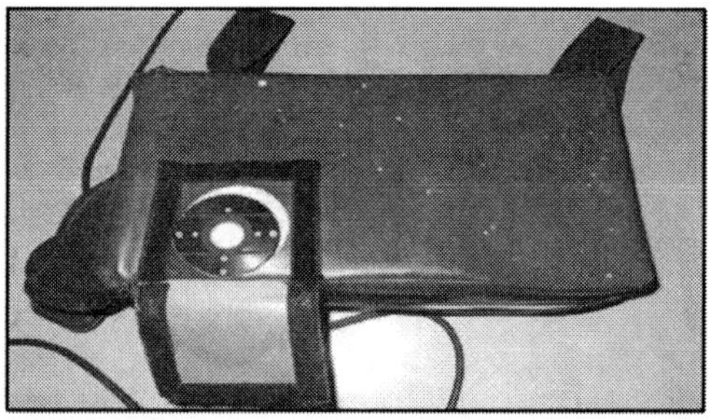

Figure 2: Spectrophotometer used for measuring Spectral irradiance

In order to check the effect of dust on the spectrum transmitted through the glass has been analyzed by using spectrometer. The measurement range of the spectrometer is 300 to 1100 nm

Soiling ratio can be calculated from the formula shown below:

$$\text{Soiling Ratio} = \frac{\text{Isc(clean)} - \text{Isc(dust)}}{\text{Isc(clean)}} * 100 \qquad (4)$$

The I_{sc} for various modules were measured using digital multi-meter and data was logged on continuous basis to a computer attached to it. We have studied the effect of spectrum and soiling losses for composite climate, quantified the effect and make it PV technology specific. Spectral transmittance depends on the absorbance property of dust particles and it can lead to the change in the useful fraction of the spectrum.

3. RESULT AND DISCUSSION

The dust effect is studied in terms of spectral transmittance which is taken under consideration. The light transmittance occurred due spectral variation of clean and unclean module observed in the Fig (3).The variation of dust with soiling ratio and wavelength is calculated and highest variation is found in lower tilt of module.

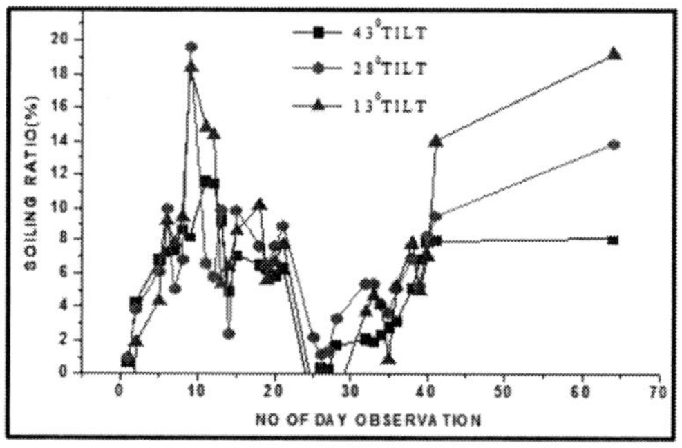

Figure 3: The Soiling ratio of glass at different Tilt Angle

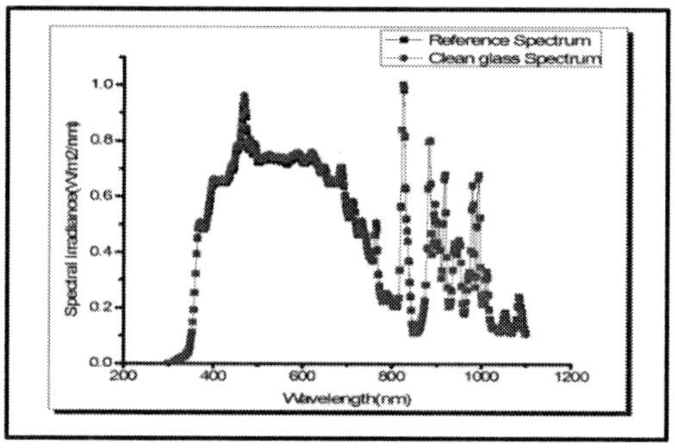

Figure 4: Spectrum clean glass respect with reference spectrum

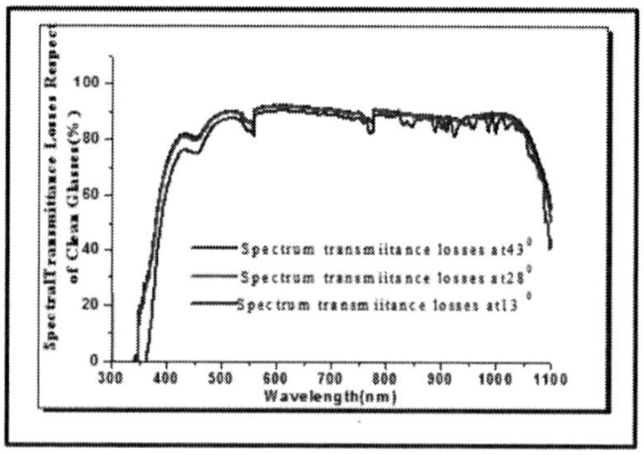

Figure 5: The transmittance losses of the Spectrum for different tilt angle

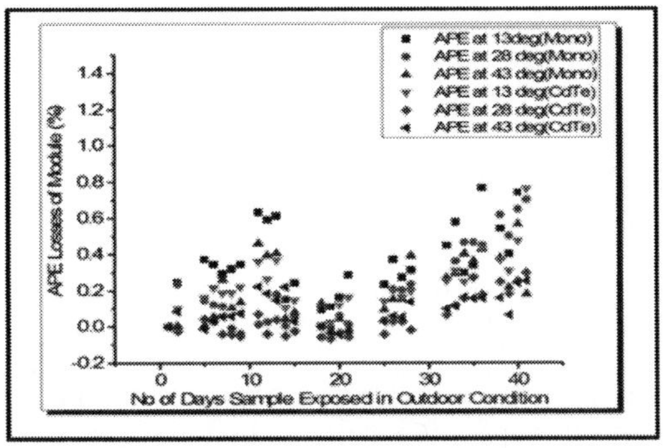

Figure 6: APE losses for diffrent tilt angle

The transmittance of dust particle has been estimated by using glass slides exposed in outdoor conditions. The accumulation of dust particle absorbs the fraction of light at particular wavelength. The spectrum with clean glass is taken as a reference spectrum to calculate the spectral transmittance. In the fig (4) the reference spectrum from the source and clean glass is measured and compared. The figure (5) shows the scattering observed in the IR range occurred due to absorbance of the dust particle in lower tilt. Spectral variation is not observed significantly in the outdoor condition when tested at zero degree angle of incidence [7].The variation need to study at different tilt angle. Fig 6 and 7 describe the APE and MMF losses for diffrent tilt. The losses is verified depending on the seasonal effect which effect the losses more in summer as compared to the winter.

Figure 7: MMF losses for diffrent tilt angle

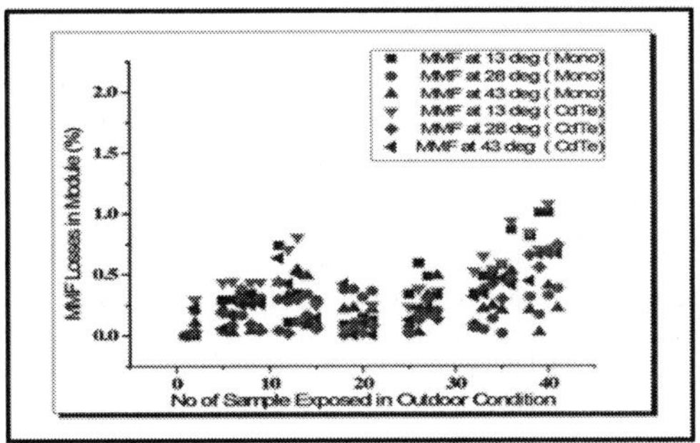

The spectral impact is the energetic gain or loss in effective irradiance available to a specific PV technology compared to total irradiation measured by a pyranometer. Spectral mismatch factor is calculated according to IEC 60904-7. MMF essentially expresses how more or less irradiance the device under test sees change in current as compared to the reference device's indication. MMF can be rewritten as the ratio of I_{SC} (shortcircuit current) of the module under investigation, and I_{SC} of the reference device as irradiance is directy influnced the current change . MMF>1 signifies the gain in spectrum power as compared with reference devices where MMF<1 determine the looses in the spectral power as compared with reference device.

4. CONCLUSION

The study conclude the transmittance losses is more in low tilt as compared to high tilt. It depends on the concentration of dust deposited in the particular area. Due to dust composition the particular region of absorption wavelength is failed to absorb the region and spectrum is shifted towards red shift. CdTe is showing high MMF as compared to Monocrystalline module with change in dust accumulation.Quantification of dust will help further to select the technology specific condition for diffrent climatic zone of India.

ACKNOWLEDGEMENT

This research is based upon work supported in part by the Solar Energy Research Institute for India and the U.S. (SERIIUS) funded jointly by the U.S. Department of Energy subcontract DE AC36-08G028308 (Office of Science, Office of Basic Energy Sciences, and Energy Efficiency and Renewable Energy, Solar Energy Technology Program, with support from the Office of International Affairs) and the Government of India subcontract IUSSTF/JCERDC-SERIIUS/2012 dated 22nd Nov. 2012

REFERENCES

[1] S.R. Williams, T.R. Betts, P. Vorasayan, R. Gottschalg. D.G.Infield" Actual PV Module Performance Including Spectral Losses In The UK"

[2] IEC 60904-7: Photovoltaic devices – Computation of the spectral mismatch correction for measurements of photovoltaic devices, 2009- 12

[3] Michael Simon, Edson L. Meyer "The effects of spectral evaluation of c-Si modules" Vol19 Issue(1),D.O.I 10.1002/pip.973.

[4] PatrickD. Burton and Bruce H. King, Member, IEEE"Spectral Sensitivity of Simulated Photovoltaic Module Soiling for a variety of Synthesized Soil Types" ,IEEE Journal for Photovoltaic ,Vol4 Issue3.

[5] Ulrike Jahn Dr Wemwer"Spectral Analysis of Various Thin film module Using high Precision Spectral Response Data and Solar Spectral Irradiance" 27thEuropeanconferenceD.O.I10.4229/27thEUPVSEC2012-4BV.2.22.

[6] Paul Grunowl et .al"Actual issues on power measurement of photovoltaic modules"

[7] Modules H. Qasem*, T.R. Betts and R. Gottschalg " Effect Of Shading Caused By Dust On Cadmium Telluride Photovoltaic". Photovoltaic Specialists Conference (PVSC), 2011 37th IEEE

[8] D.B. Magare, O.S. Sastry, R. Gupta, T.R. Betts, R. Gottschalg, A. Kumar, B. Bora, Y.K. Singh, "Effect of Seasonal Spectral Variations on Performance of Three Different Photovoltaic Technologies in India," Int J Energy Environ Eng, DOI 10.1007/s40095-015-0190-0

[9] C.R. Oster Wald et .al "Photovoltaic module calibration value versus optical air mass: the air mass function." Progress in Photovoltaics Research and Applications22(5) · May 2014 DOI: 10.1002/pip.2303

IMPROVED 1500V PID RESISTANCE: ENCAPSULANTS, COVER GLASS AND ION IMPLANTED CELLS

B. Braisaz[1], B. Commault[2 3], N. Le Quang[4], E. Gerritsen[2 3], M. Joanny[2 3], D. Binesti[1], G. Goaer[4], K. Radouane[5]
[1]EDF R&D, ENERBAT, F-77250 Moret sur Loing, France,
[2] Univ. Grenoble Alpes, INES, F-73375 Le Bourget du Lac, France
[3]CEA, LITEN, Department of Solar Technologies, F-73375 Le Bourget du Lac, France
[4]EDF ENR PWT, 33, rue Saint-Honoré, F-38 307 Bourgoin-Jallieu, France
[5]EDF EN, F-92400 La Défense, France

ABSTRACT — In this paper, several solutions against PID have been tested on mini-modules then 60-cell modules. 4 encapsulants are compared at the mini-module level: two types of ethylene-vinyl acetate (EVA), a thermoplastic polyolefin (TPO) and an ionomer. Then 3 out of 4 encapsulants are compared on 60-cell modules in this study: an ethylene-vinyl acetate (EVA), a thermoplastic polyolefin (TPO) and an ionomer.
3 modules architectures have been also tested: a module with a standard glass and a TPT backsheet, a module with a special thin glass and a TPT backsheet, and a special thin double glass module using the same thin glass (only on 60-cell module).
At the end, a module with ion implanted cells, while others components were kept identical, is PID tested.
The applied voltage was 1500V in order to prepare the future increase of DC plant voltage. On mini-modules, a first PID test at room temperature (25°C and Al-foil) has been conducted before a more severe test at 70°C (Al foil), allowing us to clearly differentiate the influence of the tested materials on PID degradation of the modules.
The conditions of these tests on 60-cell modules were 192 hours 75°C 85RH. Contrary to EVA and standard glass modules, we clearly observed no degradation on thermoplastic polyolefin (TPO), ionomer, thin glass modules and module with ion implanted cells. Using one of these components is a solution against PID.
Keywords — PID, degradation, test, encapsulant, thin glass, ion implanted cells

1. STATE OF THE ART

Development and operation of solar power plants require good knowledge of the module degradation types and their sensitivities to different stresses (temperature, humidity, bias voltage stress, UV etc...).

PID (Potential Induced Degradation) is a cause of degradation. It is due to a high negative voltage between module– cells polarities and the ground (module frame). It strongly depends on the humidity and temperature, but also on:

- Cell components (junction technology, thin film materials, etc ...)
- Module components (EVA, antireflection layer, frame, glass etc ...)
- Module aging and soiling
- System (type and topology of inverters, system voltage, DC and AC grounding system etc ...)
- Mounting system (metallic or wood support, fixing plates, dielectric characteristics of rubber, frame-support contact, tightening torque etc...)
- The environment (weather, temperature, humidity, irradiance, dirtiness etc ...)

This topic, active in terms of international publications since 2010 thanks to the papers of Solon [1][2] and NREL [3], appeared several years ago when the PV system voltages increased. But the degradation was initially identified in 1978 by Jet Propulsion Labs [4]. Today, no testing standard exists to differentiate PID prone modules to "PID free" modules. Researches to find PID free solutions are currently conducted and the first modules "PID free" come on the market.

The main theory to explain the PID phenomenon on c-Si modules comes from the paper of Naumann [5] which assumes that sodium ions are responsible of the degradation. According to the paper, the electrical field (frame-cells) triggers the diffusion of sodium ions from the glass to the encapsulating material and cells. These ions are then released and accumulate at the surface of the Silicon Nitride (AR layer). Sodium ions propagate through Si stacking faults in PN junction. In the case of P-type cells, these charges generate local shunts in the PN junction: shunt resistance and Voc are strongly affected.

Inside the module, many parameters can mitigate or suppress the PID consequences: EVA resistivity and glass resistivity [6], intermediate diffusion layer, thickness of glass and EVA, AR coatings [7], cells and silicon features etc.

Several solutions are being studied currently in order to solve PID. However, these solutions are not always validated by laboratories: a module announced by the manufacturer insensitive to the PID may actually be sensitive. The degradation mechanism due to the potential is not fully understood at the material level. In addition, the power plant voltage increase to 1500V will accentuate PID stresses.

The work done before in cited publications has identified possible electrical causes (Figure 1), of degradation processes at the microscopic level for c-Si and CdTe modules, the impacts on PV plants, the testing methods used by laboratories and a panel of solutions at 3 levels (cell, module and system) ([8], [9]).

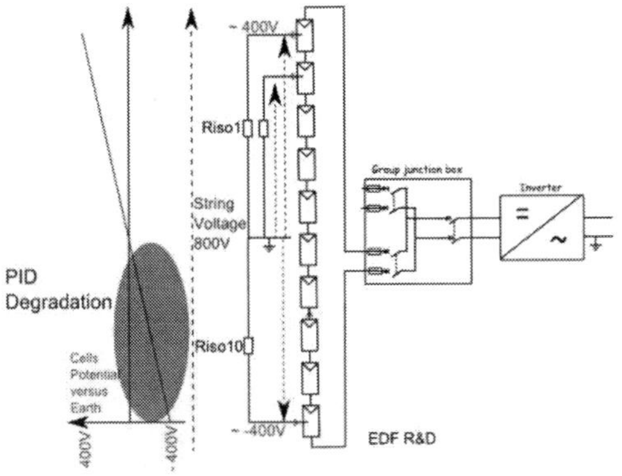

Figure 1: PID principle.

In [10], we showed that STC measurements are not sufficient for measuring PID consequences: low light behavior should be a mandatory criterion for pass/fail results in a PID test. Indeed, for example, the module tested in [10] exhibits no degradation at 1000W/m² but showed strong degradation at 200W/m² after 96hours 60°C 85RH -1000V (Draft of the 62804 standard [11]). It could have passed the standard if there had only a criterion at 1000W/m².

In [12], we defined and proposed a representative, selective and reproducible test protocol with a pass or fail criterion: **192h, 75°C, 85RH, -1500V, with 5% of maximum loss at 200W/m²**.From indoor accelerated test results and an empirical PID model, the outdoor degradation has been estimated. After several iterative simulations, an indoor test protocol with specific "pass or fail" criteria has been found in order to have limited losses after 30 years of outdoor exposition in Miami for a module at the end of a 1500V string.

2. MATERIALS AND METHODS

Mini-modules:

Two 1-cell mini-modules with a standard solar cell (p-type monocrystalline silicon with POCl3-diffused emitter, SiNx ARC, Al-BSF) have been realized with:

- 4 Encapsulants:
  - A classic ethylene-vinyl acetate (EVA),
  - A 28% of vinyl acetate EVA
  - A thermoplastic polyolefin (TPO)
  - An ionomer

Two other modules with 4 cells in series have been realized:

- 2 Glass / Backsheet assemblies:
  - Standard glass and a TPT backsheet,
  - Special thin glass and a TPT backsheet,

The module's PID sensibility on mini-modules has been tested using the following protocol: modules have been stacked together (glass-to-glass) with an aluminum foil between them. The aluminum foils have been interconnected together to the negative terminal of the voltage generator while the cell strings were connected together to the positive terminal. Thus the voltage of 1500V delivered by the generator has been set between the aluminum foil and the cells. A first test at 25°C and a relative humidity <60% has been conducted on all modules for 192h. Then a more severe test was carried out at a temperature of 70°C.

60-cell modules:

In this paper, we used our test protocol (**192h 75°C 85RH -1500V with 5% of maximum loss at 200W/m²**) on 60-cell modules to test different anti PID solutions tested before on mini-modules:

- 3 Encapsulants:
  - A classic ethylene-vinyl acetate (EVA),
  - A thermoplastic polyolefin(TPO)
  - An ionomer
- 3 Glass Backsheet assemblies:
  - Standard glass and a TPT backsheet,
  - Special thin glass and a TPT backsheet,
  - Special double thin glass module
- 1 module with implanted cells

3. MINIMODULES TESTS RESULTS

The maximum power variation of the modules during the test at 25°C is presented in Figure 2.

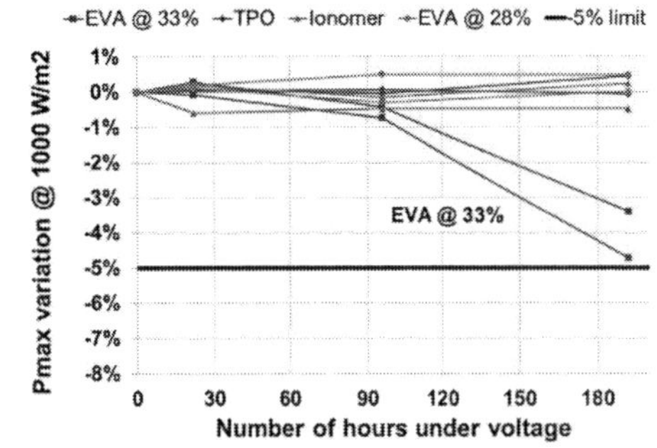

Figure 2: STC relative power evolution (%) of mini-modules with different encapsulants (2*EVA, Ionomer and Polyolefin) during PID at -1500V / 25°C / with aluminium-foil.

The maximum power of the tested modules is stable during the test except for the modules encapsulated with the EVA encapsulant at 33% of vinyl acetate.

To discriminate the impact on PID of the 3 remaining encapsulants, a PID test at 70°C has been carried out. The electroluminescence images of the mini-modules during the test are presented below.

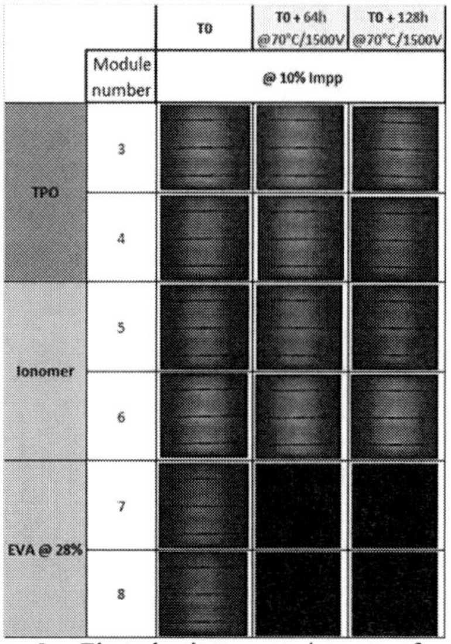

Figure 3: Electroluminescence images after PID at -1500V / 70°C / with aluminium-foil.

We have obtained a very good cell response before the test for all modules which remain stable during the test except for the modules with the EVA at 33% (Figure 2). Even if this EVA encapsulant is made with 28% of vinyl acetate it is clearly not sufficient to minimize the PID (Figure 3). On the other hand the modules with a TPO or ionomer encapsulant allow a very good and stable cell response by electroluminescence.

The evolution of the I-V curves during PID testing at 70°C for modules with the 3 different encapsulant types is presented in Figure 4).

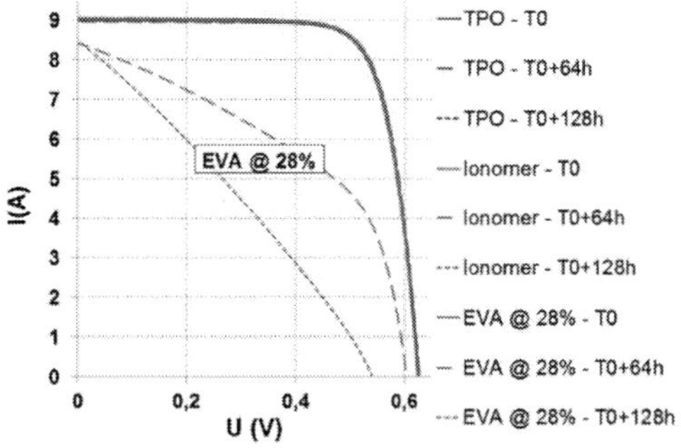

Figure 4: IV curves after PID stress test (-1500V / 70°C / with aluminium-foil).

We observed degradation for the module with the EVA at 28%, while the modules with a TPO or ionomer encapsulant do not show any degradation.

The two 4-cell mini-modules with a chemically strengthened thin glass (0.85mm thick) have been subjected to the PID tests at -1500V versus a 4-cell mini-module with a standard solar glass. The results, as relative variation of the maximum power at 1000 W/m^2 during the tests, are presented in Figure 5.

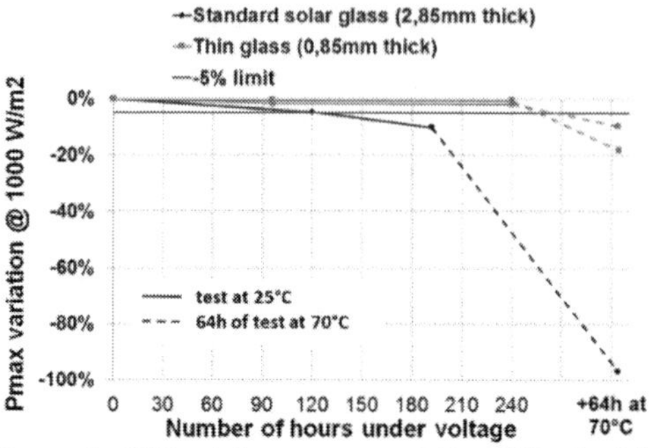

Figure 5: STC relative power evolution (%) of mini-modules with different glass during PID test (-1500V / 25°C / with aluminium-foil).

The degradation rate is limited by the use of the chemically strengthened thin glass compared to the module with a standard solar glass. Nevertheless the PID resistance induced by the chemically strengthened thin glass is not enough to have a degradation of the initial module power less than 5%.

4. STANDARD SIZE MODULES RESULTS

Encapsulant tests:

On Figure 6, we can see the module efficiency as a function of the irradiance. We observed no losses at all for Polyolefin and Ionomer modules, even at low irradiances of 200W/m². These modules slightly increased their power contrary to the EVA module which lost 66% at STC conditions and 88% at 200W/m² (too low to be displayed on Figure 6). We can see the EL images on Figure 9.

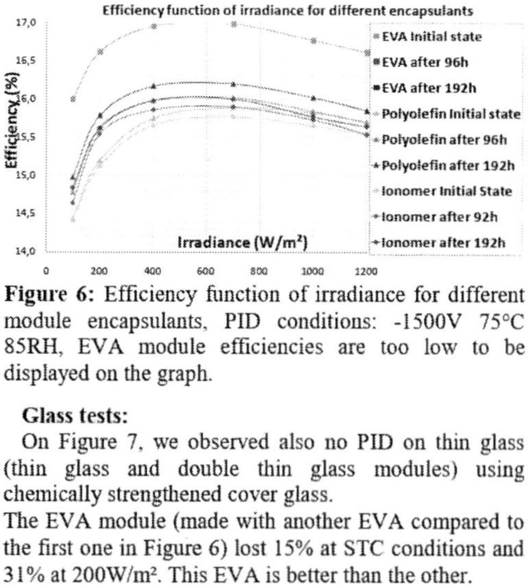

Figure 6: Efficiency function of irradiance for different module encapsulants, PID conditions: -1500V 75°C 85RH, EVA module efficiencies are too low to be displayed on the graph.

Glass tests:

On Figure 7, we observed also no PID on thin glass (thin glass and double thin glass modules) using chemically strengthened cover glass.

The EVA module (made with another EVA compared to the first one in Figure 6) lost 15% at STC conditions and 31% at 200W/m². This EVA is better than the other.

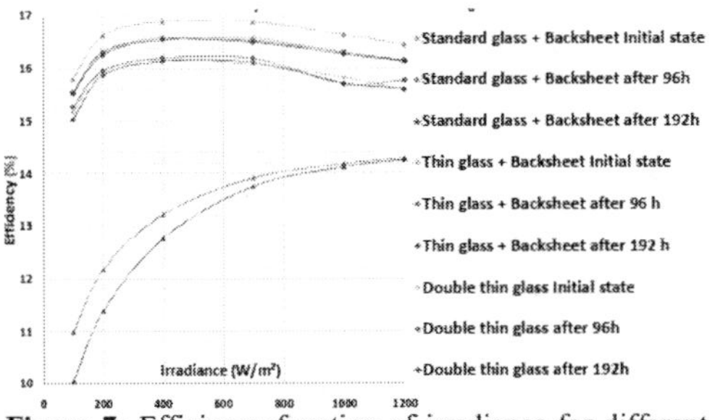

Figure 7: Efficiency function of irradiance for different module glass, PID conditions: -1500V 75°C 85RH.

Temperature coefficients:

We also measured the temperature coefficients after PID stress 192h -1500V 85RH 75°C on an EVA module. It appeared that the power was very low (-66%) but not any more dependent on temperature (Figure 8).

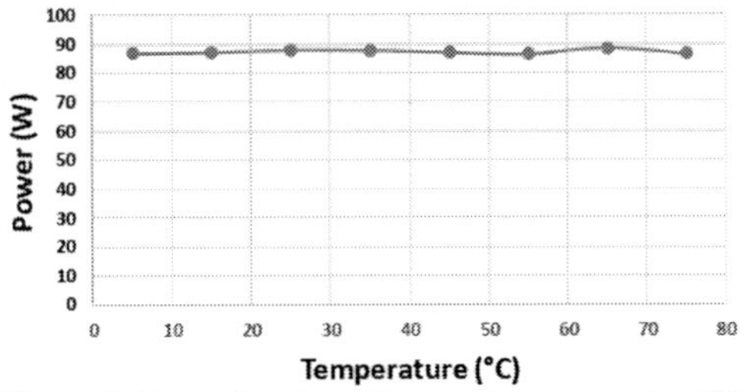

Figure 8: Power function of temperature curve after PID stress: no temperature dependency.

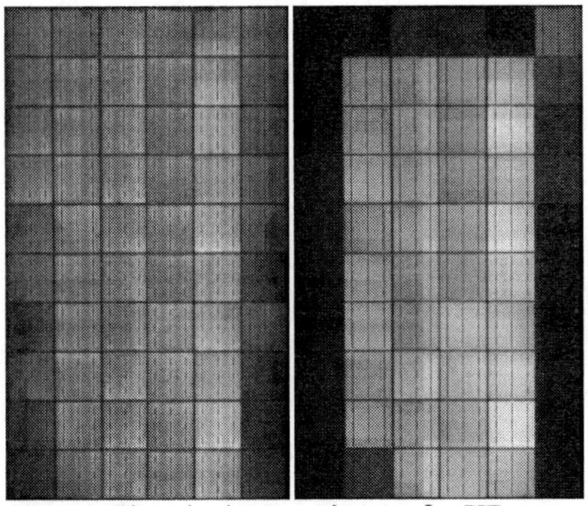

Figure 9: Electroluminescence images after PID stress - 1500V 75°C 85RH 96h (9A on the left and 0.9A on the right).

Ion implanted cells:

We tested a 60-cell module with ion implanted cells (p-type monocrystalline silicon with SiO2/SiNx stack on front side, ion implanted emitter, Al-BSF) versus a 60-cell module with standard diffused cells. We observed no degradation for the module with EVA and ion implanted cells. It seems that ion implanted cells are PID resistant. Further investigations need to be done in order to validate this interesting result also published in [19-21].

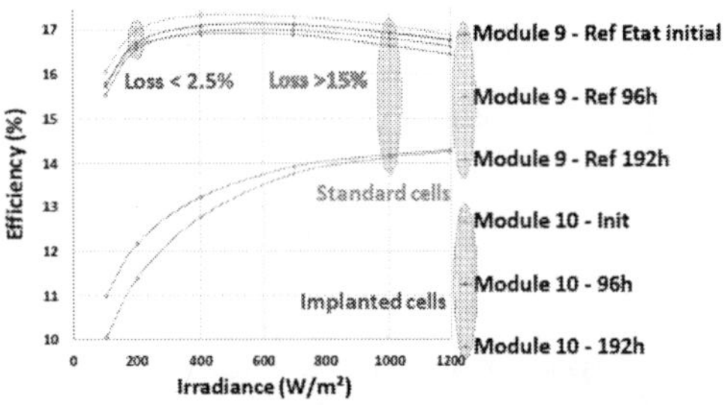

Figure 10: Efficiency function of irradiances for a module with implanted cells versus a module with diffused cells, PID conditions: -1500V 75°C 85RH.

5. CONCLUSIONS

In this paper, we demonstrated that solutions giving module resistance to PID are: TPO, ionomer, thin PID free glass with lower Na content (chemically strengthened glass), and ion implanted cells.

The most promising solution, in terms of cost, is the ion implanted cells.

6. ACKNOWLEDGEMENTS

The authors wish to thank ADEME (Agence française pour l'environnement et la maitrise de l'énergie) for funding this work by financial support. This work has been done in the frame of the ADEME project called "PID1500" in which the goals are to define an adequate PID test protocol for 1500V systems and to develop solutions against PID at the cell and module levels.

7. REFERENCES

[1] S. P. e. al., "Potential induced degradation of solar cells and panels," *IEEE*, 2010.

[2] S. P. e. al., "Potential Induced Degradation of Solar Cells and Panels," *PVSEC*, 2010.

[3] P. H. e. al., "Test-to-Failure of Crystalline Silicon Modules," *IEEE*, 2010.

[4] Hoffman, "Environmental Qualification Testing of Terrestrial Solar Cell Modules," *IEEE*, 1978.

[5] Naumann, "Explanation of potential-induced degradation of the shunting type by Na decoration of stacking faults in Si solar cells," *Solar Energy Materials & Solar Cells*, 2013.

[6] Pingel, "The local potential distribution as driver of PID," *PVSEC*, 2014.

[7] Berghold, "PID effects on crystalline silicon cells with various AR coatings," *Photovoltaic*

international, 2012.

[8] Berghold, "PID and its correlation with experience in the field PV international," *Photovoltaic International*, 2013.

[9] Kastl, "Dealing with high voltage stress," *PV magazine*, 2011.

[10] Braisaz, "PID Results at low irradiances on c-Si modules," *IEEE*, 2014.

[11] IEC, "Test methods for detection of potential-induced degradation of crystalline silicon," *62804*, 2014.

[12] Braisaz, "Evaluation of "PID 1500V Ready" modules: a new test protocol," *IEEE*, 2015.

[13] H. &. Koehl, "Modeling the leakage current for Potential Induced Degradation," *Progress in PV*, 2012.

[14] B. Dupeyrat, "Improved method for the extraction of PV module parameters, PVSEC 2013," *PVSEC*, 2013.

[15] Braisaz, "An ageing model of PV modules and a link between indoor accelerated tests an outdoor degradation," *PVSEC*, 2012.

[16] Hacke, "System Voltage Potential-Induced Degradation Mechanisms in PV Modules and Methods for Test," *NREL*, 2011.

[17] Lechner, "PID-Behaviour of Thin-Film and c-Si PV-Modules," *PVSEC*, 2013.

[18] Hattendorf, "Potential induced degradation in mono-crystalline silicon based modules: an acceleration model," *PVSEC*, 2012.

[19] Zhu, "Cost-effective industrial n-type bifacial and IBC cells with ENERGi™ P and B ion implantation" SNEC, 2014.

[20] Basame, "Ion implanted high efficiency PID free cells for today's solar market" PVSEC, 2013.

[21] Han, "Ion implantation - an effective solution to prevent c-Si PV module PID" PVSEC, 2013.

DURABILITY OF BIFACIAL SOLAR MODULES UNDER POTENTIAL INDUCED DEGRADATION: ROLE OF THE ENCAPSULATION MATERIALS

Marco Barbato[1,2,*], Matteo Meneghini[1], Andrea Cester[1,2], Alessandro Barbato[1], Giacomo Tavernaro[3], Mark Rossetto[3], Gaudenzio Meneghesso[1,2]
[1] University of Padova, Department of Information Engineering, Via Gradenigo 6/B, 35131, Padova, Italy.
[2] Centro Levi Cases, University of Padova, Via Marzolo 9, 35131, Padova, Italy.
[3] MegaCell S.r.l., Via Postumia Vecchia 9, 35010, Carmignano di Brenta, Padova, Italy.
* Phone: +39 049 8277719, Fax: +39 049 8277699, E-mail: barbatom@dei.unipd.it

ABSTRACT: In this paper we investigate the effects of the potential induced degradation (PID) in bifacial N-type solar cells (BSC) using different encapsulation materials. Three different approaches were used for the encapsulation: (i) two sheets of polyolefin elastomer type I (POE-I) in the front and one sheet of POE-I in the back of the solar cell (A-type), (ii) two sheets of POE-I in the front and one sheet of ethylene-vinyl-acetate (EVA) in the back of the solar cell (B-type) and (iii) one sheet of polyolefin elastomer type II (POE-II) in the front and one sheet of ethylene-vinyl-acetate (EVA) in the back of the solar cell (C-type). Within this work, we show how the use of different encapsulation structures may strongly affect the robustness to PID. Moreover the results demonstrate that, after 55 hours of PID, the A-type and B-type configurations show similar degradation of the maximum power (26.5%) whereas the C-Type configuration shows a considerable improvement in the PID reliability (maximum power degradation of 1.5%). The analysis of the recovery phase shows that PID degradation is mainly recoverable and that it can be recovered by applying a reverse voltage with respect to the PID stress.
Keywords: Potential Induced Degradation, Reliability, N-type solar cells, Degradation, Module.

1 INTRODUCTION

Monocrystalline bifacial solar cells (BSC) are a commercially important technology for manufacturing of solar cells. However, the efficiency of BSC is limited by the potential induced degradation (PID). It is well know that PID causes the migration of Na+ ions [1]-[3] of the external glass inside the solar cell structure or surface polarization effects [4]. This is a big issue for the solar cell reliability, since it causes the lowering of the fill factor and the conversion efficiency.

The understanding of the influence of PID in N-type BSC performance is a relevant issue in the photovoltaic community. PID can cause a relevant degradation of the efficiency and fill factor during the operation of the solar module [5] and can cause the lowering of the shunt resistance [6-7] with consequent problems in the string of solar cells. It is very important to find the degradation kinetics in order to understand the impact of PID on the solar cell efficiency during time.

The approach used in this work can be used to find the optimum encapsulation material and to study the degradation law of the PID. Moreover the study of different encapsulation materials allows to find the better solution considering both the robustness during time and the costs of the encapsulants. We propose the combined use of different materials to avoid and reduce the degradation induced by the PID stress in bifacial solar cell technology.

The purpose of this work is to demonstrate the capabilities of different encapsulation structures to limit the PID and to mitigate the ions or charges migration in bifacial solar cells. First, we assembled different solar panels with the three following structures: (i) two sheets of polyolefin elastomer type I (POE-I) in the front and one sheet of POE-I in the back of the solar cell (A-type), (ii) two sheets of POE-I in the front and one sheet of ethylene-vinyl-acetate (EVA) in the back of the solar cell (B-type) and (iii) one sheet of polyolefin elastomer type II (POE-II) in the front and one sheet of ethylene-vinyl-acetate (EVA) in the back of the solar cell (C-type). Then, we have subjected these bifacial solar panels (BSP) to 55 hours of PID @ 65 °C: an Al sheet covers the front of the module, and a constant voltage stress of -600 V was applied to the solar cells for 55 hours by using a dc power supply (Keithley 2410). The entire stress procedure was periodically interrupted to characterize the solar cells with time steps of 1h, 3h, 7h, 15h, 39h and 55h. The analysis was carried out by using both electrical and optical (Electroluminescence and External Quantum Efficiency) measurements. For the electrical characterization, two types of measurements were considered: dark I-V characterization, and I-V curves at one sun using a white LED-based solar simulator. After the 55 hours of negative bias, a +600 V PID @ 65 °C was applied for 48h to induce the recovery of the solar panel properties.

2 DESCRIPTION OF BIFACIAL SOLAR CELLS END ENCAPSULATION TECHNIQUE

Bifacial N-type solar cells are manufactured in MegaCell using 4 busbars. MegaCell produces high efficiency solar cells up to 21 % efficiency, with a bifacial factor > 88 % (ratio between rear and front efficiency), using N-type silicon wafers. The cell architecture is represented in Figure. 1.

The cells have been soldered manually using standard ribbon SnPb 60/40 with dimensions 1.2 mm x 0.2 mm. In the front an extra clear 2 mm glass and in the back a transparent back sheet have been used. For the lamination a one-step laminator has been used.

The BSCs are encapsulated using different encapsulants (example in Figure 2) using three different strategies: (i) two sheets of polyolefin elastomer type I (POE-I) in the front and one sheet of POE-I in the back of the solar cell (A-type), (ii) two sheets of POE-I in the front and one sheet of ethylene-vinyl-acetate (EVA) in the back of the solar cell (B-type) and (iii) one sheet of polyolefin elastomer type II (POE-II) in the front and one sheet of ethylene-vinyl-acetate (EVA) in the back of the

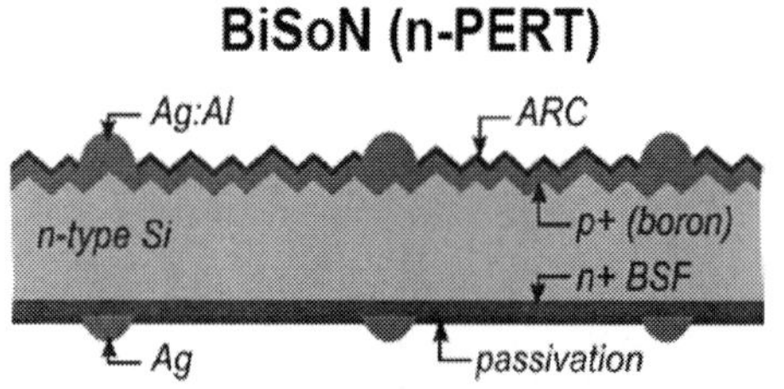

Figure 1: Schematic diagram of a Bifacial Solar Cell (BSC).

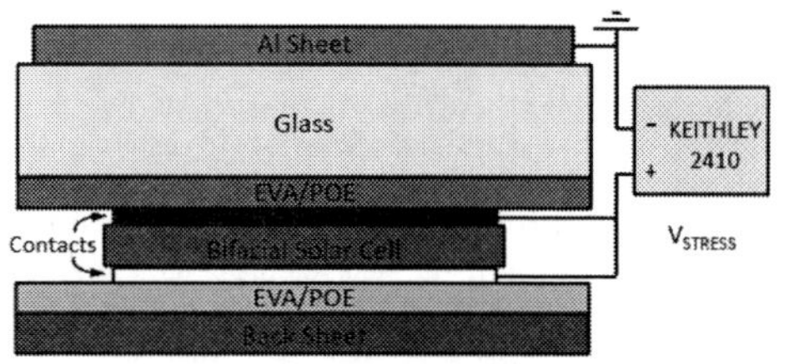

Figure 2: Scheme of the structure.

solar cell (C-type). The polyolefin elastomer type I (POE-I) and polyolefin elastomer type II (POE-II) have different physical properties. In fact the water vapor transmission rate is 2.5 [g·mm/m^2 day] for POE-I and 0.6 [g·mm/m^2 day] for POE-II.

3 EXPERIMENTAL DETAILS

In order to highlight the degradation induced by the PID stress, we take four representative samples of BSP with different encapsulation structure, respectively. The main solar cell parameters are obtained by using four typologies of measurements (named "Off line measurements"): *Dark I-V characterization*, *Light I-V characterization* using a white LED solar simulator, *Electroluminescence characterization* and *External Quantum Efficiency*. In the following we describe our measurement setups. We carried out the dark I-V characterization of the different cells by using a source meter (Keithley 2651A). The measurements were carried out with a four wire technique in order to minimize the effect of the parasitic impedance of the cables.

In order to extrapolate light I-V curves, we used a LED based solar simulator that can produce an arbitrary illumination level between 0.2 and 4 suns combined with the source meter Keithley 2651A. The solar simulator is based on a white LED array calibrated at 1 sun. The advantage of using a LED-based solar simulator is the possibility to adjust the irradiance level. In Figure 3a and Figure 3b the I-V characteristics obtained on a representative BSC are shown. The main extracted parameters are: the open circuit voltage 0.66 V (V_{OC}), the short circuit current 9.55 A (I_{SC}), the maximum power 4.68 W (P_{MAX}), the fill factor 0.74 (FF), the efficiency 19.4 (η), the shunt resistance > 4500 ohms (R_{SH}) and the series resistance < 10 mohms (R_S). The series resistance, and consequently the efficiency and the fill factor, can be improved since the solar cells have been soldered manually.

We carried out electroluminescence measurements in forward bias (+ 7 A) using an InGaAs camera (Xeva) in

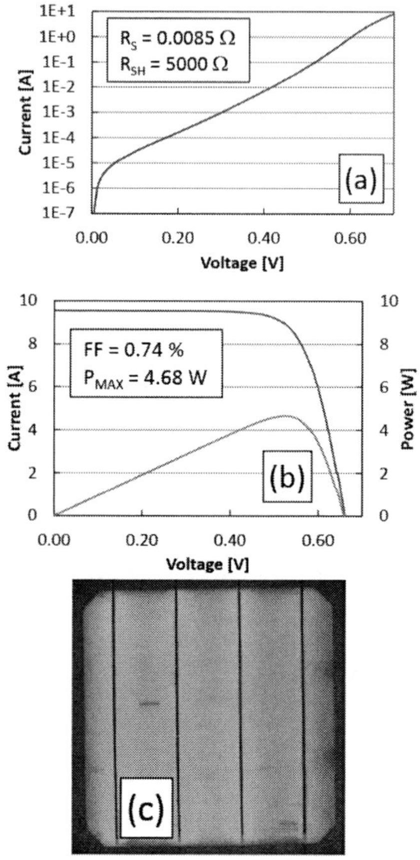

Figure 3: (a) Dark and (b) Light IV curves of a representative Bifacial Solar Cell. (c) EL @ + 7A.

order to evaluate the presence of defects and weak regions in the solar cell area. Moreover different External Quantum Efficiency (EQE) measurements were carried out during the PID stresses using the commercial instrument LOANA (pvTools) system. Figure 3c shows a representative electroluminescence image on a fresh BSC extracted with a forward current of +7 A. The PID test was performed inside an oven at 65 °C.

Figure 4 shows a schematic block diagram of the stress test: after the preliminary characterization at 25 °C, the BSP was subjected to a 1 hour PID test @ 65 °C. Then the system was cooled down at 25 °C and the "Off line measurements" were carried out. This scheme was repeated for different steps with this cumulative temporization: 1 hour, 3 hours, 7 hours, 15 hours, 39 hours and 55 hours. After the 55 hours of negative biasing, a +600 V PID @ 65 °C was applied for 48h to study the recovery of the solar panel properties.

4 PID RESULTS

Figure 5 shows the electrical characterization of one representative A-type sample subjected to 55 hours of PID @ -600 V and 48 hours of recovery @ +600 V. Figure 5a and Figure 5b respectively show the dark I-V curves and the light I-V curves. From the dark I-V characterization one can clearly see the decrease of the

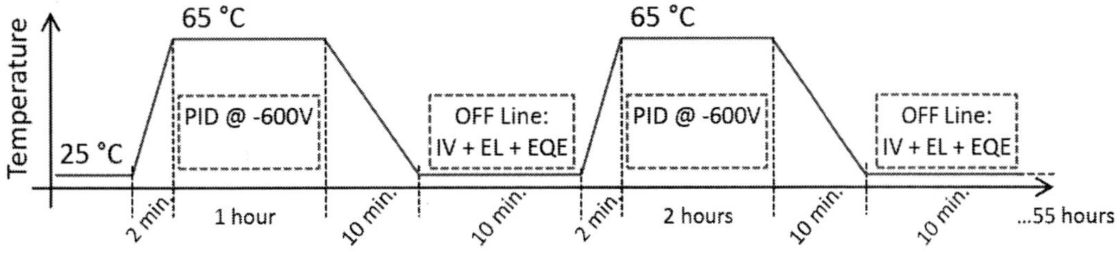

Figure 4: Scheme of the PID step stress test.

shunt resistance. Moreover the light I-V characterization highlights the lowering of the short circuit current, and the lowering of the open circuit voltage. The comparison of the different solar cells is visible in Figure 6 that shows the evolution of the normalized short circuit current (I_{SC}), the normalized open circuit voltage (V_{OC}), the normalized maximum power (P_{MAX}) and the normalized fill factor (FF) during the PID for four representative solar cells (A-type, B-type and C-type). The A-type and B-type configurations show similar results with an average decrease of the maximum power of 26.5%, conversely the C-type shows a decrease of barely 1.5%. The A-type differs from the B-type for the fill factor at the steps 39 h and 55 h. This can be related to the different encapsulation material of the rear side of the solar cells.

The different degradation of the solar cells encapsulated with polyolefin elastomer type I (POE-I) and polyolefin elastomer type II (POE-II) can be related to the different properties of the two materials. In fact the water vapor transmission rate is 2.5 [g·mm/m^2 day] for POE-I and 0.6 [g·mm/m^2 day] for POE-II.

Figure 5 and Figure 6 show also the PID recovery @ + 600 V for 48 hours. The measures indicate that the

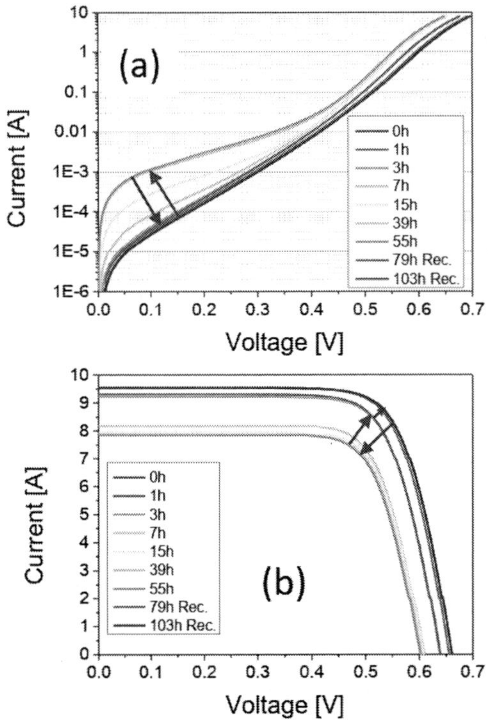

Figure 5: Dark and light I-V curves during a PID stress.

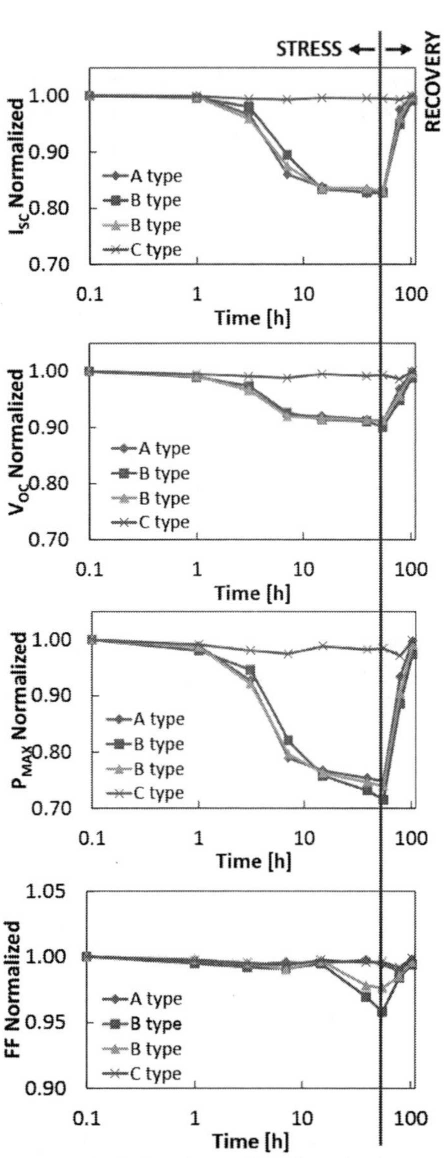

Figure 6: Evolution of normalized short circuit current, normalized open circuit voltage, normalized maximum power and normalized fill factor for the four solar panels.

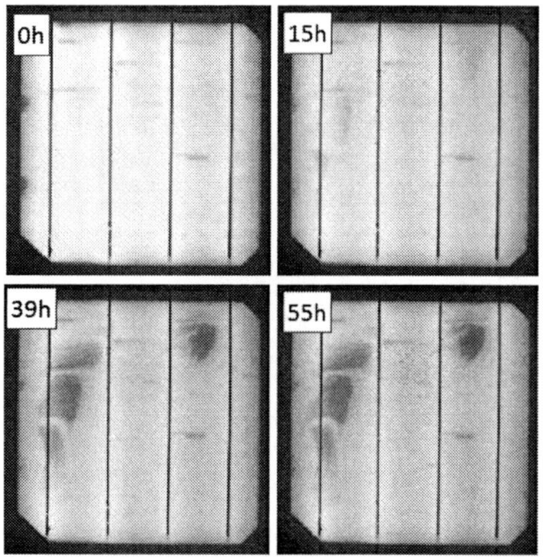

Figure 7: EL images extracted during the PID for the B-type configuration.

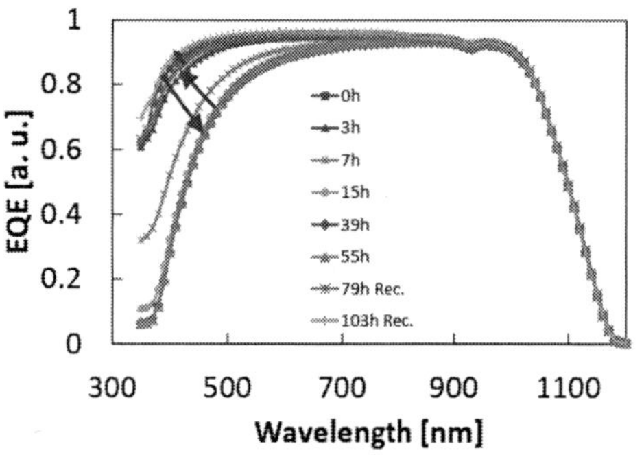

Figure 8: EQE extracted during the PID for the B-type configuration.

recovery is almost complete (99%) for all the solar cells. This result supports the hypothesis that the degradation of the solar cell is related to the migration of the Na+ ions inside the solar cell structure or to a polarization effect (charge trapping). In fact both these effects are reversible inverting the stress voltage.

In Figure 7 the Electroluminescence images of a B-type solar cell during the PID stress are reported. The images @ 39 h and @ 55 h show some dark regions on the solar cell area. These dark regions can be related to the lowering of the shunt resistance of the solar cell that is visible in the dark I-V characterization (see Figure 5). Moreover the EL images clearly reveal the weak areas of the solar cell structure that are responsible of the degradation.

Figure 8 shows the EQE during a PID stress extracted with a biasing light of 0.3 sun and a spot size of 2 cm x 2 cm on an corner of a B-type solar cell. The results show a degradation of the conversion efficiency in the UV-blue spectral range. This can be ascribed to the change of the absorption properties of the surface layers and this explains the lowering of the I_{SC} presented in Figure 5. This phenomenon is completely reversible, in fact the EQE signal recovers to the initial value after the 48 hours of positive PID at +600 V (see Figure 8).

In particular the optical investigations (EL and EQE) show that the degradation is mainly caused by a reversible phenomenon that occurs in the surface layers of the solar cell. The degradation can be caused by the Na+ ions migration inside the pn junction [8] or by a polarization effect [4] that induces an increase of the surface recombination velocity and a reduction of the absorption. Another cause can be a real shunting of the solar cell [9].

5 CONCLUSION

In this work, we show the PID effects on N-type bifacial solar cells using a step stress test strategy. This approach allows to study the evolution of the degradation and to find the weak areas in the solar cell where the PID degradation is critical (i.e. fingers, busbars, edges). We adopted three different encapsulation strategies and we show different degradation kinetics for the three configurations. In particular, the POE-II solution shows a PID degradation that is limited to 1.5% after 55 hours of continuous stress.

The electrical characterization shows that PID induces a shunting effect in the solar panel. On the contrary the FF decrease is very low for the A-type (0.6% at 55h) and C-type (0.4% at 55h) configurations. Moreover the optical (EL and EQE) analysis highlights that the main cause of the degradation can be ascribed to the change of the absorption properties of the surface layers where the charges or ions migration is higher. In particular the EQE reveals that the absorption mainly changes in the UV-blue spectral range. The recovery phase indicates that the degradation is caused by a recoverable phenomenon and that the recovery is almost complete. In fact after 48h of PID at +600 V the recovery is higher than 99% for all the four solar cells analyzed.

Future investigations can be carried out to investigate the physical nature of the degradation.

ACKNOWLEDGMENTS

This work was supported in part by Regione Veneto: project POR - parte FESR "Fondo Europeo Sviluppo Regionale" 2007-2013 azione: 1.1.1 Cod. SMUPR n. 4148., and by ENIAC JU/CALL 2010/270722-2 ERG "Energy for a green society: from sustainable harvesting to smart distribution. Equipments, materials, design solution and their applications".

REFERENCES

[1] J. Kapur, K.M. Stika, C.S. Westphal, J.L. Norwood, B. Hamzavytehrany, "Prevention of Potential-Induced Degradation With Thin Ionomer Film", in IEEE Journal of Photovoltaics, vol. 5, no. 1, pp. 219-223, Jan. 2015.
[2] S.C. Pop, et al., "Development of an ion-barrier film on solar panel glass," in 2015 IEEE 42nd Photovoltaic Specialist Conference (PVSC), pp. 1-6, 14-19 June 2015.
[3] O. Jaewon, S. Bowden, G. Tamizhmani, "Potential-Induced Degradation (PID): Incomplete Recovery of Shunt Resistance and Quantum Efficiency Losses", IEEE Journal of Photovoltaics, vol. 5, no. 6, pp. 1540-1548, Nov. 2015.
[4] Kohjiro Hara, Sachiko Jonai, Atsushi Masuda, "Potential-induced degradation in photovoltaic modules

based on n-type single crystalline Si solar cells", Solar Energy Materials and Solar Cells, vol. 140, pp. 361-365, September 2015.
[5] S. Pingel, O. Frank, M. Winkler, S. Daryan, T. Geipel, H. Hoehne, J. Berghold, "Potential Induced Degradation of solar cells and panels", in 2010 IEEE Photovoltaic Specialists Conference (PVSC), pp. 2817 2822, 20-25 June 2010.
[6] M. Barbato, M. Meneghini, A. Cester, G. Mura, E. Zanoni, G. Meneghesso, "Influence of Shunt Resistance on the Performance of an Illuminated String of Solar Cells: Theory, Simulation, and Experimental Analysis," in IEEE Transactions on Device and Materials Reliability, vol. 14, no. 4, pp. 942-950, Dec. 2014.
[7] M. Barbato, et al., "Effect of shunt resistance on the performance of mc-Silicon solar cells: a combined electro-optical and thermal investigation", in 2012 38th IEEE Photovoltaic Specialists Conference (PVSC), pp. 1241-1245, 3-8 June 2012.
[8] V. Naumann, et al., "Explanation of potential-induced degradation of the shunting type by Na decoration of stacking faults in Si solar cells," Solar Energy Materials and Solar Cells, vol. 120, pp. 383–389, 2014.
[9] V. Naumann, et al., "Potential-induced Degradation at Interdigitated Back Contact Solar Cells", Energy Procedia, vol. 55, pp. 498-503, 2014.

LIFETIME WARRANTY TEST METHOD CONSIDERING POTENTIAL INDUCED DEGRADATION RECOVERY BEHAVIOR

Kyungchan Kang*, Byongsu Kim, Sanghwan Park, Sungho Chang
LG Electronics, 191-1, Gong-dan dong, Gumi-si, Gyeongsangbuk-do, Korea
*Corresponding Author: Phone: +82 54 469 0407, E mail: kcpv.kang@lge.com

ABSTRACT: In recent years several failure modes of PV modules operated under high potentials were observed. Most dominant degradation mechanism is called "potential induced degradation (PID)" of crystalline PV modules [1]. PID test as proposed by the IEC working group does not reflect on particular modules. As a consequence, they do not fulfill the requirements of a lifetime warranty test [2]. Especially, it is not reflect the recovery characteristics of the solar cell. In this study we have focused on the lifetime test method considering the recovery characteristics of the n-type mono crystalline solar cell. The sensitivity of the modules regarding potential induced power degradation was investigated with respect to the recovery behavior under illumination. Positive as well as negative potential to ground was considered. The test was divided in two parts. First, a PID test was performed according to IEC TS 62804 ed. 1, method a. Then, a special PID-/ light-recovery sequence was performed.
Keywords: PV module, Lifetime, Degradation

1 INTRODUCTION

After PID has been found in the field, several institutes released the test standard to priory predict the PID degradation on PV Modules. And it passed the test in cell / module / system level, and being release the solution that can prevent the actual occurrence. But the question is whether these tests are able to represent the actual environment and also several types of cells. PID is caused by corrosion, ion migration, polarization, however it is observed only polarization degradation and shows reversible characteristic on particular N-type module. Present PID tests as proposed by several institutes do not reflect the reversible characteristic. In our study, we figure out the recovery characteristic and PID is not observed in real environment on the particular modules. And we suggest the test method considering recovery characteristics.

2 LG NeON® MODEL PID CHARACTERIZATION

There are two causes for PID to occur in a PV Module: Electrochemical Corrosion and Polarization Effect. (See Figure 1.)
First, Electrochemical Corrosion can be verified with a Corona Discharge Test, in which the cause of PID is not dependent on the type of ion. One prevalent ion found in the glass to cause PID is the sodium ion (Na). Because the Na carries a positive ion (+), the electrochemical Corrosion caused by the Na+ ion creates a destructive effect on the P-N Junction of the solar cells, resulting in power degradation of the solar module. This degradation is irreversible.

However, the + ion cannot shunt LG n-type NeON® cells, which have a p-type emitter on its top surface, as opposed to the n-type emitter that most p-type modules have on their top surfaces. As a result, Electrochemical Corrosion cannot be found on LG n-type NeON® modules.

Second, Polarization cumulates charges around the SiNx ARC and can disrupt the minority carrier movement and the junction electrical field, resulting in the power degradation of the cells. [3]

However, the polarization effect is a reversible reaction, and modules can recover from polarization. The solar cells naturally recover to its normal condition after PID had been occurred, and recovery is accelerated while under high temperature or opposite polarity.

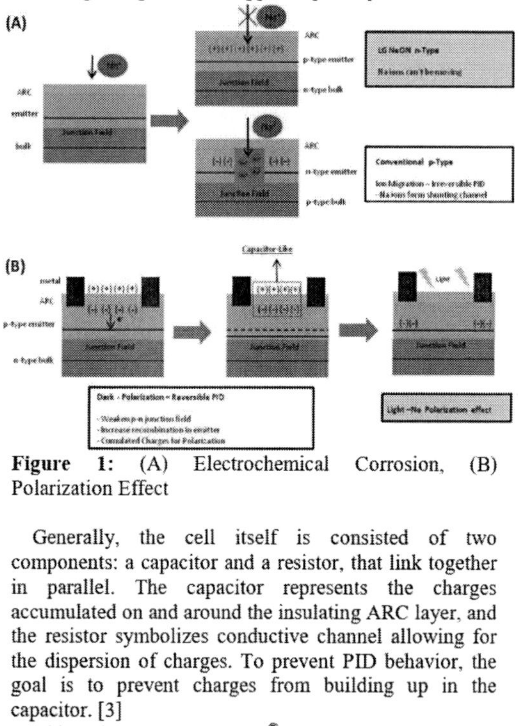

Figure 1: (A) Electrochemical Corrosion, (B) Polarization Effect

Generally, the cell itself is consisted of two components: a capacitor and a resistor, that link together in parallel. The capacitor represents the charges accumulated on and around the insulating ARC layer, and the resistor symbolizes conductive channel allowing for the dispersion of charges. To prevent PID behavior, the goal is to prevent charges from building up in the capacitor. [3]

Although LG n-type NeON® cells do show the power degradation under normal PID Test condition, they naturally prevent the charge from building up while under irradiance, eliminating any effects from polarization.

As a result, LG n-type NeON® cells do not show any effects from PID under real environmental conditions. We would like to test and verify n-type cells' performance under PID in this paper.

3 TEST SETUP

3.1 Step 1: PID Test acc. to IEC TC 62804 ed. 1

The PID test was performed by connecting the modules to a positive / negative potential between cells and frame and exposing them to warm and humid climate in a climate chamber. The test conditions were taken from the IEC TS 62804 ed. 1 – 'System voltage durability test for crystalline silicon modules'.

The test conditions are:

- Application of a voltage of 1000 V (-/+) between cells and frame
- Ambient temperature of 60 °C
- Ambient humidity of 85 %
- Test time: 96 hours

3.2 Step 2: PID- / Light-Recovery Sequence

Potential induced power degradation in a PV module can be reversible to a certain extent. Experience shows, that power, lost due to PID, can partly be regained due to changing ambient conditions.

During the PID-/ light-recovery sequence, ten cycles of changing ambient conditions are subsequently completed. During the first five hours of each cycle, the front side of the modules is exposed to an illumination of roughly 1000 W/m² at ambient conditions of 20°C and 50% relative humidity in the climate chamber (similar to daylight conditions). During the following 3 hours the ambient conditions are set to 40°C and 90% relative humidity without illumination (similar to night conditions).

During the ten cycles, a voltage of -1000 V is applied on modules M03 and M04 and +1000 V on modules M05 and M06. Several power measurements are conducted during the time of illumination.

Figure 2 and 3 shows the graph of the actual and set ambient conditions in the climate chamber.

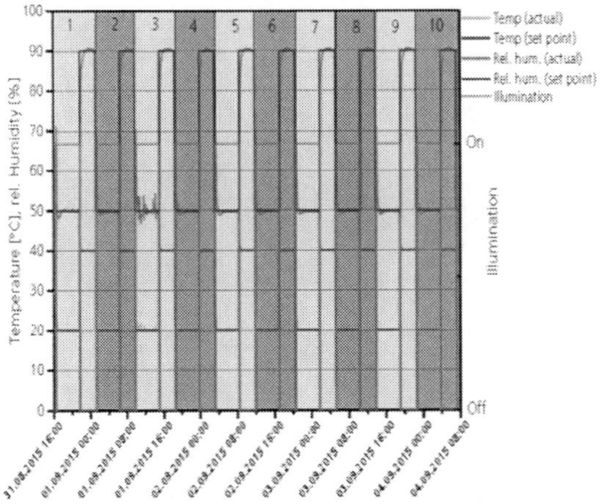

Figure 2: Climate chamber characteristic – modules M03, M04

3.3 Step 3: PID- / With Light Source

In order to verify PID effect under illumination, Step 1 was tested under a light source. There was tested under irradiance with intensity as high as 1000 W/m² and as lows as 200W/m² to simulate conditions on a fine sunny day and a cloudy day, respectively.

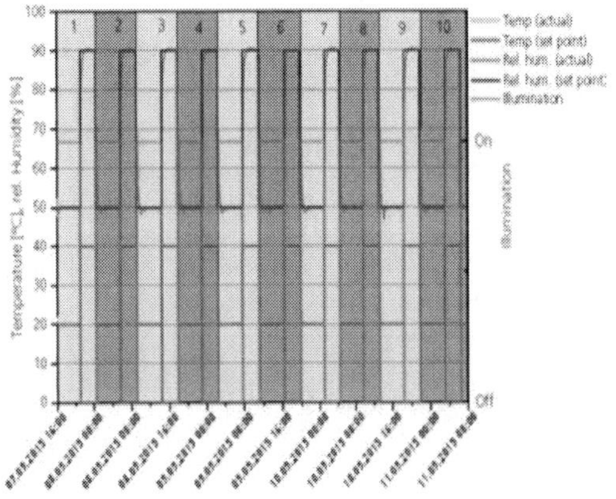

Figure 3: Climate chamber characteristic – modules M05, M06

3.4 Step 4: Leakage Current Monitor

Figure 4 shows the scheme of PID test to observe the leakage current. The module is placed in an environmental chamber so as to control module temperature, relative humidity and irradiance. The leakage current is compared in dark and light state, respectively.

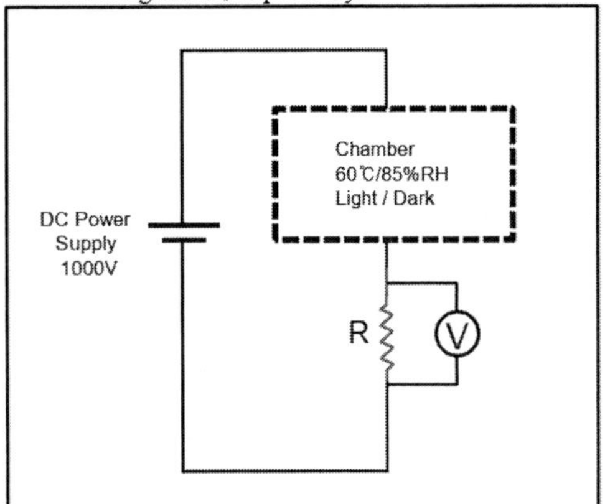

Figure 4: Schematic of leakage current measurement system

4 RESULT AND DISCUSSION

4.1 Step 1,2 Performance Measurement

The results of the performance measurements at STC are given. Table 1 and 2 list the results of the measurements before and after the PID test and the final measurement after the recovery procedure. The deviation from the initial measurement is also given.

4.2 Step 1,2 Performance Measurement

In case of PID, either the pn-junction is shunted (ion migration) or an increased surface recombination (polarization) occurs. Both cases cause a decrease of shunt resistance (Rsh) of the solar cell and module. A low value Rsh on the other hand causes power loss by providing an alternate current path through the solar cell junction and reduces the voltage from the solar cell. The effect of Rsh is particularly severe at low light levels, since there will be less light-generated current and a lower open circuit voltage (Voc). In an EL image, a change in Rsh after a PID test can be very well observed when applying a much lower current than ISC, which represents a low light level.

The PID affected area of a solar cell or even the whole cell appears dark in the image and PID affected cells can be identified. Figure 5 to Figure 6 show the EL images of the tested modules before and after the PID test and after the PID-/ light-recovery sequence.

Table I: Performance measurement at STC, PID test with -1000 V

M03

	Isc [A]	Voc [V]	Imp [A]	Vmp [V]	Pmax [W]	FF [%]	eff [%]
Initial Measurement	9.93	40.64	9.32	33.11	308.68	76.49	18.82
PID Test 96 hours at -1000 V							
After 96 hrs PID Test	9.76	40.14	9.16	32.34	296.24	75.59	18.06
Deviation to initial [%]	-1.70	-1.20	-1.80	-2.30	-4.00	-1.20	-4.00
Recovery Procedure at -1000 V							
After Recovery	9.99	40.64	9.34	33.00	308.26	75.91	18.80
Deviation to initial [%]	0.60	0.00	0.20	-0.30	-0.10	-0.80	-0.10

M04

	Isc [A]	Voc [V]	Imp [A]	Vmp [V]	Pmax [W]	FF [%]	eff [%]
Initial Measurement	9.99	40.66	9.36	33.00	308.80	76.06	18.83
PID Test 96 hours at -1000 V							
After 96 hrs PID Test	9.87	40.39	9.24	32.64	301.73	75.70	18.40
Deviation to initial [%]	-1.20	-0.70	-1.20	-1.10	-2.30	-0.50	-2.30
Recovery Procedure at -1000 V							
After Recovery	10.00	40.70	9.38	33.05	310.00	76.20	18.90
Deviation to initial [%]	0.10	0.10	0.20	0.20	0.40	0.20	0.40
Measurement Uncertainty [%]	± 1.9	± 0.6	± 2.2	± 1.2	± 2.1	± 1.6	± 2.2

Table II: Performance measurement at STC, PID test with +1000 V

M05

	Isc [A]	Voc [V]	Imp [A]	Vmp [V]	Pmax [W]	FF [%]	eff [%]
Initial Measurement	9.98	40.60	9.38	33.16	311.19	76.82	18.98
PID Test 96 hours at +1000 V							
After 96 hrs PID Test	9.93	40.46	9.36	32.90	307.99	76.64	18.78
Deviation to initial [%]	-0.50	-0.30	-0.20	-0.80	-0.10	-0.20	-1.00
Recovery Procedure at +1000 V							
After Recovery	9.98	40.53	9.41	32.98	310.19	76.65	18.91
Deviation to initial [%]	0.10	-0.20	0.20	-0.50	-0.30	-0.20	-0.30

M06

	Isc [A]	Voc [V]	Imp [A]	Vmp [V]	Pmax [W]	FF [%]	eff [%]
Initial Measurement	9.94	40.67	9.39	33.22	311.87	77.13	19.02
PID Test 96 hours at +1000 V							
After 96 hrs PID Test	9.94	40.52	9.38	33.00	309.40	76.78	18.87
Deviation to initial [%]	0.00	-0.40	-0.10	-0.70	-0.80	-0.50	-0.80
Recovery Procedure at +1000 V							
After Recovery	9.96	40.58	9.40	33.06	310.91	76.91	18.96
Deviation to initial [%]	0.20	-0.20	0.20	-0.50	-0.30	-0.30	-0.30
Measurement Uncertainty [%]	± 1.9	± 0.6	± 2.2	± 1.2	± 2.1	± 1.6	± 2.2

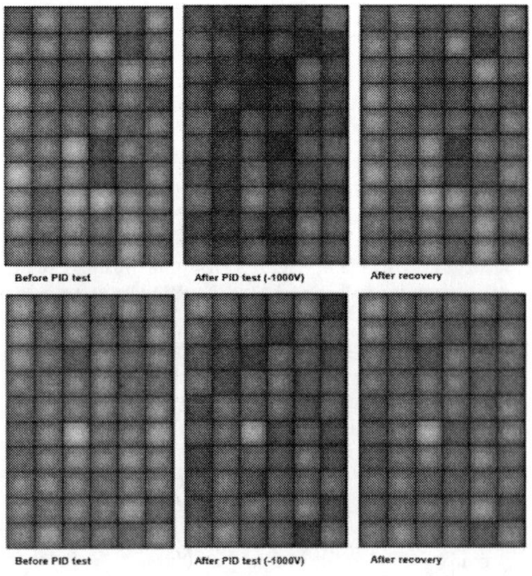

Figure 5: Electroluminescence images of module M03, M04

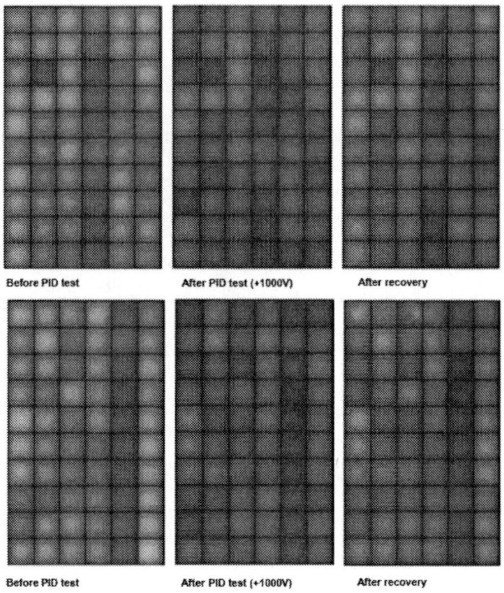

Figure 6: Electroluminescence images of module M05, M06

4.3 Step 3 Performance Measurement and EL Imaging

The results of the performance measurements at STC are given. Table 3 list the results of the measurements before and after the PID test. The deviation from the initial measurement is also given. Tables 8 to Table 9 show the EL images of the tested modules before and after the PID test.

Table III: Performance measurement at STC, PID test with -1000 V with light source

M07

	Isc [A]	Voc [V]	Imp [A]	Vmp [V]	Pmax [W]	FF [%]	eff [%]
Initial Measurement	10.07	40.52	9.56	32.76	313.24	76.76	19.10
PID Test 96 hours at -1000 V with light (200W/m²)							
After 96 hrs PID Test	10.09	40.43	9.65	32.46	313.30	76.76	19.10
Deviation to initial [%]	0.21	-0.21	0.94	-0.92	0.02	0.00	0.00

M08

	Isc [A]	Voc [V]	Imp [A]	Vmp [V]	Pmax [W]	FF [%]	eff [%]
Initial Measurement	10.07	40.48	9.57	32.75	313.26	76.86	19.10
PID Test 96 hours at -1000 V with light (1,000W/m²)							
After 96 hrs PID Test	10.04	40.41	9.48	32.85	311.36	76.76	19.00
Deviation to initial [%]	-0.33	-0.16	-0.92	0.30	-0.61	-0.13	-0.53

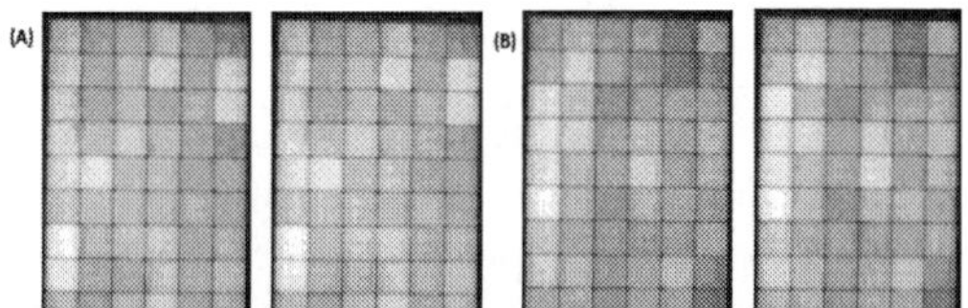

Figure 7: Electroluminescence images of module (A)M07@200W/m^2, (B)M08@1,000W/m^2

4.4 Leakage Current Comparison

Examining the leakage current during PID test, a current profile of the dark state's module similar to that of a capacitor is observed, on the other hand a current of the light state's module is not observed. (see Figure 8)

4.5 Summary and Discussion

The results given in this report are valid for the case of a potential of 1000 V (-/+) lying on the cells, while the module frame is grounded.

After the first PID test (acc. to IEC TS 62804 ed.1), a negative deviation of output power was recorded for the four modules. While for modules M03 and M04 the power loss was -4.0 % and -2.3 %, for modules M05 and M06 it was only -1.0 % and -0.8 % and therefore nearly negligible.

All modules are not rated as sensitive to PID assuming a pass/fail criterion of max. 5 % power loss. However, exposed at negative potential, the power loss is slightly higher than at positive potential, which indicates a slight sensitivity to negative ground potential. The EL images support this impression, as some slightly darker cells can be found on modules M03 and M04 after the first PID stress treatment.

During the PID-/ light-recovery sequence, all four modules regained power fast after first light exposure, although the 'critical' ground potential was continuously applied to the modules. The modules remained on a high level during the dark-light phases in PID-light recovery treatment. The power measurements conducted after the PID-/light sequence show, that the initial power could be fully regained. (see Figure 9)

And the second PID test (acc. To IEC 62804 ed.1 with light source), a negative deviation of output power was recorded for the two modules. While for modules M07 and M08 it was only 0.02% and -0.61%.

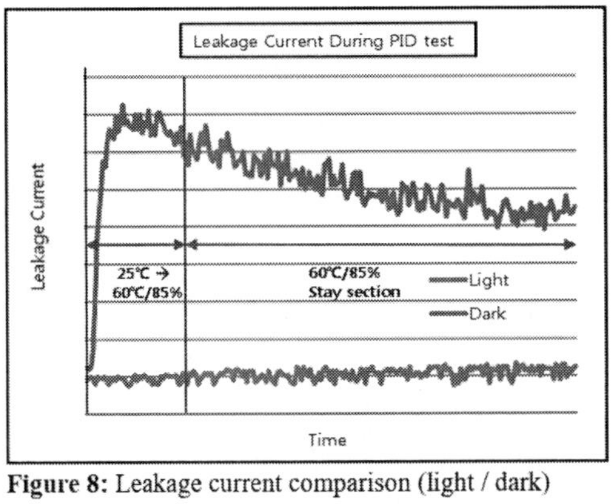

Figure 8: Leakage current comparison (light / dark)

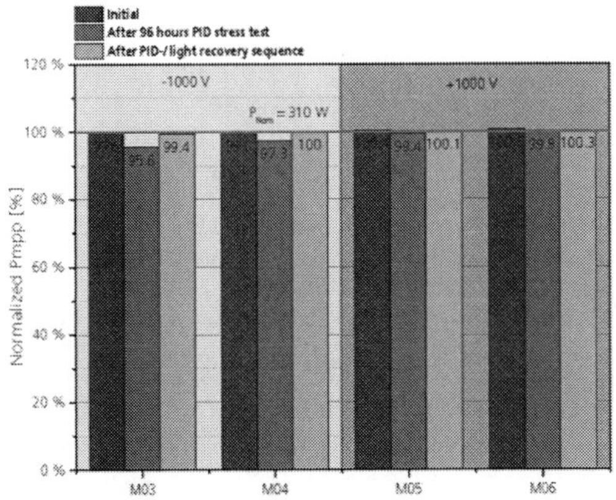

Figure 9: Normalized initial, intermediate and final PMAX (STC measurements)

5 CONCLUSION

The behavior described above shows clearly that the modules recover fast when exposed to light and that a simulated day-night sequence, as applied in the scope of this study, does not lead to a continuous decrease in power. And we confirmed that 'Polarization Effect' does not show in light state according to cell type, result from the measurement of leakage current, EL and module power after test. The test result indicates that the PID effect will be equalized in the field to the largest extent. The modules show different results depend on cell types, it requires new lifetime warranty test method considering potential induced degradation recovery behavior.

REFERENCES

[1] S. Pingel et al. "Potential Induced Degradation of Solar Cells and Panels", Conference Paper 35th IEEE, 2010

[2] H. Nagel et al. "Lifetime Warranty Testing of Crystalline Silicon Modules for Potential-Induced Degradation", 27th EUPVSEC 2012

[3] Chen-Wei et al., "Potential Induced Degradation Mechanism Observed at Module Level", 28th European Photovoltaic Solar Energy Conference and Exhibition

Does the new IEC 62804-2 PID Test Procedure cover a Service Life of CIGS PV Modules?

Peter Lechner[1], Jonathan Schnepf[1], Dieter Geyer[1],
Raymund Schäffler[2], Rolf Wächter[2], Tobias Repmann[2]
[1] Zentrum für Sonnenenergie- und Wasserstoff-Forschung Baden-Württemberg (ZSW),
Industriestraße 6, 70565 Stuttgart, Germany
[2] Manz CIGS Technology GmbH,
Alfred-Leikam-Straße 25, 74523 Schwäbisch Hall, Germany

ABSTRACT: A new work item IEC NP 62804-02 TS was submitted in order to define a procedure for evaluating effects of potential-induced degradation (PID) on thin-film PV modules. This work seeks to answer the question of how many years of outdoor service life are covered by the proposed IEC PID test procedure, applying the assumption that an equivalent transferred charge implies an equivalent severity of the PID stress. The leakage currents and transferred charge from cell to ground for a standard CIGS module configuration and a leakage current suppression (LCS) CIGS module configuration were evaluated in an environmental chamber test and in the outdoor field in Germany. The suppression of currents with the LCS configuration works very efficiently at elevated temperatures (also for the IEC test condition) but can be neglected for cool and extremely wet outdoor conditions. Comparing transferred charge in the field, based on a selected and not necessarily representative day, with the transferred charge in the 1,000 h IEC test procedure we conclude that the IEC test covers more than 50 years of service life in Germany for the standard system, but less than 2 years for the LCS configuration. Thus no generally valid equivalence between the severity of the PID stress induced by the new IEC test and the stress during a full service life can be established.
Keywords: CIGS Module, IEC 62804, PID, Leakage current

1 INTRODUCTION

Some types of c-Si [1] and thin film PV modules suffer from a substantial reduction in power after having been subjected to high voltage (HV) stress versus ground (potential-induced degradation, PID). A leakage current through the front or rear glass into the cell is the driving force for the degradation effects in all PV technologies. For c-Si modules, a thermally activated regeneration of power can be observed after relaxation of the PID stress [2, 3]. CIGS modules at a negative potential versus ground can experience HV stress [4] from sodium diffusion from the glass into the active layers [5] and lose power as a result. For CIGS cells, recovery of PID-related degradation could be observed up to a certain level [5]; conversely, degradation of modules may be irreversible [6].

Recently a proposal for a new work item IEC NP 62804-02 TS was submitted with the aim to define a procedure for evaluating effects of potential-induced degradation (PID) on thin-film PV modules. However, one crucial question has not yet been addressed: How many years of outdoor service life of a CIGS module does the proposed IEC test correspond to? This work seeks to give an answer by a simplified approach, using leakage currents and cumulated transferred charge as an indicator for the severity of the PID stress. As a worst-case scenario we assume that thermal regeneration does not play a role during outdoor operation.

Furthermore an improved CIGS module configuration (Leakage current suppression, LCS) is introduced where leakage currents are suppressed significantly compared to the standard system. For this application we evaluate the reduction of transferred charge and quantify the improvement in PID resistivity for both the IEC test procedure and real outdoor operation.

2 EXPERIMENTAL APPROACH

We tested two different Manz CIGS module system configurations, the standard configuration and the LCS configuration which should suppress the leakage currents. A bias voltage of -1,000 V from cell to ground was applied to several specimen of both module configurations and the resulting leakage currents were measured in a "matrix chamber test" in an environmental chamber at different temperature values (15, 25, 40, 60 and 85°C) and at different relative air humidity values (20, 50, 85 and 95% RH). At every temperature/humidity step several hours (5 to 10 h) waiting time are required to measure more or less stabilized leakage current values. Current measurements were performed with a calibrated high precision picoamperemeter with a 1,000 V voltage sourcing capability.

After this procedure the modules were installed at the ZSW test field in Widderstall, Southern Germany, where several PID test beds are available for monitoring module leakage currents and relevant meteorological data. The bias voltage of -1,000 V between cell and ground was triggered by the solar irradiance with a switching on threshold of 5 W/m².

3 RESULTS AND DISCUSSION

3.1 Environmental chamber measurements

The leakage current results from the matrix chamber test with different temperature/humidity combinations are displayed for both module configurations, standard and LCS, in an Arrhenius plot (Figure 1). Note that leakage currents vary over more than four orders of magnitude in our experiments, showing that they are highly dependent on temperature, humidity and module configuration. It becomes evident that there is not a constant leakage current reduction for the LCS configuration relative to the standard configuration.

In Table I we summarize the leakage currents and their ratio for some selected temperature/humidity combinations. For a 60°C and 20% RH chamber condition, which represents the case of a PV module in the field under full sunshine in a dry environment, the leakage current is significantly reduced in the LCS configuration by a factor of 38 compared to the standard case. At the IEC test condition (85°C, 85% RH) the suppression factor LCS versus standard configuration is about 25.

At chamber conditions with high humidity (95% RH) and low module temperature (15 to 25°C) the leakage currents for the LCS system are reduced only by a factor of 1.6 to 2.4; the leakage current suppression works less efficiently than at higher temperature.

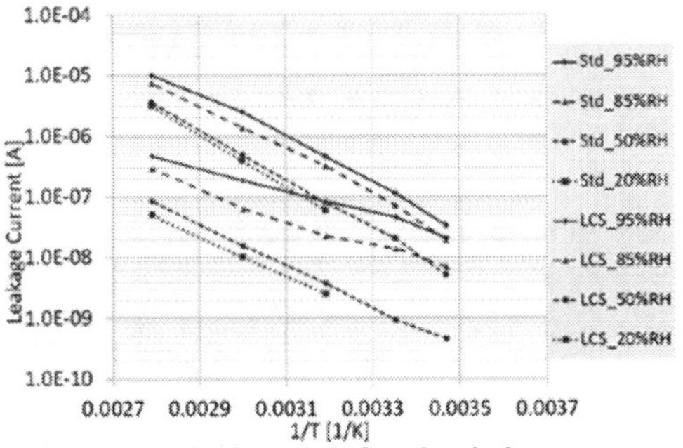

Figure 1: Arrhenius plot for the leakage currents measured in the chamber for the standard and the LCS module configuration

Table I: Leakage currents and their ratio for the standard and the LCS configuration at selected T-RH combinations

Site	T_{mod} [°C]	RH [%]	Situation	Leak. current Standard [A]	Leak. current LCS [A]	Leak. current Ratio Std/LCS
Chamber	15	95	cool/humid	3.4E-8	2.1E-8	1.6
Chamber	25	95	moderate/ humid	1.1E-7	4.7E-8	2.4
Chamber	60	20	hot/dry	3.9E-7	1.0E-8	38
Chamber	85	85	IEC test	7.5E-6	3.0E-7	25

3.2 IEC test

Several modules with standard and LCS configuration were subjected to the new IEC 62804-2 test procedure at -1,000 V bias voltage, while the leakage currents were monitored over the full test period of 1,000 h. It is important to follow the test instructions and allow the 12 to 24 h period for the stabilization of the environmental condition before starting the bias voltage. For the standard configuration modules a total cumulated charge of 33 As was transferred during the IEC test, for the LCS configuration modules the total charge was only 0.74 As (Table II). The suppression factor for the transferred charge is around 45 and therefore better than the factor 25 found for suppression of leakage currents from the previous chamber experiment. The reason for the difference obviously lies in the long term course of the leakage current during the IEC test. For the standard configuration the leakage currents are, after the initial phase, more or less constant, whereas for the LCS configuration the leakage current is consistently falling almost over the full IEC test period.

3.3 Outdoor test

From our outdoor test site in Widderstall we received detailed leakage current data for a variety of module types. As an example, Figure 2 reveals the course of the leakage currents for both module configurations, standard and LCS, during a day in May 2016. After a rainy night the modules are cold and wet in the morning and the leakage currents are high for both configurations. Later, when the modules get dry and warm, the leakage currents for the LCS system are significantly suppressed. Later in the afternoon the weather becomes rainy with thunderstorms and the leakage currents increase to similarly high values for both configurations comparable to the situation in the morning. Obviously, the LCS system does not reduce leakage currents under extremely wet environmental conditions.

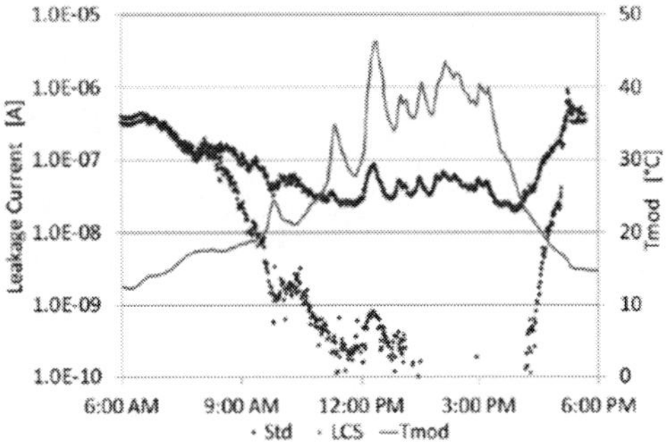

Figure 2: Leakage currents for the standard and the LCS configuration and module temperature over the course of a day in May 2016

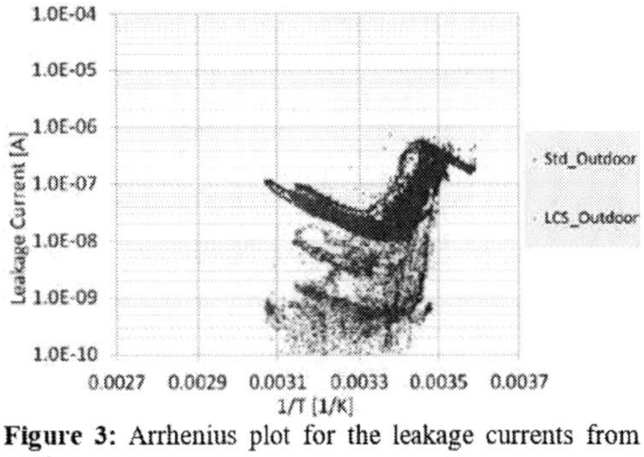

Figure 3: Arrhenius plot for the leakage currents from outdoor measurements

The leakage currents logged from an outdoor exposure period of 2 weeks in May 2016 are displayed in an Arrhenius plot in Figure 3. The "upper limit" of the leakage currents at low module temperatures (i.e., high 1/T) is more or less identical for both module configurations; the LCS system does not reduce leakage currents here. These data points are associated with situations like those occurring in the morning hours of the example day shown in Figure 2, periods where the modules are extremely wet. At higher module temperature (i.e., low 1/T) we notice that the LCS configuration shows significantly reduced leakage current values.

3.4 Lab and field comparison of leakage currents

In Figure 4 and Figure 5 the leakage current values

measured with the outdoor test setup are superimposed in the Arrhenius plot with the values gained from the matrix chamber experiments. The most interesting finding is that the outdoor leakage current "upper limit", being associated with wet and cold modules, is 1 to 1.5 orders of magnitude higher than the leakage current values measured in the matrix chamber test at low module temperature and 95% RH, which represents a situation where the chamber humidity is extremely high, but not yet condensing.

Also we can see for the LCS configuration that, unlike the standard configuration, the leakage current at IEC PID test condition is lower than at wet and cold outdoor conditions. This means that for the LCS configuration the IEC test does not represent an accelerating stress test in terms of transferred charge.

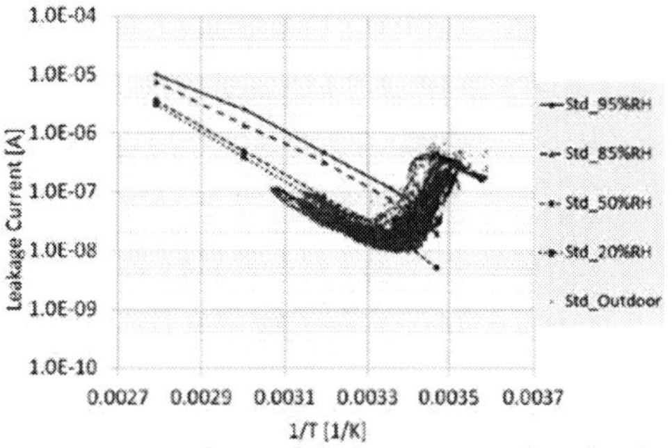

Figure 4: Superimposed leakage current data for the standard system taken from chamber and outdoor tests

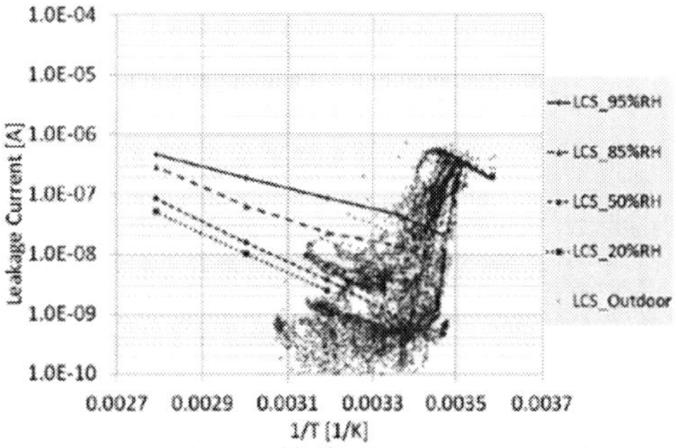

Figure 5: Superimposed leakage current data for the LCS system taken from chamber and outdoor tests

Table II: Transferred charge in different situations

Site	T_{mod} [°C]	RH [%]	Situation	Transferred Charge Standard [As]	Transferred Charge LCS [As]
Chamber	85	85	IEC test 1000h at -1000V	33	0.74
Field, Germany			Spring day 29.05.2016 at -1000V	4.3E-3	4.1E-3

Table III: Time covered by IEC test and IEC test time to cover a full service life

Item	Situation	Standard	LCS
Time to transfer IEC charge	spring day at -400V	~52 yrs	~1.2yrs
IEC test time to cover 25yrs field	spring day at -400V	~480h	~20,000h

The consequence of this situation becomes even clearer if we compare the charge transferred in the outdoor field during May 29, 2016 (the day might not necessarily represent a typical day in Germany): 0.0043 As for the standard and 0.0041 As for the LCS, while during the IEC test transferred charges are 33 As and 0.74 As respectively (Table II).

Based on the principle of equivalent transferred charge, we want to estimate the field operation time covered by the IEC test. We assume that in a non-grounded PV system with string inverter and a high system voltage of 800 V the potential between cell and ground at the negative end of the string is -400 V, unlike the -1,000 V potential used in our experiments. Furthermore we assume a reference year based on a single day as given in Table 2, knowing that this could imply significant uncertainties. Under these assumptions we find that the IEC test would cover 56 years service life for the standard configuration, but only 1.2 years for the LCS configuration (Table III). Or, vice versa, the IEC test time required to cover the transferred charge of 25 years service life would be only 480 h for the standard configuration but about 20,000 h for the LCS system.

4 SUMMARY

Leakage currents under high bias voltage from cell to ground measured at CIGS modules can be partially suppressed by modifying the module configuration from standard to LCS. From chamber tests we see a significant current reduction at elevated temperatures and both low and high relative air humidity conditions, including the IEC TS 62804-2 Draft test condition 85°C/85%RH. However, for low temperatures and high humidity chamber test conditions the leakage current suppression by the LCS configuration is found to be less efficient. Finally, for outdoor field conditions where the modules are wet and cool, we do not find any leakage current difference between both configurations. Following the principle of equivalent PID stress by equivalent transferred charge for the IEC test and cumulated charge transferred during field operation, the IEC 62804-2 test covers more than 50 years based on a selected and not necessarily representative day in Germany and therefore more than the expected module service life, but less than2 years or a fraction of the module service life for the LCS module configuration. As a consequence, no generally valid equivalence between the PID stress induced by the IEC 62804-2 test and the PID stress from real outdoor field operation can be established. This correlation must be individually established for each module design and configuration.

5 ACKNOWLEDGEMENT

We would like to gratefully acknowledge the support of a part of this work by the Federal Ministry for Economics and Technology (BMWi), contract number FZ 0325715.

6 REFERENCES

[1] S. Pingel, O. Frank, M. Winkler, S. Daryan, T. Geipel, H. Hoehne, J. Berghold; "Potential Induced Degradation of Solar cells and Panels", Proceedings 35th IEEE PV Specialists Conference, 2010, 2817

[2] J. Berghold, O. Frank, H. Hoehne, B. Richardson, M. Winkler; "Potential Induced Degradation of solar cells and panels", 25th European Photovoltaic Solar Energy Conference and Exhibition, 2010, 3753

[3] P. Lechner, S. Hummel, D. Geyer, H.-D. Mohring, "Evaluation of Recovery Methods after Potential Induced Degradation of PV Modules", 30th European Photovoltaic Solar Energy Conference and Exhibition, Hamburg, 2015,

[4] J. Springer, D. Geyer, R. Schäffler, 22th European Photovoltaic Solar Energy Conference and Exhibition, 2007, 2262

[5] V. Fjällström, P.M.P. Salome, A. Hultqvist, M. Edhoff, T. Jarmar, B. G. Aitken, K. Zhang, K. Fuller, C. Kosik Williams, IEEE Journal of Photovoltaics, 2013, 1090

[6] P. Lechner, D. Sanchez, D. Geyer, H.-D. Mohring, 27th European Photovoltaic Solar Energy Conference and Exhibition, 2012, 3152

An Investigation of Factors Contributing to Potential-Induced Degradation (PID) and Its Countermeasures

Xusheng Wang*, Songbo Wan, Albert Fu, Guoqiang Xing
Canadian Solar Inc., 199 Lushan Road, SND, Suzhou, Jiangsu, China
* Corresponding author: Phone: +86 512 66908088; Fax: +86 512 66908089; Email: Xusheng.wang@canadiansolar.com

ABSTRACT: This paper introduce the Potential induced degradation (PID) phenomenon, analyzes the PID forming mechanism, studies factors contributing to PID: temperature of the environment, relative humidity and differences in voltage potential between the solar cells in the PV module and the frame. Based on our studies, we have developed solutions designed in cell level, module level and system level to prevent PID.
Keywords: Potential induced degradation, factor, solution, cell, module, system

1 INTRODUCTION

Solar energy has wide applications. Photovoltaic (PV) modules convert sunlight directly into electricity and have been installed at a large scale worldwide. As PV systems are installed in harsh environments (humid, hot or even salty, Fig. 1), a phenomenon called potential-induced degradation (PID) is becoming a noteworthy topic. This paper studies factors contributing to PID: temperature of the environment, relative humidity and differences in voltage potential between the solar-cells in the PV module and the frame. Based on our studies, we have developed solutions designed to prevent PID.

Figure 1: Harsh environment for PV application

2 BACKGROUND

2.1 PID phenomenon

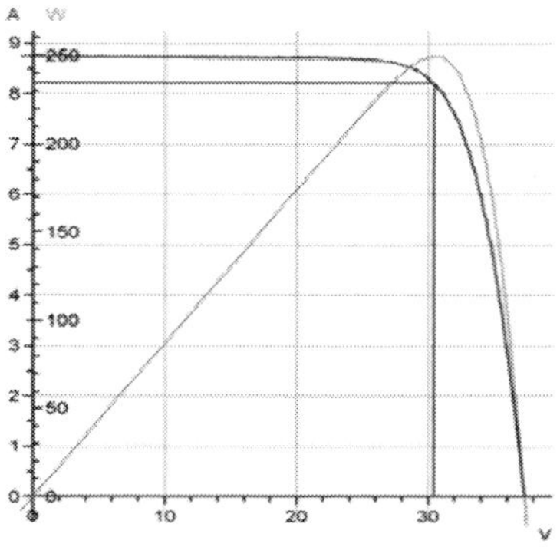

BEFORE

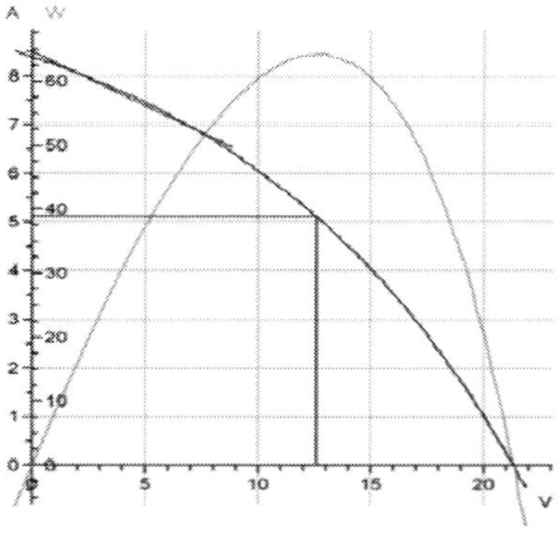

AFTER

Figure 2: I-V curves before and after PID test

Potential induced degradation (PID) is a phenomenon that occurs when the photovoltaic (PV) modules are connected in series and the solar cells are under high voltage stress. When PID happens, the module's power degrades as

Fig.2 shows, and EL (Electroluminescence) imaging shows darkened cells (Fig.3).

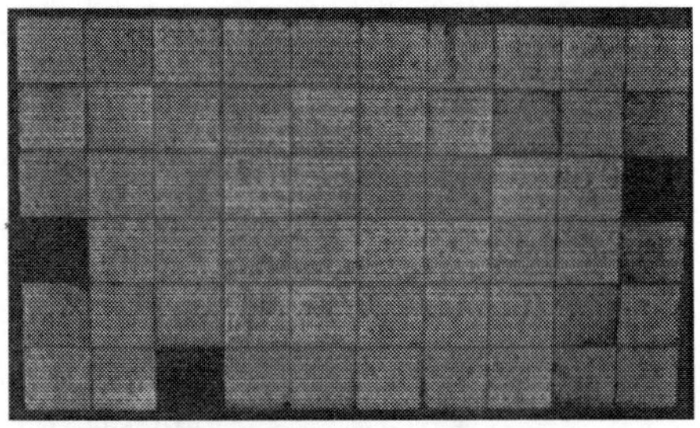

Figure 3: EL image of the module with PID

2.2 PID forming mechanism

There is a voltage difference between solar cells in the PV module and the frame connected to the ground. In a PV power plant, this voltage difference can be as large as 600V in USA and 1000V in Europe. In regards to, P-type modules operating at a negative voltage with respect to the earth ground, positive charge will accumulate on top of the anti-reflective coating of the solar cells. In essence, the charged surface attracts electrons from the cell, thereby forming leakage current and destroying the p-n junction in the cell. High humidity, high temperature and high system voltage will accelerate this problem. For applications in salty/marine environments, sodium ions (existent in the glass exterior originally) interact with the environment, causing PID effects to become even more severe. When PID occurs, the PV module's power may degrade significantly as some cells appear darker under electroluminescence examination as Fig.3 shows.

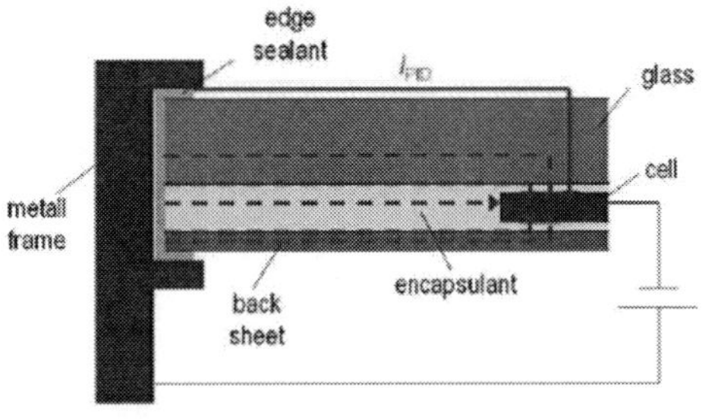

Figure 4: The speculated forming mechanism of PID effect

As shown in Fig.4, when a voltage difference occurs between the grounded metal frame and the cells, electrical leakage is also formed between the frame and the cells. The magnitude of the leakage is also dependent on the temperature of the environment, humidity, resistivity of the material that encapsulates the solar cells and surface conductivity of the modules. Recent studies suggest that sodium ions tamper with the electrical processes inherent in solar cells, resulting in a decrease in efficiency.

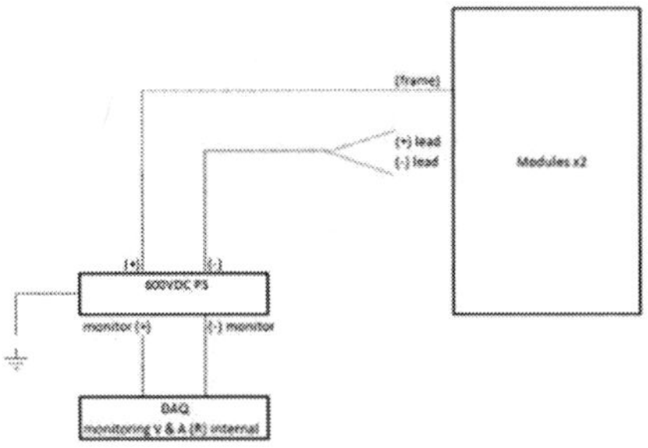

Schematic diagram of PID testing

3 FACTORS INFLUENCING PID

3.1Experiment Design

The experiments were conducted to understand the influence that temperature, humidity and bias potential have on the PID effect. For the modules with P-type cells, we used two methods of accelerated testing. The first method was conducted in a high temperature and humid environment; this method applied a negative voltage between the internal circuit (solar cells) and the module frame. The second method was conducted in a high temperature environment; this method covered the module's front surface with copper foil and applied a negative voltage between the internal module circuit and the module frame. Covering the module's front surface with copper foil provides much higher conductivity on the module's front surface in comparison to the conductivity derived from simply a high temperature and humid environment. Such accelerates the occurrence of PID. To apply a negative potential, we connect both the positive and negative poles of the module to the negative terminal of the electrical source. The frame is subsequently connected to the positive terminal of the same electrical source.

3.2 Temperature

Within the experiment, we took two modules A and B with identical encapsulation materials. The glass surfaces were covered with copper foil. The copper foil covered the glass surface in its entirety and extended so that it was also touching the frame. The independent variable was temperature. In that vein, both modules were given a negative 1000v charge. However, Module A underwent the charge in 35℃ and Module B 85℃. After testing, they were taken out of the chamber and measured for performance. Differences in maximum-power output after testing are shown in Fig.5. In summary, PID effect becomes more severe under higher temperature, as evidenced by the data noted below.

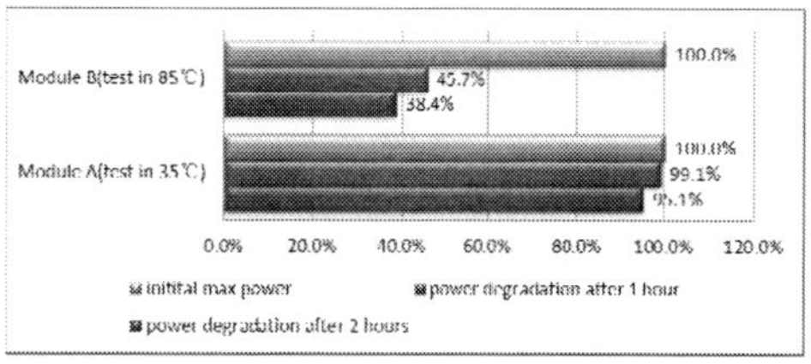

Figure 5: Power degradation of modules after accelerated testing

3.3 Humidity

With this test, we used two modules C and D with the same encapsulation materials. Unlike the previous test (2.2), there was no copper foil covering the glass surface. In this experiment, temperature was controlled and humidity altered. Module C was given a negative 1000V for one hour in 30℃ and 35% relative humidity (RH), while module D was in 30℃ and 85% RH. After testing, they were taken out of the chamber and measured for performance. Subsequently, 66 hours of additional accelerated testing under the same conditions was performed and the electrical performance was once again measured. Changes in power degradation are shown in Fig. 6. In summary, PID becomes more severe under higher humidity, as evidenced by the data.

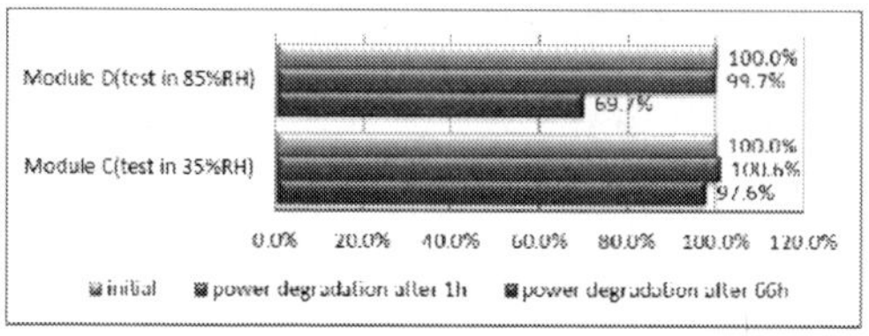

Figure 6: Power degradation of modules after accelerated testing

3.4 The Bias Voltage

(1) Testing conducted under negative bias voltage stress

In this experiment, we took two modules E and F with the same encapsulation materials. There was no copper foil covering the glass surface. Module E was supplied a negative 500V for an hour and a half in 60℃ and 85% relative humidity (RH), while module F was supplied 1500 V under the same conditions. After testing, both modules were taken out of the chamber and measured the performance. Changes in the power degradation after the accelerated testing are shown in Fig.7. Test data indicate that the PID effect becomes more severe under higher bias voltage.

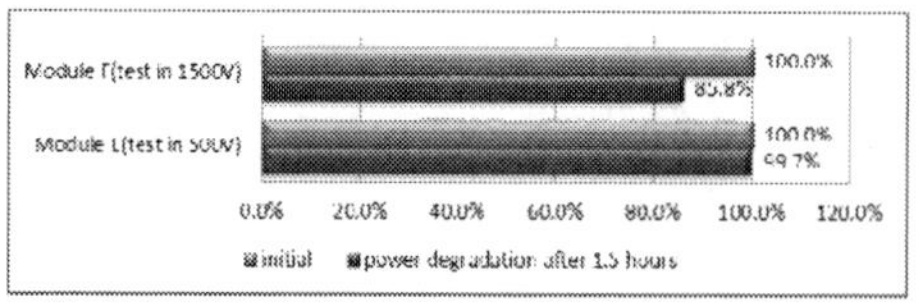

Figure 7: Power degradation of modules after accelerated testing

(2) No degradation observed when modules under positive bias voltage.

Five modules (regular type) went through 1000 hours of testing at positive 1kV bias under 85℃&85%RH environment (damp heat test).Less than 2% total power degradation was observed according to the data shown in table I. For P-type modules, positive voltage bias won't cause PID, supporting the results of other publications [1, 2].

Table I: Performance change after positive bias under DH condition

Power Difference-Post-DH1000 Relative to Initial						
Sample #	Bias	Pmax[%]	Voc[%]	Vmp[%]	Isc[%]	Imp[%]
1	+1kv	-1.4	0.2	0.1	-2.9	-1.4
2	+1kv	-0.6	0.5	1.1	-2.3	-1.7
3	+1kv	-1.4	0.1	-0.1	-3.3	-1.3
4	+1kv	-1.8	-0.1	0.9	-2.5	-2.7
5	+1kv	-1.7	-0.1	1.1	-2.2	-2.8

Table II: Sample testing conditions proposed

Agency	Foil cover	Relative Humidity	Temperature (℃)	Voltage biased	Time ([illegible])
PVEL	none	85%	85	+1kv/-1kv	6[illegible]
Fraunhofer CSP	Al foil	50%	50	-1kv	4[illegible]
PI-Berlin	Al foil	Not controlled	25	-1kv	1[illegible]
IEC	none	85%	60	-1kv	9[illegible]
TUV SUD	none	85%	85	-1kv	9[illegible]

3.5 PID Testing Conditions and Qualification

It must be noted that PID has not yet been identified as a type of degradation in the widely accepted IEC 61215 standard [3]. In other words, there is no widely accepted official standard on PID yet (as of July 2013). Different testing conditions have previously been proposed (table II) and the IEC conditions, proposed in draft standard IEC 62804, are gaining recognition [4].

Our modules passed the PID testing put forth by PVEL, PI-Berlin in 2012 and Franuhofer ISE (IEC conditions) in 2013. The three organizations mentioned above are some of the top solar-module testing laboratories in the world. To that end, we hope our customers continue using our products with confidence.

4 METHODS OF PREVENTING PID

4.1 Cell Level

It is well known that solar cells are key to the PID effect. The severity of the PID effect is dependent on how the PN junction of the cells is influenced. Preventing sodium ions of the glass from migrating to PN junction is an effective method to anti PID effect (Fig.8). One way is to grow a thin silicon dioxide film on the wafer surface. Another way is to alter the make-up of the anti-reflective coating (thickness, reflective properties, ratio of silicon and nitrogen). As such, we have successfully developed cells of high anti-PID capability. In the experiments, we made module samples with anti-PID cells and conventional cells. All the other materials remained the same. These modules were subjected to PID accelerated testing (85 ℃, 85%RH and negative 1000V). The power degradation of these modules after testing is shown in Fig.9. Module with anti-PID cells shows very little degradation, which is much better than module with conventional cells. All modules displayed power loss of less than 2% after 168 hours PID stress.

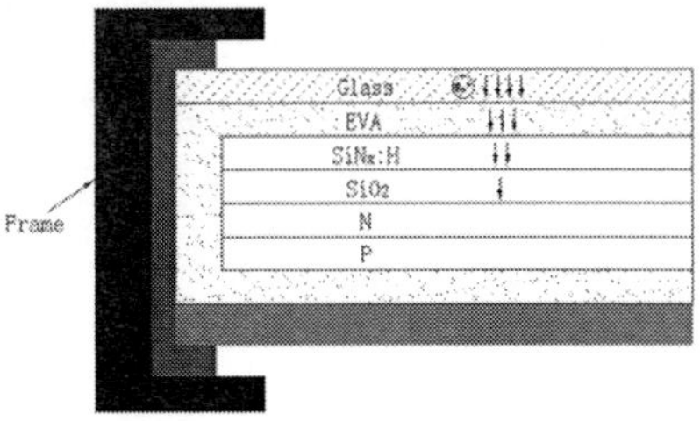

Figure 8: Migrating ways of sodium ions

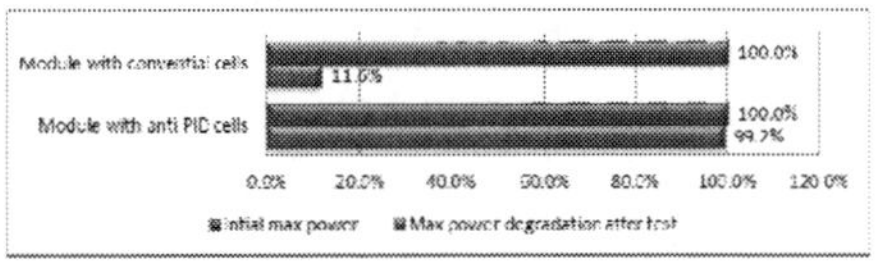

Figure 9: Power degradation of modules after accelerated testing

4.2 Module Level

On the module level, good encapsulation materials are critical in countering PID. We focus specifically on the encapsulation materials and frames, analyzing their influences on the PID effect.

In one test, conventional cells without anti-PID features were covered with encapsulation materials of differing volume resistivity. All the other materials in the modules remained the same. These modules were subjected to PID accelerated testing. We realized that increasing the volume resistivity (from about 10^14Ωcm to about 10^16Ωcm) eased the PID effect. In our tests, modules with encapsulation materials #1,2,3 achieved strong anti-PID performances. Module power loss was less than 5% after PID stress. The power degradation of these modules after testing is shown in Fig. 10. This suggests that appropriate encapsulation materials could counter the PID effect to varying degrees. For those encapsulation materials without good PID resistance, it is necessary to incorporate anti-PID cells to protect the module from PID stress.

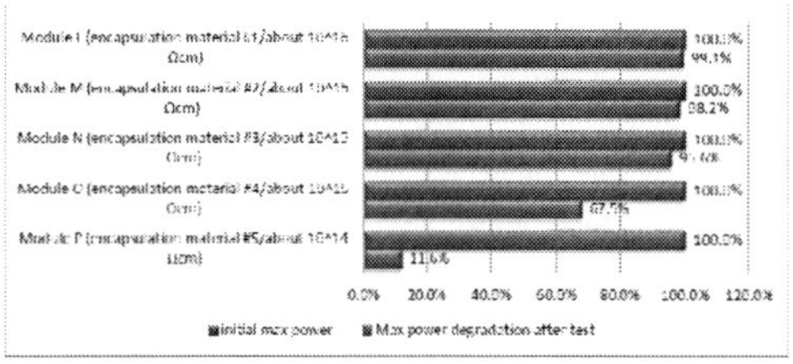

Figure 10: Power degradation of modules after accelerated testing

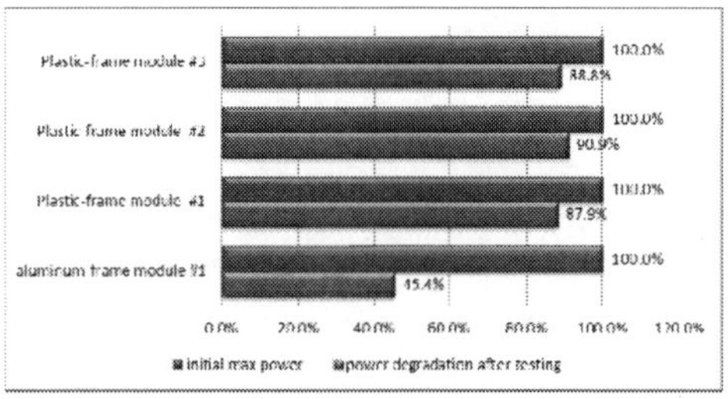

Figure 11: Power degradation of modules after accelerated testing

In another test, aluminum frame and plastic frames were used to make modules. These cells were conventional cells without anti-PID features (all the other encapsulation materials remained the same). By accelerated testing(60℃, 85%RH and negative 1000V), the PID effect was observed. Among all the modules, those with plastic frames performed better than those with aluminum-frames. The power degradation of these modules is shown in Fig. 11.

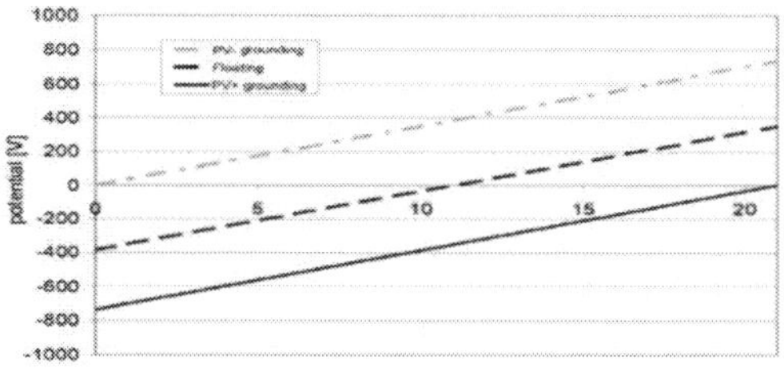

Figure 12: PV system grounding modes (PV+/PV- and floating potential schemes)

4.3 System Level

There are three possible methods of grounding in PV systems (Fig. 12).7

a. Floating

Floating mode means neither pole of the inverter is grounded. In this system, half the modules are negative biased and half modules positive biased towards earth ground.

b. Negative grounding (PV- grounding)

Ground the negative pole of inverters. All modules in the system are positive bias towards earth ground. This method can prevent sodium ions from diffusing on the cells. Testing results reported in section 2.3 show that P-type modules won't degrade when modules have positive bias towards the earth ground. Negative grounding is recommended for combating the PID effect. In comparison to other methods, negative grounding is the simplest and most cost-effective. We have a few projects in Southeast Asia, where it is both very humid and hot. Additionally, the project land is salty. In the beginning, floating mode modules characterized the system. However, after our clients complained of PID, we recommended that they retrofit their systems from floating to negative grounding. Module performance recovered and the systems are functioning as expected. The retrofit expense was minimal.

c. Positive grounding (PV+ grounding)

Ground the positive pole of inverters. All modules in the system are negative bias towards earth ground. Positive grounding is recommended for N-type modules. N-type module supplier, SunPower actually requires positive grounding [5].

Both negative and positive grounding can be easily implemented for transformer-based inverters. To minimize costs and increase inverter efficiency, more transformer-less inverters are being marketed and used for PV systems. Modules will be of floating mode if transformer-less inverters are used in the system. Neither P-type nor N-type modules want to see this in order to achieve optimal performance.

In accordance with varying local policies, different regions have different requirements on system grounding. For instance, before 2005, the NEC (National Electrical Code) required all PV systems to have one DC circuit conductor grounded whenever the maximum system voltage exceeded 50 volts. After 2005, however, ungrounded PV circuits when use of utility –interactive inverters without isolation between DC input and AC output. And in European countries, only the system into the building requires compulsory system grounding, but equipment grounding (functioning grounding) when there are any exposed metal or conductive surfaces that may become energized.

We recently offered a proposal to IEC TC82 Work Group 3, experts on PV systems standards. We believe that anti-PID solutions through proper system grounding is simple and cost-effective, in contrast to cell level and module level solutions [6]. From a systems level, we urge PV installers and module makers to adopt a common method of combating PID, namely the negative grounding method for P-type modules.

5 CONCLUSIONS

In this paper, experiments were designed and conducted to study the factors contributing to PID. We found that high temperature and humidity would trigger PID effect if modules did not have the proper voltage bias with respect to earth ground. There are realistic and applicable solutions to PID on both the cell level and module level. However, applying a proper voltage bias to earth ground is most effective solution and can prevent PID from occurring for PV modules. This solution is simple and cost-effective, in contrast to less cost-friendly cell and module level solutions. In humid and hot areas, negative pole grounding for P-type modules and positive pole grounding for N-type modules are recommended, as systems built under the mentioned specifications will achieve optimal system performance. In such cases, transformer-based inverters should also be utilized

5 REFERENCES

[1] Peter Hacke, "System voltage potential-induced degradation mechanism in PV modules and methods for test," the 37th IEEE PVSC, Seattle, USA.

[2] Mani G. Tamizh-Mani, 12–18 Year-Old PV Power Plants in Arizona: Potential Induced Degradation Analysis of 1900 Individual Modules, PV Module Reliability Workshop, Golden, CO, Feb 2012.

[3] IEC 61215, "Crystalline Silicon Terrestrial Photovoltaic Modules – Design Qualification and Type Approval" edition 2, IEC TC82 WG2, 2005.

[4] IEC 62804draft version, "System voltage durability test for crystalline silicon modules", IEC TC82 WG2, 2013.

[5] SunPower, "Handbook for Positive Grounded Batteryless Grid Connected PV Systems".

[6] Canadian Solar, "Anti-PID through proper system grounding", a proposal to IEC TC82 WG3, 2013".

[7] S.Pingel, O.Frank, M.Winkler, S.Daryan, T.Geipel, H.Hoehne, J.Berghold, Potential induced degradation of solar cells and panels.

[8] C.R.Osterwald et.al, Sol.energy.mat Sol.cells, 79, 2003

POTENTIAL-INDUCED DEGRADATION (PID): FIELD MONITORING DATA EVALUATION AND CLIMATIC CHAMBER TESTS WITH FIELD MODULES

G. Mathiak, N. Bogdanski, F. Reil
TÜV Rheinland Energy GmbH
Am Grauen Stein, 51105 Cologne, Germany
Email: Gerhard.Mathiak@de.tuv.com

ABSTRACT: Potential-induced degradation (PID) can cause high yield losses in PV systems. Anti-PID measures shall be started as fast as possible to avoid unrecoverable modules and energy yield loss. In the presence of PID, the efficiency of module strings will decrease, especially at low irradiation, where the relative power loss is higher. An approach to detect PID based on monitoring data of PV strings is evaluated.
In the field, degraded modules are typically regenerated by anti-PID boxes which apply a positive potential during night time. Climatic chamber tests with PV modules which were PID-degraded in the field showed that some c-Si module types cannot be completely regenerated after severe degradation and that thin film CIGS modules can have reversible PID effects.

Keywords: PID, Monitoring, Regeneration

1 INTRODUCTION

Potential-induced degradation (PID) can cause high yield losses in PV systems. The degree of PID degradation of a PV module can be measured quantitatively by maximum power determination at standard test conditions. In the field, infrared thermography and open-circuit voltage measurements may be used for quick PID detection [1]. In the presence of PID, the efficiency of module strings will decrease, especially at low irradiation, where the relative power loss is higher. In this paper we present an approach to detect PID based on monitoring data.

PID is a reversible phenomenon; degraded modules can be regenerated through electrical or heat treatment. In the field, degraded modules are typically regenerated by anti-PID boxes which apply a positive potential during night time. The regeneration duration is expected in a time scale of several months. In the following we report the results of climatic chamber tests with PV modules which were PID-degraded the field.

2 FIELD MONITORING DATA EVALUATION

To optimize PID detection, monitoring data were analyzed. The efficiency of randomly chosen strings in dependence of the irradiance was calculated for different days using the two-minute averages of string output power and in-plane irradiance. Especially the comparison of clear sky days in different years allows the calculation of yearly degradation. For this method irradiance sensors and modules have to be aligned accurately. A misalignment of the sensor of several degrees in the azimuth can induce a time shift of several minutes between power and irradiance signal.

An example of a string efficiency over irradiance curve is shown in Fig. 1. The irradiance and the string efficiency were calculated for every 18 July of 4 consecutive years. A high yearly degradation rate of 4 %/a was found. The degradation at low irradiance was slightly higher than at high irradiance. Degradation rates above 1 %/a are a hint for unusual high performance loss. Other failure mechanisms can cause similar curves. For example, soiling with the growth of moss, corrosion or thermomechanical fatigue of the connectors can lead to major power losses.

Module monitoring which is implemented in some module optimizers allows PID detection by the comparison of the modules at the negative and the positive pole of the module string.

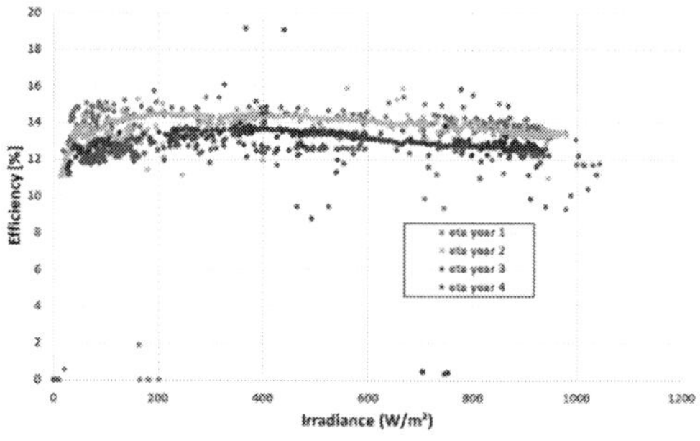

Fig. 1: String Efficiency plotted versus in-plane irradiance [W/m²] at the same day (18 July) of four consecutive years with clear sky conditions. The resulting yearly degradation is displayed in Table 1.

Irradiance	Efficiency				Degradation
	Year1	Year2	Year3	Year4	
200 W/m²	14.5%	14.4%	13.5%	12.7%	-4.3%/a
800 W/m²	13,8%	13,7%	12,7%	12.4%	-3.8%/a
900 W/m²	13.8%	13.4%	12.7%	12.2%	-4.0%/a

Tab. 1: String Efficiency at the same day (18 July) of four consecutive years. The relative degradation rate in the order of 4 %/a is clearly higher than the typical power warranty of many manufacturers (1 %/a).

3 CLIMATIC CHAMBER TESTS WITH FIELD MODULES

PID is a reversible phenomenon; degraded modules can be regenerated through electrical or heat treatment (see Fig. 2). A positive potential were applied with the PVO-box made by SMA during night time to a string with PID-degraded modules at the negative end. The PVO-box is installed parallel to the module string.

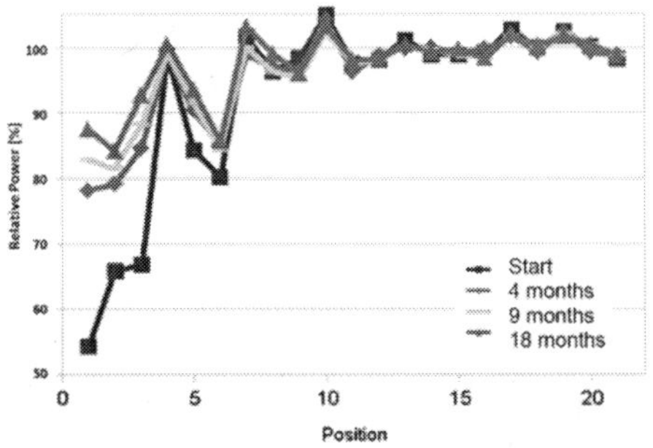

Fig. 2: Relative power [%] over position in the module string in the course of 18 months of regeneration.

Climatic chamber tests with PV modules which were PID-degraded in the field were performed to find out if PID-degraded modules can be completely regenerated.

3.1 Saturation of Regeneration of c-Si Modules

An experimental study was performed with c-Si modules from the field (Fig. 3). The modules showed PID in the field. After severe degradation in a climatic chamber (V = -1000V, t = 96h, T =60°C) the maximum power could be increased by factor 10 in a first regeneration run (V = +1000V, t = 96h, T = 60°C). Further regeneration runs in the climatic chamber induced only slight power increases, while a final degradation run (V = -1000V, t = 96h, T = 60°C) with 2 modules induced high power loss

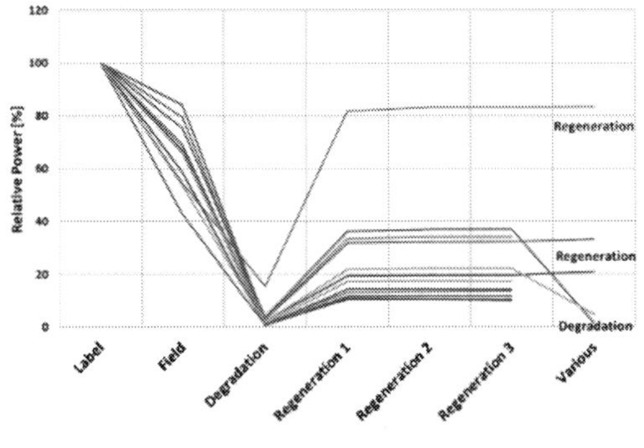

Fig. 3: Power degradation and regeneration of 12 c-Si modules. Modules from the field, degraded to 40% to 85% were selected for a further degradation run. These severely degraded modules where regenerated in the climatic chamber. Following regeneration processes with V = +1000 V lead only to minor power gains. A final run was performed with modules which were covered by adhesive Aluminum foil.

These experiments showed that this module type cannot be completely regenerated after severe degradation with the same voltage which was used for degradation. This regeneration saturation can be an effect of the sodium ions in the vicinity of the p-n junction. The effect is not attributed to glass conductivity effect since further degradation is possible and leakage currents in both directions are similar.

3.2 Degradation and Regeneration of CIGS Modules

CIGS modules showed PID in the field [3]. Normally the cause for PID of thin film modules is irreversible TCO corrosion. In this case it was possible to regenerate the modules as it is typical for c-Si modules (Fig. 4).

Six modules were covered with adhesive aluminum foil and stressed in a climatic chamber (96 h, 60°C) with the test voltage. The voltage of ±1000 V between frame and cells induced regeneration resp. degradation. Two modules without voltage had minor power losses after the climatic chamber test.

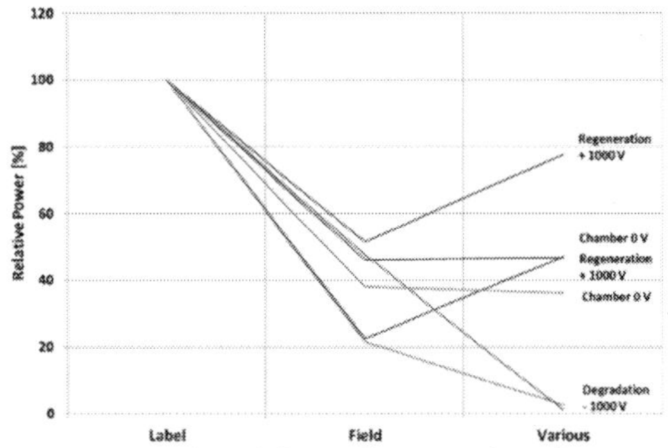

Fig. 4: Power degradation and regeneration of 6 CIGS modules. Two modules were regenerated with +1000 V, two modules were without electric tension, and two modules were severely degraded by -1000 V.

A possible reversible effect of thin film modules must be considered for the development of PID testing of thin film modules. Severe TCO corrosion testing can lead to overstress since reversible degradation can be regenerated under natural conditions. Potential reversible effects are to be considered in the development of the PID test technical specifications for crystalline and thin film modules [4, 5].

4 CONCLUSIONS

- The evaluation of field monitoring data allows determining yearly power losses and can give hints for PID. A distinct evidence for PID could not be given by string data analysis. Site inspections including on-site measurements are still necessary.
- Experiments with c-Si modules showed that some module types cannot be completely regenerated after severe degradation. This saturation is not attributed to glass conductivity effect since further degradation is possible.
- Typically thin film modules can suffer from irreversible TCO corrosion when high system

voltage is applied. We showed that CIGS modules can also have reversible PID effects.

- Anti-PID measures (Anti PID-boxes or grounding) shall be started as fast as possible to avoid unrecoverable modules and energy yield loss.

5 REFERENCES

[1] G. Mathiak, S. Kämmer, M. Schweiger, W. Herrmann: "Potential-induced Degradation (PID) – Comparison of Different Test Methods, Shunt Resistance Measurements of Individual Cells inside a Module and Infrared Imaging of PV modules". Proc. 28 EU PVSEC, Paris 2013, 3332-3335.

[2] G. Mathiak, M. Schweiger, W. Herrmann, M. Stalder, J. Laschinski: „Potentialinduzierte Degradation (PID) bei kristallinen PV-Modulen – Analyse und Gegenmaßnahmen", 28. Symposium Photovoltaische Solarenergie, Bad Staffelstein (GER), 08.03.2013.

[3] C. Dürschner: "Potential-induzierte Degradation an einer Megawatt-Freiflächen-PV-Anlage mit GIGS-Modulen", Verlag "Solare Zukunft", 2015, Erlangen.

[4] IEC/TS 62804-1:2015 : "Photovoltaic (PV) modules - Test methods for the detection of potential-induced degradation - Part 1: Crystalline silicon", published 2015-08-06.

[5] IEC/TS 62804-2: "Photovoltaic (PV) modules - Test methods for the detection of potential-induced degradation - Part 2: Thin-film". Draft ANW.

6 ACKNOWLEDGMENT

The work presented has been elaborated in the frame of the project 'Fidelitas' (FKZ: 0325735A) supported by the Federal Ministry for Economic Affairs and Energy

ANALYSIS OF PHOTOVOLTAIC MODULE AFTER POTENTIAL INDUCED DEGRADATION

Jiří Vaněk[1], Josef Hylský, Dávid Strachala, Martin Šturm, Pavel Čudek
Brno University of Technology, Faculty of Electrical Engineering and Communication Technologies,
Department of Electrotechnology
Technicka 10, 616 00 Brno, Czech Republic

ABSTRACT: This work deals with the phenomenon called Potential-Induced Degradation of PV modules (PID shortly). Currently is this problem solved in a lot of scientific institutes where is analyzed the physical description of the process that occurs due to PID. The cause of PID degradation is in a migration of positive sodium ions from glass (that protects PV modules from the environmental influences) to the PV cell which leads to a negative effect in a pn junction of photovoltaic cell. This article is focused on the entire photovoltaic module. PV module is firstly degraded and then regenerated through an applied voltage of 600 V between the aluminum frame and PV cells of module. The degradation and regeneration process was locally supported by increasing of the temperature of PV module to 80 ° C. In this way was possible to observe the influence of temperature in a PID affected PV module.
Keywords: Potential-Induced Degradation, PV modules, solar cells.

1 INTRODUCTION

Degradation of photovoltaic power plants due to the PID (Potential Induced Degradation) is one of the main degradation mechanisms, which is responsible for a reduction of efficiency of the photovoltaic systems. The problems with PID were already known in the 1980 and currently is this topic widely researched through many scientists [1].

With the gradual increase of the production of photovoltaic modules and pressure to lowering the manufacturing costs often to the detriment of the quality of used materials, began this undesirable phenomenon more frequent. The first company which warned about its negative effects was SOLON in 2010 and this has led to increased awareness of PID [2]. A detailed study about the PID in photovoltaic modules has released 2 years later the Fraunhofer Institute in Germany. Fraunhofer Institute identified PID in 46% of tested modules from 96 different manufacturers [3].

Reason of performance decrease through a PID degradation is caused by several technical issues. Especially because of the large number of photovoltaic modules connected in the series in string of PV system, then in the grounding of photovoltaic module frame (safety reasons) and by using an inverter without galvanic isolation (it is not possible to grounded the negative pole of string). This leads to a high voltage potential between the grounded frame and the solar cell of PV module at the negative pole of the string. High negative voltage may cause leakage currents and transfer of electrons from the photovoltaic cell into the aluminum frame. The movement of ions from the materials used in the construction of photovoltaic module (protective glass, foil, insulation of panels, etc.) is generated into the PV cell structure. In the end, these processes affect the final parameters of photovoltaic modules.

PID is further caused by environmental factors in the form of increased temperature and humidity in the area of photovoltaic modules (cells), that accelerate degradation and mobility of charge carriers.

2 REGENERATION OF PV MODULE

Analysis of PV module regeneration start from a study of the Research Center for Silicon Photovoltaics [4]. This study deals with the issue of photovoltaic cell regeneration affected by PID degradation, where through a higher temperature is decreased the concentration of sodium ions Na^+ in active part of the photovoltaic cells.

Through the reverse process by "re-diffusion" of sodium ions Na^+ is reduced recombination in the emitter of PV cell. This again increases the shunt resistance R_{SH} what increase the efficiency of the photovoltaic cells (module).

Regeneration was carried out on photovoltaic module ES - A - 205 EVERGREEN. Evergreen module has been uninstalled from the negative pole of photovoltaic plant affected by PID degradation. PV module parameters given by the manufacturer are presented in the following Table 1.

Table I: Parameters of PV module given by the manufacturer

P_{MPP} (W)	Eff (%)	U_{MAX} (V)	I_{MAX} (A)	U_{OC} (V)	I_{SC} (A)
205	13.1	18.2	11.2	22.7	11.9

Parameters PID degraded module before the regeneration process are shown in Table 2. The measurements were done in certified laboratories CV-LAB at Brno University of Technology.

Table II: Parameters of PV module affected by PID

P_{MPP} (W)	Eff (%)	U_{MAX} (V)	I_{MAX} (A)	U_{OC} (V)	I_{SC} (A)
179.9	11.4	17.3	10.5	22.3	11.9

The maximum power decrease of photovoltaic module was calculated to 12.4%. After five years of operation of photovoltaic module is this decrease of power attributed to the PID degradation. This is confirmed by the measured parameters of the PV module from the positive pole of a string, see Tab. III, where the reduction in power is only 3.9 %.

If the parameters are compared with the parameters of the PV module from negative string, it is quite evident that the PID module from the positive pole is not affected by the PI.. This confirms the fact that the PV modules from a positive pole of a string are not PID degraded.

Table III: Parameters of PV module from the postitive pole of string

P_{MPP} (W)	*Eff* (%)	U_{MAX} (V)	I_{MAX} (A)	U_{OC} (V)	I_{SC} (A)
197,2	12,6	17,9	10,9	22,4	11,9

Visible differences caused by degradation of PID among PV modules from the positive and negative poles are captured in images from electroluminescence measurements Fig. 1.

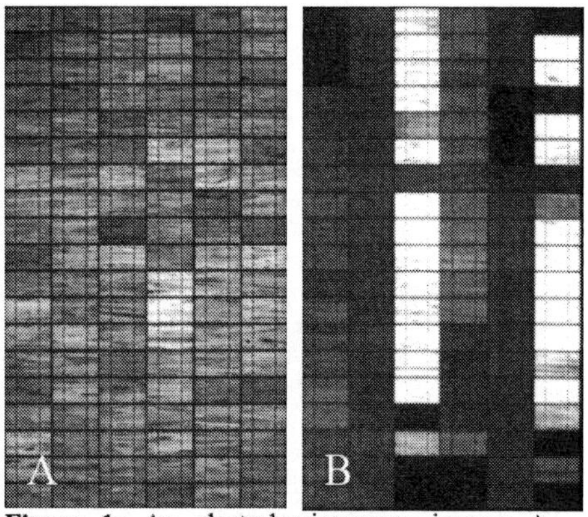

Figure 1: An electroluminescence images A - PV module from a positive pole of string, B - PV module from the negative pole of string.

3 WORKPLACE FOR REGENERATION OF PV MODULE

Workplace for the regeneration of the photovoltaic module has been created in a dark room at the Institute of Technology, BUT, see Fig.2. Photovoltaic module was placed on the aluminum plate, which was under the entire surface of the module.

A negative terminal of the DC power supply was attached on the aluminum plate and positive pole was connected to the terminals of the module. Through this connection was achieved an electric field in the opposite direction upon the occurrence in PID affected photovoltaic power plant (solar cells was on the positive potential). Because of surface of aluminum sheet under the entire module has been achieved a homogeneous electric field. A DC power was set to 600 V, which was chosen because of maximum value in the string according the manufacturer's manual. Local heating of the photovoltaic module was performed through the halogen lamp. The distance of the lamp was adjusted in the way that the temperature of PV module was 80 °C. Precise adjustment of temperature was measured using a thermocouple on the surface of the of the PV module.

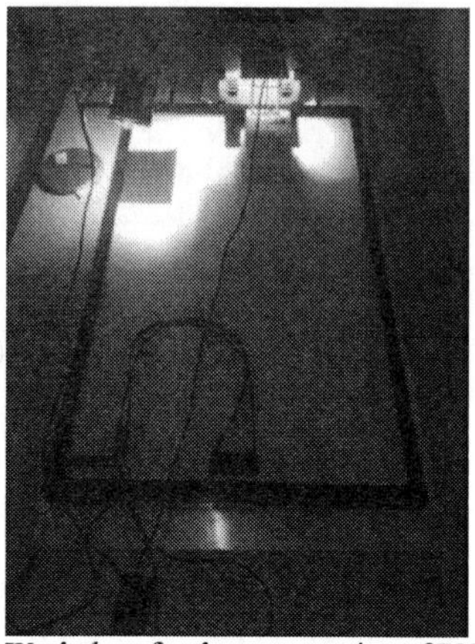

Figure 2: Workplace for the regeneration of PV module

Firstly was photovoltaic module regenerated only by an electric field and after several time intervals was added a local heating on the most affected solar cells. After each time interval was done a luminescence measurement and current-voltage characteristic was also measured.

4 RESULTS AND DISCUSSION

The regeneration of photovoltaic module was done by eight steps. In a four of them was applied just an electric field of 600 V and in a remaining steps an electric field with local heating by 80 °C. These areas were located by an electroluminescence. Figure 3. summarizes the first four PV modules gradually regenerated by an electric field (600 V) in the 22 hour steps.

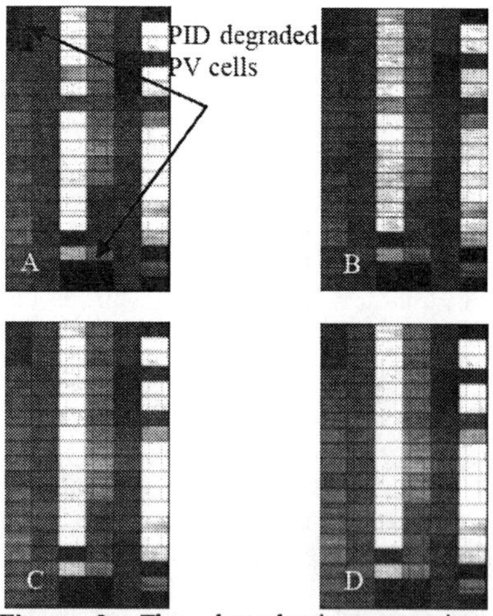

Figure 3: The electroluminescence images of the regenerated PV module by an electric field (600V) at time intervals: A - initial measurements, B - regeneration time 22 h,C - regeneration time 44 h, D - regeneration time 66 h).

During regeneration through the electric field was the increase of the maximum power insignificant, as can be seen in figure 3. Degraded PV cells are still noticeably dark. After this part of the regeneration was applied local heating of the affected areas (80 ° C). Places where the temperature is increased are indicated by the red box. The next four steps of regeneration captured by an electroluminescence is displayed in figure 4.

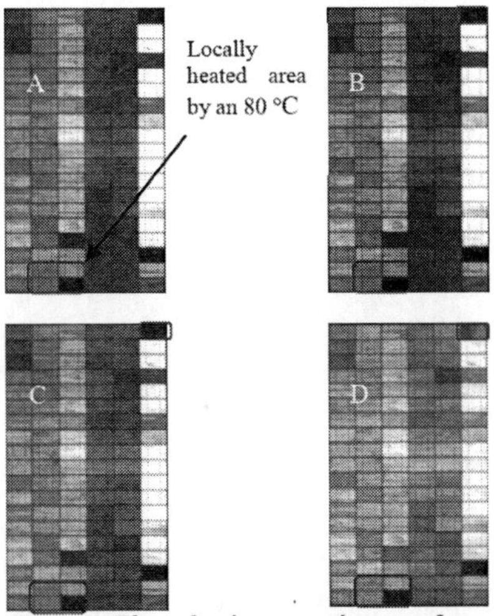

Figure 4: Electroluminescence images of regenerated PV module by an electric fielde (600 V) and with a local temperature of 80 °C: A - regeneration time 163 , B - regeneration time 187 h, C - regeneration time 210 hours, D - regeneration time 306 h.

Comparing the locations of the PV module affected by PID degradation which has been heated to 80 ° C with other PID degraded cells shows that the temperature accelerates regeneration. For steps B, C, D in Fig. 4 were added another place of local heating in the upper right corner of the PV module. Fig. 4 (D) shows the final state of the PV module after 306 h of regeneration. We are assuming that that two dark cells are because of high degree of PID degradation, which is not reversible and therefore the regeneration process is ineffective.

The characteristic of the PV module parameters on the time of regeneration are shown in the following figures Fig. 5 and Fig. 6.

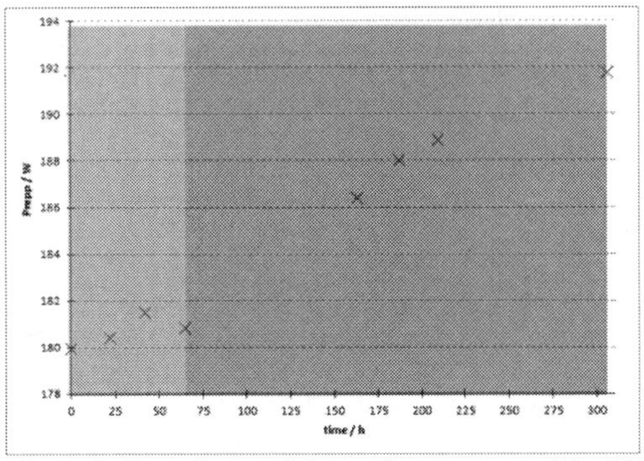

Figure 5: Dependence of the maximum power of the PV module at the time of regeneration (the blue area - regeneration of 600 V, the red area - regeneration of 600 V, 80 ° C).

The characteristic of PV module regeneration in Fig. 5 shows that the electric field has much less potential of regeneration versus regeneration with the areas heated to 80 ° C. This confirms the theory that heating of the affected areas with the temperature noticeably accelerated regeneration process. Temperature has increased the diffusion of sodium ions in the structure of a solar cell.

In FIG. 6 are shown the values of a shunt resistance at the time of regeneration. The shunt resistance of solar cell is a parameter which is extremely sensitive to the PID degradation

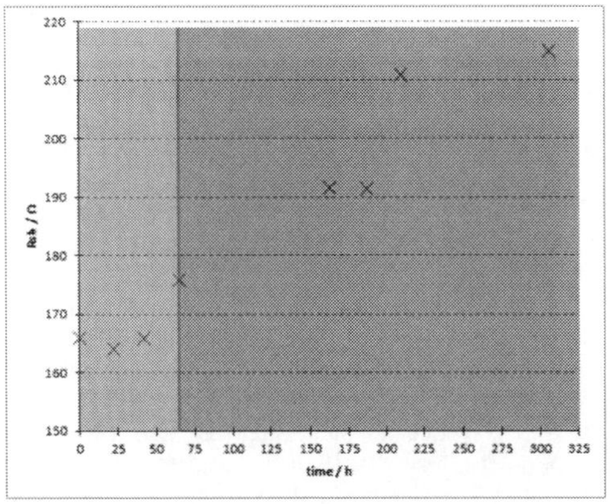

Figure 6: Dependency of shunt resistance at the time of regeneration (the blue area - regeneration by 600 V, the red area - regeneration by 600 V, 80 ° C).

During the regeneration of the PV module the value of the shunt resistance has increased by 49 Ω. Also this dependenceshows a higher gradient rise in the red area of the graph. All PV module parameters measured during the regeneration process are summarized in Table IV.

Table IV: Measured parameters of PV module during the regenration process

t (h)	P_{mpp} (W)	*Eff* (%)	*FF* (%)	ϑ (°C)	R_{sh} (Ω)
0	179.96	11.48	68.75	25	166.1
22	180.39	11.51	69.22	25	164.1
42	181.47	11.58	69.44	25	165.9
65	180.81	11.54	69.44	25	175.9
163	186.41	11.89	71.06	80	191.6
187	188.21	11.99	71.46	80	191.5
210	188.91	12.05	71.72	80	210.9
306	191.77	12.23	72.67	80	215.1

4 SUMMARY/CONSLUSION

This article deals with the regeneration of the PV module by an electric field (600 V) and with a local heating of a PID affected areas with the temperature of 80 ° C. The value of 600 V was chosen according the maximum voltage given by the PV module manufacturer manual. Tested PV module (Evergreen) with the performance of 205 W has been uninstalled from the negative string of power plant. The module was PID degraded and regenerated again. It was found that the temperature rapidly accelerates the process of regeneration, as can be seen from the dependencies in FIG. 5 and FIG. 6. Because the temperature was

increased locally at certain points of the PV module, it was possible to compare these areas with the areas regenerated only by an electric field.
On the other hand, this analysis cannot determine how the temperature can accelerate the process of regeneration for the entire PV systems without an attached reverse voltage potential. This will be observed in the follow-up work.

4.1 References

[1] Lahký, Martin: Pokles výkonu fotovoltaických elektráren v důsledku vysokonapěťového stresu Další rána pro majitele solárních elektráren., oze.tzb-info.cz[online]. 2015 [cit. 2016-5-18]. Dostupné z: http://oze.tzb-info.cz/fotovoltaika/12260-pokles-vykonu-fotovoltaickych-elektraren-v-dusledku-vysokonapetoveho-stresu

[2] Schuetze, et al, Laboratory Study of Potential Induced Degradation of Silicon Photovoltaic Modules, Q-Cells SE, Sonnenallee 17-21, 06766 Bitterfeld-Wolfen

[3] Heriber t Schmidt and Bruno Burger, Interactions between Solar Panels and Inverters, Fraunhofer-Institut für Solare Energiesysteme (ISE), December 2010

[4] LAUSCH, D., V. Naumann, A. Graff, A. Hahnel, O. Breitenstein, C.Hagendorf and J. Bagdahn,"Sodium outdiffusion from stacking faults as root cause for the recovery proces of potential-induced degradation (PID), Energy Procedia 55 (2014) 486 –493.

4.2 Acknowledgement

This research work has been carried out in the Centre for Research and Utilization of Renewable Energy (CVVOZE). Authors gratefully acknowledge financial support from the Ministry of Education, Youth and Sports of the Czech Republic under NPU I programme (project No. LO1210).

TOPIC 5.4 PV Cells and Modules - SUBMISSION FOR ORAL PRESENTATION

YIELD LOSSES OF PID-AFFECTED PV SYSTEMS - SIMULATION OF YIELD LOSSES BEYOND POWER LOSS

J. Arp[1], B. Jaeckel[2] and J. Behrschmidt[3]

[1] PV LAB Germany GmbH, Gartenstrasse 36, 14482 Potsdam, Germany,
Tel.: +49-172-1985686, j.arp@pv-lab.de, www.pv-lab.de

[2] UL International GmbH, Admiral-Rosendahl-Straße 9, 63263 Neu-Isenburg (Zeppelinheim), Germany,
Tel.: +49-162-2621901 bengt.jaeckel@gmail.com, Internet: www.ul.com

[3] 8.2 Obst & Ziehmann GmbH, Brandstwiete 4, 20457 Hamburg Germany,
Tel.: +49-40-1812604-00, joerg.behrschmidt@8P2.de, Internet: www.8P2.de

ABSTRACT:
Target of this work is to estimate the extent of yield losses of potential induced degradation (PID) affected PV systems and to compare these with the STC power losses. It is obvious that a PV system which faces a loss of STC power will face at least a yield loss, which often assumed to be in the same range like STC power loss. In fact, the yield loss will be much higher for PID since it has a significantly stronger impact on the performance at low irradiance. This effect becomes relevant already in the initial phase of PID when the STC power is hardly affected.
We simulate three PV systems with two different sets of irradiation data for a Mediterranean area (Greece) and a moderate climate (Germany). For the simulation we used PVSyst with a fixed tilted, open rack mounting structure without any shading and weather data from Meteonorm. The PVSyst simulations are resulting in different specific energy production as a function of location and performance of the string. This is further folded with the STC power loss to calculate the total relative energy yield loss of a system which is also a loss in revenue.

Keywords: PID, Degradation, Yield loss, power loss, revenue and income loss

1. INTRODUCTION/MOTIVATION

Potential Induced Degradation (PID) is a topic of extensive research since the degradation has first been published by Pingel in 2010 [1].
Authors like Berghold [2][3], Hacke [4], Nagel [5], Naumann [6][7][8], Lechner [9][10] and Pingel [1][11], published valuable information about degradation mechanism. Whereas Taubitz [12][13][14] also discusses the recovery mechanisms on cell and module level. All publications mainly deal with the topic from a research and laboratory point of view. Influences of weather on degradation are described by several authors e.g. Hoffmann [15][16] and help to develop ideas to simulate the progression of PID in the field. Different approaches are discussed to combine degradation [5][11][17][18] and recovery to best simulate the field but also on how to test modules for PID degradation and recovery [10][19][20].
Degradation and recovery mechanisms are described and it is well understood that PID affects initially the low light performance of the PV module due to its shunting mechanism of the silicon solar cell. Laboratory results are therefore great and can help to get a feeling of the severness.
The scope of this work is to look one step further. That means to look into the yield loss, which is equivalent to a loss in income of the PV asset owner.

2. APPROACH

In the first step (step #1) a set of 12 representative modules are picked from a larger batch of installed modules five years in operation and characterized using a Pasan AAA flash sun simulator. Electrical performance characterization is done at 100, 200, 400, 700 and 1000 W/m² (IV curves for 1000W/m² see Figure 1). Those modules are representative from non-degraded to highly PID degraded modules. The module's relative power loss after 5 years of operation is between 3.2% (normal degradation) and over 60% (significant PID). For further processing all IV curves are interpolated to show equi-spacing steps of 0.01A. These equi-spacing steps are needed for step #3.

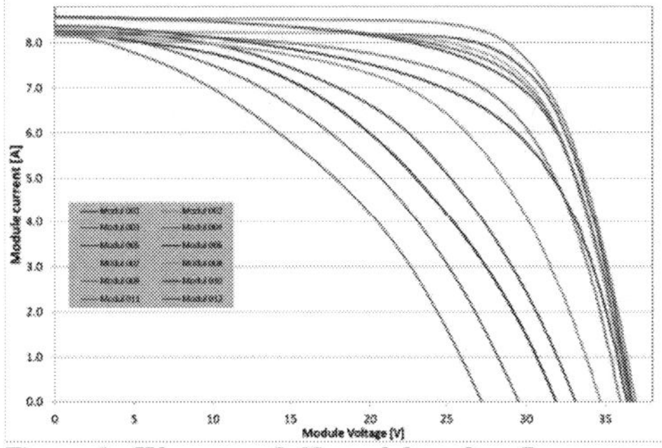

Figure 1: IV-curves of 12 modules after five years of operation showing almost no degradation to severe PID degradation.

In the second step (step #2) three degraded PV systems, one in Greece and two in Germany, are disassembled and the STC power of all PV modules is measured in a laboratory with a flash-type xenon sun simulator. The PV system in Greece is significantly degraded. One of the two PV systems located in Germany is strongly, the other one just slightly degraded. All three PV systems have been equipped with 3-phase transformerless inverters.

The distribution of the degradation in the string of these three PV systems is used for preparation of the yield simulation as described in step #3.
Step #3: Based on the IV curves generated in step #1 one PV string was synthesized for each of the three locations (one Greece, two Germany) in such a way that they match the characteristics of the field degraded strings measured in step #2 (example see Figure 2).

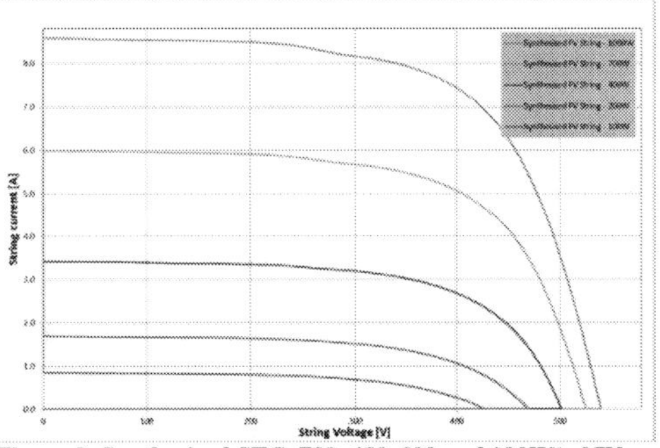

Figure 2: Synthesized STC, 700, 400, 200 and 100W/m² IV curves for the PV installation in Greece. Those curves served as base for PVsyst simulation.

Step #4 Simulation: The synthesized strings from step #3 are used for the energy yield calculation with PVsyst and Meteonorm weather data. A summary of the irradiance levels is given in Figure 3

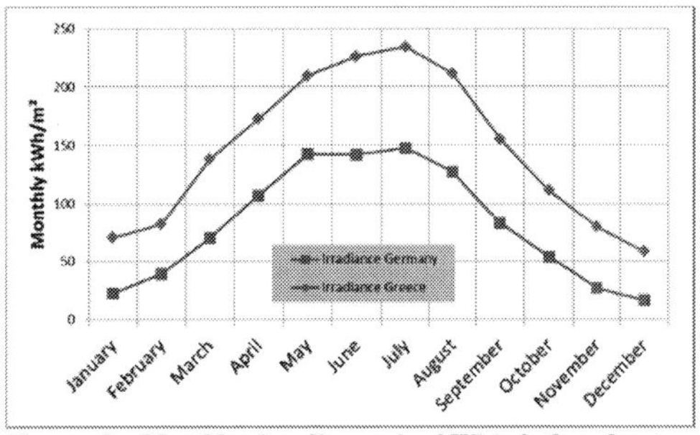

Figure 3: Monthly irradiance in kWh/m² for the two different sites. Clearly the irradiance in Greece is significantly higher compared to the German sites.

To increase comparability we simulate identical PV systems with and without PID degradation for every location. The system was assumed to be a fixed tilted (optimal for the location), open rack mounting structure without any shading and a state of the art inverter.

3. SIMULATION RESULTS

For all three sites a reference calculation is based on a naturally (without PID) degraded module. This result was set to 100% meaning optimal performance and energy production (yield) as calculated.
The three PID-degraded systems were simulated and all the data is summarized in Figure 4.
The middle part of the table summarized the measured and calculated power losses at STC conditions with respect to the normal-degraded system. PID leads to a string power loss of 4.2% for the low and 18.3% of the high PID degraded, respectively.
The left part shows the calculated specific energy production based on the IV curves from the synthesized PID affected strings with the reference of the normal string. The low PID degraded system showed a specific loss in energy production of 6.8% and the more degraded one of 12.4%. The system in Greece only showed a reduction of specific energy production of 5.4%.
The most right column shows the total energy loss caused by PID for the different loctions. The results are calculated by multiplication of STC power loss and loss of specific energy production.

4. DISCUSSION AND CONCLUSION

PID can be observed both in Germany and in Greece. In Greece there is much more sun (less moisture in the laminate), higher temperatures (higher regeneration rate, but also higher degradation rate) and especially on islands potentially a thin salty layer on the module. Germany is the more humid, rainy and colder location. Both locations have their own specific risk factors.
By comparing e.g. the string STC power loss of 13.8% of the creek PV system versus the 4.2% of the slightly degraded German PV system the specific yield loss in Greece is lower than in Germany (5.4% vs. 6.8%). This is due to the different shares of low light in the annual global irradiation (see also Figure 3). In Germany the specific yield has a large low light share whereas in Greece the yield is achieved by much higher irradiation.
Measured power loss of a string or a set of modules does not describe the real value loss of a PV asset. The total loss caused by PID (here 18.3 vs. 28.5%) is depending on the location of the PV system. The yield loss is in general higher compared to the power loss.

Summary: The losses of a PV system affected by PID occur on three levels. First the PV modules on the negative side (for p-type Si) of the string suffer a STC-power loss. Second due to the increased mismatch of the PV modules within a string the STC power of the string is below the sum of the STC power of the PV modules which are degraded to a different extend (not shown in detail herein but was considered in the simulation by the synthesized strings). And finally most important the third aspect that we are focusing on is the combination effect of power loss and specific yield loss due to poor weak light performance. This will lead to a yield loss which significantly exceeds the power loss (see Figure 4 and Figure 5). At irradiances below the STC level the PID effect is increasing with decreasing irradiance. As a result the yield loss exceeds the power loss. The extent on which this is visible is depending on the location of the PV system. A PV system at a location where a large share of the total energy production over a full year is generated at low irradiance levels (e.g. Germany) PID will cause a relatively higher energy and revenue loss compared to a location with high irradiance levels (e.g. Greece or desserts)

Germany	[kWh/kWp/year] [Germany]	%	%-loss [Germany]
Specific production normal	974	100.0%	0.0%
Specific production low PID	908	93.2%	6.8%
Specific production high PID	853	87.6%	12.4%

Greece	[kWh/kWp/year] [Greece]	%	%-loss [Greece]
Spezifischer Ertrag	1678	100.0%	0.0%
Specific production high PID	1587	94.6%	5.4%

Germany	Number of modules	kWp - defined	Wp per module defined	kWp as is	average Wp as is per module	%	%-loss
Power of string normal	15	3462	230.8	3462	230.8	100.0%	0.0%
Power of string low PID	14	3231	230.8	3095	221.1	95.8%	4.2%
Power of string high PID	20	4616	230.8	3769	188.5	81.7%	18.3%

Greece	Number of modules	kWp - defined	Wp per module defined	kWp as is	average Wp as is per module	%	%-loss
Power of string normal	15	3462	230.8	3462	230.8	100.0%	0.0%
Power of string high PID	15	3462	230.8	2984	198.9	86.2%	13.8%

Real Yield loss [Germany]
0.0%
10.7%
28.5%

Real Yield loss [Greece]
0.0%
18.5%

Figure 4: Summary of all measured/synthezied and calculated values. Left: Specific energy production/yield determined by PVsyst. Middle: measured/synthezied STC data of the three locations plus the reference system for each location. Right: Real resulting loss for the owner.

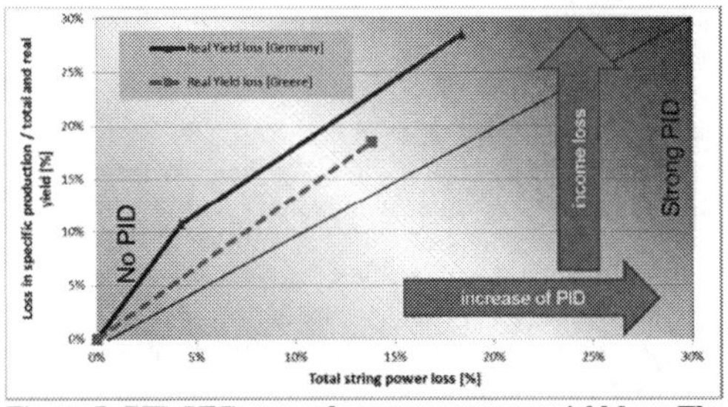

Figure 5: PID STC power loss versus energy yield loss. The line in the middle shows a 1:1 correlation. Clearly visible the loss in yield is higher compared to the STC power.

5. REFERENCES

* Thank you to Dr. Rakefet Zalashik for editing this article.

[1] S. Pingel, O. Frank, M. Winkler, S. Daryan, T. Geipel, H. Hoehne, and J. Berghold, Conference Paper 35th IEEE, (2010), „Potential Induced Degradation of solar cells and panels"

[2] J. Berghold, P. Grunow, P. Hacke, W. Hermann, St. Hoffmann, St. Janke, B. Jaeckel, S. Koch, M. Koehl, G. Mathiak, S. Pingel, L. Poehlman, P. Reinig, and A. Ukar, 28th European Photovoltaic Solar Energy Conference, (2013), "PID test round robins and outdoor correlation"

[3] J. Berghold, S. Koch, A. Böttcher, A. Ukar, M. Leers and P. Grunow, Photovoltaic International, (2013), "Potential-induced degradation (PID) and its correlation with experience in the field"

[4] P. Hacke, K. Terwilliger, St. Glick, D. Trudell, N. Bosco, St. Johnston, and Sarah Kurtz, Conference Paper 35th IEEE, (2010), "Test-to-Fail of crystalline silicon modules"

[5] H. Nagel, R. Pfeiffer, A. Raykov, and K. Wangemann, 27th European Photovoltaic Solar Energy Conference, (2012), "Lifetime warranty Testing of crystalline silicon modules for Potential induced degradation"

[6] V. Naumann, Chr. Hagendorf, St. Grosser, M. Werner, and J. Bagdahn, Energy Procedia (2012), "Micro Structural Root Cause Analysis of Potential Induced Degradation in c-Si Solar Cells"

[7] V. Naumann, D. Lausch, A. Graff, M. Werner, S. Swatek, J. Bauer, A. Hähnel, O. Breitenstein, St. Großer, J. Bagdahn, and Chr. Hagendorf, Physic status Solidi RRL 1-4, DOI 10.1002/pssr.201307090 (2013), "The role of stacking faults for the formation of shunts during potential-induced degradation of crystalline Si solar cells"

[8] V. Naumann, D. Lausch, A. Hähnel, J. Bauer, O. Breitenstein, A. Graff, M. Werner, S. Swatek, St. Großer, J. Bagdahn, and Chr. Hagendorf, Solar Energy Materials and Solar Cells, DOI: 10.1016/j.solmat.2013.06.015 (2013), "Explanation of Potential-Induced Degradation of the Shunting Type by Na Decoration of Stacking Faults in Si Solar Cells"

[9] P. Lechner, D. Sanchez, D. Geyer, and H.-D. Mohring, 27th European Photovoltaic Solar Energy Conference, (2012), "Estimation of Time to PID-failure by Characterization of Module Leakage Currents"

[10] P. Lechner, S. Hummel, and J. Schnepf, 31st European Photovoltaic Solar Energy Conference, (2015), "Evaluation of Recovery Methods after Potential Induced Degradation of PV Modules"

[11] S. Pingel, S. Janke, and O. Frank, 27th European Photovoltaic Solar Energy Conference, (2012), „Recovery methods for modules affected by Potential induced Degradation"

[12] Chr. Taubitz, M. Schütze, and M. B. Koentopp, 27th European Photovoltaic Solar Energy Conference, (2012), „Towards a kinetic model for Potential induced degradation"

[13] Chr. Taubitz, M. Kröber, M. Schütze, and M. B. Koentopp, 28th European Photovoltaic Solar Energy Conference, (2013), „Kinetic Description and Modelling of Potential Induced Degradation"

[14] Chr. Taubitz, M. Schütze, M. Kröber, and M. B. Koentopp, 29th European Photovoltaic Solar Energy Conference, (2014), „Potential Induced Degradation: Model Calculations and Correlation between Laboratory Tests and Outdoor Occurrence"

[15] S. Hoffmann, and M. Koehl, 28th European Photovoltaic Solar Energy Conference, (2013), "Modelling the leakage current for potential induced degradation"

[16] S. Hoffmann, and M. Koehl, 27th European Photovoltaic Solar Energy Conference, (2012), "Influence of Humidity and Temperature on the potential induced degradation"

[17] A. Raykov, H. Nagel, D.-J. Amankwah, and W. Bergholz, 27th European Photovoltaic Solar Energy Conference, (2012), "Climate model for Potential Induced Degradation of crystalline Silicon"

[18] J. Hattendorf, W.-M. Gnehr, R. Loew, T. Roth, D. Koshnicharov, and M. Zentgraf, 28th EU Photovoltaic Solar Energy Conference, (2013), „Potential-Induced Degradation and Temperature-driven Regeneration: A Realistic Simulation"

[19] B. Jaeckel, M. Cosic, and J. Arp, Conference Paper 40th IEEE, (2014), , "Investigation of c-Si Modules Degradation and Recovery Effect under High Potentials: CV-PID"

[20] B. Jaeckel, M. Cosic, and J. Arp, 29th EU Photovoltaic Solar Energy Conference, (2014), "Investigation of c-Si modules Degradation and Recovery Effect under High Potentials"

PID- AND UVID-RESISTANT N-TYPE SOLAR CELLS AND MODULES

Maciej K. Stodolny[a], Gaby J.M. Janssen[a], Bas B. Van Aken[a], Kees C.J.J. Tool[a],
Machteld W.P.E. Lamers[a], Ingrid G. Romijn[a], Peter R. Venema[b], Marten R. Renes[b],
Olga Siarheyeva[c], Ernst H.A. Granneman[c], Jianming Wang[d], Jikui Ma[d], Jingguang Cui[d], Fang Lang[d], Zhiyan Hu[d] and
Jochen Löffler[a]

[a] *ECN Solar Energy, PO Box 1, 1755 ZG Petten, the Netherlands*
[b] *Tempress Systems, Radeweg 31, 8171 MD Vaassen , The Netherlands*
[c] *Levitech BV, Versterkerstraat 10, 1322 Almere, The Netherlands*
[d] *Yingli Green Energy Holding Co., Ltd, 3399 North Chaoyang Avenue, Baoding 071051, China*

ABSTRACT: In this paper we report on the high stability of our n-type front junction solar cells (n-PERT) exposed to potential-induced degradation (PID) and UV-induced degradation (UVID) conditions. These intrinsically stable n-Pasha cells enable PID- and UVID-free modules even with industrially low-cost standard EVA encapsulant, independent of system grounding and system voltage. Based on intentional modifications of the Boron emitter and/or the dielectric layer in the PID-free and UVID-free n-Pasha solar cells, we are able to replicate reported degradation effects and study the mechanisms behind it. A combination of altering the boron profile and the dielectric properties together with increasing the interface defect density D_{it} is detrimental for the stability. Applying our standard optimal B-diffusion and passivation scheme assure that the UV radiation and system voltage have virtually no effect on our n-Pasha cell and module performance.
Keywords: PID, Potential Induced Degradation, UVID, UV Induced Degradation, n-PERT, stability, degradation

1 INTRODUCTION

In this paper we present potential-induced degradation (PID) and UV-induced degradation (UVID) resistant n-type c-Si solar cells enabling PID- and UVID-resistant modules even with common ethyl vinyl acetate (EVA) encapsulant, and independent of system grounding and system voltage.

The PID effect is a power loss of photovoltaic modules due to applied high system voltage. It has mainly been observed in standard front junction p-type c-Si modules [1]. The degradation mechanism has been attributed to massive shunting (*FF* loss) caused by Na^+ ions bridging across the emitter [2]. PID was also observed in rear-contact IBC n-type cells and related to polarization effects leading to a passivation loss [3]. Although our n-Pasha (Passivated all sides H-pattern) cells and modules are resistant to PID and UV exposure, it has recently been reported that front junction (FJ) n-type based modules can suffer from PID [4, 5]. In contrast to p-type, no *FF* reduction but I_{sc} and V_{oc} loss was observed. The degradation was attributed to increased surface recombination [4] though the exact mechanism was still unclear. Similarly, a passivation loss was observed while exposing n-PERT (Passivated Emitter, Rear Totally Diffused) cells [6] to UV radiation resulting in UV-induced degradation (UVID).

As PID occurs at negative voltages for p-type cells [2] and at positive voltages for n-type IBC (Interdigitated Back Contact) cells [3] or n-PERT cells [5,7], PID can be prevented by grounding the system, respectively, at the negative and the positive connector/pole, thus preventing the degradation inducing system voltages. However, this solution is not possible with 'floating' PV systems and halves the possible voltage range for grounded systems, thus increasing the balance of system costs.

The amount of PID in PV modules can be strongly reduced by applying encapsulants with higher electrical resistance thus reducing the polarization effects [5,7]. However, the cost of these encapsulants, like e.g. polyolefin, is higher than for commonly used EVA.

During the development of our n-Pasha cells over the past years, a thorough understanding has been built up on why our n-type c-Si solar cells are both PID- and UVID-free. We present results on cells with intentionally modified dielectrics and emitters to demonstrate the effect dielectrics and emitter profiles can have on PID and UVID. We also present modelling studies showing the degradation effects are consistent with deteriorating surface passivation. Our PID- and UVID-free cells presented in this paper can be integrated in modules without the necessity of using non-standard (and more expensive) encapsulants, and without restrictions on the system voltage range and on the choice or absence of grounding, providing a low-cost solution to manufacture UVID- and PID-resistant modules.

2 EXPERIMENTAL

2.1 Cell preparation

The n-Pasha cells (Cz Si, 156x156 mm^2 semi-square) are prepared according to standard industrial processes on industrial tools. The cross-section of such n-Pasha cell is presented in Fig. 1. Intentionally introduced modifications into the processing of the Boron (B)-emitter and the dielectric front side layer were applied to reproduce degradation effects that have been observed by others [4,5]. Table I summarizes the modifications for the different samples. These variations did not significantly affect initial cell efficiencies. For PID testing, cells with soldered tabs are laminated into (single-cell) mini-modules using a commercial, fast-cure EVA encapsulant. A selection of non-laminated n-Pasha cells, produced together with the cells for PID tests, undergoes UVID testing.

2.2 PID testing

The mini-modules are exposed to high-voltage stress conditions in a dedicated climate chamber (-1000V, 60°C, 85% relative humidity (RH)) for 20 h steps. The *I-V* characteristics are measured before the exposure and after each step to monitor the evolution of the module stability. PID exposure is finalized after a total of 100 h to be in line with the method discussed by NREL [8].

Table I: Overview of the intentional modifications and performed tests for the standard and experimental groups.

	Standard	Set 1	Set 2	Set 3	Set 4
Boron emitter	stable	modified	modified	stable	modified
Dielectric	stable	stable	modified A	modified B	modified B
PID test	yes	yes	yes	yes	yes
UVID test	yes	yes	no	no	yes

Figure 1: n-Pasha cell architecture featuring PID- and UVID-resistant B-emitter and dielectric layer –denoted as anti-reflective coating (ARC).

2.3 UVID testing

For the exposure to UV radiation an ATLAS UV-tester (351 nm, 1.55 W/m^2/nm, total UV irradiance 62 W/m^2) is applied. The UV exposure is carried out in a few steps until 320 h is reached. The *I-V* characteristics are measured before the exposure and after each step to monitor the evolution of the cell stability. The 320 h of direct UV exposure of n-Pasha cells in the ATLAS UV-tester corresponds roughly to a direct continuous 20 kWh/m^2 UV radiation at cell level. In modules with UV-absorbing encapsulants and solar glass the yearly UV dosage is significantly reduced and 20 kWh/m^2 direct indoor UV radiation at cell level can be only roughly approximated to circa 1 year outdoors in Central Europe.

2.4 Modelling

The effect of deteriorating surface passivation of the B-emitter in n-Pasha cells was investigated by numerical simulations. The Atlas device simulation package by Silvaco [9] was used with the parametrization and physical models for the n-Pasha cell described elsewhere [10]. The effective surface recombination velocity parameter S of the emitter was varied between $2\cdot10^3$ cm.s^{-1} and $1\cdot10^7$ cm.s^{-1}. S represents the surface recombination velocity S_n caused by surface defects (thus related to D_{it}) corrected for changes of the minority concentration induced by charges at the surface. The value of $S = 2\cdot10^3$ cm.s^{-1} represents a well passivated B-emitter surface. Increasing positive charge density at the interface will result in depletion which would lead to an increase of the effective S. At $S > 1\cdot10^7$ cm.s^{-1} surface recombination for this emitter is transport limited.

3 RESULTS

3.1 PID stability

Fig. 2 shows the time evolution of the mini-module relative efficiency loss, for various combinations of B-

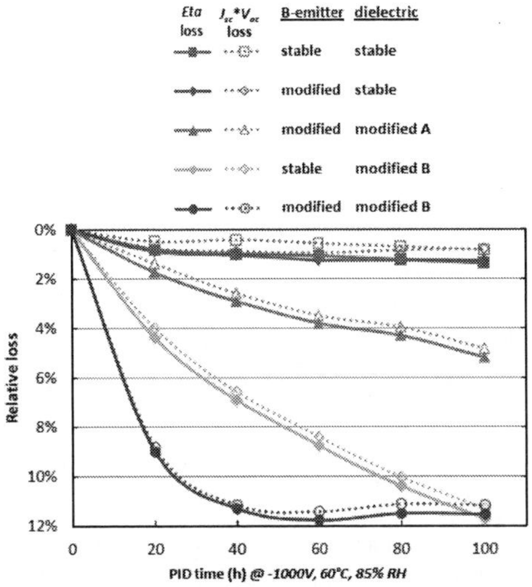

Figure 2: Time evolution of the relative efficiency compared to initial values for n-Pasha single cell laminates (mini-modules) tested till 100 h PID exposure (-1000V, 60°C, 85% RH) plotted with solid symbols. Additionally the relative J_{sc}*V_{oc} loss is plotted with open symbols. Solid and dotted lines are to guide the eye.

emitters and dielectric coatings exposed to PID conditions (-1000 V, 60°C, 85% RH). Our typical n-Pasha cells with the stable B-emitter and the stable dielectric show very high PID-stability (~1% relative efficiency loss after 100 h). However, for the combination of intentionally adjusted processing for the modified B-emitter and the modified dielectric of type-B the PID loss is reaching a saturation level of circa 12% already after ~40 h. In the case of the stable B-emitter, in combination with the modified dielectric B, deterioration is slower but after 100 h the same level of degradation is observed. For the modified B-emitter combined with the modified dielectric of type-A (different modification compared to dielectric B) the power loss is much smaller, reaching ~5% after 100 h, which is close to the pass/fail level of degradation as proposed by NREL [8], although no saturation of the efficiency loss is observed. These data show that although the combination of the modified emitter and the modified dielectric (A or B) is sensitive to fast PID degradation, the amount of PID loss is much lower than reported for n-type IBC [3] or p-type FJ solar cells [1]. Since modifications in the dielectric layer processing show the largest effect on PID, it suggests that the underlying mechanism is most sensitive to properties of the dielectric layer. This is further confirmed as the ultimate PID-solution in the form of the stable dielectric coating can be combined with either stable or modified B-emitter. Both combinations show similarly high PID-stability after a full PID-exposure test. A stability overview of the discussed combinations of stable/modified B-emitter/dielectric is presented in Table II.

Fig 2. also presents the time evolution of a relative efficiency loss expressed as J_{sc}*V_{oc} loss which is closely matching with the time evolution of the relative efficiency. The origin of these PID losses, manifested mainly by J_{sc} and V_{oc} deterioration and not by *FF* loss, will be discussed in section 4.1.

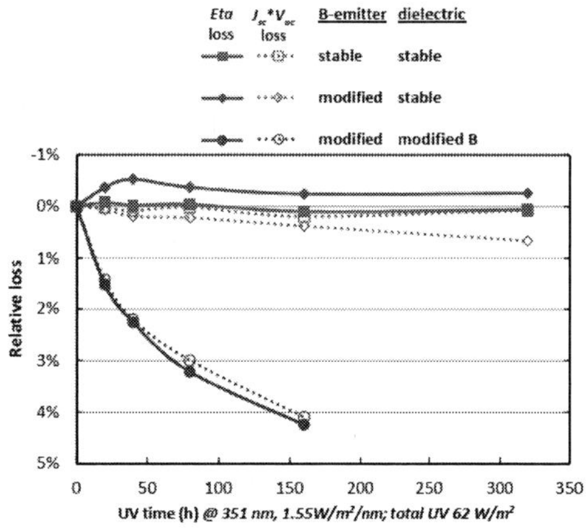

Figure 3: Time evolution of the relative efficiency compared to initial values of selected n-Pasha cells exposed to UVID plotted with solid symbols. Additionally relative efficiency $J_{sc}*V_{oc}$ loss is plotted with open symbols. Solid and dotted lines are to guide the eye.

3.2 UVID-stability

Fig. 3 shows the time evolution of the relative efficiency loss for n-Pasha cells with various combinations of B-emitters and dielectric coatings exposed to direct UVID. Our stable n-Pasha cells show high UVID-stability (<<1% loss after 320 h). Similarly for the modified B-emitter combined with the stable dielectric layer the power loss is also negligible (the minor improvement of the relative efficiency is within the measurement error). However, for the intentionally modified B-emitter and the modified dielectric layer (type B) a large UVID loss does occur reaching circa 4% loss already after 160 h UV exposure. This suggests that the underlying UV induced degradation mechanism is most sensitive to properties of the dielectric layer. A stability overview of the discussed combinations of stable/modified B-emitter/dielectric is presented in Table II.

Table II: Stability overview of the investigated combinations of B-emitters and dielectrics.

	stable dielectric	modified dielectric B
stable B-emitter	PID: ☺	PID: ☹
	UVID: ☺	UVID: N.A.
modified B-emitter	PID: ☺	PID: ☹☹
	UVID: ☺	UVID: ☹

4 DISCUSSION

4.1 Discussion on PID mechanism

Fig. 4 shows a schematic cross-sectional view of an n-type module exposed to a high system voltage. A strong cell potential bias (with respect to the frame of the module) induces a small leakage current j_D through the module. In the case of a negative bias, this results in accumulation of positive charges at the front side of the cell. These positive charges are either electronic in nature, i.e. an oxidized state in the nitride or oxide, or can

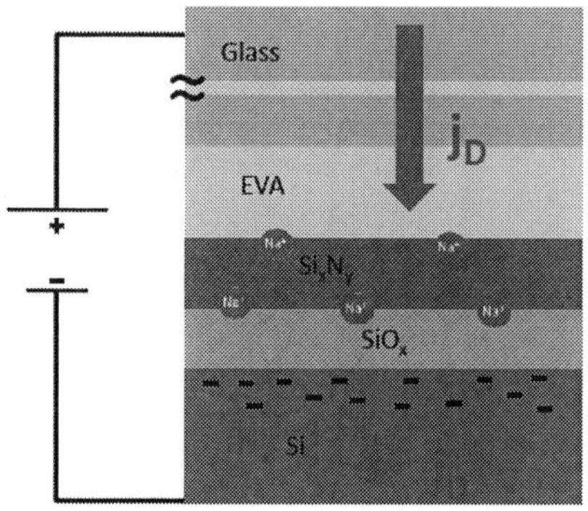

Figure 4: Schematic drawing of proposed PID mechanisms. In the case of a negative bias, positive charges accumulates at the front side of the cell. These positive charges are either electronic in nature or can result from mobile Na^+ ions present in the module materials.

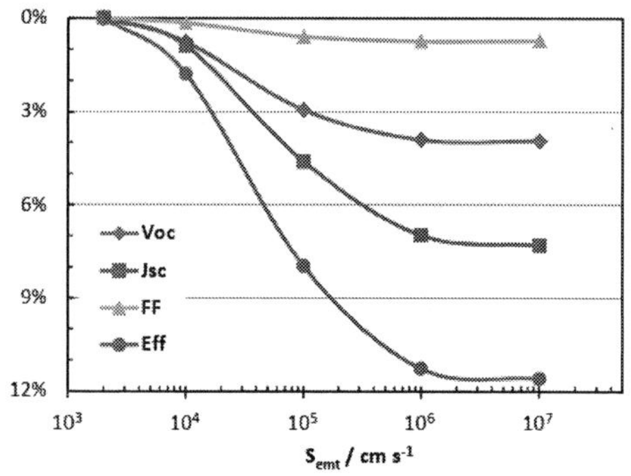

Figure 5: Modelled effect of passivation loss of the B-emitter on the cell performance.

result from mobile Na^+ ions present in the module materials. The positive charges accumulated at the front of the cell destroy the surface passivation of the B-emitter as they attract electrons (minorities) to the emitter surface and the SRH recombination (S) increases. Here S represents the surface recombination velocity S_n caused by surface defects, corrected for changes of the minority concentration induced by charges at the surface, i.e. $S=S_n \cdot n_s/n_d$. Here n_s is the electron (minority) concentration at the surface in the presence of charges, and n_d is the electron concentration without charges. As the charge increases, the surface concentrations of the charge carriers move towards inversion due to band bending [11], i.e. n_s and therefore S is expected to change by orders of magnitude. This so-called polarization effect will have detrimental effects mainly on J_{sc} and V_{oc} as modelled and depicted in Fig. 5. The saturation of the degradation effect at high values of S indicates that the surface recombination is becoming transport limited.

Modelled degradation values for $S=10^5$ cm.s^{-1}, $S=10^6$ cm.s^{-1} and $S=10^7$ cm.s^{-1} are shown in Fig. 6, together with the experimentally observed degradation. For the considered case of the modified B-emitter and the modified dielectric B (see Fig. 2), the experimentally observed degradation as presented in Fig. 6 agrees with a strong increase of the effective S. Thus the proposed model is in line with the observed phenomenological degradation.

PID can be thus avoided by i) preventing transport of positive charges or by ii) reducing the impact of accumulated positive charges. On cell level only the latter can be achieved by a proper configuration of e.g.: a) the dielectric layer properties, b) the interface properties and c) the B-emitter profile. Our stable n-Pasha cells are thus PID-resistant.

4.2 Discussion on UVID mechanism

The UV-radiation may affect both chemical and field-effect passivation. Interface defect states can be activated by UV-radiation leading to an effective increase of their density (D_{it}). This is causing a deterioration of the chemical passivation. The fixed charges Q_f at the interface may be affected by the UV-radiation leading to a deterioration of the field-effect passivation. Both effects can result in efficiency loss due to deterioration of B-emitter surface passivation, as shown in Fig. 3, consistent with modelling (Fig. 5). UVID can be thus avoided by e.g. a) reducing initial D_{it} and b) optimizing the boron profile of the emitter. Our stable n-Pasha cells are thus UVID-resistant.

The same mechanisms play a role in our IBC cells with front-floating emitter, which have also demonstrated stable PID and UVID behaviour (data not shown here).

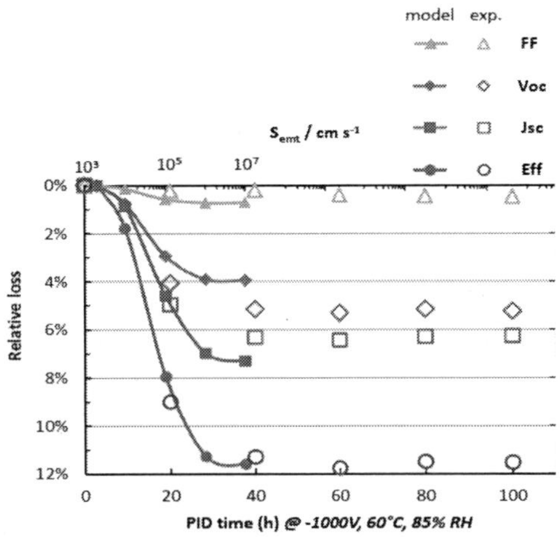

Figure 6: Comparison of relative losses in *FF*, V_{oc}, J_{sc} and efficiency observed experimentally (PID exposure, open symbols) with modelled passivation loss of the B-emitter (closed symbols). Solid lines are to guide the eye.

5 CONCLUSIONS

We report on the high stability of our n-type front junction solar cells (n-Pasha) exposed to potential-induced degradation (PID) and UV-induced degradation (UVID), with a power loss of only ~1% and <0.5% for NREL's proposed PID test [8] and ~20 kWh/m^2 direct UV exposure, respectively. Our n-Pasha cells are PID-free and UVID-free owing to a controlled processing based on a thorough understanding of the interaction between induced degradation and cell production. Based on intentional modifications of the B-emitter and/or the dielectric layer in our PID-free and UVID-free n-PERT solar cells we are able to reproduce degradation effects reported by others, and study the mechanisms behind it. A combination of tuning the boron profile and the dielectric properties together with decreasing the interface defect density D_{it} is beneficial for reducing the degradation. Therefore, the UV radiation and polarisation have negligible effect on our n-PERT cell performance.

5.1 References

[1] Berghold J, Frank O, Hoehne H., Pingel S, Richardson B, Winkler M. Potential induced degradation of solar cells and panels. Proc. 25th EUPVSEC, Valencia, Spain, 2010, p. 3753-3759.

[2] Naumann V, Lausch D, Graff A, Werner M, Swatek S, Bauer J, Hähnel A, Breitenstein O, Großer S, Bagdahn J, Hagendorf C. The role of stacking faults for the formation of shunts during potential-induced degradation of crystalline Si solar cells. Phys. Status Solidi RRL 2013; 7:5-315.

[3] Swanson R, Cudzinovic M, DeCeuster D, Desai V, Jürgens J, Kaminar N, Mulligan W, Rodrigues-Barbarosa L, Rose D, Smith D, Terao A, Wilson K. The surface polarization effect in high-efficiency silicon solar cells. Technical Digest of the 15th PVSEC, Shanghai, China, 2005, pp 410-411.

[4] Pingel S, Janke S, Frank O. Recovery methods for modules affected by potential induced degradation (PID). Proc. 27th EUPVSEC, Frankfurt, Germany, 2012, p. 3379-3383.

[5] Hara K, Jonai S, Masuda A. Potential-induced degradation in photovoltaic modules based on n-type single crystalline Si solar cells. Solar Energy Materials & Solar Cells 2015; 140 – 361.

[6] Zhao J, Schmidt J, Wang A, Zhang G, Richards BS, Green MA. Performance instability in n-PERT silicon solar cells. Proc. 3rd World Conference on Photovoltaic Energy Conversion, 2003; pp 1- 923.

[7] Halm A, Schneider A, Mihailetchi VD, Koduvelikulathu LJ, Popescu LM, Galbiati G, Chu H, Kopecek R. Potential-induced Degradation for Encapsulated n-type IBC Solar Cells with Front Floating Emitter. Energy Procedia 2015; 77 - 356.

[8] Hacke, P., Smith, R., Terwilliger, K., Perrin, G., Sekulic, B. and Kurtz, S. Development of an IEC test for crystalline silicon modules to qualify their resistance to system voltage stress. Prog. Photovolt: Res. Appl. 2014; 22 - 775.

[9] Atlas Device Simulator, Silvaco Int. Software., Santa Clara, CA, USA, 2014.

[10] Fell A, McIntosh KR, Altermatt PP, Janssen GJM, Stangl R, Ho-Baillie A, Steinkemper H, Greulich J, Muller M, Byungsul M, Fong KC, Hermle M, Romijn IG, and Abbott MD. Input Parameters for the Simulation of Silicon Solar Cells. IEEE Journal of Photovoltaics 2015; 5 - 1250.

[11] Aberle AG, Crystalline Silicon Solar Cells; Advanced Surface Passivation and Analysis. Center for Photovoltaic Engineering, University of New South Wales, Sydney, Australia; 1999.

RECOVERTY METHOD FOR SOLAR MODULSES AFFECTED BY POTENTIAL INDUCED DEGRADATION IN UTILITY-SCALE SOLAR PLANTS

Yayi Hu, Paul Ni, Qingzhu Wei, Chunhui Li, Zhangzhu Mou, Leizhen Hu, Yu Yan, Chenxi Liu, Junyu Lu, Chenyang Wu
Zhongli Talesun Solar Co. Ltd.
No.1 Talesun Road, Shajiabang, Changshu, Suzhou, Jiangsu Province, China

ABSTRACT: Potential induced degradation(PID) of PV solar modules has been observed and in the discussion for years because of the significant power loss it has contributed to. The main reason leads this is that current modules qualification testing and standards have short in ability to evaluate modules' long term durability under a large system bias in the fields. PV systems and the ambient working environment interact to cause PID, which is also attributed to factors on cells, modules and systems. In this paper, we focus on researching and verifying the application effects of PV Offset Box developed by SMA on recovering PID in the utility-scale PV plants in the region with high temperature and humility. One plant is located in Kumamoto, Japan and another one is built above the fishpond in the east of China. Modules from both plants have been found to be affected by PID.
Keywords: degradation, durability, large grid-connected PV system

1 Background

The principle and causative factors have been investigated but still can not be reached an aggreement on by the industry till now [1-2]. According to previous work, it is widely known that PV system and the ambient working environment interacts to cause PID, which is also attributed to factors on cells, modules and systems. With respect to Environment factors, both of humility and temperature have adversely impacts on the stability of module performance [3-5]. Apart from that, accumulation of sodium ions on the surface of cells also plays a significant role [6-8]. Basically, the working environment is constant and cannot be changed. Hence, efforts are suggested to be made on the level of cell, module and system.

2 Motivation

2.1 cell level

In the light of the EL images of PID affected modules (figure 1), it is observed that most of black cells are located along with the frame with non-uniform damage.

Some cells seem to be normal, while some cells have degraded heavily even from diode behavior to ohmic behavior [9]. Accordingly, it can be found cells have different levels of stability to PID. It is effective methods extensively applied by the PV industry for preventing solar modules from the incurrence of PID by adjustments of materials or properies on the level of cells and modules, such as SiNx refraction index adjustment, surface oxidation treatment or high resistance encapsulant materia[10-11].

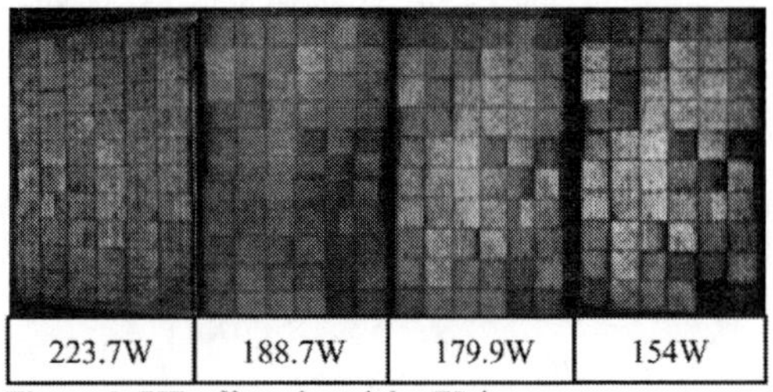

Figure 1: PID affected modules EL image

2.2 module level

The approach of mitigation of PID in the level of modules is to use the encapsulating materials with high resistivity to reduce and prevent the current leakage from cells to the frame of modules[12-13].

2.3 system level

On the system level, the degradation induced by voltage bias mainly depends on the polarity and the level of the potential between cells and ground. This system bias can trigger leakage current from solar cells and ground through encapsulating materials, i.e. glasses, EVA and backplanes, and frame, especially in damp heat environment, which further leads to accumulation of trapped chargers above cells. These charges interact with emitter and depletion region, disrupt their function and finally cause cells' out power loss (figure 2). To prevent PID, the system must not have a negative potential towards ground [14]. However, considering the modern process technology development of cells and modules, most efforts have been made is to pursue the highest efficiency by optimization of various parameters of cells and modules. Changing parameters or encapsulating materials may lead to the reduction of output performance of modules.

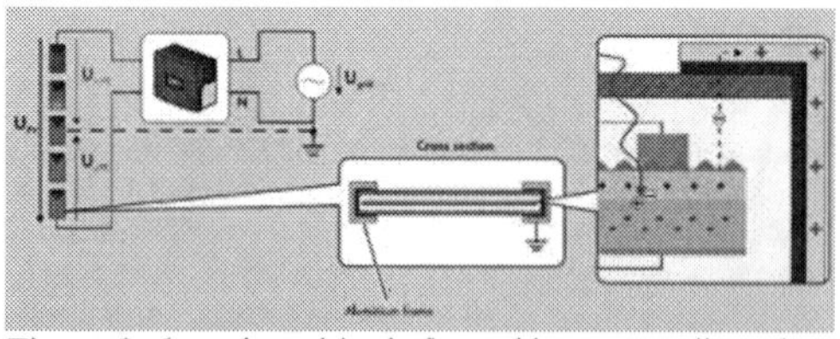
Figure 2: the voltage bias is formed between cells and frames in a PV system without grounding

Hence, resolving the PID problems in system level may be the most practical and economical counter measure currently because its financial cost is relatively lower than exchanging modules or changing system concept. Grounding the negative pole of the PV array can solve this problem. But as a result of the development of inverter technology, transform-less inverters with higher efficiency become predominant, in the case of which the PV system cannot be negatively grounded. Therefore, PV

Offset Box developed by SMA can be used as an alternative [15]. The principle and configuration are showed in figure 3.

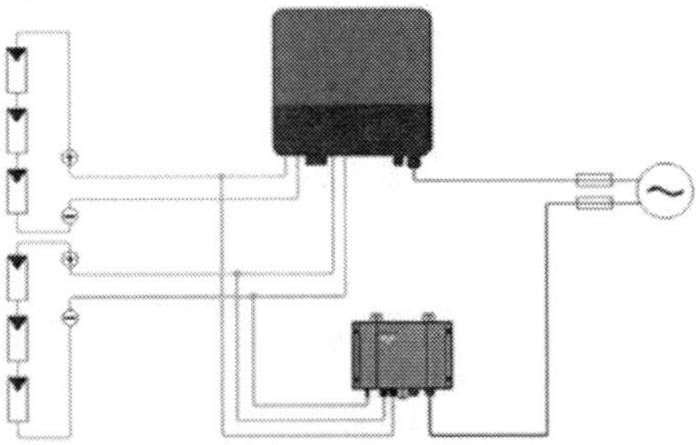

Figure 3: Principle of a PV plant with SMA PV Offset Box and multi-string inverter

3 EXPERIMENT

In this paper, we focus on researching and verifying the application effect of PV Offset Box developed by SMA on recovering PID in the utility-scale PV plants in the region with high temperature and humility.

One plant is located in Kumamoto, Japan and another one is built above the fishpond located in the east of China. Modules from both plants have been found to be affected by PID.

In the Japanese solar plant, the PV offset box was installed in Array A with 500kW capacity, which Array B with same capacity stayed the same simultaneously. Array A was under the influence of more undesirable shading than Array B due to the trees in the west. The one-and-half-month data of output electricity of Array A and B was monitored.

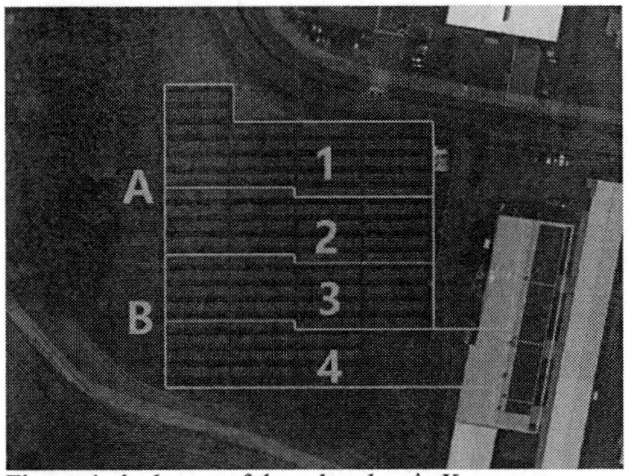

Figure 4: the layout of the solar plant in Kumamoto, Japan

The second solar plant used for testing is located in the east of China. Fifteen PID affected modules were serially interconnected and installed intentionally where the humility of environment was very high. In this solar plant, a large scale of PID happened on modules, most of which had been exchanged. The PV offset boxes were installed for 500kW PV array recovery. These PID affected modules would have been transported back to production line in factory for flash-test three times in 140 days' period.

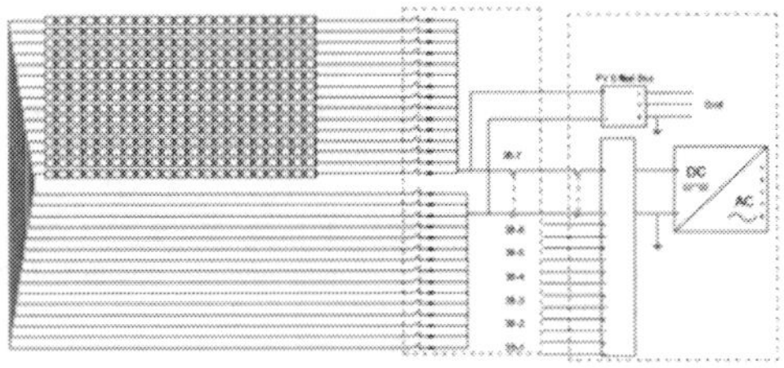

Figure 5: The connection position of PV offset Box in the solar plant, China

4 EXPERIMENT RESULT

In the Japanese solar plant, the power output of Array A was less than Array B from 1st September mainly because of trees' shadow in the afternoon. However, since 29th Sep, the daily accumulated power of Array A had surpassed that of Array B with around 10-20kW. In November, the stable 50kW more power was generated by Array A than done by Array B. In accordance with, it states that the PV offset box worked very well and had a very good repairing ability of PID with restoring around 12%-15% rated power in 78 days.

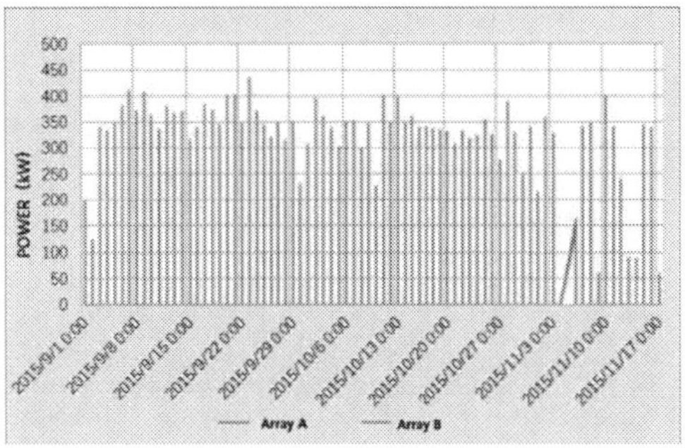

Figure 6: the 78 days' electricity output of Array A and Array B

However, it is still unknown that how fast the recovery rate is. Hence, we continued to conduct the next experiment in the solar plant in the east of China.

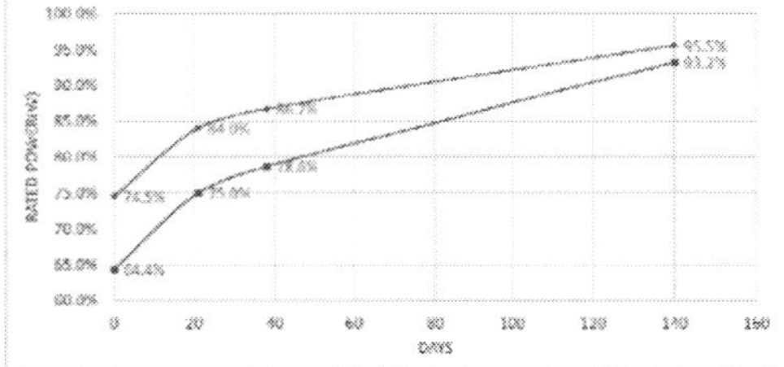

Figure 7: Fifteen PID affected modules recovery data in 140 days

The experiment was divided into four stages in the Chinese solar plant, which was last 140 days. During this period, the modules had been tested four times in 1st day, 21st day, 38th day and 140th days. At the beginning, the power of PID affected panels were measured to be from 123.52W to 229.35W with different levels of degradation. In the first stage of experiment only two modules (module a and module b) were selected and tested. The next 2 stage of testing, all 15 piece of panels (module a~o) were tested. In the light of figure 7, it is illustrated that module b with lower initial power after the incurrence of PID, had a slightly higher recovery speed up to 10.6% and 3.6% of increment separately while comparing with module a, 9.5% and 2.7% in the first and second stage experiments. In the final testing, it was found that the two modules were restored to 95.5% to 93.2% respectively. the recovery rates of module a and b are showed in table 1.

Table 1: the recovery rate of module a and b in 3 individual stage of experiments

	stage 1 (%/day)	stage 2 (%/day)	stage 3 (%/day)
Module a	0.45	0.16	0.09
Module b	0.51	0.22	0.14

As a consequence of low resistance between cells and ground, module b had a higher recovery speed. Similarly, low resistance also is a typical characteristic of modules affected by PID, which means a high degradation module has a higher recovery speed.

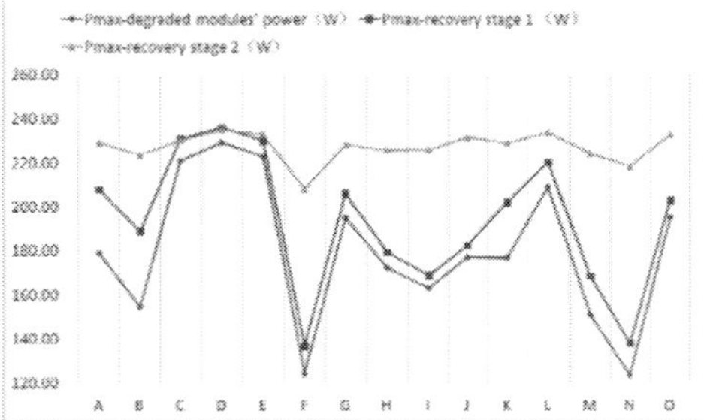

Figure 8: Fifteen PID affected modules recovery data in 140 days

The figure 8 shows that the all modules could be recovered somewhat extent of degradation. But the modules seem to be repaired back to the expected power with consideration of anual LID. That means both degradation mechanism happened in these modules including polarization (reversible) and electro chemical corrosion (irreversible).

5 CONCLUSION

Based on analyzing, Recovering PID affected modules in the level of system is the most economic method. We used PV offset boxes in two larger scale solar plants in Japan and China to repair modules and will carry on this experiment for long term. Because the damp heat environment in these two places cannot be changed, the modules are prone to PID despite they are under restoration. The experiments declare that PV offset boxes have good capability of recovering PID affected modules in 3 months. The initial recovering rate is high and will become slow at later period. The recovering rate also depends on the degradation levels of modules.

6 REFERENCE

[1] P. Hacke, K. Terwilliger, R. Smith, S. Glick, J. Pankow, M. Kempe and S. Kurtz, "System Voltage Potential-Induced Degradation Mechanisms in PV Modules and Methods for Test," 37th IEEE Photovoltaic Specialists Conference, 2011.

[2] H. Nagel, A. Metz and K. Wabgemann, "Crystalline Si Solar Cells and Modules Featuring Excellent Stability Against Potential-Induced Degradation," 26th European Photovoltaic Solar Energy Conference, p.3107,2011.

[3] Berghold et al.: "Potential Induced Degradation of solar cells and panels", 25th EU PVSEC (2010)

[4] Nagel et al., "Crystalline Si solar cells and modules featuring excellent stability against potential-induced degradation", 26th EU PVSEC (2011)

[5] Pingel et al.: "Potential Induced Degradation of Solar Cells and Panels", 35th IEEE (2010)

[6] Koch et al.: "Polarization effects and tests for crystalline silicon solar cells" 26th EU PVSEC (2011)

[7] Hacke et al.: "Characterization of Multicrystalline Silicon Modules with System Bias Voltage Applied in Damp Heat", 25th EU PVSEC (2010)

[8] Bauer et al.: "On the mechanism of potential-induced degradation in crystalline silicon solar cells", pssr (2012)

[9] Pingel, S., et al. "Potential Induced Degradation of solar cells and panels." 25th EU PVSEC (2010)

[10] Yi-Cin Chen, et al. "Improvements on PID for c-Si based solar cells" 37th IEEE (2012)

[11] Koch, Simon, et al. "Potential induced degradation effects on crystalline silicon cells with various antireflective coatings." Proceedings of 27th EUPVSEC, Frankfurt, Germany (2012): 1985-1990.

[12] Krauter, Stefan. "Identification and elimination of potential induced degradation (PID) at photovoltaic power plants." Proceedings of the 6th WCPEC, Joint Conference of the 24th International PVSEC-24, 41st IEEE Photovoltaic Specialists Conference, the 30th EU PVSEC, Kyoto, Japan. 2014.Conference (PVSC), 2010 35th IEEE. IEEE, 2010.

[13] Zhanga, YingBin, et al. "High-reliability and long-durability double-glass module with crystalline silicon solar cells with fire-safety class a certification."

[14] Pingel, S., et al. "Potential induced degradation of solar cells and panels."Photovoltaic Specialists

[15] SMA: "PV Offset Box PVO-11", downloaded from www.sma.de (15.09.2012)

PERFORMANCE EVALUATION OF PV MODULES AFTER ACCELERATED TESTING FOLLOWED BY FOUR YEARS OF FIELD EXPOSURE IN HOT-HUMID CLIMATE OF FLORIDA

Vivek Gade, Narendra Shiradkar, Shesh Vaishnav and Jared Opalewski
Jabil Solar and Environmental Test Center
2007 Gandy Blvd, Suite 100, St. Petersburg, Florida, USA- 33702
Phone: (727) 803-3988, e-mail: Vivek_Gade@jabil.com

ABSTRACT: Development of extended accelerated test protocols that don't introduce superfluous failure modes is necessary to understand long term PV module durability. Extended Damp Heat (DH) testing beyond 2000 hours is often performed to understand long term durability of PV modules. However, such type of harsh testing is known to introduce failure modes not seen in field. This work describes a non-traditional approach taken by JSEC which combines DH and outdoor exposure in Florida. Modules procured from market from 3 different manufacturers having glass-EVA-backsheet design were used in this study. The modules were first passed through DH testing for 1000, 1500 and 2000 hours for 3 manufacturers respectively. Consequently, the modules were deployed on a rooftop array in Florida for 4 years. Module characterization including Electroluminescence (EL) and flash I-V was performed initially, after DH exposure and after outdoor exposure to understand failure mechanism. It was deduced that acetic acid formed during DH exposure accelerated the corrosion of silver grid with the availability of heat, humidity and UV during outdoor exposure. However, the presence of cell cracks mitigated the corrosion in the local areas as acetic acid formed at the front of the cell could escape through backsheet.
Keywords: Reliability, PV Module, Electroluminescence

1 INTRODUCTION

Most of the reports regarding module reliability published in the literature deal with the evaluation of fresh modules after either accelerated testing or field exposure. It is important to understand the wear out failure modes and mechanisms in PV modules in order to reduce the uncertainty in predicting the PV power throughout the module lifetime [1]. Increasing the duration or the amount of stress during accelerated testing beyond certain limit to simulate wear out failure modes is not recommended because it often introduces extraneous failure modes not observed in field. For example, some degradation modes/mechanisms reported for damp heat (DH) testing beyond 2000 hours have never been observed in field [2]. On the other hand, it can take decades for wear out failure modes to show up in modules undergoing field exposure, thus making field testing to understand wear out in new PV modules impractical. Climate-specific degradation modes in PV modules make this process even more complicated [3]. In order to resolve this deadlock, a test protocol that would combine the benefits of accelerated and field testing was developed at Jabil Solar and Environmental Test Center (JSEC).

The test plan involved passing the modules through slightly extended versions of accelerated tests and sequences provided in IEC 61215 followed by outdoor exposure in hot and humid climate of Florida. The accelerated test sequences involved one or more tests such as extended damp heat (85 °C/85% Relative humidity), extended thermal cycling, humidity freeze, UV exposure and mechanical load tests. The idea was to utilize these accelerated tests to accelerate the degradation mechanisms in PV modules without introducing extraneous failures. Consequently, the PV modules that have gone through the accelerated test would potentially experience more degradation during the same amount of outdoor exposure period as compared to the fresh modules directly deployed outdoors. Since outdoor exposure in a place like Florida exposes the module to a combination of stresses such as relative humidity, heat, UV radiation and voltage, it was also goal of this study to investigate how a particular failure mechanism accelerated by the accelerated test progresses in the presence of combination of field stressors.

2 EXPERIMENTAL DETAILS

This paper presents the results from mc-Si and c-Si modules procured from market by JSEC that were passed through extended versions of damp heat testing followed by outdoor exposure. Fig. 1 shows the experimental details. Based on the initial, post DH testing and post outdoor exposure flash I-V measurements, the percentage power lost by the module with respect to initial measurement was recorded for (i) DH exposure (ii) Outdoor exposure (iii) Total of the two exposures. Percentage changes in Fill Factor (FF), Open Circuit Voltage (V_{oc}) and Short Circuit Current (I_{sc}) were also recorded. The modules were cleaned post outdoor exposure to eliminate the effects of soiling.

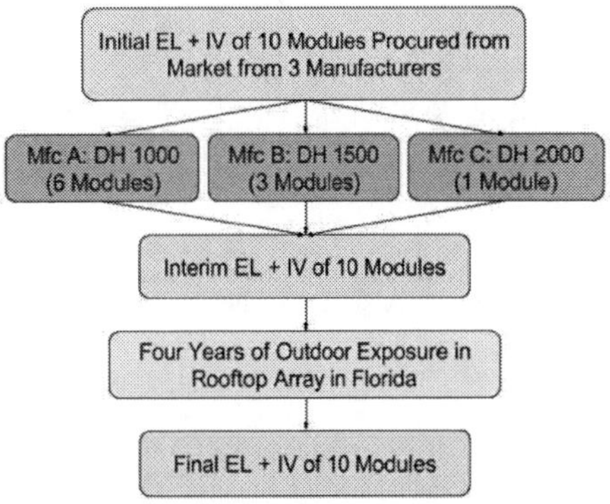

Figure 1: Experimental design for the current study involving 10 modules from 3 manufacturers

3 EXPERIMENTAL RESULTS AND DISCUSSION

3.1 Performance degradation

Fig. 2 shows the performance degradation in the PV modules only due to DH testing. The different colors show the absolute percentage contribution by FF, V_{oc} and I_{sc} towards the total power degradation. The numbers at the top of bars indicate total power degradation at the end of respective versions of DH testing. Occasionally, the value of parameter increases after testing, this is indicated by negative numbers (For example, I_{sc} of module A2).

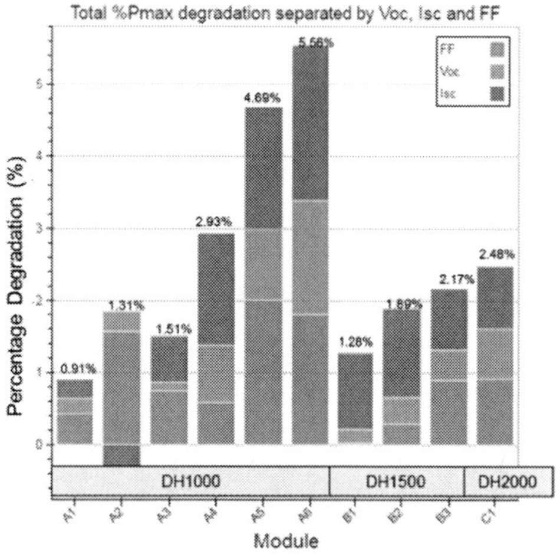

Figure 2: Performance degradation observed during only damp heat test sequences.

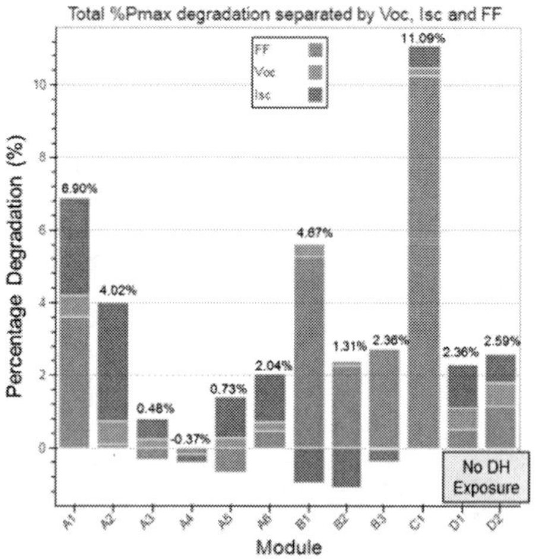

Figure 3: Performance degradation observed during only four years of outdoor exposure in Florida.

It is seen that FF and/or I_{sc} degradation is main source of power degradation. Modules from manufacturer A itself show wide variation in degradation. Despite of going through more extended versions of DH testing as compared to modules from Manufacturer A, the modules from manufacturer B and C show less degradation. Module A6 degraded more than 5%, after DH 1000 indicating that it would have failed the IEC 61215 DH qualification test. Overall, it can be said that the modules from different manufacturers have varying degree of resistance against DH degradation.

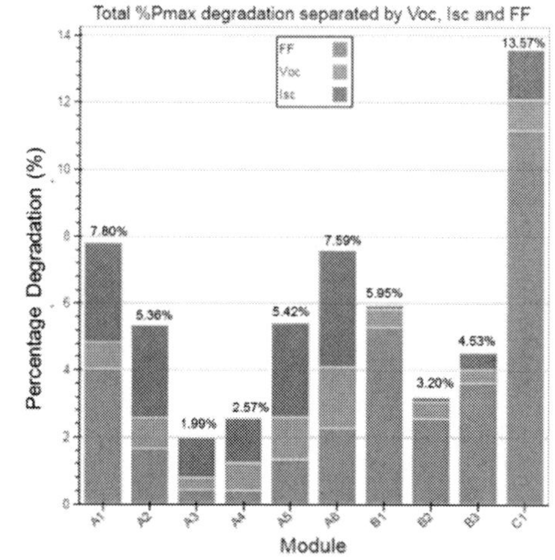

Figure 4: Total performance degradation observed during after damp heat test sequence followed by four years of outdoor exposure in Florida.

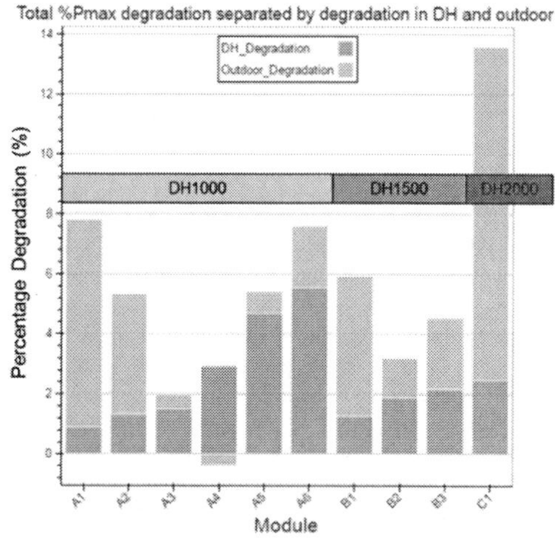

Figure 5: Total power degradation separated by degradation in damp heat test sequence and degradation observed during four years of outdoor exposure in Florida.

Fig. 3 shows the performance degradation only due to four years of outdoor exposure in Florida, shown with respect to initial values. Performance degradation of modules from Manufacturer D that did not have prior DH exposure is shown for reference. FF degradation is main source of power degradation for B & C. Module C1 shows excessively high FF degradation after outdoor exposure. Modules A3-6 had less degradation than D1-2 despite of going through prior DH exposure. For almost 2/3rd of the modules, the average annual degradation during outdoor exposure was more than 0.5% / year.

Fig. 4 shows the total degradation in modules (after DH and outdoor exposure) with respect to initial values. FF and/or I_{sc} degradation are main sources of power degradation at the end of test sequence. Fig. 5 shows the degradation separated by DH and outdoor exposure. It can be seen that some modules (A4,A5,A6) are more sensitive to degradation in DH while others (A1,B1,C1) degrade more during outdoor exposure that followed. For some modules, (A1,B1,C1) outdoor degradation post damp heat can be a dominating component in the overall degradation. This underlines the importance of outdoor testing in hot and humid climate post damp heat testing.

3.2 Degradation mechanisms

During DH testing, the moisture enters the laminate from breathable backsheet and then slowly creeps in towards the center of the cell starting from the edges [4].

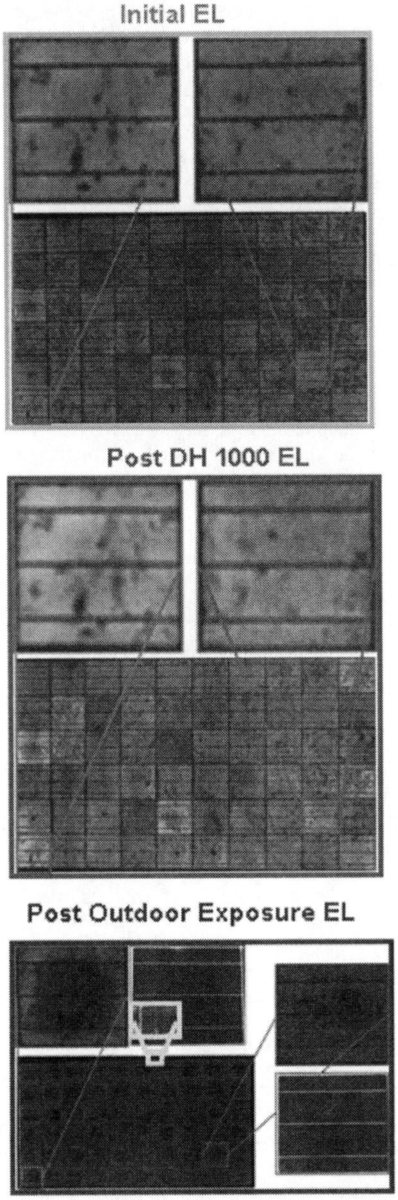

Figure 6: EL images of module A1 as it went through the test sequence involving damp heat exposure and outdoor exposure. Post outdoor exposure image also shows visible images of two cells within the module.

PV modules with EVA encapsulants are known to form acetic acid when exposed to high temperatures and humidity [5]. The acetic acid reacts with front silver grid to cause corrosion. Therefore, in highly extended damp heat testing (~ 3000 hours), the corrosion pattern first forms at the edges and then moves towards the center.

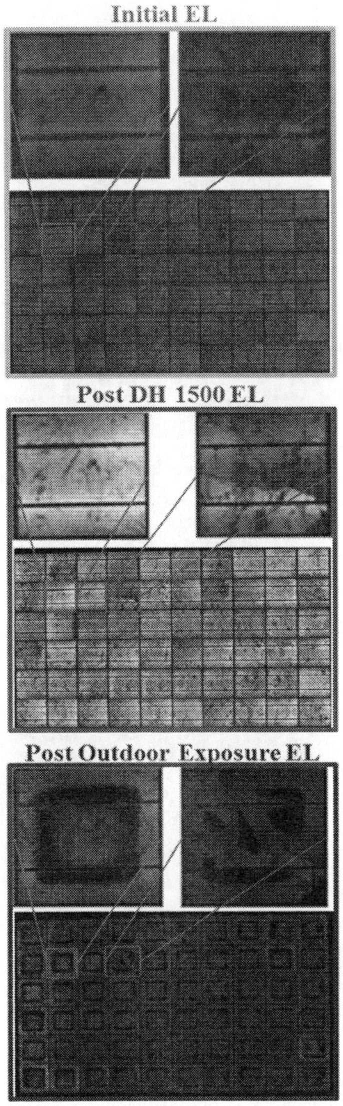

Figure 7: EL images of module B1 as it went through the test sequence involving damp heat exposure and outdoor exposure.

The rate of corrosion depends on various factors such as cell properties, temperature, humidity, availability of catalysts such as UV and possibility of acetic acid escape through the backsheet. In case of breathable backsheets, the acetic acid formed at the backsheet-EVA interface escapes out during diurnal temperature and humidity cycle experienced by the PV modules. However, since the solar cell is a barrier for acetic acid and moisture, the acetic acid formed at the front end of the solar cell can't escape out from the backsheet easily, unless the mechanical structure of the solar cell itself is compromised.

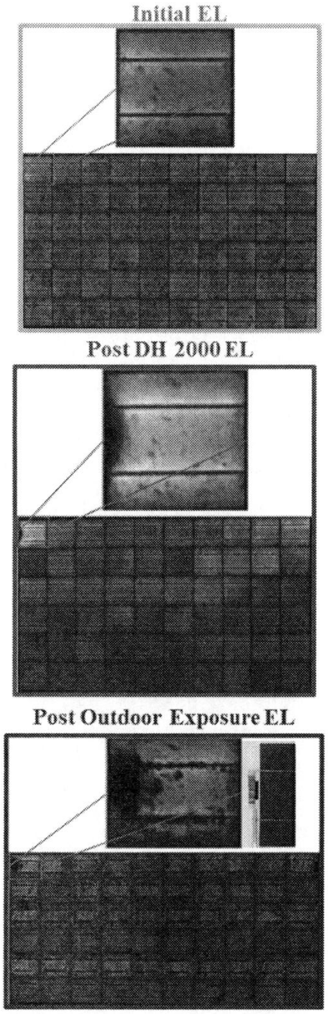

Figure 8: EL images of module C1 as it went through the test sequence involving damp heat exposure and outdoor exposure. Post outdoor exposure image also shows degraded serial number sticker in the inset.

Fig. 6 shows the EL images of module A1 as it went through the test sequence. The images of two cells are shown in inset to illustrate the effect of cell crack on the corrosion of front silver contacts. No significant corrosion was seen after DH 1000 exposure. For cell outlined in Blue, and almost all the cells in the module, significant corrosion of silver grid is visible in the EL image and visible image after outdoor exposure. The corroded fingers are seen to have turned brown in the visible image shown in inset. This indicates the outdoor exposure has accelerated the corrosion. Most likely cause is that the UV radiation during outdoor exposure catalyzed the corrosion of silver in the presence of acetic acid [6]. Acetic acid formed near the edges of the cell escaped through the backsheet, hence less corrosion is seen near the edges of the cells. For cell outlined in Green, a pin-hole crack through the backsheet / cell was observed. This probably caused the acetic acid formed at the front of the cell to escape. Hence, no corrosion seen at the circular area around the crack.

Fig. 7 shows the EL images of module B1 as it went through the test sequence. No significant corrosion seen in EL image post DH 1500 exposure. However, a peculiar corrosion pattern emerges after outdoor exposure on almost all of the cells. This again underlines the importance of outdoor stressors such as UV, heat and humidity to accelerate the acetic acid induced corrosion. Due to the nature of EVA / backsheet, the moisture, and therefore acetic acid had probably not reached until the center of the cell, therefore, the relatively less corrosion is seen near the center of the cell. On the other hand, acetic acid near the edges might have escaped through the backsheet. A cell with crack probably caused the acetic acid generated after damp heat near the center of the front of the cell to escape through backsheet. This resulted in reduced corrosion in the nearby areas of the cell crack.

Fig. 8 shows the EL images of module C1 as it went through the test sequence. No significant corrosion seen in EL image post DH 2000 exposure. However, significant corrosion is seen after outdoor exposure on almost all of the cells. Since the corrosion is seen to originate predominantly at the cell busbars, the residual flux induced corrosion seems to be the likely cause. In the inset, and EL image of a cell next to the serial number label can be seen. The adhesive used to stick the serial number label had probably resulted in enhanced degradation of the adjacent cell after DH 1500. This effect was exacerbated after outdoor exposure. Degradation around the serial number label can be seen in visible image. This underlines the importance of benign adhesives that will not affect the reliability of cells during module service life.

The modules that only experienced the outdoor exposure, without going through the damp heat test did not show any indication of corrosion. This indicates that significant corrosion was seen after outdoor exposure only when significant amount of acetic acid was formed at the front side of solar cells in the laminate due to damp heat testing.

4 CONCLUSIONS

Damp heat testing results in the formation of acetic acid within the PV modules with EVA encapsulants. Insignificant corrosion of the front silver contacts was seen just after damp heat exposure. However, significant corrosion of the front silver contacts was seen after outdoor exposure resulting in reduction of FF and module power. This effect is attributed to the UV enhanced corrosion of silver electrodes in the presence of acetic acid. Whenever acetic acid generated at the front of cells found a way to escape through backsheet, the rate of corrosion in the localized area was found to have reduced. Modules that only experienced outdoor exposure without first going through DH testing showed no corrosion probably due to insufficient acetic acid generation at the front of the solar cell. This study shows that DH followed by UV / Outdoor exposure is a much better accelerated test than simply extending the DH testing duration beyond 2000 hours for claiming long term durability. Differences in cell manufacturing processes and raw materials (silver paste, flux, encapsulant, backsheet) may have significantly impacted the final performance. These differences in processes and materials may be minor but have the potential to significantly impact the accelerated test and outdoor performance of the modules.

References

[1] J. Wohlgemuth, "IEC 61215: What it is and isn't," PV Module Reliability Workshop, 2012.
[2] J. Wohlgemuth, "Group 3: Humidity, Temperature and Voltage", PV Module Reliability Workshop, 2013.
[3] J. Wohlgemuth and S. Kurtz, "International PV QA Task Force's Proposed Comparative Rating System for PV Modules", Proc. SPIE 9179, Reliability of Photovoltaic Cells, Modules, Components, and Systems VII, 917902, 2014.
[4] M. Kempe, A. Dameron and M. Reese, "Evaluation of moisture ingress from the perimeter of photovoltaic modules", Progress in Photovoltaics, Volume 22, Issue 11 (2014) Pages 1159–1171
[5] T. Shioda, "Acetic acid production rate in EVA encapsulant and its effect on PV module performance", 2nd Atlas/NIST PV Materials Durability Workshop, 2013.
[6] Z. Chen *et al.*, "Influence of UV irradiation and ozone on atmospheric corrosion of bare silver", Corrosion Engineering, Science and Technology Vol 45 No 2 (2010)

DEVELOPMENT OF A MATLAB BASED SIZING AND SIMULATION TOOL FOR SOLAR PHOTOVOLTAIC PUMPING SYSTEM (PVPS)

Rakibul Hasan[(1)], Mike Zehner[(1)], Oliver Mayer[(2)], Ralph Gottschalg [(3)]

[(1)]University of Applied Sciences Rosenheim, Hochschulstrasse 1, 83024 Rosenheim, Germany
Phone: +49 (0) 8031 805-2400, Fax: +49 (0) 8031 805-2402, rhr.eee@gmail.com

[(2)]GE Global Research, Freisinger Landstraße 50, 85748 Garching, Germany

[(3)]Centre for Renewable Energy Systems Technology (CREST), Department of Electronic and Electrical Engineering, Loughborough University, UK

ABSTRACT: The use of renewable energy like solar PV stand-alone systems to produce electricity for water pumping, represents a good solution both economic and technical especially in rural/off-grid area. But both the natural resources water and solar are non-static and volatile. This makes the PVPS design most challenging because there is no fixed operational or duty point for motor-pump set like the conventional electric systems. Hence the focus of this research is to develop a sizing tool which can simply estimate the optimized PV generator size of a solar PV pumping system (PVPS). The developed sizing tool is named as **'DASTPVPS-Reloaded'** and the notable features of this tool are high user friendliness, minimum need of inputs, simple data base formate and interactive graphical representaiton of outputs. It can also provide different simulation outputs and give a technical guidelines to select suitable components and system layouts for PVPS.

Keywords: Solar PV generator, Array sizing, PVPS, DASTPVPS-Reloaded, Total dynamic head, Flow rate, MATLAB, Simulation.

1 INTRODUCTION

Fresh Water and electricity are two important preconditions for human development and urbanization. As time goes by more and more regions become arid or underground water level is going down due to immense use of water, especially in Africa. For the huge demand of water in domestic purpose, livestock watering and irrigation PV powered water pumping system (PVPS) has become the most popular technology now a days. It also represents a good economic, technical and environmental solution mainly for rural off-grid areas. The simplest way to design PVPS is by some developed formula or nomograms provided from different pump manufacturing company. But they cannot give real time values; are not so practical and must be calculated for some fixed values. The most accurate approach to the design of PVPS is to compute the system's configuration with the aid of simulation software/tool. Unfortunately there are only few such simulation software tools ([1], [2]) and they are not free to use. Moreover they need lots of user inputs and data base which may be very complex for general end user. Thus, the main idea of this research is to develop a sizing and simulation tool for PVPS which would have user friendliness and general people can easily access to it. The tool takes the user inputs and gives corresponding simulation results with graphical representations to obtain size of the PV array and the relative pump system. With this data and characteristic curves in every stage of simulation, the designer can decide about the most suitable components and layout of a system.

2 ALGORITHM FOR TOOL DEVELOPMENT

2.1 Basic Algorithm

The basic algorithm ([3] to [6]) to estimate PV generator size is, first the assessment of required hydraulic energy (i.e. product of developed head and water volume) and solar irradiance for a particular place. From there with the aid of equations [5] calculate the required solar generator size. The main loads for PVPS are volume of water and dynamic head from water source to the tank inlet. Therefore one of the main preconditions to design an optimal PVPS is the meticulous assessment of "Minimum Water Demand" ([7] to [9]). The electrical energy produced by the solar PV generator must be equal to the required hydraulic energy to pump water. Solar energy converts into hydraulic energy in three steps. Solar irradiance to electrical energy, then converts to mechanical energy and finally transform to hydraulic energy. In each step of energy conversion, efficiency factor must be considered to get the actual required energy.

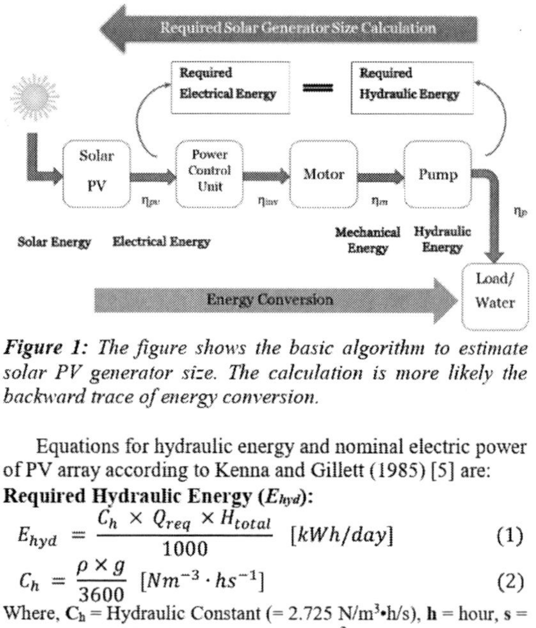

Figure 1: *The figure shows the basic algorithm to estimate solar PV generator size. The calculation is more likely the backward trace of energy conversion.*

Equations for hydraulic energy and nominal electric power of PV array according to Kenna and Gillett (1985) [5] are:

Required Hydraulic Energy (E_{hyd}):

$$E_{hyd} = \frac{C_h \times Q_{req} \times H_{total}}{1000} \quad [kWh/day] \tag{1}$$

$$C_h = \frac{\rho \times g}{3600} \quad [Nm^{-3} \cdot hs^{-1}] \tag{2}$$

Where, C_h = Hydraulic Constant (= 2.725 N/m³•h/s), **h** = hour, **s** = sec, **g** = Gravitational constant (= 9.81 m/s²), **ρ** = Water density (= 1000 kg/m³), Q_{req} = Daily water demand (m³/day), **H** = Total developed head (m).

Required Solar Generator/PV Array Size (P_{pv}):

$$P_{pv} = \frac{E_{hyd} \times 1000}{\eta_{pv} \cdot \eta_{mp} \cdot \eta_{inv} \times T_f \times \overline{G}} \quad [W_p] \tag{3}$$

Where,

1000 = Constant for unit conversion, P_{pv} = Solar generator size (Watt-peak), $\overline{G}$ = Daily average insolation on horizontal surface/GHI (kWh/m²/day), T_f = Tilt factor , η_{mp} = Motor-pump set efficiency (%), η_{inv} = Inverter efficiency (%), η_{pv} = Efficiency of PV array under operating conditions(%).

The System Curve or frequently known as total dynamic head (TDH) is the actual head required to pump the water from the source to the storage tank inlet. Most pumping systems are designed without considering right TDH which is very important. It is the sum of three components-

• Static Head (H_{static}): This is the total vertical height from static water table to top of the water tank. According to fig. 2; it's the sum of (a) and (b).

• Friction Head ($H_{friction\ loss}$): Loss due to friction and bend in the piping.

• Drawdown (H_d): A cone of depression in the well will create as an effect of the pumping operations.

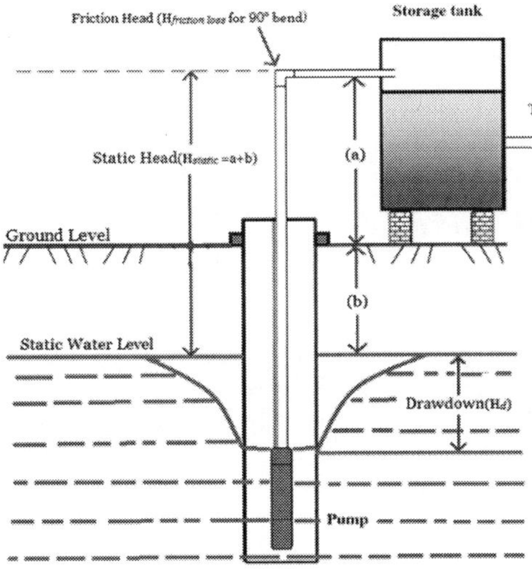

Figure 2: Understanding the total dynamic head composition.

2.2 Distinction in Algorithm with Conventional System

To get a clear idea about how PVPS are working (without battery backup), it is essential to understand the conventional pumping system. In a conventional system for the whole operational period is only one single duty point because of single power line supply or frequency (e.g. in Europe or most of Asia has power line frequency of 50*Hz* with 220–240*V*). Once the water requirement determined, technician select a pump in such a way that system works at its best efficiency point (BEP) or near to BEP of the pump. The more the duty point is deviant from the BEP the less is the system efficiency.

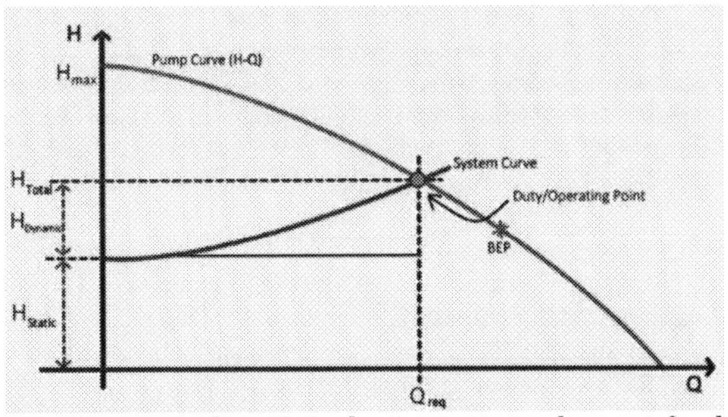

Figure 3: *Typical pump performance curve of a centrifugal pump powered by conventional power supply. The pump will only operate on the system characteristic curve. It is desirable that the pump operates at BEP or pump pick efficiency.*

On the other hand the solar power supply is unpredictable and varying throughout the day. Therefore the operating point i.e. flow rate of the pumping system is not fixed and vary along with solar irradiance (without battery system). It supplies low frequency power at early morning and late afternoon (like 20*Hz* to 40*Hz*), whereas at noon power frequency could reach up to 70*Hz* in a clear sunny day. Therefore motor synchronizes accordingly and produces multiple pump performance curve. Thus comes multiple operating points at different time. After getting head versus flow rate curve (H-Q) for different frequencies, efficiency must be imposed on it. Then the complete pump performance curve should look like ***Fig. 4***. If the hydraulic system curve ([6] and [11]) superimposes on it, the total dynamic head (TDH) intersect the ISO efficiency curve at different flow rate and head. Hence different operating points can be obtained at different pump efficiencies. For each operating points required power should be calculated according to equation (3). Motor and inverter efficiency are kept constant with change of flow rate to avoid complexity.

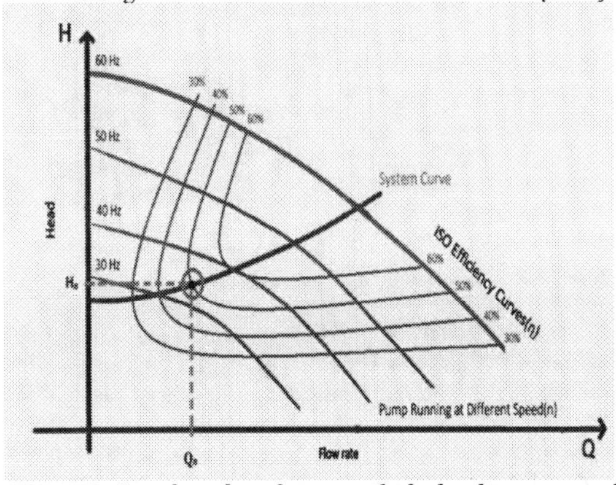

Figure 4: *Complete chart for a specific hydraulic system and pump performance curve. Green circle indicates the duty point while pump efficiency is 50%. H_n and Q_n are head and flow rate at 35Hz respectively for that instantaneous operating point.*

3 TOOL DEVELOPMENT AND FEATURES

MATLAB GUI (also known as graphical user interfaces or UIs) has been used in order to develop a simulation tool for PVPS named as **'DASTPVPS-Reloaded'**. The idea developed in section 2.1 and 2.2, could be shown as a whole in a graphical aspect (***Fig. 5***) [10]. This chart is the basis of all simulation and design tool. It makes it easier to understand and develop the tool. Hydraulic energy characteristic curve is placed in a Cartesian plane's quadrant 1. In quadrant 2, normalized power characteristic curve is drawn with respect to irradiance. By equalizing the energy of these two quadrant, flow rate curve can be estimated with respect to time in quadrant 4.

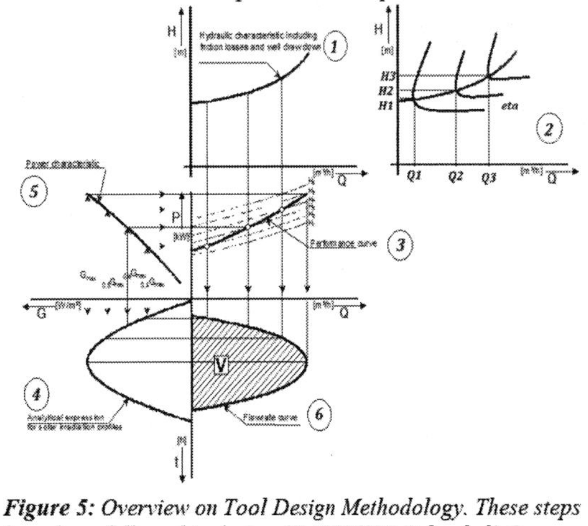

Figure 5: *Overview on Tool Design Methodology. These steps have been followed to design 'DASTPVPS-Reloaded'. Source: Modified [10].*

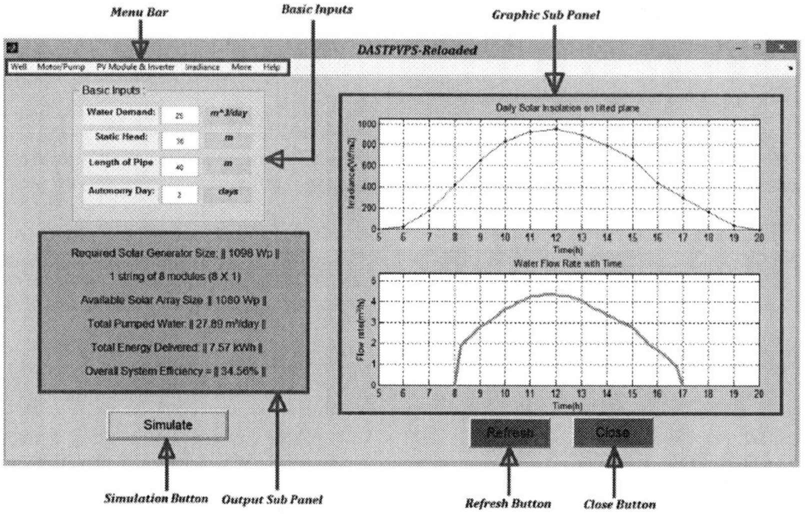

Figure 6: Tool working panel at a glimpse.

The tool's GUI comprises (***Fig. 6***) with a Menu bar, couple of sub panels and few buttons. The **'DASTPVPS-Reloaded'** starts with the sub panel "Basic Inputs". The user has to give at least 4 inputs which are minimum requirement for an approximate calculation. Other required data inputs are placed within the menu bar. With each selection of the menu bar a particular operation will be executed. Two kind of data base are needed in this tool, one is for pump specifications and other is the solar irradiance data for a particular place. This data must be created in a compatible format in Microsoft Office Excel spread sheet (***Fig. 7***). The "Outputs" sub panel shows results or the selection of parameters in every step of the simulation. Like the "Outputs" sub panel "Graphic" sub panel shows relevant characteristic diagrams/graphs to provide user a better understanding about PVPS behaviour. The "Simulate" button is for final outputs and graphs. The "Refresh" button is to run simulation from the beginning and the "Close" button is to exit/quit from the tool.

	A	B	C	D
1	Hx[m]	Qx[m3/h]	eta1 [%]	eta_pump[frac.]
2	51	0	0	0.5
3	49.5	3.5	52	0.58
4	49	4.5	58	0.61
5	49	5	62	0.64
6	48.5	5.5	64	0.67
7	47	6.5	67	0.7
8	42	8.5	70	
9	34	10.5	67	
10	31	11	64	
11	28	11.5	61	
12	24	12	58	
13	18	12.5	50	

	A	B
1	Time[h]	Irradiance[w/m2]
2	5	0
3	6	26
4	7	167
5	8	400
6	9	620
7	10	780
8	11	875
9	12	910
10	13	876
11	14	785
12	15	656

Figure 7: *Input data base for DASTPVPS-Reloaded. Irradiance data format and Pump performance data format in Microsoft Office Excel spread sheet.*

An additional feature of 'DASTPVPS-Reloaded' is the graphical representation of the idle solar energy which is lower than motor-pump operating threshold. It shows the probable time when the water flow will start and stop. It depends on how intense is the solar irradiance. This idle energy can be used for small devices like data logger, mobile phone or emergency light charging and so on.

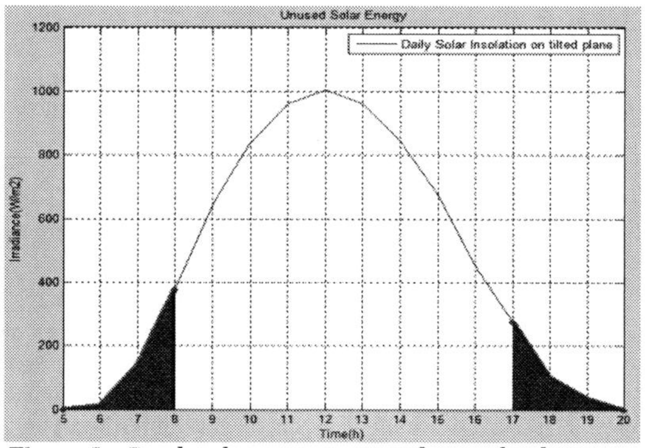

Figure 8: *Graphical representation of unused solar energy. The blue filled colour shows the energy that is not being used in water pumping.*

4 RESULTS AND VALIDATION

The Tool model validation has been accomplished by comparing the output results with two different cases. First case with the pioneering simulation tool 'DASTPVPS' and second with a real case study which is followed by an Ms Excel Model for sizing PVPS. In both reference cases the tool inputs and data base have been carefully taken for accurate simulation.

4.1 First Case: Justify with 'DASTPVPS'

In an African arid region Kampala, Uganda; a PVPS with minimum water demand of 20 m³ per day (assumed) which has 20m of static head. All specifications have been simulated in both tools and the outputs can be shown in table 1.

Table 1: *Comparison of simulation outputs for a 700 W_p system in Kampala, Uganda.*

Item	DASTPVPS	DASTPVPS-Reloaded	Difference
Array Size (ser X para)	700 W_p (7X2)	700 W_p (7X2)	_
Accumulated water	16.4 m^3	15.26 m^3	7%
TDH at max. flowrate	23.1 m	21.3 m	- 8%

Considering the rough calculation and simple input methods it hasn't resulted so different than the reference value. This could happen when there is slight difference in calculation methodology or when some input parameters (like solar irradiance) and constant values (like pipe roughness factor, fittings and valves) are not considered exactly. However the focal point of this PV sizing tool is to estimate how many panels are needed for a pumping system. The PV generator is consists of two PV string of 7 modules (50 Wpeak each) with array configuration (7X2). In that case array size is equal with the DASTPVPS and hence this is a positive approch for tool validation.

5.2 Second Case: Compare with an Ms Excel Model for sizing PVPS

Purpose of the study was to reach an optimum PVPS installation design for a farm located in Sebseb village, at 60 km southwest of Ghardaia, Algeria. This PVPS needed to be supplied a daily average water of $50m^3$. This article data has been carefully used to validate an Ms Excel Design Model [12]. Outputs from the both simulation are shown in table 2.

Table 2: *Comparison of software based simulation outputs for a SPWPS system in Ghardaia, Algeria.*

Item	Ms Excel Model	DASTPVPS-Reloaded	Diff.
Required PV Power	2537 Watts	2326 Watts	-8%
Array Size (ser X para)	$2430W_p$(9X1) (270 W_p each)	2430 W_p (9X1)	–
Accumulated water	50 m^3 (expected)	59.91 m^3	-19.8%
TDH at max. flow rate	42.75 m	40 m	-6.4%

Required Solar Power and TDH values are close (<10%). Although total pumped water volume is pretty high (-19.8%) than required. But in reference [12] for total pumped water there is only a consideration for expected volume of water. No such simulation has been done there that can predict the total pumped water. 'DASTPVPS-Reloaded' calculates approximate possible output of water in a clear sunny day which could be less or more than required water volume. As far as concerned about solar generator size or array designing, both outputs are same for the particular solar module and inverter. The overall output difference range of the two cases are around 6%-8%, which is conducive to validate this kind of simulation tool.

5 CONCLUSION

Though there is a demand for large number of inputs and assumptions for this kind of simulation tool; 'DASTPVPS-Reloaded' user interface has been tried to keep very simple and user friendly. The simulation results are just few clicks away from its initialization. The prompt outputs and graphs in each stage of inputs, allows the user to change the specific input immediately right there itself. User can also compare or try the simulation outputs for different system layout (e.g. different motor-pump set or solar panel size) quickly. In this tool total dynamic head or TDH is calculated as a function of irradiance or time. For every irradiance value the TDH will be varying, hence comes the actual required hydraulic energy and accumulated pumped water. PV generator array sizing (i.e. series-parallel connection of solar panels) is one of the main outputs of this tool. In addition there has been a menu added for calculating idle solar energy or the energy that is not being used in pumping water. Since idle power range is very low, this could be used for small electronic devices.

It can be stated, one can design and have an idea about PVPS with simple simulation of 'DASTPVPS-Reloaded'. This tool can be developed further i.e. more input/output features can be added in next versions. Data base format or source code could be modified for more accurate results. One future work on this research could be mobile/tab application (app) development as smartphones are getting smart enough to process large files. Also that app will be able to interact with the geographical position and meteorological data where the PVPS supposed to be installed.

6 REFERENCES

[1] Design of PVPS in the program PVSYST ver. 6.39, www.pvsyst.com

[2] PV-DesignPro-P ver. 6.0, a program for the design of water pumping systems, www.mauisolarsoftware.com

[3] R. Barlow, B. McNelis and A. Derrick, Solar Pumping. An introduction and Update on the technology, performance, costs and economics, World Bank Technical Paper Number 168, IT Publications, Washington D.C., 1993

[4] W. Ratterman, J. Cohen, A. Garwood, Solar Pumping Systems (SPS), Introductory and Feasibility Guide, Green Empowerment, 2007

[5] Z. Glasnovic, J. Margeta, A model for optimal sizing of photovoltaic irrigation water pumping systems, Solar Energy 81: p. 904–916, 2007

[6] O. Mayer, Photovoltaic Pumping system (PVPS), Ing.-Büro Mayer, Munich, 2000

[7] B.Reed and B. Reed, "*How much water is needed in emergencies*", WHO, WEDC, 2013.

[8] K. Watkins, "*Beyond Scarcity: Power, Poverty and Global water crisis*", Human Development Report, ISBN 0-230-50058-7, UNDP, 2006.

[9] W. Ratterman, J. Cohen and A. Garwood, "*Solar Pumping Systems (SPS) – Introductory and Feasibility Guide*", Green Empowernment (1-64), 2007.

[10] O. Mayer, Overview on Design Methods and Tools for PVPS, PVPS Seminar GTZ, Ing.-Buro Mayer, 2004

[11] M. Milnes, The Mathematics of Pumping Water, AECOM Design Build-Civil, Mechanical Engineering, The Royal Academy of Engineering

[12] M. Marlini, "*Implementation and validation of a design model for photovoltaic water pumping systems*", KTH Energy Process, 2012.

TERMOVISION TESTING OF THE SOLAR POWER PLANT LIFETIME IN THE CZECH REPUBLIC

Jandova Kristyna, Vanek Jiri
Brno University of Technology, Faculty of Electrical Engineering and Communication Technologies, Department of Electrotechnology, Technicka 10, 616 00 Brno, Czech Republic
jandovak@stud.feec.vutbr.cz, vanekji@feec.vutbr.cz

ABSTRACT: In this article we are comparing the results of analyses of photovoltaic power systems by using of thermal imager. Fifty different-sized photovoltaic power plants comprising more than twenty thousand photovoltaic modules were analyzed. The analysis was carried out over a period of one year and under various conditions and it was performed by three service companies. The results of the analysis are delivered with the recommendations, under what best conditions the measurements are to be performed.

Keywords: termovision testing, solar power plant, Czech Republic

1 INTRODUCTION

It is possible to use the method of thermovision to see the distribution of surface temperatures in a photovoltaic module. This analysis is based on a defect detection due to dissipated heat generated by the consumption of electricity produced by faulty photovoltaic cells. These generate less current than other elements of the panel and therefore do not contribute to the production of energy and they consume some energy produced by other cells. This reduction in current can be caused by a defect in the element or partial shading of the solar cell surface. Even in the case of a short term shading of the cell an increase in its temperature can be detected.

Temperature leveling is a gradual process and accuracy of the described method that can be very affected. This method can also detect heat loss generated on conductive paths or excessive loading of the bypass diodes in the wiring boxes. Because of this the method of using the thermocamera in the detection of defects on the photovoltaic modules has several advantages. Unlike most other methods, diagnosis with a thermal camera can be used to analyze already installed in-service modules without disconnecting them. Another advantage is the speed of the measurement.

2 THERMOVISION DIAGNOSTICS

While using the thermocamera for diagnosis it is important to reach the minimum necessary intensity of the radiation flux. Therefore, the recommended irradiance is at least 500 W/m2 [1]. Thermal imaging diagnostics can be performed from both the front and back of the PV module.

From the front of the imager shows particularly the temperature distribution on the surface of the cover glass panel. With low emissivity glass this diagnosis is very difficult. Glass reflects from its surface heat flux emitted by other objects, which can cause incorrect display of temperature and a defect on the thermogram. To avoid displaying of the reflected temperature readouts on the thermogram, it is needed to measure at an angle of 5-60 °, so that the reflection of the interfering objects is avoided.

In most cases of ground-based installations the diagnostics can be performed from the backside of the photovoltaic panels. The temperature measured on the backside may be higher due to higher heat flux because of the smaller thermal resistance of the backside. Measuring from the rear side minimizes interference with surrounding heat sources because the reflectance of the backside is not as high as the reflectivity of the cover glass of the front. Unfortunately, even when measured from the back an artifact of another heat source radiating through the panel may appear.

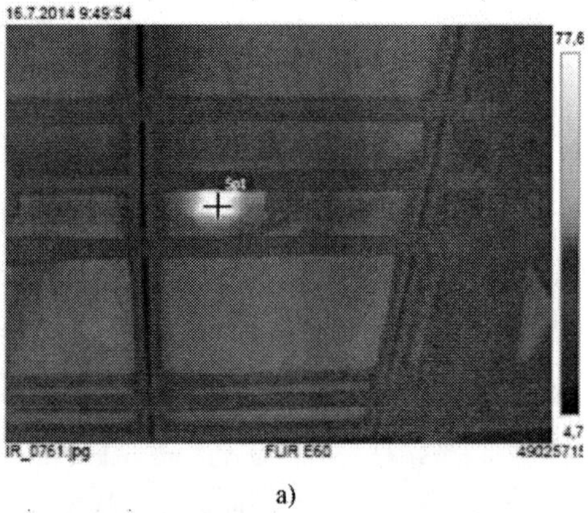

a)

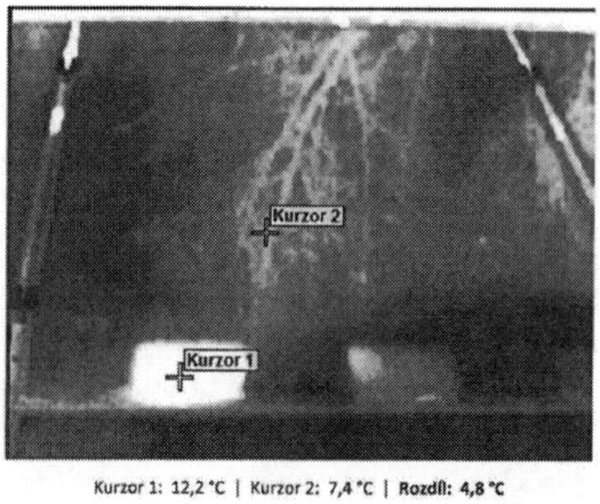

b)

Figure 1: Examples of thermographic analyses from measurements a) from the back side and b) from the front.

In Fig. 1 a) is an example of an image of the thermal camera obtained from the rear of the photovoltaic module, where the thermograph in addition to showing significant defect observed with higher temperature, it also shows the low temperature of the structure, which affects certain areas of the thermograph. The overall temperature of the scanned area is also affected by IR heat passing through the test panel. In Fig. 1 b) it is possible to see the reflected picture of the camera itself captured from the front at low ambient temperatures. It is seen as a defective area ("Kursor 1") as well as the heat reflected from the surrounding trees, which at this time have a much higher temperature than ambient air. This reflection will strongly influence the overall evaluation of the image ("Kursor 2").

3 EVALUATED POWER PLANTS

Within the service inspection were analyzed 114 differently sized photovoltaic power plants, consisting in total of more than twenty thousand photovoltaic modules. The analysis was carried out over a period of one year under various conditions and it was carried out by four service companies.

Table 1: The thermocameras used by particular service companies

Service Company	S1, S4	S2	S3
Thermocamera used	FLIR E60	FLIR E8	Irisis 4010

The nature and size of temperature anomalies enabled the division into three categories:

Category I. 0 - 10 °C – PVM does not show anomaly requiring repair; it is a reason for a follow-up at the next inspection,

Category II. 10 - 20°C – PVM shows temperature anomaly requiring a repair prior the next scheduled inspection,

Category III. > 20°C – PVM shows temperature anomaly requiring immediate attention to address the faulty state.

A fault was determined to be a measured increase in temperature critical point in comparison with a reference temperature (a reference point) of the device.

Most of the detected thermal anomalies have been found in the photovoltaic system in the town of Pilsen with a nominal power of 992 kWp in 4320 of the tested panels 4320, of which 50 panels were in Category III - Thermal anomaly requiring immediate solutions of the defective condition.

Table 2: Numbers of modules

	Number of modules	Found defects of type II	Found defects of type III
polycrystal	20592	31	94
monocrystal	2106	8	13
other types	1346	3	5
A roof installation	8973	32	97
terrestrial installation	14447	10	66

Overall, in the analysis of the four companies that analyzed 24044 panels, they found 41 panels in Category II and 112 panels in Category III. In Category I, some service organizations included all the panels, which showed undetected temperature anomaly and moderate temperature changes and some companies these two cases distinguished (in the table is the number of the panels with weaker anomalies in brackets).

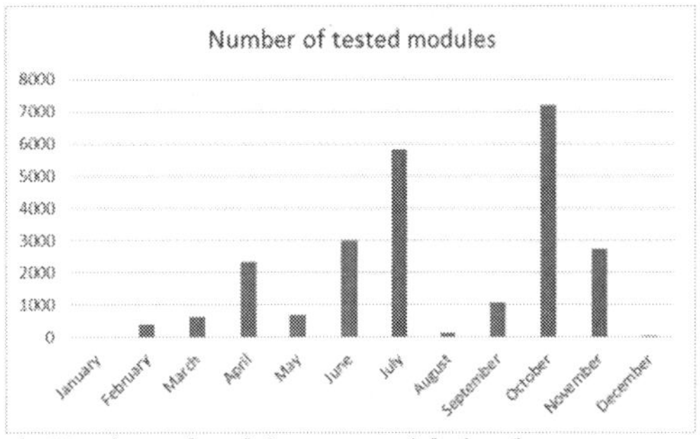

a) Numbers of modules measured during the year

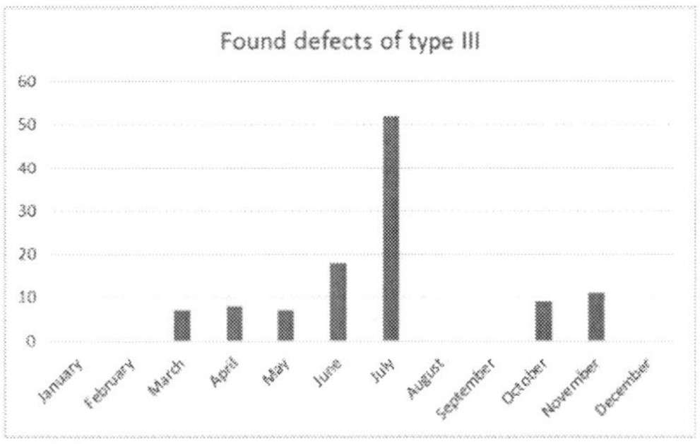

b) Numbers of detected defect modules with Type III

Figure 2a), b): An overview of individual monitored quantities as a function of the season

From the above charts it is possible to see the correlation between temperature measurements and the number of anomalies found. This feature is not related so much to the specified temperature, but rather to the seasons and the position of the sun in the sky. Changing the intensity of the incident radiation depending on the angle of incidence is given by the cosine function of the incident angle and the intensity of radiation incident perpendicular to the detector according to the following relation:

$I_{final} = \cos \varphi . I_{measured}$ (1)

Therefore, even if the measurements were made at a measured intensity 850 W/m2, the actual intensity of the solar radiation incident on the photovoltaic panel did not even have to reach the recommended minimum value of 500 W/m2 during the winter months.

As described in the article [1] the temperature difference of the analyzed region of the defective solar cell is directly proportional to the intensity of the incident radiation, see Fig. 2.

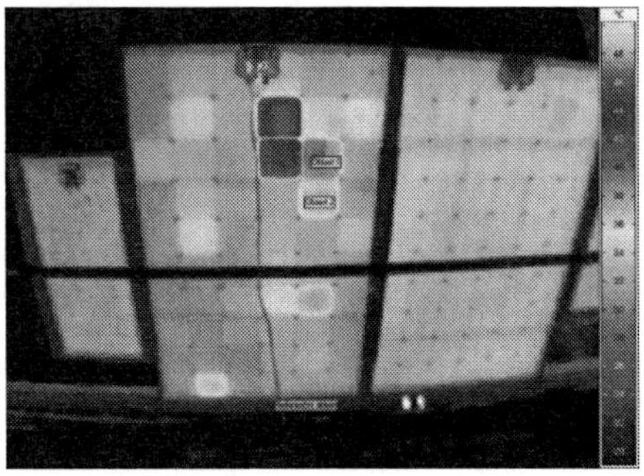

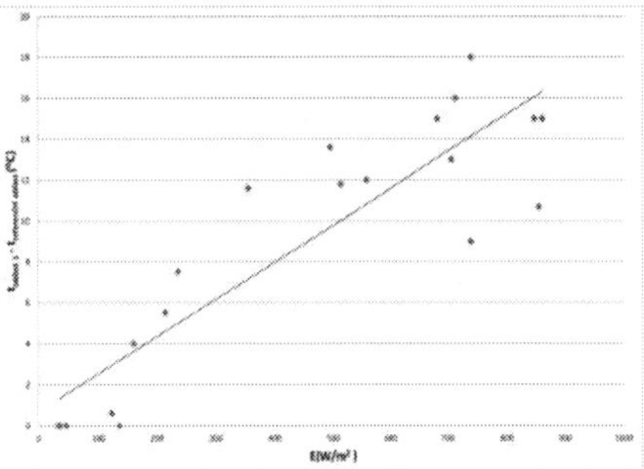

Figure 3: The effect of intensity of illumination on the temperature difference between the reference plane and a defective cell

4 CONCLUSION

The analysis of the measured results showed that most of the image recording of the photovoltaic modules of the listed power plants took place under conditions of insufficient solar radiation intensity. Therefore, it would be good to review the classification of the modules in each category. In case of insufficient intensity at the time of measurement the panels should of could move Category I to Category II and from Category II to Category III. This would therefore increase the number of modules required immediate replacement.

Table 2: Summary of results

Number of modules	**Installed power (kWp)**	**Cat. I 0 - 10°C pcs**	**Cat. II 10 - 20°C pss**	**Cat. III > 20°C pcs**
24 044	5 143,741	218	41	112
		0,9 %	0,1 %	0,5 %

5 ACKNOWLEDGEMENT

This publication was possible thanks to a support for the infrastructure of the "Centre for Research and Utilization of Renewable Energy" in the project No. LO1210 - "Energy for Sustainable Development" and the project FEKT-S-14-2293 "Materials and Technologies for Electronics II". We also thank the company Asekol Solar s.r.o. for their helpful attitude and willingness for cooperation.

6 REFERENCES

[1] STRNADEL, J.; GVRITISHVILI, R.; VANĚK, J. COMPARISON BETWEEN VISUAL AND THERMOVISION INSPECTION OF PV FIELD. In Proceedings of 28th European Photovoltaic Solar Energy Conference and Exhibition. Paris, France.: WIP, 2013. p. 3953-3956. ISBN: 3-936338-33- 7.

[2] VANĚK, J.; STRNADEL, J. Long time monitoring of heat distribution in photovoltaic module by the thermovision. In ECS Transaction. ECS Transactions. San Francisco, USA: ECS, 2014. p. 199-206. ISBN: 978-1-62332-031- 7. ISSN: 1938- 5862.

[3] KADLEC, M., VANĚK, J. Analýza FV instalace termokamerou a zjistitelné dopady metody na budoucí výkon, životnost, problémy. FF2014. Plzeň.

[4] Váry, Michal - Perný, Milan - Šály, Vladimír - Firický, Eduard: Monitoring of Grid Tied Photovoltaic System Efficiency in Operating Conditions. In: Electrotehnica, Electronica, Automatica. - ISSN 1582-5175. - Vol. 60, nr. 2 (2012), s. 9-12

[5] P. Tauš, M. Taušova, Economical analysis of PV power plants according installed performance, Acta Montanistica Slovaca, 14 (2009), pp. 92–97

[6] Z. Dostál, M. Ďulík, M. Koščová, And P. Šimon, Monitoring of degradation of photovoltaic panels using infrared method. Acta Montanistica Slovaca, 15(SPEC.ISSUE 2), pp. 111-115. (2011)

SIMPLE AND ACCURATE MONITORING OF EXPECTED PV POWER GENERATION BY USING MINI-PV MODULE

Kimihiko Saito[1], Jiro Yamazaki[2], Daishi Yoshino[2], Noboru Higuchi[3], and Michio Kondo[1,4]
[1] Faculty of Symbiotic Systems Science, Fukushima University, Kanayagawa 1, Fukushima, Fukushima, 960-1296, Japan
[2] Research Center for Advanced Imformation Science and Technology, The University of Aizu, Ikkimachi Tsuruga, Aizu-Wakamatsu, Fukushima, 965-8580, Japan
[3] Fukushima National College of Technology, Taira Kamiarakawa Nagao 30, Iwaki, Fukushima, 970-8034, Japan
[4] Fukushima Renewable Energy Institute, National Institute of Advanced Industrial Science and Technology (FREA), Machiikedai 2-2-9, Koriyama, Fukushima, 963-0215, Japan

ABSTRACT: The evaluation method of the expected photovoltaic (PV) power and energy generation by using the PV mini-module is proposed as a simple, cost-effective, and accurate monitoring and is examined by the comparison with the actual power and energy generation of the PV power system in the field. By monitoring of the deviation ratio between the daily amount of actual energy generation in DC (DEAG-DC) and of the expected energy generation estimated from the mini-module output power (DEEG-MM) on the days picked out above the appropriate threshold of its DEAG-DC, the potential of this method for highly accurate estimation is demonstrated, where it can detect the decrease of the PV power system as small as 1% in terms of the number of faulty panels with the average of the deviation ratio owing to its quite small offset less than 1% and with its small standard deviation less than a few %.
Keywords: Monitoring, c-Si, PV System, System Performance

1 INTRODUCTION

For the long term operation of photovoltaic (PV) power systems, monitoring of the slow degradation such as aging within a few % per year as well as the sudden failure accompanied with a few ten % decline of those output power is important for forecasting a fault or a latent fault of the systems and the simple and cost-effective method is desired for this purpose. One of the effective methods for the evaluation of slow degradation is the comparison between the actual energy generation (AEG) and the expected energy generation (EEG) of the PV power system estimated by the solar irradiance and the module temperature measurements, where the high accuracy of EEG estimation is essential. In this study, we demonstrate the simple and accurate estimation of EEG by using the PV mini-module as a reference which can reflect the solar irradiance and the module temperature in the power generation output at the same time.

2 EXPERIMENTS

2.1 Preparation of mini-module

We prepared the PV mini-module (Fig.1(a)) as a reference module, composed of the series connection of 10 c-Si solar cells of 125 mm x 15 mm strip, the tempered glass plate with the size of 240 mm square, the PET-based back sheet, and the Al-based frame. They were encapsulated not by the commonly used ethylene-vinyl-acetate (EVA) but by the silicone resin [1] to keep the longer power output stability than that of the conventional PV modules. This mini-module was connected to the outdoor-use current-voltage (IV) measurement circuits (Fig.1(b)). The maximum output powers of the mini-module measured under the standard test condition (with AM1.5G irradiance of 1kW/m^2 at 25°C) using the solar simulator (WACOM WXS-163S-10) by the commercial IV measurement system (WACOM IV-10020-AD1) and by this outdoor-use IV measurement circuits were 3.648W and 3.681W, respectively.

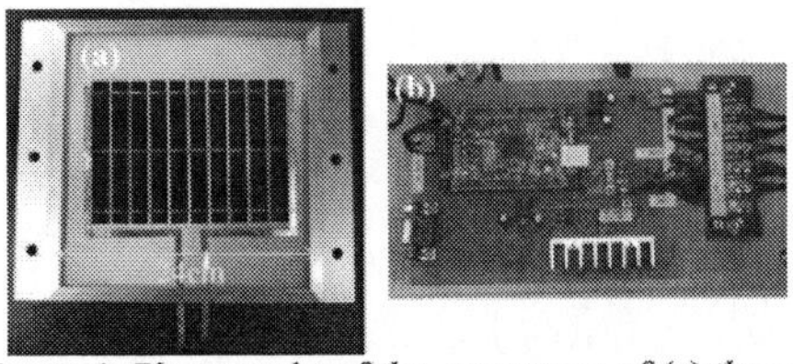

Figure 1: Photographs of the appearance of (a) the mini-module and (b) the outdoor-use IV measurement circuits

2.2 Field test

This mini-module with the outdoor-use IV measurement circuits was installed in Feb. 2015 nearby the c-Si PV power system with the rating of 31.152 kW (consisting of 132 panels with the rated power of 236W$_p$ per each) operating from Mar. 2014 on the roof of the school building in the Fukushima National College of Technology (Fig. 2).It has the same tilt angle as the PV panels of the PV power system and the outdoor-use IV measurement circuits were set up to measure the IV curve of the mini-module every 1 minute and to save the values of maximum output power obtained from it in the storage server placed at the University of Aizu through the internet connection. This PV power system has the pyranometer attached at the same tilt angle as its PV panels and its measurement output of the solar irradiance and the DC and AC output powers of the PV power system were monitored every 1 second.

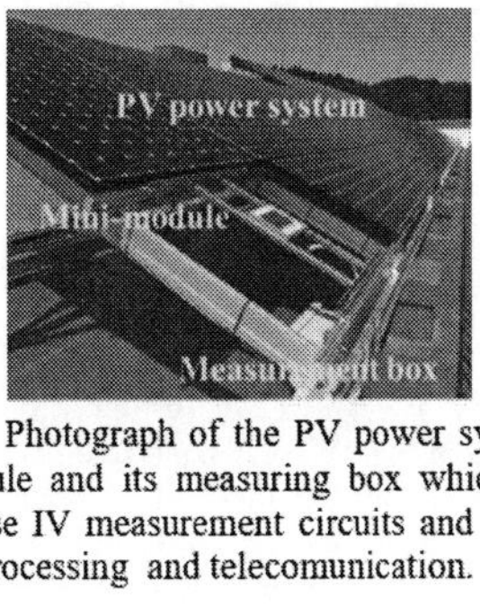

Figure 2: Photograph of the PV power system with the mini-module and its measuring box which contain the outdoor-use IV measurement circuits and the device for the data-processing and telecomunication.

2.3 Estimation of expected power and energy generation

In this study, the expected power generation outputs of the PV power system estimated from the solar irradians measured by pyranometer (EPG-SI) and from the power generation output of the mini-module (EPG-MM) were defined by the following equations, respectively.

$$[EPG - SI](kW) = \frac{[solar\ radiation\ intensity](kW/m^2)}{1\ (kW/m^2)} \times 31.152(kW) \quad \cdots(1)$$

$$[EPG - MM](kW) = \frac{[power\ generation\ output\ of\ mini-module](W)}{[rated\ power\ of\ mini-module\ (3.681)](W)} \times 31.152(kW) \quad \cdots(2)$$

Therefore, the daily amounts of the expected energy generation by the solar irradiance (DEEG-SI) and by the mini-module output power (DEEG-MM) can be expressed by the following equations, respectively.

$$[DEEG - SI](kWh) = \frac{[daily\ amount\ of\ solar\ radiation\ energy](kWh/m^2)}{1\ (kW/m^2)} \times 31.152(kW) \quad \cdots(1)'$$

$$[DEEG - MM](kWh) = \frac{[daily\ amount\ of\ energy\ generation\ of\ mini-module](Wh)}{[rated\ power\ of\ mini-module\ (3.681)](W)} \times 31.152(kW) \quad \cdots(2)'$$

3 RESULTS AND DISCUSSIONS

3.1 Comparison of power generation outputs

Figure 3 shows the example of the daily actual power generation chart of the PV power system in DC (APG-DC), the EPG-SI and the EPG-MM. Around noon, the EPG-SI overestimated the APG-DC owing to its evaluation without the consideration of the output power decrease by the increase of PV panel temperature. However, the EPG-MM demonstrates the good agreement with the APG-DC probably due to the inherent reflection of the influence of the temperature on the mini-module output power in the evaluation. Furthermore, this fact suggests no significant difference of the temperature coefficients of the output powers between the PV panels and the mini-module as a result of the usage of the conventional c-Si cells and module structure of the mini-module.

3.2 Comparison of daily energy generation

As discussed in section 3.1, the EPG-MM has enough potential for monitoring the small change of the power generation of the PV power system by the comparison with the APG-DC. However, this real-time monitoring could be easily affected by the errors attributed to the partial shading of the PV panels and so on. Therefore, we evaluated and monitored the daily amount of energy generation as well.

Figure 4 shows the timeline comparison between the daily amount of actual energy generation in DC (DAEG-DC), the DEEG-SI, and the DEEG-MM from Apr. to May in 2015. As can be seen in Fig. 4, the DEEG-MM shows the good consistency with the DAEG-DC in contrast to the large overestimation of DEEG-SI, especially on the high energy yeild days due to the reason mentioned in the section 3.1. In the monitored period from Mar. 2015 to Apr. 2016 (total 308 days under the normal operation except for the downtime for the mentainance, etc. of the PV power system and data server and also except for the period of the pseudo degradation test), the gross of the DEEG-MM underestimates against the gross of the DAEG-DC by only 1.6% in contrast to the 9.9% overestimation of the DEEG-SI.

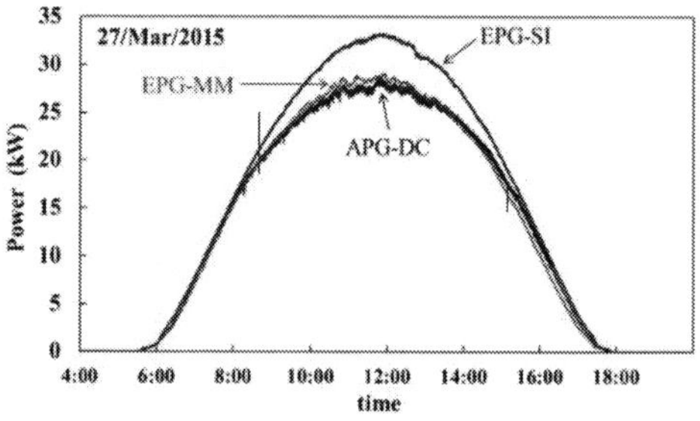

Figure 3: Comparison between the actual power generation of the PV power system in DC (APG-DC) and the expected power generation output (EPG) in the daily chart.

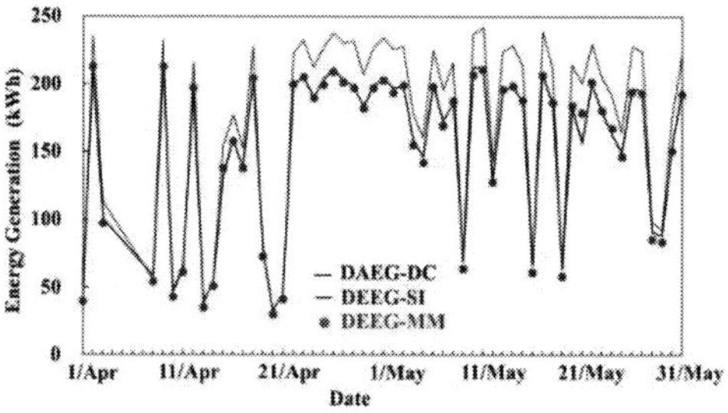

Figure 4: Timeline comparison between the daily amount of actual energy generation of the PV power system in DC (DEAG-DC) and the daily amount of the expected energy generation (DEEG) from Apr. to May in 2015.

Figure 5 illustrates the histogram of the deviation ratio of the DAEG-DC to the DEEG-MM (DR-MM) calculated by ([DAEG-DC] – [DEEG-MM])/[DEEG-MM] for the days from Mar. 2015 to Apr. 2016 (total 308 days under the normal operation as commented above) and for the days when the DAEG-DC is 100kWh and over (total 207 days). As shown in this figure, the DEEG-MM shows the small difference from the DAEG-DC on the relatively high energy yield days (100 kWh and over), however, it underestimates the DAEG-DC largely on the low energy yield days. One of the possible reasons is that it comes from the combination of the irradiation spectrum under the cloudy or rainy climate and the difference of the spectral responses between the PV panels and the mini-module which originates from the difference of the encapsulation materials [2]. By picking out the days when the DAEG-DC is 100kWh and over in order to remove the low energy yield days which cause a large error, the average of the DR-MM and its standard deviation decreased to 0.9% and 2.5%, respectively.

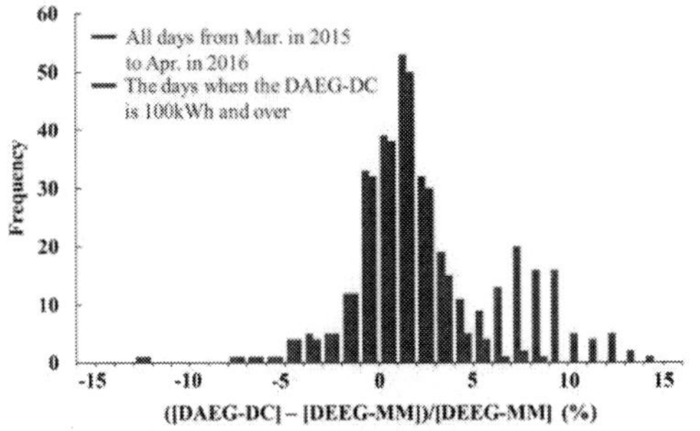

Figure 5: Histogram of the deviation ratio between the DAEG-DC and the DEEG-MM calculated by ([DAEG-DC] – [DEEG-MM])/[DEEG-MM]

Furthermore, in order to take into account the change of the length of daytime depending on the seasons, we tried not to use 100kWh/day but to use 70% of the daily amount of ideal energy generation under a clear blue sky as the threshold of the effective day for the comparison between the DAEG-DC and DEEG-MM, where the ideal energy generation is deduced from the solar irradiance on the tilted surface simulated by the calendar, the place of the PV power system and the tilt angle and mounting direction of the PV panels [3]. The decrease of ideal energy generation by the increase of PV panel temperature is estimated assuming the constant factor of 10%. In this case, the average of the DR-MM and its standard deviation are decreased to 0.7% and 1.7%, respectively, for the 151 effective days. One of the key factors for this underestimation of the average 0.7% should be the difference between the rated power and the actual generation power capacity of the PV panels, however, it is necessary to pay attention to the fluctuation not only of the average but also of the standard deviation which should depend on the seasonal change of the solar spectral irradiance considering the difference of the spectral response between the PV panels and the mini-module as discussed above.

3.3 Pseudo degradation test

In the above discussions, we have investigated the appropriateness of evaluation of the EEG by using the mini-module under the assumption of no degradation of the PV power system. For its diagnosis application to the slow degradtion of the PV power system, its detection capability of the small energy-generation decline of the PV power system should be investigated as well. Accordingly, the pseudo degradation tests were carried out twice from Nov. 20th to 30th in 2015 and from Feb. 26th to Mar. 25th in 2016 by covering one of the 132 PV panels consisting of the parallel connection of the 12 strings of the 11 panels in series with the black rubber sheet (Fig. 6).

Figure 7 shows the timeline chart of the DR-MM and its 5 day moving average on the days when the DAEG-DC was not less than 70% of the daily amount of ideal energy generation under a clear blue sky deduced from the simulated solar irradiance on the tilted surface as described in section 3.2 from Nov. 2015 to Apr. 2016 (total 69 days). In this figure, the 5 day moving averages plotted in the beginning of Nov. are evaluated from the DR-MM in the end of the previous month and the timeline chart is treated continuously without the consideration of the down period of the PV power system from Dec. 10th, 2015 to Feb. 6th, 2016. In spite of the 1.7% standard deviation of the DR-MM as mensioned in section 3.2, the decreases of the DR-MM in the periods of the pseudo degradation test, where the 0.8% decrease of the generation power of the PV power system is expected from the number of masked panels (= 1/132), are clearly observed especially by applying the moving average method. This result suggests that the EEG monitoring by using the PV mini-module has a potential for the detection of the small energy-generation change of the PV power system caused by its degradation and so on.

Figure 6: Photograph of the setup for the pseudo degradation test

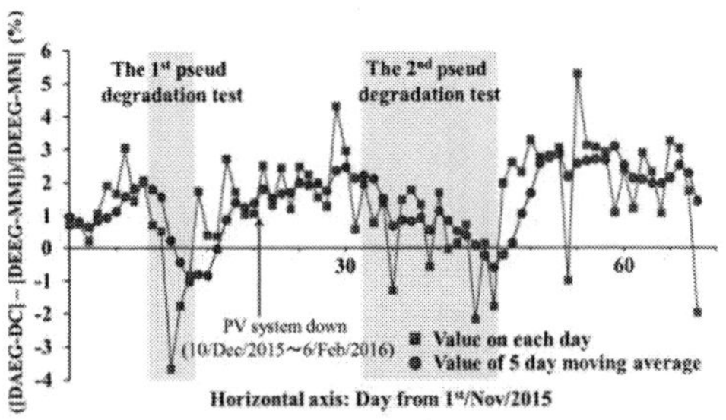

Figure 7: Timeline chart of the deviation ratio between the DAEG-DC and the DEEG-MM calculated by ([DAEG-DC] – [DEEG-MM])/[DEEG-MM] and its 5 day moving average on the days when the DAEG-DC was not less than 70% of the daily amount of ideal energy generation under a clear blue sky from Nov. 2015 to Apr. 2016

4 SUMMARY

The accurate estimation of the expected power and energy generation of the PV power system through the monitoring of the maximum output power of the PV mini-module was demonstrated by the comparison with the actual power and energy generation of the PV power system with the rating of 31.152 kW. By using the PV mini-module as a reference which reflects the influence not only of the irradiance but also of the panel temperature on the output power, the gross of the daily amount of expected energy generation (DEEG-MM) through the 14 months with the effective monitoring of 309 days was almost consistent with the daily amount of actual energy generation in DC (DAEG-DC) with the difference of 1.6%. Furthermore, by introducing the data-handling condition such as picking out the days when the DAEG-DC was not less than 70% of the daily amount of ideal energy generation under a clear blue sky deduced from the simulated solar irradiance on the tilted surface, the average of the deviation ratio between the DAEG-DC and the DEEG-MM defined as ([DAEG-DC] – [DEEG-MM])/[DEEG-MM] and its standard deviation were improved to 0.7% and 1.7%, respectively. In addition, from its moving average, the small decline of the generation power of the PV power system was detected in the pseudo degradation test where one of the 132 PV panels was covered with the black rubber sheet. These results suggest that the evaluation method of the expected energy generation of the PV power system by using the simple equipment with the PV mini-module enables the accurate monitoring for the slow degradation of the PV power system. For the improvement of its accuracy and reliability, the error analysis and the further optimization of the statistic treatment are in progress and the stability of the mini-module and of the outdoor-use IV measurement circuits will be tested.

5 ACKNOWLEGEMENTS

We thank to the Fukushima Renewable Energy Institute (FREA) of AIST, Nihon Rokki Ind. Co., Ltd., and Silicone-Electronics Materials Research Center of Shin-Etsu Chemical Co., Ltd. for their support of the preparation for the demonstration equipment. This work was supported by MEXT Japan in the Fukushima Regional Innovation Strategy Promotion Program.

6 REFERENCES

[1] S. Sakamoto, A. Ito, H. Ohwada, T. Furihata, K. H-Bae, N. Yamakawa, A. Yaginuma, T. Imataki, and M. Watanabe, Proceedings 6th World Conference on Photovoltaic Energy Conversion (Kyoto), 7WePo.10.18 (2014)

[2] H. Ohwada, J. Tsukada, T .Furihata, N. Yamakawa, A. Yaginuma, K. Hara, and A. Masuda, Proceedings of Annual Meeting of Research Center for Photovoltaics in National Institute of Advanced Industrial Science and Technology (AIST), in Japanese (2015)

[3] D. G. Erbs, S. A. Klein, and J. A. Duffie, Solar Energy Vol.28, No.4 (1982) 293

AUTOMATIC DETECTION OF DEFECTIVE SOLAR MODULES BY THERMOVISION

Vaněk Jiří, Repko Ilja, Klíma Jakub, Peroutka Tomáš
Department of Electrical and Electronic Technology,Faculty of Electrical Engineering and Communication
Brno University of Technology, Czech Republic

ABSTRACT: This paper deals with the possibility of automated detection of defective solar modules by thermovision camera mounted on a dron or on the car roof. During thermovision imaging analysis of large photovoltaic power plants there are captured large numbers of images. These images are analyzed by human operators. Given the vast amount of images, which are sometimes very similar, and given of the specificity of some defects, the work of operators can be replaced by the automation recognizing of anomalies software. Two methods for the automatic detection of defects in thermal imaging pictures were developed and validated. First one method is based on the geometric information of tested modules and on the assumption that defective cell has an increased temperature along its whole surface and therefore will appear as a regular geometric shape which is recognizable by geometric comparisons. The second method does recognition by usage of trained artificial neural network.
Keywords: diagnostic, thermovision, recognition, automatic

1 INTRODUCTION

Usage of thermography measurements during production of power can show the distribution of temperature on the photovoltaic modules surface. This temperature distribution corresponds to the heat delivered by sunlight and to the dissipation heat produced by energy lost on the defects of solar modules. Therefore, the use of thermovision camera to detect defects on photovoltaic modules has several advantages and disadvantages.

It is quite easy to detect the anomaly - unlike most other diagnostic methods which need some interventions in the system. Thermal anomaly is show on thermocamera screen and it is possible to save it to camera/computer memory. There is no need to disconnect tested modules, the thermovision inspection even must be measured without disconnecting of the PV modules. Another advantage is the speed of measurements and predetermined shape of thermal anomaly corresponding to a type of defect.

Disadvantage of this method is that the thermovision camera shows the temperature distribution on the surface of the cover glass which is combination of inner heat radiation, reflection on surface of glass and the IR radiation passing through the module. And due to the low emissivity of glass it is quite difficult to do the proper diagnosis. From its surface, glass reflects emitted by surrounding objects, which can cause incorrect display of temperature and defects on the thermogram (Fig. 1.).

Figure 1: Thermovision image of photovoltaic modules with all type of interfering signals – reflection of clouds and passage of the IR signal though modules.

Evaluation of the themovision images is therefore difficult for operator's attention and during the analyses of a large photovoltaic power plant may be easily overlooked an obvious defect. Therefore, it is extremely topical this process at least partially automate for types of defect that is it possible. There is possible to automatically detect the single hotspot (Fig. 2), multiple regular hotspots (Fig. 3), bypassed single string of module (Fig. 4), bypassed multiple parts of module and whole bypassed module. In this case the partly shadowed module (Fig. 5) should be detected as the multiple hotspot or in higher intensity as bypassed module.

Figure 2: Thermovision image of photovoltaic module with single hotspot created by the dissipation heat effect of single defective solar cells.

Figure 3: Thermovision image of photovoltaic module with multiple hotspot created by the dissipation heat effect of multiple defective solar cells.

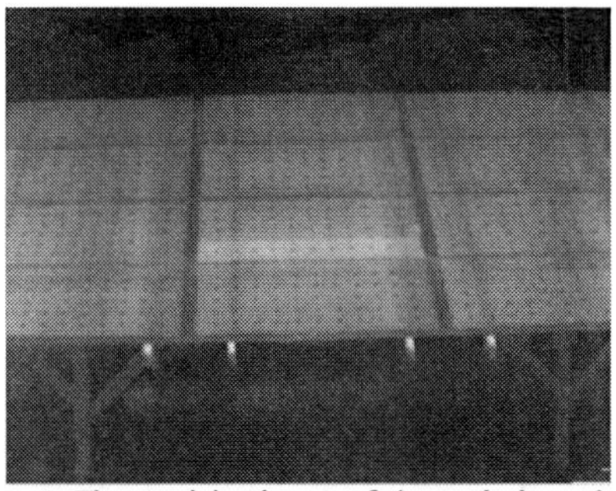

Figure 4: Thermovision image of photovoltaic module with bypassed single string of solar cell. The heat is the accumulated heat made by sunlight. There is also possible detect the heat passing through the module from junction box produced by dissipation heat of bypassed diod.

Figure 5: Thermovision image of photovoltaic module with bypassed multiple part of solar module. The heat is the accumulated heat made by sunlight. There is also possible detect the heat passing through the module from junction box produced by dissipation heat of bypassed diod.

Figure 5: Thermovision image of whole bypassed photovoltaic module. The heat is the accumulated heat made by sunlight.

Figure 6: Thermovision image of photovoltaic module with multiple hotspot created by the dissipation heat effect of multiple shadowed solar cells.

All these defects are characterized by a definable geometrical shape easily distinguishable from other thermal anomalies and therefore easily recognizable by computer sorftware. A human operator using the sofware will only need to check those images where it will occur undefined shape anomaly.

2 IMAGE ANALYZING

Thanks to the matrix arrangement of cells in the module the defect always have the same geometrical shape detectable by a mathematical method or by a cybernetic method. In first case we need to transform the input image to the shape corresponding to parameters of tested module and compare the temperature of area of single solar cell to the medium temperature of whole module or whole photovoltaic field. In the second method we need the large training set of solar modules images with predefined defects types for the training of artifitial neural network which will be then capable to recognize by the information about known shapes the similar shapes.

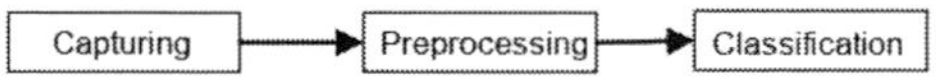

Figure 7: Basic steps of computer vision

Mathematical method.

First steps are the graphical image pixel to the temperature matrix cell transformation, intensity transformation and thresholding. During the thresholding process, individual data's point (pixel in an image) are marked as "object" pixels if their value is greater than some threshold value (assuming an object to be brighter than the background) and as "background" pixels otherwise. This convention is known as threshold above. Variants include threshold below, which is opposite of threshold above; threshold inside, where a pixel is labeled "object" if its value is between two thresholds; and threshold outside, which is the opposite of threshold inside. Typically, an object pixel is given a value of "1" while a background pixel is given a value of "0." Finally, a binary image is created by coloring each pixel white or black, depending on a pixel's label.This method is oftern used in computer preprocessing for object border finding.

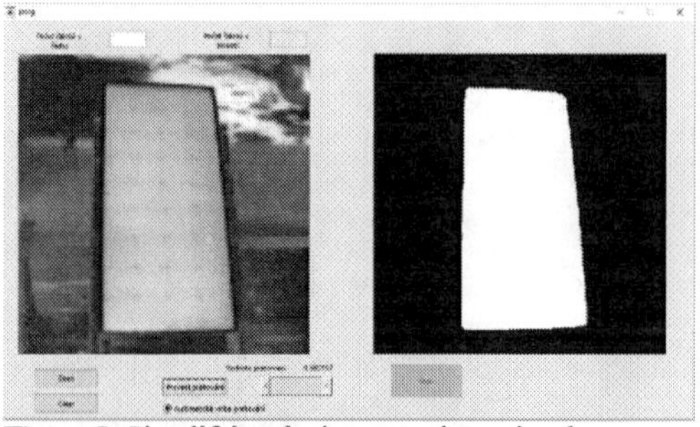

Figure 8: Simplifying the image to determine the boundaries of the module by thresholding.

Next operations will be performed for morphological

tuning of the image, in particular, erosion, extension and removal of objects having a size of less than 20% of the pixel area of the image.

Next functions of the following steps is: finding the lines limiting shaped of solar module on the image, calculating the corner points for geometric transformations, geometric transformations to template schape, and evaluate the resulting image the solar panel on the subject defects.

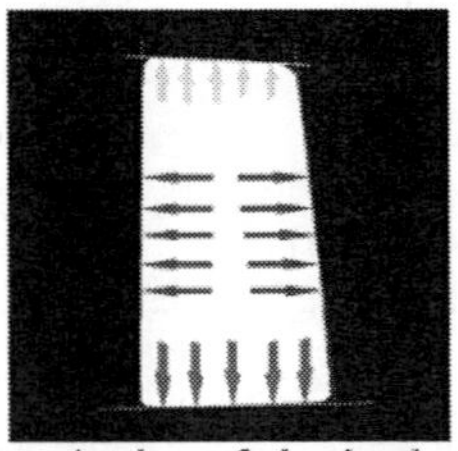

Figure 9: Determination of the border lines and the corner points of the module.

The geometric transformation is based on an affine transformation of the geometric shape defined by the four points into a rectangular shape defined by the template corresponding of photovoltaic module obtained from manufacturer datasheet.

Projection transformation matrix of homogeneous coordinates point X1 to the point X2 is:

$$P = \begin{bmatrix} a_{11} & a_{12} & d_x \\ a_{21} & a_{22} & d_y \\ p_1 & p_2 & s \end{bmatrix}$$

This matrix consists of eight elements (scale-rotation (a_{ij}) shift (d_x, d_y), perspective (p_1, p_2) and scale coefficient s. So as to clearly define this transformation, we have eight equations in relation to these values. Four points, each of them has two coordinates, determine precisely these equations, assuming that any three of them are not collinear.

We have four corner points (x1, y1), ..., (x4, y4) coordinates of the starting image. By use of affine transformation:

$$\begin{pmatrix} x' \\ y' \\ 1 \end{pmatrix} = \begin{pmatrix} a_{11} & a_{12} & d_x \\ a_{21} & a_{22} & d_y \\ 0 & 0 & s1 \end{pmatrix} \cdot \begin{pmatrix} x \\ y \\ 1 \end{pmatrix}$$

we get system of equations with results of P matrix used for transformation of the perspective shaped photovoltaic module.

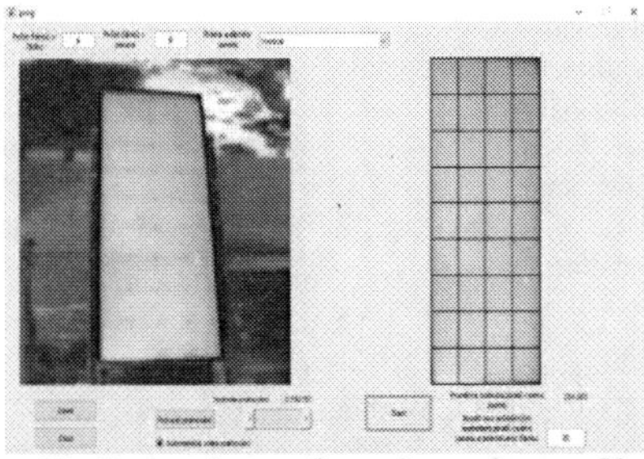

Figure 10: Geometric transformation and crop of images.

The last step is to crop the area of photovoltaic module and in an area corresponding to individual cells there is calculated the medium temperature and evaluated (Fig. 11).

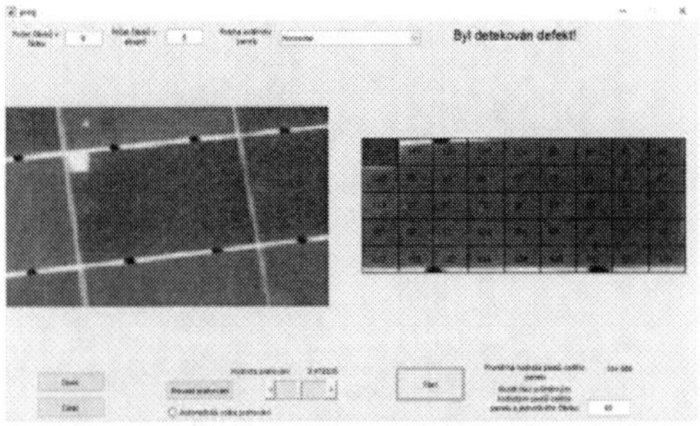

Figure 11: Evaluation of the result image by calculation of medium temperature of areas corresponding to cells areas and compering with whole medium temperature.

Cybernetic method.

Cybernetic method simulates human operator decisions. It is searching for already familiar shapes in the image and assigning a value corresponding to the probability that it is a search shape. The used method is called an artificial neural network and its success depends on the initial setup and training data, which should include all possible cases that the decision-making process may encounter.

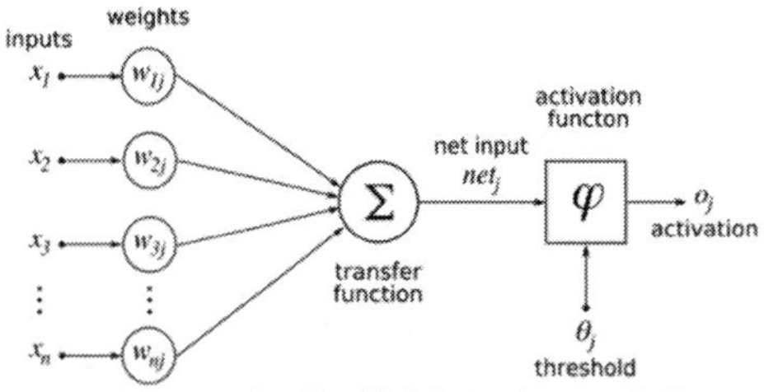

Figure 12: Example of artificial neural network diagram

Preprocessing of cybernetic method in our experimnet we chosed the simplest method. We skipped convertion of RGB temperature scale to gray shades scale. We directly applied the thresholding in the green region of spectrum to the original image. Thus we speeded upt the overall image analizing.

Figure 13: The Original image

Then we sharpened and filtered image from the grafical noise.

Figure 14: Thresholding of pixels having the green color value below its median.

Figure 15: Sharpening of filtered image

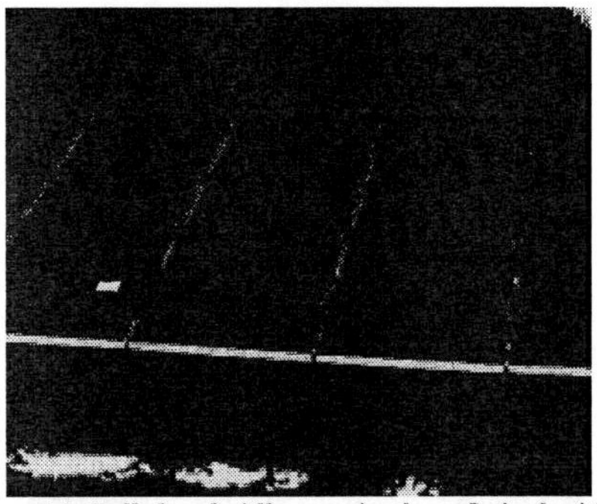

Figure 16: Full thresholding - selecting of pixel with the highest intensity.

The last step of preprocessing we re-thresholdign, which highlighted places with highest temperature. This preprocessing image was prepared for segmentation by neural network

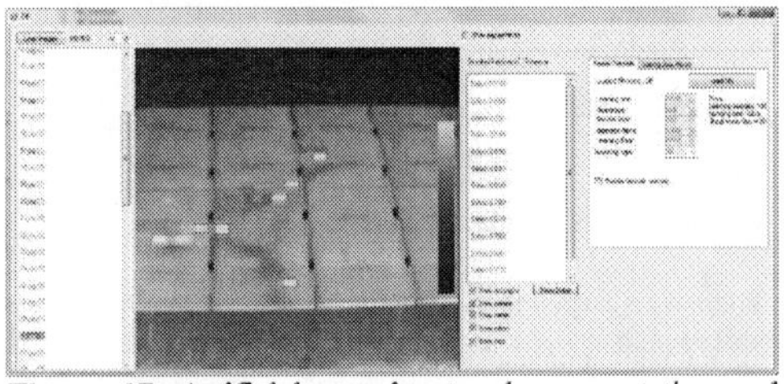

Figure 17: Artificial neural network segmentation and classification of defect places.

3 RESULTS

Both method we tested on the different sets of thermovisions images.

Table I. Results of the mathematical method.

Name of set	Correctly identified	Number of defects	Accuracy of determination
	[pcs]	[pcs]	[%]
Set 1	15	13	82
Set 2	30	70	68
Set 3	24	3	100

Table II. Results of the mathematical method.

Name of vector	Correctly identified	Number of defects	Accuracy of determination
	[pcs]	[pcs]	[%]
Vector 35	43	34	72
Vector 7	45	17	75
Vector 22	46	18	77
Vector 10	48	15	80
Vector 56	47	16	78

Both method were similar successful in defect determination. The cybernetic method is dependent on training data set but the process is much faster.

4 CONCLUSION

Both cognitive methods are possible to use for automatization of defect detection. By the averaging of temperature of the separated solar cells we are able to recognize defected solar cells or other sources of heat dissipation. By the adaptive thresholding of particular solar cell image and comparison with defect database we are able to recognize what type of defect of the solar module was detected and eliminate the manifestations of heat sources in the nearby area reflections. Usage of artificial neural network allows versatile methods primary dependent on the quality of training data.

5 ACKNOWLEDGEMENT

This publication was possible thanks to a support for the infrastructure of the "Centre for Research and Utilization of Renewable Energy" in the project No. LO1210 - "Energy for Sustainable Development" and the project FEKT-S-14-2293 "Materials and Technologies for Electronics II". We also thank the company Asekol Solar s.r.o. for their helpful attitude and willingness for cooperation.

6 REFERENCES

[1] REPKO, I. Automatické vyhodnocování termovizních snímků fotovoltaických panelů. Brno: Vysoké učení technické v Brně, Fakulta elektrotechniky a komunikačních technologií. Ústav elektrotechnologie, 2016. 30 p.

[2] KLÍMA, J. Automatické vyhodnocování termovizních snímků fotovoltaických panelů.

Brno: Vysoké učení technické v Brně, Fakulta elektrotechniky a komunikačních technologií, 2016. 66 p.

[3] HORÁK, K., KALOVÁ, I., PETYOVSKÝ, P., RICHTER, M. Počítačové vidění. Elektronické přednášky. Brno: FEKT VUT v Brně.

[4] DE VRIES, J. Object Recognition: A Shape-Based Approach using Artificial Neural Networks. Utrecht, 2006. University of Utrecht..

[5] TOSHEV, A, Ben TASKAR a Kostas DANIILIDIS. Object Detection via Boundary Structure Segmentation. Philadelphia, 2010. University of Pennsylvania.

[6] HANSEN, A. Neural Networks and Shape Identification. Jefferson City, 2010.

[7] STRNADEL, J.; GVRITISHVILI, R.; VANĚK, J. COMPARISON BETWEEN VISUAL AND THERMOVISION INSPECTION OF PV FIELD. In Proceedings of 28th European Photovoltaic Solar Energy Conference and Exhibition. Paris, France.: WIP, 2013. p. 3953-3956. ISBN: 3-936338-33- 7.

[8] VANĚK, J.; STRNADEL, J. Long time monitoring of heat distribution in photovoltaic module by the thermovision. In ECS Transaction. ECS Transactions. San Francisco, USA: ECS, 2014. p. 199-206. ISBN: 978-1-62332-031- 7. ISSN: 1938- 5862.

[9] VANĚK, J.; VESELÝ, A.; BAŘINKA, R. Automatic analysis of defective solar cells in the solar module. In Proceedings of 25th European Photovoltaic Solar Energy Conference and Exhibition. Valencia: WIP, 2010. s. 1682-1685. ISBN: 3-936338-26- 4.

[10] Váry, Michal - Perný, Milan - Šály, Vladimír - Firický, Eduard: Monitoring of Grid Tied Photovoltaic System Efficiency in Operating Conditions. In: Electrotehnica, Electronica, Automatica. - ISSN 1582-5175. - Vol. 60, nr. 2 (2012), s. 9-12

[11] P. Tauš, M. Taušova, Economical analysis of PV power plants according installed performance, Acta Montanistica Slovaca, 14 (2009), pp. 92–97

[12] Z. Dostál, M. Ďulík, M. Koščová, And P. Šimon, Monitoring of degradation of photovoltaic panels using infrared method. Acta Montanistica Slovaca, 15(SPEC.ISSUE 2), pp. 111-115. (2011)

EVALUATION OF A DETAILED ELECTRO-THERMAL PV MODEL ON A 62.5 KWP INSTALLATION

D. Anagnostos[1] (anagnostos.d@microlab.ntua.gr) , K. M. Paasch[2], H. Goverde[3 4], F. Catthoor[3 4], D. Soudris[1]
[1]National Technical University of Athens (NTUA),
Heroon Polytechneiou 9, Zographou Campus 157 80 Athens – Greece
[2]University of Southern Denmark, Alsion 2, DK-6400 Sønderborg
[3]imec, Kapeldreef 75, 3001 Heverlee , Belgium
[4]Katholieke Universiteit Leuven (ESAT), Kasteelpark Arenberg 10, 3001 Heverlee, Belgium

ABSTRACT: The ongoing growth of renewable energy installations is slowly shifting the energy mix for grids worldwide. Photovoltaic (PV) installations in specific have increased their share and PV currently covers more than 6% of the peak electricity demand in Europe. These growing numbers give hope for a sustainable future but also showcase the increasing challenges that the PV industry will have to face. With such a big number of PV installations on the grid, each with different size and characteristics, it is crucial to develop tools and methods for efficient monitoring and problem detection. The kind of dynamism, which occurs in time scales much shorter than 1 hour, requires PV models that inherently incorporate the dynamic effects of the PV system, like thermal transients and response time constants, also scalable both spatially and temporally. State of the art parametric models are unsuitable for fine time resolutions or non-uniform environmental conditions. In this work a novel PV model is presented, suitable for non-uniform, fast varying conditions on small time scales, for online monitoring and control of the PV plant. The model is validated on a 62.5 kWp installation for a period of 4 months and retains a root-mean-square (RMS) error of less than 5% even for time periods down to 10 seconds.
Keywords: Simulation, Modelling, PV System, Energy Performance

1 INTRODUCTION

The ongoing growth of renewable energy installations is slowly shifting the energy mix for grids worldwide. Photovoltaic (PV) installations in specific have increased their share and PV currently covers more than 6% of the peak electricity demand in Europe. The total capacity is expected to double by 2019, to 400 GW[1]. These growing numbers give hope for a sustainable future but also showcase the increasing challenges that the PV industry will have to face.

With such a big number of PV installations on the grid, each with different size and characteristics, it is crucial to develop tools and methods for efficient monitoring and problem detection, especially since the power grid is starting to become more active than passive, towards the smart grid era [2].

That kind of dynamism, which occurs in time scales much shorter than 1 hour, requires PV models that inherently incorporate the dynamic effects of the PV system, like thermal transients and response time constants. On the contrary, the methods that are considered state of the art at present make use of simplified parametric models, unsuitable for fine time resolutions or non-uniform environmental conditions [3][4][5].

In this work a novel PV model is presented, suitable for non-uniform, fast varying conditions on small time scales, for online monitoring and control of the PV plant. The model is validated on a 62.5 kWp installation for a period of 4 months and retains a root-mean-square (RMS) error of less than 5% even for time periods down to 10 seconds.

2 MOTIVATION

Current state of the art PV models are usually parametric, extracting coefficients from the I-V curves and then solving a set of equations to locate the operating point of the system. However, these models do not include the effects of time in their equations (steady state), ignoring system states and intrinsic time constants, mainly thermal, of the system. Therefore, when the model evaluates the energy production during a non-uniform day, and/or on short time scales, errors due to the steady state assumption begin to manifest.

For example in [3], which presents a typical parametric model, when the total RMS error is calculated for 4 days on a 1 minute resolution, even if 2 of the days are clear – sky, its value is larger than 5%. Furthermore, the normalization is performed to the maximum output power whereas in this publication all errors are normalized to the average daily

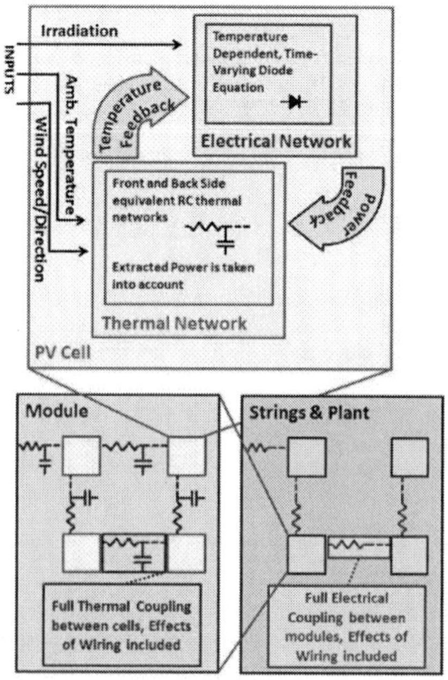

Figure 1: Proposed Model

output. As a result, the error from [3] translated to this work would be larger than 10%.

In [5] hourly RMS error values are presented with ranges from 6.5% to 11%, again with no scalability to shorter time scales.

The purpose of this work is to describe a plant level model that can bridge the gap between hourly/daily and minutes/sub-minute yields calculation and verify its accuracy on a real installation.

The model is presented in section 3 and the installation in section 4. Results are then reported in section 5 along with some conclusions in the final section.

3 MODEL DESCRIPTION

The proposed PV model, illustrated in Fig. 1, includes the optical-thermal-electrical and spatial properties of a PV module, including temporal effects. The model is constructed by coupling two electrical equivalent circuits, one describing the electrical part and one circuit for the thermal part of the PV module. This coupling, along with the inclusion of thermal capacitances ensures that the state memory of the system is also suitable for transient simulations [6].

The parameters of the optical, thermal and electrical parts of the model are determined from experimental data and from Finite Element Method (FEM) models. It has been extensively validated with measurements and FEM simulations and it is the first known model to encapsulate dynamic thermal effects, especially wind, for energy yield calculations.

The treatment of wind input is an ongoing research because it can drastically alter the thermal behavior of a PV module and therefore its power production. An improvement of over 1% RMS is achieved if the wind speed is given weights depending on the incoming direction and Fig. 2 illustrates the current iteration of these coefficients.

Previous iterations of the model have been verified for individual models with great success and even utilized in short term forecasting [7][8]. utilized in short term forecasting [7][8]. Here we evaluate the use of this model when we add the interconnection structure between a large set of modules, and when the measurement data to calibrate it are at a much coarser granularity, namely at the level of the string convertors. In section 5 it will be shown that the good accuracy can be maintained then due to the specific bottom-up physics basis of the model.

4 INSTALLATION

The 62.5 kWp PV plant at the Mads Clausen Institute at the University of Southern Denmark in Sønderborg (54°54′47′′N, 9°46′44′′E) is situated on the roof of a 4 storage building, housing a concert hall in connection with the University.

The installation is showcased in Fig. 3. The plant consist of 250 polycrystalline 250W panels of type Simax New Green Energy (SM660-250) mounted in portrait orientation with a tilt of 15° and an 45° orientation (SouthWest). The 45° orientation is due to the building layout. The panels are connected to 5 Danfoss Solar TLX

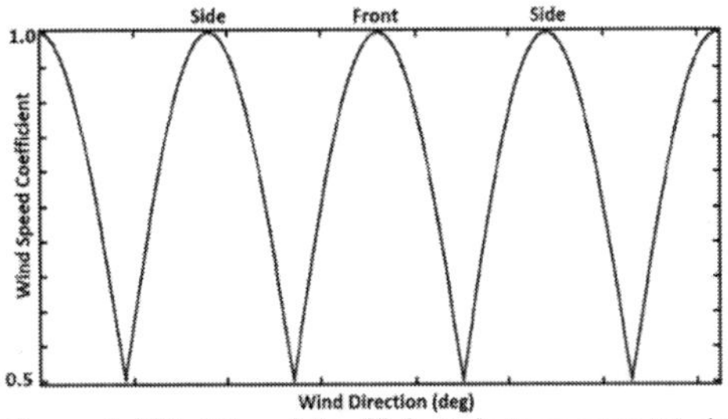

Figure 2: Wind Speed Coefficients in Respect to Module Front Face

Figure 3: Overview of the Rooftop Installation

type inverters with the strings organized in 13 string of 18 panels and 1 string of 16 panels, each having its own Maximum Power Point Tracker (MPPT). The inverters are mounted below/inside.

The panels are fixed on an open metallic structure bolted to the roof of the building, allowing the wind to pass around the panels and are free from any shadows from surrounding trees, masts etc. The plant is equipped with a sensor kit (irradiation, ambient and panel temperature) and an external weather station nearby. Data from a nearby weather station as well as a set of 31 parameters from each inverter are recorded each second since September 2013, giving a total of 184 parameters each second. These parameters include weather data (wind speed, direction, rain, irradiation, temperatures, DC string voltages, DC string currents, AC voltage pro phase, AC current pro phase, power, frequency etc.). The data are read and stored by a field-PC and each night streamed to the university server for storage [9].

The rated annual energy delivered to the grid is 52530 kWh. Yield reduction due to inter-row shading is estimated to 4.8%. The specific annual yield is 839.5 kWh/kWp.

5 RESULTS

For the purposes of this work the months of April to August 2014 are selected. Input data, namely irradiation, ambient temperature and wind speed/direction, is collected and run through the model. Visual inspection of the days reveals that most of them are infested with broken clouds and sudden changes, whereas only few can be characterized clear sky or overcast days.

As mentioned in the introduction, the main metric of accuracy in this work is the normalized RMS error, expressed as

$$RMSE(\%)=\frac{\sqrt{\frac{\sum_{1}^{N}(P_{mi}-P_{si})^2}{N}}}{P_{day}} \quad (1)$$

where N is the number of segments in each day based on the time resolution, Pmi is the DC measured power for this segment, Psi is the DC simulated power and the denominator is the average production of each day. This way the error always has an equal significance across all days.

Additionally, in order to showcase the accuracy of the model, the RMS of the individual errors for each time segment are also presented. The bias is also given as a measure of good fit with no post processing of the parameters of the model.

The average bias and RMS error for the whole installation, for all days and for different time resolutions are provided in Table 1. As expected, since the model is a bottom-up approach, the bias remains almost unchanged as the time segments are increasing in length. An offset of 1.3-1.4% for a total of 250 modules can be explained by many reasons, including wiring, mismatches, MPPTs etc. Generally it can be used for a second step of normalization which brings the offset to 0 and further improves the RMSE.

The RMS values decrease as the time periods are growing, since there is an aggregation of smaller variations and values fluctuate less around the average. Nonetheless, the performance is deemed highly satisfactory as no error surpasses 5%. In comparison to [3] and [5] there is significant improvement. Interestingly enough, even the RMS per section retains an average value below 6% for all time resolutions.

Additionally the modularity of the model allows for reports for separate sub-strings, as shown in Table 2. The bias in this case is lower as the extracted parameters are coming from these modules.

Another important result is the performance depending on the non-uniformity of the day. By using the RMS-to-mean ratio, the days are sorted from clear-sky (lower values) to highly non-uniform (higher values). The 5 most representing days for each case are presented in Fig. 4 and 5. The average RMS error for these two groups differs only by 2.8% for a time resolution of 10 seconds (4.5% to 7.3%), signifying that this dynamic model is suitable for capturing even such level of stochastic conditions.

TABLE 1: AVERAGE NORMALIZED ERRORS FOR THE 4 MONTHS PERIOD FOR WHOLE INSTALLATION

	Error per Daily Average					
	10s	1min	5min	10min	15min	1h
Bias (%)	1.36	1.37	1.38	1.40	1.40	1.39
Rms (%)	4.80	4.25	3.84	3.68	3.60	3.33

TABLE 2: AVERAGE NORMALIZED ERRORS FOR THE 4 MONTHS PERIOD FOR ONE SUB-STRING

	Error per Daily Average					
	10s	1min	5min	10min	15min	1h
Bias (%)	0.20	0.21	0.23	0.25	0.24	0.23
Rms (%)	5.18	4.45	4.06	3.87	3.80	3.47

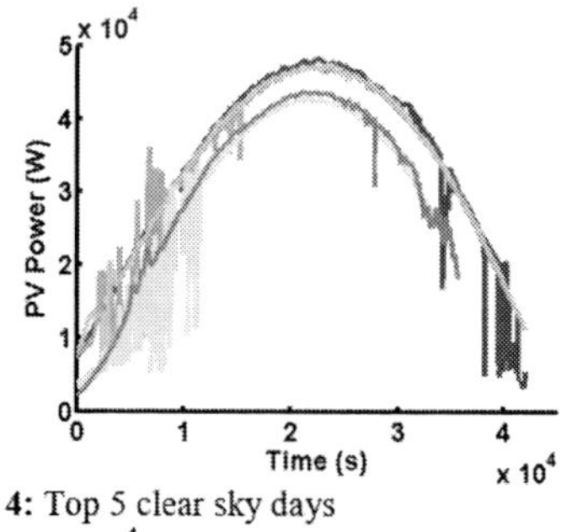

Figure 4: Top 5 clear sky days

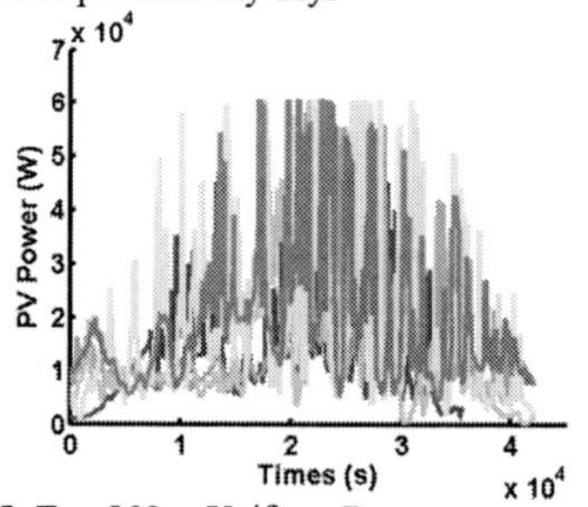

Figure 5: Top 5 Non-Uniform Days

6 CONCLUSIONS

Summing up, the proposed model is suitable for detailed assessment of energy yield for PV installations, scalable in time and space. The accuracy remains satisfactory for the whole range of time resolutions because of its inherent dynamic operation. The analysis presented in this work concerns measurements covering 4 months, with a high variability of weather conditions. The normalized RMS remains under 5% for time resolutions even down to 10 seconds, providing unparalleled possibilities of dynamic control and forecasting.

7 REFERENCES

[1] G. Masson , S. Orlandi, M. Rekinger, "Global Market Outlook for Photovoltaics 2015-2019", Solar Power Europe, available online.

[2] M. Delfanti, M. Merlo, G. Monfredini, V. Olivieri, M. Pozzi, A. Silvestri, "Hosting Dispersed Generation on Italian MV networks: Towards smart grids," 14th International Conference on Harmonics and Quality of Power (ICHQP), Sept. 2010.

[3] A. Chouder, S. Silvestre, N. Sadaoui, L. Rahmani, "Modeling and simulation of a grid connected PV system based on the evaluation of main PV module parameters", Simulation Modelling Practice and Theory, Volume 20, Issue 1, January 2012.

[4] F. Bizzarri et. al., "Model of Photovoltaic Power Plants for Performance Analysis and Production Forecast", IEEE Transactions on Sustainable Energy, vol. 4, no. 2, April 2013.

[5] C. P. Cameron, W. E. Boyson, D. M. Riley, "Comparison of PV System Performance - Model Predictions with Measured PV System Performance", 33rd IEEE PVSC, San Diego, CA 2008.

[6] H.Goverde et al., "Model requirements for accurate short term energy yield predictions during fast-varying weather conditions", 31st European Photovoltaic Solar Energy Conference (PVSEC), Hamburg 2015.

[7] D. Anagnostos, , T. Schmidt, H. Goverde, J. Kalisch, F. Catthoor, D. Soudris, "PV energy yield nowcasting combining sky imaging with simulation models", 31st European Photovoltaic Solar Energy Conference (PVSEC), Hamburg 2015.

[8] D.Anagnostos, H.Goverde, B.Herteleer, F.Catthoor, D.Soudris, J.Driesen, J.Poortmans, Demonstration and validation of an energy yield prediction model suitable for non-steady state non-uniform conditions", 6th World Conference on Photovoltaic Energy Conversion, Kyoto, 2014.

[9] K.M. Paasch, M. Nymand and F. Haase, "Sensor System for Long-term Recording of Photovoltaic (PV) IV-curves", PVSAT-9 conference, 2013, Swansea, UK.

SKELION: THE 3D SIMULATION TOOL FOR PV SYSTEMS

Juan Pons Alemán
Skelion SL
e-mail: j.pons@skelion.com
Bernabé Marí Soucase
Universidad Politécnica de Valencia,dept. applied physics
e-mail: bmari@fis.upv.es
Inmaculada Guaita
Universidad Politécnica de Valencia, dept. economy and social science.
e-mail: iguaita@esp.upv.es

ABSTRACT: This paper presents a 3D simulation software for solar systems, which is being developed under the support of the UPV Valencia university. This software is integrated inside SketchUp, a famous 3D software. This allows the users to draw, easily and with big accuracy, complex solar 3D scenes with all the tools of a professional 3D software. Software can import NREL TMY2 and TMY3 meteorological data. Using this data, software is capable to calculate shading losses and plot surface orientation factor graphs. It is possible to work with dome east-west solar panels or racking systems for ground power stations. Components can be inserted in a 3D mesh or in a plane. Also it is possible to simulate tracking and backtracking of 3D solar components with one or two axis in any plane, and to view the movement of the component instantly. Yield simulations are done using the well-known online services PVGYS and PVWATTS.
Keywords: PV System, Roofing Systems, Shading, System, Tracking.

1 INTRODUCTION

Skelion is a simulation software tool that allows design different types of solar systems, from roofs to large ground power plants, with fixed or tracking racking elements.

Skelion links different software, like PVGYS, PVWATTS and SketchUp. Also it makes some calculations like shading losses and surface orientation factor plots taking into account the shadows and the effects of the weather in the orientation.

The following sections describe the procedures to make a simulation of a solar system.

2 RECOMMENDED AREA TO INSERT SOLAR COMPONENTS

Two methods have been developed to determine the area with negligible shadow losses, in which solar panels can be inserted.

The first method is based in the well-known solstice criteria, used to determine the row spacing in a solar system, with negligible shadows [1]. The same criteria is used to project shadows of surrounding objects over the area of interest. Solar panels are inserted in the part of the area without shadows automatically by the software.

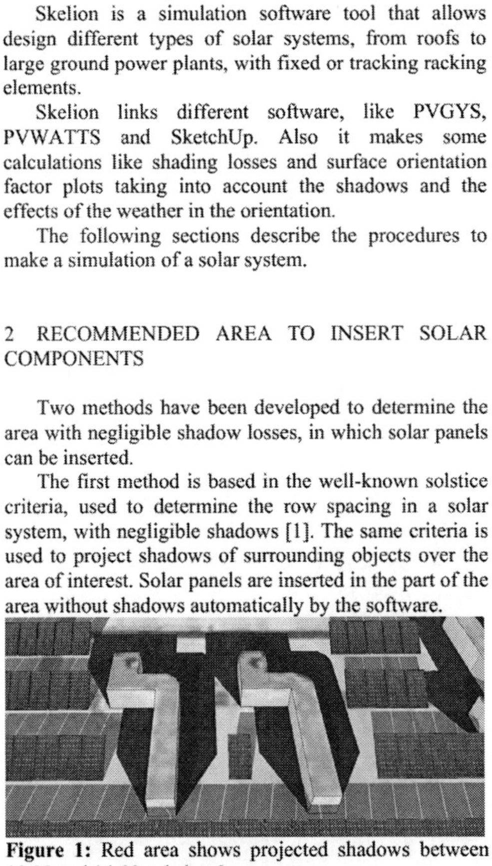

Figure 1: Red area shows projected shadows between 10:00 and 14:00 solstice day.

The second method proposed, insert all the solar panels in the area of interest. After, Skelion calculates the shading losses for each panel, taking into account all the 3d objects like trees or buildings. The panels with non-negligible shading losses are erased. In the next section is explained the method to obtain shading losses.

Figure 2: Panels with no negligible shading losses percentage will be erased.

3 SHADING LOSSES

Shading losses is defined as the total irradiance received on a shaded PV generator $G_{t,sh}$ divided by the irradiance received on unshaded PV generator G_t.

Following Perez et al. [2] the total solar irradiance on a tilted plane is:

$$G_t = B_t + D_t + R_t ;(1)$$

where B_t is the direct beam irradiance, D_t sky diffuse irradiance and R_t is the ground-reflected component.

Diffuse irradiance is decomposed in:

$$D_t = D_{cs} + D_{hz} + D_{iso}$$

where D_{cs} is the circumsolar diffuse component, D_{hz} the horizon brightening diffuse component and D_{is} the isotropic diffuse component.

Typical meteorological data TMY2 [4] or TMY3 [5] of the location are used to know the direct irradiance on horizontal plane B_h and the the diffuse irradiance on the horizontal plane D_h. Following [2] B_t and D_t are obtained from B_h and D_h. Ground-reflected component also is calculated:

$$R_t = (2 - \cos\beta) \cdot \frac{\rho}{2}$$

where ρ is the ground albedo, that is common to consider 0.2 when it is unknown.

To calculate the total irradiance received on a shaded PV generator $G_{t,sh}$ we follow a similar method used by Drift [3]. When panel is shaded, we assume direct irradiance, B_t and circumsolar component, D_{cs} are zero. Also we assume reflected is not affected.

The difference with Drift method is that we make a shadow point test. Software makes a ray-test, in an array of mxn points, in each panel, during all the year, to know if the point is shaded or not. Average of the points returns the $G_{t,sh}$.

For example, for one solar panel, in a determined instant, will have z points not shaded. Then the $G_{t,sh}$ for this panel, in that instant will be:

$$G_{t,sh} = B_t \cdot \frac{z}{mxn} + D_{cs} \cdot \frac{z}{mxn} + D_{hz} + D_{iso} + R_t$$

Software calculations has been compared with software PVSYST 6.43 using the example model called "DEMO grid-connected system at Geneva", with meteonorm 7 global and diffuse meteorological data. PVSYST returns after transposition a irradiance Gt=1403.7 kWh/m², global horizon Ghrz = 1387.9 kWh/m² (1.1%) and $G_{t,sh}$ = 1378.2 kWh/m2 (0.7%). Skelion returns Gt=1409 kWh/m² and $G_{t,sh}$ = 1399.2 kWh/m2 (0.7%). Horizon effect didn't have any appreciable effect in D_{cs} and B_t.

4 SURFACE ORIENTATION FACTOR

Surface orientation factor (SOF) [6] is equal to the ratio of annual incident solar radiation for a particular orientation I_{ann}, to annual incident solar radiation for an optimally oriented surface $I_{ann,max}$ that is found using Nelder-Mead method [7].

$$SOF = \frac{I_{ann}}{I_{ann,\max}}$$

Software calculates SOF using $G_{t,sh}$ calculated as mentioned above, for each point selected by the user, making an average. Thanks to this, it is possible to know what is the best orientation and the SOF plot for a specific shaded area (fig. 3 and fig. 4).

Thanks to the use of TMY, SOF plot will show effects like morning/afternoon cloudiness, with a shift east or west in the plot, or cloudy/clearly winters, with a shift in the tilt.

Software allows the user to multiply any hourly $G_{t,sh}$ with a value between 0 and 1 to calculate SOF plots useful for water heating systems or for pv systems in which electric tariff is cheaper in the afternoons.

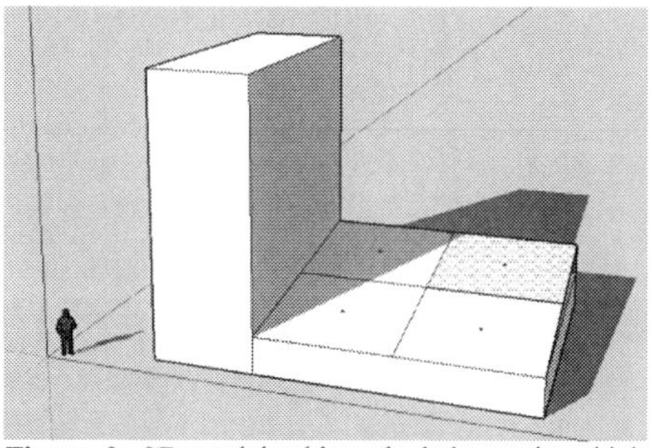

Figure 3: 3D model with a shaded area in which SOF will be calculated using 4 points average.

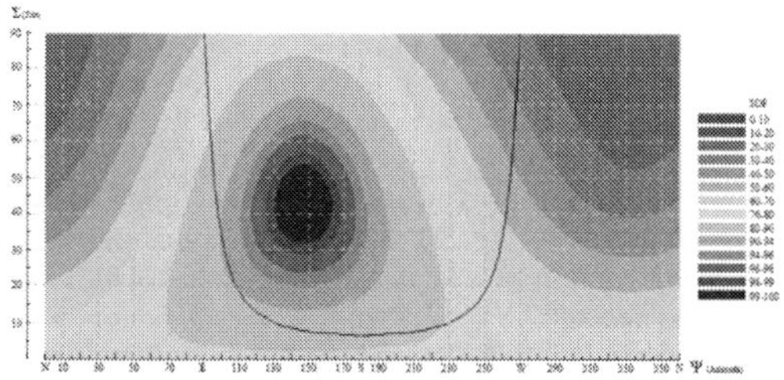

Figure 4: SOF plot of the area. Red line represents all the possible orientations of a plane with rotation axis perpendicular to the blue line of figure 3.

In the figure 4 we can see that the best orientation of the area (located in northern hemisphere) is close to an azimuth of 150°, about 30° due East from south. Area is shaded in the afternoon, that is the reason that the best azimuth is displaced from south to the east. Also we can see the best orientation of the panel if we want to orient the panel following building axis and looking to the west. In that case line of figure 4 represents all the possible orientations of the panel, and we can see that the best orientation with this criterias is around an azimuth of 100° and a tilt of 25°.

5 INSERTING SOLAR COMPONENTS

An algorithm to insert solar components automatically in a 3D mesh (fig. 5) and also in a plane has been developed. It is possible to insert dome components, with two panels in opposite orientation (fig. 6).

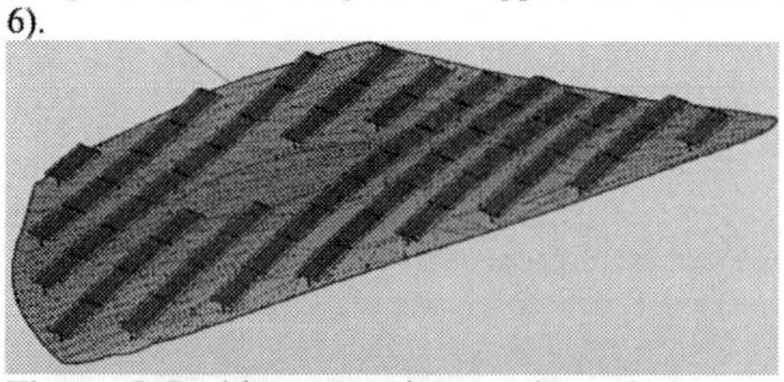

Figure. 5: Racking system in a terrain mesh.

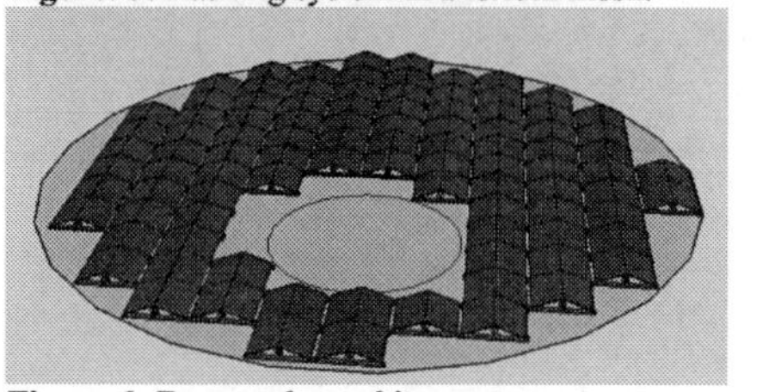

Figure 6: Dome solar racking system

6 TRACKING AND BACKTRACING

A tool to view the movement of the tracking components and the shadows changing the time with a slider has been developed. This tool solves all the possible cases for tracking and backtracking (single horizontal axis, single tilted and deviated axis (fig. 7), single vertical axis, two axes, primary vertical, two axes, primary horizontal) over any plane. Single axis components follow equations of [8] and [9] but for vertical axis, a numeric method has been developed to solve the problem in any plane orientation and not only in a horizontal plane.

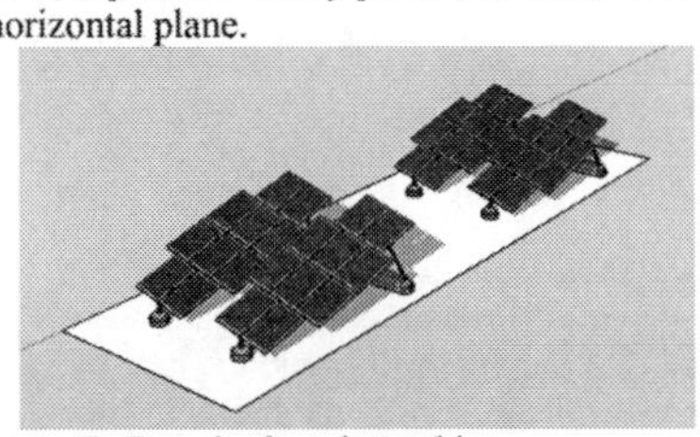

Figure 7: One single axis tracking

7 CONCLUSIONS

This paper has described the characteristics of the Skelion tool, a simulation software tool that allows to plot a 3d solar system and links PVWATTS and PVGYS to obtain the yield. Also is described how Skelion calculates shading losses and surface orientation factor plots. Terrain power plants, and tracking and backtracking features has been described briefly.

8 REFERENCES

[1] J. Appelbaum, J., Bany. Shadow effect of adjacent solar collectors in large scale systems. Solar Energy 1979, Vol. 23, pp. 497-507

[2] Perez R, Deals R, Ineichen P, Stewart R, Menicucci D. A new simplified version of the Perez diffuse irradiance model for tilted surfaces. Sol Energy 1987;39(3):221–31.

[3] M. Drif, P.J. Perez, J. Aguilera, J.D. Aguilar. A new estimation method of irradiance on a partially shaded PV generator in grid-connected photovoltaic systems. Renewable Energy 2008; 33:2048–2056

[4] National Renewable Energy Laboratory. User's Manual for TMY2s typical meteorological years. NREL 1995.

[5] S. Wilcox and W. Marion. User's Manual for TMY3 Data Sets. Technical Report NREL/TP-581-43156 2008.

[6] Craig B. Christensen, Greg M. Barker. Effects of tilt and azimuth on annual incident solar radiation for united states locations. Proceedings of Solar Forum 2001: Solar Energy: The Power to Choose. April 21-25, 2001

[7] John H. Mathews and Kurtis K. Fink. Numerical Methods Using Matlab,2004, 8-430

[8] E. Lorenzo, L. Narvarte and J. Muñoz. Tracking and back-tracking Prog. Photovolt: Res. Appl. 2011; 19:747-753

[9] E. Lorenzo, L. Narvarte. Tracking and Ground Cover Ratio. Prog. Photovolt: Res. Appl. 2008; 16:703–714

AUTOMATIC COMPUTATION OF SHADING MASK ON A PV FILED BASED ON PRODUCTION DATA

J. Dupas[1], M. Joos[2], S. Fraisse[2], B. Gaiddon[1]
[1]Hespul, 14 place Jules Ferry, 69006 Lyon France
[2]Epices Energie, 14 place Jules Ferry, 69006 Lyon France

ABSTRACT: In this paper, we propose an algorithm to compute the shading mask of any PV field based on its production data, GPS coordinates, and global solar irradiation for the site. The model is based on hourly production data, but could be extended to sub-hourly data. The detection of the shading mask through the production curve itself allows the operator to anticipate the production curve due to shading – and eliminate shading impacts in failure detection routines that can then concentrate on component (inverter, string, …) failure, so that operators can responde appropriately.
Keywords: Monitoring, Performance, Shading

1 INTRODUCTION

Today solar production of energy is increasing and operators need to predict and know the real power produced for every PV plant in order to respond to the supply and demand electricity markets. Ground mounted solar plants don't have any shading masks, as they are built in a controled environment. Building mounted systems, in a industrial, commercial or residential environment may have existing shading masks or develop shading masks as local constructions and vegetation change.

Knowing the shading mask gives a precise estimation of theoretical production to improve operational diagnostic and maintenance, and knowledge of PV field behaviour within its environment allows for better forecasting at the PV system scale., Automatic mask detection can gain operators time due to fewer entries needed to specify the PV system characteristics required for adequate monitoring.

The purpose of this work was to design an algorithm to compute the shading mask of any PV field based on its production data, GPS coordinates and horizontal global solar radiation.

2 DATA SET

The algorithm needs at least 6 months of hourly real production data and hourly theoretical production data computed without taking into account any shading mask; this dataset covers the spectrum of possible incidence angles of the sun on the PV field and generates a coherent mask. The computation of the theoretical production data using the global horizontal solar radiation and the GPS coordinates uses a well know algorithm developed by Arias[1] and John E. Hay[2].

The monitoring software Epices[3] is used to obtain a complete set of hourly points, data are then trimmed to include only points with a majority of direct solar radiation. Cloudy periods do not project clear shading masks and hence are not adapted to this task.

The clear sky index, which is equal to 1 in a clear sky situation and close to 0 when in a cloudy sky situation, is used to detect data points that require elimination because of insufficient direct solar radiation.
The clear sky index threshold is adapted to have a good compromise between the number and the qualitiy of data points. For example, in winter the threshold is set lower than in summer in order to retain a sufficient number of points.

The final shading mask will only be as precise as the least precise data entry, e.g. with hourly production data in entry, only a shading mask at a granularity of 15° can be obtained.

3 MODEL

The model is physical, as the algorithm searches for a relationship between sun height and production loss, as compared to shade-less production data.

3.1 Comparison as a function of the incident angle

Real production data is compared with theoretical production data as a function of the incidence angle between the sun and the normal to the PV panel.

Computation of the incidence angle :

$$\cos(\theta) = \sin(\gamma)\cos(\beta) + \cos(\gamma)\sin(\beta)\cos(\alpha_s - \alpha)$$

With :
θ : incidence angle
γ : sun height
β : tilted surface
α : surface azimut
α_s : sun azimut

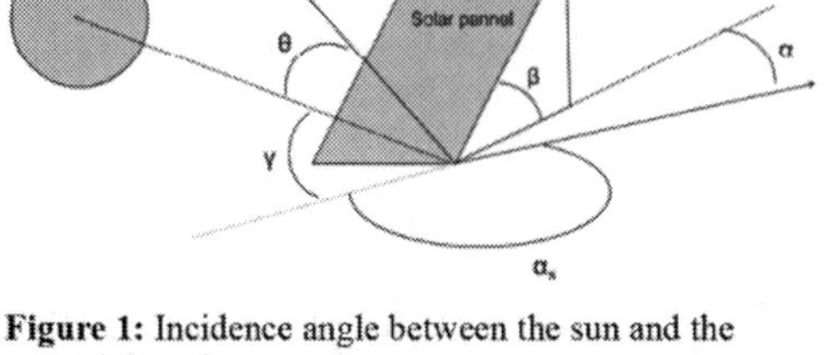

Figure 1: Incidence angle between the sun and the normal the solar pannel

Morning hours and afternoon hours are treated

separately as we get the same incidence angle for positive and negative azimut, 0° being midday.

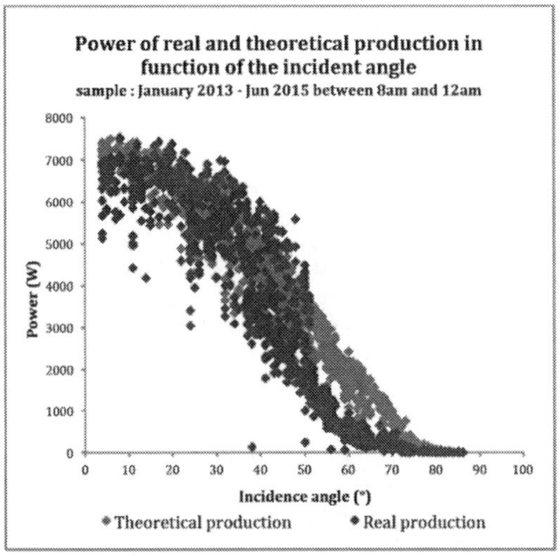

Figure 2: Power of real and theoretical production in function of the incident angle

The difference between theoretical production and real production is standardised by the maximal power of the day, then the curve is smoothed to eliminate outliers.

Every point under the mean of the curve is considered as a potential shadowed point.

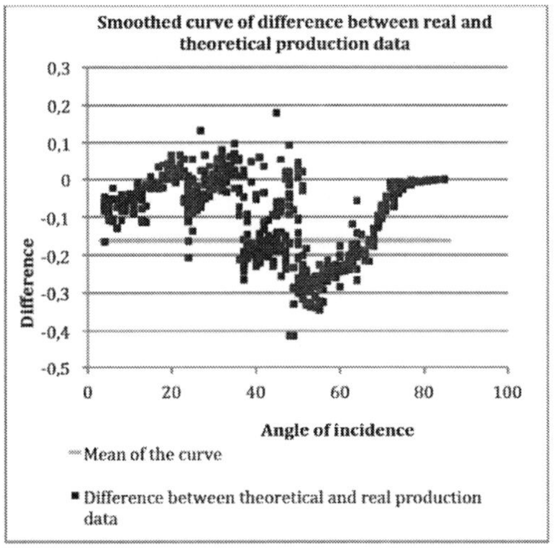

Figure 3: Smoothed curve of difference between real and theoretical production data

3.2 Analyse of the potential shadowed point

The data set of all potential shadowed points is analysed : first each points real production is compared to the theoretical production to see if the drop of production is coherent with a shading effect and then the coherence between shadowed points is tested.

If a solar panel is entirely shadowed, the string will be bypassed[4], therefor the drop in production should be at least equal to one string bypassed and it should appear more than once in the same area (same azimuth rounded to the nearest degree).

Because potentially shadowed points can not have any non-shadowed points at the same or lesser sun elevation in this model, floating shading objects, such as publicity panels for exemple, are not considered.
The assumption was made that floating shading objects would not create masks that would impact the PV production over only a short period of time, but would, if they were to have a real impact on the production, impact the production for the entire degree of azimuth, whatever the sun elevation.

3.3 Drawing the shading mask

With hourly points the granularity of the shading mask is 15°. To draw the mask, the highest shadowed points for every sequence of 15° azimuth from -120 to 120 was considered.

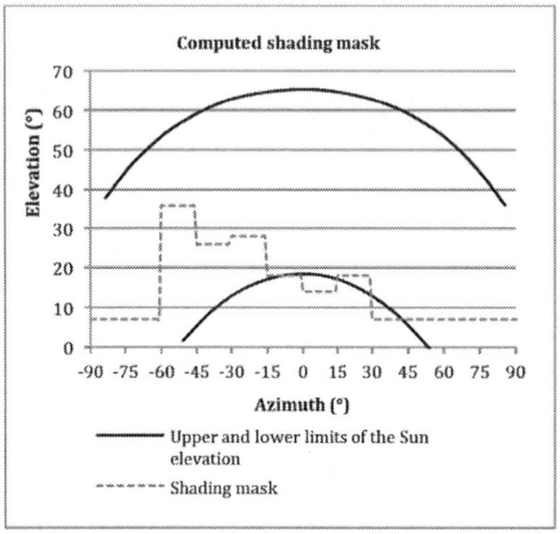

Figure 4: Computed shading mask

3 RESULTS AND COMPARISON

Several shaded PV systems have been studied in order to validate the algorithm. The one used to illustrate the methodology here is located in the west of France in Cenon-sur-Vienne. To appreciate the efficiency of the computed shading mask, the theoretical data are computed once more taking into account the shading mask and compared with the corresponding real production data.

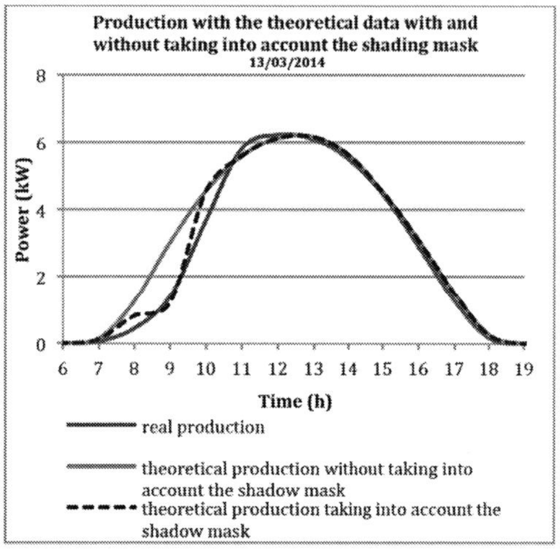

Figure 5: Production with the theoretical data with and without taking into account the shading mask

Analysis of error shows that taking into account the shading mask with the developed algorithm improves theoretical production data computation (table 1)

To compare the two data sets common indicators have been used: Normalized Root Mean Square Error (NRMSE), Normalized Mean Average Error (NMAE) and Normalized Mean Bias Error (NMBE).

Theses indicators have been normalized by the average production over a period of one year and only real production data points greater than zero were considered.

The error used as reference is :

error = (real production – theoretical production)

Table 1 : Table of errors with and without the computed shading mask

	NMAE(%)	NRMSE(%)	NMBE(%)
Without any shading mask	22.91	0.34	-10.79
Taking into account the computed shading mask	21.34	0.33	-8,36

This table illustrates improvement of theoretical production data computation as NMAE and NRMSE values decrease when taking into account the shading mask. As it is less negative, NMBE value shows that there is less over-estimation of the production when taking into account the shading mask.

Part of the remaining errors is due to the under-estimation of the production by the algorithm when in a shaded area. Indeed, the algorithm developed to compute the shading mask does not compute precisely the attenuation of irradiation as a result of shading. Instead, it assumes by default that the entire PV field is shaded when only part of it could be affected.

4 CONCLUSION

The algorithm generates a shading mask based on physical analysis of the production data, allowing fault detection routines to avoide confusing component defaults with shading effects. Results show that taking into account the calculated shading mask improves the monitoring of installations as the theoretical data are closer to the real production data.

Instead of a global mask for the entire PV field, the algorithm could be use at the string level to generate a mask for each string. It would increase the accuraty of theoretical data as the impact of shading objects on each string would be considered. As data at the string level are not available yet for a sufficient number of installation we could not mesure the improvement in this particular experiement.

8 REFERENCES

[1] J.A. Ruiz-Arias, H. Alsamamra, J. Tovar-Pescador, and D. Pozo-Vázquez. Proposal of a regressive model for the hourly diffuse solar radiation under all sky conditions.Energy Conversion and Management, 51(5) :881 – 893, 2010
[2] John E. Hay. Calculating solar radiation for inclined surfaces : Practical approaches.Renewable Energy, 3(4–5) :373 – 380, 1993. Solar radiation, environment and climate change.
[3] http://www.epices-energie.fr/ : software which compares theoretical data to production data to detect fault. It computes theoretical production based on horizontal global solar radiation, the tilt angle, the orientation and the GPS coordinates of the PV field.
[4] Ralf Haselhuhn. Le guide du photovoltaïque. DGS, Deutsche Gesellschaft für Sonnenenergie,2012.

CALCULATION- AND VISUALIZATION-TOOL (CVT) FOR PARTIAL SHADING OF PHOTOVOLTAIC SYSTEMS

Franziska Kuonen, Daniel Gfeller, Thomas Schott, Eva Schüpbach, Horst Heck, Urs Muntwyler
Bern University of Applied Sciences BFH, Engineering and Information Technology, Photovoltaic Laboratory (PV LAB)
Jlcoweg 1, CH-3400 Burgdorf, Switzerland
Tel. +4134 426 68 37, Fax +4134 426 68 13, urs.muntwyler@bfh.ch

ABSTRACT: Partial shading of photovoltaic systems can lead to disproportionate power losses. To visualize and precisely calculate the characteristics of a partial shaded PV system, a new simulation tool has been developed. Compared to existing simulation tools that address a PV module as a whole, our new Calculation- and Visualisation Tool (CVT) for partial shading of photovoltaic systems is able to provide quantifications of power losses at individual photovoltaic cell level. Using the CVT, comparisons of calculations and measurements of the whole PV generator are possible. This is especially an advantage to analyse partially shaded situations of a PV generator.

Keywords: Photovoltaics, Shading, Power Losses, Simulation Tool

1 INTRODUCTION

This study addresses the partial shading effects on photovoltaic systems and was a bachelor-thesis in spring 2015 [2]. Partial shading of photovoltaic systems can lead to disproportionate power losses. Due to the nonlinear behaviour of the photovoltaic cells, the quantification of such power losses is challenging. So far, the determination of power losses is carried out using oversimplified algorithms.

Most simulation-tools to calculate power losses calculate the losses at photovoltaic module level rather than for individual photovoltaic cells. This is a major disadvantage because of the internal connections between the photovoltaic cells and the bypass diodes (see Figures 1a-b).

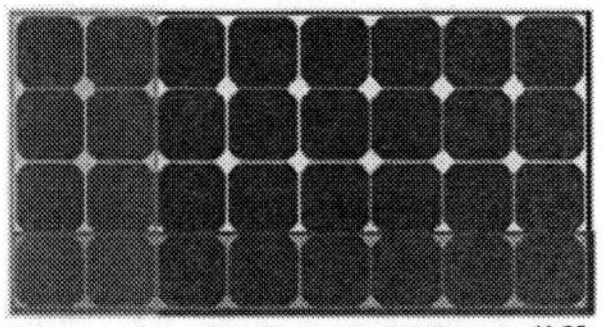

Figure 1a: SunPower SPR-95 differently shaded.

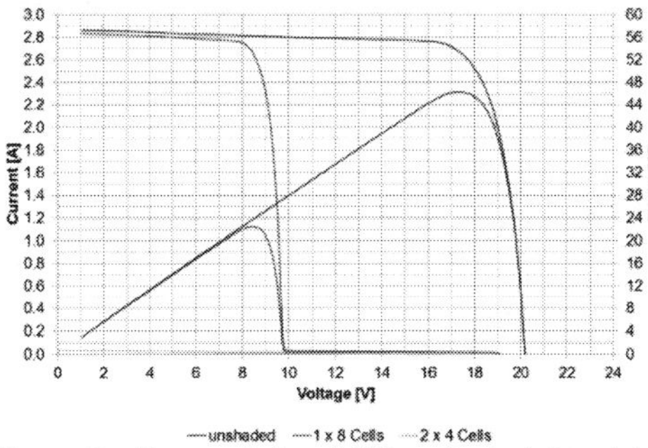

Figure 1b: Characteristics of differently shaded modules.

In order to remedy this gap, we developed a new Calculation- and Visualisation Tool (CVT) for partial shading of photovoltaic systems. The CVT considers the internal structure of the photovoltaic module and calculates the characteristic of each photovoltaic cell. Our tool consists of a drawing area, a setting area, a diagram and a table (see section 2) that are described below.

2 DESCRIPTION OF THE TOOL

2.1 Structure

The main task of the CVT is to create and calculate the photovoltaic system. The different tasks of the tool are carried out in different Tabs.

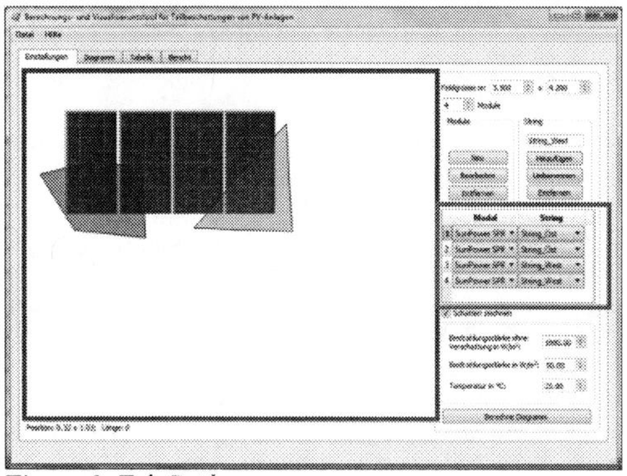

Figure 2: Tab Settings.

The first tab (see Figure 2) includes all important settings to create a photovoltaic system. Here, the user is able to vary the count of modules for the photovoltaic system.

With the aid of another window (see Figure 3), it is possible to create different module types. In this window, the count of photovoltaic cells and bypass diodes within one module are set. Different calculation models for the photovoltaic cell and the bypass diode can also be chosen.

Once a module is created it is possible to save it or to load it for another calculation later. The user is thus enabled to create an own list of photovoltaic modules.

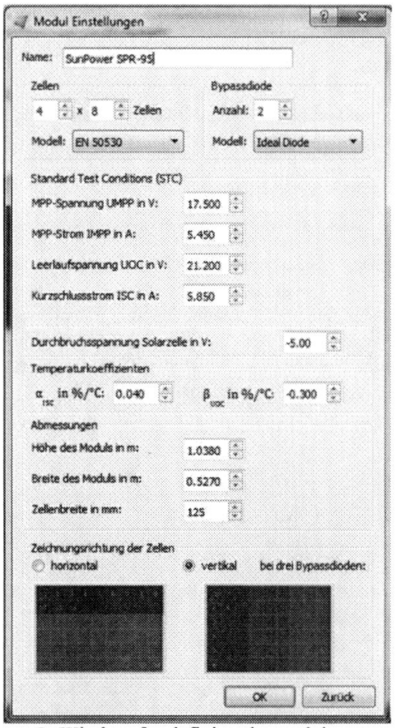

Figure 3: Window for defining the module properties.

Figure 2 shows a table (see Figure 2 green rectangle) which defines the selection of the type of module and which module belongs to which string.

Because of the calculation method, it is only possible to create crystalline modules at this stage. Also, the temperature can be defined for the whole photovoltaic system. The irradiation of each photovoltaic cell will be calculated separately.

The red rectangles on the left side of Figure 2 display a free area in which each module can be placed. The shadow can be painted with the mouse. Before the characteristics are calculated, and as a last step, the movement of the shadow in function of time is set.

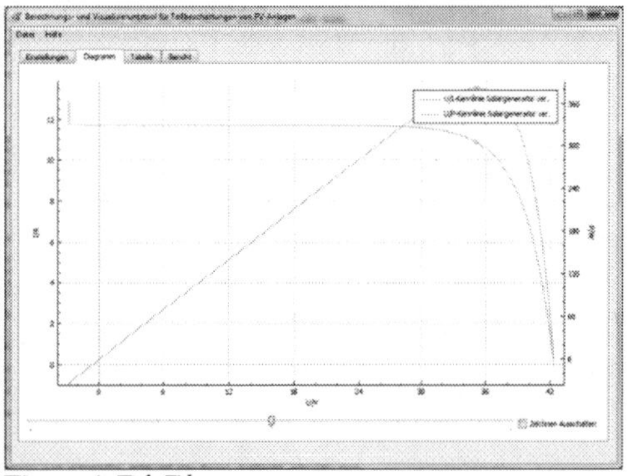
Figure 4: Tab Diagram.

In the second tab (see Figure 4), the characteristics can be displayed (e.g., of the whole photovoltaic system, of each module-string, of each module, of each cell-string and of each photovoltaic cell). The user can select from four different types of characteristics, being: V/I-characteristic (shaded or unshaded) and a V/P-characteristic (shaded or unshaded). Power losses can be visualised better when a shaded characteristic is compared with an unshaded characteristic. Inserted underneath the diagram in Figure 4 is a slider which serves to visualise the different time steps.

The calculated data can be exported into csv-files.

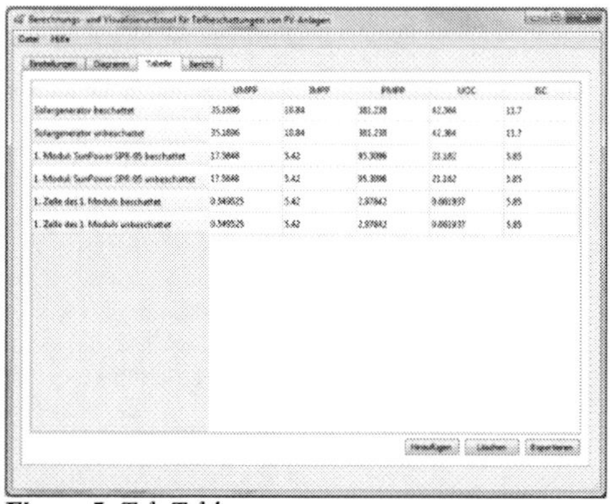
Figure 5: Tab Table.

The third tab (Figure 5) is a table containing the most important values, being: the maximum power point (MPP), the MPP-voltage, the MPP-current, the open circuit voltage and the short circuit current. As in the second tab diagram (Figure 4), it is possible to include the data from all characteristics, from the photovoltaic system to the cell level.

2.2 Design

Figure 6 shows the UML-diagram (Unified Modeling Language) of the CVT for partial shading of photovoltaic systems. The diagram reads from top to bottom and is described below:

A photovoltaic system consists of multiple module-strings. These module-strings are connected in parallel. A module-string is composed of multiple serial connected modules. A module is made up of multiple serial connected cell strings. One cell-string consists of multiple serial connected photovoltaic cells and one bypass diode. To choose between different calculation methods for the bypass diode and the photovoltaic cell, the bypass diode and the cell are implemented as abstract classes. An abstract class can be used like a model, and it is possible to implement multiple classes for both bypass diode and cell.

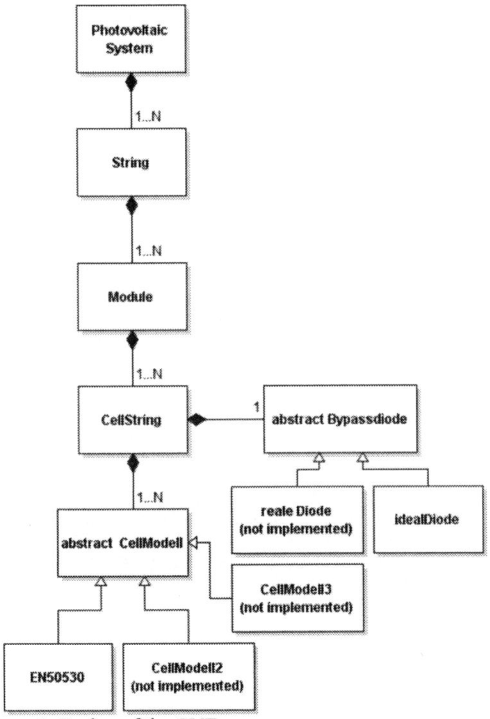

Figure 6: Design of the CVT

2.3 Calculations

There are multiple ways to calculate a characteristic of a photovoltaic cell. The best known calculation methods are the simplified model, the one-diode model and the two-diode model. In our tool, another model is implemented, namely a model from the standard EN50530 [1], with the formula:

$$I = I_{SC} - I_0 \cdot \left(e^{\frac{U}{U_{OC} \cdot C_{AQ}}} - 1 \right)$$

in which

$$C_{AQ} = \frac{FF_U - 1}{\ln(1 - FF_I)}$$

I_{SC}:	short circuit current
I_0:	irradiation depending current
U_{OC}:	open circuit voltage
FF:	fill factor

After calculating the photovoltaic cell, the voltages of the photovoltaic cells are summed up to calculate the characteristic of the cell string. As the module strings are connected in parallel, the currents are added.

The radiation is calculated for each photovoltaic cell individually.

3 COMPARISION CVT AND MEASUREMENT

Figure 7 shows the difference of the calculated values from the CVT and the measured values from the flasher of the Photovoltaic Laboratory of the BFH. The module SPR-95 from SunPower was shaded in two different ways (see Figure 1a).

As Figure 7 shows the match of the characteristics is very height.

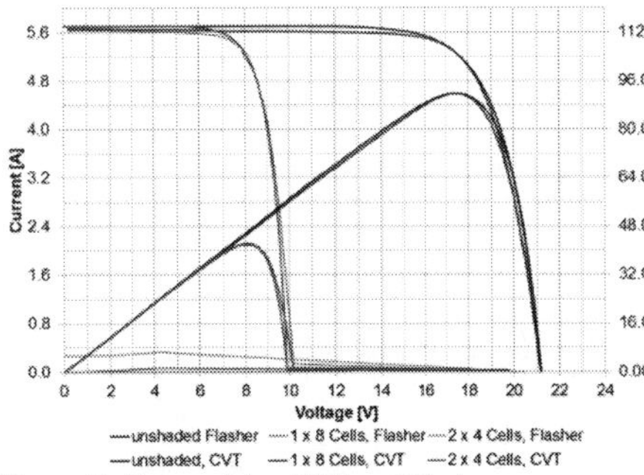

Figure 7: Comparision CVT and Flasher

4 SUMMARY AND FURTHER AVENUES

A Calculation- and Visualization-Tool (CVT) for partial shading of photovoltaic systems was developed. The CVT is able to simulate a photovoltaic system with modules arranged in a plane and consisting of crystalline cells. Different shadows can be defined and each module can be placed on a location within the drawing area in the tool. The radiation of each shadow can be set individually. The shadow can be moved to a specified direction for a predetermined number of steps. When each step is calculated it can be visualised by moving a slider. A table displays the most important values.

The tool allows different calculation methods for the characteristic of the photovoltaic cell and the bypass diode. From the results provided by the calculation, it is possible to select and visualise the characteristics for each individual cell, cell-string, module, module string and the complete photovoltaic system. The simulation results may serve e.g., as a reference for measurements of inverters that are under development.

The innovation of the CVT as compared to currently available tools is the quantification of the characteristics at individual PV cell level rather than at PV module level.

Among the planned extensions of the CVT are the implementation of various calculation models for the calculation of the solar cell and the bypass diode.

5 RESOURCES

[1] Standart EN50530

[2] F. Kuonen, bachelor-thesis: Berechnungs- und Visualisierungstool für Teilbeschattungen von Photovoltaikanlagen

THREE YEAR FIELD PERFORMANCE OF ANTI-SOILING COATINGS AT MULTIPLE LOCATIONS

Brenor Brophy[1], Karen Schexnaydre[2]
1. Enki Technology Inc., 2192 Bering Drive, San Jose, CA 95131, USA
2. SunEdison Inc., 600 Clipper Drive, Belmont, CA 94002, USA

ABSTRACT: As solar panels are being deployed at GW scales across diverse climatic regions, and as the solar industry must increasingly compete on a leveled cost of energy (LCOE) basis, the effects and cost of soiling on panel output is becoming more acute. Numerous studies have investigated soiling losses in different locations and under different climatic conditions. Studies of various anti-soiling coatings have shown that coatings can exhibit an anti-soiling effect in many locations and applications. However, the data reported typically covers only a single season or just part of a year. Additionally, it is likely that local effects play a significant role in soiling factors, but the data may lack sufficient geographic coverage to resolve such local effects. Here, we present an analysis of 3 years of field data collected at 10 locations throughout the Southwestern United States comparing the performance of a durable hydrophobic anti-reflective coating to uncoated PV glass. We report on soiling rates over the study period. We conclude that hydrophobic coatings have a significant anti-soiling effect at most arid sites in relation to the performance of the uncoated glass controls.
Keywords: Environmental Effect, Optical Properties, PV Module

1 INTRODUCTION

Soiling of solar panels can reduce their output up to tens of percent a year, depending on the season, location and cleaning schedules [1-3]. Regions with long dry periods such as Mediterranean climates can see losses of 15% over the summer, with further losses in the case of severe soiling events such as sandstorms. Rain is a natural and free way to clean most of the soiling deposits from a glass surface; however, some dirt can strongly adhere to the glass over time.

Various studies have followed the soiling rates on solar panels in specific locations for a number of years [4], and have furthermore tested the use of anti-soiling coatings on the glass [4-7]. These coatings can be applied to the outer layer of the glass, and can be hydrophilic (meaning, a water contact angle lower than bare glass, which is typically 25-40°) or hydrophobic (water contact greater than about 70°). Studies have shown anti-soiling effects and self-cleaning for both hydrophobic and hydrophilic coatings [6, 9] under different soiling conditions or at different times of the year. When outdoor studies are conducted [7] it appears that some locations may favor one coating or the other and that coating surface functionality is an important performance consideration. Many early tests and much of the PV industry's experience with glass coating anti-soiling performance and durability derive from climatically benign European regions.

In the case of hydrophobic coatings, surface functionalization may be achieved by the presence of organic molecules such as fluorocarbons and/or hydrocarbons. There is extensive industry experience regarding durability with these molecules in the form of polymers, but, there is less experience incorporating them into photovoltaic applications, which can present a very challenging environment due to high temperatures, high UV flux and weather exposure. Questions have been raised regarding the persistence and durability of the anti-soiling effect for functionalized coatings.

Here, we present three years of anti-soiling performance data from an ongoing study of a robust dual-function anti-reflective and anti-soiling coating that is dense, homogenous and intrinsically hydrophobic (Enki Technology CleanARC® coating) sited at 10 large-scale commercial solar installations throughout the USA installed and managed by SunEdison. These coatings, reported on previously [7, 8] after two years of outdoor exposure, continue to perform at the end of a third year.

2 EXPERIMENTAL

The field study originally consisted of 14 sites spread over 6 states and 4 different Köppen climate classifications. In 2015, data collection at 4 sites was discontinued, leaving 10 that are reported on in the present paper. Most sites began data collection in June or July of 2013, with one site starting about 1 year later in May 2014. Table I below summarizes the locations, which were chosen based upon availability and access and the start date. Table II summarizes the four climate classifications. Data collected up to May 31st, 2016 is used in this paper.

Table I: Field study sites

Site	Köppen	Tracker	Start
AZ-2	BWh	Single Axis	July 2013
CA-1	BWh	Single Axis	July 2013
AZ-1	BSk	Single Axis	June 2013
CA-2	BSk	Fixed	July 2013
CA-3	BSk	Single Axis	July 2013
CA-4	BSk	Single Axis	July 2013
CA-5	BSk	Fixed	June 2013
NM-3	BSk	Single Axis	June 2013
NV-1	BWk	Single Axis	May 2014
TX-2	Cfa	Single Axis	June 2013

Table II: Field study site climate classifications

Köppen	Description
BWh	Arid – Desert – Hot arid
BWk	Arid – Desert – Cold arid
BSk	Arid – Steppe – Cold arid
Cfa	Warm temperate – Fully humid – Hot summer

Each test site uses two Atersa calibrated irradiance modules, each consisting of two reference cells housed in a small module that provides a 0-65mV output corresponding to temperature compensated insolation of 0-1000W/m^2 for each cell. The modules are constructed in a similar manner to standard, glass covered crystalline modules.

Figure 1: Atersa irradiance modules installed at CA-4 test site. One module is coated with CleanARC coating and one is uncoated. Half (area over one cell) of each sensor is cleaned periodically. Irradiance is logged every minute.

One module is coated with Enki CleanARC hydrophobic anti-reflective coating and the other is uncoated. The glass above one reference cell of each module is periodically cleaned, allowing us to measure cell output for cleaned and naturally soiled conditions for each coated and uncoated glass.

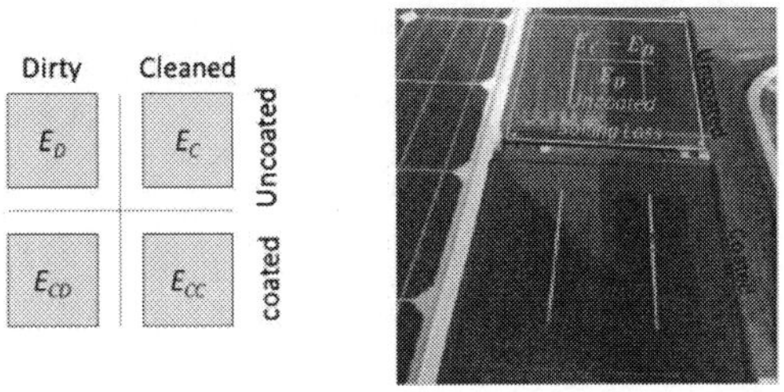

Figure 2: 2x2 matrix of coated/uncoated cleaned/dirty allows us to isolate the anti-soiling and anti-reflective properties of the coatings.

$$U_{SL} = (E_C - E_D)/E_D \quad (1)$$
$$C_{SL} = (E_{CC} - E_{CD})/E_{CD} \quad (2)$$

$$AS = U_{SL} - C_{SL} \quad (3)$$

The soiling loss measured by each sensor (uncoated U_{SL} and coated C_{SL}) is the difference of the energy yields of the dirty cell and the clean cell normalized to the energy yield of its respective dirty cell. The anti-soiling gain (*AS*) is the difference in the coated and uncoated soiling loss.

The Atersa module irradiance output for each reference cell is logged every minute and 15-minute averages of this data are used in this analysis. The irradiance data is summed over the day for each reference cell to calculate daily energy production (kWh). Precipitation data is collected for each site using the daily summary dataset from the NOAA CDO database available at http://www.ncdc.noaa.gov/cdo-web.

A practical limitation of this test protocol is that it relies on periodic manual cleaning of the cells. One issue with this is that the "clean" cell accumulates soiling between manual cleanings which leads to an underestimate of the true soiling losses at the test site. Another issue is inconsistency with manual cleaning. The clean and dirty references can reverse if the operator sporadically cleans the wrong side, which can cause a "negative" soiling loss (i.e. the 'dirty' cell is producing more energy than the 'clean' cell). In this case, it is reasonable to assume that both modules were cleaned in the same manner so the swap occurs for both at the same time. Therefore, by looking at the difference between the two we still can calculate the AS gain irrespective of which cells are designated clean and dirty.

Another issue we identified as time progressed was a small, slowly varying energy output offset between the two reference cells in the Atersa module. Although this offset error was within the ±2% accuracy limits of the modules, this error manifests as an observable difference in reported irradiance between the two cells when the glass above both is equally clean and the irradiance measure from each cell should be identical. Given the significance that any measurement error could have on our reported results, we used an offset correction algorithm to minimize this error in our final analysis.

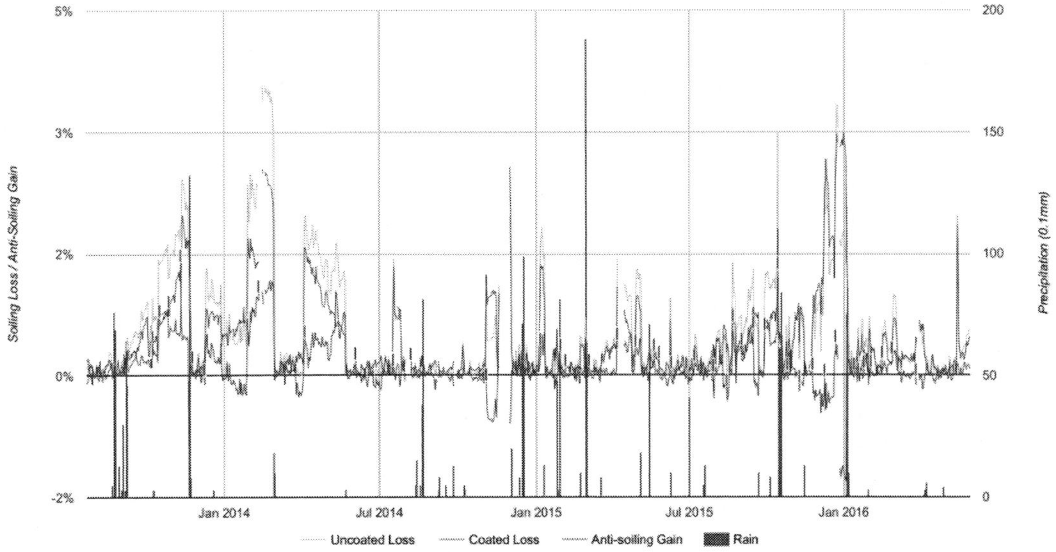

Figure 3: Processed sensor data from site CA-1 showing Uncoated Loss U_{SL}, the Coated Loss C_{SL} and the Anti-soiling Gain AS. Green bars display local precipitation. Red ticks indicate days when the designated clean reference cell was cleaned.

3 ANALYSIS METHODS

The dataset reported below incorporates data up to May 31st, 2016 for all sites. 15-minute average irradiance data was summed to daily energy yield for each reference cell. Then, following the calculations outlined previously, the uncoated and coated soiling losses and the anti-soiling gain were calculated for each day. This initial result was plotted against precipitation measured at the closest available NOAA weather station. Figure 3 above shows an example plot for site CA-1. Using these plots for the 10 test sites, 180 "reference dates" were manually identified. A reference date is a day on which both the NOAA weather station recorded precipitation and there was an observed response in the Atersa irradiance data consistent with a cleaning event caused by rain. We assume that the precipitation cleans the modules so that the glass above both reference cells is equally clean and so any measured difference between the cells on this date is due to the small, slowly varying energy output offsets mentioned previously. A correction coefficient is then calculated based on the irradiance measurements for the succeeding 7 days after the reference date (during which it is assumed minimal soiling occurs). For example, eq. (4) shows how the uncoated correction coefficient is calculated and eq. (5) shows how eq. (1) is modified to use this coefficient.

$$C_U = \sum_r^{r+7} E_D \,/\, \sum_r^{r+7} E_C \qquad (4)$$

$$U_{SL} = (C_U E_C - E_D)/E_D \qquad (5)$$

Note that a value of 1.000 means no correction is needed. The mean correction coefficient for two Atersa modules per site and all 180 reference dates was 0.9995, indicating no significant systematic positive or negative bias in the corrections. The standard deviation was 0.0082 which is small in relation to the ±2% accuracy limits of each module.

To correct for the small, slowly varying energy output offsets, the correction coefficients were linearly interpolated between reference dates and applied to the daily soiling loss results. Approximately 3.7% of days in the full dataset had soiling losses > 4 standard deviations from the mean and so were filtered out.

As the "clean" reference cells are only cleaned periodically, the calculated anti-soiling gain declines between cleaning dates as the supposedly "clean" reference cell itself becomes soiled. To determine the soiling present and the extent of any improvement in soiling performance attributable to the coating with the highest degree of accuracy and certainty, we limit the measurement readings of anti-soiling gain to just the dates directly after the clean reference cell was cleaned. For each site we then plot the uncoated soiling loss (U_{SL}) against the coating's anti-soiling gain (*AS*). We select the next day because we do not capture the time of day that a cleaning is performed and so the most reliable measurement option is to use the next full day after and assume that very little soiling has occurred overnight. Figure 4 shows the plot for site NV-1.

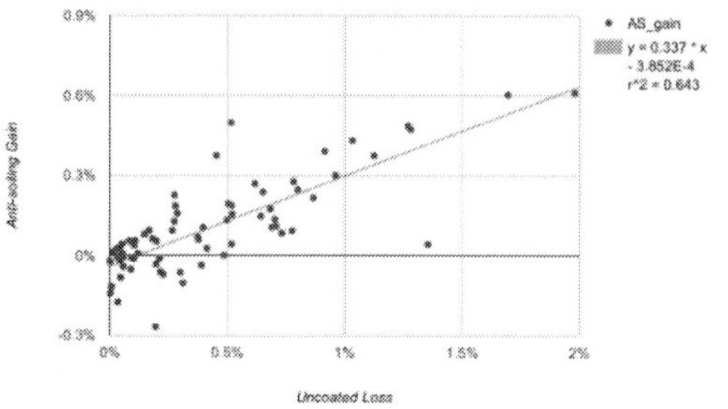

Figure 4: Scatter (XY) plot of anti-soiling gain (*AS*) vs uncoated loss (U_{SL}) with a linear fit model.

The example results from the NV-1 site show correlation between soiling losses of uncoated glass and recovery of those soiling losses on the coated glass. The slope of the linear fit model represents the average anti-soiling recovery rate due to the coating. In other words, for this example site, on average 33.7% of the measured soiling loss was recovered and attributable to the coating.

The experiment cannot measure total soiling losses at a site because the losses between cleanings are unknown; as the "clean" cell becomes soiled, actual soiling losses are masked and under reported. However, eq. (6) can be used to calculate a lower bound of total soiling loss. Using this method, we report the lower bound of soiling loss for each site over the full three years of data collection.

$$SL = \sum(E_C - E_D)/\sum E_D \qquad (6)$$

In the final step of the analysis we plotted the soiling recovery rate versus soiling loss for all sites.

4 RESULTS AND DISCUSSION

Figure 5 is a summary chart of all test sites that plots the soiling recovery rate against the lower bound of soiling loss at the site. Generally, soiling losses are low, at about 0.5% over all sites. Two California sites, CA-2 and CA-5 that are located in California's Central Valley near agricultural areas show higher soiling losses of 2.5% and 4% respectively. Eight of the ten sites show positive soiling recovery rates and two, AZ-2 and CA-3 showing negative recover rates (that is, the coated glass performed worse than uncoated glass). The average anti-soiling recovery rate for all sites is 15%.

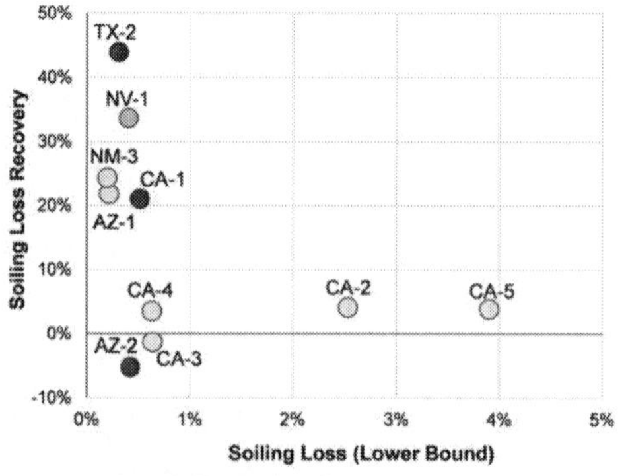

Figure 5: Plot of all test site of recovery rates vs the lower bound of soiling loss.

5 CONCLUSION

We have completed three years of soiling performance data collection at 10 sites in the USA. After careful analysis and error correction, we observe that in general these sites do not have severe soiling problems and that at most sites the hydrophobic anti-soiling coating has a positive effect, recovering on average about 15% of the losses with respect to uncoated glass.

6 REFERENCES

[1] A. Kimber, L. Mitchell, S. Nogradi, and H. Wenger, "The Effect of Soiling on Large Grid-Connected Photovoltaic Systems in California and the Southwest Region of the United States," in the 2006 IEEE 4th WPEC.

[2] F. A. Mijia and J. Kleissl, "Soiling Losses for Solar Photovoltaic Systems in California" Solar Energy, 95, pp. 357-363, 2013.

[3] M. Gostein, B. Littmann, J. R. Caron and L. Dunn, " Comparing PV Power Plant Soiling Measurements Extracted from PV Module Irradiance and Power Measurements", in the 39th IEEE PVSC, 2013.

[4] C. Cañete, R. Moreno, J. Carretero, M. Piliougine, M. Sidrach-de-Cardona, J. Hirose, S. Ogawa, "Effect Of The Self-Cleaning Coating Surface In The Temperature And Soiling Losses Of Photovoltaic Modules", in the 27th EUPVSEC, 2012.

[5] E. Klimm, T. Lorenz, K.-A. Weiss, "Can anti-soiling coating on solar glass influence the degree of performance loss over time of PV modules drastically?" in the 28th EUPVSEC, 2013.

[6] K. Midtdal, B. P. Jelle, "Self-cleaning glazing products: A state-of-the-art review and future research pathways", Solar Energy Materials & Solar Cells, 109, pp. 126–141, 2013.

[7] Z. R. Abrams, P. Gonsalves, B. Brophy, J. Posbic, "Field Performance and Persistance of Anti-Soiling Coatings on Photovoltaic Glass" in the 29th EUPVSEC, 2014

[8] B. Brophy, Z. R. Abrams, P. Gonsalves, K. Christy, "Field and Lab Verification of Hydrophobic Anti-Reflective and Anti-Soiling Coatings on Photovoltaic Glass" in the 31st EUPVSEC, 2015

[9] A. R. Hoffman, C. R. Maag, "Photovoltaic Module Soiling Studies May 1978 - October 1980" JPL PUBLICATION 80-87.

Evaluation of remote diagnoses performance by using Operating Performance Index at different measurement intervals for residential PV systems

Ajisaka Masato[1], Yuzuru Ueda[1], Ryuuji Yamada[2], Tomoyoshi Yokota[2]
[1]Tokyo University of Science
6-3-1, Niijuku, Katsushika-ku, Tokyo, 125-8585, Japan
[2]KYOCERA Corporation
3-32-42, Higashishinagawa, Shinagawa-ku, Tokyo, 140-8820, Japan

ABSTRACT: Output power of a residential PV system is reduced by many factors such as shading, a malfunction, and a breakdown. A remote diagnoses method for residential PV systems by using Operating Performance Index (OPI) is developed in this research. OPI is defined as measured power generation divided by estimated power generation. In order to vilify the effect of time resolution on accuracy of OPI calculation, *OPI(t)* was calculated in the following measurement intervals: 1, 5, 10, 30, and 60 min. We also divided the celestial sphere into the following azimuth ψ [deg.] and solar height h [deg.] resolutions: ψ = 5, 10, 15, and 20, h =5, 10, and 15. $OPI_m(\psi,h)$ is the *OPI(t)* frequency value of each sector of ψ and h. OPI_{Mp} is the mode value of $OPI_m(\psi,h)$ and it shows power generation performance of the system. We have evaluated OPI at different measurement intervals with various celestial sphere resolutions.
Keywords: Operating Performance Index (OPI), PV system, Monitoring

1 INTRODUCTION

In Japan, since the launch of Feed-In Tariff (FIT) in 2012, PV system has been installed in large quantities. PV was considered to be maintenance-free, however, PV output power could decrease by a malfunction or a breakdown. The owner of a residential PV system needs an expert knowledge to detect a malfunction or a breakdown of the system. If the installer is not an expert, the installer may not be aware of a malfunction or a breakdown. In this study, we propose a method using Operating Performance Index (OPI) [1] for a remote diagnosis, and evaluate OPI at different measurement intervals and celestial sphere resolutions.

2 OPERATING PERFORMANCE INDEX (OPI)

Operating Performance Index (OPI) is defined as measured power generation e_m divided by estimated power generation E_{EST}. OPI at time t $OPI(t)$ is calculated by the following[2]:

$$OPI(t) = e_m(t) \,/\, E_{EST}(t) \quad (1)$$

where, $e_m(t)$ is the measured power generation at time t [kWh], $E_{EST}(t)$ is the estimated power generation at time t [kWh]. Generated PV output power was estimated by using plane of array (POA) solar irradiation, air temperature, tilt and azimuth angle of the array, and the rated capacity of the system [2]. Estimated value of generation power $E_{EST}(t)$ is calculated by the following equation:

$$E_{EST}(t) = P_{AS} \cdot K_T(t) \cdot K_D(t) \cdot K_O \cdot H_{Ag}(t) \,/\, G_S \quad (2)$$

where, P_{AS} is the rated capacity [kW], $K_T(t)$ is the temperature loss coefficient at time t, $K_D(t)$ is the aged degradation coefficient at time t, K_O is the product of other loss coefficients (K_O = 0.81), $H_{Ag}(t)$ is the POA solar irradiation at time t [kWh/m^2], and G_S is the reference solar irradiance (G_S = 1kW/m^2). Global horizontal irradiation was converted into POA irradiation [kWh/m^2] by using Erbs model [3] and Perez model [4], respectively.

$K_T(t)$ is calculated by the following equation:

$$K_T(t) = 1 + [\alpha_{pmax} \cdot (T_{PA} - 25)] \quad (3)$$

where, α_{pmax} is the temperature coefficient. For example, α_{pmax} of a silicon module is -0.45[% / °C]. T_{PA} is the module temperature.

T_{PA} is calculated by the following equation[5]:

$$T_{PA}(t) = T_A(t) + \left(\frac{A}{B \times V(t)^{0.8} + 1} + 2\right) \times H_{Ag}(t) - 2 \quad (4)$$

where, $T_A(t)$ is the ambient temperature at time t [° C], A and B are constants, $V(t)$ is the wind speed at time t [m/s], and $H_{Ag}(t)$ is POA solar irradiation at time t [kWh/m^2]. A and B becomes 46 and 0.14, respectively, if PV array is installed on a stand, while A and B becomes50 and 0.16, respectively, if installed on a roof.

$K_D(t)$ is calculated by the following equation:

$$K_D(t) = 0.995^{year} \quad (5)$$

where, *year* is the elapsed year of PV systems. If $OPI(t)$ is around 1.0, it indicates that the PV is generating electric power with an intrinsic performance. Conversely, if OPI is zero, PV is not generating electric power at all.

3 DIVIDED CELESTIAL SPHERE

We divided the celestial sphere into the following azimuth ψ [deg.] and solar height h [deg.] resolutions: ψ = 5, 10, 15, and 20, h =5, 10 and 15. In this paper, the celestial sphere is divided as shown in Fig.1. $OPI(t)$ was stored in each sector, which is the celestial sphere divided based on solar azimuth and solar height information. $OPI_m(\psi,h)$ is the frequency of $OPI(t)$ in each sector. If the celestial sphere was divided into ψ = 10[deg.] and h = 10[deg.], $OPI_m(\psi,h)$ contains 140 sectors. After $OPI_m(\psi,h)$ was calculated this way, OPI_{Mp} was calculated based on $OPI_m(\psi,h)$. OPI_{Mp} is the mode of all $OPI_m(\psi,h)$. OPI_{Mp} of 1.0 indicates normal operation of a PV system. The flow chart of the proposed method is shown in Fig.2.

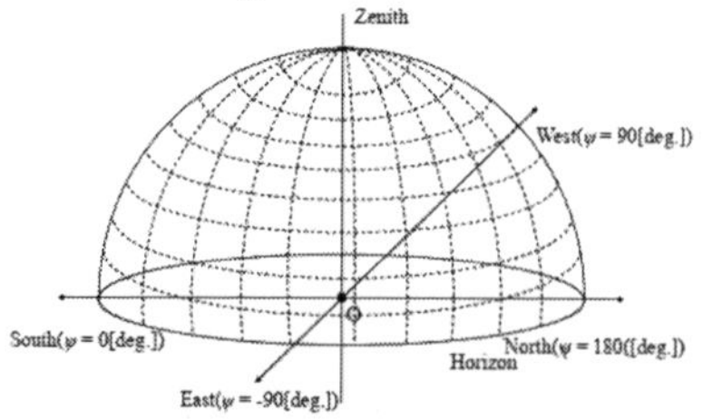

Fig.1. Divided the celestial sphere

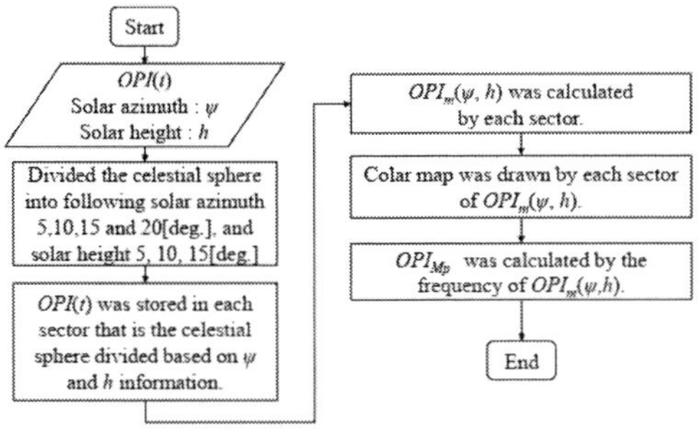

Fig.2. Flow chart of OPI_{Mp} calculation

4 ANALYSED PV SYSTEMS

Table.1 shows four sets of PV systems configuration in Ota, Gunma, Japan. Data from January 1st, to December 31st, 2007 was used in the analysis. We defined the azimuth to be -90, 0, and 90 [deg.] when the sun is in the east, south, and west, respectively. We used the data of residential PV system configuration from [6]. Overviews of the systems are shown in Table.1 and Fig.3.

Table.1 System configuration

System Number	1	2	3	4
Tilt Angle[deg.]	38.66	26.6	21.8	30.96
Azimuth[deg.]	-7	0	-6	15
P_{AS} [kW]	3.49	4.48	3.98	3.77

(a). System 1 (b). System 2

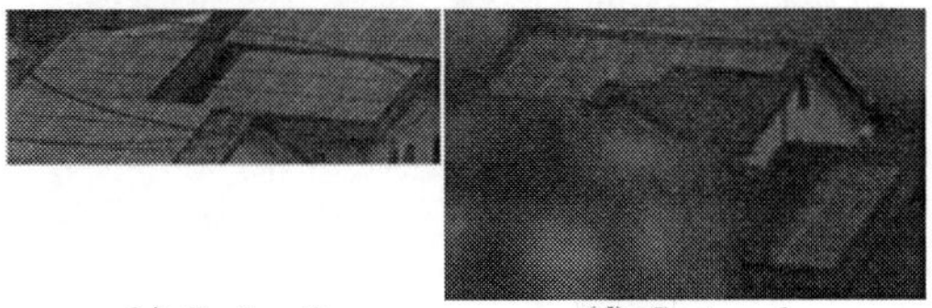

(c). System 3 (d). System 4

Fig.3. Overview of PV system

Onsite measurement data of solar irradiation, ambient temperature and wind speed at Maebashi, Gunma, Japan were used for the calculation. These data were measured at 1 minute intervals. We converted 1 minute values into 5, 10, 30 and 60 minute values. For example, when creating 5 minute values, 5 minute values were determined by taking 5 minutes average of the 1 minute values. If measurement data was lacking some data, 5 minute values were calculated as average excluding the missing data. We calculated 10, 30, and 60 minutes values in a similar manner.

5 ANALYSIS RESULT

5. 1 COLOR MAP OF $OPI_m(\psi,h)$

Color maps of $OPI_m(\psi, h)$ values are shown in Fig.4. Fig.4 (a) ~ (d) show the celestial sphere divided into different resolutions: ψ = 5[deg.] and h = 5[deg.], ψ = 10[deg.] and h = 10[deg.], ψ = 20[deg.] and h = 15[deg.], ψ = 5[deg.] and h = 5[deg.], respectively. Also, Fig.4 (a) ~ (d) were analyzed by using 1, 5, 10 and 60 minute values, respectively.

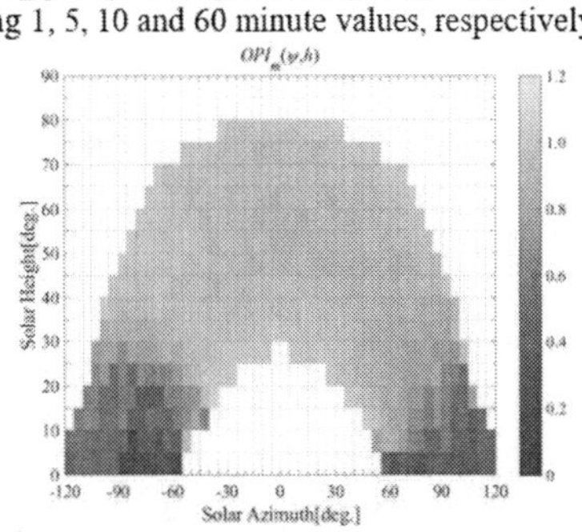

(a). System 1 (ψ = 5[deg.] , h = 5[deg.] and 1[min])

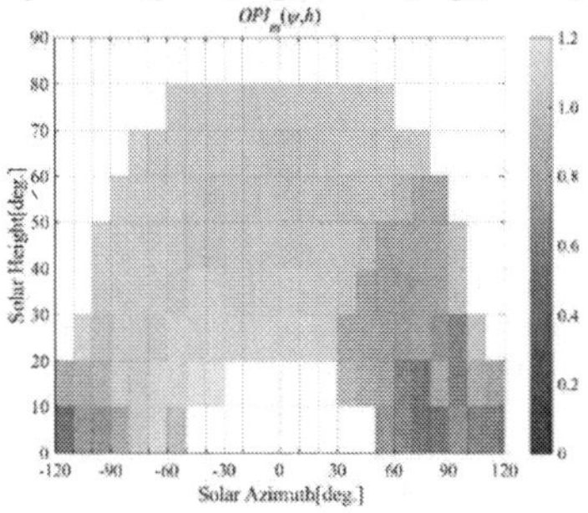

(b). System 2 (ψ = 10[deg.] , h = 10[deg.] and 5[min])

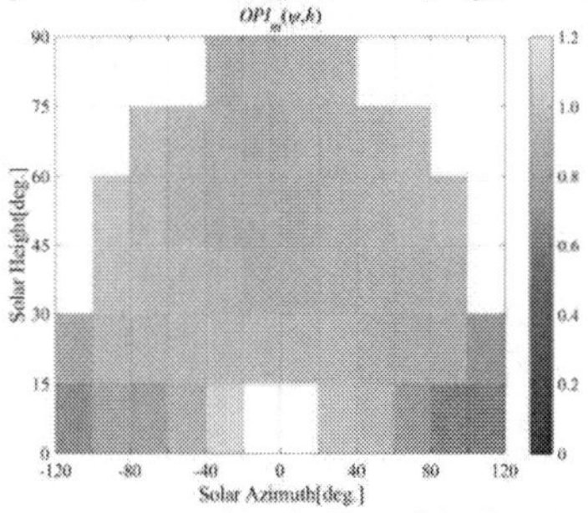

(c). System 3 (ψ = 20[deg.] , h = 15[deg.] and 10[min])

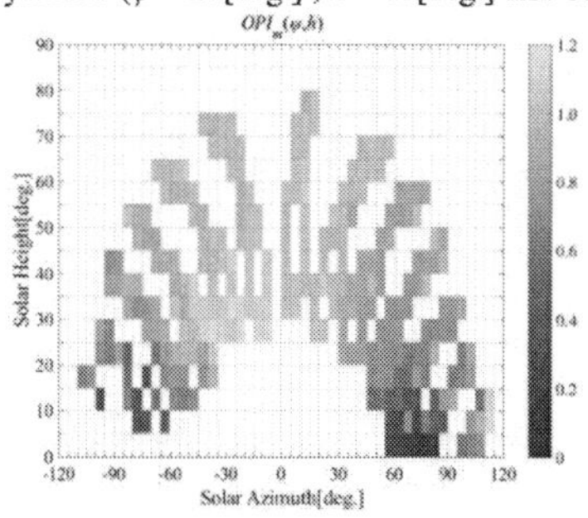

(d). System 4 (ψ = 5[deg.] , h = 5[deg.] and 60[min])

Fig.4. Color map of $OPI_m(\psi, h)$

These PV system were installed in the south side of the roof and these systems have stepped roof. In system 1, when the sun is in the east, the west side of the roof was covered by shade and left side of color map became red as shown in Fig.4 (a). Since the west side roof is covered by a shadow in the morning, PV output decreases. In other PV systems, the west side roof is higher than east side of the roof. Therefore, PV systems were covered by shadows in the evening and the right side of color map became red as shown in Fig.4 (b) ~ (d).

Blank sections appeared in Fig.4 (d). Blanks sections appeared when using 60 minute values and celestial sphere divided in a small resolution. Since the earth is moving 15[deg.] every an hour, some sectors did not contain any values. Hence, blank sector appeared when dividing the celestial sphere into small sector size.

5.2. FREQUENCY OF THE $OPI_m(\psi, h)$

Fig.5 shows frequency of respective $OPI_m(\psi, h)$ in Fig.4. Vertical axis shows the number of data and horizontal axis shows the $OPI_m(\psi, h)$ values.

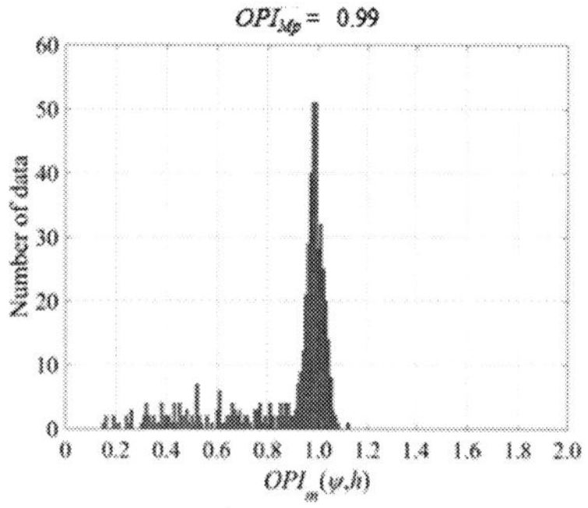

(a). System 1 (ψ = 5[deg.] , h = 5[deg.] and 1[min])

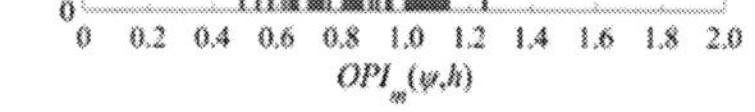

(b). System 2 (ψ = 10[deg.] , h = 10[deg.] and 5[min])

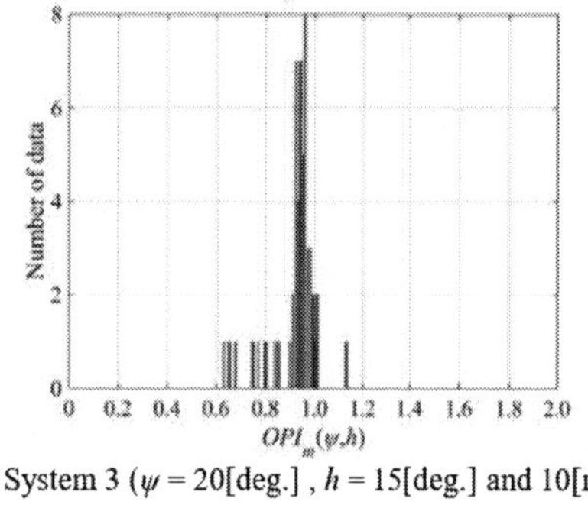

(c). System 3 (ψ = 20[deg.] , h = 15[deg.] and 10[min])

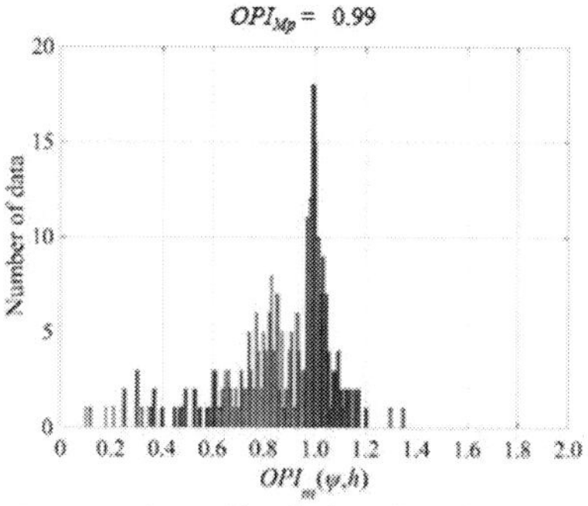

(d). System 4 (ψ = 5[deg.] , h = 5[deg.] and 60[min])

Fig.5. OPI_{Mp} histograms

These results show that OPI_{Mp} is about 1.0, which means that PV module is operating normally. Also, in Fig.5 (b) and (d), we detected the second maximum OPI_{Mp} value of approximately 0.8. We thought that the second maximum values was created by shadows. This method can detect an output reduction due to a malfunction or a breakdown without being affected by shadows.

5.3. EVALUATION OF OPI_{Mp}

Table.2 shows OPI_{Mp} analysis results of different interval values and celestial sphere division into different ψ and h. We assumed that OPI_{Mp} calculation using 1minute values and celestial sphere division into ψ = 5[deg.] and h = 5[deg.] will obtain the most accurate value of OPI_{Mp} . Hence, we defined this resolution as the standard OPI_{Mp}. When OPI_{Mp} is smaller than the standard OPI_{Mp}, OPI_{Mp} values were colored red in Table.2. Conversely, when OPI_{Mp} is larger than the standard OPI_{Mp}, OPI_{Mp} values were colored green.

Table.2. Result of several OPI_{Mp}

(a). System 1

ψ[deg.]×h [deg.]	1[min.]	5[min.]	10[min.]	30[min.]	60[min.]
5×5	0.99	0.98	0.99	1.00	1.00
10×5	0.98	0.98	0.99	1.00	1.00
15×5	0.98	0.99	0.99	1.00	1.01
20×5	0.99	0.98	0.99	1.00	1.00
5×10	0.98	0.99	0.99	0.99	1.00
10×10	0.98	1.00	0.99	1.00	1.00
15×10	0.98	0.99	0.99	1.00	1.00
20×10	0.99	0.99	0.99	1.00	1.01
5×15	0.99	0.99	0.99	1.00	1.00
10×15	0.98	0.99	0.99	1.00	1.00
15×15	0.98	0.99	0.99	1.00	1.00
20×15	0.99	0.99	0.99	1.00	1.02

(b). System 2

ψ[deg.]×h [deg.]	1[min.]	5[min.]	10[min.]	30[min.]	60[min.]
5×5	1.05	1.06	1.05	1.05	1.04
10×5	1.05	1.05	1.05	1.05	1.05
15×5	1.05	1.05	1.05	1.05	1.05
20×5	1.05	1.05	1.05	1.05	1.05
5×10	1.06	1.06	1.05	1.05	1.04
10×10	1.05	1.05	1.05	1.06	1.05
15×10	1.06	1.05	1.05	1.05	1.05
20×10	1.06	1.05	1.05	1.06	1.05
5×15	1.05	1.06	1.05	1.05	1.04
10×15	1.05	1.05	1.05	1.05	1.04
15×15	1.05	1.05	1.05	1.05	1.05
20×15	1.06	1.05	1.05	1.05	1.04

(c). System 3

ψ[deg.]×h[deg.]	1[min.]	5[min.]	10[min.]	30[min.]	60[min.]
5×5	0.95	0.95	0.94	0.95	0.96
10×5	0.95	0.95	0.95	0.95	0.96
15×5	0.95	0.98	0.94	0.92	0.99
20×5	0.95	0.93	0.94	0.93	0.96
5×10	0.95	0.94	0.94	0.95	0.95
10×10	0.95	0.94	0.94	0.95	0.96
15×10	0.95	0.94	0.94	0.98	0.99
20×10	0.95	0.93	0.94	1.00	0.96
5×15	0.93	0.95	0.95	0.96	0.94
10×15	0.95	0.95	0.95	0.95	0.98
15×15	0.95	0.98	0.94	0.92	0.96
20×15	0.95	0.93	0.96	0.97	1.00

(d). System 4

ψ[deg.]×h[deg.]	1[min.]	5[min.]	10[min.]	30[min.]	60[min.]
5×5	1.04	1.03	1.03	1.01	0.99
10×5	1.03	1.03	1.02	1.01	0.98
15×5	1.04	1.03	1.02	1.01	0.98
20×5	1.03	1.03	1.03	1.01	0.99
5×10	1.03	1.03	1.03	1.00	0.99
10×10	1.04	1.03	1.03	1.01	1.00
15×10	1.03	1.03	1.03	1.01	0.99
20×10	1.03	1.03	1.02	1.01	0.99
5×15	1.03	1.03	1.03	1.00	0.99
10×15	1.04	1.03	1.02	1.01	0.98
15×15	1.04	1.03	1.02	1.01	0.98
20×15	1.03	1.03	1.02	1.01	0.99

As shown in Table.2, in the case of calculation using all interval values, OPI_{Mp} is scattered within 4% ($0.98 \leq OPI_{Mp} \leq 1.02$) in system 1, and 2% ($1.02 \leq OPI_{Mp} \leq 1.04$) in system 2. On the other hand, calculation using 1, 5, and 10 minute values in system 3, OPI_{Mp} scattered within 5% ($0.93 \leq OPI_{Mp} \leq 0.98$), however, calculation using 30 and 60 minute values, several OPI_{Mp} were over 5% larger than the standard OPI_{Mp} (1[min], ψ = 5[deg.], h = 5[deg.]). In system 4, OPI_{Mp} was highly accurate in calculation using 1, 5, 10, and 30 minute values, however, calculation using 60 minute values, average OPI_{Mp} was lower by about 5% than the standard OPI_{Mp}. From the above results, it can be seen that OPI_{Mp} error of 5% from the standard OPI_{Mp} is a valid cutoff value. Therefore, we conclude that using 60 minute values for OPI analysis is difficult.

6. CONCLUSION

In this paper, a remote diagnoses method to evaluate PV system performance, which uses OPI to exclude shading effect, was proposed. OPI_{Mp} of normally operating PV systems were about 1.0 and OPI_{Mp} of poorly operating PV systems were lower than 1.0. When a PV system was covered by a shadow, we were able to detect the shadow by OPI_{Mp} histograms as in Fig.4 (b) and (d). In terms of interval values, OPI_{Mp} calculation result of long interval values tended to scatter. However, in order to perform a highly accurate diagnosis of a PV system, OPI_{Mp} needs to be scattered within 5% from the standard OPI_{Mp}. Therefore, we concluded that 60 minute value is not suitable for the proposed method. In terms of celestial sphere division, if improper interval resolutions of solar azimuth and height is selected for the chosen time interval, blank sections in the color map of OPI_m appeared as in Fig.4 (d). Dividing the celestial sphere in small intervals increases the number of $OPI_m(\psi, h)$, but it also reduces the number of the data in each sector. Therefore, optimal celestial sphere resolution needs to be selected for chosen time interval.

7. ACKNOWLEDGEMENT

We would like to acknowledge the financial support of NEDO.

REFERENCES

[1] M.Ajisaka et al., "Development of the operating performance index calculation method by using residential PV system information" IEEJ national convention, Vol7, pp.50-51 (2016) (In Japanese)

[2] Y.Ueda et al.,"Development of the web application for the self diagnosis of the PV systems "IEEJ Trans. Power and Energy,Vol.127,No.6,pp.11-12 (2010) (In Japanese)

[3] D.G.Erbs et al., "Estimation of the Diffuse Radiation Fraction for Hourly, Daily and Monthly-Average Global Radiation" Solar Energy Vol.28 No.4, pp.293-302(1982)

[4] R.Perez et al., "Modeling Daylight Availability and Irradiance Components from Direct and Global Irradiance" Solar Energy Vol.44, No.5, pp271-289(1990)

[5] Japanese Standards Association, JIS C8907, (2005)

[6] Kandenko Co., Ltd., "Demonstrative Research on Clustered PV systems" NEDO report (2008) (In Japanese)

32nd European Photovoltaic Solar Energy Conference and Exhibition - Munich, Germany, 20-24 June 2016
TOPIC 5: Operation, Performance, Reliability and Sustainability of Photovoltaics - 5.2 Operation of PV Systems

COMPARISON OF VARIOUS MODELS FOR THE ESTIMATION OF THE PERFORMANCE LOSS RATE OF SEVEN PV TECHNOLOGIES OVER FIVE YEARS IN ALPINE CLIMATE

P. Ingenhoven[1*], G. Belluardo[1], D. Moser[1]
[1]Institute for Renewable Energy, EURAC Research, Viale Druso 1, 39100 Bolzano (BZ), Italy
[*]Phone: +39 0471 055634; Fax: +39 0471 055699; E-mail: philip.ingenhoven@eurac.edu

ABSTRACT: The interest in methodologies for the estimation of the performance loss rate (PLR) using data from PV systems arises with the aging of an increasing amount of PV installations at different climates. These methodologies must be simple, i.e. involve few quantities as input. Furthermore, the associated uncertainty must be kept low in order to be a reliable tool for the verification of module manufacturers' warranties.
This study compares the PLR of seven PV systems based on different technologies using four different methodologies that are presented and discussed. Results are compared on the basis of the PLR and the respective regression uncertainty. We find that, especially for technologies with a strong seasonal response, the error of the estimated PLR can be significantly reduced using seasonal decomposition methods such as a simple moving average. Other more physical approaches like the PVUSA method to reduce seasonality reduce the error as well, however remaining oscillations that are not captured by the physical description enlarge the error.

Keywords: performance loss rate, degradation, PVUSA, performance ratio, moving average.

1 INTRODUCTION

In this paper we calculate the PLR, expressed as the percentage of the yearly performance loss, of seven PV systems installed at the Airport Bolzano Dolomiti (South Tyrol, North Italy), namely: mono- and polycrystalline Silicon (mc-Si and pc-Si), amorphous Silicon (a-Si), micromorph Silicon, Copper-Indium-Gallium-Selenide (CIGS), Cadmium Telluride (CdTe) and heterojunction with intrinsic layer (HIT). DC-power and weather data are filtered and treated before calculating monthly values of performance metrics. The methodologies used can be summarized as: Performance Ratio (PR) filtered for shading, PR evaluated only at high irradiances G>800W/m^2, PVUSA also evaluated only at high irradiances and the moving average of the shading-filtered PR. A linear regression is performed on a 5-year series of monthly values, in order to estimate the PLR and the associated regression uncertainty. The detailed analysis of the performance loss with different methods is an important step not only to understand the aging behavior of different PV technologies, but more importantly to understand the effect that different computation methodologies have on the value and the accuracy of the performance loss estimation. This information will serve as a valuable source in defining standard of performance loss measurements. Standards are important so that values given by the producer's warranties are verifiable in long term measurements. This in turn will make the economic analysis and the life time yield analysis of a given PV plant more accurate and will give potential investors a better idea of the risk involved in PV investment.

2 METHODS

The purpose of this work is the comparison of four different methods to calculate the performance loss rate (PLR) of seven different PV systems.
In cooperation with the Airport of Bolzano Dolomiti (ABD), EURAC research runs a multi-technology photovoltaic test facility for the detailed performance evaluation of different module technologies and mounting systems.
The site includes 24 technologies and 4 mounting systems. Power monitoring started n summer 2010. The test site has an overall nominal power of 724 kWp. The field orientation is 8.5° West of South, and the modules have a 30° tilt angle. Irradiance variables are monitored by a dedicated weather station. Five years of monitored PV system data are taken into consideration, ranging from January 2011 to December 2015. The DC-power data of the seven considered PV systems was measured by commercial inverters. Data of plane-of-array irradiance, wind speed and ambient temperature were measured by a weather station close to the PV systems and are available as 15-minutes based values. The methods involve the calculation of monthly average values of the performance estimators PR or PVUSA.
The Performance Ratio (***PR***) metric is derived from IEC61724:1998 [1] and is calculated as [2]:

$$PR = \frac{Y_a}{Y_r} = \frac{\sum P_{dc}(i)}{\sum G(i)} \frac{G_{STC}}{P_{STC}} \tag{1}$$

where P_{dc} is the DC-power of the PV system at time i, $G(i)$ is the irradiance measured in the plane of the modules at time i, while G_{STC} is the irradiance under standard test conditions (STC) and P_{STC} is the nominal power of the PV system under STC. The same formula can be written as the ratio of array yield Y_a, i.e. the generated DC-energy per kW of installed PV array, and reference yield Y_r, i.e. the ratio of the irradiation measured on the plane of the modules and irradiance at STC.
The moving average method $\boldsymbol{PR_{mov}}$ involves the use of a monthly value of a symmetrical moving average of the PR over 12 months:

$$PR_{mav}(t) = \frac{1}{2}\left(\frac{1}{12}\sum_{i=t-6}^{t+5} PR(i) + \frac{1}{12}\sum_{i=t-5}^{t+6} PR(i)\right) \tag{2}$$

This method is used in order to decrease the uncertainty associated to the estimation of the PLR as it smooths the seasonal oscillation of PR metric. On the other hand, the values of PR calculated with this technique are computable from the sixth month of data series start, up to the sixth month before the data series end, thus leaving some initial and final gaps in the series used for regression.

PVUSA (Photovoltaics for Utility Scale Applications) metric [3] consists in calculating the best-fit correlation on

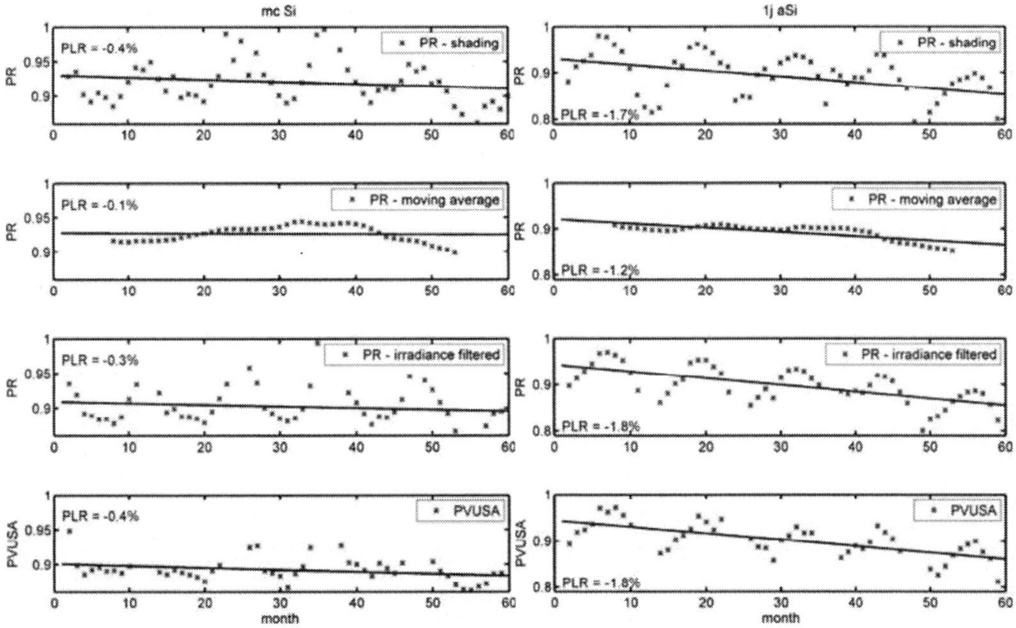

Figure 1 Monthly values of PR and PVUSA according to the four considered methodologies (1 = January 2011; 13 = January 2012; 25 = January 2013; 37 = January 2014, 49 = January 2015): (right) amorphous silicon PV system; (left) monocrystalline silicon PV system. Regression line as well as PLR value are reported. PVUSA values are normalized for the array power at STC, P_{STC}

a monthly base between 15-minutes based values of measured DC-power P_{dc} and 15-minutes based values measured irradiance on the plane of the modules G, wind speed W, and ambient temperature T, according to the following parametrized equation:

$$P_{dc}(i) = G(i)[a + b * G(i) + c * W(i) + d * T(i)] \quad (3)$$

Parameters a, b and c are then re-fed into Equation 4 in order to estimate the power at Performance Test Conditions (PTC) (G = 1000 W/m^2, T = 20 °C, W = 1 m/s):

$$PVUSA = P_{dc}(PTC) \quad (4)$$

Different filters are applied to the metrics introduced above. Overall, four different methodologies are applied to the measured data. The methods can be summarized as follows:

- Method 1: PR (Equation 1) filtered only for shading using shading diagram for each array to exclude effects due to close objects which cause the irradiance conditions to be different on the array and the weather station.
- Method 2: moving average of the PR. Equation 2 is applied on the PR values obtained in method 1.
- Method 3: PR (Equation 1) at high irradiances. PR outliers (15-minutes based) are first filtered out using a filter $\overline{PR} \pm \sigma$, where $\overline{PR}$ is the average monthly value of PR and σ is the corresponding standard deviation. Data are further filtered in order to keep only 15-minutes based values of PR corresponding to irradiance values higher than 800W/m^2.
- Method 4: PVUSA: (Equation 3 and 4) at high irradiances. The same filtering technique as Method 3 are used. Here however the PVUSA metric is used instead of PR.

Once the monthly values of the metric are calculated according to the four selected methodologies, a linear regression is calculated and the data is fitted to the linear expression:

$$\bar{P} = at + b \quad (5)$$

where $\bar{P}$ is the monthly value of the metric PR or PVUSA, t the month number, and a and b are the regression coefficients. The PLR, expressed as the percentage of performance loss in one-year period, is then calculated as:

$$PLR = \frac{12a}{b} \quad (6)$$

The uncertainty of the linear regression is also calculated according to the procedure proposed by the Guide to the Expression of uncertainty measurement [4] and [5]. The uncertainty of each measurement (y = PR, PR_{mov}, PVUSA) is given as:

$$\sigma_y = \sqrt{\frac{1}{N-2}\sum_{t=1}^{N}(y_t - a\,t - b)^2},$$

where N is the total number of months considered and y stands for the respective performance metric. The uncertainty of the gradient is then given as

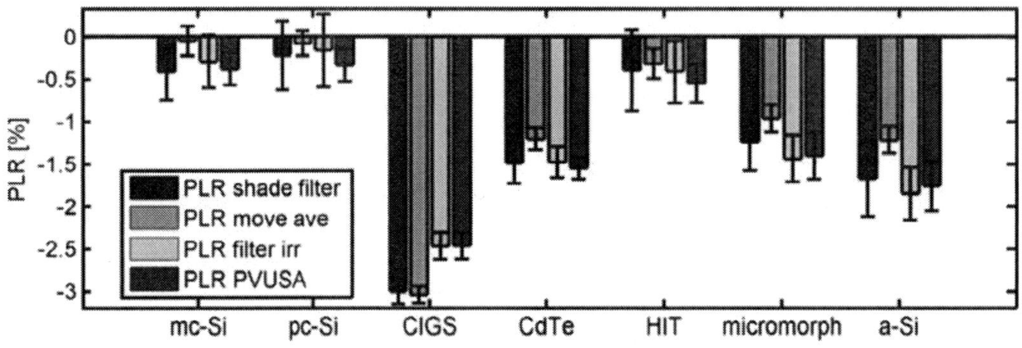

Figure 2 performance loss rates of the 7 technologies with the error for the different methods used: (dark blue) Method 1: PR filtered only for local shading was used for the evaluation, (light blue)Method 2: PR from Method 1 was smoothed using a symmetric 12 month moving average, (yellow) Method 3 PR filtered for high irradiance only, (red) Method 4: the PVUSA performance metric was used

$$\sigma_a = \sigma_y \sqrt{\frac{N}{\Delta}},$$

$$\sigma_b = \sigma_y \sqrt{\frac{\sum t^2}{\Delta}},$$

where Δ is given as

$$\Delta = N \sum t^2 - \left(\sum t\right)^2.$$

The uncertainty for the PLR is then given as

$$\sigma_{PLR} = \sqrt{\left(\frac{12}{b}\right)^2 \sigma_a^2 + \left(\frac{12a}{b^2}\right)^2 \sigma_b^2}.$$

3 RESULTS

The rates were computed with the methods described above. Figure 1 shows the monthly values of the metrics obtained by the 4 different methods. For space reasons only two of the technologies, namely amorphous silicon (right) and mono-crystalline (left) silicon are shown here in detail. The four graphs from top to bottom show method 1-4 (PR-shading), PR-moving average, PR-irradiance >800W/m^2 and PVUSA). The regression line is plotted for each method and technology and the values of the PLR are given in the figures. For a detailed results the PLR and the corresponding uncertainties are reported in Figure 2 and Table 1.

In general, Methods 1, 3 and 4 present a strong seasonality. They give similar results for PLR. However, method 3 and 4 present gaps in the metric series, since filtering data for irradiance lower than 800W/m^2 tends to exclude all points in winter. On the other hand, Method 2 shows different values of PLR, lower in absolute value. This methodology has a smoothing effect on the metric series, even though it presents a 6-months gap at the beginning and end of the data series.

For all methods used the largest performance loss was measured for the CIGS technology (~2.5-3 %/year), followed by the 1j amorphous silicon (~1.2-1.9 %/year), CdTe (~1.2-1.5 %/year) and the micromorph silicon (~1.0-1.4 %/year). After 5 years of monitoring the thin film show a much larger performance loss than the crystalline technologies such as poly-crystalline (~0.1-0.3 %/year), mono-crystalline silicon (~0.1-0.4 %/year) and HIT (~0.3-0.5 %/year). It is important to mention that the performance loss we found for the CIGS array is unusually high for this technology due to manufacturing problem of this specific product. Typical degradation rates found for CIGS are comparable to those of other thin film technologies [6]. In general the precision of all 4 methods is high taking the error bars into account.

Comparing the uncertainties of the various methods we find the largest uncertainty is found for the PR measurements that are only filtered for shading. Filtering the PR from irradiance larger than 800W/m^2 has no large effect on the error bar. A maximal reduction of the uncertainty to 70% (average reduction to 86%) of that of the former can be achieved. The PVUSA metric reduces the uncertainty by up to about 50% (average reduction to 63%) but not consistently as the physics used by in the computation of the PVUSA metric is not equally accurate for all technologies. The best result in terms of reduction of the uncertainty is achieved using the moving average of the shading-filtered PR. Here the uncertainty can be reduced to up to 33% (average reduction to 46%) of the untreated PR measurements.

4 CONCLUSIONS AND FUTURE WORK

The results found here confirm the general trends in performance loss rates that crystalline technologies age more slowly than thin film technologies. However, the main focus of this work was not the values of PLR but an analysis of various methods of computing them. In previous publication different methods of computing the PLR were explored, see [7] and [8]. In [7] more physical properties were used to reduce the uncertainty related to the PLR computation, taking into account the module temperature and correct for it using the temperature coefficients found in the data sheet. Further [7] presents a study of the influence of the spectral influence on the PLR computations. In contrast to that [8] follows a purely statistical path and uses the XARIMA time series decomposition. Here the PR time series are analyzed and the long term trend is decoupled form seasonal oscillations and a remaining random part.

Our comparison here shows that for the seven technologies under investigation here, simple statistical method based

Table 1 performance loss rates of the different technologies computed with the 4 methods. Lowest error values are found for the PVUSA and the moving average methods. The statistical error of the moving average method is overall lowest, however in contrast to the other methods does not take any physical properties into account to counter balance the fluctuations with cause the errors.

	PLR no filter [%/year]	Error [%/year]	PLR moving average [%/year]	Error [%/year]	PLR filter irradiance [%/year]	Error [%/year]	PVUSA [%/year]	Error [%/year]
mcSi	-0.41	0.32	-0.05	0.18	-0.29	0.31	-0.38	0.18
pcSi	-0.22	0.41	-0.08	0.15	-0.16	0.43	-0.34	0.19
CIGS	-3.00	0.16	-3.04	0.10	-2.46	0.16	-2.46	0.15
CdTe	-1.48	0.25	-1.20	0.13	-1.47	0.18	-1.54	0.13
HIT	-0.40	0.48	-0.32	0.17	-0.41	0.37	-0.55	0.22
micrm	-1.23	0.34	-0.96	0.16	-1.43	0.28	-1.40	0.28
1jaSi	-1.67	0.45	-1.21	0.15	-1.85	0.31	-1.76	0.30

on moving average shows the best results in terms of uncertainty reduction. More advanced methods like ARIMA or LOESS [9] could even strengthen the reduction. Future work will include studies of the effect of ARIMA and LOESS methods alone and in combination with the more physics based approaches described in [7].

ACKNOWLEDGMENTS

The research leading to these results has received funding from the European Union's Horizon 2020 research and innovation programme under the grant agreement No 649997 (Solar Bankability).

REFERENCES

[1] IEC61724:1998, 1998. Photovoltaic System Performance Monitoring Guidelines for Measurement, Data Exchange, and Analysis.

[2] Woyte, A., Richter, M., Moser, D., Reich, N., Green, M., Mau, S., Beyer, H., 2014. Analytical Monitoring of Grid-Connected Photovoltaic Systems – Good Practices for Monitoring and Performance Analysis. Tech. rep., IEA-PVPS T13-03:2014.

[3] Whitaker, C., Townsend, T., Newmiller, J., King, D., Boyson, W., Kratochvil, J., Collier, D., Osborn, D., 1997. Application and validation of a new PV performance characterization method. In: 26th IEEE Photovoltaic Specialists Conference 1997, Anaheim, California, pp. 1253–1256.

[4] GUM - JCGM 100:2008, 2008. Evaluation of Measurement Data Guide to the Expression of Uncertainty in Measurement.

[5] J. R. Taylor, An Introduction to Error Analysis: The Study of Uncertainties in Physical Measurements, 2nd edition. Sausalito, Calif: University Science Books, 1996.

[6] Jordan, D.C., Kurtz, S.R., 2013. Photovoltaic degradation rates-an analytical review. Prog. Photovolt.: Res. Appl. 21 (1), 12–29.

[7] G. Belluardo, P. Ingenhoven, W. Sparber, J. Wagner, P. Weihs, and D. Moser, "Novel method for the improvement in the evaluation of outdoor performance loss rate in different PV technologies and comparison with two other methods," Solar Energy, vol. 117, pp. 139–152, Jul. 2015.

[8] A. Phinikarides, G. Makrides, B. Zinsser, M. Schubert, and G. E. Georghiou, "Analysis of photovoltaic system performance time series: Seasonality and performance loss," Renewable Energy, vol. 77, pp. 51–63, May 2015.

[9] R. B. Cleveland, W. S. Cleveland, J. E. McRae, and I. Terpenning, "STL: A Seasonal-Trend Decomposition Procedure Based on Loess," Journal of official statistics, vol. 6, no. 1, pp. 3–33, Jan. 1990.

5BV.2.34 DRONE-BASED ASSESSMENT OF CLEANING EFFECTS ON PV INSTALLATIONS

Manuel Lanz, Eva Schuepbach, Urs Muntwyler
Bern University of Applied Sciences BFH, Department of Engineering and Information Technology
Institute for Energy and Mobility Research, Photovoltaic Laboratory (PV LAB)
Jlcoweg 1, CH-3400 Burgdorf, Switzerland
Tel. +4134 426 68 37, Fax +4134 426 68 13, urs.muntwyler@bfh.ch

ABSTRACT: Monitoring the performance of PV installations can be very labour intensive. The efficiency can significantly be increased by using a thermal imaging drone system. The Photovoltaic Laboratory (PV LAB) at Bern University of Applied Sciences BFH in Burgdorf, Switzerland, hence developed and built its own thermal imaging drone. Drone-based inspection of several Swiss PV installations operated by the PV LAB was conducted and degradation (k_G) factors were calculated. Comparison of the k_G-factors evidenced that cleaning the PV module surfaces may increase the energy yield production up to 9%. The drone-based thermal images offered a real advantage for the detection of "hot spots" in PV modules and associated differences in solar cell temperatures, indicating power decreases. However, the quantification of extra energy yields gained from cleaning homogeneous environmental pollution on PV module surfaces needs to be based on measurements.

Keywords: Monitoring, Performance, PV System, Soiling, Thermography

1 INTRODUCTION

Reductions of electrical output performance of PV modules due to environmental contamination (e.g., dust) can be significant [1,2,3]. Irregular and inhomogeneous contamination of PV modules may generate a mismatch of the cells in the substring. The affected cells may heat up and cause thermal stress, which can result in potential defects.

As the inspection of such defects, as well as the overall monitoring of the performance and quality of (especially large) PV installations are time-consuming, unmanned aerial drone vehicles (UAV) can provide an added value. An infrared multicopter drone was hence developed and built at the Photovoltaic Laboratory (PV LAB) at Bern University of Applied Sciences BFH in Burgdorf-Switzerland. Using this drone, several PV installations in Switzerland were inspected with regard to performance quality. The inspected PV installations are part of the monitoring network operated by the PV LAB at BFH since the 1990es. At some specific PV installations, the energy yield output was quantified before and after cleaning the PV module surfaces using installed temperature and irradiation sensors as input data, and by measuring transformers at the PV system.

2 THE IR-MULTICOPTER DRONE

In order to facilitate the maintenance, the drone system was assembled from commercially available components. The measuring device is a thermal imaging camera with a resolution of 382 x 288 pixels. Combined with a 62°x49° wide-angle optics, the viewing angle is 3.14 milliradian, i.e., the smallest measurable spot describes a square with 31.4 mm edge length in 10m distance. A fully radiometric video can be recorded, as reflections are easier recognizable on the video than on thermal images. Reflections are also more distinguishable from the thermal abnormalities in the video. Hence, a live video stream is sent to the pilot to check the image. As the pilot can switch between two video signals for easier flight navigation, rapid overflight of a PV installation is possible. A digital compact camera is mounted for capturing comparison shots. More technical and operational details of the IR-multicopter drone system can be found in [4]. Figure 1 shows the drone ready for take-off.

Figure 1: The IR-multicopter drone assembled by the Photovoltaic Laboratory (PV LAB) at Bern University of Applied Sciences BFH in Burgdorf, Switzerland (model S1000 octocopter with camera-system). Photo: BFH-TI.

Figure 2 shows a video capture of the IR camera mounted on the drone. At the left hand side in Figure 2, a Siemens M55 module of the PV LABs own PV installation displays a thermal abnormality ("hot spot").

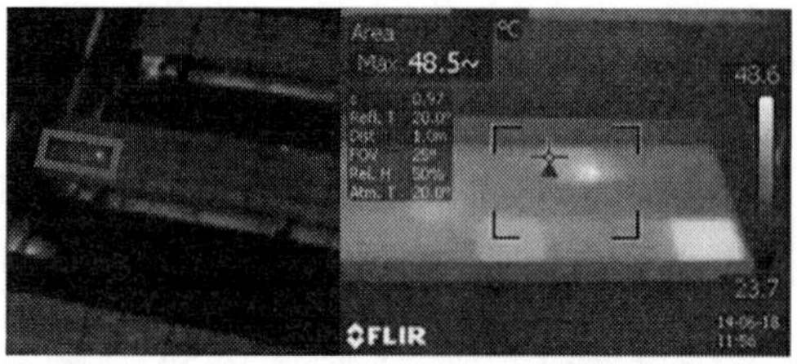

Figure 2: IR video capture, magnified (left) and close-up of marked module (right). Photo: BFH-TI.

The right hand side in Figure 2 shows a close-up of the same module and reveals that the module has a bad solder joint (at the top of the module) and a short circuit in a cell (bottom right). The deficient module was removed from the PV installation and measured in the lab. The measurements were compared with data of a new Siemens M55 module (Figure 3) and evidenced a power loss of about 20 percent in the defective module.

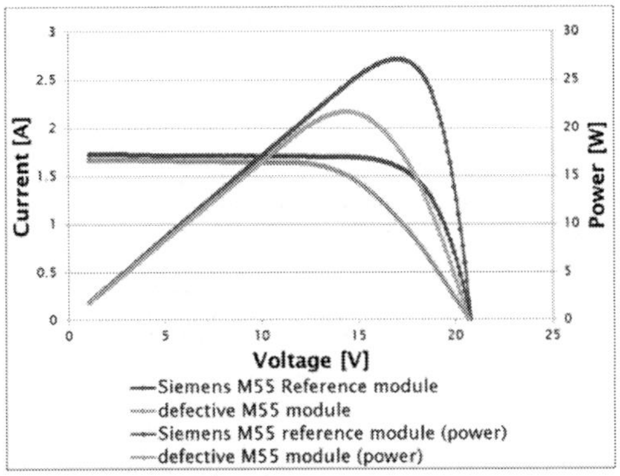

Figure 3: Characteristics of the defective module (marked Siemens M55 module in Fig. 2).

It is hence demonstrated that thermography from drone-based inspection of PV installations can identify heated dirt and heated solar cells in PV modules and thus indicate associated power losses.

3 QUANTIFYING CLEANING EFFECTS

As evidenced in Section 2, "hot spots" due to environmental pollution can be determined using thermography. In order to examine whether or not thermography can also be used to isolate effects of homogeneous environmental pollution, field-tests and power relevant measurements (explanations see below) were conducted in summer 2015 at several Swiss PV installations operated by the PV LAB.

3.1 Theory

A theoretically achievable generator yield (temperature corrected radiation yield Y_T) was calculated. With the effective yield (array yield Y_A), a quotient can be formed. The resulting generator correction factor (k_G) provides information (in percentage) about the efficiency of the plant. The following formulas describe the mathematical relationship [5].

$$k_G = \frac{Y_A}{Y_T} = \frac{Array-yield}{temp.-corrected\ radiation\ yield}$$

$$Y_A = \int_0^T y_A * dt$$

$$Y_T = \int_0^T y_T * dt$$

$$y_T = y_R[1 + c_T(T_C - T_0)]$$

$$y_R = \frac{G_G}{1\ kW/m^2}$$

y_A	normalised PV array yield
Y_A	daily PV array yield
y_R	standardised irradiation yield (reference yield)
y_T	temperature corrected irradiation yield
Y_T	daily temperature corrected irradiation yield
G_G	global irradiance in PV array level
c_T	temperature coefficient of MPP power of PV arrray
T_C	cell temperature of the PV array
T_0	STC reference temperature (25°C)
k_G	array correction factor

Formula 1: Calculation of generator correction factor k_G.

The input data were taken from PV installations in the network operated by the PV LAB at BFH [6]. The parameters included in the formula are: voltage, current and temperature of the PV modules and solar radiation in PV array level. Temperature is captured with reference cells and irradiation is measured with a Kipp & Zonen CMP 22 pyranometer. DC voltage and DC current are measured with instrument transformers.

3.2 Cleaning effects in long-term records (1994 to 2016)

The PV LAB's own PV installation (see Figure 2 in Section 2), mounted on the roof of the building, has a capacity of 50 kWp. It has been operated since the 1990s and consists of Siemens M55 modules that are regularly cleaned every four years. Figure 4 illustrates the degeneration of this PV installation due to environmental contamination and the effects of cleaning the PV module surfaces. Calculation of the k_G-values using Formula 1 evidences that about 5-8% of the produced electrical power can be regained with regular cleaning.

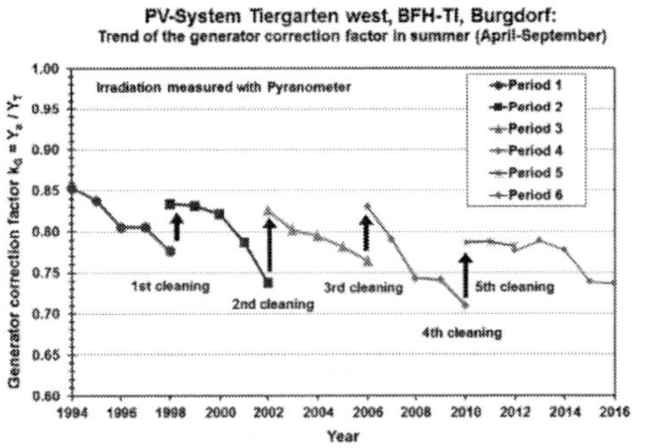

Figure 4: Energy yield increase (1994-2016) due to cleaning. Data are taken from the PV LABs own PV installation at the BFH building in Burgdorf -Tiergarten, Switzerland.

3.3 Cleaning effects at large PV installation

The PV installation on the «Stade de Suisse» football stadium in Bern, Switzerland (Fig. 5) has a total capacity of 1347 kWp. Using the IR-multicopter drone, the entire PV module surface was mapped with thermal images before and after cleaning in summer 2015. It took about 30 minutes to IR-map the 7930 modules to identify defects. 32 thermal anomalies were isolated; these had disappeared after cleaning and were, therefore, attributed to environmental pollution.

Figure 5: «Stade de Suisse» football stadium in Bern, Switzerland. The total capacity of 1347 kWp is divided into 11 subsystems with different module inclinations. The inner roof surfaces have 7 degrees, the external roof surfaces 20.5 degrees of inclination.

The thermal abnormalities were identified by comparing the images recorded before and after cleaning. The temperature differences to the "normal" (not affected) cells were between about 3 and 17 Kelvin.

To quantify the effects, the variables listed in Section 3.1 (DC current, DC voltage, module temperature and solar radiation) were measured. The k_G-values were calculated separately for each unit using Formula 1.

Figure 6 illustrates the percentage change in the performance of the units from 2010 to 2016.

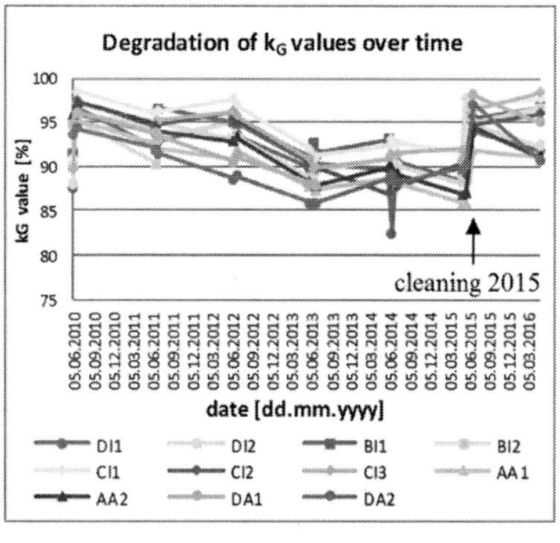

Figure 6: Degeneration of the kG values at the "Stade de Suisse" in Bern, Switzerland, from 2010 to 2016. The increase of the kG value due to cleaning in summer 2015 is clearly visible.

Table 1 lists the kG-values and evidences power gains from 4-5 percent for the subsystems with an inclination of 7 degrees that were cleaned after 5 years of operation. The subsystems with an inclination of 20.5 degrees were cleaned after 8 years of operation and display power increases between 6-9 percent.

Subsystems	Inclination	Increase in %	Notes/year
BI1	7 degrees	4.0 → ~4	2nd cleaning 10/15
BI2	7 degrees	6.4 → ~6	2nd cleaning 10/15
CI1	7 degrees	5.7 → ~6	2nd cleaning 10/15
CI2	7 degrees	4.3 → ~4	2nd cleaning 10/15
CI3	7 degrees	4.3 → ~4	2nd cleaning 10/15
DI1	7 degrees	5.6 → ~6	2nd cleaning 10/15
DI2	7 degrees	6.4 → ~6	2nd cleaning 10/15
E1 (AA1)	20.5 deg.	6.4 → ~6	1st cleaning 07/15
E2 (AA2)	20.5 deg.	8.5 → ~9	1st cleaning 07/15
F1 (DA1)	20.5 deg.	7.8 → ~8	1st cleaning 07/15
F2 (DA2)	20.5 deg.	7.5 → ~8	1st cleaning 07/15
Total		**6.2 → ~6**	

Table 1: Energy yield increase at the football stadium "Stade de Suisse" in Bern, Switzerland, after cleaning the PV modules in summer 2015.

3.4 Discussion

It is rather unlikely that the 32 abnormalities, identified with thermography (in Section 3.3) cause 80 kilowatts of power reduction (corresponding to 6% increase in performance). Rather, it is the homogeneous contamination of the entire PV module surface that causes the significant power loss. This environmental contamination cannot be recognised in the thermal images as it does not appear as "thermal abnormalities" or "hot spots" in the PV modules.

4 CONCLUSIONS AND FURTHER AVENUES

Drone-based thermography is a highly efficient method for the periodic inspection of PV modules. A thermal imaging drone can detect smallest power losses (see Section 2) in PV modules ("hot spots"). These have a negative impact on the overall performance of a PV installation.

But can a robust relationship be established between drone-based thermal images and the measured and calculated power increase after cleaning the PV module surfaces? Field-measurements (see Section 3) evidenced that cleaning of PV modules resulted in a 4 to 9% performance increase with regard to energy yield. Homogeneous environmental pollution on the PV module surfaces was, however, difficult to discern using drone-based thermal images (unless the pollution appears at the bottom of framed modules). Pollution impacts upon energy yields are thus not easy to determine, and using thermography to quantify the effects from cleaning has not been achieved yet in this study.

Further research and field-tests will be carried out by adopting thermal imaging to the long-term PV installations in the Swiss PV network operated by the PV LAB at BFH Burgdorf, Switzerland. Increasingly, thermal drone imaging is requested by individual home owners as a service for economic inspection of their own PV installations. On the technical side, an integrated GPS waypoint navigation (automatic flight via GPS coordinates) in the flight control module of the copter is planned.

Acknowledgements

We gratefully acknowledge financial support from BFH and the Swiss Commission for Technology and Innovation (CTI). This project is carried out in the frame of the Swiss Centre for Competence in Energy Research on the Future Swiss Electrical Infrastructure (SCCER FURIES) [7] and with Swiss industry partners. Here, we particularly like to acknowledge the BKW group (www.bkw.ch) and Gebäudeversicherung Bern (GVB).

References

[1] R. Appels, B. Lefevre, B. Herteleer, H. Goverde, A. Beerten, R. Paesen, K. De Medts, J. Driesen, J. Poortmans, Solar Energy 96 (2013) 283-291.

[2] F. A Mejia, J. Kleissl, ISES Solar Energy Journal 95 (2013) 357-363.

[3] E. Klimm, T. Kaltenbach, D. Philipp, M. Masche, K.-A. Weiss, M. Koehl, Proceedings 31th European Photovoltaic Solar Energy Conference (2015) 2521-2523.

[4] U. Muntwyler, E. Schuepbach, M. Lanz, Proceedings 31th Photovoltaic Solar Energy Conference (2015) 1804-1806.

[5] H. Häberlin, Photovoltaik - Strom aus Sonnenlicht für Verbundnetz und Inselanlagen (2010) 494-495, ISBN VDE 978-3-8007-3205-0.

[6] E. Schuepbach, U. Muntwyler, M. Jost, T. Schott, Proceedings 29th European Photovoltaic Solar Energy Conference (2014) 2689-2691.

[7] http://sccer-furies.epfl.ch/

FLOATING PHOTOVOLTAIC INSTALLATIONS IN MALTESE SEA WATERS

Martin Grech, martin.b.grech@um.edu.mt
Luciano Mule Stagno, luciano.mule-stagno@um.edu.mt
Matteo Aquilina, matteo.aquilina@um.edu.mt
Institute for Sustainable Energy, University of Malta, Msida MSD 2080, Malta

Marco Cadamuro, Cadamuro@generalmembrane.com
General Membrane Spa, Via Venezia 538 - 30022 Ceggia - Venice - ITALY

Udo Witzke, udo.witzke@pandiaenergy.com
Pandia Energy (Malta) Ltd, The Duke Block A Flat 6, Triq ir-Repubblika, Victoria VCT 1017, Gozo, Malta

ABSTRACT: Photovoltaic (PV) installations require about 7-10m^2 of area per Kilowatt Peak (kWp) of power. This is quite a considerable area where space is scarce, such as regions with high population densities, cities or small islands. In such cases, large scale photovoltaic installations are not viable on land. However, if one can develop a viable and economically feasible solution to install photovoltaic installations at sea, a much greater area can be utilized for renewable energy. The aim of this study was to build and test prototypes of floating PV installations suitable for marine environment deployment. Different floating devices, mounting techniques and PV modules technologies were used in this study. The installations were deployed at sea, and data was recorded. The data,collected over an extended period of time,was compared with that of a similar installation constructed on land. . The costing and feasibility of the marine installation was also modelled and studied. Sea-based modules performed better than land-based similar systems, and a design that meets both the durability and the cost requirement appears possible. The lessons learnt from these prototypes may be carried forward in designing future floating installations.
Keywords: Floating PV installations, Marine, PV System

1 A SMALL ISLAND SCENARIO

Malta is a small island having one of the highest population densities in the world, where countryside is very limited and the price of land is high. Thus, large scale photovoltaic installations are not viable on land except on rooftops and a limited number of locations, which are already close to being exhausted.

It has already been concluded that around the Maltese Islands, the wind potential is limited and the wind resource is barely enough to make a deep offshore wind farm profitable [1]. This makes floating photovoltaic installations in a strong position to be considered as an alternative method for generating renewable energy. But floating photovoltaic installations require some form of floating device as a supporting paltform. If the cost of a floating structure will be competitive with the cost of leasing or buying land, then one should consider the option of large floating PV installations seriously.

Most of the works and studies carried out so far on floating PV installations were done on closed fresh water bodies, such as reservoirs or lakes. In Japan, two such large installations producing 3300 MWh annually were recently inaugurated in Nishihira and Higashihira reservoirs [2], and the largest floating installation is planned to start operating in 2018 on Yakamura Dam [3]
.

Open sea installations offer more challenges due to water salinity which increases the probability of corrosion, salt deposits on the PV modules themselves (shading the PV panels), waves, sea currents and generally harsher environmental conditions.

2 LITERTURE REVIEW

Most of the floating PV installations designed so far were intended for calm waters – lakes, dams or reservoirs – where enviromental and weather issues are not severe [4]. Such systems have the advantage of conserving land space, reduce evaporation and algae growth and improved efficiency. The Water Corporation of Korea installed two floating PV installations on two different dams, using standard PV technology. It was reported that the two systems gave 10.3% and 13.5% more electrical output when compared with a land based system. However, there were difficulties in anchoring the floating structures firmly [5]. In another case, PV panels installed at optimum angle on land (at 30°) gave better system yield than PV installations installed on Lake 'Aichi ike' (Japan), inclined at 1.3° and 10° to the horizontal. Spraying water on the floating PV systems improved their output by 4% in winter and 8% in summer. However, regular cleaning of the PV modules surfaces was required to keep this performance advantage. Water staining and bird droppings were the largest sources of dirt on the modules [6].

Many other methods have been studied to improve the efficiency of floating PV installations. Tracking mechanisms [7], cooling and radiation concentrators were all considered to increase the energy gain [8]. Submerging the PV modules under a maximum of 5cm of water also improved the output by nearly 10% [9].

The cost of floating PV installations is of great importance to determine whether such installations are economically feasible. Kim Trapani et al estimated that the capital costs would be €2554/kWp and €1770/kWp of installed crystalline and thin film modules respectively. This included floating structure and mooring. The running costs were estimated at €2.0/kWp and €1.36/kWp respectively but system costs have come down cosniderably since then[4].

3 AIM OF THIS STUDY

The study at hand is aimed at exploring the feasibility

of PV installations in Maltese sea waters. Issues such as the location of the installation, design of the floating platforms to support the installations, anchoring and costs were considered.

3.1 Location of the installation

The location of the floating PV installation had to be chosen in such a way that the floats will be exposed to considerable waves and wind, but not in very deep sea, otherwise anchoring would be very costly. At the same time, interference with any marine activity such as location near ports, fish farms or swimming/touristic areas was avoided. The distance from the power grid was also considered, so as to reduce the length and cables cost to eventually export the generated energy to the grid.

It was also planned to install a similar installation on the ground for comparison purposes. So the floating site had to be near a suitable land site available for land-based installation.

3.2 Design of the floating platform

A number of raft designs were considered and analyzed. First, different types of wood were considered. However, as it was deemed that this kind of material will require considerable maintenance throughout its life time in a Mediterranean climate, resulting in shoring the rafts and long down times, it was not considered further.

Hollow concrete cuboids shapes were also considered. To avoid the possibility of water entering in the concrete block if it cracked, the interior void was designed to be filled with expanded polystyrene (EPS). To reduce the cost, plastic (high density polyethylene) drums may replace the EPS. To reduce the weight, aerated concrete was also featured.

One other design was to build an EPS structure on a steel frame. The EPS required special coating to make sure that it will not crumble away when faced with the harsh marine environment.

Another design was to have a metal box suspended by floatation devices. Such a design was eventually improved to a metal pipe structure supported on plastic drums floats. Such rafts can be built in modules using off-the-shelf products, and the installation can be assembled to the required size with ease. The materials used for such construction render the build very straight-forward such that no specialized or particular training is required. It would also be relatively inexpensive.

Other self floating photovoltaic devices were also part of the study. These consisted of spherical shapes, with the PV modules in-built in them (self-floating panels).

3.3 Costing

The floating platform will increase the cost of the installations. However, if the cost of the raft is offset against the purchase or long term lease of land, the floating installations may be economically feasible. Thus, one main criterion for the design of the platform were that cost of the raft per square metre was of the same order or less that the cost of purchasing or leasing land for long term PV installation. In Malta, the cost of land for such use is about €33.50 per square metre [10].

From the designs listed above, some were below this benchmark cost. Mahogany and Scott pine wood are viable, though one has to consider the lifetime of the latter in the sea. Aerated concrete encapsulating plastic drums and metal pipes structures on drum floats achieved this criterion as well.

Table I shows the approximate material costing for several types of platforms considered in relation with this study. The labour cost depends on the scale of platform, salaries and available production facilities.

Table I: Cost per unit area of various platforms

Type of floating platform	Cost €/m^2
Steel cuboid	€35.91
Steel platform with HDP drums	€19.59
Steel pipes with drums	€7.31
Wood mahogany	€17.50
Wood Scotts pine	€8.00
Concrete with EPS	€80.40
Concrete with HDP drums	€21.24
Concrete platform with drums	€17.24

4 CONSTRUCTED RAFTS

A number of rafts and floats were constructed for this project, and what follows is a brief description of each model. The orientation of the rafts on the sea surface was in such a way that the panels faced South. This orientation was ensured through the mooring system used in each case.

The floating structures were tied to concrete blocks (sinkers) placed on the sea bed. There were 12 blocks arranged in a 4 by 3 matrix. Each block measured 1.2 x 1.2 x 0.4 metres, costing €265 each, including haulage and diving charges. The sinkers were designed with such a wide footprint so that they will be less susceptible to movements by water currents and foundation movements. The floats were tied to one or more sinkers, depending on its size and shape.

It should be noted that the PV panels used in these systems were not connected to the main electricity grid. The power generated was dissipated in resistors on the rafts themselves to avoid the complication, permits and added cost of having a submarine cable connection to the grid. Such a measure is justified when considering that this project was intended to serve as a proof of concept.

4.1 Reinforced expanded polystyrene

Two rafts, measuring 7.5 metres by 1.2 metres each, were designed to keep the modules and data logging equipment. The rafts were also covered by a sheet of thermoplastic polyolefin (TPO) material to give the float more durability. The rafts housed two metal compartments which were used to store the data logging equipment, electrical loads to dissipate the energy generated and any other electronic equipment as necessary.

Each raft had three 144 Wp PV modules connected in series. Each module consisted of 22 triple junction amorphous silicon solar cells in, encapsulated in a highly transparent polymer resistant to UV radiation and weathering. A similar system was installed on land, consisting of two single PV modules, dissipating the energy in an identical resistor. It was located about 500m away from the marine rafts, so that the solar radiation and cloud coverage will be very similar on both installations.

4.2 PV module floats

The second prototype consisted of polycrystalline PV modules suitable for marine environment mounted on simple polystyrene floats. The PV modules were semi-flexible and had aluminium backing plate. On the float, there were two 40Wp modules, electrically separated. The resistors and the data logger were housed on the bigger raft described in 4.1 above, connected by sub-surface cable. One module was further protected with lacquer spray, while the other was left as supplied. A similar setup was installed on land, near the first one, for comparison purposes. Table II shows the results of the two sets of PV modules.

Table II: Daily average output of PV module floats per Wp installed

Module	Watt hour	Total Wh
Float Module 1	3.70	7.48
Float Module 2	3.78	
Land Module 1	3.30	6.73
Land Module 2	3.43	

4.3 Self floating photovoltaic devices

The third device consisted of 500mm diameter transparent acrylic self-floating hemispheres. Inside these 'domes', two types of small PV cells were installed, connected in independent systems. Twenty epoxy based 0.3W crystalline silicon modules were connected in series to form one system. The other type consisted of five flexible amorphous silicon strips, also connected in series.

The domes were connected, by means of a cable, to the main raft, described in 4.1 above, were the data loggers and resistive loads were housed.

4.4 Metal pipes rafts

Two other rafts made of steel pipes with drum floats were constructed, measuring 2.2 by 2.6 metres. The drums were reinforced from the inside to withstand the water pressure. Two PV modules and the necessary data loggers and electrical loads were installed on each; thus each raft formed an independent system.

One raft had two 260Wp polycrystalline modules on an aluminium frame, while the other raft had two 155 Wp polycrystalline cells framed in glass-glass sandwich configuration. Both sets of modules were inclined at 20° facing South on aluminium supports. These systems would also be compared against standard land-based systems.

5 RESULTS

5.1 Reinforced expanded polystyrene with amorphous PV modules

The two rafts have performed very well in the marine environment. They endured gale force winds and waves having a swell up to 2 metres high. The raft materials and the PV modules did not show any sign of corrosion or weathering during the 18 months at sea so far.

The sea salt deposits on the PV modules of the raft were found to be negligible and much less dust was deposited on them, when compared with the land based installation.

The electrical output of the modules on the raft was 4.2% more than those installed on land.

The PV modules on the rafts were attached to the solid flat surface underneath, in such a way that the module on the rafts did not gain any sea cooling advantage.

a) Reinforced EPS with thin film PV modules

b) Crystalline PV module float

c) Steel pipes rafts with HDP drums

d) Floating dome

Figure 1 (a – d): Floating PV installations

Table III: Quarterly average output per Wp installed of the thin film PV module

Quarter	Raft Module	Land Module
2015 Q2	409	408
2015 Q3	378	332
2015 Q4	134	137
2016 Q1	199	203
2016 Q2*	321	304
Total	**1441**	**1384**

** Till 16 June 2016*

Table III compares the outputs of the raft and land installations. During the spring and summer time, the sea installations perform better.

As a side observation, two 20Wp polycrystalline modules were installed at 30° inclination on the raft to charge the equipment battery. The modules were constructed on glass sheet, ethylene vinyl acetate (EVA) films, standard back sheet and framed in aluminium. They have performed very well and did not show any sign of deterioration during the 15 months sea trial.

5.2 PV module floats

The average electrical output of the two PV modules installed at sea was 11% higher than the land based pair of modules.

The measurements were taken during August – September 2015, when the air temperature was about 33°C. The surface temperature of the modules on land is higher than this. However, the sea temperature was 24°C and it effectively cooled the floating PV modules, being installed near the sea surface and having aluminium back support. Thus most of the efficiency gain may be attributed to the cooling effect.

Though the PV modules were claimed to be suitable for marine environment, the aluminium edges were not suitably protected, and corrosion started to develop after about one month. This eventually led to the failure of the PV modules. However it is probable that this setup could be used long term if the panels are better encapsulated.

5.3 Self floating photovoltaic devices

The hemi-spherical design of these devices implies that some of the modules will not be optimally exposed to solar radiation. The epoxy based crystalline modules output was severely affected by this and the output was low. The amorphous silicon flexible strips gave an average of 4 Wh output. Some condensation on the inside was noted, but most cleared in warmer temperatures.

Given the small scale of the model, this design may be applicable to situations where the power needed is small. But to extract significant amount of energy, the cost of cabling to connect each dome and the individual mooring could offset the energy gains. Moreover the original design was envisioned to be a hemisphere or inverted saucer shaped container with semi-transparent thin film module all over the inner surface. This would ensure some level of production even from the side opposite the sun. In addition the extra cost of individual mooring could be comparable or less in cost to a raft in shallow waters.

5.4 Metal pipes rafts

These structures have been launched only recently (April 2016), but still have experienced some rough weather. Structurally the rafts performed as expected without any issues. However, the HDP drums had to be reinforced from the inside, as some collapsed from side pressure and forces. Output is being measured and compared to land-based systems.

6 CONCLUSIONS AND DISCUSSION

6.1 Energy output

From the designs and results presented above, it can be shown that it is possible to have PV installations at sea, but the success and duration of installation depends on the suitability of materials and the preventive measures taken to withstand the marine environment.

The reinforced polystyrene raft gave good results in terms of electrical output, reliability and marine worthiness. Such a construction was demonstrated to have a long life time.

The PV floats had good efficiency but due to lack of adequate edge protection, the modules failed after a number of weeks. With better module encapsulation, this design could be further tested and improved. It is expected that positive results could then be achieved.

The small domes can be very effectively used to power low demand nearby equipment or in large numbers in shallow waters where the extra costs associated with mooring would be offset by the absence of a raft. Moreover the design would be such that the inside of the domes (or inverted saucers) would contain the semi-transparent thin film PV material – or the panel could sit flat at the bottom of the container. Further, keeping in mind that the floats have to be placed at a safe distance from each other, the area needed will also be higher.

It was observed that generally the installations on the sea benefited from the cooler and cleaner environment, while the salt deposits do not appear to be an issue.

6.2 Cost of rafts

The cost of the rafts' construction is crucial for the economic viability of the whole project. Other studies have been conducted about the costs of different types of rafts/platforms suitable for marine PV installations. Costs vary considerably, as they are dependent on the labour cost, scale of platform undertaken and availability and cost of materials needed. However, it was shown that some types of platforms can be built with costs close to or cheaper than purchase or long lease of land for PV installations, as seen in Table I.

7 ACKNOWLEDMENTS

This study was part of the SolAqua project (R&I 2012-041) funded by the Malta Council for Science and Technology. The partners in the project were General Membrane Spa (Italy), Pandia Energy (Malta) and the University of Malta.

8 REREFENCES

[1] Mott MacDonald, *Feasibility Study For Increasing Renewable Energy Credentials, January 2009.*
[2] http://www.sciencealert.com/two-enormous-floating-solar-power-plants-have-just-opened-in-japan?utm_source=Article&utm_medium=Website&utm_campaign=InArticleReadMore
[3] http://www.sciencealert.com/the-world-s-largest-floating-solar-power-plant-is-being-built-in-japan?perpetual=yes&limitstart=1
[4] K Trapani, D Millar, *Proposing Offshore Photovoltaic (PV) Technology To The Energy Mix Of The Maltese Islands*
[5] Young-Kwan Choi, Nam-Hyung Lee, Kern-Joong Kim, *Empirical Research On The Efficiency Of Floating PV Systems Compared With Overland PV Systems, 2013*
[6] Yuzuru Ueda, Kosuke Kurokawa, Makoto Konagai, Sadayuki Takahashi, Akito Terazawa, Hiroyuki Ayaki, *Five Years Demonstration Results Of Floating PV Systems With Water Spray Cooling, 2013*
[7] R. Cazzaniga, M. Rosa-Clot, P. Rosa-Clot, G. M. Tina, *Floating Tracking Cooling Concentrating (FTCC) System*
[8] http://phys.org/news/2012-02-italian-solar-panels.html
[9] Marco Rosa-Clot, Paolo Rosa-Clot, *Submerged Photovoltaic Solar Panel: SP2,* August 2008
[10] E Vassallo, Luciano Mule Stagno, *Assessing The Design Options for The Optimization of A 2 MWp Ground Mounted System in Malta, 2014*

DEVELOPMENT, APPLICATION AND VALIDATION OF A COMPACT, PORTABLE SOLAR CELL CHARACTERIZATION DEVICE UTILIZED FOR BIPV ANALYSIS

Dominik Holzmann[1], Christoph Mayer[1], Lukas Neumaier[1], Christina Hirschl[1]
CTR Carinthian Tech Research AG, Europastraße 12, 9524 Villach, Austria
dominik.holzmann@ctr.at, christoph.mayer@ctr.at, lukas.neumaier@ctr.at, christina.hirschl@ctr.at

ABSTRACT: During this work a portable solar cell characterization device as well as its application in a building integrated photovoltaic (BIPV) test system is shown. The solar cells are characterized by measuring their current-voltage curve (I-V-curve) in user-defined intervals. A test setup was build that allows the application of various types of façade elements and solar cells. Up to four solar cells, oriented at the four cardinal directions, can be measured simultaneously over a whole day, which allows a comprehensive comparison of them. Besides additional energy yields also advantages of a more consistent/user adapted energy distribution over a day, resulting from façade-mounted solar cells/modules, can be evaluated. To enable a more detailed data analysis the local irradiance as well as the temperature of each cell and the specific façade panel is determined. Outdoor measurements under the influence of natural sunlight and various weather conditions were conducted with crystalline cells and thin film (CZTS) modules, mounted on exchangeable façade elements.
Keywords: solar cell, module, BIPV, façade, solar cell characterization device

1 INTRODUCTION

Building integrated photovoltaic (BIPV) aims to make use of the building façade areas to generate electricity. To realize architectural highlights and enhance the energy yield, BIPV can be integrated on sidewalls or roofs of buildings. To evaluate the performance of BIPV-elements in dependence of directional effects (temperature variations, reflection effects and irradiance), a portable test setup was developed. This setup consists of a quadratic building like structure, where various types of façade elements equipped with mini PV modules and sensors can be attached. The test system can be placed anywhere outside to allow solar cell measurements under realistic, natural conditions. With this characterization device long term tests of solar cells on façade elements were performed. Based on the test results, statements about the cost effectiveness and benefits of various cell and façade types can be approximated.

2 SOLAR CELL CHARACTERIZATION DEVICE

The custom single cell characterization device (SCCD) shown in Figure 1 was developed and validated by utilizing a commercial solar cell flash tester [1]. The SCCD enables solar cell characterization as well as long term monitoring under realistic conditions. Besides the measurement of the current-voltage curve other relevant parameters, like the cell and the façade temperatures as well as the solar irradiance are measured, which are used to calculate the STC parameters. The current-voltage curve is obtained by applying a variable load and a shunt resistor to the solar cell and measure the resulting current and voltage. From the I-V-curve various important cell parameters like the short circuit current i_{sc}, the open circuit voltage v_{oc} and the maximum-power-point (MPP) can be obtained [2]. Especially the field of research and development on solar cells has interest in measuring the current voltage curve of single solar cells on site.

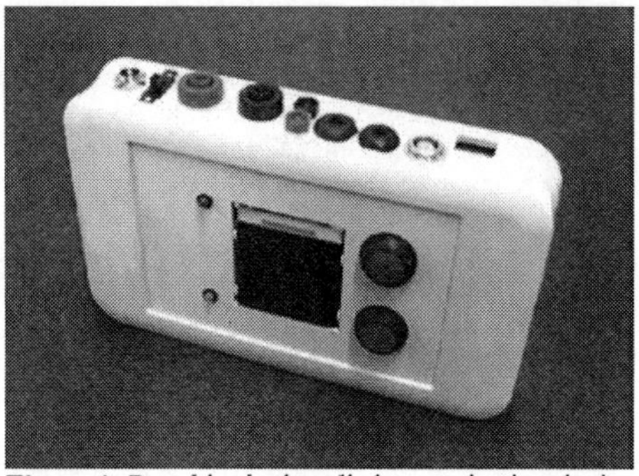

Figure 1: Portable single cell characterization device (SCCD)

Especially for BIPV monitoring applications, two temperature sensors are available to allow measurements of the façade interspace as well as the backside temperature of the PV cell/module. A typical measurement setup utilizing the portable SCCD is shown in Figure 2.

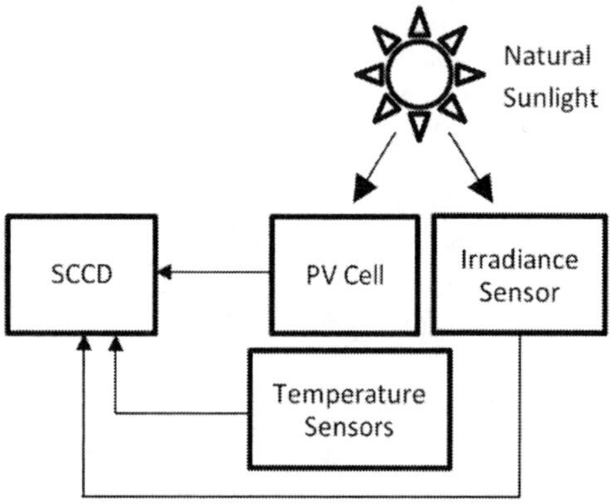

Figure 2: Schematic measurement setup

Since crystalline and thin-film solar cells differ regarding their maximum cell voltage and current, the SCCD can be configured according to the respective cell technology. Currently the two used measurement ranges are shown in Table I.

Table I: Measurement ranges of the SCCD

Parameter	Range	Accuracy
Cell Voltage [V]	0 – 4 / 0 – 15	-
Cell Current [A]	0 – 10 / 0 – 0.2	-
Temperature [°C]	-20 to +400	±2
Irradiance [W/m²]	0 – 1400	±10 %

An integrated logging function (adjustable interval) enables long term measurements and furthermore allows the comparison and validation of the performance of BIPV façade elements mounted at different orientations. Such long term measurements were performed for different solar cell types on the façade test system.

3 FAÇADE TEST SYSTEM

To compare different façade systems and solar cell technologies for BIPV applications the experimental setup shown in Figure 3 was developed [3]. It imitates a building with four exchangeable façade panels where mini BIPV modules and irradiation- as well as temperature sensors are integrated. The façade elements are made of composite wood and have a width of 30 cm, a height of 50 cm and a thickness of 1.2 cm. They are mounted using a standard aluminum façade mounting structure (horizontally aligned profiles and clasps). One temperature sensor is placed directly behind the solar cell. The second temperature sensor is mounted at the backside of the façade element (air gap between "wall" and façade panel). The knowledge of the façade interspace temperature might be useful to find façade systems that are best suitable for BIPV integration regarding a possible cooling effect, which would positively influence the PV module's performance. A lower temperature in the façade interspace could subsequently lead to a lower cell temperature and therefore an increased power output. For conducting the measurements, each façade panel is equipped with a PV module which is furthermore connected to one of the SCCDs. The experimental setup is placed outside with one façade element oriented respectively to one of the four cardinal directions (north, south, east and west). The north façade points towards a building (distance ~ 10 m) to measure the impact of indirect/scattered light due to reflections of surrounding structures.

Figure 3: BIPV test system – outdoor measurement

Two different types of solar cells were mounted on the exchangeable façade elements shown in Figure 4. Both cell types were tested outside with the experimental setup for several days. The recorded data is used to evaluate the advantages and disadvantages of both solar cell technologies when utilized in BIPV.

Figure 4: Façade panel equipped with crystalline solar cell

4 RESULTS

Crystalline solar cells and thin-film modules were chosen for a comparative test regarding their qualification in BIPV applications. Typically crystalline solar cells/modules are oriented in south direction since their good performance at high irradiation levels. Thin film solar cells show a higher sensitivity for indirect light/diffuse irradiation compared to crystalline cells. This characteristic makes them interesting for BIPV where an optimal orientation of the solar modules might not be possible due to structural conditions, aesthetic aspects or statutory provisions.

4.1 Crystalline cells

Standard crystalline solar cells with an active area of 15.6 cm times 15.6 cm and based on three bus-bar technology were used. Their short circuit current is approx. 9 A and the open loop voltage is approx. 0.6 V at STC. Measurement results for crystalline cells recorded with sunshine in the morning hours and upcoming clouds starting at noon are shown in Figure 5. The power output (maximum power point) of the solar cell was measured and evaluated with the SCCD for each cardinal direction. Under these conditions, the BIPV cells oriented to the south and east plane produced the highest amount of energy (4.93 Wh, 6.44 Wh, respectively). Nevertheless a noticeable electricity production was also measured for the PV cells oriented west and north (1.05 Wh, 0.49 Wh).

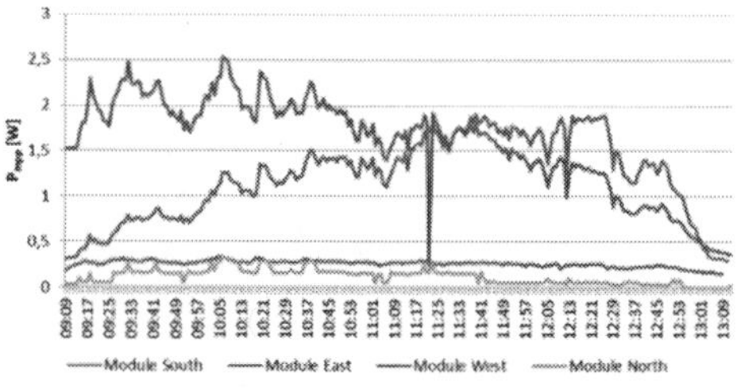

Figure 5: Crystalline solar cells, power output

The local irradiance, shown in Figure 6, was recorded during the tests utilizing a high precision pyranometer (Kipp&Zonen CMP11). The local irradiance data directly correlates to the power output measured at the solar cells.

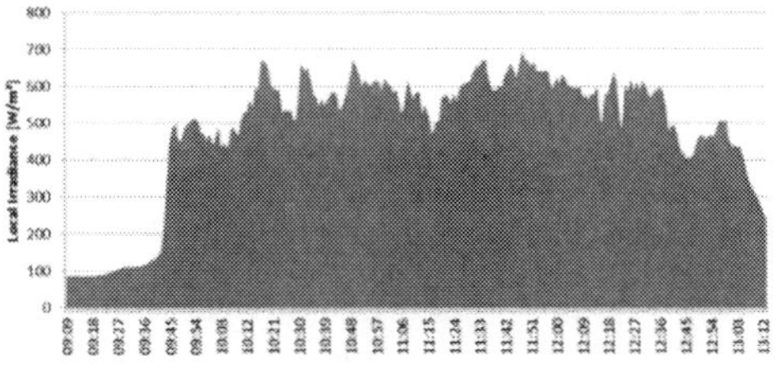

Figure 6: Crystalline solar cell test - local irradiance

The exemplary test with the crystalline cells was conducted on the 20th of November 2015 at an average ambient temperature of 8 °C. Figure 7 shows the measured temperatures on the south side solar cell and façade as well as the maximum power of the module. A higher warming of the cell compared to the backside of the façade is clearly visible.

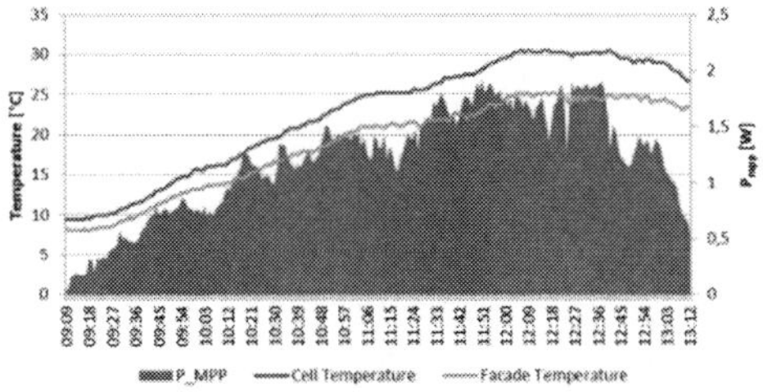

Figure 7: South side with crystalline solar cell - cell and façade temperatures

The maximum measured temperatures on the respective cell and façade panel are shown in Table II.

Table II: Maximum cell and façade temperatures - crystalline solar cell test

Parameter	East	South
Max. Temperature Cell [°C]	28.75	30.50
Max. Temperature Façade [°C]	22.75	25.25

During the test, the local irradiance as well as the cell temperature did not meet the standard test conditions (STC, 25 °C, 1000 W/m²). The recorded data represents realistic results under the influences of temperature, irradiance, weather effects (e.g.: cooling wind) and diffuse light (light scattering on surrounding objects). Nevertheless corrective STC calculations are possible to determine the current-voltage characteristics of the respective solar cell under standardized conditions.

4.2 Thin-film CZTS modules

The thin-film CZTS modules used for this test have an active area of 9 cm times 11 cm. The short circuit current is approx. 64 mA, the open circuit voltage is approx. 12.3 V and the maximum power is 0.395 mW. The measured power output for the CZTS modules is shown in Figure 8. As expected, the highest amount of energy is produced by the module oriented to the east during the morning hours. Due to the alternating position of the sun, the energy produced by the module on the south façade rises until noon, whereas the energy yield of the eastern module declines. The measurement data for the south and east modules are correlated to the changing position of the sun during the morning. The highly illuminated east and south modules produced 1.24 Wh 0.78 Wh respectively during the measurement interval. The indirect lighting on the north and west module lead to 0.17 Wh and 0.22 Wh respectively.

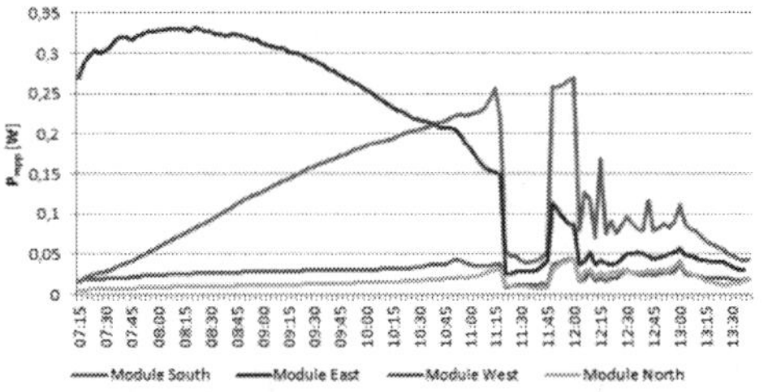

Figure 8: Thin-film CZTS modules, power output

The measured maximum power as well as the local irradiance which is shown in Figure 9, indicating an energy drop at noon due to a large cloud. After 12 o'clock increasing cloudiness leads to a lower local irradiance and energy production of the modules.

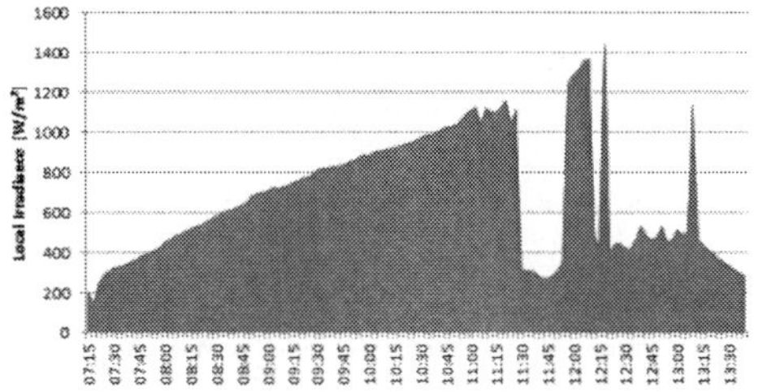

Figure 9: Thin-film CZTS module test - local irradiance

The thin-film CZTS modules were tested on the 17th of May 2016 at an average ambient temperature of 20°C. The module and façade temperature as well as the maximum power for the south side module is shown in Figure 10. The module temperature reaches again distinct higher values than the façade.

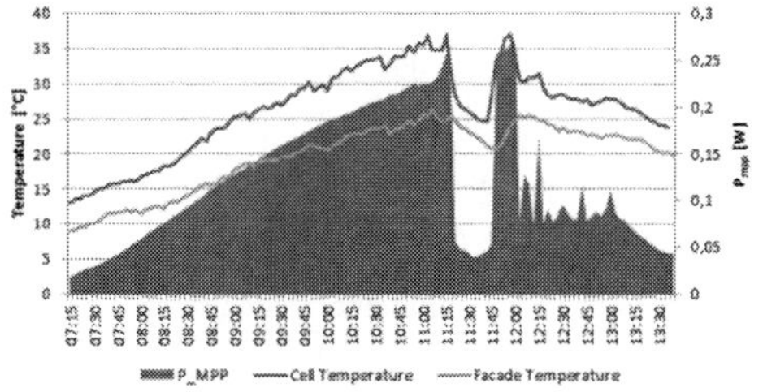

Figure 10: South side thin-film CZTS module, module and façade temperature

The maximum module and façade temperatures are shown in Table III.

Table III: Maximum module and façade temperatures - CZTS module test

Parameter	East	South
Max. Temperature Module [°C]	39.75	37.0
Max. Temperature Façade [°C]	28.75	26.5

4.3 Practical Benefit of BIPV

An example for a standard load profile for a household (H0) on a winter day is shown in Figure 11. For the shown timespan the energy demand is nearly constant. Furthermore the maximum power on a winter day (11th February 2016) for crystalline cells mounted on the façade test system in all directions is displayed. Using solar cells on all façades instead of only using them on the south of the test system provides a nearly constant energy yield through most of the day. This allows a good coverage of a standard household energy demand.

PV systems oriented to the south, generate high energy peaks on clear days at noon. The standard load profile shows that the high amount of energy, which is available at noon, is not consumed in most households. Therefore significant improvements concerning economic energy usage are possible by adapting the configuration/orientation of the BIPV system.

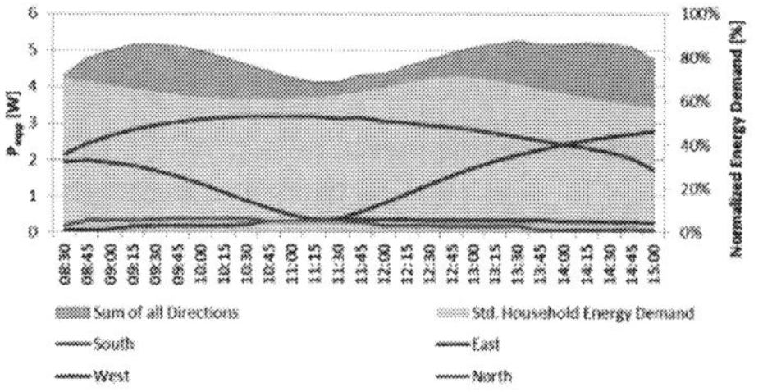

Figure 11: Power output from crystalline solar cells in comparison to the energy demand of a standard household [4]

5 SUMMARY AND OUTLOOK

Within this work a portable single solar cell characterization device and a test structure for BIPV façade elements were developed. Façade elements with crystalline cells and thin-film modules were evaluated under natural conditions for whole days.

Typically solar modules on buildings are mounted on the direction with the highest local irradiance, the south façade. Using BIPV on all outer walls of a building instead of only one leads to a uniform appearance and increases the amount of energy produced. Further the produced energy is distributed more regular through the day. This corresponds very well to the energy demand of standard households and public buildings respectively.

Based on our measurements the additional energy benefit from using solar cells/modules on all façade directions compared to only use them on the south side is around 262% for crystalline- and 194% for thin-film CZTS technology.

During the practical tests of the different cell types, potential ideas for improving the SCCD were identified: Currently each device is able to measure one solar cell. Therefore four devices are needed for our BIPV façade test system. It is desirable to develop a single system that is able to measure four or more solar cells simultaneously. Furthermore, the short circuit current and open circuit voltage of different solar cell technologies are differ widely. To measure all kinds of solar cell technologies, a user friendly adaption of the measurement range will be implemented in the future.

6 ACKNOWLEDGEMENT

The project PV@Fassade is funded by the Austrian Climate and Energy Fund and conducted under the "Energy Mission Austria" program.

7 REFERENCES

[1] C. Mayer, D. Holzmann, CIEEE, New Approach of designing a Solar Cell Characterization Device, (2015).
[2] International Electrotechnical Commission. Standard IEC 60904-1. Photovoltaic Devices. Part 1: Measurement of Photovoltaic Current-Voltage Characteristics.
[3] L. Neumaier, D. Holzmann, W. Mühleisen, Ch. Hirschl, G. C. Eder, 31. Symposium Photovoltaische Solarenergie, Bad Staffelstein, Ertragssimulationen gebäudeintegrierter Photovoltaik mit unterschiedlichen Modultechnologien und Konzepten, (2016).
[4] T. Karner, Bachelor Thesis, Technical University of Vienna, Synthetische Haushaltslastprofile, (2014).

ADVANCED PERFORMANCE MONITORING SYSTEM FOR IMPROVED RELIABILITY AND OPTIMIZED LEVELIZED COST OF ELECTRICITY

George Makrides*[1], Alexander Phinikarides[1], Juergen Sutterlueti[2], Steve Ransome[3] and George E. Georghiou[1]
[1]FOSS Research Centre for Sustainable Energy, Photovoltaic Technology Laboratory,
Department of Electrical and Computer Engineering, University of Cyprus
75 Kallipoleos Avenue, P.O. Box 20537, Nicosia, 1678, Cyprus
[2]Gantner Instruments GmbH, Montafonerstraße 4, 6780 Schruns, Austria
[3]SRCL, Kingston upon Thames, United Kingdom
*corresponding author George Makrides (phone: +357 22 894397, Fax +357 22 895370, makrides.georgios@ucy.ac.cy),
Alexander Phinikarides (phinikarides.alexander@ucy.ac.cy), Juergen Sutterlueti (j.sutterlueti@gantner-instruments.com),
Steve Ransome (steve@steveransome.com), George E. Georghiou (geg@ucy.ac.cy)

ABSTRACT: The key factor that will enable and enhance the further increase of the uptake of photovoltaic (PV) technology globally, is the reduction in PV electricity costs by increasing the lifetime output of PV systems. This can be achieved by improving the reliability and service lifetime performance through constant, solid and traceable PV plant monitoring of installed systems. In this way, the investment cost, levelised cost of electricity (LCoE) and in general PV competitiveness can be enhanced positively. It is with this background that the project "Innovative Performance Monitoring System for Improved Reliability and Optimized Levelised Cost of Electricity" (IPERMON) funded by the SOLAR-ERA.NET Transnational Calls PV3 and CSP3, has been initiated between Gantner Instruments (GI) GmbH and the PV Technology Laboratory of the University of Cyprus (UCY), in order to develop an innovative monitoring system with enhanced capabilities, far beyond the state-of-the-art. The proposed monitoring system will be capable to detect performance losses and failures, determine the degradation rate and provide accurate hourly and day-ahead power production forecasts. The scope of this work is to present the main scientific, technological and commercial objectives of the project in the field of PV performance monitoring systems and to present first results for the formulation and benchmarking of innovative guidelines and algorithms for the detection and diagnosis of performance losses, degradation and failures.
Keywords: c-Si, degradation, forecasting, energy performance, lifetime, PV system.

1 INTRODUCTION

A key factor that will enable the further increase of the uptake of photovoltaic (PV) technologies is the reduction of PV electricity costs by increasing the lifetime output. In particular, this can be achieved by improving the reliability and service lifetime performance through constant, solid and traceable PV plant monitoring of installed systems. In this respect, safeguarding the optimal operation of PV systems directly impacts positively the investment cost, levelised cost of electricity (LCoE) and in general the technological competitiveness.

A main challenge in the quest for ensuring quality of operation especially for grid-connected PV systems is to safeguard reliability and optimal performance by identifying and quantifying accurately the factors behind the various loss mechanisms, minimising them by best design practice, while also detecting and diagnosing potential failures at early stages or before occurrence, through robust monitoring, fault detection and preventive maintenance.

At present, the most common guidelines or standards for PV system performance monitoring include the IEC 61724 standard [1], the guidelines of the European Joint Research Centre in Ispra, Italy [2-3] and the National Renewable Energy Laboratory (NREL) [4]. In addition, many academic institutions and organisations are particularly active in the development of sophisticated failure detection routines (FDR), degradation rate determination and accurate forecasting methodologies that are necessary for improved production [5-9].

At the moment, as most approaches for performance loss and failure identification are yet at an initial research stage and not fully tested and verified on a large amount of field data, there is as yet no complete monitoring platform for investors and in general end users to assess and diagnose performance issues of PV systems. As a result, there is a strong market need for a unified independent baseline platform for effective and automated PV plant management, which will be vendor-independent, and that will offer reduced risk for PV plant owners, preventive maintenance and improved financial key performance indicators (KPI) for investors.

The purpose of this work is to present the main scientific, technological and commercial objectives of the IPERMON project in the field of PV performance monitoring systems and to present first results for the formulation and benchmarking of innovative guidelines and algorithms for the detection and diagnosis of performance losses, degradation and failures. The end-product of the project will be an advanced monitoring system, which will act as a high level watchdog in ensuring reliability and operational quality of PV power plants and eventually yielding increased lifetime output.

2 PROJECT WORK PLAN

The project will be executed by a transnational consortium from two countries, Gantner Instruments (GI) GmbH and the PV Technology Laboratory of the University of Cyprus (UCY) and the duration is 36 months. More specifically, the methodology and technological process that will be utilised for this work includes the following steps:

1) Analytical review and benchmarking of data acquisition practices and guidelines.
2) Development of failure detection routines (FDR) to analyse sub- five minute meteorological and

PV operational data-sets against profiles of different failures and comparisons with predefined profiles (energy loss, deviation, behavioural change, spatial/neighbour PV array dimension, duration and modelled desired output correlations) of common failure modes.

3) Application of filtering steps and advanced statistical analysis techniques such as the Autoregressive Integrated Moving Average (ARIMA), regression model with ARIMA (regARIMA), robust Principal Components Analysis (rPCA) [10-13] and Loss Factors Model (LFM) [14] for the assessment of data acquisition quality, outages and performance assessment.
4) Application of trend-extraction methods for the algorithm formulation of performance losses, failures and long term degradation.
5) Implementation of advanced pattern classification and recognition algorithms employing artificial neural networks (ANN) for the production forecasting of grid-connected PV systems.
6) Emulation of failures and validation of the algorithms.
7) Use of web-based application programming to integrate the developed algorithms to an advanced web-portal.
8) Testing of the performance and accuracy of the tool against real operating PV systems.

The developed monitoring solution will be hosted as a plug-in to the already implemented web-portal of GI (see Fig. 1).

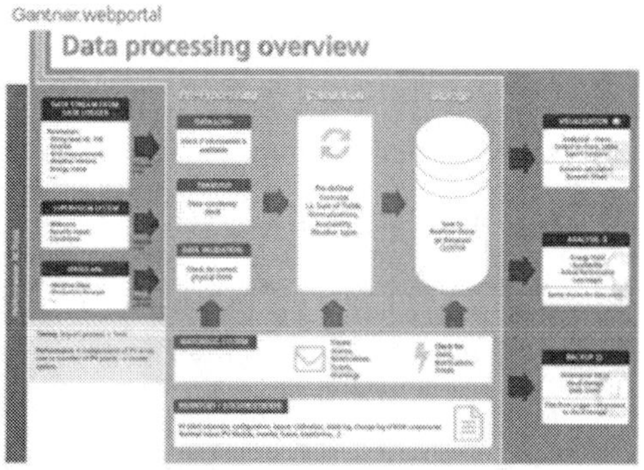

Figure 1: GI webportal functionality outline.

3 METHODOLOGY

3.1 Data-acquisition guidelines

Data-acquisition through advanced monitoring systems is necessary to measure and assess the performance of PV systems and also to identify design faults and failures. For many large scale grid-connected PV systems, performance analysis on acquired data is further performed in order to identify and prevent economic losses due to operational problems and timely detection.

The first part of the project is the formulation of monitoring guidelines on the acquisition and analysis of PV performance data-sets for the development of innovative algorithms for the diagnosis of performance losses and FDR. In particular, it is imperative that in order to detect failures and performance losses the data-sets sampled and acquired will be at a high resolution (1-5 minutes), thereby extending the measurement requirements beyond the recommendations of the IEC 61724 standard [1], for measurement, data exchange and performance analysis of PV systems. Typical measurements acquired at the developed testing sites will include with respect to the meteorological data, the total irradiance in the plane of the array (E_{POA}), ambient temperature (T_{amb}), module temperature (T_{mod}), wind speed (W_s) and wind direction (W_d). Additionally, with respect to the PV operational system side data acquired include (DC voltage, current and power) and utility side (AC grid voltage, current, power, outage duration period). Figure 2 shows an outline of the proposed advanced monitoring system installation including sensors, data acquisition devices, storage systems and portals for large scale PV plants.

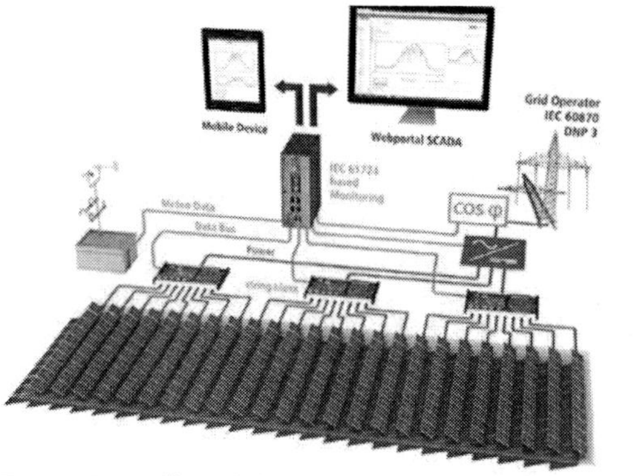

Figure 2: Outline of the advanced monitoring system installation for large scale PV plants.

In addition, an important consideration before analysing any time series data set is the quality of data. In this sense, the service time i.e. the duration of monitoring without outages, is an important parameter to be known before commencing the analysis of the data and quantification/qualification of losses. In particular, attention should be given to the following details [4]:

- The data must be carefully examined for missing or erroneous data.
- Inconsistencies in the frequency of data collection and/or duplicate records should be identified and addressed.
- Nonzero irradiance and electrical measurements that might affect the expected or measured energy should be removed. Night time data should only be used for meteorological data such as temperature, precipitation and windspeed analysis.

Throughout the progress of the project different methods will be investigated on how to analyse the time series for outages and what data filtering techniques and algorithms can be applied to reconstruct the time series. The influence of outages on the degradation rate estimation and energy yield production will also be studied.

3.2 Degradation rate determination

Based on previous work performed at the UCY in the field of degradation rate determination [10-13], the developed algorithms will make use of data sets acquired for the irradiance, ambient/module temperature and DC/AC maximum power (P_{mp}), in order to first calculate the performance ratio. Once the performance ratio (PR) is calculated on a daily and monthly basis then suitable statistical techniques such as seasonal autoregressive integrated moving average (SARIMA) and principal component analysis (PCA) are applied on the implemented time series. The following flow chart describes the general principle of the degradation rate algorithms (see Fig. 3).

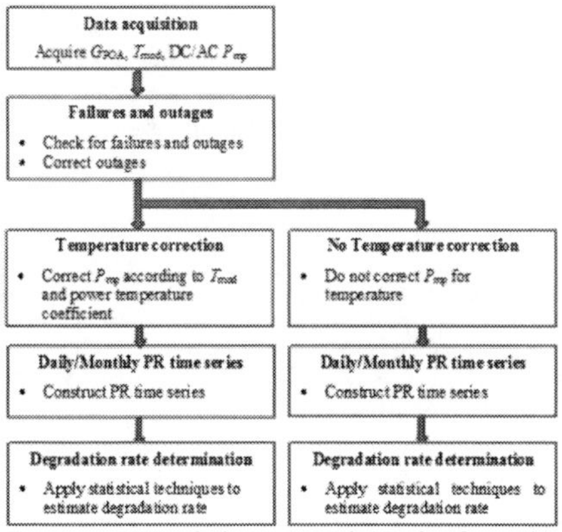

Figure 3: Degradation rate determination algorithm procedural flow chart.

3.3 Failure detection routines (FDR)

Based on the proposed guidelines different algorithms will be developed in order to identify and quantify system performance losses due to thermal and irradiance behaviour, soiling, shading, string mismatch, bypass diode effects, inverter failures and insulation faults, non-optimal maximum power point tracking (MPPT) and unexpected trends. The algorithms will perform real-time comparisons against known failure profiles and performance effects, threshold levels and modelled expected performance in order to first identify the loss and then quantify the effect.

More specifically, the FDR will operate on acquired data sets by producing the daily power production profile which will be tested against known failure/malfunction profiles, determining thus possible causes. The important parameters considered for the verification of the failure include:

- Comparison with simulated energy yield models.
- Magnitude of the fault.
- Parameters affected and correlated with the fault.
- Duration of the fault (day to month).
- Cross reference with other healthy parameters.

For the development of the FDR the system will comprise of the failure detection tool and the failure profiling and footprint methods. The routines will evaluate patterns of energy loss by creating profiles and comparisons with predefined profiles of frequently known failures such as hot spots, bypass diodes, inverter faults, grid outages and others. The operational verification of the algorithms will be performed on an experimental test-setup at the UCY and the outdoor test facility (OTF) of GI against emulated faults and measured performance effects (see Fig. 4).

Figure 4: Outdoor test facilities of a) GI in Tempe, Arizona and b) UCY in Nicosia, Cyprus.

The flow chart below describes the procedural flow of the FDR (see Fig. 5).

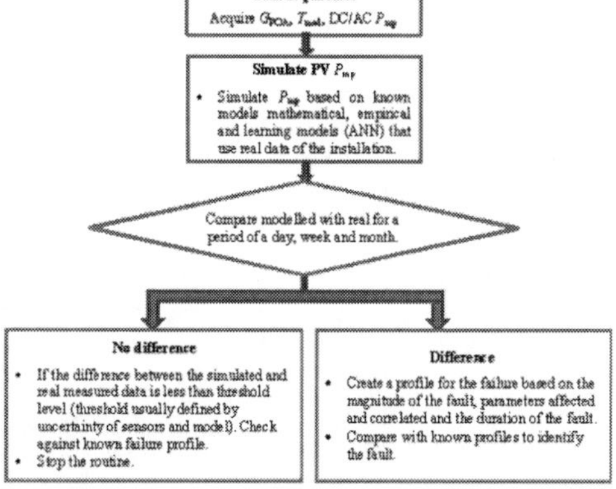

Figure 5: FDR procedural flow chart.

3.4 Integration of the Loss Factors Model (LFM) into the analysis

The Gantner Instruments/SRCL Loss Factors Model (LFM) (see Fig. 6) is being integrated into this analysis to provide an even better understanding of the performance and degradation modes.

The LFM has 6 normalised orthogonal parameters associated with I_{SC}, R_{SC}, I_{MP}, V_{MP}, R_{OC} and V_{OC}. The product of all 6 gives the PR_{DC}. Using normalised coefficients means:

- An easy sanity check for bad data if not ~1.
- It is easy to tell good from bad values e.g. 98% or 90%.
- It can differentiate PV technologies – e.g. thin Film usually has a higher loss due to R_{OC} (from thin conductive oxide resistivity).

The system can plot LFM values vs. the parameters

below to determine the following [15]:

- vs. Time → Degradation or annealing rates
- vs. Irradiance → Low light performance behaviour
- vs. Module temperature → Thermal coefficients

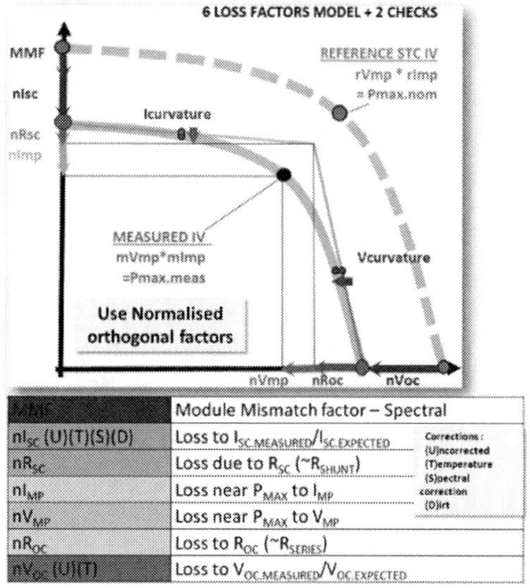

Figure 6: Gantner Instruments/SRCL LFM.

Figure 7 analyses degradation rates as functions of LFM coefficients for a CIGS module in Arizona. Most degradation rates reported in the literature are only for standard test condition (STC) corrected efficiency which is limited by any inaccuracy in the I_{SC} value (this could be caused by soiling, snow, spectral, angle of incidence or irradiance sensor changes with time). Because the LFM separates out the degradation into almost independent coefficients the cause of the degradation can be seen – Figure 7 clearly shows the residual (measured – predicted) degradation is dominated by nV_{OC} , nI_{MP} and nR_{SC} with a smaller drop in nR_{OC}.This implies that the shunt resistance is falling as is the current near I_{MP}.

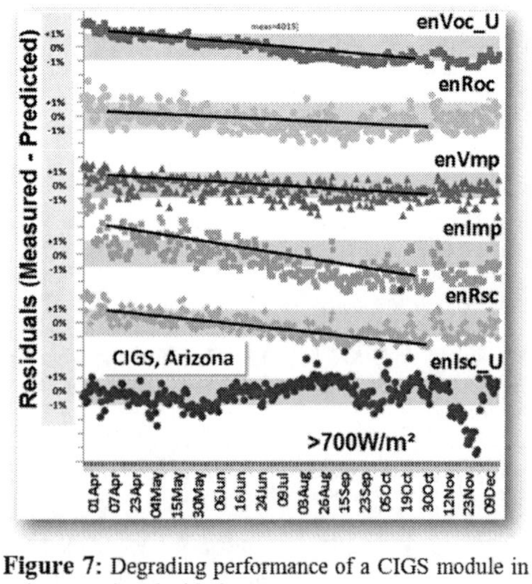

Figure 7: Degrading performance of a CIGS module in Arizona analysed with the LFM.

4 RESULTS

Routines and algorithms for performance loss rate estimation due to high temperature, soiling and failure detection of modules/arrays and sensors have already been implemented by the PV Technology Laboratory UCY [10-13]. As a snapshot of the undertaken work, the results of the degradation rates of different systems applying the developed long term degradation algorithms and the results of indoor techniques using a solar simulator and electroluminescence (EL) imaging, demonstrated that the estimated annual PLR of crystalline Silicon (c-Si) arrays with no visible defects were comparable with outdoor time series analysis and indoor STC maximum power determination results. Especially strong agreement was exhibited between STC and regARIMA modelling results (see Fig. 8) [10]. Some of the higher degradation rates that have been observed through the indoor STC testing were attributed to failures as also identified by EL imaging.

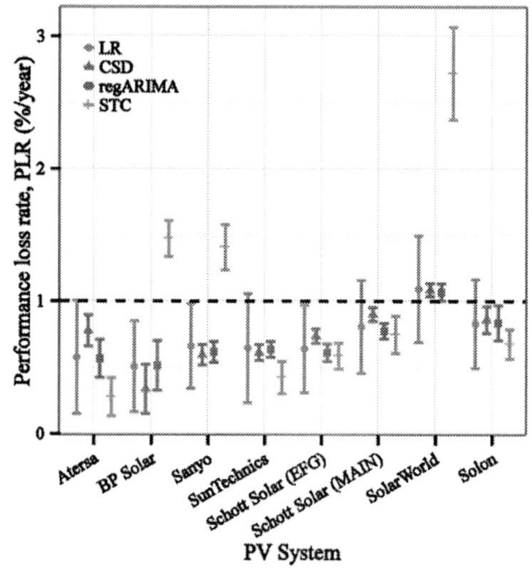

Figure 8: Annual performance loss rates evaluated through outdoor time series analysis for the evaluation period June 2006 – October 2014 and indoor testing at STC. The 1 %/year line is used as a visual aid [10].

Finally, preliminary FDRs developed to detect Potential Induced Degradation (PID), string and inverter defects, shading and hot spots are currently tested and benchmarked on historical data sets acquired for different technology PV systems. First results showed that the FDRs can detect malfunctions in the normal operation of PV systems. More specifically, the routine recognized that the power loss (difference between the measured and modelled power) was outside predefined limits and suggested an inverter or grid outage failure (see Fig. 9). As the routines are yet in the initial development stage an important step to ensure correct detection of specific faults is to include predefined failure patterns. Depending on the degree of coincidence between the failure profile and the predefined profiles along with inputs from other measurements the routines will be able to distinguish amongst the failures.

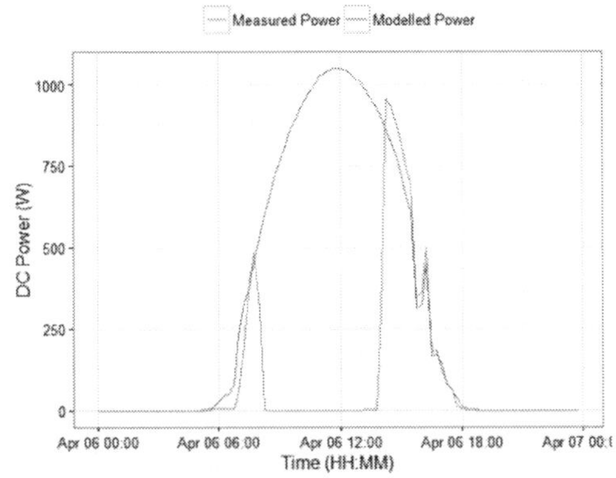

Figure 9: Measured and modelled power of a PV system during the FDR testing phase.

5 CONCLUSIONS

In this work the main scientific, technological and commercial objectives of the project "Innovative Performance Monitoring System for Improved Reliability and Optimized Levelised Cost of Electricity" (IPERMON) funded by the SOLAR-ERA.NET Transnational Calls PV3 and CSP3, in the field of PV performance monitoring systems were presented.

Specifically, the expected results of the proposed work, which include the development of performance loss, failure and degradation rate detection algorithms, improved forecasting and a reliable solution for PV monitoring, will create new knowledge which will advance the field of PV reliability. The advanced monitoring system, which will be the main outcome of this project, will offer reduced risk for PV plant owners, preventive maintenance and improved financial KPI for investors.

Finally, it is anticipated that the successful completion of the project will push a highly innovative end-product straight to the market that will act as a high level watchdog for ensuring reliability and operational quality of PV power plants.

6 ACKNOWLEDGEMENTS

The project is co-financed by the SOLAR-ERA.NET Transnational Calls of the European Union under the grant agreement number SOLARERANET3-147. The authors would also like to acknowledge the financial support from the Cyprus Research Promotion Foundation.

7 REFERENCES

[1] IEC 61724 Standard, Photovoltaic system performance monitoring guidelines for measurement, data exchange and analysis, IEC 1998.

[2] G. Blaesser and D. Munro, Guidelines for the assessment of photovoltaic plants Document A Photovotlaic system monitoring, Commission of the European Communities, Joint research Centre, Ispra, Italy, EUR 16338 EN, 1995.

[3] G. Blaesser and D. Munro, Guidelines for the assessment of photovoltaic plants Document B Analysis and presentation of monitoring data, Commission of the European Communities, Joint research Centre, Ispra, Italy, EUR 16339 EN, 1995.

[4] S. Kurtz, E. Riley, J. Newmiller, T. Dierauf, A. Kimber, J. McKee, R. Flottermesch and P. Krishnani, Analysis of photovoltaic system energy performance evaluation method, NREL Technical Report, NREL/TP-5200-60628, 2013.

[5] S. Stettler, P. Toggweiler, E. Wiemken, W. Heydenreich, A. C. de Keizer, W. van Sark, S. Feige, M. Schneider, G. Heilscher, and E. Lorenz, Failure detection routine for grid-connected PV systems as part of the PVSAT-2 project, *in: 20th EUPVSEC*, 2005, pp. 2490- 2493.

[6] A. Drews, A. C. De Keizer, H. G. Beyer, E. Lorenz, J. Betcke, W. Van Sark, W. Heydenreich, E. Wiemken, S. Stettler, and P. Toggweiler, Monitoring and remote failure detection of grid-connected PV systems based on satellite observations, *Solar Energy*; 2007; 81: 548-564.

[7] T. Oozeki, T. Izawa, K. Otani, and K. Kurokawa, An evaluation method of PV systems. *Solar Energy Materials and Solar Cells*; 2003; 75: 687-695.

[8] S. Mau and U. Jahn, Performance analysis of grid-connected PV systems, *in: 21st EUPVSEC*. 2006, pp. 2676-2680.

[9] S. J. Ransome, J. H. Wohlgemuth, S. Poropat, and E. Aguilar, Advanced analysis of PV system performance using normalised measurement data, *in: 31st IEEE Photovoltaic Specialists Conference*, 2005, pp. 1698-1701.

[10] A. Phinikarides, G. Makrides and G. E. Georghiou, Estimation of annual performance loss rates of grid-connected photovoltaic systems using time series analysis and validation through indoor testing at standard test conditions, *in: 42nd IEEE Photovoltaic Specialists Conference*, 2015, pp. 1-5.

[11] A. Phinikarides, G. Makrides, M. Schubert and G. E. Georghiou, Analysis of photovoltaic system performance time series: seasonality and performance loss Renewable Energy, *Renewable Energy*; 2015; 77: pp. 51-63,

[12] G. Makrides, B. Zinsser, M. Schubert and G. E. Georghiou, Performance loss rates of different photovoltaic technologies, *Solar Energy*; 2014; 100: 28-42.

[13] A. Phinikarides, N. Kindyni, G. Makrides and G. E. Georghiou, Review of photovoltaic degradation rate methodologies, Renewable and Sustainable Energy Reviews, 2014; 40: 143-152.

[14] J. Sutterlueti, S. Ransome, R. Kravets and L. Schreier, Characterising PV modules under outdoor conditions: What's most important for energy yield, *in: 26th EUPVSEC*, 2011, pp. 3608-3614.

[15] S.Ransome and J. Sutterlueti, Degradation Analysis of PV Technologies Using NREL and Gantner Instruments Outdoor Data, *in: 43rd IEEE Photovoltaic Specialists Conference*, 2016.

A USE OF ARTIFICIAL INTELLIGENCE FOR IMPROVING PV ARRAY PERFORMANCE (EMPIRICAL APPROACH)

André Macq *(1), Loïc Mercier des Rochettes (1), Laura Martin-Carron (1), Nicolas Cristi(1), Marie-Pierre Gleizes (2), Carole Bernon (2) and Pierre Glize (2)
(1) SUNiBRAIN S.A.S.,
52 Blv. Gabriel Koenigs, F-31300 Toulouse, France
(2) IRIT, Université de Toulouse, Toulouse, France
(*) corresponding author e-mail: andre.macq@sunibrain.com

ABSTRACT: This paper reflects research to design an Artificial Intelligence (AI) platform for controlling the cooling of Si-based PV modules in commercial solar arrays. The AI technology used for this purpose is based on Adaptive Multi-Agent Systems (AMAS). The information fed to the control platform combines locally-collected and exogenous sources, such as grid data and weather forecasts. Elaborating on an existing PV modules cooling equipment driven by traditional industrial algorithms, the AI control platform substantially improves the use of information to make or recommend operational decisions. Thanks to AMAS techniques, the platform can anticipate events by analysing a broader range of information, thereby improving the economics of PV modules cooling. Introducing Artificial Intelligence in this and related areas of solar plant operations has the potential of further improving resource utilization and maintenance efficiency.
Keywords: Artificial Intelligence, Multi-agent systems, Photovoltaics, PV Cooling

1 INTRODUCTION

This work derives from year-long software development efforts to precisely control cooling/cleaning equipment fitted to regular commercial Si-based PV arrays situated in temperate and tropical climates [1,2]. The software controls the equipment with the aim to maximise the power output of entire PV systems based on a possibly very limited water resource. In this paper, we examine the cooling of the PV arrays based exclusively on rainwater collected at site.

Using Artificial Intelligence techniques for taking optimal decisions about when to automatically cool PV modules avoids creating *ad hoc* models depending on PV types or their geographical location. Learning abilities have then to be integrated in such a decision support tool, in order to identify optimal production conditions for these panels.

Interesting results were already possible with classical industrial controller algorithms, but the addition of AI to the software aims to attain the most efficient and cost-effective cooling approach.

2 MEASUREMENTS / SOFTWARE APPROACHES

Both traditional algorithms and Artificial Intelligence may be used to control industrial processes. The aim of this research is to compare the performance of the two approaches, in terms of cooling effect and/or addition power output from a PV array. Cooling PV panels produces a significant increase in power (primarily in voltage), almost directly proportional to the cooling intensity [3]. This increase is the translation into practice of the PV modules thermal coefficient, as specified in manufacturer datasheets.

To measure the performance of cooling software, we compare the power output differential of the two PV fields and the quantity of water, expressed in % [1]. The PV Performance Ratio (PR) was not used here, because the effectiveness of a rainwater-based cooling system relates to weather variables other than solar irradiation (seasonal precipitations, humidity, wind characteristics).

Figure 1: The demonstrator is located on the farm in Saint- Marcel . The photovoltaic plant has a power of 777 kWp for a surface of 7000 m² .

In order to be able to representatively make these calculations and to statistically analyze the determinants of any output gain, it is necessary to use twin solar (Fig.1) generators instrumented with relevant measurement instruments (separate electric meters at the grid injection point, multiple sensors for PV temperature, wetting, irradiance, weather parameters, etc.).

The initial research uses a twin solar system (specially built in 2011 to offer the possibility of real-time comparison). The installation consists of a steel roof with mounting systems and PV modules with two identical solar generators of which one only is equipped with rainwater-cooling. In 2015 a second twin generator of 250 kWh was equipped in a region benefitting from different weather patterns (Figure 2).

Figure 2: Twin generator of 250 kWh.

The same industrial control software was used on all comparators [1]. This software was based on imperative programming and traditional optimization models.

We compare the outcome of this software with an alternative optimisation approach offered by an Artificial Intelligence technique (Fig.3): called Adaptive Multi-Agent Systems (AMAS). The AMAS approach has been successfully used in many domains of applications for controlling, in a real-time manner, systems such as car heat engines [4], bioprocesses [5], videogames [6], robots [7] or ambient systems [8].

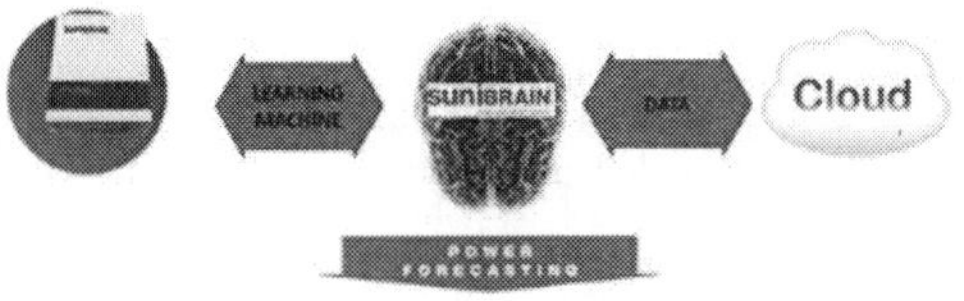

Figure 3: The development industrial control software with an optimisation approach offered by AMAS.

3 INDUSTRIAL ALGORITHMS vs. ARTIFICIAL INTELLIGENCE

3.1 Industrial Algorithms

The algorithmic approach showed an ability to calculate the cooling opportunities, i.e. the moments when it is economically useful to cool the solar field. This was based on static models that are predetermined by engineers.

The static rules are based on logical deductions and previous man-made observation of a prototype system, which are then implemented in imperative programming within a fixed core program. This is generally efficient but this kind of solution is unable to evolve by itself. If an unpredicted situation appears, algorithmic systems will end up with a cooling configuration as appropriate as it can, according to the logic insufflated initially (i.e. not necessarily an optimum configuration).

There are two areas where the algorithms, in spite of 5 years of R&D, failed to produce excellence:

1. The first area is related to errors in assessing data: a traditional industrial algorithm is fixed, it cannot be adjusted frequently (and by definition it does not adjust by itself). It will assess data based on hard-coded formula calculations. On the contrary, the AI software adjusts to find optimal ways to use the data.
2. The second area is related to cooling strategies: testing and choosing among different cooling approaches (for example continuous water sprinkling) vs. a traditional industrial algorithm is fixed, it cannot be adjusted frequently (and by definition it does not adjust itself).

3.2 Artificial Intelligence

The AI technology used for this control platform is based on Multi-Agent Systems (MAS) and more specifically Adaptive ones (AMAS). A Multi-Agent System is a system composed of a set of interacting agents [9]. An agent is an autonomous entity which evolves in an environment from which it has only a local and incomplete perception. An agent possesses skills that enable it to carry on a behaviour which is generally dictated by a local objective [12].

MAS are a recognized paradigm to deal with complex problems and the AMAS approach, used here, is focused on the cooperative behaviour of the agents composing a MAS. The functional adequacy theorem ensures that the global function performed by any kind of system is the expected one if all its parts interact in a cooperative way [10]. Cooperation, as intended in AMAS, is not only a mere resource or task sharing, but truly a behavioural guideline. This cooperation is considered in a proscriptive way, implying that agents have to avoid or solve any Non Cooperative Situation (or NCS) encountered. Therefore, an agent is considered as being cooperative if it understands perceived signals without ambiguity, if received information is useful for its reasoning and if its reasoning leads to useful actions toward other agents. Once an agent detects a NCS it takes actions for coming back to a cooperative state by changing its internal state (tuning) or its relationships with others (reorganization) or by creating new agents or removing itself (evolution). These changes therefore make the structure of the global system self-organize and the global function performed by this system emerges from the interactions between agents. This AMAS is thus able to react to changes coming from the environment and therefore becomes (self-) adaptive.

The AMAS involved in the control platform performs a contextual learning by identifying all the specific environmental conditions (temperature, light, sunshine, wind direction, hygrometry…) which may influence the functioning of the panels. These environmental conditions are embedded into specific agents, called context agents, which are dynamically created and self-adapt on-the-fly. A context agent is composed of three parts: a context, an action and an appreciation. The context is a set of validity ranges – one per percept – that is used by the agent to determine if the context it represents is valid (when all the ranges are valid). The action is the modification of the environment proposed by the agent when it considers itself valid. The appreciation represents the estimation the agent has about the effects of the action on the environment; this appreciation is constantly adjusted by the agent depending on what it perceives in its environment [11].

4 RESULTS AND DISCUSSION

Traditional industrial algorithms operating the cooling equipment over long periods were able to increase by around 6 percent the power output of the PV array. Power output gains on each of the twin solar generators (full-year 2013 on Comparator 1 and 6 months in 2015H2 on Comparator 2) were respectively 6.24% and 5.5%. However, operation of the cooling system was less than perfect, as shown by our statistics on the first 6 months of the second twin comparator:

- Days in the analysis period: 184
- Days when cooling system was operated: 106
- Equivalent full days of non-performance due to errors: 27
- Days when cooling system was not or sub-optimally operated: 78

It is estimated that AI techniques will enable to gain more than 20% additional productivity over the cooling equipment's performance using traditional industrial algorithms.

Difficulty when designing models to control systems resides in the fact that these models depend on the situation. Using an AMAS enables to ignore these situations, a learning phase during a period of some months enables to calibrate this control system before actually using it in production phase. The learning is automatically done, in real-time, customized and able to adapt to situations that were not previously encountered by the control system.

4.1 Principle and Architecture of the AI System

The core of the AI system is based on learning the maximum power of photovoltaic panels. For this, a multi-agent of context agents has to find the maximum power produced by a set of panels (Fig.4) (typically one or more strings of PV panels, but to simplify things we will call each of these agents a "panel agent"). This optimality depends on environmental conditions (temperature, wind, irradiance...) and endogenous conditions (when internal temperature and cleanliness of panels are optimal). Feedbacks are required to make the system learn and the variation in photovoltaic production plays this role: when it increases, the decision is considered as good; otherwise, context agents reconsider their decision by modifying their validity range.

Each set of panel is "agentified" in the AI system and these panel agents use the multi-agent context system in order to know the maximum power they are able to obtain under the current environmental conditions. Thus the criticality of an agent panel corresponds to the difference between this maximum power value and the actual power.

The second important software component is the scheduler whose role is to decide which action to perform. This scheduler receives the criticality of the available water (actual rainwater collected and the rain expected by the meteorological forecast) and all the criticalities of the panel agents. According to these criticalities, the scheduler decides or not to cool (or clean up) a given set of panels.

The initial MAS calibration is obtained from two data sets: firstly, the system learns the behavior of the currently used industrial tool, and secondly, this behavior is tuned using real historical data obtained on the target site. When installed on the site, the AI system learns from real time data.

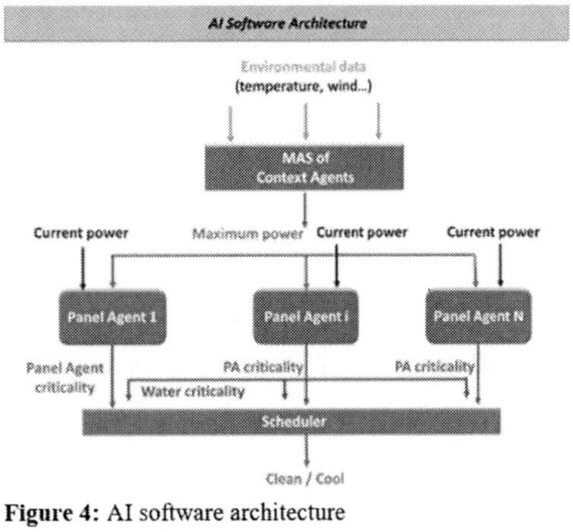

Figure 4: AI software architecture

The standard MAS context is currently used in several applications (singularity detections in a Building Management System [11], heat engine calibration [4], robotic learning from demonstration [8], ambient systems [7], bioprocess control [5]) and is currently in the integration phase for large-scale agricultural project.

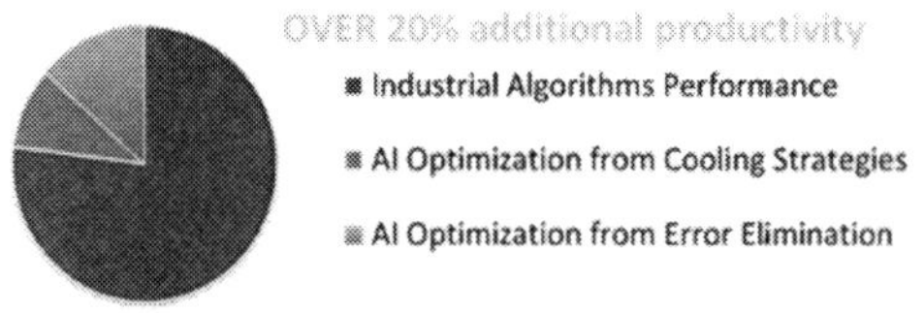

Figure 5: AI compared to traditional algorithms.

The figure above (Fig.5) shows that substantial improvements are possible using AI compared to traditional algorithms. The gain of power output using traditional algorithms is shown in red. Additional gains from AI are expected from reducing errors and improving the cooling strategies.

5 FUTURE WORK AND CONCLUSIONS

This work derives from year-long software development efforts to precisely control cooling/cleaning equipment fitted to regular commercial Si-based PV arrays situated in temperate and tropical climates. Interesting results were already possible with classical industrial controller algorithms based on logical deductions. It is expected that a control system based on an AMAS (multi-agent system with adaptive abilities) will enable to cool and clean panels in a more effective manner than more classical approach.

The AMAS approach will be applied to additional areas of PV O&M. Once this control system is deployed on test sites, it is also expected that this AMAS will enable to be used for signaling problems that may occur during the functioning of the set of panels such as failures.

A new test bench is being built (2016H1) on the premises of the University Paul Sabatier in Toulouse, France, as part of the University's neOCampus project.

6 ACKNOWLEDGEMENTS

The authors are indebted to the FEDER / FSE program of the European Union managed by the Languedoc-Roussillon Midi-Pyrénées Region, and to BPI France for co-funding this research.

7 REFERENCES

[1] Nicolas Cristi, André Macq and Laura Martin-Carron, Evaporating pure rainwater to increase the yield of commercial-size pv arrays, Proceedings 32nd European Photovoltaic Solar Energy Conference and Exhibition (to be published), 2016

[2] Laura Martin-Carron, Roland Becker, Didier Graebling, Robert Luce, Daniel Ugarte, André Macq and Nicolás Cristi, Air cooling of photovoltaic panels: a numerical approach, Proceedings 32nd European Photovoltaic Solar Energy Conference and Exhibition (to be published), 2016

[3] E Skoplaki, JA Palyvos, On the temperature dependence of photovoltaic module electrical performance: A review of efficiency/power correlations, Solar Energy 83 (2009) 614–624

[4] J. Boes, F. Migeon, and F. Gatto. Self-Organizing Agents for an Adaptive Control of Heat Engines, International Conference on Informatics in Control, Automation and Robotics (ICINCO 2013), Reykjavik,

SciTePress, p. 243-250, July 2013.
[5] S. Videau, C. Bernon, and P. Glize. Towards Controlling Bioprocesses: A Self-adaptive Multi-agent Approach, Journal of Biological Physics and Chemistry, AMSI & Collegium Basilea, Switzerland-Georgia, Vol. 10 N. 1, p. 24-32, March 2010.
[6] L. Pons, and C. Bernon. A Multi-agent System for Autonomous Control of Game Parameters, IEEE International Conference on Systems, Man and Cybernetics, Manchester, IEEE, p. 583-588, January 2014.
[7] N. Verstaevel, C. Régis, M-P. Gleizes, and F. Robert. A Distributed User-Centered Approach For Control In Ambient Robotic, European Congress on Embedded Real-Time Software (ERTS 2015), p. 102-111, Jan. 2015.
[8] V. Guivarch, V. Camps, A. Péninou, and P. Glize. Self-adaptation of a learnt behaviour by detecting and by managing user's implicit contradictions, IEEE/WIC/ACM International Conference on Intelligent Agent Technology (IAT 2014), Warsaw, Poland, IEEE Computer Society, p. 24-31, August 2014.
[9] M. Wooldridge, An Introduction to Multi-Agent Systems. Wiley, 2002.
[10] M-P. Gleizes, V. Camps, and P. Glize. A Theory of Emergent Computation based on Cooperative Self-organization for Adaptive Artificial Systems, 4th European Congress of Systems Science, Valencia Spain, 1999.
[11] J. Boes, J. Nigon, N. Verstaevel, M-P. Gleizes, and F. Migeon. The Self-Adaptive Context Learning Pattern: Overview and Proposal, International and Interdisciplinary Conference on Modeling and Using Context (CONTEXT 2015), Larnaca, Cyprus, Springer, LNAI 9405, p. 91-104, 2015.
[12] J. Ferber, Multi-Agent Systems: An Introduction to Distributed Artificial Intelligence, Addison-Wesley Longman Publishing Co., Inc., 1999.

FLOATING PV POWER SYSTEM EVALUATION OVER FIVE YEARS (2011~2016)

Dr. Chang-sub Won, Waithiru Lawrence, Dongchan Kim, Boram Kang, Kwangwook Kim, Gunhyun Lee
Electronics R&D Center, LSIS Corporation, Anyang, Republic of Korea
431-831 LSIS R&D Campus, LS-ro 116 beongil 40, Dongan-gu, Anyang-si,
Telephone: +82-31-450-Email: cswon@lsis.com

ABSTRACT: We introduce sections of an ongoing reliability research study on a test bed 100kW and commercial 500kW Floating PV systems installed in 2011 and 2012 repectively. Based on periodic testing field samples, and Manuel and Ignacio [1] module degradation model as reference, initial predictions are made on module useful life is made. When 2.4% initial LID loss is factored, an annual degradation rate of 0.45%/year is deduced. PO when implemented in combination with other high performance components like junction box, edge sealant materials, significantly reduces water ingress induced degradation, improving long term PV module reliability, ultimately increasing solar technology viability, durability and bankability).
Keywords: Polyolefin (PO) encapsulation, Floating Photovoltaic System, Light induced degradation (LID), Anti-Damp Proof Module (ADPM).

1 INTRODUCTION

PV module durability and reliability concerns are the major concerns for competitiveness of solar technology. To develop an Anti-Damp Proof Module (ADPM), LS Industrial Systems PV module experts bench-marked both a 45kW island site (Pyeongsa Island Korea) installed in 1992, and a floating buoy PV system, for module failure analysis. The rationale for power loss was analyzed. Corrosion related water ingress through the frame edges and rear side of the modules was investigated for understanding of failure mechanism [2]. Cell electrode, frame and junction box showed severe humidity ingress related corrosion. Acetic acid triggered EVA yellowing was also observed in study. Pb-free ADPMs were developed with higher IP and insulation grade and installed on two floating systems (Fig.1). Data was collected and analysed over successive years for module degradation and system performance. Light Induced Degradation (LID) experiments were undertaken to ascertain initial module power loss.

Our focus here is to briefly introduce module field data, and make preliminary prediction for useful life of based on five year data.

2 FLOATING PHOTOVOLTAIC MODULE AND SYSTEM

2.1 Pb-Free PVM with Polyolefin Encapsulant

Initial research work focused on developing an eco-friendly and durable ADPM with Pb-free materials, other high performance BOM including Polyolefin encapsulation for a dam or water reservoir application. Compliance with local authority tests, regulations was successful. Deployed PV modules received VDE certification and IP grade conformity tests at DeT Norske Veritas in Norway, higher than normal wet leakage current and insulation strength.

2.2 Floating System Overview

The Floating PV system site information is briefly outlined below in Table1 and Fig.1 below

Table I: FLOATING PV SITE INFORMATION

SITE	TEST BED	COMMERCIAL
Site Name	Hapcheon Dam	Hapcheon Dam
Site coordinates	N 35.5° 33'36"	E 128° 02' 26"
Inst Capacity	100kW	500kW
Installation Date	2011 Nov.	2012 June
Module Slope	30°	30°
Module type	c-Silicon	c-Silicon;
Module	417pcs	1,667pcs
Mounting	Alum., Steel, FRP	Aluminium,
Mounting type	Fixed	Fixed
Water Depth	20 meters	40 meters
Annual Energy	144MWh	714MWh

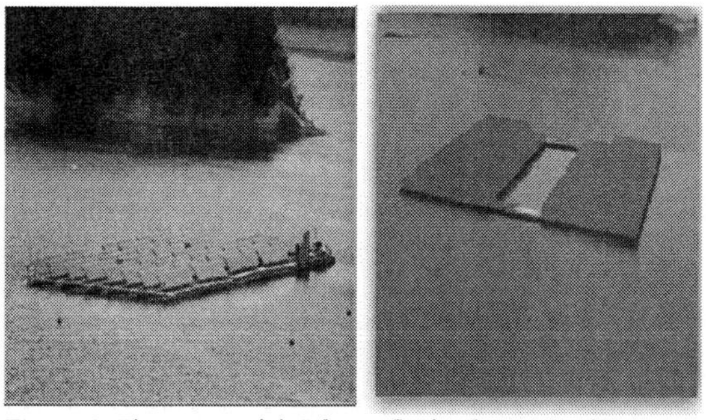

Figure 1: Fig. 1. Aerial View of Floating PV Sites; 100kW (Left), 500kW (Right)

3 EXPERIMENT AND APPROACH

3.1 Experiment Data Acquisition

Material composition of modules installed in both 100kW and 500kW sites were identical. One hundred (100) field modules were tested for power output at STC [3] bi-annually for power output characteristics in a lab environment between 2011 and 2015. The procedure is shown below pictorially in Fig.2.

Figure 2: Data acquisition processes

3.2 Data Modeling

To predict module long-term performance, reference was made to Va´zquez and Rey-Stolle Time Degradation model [1] . Fig. 3 is a quantitative illustration of time evolution of a module.

$$p(P,t)=\frac{1}{\sqrt{2\pi}(\sigma_O+Bt)}\exp\left\{-\frac{1}{2}\left[\frac{P-(P_O-At)}{\sigma_O+Bt}\right]^2\right\} \quad (1)$$

The time based model defines module power degradation based nominal power, standard deviation and parameters A (% yearly decrease in nominal module power) and B (% yearly increase in standard deviation). $P_0 - At$ represents linear decrease in nominal power, while $\sigma_O + Bt$ linear increase in standard deviation. The red circle indicates a critical limit where module power is above 80% of nominal power at warranty time (t_W).

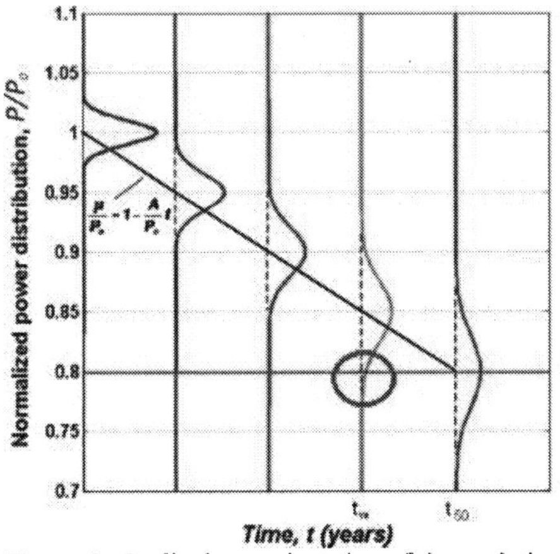

Figure 3: Qualitative reorientation of the evolution over time of module

4 RESULTS

4.1 Field Data

Floating PV module data is presented next in Table 2, and normal distribution shown in Figures 4 below.

Table II: Floating PV Pmax Data (100 samples)

FLOATING	Yearly PV (@STC)		
YEAR	Yr 0	Yr 2	Yr 4
PVM [W]	242.57	242.19	241.81
STDEV(%)	0.5106	0.9163	1.1065

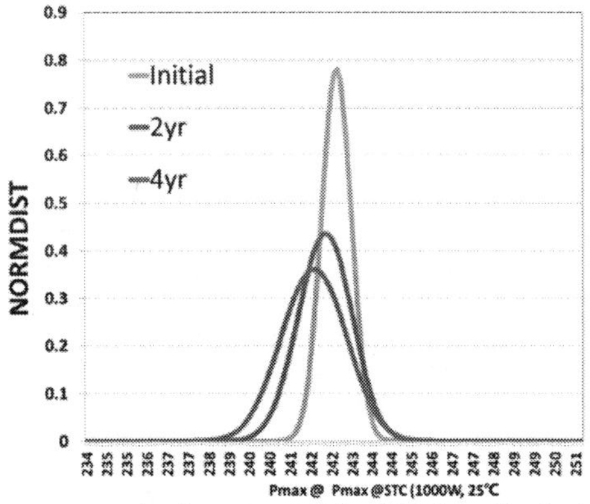

Figure 4: Floating PV Module field data distribution (@STC); 100 samples.

4.2 Module Life Prediction

A simulation is run to predict floating PV module lifetime performance, with parameters A and B in equation (1) deduced from average field data in Fig. 5. Results show module nominal power falling slightly by 13.9% after twenty year (25yrs).

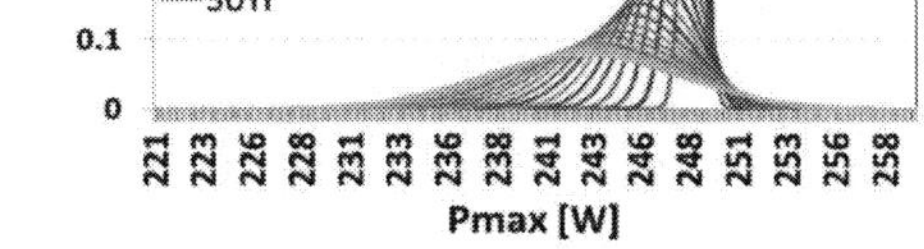

Figure 5 Prediction of Floating PV module Distribution using 100pcs polyolefin module Field Data

5 CONCLUSION AND FUTURE WORK

LS Industrial Systems successfully introduced new polyolefin-encapsulated PV modules with specialized junction box and sealing in highly humid 100kW and 500kW floating PV sites. Visual inspection after 5 years shows very positive results, with no delamination. Separate research work on 100kW has the system producing 10% more power than compared to conventional systems [4]

By March 2013, after concerted research work, LSIS successfully replaced 300MW EVA capacity equivalent with polyolefin in PV module manufacturing line.

Investigation of the 500kW is continuing and will be share in future work. We trust that with improving floating module and system performance, system LCOE will be at par with conventional systems given rapid material advancements and favorable government policies.

5 ACKNOWLEDGMENT

This work was supported by the Korea Institute of Energy Technology Evaluation and Planning(KETEP) and the Ministry of Trade, Industry & Energy(MOTIE) of the Republic of Korea (No. 20153010012060).

5.1 References

[1] Manuel Va´zquez and Ignacio Rey-Stolle, "Photovoltaic Module ReliabilityModel Based on Field Degradation Studies", Progess in Photovoltaics Research and Application, Prog. Photovolt: Res. Appl. 2008; 16:419–433

[2] Gi-Hwan Kang , Kyung-Soo Kim, Chi-Hong Park, L Waithiru, Gwon-Jong Yu, Hyungkeun Ahn, Deuk-Young Han, "Current-Voltage Characteristics with Degradation of Field-Aged Silicon Photovoltaic Modules"

[3] Young-Kwan Choi, "A Study on Power Generation Analysis of Floating PV System Considering Environmental Impact", International Journal of Software Engineering and Its Applications Vol.8, No.1 (2014), pp.75-84

[4] IEC Standard 61724, Photovoltaic system performance monitoring Guidelines for measurement, data exchange and analysis, 1993.

MODELLING AND ENERGY MANAGEMENT OPTIMISATION OF BATTERY ENERGY STORAGE SYSTEM (BESS) BASED PHOTOVOLTAIC CHARGING STATION (PV-CS) FOR UNIVERSITY CAMPUS

Ayda Esfandyari* [1], Agata Świerc [3], Brian Norton[1], Michael Conlon[1], Sarah J. McCormack [2]

[1] Dublin Energy Lab, Greenway Hub (ESHI), Dublin Institute of Technology, Grangegorman, Dublin 7, Ireland
ayda.esfandyari@dit.ie , president@dit.ie, michael.conlon@dit.ie

[2] Department of Civil, Structural & Environment Engineering, Trinity College Dublin, Dublin 2, Ireland
mccorms1@tcd.ie,

[3] Department of Heating, Ventilation and Dust Removal Technology, Silesian University of Technology, Gliwice, Poland
agata.swierc@polsl.pl,

*Corresponding author: ayda.esfandyari@dit.ie

ABSTRACT:

As utilization of Photovoltaic Charging Stations (PV-CS) that generate clean electricity from the sun increase, Dublin Institute of Technology (DIT) adopts this application for accommodating the required charge of small campus Battery Electric Vehicles (BEVs). This paper presents the virtual simulation of the 10.5 kW Battery Energy Storage System (BESS) based PV-CS model. Transient System Simulation (TRNSYS) built-in climatic data and modular structure properties were adopted to replicate the experimentally proposed PV-CS, where special attention was paid to the electrical measurments and energy flow signals. The objective was set to model the PV-CS system, formulate an energy managment optimisation and justify the ideal value and or potential range of the equivalent battery size. The primary assessment for energy management of the charging infrastructure was performed through the formulation of analytical energy balancing optimization. The energy balancing approach adopted the Simple Payback Period (SPP) method in order to investigate the acquired positive gains (Gain-1 and Gain-2) by BESS unit. The key variables for tuning the BESS capacity were load profile and size of BESS. The resultant measurement signals from TRNSYS were monitored and compared to their analytical equivalents, where verification and conclusion on accuracy improvements for BESS capacity and reliable system performance were drawn.

Keywords: PV System, PV Charging Station (PV-CS), Battery Electric Vehicle (BEV), Battery Energy Storage System (BESS)

1 INTRODUCTION

Almost 95% of the world's transportation energy comes from petroleum-based fuels; largely CO_2 emitting gasoline and diesel that are likely to be depleted by 2049 [1]. As the result, Green-House Gas (GHG) emissions from the transport sector are substantial and hence, Europe is accounting for nearly a quarter GHG emissions [2].

Battery Electric Vehicles (BEV) powered from renewable generated electricity are an alternative to the conventional vehicles and ideal solution for reversing these trends [3]. BEVs can offer significant benefits, i.e, such as lowering CO_2 emissions and achieving an environmentally sustainable future, when charged with PV generated electricity. This can result in zero emissions at both generation and usage points [4].

Solar energy is an abundant source of renewable energy with enormous potential for delivering clean and reliable electricity. Photovoltaic (PV) technologies are advancing at a rapid rate on the global scale, and as predicted by International Energy Agency (IEA) PV will account for 11% of global electricity generation by 2050 [5]. PV can be harnessed in the paradigm of PV charging station (PV-CS) on both urban and remote scales, to charge batteries and accommdate load demand.

PV Charging Station (PV-CS) that charge batteries with zero emissions at the generation point, is the focus of this research [6]. The proposed PV-CS is designed to supply the required charge of the campus BEVs load. The campus fleets are two CarryAll 13.76 kWh vehicles, which are referred to as Light weight Electric Vehicles (LEVs). These vehicles are utilised for short distance commutes at day/night time around Dublin Institute of Technology's (DIT) expanding campus at Grangegorman, Dublin 7, thus fulfilling the campus's sustainability targets systems.

2 OBJECTIVE

The objective of this paper is: to develop the TRNSYS based PV-CS modelling and verify BESS unit capacity. The specific objectives are: development of the numerical model for performance of PV-CS (Option that incroprates BESS unit) via relevant TRNSYS components and input-ouput mapping, evaluation of BESS sizing curves for optimal size and energy managements. The optimization problem was formulated with an objective to select the BESS capacity while meeting the required daily load demand.

This paper is organized as following: Section 3 provides the design and system development of PV-CS. Section 4 details the modelling procdure of BESS based PV-CS in TRNSYS platform. The energy managment optimisation formaulation is described in section 5. Comparison of the TRNSYS and theoretical energy management signals and numerical simulation results for BESS capacity sizing are discussed in section 6. Finally, the conclusion on TRNSYS performance and verified BESS size is presented in section 7.

3 PV-CS

3.1 Design option

PV based charging stations have been considered in numerous applications on both urban and remote scales system. The system on the high level can include various components such as: PV and system components; controller, energy storage system and load [7].

Likewise, several prominent design options exist such as: coupled PV-grid without storage where PV is used as the primary source of generation and any insufficiencies is accommodated via the grid [8-9]. The addition of storage as part of the design is denoted as coupled PV-grid with storage [10-11].An alternative option can exclude the grid that is referred to as standalone PV charger or decoupled PV. This is an applicable choice in remote areas where grid extension is unavailable or costly [12]. Standalone configuration can also include auxiliary storage or decoupled PV with storage, [13-14].

To consistently utilize the solar generated electricity in Grangegorman where solar radiation and charging habits of the users fluctuate, the designated PV-CS design will incorporate a Battery Energy Storage System (BESS). In this case, PV generated electricity during daylight will be exploited and stored in the BESS unit. Thereafter, BESS would act as the primary supply to accommodate the LEVs demand. The appropriate BESS capacity is the prevailing factor to accomplish the charging scheme [15]. Hence, based on the chosen BESS capacity once the unit is fully charged, the surplus generation from the PV can be spilled in the utility grid. Besides, the grid can work as the back-up for charging the load, particularly when the stored charge in the batteries is low and close to the Depth of Discharge (DoD) capacity limit.

3.2 BESS unit

Sizing of a BESS based system, when excluding the grid option, can be undertaken by taking into account load estimation, battery sizing and PV sizing as well as solar resource estimation, charge controller specification and system wiring [16]. Lee et al [17] monitored the load profile of various appliances and the BESS storage size capacity estimated with respect to 95% and 99% system availability via Peak Sunshine Hour (PSH). PSH is defined as "the equivalent number of hours per day, with solar irradiance equalling 1000 [Watt /m^2]". Consequently the size of the PV array was estimated based on inverter losses, system voltage, PSH and temperature coefficient. Based on the estimated values, costs of the various systems were estimated, (99% tilted, 95% tilted, 99% axis tracking, 95% axis tracking). These included the cost of: PV panels, inverter, charge controller, BESS, O&M costs and labour cost. The most viable option was identified as 99% tilted axis with the estimated cost value of $ 89,000.

Likewise, the campus BESS capacity was evaluated based on: Peak Sunshine Hour (PSH) and correlation between generation and load demand [6-15]. In the first case, the days of autonomy that determine the self governing properties of the batteries were calculated according to PSH of 2.83 hours for a typical day in Dublin. This resulted in 72 or 32 Trojan compatible batteries for summer and winter load profiles respectively. In the second approach, the DC outputs of the array were normalized, grouped into specified generation categories and correlated with the load demand profile of the LEVs, where the optimal capacity was obtained to fall in the range of 6-8 kWh.

4 MODELLING PROCEDURE

4.1 TRNSYS

TRNSYS is a modular, open-source, Fortran-based energy modelling program, that is currently adopted for a variety of thermal and photovoltaic system modelling [18]. For effective simulation of PV-CS, TRNSYS modular structure requires identification of the applicable system components. The components in TRNSYS system are referred to as TYPEs. Each TYPE has a set of time-independent parameters as well as time-varying inputs that allow estimation of the matching outputs.

TRNSYS simulation tool has been used in several studies for system sizing [19-20-21]. A standalone PV system model for residential house lighting was implemented [20] using TRNSYS 16. The model incorporated a battery unit and was tested using the Ipoh city (Malaysia) weather data. The sizes of the PV and battery components were determined via the PSH method. The simulated results were further correlated with the experimentally measured data [19], and TRNSYS simulation model gave comparable results to the actual measurements [21] conducted an analysis on the system performance of battery based standalone solar home systems in Malawi for rural electrification. The equivalent model in TRNSYS considered low and high isolation profiles and was further compared to the experimentally designed system. The results showed the system was more reliable in high insolation areas.

4.2 Model components

The PV-CS model was developed to replicate the experimentally proposed model in [9].The model created through the connection of selected components in TRNSYS. The flow diagram for this model and the components utilised are shown in Table 1.

Component	TYPE	Measured
Solar radiation processor	15	Irradiance
PV panels	94a	Power, V I
BESS	47	Power/SOC
Charge controller	48	Energy balancing
Unit converter	57	-
Printer	65	-
Online plotter	25	-

Table 1 PV-CS components TRNSYS TYPES

In order to replicate the experimentally proposed PV-CS the climatic scenarios of the installation site were considered. The weather conditions in TRNSYS was obtained using TRNSYS' built in meteorological library (TYPE 15). A four parameter model of PV system i.e., TYPE 94a, was used to present an empirical equivalent circuit model and predict the current-voltage characteristics of the 10.5 kW crystalline PV array. The battery/BESS (TYPE 47) that is currently limited to lead-acid type utilised power as an input. Subsequently the capacity was adjusted and the variations of BESS State of Charge (SOC) and power over time were considered given the rate of charge or discharge. For the purpose of this work it was assumed that the SOC of battery is initially at its minimum limit. The charge controller (TYPE 48) operated as a power conditioning device on maximum power tracking system [18].The limit on fractional SOC in controller was arranged to 0.4 and higher than lower limit of 0.1 on SOC. This was to make sure that charging the BESS by PV had priority over serving the load. Hence, the PV generated electricity would primary be stored in BESS and used at night time (when the solar energy is unavailable) to charge the load. Finally, TYPE 14 served the purpose of defining a time-dependent load profile for the campus LEVs. The pattern of forcing functions was established by a set of discrete points indicating the value of the function at various times throughout one day cycle.

4.3 Input-output mapping

The manner in which the input and output components were connected influenced the system control and its convergence. As illustrated in Figure 1, for input-output mapping, weather component/solar insolation, i,e, the total solar radiation at the slope angle of 10 degrees, was used to represent the solar radiation of the site. The radiation values were fed as inputs into crystalline PV panels, and the irradiance and DC output power of the panels were derived. PV's Maximum Power Point (MPP) was converted from Watts to KJ/hr and linked as inputs to the charge controller. The charge controller, with an energy balancing capability, along with the BESS system represented the key components for overall energy management relating to the charging infrastructure. Thus, it was important to ensure the chosen controller TYPE was matched with BESS unit. Hence, Mode 1 of TYPE 48 was coupled with Mode 1 of BESS TYPE 47. The fractional SOC and power of BESS were linked to the charge controller as inputs and outputs. The charge controller also measured the daily load power profile of the site which was converted to KJ/hr. This was mainly to establish the overall system control and correlate the charge/discharge between generation, BESS and load demand. All the output finals such as irridaince, energy balancing parameters and BESS unit values were converted back to their kWh for energy managment analyses.

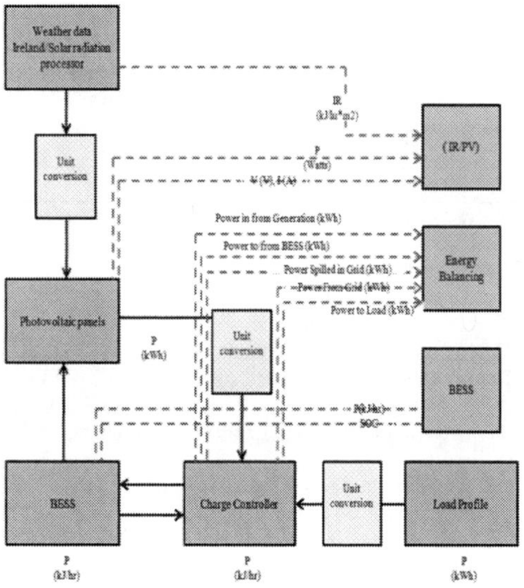

Figure 1 PV-CS TYPES input-output connections

Figure 4 presents the complete implemented PV-CS model in TRNSYS with all the system components expalined above.

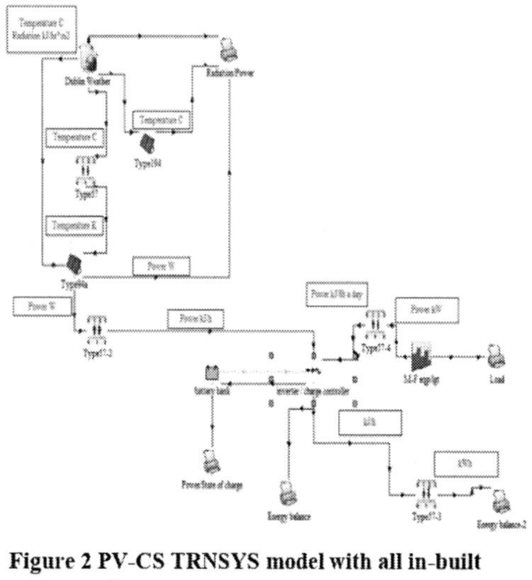

Figure 2 PV-CS TRNSYS model with all in-built components

As a consequence, the primary assessment for energy management of the charging infrastructure was performed through formulation of theoretical energy balancing optimization. The size of the lead-acid based BESS unit in TRNSYS was subsequently matched to the identified values obtained from this energy balancing approach. Moreover the resultant measurement signals from TRNSYS were monitored and compared to their theoretical equivalents.

4 ENERGY MANAGEMENT

4.1 Empirical

Accordingly, the energy balancing optimization adopted the Simple Payback Period (SPP) method in order to investigate the acquired positive gain by BESS unit, while the key variables for tuning the theoretical BESS capacity were load profile and size of BESS. The parameters used for energy appraisal are listed in Table 2. At this point, battery losses were not considered into system modelling.

Energy balancing parameter
Energy at the beginning of the day in BESS (Start cycle)
Energy stored at the end of the day (Theoretical)
Energy stored at the end of the day (Actual)
Load covered from the battery
Energy bought from the grid
Energy remained for the next day
Energy spilled into the grid

Table 2 Energy balancing control parameters

Weather and solar radiation data from TRNSYS model was utilised on hourly basis for annual period. PV DC output data was merged and the total daily output generation of the system was acquired. To illustrate a realistic load demand scenario, the maximum capacity specification of 13.73 kWh for one LEV, provided from the manufacturer was considered. Moreover, it was vital to examine the incremented BESS volume for a range of load values. As a result, the maximum load (100%) was iterated by an established factor of 20% inspired by 80%-20% Pareto principal [22], and the relationship between load covered from the battery and the overall demand was perceived. The basic method of load iteration and BESS tuning is listed in the Table 2.

Variables	
Load	100%-80%-60%-40%-20%
BESS	7kWh-14 kWh-21 kWh

Table 3 Load iteration factor and BESS capacity tuning

As shown in Figure 3, in order to implement the empirical method the following steps were considered. Initially the energy stored at the beginning of the procedure, at day 1 in BESS (Start cycle) was assumed to be zero. Nonetheless, this value was determined for each individual day of year with respect to the energy that remained for the next day. Energy available at the end of the day to BESS (Theoretical) was sum of the following parameters: start cycle + energy remained from the previous day/for the next day + daily solar generated energy at the end of the day. Energy stored at the end of the day (Actual) was assumed to be equal to the total energy available at the end of the day, unless: total energy available at the end of the day > Maximum possible capacity of BESS. In this case, Actual value = Maximum possible capacity of BESS. Load covered from the battery was arbitrated according to minimum value between Actual value and Maximum possible value of Load. Hence, Load covered from the battery would be equal to Energy stored at the end of the day (Actual), unless Energy stored at the end of the day (Actual) > Maximum value of Load. Only under this scenario, Load covered from the battery was capped to be equal to Maximum daily load demand. Energy bought from the grid was estimated based on; the supplementary energy needed for accommodating the remaining load that was not covered via BESS. When the Energy stored at the end of the day (Actual) > Load covered from the battery, this energy was carried over the next day as part of Energy remained for the next day. Finally, Energy spilled into the grid was obtained by the difference between Energy available at the end of the day and Energy stored at the end of the day (Actual). In other words, the slurps PV generated energy was spilled into the utility grid.

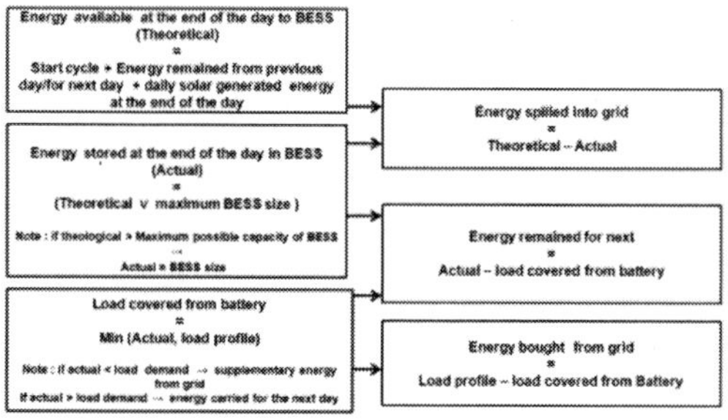

Figure 3 Logical description of energy control management

Moreover, as tabulated in Table 4, the aforementioned energy balance values were adjusted to their cost equivalents. Annual savings were the energy costs avoided annually consisted of annual energy savings multiplied by the cost per unit. These were obtained from operation of PV-CS and referred as Gain-1 and Gain-2. In other words, these gains were defined as Load covered from the battery + Energy spilled into the grid. Energy spilled into the grid for Gain-1 was assumed to be dumped energy that made zero profit, as there is currently no feed-in-tariff for selling the electricity back to the grid in Ireland. On the other hand, for Gain-2 the boundary was defined in such a way that surplus energy spilled into the grid would be used to serve the other campus's energy demand as the PV-CS is collected within a civic area. Thus, Energy spilled into the grid was set to be equal to half of electricity tariff/ Energy bought of 0.18 cents per kWh. However, this price was versatile which can be altered to different scenarios as part of the future work. Finally, SPPs were altered with respect to the inventory gains influences. SPP was defined as the minimum amount of time in years that would be required for the positive cash flow to surpass the initial assessment of PV-CS, excluding the time value of money. In other words, the payback period was defined as the ratio of capital cost investment for building and deployment of PV-CS over the annual cost savings resulted from BESS incorporation.

6 Results

6.1 Empirical results

The energy for Case 1&2 sample load ratios and BESS unit sizes, i.e. case 1 : 7 kWh, case 2 : 14 kWh and case 3: 21 kWh, for total yearly period. as it was specified in Table 3, is shown in Figure 4 and 5.

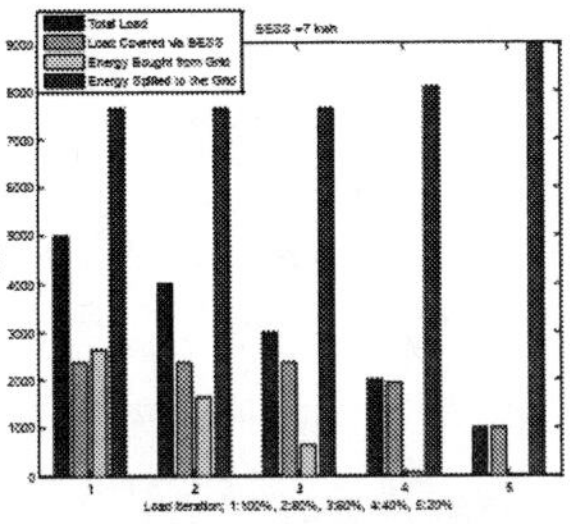

Figure 4 Load iteration for 1 unit of 7 kWh BESS

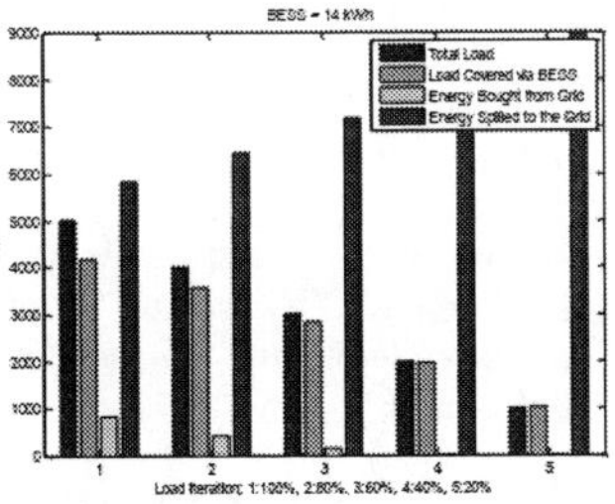

Figure 5 Load iteration for 2 unit of 7 kWh BESS

The aforementioned energy balance tables were adjusted to their cost equivalents, where the average cost of the system components were extracted from [15]. These results are summarised in Figure 7 and 8.

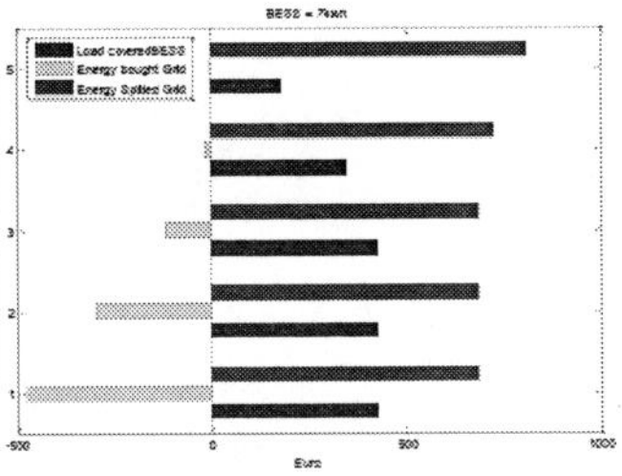

Figure 7 Positive/ negative cash flows 7 kWh BESS

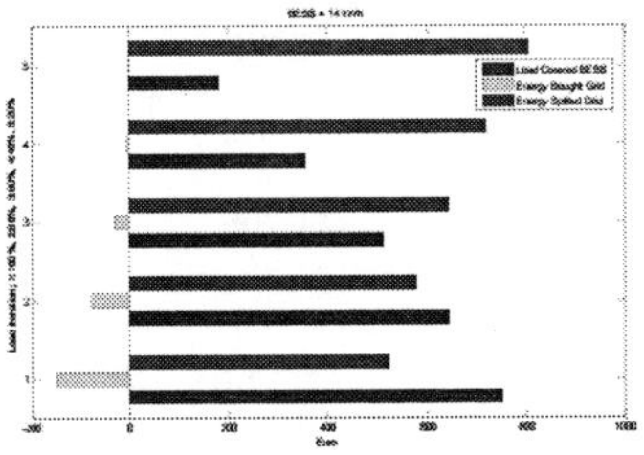

Figure 8 Positive/ negative cash flows 14 kWh BESS

When deciding between the optimal BESS capacity for PV-CS project, the decision was made to accept the case with the shorter payback. Figures 9-10 present (a) Gain-1, Gain-2, and (b) SPP1 and SPP2

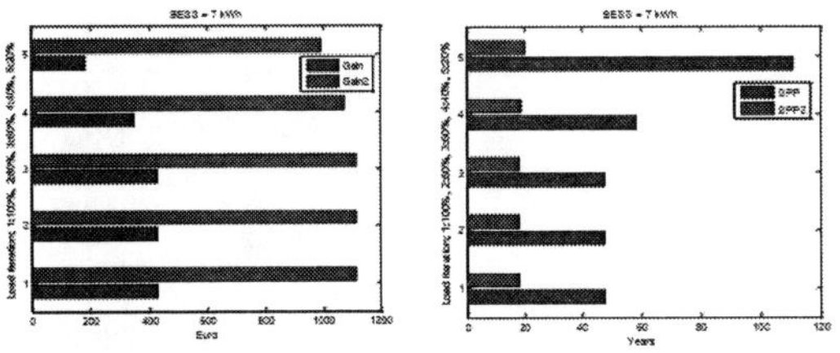

Figure 9 Economic analyses for Case 1: 7 kWh BESS unit

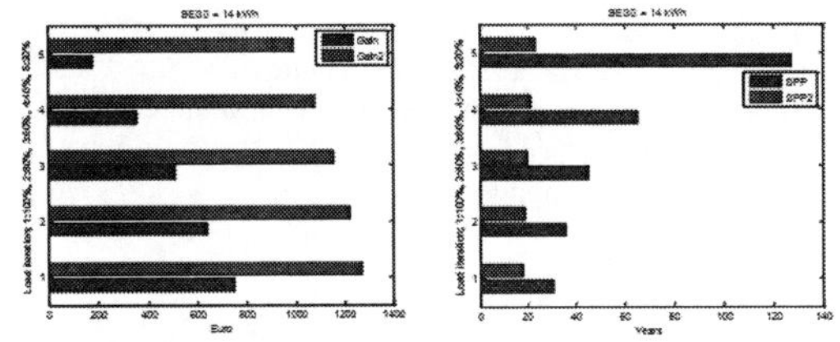

Figure 10 Economic analyses for Case 2: 14 kWh BESS unit

Since there were no constraints applied at this point in the system study, the price of energy spilled influenced the optimal choice. As the optimization was performed for 3 different BESS sizes (Case 1, 2, 3), it gave each case 33% chance of captivation. The difference between Case 2-3 was negligible, thus the optimization was focused on the assessment between Case 1 and 2. The values of SPPs (SPP1 – SPP2) are summarized in Table 5 and 6. By looking at the green fields in the energy tables for SPP of Case 1 and 2, it was evident that for 60% - 100% of the load demand, SPP value of Case 2 was shorter which subsequently led to a more desirable choice. On the other hand, when the load demand is smaller and in the range of 20%-40%, Case 1 was identified to be more viable. With SPP2, the results were quite similar and the differences between values were insignificant. This was mainly due to the cost of selling electricity back into the grid.

The concept of 80%-20% which is referred as Pareto principal was developed by Pareto an Italian economist could be applied to for finalising the scenarios. The principal states; "80% of the output is a direct result of about 20% of the input".While Pareto optimization is a widely accepted rule of thumb, the validity of this approach in business field and software is particularly indispensable. In this case it could be referred to 33% - 60% and 33% -40% for the explained scenarios above.

Variables	SPP_C1	SPP_C2	SPP_C3
100%	47.58	31.17	34.11
80%	47.58	36.43	40.48
60%	47.58	45.75	51.25
40%	58.36	65.70	74.20
20%	110.99	127.71	147.33

Table 5 Scenario comparison for BESS sizing for Case1, 2 and 3 for Gain-1

System	SPP_C1	SPP_C2	SPP_C3
100%	18.22	18.36	20.59
80%	18.22	19.17	21.62
60%	18.22	20.76	22.91
40%	18.89	21.72	24.62
20%	20.47	23.63	26.86

Table 6 Scenario comparison for BESS sizing for Case1, 2 and 3 for Gain-2

However, if BESS was tuned for more numbers, the 33% value could be varied to a difference percentage. In summary, if the fleet are to be enlarged in the future, there would be the probability of serving more cars from the installation, thus designing PV-CS for Case 2 could be a more reliable option.

4.2 TRNSYS

In order to verify the obtained energy performance values from the emprical optimisation approach, the results were compared with their equivalent values obtained from the output performance of TRNSYS model. As explained in Figure 4, in the TRNSYS based approach, the components were adjusted once more with respect to the Table 3 methodolgy. Hence, in this case the *n* number of BESS capacity and load values were iterated. This energy and cost data generated via this approach were very comparable to their empirical equivalents. In other words, TRNSYS provided similar range of data and thus same conclusion could be drawn for choosing the BESS capacity.

7 CONCLUSION

The main motivation behind campus PV-CS scheme is to harvest electricity, thus accommodating the load requirements of campus LEV fleet. This paper presented a deterministic method for sizing of the BESS based via energy managment optimization and TRNSYS electrical signals. The obtained sizing curve from the two approches were compared the result were similar. The recommeded size of the BESS unit would mainly depend on the load demand of the buggies and the energy tariffs for spilling the electricty into the grid.

8 ACKNOWLEDGMENTS

Authors would like to acknowledge the Fiosraigh scholarship (DIT), Grangegorman Development Agency, Dublin Institute of Technology, and Trinity College Dublin for all the support and facilities.

9 REFERENCES

[1]. BP, "BP Statistical Review of World Energy", 2015
[2]. European Commission, "A 2030 framework for climate and energy policies,", 2013
[3]. W. Smith, "Can EV electric vehicles address Ireland's CO2 emissions from transport?, *Energy*, vol. 35, no. 12, pp. 4514–4521, 2010
[4]. SEAI, "A Guide to Electric Vehicles", 2012
[5]. IEA, "Technology Roadmap Solar photovoltaic energy," *Int. Energy Agency*, pp. 1–42, 2010
[6]. A. Esfandyari, B. Norton, M. Conlon, and S. J.Mccormack, "The Battery Energy Storage System (BESS) Design Option for On-Campus Photovoltaic Charging Station (PV-CS)", Solar World Congress (SWC15), ISES, Daegu, Southkorea, 2015
[7]. A. Esfandyari, K.OFarell, B. Norton, and S. J.Mccormack, "Electric Bike Photovoltaic Charging Station (PV _ CS):Experimental Design for fault Investigations,", Irish Transport Research Conference (*ITRN*), Limerick, Ireland,2014
[8]. E. Flood, K. McDonnell, F. Murphy, and G. Devlin, "A Feasibility Analysis of Photovoltaic Solar Power for Small Communities in Ireland," *Open Renew. Energy J.*, vol. 4, no. 1, pp. 78–92, 2011
[9]. P. Kadar and A. Varga, "PhotoVoltaic EV charge station," *SAMI 2013 - IEEE 11th Int. Symp. Appl. Mach. Intell. Informatics, Proc.*, pp. 57–60, 2013
[10]. P. Goli and W. Shireen, "PV powered smart charging station for PHEVs," *Renew. Energy*, vol. 66, pp. 280–287, 2014
[11]. A. Al Bdwawi, H. Al Ahbabi, S. A. Al Shamsi, and A. a. Hussein, "Modular Photovoltaic Charging Station for ARE University Golf Carts," *Energy Procedia*, vol. 75, pp. 314–318, 2015
[12]. S. J. Tong, A. Same, M. a. Kootstra, and J. W. Park, "Off-grid photovoltaic vehicle charge using second life lithium batteries: An experimental and numerical investigation," *Appl. Energy*, vol. 104, pp. 740–750, 2013
[13]. D. a G. Redpath, D. McIlveen-Wright, T. Kattakayam, N. J. Hewitt, J. Karlowski, and U. Bardi, "Battery powered electric vehicles charged via solar photovoltaic arrays developed for light agricultural duties in remote hilly areas in the Southern Mediterranean region," *J. Clean. Prod.*, vol. 19, no. 17–18, pp. 2034–2048, 2011
[14]. E. K. Mueller and O. M. Mueller, "Off-Grid , Low-Cost , Electrical Sun-Car System for Developing Countries," *Glob. Humanit. Technol. Conf.*, pp. 14–17, 2014
[15]. A. Esfandyari, B. Norton, M. Conlon, and S. J. Mccormack, "Essentials for on-Campus Photovoltaic Charging Station(PV-CS): Grangegorman", Irish Transport Research Conference (*ITRN*), Galway, Ireland, 2015
[16]. B. B. Bafrouei, "Modeling and Optimization of Off-Grid Solar Street Lighting System", 2014
[17]. K. H. Lee, K. Malmedal, and P. K. Sen, "Conceptual design and cost estimate for a stand-alone residential photovoltaic system," *2012 IEEE Green Technol. Conf.*, 2012
[18]. TRNSYS, "Trnsys Manual,", 2014
[19]. D. Firmanda, A. Riza, S. I. H. Gilani, and M. S. Aris, "Measurement and simulation of a standalone photovoltaic system for residential lighting in Malaysia," vol. 2, no. 1, 2011
[20]. D. Firmanda, A. Riza, and S. I. H. Gilani, "Standalone Photovoltaic System Sizing using Peak Sun Hour Method and Evaluation by TRNSYS Simulation,"2014
[21]. L. Nkhonjera and J.-C. Wu, "Performance Analysis of Battery Based Stand Alone Solar Home Systems in Malawi", 2015
[22]. Y. S. Chen, P. P. Chong, and M. Y. Tong, "Mathematical and computer modelling of the Pareto principle," *Math. Comput. Model.*, vol. 19, no. 9, pp. 61–80, 1994

OUTDOOR PERFORMANCE AND MODELLING OF INNOVATIVE CRYSTALLINE SILICON PHOTOVOLTAIC MODULES UNDER HOT CLIMATIC CONDITIONS

George Makrides*[1], Alexander Phinikarides[1], Evelyn Herzog[2], Matthias B. Strobel[2] and George E. Georghiou[1]
[1]FOSS Research Centre for Sustainable Energy, Photovoltaic Technology Laboratory,
Department of Electrical and Computer Engineering, University of Cyprus
75 Kallipoleos Avenue, P.O. Box 20537, Nicosia, 1678, Cyprus
[2]Hanwha Q CELLS GmbH, Sonnenallee 17-21, 06766 Bitterfeld-Wolfen OT Thalheim, Germany
*corresponding author George Makrides (phone: +357 22 894397, Fax +357 22 895370, makrides.georgios@ucy.ac.cy),
Alexander Phinikarides (phinikarides.alexander@ucy.ac.cy), Evelyn Herzog (E.Herzog@q-cells.com),
Matthias B. Strobel (M.Strobel@q-cells.com), George E. Georghiou (geg@ucy.ac.cy)

ABSTRACT: The outdoor assessment of photovoltaic (PV) systems provides useful data to validate performance modelling approaches. The scope of this work is to analyse and model the outdoor performance of different grid-connected poly-crystalline Silicon (poly-c Si) PV systems installed at the PV Technology Laboratory of the University of Cyprus (UCY). These systems provided by Hanwha Q CELLS were installed side-by-side at a fixed-plane and monitored for a period of one year. Apart from PV system operational data, meteorological data were also extracted. Specifically, the focus of this ongoing analysis is to compare the performance of different modelling methods against real outdoor energy measurements and to investigate the impact of using real meteorological data for the simulation using the PVWATTS model and the PVSYST simulation software. The annual DC Performance Ratio (PR_{DC}) results using the PVWATTS model and calculated using the PVSYST software showed that the prediction error (PE) was below 2 % for the PVSYST simulation, whereas for the PVWATTS model PE differences of up to 8 % were observed when compared to the measured PR_{DC}.
Keywords: energy yield, modelling, photovoltaic, simulation, system performance.

1 INTRODUCTION

The introduction of new innovative material and fabrication techniques associated with crystalline-Silicon (c-Si) photovoltaic (PV) modules renders the evaluation of such modules under real outdoor conditions imperative. In this sense, outdoor field performance assessments provide useful benchmarks for modelling the performance of the technology when exposed to a specific environment. With the advent of new technologies and industrial methods in the quest to achieve higher efficiencies, the need for a better understanding of how the innovative technologies perform under the operating conditions they are exposed to during their lifetime and improving the energy yield prediction through the development of various simulation tools, becomes of crucial importance [1-6].

To this extent, different mathematical, empirical and electrical models for simulating the performance and output of PV technologies have been developed [7-12]. Additionally, over the years there has been a concerted effort to simplify and improve these models in order to provide more accurate predictions [13-17].

The purpose of this work is to assess and compare the outdoor performance of c-Si modules with different temperature coefficients under real outdoor prevailing conditions, with the results of two major performance models. More specifically, three special technology types of Hanwha Q CELLS poly-crystalline Silicon (poly-c Si) PV modules were evaluated outdoors by assessing their performance under real outdoor exposure at the hot, high irradiance conditions of Cyprus. The PV modules were mounted side by side and grid-connected in such a way so that all systems experience identical environmental conditions providing in this way the same exposure. The PV systems were monitored since June 2014 and both climatic data and PV operational measurements were acquired through the same data acquisition measuring platform at a resolution of one second and stored as fifteen-minute average data-sets.

Finally, the measured data sets were used to validate the power output predictions of two modelling approaches – PVWATTS [18] and PVSYST [19].

2 EXPERIMENTAL SETUP

2.1 Outdoor test facility description

The outdoor test facility of the University of Cyprus (UCY) includes a fixed plane infrastructure for outdoor performance assessments at both the module and system level. The different c-Si PV module types were mounted in landscape arrangement on aluminium mountings, at the optimum annual energy yield plane-of-array (POA) angle for Cyprus of 27.5°. The modules were installed in such a way so that identical conditions were applied to all systems and that comparison between the performance of the systems could be directly performed.

Each PV system installed in Cyprus, was connected to the data-acquisition platform of the PV Technology Laboratory of the UCY and used for the monitoring and storage of meteorological and PV system real outdoor operational data. The platform comprises of meteorological and PV operational measuring sensors connected to a central data acquisition system. The performance of the different PV systems and the prevailing meteorological conditions were monitored at a resolution of one second and stored as fifteen-minute average data-sets. Specifically, the meteorological measurements include the incident global irradiance (G_I), wind direction (W_a), wind speed (W_s), as well as ambient temperature (T_{amb}). The PV system operational measurements include maximum power DC current (I_{mp}), voltage (V_{mp}) and power (P_{mp}), as measured at the output of each PV system (DC side).

2.2 PV system description

Each of the PV systems comprises of five poly-c Si PV modules, each from 255 ~ 260 Wp as depicted by the manufacturer's datasheet. The modules of each system are connected in series to form a PV string at the input of the respective identical inverters. The same inverters are used to exclude the influence of different maximum power point tracking (MPPT) methods to the DC yield. Furthermore, it is worth noting that during the evaluation period the systems had been affected by partial shading during the very late afternoon hours of the winter season and were also dismounted on a regular basis for standardized indoor testing assessment. For a fair comparison, such data was rejected in the analysis. The main technical specifications of the three different PV system types are summarised in Table I.

Table I: Installed PV system technical characteristics.

System 1 (ucy1-11)	
Modules	5 × Hanwha Q CELLS 260Wp
System power (manufacturer's datasheet)	1300 W
Installation date	19/07/2013
Efficiency (reference to module area)	15.50 %
System 2 (ucy1-04)	
Modules	5 × Hanwha Q CELLS 255Wp
System power (manufacturer's datasheet)	1275 W
Installation date	19/06/2014
Efficiency (reference to module area)	15.35 %
System 3 (ucy1-08)	
Modules	5 × Hanwha Q CELLS 260Wp
System power (manufacturer's datasheet)	1300 W
Installation date	19/06/2014
Efficiency (reference to module area)	15.71 %

2.3 Uncertainty analysis

The main sources of uncertainty in the outdoor performance evaluations due to the instrumentation are presented in this section. In particular, the total irradiance was measured using two thermopile pyranometers (pyranometer 1 and 2) in the spectral range of 285 - 2800 nm, with a ±2 % and ±5 % expected daily uncertainty, respectively. In addition, the response time of the installed pyranometers is < 5 s (for 95 % response) which is longer than the PV module response time and the data acquisition interval (of 1 s), and this introduces an additional uncertainty contribution in the measurements.

Furthermore, parameters including MPP voltage and current were obtained using an active sensor board and were then connected to the data acquisition analogue channels. Both DC voltage and current were measured with an accuracy of ±0.5 % over the full range of measurement scale.

In addition, the AC power and energy yield were measured using an AC meter which is primarily associated with a metering measurement uncertainty of ±1 % and an additional ±1 % uncertainty that accounts for differences caused by the inverters (efficiency and MPPT accuracy). DC power and energy were also calculated from the measurements of DC voltage and current and are therefore associated with an uncertainty of ±1 %.

The installed data acquisition equipment, sensors and associated accuracy are summarized in Table II. Periodic calibrations and inspections of the sensors installed were performed to ensure good quality measurements and reveal any departures from the real measurements. The systems and sensors were regularly cleaned.

Table II: Summary of installed data acquisition equipment and sensors including manufacturer type and accuracy information.

Equipment/Sensor	Manufacturer type	Accuracy
Data acquisition	Delphin TopMessage	±0.01 %
Total irradiance POA	Kipp Zonen CMP11	±2 %
Total irradiance POA	Hukseflux SR11	±5 %
Ambient temperature	Rotronic HC2S3	±0.1 °C at 25 °C
Relative humidity	Rotronic HC2S3	±0.8 % at 25 °C
Wind speed	NRG 40C	± 0.45 m/s at 10 m/s
Wind direction	NRG 200P	±1 %
Module temperature	Heraeus (Pt100)	± 0.75 %
DC Voltage sensor	Muller Ziegler Ugt	±0.5 %
DC Current sensor	Muller Ziegler Igt	±0.5 %
AC Power	Muller Ziegler EZW	±1 %

3 METHODOLOGY

The performance of each PV system and the prevailing meteorological conditions were monitored at a resolution of one second and stored as fifteen-minute average data-sets, over the evaluation period. The energy yield at the DC side of the installed PV systems was predicted by using the mathematical model PVWATTS [18] and the simulation software PVSYST [19].

3.1 PVWATTS

The PV production model, PVWATTS, relates both the effective irradiance and the PV operational cell temperature to the produced maximum power (P_{mp}). The equation that forms the basis of the PVWATTS model is shown below [20]:

$$P_{mp} = \frac{E_e}{E_0} P_{mp0}\left[1 + \gamma\left(T_c - T_0\right)\right] \quad (1)$$

where, E_e is the effective irradiance adjusted for angle of incidence losses, soiling and spectral mismatch, E_0 is the reference irradiance of 1000 W/m^2, P_{mp0} is the maximum power at STC, T_0 is the reference PV cell temperature of 25 °C and γ is the maximum power temperature coefficient.

Additionally, the operational cell temperature for each system was obtained by employing the Faiman temperature model (also used in the PVSYST software tool) and given by the following equation [21]:

$$T_c = T_a + \frac{\alpha E_{POA}\left(1 - eta_m\right)}{U_0 + U_1 W_s} \quad (2)$$

where, T_c is the operating cell temperature (°C), T_a is the ambient air temperature (°C), α is the adsorption coefficient of the module (common default value is 0.9), E_{POA} is the irradiance incident on the plane of the module or array (W/m^2), eta_m is the efficiency of the PV module, U_0 is the constant heat transfer component (W/m^{2}K), U_1 is the convective heat transfer component (W/m^{3}sK) and W_s is the wind speed (m/s).

3.2 PVFORM

An enhancement to the PVWATTS mathematical model used in this investigation is PVFORM [22], which provides an improvement at low irradiance levels. In particular, for POA solar irradiance $\leq$ 125 W/m^2 the mathematical model is improved by incorporating the equation below:

$$P_{mp} = \frac{0.008(E_e)^2}{E_0} P_{mp0}\left[1+\gamma(T_c - T_0)\right] \qquad (3)$$

3.3 PVSYST

PVSYST is an energy modelling tool that can be used to simulate the energy production of grid-connected PV systems (details on simulation can be found in [2][23]). It is important to note that PVSYST contains parameters that can be customized based on the PV technology and module, with the additional functionality utilised in this investigation of importing meteorological data-sets as inputs to the simulation.

More specifically, the simulations were performed using PVSYST v6.43 with imported meteorological data-sets. The arrays were oriented in a way to emulate the real installation and configured with 5 modules per string. Furthermore, the module and inverter data-sets were selected as provided by the official manufacturer data sheets, as given at the PVSYST database. Finally, with respect to the simulation losses the following were considered:

- Ohmic losses: AC and DC cabling losses of 1 %.
- Mismatch losses of 0.3 %.
- Module quality loss of 0 % (no positive sorting or quality loss is applied).
- Soiling loss of 1 % since the modules are cleaned periodically therefore a low soiling loss is expected.
- Field thermal loss factor of 20W/m^{2}K (average circulation behind the modules).
- No shading losses

4 RESULTS

4.1 Meteorological data analysis

The annual total irradiation in the POA measured in Nicosia, Cyprus was 2089 kWh/m^2. In addition, the monthly total irradiation in the POA exhibited a seasonal pattern with the highest irradiation occurring during the summer and lowest during the winter (see Fig. 1).

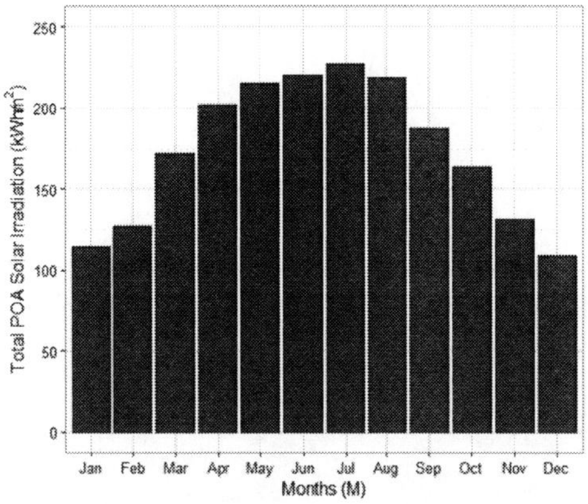

Figure 1: Monthly total irradiation in the POA over the period June 2014 – June 2015 in Nicosia, Cyprus.

To further investigate the contribution of the solar irradiance levels to the total irradiation, the solar irradiance was categorized into 100 W/m^2 bins. The plot of the total POA irradiation against the solar irradiance bins clearly shows that the low irradiance level ($\leq$400 W/m^2) presents a low share of 14 % to the yearly total irradiation, while the highest share of 21 % was for the 900 - 1000 W/m^2 bin (see Fig. 2).

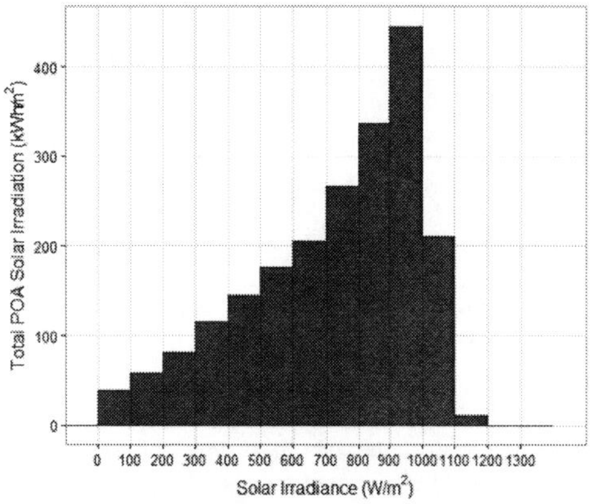

Figure 2: Total POA irradiation share per solar irradiance bins (100 W/m^2) over the period June 2014 – June 2015 in Nicosia, Cyprus.

Similarly, the monthly average ambient temperature ranged between 9 – 29 °C, with temperatures reaching maximum values of 40 °C during the summer (see Fig. 3).

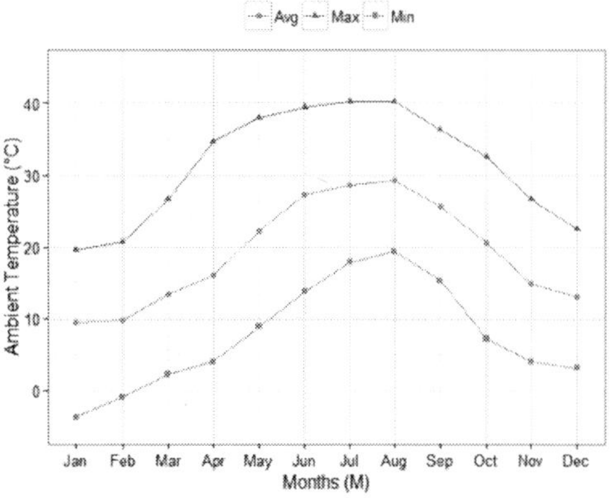

Figure 3: Monthly maximum, average and minimum ambient temperature over the period June 2014 – June 2015 in Nicosia, Cyprus.

4.2 System analysis

Table III summarizes the annual energy yield normalized to the manufacturer's rated power (E_{DC}) at the DC side of the three systems installed in Nicosia, Cyprus. Over the period of a year the highest energy yield normalized to the manufacturer's rated power was 1781 kWh/kWp produced by System 3 (without considering shading and outage corrections).

Table III: Measured annual DC energy yield normalized to the manufacturer's rated power (E_{DC}) over the period 19/06/2014–19/06/2015.

System	E_{DC} (kWh/kWp)*
System 1 (ucy1-11)	1744
System 2 (ucy1-04)	1761
System 3 (ucy1-08)	1781

*The systems were affected by partial shading during the late afternoon hours of the winter season and outage occurrences throughout the evaluation period because of indoor tests performed.

Additionally, the monthly PR_{DC} of each system demonstrated that the seasonal variation ranged between 79 % – 94 % (see Fig. 4) including power loss from shading and incomplete days where the modules were measured indoors.

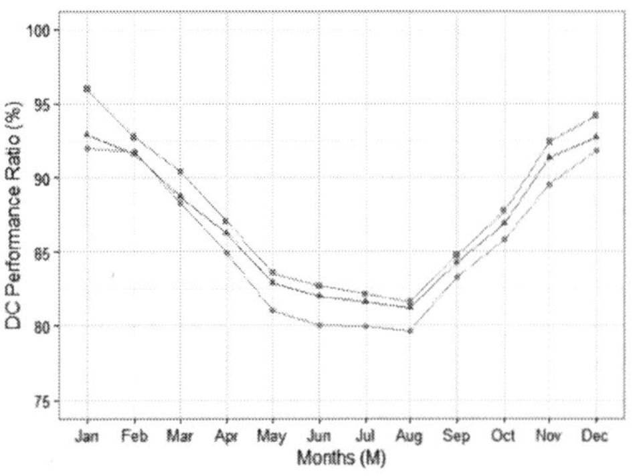

Figure 4: Monthly PR_{DC} of the installed systems over the period June 2014 – June 2015 in Nicosia, Cyprus.

In order to provide further insight into the dependence of the produced power with respect to the solar irradiance and operating temperature, the quantity $\Delta\eta_{rel}(G,T)$, which is the deviation of the module efficiency at any given irradiance and temperature conditions relative to the value at the corresponding STC efficiency (η_{STC}) was constructed. Periods when the systems were affected by partial shading during the late afternoon hours of the winter season and outage occurrences throughout the evaluation period were filtered out. The following plots show that for all systems the module efficiency varies predominantly between 80 – 100 %, exhibiting in parallel lower efficiencies at low light and high temperature conditions (see Fig. 5).

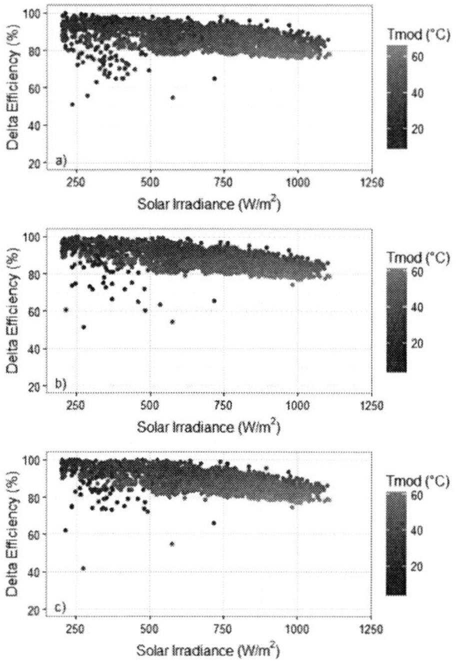

Figure 5: Deviation of the module efficiency at given irradiance and temperature conditions relative to the value at the corresponding STC efficiency ($\Delta\eta_{rel}$) over the period June 2014 – June 2015 in Nicosia, Cyprus for a) System 1, b) System 2 and c) System 3.

4.3 Comparison of measured to modelled array output

The annual PR_{DC} for the three systems, predicted using the PVWATTS model and calculated using the PVSYST software, demonstrated that the simulated PVSYST results showed good agreement when importing the real meteorological data-sets as inputs to the simulation (see Fig. 6). The prediction error (PE) of the PR_{DC} was calculated by subtracting the predicted PR_{DC} from the measured PR_{DC}, given by:

$$PE = Measured\ PR_{DC} - Predicted\ PR_{DC} \quad (4)$$

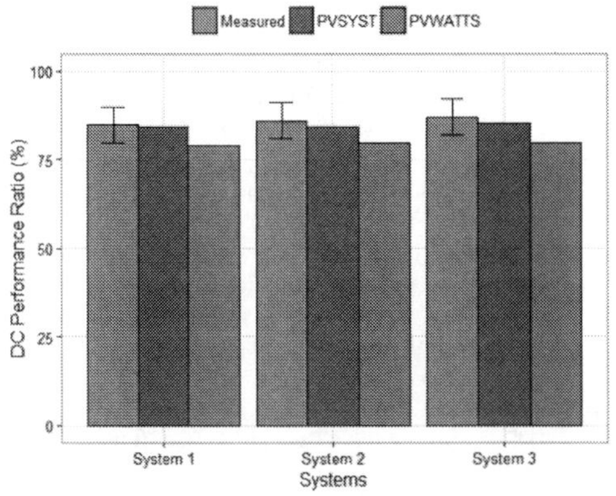

Figure 6: Comparison of measured and modelled annual PR_{DC} in Nicosia, Cyprus.

Table IV summarizes the *PE* of the measured and predicted (PVSYST and PVWATTS) annual PR_{DC}, for each of the three systems installed in Nicosia, Cyprus. In particular, the prediction *PE* was below 2 % for the PVSYST simulations whereas for the PVWATTS model the *PE* ranged from 6 % – 8 %.

Table IV: *PE* of annual PR_{DC} (PVSYST and PVWATTS) for each of the three systems installed in Nicosia, Cyprus, over the period June 2014 – June 2015.

System	Annual PR_{DC} *PE* PVSYST (%)	Annual PR_{DC} *PE* PVWATTS (%)
System 1	0.5	5.6
System 2	1.8	6.5
System 3	1.7	7.5

Useful information on the performance of the simulation models was obtained from the plots of the *PE* of the measured and predicted (PVSYST and PVWATTS) monthly PR_{DC}, for each of the three systems over the evaluation period (see Fig.7). The results clearly demonstrated that the highest *PE* was exhibited during the winter season and this could be associated with simulation model limitations at low light conditions.

To further provide insight into the performance of each model, the measured, PVSYST simulated and PVWATTS predicted PR_{DC} of System 1 was plot against the measured solar irradiance in the POA as a function of module temperature. More specifically, the comparison of the plot of Figure 8a and Figure 8b shows that PVSYST tends to underestimate the PR_{DC} at low irradiance levels. On the other hand, the PVWATTS model tends to underestimate the predicted PR_{DC} at both low and high irradiance levels, as depicted in Figure 8c.

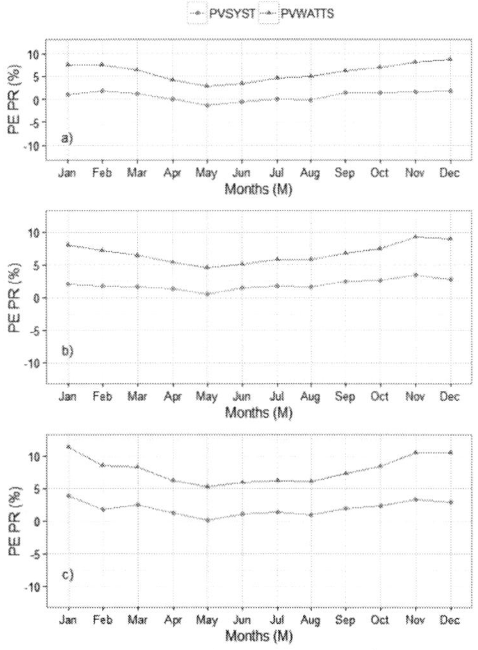

Figure 7: *PE* between the measured and the predicted monthly PR_{DC} for a) System 1, b) System 2 and c) System 3.

Improvements to both modelling approaches could be made by considering the spectral losses, low-light efficiency losses (which both respect to STC depend on the PV module parameters of series (R_{series}) and shunt resistance (R_{shunt})), availability and module quality losses.

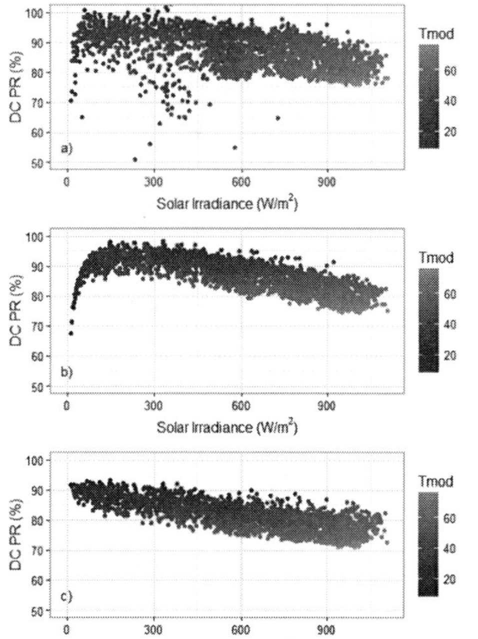

Figure 8: Solar irradiance plot of System 1 against a) Measured PR_{DC}, b) Simulated PVSYST PR_{DC} and c) Predicted PVWATTS PR_{DC}.

5 CONCLUSIONS

In this work the outdoor performance of different grid-connected poly-c Si PV systems installed side-by-side at the PV Technology Laboratory of the UCY and provided by Hanwha Q CELLS was analysed. In particular, over the period 19/06/2014 – 19/06/2015, the highest energy yield normalized to the manufacturer's rated power was 1781 kWh/kWp produced by System 3.

The results obtained using the PVWATTS model and calculated using the PVSYST software showed that the installed systems produced higher energy compared to the predicted results. Specifically, the annual PR_{DC} prediction error was below 2 % for the PVSYST simulation, whereas for the PVWATTS model prediction error differences of up to 8 % were observed when compared to the measured annual PR_{DC}.

Finally, the analysis showed that by importing the meteorological data-sets as inputs to the simulation, PVSYST tends to underestimate the power at low irradiance levels. On the other hand, the PVWATTS model tends to underestimate at both low and high irradiance levels.

6 REFERENCES

[1] D. Thevenard, G. Leng, and S. Martel, "The RETScreen model for assessing potential PV projects," in Conference Record of the 28th IEEE Photovoltaic Specialists Conference, 2000, pp. 1626–1629.

[2] A. Mermoud and T. Lejeune, "Performance assessment of a simulation model for PV modules of any available technology," in 25th European Photovoltaic Solar Energy Conference, 2010, no. September, pp. 4786–4791.

[3] D. L. King, J. K. Dudley, and W. E. Boyson, "PVSIM: a simulation program for photovoltaic cells, modules, and arrays," in Conference Record of the 25th IEEE Photovoltaic Specialists Conference, 1996, pp. 1295–1297.

[4] C. P. Cameron, W. E. Boyson, and D. M. Riley, "Comparison of PV system performance-model predictions with measured PV system performance," in 33rd IEEE Photovoltaic Specialists Conference, 2008, pp. 1–6.

[5] H. Matsukawa, P. S. Pimentel, T. Izawa, S. Ike, H. Koizumi, and K. Kurokawa, "An integrated design software for photovoltaic systems," in Proceedings of 3rd World Conference on Photovoltaic Energy Conversion, 2003, pp. 1930–1933.

[6] G. Becker, B. Schiebelsberger, W. Weber, J. Schumacher, M. Zehner, G. Wotruba, and C. Vodermayer, "Energy yields of PV systems - Comparison of simulation and reality," in 23rd European Photovoltaic Solar Energy Conference, 2008, pp. 3184–3186.

[7] G. Makrides, B. Zinsser, M. Schubert and G. E. Georghiou, Energy yield prediction errors and uncertainties of different photovoltaic models. Progress in Photovoltaics: Research and Applications, 2011: 21: 500–516.

[8] F. Mavromatakis, G. Makrides, G. E. Georghiou, A. Pothrakis, Y. Frangiadakis, E. Drakakis and E. Koudoumas, Modelling the Photovoltaic Potential of a site, Renewable Energy, vol. 37, no. 7, pp. 1387-1390, Jul. 2010

[9] M. D. Siegel, S. A. Klein, and W. A. Beckman, "A simplified method for estimating the monthly-average performance of photovoltaic systems," Sol. Energy, vol. 26, no. 5, pp. 413–418, Jan. 1981.

[10] U. Eicke, D. Pietruschka, J. Schumacher, J. Fernandes, T. Feldmann, and E. Bollin, "Improving the energy yield of PV power plants through internet based simulation, monitoring and visualisation," in 20th European Photovoltaic Solar Energy Conference, 2005, pp. 2670–2673.

[11] S. Ransome, "Comparing PV simulation models and methods with outdoor measurements," in 35th IEEE Photovoltaic Specialists Conference, 2010, pp. 2306–2311.

[12] Y. Sukamongkol, S. Chungpaibulpatana, and W. Ongsakul, "A simulation model for predicting the performance of a solar photovoltaic system with alternating current loads," Renew. Energy, vol. 27, no. 2, pp. 237–258, Oct. 2002.

[13] S. J. Ransome, J. H. Wohlgemuth, S. Poropat, and E. Aguilar, "Advanced analysis of PV system performance using normalised measurement data," in Conference Record of the 31st IEEE Photovoltaic Specialists Conference, 2005, pp. 1698–1701.

[14] W. De Soto, S. A. Klein, and W. A. Beckman, "Improvement and validation of a model for photovoltaic array performance," Sol. Energy, vol. 80, no. 1, pp. 78–88, Jan. 2006.

[15] S. Elies, M. Hermle, and B. Burger, "Improved two-diode model for more detailed simulation of I-U-curves for solar cells and modules," in 24th European Photovoltaic Solar Energy Conference, 2009, pp. 3737–3740.

[16] S. Singer, B. Rozenshtein, and S. Surazi, "Characterization of PV array output using a small number of measured parameters," Sol. Energy, vol. 32, no. 5, pp. 603–607, Jan. 1984.

[17] M. Akbaba and M. A. A. Alattawi, "A new model for I–V characteristic of solar cell generators and its applications," Sol. Energy Mater. Sol. Cells, vol. 37, no. 2, pp. 123–132, May 1995.

[18] W. Marion and M. Anderberg, "PVWATTS-An online performance calculator for grid-connected PV systems," in ASES Solar Conference, 2000.

[19] "PVsyst photovoltaic software." [Online]. Available: http://www.pvsyst.com.

[20] "PV performance Modelling Collaborative" [Online]. Available: https://pvpmc.sandia.gov/.

[21] D. Faiman, "Assessing the outdoor operating temperature of photovoltaic modules," Prog. Photovoltaics Res. Appl., vol. 16, pp. 307–315, 2008.

[22] D. F. Menicucci and J. P. Fernandez, "User's Manual for PVFORM: A Photovoltaic System Simulation Program for Stand-Alone and Grid-Interactive Applications," 1088.

[23] B. Wittmer, A. Mermoud, and T. Schott, "Analysis of PV grid installations performance, comparing measured data to simulation results to identify problems in operation and monitoring," in 30th European Photovoltaic Solar Energy Conference and Exhibition, 2015, pp. 2265 – 2270.

EVALUATION OF SOILING DURING A 2-MONTHS DROUGHT AND CONSTRUCTION WORKS NEAR A PV TEST FACILITY IN NORTH-EAST OF ITALY

G. Belluardo[1*], P. Ingenhoven[1], D. Moser[1]
[1]Institute for Renewable Energy, EURAC Research, Viale Druso 1, 39100 Bolzano (BZ), Italy
*Phone: +39 0471 055626; Fax: +39 0471 055699; E-mail: giorgio.belluardo@eurac.edu

ABSTRACT: The accumulation of dust on PV modules surface results in a loss of performance. PV plants installed in areas prone to pollution, or to the presence of sand or sea salt in the atmosphere are more affected by soiling. The process is also favored by the scarcity of rainfalls. The South Tyrol region (North-East of Italy) is not particularly affected by this process. However, in the last two months of 2015 a dry period with no rainfalls was registered in the area. This coincided with construction works on the extension of the runways at Airport Bolzano Dolomiti, where a multi-technology PV test facility is installed.
This paper calculates the performance losses, expressed as daily percentage, of 13 crystalline-silicon based PV technologies. Comparisons are carried out considering different glass textures (smooth vs structured), orientations of the modules within the array (portrait vs landscape) and the presence of a frame.
Keywords: soiling; performance loss; crystalline-silicon

1 INTRODUCTION

Soiling is the accumulation of dust that causes a decrease of transmission of the solar spectrum through the front glass of a photovoltaic (PV) module, and therefore a decrease of PV generated power. Plenty of studies have focused on this topic, especially in climates more prone to this effect, such as desert and coastal areas. An overview of the most significant studies on soiling is reported in [1]. In particular, Ghazi et al. [2] found that the problem of soiling is more severe in the Middle East and North Africa. If we consider that many large-scale PV plants are installed in these parts of the world due to favorable irradiance conditions, the importance of correctly studying the mechanism of soiling as well as providing reliable assessment tools raise.

Soiling of PV modules depends on two main factors: environmental conditions and PV module and system characteristics. In addition to how much an area tends to accumulate dust, also the properties of dust (i.e. particle size, shape and weight; chemical, biological and electrostatic property) have been demonstrated to play a role [3]. Wind is another important factor, since it can increase or decrease soiling effects depending on the speed and the degree of turbulence [4]. From the PV module side, the most important influence is exerted by the surface of the front glass. Textured glass tends to accumulate dust more, but also smooth glass presents different levels of soiling depending on the presence of special coatings, such as anti-reflection (AR) or self-cleaning (SC) coatings. In particular, Appels et al. [5] observed that a combination of AR and SC coatings decrease the reduction of transmittance due to soiling more than the application of just one of them. Furthermore, the effect of soiling is higher the less the PV module is inclined.

From 25th August to 30th November 2015 construction works were carried out at the Airport Bolzano Dolomiti (South Tyrol, North-East of Italy) that released dust in the air. The PV test facility installed in August 2010 and monitored by EURAC is just few meters away and was subjected to the deposition of this dust. This process was increased by the fact that during the period between 31st October 2015 and 2nd January 2016 no rainfall was registered in the area. The purpose of this study is to quantify the mean loss of performance, expressed in *%/day*, occurred during this rainless period. Since the considered PV systems have different glass surface textures, orientations (portrait or landscape) and have or not a metal frame, results are discussed considering these properties.

2 METHODOLOGY

Thirteen silicon-based photovoltaic groups (mono- and polycrystalline silicon, ribbon) ranging between 1.2 kWp and 4 kWp are considered in this study. They are part of the multi-technology test facility of Airport Bolzano Dolomiti (North-East of Italy) [6], and are installed on fixed racks with a tilt of 30° and an orientation angle of 8.5° West of South. Even though the multi-technology test facility consist also of thin-film technologies, the choice of focusing only on silicon-based technology was driven by the fact that they are affected by spectral effects in a similar way, having similar spectral responsivities.

Electrical parameters, such as DC-current and DC-voltage are acquired by commercial inverters with a 15-minutes frequency. A weather station is installed in the proximity, with a secondary standard pyranometer providing measurements of global irradiance on the arrays' plane and Pt100 sensors providing measurements of back-of-module temperature.

The works at Bolzano Airport started on the 25th August 2015 and ended on 30th November 2015. The period of drought, i.e. without rain, lasted from 31st October 2015 to 2nd January 2016. During that period, the pyranometer was regularly cleaned. Cumulative daily records of rainfalls were acquired from a meteorological station run by the Meteorology Department of the Autonomous Province of Bolzano, around 3 km away from the test site.

In a first instance, a filter for $PR^*-\sigma<PR^*<PR^*+\sigma$ was applied, where PR^* is the mode (most probable value) of the distribution frequency of the 15-minutes based values of Performance Ratio, PR, in a month and σ is the standard deviation of all 15-minutes based values of PR from the value PR^* in a month. The metric PR is defined by IEC61724:1998 [7]. On the DC-side, it can be expressed as [8]:

$$PR = \frac{\sum P_{dc}(i)}{\sum G(i)} \frac{G_{STC}}{P_{STC}} \quad (1)$$

where P_{dc} is the DC-power of the PV system at time i, $G(i)$

is the irradiance measured in the plane of the modules at time i, while G_{STC} is the irradiance under standard test conditions (STC) (corresponding to 1000 W/m^2) and P_{STC} is the nominal power of the PV system under STC, given by the manufacturer. The purpose of this filter is to exclude performance values corresponding to inhomogeneous conditions of irradiance as detected by a PV system and by the irradiance sensor. This happens for example for mountain shading at sunrise and sunset, as well as for mutual shading between consecutive PV arrays, which is stronger especially at winter time with lower sun elevation.

Another filter was set for irradiance $G > 500$ W/m^2, in order to exclude performance values corresponding to cloudy-sky conditions (diffuse component of irradiance prevailing) and situations when sun is behind the surrounding mountain. The selected value has been verified to be a good threshold in the specific location of Bolzano airport, both for summer and winter time.

The metric used to evaluate the performance of the considered PV systems during the period of drought is the Performance Ratio calculated on DC-current on a daily base, PR_{Impp}:

$$PR_{Impp} = \frac{\sum I_{dc}(i)\ G_{STC}}{\sum G(i)\ \ I_{STC}} \quad (2)$$

where I_{dc} is the DC-current of the PV system at time i. PR_{Impp} differs from the Performance Ratio PR defined by Equation 1 for the use of current in place of power. The choice of PR_{Impp} in place of PR is motivated by the fact that it has been verified that the soiling occurred on the considered PV systems acted mainly on the current of the PV modules. Furthermore, DC-current is less affected by temperature than DC-power or DC-voltage, and therefore the seasonality of PR_{Impp} is lower than simply using PR.

However, also in this case still a temperature effect is affecting the PV system performance. Therefore, a further temperature correction was performed using the following expression:

$$PR_{Impp\_Tcorrected} = PR_{Impp} \frac{1}{1+\alpha(T_{cell}-T_{STC})} \quad (3)$$

where α is the temperature coefficient in maximum power point current (%/°C), given by the manufacturer, T_{cell} is the cell temperature, considered in a first approximation equal to the back-of-module temperature, and T_{STC} is the module temperature at STC (corresponding to 25°C). In the prosecution of this paper, we will always refer to the term PR_{Impp} in the meaning of Performance Ratio calculated on DC-current and corrected for temperature to STC.

The losses of performance of each PV system, expressed in *%/day*, is calculated by a linear regression performed on the daily averages of PR_{Impp}, starting from the first day (31st October 2015) and ending on the last day (2nd January 2016) of the dry period.

3 RESULTS AND DISCUSSION

The presented method was applied to a data series ranging from July 2015 to March 2016, encompassing the period of drought. The daily averages of PR_{Impp} for one technology are reported in Figure 1. In order to have a clearer view of the performance pattern, a smoothing average was calculated on the daily values, using a central average of 7 days. A clear decreasing trend is evident after the last rainfall registered on the 30th October 2015, while a substantial recovery of performance occurs right after the first rainfall after the dry period, occurred on 3rd January 2016.

Figure 2 zooms-in the period of drought, and shows the regression line. In particular, its angular coefficient corresponds to the loss of performance occurred in one day due to soiling. Actually, this value includes also the performance loss normally occurring in PV systems due to the material degradation and to a non-optimal electric performance caused by the environmental conditions. However, considering that the highest performance loss calculated for the considered technologies is -0.6%/year [9], the corresponding daily impact is not significant (0.0016%/day) and can be therefore neglected.

The values of daily losses as calculated by the regression shown in Figure 2 are reported in Table 1. In order to highlight possible influences of frame presence, module orientation and glass texture, this information has also been reported. Performance losses due to soiling range from a minimum of -0.03%/day to a maximum of -0.08%/day. Considering the 62-days period of drought, these correspond to a minimum of -1.56% and a maximum of -5.10% losses. The highest loss occurs for the PV system with textured glass modules, probably due to the higher tendency of dirt and dust to be trapped in this kind of structure. A similar high value of loss occur for the PV system with landscape-oriented modules and frame. Despite many studies focus on the difference between performance losses due to sharp shading in landscape and

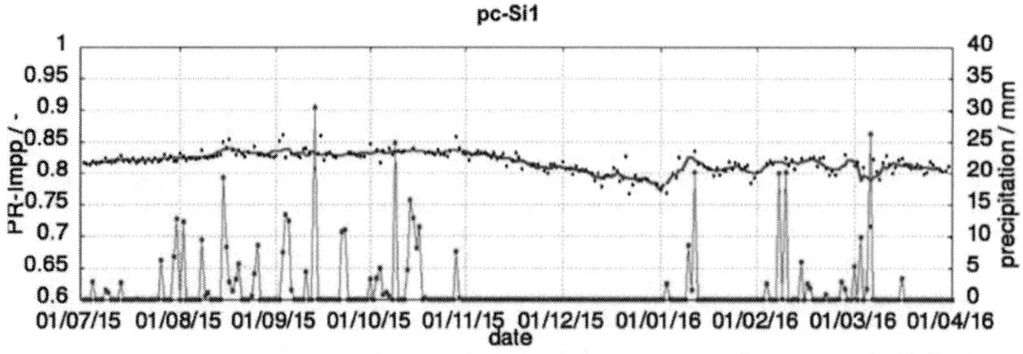

Figure 1: Daily averages of PR_{Impp} (black dots) for the technology pc-Si1 and corresponding smoothing average for three periods: before (green line), during (red line) and after (blue line) the period of drought (31st October 2015 - 2nd January 2016). A secondary y-axis shows the corresponding daily cumulative precipitation.

portrait oriented modules, fewer focus on the effect of soiling, especially of the non-homogeneous type as in our case. In particular, Appels et al. [5] have shown that a landscape oriented module could be affected by higher losses than portrait-oriented when subjected to non-homogeneous soiling.

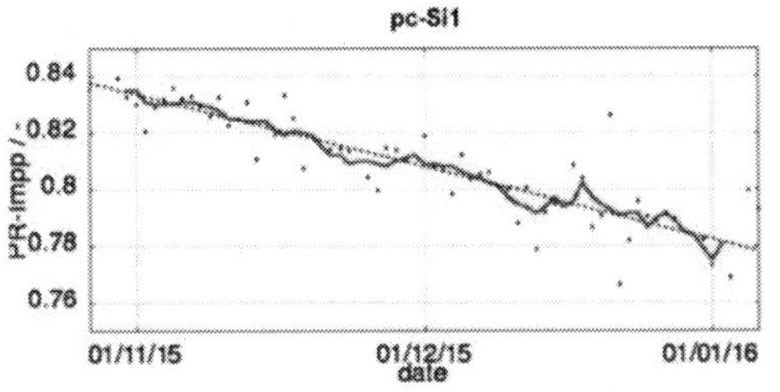

Figure 2: Daily averages (black dots) of PR_{Impp} for the technology pc-Si1 during the period of drought between 31st October 2015 and 2nd January 2016; corresponding smoothing average (red line) and regression line (black dashed line).

A possible explanation for this is given by Maghami et al. [1]. When dust deposits on a portrait-oriented module, this tends to cause a partial shading that is different from string to string, and is higher on the bottom string. In such a condition, the different strings deliver different voltages to the inverter, and in turn several power peaks in the I-V curve may raise. As a consequence, the inverter could track the wrong optimum point, i.e. a lower power local peak.

The other PV systems composed of landscape-oriented modules, mc-Si1 and mc-Si2, do not seem to be affected by a similar amount of soiling loss. However, in this case the modules have no frame, which constitute an element for the accumulation of dust and dirt at the bottom of the module. In this sense, the absence of frame in these two systems may counteract the negative effect of their landscape orientation.

Table I: main characteristics and daily performance loss due to soiling of the considered technologies.

Tech	type of glass	frame	orientation	soiling performance loss (%/day)
mc-Si1	smooth	no	landscape	-0.03
mc-Si2	smooth	no	landscape	-0.03
mc-Si3	smooth	yes	portrait	-0.07
mc-Si4	smooth	yes	portrait	-0.03
mc-Si5	smooth	yes	portrait	-0.05
pc-Si1	textured	yes	portrait	-0.08
pc-Si2	smooth	yes	portrait	-0.03
pc-Si3	smooth	yes	landscape	-0.08
pc-Si6	smooth	yes	portrait	-0.03
pc-Si7	smooth	yes	portrait	-0.04
pc-Si8	smooth	yes	portrait	-0.06
pc-Si9	smooth	yes	portrait	-0.05
ribbon1	smooth	yes	portrait	-0.05

4 CONCLUSIONS

The loss of performance due to soiling occurred during a period of drought between 31st October 2015 and 2nd January 2016 has been assessed on thirteen crystalline-based photovoltaic systems. They are installed at the multi-technology test facility of Bolzano Airport (North-East of Italy) and are characterized by different orientations (landscape vs portrait), surface glass texture (smooth vs textured), and by the presence or absence of metal frame. Losses were calculated in terms of *%/day* from a regression of daily averages of Performance Ratio calculated on DC-current, PR_{Impp}, and corrected for temperature to STC. Results show that soiling losses range from -0.03%/day to -0.08%/day, corresponding to losses between -1.56% and -5.10% on the whole 62-days period of drought. Higher soiling losses are calculated on the PV system composed of modules with textured surface glass, as well as on the PV system composed of modules with landscape orientation. This last finding might be related with a miss-tracking of the inverter due to the different voltages delivered by the differently-soiled strings of the module.

5 ACKNOWLEDGEMENTS

The research leading to these results has received funding from the European Union's Horizon 2020 research and innovation programme under the grant agreement No 649997 (Solar Bankability). The authors wish also to acknowledge the Stiftung Südtiroler Sparkasse for financing the institute's participation in the IEA Task 13.

6 REFERENCES

[1] M.R. Maghami, H.Hizam, C.Gomes, M.A. Radzi, M.I. Rezadad, S. Hajighorbani, 2016. Power loss due to soiling on solar panel: A review. Renewable and Sustainable Energy Reviews 59, 1307–1316

[2] S. Ghazi, A. Sayigh, K. Ip, 2014. Dust effect on flat surfaces – a review paper. Renew Sustain Energy Reviews33, 742–751.

[3] M. Mani, R. Pillai, 2010. Impact of dust on solar photovoltaic (PV) performance: Research status, challenges and recommendations. Renew Sustain Energy Rev 14, 3124–3131.

[4] D. Goossens, Z. Offer, A. Zangvil, 1993. Wind tunnel experiments and field investigations of eolian dust deposition on photovoltaic solar collectors. Solar Energy 50, 75–84

[5] R. Appels, B. Lefevre, B. Herteleer, H. Goverde, A. Beerten, R. Paesen, K. De Medts, J. Driesen, J. Poortmans, 2013. Effect of soiling on photovoltaic modules. Solar Energy 96, 283–291

[6] G. Belluardo, P. Ingenhoven, D. Moser, 2013. Medium-term degradation of different photovoltaic technologies under outdoor conditions in alpine area. Proc. 28th EU PVSEC, 4318 - 4321.

[7] IEC61724:1998, 1998. Photovoltaic System

Performance Monitoring Guidelines for Measurement, Data Exchange, and Analysis.

[8] A. Woyte, M. Richter, D. Moser, N. Reich, M. Green, S. Mau, H.G. Beyer, 2014. Analytical Monitoring of Grid-Connected Photovoltaic Systems – Good Practices for Monitoring and Performance Analysis. Tech. rep., IEA-PVPS T13-03:2014.

[9] G. Belluardo, P. Ingenhoven, W. Sparber, J. Wagner, P. Weihs, D. Moser, 2015. Novel method for the improvement in the evaluation of outdoor performance loss rate in different PV technologies and comparison with two other methods. Solar Energy 117, 139–152.

GLOBAL METHOD FOR CALCULATING LOCATION SPECIFIC MPP TRACKING LOSSES USING AVAILABLE WEATHER STATISTICS

Matthias Egler, Sam Gordon, Pascal Yim
OST Energy Ltd, Nile House, Second Floor, Nile Street, Brighton (UK), BN1 1HW
Phone: +44 (0)1273 819 429, e-mail: matthiasegler@ostenergy.com

ABSTRACT: The fraction of site-specific annual static irradiation conditions varies significantly between different locations and subsequently leads to variations in annual inverter Maximum-Power-Point-Tracking (MPPT) losses. However, there is a lack of a global method to estimate this loss in a project specific manner within PV energy yield studies.

On the basis of EN 50530:2010 MPPT efficiency definitions and worldwide weather statistics obtained from Baseline Surface Radiation Network (BSRN) records, a linear correlation between the annual ratio of Diffuse to Global Horizontal Irradiation (DHI/GHI) and static irradiance conditions has been found. DHI and GHI are standard information in state-of-the-art yield calculations.

Using DHI/GHI, site- and product-specific annual MPPT losses are estimated assuming MPP Tracking efficiencies corresponding with different levels of industry standard inverters. The results display how PV energy yield studies can become more site- and product-specific.

Keywords: Modelling, Simulation, Inverter, MPPT Performance, Solar Radiation

1 INTRODUCTION

In energy yield studies for PV systems, typically overall MPPT losses due to the inverters control algorithm are either neglected or estimated using a default assumption for static conditions irrespective of location. In 2015, estimates of site-specific variations of annual MPPT losses have been presented by the authors [1]. In this study, inverter MPPT efficiencies were used as defined in EN 50530:2010 [2] along with corresponding average inverter performance data and with weather statistics obtained from Baseline Surface Radiation Network (BSRN) [3] records of various locations worldwide. The BSRN data show a site-dependent range of annual static conditions of between 40% and 90% of GHI and subsequent yearly MPPT losses ranging from approximately 0.9% to 0.4% have been revealed. These results highlighted the need for a globally applicable approach to estimate project specific MPPT losses within yield calculations. Furthermore a linear correlation between annual DHI/GHI ratios and the fraction of annual static irradiance conditions has been observed.

Following the methodology outlined in the 2015 paper [1] the meteorological database has been expanded by two locations. It now includes high-quality records and subsequent weather statistics obtained at 17 BSRN weather stations located in Africa, America, Asia, Australia and Europe. The purpose of this follow-up paper is to present the global correlation between annual DHI/GHI ratio and the amount of annual static weather conditions (the variability factor), i.e. a worldwide basis to estimate project-specific MPPT losses by using readily available site specific weather data. In addition, annual MPPT losses determined for different inverter efficiencies are shown to present the relevance of a site- and product-specific approach for PV energy yield studies. Finally, the uncertainty associated with the proposed global approach to estimate annual MPPT losses is evaluated.

2 WORKING BASE

2.1 European standard EN 50530:2010

EN 50530:2010 defines irradiance slopes in order to characterise the MPPT performance of PV inverters under dynamically changing conditions [2]. The defined slopes are varying from 0.1W/m²/s to 100W/m²/s and can be used to differentiate static and dynamic conditions. Information from corresponding inverter testing combined with weather statistics can be used to estimate MPP Tracking losses, as outlined in [1]. Since the definition of the irradiance slopes refers to second time series, which are usually unavailable or limited to single sites, the gradients are adapted to more common resolutions (see Table I). Note that converted gradients exceeding 1300W/m² are beyond the physical possible range.

Table I: Irradiance slopes given by EN 50530:2010 and converted for 1min, 2min and 3min time series

EN 50530:2010 (W/m²/sec)	Converted 1min	Converted 2min (W/m²/Xmin)	Converted 3min
0.1	6	12	18
0.5	30	60	90
1	60	120	180
2	120	240	360
3	180	360	540
5	300	600	900
7	420	840	1260
10	600	1200	n/a
14	840	n/a	n/a
20	1200	n/a	n/a
30	n/a	n/a	n/a
50	n/a	n/a	n/a
100	n/a	n/a	n/a

Assuming that the rate of change is more relevant to MPPT efficiency than the time over which the drop or increase occurs, each gradient over a second has to be multiplied by 60 or a corresponding multiple to arrive at its adopted minute resolution value.

2.2 Irradiation data

The irradiation data used in this work have been obtained through the Baseline Surface Radiation Network. The objective of the BSRN is to provide measurements of the best possible quality and with a high temporal resolution (1 minute to 3 minutes) in a uniform and consistent way. The network currently includes met stations at 58 locations spread over the globe in contrasting climatic zones [3].

Overall data of 17 BSRN locations are used in this study, which consist predominantly of 1 minute records, except for two sites with 2 minute and 3 minute resolution. The number of years included varies from 2 years to 18 years driven by the total available years in the BSRN archive [4].

Table II presents latitude (LAT) and longitude (LON) of the 17 locations included in this study, which cover different climatic zones worldwide.

Table II: BSRN location information

Location	LAT (°North)	LON (°East)
Izana, ESP	28.309	-16.499
Cener, ESP	42.816	-1.601
Carpentras, FRA	44.083	5.059
Palaiseau, FRA	48.713	2.208
Camborne, UK	50.217	-5.317
Lindenberg, GER	52.210	14.122
Alice Springs, AUS	-23.798	133.888
Fukuoka, JP	33.582	130.375
Xianghe, PRC	39.754	116.962
Florianopolis, BRA	-27.533	-48.517
Brasilia, BRA	-15.601	-47.713
Desert Rock, USA	36.626	-116.018
De Aar, RSA	-30.667	23.993
Gobabeb, NAM	-23.561	15.042
Ilorin, NIG	8.533	4.567
Solar Village, SA	24.910	46.410
Sede Boqer, ISR	30.905	34.782

Note that the equipment to record GHI and DHI has been exchanged regularly over the years.

2.3 Inverter MPPT efficiency data

MPPT efficiency quantifies the effectiveness with which the inverter tracks the MPP of the PV array (or module) and depends on the field conditions. Considering good quality inverters, during static input conditions, MPPT efficiency should perform similarly and produce negligible differences in energy loss due to MPPT. However when considering dynamic input conditions the variation in the efficiency of different MPPT methods implemented in the inverter may produce considerable variation in performance.

Based on information provided confidentially by a number of inverter manufacturers, and some results published in research, this study has collected static and dynamic MPPT performance measurements from PV inverters in the current market place that have been measured according to [2]. The results reflect a number of inverter sizes from single phase string inverters to large three phase centralised units of greater than 500kW, we have not noted any pattern in the results relating to inverter rating. A summary of the tested MPPT efficiencies obtained can be found in Table III below;

Table III: MPPT efficiencies

	MPPT efficiencies		
	Min.	**Mean**	**Max.**
Static MPPT efficiency ($\eta_{mppt,stat}$)	99.3%	99.7%	99.9%
Dynamic MPPT efficiency ($\eta_{mppt,dyn}$)	96.6%	98.6%	99.5%

3 ANALYSIS

3.1 Dynamic and static annual irradiation conditions

BSRN provides meteorological time series saved in monthly files, with availability of data varying from location to location and time steps of 1 minute to 3 minute. A regional classification of the selected sites, the years included in this study and the underlying temporal resolution of the records is summarized in Table IV.

Table IV: Regional classification and years included in the analysis as well as time step information

Location	Region	Years	Time step
Izana, ESP	Europe	2010-2014	1
Cener, ESP	Europe	2010-2012	1
Carpentras, FRA	Europe	1997-2014	1
Palaiseau, FRA	Europe	2006-2013	1
Camborne, UK	Europe	2001-2007, 2014	1
Lindenberg, GER	Europe	1995-2006	1
Alice Springs, AUS	Asia & Oceania	1995, 1997-98, 2002-09, 2011-12	1
Fukuoka, JP	Asia & Oceania	2011-2013	1
Xianghe, PRC	Asia & Oceania	2005-2010	1

Location	Region	Years	Time step
Florianopolis, BRA	Ameri-cas	1996*, 1998*, 2014	1 / 2*
Brasilia, BRA	Ameri-cas	2009-2011	1
Desert Rock, USA	Ameri-cas	1999-2008	3
De Aar, RSA	Africa	2002, 2004	1
Gobabeb, NAM	Africa	2013-2014	1
Ilorin, NIG	Africa	1996-1997, 2000-2001	1
Solar Village, SA	Africa	1999-2002	1
Sede Boqer, ISR	Africa	2003-2011	1

Most of the sites have been already analysed in [1]. Two additional BSRN locations are included in this work, Alice Springs, AUS and Desert Rock, USA, both have been quality checked and studied accordingly (see [1]).

On the basis of the yearly datasets considered in this paper, a range of fraction of dynamic conditions from 9.9% (Gobabeb, NAM; 2013) to 63.0% (Camborne, UK; 2002) has been obtained. For the additional sites, the resulting long-term numbers are 23.3% (Alice Springs, AUS) and 18.0% (Desert Rock, USA), with single years varying between 19.9% and 29.4% at the Australian and 14.8% to 20.5% at the USA location. The dynamic fractions found in this work are in line with other industry research findings which indicate for Central Germany and Central Europe values of ca. 32.5% [5] or between 40% and 65%, respectively [6].

In order to evaluate the effect induced by the minute resolution of the time series, the data of some BSRN locations are averaged to time steps of 2 minutes to 5 minutes as well as 15 minutes. Using the relevant converted irradiance gradients, respective variability factors for each temporal granularity are obtained.

Figure 1 presents the time step dependent fraction of annual static or constant GHI conditions found for Camborne (UK), Florianopolis (BRA), Fukuoka (JP) and Solar Village (SA). Subsequently, using a 2nd order polynomial function for each BSRN location, the variability factor for 1sec resolution can be estimated.

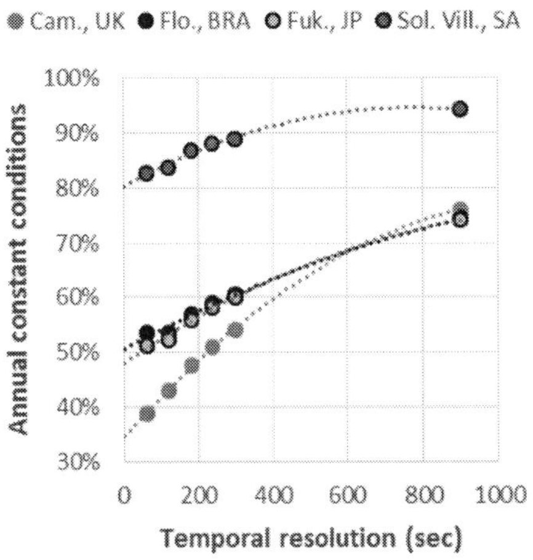

Figure 1: Time step dependent annual constant conditions

For the four sites, the resulting difference between second and minute resolution varies between -2.3% (Solar Village, SA) and -4.3% and seems to feature an approximately linear correlation with irradiance variability, as displayed in Figure 2.

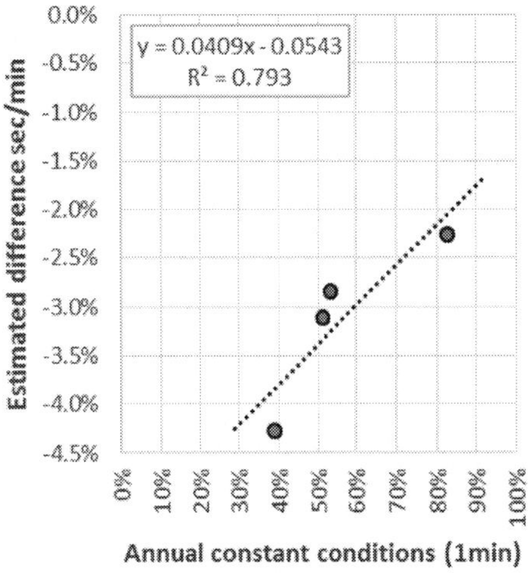

Figure 2: Difference induced by temporal resolution depending on annual constant conditions

The estimated difference is in the same order of magnitude as the year on year variation observed in the 1min BSRN records for the discreet sites.

3.2 Static irradiation fraction correlation with DHI/GHI

In previous work, evidence for a linear correlation between site-specific annual static GHI fraction and annual DHI/GHI ratio has been noted (c.f. [1]). Figure 3 presents the data points of each site and each included year as well as the resulting linear trend line.

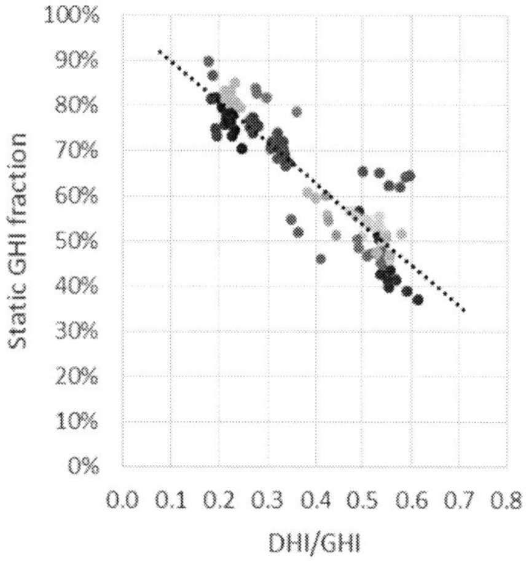

Figure 3: Global correlation of annual static irradiation fraction against annual DHI/GHI ratio (Color code of data points: Europe = blue; Asia & Oceania = green and black; Americas = grey; Africa = orange/red)

The coefficient of determination, R^2 of the displayed global linear regression is 0.820. The results considering global as well as regional trend lines along with R^2, Mean Bias Error (MBE) and Root Mean Square Error (RMSE) are summarized in Table V. Note that the number of sites and data points for each regression varies depending on the includable BSRN locations for each region.

Table V: Linear regression coefficients and accuracy indices

Region	a	b	R^2	MBE	RMSE
Global	-0.904	0.990	0.820	1.0%	10.1%
Global*	-0.879	0.977	0.816	1.0%	10.0%
Europe	-1.019	1.022	0.942	0.4%	5.9%
Africa	-0.904	1.013	0.825	0.6%	7.1%
Asia & Oceania	-0.494	0.875	0.764	0.5%	7.8%
Americas	-1.515	1.152	0.903	0.5%	8.0%
Americas*	-1.336	1.076	0.901	0.4%	7.2%

The accuracy metrics MBE and RMSE have been applied as defined in [7] with each error calculated between the estimated annual static GHI fraction on the basis of the linear regressions and the corresponding values determined from the BSRN records.

In order to compensate in the regression analysis for the BSRN datasets coming in 2 minute and 3 minute resolution (Florianopolis, BRA and Desert Rock, USA), results from the temporal effect analysis have been used. In particular, the variability factors of Florianopolis, BRA obtained for 1996 and 1998 have been adjusted by 0.7% and all annual static GHI fractions of Desert Rock, USA by 3.7%. No sufficient estimate can be drawn from the 3 minute time step data of the US location, in this case the temporal effect findings of Solar Village, SA have been used, as this site features similar DHI/GHI and variability factor. The subsequently received adjusted set of data points are reflected in the regressions marked with "*" in previous Table V, Figure 4 and Figure 5 below.

3.3 Estimation of annual MPPT losses

As presented in Table III above, data corresponding with various levels of industry standard inverter MPPT quality will be used to evaluate the product-dependent influence on annual MPPT losses. The following expression along with weighting factors for dynamic (DYN) and static (STAT) irradiance conditions for the given location as well as information on static and dynamic MPPT efficiency ($\eta_{mppt,stat}$ and $\eta_{mppt,dyn}$) enable the estimation of annual MPPT losses:

$$\text{MPPT losses} = 100\% - (\eta_{mppt,dyn} \cdot DYN + \eta_{mppt,stat} \cdot STAT)$$

Using the methodology provided in this research, DYN and STAT coefficients can be found for any given location from the DHI/GHI ratio.

4 RESULTS

4.1 Estimation of static irradiation fraction

Application of the regression coefficients determined in previous section leads to the trend lines over DHI/GHI displayed in Figure 4.

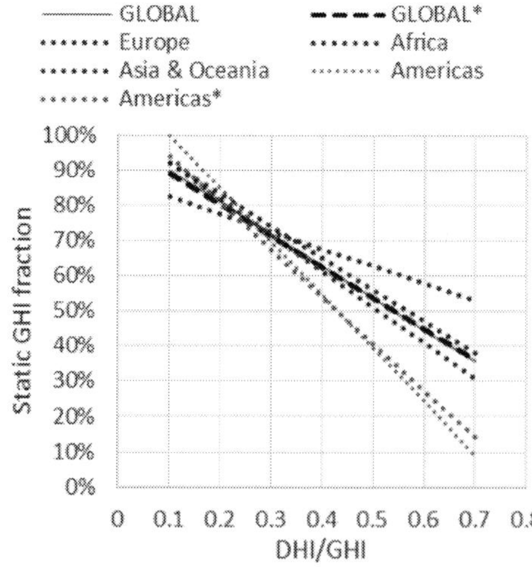

Figure 4: Annual static GHI fraction over annual DHI/GHI ratio estimated using global and regional regression coefficients

The global and regional linear regressions lead to relatively similar results for DHI/GHI ranging from 0.2 to 0.3, but show increasing deviations below and above this range. However, both global (Global and Global*) as well as the European and African trend lines, which are the datasets with the most sites and data points available, agree well in all cases. This means that the variations found for other regions could be caused by regional differences, but also by other factors, such as an insufficient basic set of data. Considering the worst case, at a DHI/GHI ratio of 0.8, the maximum difference between the Global* estimate and the highest deviation (Americas) determined here is approximately 27%.

4.2 Estimation of annual MPPT losses

Using DYN and STAT weighting factors derived from the regressions presented above, and the average MPPT efficiency values of 99.7% for $\eta_{mppt,stat}$ and 98.6% for $\eta_{mppt,dyn}$, the resulting total MPPT efficiencies as displayed in Figure 5.

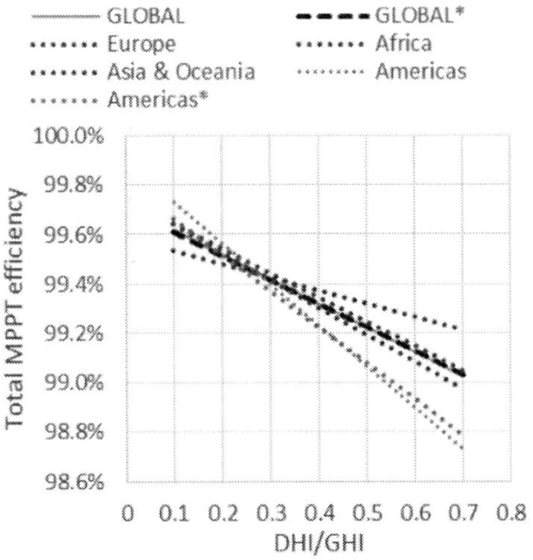

Figure 5: Total MPPT efficiency over annual DHI/GHI ratio estimated using average inverter MPPT efficiency characteristics as well as global and regional regression coefficients to estimate static GHI fraction

Since the input derived from the different regression lines changes linearly, estimated annual total MPPT efficiency changes linearly too. This means that the representativeness of the trend lines determines the accuracy of the resulting annual MPPT losses. In the worst case, the maximum difference between using regional regressions and Global* is ca. 0.3%.

Considering the small annual MPPT loss variation induced by using different regression lines as well as the greater confidence with the global trend lines, we consider Global* a fair estimate for worldwide variability factor estimation.

Using the min. and max. MPPT efficiencies as displayed in Table III, annual MPPT losses of 1.7% and 0.2% for a DHI/GHI ratio of 0.4 can be obtained. The corresponding value for above applied average MPPT efficiencies is 0.7%. This result highlights the relevance of product-related values in project-specific PV energy yield studies and underlines the need for that information.

4.3 Uncertainty of annual MPPT losses

An uncertainty calculation for the suggested approach to estimate annual MPPT losses has considered:

- The influence associated with the proposed use of Global* regression
- Effects from inaccurate MPPT efficiency information
- Temporal resolution effects

Using the variability factors provided by the different regression lines as displayed in Figure 4 or the total MPPT efficiencies as shown in Figure 5, respectively, the resulting standard deviation over all DHI/GHI ratios and trend lines is 0.23%. Similarly, assuming the worst case, which is applying $\eta_{mppt,stat}$ of 99.9% and $\eta_{mppt,dyn}$ of 96.6%, an additional standard deviation value of 0.47% is determined. The uncertainty associated with the temporal resolution has previously been calculated (see [1]). The resulting average effect on this basis is 0.03%. Table VI summarizes the different uncertainty positions and presents the total uncertainty of the estimated annual MPPT losses. The uncertainties are combined using the standard error approach.

Table VI: Uncertainty calculation for estimation of annual MPPT losses

Position	Uncertainty
Application of Global* regression coefficients instead of region specific	0.23%
Uncertainty from inaccurate MPPT efficiency information	0.47%
Temporal resolution effect	0.03%
TOTAL	**0.53%**

The total uncertainty of the annual MPPT losses induced by approximations made in the modelling of the relevant weather statistics and the MPPT performance of the inverter is determined with 0.53% globally.

5 CONCLUSION

Annual DHI/GHI ratios and static GHI conditions have been determined from BSRN records of locations spread over the globe. The relation of both variables provide linear regressions, which can be used for estimation of annual MPPT losses. Furthermore, the analyzed BSRN records along with the regression lines give indication of the uncertainty associated with the proposed approach. Overall, the findings form the basis of a global approach to estimate site-specific annual MPPT losses in PV energy yield studies, as GHI and DHI are available information in state-of-the-art yield calculations. Furthermore, combining the estimation of site-specific variability factor with the inverters variable MPP Tracking efficiencies found in the market shows that there is also a product-dependent aspect. This means that following this approach both site- and product-specific annual MPPT losses can be determined for PV energy yield studies.

[1] Egler, M.; Gordon, S.: Estimating Overall MPP Tracking Losses Using Weather Statistics; 31st European Photovoltaic Solar Energy Conference and Exhibition, Hamburg / Germany, 14-18 September 2015

[2] EN 50530:2010, Overall efficiency of grid connected photovoltaic inverters; 31 May 2010
[3] Baseline Surface Radiation Network, BSRN; http://bsrn.awi.de
[4] BSRN Data Retrieval via Publishing Network for Geoscientific & Environmental Data (PANGAEA); http://www.pangaea.de/PHP/BSRN_Status.php
[5] Schäfer, T.: Bachelorarbeit – Berücksichtigung des dynamischen Verhaltens von PV-Wechselrichtern bei netzgekoppelten PV-Anlagen mit c-Si Modulen; Fachhochule Köln; 31. August 2011
[6] Bründlinger, R.; Bettenwort, G.; Bletterie, B.; Laschinski, J.: 99,9% MPP-Tracking Performance – nur die "halbe Wahrheit"?; 23. Symposium Photovoltaische Solarenergie; 5.- 7. März 2008
[7] Pelland, S.; Remund, J.; Kleissl, J.; Takashi, O.; De Brabandere, K.: Photovoltaic and Solar Forecasting – State of the Art; IEA PVPS Task 14, Subtask 3.1, Report IEA-PVPS T14-01:2013, October 2013, ISBN 978-3-906042-13-8

CELL TO MODULE LOSSES OF AN MWT MODULE

L.H. Slooff, E.E. Bende, M.J. Jansen, L.A.G. Okel, F.J.K. Danzl, P. Manshanden,
[1]ECN Solar Energy, P.O. Box 1, 1755 ZG Petten, The Netherlands
Phone: +31 88 515 4314, Fax: +31 88 515 8214, E-mail: slooff@ecn.nl

ABSTRACT: The efficiency of photovoltaic modules is for a large part determined by the solar cell efficiency, but the electrical losses from current transport between the cells and to the module contacts, and optical losses also play an important role. In order to investigate the losses in the module the Cell to Module (CtM) ratio is often used. It is determined by the ratio in performance between the module and the cell and gives an indication of the electrical and optical losses in the module. In principle this seems a straightforward approach. However, a good CtM measurement is not as trivial as it seems. Many factors play a role. Not only the way the measurement is done, but also which parameters are used for the cell in the comparison. After all, a module consists of many cells that are all slightly different. The various aspects in measurement and their influence on the CtM ratio will be discussed and suggestions for a correct measurement protocol to determine absolute CtM ratios will be given.
Keywords: Cell to module ratio, CtM ratio, PV module, MWT

1 INTRODUCTION

Research in the field of photovoltaic (PV) modules is always focused on increasing the efficiency and reducing the costs. The efficiency of a PV module is for a large part determined by the efficiency of the cell. But the final module efficiency is also determined by the losses that occur when the cells are put together in the module. To minimize these losses and to optimize the module efficiency it is important to know the cell to module loss, or CtM ratio. CtM losses are always expressed as the ratio between the module output and the cell output, where the output can be the power, short circuit current (Isc), fill factor (FF), open circuit voltage (Voc) or efficiency (η). This trivial definition is not as simple as it seems. The numbers for both the cell and module output are influenced by many factors in the measurements itself. As a result, the CtM ratio is related to the way the measurements were done and thus CtM ratios can only be compared if the measurements were done in a similar way. Besides that it is also affected by the cell parameter that is used for the comparison. As there are many cells in the module it is not straightforward which one should be used. In most of the publications on CtM these issues are not or only partly addressed[1,2,3,4] although there are a few exceptions.[5] This paper first addresses the importance of differentiation between an absolute and relative CtM ratio. Next the different factors that influence the measurements are discussed. Subsequently an example is given of relative CtM ratios that are based on three different approaches for the input of the cell parameters. Finally a measurement and calculation procedure is proposed that enables the comparison of CtM ratios between different institutes/companies.

2 RELATIVE AND ABSOLUTE RATIOS

The uncertainty in the measured current of a cell or module is determined by many factors. It depends for example on the reference cell that is used, the reproducibility in adjusting the irradiation of the solar simulator and the spectral mismatch between the spectrum of the solar simulator and the AM 1.5 spectrum. But it also depends on the uncertainty in the electrical measurements. However, the effects on the current are expected to be much smaller than the effect of the spectral mismatch. For this reason we will focus here on the effects resulting from the spectral mismatch. If a cell is measured on two different solar simulators, the measured currents will be different as the lamp spectra will be different. To obtain similar currents, the current needs to be corrected for the differences in the lamp spectra. This is done by measuring the spectral response (SR) of the cell and the lamp spectrum. I-V measurements of PV cells are often performed on a different setup than the module measurements and thus the currents cannot be directly compared. If the currents are not corrected for the spectral mismatch, only a relative comparison can be done. For an absolute comparison, the mismatch (MM) between the cell and module measurement needs to be determined. For this, the SR of both the cell and the module must be measured as well as the lamp spectra of both sun simulators. Then the MM factor can be derived using the following formula:

$$MM = \frac{I_{ref,AM1.5} * I_{test,lamp}}{I_{test,AM1.5} * I_{ref,lamp}}$$

Where $I_{ref,Am1.5}$ is the current from the calibrated reference cell/module for AM1.5 spectral conditions, is the current of the test cell for AM1.5 spectral conditions, $I_{ref,lamp}$ is the current of the reference cell as measured in the solar simulator and $I_{test,lamp}$ is the current of the test cell as measured in the solar simulator. $I_{ref,Am1.5}$ and $I_{test,Am1.5}$ are derived from the known AM1.5 spectrum and the measured spectral response.
The MM is determined with respect to the AM1.5 spectrum. This means that the current of both the cell and the module must be corrected before they can be compared. The correct current $I_{test,cor}$ can now be estimated from the measured current $I_{test,meas}$ and the mismatch factor as follows:

$$I_{test,cor} = \frac{I_{test,meas}}{MM}$$

From the above it is clear that when comparing CtM ratios between institutes or companies an absolute CtM must be used. Even within an institute absolute CtM is needed when different cell technologies are used, as again the SR will be different for the different cell technologies.
A relative CtM can be used only in the following situations:

- To check if a module line is performing according to specs. A CtM is determined for the specific module and trends in CtM can be followed.
- To study cell processing or module fabrication improvements that do not affect the SR of the module. For example metallization or tabbing, different and minor changes in cell processing.

3 MEASUREMENT FACTORS

The basics in performing measurements for determining the CtM ratio is that the cell measurement must be as similar to the module situation as possible. This means that the following points should be considered for each cell type and situation.

- Cell contacting
- Chuck reflections
- Reference cell
- Reference module
- Bi-facial cells

Cell contacting

If an H-pattern cell is used, the number of contact pins on the cell can have an influence on the FF of the cell measurement. If in the module the tabs are soldered at more points than there are measurement probes in the cell measurement, a relatively higher FF will be measured. In that case a CtM gain in FF can be obtained whereas in reality the CtM will be lower or can even be a loss. Or the other way round. Especially if lower conducting or floating busbars are used this could have a large impact. Back contacted cells like IBC and MWT have fixed contact points, so for these cell types fill factor changes due to different contact points should not occur.

Chuck reflections

The measurement chuck on which the cell is positioned will reflect light that can eventually reach the solar cell and contribute to the current of the cell. For mono-facial cells this concerns mainly light that hits the chuck outside the cell, but for bi-facial cells this also holds for light that is not absorbed by the cell and is reflected from the chuck back to the cell. Therefore, the area outside the cell should be covered with non-reflecting material and for bi-facial cells between the chuck and the cell as well.

Reference cell

For accurate cell measurements a calibrated reference is used and the measured cell data are corrected for spectral mismatch. This means that the spectral response (SR) of the reference cell and test cell must be known as well as the spectrum of the lamp of the solar simulator. Preferably, the reference cell should be of the same type and size as the test cell. The size will be of influence if the irradiance is not homogeneous over the area.

Reference module

Module measurements also use a reference module. Like with the cells, the SR of both the reference and test module and the spectrum of the lamp of the flash tester must be known for spectral mismatch correction. Here also the type and size of the module are preferably similar to the test module.

Bi-facial cells

Besides the chuck reflection there are several other aspects that need to be considered when measuring bi-facial cells. A committee is working on a standardized norm for the measurements of bi-facial cells or modules[6], but at the moment there is no norm available. However it is suggested to measure a bi-facial cell [7] under front and rear side illumination and determine the bi-faciality ratio: bi-faciality=$I_{sc,rear}/I_{sc,front}$

Then the compensated current density can be derived:

$$I_{sc,com} = (1 + 0.2 * bifaciality) * I_{sc,front}$$

4 APPROACH

I-V measurements were performed on a series of 60 MWT cells that were processed at ECN. For each cell the I-V curve was subjected to a 2-diode model fit.[8,9] These fit parameters were put into a model that calculates the I-V characteristics of a series connection of cells. The module contained 3 individual strings as in a typical MWT module. For each string the series connection was calculated, as well as the series connection for the complete module.[10] This apprach will be called String to Module (StM). The measured currents were not corrected for MM so the results show only relative StM and CtM ratios. The I-V curve summing result was compared to more commonly used approaches for determining the CtM ratio:

1. The minimum Isc of the cells, the sum of the Voc's and an average FF: this allows for determining a CtM ratio for current, Voc, FF and Pmp
2. The average Isc of the cells, the sum of the Voc's and an average FF: this allows for determining a ratio for CtM current, Voc, FF and Pmp
3. Summing the powers from the measured individual cells: allows for determining a CtM ratio for Pmp only.

3 RESULTS

In Table I the calculated cell parameters for each string and the calculated parameters for the module are given. The table does not show CtM or StM ratios. It shows the string or module parameters that can be used as input for the CtM or StM ratio. For a CtM or StM ratio, also the module should be measured, which was not done yet. The overall difference in power between the different approaches can be more than 1%. It is clear that a CtM ratio can only be compared if it is based on the same approach for the cell or string input. The difference is mainly due to a difference in Isc. Often the minimum current of the cells in a string is used to obtain a CtM ratio for the current, as the cell with the lowest current determines the current in a series connected string. However, in reality this is not true as it also depends on the FF of the cells. In the examples below it is shown that the average current of the cells gives a much closer match to the current of the strings and the module. The sum of the Vocs and the average FF also are a good approach to determine CtM ratios.

Table I: Cell, string and module parameters for the various approaches. Percentage are relative to the sum of the I-V curves of the cells in the string.

string 1	Pmp[W]	Isc[A]	Voc[V]	FF[%]
∑ I-V of cells StM	100.24	9.188	14.00	77.96
min I_{sc}, ∑ V_{oc}, av FF	99.64	9.137	13.98	77.98
Δ wrt ∑ I-V	-0.60%	-0.56%	-0.08%	0.03%
av I_{sc}, ∑ V_{oc}, av FF	100.21	9.190	13.98	77.98
Δ wrt ∑ I-V	-0.03%	0.01%	-0.08%	0.03%
∑cell powers	100.36			
Δ wrt ∑ I-V	0.11%			

string 2	Pmp[W]	Isc[A]	Voc[V]	FF[%]
∑ I-V of cells StM	95.56	9.191	13.36	77.81
min I_{sc}, ∑ V_{oc}, av FF	94.85	9.122	13.35	77.86
Δ wrt ∑ I-V	-0.76%	-0.75%	-0.06%	0.06%
av I_{sc}, ∑ V_{oc}, av FF	95.59	9.194	13.35	77.86
Δ wrt ∑ I-V	0.03%	0.03%	-0.06%	0.06%
∑cell powers	95.70			
Δ wrt ∑ I-V	0.14%			

string 3	Pmp[W]	Isc[A]	Voc[V]	FF[%]
∑ I-V of cells StM	78.18	9.256	10.83	78.02
min I_{sc}, ∑ V_{oc}, av FF	77.40	9.169	10.82	77.99
Δ wrt ∑ I-V	-1.02%	-0.95%	-0.04%	-0.03%
av I_{sc}, ∑ V_{oc}, av FF	78.20	9.265	10.82	77.99
Δ wrt ∑ I-V	0.02%	0.09%	-0.04%	-0.03%
∑cell powers	78.27			
Δ wrt ∑ I-V	0.11%			

Module	Pmp[W]	Isc[A]	Voc[V]	FF[%]
∑ I-V of cells StM	273.95	9.205	38.18	77.94
min I_{sc}, ∑ V_{oc}, av FF	271.34	9.122	38.16	77.94
Δ wrt ∑ I-V	-0.96%	-0.90%	-0.06%	0.00%
av I_{sc}, ∑ V_{oc}, av FF	274.12	9.216	38.16	77.94
Δ wrt ∑ I-V	0.06%	0.12%	-0.06%	0.00%
∑cell powers	274.33			
Δ wrt ∑ I-V	0.14%			

5 CtM PROTOCOL

The paragraphs above show that both the measurements procedure as well as the calculation procedure influence the resulting CtM ratios. For this reason it is important that similar procedures are used when comparing data between institutes or companies. We propose the following protocols:

Measurement protocol.

- measure all cells in the solar simulator: area outside the cell should be masked to avoid chuck reflections
- measure the SR of at least 3 typical cells
- calculate the MM factor for the cells using the average SR, the solar simulator lamp spectrum, the SR of the reference cell and the AM1.5 spectrum
- Correct cell power and current with the MM factor
- Measure the module in the flash tester: unmasked
- Measure the SR of the module
- Determine the MM factor using the SR of the module, the lamp spectrum of the flash tester, the SR of the reference module and the AM1.5 spectrum
- Correct the module power and current with the MM factor

For bi-facial cells and modules the protocols are rather similar, but now the cell should be measured on a black background to avoid chuck reflection from underneath the cell and the module should be measured under front and rear side illumination. Furthermore the proposed measurement standard for bifacial power measurement, as described above should be followed.

Calculation protocol.

CtM ratio of the power:

- Take the average MM corrected Isc of the individual cells, the sum of the Voc's of the cells and the average FF of the cells.
- Calculate the expected module power from these numbers.
- Take the ratio of the measured MM corrected module power over the expected module power.
- For a CtM ratio for the efficiency, also the cell and module area become important. Here total cell area and total module area, including frame, should be used.

6 CONCLUSIONS

Various aspects in PV cell and module measurement and their influence on the CtM ratio have been discussed and suggestions for a correct measurement protocol to determine absolute CtM ratios will be given. The importance of absolute CtM ratios for comparisons between companies and institutes is explained and the influence of the cell parameters are shown. For determining CtM ratios the average Isc, sum of Voc and average FF give a good approximation with respect to the summing of the individual IV curves. For that approximation the difference is well below 0.1% whereas summing of the power of the individual cells gives an error of 0.1% to 0.15%. It is therefore recommended to either sum the individual IV curves or to take the average Isc, sum of Voc and average FF as cell parameter input in CtM ratios.

8 ACKNOWLEDGMENTS

This work has been supported by the Dutch Government.

9 REFERENCES

[1] I. Haedrich, M. Wiese, B. Thaidigsman, D. Eberlein, F. Clement, U. Eitner, R. Preu, H. Wirth, Energy Procedia 38 (2013) 355

[2] I. Haedrich, U. Eitner, M. Wiese, H. Wirth, Sol. En. Mat & Sol. Cells 131 (2014) 14

[3] S.M. Dasari, P. Srivastav, R. Shaw, S. Saravanan, P. Suratkar, Renew. Energy 50 (2013) 82

[4] T-H. Jung, H-E. Song, H-K. Ahn, G-H. Kang, Solar Energy 103(2014) 253
[5] J.P. Singh, Y. S. Khoo, J. Chai, Z. Liu, Y. Wang, Photovoltaic International 31 (2016) 90
[6] IEC/TS 60904-1-2 Ed. 1.0 Photovoltaic devices - Part 1-2: Measurement of current-voltage characteristics of bifacial photovoltaic (PV) devices
[7] https://pvpmc.sandia.gov/pv-research/bifacial-pv-project/bifacial-pv-characterization-and-rating-standards/
[8] A.R. Burgers, J.A. Eikelboom, A. Schönecker, and W.C., Sinke. Improved treatment of the strongly varying slope in fitting solar cell I-V curves. In Proceedings 25th IEEE PVSC Conferenc (1996) 569
[9] In-house built model ivfit. Commercial available via **pvwebtools@ecn.nl**
[10] G. Friesen, S. Dittmann, S. Williams, R. Gottschalg, H.G.Beyer, A. Guerin de Montgareuil, N.J.C.M. van der Borg, A.R. Burgers, R. Kenny, T. Huld, B. Müller, C. Reise, J. Kurnik, M. Topic. 24th European Photovoltaic Solar Energy Conference (2009) 3189

Annual Yield Comparison of Module Level Power Electronics and String Level PV Systems with Standard and Advanced Module Design

K. Sinapis[1], T.T.H. Rooijakkers[1,2], C. Tzikas[1], G.B.M.A. Litjens[2], M.N. van den Donker[1], W. Folkerts[1], W.G.J.H.M. van Sark[2]
[1]Solar Energy Application Centre, Eindhoven, The Netherlands
[2]Utrecht University, Copernicus Institute, Utrecht, The Netherlands

ABSTRACT: This study focuses on the partial shade-mitigating effects related to the insertion of additional ideal by-pass diodes in residential-scale photovoltaic (PV) systems. For this purpose, quantification of the resulting energy yield benefits is carried out in a representative residential environment. It is widely recognized that partial shading inflicts disproportional losses to the energy output of PV systems. Increased granularity levels in cell groups are perceived as a potentially promising measure to increase the shade-tolerance of photovoltaic devices. The past years, introduction of module level electronics promise to reduce further shading losses. The developed model includes a shading evaluation of the installation with means of 3D modeling, insertion of additional by pass diodes resulting in smaller cell groups, irradiance calculations, PV cell modelling and finally an empirical power conversion model. Results suggest that in the reference case of 3 by pass diodes the micro inverter system is performing the best under partial shading. By increasing the cell group granularity the string inverter systems seems to benefit due to the wide maximum power point voltage window.
Keywords: MLPE, BIPV

1 INTRODUCTION

Penetration of solar photovoltaic (PV) systems in the Netherlands and worldwide has remarkably increased the past years and it is forecasted to keep growing in the future [1]. Particularly the application of BIPV and BAPV systems are projected to thrive in the following years as a result of increasing electricity prices for the residential sector and decreasing PV component costs. Residential and small commercial PV systems are typically installed in an urban environment. Roofs and terraces are often affected by shade coming from the close proximity of buildings, poles, antennas, dormers etc and thus introduce electrical and thermal mismatch losses between cells and modules. These are generally caused by manufacturing tolerance, heterogeneous irradiation conditions which are especially important for larger systems, panel degradation and thermal mismatch of the solar panels. Solar panels are connected in series and thus sharing the same current in a string. This topology is prone to power losses if the solar cells in the panel are not operating under the same conditions thereby reducing the current of the panel and consequently of the whole string. Partially shaded solar cells may become reverse biased because of the series connection and thus act as a load consuming the power that is generated by the unshaded cells. Two negative effects occur from partially shaded operation of a PV system: power loss and increased temperature of the shaded cells (hot-spot). By-pass diodes (BPD) have been applied in solar panels to prevent power consumption from shaded cells and to prevent hot-spots by by-passing the shaded substrings of the solar module. Most of the solar modules include one by-pass diode connected anti parallel per 16-24 cells [2]. Further increase of the granularity could potencialy result in additional power harvesting. On the other hand manufacturability of such module lay-outs is technically challenging and costly.

The use of module level power electronic devices (MLPE) has been proposed to mitigate electrical and thermal mismatch losses [3-5] in the field by tracking the maximum power point of individual modules. In general MLPE devices consist of two main categories: micro inverters and power optimizers. In this paper micro inverters and boost power optimizers are considered.

Although modelling tools have been developed based on a variety of software platforms, most of them don't consider the system architecture. There are many available models which can be different in terms of mathematical sub-models and assumptions. Some models lack transparency and as a result project developers are expressing concerns regarding PV performance validity forecast, especially when shading is present. The key challenges of partial shading PV models is therefore to generate accurate yield predictions under heterogeneous irradiance conditions with reduced simulation time. In this paper a model is presented that considers cell shading fractions determined by a 3D model and applies an irradiance model to determine the effective irradiance on a partially shaded cell. Moreover, the model takes into consideration the cell layout of the module and the number of BPDs. The PV system architecture and associated power electronics efficiency losses are also taken into account.

2 YIELD MODEL

The complete yield model includes 5 different models integrated into one. Namely, it includes a 3D SketchUp model, a shade detection model, a radiation model, a DC and an AC simulation model. All the model inputs used in the complete model and the flow of simulation processes are shown in Figure 1.

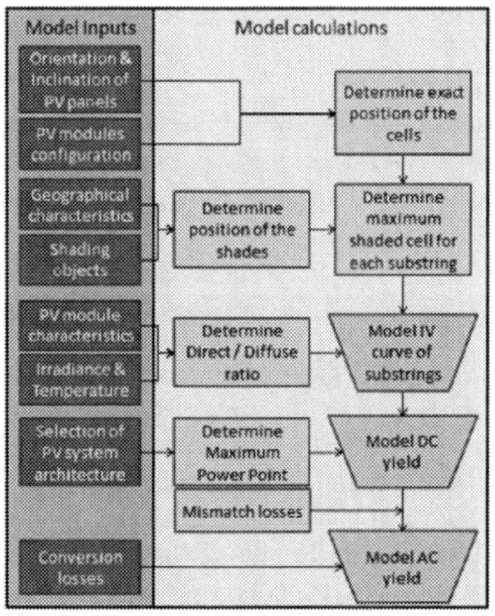

Figure 1: Building blocks of the simulation model presented in PVSEC 2015 with an oral presentation at session 5.AO 8.4

2.1 Shading Scenario

The partial shading conditions opted for in this research are set to match typical shading objects on a representative "reference" rooftop. The rooftop reference is determined for The Netherlands because the Dutch documentation on the housing stock is extensive. In terms of sensitivity, both the individual and aggregate effects of the shading object types are investigated. As this study is set to explore a realistic high-end range of the by-pass energy recovery potential under partial shading conditions, and by absence of average Dutch dormer dimensions, the shading scenarios are set relatively aggressively in the dormer-incorporating cases. Further chimney and exhaustion poles are situated in representative locations as seen in Figure 2.

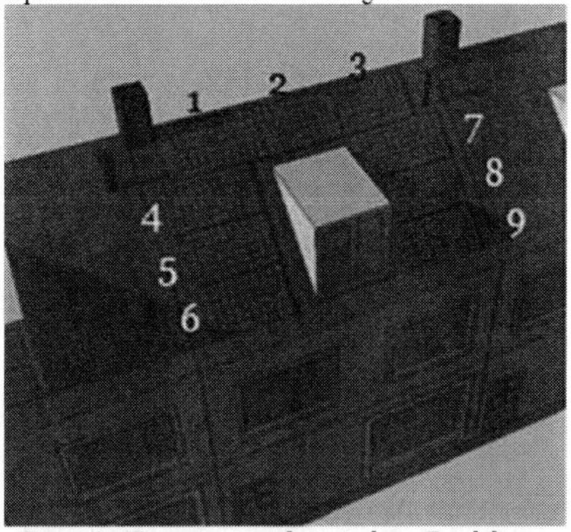

Figure 2: 3D impression of a typical row Dutch house. Dormers, chimneys and poles are usually present, causing obstruction shading to parts of the PV system

2.2 Module layout

A selection of 3 (reference), 6, 12 and 30 horizontally aligned BPD substring groups along with 5 and 10 vertically aligned BPD groups and the cell-wise case of 60 BPDs is assumed to be representative. These substring group selections can be seen in **Error! Reference source not found.**. This incremental approach is set out to evaluate the performance effect of stepwise BPD additions. As this research is set out for exploratory purposes, the BPDs modelled are assumed to be ideal smart by-pass diodes. This means that current leakages are neglected and that no voltage losses are assumed to occur when substrings are bypassed.

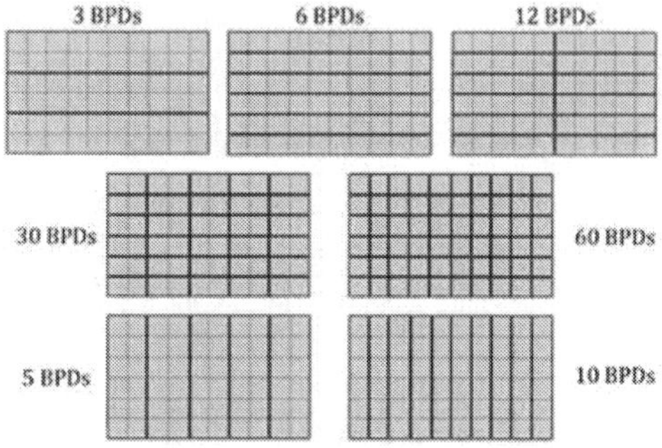

Figure 3: Substring granularities or variation of per-module BPDs investigated in this research. The 3, 6, 12 and 30 BPD cases are referred to as horizontally aligned substrings; the 5 and 10 BPD cases as vertically aligned substrings. Bypass diodes per cell are assumed in the 60 BPD case

2.3 Irradiance Model

The effective irradiance on a cell can be expressed as the area-weighted accumulation of its unshaded ($G_{U,k}$) and shaded ($G_{diff,k}$) irradiance-receiving fractions (where σ_P represents the cell shade percentage):

$$G_{cell,eff} = \left[(100-\sigma_P)\cdot G_{U,k} + \sigma_P \cdot G_{diff,k}\right] / 100 \qquad (1)$$

The effective irradiance on a substring (connecting cells in series) depends on the irradiance received by the cell receiving the least irradiance – the cell shaded most heavily in the case of this study:

$$G = G_{eff} = min(G_{cell,eff}) = G_{cell,eff}(max(\sigma_P)) \qquad (2)$$

Figure 4: Principle of deriving the effective cell irradiance from its shaded and unshaded fractions. In this example, σP = 50%. Modified from previous similar work (Sinapis, et al., in press; Tzikas, 2015)

2.4 PV cell and conversion model

A mono-crystalline cell can be modelled with the equivalent electric circuit of a simplified double diode model developed by Ishaque [6] and shown in Figure 5.

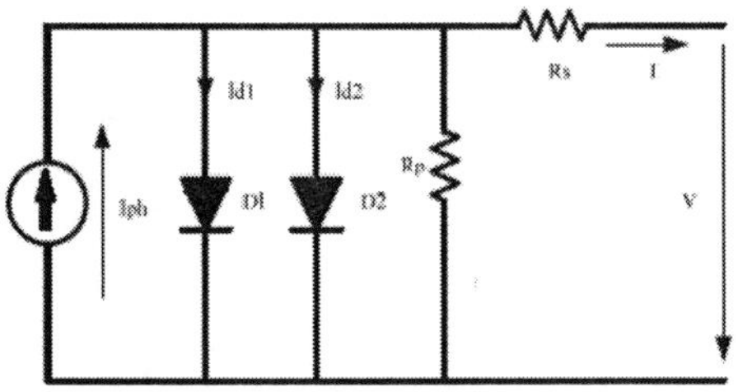

Figure 5: Double diode equivalent circuit for a PV cell

Nearly all modern inverters have more than 99% MPPT efficiency. While Perturb and Observe (P&O) is the most used algorithm new hybrid algorithms have been implemented by inverter manufacturers to boost performance at partial shading conditions [7-8]. This is achieved by frequent scans of the P-V curve of the solar modules which ensure that the inverter will detect the MPP even in the case of lumpy P-V curves. In this study the MLPE devices are using the hybrid P&0 algorithm while the string inverter system has the option to activate it. Note that the string inverter is delivered from the manufacturer with the shadow mode deactivated. The model assumes that the MPP of the solar modules is always found and kept when the hybrid algorithm is used, however the string inverter is modeled with the hypothesis that when the shadow mode is deactivated the solar modules are operated at a local maximum when partial shading is present.

3 SIMULATION RESULTS

By using typical meteorological year's irradiation data by Meteonorm [9], a full year simulation for unshaded and partially shaded scenarios has been performed. Meteonorm provides measured irradiance data for a variety of locations. Moreover, the data can be decomposed and trans-positioned by using known irradiance models. A constant albedo factor of 0.15 has been used for the simulations.

In Figure 6 the annual irradiance reduction due to shading can be seen. Modules around the dormer suffer the most while the chimney and poles have a small effect on the annual irradiance received by the cells.

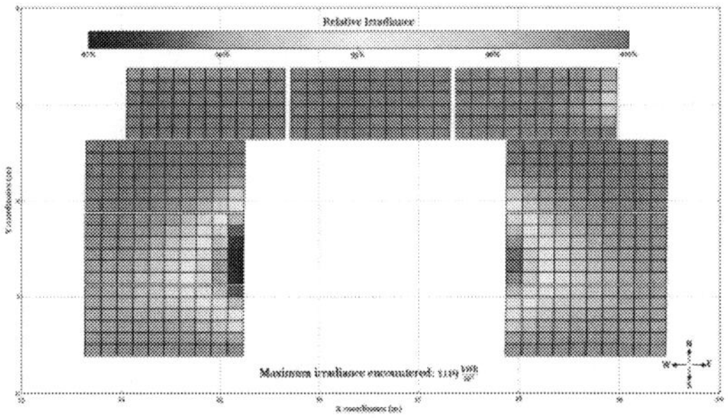

Figure 6: Annual irradiance reduction due to shading for the modelled roof

The effect of a triuangular shade on a PV module caused by the dormer situated in close proximity can be seen in *Figure 7* and *Figure 8.*

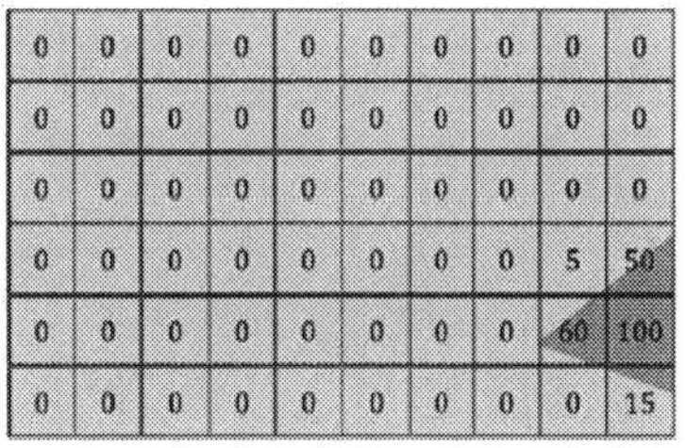

Figure 7: Partially shaded PV module with shaded cell percentages

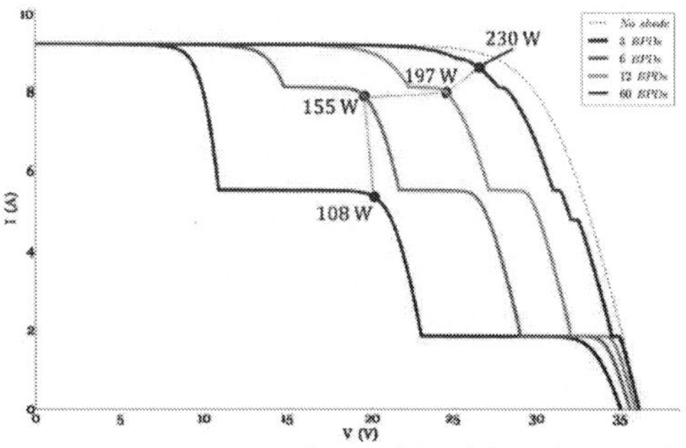

Figure 8: IV responses of example module under partial shading as in Figure 7. The cases of 3, 6, 12 and 60 per-module BPDs are shown. All increases lead to increases in extractable power, as can be seen from the highlighted MPP-values. The grey dashed line represents the IV curve in unshaded conditions. Total irradiance: 1000 W/m2; diffuse irradiance: 200 W/m2; Tmod: 40 °C

3.1 Annual Yield Results per system (NL)

The MI-type system provides the highest absolute system output if the amount of BPDs per panel is 3 or 6. However, the SI system with active shadow function (SI+) surpasses the specific yield performance of the MI-type system for other per-panel BPD amounts. The fact that either the MI or the SI+ system types come out on top can be rationalized. The reason is that SI and MI only require single-step conversion whereas the PO system requires two conversion steps: DC-to-DC followed by DC-to-AC. This two-step conversion leads to lower overall conversion efficiencies and therefore lower specific yield figures for the PO-type system. The MI performance advantage at low BPD amounts can be attributed to MPP optimization per panel instead of system-wise for the SI-type system. This leads to a flexibility advantage of MI-type systems over SI-type systems at low amounts of BPDs per module. The beneficial efficiency of the SI system combined with increased bypass flexibility of shaded module parts within the system makes it the best-performing system if the system bypassing flexibility is increased.

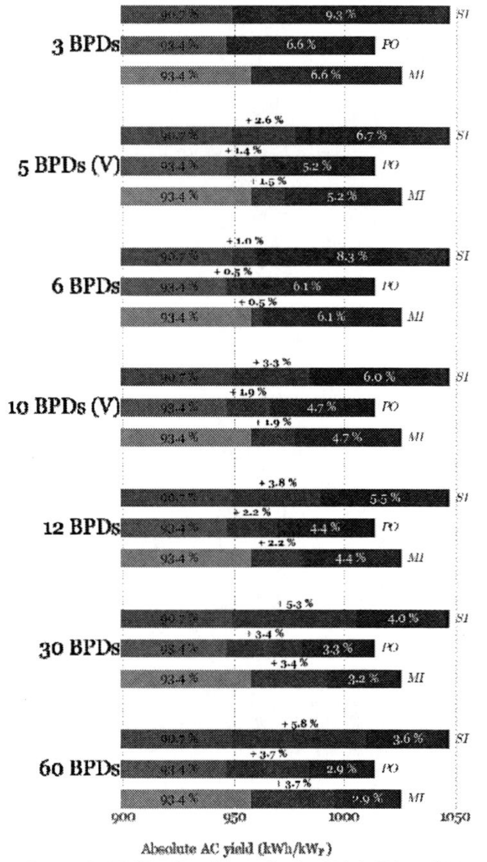

Figure 9: TMY AC yield results for SI(+), PO and MI systems. Shading index values and potential benefits when increasing the amount of BPDs per module in a stepwise fashion from 3 to 60 are illustrated

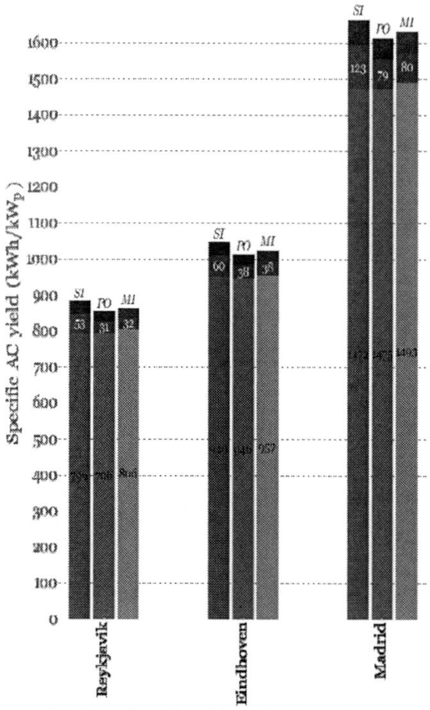

Figure 10: Specific AC yield results for SI(+), PO and MI systems placed in Reykjavik, Eindhoven and Madrid. The red, blue and light-green bars and the corresponding black text indicate the 3-BPD reference specific AC yield in kWh/kWp. The dark green bars and the corresponding white text indicate the potential benefits in kWh/kWp if 60 BPDs per module are used instead of three

3.2 Geographic location sensitivity

Another crucial sensitivity component is the geographic location. This is because incident solar angles and direct-diffuse radiation ratios vary among different places on Earth. For this purpose an evaluation of three different representative locations have been investigated. The figure below presents the specific yield and the potential benefits in kWh/kW$_p$ for the reference case of 3 BPD and the potential improvement with 60 BPD. In Reykjavik and Eindhoven simulation runs, the MI system output exceeds those of the other system architectures when the amount of BPDs is 3. When the granularity of the cell groups increase to 60, the string inverter is able to recover significantly more yield.

4 CONCLUSIONS

In this paper we investigated the benefits of module level electronics and string inverters with various numbers of cell groups. Results suggest that the increased granularity can lead to higher performance under partial shading in almost all cases and locations. While the MI system is marginal better in the reference case of 3 BPD, the string inverter system is benefited the most when the granularity increases.

5 REFERENCES

[1] IEA PVPS Annual Report 2014 http://iea-pvps.org/index.php?id=6&eID=dam_frontend_push&docID=2040 (accessed 16 January 2015)

[2] Silvestre S., Boronat A., Chouder A., 'Study of by-pass diodes configurations on PV modules' Applied Energy 2009 Volume 86, Issue 9 p. 1632-1640

[3] Poshtkouhi S., Palaniappan V., Fard M., Trescases O., 'A general approach for quantifying the benefit of distributed power electronics for fine grained MPPT in photovoltaic applications using 3-D modelling' Proc IEEE on Power Electronics vol.27, No 11 page 4656-4666, 2012

[4] Woyte A., Nijs J. , Belmans R. 'Partial shadowing of photovoltaic arrays with different system

configurations: literature review and field test results' Solar Energy 74 (2003) p. 217-233)
[5] Hanson A.J, Deline C, A, MacAlpine S. M., Staruth JT, Sullivan C. R., "Partial-Shading Assessment of Photovoltaic Installations via Module-Level Monitoring" IEEE J. Photovoltaics, Vol. 4 No. 6 November 2014
[6] K. Ishaque, Z. Salam and Syafaruddin, "A comprehensive MATLAB Simulink PV system simulator with partial shading capability based on two-diode model," *Solar Energy,* vol. 85, no. 9, p. 2217–2227, September 2011.
[7] Chih-Tang, S., Noyce, R.N., et al., 1957. Carrier generation and recombination in P–N junctions and P–N junction characteristics. Proceedings of the IRE 45 (9), 1228–1243.
[8] X. Jinhui, Y. Zhongdong, W. Bingbing and P. Jun, "Design of PV array model based on MTDC/PSCAD", Power and Energy Engineering Conference, 2009.
[9] http://meteonorm.com/

IR-imaging a tracked PV-plant using an unmanned aerial vehicle

Claudia Buerhop[1], Hans Scheuerpflug[1], Tobias Pickel[1], Christian Camus[1], Jens Hauch[1] and Christoph J. Brabec[1,2]
[1] Bavarian Center for Applied Energy Research (ZAE Bayern), Haberstraße 2a, D-91058 Erlangen, Germany
[2] i-MEET, FAU Erlangen-Nürnberg, Martensstraße 7, D-91058 Erlangen, Germany
email: claudia.buerhop-lutz@zae-bayern.de,

ABSTRACT: The on-site inspection of solar farms with infrared cameras from the air using unmanned aerial vehicles UAVs is called aIR-PV-check. A 2.4 MW PV-tracking system with more than 200 arrays was IR-inspected. 75% of the trackers had defective modules, which corresponds to more than 8% of the modules. Fairly high temperatures above 80°C were often measured at defects. The spatial distribution of the modules within the tracker plane was studied. Statistical data evaluation shows that more defective modules are found in the upper half. For the vertical failure distribution one maximum exists in the center, whereas for the horizontal dimension two maxima are present. One explanation may be the considerably high wind loads at the installation spot in context with the stiffness of the tracker system construction. It is shown that IR-inspection is a suitable method to reveal failures in tracking PV-systems where monitoring data fail for various reasons.

Keywords: thermography, defects, IR-imaging, tracked PV modules.

INTRODUCTION

Industrial inspection of PV-plants gains importance. IR-thermography is a valuable tool for quality control of stationary PV-plants [1, 2]. There are several advantages. IR-imaging is contact-free, nondestructive, provides two-dimensional images, and can be carried out during operation.–Reliable images require certain measurement conditions, high solar irradiance, no wind, adequate resolution and avoidance of reflections [3]. Using aerial, drone-mounted camera systems facilitates the inspection [4, 5]. We call the method aIR-PV-check. Irregularities in the IR-images of PV-modules are a strong indication for defective sites, as described elsewhere [6]. Early outdoor performance evaluation of tracker systems is described by Quintana et al. [7].

Fig. 1: Aerial photography of the PV-plant with tracked module tables, 2.4 MW, Germany

This paper gives a first impression of the IR-inspection of tracked PV-modules, the occurrence and visualization of defects and the correlation with monitoring data.

EXPERIMENTAL PROCEDURE

A 2.4 MW PV-plant with two-axial tracked PV-arrays was inspected using IR-imaging. The solar park, as seen in Fig. 1, consists of more than 200 biaxial, automatically tracked arrays on an area of 7.3 ha. Each tracker has three strings with 15 modules. These crystalline modules have a nominal power of 244 W in average. The operation period is about 5 years. Due to the high packing density in the solar plant temporal shading is inevitable. The PV-plant is located in a region where high wind velocities and high wind loads occur.

For investigating the installed PV-modules or PV-plants aerially, drone mounted IR-imaging systems are used. The measurement system, as shown in Fig. 2, consists of an unpiloted drone (Multicopter), a lightweight IR-camera PI 450 (optris), a visible camera GoPro and equipment for navigating. Solar irradiance E and ambient temperature T_{amb} are recorded, too. The measurements were carried out on a sunny day in fall with a solar irradiance $E \approx 670$ W/m² and an ambient temperature of $T_{amb} \approx 15$°C.

Fig. 2: Measurement system for aIR-PV-check of PV-plants

Because the module tables move according to sun orientation, the hilly ground and the closeness of packing, the systematical routing through the solar generator is challenging. Therefore, obtaining highly resolved IR-imaging of every table is fairly time-consuming. In particular, the identification of the tables is more challenging than for non-tracked field installations.

The evaluation of the monitoring data, e. g. the string energy output and daily output per tracker table, will complete the investigations. Statistical evaluation of the IR-images shed some light on the possible design-related root-cause of the defects.

RESULTS

An overview IR-image of several module arrays is shown in Fig. 3. In the foreground tables with defective modules are visible. These modules appear rather hot compared to the apparently "cooler" modules in the background. This is due to the perspective, the view angle. It becomes obvious, that the highly resolved modules show the clearest and best evaluable thermal features. Close-up IR-images of several module tables follow for better defect inspection.

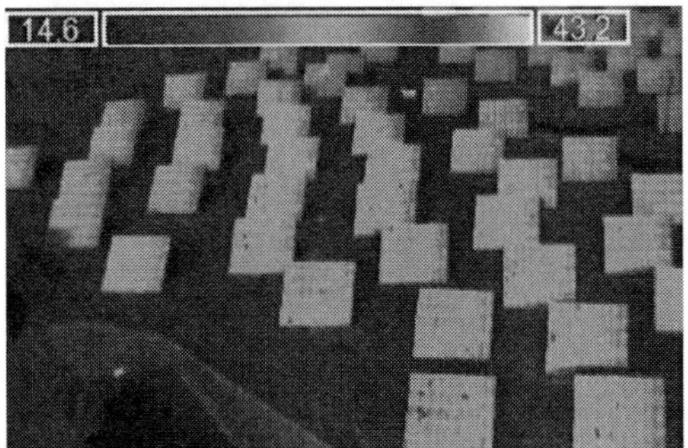

Fig. 3: IR-overview of the tracked PV-module tables, showing defective modules in the foreground, $E \approx 670$ W/m², $T_{amb} \approx 15$°C.

About 25% of the tables are defect-free and do not show any defective modules with irregularly heated cells in the IR-inspection. One of these tables is shown exemplarily in Fig. 4 This table is one with the highest daily yield *Y*.

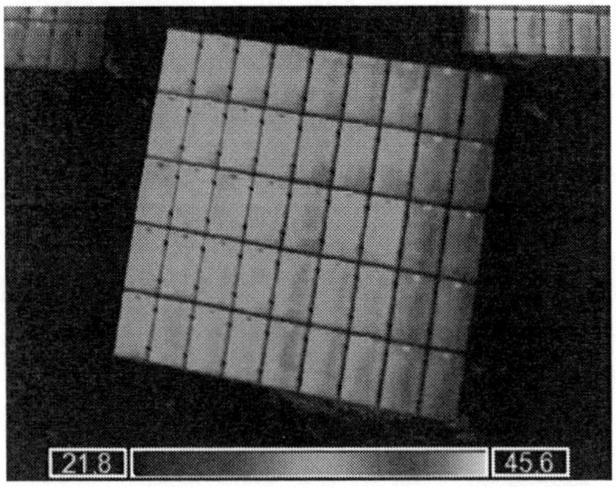

Fig. 4: Tracker 42c, tracker without defective modules and maximum energy output $Y = 44$ kWh, $E \approx 670$ W/m², $T_{amb} \approx 15$°C

Several tables are exhibiting extremely inferior performance. One of these tables is presented in Fig. 5. Defective modules can hardly be seen. Suspicious is the large shading area of the lowest module row during the inspection time. Presumably that causes the extremely poor energy yield.

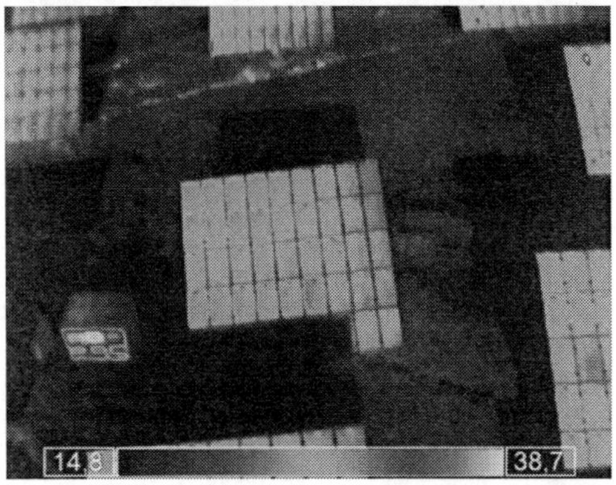

Fig. 5: Tracker 40b, tracker with severe shading effects and no defective module, energy output $Y =$ 33.5 kWh, $E \approx 670$ W/m², $T_{amb} \approx 15$°C

The majority of the tables exhibit both local shading of differing size and defective modules. Several examples are shown in Fig. 6 to fFig. 8. Partial shading of modules in the lowest row is inevitable since the PV-plant has a high packing density of the tracking systems. The influence on the energy yield will be studied later.

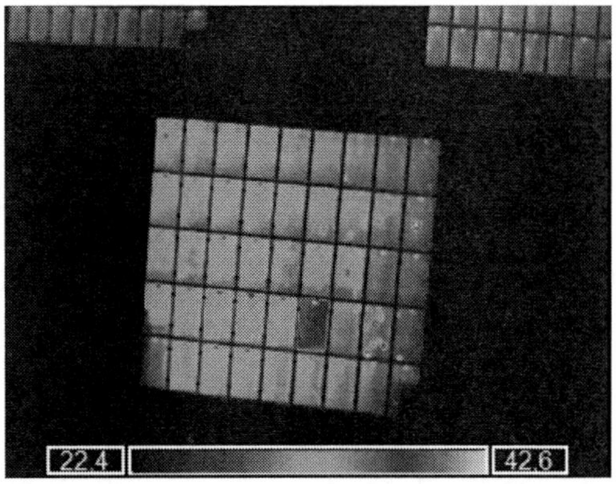

Fig. 6: Tracker 8c, energy output Y = 31.9 kWh, E ≈ 670 W/m², T_{amb} ≈ 15°C

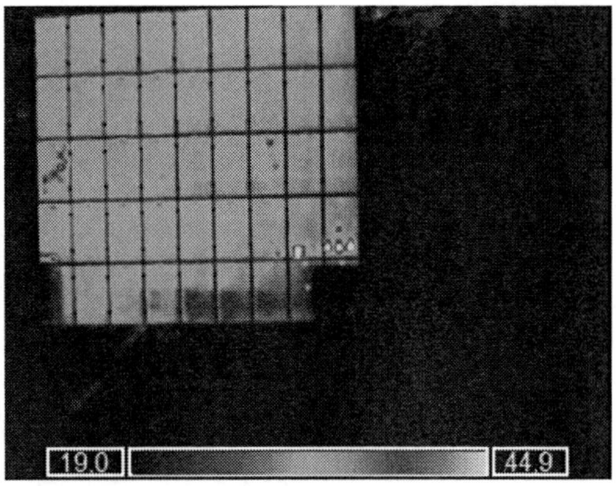

Fig. 7: Tracker 21a, energy output Y = 32.5 kWh, E ≈ 670 W/m², T_{amb} ≈ 15°C

The percentage of more than 8% defective modules in this PV-plant is fairly high. Based on the IR-pattern it is assumed, that almost half of the defective PV-modules contain fractured cells [6], see Fig. 6, Fig. 7, Fig. 8. Also, string failures, bypassed substrings (Fig. 6, module with homogeneously elevated temperature) and defective bypass (Fig. 7) are detected.

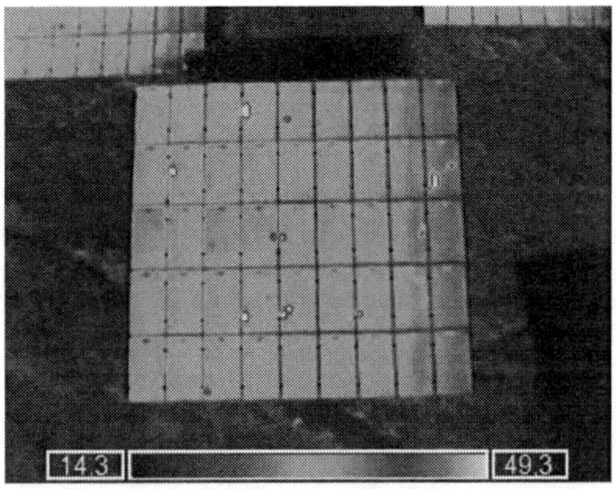

Fig. 8: Tracker 58c, energy output Y = 37.0 kWh, E ≈ 670 W/m², T_{amb} ≈ 15°C

Because of the rather high occurrence of defective modules in the IR-inspection [8], the data are analyzed in detail. This covers the number of appearances per array (table), the distinction between horizontal rows and vertical columns as well as resulting defect temperature.

The number of defective modules per array (45 modules each) reached up to 15 modules per array, as displayed in Fig. 8. In average an array has 2.5 defective modules.

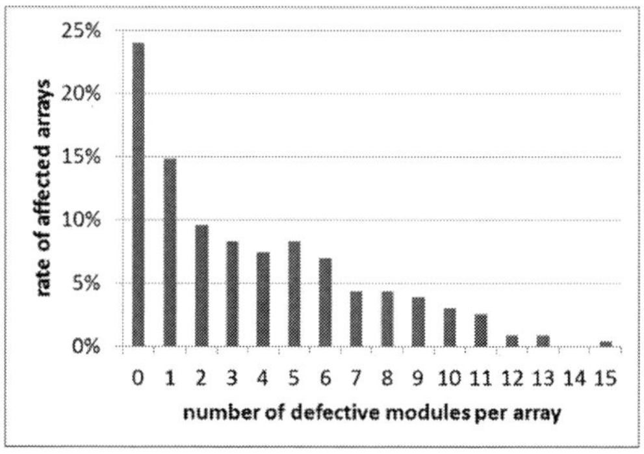

Fig. 9: Statistical distribution of the number of defective modules detected per array of 15 modules

The location of these detected defective modules is summarized in fig. 10 and fig. 11. Fig. 10 and 11 display the rate of defective modules with respect to the horizontal row and vertical column in an array, respectively. The data are evaluated over all arrays.

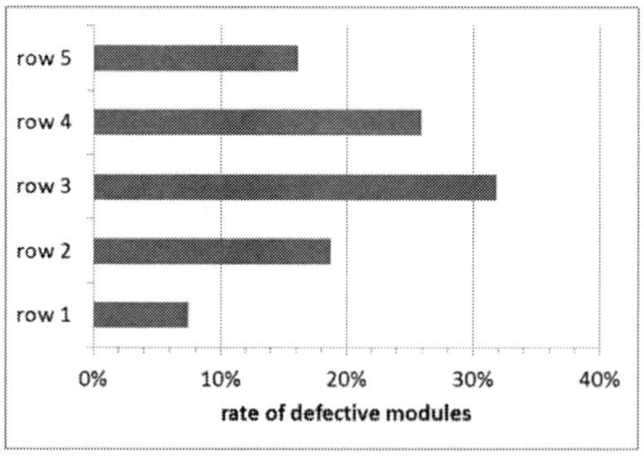

Fig. 10: Statistical distribution of detected defective modules in terms of horizontal module rows

The data reveal that the rate of defective modules accumulates in the upper three rows to 74% (roughly 6% of all installed modules). The majority of defects are present in the middle module row. Thus, one pronounced maximum of defective module exists in the middle row.

The same data evaluation for the vertical orientation (fig. 11) clearly depicts two maxima and a minimum in the array center. The maxima positions correspond to the second and forelast column.

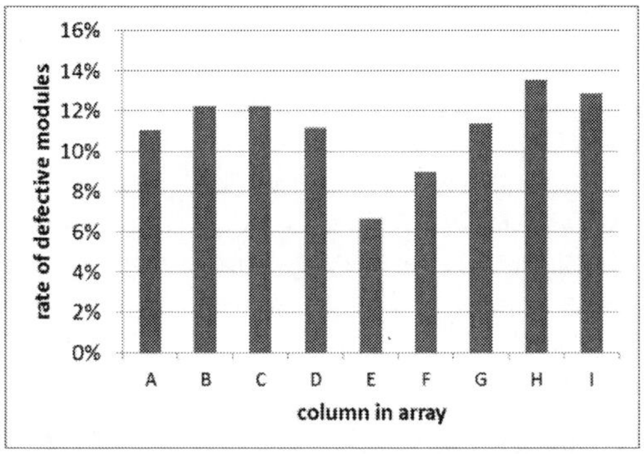

Fig. 9: Statistical distribution of detected defective modules in terms of vertical module columns

These phenomena might be explained by high wind loads, especially at this particular installation, in combination with the extended array dimensions of 9 x 8 m² together with the own weight of 2187 kg of the tracking system. The horizontal carrier is located in the middle of the array; the two vertical supports are position corresponding to B and H (in fig. 11). If high wind load are present, considerable bending moments and stresses can be induced in the modules depending on the position in the array. Therefore, the modules in the upper row show less failure, because the construction does not seem as stiff as for the center row. Here, strong and heavy steel truss have to carry the load of the array and limit the mechanical reaction of the system. Since these steel trusses are located under the modules of column B and H, they hinder the modules to react elastically to wind loads. Due to the large clearance of the truss of about 7 m, the modules in the middle, column E, can react to the wind load and have the lowest number of defects.

Characterizing the defects by their defect temperature shows fig. 12. Defect temperatures from 40°C up to temperatures above 100°C were measured for defective modules. Half of the affected arrays had defective modules with defect temperatures above 80°C. These high local temperatures indicate severe failure and are likely to lead to accelerate aging, e. g. change of encapsulant properties [9].

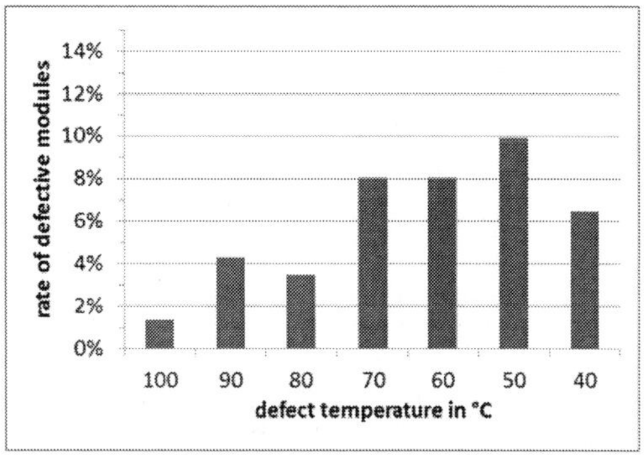

Fig. 10: Statistical distribution of maximum temperatures measured at defects

Analyzing the influence of the wind on the rate of defective modules, the PV-plant could be separated in a wind–rich and a wind-poor region. In the wind-rich area the tables often have to expect heavy wind loads, whereas in the wind-poor area the wind load plays a minor role. The data reveal that in the wind rich area the probability for IR-suspicious modules is 2.5 times higher than in the other area.

Possible explanations for the high defect rate may be the tracker motion that may cause resonance vibrations of the modules. Another reason are probably high wind loads to the system [10]. Considerable visible bending of the entire array was observed during heavy wind loads.

For evaluating the monitoring data the day-based energy yield is plotted for one set of arrays in comparison to the expected yield for two different days, see fig. 13. Depending on the weather conditions energy yields of about 50 kWh and 120 kWh are expected. Almost all module tables stay behind these expectations. The monitoring data scatter below the expected horizons. In average the arrays yield 38 kWh and 101,5 kWh, respectively. Even the "good" arrays with healthy modules hardly reach the values. For a better understanding the data of the depicted tracker systems in fig. 4 to fig. 8 are highlighted in the graphs. It is obvious, that the yield data are strongly influenced by seasonal effects (e. g. solar irradiance, tracker orientation, inclination) and the high packing density of the trackers resulting in shading issues.

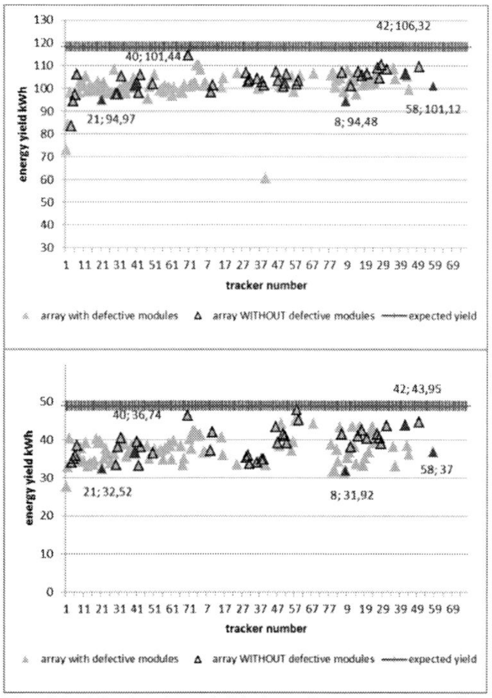

Fig. 11: Monitoring yield data for a summer day May 2015, E_{max} = 1000 W/m^2, T_{amb}= 32°C (upper image), and a day in autumn, October 2015 E_{max} = 720 W/m^2, T_{amb}= 15°C. The triangles mark the daily energy yield of a single tracker. Black framed triangles mark the trackers, which were identified as defect-free by the IR-imaging. Red triangles refer to trackers shown in fig. 4 to fig. 8 (first number = tracker, second number =yield).

Therefore, the monitoring data of the energy yield is in this case not very helpful to locate defect dominated power losses. As the energy yields of defect-free trackers point out, values between 35 and 49 kWh and respectively 83 and 114 kWh are realized. Here, shading phenomena due to the high packing in the PV-plant and seasonal and daily affects may have a significant influence on performance. Furthermore, the module temperature can differ significantly due to pronounced convective heat losses of distinct arrays at wind exposed sites. The monitoring data of distinct trackers are hard to compare due to differing shading and cooling situations for different periods and dates.

Tab. 1: Yield data defect-free (df) tracker and tracker with defective (d) modules

	$Y_{df,mean}$ kWh	$Y_{d,mean}$ kWh	$\Delta Y/Y_{df,mean}$
May 2015	103,2	100,8	2,3%
Oct.2015	39.3	37.3	5,2%

The comparison of the mean energy yield for defect free arrays and defective arrays are shown in tab. 1. The data shows a difference of approximately 2 to 5%. This allows a rough estimation of the severe impact on the power output of the defective modules detected by IR-inspection. Independent of shading, seasonal effects such power reducing failures can be detected using IR-inspection.

SUMMARY

Biaxially tracked PV-systems can be inspected using aIR-PV-check. Autonomous routing throughout the scattered PV-trackers is not possible. Manual adjustments of the flight path and corrections of the camera orientation with respect to the module surface are necessary. Remarkable is the high rate of defective modules and their spatial distribution in the tracker system. This power reducing failures are hard to detect in monitoring data, whereas it can be revealed by IR-inspection.

REFERENCES

[1]D. L. King, et al., Photovoltaic Specialists Conference, 2000. Conference Record of the Twenty-Eighth IEEE (2000) 1487.

[2]U. Jahn, et al., ep Photovoltaik aktuell (2008) 32.

[3]C. Buerhop, et al., 26th EU PVSEC (2011) 3413.

[4]C. Buerhop and H. Scheuerpflug, 29th EU-PVSEC (2014) 2975.

[5]C. Buerhop and H. Scheuerpflug, Third Southern African Solar Energy Conference SASEC (2015) 224.

[6]C. Buerhop, et al., Solar Energy Materials and Solar Cells 107 (2012) 154.

[7]M. A. Quintana, et al., Photovoltaic Specialists Conference, 1993., Conference Record of the Twenty Third IEEE (1993) 1197.

[8]C. Buerhop, et al., SPIE (2016), to be published.

[9]C. Camus, et al., 32th EU-PVSEC (2016), to be published.

[10]Solar-Prinz, 2010, http://www.solar-prinz.de/tracker-systeme-mover-nachgefuhrte-solaranlagen/341.

aIR-PV-check of thin-film PV-plants - Detection of PID and other defects in CIGS modules

Claudia Buerhop[1], Tobias Pickel[1], Hans Scheuerpflug[1], Christian Dürschner[2], Christian Camus[1], Jens Hauch, Christoph J. Brabec[1,3]
[1] Bavarian Center for Applied Energy Research (ZAE Bayern), Haberstraße 2a, D-91058 Erlangen, Germany
[2] Photovoltaik Ing.-Büro Dürschner, Anna-Rosenthal-Weg 21, D-91052 Erlangen, Germany
[1] i-MEET, FAU Erlangen-nürnberg, Martensstraße 7, D-91058 Erlangen, Germany
email: claudia.buerhop-lutz@zae-bayern.de, Author for correspondence: Claudia Buerhop

Abstract: The on-site inspection of solar farms with infrared cameras from the air using unmanned aerial vehicles (UAVs) is called **aIR-PV-check**. Applying this method with a sophisticated image processing enables the quality control of installed thin-film PV-modules. Various module defects can be detected and visualized based on their elevated temperature. There are modules with cracks and ones affected by potential induced degradation PID. The IR-imaging data correspond well with EL-images and electrical data. Thus, modules with slightly elevated temperature - typical for PID modules - appear damaged in the EL-images as well as in their electrical behavior. Furthermore, modules with different power classes can be located. Therefore, IR-imaging is a fast and suitable technique to detect PID and other features in installed CIGS modules.

Keywords: thermography, on-site inspection, CIGS, defect localization, PID.

INTRODUCTION

The IR-inspection of installed crystalline PV-modules is known quite well [1-3]. There are several advantages: IR-imaging is contact-free, nondestructive, provides two-dimensional images, and can be carried out during operation. Using unmanned aerial vehicles (UAV) as a carrier for an IR-camera even facilitates the inspection [4, 5]. At ZAE Bayern, a new method for aerial inspection of PV-plants using IR-cameras mounted to an UAV has been developed - the **aIR-PV-check**. This method takes advantage of the mobility and navigability of UAVs and the fast and contactless detection of defective PV-modules. Using aIR-PV-check even large and non-accessible solar parks can be easily inspected. Irregularities in the IR-images of PV-modules are a strong indication for defective sites, as described elsewhere [6]. Current measuremnts support the location of defective module strings [7].

Fig. 1: Aerial photography of PV-plant with thin-film modules made with aIR-PV-check, 1 MWp, Germany

However, inspecting thin-film modules under outdoor conditions is quite challenging using IR-imaging. The optical properties of the used glasses differ. The expected defects are different from those known from crystalline modules. Because the defects are much smaller, the heat generation at defect sites is lower. Thus, both the lateral and the radiometric resolution of the IR-camera are important for the detection of small defect sites [4].

Beside shunts other defects, like potential induced degradation (PID), are discussed in [8]. PID decreases the open circuit voltage and the output power of CIGS modules, if they are at high negative voltage versus ground. In electro-luminescence (EL) images PID affected module areas are dark corresponding to reduced electrical activity. The intensity of the EL-signals corresponds to the advanced state of PID deterioration of the module [9-11].

This paper gives a first overview of findings in PV-plants with CIGS modules using IR-inspection. The visualization of PID affected modules and their differentiation to other defective modules will be described.

EXPERIMENTAL PROCEDURE

A 1 MWp PV-plant with thin-film modules in Germany was inspected using IR-imaging. There are 13320 modules installed, connected in 1110 strings with 12 modules each and arranged in 53 rows. The CIGS-modules had a nominal power of 75 W in average. Installed in 2009 the operation period of the modules was about 6 years. In several maintenance periods quite a large number of defective modules showed cracks and/or TCO corrosion and had to be replaced by other modules.

Since the original module type was not available anymore, modules of higher power classes were installed instead.

Fig. 2: Measurement system for aIR-PV-check of PV-plants

For investigating the installed PV-modules or PV-plants the **aIR-PV-check** system was used [5]. The measurement system, as shown in Fig. 2, consists of an unpiloted drone (Multicopter), a lightweight IR-camera PI 450 (optris), a visible camera GoPro and equipment for navigating. Solar irradiance E and ambient temperature T_{amb} are recorded, too.

The hilly ground, as seen in Fig. 1, and the vicinity to a highway make the inspection ambitious at this site.

For verifying the findings visual inspection and IV-curves of selected individual PV-modules and entire module strings were performed with a PVPM1000C40 (peak power and I-V-curve measurement device) [12]. The IV-curves were measured at nearly ideal conditions: mid-day (11:00 am to 3:00 pm, European daylight saving time) at very sunny days only ("clear sky days") with a solar irradiance between 850 W/m² and 1000 W/m². Air temperature was about 20-25°C and module temperature between 35-50°C (both temperatures gradually rising during day but steady during each IV-curve measurement).

Additionally electroluminescence images of all strings were taken with pvVision (a modified single lens reflex NIKON camera to get EL images) combined with pvServe (a device to apply back current into the strings during the night) [13]. The EL images were taken during nighttime (after sundown, 9:00 pm to 3:00 am). The back current for the modules was chosen close to the Isc-current during daytime.

RESULTS

The aIR-PV-check of the thin-film PV-plant was carried out on a summer day with high solar irradiance $E \approx 1000$ W/m², high ambient temperatures $T_{amb} \approx 35$°C but fleecy clouds. Fig. 3 shows an IR-image of a large number of module tables. Although the lateral resolution is pretty poor for identifying defects in the modules, one site is quite eye-catching (marked with an arrow in Fig. 3, upper image). This module position is much colder than the surrounding modules. The visual inspection revealed severe module deterioration: the fractured module parts dangled from the rack, a gap remains, compare fig. 3, lower image.

Large temperature increases result from the clouds. This demonstrates the importance of optimal measurement conditions, especially for the inspection of thin-film modules with highly reflecting surfaces.

Fig. 3: upper image: IR-overview of the thin-film PV-plant, showing a distinct site with significantly lower temperature, $E \approx 1000$ W/m², $T_{amb} \approx 35$°C. The inhomogeneous temperature distribution is due to the perspective view the movement of the fleecy clouds. Lower image: Visual image of the detected free space with the dangling module.

aIR-PV-check inspections show that various defects in thin-film PV-modules can be detected. Three interesting observations will be presented here.

Detection of cracks

For inspecting defective modules, images with a higher lateral resolution are necessary. A frequently observed pattern is an one-dimensional temperature increase. This is usually observed at a module edge parallel to the cells. Exemplarily, two findings of such defective modules are shown in Fig. 4. Further inspections reveal cracks in the glass as well as in the semiconductor as the root-cause interrupting the continuous current flow.

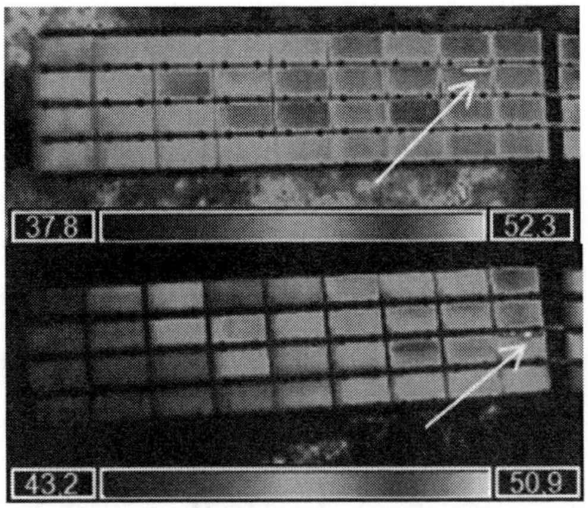

Fig. 4: IR-images showing locally heated thin-film modules, $E \approx 1000$ W/m², $T_{amb} \approx 35$°C

Detection of PID

A second phenomenon observed after careful image evaluation is a slightly increased temperature of lower module rows. Typical temperature differences measured between modules mounted in the upper and the lower row are 1,5 – 2 K, see Fig. 5. The temperature increase occurs always at the lower rows as the six module strings on two exemplary tables illustrate. Since this corresponds well to the negative voltage at the string end (string is indicated by the white lines), that might be an indication for PID in CIGS modules.

For evidence further analyses were carried out. IV-curves were measured, EL-images taken and monitoring data evaluated and compared with the findings of the IR-images

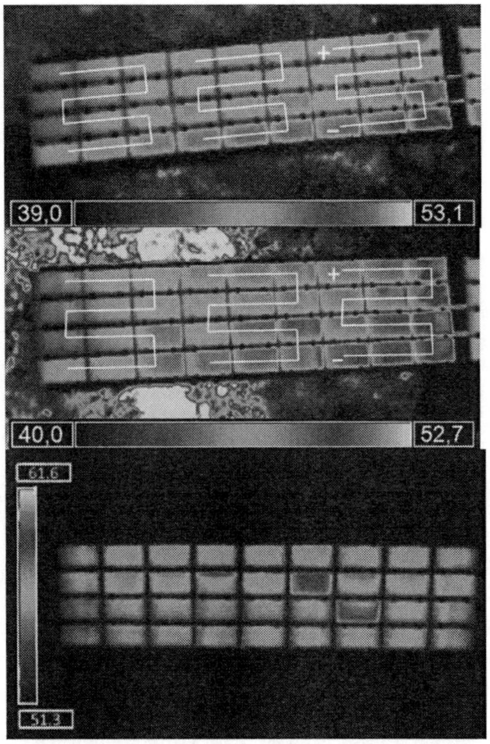

Fig. 5: IR-images showing PID at CIGS modules by a slight temperature increase in the lower row at two tables (upper and middle images), $E \approx 1000$ W/m², $T_{amb} \approx 35$°C. White lines indicate the serial connection of the strings. The temperature gradient (approx. 4 K) from the left to the right is due to the locally differing temperature transport by convection. Lower IR-image showing a table without PID-affected modules but three with other faults.

First, the two sets of daily monitoring data from the year 2012 and 2015, see fig. 6, are analyzed. The string current data from 2012 reveals the typically shaped daily current production. All strings behave quasi identically. (Height during mid-day is nearly the same, offset in time is caused by hilly ground, so that effective orientation of some modules is not south but south east or south west.) In 2015, the situation is totally different. There is a broad variety of different current production visible in the graph. Many curves are flattened. Thus, the current during the day is extremely lowered depending on the health state of the string.

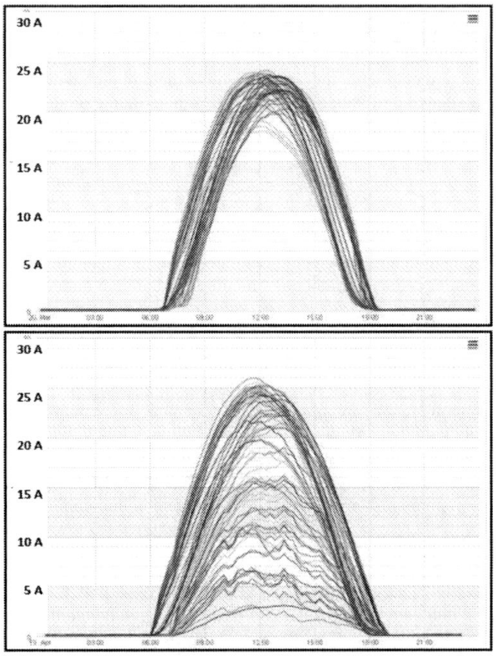

Fig. 6: Monitoring data of the string current during daytime of 62 string groups at high solar irradiance, 20th March 2012 (upper graph), monitoring data of the same string current, 19th April 2015 (lower graph)

For a better understanding the module IV-curves of a healthy, normally heated string and of one suspiciously heated are measured, see fig. 7. The IV-curves of the regularly heated modules are almost identical (fig. 7, upper graph). The measured deviations are likely to be due to normal fluctuations of the manufacturing process. The lower graph in fig. 7 shows the IV-curves of modules of an irregularly heated string. These IV-curves differ strongly. Three stages of degradation can be stated. Stage 1: no degradation is recognizable. Stage 2: the open circuit voltage V_{oc} decreases. As the degradation proceeds, stage 3 shows additionally a reduction of the short circuit current I_{sc}.

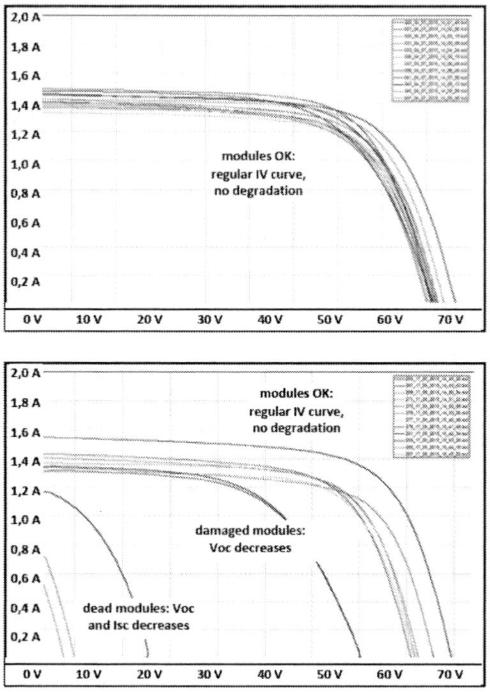

Fig. 7: IV-curves of 12 PV-modules of two different strings, upper graph: modules without PID, lower graph: PID-affected modules with PID-damage at different stages

EL-images of a string with 12 modules at different degradation stages and different grey states are shown in fig. 8. Modules no. 1 and 3 to 5, respectively the four "black" modules in the two lower rows, are not operating anymore. These modules do not emit any electroluminescence. Their IV-curves show a decrease of voltage and current. Module no. 6 appears half inactive (black) half active (bright). This module shows beginning degradation and as the IV-curve point out considerable voltage reduction. Module no. 2 and 7 to 12 are still in good shape, as concluded from the homogenous, bright shining EL-images.

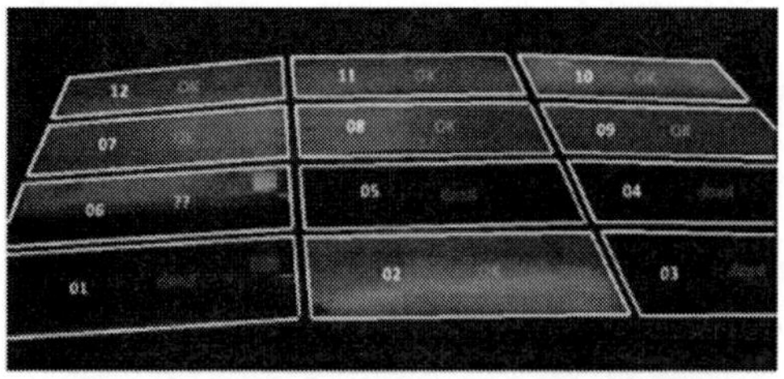

Fig. 8: EL-image showing an entire module string (compare fig.5, lower image, right string), where the four framed modules are not operating and three other modules show severe inactive, black areas.

In general: The more a module is positioned to the negative pole of the string, the bigger is the damage of the module. Usually module no. 2 also would be expected to be dead, but it is not the original module but has been installed as a replacement of the original one.

Table 1: measured power of the modules visible in fig. 8

78,3 Wp	79,8 Wp	81,7 Wp
79,5 Wp	82,7 Wp	82,6 Wp
56,9 Wp	28,8 Wp	14,7 Wp
27,3 Wp	88,1 Wp	9,5 Wp

Table 1 shows the measured power of the modules shown in fig. 8. (The data are the average of three measurements with the uncertainty is approx +/- 5%). Modules no. 2 and 7 to 12 still are within specification, module no. 6 is damaged (decrease of voltage only), modules no 1 and 3 to 5 are (near to) dead (decrease of voltage and current).

The additional measurements on module and string level give a strong indication for PID at the detected, irregularly heated modules with IR-imaging. Measured IV-curves, module output power and EL-images reveal clearly the PID characteristics. The modules at the string end (in the lowest row) have the lowest V_{oc} and lowest power output (less than 30%) compared to the unaffected modules at the string beginning. Thus, elevated temperature in IR-image, dark EL-images and reduced voltage indicate PID and correspond well for the investigated strings. Nevertheless, the monitoring data record the poor performance of the PID affected modules with an advanced state of damage.

Detection of power mismatch

However, not all temperature increases may be caused by defective modules, as fig. 9 illustrates. During the operation period several damaged modules have been replaced by substitutes, usually modules of a higher power class. Due to that fact the modules in one string do not have the same current and voltage at their respective maximum power point (mpp) - their individual P_{mpp}s do not match perfectly. Under operating conditions the string is at its mpp. That means a shift of the individual I_{mpp} and U_{mpp} of the modules. Accordingly less absorbed solar radiation is converted to power output and more heat loss occurs in particular modules depending on the string configuration. The IR-image displays clearly that the "old" modules have a much higher temperature than the "new" ones. The temperature varies with the position in the table because of heat convection. Typically temperature differences of $\Delta T \approx 1.4 - 2.5$ K can be measured between two modules. That means in this case, the "old" modules are not working at their right working point because an averaged current is flowing through the string.

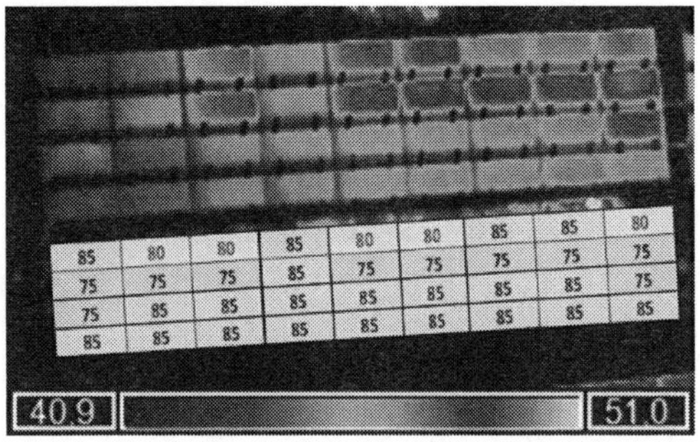

Fig. 9: IR-image showing different temperature distributions due to modules of different power classes (data in the table), $E \approx 1000$ W/m², $T_{amb} \approx 35$°C. Here, no PID affected modules due to previous replacement are present.

The analysis points out that different effects may overlay in a PV plant. In order to get to a proper conclusion a careful evaluation of all data is necessary. Here, e. g. the replacement of modules, their power class as well as the string configuration has a significant impact on the IR-images, and may lead to misinterpretation if not taken into account.

SUMMARY

IR-inspection is a fast and suitable method to detect PID under operation conditions. Weather conditions, especially a clear, cloud-free sky, are important for the inspection of thin-film PV-plants. Under optimal measurement conditions, and sophisticated image post-processing irregularities can be identified and located using aIR-PV-check as in the case of crystalline silicon PV modules. Defective modules with glass fracture and cracks as well as with PID can be detected in IR-images. The PID findings correlate to the EL-images and electric data. Different stages of PID can be recognized. However, for a reliable data interpretation the knowledge of the solar plant is necessary.

In conclusion, IR-imaging is a valuable tool to identify irregularities in thin-film PV-plants fast and without operation interruption.

REFERENCES

[1]E. Molenbroek, et al., Photovoltaic Specialists Conference, 22nd IEEE (1991) 547.

[2]J. Wohlgemuth and W. Herrmann, Photovoltaic Specialists Conference, 2005. Conference Record of the Thirty-first IEEE (2005) 1062.

[3]U. Jahn, et al., ep Photovoltaik aktuell (2008) 32.

[4]C. Buerhop and H. Scheuerpflug, 29th EU-PVSEC (2014) 2975.

[5]C. Buerhop and H. Scheuerpflug, Third Southern African Solar Energy Conference SASEC (2015) 224.

[6]C. Buerhop, et al., Solar Energy Materials and Solar Cells 107 (2012) 154.

[7]M. Diehl, 2015, http://www.photovoltaikbuero.de/pv-know-how-blog/fehlende-modulstraenge-an-duennschicht-photovoltaikanlagen/.

[8]Y. Seira, et al., Japanese Journal of Applied Physics 54 (2015)

[9]P. Lechner, et al., 28th EU-PVSEC (2013) 2810.

[10]P. Lechner, et al., 28th PVSEC (2014) 2342.

[11]C. Dürschner, PID@CIGS (2015), photovoltaic meetings

[12]peak power and IV-curve measuerment equipment www.pv-engineering.de.

[13]electroluminescence (EL) imaging and back current equipment www.pvserve.de.

IR-images of defective PV-modules influenced by short-time changes of the electric system

Claudia Buerhop[1], Johannes Dettelbacher[1], Tobias Pickel[1], Christian Camus[1], Jens Hauch[1], Christoph J. Brabec[1,2]
[1]Bavarian Center for Applied Energy Research (ZAE Bayern), Haberstraße 2a, D-91058 Erlangen, Germany
[2]i-MEET, FAU Erlangen-Nürnberg, Martensstraße 7, D-91058 Erlangen, Germany
email: claudia.buerhop-lutz@zae-bayern.de, Tel.: (+49)9131-9398-177, Fax: (+49)9131-9398-199
Author for correspondence: Claudia Buerhop

Abstract: IR-thermography is a useful tool for the inspection of installed PV-modules. It is well-known that ambient conditions like irradiance and wind velocity have an influence on the measured temperatures. In this investigation, we study the influence of short-time changes of the electrical system, caused by the global maximum power point search algorithm of a grid-tie inverter, on IR-images. Considerable changes of the temperature difference at defects within seconds were detected. Therefore, we perform an experimental study with different types of inverters on a test facility at the ZAE Bayern and thermal simulations of the behavior of different types of module defects. Furthermore, we discuss the impacts on the defect temperature and make a proposal to avoid misinterpretation of IR-images of PV-modules affected by short-time changes of the electric system.

Keywords: thermography, installed PV-modules, defect temperature, grid-tie inverter.

INTRODUCTION

Thermography is a valuable tool for industrial inspection of installed photovoltaic (PV) modules. Infrared (IR) -measurements are contact free, nondestructive and can be carried out under operating conditions whilst providing two dimensional images [1]. The evaluation of the module quality due to IR-images is usually done by comparing the temperature of defective module sites – hot cells or hotspots – with non-defective ones [2]. It is well-known that the temperatures of defects are affected by environmental conditions like solar irradiation or ambient temperature. Furthermore, the configuration of a string of the module investigated can influence the temperature of defects [3-6]. But also changes in the electrical system can modify the thermal behavior of defective PV-modules, e. g. inverter characteristics.

This paper shows how short-time temporal changes of the current flow caused by an inverter, affect the defect temperature.

EXPERIMENTAL PROCEDURE

For investigating the influence of short term changes in the electric system on the temperature distribution of defective modules, distinct modules and two different inverters were installed at the ZAE test facility, see **Fehler! Verweisquelle konnte nicht gefunden werden.**. Here, the modules were connected in series in a string and held at the global maximum power point (GMPP). The data recording included: weather data like solar irradiance, ambient temperature, wind velocity, and module data as temperature at certain spots and the operating point (current and module voltage). The temperature was measured with 1-wire DS18S20 sensors attached at several sites on the module and with a lightweight IR-camera PI 450 (optris). The IR-camera has a resolution of 382 x 288 pixels and a frame rate of 80 Hz. Using the IR-image, the temperature distribution across the module could be recorded, too.

The inverters used differed in their procedure to find the GMPP of the string. Inverter S performs a periodic scan every 30 minutes for about 2 minutes until the GMMP is reached. The other inverter F uses a more sophisticated approach to detect the GMPP and performs a periodic scan every 10 minutes in less than one second [7].

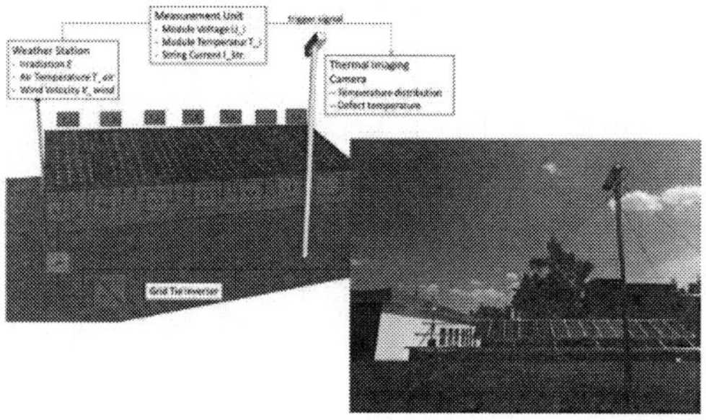

Fig. 1: Test facility at ZAE Bayern for analysis of defective modules at conditions close to real operating conditions with extended data logging, left: schematic sketch of the data monitored, right: photograph showing the PV-modules installed and the continuously monitoring IR-camera

Different pre-characterized defective modules were mounted at the test facility. For the analysis modules with cell fracture and contact failure were selected. The string consists of 10 modules, one defective one and nine non-defective ones. All modules were characterized thoroughly in the lab with regard to defect type and output power (STC). EL-imaging was carried out with high resolving Si-camera CoolSamBa HR-830 with 8.3 MPixel (Sensovation). The format of the detector was 3326 x 2504 pixels. The modules were excited with an external power supply. The applied current was close to the short circuit current. IV-curves were measured using a SPIRE sun simulator 4600 SLP.

For better understanding of the findings, the transient temperature development is modelled, considering temporal changes of the heat source and heat transfer.

RESULTS

Experimental results

IR-images and electrical data were evaluated for quasi-constant weather conditions. IR-images with a time interval of 30 seconds are shown in Fig. 2. The recorded temperature difference is about 10 K. Since the temperature drop is not caused by altering ambient conditions, the reason may be due to changes in the electrical system.

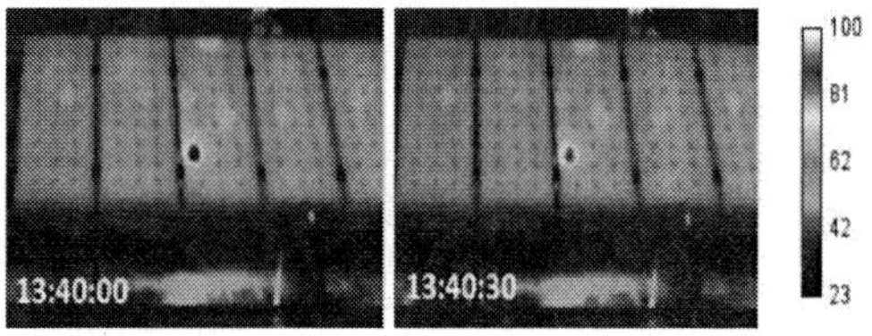

Fig. 2: IR-image of the same scene at two instants of time under unchanged weather conditions, at 13:40:00 T_{defect} = 91°C (left), at 13:40:30 T_{defect} = 80°C (right), E = 940 W/m², T_{amb} = 28°C, inverter S

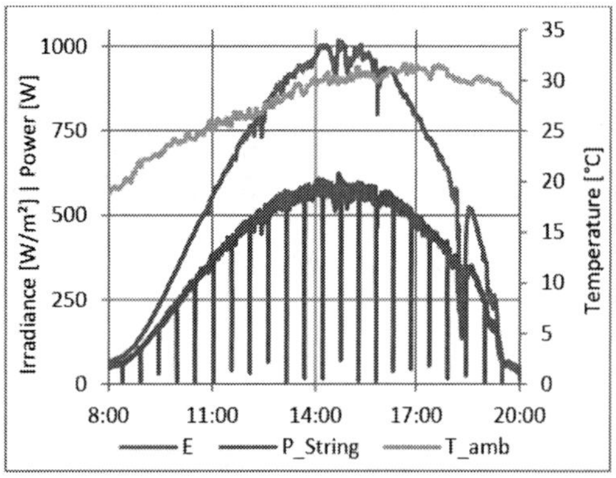

Fig. 3: Monitoring data for June 12th, 2015 at ZAE test site showing solar irradiance E, ambient temperature T_{amb}, and string power P_{String}

The monitored electrical data and the measured solar irradiance E and ambient temperature T_{amb} are plotted in Fig. 3 for one day. It was a nice summer day with quite high ambient temperature and high solar irradiance. The string power is mainly determined by the irradiance during the day. The periodic, short-time power drops every 30 min are striking. A closer look at the string current and voltage reveals that the inverter is scanning for the GMPP (Fig. 4). After 2 min the procedure is finished and the GMPP is reached.

How the cell and defect temperatures are affected by the scanning procedure is shown in Fig. 4. The data are plotted for quasi-onstant solar irradiance and ambient temperature, the monitored current is fairly constant except for the scanning time intervals. No influence of the scanning is observed for the reference temperature T_{ref}. In contrast, for the temperature at the defect site T_{def} a remarkable temperature drop is recorded which is synchronous with the current decrease. Significant temperature differences up to -10 K or more could be registered. While the current recovers instantaneously after the scan ended, the recovery of the defect temperature is rather slow.

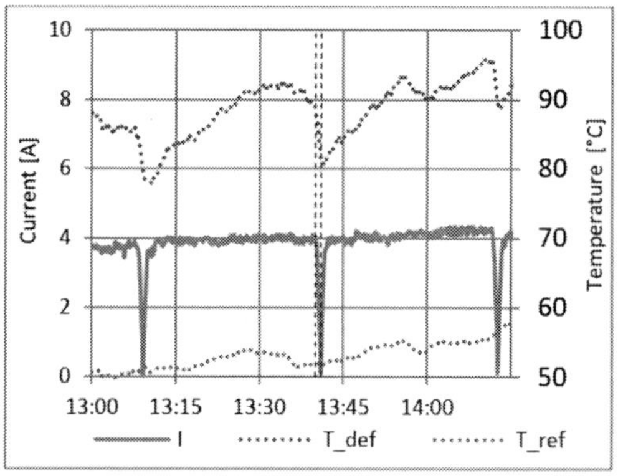

Fig. 4: Highly temporally resolved, measured data for the current I, defect temperature T_{def}, and temperature of a reference cell T_{ref}, inverter S

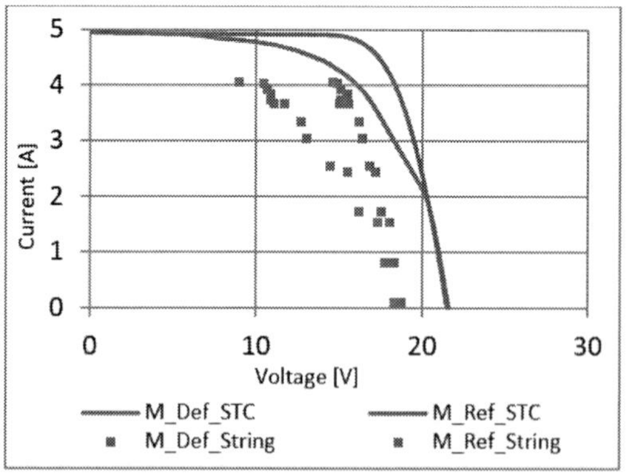

Fig. 5: IV-curves of the modules of interest measured under STC (lines), operating points of these modules in the string (dots)

The IV-curves for the defective modules and exemplarily for one defect-free reference module is shown in Fig. 5. The measured operating currents and voltages during the scanning are presented by the dots. The offset to the STC-IV-curves is caused by the increased temperature of the modules at operating conditions.

Under comparable ambient conditions the other inverter F does not show the periodic features. Fig. 6 shows the monitored string power, E, and T_{amb} plotted for a sunny day. The module's power and temperature as well as the defect temperature are not affected by the scanning procedure.

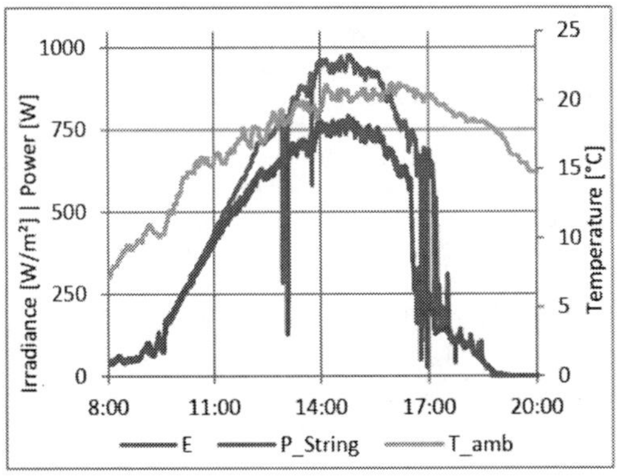

Fig. 6: Monitoring data for September 21th, 2015 at ZAE test site showing E, T_{amb}, and P_{String} using Inverter F

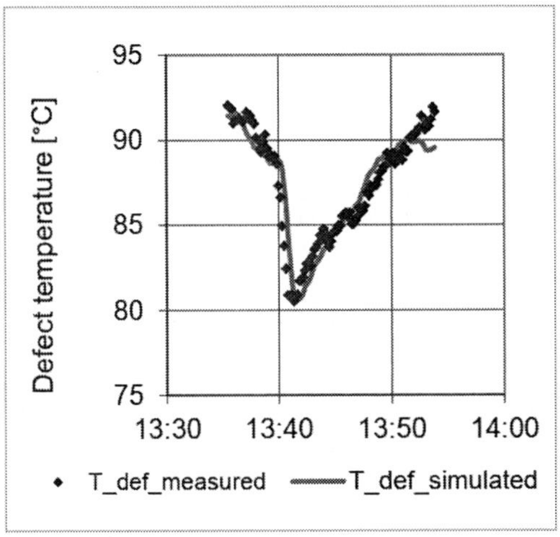

Fig. 7: Comparison of measured and simulated data

Modelling results

Short-term changes in the PV-system lead to an instantaneous change of the electrical parameters. However, the thermal response is rather slow. Since the calculated Biot-number of the system is $Bi = 0.08 < 0.1$ [8] the heat transfer from the surface to the surrounding is the governing factor.

The transient temperature is therefore a function of the heat transferred and the electrical power produced. Knowing the material properties, a module heat capacity of $C_{module} = 7500$ J/kg can be determined. Assuming a heat transfer coefficient of $h = 25$ J/K/m², a relaxation time of $\tau = 12$ min is estimated for reaching 90% of the thermal equilibrium.

The differential equation was solved stepwise using the Euler procedure. The fit to the measured data is shown in Fig. 7. Using the adapted parameter a real relaxation time of $\tau_{real} = 8$ min is obtained.

CONCLUSION

Concluding, the scanning procedure is relevant for the defect temperature and power of defective modules. 8 min after a short-time electric change in the system, the temperature should have reached the thermal equilibrium. Thus, for IR-inspection of PV-plants the scanning procedure of the inverter should be checked. In case the scanning procedure is unknown the current flow of a PV-string and the irradiance during the measurement should be logged, otherwise the defect temperature might be underestimated.

Similar situations may be caused by instant weather changes and regulations by the net provider. Here, IR-imaging needs more attention.

ACKNOWLEDGEMENT

We gratefully thank the German Federal Ministry for Economic Affairs and Energy (BMWi) for financial funding this project.

REFERENCES

[1]U. Jahn, C. Buerhop, U. Hoyer, B. Lerche and S. Wittmann, ep Photovoltaik aktuell (2008) 32.

[2]C. Buerhop, D. Schlegel, M. Niess, C. Vodermayer, R. Weißmann and C. J. Brabec, Solar Energy Materials and Solar Cells 107 (2012) 154.

[3]R. Moretón, E. Lorenzo and L. Narvarte, Solar energy 118 (2015)

[4]C. Buerhop, T. Pickel and H. Scheuerpflug, 31. Symposium Photovoltaische Solarenergie (2016).

[5]C. Buerhop, H. Scheuerpflug and T. Pickel, 31th PVSEC (2015) 1692.

[6]M. Köntges, I. Kunze, S. Kajari-Schröder, X. Breitenmoser and B. Bjørneklett, Solar Energy Materials and Solar Cells 95 (2011)

[7]S. Lyden and M. E. Haque, Renewable and Sustainable Energy Reviews 52 (2015)

[8]W. Polifke and J. Kopitz, Wärmeübertragung

SELECTION CRITERIA OF PV TECHNOLOGY BASED ON LOCATION

Gopal Kumar [1], Rajesh Kumar[1], PK Dash[1] , Rahnuma Siddiqui[1], Ganesh Gowri[1], Manoj Morampudi[1] , Sneh Lata[1],Sahan Raghava[1], Pragati Rajput[1],
[1]National Institute of Solar Energy, 19th milestone, Gwalpahari, Gurgaon-Faridabad road, Haryana-122003, India.
Email: gopaljha1987@gmail.com

ABSTRACT: The suitability of a PV technology for a particular site depends on five major factors like annual solar intensity distribution, annual temperature distribution, efficiency of PV module, module temperature coefficients, variations in the Sun spectrum distribution and rate of power degradation in the PV modules with time. The PV technologies which are used in this study are mono, multi crystalline silicon, amorphous silicon single junction, micro crystalline amorphous silicon, CdTe, and Interdigitated Back contact (Maxeon) and Hetero-Junction with thin Intrinsic (HIT) layer technologies. A detailed comparison of PV technologies has been done by using indoor and outdoor conditions. The outdoor performance of these modules are evaluated by using the outdoor test bed established in NISE and the efficiencies are estimated as a function of intensity under indoor conditions by using Endeas, Quicksun 700A single pulse sun simulator. This paper presents the details of the work carried out and the understanding of the performance of the above technologies in terms of performance ratio and utilization fraction.
Keywords: a-Si, back Contact, c-Si, CdTe, CIGS, Heterojunction, Micromorph, performance ratio, Solar cell Efficiencies,

1 INTRODUCTION

The module performance is rated under Standard Test Condition (STC) i.e. Air mass 1.5 spectral distributions, irradiance level of 1000 W/m2 and module temperature of 25 ^{0}C at 0 angle of incidence as per IEC 61215 [1]). These conditions are seldom visualized in the actual field in most of the sites/locations. In real field conditions, modules are experienced wide range of irradiance and its spectrum, temperature and these operating conditions varies with the time of the day, day of a year, season and location. In real field conditions PV Modules performance are dependent on site temperature, intensity and spectrum. The field performance can be judged through indicators such as yield (Y_i), Performance Ratio (PR) and Utilization Fraction (UF). In the present study, we attempted to understand variations in cell (in module) efficiency, for different PV technologies with respect to Irradiance. India has at least seven different climatic zones, with whose field a condition varies with the time of the day, day of a year, season and location. Recently, India has revised the target of deploying grid connected solar power plants under Jawaharlal Nehru National solar mission from 20 GW of to 100 GW by 2022. In view of such studies are important to know how different PV technologies achieve their best energy yields at different sites.

2 EXPERIMENTAL SET UP:

An experimental test facility for PV modules in indoor lab at NISE (National Institute of Solar Energy), Gurgaon as shown in the Fig 1. is the photo of a large area (2m X 2m) single pulse sun simulator model QuickSun 700A, from M/s Endeas, Finland. The lab has established quality system as per ISO 17025, for qualification testing of PV modules as per IEC61215 or IEC 61646 standards. This laboratory is the first laboratory in India to acquire accreditation for testing PV modules by NABL, as per ISO 17025 quality systemVolume of paper This laboratory is the first laboratory in India to acquire accreditation for testing PV modules by NABL, as per ISO 17025 quality system. Figure.2 depicts the details of outdoor test bed(s) set up in NISE for performance evaluation of different technology PV modules. The test bed has all the PV technology module arrays, including mono & multi crystalline silicon, Maxeon®cell technology from M/s Sun Power, HIT® technology modules from M/s Sanyo, latest CdTe modules, amorphous silicon single junction and triple tandem module. PV arrays with CIGS and micro-crystalline silicon are under installation. The power generation from these arrays is connected to a localized mini-grid through MPPT based high quality String inverters. The performance data as energy yields, radiation on module plane, ambient temperature module temperature recorded once in every minute.

Fig. indoor lab Facility at NISE

2.1.Result & Conclusion

The global radiation data on the module plane and the horizontal are recorded through a dedicated weather platform established at outdoor test beds. Initially before installation modules are tested for their STC wattage. Further few representative modules from each technology are tested for the efficiency under varying intensities, from 200 W/m^2 to 1100 W/m^2. The data is shown in figure 3. Further the details are also tabulated in the table 1. From the data it is clear that different technology behave differently under varying intensity conditions[5] [6]. Silicon based mono, multi, amorphous Si, and HIT cells efficiency is lower under low intensity and the efficiency increases with intensity and saturates after 600 W/m^2 radiation. On the contrary thin film based technology modules CdTe & CIGS showed better efficiencies under low intensities and at higher intensities the efficiencies decreased [7]. The magnitude of the variations can be obtained from the data at the table1. The distribution of the energy at different intensities ranges are plotted as bar graph in the figure.4. This graph is clearly showing the distribution of radiation, indicating maximum intensities are in the range < 800 W/m^2, only 15% of the radiation energy available in the intensity range above 800 W/m^2. In figure.5 a bar graph is plotted between the temperature distribution for crystalline silicon and thin film technology module during the years 2010 and 2011. These plots show around 60% of energy from the module is generated in the temperature range between 35°C to 65°C. In figure.6 the data of Performance ratio (PR) is plotted in bar graph for different technology PV modules installed at NISE in the outdoor test bed [8]. From the bar graph it is evident that the PR is higher (about 0.9) for those technologies showing less variation in cell efficiency with intensity. These includes Sun Power's Maxeon® cell technology, HIT, CdTe (new) and amorphous silicon triple tandem, other technologies such as multi crystalline silicon , CIS, CdTe (degraded) are showing lower PR values(<0.8) few .

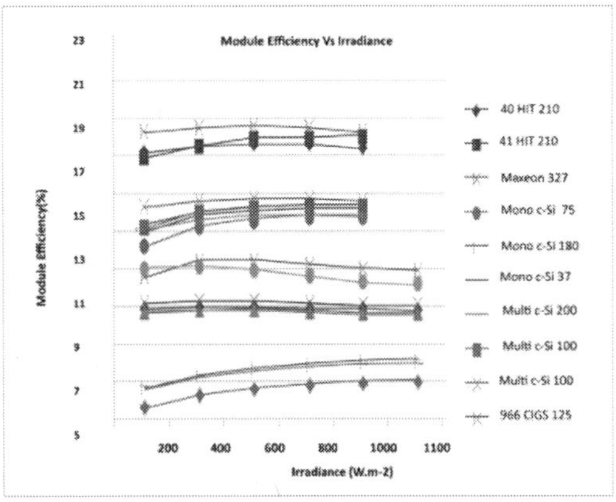

Figure.3 (Variation of efficiencies of different PV technologies with intensity)

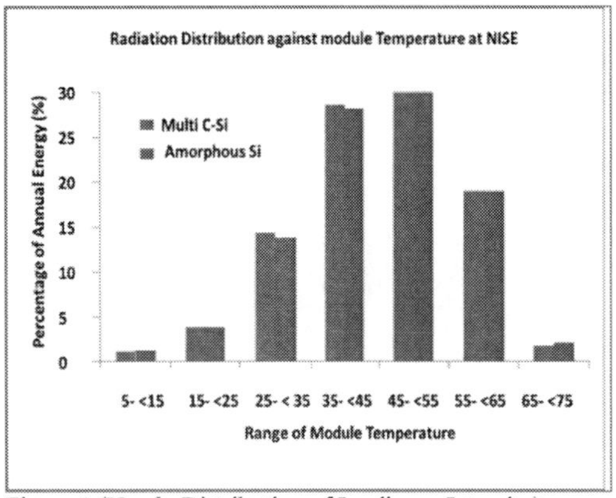

Figure.4 (Yearly Distribution of Irradiance Intensity)

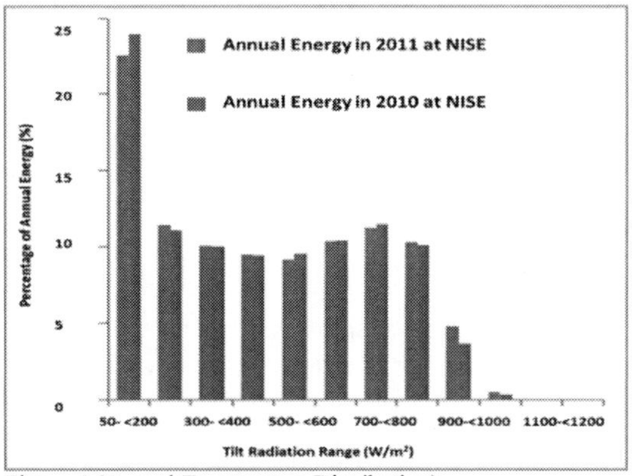

Figure.5 (Annual Temperature Distribution)

Type of Technology		Sample name	Pmax at STC	Irradiance (W/m2) for maximum cell efficiency	Irradiance (W/m2) for Minimum cell efficiency	Average variation of efficiency
Crystalline	Mono C-Si	Mono 37	44.7	800-1000	200	1.12 %
		Mono 75	84.6	800-1000	200	
		Mono 180	182.5	800-1000	200	
	Multi C-Si	Multi 200	230	600-800	200	0.50 %
		As Multi 100	107.9	800-1000	200	
		V Multi 100	106	600-800	200	
Thin film	a-Si	96 a-Si	98.2	1100	200	1.34 %
		92 a-Si	98.3	1100	200	
	CdTe	180 CdTe 80	67.2	600	200.1000-1100	0.29 %
		201 CdTe 80	69.5	400-600	1000-1100	
		526 CdTe 80	67.9	400-600	1100	
		523 CdTe 80	66.1	200-600	1000-1100	
	CIGS	966 CIGS 125	125.9	400-600	200	0.69 %
		931 CIGS 125	119.1	400	1100	
HIT Technology		40 HIT 210	207	600-800	200	0.36 %
		41 HIT 210	213	1000	200	
Maxeon Cell Technology		327	300	600	200,1000	0.10 %
Micromorph, micro crystalline /a-Si		170 MM 139	117.2	1100	200	1.09 %
		460 MM 139	120	1100	200	

Table .1 (Efficiency Variation at different Irradiance)

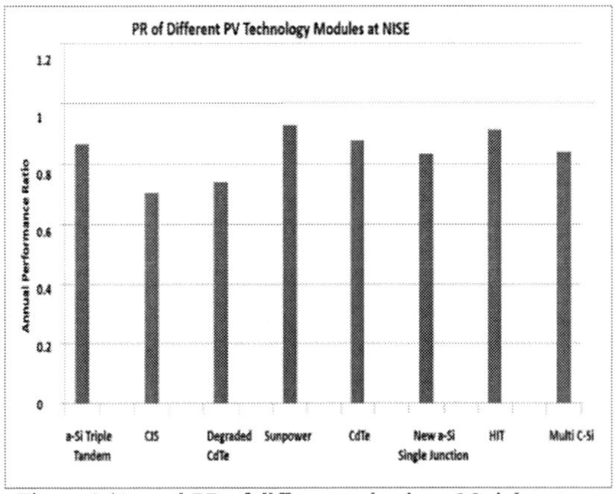

Figure.6 (Annual PR of different technology Modules at NISE

3. Conclusion

The efficiency variation with intensity of different PV technologies has been found to affect the PR values of these modules for the specific site (NISE). Similarly, variations of module temperature and temperature co-efficient estimated at different irradiance level will further append on the performance information. The above study is useful to estimate energy yield and PR for different PV technologies under various climatic zones. This study is most useful to predict energy yield of PV technology module for a specific site.

3.1 Acknowledgement:

The authors would like to acknowledge support and help provided from National Institute of Solar Energy, Gurgaon, An autonomous institute under Minister of New and Renewable Energy

4. References:

1. IEC 60904-9: Solar Simulator Performance Requirement.
2. IEC 61215:2005 International Standard Crystalline Silicon Terrestrial Photovoltaic Modules- Design Qualification and type approval.
3. IEC 61646: International Standard Thin Film Terrestrial Photovoltaic Modules- Design
4.O.S.Sastry et al; Performance Comparison of three Technology Modules under similaroutdoor condition; 26th EU PVSEC , Hamburg
5. Y. Ueda et al; comparison between I-V measurement and the performance in various kinds of PV technologies; 25th EU PVSEC, Valencia.
6.O.S.Sastry et al; Degradation Analysis of different PV modules after prolonged field Operation; 26th EU PVSEC , Hamburg .
7. Ulrike Jahn et al; Comparison of different thin film Technologies- performance characteristics obtained from laboratory and field test; 25th EU PVSEC, Valencia
8. O.S.Sastry et al; Degradation in PR and yields of exposed modules under arid condition; 26thEU PVSEC, Hamburg.

FORECASTING DEGRADATION RATES OF DIFFERENT PHOTOVOLTAIC SYSTEMS USING ROBUST PRINCIPAL COMPONENT ANALYSIS AND ARIMA

Andreas Kyprianou[1], Alexander Phinikirades[2], George Makrides[2] and George E. Georghiou[2]
FOSS Research Centre for Sustainable Energy
[1]Department of Mechanical and Manufacturing Engineering, University of Cyprus
[2]Photovoltaic Technology Group, Department of Electrical and Computer Engineering, University of Cyprus
75 Kallipoleos Avenue P.O. Box 20537, Nicosia, 1678, Cyprus

ABSTRACT: Degradation rates based on forecasting of performance ratio (PR), R_p, time series are computed and compared with actual degradation rates. A three year forecasting of monthly PR, measured from PV connected systems of various technolgies is performed using the seasonal ARIMA (SARIMA) time series model. The seasonal ARIMA model is estimated using monthly PR measured over a 5 year period and based on this model forecasting is implemented for the subsequent three years. The degradation rate at the end of the forecasting period, eighth year, is computed using a robust principal component analysis (RCPA) based methodology. The degradation rates obtained for various (PV) systems are then compared to the ones obtained using the actual eight year data.
Keywords: degradation, experimental methods, PV modules

1 INTRODUCTION

As the technology of photovoltaics advances and through grid connection are embedded in urban areas [1] the need for forecasting of their power output and performance becomes vital for both the stability of the grid system and the urban energy planning and policy making [2]. Over the years significant research has taken place in forecasting the power output, where many time series based forecasting methodologies ranging from neural networks to ARMA and ARIMA have been proposed. A comparison of time series based power output forecasting techniques can be found in [3]. A characteristic of the techniques compared in [3] is that they do not utilize exogenous inputs. Recently the use of ARMA with exogenous input, ARMAX, as a means of imroving forecasting by using verious metereological variables as exogenous inputs has been proposed [4].

An important factor that has not to be taken into consideration in forecasting power output is that of module PV degradation rate. A reason for this was both the lack of a well defined index of degradaion rate and methodology of its computation. A step towards this end using the robust principal component analaysis (RPCA) was taken in [5].

PV system degradation is due to an inexplicable combination of many environmental factors and parameters relating to their constituent instruments. Forecasting of PV degradation rates poses two serious challenges, that of forecasting the future PRs and the uncertainty associated with PR measurements.

The purpose of this work is to combine the seasonal ARIMA (SARIMA) model with the RPCA based methodology, proposed in [5], of computing PV degradation rates. The SARIMA model is used to forecast the PR values whereas the RPCA based methodology is used to deal with the uncertainty in the data.

The rest of the article is organized as follows. Section 2 at the beginning discusses how the data should be organized for forecasting and RPCA. Then it outlines the SARIMA and RPCA theory. In section 3 the degradation rate definition is given and based on this definition forecasting is carried out of which the results are also given. A discussion and the conclusions on these results are given in section 4.

2 THEORY AND METHODOLOGY

2.1 Data organization

The monthly PR time series of an eight-year period of eleven grid connected PV systems was used to forecast their degradation rates. Initially the 5 year monthly PR time series were used to estimate SARIMA models for each of the eleven PV systems. The estimated SARIMA models were then used to generate monthly PR forecasts for the next three years. The forecast time series are then appended to the 5 year monthly PR time series to generate eight-year time series of monthly PRs for each PV.

Then the RPCA methodology proposed in [5] was used to obtain the degradation rates at the end of the eighth year. In order to implement the RPCA methodology the 96-month PR time series for each photovoltaic module was divided into five 12-month time series. Subsequently the 12-month time series of each PV system are collected in a matrix **D** of dimension 8x12 called data matrix to be analyzed RPCA.

2.2 Seasonal Auto Regressive Integrating Moving Average Model (SARIMA)

SARIMA is composed by three different basic time series operators, namely the autoregressive, the moving average and the integrating operators. These operators are founded on the basic concepts of translating and differencing a time series. For a time series x_t they are respectively defined by [5,6],

$$\mathbf{T}x_t = x_{t-1} \quad (1)$$

and

$$\nabla x_t = (1-\mathbf{T})x_t \quad (2)$$

The autoregressive operator models the current value of a time series as sum of a linear superposition of its previous values and a random value e_t as follows,

$$x_t = r_1 x_{t-1} + r_2 x_{t-2} + ... + r_p x_{t-p} + e_t \quad (3)$$

which can be written in terms of a polynomial of the translation operator (1),

$$r(\mathbf{T})x_t = e_t \quad (4)$$

where,

$$r(\mathbf{T}) = 1 - r_1\mathbf{T} - r_2\mathbf{T}^2 - \ldots - r_p\mathbf{T}^p \quad (5)$$

is a pth order polynomial in the operator **T** known as the autoregressive operator.

On the other hand the moving average operator models x_t by a superposition of q previous values of e,

$$x_t = e_t - m_1 e_{t-1} \ldots - m_q e_{t-q} \quad (6)$$

which can be expressed in operator form,

$$x_t = m(\mathbf{T})e_t \quad (7)$$

where the qth order polynomial,

$$m(\mathbf{T}) = 1 - m_1\mathbf{T} - m_2\mathbf{T}^2 - \ldots - m_q\mathbf{T}^q \quad (8)$$

is known as the moving average operator.

The seasonal ARIMA model of seasonal data x_t of period s is obtained by combining the operators (2), (4) and (7) in the following manner [6],

$$r(\mathbf{T})r_s(\mathbf{T})\nabla^d \nabla_d^{d_s} x_t = m(\mathbf{T})m_s(\mathbf{T}^s)e_t \quad (9)$$

where $r_s(\mathbf{T})$ and $m_s(\mathbf{T})$ are autoregressive and moving average polynomials in $\mathbf{T}^s$ respectively, $\nabla^d = (1-\mathbf{T})^d$ is a non-seasonal differencing operator and $\nabla^{d_s} = (1-\mathbf{T}^s)^{d_s}$ a seasonal differencing operator. The values of the parameters defining the operators are estimated from the data.

Simulated realizations from year 5 to year 8 of the estimated SARIMA model for the Schott multi-c-Si system shown in the top plot of figure 1. The bottom plot of figure 1 depicts the forecast PR values of the SARIMA model for the same PV grid-connected system.

The forecast curve in the bottom plot of figure one corresponds to the mininum square error. It is curve corresponding to predicted PR values that, given the 60 measured PRs and model parameters, minimizes the square error.

It is shown in [6] that the forecast value, $\hat{x}_t(h)$, at an instat of h time steps ahead from a time instant t is the conditional expectation,

$$\hat{x}_t(h) = \mathrm{E}[x_{t+h} | \mathbf{m}, \mathbf{r}, \mathbf{m}_s, \mathbf{r}_s, x_t, x_{t-1}, \ldots] \quad (10)$$

where x_{t+h} is the random variable generated by the SARIMA model (9) at the future time instant $t+h$ **m**, **r**, $\mathbf{m}_s$, $\mathbf{r}_s$ are the coefficients of the polynomials $m(\mathbf{T})$, $r(\mathbf{T})$, $m_s(\mathbf{T})$, $r_s(\mathbf{T})$ respectively, and x_t, x_{t-1},... are actual PRs up to the time t.

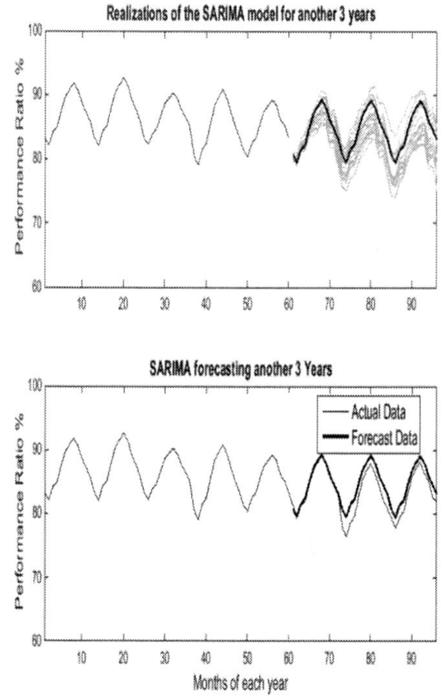

Figure 1: SARIMA model of the Schott Main multi-c-Si PV system. *Top* 20 realizations for the next three years of the estimated model. *Bottom* Forecast and actual PR Values for the next three years.

2.3 Robust Principal Component Analysis (RPCA)

The forecast data of each of the PV systems is appended to the original PR data for a five year period, bottom plot of figure 1and collected in a data matrix **D**. Then an RPCA analysis on **D** is carried out. In the context of the Robust PCA the data matrix **D** is expressed as:

$$\mathbf{D} = \mathbf{K} + \mathbf{E} \quad (11)$$

where, **K** is a low rank data matrix and **E** an unknown sparse perturbation matrix. Robust PCA extracts **K** by solving the following optimization program:

$$\begin{aligned} &\text{minimize} \quad \|\mathbf{K}\|_s + \mu\|\mathbf{E}\|_1 \\ &\text{subject to} \quad \mathbf{D} = \mathbf{K} + \mathbf{E} \end{aligned} \quad (12)$$

where, $\|\mathbf{K}\|_s$ is a norm defined by the sum of the singular values of **K** and $\|\mathbf{E}\|_1$ a norm defined by the summation of the absolute values of **E** and μ is a weighting parameter that trades off between the two norms. The robust PCA program was solved using the method of augmented Lagrange multiplier [8]. The robust data for each year is shown in Figure 2.

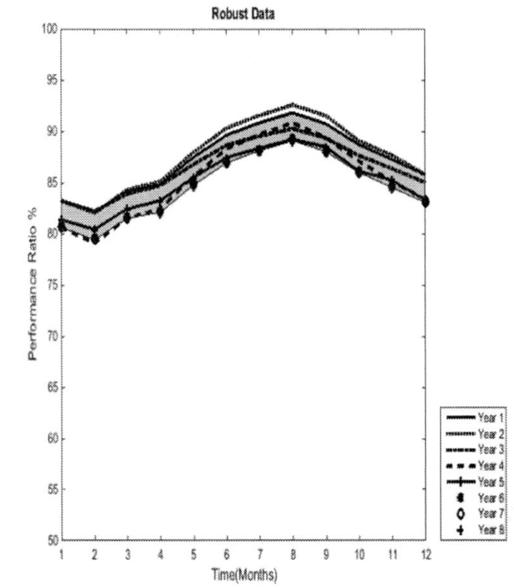

Figure 2: The yearly robust data

3 DEGRADATION RATE DEFINITION AND RESULTS

The robust PCA data is used to define the degradation rate of the PV system. The grey region in Figure 2 represents the eight-year PR decay of the forecast data. The percentage eight-year degradation rate is defined as the area of this region divided by the area beneath the curve of the robust data of the first year. The average annual percentage degradation rate is then defined as the percentage eight-year degradation rate divided by five. The computed forecast average annual degradation rate for the Schott main-c-Si system is 0.54 %/yr whereas the one computed in [5] using the actual data is 0.59. The forecast average annual degradation rates for all the different technology PV systems are listed in Table I together with those computed using the actual data [5]. The PV systems installed range from mono-c-Si and multi-crystalline Silicon (multi-c-Si), Heterojunction with Intrinsic Thin layer (HIT), Edge defined Film-fed Growth (EFG), Multi-crystalline Advanced Industrial cells (MAIN) to amorphous-Silicon (a-Si), cadmium telluride (CdTe), copper indium gallium diselenide (CIGS) and other PV technologies from a range of manufacturers.

Table I: Forecast degradation rates per PV system and the actual degradation rate [5].

PV System		Actual (%/yr)	Forecast (%/yr)
Solon *	multi-c-Si	0.83	0.56
Sanyo	mono-c-Si	0.76	0.97
Atersa	mono-c-Si	0.40	0.47
Sunpower	mono-c-Si	0.75	0.87
Schott EFG	multi-c-Si	0.59	0.54
BP Solar*	mono-c-Si	1.02	0.70
Solar World	multi-c-Si	1.03	0.93
Schott MAIN	multi-c-Si	0.73	0.68
Würth Solar	CIGS	2.33	1.94
First Solar	CdTe	1.96	2.10
MHI	a-Si	1.38	1.71

* The Solon multi-c-Si and BP Solar mono-c-Si systems were partially shaded during the 2nd, 3rd and 4th years

4 CONCLUSIONS

The study presented in this paper combines two methods to achieve degradation forecasting by at the same time dealing with the uncertainty that is incorporated in the measurement and computation of PRs. In order to circumvent the need of estimating uncertainty and then assessing its effect on SARIMA the RPCA methodology was employed. RPCA does not require any prior knowledge about the uncertainty that for example the environmental conditions and instrument tolerances generate. RPCA combined with SARIMA based forecasting give a methodology of degradation rate forecasting.

The degradation rates obtained from the forecast data were comparable with those obtained from the real data. Significant deviations between the forecast degradation rates and the ones obtained from the actual data occurred for those PV systems that were partially shaded for a long period during their operation. The forecast values for these PV systems were lower than the actual ones reflecting the fact that shading reduces the measured PR of the PV systems.

5 REFERENCES

[1] A. Strzalka, N. Alam, E. Duminil, V. Coors and U. Eicker, Applied Energy, vol. 2(1), pp. 413–421, 2012.

[2] A. Woyte, V. V. Thong, R. Bellmans and J. Nijs, IEEE Transactions on Energy Conversion, vol. 21(1), pp. 202-209, 2006.

[3] H. T. C. Pedro and C. F. M. Coimbra, Solar Energy, vol. 86, pp. 2017–2028, 2012.

[4] Y. Li, Y. Su and L. Shu, Renewable Energy, vol. 66, pp. 78–79, 2014.

[5] A. Kyprianou, A. Phinikarides, G. Makrides and G. E. Georghiou, IEEE Journal of Photovoltaics, vol. 5, pp. 1698–1705, 2015.

[6] G. E. P. Box and G. M. Jenkins, Time series analysis, forecasting and control, Prentice Hall, 1976.

[7] J. D. Hamilton, Time series analysis, Princeton University Press, 1994.

[8] L. Zhounchen, M. Chen, and Y. Ma, arXiv preprint, arXiv:1009:5055.

PERFORMANCE OF A MODULE AND DEFECT DETECTION ALGORITHM FOR AERIAL INFRARED IMAGES AS A FUNCTION OF THE FLYING ALTITUDE

Manuel Dalsass[1]*, Sergiu Deitsch[2], Philipp Luchscheider[1], Florian Gallwitz[2], Christoph J. Brabec[3,4]
[1]ZAE Bayern, Unterkotzauer Weg 25, 95028 Hof, Germany
[2]TH Nürnberg, Department of Computer Science, Hohfederstraße 40, 90489 Nürnberg, Germany
[3]ZAE Bayern, Haberstraße 2a, 91058 Erlangen, Germany
[4]FAU Erlangen-Nürnberg, i-MEET, Martensstraße 7, 91058 Erlangen, Germany
*Corresponding author, phone: +49 9131 9398 406, fax: +49 9131 9398 199, email: manuel.dalsass@zae-bayern.de

ABSTRACT: In recent years, remote-controlled unmanned air vehicles (UAV) have been used increasingly to attain aerial infrared (aIR) recordings of large photovoltaic (PV) power plants with which thermally suspicious PV modules can be identified and localized. But so far, aIR recordings are still analyzed mostly manually – a process which can be extremely time-consuming and error-prone. In this context, the performance of a module and defect detection algorithm for aIR images is evaluated as a function of the flying altitude. For this purpose, aIR recordings of a PV power plant are generated from six flying altitudes from 5 m to 30 m above ground level by means of a drone measurement system and used as data input for the image processing tool. The results indicate optimal flying altitudes for both module and defect detection. However, the most suitable flying altitude may be a compromise with further conditions. The findings of this work will advance future automatic monitoring and early detection systems.
Keywords: PV module defects, aerial thermography, automatic detection, monitoring

1 INTRODUCTION AND MOTIVATION

The market for photovoltaic (PV) systems has been growing tremendously in the last years. In Germany alone, about 1.5 million PV systems with an overall nominal power of approximately 40 GW have been installed at the end of 2015 [1]. Considering the necessary continuous maintenance of these PV systems, efficient quality assessment tools are needed.

In this context, infrared (IR) thermography is a contactless, non-destructive, and fast assessment tool which can be used to evaluate the quality of PV modules under real operating conditions. In IR images of PV modules, thermal abnormalities such as hot spots or hot areas indicate irregularities which are probably caused by defects [2-5]. In recent years, remote-controlled unmanned air vehicles (UAV) have been used increasingly to attain aerial IR (aIR) recordings of large PV power plants with which thermally suspicious PV modules can be clearly localized [6-7]. But despite the technical progress, aIR recordings of aIR-PV-checks of large PV power plants are still analyzed mostly manually – a process which can be extremely time-consuming and error-prone.

With regard to future reliable monitoring and early detection systems for PV power plants, this work aims at enhancing the evaluation software for aIR-PV-checks presented in [8] which detects thermal abnormalities in aIR recordings. This work is to combine the software implementation approach with practical experiences of the manual interpretation of thermal abnormalities.

2 EXPERIMENTAL PROCEDURE

2.1 Test Object

The test object for this study is a PV power plant located in northern Bavaria, Germany, and has a simple layout. On a spacious flat terrain, 6,560 polycrystalline PV modules with a nominal power of 235 W_p each are installed in parallel rows having a tilt angle of 25° in southern direction.

One PV module has 60 solar cells and three bypass diodes (each for 20 solar cells on the left side, the center, and the right side). One string consists of 20 modules connected in series.

2.2 Data Acquisition

For the generation of the aIR recordings, a remote-controlled octocopter with an extensive navigation system (DaVinci Copters Scarabot X8) is used (see Figure 1). Two lightweight cameras are installed beneath the octocopter, the IR camera Optris PI450 and the RGB camera GoPro Hero3+.

The observation angle of the cameras is almost perpendicular to the flat terrain. Since the tilt angle of the PV modules is 25°, the observation angle of the cameras towards the glass surface of the PV modules is smaller than 30° as recommended in [9].

To achieve not only usable, but also challenging aIR data for the detection algorithms, the measurement conditions do not need to be optimal. Thus, lower irradiation values down to 400 W/m², maybe due to a rather cloudy sky, are permitted, too.

Figure 1: The remote-controlled octocopter with integrated IR and RGB cameras used for the generation of the aIR data.

2.3 Approach

Altogether, 20 radiometric aIR data sets of different spots of the PV power plant are generated from six flying altitudes above ground level (5 m, 10 m, 15 m, 20 m, 25 m, and 30 m) using the remote-controlled drone. The selected spots for the flying altitude analysis are to represent typical aIR images of PV power plants that have a layout similar to the test object. The spots include central areas and marginal areas with both healthy and defective modules. The orientation of the rows within an aIR image varies between the three alignments: horizontal (about 0°), diagonal (about 45°), and vertical (about 90°).

The generated aIR images are used as input data for the module and defect detection algorithms presented in [8]. Subsequently, the performances of these algorithms are evaluated by comparing their results with the results of the manual evaluation of the same data set. Finally, mandatory and practical measurement conditions are taken into account and discussed.

3 RESULTS AND DISCUSSION

3.1 Measurement Conditions

The aIR images were recorded on June 4th, 2016, around noon when the solar irradiation values ranged from 450 W/m² to 850 W/m² due to a scattered cloud cover (see Figure 2). There was almost no wind and the ambient temperature was about 23°C.

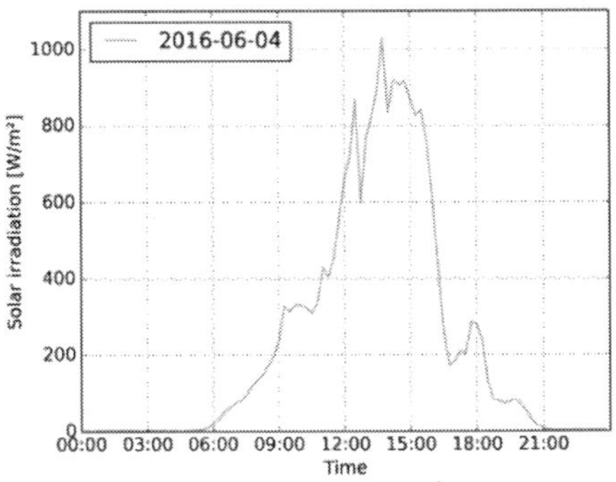

Figure 2: Solar irradiation on June 4th, 2016.

3.2 Automatic Detection of Modules

Exemplary for the aIR data sets, Figure 3 shows the aIR images and the results of the module detection algorithm of a marginal area of the PV power plant from the six flying altitudes. The inactive substring on the right side of the central module is noticeable from all six flying altitudes. Three tendencies can be noticed.

Firstly, the algorithm clearly separates the background from the PV modules. Thus, modules in marginal areas are detected reliably. As seen in Figure 4, the results of the module detection algorithm are appropriate even for areas between two segments of the PV power plant (left image) as well as for areas with other rectangular objects nearby the module rows (right image).

Secondly, the results of the algorithm strongly depend on the quality of the aIR images. If parts of a module row are too blurred or distorted in the aIR image, the algorithm detects multiple modules as one module (see Figure 3). In these cases, the separating area between two neighboring modules cannot be identified. The reason for blurred or distorted areas in the aIR image could be the quality of the optical lens of the IR camera. In the analyzed aIR images generated with the IR camera Optris PI450, the image areas on the right side seem to be less focused.

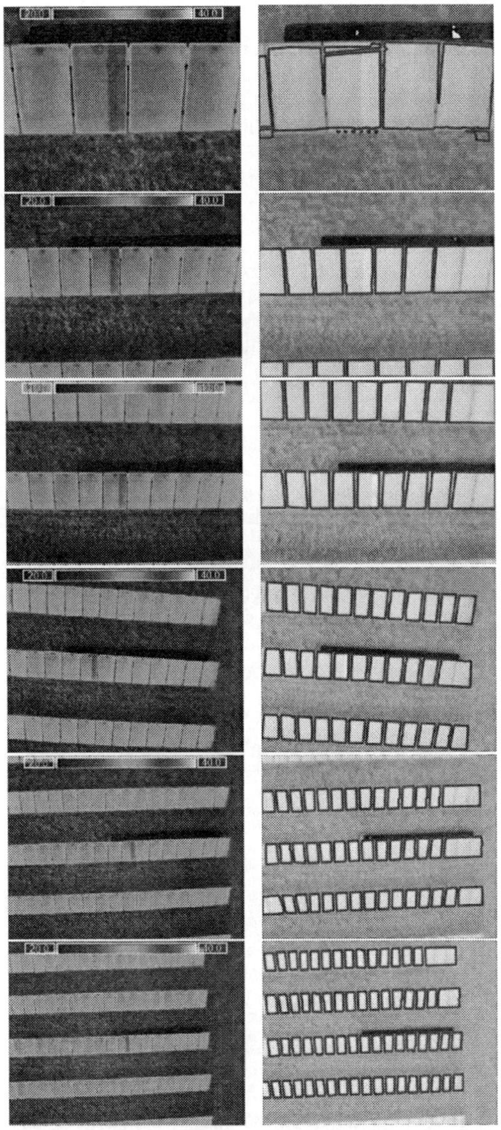

Figure 3: The aIR images displayed in false color (left column) and the results of the module detection algorithm (right column) of a marginal area of the PV power plant from the six flying altitudes (from top to bottom: 5 m, 10 m, 15 m, 20 m, 25 m, 30 m; horizontal alignment). The inactive substring is noticeable from all flying altitudes.

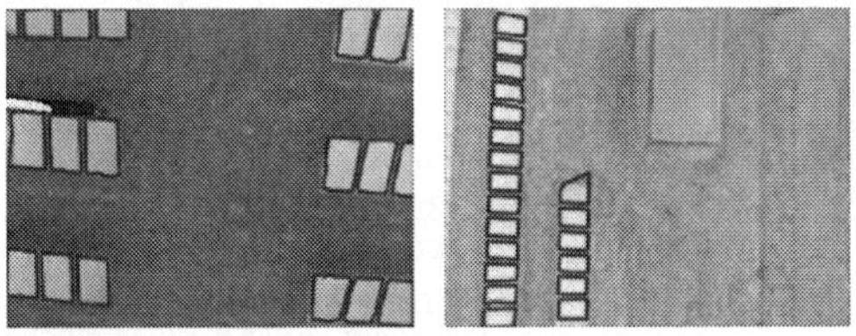

Figure 4: Special cases of marginal areas of the PV power plant – module detection results of an area between two segments of the PV power plant (left image; flying altitude 15 m; horizontal alignment) and of an area with one other rectangluar object nearby the module rows (right image; flying altitude 30 m; vertical alignment).

Thirdly, the performance of the algorithm is not satisfactory for the aIR image taken from a flying altitude of 5 m. Although the aIR image is of excellent quality, none of the modules is detected correctly. The reason for this behavior is probably that the separating areas between two neighboring modules are too large for the implemented separator function in these cases. However, a modification of the separator function could improve the performance of the algorithm for low flying altitudes.

For the evaluation of the performance of the module detection algorithm, the aIR images are first analyzed manually in terms of identifying all PV modules with a visible module area of greater than 95 %. By this means, all modules not suitable for a solid thermographic analysis are discarded at this point. Considering only the suitable PV modules, the results of the module detection algorithm are compared to the results of the manual analysis. To give an example, Table I summarizes the evaluation of the performance of the module detection algorithm for the six aIR images shown in Figure 3.

Table I: Evaluation of the performance of the module detection algorithm for the six aIR images shown in Figure 3. Only modules with a visible module area of greater than 95 % are considered.

Flying altitude	Manual detection	Automatic detection		
		Correct	Incorrect	Rate
5 m	4	0	4	0 %
10 m	7	5	2	71 %
15 m	20	16	4	80 %
20 m	36	34	2	94 %
25 m	47	38	9	81 %
30 m	68	63	5	93 %

Figure 5 illustrates the detection rates of the module detection algorithm for all 20 aIR data sets as a function of the flying altitude. The data sets are subdivided into the three alignment groups of the module rows in the aIR images (horizontal alignment: 8 data sets; diagonal alignment: 5 data sets; vertical alignment: 7 data sets).

At first glance, it can be noticed that the detection rates for flying altitudes of 5 m, 10 m, and 15 m range from 0 % to 100 %. In the case of the flying altitude of 15 m, it can be assumed that the detection rate of 0 % is an exception because all other 19 detection rates for that flying altitude are greater than 70 %. For the flying altitudes of 5 m and 10 m, however, 27 of 40 detection rates are smaller than 70 %. Thus, the performance of the algorithm for aIR images taken from relatively low flying altitudes is not reliable and needs to be improved. In contrast, detection rates greater than 75 % are achieved for aIR images taken from the flying altitudes of 20 m and 30 m. In the cases of the aIR images taken from a flying altitude of 25 m, the detection rates are even greater than 80 %. Thus, the performance of the module detection algorithm for aIR images taken from relatively high flying altitudes seems to be reliable.

In Table II, the average detection rates of the module detection algorithm for the 20 aIR data sets are listed for the six flying altitudes as well as for the three alignment groups. For the given measurement setup, exceptionally high detection rates of 99 % and 98 % are achieved for aIR images with vertically oriented module rows generated from flying altitudes of 20 m and 25 m, respectively. Consequently, these parameters can be recommended for similar aIR-PV-check measurement setups.

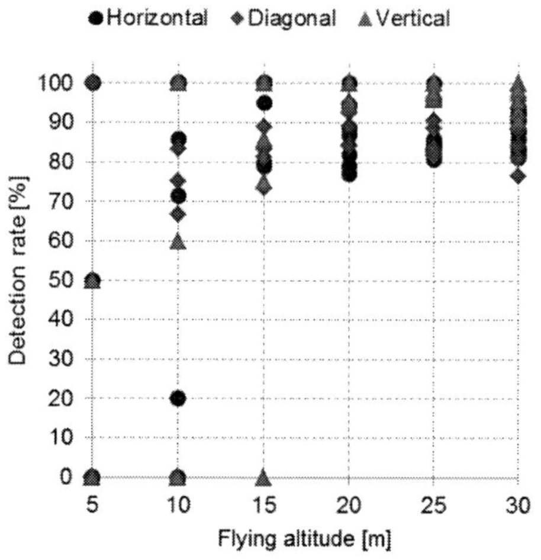

Figure 5: The detection rates of the module detection algorithm for all 20 aIR data sets as a function of the flying altitude. The data sets are subdivided into the three alignment groups (horizontal, diagonal, and vertical).

Table II: Average detection rates of the module detection algorithm for the six flying altitudes and for the three alignment groups (horizontal alignment: 8 data sets; diagonal alignment: 5 data sets; vertical alignment: 7 data sets). Again, only modules with a visible module area of greater than 95 % are considered.

Flying altitude	Automatic detection of modules Average detection rates			
	Horizontal	Diagonal	Vertical	Overall
5 m	19 %	20 %	7 %	15 %
10 m	60 %	65 %	51 %	58 %
15 m	94 %	85 %	80 %	87 %
20 m	88 %	89 %	99 %	92 %
25 m	87 %	85 %	98 %	91 %
30 m	87 %	84 %	96 %	90 %
Overall	72 %	72 %	72 %	72 %

3.3 Automatic Detection of Thermal Abnormalities

Based on the results of the module detection algorithm, all identified modules within an aIR image are extracted and saved in one data container which is used as input for the abnormality detection algorithm. All aIR images taken from a flying altitude of 5 m are not considered any further since the detection results of the module detection algorithm are unusable.

The purpose of the abnormality detection algorithm is the reliable identification of modules with any significant temperature abnormality regardless of its spatial pattern. A classification of the temperature abnormality as summarized in [10] is not considered at this point. Regarding the two aIR images in Figure 6, the algorithm should label exactly one module in each aIR image as thermally abnormal – the module with the inactive substring in the left image (central module in the left module row) and the module with the hot spot in the right image (central module in the second module row from the left side).

Altogether, 12 aIR data sets contain thermal abnormalities (3 modules with inactive substrings and 9 modules with hot spots). Accordingly, in each of the 60 aIR images there is exactly one module with a thermal abnormality which is to be identified by the algorithm.

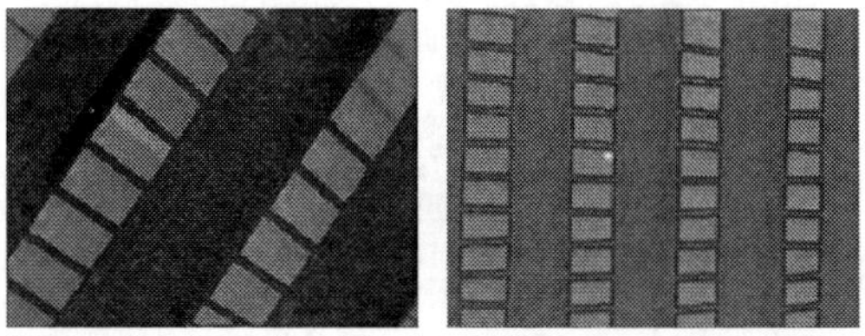

Figure 6: Examples for thermally abnormal modules which need to be detected by the abnormilty detection algorithm: module with an inactive substring (left image; flying altitude 10 m; diagonal alignment) and module with a hot spot (right image; flying altitude 20 m; vertical alignment).

The evaluation of the performance of the abnormality detection algorithm is based on the F_1 score. Defined as a weighted average between precision and recall, the F_1 score is a measure of accuracy which can be used as a single performance test for positive classifications. Its best value is 100 %, its worst value 0 %.

The calculated F_1 scores for the remaining five flying altitudes and for the three alignment groups are summarized in Table III. In all but four cases, the F_1 scores are exceptionally high (greater than 99 %). These cases also include the recommended parameters for similar aIR-PV-check measurement setups with regard to results of the module detection algorithm, namely aIR images with vertically oriented module rows taken from a flying altitude from 20 m and 25 m. The four cases for which the F_1 score is smaller than 99 % occur only for the flying altitudes of 10 m and 15 m. Hence, it can be concluded that the algorithm performs best for aIR images taken from flying altitudes ranging from 20 m to 30 m.

Table III: F_1 scores of the abnormality detection algorithm for the remaining five flying altitudes and for the three alignment groups (horizontal alignment: 8 data sets; diagonal alignment: 5 data sets; vertical alignment: 7 data sets).

Flying altitude	Automatic detection of thermal abnormalities F_1 scores		
	Horizontal	Diagonal	Vertical
10 m	94.4 %	100.0 %	97.6 %
15 m	97.5 %	97.6 %	100.0 %
20 m	100.0 %	100.0 %	99.3 %
25 m	99.6 %	99.4 %	99.6 %
30 m	99.7 %	99.5 %	99.7 %
Overall	99.4 %		

3.4 Mandatory and Practical Measurement Conditions

The results of the detection algorithms indicate that there are favorable flying altitudes for the module detection (20 m to 25 m) and for the abnormality detection (30 m). Nevertheless, mandatory and practical measurement conditions have to be taken into account as well so that the most suitable flying altitude of the drone measurement system may be a compromise with further conditions.

As for mandatory measurement conditions, local air traffic laws and rights have to be obeyed in any case. In Germany, for example, general flight permissions can be granted for commercial purposes – however, exceptional flight permissions for both private and commercial purposes are necessary if the weight of the UAV exceeds 5 kg or if the PV power plant is located nearby airports or public places [11]. A valid liability insurance is required by law for any UAV.

As for practical measurement conditions, it must be noted first that, especially with regard to legal procedures, the geometric resolution of a solar cell in an IR image should be at least 5 x 5 pixels as recommended in [12]. Since the used IR camera has a geometric resolution of 382 x 288 pixels and a field of view (FOV) of 38° x 29°, the validity of the aIR images taken from a flying altitude of 30 m is limited for this reason. For the given measurement setup, the minimum geometric resolution is ensured for flying altitudes up to about 25 m. However, it can be assumed that IR cameras with higher geometric resolutions facilitate the generation of valid aIR recordings taken from flying altitudes greater than 25 m.

Although the geometric resolution of a solar cell in an aIR image inherently increases with shorter observation distances, aIR-PV-checks performed at flying altitudes smaller than 10 m above ground level become challenging.

In the case of programmed way point flights, the actual flight path greatly depends on the spatial inaccuracies of the used external navigation system. The standard positioning service (SPS) of the global positioning system (GPS), for example, achieves a currently guaranteed horizontal accuracy of about 8 m respectively 4 m root mean square (RMS) [13]. Therefore, even a precisely programmed way point flight does not ensure that the module rows are overflown as requested to generate suitable aIR recordings.

Also, the spatial orientation within large PV power plants with monotonous layouts is complicated for flying altitudes smaller than 10 m and the given FOV of the IR camera because the number of distinctly detectable features in the aIR images decreases. This fact concerns both the manual analysis and automatic analysis tools.

Furthermore, the thermal reflection of the drone measurement system on the module surfaces is more likely for shorter observation distances. Although these hot spots are normally moving on the module surfaces in aIR videos, they can lead to misinterpretations when analyzing single aIR images. Therefore, any thermal reflections of the measurement system should be avoided.

In contrast, a flying altitude ranging from 20 m to 25 m has one important advantage. The flight planning is very flexible since at least two or even three module rows are captured completely in the aIR recordings. For example, this circumstance can be used to shorten the flight path to a certain extend (one-way flight for two or even three module rows). Thus, flight path and flight speed can be easily adapted to the layout of the PV power plant. By this means, the overall measurement time can be reduced significantly.

As described in [14], it should be noted at this point that the temperature values of thermal abnormalities detected with an IR camera decrease for greater observation distances as a result of the decreasing geometric resolution. For observation distances smaller than 30 m, this effect can practically be neglected.

4 SCIENTIFIC RELEVANCE

This work provides a recommendation for a suitable flying altitude of drones with regard to the development of fully automatic aIR-PV-checks of large PV power plants. The consideration of given measurement conditions as well as practical experiences of manual analyses facilitates software implementations to produce reliable results for maintenance services.

5 SUMMARY AND OUTLOOK

With regard to future automatic analyses of aIR recordings of large PV power plants, the performance of the module and defect detection algorithm presented in [8] is evaluated as a function of the flying altitude. In this context, the spatial orientation of the module rows in the aIR images as well as mandatory and practical measurement conditions are taken into account. For aIR-PV-checks with similar measurement setups, the detection algorithms perform best for aIR images with vertically oriented module rows which are generated from flying altitudes from 20 m to 25 m.

Although the detection algorithms provide reliable and satisfactory results for the given measurement conditions, their performances need to be evaluated for PV power plants with different layouts and for IR cameras with higher geometric resolutions. Also, the classification of thermal abnormalities may be integrated.

6 ACKNOWLEDGEMENTS

The research project "Smart Grid Solar" is co-financed by the European Union through the European Regional Development Fund and by the Free State of Bavaria.

This work was partly supported by the Research Training Group 1773 "Heterogeneous Image Systems" funded by the German Research Foundation (DFG) and by the Energie Campus Nürnberg (EnCN).

The authors thank the company IBC Solar AG for the possibility to use one of their PV power plants as a test object for this study.

7 REFERENCES

[1] Fraunhofer-Institut für Solare Energiesysteme ISE, Aktuelle Fakten zur Photovoltaik in Deutschland – Fassung vom 25.12.2015 (2015), in German.

[2] E. Molenbroek, D. W. Waddington, K. A. Emery, Photovoltaic Specialists Conference 1991, Conference Record of the 22nd IEEE (1991) 547.

[3] W. Herrmann, W. Wiesner, W. Vaassen, Photovoltaic Specialists Conference 1997, Conference Record of the 26th IEEE (1997) 1129.

[4] J. Wohlgemuth, W. Herrmann, Photovoltaic Specialists Conference 2005, Conference Record of the 31st IEEE (2005) 1062.

[5] C. Buerhop, D. Schlegel, M. Niess, C. Vodermayer, R. Weißmann, C. J. Brabec, Solar Energy Materials and Solar Cells 107 (2012) 154.

[6] C. Buerhop, R. Weißmann, H. Scheuerpflug, R. Auer, C. Brabec, Proceedings 27th European Photovoltaic Solar Energy Conference (2012) 3370.

[7] C. Buerhop, H. Scheuerpflug, Proceedings 29th European Photovoltaic Solar Energy Conference (2014) 2975.

[8] S. Dotenco, M. Dalsass, L. Winkler, T. Würzner, C. Brabec, A. Maier, F. Gallwitz, Automatic Detection and Analysis of Photovoltaic Modules in Aerial Infrared Imagery, 2016 IEEE Winter Conference on Applications of Computer Vision (WACV) (2016).

[9] ZAE Bayern e. V., Abschlussbericht der Machbarkeitsstudie zur Überprüfung der Qualität von Photovoltaik-Modulen mittels Infrarot-Aufnahmen (2007), in German.

[10] IEA-PVPS, Review of Failures of Photovoltaic Modules (2014).

[11] Bundesministerium der Justiz und für Verbraucherschutz, Luftverkehrs-Ordnung vom 29. Oktober 2015 (BGBl. I S. 1894) (2015), in German.

[12] Bundesverband für Angewandte Thermografie e. V., VATh-Richtlinie Elektrothermografie, Teil B: Niederspannung (Fassung Mai 2014) (2014), in German.

[13] United States of America, Department of Defense, Global Positioning System Standard Positioning Service Performance Standard (4th Edition) (2008).

[14] C. Buerhop, H. Scheuerpflug, Proceedings 29th European Photovoltaic Solar Energy Conference (2014) 3260.

A SIMULATION BASED OPTICAL AND ELECTRICAL APPROACH TO ESTIMATE ENERGY YIELD OF VARIOUS DESIGNS OF CURVED MODULES

Hamed Hanifi[1,2]*, Charlotte Pfau[1], Jens Schneider[1], Joerg Bagdahn[1,2]
[1]Fraunhofer-Center for Silicon Photovoltaics CSP, Otto-Eissfeldt-Str. 12, 06120 Halle (Saale), Germany
[2]Anhalt University of Applied Sciences, Bernburger-Str. 55, 06366, Koethen, Germany
*Corresponding author: +49 345 5589 5515, hamed.hanifi@csp.fraunhofer.de

ABSTRACT: The concept of sustainable structures is tightly involved with the architectural designs. Photovoltaic is the cheapest way to generate small amount of electricity anywhere. When PV follows functional or aesthetic designs, curved shapes can become desirable. PV modules with curved shapes can follow different forms over the structures and generate electrical power. In this paper, we have simulated the radiation power over 60-cell modules with different angle of curvature by considering the location and the elevation of the modules. In the next step, we have simulated two different module interconnection layouts to investigate the performance of each module at specific levels of curvature. We show that after a range of curvature, it is better to use parallel coupled strings and change the interconnection of the modules to reduce current mismatch over the strings. In this study, we show that for the modules with the 8 degrees tilt angle between the cells, it is better to use standard series connected cells. For higher levels of curvature it is necessary to rearrange the interconnections. The module with parallel connected strings does not have bypass diodes and it can save up to 125% energy compared to the conventional one in high levels of curvature.
Keywords: Modelling, Simulation, Curved Modules, Building Integrated PV (BIPV), PV Module, Module Technology, Module Manufacturing, Module Interconnections, LTSPICE

1 INTRODUCTION

New visions in architecture look to have more sustainable buildings beside eye appealing designs. Sustaining the building is important due to high consumption of energy in buildings. Higher tax rates or energy price in different countries over fossil based fuels and emerge of renewable energies as a sustainable and decentralized source of energy affect the market to put this source of energy under attention [1]. Photovoltaic is a promising technology to integrate energy generators on obstacles such as buildings or electrical cars to have higher sustainability.

Previous researches explain that a PV system is an effective solution to reduce electricity cost and reduce carbon dioxide emission [2]. PV generators can be installed as facades or modules in buildings or structures for power production purposes beside architectural innovations. Shapes and sizes of structures vary upon their applications. PV cells can be integrated in large size structures such as football stadiums as well as small scale applications such as electrical cars. However, Integration of PV modules in structures should follow different shapes and designs. Curved modules are a solution to overcome non-linear shapes in buildings or other structures. However, fabrication and performance of such modules should be taken into account. As an example, the integrated modules in conventional car- or bike-ports are only located on the rooftop of the structure. This shape limits architectural designs and the active area of the integrated PV generator. However, using curved modules can increase the generated energy by the PV generators as well as increasing eye appealing structural designs. **Figure 1**-2 show a golf car with curved rooftop and a bike port with curved PV modules.

The challenges of curved PV modules can be divided to the fabrication process and the electrical performance. Production process of curved modules should be well adjusted to avoid cell breakage due to bending angles and increase the reliability of the module. Furthermore, conventional connection of PV modules might not be a proper design for different structures. The integrated solar cells in the module receive different level of irradiation depending on the location and the curvature angle between different integrated cells. This mismatch between the current generated by solar cells in a string limits the output power of the modules considerably. By novel module technologies and rearranging the module designs, optimum power can be harvested from the curved module as well as epic architectural design.

Figure 1: A golf car with curved module over the rooftop by Solarcar

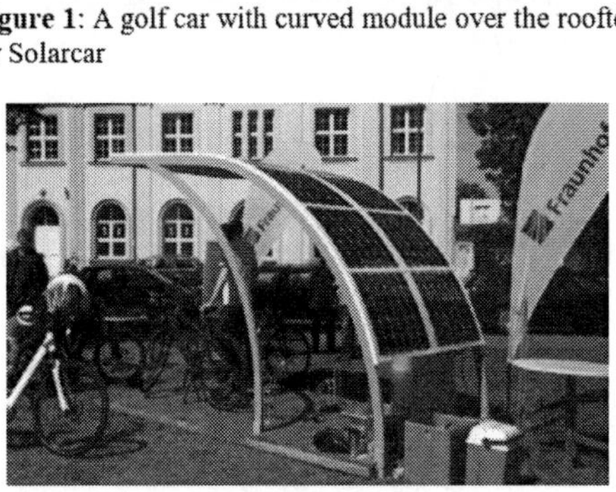

Figure 2: Curved shape bike port by Fraunhofer CSP presented in Saxony-Anhalt days, Koethen

In this work, we have simulated optical and electrical properties of curved modules with different angle of curvature to provide some general rules for designers.

THEORY AND SIMULATION

1.1 Optical simulation

An analytical model is developed to simulate the optical characteristics of curved modules depending on their location. The calculations are for 60 cell modules with two different interconnection designs and different levels of curvature over the 10 cell axis. ω represents the tilt angle between adjacent cells as indicated in **Figure 3**. The module bending angle is varied between 0 degrees (ω = 0°) for the flat module to 144 degrees (ω = 16°) for the extreme case curved module.

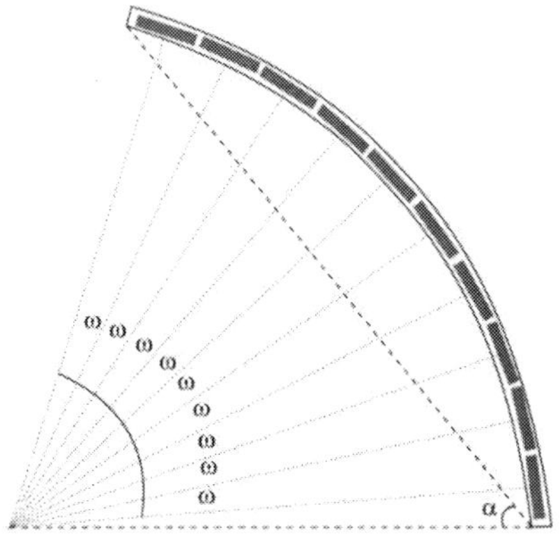

Figure 3: An example for curved module with ω = 8° tilt per cell. α is the tilt angle between the module and the earth.

The calculations are done for equinox (around 21st of March and September) at a place with latitude of 50°N. All modules are lifted up by α = 50° towards south, thus the flat module or the base of the curved module faces the sun perpendicularly at solar noon which is the optimum tilt angle. Radiation power per unit area is determined for relative sun position at different time, location and individual cell tilt/orientation by considering atmospheric and projection losses as well as the variation of transmissivity of the glass/encapsulation with the incident angle.

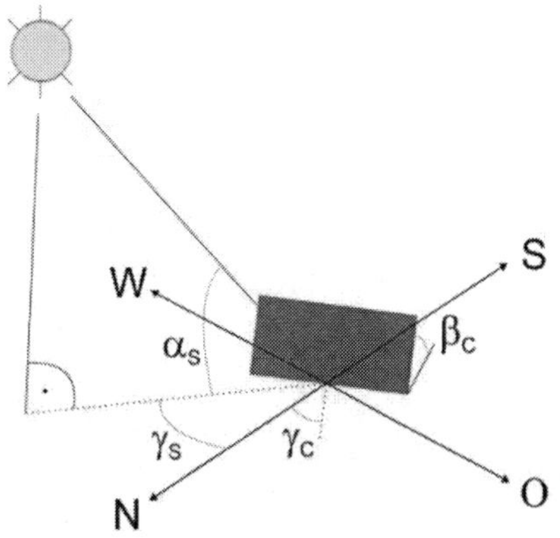

Figure 4: Apparent position of the sun relative to a point on a cell by considering the orientation and tilt angle.

Figure 4 demonstrates the positioning of the cell to the sun where α_s and γ_s are the elevation and azimuth of the sun, respectively, and they are dependent to the location, date and the time. β_c and γ_c represent the cells tilt and azimuth subsequently. By definition, the elevation is 0° for the horizontal case and 90° for the vertical case. The azimuth is 0° for the north direction, 90° for east, 180° for south and 270° for west.

To determine the radiation power over cells, solar power above the atmosphere is reduced by the factors mentioned above:

$$S_{cell} = S_0 \cdot (f_{AM} \cdot f_{Projection} \cdot f_{Glass/encap}) \tag{1}$$

where S_{cell} is the radiation power density reaching the solar cell, S_0 is the radiation power density hitting the outer surface of the earth's atmosphere, f_{AM} is the air mass loss factor, $f_{Projection}$ represents the projection loss and $f_{Glass/encap}$ gives the loss in the module front glass and encapsulation. Here, the value of the solar constant is used for the initial radiation power, i.e. $S_0 = 1353\ W/m^2$.

The air mass AM quantifies the power loss of light as it passes the earth atmosphere due to absorption by air and particles. Considering the curvature of the atmosphere, the air mass is calculated by [3])

$$AM = \frac{1}{\cos(90°-\alpha_s)+0.50572\,(96.07995-(90°-\alpha_s))^{-1.6364}} \tag{2}$$

The power reduction in the atmosphere can be calculated from the air mass using [4]:

$$f_{AM} = 0.7^{AM^{0.678}} \tag{3}$$

When the cell surface and the incident light are not perpendicular to each other, the power density that hits the surface is further reduced, since the incoming light is spread over a larger area. For arbitrary sun position, cell orientation and tilt, the power density on the cell surface can be determined via [5]:

$$f_{Projection} = [\cos(\alpha_s)\sin(\beta_c)\cos(\gamma_c - \gamma_s) + sin(\alpha_s)\cos(\beta_c)] \tag{4}$$

The power of the radiation hitting the module surface is further reduced by optical reflection and absorption losses in the front materials of the solar module stack, i.e. the front glass and encapsulation. The resulting transmissivity decreases with increasing incident angle. Using the transmissivity τ of a typical front stack as a function of the incident angle α_{in}, the resulting power loss factor $f_{Glass/encap}$ is calculated for arbitrary relative sun positions, cell orientations and tilts.

$$f_{Glass/encap} = 1 - \tau(\alpha_{in}) \tag{5}$$

where α_{in}, is given by α_s, γ_s, β_c and γ_c.

1.2 Electrical simulation

An electrical model is developed in SPICE to investigate the performance of curved modules with different irradiation level for each degree of curvature between the cells. The model is based on two diode model considering the resistance of string connections in the module [6, 7, 8]. Depending on the level of curvature and orientation of the module, irradiation is set based on the optical simulation results.

The electrical parameters used for the model, are derived from a mono-crystalline solar cell with three busbars. The measured parameters are given in **Table 1**, where A is the area of the cell, J_{ph} is the ideal photo-current density, n_1 and n_2 represent the quality factors of the diffusion and the recombination diodes, J_{01} and J_{02} are the dark saturation current densities of the diffusion and the recombination diodes, R_s and R_{sh} are the series and shunt resistance densities subsequently, V_{oc} is the open-circuit voltage and finally FF is the fill factor of the cells [7].

Table 1: The electrical parameters of the mono-crystalline solar cells according to the electrical measurements [7]

Cell parameters	Value
A	243.36 cm^2
J_{sc}	0.036 A/cm^2
n_1	1
n_2	2
J_{01}	6.20×10^{-13} A/cm^2
J_{02}	7.30×10^{-9} A/cm^2
R_s	0.750 $\Omega\cdot cm^2$
R_{sh}	100 $k\Omega\cdot cm^2$
Voc	0.641 V
FF	78.30 %

String connections are copper leaded by SnPbAg with the resistivity of $\rho = 1.53\times10^{-5}$ $\Omega\cdot$mm and cross section of 0.2×5 mm^2 [7].

We have compared two different module designs for the curved modules. The modules layouts are shown in **Figure 5**. Each block in module layouts represents one string with series connected solar cells and defined polarization. **Figure 5.(a)** demonstrates the electrical interconnection of a curved module with standard series interconnection with 6 strings, 10 cells per string and three bypass diodes. The cells within each string are elevated with an angle to the neighboring cell. **Figure 5.(b)** shows the electrical interconnection of a module with series and parallel connection. The module has 10 strings, each with 6 cells. The strings are connected to each other two by two in series and the series connected strings are coupled in parallel. The outgoing string connectors sum up in middle to prevent high current flow and reduce the electrical losses. All cells in a string have the same angle of curvature and each string is elevated with an angle to the neighboring string. The parallel connection of strings leads to independent performance of each series strings and allows the module to perform without bypass diodes.

Finally, both modules are simulated under different irradiation levels over the integrated cells. To have a better understanding of module performance over the day, the energy yield of each module is calculated during the day on 21st of March by **Equation 6**. For simplification purposes, the effect of temperature for energy yield calculation is neglected.

$$EY = \sum P_{module} \cdot t \qquad (6)$$

Where EY is the energy yield in kWh, P_{module} is the module power in kW and t is the period of the time in hours that module receives solar radiation. On 21st of March sun rises and sets at 6 a.m. and 6 p.m., respectively, defining this period..

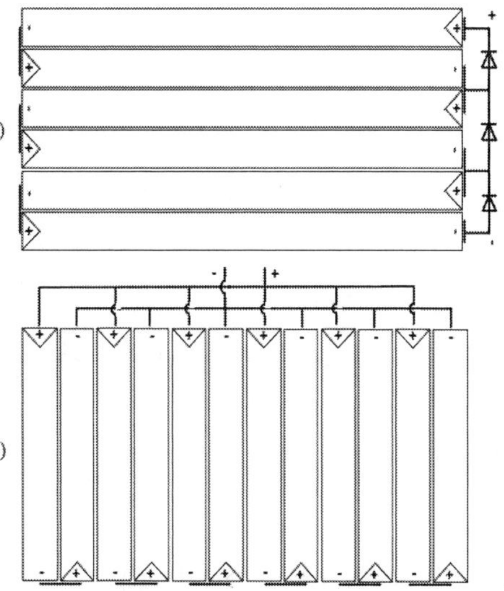

Figure 5: (a) Module layout with standard series connections and three bypass diodes. A row of cells with the same angle are distributed in all strings. (b) Module layout with parallel and series connection without bypass diode. Each row of cells with the same angle are located in one string. The string connectors are gathered in the center to avoid high power losses due to high current passing through the connectors.

2 RESULTS

As explained in theory section, the radiation power of the sun decreases by different factors. These factors are calculated for each solar cell in the module with specific tilt angle.

Figure 6 demonstrates the reduction factors by atmosphere, projection and the reflection in air/glass interface for a cell with $\omega = 8°$ tilt angle in a module facing south with $\alpha = 50°$ on 21st of March.

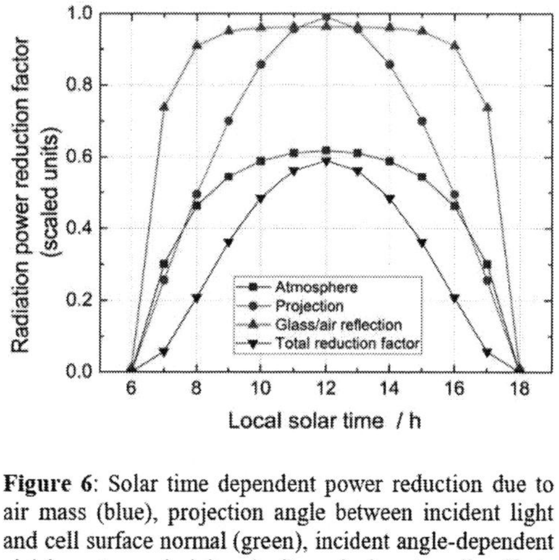

Figure 6: Solar time dependent power reduction due to air mass (blue), projection angle between incident light and cell surface normal (green), incident angle-dependent air/glass transmissivity (red) and due to all effects together (black) calculated for an individual cell which is tilted by $\omega = 8°$ relative to the south facing module with $\alpha = 50°$(i.e. $\gamma_c = 42°$, β_c=180°). $f = 1$ corresponds to 1353 W/m^2.

By deriving the reduction factors per hour for each solar cell with different tilt angle in the module, the radiation power over the module is calculated.

In the next step, the electrical characteristics of both module layouts shown in **Figure 5** are determined. The detailed information of the module characteristics under standard test conditions (STC) are shown in **Table 2**, where P_{mpp}, V_{mpp} and I_{mpp} are the power, voltage and current of PV modules at maximum power point respectively. V_{oc} is the open-circuit voltage, I_{sc} is the short-circuit current and finally, FF represents the fill factor of the modules.

Table 2: Electrical characteristics of series and parallel connected PV modules without curvature (flat) at standard test conditions

STC	Series connection	Parallel connection
P_{mpp}	266.97 W	250.94 W
V_{oc}	38.60 V	7.72 V
I_{sc}	8.76 A	43.08
FF	0.78	0.74
V_{mpp}	32.19 V	6.09 V
I_{mpp}	8.29 A	41.20 A

The current passing through the cell connectors and string connectors is the same for both modules. However, the flat module with parallel connected strings (see **Figure 5.(b)**) has a main string connector which conducts 4.9 times more current, which leads to higher electrical losses. These losses decrease the output power of the parallel connected module by 6% compared to the flat module with series connected strings at standard test conditions.

Figure 7 demonstrates the energy yield of the parallel connected and series connected strings. For the level of curvature below 8 degrees, the module with series connected strings shows better performance compared to the module with parallel connected strings. The short-circuit current of each string in the module with series connected strings is limited by the top and bottom cells, which receive the minimum irradiation compared to the other rows. In this range of curvature, the amount of irradiance distributed over the module and received by each row of the cells is close to the central row which leads to slight decrease in power ratio by almost 15% of the related flat module (see **Table 3**). However, with higher degree of curvature between the rows, bottom and top rows receive considerably lower irradiation, which limits the total short-circuit current of the strings considerably and leads to significant power drop. This power drop for a curvature level of 16 degrees is almost 73.5% of the related flat module (see **Table 3**).

Table 3: Energy yield ratio of curved modules with different interconnection to the related flat module

Angle of curvature 'ω' [°]	Energy yield ratio [%]	
	Parallel	Series
4	97.88	97.44
8	91.03	85.56
12	78.78	61.15
16	62.35	26.55

On the other hand, for the range of curvature below $\omega = 8°$, the output power of the curved module with parallel coupled strings is slightly lower than the series connected strings layout due to higher electrical losses in the junction tab. This power loss is 3.8% in case of flat modules. However, for the curvature range over $\omega = 8°$, the parallel coupled module considerably better performance. For the extreme case with $\omega = 16°$, almost 125% higher energy can be gained (see **Table 4**). Connection of strings in parallel and rearranging the interconnection of the cells reduces the limitation of the short-circuit current by the less illuminated cell rows.

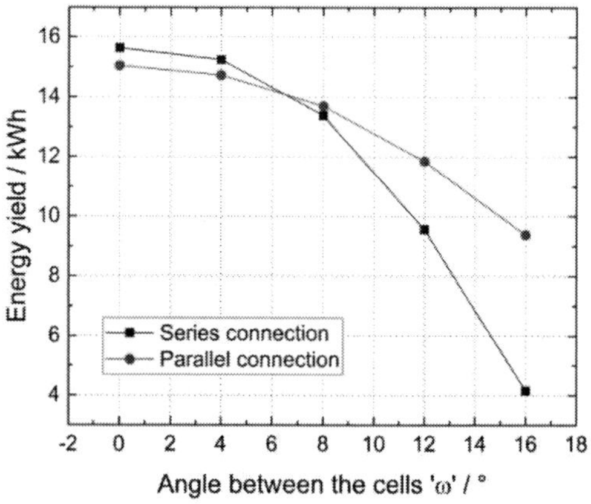

Figure 7: Absolute energy yield values of series and parallel connected modules with different angle of curvature between the cells

Table 4: Power loss and gain of the module with parallel connected strings over the module with series connected strings

Angle of curvature 'ω' [°]	Energy loss and gain ratio [%]
0	-3.8
4	-3.38
8	2.34
12	23.92
16	125.83

3 DISCUSSION

The simulations are done for curved modules with different angle of curvature. We have investigated only the electrical and optical properties of such modules. The feasibility of construction and mechanical properties of such modules should be discussed further.

The properties of junction point and the properties of the output tabs in such modules are important issues. The resistance of output tabs can be enhanced by using different geometries to reduce the electrical losses.

4 CONCLUSION AND OUTLOOK

Modules with curved shapes can be matched with different architectural designs and help the sustainability of the structures. The design, technology and location of such modules matters. We have developed and optical and electrical approach to estimate the energy yield of various designs of curved modules. In modules with a range of curvature below 8 degrees, the conventional series interconnection is a proper design choice. For the level of curvature over 8 degrees, the series connected modules lose their power dramatically. This reduction can reach up to 73.5% for 16 degrees angle between the cells. However, by changing the strings connection from series to parallel, more flexibility in design and higher energy can be achieved. Parallel connected modules do not need bypass diodes which lead to reduction of material consumption and module price.

The optimum case to harvest maximum power from curved modules is to couple all strings in parallel. However, this connection leads to very high current and electrical losses consequently. By cutting solar cells in half and connecting the cells in parallel, the electrical losses in the output tab reduce to 25% of its full size design and the module will be less sensitive to the current mismatch.

5 ACKNOWLEDGEMENTS

We thank the Federal Ministry of Economic Affairs and Energy (BMWi) of the German Government for the financial support of this work within the "AdmMo" project (FKZ 0325775E).

6 REFERENCES

[1] M. Salvador und S. Grieu, „Methodology for the design of energy production and storage systems in buildings: Minimization of the energy impact on the electricity grid," *Energy and Buildings,* Bd. 47, pp. 659-673, 2012.

[2] L. G, M. Rasul, M. Amanullah und M. Khan, „Techno-economic simulation and optimization of residential grid-connected PV system for the Queensland climate," *Renewable Energy,* Bd. 45, pp. 146-155, 2012.

[3] A. Meinel und M. Meinel, Applied Solar energy, Addison Wesley Publication Co., 1976.

[4] F. Kasten und A. Young, „Revised optical air mass tables and approximation formula," *Applied optics,* pp. 4735-4738, 1989.

[5] PVEDUCATION, „Arbitrary Orientation and Tilt," [Online]. Available: http://www.pveducation.org/pvcdrom/properties-of-sunlight/arbitrary-orientation-and-tilt.

[6] J. Gow und C. Manning, „Development of a photovoltaic array model for use in power-electronics simulation studies," *IEE Proceedings,* Bd. 146, 1999.

[7] H. Hanifi, J. Schneider und J. Bagdahn, „Reduced Shading Effect on Half-cell Modules – Measurement and Simulation," in *31th European Photovoltaic Solar Energy Conference and Exhibition*, Hamburg, 2015.

[8] C. Luis und S. Silvestre, Modelling Photovoltaic Systems Using PSpice, John Wiley & Sons Ltd, 2002.

A SOFTWARE SUITE FOR SIMULATION AND DESIGN OF PV PLANTS

Ismaël Lokhat, Stéphane Boussac, Benoît Lelong
CYTHELIA ENERGY
350 route de la Traverse, 73000 Montagnole, France

ABSTRACT: The drop of the prices of the PV modules and other BOS components lead to an increasing part of the 'soft' costs, which include the design and the engineering of the system. The decrease of these soft costs has thus become one of the challenges the PV industry has to tackle to achieve further global systems costs reduction. During the design of a PV power plant, several software are used, and most of the time, these software are independent. As the design purpose is mainly the quest for an optimum between the yield and the investment cost, the process is iterative and can be very time-consuming if there is no straightforward communication between the different tools. The suite of three compatible software called Archelios Suite, is based on a unique Photovoltaic Information Model (PVIM) which allows a fast and efficient interaction between the software and therefore the reduction of the time dedicated to the design of the plant.
Keywords: Design, Safety, Software, Simulation, System Performance

1 INTRODUCTION

Pre-construction phases of a photovoltaic power plant project often require several tools or software, which are autonomous and poorly able to communicate between themselves. It is especially the case with the electrical design software (which enables to calculate the characteristics of the electric components and the compatibility between them, e.g. the compatibility between the photovoltaic array and the inverters) and the expected yield calculation software. During the engineering process, the design engineer has to compare several solutions or configurations, and each of these options has an impact on the sizing of the plant and its yield. When optimizing the yield (annual or seasonal depending on the context), the design engineer has to make sure that the options are still technically realistic and economically viable. Therefore, a design process aimed at optimization is iterative and is time-consuming when two (or more) independent software are used. Adopting a single tool, or at least tools that can interact, helps companies to diminish the soft costs of PV plant projects, by reducing the time for the design studies. This principle was the guiding thread of the development of the Archelios Suite.

The Archelios Suite is composed of three software: Archelios PRO, which main purpose is the yield assessment, Archelios CALC which aim is to check the compliance of the PV system with grid-connected photovoltaic systems standards, and Archelios MAP, which is used for the generation of Solar Cadastre.

These software are able to communicate between each other thanks to a single file, called the Photovoltaic Information Model, or PVIM.

2 THE Archelios SOFTWARE

2.1 Archelios PRO

Since its first release 2008, the Archelios PRO software has been recognized by third-parties evaluation as one of the most accurate software for PV yield prediction [1], [2].

In 2014, the desktop version was migrated to a full online application, which enables the yield prediction, but also the sizing and the economic evaluation of the plant.

Thanks to a SketchUp plugin, a 3D scene can be easily defined in order to take into account the near shadings in the yield computation.

The PV modules layout can be performed within SketchUp, for both roof-integrated and sheds (and also ground-mounted) configurations.

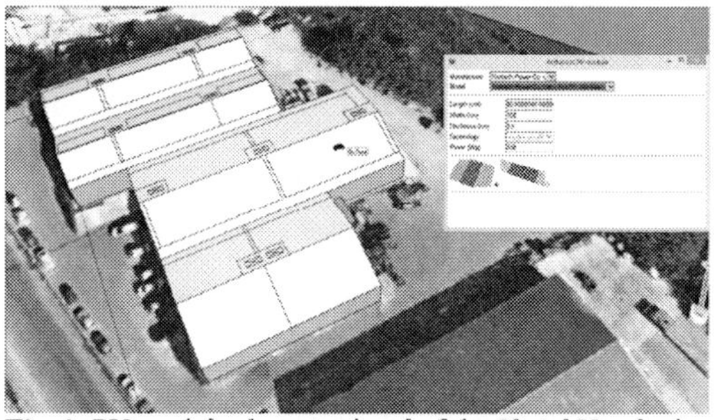

Fig. 1: PV modules layout wizard of the SketchUp plugin

The plugin also offers some features like a wizard for the PV modules layout and an irradiation assessment tool, which is particularly useful to identify the areas that are mostly affected by the surrounding obstacles.

The string definition and cabling is performed online within Archelios PRO, thanks to a specific 3D tool designed for a simple and efficient input of several default cabling patterns.

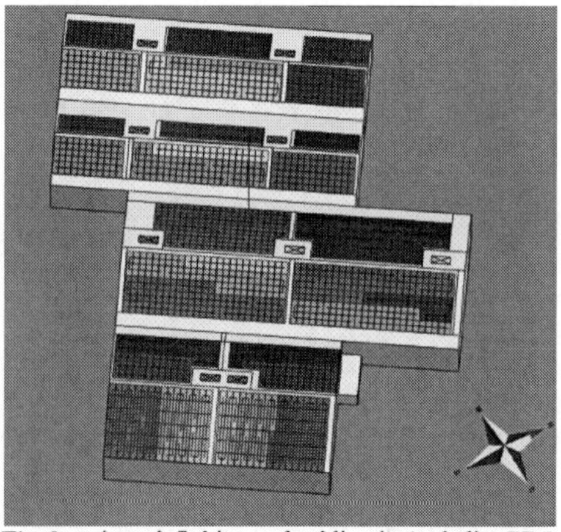

Fig. 2: strings definition and cabling in Archelios PRO

At each time step, the software computes the incident irradiance for each module, and then their I-V curves. As the I-V curves are added up according to the string definition, the software allows the comparison between the cabling options, or between string/central inverters and micro-inverters or optimizers.

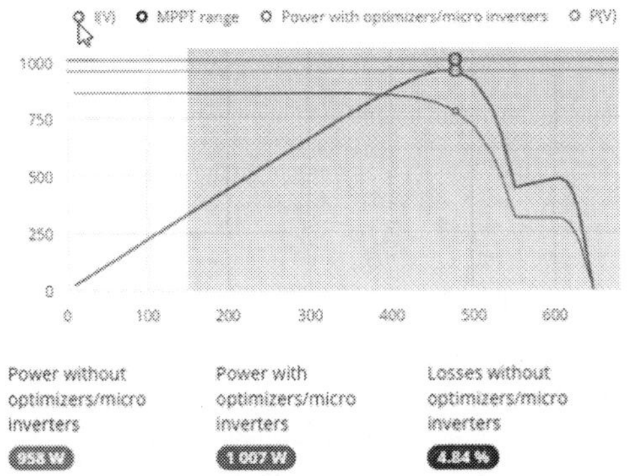

Fig. 3: I-V curve of a PV array

Actually, as the diffuse irradiance is also modified by the obstructions, two adjacent or nearby PV modules may have different incident irradiance even if they are not shadowed. For small PV systems, the computation time may be acceptable, but for large-scale PV plants with tens of thousands of PV modules, the duration would easily exceeds one hour, which is not bearable from a user perspective. The optimization basically consists in grouping the PV modules which have close incident irradiances and temperatures. Thanks to a major optimization work to drastically reduce computation time, which allows to maintain an acceptable loss of accuracy (< 0.5%), the operability of the software for up to 50 000 PV modules plants has been validated. Validation for larger plants is currently ongoing.

Besides the sizing and the yield computation, Archelios PRO also offers economic evaluation features, which are able to deal with the most common support mechanisms like feed-in-tariff, net metering or self-consumption. This ability is crucial in any optimization process, along with the variant comparison function.

2.2 Archelios CALC

Archelios CALC main purpose is to ensure that the proposed design complies with the applicable standards. The software manages the international IEC 60364 "Electrical Installations for Buildings" standard and the French C 15-712 and C 15-100 standards.

From the PV modules to the grid, Archelios CALC helps the designer to check the sizing of all the electrical equipment including the cables and the protections devices: fuses, breakers, surge arresters, etc.

Once the system's design is completed, the user can export calculation reports and electrical synoptic of the plant.

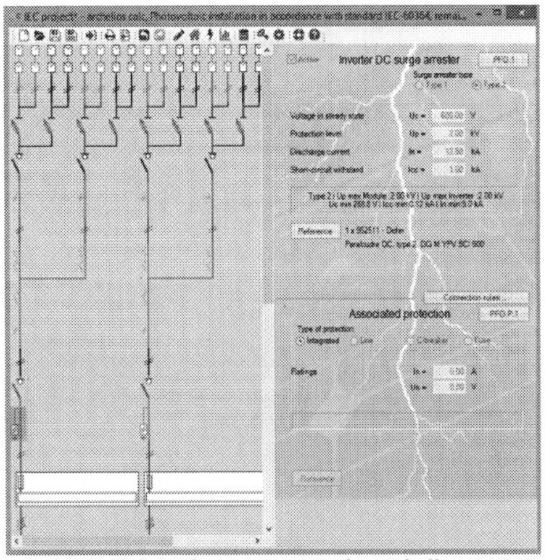

Fig. 4: Sizing of surge arresters in Archelios CALC

2.3 Archelios MAP

From a 3D model of a city, Archelios MAP is able to compute the annual irradiation on each roof or façade of the city, thanks to the very same solar resource algorithms used in Archelios PRO. The results can then be integrated into the Geographic Information System (GIS) of the city, and be made accessible to the public, who can assess the solar potential of buildings.

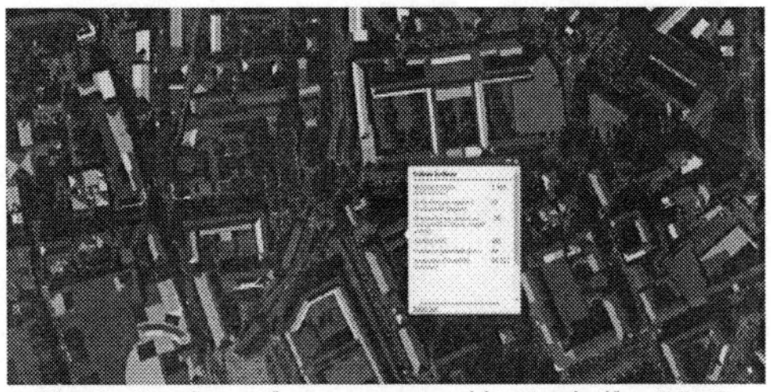

Fig. 5: A solar cadastre generated by Archelios MAP.

3 The PHOTOVOLTAIC INFORMATION MODEL (PVIM)

Following the BIM (Building Information Model) concept, the previous 3 (4 with the Sketchup plugin) software of the Suite have the ability to communicate through a unique PVIM (PhotoVoltaic Information Model) file.

Many definitions of BIM can be found, but it can be simply defined as a mean, by which every stakeholder of the building's design, construction and operation process can easily interact between each other. Concretely, the BIM is a digital model, which includes physical and functional characteristics of the building, founded on open standards for interoperability.

As the need for a BIM approach in PV projects mostly exists for new building projects, it seems relevant to have for the other projects (on existing buildings and ground-mounted plants) a BIM-like standard for photovoltaic systems only. PV professionals dealing with such projects would have access to the features offered by the BIM, without investing in costly BIM software available on the market. The PVIM could also be a first step towards a full BIM approach as described for instance in [3].

Like the BIM for buildings, the PVIM file is at the centre of all the different stages of the life-cycle of PV plant. Thanks to the SketchUp plugin or Archelios MAP, the 3D physical digital is easily created, and then enriched with more specific data from PRO and CALC.

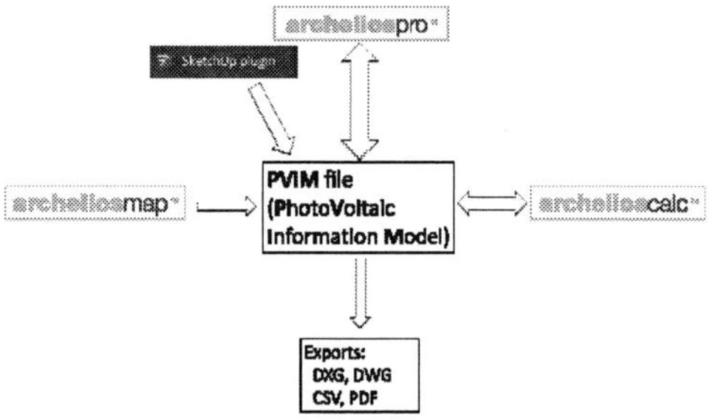

Fig. 6: PVIM within Archelios Suite

To ensure a perfect compatibility, the software must also share a common database which contains data that are useful for the overall Suite, but not for each software. For example, costs data will be useful in PRO, which deals with the economics of the project, but not in CALC.

So far, only conception, engineering and construction (indeed, the PVIM created during pre-construction phases can be updated to take into account any modifications or discrepancies that may occur) are addressed by the software, but any other operation (or monitoring) software could be interfaced with the PVIM.

4 CONCLUSION

The PVIM, even if not as thorough as the BIM, allows saving time during the optimization-oriented conception and engineering of PV plants. With the Archelios Suite software, the user has access to every feature needed for an optimized design of its PV systems.

Integration to the BIM of the Archelios software would be part of future developments.

5 REFERENCES

[1] Axaopoulos *et al.*: Accuracy analysis of software for the estimation and planning of photovoltaic installations. International Journal of Energy and Environmental Engineering 2014, 5:1

[2] Photon Profi, April 2011

[3] Gupta *et al.*: A conceptual framework to support solar PV simulation using an open-BIM data exchange standard. Automation in Construction 37 (2014) 166-181

SPECTRAL STUDIES INVESTIGATING THE INFLUENCE OF DUST ON SOLAR TRANSMITTANCE

KHADIJA K. KHANUM, MONTO MANI, PRAVEEN C. RAMAMURTHY
Indian Institute of Science, Bangalore: 560012, INDIA
Email: monto_mani@hotmail.com

ABSTRACT: Photovoltaic (PV) performance is significantly impacted due to dust settlement on the glass covering the panel, fundamentally understood as transmittance loss. Maintaining a dust free PV panel requires a minimum of 0.5 L/m2 of water daily in most tropical conditions where dust settlement is common. Most commercially available cells are Si based which harnesses as electricity most of the visible part of the solar spectrum, while the rest of the spectrum is absorbed as heat. However given the fact that the cross-section of a typical solar panel is multi-layered, with the top being a hi-transmittance glass, it is crucial to understand influence of dust on this primary layer. Dust though universally present is not universally uniform and varies depending on the atmospheric pollution, terrain (natural and anthropogenic) aggravations and vegetation. The composition is thus extremely varied. Little is understood on the geometry of settled dust, its composition and spectral absorptivity in response to incident direct and diffused radiation and its angle of incidence.

The current study experimentally investigates the spectral response (Spectroradiometer) of dust characteristic to Bangalore for varying densities from mild soiling (~1g/m2) to heavy soiling (>5g/m2) and its elemental composition. The spectral absorptivity has also been compared for near direct solar incidence vs diffused. Elemental composition (EDAX) of the dust samples partially explained their spectral absorptivity. The results of the study provides critical insight into possible benefit of dust, at specific densities and compositions, in reducing the heat gain of solar cell by absorbing IR at the surface of the glass.
Keywords: Photovoltaic, Dust, Spectral Response, Absorption, Solar Radiation, Dust

1 INTRODUCTION

Studies into dust impact on solar devices has been traced to the 1940's [1]. Dust, though ubiquitous, is extremely complex and can vary depending on the PV system configuration and geometry, ambient parameters, wind velocities and its properties (see Figure 1). Dust not only settles on the PV panel but can also affect visibility in conjunction with humidity, rising and suspended dust [3]. Dust on PV panel and interferes with illumination quality by attenuating and scattering light depending on the constitution, density and size distribution of dust [2] citing [3]. Dust setting on the PV panel adds another layer, like an optical filter, and can vary on its spectral response, i.e. varies in the transmittance, absorption and scattering varying across the solar spectrum (wavelength). Earlier studies cited in [4] have restricted their observation on impact of dust to wavelengths required for the PV cell and have overlooked the remainder of the wavelength which could very well be absorbed by the dust. The current study is a preliminary investigation into the spectral response of dust settlement at varying angles of inclination and dust densities.

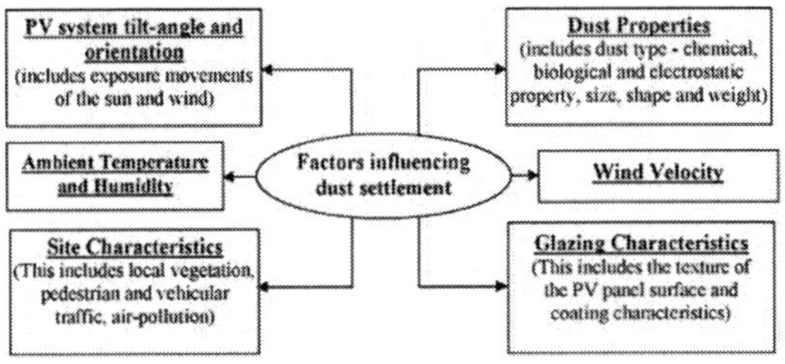

Figure 1: Factors influencing dust settlement [5]

Earlier studies [6] reported that traditional broom was able to achieve more than 90% cleaning efficiency as compared to water-based cleaning normally requiring 0.5 L/m^2/day. Efficiency improvement by a water-cleaned panel was in the range of 0.2~0.4% over the broom cleaned surface. Less than 50 ml/m^2 of water was required to completely clean the surface using a micro-fiber cloth. The dust deposition density was found to range between 0.68~1.11 mg/sq.cm for Poly-Si (25-30% output loss) collected (under natural outdoor conditions) over nearly four months.

2 SALIENT OBSERVATIONS

Four glass coupons (slides) were placed outdoor at four angles of inclination (0, 15, 30, 45 deg) next to three pairs of 37W$_p$ mono-crystalline Si panels. Each pair was maintained at 15, 30 and 45 deg slope to understand the severity of dust settlement and associated degradation. One panel was maintained clean as a reference for the adjoining (soiled) panel.

Surface temperatures have been reported to be higher with dust settlement [1], but the explanation for the same remains unclear, and could be attributed to the higher absorption of IR by dust on the surface [7]. Bird dropping and stains too register a higher thermal footprint when observed using a thermal imager (see Figure 2).

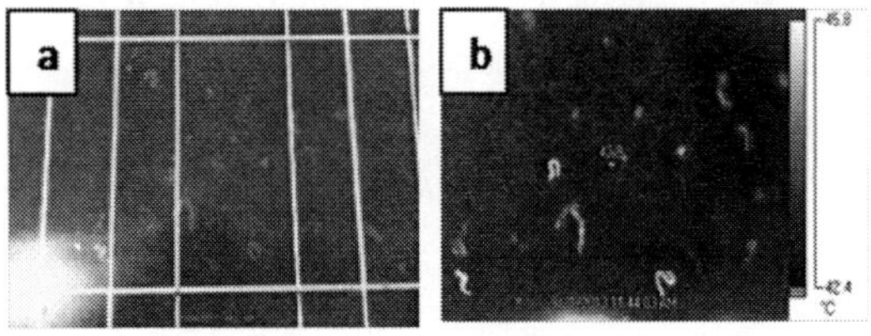

Figure 2: Bird dropping register higher thermal signature [7]

FESEM imaging on dust collected at various angles of inclination provides an insight into their elemental composition (Figure 3). However, these show little variation for the various slopes (Figure 4). This was verified to check the possible influence of wind and moisture on altering the pattern of dust settlement. The results provide critical insight into possible benefit of dust in absorbing IR at the surface of the glass. C and Si are among the major elements in the dust. It is known that organic elements and Si in oxide form absorbs in the IR range [8][9]. The thresholds for dust settlement, the nature of dust and its spectral response needs further investigated.

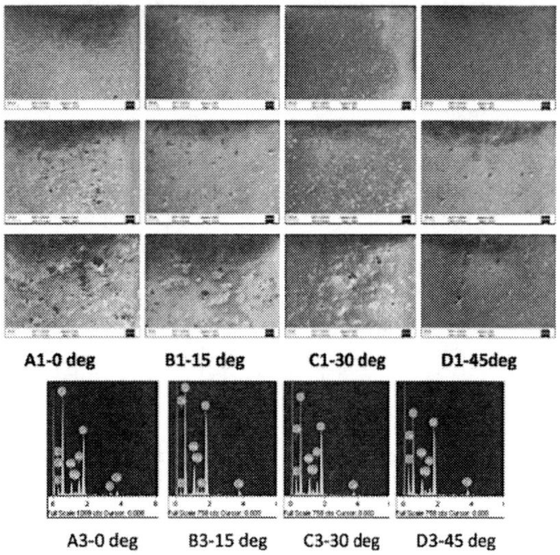

Figure 3: FESEM images of dust samples collected at various tilt angles

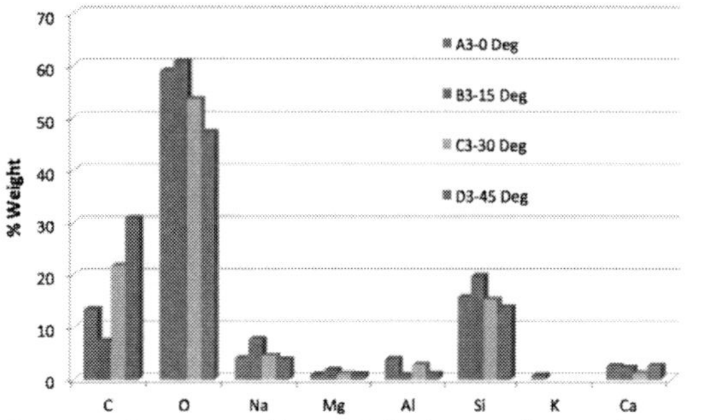

Figure 4: Dust composition assimilation at various angles

Figure 5 provides a comparative illustration on the actual degradation observed for dust collection and densities at various angles of inclination.

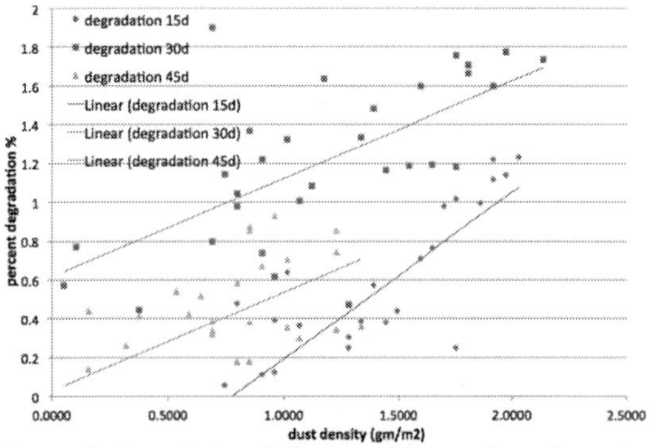

Figure 5: Degradation (PV output %) at various tilt angles and settlement densities

Figure 6 illustrates the spectral response studies on dust coupons maintained at various angles of inclination. An OceanOptics instrument was used for this study. There is noticeable variation in the % transmittance in the visible part of the spectrum, but comparable absorption at the IR range. The instrument was however unable to record transmittance loss beyond 950 nm.

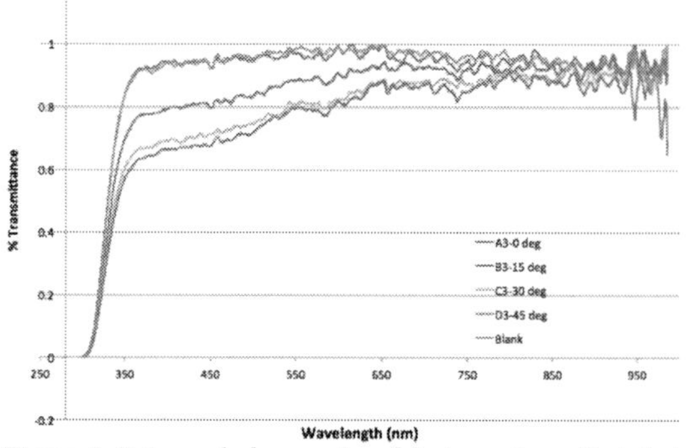

Figure 6: % transmission spectra of dust samples collected at various panel tilt angles

Two dust samples (Figures 7 & 8) were further analyzed using the FieldSpec Spectroradiometer. The results substantiate the fact that for varying dust characteristics and/or dust densities the spectral response (transmittance/absorption/scattering) could vary significantly.

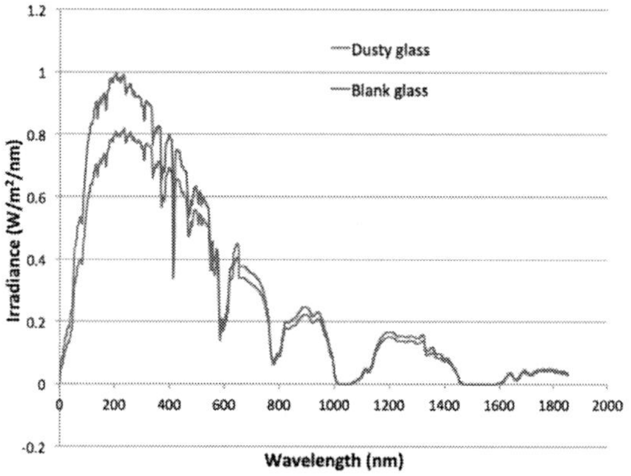

Figure 7: Dust sample indicating higher absorption at visual spectrum

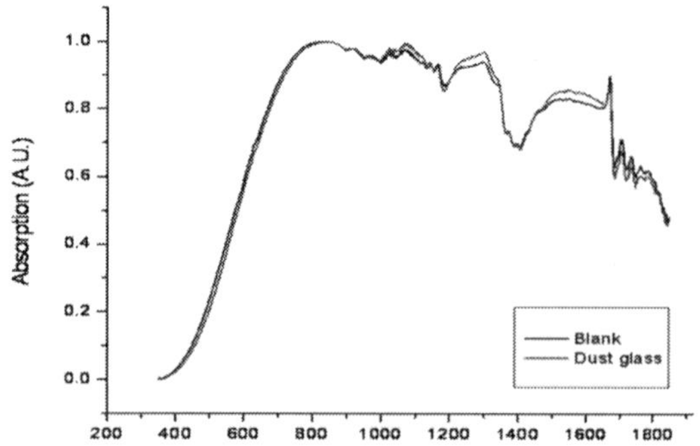

Figure 8: Dust sample indicating higher absorption at IR

Table 1 summaries the % transmittance loss observed for the three broad spectral bands (UV, visible, IR) and notes that the transmittance loss is higher in the IR range. There is also interesting variation in % transmittance loss for dust settlement observed at various angles. This requires careful investigation and scrutiny.

Table 1: % transmittance loss for varying dust densities

Slope	Dust density mg/cm^2	%Transmittance loss (UV) 300-985 nm	%Transmittance loss-Visual 390-700 nm	%Transmittance loss - IR 700-986 nm
0°deg	0.036	1	1	2
15°	0.007	6	10	2
30°	0.026	12	18	4
45°	0.041	13	20	6

3 CONCLUSIONS

- Dust characteristics selectively impacts spectral absorption
- With variation in dust density IR spectral absorption increases compared to visible region
- IR absorption for specific dust type and density could be beneficial to PV output (and reducing cleaning requirement)
- Study into elemental composition of dust and their spectral absorption is crucial

4 ACKNOWLEDGMENT

This (paper) is based upon work supported in part under the US-India Partnership to Advance Clean Energy-Research (PACE-R) for the Solar Energy Research Institute for India and the United States (SERIIUS), funded jointly by the U.S. Department of Energy (Office of Science, Office of Basic Energy Sciences, and Energy Efficiency and Renewable Energy, Solar Energy Technology Program, under Subcontract DE-AC36-08GO28308 to the National Renewable Energy Laboratory, Golden, Colorado) and the Government of India, through the Department of Science and Technology under Subcontract IUSSTF/JCERDC-SERIIUS/2012 dated 22nd Nov. 2012. This work is partially supported by the Robert Bosch Centre for Cyber Physical Systems (RBCCPS) at Indian Institute of Science, Bangalore.

5 REFERENCES

[1] Ghazi, S., Sayigh, Ali, & Ip Kenneth, 2014. Dust effect on flat surfaces – A review paper. Renewable and Sustainable Energy Reviews, 33, pp. 742–751.

[2] Qasem, H., Betts, T.R., Mullejans, H., AlBusairi, H., & Gottsschalg, R., 2014. Dust-induced shading on photovoltaic modules. Progress in Photovoltaics: Research and Applications, 22, pp. 218–226.

[3] Al-Hasan, A.Y., 1998. A new correlation for direct beam solar radiation received by photovoltaic panel with sand dust accumulated on its surface. Solar Energy, 11, pp. 323–333.

[4] Sayyah, A., Horenstein, M. N., Mazumder, M. K., 2014. Energy yield loss caused by dust deposited on photovoltaic panels, Solar Energy, 107, pp. 576-604.

[5] Mani, M., & Pillai, R., 2010. Impact of dust on solar photovoltaic (PV) performance: research status, challenges and recommendations. Renewable and Sustainable Energy Reviews, 14, pp. 3124–3131.

[6] Khanum, K.K., Soni, S., Praveen Ramamurthy, Mani, M., "Evaluating effectiveness of non-water based cleaning mechanisms for PV systems" in Third Southern African Solar Energy Conference, 11-13 May 2015, pp. 29 – 32.

[7] Khanum, K.K., Rao, A., Balaji, N.C., Mani, M., & Ramamurthy, P.C., 2016. Performance evaluation for PV systems to synergistic influences of dust, wind and panel temperatures: spectral insight. Poster presented at the 43rd IEEE PV Specialists Conference, 10-15 June 2016, Portland, OR.

[8] Kitamura, R., Pion, L., Jonaaz, M., 2007. Optical constants of silica glass from extreme ultraviolet to far infrared at near room temperature. Applied Optics, 46, pp. 8118 – 8133, 2007.

[9] Sheibley, D.W., Fowler, M. H., 1966. Infrared spectra of various metal oxides in the region of 2 to 26 microns, NASA Technical Note D-3750.

ANALYSIS OF DIFFERENT SHADING PATTERN ON THE TOTAL CROSS TIDE CONNECTED CONFIGURATION OF SOLAR PV POWER PLANT

Dharmveer Singh[1], Birinchi Bora[2], Yogesh Kumar Singh[2], Rashmi Singh[2], Supriya Rai[2], B. Pradhan[1], Manandar Bangar[2], Renu[2], Ashish Sharma[1],Kaushik Saikia[1], O. S. Sastry*
1. Central University of Jharkhand, Brambe – 835205
2. National Institute of Solar Energy, Gurgaon -122003
*sastry284@gmail.com Tel: +91-0124-2579213, Telefax No. : +91-124-2579207

ABSTRACT: The aim of this study is to investigate the detrimental effects of partial shading on total cross tide connected solar PV modules, and compare their performance in terms of output power. In order to find optimum pattern least susceptible to partial shading effects, an array of 20 modules under variety of partial shaded conditions has been studied to test several shading profiles and results are presented. It is found that the effect of parasitic series resistance increases with increase in the area of partial shading. It concluded that power output is 4 % more in total cross tide connection, as compared to conventional series parallel connections.
Keywords: Partial shading, Power, Series and Parallel connection, TCT connection and Resistance.

1 INTRODUCTION

Use of solar PV technology for production of electricity is growing day by day. It is preferable to produce electricity by using large size power plant to reduce the cost of power generation. A number of series and parallel connected PV modules are used to form a solar array for a desired voltage and current level in solar PV plant. Performance of a series connected string of solar cells is adversely affected if all its cells are not equally illuminated (partially shaded). In a solar array spread over vast area, it is likely that partial shadow may fall over some of its cells or modules due to tree leaves falling over it, tree near it, bird drops on the array, non-uniform dust deposition in a string or in different strings, passing cloud, shade of a neighbouring building etc. Nowadays there is an increasing trend to integrate the solar PV arrays at the design level in the building itself. In such cases it is tough to avoid shading of array due to neighbouring buildings throughout the day in all the seasons. In a series connected string of cells, all the cells carry the equal amount of current. Even though a few cells under shade produce less photon current but these cells are also forced to carry the same current as the other fully illuminated cells. The shaded cells may get reverse biased, acting as loads, draining power from fully illuminated cells. If the system is not correctly protected, hot-spot problem [1], cell damaging can arise and in several cases, the system can be irreversibly damaged. This makes the study of partial shading of modules an important issue. In recent years, the impact of partial shading on the PV array performance has been widely discussed in a number of paper [2, 3, 4, 5]. From the literature review, it is found out that for every configuration of PV plant impact, of partial shading is different. Partial shading also have a great impact on the increase in series resistance. Series resistance is a parasitic phenomenon which leads to power and fill factor decrement and its long term effect tends to degradation of the modules [6, 8]. People have already reported about the different connection of array in power plant such as simple series (SS), series parallel (SP), bridge link (BL), honey comb (HC), total cross tide (TCT)[7].

2 EXPERIMENTAL APPROACH

To study the effect of partial shading on different shading pattern of PV plant and observe the least affected pattern, a power plant of Mono-Crystalline Technology has been installed at the premises of National Institute of Solar Energy (NISE), Gurgaon (Latitude 28°37' N, Longitude 77° 04'E). The site is located at an altitude of 215 m (above mean sea level) having composite climate. These modules are tilted at 28°. The mono crystalline test bed comprises 20 (twenty) modules. Using different number of modules different configuration of the array are made. These module having a total of 36 cells having peak power of 70 W. In this study, PVPM 2540C, PV Engineering Germany IV tracer, Pt 100 temperature sensor and irradiance sensor is used. The accuracy of the instrument is about ±0.5% .

Figure 1 Test Bed for partial shading of Monocrystalline module.

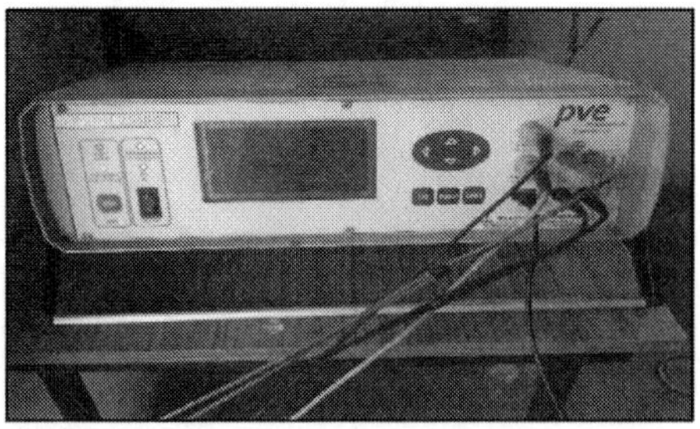

Figure 2 PVPM array tester 2540C

3 RESULT AND DISSCUSSION

People studied different type of connection configuration of PV module power plant. TCT configuration is one of them which gives more power. Connection of total cross tide of modules in the array are developed and performed to check the electrical parameter of the PV plant. Twenty PV modules are taken which are divided in four segments, comprising each string of five modules. 41 different shading pattern are developed and some of them are used in experiment using mesh and cardboard. The abbreviation for those shading pattern are given as follows

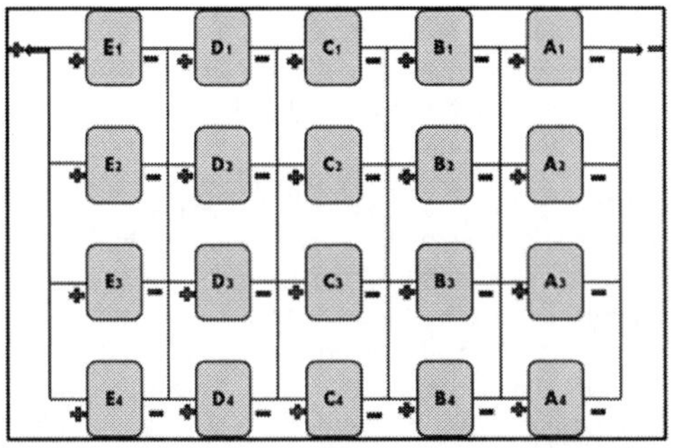

Figure 3: Total Cross Tide connection of 20 Module in (5x4) array

O - Open (without shade) **S** - Shade with single mesh **T** - Shade with double mesh **D** - Dark shade using cardboard. NREL method is used to convert the outdoor Power measurment into STC conversion

$$P = (P_{MAX} * \frac{1000}{I}) + (\frac{1}{(1 + \gamma \times (T - 25))}$$

$P_{MAX} = Test\ measured\ at\ STC\ Condition$
I = Irradiation
γ = Power temperature cofficient

Performance of the power plant is tested at different shading conditions by using mesh. It is observed that there is no significant change in the spectrum of the incident light however the intensity changes. (Figure 4.) Covering the PV module with different meshes allows only 50%, 30%, 20% and 10% light intensity due to which the current get decreased. The shading pattern used in the study of the TCT connection is given below in Table 1. As already mentioned above, the power plant is divided in 4 strings, the power output with different shading pattern are given below. The power output of a particular condition is translated to STC by using procedure 1 of IEC 60891 standard and find out the maximum occurring power output by using histogram method as shown in figure 5. For the array of total cross tide configuration of PV module.

Table 1 The Shading pattern and respective STC power with series and shunt resistance

Shading Pattern in TCT Connection	STC Power at 1000 W/m²	Series Resistance (R_s) Ω	Shunt Resistance (R_{sh}) Ω
OOOO	845	3.26	368
SOOO	720	3.46	738
SSOO	680	2.81	685
SSSO	660	3.18	633
SSSS	555	3.46	481
STOO	585	3.00	605
SSTO	530	3.51	643
DOOO	625	3.50	567
DDOO	450	4.61	785
SDOO	570	3.70	658
TDOO	510	3.66	826

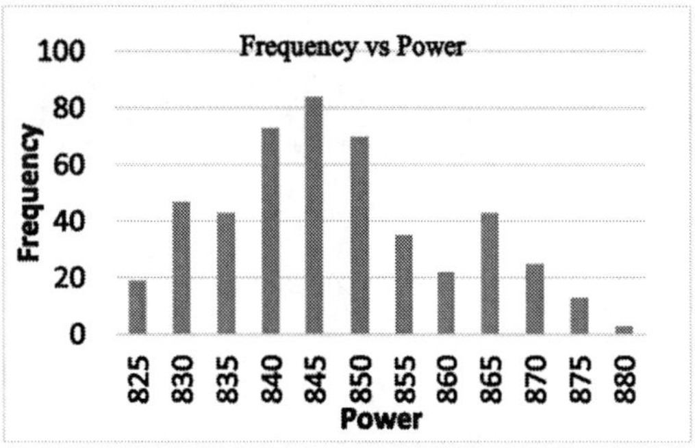

Figure 4 Frequency of STC power in TCT connection without shade

Table 2 The Shading pattern comparision of TCT power with Series Parallel power

S. No.	Total Cross Tide Power (W)	Series-Parallel Power (W)	Percentage (%)
1	785	750	4.45
2	830	770	6.02
3	720	710	1.38
4	560	545	2.67
5	505	475	5.94
6	635	605	4.72

Figure 4. Describe the frequency occurance of Power at diffrent TCT Configration while Figure 5. Describes the change of irradiance due to diffrent mesh configrations

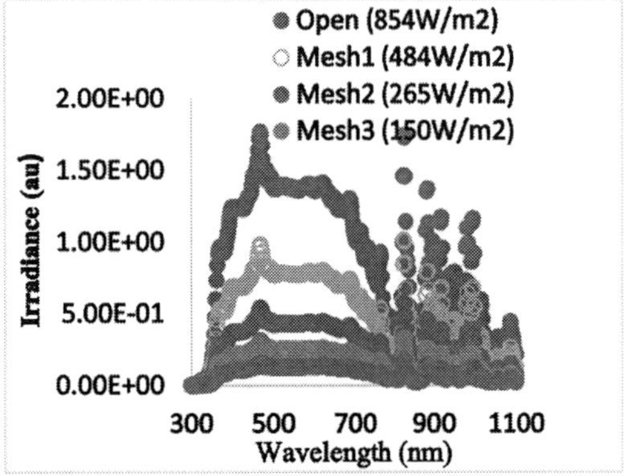

Figure 5: Spectrum of incident light using meshes of different transmissivity

The series resistance (R_s) increases and shunt resistance (R_{sh}) decreases with increase in shading. The power output of conventional series parallel connection and total cross tide connection are compared and 4 -6% more power is found in TCT connection.

4. CONCLUSION

In this paper eleven different configuration of shading is studied and comparison of power output of conventional series parallel connection with total cross tide connection is done. It is found that TCT is producing 4% more power as compared to conventional connections. Additionally, it is been observed that there are variation in series and shunt resistance with different shading pattern and connection. It is concluded that TCT connection is more suitable connection.in adverse condition due to partial shading in large area of power plant. The power output for TCT connection will be economically good.

6.ACKNOWLEDGEMENT

This research is based upon work supported in part by the Solar Energy Research Institute for India and the U.S. (SERIIUS) funded jointly by the U.S. Department of Energy subcontract DE AC36-08G028308 (Office of Science, Office of Basic Energy Sciences, and Energy Efficiency and Renewable Energy, Solar Energy Technology Program, with support from the Office of International Affairs) and the Government of India subcontract IUSSTF/JCERDC-SERIIUS/2012 dated 22nd Nov. 2012.

7. REFERENCES

[1]Quaschning.V, Hanitsch.R. "Numerical Simulation of Current-Voltage characteristics of photo voltaic systems with shaded solar cells", Solar Energy, pp 56,513-520. (1996).

[2]W. Herrmann, W. Wiesner, W. Waassen. "Hot spots investigations on PV modules—new concepts for a test standard and consequences for module design with respect to by-pass diodes", Proceedings of the 26th IEEE Photovoltaic Specialists Conference, pp. 1129–1132, (1997).

[3]Kaushika. N.D., Gautam.N.K. "Energy yield simulations of interconnected solar PV arrays". IEEE Transactions on Energy Conversion, 18(1), pp 127-134, (2003).

[4]M. Klenk, S. Keller, L. Weber, C. Marckmann, A. Boueke, H. Nussbaumer, P. Fath, R. Burkhart. "Investigation of the hot-spot behaviour and formation in crystalline silicon POWER cells", PV in Europe, From PV technology to energy solutions, Proceedings of the International Conference, pp. 272–275. (2002).

[5]Woyte.A., Nijs.J., Belmans.R. "Partial shadowing of photo voltaic arrays with different system configurations: literature review and field test results", Solar Energy 74, pp 217-233, (2003).

[6]Martin Wolf and Hans Rauschenbach, "Series resistance effects on solar cell measurements", Advanced Energy Conversion. Vol. 3, Pp. 455-479. Pergamon Press, (1963).

[7]F. Belhachat, C. Larbes, "Modeling, analysis and comparison of solar photovoltaicarray configurations under partial shading conditions". Solar Energy 120, 399–418(2015)

[8]Rashmi Singh, Birinchi bora, O.S. Sastry, Supriya Rai, Manander Banger, Renu, Kaushik Saikia, Effect of Series and Parallel Combination of Solar PV Modules on series resistance, Indian Science congress Association -2016, Mysore.

DESIGN AND ANALYSIS OF 10 MWP GRID CONNECTED PV SYSTEM INSTALLED WEST KUWAIT

Abdullah Heba[1], Kamel Rashad[1,2], El-Sayed M.[1,3]
[1]Electrical Engineering Department College of Engineering and Petroleum, Kuwait University, Kuwait
[2]Electrical Engineering Department, PhD, Assuit University, Assiut, Egypt
[3]Electrical Engineering Department, Phd, RWTH University, Aachen, Germany
habdullah_ee@yahoo.com, rmkamelm@gmail.com, elsmah@hotmail.com

ABSTRACT: A 10MWp PV grid connected system is modelled for Kuwait Oil Company (KOC). The system will contribute in the reduction of the peak demands that happen in summer. This research studies the effect of Kuwait temperature on the performance of photovoltaic power plant to provide guidelines for future projects in Kuwait. System performance is evaluated for different array orientations by analyzing the monthly and annual energy and losses for each orientation. The optimal tilt angle is obtained for each month. Having a seasonal tilt angle will do the job of a tracker with less performance by 1.4%. The annual capacity factor for the system with a tracker is 19.48% and the yield factor is 1697 kWh/kWp/year. PV system is modelled in matlab and load flow analysis is performed for KOC grid utility before and after connecting the PV plant. The plant is studied at high PV penetration with different voltages and loading conditions. PV inverter var potential and control is demonstarted. Volatge regulation is better achieved at conditions of high upstream voltages than lower. The proposed PV plant contribution to KOC's annual consumption is 29%. Such work is helpful for PV technology utilization and integration in Kuwait.

Keywords: Photovoltaic – Grid-Connected – Inverter

1 INTRODUCTION

This research represents the first milestone in utilizing solar energy in Kuwait and outlines the guidelines for future PV integrations to the existing distribution systems. A study case has been chosen to model and simulate a 10MWp PV grid connected power system to one of Kuwait Oil Company's power grids which feeds and distributes electricity to Electrical Submersible Pump (ESP) systems at the Umm-Gudair field, West Kuwait. Even though Kuwait location provides sufficiently high levels of irradiance, the temperatures are high as well which effects the performance of PV dramatically. The paper proceeds in the following manner. First a brief review of the literature regarding PV estimation and evaluation. Second, the PV plant design procedure is explained in detail. Then, the PV system is modelled and analyzed hourly under temperature and irradiance variation using site meteorological data. An hourly temperature loss is calculated to be used for future PV design guidelines, which has never been presented before and specifically for Kuwait. PV system performance evaluation is done for different tilt angles to find the optimum angle for every month. Next, KOC's distributed system is modelled and load flow is analyzed before and after connecting the proposed PV plant using Newton-Raphson method [13]. The impact of connecting the PV system on the upstream losses is demonstarted as well as the potential Var control the PV inverter can deliver for voltage regulation. Finally theoretical implication is discussed and research findings are presented.

PV efficiency is measured by standard test conditions which are rarely the case in outdoor conditions [1].The Daily Solar Radiation on horizontal surface in Kuwait increases rapidly from 3786.5 Wh/m^2 on 15th of January to 7984.9 Wh/m^2 on 15th of June and then decreases very rapidly to 3691.6 Wh/m2 on 15th of November. It reaches its peak values during months from April to August [2]. The annual based optimum tilt angle of a PV array is approximately equal to latitude of the site location [3]. The loss of energy when using the yearly average fixed tilt angle is around 8% compared with the optimum tilt for each month at Madinah [3]. Literature shows the effect of temperature power losses annually rather than hourly. A comprehensive study by [4] has identified seven factors that influences the annual performance of PV modules which are; cumulative solar irradiance, Module power rating, operating temperatures (cause losses from 2% to 10%), maximum power point voltage dependence on irradiance levels, and variation in solar spectrum. The temperature dependence of solar cell parameters including other materials has been discussed [5]. PV simulation PVsyst® software package can be used to evaluate PV performance with relative error of 5% between estimated and measured values [6, 7]. Annual performance ratio for grid-connected systems is 60% to 80% and it is higher standalone systems [8]. Two sites in Kuwait with Grid-Connected Photovoltaic System have been studied in [9]. Kuwait yearly peak load in 2014 occurred in Jun 11th at 2:30 pm [10]. A 10MWp PV power plant with fixed installation can generate 16240 MWh/yr in Kuwait with a capacity factor of 18.5% as studied by [16].

2 PV PLANT DESCRIPTION & DESIGN PROCEEDURE

The location of the investigated PV grid connected plant is inside a green field owned by Kuwait Oil Company (KOC). The site is at latitude: 28.97 and longitude: 47.62.The objective of connecting the PV plant to KOC grid is to feed the distributed system which feed ESP pumps in parallel with the electricity provider "Ministry of Electricity and Water" (MEW) to contribute in the reduction of summer peak demands. The running load at the distributed system 11kV bus is 9.09 kVA and power factor PF=0.88 (see Fig. 1). The upstream bus of the substation is connected directly transformer less to MEW which makes the substation vulnerable to voltage drops caused by MEW load profile during summer peaks. Metrological data for project PV site is based on averaging satellite data from 1st January 1994 till 31st December

2010. The data is analyzed hourly by excel spread sheet. Minimum and maximum solar radiations are obtained to size the capacity of the PV plant. Different tilt angles are taken into consideration to find optimum solution. System losses are estimated using PV simulator PVsyst and electrical equipment efficiencies provided in data sheets. PV modules are compared according to smallest installation area possible with largest potential of energy production. This is mostly achievable by polycrystalline technology where the efficiency is higher per 1 m^2 compared with thin film in addition higher power ratings are available in the market[14]. PV module Trina solar TSM-310PC14 with 310 W_p is chosen. Number of modules is 32452 and total collector plane is 62968 m^2. The plant is configured to have 12 central inverters with 6 step up transformers, "plant stations", to reduce dc cabling, ohmic losses and voltage drops. Number of stations are chosen to be six and each station has two inverters, Sunny Central 720CP XT, and one three winding transformer sized 1400 kVA and 1600 kVA respectively.

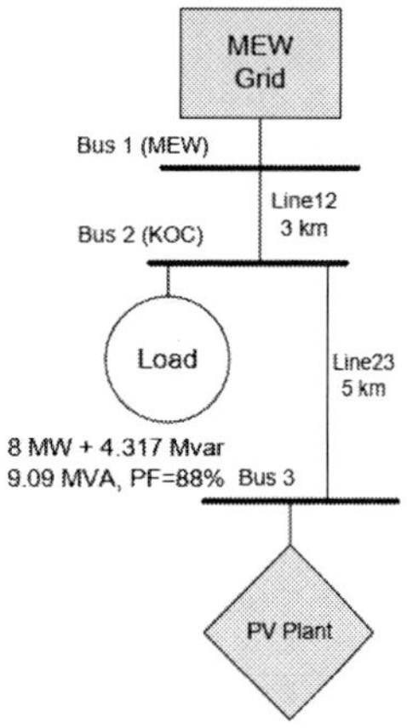

Figure 1: PV grid connected system configuration

3 PV MODEL & SIMULATION

The PV array consists of 140 number of strings connected to each inverter. Each string consists of 19 modules connected in series. There are two sub array configurations. The inverter stations 'Plant 01' to 'Plant 04' are identical, each consist of 20 sub-arrays (10 for each inverter) totaling a number of 5320 PV modules at each station. The inverter stations 'Plant 05' and 'Plant 06' are identical, each consist of 21 sub-arrays (10 into one inverter and 11 to another) totaling a number of 5586 PV modules at each station. Hence, the overall PV project consists of 3242 PV modules. PV generation depends on both irradiance and temperature and both of these factors are studied in detail.

3.1 Solar energy potential in Kuwait

The simulation involves studying different scenarios for the PV array inclination or tilt angle and results are compared. The purpose of this section is to study the solar energy potential in Kuwait. Knowing that solar plants are being installed in the country for different application purposes it is necessary to understand the monthly solar potential accordingly as well as peaking and daylight hours. PVsyst® software package is used to model and simulate the PV plant and estimate energy generation. The plant capacity is studied by obtaining hourly peak values. Daylight hours is used for averaging the power output. Overall system components power losses are modeled in PVsyst® software. The orientation of PV array is adjusted for 20, 30, 40, and 50 degrees tit and simulated to find optimum angles with highesst solar energy collection per month. To have maximum annual irradiation, three options are being considered to go under study. Having a W-E tracker, seasonal tilt with two angles (ST2) and seasonal tilt with four angles (ST4). Having a seasonal tilt of two angles means that there will be two options of positions to choose throughout the year that is to collect monthly maximum irradiation, in this case for ST2 it is 20° in summer (Apr. to Sep.) and 50° in winter (Oct. to Mar.). For ST4 a four season tilt is to have four options of tilt angles in this case 20°, 30°, 40°, 50° each month has the tilt angle that provides the maximum irradiance.

3.2 Calculating temperature losses

The ambient temperature T_{amb} range during summer is between 26°C and 49°C obtained from meteological data. Effective irradiance is obtained G_{EFF} by degrading global irradiance with soiling and Incident Angle Modifier (IAM) loss fators, which are assumed 3% and 3.5% respectively. The output power generated by the PV is dependent on the PV's performance which is governed by its efficiency η_{ref}. With hourly meteorological data, the operating efficiency of the PV η_i is obtained for every hour i, PV array power $P_A(i)$ and hourly Temperature Derating Percentage P_{i_TDP}% are obtained by applying the following equations:

$$G_{EFF} = G(1 - \text{Losses\%}) \quad (1)$$

$$T_{cell}(i) = T_{amb}(i) + \left[\frac{NOCT-20}{800}\right] G_{EFF}(i) \quad (2)$$

$$\eta_i = \eta_{ref}(1 - \gamma(T_{cell}(i) - T_{ref}) \quad (3)$$

$$P_{ref}(i) = G_{EFF}(i) \times \eta_{ref} \times A \quad (4)$$

$$P_A(i) = G_{EFF}(i) \times \eta_i \times A \quad (5)$$

$$P_{i\_TDP}\% = \left[1 - \frac{P_A(i)}{P_{ref}(i)}\right] \times 100 \quad (6)$$

Monthly lossess due to temperature are calculated by obtaining the sum of the reference power obtained in equation (4) for each month while having the temperature coefficient zero and comparing it with the sum of the array output at operation conditions obtained in equation (5).

3.2 PV plant performance evaluation

The capacity factor evaluates the PV plant while ignoring the environmental conditions. It is the ratio of the generated energy into the grid to the rated DC annual energy of the PV, and is given by the following equation:

$$CF = \frac{E_{Grid}[\frac{kWh}{yr}]}{P_o[kW] \times 8760\,[\frac{h}{yr}]} \quad (7)$$

Yield factor is the ratio between the net ac generated outputs of the system to the peak power of the installed PV array. The performance ratio PR is a key indication independent of the orientation and inclination that can be used to compare different PV plants at different locations.

This is obtained using Pvsyst solar simulation software. The ratio includes all system losses and it can be used to estimate the PV energy output from a PV plant.

The design criteria of this project is to generate a minimum of 20 MWh during 4 hours from 10:00am to 2:00 pm in the months starting from 1st Jun. till 20th Sep. where Kuwait's peak load demand occurs [16]. Thus an average of 2 MWe is estimated for those hours. This means that the required design rating for the PV plant is 10 MWp with an approximation of 20% capacity factor.

3.3 Enhancement of grid performance

PV system and KOC substation are both modeled in Matlab. Load flow analysis before and after connecting the PV plant is obtained and analyzed. The effect of PV connection to the power grid is studied by presenting the grid load profile before and after PV connection, upstream line losses and the voltage profile. The line losses P_{line} is calculated by the line current I and impedence R using the following equation

$$P_{line} = I^2R \quad (8)$$

The line losses are calculated in Matlab by the following equation

$$P_{loss} = \frac{I_{12}^2 R \times I_{base} V_{base}}{P_{base}} \quad (9)$$

P_{loss} is calculated to find the power losses in p.u. value. Because the values of I_{12} and R are in p.u., they are multiplied by the base values $I_{base} = 701.8$ A and $V_{base} = 11000$V then divided by P_{base} =10 MW which represents the rated value of the PV plant. The plant maximum apparent capacity is S=9.504 MVA. For every hour, the PV array potential power output from the system P_{inv} with a tracker is obtained and accordingly the available reactive power potential at the inverter output Q_{inv} is calculated using the following equation:

$$Q_{inv} = \sqrt{S^2 - P_{inv}{}^2} \quad (10)$$

Calculations are done in a spread sheet using excel. The max. power that can be generated at the inverter output is equal to the max. power rating of the inverter. As the PV system connected to Bus3 in Fig.1 has twelve inverters, each has a rated output ac power of 792 kVA. This means that the PV station can have max. 9.504 MVA at the inverter output. Thus, the PV plant connected to Bus3 will be assumed to be 9.5 MW. Further more, the transformer losses and auxilliary loading are -0.8% and -1.4% respectively which results with 9.3 MVA as the final power that can reach the grid at Bus2.

PV penetration is effected by the voltage variations from the upstream power grid delivered by MEW. The limitation transient condition of the voltages at KOC grid is to be at the range of 0.95 p.u to 1.05 p.u. and the line currents at 1 p.u., this is where the PV penetration levels are limited accordingly. The effect of voltage variations at the upstream grid is demonstared by load flow analysis and Inverter Var control is used to increase the penetration limited by the voltage condition [12].

During night the inverter can operate as a voltage controller which is an enhancing tool for the system voltages knowing that there is a voltage swing at the 11kV from the upstream grid which cannot be fixed by transformer tapping. From site measured data during summer, the voltage at Bus2 is 0.972 pu and 1.027 pu in winter. The voltage drops in summer is due to the insufficient reactive powr available during peeking hours in summer that is why reactive power injection to the grid is required to compensate for the motor loads. In winter the voltage rises because of the increaese in the available reactive power thus the inverter is used to consume the extra reactive power floating in the grid [11].

4 RESULTS & DISCUSSION

The average daylight hours per month in Kuwait have been obtained and presented in Table I.

Table I: Average Daylight hours in Kuwait 1994-2010

Month	Daylight hours
Jan	334
Feb	330
Mar	373
Apr	390
May	429
Jun	420
Jul	434
Aug	424
Sep	372
Oct	363
Nov	330
Dec	336

Daylight hours in Kuwait is 4535 and is used to calculate the energy generated for yhe PV plant throughout the day. Each month has an optimum tilt angle that will result with the highest value of monthly irradiation comparing with the other orientation options. Optimized tilt angles for max irradiance is obtained for every month and shadawed in Table II.

Table II: Monthly Optimized tilt angles for max. irradiance

Month	GHI	20°	30°	40°	50°
Jan	101.5	129.4	138.9	145	147.5
Feb	109.8	132	138.4	141.5	141.1
Mar	166.6	186.7	190.1	188.9	183.1
Apr	160.3	166.1	163.2	156.6	146.4
May	208.9	205.6	196.7	183.4	166.1
Jun	227.1	217	204.3	186.9	165.6
July	226.4	219.4	208.2	192.1	172
Aug	214.6	217.9	211.7	200.5	184.6
Sep	186.5	203.2	204.2	200.2	191.3
Oct	143.8	166.3	171.8	173.1	170.3
Nov	102.1	127.9	136.4	141.6	143.3
Dec	94.1	124.1	134.7	142	145.7
Year [kWh/m²]	1941.7	2095.6	2098.8	2051.9	1957.2

The optimal angles are suggested for seasonal tilt "S2T" twice a year and for "S4T" four times a year as suggested in Table III. The annual global incidence of a seasonal tilted plane SL2 and SL4 compared with a W-E tracker is less by 2.1% and 1.67% respectively. It depends on the purpose of the PV installation to decide whether to have a tracker or not. Having the load demand peaking in summer, the optimum choice for installing the PV modules is 20° tilt angle which will help in reducing peak demands in Jun. But if the purpose of PV design is to reduce annual demand, then 30° tilt angle is the optimal choice where the annual global incidence is the highest 2098.8 kWh/m². It is necessary to account for soiling factor while coming to decide the tilt angle. Dust accumulation decreases with higher tilt angles [15]. More dust accumulations means less light reaching into the PV panel thus reduction in its performance.

Table III: Average Daylight hours in Kuwait 1994-2010

Month	GHI	(W-E)	SL2	SL4
Jan	101.5	147.7	147.5	147.5
Feb	109.8	142	141.1	141.5
Mar	166.6	190.7	183.1	190.1
Apr	160.3	167.3	166.1	166.1
May	208.9	212.3	205.6	205.6
Jun	227.1	230.9	217	217
July	226.4	229.5	219.4	219.4
Aug	214.6	221	217.9	217.9
Sep	186.5	205.1	203.2	204.2
Oct	143.8	173	170.3	173.1
Nov	102.1	143.2	143.3	143.3
Dec	94.1	145.8	145.7	145.7
Total	1941.7	2208.5	2160.2	2171.4

The PV output is calculated hourly by the corresponding temperature and irradiance using equations (1) to (6). The annual energy E_A from the array is the sum of all power generated for 8760 hours (during night P_A=0). The calculations have been performed and E_A is calculated 17208.41MWh/yr at horizontal surface with the account of irradiance and temperature degradations only without yet considering soiling or mismatching losses and it is 16188.617 MWh/yr after considering the losses at the PV module. Thus the average power output from the PV array is approximately 2 MW which is calculated over 8760 hrs and 3.56MW calculated by 4535 daylight hours . This is very low compared with the peak rating 10 MW_p and this is because during night, the plant generates zero power which leads to a low capacity factor for grid connected PV plants. Maximum PV array output is obtained and is 8.13 MW on April 26th. The sum of the reference power is 18290.558 MWh/yr and the losses due to temperature is 10.825%. The losses increase significantly in summer months. In summer (May. to Aug.) the losses reach 15.8% and exceeds 20% in Jun. The results are summarized in Table IV and a plot has been prepared using Matlab to present the PV power losses as a function of ambient temperature and global insolation as shown in Fig. 2. The PV temperature derating is max in summer exceeding 20% and negative values occur in winter as shown in Fig.3.

The calculated annual losses are validated using PVsyst simulation some variation occur, see Fig. 4, because PVsyst does not use NOCT approach but rather uses an energy balance between ambient temperature and cell heating giving consideration for mounting mode. Monthly temperature losses for different tilt angles is shown in Fig. 5. The orientation which provides more irradiance experiences more temperature losses because of longer operating duration dependent on radiation level. The overall system losses is obtained for different tilt angles as shown in Fig. 6. The system with a tracker will have more system losses but even though, it will slightly have less system performance ratio compared with installed systems of fixed tilt or seasonal tilt angles. This variation is very small and can be neglected.

The performance ratio of recommended system with a tracker is 76.8% as shown in Fig. 8 showing less performance in summer than in other months. The 10MWp system with a tracker produces 1707MWh/yr system (see Fig. 8) and the capacity factor is thus 19.48% and it is the highest rate achieved compared with the other orientation options suggested in this research.

Table IV: PV array power and losses due to temperature.

Month	E_A	E_{ref}	Losses %	Max loss%
Jan	948.18	948.754	0.06	5.75
Feb	1007.13	1036.990	2.88	8.49
Mar	1434.82	1547.939	7.31	13.16
Apr	1360.34	1498.833	9.24	15.82
May	1689.56	1953.271	13.5	18.71
Jun	1789.91	2123.366	15.704	20.36
Jul	1777.46	2116.47	16.017	20.52
Aug	1688.24	2006.09	15.844	20.13
Sep	1507.04	1743.83	13.579	18.11
Oct	1219.01	1356.34	10.125	14.36
Nov	903.11	945.75	4.509	10.83
Dec	863.79	876.93	1.498	6.43
Total	16188.62	18154.56	10.829	-

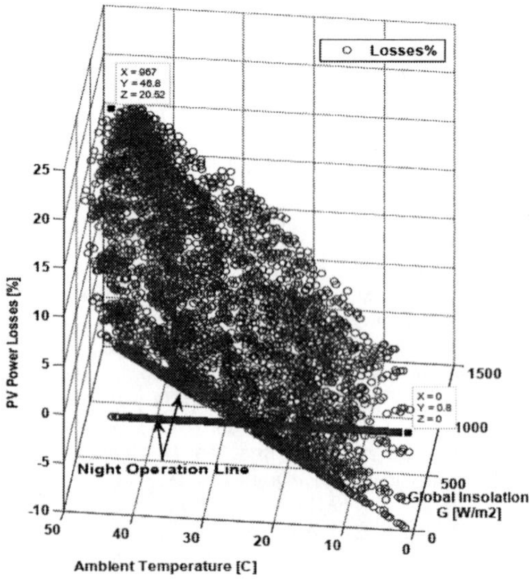

Figure 2: PV power losses with temp & irradiance

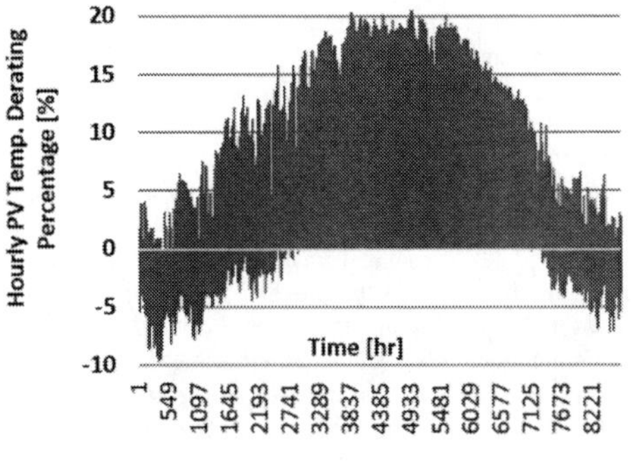

Figure 3: Hourly PV Temperature Derating Percentage

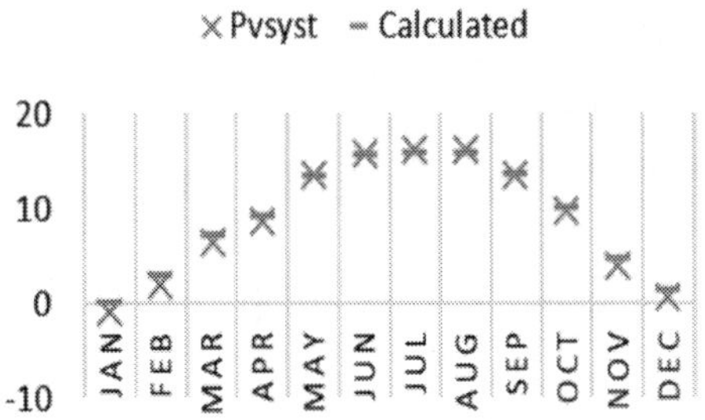

Figure 4: PV power losses validation with PVsyst

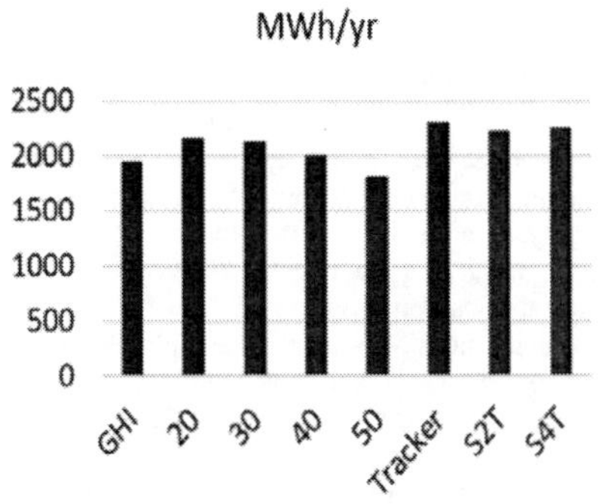

Figure 5: PV power losses due to temperature for different inclination options

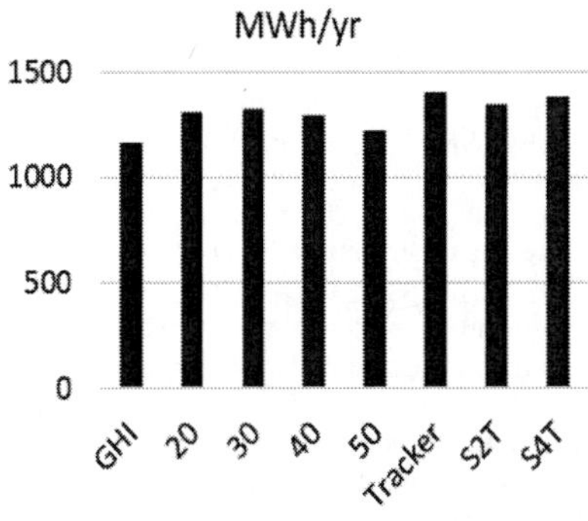

Figure 6: System losses for different inclination options

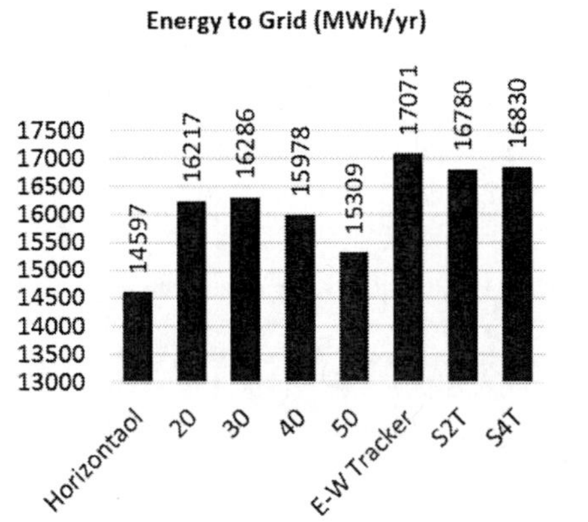

Figure 7: Annual PV Energy Injected to the Grid for different inclination options

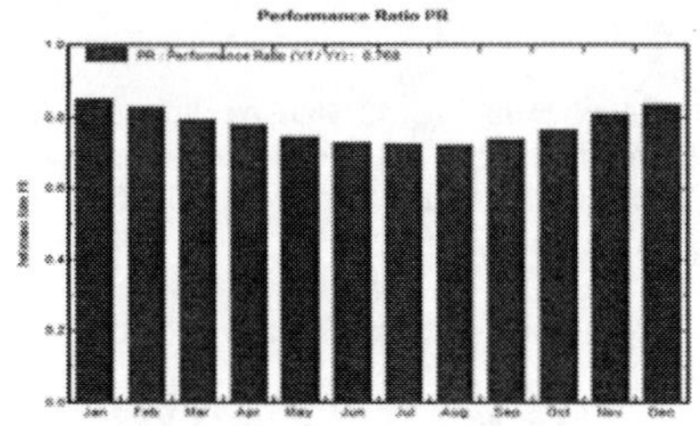

Figure 8: PV plant Performance with tracker

Checking the design requirement which is to have to deliver 20 MWh to the grid from 1st Jun to the 20th Sep during the four hours of 10:00 am to 2:00 pm for the system with an E-W tracker. The generated power at these months at the required times are obtained by simulation and plotted. Also, the generated power during the required time is summed to check if the requirement of 20 MWh is achieved or not. The power injected to the grid for the month of Jun to the 20th of Sep are plotted in Fig.9 to Fig.12 and are compared to study to obtain the day that has less with the least power generation and to check if the design requirement is satisfied.

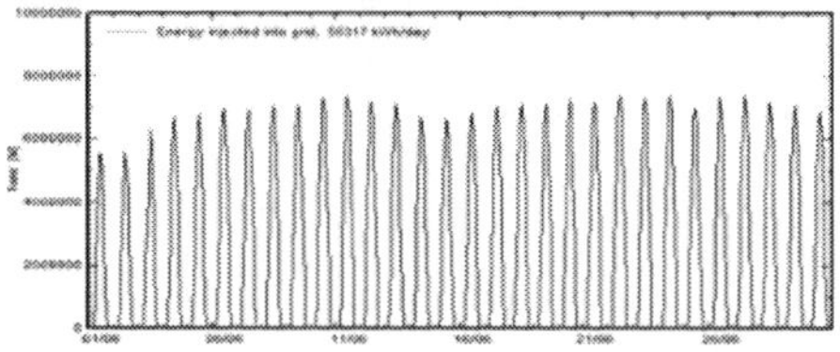

Figure 9: Hourly power injected to the grid (Jun).

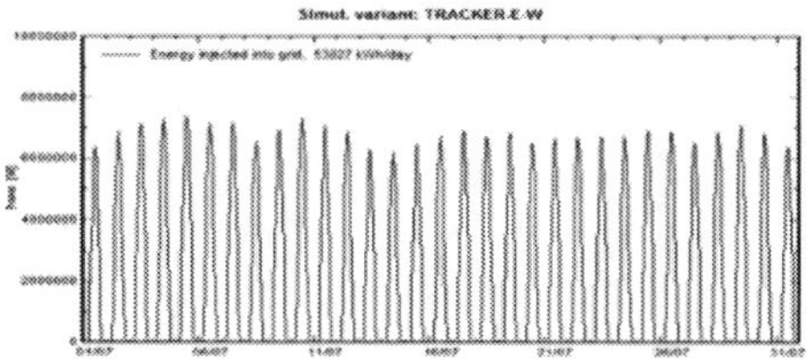

Figure 10: Hourly power injected to the grid (Jul).

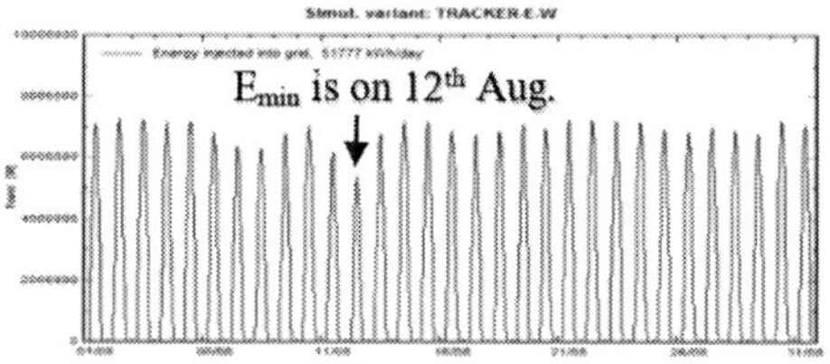

Figure 11: Hourly power injected to the grid (Aug).

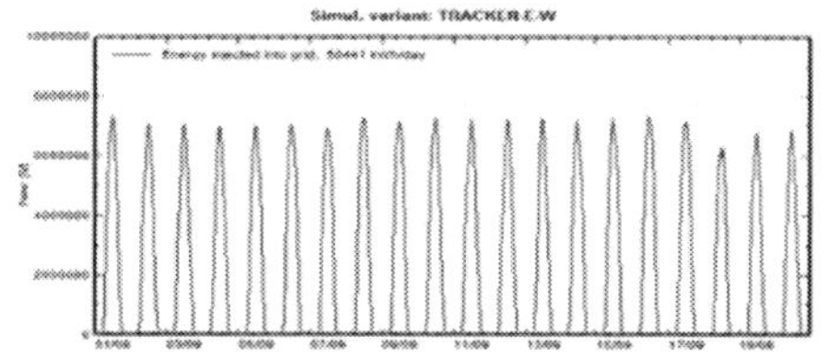

Figure 12: Hourly power injected to the grid (Sep 1st to Sep 20th).

It is found from Fig. 11 that on Aug 12th the power injected into the grid is the least power generation during the period under study. As shown in Table V, the design condition is satisfied.

Table V: Hourly power injected to grid on 12th Aug.

Hour	P_{Grid} [MW]
10:00	4979.4
11:00	5427.3
12:00	5361.4
13:00	4801.1
Total	20.5 MWh

The case is studies as well for the PV plant with different tilt angles and it is found that all cases achieve 20MWh in the 4 hours between 10 am to 2 pm from 1st Jun to 20th Sep, see Table VI. This means any of these tilt angles can be chosen where in any of these cases, the PV plant will meet the design criteria.

Table VI: Design criteria satisfied for different panel tilt angles

Day	20°	30°	50°	E-W tracker
12th Aug	20.5 MWh	20.1 MWh	20.1 MWh	20.5 MWh

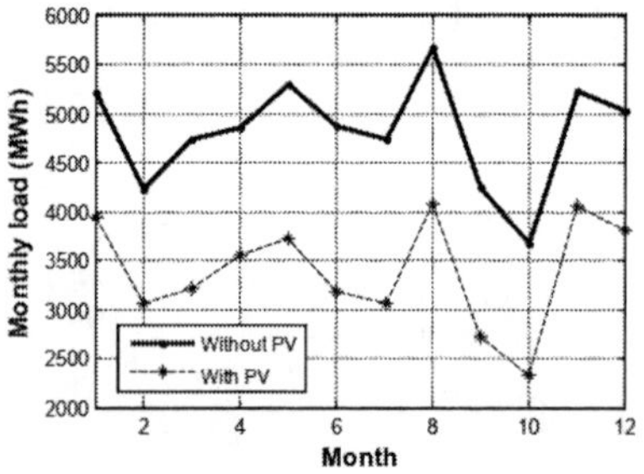

Figure 13: Distributed system load profile

The average inverter output from horizontal oriented plant, considering day hours is 3.9 MW and reaches the grid at an average of 3.29MW, which is caused from 2.93% losses by the 5 km ac cables, transformers and auxiliary loads. The impact of connecting the PV system to KOC substation is to intraduce a reduction in the annual load consumption from MEW by 19.3% for horizontal surface and 29% for the PV system with a tracker. The substation load profile before and after connecting the proposed PV plant with a tracker is shown in Fig. 13. The bus voltages are enhanced to desirable levels by supplying or consuming kVAR. A plot between the available power at the PV array output and the potential of available reactive power at the inverter output is shown in Fig. 14. Maximum supply of reactive power is 9.504 MVAR at night. The annual average power potential of the inverter is 8.8 MVAR. The available VAR that can be used for voltage regulation during peeking hour in Jun 11th at 2:00 pm is 7.56 MVAR which can be available for the purpose of voltage regulation.

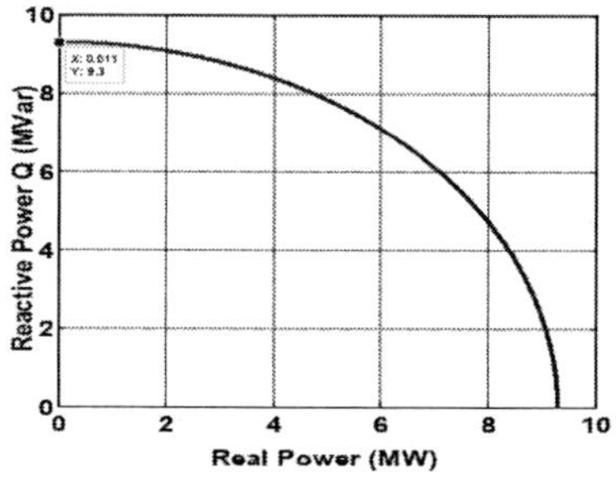

Figure 14: Annual Inverter kVAR Potential

Power losses in the line increases proportionally with the load and drawn current. Hence there are three cases of loading conditions (minimum, maximum and average loading) which are developed to study the effect of changing the load on the losses in Line12 . The losses in the upstream line is demonstrated and plotted for different loading conditions as a function of the PV power injected to the grid at Bus3 (see Fig. 15). For max. loading on Bus2, the PV will always reduce the losses in the line no matter what value of the injected power from the PV plant was. For min loading, once the PV power exceeds approximately double the loading power the losses in Line12 increases and reaches very high levels at PV max. penetration.

The PV max penetration limit is obtained and plotted in Fig. 16 and is 28.9 MW if limited by the voltage of Bus3 only. But this is not the case where the ampacity of the PV plant cable Line23 must be taken into consideration. At maximum penetration limited by voltage, the current in Line23 is 2.059 p.u. which exceeds the cable maximum handling capacity. The base apparent power of the PV plant is S_{base}=13.37 MVA and is related to the line ampacity base value I_{23}=701.79 A. The result of limiting the maximum penetration by the ampacity in Line23 with the condition not to exceed 1 p.u. is 13.6 MW as shown in Fig.17. The cable ampacity is 13.3 MW and the line losses is 0.3 MW. Varyng the load condition does not effect the PV max penetration limit as shown in Fig. 18. The effect of changing the upstream voltage on PV max. penetration is shown in Table VII. When the upstream voltage reaches 1.05 p.u. (high) the PV Max. penetration is reduced to its minimum at 2.7 MW. It is limited by the voltage of PV bus (Bus3). For this case, at high upstream voltages the PV max. penetration is limited by the voltage as shaded. The max PV penetration is increased from2.7 MW to 11.3 MW for the case of V1=1.05, that is when the inverter injects 7.5 Mvar into the grid. At V1=1.04 when the inverter injects 4 Mvar the PV penetration increases to 12.4 MW after it was just 7.8. The effect of PV VAR control on the PV maximum penetration is shown in Fig.19. where the maximum penetration was limited by the voltage before injecting reactive power, then it has become limited by the current in Line23 after injecting reactive power using the PV invererter as shown in Fig. 20.

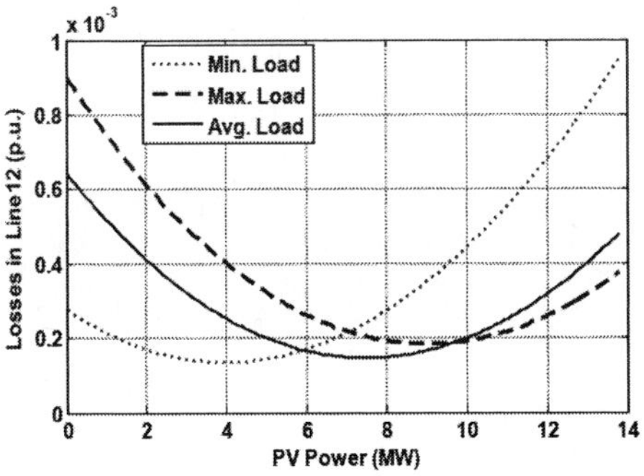

Figure15: Losses in feeder Line12 for different loading

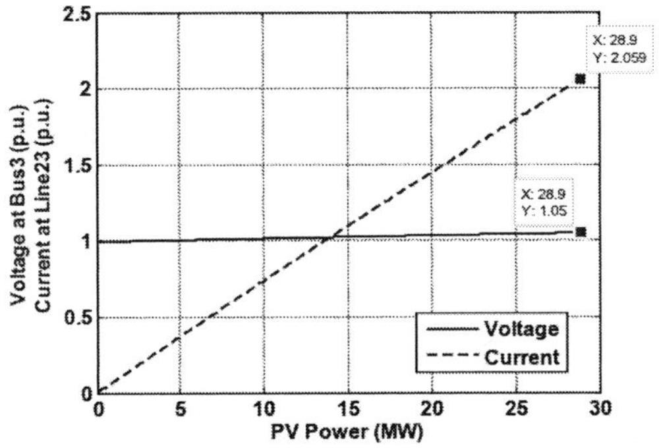

Figure16: Max. PV penetration limited by voltage

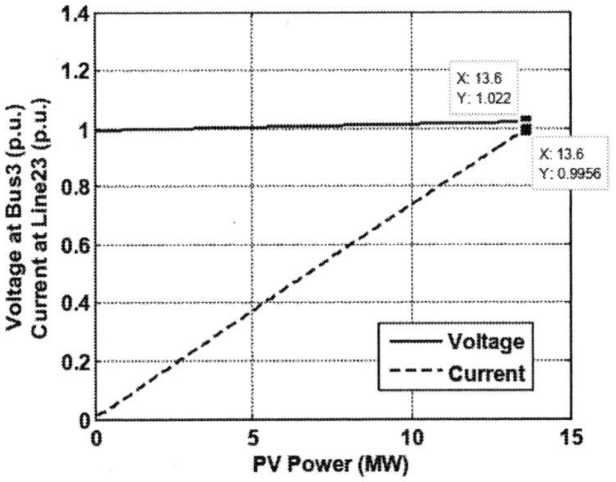

Figure 17: Max. PV penetration limited by voltage and current

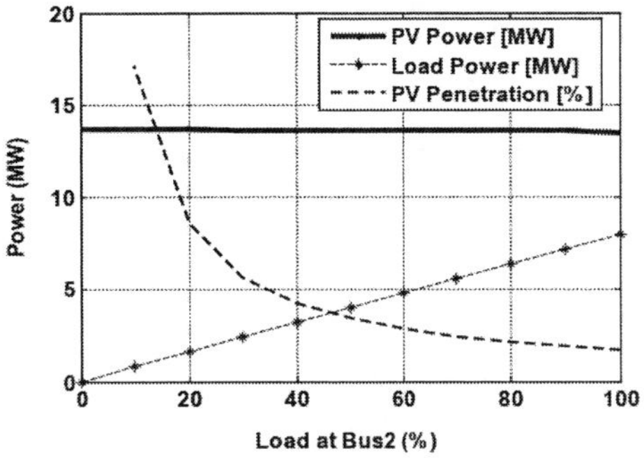

Figure 18: PV max as a function of loading condition

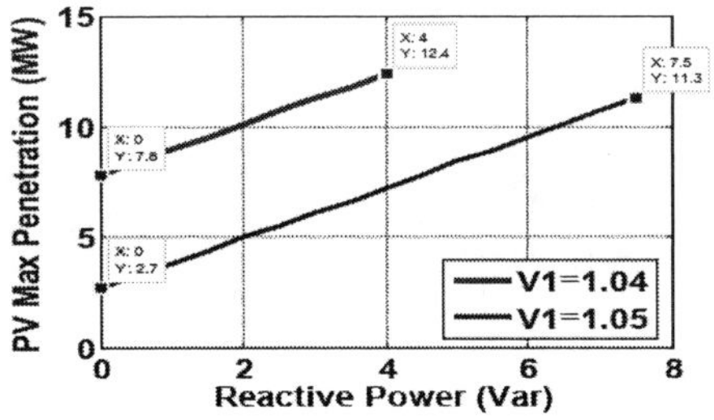

Figure 19: Effect of PV VAR control on PV maximum penetration

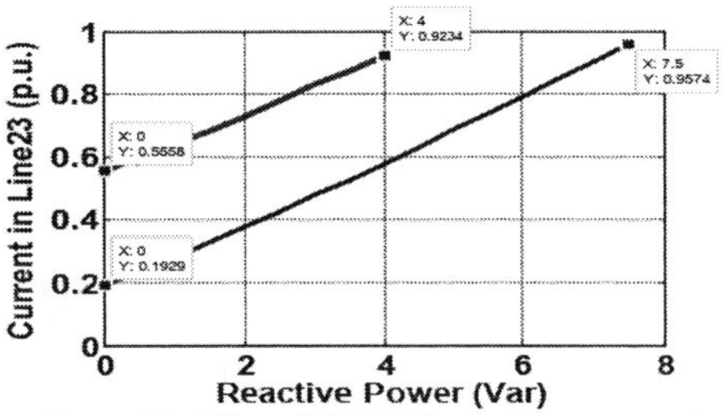

Figure 20: Effect of PV VAR control on current in Line23

Table VII: Effect of changing upstream voltage on PV Max. Penetration limited by voltage and current (voltage and current are in p.u., power is in MW)

Without PV		With PV at Bus3		
V_1	V_3	V_3	I_{23}	P_{pv}
1.050	1.045	1.050	0.193	2.700
1.040	1.034	1.050	0.556	7.800
1.030	1.024	1.050	0.926	13.000
1.020	1.014	1.042	0.998	13.900
1.010	1.004	1.032	0.993	13.700
1.000	0.994	1.022	0.996	13.600
0.990	0.984	1.012	0.998	13.500
0.980	0.974	1.001	0.993	13.300
0.970	0.964	0.991	0.996	13.200
0.960	0.954	0.981	0.998	13.100
0.950	0.944	0.971	0.993	12.900

At night the operaion of the PV palnt will either inject or consume reactivepower depending on the grid requirement. The reactive power available at Bus2 is 9.3 Mvar. Table VIII shows the voltage levels V2 at Bus2 and V3 at Bus3, effected by the upstream voltage V1 at Bus1. The impact of inverter control to regulate the voltages at Bus2 and Bus3 is demonstrated in Fig. 21. An improvement of 2% of voltage regulation is achieved at Bus3 and 0.5% at Bus 2 at voltage lower than 1 pu, and an improvement of 3% at Bus3 and 1.4% at Bus 2 is achieved for busses with high voltages (mote than 1 pu).

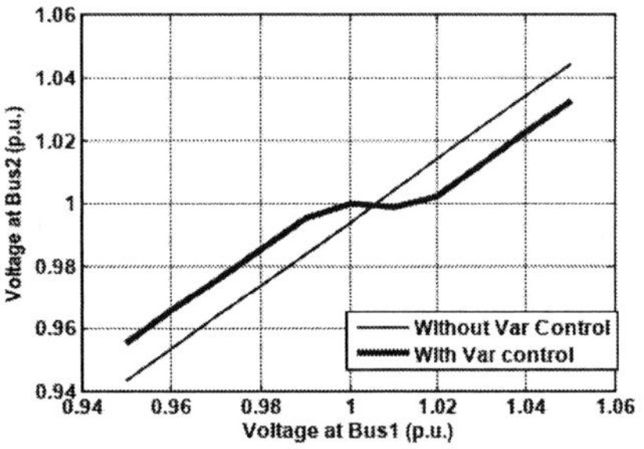

Figure 21: Voltage regulation using PV inverter

Table VIII: Voltage profile effecter by var control

Without PV Inverter Var Control				With PV Inveretr Var Control		
V1	Q	V2	V3	V2	Q	V3
0.95	0	0.9437	0.944	0.949	-9.3	0.965
0.96	0	0.9537	0.954	0.959	-9.3	0.975
0.97	0	0.9638	0.964	0.969	-9.3	0.985
0.98	0	0.9739	0.974	0.979	-9.3	0.995
0.99	0	0.9839	0.984	0.989	-9.3	1.005
1	0	0.994	0.994	0.999	-9.3	1.015
1.01	0	1.0041	0.994	0.989	+6.6	0.972
1.02	0	1.0141	1.004	0.999	+6.7	0.983
1.03	0	1.0242	1.014	1.009	+6.9	0.993
1.04	0	1.0342	1.024	1.019	+7.1	1.003
1.05	0	1.0443	1.034	1.029	+7.3	1.014

5 CONCLUSION

The estimation of the amount of sunlight that will fall on a given surface at a given angle and time in a given place cannot be predicted exactly, that is why photovoltaic power systems is dependent on the use of data based on measurements averaged over a long time. For photovoltaic projects located west Kuwait, having to decide whether to have the panels installed on a E-W tracker or have them at a fixed tilt depends on the purpose and need of the solar power system. It has been concluded that the optimum fixed tilt is 30° to generate maximum annual energy which is consistent with literature reading that proposes tilt angles to be equal to the longitude angle of the PV site. If more solar energy is required then using a 1-axes E-W tracker will increase the Global irradiance up to 13% of GHI. If peak loads are the main concern, then having the fixed tilt angle at 20° is chosen rather than 30° for the months April to August where Kuwait's peak loads occur. Overall, for highest Global Irradiance, installing a tracker will be the first choice then comes the seasonal tilt option and at last fixed tilt at 30°. Environmental assessment as well as economical evaluations must come into account and study for optimum solution.

There is a 10.8% drop in annual energy production due to temperature effect on PV performance with a PV system on a horizontal orientation. Predicting the annual energy using the STC given ratings will give an overestimate of the total annual produced energy. The PV plant peaks in April and not June. Even though the maximum irradiance occurs in June, high temperature causes significant losses in PV performance. The annual losses due to temperature obtained were 10.8% and this underestimates the actual losses that occur in summer. For summer condition in Kuwait, it is necessary to highlight that summer months temperature losses exceeds 20% in June.

The hand calculation for estimating the annual energy produced by the PV plant has been validated by PV simulator. The average inverter output from horizontal oriented plant is 1.76 MW and reaches the grid at an average of 1.7 MW which is caused from 3% losses by aux loads, 5 km ac cables and six transformers. The average inverter output from horizontal oriented plant, considering day hours only is 3.9 MW and reaches the grid at an average of 3.29 MW. Having an E-W tracker increases the average power injected to the grid by 14% which makes it 3.76 MW. The minimum power requirement of 20MWh to be injected to the grid during the period from 1st Jun to 21st Sep and from 10:00 AM to 2:00 PM is satisfied with E-W tracker, 20° tilt, seasonal tilt and 50° tilt. The impact of connecting the PV system to KOC grid is a reduction in the annual load consumption from MEW by 19.3% for horizontal surface and 29% with a tracker.

It is found that a 20° tilt angle is optimized for peak load reduction during summer. For maximum annual energy, 30° tilt is chosen for optimal solution. For maximum annual energy, E-W tracker then seasonal tilt and last a tilt angle of 30 is chosen for optimal solution. Choosing one of these options depends on other factors such as economical and environmental evaluations left for future work. If the PV plant purpose is to inject energy to the loads through the whole year, then the angle must be optimized for the months with the lowest energy production Nov., Dec. and Jan. so the tilt angle will be 50°. Having the tilt angle at 50° decreases the annual energy produced by 6% while the tracker improves it by 4.8% in comparison with a 30° tilt angle. Choosing a higher inclination angle will reduce dust accumulation.

The months with most energy production potential are May to August. Kuwait has a high potential for solar energy with a daily global irradiance of 5.319 kWh/m^2 and an annual global horizontal irradiance of 1941 kWh/m2. The proposed system with E-W tracker shows high energy productivity with an increase of average power injected to the grid by 14%. Seasonal tilt can do the job of a tracker with less energy generation by 1.6% on the other hand it is cheaper, less maintenance and power consumption. An economical evaluation comparison between having a tracker and having a seasonal tilt angle operation is kept for future work.

The performance ratio of the PV system with a tracker is 76.8%. The system with a tracker will have more system losses but even though, it will slightly have less system performance ratio compared with installed systems of fixed tilt or seasonal tilt angles. This variation is very small and can be neglected. The 10 MWp system with a tracker produces 1707 MWh/yr system and the capacity factor is thus 19.48% and it is the highest rate achieved compared with the other orientation options suggested in this research. Such work is helpful for PV technology utilization in Kuwait.

At max. PV penetration, the losses in the upstraem line is the lowest in the case for max. loading and highest for min. loading. At max. loading the line losses are always reduced no matter what the PV plant rating was. At min. loading, the upstream Line12 losses start to reduce once PV panels are connected until the PV rating exceeds almost double the load power rating, then the losses in the line is always higher than the losses before installing the PV plant.

PV penetration limit is unaffected by the loading condition but highly reduced at high voltages. The PV inverter was used to increase the max. PV penetration and it was achieved where the limitatiion factor on the penetration limit was the votage.

Finally, voltage regulation during night can be achievedv to a limit due to the fact that the reactive power available by the PV inverters is not sufficient enough to overcome the voltage drops. But even though, voltage regulation has been domonstrated effectively at high voltage rises.

6 REFERENCES

[1] Mattei M, Notton G, Cristofari G, Muselli M, Poggi P. Calculation of the polycrystalline PV module temperature using a simple method of energy balance. Renew Energ 2006; 31: p. 553-567.

[2] F. Q. Al-Enezi and J. K. Sykulski, Modeling of a Photovoltaic Module Considering the Solar Energy Available from Horizontal Surfaces over Kuwait Area, JOURNAL OF ELECTRONIC SCIENCE AND TECHNOLOGY, VOL. 10, NO. 2, JUNE 2012.

[3] M. Jamil Ahmad and G.N., Optimization of Tilt Angle for Solar Collector to Receive Maximum Radiation, Tiwari Centre for Energy Studies, Indian Institute of Technology Delhi, Hauz Khas, New Delhi-11 00 16, India.

[4] M. Benghanem, Optimization of tilt angle for solar panel: Case study for Madinah, Saudi Arabia, Department of Physics, Faculty of Sciences, Taibah University, P.O. Box 344, Madinah, Saudi Arabia.

[5] King, D. I., Boyson, W. E. and Kratochvil, J. A 2002. Analysis of factors influencing the annual energy production of photovoltaic systems. Proc. 29th IEEE Photovoltaic Specialist Conf., New Orleans, pp. 1356-1361.

[6] Fan, J. C. C., 1986. Theoretical temperature dependence of slar cell parameters. Solar Cells, Vol. 17, pp. 309-317.

[7] Group of Energy, PV syst V6.32, November 14th, 2011, Institute of the Sciences of the Environment, University of Geneva, Switzerland.

[8] Ali Hamzeh et al., International Journalvof Renewable Energy Reaserch Vol.5, No.4, 2015 .

[9] Alsema E. Fraile, D. Frischknecht, R. Fthenakis, VM. Held, M. Kim, HC. Polz, W. Raugei, \m. De Wild Scholten, M. (2009). Methodology Guidlines on Life Cycle Assessment of Photovoltaic Electricity, Subtask 20 LCA, IEA PVPS Task 12.

[10] Statistic Dept. & Information Center, Ministry of Electricity & Water, Electrical Energy Statistic Year Book 2015.

[11] "Stott and O. Alsac, "Fast decoupled load flow" IEEE Trans. on PAS, vol. 93, no. 3, pp. 859-869, May/June 1974".

[12] R. Tonkoski and L. Lopes. "Voltage regulation in radial distribution feeders with high penetration of photovoltaic," in Proc. IEEE Energy 2030, Atlanta, GA, 2008.

[13] Hadi Sadaat,"Power System Analysis", book, ISBN 0-07-561634-3

[14] Tomas Markvart, "Solar Electricity", book, ISBN: 0-471-98852-9.

[15] Al Busairi, H.A., Moller, H.J., 2010. "Performance evaluation of CdTe PV modules under natural outdoor conditions in Kuwait". In: 25th European Photovoltaic Solar Energy Conference and Exhibition/ 5th World Conference on Photovoltaic Energy Conversion, 6-10 September. Valencia, Spain, pp. 3468-3470.

[16] Alhajraf Salem, Kuwait Institute of Scientific Research, Science & Technology Magazine, issue 213, pages 12-15, October 2012.

Alsayegh, Osamah A., Fairouz, Fatma A., "Renewable Energy Supply Options in Kuwait", International Journal of Environmental, Chemical, Ecological, Geological and Geophysical Engineering Vol:5, No:12, 2011

PERFORMANCE ANALYSIS OF DIFFERENT THIN FILM MODULE TECHNOLOGY IN INDIAN CLIMATIC CONDITION

Yogesh Kumar Singh, Birinchi Bora, Ramayan Singh, Suparva Chakravarty, O S Sastry*, Rashmi Singh, Supriya Rai,
National Institute of Solar Energy 122003, Gurgaon, India
*sastry284@gmail.com, Tel: +91-0124-2579213, Telefax No. : +91-124-2579207

ABSTRACT: The potential of solar power across the country is more than 740 GW. Performance of module technology also depends on the different parameters such as irradiance, temperature, wind speed, humidity and rain fall. The photovoltaic simulation tools are useful for energy estimation of solar systems. This paper presents the performance assessment of small capacity grid connected solar photovoltaic system (GCSPVS) using PV* SOL software [1, 2]. Simulation software are used to estimate the performance of GCSPVS by calculating the performance ratio using PV* SOL software. This paper present's PR estimation of three thin film modules CdTe, CIGS and a-Si-H modules for different climatic zones. The performance modelling data is compared with the measured data to compare the estimation error.
Keywords: Solar photovoltaic, Performance Ratio, climatic zone, PV* SOL software

1 INTRODUCTION

In India the major player in thin film is CdTe. CdTe installed capacity more than 350 MW in across India. Beside the CdTe there is largest installation in CIGS and a-Si-H modules in various parts in India. The band gap of CdTe module is around 1.5 eV which is well matched to the spectrum range falling in rainy seasons of India. CIGS band gap around 1.2 eV and a-Si-H band gap is around 1.7 eV. The behaviors of all technology are different in different climatic zones. With the help of estimated energy yield and reference yield the particular technology can be selected for deployment. Characteristics of indian climatic condition is given in table 1. In a hot and dry condition the temperature range is greater than 30 and relative humidity is less than 55. Warm and humid zones temperature zone variance greater than 25-30 and relative humidity behavior range around greater than 55 to 75, temperate climatic zone the temperature range is 25 to 30 and humidity range is less than 75, cold climatic zone temperature less than 25 and humidity in all value and composite climatic zones temperature and humidity condition are different.

Table 2: Maximum mean Temperature and Mean relative humidity of different climatic zones of India

Climate	*Mean monthly maximum Temperature (deg. C)*	*Mean Relative humidity (%)*
Hot and Dry	>30	<55
Warm and	>30	>55
Humid	>25	>75
Temperate	25-30	<75
Cold	<25	All values
Composite	This applies when six month or more do not fall within any of above categories	

Five different representative locations has been selected as per climatic zones of India are Gurgaon, Jodhpur, Bengaluru, Chennai and which are shown in table 2.

Table 2: Examples of each climatic range in India

Climate	Region In India
Composite	Gurgaon
Hot and Dry	Jodhpur
Temperate	Bengaluru
Warm and Humid	Chennai
Cold	Leh

2 METHODOLOGY

With the help of PV* SOL simulation software, performance of three different thin film module technology: CdTe, CIGS and a-Si-H are modelled. The details are given in table 3.

Table 3: Electrical output of the three different technology

Technology	*Pmax*	*Voc*	*Isc*	*Vm*	*Im*
CdTe	80	60.8	1.88	48.5	1.65
CIGS	120	55.2	3.32	42.7	2.81
a-Si-H	75	91.8	1.4	67	1.12

Figure 1: Test-bed set-up of three Technologies

The test bed of different thin film modules CdTe, CIGS and a-Si-H are shown in fig1. Three technologies are studied on the basis of performance ratio. The performance ratio depends on the array yield and reference yield. The ratio of the array yield to reference yield is called the Performance ratio. The performance ratio [3] is defined as below:

2.1 Array yield (YA)

Array yield (YA) is defined as the total energy generated by the PV array for a defined period (day, month or year) divided by the rated output power of the installed PV array and is given by:

$$Ya = EDC/Ppv\ rated$$

2.2 Reference yield (YR)

Reference yield is defined as the ratio of total in plane solar insolation Ht (kWh/m^2) to the reference irradiance G (1 kW/m^2).This parameter represents an equal number of hours at the reference irradiance and is given by:

$$r = Ht(kWh/m2)/G(kW/m2)$$

2.3 Performance ratio (PR)

Performance ratio is defined as the ratio of the array yield (Ya) to the reference yield (Yr). This normalizes performance parameter with respect to the incident solar radiation is a dimensionless quantity and provides important information on the overall effect of losses. This parameter is used to evaluate the long-term changes in the performance and decreasing year wise PR values are indicative of loss in the performance:

$$PR = Ya/Yr$$

3 RESULTS AND DISCUSSION

Simulation study is used for forecasting the PV system output would useful in enhancing system performance, stability and reliability. With the help of simulation software analysis the thin film module technology, performance ratio in different climatic zones in India has been analyzed.

With the help of PV* SOL simulation result the data are obtained in the monthly wise PR format. In the figure PR is plotted with monthly distribution for a year. In the most cases the performance lies 0.79 to 0.9. In fig 2 CdTe in composite location performance is same in different seasons. But in the cold region the PR is high in winter season, but in the rainy season PR is low. In Hot and dry zones the PR is going down in the summer session, rest climatic zones patterns are very same to each other. In fig 3 CIGS technology PR range lies 0.79 to 0.9. In cold region the performance of CIGS is more than the other climatic zones. In temperate climatic zones CIGS module's behavior varies from winter to summer season. In hot and dry location PR is lower in the summer season. In the month of September all PR are same for different Climatic zones.

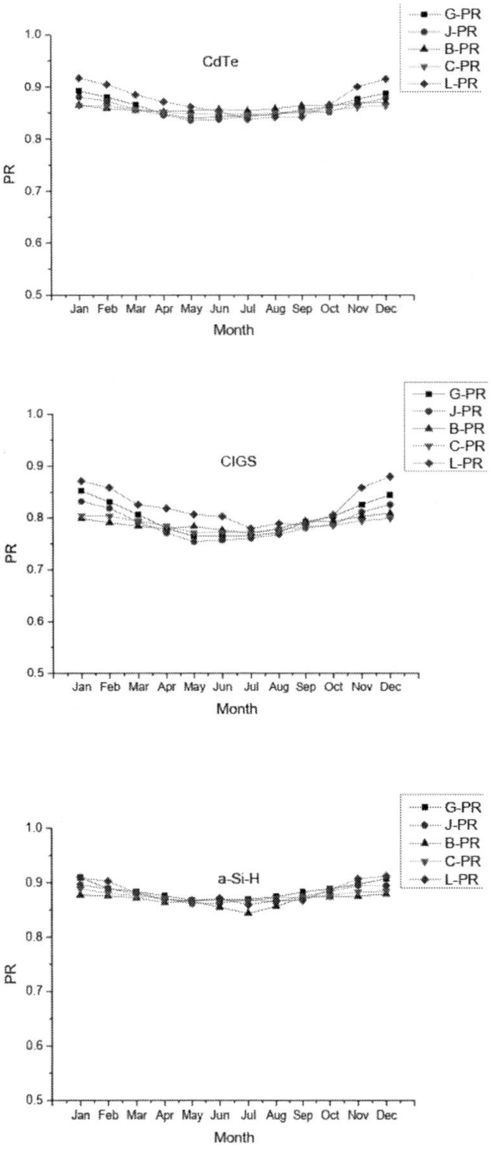

Figure 2: Monthly PR of the Three SPV technology

In temperate climatic zones CIGS module's behaviour varies from winter to summer session. In hot and dry location PR is lower in the summer season. In fig 2 a-Si-H PR values lie in 0.84 to 0.93.

The deviation of the measured value in real time data and the predicted value from the PVSOL software is also done. The relative error range for CdTe technology is minimum 2%, for CIGS technology minimum error is 4% and for a-Si-H minimum value is 1%.

4. CONCLUSION

In this paper PV* SOL software are used to estimate the power output for different climatic zones. With the help of this study it can be observed that thin film PV technology have bright future in different Indian climatic zone.

ACKNOWLEDGEMENT

This research is based upon work supported in part by the Solar Energy Research Institute for India and the U.S. (SERIIUS) funded jointly by the U.S. Department of Energy subcontract DE AC36-08G028308 (Office of Science, Office of Basic Energy Sciences, and Energy Efficiency and Renewable Energy, Solar Energy Technology Program, with support from the Office of International Affairs) and the Government of India subcontract IUSSTF/JCERDC-SERIIUS/2012 dated 22nd Nov. 2012.

REFERENCE

[1] Stand-Alone Photovoltaic (SAPV) System Assessment using PVSYST Software Y.M.Irwana*, A.R.Ameliaa , M.Irwantoa , Fareq.Ma , W.Z.Leowa , N.Gomesha , I.Safwati Science Direct Energy Procedia 79 (2015) 596 – 603.

[2] PV* SOL Premium software

[3] Performance Parameters for Grid-Connected PV Systems B. Marion, J. Adelstein, and K. Boyle National Renewable Energy Laboratory31st IEEE Photovoltaics Specialists Conference and Exhibition Lake Buena Vista, Florida January 3-7, 2005

PERFORMANCE CMPARISION OF PV MODULE BASED OF TEMPERATURE COFFICIENT IN INDOOR AND OUTDOOR CONDITIONS AS PER IEC 61853-1 .

Manoj Morampudi[1], Birinchi Bora[1], Gopal Kumar[1], Rajesh Kumar[1], Rahnuma Siddiqui[1], Sneh Lata[1], Ganesh Gowri[1], Bharat Dubey[1], Pragati Rajput[1], Sahan Raghava[1], Mayank Singh[1], Jithender Vyas[1]

[1]National Institute of Solar Energy,
Gurgaon, India, 122003

ABSTRACT: In real world energy scenario, photovoltaic module power generation from laboratory test conditions to field conditions vary irrespective of capacity derived from energy rating of module as per IEC 61853-1 standard. Performance consistency over linearity of module at different irradiance levels enables the precise comparison of measured data at both indoor and outdoor test conditions. Hence the current research on irradiation and thermal impact over photovoltaic module is intended to analyse through indoor simulated and outdoor realistic performance based on 23 test conditions of IEC 61853-1" with deviation in terms of percentage relative temperature coefficients.
Keywords: Photovoltaic Module, Crystalline silicon, IEC 61853-1

1 INTRODUCTION

STC power outcome in laboratory conditions remained still an anticipation to attain matching results with PV module mounted at real field. Furthermore to estimate PV energy yield accurately from individual module level IEC 61853-1[3] standard is equally important. Consequently laboratory testing of each module sample by simulating the level of intensity and temperature ranges at lab is cost effective. Hence testing of PV module at field conditions might be a most feasible and economical solution. In a country with wet and dry climate like India, it becomes almost important to draw a comparison in the performance based on irradiance and temperature as its variation is high. Hence it is intended to analyse the comparison of indoor simulated and outdoor realistic performance at 23 test conditions as per IEC 61853-1 with deviation in terms of percentage relative temperature coefficients.

PV module performance varies from simulated to outdoor field condition depending upon temperature and intensity variations. To express the conformity, a Standard Test Conditions (STC) norm was being followed and defined as a 'typical' sunny day with an irradiance of 1000W/m^2, an Air mass (AM, an indicator for the composition of the light spectrum, defined in IEC norm 60904-3 (2008)) of 1.5 and a test specimen temperature of 25°C (IEC-I 2008 and Gueymard et al. 2002)[1]. Also, three different procedures were developed to correct (field) measured IV characteristics to the STC, making outdoor comparison between different PV-panels possible (IEC 2009)

The main object of this study is to determine the temperature coefficients of short-circuit current Isc (α), open-circuit voltage Voc (β), and maximum power Pmax (δ) of PV module with respect to various temperature and intensity levels to simulate energy yields of the test specimen at all intensity ranges to conclude coefficients influence. As per IEC 61215[4] for crystalline silicone PV modules the coefficients so determined are only valid at the irradiance at which the measurements were made. As the influence of coefficients either in current and voltage is non-linear in real time fields due to variation in intensity and temperature. For modules with known linearity over a certain irradiance range according to IEC 60891[5], the calculated coefficients can be considered valid over that irradiance range.

2 Experimental setup:

Indoor test setup: Make: QuickSun 700A Flash System Class AAA solar simulator according to IEC 60904-9 Xenon flash tube with AM1.5G filter - +/- 2 % irradiance uniformity with proprietary optics.

Figure 1: Indoor sun simulator test setup, Endeas quick sun 700A

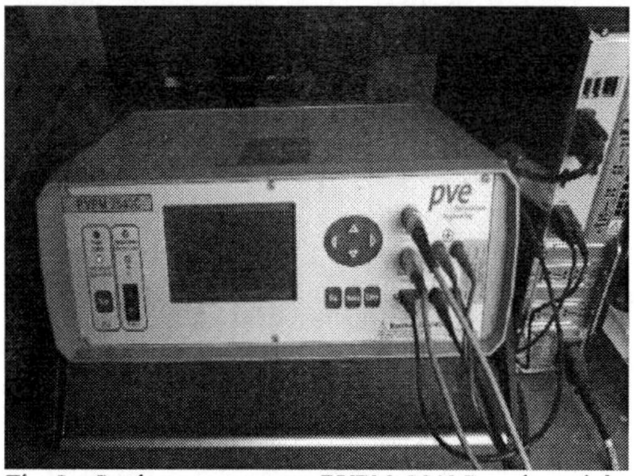

Fig 2: Outdoor test setup, PVPM 2540C and module mounting

Outdoor test setup equipped with measurement and evaluation unit of industrial class PVPM 2540C model with measuring accuracy for the I-V characteristics, having four-wire-measurement leads to avoid systematic errors in voltage measurement. Irradiation reference sensor (Phox) with integrated Pt100 temperature sensor to measure back surface temperature of the module under test.

Fig 3: Module mounting at outdoor test setup

3 METHODOLOGY

The electrical output performance of multi crystalline silicone PV module is determined under simulated conditions at indoor facility using sun simulator, current versus voltage (I-V) measurement sweeps at various intensity levels and desired temperature ranges are simulated using climatic chambers. The module is mounted on a manual single axis tracker at a fixed tilt with reference cell. I-V curve is sweep using PVPM 2540C having a radiation sensor of phox and an integrated temperature sensor of pt100 with lead are attached to the module back sheet to record temperature simultaneously. Actual outdoor test conditions does not necessarily match the particular irradiance and temperature ranges defined as per IEC 61853-1 standard; so the translation procedures based on extrapolate and interpolate to obtain desired conditions are needed to translate the measured data from the actual test conditions to the reported rating conditions[2]. The measured current-voltage characteristic corrected to standard test conditions or other selected temperature and irradiance values by applying the following equations

$$I_2 = I_1 + I_{SC}.\left(\frac{G_2}{G_1} - 1\right) + \alpha.(T_2 - T_1)$$

$$V_2 = V_1 - R_S.(I_2 - I_1) - K.I_2.(T_2 - T_1) + \beta.(T_2 - T_1)$$

Where

I_1, V_1 are coordinates of points on the measured characteristics;

I_2, V_2 are coordinates of the corresponding characteristics;

G_1 is the irradiance with the reference device;

G_2 is the irradiance at the standard or other desired irradiance;

T_1 is the measured temperature of the test specimen;

T_2 is the measured temperature of the test specimen

I_{sc} is the measured short-circuit current of the test specimen at G_1 and T_1

α and β are the current and voltage temperature coefficients of the standard or target irradiance for correction and within the temperature range of interest.

R_s is the internal series resistance of the specimen;

K is a curve correction factor

Obtained coefficients are expressed in terms of relative percentages by dividing the calculated α and β by the values of Isc, Voc and Pmax at 25°C (1000W/m²) [2], extended to determine relative temperature coefficients at 15°C, 50°C and 75°C at both indoor simulated and outdoor real time conditions. The values of Isc, Voc and Pmax are plotted as functions of temperature for each set of data. Given a certain irradiance level, it is to be noted that β(for Voc) and δ(for Pmax) are the two most sensitive to temperature changes. And the data is compared against separately processed and analysed data from indoor lab testing using sun simulator that I-V characteristics of the PV device are to be measured in accordance with IEC 60904-1.

These parameters are determined from measurements in simulated sunlight. The coefficients determined are valid at the irradiance at which the measurements were made. The temperature of the module was measured at the four positions and their values averaged. Increased the PV module to the temperature of interest until its temperature is uniform within ±2 °C till the module temperature has stabilized, set the irradiance to the desired level, using the reference device (IEC 60904-2). Record the current-voltage characteristic and temperature of the specimen and take the values of ISC, Voc and Pmax with respect to temperature of the module in steps of 75°C, 55°C, 25°C, 15°C and repeat with respect to intensity range of 200 W/m², 400 W/m², 600W/m², 800 W/m², 1000 W/m², and 1100W/m². Plotted the values of Isc, Voc and Pmax as functions of the device temperature and construct a least-squares-fit curve through each set of data.

Table I: 23 sets of irradiance and temperature conditions

Irradiance (W/m²)	Module Temperature °C			
	15	25	50	75
1100	NA	1	2	3
1000	4	5	6	7
800	8	9	10	11
600	12	13	14	15
400	16	17	18	NA
200	19	20	21	NA
100	22	23	NA	NA

5 Results and discussion:

The performance of PV module tested in indoor condition and in outdoor condition. The comparison table of Pmax in outdoor and indoor conditions are given below table

Table II:Pmax at indoor simulated conditions

Irradiance (W/m²)	Module Temperature °C			
	15	25	50	75
1100	NA	102.14	93.69	86.58
1000	95.35	92.91	86.39	79.99
800	76.23	74.25	68.23	63.24
600	56.84	55.48	50.26	46.67
400	37.18	36.41	33.23	NA
200	17.72	17.43	15.81	NA
100	8.98	8.95	NA	NA

Table III: Pmax at outdoor real time conditions

Irradiance (W/m²)	Module Temperature °C			
	15	25	50	75
1100	NA	91.88	84.61	79.14
1000	89.17	86.6	80.08	73.43
800	71.39	69.4	64.32	59.11
600	55.62	54.18	50.49	46.68
400	37.93	37.07	34.83	NA
200	18.14	17.87	17.36	NA
100	8.14	8.17	NA	NA

Table IV: Relative temperature coefficient of Pmax at indoor and outdoor w.r.t intensity

Relative temperature coefficient of Pmax w.r.t intensity							
Intensity Isc	100	200	400	600	800	1000	1100
Indoor at 15°C	0.3390	0.1695	0.0840	0.0557	0.0417	0.0333	0.0302
Outdoor at 15°C	0.3333	0.1570	0.0779	0.0516	0.0394	0.0322	0.0293
Indoor at 25°C	0.3279	0.1667	0.0833	0.0554	0.0416	0.0332	0.0302
Outdoor at 25°C	0.3276	0.1545	0.0772	0.0514	0.0393	0.0321	0.0292
Indoor at 50°C	0.3030	0.1639	0.0831	0.0550	0.0413	0.0327	0.0297
Outdoor at 50°C	0.3016	0.1484	0.0757	0.0508	0.0389	0.0319	0.0290
Indoor at 75°C	0.2817	0.1574	0.0814	0.0542	0.0408	0.0325	0.0295
Outdoor at 75°C	0.2794	0.1429	0.0745	0.0501	0.0385	0.0316	0.0287

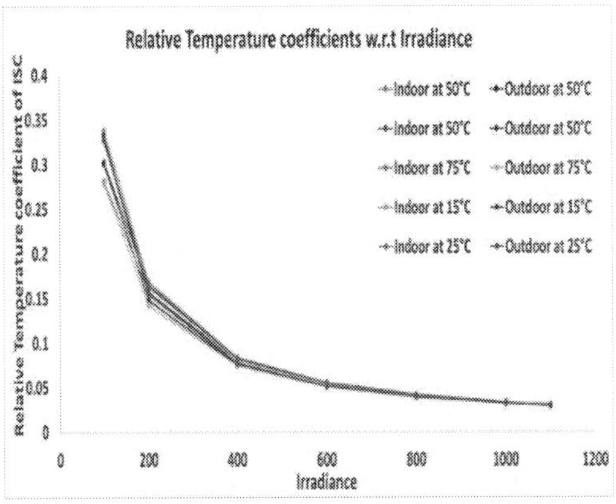

Irradiance (W/m^2)	Relative temperature coefficient: Indoor Simulated conditions	Outdoor real time conditions	Deviation
1100	-0.0033	-0.0030	8.0502
1000	-0.0028	-0.0030	-9.9524
800	-0.0023	-0.0024	-1.4555
600	-0.0018	-0.0017	5.6597
400	-0.0011	-0.0011	4.1026
200	-0.0005	-0.0003	32.9983
100	-0.0001	0.0000	75.5039

Table5: Deviation in terms of Percentage relative temperature coefficients of Pmax

The performance measurement comparison has been done for outdoor and indoor testing conditions are calculated by using single temperature coefficient.

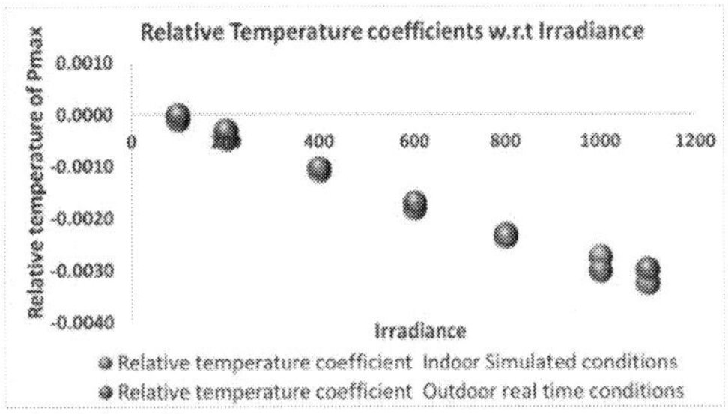

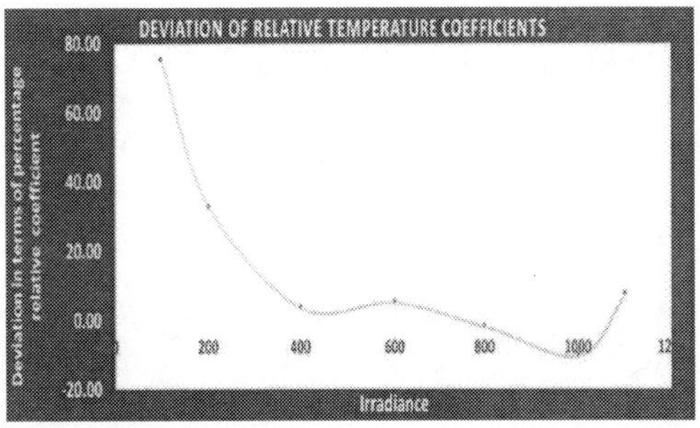

6 Conclusions:

To determine the energy rating of the PV module as per IEC 61853-1, at 23 conditions are performed under simulated conditions using simulator is not enough in terms of accuracy. Four different temperature ranges with respect to intensity variation has proven that the response over module in terms of coefficientsα, β and δ are varying with reference to test patterns of indoor with outdoor conditions.

To get realistic energy estimation, have to mandate the necessity to determine coefficients should be consider both indoor for manufacturing compliance followed by outdoor uncontrollable measurement for real time accurate energy estimation.

In tropical country like India maximum of intensity fall between in the range of 600-800nm, form the current analysis concluded that the variation for Pmax is in the range of -1.46% to +5.66% of measurement deviation. So it should be necessary to prefer measurement of temperature coefficients to estimation the optimum energy rating of a PV module other than traditional indoor simulated testing under controlled conditions.

7 Acknowledgement:

The authors would like to acknowledge support and help provided from National Institute of Solar Energy, Gurgaon, An autonomous institute under Minister of New and Renewable Energy

8 References

[1] C.A. Gueymard et al. Proposed reference irradiance spectra for solar energy systems testing

[2] Paghasian, K. ; TamizhMani, G. Photovoltaic module power rating per IEC 61853–1: A study under natural sunlight

[3] IEC 61215

[4] IEC 61853-1

[5] IEC 60891

CONTROL STRATEGY OF A PHOTOVOLTAIC MODULE EMULATOR BASED ON HILL-CLIMBING AND SINGLE-DIODE MODEL

Juan D. Bastidas-Rodríguez[1], Brian Ospina[2], Jhoan S. Parra[3], Edinson Franco-Mejía[4]
[1]Universidad Industrial de Santander (Colombia), Carrera 27 Calle 9 Office LP 211, Bucaramanga – Colombia, +57 6344000 Ext. 1466, jdbastir@uis.edu.co. [2]Universidad del Valle (Colombia), brian.ospina@correounivalle.edu.co.[3]Universidad del Valle (Colombia), jhoan.parra@correounivalle.edu.co.[4]Universidad del Valle (Colombia), edinson.franco@correounivalle.edu.co.

ABSTRACT: This paper introduces a control strategy of a photovoltaic (PV) module emulator based on the single-diode model and a hill-climbing (HC) algorithm. A PV module emulator (PVME) is implemented by using a power converter whose output current and voltage match the current vs. voltage (I-V) curve of a PV module for a given load. The control strategy is divided into two parts: the definition of the operating point, and the setting of the power converter duty cycle. The PVME operating point is the intersection between the I-V curve of the PV module (given by the single-diode model) and actual load, which can be found by using Newton-Raphson (NR) method. The duty cycle of the power converteris tracked by an HC algorithm, which perturbs the duty cycle in different directions and evaluates the error until the desired operating point is reached. The proposed control strategy is validated with simulations and experiments taking an 85 W PV module as a reference. The relative errors obtained in the I-V curve reproduction are less than 4% and 5% for the simulation and experiments, respectively.
Keywords: Photovoltaic module emulator, single-diode model, hill-climbing algorithm, DC-DC power converter.

1 INTRODUCTION

In the last years, photovoltaic (PV) systems have taken a relevant place in the electric power generation to mitigate the climate change and supply the growing energy demand [1]. However, there are some challenges that motivates research on PV systems, particularly the increment of the efficiency of the whole PV system, the PV panels and the PV cells.

The research activities focused on the improvement of PV system's efficiency include the evaluation of new Maximum Power Point Tracking (MPPT) techniques, new structures and architectures of power converters, new techniques to estimate the PV system energy production, new design procedures of the PV systems, among others. Those research topics require experimental analysis and evaluations of the whole system and its components to compare the proposed solutions with the ones available in literature and quantify the advantages and disadvantages.

Comparing two or more MPPT techniques or control strategies is not simple, since the variability in the operating conditions of the PV panels (i.e. irradiance and temperature) difficults the repeatability of the experiments. Moreover, operating conditions of the PV panels is limited by the location and season where the panels are installed. That is why, a PV module emulator is a useful tool not only for research activities but also for teaching purposes.

A PV module emulator (PVME) is a power converter with a control strategy whose output voltage and current reproduce the electrical behavior of a PV module. In literature there are different hardware structures to implement a PV module emulator. The main hardware structures of PVME are briefly described below:

- PVME based on power converters [2]–[7].
- PVME based on the equivalent circuit and composed by a current source, diodes and resistors [8].
- PVME based on the voltage and current amplification of a PV cell or a photodiode [9], [10].
- PVME based on the variation of the series resistance in a voltage source [11], [12].

PVME implemented with power converters is the widely used structure in literature, since it provides flexibility in the voltage, current and power of the PV module to be emulated. Additionally, the control of the power converters can be implemented on embedded systems, which allow the modification of the control parameters or the emulated environmental conditions through the communication with a computer or other devices.

The challenge in PVME implemented with power converters is to develop a control strategy that allows a precisely reproduction of the nonlinear behavior of a PV module under different irradiance and temperature conditions.

In [6] and [2] the authors use microcontrollers to control a power converter used to implement the PVME. On one hand, in [6] the authors use a Buck-Boost converter and a low cost microcontroller sets the operating point by using a piecewise linear approximation of the analytical model of a PV module. On the other hand, the authors in [2] use a Buck converter and implement the single diode model of a PV module on a microcontroller. Such a microcontroller generates the reference of a PI compensator that regulates the output voltage of the Buck converter.

PVMEs implemented on a step-down DC-DC converter are described in [4] and [5]. In [4] the DC-DC converter is controlled by an embedded system based on an FPGA. The embedded system has an algorithm to calculate the current vs. voltage (I-V) curve of the PV module, which is used to calculate the reference of the DC-DC converter output voltage. In [5] the power converter is controlled by a DSP and the irradiance and temperature conditions are defined by a potentiometer connected to the ADC channels of the DSP. However, the irradiance and temperature conditions are predefined and stored in a look-up table; therefore, the PVME cannot reproduce the PV module behavior in a continuous range of irradiances and temperature.

This paper presents a hill-climbing (HC) based control strategy for a PVME implemented with a power converter. The proposed control strategy calculates the theoretical operating point, i.e. the intersection of the I-V curve and the load line, by solving a nonlinear equation constructed from the single diode model and the information of the current load (converter output voltage and current). Such a nonlinear equation is solved by the Newton-Raphson (NR) method and provides the reference for a HC algorithm that searches the duty cycle that minimizes the error between the theoretical and power converter operating points.

The rest of the paper is organized as follows, the single diode model of a PV module is presented in Section 2; the hardware structure used to implement the proposed PVME is presented in Section 3; Section 4 describes the proposed HC based control strategy; Section 5 presents the simulation and experimental results; and finally Sections 6 shows the conclusions.

2 PHOTOVOLTAIC MODULE SINGLE-DIODE MODEL

The single diode model is widely used in literature to represent a PV module, since it provides a trade-off between accuracy and complexity. The circuital implementation of the single diode model is shown in Fig. 1 [13], where the current source (I_{ph}) represents the current produced by the photovoltaic effect, the diode represents the nonlinear behavior of the PV cells, and the resistors in parallel (R_h) and series (R_s) represent the losses produced by the leakage currents and the ohmic contacts, respectively.

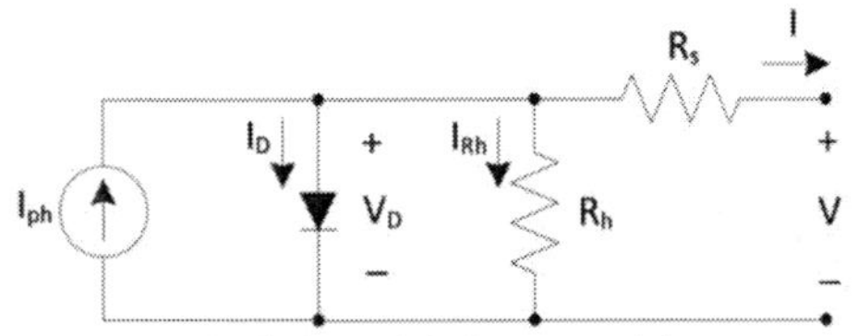

Figure 1: Single diode model of a PV module.

The output current of a PV module composed by N_s series connected cells (I) is shown in (1), where V is the PV module output voltage, I_{sat} is the inverse saturation current of the diode, η is the ideality factor, and V_T is the thermal voltage. V_T is proportional to the cells temperature in Kelvin (T) as shown in (2), where k and q are the Boltzmann constant and the electron charge, respectively. From (1) it is also possible to express the voltage as a function of the current as shown in (3).

$$I = I_{ph} - I_{sat} \cdot \left(e^{\frac{V+R_S \cdot I}{N_S \cdot \eta \cdot V_T}} - 1\right) - \frac{V+R_S \cdot I}{R_h} \quad (1)$$

$$V_T = \frac{k \cdot T}{q} \quad (2)$$

$$V = R_h \cdot I_{ph} - R_h \cdot I_{sat} \cdot \left(e^{\frac{V+R_S \cdot I}{N_S \cdot \eta \cdot V_T}} - 1\right) - (R_h + R_s) \cdot I \quad (3)$$

3 PVME STRUCTURE

In this paper a PVME implemented with a power converter is selected, since this structure is more flexible than other structures presented in [14] with respect to both, the PV modules electrical characteristics and the operating conditions to be emulated.

The diagram of Fig. 2 shows the structure of the PVME used in this paper, which is formed by four sub-systems: a DC-DC power converter, a control unit, a current and voltage measurement system, and an application running on a PC. The irradiance, temperature and electrical characteristics of the PV module are defined by the user in the PC application and it calculates the single diode model parameters, which are sent to the control unit. The control unit, based on a microcontroller, takes the information from the measurement system and uses the proposed control strategy (described in the following Section) to modify the duty cycle of the DC-DC converter in order to reproduce the PV module behavior in the DC-DC output.

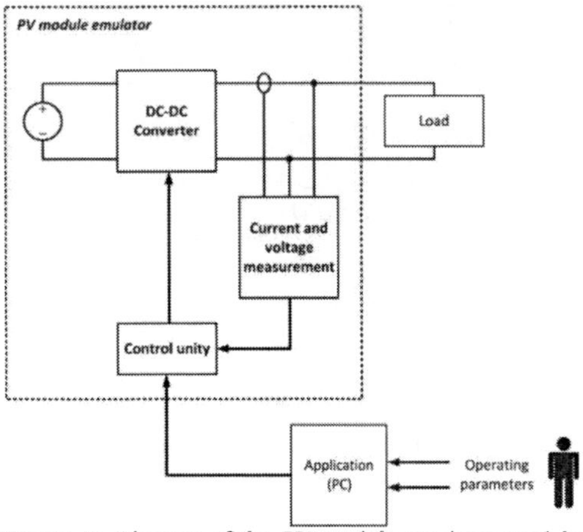

Figure 2: Diagram of the PV module emulator used in this paper.

4 PVME CONTROL STRATEGY

The objective of the control strategy is to set the operating point of the DC-DC converter (i.e. output current and voltage) in the intersection of the emulated I-V curve and load line (see Fig. 3). Moreover, the control strategy needs to recalculate the operating point when there is a change in the PV module operating conditions (irradiance and temperature) or in the load line.

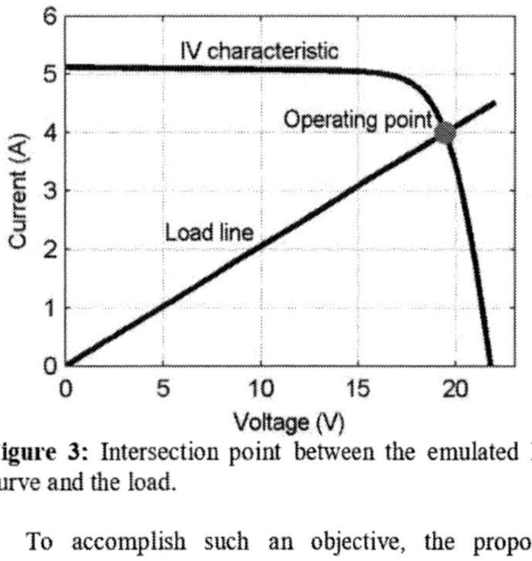

Figure 3: Intersection point between the emulated I-V curve and the load.

To accomplish such an objective, the proposed

control strategy is divided in two stages, the first one is the calculation of the desired operating point, and the second one is the search of the duty cycle that makes the DC-DC converter to operate in the desired point.

4.1 Definition of the operating point

For a resistive load the current and voltage relationship is defined by Ohm's law as show in (4), where G_L is the inverse of the load resistance and V_L and I_L are the load voltage and current, respectively. For a PV module the relationship of current and voltage is given in (1).

To calculate the intersection of the PV module I-V curve and the load line, the voltage and currents must be equated, which is performed by replacing (4) in (1). Then, the root of the resulting equation, i.e. (5), must be solved to obtain the voltage of the intersecting point. Finally, the intersecting point current can be calculated by evaluating (4) for the intersecting point voltage.

$$I_L = G_L \cdot V_L \tag{4}$$

$$I_{ph} - I_{sat} \cdot \left[e^{\frac{V_L \cdot (1 + R_S \cdot G_L)}{N_S \cdot \eta \cdot V_T}} - 1 \right] - \frac{V_L \cdot [1 + G_L \cdot (R_s + R_h)]}{R_h} = 0 \tag{5}$$

It is worth noting that the root of (5) is between zero volts and the open-circuit voltage (V_{oc}) of the PV module; moreover, it has only one solution since the derivative of (5) is definite negative. Therefore, (5) can be solved by a simple method, like NR, with a reduced number of iterations; additionally, using the knowledge of the equations it is possible to define an initial value close to the solution, which reduces even more the number of iterations [15].

The basic equation for a new iteration of the NR method is shown in (6), where x_i and x_{i+1} are the present and the next approximations of the solution, respectively, while f and f' are the function to be solved and its derivative, respectively. Replacing (5) in (6) results in (7), which is the equation used by the NR method.

$$x_{i+1} = x_i - \frac{f(x_i)}{f'(x_i)} \tag{6}$$

$$V_{Li+1} = V_{Li} + \frac{I_{ph} - I_{sat} \cdot \left[e^{\frac{V_{Li} \cdot (1 + R_S \cdot G_L)}{N_S \cdot \eta \cdot V_T}} - 1 \right] - \frac{V_{Li} \cdot [1 + G_L \cdot (R_S + R_h)]}{R_h}}{I_{sat} \cdot \left(\frac{1 + R_S \cdot G_L}{N_S \cdot \eta \cdot V_T} \right) \cdot e^{\frac{V_{Li} \cdot (1 + R_S \cdot G_L)}{N_S \cdot \eta \cdot V_T}} + \frac{1 + G_L \cdot (R_S + R_h)}{R_h}} \tag{7}$$

The algorithm used to implement the NR method is introduced in Fig. 4. The algorithm starts defining a guess solution for the unknown voltage (V_0), then, the load conductance (G_L) is calculated from the voltage and current measured at the output of the power converter (V_{op} and I_{op}), such information is used to calculate the next solution (V_{Li+1}). If V_{Li+1} is less than zero, then V_{Li+1} is set to zero, otherwise the relative difference between the next (V_{Li+1}) and the present (V_{Li}) solution is calculated and stored in the variable e_r. After, the actual solution is updated ($V_{Li}=V_{Li+1}$) and the stopping criteria (i.e. maximum number of iterations (i_m) and minimum variation of the next solution (ie_a)) are evaluated. Finally, the voltage operating point (V_L) is defined as the value of the present solution (V_{Li}).

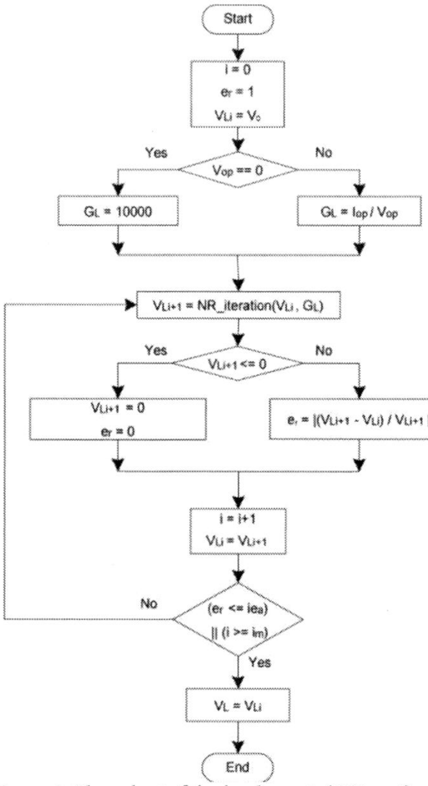

Figure 4: Flow chart of the implemented NR method.

It was found, experimentally, that NR method converges slowly when the solution is a less than the maximum power point voltage (V_{mpp}) of the emulated I-V curve, which may cause the premature termination of the NR algorithm (for exceeding i_m) and a significant error in the solution. To solve this situation, the I-V curve is divided in two regions (see Fig. 5) by the load at the maximum power point $G_{mpp}=I_{mpp}/V_{mpp}$, where I_{mpp} is the maximum power point current. In the voltage region the operating point is obtained following the procedure described in Fig. 4, i.e. solving a nonlinear equation to obtain the operating point voltage (V_L), while in the current region a similar procedure is performed to obtain the operating point current (I_L).

To obtain I_L an equation similar to (5), but as function of the operating point current, is required; such an equation (8) is obtained by solving V_L from (4) and replacing it on (3).

$$I_{ph} - I_{sat} \cdot \left[e^{\frac{I_L \cdot (\frac{1}{G_L} + R_S)}{N_S \cdot \eta \cdot V_T}} - 1 \right] - \frac{I_L \cdot \left[\frac{1}{G_L} + R_S + R_h \right]}{R_h} = 0 \tag{8}$$

The operating point current is obtained by solving (8), taking into account that its derivative is definite negative and the solution is between zero and the module short-circuit current (I_{SC}). Hence, a NR method, similar to the one presented in Fig. 4, can be used to find I_L. The basic equation used by the NR method to calculate the operating point current is shown in (9).

$$I_{Li+1} = I_{L_i} + \frac{I_{ph} - I_{sat} \cdot \left[e^{\frac{I_{Li} \cdot (\frac{1}{G_L} + R_S)}{N_S \cdot \eta \cdot V_T}} - 1 \right] - \frac{I_{Li} \cdot \left[\frac{1}{G_L} + R_S + R_h \right]}{R_h}}{I_{sat} \cdot \left(\frac{\frac{1}{G_L} + R_S}{N_S \cdot \eta \cdot V_T} \right) \cdot e^{\frac{I_{Li} \cdot (\frac{1}{G_L} + R_S)}{N_S \cdot \eta \cdot V_T}} + \frac{\frac{1}{G_L} + R_S + R_h}{R_h}} \quad (9)$$

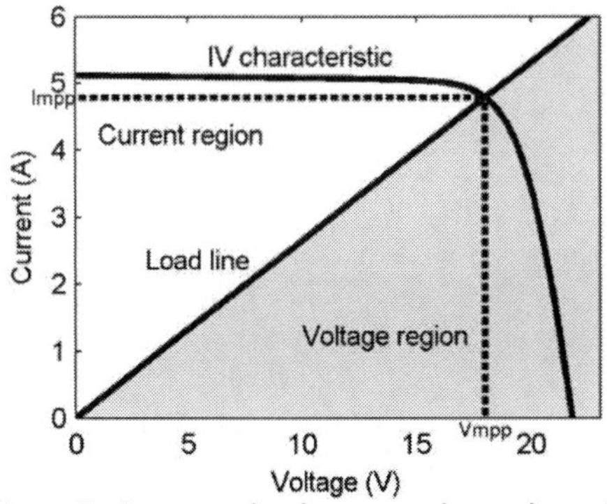

Figure 5: Current and voltage operating regions of the PVME. The regions are divided by the load line at the MPP: $G_{mpp}=I_{mpp}/V_{mpp}$.

The expression presented in (10) is used to establish the region where a given operating point is located. If y is less than zero ($y<0$) G_L is greater than G_{mpp} ($G_L>G_{mpp}$), if y is greater than zero ($y>0$) G_L is less than G_{mpp} ($G_L<G_{mpp}$), and if y is zero ($y=0$) G_L is equal to G_{mpp} ($G_L=G_{mpp}$). Moreover, a hysteresis band was defined to avoid oscillations in the definition of the operating point region as shown in Fig. 6, where the limits (d_1 and d_2) are defined as follows, $d_1 = 0.04 \cdot V_{oc} \cdot I_{mpp}$ and $d_2 = 0.08 \cdot V_{mpp} \cdot I_{sc}$[16].

$$y = V_{op} \cdot I_{mpp} - V_{mpp} \cdot I_{op} \quad (10)$$

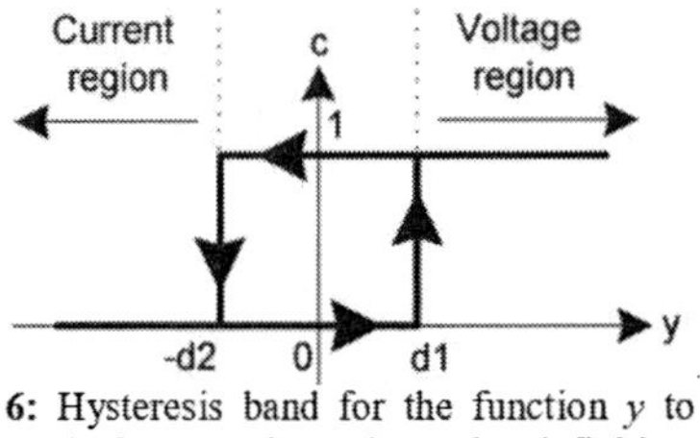

Figure 6: Hysteresis band for the function y to avoid oscillations in the operating point region definition.

4.2 Converter control based on hill-climbing

The proposed control strategy is illustrated in the flow chart of Fig. 7, which is divided in two blocks, the definition of the operating point and the control of the power converter based on HC method. In the proposed control strategy a Buck converter is used as reference, however, it could be easily adapted for other power converters.

The definition of the desired operating point starts with the identification of the operating point region by evaluating (10) and hysteresis band defined at the end of Section 4.1. If the operating point is in the voltage region ($c=1$), then (5) is solved with the NR method to obtain the reference operating point voltage (V_L), otherwise (i.e. $c=0$) (8) is solved to obtain the reference operating point current (I_L).

There are two options to modify the duty cycle of the power converter, the first one using the reference V_L and second one using the reference I_L. The duty cycle modification in both options is based on the HC procedure, which is to identify the sign of the error and increase or reduce the duty cycle by a fixed quantity in order to reduce the error.

If V_L was calculated in the previous stage, the error in the operating voltage is evaluated as $E_{Vop}=V_L-V_{op}$. If the absolute value of E_{Vop} is greater than a threshold (e_{vth}) the present duty cycle (d_k) is perturbed by a fixed quantity (Δd_v), otherwise, the duty is not modified. Moreover, if the E_{Vop} is positive and greater than e_{vth}, the output voltage of the power converter is less than the desired operating voltage, therefore, the duty cycle of the power converter is increased; otherwise, if E_{Vop} is negative and less than $-e_{vth}$, the duty cycle is reduced to reduce the output voltage of the power converter.

If I_L is calculated in the first block of the control strategy, the error in the operating point current is evaluated as $E_{Iop}=I_L-I_{op}$. A threshold value is also considered to modify the duty cycle (e_{ith}); then, if $E_{Iop}>e_{ith}$ or $E_{Iop}<-e_{ith}$, the duty cycle is increased or reduced by Δd_i, respectively, since an increment in the duty cycle produces an increment in the output current of the power converter and vice versa. Finally, a saturation function is implemented to force the duty cycle to be within its maximum (d_{kmax}) and minimum (d_{kmin}) values.

It is worth noting that both, Δd_v and Δd_i, need to be adjusted to obtain a tradeoff between the convergence speed and oscillations around the solution. Additionally, the definition of the voltage and current error thresholds (e_{vth} and e_{ith}) reduce the oscillations of the operating point. In this paper the parameters Δd_v, Δd_i, e_{vth} and e_{ith} were adjusted experimentally by observing the system behavior.

5 RESULTS

The proposed control strategy was validated with simulations and experiments. The PV module to be emulated is an ERDM 85 [17], which is a monocrystalline PV module composed by 36 series connected cells. The power stage used is a Buck converter with a switching frequency of 200 kHz.

5.1 Simulation results

A Buck converter and the proposed control strategy were implemented in PSIM software as shown in Fig. 8. The power converter is composed by a mosfet (Q_1), an inductor with losses ($L_1 = 47\ \mu H$ and $R_{L1} = 91.2\ m\Omega$) and a capacitor ($C_o = 333\ \mu F$); moreover, the Buck converter is fed by a $30\ V$ DC voltage source and the load is implemented with a variable resistor (R_l). The control strategy is programmed in *C* language and implemented in PSIM using a *dll* block, whose inputs are voltage and current of the Buck converter, and its output is a voltage proportional to the duty cycle. The duty cycle is generated with a comparator and a sawtooth voltage source.

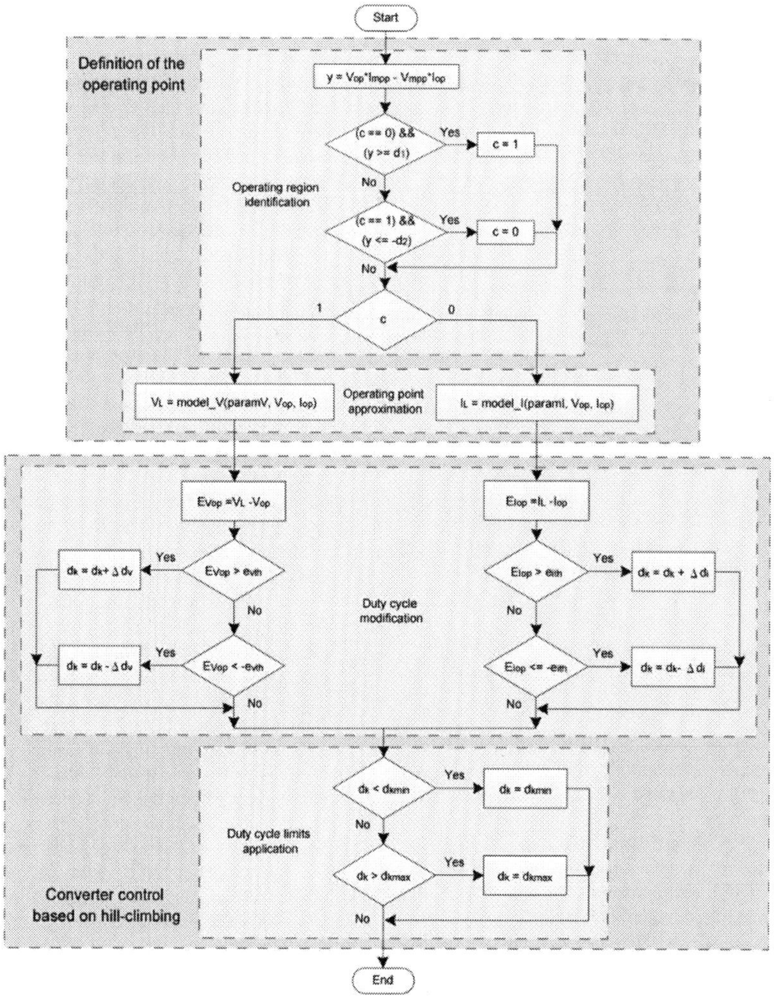

Figure 7: Flow chart of the proposed control strategy: definition of the desired operating point and modification of duty cycle of a Buck converter.

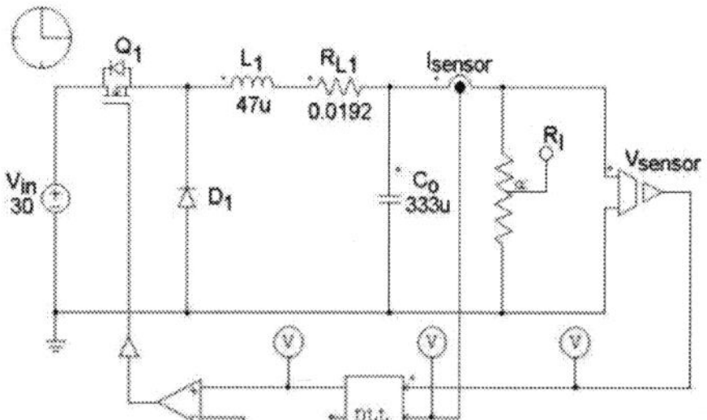

Figure 8: Schematic diagram of the Buck converter and the proposed control strategy implemented in PSIM.

The proposed control strategy require the single-diode model parameters of the ERDM 85 PV module, hence, they are calculated using the procedure proposed in [18]. Additionally, the single-diode model parameters are used to implement a circuital model of the ERDM 85 PV module to validate the results obtained by the PVME with the proposed control strategy.

The I-V and power vs. voltage (P-V) curves of the ERDM 85 PV module are obtained for different operating conditions (i.e. different values of irradiance (G) and cell temperature (T)) using both, the proposed PVME and the circuital model. Those curves are obtained by modifying the load resistance from a maximum value ($R_{lmax}=100\ \Omega$) to a minimum value ($R_{lmin}=1\ \Omega$) in both, the proposed PVME and in the circuital model.

An example of the I-V and P-V curves obtained, for $G = 900\ W/m^2$ and $T = 35\ °C$, are presented in Fig. 9, where the single-diode model parameters for those operating conditions are: $I_{ph} = 4.629\ A$, $I_{sat} = 2.029\ nA$, $\eta = 1.016$, $R_s = 0.203\ \Omega$, and $R_h = 226.141\ \Omega$, using the procedure proposed in [18]. The average relative error for the current (E_{arI}) and power (E_{arp}) are 0.279% and 0.560%, respectively.

Other 4 operating conditions are evaluated obtaining average relative current and power errors under 4% (see Table I). Such results show the proposed control strategy reproduces, satisfactorily, the I-V and P-V curves of a PV module.

The main causes of the current and power errors are the stopping criteria of the implemented NR method (i.e. minimum error and maximum number of iteration) and the threshold values used in the duty cycle modification (e_{vth} and e_{ith}). The stopping criteria of the NR method induce errors in the estimation of the reference operating point voltage or current, while the thresholds used in the modification of the duty cycle induce steady state errors in the output voltage or current of the power converter.

Table I: Summary of the simulation results

Test	G (W/m^2)	T (°C)	E_{ari} (%)	E_{arp}(%)
1	900	35	0.279	0.560
2	900	62	0.213	0.426
3	650	52	0.378	0.756
4	460	28	0.886	1.769
5	290	40	1.756	3.529

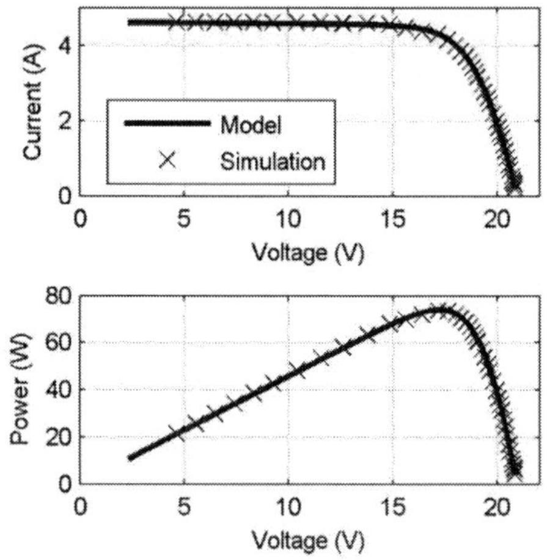

Figure 9: I-V and P-V curves of the PVME controlled by the proposed strategy (Simulation) and the circuital model (Model) of an ERDM 85 PV module for G=900 $W/m^2$and T = 35 $°C$.

5.2 Experimental results

To evaluate experimentally the proposed control strategy, the Buck converter shown in Fig. 8 was build and the proposed control strategy was implemented on a microcontroller-based development kit (Tiva C Series TM4C123G). Moreover, the variable load was implemented with an electronic load BK 8514.

The experimental evaluation is performed by comparing the I-V and P-V curves obtained with the implemented PVME and the curves provided by the single-diode model of an ERDM 85 PV module.

The first step is to calculate the model parameters for an ERDM 85 module for a given operating condition (*G=623 W/m^2* and *T=40 °C*) by using the procedure proposed in [18]. In this case the parameters obtained are: I_{ph} = *3.208 A*, I_{sat} = *4.176 nA*, η = *1.016*, R_s = *0.203* Ω, and R_h = *226.141* Ω.

The PVME is configured with those parameters and the I-V and P-V curves are extracted by modifying the electronic load between a maximum (R_{lmax}=*100* Ω) and a minimum (R_{lmin}=*2* Ω); such curves are marked as *Emulator* in Fig. 10. Finally, the theoretical characteristic curves (I-V and P-V) are calculated by solving (1), for voltages between zero and V_{oc}. Those curves are introduced in Fig. 10 and marked as *Model*.

The results of Fig. 10 show a satisfactory reproduction of the theoretical characteristic curves of the emulated PV module, since the average relative current and power errors obtained are 2.829 % and 3.018 %, respectively. Other operating conditions are evaluated to quantify the errors in the characteristic curves for different values of irradiance and cell temperature.

The results of those tests are presented in Table II, where it can be observed that the experimental current and power errors are below 5%. Those results show that the proposed PVME is able to reproduce the characteristic curves of a PV module.

It is worth noting that the operating points of the PVME have a maximum and minimum voltage limited by the maximum and minimum limits of the power converter duty cycle. Therefore, short-circuit current and open-circuit voltage cannot be reached. The current and voltage ranges obtained with the proposed PVME are from *0.7 A* to *5.5 A*, and from *5 V* to *23 V*, respectively.

The experimental current and voltage errors are close but greater than the simulation errors as expected. The source of the experimental errors are the same as those explained at the end of Section 5.1.

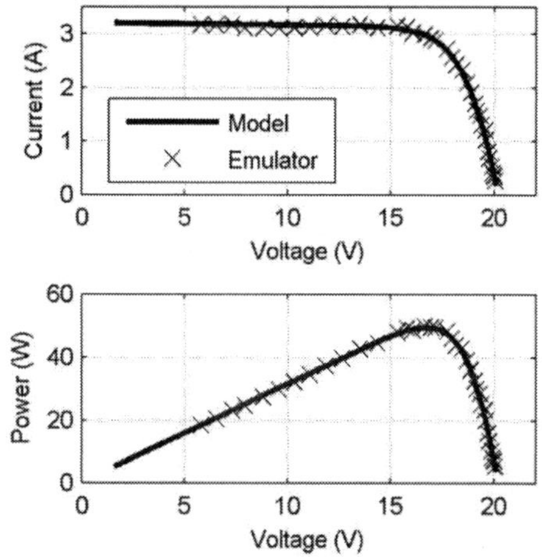

Figure 10: I-V and P-V curves of the PVME controlled by the proposed strategy (Experimental) and the single-diode model (Model) of an ERDM 85 PV module for *G=623*W/m^2 and *T = 40°C*.

Table I: Summary of the experimental results

Test	G (W/m^2)	T (°C)	E_{ari}(%)	E_{arp}(%)
1	623	40	2.829	3.018
2	900	62	2.664	3.035
3	900	35	2.870	3.105
4	589	44	3.716	4.182
5	814	57	3.743	4.161

6 CONCLUSIONS

A PVME control strategy based on HC algorithm has been proposed and validated. The control strategy is divided in two parts, the first one is to indentify the desired operating point using the single-diode model, and the second part is the modification of the duty cycle of a power converter in order to reduce the operating point error of the PVME.

The proposed control strategy was validated with simulation and experimental results. In both cases the emulated PV module was an ERDM 85 PV module and the power converter used was an asynchronous Buck.The average relative errors (in current and power) obtained in the simulation and experimental results were below 4 % and 5 %, respectively, which shows the effectiveness of the proposed PVME control strategy to reproduce the characteristic curves of a PV module.

The proposed control strategy offers a high flexibility in the definition of the PV to be emulated, the irradiance and temperature conditions, and the power converter used for the PVME implementation. Moreover, the control strategy is based on simple algorithms that can be implemented in a low-cost microcontroller-based embedded system.

The proposed control method can be improved by using a high-switching-frequency power converter in order to obtain a setting time closer to the one of a PV module. Moreover, it would be interesting to extent the proposed control strategy for PV arrays, which would allow the emulation of arrays with different sizes and configurations.

7 ACKNOWLEDGMENT

The implementation of the photovoltaic module emulator was supported by the program Apoyo a Semilleros de Investigación of the Facultad de Ingeniería of Universidad del Valle. This work was also supported by the Departamento Administrativo de Ciencia, Tecnología e Innovación de Colombia (COLCIENCIAS) under the project 1118-669-46197 and the Universidad Industrial de Santander.

8 REFERENCES

[1] IEA Photovoltaic power systems programme, "Trends 2015 in photovoltaic applications: Survey report of selected IEA countries between 1992 and 2014," 2015.

[2] a. Vijayakumari, a. T. Devarajan, and N. Devarajan, "Design and development of a model-based hardware simulator for photovoltaic array," *Int. J. Electr. Power Energy Syst.*, vol. 43, no. 1, pp. 40–46, Dec. 2012.

[3] K. Khouzam and K. Hoffman, "Real-time Simulation of Photovoltaic Modules," *Sol. energy*, vol. 56, no. 6, pp. 521–526, 1996.

[4] E. Koutroulis, K. Kalaitzakis, and V. Tzitzilonis, "Development of an FPGA-based system for real-time simulation of photovoltaic modules," *Microelectronics J.*, vol. 40, no. 7, pp. 1094–1102, 2009.

[5] L. Yuan, T. Lee, F. Z. Peng, and L. Dichen, "A hybrid control strategy for photovoltaic simulator," *Conf. Proc. - IEEE Appl. Power Electron. Conf. Expo. - APEC*, pp. 899–903, 2009.

[6] D. D. C. Lu and Q. N. Nguyen, "A photovoltaic panel emulator using a buck-boost DC/DC converter and a low cost micro-controller," *Sol. Energy*, vol. 86, no. 5, pp. 1477–1484, 2012.

[7] C. C. Chen, H. C. Chang, C. C. Kuo, and C. C. Lin, "Programmable energy source emulator for photovoltaic panels considering partial shadow effect," *Energy*, vol. 54, pp. 174–183, 2013.

[8] H. Votzi, F. A. Himmelstoss, and H. Ertl, "Basic linear-mode solar-cell simulators," in *Proceedings - 35th Annual Conference of the IEEE Industrial Electronics Society IECON 2009*, 2009, pp. 261–265.

[9] S. Armstrong, C. K. Lee, and W. G. Hurley, "Investigation of the harmonic response of a photovoltaic system with a solar emulator," *2005 Eur. Conf. Power Electron. Appl.*, vol. 9, pp. 1–8, 2005.

[10] H. Nagayoshi, S. Orio, Y. Kono, and H. Nakajima, "Novel PV array/module I-V curve simulator circuit," *Conf. Rec.. Twenty-Ninth IEEE Photovolt. Spec. Conf. 2002.*, pp. 1535–1538, 2002.

[11] S. H. Lloyd, G. A. Smith, and G. D. Infield, "Design and construction of a modular electronic photovoltaic simulator," in *Eighth International Conference on Power Electronics and Variable Speed Drivers*, 2000, no. 475, pp. 120–123.

[12] a. K. Mukerjee and N. Dasgupta, "DC power supply used as photovoltaic simulator for testing MPPT algorithms," *Renew. Energy*, vol. 32, no. 4, pp. 587–592, Apr. 2007.

[13] J. D. Bastidas-Rodriguez, C. A. Ramos-Paja, and E. Franco-Mejía, "Modeling and parameter calculation of photovoltaic fields in irregular weather conditions," *Ingeniería*, vol. 17, no. 1, pp. 37–48, 2012.

[14] M. C. Di Piazza and G. Vitale, "PV Emulators: Concepts and Realization," in *Photovoltaic Sources: Modelig and Emulation*, London: Springer-Verlag, 2013, pp. 175–191.

[15] S. C. Chapra and R. P. Canale, "Método de Newton-Raphson," in *Métodos numéricos para ingenieros*, 5th ed., México D.F.: McGraw-Hill Interamericana, 2006, p. 1001.

[16] G. Escobar, M. J. Lopez-Sanchez, I. E. Llanez, E. E. Ordonez, and M. Hernandez-Gomez, "A combined controller for a PV simulator," *2014 Int. Conf. Electron. Commun. Comput.*, pp. 65–70, 2014.

[17] ERDM SOLAR, "ERDM85 75 - 85 Watt panel solar monocristalino - Línea standard." pp. 1–2.

[18] J. Accarino, G. Petrone, C. a. Ramos-Paja, and G. Spagnuolo, "Symbolic algebra for the calculation of the series and parallel resistances in PV module model," *4th Int. Conf. Clean Electr. Power Renew. Energy Resour. Impact, ICCEP 2013*, pp. 62–66, 2013.

OPTIMUM SIZING AND EXPLOITATION OF NDEM'S SOLAR POWER PLANT PRODUCTION

Serigne Ndiangue LEYE[1], Senghane MBODJI[1,2], Fakoro Souleymane DIA[2], Amadou DAIO[2] and Gregoire SISSOKO[2]
[1]Department of physics, University Alioune DIOP of Bambey, P.O Box 30 Bambey, SENEGAL;
[2]Laboratory of Semiconductors and Solar Energy, Department of Physics, Faculty of Science and Technology, Cheikh Anta Diop University, Dakar, SENEGAL.
E-mail: senghane.mbodji@uadb.edu.sn; ndiangueleye@hotmail.com
Adresse: PO. Box 30, Bambey

ABSTRACT: The electrical power is not continuously available in several places. The village of Ndem, in Senegal, lived this situation until December 2012. In these places some activities may not be made for lack of electricity. That is why, electricity generated through photovoltaic panels was used to drive the training and textile centers of this Senegalese village.
In this work, a parameter study is carried out to find out the best combination of the PV panel wattage, battery capacity, regulator and inverter.
Hence, it is conclude that it is feasible to operate the two centers on solar photovoltaic without grid power with 6.4 KWp of 30 panel arrays and 16 lead batteries with a capacity of 440 Ah.
We present the monthly and daily production of the photovoltaic central and the efficiency of the inverter in 2013.
We also present the radiance and the temperature effect on the daily and monthly production. The study also carries out the utility of maintenance.

Keywords: Solar Photovoltaic, Radiance, Temperature

INTRODUCTION

The increase of oil price pushes policymakers and researchers to move more and more on alternative energies. Renewable energies are one of the main development challenges and may allow more or less long term to reduce energy costs but also reduce the emission of greenhouse gases.
Indeed, renewable energy in general and solar energy in particular are sustainable and comprehensive development tool for sunny countries [1, 2].
Therefore, many actions are defined and implemented to increase the use of solar energy, which appears in various forms: solar photovoltaic, solar thermal...
Regarding Senegal, context is marked by untimely power cuts which can last several hours. This slows down the work of several sections of the population whose activities depend on the electric current.
To address this problem, several solutions were considered including the installation of photovoltaic power plants.
To this end, the Non-Governmental Organization MAAM SAMBA decided to make self-powered textile center and training center in the village of NDEM by installing a photovoltaic solar plant.
Our study focuses on environmental conditions, the sizing of the solar plant of training and textile centers, monthly and daily production of year 2013 and the effect of temperature on the panels.

1 ENVIRONMENTAL CONDITIONS OF THE TRAINING AND TEXTILE CENTERS OF NDEM VILLAGE.

Figure 1 shows the complete radiation data of 2012 year.
The monthly sum of the global irradiation is maximal for the months of March, April and May with respectively 203, 212 and 213 kWh/m^2. It is also noted, during these months, the most elevated daily global irradiations. We remarked a regular increase of the temperature and the peaks are noted during the months of June to September.
The lowest radiation is observed in December [3]. On the other hand the most elevated radiations are observed between March and May.
It is noted, according to [4], that the minimal temperature in the local is 10 ^{0}C to the sunrise. The temperature can reach to a maximum of 45 ^{0}C.

Table I: radiation data in 2012 year [5].

Month	Gh_m (kWh/m^2)	Gh_d (kWh/m^2)	Dh_d (kWh/m^2)	T_{24} (^{0}C)
January	163	5.26	2.20	23.7
February	163	5.84	2.50	25.2
March	203	6.56	2.83	26.1
April	212	7.06	2.96	26.3
May	213	6.87	3.03	27.1
June	189	6.30	3.24	28.6
July	181	5.84	3.14	29.4
August	171	5.52	2.93	29.4
September	169	5.62	2.78	29.8
October	176	5.68	2.54	29.5
November	158	5.25	2.26	27.3
December	151	4.88	2.05	25.0
Year	2149	5.89	2.70	27.3

2 SIZING OF SOLAR POWER PLANT OF TRAINING AND TEXTILE CENTERS OF NDEM VILLAGE.

The solar power plant sizing of the training and textile centers of the Non-Governmental Organization MAAM SAMBA, in Ndem village, is made starting from the corresponding threshold of 300W corresponding to a one sewing machine power. Therefore, it is considered that, if the power goes up to 300W one machine is on and when the power goes up to 600W both two machines operate simultaneously. If n machines operate simultaneously it has a total output of 300 nW.

Furthermore the sewing machines used by the dressmakers are relatively old and use cold or function with DC current. For a nominal power of 250W, the power of the machine when it turn on is between 250 and 260 W and when the pedal is engaged the power of the machine is between 270 and 280W and finally, at the moment of ignition, the power is between 800 and 1100W.

Solar power plant sizing also take into account:

- the inability to take into account the sharp fluctuations of the energy consumed;
- the need to increase the amount of energy for the operation;
- the energy consumed during the work time is equal to the product of the work of the power of sewing machines;
- calls power which are three times superior to the nominal power.

For the analysis of needs, the artisans of the textile center work six days a week from 9h to 18h with a break of 2 hours between 12h and 14h. The workshop consists of 8 electric sewing machines but in fact currently only 6 machines are used simultaneously.

Hence, the solar power plant system must provide a minimum daily energy of 21.88 kWh.

2.1 Storage battery

The PV panels charge a field of 16*12Vx110Ah lead acid batteries connected in a game of 4x4 in series/shunt to give a nominal output of 48V. They have a life of six years, resist the temperature and are especially conceived for the PV. Considering the peaks of temperature in NDEM, the lead batteries of mark Victron Gel Deep cycle with a C20 capacity of 440 Ah are best suited as recommended by [1,6].

2.2 Photovoltaic panels

The solar power plant is composed of 30 photovoltaic panels made by silicon polycrystalline solar cells and manufactured by Solar Direct. The output of one module is understood between 210 and 245 Wp. It has a power of 6.4 kWp.

The photovoltaic panels are arranged as a set of five panels in series with six panels in parallel combinations to supply the sufficient voltage and array current to charge the batteries.

The orientation of the panels take account to the geographical position of the village for which the panels must be oriented full south and tilted a constant angle of 15°. This angle permits us to have an optimum output.

If in Ndem, the tilt of panels is fixed to 15°, in Yenagoa, in Nigeria, solar panels can be tilted of an angle understood between 15° to 45° [7].

This solar power plant is deprived of tracker system contrarily to those doted of an axis tracking systems [8].

2.3 Technical area and functional diagram of our solar power plant installation

The inverters and the batteries are installed in the technical area located on the part northeast of the purchasing group building. The technical area has main functions of:

- protecting the devices and the batteries of the dust and the direct radiance of the sun;
- assuring adapted ventilation and the working of these components to acceptable temperatures (<25°C);
- making inaccessible the access to these components to non-authorized people with an average temperature in the local between 20 and 30 °C.

In spite of the presence of the network, it has been preferred to install photovoltaic modules rather than a unique battery park. It has been decided to maximize the photovoltaic production, in order to minimize the need in batteries that are the most expensive components in the photovoltaic installation; while knowing that it is always necessary to have some batteries to control the fluxes of power, and to assure the working to the dawns and to the twilight.

In order to minimize the electric losses, the coupling between the photovoltaic panels and the network has been chosen therefore by means of connected inverters. The energy produced solely by the modules pass through these inverters, that have elevated outputs (>95%) and the batteries are only used to assure the energy at the end and in beginning of day.

2.4 Remote monitoring system

The maintenance and the surveillance of the solar power plant station are fundamental [9] if one wants to maintain the operating time of 20 to 30 years. That is why, our system is endowed with a remote monitoring system. The simple indicators (types light red/light green) give the state of discharge of the batteries. They allow users to stop consuming if the batteries are unloaded or if the needed current is too important in relation to the current threshold. If the light is green, it shows that the extracted power is generated only by the panels. If it is orange, the extracted power is generated by the panels and the batteries. Therefore it can have a too important discharge risk on the batteries. Finally if the light is red, the batteries are discharged and must reload themselves so that they can continue their working.

3 RESULTS

3.1 Monthly and daily production of year 2013.

The figure 1 shows that the month of April has a production estimated to 500 kWh against 900 kWh of forecasting. This month gives the best annual production. The weakest monthly production is observed during the month of March 2013 with less than 200 kWh, where it was foreseen 900 kWh.

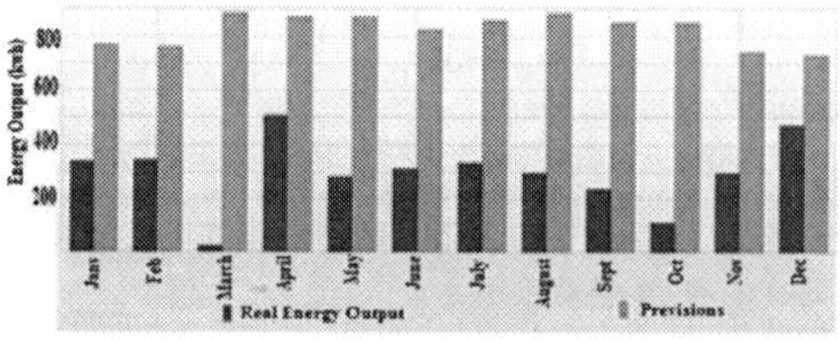

Figure 1: Yearly production of the solar power plant in year 2013.

The month of April is characterized by a very strong sunshine. It is then normal that the best production corresponds to this month. What is curious is that the month of March gives lower production. This month is hotter than months of the December, January and February. The gap of temperature between the months of March and April, when one trusts data of Solargis [5], is

0.2 °C; what is negligible.
More, if we compare the monthly irradiation, one sees that the one of the month of March is more important than those of every month of the year 2013, except those of the months of April and May. One sees besides that the gaps of monthly irradiation of March, April and May are negligible. All this makes us conclude that there is an incoherence concerning the production of the month of March.
Globally, one notes that there are a lot of losses that would be bound to the recombination [10, 11], the meteorological conditions, the diffusion of the irradiation, the temperature, etc.

Figures 2, 3, 4 and 6 give the daily production of January, February, March, August and September 2013.

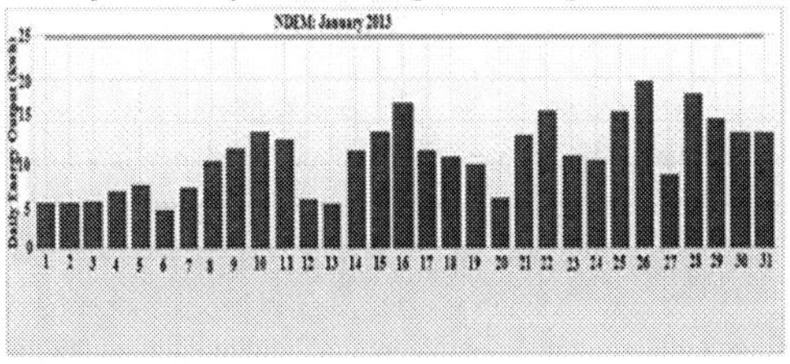

Figure 2: Daily production of the solar power plant of the month of January 2013

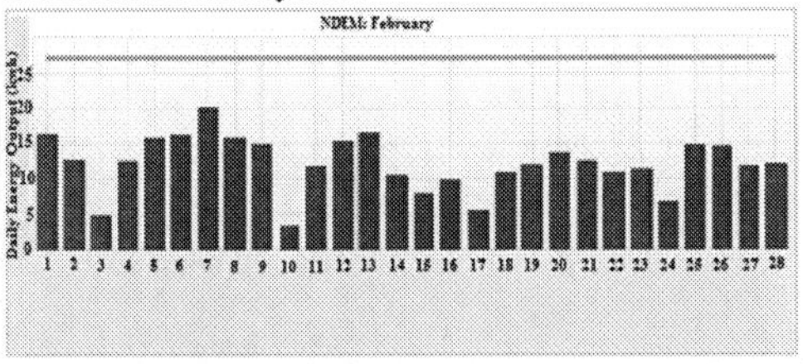

Figure 3: Daily production of the solar power plant of the month of February 2013

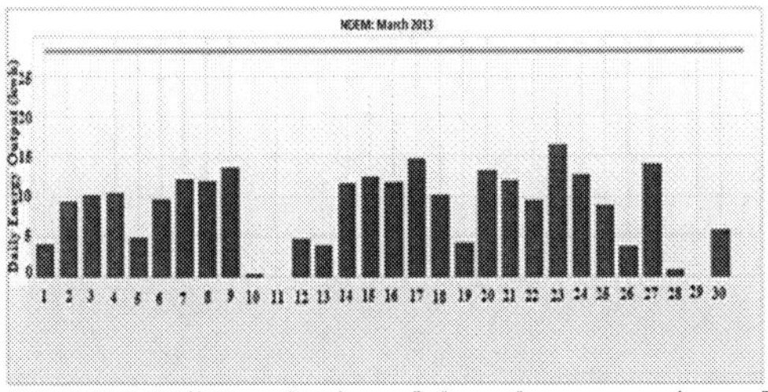

Figure 4: Daily production of the solar power plant of the month of May 2013

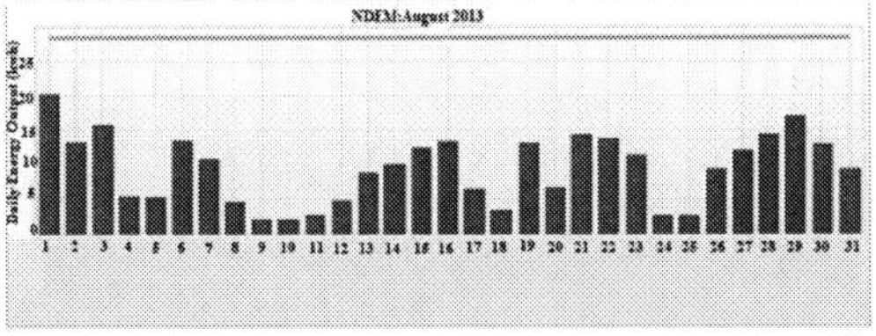

Figure 5: Daily production of the solar power plant of the month of August 2013.

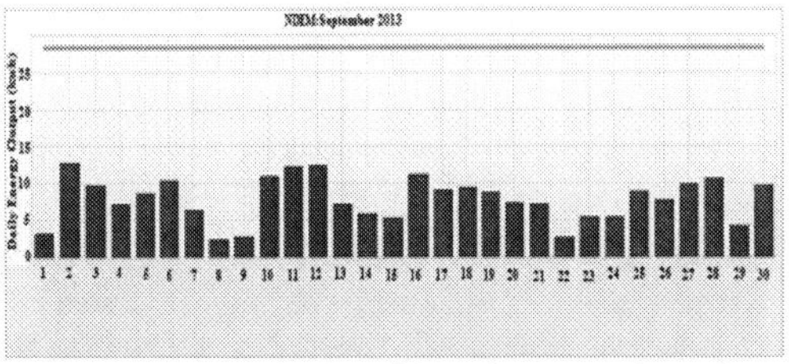

Figure 6: Daily production of the solar power plant of the month of September 2013.

The daily production presents also disparate behaviors. Let's consider the month of August 2013 for example. For this month, the best productions are observed the 1st, 29, 3, 28, 21, 22, 16, 6, 15, 7 and August 2, respectively. The weakest productions have been remarked for 9, 10, 11, 24, August 25 and 18, 2013, respectively.

The disparities can be due to the variation of the daily global irradiation and temperature.
On the other hand, one sees that the 9 and 10 august, have the same productions. These days are certainly characterized by the same temperature and the same daily global irradiations.

We noted for the month of March 2013 an anomaly: the daily productions of the 11, 29 and 31 May are null and those of 10 and 28 May negligible. The solar power plant didn't work. Is the system down? Are the climatic conditions favorable?

That is why, it would be necessary to have a set of measures of the years 2013, 2014 and 2015 to make the analysis of this anomaly.

3.2 Radiance and temperature of the panels

We present in figure 7 the monthly solar radiance of the solar power plant NDEM in 2013.

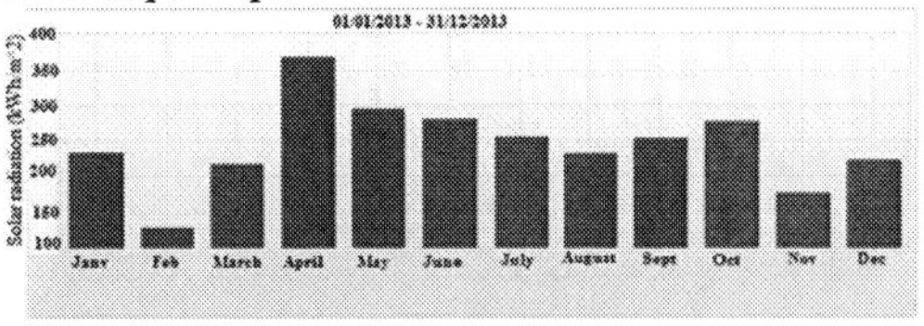

Figure 7: Solar radiance of year 2013.

Figure 8 shows the local and solar panels temperatures.

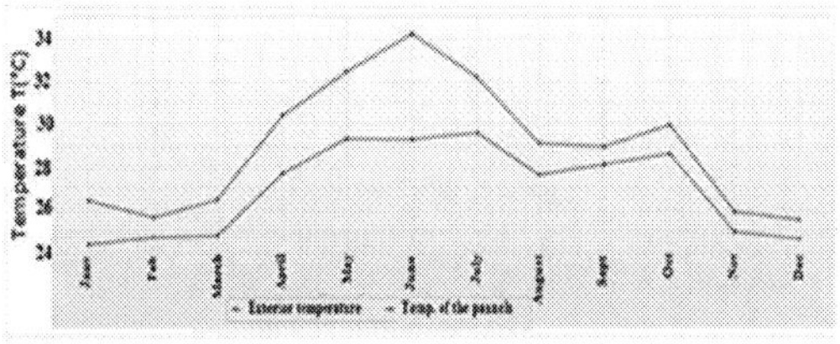

Figure 8: Temperature of the solar panels of year 2013.

Through the figure 8, we see that the external temperature is always lower to the one of the solar panels.

The temperature influence the short-circuit phocurrent and open circuit photovoltage. While influencing these two parameters, the temperature contributes to the modification of the I-V characteristic, the fill factor (FF) and the solar cell's power output. The temperature is besides one of the main parameters that influences photovoltaic conversion output.

Figure 8 also show that, the temperature of the solar panels is superior to the external temperature, they warm up. This warming up decreases the output of production of the panels according to [12, 13].

Based on the conclusion of [13], the decrease of the production of the solar panels is very quickly caused by a warming up of the panels responsible of the losses and decrease of solar power plant's efficiency.

One also sees that, in year 2013, the temperature of solar panels decreases slightly between January and February. It increases until June before decreasing until September and soars again until October. It falls from October to December 2013.

One notes a strong increase of the temperature of the solar panels during May to July 2013 due, certainly, to a very important radiance.

These last entail an absorption of the solar radiance that generates an increase of the temperature of the modules on its turn, provoking a relative decrease of the open circuit photovoltage as shown in [14].

Also, according to [13], the variation of the temperature has a direct effect on the operating voltage of converter DC/AC and the solar panels leading to the decrease of the output of the solar panels' production.

CONCLUSION

Our work has shown that the best solar production is observed during the month of April 2013 whereas the weakest solar power plant production is noted in March 2013.

The analysis of the solar power plant production has shown that there are a lot of losses that would be bound to the recombination, the meteorological conditions, the diffusion of the irradiation, the warming-up of the solar panels, etc.

NOMENCLATURE

Table II: Nomenclature

Symbol	Designation	Unit
Gh_m	Monthly sum of the global irradiation	kWh/m^2
Gh_d	Daily sum of the global irradiation	kWh/m^2
Dh_d	Daily sum of the diffuse irradiation	kWh/m^2
T_{24}	Daily temperature (24h)	°C

ACKNOWLEDGMENT

The authors would like to thank the Non-Governmental Organization MAAM SAMBA, in Ndem, Senegal, and the "Centre d'Excellence en Mathématiques, Informatique et TIC (CEA-MITIC)" in Gaston Berger University, Senegal.

CEA-MITIC sponsored the participation of the authors in EUPVSEC 2016.

References

[1] A. Bennouna, D. Zejli and R. Benchrifa, Revue des Énergies Renouvelables, Vol. 7 (2007) 1.

[2] F. Gama-Yettou, A. Gama, A. Malek, L. Serir and H. Zeraia, Revue des Énergies Renouvelables, Vol. 7 (2007) 141.

[3] K. Namrata, Dr. S. P. Sharma and Dr. S.B.L. Saksena, International Journal of Science and Advanced Technology, Vol 2 (2012) 138.

[4] Team Panels Partners, Projet d'électrification photovoltaïque du centre d'artisanat du village de NDEM, Sénégal (2011) 1.

[5] Site de Solargis : www.solargis.info/pvplanner

[6] A. Labouret, P. Cumunel, J. Braun and B. Faraggi, Livre 5e édition, ETSF (2010).

[7] K. P. Ainah and J. T. Afa, Turkish Journal of Physics, Vol. 37 (2013) 242.

[8] S. Bazyari, R. Keypour, S. Farhangi, A. Ghaedi and K. Bazyari, Vol 2 (2014) 718.

[9] M. Hatti, Green Energy and Technology, (2014) 61.

[10] S. Mbodji, B. Mbow, F. I. Barro and G. Sissoko Turkish Journal of Physics, Vol 15 (2011) 281.

[11] S. Mbodji and G. Sissoko, Int. J. Pure Appl. Sci. Technol., Vol 6(2011) 103.

[12] A. Yusoff, N. T. R. N. Kumara, A. Lim, P. Ekanayake and K. U. Tennakoon, Journal of Biophysics, Vol. (2014) 1.

[13] M. A. K. El-Adawi1 and I. A. Al-Nuaim, Journal of Optics and Photonics, Vol. 4 (2014) 219.

[14] M. A. Hamouda, M. Saïdi, A. Louchene, C. Hamouda and A. Malek, Revue des Énergies Renouvelables, Vol. 14(2011)187.

LOWCOST-OUTDOOR-ELECTROLUMINESCENCE: SIGNIFICANT IMPROVEMENTS OF THE METHOD

K. Mertens, A. Arnds, G. Behrens*), A. Domnik*)
Münster University of Applied Sciences, Electrical Engineering, Testing Lab for Photovoltaics
Stegerwaldstraße 39, D-48565 Steinfurt, +49 2551/962-111, mertens@fh-muenster.de
*) University of Applied Sciences Bielefeld, Minden, Germany, grit.behrens@fh-bielefeld.de

ABSTRACT:

We show methods to use the video mode of modified DSLR cameras for outdoor-EL measurements. This by far simplifies the analysis of module failures onsite. As a result, the Low-Cost-Outdoor-EL method will continue its way to become a standard inspection method besides thermography to analyze PV plants onsite.

1 INTRODUCTION

Electroluminescence pictures (EL pictures) of solar modules are a **powerful tool** for **quality analysis**. Recently, it was shown that with modified inexpensive DSLR cameras and suitable blocking filters even onsite-outdoor measurements can be realized. This **LowCost-Outdoor-Electroluminescence method** facilitates a **detailed error analysis of whole strings** with respect to micro cracks, potential induced degradation (PID), defective bypass diodes etc. [1-5].

A **disadvantage** of the hitherto existing method lies in the relatively **long exposure time** which enforces the **use of a tripod**. At least for a quick first inventory of the PV plant it would be desirable to find the string set under current on the roof with the **video mode** of the camera and to already roughly **examine it for peculiarities**.

This paper describes a **further development** of the LowCost-Outdoor-EL-method which also facilitates the **creation of EL-video exposures**. First cheap **consumer cameras** are considered. By **manipulating** the **camera** operation system **higher single frame brightness** can be achieved **in video modus**. After that, possibilities of **semi-professional cameras** will be demonstrated which for the **first time** offer the possibility of **EL examinations with copters** (UAVs). Finally, we describe methods for **video enhancements** by **suppression of extraneous light.**

2 DEPLOYMENT OF CONSUMER CAMERAS

With modified DSLR cameras good EL-pictures can be taken with **minimal exposure times** of about **0.5 seconds**. The reason lies in the **poor sensivity of** the **CMOS sensor** for infrared light. However, the **frame rate** in video mode of typical cameras is in the order of **25 or 50 fps** (frames per second). This leads to exposure times of 20 and 40 ms, respectively.

The camera deployed in our lab is a Canon EOS 700D. Here the frame rate can only be adjusted to 25 or 30 fps, respectively. However, for Canon models the **alternative camera operation system *Magic Lantern*** exists which is an **enhancement software** of the original firmware, that offers increased functionality with access to a variety of normally blocked functions.

For illustration, Figure 1 shows a screenshot of the original software of the EOS 700D as well as the view after implementing the enhancement software.

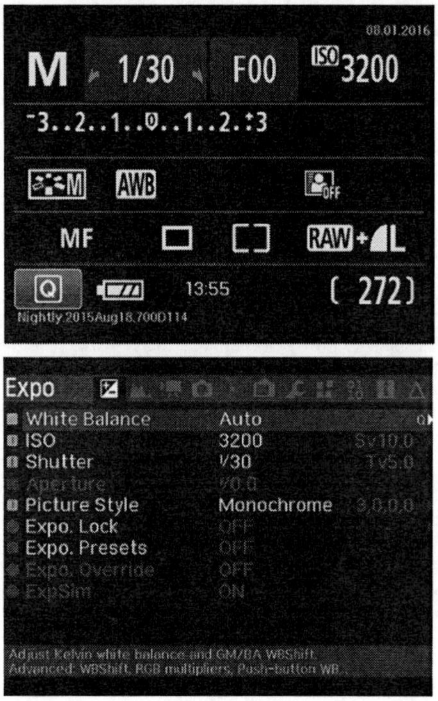

Figure 1: Screenshot of the original Canon (upper) and the *Magic Lantern* software (lower)

With the new software, the **frame rate can be adjusted** to even almost arbitrary values. As an example, **Figure 2** shows the comparison of two exposures (15 fps and 5 fps resp.). Clearly visible the brightness of the first picture is not sufficient whereas the **second picture** also **shows details of the module.** In practice to make a **first quick inventory of the PV plant** a **frame rate of about 7 fps** has been well-tried. Here the brightness of the video is sufficient and at the same time an about-turn of the camera results in a smooth video representation. Moreover, the **video mode helps** a lot **to find** the **correct**

focus adjustment which is quite problematic in picture mode.

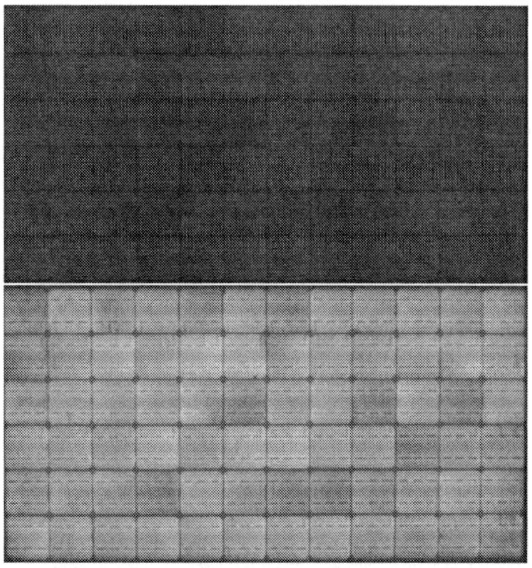

Figure 2: Depiction of two single video frames at different frame rates: Only the smaller frame rate results in a sufficient brightness of the module (upper: 15 fps, lower: 5 fps)

3 DEPLOYMENT OF SEMI-PROFESSIONAL CAMERAS

Only shortly, **compact system cameras** (price smaller than 2000 €) are on the market which show **excellent detector sensivities**. E.g. one model of the Sony Alpha series features **ISO values** up to **400.000**. This is combined with a relatively high sensivity also in the for EL measurements relevant IR spectrum. Therefore, this **camera is suitable** (after modification) **to take EL videos** at high frame rates with meaningful results of PV plants.

As an example, **Figure 3** (upper) shows a single frame of a video which was taken with 25 fps. Obviously the **brightness is sufficient** to see details and to detect errors.

With this type of camera, the **whole inspection of the PV plant can be done in video mode** without taking single pictures. This extremely **speeds up the inspection** of the plant. The EL video technique even facilitates the **deployment of copters** with attached EL camera.

As a copter has only restricted battery time it is desirable to subsequently energize the separate strings of a plant. Figure 3 (lower) shows a **developed string-switch-box**. This box can **subsequently switch** the **current** of a high voltage power supply **to 8 different strings**. The control can be done by selection switches or by a **wireless remote control** device.

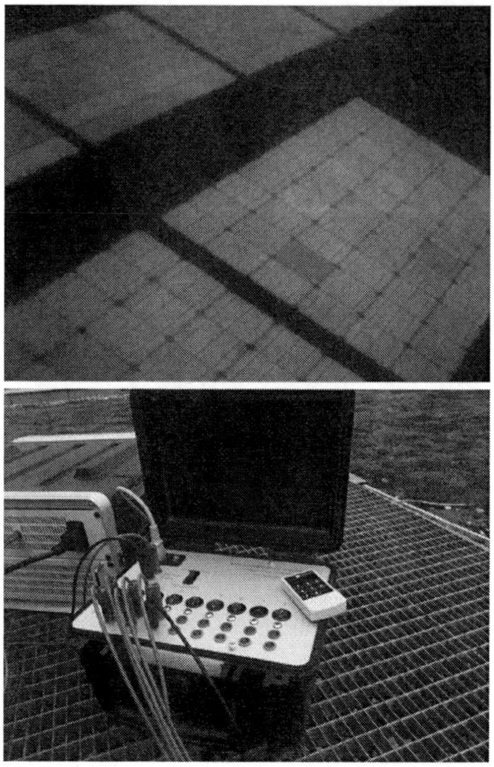

Figure 3: Upper: Single frame of a video that was made with 25 fps (Sony Alpha 7s): The picture shows all relevant details
Lower: String switching box with remote control to quickly switch between different module strings

Figure 4 shows two examples of EL pictures taken from a copter. The upper one shows a module with a **defective bypass diode** (short circuit). In the lower picture the **typical EL pattern of PID** infected modules is clearly visible.

4 SUPPRESSION OF EXTRANEOUS LIGHT

With the blocking filters mentioned above already a large amount of the extraneous light can be suppressed. However, in the IR emission range **around 1100 nm sunlight still has large spectral contributions** [3]. Therefore, EL measurements are **typically** done **in the night**.

In order to make high-contrast **EL exposures also at dawn** first a picture with EL current and then one without EL current is taken. By forming the **difference of both pictures** already **large contrast improvements** can be attained.

In case of **EL videos** the reverse feeding **current must be modulated**. In the simplest case it is simply switched on and off. Afterwards the frames of the "dark cycle" can be subtracted from the "bright cycle". An example of such an extraneous light suppression is shown in **Figure 5**. In the upper picture the EL characteristics are only poorly to be seen. After subtraction of the dark picture ("without current") from the original one a **much better EL contrast is achieved** (lower picture).

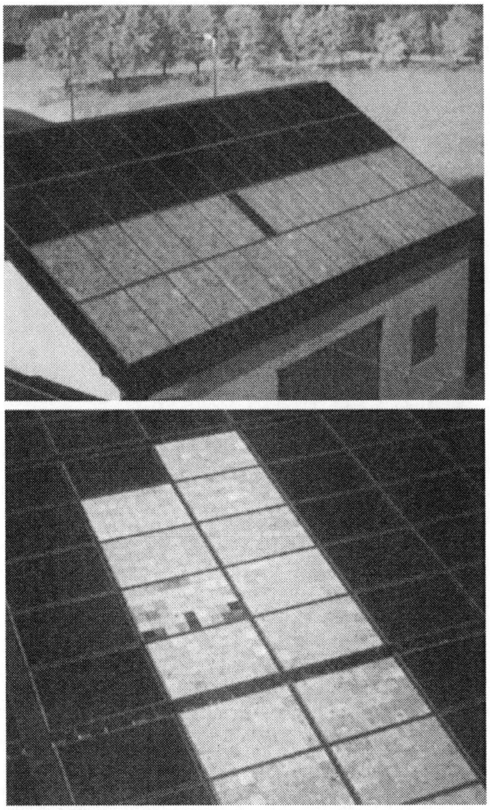

Figure 4: Examples of EL exposures taken with a copter:
Upper: Module with a defective bypass diode
Lower: Typical EL pattern of PID infected modules
(Source: Fladung Solartechnik, Aachen, Germany)

5 CONCLUSION

With suitable measures, the **Low-Cost-Outdoor-EL method** can be **extended to the possibility to take EL videos of whole PV plants**. With this extension the **capabilities** of the method are **significantly improved and expanded**.

Besides facilitating a **quick overview of the PV plant** to whole **detailed plant inspections** the video mode offers a lot of **possibilities** that were **not possible up to now**.

As a result, the **Low-Cost-Outdoor-EL method** will continue its way to **become a standard inspection method besides thermography** to analyze PV plants onsite.

Figure 5: Reduction of the unwanted extraneous light at dawn: The middle picture (without current) is subtracted from the upper picture (with current). As a result, the lower picture ("difference") shows a drastically improved contrast.

REFERENCES

[1] Mertens, K.; Stegemann, T.; Stöppel, T.: LowCost EL - Erstellung von Elektrolumineszenzbildern mit einer modifizierten Standard-Spiegelreflexkamera, 27. Symp. Photovoltaische Solarenergie, 2012

[2] Mertens, K., Pascual Gonzales, D., Diehl, M.: „LowCost-Outdoor-EL: Kostengünstige umfassende Vorort-Qualitätsanalyse von Solarmodulen", 30. Symp. Photovoltaische Solarenergie, 2015

[3] Mertens, K., Kösters, H., Diehl, M: „Low-Cost-Outdoor-EL: Cost-Efficient Extensive on-Site Quality Analysis of Solar Modules", Proceedings of 31st EU PVSEC, Hamburg, 2015

[4] Mertens, K.: Photovoltaik - Lehrbuch zu Grundlagen, Technologie und Praxis; 3., neu bearbeitete und aktualisierte Auflage, Carl Hanser Verlag, München, 2015

[5] Mertens, K.: Photovoltaics - Fundamentals, Technology and Practice, John Wiley & Sons Ltd, London, 2014

INNOVATIVE SEMI-AUTOMATIC CLEANING TECHNIQUE FOR HIGH CONCENTRATION PHOTOVOLTAIC PANELS

*D. Dahlioui, Y. El Fatimy, A. Benazouz, A. Barhdadi**
Semiconductors Physics and Solar Energy Research Team (PSES)
Ecole Normale Supérieure, University Mohammed V of Rabat
PO Box 5118-Takaddoum, Rabat, Morocco
*Senior Associate of the International Centre for Theoretical Physics, Trieste, Italy
&
G. Borelli, M. Carpanelli, D. Verdilio
BECAR s.r.l, Beghelli Group, Vialedella Pace, 1 - 40050 Monteveglio, Bologna, Italy

ABSTRACT: In this work, we have first demonstrated that soiling has a significant effect on the high concentrated photovoltaic system productivity. Next, we presented the integrated cleaning technique we have designed and we currently try to implement on the system. This innovative technique allows first reducing soiling of the HCPV modules, avoid micro-scratches on their glass surfaces and improve their energy productivity. The proposed technique is based on the use of compressed air flux. This study can lead to a final prototype achievement that we propose in terms of perspectives, such as the integration of a dirt sensor and the allocation of automatic control of all the system. These additional improvements would be useful to develop the first Moroccan proper cleaning solution for high concentration photovoltaic panels.
Keywords: Photovoltaic, concentration, soiling, design, air, motor, contactors, control.

1 INTRODUCTION

The Concentrated Photovoltaic Technology (CPV) is a promising alternative to both flat PV and Concentrated Solar Power (CSP) technologies, especially for applications in sunny areas where high temperatures decrease the flat PV performances on one hand and the lack of water prevents the easy deployment of CSP on the other hand [1]. CPV delivers the outstanding efficiencies at all levels (cell, module, and system) and the highest energy yield with comparison to the other solar technologies. It has also a compelling cost advantage because of its leading LCOE (*levelled cost of energy)* in the target market [2]. Moreover, CPV is a proven, certified, predictable technology with rapid deployment, scalability, and a strong solar market forecast [1].

At home university, we have recently installed an innovative grid-connected HCPV (High Concentration Photo-Voltaic) power system [3, 4] and three others grid-connected PV micro plants, based on three different silicon technologies. These installations are part of the small RDI solar energy platform we are progressively developing at the university to provide the researchers and PhD/Master's degree students with the necessary equipment and materials to develop specialized training sessions and to perform proper research activities on emerging CPV and PV systems.

The innovative HCPV power generation system we have installed is composed by: (a) a sun tracker (heliostat), which hosts a set of 48 HCPV single modules of 64 cell each, (b) the DC/AC converters (inverters) and (c) the control electronics. The system's nominal peak power is 8160 W at 1 kW/m^2 DNI and standard cells temperature of 25 °C. The system total weight is 3900 kg and its size is 7.670 x 5.560 m^2 [5]. The system, baptized "Life Tree", is well furnished by the necessary meteorological measurement tools and operating as any small PV power plant. This achievement has been performed in the framework of a scientific and technical cooperation agreement the European Beghelli Group aiming at improving the system performances in terms of energy efficiency, technology, and cost-effectiveness.

In this paper, we start by demonstrating the effect of soiling on HCPV power system panels before presenting the integrated cleaning technique we have designed and we currently try to implement on the system. This innovative technique allows first reducing soiling of the modules, avoid micro scratches on their glass surfaces and improve their energy productivity.

2 CONTEXT OF STUDY

2.1 Necessity for panels cleaning

Various studies show that solar panels lose 3 to 16% of their effectiveness due to the accumulation of dirt and dust that settle on the panels and on the edges of the frame. There are various sources of pollution among them urban and agricultural pollution, chimneys ashes, bird droppings, pollens, salt deposit due to spray and rains sand. In order to demonstrate that solar panels lose their effectiveness due to soiling, we have first monitored the productivity of the generator during one year. The monitoring results are reported in Figure 1. We noticed that the productivity of the photovoltaic panel increases due to a cleaning action.

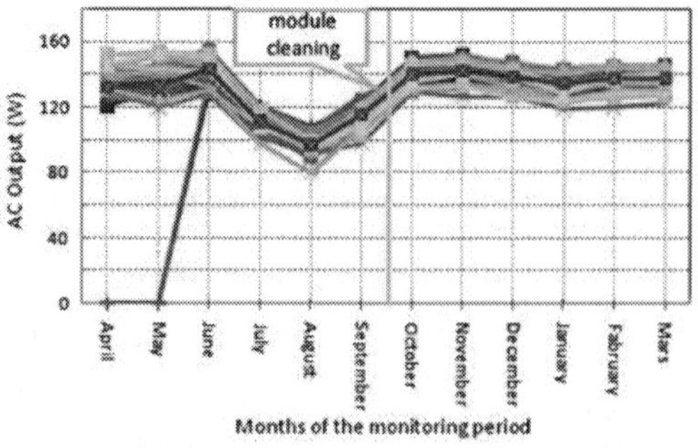

Figure 1: Graphical analysis of the maximum AC power provided by each HCPV module

Thereafter, we have chosen the comparison between two adjacent modules (M1 and M2), one of which is regularly cleaned, and the other of which naturally accumulates environmental contaminants as described in Figure 2, then we have compared in real time the differential output between the two inverters in order to illustrate that HCPV modules soiling, is a significant source of lost potential power generation for HCPV installations.

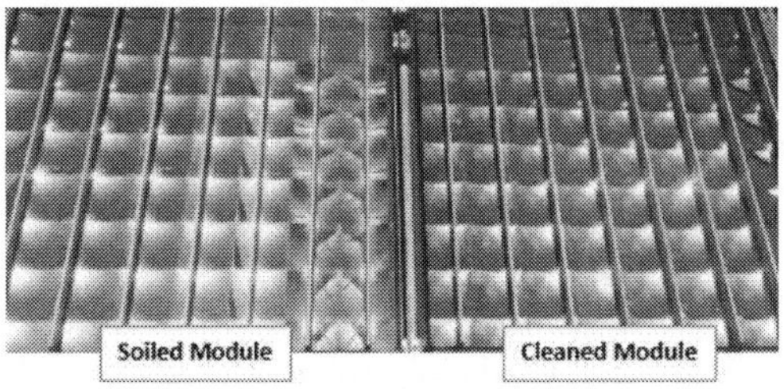

Figure 2: View of cleaned HCPV module in comparison with the soiled one that has not been cleaned within twenty days.

As shown in Figure 3, the gain of the power generated by the cleaned module is significant. It can reach 3.270 kWh per day compared to the soiled. So we are led to think about an innovative technique to rid appropriately and efficiently the photovoltaic panels of the dust [6].

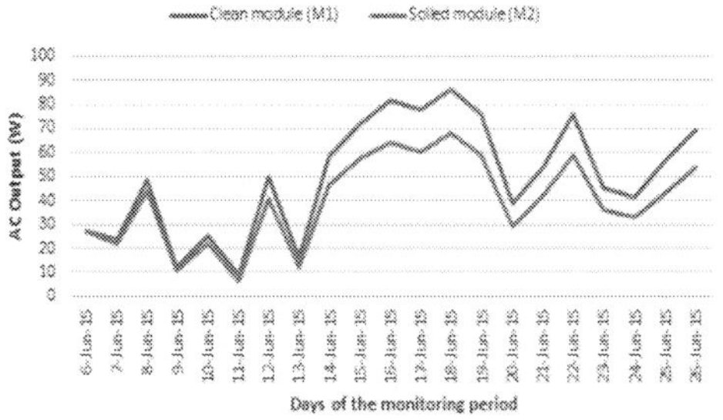

Figure 3: Graphical analysis and evolution of the power produced by the cleaned the soiled HCPV modules.

3 DESCRIPTION OF THE PROPOSED SOLUTION

The proposed technique is based on the use of compressed ambient air flux. The process consists on transporting the compressed air, coming from a compressor, to the air moving duct on which we have already created slots to allow the air flow on the surface of panels via a flexible. The compressed air duct will perform a translational motion on the panel by means of a translation system driven by a motor controlled by contactors. So in this study, we propose a rigorous design using catalogs, for performing the selection of key controls and training system by taking into account several technical and economic criteria.

3.1 Presentation of the cleaning system

Our cleaning system consists of the following components as shows in Figure 4: Compressor, compressed air duct, pulley-belt system and motoreducer.

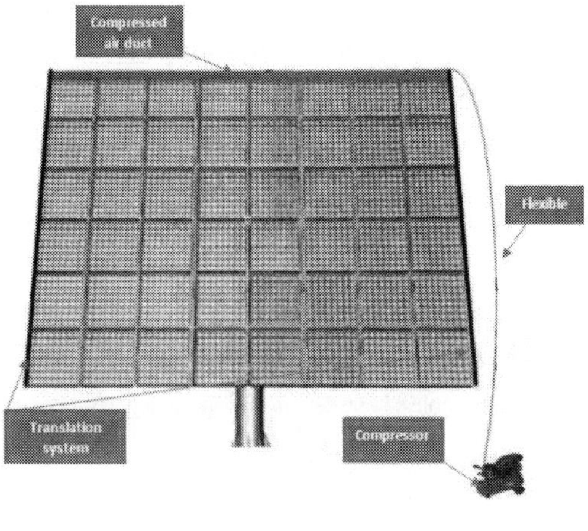

Figure 4: Simplified architecture of dust removal technology of the HCPV power panel.

4 DETAILS OF THE CLEANING TECHNIQUE

4.1 Tender specifications

We desire design semi-automatic cleaning technique for very high concentration photovoltaic panels connected to the low voltage network for this, table 1 summarizes a set of requirements that must be taken into account.

Table I: Tender specifications

Specifications	
Specifications of the system	Using air Daily cleaning action Period: 15 min
Characteristics of the load	Constant load Negligible weight
Characteristics of the motor	Constant speed
Characteristics of the control	By contactors
Ground clearance	Distance: 3130 mm

4.2 Dimensioning the compressed air network

We started by dimensioning the air flux. The duct system must ensure the different air flux [7]. Once known flux rates, it is necessary to determine the characteristics of all its components (compressor, piping sections...) summarized in Table 2.

Table 2: Data related to HCPV grid-connected system.

Data	
Module surface to clean	$S = a^2$, $a = 918$ mm
Average thickness of dust to remove out from the panel	$e = 0{,}5$ mm
Density of the glass	2500 $kg.m^{-3}$
Allowable transport rate	$C = 5\ m^3.h^{-1}.kg^{-1}$
Number of modules dusted	$N = 48$
Production dust in kg per hour	$a^2.\rho.e.n$

Because the regulation prohibits the use of suction networks beyond 0,2 $kg.m^{-3}.h^{-1}$ or 5 $kg.m^{-3}.h^{-1}$, the air flux expected, with a safety factor of 1.25, is about :

$$Q = 1{,}25 . C . a^2 . \rho . e . n \qquad (1)$$

Then the air flux that we expected is about 316 m^3h^{-1}.

For the compressed air flux rate, we avoid using excessively high speeds in order to minimize energy losses and air flux noise and the choice is a compromise between the cost and the facility of its operation. The following table shows the air flux speed depending on the flux for the "low pressure" equipment.

Table 3: Air flux rates

Speed transit in ducts	Maximum speed
Flux maxi < 300 [m^3/h]	2.5 [m/s]
Flux maxi < 1000 [m^3/h]	3 [m/s]
Flux maxi < 2000 [m^3/h]	4 [m/s]
Flux maxi < 4000 [m^3/h]	5 [m/s]
Flux maxi < 10000 [m^3/h]	6 [m/s]

With regard to the air flux that we have previously calculated, a speed of 3 $m.s^{-1}$ is the most optimal.

Regarding the choice of compressed air duct, having chosen the constant speed dimensioning method, we calculate the diameter of the duct to maintain a constant speed throughout the installation.

$$Q = S . V = \frac{D^2}{4} . V . \pi \qquad (2)$$

We opted for the choice of the aluminum line since this metal has a combination of attractive properties (Lightness, high stability, high malleability, easy machining, high corrosion resistance and easy to pierce). Note: We created slots on the duct to allow the flux of air on the photovoltaic panel as illustrated in Figure 5.

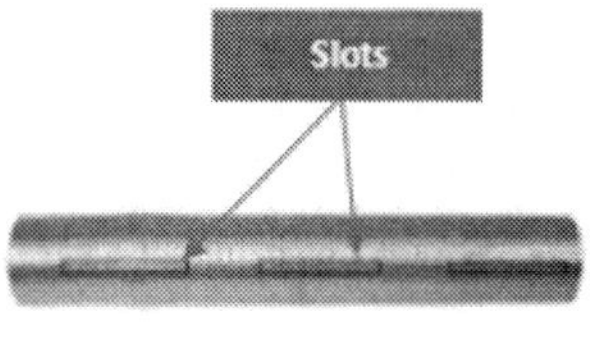

Figure 5: Compressed air duct

While the transport of air under pressure, it decreases in the presence of obstacles such as restrictions, bends, narrow passages etc. This pressure reduction is called loss. The production of compressed air is always accompanied by a loss, and the losses increase with the length and narrowness of the supply circuit. It is the air path of our system as shown in Figure 6. This method involves setting the linear pressure loss in the branch of the strongest network (a priori, the longest), for example to a value of 1 AP/m (current value of compromise between the acoustic problems at speeds too high air and the investment in the duct size). For the pressure losses, the calculation results are shown in table 4.

The compressor chosen must provide a higher or equal flux rate of 316 $m^3.h^{-1}$ taking into account the loads in terms of pressure loss, which will be estimated to 22.91 Pa.

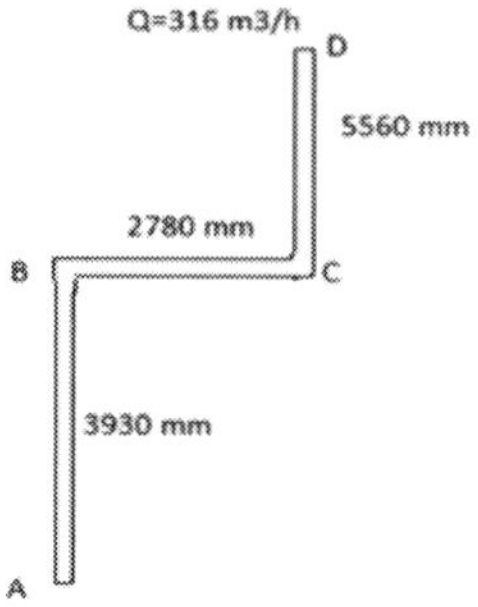

Figure 6: Path traveled by the flexible into the compressed air duct

Table 4: Pressure losses

Section	Flux [m^3/h]	Length [m]	Dp linear [Pa/m]	Dp [Pa]	S Dp [Pa]
A-B	316	3.930	1	3.93	3.93
B-C	316	2.780	1	2.78	6.71
C-D	316	5.560	1	5.56	12.2
D	316	---	1	12	22

We should outline that air debit supported by HCPV panel front glass is a priority for our study. For this reason we have performed a dimensioning of the compressed air network that led us to choose the appropriate compressor able to generate the air debit previously determined.

4.3 Dimensioning of the motor

For the electric motor type selection, we opted for a three-phase cage induction motor, which we have seen its benefits as its low cost, standard Sealed motor and easy to realize a direct starting on the network and list goes on.

The power for load movement has been estimated to define the range in which the motor will be selected [8]. For the displacement objective, we first calculated the compressed air duct translation speed. For this we determined the cleaning period that will be performed for 15 min in traveling a distance of d = 2 x 5.560 m. The air mobile conduct will make a back-and-forth movement; so it will cover a distance of twice the length of the HCPV panel. All dimensioning parameters, details, assumptions and calculations are now available in our published scientific report [9]. We focused on smaller models of three-phase cage induction motors, supplied by Leroy-Somer, with 0.06 kW power output which is slightly higher than we have calculated to better respect the safety margin. Motor characteristics are shown in table 6.

Table 6: Motor characteristics

Motor type	LS 56 M
Nominal power	0,06 kW
Nominal torque	0,4 Nm
Number of pole	4
Nominal speed	1380 rpm

4.4 Dimensioning of the movement transmission chain

In order to ensure the displacement of the duct, it was necessary to pre-define the range in which the motoreducer will be chosen. So, we have applied an accurate approach detailed in the TechnoMeca catalog [8]. Figure 7 illustrates the transmission of motion that occurs by the motoreducer and the pulley-belt system. We want a low-speed output, so a reduction ratio was calculated based on the displacement speed of the compressed air duct and the radius of the pulleys.

$$R = \frac{W_1}{W_2} = \frac{n_1}{n_2} \qquad (3)$$

where n_1 and W_1 are the motor output speed in rpm and in rad.s^{-1} respectively. Also, n_1 and W_1 are the desired speed output in rpm and rad.s^{-1} respectively.

The reduction ratio obtained is relatively important. Therefore the use of a reducer is essential. To do this, our choice rested on a coaxial motoreducer from Leroy-Somer perfectly compatible with the motor chosen. The reduction ratio is slightly below the reduction ratio of the cinematic chain.

Based on the system architecture, we have to move the compressed air duct and this will be done using a transmission system of the motor rotation into a translation movement, we opted for a pulley-belt system for its benefits such as flexible transmission, ability to vary the center distance, no lubrication, purchase cost and reduced installation, silent and light weight. The pulleys are of the same diameter seen that we want to transmit the movement without any changes.

Since a motor has to be controlled remotely or automatically and its use needs to be secure and safe, manual control is replaced by a control contactor. To order the system, a semi-automatic control will be developed using three (Forward, Reverse and Off) buttons and integrated in the HCPV plant control interface. All considerations and calculations including electrical, mechanical and economical concerns of theses issues are detailed and commented in [9].

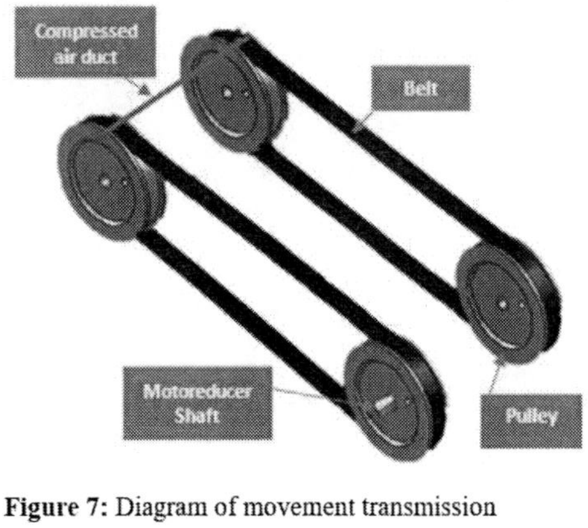

Figure 7: Diagram of movement transmission

5 RESULTS AND DISCUSSION

The first results we have obtained are shown in table 10. We have confirmed by a comparison of monthly consumption as shown in Table 7, that the traditional manual cleaning costs more than the proposed cleaning solution. On the other hand, based on the power graph of photovoltaic modules, we found that the estimated monthly gain in terms of energy is around 1881 kWh. Our proposed and designed cleaning technique is therefore cost efficient.

Table 7: Calculation of monthly consumption

Energy receiver type	Energy consumption per day (kWh)	Energy consumption per month (kWh)
Compressor	7.5	195
Motoreducer	0.015	0.4
Total monthly energy consumption of the system (kWh)		195,4
Cost of the total monthly energy consumption of the system (MAD)		141
The cleaning agent expenses (MAD)		1300

6 CONCLUSION AND PERSPECTIVES

The different experiments we have performed have clearly shown the effect of soiling on the HCPV system productivity. For this reason we thought of a development of an innovative technique for cleaning the system modules. The proposed cleaning solution has been proved that it costs less than the traditional manual cleaning. On the other hand, based on the power graph of photovoltaic modules, we found that the estimated monthly gain in terms of energy is around 1881 kWh. Our proposed and designed cleaning technique is therefore efficient.

This study can lead to a final prototype achievement that we propose in terms of perspectives, such as the integration of a dirt sensor, the electrostatic precipitator, the environment protection from the removed dust out of the panel, and the allocation of automatic control of all system. These additional improvements would be useful to develop the first Moroccan proper cleaning solution for high concentration photovoltaic panels.

ACKNOWLEDGMENT

This work has been supported by IRESEN (Institut de Recherche en Énergie Solaire et Énergies Nouvelles) Moroccan Institute, in the framework of SOLEIL Inno-PV Project. Its presentation at EU-PVSEC 2016 has been supported by GIZ

REFERENCES

[1] M. Crowley, "Why CPV ? The CPV value proposition", CPV Consortium, 2012
[2] B. Prior, "Cost and LCOE by Generation Technology", 2009-2020, GTM Research, November 2011
[3] A. Barhdadi, A. Benazzouz, B. Fabrizio, D. Verdilio "Implementation of high concentration photovoltaic grid connected power plant for training, research, innovation and solar electricity production" Journal of Physics Conference Series 605, 2015, 012042, pp. 1 – 8
[4] A. Barhdadi, G. Borelli, M. Carpanelli, and D. Verdilio, "New high concentration photovoltaic power plant for training, research and electricity production" Proceedings of the 29th European PV Solar Energy Conference and Exhibition, 22 - 26 Sep. 2014, Amsterdam, Netherlands, pp. 2125-2127.

[5] A. Barhdadi, "New high concentration photovoltaic power plant for training, research, innovation and solar electricity production at the university" ANSOLE News, Issue 3, 2014
[6] Y. EL Fatimy "Comparison of electric productivity between PV and CPV panels: Soiling effect" Master degree thesis, pp.48-53. 2015.
[7] S. Tachoula Tsogno, "Conception d'un système d'aspiration de poussière au poste de ponçage de carreaux de la Mipromalo", Polytechnic National School of Yaounde, Industrial Engineering 2005.
[8] F. Binet, "TechnoMeca: Dimensionnement d'une motorization d'axe", pp. 27 – 33, Nov. – Dec. 2007.
[9] D. Dahlioui, "Conception d'une technique de dépoussiérage semi-automatique d'une installation photovoltaïque à très haute concentration connectée au réseau de basse tension" Scientific Report of Assistant Engineer, 2014, INSAK, Morocco

PV MODULE AGEING IN SOUTHERN EUROPE – HOT SPOTS AND IMPACT ON YIELD

Matthias Grottke*, Oliver Beck*, Francisco Espín**
*WIP Wirtschaft und Infrastruktur GmbH & Co Planungs-KG, Sylvensteinstrasse 2, 81369 Munich, Germany,
Phone: +498972012-737, Fax. -791, matthias.grottke@wip-munich.de;
**Efficiency Services Consulting, Bullas, Spain; francisco.espin@efficiencyconsulting.es

ABSTRACT: By 2006 the common expectation was that crystalline PV modules produced in Europe by a leading manufacturer (and certified according to IEC 61215 and IEC 61730) were designed for long-term outdoor operation, including in challenging ambients. However, monitoring results of systems using large-area, Si-cell based Solon modules have shown another reality. Operation in an ambient with high irradiance, high temperatures and high temperature variations - namely on two-axis trackers in Southern Spain - resulted in accelerated module ageing. By 2016, after ten years of operation, every second PV module of a PV park with such trackers shows degradation effects. The peak capacity of sub-systems is partially reduced to below 80% of the initial capacity. A variety of defect mechanisms is observed. Countermeasures to remedy the situation had to be developed and implemented to keep systems operational.
Keywords: Crystalline PV Modules, Degradation, Energy Performance, Tracking

1 SOLAR TRACKING TECHNOLOGY UNDER SCRUTINY

The Spanish PV boom began in 2007 with the implementation of the Royal Decree 661/2007. As a consequence, PV system installations jumped from 19 MWp per year in 2005, to 103 MWp in 2006 and 544 MWp in 2007 to 2,708 MWp in 2008. This significant market growth was in line with significant growth in other key European markets. PV module costs were comparatively high and dual-axis tracking systems promised to be an economical solution for markets with high direct solar radiation levels.

A premium European PV module manufacturer, the company Solon, had developed a mechanically robust dual-axis tracking PV system using the latest PV cell technology. One of the first PV farms in Spain to implement Solon dual-axis trackers was the Solar farm in the region of Murcia which was installed in 2006 and connected to the grid in 2007. It has an overall plant capacity of 230 kWp and is the focus of this work. This PV farm uses two different versions of the Solon dual-axis tracker "Mover L Poly" (28 units).

Figure 1. Dual-axis tracker "Solon Mover L Poly" at the Murcia pilot plant.

Solon manufactured two-axis solar trackers in two configurations for the Spanish market: (a) the Solon Mover L with 6.3 to 9.5 kWp and (b) the Solon Mover XL with 14.2 – 19.8 kWp. The Solon Mover L was the first of the two Solon trackers available on the market and was offered in three versions per **Table 1**.

The Solon Mover L trackers "Poly" and "Mono" carry 12 PV modules. Always four modules of a line are connected in parallel to one inverter.

Soon many solar farms with Solon Movers followed in Spain. Solon Movers with a capacity significantly exceeding 40 MWp were installed throughout Spain. **Table 2** below lists a selection of ten relevant size installations. System implementation was verified by the author.

Table 1. Versions of the "Solon Mover L".

	Capacity [kWp]	Module Characteristic
1 Poly	6.32 - 7.49	160 polycrystalline Si cells @ 6.2"
2 Mono	7.05 - 7.53	160 monocrystalline Si cells @ 6.2"
3 High Performance	8.37 - 9.45	260 monocrystalline Si cells @ 5.0"

Table 2. Selection of larger PV plants in ES using Solon technology 2-axis tracking systems.

		Capacity [MWp]
1	Jerez de los Caballeros	10,4
2	Flix, Ribera d'Ebre	8,2
3	Lebrija, Guadalquivir	6,6
4	Llerena, Badajoz	4,0
5	Puerto Lápice	3,5
6	El Redondo, Jorqueta	2,2
7	Don Benito	2,0
8	Galera, Alqueria	1,0
9	Cúllar, Ventanas	1,0
10	Cabanillas	1,0
	Total	**39,9**

At the pilot PV plant in Murcia, priority was given from the beginning to an effective O&M and an early identification of system malfunctions.

However, nine years after PV system installation, by January 2015, a significant number of PV modules exhibited cells with optical discoloration. This was the key reason to initiate a module-level state-of-health analysis. It was not an obvious reduction in power output of the system.

Annual yield figures of the PV system did not yet

provide direct evidence to take action per **Figure 2**.

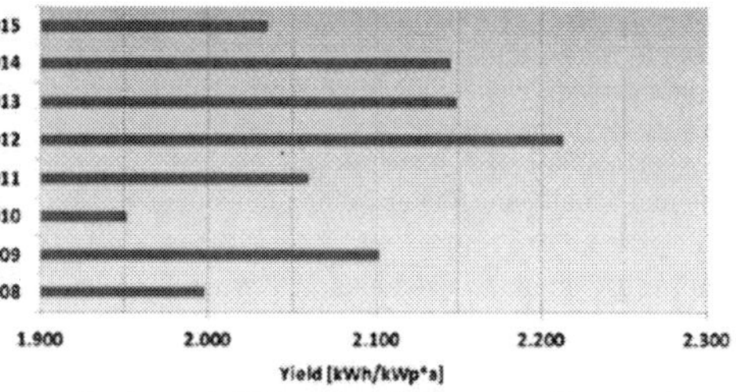

Figure 2. Annual PV park yield on medium voltage level.

This paper presents (a) strategies developed for detecting system degradation at a plant using basic electricity meter level monitoring, (b) module defects identified via IR Thermography and inverter level power monitoring, and (c) solutions tested in the field to remediate key defects.

2 IDENTIFYING SYSTEM BASED DIFFERENCES IN PERFORMANCE

As previously mentioned, basic annual yield figures of the PV system did not show underperformance. Average monthly yield figures per **Figure 3** did not show evidence either. Monthly average figures for 2015 are not generally low (green marks) and do show peaks for January, June and November.

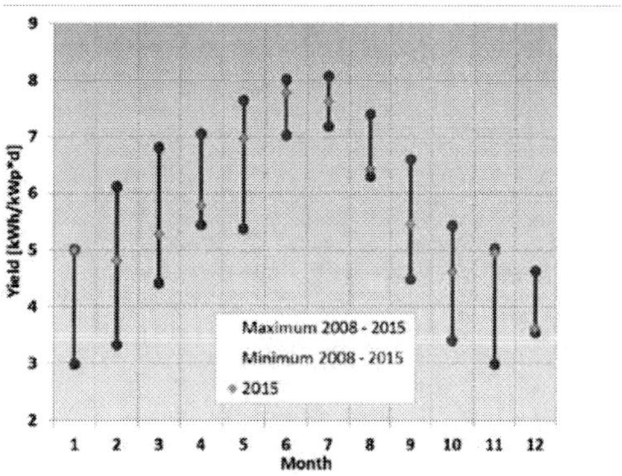

Figure 3. Monthly average yield per day on medium voltage level for different years (between 2008 and 2015).

The figures change when focusing exclusively on "clear-sky" days of every month and year. The respective results are presented in **Figure 4**. This graph suggests that the peak performance of this park is decreasing: Recent daily peak yields per month on medium voltage level are the lowest in most months in the last year as compared to the period 2010-2016 (exceptions are January, March and September). The recorded difference between the latest figures and the "ever peak" is significant: it varies between -4.4% and -11.7%.

The observation is confirmed when having a closer look to higher resolution hourly data as per **Figure 5**. The profile for 2015 is significantly below the corresponding figures for 2010 and 2011 (-2% to -16% during peak generation periods). Note that shading by green is not an issue at this plant and PV modules are periodically cleaned.

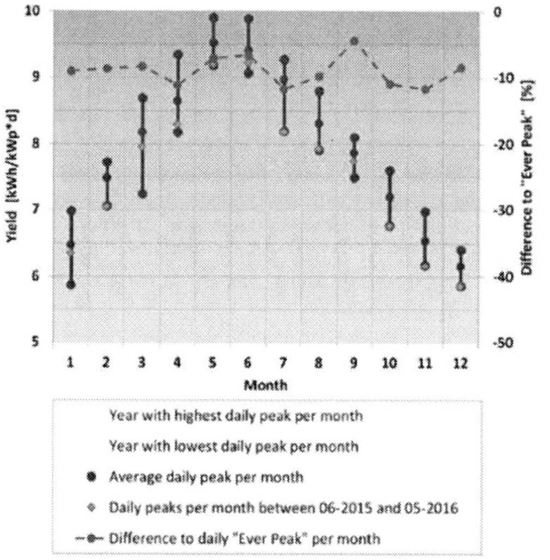

Figure 4. Daily peak yields per month on medium voltage level for different years (between 2010 and 2016).

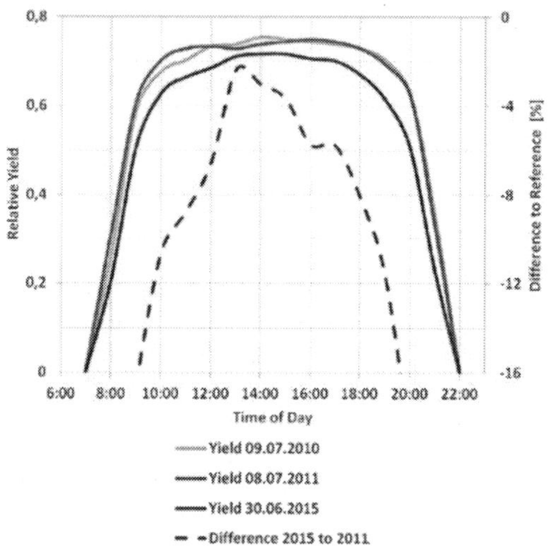

Figure 5. Daily relative yield on clear-sky days for three years (2010, 2011, 2015) on medium voltage level.

This type of graph forms part of a protocol which is automatically generated on a daily basis to support the O&M services. The side by side comparison of such graphs on the tracker level directly permits the quick identification of faulty system behavior per **Figure 6** [1].

Figure 6 shows that the output of Inverter #1 of Mover #28 is 25% lower than the output of the two other inverters of this Mover. It was found that one of four modules connected in parallel to the inverter was defective. Also, the performance of the second inverter sub-system is lower for a defined period.

Most of the corresponding module level defects were identified in detail via IR thermography.

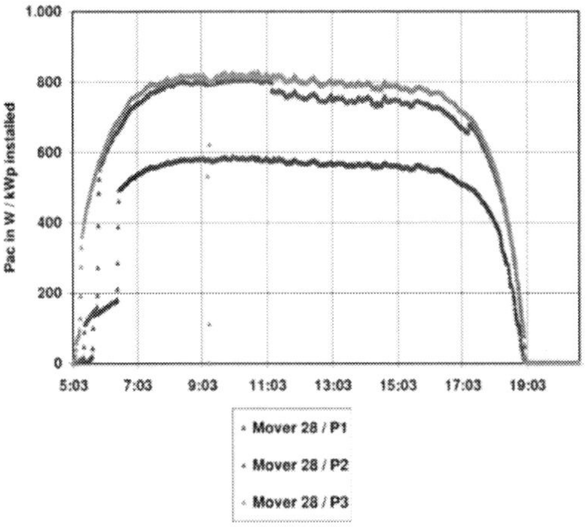

Figure 6. Daily relative performance of the three independent PV fields/ inverters of one Mover (16.05.2015).

3 IDENTIFYING MODULE LEVEL DEFECTS VIA IR THERMOGRAPHY

By January 2015, a significant number of PV modules exhibited cells with optical discoloration, mostly browning. Consequently an IR thermography and closer technical analysis was done on the module level by mid 2015. This work revealed that 149 of 336 PV modules, or 44% of the PV modules, show defects (**Figure 7**). Many of these defects were visualized only via IR thermography. Some of the identified visible defects/ discolorations did not result in measurable hot spots.

	Mover #	Total	Mover #	Total
	1	5	15	5
	2	3	16	8
	3	3	17	7
	4	10	18	7
	5	8	19	0
	6	3	20	3
	7	4	21	0
	8	5	22	5
	9	5	23	3
	10	7	24	4
	11	8	25	9
	12	5	26	3
	13	8	27	7
	14	5	28	9
Total defective modules		79		70
Total defective modules [%]		**47**		**42**
Overall number of modules		168		168

Mover w/o defective module (2 movers)
Mover with defective modules (visible defects; 4 movers)
Mover with defective modules (measurable defects; 7 movers)
Mover with defective modules (significant defects; 6 movers)
Mover with defective modules (significant defects; 9 movers)

Figure 7. Results of module level IR thermography in combination with inverter level power output analysis and, in specific cases, IV curve measurements.

Per **Figure 8**, twelve of the 336 PV modules (4%) showed hot spots that can be classified as severe, in line with the classification scheme used by R. Moretón et al. **[3]**.

In this case, the ΔT between the PV cell with a hot spot and non-defective cells in the vicinity is > 30°C. In the worst case, the ΔT was 88°C at operation conditions with G > 700 W/m² (hot spot with 155°C). Further 21 modules (6%) showed hot spots with ΔT of 20°C to 30°C.

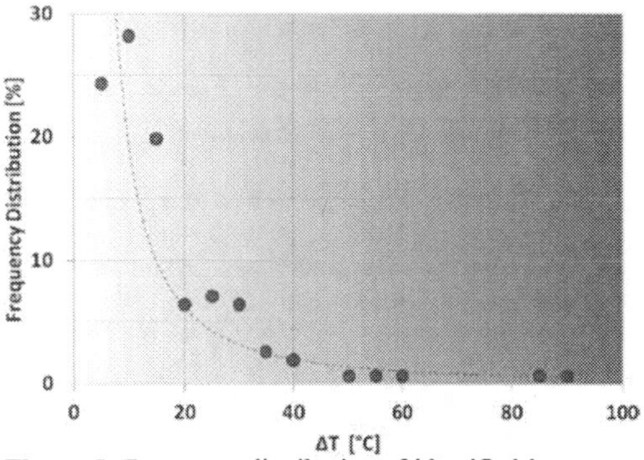

Figure 8. Frequency distribution of identified hot spots. ΔT is the temperature difference between the PV cell with a hot spot and non-defective cells in the vicinity.

Figure 9. A micro-crack divides the above cell in two parts (blue arrow). The hot spot on the remaining cell part reaches temperatures of 133°C (red arrow). The respective cell had to be bridged from the modules back side to avoid further module degradation.

Defects identified include:

- burn marks at the back sheet
- burn marks at the front,
- discoloration of the encapsulant associated with overheating along the metallic interconnections
- discoloration of the solar cell
- browned EVA in the center of the cell, including severe discoloration
- contact problems at the PV module junction box
- single rows that are warmer than others
- etc.

Some of these defects had to be repaired on site, such as (a) the reconnection of defective junction boxes or (b) the bridging of defective cells exhibiting high temperatures. In the mid-term selected defective PV modules will have to be replaced by new modules with similar characteristics.

Note that by 2006 a fixed installed PV system with 31 kWp comprising 135 PV modules of Solon, type P220/6+, was implemented and is still operated next to the dual-axis tracking systems. The modules are smaller but use same-type PV cells (60 cells/ 1.64m² vs. 160 cells/ 4.39m²). None of these modules show any discoloration or visible defects.

4 CONCLUSIONS

Sun tracked PV modules are normally installed in regions with a high direct solar radiation level. These modules are subject to longer periods of elevated module temperatures and to higher module temperature variations as compared to fixed installed systems. In Murcia (ES) this resulted in accelerated module ageing of the tracked PV modules. The fixed tilted PV modules using same-type PV cells do not yet show degradation effects. Nine years after system installation, 44% of the tracked modules are defective. Most of them show cell discoloration. In the case of the Murcia PV plant, this results in a loss of overall generation capacity.

Typical PV system operation times do exceed 20 years. Thus system design should allow for component replacement (modules/inverter), including from alternative manufacturers.

O&M services should be capable of identifying module defects, classifying defects by severeness-level and to remedy the most critical defects thus preventing accelerated system degradation.

5 REFERENCES

[1] M. Grottke et al., Performance of a 230 kWp Solar Park in Spain with Two-Axis Trackers from a European Market Leader; Proceedings of the 24th EUPVSEC (2009), Hamburg, Germany, p.4080 – 4082.

[2] R. Moretón et al., Dealing in Practice with Hot-Spots, Proceedings of the 29th EUPVSEC (2014), Amsterdam, The Netherlands, p.2722 – 2727.

[3] M. Köntges et al., Performance and Reliability of PV Systems, Subtask 3.2: Review of Failures of Photovoltaic Modules, IEA PVPS Task 13, External final report IEA-PVPS, March 2014.

COMPARATIVE PERFORMANCE INVESTIGATION OF MONO, POLY-CRYSTALLINE AND AMORPHOUS SILICON PHOTOVOLTAIC MODULES UNDER NORTHERN MOROCCO CLIMATE CONDITIONS

K. Attari, A. El yaakoubi, Z. Leemrani, A. Asselman
Optic and photonic team, Abdelmalek Essaadi university, MOROCCO
Email: attari.kamal@gmail.com ali.elyakoubi@gmail.com

ABSTRACT: In this paper, a PV system installed in Tetouan's faculty is investigated , its total capacity is 5.86 Kw_p and comprises three types technologies modules, monocrystalline (m-Si) , polycrystalline (p-Si) and amorphous (a-Si). The purpose of this study is to give real and simulated results, real results based on climate data and by monitoring the PV system during the year 2015 and simulated data by the PVsyst software. Those results can be used to make more appropriate technology choice in similar climate.

Keywords: grid-connected photovoltaic, PVsyst software, generated energy, final yield

1. INTRODUCTION

In the recent years, solar energy has expanded in Morocco where the country seeks to become leader in the field of renewable energies and particularly in photovoltaic energy by developing research centres and studies in all the kingdom and by the integration of the solar system on most faculties. Various studies have been conducted on the performance parameters of installed PV power plants in different geographical locations and different climatic conditions [1–6]. Eke and Demircan [7] are prepared performance analysis of a multi crystalline Si photovoltaic module under Mugla climatic conditions in Turkey. Grid-connected PV power plants studies are based on the recent experimental studies and there is still a need for detailed performance analysis about mc-Si and p-Si PV module. The paper [8] compare the experimental behavior of these two common PV module technologies (m-Si and p-Si). In the present paper detailed study is given with a comparison between the three technologies; m-Si, p-Si and a-Si. An estimation of the energy produced is calculated depending on real and simulated data. In the first part of this study, a presentation of the photovoltaic technologies and the monitoring station where data comes from are given, then a presentation of the site characteristic is introduced as well. In the second part of the study, a mathematical formulation of the site and the PV technology parameters are provided to validate the collected data, then a numerical simulation with PVsyst software also will be given. At the end of the work, the results and a discussion will be presented.

2 DESCRIPTION OF PV INSTALLATION

Three types of PV system are under consideration of this study: Monocrystalline silicon PV panels, Polycrystalline silicon PV panels and Amorphous silicon PV panels. These technologies are installed in a public establishment in Tetouan north of Morocco. The capacity of this connected grid installation is 5.86 Kw_p. The installation is linked with the web, by means of a monitoring station, in order to supervise the production of the installation and the metrological conditions of the site. The data that we have used through this work are obtained from this station. The study and the analysis are done for the years 2015. The study is done on daily, monthly and annually basis. A Technical data of the PV modules are given in Table 1

Table I: PV modules specifications.

PV module	mc-Si	p-Si	a-Si
STC Power Rating	255 W_p	255 W_p	155W_p
Peak Efficiency	15.21%	15.21%	9.9%
Vmp	30.6V	30.9V	65.2V
Isc	9.05A	8.88A	2.56A
Imp	8.43A	8.32A	2.38A
Voc	38.7V	38V	85.5V

Figure 1: The PV system in Tetouan University

3 MATHEMATICAL FORMULATIONS

The mathematical formulation of the site and the system technology parameters are:

Energy output

The total energy generated by the PV system is determined from [9]:

$$E_{AC,d} = \sum_{d=1}^{24} E_{AC,h} \quad (1)$$

Nomenclature

Tamb ambient average temperature
GlobInc global irradiation in the collector plan
GlobEff effective global irradiation the collectors
E_{array} energy produced by the PV array
E_{Grid} energy injected into the grid
EffArrR PV array effective Earray related to the irradiation on the collector's total area
EffSysR system efficiency EGrid related to the irradiation on the collector's total area
GlobHor global irradiation in the horizontal plane
a-Si amorphous
m-Si monocrystalline
p-Si polycrystalline
PV photovoltaic

And

$$\sum E_{AC,m=}\sum_{d=1}^{N} E_{AC,d} \quad (2)$$

Where $E_{AC,h}$ is AC energy output at hour h, $E_{AC,d}$ is the daily AC energy output E_{AC}, m is the monthly AC energy output and N is the number of days in a month.

Array yield

The array yield (Ya) is defined as the energy output from a PV array over a defined period (day, month or year) divided by its rated power and is given as [10]:

$$Y_A = \frac{E_{DC}}{P_{PV}, rated} \quad (3)$$

Where Ya is the array yield and E_{DC} is the DC energy output (kW h)

Final yield

The final yield is defined as the annual, monthly or daily net AC energy output fromthe inverter divided by the rated power of the installed PV array, is given as [11,12]

$$Y_{F,d} = \frac{E_{AC}}{P_{pv}, rated} \quad (6) \qquad Y_{F,m} = \frac{1}{N}\sum_{d=1}^{N} Y_{F,d} \quad (7)$$

System efficiencies

Main groups of PV system efficiencies are: PV module efficiency, inverter efficiency and system efficiency. The PV module efficiency is calculated as [13]

$$\eta_{PV} = \frac{100 * E_{DC}}{H_t * S} \ (\%) \ (13)$$

The monthly PV module efficiency ($\eta_{PV,m}$) is calculated as [14,15]

$$\eta_{PV,m} = \frac{\sum_{i=1}^{n} E_{DC}}{S * \sum_{i=1}^{n} H_t} \ (\%) \ (14)$$

Where E_{DC} is a total energy produced by the solar PV modules, n is the number of days in a month and S (m^2) the total area occupied by the PV modules. The inverter efficiency is given as :

4 RESULTS AND DISCUSSION

The results are given in form of tables and figures to evaluate the performances of the present study. The tables represents global irradiation, energy produced by the modules, energy injected into the grid, ambient temperature, PV module efficiency and system efficiency for each technology over the monitored period.

Figure 2 shows the horizontal irradiation and power for the three technologies: m-Si, p-Si and a-Si in a winter day. It can be observed that the p-Si modules are greater than the others in particular in the hours when the irradiation is higher. The maximum irradiation was found 729 wh/m^2 and the maximum power for the P-Si was 1925 w, for the m-Si it was 1892 w and for the a-Si it was 1651 w.

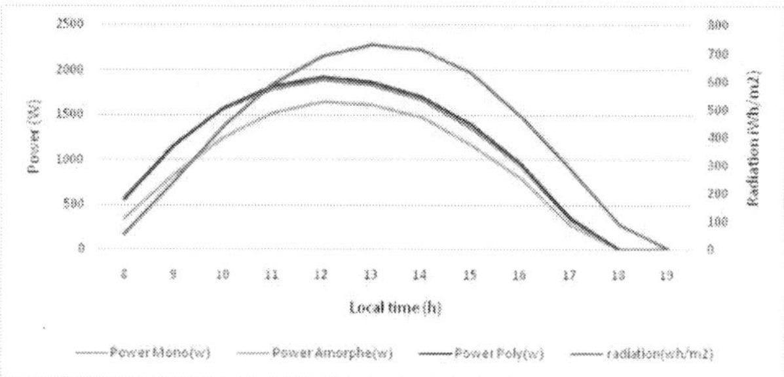

Figure 2: Power daily total PV module output in winter

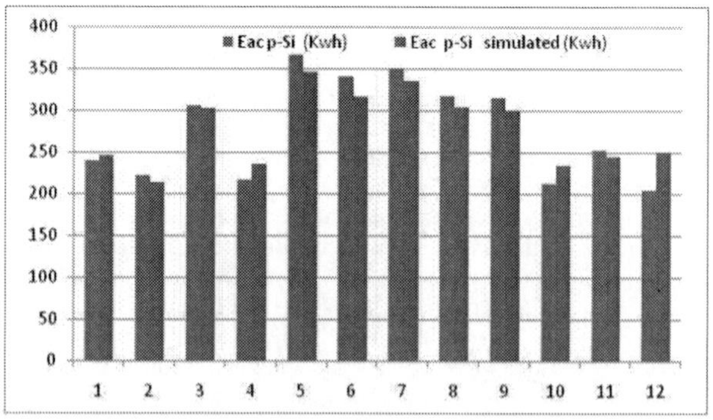

Figure 3: Monthly produced energy and simulated energy for P-Si

Figure 3 shows the energy produced by the inverter during a period of one year from p-Si panels and simulated values are also given, it can be observed that the measured values are higher in most months except January, April, October and December. The maximum

Table II: PVsyst parameters at a tilt angle of 32° for Amorphous Modules

	GlobHor (kWh/m²)	T amb (°C)	GlobInc (Kwh/m²)	GlobEff (Kwh/m²)	EArray(Kwh)	E_Grid (Kwh)	EffArrR (%)	EffSysR(%)
January	92.7	12.5	144.8	141	252.2	225.3	8.63	8.26
February	96.3	13.2	126.3	122.8	222.7	197.5	8.69	8.3
March	155	15.2	185.1	180.1	318.7	290.9	8.72	8.34
April	140.1	16.6	143.8	139.2	252.3	225.3	8.74	8.32
May	232.8	19.1	217.3	210.2	367.3	338.6	8.65	8.27
June	227.7	22.5	202.3	195.5	337.3	312.7	8.59	8.2
July	241.8	25.2	218.7	211.6	355.5	335.8	8.53	8.15
August	196.8	25	195.8	190.1	322.1	301.1	8.54	8.16
September	168.3	23	191.6	186.2	315	295.3	8.55	8.18
October	114.4	19.9	142.7	138.8	245.4	221	8.61	8.22
November	96.9	16.4	149	145.1	250.8	227.4	8.47	8.1
December	86.8	13.9	147.1	143.2	251.9	232.4	8.5	8.13
Year	1849.6	18.57	2064.4	2003.9	3491.2	3196.1	8.6	8.22

Table III: PVsyst parameters at a tilt angle of 32° for Polycrystalline Modules

	GlobHor (kWh/m²)	T amb (°C)	GlobInc (Kwh/m²)	GlobEff (Kwh/m²)	EArray(Kwh)	E_Grid (Kwh)	EffArrR (%)	EffSysR(%)
January	92.7	12.5	144.8	141	260.2	247.4	13.81	13.13
February	96.3	13.2	126.3	122.8	226.4	214.9	13.77	13.07
March	155	15.2	185.1	180.1	320	303.6	13.28	12.6
April	140.1	16.6	143.8	139.2	250.1	236.3	13.37	12.63
May	232.8	19.1	217.3	210.2	366.3	346.8	12.95	12.26
June	227.7	22.5	202.3	195.5	335.6	317.6	12.75	12.07
July	241.8	25.2	218.7	211.6	355.7	336.6	12.5	11.83
August	196.8	25	195.8	190.1	322.5	305	12.65	11.97
September	168.3	23	191.6	186.2	317.6	300.8	12.74	12.06
October	114.4	19.9	142.7	138.8	247	234.1	13.3	12.61
November	96.9	16.4	149	145.1	259.3	246.1	13.37	12.69
December	86.8	13.9	147.1	143.2	263.2	250.2	13.75	13.07
Year	1849.6	18.57	2064.4	2003.9	3523.8	3339.3	13.12	12.43

Table IV: PVsyst parameters at a tilt angle of 32° for Monocrystalline Modules

	GlobHor (kWh/m²)	T amb (°C)	GlobInc (Kwh/m²)	GlobEff (Kwh/m²)	EArray(Kwh)	E_Grid (Kwh)	EffArrR (%)	EffSysR(%)
January	92.7	12.5	144.8	141	260.2	247.8	13.81	13.15
February	96.3	13.2	126.3	122.8	226.5	215.4	13.78	13.1
March	155	15.2	185.1	180.1	319.6	303.9	13.26	12.61
April	140.1	16.6	143.8	139.2	250.8	237.8	13.4	12.71
May	232.8	19.1	217.3	210.2	366.2	348	12.95	12.3
June	227.7	22.5	202.3	195.5	335.6	318.6	12.75	12.1
July	241.8	25.2	218.7	211.6	355.6	337.6	12.49	11.86
August	196.8	25	195.8	190.1	322.2	305.8	12.64	12
September	168.3	23	191.6	186.2	317.2	301.4	12.72	12.09
October	114.4	19.9	142.7	138.8	246.9	234.5	13.3	12.63
November	96.9	16.4	149	145.1	259	246.4	13.36	12.71
December	86.8	13.9	147.1	143.2	263.2	250.7	13.75	13.1
Year	1849.6	18.57	2064.4	2003.9	3523	3348	13.11	12.46

measured values were found in May with 367.84 Kwh and 346.8 Kwh for the simulated values. The minimum measured value was found 204.3 Kwh in December and 214.9 Kwh in February for simulated values. Figure 5 shows the results of measured and simulated values from m-Si panels, for this technology both values are similar for February with 218 kwh for measured values and 215.4 kwh for simulated values, for March 303.3 Kwh for measured values and 303.9 Kwh for simulated value and November with 245 for measured values and 246.4 for simulated values, for the other months the values were very close. The maximum value was found in May with 362.42 Kwh for measured values and 348 for the simulated values

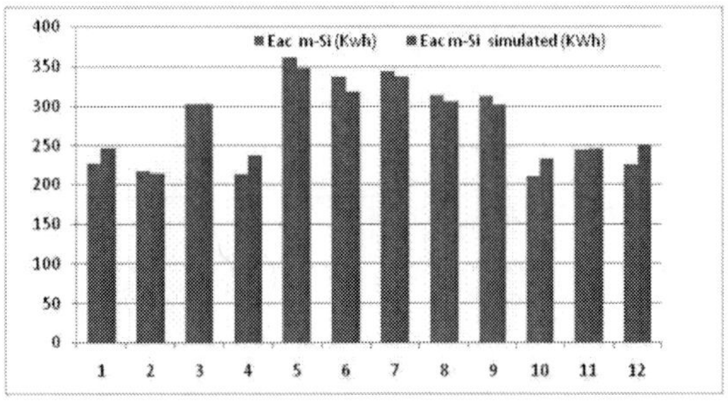

Figure 4: Monthly produced energy and simulated energy for m-Si modules

. Figure 5 gives results of monthly produced energy during 2015 from the a-Si panels, for this technology almost all months are close to the simulated values except December with 161.8 Kwh for measured values and 225.4 Kwh for simulated values. The maximum value was found in May with 322 Kwh for measured values and 338 Kwh for the simulated values. The minimum value was found in December.

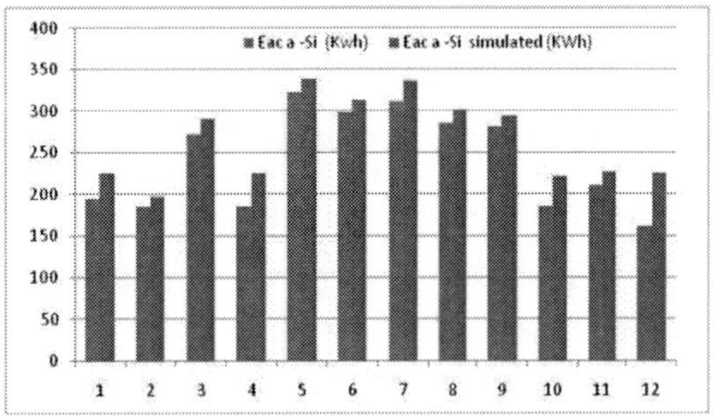

Figure 5: Monthly produced energy and simulated energy for a-Si modules

The table1 describe the simulation results of amorphous modules array and system efficiencies. The annual module efficiency was 8.6% and the annual system efficiency was found 8.22%. The array energy ranged from 206.9 Kwh in February to 354.1 Kwh in May despite july has a higher irradiation 241.8 Kwh/m^2, thisis due to the combination of high irradiance and lower temperature in May, however July has a higher temperature 25.2 °C against 19.1°C in May as shown in Table 1.

The table 3 and 4 shows the results for p-Si and m-Si modules, the annual array efficiencies were found 13.12% and 13.11% respectively. The annual system efficiencies are 12.43% for the p-Si panels and 12.46% for the m-Si panels. The simulated annual energy array output was found 3523.8 Kwh for the p-Si panels and 3523 Kwh for m-Si panels. The total energy injected into the grid was 3339.3 Kwh for p-Si panels and 3348 Kwh for the m-Si. Those simulated values are close to the measured values: 3351.24 Kwh for p-Si modules and 3315.69 Kwh for m-Si modules but the measured values shows that the p-Si are better than the m-Si and a-Si modules.

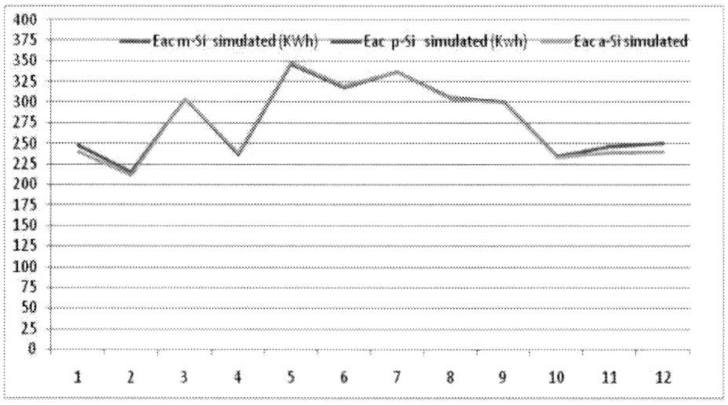

Figure 6: Monthly simulated energy output for the three technologies

Figure 6 shows the Monthly energy simulated injected into the grid by the three technologies, in this simulation the 155 Wp amorphous panels are replaced by an equivalent module with the same power and characteristics of the other two technologies .The results demonstrate that the m-Si and p-Si modules are better in winter months: October, November, December, January February as shown in Figure 6

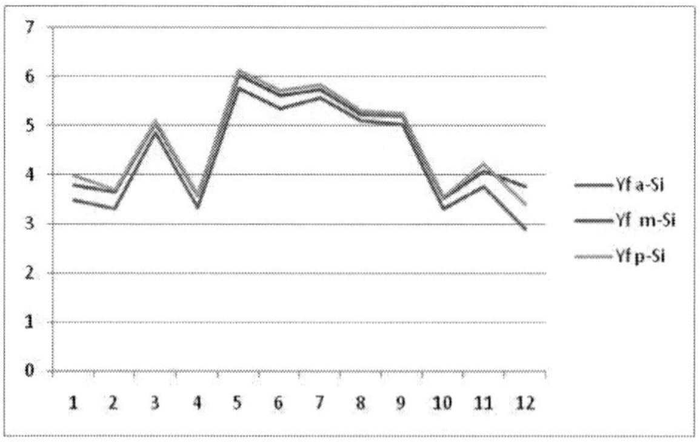

Figure 7: Comparison of final yield for the three technologies

Figure 7 shows the monthly average final yield for the three technologies depending on real measured data, the p-Si technology has the higher value with an annual average of 4.65 kwh/kw$_p$/day, the second is the m-Si with a close annual avearage value of 4.60 kwh/kw$_p$/day and finally the last one was the a-Si technology with avearage annual value of 4.31 kwh/kw$_p$/day.

CONCLUSION

Throughout the presented results,the experimental parameters of the PV system coincide well with the estimated one by the software, the result obtained showed a significant improvement of the presented site in term of produced energy between the three technologies of the PV system used in the investigation. The p-Si has the better average annual final yield with 4.65 kwh/kw$_p$/day and the higher mesuread annual energy injected into the grid with 3351.24 Kwh. This study reveals the potential in term of investment of this technology in the country and similar climate areas.

Acknowledgements

The authors acknowledge l'Institut de Recherche en Energie Solaire et Energies Nouvelles (IRESEN) in Morocco .

References

[1] Zhou W, Yang H, Fang Z. A novel model for photovoltaic array performance prediction. Appl Energy 2007;84:1187–98.
[2] Kim J-Y, Jeon G-Y, Hong W-H. The performance and economical analysis of grid-connected photovoltaic systems in Daegu,Korea.Appl Energy 2009;86:265–72.
[3] Li DHW, Cheung KL, Lam TNT, Chan WWH. A study of grid-connected photovoltaic (PV) system in Hong Kong. ApplEnergy 2012;90:122–7.
[4] Su Y, Chan L-C, Shu L, Tsui K-L. Real-time prediction models for output power and efficiency of grid-connected solar photovoltaic systems. Appl Energy 2012;93:319–26
[5] Sharma V, Kumar A, Sastry OS, Chandel SS. Performance assessment of different solar photovoltaic technologies under similar outdoor conditions. Energy 2013;58:511–8.
[6] Padmavathi K, Daniel SA. Performance analysis of a 3 MWp grid connected solar photovoltaic power plant in India. Energy Sustain Dev 2013;17:615–25.
[7] Eke R, Demircan H. Performance analysis of a multi crystalline Si photovoltaic module under Mugla climatic conditions in Turkey. Energy Convers Manage 2013;65:580–6.
[8] Saeed Edalati ,MehranAmeri , MasoudIranmanesh.Comparative performance investigation of mono- and poly-crystalline silicon photovoltaic modules for use in grid-connected photovoltaic systems in dry climates.Applied Energy 2015: 160;255–265
[9] Ayompe LM, Duffy A, McCormack SJ, Conlon M. Measured performance of a 1.72 kW rooftop grid connected photovoltaic system in Ireland. Energy Convers Manage 2011;52:816–25.
[10] Padmavathi K, Daniel SA. Performance analysis of a 3 MWp grid connected solar photovoltaic power plant in India.
Energy Sustain Dev 2013;17:615–25.
[11] Mondol JD et al. Long term performance analysis of a grid connected photovoltaic system in Northern Ireland. Energy Convers Manage 2006;47(18– 19):292547.
[12] Al-Otaibi A. Performance evaluation of photovoltaic systems on Kuwaiti school's rooftop. Energy Convers Manage2015 ;95 :110-119
[13] Adaramola M, Vagnes E. Preliminary assessment of a small-scale rooftop PV grid tied in Norwegian climatic conditions. Energy Convers Manage 2015;90:458–65.
[14] Filippo Spertino, Fabio Corona. Monitoring and checking of performance in photovoltaic plants: a tool for design, installation and maintenance of grid connected systems. Renew Energy 2013;60:722–32.
[15] Chaiyant Boonmee, Boonyang Plangklang, Napat Watjanatepin. System performance of a three-phase PV-grid-connected system installed in Thailand:data monitored analysis. Renew Energy 2009;34:384–9.

INVESTIGATION AND DIAGNOSTIC TOOL COMPARISON: INFRARED THERMOGRAPHY VS ELECTROLUMINESCENCE

Dario Bertani[1], Claudio Liciotti[2], Salvatore Guastella[1], Cinzia Camilloni[2]
1 Ricerca sul Sistema Energetico – RSE S.p.A., TGM - Power Generation Technologies and Materials,
Via Rubattino, 54 - 20134 Milano (MI) – Italy
Phone: +39 02 3992 4719, e-mail: dario.bertani@rse-web.it
2 KB Development srl, Via A.Volta 13, San Zeno Naviglio (BS) – Italy
Phone: +39 3475354967, e-mail: cli@kbdev.it

ABSTRACT: This paper presents the comparison of two PV modules diagnostic/inspection methods in the field. The comparison of infrared thermography and electroluminescence has been performed with the practical experience from the measurements. The committee draft 82/1109/CD (IEC/TS 60904-12 Ed. 1.0 Photovoltaic devices - Part 12: Infrared thermography of photovoltaic modules) and committee draft 82/1062/CD (IEC/TS 60904-13 Ed. 1.0 Photovoltaic devices - Part 13: Electroluminescence of photovoltaic modules) were taken into account. The aim of this work is to identify the best diagnostic tool for assessing the performance of photovoltaic modules and identifying the most commons failure causes in outdoor and indoor conditions.

Keywords: Diagnostics, Electroluminescence, PV Modules, Failure Causes, Thermography.

1 INTRODUCTION

Infrared (IR) thermography could be an important diagnostic tool for assessing the performance of photovoltaic panels [1]. It is relatively fast, simple, inexpensive and useful to identify failures on the modules. The testing of Photovoltaic (PV) systems by infrared thermography has been carried out for more than a decade and is increasingly becoming an important standard. Capture thermal images of photovoltaic modules by infrared thermography, process images to obtain metrics about the image taken in quantitative terms, and provide guidance to qualitatively interpret the images are the scopes of an IEC committee draft (82/1109/CD) [2]. Electroluminescence (EL) imaging is a common indoor measurement technique [3] used to identify the damage and the failure cause in PV modules. Capture electroluminescence images of photovoltaic modules, process images to obtain metrics in quantitative terms and provide guidance to qualitatively interpret the images are the scopes of an IEC committee draft (82/1062/CD) [4]. EL could be also used in outdoor conditions. Both techniques are going to be standardized and could be used by O&M providers to analyze the status of PV installations and to identify typical issues on PV modules.

The aim of this work is to identify the best diagnostic tool for assessing the performance of photovoltaic panels and identify the most commons failure causes in outdoor and indoor conditions.

2 TESTING PROCEDURES

Some damaged modules were identified in some PV installations. These modules were analyzed with infrared thermography, during a sunny day; electroluminescence was attempted during the following night. Both techniques were repeated in indoor conditions to better understand the effects of the environmental conditions on the tests.

Other c-Si modules (modules manufactured with different technology and modules installed more than 20 years ago) were also tested to validate indoor analysis techniques.

3 OUTDOOR IR

Thermography is a safe, non-contact analysis technique to check installed photovoltaic modules. The emissivity of silicate glass depends strongly on the reflection angle [5]. Usually glass has an emissivity of 0,85 in the 8–14 μm waveband [6]. In order to perform meaningful measurements a high emission is required and the camera image shall be taken as perpendicular to the PV module surface as possible [7]. However, taking images perpendicularly to the PV module surface implies the imaging of self-reflection of measuring personnel and IR-camera apparatus and reflection of heated objects like sun, near-by buildings and trees. In order to avoid reflection a viewing angle of 5–30° is a good compromise (where 0° is perpendicular).

To achieve sufficient thermal contrast a solar irradiance of 700 W/m^2 or higher is advisable [6]. During thermography analysis the environmental conditions and DC load (i.e. the level of the current fed into the PV module) should be monitored.

IR images were captured using IR-cameras, in compliance with the minimum requirements for IR-cameras used for inspecting PV plants recommended by 82/1073/CD [7], as perpendicular to the PV module surface as possible, avoiding self-reflection of measuring personnel and IR-camera. The IR-camera emissivity was set to 0,85.

Table I: Thermography analysis conditions

Location	Milan
Type of mounting	Tilted ground mounted system (tilt 30° - azimuth 0°)
Air temperature [°C]	27
Wind speed [m/s]	0,9
P.O.A. Irradiance [W/m^2]	960
PV modules type	60 polycrystalline solar cells module
PV modules eff. [%]	12,56
PV module connection	4 modules in series; short circuit condition

A photovoltaic reference device compliant with IEC 60904-2 with an estimated uncertainty of 3 % was used to measure the solar irradiance (W/m^2) in the plane of the array of modules. The analyzed modules string was set up in short circuit condition.

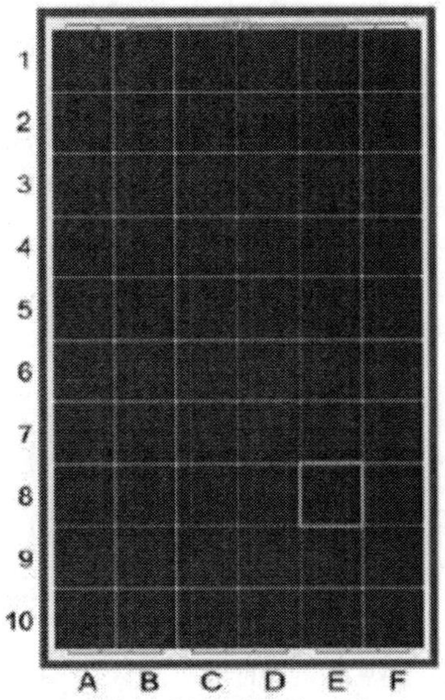

Figure 1: Matrix for cell identification [7].

For PV cells identification, the matrix reported in figure 1 was taken as reference.

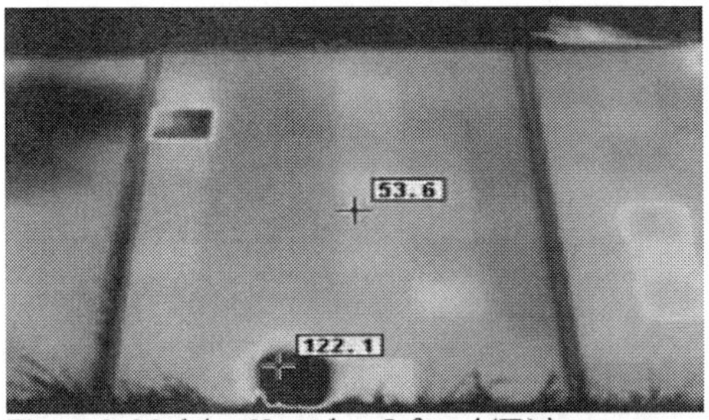

Figure 2: Module n°2 outdoor Infrared (IR) image.

Module number 2 presents some thermal abnormalities. Two cells (A3, C10) are warmer than others and present a temperature > 40 K than the average value over the cell area. According to annex C of IEC 62446-3 TS [7], this kind of thermal abnormality allocates this module in Class of Abnormality (CoA) 3 (Safety relevant thermal abnormality).

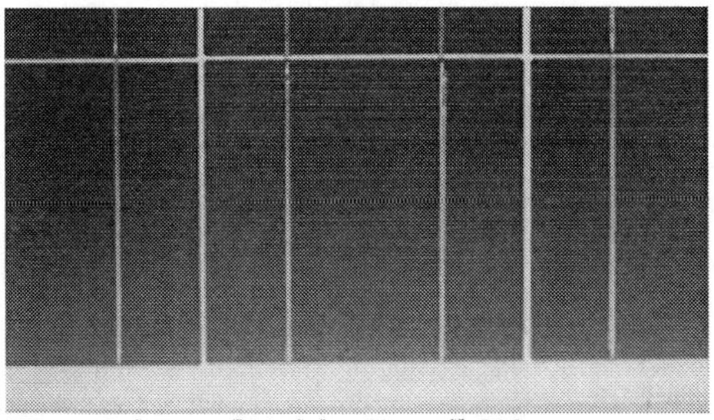

Figure 3: image of Module n°2 – cell C10.

Visual inspection was performed on the examined module but evident defects were not recognizable (see Figure 3 relative to the most overheated cell, C10).

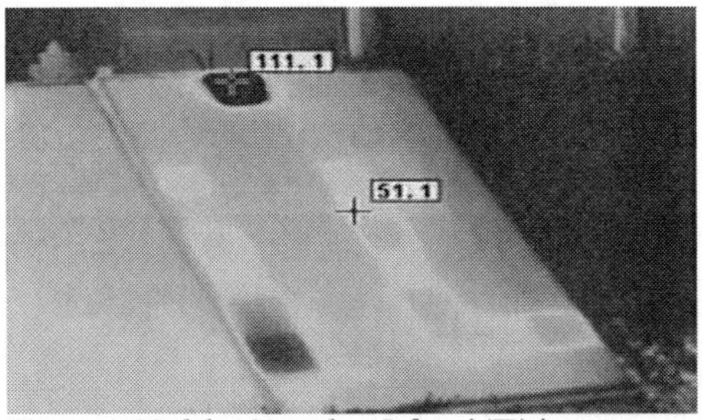

Figure 4: Module n°4 outdoor Infrared (IR) image.

Sample number 4 presents some thermal abnormalities as well. Two cells are sensibly warmer than others and present a temperature > 40 K average value over the cell area. Other cells show some less sever overheating (e.g. sub-string D). According to the annex C of IEC 62446-3 TS [7], this kind of thermal abnormality allocates this module in Class of Abnormality (CoA) 3 (Safety relevant thermal abnormality).

Hot areas are frequently associated with Joule heating, when current is concentrated due to obstacles in normally available paths. Heating of cells may also occur according to mismatch in current generated by particular cells in series strings. These cells will dissipate power driven by the current of normally producing cells.

Also on the module n°4 visual inspection was performed and visible defects were not recognizable.

3 INDOOR IR

Indoor IR imaging requires that the modules are dismounted from their ground mounting system and brought into a laboratory. Isothermal environment conditions are required in the laboratory, in order to avoid localized effects of air conditioning and ventilation. The imaging laboratory temperature shall be between 18 °C and 25 °C [2]. After placing the module in the correct position for imaging, glass/polymeric modules may be imaged on the front (normally sun-facing side) and on the back (backsheet side).

To perform the indoor testing phase, the analyzed modules were moved in a laboratory room (room temperature of 23 °C). The modules were placed in a stable position and forward biased; a current equal to the short circuit (Isc) value was applied to the modules.

Figure 5: image of Module n°2 placed in a stable position inside to the laboratory room.

Image at time zero, before biasing of the module, was captured for each module. IR Images were taken at 2 minutes intervals.

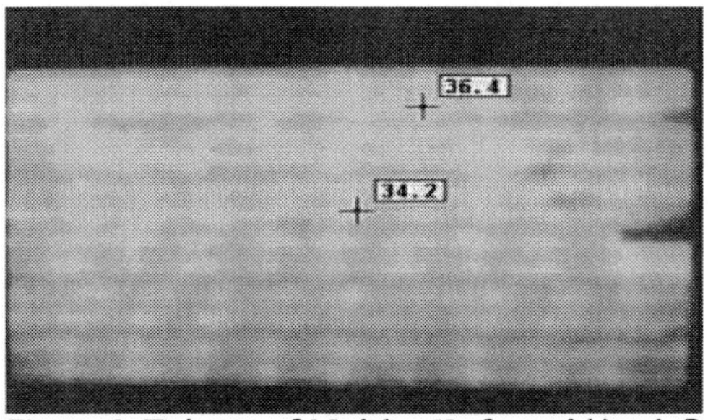

Figure 6: IR image of Module n°2, forward biased @ Isc, placed in a stable position inside the laboratory room after 4 minutes.

Current doesn't flow or flows less than normal through a series string of cells in the module because of areas of broken cells disconnected from the circuit (e.g. cell C10). The presence of areas of carrier recombination or shunt paths through the cell in a module where the I^2R product is low, usually where the low resistance path is distributed over a large area, may also cause the cell to be cooler because of reduced power dissipation.

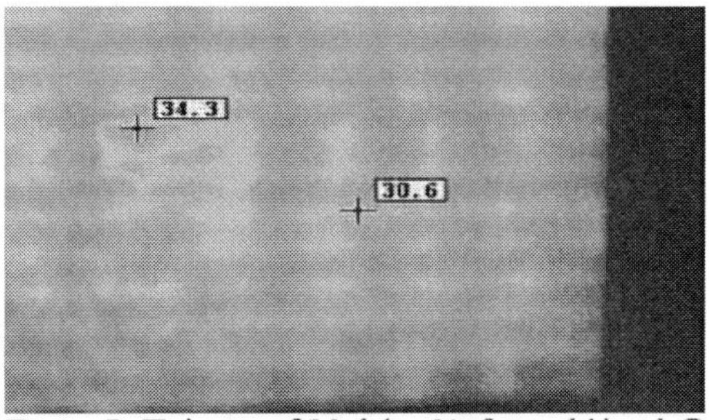

Figure 7: IR image of Module n°4, forward biased @ Isc, placed in a stable position inside the laboratory room after 4 minutes.

Usually, heat transfer through the front glass will generally follow a longer and more variable path than through thin polymeric backsheet [2]. In order to achieve better indoor IR images is then recommendable to capture images of the backside of the modules (backsheet side). Emissivity values should be modified accordingly: glass has an emissivity of 0,85 in the 8–14 μm waveband [6] while polymeric backsheet has an emissivity of 0,95 in the 8–14 μm waveband [6]. Non isothermal environment conditions could afflict the images capture: isothermal environment conditions are strictly recommended in order to avoid localized effects of air conditioning and ventilation.

Indoor IR imaging was also applied to a 36 cells (5" polycrystalline cells) module, operating for almost 20 years. It allows to identify clearly the degradation effects due to the prolonged exposure to the atmospheric agents. The inactive cells parts look cooler that the other parts of the module. The temperature differences between defective module parts and normal operating device are sensibly lowered with indoor IR imaging, when compared with outdoor IR analysis.

Figure 8: 20 years old 36 cells module placed in a stable position inside the laboratory room.

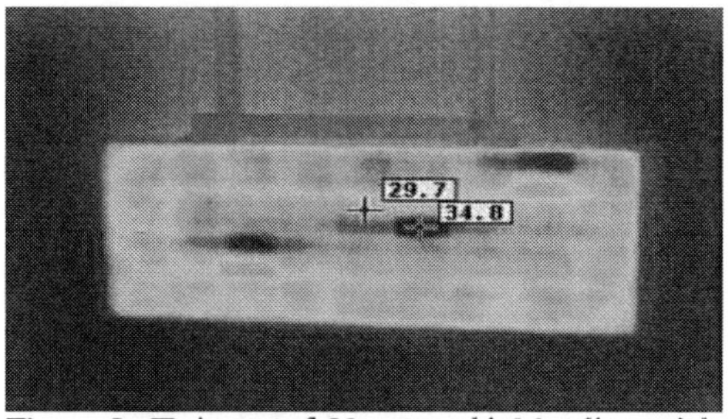

Figure 9: IR image of 20 years old 36 cells module forward biased @ Isc and placed in a stable position inside the laboratory room, after 4 minutes.

4 INDOOR EL

Indoor EL imaging technique requires that the modules are dismounted from their ground mounting system and brought inside a dark chamber in order to perform the analysis. Modules are imaged on the front side. The camera should be placed in front of the module and the angle of view relative to the surface should be normal. The distance between camera and module should be set according to focal length and viewing angle of the used lens. The contrast range for the camera is optimized by adjusting the exposure time, f-aperture, integration time, or by gain after a survey of examined modules. All necessary preparations, including preparations for camera settings, focusing, angle, image framing and darkened environment were performed in a laboratory dark room (room temperature 23 °C). The plus and minus modules poles were connected to DC power supply (AIM-TTI INSTRUMENTS - CPX400S) matching the plus pole of the power supply with the plus of the sample.

4.1 Cameras sensors, lenses and optical filters

Two different camera setup were used in order to improve the EL image capture.

Setup 1:

- Sensor Canon CMOS APS-C 12,2 Megapixel;
- Lens 50mm f/1.4;
- Infrared-passing filters 950nm.

Set up 1 camera advantage is the high resolution. Set up 1 camera disadvantage is the long shooting time.

Setup 2:

- Sensor CMOSIS CMV4000-3E12M1PP 4 Megapixel;
- Lens 16mm f/1.8;
- Infrared-passing filters 780nm.

Set up 2 camera advantage is the short shooting time while its disadvantage is the lower resolution. Main camera settings used with the two setups are reported in Table II.

Table II: Camera settings for EL experimental setups

Setup 1	**Setup 2**
ISO 3200	19 fps
F 3,5	F 1,8
Exposure time: 30s	Exposure time: 0,5s
IR Long-pass filter 950nm	Near IR Long-pass filter 780nm

Images acquired with the modified reflex camera (Setup1) are in false colors, they were then converted, during the post-processing phase, in a grey scale in order to be compared with the images taken with the "industrial" camera (Setup 2). Level range adjustments were performed in order to obtain more information from the acquired images.

4.2 Biasing and imaging

Images were taken at Isc bias; they are useful to get an overview on the module and make a diagnostic analysis on it.

4.3 Images at Isc Bias Module n° 2

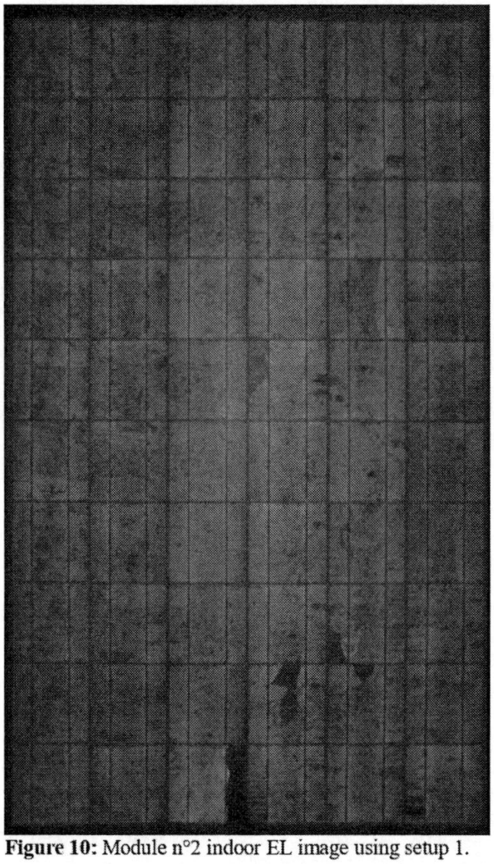

Figure 10: Module n°2 indoor EL image using setup 1.

Module n°2 showed:

- Several cracked cells that indicate partially or totally electrically disconnected regions from the remaining module electrical circuit. According to [4], some of these cracks are classified as mode A (B8, C5), since they appear as line defects and no cell areas appear electrically disconnected. Others look more critical, and are classified as mode B (B10, D5, E4), since they delineate partially electrically disconnected regions, or as mode C (C10; D9; E8; E9);
- Broken or defective grid fingers;
- Wafer/cell contamination (e.g. D1; D2; D3; D4; etc.); these localized areas of low EL signal, especially when following a similar pattern from cell to cell, are likely caused by cell contamination by handling equipment on the production lines.

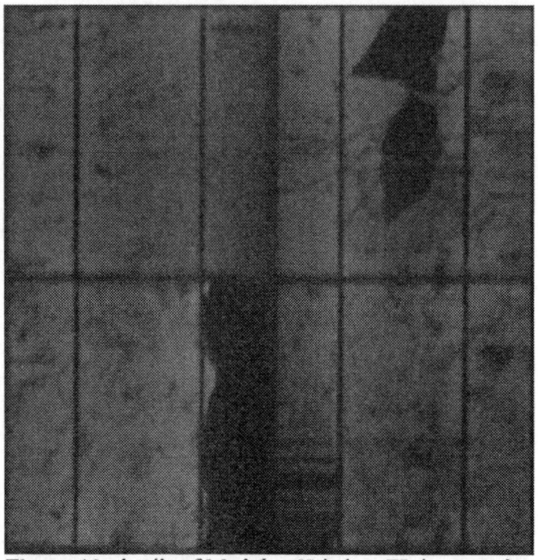

Figure 11: details of Module n°2 indoor EL image using setup 1.

Figure 12: details of Module n°2 indoor EL image using setup 2.

The comparison with the outdoor IR image of the same module showed a correspondence of a mode C crack (which caused the interruption of a busbar) with strong overheating of the same cell (C10).

Through both setups it was possible to obtain good quality images of the whole module, with sufficient resolution to perform a detailed diagnostic analysis. Sample's EL luminescence detectable with setup 2 is sensibly lower (mainly due to the lower exposure time), but still allows the operator to detect and evaluate each of the over-mentioned defects. The effects of the lower resolution and exposure time of setup 2 can be clearly observed in the detail image (Figure 11 and 12).

4.4 Images at Isc Bias Module n° 4

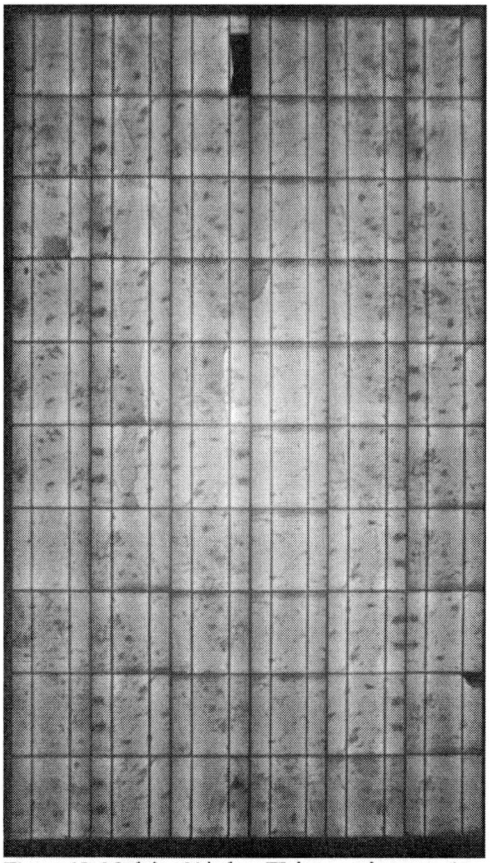

Figure 13: Module n°4 indoor EL image using setup 1.

Module n°4 shows:

- Several cracked cells classifiable, according to [4], as mode A (B2, B5, C5, D2), mode B (A3, B6, D4) and mode C (C1).
- Wafer/cell contamination (ex. C2; C3; C4; C6 ecc.); these localized areas of low EL signal, are systematically present in several cells and are therefore assignable to cell contamination by handling equipment.
- Lower EL luminescence on several cells' edges (e.g. B5, B7, B8, D1 to D9, etc.); due to elevated defects or impurity concentration in the ingot portion used for their production.

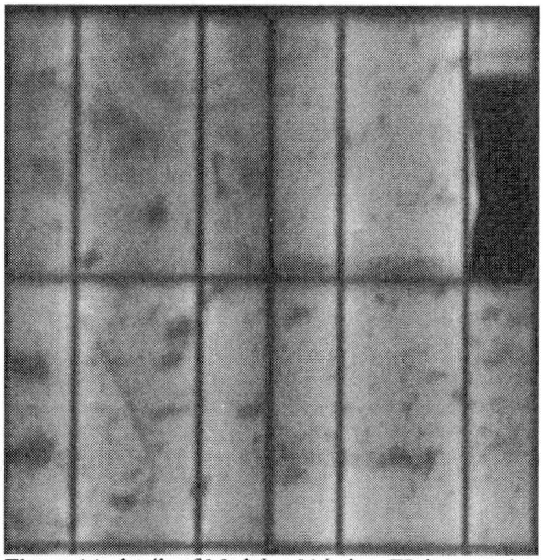

Figure 14: details of Module n°4 indoor EL image using setup 1.

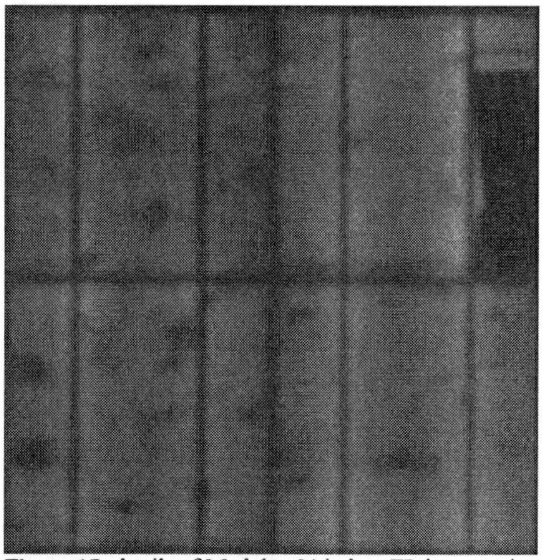

Figure 15: details of Module n°4 indoor EL image using setup 2.

4.5 Images of other c-Si modules type

Acquiring EL images with high resolution allows the identification of some details on the cells that would not be recognizable with low resolution camera.

Heterojunction technology is also analysable using EL imaging. As showed in Figure 17 no major defects were located, the typical chess pattern of the cells' EL luminescence is observable. Using high resolution camera, it was possible to recognize also contamination by handling equipment in cells production line. EL is suitable also with back contacts module technology. The wafers/cells integrity is absolutely recognizable.

Several cracks with no disconnected cell parts were detected, together with contamination by handling equipment and other dark areas.

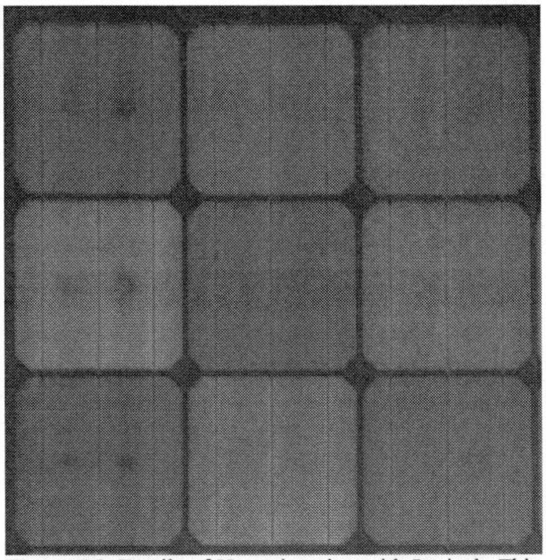

Figure 16: Details of Heterojunction with Intrinsic Thin layer module indoor EL image using setup 1.

Figure 17: Heterojunction with Intrinsic Thin layer module indoor EL image using setup 1.

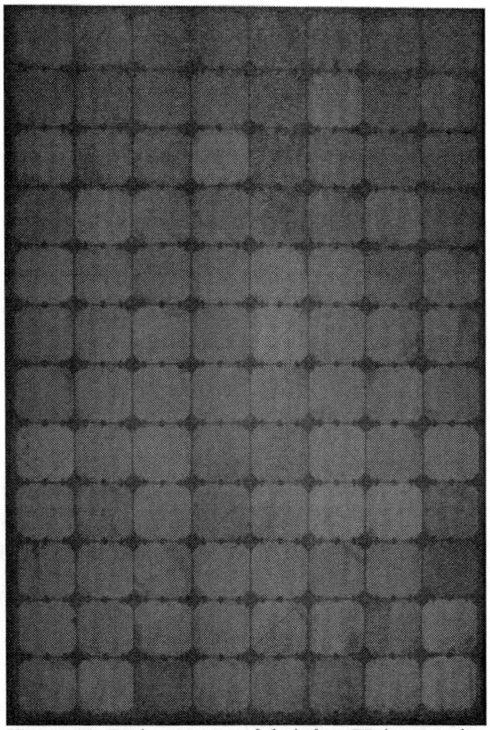

Figure 18: Back contact module indoor EL image using setup 1.

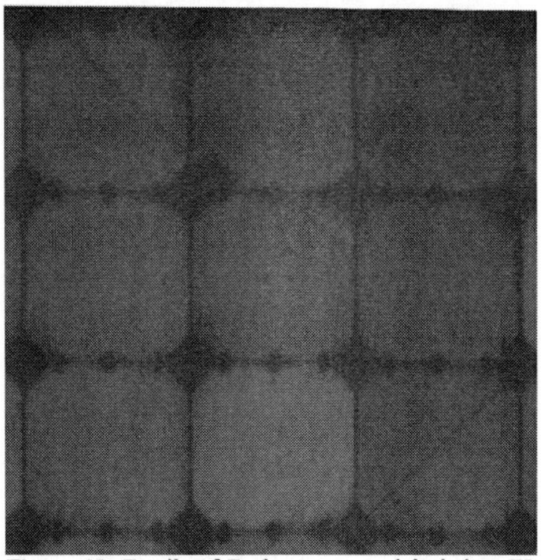

Figure 19: Details of Back contact module indoor EL image using setup 1.

5 OUTDOOR EL

Outdoor EL imaging technique allows to capture modules EL images directly on their ground mounting system. Also in this imaging technique, the camera should be placed in front of the module and the angle of view relative to the surface should be normal.

PV modules were forward biased using a DC power supply able to provide sufficient voltage to the string and achieve I_{SC} (Sorensen SGA600X17 600 Volt, 17 Amp, High Power Modular DC Power Supply).

Setup 1 was used in order to capture the EL image

outdoor. Main camera settings used with the Setup 1 are reported in Table III.

Table III: Camera settings for EL experimental setup 1

Setup 1
ISO 800
F 4,5
Exposure time: 30 s
Infrared-passing filters 950nm

In order to eliminate the reflections on the module glass due to various environmental lights (city lights, moon, etc.) also a Circular Polarizer (CPL) filter was placed in front of setup 1 camera lens.

EL images were acquired after the sunset (21:14 CEST at the measurement location) at regular time intervals in order to evaluate the effects of light pollution.

Figure 20: EL images using setup 1 recorded at different time and different light pollution.

The two EL images in Figure 20 were recorded at 21:30 CEST and 21:54 CEST respectively. Due to the relatively high level of luminosity still present at the time of acquisition of the first image, most of module's EL luminescence is not detectable, and only the most severe defect (cracks in cell C6) is barely visible. In the second image, acquired 24 minutes later, EL luminescence is clearly visible and a complete diagnostic analysis is possible. The obtained image quality is comparable with indoor EL images.

As soon as a sufficiently dark condition was reached (22:05 CEST), an EL image of the whole string of 4 modules was captured.

Figure 21: EL image of the whole string of 4 modules using setup 1.

6 CONCLUSION

Outdoor IR imaging is a useful investigation tool to quickly identify some defects over an extended number of PV modules, directly on field. A macroscopic investigation could be conducted also applying IR camera on drones.

Indoor IR imaging didn't give relevant additional information, while indoor EL imaging was proven to be the best diagnostic tool to precisely identify the nature of each failure mode.

Experimental results showed that outdoor EL analysis can be performed with low-cost cameras as well, if appropriate precautions are taken. This technique is confirmed as a valid tool to perform a detailed on field diagnostic analysis, avoiding the need of dismounting PV modules from their support structure and transporting them inside a dark room.

7 FUTURE WORK

In order to further evaluate PV modules' failure modes, several modules failures will be analyzed and diagnosed using indoor and outdoor EL imaging to provide a guidance of the failures.

8 ACKNOWLEDGEMENT

The authors wish to thank RSE and KB Development technicians for their help in the preparation of experimental setups and the retrieval of the samples.

9 REFERENCES

[1] G.S. Spagnolo, P. Del Vecchio, G. Makary, D. Papalillo, A. Martocchia: "A review of IR thermography applied to PV systems", 11th International Conference on Environment and Electrical Engineering (EEEIC), 2012, Venice, Italy, 18-25 May 2012

[2] 82/1109/CD - IEC/TS 60904-12 Ed. 1.0 Photovoltaic devices - Part 12: Infrared thermography of photovoltaic modules

[3] P.Bentz, "Practical aspects of Electroluminescence" Proceedings of SOPHIA-workshop 'PV-Module Reliability', organized by Fraunhofer ISE and JRC in Freiburg, Germany (2014).

[4] 82/1062/CD - IEC/TS 60904-13 Ed. 1.0 Photovoltaic devices - Part 13: Electroluminescence of photovoltaic modules

[5] C. Buerhop, H. Scheuerpflug, R. Weißmann, "The role of infrared emissivity of glass on IR-imaging of pv-plants", 26th European Photovoltaic Solar Energy Conference, Germany (WIP, Hamburg, 2011)

[6] M. Köntges, "PV Module Reliability Assessment Using IR and EL Imaging Techniques", 29th European Photovoltaic Solar Energy Conference, Netherlands (WIP, Amsterdam, 2014)

[7] 82/1073/CD - IEC/TS 62446-3 Ed. 1.0 Photovoltaic (PV) systems - Requirements for testing, documentation and maintenance - Part 3: Outdoor infrared thermography of photovoltaic modules and plants.

EVALUATION OF A PV-PANEL VIA LONG TERM HIGH SPEED RECORDING OF IV-CURVES

Kasper M. Paasch[1]*, Morten Nymand[1], Cristina Cornaro[2], Marco Pierro[2] and Francesco Bucci[2]
[1]University of Southern Denmark, MCI/MMMI, Odense, Denmark
[2] Department of Enterprise Engineering, University of Rome "Tor Vergata", Rome, Italy
* Corresponding Author

ABSTRACT: This paper presents results from a high speed photovoltaic sweeper system measuring the IV-characteristics once per second of a solar panel under test at the ESTER PV test laboratory at the Tor Vergata University in Rome and in Denmark. The high temporal resolution of the scans shows the detailed temperature characteristics of the primary panel parameters compared to the reference irradiation/spectral data obtained via the ESTER instrumentation. An analysis of the irradiance levels shows that the STC and NOCT conditions in real life are extremely rare events. Analysis show that the temperature coefficient is depending on the irradiance level.
Keywords: PV module, Thermal Performance, Qualification and Testing

1 INTRODUCTION

The characteristic of a solar panel is measured under fast changes of irradiation by applying a novel IV-recording system. The IV-curve tracer system is called PVSweeper. The IV-curve is swept every second and the panel has been monitored for a period extending 6 months. The use of a high speed IV-curve tracer enables the identification of the real maximum-power-point without the use of any solar inverter with build-in -tracking circuitry. The irradiance measurements during the solar equinox of March 20, 2016 are included as an example.

2 MOTIVATION

Current state of the art PV models are usually parametric, extracting average coefficients from the I-V curves. The aim is via high speed recordings to monitor fast changes in the characteristic of solar panels with respect to temperature.

3 IV-TRACER DESCRIPTION

The main building blocks of the developed IV-curve tracer is shown in figure 1. The system is optimized for sweeping a panel up to 250W each second with a sweep time below 100 ms [1, 2].

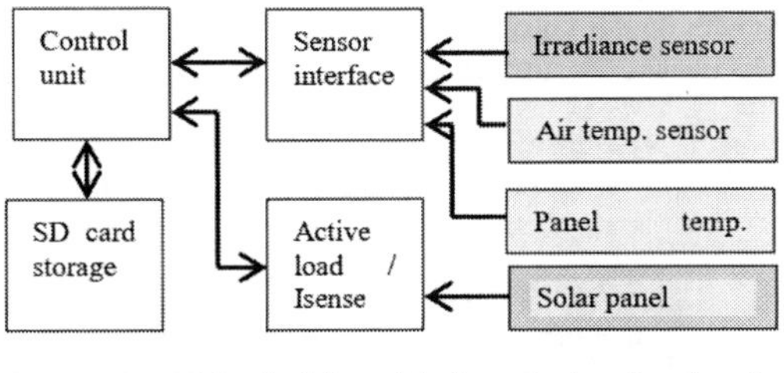

Figure 1: Main building blocks of the developed PVSweeper system.

4 INSTALLATION

A 225W polycrystalline panel was monitored for approximately 6 months at the outdoor ESTER PV monitoring station in Rome as a larger variation in irradiation level and temperature was expected there than in Denmark. The panel was monitored with the developed IV-curve tracer instrument to obtain IV-sweep data every second from March 2015 to October 2015. The ESTER system records intensity measurements from a range of sensors (a polycrystalline reference cell, a monocrystalline reference cell and a pyranometer) as well as the optical spectrum [3-5]. See figure 2. The ESTER irradiance data are recorded once per minute, where the PVSweeper sweeps recorded once per second. The uncertainty of the measured irradiation levels has been addressed in [3, 4]. Deviations between the ESTER sensors are in the range of ±4%@1000 W/m^2 are documented and have also been observed by the tests performed.

Figure 2: Left: ESTER PV test platform with the sensors mounted. Right: Close-up.

5 COMPARISON OF THE ESTER IRRADIANCE SENSORS

The relationships between the irradiation levels recorded by the PVSweeper and the 3 ESTER sensors are shown in figure 3, together with the calculated polynomial fit of 1st order.

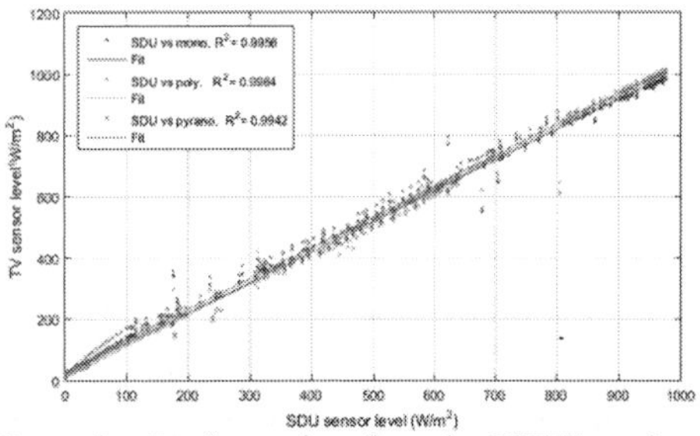

Figure 3: Irradiance data from the ESTER station relative to the PVSweep irradiance measurements.

All linear fits shows a R2-value >0.99. It is

concluded that the intensity measurement of the PVSweeper has an acceptable degree of correlation with the sensor systems mounted at ESTER. The relationship between the ESTER mono sensor and the other ESTER sensors for March 21, 2015 (a very clear day) is shown in figure 4.

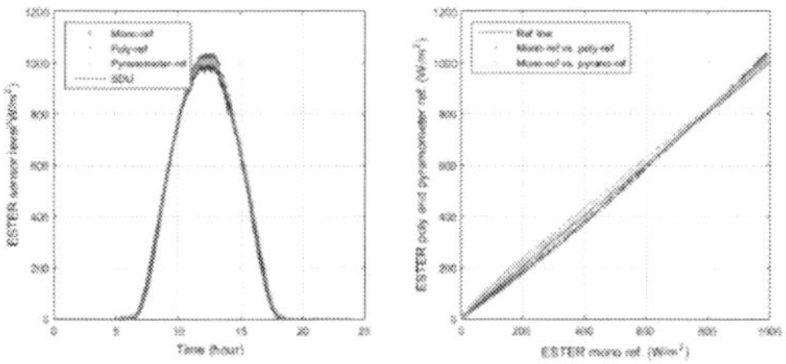

Figure 4: Recorded differences between ESTER sensors and the sensor attached to the SDU PVSweeper unit for March 21, 2015 in Rome.

It can be observed that the pyranometer gives a higher reading than the mono sensor and the poly crystalline sensor for all irradiance levels and that the polycrystalline sensor for levels below 800 W/m^2 gives a lower reading, but a higher reading above that level. The irradiation measurements from the SDU irradiance sensor were used to scale the sensor output to the ESTER mono crystalline reference cell, due to the same crystal type and comparable performance.

The recorded values for March 20, 2015, during a solar equinox, are shown below for the ESTER sensors and the SDU irradiation sensor. It can be observed that the 3 ESTER reference sensors differ in amplitude by a few percent between approximately 8 and 16 o'clock, see figure 5.

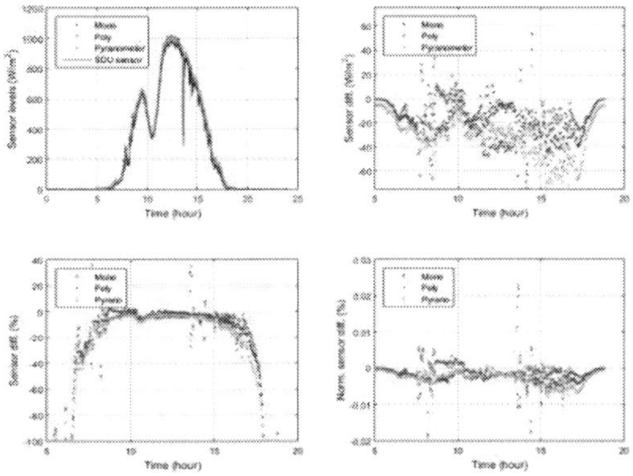

Figure 5: Top left: Recorded irradiance levels for March 20, 2015. The SDU sensor level is scaled to the ESTER mono ref. cell. Top right: Sensor difference in W/m^2. Bottom left: Sensor difference in %. Bottom left: Sensor difference in % between normalized levels. Bootom right: Difference plots.

The temporal variation of the ratio between the normalized sensor levels (figure 5, bottom right) indicates a small spectral dependency, which is further analyzed in the next section.

6 RESULTS OF PANEL TEST

The raw current and voltage values recorded by PVSweeper are compensated for the influence of ambient temperature and nonlinearity of the current measurement circuit. The panel under test is a 225W polycrystalline panel of type ND-225R1J from the company SHARP. The main panel data are listed in table1.

Table I: Panel data

Panel Type	SHARP Solar Module ND-225R1J		
Serial no.	134459494		
	STC	**NOCT**	**Unit**
Nom. power	225	162.1	W
V_{OC}	36.6	35.8	V
I_{SC}	8.28	6.68	A
V_{MPP}	29.3	26.1	V
I_{MPP}	7.68	(6.21)[1]	A

[1] Calculated value.

The developed PVSweeper unit records a scan consisting of 100 measurement points every second, as well as the panel temperature, ambient temperature and solar irradiation level. Via data post processing the P_{MPP}, V_{MPP} an I_{MPP} values are extracted . As an example the recorded PV-curves during the solar equinox on March 20, 2015, are shown in figure 6.

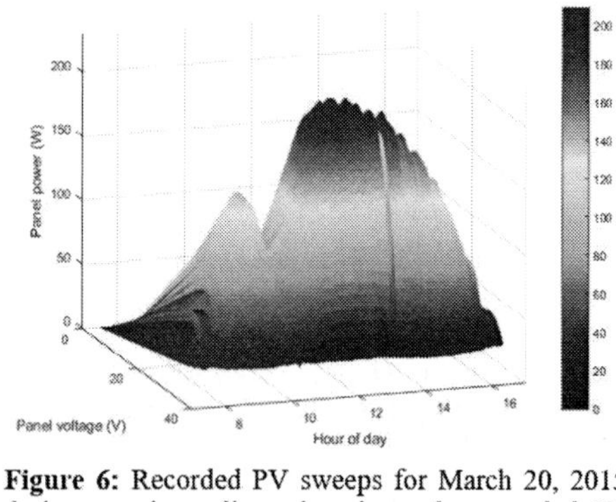

Figure 6: Recorded PV sweeps for March 20, 2015, during a solar eclipse, based on the recorded IV-values. 1 minute averaging is applied.

The distribution of the recorded V_{MPP}, I_{MPP} and P_{MPP} values for the test period is shown in figures 7 and 8.

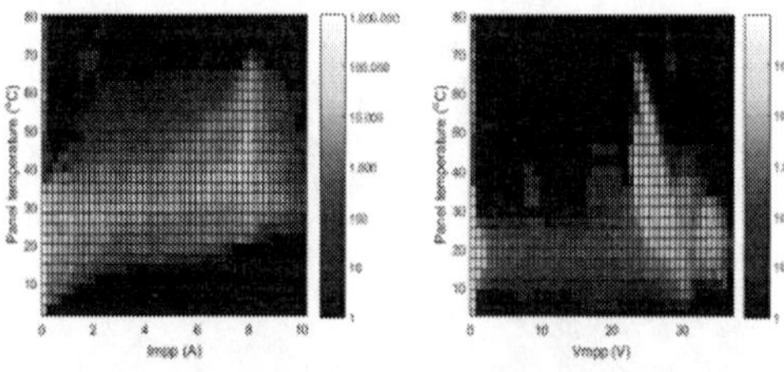

Figure 7: Distribution of panel temperature and I_{MPP} level (left) and V_{MPP} level (right) for day 30-100 of the observation period. A total of approximately 13 mill. measurements pairs are presented in each plot.

The occurrence of combinations of P_{MPP} level and panel temperature for the tests performed with the developed PVSweeper at the ESTER Test facility in Rome is shown in figure 8. It can be observed that the STC condition is outside the main area of operation but also that the irradiation/temperature distribution is different from the other test plants as more irradiation around the level of 1000 W/m² is present.

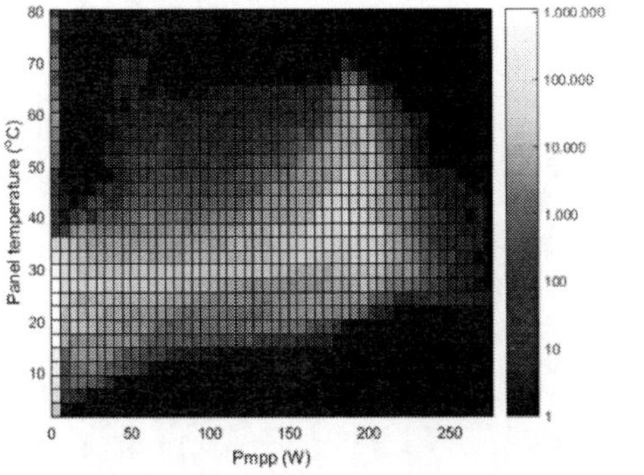

Figure 8: Distribution of panel temperature and P_{MPP} level for day 30-100 of the observation period. A total of approximately $13x10^6$ measurement pairs are presented in the plot.

An abrupt change in irradiation will either let the panel cool down or warm up, depending on the direction of the change. The thermal mass of the panel will delay the panel temperature to adapt to the new thermal equilibrium for the panel, resulting in a power deviation of the power production under MPPT relative to the instant irradiance. The thermal effect has been addressed in the literature, i.e. [6], stating a thermal time constant in the range of 7-10 minutes for a laminated (85W) PV panel and even longer time constants for larger panels with thicker back plates, up to twice as long. This thermal effect has been observed in the recorded IV-curves, see figure 9 and 10.

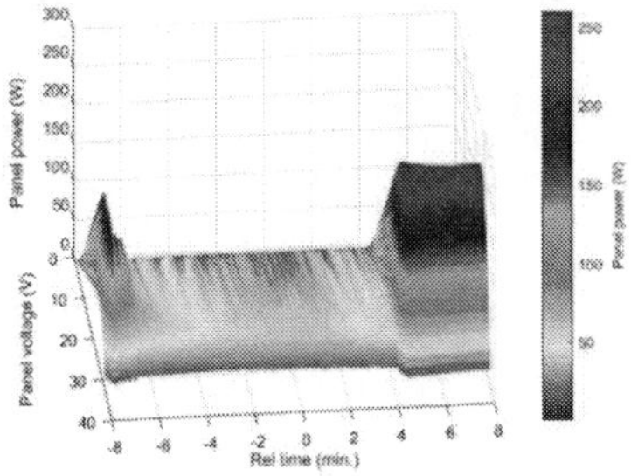

Figure 9: IV scans around an event with fast change in irradiation level, recorded April 17, 2015.

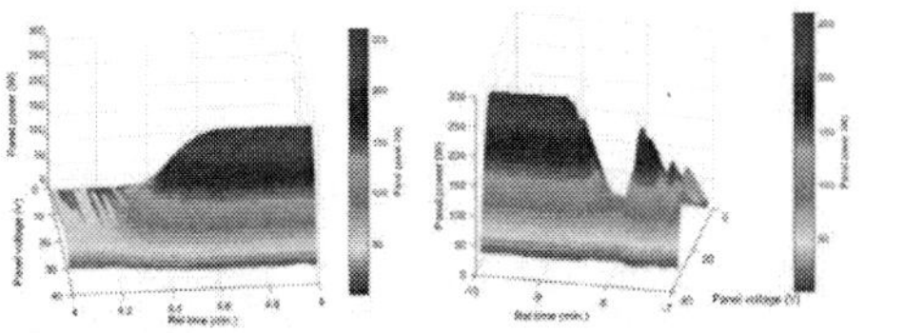

Figure 10: IV scans around the transitions areas. Left: Downwards. Right: Upwards.

The extracted P_{MPP} values and the irradiation levels are shown in figure 11.

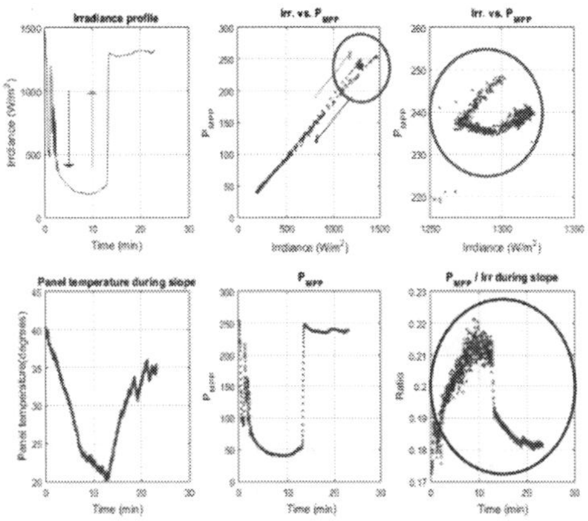

Figure 11: Extracted plots with irradiation and P_{MPP} data.

The irradiance profile is shown top-left. The arrows indicate the direction of the change. The plot located top-center shows the decline from 236 W power level @ 1300 W/m² to around 40 W @ 200 W/m² and back up to 248 W @1300 W/m² (green circle). The panel cools down from 40 °C to 20 °C during the shadowing and then warms up during the next 10 minutes. The temperature effect can be observed in the ratio between the P_{MPP} and the irradiation, shown in the plot bottom-right (red circle). An estimate for the time constant of the PV panel is 10 minutes.

The dynamics of cloud induces fast shadowing has recently been addressed in various publications, addressing the need for better modelling of the performance of PV-systems under varying atmospheric conditions, i.e. [9-12].

7 PANEL TEMPERATURE ANALYSIS

The temperature coefficient τ for a the specified series of irradiance levels *Irr* can be determined by fitting a series of suitable functions to the power and irradiance data measured at a specific temperature *T*. A power law function of the form given in eq. 1 was chosen.

$$P(Irr,T) = 10^{k(T)} * (Irr)^{v(T)} \qquad (1)$$

A set of (k,v) coefficients are calculated for each chosen temperature *T*. In figure 12 the extracted P_{MPP} values are shown as a function of the air mass range 1.0-3.0 as well as the fitted power law function for each temperature. The solar elevation and corresponding air mass factor were calculated for each sample time, using the actual time stamp of each measurement and applying the modified procedure introduced by [7]. The air mass interval of 1.0-3.0 has been accepted to obtain sufficient data points, even if the variation -in solar elevation / air mass introduces a potential spectral dependency due to varying atmospheric absorption spectra. The STC location is marked with a red dot.

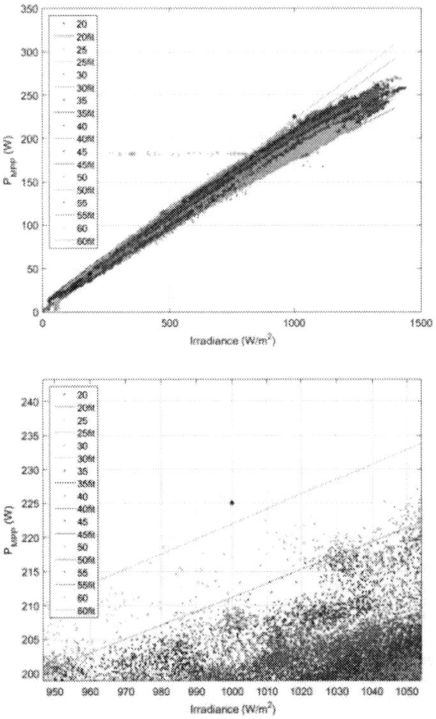

Figure 12: Top: Calculated P_{MPP} levels as a function of solar irradiance in steps of 5°C with fitted polynomials of 1st order. Data from day 30-100 applied, AM interval 1-3. A total of approximately 418,000 out of 13 mill. data points are shown. Bottom: Zoomed. The red dot indicated the location of the STC regarding irradiance, temperature and panel MPPT-rating.

Environmental conditions at or close to STC regarding temperature (25±0.5°C, irradiation 1000±2W/m^2, air mass AM1.5±-0.1) are not present in the recorded dataset. For an extended interval around STC (25±1°C, 1000±5W/m^2, AM1.5±0.2) only 4 sets of data are found. The average P_{MPP} value of these data sets was 218W.

The linear fit for T_{panel}=25°C for the interval 100-1400 W/m^2 shown in figure 12 indicate that for an irradiation level of 1000 W/m^2 a P_{MPP} level of 222W for the monitored panel is fitted. This is 1.4% below the nominal P_{MPP} level of 225.0 W, but the nominal power loss due to aging of the panel is not known and the data sheet guaranties max. 4% loss during the first year of operation. The panel under test has been in operation for approximately 1.5 year in total.

The fitted functions are used to calculate the temperature coefficients τ for each temperature combination (T_a,T_b) and irradiance level *Irr* via eq. 2.

$$\tau_{T_a,T_b}\left(Irr,\Delta T_{a,b}\right)=\frac{P_{T_a}(Irr)-P_{T_b}(Irr)}{P_{T_b}(Irr)\,\Delta T_{a,b}} \qquad (2)$$

for a=1:n and b=1:n, n being the number of temperature intervals.

The analysis is performed for T=20°C, 25°C ... 60°C. An average temperature coefficient τ_{Irr} for each selected irradiance level *Irr* was found by averaging all valid calculated combinations for said level *Irr*. The calculated values are shown in figure 13. Temperature coefficients obtained at irradiance values below 200 W/m^2 were discarded due to a low number of actual measurements at some temperatures.

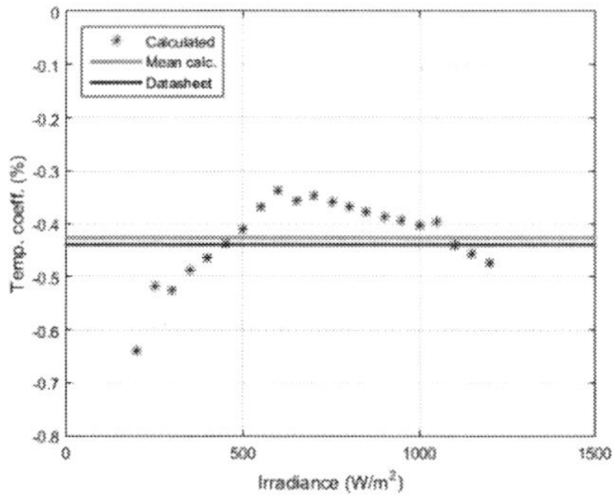

Figure 13: Calculated averaged temperature coefficients for the tested SHARP PV panel as a function of irradiance level.

The average temperature coefficient, estimated on the basis of the fitted functions for the interval 200-1200 W/m^2, is -0.43±0.07%, close to the data sheet value of -0.440%. An apparent dependency of the irradiation level is observed which is to be expected as P_{MPP} is the product of I_{MPP} and V_{MPP}, both of which vary with temperature and irradiance, acc. to (eq. 3) [8].

$$\frac{dP_{MPP}}{dT}=V_{MPP}\frac{dI_{MPP}}{dT}+I_{MPP}\frac{dV_{MPP}}{dT} \qquad (3)$$

The measured values are average values over all temperatures for a given irradiation interval. The data sheet information of a temperature coefficient of -0.440%/°C is only to be seen as an average coefficient. This observed dependency is subject of an ongoing investigation as more data becomes available from the running measurement program in Rome.

8 CONCLUSION

Results from a long-term measurement campaign with a novel high-speed IV curve tracer is presented, showing the fast dynamics induced by moving clouds as well as the detailed temperature response of a polycrystalline PV panel and the occurrence of V_{MPP} as function of temperature and irradiation level. A temperature coefficient dependence on the irradiation level is identified for the tested panel.

6 REFERENCES

[1] K.M. Paasch, M. Nymand and F. Haase, "Sensor System for Long-term Recording of Photovoltaic (PV) IV-curves", PVSAT-9 conference, 2013, Swansea, UK.

[2] K.M. Paasch, M. Nymand and S.B. Kjær, "Distributed measurement system for long term monitoring of clouding effects on large PV plants", EPE'13 ECCE Europe 15th European Conference on Power Electronics and Application, Lille, France, 2013.

[3] A. Spena et al., "DATA VALIDATION AND UNCERTAINTY EVALUATION OF THE ESTER OUTDOOR FACILITY FOR TESTING OF PHOTOVOLTAIC MODULES", 24th European Photovoltaic Solar Energy Conference, 21-25 September 2009, Hamburg, Germany. DOI: 10.4229/24thEUPVSEC2009-4AV.3.89

[4] Cristina Cornaro, Davide Musella, Domenico Chianese, Gabi Friesen, Sebastian Dittmann, "OUTDOOR PV MODULE PERFORMANCE COMPARISON AT TWO DIFFERENT LOCATIONS", Conference Proceedings of AMSE-ATI-UIT 2010 Conference on Thermal and Environmental Issues in Energy Systems, Sorrento, Italy.

[5] Cristina Cornaro and Davide Musella, Comparative Analysis of Crystalline and Double-Junction Amorphous Silicon Modules Performance in Outdoor Conditions J. Sol. Energy Eng. 135(3).

[6] S. Armstrong and W.G. Hurley, "A thermal model for photovoltaic panels under varying atmospheric conditions", Applied Thermal Engineering 30 (2010), pp. 1488-1495.

[7] Keith A, Pickering, "The Southern Limits of the Ancient Star Catalog and the Commentary of Hipparchos", DIO The International Journal of Scientific History, Vol. 12, September 2012. ISSN: 1041-5440.

[8] David L. King et al., "Temperature Coefficients for PV Modules and Arrays: Measurement Methods, Difficulties, and Results", 26th IEEE Photovoltaic Specialists Conference, September 29-October 3, 1997, Anaheim, California.

[9] H.Goverde et al., "Model requirements for accurate short term energy yield predictions during fast-varying weather conditions", 31st European Photovoltaic Solar Energy Conference (PVSEC), Hamburg 2015.

[10] D. Anagnostos, , T. Schmidt, H. Goverde, J. Kalisch, F. Catthoor, D. Soudris, "PV energy yield nowcasting combining sky imaging with simulation models", 31st European Photovoltaic Solar Energy Conference (PVSEC), Hamburg 2015.

[11] Anssi Mäki and Seppo Valkealathi, "Mismatch losses in photovoltaic power generators due to partial shading caused by moving clouds". 27th European Photovoltaic Solar Energy Conference and Exhibition. EUPVSEC 2012.

[12] D.Anagnostos, H.Goverde, B.Herteleer, F.Catthoor, D.Soudris, J.Driesen, J.Poortmans, "Demonstration and validation of an energy yield prediction model suitable for non-steady state non-uniform conditions", 6th World Conference on Photovoltaic Energy Conversion, Kyoto, 2014.

LESSONS LEARNED FROM THE DESIGN AND OPERATION OF A 300 KWP PV SYSTEM WITH FULL SELF-CONSUMPTION OF THE ENERGY PRODUCED

Gaiddon, B.[1], Joos, M.[1], Thebault, A.[2], Derobert, C.[2], Debray, N.[3], Dupret, M.[4], Rozel, B.[4]
[1]Hespul, 14 place Jules Ferry, 69006 Lyon, France
email: bruno.gaiddon@hespul.org , marine.joos@hespul.org
[2]Enercoop, 16-18 quai de la Loire, 75019 Paris, France
email: adrien.thebault@enercoop.org , camille.derobert@enercoop.org
[3]Enercoop Bretagne, 6 Bis avenue Louis Barthou, 35000 Rennes, France
email: nicolas.debray@enercoop-bretagne.fr
[4]Enertech, 26160 Félines sur Rimandoule, France
email: dupret@enertech.fr , rozel@enertech.fr

ABSTRACT: This article presents the lessons learned from the design and operation of a 300 kWp PV system installed, during spring and summer 2015, on the roof of a new logistic building located in western France. As this building is equipped with refrigerating units, thus with a high power consumption during hot and sunny days, it has been decided to consume the energy produced on-site without any regulated feed-in tariff but with a B-to-B contract, a first of its kind in France. In order to not inject power into the local MV distribution grid, the active power management feature of PV inverters has been implemented, also a first of its kind in France. This paper gives a summary of the detailed on-going monitoring campaign that aims to check if the actual power consumption of the site matches the consumption forecast and if the active power management feature of PV inverters has a significant impact on the total energy generation of this PV system.
Keywords: PV system, Battery Storage and Control, System Performance

1 INTRODUCTION

Launched by ecological and ethical business organizations in 2005, Enercoop is a French cooperative that supplies green power [1]. Members of Enercoop are producers, consumers and other related partners of green electricity. Enercoop has an annual turnover of EUR 20 million and employs over 80 staff members. It supplies 100 GWh to 23 000 customers in France. Enercoop has a network of 10 regional cooperatives who actively develop renewable energy projects and promote rational energy use.

Mid 2013, Enercoop started to study the possibility to design and build a PV system on the roof of a new logistic building located in western France. This building, owned by Biocoop [2], an organic foodstuff cooperative, was necessary to temporarily store foodstuffs before delivering them to local food stores. The access to a feed-in tariff for medium-size PV systems based on solar-auctions was judged too complex and to uncertain, especially for PV systems located in the northern part of the country with a lower solar irradiation. Also, as this new building was planned to be equipped with refrigerating units, it was assumed that its power consumption could fit well with the power production of a PV system. Therefore Enercoop decided to study the possibility to consume on-site the energy produced by the PV system without any power injection into the local MV distribution grid and to experiment a new business model without any regulated feed-in tariff.

2 PRELIMINARY DESIGN

2.1 PV potential of the roof

The site owned by Biocoop is located close to Rennes in Brittany in western France. It was originally composed of an office building and a logistic building of 6.000 m2, but theses premises were too small and Biocoop needed more storage surface and started to work on a 4.000 m2 extension (see figure 1).

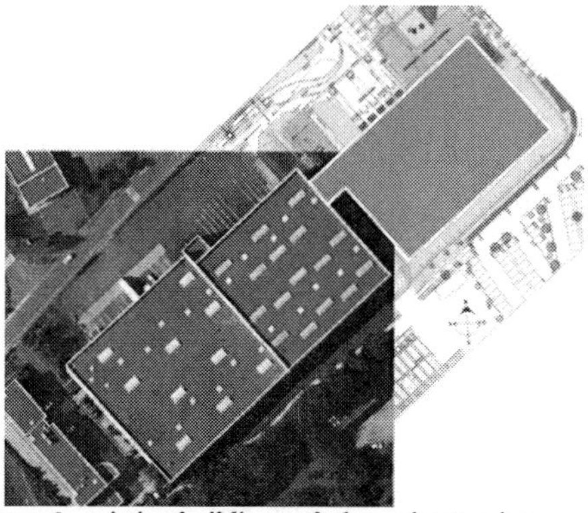

Figure 1: existing building and planned extension

This new building has a roof area of 3.500 m2 made from multi-layer membrane with a slope of 3% and a southeast orientation (-37°). With the use of standard multi-crystalline PV modules with an efficiency of 15%, the PV potential of this roof was assessed at 330 kWp (see figure 2).

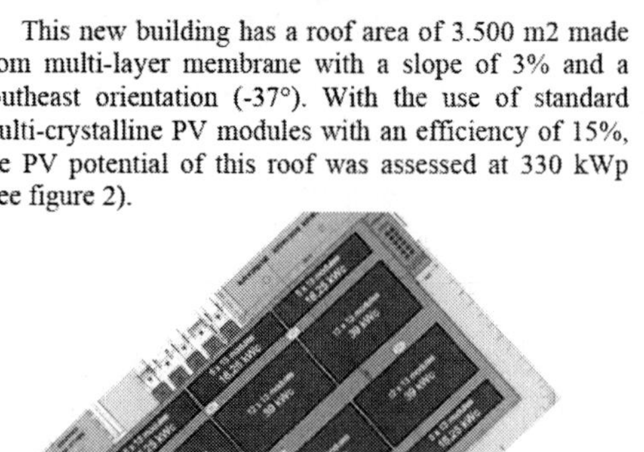

Figure 2: preliminary design with a PV potential of 330 kWp

2.2 Building load profile

The main challenge to study the possibility to consume on-site the energy produced by this PV system is to have a good quality load profile of the building, especially for new buildings. In this case, by chance, the site had an existing logistic building of 6.000 m2 that had been monitored during September 2013. Mean daily load profiles have been generated as a direct outcome of this monitoring campaign (see figure 3).

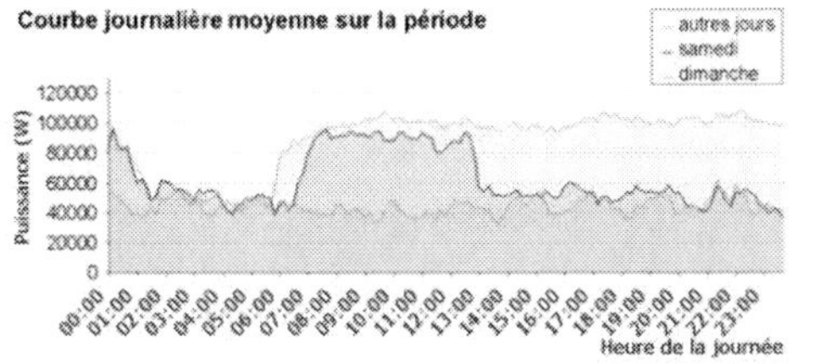

Figure 3: daily average power consumption of the existing site (yellow: days of the week; purple: Saturday; green: Sunday)

In order to scale-up these results to the new building under construction, a detailed analysis of the power of new devices planned has been undertaken and an hourly load profile for 2012 has been generated for the whole site including the new building.

2.3 PV production profile

A detailed self-consumption study also requires a good quality PV production profile. For this study, a specialized company called EPICES Energie [3] generated an hourly production profile for this site also for 2012 based on satellite data and a dedicated numerical PV model.

2.4 Self-consumption study

The set of hourly consumption data has then been compared to the set of hourly production data. It appears, as expected, that due to the fact that this building is equipped with refrigerating units, the expected power consumption fits well with the simulated power production of the 330 kWp PV system. This is true almost every day of the year of reference (2012) expect during sunny Sundays and sunny Saturday afternoons (see figure 4).

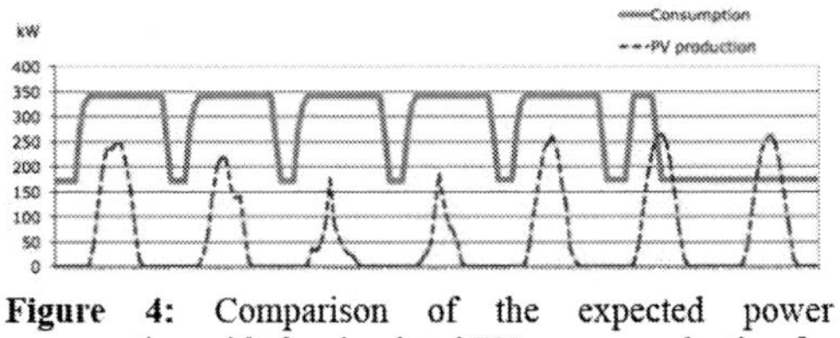

Figure 4: Comparison of the expected power consumption with the simulated PV power production for a week of March 2012 (19/03/2012-25/03/2012).

Thus, the conclusion of the preliminary study was that 98,6% of the energy produced by a 330 kWp PV system could be consumed on-site without any storage.

Finally, in order to limit financial risk and to take into account the various assumptions made during the preliminary deign phase, Enercoop decided to limit the PV system peak power to 300 kWp (90% of the PV potential) and to limit the active power of the system to 250 kVA in order to stay under the low-voltage power range of the French regulation (under sizing ratio of inverters of 83%).

3 DETAILED DESIGN AND INSTALLATION

Enercoop then launched on March 2014 a call for tender to select a system supplier and installer and three companies sent a detailed offer to design and build this PV system. After a technico-economical analysis of the offers received, Enercoop selected Armorgreen [4], a local company with very good PV references. The PV system was designed with the following components:

- PV modules: Sillia 60P255-V (255Wc) – 1176 units,
- Support structure: Inova PV from EPC Solaire,
- Inverters: Kaco Powador 60.0 TL3-M – 5 units,
- Monitoring: Meteocontrol Weblog Pro GPRS.

In order to further guarantee that no power will be injected into the MV grid, Enercoop decided to also use the monitoring device as an active power controller.

Armorgreen did the installation work during spring and summer 2015 and commissioned the PV system in September 2015 (see figure 5). The system price was approximately 1.35 €/Wp.

Figure 5: overview of the 300 kWp PV system installed on the roof of the new logistic building

4 OWNERSHIP AND B-TO-B PURCHASE AGREEMENT

Although the energy produced is physically consumed on-site, the PV producer and the power consumer are different entities. The PV producer and thus the owner of the PV system is a special purpose company (SPV) called Soleil du Grand-Ouest (SGO) and owned by the regional cooperative Enercoop Bretagne, a citizen RES investment fund called Energie-Partagée [5] and Biocoop, who is also the site owner and the power consumer.

Two contacts bind SGO and Biocoop. First, a contract that allows SGO to use the roof to operate the PV system installed on the building owned by Biocoop. Second, a B-to-B power purchase agreement to commit Biocoop to purchase from SGO over the long term the energy produced by the PV system. The price of the energy consumed was set at 185 €/MWh for a period of 20 years with the possibility to extend the contract by 10 additional years.

5 OPERATION

This PV system was commissioned in September 2015 and the validation of the main assumptions made during the preliminary design phase took place on May 2016 on the basis of April 2016 data.

The first task that has been done was to collect from ERDF (now Enedis), the Distribution Network Operator (DNO), the consumption profile of the entire site given that the DNO just sees the net consumption, i.e. the total consumption minus the PV production. In case of pure self-consumption, this consumption profile shouldn't have negative value, as no power should be normally injected into the medium voltage (MV) grid. The 10 minutes data provided by the DNO shows that the minimum power consumption value measured during April 2016 was 12 kW at 15:40 on Monday 18 April 2016 (see figure 6). This means that all the energy produced during this month was consumed on-site with no power flow back to the grid.

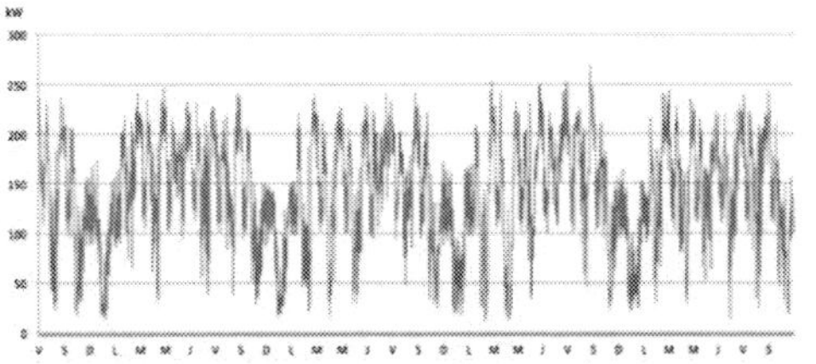

Figure 6: net consumption measured by the DNO during April 2016 with no negative value and thus evidence of pure self-consumption of the energy produced

A deeper analysis of the situation was then undertaken for one week, from Monday 18 April to Saturday 21 April 2016. This highlighted the fact that the power regulation was too efficient and led to production losses. Thus, when the instantaneous power measured by the DNO reaches 0, the power controller then reduces the power delivered by the PV system down to 10% of the production power (see figure 7).

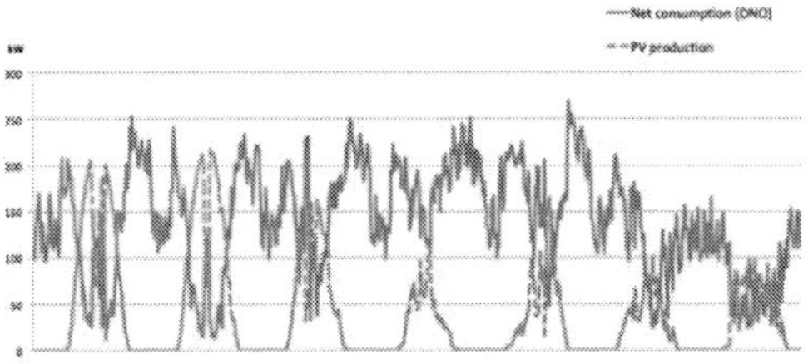

Figure 7: net consumption and PV production for one week of April 2016 with evidence of production losses due to inefficient regulation, especially during the two first days of the week

The correction of the power regulation factor can be improved with the analysis of the gross consumption during Sundays, when the consumption load is the smallest. The gross consumption profile was obtained by summing the net consumption profile and the PV production profile. This profile is nonetheless not very accurate since these profiles are measured by 2 different meters that have not been synchronized (DNO's meter and PV data logger). But, the gross consumption profile of the site shows that a base load of approx. 100 kW is consumed during the weekend (see figure 8). Therefore, the power regulation factor could be set at 40% (100 kW/250 kW) without any risk to inject power into the distribution grid but with reduced PV production losses.

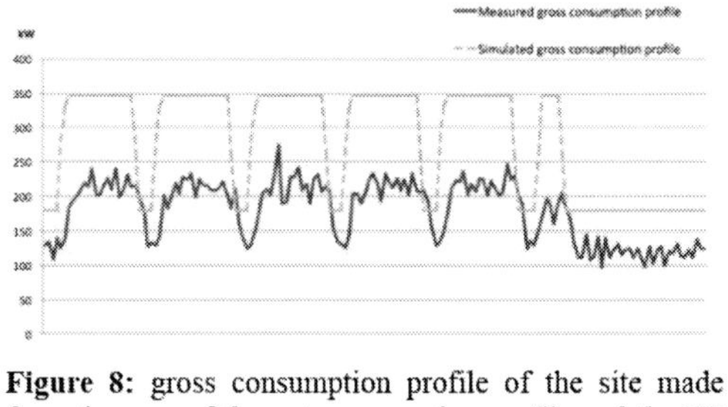

Figure 8: gross consumption profile of the site made from the sum of the net consumption profile and the PV production profile

This figure also shows the difficulty to forecast consumption data as there is a significant gap between the measured gross consumption profile from 2016 and the simulated gross consumption profile from 2012 despite the fact that the mean temperature of Rennes for April 2012 and April 2016 is strictly the same (9,4°C).

5 SUMMARY OF MAIN FINDINGS

- Logistic sites are suited for self PV consumption as their consumption profiles generally fit well with PV production profile and require no battery,
- Main challenge during design phase is the quality of the consumption load profile which may be complex in the case of new buildings,
- Necessity to use the same data logger or two synchronized data loggers for net consumption and PV production in order to correctly calculate the gross consumption profile,
- In case of PV power regulation, necessity to correctly set the power regulation factor in order to avoid unnecessary production losses.

6 ACKNOWLEDGEMENTS

Authors wish to thank Armorgreen, the PV system supplier, and Biocoop, the site owner, for their strong involvement and for having made this project possible.

7 REFERENCES

[1] www.enercoop.fr
[2] www.biocoop.fr
[3] www.epices-energie.fr
[4] www.armorgreen.fr
[5] www.energie-partagee.org

A HIGH SPEED GLOBAL MAXIMUM POWER POINT TRACKING ALGORITHM FOR PV SYSTEMS

Mustafa Engin Başoğlu, Bekir Çakır
Department of Electrical Engineering, Kocaeli University
Campus of Umuttepe, Kocaeli, Turkey
mengin.basoglu@kocaeli.edu.tr, bcakir@kocaeli.edu.tr

ABSTRACT: Partial shading is one of the unfavourable conditions for photovoltaic modules since available power of them decreases. Besides that, maximum power point tracking becomes difficult in that kind of condition. Multiple maximum power points occur on the voltage - power curve of photovoltaic modules, if modules have bypass diodes. Hence, this paper proposes a new algorithm which provides fast and accurate tracking in partial shading conditions. It is shown that values of duty ratios change in limited range in partial shading which is approximately 40% in buck-boost converter topologies which is the major idea of the proposed algorithm. A few duty ratios, which are based on the number of bypass diodes, are calculated by using equivalent resistance of photovoltaic module which is ratio of voltage and current of a module. Then, these duty ratios are applied to the DC-DC converter. A single ended primary inductance converter (SEPIC) is modelled in MATLAB/Simulink, in order to validate the performance of the proposed algorithm. Simulation results show that proposed algorithm has capability of tracking global maximum power point in all partial shading conditions and it has simple structure which can make this algorithm applicable for commercial products.

Keywords: DC-DC converter, shading, photovoltaic, PV module, maximum power point tracking

1 INTRODUCTION

Due to the limited reserves of fossil energy resources, environmental concerns and public activities, kind of resources used in energy generation has become important in recent years. According to the British Petroleum's report, large share of fossil sources in primary energy consumption is 86.3% in 2014. On the other hand, solar energy, which is one of the renewable energy sources, is an important alternative for meeting the growing energy demand. In 2014, renewable share in primary energy consumption is 9.25%. Besides that, share of photovoltaic (PV) energy in electricity generation is 0.79% in the same year [1].

PV modules are the components generated electrical power in solar power systems. Furthermore, PV modules are nonlinear direct current (DC) power source whose voltage-power (V-P) and voltage-current (V-I) curves are logarithmic structure. This nonlinearity results in difficulties in maximum power extraction, since there is only single maximum power point (MPP) on the V-P curve in uniform irradiance condition. Conventional maximum power point tracking algorithms (MPPT) can track MPP accurately in that kind of condition [2-6]. On the other hand, if a partial shading condition (PSC) occurs, this curve gets more complex compared with the curve in uniform irradiance condition which causes to the fault operation in conventional techniques and algorithm. In order to track global maximum power point (GMPP), several approaches have been proposed [7-15] which are classified into two groups. These are analytical approaches (PV characterization based) and hybrid methods which generally consist of one of the soft computing methods such as ant colony, cuckoo search, genetic algorithm etc. and conventional algorithms (P&O, Incremental conductance). Main features of the soft computing based methods are the high computational requirement and complexity. Besides, their applicability for commercial purposes may be limited. Therefore, simple approaches are fashionable in MPPT under PSCs.

Commercial PV modules have low efficiency which decreases the capacity factor in solar power systems. In order to obtain high capacity factor, highly efficient MPPT algorithms are required which can track MPP in PSC and rapidly changing environmental conditions. These algorithms should have high speed (low response time), low fluctuation at MPP and tracking capability in PSCs. In this study, a high speed global MPPT (GMPPT) algorithm is proposed. It has simple structure and high speed properties due to its scanning feature in limited range. Besides that, tracking of the GMPP is guaranteed in all PSC cases due to its scanning structure. The remains of the paper continue as follows: In Section 2, Methodology of the proposed algorithm is briefly explained. Section 3 presents the simulation results that validate the performance of the proposed algorithm given in Section 2. In the last section main contribution of this study is summarized.

2 METHODOLOGY

Maximum power extraction from PV modules is an important subject because of low power conversion efficiency of modules. In order to increase the capacity factor of a PV module, an efficiently operated MPPT algorithm is required in any irradiance conditions. In this context, V-P curves based algorithm is introduced in this study. First, it is required to observe some V-P curves of PV array for certain conditions.

2.1 Voltage - Power Curve of PV Arrays in PSC

Even if PV modules have one MPP in uniform irradiance condition, they have multiple MPPs in PSCs. So as to find out change of voltage and power in different PSCs, V-P curves should be generated by making some assumptions for irradiance. Assume that maximum, minimum and increment value of irradiances are $1000W/m^2$, $100W/m^2$ and $100W/m^2$, respectively [16]. According to these assumptions, number of MPPs can be formulated for a PV array which consists of three series connected PV modules as in (1). This formula is used, if number of MPPs is the maximum for a PV array. Block diagram of the partially shaded PV array is shown in Figure 1.

$$c = \frac{m!}{(m-n)!n!} \tag{1}$$

where c is the number of shading cases, m is the ratio of maximum and minimum value of irradiance, n ise the number of bypass diodes in a PV array. For given assumptions, there may be 120 different shading patterns. All V-P curves of the PV array given in Figure 1 are presented in Figure 2. It is clear that voltage variations are very low in these curves. It can be said that MPPs are clustered in certain regions on the V-P curve.

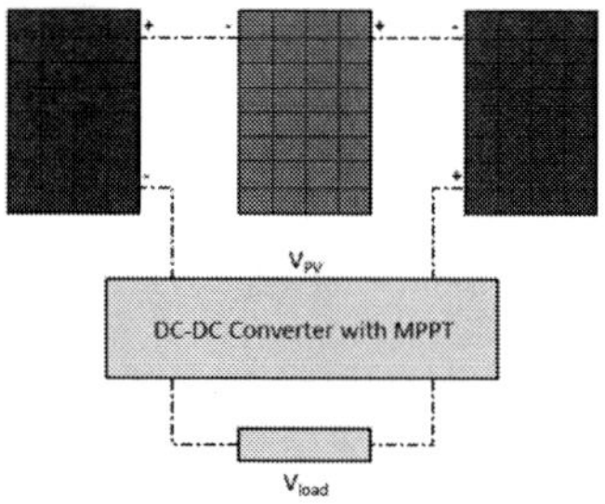

Figure 1: Block diagram of the PV array

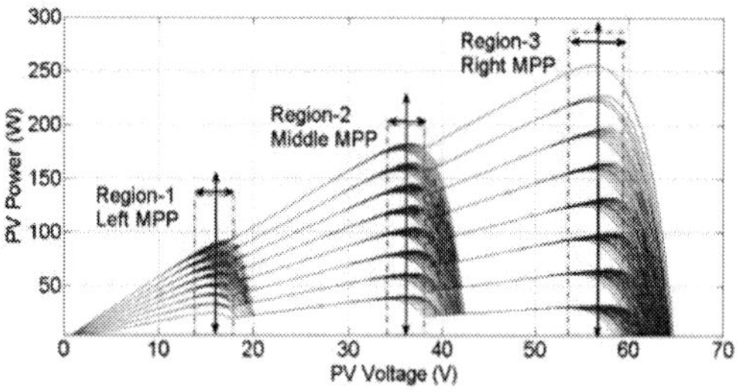

Figure 2: V-P curves of 120 different shading patterns

2.2 Duty Ratio Calculation

Proposed algorithm uses limited scanning approach which is based on the number of bypass diodes included in a PV array. According to the proposed configuration, there may be three MPP regions on the V-P curve as shown in Figure 2. For these MPP regions, average equivalent resistance (ER) of the PV array is calculated as in (2).

$$R_{PVavg} = \frac{\sum_{x=1}^{c} \frac{V_{mpp}}{I_{mpp}}}{c} = \frac{\sum_{x=1}^{c} \frac{V_{mpp}^2}{P_{mpp}}}{c} \tag{2}$$

In (2), R_{PVavg} is the average ER of the PV array, V_{mpp} and I_{mpp} are the voltage and current of this PV array at MPP, respectively. For each MPP regions, duty ratio is calculated as in (3). As is clear in this equation, duty ratio, which is abbreviated as D, is only based on the value of load resistance, since R_{PVavg} is assumed as constant.

$$D = \frac{1}{1+\sqrt{\frac{R_{PVavg}}{R_{load}}}} \tag{3}$$

2.3 Proposed Algorithm

In order to determine whether shadowing occurs, polarities of voltages of the bypass diode are checked. If one of the bypass diodes has negative voltage in a PV array for certain condition ($I_{PV} \approx I_{SC}$), it means to partial shading. As it is clear in Figure 3, if PSC occurs, one of the modules has negative voltage at least. This condition is formulated in (4).

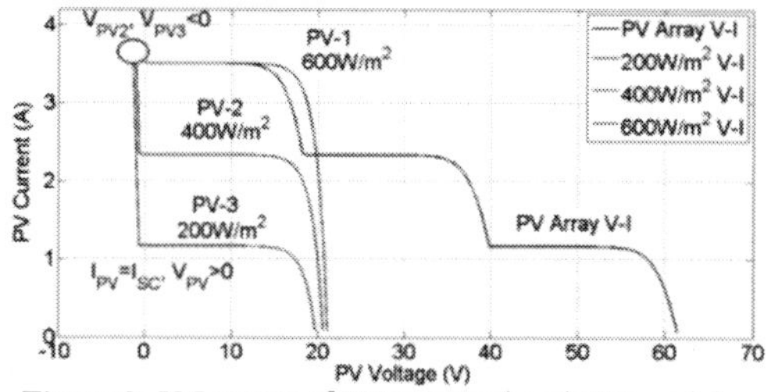

Figure 3: V-I curves of PV array and each PV modules

$$I_{PV} \cong I_{SC} \rightarrow V_{d1}, V_{d2}, \ldots, V_{dn-1} < 0, V_n > 0 \tag{4}$$

In order to ensure $I_{PV} \approx I_{SC}$, duty ratio is set to value (0-100%) which depends on the type of power processing converter. For a SEPIC, (5) can be obtained from the basic equations of buck-boost converter topology.

$$R_{PV} = R_{load} \frac{(1-D)^2}{D^2} \tag{5}$$

This equation can be written as in (6), if ratio of voltage and current of PV array is written instead of R_{PV}. According to (6), duty ratio has to be set to minimum value, so as to provide $I_{PV} \approx I_{SC}$.

$$\frac{0}{I_{SC}} = R_{PV} = R_{load} \frac{(1-D)^2}{D^2} \tag{6}$$

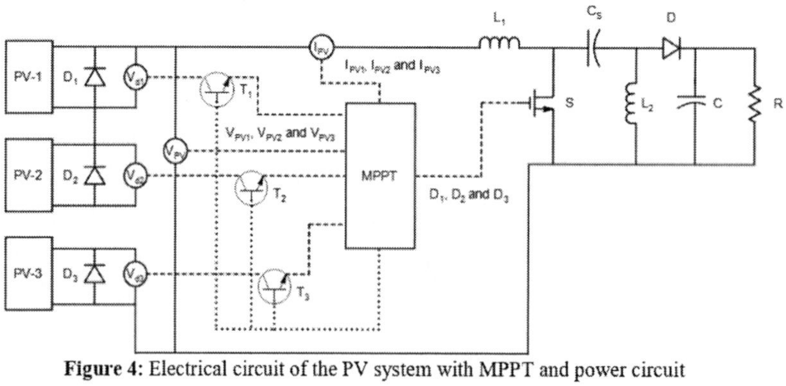

Figure 4: Electrical circuit of the PV system with MPPT and power circuit

Figure 5.a shows the flowchart of the proposed algorithm. It is understood whether PSC occurs by checking the voltage of bypass diodes. If PV array is partially shaded, duty ratios calculated via (2) and (3) are applied to SEPIC for user defined time. This time determines the power scanning time for PSCs. In this time interval, voltage and current of PV array are measured. Then, power of the PV array is calculated and saved. At the end of these intervals, maximum value of power is determined. For a PV array presented in Figure 1, there are three bypass diodes. Therefore, there are three duty ratios and three time intervals. It is sure that stored powers are compared, so as to initiate the value of duty ratio. Initial value of the duty ratio is set based on time interval which maximum power is extracted from PV array as illustrated in Figure 5.b.

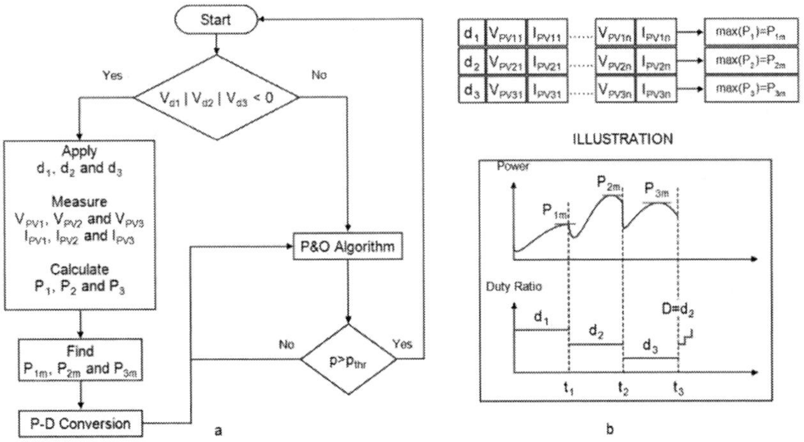

Figure 5: Electrical circuit of the PV system with MPPT and power circuit

3 SIMULATION RESULTS

In order to validate the performance of the proposed algorithm, a SEPIC has been designed and modelled in MATLAB/Simulink. Some specifications of the SEPIC and shading patterns are listed in Table 1.

Table I: Simulation scenarios and results

Case	Irradiance (kW/m^2)	Response Time (ms)	Efficiency at MPP (%)
1	0,4 - 0,5 - 1	100	99.5
2	0,2 - 0,4 - 0,6	80	98.1
3	0,1 - 0,2 - 1	80	88.6

In the first case, PV array receives three different irradiance levels which are $400W/m^2$, $500W/m^2$ and $1000W/m^2$. Therefore, there are three MPP regions for this case. As is clear in Figure 6.a, duty ratio is set to 5% (minimum value), in order to detect whether PSC occurs. Since two of the voltages of bypass diodes have negative voltages which are presented in Figure 7, duty ratios calculated are applied to SEPIC one by one. During this process, variations in power is saved until t=60ms (the time can be adjusted by user). Since value of maximum power is obtained in case of $d=d_3$, duty ratio is initiated as d_3. Then, P&O is used, in order to track the GMPP in the local MPP regions. Power scanning and convergence takes about 100ms in this case. Tracking efficiency at steady state (MPP) is 99.5%.

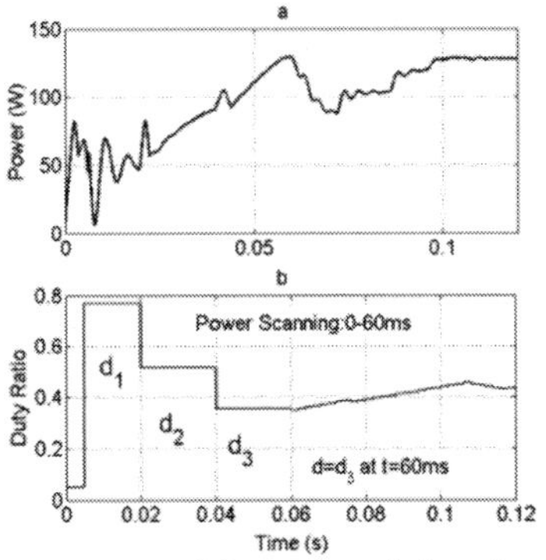

Figure 6: Power and duty ratio variations for case 1

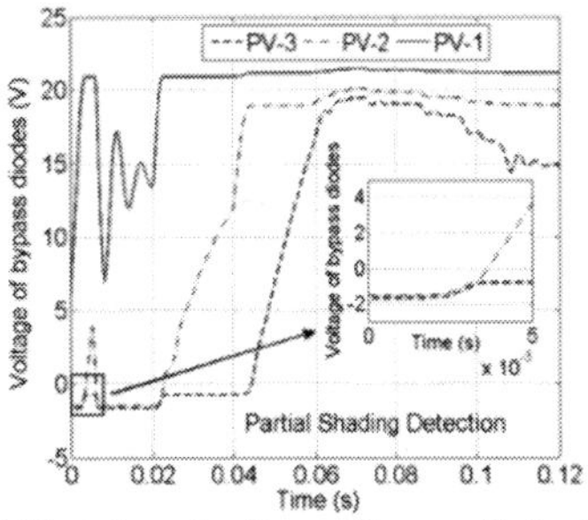

Figure 7: Variation of voltage of bypass diodes in case 1

GMPP may be on the left, middle or right of the three MPP regions for a PV array included three PV modules. In the second case, GMPP is on the middle of the V-P curve. Therefore, d_2 must be initialized, in order not to be failed in local MPP. Case 2 refers to this condition. Values of solar irradiances are $200W/m^2$, $400W/m^2$ and $600W/m^2$ in this case. As shown in Figure 8, duty ratio is initiated properly which results in detection of right MPP region. As a summary, GMPPT is realized in 80ms in this case. Tracking efficiency at steady state is 98.1%.

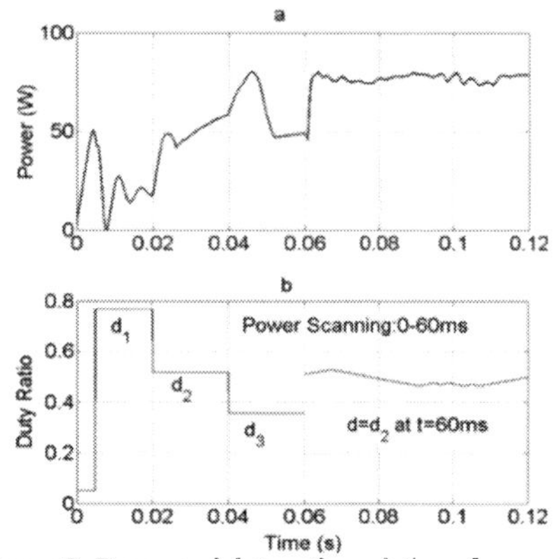

Figure 8: Power and duty ratio variations for case 2

Left, middle and right MPP regions have already been given in Figure 2. In the last case, GMPP is on the left of the MPP regions which can be understood from initialization value of duty ratio at MPP in Figure 9.b. When value of duty ratio equals to d_1, maximum value of array power is stored as can be understood from Figure 9.a. d_1 is initiated at the end of the power scanning. Then, P&O is applied to the SEPIC. Convergence time is about 80ms in this case. Tracking efficiency is 88.6%.

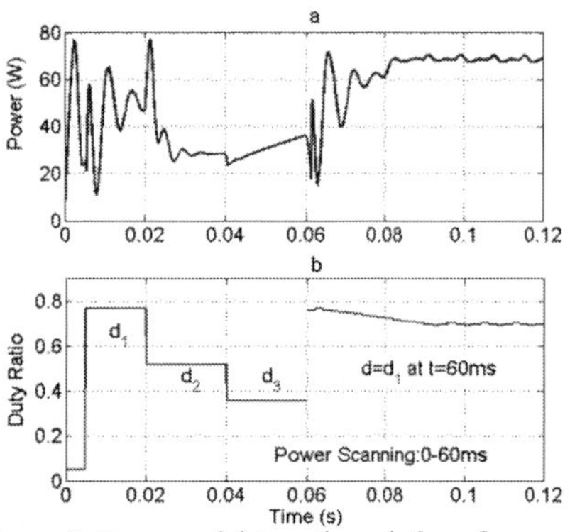

Figure 9: Power and duty ratio variations for case 3

4 CONCLUSIONS

This paper presents a new GMPPT algorithm which manages to efficient tracking in PSCs. Advantages of the proposed algorithm can be summarized as follows.

- Due to the scanning procedure, it is impossible to choose the wrong MPP region in the propose algorithm.
- This algorithm uses the relationship between voltage and current of PV array which is much known and does not require additional information about the PV system.
- Partial shading detection is provided by the voltage value of the bypass diodes included in a PV array. There is no need to have additional component.
- It is verified that duty ratios vary in limited range for all MPP regions which ensures the low scanning time as validated in simulation studies.
- This algorithm has simple structure and it can be used in commercial products in the market.

In further research, a prototype will be implemented for experimental validation of these algorithms.

5 CONCLUSIONS

[1] British Patreloum, BP Statistical Review of World Energy, 64th edition, June 2015.

[2] T. Esram, P.L. Chapman, "Comparison of photovoltaic array maximum power point tracking techniques", **IEEE T ENERGY CONVER,** 22(2), 439-449, (2007).

[3] A.R. Reisi, M.H. Moradi, S. Jamasb, "Classification and comparison of maximum power point tracking techniques for photovoltaic system: a review" **RENEW SUST ENERG REV**, 19, 433-443, (2013).

[4] F. Lui, S. Duan, F. Lui, B. Liu, Y. Kang, "A variable step size INC MPPT method for PV systems" **IEEE T IND ELECTRON,** 55(7), 2622-2628, (2008).

[5] T. Noguchi, S. Togashi, R. Nakamoto, "Short-current pulse-based maximum-power-point tracking method for multiple photvoltaic-and-converter module system", **IEEE T IND ELECTRON**, 49(1), 217-223, (2002).

[6] M.A.S. Masoum, H. Dehbonei, E.F. Fuchs, "Theoretical and experimental analyzes of photovoltaic systems with voltage and current based maximum power point tracking", **IEEE T ENERGY CONVER,** 17(4), 514-522, (2002).

[7] K. Ishaque, Z. Salam, "A deterministic particle swarm optimization maximum power point tracker for photovoltaic system under partial shading condition," **IEEE T IND ELECTRON,** 60(8), 3195-3206 (2013).

[8] M. Sarvi, S. Ahmadi, S. Abdi, "A PSO-based maximum power point tracking for photovoltaic systems under environmental and partially shaded conditions", **PROG PHOTOVOLTAICS**, 23(2), 201-214, (2015).

[9] S.A. Rizzo, G. Scelba, "ANN based MPPT method for rapidly shading conditions", **APPL ENERG**, 145, 124-132, (2015).

[10] K. Punitha, D. Devaraj, S. Sakthivel, "Artificial neural network based modified incremental conductance algorithm for maximum power point tracking in photovoltaic system under partial shading conditions", **ENERGY**, 62, 330-340, (2013).

[11] B.N. Alajmi, K.H. Ahmed, S.J. Finney, B.W. Williams, "A maximum power point tracking technique for partially shaded photovoltaic systems in microgrids", **IEEE T IND ELECTRON**, 60(4), 1596-1606, (2013).

[12] S. Daraban, D. Petreus, C. Morel, "A novel MPPT (maximum power point tracking) algorithm based on a modified genetic algorithm specialized on tracking the global maximum power point in photovoltaic systems affected by partial shading", **ENERGY**, 74, 374-388, (2014).

[13] N. Gokmen, E. Karatepe, G. Urganli, S. Silvestre, "Voltage band based global MPPT controller for photovoltaic systems", **SOL ENERGY**, 98, 322-334, (2013).

[14] A. Kouchaki, H. Iman-Eini, B. Asaei, "A new maximum power point tracking stregy for PV arrays under uniform and non-uniform insolation

conditions ", **SOL ENERGY**, 91, 221-232, (2013).
[15] A. Murtaza, M. Chiaberge, F. Spertino, D. Boero, M.D. Giuseppe, "A maximum power point tracking technique based on bypass diode mechanism for PV arrays under partial shading", **ENERG BUILDINGS**, 73, 13-25, (2014).
[16] J. Qi, Y. Zhang, Y. Chen, Y "Modeling and maximum power point tracking (MPPT) method for PV array under partial shading conditions", **RENEW ENERG**, 66, 337-345, (2014).
[17] R. Kotti, W. Shireen, "Efficient MPPT control for PV systems adaptive to fast changing irradiation and partial shading conditions", **SOL ENERGY**, 114, 397-407, (2015).
[18] M.E. Başoğlu, B. Çakır, "Comparisons of MPPT performances of isolated and non-isolated DC-DC converters by using a new approach," **RENEW SUST ENERG REV**, 60, 1100-1114 (2016).

ANALYZING THE PERFORMANCE OF COMMERCIAL PV MODULES UNDER FIELD CONDITIONS

Jong-Keun Lim, Jin Hyung Ahn, Moonseok Kim, San-Il Yoon, Myung-Ick Hwang, and Eun-Chel Cho
Hyundai Heavy Industries Co., Ltd.
17-10 Mabuk-ro 240beon-gil, Yongin 16891, South Korea

ABSTRACT: The performance of commercial PV modules has been compared under field conditions at HHI's outdoor testbed, located in South Korea. We have established the testbed with the capability of PV monitoring system in which can be collected the individual PV modules' electrical data and meteorological data including solar irradiation, ambient temperature, humidity, wind speed etc. This study was characterized outdoor module performance of several commercial modules, which were purchased from market. In this experiment, it is found that the light-induced degradation (LID) and the cell structure are the significant factors to determine the performance ratio (PR) of the modules under same field conditions. The performance ratio of LID-resistant mono-PERC module has the highest value due to the lower LID and the better cell structure. Additionally, it is investigated that multi-wire busbar can improve the PR of module by reducing the electrical loss of the module.
Keywords: PV module, field conditions, outdoor testbed, performance ratio, PERC, cell structure, LID, electrical loss, multi-wire busbar

1 Introduction

It is not easy to investigate the performance of the commercial modules under field conditions. Power generation of the modules is affected by the various factors like measurement error, the difference between rated power and real power of modules, temperature coefficient, light-induced degradation (LID), the I-V characteristics by intensity of irradiance etc. [1] [2] In this study, the objectives is to find the significant factor to determine the performance ratio of the tested modules under field conditions. The performance ratio of the commercial PV modules will be analyzed based on the PV modules and weather data obtained at our outdoor testbed.

2 Outdoor testbed setup

We have established our outdoor testbed to investigate the performance of the individual PV modules. Table I shows the components of our outdoor test-bed facility. Module optimizer (Tigo company 2ES50XL model) has been used to check the electrical performance. It is possible to collect the electrical data of each PV modules by optimizing each modules individually regardless of the module type or rated power. And the meteorological data (including solar irradiation, ambient temperature, humidity, wind speed etc) can be collected through weather station synchronized with each PV module's data (Vmpp, Impp, Pmpp). Fig. 1 shows the whole view of our outdoor testbed and PV monitoring system configuration.

Table I: Components of Outdoor test-bed

Location	Eum-seong , South Korea
Capacity	10kW (5kW x 2)
The number of Available PV module	32ea (16ea x 2)
Inverter	ABB (PVI series model)
The system to measure module's characteristics	Tigo optimizer (2ES50XL model) + Monitoring S/W
Meteorological sensor	Lufft weather station (WS501-UMB) Kipp & John Pyranometer (SMP3)

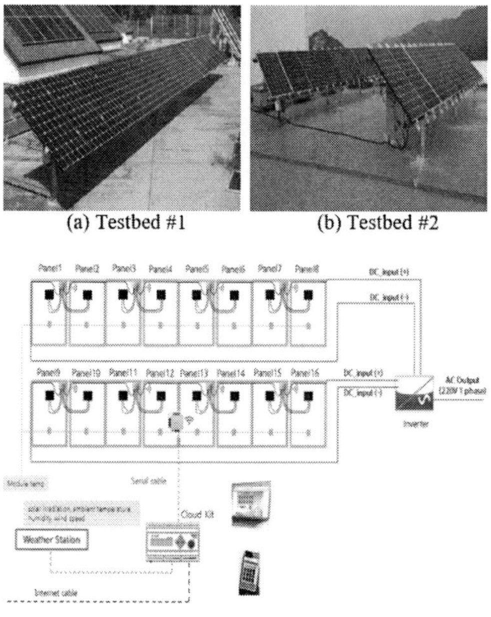

(a) Testbed #1 (b) Testbed #2

(c) PV monitoring system configuration

Figure 1: (a) testbed #1 (b) testbed #2 (c) PV monitering system configuration

3. Comparison of Performance ratio of PV modules

3.1 The effect of cell type & structure

In order to check the difference of module's performance ratio by the cell type and structure under outdoor field condition, we have compared the performance ratio of the modules ; (A) 260W multi module, (B) 275W mono module, (C) 285W mono-PERC module. Table II shows the specification of 3 modules. In this experiment, we have classified STC Power, Measured Power, Stabilized power. STC power is the power presented at label of commercial module. Measured power is the power measured at our solar simulator under STC condition. Stabilized power is the power measured at our solar simulator under STC condition after operating for 4 weeks under field conditions. It is found that PERC module(C) have a higher PR value than conventional modules (multi, mono) in spite of higher LID (4.44%).

Table II: The specification of tested modules to check the PR of module according to the cell type & structure

Module	A	B	C
Cell type	Multi	mono	mono
Cell structure	Full BSF	Full BSF	PERC
LID- resistant	No	No	No
STC power [W]	260	275	285
Measured Power [W]	261.2	275.7	283.8
Stabilized power [W]	260.4	269.1	271.2
LID [%]	0.33%	2.4%	4.44%
Performance Ratio	0.89	0.887	0.897

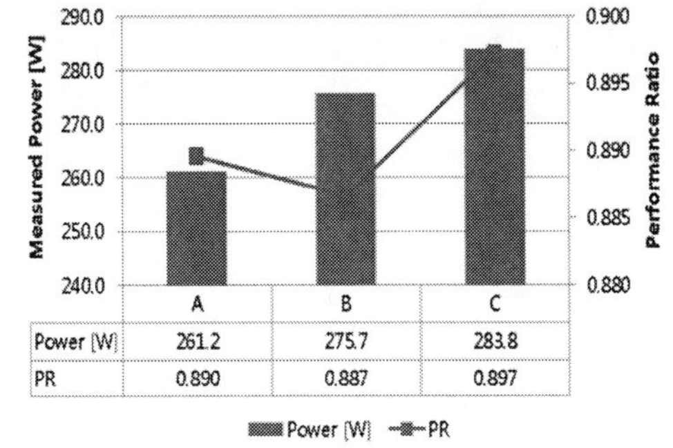

Figure 2: The performance ratio of tested modules accoridng to the cell type & structure

3.2 The effect of LID-resistant

LID can be reduced by applying proper heat & light to cell. [3] So many module manufacturers are recently promoting their LID-resistant modules. In order to check the difference of module's PR according to the LID under field condition, we have compared the performance ratio of the modules ; (C) 285W mono-PERC module, (D) 285W LID-resistant mono-PERC module. Table III shows the specification of 2 modules. In this experiment, it is found that the PR of LID-resistant PERC module is improved by 3.2% due to the reduced LID (4.44% →0.44%).

Table III: The specification of tested modules to check the PR of module according to the LID resistant processing

Module	C	D
Cell type	mono	mono
Cell structure	PERC	PERC
LID- resistant	No	YES
STC power [W]	285	285
Measured Power [W]	283.8	285.9
Stabilized power [W]	271.2	284.6
LID [%]	4.44%	0.44%
Performance Ratio	0.897	**0.929**

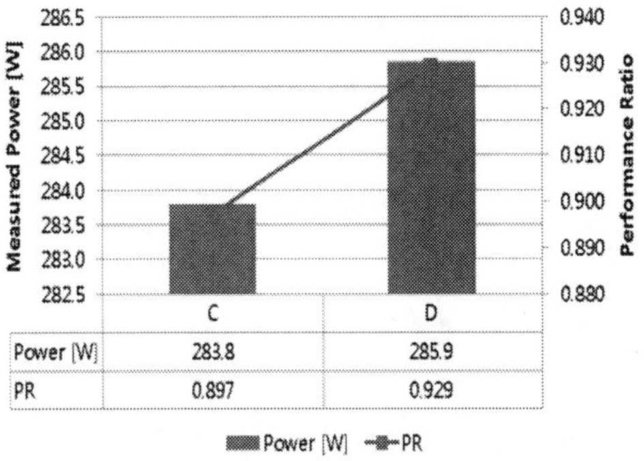

Figure 3: The performance ratio of tested modules accoridng to LID resistant processing

3.3 The effect of multi-wire busbar

In order to check the difference of module's PR by the busbar under field condition, we have compared the performance ratio of the modules ; (C) 285W mono-PERC module with 3 busbar, (E) 285W mono-PERC module with multi-wire busbar. Table IV shows the specification of 2 modules. In this experiment, it is found that PERC module(E) with multi-wire have 3.1% higher PR than PERC module(C) with 3 busbar due to the reduction of electrical loss.

Table IV: The specification of tested modules to check the PR of module according to the busbar type (3busbar, multi-wire busbar)

Module	C	E
Cell type	mono	mono
Cell structure	PERC	PERC
LID- resistant	No	No
Busbar type	3 busbar	Multi-wire (12 busbar)
STC power [W]	285	295
Measured Power [W]	283.8	295.5
Stabilized power [W]	271.2	282.5
LID [%]	4.44%	4.4%
Performance Ratio	0.897	**0.928**

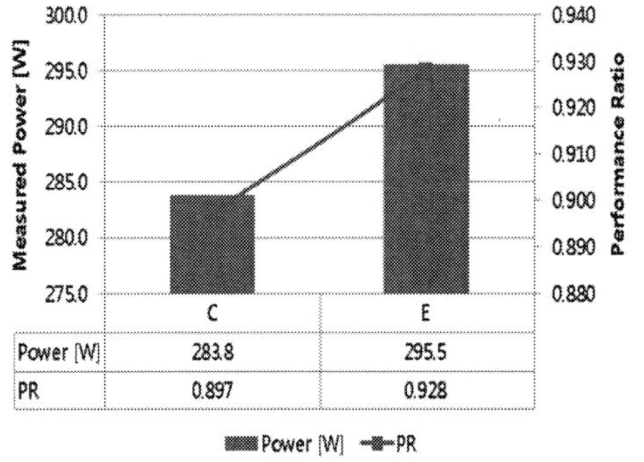

Figure 4: The performance ratio of tested modules accoridng to busbar type (3busbar, multi-wire busbar)

4 Conclusion & Discussion

We have established our testbed with the PV monitoring system, in which can be obtained the individual PV module's electrical and meteorological data. The PERC cell structure have about 1% higher PR than the conventional cell structure with full BSF in spite of relatively higher LID. By applying the LID-resistant process, the PR of PERC module can be improved by about 3.2%. In addition, the PR of PERC module with the multi-wire busbar can be improved by about 3.1%. In our testbed, the PR of LID-resistant PERC module with multi-wire busbar will be evaluating on June, 2016 to achieve the maximum value of PR.

ACKNOWLEDGMENTS

This work was partly supported by the New & Renewable energy Core Technology Program of the Korea Institute of Energy Technology Evaluation and Planning (KETEP), with the financial grant provided by the Ministry of Trade, Industry & Energy, Republic of Korea (No. 20133010011780).

References

[1] Wilfried G.J.H.M. van Sark1, Atse Louwen1, Arjen C. de Waal1,2, Boudewijn Elsinga1, Ruud E.I. Schropp2, UPOT: THE UTRECHT PHOTOVOLTAIC OUTDOOR TEST FACILITY, 27th European Photovoltaic Solar Energy Conference and Exhibition

[2] N.H. Reich, B. Mueller, A. Armbruster, W.G.J.H.M. van Sark, K. Kiefer, and C. Reise, Performance ratio revisited: is PR > 90% realistic?, Progress in Photovoltaics: Research and Applications 20 (2012) 717-726

[3] Kyumin Lee, Moonseok Kim, Jong-Keun Lim, Jin-Hyung Ahn, Myung-Ick Hwang, and Eun-Chel Cho, natural recovery from lid: regeneration under field conditions?, 31st European Photovoltaic Solar Energy Conference and Exhibition

ANALYSIS OF THE PERFORMANCE OF PV MODULES WITH CELL-STRING LEVEL OPTIMIZERS FROM A LCOE PERSPECTIVE

Shengcheng Zhang, Peng Quan, Shifeng Deng, Esther Lee, Jun Yu, Mingdong Wu, Zhen Zhang, Pierre J. Verlinden, Zhiqiang Feng
State Key Laboratory of Photovoltaic Science and Technology, Trina Solar
No. 2 Trina Road, Trina PV Park, New District, Changzhou, Jiangsu Province, China 213031

ABSTRACT: This paper reviews and predicts the PV system component cost from 2013 to 2020. The added cost of cell-string smart module is also summarized by technology and geographical region. The benefit of a PV module with cell-string optimizers (or sub-string optimizers) is analyzed. The energy yield of systems with and without smart module is simulated taking into account the ground coverage ratio (GCR), panel mismatch, soiling, row-to-row shading, and the power degradation. Incorporating all information provided, a template of LCOE calculation is proposed. The LCOE of a conventional system and a smart module system is calculated by the region such as the USA, Europe, China and Japan. The present value of LCOE is obtainable, as well as its future values up to the year 2020 based on the system and smart module price roadmap. The analysis predicts that, if the added cost of smart PV modules gets reduced to approximately \$0.06(USD)/Wp, the installation of smart PV modules for residential, commercial rooftop projects and utility-scale projects will be more feasible. In the case of rooftop distributed systems in China, the application segment which has been growing rapidly in recent years, the LCOE of cell-string level smart PV systems is already approximately 2~5% less than that of conventional PV systems.
Keywords: smart module, LCOE, optimizer, Energy gain

1 INTRODUCTION

Over the past several years, DC optimizers have become an important component in many of the residential, commercial and utility-scale solar PV systems. Incorporating DC optimization technology into the system enables system designers to minimize the power loss from shading, uneven soiling or debris and module mismatch within a string.

The cell-string smart module is a smart PV module containing cell-string level optimizers, i.e. one DC-DC converter per each of the sub-strings in a module. It offers better energy optimization and partial shading tolerance than module-level optimizers. It is a highly-integrated power regulator in each cell-string of a PV module. In this paper, the benefits of the cell-string level optimizer is discussed, and how systems equipped with this technology can be expected to have higher energy density and yield with enhanced reliability at a low installation cost.

There have been many papers on the LCOE analysis of systems with conventional PV modules, and the technical analysis of systems with smart PV modules. However, not many papers have been dedicated to the LCOE analysis of systems with smart PV modules. In this paper, the performance of a cell-string level smart PV module, a low-cost power optimization solution, is analyzed from the LCOE perspective.

2 PERFORMANCE OF SMART PV MODULE

2.1 Energy density and land coverage

A typical conventional PV system will be limited in size, to avoid objects and row-to-row shading. The ground coverage ratio (GCR) is a ratio of a module area to the total area, and is set to minimize losses due to row-to-row shading within the system array—typically to limit annual energy loss to below 3%. By independently optimizing cell strings, rather than bypassing affected cell strings, the module with cell string optimizers continues to generate the maximum possible energy from shaded cell strings, enabling increased density and expanded system size.

Cell-level optimization allows a system designer to configure a system with higher tolerance to inter-row shading, resulting in up to 20% higher energy density than a system with conventional modules. Effectively, the system can maintain the same kWh/kWp as a conventional system, but with higher ground coverage ratios. Using cell-level smart modules on a commercial rooftop project, a system designer/developer can save up to 10% of land cost [1].

2.2 Panel mismatch loss

The cell-string level optimizer finds an optimal working point for each cell-string and boosts the current of underperforming cell-strings to match that of better performing cell-strings. This minimizes the energy loss due to string mismatch within a module.

2.3 Soiling loss

Soiling losses are always factored into modeling and simulation. The soiling pattern, common to all modules, consists of relatively uniform soiling from top to bottom of a module with heavier soiling close to the bottom edge of the module, and bird droppings. The soiling with uniform pattern cannot usually be addressed by a simple panel-level optimizer. In such case, there is no mismatch among the panels, and the performance of a conventional module and of a module-level optimizer is more or less the same. A module with cell string optimizers, however, can act as an insurance policy to mitigate performance loss from dust and debris. If about 40 percent of the bottom row (long side) is affected, the bypass diode turns on, thereby reducing panel output by one third (see Figure 1). In the absence of module-to-module mismatch, module-level optimizers provide no benefit. Only by removing the bypass diode, isolating the cell strings, and continuing to generate energy from the shaded cell string can the impact of this effect be reduced.

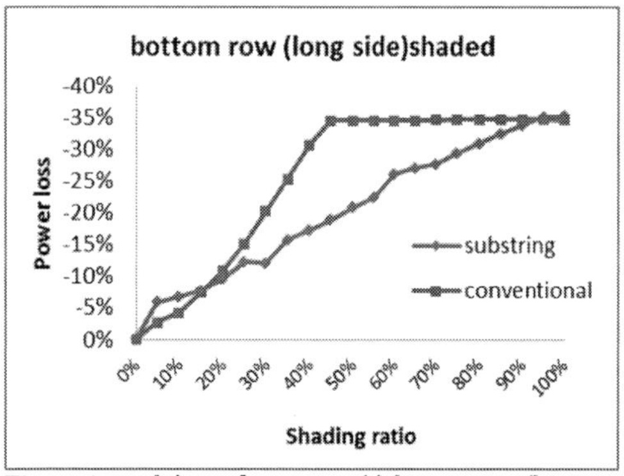

Figure 1: module performance with bottom row (long side) shaded

In the study of rooftop systems with bottom row soiling, systems with cell-string optimizers experienced 2.4% power loss due to soiling, while conventional systems experienced 16.4% power loss due to soiling.

2.4 Row-to-row shading loss

In a module with cell-string level optimizers, the bypass diodes are replaced with MPPT integrated circuits (ICs). When a cell-string in a module is underperforming, say due to partial shading, the smart IC chip will maintain the sub-string at maximum power point (MPP) and perform a DC-DC conversion to artificially increase the output current of the cell-string (and lower the voltage) to match that of unshaded neighboring cell-strings. Subsequently, the shaded cell-string will continue to contribute with the maximum output power that it can provide, instead of being by-passed by a diode if in conventional modules. The monthly energy output comparison between shaded modules with cell-string optimization and shaded conventional modules without optimizers is shown below (see Figure 2).

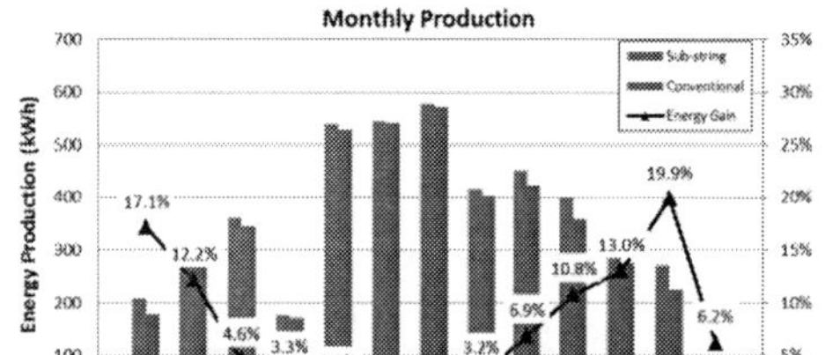

Figure 2: Comparison between a conventional module (red) and a module with cell-string optimizers (blue). Location: Fremont.

2.5 Reduce annual energy degradation

Over the course of its lifetime, a module can degrade due to various issues, such as thermal cycling, fatigue of interconnections, hot-spots, micro-cracks or potential induced degradation (PID), all of which can reduce its energy production. Cell-string optimization can prevent and mitigate problems arising from partial shading and mismatch between cell-string and between modules.

Hot-spot in a module occurs when its operating current exceeds the short-circuit current of a shaded or a faulty cell(s) within it. When such a condition occurs, the affected cell(s) is forced into a reverse bias mode and the power dissipated as heat causes overheating of the cell. Trinapeak, Trina solar's proprietary module with cell-string optimizers, eliminates the reverse bias and prevents the hot-spot formation. Figure 3 shows the IR image of a single shaded cell in a conventional module and a Trinapeak module. It is clear that a hot-spot has developed only in the conventional module.

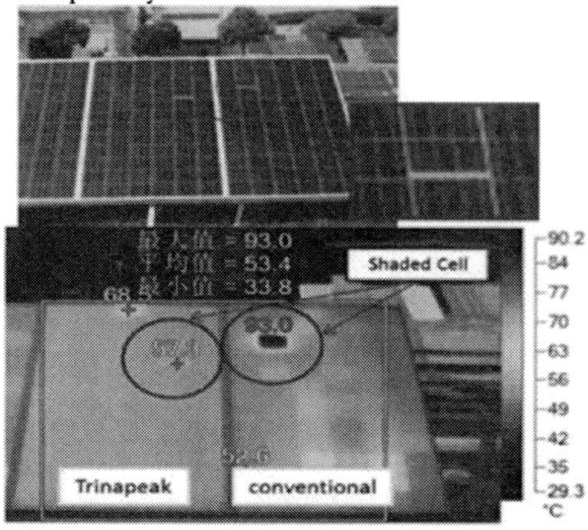

Figure 3: Hot spot test with a single cell shaded

The temperature across a single shaded cell in both conventional and Trinapeak modules was measured. There was a temperature difference of 51°C between the hottest and the coldest points in the conventional module while the Trinapeak module exhibited only 18°C difference (see Figure 4). Modules with Trinapeak technology have a lower "hot-spot" operating temperature, and in turn, these modules are protected from the long term damage and are expected to have a better durability.

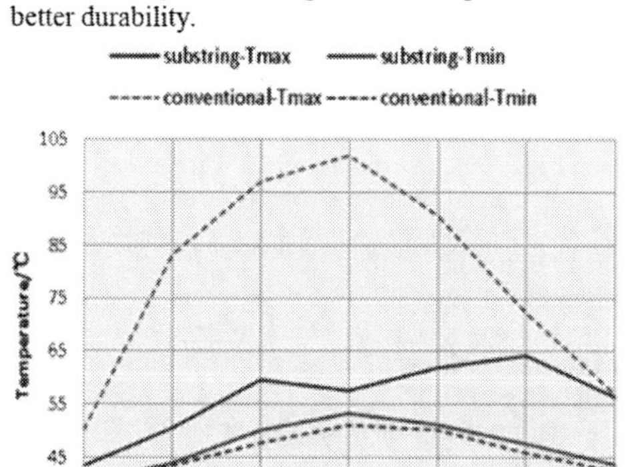

Figure 4: Module Temperature in Hot spot test

3 COST TRENDS OF MODULE AND SYSTEM

3.1 Global standard PV module cost trends 2014~2020

With technical progress and reduction in raw material cost, the PV module price will decrease constantly in the coming years. According to statistics and forecast of GTM, the price of global PV modules without optimizers will reduce by 20%, from \$0.64(USD)/W in 2014 to \$0.48(USD)/W in 2020.

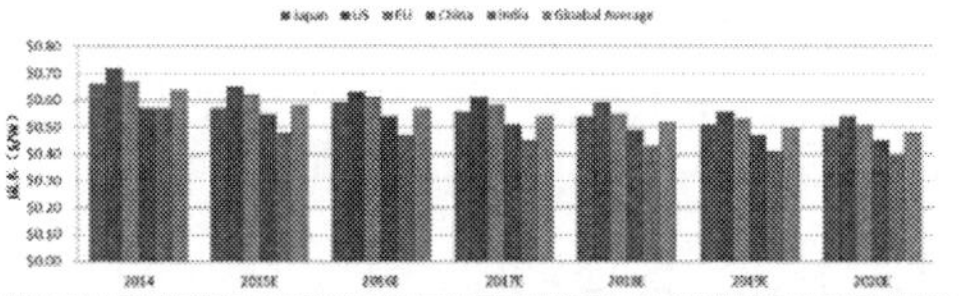

Location	2014	2015E	2016E	2017E	2018E	2019E	2020E
Japan	$ 0.66	$ 0.57	$ 0.59	$ 0.56	$ 0.54	$ 0.51	$ 0.50
US	$ 0.72	$ 0.65	$ 0.63	$ 0.61	$ 0.59	$ 0.56	$ 0.54
EU	$ 0.67	$ 0.62	$ 0.61	$ 0.58	$ 0.55	$ 0.53	$ 0.51
China	$ 0.57	$ 0.55	$ 0.54	$ 0.51	$ 0.49	$ 0.47	$ 0.45
India	$ 0.57	$ 0.48	$ 0.47	$ 0.45	$ 0.43	$ 0.41	$ 0.40
Gloabal Average	$ 0.64	$ 0.58	$ 0.57	$ 0.54	$ 0.52	$ 0.50	$ 0.48

Figure 5: Global standard module cost trends 2014~2020

3.2 Global DC smart module cost trends 2014~2020

Smart module pricing is primarily dependent on the price of PV modules and electronics components. As the market matures and expands, it is expected that the optimizer cost will decline steadily, from $0.14(USD)/W in 2015 to $0.03~0.05(USD)/W in 2020.

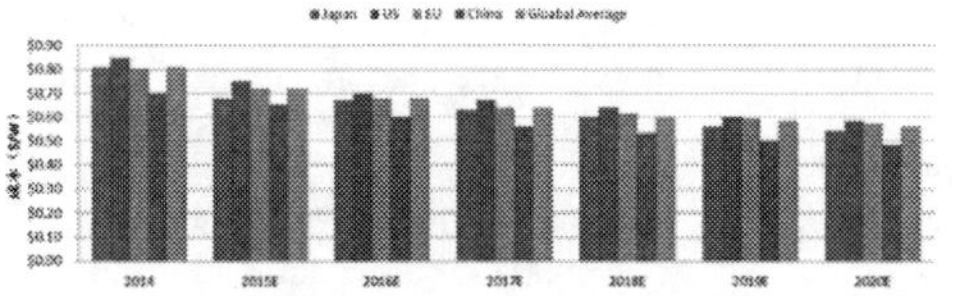

Location	2014	2015E	2016E	2017E	2018E	2019E	2020E
Japan	$ 0.81	$ 0.68	$ 0.67	$ 0.63	$ 0.60	$ 0.56	$ 0.54
US	$ 0.85	$ 0.75	$ 0.70	$ 0.67	$ 0.64	$ 0.60	$ 0.58
EU	$ 0.80	$ 0.72	$ 0.68	$ 0.64	$ 0.61	$ 0.59	$ 0.57
China	$ 0.70	$ 0.65	$ 0.60	$ 0.56	$ 0.53	$ 0.50	$ 0.48
Gloabal Average	$ 0.81	$ 0.72	$ 0.68	$ 0.64	$ 0.60	$ 0.58	$ 0.56

Figure 6: Global smart module cost trends 2014~2020

3.3 Global PV system (with standard modules) cost trends 2014 VS 2020

System price varies across market segments and geographies. Residential markets continue to depend heavily on subsidy programs that limit the attractiveness outside major feed-in tariff markets. The U.S.A, Japan, Germany, Australia, and the U.K. will comprise two-thirds of the global residential market. Japan and the U.S.A are expected to have the largest cost reduction, with costs dropping by 40%-45% by 2020. Meanwhile, cost in Europe is expected to fall by 20%-25% by 2020.

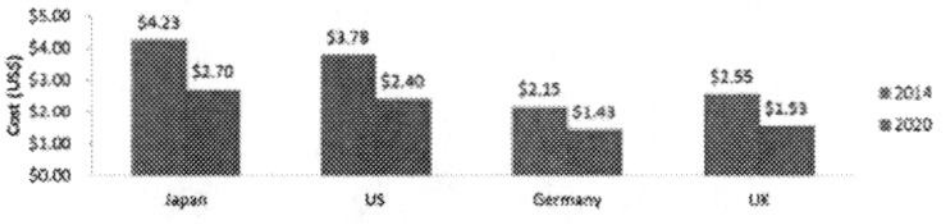

Figure 7: Global residential system cost

Much like residential systems, global commercial systems are subject to various local influences that create significant disparities in project pricing. Compared to residential systems, module and labor costs play the most significant role in determining price for commercial flat-roof systems.

Figure 8: Global commercial system cost

Ground-mount projects are estimated to make up nearly 70% of the global market - exceeding 350 GW_{DC} - between 2015 and 2020. Module pricing variability has narrowed significantly over the past few years, and the most module vendors are now selling at similar prices globally. The BOS price is extremely susceptible to local market conditions.

Figure 9: Global utility scale system cost

4 LCOE CALCULATION AND RESULTS

4.1 LCOE definition

$$LCOE = \frac{\text{Initial Cost} + \text{Annual Costs} - \text{Residual Cost}}{\text{Lifetime Energy Produced}}$$

Where,

Initial Cost: Modules, BOS, contractor work (i.e. siting), inverters.

Annual Costs: Fixed O&M, debt/other financing structures.

Residual Cost: Post-lifetime system value (resale/repurposing of parts)

4.2 LCOE calculation

Some system parameters were assumed in the calculation (see Table I).

Table I Assumed system parameters in LCOE calculation

Parameter	Unit	Value
Irradiation	kWh/m^2.day	4
Annual power degradation	%	0.7
System Efficiency	%	77.50
Lifetime	years	25
Capital Ratio	%	30
Annual Interest rate	%	5
Length of Maturity	years	20
Tax rate	%	8.5

Generally the application of a smart module can bring about 2% to 6% of energy gain depending on the environmental conditions. In LCOE calculation of smart module systems, we set an energy gain of 6% for residential and commercial systems, and 4% of energy gain for ground mounted systems. The added-on cost of a smart module is also considered in calculation. For residential and commercial systems, the added-on cost is $0.07(USD)/W. For ground mounted systems, it is $0.06(USD)/W. A smart module can also reduce the operating and maintenance costs of system, due to its higher module reliability and lower soiling loss. The assumed factors of a smart module are summarized in Table II.

Table II: Smart module assumption in LCOE calculation

System	Energy gain	Added on Cost	Reduction in O&M
Residential	6%	US$0.07/W	20%
Commercial	6%	US$0.07/W	20%
Ground-mounted	4%	US$0.06/W	10%

The LCOE of residential, commercial and ground-mounted systems are calculated respectively. The LCOE of conventional systems and smart module systems is calculated by the region – the USA, Germany, China and Japan (see Table III).

Table III: LCOE of residential, commercial and utility scale systems (standard modules and smart modules)

Residential		Conventional module				Smart module			
		US	Japan	Germany	China	US	Japan	Germany	China
Capacity	kW	5	5	5	5	5	5	5	5
Module price	$/W	0.69	0.60	0.65	0.53	0.76	0.67	0.72	0.60
System price	$/W	3.54	4.23	2.15	1.58	3.61	4.30	2.22	1.65
Yield at fist year	kWh	7300	7300	7300	7300	7738	7738	7738	7738
Total yield	kWh	167170	167170	167170	167170	177200	177200	177200	177200
Total investment	$	17700	21150	10750	7890	18050	21500	11100	8240
Annual O&M	$/W	0.05	0.05	0.05	0.02	0.04	0.04	0.04	0.016
Feed-in tariff	$/kWh	0.39	0.36	0.31	0.14	0.39	0.36	0.31	0.14
LCOE	$/kWh	0.232	0.261	0.162	0.099	0.217	0.244	0.150	0.094

Commercial		Conventional module				Smart module			
		US	Japan	Germany	China	US	Japan	Germany	China
Capacity	kW	200	200	200	200	200	200	200	200
Module price	$/W	0.69	0.66	0.65	0.53	0.76	0.67	0.72	0.60
System price	$/W	2.33	2.80	2.00	1.31	2.40	2.87	2.07	1.38
Yield at fist year	kWh	292000	292000	292000	292000	309520	309520	309520	309520
Total yield	kWh	6686800	6686800	6686800	6686800	7088008	7088008	7088008	7088008
Total investment	$	466000	580000	400000	262985	480000	594000	414000	276985
Annual O&M	$/W	0.05	0.05	0.05	0.02	0.04	0.04	0.04	0.016
Feed-in tariff	$/kWh	0.22	0.32	0.28	0.15	0.22	0.32	0.28	0.15
LCOE	$/kWh	0.162	0.197	0.152	0.088	0.150	0.183	0.141	0.084

Utility		Conventional module				Smart module			
		US	Japan	Germany	China	US	Japan	Germany	China
Capacity	kW	5000	5000	5000	5000	5000	5000	5000	5000
Module price	$/W	0.69	0.60	0.65	0.53	0.75	0.66	0.71	0.59
System price	$/W	1.45	2.45	1.35	0.99	1.51	2.51	1.41	1.05
Yield at fist year	kWh	7300000	7300000	7300000	7300000	7592000	7592000	7592000	7592000
Total yield	kWh	167170000	167170000	167170000	166819245	173856800	173856800	173856800	173596015
Total investment	$	7250000	12250000	6750000	4950966	7550000	12550000	7050000	5230966
Annual O&M	$/W	0.05	0.05	0.05	0.02	0.045	0.045	0.045	0.018
Feed-in tariff	$/kWh	0.2	0.32	0.22	0.14	0.2	0.32	0.22	0.14
LCOE	$/kWh	0.121	0.176	0.118	0.072	0.116	0.170	0.113	0.071

4.3 Results

The LCOE of a smart module system is compared to that of a standard module system, taking system scale and geographic variations into account.

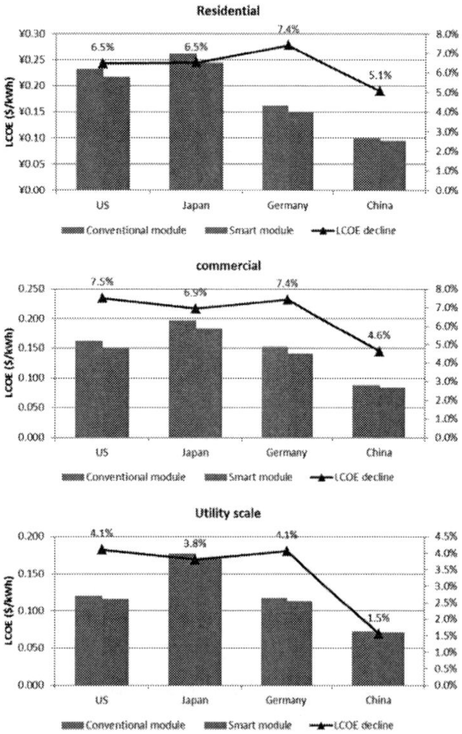

Figure 10: Comparisons of LCOE of a standard module system and a smart module system

For residential and commercial systems in the USA, Japan and Germany, by employing smart modules, LCOE can be reduced by 6.5%~7.5%, while for a system in China, 5.1% LCOE reduction can be expected. For utility scale systems with smart modules, 4% LCOE reduction could be expected in the USA, Japan and Germany. Even in China with the lowest system price available in the world, the adoption of smart PV modules would still benefit developers and investors.

The comparative LCOE analysis of residential and commercial rooftop PV projects using current system costs shows that to justify expected LCOE reduction, there needs to be an additional 6% energy production for a smart PV module. For an utility scale PV project, there needs to be 4% additional energy production. If all types of system losses such as the mismatch loss caused by dust and birds droppings (2%), and the power degradation loss (1%) and inter-shading loss of the front-rear rows (2~3%) were taken into account, the adoption of smart modules would result in at least 4~6% energy gain for residential, commercial and utility scale systems and projects. The inclusion of cell-string level DC optimizers would be easily justifiable for any scale projects.

5 CONCLUSION

As the $/Wp value between the smart PV modules and the conventional modules narrows, the applicability of smart PV modules for residential, commercial rooftop systems and utility-scale projects increases. The comparative LCOE analysis of PV systems using modules with sub-string optimizers and conventional PV modules, for residential, commercial and utility-scale applications has been presented. The analysis is long and complex because many different cases of mismatches or partial shadings have to be considered, for example, local degradation, bird dropping, row-to-row shading, and soiling etc. In a very practical case of partial shading, we showed that a PV system using modules with cell-string optimizers can generate 19.9% more energy than the same PV system with modules without cell-string optimizers. The usage of cell-string optimizers also reduces hot-spot problems, improves reliability and reduces LCOE.

REFERENCES

[1] Chris Deline, Bill Sekulic, Josh Stein, Stephen Barkaszi, Jeff Yang, Seth Kahn. Evaluation of Maxim Module-Integrated Electronics at the DOE Regional Test Centers. IEEE 40th Photovoltaic Specialist Conference (PVSC), June 2014.
[2] Alonso M C, Ruiz J M, Chenlo F. Experimental study of mismatch and shading effect s in the I-V characteristic of a photovoltaic module[J]. Solar Energy Materials & Solar Cells, 2006, 90(3): 329~340.
[3] Sharma A K, Dwivedi R, Srivastava S K. Performance analysis of a solar array under shadow condition[J]. IEE Proceedings G, 1991, 138(3): 301~306.
[4] Scott Moskowitz, MJ Shiao. Smart and AC PV Modules 2015-2020. GTM research, November 2015.
[5] MJ Shiao. PV Balance of Systems 2015: Technology Trends and Markets in the U.S. and Abroad. GTM research, August 2015.

TESTING OF SMART PV MODULES

Daniel Gfeller, Urs Muntwyler, Christian Renken, Luciano Borgna
Berne University of Applied Sciences (BFH), Engineering and Information Technology
Photovoltaic Laboratory (PV-Lab), Jlcoweg 1, CH-3400 Burgdorf, Switzerland
Phone +41 (0)34 426 68 11, Fax +41 (0)34 426 68 63, urs.muntwyler@bfh.ch, www.pvtest.ch

ABSTRACT: There are more and more PV modules with integrated electronic devices on the market. These smart PV modules have their electronic devices either directly inside the junction box or in a box in close proximity to the module. Common applications for smart modules are MPP tracking on module level (with microinverters or power optimizers), monitoring of the module's electrical data or safety features such as disconnecting the module in case of fire. Standard tests for PV modules are not intended for modules containing electronics other than bypass diodes in their junction box. If there are for example large capacitors in parallel to the cells, it is no longer possible to measure the module's I/V curve with a flashed solar simulator. In case of microinverters or power optimizers in the junction box, the module's DC terminals might not even be accessible, especially when the junction box is potted. This paper proposes methods for classifying and testing of different types of smart PV modules.

1 INTRODUCTION

Most PV modules simply have bypass diodes in their junction box. However, there are also modules with more sophisticated electronics on the market. These smart PV modules can have all kinds of functions (e.g. MPP tracking, monitoring or safety). Accordingly, there are many possible topologies and configurations of these electronics. Proper testing of these modules is sometimes difficult, because standard test procedures for PV modules do not foresee complex electronics in the module. In this paper, some typical problems at testing of smart PV modules are being discussed. In addition, a classification scheme for smart PV modules is presented and some possible testing methods are proposed. The focus lies on the characterization of the electrical data of the module (especially the MPP power) when the latter is considered to be a black-box. Mechanical stress tests are not considered. Moreover, even though the authors are somewhat skeptical concerning complex electronics in a rough environment such as the back of a PV module, possible problems with these devices in the field (e.g. endurance or reliability) are not discussed in this paper. An exception of this are the lightning current tests under section 8.

2 COMMON APPLICATIONS FOR SMART PV MODULES

Microinverters

Microinverters are one of the most popular reasons for electronics in a PV module. If each PV module has its own inverter, mismatch losses of the PV plant can be minimized. Also, the design of the plant becomes very flexible, because the modules have no mutual dependencies.

Power optimizers

Power optimizers basically aim for the same goal as microinverters. That is the MPP tracking on module level. Unlike microinverters, power optimizers just perform a dc-dc-conversion. Thus, there is still an inverter needed to feed-in the power into the mains.

Monitoring

There are some products on the market which allow monitoring of the module's electrical data. With this, defective modules and other flaws in a PV array can easily be detected. The transmission of the monitored data is usually performed by powerline or radio communication.

Safety circuit breakers

To place safety circuit breakers directly into the PV modules allows a complete shutdown of the PV array, e.g. in case of fire. By this, dangerous voltages, which might put the firefighters at risk, can be prevented. It is also possible that circuit breakers can autonomously detect faults, such as high temperatures (fire) or electrical arcs.

Active bypass diodes

Most PV modules use schottky diodes as bypass diodes. However, the power dissipation of these diodes is critical if the module is in full bypass operation (i.e. when all diodes carry the full string current). An alternative are active bypass diodes that emulate the function of normal diodes but with much better characteristics (very low forward voltage and negligible leakage current).

Combined functions

The applications listed so far can also be combined. Monitoring functions for example are also included in most microinverters and power optimizers.

3 POSSIBLE PROBLEMS AT TESTING OF SMART PV MODULES

Power consumption

The electronics in smart modules need a power supply. This leads to the question of how the MPP power shall be measured and specified – with or without the losses caused by the module's own consumption. The fairest assessment would be to include the own consumption into the specifications because that is the real life situation. However, there is the problem that this power consumption can be time-variant and therefore in a black-box test not deterministic.

Unknown components in parallel to the solar cells

Often, smart PV modules contain electronic components in parallel to the solar cells. Of course this has an impact on the electrical characteristics of the module, especially but not exclusively in transient operation. One example are capacitors in parallel to the solar cells who slow down the transient response of the module. This problem is explicitly discussed under section 5. However, in black-box tests these electronic circuits are unknown. Any attempt of measuring the electrical characteristics at the connections of the junction box might lead to undefined test results.

Missing accessibility of the DC terminals

Especially at microinverters and power optimizers, the DC terminals of the actual module (without electronics) might not even be accessible. If the electronics cannot be detached, it is virtually impossible to measure the I/V characteristics and with that the MPP power of the module. In some cases it is possible to open the housing of the electronics, disconnect them from the cells and make a direct connection (bye-bye black-box testing). However, if the housing is filled with potting mass, this becomes very difficult.

4 CLASSIFICATION SCHEME FOR SMART PV MODULES

Figure 1 shows a simple classification scheme for smart PV modules with respect to black-box testing. This means that on the module only manipulations are needed that are intended to be done by a normal user as well. For white-box testing the scheme also applies, as long as the module under test can be modified to the testers needs (destructive or nondestructive).

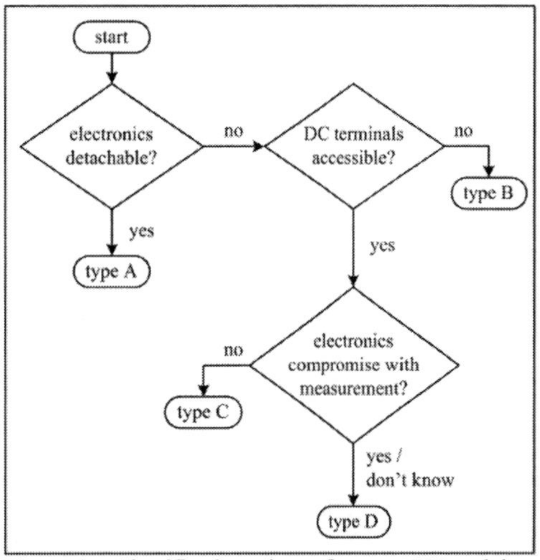

Figure 1: Classification scheme for smart PV modules

This classification scheme differentiates between four types of smart PV modules (A to D). Which tests are suitable for which type of module will be discussed under section 9. In most cases, the common flasher-tests are not performable with smart PV modules. As an alternative, the smart PV module can be characterized as a complete system by its energy output over a certain amount of time (see section 6 and 7). However, this kind of test is time consuming and therefore only suitable for sample testing.

5 CURRENT MEASUREMENT ERROR BECAUSE OF CAPACITORS IN PARALLEL TO THE SOLAR CELLS

Usually, PV modules are being measured with flashed solar simulators (so called flashers). With this kind of measurement, the solar cells are only shortly exposed to the simulated sunlight. During this time, there is a voltage sweep between zero and the module's open circuit voltage at the module's DC terminals, allowing a measurement device to scan the I/V curve. However, if there are capacitors in parallel to the solar cells, they are being charged or discharged during the sweep (open circuit to short circuit or vice versa). The total charge of the capacitor is the product of the capacitance (C) and the module's open circuit voltage (V_{OC}). The charging process leads to an error current (I_{ERROR}) that compromises the measurement. I_{ERROR} is reciprocally proportional to the sweep time (t_{SWEEP}). It is calculated by formula I.

$$I_{ERROR} = \frac{C \cdot V_{OC}}{t_{SWEEP}}$$

Formula I: Calculation of the error current

Please note that formula I is only valid, if the voltage sweep is linear (constant dV/dt). If the sweep is non-linear, I_{ERROR} calculated by formula I is the mean value of the actual, non-constant error current. Assuming a 72-cell module with V_{OC} = 45 V and a sweep time of 10 ms (linear), the current measurement error will be 4.5 mA per microfarad capacitor size. Assuming an MPP power of 300W, the measurement error because of the capacitor will be about 0.54‰ per microfarad. As capacitors for electronic devices may have several hundred microfarads, the measurement error can easily become unacceptable. By the way: The cell itself has a parasitic parallel capacitance. The sweep time for the measurement should therefore not be too low. With some cell types (mostly thin film) this can already be a problem with a sweep time of 10 ms. If the sweep is linear, the expected error current is not too high (e.g. <5% of I_{MPP}) and the size of the capacitor is known, the error caused by the capacitor can be corrected mathematically.

6 CHARACTERIZATION OF SMART PV MODULES USING A CONTINUOUS SOLAR SIMULATOR

With a continuous solar simulator it is possible to test most types of smart PV modules. Due to the constant irradiation of the solar cells, the potentially time-variant behavior of the electronics attached to the module is less troubling. For instance, the problem with capacitors in parallel to the solar cells discussed under section 5 can be avoided with a continuous solar simulator, because the voltage sweep for the I/V curve measurement can be very slow (I_{ERROR} then becomes negligible). Even if the electronic device attached to the module has fluctuating power consumption, a test with a continuous solar simulator is possible because if the measurement period is long enough, the fluctuations are being time-averaged. Thus, the average value of the measurement is representative for real operation. Especially for tests of smart modules with microinverters or power optimizers continuous solar simulators are congenial because they allow reproducible and stable test conditions for the complete system. If the simulated irradiation of the solar simulator is sufficiently accurate and stable, the power delivered to the module can be determined by calculating the product of irradiation (G) and module area (A_M). The efficiency of the system (η) is simply the ratio of the electric output power (P_{EL}) to the power delivered to the module. If P_{EL} and G are constant (static test), the efficiency is calculated by formula II (instantaneous value).

$$\eta_{STATIC} = \frac{P_{EL}}{G \cdot A_M}$$

Formula II: Calculation of the static efficiency

However, with a real inverter or power optimizer P_{EL} will not be perfectly constant because of the MPP tracking. Moreover, there is also the possibility for dynamic test patterns were G is time-variant. In this general case, the efficiency is the ratio between the electrical output energy from the Module to the radiation energy to the module. The energies again are the integrals of the powers. The static values G and P_{EL} are being replaced by their instantaneous values g(t) and $p_{EL}(t)$. This leads to formula III.

$$\eta_{DYNAMIC} = \frac{\int_{t1}^{t2} p_{EL}(t)dt}{A_M \cdot \int_{t1}^{t2} g(t)dt}$$

Formula III: Calculation of the dynamic efficiency

The test profiles could be defined in analogy to EN 50530, the European standard for testing of the overall efficiency of PV inverters [1]. This standard comprises both, static and dynamic test patterns. For the static measurements, the static efficiency should therefore be measured at an irradiance G of 50, 100, 200, 300, 500, 750 and 1000 W/m^2. This would allow calculating an average efficiency analog to the European efficiency in EN 50530. This would give information about the overall efficiency of the system under realistic conditions. For the dynamic measurements, the test profiles of EN 50530, where the irradiation is following linear ramps between 100 and 500 W/m^2 as well as between 300 and 1000 W/m^2 can also be adapted. However, the temperature of the solar cells during these long tests is virtually uncontrollable (that's one of the nice features of flash-tests, because there the heating of the solar cells during the short measurement period is negligible). To ensure that every tested module is being treated equally, the physical test setup has to be defined explicitly (e.g. ambient temperature of 25 °C, vertical mounting of the module and an air gap of 10 cm on both sides of the module). Moreover, the temperature of at least one cell should be measured and recorded as well during the tests.

7 OUTDOOR TESTS FOR SMART PV MODULES

A good characterization of smart modules under real life conditions can be achieved with outdoor tests. However, in an outdoor test stand, precise and reproducible measurements similar to laboratory measurements are not possible, because on the outside there are uncontrollable interferences like weather, dirt on the modules or even partial shading (although a good test array shouldn't have any near shading). Therefore, it is impossible to determine an absolute and representative value for the device's performance (even though it is attempted sometimes). But with a clever test setup, it is possible to make a fair comparison of different systems. The tricky thing about this is that every module in the test setup should be treated equally. Because of the volatile environmental conditions on the outside, this is not entirely possible. However, with a sufficiently large number of modules of each type (sample size) statistical correlations can be found between module type and energy yield. But to achieve good, neutral data, the physical arrangement of the modules must be well-thought-out. Figure 1 shows an optimized layout for a test array with three different module types (numbered 1-3), each of them having a sample of three modules. In this case, the modules are being positioned in a single row.

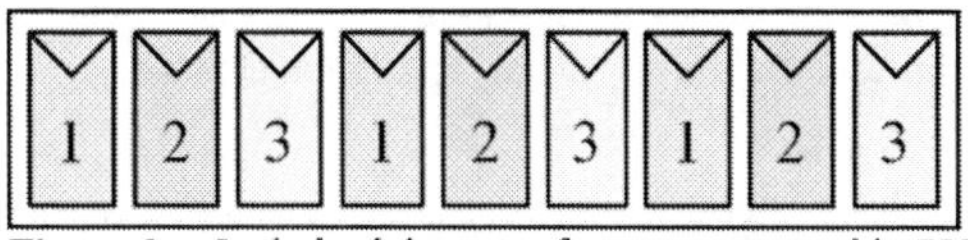

Figure 2: Optimized layout of a test array with PV modules in single row alignment

Due to the alternating placement of the modules, interference factors (partial shadings, inhomogeneous thermal characteristics, etc.) are being distributed as evenly as possible between the different types of modules. However, a sample of three specimens per module type is a bit low. Of course this layout can be extended to an arbitrary number of modules, but in most cases it will be more useful to align the modules in a matrix. In this case, the alternating placement of the modules should be performed in both columns and rows. Figure 3 shows an example of such an array.

Figure 3: Optimized layout of a test array with PV modules in matrix alignment

The PV lab of BFH runs s smart PV module test stand for 20 PV modules in single row alignment, mounted on a bike shed. Due to the low mounting height, the modules are easily accessible for manipulations and measurements. So far, we made very good experiences with this outdoor test stand, which is shown in figure 4.

Figure 4: Smart PV module test stand of BFH's PV lab

To compare the different types of smart modules, the Energy output of each individual module has to be monitored over a certain amount of time (ideally one year or longer). The energy has then to be normalized by dividing it by the module's nominal power. This way, the final yield Y_f [4] of each module can be determined. Each of these yields represents a single, independent measurement. With this data, it is possible to make a statistical analysis for each type of smart module. This can be done by calculating the expected value, the standard deviation and the confidence interval of the energy yield for each module type. The results of the different types can then be compared. A visualization of the test results, e.g. with boxplots, is also possible.

8 LIGHTNING CURRENT TESTS FOR SMART PV MODULES

Because of their large area, solar cells are quite immune against currents surges. In a standard PV module, usually the bypass diodes are the weakest parts [2, 3]. But even they are power electronic devices and as such to some degree robust. Things change when there are more sophisticated electronic devices inside the junction box, especially when there are parts based on CMOS technology. These circuits can already be damaged because of electrostatic discharges from a human body (ESD). Consequently, they can easily be destroyed by atmospheric discharges such as lightning currents. As smart PV modules may contain complex electronic circuits and are self-evidently exposed to the weather, a test of the immunity against lightning currents is reasonable. Before we are discussing a possible test setup, let's make one thing clear: An all-embracing protection against lightning currents is nearly impossible (or at least unpayable). If the lightning hits the module or the nearby DC lines directly, a damage of the module and any electronics that comes with it is almost certain. But as the average density of lightning strokes is quite low, the risk of a direct hit is small. Nearby lightning strokes however, may occasionally happen. A (smart) PV module should not be damaged by them. The test setup used by BFH's PV lab is simple: A simulated lightning current I_L (10/350 μs) is generated in a pulse current generator. The module under test is being placed in a certain distance (e.g. 20 cm) to the lightning current conductor. Because of I_L, an alternating magnetic field (B) develops around the conductor. This magnetic field induces a voltage (V_I) inside the module's cell loop that leads to a potentially destructive current surge (I_I). This test setup is shown in figure 5.

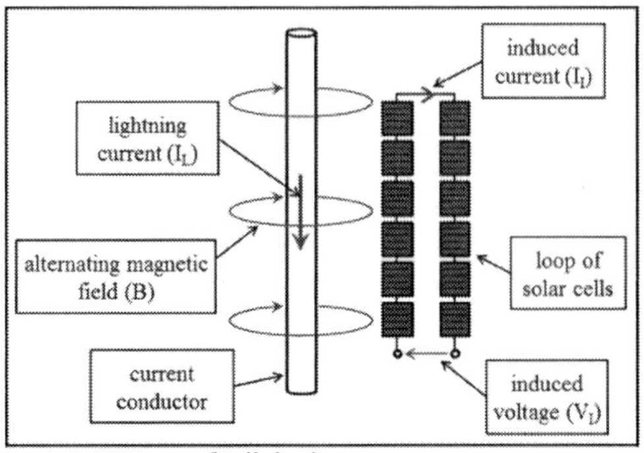

Figure 5: Setup for lightning current tests

Similar to the tests discussed under sections 6 and 7, with this setup the complete system (module and electronics) is being tested (as systems engineers we like to do that). With the pulse current generator in BFH's high voltage lab, simulated lightning currents with a peak value of up to 110 kA can be generated. The induced current surges can easily go up to a few kiloamperes. However, because of the short pulse duration (few microseconds), the induced charge is limited. Therefore, previous tests of smart PV modules (i.e. with microinverters) showed pleasantly good results, because the capacitors in parallel to the solar cells absorbed most of the current surge and by doing so protected the electronic circuits. In some cases, smart PV modules withstood higher simulated lightning currents than standard PV modules (damage of the bypass diodes). Figure 6 shows a test setup with a 60-cell PV module inside of BFH's pulse current generator. The tube on the left side is the lightning current conductor. At the bottom of the module, there is a receptacle for a microinverter, which for this test has been replaced with the measurement lines for the induced voltage and current.

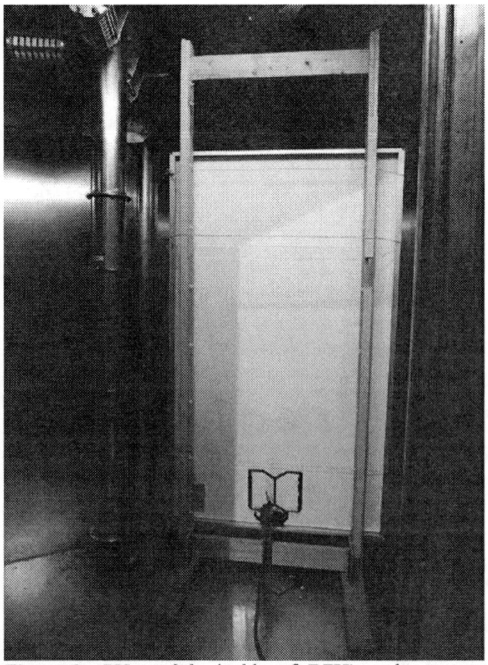

Figure 6: PV module inside of BFH's pulse current generator

9 CHOICE OF THE RIGHT TESTS FOR THE RIGHT TYPE OF SMART PV MODULE

The classification of the smart PV modules (type A to D) in this section refers to the classification scheme under section 4. The lightning current tests under section 8 are suitable (and recommended by the authors) for each of the four types.

Module type A

If the "smart" part of the module can be detached, it is possible to simply test the bare module and the electronics separately. In most cases, this allows a good estimation of the behavior of the complete system. The bare module can be flash-tested like any normal PV module. The electronics have to be tested depending on their function (e.g. tests of EN 50530 for a microinverter). However, the whole truth (that is the behavior of the complete system under real conditions) can only be tested with an outdoor test, e.g. as proposed under section 7. Tests with a continuous solar simulator as proposed under section 6 might also prove to be useful, because they also characterize the complete system, but are much less time-consuming than an outdoor measurement.

Module type B

If the DC terminals are not accessible, measurement of the module's I/V characteristic is impossible in a black-box test. This is usually the case when a microinverter or power optimizer is permanently fixed to the module (if it's a power optimizer, there is a DC terminal, but it is not directly connected to the solar cells). These modules are best being characterized with a continuous solar simulator (section 6) and an outdoor performance test (section 7).

Module type C

This type of module is an exception, because in a black-box test it actually is impossible to tell how the (unknown) electronics will or will not compromise with the measurement. It is however possible that the manufacturer of the smart module reveals the module's inner life or specifies that this inner life does not or only negligibly influence the electrical data of the solar cells. Active bypass diodes can fall under this category. Such a module can be tested like any conventional PV module (i.e. flash testing).

Module type D

If the electronics do compromise with the measurement or it is unknown if they do so, measuring the I/V characteristic will be impossible in most cases. The only exception is when the behavior of the electronics is known with sufficient accuracy and their influence can be compensated. For instance, if the only influence is caused by (small) capacitors in parallel to the solar cells, the current measurement error in flash tests can be corrected mathematically (see section 5). However, in a black-box test, the required information will usually not be available or not with the accuracy required (e.g. if there is an electrolytic capacitor in parallel to the solar cells, it may have a tolerance of +/-20%). Therefore, these modules can only be characterized as complete systems under real-life conditions, namely with a continuous solar simulator (section 6) and an outdoor performance test (section 7).

10 CONCLUSIONS

Smart PV modules can perform many tasks. However, in most cases, smart PV modules cannot be tested like normal modules. If and how they can be tested depends on the configuration and the function of the module. In this paper, a simple classification scheme is presented. Especially the well-known flash tests cannot be performed with most smart PV modules. As an alternative, I/V curve measurement with a continuous solar simulator is possible in some cases. Such a simulator can also be used for entire test profiles for smart PV modules with a microinverter or power optimizer. A very good characterization can be achieved with an outdoor performance test. A possible test setup is proposed in this paper. Also, the authors think that the immunity against nearby lightning currents should be tested, because the compelex electronics inside of smart modules might be sensitive to surge currents.

DISCLAIMER

ACKNOWLEDGEMENTS

Thank goes to all the institutions that gave us financial support, particularly the Swiss Federal Office of Energy (SFOE), the Swiss Competence Centers for Energy Research (SCCER) and the Building Insurance Berne (GVB).

REFERENCES

[1] R. Bruendlinger, N. Henze, H. Haeberlin, B. Burger, A. Bergmann, F. Baumgartner, "prEN 50530 – The new European Standard for Performance Characterisation of PV Inverters", 24th European Photovoltaic Solar Energy Conference and Exhibition, Hamburg, Germany, 2009.

[2] H. Haeberlin, M. Kämpfer, "Measurements of Damages at Bypass Diodes by Induced Voltages and Currents in PV Modules Caused by Nearby Lightning Currents with Standard Waveform", 23rd European Photovoltaic Solar Energy Conference and Exhibition, Valencia, Spain, 2008

[3] H. Haeberlin, "Damages at Bypass Diodes by Induced Voltages and Currents in PV Modules Caused by Nearby Lightning Currents", 22nd European Photovoltaic Solar Energy Conference and Exhibition, Milano, Italy, 2007

[4] H. Haeberlin, C. Beutler,"Normalized Representation of Energy and Power for Analysis of Performance and On-line Error Detection in PV-Systems", 13th European Conference on Photovoltaic Solar Energy Conversion, Nice, France, 1995

All papers can be downloaded under www.pvtest.ch.

WEIGHTED EFFICIENCY OF SPV POWER CONVERTERS/INVERTERS IN INDIAN COMPOSITE CLIMATE

Kamlesh Yadav[AB], O.S. Sastry[A], B.Bora[A], M.kumar[A], R.Singh[A], R.Parmar[A], Atul Kumar[B], B.prasad[B]
[A]National Institute of Solar Energy (NISE), Gurgaon, India
[B]The Energy and Resources Institute, New Delhi, India
*sastry284@gmail.com, Tel: +91-0124-2579213, Telefax No. : +91-124-2579207

ABSTRACT: Indian climate is broadly divided into 4 climatic zones, Hot & dry, Hot & humid, Composite and cold & Cloudy. Most of the area comes under composite climate [1]. Further the performance of SPV power plants depends on the site, PV Module technology and load wise inverters/power converters efficiency. So with the change of the technology and site, weighted efficiency of the inverter also varies. This paper coins the new efficiency term as Indian© efficiency based on irradiation distribution similar as European and Californian efficiency which is suitable to Indian climatic condition. To showing technology and site dependency the study also compares the Indian© weighted efficiency of a sample inverter with weighted efficiency using the generation energy weighted factors of a-si, multi-si, HIT and CdTe technology module. It is found that same inverter performs best with CdTe Modules in the composite climate zone of India.
Keywords: SPV, European Efficiency; Californian Efficiency; Indian efficiency, weighted factor; composite climate.

1 INTRODUCTION

The total electricity generation capacity installed as on March 2015 in India is 263.66 GW, out of it the renewable energy installed is 34.5 GW, which is about 13% of the total installed capacity [2]. The increase in solar cell conversion efficiency and improvements in the long-term performance reliability of PV module are some of the prime factors driving solar PV technologies. The major impact of these improvements resulted in setting up of Giga-watt (GW) size solar power parks in different states of the union. These solar parks feed power into high voltage grid. It has also resulted in multi fold increase in number of small capacity Off-grid system for various applications. Thus, SPV technologies have become boon to the energy starved states and especially to the island regions of the country, which hitherto relied extensively on diesel powered generators for day to day requirements of power. Realizing the abundance potential of solar energy as a resource, there is a very ambitious program for installation of 100 GWp solar power plants by 2022 in India [2]. To achieve such an ambitious goal in realistic manner, it is imperative to monitor the SPV products by strict enforcement of quality standards and installation practices suited to Indian condition which differs vastly from that of the nations of Europe.

Performances of Photovoltaic technologies depend upon radiation, air-mass, temperature, humidity and wind speed; which vary from place to place. Each solar photovoltaic technology has different spectral response [3-5]. Hence there is need to choose appropriate SPV technologies for different places. Module temperature which is also site specific is another major contributing factor affecting the performance of PV Technology. With rise in module temperature, voltage and power decreases, current will have little gain but overall power decreases with increase in module temperature [6-7]. One of the main components in solar PV system is its maximum power tracking. MPP tracking is more challenging in case of varying climatic conditions and partial shading [6]. The estimation of the payback time and the profitability of grid tied photovoltaic (PV) systems require information about performance of PV modules and inverters. For this reason, the strong attention should be given on the efficiency. Further this paper focuses on the inverter efficiency. In European and American country there is concept of European and Californian efficiency for the inverters on which the inverter will operate in field condition.

European efficiency is calculated for more than five different loads with different weighted factor [8]. The main reason of taking weighted factor is to relate the influence of the irradiation distribution depending on location.

$$\eta_{euro}=0.03\times\eta_{5\%}+0.06\times\eta_{10\%}+0.13\times\eta_{20\%}+0.1\times\eta_{30\%}+0.48\times\eta_{50\%}+0.2\times\eta_{100\%}$$

CEC efficiency is also a weighted inverter efficiency the difference between Euro and CEC efficiency is weighted of different input level. Efficiency defined by the Californian Energy Commission (CEC) [8]:

$$\eta_{CEC}=0.04\times\eta_{10\%}+0.05\times\eta_{20\%}+0.12\times\eta_{30\%}+0.21\times\eta_{50\%}+0.53\times\eta_{75\%}+0.05\times\eta_{100\%}$$

European efficiency and CEC efficiency can provide a better idea about the inverters operating profile throughout a day in the field condition. But it is not principled for Indian Climatic condition as India has different intensity and ambient temp reason throughout the country. Here main aim of the study is to define a new inverter efficiency (as Euro and CEC) named as Indian© efficiency for composite climatic zone. The paper also presents a detailed comparison of the generated weighted factor for four different technology and incidence energy weighted factor of the site. Further comparison is done with weighted efficiency; it is highest with CdTe module 93.2% and lowest with the m-si modules 91.8%.

2 EXPERIMENTAL PROCEDURE

In order to study Comparative analysis of efficiency of three different types of technology, PV array test bed Amorphous silicon, HIT and multi crystalline silicon PV modules with capacity 1.2, 1.68 and 1.6 kWp respectively were installed in the year 2009 at the premises of NISE as a joint project between NISE and AIST, Japan. The modules are fixed at tilt of 28°, facing towards south. The performance data of each technology along with the environmental parameters are recorded in a certain interval of time. The 3.2 kWp test bed of CdTe technology based PV modules was installed at a fix tilt (20°), by M/s First Solar, USA. A pyrometer (Make: Kipp & Zonen, CMP11, error <10 W/m^2) and T-type thermocouples are used to record individual PV module temperature.

A year's data of radiation has been analyzed for NISE by using contour map for all four technologies. The irradiation has been plotted as function of intensity and ambient temperature. We made two by two matrices of falling index of incidence energy by intensity and ambient temperature for a particular range of intensity and temperature. A year's data has been summed up for the incidence energy.

Further A sample inverter sample efficiency data has been used. The efficiency of the inverter is taken at different load and voltage in the power electronics lab with Solar array simulator make: Terra SAS (ETS600X), 15 KVA capacities and power measurements, are done with a power analyzer make: Zimmer (LMG-450).

3 RESULT AND ANALYSIS

Major part (37.14 %) of energy falls in the range of 700-800 W/m^2 at NISE as can be seen from the figure 1. It has been found from the radiation distribution with respect to ambient temp at NISE that 23.23 % of energy falls in the range of 30-35 ^{0}C. Weighted factor of irradiation is presented in table 1. Based upon the weighted factor the proposed inverter efficiency for Indian composite climate is:

$$\eta_{Indian©}=0.248\times\eta_{10\%}+0.121\times\eta_{20\%}+0.092\times\eta_{30\%}+0.159\times\eta_{50\%}+0.215\times\eta_{75\%}+0.162\times\eta_{100\%}+0.003\times\eta_{120\%}$$

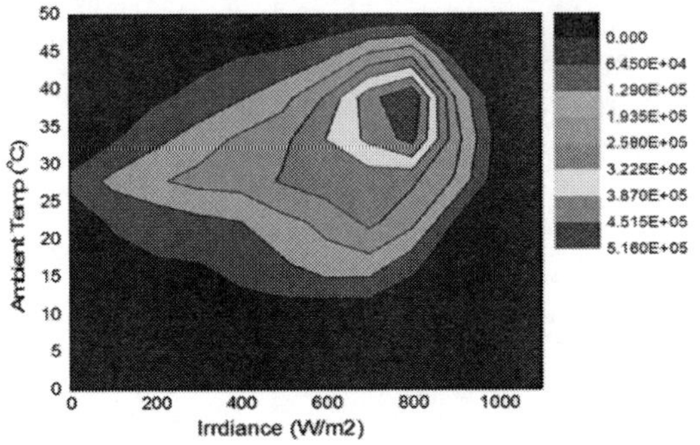

Figure 1: Energy yield distribution with respect to irradiance and ambient temp

In figure 2, it is shown that for a-Si the maximum yield is in the intensity range of 750-800 W/m^2 at module temp of 55 ^{0}C, for HIT it is 800 W/m^2 at module temp of 55 to 60 ^{0}C, for m-Si it is 700-800 W/m^2 at module temp of 55 ^{0}C and for CdTe it is 800 W/m2 and module temp

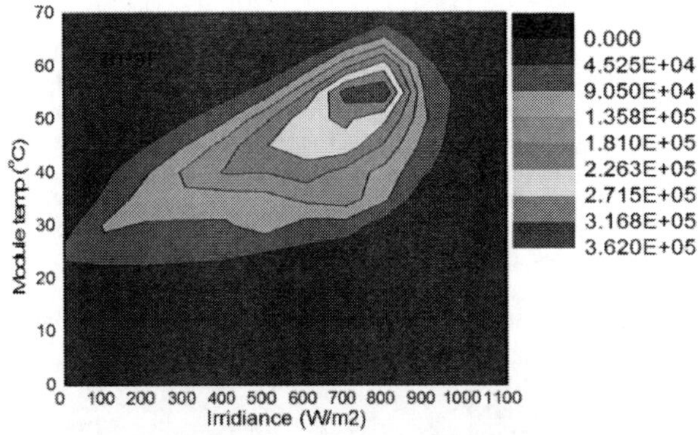

in the range of 55-60 ^{0}C. Corresponding weighted factors

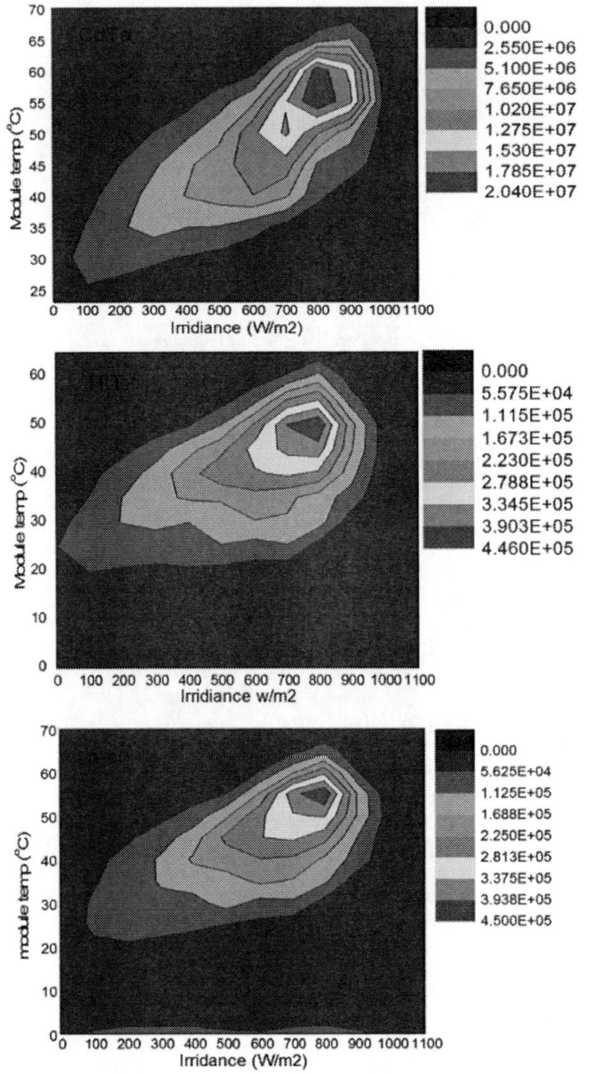

are given in table 1.

Figure 3: Technology wise yield distribution with respect to irradiance and Module temperature

Further a string inverter sample of 3 KVA is tested at different load at three different voltage: 150, 360 and 450V. On 360 V this sample inverter have lowest efficiency 89.4 at 10% load and peak efficiency 95.2 at 100% load. Here again weighted efficiency of the sample is calculated which gives the actual inverter field efficiency over hole day. The highest weighted efficiency is coming with CdTe modules and lowest with m-si modules.

Table 1 Different Weighted Factors and weighted efficiency

Load %	Sample inverter Efficiency at different loads	Weighted factor				
		a-Si	m-Si	HIT	CdTe	India ©
10%	89.4	0.385	0.48	0.359	0.231	0.248
20%	92.01	0.103	0.086	0.1	0.105	0.121
30%	93.6	0.083	0.072	0.082	0.168	0.092
50%	94.8	0.146	0.127	0.142	0.141	0.159
75%	95.1	0.191	0.19	0.202	0.273	0.215
100%	95.2	0.091	0.044	0.114	0.082	0.162
120%	94.8					
Weighted eff		**92.3**	**91.8**	**92.5**	**93.2**	**92.8**

4 CONCLUSION

This study proposes India© efficiency:

$$\eta_{India©}=0.248\times\eta10\%+0.121\times\eta20\%+0.092\times\eta30\%+0.159\times\eta50\%+0.215\times\eta75\%+0.162\times\eta100\%+0.003\times\eta120\%$$

Further it derived that best inverter efficiency can be achieved with the CdTe modules in Indian composite climate. For selected sample inverter peak inverter efficiency is 93.2% with CdTe and lowest 91.8 with m-si. Hence the study clearly establishes that the performance of SPV based power converters/inverters is also dependent on site and selected PV module technology.

ACKNOWLEDGEMENT

This research is based upon work supported in part by the Solar Energy Research Institute for India and the U.S. (SERIIUS) funded jointly by the U.S. Department of Energy subcontract DE AC36-08G028308 (Office of Science, Office of Basic Energy Sciences, and Energy Efficiency and Renewable Energy, Solar Energy Technology Program, with support from the Office of International Affairs) and the Government of India subcontract IUSSTF/JCERDC-SERIIUS/2012 dated 22nd Nov. 2012

REFERENCES

[1] http://mnre.gov.in/solar-energy/ch2.pdf

[2] http://cseindia.org/docs/photogallery/ifs/Renewable%20Energy%20in%20India%20Growth%20and%20Targets.pdf

[3] Effect of Seasonal Spectral Variations on Performance of Three Different Photovoltaic Technologies in India, D. B. Magare, O.S. Sastry, R. Gupta, T.R. Betts, R. Gottschalg, A. Kumar, B. Bora, and Y.K. Singh, International Journal of Energy and Environmental Engineering (IJEE) Performance modeling of three PV module technologies based on clearness index and air-mass using contour map, Birinchi Bora, O.S. Sastry, Arun Kumar, Renu and Manander Bangar, 42nd IEEE PVSC].Spectrum changes with the seasonal variation

[4] Analysis of Spectrum Variations in different Seasons in Composite Climate zone of India, Manander Bangar, Birinchi Bora, Renu, Arun Kumar and O.S. Sastry, 42nd IEEE PVSC

[5] Effect of module temperature and seasonal variation analysis of the outdoor performance of CdTe at composite climate of India, Birinchi Bora, O.S. Sastry, Yogesh Kumar Singh, Ramayan Singh, Kamlesh Yadav, Vikrant Sharma, Manander Bangar, Mithilesh Kumar, Invertis Journal of Renewable Energy, Vol. 5, No. 3, 2015

[6] Comparison of Photovoltaic Array Maximum Power Point Tracking Techniques, Trishan Esram, Student Member, IEEE, and Patrick L. Chapman, Senior Member, IEEE, IEEE TRANSACTIONS ON ENERGY CONVERSION, VOL. 22, NO. 2, JUNE 2007

[7] O. Wasynczuk, "Dynamic behavior of a class of photovoltaic power systems,"IEEE Trans. Power App. Syst., vol. 102, no. 9, pp. 3031–3037,Sep. 1983.

[8] Sandia_Guideline_2005.pdf document, p17.

TESTING OF MULTI-MPPT PV INVERTERS: APPROACH AND TEST RESULTS

Daniel Gfeller, Urs Muntwyler, Luciano Borgna
Berne University of Applied Sciences (BFH), Engineering and Information Technology
Photovoltaic Laboratory (PV-Lab), Jlcoweg 1, CH-3400 Burgdorf, Switzerland
Phone +41 (0)34 426 68 11, Fax +41 (0)34 426 68 63, urs.muntwyler@bfh.ch, www.pvtest.ch

ABSTRACT: The European standard EN 50530 defines procedures for measuring the conversion and MPPT efficiency of PV inverters. The standard has been released in 2010 when multi-MPPT PV inverters were not yet widely-used. Therefore, the scope of EN 50530 is limited to PV inverters with only one MPP tracker. Today however, multi-MPPT inverters have become a market standard. The question is now what tests are necessary to obtain a good characterization for these devices. The easiest approach would be to simply use the test profiles defined in EN 50530 on each of the inverter's inputs simultaneously. But by doing so, one would disregard the main purpose of the multi-MPPT technology: to achieve a good performance under inhomogeneous conditions. Therefore, measurements with different test profiles on each MPP tracker should be performed as well. This work proposes some modifications for existing test standards and new methods for testing of multi-MPPT PV inverters. It also presents actual measured data of real inverters, which have been tested on BFH's multi-MPPT inverter test bench.
Keywords: Inverter, Qualification and Testing, MPPT Efficiency

1 INTRODUCTION

Even though the PV inverter is one of the most important parts of a PV power plant, proper testing and characterizing of such devices has been neglected for a long time. In the past, the PV inverter's only quality feature that has been paid some attention to was the conversion efficiency. Other characteristics, namely the MPP tracking efficiency, have either been neglected or assumed to be ideal – an assessment that in many cases is far too optimistic. Because of this, many flaws in such devices have not been localized and have caused a considerable loss in energy yield. This was at a time, when the price per watt of a PV array was a multiple of what we pay today. Only in 2010, the European standard EN 50530 has been released [1]. This standard defines test procedures for the overall efficiency of grid connected PV inverters including conversion and MPPT efficiency with both static and dynamic test profiles. When EN 50530 was first released, multi-MPPT PV inverters were not yet very popular. Consequently, the scope of this standard does not include multi-MPPT inverters. Today however, many modern PV inverters have at least two MPP trackers. Technically, multi-MPPT inverters are now in the same position as single-MPPT inverters have been before the release of EN 50530. Of course it is possible to adapt the existing normative test profiles and use them on each MPP tracker simultaneously (which actually is the current standard procedure). But this allows only a partial characterization of the inverter. The main reason for the multi-MPPT technology is to achieve a good performance under inhomogeneous conditions (e.g. in a plant with multiple module orientations, different number or modules per string or partial shading). This very feature cannot be tested with the existing normative test procedures.

2 PV INVETER TESTS AT BFH'S PV-LAB

The PV-Lab of BFH is one of the first and most experienced testing centers for PV inverters in Europe. Already in 1994, first tests on grid connected PV inverters were performed. Compared to the devices we have today, these early PV inverters were downright primitive. In these first years, PV inverter tests at BFH have been performed with an on-side PV array of 60kWp. However, it soon became clear that a good qualification of a PV inverter (especially of the MPPT performance) is only possible under highly stable and reproducible test conditions. With a real PV array, this is not possible. Therefore, since 1999 tests have been performed with PV array simulators. After having made bad experiences with a commercially available simulator, BFH's PV-Lab began developing its own PV array simulators. Today, the PV-Lab has a well-equipped inverter test stand with two single-string simulators (20 kW & 100 kW) and one multistring simulator (3 x 11.5 kW). The accreditation of the test stand is underway [2, 3].

3 EXISTING TEST PROCEDURES

EN 50530 defines test profiles for both static and dynamic conversion and MPPT efficiency measurements. The aim of the static measurements is to determine the inverters conversion and tracking performance under steady-state conditions. For this, the inverter is being measured while operating at different voltage and power levels. The essence of these tests is the European efficiency (η_{EU}) which is a weighted average of the individual measurements and is a good approximation for the device's average performance in the central European climate. In the same manner but with different weighting factors, the CEC efficiency (η_{CEC}; California energy commission) is calculated. These measurements are performed at three voltage levels (namely at the minimum, the rated and the maximum MPP voltage). The conversion efficiency of many PV inverters has a high dependency on the device's operating voltage. Sadly, many manufacturers specify the European efficiency at the optimum MPP voltage only – and often they do generously round up the value (+0.2% or so are standard).

The aim of the dynamic measurements is to show how good the inverter's MPP tracking algorithm can adapt the device's operating point to a non-static maximum power point. For this, the inverter is being measured while operating with test profiles with a time-variant simulated irradiance (the cell temperature is assumed to be constant). The simulated irradiance describes linear ramps between different irradiance levels and with different slopes. The goal of this is to simulate both slow and fast variations in irradiation. Tests at BFH's PV lab show that even modern PV inverters occasionally have problems following the MPP at several slopes, leading to a loss in tracking efficiency.

4 EXISTING PROBLEMS AT MEASURING THE EUROPEAN AND CEC EFFICIENCY AND PROPOSED SOLUTION

If the MPP power of the simulated PV array is higher than the inverter's rated power, the inverter usually reduces its input power by moving the operating point out of the MPP. As the operating point is then no longer in the MPP, the MPPT efficiency becomes very poor. In the measurements required for the calculation of η_{EU} and η_{CEC}, one measurement must be performed at the inverter's rated power. If for some reason the MPP power of the simulated PV array is slightly higher than the inverter's rated power, the device will limit its input power. This has a negative impact on the MPP tracking efficiency. This can easily happen if for an instance the PV array simulator cannot be programmed with the accuracy required. If the inverter's manufacturer specifies the rated AC power only (this is very common) the rated DC power must be estimated using the specified conversion efficiency. In such cases the MPP power of the PV array simulator can easily be 1-2% too high. This will lead to a very low MPP tracking efficiency and an unfair test result. To prevent this systematic error, the highest power level in the measurements required for η_{EU} and η_{CEC} should not be at 100% of the inverters rated power, but rather at 95%. This modification would only have a minimal effect on η_{EU} and η_{CEC} under normal conditions. A similar problem can occur because of the MPP voltage used for the tests. Some of the measurements have to be performed at the inverter's minimum or maximum MPP voltage. Because it is not a good practice to characterize a device at its limits, these measurements should be performed at e.g. 105% of the minimum and 95% of the maximum MPP voltage. Moreover, if the the inverter's maximum MPP voltage is too close to its maximum DC voltage, technically the test cannot be performed. According to EN 50530, the simulated PV array has a ratio of 0.8 between the MPP voltage and the open circuit voltage. If the inverter's maximum DC voltage is not at least 25% higher than its maximum MPP voltage and the MPP of the simulated PV array is at the inverter's maximum MPP voltage, the open circuit voltage of the PV array is higher than the inverter's maximum DC voltage. This might damage the inverter. The author's proposition in such cases is to choose the maximum MPP voltage for the static measurements no higher than 75% of the inverter's maximum DC voltage. Frankly speaking, no reasonably designed PV plant should have its nominal MPP voltage at more than about 75% of the inverter's maximum DC voltage, because the inverter should also survive the array's open circuit voltage at cold temperatures. Unfortunately, many PV inverters have a specified ratio between maximum MPP voltage and maximum DC voltage that is much higher. Some manufacturers even specify these two voltages to be identical (which is nonsense). This might mislead overeager plant designers to design the PV array with voltages too high for the inverter in use [4].

5 MISSING TEST PROCEDURES IN EXISTING STANDARDS AND PROPOSAL FOR A NEW TEST WITH A PARTIALLY SHADED PV ARRAY

The existing test procedures described in EN 50530 allow a good characterization of a single-MPPT PV inverter as long as it is connected to a completely unshaded PV array. Such an array has a P/V characteristic that is a continuous curve – just like the normative test curves in EN 50530. However, when a PV array is partially shaded, the array's P/V characteristic can have one or more local maxima aside from the actual MPP. In this situation it can happen that the inverter's operating point is stuck in one of these false peaks. It is even possible that due to changes in the partial shading, the MPP can change from one local maximum to another. Some inverters have problems to follow these changes of the MPP's position. They keep their operating point in the formerly correct but now false maximum. In such cases, considerable tracking losses (>>10%) for time periods of up to several hours are possible. Despite these potentially severe losses, this scenario is not covered by the existing test procedures in EN 50530. Here, a new test is proposed. This test is based on a simulated PV array with a fill factor of 72% (model according to EN 50530). Open circuit voltage and short circuit current of the array are set so that the MPP (under STC conditions) is at the inverters rated voltage and DC power. At the beginning of the test, the entire PV array is irradiated with 160 W/m^2. This is a realistic value for shaded PV modules on a clear day. After a settling time, which allows the inverter to start up and find the MPP, the irradiation on 80% of the modules (which are in series to the remaining 20%) begins to rise with a slope of 0.4 W/m^2 per second. After 1'600 seconds the irradiation on this (now unshaded) part of the array reaches 800 W/m^2. Then the test is complete. After 45 seconds, when the simulation on the unshaded part of the array reaches 178 W/m^2, a second local maximum forms on the P/V curve. When the simulated irradiation on the unshaded part of the array is about 231 W/m^2 – that's 178 seconds after beginning of the slope – the MPP makes a jump from the original maximum to the new one. A good MPP tracker should be able to quickly change the inverter's operating point to the new MPP, whose voltage is about 25% lower than the former MPP's voltage. After the jump of the MPP, the irradiation on the unshaded part of the array continues to rise for 1422 seconds, which is somewhat more than 23 minutes. That is enough time for a good MPP tracker to find the new MPP. However, an MPP tracker that remains in the false local maximum will harvest less than 48% of what would be possible. Figure 1 shows some P/V characteristics of the proposed test array. The values are normalized to the open circuit voltage and short circuit current at STC. Due to this normalization, if the array was under STC, short circuit current and open circuit voltage would be one and the MPP power would be equal to the fill factor. Three curves are plotted. The dark blue curve is the P/V characteristic at a homogeneous irradiation of 160 W/m². This curve is continuous and has only one maximum. That is the situation during the settling time and at the beginning of the test. The red curve represents the characteristic when the irradiance on the unshaded part of the array is 231 W/m². Here, both maxima have the same value. This is the moment when the MPP jumps from the right to the left peak. The purple curve shows the characteristic at the end of the test, when the irradiation on the unshaded part of the array is 800 W/m². Pay attention to the power difference of the two maxima. An inverter with its operating point stuck in the false peak would lose about 69% of power in this final situation.

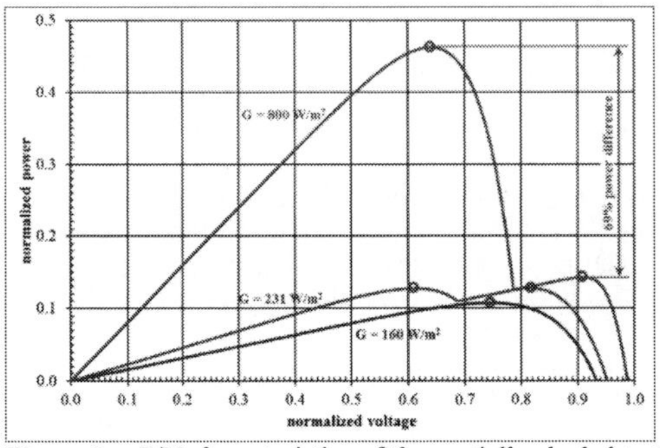

Figure 1: P/V characteristics of the partially shaded test array with different irradiations

Figure 2 shows power and voltage of the two maxima of the P/V curve as a function of the time. The values are normalized in the same manner as in Figure 1. The solid lines represent voltage and power of the actual MPP. The dashed lines represent voltage and power of the other peak. Remarkable is the drop of the MPP voltage at 178 seconds, when the MPP jumps from one peak to the other. Finding the actual MPP again is a demanding task for the device under test.

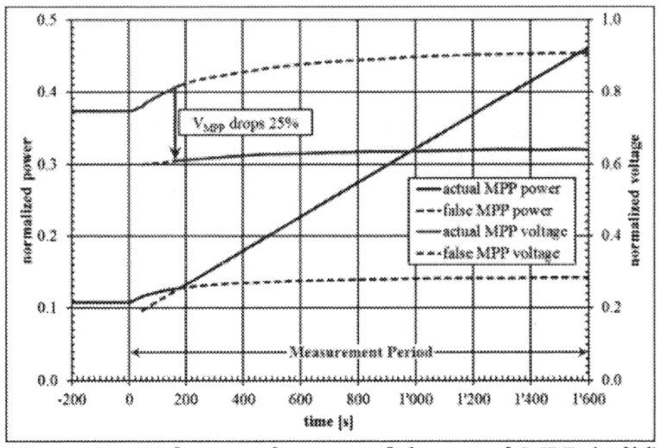

Figure 2: Voltage and power of the actual MPP (solid) and the false local maximum (dashed)

One might criticize that this test is not realistic, because a linear ramp of irradiance on the unshaded part of the PV array does not occur in reality (especially when the irradiation in the shade is constant). In reality, a shadow usually moves over the array. So it's rather the number of modules that is shaded and not a variation of the irradiance that influences the resulting P/V characteristics. However, to keep the test simple but fair, a compromise must be made (this is also done in some of the tests in EN 50530 and many other standards). A major problem of scenarios with moving shadows on a PV array is that the MPP voltage can fall to very small values (e.g. 10-20% of the MPP voltage at STC). This is outside of the tracking window of most PV inverters. Consequently, the device under test has no chance to properly track the MPP and achieve a good test result. Such a test is unfair and the result is not representative for the device under test. Of course, this is exactly what can happen in reality. But then the error rather lies in the design of the PV array and is not a flaw of the inverter (granted, modern PV inverters are good, but you cannot expect them to perform "magic"). It is the belief of the authors that this test is close enough to the reality, but still strict enough to unveil problems in MPPT algorithms. It is a chance to localize one of the last major sources of errors in PV systems with both single- and multi-MPPT PV inverters.

6 EQUIPMENT NEEDED FOR TESTING OF MULTI-MPPT PV INVERTERS

Compared to single-MPPT PV inverters, much more equipment is needed for testing multi-MPPT PV inverters. Two main components are needed to determine the MPPT performance: A high precision power analyzer and a highly stable PV array simulator. Measuring the MPPT efficiency means to compare how much energy a PV inverter draws from a PV array (or simulator) in a certain amount of time to the amount of energy that would have been available in said amount of time. The latter is the integral of the MPP power over the measurement period. Because of this, the MPP power of the simulated PV array must be known exactly at any time of the test. Consequently, accuracy and stability (mainly the thermal drift) of the PV array simulator set a limit to the uncertainty of the measurement. Therefore, a highly stable PV array simulator is essential for this measurement (drift < 0.1%). However, such a high stability is a very strict requirement for a power electronic device like a PV array simulator. For testing multi-MPPT PV inverters, one such simulator is required on each MPP tracker. Moreover, each of these inputs must be measured with a separate channel of a precision power analyzer. As most multi-MPPT PV inverters are three-phase devices, three more power measurement channels are needed for measuring the AC power. The measurement periods of all these devices must be synchronized exactly.

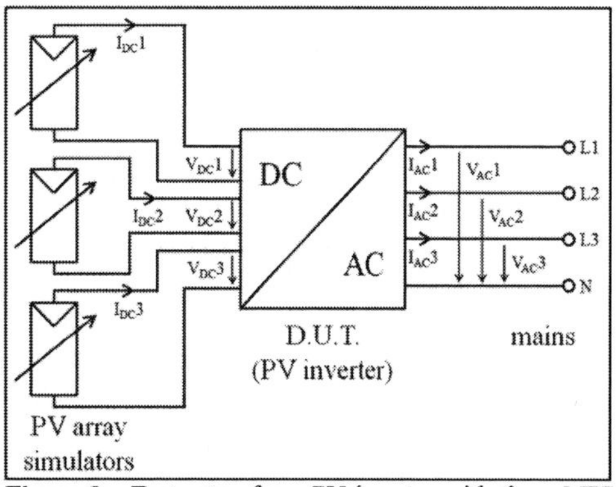

Figure 3: Test setup for a PV inverter with three MPP trackers and a three-phase output

Figure 3 shows an exemplary test circuit for a three-phase PV inverter with three MPP trackers. In total, six currents and six voltages must be measured. Thus, six power analyzer channels are required. If the inverter had only two MPP trackers, it would still take five channels to perform the measurement. It is essential that the cable losses are compensated. For this, all voltages must be measured directly at the inverter's terminals. However, the cable losses also have a slight impact on the MPP's position. Therefore, the calibration of the PV array simulators must be performed at the end of the DC cables. Ideally, this is done with the same power analyzer that is used in the actual measurement. By this, any bias of the power analyzer is being compensated.

7 PROPOSED NEW PROCEDURE FOR TESTING OF MULTI-MPPT PV INVERTERS

To show the inverter's performance under inhomogeneous conditions, a new dynamic test profile is proposed. In this profile, the simulated irradiance follows linear ramps of 150 seconds duration between 100 and 800 W/(m^2*s), but the slopes on the different MPP trackers are time staggered. Again, the simulated PV arrays are modeled according to EN 50530 with a fill factor of 72%. The simulated curves are scaled so that the MPP at STC would be at the rated voltage and power of the corresponding MPP trackers. The cell temperature is assumed to be constant at 25°C. Figures 4, 5 and 6 show these test profiles for inverters with two, three and four MPP trackers.

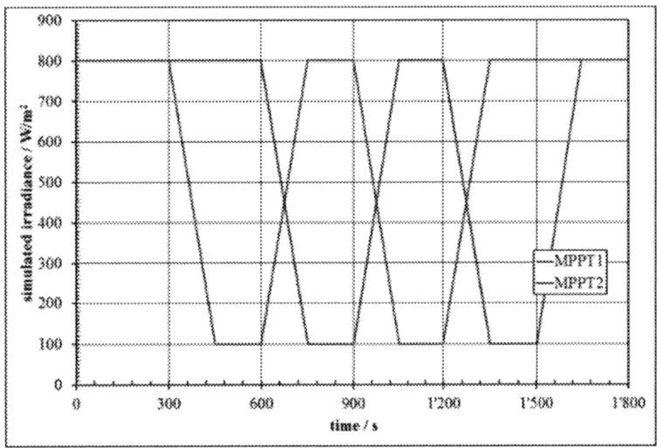

Figure 4: Test profile for two MPP trackers

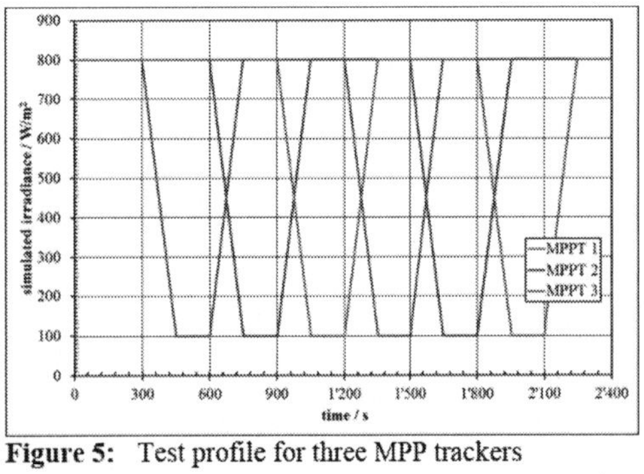

Figure 5: Test profile for three MPP trackers

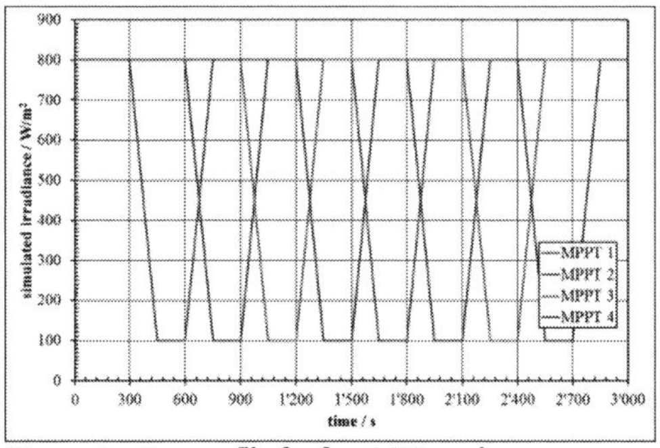

Figure 6: Test profile for four MPP trackers

If the inverter has more than four MPP trackers, the test profile of figure 6 is being used with the fifth MPP tracker having the same profile as the first, the sixth MPP tracker having the same profile as the second and so on. The test can therefore be performed with an arbitrary number of MPP trackers. Indeed, there are no PV inverters with more than four MPP trackers on the market right now (at least none that the authors knew about). Perhaps in future, there will be multi-MPPT central inverters with a dozen or so MPP trackers. With the profiles proposed here, the tests would be prepared for any such development (of course, testing of such devices would require a lot of test equipment). As the size of the simulated PV array is defined in relation to the power of the MPP tracker, devices with MPP trackers of different sizes also would not be a problem. These tests are quite simple and short, but they still allow a good characterization of the inverter under conditions, when one MPP tracker runs with a power much lower than the others.

8 SUGGESTED MODIFICATIONS IN FUTURE STANDARDS FOR TESTING OF PV INVETERS

Sooner or later, the standards for testing the overall efficiency of PV inverters (i.e. EN 50530) must be upgraded, so that their scope includes multi-MPPT inverters (note of the authors: sooner would be better than later). In this revision, the authors propose that the test procedures should be set as follows:

- Measuring the static total efficiency [5] with different voltage and power levels. The test can be adopted from EN 50530. The parameters should be adjusted according to section 4 of this paper. For multi-MPPT inverters, the tests should be performed on each MPP tracker simultaneously.
- Measuring the dynamic total efficiency. These tests can be adopted from EN 50530. For multi-MPPT inverters, the tests should be performed on each MPP tracker simultaneously. Further changes are not required.
- Measuring the total efficiency with a partially shaded PV array (see section 5). For multi-MPPT inverters, the tests should be performed on each MPP tracker simultaneously. This test is new.
- Multi-MPPT inverters only: Measuring the dynamic total efficiency under inhomogeneous input conditions (see section 7). This test is new.

9 MEASURED DATA OF ACTUAL MULTI-MPPT PV INVERTERS

Figure 7 shows the static efficiency of a PV inverter (inverter A) with three MPP trackers and a rated AC power of 15 kW. The device is manufactured in Europe. The manufacturer specifies a European efficiency of 97.5%. From this specifications, a maximum DC power of 15 kW / 0.975 = 15.385 kW can be assumed. This corresponds to 100% on the horizontal axis.

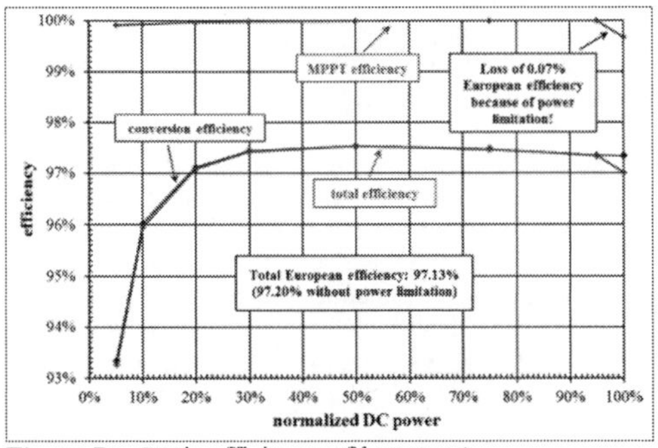

Figure 7: Static efficiency of inverter A

As you can see, the MPPT efficiency is virtually 100%. Thus, the total efficiency [5] is nearly identical to the conversion efficiency. However, in this measurement, the problem mentioned under section 4 occurred. The DC power of 15.385 kW is slightly too high for this device. Because of this, the inverter moves its operating point out of the MPP. The AC power is then limited to about 14.817 kW which is even 1.22% below the device's power rating. As a result, the MPPT efficiency (red line) drops 0.33% between 95% and 100% DC power. The total European efficiency is 97.13%. If the power limiting problem didn't occur, it would be 0.07% higher. This could easily be achieved, if for the calculation of the European efficiency the measurement point at 95% instead of 100% DC power would be used (as it is proposed under section 4). The conversion efficiency would not measurably be affected by this modification.

Figure 8 shows the same chart of inverter B. This device is manufactured in China. It has a rated AC power of 23 kW and three MPP trackers. The manufacturer also specifies a maximum usable DC power of 23.6 kW (which is 100% on the horizontal axis).

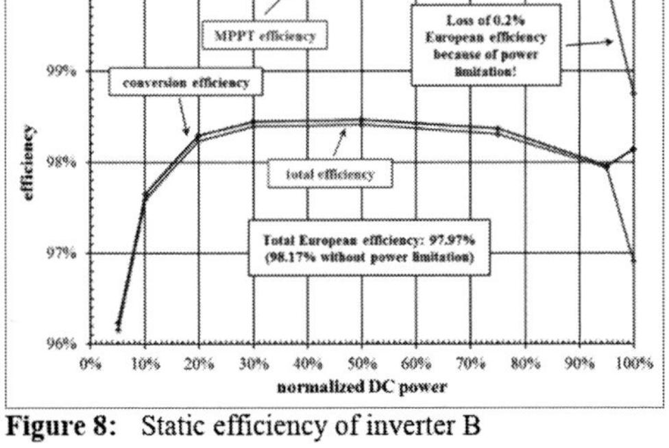

Figure 8: Static efficiency of inverter B

In this case, the loss of MPPT efficiency because of the power limitation is even higher. Even though the manufacturer explicitly specifies a DC power of 23.6 kW, the device's operating point is more than 290 W lower. Because of this, inverter B loses about 0.2% of European efficiency. Still, the measured European efficiency of 97.97% is very good. Here also, the use of the measurement point at 95% of the rated power for calculating of the European efficiency would allow a better characterization of the device. Remarkable is the rise of the conversion efficiency between 95% and 100% of normalized power. The reason for this is probably that the inverter increases the operating voltage on all three MPP trackers by about 35V to reduce the input power. At this operating point, the conversion efficiency is about 0.18% higher compared to the previous measurement point. However, this gain is far too low to compensate the MPP tracking losses caused by the power limitation. All in all, this leads to a drop of the total efficiency of more than 1% between 95% and 100% of the rated power.

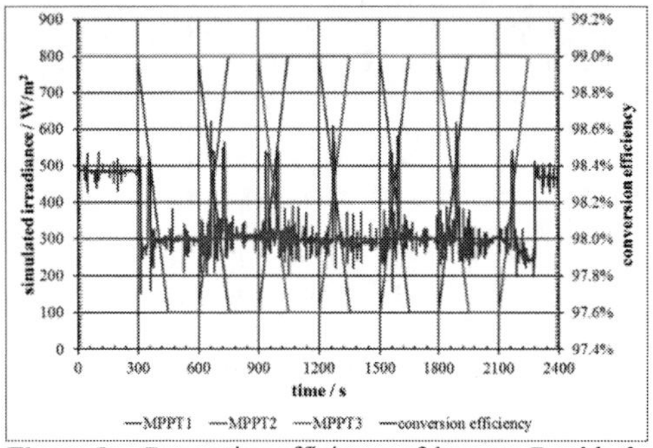

Figure 9: Conversion efficiency of inverter B with the proposed new test profile

Figure 9 shows the results of the new test proposed under section 7, performed with inverter B. As you can see, as soon as the input conditions become inhomogeneous, the conversion efficiency (purple line) drops about 0.4%. A further investigation of this problem showed that under homogeneous input conditions, all three DC inputs have the exact same voltage (except for the measuring noise). We assume that in this case, the inverter's input stages (probably boost converters) are being bypassed and all three inputs are connected directly to a common DC link. By this, the losses of the input stages can be avoided. In figure 10 you can see these three DC voltages. The picture shows a zoom of the first ramp on MPPT1.

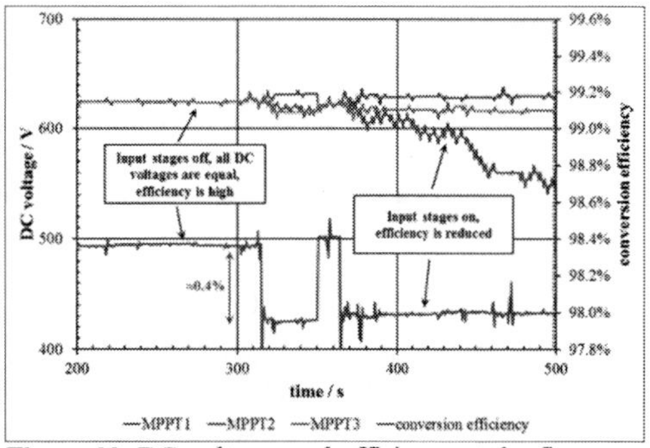

Figure 10: DC voltages and efficiency at the first ramp of MPP tracker 1

The measurements of pictures 9 and 10 show the benefit of this new proposed test profile. With the existing tests (e.g. figure 8), the loss of efficiency at inhomogeneous conditions would not have been detected. Intentionally or not, the manufacturer of inverter B makes a benefit of the current standards. But as a multi-MPPT inverter should also perform well under inhomogeneous conditions, the characterization of this device using current normative tests is not entirely representative for the operation in an actual PV array.

7 CONCLUSIONS

The test procedures in EN 50530 allow quite a good characterization of a single-MPPT PV inverter. What's missing is a test to check if the inverter has the ability to find the actual MPP on a partially shaded PV array, where the P/V curve has more than one peak. A simple test for this situation is proposed in this paper. Also, the test procedures for the static efficiency should be modified, so that no tests are being performed at the very limit of the device's specifications (voltage and power). The modifications proposed in this paper would only have a minimal impact on the test results, but they would prevent the inverter from operating in a non-representative state, e.g. with power limitation. Current standards do not include tests that are specifically designed for multi-MPPT PV inverters. Simply adapting the tests from EN 50530 and perform them on each of the inverter's MPP trackers simultaneously does not allow a good characterization of these devices, which are deliberately designed to achieve a good performance under inhomogeneous input conditions. In this paper, a simple but effective test method with inhomogeneous input conditions is being proposed. Measurements of two multi-MPPT PV inverters show that if the measurement of the static efficiency (and the calculation of the European or CEC efficiency) is performed strictly according to EN 50530, the inverter might reduce the input power at the last measurement point. This leads to a poor and non-representative MPP tracking performance. The measurement with the proposed test profile for multi-MPPT PV inverters shows that the conversion efficiency under inhomogeneous conditions can be considerably lower compared to conventional tests. Therefore, it would make sense to include this or a similar test in upcoming standards for PV inverters.

DISCLAIMER

Information contained in this paper is believed to be accurate. However, errors can never be completely excluded. Therefore, any liability for correctness and completeness of the information or from any damage that might result from its use is formally disclaimed

ACKNOWLEDGEMENTS

Thank goes to all the institutions that gave us financial support, particularly the Swiss Federal Office of Energy (SFOE), the Swiss Competence Centers for Energy Research (SCCER) and the Building Insurance Berne (GVB).

REFERENCES

[1] R. Bruendlinger, N. Henze, H. Haeberlin, B. Burger, A. Bergmann, F. Baumgartner, "prEN 50530 – The new European Standard for Performance Characterisation of PV Inverters", 24th European Photovoltaic Solar Energy Conference and Exhibition, Hamburg, Germany, 2009.

[2] U. Muntwyler, D. Gfeller, L. Borgna, E. Schüpbach, "Ultrafast Multi-MPPT PV Inverter Test Bench", 29th European Photovoltaic Solar Energy Conference and Exhibition, Amsterdam, Netherlands, 2014.

[3] H. Haeberlin, L. Borgna, D. Gfeller, M. Muenger, P. Schaerf, "New PV Array Simulator of 100 kW: Results of first Tests at a PV Inverter of 100 kW", 24th European Photovoltaic Solar Energy Conference and Exhibition, Hamburg, Germany, 2009.

[4] H. Haeberlin, "Optimum DC Operating Voltage for Grid-Connected PV-Plants", 20th European Photovoltaic Solar Energy Conference and Exhibition, Barcelona, Spain, 2005.

[5] H. Haeberlin, L. Borgna, M. Kämpfer, U. Zwahlen, "Total Efficiency η-tot – A new Quantity for better Characterisation of Grid-Connected PV Inverters", 20th European Photovoltaic Conference, Barcelona, Spain, 2005.

All papers can be downloaded under www.pvtest.ch.

COMPARING THE IMPACT OF THE OFF-GRID SYSTEM AND ON-GRID SYSTEM ON A REALISTIC LOAD

Ali Algaddafi,[1*] Jubran Alshahrani,[2] Sabir Hussain,[3] Khalifa Elnaddab,[4] Eljaroshi Diryak,[5] Ibrahim Daho[6]
[1*, 2, 3] Institute of Energy and Sustainable Development, De Montfort University
The Gateway, Leicester, UK, LE1 9BH
[4, 5, 6] Department of Engineering, Sirte University, Sirte, Libya
alisirte@yahoo.com,[1] jubran.2050@gmail.com,[2] shussain95@yahoo.com,[3] alnar_28@yahoo.com, [4] aljaroshi1966@gmail.com,[5] ibrahim.lameen.daho@gmail.com[6]

ABSTRACT: The main features of the power electronic inverter are used to facilitate the integration of different renewable energy sources; the management of power flow between the generator, the load and the grid; and to provide a reasonable power quality to the grid or load. However, where the inverter output of the renewable energy resource fluctuates, such as the impact of cloudy weather on the solar system, this may lead to degradation in the efficiency and distortion of the power quality. Therefore, enhancing the power quality and designing an optimal solar system controller to improve the reliability, while taking into account the cost and output power are explored. The off-grid and on-grid systems were configured in the laboratory, with this paper studying the impact of the off-grid system on the non-linear load when compared with the on-grid system in terms of the power quality and output current distortion. The results reveal that the off-grid system has a lower effect on a load than the on-grid system. Therefore, the performance of the off-grid system is superior to the on-grid system. To evaluate the proposed system improvements achieved in the AC Mini-Grid system, the experimental works were implemented and validated by real-time measurements.
Keywords: AC Mini-Grid System, On-Grid System, Off-Grid System, Network, System Protection

1 INTRODUCTION

Electrical energy production has been increasing significantly over recent years to satisfy the energy demands associated with the rising global population and strong economic growth. According to Tagare et al [1, 2], global electricity demand is predicted to increase by 2.20% per annum between 2008 and 2035. Meanwhile, 87% of international power generation is dependent upon the burning of fossil fuels, which increase greenhouse gas emissions and pollution [3]. Thus, solar energy has attracted the attention of researchers as an important component of renewable energy. Solar energy can be utilised to support the grid or supply isolated areas where extending the grid is not possible or economical, with the storage batteries utilised as a backup system that enables an emergency response to any disturbance or contingencies and provides an enhanced overall response. Frequency Disturbance Recorders (FDRs) were deployed by Guo et al [4]. Although the FDR can measure the voltage magnitude, angle, and frequency, the calculated measurements were at 0.1 s intervals. This requires new techniques such as the self-adaptive threshold to enhance and achieve the calculation in a short period of time. A ripple control based control system was presented in [5] to manage specific loads, by controlling the power droop control in order to limit the grid voltage. While this also allows an increase of the photovoltaic (PV) hosting a capacity of an electrical power system, the weakness was the voltage rising above 10%, with the measurements taken at 10 min intervals. Therefore, this should be measured more quickly to avoid the issue of rising voltage. Furthermore, the communication link to the grid code should be both reliable and able to handle missing data.

There are medicines that should be stored at low temperatures in the fridge, such as vaccines, insulin, and some antibiotics. The fridge thus requires a good quality power and the flexibility to operate in an off-grid system. Moreover, Magnetic Resonance Imaging (MRI), renal dialysis, computerised axial tomography, cardiac monitoring devices and pulse oximeters require a high quality power to avoid damage and ensure their continuing operation. For example, an MRI machine can cost from (USD) $150,000 to $1.2 million [6]. Therefore, in this paper, the testing of the electrical power quality that supplies a resistive load that is located in the Laboratory of Engineering and Renewable Energy at the University of Leicester in the Department of Engineering is evaluated and compared between the off-grid and on-grid systems. This paper aims to describe the configuration and compare the impact of the off-grid system and on-grid system on the realistic load and is organised as follows. Section 2 presents the system description, including the procedures for establishing the on-grid system, where protection for the network is required. The configuration of the off-grid system is included in the same section. Section 3 features a comparison between the off-grid and on-grid systems, while also considering the general measures taken to conserve battery life and highlighting the limitations and challenges of the AC Mini-Grid system. Section 4 discusses the acquired results, before the final section of this paper summarises the works that have been achieved.

2 SYSTEM DESCRIPTION

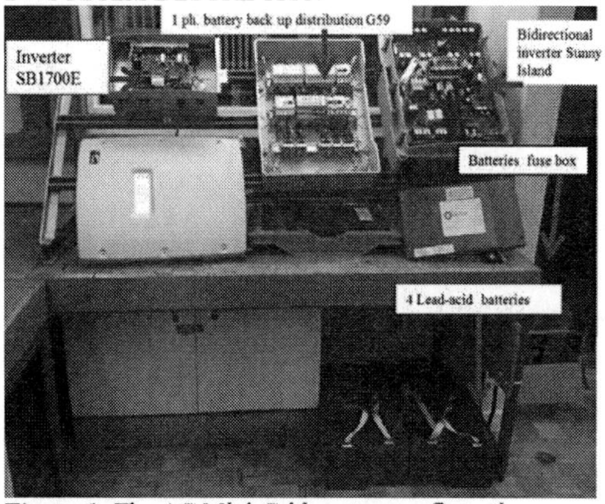

Figure 1: The AC Mini-Grid system configuration

The system is configured as illustrated in Figure 1, which consists of a Sunny Island (SI) 6.0H inverter, storage batteries, and a Network and System (NS) protection box. Figure 2 presents the AC Mini-Grid system diagrammatically, which can integrate with the existing PV system and protect the network. However, as the schematic diagram of the automatic transfer switch is not adequately explained, the PSpice software is utilised to describe the automatic transfer switch (NS protection), as shown in Figure 3.

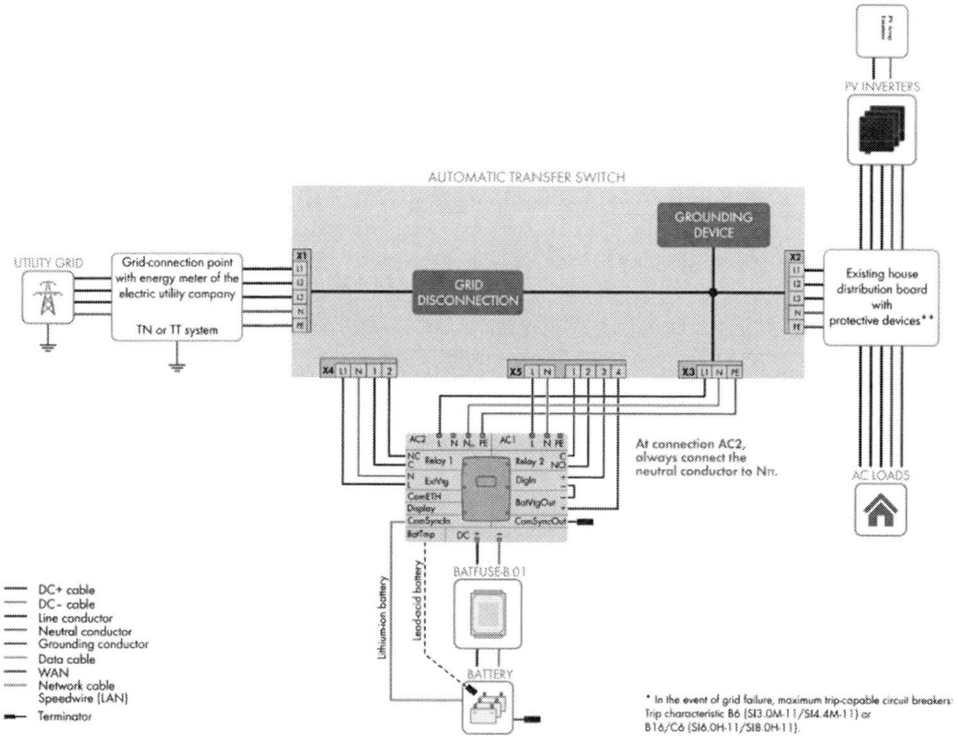

Figure 2: Schematic of the SI inverter storage system with battery backup function [7]

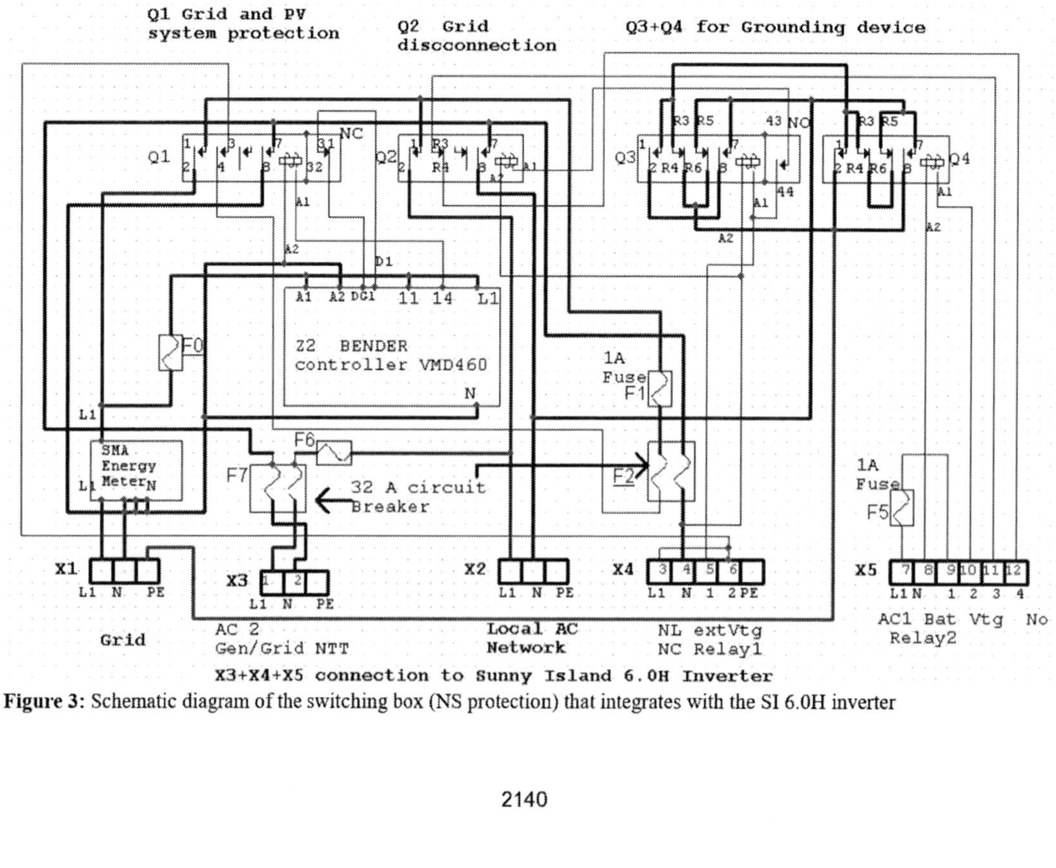

Figure 3: Schematic diagram of the switching box (NS protection) that integrates with the SI 6.0H inverter

When contactor Q3 is switched on, contactor Q2 can be turned on. This allows the connection between the network grid and the generator power, such as a PV system. The relay Q1 is controlled by a Bender VMD460, which provides protection for the network. Relays Q3 and Q4 are utilised to create grounding devices while relay Q2 is implemented to grid disconnect when the grid code does not match the generator power.

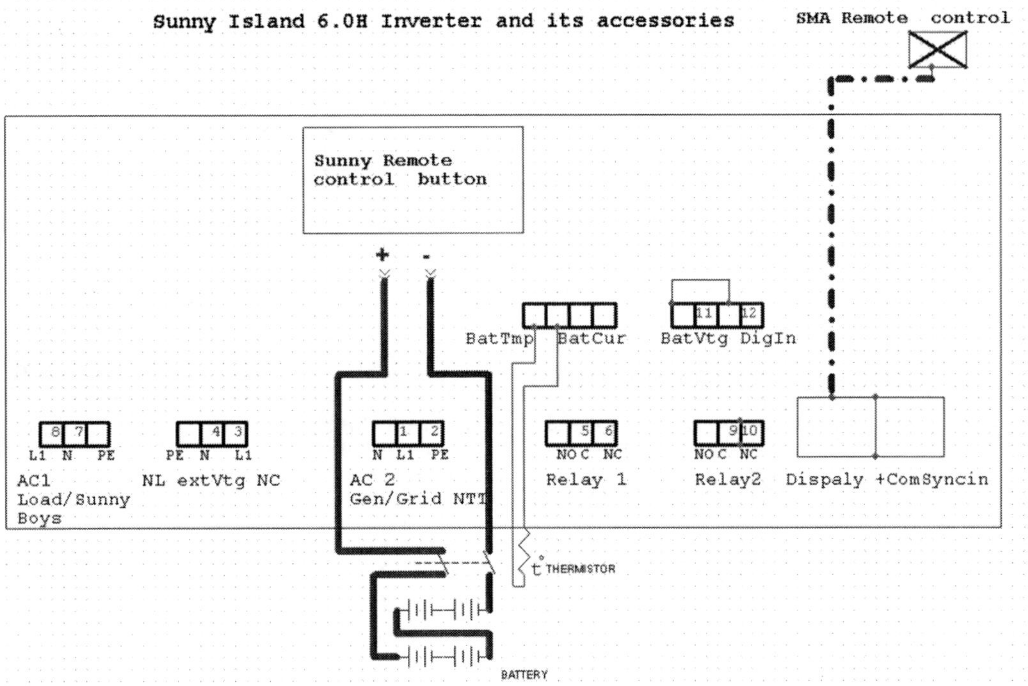

Figure 4: Schematic diagram of the SI 6.0H inverter with storage system as a battery backup function

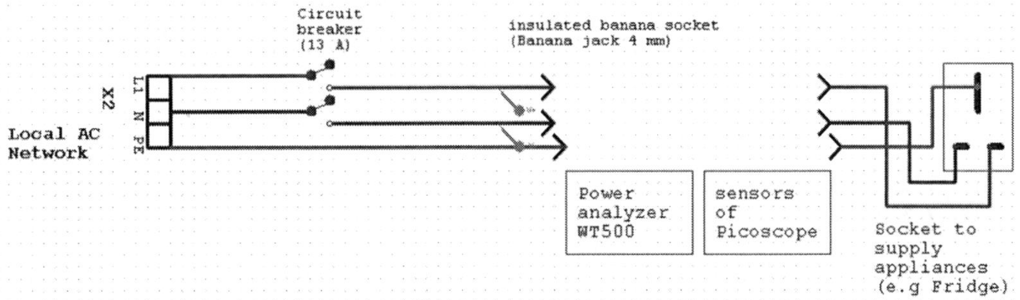

Figure 5: Schematic diagram of the testing impact of the off-grid and on-grid systems with realistic loads as future work

The diagram of the SI 6.0H inverter is shown in Figure 4, which can help us to comprehend the physical connection between the NS protection and the SI 6.0H inverter. Furthermore, the essential accessories of the SI 6.0H inverter such as SMA Solar Technologies remote controller and battery temperature sensor are also presented.

2.1 The On-Grid System

The procedure for the configuration of the SI 6.0H inverter in the grid-connected system is as follows:

1- All wires must be checked according to the SI installation manual. All components must be locked, except for the battery fuse.
2- The battery fuse should be locked as the last component to activate the SI 6.0H inverter.
3- By pressing and holding the control button on the Sunny Remote Control, the screen will be displayed.
4- The instructions that appear on the screen should be followed in order to set up the parameters according to the Installation Quick Reference Guide [7].

As the battery storage can be damaged by incorrect setting of the battery parameters including the battery type, nominal voltage, and capacity, the battery storage parameters should thus be selected with care. The grid-connected inverter is considered as a main element in the on-grid system and is commonly utilised when the grid is available as it offers many features such as Maximum Power Point Tracking (MPPT), anti-islanding, power droop and fault ride through.

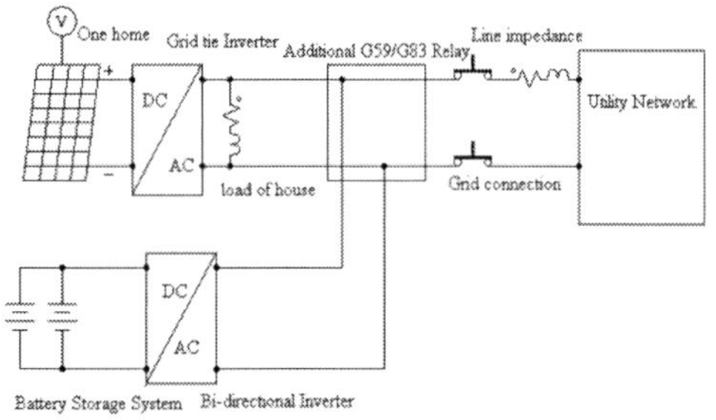

Figure 6: Circuit diagram of the AC-Mini-Grid system

The AC Mini-Grid is configured and mounted on the assembly frame based on the circuit diagram presented in Figure 6 and the physical configuration shown in Figure 1.

2.2 The Off-Grid System

The off-grid system is pertinent, since there are more than one billion people living globally without access to electricity. The high cost of expanding the public grid alongside lower power demand in isolated areas may cause the connection of remote areas to be deemed uneconomical. Thus, a stand-alone system represents a logical alternative solution. The advantages of the SI 6.0H inverter as a stand-alone system are described as follows:

1- The inverter features the flexibility of connection as a single phase or three phase.
2- An optimal battery management and state of charge can be achieved to guarantee long battery life.
3- Low-cost integration of standard AC consumers, generators, and a renewable energy source is applicable.
4- The system employing the SI 6.0H inverter does not require any additional controller monitoring unit.

A stand-alone solar system or off-grid system consists of four basic elements: a power generator, a charge controller, a storage battery and an inverter.

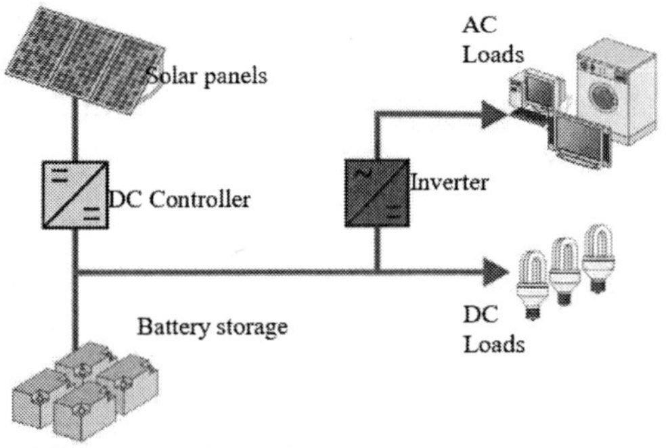

Figure 7: Stand-alone solar system

A grid-connected inverter or AC coupling with SI 6.0H inverters also features a number of advantages, such as the compatibility to connect with the public grid and simplicity of installation. The off-grid system offers a highly reliable system structure. An SI 6.0H inverter is connected to the battery to form the AC grid system, where the voltage and frequency of the output can be controlled, while the power can be consumed directly by the demand load. Surplus power is then stored in a battery for use when there is insufficient supply to satisfy the demand, as described in Figure 8.

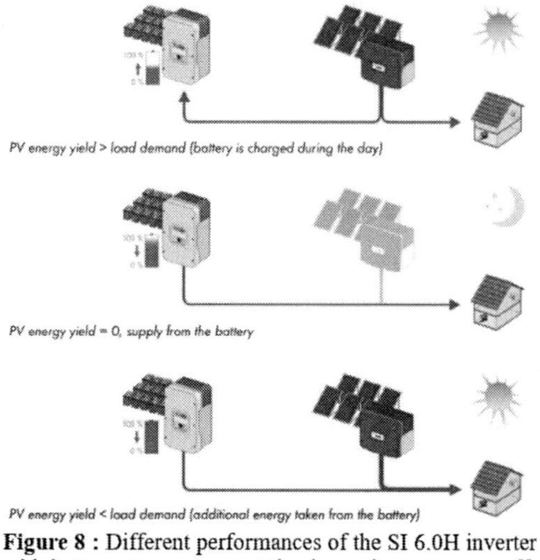

Figure 8 : Different performances of the SI 6.0H inverter with battery storage to meet the demand power as an off-grid system [8]

The off-grid system is configured as per the procedure in the on-grid system. However, two steps should be carefully considered when setting the parameters:

1- The grounding in off-grid systems differs from the grid, where the distribution can be designed differently. Personal protection is the purpose of grounding in all grid configurations, and in a stand-alone scenario the neutral point of the source is grounded [8].
2- Instead of selecting the on-grid, the off-grid should be selected in the remote control of the SI 6.0H inverter and the display instructions then followed.

3 COMPARISON BETWEEN THE OFF-GRID AND ON-GRID SOLAR SYSTEMS

The on-grid system is known variously as grid connected, grid tie, grid-intertied, utility-interactive, grid-direct, or integrated grid systems. On-grid systems are connected to the grid and enable the utilisation of both solar energy and electricity from the grid. Unlike the off-grid system, the electricity demand produced by the on-grid system is not required to be 100% of the demand Moreover, the surplus power can be sent to the grid. The on-grid system will not continue to operate when the grid fails, since this can provide protection for the utility workers.

The off-grid system is known as a backup system. This system is independent of the grid, but requires that the PV panels are able to produce sufficient electricity to cover all demand. Thus, the off-grid system requires storage batteries or a backup source of energy such as a generator. These characteristics make the off-grid system more expensive, more complex and less flexible than the on-grid system. It is commonly used to supply homes, buildings, boats, remote agricultural pumps, gates, and traffic signs.

In general, the solar system could either be part of an

on-grid or off-grid system. The on-grid system is typically utilised in urban zones with those houses that have utility networks. The main advantage of the on-grid system is that battery storage is not required, as the excess power can be exported into the grid. Therefore, the cost of the battery and regulator is avoided in the on-grid system. Nevertheless, the on-grid system cannot provide power when there is a grid outage.

The off-grid system in this paper is a solar system that is fully independent of the national grid, and thus requires battery storage as a backup system to offer supply during night periods. Although off-grid systems have many advantages, there are also drawbacks such as the batteries being expensive and not being particularly environmentally friendly. Nevertheless, the off-grid system can represent an affordable solution for isolated areas where the grid is unavailable and a critical load must be consistently supplied. The battery storage in the off-grid system requires ongoing maintenance and periodic replacement. Table I lists a comparison between the off-grid system (stand-alone inverter) and on-grid system (PV inverter).

Table I: Comparison between the PV and stand-alone inverters [9]

Type of comparison	PV inverter	Stand-alone power inverter
Direction of energy flow	Unidirectional inverter	Bidirectional converter
Function	MPPT sinusoidal grid current	Management of batteries, generators, and loads; sinusoidal grid voltage
Overload capacity	Approx. 110 %	Approx. 300 % (short-circuit proof)
Active / reactive power	Feed-in of active power	Loads with any Power Factor (PF)
Typical DC voltage	125–750 V (string technology) Time [4ms/Div.]	12 V, 24 V, 48 V

The proposed AC-Mini-Grid system can be operated in both off-grid and on-grid systems. It can provide virtually an Uninterruptible Power Supply (UPS) function. A commercial UPS is used to respond rapidly with secure power provision for small machines such as computers and printers. However, for large loads, the fast response of the UPS function is challenged. For example, the AC Mini-Grid takes five to seven seconds to respond to change from on-grid to the off-grid system [7].

3.1 Limitations of the AC Mini-Grid system and general advice for preserving battery life

This paper proposes that the AC Mini-Grid can be utilised in rural areas in Libya, or any such isolated areas in the world. As with any system, this one has its advantages as well as limitations. A current ripple that has twice the fundamental frequency of the power supply appears at the DC side of the SI 6.0H inverter. This ripple current is the main disadvantage of this system as it may reduce the lifetime of the batteries, fuel cells, and electrolytic capacitors. Thus, to extend the lifetime of these components, the ripple current should be reduced in the grid interconnection converter [10, 11, 12, 13] and a high-speed control employed to reduce the current ripple, although an energy buffer requires a larger capacitor. An active filter was utilised in [14, 15] as opposed to a large capacitor. However, this method requires a number of switching devices, which increases the overall cost. A centre tap of the isolation transformer was used by Itoh and Hayashi [15], where the primary-side inverter in the DC/DC converter was regulated by the differential voltage and common-mode voltage. Nevertheless, this method is only suitable for high-power prototypes, and the optimisation of the transformer should be considered.

Figure 9 describes the charge battery current that arrives from the PV inverter, where it has a harmonic component from the PV inverter and is then converted by a bidirectional converter to battery storage.

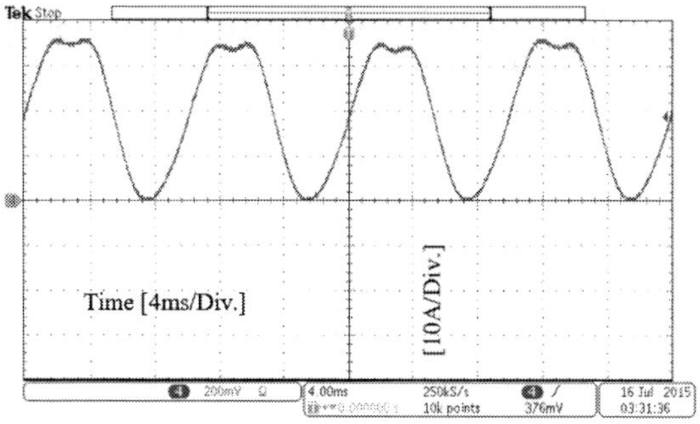

Figure 9: Charge current from the grid [50A/Div.]

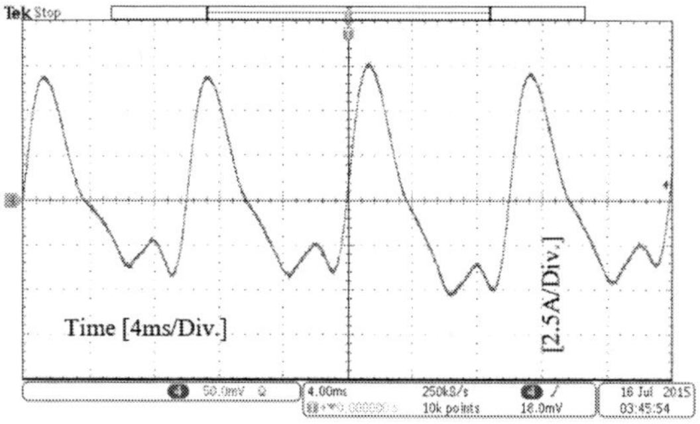

Figure 10: The waveform of the battery current [50A/Div.] when the batteries are close to full charge

When the battery is fully charged, the current charge produces the waveform seen in Figure 10. Since the battery voltage is a constant 48 V, the power should, therefore, be zero as the current charge has a positive half period, while the other half period is almost equal to the first period but in negative. This explains why the power is shown as zero in the screen display in the remote control of the SI 6.0H inverter.

3.2 General advice and suggestions for maintaining battery life

The issues of battery storage and consumption of energy are essential concerns for the solar system since the majority of systems work in conjunction with battery storage. Therefore, the efficiency of those systems is associated with the efficiency and performance of the battery storage. Some advice and suggestions for maintaining the battery life are presented as follows:

1- Switch off the battery usage when there is an alternative source of energy such as the wind, solar or the public network. This increases the time between charging periods.
2- Avoid discharging a full battery to zero, because when completely discharged the battery may malfunction.
3- Avoid fully charging the battery. Leave a small cap for a full charge, such as only charging to 95% of the battery's storage capacity.
4- Do not expose the battery to high temperatures, such as direct sunlight.
5- Ensure appropriate and suitable charging of the battery, such as partially charging the battery and then stopping when its temperature is high.
6- Where the battery is not being used, separate it from the charging system.
7- Clean the battery poles when the battery is not being used for a period of time.

4 RESULTS AND DISCUSSION

The output voltage and current between the SI 6.0H inverter and PV inverter are measured, together with the current injected into the battery, where the voltage is fixed at 48 V; hence, the battery voltage is not included, as shown in Figure 11. We notice that the AC Mini-Grid system operates at unit PF.

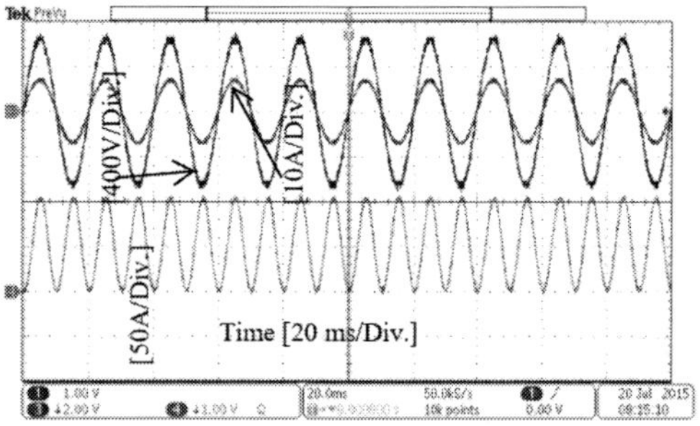

Figure 11: Operating AC Mini-Grid systems at unit PF; the output voltage is trace 1, the current is trace 3 and trace 4 is the charger current

Figure 12 presents the response of the AC Mini-Grid during an interruption in the network.

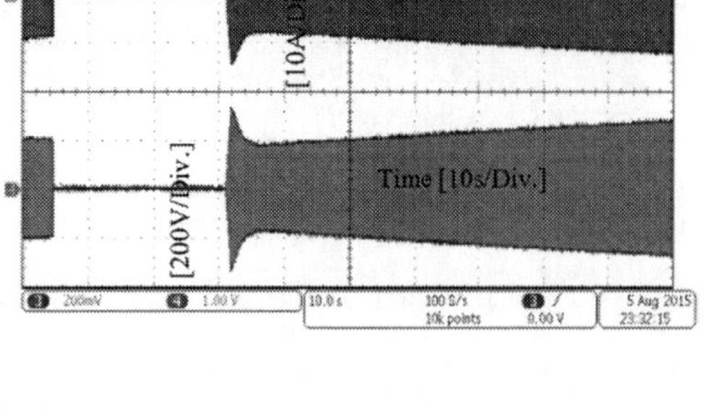

Figure 12: Response of the AC Mini-Grid when transferring from on-grid to off-grid; trace 4 is the output voltage and trace 3 is the output current

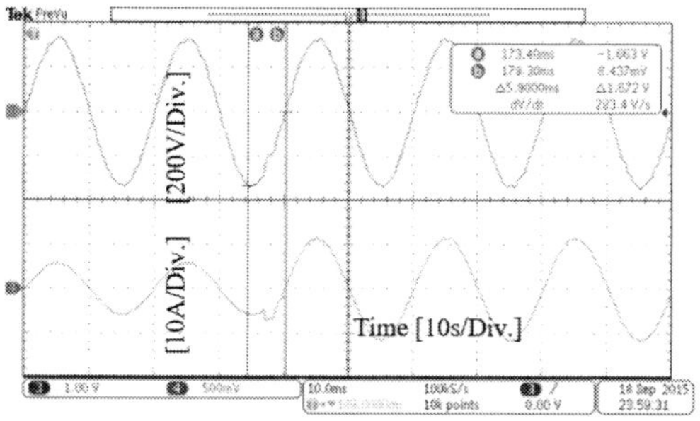

Figure 13: Response of the output current of the Mini-Grid in trace 4 and voltage in trace 3

The response of the output current to changing the resistive load is shown in Figure 13. This paper only tests the impact of varying a resistive load, due to the ease of understanding the behaviour. In our next article different loads such as a fridge, washing machine and the fluorescent lamp will be considered.

In order to evaluate the output current, the Total Harmonic Distortion (THD) and a Fast Fourier Transform (FFT) algorithm are used to analyse the output current, as seen in Figure 14.

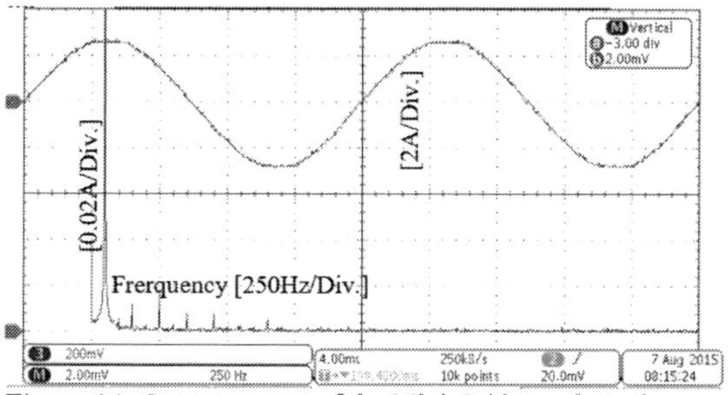

Figure 14: Output current of the Mini-Grid travels to the domestic load [10A/Div.], with the THD, found to be 0.64% [16]

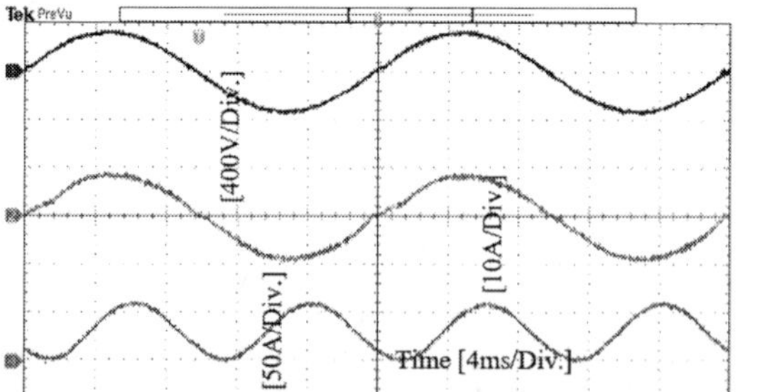

Figure 15: Trace 1 is the grid voltage, trace 2 is the current injected into the battery from the solar system, and trace 4 is the ripple current injected into the storage battery that has DC and AC components

It can be noticed that the battery charge current has two elements: AC components and a DC offset. The charge current should be DC with a small ripple. This ripple current has a frequency twice that of the grid, and thus may cause damage to the battery and also reduce its lifespan. This is considered to be the main limitation of the AC Mini-Grid.

This paper aims to investigate the solar system with energy storage to enhance the system's reliability. The AC Mini-Grid not only provides backup power to improve the supply security but also offers optimal self-consumption. Optimal self-consumption facilitates the management of the battery in summer and winter over the entire year, allowing the battery to be charged to a high degree in the winter months, while during the shorter summer nights a smaller portion of the battery must be reserved for backup power. Therefore, the AC Mini-Grid has the main function of reducing the battery costs by increasing the self-consumption and battery service life.

4 CONCLUSION

This paper introduces a comparison between the off-grid system and on-grid system, while the limitations of the system are also highlighted, with the main identified challenge being the ripple current of the battery storage, which reduces the battery life. The off-grid system does not require a source of reserve power, as this system relies on a generator or storage battery to provide electricity during cloudy days or at night.

The solar system described in this paper is a solution for increasing self-consumption and reliable backup power. It can be integrated with an existing system and ensures continuous power supply to all consumers, which guarantees the security of electricity as when the grid fails, the backup power is harnessed. The schematic diagram and circuit of the AC Mini-Grid are presented in detail, together with the testing of the system's response for the variation of the resistive load being investigated. The system response was found to be very fast at less than six ms. Future work will consider the system with different household appliances.

5 REFERENCES

[1] D.M. Tagare, Electricity Power Generation: The Changing Dimensions, Vol. 56 (2011) John Wiley and Sons.

[2] Anon., World Outlook Energy 2015, Available: https://www.iea.org/Textbase/npsum/WEO2015SUM.pdf [Accessed: 03/01/2016].

[3] B.K. Bose, Global Warming: Energy, Environmental Pollution, and the Impact of Power Electronics, Industrial Electronics Magazine, IEEE, Vol. 4 (2010) 6–17.

[4] J. Guo et al., "Design and implementation of real-time off-grid detection tool based on FNET/GridEye", PES General Meeting| Conference and Exposition, IEEE, (2014) 1–5.

[5] D. Müller, A. Solvatec, Ripple control based control system for decentralised photovoltaic power plants, 31st European Photovoltaic Solar Energy Conference and Exhibition (2015).

[6] Anon., A single scan runs $2,600 on average (before your insurance kicks in). Here's what makes this common diagnostic procedure so expensive, Available: http://time.com/money/2995166/why-does-mri-cost-so-much/ [Accessed: 27/04/2016].

[7] Anon., SMA flexible storage system with battery backup function, Available: http://www.windandsun.co.uk/media/304189/Flexible-Storage-Battery-Backup-Planning-Guide.pdf [Accessed: 01/11/2014].

[8] Anon., Solar Stand-Alone Power, and Backup Power Supply, Available: http://files.sma.de/dl/10040/INSELVERSOR-AEN101410.pdf [accessed: 06/01/2016].

[9] Anon., World Outlook Energy 2015, Available: http://files.sma.de/dl/7910/SI-Verteilsystem-TI-en-10.pdf [Accessed: 07/02/2016].

[10] D. Polenov, H. Mehlich, J. Lutz, Requirements for MOSFETs in Fuel Cell Power Conditioning Applications, Power Electronics and Motion Control Conference, EPE-PEMC. 12th International (2006) 1974–1979.

[11] G. Fontes, T. INPT-ENSEEIHT, C. Turpin, S. Astier, T.A. Meynard, Interactions Between Fuel Cells and Power Converters: Influence of Current Harmonics on a Fuel Cell Stack, Power Electronics, IEEE Transactions, Vol. 22 (2007) 670–678.

[12] L. Changrong, L. Jih-Sheng, Low Frequency Current Ripple Reduction Technique With Active Control in a Fuel Cell Power System With Inverter Load, Power Electronics, IEEE Transactions, Vol. 22 (2007) 1429–1436.

[13] S.K. Mazumder, R.K. Burra, K. Acharya, A Ripple-Mitigating and Energy-Efficient Fuel Cell Power-Conditioning System, Power Electronics, IEEE Transactions, Vol. 22 (2007) 1437–1452.

[14] J.I. Itoh, F. Hayashi, Ripple Current Reduction of a Fuel Cell for a Single-Phase Isolated Converter Using a DC Active Filter With a Center Tap, Power Electronics, IEEE Transactions, Vol. 25 (2010) 550–556.

[15] M. Saito, N. Matsui, Modeling and control strategy for a single-phase PWM rectifier using a single-phase instantaneous active/reactive power theory, Telecommunications Energy Conference, INTELEC '03. The 25th International (2003) 573–578.

[16] A. Algaddafi, N. Brown, R. Gammon, J. Al-Shahrani, Modelling a Stand-Alone Inverter and Comparing the Power Quality of the National Grid with an Off-Grid System, IEIE Transactions on Smart Processing & Computing (2016).

DESIGN AND PIL SIMULATION OF AN AEKF FOR REAL TIME BATTERY SOC ESTIMATION USING ARM BASED CORE

[1,*]Ahmed GAGA, [2]Omar DIOURI, [3]Youssef CHEDDADI, [4]Fatima ERRAHIMI, [5]Najia ES-SBAI
[1, 2,3,4,5] Faculty of Sciences and Technologies, Laboratory of Renewable Energies and Intelligent Systems,
Sidi Mohamed Ben Abdellah University of Fez, Morocco
Road Imouzzer, BP 2202, Fez, Morocco
*Email: gaga.ahmad@gmail.com *Phone: +212602286735

ABSTRACT: Battery management system (BMS) is an electronic system responsible for all control and management operations of various battery parameters, especially SOC. The battery SOC estimation process is one of the most complicated techniques due to the fact that battery chemistry is a time variant system with a strong non linearity. An EKF algorithm is adopted to estimate internal states of complex systems. In this paper we have chosen to dynamically estimate the SOC by decreasing divergence due to parametric uncertainty of battery model, measurement and process noise. To further enhance this algorithm an adaptive calculation of noise covariance is proposed to combine between a better convergence and robust results. For the implementation one of the powerful microcontrollers in the market is used, the STM32F4 ARM based core combining low cost and high performances with DSP and FPU capability. To validate the simulation results for real implementation, a PIL configuration is made to use a real hardware target within a simulation process. The obtained results demonstrate that the AEKF and ARM based core processor are the best hardware and algorithm combination to estimate an accurate battery SOC with the best real time constraint targeted electrical storage market especially EV and autonomous photovoltaic systems.
Keywords: KALMAN Filtering, Adaptive EKF, Battery Storage and Control, PIL simulation, STM32F4, Thevenin RC network, SOC Estimation, Plant Control, Modeling.

1 INTRODUCTION

In the last few years, several energy storage techniques have appeared and been developed especially electrical storage represented by the device called battery. Batteries are devices which can ensure the conversion between both; electrical and chemical energy types to store or supply electricity. Among the most battery technology used in the industries we found Lithium Ion and Lead Acid batteries which are presently considered as the practical energy storage solution for several applications area like an autonomous solar installations, electric and hybrid vehicles (HEV) and so on. The difference between batteries technology is due to several parameters such as their high energy density, length of cycle-life, high specific energy and power… [1]. For most electrical battery applications, the internal states determination of the battery like state of charge (SOC) and state of health (SOH) are critical parameters as their mirrors the remaining capacity in the storage device and if the actual battery conditions much the ideal specification.

These two battery parameters are often used to design the optimum control for the battery management processes. Therefore, in order to have an accurate information and efficient management system; an accurate battery SOC estimation technique is a dominant importance.

In the battery internal states estimation, a number of aspects can influence the accuracy of SOC results, such as high non linearity between an open circuit voltage (OCV) and (SOC), an hysteresis phenomena between charge and discharge curves, the noise that affect measurement data which is linked to the sensors accuracy, the process noise that affect directly the internal state due to the battery chemistry nature [2]. Inaccurate battery SOC evaluation can cause several problems such as undesirable overcharge or over discharge of the battery. Both of these drawbacks lead to the reduction of battery life time and the decrease of battery use efficiency [3].

This work proposes the complete process to extract a battery SOC using one of the commonly known algorithms for dealing with the internals and non-measurable states estimation of nonlinear system based on the KALMAN filtering. The aim of this work is a design and implementation of a mixture between a discrete extended KALMAN filter (DEKF) and the traditional coulomb counting technique [4] to estimate SOC for several batteries chemistry, especially battery technology used in solar applications. An equivalent electrical circuit model based on second order Thevenin RC network is presented, the order of network is chosen to balance between parameters identification complexity and model fidelity. In order to identify battery parameters, a battery block is built using an equivalent electrical circuit, then an automating battery parameters estimation using Simulink design optimization tools is established to correlate model parameters to experimental data based on nonlinear mean square method.

2 BATTERY MODELING

2.1 Electrical model and state space equations

One of the most important information in battery management systems is the remaining capacity level which is called State Of Charge. SOC is an internal state of battery which cannot be measured directly during battery operation by using physical sensors [5]. Therefore, a virtual sensor or an estimator represents the only manner to obtain information about its SOC value. To do that, an electrical battery model is needed. A range of battery models are designed to capture battery dynamic for various test cases, among them we found the electrochemical equivalent circuit and electrical circuit models which are widely used in several battery applications studies. The electrical circuit models use equivalent electrical circuits to reflect the I-V characteristics of real batteries by using current and voltage sources, and passive elements like resistors and capacitors.

Due to the notable relaxation effect of the wide range of battery technologies, especially lead acid and Li-ion types, a reliable and accurate model is needed to cover different realistic conditions. Therefore, a second order electrical Thevenin model is selected to further reflect the battery behaviors as shown in Fig. 1 [6].

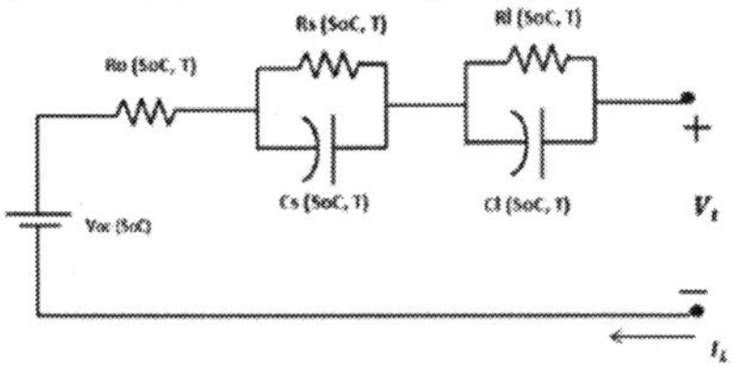

Figure 1: Thevenin second order Battery Electrical Model

In this model, the electrical element R_0 models the surface layer impedance, accounting for the electric conductivity of the electrolyte, electrodes and separator, generally it represents the internal bulk resistance. The concentration and activation polarization are modeled respectively by R_l, C_l and R_s, C_s.

The applied model in Fig. 1 can be expressed the following electrical behavior:

$$V_t = V_{oc} - V_{R0} - V_s - V_l \tag{1}$$

$$\dot{V_s} = -\frac{1}{R_s C_s} V_s + \frac{1}{C_s} I_L \tag{2}$$

$$\dot{V_l} = -\frac{1}{R_l C_l} V_l + \frac{1}{C_l} I_L \tag{3}$$

Where V_t represents the battery terminal voltage, V_{oc} is the battery open circuit voltage (OCV) and I_l is load current. U_s and U_l are the short and long time transient voltage responses for charging and discharging respectively.

2.2 Parameters identification

In the battery electrical model, two sets of parameters are needed to be identified, namely the relationship between OCV and SOC and also the parameters, R_s, R_l, C_s, C_l, R_0.

2.2.1 OCV=*f*(SOC) function identification

In many applications, it is advisable to keep the battery SOC limited within a given operation area, generally between 20% and 80% or 10% and 90% range, it depends on battery technology and actual application. A battery SOC which verifies the above condition protects a battery from any damage like a deep discharge and overload and also enhances its life time.

In this work, the nonlinear least square method is applied to identify the parameters of function OCV=f(SOC) with the condition that the SOC must be between 0% to 90%.

The main least square method application is data fitting; in this part we use it to find the best fit in the least square sense to minimize a quadratic residual in the objective function (Equation 5). This function is formed based on the error between OCV measurement and the approximation function which is a 6th order polynomial function as shown in Equation 5:

$$P(x) = a_0 + a_1 x^1 + a_2 x^2 + \cdots + a_n x^n \,; n = 6 \tag{4}$$

The following equation is set to obtain the parameters: $a_n \,; n \in (1,2,\ldots,6)$

$$\varepsilon(a_0, a_1, \ldots, a_n) = min \sum_{i=0}^{m} \left(y_i - \sum_{j=0}^{n} a_j x_i^j \right)^2 \tag{5}$$

By using Matlab optimization tools and according to the equations (4) and (5), the parameters of OCV=*f*(SOC) can be identified based on the equations (7) and (8), the Table 1 shows the parameters identification results.

$$OCV = 3.5 + 0.46(SOC)^1 + 2.6(SOC)^2 - 17(SOC)^3 + 40(SOC)^4 - 37(SOC)^5 + 13(SOC)^6 \tag{6}$$

Table I: OCF = f(SOC) Parameters identification results

Parameter	a_0	a_1	a_2	a_3	a_4	a_5	a_6
Value	3.5	0.46	2.6	-17	40	-37	13

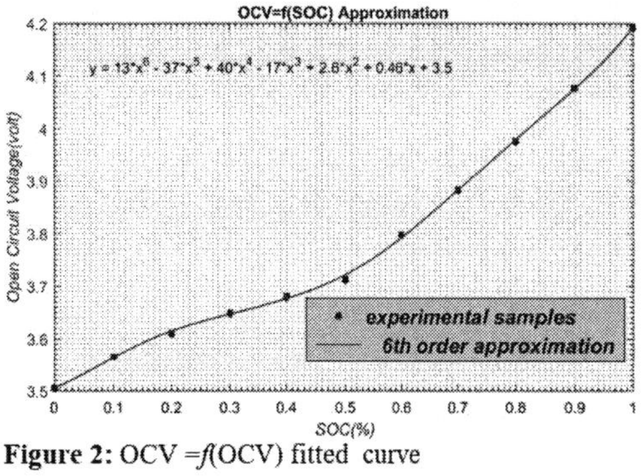

Figure 2: OCV =*f*(OCV) fitted curve

2.2.2 Electrical parameters identification

One of the most important effects which characterize the battery dynamic is the hysteresis phenomenon. This effect was observed in the battery voltage response due to the internal parameters. However, when the battery switches from discharge or charge state, the unexpected change of current causes the large magnitude change in the voltage, this change is caused by the internal Ohmic resistance. When the battery is not connected, an electrochemical effect of the internal battery imposes a slow voltage change.

So, we can exploit this response phenomenon based on the experimental data to identify the internal parameters of the battery, like an Ohmic and the polarization resistance. The HPPC voltage response test is used to accomplish this task as shown in Fig. 3:

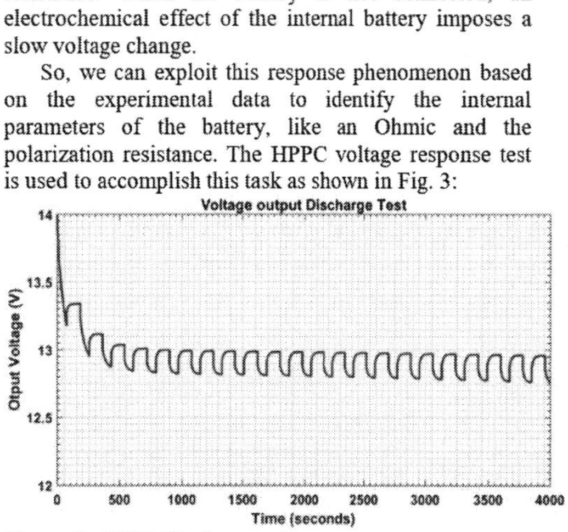

Figure 3: HPPC Discharge voltage response

A genetic algorithm is used to identify the battery parameters by finding the optimal value of $\boldsymbol{R_l}, \boldsymbol{C_l}; \boldsymbol{R_s}; \boldsymbol{R_s}$. The objective function of the used genetic algorithm is given as shown in Equation (7): [7]

$$\begin{cases} min\left\{f\left(\hat{\mathcal{X}}_k^{(g)}\right)\right\} \\ f\left(\hat{\mathcal{X}}_k^{(g)}\right) = \frac{1}{N}\sum_{i=1}^{N}\left(V_{t,k} - \hat{V}_{t,k}\left(\hat{\mathcal{X}}_k^{(g)}\right)\right)^2 \end{cases} \quad (7)$$

Where $\hat{\mathcal{X}}_k^{(g)}$ is the estimation value of current population $\mathcal{X}_k$ at generation(g); $\mathcal{X}_k$ is the current individual k of the population $\mathcal{X}$, where: $\mathcal{X} = \begin{bmatrix} \tau_l & 0 \\ 0 & \tau_s \end{bmatrix}$. $\hat{V}_{t,k}$ is the estimation value of V_t at the individual k; N is the estimation length.

So, for parameters identification; the appropriate polarization time constants $\tau_l = R_l C_l$ and $\tau_s = R_s C_s$ need to be given in advance based on the battery characteristics: the following regression equation is built:

$$\begin{cases} V_{t,i} = ocv - R_0 I_{L,i} - R_l I_{l,i} - R_s I_{s,i} \\ I_{R_l,i} = \left(1 - \frac{\left(1 - exp^{-\frac{\Delta_t}{\tau_l}}\right)}{\frac{\Delta_t}{\tau_l}}\right) I_{L,i} + \\ \left(\frac{1 - exp^{-\frac{\Delta_t}{\tau_l}}}{\frac{\Delta_t}{\tau_l}} - exp^{-\frac{\Delta_t}{\tau_l}}\right) I_{L,i-1} + I_{R_l,i-1} exp^{-\frac{\Delta_t}{\tau_l}} \\ I_{R_s,i} = \left(1 - \frac{\left(1 - exp^{-\frac{\Delta_t}{\tau_s}}\right)}{\frac{\Delta_t}{\tau_s}}\right) I_{L,i} + \\ \left(\frac{1 - exp^{-\frac{\Delta_t}{\tau_s}}}{\frac{\Delta_t}{\tau_s}} - exp^{-\frac{\Delta_t}{\tau_l}}\right) I_{L,i-1} + I_{R_s,i-1} exp^{-\frac{\Delta_t}{\tau_s}} \end{cases} \quad (8)$$

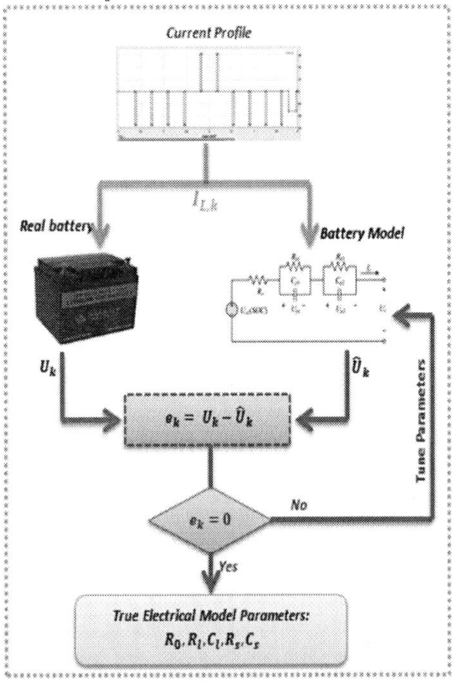

Figure 4: Workflow of Identification Principle

3 KALMAN FILTERING PROCEDURE

3.1 KALMAN Filter Principle

KALMAN Filtering technique or simply KALMAN Filter (KF) also known as Linear Quadratic Estimator was introduced in 1960 [8].This filter provides an optimal solution for estimating internal states variable of a dynamic system using series of measurement over time, including inaccuracies and statistical noise. However, KALMAN filter algorithm is applied for any dynamic nature; if the system is linear, an algorithm is applied directly. For the nonlinear systems, a linearization process is assumed at each time step to approximate the nonlinear system with a linear time varying (LTV) system using Taylor series expansion. This approximation technique leads to a modified KALMAN filtering "Extended KALMAN filter (EKF)" which is used on a real nonlinear system [9-13]. The dynamic modeling for both linear and nonlinear system is given in the following equations:

$$\begin{cases} x_{k+1} = f(x_k, u_k) + w_k \\ y_k = g(x_k, u_k) + v_k \end{cases} \quad (9)$$

Where equations (19), (20), (21), (22) represent the state space equations, they express the system dynamic and the output equation respectively. Functions $f(x_k, u_k)$ and $g(x_k, u_k)$ are nonlinear transition and nonlinear measurement functions. Process and measurement noise are modeled by a zero-mean white Gaussian stochastic process with covariance process noise matrix Q of w_k vector and covariance measurement noise matrix R of v_k vector.

The w_k and v_k must satisfy the equation (10):

$$\begin{cases} E(w_k) = q_k; Cov(w_k, w_j) = Q_k \delta_{k,j} \\ E(v_k) = r_k; Cov(v_k, v_j) = R_k \delta_{k,j} \\ Cov(w_k, v_j) = 0 \end{cases} \quad (10)$$

At each time step, both $f(x_k, u_k)$ and $g(x_k, u_k)$ functions are linearized around to the operation point based on the first order Taylor-series expansion, the rest of the series is neglected. Assuming that $f(x_k, u_k)$ and $g(x_k, u_k)$ are differentiable at all operate points, we can construct new dynamic matrices A_k and $C_k$from a given nonlinear function as follow; equations (11) and (12):

$$A_k = \frac{\partial_f}{\partial_x} \quad (11)$$

$$C_k = \frac{\partial_g}{\partial_x} \quad (12)$$

EKF starts filtering using the available information on the initial state $(\hat{X}_0)$ and error covariance $(\hat{P}_0)$ as shown in equation (13).

$$(\hat{X}_0) = E(X_0)\,;\,(\hat{P}_0) = E\left[(X - \hat{X}_0)(X - \hat{X}_0)^T\right] \quad (13)$$

An illustration of the EKF algorithm is given in Fig. 6 [9, 16].

3.2 Battery SOC estimation using AEKF

Battery state of charge or simply SOC is a relative quantity that describes the ratio of the remaining capacity and the maximum available capacity, the SOC formula is given by equation (14):

$$SOC_k = SOC_0 - \frac{1}{C_n}\int_0^k \eta_i I_{L,t} dt \quad (14)$$

Where SOC_k is the State of Charge at the kth sample time and SOC_0 is the initial SOC, $I_{L,t}$ is the instantaneous charge or discharge current; η_i is the Coulomb efficiency, which is a function of temperature and current. C_n is the maximum battery capacity. The available SOC range varies between 0% and 100%.

The discretization of Equation (14) is given by the following Equation:

$$SOC_k = SOC_{k-1} - \frac{\eta_i I_{L,k}\Delta_t}{C_n} \tag{15}$$

Where Δ_t represents the sampling time. Equation (15) is the base of the iterative calculation of actual SOC, it uses numerical coulomb metric counting [13-15].

The extended KALMAN filter has been widely used for parameter identification and state estimation in battery systems and precisely to extract the SOC [17-19]. However, its performance is depends deeply on the accuracy of the determined noise matrices. Consequently, due to the complex and various operation environments of battery applications, the EKF has not been used in practice. To overcome this problem, an adaptive variant of EKF is used. An AEKF approach employing the process and measurement covariance noise matrix matching is applied [20].

So, to apply an adaptive EKF for the SOC estimation, we must first have a system model in a state-space representation in discrete form. However, we assume a general structure for discrete-time dynamic systems.

$$\begin{cases} x_k = Ax_{k-1} + Bu_{k-1} + w_{k-1} \\ y_k = Cx_k + Du_k + v_k \end{cases} \tag{16}$$

Where x_k is the state vector at the kth sampling time. w_k is the unmeasured "process noise" that affects directly the internal system states and v_k is the measurement noise which is reflected in the system output estimation y_k. w_k and v_k are assumed to be Gaussian white noise with zero mean and covariance Q_k and R_k respectively. The dynamic of the system is described by the matrices A, B, C and D, and are time varying and determined by internal battery parameters. An implementation flowchart of the AEKF algorithm is shown in Fig. 7. The Adaptive EKF allows the parameters Q and R to be estimated and updated iteratively.

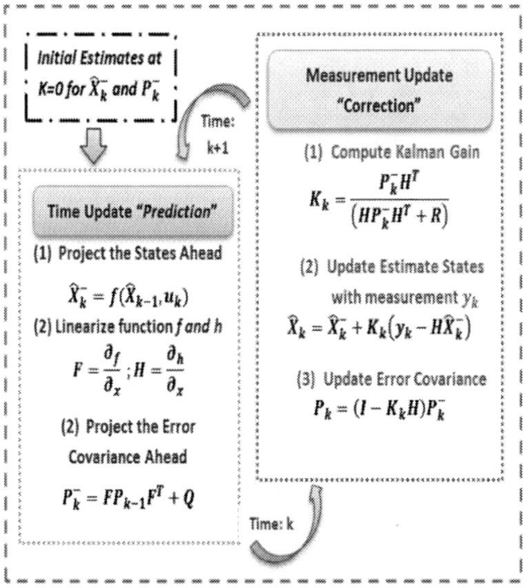

Figure 6: Extend KLMAN Filter Algorithm

The deterministic discrete-time state equations of the battery Electrical Circuit Model can be described by the equation (17):

$$\begin{cases} SOC_k = SOC_{k-1} + \frac{\eta_i I_{L,k}\Delta_t}{C_n} \\ U_{R_l,k} = U_{R_l,k-1}\exp^{-\frac{\Delta_t}{\tau_l}} + I_{L,k-1}R_l\left(1 - \exp^{-\frac{\Delta_t}{\tau_l}}\right) \\ U_{R_s,k} = U_{R_l,k-1}\exp^{-\frac{\Delta_t}{\tau_s}} + I_{L,k-1}R_s\left(1 - \exp^{-\frac{\Delta_t}{\tau_s}}\right) \\ V_{t,k} = ocv(SOC_k) - R_0 I_{L,k} - U_{R_l,k} - U_{R_s,k} \end{cases} \tag{17}$$

A state vector is defined as follows:

$$X = \begin{bmatrix} SOC \\ U_{R_l} \\ U_{R_s} \end{bmatrix}$$

The initial state mathematical expectation, error covariance, initial process noise covariance and initial measurement noise covariance are respectively shown in equation (18):

$$\begin{aligned} &\hat{X}_0 = E(X_0); \\ &P_0 = var(X_0) = E\left[(X_0 - \hat{X}_0)(X_0 - \hat{X}_0)^T\right]; \\ &Q_0; R_0 \end{aligned} \tag{18}$$

Finally, based on the equations (17) we can extract the state space matrices of the discrete system as follows:

$$A_k = \begin{bmatrix} 1 & 0 & 0 \\ 0 & \exp^{-\frac{\Delta_t}{\tau_{l,k}}} & 0 \\ 0 & 0 & \exp^{-\frac{\Delta_t}{\tau_{s,k}}} \end{bmatrix} \tag{19}$$

$$B_k = \begin{bmatrix} \frac{\eta_i \Delta_t}{C_n} \\ R_{l,k}\left(1 - \exp^{-\frac{\Delta_t}{\tau_{l,k}}}\right) \\ R_{s,k}\left(1 - \exp^{-\frac{\Delta_t}{\tau_{s,k}}}\right) \end{bmatrix} \tag{20}$$

$$C_k = \begin{bmatrix} \frac{\partial_{ocv}}{\partial_{SOC}} & -1 & -1 \end{bmatrix} \tag{21}$$

$$D_k = -R_0 \tag{22}$$

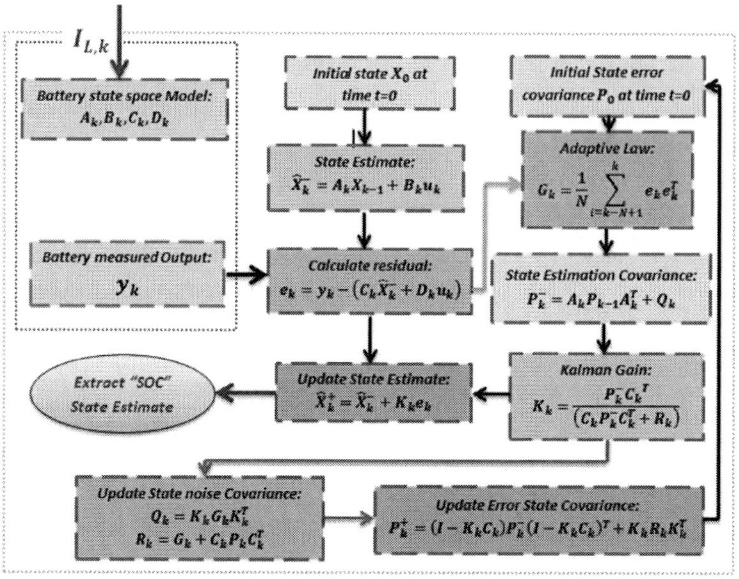

Figure 7: Flowchart of the AEKF algorithm

4 PIL SIMULATION

4.1 Processor In the Loop simulation

As the microcontroller market has rapidly developed, researchers have proposed several approaches to design implement and test algorithms for digital control plants. A novel procedure, called Processor-In-the-Loop (PIL), allows developers to verify the actual control software running in a targeted microcontroller device which controls a virtual prototype of a system. In this part, we focus on the demonstration of how to implement PIL technique in a virtual battery system with a high performance processor ARM based core control platform.

To do that, we use MATLAB environment both for system design and code generation to perform a PIL co-simulation with a targeted hardware. During the **Processor-In-the-Loop** (PIL) phase, the Adaptive EKF control algorithm is compiled and downloaded into an embedded target processor, as a processor we have used the STM32F429 ARM based core microcontroller. After downloading a compiled algorithm in the hardware, it communicates directly with the virtual battery model through serial communications. Fig. 8 illustrates well the PIL simulation procedure:

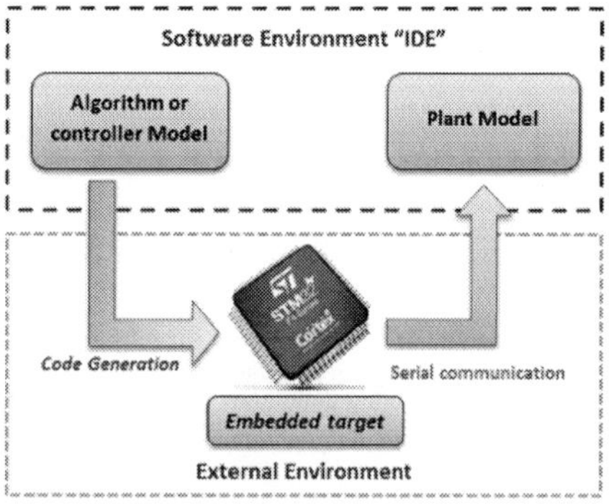

Figure 8: PIL simulation principle

Firstly, we must verify code of our AEKF algorithm before validation in hardware target with PIL simulation and external mode. The chosen target is STM32F4-Discovery Board with ARM Cortex-M4 based core. However, an Embedded Coder Support Package for this board is needed to enable hardware and software communication.

In a PIL simulation, the generated code by a suitable Embedded Coder runs on the STM32F4-Discovery Board. The PIL simulation results are transferred to Simulink to verify the numerical similarity between the AEKF model control algorithm and the generated C code, this verification process is an important part of the development cycle to guarantee that the performance of the deployment code matches well design requirement. To enable tuning parameters capability and real time data logging while running AEKF algorithm model in the target, we can use Simulink External mode.

5 DISCUSSION AND EXPERIMENTAL RESULTS

For simulation, we have used Matlab/Simulink. The Adaptive EKF algorithm is compared with the real coulomb metric results. The initial values used in the simulation are as follows:

$$X_0 = [0.1 \quad 0 \quad 0]^T\,;\; R_0 = 0.1\,;\; Q_0 = 0.1diag[3]$$

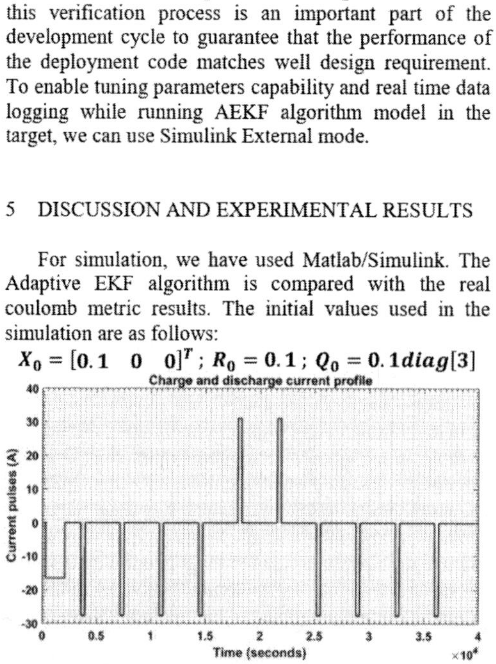

Figure 9: Charge and Discharge Current Profile

The simulation results are shown in Fig.10, in which the purple line represents the real SOC state and yellow line represents the Adaptive EKF for SOC estimation results. The Y axis represents the SOC values witch vary from 0% to 100%. The X axis represents the simulation time from $0\ sec$ to $4 \times 10^4 sec$. The battery was charged and discharged by current profile shown in Fig. 9.

The simulation results of used Adaptive EKF methods is given in Fig. 10, while zoom figures for the zones C, A and B in this figure are shown in Fig. 11, Fig. 12 and Fig. 13 respectively. The analysis of the results in these graphs shows that the error in the used Adaptive EKF algorithm is less than 2% with an imposed initial error greater than 90%.

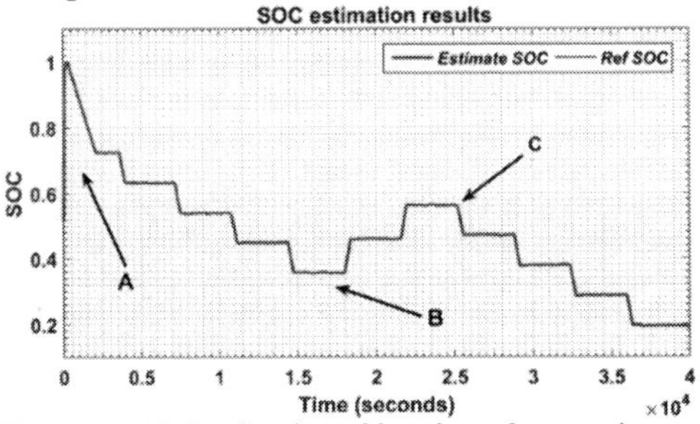

Figure 10: SOC estimation with a given charge and discharge scenarios

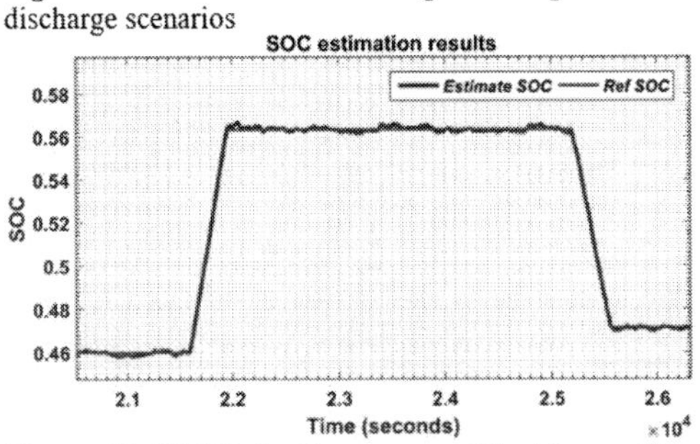

Figure 11: SOC Estimation accuracy (Fig. 10 zone C)

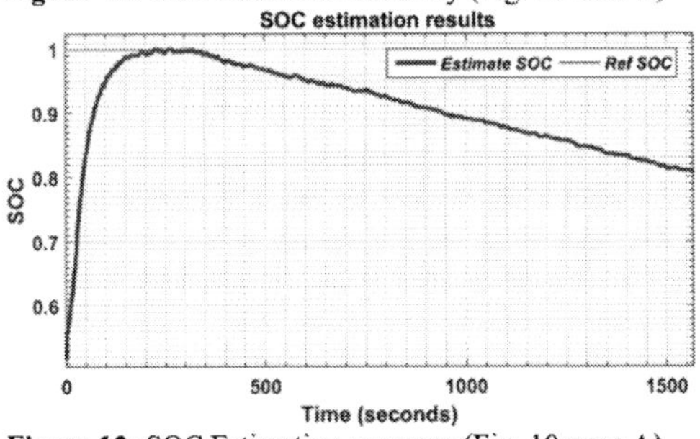

Figure 12: SOC Estimation accuracy (Fig. 10 zone A)

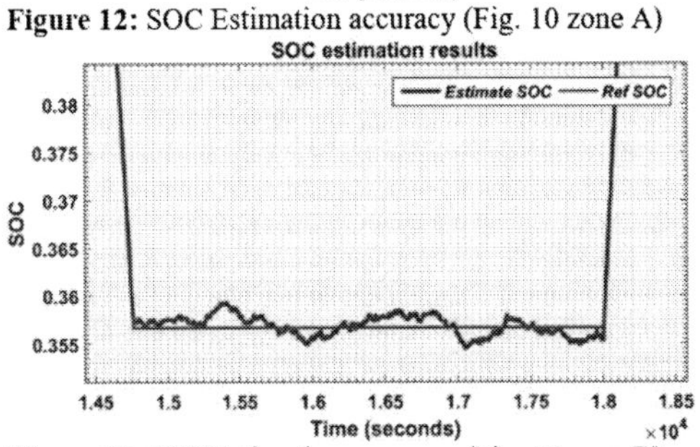

Figure 13: SOC Estimation accuracy (Fig. 10 zone B)

A random current profile is chosen Fig. 9 to verify the good response of the used Adaptive EKF algorithm, the estimate vector state is has three components, which is simply the internals state modeled by the electrical model. Fig. 14 shows respectively the error convergence of estimate SOC, RC network voltage and estimated output voltage.

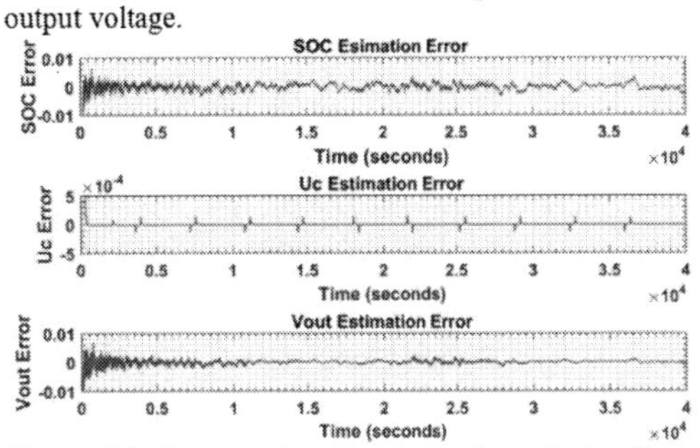

Figure 14: Battery internal states and terminal voltage errors convergence

Finally, Fig. 15 focuses on the SOC estimation accuracy and shows by zooming on the error convergence results in Fig. 14 that the error peak is less than 1% with a fast convergence time (less than 150 sec).

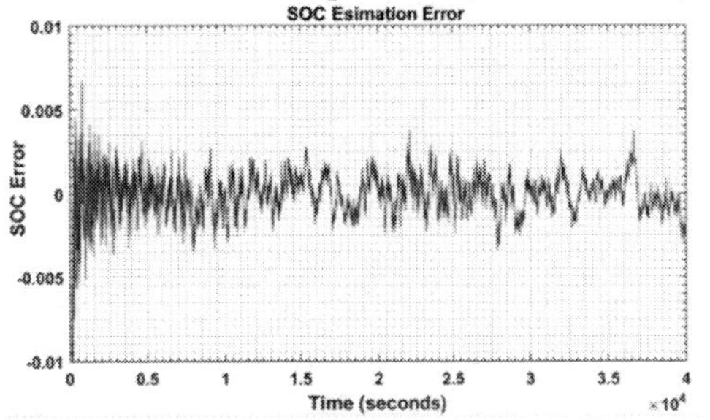

Figure 15: SOC Error peak (less than 1%)

6 CONCLUSION AND FUTURE WORK

This work presents the Processor in the Loop simulation of an Adaptive EKF algorithm for battery SOC estimation. The presented EKF introduces a higher SOC estimation accuracy by adaptively adjusting the covariance matrices of the process and measurements noise in real time based on a series of measurable output voltage. A second-order Thevenin electrical battery model based on RC network was used to reproduce the behavior of several battery types; especially Li-Ion. The model order is chosen to balance between the computational complexity and model accuracy. The HPPC battery test and experiment data were obtained to exploit response behavior to perform parameters identification like hysteresis effect and relaxation time. Simulation results showed that the AEKF algorithm can estimate the battery SOC with higher accuracy compared to the existing technique with error less than 2%. For the implementation one of the most powerful microcontrollers in the market is used, the STM32F4 ARM based core combining low cost and high performances with DSP and FPU capability. To validate the simulations results for real implementation, a PIL configuration is made which provides capability to use a real hardware target within a simulation process.

The obtained results demonstrate that the AEKF and ARM based core processor are the best hardware and algorithm combination to estimate an accurate battery SOC with the best real time constraint targeted electrical storage market especially EV and autonomous photovoltaic systems.

Future work will focus on developing a stand-alone system based on the same processor used for PIL estimation, without need to connect with any external environment. The estimates results will be communicated to the user using a local network based on CAN standard, and exploit the internal FPU unit in the used microcontroller to perform online battery parameters identification which extends greatly the performance and accuracy of the battery management system.

7 REFERENCES

[1] M. Dubarry, B.Y. Liaw, M.S. Chenb, S.S. Chyanc, K.C. Hanc, W.T. Siec, S.H. Wub,J. Power Sources 196 (2011) 3420-3425.

[2] D. Linden, T. Reddy, Handbook of Batteries, third ed., McGraw-Hill, New York.

[3] M. Matsuura, IEEE Power. Energy Mag. 26 (2011) 1172-1180.

[4] M. Dubarry, B.Y. Liaw, N. Vuillaume, J. Power Sources 186 (2009) 500-507.

[5] G. Marangoni, Battery Management System for Li-Ion Batteries in Hybrid Electric Vehicles, University of Padova, 2010.

[6] L. Lam, P. Bauer, E. Kelder, in: Telecommun. Energy Conf, (INTELEC), 2011, pp.1-9.

[7] Hongwen He, Rui Xiong, Jinxin Fan, Energies 4 (2011) 582-598.

[8] W.X. Shen, Energy Convers. Manag. 48 (2007) 433-442.

[9] W.X. Shen, K.T. Chau, C.C. Chan, IEEE Trans. on Veh. Technol. 54 (2005) 1705-1712.

[10] F. Zhang, G. Liu, L. Fang, in: IEEE ICRA, 2009, pp. 1863-1868.

[11] G.L. Plett, J. Power Sources 134 (2004) 277-292.

[12] W. Junping, G. Jingang, D. Lei, Energy Convers. Manage. 50 (2009) 3182-3186.

[13] J. Li, J.K. Barillas, C. Guenther, M.A. Danzer, J. Power Sources 230 (2013) 244-250.

[14] Lijun Gao, Shengyi Liu, Roger A. Dougal, IEEE Trans. Comp. Packag. Technol. 25 (3) (September 2002).

[15] Sungwoo Cho, Hyeonseok Jeong, Chonghun Hana, Shanshan Jin, Jae Hwan Lim, Jeonkeun Oh, Comput. Chem. Eng. 41 (2012) 1-9.

[16] Y. Morita, S. Yamamoto, S.H. Lee, N. Mizuno, Asian J. Control 8 (2006) 268-273.

[17] G.L. Plett, J. Power Sources 134 (2004) 252-261. J. Wang, J. Guo, L. Ding, Energy Convers. Manage. 50 (2009) 3182-3186.

[18] M. Smith, A. Miller, Proceedings 17th European Photovoltaic Solar Energy Conference, Vol. I (2002) 903.

[19] Saeed Sepasi, Reza Ghorbani, BorYann Liaw, J.Power Sources 245 (2014) 337-344.

[20] G. YE, P.P. SMITH, J.A. NOBLE, Ultrasound Med. Biol. 36 (2010) 234-249.

Performance Comparison of three Inverters with different Transformer topology

Mithilesh Kumar, O.S. Sastry*, Kamlesh Yadav, Richa parmar, Rashmi Singh, Birinchi Bora.
National Institute of Solar Energy (NISE), Gurgaon, India
*sastry284@gmail.com, Tel: +91-0124-2579213, Telefax No. : +91-124-2579207

ABSTRACT: This paper compares the performances of three standalone inverters used in SPV applications, having three output configurations, viz. transformer less (TL), EI core and toroid core (TC) transformers. The performance is analyzed in terms of efficiency, harmonics generation, crest factor (CF), power factor (PF) and weighted factor for efficiency for linear (LN) and non-linear (NL) type of loads. It is observed that for linear loads the weighted factor of efficiency more than 87%, for TL, EI and TC are 0.85, 0.75 & 0.55 correspondingly whereas for Nonlinear load weighted factor for efficiency more than 85% are 0.80, 0.75 and 0.35 respectively. It has been also observed that in TL inverter distortion in voltage (U_{thd}) and variation in crest factor (CF) is comparatively less than rest of two other configurations. Further the study also gives the approach for selection of suitable topology of inverter based upon loads requirement.
Keywords —Photovoltaic systems; DC-AC Conversion; transformer based inverters, toroid core; EI core; transformer-less inverters.

INTRODUCTION

The demand of solar photovoltaic(SPV) generated power is increasing rapidly [1] as climatic conscious users are demanding for the green power. The power generated from the SPV power sources is normally DC, and most conventional applications require AC power. Therefore, there is a need of DC-DC and DC-AC converters as an intermediary, which are based on high efficiency power electronics [3]. Further, utilization of generated energy by PV modules depends on the conversion efficiency and power quality from these inverters, on which performance and operational life of loads are also dependent. In off grid mode, quality of power in the networks depends upon the single source which can be optimized according to the user need. This paper presents the selection approach of suitable off grid systems based upon best working range of load percentage and power quality.

2 EXPERIMENTAL PROCEDURE

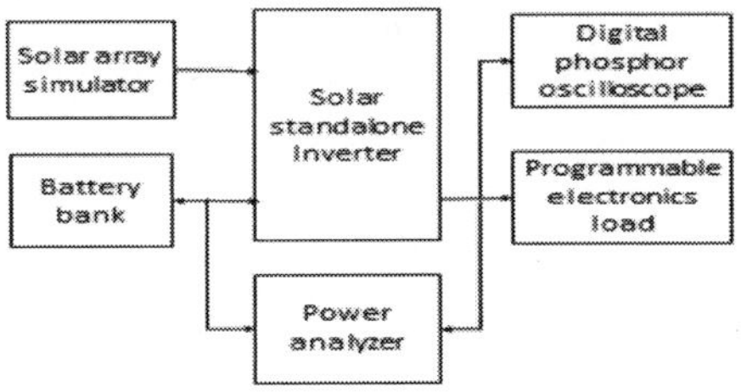

Figure 1: Experimental test setup

National Institute of Solar Energy (NISE) having testing lab for inverters. Solar array simulator make: Terrasaas (ETS600X), 60 KVA capacities for simulation of SPV array, and for power measurements power analyser make: Zimmer (LMG-450) are used for efficiency and power quality measurement. A DSP oscilloscope (DPO 4034) Tektronics make is used for analyzing the wave forms. To simulate different types of load condition Programmable AC/DC electronics load model 63804 is used, battery bank to supply input DC power & programmable AC source for grid simulation: model 61704. Among the sample of different manufactures, sample of similar capacity of three different topologies (EI, TC and TL) are selected randomly for the experiment purpose. Incase of transformer based topologies (EI and TC), inversion(DC to AC) is done first and then stepping up the AC voltage to the desired level by using 50Hz transformer. However in transformer less topology first DC voltage boost up and then inversion is done. In this study three different types of inverters of similar rating are studied. All three inverters are coupled with 24 V battery bank and solar Photovoltaic (SPV) panel in parallel mode. The power output from these inverters is in single phase 230 V, 50 Hz. Weighted factor of efficiency is meassured by varying loads from 5% to 100% with total 20 nos. of observations. For each load the best efficiency obsevation is selected and it is divided by total nos of observations.

3 Result And Discussion

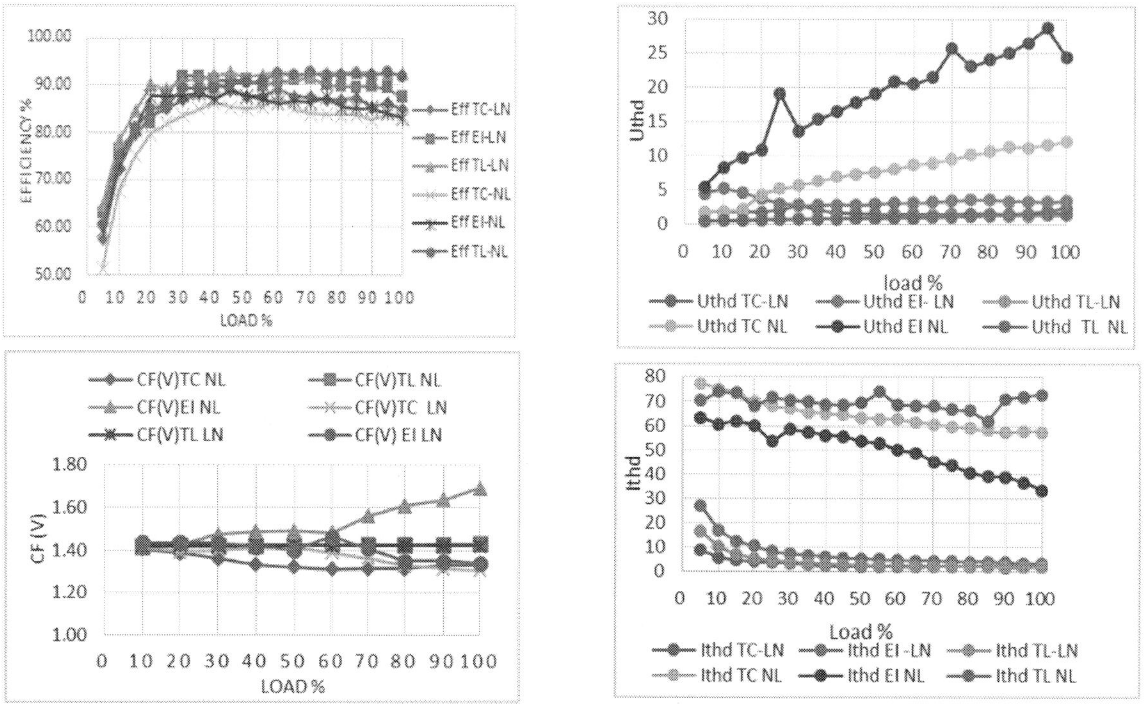

Figure 2: Parameter comparison: efficiency, U_{THD}, I_{THD} and Crest factor for Voltage

From figure 2 it has been observed that the efficiency of transformer-less inverter is always higher than transformer based inverter for any type of load application. It is observed in TL inverter the threshold load values for peak efficiency is 20% and 30% for linear and non-linear loads respectively. Whereas for TC transformer based inverters, the peak efficiency is in the range of 35% to 85%, 40% to 70% for linear and nonlinear loads respectively. EI transformer based inverters these values are 25% to 95%, 20% to 90% for linear and nonlinear loads respectively. For all type of loads and ranges, it is found that the efficiency of TL inverter is better than that of the any transformer based inverter, whereas with in the transformer based inverters EI cores is better then and TC Inverters .Total Harmonics Distortion (THD) of current and voltage waveforms is changing when inverters goes from linear loads to nonlinear loads. The EI core transformer based inverter generating higher distortion in voltage than toroid core. However, TL inverter generates least harmonics in voltage. With the peak range of the efficiency with nonlinear load application for toroid core U_{THD} varies from 6.98% to 9.58% for EI core U_{THD} varies from 10.88 to 25.72 % and for transformer less inverter it is varies from 0.68 to 1.60.

The higher value for THD for voltage of any inverter will degrade the network quality [4] or may harmful for that loads which are actually not cause of harmonics generation in network [2]. So the transformer-less inverter may be better to use in local electrical network. However current harmonics generated in case of nonlinear load for all three inverters are very high, but higher current harmonics is harmful only for those loads which are cause for it [5]. The study also shows that inverter with transformer based that means transformer may be responsible for increase of U_{THD}.

4 Conclusion

The transformer less inverter has wide range of weighted factor for better efficiency of operation and showing comparatively least THD in voltage, than transformer based inverters. The transformer based inverters giving medium range of weighted factor for best efficiency of operation and toroid core transformer inverter having narrow band. Weighted factor for best efficiency range suggest the user to operate in this range to save more power and can get more back-up from battery storage. It also suggest that for the designers of Balance of System (BOS) who are designing solar charge controller for existing inverters, or interface unit to convert home UPS in to solar inverters, should study loss of power during average efficiency of operation range particularly for nonlinear loads and then design the actual size of interface unit or solar charge controller. From the study of voltage harmonics and crest factor the transformer less standalone inverter is better for all types of loads.

ACKNOWLEDGEMENT

This research is based upon work supported in part by the Solar Energy Research Institute for India and the U.S. (SERIIUS) funded jointly by the U.S. Department of Energy subcontract DE AC36-08G028308 (Office of Science, Office of Basic Energy Sciences, and Energy Efficiency and Renewable Energy, Solar Energy Technology Program, with support from the Office of International Affairs) and the Government of India subcontract IUSSTF/JCERDC-SERIIUS/2012 dated 22nd Nov. 2012

REFERENCES

[1] MNRE target of 100GW SPV project, www.mnre.gov.in

[2] A.E.Emanuel, J.janczak,D.J.pileggi, E.M., Gulachenski, M.Breen, T.J. Gentile. Voltage distortation in distribution feeders with nonlinear load application.

[3] F. Ballabjerg, F. Lov, T. krekes, R. Teodorescu, Trends in power electronics and control of renewable energy system. 14th International power electronics and motion control conference, EPE-PEMC 2010.

[4] Kamlesh Yadav, O.S. Sastry, R. Wandhare, N. Sheth, M.kumar, B.Bora, Rashmi Singh, Renu, A. Kumar, Performance comparison of controllers for solar PV water application. Solar energy 119(2015) 295-202.

[5] Alxander E.Emanual, John A.orr, David cyganski, Edward M.Gulachenski, A survey of harmonics and currents at the customers Bus. IEEE transactions on power delivery vol. 8 no .1 January 1993.

A NOVEL SUITABLE RESONANT FILTER TO IMPROVE THE THD FOR A PV INVERTER

Rachid El Bachtiri Maha Khanfara, Karima El Hammoumi
REEPER Group, PERE Lab, EST school; CED STI, FST; USMBA University; Fez
rachid.elbachtiri@usmba.ac.ma, maha.khanfara@usmba.ac.ma, karima.elhammoumi@usmba.ac.ma

ABSTRACT: This paper presents a simple resonant filter for a PV inverter. We have considered a five level inverter to reduce THD of the output voltage. We have used an inverter control allowing the elimination of third and fifth harmonics. We have already shown, in a precedent study, that low pass filter is better than the tuned passive filter accorded to the seven harmonic. In this paper, we will compare performance of the low pass filter and the resonant series converter. Indeed, a resonant converter ensures a soft communication for electronic switch when it is adequately controlled. For comparing different cases, we have simulated the operation with PSIM and discussed the corresponding Total Harmonic Distortion (THD).
Keywords: PV inverter, Resonant filter, Harmonics.

1 INTRODUCTION

Photovoltaic systems connected to the grid usually use static converters especially inverters that generate voltage harmonics. To limit the level of distortion at the point of common coupling to the network, a various harmonic standards and engineering recommendations are employed [1]. European regulation defines all the control parameters and quality of an electrical distribution network, especially concerning electrical disturbances with a Total Harmonic Distortion coefficient THD < 5% [2].

To comply with these harmonic standards, PV installations often use filters to reduce the level of harmonic pollution in electrical network [3].

In a precedent study, we have already shown that low pass filter is better than the tuned passive filter accorded to the seven harmonic in order to have a better THD [4]. In this paper, we will compare performance of the low pass filter and the resonant series converter. The LC cell is mounted in series with the inverter and it was calculated to have the resonant frequency equal to the grid frequency. We obtain a resonant series converter which can insure a soft commutation and improve the system efficiency. We have calculated spectrum and THD in different situations. To validate our results we have used some simulations with Psim.

2 STUDIED SYSTEM

The studied system consists of a PV module connected in cascaded with H-bridge five-level voltage inverter, coupled to the grid through a filter to limit the network harmonic pollution as it is shown in Figure 1.

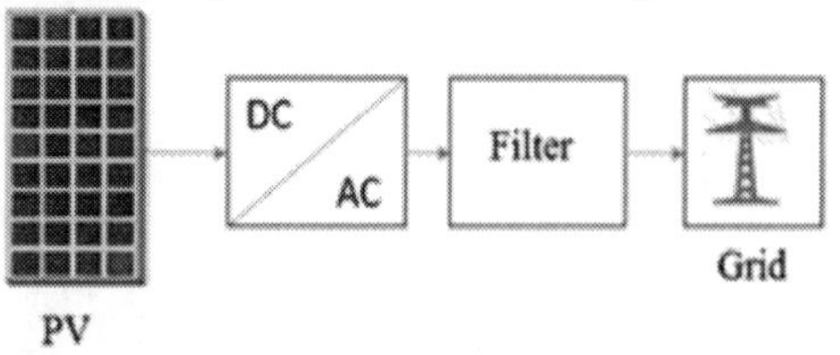

Figure 1: Block diagram of the studied system

The inverter considered in this study is a cascaded H-bridge five-levels with two H-bridge cells, each one is fed by a solar panel. Each one of the cells has four semiconductor switches which can be MOSFETs, GTOs or IGBTs mounted in parallel with inversed diode.

The output voltage of the inverter Vs is the sum of voltage generated by each H-bridge cell [5-6]. The circuit configuration and waveform of our inverter are shown in Figure.2 and Figure.3.

E1 and E2 are the DC voltage from PV generators.

α and β are the control angles at different level (Figure.3).

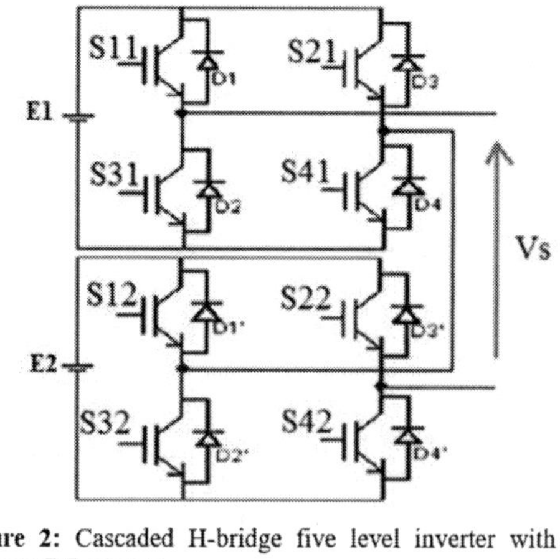

Figure 2: Cascaded H-bridge five level inverter with separate DC sources

S11, S21, S31 ...S42, represent the gating signals used for controlling semiconductor switches.

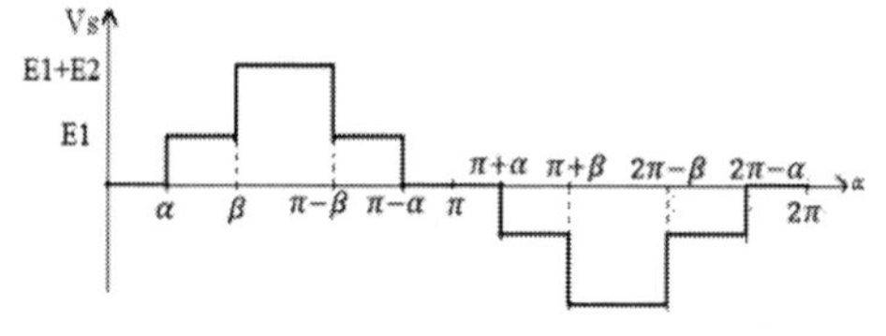

Figure 3: Five level inverter voltage waveform

In order to eliminate the third and the fifth harmonics, the multilevel control used, is covered by two switching angles α and β such :

$$\alpha = 12° \quad (1)$$

$$\beta = 48° \quad (2)$$

3 SIMULATION RESULTS

To validate the analytical result, we have simulated the single phase inverter and different filters using Psim software.

3.1 Five-levels inverter

Figure. 4 shows the proposed topology of the photovoltaic inverter

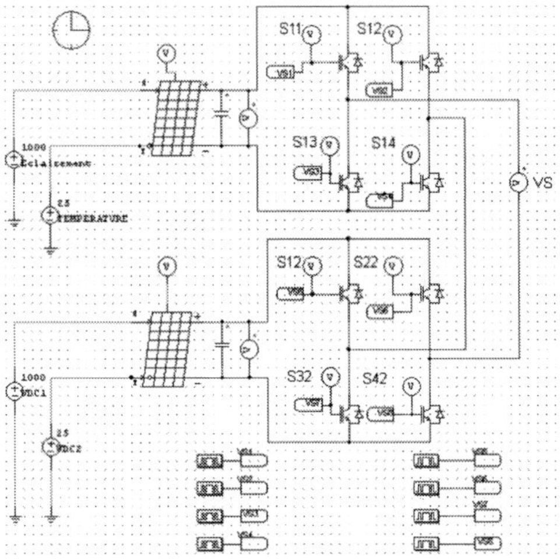

Figure 4: Cascaded H-Bridge five levels inverter

We obtain the waveform of the output voltage Vs illustrated on Figure 5.

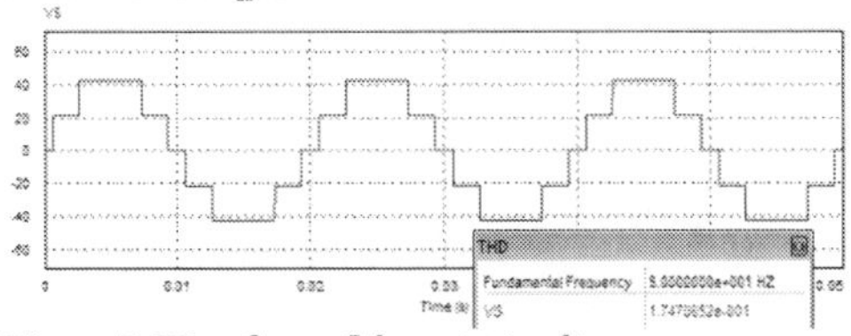

Figure 5: Waveform of the output voltage

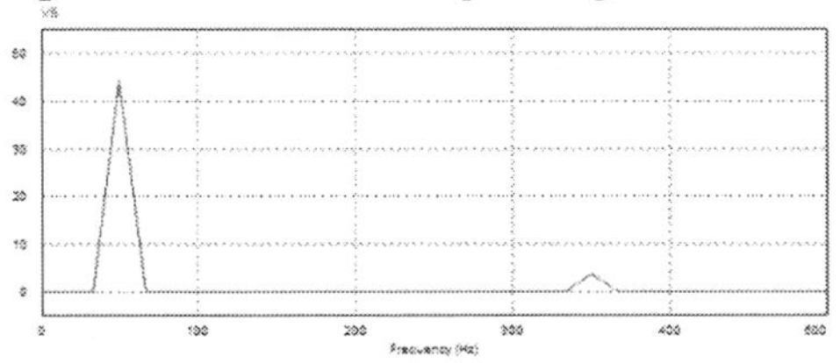

Figure 6: FFT analysis of the inverter

From the FFT analysis of above result, it can be seen that, 3rd and 5th harmonic are removed, but the THD > 17% due to the harmonics over than 5 still exist.

3.2 Low-pass filter

To eliminate or mitigate any type of harmonic frequencies over a threshold frequency, usually we use low pass filter because they block the unwanted harmonics and allow a certain frequency range to be transmitted.

We chose a simple passive LC filter shown below:

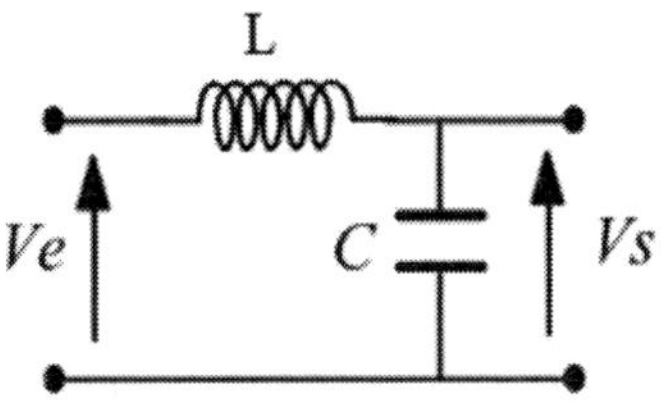

Figure 7: Low-pass filter

The resonant frequency of the filter is :

$$f_c = \frac{1}{2\pi\sqrt{L_f C_f}} \quad (3)$$

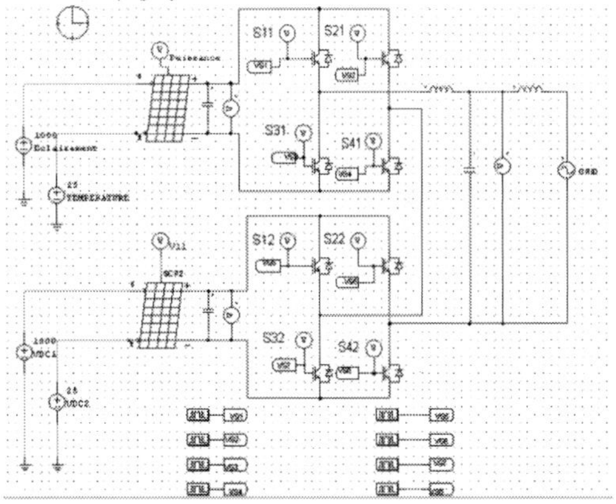

Figure 8: Five levels inverter with low-pass LC filter

The photovoltaic inverter with the low-pass LC filter connected to grid is shown in Figure. 8

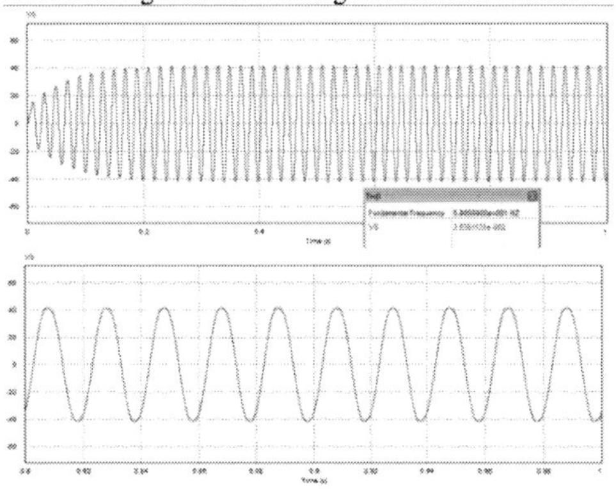

Figure 9: Waveform of the inverter with low-passfilter output voltage

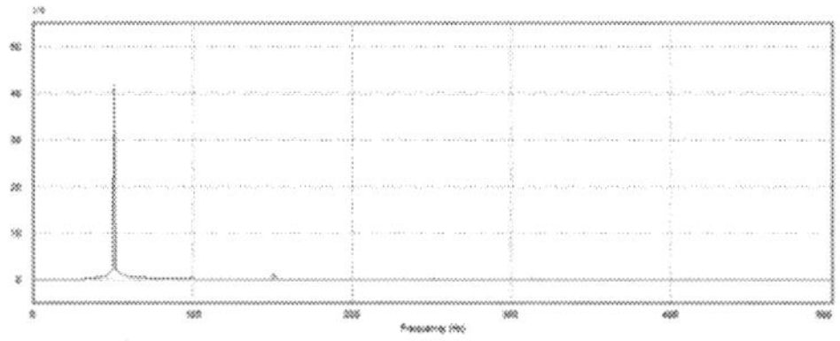

Figure 10: FFT analysis of the inverter with low-pass LC filter

We note that the LC low-pass filter removes all frequencies above the fundamental (50 Hz), except the 3rd harmonic with a small value.

3.3 Resonant series filter

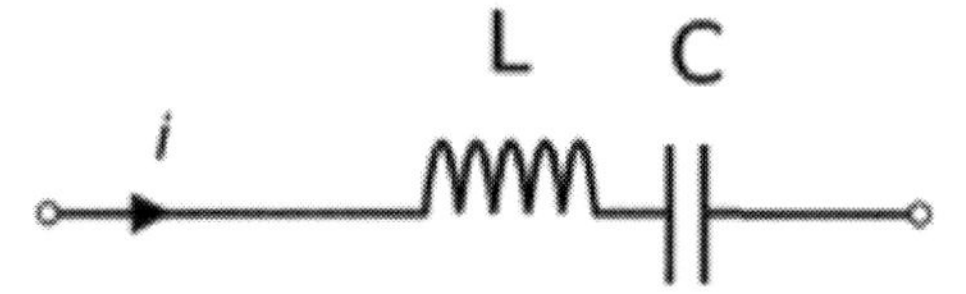

Figure 11: Resonant series filter

This filter presents very low impedance for 50Hz, so we have:

$$Z_L + Z_C = 0 \quad (4)$$

$$jL\omega + j\frac{1}{C\omega} = 0$$

$$\|Z_T\| = L\omega - \frac{1}{C\omega} = 0$$

$$\Rightarrow L\omega = \frac{1}{C\omega} \quad (5)$$

So $\begin{cases} \omega = \omega_0 = \frac{1}{\sqrt{LC}} \\ f_0 = \frac{\omega_0}{2\pi} = \frac{1}{2\pi\sqrt{LC}} \end{cases}$ (6)

The figure below shows the simulation scheme of photovoltaic inverter with resonant series filter at 50Hz.

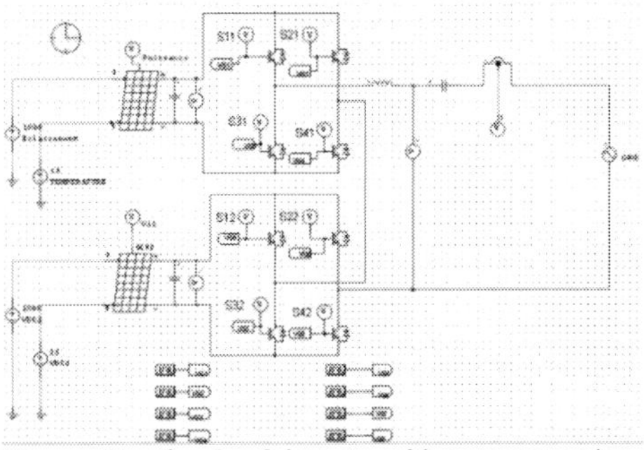

Figure 12: Five level inverter with resonant series filter at 50Hz

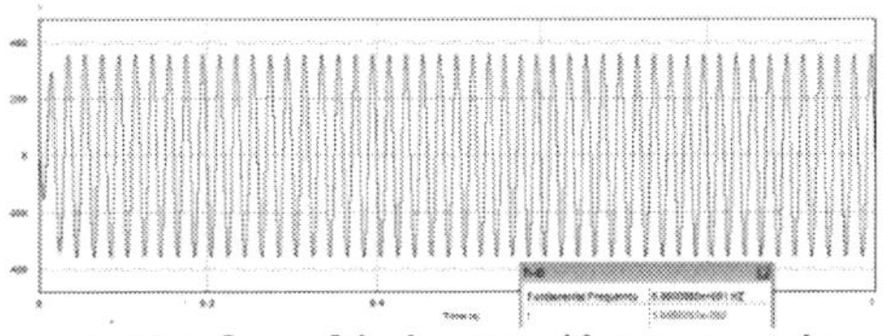

Figure 13: Waveform of the inverter with resonant series filter at 50Hz output current

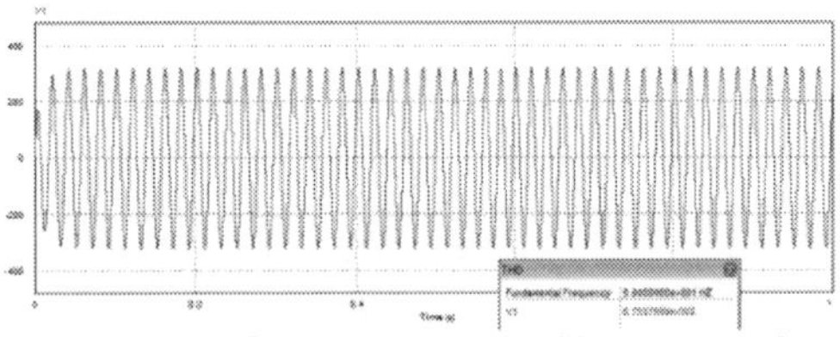

Figure 14: Waveform of the inverter with resonant series filter at 50Hz output voltage

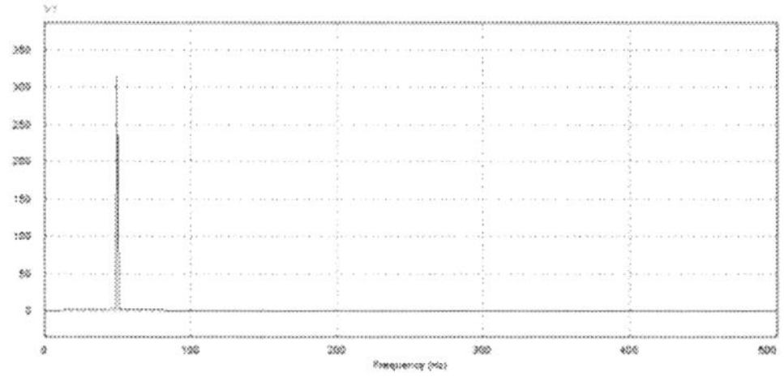

Figure 15: FFT analysis of the inverter with resonant series filter at 50Hz

There is a tremendous reduction in THD voltage from 3.83% given by low pass LC filter to 0.66% by the resonant series filter at 50Hz.

4 CONCLUSION

The simple filter proposed gives good results. First, we have modeled and simulated a single phase five-levels inverter using pre-calculated switching angles for eliminating third and fifth harmonics. Two filters have been used, a low-pass filter and a resonant series one. Simulation results obtained shows that pre-calculated switching technique and low-pass filter has a good sine wave output voltage with THD = 3.8 %. However the proposed resonant series filter improve output voltage THD consequently and it reaches 0.66 %.

4.3 References

[1] Hussein A. Kazem " Harmonic Mitigation Techniques Applied to Power Distribution Networks" Hindawi Publishing Corporation Advances in Power Electronics Volume 2013, Article ID 591680, 10 pages http://dx.doi.org/10.1155/2013/591680*Review* January **2013**

[2] WB / ORD-TE" référentiel technique : harmoniques présentation, prise en considération, solutions pour limitations".Vialis Paris .version v0 du 01.09.2011

[3] Boualem BOUKEZATA, Abdelmadjid CHAOUI, Jean Paul GAUBERT et Mabrouk HACHEMI "Système solaire photovoltaïque connecté au réseau électrique et associe a un filtre actif parallèle" SYMPOSIUM DE GENIE ELECTRIQUE (SGE'14) : EF-EPF-MGE 2014, 8-10 JUILLET 2014, ENS CACHAN, FRANCE

[4] Maha KHANFARA, Rachid ELBACHTIRI ,Karima ELHAMMOUMI, Mohammed BOUSSETTA"A comparison between two filters for PV inverter controlled to eliminate firsts harmonics" will be presented at The 2nd International Conference on Information Technology for Organizations Development March 30 - April 1st, 2016

[5] A. Fri, R. El bachtiri and A. El Ghzizal "Comparative study of H-bridge multilevel inverters dedicated to PV," IEEE International Renewable and Sustainable Energy Conference, 07th-09th March 2013, Ouarzazate Morocco.

[6] A. Fri, R. El Bachtiri and A. El Ghzizal, "A Comparative Study of Various Multicarrier PWM Strategies Single Phase Cascaded Inverter Symmetric 5L For PV Systems" WOTIC'2013, 26th - 27th December 2013 ; Fez, Morocco.

PHOTOVOLTAIC MODULES MONITORING SYSTEM USING A WIRELESS SENSOR NETWORK

E. Ortega [1], G. Aranguren [1], M.J. Sáenz[2], R. Gutiérrez[2] and J.C. Jimeno [2]
[1] Electronic Design Group. University of the Basque Country. Alameda Urquijo S/N, 48013-Bilbao, Bizkaia, Spain.
[2] Technological Institute of Microelectronics. University of the Basque Country. Technological Park, Building 105. 48170-Zamudio, Bizkaia, Spain.

ABSTRACT: In this work the first approach to a wireless sensor network for photovoltaic plant monitoring is presented. This network is designed to detect faults in each individual photovoltaic module of the plant and to localize physically without disconnecting it from the rest of the system. There will be one electronic circuit in parallel for each photovoltaic module and they will be connected to a central control unit with a Wireless Sensor Network. In this work the general wireless network, a first prototype of the electronic circuit for a single photovoltaic module monitoring and several experimental tests with this prototype are presented. The system was designed in terms of easy implementation and low cost, without any extra sensors.
Keywords: PV module, PV system, performance, monitoring, wireless sensor network.

1 INTRODUCTION

Due to the strong increase in the global energy consumption and the future growing predictions it is impossible to generate all the demanded energy from non renewable sources. These resources, in addition of being very polluting have a very slow regeneration speed so the depletion risk is elevated. Because of these, the weight of renewable energies in the energy production is growing sharply. Among them, the solar energy is growing and the number of photovoltaic (PV) plants and the number of PV modules in each of them is rising. Each PV plant could include thousands of PV modules connected in arrays and the failure of a single module usually implies a loss on energy greater than the energy of a single module, sometimes, as much as the energy of all the PV string.

There are a large variety of facts in the origin of these faults such as dust [1], growth of vegetation [2] or wire worsening among others which many times can be easily sorted out if they are detected.

To detect these failures and to improve the PV plant operation monitoring is admitted as an effective path, with measurable effects on terms of the performance ratio, PR.

There are some ways of monitoring the whole PV plant, in some cases adding new sensors of temperature and irradiance [3] which measure the power generated by the PV plant to detect energy losses or through statistical analysis [4] to predict the PV plant behaviour. With these methods, even if it is possible to detect a fault there is not a clear indication of which is the PV module causing that fault. In other works, previously disconnecting the PV modules from the system, it is possible to individually measure them [5].

These two methods imply a loose of productivity, in the first case because it takes more time, when a fault is detected, to detect which is the PV module with the fault and in the second one because the PV module is not producing energy during the monitoring process.

In this work a new approach based on previous work [6] is presented. Every PV module failure has an influence on the I-V characteristic of the module which could be used to detect any malfunctioning. The proposed system is able to detect a fault in each individual PV module without disconnecting it from the array. This approach consists of multiple electronic circuits connected in parallel with each module of the PV System. These circuits are able to measure the voltage in the module and the current injected, obtaining this way the I-V characteristic of each module without modifying the modules circuitry and without interrupting the system operation. This system will be controlled from a central control unit which will be connected to the electronic circuits of each module through a Wireless Sensor Network (WSN).

In addition, a first prototype of the system, with the electronic circuit for a single module and the results of a field test with a PV-module will be presented.

2 DESCRIPTION OF THE SYSTEM

The proposed system consists on the Wireless Module Interconnection Circuit, WMIC, and the Wireless Control System, WCS. There is one WMIC for each PV module and one WCS for the entire PV plant. This system was designed in terms of low cost, reliability, resistance to extreme weather conditions, easy installation and minimum maintenance. The WMIC-WCS connection was made wirelessly to save wiring and electrical protections when connecting the circuits between them, which are at different voltages.

The WMIC consist of an electronic circuit connected in parallel to each PV module. This module has an electronic circuit to monitor the PV module and a digital part, with a control unit and a radio frequency (RF) device for the wireless communication with the WCS. The circuit is powered directly from the PV-module. Furthermore, the WCS consist of a control unit and a RF device and acts as the sink node of the network.

The interconnections of the system can be shown in Fig. 1, were it can be seen several WMIC, one for each PV-module and one WCS for all the system.

In order to save energy during the not operation time the WMIC modules are in low power mode (sleep) and when a monitoring cycle is required the WCS awakes the WMIC modules and they establish the WSN to start with the monitoring process. During the network establishment there are some limitations due to the range of the RF devices so all the WMIC modules cannot connect directly with the WCS module as receptors. Because of that some WMIC nodes act both as receptors and as transmitters spreading the WCS signal. When the network is established, the WCS start ordering sequentially to all the WMIC modules to check their attached PV-module status and the WCS reports the results to a computer. When a PV-module has any fault

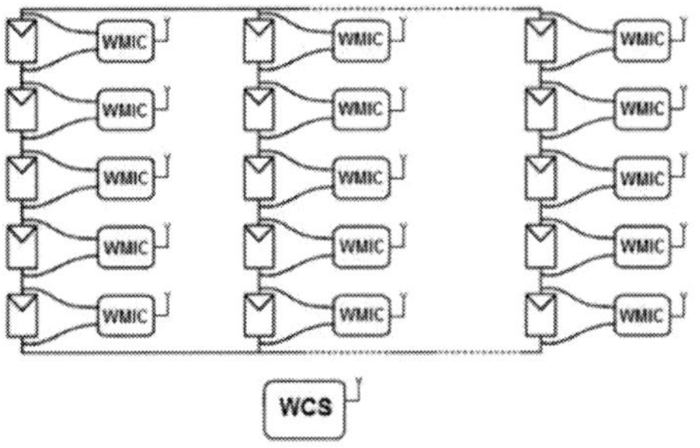

Figure 1: Electrical circuit of the PV array including the Wireless Monitoring System.

the system localizes the attached WMIC module physically in the plant so it can be fixed. After the monitoring process is finished the WMIC modules switch off until next wake up.

The issues related with the network establishment, the WMIC physical localization and the monitoring system development will be discussed in future works.

3 CIRCUIT DEVELOP

For this study, the prototype of the circuit developed has the functionality shown in Fig. 2. Once the WSN is established, the WCS module orders one by one to the WMIC modules to check the status of their PV modules. It is in that moment when the functionality previously mentioned starts operating. The WMIC connects an electronic circuit in parallel to the PV module and produce small variations in the operating point of the modules. From that they obtain 5 simple parameters from the module and transmit these to the computer. It is possible to find the simulated I-V characteristic, the operating point and the operation conditions in terms of illumination level and temperature. From each module a couple of numbers are obtained: the distance from the simulated maximum power point to the estimated power from this module at this operating conditions, what is a measure of the module degradation and the distance from the operating point and the estimated maximum power point, what is a measure of the misfit in the operation point. Checks if the I-V characteristic is correct, otherwise, depending on the value of the two previously explained numbers the system takes the decision about if any module will not be working correctly and it will have a fault.

For the develop of the electronic circuit, beginning from the previous studies [6] which define the expected voltage and current as well as the way to obtain them, a circuit was developed, which diagram is shown in Fig.3.

In the WMIC module, from left to right we have first the electronic circuit connected in parallel to the PV module. That is a simple two capacitors circuit that allows moving the output current from the module to small distances around their operating point. Figure 3 shows a simplified schematic of this circuit with the two capacitors and six switches: 4 externally controlled (transistors) and 2 only controlled by the direction of the current (diodes).

To control this circuit, a microcontroller (MCU) could carry out all the functionalities needed with low size, cost and consumption. Taking into account that, there is not a high volume of data and processing and that the number of nodes in the WSN is very high a FPGA as

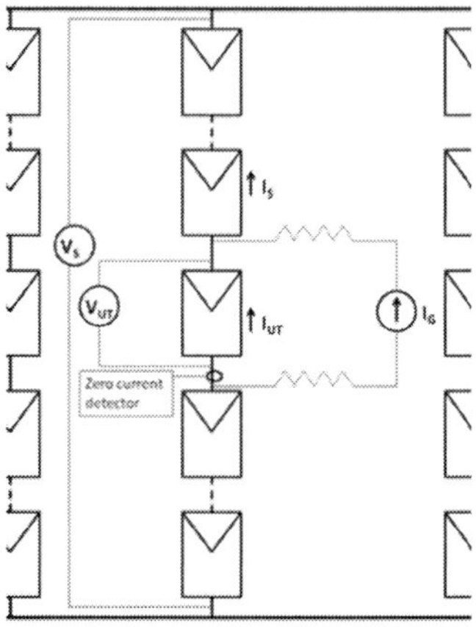

Figure 2: Setup used for the module under test. The IG current generator and the VUT Voltmeter are connected to the module trough their WMIC

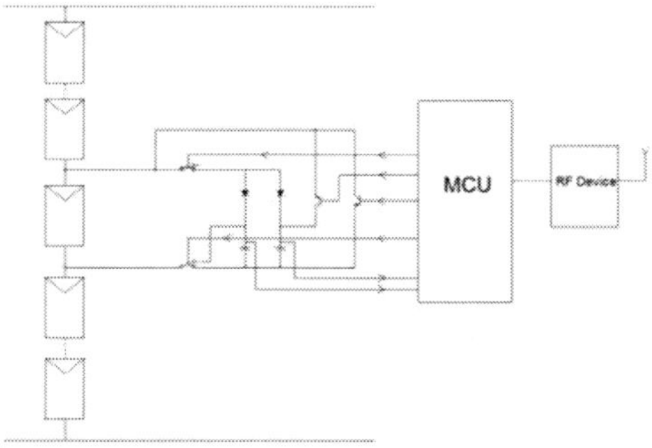

Figure 3: WMIC Module. From left to right: the PV array, the electronic circuit connected in parallel to the PV module, the MCU and the RF device for the wireless communication.

in other studies of WSN [7] was not appropriate. This MCU will stay in sleep mode, with minimum power consumption, until it receives the wake up order from the WCS. In that moment, after the WSN is established the MCU switches the circuit to connect it with the PV module. With the circuit some analogical signals are obtained and the MCU reads them, process them, determines the PV module status and sends the results to the WCS trough the RF device.

For the WSN, several alternatives were considered according to other studies: Bluetooth, also for PV modules monitoring [8], ZigBee, for vineyards [9] or railway [10] monitoring, 6LOWPAN, WIFI and the sub 1 GHz band. Finally, the sub 1GHz was chosen, with technology of Texas Instruments, due to the low data rate to transmit and the high number of nodes in the network, with the objective of minimizing the cost per module.

To carry out the test of this system a prototype of the WMIC module was developed as it can be seen in figure 4.

4 EXPERIMENTAL RESULTS

Using a couple of PV systems an experiment has been realized to test the prototype of the WMIC module. One of the systems is shown in figure 5, is a 1.2 kW system formed by modules with different number of cells

Figure 4: WMIC Prototype

Figure 5: 1.2 KW PV experimental system.

Figure 6: 11 KW partially shadowed PV system for testing

and with an open electrical configuration in order to simulate different malfunctions behaviors. The other, shown in figure 6, is an 11 kW PV system subjected a large partial shadows form the surrounding buildings.

With these two systems, the WMIC prototype of figure 4 has been connected in parallel to a PV module. The microcontroller and the digital part are powered from the PV module through a DC/DC circuit and the analogical part directly from the module.

Switching the circuit with the microcontroller to different states, the capacity of the circuit to inject current on the PV module and to measure that current and the changes on the PV module voltage has been tested. With these measures, in the future it will be possible to develop an algorithm on the microcontroller which will rebuild the I-V characteristic of the PV module. In addition, it will measure the distance between the theoretical operating point and real working point of the PV module.

The switching process of the WMIC has three steps, controlled by three signals of the microcontroller. The first one, called S1, is the idle state. During the normal working of the WMIC the circuit is in this state, with minimum power consumption. When the monitoring process starts, the microcontroller changes the circuit to the next state, S2, where current is injected to the PV module in one direction and during the next state, S3, in the opposite direction.

The prototype has three measure points. The first one for the module voltage through a voltage divider, to adapt the module voltage to the microcontroller voltage (5V). The other two measure points are for the current, one in each direction. The value of the current is obtained measuring the voltage in a 1 ohm resistor.

When the switching process starts, the microcontroller also starts sampling these measuring points. In order to avoid high frequency noise during the acquisition process a 1 kHz low-pass filter was added at the output of the analogical circuit, before the ADC inputs of the microcontroller.

In figure 7 it can be seen the three steps of the circuit controlled by three signals of the microcontroller and the variations of the module voltage and the injected current on the module (in the two directions).

In figure 8 a zoom of the signals in the transition from the S1 state to the S2 state is shown. As it can be seen, the injected current causes a peak also in the module voltage of the same sign of the current. With an injected current of negative 120mA (signal 2) a negative peak of more than 300mV of amplitude is induced on the PV module with a pulse length of 5ms approximately.

In figure 9 is shown a zoom of the transition from S2 to S3, with current injected on the other way. As it can be seen a voltage peak of similar amplitude but opposite sign to the generated on the transition S1 to S2 appears on the module voltage. With an injection of positive 120mA a positive peak of 200mV of amplitude and 1.5ms of length is induced on the PV module.

Lastly, in figure 10 the transition from S3 back to S1 is shown. This transition is to come back to the idle state, it generates a smaller disruption on the PV module but it has no relevance for this study.

With these two current injections (from S1 to S2 and form S2 to S3), one positive and the other one negative,

the PV module has been disrupted. Modifying the output current form of the module to small distances around his working point. With this currents and their corresponding module voltage it is possible to reconstruct in the microcontroller the I-V characteristic of the PV module and to measure the degradation level of the module.

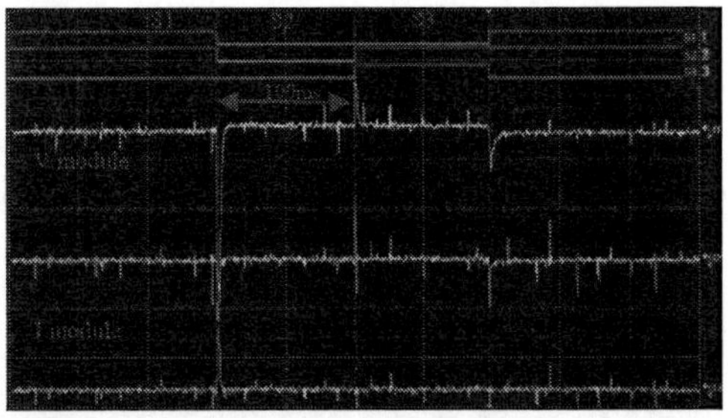

Figure 7: Switching states and signals of the module.

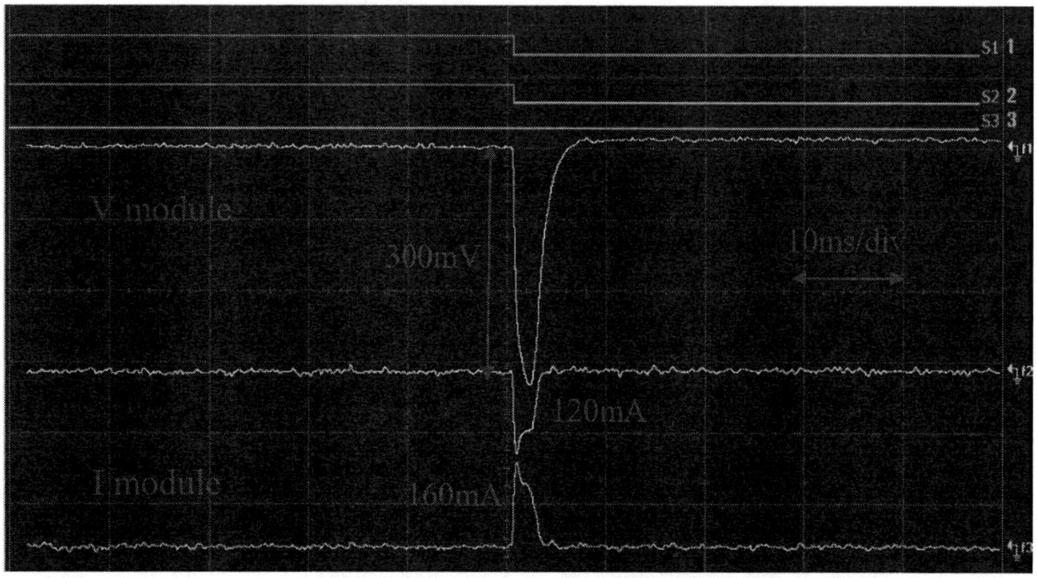

Figure 8: Signal in the transition from S1 to S2

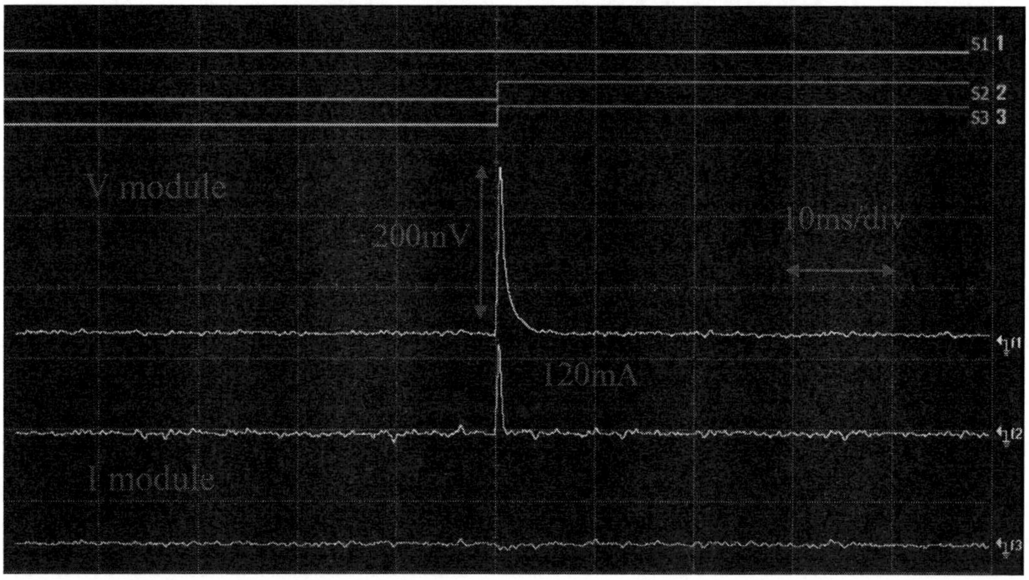

Figure 9: Signal in the transition from S2 to S3

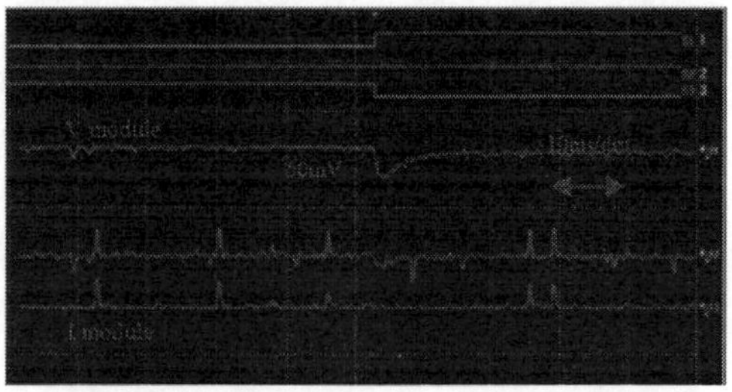

Figure 10: Signal in the transition from S3 to S1

6 CONCLUSIONS

A novel system for wireless PV plant monitoring and a first prototype of the system for a single PV module have been presented. The designed monitoring circuit allows to individually measure each PV module and to detect individual faults in the PV plant which increases the PV plant productivity and decrease economic losses. The wireless monitoring system is based in two circuits, one WMIC for each PV module and one WCS for all the PV system. The WMIC has an electronic circuit and a digital part for processing and communications and the WCS is the central node of the network. The sub 1 GHz band chosen for the WSN wireless communication offers a low cost solution for wireless PV modules monitoring systems in comparation with other typical comercial bands.

This work has proven that tinitial approach for the wireless PV modules monitoring system, which consisted on producing small variations in the operating point of the modules by modifying its output current and to measure these variations without any extra sensors, is valid. In addition, the circuit has been proven with a prototype of the PV module monitoring circuit being able to observe the voltage variations with the current injections.

With the goal of minimizing energy losses a new prototype will be developed with which it will be possible to shortcut failing PV module to allow to the rest of the array to continue working. In addition, with the improvement of this new prototype we pretend to minimize the electronic components costs and the energy consumption.

Afterwards, an algorithm to rebuild the I-V characteristic curve of the PV module from the obtained data will be developed. Comparing the obtained I-V curve with the theoretical the degradation of the PV module will be determined.

With the aim of connecting the entire system, a dynamic and multi-hop network is going to be developed communicating each WMIC with the WCS and to detect WMICs physical position on the PV plant when a module failures we are going to work in different techniques.

Finally, we are going to work on the WCS, to create a graphic interface from where all the PV systems could be monitored from a central point.

7 REFERENCES

[1] M. Seyedmahmoudian, S. Mekhilef, R. Rahmani, R. Yusof y E. Taslimi Renani, «Analytical Modeling of Partially Shaded Photovoltaic Systems,» *MDPI. Energies,* vol. 51, pp. 128-144, 2013.

[2] M. Catelani, L. Ciani, L. Cristaldi, M. Faifer, M. Lazzaroni y M. Rossi, «Characterization of photovoltaic panels: The effects of dust,» *2nd IEEE Energycon. Conference & Exhibition. Advances in Energy Conversion Symp,* 2012.

[3] A. Chouder y S. Silvestre, «Automatic supervision and fault detection of PV systems based on power losses analysis,» *Energy Conversion and Management,* vol. 51, pp. 1929-1937, 2010.

[4] A. Carretero y H. Hoffmann, «Where have my watts gone? Five years of utility scale PV power plant monitoring,» *28th European Photovoltaic Solar Energy Conference and Exhibition,* pp. 4045-4050, 2013.

[5] Y. Hirata, S. Noro, T. Aoki y A. Shiwa, «Simple method to measure I-V curve of photovoltaic failures compared with other conventional method,» *28th European Photovoltaix Solar Energy Conference and Exhibition,* pp. 3695-3698, 2013.

[6] A. Ulasenka, J. Jimeno, G. Obieta, R. Gutierrez y M. Saenz, «Module to module monitoring system, M3S, a new strategy for PV-System monitoring,» *42th IEEE Photovoltaic Specialist Conference,* 2015.

[7] D. Manoranjan y P. Banoj Kumar, «Prototyping a wireless sensor node using FPGA for mines safety application,» *International Journal on Electrical & Power Engineering,* vol. 2, nº 2, 2011.

[8] M. Methfessel, S. Lange, R. Kraemer, M. Zessack, P. Kollermann y S. Peter, «Real-Life deployment of bluetooth scatternets for wireless sensor networks,» *Real-World wireless sensor networks,* vol. 1, pp. 43-51, 2014.

[9] C. Alippi, G. Boracchi, R. Camplani y M. Roveri, «Wireless sensor networks for monitoring vineyards,» *Methodologies and Technologies for Networked Enterprises,* pp. 295-310, 2012.

[10] H. Wang y J. Yu, «The design and implementation of smart monitoring system for large-scale railway maintenance equipment cab based on ZigBee wireless sensor network,» *Sensors & Transducers Journal,* vol. 173, pp. 177-188, 2014.

EXPERIMENTAL EVALUATION OF THE SOLAR RADIATION GAINS OVER PHOTOVOLTAIC CELLS DUE TO THE USE OF TiO_2 TREATED SURFACES. APPLICATIONS TO PHOTOVOLTAIC SYSTEMS WITH MICRO-INVERTERS

Isidoro Lillo[1], Rocío Domínguez[2], Miguel Larrañeta[2] and Manuel Silva[1]
1 Department of Energy Engineering. University of Seville. Spain
2 Andalusian Association for Research and Industrial Cooperation. Group of Thermodynamics and Renewable Energy.
Camino de los Descubrimientos s/n, 41092, Seville, Spain. Phone/Fax number: (+34)954487236/(+34)954487233.
E-mail: isidorolillo@etsi.us.es

ABSTRACT: In building integrated photovoltaic (PV) systems, the reflection of the solar radiation by the walls of the buildings may cause a non-uniform distribution of the incident radiation on the different PV modules. However, by using micro-inverters at strings levels, electricity production may increase instead of provoking losses by mismatching. For this purpose, the maximum possible solar gain due to reflections is required.
The annual profit of the solar radiation depends on factors such as climate, latitude and the relative position of the cell to the reflective surface. In the location of Seville (Spain), the gain on a PV cell generation due to radiation reflected by a TiO_2 treated perpendicular surface reaches a 13% for an entire year. Therefore, it is well demonstrated that using micro-inverters not only involves a loss reduction of a solar installation but also to obtain a profit from the solar radiation that cannot be achieved with centralized inverters.
Keywords: TiO_2, Solar radiation, Photovoltaic.

1 INTRODUCTION

In building integrated PV systems, the solar radiation distribution on the different PV modules can be inhomogeneous. This is due to the presence of partial shadows, dirt and/or PV modules with very different orientations and inclinations.

The inhomogeneous distribution of the solar radiation on a PV module usually produces significant reductions on the electrical production of the installation due to mismatching. A way to reduce these energy losses due to partial shadows and dirt is using micro-inverters at cell, string or module levels [1-7]. The benefits of micro-inverters against of centralized inverters have been extensively analyzed in the scientific literature [8-20].

Nevertheless the use of micro-inverters can encourage another phenomenon that occurs in the building sector. Sometimes due to structural elements of the building and its surrounding the reflection of the solar radiation occurs on the PV modules. These reflections of radiation produce a gain of the global total radiation on cells, strings or modules. If mismatching losses between cells, string or modules are avoided by using micro-inverters, the electric production of the PV system can be increased due to its increased solar radiation gains.

To achieve the maximum reflections is necessary to dispose of superficially treated surfaces with highly reflective materials. Among the most common reflective materials used on the exterior of the buildings include aluminum, galvanized steel, white cement or coatings finished with white paint highlight.

One of the pigments that have better reflective qualities is the TiO_2. This is the whitest pigment available in the market. With a low price, it has very good reflective properties in the visible range, reflecting around 90% of the incident radiation. It has a high refractive index, making it a material with a high opacifying power. TiO_2 also has a wide range of application in nano-technology field, as it shows photocatalytic and super-hidrophilic properties that make it a self-cleaning material, used in plenty of solar applications like PV or building integration [21-35].

The goal of this study is to evaluate the use of TiO_2 as a reflective pigment to increase the gain of reflective incident irradiance on PV modules that have micro-inverters at string or module level. To this the solar radiation gain of a cell that receives reflection respect of other that not receives has been experimentally measured. Besides this gain, the spectral distribution of reflection radiation by the TiO_2 treated surface is analyzed.

As Figure 1 shows, it has been dealt a vertical surface facing with TiO_2 towards south over a horizontal monocrystalline silicon calibrated solar cell (cell 2) located outside. Simultaneously another identical calibrated cell (cell 1) was placed on a horizontal surface without reflection. We assessed the gain of solar radiation that receives the reflected cell (cell 2) versus the non-reflected cell (cell 1) in different types of days.

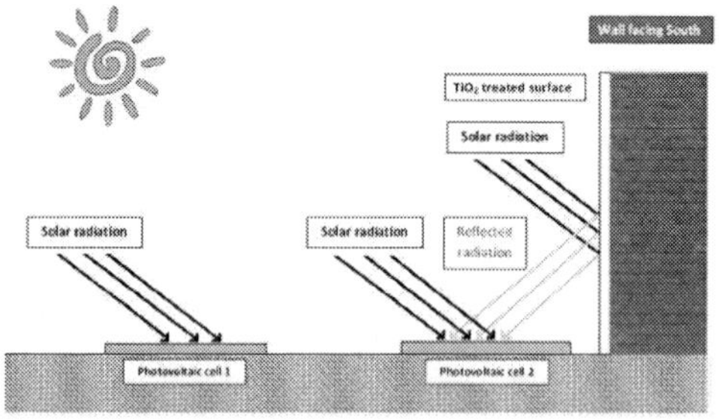

Figure 1. Experimental installation.

2 RESULTS

The solar radiation gain received by the cell depends on geometrical considerations defined by the relative position between the sun, the reflecting surface and the solar cell. Climatic considerations as the level of irradiance at each time, the value of the direct solar radiation and the composition of the atmosphere also affect to the value of the solar radiation gain. The existing combinations are limitless.

By way of example, Figure 2 and Figure 3 show the incident solar irradiance between cell 1 and cell 2 along a cloudy day and another clear day. The vertical plane of the reflecting surface is facing south.

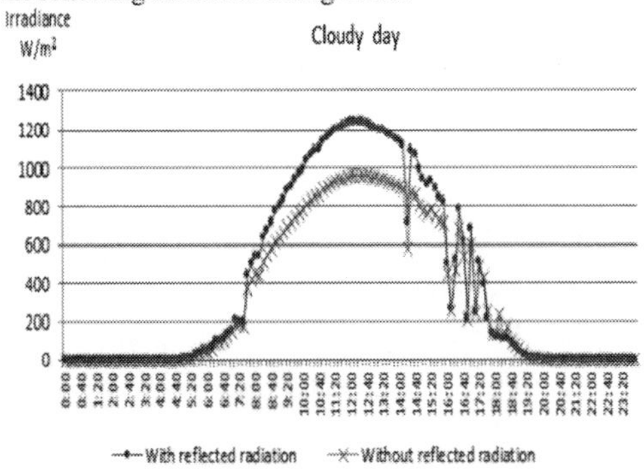

Figure 2. Solar radiation on the reflected and non-reflected cell for a cloudy day.

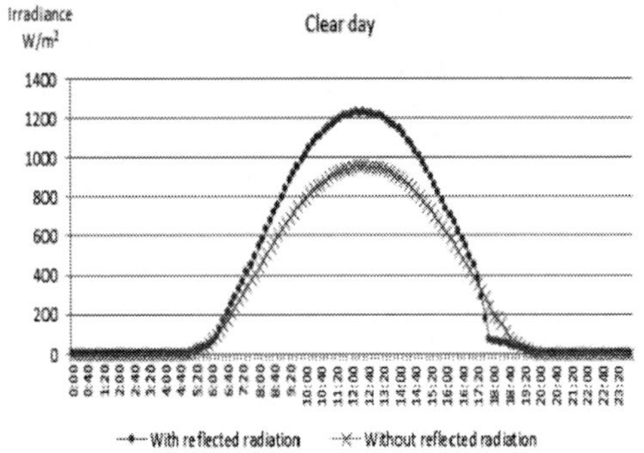

Figure 3. Solar radiation on the reflected and non-reflected cell for a clear day.

On the basis of the results obtained of solar radiation gain for each relative position, the results have been extrapolated for a typical meteorological year of Seville (Spain). It is regarded that the analyzed cell is in a horizontal position and the reflecting surface of TiO_2 is facing south.

Furthermore, it has been experimentally established that in the Northern Hemisphere for solar altitudes between 10° and 35°, in the morning and with vertical reflective surfaces south-east facing, the gains reach maximum values of 150%.

Solar radiation gain not only has been measured in absolute value, since reflected radiation presents a specific spectral distribution in function of the reflective material. Figure 4 ans Figure 5 show, in function of waveleength, the behaviour of the reflecting surface treated with TiO_2.

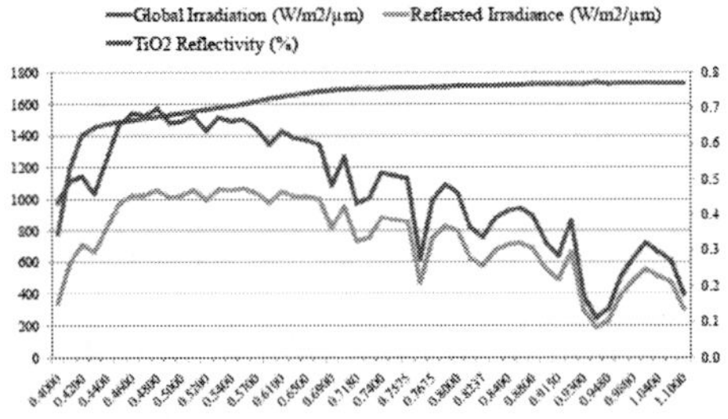

Figure 4. Spectral reflectivity of the surface treated with TiO_2.

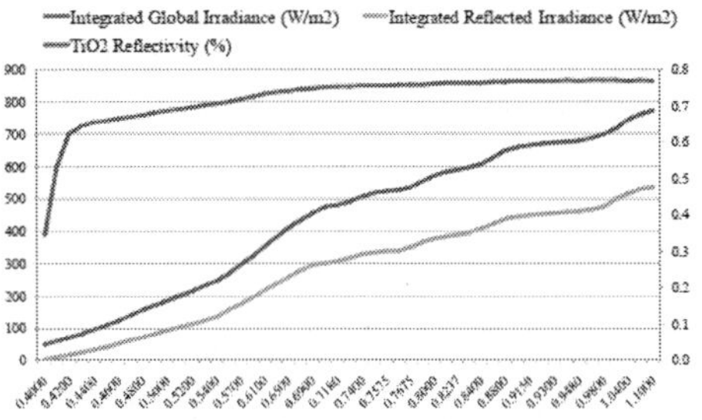

Figure 5. Accumulated values of solar irradiance and reflected irradiance of a surface treated with TiO_2.

It is noted as above a wavelength of 700nm the reflectivity of the surface is between 75% and 80%. For wavelengths of 400nm the reflectivity decreases in order of 35%, that is, the reflectivity of the reflecting surface improves as the wavelength increases.

On the other hand, each PV technology has its own spectral response and thus their behaviour varies depending on the spectral distribution of the incident solar radiation, both radiation reflected and radiation directly reflected.

Figure 6 shows a comparative analysis of the intensity that would be generated for the same reflected solar radiation for five different PV technologies, monocrystalline, polycristalline, amorphous, CdTe and CIGS due to the spectral response of each technology in function on the spectral distribution of the incident solar radiation.

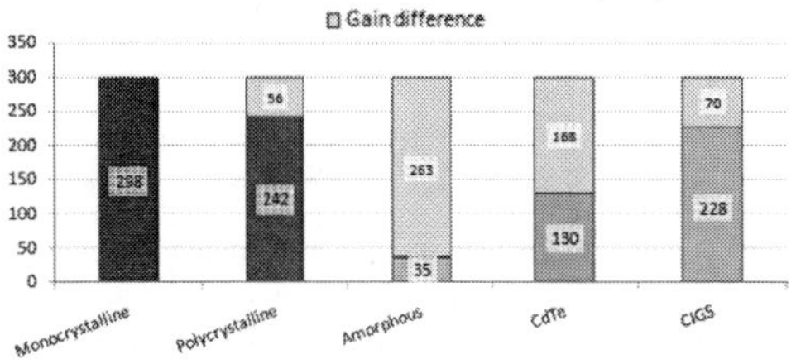

Figure 6. Comparative analysis of the intensity produced by each technology to the same spectral radiation emited by a surface treated with TiO_2.

Figure 6 shows that the monocrystalline silicon PV technology presents the best response. In contrast, the amorphous silicon PV technology presents the worst response with a decrease of 88% of net gain in the conversion of incident solar radiation into electrical production.

3 CONCLUSIONS

The results show that the use of these painted surfaces can significantly increase the PV production in buildings. The use of TiO_2 pigment as reflector achieves an homogeneous distribution of the additional solar radiation reflected with appreciable results and without incurring in huge costs.

The annual profit in solar radiation depends on the relative sun-surface-cell position and the amount of direct solar radiation and weather conditions.

The electrical production gain at the output of the module depends on the solar radiation gain and the spectral response of the PV technology.

Monocrystalline silicon technology obtains higher electrical production gains compared to other PV technologies.

Therefore, it is well demonstrated that using micro-inverters not only involves a loss reduction of a PV solar installation, but also allows to obtain a profit from the solar radiatoin that cannot be achieved with centralized inverters.

4 REFERENCES

[1] Micro-inverter improves array reliability. Limpscombe, Gail. 2008, Solar Today, Volume 22, Issue 7, p. 58.

[2] Performance comparison between micro-inverter and string-inverter Photovoltaic Systems. Fabio Famoso, Rosario Lanzafame, Simone Maenza, Pier Francesco Scandura. 2014. 2014, Energy Procedia,Volume 81, pp. 526-539.

[3] Control and circuit techniques to mitigate partial shading effects in photovoltaic arrays . Ali Bidram, Ali Davoudi, Robert S. Balog. 2012. 2012, IEEE Journal of Photovoltaics, 2 (4), pp. 532-546.

[4] Micro-inverters Promising solutions in solar photovoltaics. Hadeed Ahmed Sher, Khaled E.Addoweesh. 2012. 2012, Energy for Sustainable Development 16 , pp. 389-400.

[5] A comprehensive techno-economic review of microinverters for Building Integrated Photovoltaics (BIPV). Hanumath Prasad Ikkurti, Suman Saha. 2015. 2015, Renewable and Sustainable Energy Reviews 47, pp. 997-1006.

[6] Microinverters and power optimizers: Power conversion migrates down to the PV panel. Conner, Margery S. 2009. 2009, EDN Volume 54, Issue 24, pp. 24-27.

[7] A Review of Single-Phase Grid-Connected Inverters for Photovoltaic Modules. Soeren Baekhoej Kjaer, John K. Pedersen, Frede Blaabjerg. 2005. 2005, IEEE Transactions on Industry Applications,41 (5), pp. 1292-1306.

[8] "Partial Shade Evaluation of Distributed Power Electronics for Photovoltaic System," Deline, J. Meydbray, M. Donovan, and J. Forrest, 2012. IEEE Photovoltaic Specialists Conference Austin, Texas. NREL/CP-5200-54039.

[9] A performance and economic analysis of distributed power electronics in photovoltaic systems, C. Deline, B. Marion, J. Granata, S. Gonzales, NREL Technical Report, NREL/TP-5200-50003, January 2011.

[10] "Analysis of potential for mitigation of building integrated PV array shading losses through use of distributed power converters," S.M. MacAlpine, M.J.Brandemuehl, R. W.Erickson, in Proc. 4Th Int. Conf.on Energy Sustinability, ES2010, May 17-22. 2010.

[11] "Characterizing Shading Losses on Partially Shaded PV Systems, C. Deline, 2010. PV Performance Modeling Workshop. NREL/PR-520-49504.

[12] "Photovoltaic Shading Testbed for Module-Level Power Electronics," C. Deline, J. Meydbrey, M. Donovan and J. Forrest, NREL Technical Report NREL/TP-5200-54876 (2012).

[13] "Analaysis of potential for mitigation of building-integrated PV array shading losses through use of distributed power converters," S.M. MacAlpine, M.J.Brandemuehl, R.W.Erickson, in Proc. 4Th Int. Conf. on Energy Sustainability, ES2010, May 17-22.2010.

[14] Performance of Mismatched PV Systems with Submodule Integrated Converters," C. Olalla, C. Deline, D. Maksimovic, "IEEE J. Photovoltaics, 2013.

[15] "Potential for power recovery: simulated use of distributed power converters at various levels in partially shaded photovoltaic arrays," S.M. MacAlpine, M.J.Brandemuehl, R.W.Erickson, in Proc. PVSEC 2011.

[16] "Control of Submodule Integrated Converters in the Isolated - Port Differential Power Processing Photovoltaic Architecture," Y. Levron, D. Clement, B. Choi, C. Olalla, D. Maksimovic, and S. Member, IEEE J. Emerg. Sel. Top. Power Electron., 2014.

[17] "Architectures and Control of Submodule Integrated DC – DC Converters for Photovoltaic Applications," C. Olalla, D. Clement, M. Rodriguez, and D. Maksimovic, IEEETrans. Power Electron., vol. -, no. -, pp. 1–18, 2012.

[18] "Generation Control Circuit for Photovoltaic Modules," T. Shimizu, M. Hirakata, T. Kamezawa, and H. Watanabe, IEEE Trans. Power Electron., vol. 16, no. 3, pp. 293–300, 2001 K. Kim, P. Shenoy, and P. Krein, "Converter Rating Analysis For Photovoltaic Differential Power Processing Systems," IEEE Trans. Power Electron., vol. 8993, no. c, pp. 1–1, 2014.

[19] "Differential Power Processing for Increased Energy Production and Reliability of Photovoltaic Systems," P. Shenoy, K. A. Kim, B. B. Johnson, and P. T. Krein, IEEE Trans. Power Electron., vol. 28, no. 6, pp. 1–19, 2013.

[20] "Performance of Power Limited Differential Power Processing Architectures in Mismatched PV Systems," C. Olalla, C. Deline, D. Clement, S.Member, Y. Levron, M. Rodriguez, D. Maksimovic, and S. Member, IEEE Trans. Power Electron., pp. 1–13, 2014.

[21] TiO2 pigment technology: a review. Juergen H.Braun, Andrejs Baidins, , Robert E.Marganski. 1992. 1992, Progress in Organic Coatings 20, pp. 105-138.

[22] Mini review on photocatalyst of titanium dioxide nanoparticles and their solar applications. Yucheng Lan, Yalin Lu, Zhinfeng Ren. 2013, Nano Energy , págs. 1031-1045.

[23] Nanostructured Titanium Dioxide Materials: Properties, Preparation and Applications. Alireza Khatase, G.Ali Mansoori. 2011, World Scientific.

[24] Titanium Dioxide Photocatalysis: present situation and future approaches. Akira Fujishima, Xintong Zhang. 2006. 2006, C.R. Chimie 9, pp. 750-760.

[25] TiO2 photocatalysis: Design and applications. Kazuya Nakata, Akira Fujishima. 2012. 2012, Journal of Photochemistry and Photobiology C: Photochemistry Reviews 13, pp. 169-189.

[26] Influence of TiO2 nanoparticle synthesis on the properties of thin coatings. Nina Barth, Mandy Zimmermann, Andre Erek Becker, Tobias Graumann, Georg Garnweiter, Arno Kwade. 2015. 2015, Thin Solid Films, pp. 20-27.

[27] TiO2 photocatalysis and related surface phenomena. Akira Fijishima, Xintong Zhang, Donald A.Tryk. 2008. 2008, Surface Science Reports 63, pp. 515-582.
[28] TiO2 photocatalysis Design and application. Kazuya Nakata, Akira Fujishima. 2012. 2012, Journal of Photochemistry and Photobiology C: Photochemistry Reviews 13, pp. 169-189.
[29] Photocatalytic activity of TiO2–SiO2 nanocomposites applied to buildings. Luis Pinho, Maria J.Mosquera. 2013. 2013, Applied Catalysis B: Environmental 134-135, pp. 205-221.
[30] The effect of process parameters on the structure, photocatalytic and self-cleaning properties of TiO2 and Ag-TiO2 coatings deposited using reactive magnetron sputtering. P.Navabpour, S.Ostovarpous, J.Hampshire, P.Kelly, J.Verran, K.Cooke. 571, Manchester : s.n., 2014, Thin solid films 571, págs. 75-83.
[31] Studies on photokilling of bacteria on TiO2 thin film. Kayano Sunada, Toshiya Watanabe, Kazuhito Hashimoto. 2003. 2003, Journal of Photochemistry and hotobiology A: Chemistry 156, pp. 227-233.
[32] Preparation of a highly-reflective TiO2/SiO2/Ag thin film with self-cleaning properties by magnetron sputtering for solar front reflectors. Y.J.Xu, J.X.Liao, Q.W.Cai, X.X.Yang. 2013, Solar Energy Materials & Solar Cells 113, págs. 7-12.
[33] Comparison of self-cleaning properties of three titania coatings on float glass. Minna Piispanen, Leena Hupa. 2011. 2011, Applied Surface Science 258, pp. 1126-1131.
[34] The effects of TiO2 pigmented coatings characteristics on temperature and brightness of a coated black substrate. Mehdi Baneshi, Shigenao Maruyama, Atsuki Komiya. 2012. 2012, Solar Energy 86, pp. 200-207.
[35] Applications of TiO2 nanotube arrays in environmental and energy fields: A review. Qingxiang Zhou, Zhi Fang, Jing Li, Mengyun Wang. 2015. 2015, Microporuos and Mesoporous Materials 202, pp. 22-35.

NON-UNIFORMITY MEASUREMENTS OF A STEADY STATE SOLAR SIMULATOR USING THE HISHIKAWA-HASHIMOTO METHOD AND SUBSEQUENT IMPROVEMENT

Bernd Doll[1], Ulrich Hoyer[1], Markus Hofer[1], Tobias Pickel[1], Claudia Buerhop-Lutz[1] Christian Camus[1], Jens Hauch[1], Christoph Brabec[1,2]

[1] Bavarian Center for Applied Energy Research, Haberstraße 2a, 91058 Erlangen, Germany
[2] Materials for Electronics and Energy Technology (i-MEET), Friedrich-Alexander-Universität Erlangen-Nürnberg, Martensstraße 7, 91058 Erlangen, Germany
e-mail: bernd.doll@zae-bayern.de

ABSTRACT: A steady state solar simulator has been set up at the ZAE Bayern. The system is meant to be used for the characterization of photovoltaic (PV) of various technologies. In particular, it can be used for pre-conditioning of thin film modules in order to enable reliable power and *I-V* measurements. Furthermore, it allows for determining the temperature coefficient of PV modules of various technologies. The latter application will also be discussed in this work. Here, the optimization of the steady state solar simulator based on a method originally introduced by Hishikawa et al. for pulsed solar simulators is described and the results of this optimization are presented.

Keywords: Characterization, Experimental Methods, Light-Soaking, Qualification and Testing

1 INTRODUCTION

A typical steady state solar simulator consists of various components that influence the distribution of the light: usually several lamps, mirrors, reflectors and other optical components. Therefore, the task to optimize the solar simulator towards good uniformity of the irradiation is very time consuming because traditionally the measurement is done using the short circuit current values of one or more solar cells moved over the entire area.

In 2008 Hishikawa et al. [1] presented a method to measure the lateral homogeneity of pulsed solar simulators using a camera based system in very short time. In this manuscript, we present the use of this technique to measure and improve the lateral non-uniformity of a steady state solar simulator for an accelerated adjustment of the lamps of the solar simulator and other optical components. The raw measurements of the camera are corrected for the optic's influences and the measurement method is validated using the traditional method of scrolling a reference cell over the one sun plane.

Using this technique, a steady state solar simulator for full-size PV modules is built up with a low non-homogeneity value, resulting in good measurement and test results.

2 TECHNICAL DETAILS

The steady state solar simulator illuminates an area of 200 cm x 170 cm at an intensity of approximately 1 sun. It constitutes of six halogen-metal vapor lamps of 2000 W each, which are operated via electronic ballasts. The system runs at 32 A three phase current, which is distributed to the electronic ballasts and the cooling fans via an automatic circuit breaker. The setup also consists of an air ventilation system for heat removal and two mirrors at the sides of the setup for adjusting the homogeneity of the illumination.

For safety reasons, the UV part of the irradiation was artificially cut off. The setup is built in a mobile version so that it can be moved freely within the building.

3 EXPERIMENTAL DETAILS

The uniformity measurement was done after the Hishikawa-Hashimoto method reported in [1]. The technique consists of measuring the irradiance distribution in the plane of one sun two dimensionally using a Panasonic DMC-FZ45 digital camera with consecutive analysis with a computer and classification according to the standard IEC 60904-9 with respect to the desired class A, B or C. The analyzed areas of one sun with an irradiance of 1000 W/m² (Fig. 1) are either 160 cm x 150 cm or 120 cm x 120 cm large.

In order to characterize the homogeneity of the distribution of the steady state solar simulator's irradiation a homogeneously colored grey cloth was fixed in the illuminated module plane. The intensity distribution obtained from this illuminated cloth served as basis for further analysis of the irradiation emitted by the lamps.

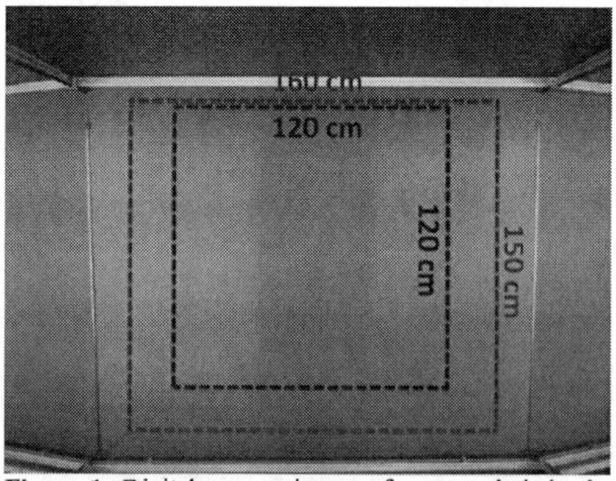

Figure 1: Digital camera image of a grey cloth in the module plane with non-optimized optical configuration and two different analyzed one sun areas with edge length of either 160 cm x 150 cm or 120 cm x 120 cm.

To show the power of the technique the raw measurement image of a non-optimized configuration is shown in Fig. 2 a) (Fig. already converted into W/m²). Clearly brighter horizontal areas can be observed alternating with less bright areas, which are caused by the three levels of the lamps. Also the reflection of the mirrors on the left and right side can be seen. In this configuration, the mirror reflections cause a darker stripe in the center.

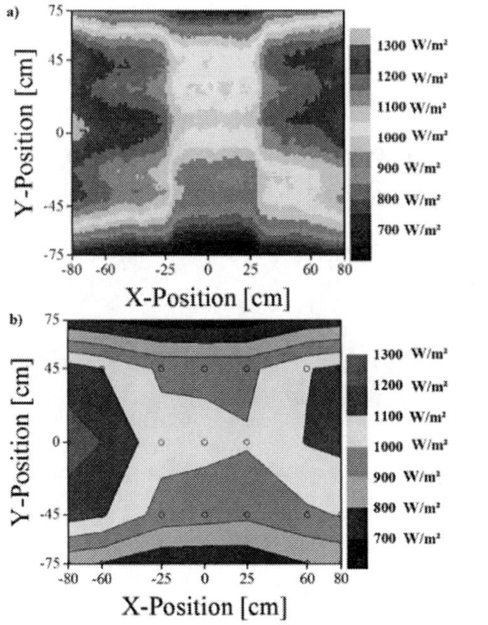

Figure 2: a) Corrected converted and calibrated image with the Hishikawa-Hashimoto method of the irradiance distribution. b) Contour-plot of the irradiance distribution generated by traditional measurements with a silicon reference cell scrolling along the measuring points.

In the next step, the image is corrected for noise and vignetting and calibrated. The camera measurements were corrected in three steps with respect to the influences of the optics by dividing the raw measurements by an image of a uniformly illuminated area: 1) Reducing the noise, by reduce the size of the image and the amount of pixels; 2) flat field correction for higher linearity of the camera intensities [2] and 3) calibration of the camera Pixels with a Si-cell. Therefore, traditional measurements are done using a crystalline silicon solar cell: the cell is moved over the illuminated area and the resulting short circuit current is measured. Since the short circuit current is proportional to the intensity of the irradiation the intensity can be calculated using a known current value at 1000 W/m². The calibration to irradiation in the units of W/m² is done by comparing the minimum and maximum values of the two different methods with a pixel size matching the reference cell area. Thus, the digital values of the camera can be assigned to irradiation values. The result for the digital image and reference cell scrolling is shown in Fig. 2 a) and b). The comparison of the different methods shows a very good agreement, whereas the camera image exhibits a much better resolution. The darker stripe in the middle of Fig. 1 corresponds to a decrease of the irradiation of approximately 100 W/m².

With the validated Hishikawa-Hashimoto technique the effects of different optical configurations on the distribution of the irradiation intensity distribution can be measured very quickly. In Fig. 3, the effect of different mirror angles is shown. In Fig. 3 a) the mirrors are too steep, leaving a darker gap in the center. When the mirrors are flatter in Fig. 3 b) the intensity gap closes and more light is reflected towards the middle. In the extreme case in Fig. 3 d) the mirrors act as concentrators for the light, resulting in a very inhomogeneous distribution.

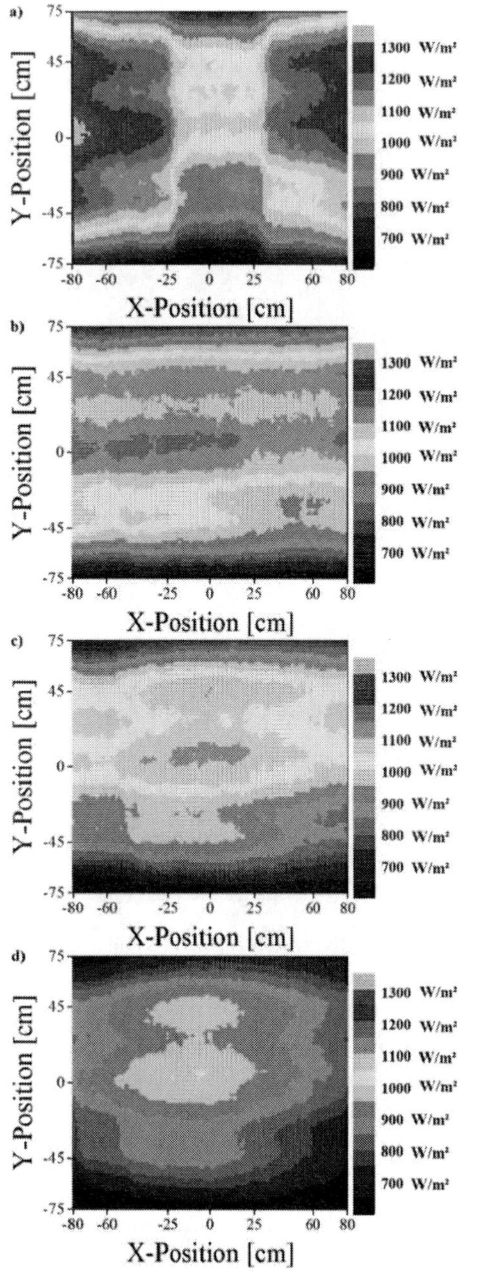

Figure 3: Irradiance distributions for mirror position optimization with no lamp tilting and both mirror at the same position for a) mirror position 0 cm, b) mirror position 20 cm, c) mirror position 34.5 cm, d) mirror position 45 cm.

The most homogenous irradiation distribution setup was found with a lamp tilt angle for the upper and lower lamps of 2.62° and mirror positions for the left one 25.5 cm and 20 cm for the right one and is reported together with the deviation to one sun in Fig. 4. The measurements only exhibit deviations slightly higher than 10% in the lower left corner. In all other areas deviation below 10% were measured.

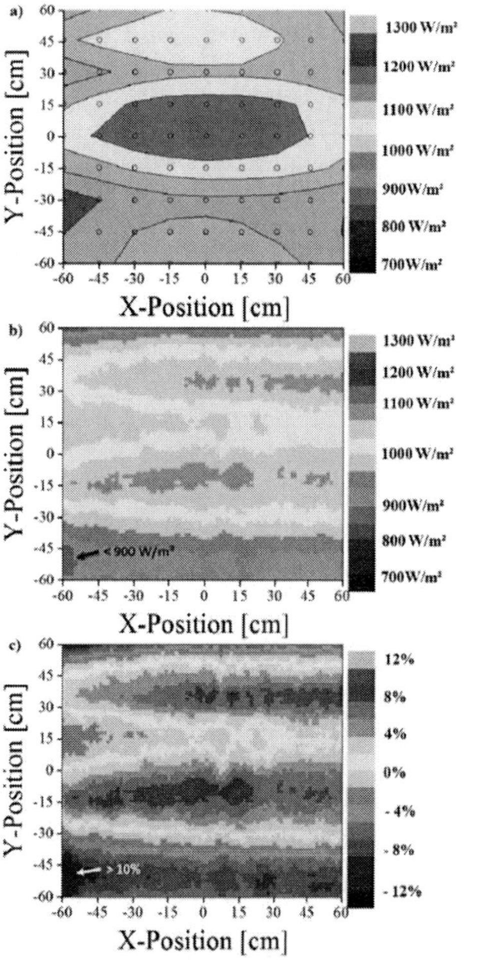

Figure 4: Optimized adjustments, upper and lower lamps 2.62° tilted and left mirror position 25.5 cm and 20 cm for the right mirror. a) Irradiance distribution measured with the traditional method, b) Irradiance distribution measured with the camera, c) calculated deviation to 1000 W/m².

4 APPLICATION

Upon optimization the steady state solar simulator was applied for illuminating PV modules in order to correlate the module performance to the module temperature.

Firstly, the influence of the homogeneity of the solar simulator's irradiation on the module temperature was investigated. A 36-cell monocrystalline silicon PV module was fixed in the module plane of the solar simulator after a lamp stabilization period of one hour. The module was held in short circuit conditions. The temperature of the PV module was measured using a MIDAS infrared (IR) camera. In the initial configuration (Fig. 5a) homogeneity of the irradiation was not optimized, whilst it was optimized for optimum homogeneity in the module area using the Hishikawa-Hashimoto method afterwards (Fig. 5b). For comparison, an electroluminescence (EL) image of the module is also shown (Fig. 5c).

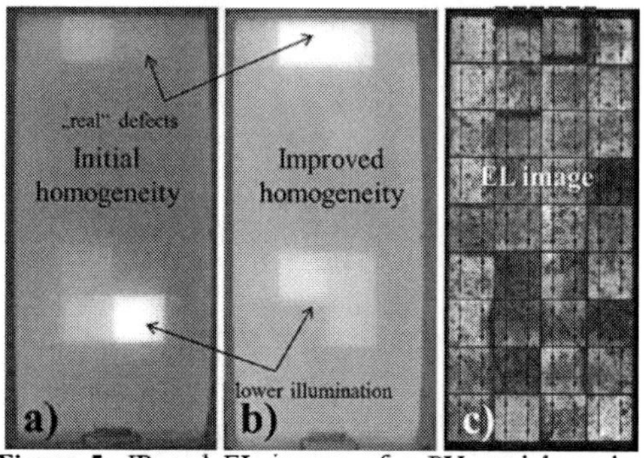

Figure 5: IR and EL images of a PV module under illumination in the steady state solar simulator: a) IR image under non-optimized illumination homogeneity; b) IR image after homogeneity optimization in the module area; c) EL image of the same module with 2 defective cells marked by the red dashed line.

It can be seen that for non-optimized conditions the highest temperatures are obtained in the central lower part of the module (Fig. 5a). This position coincided with the location of the lowest illumination intensity in this configuration. After optimization of the homogeneity (Fig. 5b) the cells in this area only show a slightly increased temperature, whereas the highest temperatures are obtained at the upper edge of the modules. The EL image of the same module reveals that these cells indeed exhibit non-active cell area. Thus, the increased temperature in the IR image is a real effect of the defective cells, whereas the increased temperature in the non-optimized case is a direct consequence of the inhomogeneous illumination. These results show the importance of a homogeneous illumination of the steady state solar simulator – in particular, if the solar simulator is to be used for IR camera measurements.

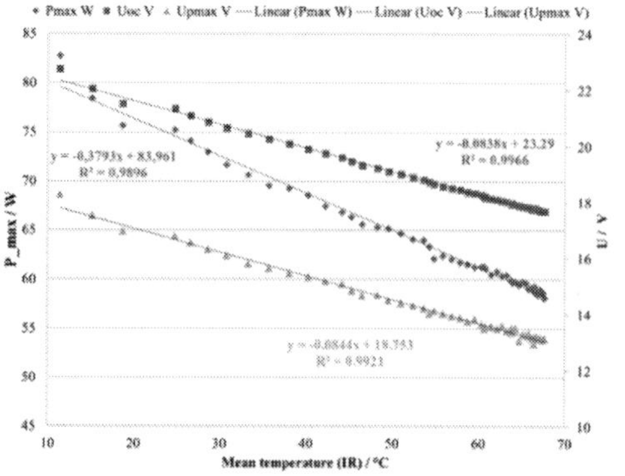

Figure 6: Correlation of the module parameters P_{max}, U_{OC} and U_{Pmax} and the mean module temperature of a PV module illuminated in the steady state solar simulator.

In a further investigation a PV module of the same kind was mounted in the solar simulator in order to investigate the correlation between module performance and module temperature. Again, the module temperature was monitored with an IR camera during illumination. Additionally, a full I-V curve was recorded every 15 seconds. Measurements were continued until the mean module temperature stabilized at a level of approximately 70°C. Based on these measurements, the temperature coefficients of the module were determined. In Fig. 6, the correlation of the module parameters P_{max}, U_{OC} and U_{Pmax} and the mean module temperature are shown. The temperature coefficients of for P_{max}, U_{OC} and U_{Pmax} were determined as -0.38 W/K, -0.08 V/K and -0.08 V/K, respectively. These values were found to be in good agreement with the values stated in the module data sheet.

SUMMARY

In this work, a steady state solar simulator was characterized and optimized based on a modification of the Hashimoto-Hishikawa method. Thus, a rapid and accurate method was introduced for optimizing the homogeneity of the illumination distribution of a steady-state solar simulator. Upon, optimization the distribution of the illumination homogeneity on the module temperature distribution was investigated. Furthermore, the optimized solar simulator was used for determining the temperature coefficient of a PV module. The temperature coefficients were in good agreement with the data sheet values of the module. Therefore, it has been shown that the solar simulator can be employed as a versatile tool for studying the temperature behavior of PV modules

ACKNOWLEDGEMENT

Parts of this work were conducted as part of the research project Smart Grid Solar and were co-financed by the European Union through the European Regional Development Fund and by the Free State of Bavaria. Parts of this work were funded by the Federal Ministry of Economics (AQUAM, Förderkennzeichen 0325807A).

REFERENCES

[1] Y. Hishikawa, H. Shimura , Y. Nakanishi , and T. Hashimoto, Proceedings 23rd European Photovoltaic Solar Energy Conference (2008) pp. 2974-2977

[2] P. Kask, K, Palo, C. Hinnah, T. Pommerencke, Journal of Microscopy, Vol. 0 (2016), pp. 1-13

MEASURING UNIFORMITY UNDER SIMULATED SUNLIGHT

F. Plag*[1], F. Haas[2], K. Ramspeck[2] and S. Winter[1]

[1]Physikalisch-Technische Bundesanstalt (PTB), Bundesallee 100, D-38116 Braunschweig, Germany
[2]h.a.l.m. elektronik GmbH, Burgstraße 106, D-60389 Frankfurt am Main, Germany

*corresponding author: Tel. +49 531 592 4148, fax: +49 531 592 694148, e-mail: Fabian.Plag@ptb.de

ABSTRACT: To minimize the measurement uncertainty of indoor measurements of PV module efficiency, the accurate determination of the spatial distribution of irradiation is essential. The international standard IEC 60904-9 (2007) describes procedures to classify the basic optical properties of solar simulators, including the measurement of the spatial distribution of irradiation [1]. The development and characterization of high quality light sources for solar simulation with lowest non-uniformity requires adequate means and methods to determine repeatable and reliable values of the irradiance non-uniformity.
We present different methods for measuring pulsed solar simulators irradiance uniformity on PV module scale and discuss the results obtained with the different methods. Herein, special emphasis is laid on the characterization and evaluation of the measurement equipment used and the impact on measurement uncertainty for the irradiance non-uniformity. According to our observations, it is required to consider local temporal stability of the irradiance distribution for precise and comparable measurements. In our investigation, we are also involving optical properties of typical crystalline silicon c-Si device under test modules to be part of the uncertainty analysis of non-uniformity measurements.
We discuss the requirements for equipment and methods to report reliable values on solar simulators irradiance non-uniformity below 1 % and the corresponding measurement uncertainty affecting PV module efficiency measurements.
Keywords: module, qualification and testing, solar simulator, characterization, traceability, experimental methods

1 INTRODUCTION

The measurement of solar simulators irradiance uniformity is vital for precise and traceable determination of PV module efficiency, derived from *I-V* measurements. In the last years, requirements for solar simulator performance become more crucial due to increasing PV module size.

Improvements in solar simulation capabilities in the past years require also improvements in characterization tools for simulators. To address the technical possibilities to modify a solar simulators spectrum and total irradiance our study covers the investigation of: spectral non-uniformity on a large area, temporal stability of irradiance including spatial information and a separate evaluation of standard testing conditions- (1000 W/m² STC) and low light uniformity (e.g. 200 W/m²).

Measurements with a high scanning resolution and small detector size allow us to evaluate local non-uniformity on below usual cell size scale.

This work focuses on a beyond the state-of-the-art analysis on two non-uniformity measurement methods for determining light sources irradiance properties and their significance for its measurement uncertainty. The analysis involve also the determination of spectral- and angular mismatch effects and their impact on non-uniformity measurements below values of 1% with respect to typical c-Si PV modules.

The irradiance *non-uniformity* in percent is defined in the international standard IEC 60904-9 Ed.2 (2007) as follows:

$$non\text{-}uniformity = \frac{E_{\max} - E_{\min}}{E_{\max} + E_{\min}} \cdot 100\% \qquad \text{eq. 1.1}$$

2 REQUIREMENTS ON THE INSTRUMENTATION FOR MEASURING UNIFORMITY

Generally, three methods for the determination of irradiance non-uniformities under simulated sunlight are known:

2.1 Single scanning detector

A single detector can be used to scan the designated test area subsequently. Often encapsulated (reference) solar cells or photodiodes are used. Due to the sequential measurement procedure, it is recommended to correct for irradiance fluctuations at each measured position, particularly when using pulsed light sources.

Advantages:
- A single detector is sufficient.
- Moderate requirements for the data acquisition system and scanning procedure, if fully automated.
- Usually the *I-V*-data acquisition system (Curvetracer) can be used for determination of the irradiance distribution $E(x,y)$.

Disadvantages:
- The temporal fluctuation of $E(x,y)$ within one flash pulse is not provided due to the sequential measurement with the irradiance correction of a single monitor solar cell.
- Reflectance conditions may differ from those of a real PV module measurement

2.2 Array of detectors

Instead of a single scanning detector an assembly of multiple (reference) solar cells can be used, e.g. a modified PV module with individually contacted single solar cells.

Advantages:

- Simultaneous, thus fast measurement of $E(x,y)$.
- In case of a high sample rate, local temporal non-uniformities can be detected with a reasonable effort.
- Properties like reflections of the array can be designed comparably to the PV modules that shall be tested with the simulator.

Disadvantages:

- Comprehensive characterization and calibration of each detector is necessary.
- Detector size and step width are fixed with the consequence of limited flexibility in resolution.
- For measurements under low light conditions the non-linearity of the individual detectors becomes crucial.

2.3 Camera based methods

For a simultaneous measurement of the irradiance distribution, a CCD camera can be used. Hishikawa et al. first introduced this method for solar simulators in September 2008 [2] and after that, Winter et al. published a technique for monochromatic light in October 2008 [3]. The camera is viewing at a white screen in place of the PV device, which is irradiated by the light source. The screen acts as standard for hemispheric reflectance, hence it has to be homogeneous and perfectly planar.

After calibrating the CCD camera including lens and aperture setup, non-uniformity can be derived from the pictures.

However, this method requires an extensive characterization of the CCD camera system (including optical effects of lens and screen) and is used for applications with smaller areas and moderate uncertainty requirements (e.g. for quick live metering and online optimization of the beam shape) [3].

In addition, the use of a large white screen is expected to cause additional reflections between the designated test area, the optics of the simulator and the surroundings which leads to different radiative situation compared to a setup with a PV module. To measure non-uniformity values of below 1 % this characterization procedure appears to be extremely demanding. Therefore, this method was not further examined in our comparison of uniformity measurements on module scale.

3 MEASUREMENTS AND CHARACTERIZATION

We applied the two different methods described in 2.1 and 2.2 for measuring the total irradiance uniformity of a pulsed solar simulator. Therefore, we scanned the designated measurement area (2000 x 1000 mm²) both with an encapsulated 5-inch (125 x 125 mm²) multi-crystalline-Si solar cell and a smaller cell with the size of 50 x 50 mm². In case of the array method, we captured the measurement area simultaneously with a multichannel PV module consisting of 60 individually calibrated 6-inch (156 x 156 mm²) solar cells.

We varied both methods systematically in order to evaluate their sensitivity to potential influences in order to derive a related measurement uncertainty. The variation includes a decreasing step width for scanning (resulting in a higher spatial resolution), testing rotational symmetry (180°) of the multichannel instrument and repeatability of the measurements for both methods.

In order to surpass the characterization beyond the state-of-the-art specified by the standard IEC 60904-9 Ed.2 we measured the spatial distribution of spectral irradiance by additionally scanning the area with an array spectroradiometer [4]. Then we calculated the spatial distribution of the spectral mismatch for an exemplary case of c-Si module spectral responsivities.

Local temporal stability of the irradiance of a pulsed solar simulator was investigated by using the multichannel PV module combined with a high-speed data acquisition.

In addition, we examined the influence on total irradiance uniformity when using neutral density filters for light attenuation (down to 200 W/m²). The attenuating filter is located in the lamp housing

3.1 Reproducibility of the scanning method

To validate the long-term reproducibility of the scanning method we compared two measurements of the non-uniformity. One was performed before and one after the re-assembly of the basic optical parts of the simulator, including the alignment lamp housing and the uniformity filter glass.

The compensation of irradiance fluctuation is performed by using a synchronized monitor detector at a stationary position.

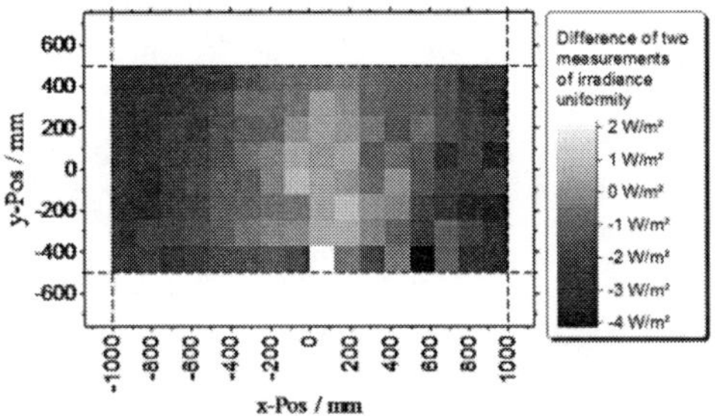

Figure 1: Difference in the non-uniformity of irradiance. The optical parts of the simulator system were detached and stored almost one year before remounted in a PV module laboratory to be tested for reproducibility. In 2014 a non-uniformity of 0.74 % and in 2015 a non-uniformity of 0.60 % was determined according to Equation 1.1.

It turns out that the standard uncertainty (derived from the standard deviation) of seven repeated scans without re-assembling of the setup is 0.01 % (k=1). Tolerances (considering deviations including alignment and aging effects of lamp or filters) could explain the responsible difference of 0.14 % in the non-uniformity.

Systematic effects (e.g. spectral and angular mismatch [5]) cancel out when using particularly the same measuring equipment. If these properties are differing to those of the device under test PV module, then they have to be treated as contribution to uncertainty of the PV module efficiency measurement (see section 3.5).

3.2 Enhancement of the resolution

To examine the impact of local non-uniformity a smaller detector size and step width was chosen for scanning the test area (see Figure 2). The detectors optical properties (responsivity and angular response) are assumed to be identical compared to the large area detector.

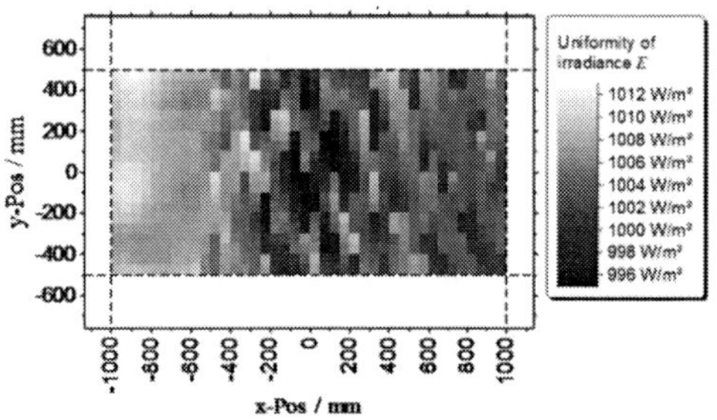

Figure 2: High-resolution map of the investigated solar simulators irradiance non-uniformity resulting in a value of 0.84 % according to Equation 1.1. A step width of 50 mm was chosen to scan the test area of 2000 x 1000 mm² with a 50 x 50 mm² sized and encapsulated solar cell.

In order to compare measurements of 50 x 50 mm² cells to those performed with 156 x 156 mm² cells we binned nine adjacent measurement points of the high-resolution map accordingly. This approach implies the assumption for a lateral uniform responsivity of the scanning detector $s_{\text{ref}} \neq f(x,y)$. The calculated non-uniformity of approximately the same area (1950 x 960 mm²) is then 0.72 % according to Equation 1.1. Differences to the uniformity of the rather low-resolution scan with the 5-inch solar cell are taken into consideration for our uncertainty analysis (see section 3.5).

3.3 Simultaneous measurement using an array of detectors

For the simultaneous irradiance uniformity measurement a customized 60 cell c-Si module with a cell size of 6-inch (156 x 156 mm²) was used as an array of detectors. As opposed to a conventional c-Si PV module each cell is individually contacted to a shunt resistor for voltage measurement. The analog signal is then converted by 12 bit AD-electronics respectively. The readout electronics is microcontroller based and performs at a maximum sample rate of 2 kHz. The instrument is equipped with a trigger system for the simultaneous and temporal resolved non-uniformity data acquisition. Detector and AD-electronics are henceforward summarized as one channel.

For using this device as an equipment for temporal resolved non-uniformity measurements, substantial characterization steps are required. These are at least:

3.31 Irradiance calibration of each channel

Each channel has an individual sensitivity. In order to derive homogeneous measurements with the detector array, the sensitivity of each of the 60 channels has to be determined under the same irradiance condition.

We used a Class AAA (according to [1]) single lamp xenon solar simulator of flash type designed for solar cells at short distance for the calibration. The method is comparable to the indoor procedure introduced by Schinke 2015, but without temperature variation [6]. A mask was used to shade all channels but one to minimize multiple reflections.

An encapsulated reference solar cell was used to adjust the irradiance of the solar simulator to 1000 W/m². This was repeated for each channel while the light source was moved across the module area. A monitor solar cell measures the average irradiance signal during calibration of the monitor solar cell I_{mon} at each flash pulse. The temperature T was also measured at the position of the monitor solar cell in order to correct for deviations to 25 °C with the temperature coefficient α.:

$$AD_{i,\text{corr}} = AD_{i,\text{meas}} \cdot \frac{1000\frac{\text{W}}{\text{m}^2}}{I_{\text{mon}}} \cdot (1 + \alpha(25°C - T)), \qquad \text{eq. 3.1}$$

$i = 1 \ldots 60$.

For a homogenization of all channels the measured AD-Counts $AD_{i,\text{meas}}$ can be corrected to $AD_{i,\text{corr}}$. To evaluate the reproducibility of each channel a set of ten acquisitions was taken for averaging. The derived standard uncertainty (k=1) is 0.03 % in average for the 60 channels. Outliers exhibit values of up to 0.16 % in the standard uncertainty.

We found out that the digital output at 1000 W/m² is in average at approximately 2060 counts. This leads to the conclusion that a resolution of 11 bits is usable for measurements at 1000 W/m² with quantization steps of $u_{\text{ADCstep}} = \pm\, 0.25$ W/m². With a noise level of three counts a signal to noise ratio SNR of 1:700 can be realized. This value corresponds to $u_{SNR} = \pm\, 0.73$ W/m².

The limitation in resolution due to digital quantization steps and the SNR results in a combined standard uncertainty of the data acquisition $u_{\text{DAQ}}(E_i) = 0.45$ W/m² with $k = 1$, when using Equation 3.2:

$$u_{\text{DAQ}}(E_i) = \sqrt{\frac{u_{SNR}^2}{3} + \frac{u_{\text{ADCstep}}^2}{3}} \qquad \text{eq. 3.2}$$

To analyze the linearity of the array detector in terms of irradiance we varied the irradiance of a PV module solar simulator setup. During the measurements, the array of detectors was thermally stabilized in a climate chamber providing a front glass window. It was found, that for irradiances in close range of 1000 Wm² ± 100 W/m² the effect of non-linearity on different channels is negligibly small. It is further assumed that non-linearity is not affecting the final non-uniformity measurement for irradiances within the investigated irradiance range.

3.32 Rotational symmetry for the determination of systematic effects

To validate the non-uniformity measurements performed with the detector array we measured the non-uniformities of seven flash pulses for each of two rotational orientations (0° and 180° around the surface normal of the array). For better comparability, we normalized both averaged non-uniformities. Systematic deviations of up to 0.35 % were identified for the 60 channel array (see Figure 3) when comparing both situations.

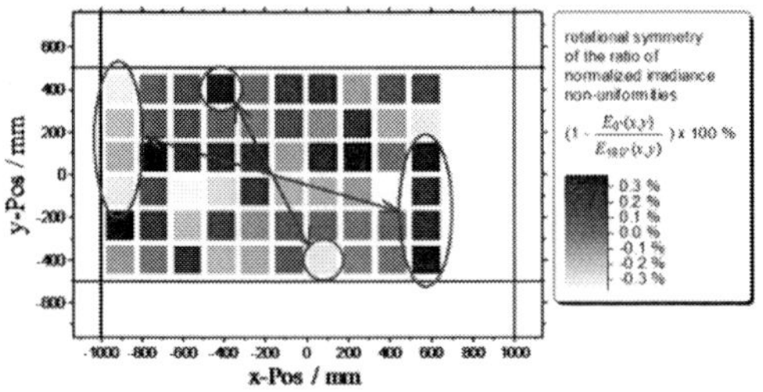

Figure 3: Rotational symmetry of the 60 channel array of detectors for the determination of systematic effects. For each of the two orientations an average of seven repeated flash pulses was taken for minimizing statistical effects.

In a first approach, the maximum deviation was used as uncertainty (rectangular distribution) for the calibration and linearity of the device. The standard uncertainty is then calculated as follows:

$$u_{CAL}(E_i) = \sqrt{\frac{(0.35\ \%)^2}{3}} \qquad \text{eq. 3.3}$$

3.33 Temporal resolution and investigation of stability

A maximum acquisition rate of 2 kHz allows us to measure fluctuations at a time interval of 0.5 ms. We investigated a single flash pulse of a solar simulator with a total duration of 35 ms.

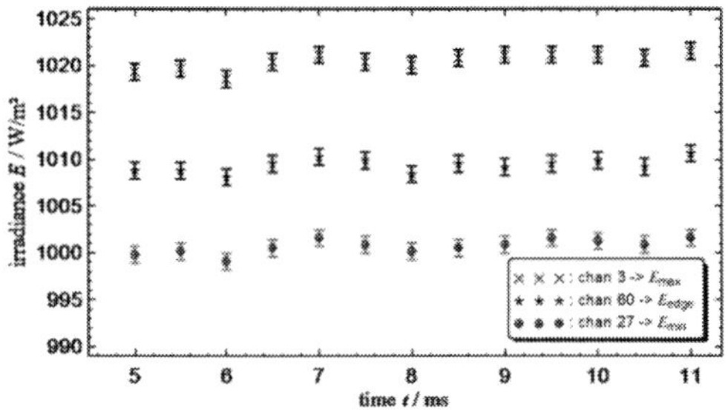

Figure 4: Temporal stability of the three representative channels during a time interval of 10 ms within a flash pulse of 35 ms duration. The uncertainties in this graph are representative for the array of detectors data acquisition $U_{DAQ}(E_i)$ with $k = 2$.

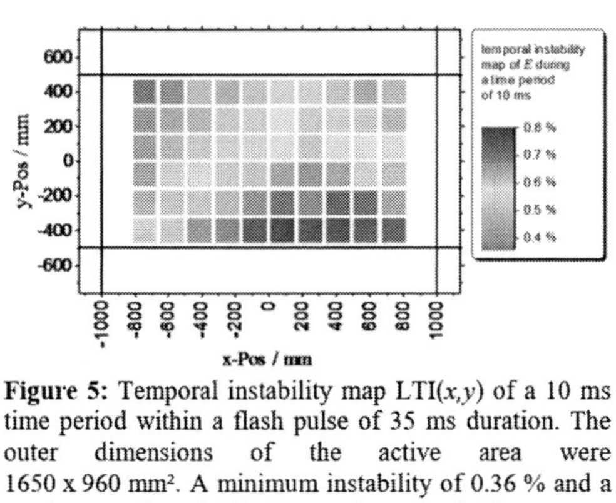

Figure 5: Temporal instability map LTI(x,y) of a 10 ms time period within a flash pulse of 35 ms duration. The outer dimensions of the active area were 1650 x 960 mm². A minimum instability of 0.36 % and a maximum value of 0.77 % was determined. The average instability is 0.57 %.

The temporal long-term instability (LTI) according to [1] shows also a spatial dependency (see Figure 5). Differences of more than a factor of two were observed.

The laterally distributed irradiance fluctuations $E(t,x,y)$ cannot be compensated for PV module performance measurements when using a single monitor solar cell positioned in plane with the module. In order to study the impact on the corrected output current of a PV module $I_{STC,\,DUT}$ we assume that average fluctuations of 0.4 % can be detected with the monitor solar cell I_{mon} at an appropriate position close to the module area (see Equation 3.3). Irradiance instabilities at differently located positions on the module contribute directly to the uncertainty of the irradiance correction. In our case, this uncertainty is up to 0.37 % in combination with a c-Si PV module (DUT) with same cell dimensions and without bypass diodes.

$$I_{STC,DUT} = I_{STC,ref} \cdot \frac{I_{meas,DUT} / I_{mon,DUT}}{I_{meas,ref} / I_{mon,ref}} \cdot \prod_i f_i \qquad \text{eq. 3.3}$$

$I_{mon,\,ref}$ is the irradiance signal of the monitor solar cell during the adjustment of the simulators irradiance with a reference solar cell. $I_{mon,\,DUT}$ is the signal of the monitor at the time of the PV module (DUT) measurement. f_i are the correction factors for other systematic effects.

3.34 Low light conditions at 200 W/m²

To check how both methods perform under low light conditions we measured the non-uniformity at an irradiance of 200 W/m² (see Figure 6 and 7). The solar simulators light source was attenuated with neutral density filters to prevent changes in the spectral distribution.

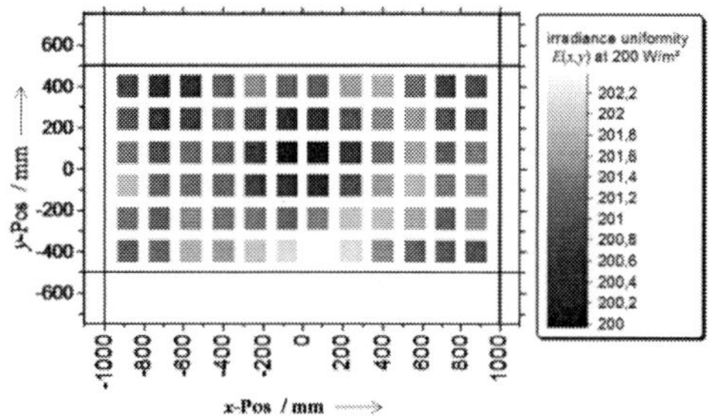

Figure 6: Single detector scan measuring non-uniformity of the investigated solar simulator at low light conditions (200 W/m²). To reduce type A uncertainties a set of seven repeated measurement was taken and an average irradiance distribution was determined. The value of the non-uniformity of irradiance over an active test area of 1920 x 960 mm² is 0.65%.

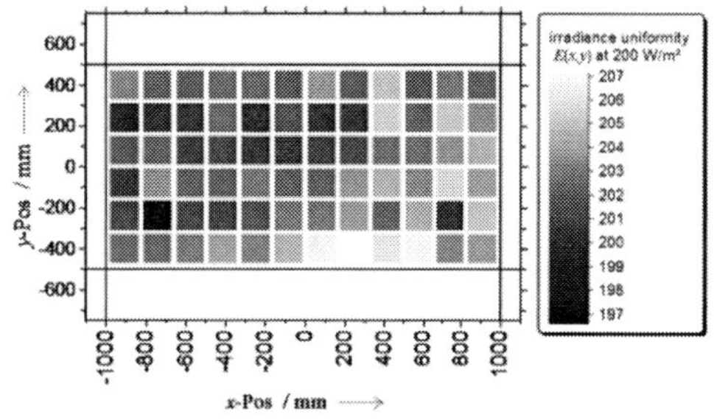

Figure 7: Array of detectors measured non-uniformity of the investigated solar simulator at low light conditions (200 W/m²). To reduce type A uncertainties a set of seven repeated measurement was taken and an average irradiance distribution was determined. To cover the designated test area properly we shifted the smaller module with the outer dimensions of 1600 x 1000 mm² once 160 mm to the left and once 160 mm to the right for each run. The value of the non-uniformity of irradiance over an active test area of 1950 x 960 mm² is 2.6% for this case.

Obviously, the measurement with the array of detectors shows a large deviation to the results obtained with the scanning method. A difference of almost 2 % in the low light non-uniformity was identified.

We suppose that non-linearities of each individual channel (solar cell and AD-electronics) are responsible for this extensive deviation.

3.4 Consideration of the lateral spectral irradiance distribution

The influence of non-uniformity of the spectral irradiance was investigated by creating a spectral mismatch map $M(x,y)$ for an exemplary combination of reference and the device under test module. This procedure was already introduced in 2014 an can be looked-up in [4]. For our calculation, the spectral responsivities s_{ref} (uniformity mapping or reference device) and s_{DUT} (device under test DUT module, which is measured afterwards) were used (see Figure 8). The mapping detector is assumed to consist of typical multi-crystalline BSF material and the DUT assumed to be a module made of c-Si PERC cells.

We performed a mapping of the lateral spectral irradiance distribution using an array spectroradiometer with wavelength coverage from 300 nm - 1100 nm.

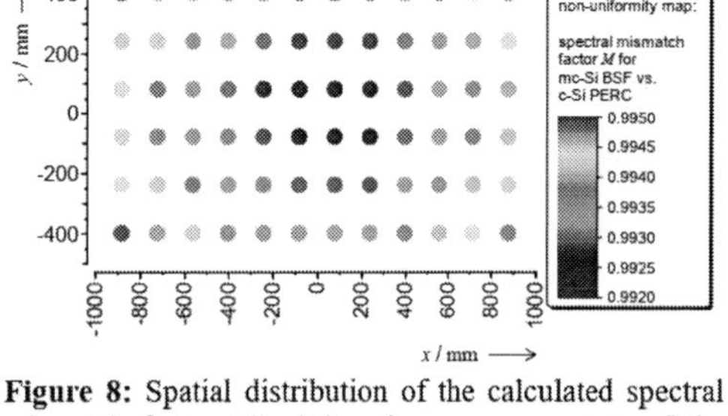

Figure 8: Spatial distribution of the calculated spectral mismatch factor $M(x,y)$ based on measurements of the simulators spectral irradiance and the spectral responsivities of a mc-Si BSF and a CZ-Si PERC solar cell. A spatial variation < 0.3 % in spectral mismatch is observed for this combination.

3.5 Comparison of scanning and array method at 1000 W/m² and measurement uncertainty

For comparison reasons we performed scanning measurements in a pulsed xenon solar simulator with a single solar cell at the same positions where the single channels of the array of detectors are located. The same simulator was then measured with the calibrated array of detectors. Both measured non-uniformities are displayed in Figure 9 for rating the solar simulators performance.

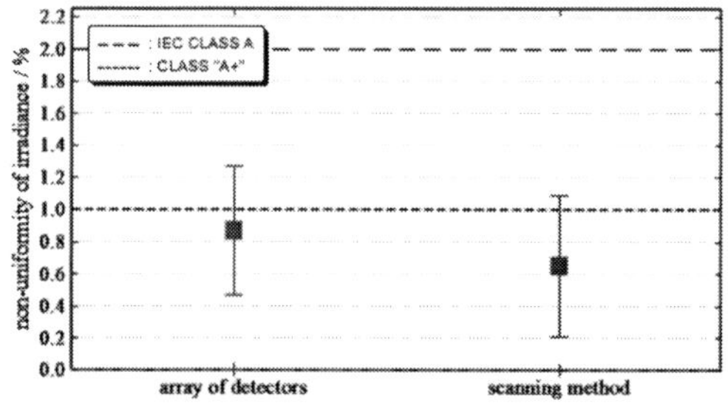

Figure 9: Measured non-uniformities of irradiance at 1000 W/m². Corresponding expanded measurement uncertainties (k=2) for the non-uniformity are shown in red.

For the determination of the measurement uncertainty we considered the following quantities influencing the result:

- Differences in the detector size/changes in the resolution may have effect on the measured quantity. In our case a 125 x 125 mm² (five-inch) solar cell was used for the scanning method. The difference in the non-uniformity compared to measurements with six inch solar cell which was used in the array of detectors was estimated to be 0.2 %.
- Local temporal stabilities which cannot be compensated by a monitor solar cell need to be considered when scanning the irradiated area subsequently. They do not contribute to the non-uniformity measured with an array of detectors, because of the real-time acquisition.
- In case of an array of detectors, multiple channels have to be calibrated and investigated in terms of their irradiance linearity. In our case the array performed very poor at low light conditions around 200 W/m².
- Spectral mismatch effects due to non-uniformities of the spectral irradiance need to be considered, when the spectral responsivities of the reference device and the DUT are not identical.
- Reflections from sidewalls, floor, baffles, and the optics of the solar simulator need to be considered, when the reflective properties of the detector, dimensions and the configuration of the experimental setup is differing from a usual PV module measurement. E.g., an automated x-y-linear-stage for scanning may cause different reflections than an array detector shaped like a PV module.
- Many solar simulators designed for PV module measurements provide angular dependencies of their irradiation field. The angle of

incidence often changes for each position on the test area. Differences in the angular responsivities of the detector(s) used for non-uniformity measurement to the actual DUT have to be considered. The Procedures are described in [4].

- Finally, the data acquisition system with corresponding *SNR* and AD-quantization should be taken into account.
- Temperature effects are regarded as negligibly low when measuring pulsed light sources.

3.51 Uncertainty budget for the scanning method irradiance measurements

Table I: Uncertainty budget for irradiance measurements using the scanning method with a single solar cell characterizing a pulsed xenon solar simulator subsequently at 1000 W/m².

quantity	distrib.	type	value %
different detector size	rect.	B	0.2
local temporal instability	rect.	A	0.37
spectral mismatch	rect.	B	0.3
reflection	rect.	B	0.1
angular mismatch	rect.	B	0.1
data acquisition	gauss.	A	0.01

Scan: comb. uncertainty $u_{scan}(E(x,y))$ k=1 <u>0.31</u>

3.52 Uncertainty budget for the array of detectors

Table II: Uncertainty budget for the <u>irradiance</u> measured with the array of detectors characterizing a pulsed xenon solar simulator simultaneously at 1000 W/m².

quantity	distrib.	type	value %
cal. and linearity of array	rect.	B	0.35
spectral mismatch	rect.	B	0.3
reflection	rect.	B	0.1
angular mismatch	rect.	B	0.1
data acquisition	gauss.	A	0.04

Array: comb. uncertainty $u_{array}(E(x,y))$ k=1 <u>0.28</u>

The influence of the calibration and linearity significantly rises when the simulators non-uniformity is measured at deviating irradiance levels than the array of detectors is designed and calibrated for (e.g. 200 W/m²).

Measurement uncertainty contributions of type A can be further reduced by repeating the sequential measurement. As a consequence the standard uncertainty of the scanning method can be reduced to a minimum of $u_{scan}(E(x,y)) = 0.23$ % ($k = 1$).

3.53 Uncertainty propagation and impact on non-uniformity measurements

To calculate the combined standard uncertainty for the non-uniformity u(*non-uniformity*) we applied the Gaussian law for uncertainty propagation regarding the input quantities as uncorrelated (see Equation 3.4)

$$u_{scan\,/array}\,(non\text{-}uniformity) = \sqrt{\left(\frac{\partial f}{\partial E_{max}}\right)^2 u^2(E_{max}) + \left(\frac{\partial f}{\partial E_{min}}\right)^2 u^2(E_{min})} \quad \text{eq. 3.4}$$

For calculating the sensitivity coefficients $c_{max/min}$ the partial derivatives are taken from our modeling Equation 1.1 which is here $f(E_{max},E_{min})$:

$$\frac{\partial f}{\partial E_{max}} = c_{max} = \frac{2\,E_{min}}{(E_{max}+E_{min})^2} \approx 0.5$$
$$\frac{\partial f}{\partial E_{min}} = c_{min} = \frac{-2\,E_{max}}{(E_{max}+E_{min})^2} \approx -0.5 \quad \text{eq. 3.5}$$

when $E_{max/min}$ are normalized and their values are very close to 1.
Assuming further that

$$u^2(E_{max}) = u^2(E_{min}) \quad \text{eq. 3.6}$$

for each method respectively we receive with Equation 3.4:

$$u_{scan}\,(non\text{-}uniformity) = \mathbf{0.21\,\%} \quad \text{eq. 3.7}$$
$$u_{array}\,(non\text{-}uniformity) = \mathbf{0.20\,\%} \quad \text{eq. 3.8}$$

as combined standard uncertainties (k = 1) for the non-uniformity measurements using two different methods.

To verify the impact of correlated quantities in Equation 1.1 we applied a Monte Carlo simulation. Our analysis lead to the conclusion that correlated parts contributing to the combined standard uncertainty $u_{scan/array}$(*non-uniformity*) are negligibly small.

4 DISCUSSION OF THE RESULTS

The results of our comprehensive study on the measurement equipment used for non-uniformity measurements have shown that several different effects are of importance to achieve precise and reliable non-uniformity information. For traceable PV module performance measurements with high accuracy, we do not recommend to characterize the solar simulators performance independently from its intended use. This means that systematic deviations from the real measurement situation of the DUT PV module have to be considered.

The impact on uncertainty due to different optical properties (uniformity mapping device vs. DUT) should at least be estimated. This includes the well-considered selection of a suitable detector (size, spectral- and angular responsivities) and the experimental setup itself.

To detect local temporal instabilities a multi-channel array of detectors is mandatory. An array of detectors with high-speed data acquisition can identify two coupled effects simultaneously (temporal- and lateral irradiance distribution). Although an extensive calibration and characterization is required, a detector array raises the effectiveness of the characterization and optimization of a light source.

For simulator concepts, applying multiple light sources and optical components for the optimization of the light field, the spectral irradiance uniformity should be taken into consideration additionally. Possible non-uniformities of the spectral irradiance distribution can be embedded into a spatially resolved mismatch correction when determining the non-uniformity of irradiance. Therefore a calibrated array spectroradiometer and the use of a reference solar cell as a scanning device or a "reference array of detector" PV module is needed to adjust and correct the solar simulators spatial irradiance.

For the uncertainty budget of precise efficiency measurements of PV modules, all quantities have to be taken into account in particular. Our uncertainty analysis indicates the challenge to prove that non-uniformity of the irradiance is certainly below 1%. Even if the solar simulators non-uniformity was characterized comprehensively with state of the art equipment and methods described in this work a value below 0.6 % has to be determined to certainly classify a simulator as "CLASS A+" for irradiance non-uniformity.

5 CONCLUSION AND OUTLOOK

Two methods for measuring the uniformity of the irradiance emitted by a solar simulator we compared comprehensively. We measured the uniformity of a flash pulse xenon solar simulator on module scale both with a single scanning detector and an array of detectors. The evaluation shows a good agreement of both methods within their uncertainties.

We observed that for traceable measurements PV module efficiencies, a stand-alone characterization of the simulators irradiance properties independent from the device under test (DUT) is not sufficient. We brought spectral and angular dependent properties of the DUT into interaction with the radiant source properties to achieve a reliable informative value of the non-uniformity of irradiance with corresponding uncertainties. The combined standard uncertainties of both methods are in the range of $u(non\text{-}uniformity) = 0.2\ \%$ ($k = 1$). With this critical view, a solar simulator performance classification according to Equation 1.1 for non-uniformity values below 1 % is challenging.

Measurements with an array of detectors benefit from the simultaneous acquisition of temporal and spatial information, which affects the accuracy of PV module efficiency measurements. The major contribution to uncertainty using the array method is the irradiance calibration of individual detectors. According to our investigation, we suppose that there is room for improvements for decreasing uncertainty of the calibration by revising the procedure. Outdoor calibration procedures, Temperature dependencies, and measurements under low light conditions are subject to further studies.

6 ACKNOWLEDGEMENTS

The authors acknowledge financial support of the MNPQ-Programme provided by the German Federal Ministry of Economic Affairs and Energy. This work is supported by the Braunschweig international graduate school of metrology (igsm).

7 REFERENCES

[1] IEC 60904-9 Edition 2. Photovoltaic devices - Part 9: Solar simulator performance requirements. Geneva, 2007

[2] Hishikawa, Y., et al., Digital Camera-Based Quick and Accurate Measurement of Solar Simulator Uniformity. Valencia: 2008. Proceedings of 23rd European Photovoltaic Solar Energy Conference and Exhibition. pp. 2974 – 2977

[3] Winter, S., et al., Camera-based Uniformity Measurements of Irradiation Fields. Daejeon : Proceedings of NEWRAD 2008, 2008. S. 189 - 191.

[4] Plag, F., et al., Comprehensive Analysis of a Pulsed Solar Simulator to Determine Measurement Uncertainty Components. Amsterdam : 2014. Proceedings of 29th European Photovoltaic Solar Energy Conference and Exhibition. pp. 2435 - 2442.

[5] Monokroussos, C., et al., Effect of Glass/EVA/Backsheet encapsulation Packadge on the industrial Rating of c-Si PV Modules. Hamburg : 2015. Proceedings of the 31st European Photovoltaic Solar Energy Conference and Exhibition. pp. 1883 - 1887.

[6] Schinke, C., et al., Messunsicherheitsanalyse zur präzisen Bestimmung der lateralen Bestrahlungsstärkeverteilung von Sonnensimulatoren für PV-Module. Bad Staffelstein, Deutschland : Tagungsband 30. Symposium Photovoltaische Solarenergie (04.-06.03.2015), 2015. ISBN 978-3-943891-45-4.

INFLUENCE OF LOW CONCENTRATION ON THE ENERGY HARVEST OF PV SYSTEMS USING BIFACIAL MODULES

Hartmut Nussbaumer*, Giordano Petrzilek, Steve Schartinger, Markus Klenk, Nico Keller, Thomas Baumann, Fabian Carigiet, Franz Baumgartner
Zurich University of Applied Science, SoE, Institute of Energy Systems and Fluid Engineering
Technikumstrasse 9, 8401 Winterthur, Switzerland
*phone.: +41 58 934 4799, *e-mail: hartmut.nussbaumer@zhaw.ch

ABSTRACT: PV systems using bifacial solar modules have shown an increased energy harvest compared to systems with standard, monofacial panels. Due to bifaciality of the modules, the tilt angle, the installation height, the distance between the modules and in particular the ground reflectance have a significant influence on the energy harvest. For bifacial module arrays reflective elements in between the rows might be an option in order to achieve a low light concentration resulting in a further increased energy harvest. A simulation of such conditions is hardly possible; measurements at real systems would require considerable effort. In this work we present results obtained by a measuring setup with rotating mini modules. In particular the influence of the ground reflection was studied by using various grounds and reflecting elements with different shapes. Prior to the analysis of the reflector elements the measurement setup is described and the assignability of results from the miniaturized test array to a corresponding large scale system is demonstrated.

Keywords: Bifacial, Energy yield, Low Concentration

1 INTRODUCTION

Currently bifacial technology attracts considerable interest in the PV community [1, 2]. Due to technical progress, such as improved bifacial cell concepts or the availability of thin solar glass, the technology gets increasingly attractive. Also, some of the PERC solar cell technologies, which are currently implemented in industrial production, allow a comparatively simple adaption to a bifacial lay-out [3]. The general trend towards glass/glass-modules with superior reliability, as well as the interest in "peak shaving" and customized solutions for specific applications, further supports the development towards bifacial technology.

Nevertheless, the installed capacity of bifacial systems is still negligible. The lack of reliable field data still deters possible investors. Even in the PV community there is considerable uncertainty about the real benefit due to bifaciality, as reflected by the numerous publications dealing with this issue.

The potential for an improved module power output was repeatedly demonstrated by simulations [4, 5, 6, 7] and measurements on single modules [8, 9, 10, 11, 12, 13, 14, 15] or installations [16] in various orientations. In real, extended systems however, the arrangement of multiple modules will have additional effects, such as direct shading by modules in adjacent rows or indirect shading of the modules surrounding, which results in a reduced albedo. Data of larger systems are rare [17, 18, 19, 20], and the results are linked to a specific mounting and / or the respective orientation. Thus, presently not only the actual yield gain of bifacial installations but also the optimized installation conditions or module orientations are unknown.

2 OBJECTIVES AND APPROACH

At the Zurich University of Applied Science, ZHAW, we currently put an array for the systematic measurements of bifacial systems with differing mounting conditions to the test. The array is based on large, commercially available, 60-cell modules. There is a continuous, automated variation of the tilt angle. Other parameters like the distance between adjacent rows, the installation height or the ground reflectivity are manually adjusted. The details of this measurement device will be reported elsewhere [21].

While this large measurement array delivers important results in continuous measurements, the manual change of the parameters will need some time due to the considerable dimensions.

In several regard it is be beneficial to have a measurement device which allows a fast and flexible variation of the installation parameters. A fast variation of the installation parameters enables a direct comparative analysis at virtually unchanged lighting conditions. The smaller dimensions and the lower cost may also enable the use of similar systems at different locations. Another option is the direct comparison of similar systems with differently adjusted parameters which are used in parallel. First however, it has to be proven that the results can be correlated to measurements at large, real systems. This can be obtained by a direct comparison of the results from a large system and a similar but miniaturized device. It is therefore our objective to compare the results of the described test rig with real modules, continuously changing tilt angle and variable mounting conditions, to a miniaturized version.

3 MEASUREMENT SET-UP

The basic concept of the large as well as of the miniaturized test rig is shown in figure 1. Three rows of modules, or solar cells for the miniaturized version, with manually adjustable distance between the rows, are mounted on vertically adjustable pillars in order to position the modules or solar cells in different heights. An important feature of the measurement set-up is the automatic variation of the tilt angle in certain steps. For each step an I/V- curve is measured in order to get the power as a function of the tilt angle of the modules. All panels change their tilt angle continuously and coordinated with the central row.

Due to a change of the orientation of the whole measurement test array one can measure the power output of a typical flat roof or ground mounted, free field PV system, mainly oriented to the south (northern hemisphere) or a tracking system changing the orientation from east to west. In the latter case, the measurement system can be used as tracking system itself. In addition the influence of different reflective grounds can be analyzed.

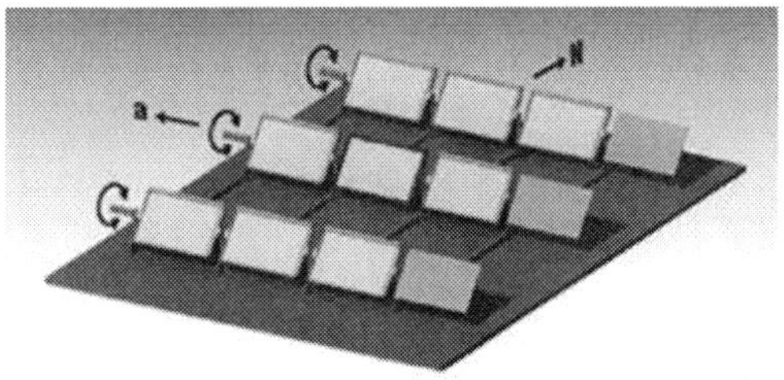

Figure 1: Measurement setup with permanently revolving modules. Height, distance between the rows and reflecting ground can be changed manually. Also the global orientation of the array can be varied. The most relevant module in the center, which is best suited to represent the actual conditions in real installations, is marked red.

Obviously for the miniaturized test array the effort to change the reflecting ground, the orientation, the installation height or the row distance is significantly reduced as compared to changes on the large test rig.

The objective of both test rigs is to measure the potential power output and energy harvest of bifacial PV modules under real conditions, also considering effects which occur in extended arrays, e.g. shading due to adjacent modules. This is of particular interest for some orientations which are relevant for bifacial modules, such as vertical installations. Because of the bifaciality, the PV modules are also sensitive to indirect shading due to reduced diffuse reflectance from the ground. Therefore panels in the center of such an array are best suited to reflect the general conditions for a typical module in an extended bifacial power plant. By analysis of the other modules in the array the effect due to the limited extension and the validity of simulation tools can be tested.

As shown previously [4], a matrix of 3x3 modules may not be sufficient to reflect the shading effects in extended arrays for specific installation conditions, such as vertical installation. This is particularly true for small distances between the rows.

Therefore additional shading elements are respectively applied to one side of the rows, a further extension in the large array is not feasible due to the limited available space on the roof of the building.

Figure 2 shows the large test rig, equipped with commercially available bifacial modules from the cell & module manufacturer Megacell (Italy).

Figure 2: Installation on the roof of the ZHAW in Winterthur. Details of this measurement device will be reported elsewhere [21].

The lay-out of the miniaturized test device is a downscaled version of the large test array; also the same BiSoN solar cells were used. In order to keep the proportions of real modules, bifacial solar cells were correspondingly cut to the size. Thus, the relative dimensions, such as the installation height over the ground, can be maintained. Accordingly it can be assumed that also the influence of these factors will show a comparable dependency. It should thus also be possible to analyze the impact of the installation conditions and their interrelation (height, albedo, tilt angle,..), also for different insolation situations. The validation of the results is then obtained by a comparison with the large system.

Provided that there is a reliable correspondence, the small system then allows a fast scanning of multiple differing set-ups, which are easily established by a readjustment of the device. For instance, the effect of various reflecting grounds or installation heights can be studied effectively. A great advantage with regard to the outdoor measurements is the fast recording of measurement data at comparatively unchanged radiation conditions.

Obviously not all parameters of a large system can be reflected by the miniaturized device, especially due to the fact that the solar cells will show a different behavior compared to full-scale solar modules, which are a series connection of multiple cells grouped in strings. This is particularly true for the response to direct shading, which typically results in the activation of the bypass-diodes in real modules. Nevertheless, at least indirect shading effects, which means reduced, reflected light from the ground, should be correctly represented. Therefore, as in the large device [21], not only a single element, but multiple elements are installed in a downscaled arrangement. Accordingly there is not only a single rotating element but also one row in front and one row behind of it.

Even if it is attempted to reflect the conditions of the large test array as good as possible, it is essential to supervise the respective shading conditions during the experiments. The frame structure of this prototype is still quite coarse and might be improved in future rigs, but in the current version increased shading effects especially for low sun positions have to be expected.

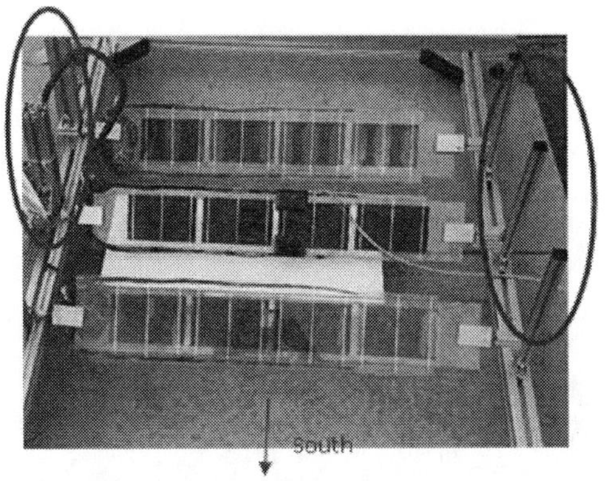

Figure 3: Measurement setup (miniaturized test array) with revolving mini modules. The arrangement is the downscaled version of the large test array with commercial modules. Installation height, distance between the rows and reflecting ground can be changed manually and considerably faster than for the large test rig with real modules. The vertically adjustable mounting elements result in unwanted shading effects in the morning and evening.

4 SMALL AND LARGE ARRAY - COMPARISON

A prerequisite for meaningful measurements on the small test rig is a sufficient correlation of the measurement results of both test rigs. Therefore, the measured data of the miniaturized device has to be correlated to the results obtained by the large array.

In both systems the respective output versus the tilt angle is recorded for a specific set of tilt angles. The duration of a cycle is one minute, in which one cycle means the movement from the horizontal position to the vertical position (for south orientation of the front side). There are 12 tilt angles α / measurement positions in each cycle (0°, 10°, 15°, 18°, 21°, 25°, 30°, 35°, 37°, 45°, 60°, 90°). The rotation and the data collection of both systems are synchronized.

Due to the proportionality of the current to the light intensity it is reasonable to assume that the normalized short circuit current of both devices should show a good correspondence if a linear adaption factor is applied. Furthermore, for typical lighting conditions the open circuit voltage should be almost constant. However, the differing set-up of real modules and small parts of contacted solar cells may result in differing fill factors and therefore affect the normalized Vmpp, Impp, Pmpp values.

Surprisingly, the simple direct comparison of the normalized Pmpp values already showed a good correspondence for a wide time range and irradiation conditions. In figures 4 to 6 the course of the normalized Pmpp values for the large and the miniaturized device are shown for three different days and arbitrarily selected tilt angles of 37° and 35°. The measurements were carried out at a comparatively clear day with a partly cloudy phase during noon (figure 4) and at two partly cloudy days (figures 5 and 6). For the major part of the graphs in figures 4 to 6 the curves are almost congruent. As expected there are undesirable measurement artefacts due to shading and reflection effects at low sun positions for the miniaturized device (figure 4), which are due to the coarse set-up of the miniaturized test rig. There are also deviations due to issues with the data recording for the miniaturized array around noon (figures 4 and 5) and in the morning (figure 6).

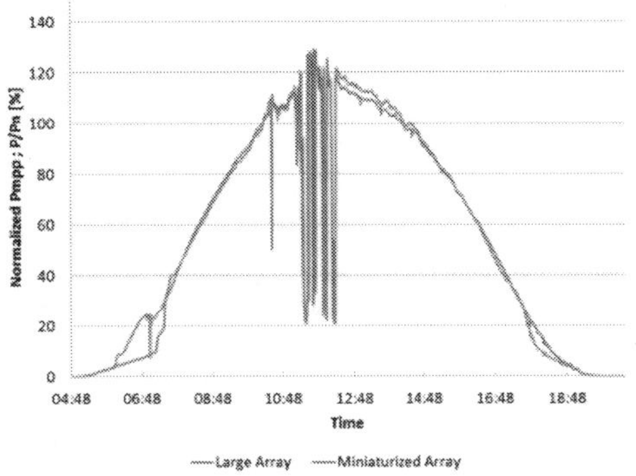

Figure 4: Normalized Pmpp values for the large and the miniaturized test array are shown for one specific, mostly sunny, day (2016/29/04) at an arbitrarily selected tilt angle of 37°.

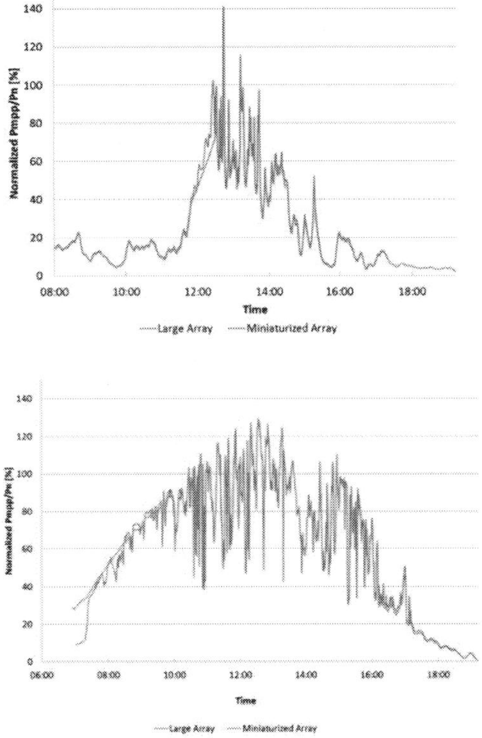

Figures 5 and 6: Normalized Pmpp values for the large and the miniaturized test array recorded at the 2016/16/05 (above) and the 2016/18/05 (below) for an arbitrarily selected tilt angle of 35°.

Nevertheless, the measurements which were carried out at the large and the miniaturized array showed a surprisingly good correspondence up to now. If critical phases with low sun positions are omitted the small rig is actually a suitable means to reproduce data of the large system. We are however aware that fill factor may show a deviating behavior for other measurement conditions with e.g. deviating temperature and varying series resistance. It is therefore necessary to analyze the generated data for a wider range of ambient conditions and it is expected that correction factors for some parameters have to be applied, once there is sufficient data available.

The data presented in the figures 4 to 6 is restricted to single tilt angles. A major feature of the test rigs is however that, within a minute, measurements for a whole range of twelve tilt angles between 0° and 90 ° are carried out. A compiling of all this data within one graph is obviously confusing, particularly for a day with quick and pronounced fluctuations of the irradiation. Procedures to perform a meaningful quantitative analysis are given in the references [21, 22]. The main objective of this work is however the application of differently shaped reflectors and to check their suitability to generate a low concentration effect.

5 MEASUREMENTS WITH REFLECTORS

The potential for an improved module power output of bifacial modules was repeatedly demonstrated by simulations [4, 5, 6, 7] and measurements on single modules [8, 9, 10, 11, 12, 13, 14, 15] or installations [16] in various orientations. The power gain due to the bifacial lay-out is inherently based on the utilization of the radiation which is impinging on the solar cells rear side. A reflecting surface is a prerequisite for a successful application, in almost all applications the reflectors are flat reflecting surfaces. Typically the surrounding ground around the module is used as reflector, but also the backsheet at the laminate rear side can be the reflective surface.

Low concentration was repeatedly proposed for bifacial applications [17, 23, 24, 25, 26, 27, 28]. While low concentration by comparatively simple reflector arrangements around the modules is well known for monofacial solar modules, there are only very limited reports for bifacial applications [17, 24]. This motivated the experimental testing of suchlike simple reflectors at the rear side of bifacial installations which may be realized in a comparatively simply manner. In total 6 reflector types were compared in the following measurements:

- black, flat 0°
- white, flat 0°
- white, flat 15°
- white, flat 30°
- white, parabolic 15°
- white, parabolic 30°

The 0° types are flat positioned foils while the 15° and 30° flat ones are reflectors with triangular and parabolic shape. The angle is measured between the reflectors contact point to the ground and the shapes cut-off. The white reflector structures are covered with a highly reflective foil (albedo factor of 0.86).

The measurements which are presented in this work were carried out by a fast replacement of the respective reflectors at a specific day at noon in order to exclude effects due to differing sun positions. Nevertheless a direct comparison of the output may be misleading, because changing lighting conditions are unavoidable in outdoor measurements. Therefore the actually measured output is related to the respective nominal power at the given insolation conditions during the measurement.

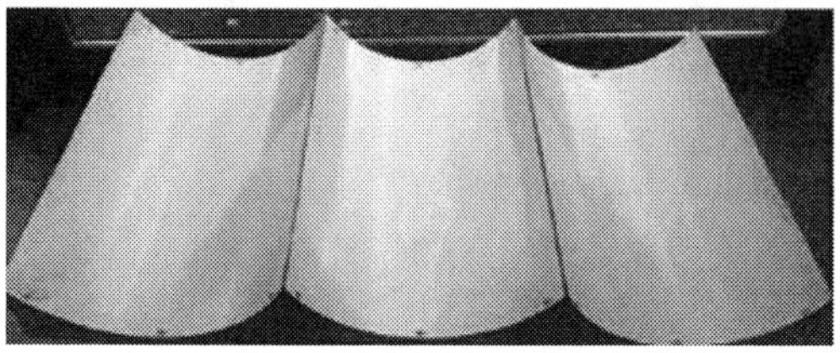

Figure 7: Simple diffuse reflector with parabolic shape which is placed beneath the rotating elements of the miniaturized array.

It is important to point out that the short-term measurements which are shown in the following graphs, do not reflect the yield gain due to the reflectors, but are a snapshot of the power output for the given conditions with high sun elevation at noon. To obtain meaningful results concerning the yield, the measurements have to be extended and need to include a larger range of sun positions and ambient conditions.

For all reflector types the measurements were carried out with permanently changing tilt angles at twelve defined position between 0° and 90 ° tilt angle. One cycle with the recording of twelve IV-curves is completed within one minute cycle time.

Figure 8 represents a comparative overview of the measurements as recorded for a specific set of installation conditions (height, row distance) with different reflectors. All but one reflector make use of the same highly reflective white reflecting foil. The differing reflector type is a black sheet on flat ground.

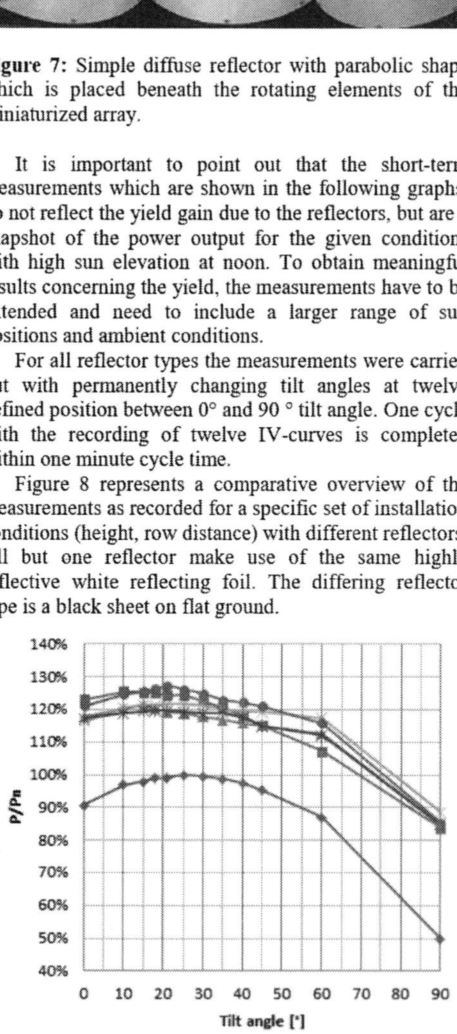

Figure 8: Comparative overview of the measurements as recorded for a specific set of installation conditions (77 mm installation height, 230 mm row distance) with different reflectors.

The difference between the flat black surface and the other reflectors with white reflecting foil material is the most obvious effect. This difference is the lowest at about the optimum tilt angle for monofacial modules. This is plausible, because for this orientation the front side has the major influence on the modules total output. Nevertheless, even at these conditions there is a considerable difference between the flat black surface and the white reflectors.

For very low and steep tilt angles the bifacial properties get more important, hence also the effect of the reflecting surrounding is increasing. Reflecting surfaces with high albedo are thus of particular importance for horizontal or vertical installations.

When comparing differing reflector shapes, covered with the same white foil, there is a roughly similar course of the output versus the tilt angle, but quantitatively there is a range of about 10%, relative to the absolute output, due to the different reflectors. The flat white reflector is well suited for low tilt angles, while the curved reflectors are superior at steeper angles around 60°. For vertical orientation the differences between all white reflector types decrease significantly. The parabolic 30° reflector seems to be slightly advantageous in the medium tilt angle range.

The results in figure 8 represent an overview of the measurements for an installation height (axis) of 77 mm and a row distance (axis to axis) of 230 mm. The installation height and the row distance were varied in several steps. According to the used scale the respective parameters correspond to the following distances in the large array:

- Installation height 77 mm ⇒ ~1.0 m
- Installation height 107 mm ⇒ ~1.3 m
- Installation height 127 mm ⇒ ~1.6 m
- Row distance 230 mm ⇒ ~2.9 m
- Row distance 275 mm ⇒ ~3.4 m

While figure 8 depicts the results for the lowest elevation and the smallest row distance, figure 9 shows the results for an installation height of 127 mm and a row distance of 275 mm, the maximum for both parameters within the range that was analyzed by us.

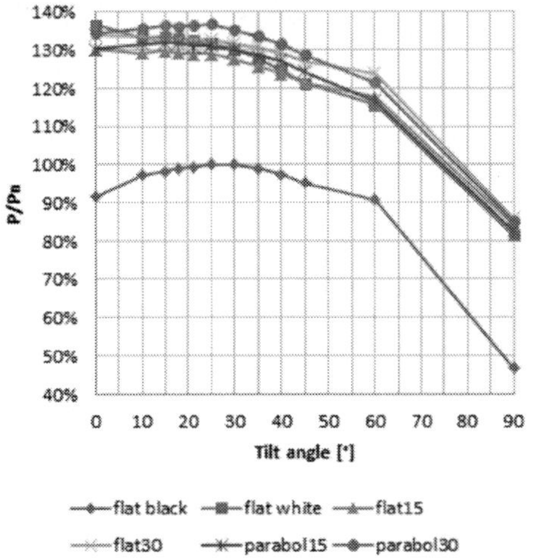

Figure 9: Comparative overview of the measurements as recorded for a differing set of installation conditions (127 mm installation height, 275 mm row distance) compared to figure X, but with the same reflectors.

The general trend and the qualitative course of the results is quite similar to the one described in figure 8. Again, all white reflectors produce a roughly comparable course of the output, within a fluctuation range of about 10%. The curved reflectors are superior to the flat white reflector at steeper angles around 60°. For vertical orientation the differences between all white reflector types almost vanish. The parabolic 30° reflector is again advantageous for the medium tilt angle range. This general course is not limited to measurements shown in figures 8 and 9, but was confirmed by all measurements with varying set-up. According to the results an output gain with increased installation height is not limited to flat reflecting surfaces, but is also observed for the curved structures.

A qualitative difference between the figures 8 and 9 is observed for very low tilt angles. The slope towards lower values, as seen in figure 9, is flattened and the values remain close to the maximum level. A shift of the optimum tilt angle towards lower values with increased installation height was calculated in simulations [29, 30,] which however concern the annual yield.

Due to the high sun elevation at noon there is no direct shading by adjacent rows. While the white reflectors are identical, an increased albedo may indirectly result from a reduced shading of the surrounding ground due to an increased row distance. According to simulations [30] the effect of an enhanced albedo may result in a lowered or increased optimum tilt angle, dependent on the installation height.

Quantitatively, there is a difference in the output for the variied installation conditions. While the results for the black reflector are virtually identical, the measured output is higher for all white reflector types, with the highest output for the largest installation height and row distance. This is reasonable and also according to simulations [29, 30] concerning the annual yield. Again, it has to be emphasized, that the shown measurements are no long term study, but only reflect very specific conditions.

The maximum absolute output with the white reflectors is increased by about 35%, compared to the maximal values for the black flat surface. This maximum absolute output is obtained with the white 30° parabolic reflector, for large installation height and row distance. The maximum gain due to reflectors is obtained for horizontal mounting conditions with a delta of about 45% for the white flat and the white parabolic 30° reflector.

6 CONCLUSION AND OUTLOOK

The miniaturized test rig is a suitable tool to predict the power output of large PV systems. Due to the chosen setup with permanently revolving modules the whole range of tilt angles is analyzed. Because of the small dimensions it is possible to quickly adjust also different installation conditions, such as module height and row distance. In this study the effect of different reflective surfaces was investigated.

While the application of a white reflection foil alone has a major impact, the use of differently shaped reflectors also has an reproduceable effect. Differently shaped reflectors are favorable for specific tilt angles. Thus, the use of such reflectors may be beneficial, depending on the application. The highest output was obtained with the parabolic shaped reflector, which could therefore be an interresting option for bifacial PV systems.

The presented results were limited to some first measurements at noon in order to check the general properties and to get comparable data. Therefore, these results do not reflect the total yield gain due to the reflectors, but are a snapshot of the power output for the given conditions with high sun elevation at noon. To obtain meaningful results concerning the long-term yield, the measurements have to be extended and need to include a larger range of sun positions and ambient conditions.

6 REFERENCES

[1] A. Kreutzmann, "Von Pandas, Bisons und Zebras", Photon 6-, pp. 22-27, 2015.
[2] J. Libal, R. Kopecek, "Bifaciality: One small step for technology, one giant leap for kWh cost reduction", PVTech, January 13th , 2015.
[3] A. Teppe, C. Gong, K. Zhao, J. Liu, S. Wang, J. Dong, S. Zhou, S. Keller, M. Klenk. I.. Melnyk, P. Fath, "Progress in the industrial evaluation of the mc-Si PERCT technology based on boron diffusion", Energy Procedia, vol. 77, August 2015, pp. 208–214, 5th International Conference on Silicon Photovoltaics, SiliconPV, 2015.
[4] H. Nussbaumer, M. Klenk, D. Schär, T. Baumann, F. Carigiet, N. Keller, F. Baumgartner, "PV installations based on vertically mounted bifacial modules evaluation of energy yield and shading effects", Proceedings to the 31th European Solar Energy Conference, Hamburg, pp. 2037 – 2041, 2015.
[5] Yusufoglu, U. A., T. M. Pletzer, L. J. Koduvelikulathu, C. Comparotto, R. Kopecek, and H. Kurz,"Analysis of the annual performance of bifacial modules and optimization methods", IEEE J. Photovolt. 5, pp. 320-328, 2015.
[6] U.A. Yusofoglu, T. H. Lee, T. M. Pletzer, A. Halm, L. J. Koduvelikulathu, C. Comparotto, R. Kopecek, H. Kurz "Simulation of energy production by bifacial modules with revision of ground reflection", Energy Procedia 55, pp. 389-395, 2014.
[7] N. Kasahara, K. Yoshicka, T. Sailoh, "Performance evaluation of bifacial photovoltaic modules for urban application" Proceedings to the 3rd World Conference on Photovoltaic Energy Conversion, vol. 3, Osaka, Japan, pp. 2455-2458, 2003.
[8] B.B. Van Aken, A. J. Carr,, "Relating Indoor and outdoor performance of bifacial modules", 40th IEEE Photovoltaic Specialist Conference, Denver, USA, pp. 1381-1383, 2014.
[9] J. Lossen, T Böschke, H-J. Krokoszinski, "Increased energy yield for Bosch's n-type solar cells in bifacial application"; Bifi PV Workshop, Konstanz, 2012.
[10] H. Nussbaumer, F. Baumgartner, T. Baumann, D. Schär, F. Carigiet, C. Comparotto, R. Harney, A. Schneider „New opportunities for a better power distribution by the use of bifacial modules in future PV systems", Bifi PV Workshop, Chambery, 2014.
[11] C. Comparotto, M. Noebels, L. Popescu, A. Edler, J. Ranmeyer, T. Klaus, V. D. Mihailetchi, R. Harney, J. Lossen, T. S. Böscke, D. Schär, H. Nussbaumer, T. Baumann, F. P. Baumgartner "Bifacial n-type solar modules: indoor and outdoor evaluation", Proceedings to the 29th European Solar Energy Conference EUPVSEC, Amsterdam, pp. 3248-3250, 2014.
[12] J. P. Singh, T. M. Walsh, A. G. Aberle "Performance investigation of bifacial PV modules in the tropics"., Proceedings to the 27th European Solar Energy Conference EUPVSEC, pp.3263-3266, 2012.
[13] A. Moehlecke, F. S. Febras, I. Zanesco, "Electrical performance analysis of PV modules with bifacial silicon solar cells and white diffuse reflector" Solar Energy, vol. 96, pp. 253-262, 2013.
[14] Z. Wang, Z. Sun, B. Yu, M. Liu, C. Wu, S. Han, G. Li, S. Tian "A study on electrical performance of N type bifacial module using transparent backsheet" Proceedings of the 28th European Photovoltaic Solar Energy Conference, Paris, pp. 3192-3195, 2013.
[15] F. Baumgartner, A. Büchel, H. Nussbaumer, T. Baumann, D. Schär, C. Comparotto, R. Harney, A. Schneider „ Bifacial modules used in cable based PV carport applications" Proceedings to the world PV Conference, Kyoto, 8WePo.9.33LN, 2014.
[16] K. Sugibuchi, N. Ishikawa, S. Obara "Bifacial PV Power Output gain in the field test using "earthON" high bifaciality solar cells", Proceedings of the 28th European Photovoltaic Solar Energy Conference, Amsterdam, pp. 4312-4317, 2013.
[17] V. Poulek, A. Khudysh, M. Libra "Innovative low concentration PV systems with bifacial solar panels". Solar Energy 120, pp. 113-116, 2015.
[18] bSolar, http://www.b-solar.com/Technology.aspx?Sel=Field Results
[19] N. Eisenberg, L. Kreinin, N. Bordin, Y. Kofman, "High efficiency industrial PERT p-type bifacial cell and field results", Bifi PV Workshop, Chambery, 2014.
[20] T. Nordmann, T. Vontobel, L. Clavadetscher, "15 years of practical experience in development and improvement of bifacial photovoltaic noise barriers along highways and railway lines in Switzerland" Proceedings to the 27th European Photovoltaic Solar Energy Conference, pp. 3843-3847, 2012.
[21] F.P. Baumgartner, G. Petrzilek, S. Schartinger, T. Baumann, F. Carigiet, N. Keller, M. Klenk, H. Nussbaumer, "Bifacial outdoor rotor tester", to be presented at the 32nd European Photovoltaic Solar Energy Conference, Munich, 2016.
[22] H. Nussbaumer, G. Petrzilek, S. Schartinger, M. Klenk, N. Keller, T. Baumann, F. Carigiet, F.Baumgartner, Miniaturized Test Array as a Means to Determine the Energy Harvest of Bifacial Installations, Proceedings of the IEEE PVSC, Portland, 2016
[23] P. Ooshaksaraei, R. Zulkifli, S.H. Zaidi, M. Alghoul, A. Zaharim, K.Sopian, Terrestrial Applications of Bifacial Photovoltaic Solar Panels, Recent Researches in Power Systems and Systems Science, p. 128-131
[24] V.Poulek, M. Libra, "New bifacial solar trackers and tracking", Solar Energy Materials and Solar Cells, February 1998
[25] P. Y. Vorobiev, J. Gonzalez-Hemhdez, Y.V. Vorobiev , "Optimization of the Solar Energy Collection in Tracking and Non-Tracking Photovoltaic Solar System", 1 st International Conference on Electrical and Electronics Engineering, p. 310-314
[26] T. Uematsu, Y.Y., K. Tsutsui, Y. Miyamura, H.Ohtsuka, T. Warabisako, T. Joge, "Design andcharacterization of fat-plate static-concentrator photovoltaic modules", Solar Energy Materials and Solar cells, 2001, 67.
[27] Ortabasi, U., "Low Concentration Photovoltaic

Module design using bifacial solar cells", proceedings of IEEE. 1988.
[28] I. Edmonds, "The performance of bifacial solar cells in static solar concentrators", Sol. Energy Mater., vol. 21, no. 2, pp. 173–190, 1990
[29] U.A. Yusofoglu, T. H. Lee, T. M. Pletzer, A. Halm, L. J. Koduvelikulathu, C. Comparotto, R. Kopecek, H. Kurz, "Simulation of energy production by bifacial modules with revision of ground reflection", Energy Procedia 55, pp. 389-395, 2014.
[30] Ismail Shoukry, PhD thesis, University of Stuttgart, October 2015

ANGULAR-DEPENDENT OUTDOOR INVESTIGATION OF BIFACIAL MODULES

Stephanie Malik [1,*], David Dassler [1,2], Jens Froebel [1], Akshayaa Pandiyan [1], Matthias Ebert [1]
[1] Fraunhofer Center for Silicon Photovoltaics CSP, Otto-Eissfeldt-Strasse 12, 06120 Halle (Saale), Germany
[2] Anhalt University of Applied Sciences, Bernburger Str. 55, 06366 Koethen, Germany
[*] Telephone +49 (0) 345 5589-5212 | Stephanie.Malik@csp.fraunhofer.de

ABSTRACT: Modules with bifacial technology show a high gain compared to monofacial modules, due to utilization of an illuminated rear side, as reported in different publications [1, 2]. The utilization of both sides leads to new opportunities for outside applications, such as open field or building integrated (BIPV) installations. Regarding this fact, bifacial module design optimization is challenging. This paper investigates the outdoor bifacial behavior for different angles of incidence (AOI) with the help of an individual-driven two-axis tracking system. With a bifacial module, the angular dependent response under outdoor conditions gets a new significance due to the additional contribution of the rear incident light. In this paper, the performances of bifacial and monofacial modules are compared to each other considering installation scenario, AOI, direct and diffuse irradiance, module temperature and backlight conditions. The paper shows the module preparation, measuring set-up and tracking sequences as well as an improved method to evaluate the relative gain.

Keywords: bifacial solar modules, bifacial gain, angle of incidence, two axis tracking system

1 INTRODUCTION

Bifacial technology is an extremely promising development that may lead the future solar market [3]. Following the current roadmap of ITRPV [4], the worldwide market share for bifacial cell technology will rise from 5 to 20% in the next 10 years.

The utilization of both module sides leads to new opportunities for outside applications, like open field or building integrated (BIPV) installations; however, the extent of the rear incident light depends on several factors including irradiance conditions, tilt angle, height, albedo and mounting structure. These factors make estimating the actual bifacial gain under real outdoor conditions more difficult.

Research to quantify the gain of the bifacial modules with short-term single module outdoor measurements is an absolutely necessary input into this inquiry, and the literature includes several outdoor studies based on fixed installations. The first investigation using a two axis tracking system was done by Egido M.A. 1986 [5] in order to verify the accuracy of his introduced theoretical model - for the collection of irradiance by a bifacial pv panel. A more recent new approach is to investigate the "angular response" of bifacial modules with an individual-driven two-axis tracking system to assign angular behavior. This enables innovative short time measurements to be taken, within the context of needs and specific conditions of producers and users and can accelerate the introduction of these modules.

For this experiment, the mono- and bifacial modules faced the sun at different orientations and tilt angles. Different scenarios of fixed orientations, sun-tracked and individual movements of the two-axis tracking system were implemented to determine the angular-dependent bifacial gain. Subsequently, the performances of bifacial and monofacial modules were compared to each other – considering installation scenario, AOI, direct and diffuse irradiance, module temperature and the back light conditions.

2 APPROACH

The approach, to determine the behavior of bifacial modules regarding the front and rear side, by applying a fast and flexible outdoor characterization tool, consisted of three parts (1)-(3) shown in Figure 1 below.

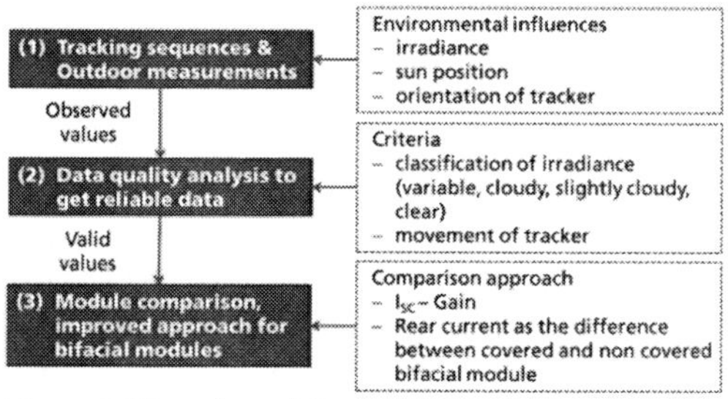

Figure 1: Flow chart of the approach

Part (1), described in chapter 3, consists of the outdoor measurements of bifacial and monofacial modules accounting for different tracking sequences. The main challenge was to define the tracker orientation in terms of irradiance conditions and sun position.

Part (2), described in chapter 4, concerns the implementation of different criteria to produce a reliable data set and to allow for diverse analysis methods. The focus was a classification of irradiance conditions in 4 classes (variable, cloudy, slightly cloudy and clear) and the detection of tracker movements, which do not relate to tracking the sun in a standard way regarding azimuth and elevation.

Part (3), described in chapter 5, investigates the differences and similarities between bifacial and monofacial modules for clear and cloudy conditions and different tracking sequences to quantify the bifacial gain.

3 TRACKING SEQUENCES & OUTDOOR MEASUREMENTS

The investigation was carried out on an individual-driven two-axis tracking system in Saxony-Anhalt (Germany) in spring and summer 2016. Table I, summarizes the properties of the solar tracker. Figure 2 presents the backside of the solar tracker, equipped with measuring equipment and bifacial and monofacial module samples.

Table I: Solar tracker properties

Property	Parameter
mast height	3.6 m
panel for module mounting	12 m²
azimuth range	35° - 325°
elevation range	0° - 90°

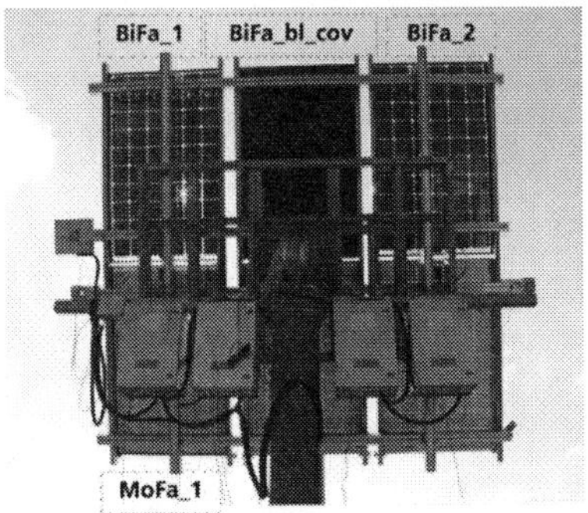

Figure 2: Bifacial (BiFa) and monofacial (MoFa) modules and measuring equipment on an individual-driven two-axis tracking system in Germany (view: rear side of the tracker)

Current-Voltage-curves, module temperature, probed at the backside, of every module and irradiance is measured synchronously and independently in a 10 seconds time interval. During these measurements each module is operated at its maximum power point (MPP). The measuring device has a current-full scale range (FSR) of 10.24 A as well as 12.5 A (dependent on the measuring place) with a resolution of 1 mA. The FSR of the voltage is 259 V with a resolution of 10 mV. The weather data acquisition system includes Si-crystalline irradiance sensors in module plane (2 on the front side & 1 on the backside of the tracker) and a pyranometer in the module plane. The global horizontal irradiance is measured independently of the solar tracker with an additional pyranometer. Wind direction and speed as well as ambient temperature, are logged using a combined device [6]. Unfortunately, the albedo of the ground, consisting of concrete, isn't measured. Notably, however, the ground was not modified during measurements.

The module samples are three bifacial modules (BiFa) and one monofacial module (MoFa) from the same module manufacturer and have the same properties for front and rear glass and encapsulation. For the purpose of examining the impact on the front and rear side of the bifacial modules, one sample is covered on the backside with a black foil (BiFa_bl_cov). The black cover was chosen to avoid the penetration of additional light due to reflection or scattering, which is a characteristic of white foil.

The performance of each module was characterized indoor under standard test conditions with a class AAA module flasher (at 1000 W/m² with an AM1.5 spectrum and module temperature of 25 °C). The measurements for the bifacial modules were carried out "one-sided", with a black rear cover for the side of the module that is not facing the light of the flasher. Table II details the indoor characterization results at STC for the front side. The electrical parameters of the bifacial modules are almost equal.

Table II: Module samples and their indoor characterization results at STC; front measurements with a black cover on backside

Module sample	V_{OC} [V]	I_{SC} [A]	P_{MPP} [W]	FF [%]
Bifa_1	38.9	9.6	283.4	75.9
Bifa_2	38.9	9.5	281.8	76.1
BiFa_bl_cov	38.9	9.5	281.5	75.9
MoFa_1	38.4	9.5	280.6	77.1

After the prior indoor characterization the modules were installed on the two-axis tracking system (see Fig. 2).

Over the course of the day, different sequences for the tracking system were applied, beginning with fixed settings, in different orientations and 2-axis tracking ("following the sun"). Furthermore, a special sequence for the bifacial modules was realized, to examine the maximum on the back side. In this sequence, the sun's azimuth was tracked on the back side. For investigating the impact of the angle of incidence (AOI), different algorithms were implemented, to cover the requirements of IEC 61853 [7]. In Table III all orientations and sequences of the solar tracker are summarized.

Regarding the cases 8 & 9 of Table III, the approach of measurement techniques for angle of incidence effects, introduced by King, Bruce H. [8], was applied. Two different tracking procedures (Table III, case 6 & 7), are shown as examples in Figure 3. The data evaluation of this paper is focused on case 5 and 8.

Table III: Summary of orientations & sequences of the solar tracker (AZ = tracker azimuth, EL = tracker elevation)

Case		Orientation / Sequence of Solar Tracker
Fixed position	1	AZ: 180° (south); EL: 0° (horizontal)
	2	AZ: 180° (south); EL: 90° (vertical)
	3	AZ: 90° (east); EL: 90° (vertical)
	4	AZ: 270° (west); EL: 90° (vertical)
Sun tracking	5	2 - axis tracking ("following the sun")
	6	AZ: tracking with focus on the backside EL: 90° (vertical) (see Fig. 3a)
Achievement of different AOI	7	AZ: changes in equal steps EL: start with tracking the sun and then remain constant during cycle (see Fig. 3b)
	8	AZ: changes in equal steps EL: sun tracking with a small offset during cycle
	9	AZ: sun tracking during cycle EL: changes in equal steps

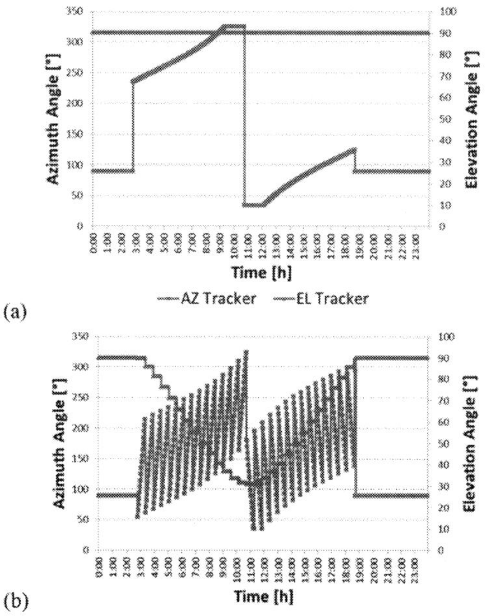

Figure 3: Two individual driven tracking sequences; see Table III case 6 (a), case 7 (b)

In Figure 3a, the elevation of the solar tracker remains at 90° (vertical) over the whole day, the azimuth (angle definition in reference to the front of the modules) follows the sun's azimuth with the backside of the modules. There is an interruption for several hours during the middle of the day, due to the limited rotational range of the solar tracker. The tracking sequence of Figure 3b starts with an optimal facing of the sun in azimuth and elevation. The elevation is then kept constant for the tracking cycle, while the azimuth is changed in equal steps, to achieve an AOI up to 90° – then a new cycle starts.

4 DATA QUALITY ANALYSIS

To evaluate the measured data in diverse ways, a differentiation of irradiance conditions was necessary. The classification of the insolation on the degree of cloudiness enabled the categorization of every measured data set. For this, an approach introduced by Tina G.M. [9], was implemented to define the insolation situation as clear, slightly cloudy, cloudy or variable. The procedure, described in detail in [9], determines an "index of serenity" k(t) (1), taking into account the measured and a calculated global horizontal irradiance following the ESRA clear sky model (CSM). In the second step the Moving Average (MA) of the "index of serenity" is calculated for 3 time steps (2). The last equation (3) sums up the absolute differences between the "moving average" and the actual "index of serenity" for 3 time steps, named as "moving function" MF. The classification is done by following a decision tree and limits for k(t) and MF displayed in [9].

$$k(t) = \frac{|x_{measured}(\Delta t) - x_{CSM}(\Delta t)|}{x_{CSM}(\Delta t)} \tag{1}$$

$$MA = \frac{1}{3}\sum_{s=0}^{2} k(\Delta t - s) \tag{2}$$

$$MF = \sum_{s=0}^{2} |MA - k(\Delta t - s)| \tag{3}$$

Table IV: Description of parameters; approach [9]

Parameter	Description
k(t)	"index of serenity"
$X_{measured}$	measured global horizontal irradiance
X_{CSM}	calculated global horizontal irradiance (ESRA clear sky model)
Δt	timestamp
s	index of accumulation
MA	moving average
MF	moving function

Figure 4 shows an example for the irradiance classification for a clear and non-clear day. The implemented algorithm works well, and was applied to the whole data set.

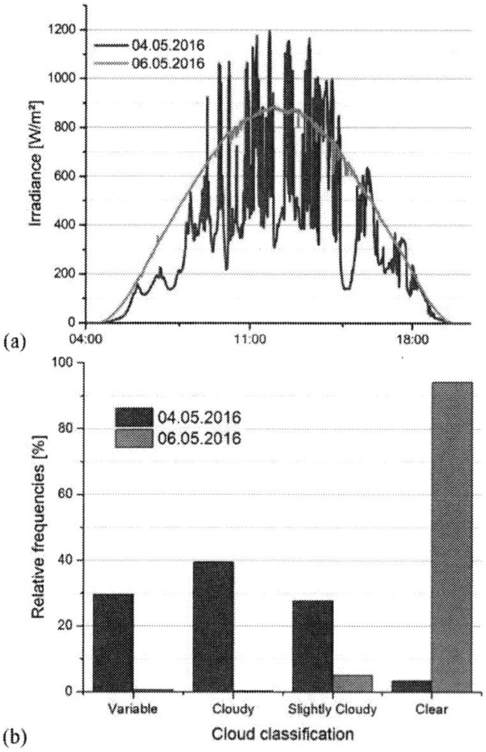

Figure 4: Measured irradiance for a clear (orange) and non-clear day (green) (a) and their corresponding relative frequencies of cloud classification (b)

5 MODULE COMPARISON RESULTS

To analyze the front and rear behavior of the bifacial modules, the rear current was calculated as difference between a non-covered and a covered bifacial module. A temperature correction of the current was not performed due to the small temperature dependence of I_{SC}. Additionally, the percentage gain in I_{SC} — determined by bifacial versus monofacial module — is used for module comparison.

In Figure 5, a clear - partially cloudy day is shown, where a two axis tracking mode was applied (case 5 in Table III). The I_{SC} of every module sample — the bifacial rear current as well as the I_{SC} gain — is displayed. The bifacial rear current, during clear conditions, is around 0.5 A. Under clear conditions the bifacial modules show an I_{SC} gain of around 4 to 5%. Under cloudy conditions, the I_{SC} gain varies widely. The expanded uncertainty for a small I_{SC} arround 1.5 A corresponds to 0.015 A.

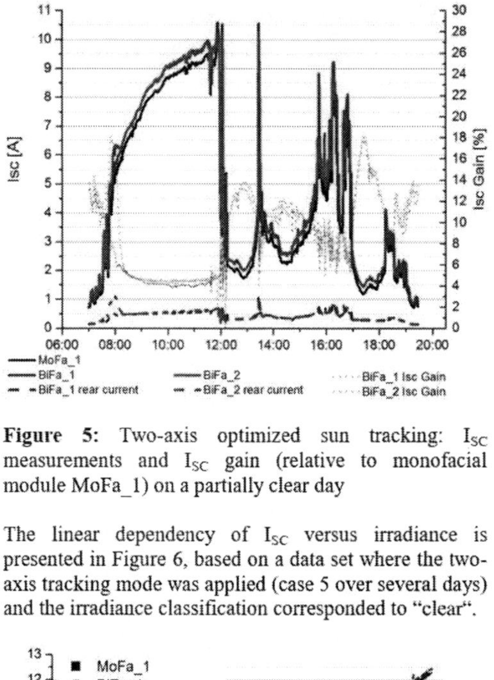

Figure 5: Two-axis optimized sun tracking: I_{SC} measurements and I_{SC} gain (relative to monofacial module MoFa_1) on a partially clear day

The linear dependency of I_{SC} versus irradiance is presented in Figure 6, based on a data set where the two-axis tracking mode was applied (case 5 over several days) and the irradiance classification corresponded to "clear".

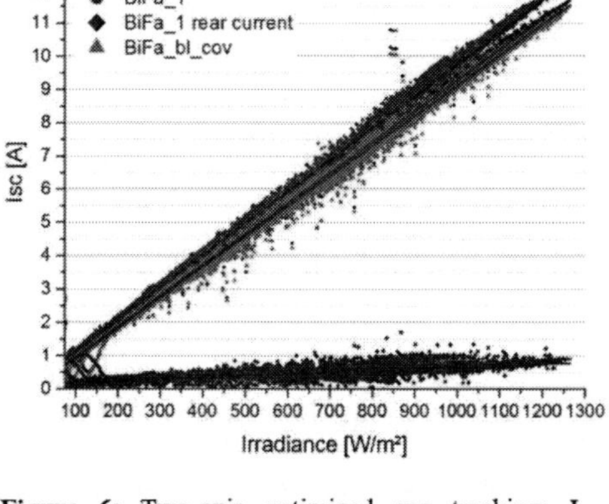

Figure 6: Two-axis optimized sun tracking: I_{SC} as function of measured irradiance for clear classification

The straights of the monofacial module and the black covered bifacial module lie almost immediately above one another. The slope of the bifacial module (BiFa_1) is larger than of the monofacial and black covered bifacial module. The relative difference between bifacial and monofacial modules is approximately 4%.

I_{SC} in dependence of AOI

Different angles of incidence were achieved by applying the tracking cycle case 8 of Table III — tracking the sun in the elevation with a small offset during whole cycle and changing the azimuth in equal steps. Figure 7 shows the results for I_{SC} in dependence of AOI of this procedure (AOI = 0° corresponds to: module is normal aligned to the sunrays). As part of the process of preparing this graph, tracker movements were detected and excluded. Only short time periods between the different steps were taken into account. Additionally, the irradiance was filtered between 800 and 1000 W/m² and the irradiance classification "clear" was chosen.

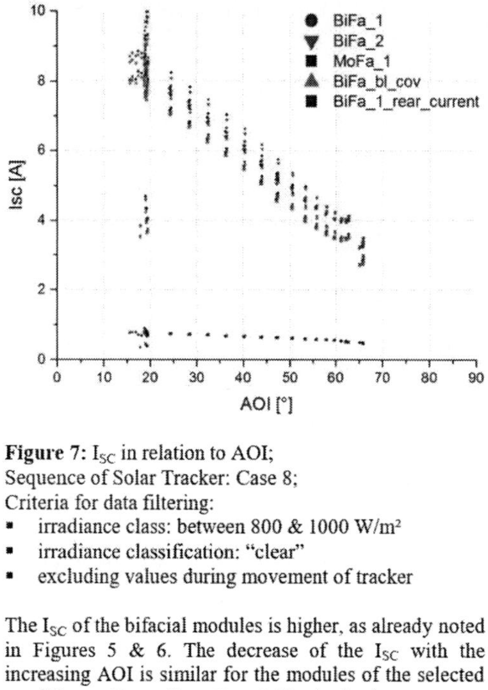

Figure 7: I_{SC} in relation to AOI;
Sequence of Solar Tracker: Case 8;
Criteria for data filtering:

- irradiance class: between 800 & 1000 W/m²
- irradiance classification: "clear"
- excluding values during movement of tracker

The I_{SC} of the bifacial modules is higher, as already noted in Figures 5 & 6. The decrease of the I_{SC} with the increasing AOI is similar for the modules of the selected conditions. Regarding the AOI, it is important to mention, that the same antireflective coated glass was used for every sample. The current of the rear side (BiFa_1_rear_current) also decreases slightly but is not affected like the current on the front side. Due to the small changes in the elevation, through the tracking the sun's elevation, and bigger steps in the azimuth, the diffuse part of the light is assumed as constant during the tracking cycle, which provides further support to the results of this experiment.

Bifaciality factor (BF) referred to Power

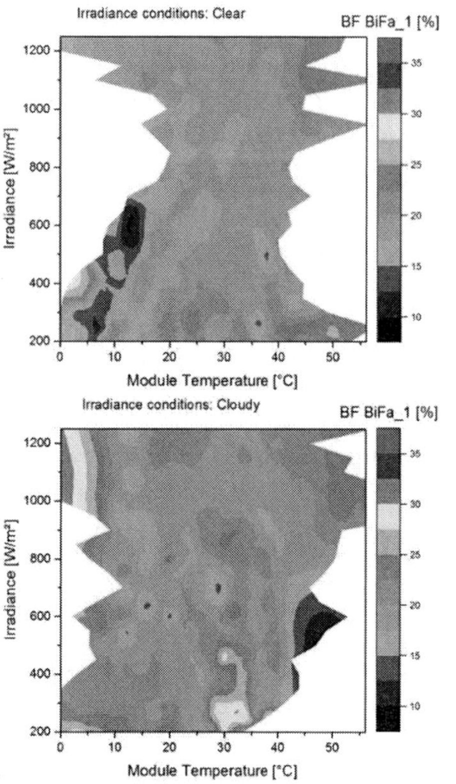

(a)

(b)

Figure 8: Bifaciality factor (BF) in relation to irradiance and module temperature class and BiFa_1;

- class width: irradiance 50 W/m²; temperature 2 °C
- basis of data set: days with two-axis optimized sun tracking

(a) under „clear" irradiance conditions
(b) under „cloudy" irradiance conditions

Beside the evaluation of the I_{SC}, the power (P) in MPP was analyzed as well. To take both sides of the bifacial module into account, the "bifaciality factor" (BF) was determined (4). The difference in power between the bifacial (BiFa_1) and the black covered bifacial module (BiFa_bl_cov) gives the power on the rear side (P_{rear}). The temperature of the covered module was slightly different to the non–covered module. For that reason the power of the covered module was corrected with the temperature coefficient for the power, taken from the module data sheet (index of Power: BiFa_bl_cov$_{corr}$).

$$BF = \frac{P_{rear}}{P_{front}} = \frac{P_{BiFa\_1} - P_{BiFa\_bl\_cov_{corr}}}{P_{BiFa\_bl\_cov_{corr}}} \quad (4)$$

In Figure 8 the mean bifaciality factor (BF) in relation to irradiance class, module temperature class and BiFa_1 is presented. The data set is based on days with two-axis optimized sun tracking (case 5) for clear (a) and cloudy (b) conditions. The average BF was determined per irradiance and temperature class. The standard deviation of BF in classes > 400 W/m² is rather small. Higher deviation can be found in values under 400 W/m².

The different shapes of the contour plots refer to available data sets for the different conditions. In general, the bifaciality factor lies between 15% and 25% for both cases and is more homogeneous as expected. A clear tendency cannot be seen neither for "clear" nor for "cloudy". It is a "snapshot" for a small data set under varying conditions at a specific location.

Unfortunately, the effect of the albedo of the ground cannot be sufficient quantified in the existing investigation. In [10] can be found estimations, to get an idea of the impact of the albedo of the height of module installation. As the module exceeds a certain height, the "view factor" (VF) from the reflective surface to the module begins to drop significantly, leading to a lesser amount of reflected radiation reaching the rear surface. "The VF, which is a geometric entity ranging from zero to one, defines the fraction of radiation emitted from a surface, which is received by another." [10] The maximum investigated height in [10] was 2.5 m, in comparison to the present case of 3.6 m (installation height of bottom edge of bifacial modules). Additionally the inhomogeneity of the irradiance reaching the rear surface of the modules is reduced significantly through a higher module installation [10]. Both cases can be applied to our measuring set-up.

But nevertheless the obtained results of this investigation are related to a specific location, measuring period and set-up. Further work is needed to set-up a universal outdoor measuring system for bifacial modules and standardized routines to analyze the data precisely.

6 SUMMARY

In this investigation, individual sequences, as well as fixed settings and optimal sun-tracking, were applied with a two-axis tracking system to determine the angular behavior of bifacial modules and the consequent derived gain. Furthermore, a variety of different criteria regarding classification of irradiance conditions in four classes (variable, cloudy, slightly cloudy and clear) and the detection and exclusion of tracker movements, were implemented to ensure that a reliable data set was collected, and to allow for diverse analysis.
The module comparison showed an I_{SC} gain of around 4 to 5% for the bifacial modules under clear conditions. The decrease of the I_{SC} with increasing AOI was similar for the bifacial and monofacial modules in the selected conditions. The bifaciality factor, evaluated for different irradiance and module temperature classes, lies in general between 15% and 25% for clear as well as for cloudy conditions and is more homogeneous as expected. A clear tendency cannot be seen with increasing irradiance or module temperature.
Finally, fast, wide tracker measurements under relative constant conditions are a big advantage for bifacial module gain characterization.

7 ACKNOWLEDGEMENT

This work was supported by the Federal Ministry for Economic Affairs and Energy (BMWi) within the project "Fidelitas" (project no. 0325735C).

8 REFERENCES

[1] Böscke, T. S.; et al, "Bifacial n-Type Cells With >20% Front-Side Efficiency for Industrial Production", IEEE Journal of Photovoltaics, Vol. 3, No. 2, 2013

[2] Van Aken, B. B.; et al, "Relation between indoor flash testing and outdoor performance of bifacial modules", 29th EU-PVSEC, Amsterdam, 2014

[3] Kopecek, R; et al, "Bifaciality: One small step for technology, one giant leap for kWh cost reduction", Photovoltaics International, Volume 26.

[4] International Technology Roadmap for Photovoltaic (ITRPV) Report 2016

[5] Egido, M. A.; Lorenzo, E., "Bifacial photovoltaic panels with sun tracking", Int. J. Sol. Energy, vol. 4, pp. 97–107, 1986.

[6] Malik, S.; et al, "Outdoor data evaluation of half-/full-cell-modules with regard to measurement uncertainties and the application of statistical methods", 29th European Photovoltaic Solar Energy Conference and Exhibition, P. 3269-3273, 2014, Amsterdam

[7] International Electrotechnical Commission (IEC), International Standard IEC 61853-2, Photovoltaic (PV) Module Performance Testing and Energy Rating Part 2, 2010

[8] King, Bruce H.; et al, "Recent Advancements in Outdoor Measurement Techniques for Angle of Incidence Effects", Photovoltaic Specialist Conference (PVSC), IEEE 42nd, 2015, New Orleans, DOI: 10.1109/PVSC.2015.7355849

[9] Tina, G.M.; et al, "Analysis of forecast errors for irradiance on the horizontal plane", The International Renewable Energy Congress IREC, 2011, DOI:10.1016/j.enconman.2012.05.031

[10] Ufuk A. Yusufoglu; et al, "Analysis of the Annual Performance of Bifacial Modules and Optimization Methods", IEEE Journal of Photovoltaics, 2015, DOI: 10.1109/JPHOTOV.2014.2364406

THE SI-TRACEABLE CALIBRATION OF SHUNTED REFERENCE SOLAR CELLS VIA DIFFERENTIAL SPECTRAL RESPONSIVITY MEASUREMENTS.

Florian Witt*[1], Ingo Kröger[1], Stefan Winter[1]
[1]Physikalisch-Technische Bundesanstalt (PTB)
Bundesallee 100, 38116 Braunschweig, Germany

*corresponding author: Tel. +49 531 592 4143, Fax: +49 531 592 694143, E-mail: Florian.Witt@ptb.de

ABSTRACT:
Shunted reference solar cells are commonly used to determine and to monitor the irradiance at outdoor solar cell testing facilities, because they allow the convenient measurement of a voltage without current-to-voltage converter. This type of reference solar cell contains a resistor with an only roughly known low resistance that is built in the housing and connected to the cell, so that a voltage value proportional to the current of the solar cell under illumination can be measured. Until now this type of reference solar cell was commonly calibrated using a sun simulator. A new significantly improved voltage amplifier allows the application of precise differential spectral responsivity calibrations of such devices. In this work the application and validation of a new calibration chain for the calibration of shunted reference solar cells with lowest measurement uncertainties is presented. By the use of the new, high precision voltage amplifier the DSR- method was transferred to shunted reference cells and a measurement uncertainty (MU) of the shunt voltage under standard test conditions below 0.8 % with k=2 was achieved. To validate the calibration results two different SI-traced measurements, a direct calibration of both types of amplifier as well as a calibration with the combination of a reference photodiode and a reference resistor were performed and compared.

Keywords: Calibration, Spectral Response,

1 INTRODUCTION

Differential Spectral Responsivity (DSR) measurements are a precise tool for the calibration of the short circuit current I_{SC} of reference solar cells which are used as irradiance standards in laboratories worldwide [1].
For the determination of the solar irradiance in outdoor applications a precise short circuit current measurement is difficult. To circumvent this problem so called shunted reference solar cells are available were the I_{SC} flows through a shunt resistor and the voltage drop across the resistor (U_{shunt}) is determined. Although voltage measurements are more comfortable performed outdoors than current measurements the commonly high calibration uncertainty of U_{shunt} under standard test conditions (U_{STC}) for such devices limits their application. In this work the application of DSR for precise calibrations of U_{STC} is presented.

2 METHODS

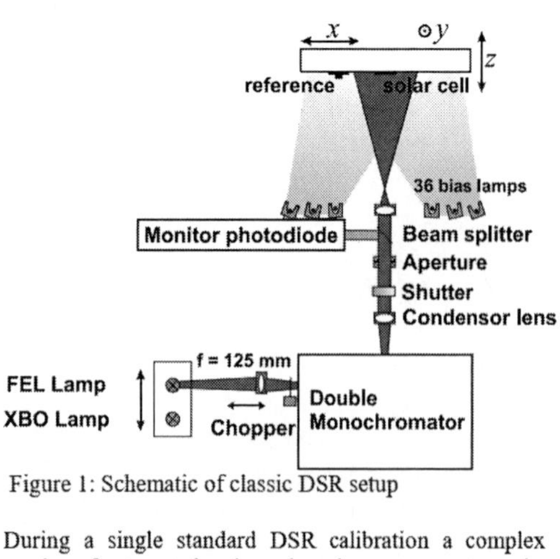

Figure 1: Schematic of classic DSR setup

During a single standard DSR calibration a complex routine of automated and monitored measurements on the device under test (DUT) (here a commercially available shunted reference solar cell in WPVS design) and reference are performed following multiple mirrored sequences of relative and absolute measurements. Here DUT and reference are connected via a relays switching system to the same measurement electronics. The absolute differential spectral responsivity is determined at few chosen wavelengths while relative spectral responsivities are determined over a broad wavelength regime at multiple bias light levels below and above 1000 W/m^2 (Fig.1). From these data the absolute responsivity at multiple irradiance levels is determined and finally the responsivity at standard test conditions (Figure 2b) and therefore the short circuit current under standard test conditions calculated from equation 1.

$$I_{SC} = \int s(\lambda) \cdot E_{AM1.5}(\lambda) \mathrm{d}\lambda \quad [s(\lambda)] = \mathrm{A} / (\mathrm{W/m^2})$$

Equation 1: Calculation of I_{SC} with the spectral irradiance responsivity at standard test conditions $s(\lambda)$ and the standard spectral irradiance at AM1.5 conditions $E_{AM1.5}(\lambda)$ for reference solar cells without additional shunt resistor.

$$U_{SC} = \int s_U(\lambda) \cdot E_{AM1.5}(\lambda) \mathrm{d}\lambda \quad [s_U(\lambda)] = \mathrm{V} / (\mathrm{W/m^2})$$

Equation 2: Calculation of the voltage drop U_{SC} with the voltage based spectral irradiance responsivity at standard test conditions $s_U(\lambda)$ and the standard spectral irradiance at AM1.5 conditions $E_{AM1.5}(\lambda)$ for reference solar cells with an integrated additional shunt resistor.

The determination of the measurement uncertainties is performed by a Monte Carlo simulation. For the shunt voltage calibration of the voltage based absolute spectral responsivity the realization of a voltage responsivity standard was necessary.

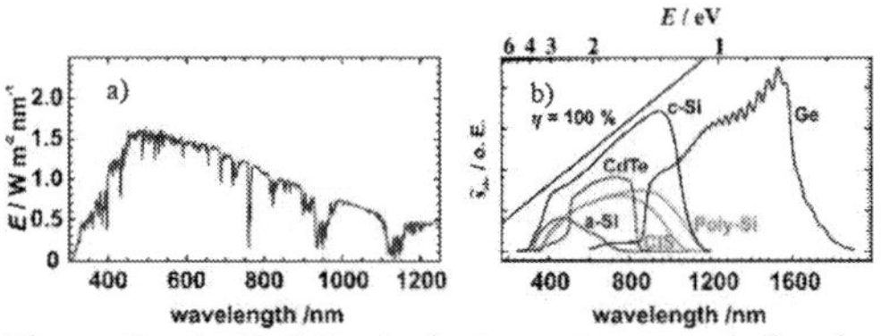

Figure 2: a) AM1.5 standard spectrum used for the determination of I_{SC} in equation 1 and U_{SC} in equation 2. b) Spectral responsivities of different solar cell devices.

Two approaches where chosen, the attachment of a calibrated resistor to the irradiance standard detector and for comparison the direct calibration of the applied current and voltage amplifiers.

In Figure 3a) the DSR primary calibration routine "a)" with calibrated photodiodes as references is demonstrated. A trans-impedance amplifier (TIA) transforms the short circuit current in a voltage and the AC part is subsequently measured with a lock-in amplifier (LIA). The device under test is measured with the same instruments as the reference so that the measurement uncertainties of the electronic equipment are canceled out. Only the linearity of the devices must be confirmed or taken into account.

To compare a shunted reference solar cell with the irradiance standard in calibration routine "b)" the photodiode is additionally shunted with a calibrated resistor (figure 3b). The respective U_{shunt} of the DUT and reference is amplified by the same voltage amplifier (VA) and measured again with LIA. Thus the VA needs not to be absolutely calibrated. Since U_{shunt} of the reference is measured in the scale of µV no additional correction due to the references IV characteristics is necessary.

To verify the calibration result obtained by method b) the reference was measured directly with a TIA while the device under test was measured via a VA. In this case (calibration routine "c)") both amplifiers need to be calibrated absolutely.

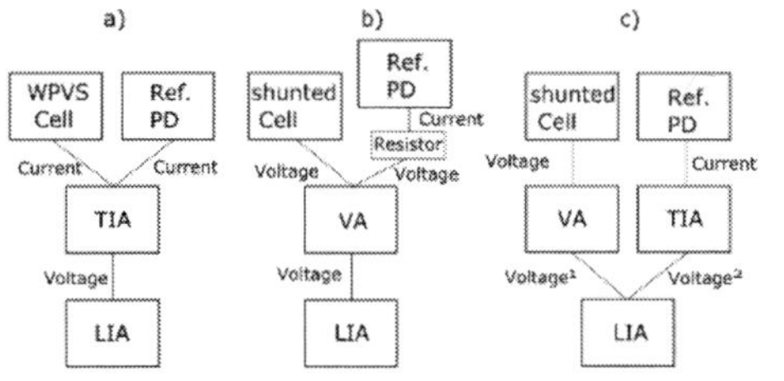

Figure 3: Scheme of the electronical interconnection in DSR calibrations of a) standard reference solar cells. Shunted reference cells are either calibrated to a shunted reference b) or alternatively with calibrated voltage and current amplifiers c).

3 CALIBRATIONS OF AMPLIFIERS

For absolute AC calibration of the applied amplifiers a stable voltage source, reference resistors and calibrated HP3458 digital multi meters (DMM) were used. The DMMs were set to a burst high speed mode were about 38.000 data points of the AC reference signal and the Amplifier output were acquired in parallel.

A Fourier transformation was applied to the data sets to determine the amplification factor at the fixed reference frequency by comparing the integrated maxima. This procedure was repeated at multiple AC amplitudes. With this method the AC amplification factors could be determined with a relative uncertainty of 0.5% (k=2).

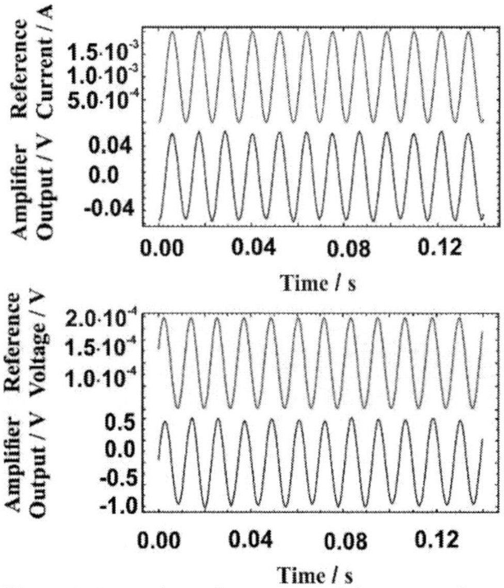

Figure 4: Comparison of an excerpt of the measured AC reference current and TIA output (top) and reference AC voltage and VA output (bottom). Total Measurement time: 7.5 s.

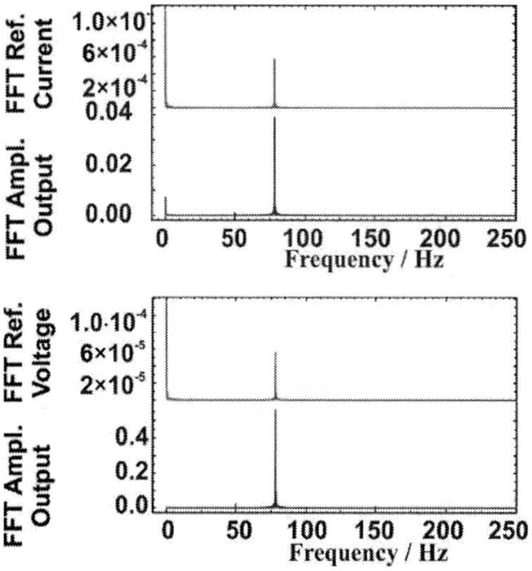

Figure 5: Fourier transformation of the measured AC signals. The integrated signals (+- 0.2 Hz) at the reference frequency are compared and the amplification factor is subsequently calculated. If the signal would be measured with the AC-mode of a DMM the 50 Hz noise that occurs only at the output signal, would alter the result.

4 RESULTS AND OUTLOOK

Calibration method b and c result in well comparable spectral responsivities and U_{STC} values. The main difference in the spectral responsivities originates from a slight upward shift of method c). In general the measurement uncertainty in method c) determined by Monte Carlo simulations is higher. This originates from the much larger measurement uncertainty of the amplifier calibration in comparison to the reference resistor calibration. The direct comparison of method b) and c) results in an E_n=0.2.

For further validations a comparison to outdoor measurements is planned in future. Also the possibility to compare the U_{STC} DSR calibration of a standard WPVS solar cell with an attached calibrated resistor to a I_{STC} DSR calibration will be investigated.

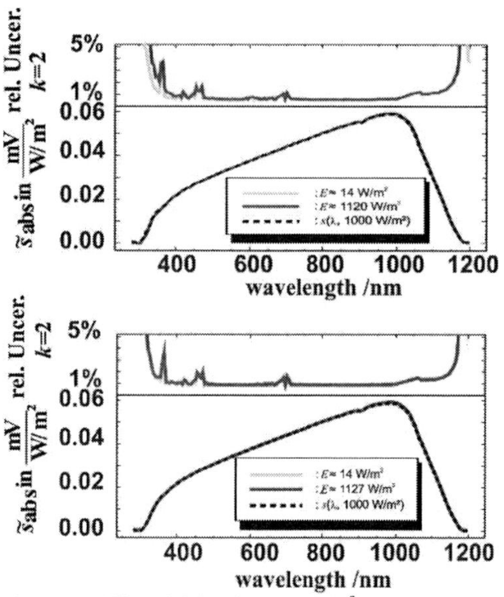

Figure 6: Differential (at about 10 W/m^2 and about 1120 W/m^2) and absolute spectral responsivity at AM 1.5 conditions obtained by method b) (top) and method c) (bottom).

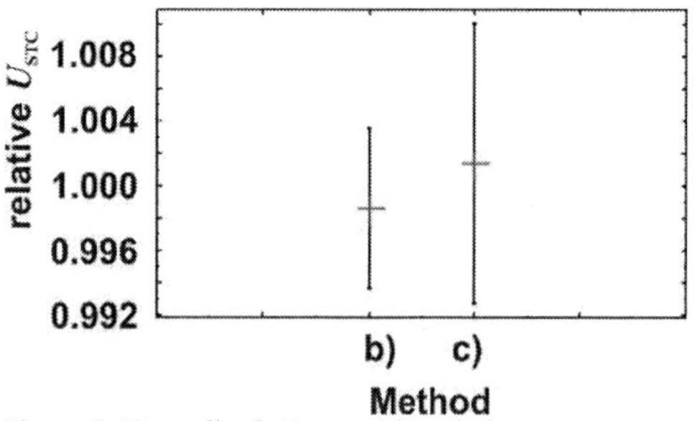

Figure 7: Normalized U_{STC} and its relative measurement uncertainty.

5 REFERENCES

[1] S. Winter et al, Proceedings. 16th European Photovoltaic Solar Energy Conf.,(2000), 4ff

INVESTIGATION OF THE INFLUENCE OF TEMPERATURE INHOMOGENEITY ON THE MEASUREMENT UNCERTAINTY OF SOLAR CELL TEMPERATURE COEFFICIENTS

Antje Schweitzer[1], Ingo Kröger[1], Stefan Winter[1]
[1]Physikalisch-Technische Bundesanstalt (PTB), Bundesallee 100, 38116 Braunschweig, Germany
Tel.: +49 531 592 4189, fax: +49 531 592 69 4183, e-mail: antje.schweitzer@ptb.de

ABSTRACT:
In this study we characterize a newly developed solar cell chuck of PTB for suitability of temperature dependent measurements. We investigate the relative temperature distribution on a tempered solar cell and the temperature offset between the solar cell and the PT100 sensor which is used for temperature regulation. We adjusted the setpoint temperature from TSP=15 °C up to 75 °C and measured the relative temperature distribution with an infrared camera. Performing temperature dependent differential spectral responsivity (DSR) measurements, we determined the spectral temperature coefficient of an exemplary silicon solar cell. With these data we can derive the influence of the temperature inhomogeneity on the measurement uncertainty. Non-irradiated we measured the absolute temperature with sixteen calibrated PT100 sensors distributed on the solar cell's surface. Applying the open circuit method (V_{OC}-method) we investigated the influence of incident irradiation on the temperature behavior of the solar cell chuck and the solar cell. We found out that the highest uncertainty contribution arises through the calibration of the integrated PT100 sensors of the chuck.

Keywords: calibration, characterization, spectral responsivity

1 INTRODUCTION

The characterization and calibration of solar cells and modules is performed, for worldwide comparability, under defined standard test conditions including irradiance, spectral distribution and temperature. These test conditions are fixed in the IEC 60904-3 standard with an irradiance of 1000 W/m^2, an AM1.5g spectrum and a device temperature of 25°C. However, for characterization under realistic conditions it is necessary to extend the standard test conditions with different irradiances, temperatures and angular dependence. These extended test conditions, fixed in the IEC 61853-1 standard, define a matrix with irradiances from 100 W/m^2 to 1100 W/m^2, temperatures from 15°C – 75°C and the according percentage partial efficiency i.e. electrical power of a PV device. If the corresponding climatic conditions are known for a particular location, it is possible to determine the module with best performance by combining the device properties with the climatic conditions.

Since the electrical parameters of a photovoltaic device depend strongly from the temperature of the pn-junction, it is important to temper the solar cell homogenously to the wished setpoint temperature. A common method for tempering of photovoltaic devices is a solar cell chuck equipped with Peltier elements for temperature regulation. According to expectations the ambient air temperature can influence the temperature distribution on a solar cell and also induce a temperature offset between solar cell and setpoint temperature. Furthermore radiation of different irradiance levels may affect the temperature adjustment.

The PTB calibrates solar cells performing the DSR method. The calibration is carried out, according to the standard, at 25°C. The spectral responsivity is usually determined for 8 different irradiance levels from E=0W/m^2-1300W/m^2. In case that the chuck heats differently under irradiation, a systematic error is expected due to the temperature offset between solar cell and chuck. In order to estimate the impact of such a front heating on a DSR calibration, we made an exemplary calculation based on real experimental data. We assumed, that the temperature offset between chuck and solar cell increases linearly with the irradiance and reaches 2K at the irradiance level of E=1000W/m^2. The influence that would be induced by a different temperature behavior is depicted in Figure 1. The differential spectral responsivity of a 2cm^2 monocrystalline solar cell at 25 °C is shown in red, the simulated one in black. The short circuit current of this solar cell would be overestimated by approximately 0.1%. The motivation of this study is to find out, if there is a deviation between the solar cell's temperature to the setpoint temperature and if it has an influence on the measurement uncertainty of temperature coefficients.

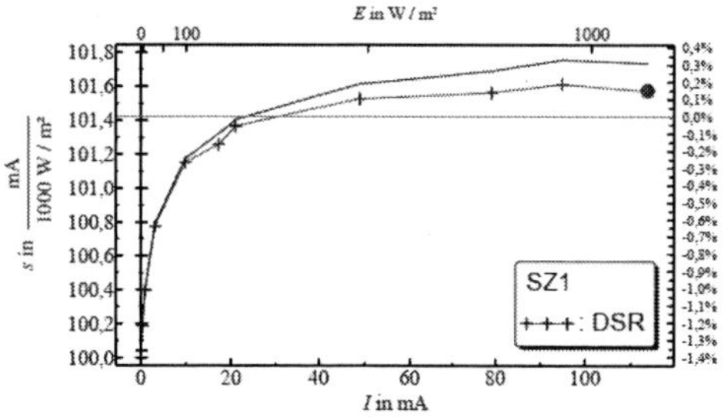

Figure 1: Measured differential spectral responsivity of a 2 cm^2 monocrystalline solar cell at 25 °C (red) and a simulated differential responsivity (black) with the assumption, that the solar cell heats up about 2K for the range of used irradiance level up to 1000W/m^2.

2 EXPERIMENTAL

For our investigations we used a Peltier element tempered solar cell chuck, which was developed by the PTB for calibration of 6" solar cells from 15°C to 75°C. The chuck has a copper plate with gold surface finishing and is equipped with nine Peltier elements and nine PT100 sensors that are incorporated in the copper plate

1 mm beneath the surface. The Peltier elements need a power supply unit which is controlled by temperature control software programmed in LabView. The readout of the PT100 sensors is ensured also via LabView software and a relaisbox. One of the nine PT100 sensors is used for temperature controlling and its readout value is delivered as input parameter to the LabView temperature control software. The whole setup is hereinafter referred to as temperature control unit.

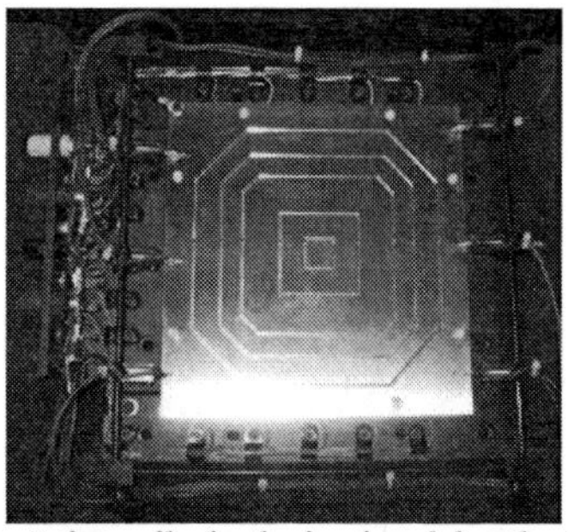

Figure 2: Solar cell chuck developed by the PTB for calibration of 6" solar cells. It allows to measure the open circuit voltage V_{oc} and short circuit current I_{sc} of a solar cell up to a size of 6" at temperatures from 15°C to 75 °C.

2.1.1 Absolute temperature measurement with PT100 sensors

Due to the strong correlation between short circuit current I_{SC} and open slope voltage V_{OC} it is essential to be aware of the absolute temperature of the pn-junction. For this reason we fixed 17 calibrated PT100 sensors on the surface of the cell. To ensure a good thermal conductivity, we used thermal paste between the solar cell and the sensors. The temperature of the chuck is determined by reading out its integrated PT100 sensors.

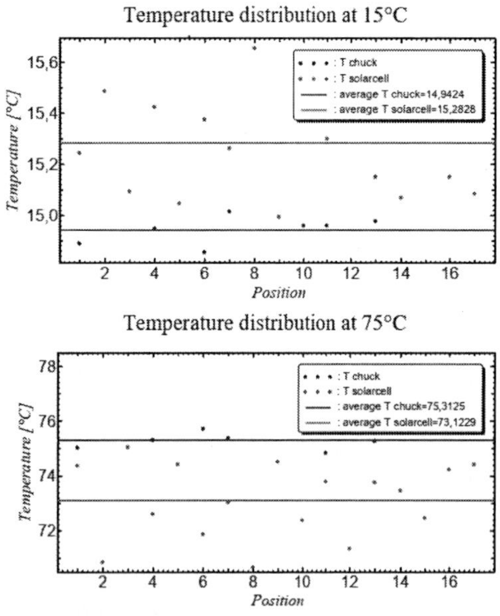

Figure 3: Absolute temperature measurement of the solar cell and the chuck with calibrated PT100 sensors at T_{SP}=15°C and 75°C.

We adjusted different setpoint temperatures of T_{sp}=15°C, 25°C, 35°C, 40°C and 75°C and read out at the same time the PT100 sensors. Exemplary the measurement for T_{sp}=15°C and 75°C is shown in Figure 3. The temperature offset between the solar cell and the solar cell chuck temperature is about -0,34 to 0,50 K for the temperature range between 15°C and 40°C, see Table 1. For higher temperatures a high offset with 2,20 K arises. The measurement uncertainty for the PT100 sensors on the solar cell was determined to $u_{T,SC}$=0,5°K and for the PT100 sensors in the chuck to $u_{T,CH}$=0,3°K (k=1). Therefore it has to be taken into account that between 15°C and 35 °C the offset lies within the order of the measurement uncertainty. The high temperature deviation at a setpoint temperature of 75°C might be caused by the non-perfect thermal contact of the PT100 sensors. We assume, that the influence of the heating through incident irradiation is higher than the temperature gradient between solar cell and ambient air temperature without radiation. Therefore we applied the V_{OC}-method for investigating temperature adjustment under irradiation.

T_{SP} [°C]	T_{SC} [°C]	$\sigma(T_{SC})$ [°C]	T_{CH} [°C]	$\sigma(T_{CH})$ [°C]	T_{CH} - T_{SC} [°C]
15	15,28	0,20	14,94	0,05	-0,34
25	25,13	0,13	25,09	0,15	-0,04
35	34,92	0,25	35,20	0,19	0,28
40	39,78	0,33	40,27	0,22	0,50
75	73,13	1,37	75,31	0,33	2,20

Table 1: Absolute temperature measured with calibrated PT100 sensors. Comparison of the absolute temperature on a non-irradiated solar cell (T_{SC}) with calibrated PT100 sensors placed on the solar cell and integrated in the chuck (T_{CH}) at ambient condition (T_{air} = 25°C) at different chuck setpoint temperatures. The calculated standard deviation σ shows the dispersion of the measured values about the average.

2.1.2 Temperature adjustment applying the V_{OC}-method

As the bandgap and charge carrier generation of a semiconductor reduces with the temperature, consequently the open circuit voltage decreases with increasing temperature. Out of that reason the V_{OC}-method is a commonly used and very precise method for temperature adjustment of photovoltaic devices. We applied the V_{OC}-method to a DSR-facility and investigated the temperature adjustment for different setpoint temperatures and under different irradiance levels. The photovoltaic device is tempered non-irradiated to a certain temperature. After an instantaneous irradiation the voltage increases and reaches within a short time range around 3ms down to a few µs [1] the maximum initial value V_{OC}. After a few milliseconds the solar cell starts to warm up and the voltages decreases. The temperature setpoint for the heat control unit has to be adjusted until the initial voltage value has been reached. The difference between the initial and adapted setpoint temperature represents the temperature offset between the junction of the solar cell and the temperature regulation sensor. We adjusted the setpoint temperature to T_{SP}=15°C, 25°C, 40°C and 75°C and irradiated the solar cell with 1019 W/m². Exemplary the measurement for 25 °C is shown in Figure 4.

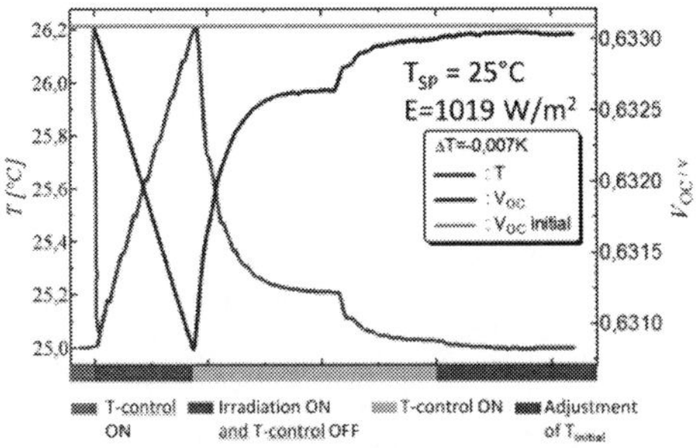

Figure 4: After tempering the solar cell to a certain temperature, the solar cell is irradiated and the temperature control is switched off. After the temperature control unit is switched on, the open circuit voltage (blue) reaches the initial value represented by the green line which indicates, that there is now significant offset between chuck and solar cell.

The open slope voltage reaches, after the temperature control is switched on again, the initial value $V_{OC,initial}$. The measurement indicates that the temperature control unit adapts the regulating parameters after irradiation, so that the solar cell reaches the same temperature as before being heated by incident irradiation. It is not required to adapt the setpoint temperature of the chuck. As the difference between the initial and final chuck temperature is smaller than 0,1K for the temperature range of T=15°C -75°C, we can conclude that the thermal contact between pn-junction and chuck is excellent and hence there is no temperature gradient across the solar cell chuck interface due to frontside irradiation.
The impact of front heating due to 1000W/m² sun-like irradiance should be much larger than the impact due to the temperature difference between the ambient temperature and setpoint temperature discussed in the previous section. Therefore we conclude that the large offset observed at 75°C probably originates from poor thermal contact of the attached PT100 sensors to the solar cell.

2.1.3 Relative temperature distribution using an infrared camera

The relative temperature distribution is determined with an infrared camera VarioCAM from infratec. It is suitable for the temperature range from -40°C to 500°C and has a measurement accuracy of ±1% for the spectral range from 7,5 µm -14 µm. We adjusted temperatures from T_{SP}=15°C, 25°C up to 75°C in steps of 10°C and after temperature equilibrium has been established we took five thermographic images from a defined distance to determine the average lateral temperature distribution. For setpoint temperatures near typical room temperatures the histogram shows a near-Gaussian temperature distribution (see Figure 5 top with T_{SP} = 25°C). A setpoint temperature far higher than the room temperature leads to increased heat flows, thus resulting in an inhomogeneous solar cell temperature, since the heat flow between solar cell and chuck, particularly in the edge regions, is not perfect. This leads to a non-Gaussian histogram with increased FWHM and a shifted mean temperature as visible in Figure 5 bottom for T_{SP}=75°C. If the short circuit current temperature coefficient is non-linear in the temperature range, measurement errors may occur. Hence we determine the linearity of the short circuit current.

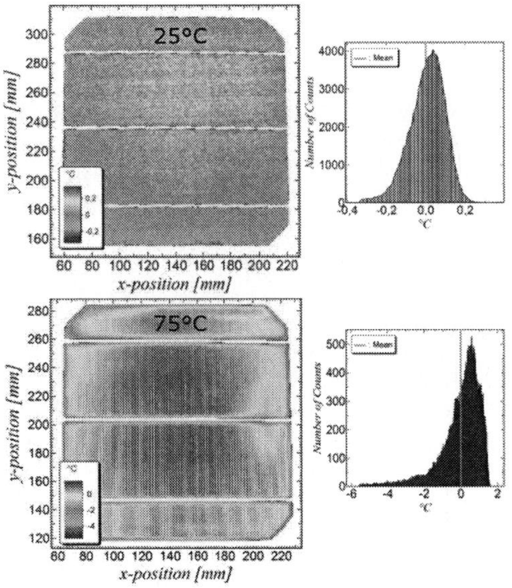

Figure 5: Thermographic images of a 156x156mm solar cell tempered at T_{SP}=15°C (top) and 75°C (bottom) which show the relative temperature distribution. The associated histogram of the solar cell's temperature distribution is depicted on the right side.

2.1.4 Temperature dependent differential spectral responsivity measurements

The spectral temperature coefficient and hence the temperature coefficient of the short circuit current is determined performing temperature dependent differential spectral responsivity (DSR) measurements. From this data the influence of the temperature inhomogeneity on the measurement uncertainty of the temperature coefficients can be derived. We tempered the solar cell at different setpoint temperatures from T_{SP}=15°C up to 75°C in 5°C steps and determined the absolute temperature dependent spectral responsivity, see Figure 6. With $I_{SC}(T) = \int s(\lambda, T) \cdot E_{AM1.5G} d\lambda$ we determine the temperature dependent short circuit current and the temperature coefficient as depicted in Figure 7. The graph shows, that the short circuit current changes linear with the temperature over the whole temperature range leading to the conclusion, that for an inhomogeneous tempered solar cell the overall generated current corresponds to the current that would have been generated at the associated mean temperature. Hence we can conclude that the temperature inhomogeneity has no significant impact on the measurement uncertainty of I_{SC}.

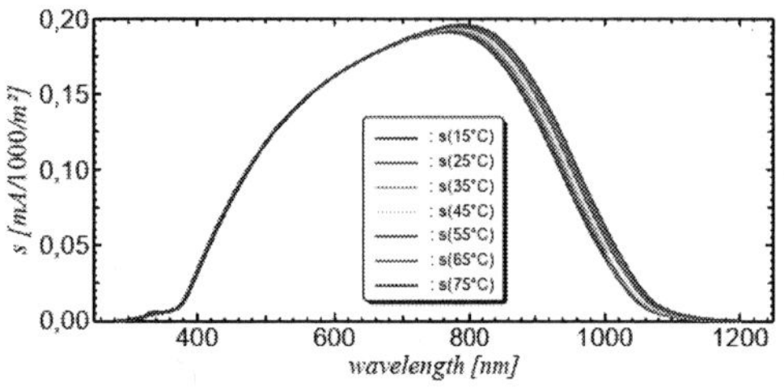

Figure 6: The absolute spectral responsivity dependent on temperature determined for a silicon solar cell using the DSR-method for the temperature range from T_{SP} =15°C up to 75°C.

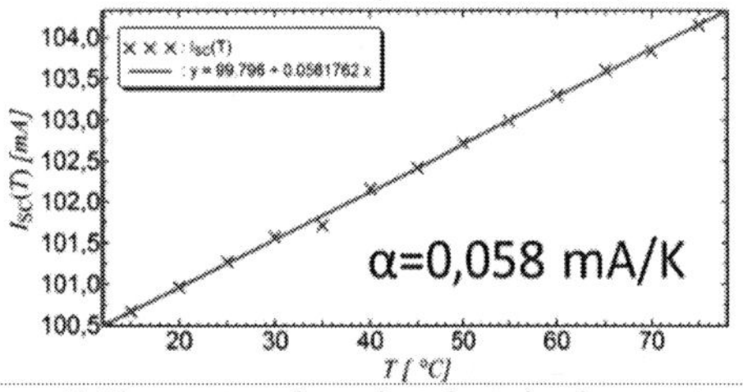

Figure 7: Temperature dependent short circuit current and temperature coefficient determined through integration and weighting of the spectral responsivity with the AM1.5G spectrum.

3 CONCLUSION

In this study we characterized a newly developed solar cell chuck for its suitability for temperature dependent measurements. We investigated the temperature behavior of the chuck and the solar cell for different temperatures and irradiances applying three different measuring methods.

For the temperature range from T_{SP}=15°C up to 75°C we measured the absolute temperature with calibrated PT100 sensors placed on the non-irradiated solar cell and integrated in the chuck at ambient condition (T_{air} = 25°C). We found out, that for higher temperatures a high offset of 2,20 K arises. The impact of front heating due to 1000W/m² sun-like irradiance should be much larger than the impact due to the temperature gradient between ambient temperature and setpoint temperature as discussed in the previous section. Therefore we conclude that the large offset observed at 75°C probably originates from poor thermal contact of the attached PT100 sensors to the solar cell.

We applied the V_{OC}-method to investigate the influence of irradiance of a DSR-facility on the temperature behavior of the cell, the chuck and the temperature adjustment of the heat control unit. The measurement shows, that the difference between the initial and the final temperature of the chuck is smaller than 0,1 K for T_{CH}=15°C -75°C and E=0W/m²-1000W/m². Thereby we can assume that an excellent thermal contact between pn-junction and chuck is given.

We determined the relative temperature distribution with an infrared camera showing a broadening of the distribution at high temperatures. Additionally we determined the temperature dependent spectral responsivity of a silicon solar cell by performing DSR-measurements and calculated an exemplary temperature coefficient of the short circuit current. The temperature coefficient shows very linear behavior; hence the temperature inhomogeneity has no significant impact on the measurement uncertainty of I_{SC}.

As a final result, the developed solar cell chuck is suitable for temperature dependent measurements under irradiation. The V_{OC}-method is the preferred method for determining the temperature of a solar cell. Since the temperature inhomogeneity has a negligible impact on the generated current and there is no expected temperature offset between pn-junction and temperature sensor within the solar cell chuck, the highest uncertainty contribution arises through the calibration of the integrated PT100 sensors of the chuck. Therefore we expect a measurement uncertainty for the temperature measurement of u_T=0,3K (k=1).

4 ACKNOWLEDGEMENT

The research work leading to this article was partly carried out within the EMRP ENG55 project “Photoclass”. The EMRP is jointly funded by the EMRP participating countries within EURAMET and the European Union.

The authors want to thank the PTB working group 4.41 *time and frequency* for the gentle providing of the infrared camera.

5 REFERENCES

[1] R. Schwarz et al, Proceedings. MRS Spring Meeting, Time Measurements and Flying Spot Technique in Microcrystalline Silicon Solar Cells.,(2000)

HIGH EFFICIENCY PHOTOVOLTAIC MODULES PERFORMANCE MEASUREMENTS USED LONG PULSE I-V SIMULATOR

Han-Chang Liu[1*], Chung-Teng Huang[1], Wen-Kuei Lee[1], Fu-Ming Lin[1], Jon-Lian Kwo[2], Yi-Ching Ou[2] and Li-Yuan Liao[2]
1. Green Energy & Environment Research Laboratories, Industrial Technology Research Institute, Hsinchu, Taiwan
2. All Real Technology Co, Kaohsiung, Taiwan
* E-Mail: itri960529@itri.org.tw

ABSTRACT: With the increasing efficiency demands of photovoltaic (PV) industry, a number of new design and high efficiency silicon solar cells were appeared. Photovoltaic module performance measurement has become more important. The traditional simulator with standard spectrum and short pulse cannot test modules performance accurately, especially in high efficiency solar modules. In this work we studied the current–voltage (I–V) performance comparison of different type crystalline silicon modules with four laboratories. In this paper inter-comparison results of short-circuit current (Isc), open circuit voltage (Voc), maximum output power (Pmax), and fill factor (F.F.) are discussed. Finally, we found the optimum solution of transient effect on the high efficiency HIT (Heterojunction with Intrinsic Thin layer) and bifacial solar modules. The solar simulator used 120msec long pulse can improve these defects for high-efficiency solar modules measurement.
Keywords: bifacial, capacitance, n-type, PV module

1 INTRODUCTION

The photoelectric conversion rate is defined as the ability to convert light energy into electrical energy. There is a set of standard test conditions (STC) that requires how to measure the photoelectric conversion rate. According IEC 60904-9 [1] the AM1.5G spectrum, 1000W/cm^2 radiance and 25℃ are required to meet before the measurement. But all companies and laboratories just use solar simulator to imitate the AM1.5G spectrum instead the light from the sun for testing the conversion efficiency of the solar modules. As Solar simulators varies, the spectrum of different simulators not the same caused wrong measurement results. The key performance parameters short-circuit current (Isc), open circuit voltage (Voc), maximum output power (Pmax) and fill factor (F.F.) were extracted from the STC-corrected curves for each high efficiency modules for comparison with each other and to the steady-state simulator control measurements [2,3]. The traditional simulator with a signal flash short time light pulse cannot test them accurately which effects caused by the solar cell technology, like bifacial solar cell and HIT (Heterojunction with Intrinsic Thin layer) solar cells [4]. In this work, except inter-comparison work, we also studied the influence of transient effect on the characterization of high efficiency solar modules by experiments and simulations. Preferable IV measurement method was provided at the end of the paper.

2 EXPERIMENTAL PROCEDURE

In this paper, the Fraunhofer ISE simulator measurements were used as the comparison baseline. All I-V data were normalized to the Fraunhofer ISE measurements. The percentage difference from this baseline for each PV parameter from each measured I-V curve was then calculated. Then inter-comparison were performed to three test laboratories include Fraunhofer ISE (Institute for Solar Energy Systems), TTC (Telecom Technology center), ITRI-GEL (Industrial Technology Research Institute-Green Energy and Environment Research Laboratories) and a solar simulator provider All Real. Those are capable of measuring the I–V performance curves of PV modules. Three single-crystalline Si modules and one multi-crystalline Si modules were used as the test specimens, as shown in table 1. In this work is also to provide a method for rapid measurement of these crystalline Si modules to eliminate electrical properties of high efficiency photovoltaic modules internal capacitance effects.

Table 1. The 4 pieces of different high efficiency solar modules used in the inter-comparison

Module Name	No. of module	Cell type	No. of cells in a module	Module Size
TS60-6M3-280S	1	P-type PERC c-Si 6 inch	6 x 10 cells	164.9 cm x 99.3 cm
TS60-6P3-265	1	P-type Poly c-Si 6 inch	6 x 10 cells	164.9 cm x 99.3 cm
PST254EarthNO60	1	N-type Bifacial, c-Si (6 inch)	6 x 10 cells	166.3 cm x 99.7 cm
VBHN240SJ25	1	N-type HIT c-Si (5 inch)	6 x 12 cells	158 cm x 79.8 cm

3 RESULTS AND DISCUSSION

The results for Pmax, Isc, F.F., and Voc are summarized in figures 1–4. The names of the laboratories are not specified, because the purpose of the present study is to confirm the present status of agreement among the laboratories and analyze the results for further improvement of accuracy. All of the measurements were carried out indoors on the basis of the IEC standards or equivalent domestic standards using various types of pulsed solar simulators, such as long-pulse simulators of up to about 120msec, short-pulse simulators of up to about 10msec, or multi-flash simulators.

3.1 Inter-comparison testing

Three single-crystalline Si modules and one multicrystalline Si modules were used as inter-comparison test and totally four laboratories participated the test. The dominant factors to affect the measurement results are not the same for all the participants. Both the hardware issues and software issues are important. All these values of Voc measured at most of the testing laboratories were within the range of about ±0.6% but the sample name "PST254EarthNO60" was within -1% by laboratory B's I-V simulator as shown in figure 1. The ranges of the values of Isc most of the testing laboratories were within +4% but only "TS60-6P3-265" sample was within -0.3% by laboratory B's I-V simulator as shown in figure 2.

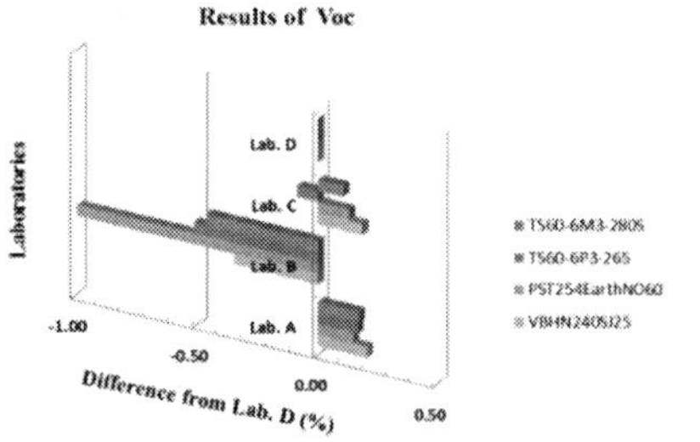

Figure 1. Comparison of the Voc rating of 4 species solar module at STC conditions by all testing laboratories. The values were normalized to the Lab. D data.

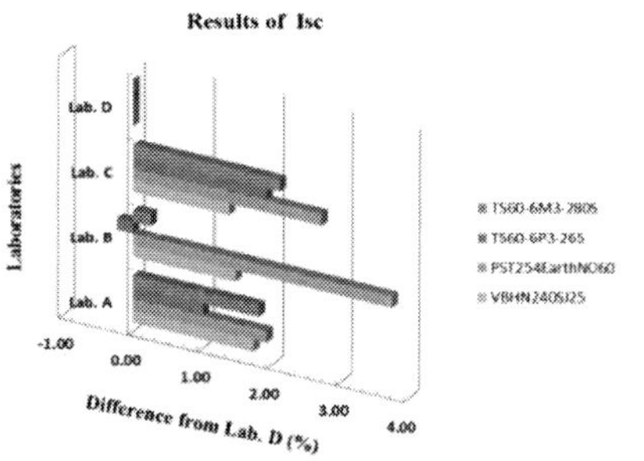

Figure 2. Comparison of the Isc rating of 4 species solar module at STC conditions by all testing laboratories. The values were normalized to the Lab. D data.

The values of F.F. most of the testing laboratories were within -1.5% but sample "PST254EarthNO60" was within +0.5% by laboratory C's I-V simulator as shown in figure 3. The values of Pmax most of the testing laboratories were within +3% but "TS60-6P3-265" and "TS60-6M3-280S" samples were within -1.5% by laboratory B's I-V simulator as shown in figure 4. In laboratory B tends to measure a slightly lower Pmax, which is correlated with Isc measurements and therefore possibly explained by the calibration of the reference device.

In high efficiency solar modules like bifacial and HIT solar modules have some measurement errors must to overcome. The measurement procedures for bifacial solar modules are not yet standardized. In a bifacial module " PST254EarthNO60" in which the reflection of transmitted IR-light from the supporting chuck back into the cell may cause measurement errors. In n-type HIT module "VBHN240SJ25" in which these cells may have a significant "internal capacitance" can result in measurement errors. These measurement errors can cause these test modules have I-V deviations.

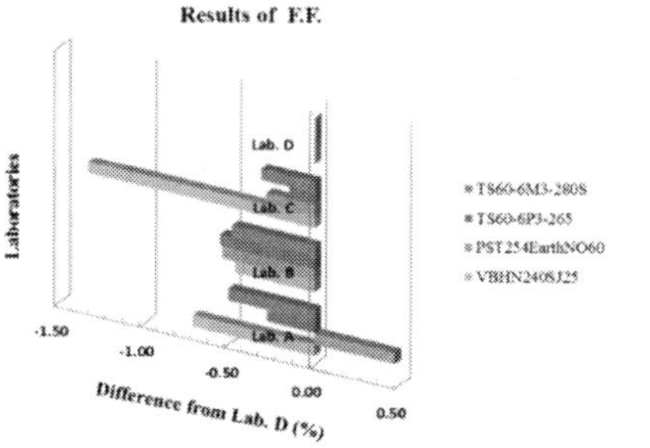

Figure 3. Comparison of the F.F. rating of 4 species solar module at STC conditions by all testing laboratories. The values were normalized to the Lab. D data.

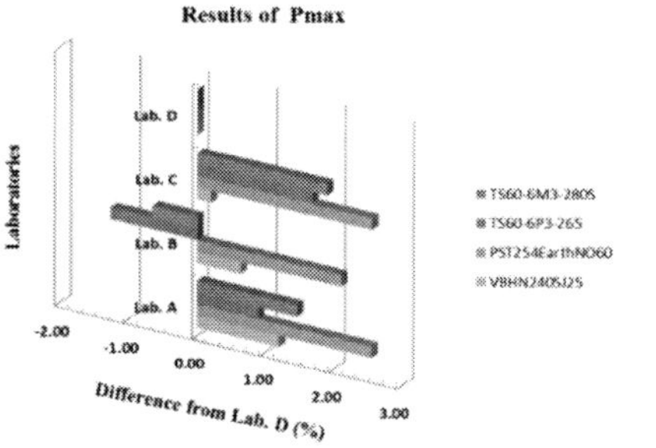

Figure 4. Comparison of the Pmax rating of 4 species solar module at STC conditions by all testing laboratories. The values were normalized to the Lab. D data.

3.2 Transient effects

High efficiency solar cells may have a significant "internal capacitance" which can result in measurement errors due to transient effects. In this section illustrates how transient effects on solar module caused by the high efficiency solar cell technology. It can be detected by comparing I-V curves measured in the forward and reverse sweep directions.

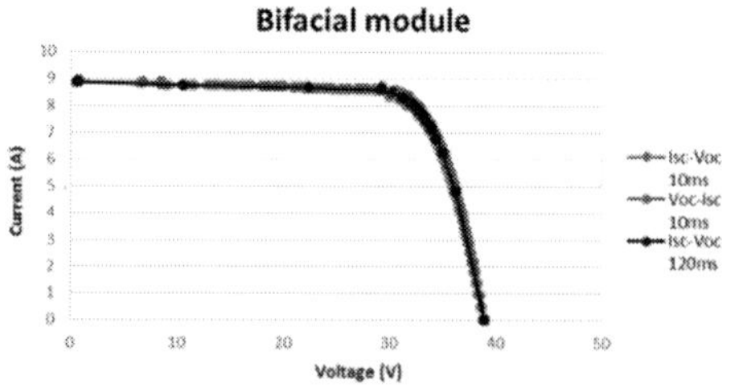

Figure 5. The I-V measurement for bifacial module: (a) short pulse forward scan (Isc to Voc); (b) short pulse reverse scan (Voc to Isc); (c) long pulse scan.

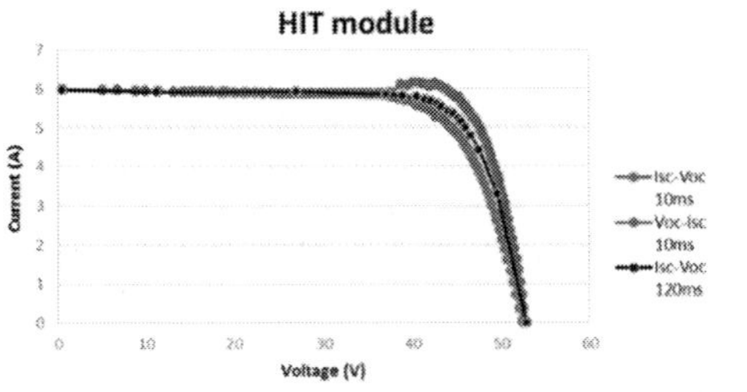

Figure 6. The I-V measurement for HIT module: (a) short pulse forward scan (Isc to Voc); (b) short pulse reverse scan (Voc to Isc); (c) long pulse scan.

In figure 5 and figure 6 shows the I-V measurements of bifacial and HIT modules with 10 ms pulse and 120 ms, respectively. These I-V curves showed mismatch with the Isc to Voc forward scan and Voc to Isc reverse scan. Because the bifacial and HIT modules have high capacitance to cause the current does not follow as the voltage increases. In the opposite direction the current overshoots as the electrons charged by the cell capacitance are released. The long pulse scan curves showed very high correlation with C laboratory which measured by a 120msec pulse. That means long pulse with 120msec simulator can solve the internal capacitance to avoid the transient effect.

4 CONCLUSIONS

The inter-comparison had been carried out with Taiwan famous laboratories and solar simulator provider. There is a multitude of different solar simulators in the test. It is clear that in the case of industrially available solar simulator, the measuring methodology is absolutely crucial. In IEC 60904-9 simulator evaluation spectral range is from 400nm to 1100nm, less than today's variety of multi-junction cells and other high efficient spectral response range. Thus A-level specification only spectrum solar simulator has been unable to meet the real test requirements. In this paper we showed the wide wavelength range from 300nm to 1300nm to meet the high efficiency PV module. On the other hand, we showed a long pulse 120msec method to solve transient effect to reduce conventional short pulse I-V measurement deviation and multi-flash method caused temperature effect in high efficiency solar cell modules I-V measurement.

ACKNOWLEDGEMENTS

The financial support provided by Bureau of Energy, Ministry of Economic Affairs in Taiwan is gratefully acknowledged.

REFERENCES

[1] IEC 60904-9 ed2.0. Photovoltaic devices – Part 9: solar simulator performance requirements, 2007.

[2] T.R. Betts, T. Zdanowicz, M. Prorok, W. Kolodenny, H. de Moor, N. v.d. Borg, D. Stellbogen, J. Hohl-Ebinger, W. Warta, G. Friesen, D. Chianese, A. Guerin de Montgareuil, W. Herrmann, J. Diaz Berrade, J. Moracho, A.B. Cueli, A.R. Lagunas R, R. Gottschalg, IEEE 4th World Conference on Photovoltaic Energy Conference V(2) (2006) 2238

[3] Y. Hishikawa, H. Liu, H. H. Hsieh, T. Inoue, K. S. Kim, C. Limsakul, S. W. Kim, R. Ninae and K. Morita, Prog. Photovolt: Res. Appl. (2012) 1181.

[4] W. Herrmann, S. Mau, F. Fabero, T. Betts, N. van der Borg, K. Kiefer, G. Friesen, W. Zaaiman, proceedings 22th European Photovoltaic Solar Energy Conference, (2007) 2506.

FAULT DETECTION OF PHOTOVOLTAIC (PV) MODULES THROUGH ANALYSIS OF REVERSE I/V CURVES

G. Vannier*, J.A. Tsanakas, N. Chaintreuil, D.L. Ha, F. Barruel, J. Merten
Commissariat à l'Énergie Atomique et aux Énergies Alternatives (CEA)
Département Technologies Solaires (DTS), Institut National de l'Énergie Solaire (INES) Laboratoire des Systèmes Photovoltaïques (LSPV)
50 Avenue du Lac Léman | F-73375 Le Bourget-du-Lac, France
E-mail : godefroy.vannier@cea.fr

ABSTRACT: This paper presents a new way of fault detection for photovoltaic (PV) modules by means of reverse I/V curves analysis. The method of work starts by selecting solar cell samples with and without faults. The first step is to measure the direct and reverse I/V curves of each solar cell under investigation. Simulations with one diode model [1] and two diodes model [2] are also performed, comparing the models to the I/V measurements. The second step is to implement three PV modules, each one consisting of four solar cells, with at least one fault in each PV module. At this step, the objective is to trace the reverse I/V curves per PV module, with the breakdown voltage. Given the results of all measurements, we can analyze, compare and discuss on the correlation between the signatures of each fault type and the resultant reverse I/V curves and the infrared (IR) thermal signature for each PV module.

Keywords: reverse I/V curve, breakdown voltage, fault diagnosis, PV module, solar cell

1 INTRODUCTION

The power generation of PV plants and the consequent payback time of investments on PV installations, are significantly dependent on the operational lifetime of PV modules and their electrical performance. The experience accumulated, through the recent decades, upon extensive quality control and qualification testing of PV modules at the manufacturing/assembling stages, had a key contribution towards minimizing the infant mortality ratio of such devices. As a result, although PV modules are characterized by limited electrical efficiency, they comprise the most robust component of a PV system, typically featuring warranty periods of up to 25 or 30 years.

However, long-term performance of PV modules and their overall reliability, can be drastically affected by faults occurring mainly under real (field) operation conditions but also during the post-manufacturing stages of transportation and installation. Such faults result in specific abnormal operation, principally characterized by reduced electrical power output, presence of hot spots, excessive thermal/mechanical stresses or even safety risk for the installation [3] [4]. Focusing on the predominant technology of PV modules, the silicon-wafer based ones, we can distinguish the following main types of faults [3]:

- Optical degradation; delamination, bubbles, discoloration of the encapsulating.
- Electrical mismatches; cell cracks/fracture, breakage of interconnection ribbons, poor soldering, snail tracks, shunts, shading.
- Non-classified; potential induced degradation (PID), defective/short-circuited bypass diode, short-circuited module.

Today, the goal of a universal monitoring tool, able to provide on-line detection and effective diagnosis over any possible fault occurring under operation in the field, remains a technical and economical challenge, especially when dealing with large-scale PV plants consisting of hundreds or thousands of PV modules. In this paper, we propose a new technique of performing fault detection in PV modules, by analyzing their reverse I/V curves, with the breakdown voltage.

2 MODELING OF REVERSE I/V CURVE

Several techniques of performing I/V curves modeling and simulations can be found in the literature [5,6]. Among them, the most common way is to generate the direct I/V curve with the model of one (five parameters for the input information) [1] or two diodes (seven parameters for the input information) [2]. In this study, we apply two models; the Bishop Model and the Quaschning Model. The next step is to add the avalanche effect in serial (Bishop Model) [5] or in parallel (Quaschning Model) [6] of the leakage current. There are two parameters to add for the reverse I/V curves analysis of the two models. To add the five or seven parameters of the electrical equivalent circuit of one or two diodes, we have three parameters for the reverse I/V curve in the two models:

- V_{br} : breakdown voltage
- n : avalanche breakdown exponent
- α : correction factor

Before describing the Bishop and Quaschning Model, we have for one diode model equation :

$$I(V) = -I_{ph} + I_{01}\left(e^{\frac{(V-I.R_s)e}{n1.K.T}} - 1\right) + \frac{V - I.R_s}{R_p} \quad (1)$$

And for two diodes model equation :

$$I(V) = -I_{ph} + I_{REC} + I_{01}\left(e^{\frac{(V-I.R_s)e}{n1.K.T}} - 1\right) + I_{02}\left(e^{\frac{(V-I.R_s)e}{n2.K.T}} - 1\right) + \frac{V - I.R_s}{R_p} \quad (2)$$

From the equations (1) and (2), we have respectively the Bishop Model and Quaschning Model. The interest of the two diodes Model is the greater accuracy in extracting equivalent parameters using the Variable illumination Method (VIM) [2].

For the Bishop [5] model, we have the equation:

$$I(V) = -I_{ph} + I_{01}\left(e^{\frac{(V-I.R_s).q}{n1.KT}} - 1\right) + \frac{V - I.R_s}{R_p}.\left[1 + a\left(1 - \frac{V - IR_s}{V_b}\right)^{-n}\right] \quad (3)$$

The equation (3) is associated to the following equivalent electrical schema:

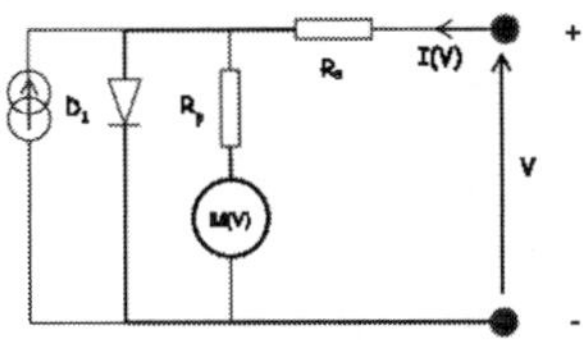

Figure 1 : Electrical equivalent schema of Bishop Model

With:

- I_{ph} :the photocurrent
- M(V) : the avalanche effect represented as a voltage

And for Quaschning :

$$I(V) = -I_{ph} + I + I_{D1}\left(e^{\frac{(V-I.R_s).q}{n1.KT}} - 1\right) + I_{02}\left(e^{\frac{(V-I.R_s).q}{n2.KT}} - 1\right) + \frac{V - I.R_s}{R_p} - \alpha(V + IR_s).\left(1 - \frac{V + IR_s}{V_b}\right)^{-n} \quad (4)$$

Equation (4) is associated to the equivalent electrical schema, as follows:

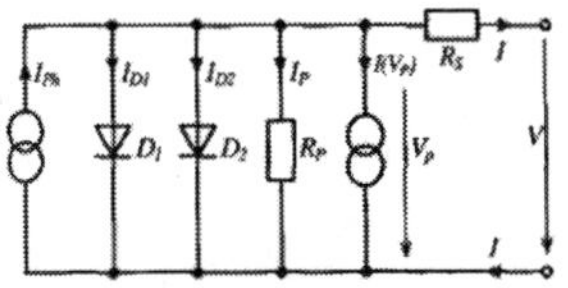

Figure 2 : Electrical equivalent schema of Quaschning Model.

With:

- I_{ph} : the photocurrent
- I_{D1} : the current through the first diode
- I_{D2} : the current through the second diode
- I_P : the leakage current
- I(Vp) : the avalanche effect

3 EXPERIMENTAL SET-UP

The experimentation consists of three steps. The first step is the selection of the solar cells to be investigated, through extraction of their equivalent parameters and reverse I/V curves analysis for each solar cell. Characterization of each solar cell was performed by a Newport solar simulator, as well as by "LEDVIM", a prototype developed by our laboratory at CEA INES. The solar simulator provides one I/V curve (under STC conditions) for each cell. On the other hand, LEDVIM provides several I/V curves in function of the irradiance level. The light source of the LEDVIM is a 7x7 power LED matrix.

In addition, for each solar cell, we also trace the corresponding reverse I/V curves, also using the LEDVIM bench, with a new software interface.

In the last step, because of the need to inject high power and voltage, on PV module scale, in order to trace the reverse I/V curve of each studied module, a power supply with a remove interface, was employed.

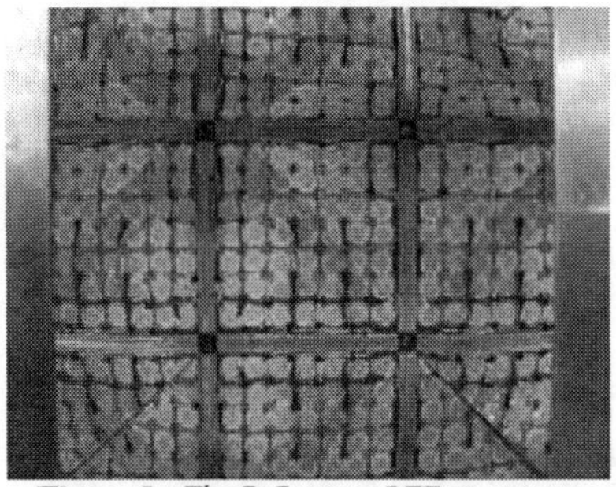

Figure 3 : The 7x7 power LED structure.

3.1 Selection of solar cells

We have the two types of solar cells:

- 15 solar cells type of PERC as "good" solar cells
- 12 solar cells type of ALU as "bad" solar cells, with high leakage current

For each solar cells, we characterizing in solar simulator in STC condition and in LEDVIM. Based on the results from characterizing from the two way of characterizing, the interested of LEDVIM is the accuracy of parameters equivalent extracting because of seven parameters extracted (two diodes models).

Also with LEDVIM we can trace the reverse I/V curves for each solar cells. For the reverse I/V curves models, for the Bishop Model, we can associate the equivalent parameters from the one or two diodes models and for the Quaschning Model, we can only associate with the two diode models.

3.2 Characterization and extraction of equivalent parameters

We selected 11 "good" solar cells and 1 "bad" solar cell. The objective is to make three 4-cell PV modules. We recapitulate the main characteristics of each PV module:

- Module 1 : reference module without fault
- Module 2 : module with a high leakage current in a solar cell
- Module 3 : Module with a cracked cell

For the reference PV module (Module 1), we select the four "good" solar cells with the lowest dispersion in equivalent parameters values and on breakdown voltage. The second module, we randomly select a "bad" solar cell in connection with three "good" solar cells. For the last module, we also select the four "good" solar cells, as for the case of the first PV module, but there is one cell which deliberately features a crack.

3.3 Reverse I/V curves tracing and IR thermography

On each PV module, using a power supply, we trace reverse I/V curves for a voltage around 100 V. Because of the expected wide voltage range, we will limit the numbers of points for each reverse I/V curves to 40. And we estimate the maximal power to inject from the power supply on each PV module to more than 500 W for reverse I/V curve tracing.
In the resulting I/V characterization, we observe the shape of reverse I/V curves.

We also performed IR thermographic tests under both forward and reverse bias, connecting a power supply to the PV module.

4 RESULTS

4.1 Extraction of equivalent parameters

As for the "good" solar cells, we have the results from two different characterization devices (solar simulator and LEDVIM). As we can see, there are some differences in the values of series and parallel (shunt) equivalent resistances. For the series resistance, such result is because the electrical contact is not the same for the two ways methods of characterization. The increase in series resistance reduces the STC efficient.

Concerning the parallel resistor, there is a big difference between the R_{shunt} values provided by solar simulation and the LEDVIM, because of the two way of extraction of parallel resistance. For the solar simulation the parallel resistance is from the slope of I/V curve in STC conditions at Isc, while for the parallel resistabce from LEDVIM, this value is deduced in very low irradiance.

Table I: Results of parameters extraction from sample I/V curve in STC conditions of "good" solar cells.

cell number	Rshunt	Rs	FF STC	STC efficient
1	26,39	0,0049	79,22%	18,70%
2	22,73	0,0050	78,56%	18,58%
3	25,32	0,0049	78,66%	18,74%
4	22,52	0,0048	78,83%	18,83%
5	28,17	0,0050	78,76%	18,72%
6	37,04	0,0048	78,82%	18,86%
7	56,50	0,0050	78,81%	18,76%
8	28,49	0,0049	78,85%	18,87%
9	27,47	0,0049	78,99%	18,86%
10	19,88	0,0048	78,78%	18,81%
11	47,39	0,0048	78,85%	18,88%
12	54,05	0,0050	78,75%	18,83%
13	31,45	0,0049	78,77%	18,81%
14	39,22	0,0050	78,65%	18,38%
maximum	56,50	0,0050	79,22%	18,88%
minimum	19,88	0,0048	78,56%	18,38%

For "bad" solar cells we will show only a part of results. The main differences for "bad" solar cells is the leakage current represented as parallel resistance of low value. We will not show all the results for each "bad" solar cells in this work.

Table II: Results of parameters extraction from several I/V curves in function of irradiance and using of variable illumination method (VIM) for "good" solar cells.

cell number	Rp	Rs	I01	I02	n1	n2	FF STC	STC efficient
1	381,28	0,011	1,72E-08	1,59E-04	1,23	3,72	71,01%	15,91%
2	331,20	0,010	6,79E-08	1,58E-04	1,31	3,52	71,21%	15,71%
3	352,96	0,012	8,35E-09	1,93E-04	1,18	3,91	69,74%	15,55%
4	674,64	0,016	1,23E-08	6,51E-05	1,21	3,10	65,26%	14,35%
5	852,67	0,010	4,01E-08	5,55E-05	1,28	3,13	71,09%	15,60%
6	889,68	0,011	1,11E-08	5,63E-05	1,20	3,21	70,72%	15,61%
7	1181,27	0,010	1,61E-08	5,23E-05	1,22	3,06	71,43%	15,89%
8	449,80	0,012	3,10E-08	8,87E-05	1,26	3,37	69,92%	15,20%
9	381,89	0,011	1,56E-08	1,42E-04	1,22	3,87	70,15%	15,30%
10	181,89	0,011	2,33E-08	4,63E-04	1,24	4,30	70,41%	15,46%
11	602,40	0,011	1,98E-08	8,66E-05	1,23	3,47	70,78%	15,47%
12	806,34	0,011	1,93E-08	5,91E-05	1,24	3,30	70,93%	15,70%
13	318,25	0,011	1,33E-08	1,45E-04	1,22	3,82	71,10%	15,74%
14	314,26	0,011	4,45E-08	1,37E-04	1,28	4,26	70,86%	15,51%
15	644,35	0,011	2,27E-08	8,39E-05	1,25	3,40	70,50%	15,61%
maximum	1181,27	0,016	6,79E-08	4,63E-04	1,31	4,30	71,43%	15,91%
minimum	181,89	0,010	8,35E-09	5,23E-05	1,18	3,06	65,26%	14,35%

Table III: Results of parameters extraction from sample I/V curve in STC conditions, for "bad" solar cells.

cell number	Rshunt	Rs	FF STC	STC efficient
3 p 259	4,95	0,0058	75,88%	16,93%
3p269	7,10	0,0073	75,89%	15,36%
3p281	6,34	0,0069	75,79%	15,65%
3p285	8,51	0,0061	76,19%	16,65%
3p291	11,86	0,0052	77,33%	17,41%
maximum	11,86	0,0073	77,33%	17,41%
minimum	4,95	0,0052	75,79%	15,36%

Table IV: Results of parameters extraction from several I/V curves in function of irradiance and using of variable illumination method (VIM) for "bad" solar cells.

cell number	Rp	Rs	I01	I02	n1	n2	FF STC	STC efficient
3 p 259	5,64	0,011	4,19E-06	9,59E-03	1,67	7,42	67,85%	14,56%
3p269	10,78	0,011	3,29E-07	6,49E-03	1,38	7,64	69,02%	14,27%
3p281	9,18	0,013	6,98E-07	7,61E-03	1,45	7,69	67,00%	13,93%
3p285	13,69	0,013	4,75E-07	4,82E-03	1,44	6,95	67,82%	14,21%
3p291	15,43	0,013	2,41E-07	4,05E-03	1,40	6,72	67,89%	14,49%
3p276	16,89	0,011	1,30E-06	3,43E-03	1,52	6,29	69,21%	14,25%
maximum	16,89	0,013	4,19E-06	9,59E-03	1,67	7,69	69,21%	14,56%
minimum	5,64	0,011	2,41E-07	3,43E-03	1,38	6,29	67,00%	13,93%

4.2 Reverse I/V curves of solar cells

For each solar cell ("good and "bad"), as well as for the cracked cell, we trace the reverse I/V curves.

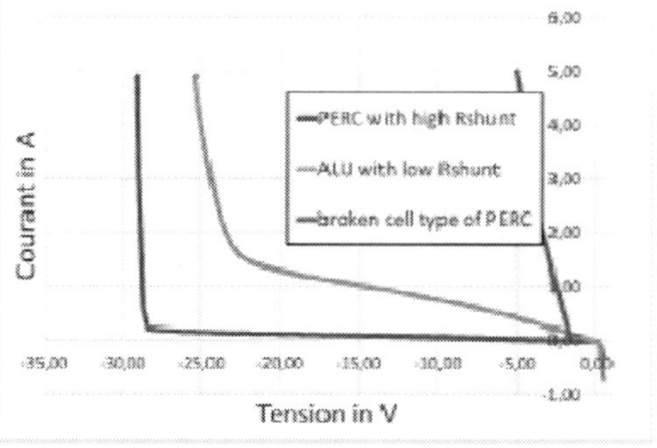

Figure 4 : Reverse I/V curves of solar cells.

We note that, for the group of "good" solar cells, the breakdown voltage of each solar cell varies around 10%. For the PERC cell, the index of exponent is very high.

For the broken cell, we can consider the breakdown voltage as very small, so we can eliminate it for the PV module. For the cell type Alu, the low parallel resistance causes the slope before the breakdown.

3.3 Reverse I/V tracing on PV modules and IR thermography

For tracing the PV modules who have a breakdown around 100V, thus, it is necessary to use a power supply. We perform the tracing I/V in both directions to check the pertinence of I/V curves.

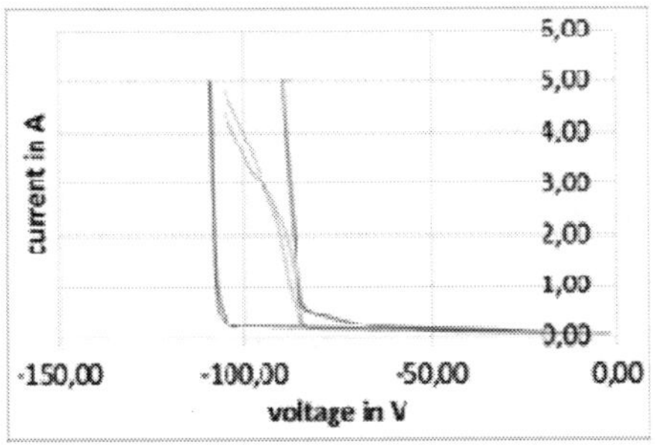

Figure 5 : Reverse I/V curves of the PV modules

In accordance to the reverse I/V curves of PV modules, we also performed thermographic tests in both forward and reverse bias, connecting a power supply to the PV modules.

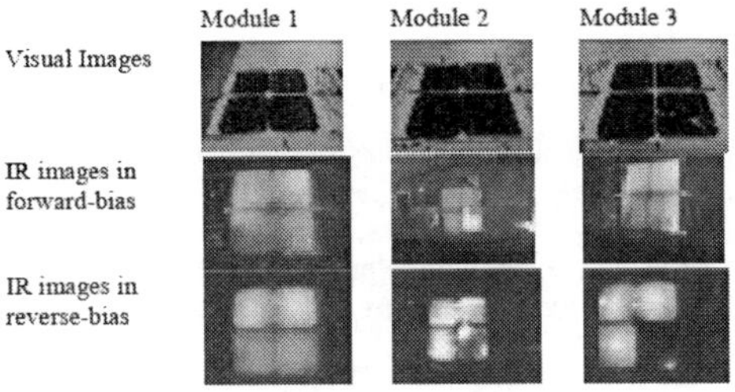

4. FURTHER DISCUSSION

With the reverse I/V curve analysis of the three PV modules, it is possible to detect certain fault modes by breakdown voltage measurements. In particular, the presence of high leakage current in the cell of Module 2, is possible to be witnessed by the shape of the corresponding reverse I/V curve. Besides, for the case of the cracked cell, the slope actually refers to a very high parallel resistance, and it may be negligible for the analysis.

Concerning the IR thermal analysis, under both forward and reverse bias, we made an effort to compare and correlate the resultant thermal signatures with the occurring fault modes. Based on the results, IR imaging is not enough "sensitive" to indicate high leakage current related faults, due to its low impact on the temperature difference, ΔT. On the other hand, cases of cracked cells, are clearly and reliably revealed under both two biasing conditions. Such abnormally high ΔT may be also indicative of the different breakdown voltages within the area of the broken cell.

5. CONCLUSIONS

A new way of fault detection and analysis, presented in this paper, extends our knowledge about the reverse I/V characteristics of defective PV modules and cells. The results show that it is possible to detect specific faults occurring either on cell or module scale, through the analysis of reverse I/V curves and indoor IR imaging under forward and reverse bias.

Extension of this work on larger scale involves certain technical obstacles and challenges. For this reason, we currently focus only on extending our understanding of the presented method and the "behavior" of reverse I/V, for other real cases of faults on PV module and solar cell scale, such as shunts, broker interconnections, poor soldering, etc.

Acknowledgments

The authors of this paper would acknowledges to the solar cell laboratory for providing the cells and the module laboratory to make the PV modules.

References

[1] J. Merten, J. M. Asensi, C. Voz, A. V. Shah, R. Platz, and J. Andreu, "Improved Equivalent Circuit and Analytical Model for Amorphous Silicon Solar Cells and Modules", *IEEE Transactions on Electron Devices*, Vol. 45(2), 1998.

[2] J. Merten, L. Sicot, Y. Delesse, A. Guerin de Montgareuil, "Outdoor evaluation of the energy production of different module technologies", Proceedings of the *23rd EU-PVSEC*, Valencia, 2008, DOI: 10.4229/23rdEUPVSEC2008-4AV.3.23.

[3] Kontges M, Kurtz S, Packard C, Jahn U, Berger KA, Kato K, Friesen T, Liu H, Van Iseghem M., "Performance and Reliability of Photovoltcaic Systems – Subtask 3.2: Review of Failures of Photovoltaic Modules", *External Final Report by International Energy Agency (IEA) for Photovoltaic Power Systems Programme (PVPS)*; 2014.

[4] B. Brooks, "The Bakersfield Fire - A Lesson in Ground-Fault Protection," *SolarPro. 4(2)*, pp. 62 - 70, 2011.

[5] J. Bishop, "Computer simulation of the effects of electrical mismatches in photovoltaic cell interconnection circuits", *Solar Cells*, Vol. 25(1), 1988, pp. 73-89.

[6] V. Quaschning, R. Hanitsch, "Numerical simulation of current-voltage characteristics of photovoltaic systems with shaded solar cells", *Solar Energy*, Vol. 56(6), 1996, pp. 513-520.

EXERGY ANALYSIS OF A SOLAR PHOTOVOLTAIC MODULE

F. Serrano-Casares and E. Zaragoza
University of Málaga. Energy Group
Campus de Teatinos E-29071 Málaga. Spain

ABSTRACT: PV energy is the direct conversion of solar radiation into electricity. In this paper, an analysis of the influence of parameters such as global irradiance or temperature in the performance of a PV installation has been carried out.

A PV module was installed in a building at the University of Málaga, and these parameters were experimentally determined for different days and different conditions of irradiance and temperature. Moreover, IV curves were obtained under these conditions to know the open-circuit voltage and the short-circuit current of the module. With this information, and using the first law of thermodynamics, an energy analysis was performed to determine the energy efficiency of the installation. Similarly, using the second law of thermodynamics, an exergy analysis is used to obtain the exergy efficiency.

The results show that the energy efficiency varies between 10% and 12% and the exergy efficiency between 14% and 17%. It was concluded that the exergy analysis is more suitable for studying the performance, and that only electric exergy must be considered as useful exergy. This exergy efficiency can be improved if heat is removed from the PV module surface, and an optimal temperature is reached.

Keywords: exergy efficiency, PV module, optimal temperature.

1 INTRODUCTION

The first application of the PV technology was in the two first geostationary satellites from USSR and USA. The aerospace industry was the first to motivate the research for more effective modules. Since then, and as a consequence of the oil crisis in 1973, the PV industry has been focused in the reduction of energy consumption and its utilization as a clean energy source.

The energy efficiency of a solar module is conventionally used to measure the efficiency of a PV module. Even though some authors have published exergy studies of PV installations before, there hasn't been enough deepening in this subject, especially considering its benefits.

Exergy is the maximum amount of work that can be obtained from a system until it reaches equilibrium with the surroundings. Unlike energy, exergy can be destroyed due to irreversibilities present in the real process. Generally, exergy analysis provides a more meaningful definition of the efficiency of processes than the energy analysis, due to the fact that exergy efficiency measures the degree of closeness to an ideal process.

2 ENERGY EFFICIENCY

Energy analysis is based on the first law of thermodynamics, and only takes into account the amount of energy used and the efficiency of the energy processes. Therefore, it ignores reductions of energy potential, which could be productively used in other physical or chemical processes. However, it is effective in those applications in which the efficiency depends only on the amount of energy used.

Energy efficiency is calculated as the ratio between the electric energy output and the solar energy input [1] [2]:

$$\eta_{en} = \frac{V_{OC} \cdot I_{SC} \cdot FF}{A \cdot G} \qquad (1)$$

Where V_{OC} is the open-circuit voltage (V), I_{SC} is the short-circuit current (A), FF is the fill factor, A is the surface area of the module (0.374 m^2) and G is the global irradiance (W/m^2). The fill factor is defined as the ratio between the real power of the PV cell and the power for maximum current and voltage.

The energy absorbed by the PV module is turned into electric and thermal energy, the latter being dissipated to the surroundings by convection, conduction and radiation.

3 EXERGY EFFICIENCY

Exergy analysis includes a consideration of the quality of the energy or capacity, which allows a more effective and efficient evaluation of the use of energy potential. For a continuous process during a finite time interval, the exergy balance of a PV module can be written as:

$$Ex_{in} - Ex_{out} = Ex_{loss} \qquad (2)$$

Ex_{out} is the maximum exergy that can be obtained from a system whose exergy input is Ex_{in}. The lower the exergy consumption, the lower the exergy losses (Ex_{loss}). This degradation in the quality of energy is called **availability loss** or **irreversibility**.

The exergy efficiency of a PV module is defined as the ratio of the exergy output and the exergy input:

$$\eta_{ex} = \frac{Ex_{out}}{Ex_{in}} \qquad (3)$$

The exergy input only includes exergy coming from the solar radiation intensity [3] [4]:

$$E_{X_{in}} = AG\left[1 - \frac{4}{3}\left(\frac{T_a}{T_s}\right) + \frac{1}{3}\left(\frac{T_a}{T_s}\right)^4\right] \qquad (4)$$

Where T_a is the ambient temperature and T_s is the Sun's surface temperature (5780 K).

The exergy output is also divided in thermal exergy (Ex_{term}), which is dissipated to the surroundings, and electric exergy (Ex_{elec}). The equations, according to [5], are:

$$E_{X_{out}} = E_{X_{term}} + E_{X_{elec}} \quad (5)$$

$$E_{X_{term}} = Q\left(1 - \frac{T_a}{T_m}\right) \quad (6)$$

$$E_{X_{elec}} = V_{OC} \cdot I_{SC} \cdot FF \quad (7)$$

Where Q is the dissipated thermal heat and T_m is the module temperature.

4 EXPERIMENT

To perform the experiments, a test bench installed on the rooftop of the building of the Andalusian Institute of Renewable Energies, at Málaga (36.73 N, 4.55 W), was used. The PV module was placed on a steel mobile structure. This structure allows an adjustment of both azimuth and inclination. A pyranometer was used to measure the global irradiance on the module plane, and a temperature sensor placed on the backside of the module measures its temperature.

A set of experiments at real sun were performed to obtain the I-V curve of the module via a hardware system specifically designed according to IEC. A software developed by the CIEMAT (Centro de Investigaciones Energéticas, Medioambientales y Tecnológicas) was used to know the open-circuit voltage, the short-circuit current, and other characteristic parameters, from the I-V curves of each measurement.

Measurements were taken every 20 minutes. If necessary, the mobile structure was adjusted in every measurement to ensure that solar radiation was normal to the module. The computer then collects the measurements trough a GPIB card to be processed later by the program.

Table I: Characteristics of the PV module.

Physical	
Type	Monocrystalline silicon
Number of cells in series	36
Number of cells in parallel	1
Nominal Operating Cell	47
Surface area (m^2)	0.426
ELECTRICAL	
Temperature (NOCT) (°C)	
Nominal voltage (V)	12
Maximum power (W)	53 ± 10%
Short-circuit current (A)	3.27
Open-circuit voltage (V)	21.6
Maximum power current (A)	3.05
Maximum power voltage (V)	17.4
Efficiency (%)	12.44

5 RESULTS AND DISCUSSIONS

The experiments were carried out in different days and different conditions:

5.1 Sunny day, August.

Measurements were taken from 11:00 to 19:00.

As shown in Figure 1, the irradiance and the exergy input follow the same profile, according to equation (4), and the bigger the irradiance, the bigger the difference between the two variables. The exergy input is considerably high due to the high surface temperature of the Sun, and can be used to produce useful work.

In Figure 2, one can see the exergy input and the exergy output, divided in electric exergy and thermal exergy. It can also be seen that the exergy loss is very high, constituting around 85% of the exergy input, due to irreversibilities in the process. This shows that current silicon modules don't exploit enough the high energy content of the solar radiation. The percentage of exergy loss is always the same, regardless of the irradiance conditions.

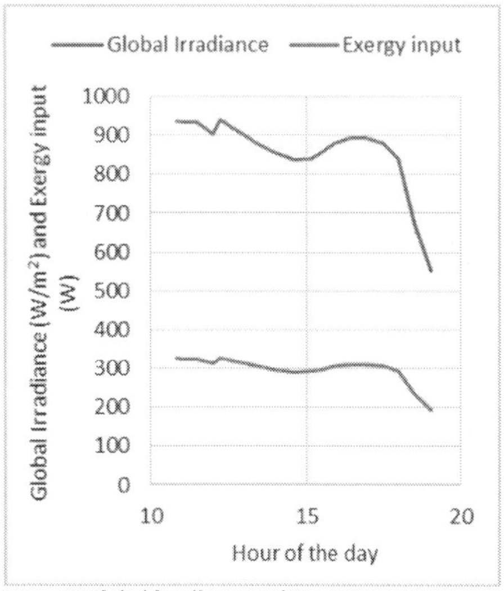

Figure 1: Global irradiance and Exergy Input on a sunny day.

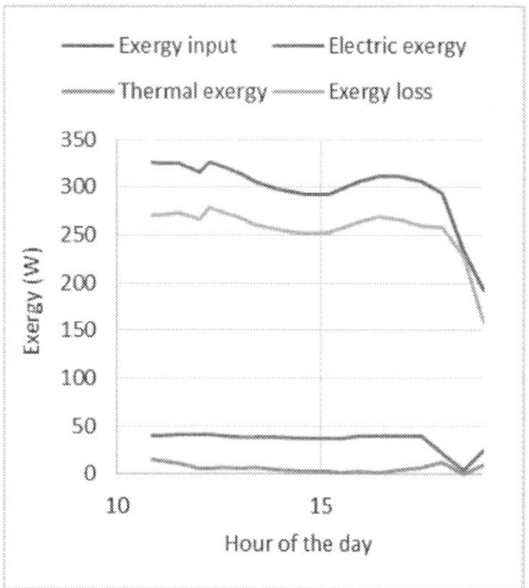

Figure 2: Exergy balance of the module on a sunny day.

The exergy efficiency, as seen in Figure 3, is higher than the energy efficiency, although both follow the same profile and are practically constant during the day. The average exergy efficiency is around 13% and the energy efficiency is around 11%.

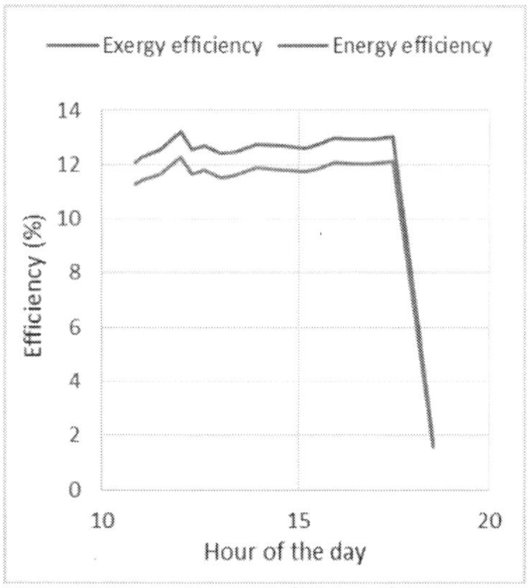

Figure 3: Exergy and energy efficiency of the module on a sunny day.

If we plot the module temperature and the ambient temperature alongside the exergy efficiency (Figure 4), it can be seen that the efficiency is proportional to the module temperature, as in equation (6). Since the ambient temperature is practically invariable in comparison with the module temperature, the exergy input doesn't have a big influence in the exergy efficiency. On the other hand, when the module temperature raises the thermal exergy raises, and so does the exergy output. Even though the open-circuit voltage decreases when the module temperature raises, and consequently the electric exergy decreases, this decrease is insignificant compared to the increase of the thermal exergy.

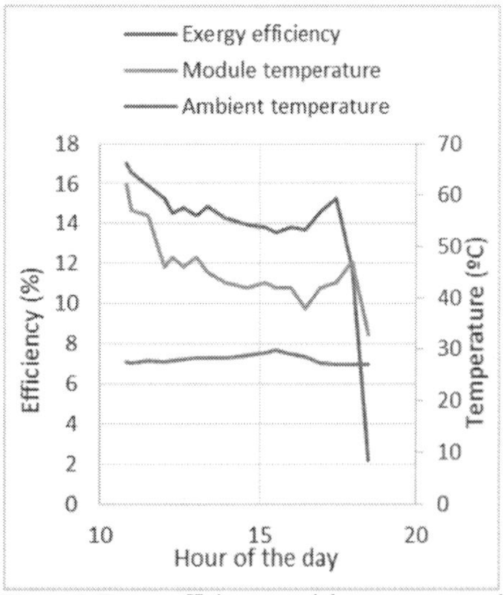

Figure 4: Exergy efficiency, module temperature and ambient temperature of the module on a sunny day.

5.2 **Cloudy day, July**

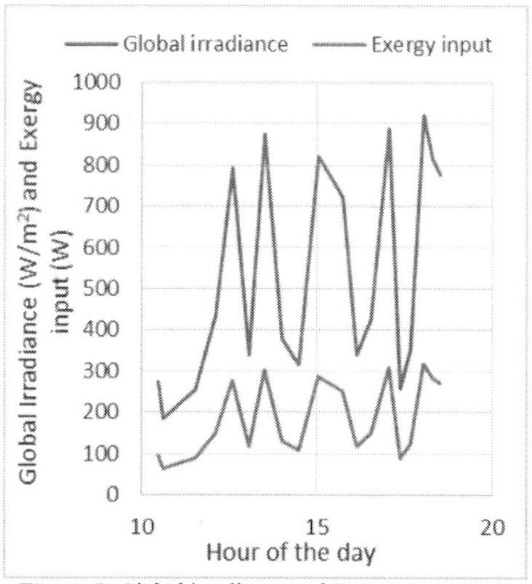

Figure 5: Global irradiance and Exergy Input on a cloudy day.

In Figure 5, one can see the abrupt variation of the global irradiance during the day, which is proportional to the exergy input, due to cloud's momentary appearances. The difference between this figure and the one corresponding to the sunny day is significant (Figure 1).

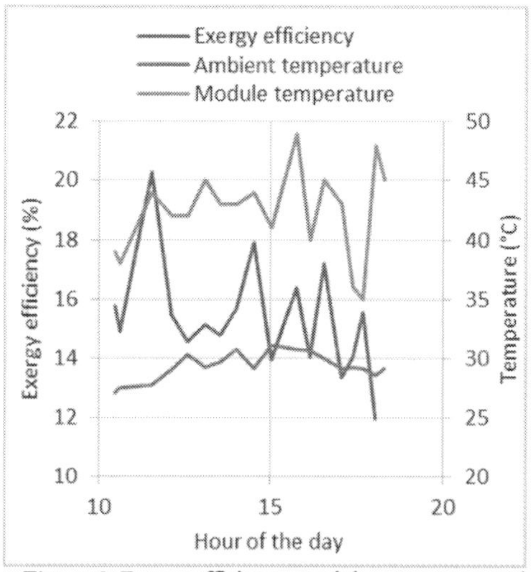

Figure 6: Exergy efficiency, module temperature and ambient temperature of the module on a cloudy day.

Exergy efficiency is also more variable than on a sunny day, and should be directly proportional to the module temperature, according to equations (3), (5) and (6). However, there is a measure at 17:40 where the module temperature raises but the exergy efficiency abruptly falls, constituting the minimum efficiency measured on that day. Neither the module temperature nor the ambient temperature are low enough to produce such a peak in exergy efficiency. This peak is mainly produced by a fall in the global irradiance, together with a raise in the module temperature. A lower global irradiance implies a lower exergy input (equation 4) and a higher module temperature implies a higher thermal exergy. This didn't happen on the sunny day because the irradiance on that day was practically invariable and high.

5.3 **Sunny day, November, changing azimuth.**

An experiment was performed to see how the exergy efficiency varies between a changing azimuth and a fixed azimuth. Since it was winter, less measurements could be taken because of the shorter daylight.

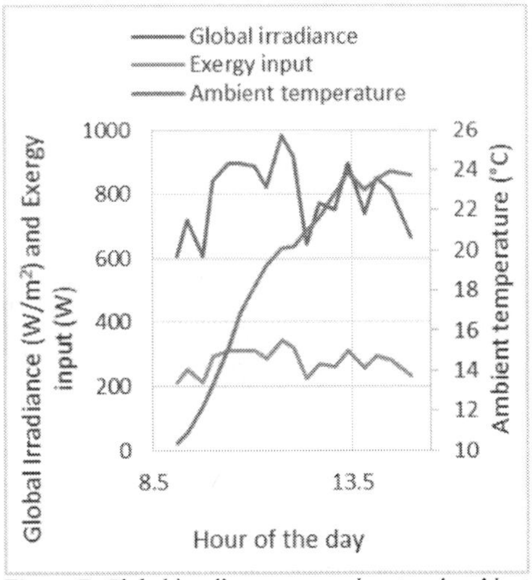

Figure 7: Global irradiance, exergy input and ambient temperature on a sunny winter day, with changing azimuth.

It is clear that it's winter because the ambient temperature never rises above 24 °C, and is around 10 °C at the beginning of the day.

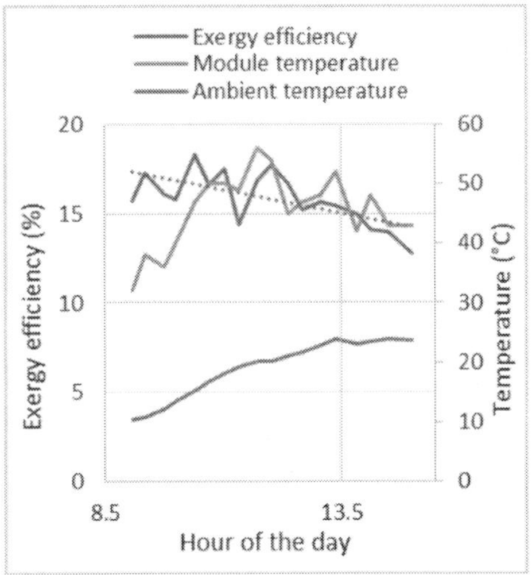

Figure 8: Exergy efficiency, module temperature and ambient temperature on a sunny winter day, with changing azimuth.

In Figure 8, it can be seen that the exergy efficiency curve has a downward trend. This is due to the increase in ambient temperature, which raises the exergy input (equations 3 and 4). It can also be seen that the efficiency increases when the module temperature increases.

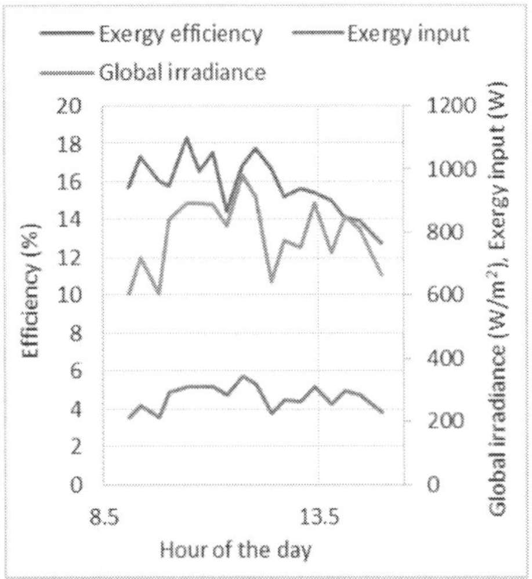

Figure 9: Exergy efficiency, exergy input and global irradiance on a sunny winter day, with changing azimuth.

Even though ambient temperature varies more than in summer, it can be seen that the exergy input is more influenced by global irradiance than ambient temperature, given that they follow the same profile. There is a fall in global irradiance at 12:10, when exergy efficiency raises, even though the module temperature decreases, which means that **global irradiance has more influence on exergy efficiency than module temperature**.

5.4 **Sunny day, November, fixed azimuth.**

These measurements were taken at the same time as the changing azimuth. Every 20 minutes, a measurement was taken with the module facing the sun, then the structure was redirected to the south to take the measurement with the fixed azimuth. They were taken to show the efficiency of a PV module with a fixed azimuth, like most of the domestic installations.

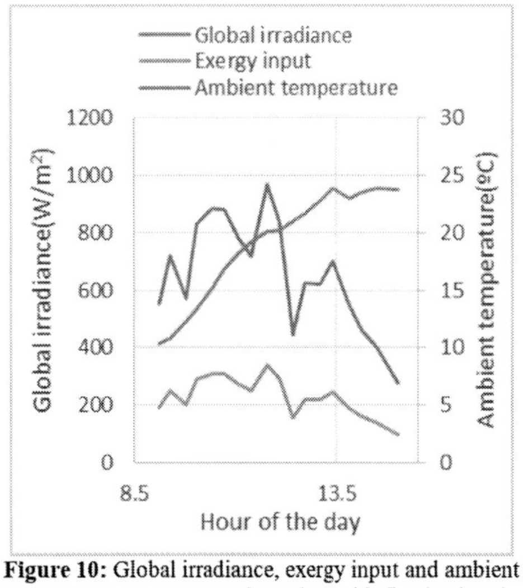

Figure 10: Global irradiance, exergy input and ambient temperature on a sunny winter day, with fixed azimuth.

As one would expect, ambient temperature is the same as the changing azimuth because it is the same day. Global irradiance, on the other hand, is practically the same until 11:45, but then it decreases more than with the changing azimuth.

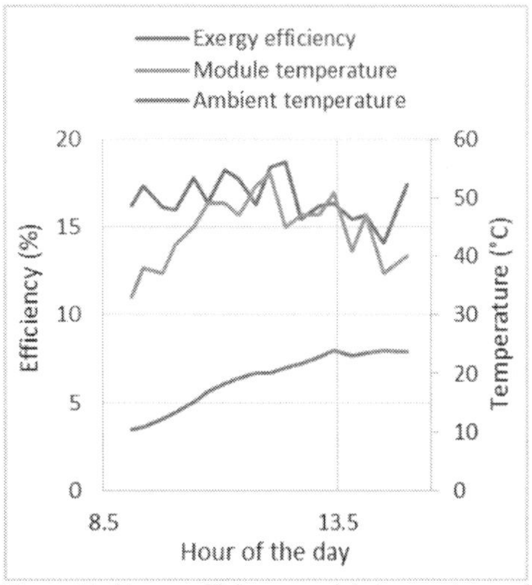

Figure 11: Exergy efficiency, module temperature and ambient temperature on a sunny winter day, with fixed azimuth.

In the same way, the module temperature is lower for a fixed azimuth than for a changing azimuth, as it is not perpendicular to the sun rays. This would make the exergy efficiency bigger than with changing azimuth, however, global irradiance is lower, as shown in Figure 12, so the exergy efficiency is lower too. A comparison of both exergy efficiencies can be seen in Figure 13.

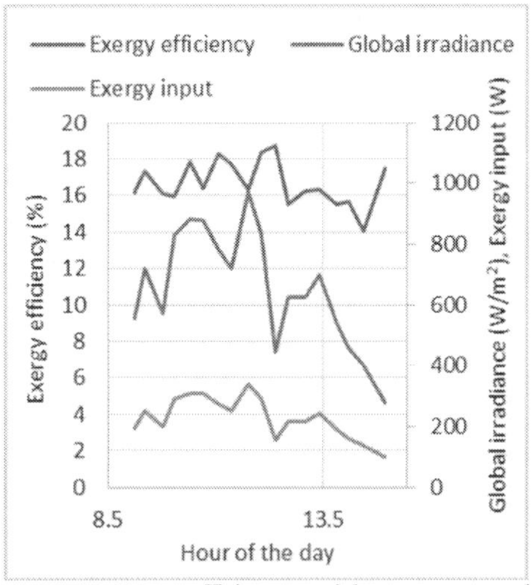

Figure 12: Exergy efficiency, module temperature and ambient temperature on a sunny winter day, with fixed azimuth.

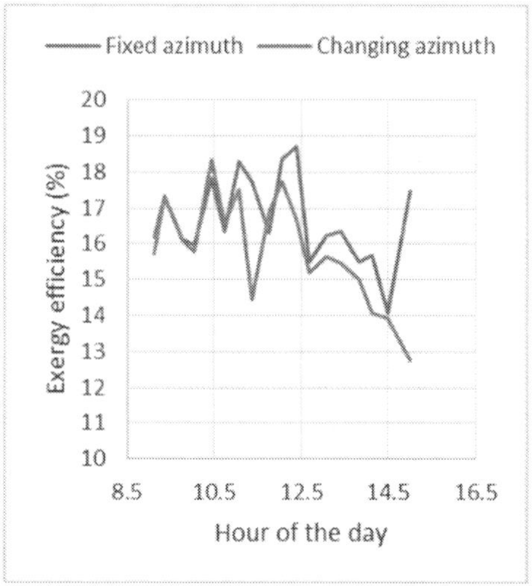

Figure 13: Exergy efficiency on a sunny winter day, for both fixed azimuth and changing azimuth.

4 CONCLUSION

- The PV module has a better efficiency on a sunny day than on a cloudy day, besides producing more electricity.
- **Exergy efficiency** is a **more suitable** definition of efficiency for a PV module, because it takes into account the energy potential and the useful energy.
- The current silicon panels are not capable of profiting all available solar energy.
- The factor that influences the module performance the most is **global irradiance**, followed by module temperature. There are more factors that influence in performance, such as wind speed, but to a lesser extent.
- In order to study the influence of module temperature in exergy efficiency, **thermal exergy cannot be considered as part of useful exergy.**

This was considered according to a paper published in the International Journal of Ambient Energy [5]. After trying to determinate the optimal module temperature, it could be seen that when the module temperature increased, thermal exergy increased, and therefore exergy efficiency increased. However, we know that a rise in the module temperature has a negative effect on the efficiency, because it reduces the electric power of the module.

Therefore, to express exergy efficiency correctly it has to be considered **only the electric exergy as useful exergy**, which is the final product. Thermal exergy is considered as an energy loss dissipated to the surroundings, and not as useful energy. Exergy efficiency equation would be:

$$\eta_{ex} = \frac{Ex_{elec}}{Ex_{in}} \qquad (8)$$

Exergy efficiency, according to its new definition, is represented in Figures 14, 15 and 16, alongside module temperature.

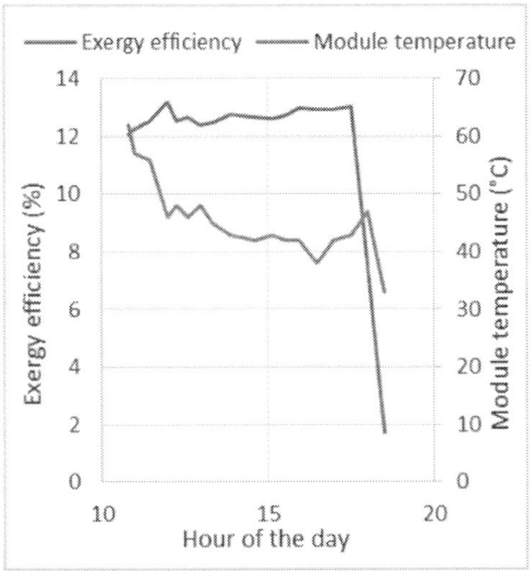

Figure 14: Exergy efficiency and module temperature on a sunny day.

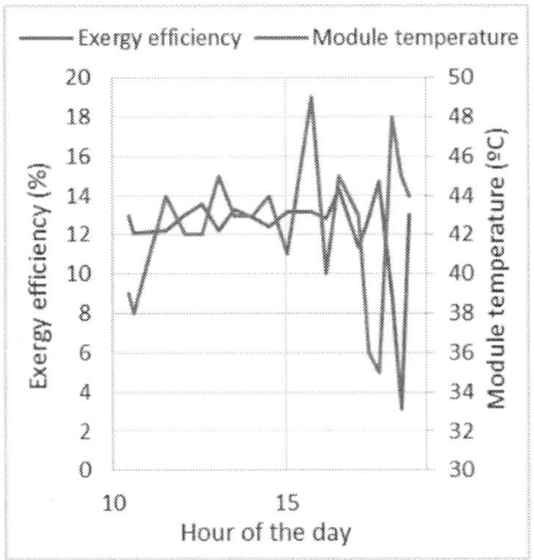

Figure 15: Exergy efficiency and module temperature on a cloudy day.

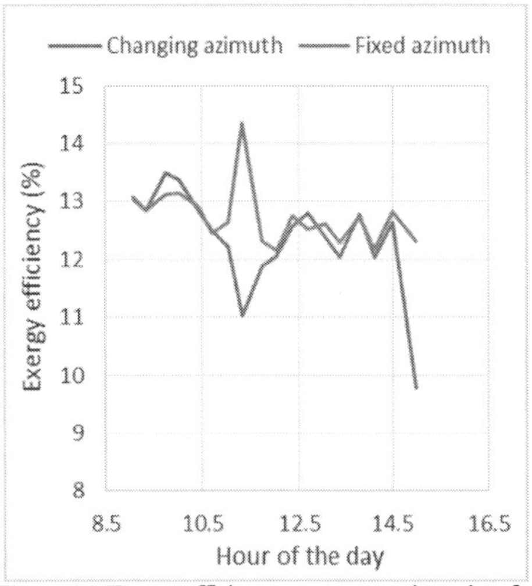

Figure 16: Exergy efficiency on a sunny winter day, for both fixed azimuth and changing azimuth.

It can be seen that, this time, exergy efficiency raises when the module temperature increases. However, the module temperature shouldn't be too low, because it can lead to a low performance of the module.

- The optimal operating module temperature ranges from 35 and 45 °C.
- A cooling system could be implemented on the surface of the module, to avoid overheating.

5 REFERENCES

[1] A. D. Sahin, I. Dincer, M. A. Rosen, Thermodynamic analysis of solar photovoltaic cell systems (2007), Solar Energy Materials & Solar Cells 91.

[2] A. Joshi, I. Dincer, B. Reddy, Thermodynamic Assessment of Photovoltaic Systems (2009), Solar Energy 83.

[3] R. Petela, Exergy of Undiluted Thermal Radiation (2003), Solar Energy 74.

[4] R. Petela, An Approach to the Exergy Analysis of Photosynthesis (2008), Solar Energy 82.

[5] K. Sudhakar, T. Srivastava, Energy and exergy analysis of a 36 W solar photovoltaic module (2014), International Journal of Ambient Energy, 35:1, 51-57.

Short Circuit Current Measurements at Clear-Sky Conditions on Photovoltaic Modules: Basic for a Reliable Self-Reference Algorithm

M. Wachter, L. Gottschalk, M. Simmler, A. Schulze+, F. Becker*, M. Sayala*, B. Hüttl

University of Applied Sciences Coburg, Department of Electrical Engineering and Computer Sciences, 96450 Coburg, Friedrich-Streib-Str. 2, Germany, Tel.: +49 (0) 9561 317 202, bernd.huettl@hs-coburg.de
+ University of Applied Sciences Rosenheim, Department of Applied Natural and Social Sciences, Germany, 83024 Rosenheim, Hochschulstr. 1
** Calyxo GmbH, Germany, 06766 Bitterfeld-Wolfen, Sonnenallee 1a,*

ABSTRACT: Outdoor or just on site analyzes of photovoltaic modules can offer some advantages in comparison to indoor Lab studies. In this way the electrical performance is monitored and logged under natural operating conditions in full range of irradiation and temperature. The entirety of monitored IV data allows deriving low-light characteristics of electric parameters, temperature coefficients and root cause analysis for possible degradation.
We apply so-called self-reference algorithm in order to increase the accuracy of analysis. The improved precision is due to application of effective acting irradiation and effective temperature instead of externally measured data. Effective irradiation is determined in self-referencing scheme: the short circuit current of the module is assigned to the irradiation. This assignment has to be calibrated, preferably at standard test conditions for irradiation.
In this contribution we demonstrate a process of precise measurement of I_{sc} under clear-sky outdoor conditions and the determination of the short circuit current for STC. The measurement method avoids errors by spectral deviations with respect to AM 1.5G spectrum, by optical reflection losses for high angles of incidence and errors by high albedo impact. In particular we investigate comparatively a CdTe module under outdoor conditions and by means a calibrated indoor Lab flasher.

Keywords: Characterization, Module, Monitoring, Performance, CdTe

1 INTRODUCTION

Accurate electrical outdoor analyses of photovoltaic (PV) modules are useful to assess their electrical parameters just on site. For this purpose we have set up an outdoor test system equipped with silicone rubber heaters on the backside of modules and with passive electric loads for logging of IV curves in high duty cycle (6 measurements per minute). In this way approximately 7000 IV measurements are performed per day within a wide range of irradiation (10 …≈ 1200 W/m²) and of temperature (ambient temperature … 90°C) due to additional heating.

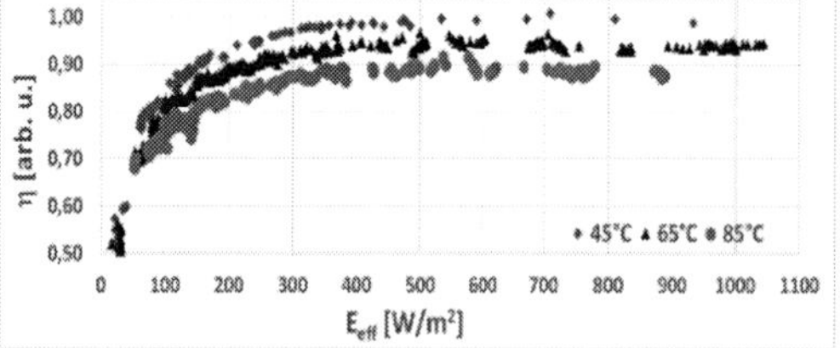

Figure 1: Low-light curves of V_{oc} and η at three different temperatures (45±2.5°C; 65±2.5°C and 85±2.5°C) measured on CdTe module

The entirety of IV data allows conclusions on temperature coefficients for open circuit voltage (V_{oc}), power at maximum power point (P_{mpp}) and for short circuit current (I_{sc}). Moreover low-light characteristics of relevant electrical and modelling parameters can be studied, and root cause analysis of power losses and possible degradation can be assessed.

For example Fig. 1 shows such low-light curves of V_{oc} and efficiency (η) for CdTe modules in a wide temperature range.

Accurate IV characterization is achieved by application of self-reference algorithm (SRA) which was introduced by [1]. SRA assigns an effective irradiation (E_{eff}) by self-referencing with I_{sc} and for effective temperature (T_{eff}) by self-referencing with V_{oc}. The effective data are at first obtained within "i" distinct irradiation intervals (ΔE_i) by distinct temperature coefficients for V_{oc} (TCV_i) and V_{oc}'s at 25°C ($V_{oc,\,25°C,\,i}$) according equation (1):

$$T_{eff} = \left(\frac{V_{oc,\,meas,\,i}\left(T_{eff}\right)}{V_{oc,\,25°C,\,i}} - 1 \right) \cdot \frac{100\%}{TCV_i} + 25°C \quad . \tag{1}$$

$V_{oc,\,meas,\,i}$'s are the real time detected V_{oc}'s in each interval ΔE_i. The TCV_i's and $V_{oc,\,25°C,\,i}$'s are determined by linear regression which is applied on all measured data pairs of $V_{oc,\,meas,\,i}$ and back side temperatures (T_{meas}) for each ΔE_i as reported elsewhere [2].

The effective irradiation E_{eff} depends on measured short circuit currents ($I_{sc,\,meas}$) according to equation (2):

$$E_{eff} = \frac{1000\,W/m^2}{I_{sc,\,STC}} \cdot \frac{I_{SC,\,meas}\left(T_{eff}\right)}{1 + \frac{TCI \cdot \left(T_{eff} - 25°C\right)}{100\%}} \quad . \tag{2}$$

Equation (2) needs knowledge of the temperature coefficient for I_{sc} (TCI) and of I_{sc} under STC conditions ($I_{sc,\,STC}$). This current can be determined by means of a calibrated Lab flasher. However, this may introduce a few flasher calibration issues. Moreover there is danger of

$I_{sc,\,STC}$ outdoor deterioration due to pollution, dust or module degradation.

In this contribution we consider a method of $I_{sc,\,STC}$ determination under clear-sky conditions. For this purpose we measure $I_{sc,\,meas}$ and control the spectral irradiation, the angle of incidence and possible albedo. If necessary, we have to correct the measured value according following equation:

$$I_{sc,STC} = I_{sc,meas} \cdot k \cdot t \cdot m \quad , \qquad (3)$$

with k as a spectral correction factor, with t for scaling the measured I_{sc} to temperature in use according temperature coefficient. The factor m corrects possible deviation of irradiation from STC.

2 TECHNICAL APPROACH

The module under test and additional control equipment are mounted on a simple tracker system (Fig. 2). We can change manually the angle of incidence (θ) by tilting in vertical direction. We use a calibrated pyranometer in module plane for detection of naturally offered clear-sky irradiation (E_{pyr}). The module temperature is controlled with a backside sensor. In addition, we use a set-up that synthesizes the spectrum of irradiation on the basic of single spectral probes. We call it "PV spectrometer", also placed on the tracker as shown in fig. 2.

Figure 2: Manual tracker system for adjustment of defined angle of incidence

It is beneficial to measure the global irradiation (in horizontal plane) too, separated into direct and diffuse components.

A precise determination of $I_{sc,\,STC}$ by simple allocation of $I_{sc,\,meas}$ to E_{pyr}, is not possible because the following distortions occur:

a) A deviations of the real irradiation spectrum E(λ) with respect to AM 1.5G results in error for $I_{sc,\,meas}$ as expressed by equation (4):

$$I_{sc,STC} = \int_{\lambda} E_{AM1.5}(\lambda) \cdot SR(\lambda)\, d\lambda \neq I_{sc,meas} = \int_{\lambda} E(\lambda) \cdot SR(\lambda)\, d\lambda \quad . \qquad (4)$$

SR(λ) in (4) is the spectral response of the module.

b) Moreover there are optical reflection losses for angled beam path irradiation (E_{dir}) which reach the solar cell or module ($E_{mod,\,dir}$) according (5):

$$E_{mod,dir}(\theta) = E_{dir}(0°) \cdot \cos(\theta) \cdot f(\theta) \qquad (5)$$

with f(θ) as an angle factor function [3, 4].

c) Albedo can cause spectral distortions of original spectrum E(λ) to $E_{alb}(\lambda)$ [5] resulting in additional errors for $I_{sc,\,meas}$ according equation (3).

d) Deviating temperature and irradiation from STC during measurement result in errors that can be corrected simply.

For handling the risk (a) there is the challenge to determine deviations of E(λ) with respect to $E_{AM1.5}(\lambda)$ and conclude a correction factor k. Relevant spectral changes for common PV materials (Si, CdTe, CIGS) are given by concentration variations of aerosols, water and ozone [e.g. 6], examples are shown in fig. 3. The solar spectra in fig. 3 were calculated with the model SMARTS [6] which needs meteorological (e.g. concentration of water, ozone and aerosols) and geographical input data.

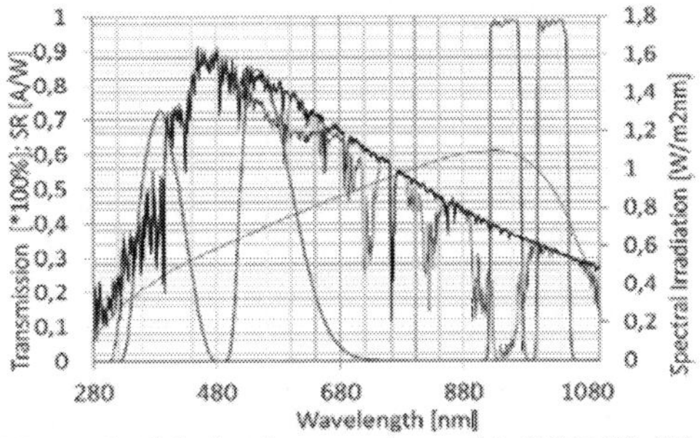

Figure 3: Calculated spectra by model SMARTS [6], black: without water and ozone absorption, red: severe ozone absorption, green: strong water absorption, grey: SR(λ) of Si-sensors, and blue: 4 wavelength characteristics of chosen optical band passes (BP1, to BP4)

Commercial available spectroradiometers offer satisfying precision to measure E(λ) under clear-sky conditions [7]. However, spectroradiometers are normally not part of standard test equipment of mobile Labs. We use for detection of spectral deviations a "PV spectrometer" which only observes four selected optical band passes (BP) of E(λ) by using Si-sensors with optical filters, as shown in fig. 3.

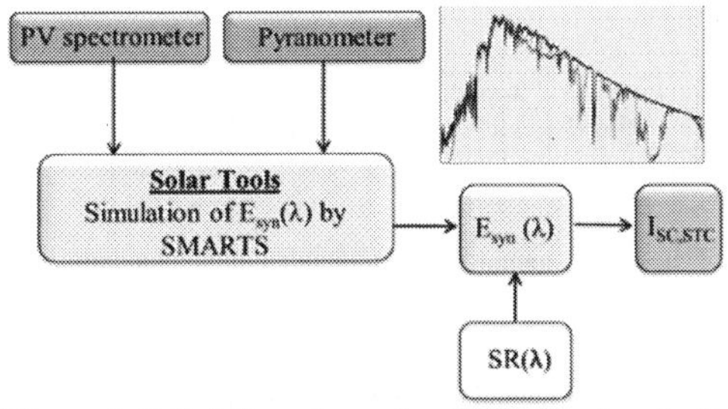

Figure 4: Calculation scheme for simulation of synthetic spectrum E_{syn} by means of detected deviations to $E_{AM1.5}$ and of simulator SMARTS [6], note: the pyranometer controls the irradiation level and confirms the success of simulation

The current signals of the four filtered sensors are calibrated by a measurement with a standard spectrum AM 1.5G. Hence the "PV spectrometer" can detect sensitively deviations from STC and can attribute them to three relevant atmospheric causes, mentioned above. The detected spectral deviations are communicated to the solar radiation model SMARTS [6] that synthesizes the actual spectrum (E_{syn}). The calculation scheme is shown in fig. 4.

We use the generated spectrum $E_{syn}(\lambda)$ to correct the measured current $I_{sc,\,meas}$ to the $I_{sc,\,STC}$ with spectral correction factor k according equation (6):

$$I_{sc,STC} = I_{sc,meas} \cdot \frac{\int_\lambda SR(\lambda) \cdot E_{AM1.5}(\lambda) d\lambda}{\int_\lambda SR(\lambda) \cdot E_{syn}(\lambda) d\lambda} = I_{sc,meas} \cdot k \quad . \quad (6)$$

We use the manual tracker shown in fig. 2 to treat the risk (b) experimentally. For this purpose we measure the $I_{sc,\,meas}$ as a function of incident angle θ. We introduce an angle characteristics $Ch_{meas}(\theta)$ for measured short circuit current:

$$Ch_{meas}(\theta) = \frac{I_{sc,meas}(\theta)}{I_{sc,meas}(0°)} \quad . \quad (7)$$

$Ch_{meas}(\theta)$ describes the optical reflection losses of solar irradiation. Moreover, $Ch_{meas}(\theta)$ turns out a maximum tilt angle θ_{max} of module (with respect to perpendicular incidence) for that we get acceptable low changes of $I_{sc,\,meas}$. In case of I_{sc} measurements with tilt angles below θ_{max} no irradiation correction is needed and no contribution to scaling factor m occurs.

For modelling this angle behavior we have to handle separately direct and diffuse irradiation, E_{dir} and E_{diff}, respectively. Tilted direct irradiation penetrates into the module according equation (5) and only a part $E_{mod,\,dir}$ reaches the active cell of module. For modelling the diffuse irradiation part that reaches the solar cell ($E_{mod,\,diff}$), an anisotropic angle model based on the concept of [8] is used. The model of [8] distributes an anisotropic diffuse irradiation source $E_{diff,\,source}$ on a hemisphere across the angels α and φ in accordance to meteorological conditions. We use for our aim only clear-sky conditions. The part $E_{mod,\,diff}$ is calculated by integration above visible hemisphere (A_{vis}):

$$E_{mod,diff} = \frac{1}{A_{vis}} \cdot \iint_{A_{vis}} E_{diff,source}(\alpha,\varphi) \cdot \cos(\theta(\alpha,\varphi)) \cdot f(\theta(\alpha,\varphi)) dA \quad . \quad (8)$$

We use in our numeric angle model the Fresnel equation for a flat glass/air interface as angle factor function f(θ) [e.g. 9]. The total irradiation that reaches the module's active cell (E_{mod}) causes $I_{sc,\,meas}$ and consists of the direct and diffuse component:

$$E_{mod} = E_{mod,dir} + E_{mod,diff} \quad . \quad (9)$$

Hence, E_{mod} should follow the function of $Ch_{meas}(\theta)$:

$$Ch_{meas}(\theta) = \frac{I_{sc,meas}(\theta)}{I_{sc,meas}(0°)} \stackrel{!}{=} \frac{E_{mod}(\theta)}{E_{mod}(0°)} = Ch_{theo}(\theta) \quad . \quad (10)$$

A comparison of measured and calculated Ch(θ) could reveal an impact of albedo (risk c). Albedo is "only" a problem for outdoor $I_{sc,\,STC}$ determination if the spectrum of reflected radiation $E_{alb}(\lambda)$ is changed with respect to $E_{syn}(\lambda)$ [5]. Note, possible spectrum changes by albedo can hardly be forecasted. According equation (4) the $I_{sc,\,meas}$ determination with respect to E_{pyr} becomes vague in presence of predominant albedo. In case of significant albedo the course of $Ch_{meas}(\theta)$ should deviate measurably from $Ch_{theo}(\theta)$. In particular $Ch_{meas}(\theta)$ has to be checked in the angle range of θ = 0° to θ_{max} on possible albedo. If albedo is significant the $I_{sc,\,STC}$ determination has to be rejected.

Deviations from STC temperature and irradiation (risk d) during measurement can be corrected easily by scaling factors.

For irradiation counts: the measured $I_{sc,\,meas}$ follows linearly to actual irradiation E_{pyr} and scaling factor m for equation (3) can be written :

$$m = \frac{1000\, W/m^2}{E_{pyr}} \quad . \quad (11)$$

The temperature correction is in accordance to the linear behavior of I_{sc}, with temperature coefficient, and the temperature correction factor becomes:

$$t = \frac{I_{sc,25°C}}{I_{sc,meas}(T)} = \frac{1}{1 + \frac{TCI}{100\%} \cdot (T - 25°C)} \quad . \quad (12)$$

3 EXPERIMENTAL RESULTS AND SUMMARY

The concept of outdoor $I_{sc,\,STC}$ determination was proven on a CdTe module. The SR(λ) characteristics has been measured in advance by a calibrated EQE system.

For outdoor $I_{sc,\,STC}$ determination we selected a day in June at local time of 11:30 am and with required clear-sky conditions. The manual tracker was oriented to azimuth of the sun and the elevation angle was set to get the incident angle of 0°.

During the measurement campaign the pyranometer in module plane detected a stable value of E_{pyr} = 1023 ± 10 W/m² for θ = 0°, and the temperature at backside was found to be 45 ± 5 °C.

For these slight deviations from STC we derive the scaling factors with equations (11) and (12):

- m = 0.994 and
- t = 0.990.

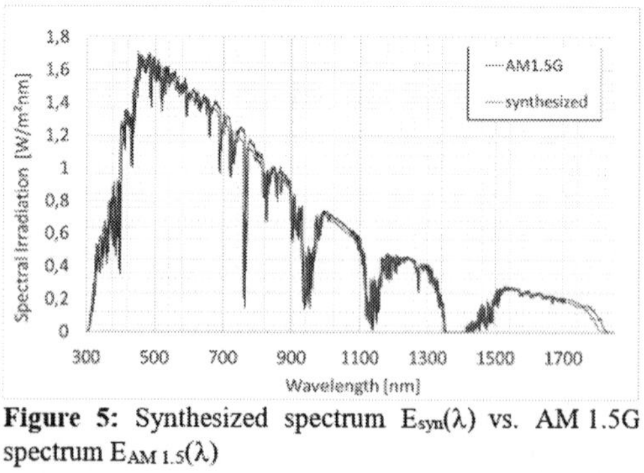

Figure 5: Synthesized spectrum $E_{syn}(\lambda)$ vs. AM 1.5G spectrum $E_{AM\,1.5}(\lambda)$

For perpendicular incidence we detected with "PV spectrometer" only slight differences of sensor signals with respect to expected values for AM 1.5G spectrum.

Hence the synthesized spectrum $E_{syn}(\lambda)$ is similar to standard spectrum as shown in fig. 5. In particular only slight concentration deviations for water (w) and ozone (u) in comparison to STC atmosphere are deduced by the spectrum synthetization with SMARTS: we found $w_{syn} = 1{,}37$ cm vs. $w_{AM1.5} = 1{,}42$ cm and $u_{syn} = 0{,}55$ DU vs. $u_{AM1.5G} = 0{,}34$ DU, respectively. The parameters describing the optical impact of aerosols (according Angstrom model with parameters α and β [6]) are changed slightly from $\alpha_{AM1.5G} = 1{,}2$ and $\beta_{AM1.5G} = 0{,}033$ to $\alpha_{syn} = 0{,}79$ and $\beta_{syn} = 0{,}029$, respectively.

By application of equation (6) the spectral correction factor is determined to be k = 1,016.

By means the manual tracker system we measured the short circuit current as a function of incident angle $I_{sc,meas}(\theta)$. The derived angle characteristics $Ch_{meas}(\theta)$ is presented in fig. 6.

The maximum acceptance angle for tilting θ_{max} was concluded by following rule: keep the variation of $I_{sc,meas}(\theta)$ below 1%. In this way we have determined an θ_{max} with 5°. Hence it is possible to have insignificant additional reflection losses by tilting the module within the angle range of ±5°.

For our measurement the right hand side of $Ch_{meas}(\theta)$ characteristics (positive angles θ) was oriented towards ground, the left hand side was tilted towards sky hemisphere. It is obvious from fig. 6 that the level of right $Ch_{meas}(\theta)$ is increased with respect to left hand side. Moreover there is a kink on the right hand side of $Ch_{meas}(\theta)$ showing an increase of $Ch_{meas}(\theta)$ too. We conclude from this increase of irradiation an impact of albedo. Our simulations support this assumption. The angle characteristics $Ch_{theo}(\theta)$ was simulated with the numeric model mentioned above. $Ch_{theo}(\theta)$ is shown in fig. 6 too, and it turns out that our model underestimates the irradiation on the right hand side. Our model does not consider albedo impact, hence albedo makes the difference. In addition we modeled the expected diffuse irradiation in module plane by simulation system SMARTS, also shown in fig. 6 ($Ch_{theo\_SMARTS}(\theta)$). This simulator takes into account albedo for calculations. By implementing an albedo level of 0.3 into SMARTS we can exactly simulate the measured $Ch_{meas}(\theta)$. The mentioned results confirm that albedo is relevant for our investigations. However, in angle range below θ_{max} the albedo can be neglected as shown by the good fits of measured and simulated Ch(θ). We conclude: $I_{sc,STC}$ determination can be performed within the acceptance angle without albedo danger.

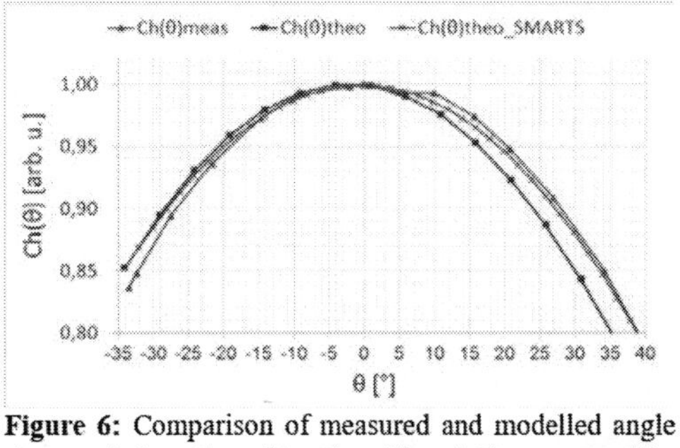

Figure 6: Comparison of measured and modelled angle characteristics, $Ch_{meas}(\theta)$, $Ch_{theo}(\theta)$ and $Ch_{theo\_SMARTS}(\theta)$

We summarize the results of the measurement campaign: With measured $I_{sc,meas} = 2.184$ A at a temperature of 45°C and an irradiation in module plane of $E_{pyr} = 1023$ W/m^2 we derive with predetermined correction factors the STC value for short circuit current:

$$I_{sc,STC} = 2{,}184A \cdot 1{,}016 \cdot 0{,}990 \cdot 0{,}994 = 2{,}168A$$

We took into account the error sources of fluctuating sensor signals of "PV spectrometer", pyranometer and cell temperature and conclude according error propagation a measurement uncertainty of 2,168A ± 0,022A or a measurement inaccuracy of acceptable 1%.

For comparison the module was also investigated by means a flasher system under STC conditions. The flasher was calibrated by WPVS cell, known flasher spectrum and spectral response of the module. We found there $I_{sc,STC,indoor} = 2.165$ A which fits well to our outdoor measurements.

The measurement concept of I_{sc} looks reasonable, further proofs also with other cell technologies are underway.

4 REFERENCES

[1] H.-D. Mohring, D. Stellbogen, D. Geyer, P. Lechner; "Comparative Assessment of Different Thin Film Technologies from Outdoor I-V Characteristics", 26th European Photovoltaic Solar Energy Conference, (2011), 4AV.2.46, pp. 3625 - 3628.

[2] D. Kaiser, M. Simmler, M. Wachter, T. Faber, F. Becker, M. Sayala, T. Kaden, A. Schulze, B. Hüttl; "Concept for Fast and Precise PV Module Outdoor Characterization", 31th European Photovoltaic Solar Energy Conference, (2015), 5AV.6.10, pp. 1956 - 1958.

[3] D.I King, J.A. Kratochvil, W.E. Boyson "Measuring Solar Spectral and Angle-of-Incidence Effects on Photovoltaic Modules and Solar Irradiance Sensors", Proc. 26th IEEE Photovoltaic Specialists Conference (PVSC), (1997), pp. 1113–1116.

[4] M. Müller; "Modellierung und Verifizierung des Jahreswirkungsgrades von Solarzellen", Diploma Thesis, University of Leipzig, private communications.

[5] R.W. Andrews, J. M. Pearce; "The effect of spectral albedo on amorphous silicon and crystalline silicon solar photovoltaic device performance", Solar Energy, Vol. 91, (2013), pp. 233-241.

[6] C. Gueymard, "A Simple Model of the Atmospheric Radiative Transfer of Sunshine: Algorithms and performance assessment", Report FSEC-PF-270-95 (1995), Florida Solar Energy Center, 1679 Clearlake Road, Cocoa, FL 32922.

[7] A. Habte, A. Andreas, L. Ottoson, C. Gueymard, G. Fedor, S. Fowler, J. Peterson, E. Naranen, T. Kobashi, A. Akiyama, and S. Takagi; "Indoor and Outdoor Spectroradiometer Intercomparison for Spectral Irradiance Measurement", Technical Report of NREL/TP-5D00-61476, (May 2014), free available at www.nrel.gov/publications.

[8] St. Darula, R. Kittler; "CIE general sky standard defining luminance distributions", Proc. Canadian conference on building energy simulation (2002)

[9] E. Hecht: „*Optics*", Oldenbourg , Munich (2005), p. 221

5 ACKNOWLEDGEMENT

The authors gratefully acknowledge the German Federal Ministry of Education and Research (BMBF) for funding the project "PV-Präzis" under support code FKZ 03FH018PX4.

ANALYSIS OF AIR MASS DEPENDENCE OF THREE PHOTOVOLTAIC ARRAYS

Huaxin Wang[1], M.A. Muñoz[1], G.P. Moreda[1], M.C. Alonso-García[2]
[1] Dep. Agrororestry Engineering. LPF-TAGRALIA. ETSIAAB, UPM
Avda. Complutense, s/n, 28040 Madrid, Spain
miguelangel.munoz@upm.es Phone: +34 913365458; Fax: +34 913365406
[2] Photovoltaic Laboratory, Renewable Energy, CIEMAT
Avda. Complutense, 22, 28040 Madrid, Spain

ABSTRACT: The solar spectrum, which is also could be described by air mass factor, has a significant relationship with the performance of photovoltaic modules. The air mass dependence has been investigated by a theoretical method as well as an actual case of three different photovoltaic arrays: polycrystalline silicon, amorphous silicon and cadmium telluride, at the EUIT Agrícola-UPM (Madrid, Spain, 40.4426°N, 3.7295°W). The calculating results show that: AM dependence of poly-Si array is positive because an elevating AM-value indicates a red shift and poly-Si module is more red sensitive; On the contrary, as a-Si module and CdTe module are more blue sensitive, their AM dependences are negative. Furthermore, the outdoor AM dependences of three arrays are corresponding to the calculating results although the thermal annealing effect could be superimposed on spectral effect in the case of a-Si.
Keywords: PV system, spectral response, polycrystalline silicon, a-Si, CdTe

1 INTRODUCTION AND OBJECTIVS

In outdoor conditions, the two main meteorological issues affecting the electricity production of a PV module are the irradiance and ambient temperature. Besides, seasonal variation in solar spectrum contributes to the seasonal changed irradiance. Therefore, seasonal variation in PV module's output could be influenced by the overlap of thermal effects and spectral changes [1–5]. Since different semi-conductor materials have different band gaps, they vary in response to radiation of different wavelengths, thus vary from each other both in efficiency and spectral response. On the other hand, the spectral components of solar radiation also have a seasonal change: they are blue-rich in summer and red-rich in winter respectively [6]. Commonly, solar spectral irradiance distribution can be characterized by three approaches: average photon energy [1], useful fraction [7], and air mass (AM) [8], which is the path-length through the atmosphere and used in this study.

In this study, the theoretical and the outdoor AM dependences of three different arrays: polycrystalline silicon (poly-Si), amorphous silicon (a-Si) and cadmium telluride (CdTe) are investigated.

2 MATERIALS AND METHODS

2.1 Air mass

The term ''air mass'', which is obtained through geometrical consideration, describes the relative path length that the sun's rays traverse through the atmosphere. It can be a specific substitute in the absence of experimental spectral measurements on solar irradiance, although it is without the consideration of the vapor, turbidity and air pollution. The simplest formula (1) for calculating the pressure corrected AM [8] is just related to the position of the sun in the sky and the site altitude. Complex mathematical expressions of the AM provided by [8] and (Gueymard 1995) show no difference between equation (2) at zenith angles less than 80° [10].

$$AM = \cos(Z_s)^{-1} * \frac{P}{P_0} \qquad (1)$$

$$\frac{P}{P_0} = \exp(-0.0001184 * h) \qquad (2)$$

Zs: sun's zenith angle, calculated based on [11]

Where:

P: local air pressure;

P0: sea level air pressure;

h: altitude, 650m of Madrid;

The solar spectral irradiance data shown in Figure 1 were simulated by the application of SMARTS (Simple Model of Atmospheric Radiative Transfer of Sunshine) developed by Gueymard. The normalized spectral responses of different technologies were provided by the accredited independent laboratory (CIEMAT-DER, Madrid).

2.2 Theoretical mathematical model:

The short circuit density of the three PV technologies could be calculated by:

$$Jsc = \int E_\lambda^{Am}(\lambda) \cdot SR(\lambda) \cdot d\lambda \qquad (3)$$

Jsc: short circuit current density of the test module;

$E_\lambda^{Am}(\lambda)$: spectral irradiance of the sunlight at different AM values;

$SR(\lambda)$: spectral response of the test module;

3 RESULTS AND DISCUSSIONS

3.1 Calculating results

The calculating results based on equation (3) and Figure1, are demonstrated in Figure 2 and 3.

Figure 2 is simulating the results obtained from a pyranometer. In this case, the AM dependence of poly-Si is positive, whereas the Isc of a-Si and CdTe would decrease with the rising AM values.

Figure 3 is simulating the results obtained from a c-Si based irradiance sensor. Consequently, there is no AM dependence in the Isc of poly-Si. However, the Isc of a-Si and CdTe decline faster with the rising AM value.

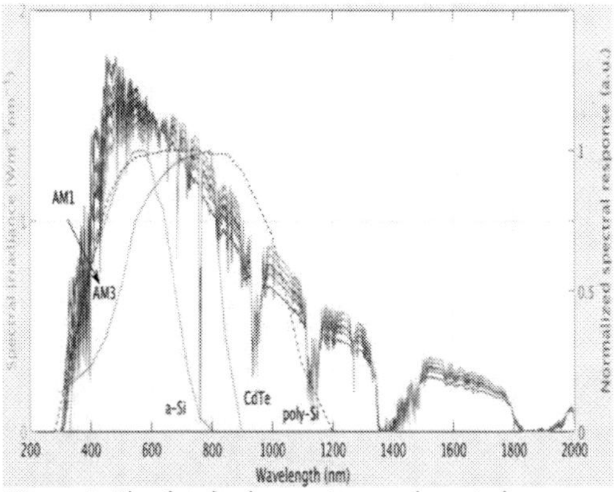

Figure 1: Simulated solar spectrum and spectral response of different technologies.

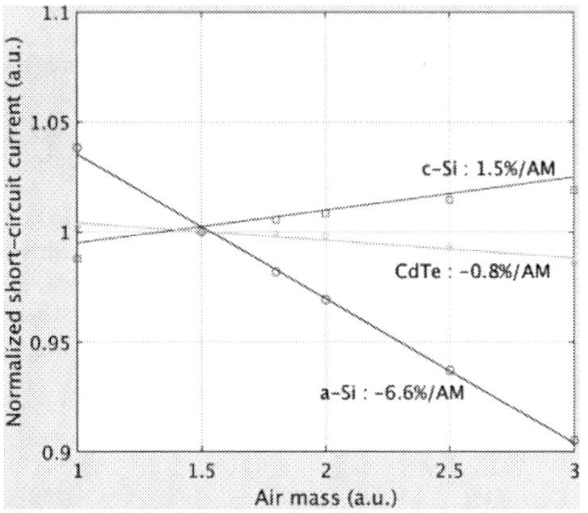

Figure 2. Calculating results of three technologies as the irradiance sensor is a pyranometer.

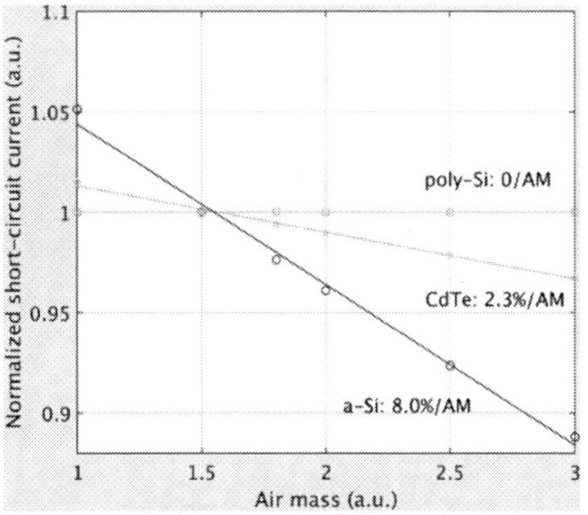

Figure 3. Calculating results of three technologies as the irradiance sensor is a c-Si based sensor.

3.2 Outdoor results

As pre-mentioned, AM is calculated by a geographical consideration and fails to contain the weather information, which also affects the spectral distribution. Thus a pre-clarification about the weather condition or a uniform weather condition should be decided. All the data shown in Figure 4 was collected under clear-sky condition and then corrected and normalized:

- Clear-sky condition: the variation of the irradiance is less than 10% from the maximum value to the minimum value recorded during the 15-min recording interval;
- Irradiance range: the reason of choosing the irradiance level at 700W/m^2 is inverter clipping happens occasionally once the irradiance is beyond 800W/m^2, subsequently the Pmp would be affected;
- Correction and normalization: the maximum power was linearly corrected to the condition (G=1000W/m^2, Tc=25°C) using the maximum power (Pmp) temperature coefficient provided by the manufacturers, and then divided by the linear fitted value at AM1.5.

Figure 4 shows the AM-dependence of the Pmp of the three arrays from December 2014 to December 2015. The higher AM values indicate the solar spectrum tending towards the redder or longer wavelength.

As depicted in Figure 4, in both the irradiance ranges, the AM-dependence of Pmp for poly-Si, CdTe and a-Si arrays increased successively. The regression line parallel to x-axis in Figure 4a indicates no clear AM-dependence of Pmp generated by the poly-Si array, owing to a c-Si based reference irradiance sensor and the determinate irradiance range. The AM-dependence of Pmp for a-Si and CdTe arrays is stronger, suffering a decreasing trend for higher AM values (red-shift). The gradient of the applied linear least square fit can also provide information about the AM-dependence. Comparing the two irradiance ranges, the AM-dependence of Pmp follow the same pattern for each array. Pmp for a-Si array varies at a higher rate ~11.22% compared to the ~4% variation of the CdTe array's Pmp, due to the narrower and more blue-sensitive spectrum response. Note that these AM-dependences for a-Si and CdTe are based on a c-Si irradiance sensor. Furthermore, as is shown in Figure 4c, the notable gap between the Pmp gathered from two periods, i.e. Apr.2015 –Jun.2015 and Jun.2015 – Sep.2015, suggests there are some other factors contributing to the seasonal variation of a-Si, besides the changeful AM. It will be explained by *thermal annealing effect* and *light soaking effect*.

4 CONCLUSION

- No detective air mass (AM) dependence of poly-Si array has been found owing to the application of c-Si based irradiance sensor, whereas the calculating result shows that poly-Si has a positive AM dependence;
- a-Si has the highest AM dependence among the three technologies, although there is an overlap of *thermal annealing effect* and spectral effect in a-Si array;
- AM dependence of CdTe array is lower compared to the value of a-Si array, due to the more red-sensitive spectrum response.

ACKNOWLEDGEMENTS

We want to acknowledge the LPF-TAGRACIA research group of Technical University of Madrid (UPM) for the funding of the project and to CIEMAT for their cooperation. Also we want to acknowledge the financial support from the program of China Scholarships Council (CSC).

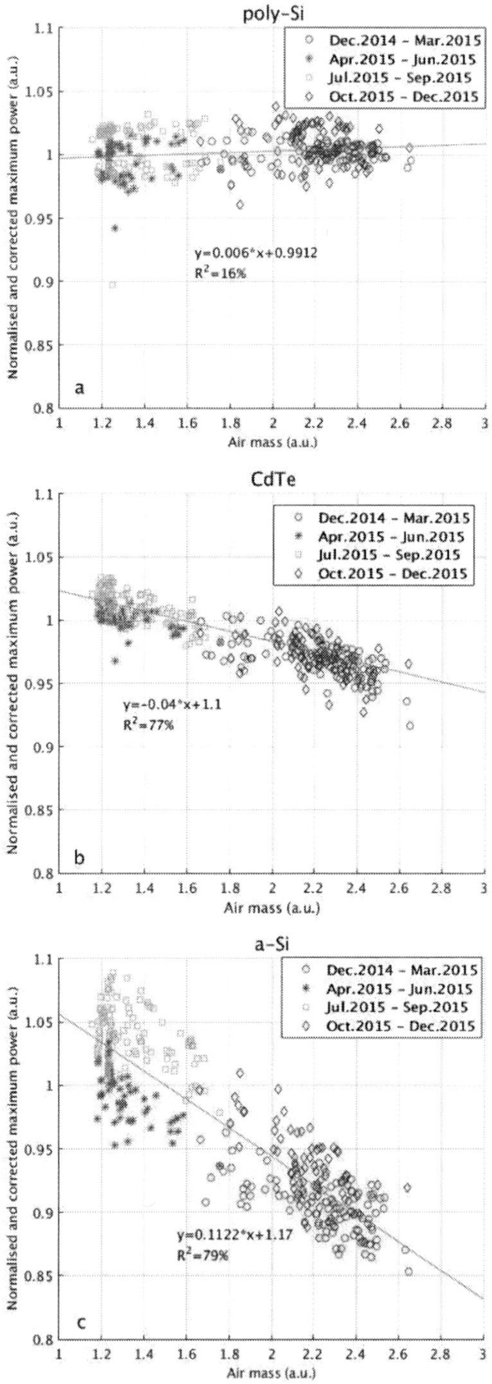

Figure 4. Seasonally classified maximum power dependence on AM-units for the poly-Si (a), a-Si (b) and CdTe (c) array under clear-sky condition.

REFERENCES

[1] N.K. A, Y.N. A, N. , Takashi Minemoto b, Hideyuki Takakura a, Estimation of irradiance and outdoor performance of photovoltaic modules by meteorological data, (n.d.).

[2] G. Makrides, B. Zinsser, A. Phinikarides, M. Schubert, G.E. Georghiou, Temperature and thermal annealing effects on different photovoltaic technologies, Renew. Energy. 43 (2012) 407–417.

[3] A. Phinikarides, G. Makrides, B. Zinsser, M. Schubert, G.E. Georghiou, Analysis of photovoltaic system performance time series: Seasonality and performance loss, Renew. Energy. 77 (2015) 51–63.

[4] N. Aste, C. Del Pero, F. Leonforte, PV technologies performance comparison in temperate climates, Sol. Energy. 109 (2014) 1–10.

[5] T. Minemoto, M. Toda, S. Nagae, M. Gotoh, A. Nakajima, K. Yamamoto, et al., Effect of spectral irradiance distribution on the outdoor performance of amorphous Si//thin-film crystalline Si stacked photovoltaic modules, Sol. Energy Mater. Sol. Cells. 91 (2007) 120–122.

[6] A. Virtuani, L. Fanni, Seasonal power fluctuations of amorphous silicon thin-film solar modules: distinguishing between different contributions, Prog. Photovoltaics Res. Appl. 22 (2014) 208–

[7] R. Gottschalg, D.G. Infield, M.J. Kearney, Experimental study of variations of the solar spectrum of relevance to thin film solar cells, Sol. Energy Mater. Sol. Cells. 79 (2003) 527–537.

[8] D.L. King, J.A. Kratochvil, W.E. Boyson, Measuring solar spectral and angle-of-incidence effects on photovoltaic modules and solar irradiance sensors, in: Conf. Rec. Twenty Sixth IEEE Photovolt. Spec. Conf. - 1997, IEEE, 1997: pp. 1113–1116.

[9] C.A. Gueymard, SMARTS2-A simple model of the atmospheric radiative transfer of sunshine: algorithms and performance assessment, (n.d.).

[10] M. Paulescu, E. Paulescu, P. Gravila, V. Badescu, Weather Modeling and Forecasting of PV Systems Operation, 2012.

[11] I. Reda, A. Andreas, Solar position algorithm for solar radiation applications, Sol. Energy. 76 (2004) 577–589.

OUTDOOR PERFORMANCE AND SEASONAL ANALYSIS OF SUNPOWER (MAXEON™) TECHNOLOGY PV MODULES IN COMPOSITE CLIMATE OF INDIA

Ashish Sharma[1], Birinchi Bora[2], Yogesh K. Singh[2], Bishnu M. Jha[1], Rashmi Singh[2], Supriya Rai[2], Manander Bangar[2], Renu[2], S. Chakraborty[2], D. Singh[1], K. Saikia[1], S.K. Samdarshi[1], O.S. Sastry[*2]
[1]*Central University of Jharkhand, Brambe, Ranchi, Jharkhand- 835205*
[2]*National Institute of Solar Energy, Gurgaon-122003*
*sastry284@gmail.com,9599935891

ABSTRACT: Performance of Sun Power based Maxeon™ technology PV module in composite climate zone of India has been analysed after 2 years of commissioning. A temperature and irradiance matrix of power for these modules as per IEC 61853 Part 1 is presented here. The seasonal performances of the modules in terms of percentage of normalised power of the array under outdoor test conditions are analysed. The time series analysis of the performance of the technology is also studied here and found a significant degradation in the performance of the tested module.
Keywords: Sun power, IEC 61853, Seasonal performance, performance ratio, degradation

1 INTRODUCTION

Outdoor performance study of the module is important to understand the viability and usability of the modules in the different climatic conditions. Reliability analysis of module enables us to ensure the commercial feasibility at a specific location. Also to understand the effects of various outdoor factors like dust and wind speed on the module that constantly changes with time. The module performance is rated under Standard Test Condition (STC) i.e. Air mass 1.5 spectral distributions, irradiance level of 1000 W/m^2 and module temperature of 25 ^{0}C at 0^0 angle of incidence as per standard IEC 61215 [1]. During the outdoor operations, module experiences a wide range of change in temperature, irradiance and spectrum with time, day, and season [2]. Due to these varied conditions, the prediction of the performance of PV modules on the basis of STC rating may not be accurate. Bora et al. reported on the seasonal performance of CdTe technology in the outdoor condition. This study has shown the effect of module temperature on performance ratio of PV module [3]. For the testing of module in outdoor condition it is very tough to get exact conditions for the study as per IEC 61853 every time because of the change in environmental conditions. So, different translation methods are used for the estimation of power output from outdoor data. IEC has formulated IEC 60891 standard which deals with 3 methods for the performance calculation [4]. Sarah et al. has reported the comparison between IEC 60891 methods and found procedure 3 to be most suitable among three procedures. However, considered data should be at high intensities for the corrections [5]. IEC 60891 have provided the procedure to find out the temperature coefficient. But the main problem is to calculate the temperature coefficient from the real time outdoor data. A method has also been devised by NREL which directly translated the I-V curve to the corresponding I-V curve at desired irradiance and temperature.

Degradation is a very common problem in the field exposed modules. With the time of exposer in the field, module performance decreases. Power degradation doesn't solely depend upon a specific factor rather it is a combined effect of various factors. Sharma et al found the degradation in three different photovoltaic technologies under similar testing conditions. It is found that degradation in each technology is to be 5.7 ± 2.5%, 0.51 ± 0.017%, 0.31 ± 0.016% in a-Si, m-C-Si and HIT modules respectively [6]. This study presents the performance of SunPower Maxeon™ technology. The performance parameters are measured or estimated under 23 different temperature and irradiance conditions as per IEC 61853 standard [7]. The study also provides percentage of normalized power in real outdoor conditions in different seasons, with respect to the rated power rating of the module. The time series analysis for a particular irradiance and module temperature condition has also been done here.

2 EXPERIMENTAL APPROACH

An experimental test-bed facility of PV modules was setup at NISE (National Institute of Solar Energy), Gurgaon (Latitude 28^0 37 N, Longitude $77^0$4 E), Gurgaon as shown in the Fig 1 to study the performance of the Sun Power technology. It consists of an array of 5 modules, an I-V tracer for continuous monitoring of power output, along with a reference cell fixed co-planer with the tilt of modules to measure in plane global radiation and a temperature sensors attached to the rear of the module to record the module temperature data.

Figure 1.a: Test bed of MAXEON™ technology

Each module has a rated power of 327 Wp, open circuit voltage of 64.9 V and short circuit current of 6.46 A. Modules are mounted on manual dual axis tracking structure and I-V data is recorded at every 15 seconds interval of time using 4 probe connectors. All modules are cleaned everyday morning before performing the measurement to avoid dust deposition or other dirt like bird dropping effect on the output.

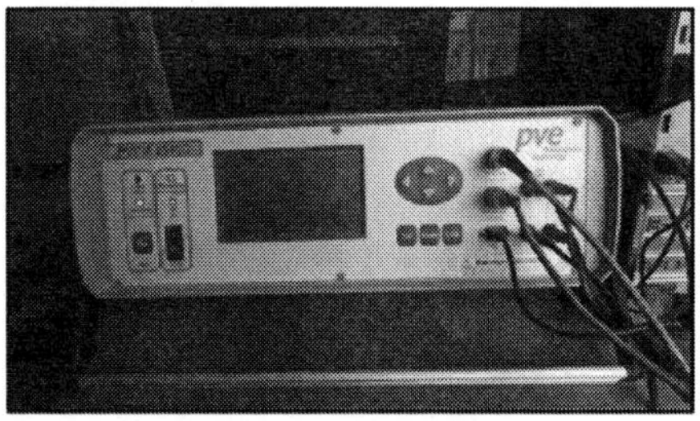

Figure 1.b: IV curve tracer PVPM

3 RESULTS AND DISCUSSION

The effect on the performance of PV module has been analyzed by using three ways:
The data has been divided seasonally into 4 seasons as winter from mid-November to mid- March, summer from mid-March to end- June, and monsoon from starting- July to mid- September, post monsoon from mid-September to mid-November. The data of the irradiance between 700 and 800 W/m^2 and temperature between 35 and 45°C has been selected for the study of time series history of the module. The conditions of temperature and irradiance is so selected that for this climatic zone the occurrence of the particular condition is highest for a particular year. For the translation of data IEC 60891 procedure 1 is used in this paper. In this procedure the current and voltage needs to find out through the I-V characteristics using the following equations:

$$I_{sc,STC} = I_{sc} + I_{sc}(\frac{1000}{G} - 1) + \alpha(25 - T)$$

$$V_{oc,STC} = V_{oc} - R_s(I_{sc,STC} - I_{sc}) - \kappa * I_{sc,STC}(25 - T) + \beta(25 - T)$$

where, α= 0.0035; β=-0.1766; κ=0.00125 are used. $I_{sc,STC}$ is the rated current, I_{sc} is the short circuit current, R_s is the series resistance, $V_{oc,STC}$ is the rated voltage, V_{oc} is the measured voltage, k is the curve correction factor, α is the temperature coefficient of the current and β is the temperature coefficient of the voltage.
For the performance measurement of the pv modules as per the IEC 61853, only clear sky data has been taken into account. In table 1, the power matrix of the Sun Power technology has been given and it has been observed that at 1100 W/m^2 and 25°C, the efficiency of the plant is found to be highest around 20% while at 800 W/m^2 and 75°C it is lowest. The study has been divided into 4 seasons namely, winter, summer, monsoon and post monsoon. The results of the seasonal performance of module is shown in Fig 2.

Table 1: Power matrix according to IEC 61853 at 23 different conditions

Irradiance(W/m^2)	Power at different temperature (W)			
	15° C	25° C	50° C	75° C
1100	NA	1741	1673	1452
1000	1362	1357	1335	1342
800	1142	1140	1100	1053
600	914	902	900	850
400	614	608	602	NA
200	312.5	308	NA	NA
100	150.5	144.4	NA	NA

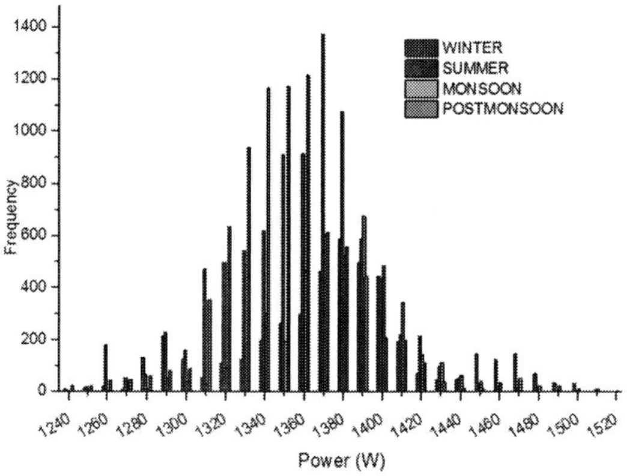

Figure 2: Seasonal performance of the array

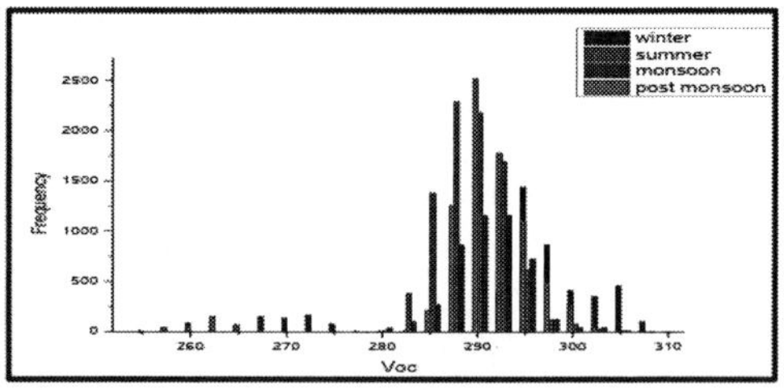

Figure 3.1: Distribution of V_{oc}

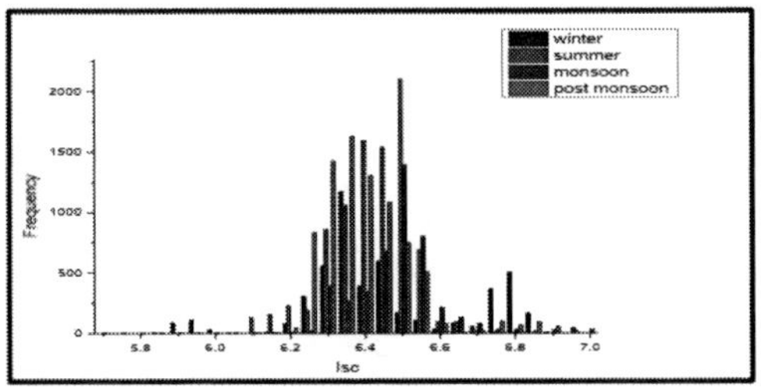

Figure 3.2: Distribution of I_{sc}

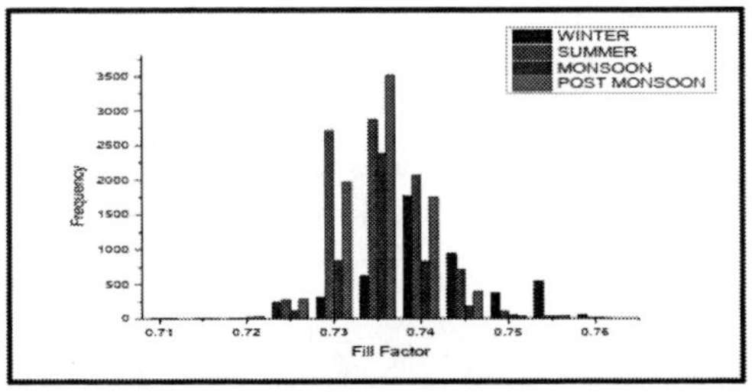

Figure 3.3: Distribution of FF

The maximum occurring power is around 1360 to 1380 Wp, while the voltage is around 290 V. The current around 6.5 A is having maximum frequency occurring in summer. Percentage of normalised power is 85% in winter and monsoon, 83% in summer and 82% in post monsoon with respect to name plate value.The time series data of performance from one year data has been plotted in figure 3 for the intensity 700 to 800 W/m^2.

It has been observed that there is 0.5%/year degradation in the performance for the particular condition of the data

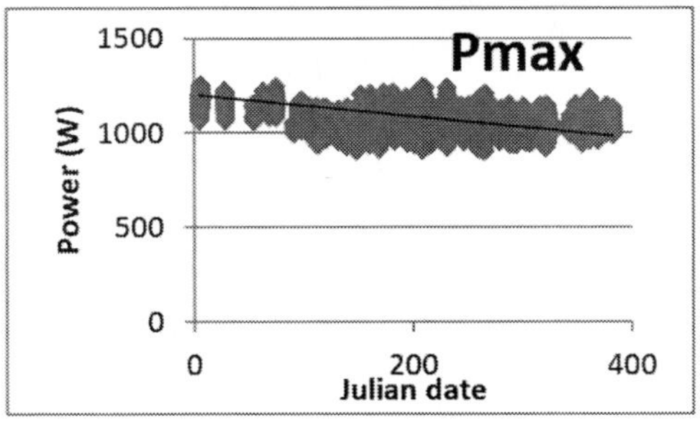

Figure 4: Time series for maximum power

Significant degradation is observed in current and fill factor however the voltage has remained almost same throughout the time period.

5. CONCLUSION

The study of SunPower technology in composite climate of India has been presented in the paper considering power output, efficiency and degradation. Higher efficiency of the plant has been observed in higher range of irradiance. Degradation study done by using the time series performance data reveals that power degrades by 0.5%/year. Over the time short circuit current and fill factor has shown significant degradation in comparison to open circuit voltage.

6 ACKNOWLEDGEMENT

This research is based upon work supported in part by the Solar Energy Research Institute for India and the U.S. (SERIIUS) funded jointly by the U.S. Department of Energy subcontract DE AC36-08G028308 (Office of Science, Office of Basic Energy Sciences, and Energy Efficiency and Renewable Energy, Solar Energy Technology Program, with support from the Office of International Affairs) and the Government of India subcontract IUSSTF/JCERDC-SERIIUS/2012 dated 22nd Nov. 2012

7. REFERENCES

[1]International Electro technical Commission 61215 Second edition, 2005-04. Crystalline silicon terrestrial photovoltaic modules – design qualification and type approval.

[2].O.S.Sastry et al. Performance analysis of field exposed single crystalline silicon modules, Solar Energy materials & Solar Cells, 94(9), 1463-1468, 2010.

[3].Effect of module temperature and seasonal variation analysis of the outdoor performance of CdTe at composite climate of India, Birinchi Bora, O.S. Sastry, Yogesh Kumar Singh, Ramayan Singh, Kamlesh Yadav, Vikrant Sharma, Manander Bangar, Mithilesh Kumar, Invertis Journal of Renewable Energy, Vol. 5, No. 3, 2015

[4].International Electrotechnical Commission. (2009). IEC 60891 (ed.2). Photovoltaic devices—Procedures for temperature and irradiance corrections to measured I-V characteristics.

[5].Sarah Sowmya Priya et al. Comparison of curve correction procedure of current and voltage as per IEC 60891for thin film technology, 42nd IEEE PVSC

[6].Sharma V. et al. Degradation analysis of a-Si, HIT and Multi-Crystalline-Si solar photovoltaic modules under outdoor conditions, Energy 72, 536-546, 2014.

[7].International Electrotechnical Commission. (2011). IEC 61853-1, Photovoltaic (PV) module performance testing and energy rating.

STUDYING THE EFFECT OF SPECTRAL DISTRIBUTION WITH SEASONAL AND IRRADIANCE VARIATIONS

Imon Kalyan Barua*[1], Birinchi Bora[2], O.S. Sastry*[2], Rashmi Singh[2], Supriya Rai[2], Manander Bangar[2], Mithilesh Kumar[2], B. Prasad[1]
[1]TERI University, New Delhi – 110070
[2]National Institute of Solar Energy, Gurgaon-122003
*sastry284@gmail.com, Tel: +91-124-2579213, Telefax No. : +91-124-2579207

ABSTRACT: This study attempts to quantify the effect of spectral variation during different seasons at different intensity values. The percentage contribution of different wave length has been studied by comparing with the AM 1.5 standard spectrum. The difference in the percentage of energy contribution in two wavelength regions in terms of blue and red shift is estimated. The technology specific nature of the spectrum and quantification of its effect on PV module in different seasons are also studied here. Finally the characteristic indicator – Mismatch Factor (MMF) is calculated to quantify the spectral variation and its impact on a photovoltaic (PV) module based on CdTe technology
Keywords: Photovoltaics, Spectral variation, Solar Spectrum, PV Module, Energy Contribution, Seasonal variation, Mismatch factor.

1 INTRODUCTION:

India can be broadly categorized into six regions of different climates - Hot and Dry, Warm and Humid, Moderate, Cold and Sunny, Cold and Cloudy and Composite. Bansal and Minke, in 1988, produced the "Climatic Zones in India Map" by evaluating the mean monthly data from 233 weather stations, and then delineating the six climatic zones. On this basis, if the defined conditions prevail in a given location for more than six months then that location is allocated to one of the first five climate zones, else they are categorized as Composite if none of these categories can be identified for six months or longer [1].The focus of the present paper is to study the effect of the spectrum variation in terms of the mismatch factor (MMF) in different seasons and in different irradiances on photovoltaic (SPV) modules based on CdTe technology. This study will be useful in quantifying the spectrum variations (MMF) of a PV module based on CdTe technology. This part of the world is known to be one of the largest monsoon systems [2, 3]. The monsoon season accounts for nearly 80% of the annual rainfall in India [4]. Spectrum is affected by scattering and absorption processes in the atmosphere. Presence of clouds and aerosols attenuate certain wavelengths of solar radiation in comparison to other and hence it will be helpful to study the effect of changes in spectrum on MMF. Besides the monsoon season, this study also presents the variation of MMF in the other seasons namely – winter, summer and post-monsoon which showed very different trends. This study of spectral distribution and measurement of spectrum was conducted at the National Institute of Solar Energy (NISE), Gurgaon, located in the state of Haryana (North India, near South-West of Delhi). We have divided the climatic conditions of the given location into four seasons: winter, summer, monsoon and the post-monsoon. The winter starts in the beginning of November and continues till the end of February. Summer starts from March and continues till the end of June. The period from July to September is the monsoon and the month of October, which is the month signifying the transit period from summer to winter, is considered to be the "post-monsoon" season [5].

A wide range of literature has been reviewed which investigate the spectral effect on different technologies at different places by different spectrum-related parameters. Most of these studies were undertaken at European and Japanese locations. In the current paper, seasonal spectral variations have been performed at an Indian site, which have very different seasonal conditions compared to other sites reported in the literature. In general, the outdoor performance of a SPV module is lower compared to its performance at Standard Test Conditions (STC). This is so because normally the outdoor conditions are not the same as considered in the STC. Therefore, it is important to know the value of the parameters and their variations, which affect the performance of SPV module in actual field conditions. Spectrum is one such important parameter which changes with locations and seasons [6].

2 MATERIALS AND METHOD

MMF is a correction factor applied to the short circuit current by taking into consideration the spectral mismatch between standard AM1.5 spectrum to the actual spectrum measured under outdoor condition at the time of measurement [7].

$$MMF = \frac{\int_a^b SR_{PVdevice}(\lambda)E_{meas}(\lambda)\,d\lambda}{\int_a^b SR_{PVdevice}(\lambda)E_{ref}(\lambda)\,d\lambda} \times \frac{\int_a^b SR_{ref}(\lambda)E_{ref}(\lambda)\,d\lambda}{\int_a^b SR_{ref}(\lambda)E_{meas}(\lambda)\,d\lambda} \quad (1)$$

where $SR_{PVdevice}(\lambda)$ and $SR_{ref}(\lambda)$ are the relative spectral response of the PV module and the pyranometer (AW^{-1}), respectively, $E_{meas}(\lambda)$ is the measured relative spectral irradiance under site condition (Wm^{-2}), $E_{ref}(\lambda)$ is the reference spectral irradiance (Wm^{-2}) according to standard IEC 60904-3 [8], a and b are wavelength (λ) ranges of spectroradiometer from 400 nm to 800 nm. MMF is the factor which expresses that quantity observed by the device under more or less irradiance at current outdoor conditions compared to the reference device [7].

Table 1: Specifications of the CdTe technology module used in test bed at STC

Open Circuit Voltage	Short Circuit Current	Max. Power	Voltage at P_{max}	Current at P_{max}
V_{OC} (V)	I_{SC} (A)	P_{max} (W)	V_{mpp} (V)	I_{mpp} (A)
60.8	1.88	80	48.5	1.65

3 RESULTS AND DISCUSSIONS

The site of testing lies under a composite climate in which a wide variation in environmental conditions is observed with changing seasons. The spectrum of radiation incident on the modules varies with seasons and to study the spectral variation in different parts of the spectrum, the entire spectrum was divided into nine different wavelength bands. The peak time of clear days was chosen from different seasons to estimate the percentage change in those five wavelength bands with respect to standard AM1.5 spectrum. Figure 1, figure 2, figure 3 and figure 4 show the percentage change in energy contribution with respect to standard AM1.5 spectrum at the site during the seasons of summer, monsoon, post-monsoon and winter at different intensities of radiations of 100 Wm-2, 200 Wm-2, 400 Wm-2, 600 Wm-2, 800 Wm-2, and 1000 Wm-2 on the selected days of 21st day of each month and separated into season wise.

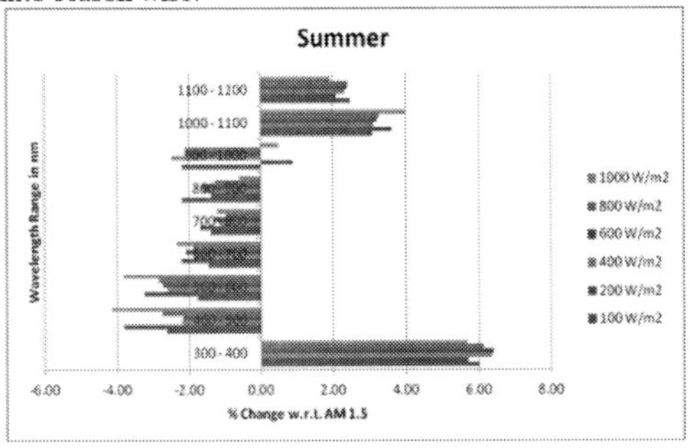

Figure 1: Percentage change in energy contribution in the six different wavelength bands with respect to standard AM1.5 spectrum at different intensities during summer

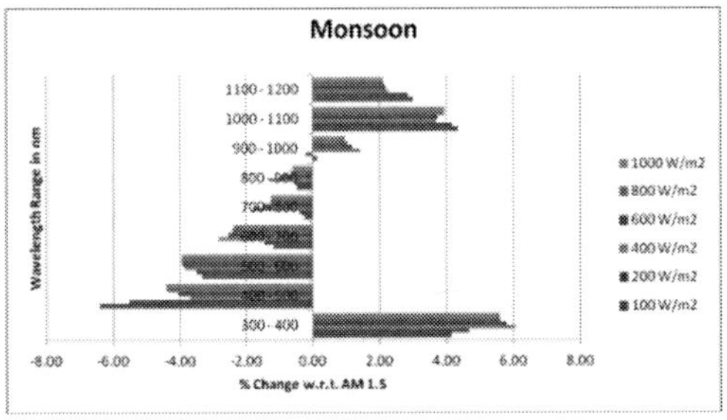

Figure 2: Percentage change in energy contribution in the nine different wavelength bands with respect to standard AM1.5 spectrum at different intensities during monsoon

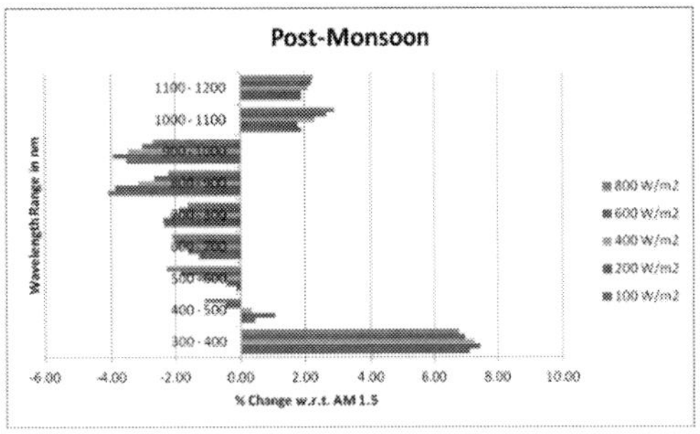

Figure 3: Percentage change in energy contribution in the six different wavelength bands with respect to standard AM1.5 spectrum at different intensities during post-monsoon

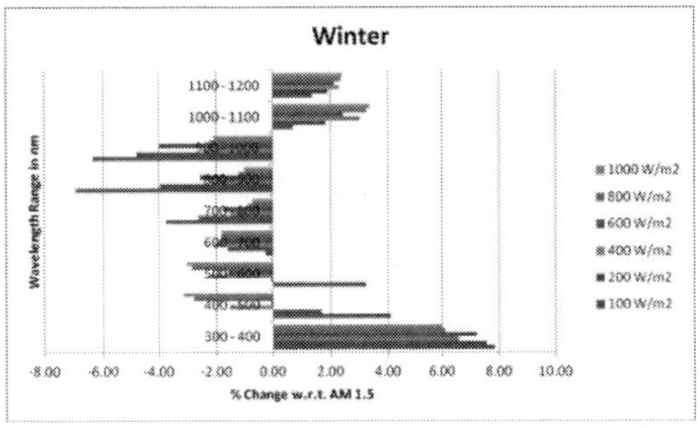

Figure 4: Percentage change in energy contribution in the six different wavelength bands with respect to standard AM1.5 spectrum at different intensities during winter

Figures 1 to 4 show that throughout the entire year, the percentage contribution of energy is discretely greater in the blue region and higher infrared region of the spectrum in comparison to the standard AM1.5 spectrum. However, in the middle region of the spectrum, we observe a drastic fall in the percentage change in energy contribution. Of all the seasons, the maximum change in the blue region is observed during the winter season and that in the infrared is observed during monsoon season. But the fall in the percentage contribution of energy with increase in wavelength till the band of 900 – 1000 nm is unanimous in all the seasons except for in monsoon where the percentage contribution begins to increase from this band of wavelength.

Since we are dealing with the CdTe technology, as the band gap of the material is 1.45 eV, it does not absorb photon having wavelength higher than 850 nm [9]. Therefore, from the perspective of CdTe technology in the range of 400 – 800 nm there is a uniform agreement in all the seasons in terms of percentage change in contribution of energy of the various wavelength bands.

The change in energy content in different wavelength range is compared with the AM 1.5 spectrum and is given in table 1 below.

Table 1: Percentage difference in Energy Contribution w.r.t. AM1.5 for different Irradiances

Shift Obs.	Seasons	Percentage Difference for Different Irradiances					
		100 Wm^{-2}	200 Wm^{-2}	400 Wm^{-2}	600 Wm^{-2}	800 Wm^{-2}	1000 Wm^{-2}
Blue	Winter	12.67	5.54	-1.33	1.54	-3.66	-4.32
	Summer	-2.11	-5.63	-2.56	-2.75	-3.64	-6.74
	Monsoon	-9.41	-8.42	-6.13	-6.72	-7.26	-7.27
	Post-Monsoon	3.90	4.30	2.26	0.53	-0.80	-
Red	Winter	- 15.71	-8.58	-1.68	-4.57	0.65	1.32
	Summer	-0.89	2.64	-0.45	-0.27	0.63	3.74
	Monsoon	6.46	5.46	3.12	3.72	4.27	4.28
	Post-Monsoon	-6.93	-7.33	-5.29	-3.55	-2.22	-

3.1 Variations in Mismatch Factor

For understanding the seasonal variations of the CdTe technology in Indian conditions, the mismatch factor (MMF) has been calculated from measured parameters at the site. Spectral mismatch depends on the difference of the measured spectral irradiance distribution (at outdoor) with respect to the standard AM1.5 spectrum. It influences the short circuit current (I_{SC}), because the spectral response of the module differs by some extent with the spectral response of the reference cell. Figure 5 shows the calculated seasonal variation of MMF for CdTe technology at different intensities. The highest MMF variation of 106.5 % was observed in winter at an irradiance of 600 Wm^{-2} whereas lowest was found to be at 102.8% during Monsoon season at an intensity of 100 Wm^{-2}.

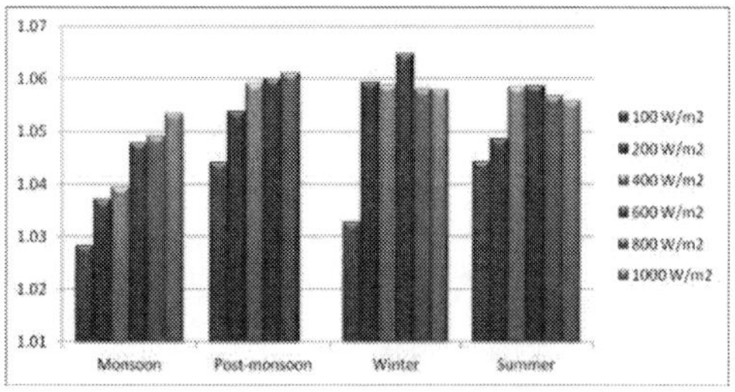

Figure 5: – Variation of MMF with seasons at different intensities of paper

4 CONCLUSIONS

The current paper presents the comparative effect of seasonal spectral variations on a PV module based on CdTe technology. The spectral effect on the above-mentioned PV technology has been quantified simultaneously in absolute value on a monthly basis and discussed qualitatively and quantitatively on a seasonal basis by MMF. It was observed that MMF variation had a near linear trend with positive slope during the season of monsoon with rise in irradiance values. During post-monsoon season a non-linear rise in the MMF values was observed till it stabilized at a value around 106%. During winter a sharp increase in the MMF was observed as intensity level rose from 100 Wm-2 to 200 Wm-2 and then it stabilized at a value slightly less than 106%. During summer, a linear rise was observed in the MMF with rise in intensity and it was observed to have stabilized at a value around 106%.

ACKNOWLEDGEMENT

This research is based upon work supported in part by the Solar Energy Research Institute for India and the U.S. (SERIIUS) funded jointly by the U.S. Department of Energy subcontract DE AC36-08G028308 (Office of Science, Office of Basic Energy Sciences, and Energy Efficiency and Renewable Energy, Solar Energy Technology Program, with support from the Office of International Affairs) and the Government of India subcontract IUSSTF/JCERDC-SERIIUS/2012 dated 22nd Nov. 2012

References

[1] "CLEAR Comfortable Low Energy Architechture," GNU Free Documentation License (GFDL), [Online]. Available:http://www.new-learn.info/packages/clear/thermal/climate/diversity/india/index.html [Accessed 28 04 2016].

[2] R. R. Kelkar, "Monsoons elsewhere in the world," Monsoon Monograph, vol. 1, pp. 96 - 119, 2012.

[3] V. Prasanna, "Impact of monsoon rainfall on the total food grain," J. Earth Syst. Sci., vol. 123, no. 5, pp. 1129 - 1145, 2014.

[4] B. Parthasarathy, A. A. Munot and D. R. Kothawale, "All India monthly and seasonal rainfall series: 1871–1993," Theor. Appl. Climatol, vol. 49, pp. 217 - 224, 1995.

[5] O. S. Sastry, S. Saurabh, S. K. Shil and P. C. Pant, "Performance analysis of field exposed single crystalline silicon modules," Solar Energy Materials & Solar Cells, no. 94, p. 1463 – 1468, 2010.

[6] L. J. Gray and e. al, "Solar influence on climate," Rev. Geophys., vol. 48, no. 4, pp. 1 - 53, 2010.

[7] D. Dirnberger, G. Blackburn, B. Müller and C. Reise, "On the impact of solar spectral irradiance on the yield of different PV technologies," Sol. Energy Mater. Sol. Cells, vol. 132, pp. 431 - 442, 2015.

[8] I. E. Commission, "Photovoltaic devices — Part 3: measurement principles for terrestrial photovoltaic (PV) solar devices with reference spectral irradiance data.IEC 60904-3 Edition 2.0," International Electrotechnical Commission, Geneva, 2008.

[9] C. S. Solanki, Solar Photovoltaics - Fundamentals, Technologies and Applications, Delhi: PHI Learning Private Limited, 2013.

INTERCOMPARISON OF PTB AND ESTI SPECTRORADIOMETERS USING SIMULATED AND NATURAL SUNLIGHT.

Ingo Kröger[1], Roberto Galleano[2], Fabian Plag[1], Harald Müllejans[2], Stefan Winter[1]
[1]Physikalisch-Technische Bundesanstalt (PTB), Bundesallee 100, 38116 Braunschweig, Germany
[2] European Solar Test Installation (ESTI), Joint Research Center, Via Fermi 2749, 21027 Ispra, Italy
Tel.: +49 531 592 4147, fax: +49 531 592 694147, e-mail: ingo.kroeger@ptb.de

ABSTRACT: JRC ESTI and PTB performed a comparison of spectral irradiance measurements on various light sources at the JRC premises. The measured sources included four standard lamps (two calibrated by PTB and two calibrated by NPL), global natural sunlight at clear sky conditions, a steady-state xenon-halogen solar simulator for solar cell calibration (Wacom) and a steady-state xenon large area solar simulator (Apollo). The PTB system is composed of three compact array spectroradiometers (CAS) equipped with Si, InGaAs, and extended range InGaAs detectors, respectively, in order to cover the wavelength range from 250 to 2150 nm. The JRC-ESTI rotating grating spectroradiometer is composed by a single stage monochromator and a Si-PbS sandwich detector in order to cover a wavelength range from 300 to 2500 nm. All instruments were calibrated in house by the respective owner institution using respective procedures and traceability chains. The measured spectra will be compared and deviations will be discussed with respect to the given uncertainties. Finally, exemplary spectral mismatch calculations will be shown in order to evaluate the impact of the deviations on PV device calibration.
Keywords: spectroradiometer, solar simulator, intercomparison, calibration, characterization, spectral responsivity

1 INTRODUCTION

For calibration and testing of photovoltaic devices solar simulators are generally used. The international standard IEC 60891 [1] defines the standard test conditions (STC) that have to be maintained during calibration: irradiance E = 1000 W/m², AM1.5G solar reference spectrum and the device temperature of 25 °C. Whenever the spectral responsivity of the device under test (DUT) differs (significantly) from the spectral responsivity of the reference device a spectral mismatch correction [2] is crucial for correcting the measured I_{STC} of the DUT. The spectral mismatch factor f_{MM} can be calculated from the spectral irradiance of the solar simulator $E_{Sim}(\lambda)$, the spectral irradiance of the AM1.5G standard spectrum $E_{AM1.5G}(\lambda)$ and the relative spectral responsivity of the reference device $s_{Ref}(\lambda)$ and the DUT $s_{DUT}(\lambda)$:

$$f_{MM} = \frac{\int E_{AM1.5G}(\lambda)\cdot s_{DUT}(\lambda)\mathrm{d}\lambda}{\int E_{Sim}(\lambda)\cdot s_{DUT}(\lambda)\mathrm{d}\lambda} \cdot \frac{\int E_{Sim}(\lambda)\cdot s_{Ref}(\lambda)\mathrm{d}\lambda}{\int E_{AM1.5G}(\lambda)\cdot s_{Ref}(\lambda)\mathrm{d}\lambda}$$

In order to apply this correction, the relative spectral irradiance of the solar simulator has to be measured using calibrated spectroradiometers. The uncertainty of the spectral irradiance measurement propagates directly into the spectral mismatch correction factor and hence into the calibration uncertainty of the DUT. For large spectral mismatch corrections (i.e. when the spectral responsivity of reference device and DUT differ significantly) this uncertainty contribution might become dominant for the calibration. Intercomparison of spectral irradiance measurements is of valuable interest in order to validate such measurements and their corresponding uncertainties.

The European Solar Test Installation of the Joint Research Center (JRC ESTI) and Physikalisch-Technische Bundesanstalt (PTB) performed a comparison of spectral irradiance measurements on various sources at the JRC premises. The measured sources included standard lamps, natural sunlight under clear sky conditions and steady-state solar simulators.

2 INSTRUMENTATION

The spectroradiometers involved (see Figure 1) were a fixed grating compact array spectroradiometers (CAS), also known as polychromator, owned by PTB (CAS), and a rotating grating spectroradiometer owned by JRC ESTI (OL750).

Figure 1: Two different spectroradiometers were compared: the OL750 of ESTI (top) and a compact array spectroradiometer system (bottom).

All instruments were calibrated in-house by the respective owner institution using respective procedures and calibration chains. Radiometric calibrations were performed using standard lamps as transfer standards for spectral irradiance. The calibration of the wavelength scale was performed using pencil style spectral calibration lamps with characteristic well-defined atomic

emission (JRC ESTI) or a tunable laser system and a Fourier-Transform Spectroradiometer (PTB) [3].
The PTB CAS system is composed of three spectroradiometers equipped with Si, InGaAs, and extended range InGaAs detectors, respectively, in order to cover the wavelength range from 250 nm to 2150 nm. The spectral bandwidth of the instrument is 3.7 nm (Si, 220 nm – 1020 nm), 9 nm (InGaAs, 780 nm – 1650 nm) and 15 nm (Ext-InGaAs, 1500 nm – 2150 nm). The JRC-ESTI rotating grating spectroradiometer is composed by a single stage monochromator and a Si-PbS sandwich detector in order to cover the wavelength range from 300 nm to 2500 nm. The spectral bandwidth of the instrument is 5 nm (Si, 300 nm – 1098 nm) and 10 nm (PbS, 1100 nm – 2500 nm).

3 EXPERIMENTAL

In the following section the comparison of spectral irradiance measurements of four different light sources are shown. The deviations will be identified as discussed.

3.1 Standard lamp

The spectral irradiance measurement of a standard lamp was performed using the JRC ESTI calibration facilities and a PTB traceable standard lamp (Acton 1000W tungsten-halogen lamp). The optical bench for alignment of lamp and entrance optics is hosted in a black painted room and the walls are at least 1.5 m away from the calibration set-up. Moreover, a baffle system composed by three thermally isolated plates and water cooled to 23 °C is used to reduce room stray light and reflection from reaching the input optic of the spectroradiometer under calibration. The calibration and working distance of the standard lamp is 70 cm and it is reproduced by interposing a calibrated 70 cm long standard gauge between the reference planes of the lamp and the DUT entrance optics.

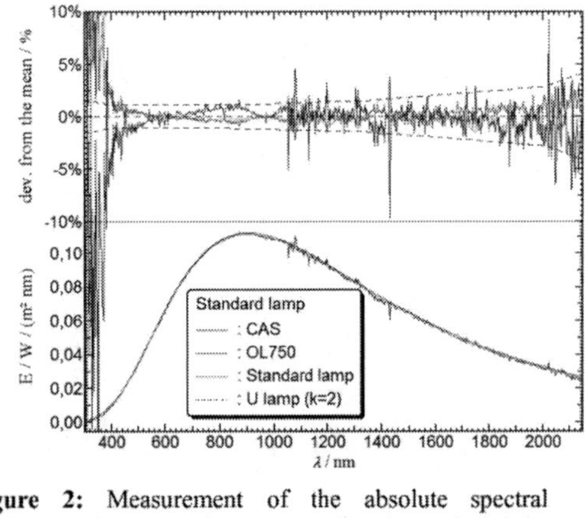

Figure 2: Measurement of the absolute spectral irradiance of PTB standard lamp using the CAS and the OL750 (bottom). The deviation from the mean including the calibration values of the standard lamp is shown in the upper graph. The dotted line denotes the expanded uncertainty of the standard lamp.

Figure 2 summarizes the results obtained during the standard lamp comparison exercise. The figure reports in the lower plot the lamp calibration value and the two spectra as measured by the CAS and OL750 spectroradiometers, and, in the upper plot, the wavelength-by-wavelength percent deviation values of the two measured spectra from the calibration value of the standard lamp. The upper plot of each figure also shows (as dashed lines) the expanded uncertainty of lamp calibration, as given by the respective calibration certificates. Most of the deviations in the wavelength range from 400 to 2100 nm provide consistency within the measurement uncertainty of the measured standard lamp. Only few points in this wavelength interval lie outside the uncertainty band limits and are clearly due to local measurement artefacts or noise. Deviation values outside uncertainty band limits are clearly visible analyzing the 300 to 400 nm wavelength band. This is probably due to the low irradiance emission from the standard lamps at these wavelengths, with a consequent worse signal-to-noise ratio or due to different internal spectral straylight behavior of the two instruments

3.2 Natural sunlight

The spectral irradiance of global natural sunlight at clear sky conditions was measured on 15th of April 2015, by the two instruments PTB CAS and ESTI OL750 in parallel, both mounted on solar tracker pointing directly at the sun.

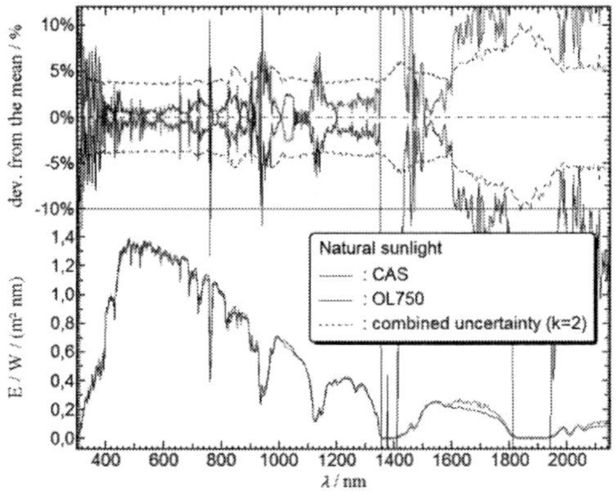

Figure 3: Measurement of the absolute spectral irradiance of natural sunlight using the CAS and the OL750. The deviation from the mean is shown in the upper graph. The dotted line denotes the combined expanded uncertainty of both instruments.

The results are shown in Figure 3. In the lower plot the measured absolute spectral irradiance is shown. The OL750 spectra were corrected in the wavelength regime between 832 nm – 1048 nm by a factor of 0.96 because of a well-known systematic instrumentation offset. In the upper plot the deviation from the mean value as well as the combined expanded uncertainty $U_{combined} = sqrt(U^2_{CAS} + U^2_{OL750})$ is shown. Below 400 nm there are significant deviations between the spectra, the CAS giving higher values than the OL750. This seems, at least partly, related to the calibration of the two instruments (see previous section). The agreement above 400 nm is much better. Below approximately 1000 nm the OL750 is higher than the CAS. At even longer wavelengths (> 1050 nm) the situation is reversed. Large deviations are visible for the absorption bands 1350 nm -1400 nm and 1800 nm – 2000 nm due to the low signal-to-noise ratio and in general above 1600 nm probably due to possible non-linearity of the Ext-InGaAs CAS. From comparing the positions of strong absorption lines, an agreement of the wavelengths scale of < 0.5 nm can be stated.

3.3 Steady state solar simulator for solar cells (Wacom)

The spectral irradiance of the Wacom was measured *subsequently* by the two instruments. The entrance optics was carefully adjusted at the same position in the light field, in order to compare the *absolute* spectral irradiance. The Wacom solar simulator was prior successfully checked for long term stability. The OL750 spectra were corrected in the wavelength regime between 832 nm – 1048 nm by a factor of 0.96 because of a well-known systematic instrumentation offset.

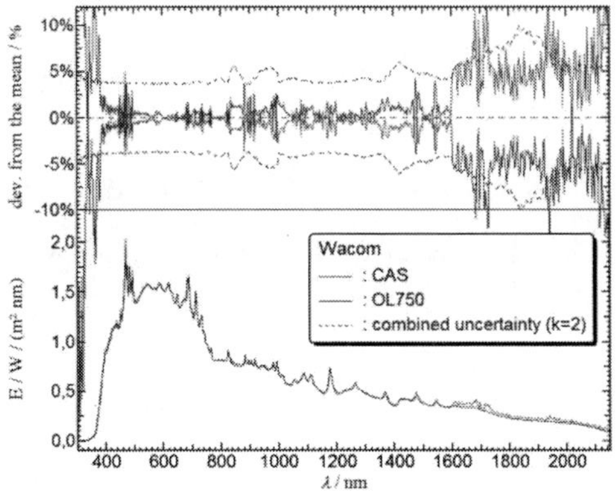

Figure 4: Measurement of the absolute spectral irradiance of the Wacom solar simulator using the CAS and the OL750. The deviation from the mean is shown in the upper graph. The dotted line denotes the combined expanded uncertainty of both instruments.

Figure 4 shows the comparison results similar to the measurements on natural sunlight. Deviations for wavelengths below 400nm seem to be consistent with difference due to instrument calibration. The agreement up to 1600 nm is good, whereas above systematic strong variations are seen as before.

The higher resolution of the PTB CAS is again evident, for example in the peak region around 450 nm. This difference in spectral bandwidth of course leads to spikes in the deviation plots wherever strong spectral lines occur. From comparing the positions of strong spectral lines, an agreement of the wavelengths scale of < 0.5 nm can again be stated.

3.4 Solar simulator for modules (Apollo)

The spectral irradiance of the Apollo was measured *in parallel* by the two instruments. The entrance optics were placed at different positions in the large light field (2 m by 2 m). Hence only the *relative* spectral irradiance can be compared.

Figure 5 shows the comparison results that are slightly different to the observations from natural sunlight and Wacom solar simulator. Generally the deviations are much larger but still within the combined uncertainties. The strongest deviations are located in the spectral regions of the strong spectral lines and are hence related to the different spectral bandwidth of the two instruments. From comparing the positions of strong spectral lines, an agreement of the wavelengths scale of <0.5nm can be stated another time.

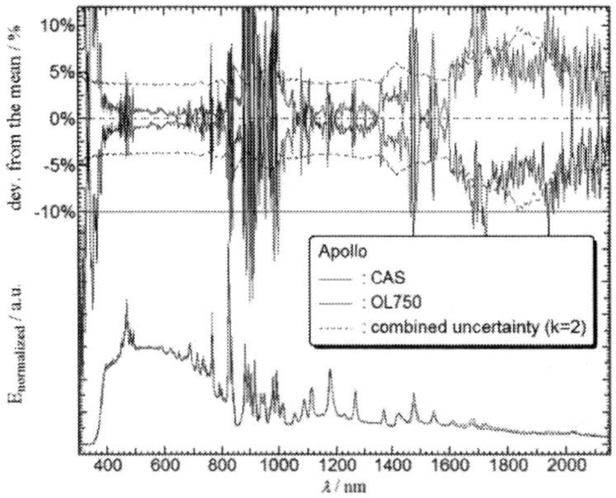

Figure 5: Measurement of the absolute spectral irradiance of the Apollo solar simulator using the CAS and the OL750. The deviation from the mean is shown in the upper graph. The dotted line denotes the combined expanded uncertainty of both instruments.

The stronger deviation can be related to a small spectral inhomogeneity of the solar simulator since it consists of 11 individually adjustable light sources with probably slightly different spectral irradiance that contribute differently to the individual spectrometer positions in the light field.

4 RESULTS AND DISCUSSION

In summary the shown comparison demonstrates generally good agreement of both instruments. Major differences are mostly well within the combined uncertainties. The agreement of the wavelength scale is better than 0.5 nm in the whole spectral range. Stronger deviations could be related to:

- A significant deviation in the UV-region from 300 nm to 400 nm that is (at least partly) related to calibration and traceability or different internal straylight behaviour
- The different spectral bandwidth that lead to deviation at narrow spectral features
- The Ext-InGaAs CAS that suffers probably from non-linearity

However, even if the agreement appears to be very good, the question arises, what would be the impact on typical spectral mismatch factors if the spectral irradiance of the light source would be measured by the two different instruments?

In the following section we performed exemplary spectral mismatch factor (f_{MM}) calculations taking the spectra from Figure 3, Figure 4 and Figure 5 and the exemplary spectral responsivities of different PV technologies from Figure 6. The reference device was chosen to be a bare crystalline silicon solar cell. For the devices under test exemplary spectral responsivities of c-Si, a-Si, CdTe and GaAs are chosen. The results of these spectral mismatch correction factor calculations are shown in Table 1.

Table 1: Calculated spectral mismatch correction factors (f_{MM}) for a c-Si reference and different devices under test (DUT) c-Si, a-Si, CdTe, GaAs using the spectra from CAS and OL750

		Ref.: c-Si DUT: c-Si	Δ / %	Ref.: c-Si DUT: a-Si	Δ / %	Ref.: c-Si DUT: CdTe	Δ / %	Ref.: c-Si DUT: GaAs	Δ / %
Natural sunlight	OL750 (ESTI)	1.0036	-0.03 %	0.9712	-0.24 %	0.9994	-0.56 %	0.9992	-0.71 %
	CAS (PTB)	1.0033		0.9688		0.9938		0.9921	
Apollo	OL750 (ESTI)	1.0092	-0.07 %	1.0184	2.39 %	1.0174	1.23 %	1.0004	0.78 %
	CAS (PTB)	1.0085		1.0423		1.0297		1.0082	
Wacom	OL750 (ESTI)	1.0126	-0.05 %	1.0001	0.49 %	0.9972	-0.05 %	0.9733	0.20 %
	CAS (PTB)	1.0121		1.0050		0.9967		0.9753	

From these results several conclusions can be drawn. The lowest deviation of the MMF is of course the case of Ref. = c-Si and DUT = c-Si, since the spectral responsivities of reference and DUT are similar. However, the deviations for this "best case scenario" show that uncertainties related to spectral mismatch should not be neglected for calibration of silicon based photovoltaic devices.

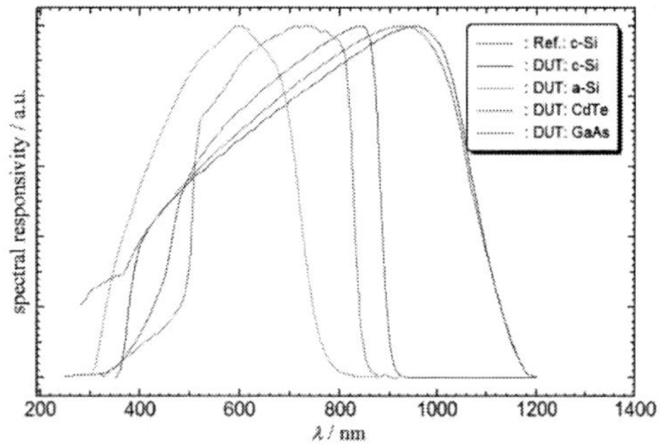

Figure 6: Relative spectral responsivity curves that were used for the exemplary spectral mismatch calculations. The red curve was taken for the reference.

Low deviation of the f_{MM} is observed for the Wacom solar simulator and natural sunlight with all devices. In this worst case scenario where spectral responsivity of reference and DUT differ significantly high deviations up to 0.71% can be expected that are only related to the measurement of the spectral irradiance of the light source. Please note that the deviation of 0.71% occurs at natural sunlight that normally can be expected to be more similar to the AM1.5G spectrum than a solar simulator. These large differences can be significantly reduced if spectrally matched reference devices are used.

Significant deviation of the f_{MM} is observed for the Apollo solar simulator for large differences in SR between reference and DUT. These differences are likely due to the fact that between 400-700 nm the CAS is systematically higher than the OL750 and between 700-1100 nm systematically lower. Therefore, in the extreme case (a-Si vs c-Si) the deviation in f_{MM} is largest, and then diminishes as the SR of the DUT becomes better matched to the reference. The differences in the measured spectra might be originated by the spectral inhomogeneity of the light beam, since it is the result of the superimposition of 11-lamp light beams, each focused on the target plane. The spectroradiometers were located at different positions in the target plane during the simultaneous spectral measurements.

5 SUMMARY

A spectroradiometer comparison of PTB and JRC ESTI was performed using different light sources at the JRC premises. Generally, a good agreement of the wavelength scale and the absolute spectral irradiance measurements within the combined uncertainties was found. For wavelengths below 400 nm there are significant deviations related (at least partly) to calibration and traceability. The ExtInGaAs-CAS (above 1600 nm) of PTB shows systematically higher measured values, which might be related to its inherent instability. The different spectral bandwidth leads to deviations at narrow spectral features. However, this could be resolved by mathematical methods.
Finally, exemplary spectral mismatch calculations were performed in order to evaluate the impact of the spectral irradiance measurements on solar cell calibration. Deviations of up to 2.39 % were reported for the spectral correction to STC.

6 ACKNOWLEDGEMENTS

The research work leading to this article was partly carried out within the EMRP ENG55 project "Photoclass". The EMRP is jointly funded by the EMRP participating countries within EURAMET and the European Union. We thank D. Pavanello and W. Zaaiman for help with the measurement on the solar simulators and the outdoor facilities.

7 REFERENCES

[1] IEC 60891 Edition 2, Procedures for temperature and irradiance corrections to measured I-V characteristics, Geneva (2009)

[2] IEC 60904-7:2008, Computation of the spectral mismatch correction for measurements of photovoltaic devices

[3] I. Kröger et al, Proceedings. 29th European Photovoltaic Solar Energy Conf. (2014), 3447ff

LOOKING AT THE ANNUAL YIELD FROM VARIOUS ANGLES: OPTICAL MODEL VERIFICATION FOR STRUCTURED GLASS.

L.H. Slooff[1], A.J. Carr[1], P.M. Sommeling[1], R. van de Voort[2],
[1]Energy research Centre of the Netherlands, P.O.Box 1, 1755 ZG Petten, The Netherlands, tel: +31 224 564314, fax: +31 224 56 8214, email: slooff@ecn.nl
[2]SCX Solar B.V., Zaanstraat 2, 5712 SN Someren, The Netherlands
Phone: +31 88 515 4314, Fax: +31 88 515 8214, E-mail: slooff@ecn.nl

ABSTRACT: In software for yearly yield calculations, normally an empirical formula is used to take the angular dependence into account. This has been shown to work nicely for standard modules. Recently light management by structures on or in the cells or module has become increasingly important. They are used to increase the path length of the light inside the module or cell, thereby increasing the response in those parts of the spectrum where the light is weakly absorbed. The response of these modules as a function of the angle of incidence is however different compared to modules without these structures. As a result, the current yearly yield calculation methods do not work for these modules. In order to derive trustworthy yearly yield estimates, the models need to be adapted. The purpose of the work is to provide a model to determine the angular dependence of output from PV modules with light management structures in the module glass. These models provide both the means to optimize light management structures and to predict performance through yearly yield calculations.
Keywords: Module, Modelling, Simulation Yearly Yield

1 INTRODUCTION

The power output of Photovoltaic (PV) modules is normally given as the power output for 1000 W/m^2 AM1.5 perpendicular illumination. In reality, a module will almost never face these conditions. The actual yield will be strongly dependent on the location and the orientation of the module. To derive an estimate for the yearly yield at a specific location and for a specific orientation, yearly yield calculation tools exist.[1,2] These tools are commonly used for flat glass modules, but the angular dependence in these tools needs to be adapted for modules containing light management structures.[3,4,5] Here we provide a model that is able to derive the angular dependence of modules with light management structures and we present a verification of the model for flat glass as well as two different types of structured glass. The verification has been made both with controlled STC measurements, and an in- an outdoor monitoring campaign. These models have then been used to derive incident angle modifiers which have in turn been applied to annual yield calculations using PVsyst for various location and orientations, including façade installations.

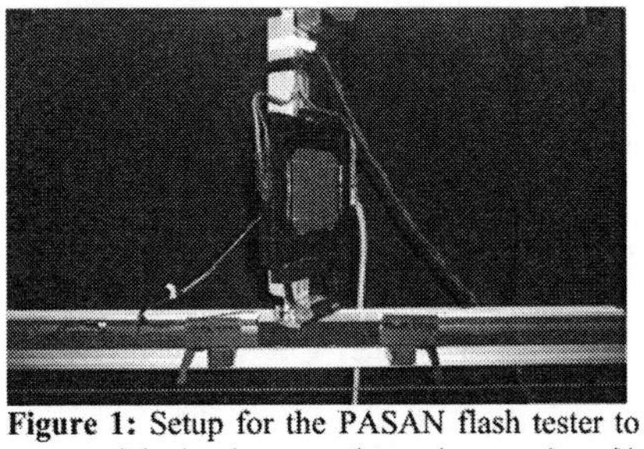

Figure 1: Setup for the PASAN flash tester to measure output of the laminates under various angles of incidence

2 APPROACH

A ray-tracing model is used to derive the response of a PV module as a function of the angle of incidence. It is especially designed to implement various light management structures in the module glass. The output is the External Quantum Efficiency for the specific modules which is converted into a current density using a specific illumination spectrum. The EQE of modules with light management structures will depend on the angle of incidence of the incoming light. This angle can be changed in the model to obtain the angular dependence of the module.

To verify the model several single cell laminates were manufactured and then measured in a PASAN flash tester under different angles. For the measurements a special setup was used to be able to change the angle of incidence, see Figures 1 and 2.

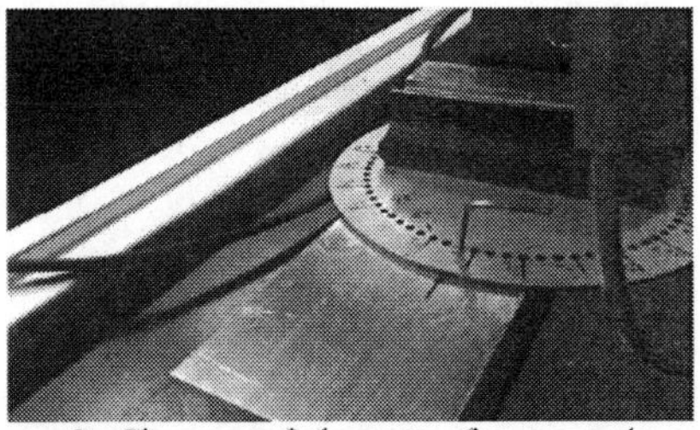

Figure 2: Close-up of the setup for measuring under various angles of incidence

A set of these laminates has also been installed outdoors, see Fig. 3., providing further verification of the modelled results. The model results are then used to estimate the influence of orientation and location on the yearly yield.

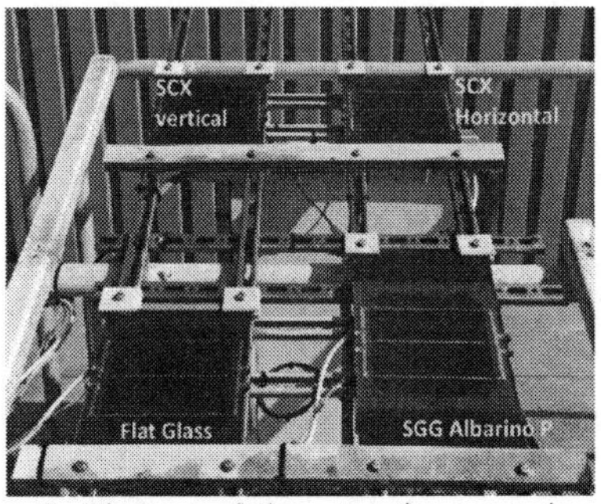

Figure 3: Close-up of the setup for measuring under various angles of incidence

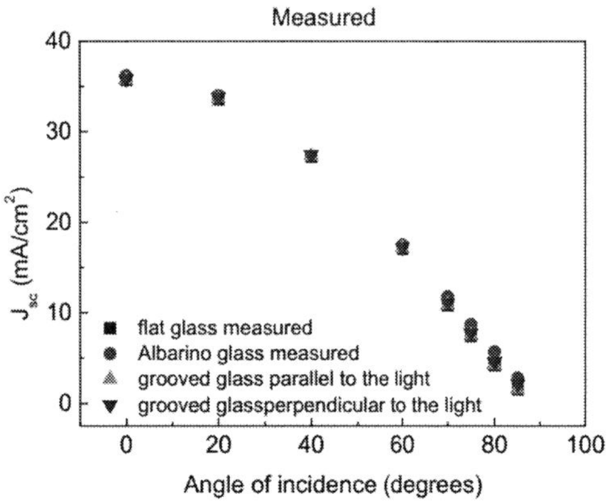

Figure 4: Measured short circuit current density of single cell laminates

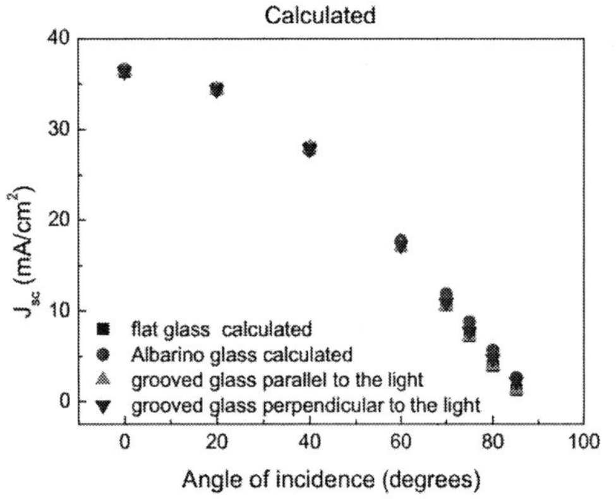

Figure 5: Calculated short circuit current density of single cell laminates

3 RESULTS

Three different glass types were measured, flat glass, random pyramid structured glass (Albarino P from Saint-Gobain), and SCX Sologlass a grooved glass from SCX Solar B.V. Per glass type there were several laminates to check reproducibility of the measurement. Figure 4 shows the results of the measurements and Figure 5 the calculation results.

The results show a very good agreement between the model and the measured data (flash tester). The flat glass shows the lowest performance, almost similar to the grooved glass with the grooves parallel to the incoming light. Perpendicular to the incoming light the grooved glass performs slightly better and the Albarino P glass outperforms all three.

Outdoor measurements also showed good correlation with the modelled results. Those results will be published in a future paper.

From the angular dependence of the current density, the incidence angle modifiers (IAM)[6,7] were calculated, i.e. the ratio between the current for a specific angle of incidence and the current at normal incidence, see Fig. 6. The IAM was subsequently implemented in PVsyst. Annual yield calculations were performed for 3 different system designs, see fig. 7, in two locations, Amsterdam (location 52 21' N) and Doha (location 25 15' N). Figure 8 shows the sun altitude and azimuth during two different days, June 21 and December 21, for both locations. Clearly the sun is much higher in the sky in Doha compared to Amsterdam. The results of the PVsyst calculation are shown in Figs 8 and 9 respectively. It shows that both the Albarino P and the SCX glass give a gain with respect to flat glass. Albarino P gives the highest gain, between 0.92% and 2.38% depending on the location and orientation. The SCX glass shows a gain between 0.29% and 0.87%. The results also show that the high angle of incidence benefits of these structures are maximized in a façade type application where the incoming light is dominated by higher angles of incidence.

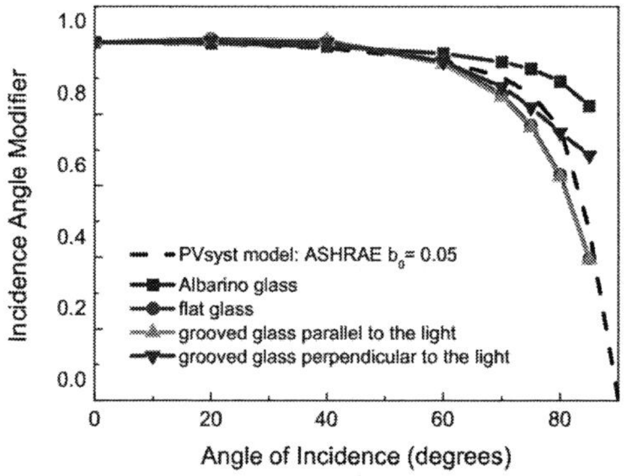

Figure 6: Incident Angle Modifiers for use of annual yield calculations. These are calculated based on the modelled results.

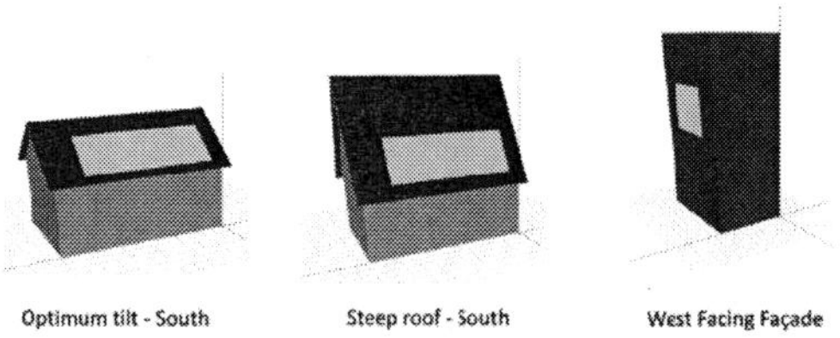

Figure 7: Three installations simulated in annual yield calculations. The annual yield calculations have been performed for a 5kW system with the shown orientation and at two locations.

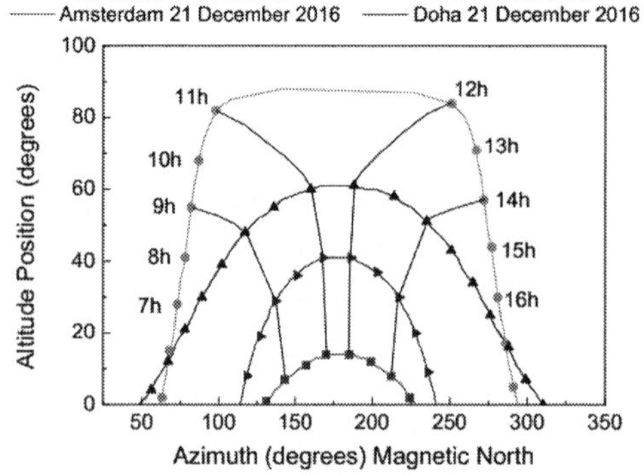

Figure 8: Sun altitude and azimuth during the day for Amsterdam and Doha on June 21 and December 21.

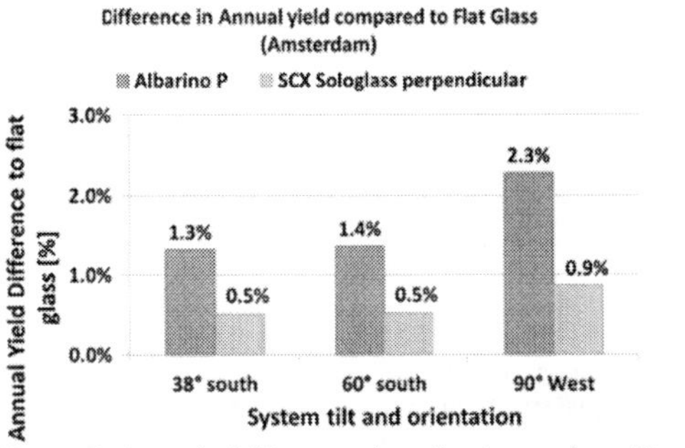

Figure 8: Annual yield comparison for Amsterdam. The results for the two glass types are relative to the flat glass.

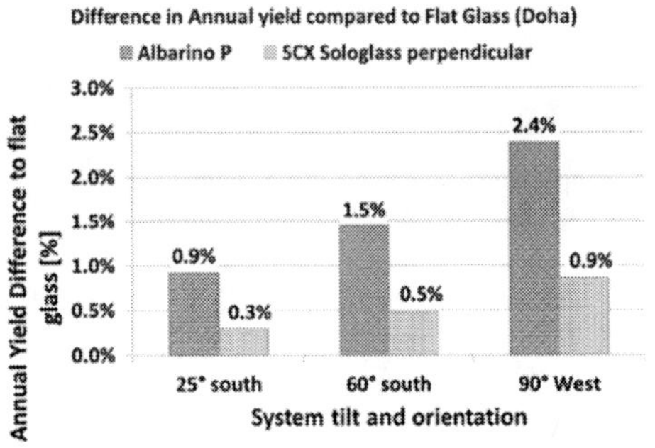

Figure 9: Annual yield comparison for Doha. The results for the two glass types are relative to the flat glass.

5 CONCLUSIONS

Single cell laminates with different types of structured glass have been made and measurements under indoor and outdoor conditions have been performed. A ray-tracing model has been made to predict the performance of the laminates as a function of the angle of incidence. The results correlate very well with the measured data. From the model the incidence angle modifier was derived and used in the PVsyst software to model annual yield for different locations and orientations. The result show that Albarino P structured glass performs best, but that the grooved glass from SCX also shows an improvement in output compared to flat glass. The effect of structured glass is highest for façade like applications where the angle of incidence in general is rather high.

8 ACKNOWLEDGMENTS

This research was funded by the Ministry of Economic Affairs of the Netherlands

9 REFERENCES

[1] http://photovoltaic-software.com/professional.php
[2] A. Mermoud, T. Lejeune, 25th EU-PVSEC (2010)
[3] M. Duell, M. Ebert, M. Muller, B. Li, M. Koch, T. Christian, R. F. Perdichizzi, B. Marion, S. Kurtz, D. M. J. Doble, 25th EU-PVSEC (2010) 3867
[4] http://www.sg-solar-glass.com/fileadmin/user_upload/broschueren/2-S_Albarino_ST.pdf
[5] P. Sánchez-Friera, D. Montiel, J.F. Gil, J.A. Montañez, J. Alonso, IEEE 4th World Conference on Photovoltaic Energy Conference (2006) 2156
[6] M.J. Carvalho, P. Horta, J.F. Mendes, M.C. Pereira, W.m. Carbajal, Proc of ISES World Congress I-V (2007) 608
[7] M. Ebert, H. Stacheit, I. Hädrich, U Eitner, 29th EU-PVSEC (2014) 3244

SEASONAL ANALYSIS OF MOST FREQUENT CONDITION AND ENERGY RATING OF PV MODULE TECHNOLOGIES

Birinchi Bora[1, 2], O.S. Sastry*[1], B. Prasad[2]
1National Institute of Solar Energy, Gurgaon-122003,
2TERI University, New Delhi, Delhi 110070
*sastry284@gmail.com, Tel: +91-0124-2579213, Telefax No. : +91-124-2579207

ABSTRACT: This paper describes the effect of irradiance and temperature on the performance of photovoltaic module technology. The contour plots of module temperature and irradiance for different seasons are plotted to find the most frequent condition for different seasons. The median performances of the three different PV technologies are studied here. The spectrum and angle of incidence distribution for a fixed tilt module are also studied in this paper.
Keywords: SPV Module, energy rating, MFC, contour plot, Spectrum.

1 INTRODUCTION

The study about the energy rating and reliability of PV module technologies is important to enhance the commercial point of view of using solar photovoltaic. Usually the PV module performance is rated under Standard Test Condition, which corresponds to AM 1.5 spectral distributions, irradiance level of 1000 W/m^2 and module temperature of 25 ^{0}C as per IEC 61215 [1]. The performance of PV modules in the real outdoor conditions cannot be predict accurately using the STC rating only, as in the real outdoor condition the operating conditions are very much wide. The performance evaluation of PV module under real weather condition is of great research interest to find out reliable energy estimation in the outdoor. The power output of PV module is mainly depends on the spectrum, incident angle, module temperature and intensity of radiation. Parameters are varies with the time of the day, day of year, location and season. So, it is difficult to estimate the energy rating of the PV module only from the STC. As per IEC 61853 standard, modules needs to be tested at 23 test conditions of intensity and temperature [2]. All these testing can be done in the indoor lab but we are concentrating in this paper about the testing conditions in the outdoor only. Lots of people have reported about the performance evaluation of in the outdoor conditions. Tamizhmani et al reported the performance evaluation of the different PV module technologies in 23 different test conditions using mesh [3]. A comparison study of the performance PV module under natural sun light and indoor condition has also been reported by Kenny et al. Somehow if actual outdoor conditions do not match the specific condition then translation or interpolation procedure can be used to estimate the power output from the module [4]. In case of long outdoor deployment of PV module almost all these conditions can be achieved.

This study was undertaken with the main objectives to validate the different irradiance – temperature data set of conditions in the real outdoor condition of India to estimate the long term performance of three different PV module technologies. This study consists of seasonal variation of different Irradiance-temperature set of conditions for the composite climate of India for three PV technologies and the performance evaluation in different conditions from three years performance data. The angle of incidence and spectrum distribution with respect to the intensity level are also studied in this paper. The estimation of the mostly occurred condition for four different seasons has also been studied to find out a methodology for estimating energy rating of PV module.

2 METHODS AND METHEDOLOGY

The experimental set up consists of three technology modules: amorphous silicon single junction (a-Si), Hetero junction intrinsic thin layer (HIT), and multi crystalline silicon (m-Si). The size of the a-Si power plant is 1.2 kWp, HIT is 1.68 kWp and m-Si is 1.6 kWp. The array of a-Si is consists of 4 modules in series and 4 in parallel and the nominal wattage of each a-Si module is 75 Wp. The array of HIT module technology consists of 4 modules in series and 2 modules in parallel and the nominal STC wattage of each module is 210 Wp. The array of m-Si power plant consists of 10 modules in series and 1 module in parallel and the nominal STC wattage of each module is 160 Wp. For each technology, one module is in open circuit condition and one module is connected individually to collect the I-V data. The test bed installed at a fix tilt (28°) equal to latitude of the site. The environmental parameters are logged through CR1000 data logger once in every minute where as I-V data of each PV technology is logged once in every ten minute. All the data have been stored to PC and down loaded to memory module once in every month. The three year (2010, 2011 & 2012) performance data of all the three technologies has been used for the analysis. For the seasonal analysis, all the data are divided into four seasons, viz. winter, summer, monsoon and the post monsoon or the transit period. The winter starts late in November and continues up to the beginning of March. The summer is from late March to till the end of June. The period from beginning of July to mid-September is the south west monsoon season. From mid-September to the end of November constitutes the post monsoon.

Figure 1: Experimental set up

The recorded data is analyzed by drawing contour graphs, using the method developed by Shingo Nagae, et al [5]. The advantage of the contour graphs is to understand the explicit behavior of the three technology based PV modules with respect to specific environment parameters. Following are the steps to draw the contour graphs:

1. Make two dimensional arrays for irradiance and output power. Columns are indexed by module temperature and irradiance.
2. A column matrix of module temperatures in the range from 0 - 80 with step of 5 °C is prepared.
3. A column matrix of Irradiance in the range from 0.05 to 1.1 kW/m² with a step of 0.05 is prepared.
4. Output power of the module and the irradiance added to the elements of each array.

3 RESULTS AND DISCUSSION

The energy rating of PV technologies is utmost important subject of study. The approach using most frequent condition of temperature and irradiance is one of the new method with which accuracy of estimation can be improved. For the energy estimation of different PV technologies, power rating needs to be checked out at 23 different conditions. However in real outdoor condition there are more variables like spectral variance and angles of incidence which are also needs to be taken into account for the energy rating. In this paper the seasonal variation of MFC and the variation of angle of incidence are analyzed to increase the estimation accuracy. The most frequent conditions analysis has been done for three PV technologies for composite climate of India. It has been found that in different seasons of the year the MFC is different for different technologies. In the figure 2, the irradiation as a function of irradiance and module temperature has been plotted for a-Si. It has been observed that for different seasons MFC is different for a-Si technologies in composite climate of India. In monsoon season there are two different conditions of MFC. In the table 1 the MFC for the three technologies has been given for the different seasons.

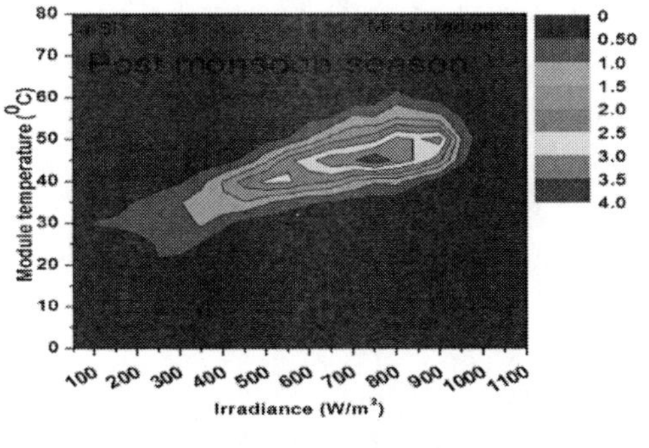

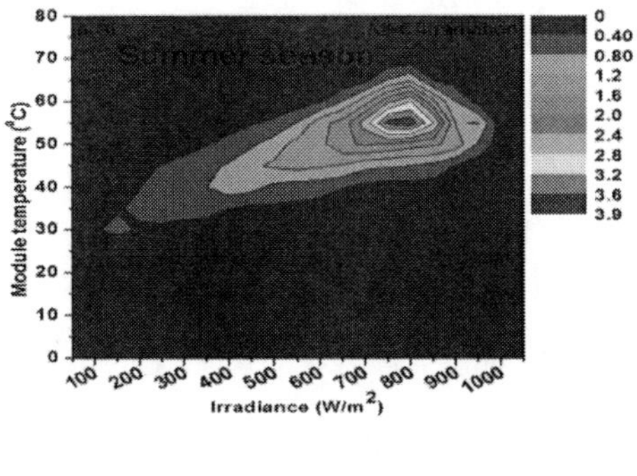

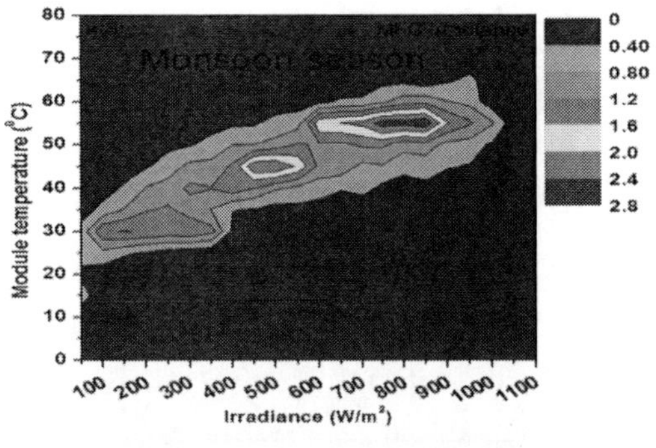

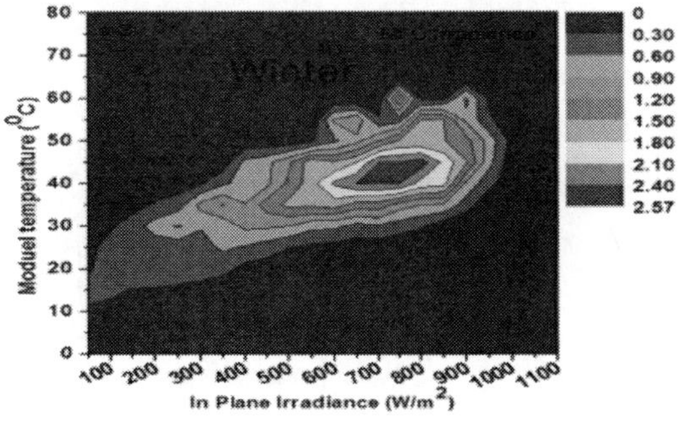

Figure 2: MFC of a-Si in different seasons

In the figure 3, Irradiation is plotted as a function of angle of incidence and irradiance for the module fixed at tilt equal to latitude. The different color pallets represent the total energy falling in plane of the module for the whole year. It has been observed that at angle of incidence (AOI) 20°, most of the radiation falls in terms of energy. The angle of incidence for the high radiation level is low. Up to 600 W/m², the highest angle of incidence is 40°. The data set designed for the energy ratings need to be incorporated this information also. Average photon energy (APE) is calculated for the tested site is calculated from the measured data of spectrum throughout a particular year. The data has been plotted in the figure 4. It has been observed that the spectrum changes with the change in the irradiance level and humidity content also. For the accurate measuring of energy rating at different irradiance level it needs to be measured the performance at different APE with different spectrum.

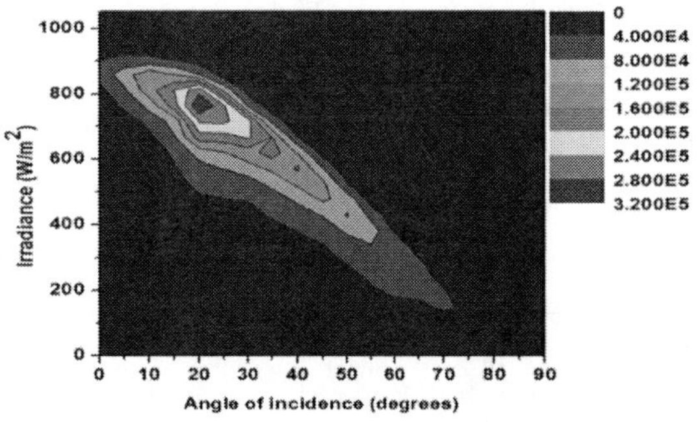

Figure 3: Irradiation as a function of angle of incidence and irradiance

Table 1 MFC for different seasons

Technology	Seasons			
	Winter	Summer	Monsoon	Post monsoon
a-Si	740 W/m^2, 42 ^{0}C	770 W/m^2, 55 ^{0}C	470 W/m^2, 45 ^{0}C; 800 W/m^2, 55 ^{0}C	740 W/m^2, 45 ^{0}C
HIT	800 W/m^2, 35 ^{0}C	800 W/m^2, 55 ^{0}C	450 W/m^2, 43 ^{0}C; 800 W/m^2, 55 ^{0}C	800 W/m^2, 50 ^{0}C
m-Si	840 W/m^2, 40^{0}C	840 W/m^2, 55^{0}C	450 W/m^2, 35^{0}C; 800 W/m^2, 55 ^{0}C	750 W/m^2, 45^{0}C

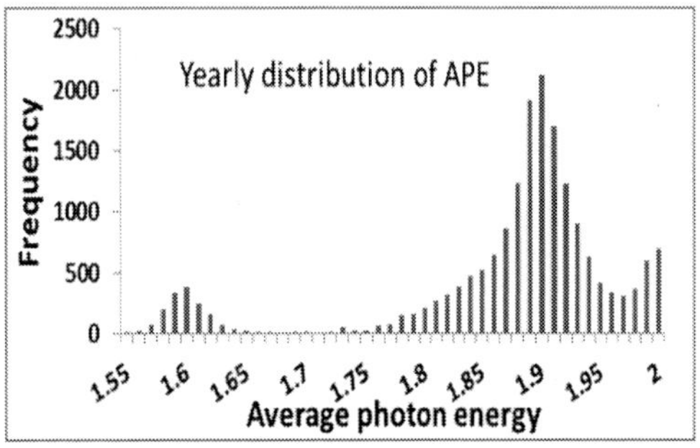

Figure 4: Yearly distribution of Average photon energy for composite climate of India

4 COCCLUSION

In this paper MFC in different seasons has been studied for three different technologies. It has been found that for different seasons the MFC are different for each PV technologies. A methodology for the estimation of energy rating using the effect of spectrum, angle of incidence, irradiance and temperature is analyzed in this paper. For each irradiance level the different range of AOI and spectrum can be incorporated for measuring the energy rating.

ACKNOWLEDGEMENT

This research is based upon work supported in part by the Solar Energy Research Institute for India and the U.S. (SERIIUS) funded jointly by the U.S. Department of Energy subcontract DE AC36-08G028308 (Office of Science, Office of Basic Energy Sciences, and Energy Efficiency and Renewable Energy, Solar Energy Technology Program, with support from the Office of International Affairs) and the Government of India subcontract IUSSTF/JCERDC-SERIIUS/2012 dated 22nd Nov. 2012

References

[1] IEC 61215

[2] IEC 61853

[3] G. TamizhMani, K. Paghasian, J. Kuitche, M. Gupta and G. Sivasubramanian: Photovoltaic Module Power Rating per IEC 61853-1 Standard – A Study Under Natural Sunlight, Solar ABCs Study, Report March 2011 (www.solarABCs.org).

[4] Power rating of photovoltaic modules including validation of procedures to implement IEC 61853-1 on solar simulators and under natural sunlight, Robert P. Kenny, Davide Viganó, Elena Salis, Giorgio Bardizza, Matthew Norton, Harald Müllejans and Willem Zaaiman, Prog. Photovolt: Res. Appl. 2013; 21:1384–1399

Nagae S, Toda M, Minemoto T, Takakura H, Hamakawa [5] Y. Evaluation the impact of solar spectrum and temperature variations on output power of Silicon based photovoltaic modules. Solar Energy Materials & Solar Cells. 2006; 90: 3568 – 75.

ANGLE RESOLVED PERFORMANCE MEAUSREMENTS ON PV GLASS AND MODULES

Line Tollund Juutilainen*[,a], Sune Thorsteinsson[a], Peter Behrensdorff Poulsen[a], Anders Thorseth[a], Mekbib Wubishet Amdemeskel[a], Stela Canulescu[a], Peter Melchior Rødder[b], Kristin Rødder[b]
[a]Department of Photonics Engineering, Technical University of Denmark, Frederiksborgvej 399, Building 130, 4000 Roskilde, Denmark, *ltju@fotonik.dtu.dk
[b]SolarLab, Gunnar Clausens vej 9, 8260 Viby J., Denmark

ABSTRACT: The angular response of PV-modules has significant impact on the energy production. This is especially pronounced in BIPV where installation angles often are far from optimal. Nevertheless, a gain in energy yield may be obtained by choosing a proper glass as superstrate. In this work we present the concept of PV balconies as a cost efficient and easy way of integrating PV into buildings. The experimental work consists of the fabrication of single cell mini modules with different glass covering, and characterizing their angular response in a custom made setup, where only the direct sunlight is used as a light source. As a special case we estimate the annual yield for each glass in the case of PV balconies for a specific geographical location and orientation of the module.
Keywords: Solar Cell, PV applications, Building Integrated PV (BIPV), Angular Performance

1 MOTIVATION

Maximizing the angular response of stationary PV-modules have a significant influence on annual energy yield since it is only a small fraction of the time the PV module is exposed to direct radiation at normal incidence [1]. Especially on BIPV applications the normal incidence can be completely absent.

Building integrated photovoltaics BIPV, is a very emerging field within the PV sector and various market surveys have predicted impressive growth rates - one of them an annual growth rate of 20 % [2]. This market is driven by 2 forces [3]:

- Energy requirements on new building and renovations projects are expressed in net energy requirements, which favor energy generating building components.
- The price of installed PV pr. square meter is approaching that of conventional building materials.

When installing PV on buildings – especially as BIPV – it is the planes on the building that determines orientation, and thus the orientation of the PV cannot be optimized with respect to energy yield, with the result that the PV during most of its operation hours is far from optimal angles.

The Technical University of Denmark in collaboration with SolarLab, is currently developing PV balcony fences. One of the main advantages of this particular BIPV application is the ease of carrying the concept of regular balconies into a PV variant, and this without a rapid increase in installation costs. Of the total price of mounting PV balcony fences the solar cell will constitute less than a one fourth. An elaboration hereof is shown in Figure 1.

PV balconies are a case where the installation angle is vertical, and this orientation results in large incident angles, with a reduction in the energy harvest.

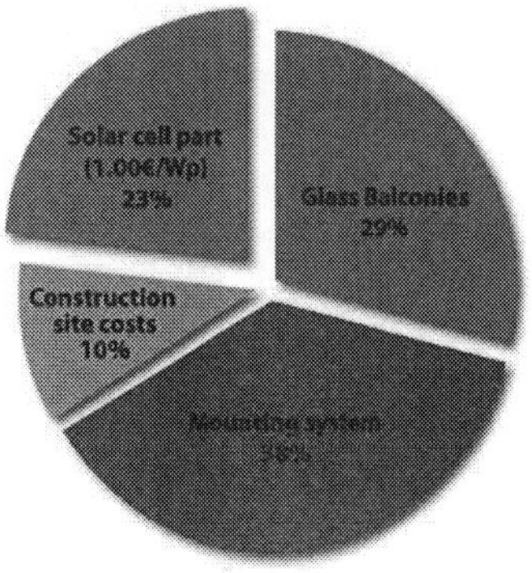

- "Glass balconies" is the market price of common fence glass for balconies.
- "Mounting system" includes all mechanical elements for attachment on the building
- "Construction site costs" includes disassembly of old balconies and construction of new ones, all inclusive.
- "Solar cell proportion" is the marginal cost for 34 kWp solar cells integrated in the fence glass, including all electrical devices needed for grid connection.

Figure 1: Cost breakdown for installed PV-balconies. The solar cell part will only constitute an increase of the total cost of 23 %.

Even though the cosine loss (projection loss) is determined by the building and can't be altered, energy gains can be found by choosing the right type of glass. Several types of antireflective coated glass (ARC-Glass), structured and deep structured are commercially available. In [4] different glass textures has been evaluated for outdoor performance. However, valid data is time consuming, and great care must be taken to ensure that only the optical losses are measured. Herman et al [5], has characterized full panels angle resolved in a sun simulator, and has previously characterized glass with solar cells at light detectors using paraffin as index matching medium between the glass and the top EVA of the PV mini module [1].

In this work, we have developed another method to characterize light coupling to solar glass based on measurements using the real sun as light source.

2 EXPERIMENTAL

2.1 Samples and preparation

5 glass samples from 3 different European suppliers were collected. Furthermore, a raw cell and a cell with EVA were prepared, i.e. without any glass cover. Their characteristics are outlined in Table I.

Table I: The different glass types used in the experiments.

Glass thickness (mm)	AR details
4	Deeply pyramidal structure
4	Low iron float glass
3.2	Matt surface + AR coating
4	Matt surface + AR coating
3.2	Prismatic surface + AR coating
-	Raw cell
-	Cell + EVA

All of the above glasses followed the exact same characterization process: All glasses were sent to a European module manufacturer, where one full cell mini modules were made. The cells were taken from the same box, so all cells came from the same binning and were most likely produced in the same batch, ensuring the cells were as similar in performance as possible. All the mini modules were flashed post lamination, and sent back to our research facility, where the angular response of the mini modules where measured using natural sunlight as described in the following section.

2.2 Angular characterization of single cell panels

An illustration and a picture of the setup used is shown in Figure 2, we named it the Solar Canon. Two tubes are directed towards the sun, letting only the direct sunlight onto the sample, which is located in a box at the end of the tube. In the box the sample is placed on a rotating sample holder, and in the other box a reference sample is placed with an orientation normal to the rays. The short circuit current is measured for the reference sample and the sample under test simultaneous together with a time stamp. The tube-lightbox assembly is mounted on axis attached to a worm gear where the tilt is adjusted manually. This shaft is mounted on a stage that can rotate around its own axis. The tubes are to prevent diffuse light from reaching the sample. The direct light from the sun differs from a perfectly collimated beam by approximately 0.5 degrees. The amount of diffuse light is determined by the diameter to length ratio of the tube, $\theta = \tan^{-1}\frac{d}{L}$, with d being the diameter of the tube and L being the length of the tube. Initially, a 5 degree angle was the target, similar to measurements of direct sunlight, and with an inner tube diameter of 18 cm the length of the tubes was close to 2 meters. However, initial experiments showed that the direction of the sun was moving too fast for this kind of manual measurement. Thus the requirement was loosened a bit.

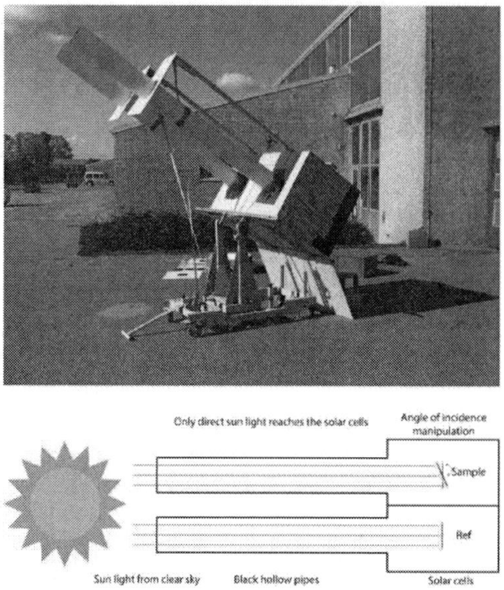

Figure 2: Illustration and picture of the Solar Canon. Two pipes leads direct sun light into boxes containing the test sample and a reference sample respectively. The test sample rotates as the angular performance is measured.

3 RESULTS

3.1 Raw data

In Figure 3 an example of the collected data is shown. The calibrated current is obtained by normalizing with respect to the current measured in the reference sample, hereby changes in the solar intensity during a measurement session is accounted for. The energy reaching the sample is the projection of the area normal to the rays, referred to as the Cosine Law. The calibrated current is expected to lie below the cosine function. The measured current values lie very close to the cosine function and even above.

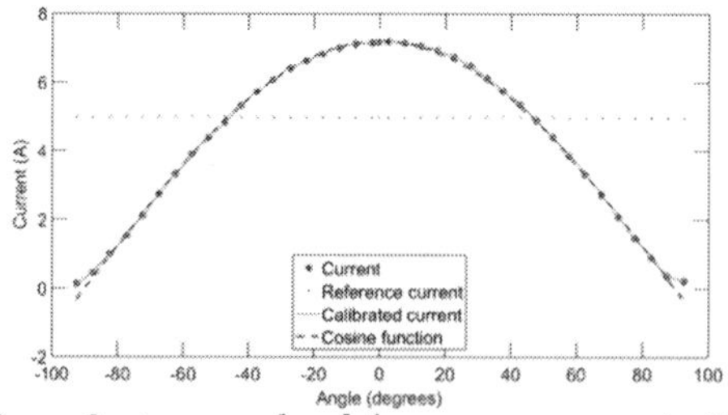

Figure 3: An example of the raw measurements. The shown case is for the sample with prismatic surface and ARC. Shown are also the calibrated current and the cosine function.

3.2 Angular response curves

The angular response is found by normalizing the calibrated current with respect to the value at normal incidence and again setting the data with respect to the Cosine Law, following the procedure by Hermann et al [5][6]. The results are shown in Figure 4.

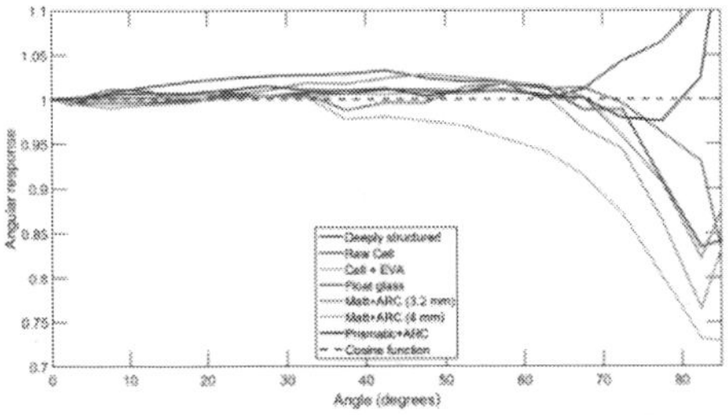

Figure 4: Angular response is obtained by normalizing with respect to normal incidence and reference to the cosine function.

Figure 4 makes it evident that the data collected gives results larger than the cosine function especially at larger angles, suggesting that more energy is transmitted onto the cell than reaching the surface. The slightly larger values than the cosine function in Figure 3, diverges into huge variances when computing the angular response. Clearly, an error is introduced. Two main issues have been identified as causing the values larger than the cosine function. Firstly, the uncertainty of the protractor adjustment has been assessed to be no more than 2 degrees, estimated through repeatability measurements. However, the ability of accuracy of the protractor becomes more and more necessary as the angle of incidence increases since the angular response curve is obtained by referencing to the cosine to the angle, hence dividing by a smaller and smaller value. An example of how the uncertainty on the angular response increases with increasing angle can be seen in Figure 5. Error bars are not applied in Figure 4 to make the data more clear.

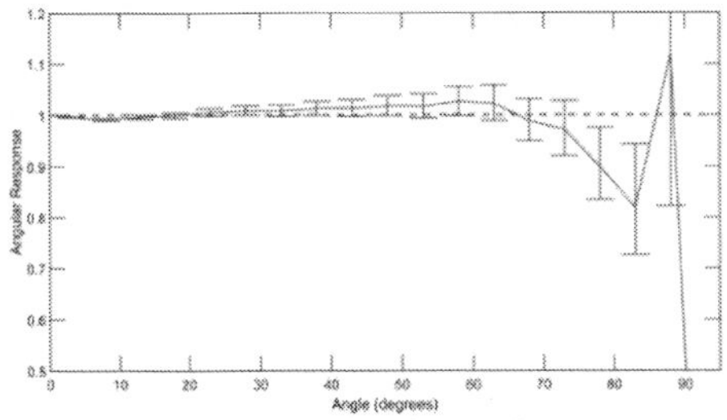

Figure 5: Errorbars applied to the angular response curve of the Matt+ARC (4 mm) sample.

Secondly, during disassembly of the protractor/sample holder part it was found that the disc where angles were read was misaligned with the protractor arm, possibly resulting in current values were measured at angles closer to normal incidence than the angles that were actually registered. This error is not present at normal incidence but becomes more evident as the angle approaches 90 degrees. Future work will aim at addressing these issues. It should be noted that in the case of highly structured glass, violation of the Cosine Law might be plausible, since the actual angle of incidence on one of the structures does not match that when viewing the surface as a whole.

3.3 Annual yield estimation

When knowing the exact location of the building where balconies are being implemented and the orientation of these, an estimation of the annual yield for the various glasses that have been investigated can be made. This allows for choosing the best suited glass giving the largest energy output for a specific case. An example of such annual energy yield estimation is shown in Figure 6. The case shown is for a south faced balcony in Roskilde, Denmark. The results are obtained with irradiance data obtained from PVgis folded with the results from the angular response curves and also considering the absolute transmission. Due to the large angular response values there is a constant overestimation of the energy yield.

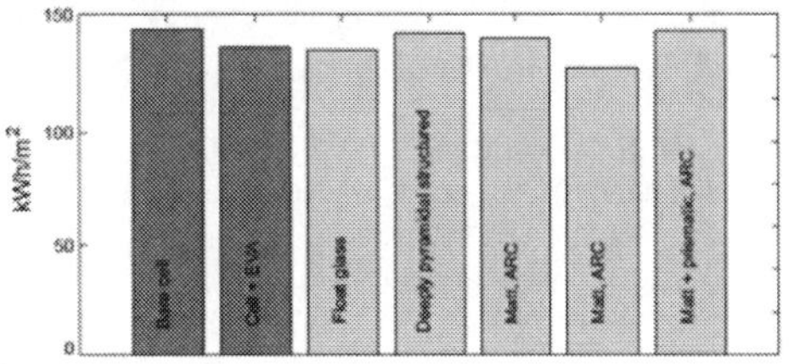

Figure 6: Estimation of annual energy yield for the different types of glasses. The highest yield is obtained when applying the prismatic structured glass with AR coating.

4 CONCLUSION AND FUTURE WORK

Within this work we have established a measurement setup for measuring the angular response on mini modules using the direct solar radiation. Initial measurement showed this principle feasible.

An analysis of the angular performance of solar glass allows for proper assessment of the cover glass to use in BIPV projects, where odd angles of incidence often is the case. An example hereof presented in this paper is PV balcony fences, which is an inexpensive and easy way to integrate PV into new and existing buildings.

We found that using the real sun as light source is realistic, since it is a rather simple procedure to align the Solar Canon with the sun and ensuring that both the reference and the test sample are entirely illuminated.

During the experiments it has become clear that the alignment of the different parts of the Solar Canon has been insufficient, e.g. the calibration of the protractor with the rotating sample resulting in angle measurements of insufficient precision. Regrettably, this error was not possible to account for and extract from the data due to additional uncertainties in the adjustment of the protractor. Moreover, the Solar Canon as it is now has proven to be insufficiently robust. As the Solar Canon is moved around the relative positions of various parts are continuously being altered. An important improvement will be the installation of an automatic and precise protractor. During the summer 2016 improvements will be implemented and new measurements will be performed. Furthermore, it is planned to additionally perform measurements with a solar simulator under more controlled conditions in the laboratory to compliment the data from the Solar Canon.

5 REFERENCES

[1] W. Herrmann, L. Rimmelspacher, and M. Reuter, "Optical characteristics of PV module front glasses," pp. 4809–4814, 1801.

[2] Transparency Market Research, "BIPV market report." [Online]. Available: http://glassmagazine.com/news-item/commercial/bipv-market-forecast-grow-187-percent-annually-through-2019-1513504.

[3] W. F. G. Verberne, P. Bonomo, F. Frontini, M.N. van den Donker, A. Chatzipanagi, K. Sinapis, "BIPV Products for Façades and Roofs: a Market Analysis," 29. Eur. Photovolt. Sol. Energy Conf. exibition, pp. 3630–3636, 2014.

[4] a. N. S. R. Williams, T. R. Betts, R. Gottschalg, D. Neumann, M. O. Prast, A. Nositschka, "Evaluating the Outdoor Performance of PV Modules with Different Glass Textures," p. 5, 2011.

[5] W. Herrmann, M. Schweiger, and L. Rimmelspacher, "Solar simulator measurement procedures for determination of the angular characteristic of PV modules," vol. 2, pp. 2–5.

[6] W. Herrmann and L. Rimmelspacher, "Angular transmittance of solar glass and impact on PV energy yield," no. October, pp. 21–22, 2014.

ENERGY RATING OF CRYSTALLINE SOLAR MODULES: INVESTIGATION OF UNCERTAINTIES DUE TO BINNING IN MASS PRODUCTION

Gerhard Kleiss[1], Holger Schülbe, Bernd Nacke
Institut für Elektroprozesstechnik (ETP), Leibniz Universität Hannover
Wilhelm-Busch-Str. 4, D-30167 Hannover, Germany

ABSTRACT: In the light of growing contribution of renewable energy (RE) to the global energy supply, the assessment of energy production of RE technologies becomes most important. In the case of electricity-related RE resources, such as photovoltaics (PV), wind, biomass, and CSP often total yearly energy production is the economic driver for RE projects and the prediction of this estimated yearly energy production is an important tool in the design of RE projects. Such judgements on energy production are obviously dependent both on the device's parameters and on factors related to installation site and system type. This study focuses on PV and the newly published "Energy Rating" standard IEC 61853, which specifies comparatively close limitations for PV module characterization. However, even within the required close set of data points serving as input for the calculations, variations of the predicted energy yield (the output) are possible. The impact on the precision of the energy yield prediction arising from production tolerances of module mass production is investigated. Statistical models for parameter variation as well as other numerical models are applied. A considerable uncertainty due to scatter in production parameter is revealed and conclusive recommendations are presented.

KEYWORDS: PV, energy yield, energy rating, diode equation, statistical parameters of module mass production

1. INTRODUCTION

It is now since at least three decades that observations were made that power and/or efficiency ratings of solar modules at STC (Standard Testing Conditions as per IEC 61215) are not corresponding to actual power and energy generation of fielded modules [1]. The performance depends on operational conditions as well as on module-related performance. With intend to create a criterion for the module-related performance under so-called "Realistic Reporting Conditions" (RRC) [2], several approaches were followed in the research community. All of these approaches target to have solar modules comparable with respect to their energy production during operation (in contrast to the rating at STC). In all cases, the dependency of the module efficiency (or MPP power) on irradiation and operational temperature is considered as a minimum input for the assessment. More sophisticated approaches consider also the non-perpendicularity of irradiance and the deviation of actual spectral distributions of the natural sunlight from the idealized AM1.5 standardized spectrum.

The International Electrotechnical Commission has been issuing the IEC-61853 series to address energy rating of solar modules. This is a clear progress and allows system designers to have access to a better data base for predicting system yield and generated electric work.

IEC 61853-1 sets a minimum of reporting conditions [3] as a base requirement for further calculation.

Table I: Five minimum reporting requirements as per IEC 61853-1

		Irradiance	Cell Temp.
1	STC (Standard Test Cond.)	1000 W/m²	25 °C
2	NOCT (Nom. Op. Cell Temp.)	800 W/m²	20 °C
3	LIC (Low Irrad. Cond.)	200 W/m²	25 °C
4	HTC (High Temp. Cond.)	1000 W/m²	75 °C
5	LTC (Low Temp. Condition)	500 W/m²	15 °C

[1] corresponding author: kleiss@etp.uni-hannover.de

2. METHOD

2.1. Simplified Model for PV modules

A single diode model is used as a base to simulate the I-V characteristics of a solar module:

$$I(V) = I_{photo} - I_o \cdot \left[e^{\frac{q(V+I\cdot R_s)}{nkT}} - 1\right] - \frac{V+I\cdot R_s}{R_p} \quad \text{(eqn. 1)}$$

This single diode model of (eqn. 1) depends initially on 5 parameters and is later simplified to 3 parameters. When a dependency on irradiance G and (cell-) temperature T is to be considered then the following set of equations (eqn. 2.1) through (eqn. 2.5) is used:

$$I_{photo}(G, T_{mod}) = \frac{G}{G_{STC}} I_{photo}(G_{STC}, T_{STC}) \cdot [1 + \alpha \cdot (T_{mod} - T_{STC})] \quad \text{(eqn. 2.1)}$$

$$I_0(T_{mod}) = I_0(T_{STC}) \cdot \left(\frac{T_{mod}}{T_{STC}}\right)^3 \cdot e^{-\frac{E_{gap}}{kT_{mod}} + \frac{E_{gap}}{kT_{STC}}} \quad \text{(eqn. 2.2)}$$

$$R_s(G,T) = const. \quad \text{(eqn. 2.3)}$$

$$R_p(G,T) \to \infty \quad \text{(eqn. 2.4)}$$

$$n = 1 \quad \text{(eqn. 2.5)}$$

An extension assumption is made with respect to the model given by (eqn. 1) and (eqn. 2.1) through (eqn. 2.5), namely that there are only three parameters considered to be influenced by cell/module technology and G or T

- $I_{photo}(G_{STC}, T_{STC})$
- $I_0(T_{STC})$
- R_s

whereas other parameters α, n and R_p are regarded as fixed and not to depend on cell/module technology, and in particular not on temperature T or irradiance G.

For calculations within this study the following numerical values are assumed: $\alpha = 0.01\ \%/K$, $n = 1$, and $R_p = 5000\ \Omega cm^2$.

This simplification seems acceptable since the target of this study is to deal with the propagation of uncertainties and no absolute results are considered (which would indeed require modeling on a more detailed level). It is well understood that for an exact prediction of absolute values more sophisticated models apply also to lumped resistances depending on G and T – compare for example [4].

Module temperature T_{mod} is calculated from meteorological data (standard south tilt of POA, plane of array) given for c-Si modules in accordance to [5] per Faiman's approach [6]

$$T_{mod} = T_{amb} + \frac{G_{POA}}{u_1 + u_2 \cdot v} \quad \text{(eqn. 3)}$$

with applicable coefficients u_1, u_2 for typical crystalline PV modules taken from [7].

Ultimately, the equations (eqn. 1) through (eqn. 3) form an implicit set of equations to be resolved for calculating the maximum power P (and thus η), so that numerical computation is needed rather than linear problem solving. Work presented here has been computed based on applicable algorithms [8].

2.2. Meteorological Data and CSER

Table II: sample data sets provided in the Committee Draft of IEC 61853-4 for different climatic regions

#	Latitude	Longitude	Type *(example in country)*
1	1°S	11°E	tropical humid *(Gabun)*
2	24°55'N	46°25'E	Desert *(Saudi Arabia)*
3	38°N	3°15'W	Mediterranean *(Spain)*
4	48°N	18°E	temperate continental *(Slovakia)*
5	56°N	4°W	temperate coastal *(UK – Scotland)*
6	34°N	83°E	high elevation *(China – Tibet)*
7	18°N	80°E	tropical semi-arid *(India)*

Table II summarizes the meteorological data sets provided by the latest CD for IEC 61853-4 [5]. These data sets contain by hourly resolution the total global irradiance G_{POA} and the ambient temperature T_{amb} that allow directly application of (eqn. 1) through (eqn. 3) for calculation of results.

Since tabulated data for G_{POA} and T_{amb} from [5] can be straightaway used to compute the results of this study, an abbreviated method for calculation compared to the upcoming IEC 61853-3 method [9] is applied: All considered installation face POA 20° towards equator and thus no correction of "beam" and no further "special" corrections per [9] were applied.

Influence of spectral effects is excluded at this time (as only crystalline devices are within the scope of this study and differences between mc-Si and cz-Si are deemed considered not to be large). The applied temperature model of (eqn. 3) [6] is exactly as used in IEC 61853-3 [9].

An important output value of the energy rating algorithm described in CD IEC 61853-3 [9] is CSER (Climatic Specific Energy Rating), as defined in clause 7.8 of [9]:

$$\text{CSER} = \frac{\sum_{1\,\text{year}} E(G_{POA}, T_{mod})}{P_{max,STC} \cdot H_p} \quad \text{(eqn. 4)}$$

In this work the time resolution for calculations is set to one hour (in accordance to [9]). Therefore, the following equation is used to calculate the energy production at a given hour of in total 8760 hours of the standard year used:

$$E(G_{POA}, T_{mod}) = P(G_{POA}, T_{mod}) \cdot 1\ \text{h} \quad \text{(eqn. 5)}$$

$P(G_{POA}, T_{mod})$ of (eqn. 5) is calculated from data for G_{POA} and T_{mod} obtained from the time series [5].

Likewise, the total yearly sum of solar energy in the plane of array (POA) H_p of (eqn. 4) is described as the sum of hourly POA irradiances times one hour:

$$H_P = \sum_{1\,\text{year}} G_{POA} \cdot 1\ \text{h} \quad \text{(eqn. 6)}$$

2.3. Model Parameter Variation and Confinement

The outcome of any mass production inevitably leads to a scatter with respect to certain parameters. In the case of solar modules power (and thus their efficiency) at standard testing conditions STC is affected in particular.

When looking at STC power, it is interesting to note that general requirements from other industries for process capability (e.g. derived from 6σ or 3σ) are rarely ever met in PV solar industry for $P_{max,STC}$ when today's standard sorting of 5 Wp classes is applied. Moreover, the measurement uncertainty of end-of-line measurement might exceed the difference between upper and lower level for sorting.

A first selection criterion for basic parameter variation to solve (eqn. 1) through (eqn. 3) is given according to the following constraints of Table III, including fixed parameters already described in section 2.1.. 60 cells are modeled in series connection.

Table III: outer limits for the multi-dimensional parameter variation approach used in this study

	variation range	variation type
I_{photo}(STC)	35.0 ... 39.0 mA/cm²	linear
I_0(STC)	1 10^{-12} ... 1 10^{-10} A/cm²	logarithmic
R_s	0.1 ... 2.0 Ohm cm²	linear
R_p	5000 Ohm cm²	- none -
n	1	- none -
α	0.01 %/K	- none -

A second limitation within this study is introduced by looking at a 5Wp bin power-class-of-modules that is fixed according to 260Wp ... 265Wp. This means looking at a typical single class of modules, exactly as a customer or a system designer would do.

A third limitation is introduced by calculating the power of all required operation parameters per IEC 61853-1 (see Tab I or [3]) and averaging these values. A maximum relative deviation of less than 2% from the average values is the selection criterion of each operating condition tabulated in Tab I.

In summary: this study assumes a threefold funnel of criteria, where:

(a) boundary values from Table III can be regarded typical to ring-fence the production output (where in particular a variation with respect to the used cell is postulated)

(b) one single typical module power class for c-Si solar modules is selected: here only (260 Wp -0/+5Wp) with 60 cells per module

(c) the characterization according to IEC 61853-1 has been done with an assumed error of (rel.) 2%.
In more general terms it can be stated, that this 2% rule is at the same time in the order of magnitude of a lower limit of (absolute) module measurement.

In our computational experiment the application of this described threefold funnel leads to a quite flat distribution of STC powers for the modules of class 260 Wp -0/+5Wp (see Fig. 1).

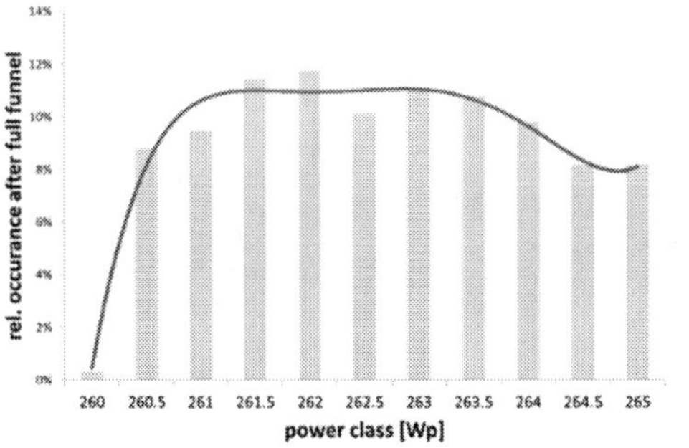

Fig. 1: (relative) histogram for STC power of modules after applying the selective funnel (a), (b), (c)

Only the very lowest bin is showing little occurrence due to cut-off criteria from (b) and (c). When looking at a real production output the latter restriction does not exist.

It is again important to note, that the distribution shown here is intended to cover the mass production with many input cell classes and not only the outcome of a distinct cell class.

3. RESULTS

3.1. Properties of solar modules with high impact on outdoor performance

Basically, the output of a solar module operated at conditions different from Standard Testing Conditions (STC) shall vary due to irradiance, (cell-) temperature, spectrum and spatial distribution of the irradiation (non-perpendicular light). It is known that for c-Si modules the impact of irradiance and temperature has the largest importance. It has been shown that for c-Si devices both effects separate and the total impact can be formulated as an analytic product of both single deviation modes for G and T [10].

The temperature impact can be characterized by looking at the normalized temperature coefficient of power (or efficiency) as defined by (eqn. 7).

$$\mathrm{TC}_\eta = \frac{\frac{d}{dT}\eta\left(G=1000\frac{W}{m^2},\ T=25°C\right)}{\eta\left(G=1000\frac{W}{m^2},\ T=25°C\right)} \qquad \text{(eqn. 7)}$$

Likewise, the module's relative performance at low light levels

$$\eta(G,\ T = 25°C) \qquad \text{(eqn. 8)}$$

affects the module's response to irradiances differing from $G = 1000\ W/m^2$.

It is important to point out that the parameter variation as described in section 2.3 does not only influence the STC power but also the key characteristics in (eqn. 7) through (eqn. 8), which are decisive for the outdoor performance (and for CIGS) of a PV module.

3.2. Observed variation in the normalized temperature coefficient

Parameter variations result in observed differences of TC_η ranging from -4211 ppm/K to -4851 ppm/K, which can be considered covering the range full bandwith of values observed in data sheets of actual sold modules. This observed span in TC_η is significant when looking at the impact on outdoor operation.

Obviously, a smaller (absolute) TC_η is favorable since most of the time solar modules are operated at temperatures higher than 25°C. An exception is the "high elevation scenario" (#6 of Table II yields to results that show a distinct advantage w.r.t. STC – see Fig. 5).

A dependency of TC_η on $I_0(\mathrm{STC})$ is shown at Fig. 2 and further work shall address this analytically. At this time it may be concluded, that modules / cells that have smaller I_0 tend to have a smaller (absolute) TC_η (with less temperature related relative losses in most climates).

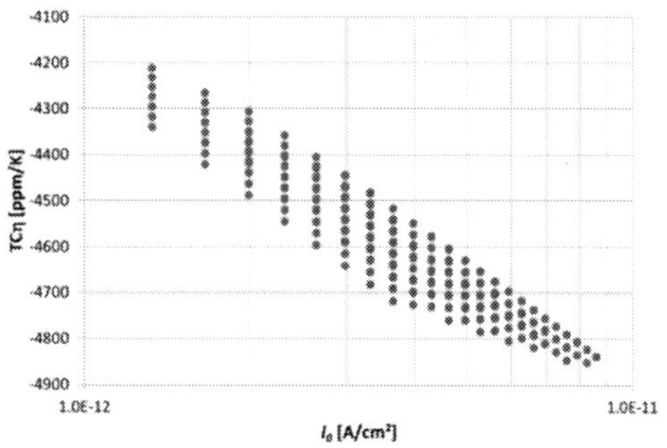

Fig. 2 correlation between TC_η and $I_0(STC)$ as result of the parameter variation with a funnel defined in section 2.3.

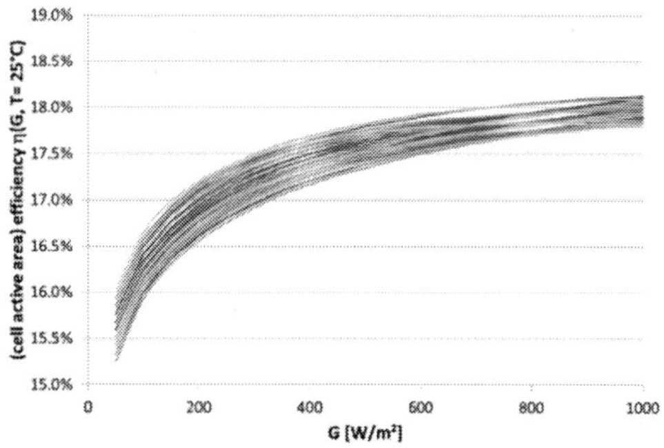

Fig. 3: performance of module efficiencies (active cell area – no power loss in cell-to-module) following parameter variation as described in section 2.3

3.3. Observed variation on impact on low-light performance

A second parameter of interest is the low-light-level performance $\eta(G,\ T = 25°C)$ as defined in (eqn. 8). Typical results are depicted in Fig. 3. Efficiency has been calculated with respect to the active cell area with (idealized) zero loss for cell-to-module conversion.

3.4. Other parameters: I_{photo}, R_S, and CSER

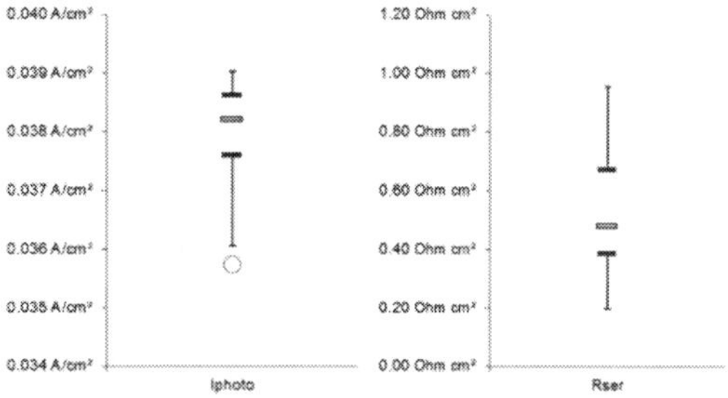

Fig. 4: outcome of parameter variation and application of funnel described in section 2.3: Box-and-whisker plots of I_{photo} and R_S (median in orange, 25% and 75% quantiles in black; only 1 out-layer for I_{photo}).

I_{photo} and R_S vary considerably and a relationship between R_S and $\eta(G,\ T = 25°C)$ can be postulated since ohmic losses limit efficiencies at higher irradiance levels.

CSER is site-dependent and results are given for the locations shown in Table II. Unless in the "high altitude" scenario all CSER are below unity reflecting that in most operation situations outdoor power generation is below to what would be expected from STC rating (see Fig. 5).

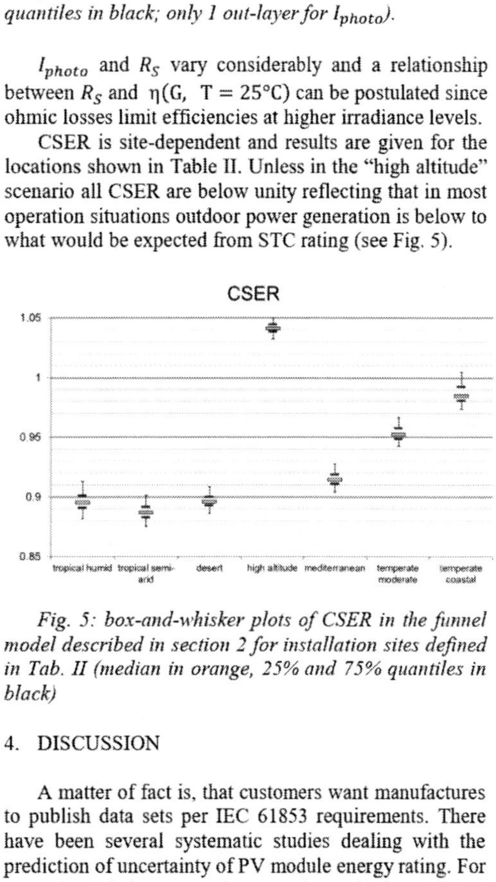

Fig. 5: box-and-whisker plots of CSER in the funnel model described in section 2 for installation sites defined in Tab. II (median in orange, 25% and 75% quantiles in black)

4. DISCUSSION

A matter of fact is, that customers want manufactures to publish data sets per IEC 61853 requirements. There have been several systematic studies dealing with the prediction of uncertainty of PV module energy rating. For example Dirnberger and co-workers [11] published studies that focus on uncertainties introduced by measurement errors during the in-door calibration of the candidate modules. Roy, Betts and Gottschalg [12] investigated predominantly the impact of uncertainties by environmental operation conditions.

The key result of this study arises from data presented in Fig. 5: The whiskers (from product scatter after application of the funnel criteria, only) largely overlap between several locations (tropical humid, tropical semi-arid, desert, and Mediterranean): The observed span for c-Si product in the suggested scenario is in the order magnitude of (min.) 3%. This said 3% uncertainty (related to production scatter, only) as such is usually not acceptable when predicting the profitability of a distinct solar PV project. It may be argued that such effects might be disappear due to the average performance of a large population of solar modules within an installation, but there are two effects that should be kept in mind:

(a) Performance of individual solar modules typically influences the performance of a larger series connected solar array always such that the "lower performance" modules have a highter impact compared to "high performance modules"
(b) There is a possibility that performance predictions according to data sets of IEC 61853 could have been done according to samples that have accidentally chosen as representatives of "very good" or "very poor" performing modules: This would be expected to reduce the said uncertainty for CSER of 3% to 1.5%.

In conclusion, it is suggested to consider a standard uncertainty in CSER for c-Si modules of 1.5% due to production scatters, only. Other technologies are expected to even show higher uncertainty margins.

5. LITERATURE

[1] R. W. Taylor, "System and module rating: Advertised versus actual capability," *Sol. Cells*, vol. 18, no. 3–4, pp. 335–344, 1986.

[2] A. Raicu, K. Heidler, G. Kleiss, and K. Bücher, "Annual and seasonal energy rating of mono-Si, a-Si and GaAs test cells for the USA by the RRC method," *Proc. 22nd IEEE PVSC, Las Vegas*, pp. 744–749, 1991.

[3] IEC, "IEC 61853-1:2011," 2011.

[4] A. Mermoud, "PV module one-diode model as implemented in PVsyst," in *1st European Workshop on PV performance modelling*, 2013.

[5] IEC, "TC82 IEC 61853-4 Ed.1.0," vol. 82/1067/CD, pp. date of circulation 2015–01–15, 2016.

[6] D. Faiman, "Assessing the outdoor operating temperature of photovoltaic modules," *Prog. Photovoltaics*, vol. 16, no. 4, pp. 307–315, 2008.

[7] M. Koehl, M. Heck, S. Wiesmeier, and J. Wirth, "Modeling of the nominal operating cell temperature based on outdoor weathering," *Sol. Energy Mater. Sol. Cells*, vol. 95, no. 7, pp. 1638–1646, 2011.

[8] W. H. Press, B. P. Flannery, S. A. Teukolsky, and W. T. Vetterling, *Numerical Recipes in Pascal (First Edition): The Art of Scientific Computing*. Cambridge University Press, 1989.

[9] IEC TC 82 WG, "CD IEC 61853-3 Ed. 1.0 Photovoltaic (PV) module performance testing and energy rating – Part 3: Energy Rating of PV Modules," *82/1067/CD*, 2016.

[10] G. Kleiss, *Energetische Bewertung von photovoltaischen Modulen auf der Grundlage der Jahreswirkungsgradmethode*. Berlin: Verlag Dr. Köster, ISBN 3-89574-246-5, 1997.

[11] D. Dirnberger, B. Müller, U. Kräling, and S. Dittmann, "Uncertainties of Laboratory Masurements for Energy Rating," in *27th European Photovoltaic Solar Energy Conference and Exhibition*, 2007, pp. 3294–3301.

[12] J. Roy, T. R. Betts, and R. Gottschalg, "Accuracy of Energy Yield Prediction of Photovoltaic Modules," *Jpn. J. Appl. Phys.*, vol. 51, pp. 20–25, 2012.

EVALUATING THE INFLUENCE OF TYPHOON ON PV MODULE RELIABILITY

Maoyi Chang, C. H Hsueh, Haomin Chen and Chienyu Chen
AU Optronics Corporation
No.1 JhongKe Rd, Taichung, 40763, Taiwan

ABSTRACT: Around 25MW PV capacity accumulated is damaged over thirty project sites after two sequential typhoons striking into Taiwan in 2015. Meteorological data for gusts of wind is carefully analyzed. Severe damage takes place at the specific location where its maximum wind speed accompanies with the wind direction of the north. 1.6% and 1.5% power degradation is respectively observed for P-multi and P-mono module in our outdoor testing system. The effective wind speed and the equivalent wind load with cycle time distribution are extracted to fit the typhoon scenario from the monitor system in every seven seconds. Dynamic mechanical load test (DMLT) is applied to imitate cell cracking resulted from the wind load during typhoon incursion, and [DMLT-TC50-HF10] sequence test with various wind load pressure and module framing is utilized to evaluate subsequent degradation in the field. Modules detaching from mounting rack due to clamping loosening are also observed in several projects. Proper mounting configuration and structure design are required to against typhoons through module service lifetime.
Keywords: photovoltaic module, reliability, typhoon.

1 INTRODUCTION

In the summer of 2015, two extreme typhoons named Soudelor and Dujuan sequentially attack Taiwan, and there are accumulated around 25MW PV deployment capacity damaged spreading out more than 30 project sites through the whole country. In this work, we start to analyze wind speed of gusts and wind direction from weather observation data in four locations. Massive damage occurs at the location during maximum wind speed period by north wind. It explains harsh PV disaster takes place in the central and the south central part of the island without the highest wind speed among the island. 1.6% and 1.5% power degradation is respectively for P-multi and P-mono module installed in Taichung.

The effective wind load in the front side and the rear side of modules is derived from gust wind speed and wind direction in our outdoor test system by every seven seconds. The typhoon equivalent wind load is acquired and simulated by dynamic mechanical load test (DMLT) with the cycling distribution in various pressure ranges [1, 2, 3, 4, 5]. [DMLT - thermal cycling (TC50) - humidity freeze (HF10)] sequence is applied for evaluating the long term reliability after wind load in PV module industry [6, 7, 8]. Split conditions of DMLT pressure and frame structure in the sequential test are performed to investigate cell cracking and to simulate the degradation as the influence of typhoon on module.

Mounting configuration is also a very crucial factor for module reliability against the impact of typhoon. During blast period, a lot of modules escape from racking system and fall down on the ground due to cramping jig loosening. Using screw bolts through mounting holes to fix between module and mounting rack is recommended for typhoon sensitive regions.

2 OBSERVATION OF TYPHOON IN THE FIELD

Tropic cyclone is named as typhoon in the northwest Pacific Ocean or as hurricane in the northeast Pacific and the northern Atlantic Ocean when the average wind speed excesses 17.5 m/sec (63 km/hr). In Figure 1 [9], it shows global tropic cyclone historical tracks from year 1985 to 2005. The geometric regions high typhoon track density are coincident with emerging PV deployment regions including East and South Asia, or north Australia and coast lines in northern American continent. The correlation between typhoon wind speed and wind load on the Beaufort scale is listed as in Table I.

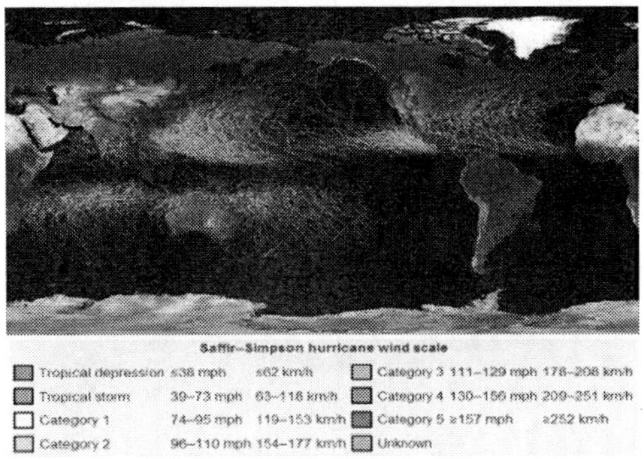

Figure 1: Global tropic cyclone tracks from year 1985 to 2005.

Table I: The typhoon wind speed and wind load on the extended Beaufort scale.

The Beaufort Scale	1-minute Sustained Wind Speed		Wind Load	
	(km/hr)	(m/sec)	kgf/m2	Pa (Nt/m2)
0-7	< 62	< 17.2	< 35	< 343
8	63 - 75	17.5 - 20.8	35-52	343 - 510
9	76 - 87	21.1 - 24.2	52-72	510 - 750
10	88 - 103	24.4 - 28.6	72-97	750 - 950
11	104 - 117	28.9 - 32.5	97-128	950 - 1254
12*	118 - 132	32.8 - 36.7	128-164	1254 - 1607
13*	133 - 149	36.9 - 41.4	164-206	1607 - 2019
14*	150 - 166	41.7 - 46.1	206-256	2019 - 2509
15*	167 - 183	46.4 - 50.8	256-312	2509 - 3058
16	184 - 201	51.1 - 55.8	312-377	3058 - 3695
17	202 - 219	56.1 - 60.8	377-499	3696 - 4890
>17	>220	> 61.1	>499	> 4890

2.1 Wind speed and wind direction

Typhoon tracks are classified into ten categories around Taiwan. Two harsh typhoons, named as Soudelor and Dujuan, sequentially hit Taiwan in the summer of 2015. The tracks of those two typhoons are quite similar to each other and classified as the third path category as in Figure 2. Four different location weather stations of the national central weather bureau are selected [10]. The maximum wind speed of gust for both typhoons takes place in Suao (the northeast part of the island) reaching 66.1 m/sec and 68.4 m/sec respectively in Figure 3. Wuqi harbor which is located 15 km away from Taichung city has the second highest wind velocity, but Taichung and Kaoshiung own much weak wind load level.

The intensity of gust deceases from the northern to the southern part of Taiwan, while severe PV system damage area is located among Taichung to Kaoshiung (from the central to the southern part) with lower wind velocity. In Figure 4, for typhoon Soudelor, Suao harbor has east direction wind load during its maximum wind velocity, while Wuqi harbor owns north direction wind load and its maximum wind velocity at the same blast period. PV system suffers the largest wind loading from the module backside during north direction gust. It well explains that rigorous module damage occurs in the central and south central part without higher wind speed than other locations of the island.

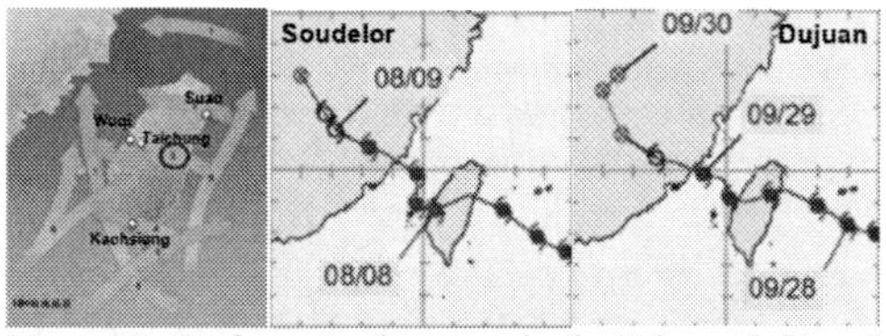

Figure 2: Typhoon track categories in Taiwan (left) and invasion paths for Soudelor (middle) and Dujuan (right).

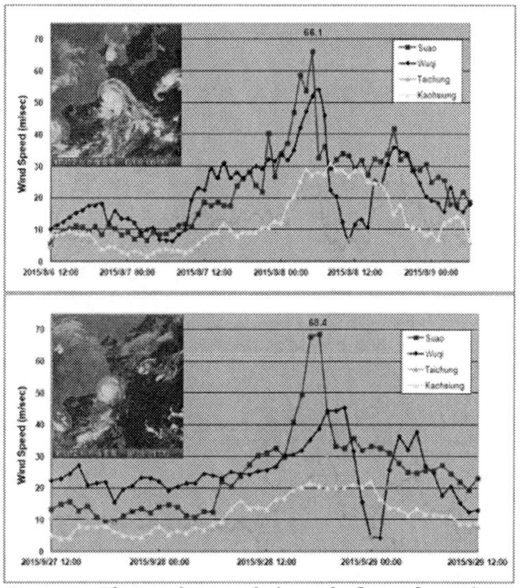

Figure 3: The maximum wind speed of gust for typhoon Soudelor (top) and typhoon Dujuan (bottom) in different observation sites.

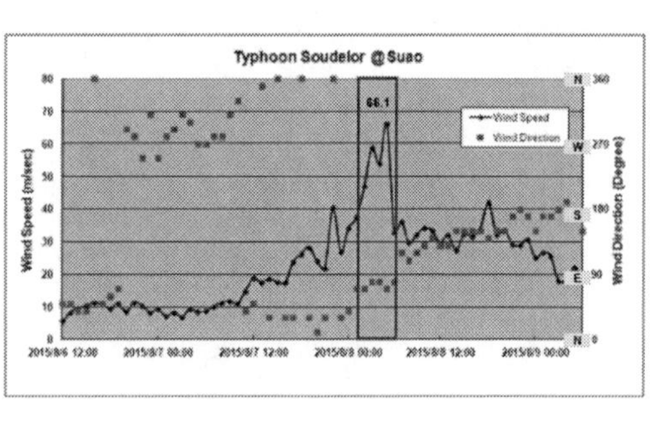

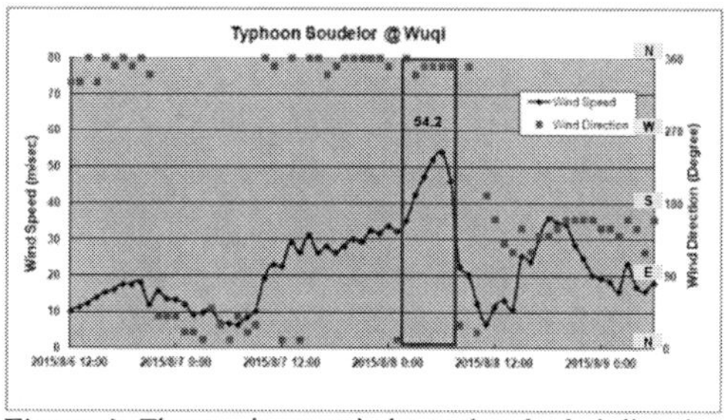

Figure 4: The maximum wind speed and wind direction of gusts in Suao (top) and in Wuqi (bottom) for typhoon Soudelor by every hour.

2.2 Power degradation and cell cracking

Our outdoor monitor system is located in Taichung. Module is periodically taken off from the system and the power output is also measured at STC condition in our testing laboratory. Power degradation after those two sequential typhoons (with the maximum wind speed: 35.7 m/sec and 36.2 m/sec) is 1.6% and 1.5% for P-multi and P-mono module respectively as shown in Figure 5. However no obvious appearance defect was found, cell cracking accompanied with dark area in EL image demonstrates the influence of wind load on modules as Figure 6. Some cell cracks originally existed in the P-mono module is due to the transportation problem 10 months before typhoon attack.

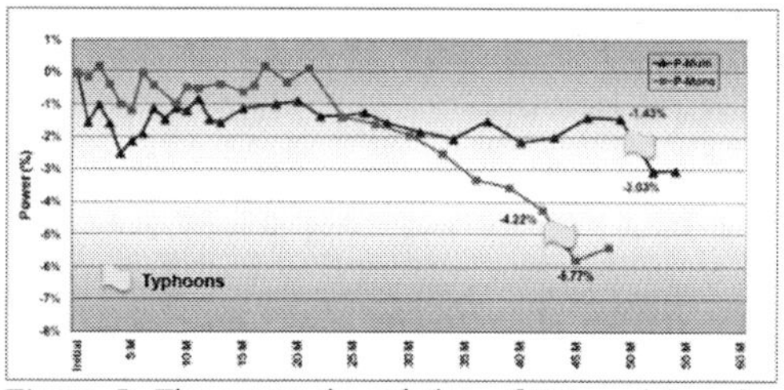

Figure 5: The power degradation of PV modules after suffering two sequential typhoons.

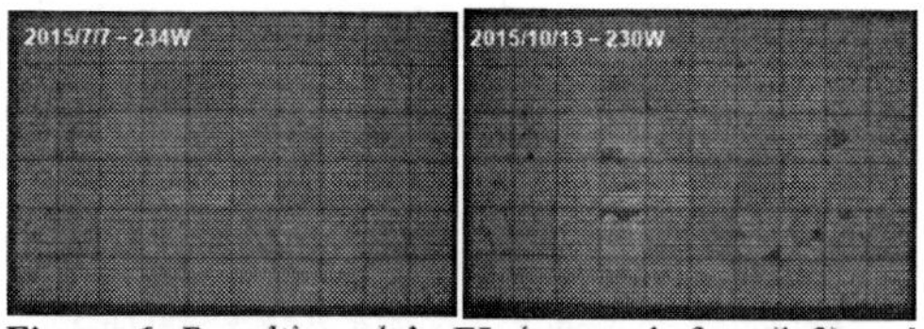

Figure 6: P-multi module EL images before (left) and after (right) two sequential typhoons.

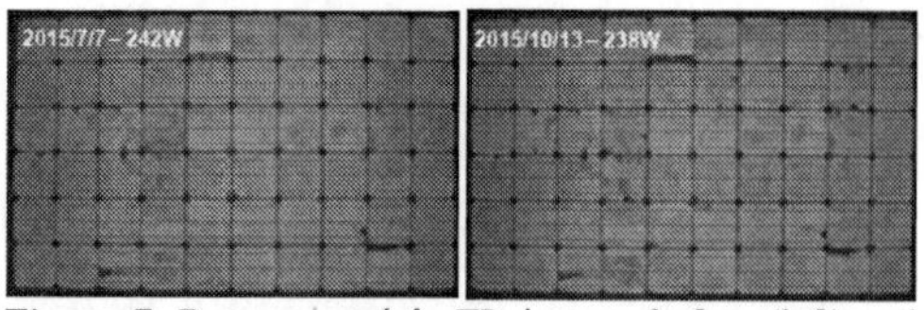

Figure 7: P-mono module EL images before (left) and after (right) two sequential typhoons.

3 THE EQUIVALENT WIND LOAD SCENARIO AND THE SEQUENCTIAL TEST

3.1 The effective wind speed and the equivalent wind

load with cycle time distribution

Wind speed and wind direction in Taichung during those two sequential typhoons are carefully inspected from the outdoor monitoring system with data collected in very seven seconds. Again, as the similar result from hourly data analysis, most high velocity gusts occur with the north direction wind at the same time for both two typhoons as in figure 8.

Module bends forward to the front (glass) side or the back side for suffering wind load. The direction of effective wind speed or loading pressure is therefore in the north or in the south direction, since the azimuth is 0° (facing to the south) for the system. The effective wind speed in Figure 9 is derived from cosine correlation with measured wind speed. Using the function of Table I, one would convert effective wind speed to the equivalent wind load in pressure scale with cycle time distribution for each typhoon scenario as in Figure 10. By collecting long-term (25 years) historical typhoon meteorological data and accumulating all of the "cycle times – effective wind load distributions", we might to be able to evaluate the influence of all typhoons on module through whole service lifetime in the specific region.

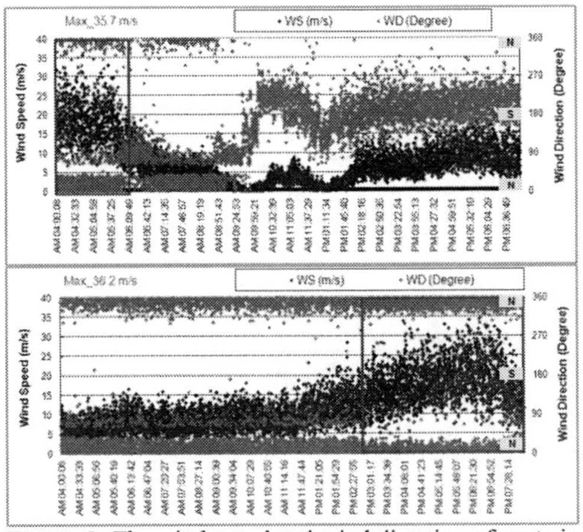

Figure 8: The wind speed and wind direction of gusts in Taichung outdoor monitor system for typhoon Soudelor (top) and Dujuan (bottom) by every seven seconds.

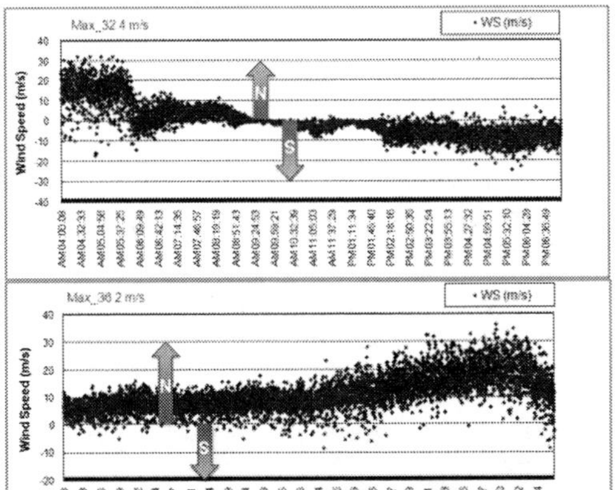

Figure 9: The effective wind speed of gusts in Taichung for typhoon Soudelor (top) and Dujuan (bottom).

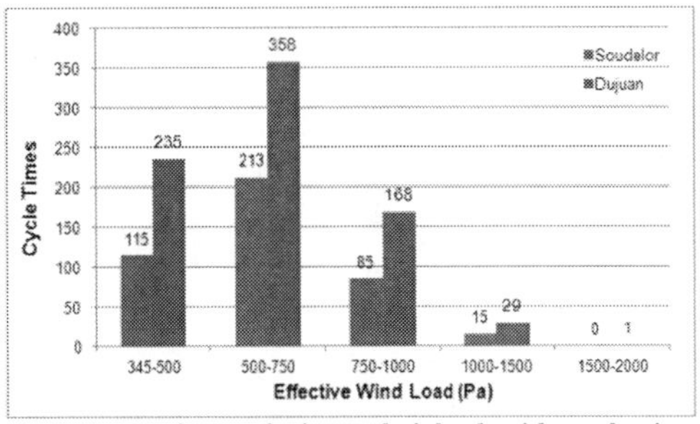

Figure 10: The equivalent wind load with cycle time distribution for each typhoon scenario.

3.2 [DMLT-TC50-HF10] sequential test

Dynamical mechanical load test (DMLT) is applied to simulate the equivalent wind load for typhoon scenario. The equipment for DMLT is as shown in Figure 11. DMLT is performed with total 1000 cycle times and three cycle times per minute according to the draft version of IEC 62782 (82/761/CD) [3]. [DMLT-TC50-HF10] testing sequence has been reported an efficient method to verify the degradation of module especially for cell cracking. Not only cell cracking due to typhoon continuously growths after the event, but also residue thermal stress due to cell interconnect soldering process in module manufacturing is prone to be released by cell cracking after thermal cycling [11].

Loading pressure from 1,440Pa (the maximum equivalent wind load of typhoon Dujuan in Taichung) to 4, 000Pa (level 17 in the extended Beaufort scale) is performed for investigating P-multi modules in the sequential test as in Figure 12. Larger DMLT pressure induces deeper degradation in following test steps. Module frame structure is also the key to against wind load. Three types of module framing structures show different power degradation result after [1440Pa DMLT-TC50-HF10] test sequence in Figure 13. EL image in each step during the sequential test is shown in Figure 14.

Figure 11: The equipment of dynamic mechanical load test for simulating typhoon wind load.

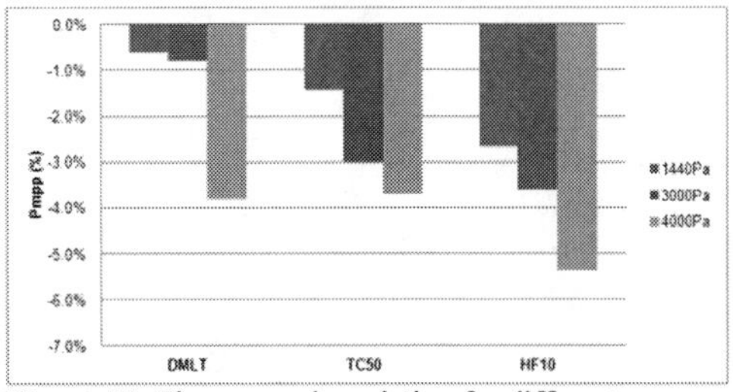

Figure 12: The power degradation for different pressure of [DMLT - TC50 - HF10] sequence.

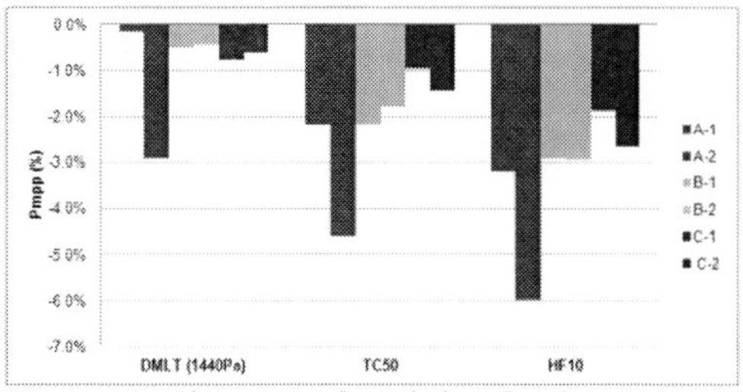

Figure 13: The power degradation for three types of module frame structure after [1440Pa DMLT - TC50 - HF10] sequence.

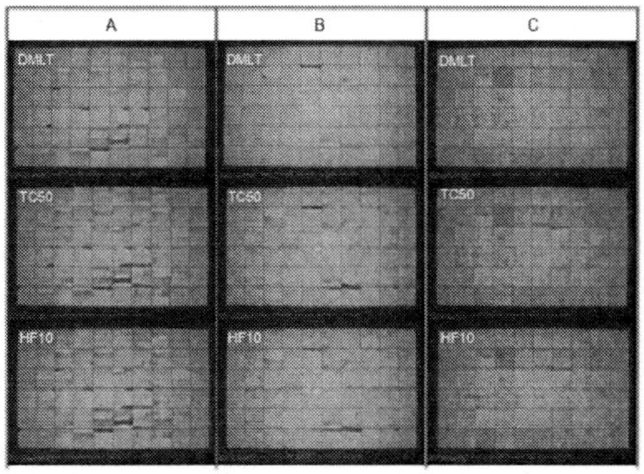

Figure 14: EL images for various types of frame structures during the sequential test.

4 MOUNTING CONFIGURATION

To sustain module reliability against extreme gusts of typhoon, the mounting configuration plays an important role for PV system in the field. Clamping jig is commonly utilized to fasten module on back rail in mounting system. Numerous modules escape from racking, collide with each other and fall down on the ground in several project sites due to cramping jig loosening or clamping kit deformation as in Figure 14.

Although output power performance is important, the safety for PV system is more crucial especially for roof mount application. Using screws or bolts through module mounting holes to fix between modules and mounting racks is one of the possible solutions to prevent this kind of issue. In addition, rack mounting or roof mounting structure should be designed strongly enough with adequate safety factor.

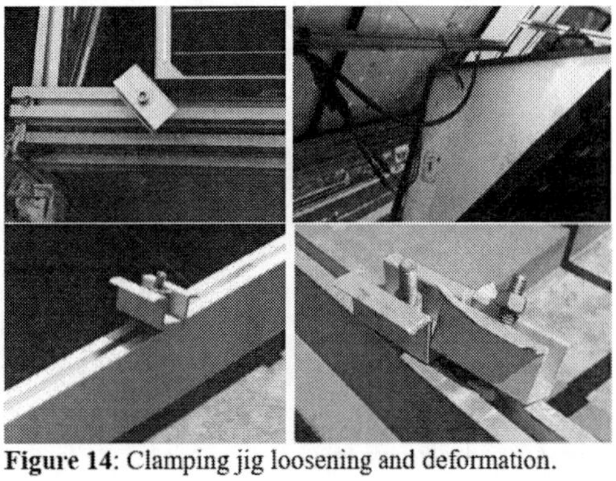

Figure 14: Clamping jig loosening and deformation.

5 CONCLUSION

In this work, we carefully inspect the weather data of two sequential typhoons attacking Taiwan in the summer of 2015. By combining wind speed and wind direction data, it concludes that disaster damage occurs at specific locations when the maximum gust wind speed of gust accompanies with wind direction from the north. Power degradation due to cell cracking for P-multi and P-mono modules are observed in the field.

The effective wind speed and the equivalent wind load with the cycle time distribution are successfully extracted in every seven seconds to build up the typhoon scenario model. This model would be used to predict the influence of accumulated wind load on PV module through whole service lifetime (25 years). [DMLT-TC50-HF10] test sequence is also applied to evaluate the influence of typhoon on module reliability with various loading pressure and different types of module frame.

Finally, module mounting configuration is discussed. Clamp jig loosening also results in module damage, and appropriate mounting structure design is required to against typhoon.

REFERENCE

[1] S. Koch, et al., "Dynamic mechanical load tests on crystalline silicon modules", 25th EU-PVSEC, 2010, p. 3998-4001

[2] IEEE 1262, "IEEE recommended practice for qualification of photovoltaic (PV) modules", 1995

[3] IEC 62782 (draft), "Dynamic mechanical load testing for photovoltaic (PV) modules", 2013

[4] N. Bosco, et al., "Evaluation of dynamic mechanical loading as an accelerated test method for ribbon fatigue", 39th IEEE PVSC, 2013, p. 3173 - 3178

[5] G. Mülhöfer, H. Berg, C. Ferrara, W. Grzesik, D. Philipp, "Influence of mechanical load at low temperatures on cell defects and power degradation at full scale PV modules", 28th EU-PVSEC, 2013, p. 2968-2971

[6] J. H. Wohlgemuth, "The effect of cell thickness on module reliability", 33rd IEEE PVSC, 2008, p. 444-445

[7] J. H. Wohlgemuth, Sara Kurtz, "Photovoltaic module qualification plus testing", 40th IEEE PVSC, 2014, p. 3589 - 3594

[8] IEC 62892-1 Ed. 1.0 (Draft, 82/984/CC), "Testing of PV modules to differentiate performance in multiple climates and applications - Part 1: Requirements for testing", 2015

[9] Douglas Inman and Scott Jenkins, "Climate patterns in the coastal zone.", Encyclopedia of Coastal Science. Springer Science & Business Media, Nov 8, 2006. pg. 246

[10] Taiwan central weather bureau, "Typhoon Database", http://rdc28.cwb.gov.tw/TDB/ntdb/pageControl/windsearch

[11] H. Chen et al., "The influence of cell busbar pattern on PV module Reliability", 28th EU-PVSEC, 2014, p. 2562-2565

FAILURE CLASSIFICATION OF DEFECTIVE PV MODULES BASED ON MAXIMUM POWER POINT ANALYSIS

Frank W. Fecher [1*], Tobias Pickel [1], Claudia Buerhop-Lutz [1], Christian Camus [1], Christoph J. Brabec [1,2]

[1] Bayerisches Zentrum für Angewandte Energieforschung (ZAE Bayern), Haberstraße 2a, 91058 Erlangen, Germany
[2] Lehrstuhl Materialien der Elektronik und der Energietechnologie (i-MEET), Friedrich-Alexander-Universität Erlangen-Nürnberg, Martensstraße 7, 91058 Erlangen, Germany
* Corresponding author: email: Frank.Fecher@zae-bayern.de, phone: + 49 9131 9398 - 170, fax: - 199

ABSTRACT: Photovoltaic (PV) modules are often affected by defects caused by transport and installation or by field degradation due to environmental conditions. If a power reduction is observed in the inverter data, a detailed error cause analysis is often very cost-intensive even when using imaging methods like thermography or electroluminescence. We present a method to classify the defect type of PV modules only by evaluating the maximum power output (P_{mpp}) for different irradiances. Thereby, low parallel resistance, high series resistance and broken cells can be distinguished from each other. Simulations and consistent explanations confirm the experimental findings. The method works particularly well for parallel resistance problems like potential induced degradation (PID). Thus, this failure type can easily be diagnosed in modules. The method should also be applicable to data of module strings and to thermography-based failure analysis in field which will be investigated in the future.

Keywords: Defects, Degradation, Photovoltaic, PV Module, Simulation

1 INTRODUCTION

Photovoltaic modules often exhibit power reduction due to failures. However, for the classification of the failures the modules have to be transported to laboratories or expensive tests have to be performed in field. In this paper, we introduce a method for classifying the defect in a PV module based on simplified *I-V* curve analysis.

2 METHODS, MODELS AND MATERIALS

The *I-V* curves of the crystalline silicon (c-Si) modules were measured at different irradiances at 25 °C with the Spire Solar's Spi-Sun Simulator™ 4600SLP (class AAA performance). The classification of the module defects was done by using the *I-V* curves and the electroluminescence (EL) images of the modules. 12 defect modules plus 4 defect-free reference modules from 4 manufactures were selected for the investigations. The EL images showed usually more than one defective cell in a module. Exemplarily the EL images and the measured *I-V* curves for different irradiances are shown for the three defect types in Figure 1.

The simulations were performed qualitatively using the circuit simulator LTspice (version 4) from Linear Technology. The equivalent circuit consisted of 60 cells modelled with the one-diode model and 3 bypass diodes for the substrings. The simulation parameters are listed in Table 1.

Table1: Simulation parameters of defect-free cells

Parameter	Value
Photocurrent I_{ph}	5 A
Series resistance R_s	5 mΩ
Parallel resistance R_p	150 Ω
Saturation current I_s	$6.2 \cdot 10^{-8}$ A
Ideality factor n	1.28

3 EXPERIMENTS

From the *I-V* curves the power outputs at maximum power point (P_{mpp}) of the modules were determined for different irradiances. The modules were classified into three defect types: 1) low parallel resistance, 2) high series resistance and 3) broken cells. The low parallel resistance R_p (e.g. due to poorly isolated cell edges) were simulated by reducing the R_p to 3.16 Ω and 1 Ω in one or all 60 cells. For the series resistance defect type, R_s of the substring was increased to 0.2 Ω and 0.32 Ω in one or all three substrings. A broken cell was simulated by connecting the broken cell part in parallel to the residual main cell via a contact resistance $R_{contact}$ as described in [1], because broken cell parts are only seldom completely disconnected in reality. The broken cell area was assumed to be 30 %. $R_{contact}$ was set to 1 Ω, 3.16 Ω and 10 kΩ in one or all 60 cells.

4 RESULTS AND DISCUSSION

Plotting the relative maximum power P_{mpp} of a defective module referred to a non-defective reference module (*relative* $P_{mpp} = P_{mpp,bad} / P_{mpp,reference}$) versus the irradiance reveals a positive slope in the case of low parallel resistance problems like PID, bad cell edge isolation or shunts (cf. Figure 2a). In our case, modules 1–5 revealed cells with bad edge isolation. For other defect types like high series resistance or cell breakage, the curves decrease with increasing irradiance (cf. Figure 2b and Figure 2c). The high series resistances in modules 6–9 are due to solder or contact failures of the busbars in the cells as diagnosed by EL images. The kink in the curve of module 6 is caused by one shunted cell in the module additionally to the cells with solder failures. Modules 10–12 were affected by broken cells.

These experimental findings could be qualitatively verified by simulations where one cell or all cells in the module were simulated as being defective (cf. Figure 3). Like in Figure 2a, the simulated curves in Figure 3a show an increase in *relative* P_{mpp} with irradiance for low parallel resistance failure. For one defect cell the deviation from one is small but visible. When all cells of the module are affected, the reduction in *relative* P_{mpp} is large for low irradiance and vanished for higher irradiances, as is seen in Figure 3a. For series resistance

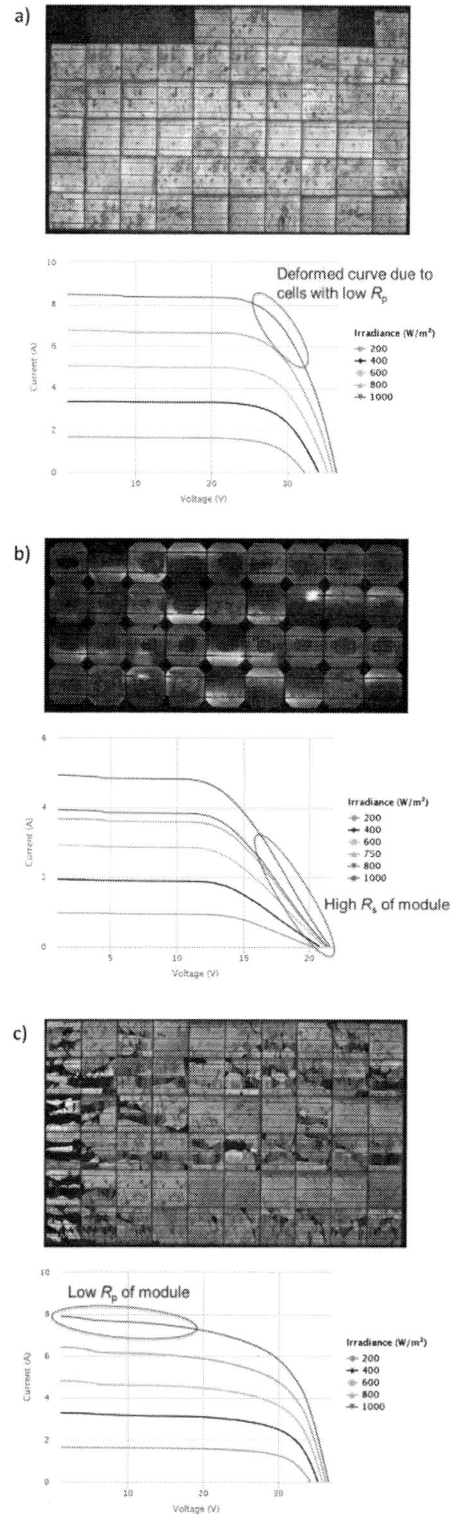

Figure 1: Electroluminescence images and I-V curves at different irradiances with a) low parallel resistance (module 3), b) high series resistance (module 8) and c) broken cells (module 10).

failures the simulated curves in Figure 3b exhibit a linear decrease for all parameters and their extrapolations meet all in *relative* P_{mpp} = 1 at zero irradiance. This trend is seen in Figure 2b, but not exactly as the modules often suffered also on other defects. Especially, the fact of one fix point for the extrapolated experimental curves cannot be seen. For broken cells in the module the simulated curves in Figure 3c show a negative slope, except for the parameter $R_{contact}$ = 10 kΩ where the broken cell part is completely isolated. If only one cell is defect, the curves decrease almost linear in the simulated range of irradiances. When all 60 cells are affected, the curves asymptotically approach the curve for $R_{contact}$ = 10 kΩ. A reduction in the slope is also seen for higher irradiances in Figure 2c.

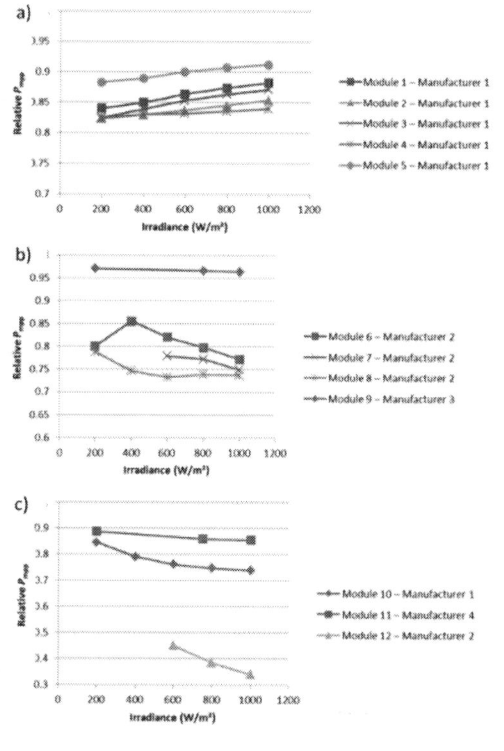

Figure 2: Experimental results: Relative module power versus irradiance for a) low parallel resistance, b) high series resistance and c) broken cells.

The results from failures due to low parallel resistance can be explained by the shunt screening effect [2]. The higher the irradiance is, the higher are the resistive losses in the electrode and the shunt's influence region is reduced [3]. Thus, the relative output power rises with irradiance. In the case of cell breakage, the broken cell parts become relatively more isolated with rising irradiance which leads to lower *relative* P_{mpp}. This is due to the higher voltage drop in the contact resistance as more current flows through the whole cell and, hence, the broken cell part. For higher irradiances (depending on the parameters, but usually > 1000 W/m²) the curves approximate to the one for an electrically isolated cell break with $R_{contact} = 10^4$ Ω and positive slope. In the case of contact problems (like soldering failures of the busbars) leading to higher series resistances, the voltage drop over the series resistance ($V = R \cdot I$) leads to a decreasing relative power output as the short-circuit current (I_{sc}) in the system rises linearly with irradiance E, whereas the V_{oc} rises only logarithmically ($I_{sc} \propto E$, $V_{oc} \propto \ln(E)$). Therefore, the findings are consistent with experimental results.

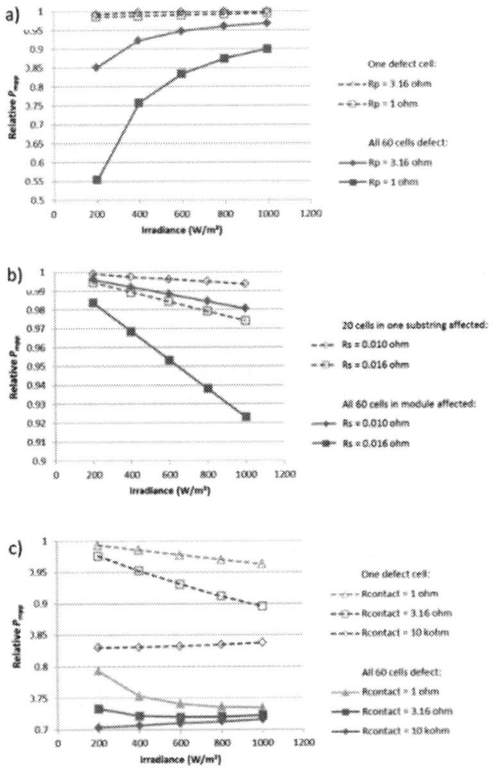

Figure 3: Simulated results: Relative module power versus irradiance for a) low parallel resistance, b) high series resistance and c) broken cells (with 30 % cell breakage).

In summary, a rising *relative* P_{mpp} for increasing irradiance indicates a low parallel resistance problem, whereas for decreasing values high series resistances or broken cells can be the reason. As seen from the simulation, the impact on the *relative* P_{mpp} is much lower in the case of a single defective cell than in the case where all cells of the module are defective. Nevertheless, the correct slope behavior is observed and the results agree qualitatively. Hence, the transfer of the evaluation method to module strings with power data from the inverter should be feasible.

5 SUMMARY

We showed that defect classification in PV modules based on simplified *I-V* curve analysis is possible. Therefore, the relative maximum power (related to power of a reference module at same irradiance) of PV modules versus irradiance was evaluated. By this simple method, the failure types 1) low parallel resistance, 2) high series resistance and 3) broken cells can easily distinguished for PV modules. Best agreements were obtained for a low parallel resistance failure. By considering the effect of module temperature on the performance, the presented method should also be applicable to data of module strings and to thermography-based failure analysis in field which will be investigated in the future.

6 REFERENCES

[1] M. Köntges, I. Kunze, S. Kajari-Schröder, X. Breitenmoser, and B. Bjørneklett, "The risk of power loss in crystalline silicon based photovoltaic modules due to micro-cracks," *Solar Energy Materials and Solar Cells*, vol. 95, no. 4, pp. 1131–1137, 2011.

[2] V. G. Karpov, G. Rich, A. V. Subashiev, and G. Dorer, "Shunt screening, size effects and I /V analysis in thin-film photovoltaics," *Journal of Applied Physics*, vol. 89, no. 9, pp. 4975–4985, 2001.

[3] F. W. Fecher, A. Pérez Romero, C. J. Brabec, and C. Buerhop-Lutz, "Influence of a shunt on the electrical behavior in thin film photovoltaic modules - A 2D finite element simulation study," *Solar Energy*, vol. 105, pp. 494–504, 2014.

6 ACKNOWLEDGEMENT

The authors gratefully thank the German Federal Ministry for Economic Affairs and Energy for financial support of this work (FKZ 0325807A).

RELIABILITY OF BONDING OF THE RAIL ATTACHMENT FIXTURE TO THE REAR GLASS OF DUAL-GLASS PV MODULES

Jing Mao , Qiangzhong Zhu, Jianmei Xu, Hui Shen, Yunhua Shu, Zhichao Ji, Pierre J. Verlinden and ZhiQiang Feng
State Key Laboratory of PV Science and Technology, Trina Solar
No.2 Trina Road, Trina PV Park, New District, Changzhou, Jiangsu, China, 213031

ABSTRACT: A rail attachment fixture bonded to the back of dual-glass modules presents several advantages compared to standard clamp systems. This paper reports on the reliability of the bonding technology for dual-glass PV modules. A detailed research of the bonding reliability of the back rail attachment fixture for dual-glass modules is presented in this paper. The bonding is based on silicone structural adhesive bonding to dual-glass modules. A full set of environmental tests from IEC61215 standard of photovoltaic industry were carried out to explore and analyze the bonding adhesive. Silicone adhesive is well known to be sensitive to water vapor and has high water permeability. It is prone to hydrolysis, which can lead to a lower strength of the back rail attachment in high humidity and high temperature environments. Damp heat (DH) testing was extended to 3,000 hours, which is 3 times of the usual duration of damp heat testing. Additional testing was carried out to evaluate the reliability of back rail attachment in harsh environments such as marine, acidic and alkaline environments. Extended experimental analysis shows that the silicone structural adhesive can meet the safety and reliability of dual-glass modules in long service life.
Keywords: rail attachment fixture, silicone structural adhesive, dual-glass module, the bonding reliability

1 INTRODUCTION

With the increasing demand of PV generation market, customers are setting higher requirements for quality and reliability of PV products. Dual-glass modules are designed for high reliability, in particular in harsh environments such as tropical, desert, marine and corrosive environments [1].

Dual-glass modules are usually provided with two strengthened glass panels and without frame. The traditional way to attach dual-glass modules to the rail structure is to use 4 clamps. A new attachment scheme is proposed, using back rail attachment fixtures that are glued to the rear glass panel via a silicone adhesive.

The aluminum fixtures are bonded on the back strengthened glass, optimizing traditional installation methods and allowing for fast installation of the dual-glass modules. This innovative mounting scheme, specifically designed for utility-scale ground-mounted PV stations reduces labor cost and ultimately LCOE. The fixture adhesion to the glass is provided by a silicone adhesive. The mechanical properties of silicone is known to be stable under ultraviolet light exposure (Si-O bond is about 420kJ/mol and 320nm UV energy is about 360kJ/mol).The structural silicone adhesive has been widely used in construction, aerospace, photovoltaics, electronics and other fields, for more than 40 years.

The aim of this paper is to present a reliability study of silicone adhesive for the bonding of back rail attachment fixture to dual-glass PV modules, in particular for harsh environments such as tropical, marine and corrosive environments.

2 EXPERIMENTAL

The structure of dual-glass PV modules with a back rail attachment fixture, a typical "clamp-less" installation of dual-glass modules and a traditional installation with clamps are displayed in Fig.1. The traditional installation with clamps is identical to the installation of thin-film modules. Rail attachment fixtures are attached in factory to the back of the modules using silicon adhesive. This installation of the modules can be completed by a single person, greatly improving the efficiency of installation and cost. In addition, there is no protruding fixture above the surface of the module, reducing the risk of partial shading, soiling and hot spots.

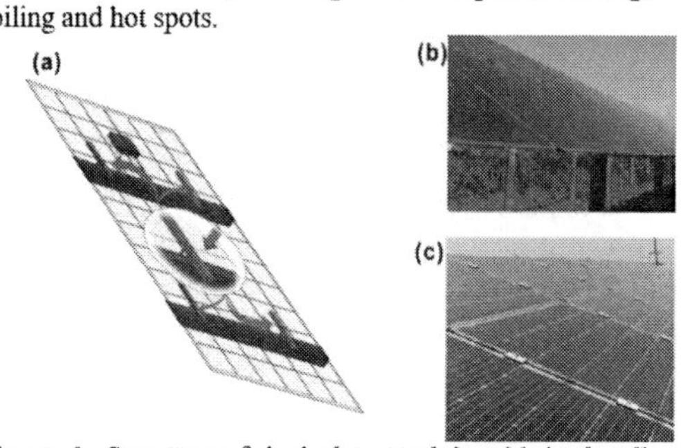

Figure 1: Structure of dual-glass module with back rail attachment fixture (a), clamp-less installation of dual-glass modules (b) and traditional dual-glass modules with installation using clamps (c)

To investigate the reliability of the bonding between the rail attachment fixture and dual-glass module with silicon adhesive in different environments, we carried out a series of reliability accelerated tests and characterizations in the State Key Lab of PV Science and Technology (PVST) at Trina Solar. We first carried out a standard set of environmental testing following the IEC61215 standard, including 1,000 hours of damp heat（85℃, 85% R.H.）, 200 thermal cycles（-40℃to +85℃）, 50 thermal cycles followed by 10 humidity-freeze cycles（+85℃ to -40℃, 85% R.H.） and UV exposure（60kWh/m2）to prove that the silicon adhesive is qualified for a conventional environment. In a second phase, we extended the damp heat (DH) test to 3,000 hours to simulate a high temperature and high humidity environment. Bonding test samples were also kept in harsh environments such as marine, acidic and alkali environments, to evaluate the stability of the mechanical properties of the silicone bonding. All samples are made according to the European standards ETAG002.e

3 RESULTS AND DISCUSSION

3.1 Type selection of silicone adhesive

During Mechanical Loading (ML) Test of PV modules under the IEC standard 61215 and the force analysis is showed in Fig. 2, the tensile strength of the structural adhesive should be qualified up to a wind pressure of 2400Pa. According to the formula: P(tensile)=N/S(hook)=S(module)*P(module)/ S(hook) ≈ 0.3MPa. Taking into account the stress in the natural environment, we need to keep some safety margin. So we choose the structural adhesive whose tensile strengthen is above 1MPa as the critical property and selection criteria.

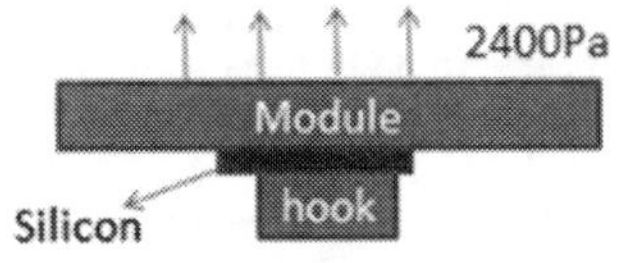

Figure 2: Force analysis of ML

According to the requirement of ETAG002.e, silicone-bonded samples were kept in a water bath at 45℃ during 21 days. This test condition was adopted to determine the adhesive properties in the selection of structural adhesive. In the development stage of dual-glass modules with back rail attachment fixtures, we enhanced the requirements of reliability testing and increased the water temperature to 55℃. The pass criterion was set at 85% adhesion level or Cohesive Failure (CF) area after 21 days in the hot water bath, which means less than 15% of the area showing loss of adhesion. We selected two kinds of structural adhesive from manufacturer A and manufacturer B. The adhesive from manufacturer A lost adhesion to the glass after the first 7 days in the water bath as shown in Fig.3(a), while the adhesive from manufacturer B met the requirements and passed the water bath test (55°C, 21 days) keeping 85% of CF area, as shown in Fig.3(b). The silicon adhesive of manufacturer B was selected for evaluation in the following reliability tests.

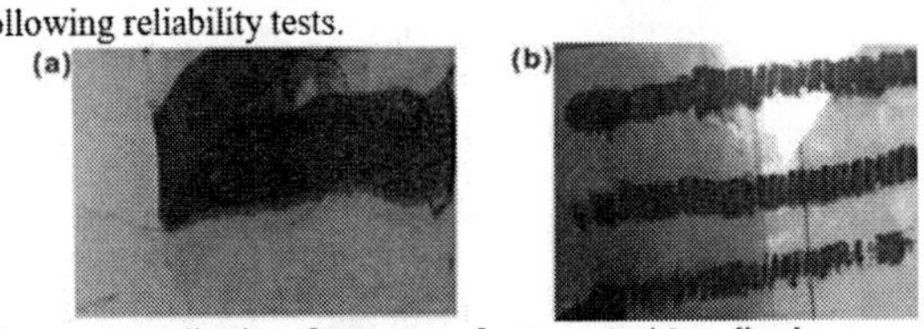

Figure 3: Adhesive from manufacturer A (a), adhesive from manufacturer B (b)

3.2 Conventional Environments

Test samples were prepared to evaluate the mechanical strength in two perpendicular directions (pull strength and shear strength) to simulate the forces applied to the rail attachment fixture during normal service life, as shown in Fig4 (a).

A standard set of environmental testing following the IEC61215 standard, including 1,000 hours of damp heat （85℃, 85% R.H.）, 200 thermal cycles （-40℃ to +85℃）, 50 thermal cycles followed by 10 humidity-freeze cycles（+85℃ to -40℃, 85% R.H.） and UV exposure（60kWh/m^2）, was set as qualification test for a temperate climate environment. Because the energy of the Si-O bond is larger than the energy of UV photons, we increased the UV test from 15kWh/m^2 to 60kwh/m^2. The selected silicon adhesive showed an excellent performance in the standard reliability test sequence, showing less than 5% degradation in mechanical strength after any environmental test as shown in Fig. 4(b).

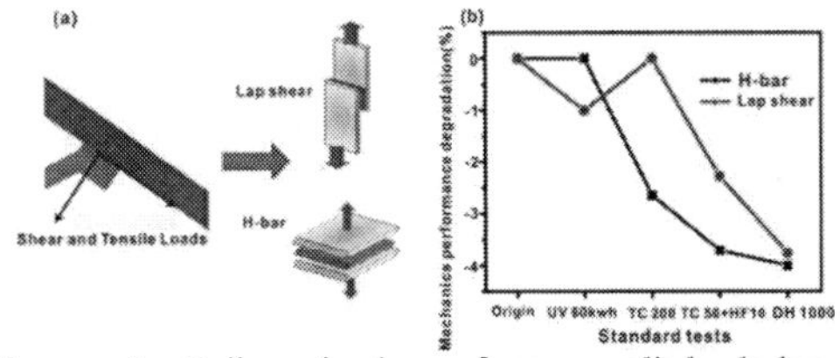

Figure 4: Pull and shear forces applied during environmental testing (a), Degradation in percent of the original strength of the pull and shear strength after environmental testing (b)

3.3 High temperature and High humidity Environment

Standard tests in IEC 61215 are not enough in real applications of dual-glass module that are specifically designed for harsh climates or environments. As shown in the standard test, the degradation of the mechanical performances is the largest in the DH test. The reason for this degradation is sensitivity of silicone to water vapor and its high water permeability. It is prone to hydrolysis, which leads to lower mechanical strength of the polymer chain. Therefore, we extended the damp heat (DH) test to 3,000 hours. As presented in Fig.5, the structural adhesive can still maintain more than 80% of its mechanical strength after DH3,000. Because the main polymer chain of the silicone structural adhesive is composed of Si- O bonds, which are not affected by the high temperature and high humidity environment, the strength of the silicone adhesive is usually quite stable.

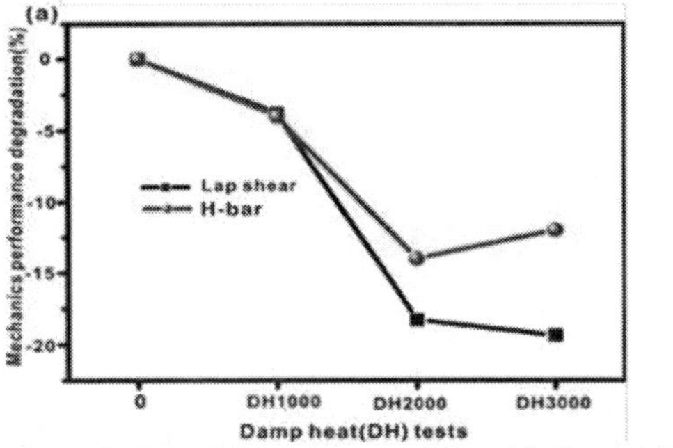

Figure 5: Degradation in percent of the original strength of the pull and shear strength after extended damp-heat testing

3.4 Marine Environment

A large amount of PV systems are located in coastal areas. Marine Environment is a severe test for PV components, particularly metal components. The coastal air contains a large amount of salt (Cl-) due to the spray or evaporation of sea water. Metal parts in PV modules, such as the aluminum frame or fixtures, can be seriously corroded by Cl- which can easily penetrate the anodic oxide protective layer. In addition, the relative humidity in marine environment is usually greater than 70% and has a great influence on corrosion. In this experiment, we studied the bonding of aluminum parts to glass with a silicone adhesive, which has a non-negligible water permeability. If seawater can reach the aluminum-silicone interface, it can produce crevices and surface corrosion. So in order to ensure that the silicone bonding of rail attachment fixture to dual-glass module can be well used in the marine environment, we designed the following marine environment test.

We used a test of immersion in a bath of artificial seawater for up to 36 days to accelerate the material corrosion and aging. We actually used two types of artificial seawater, differentiated by their pH value. In the first solution, to create acidic seawater, we mixed water with 3.5%NaCl and added H_2SO_4 to a pH of 3. In the second solution, to create alkaline seawater, we mixed water with 3.5%NaCl and added $NaHCO_3$to a pH of 8.1. The aluminum samples were made of aluminum with either an anodic oxide as protective coating or just a natural oxide with a thickness in the range of only 0.02 to1μm

3.4.1 Crevice corrosion of aluminum parts and silicone adhesive

The seawater immersion test showed two different types of failure after tensile test with natural oxide Al as shown in Fig. 6(a) and Fig. 6(b): a cohesive failure (meaning "failure with the silicone keeping a good adhesion to glass or Al") in alkaline seawater environment and an adhesive failure (meaning "failure with loss of adhesion to glass or Al) in acidic seawater environment. Experimental results show that crevice corrosion in acidic seawater is more serious than in weak alkaline seawater. The interfacial damage of the Al samples with natural oxide was inspected under an optical microscope after adhesive failure in weak acid immersion test as shown in Fig. 6(c). The electrochemical corrosion in the acidic seawater is assumed to be formed by the different density of oxygen inner and out of crevices. The oxygen out of the crevice can keep the surface passivated. In other words, it can react with a large amount of electrons, which resulted in a positive pole. To the contrary, the inner of the crevice is negative. Under the effect of Cl- and H+ in the crevice, the oxide film is damaged. It results in the activity of aluminum and accelerates the crevice corrosion.

In marine environment, the small thickness of the natural oxide film is prone to corrosion. The tensile test results show that there is a great risk of crevice formation due to corrosion on aluminum parts that are only protected by a thin natural oxide aluminum, as shown in in Fig. 6(d) and Fig. 6(e)

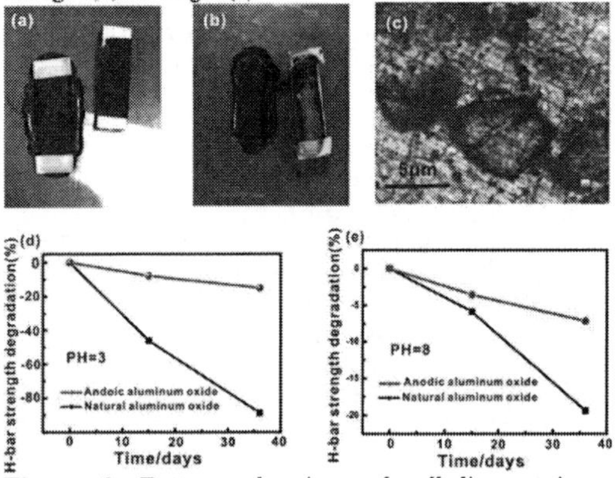

Figure 6: Test samples in weak alkaline marine environment (a) , Test samples in weak acidic marine environment (b), The interfacial damage(c), Degradation of tensile strength after exposure to weak acidic marine environment(d),or to weak alkaline marine environment(e) with two different types of aluminum oxide

3.4.2 Surface corrosion of aluminum

From the microscope analysis of surface corrosion of aluminum, it appears that pitting corrosion is present for both acidic and alkaline immersion tests. However, the surface of aluminum samples with a natural thin oxide present the most serious surface corrosion after the weak alkaline seawater immersion. The optical microscope pictures of the surface of the aluminum samples after 36 days of immersion are showed in table I. It is clear from the tensile strength measurement and optical microscope inspection that Aluminum parts must be protected by thick anodic oxide if used in marine environments.

Table I: Optical Microscope Pictures of the surfaces of the aluminum samples after 36 days of immersion with 5 μ m magnification

Samples	Initial	weak acidic marine environment	weak alkaline marine environment
anodic oxide aluminum		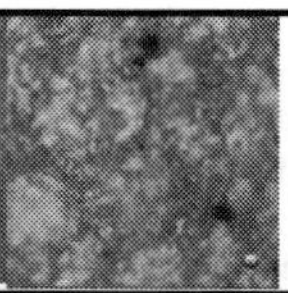	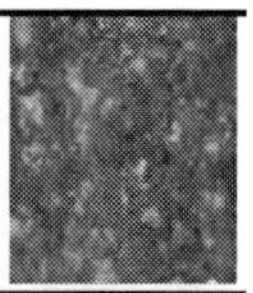
natural oxide aluminum		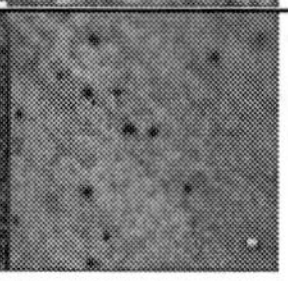	

3.5 Acidic and alkaline Environments

After immersion of glass samples with silicone adhesive in the acidic or alkaline solution for 7 days, we performed a peeling test. We first prepared a strong acid solution with 5% H_2SO_4 in water (pH=1) and a strong alkaline solution with 5% NaOH in water (pH=13). The adhesion of the silicone adhesive to glass was tested after 7days of soaking in the solution. The test results are shown in Figure 7. The Cohesive failure area in the peeling test remains 85% after soaking in either the acidic or alkaline solution.

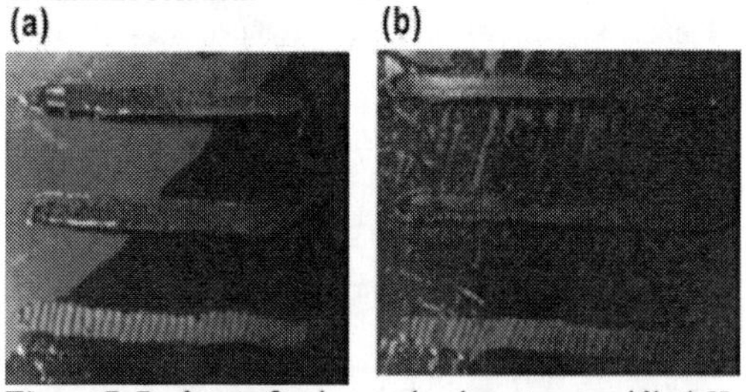

Figure 7: Peel test after immersion in a strong acidic (pH = 1) (a) or alkaline (pH=13) solution (b)

4 CONCLUSION

Silicone adhesive has over 40 years of history in the field of building, aeronautics, electronics, photovoltaics and glass curtain wall. This leads us to believe that structural silicone adhesive can be used in the field of photovoltaic for bonding double glass modules to a backside rail attachment fixture. This paper describes the beginning of a long program of reliability testing of silicone adhesive for the mounting of double-glass PV modules. Damp heat test seems to be the most difficult test to pass for silicone adhesive. Because dual-glass PV modules are specifically designed for tropical climates or other harsh climates, the damp heat test was extended to 3000 hours. It results in a degradation of less than 15% in the tensile strength of the silicone adhesive and less than 20% in shear strength. It was also shown that a thick anodic oxide is necessary to prevent the corrosion of aluminum parts in a corrosive or marine environment and to ensure a good adhesion of the silicone to Aluminum. A weak alkaline environment is more detrimental to the degradation of silicone than weak acidic environment.

References

[1] P. J. Verlinden, Y.B. Zhang, J.M. Xu, Y.H. Shu, Y. Yang, P. Quan, S. Zhang, G. Xu, J. Mao, F.L. Ping, Z.Q. Feng, "Optimized PV Modules for Tropical, Mountain and Desert Climates" – 29th European Photovoltaic Solar Energy Conference and Exhibition, Amsterdam, The Netherlands, 22-26 September 2014.

EVALUATION OF THE DURABLITY OF METALLIZATION PASTES VIA ACCELERATED AGING METHOD

Hsin-Cheng Lin*, Yi-Cin Chen, Cheng-Hua Wu, Chen-Chan Wang, Ching-Tang Tsai, Kang-Cheng Lin and Walt K.W. Huang
Gintech Energy Corporation
No. 21, Kebei 1st Rd., Hsinchu Science Park, Jhunan Township Miaoli County 350, Taiwan

ABSTRACT: An accelerated aging test with elevated temperature and humidity condition has been investigated and applied to cell and module fabrication in c-Si PV industry. Not only efficiency enhancement but also durability is the crucial criteria for metallization paste selection. In this work, a series of solar cells produced with different front-side or rear-side pastes were fabricated into 1*1 single-module samples. The samples underwent the highly accelerated stress test (HAST) with the typical test condition of 121°C and 100% relative humidity. The power decay ratio of the cells with various paste suppliers differed significantly and supposedly resulted from the decrease in fill factor. The result corresponded to the observations in the electroluminescent images where dark area emerged around soldered busbars. It is demonstrated that the cells with various metallization pastes displayed distinct ability to resist the corrosion and HAST is a promising approach to quickly identifying the paste evaluation in the cell production.
Keywords: Durability, Degradation, Metallization, PV Module

1 INTRODUCTION

It is known that reliability test plays an important role in the evaluation of bill of materials, such as cells and backsheets. Generally, damp heat test is utilized to identify the feasibility of module design and material selection. However, damp heat test typically requires thousands of hours to obtain the result, which is too time-consuming for determining the material property. In terms of this issue, highly accelerated stress test (HAST) with the relatively high hygrothermal condition is introduced to establish a quick estimation method for material selection.

2 EXPERIMENTAL

2.1 Single module preparation

1*1 single-module samples were fabricated using multicrystalline silicon solar cells, commercially available glass, ethylene vinyl acetate (EVA) encapsulant and backsheet.

2.2 Test condition and cell characterization

HAST test was performed in a climate chamber where elevated temperature and humidity condition were set as 121°C and 100% relative humidity.

I-V characteristics as well as electroluminescent images were measured to observe the degradation of single-module samples after HAST test.

3 RESULTS AND DISCUSSION

3.1 I-V performance of different front-side pastes under HAST test

We found various cells show different I-V characteristics undergo 100 or 200 hours of HAST test. Fig. 1 shows test results for cells with different front-side metallization pastes. Paste A & C had smaller decay rate than others. Fig. 2 shows peeling force of cells. Results reveal higher peeling force has smaller power decay, which is shown from less degree of FF degradation.

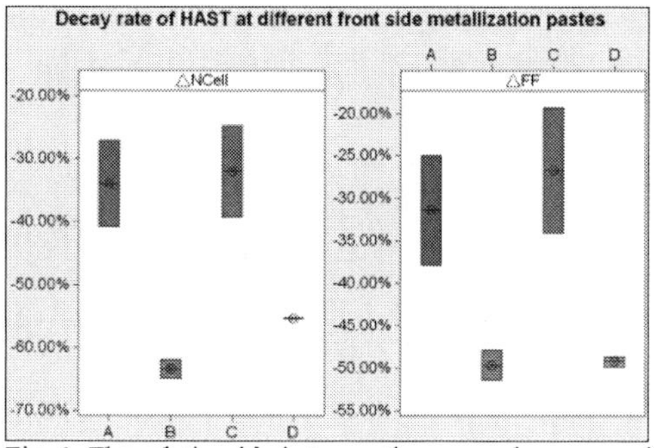

Fig. 1. The relationship between the power decay as well as fill factor ratio and the cells with various front-side pastes under HAST test.

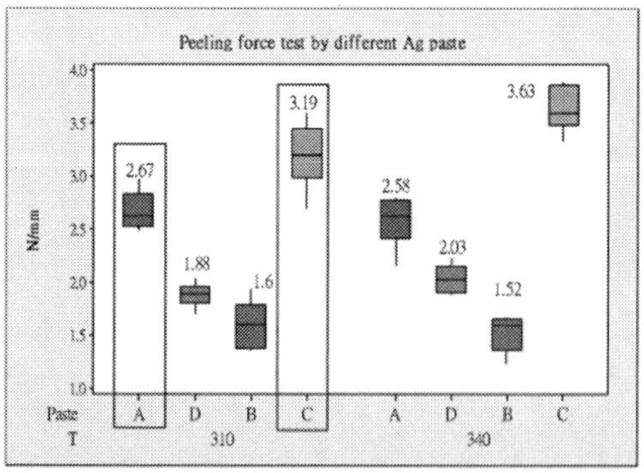

Fig. 2. The relationship between the peeling force and the cells with various front-side pastes.

3.2 I-V performance of different rear-side pastes under HAST test

Fig. 3 shows I-V characteristics from two different rear-side metallization pastes on cells. According to EL images during increasing stress time, it is observed that moisture ingresses through backsheet & encapsulation and cause corrosion around cell busbar, which makes lower peeling force (< 1N) along cell busbar, as shown in Fig. 4.

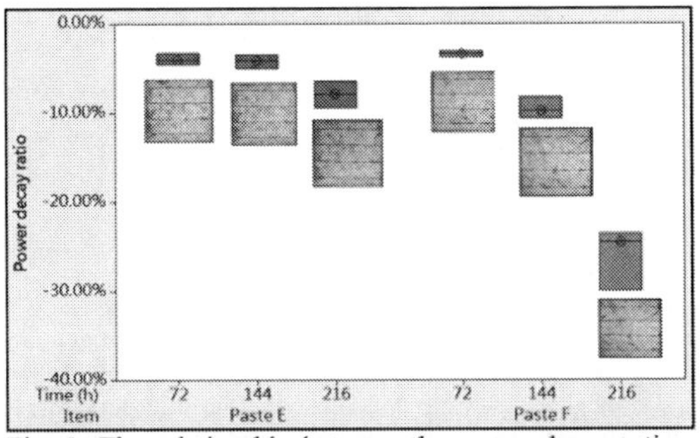

Fig. 3. The relationship between the power decay ratio and the cells with various rear-side pastes under HAST test from 72h to 216h.

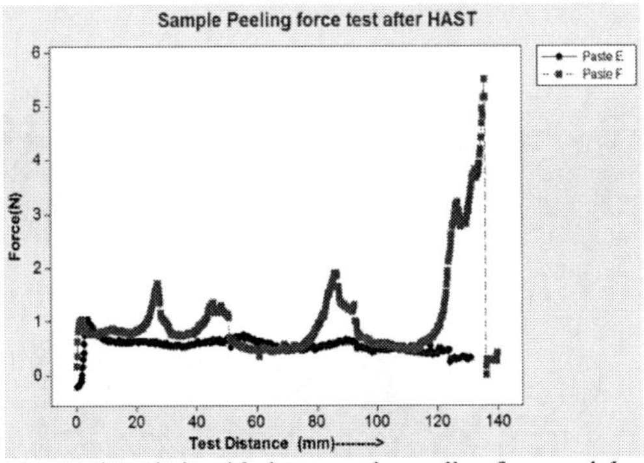

Fig. 4. The relationship between the peeling force and the cells with various rear-side pastes under HAST test for 216h.

3.3 Cell process improvement

As the effect, we design different approach to improve the durability of the PV modules after HAST. Fig. 5 shows the newly designed cell has smaller power decay than reference by 30%. In addition, we are trying to find a correlation with DH 3000. Fig. 6 is EL image of PV module tested by DH3000, which shows similar failure mechanism to HAST. Therefore highly accelerated stress test is an efficient method to help screen out weaker product.

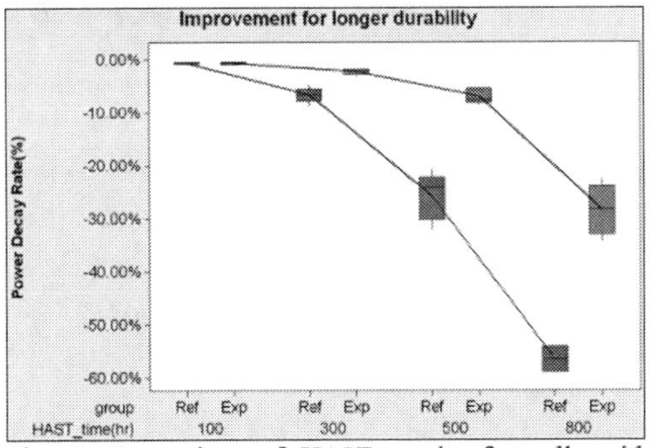

Fig. 5. Comparison of HAST results for cells with original and improved process.

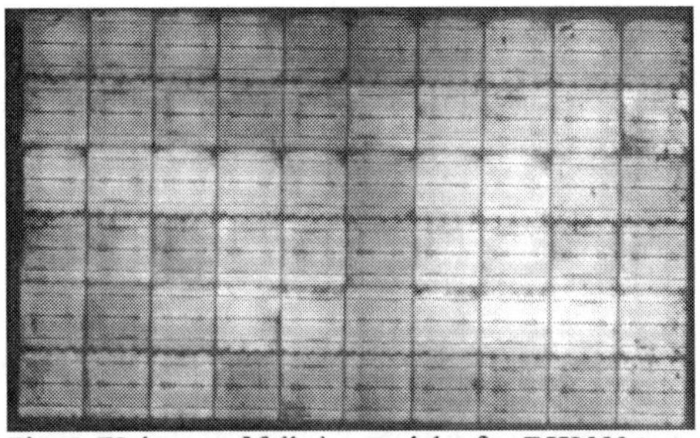

Fig. 6. EL image of full-size module after DH3000 test.

4 CONCLUSIONS

Highly accelerated stress test (HAST) has been utilized to distinguish paste durability because it considerably reduced the environmental test time compared with standard damp heat test. In this study, significant difference in power decay ratio demonstrated that the metallization pastes possess distinct ability to resist corrosion in harsh climate condition. It is concluded that HAST is an outstanding method of paste evaluation and could be further applied to other material selection in cell and module production.

5 REFERENCES

[1] M. Smith, A. Miller, Proceedings 17th European Photovoltaic Solar Energy Conference, Vol. I (2002) 903.

[2] K.W. Boer, Solar Cells 16 (1996) 591.

EFFECT OF THE REVISION OF MECHANICAL LOAD TEST IN IEC61215 CERTIFICATION STANDARD

Jin Hyung Ahn, Kyumin Lee, Moonseok Kim, Jong-Keun Lim, San-Il Yoon, Myung-Ick Hwang, and Eun-Chel Cho
Hyundai Heavy Industries, Co., Ltd.
17-10 Mabuk-ro 240beon-gil, Yongin 16891, South Korea

ABSTRACT: The IEC61215 standard has been revised to a new version since March 2016. Among many revised items, mechanical load test is one of the major revision in which the test method and the pass criteria were changed. The most important change was as follows. In the new standard, it is clearly stated that the module deformation can be restricted by underlying support rails if the mounting method of the manufacturer contains them. Also in the new standard, the power drop after the higher load (test load) is not evaluated. When we tested nearly identical modules using the test method in the old version and the new version, we found that the mechanical robustness of the rear side of the module acted as the limiting factor to decide the mechanical load specification of the module. However, in the real installation environment, the mechanical stability of the front side is still an important design point. For modules which use PERC cells, process improvement to reduce cell cracking is required to improve power drop.
Keywords: Mechanical load test, reliability, standard

1 INTRODUCTION

IEC61215 is one of the crucial certification standards to reach the commercialization of the developed PV modules. Thus, the IEC standard sets a guideline to select materials and to design the structure of PV modules. Mechanical load test is an important testing procedure which affects the dimensional design of the structural constituents of PV modules such as glass, frame, and films. As the test method and the pass criteria of mechanical load test in IEC61215 are changing, even the indentical module will show different result. In this circumstance, module manufacturers will need to react properly to the change. In this article, the effect of the standard change was analyzed by comparing the test result using the old version and the new version of the IEC61215 standard.

2 EXPERIMENTAL

2.1 Analysis of IEC61215 standard

As the first step, the change of the mechanical load test in IEC61215 was analyzed. We firstly interpreted how the clauses were performed in real test. The "old standard" (IEC61215 Ed.2) and the "new standard" (IEC61215 Ed.3) were compared in the view of real test equipment, method and pass criteria.

2.2 Test of PV modules according to the old and new standard

Closely identical PV modules were tested according to the old and new standard. The power drop and electroluminescence (EL) image were compared after the tests.

The test according to the old standard was performed without support rails, in which the gap between the module surface and the metal rail is sufficiently large that the bent module surface does not contact with any parts of the test base. The test according to the new standard was performed using a test base with support rails, in which the module frame is fixed on metal rails that the bent module surface touches the metal rails of the test base. This type of the test base simulates the module installation using support rails on the rear side of a module. Fig. 1 compares the test base for the old and new standards to test the front side load condition. When we test the rear side load condition, the test base without support rails was used for both old and new standard. The load was applied using sand bags which weigh 8kg each. The rectangular block-shaped sand bags were put on the module surface evenly on the entire module surface including the module frame.

Old standard	
New standard	

Figure 1: Comparison of test bases. To test according to the old standard, the test base with rear side support rails was used, while one without rear side support rails was used for the new standard.

3 RESULTS

3.1 Revision of mechanical load test of IEC61215

Important changes in the new standard can be summarized as below.

1) The load condition

The old standard specified the load condition as 2400Pa for wind load and 5400Pa for snow load. The new standard gives more degree of freedom to module manufacturers by the adoption of design load and test load which manufacturers can specify freely considering their products and mounting methods. The design load is determined by the module manufacturer according to its module structure design and installation method. The test load is determined as the design load multiplied by safety factor which is larger than 1 and typically set as 1.5 in UL 1703. The minimum test load has to be 2400Pa. Module manufacturers can specify the loads to front and rear surface differently.

2) Test base

The clauses in the old standard are stated somewhat ambiguously. In clause 10.16.2 (a) it states that the module may deflect freely during the test. This clause can be understood that the module deflection should not be restricted by any kind of supporting medium. On the contrary, the clause 10.16.3 (b) states to mount the module using the method prescribed by the manufacturer. This clause can be understood, if the mounting method includes a supporting rail, the supporting rail should be included in the test base. From these ambiguous clauses, some certification organizations do not have allowed the module surface to touch anything during the deflection, while other organizations do. The new standard clearly states that the module may deflect freely within the constraints of the manufacturers prescribed method of mounting, which means that the module can touch underlying support rails.

3) Test procedure

In the old standard, the higher load (5400Pa) is applied only on the front surface which is more mechanically stable than the rear side. In the new standard, the higher load (test load) should be applied on both sides.

4) Pass criteria

In the new standard, module power drop is not considered after applying the test load to both sides and frame bending is allowed as long as other parts such as glass or corner joints are not broken and the electrical insulation is maintained.

3.2 Comparison of the test result using the old and the new standard

According to the old and the new standard the same test result can be interpreted differently. Fig. 2 shows the power change and EL image of a module (not from Hyundai Heavy Industries, and its abbreviation is HHI) which was tested at 2400Pa and 5400Pa load condition according to the old standard without any underlying support rails.

After applying 2400Pa to the both sides for 3 cycles the power drop was -2.1%. The application of 5400Pa to the front side affected the module critically and the power drop reached -9.4%. According to the old standard, the module fails the mechanical load test because the power drop after the 5400Pa is over the -5% pass criterion. However, according to the new standard this result may meet the pass criteria because the result after applying the design load (2400Pa) was below -5% pass criterion and new standard does not evaluate the power drop after applying the higher load (5400Pa) .

The front side gets more resistant to the applied load when we adopt underlying support rails in the mounting method, which is clearly allowed in the new standard. Fig. 3 compares the power drop and EL image of two closely identical modules (not from HHI) which were tested at 5400 Pa to the front side after 3 cycles of 2400Pa load. One module was tested using the old standard without rear side support rails and the other was tested using the new standard with rear side support rails.

Test	Power drop	EL Image
Initial	-	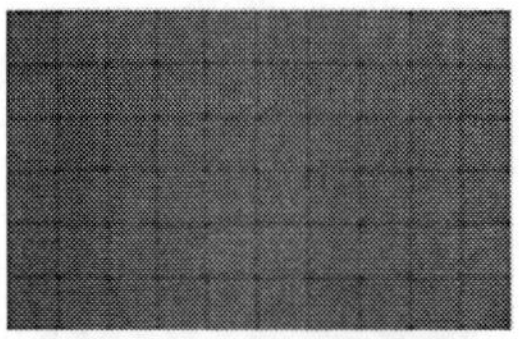
2400Pa	-2.1%	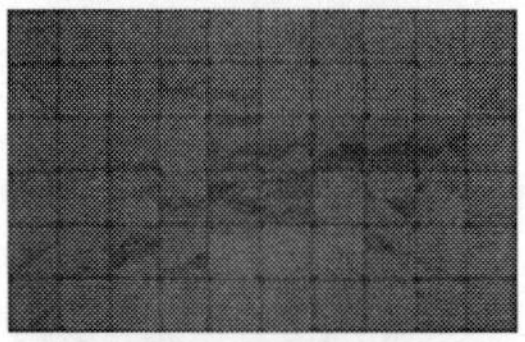
5400Pa	-9.4%	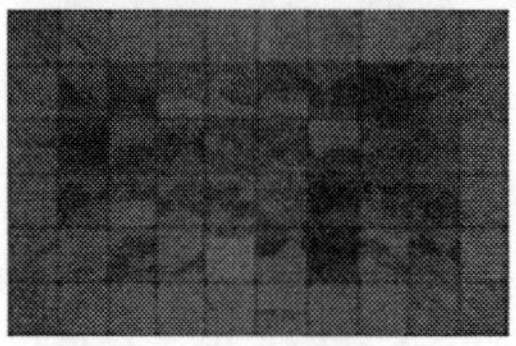

Figure 2: Power drop and EL images of a PV module after mechanical load test at 3 cycles of 2400Pa to both sides and 1 time 5400Pa to the front side. The test was performed using a test base wihtout support rails according the old standard.

With underlying support rails, the rear side module surface generally touches the underlying support rail and the module takes less stress with limited deformation by the rails. In Fig. 3 we can see the module which was test ed with the rear side support rails (new standard) suffered from notably reduced the crack formation. The power drop also was notably reduced to -3.2%, while that of the module which was tested without support rails was -9.4%. Thus, when we consider the mechanical load test to the front side only, we may increase the mechanical load specification of the module.

Test	Power drop	EL Image
Old standard	-9.4%	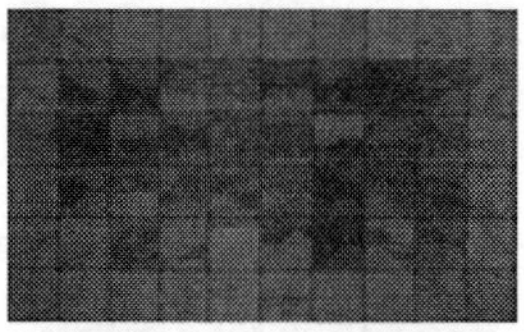
New standard	-3.2%	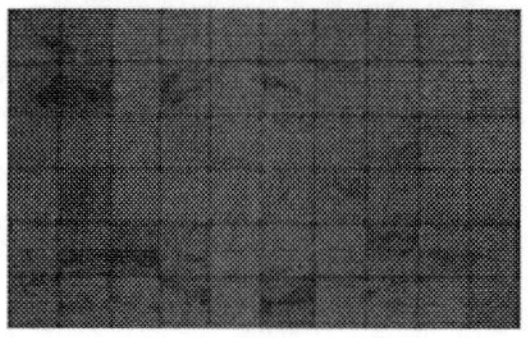

Figure 3: Comparison of power drop and EL images of a PV module after mechanical load test at 5400Pa according to the old standard (without rear side support rails) and the new standard (with rear side support rails).

However, in general, PV modules show larger deflection when we apply a load to the rear side than to the front side. This is related to the design of the module frame. The width of the frame which grabs the edge of the glass/backsheet laminate is larger on the rear side. This is because the width of the frame on the front side is limited to secure the gap between the frame and the cell. Additionally, when we apply load to the rear side of a module, the module does not see any reinforcing support from the support rails on the rear side. In Fig. 4, we can see a PV module was broken when we apply less than 5400 Pa to the rear side. The module sustained 5400Pa load test to the front side without breakage.

Figure 4: The breakage of a module during the 5400Pa load test to the rear side. The module was broken when 5163Pa was applied.

3.3 Implication to module manufacturers

In the old standard, the higher load (5400Pa) is applied on the front side only and the rear side takes just the smaller load (2400Pa). However, in the new standard, the higher load (test load) should be applied to both sides of the module. Thus, the resistance of the rear side to the mechanical load will act as a limiting factor. The new standard does not specify the load condition such as 2400Pa and 5400Pa. Instead, it allowes module manufacturers to decide freely the design load and the test load. In this circumstance, it will be necessary for module manufacturers to evaluate the resistance of the rear surface precisely and to design the module structure to improve it.

Among many phenomena of the test failure, physical breakage of the module, especially the separation of glass/backsheet laminate from the frame, which generally accompanies the breakage of the glass, will be the weakest point. According to the pass criteria of the new standard, the power drop is not measured after the rear side test using the test load and the frame deformation is not regraded as a failure. Possible solution to improve the mechanical stability of the rear side can be to reinforce the frame design to support the bent glass/backsheet laminate more strongly or to increase the adhesion strength between glass/backsheet laminate and the frame.

However, even if the mechanical stability of the rear side gets more important in the new standard than the old one, considering the real environmental condition in the field, the front side is still required to be more resistant than the rear side. Assuming the safety factor 3, 5400 Pa corresponds to the snow deposit of 0.9m, assuming the snow density 20% of water [1] and corresponds to wind speed of 195 km/h (Category 4 hurricane) [2]. As modules are generally installed on an inclined structure, the load from snow is concentrated more on the lower frame. This results in more frequent module fracture in a snow load condition than by a strong wind condition. To evaluate the resistance of PV modules to non-uniform snow load, a new IEC standard is being prepared. (IEC62938 Draft) [3].

Another point to be improved is the mechanical property of PERC modules. Recently, PERC technology has been getting popular as a high efficiency solution for both mono- and multi-crystalline Si cells. However, PERC cells tend to be more vulnerable to crack formation than conventional Al-BSF cells. Fig. 5 shows the EL images before and after 5400Pa test of modules which use mono-crystalline Si PERC cells. In Fig. 5, we can see large number of cells were cracked after the 5400Pa test and the power drop was larger than 5% which is over the pass criteria of old standard. It was reported that the laser ablation process to remove backside passivation and form contact holes generates crystalline defects on the Si wafer surface [4]. The crystalline defects act as a crack source when the module/cell is stressed by an external load.

To improve the mechanical property of PERC modules, it is required to improve the cell fabrication process to minimize the crystalline defects from the laser ablation process. When a PERC module from HHI was tested with the same method, we can see that the crack in the module was much reduced and the power drop was -1.5%, well below the pass criteria of -5%.

Module	Initial	After 5400Pa
Company A		
Company B		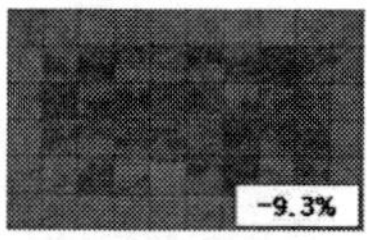
HHI		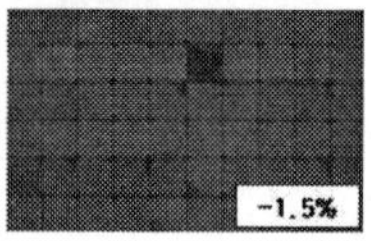

Figure 5: EL images before and after 5400Pa test of modules which use mono-crystalline Si PERC cells. The 5400Pa load was applied on the front side only.

4 CONCLUSION

We tested nearly identical modules using the test method in the old version and the new version of IEC61215. As rear side supports rails are clearly allowed in the new standard, IEC61215 Ed.3, the mechanical resistance of the module against the front side load test becomes more favorable to the module manufacturers. In the new standard, we have to apply the same higher load (test load) to the rear side as well as the front side. In this circumstance, the mechanical resistance of the rear side to the load will become a limiting factor to decide the mechanical load specification of the module.

Thus, it is necessary for module manufacturers to develop the module design or process to improve the mechanical property of the rear side. Also for PERC modules, process development to improve the mechanical stability of cells to reduce the crack formation is required.

ACKNOWLEDGMENTS

This work was partly supported by the New & Renewable energy Core Technology Program of the Korea Institute of Energy Technology Evaluation and Planning (KETEP), with the financial grant provided by the Ministry of Trade, Industry & Energy, Republic of Korea (No. 20133010011780).

REFERENCES

[1] Vivian Meløysund, Bernt Leira, Karl V. Høiseth, and Kim Robert Lisø, Meteorol. Appl. 14 (2007) 413.
[2] ASCE 7-05 Standard, Chapter 6 Wind Loads, Ameican Society of Civil Engineering, (2006) 38.
[3] IEC (International Electrotechnical Commission) TC 82 Project Files.
[4] Chi Dang, Riet Labie, Loic Tous, Richard Russell, Maria Recaman, Jan Deckers, Angel Uruena, Filip Duerinckx, Jef Poortmans, Energy Procedia 55 (2014) 649.

IN-SITU MONITORING OF MOISTURE INGRESS IN PV MODULES WITH DIFFERENT ENCAPSULANTS

Marko Jankovec[1], Gašper Matič[1], Eleonora Annigoni[2], Federico Galliano[2,3], Heng Yu Li[2,3], Laure-Emmanuelle Perret-Aebi[3], Fanny Sculati-Meillaud[2], Christophe Ballif[2,3], Marko Topič[1]

[1]University of Ljubljana, Faculty of Electrical Engineering, Laboratory of photovoltaics and optoelectronics, Tržaška cesta 25, SI-1000 Ljubljana, Slovenia
[2]École Polytechnique Fédérale de Lausanne (EPFL), Institute of Microengineering (IMT), Photovoltaics and Thin Film Electronics Laboratory, Rue de la Maladière 71b, 2000 Neuchâtel, Switzerland
[3]CSEM, PV-center, Jaquet-Droz 1, 2000 Neuchâtel, Switzerland

Email: Marko.Jankovec@fe.uni-lj.si

ABSTRACT: In 2015 we presented a new measurement method and setup for in-situ moisture ingress monitoring of PV modules for laboratory and field testing applications. The method which is based on miniature digital temperature and humidity sensors was successfully applied to different PV-like stacks with EVA encapsulants. In this work we present the test results of the experiment with sensors laminated in EVA, TPO, POE and Ionomer encapsulants. All samples were exposed to elevated humidity and temperature conditions in a climate chamber. Sensors prove to be robust to the harsh lamination temperature and pressure conditions and produce results that are consistent with the expected moisture ingress properties of the encapsulants. A one-dimensional equivalent electrical simulation model of water diffusion into the encapsulant was developed on the basis of the analogy between the diffusion process and electrical conductivity. The model was implemented in a SPICE simulation program and solved using the transient analysis. The diffusion coefficient of the EVA was obtained as a free fitting parameter and agrees well with the value obtained from an independent DVS experiment, validating our method. Although the experiment lacks the reference diffusion coefficient values of other three tested encapsulants, the fitting results show that the obtained coefficients are in the expected range according to the manufacturers' specifications.

Keywords: PV module, PV materials, Encapsulation, Reliability, Environmental effect, Humidity

1 INTRODUCTION

A key factor that determines the reliability of the PV modules is their ability to withstand challenging outdoor conditions for at least 25 years. One of the important environmental factors is also moisture ingress, which is strongly influenced by the PV module configuration, mounting location and the choice of PV encapsulants [1–4]. To model the influence of moisture on the lifetime of all kinds of PV modules, the moisture ingress phenomenon has to be well characterized in all possible materials and configurations.

With previously determined properties of moisture diffusion and solubility of the materials used it is possible to simulate the moisture ingress behavior in PV modules using analytical [5] or numerical approach [6]. However, these results have to be confirmed by measurements. Some attempts of in-situ moisture measurement methods have been reported using custom made humidity sensors in combination with the external analog readout electronics [7,8]. Although these methods provide results that are consistent with simulations they require time-consuming and inconvenient calibration procedures and despite this still hardly reach an adequate measurement accuracy. Due to the rapid progress of integrated circuits lately, miniature calibrated digital sensors are now commercially available which can be used for in-situ PV module humidity measurement applications.

Recently we developed a new in-situ measurement technique based on miniature digital humidity sensors SHT25 from Sensirion [9–11]. Sensors are soldered on a Printed Circuit Board (PCB) that is suitable for lamination in a PV module. An external digital data acquisition reader was developed which provides readout capabilities of up to 30 encapsulated sensors via a reliable digital bus. The use of a digital interface ensures that no additional noise is introduced into measurements along the transmission line, providing extremely low noise readouts. The described measurement technique proved to be a valuable tool for measuring the water vapor transmission rate (WVTR) and the diffusion coefficient of the Ethylene-vinyl acetate (EVA) encapsulant and presents a good basis for an outdoor system for monitoring the water concentration in PV modules in the field [12].

Till now the influence of EVA on the sensors' readings has already been thoroughly investigated. However, other types of encapsulants are recently gaining attention due to their superior electrical, optical and moisture-related properties. Some of them are good candidates to become dominant in the near future of the PV module construction. Hence, the purpose of this study is to validate the operation of SHT25 sensors encapsulated in Thermoplastic olefin (TPO), Polyolefin elastomer (POE) and Ionomer and compare the results with those obtained in the EVA.

2 EXPERIMENT

2.1 Humidity sensors SHT25

Humidity sensors SHT25 from Sensirion utilize capacitive sensing of dielectric properties of a polymer, which absorbs or releases water proportional to the environmental humidity. The humidity sensor is integrated together with a signal processing electronics, which provides temperature compensated and calibrated data on a digital communication interface. Sensors are packed in a package with 3x3 mm^2 footprint and 1.1 mm height and therefore introduce minimal impact into the module stack. As seen in Fig. 1, the sensitive part is exposed through the top opening while other parts are sealed.

Fig. 1: Digital temperature and *RH* sensor SHT25 [13].

To our knowledge, these are currently the only commercially available digital sensors in such a small package that offer relative humidity (*RH*) readings with typical absolute *RH* uncertainty below 2 %RH in large *RH*/*T* ranges. (Fig. 2).

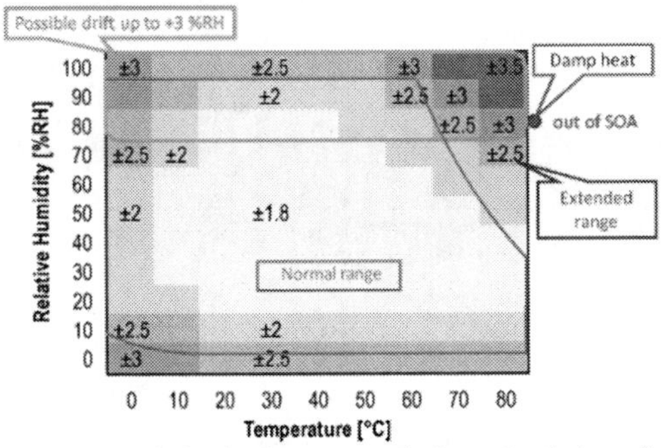

Fig. 2: Typical absolute uncertainties of relative air humidity measurements given in %RH [13].

2.2 Sensor strips

Sensor strips (Fig. 3) are designed in a form of 12 mm wide and 1 mm thick PCBs (FR-4 material) with a 35 µm thick copper plane on each side. They are coated by a solder mask for protection against corrosion. For each sensor, a square slot the size of the sensor footprint is milled out of the PCB. The sensors are placed in the slots and soldered on the bottom side to the connection lines, which are routed from the edges of the slots to the connector on the far right side of the sensor strip. Next to each sensor a small bypass capacitor is added to assure stable power supply voltage. Each sensor strip contains four slots with embedded sensors. Sensors S1-S3 are laminated between two sheets of different encapsulants while sensor S4 measures temperature and *RH* of the surrounding air. The tested encapsulants were EVA, TPO, POE and Ionomer. Thicknesses of individual sheets of encapsulants were 0.45 mm, except with Ionomer which is thinner (0.15 mm). Therefore, three sheets of Ionomer were used on each side to achieve the same overall thickness.

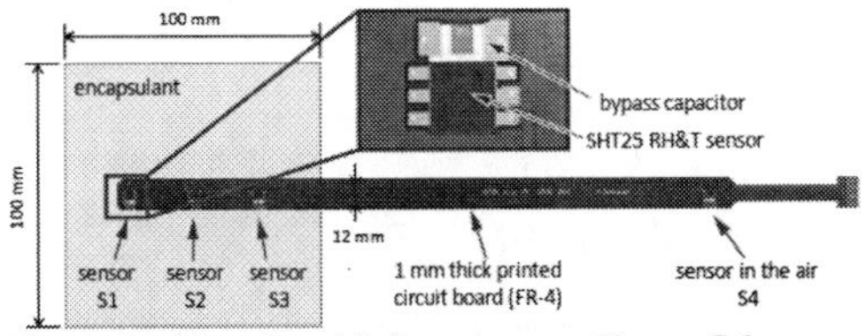

Fig. 3: Sensor strip with four sensors. Three of them are encapsulated and the fourth is exposed to the air.

In order to remove the residual water, the sensor strips were baked at 80°C in an oven for 24h prior to the lamination.

Fig. 4 represents the cross section of the sensor strip at the position of one of the sensors. The water diffusion towards the sensor predominantly follows a perpendicular path through the sheets from both sides simultaneously. Assuming much higher lateral dimensions compared to the sheet thickness other ingress paths can be neglected. However, the entrance to the active area in the sensor's housing is comparable to the sheet thickness (0.3 mm) and has to be accounted for in the overall moisture ingress path length.

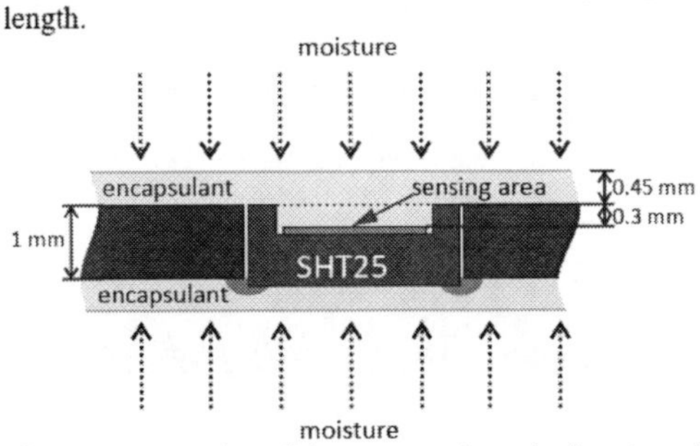

Fig. 4: Cross section of the sensor strip at the location of the sensor.

2.3 Testing procedure in a climatic chamber

All of the prepared sensor strips were put in a climatic chamber at a constant temperature T = 40 °C. Samples were preconditioned at 10 %RH until all *RH* readings of encapsulated sensors were stabilized and moisture equilibrium with the surrounding air was reached. Then ambient *RH* was varied stepwise from 10 %RH to 90 %RH in 10 %RH steps. After each change, the *RH* was kept stable for 8 h. When the last 90 %RH step was finished, the final *RH* was set to 10 %RH until all sensor readings reached initial equilibrium state. All the RH data were simultaneously sampled every 60 s.

3 MOISTURE INGRESS MODEL

Since a PCB with copper layers and protective coatings can be treated as impermeable it is reasonable to assume that the main water diffusion path is only from the top side as depicted in Fig. 5.

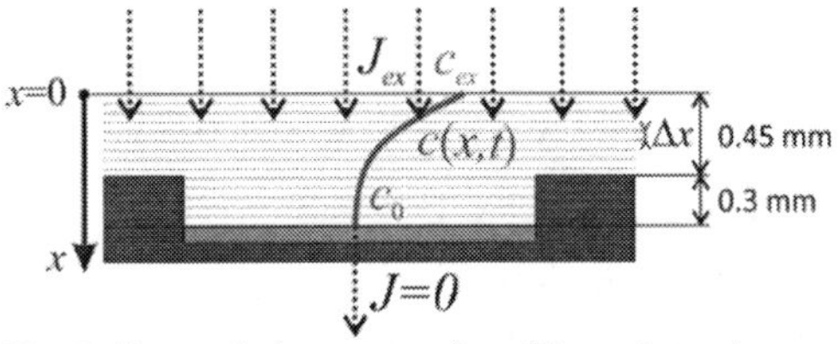

Fig. 5: Geometrical representation of the moisture ingress through the encapsulant to the active area of the sensor.

In the presented case, the moisture ingress can be modelled as a one-dimensional (1-D) Fick's diffusion assuming a constant diffusion coefficient D,

$$J(x,t) = -D\,\frac{\partial c(x,t)}{\partial x}, \qquad (1)$$

where $c(x,t)$ is the water mass density and $J(x,t)$ is the water mass density flow. If the sheet is discretized into n

layers with thickness of Δx, the Eq. (1) can be rewritten into a discrete form as

$$J(x_i,t) = -D\,\frac{c(x_{i+1},t)-c(x_i,t)}{\Delta x}. \qquad (2)$$

Using the analogy between the electricity (Ohm's law) and the diffusion (Fick's 1st law), the problem can be transformed into an equivalent electrical model and solved using SPICE electrical circuit simulator (Fig. 6).

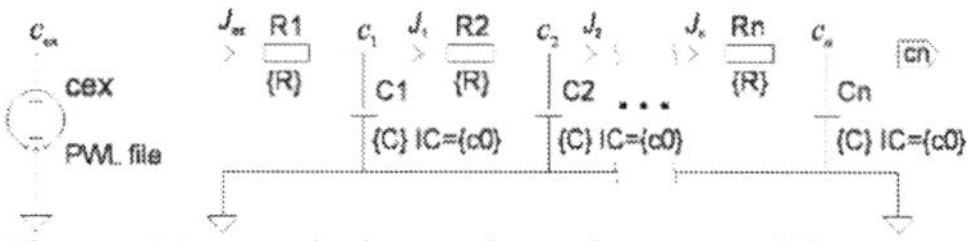

Fig. 6: SPICE equivalent moisture ingress model.

In the equivalent electrical model, resistances represent the inverse value of a diffusion coefficient in a small finite volume $\Delta x \cdot A$,

$$R_i = \frac{\Delta x \cdot A}{D}, \qquad (3)$$

while capacitances represent the volume $\Delta x \cdot A$ itself

$$C_i = \Delta x \cdot A. \qquad (4)$$

Since the resulting currents in the electrical circuit represent the moisture mass density flows $J(x_i)$, all parameters can be normalized to unit area A = 1 m^2.
The measurement results of the sensors are given in %RH. Hence, voltages of the nodes with respect to the reference potential represent the measured *RH* values

$$V_i(t) = RH(x_i,t). \qquad (5)$$

Assuming a linear relation of the water concentration to the partial water vapor pressure p_{H_2O} (Henry's law) and definition of the *RH* we can write

$$c = S \cdot p_{H_2O} = S \cdot RH \cdot p_{H_2Osat}, \qquad (6)$$

where S is the solubility of the encapsulant and p_{H_2Osat} is the partial water pressure in saturation conditions. Thus the Eq. (2) can be rewritten to become

$$J(x_i,t) = -P_{sat}\,\frac{RH(x_{i+1},t)-RH(x_i,t)}{\Delta x}, \qquad (7)$$

where P_{sat} is the permeability of the encapsulant at saturation partial water vapor pressure

$$P_{sat} = D \cdot S \cdot p_{H_2Osat}. \qquad (8)$$

For the simulation, the stack depth was divided into 160 layers (Δx = 4,6875 µm). This discretization turned to be sufficient, matching the analytical solution of the ideal case [5] with an unnoticeable error. The SPICE model was solved using transient analysis, where the driving voltage waveform signal was taken from the actual measured *RH* values in the air. The initial condition (IC) was determined from the initially measured steady state *RH* in the encapsulant at the beginning of the experiment when the system was in equilibrium with the surrounding air.

4 RESULTS

Measured and simulated *RH* data during the climatic chamber experiment are shown in Fig. 7-10. The fitting parameters used in simulations are denoted in the top right corner of the corresponding figures. In each figure the part of the first *RH* step is enlarged to demonstrate the quality of the fit. The permeability coefficients of the materials in P_{sat} were set manually as a compromise to equally match the shape of the curves and the initial dead time.

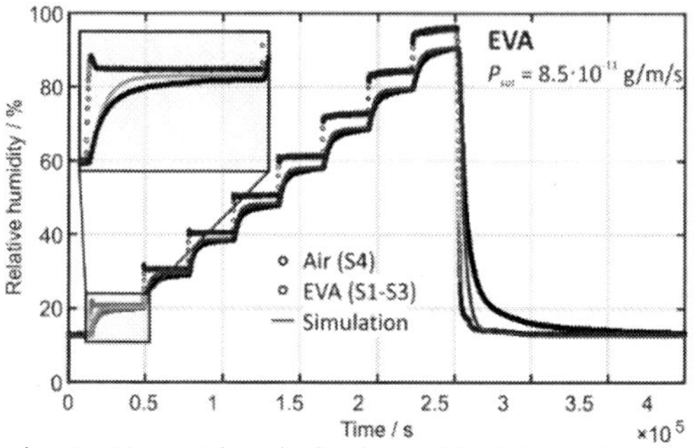

Fig. 7: Measured and simulated *RH* data during the climatic chamber experiment for EVA.

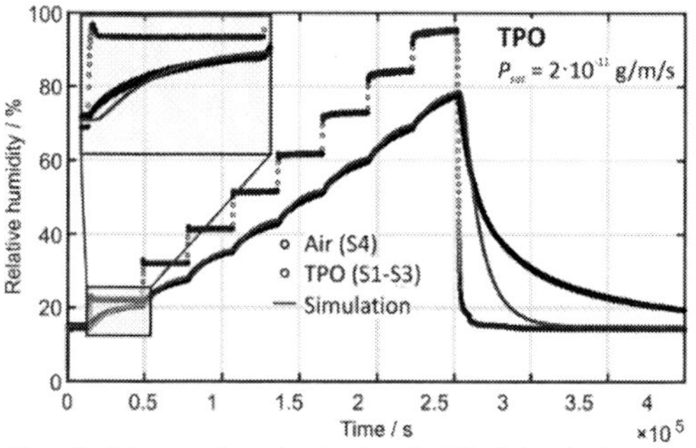

Fig. 8: Measured and simulated *RH* data during the climatic chamber experiment for TPO.

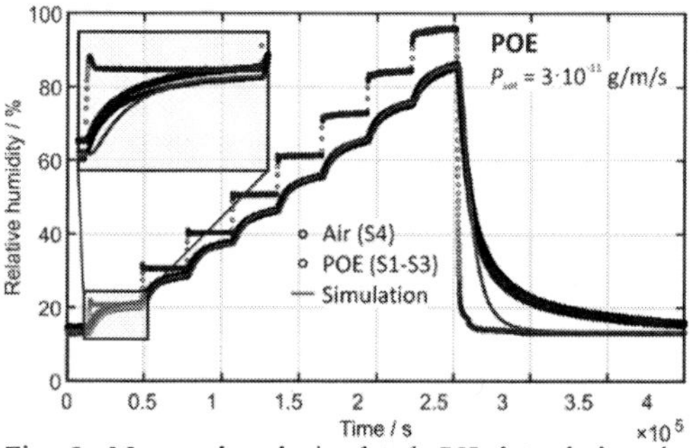

Fig. 9: Measured and simulated *RH* data during the climatic chamber experiment for POE.

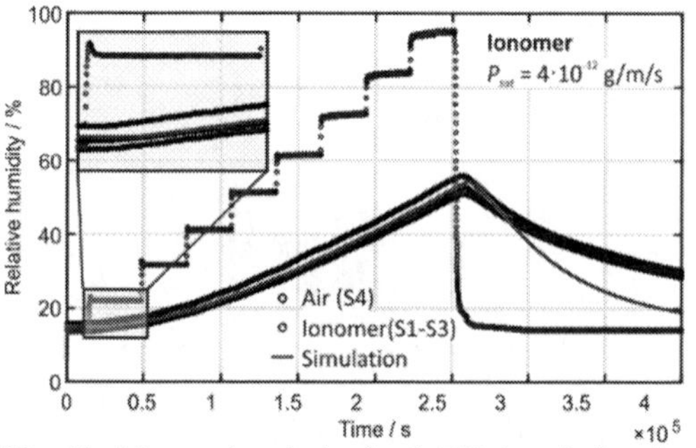

Fig. 10: Measured and simulated *RH* data during the climatic chamber experiment for Ionomer.

5 DISCUSSION

SHT25 humidity sensors prove to be robust to high temperature lamination conditions and perform reliably in all encapsulants under test. Climate chamber tests indicate a linear correlation between the water mass density c and the RH readings of the encapsulated sensors in all materials under test.

SHT25 sensors deliver measurement results recalculated to the %RH as if they are exposed to the air. However, when encapsulated their RH readings are affected by the moisture transition from encapsulant to the sensitive polymer part of the sensors, which is different in general. Thus the obtained RH results in the encapsulants were linearly scaled to account for this effect by matching the measured and simulated RH data in both equilibrium states at the beginning and at the end of the RH steps. Then in each case the permeability value was tuned to best fit the rising parts of the RH signal, where the system and corresponding model can be fairly assumed to be linear due to the relatively small RH steps. In the second part of the signal when the external RH suddenly drops, the model does not fit well anymore. This discrepancy can be accounted to the second order effects caused by the large external RH drop, including inadequacy of 1-D modelling, non-Fickian diffusion and residual water absorption in the FR-4.

Some material data, reference and fitting results are summarized in Table 1.

Material	Ref. $WVTR$ $\left(\frac{g}{m^2 \cdot s}\right)$	Ref. P $\left(\frac{g}{m \cdot s \cdot mbar}\right)$	Fit. P_{sat} $\left(\frac{g}{m \cdot s}\right)$
EVA	$14.7 \cdot 10^{-5}$ @40°C	$4.1 \cdot 10^{-9}$	$8.5 \cdot 10^{-11}$
TPO	$3.3 \cdot 10^{-5}$ @38°C	$8.8 \cdot 10^{-10}$	$2 \cdot 10^{-11}$
POE	$5.8 \cdot 10^{-5}$ @25°C	-	$3 \cdot 10^{-11}$
Ionomer	$7.6 \cdot 10^{-6}$ @38°C	-	$4 \cdot 10^{-12}$

Table 1: Reference and fitted humidity ingress related parameters of all encapsulants under test.

For EVA and TPO we obtained reference permeability values from independent DVS experiments. As a relevant comparison parameter for all materials under test we could only obtain water vapor transmission rates ($WVTR$) from the datasheets. Since from the fits we cannot obtain actual permeability values, we can only compare the ratios of these values. The obtained reference and fitted permeability ratios between the EVA and TPO are respectively 4.66 and 4.25, which is close enough to confirm the validity of our approach. Assuming the validity of the results of the other encapsulants we can conclude that the permeability values of the POE in Ionomer are respectively 2.8 and 21 times lower than that of the EVA at 40 °C. Reference $WVTR$ ratios to the EVA (4.45, 2.53 and 19.3 respectively for the TPO, POE and Ionomer) even though not stated at the same temperatures also agree well with our findings.

6 CONCLUSION

The presented measurement setup proves to be resistant to the lamination conditions and deliver moisture ingress results that are expected for the types of the materials used in the test. The obtained water diffusion properties of all encapsulants are consistent with the advertised moisture related properties from manufacturers.

However, correlation between the sensors' RH readings and the water mass concentrations in the encapsulants still has to be determined by a reference measurement method. This would allow us to measure the actual water mass density in different encapsulants from which we could directly calculate absolute values of the moisture related parameters of these materials.

SPICE simulation model of moisture ingress is easy to use and gives promising results. For our specific use case it should be extended to include a variable diffusion coefficient and a more realistic 3-D geometry description. Including all mentioned improvements, this monitoring method will be able to measure the actual water content in different PV modules with various encapsulants in laboratory and field testing applications.

7 REFERENCES

[1] N. Park, C. Han, and D. Kim, "Effect of moisture condensation on long-term reliability of crystalline silicon photovoltaic modules," *Microelectron. Reliab.*, vol. 53, no. 12, pp. 1922–1926, Dec. 2013.

[2] G. J. Jorgensen, K. M. Terwilliger, J. A. DelCueto, S. H. Glick, M. D. Kempe, J. W. Pankow, F. J. Pern, and T. J. McMahon, "Moisture transport, adhesion, and corrosion protection of PV module packaging materials," *Sol. Energy Mater. Sol. Cells*, vol. 90, no. 16, pp. 2739–2775, Oct. 2006.

[3] S. Mekhilef, R. Saidur, and M. Kamalisarvestani, "Effect of dust, humidity and air velocity on efficiency of photovoltaic cells," *Renew. Sustain. Energy Rev.*, vol. 16, no. 5, pp. 2920–2925, Jun. 2012.

[4] M. M. Rahman, M. Hasanuzzaman, and N. A. Rahim, "Effects of various parameters on PV-module power and efficiency," *Energy Convers. Manag.*, vol. 103, pp. 348–358, Oct. 2015.

[5] J. Crank, *The mathematics of diffusion*, 2d ed. Oxford, [Eng]: Clarendon Press, 1975.

[6] N. Kim and C. Han, "Experimental characterization and simulation of water vapor diffusion through various encapsulants used in PV modules," *Sol. Energy Mater. Sol. Cells*, vol. 116, pp. 68–75, Sep. 2013.

[7] G. Huyberechts and L. Frisson, "In situ formation of humidity-sensitive devices for the evaluation of solar panel encapsulations," *Sens. Actuators B Chem.*, vol. 27, no. 1–3, pp. 308–311, Jun. 1995.

[8] T. Carlsson, J. Halme, P. Lund, and P. Konttinen, "Moisture sensor at glass/polymer interface for monitoring of photovoltaic module encapsulants," *Sens. Actuators Phys.*, vol. 125, no. 2, pp. 281–287, Jan. 2006.

[9] Eleonora Annigoni, Marko Jankovec, Federico Galliano, Heng Yu Li, Christophe Ballif, and Fanny Sculati-Meillaud, "Moisture ingress into PV modules: long-term simulations and a new monitoring technique," presented at the NREL PV Module Reliability Workshop, Golden, Colorado, U.S., 2015.

[10] Marko Jankovec, Federico Galliano, Eleonora Annigoni, Heng Yu Li, Fanny Sculati-Meillaud, Christophe Ballif, Laure-Emmanuelle Perret-Aebi, and Marko Topič, "Moisture ingress in PV modules: innovative and versatile in-situ

monitoring technique," presented at the 30th European Photovoltaic Solar Energy Conference and Exhibition, Hamburg, Germany, 2015.

[11] Marko Jankovec, Federico Galliano, Eleonora Annigoni, Heng Yu Li, Fanny Sculati-Meillaud, Laure. E. Perret-Aebi, Christophe Ballif, and Marko Topič, "In-situ monitoring of moisture ingress in PV modules using digital humidity sensors," *J. Photovolt.*, vol. Accepted for publication, 2016.

[12] Marko Jankovec, Federico Galliano, Eleonora Annigoni, Heng Yu Li, Fanny Sculati-Meillaud, Christophe Ballif, Laure-Emmanuelle Perret-Aebi, and Marko Topič, "Long term stability of humidity sensors, laminated in EVA encapsulant," presented at the M I D E M 2 0 1 5 - International Conference on Microelectronics, Devices and Materials, Bled, slovenia.

[13] "SHT-25 datasheet." [Online]. Available: http://www.sensirion.com/fileadmin/user_upload/customers/sensirion/Dokumente/Humidity/Sensirion_Humidity_SHT25_Datasheet_V3.pdf.

SHADOWING INVESTIGATIONS ON THIN FILM MODULES

Stefan Wendlandt[1,2], Sarah Berendes[1], Thomas Weber[1], Juliane Berghold[1], Stefan Krauter[1,2] and Paul Grunow[1]
[1] PI Photovoltaik-Institut Berlin AG, Wrangelstr. 100, D-10997 Berlin, Germany
Phone: +49 (30) 814 52 64 205, Fax: +49 (30) 814 52 64 101, e-mail: wendlandt@pi-berlin.com
[2] University of Paderborn, Faculty of Electrical Engineering, Informatics and Mathematics, Department for Electrical Energy Technology & Sustainable Energy Concepts, Pohlweg 55, D-33098 Paderborn, Germany, e-mail: krauter@nek.upb.de

ABSTRACT: PV modules in solar plants are exposed to changing shading conditions. Due to partial module shading, cells are forced to operate in their reverse bias, which potentially leads to hot spots. In extreme situations, the extraordinarily high temperature gradient may cause in glass breakage. As a consequence, the thermal stability of pv modules is an important safety issue, especially in overhead installations. Warranty issues might be another consequence. In this paper, the thermal hot spot stability of different commercial thin film technologies will be investigated under artificial light exposure. Realistic shading conditions are simulated by shading elements of field realistic shapes that are moved step wise over the module. From ir images the failure location will be derived and correlated to other investigation technologies, such as power measurements, visual inspection and electroluminescence. This work also presents a hot spot risk benchmark of the investigated thin film modules, and their consequences regarding reliability issues. This knowledge is important for all quality assurance activities along the pv value chain, in the field and in the laboratory, as well as for the process of standardization.

Keywords: shadowing, thin film, module reliability

1 INTRODUCTION

On one hand, thin film technology (TF) provides – due to its large scale coating, integrated series circuitry, high material savings and high flexibility within the realization process of innovative module and cell concepts. Plus, it carries a high potential for cost reduction. On the other hand, beside these advantages TF-technology entails some risks. In this paper the hot spot risk of thin film modules is investigated.

It is known, that the reverse bias characteristic depends on the thickness of the i-ZnO layer at CIGSe modules [1]. *Messmann-Vera* has shown that, in reverse bias, CIGSe cells with thin i-ZnO layers have high currents at low voltages. *Herrmann & Alonso et al.* developed a hot spot test sequence for standardization [2]. *Silverman et al.* have done shading experiments and simulations [3]. He has observed a power loss of about 4% to 11% after hot spot testing at CIGSe modules. Furthermore, in his work, he presents visible defects and black spots at EL images in the shaded area after the hot spot testing. *Silverman et al.* has also published the dependency of reverse bias behaviour of CIGSe cells and illumination [4]. *Johnston et al.* have compared different imaging techniques like ReBEL, ILIT, $DLIT_{for}$, and $DLIT_{rev}$ and concluded that these methods can be used to identify hot spots [5]. Also, he has been able to identify weak diodes at hot spot locations, using a microscope. *A. Vasko et al.* has developed electrical and thermal-numerical models to simulate shadowing effects on thin film modules [6]. *Philip Mack et al.* have shown, that the reverse characteristics of CIGSe cells are affected by metastabilities, as it has been equally observed for the forward characteristics [7]. In 2009 *P-O. Westin et al.* have presented, that visual damages, which occur during reverse bias of CIGSe modules, are forced in cases where the heating produces evaporation and pore formation at the absorber/window interface [8].

When thin film modules became mass production – hot spot sensitivity represented a widespread problem [9]. The industry realized the problem and took measures like smaller cells, segmentation and improved production steps. A good description of the hot spot problem for thin films and a comparison to c-Si modules can be found here [10].

In this paper the hot spot risk of current thin film pv module technologies will be studied. Therefore, realistic shadowing conditions are simulated by shading elements of field realistic shapes that are moved step wise over the module. From ir images the failure location will be derived and correlated to other investigation technologies, such as power measurements, visual inspection and electroluminescence. This work also presents a hot spot risk benchmark of the investigated thin film modules, and their consequences regarding reliability issues.

2 EXPERIMENTAL

In this study the term "hot spot" is used for any local heating during module shadowing, observed with the ir camera.

Modules with a-Si, a-Si/µc-Si, CIGSe and CdTe semiconductors and different cell interconnection concepts were investigated and compared. Table I shows an overview of the stressed modules.

Table I: Overview of shadowed thin film modules.

Indication	Technology	Interconnection	Cell-Current
Module 1	a-Si	series	1.01 A
Module 2	a-Si	series	0.92 A
Module 3	a-Si	series-parallel	0.49 A*
Module 4	a-Si/µ-Si	series-parallel	0.38 A*
Module 5	CIGSe	series	2.54 A
Module 6	CIGSe	series	1.82 A
Module 7	CIGSe	series-parallel	0.32 A*
Module 8	CdTe	series	1.71 A
Module 9	CdTe	series	1.72 A

* Cell current related to over 60 % of the module area

IR images were analyses from field inspections. This enables to distinguish the most realistic shadowing conditions (scenarios), which have been applied on the modules (see Table II). To realise the shadowing, an opaque cardboard has been used. The shading element has been placed directly on the module.

Table II: Applied shadowing scenarios.

Scenario	Description	Schematic view
#1	Increasing shadow from the short module edge. Step size 10 cm	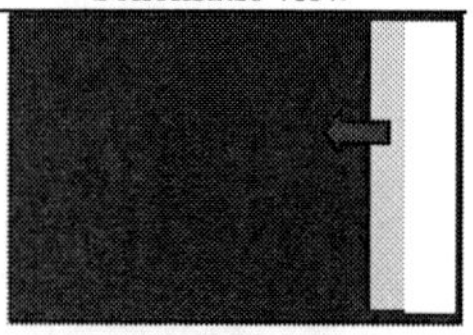
#2	Decreasing shadow at the short module edge. Shadow-width 10 cm	
#3	Shifting shadow from the short module edge. Shadow-width 10 cm	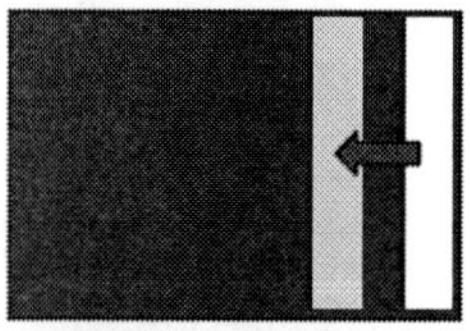
#4	Rotating shadow element from the short module edge. Shadow-width 10 cm. Rotating angle 5…10°	
#5	Diagonal shifting shadow from the short module edge. Step size 10 cm	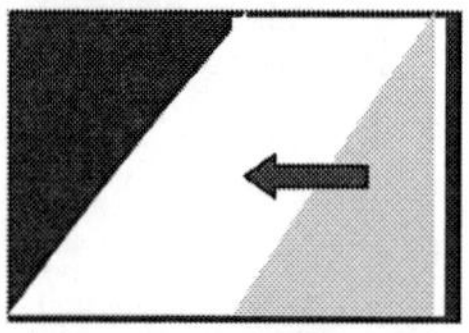
#6	Increasing shadow from the long module edge. Step size 10 cm	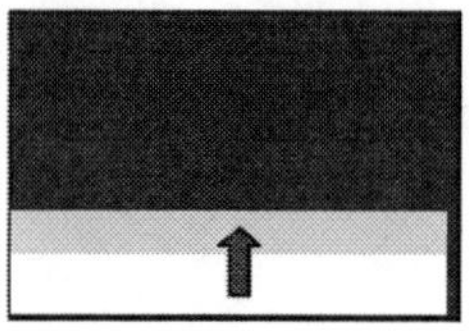
#7	Shifting shadow perpendicular from short module edge. Shadow-width 10 cm	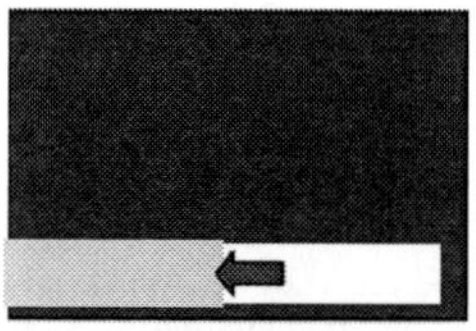
#8	Shifting shadow from the long module edge. Shadow-width 10 cm	
#9	Rotating shadow element from the long module edge. Shadow-width 10 cm. Rotating angle 5…10°	
#10	Single shadowing points	

The tests were performed in a class C solar simulator at E_e = 800 W/m². The modules were kept at the operation point of their string I_{MPP}. Additionally to these settings, the mean module temperature of the unshaded module area is set to 55°C. After a stabilization phase of around 30 min, the modules are shadowed as shown in Table II. After each shadowing exposure of 10 min an infrared image was taken to measure the temperature distribution, as well as to determine critical hot spot locations for the specific module design.

3 RESULTS

The tested modules are grouped and presented according to their respective technology types.

3.1 Investigations on CdTe modules

The ir investigations of CdTe modules, 8 and 9, show visible hot spots under various testing conditions. The temperature range of the detected hot spots varies between 70…90 °C for module 8, and 60…70°C for module 9. Hot spots of CdTe modules are mainly detected in cases, where the shading element is located in the direction of cell interconnection (#1), or in cases of diagonal shading (#5). Both scenario types cause, under certain testing conditions, the shading of a number of cells, or of a big surface area of a number of cells that are serial-connected. Therefore, hot spot occurrence is enhanced. A selection of characteristic testing results of CdTe modules can be derived from Figure 1.

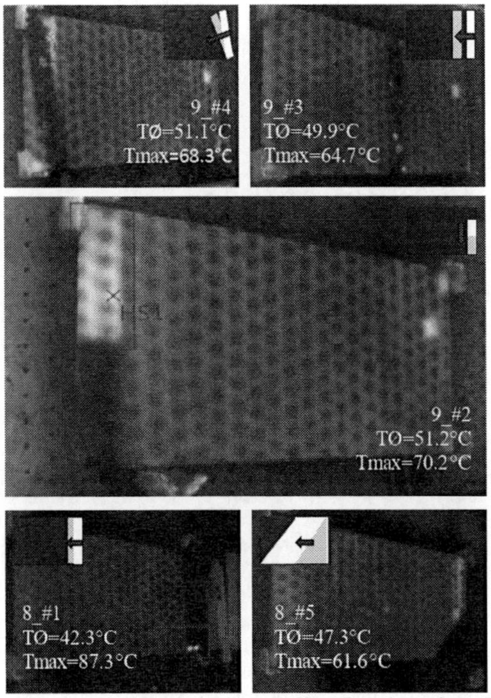

Figure 1: Selected ir images for five investigated shadowing conditions of modules 8 and 9.

Figure 1 shows module 9 after shadowing scenario #2 (shading rate of 50 %) and module 8 after shadowing scenario #5 at a low diagonal shading rate. In both cases hot spots can be observed.

Figure 2 summarizes the results for CdTe modules. The main hot spot locations were found within the shaded area (9_#3), or on its edges (9_#4), as well as on the edge, or in the corner of the module itself (8_#1, 8_#5). Another finding for module 9 is the significant temperature increase at the location of the junction box.

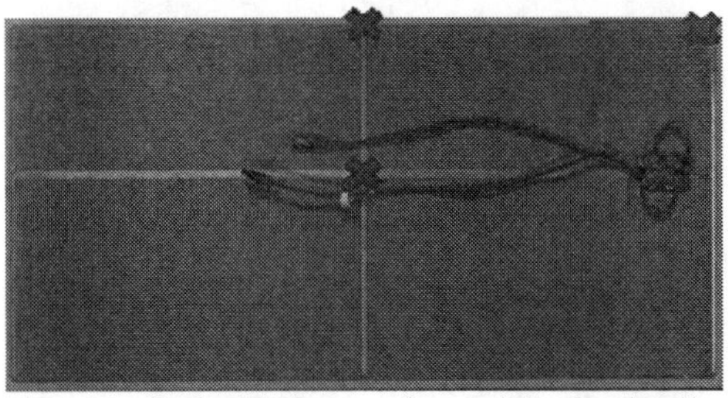

Figure 2: Back side image of a module showing the typical hot spot locations

The results of the investigated CdTe modules can be evaluated as uncritical for the integrity of the laminate and resulted under each condition and for each module tested to no further problem.

3.2 Investigations on CIGSe modules

Regarding the testing results of CIGSe technology modules, no critical hot spots can be detected for modules 5 and 6, despite of very low heating rates of small cell areas in the corner of the module.

Whereas, module 7 shows hot spots at temperature ranges between 65…80°C, which are considered to be uncritical. As can be seen from Figure 3, a shading scenario of type #6, which corresponds to shading in the direction of cell interconnection, leads to hot spot development. With an increasing shading area (up to 20 cm), the number and intensity of hot spots has been reduced. Furthermore, hot spots have emerged over the full width of the module for shading scenario #7, #8 and #9. Scenario #7 shows additional partial areal heating of unshaded cell segments.

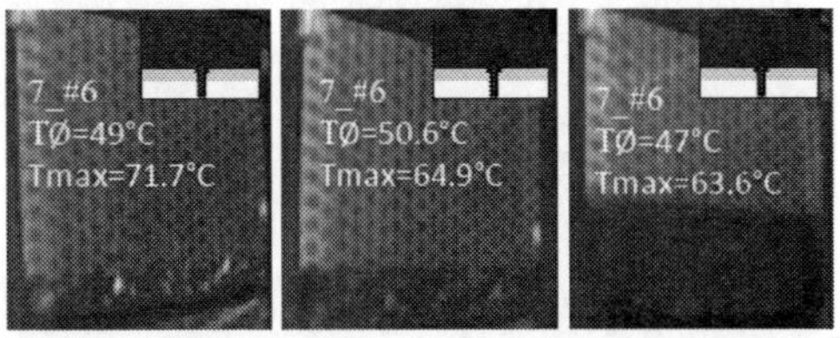

Figure 3: IR images for shading scenario type #6 of module 7, left 10 cm, middle 15 cm and right 25 cm.

For shading scenario type #9, three heating characteristics can be found (see Figure 4). While in case 1, hot spots solely appear within the shading area (0…20 cm shift of the shading element), case 2 shows additional heating on the edges of the shading element at regular distances (20…30 cm shift). In case 3, no more hot spots emerge within the shaded area, solely regular edge heating can be found (30 cm module width shift). The effect of edge heating can be observed for scenario #10, as well.

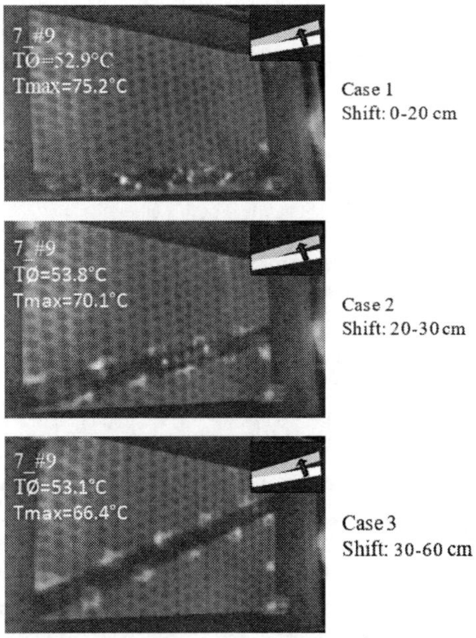

Figure 4: IR images for shading scenario type #9 of module 7 (CIGSe).

Due to its regular distances and its homogenous heating rate, the hot spot development on the edges of the shading element, is most likely to be a result of segmentation. As figured out from visual inspection (VI), edge heating mainly occurs at the cell boundaries.

Comparing ir images with VI inspection results, one can see that hot spots are most likely to be found at the cell interconnections or in the range of edge insulation (see Figure 5). As can be found in Figure 5, further hot spots have been detected at segment boundaries, on the edges, as well as, in the corners of the module.

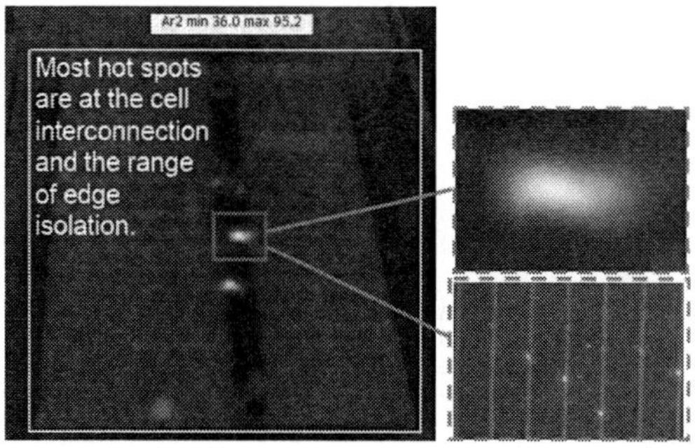

Figure 5: Front-side image and IR image showing hot spot locations

CIGSe technology modules show uncritical temperatures for all heating scenarios.

3.3 Investigations on a-Si modules

For the modules 1 and 2 hot spots have been found for shading scenario #6, which, corresponds to the direction of cell interconnection. Also the number and the temperature of hot spots decrease with an increased shading area. Module 1 shows hot spots for scenario type #8 (over the whole width of the module) and for scenario #9 (10 cm shift), with a temperature range of 65...80°C. Regarding diagonal shading scenarios, hot spots and partial areal heating on the edges of the module, as well as, on the edges of the shading element, become visible for module 3. No critical heating characteristics have been found for module 2.

Module 2 shows hot spots for scenario types #8 over the full width of the module. For this tape of scenario, the shading direction corresponds to the cell interconnection. In this case, a very critical hot spot temperature up to 130°C has been reached. The corresponding ir image can be seen in Figure 6.

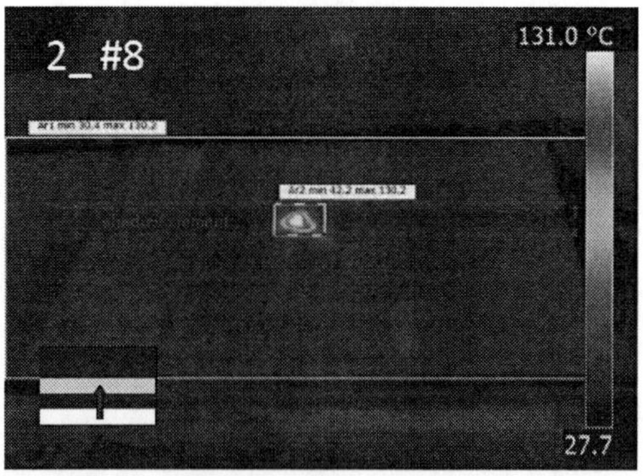

Figure 6: IR image for scenario type #8 of module 2 (a-Si) reaching a very critical temperature.

At high temperatures, the risk of cell damage, delaminations and glass breakage is enhanced. Figure 7 shows that hot spot locations of module 2 have been observed, as so called "white spots" after testing.

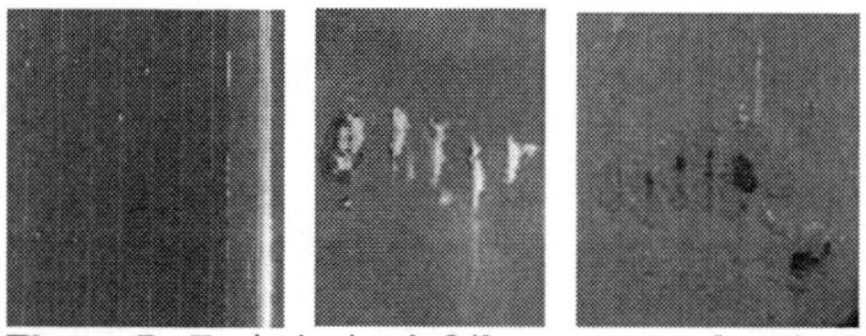

Figure 7: Typical visual failure pattern of a-Si after shadowing. Left: weak delaminations "white spots"; middle: strong delaminations; right: strong delaminations with carbonization.

3.4 Investigations on a-Si/µc-Si module

Some module types of a-Si/µc-Si modules seem to be very sensitive to hot spots. Field returns in the laboratory and field inspections revealed a visible delamination of the semiconductor layers showing white / bright, large cell areas in the module centre or at the edges. Figure 8 shows delamination of an affected field module (middle, left image). The performed hot spot test with scenario #3 (shifting shadow from the short module edge) lead to new hot spots of 57°C in maximum, resulting in new delaminations, as can be seen in the visual images that are taken after the test (middle right images). These delaminations correlate locally with the hot spots found in the ir image and dark cell parts in the electroluminescence image. Therefore, a correlation with power losses must be assumed but could not be proven in this case, due to superimposed degradation effects and high measurement uncertainties for thin film modules.

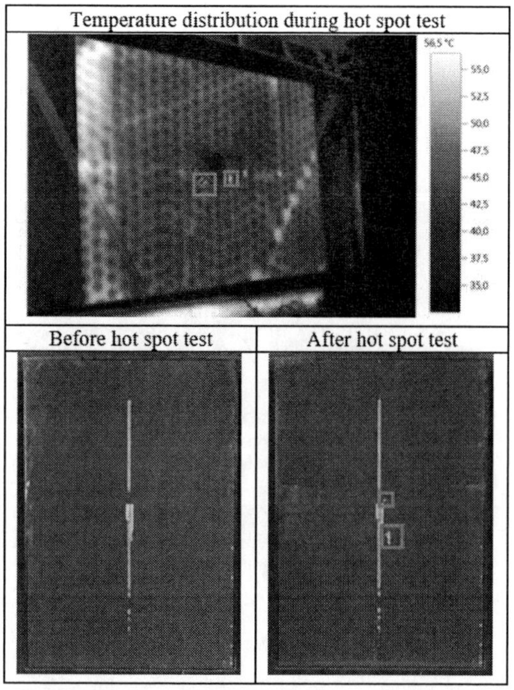

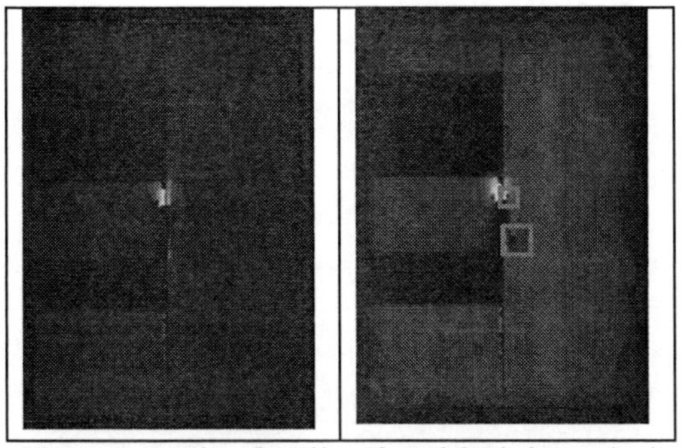

Figure 8: Results of a-Si/μc-Si module: ir image (top), visual (middle) and electroluminescence (bottom) before and after the performed hot spot test.

An important question that arises when evaluating the presented hot spot results would be the differentiation to another common failure mode of Si-TF modules. Negative bias conditions and water ingress in the module lead the so called "TCO corrosion". In this work, this failure mode could be provoked due to TCO corrosion testing and a comparison of the failure mode appearance could be derived.

The TCO corrosion tests are described elsewhere in detail and were performed under 100h of biased damp-heat testing (-1 kV against ground, 85°C, 85%RH) [11]. Figure 9 shows a hot spot field effected module before (left) and after TCO corrosion test (right). The coloured framed images show enlargements of delaminations due to hot spots (blue) and TCO corrosion (orange). The same blue framed module part is depicted in Figure 11, again, showing the comparison before and after TCO corrosion test. One can see the hot spot delamination before the test on the left side. These delaminated parts appear homogenous bright and thick worm like. The TCO corrosion in the right-site images appear superimposed to the hot spot one, appear thin worm like and less bright. It's clear visible, that the TCO corrosion is starting from the edge (due to the water ingress) whereas the hot spot delamination can occur everywhere.

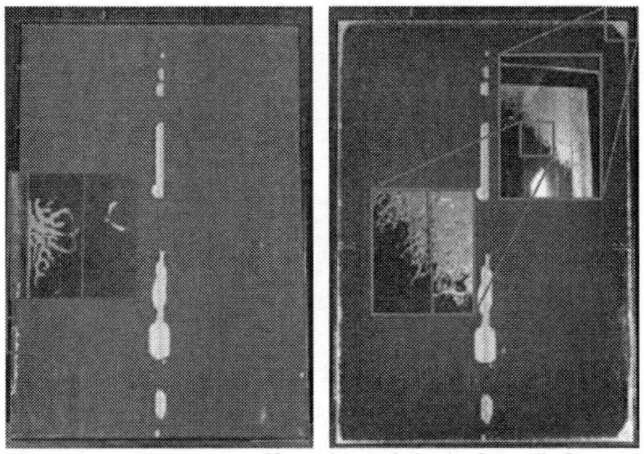

Figure 9: Hot spot effected module before (left) and after TCO corrosion test (right). The coloured, framed images show enlargements of delaminations due to hot-spots (blue) and TCO corrosion (orange).

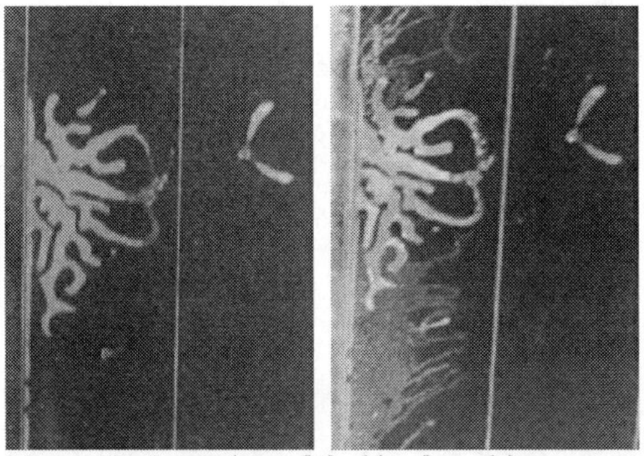

Figure 10: Comparison of the blue framed images part from Figure 9. Left: showing the area before TCO corrosion test (showing only hot spot delamination). Right: showing the same area after the TCO corrosion test with the new failure mode (TCO corrosion) superimposed.

4 SUMMARY AND CONCLUSIONS

The hot spot risk associated with thin film modules arise (among other reasons) due to production-related small-grain crystal structures producing poorer diode characteristics in thin-layer cells with reverse current, thus producing a lower blocking effect.

Thin film modules should be designed to withstand critical shading and resulting thermal stress due to hot spots, whereby each technology / product has its own optimal parameter. The cell current (high current → high risk), cell interconnection design (the shading scenarios, where parts of serial connected cells are covered lead to current gradients within the module), parameter of interconnection process (laser speed, laser type, laser intensity) and glass properties could be identified as important risk factors. Due to the fact, that high cell currents are more likely causing higher temperature gradients, it should be discussed, whether a serial-parallel interconnection should be favoured over a serial interconnection of cells. A serial parallel connection would, on the one hand, lead to reduced cell currents, whereas, on the other hand, the hot spot risk at the cell boundaries will be increased. In order to classify the optimal parameters, further investigations are necessary. Next to it the material side like the stoichiometry relation of semiconductors layers and its thickness are also important to reduce the risk.

From all shadowing experiments done, the following summery is showing the risk benchmark (see Table III & IV). The following legend is used: + high/often, +/- medium, - low/rarely. For the modules under investigation, Si thin-film modules exhibit a high risk to hot spots resulting likely in cell and glass breakage and in consequence power and yield loss. The investigated CIGSe modules perform quite better. They are showing frequently hot spots and cell damages during reverse bias but always in uncritical temperature ranges. The CdTe modules investigated performed best and showed no problems.

These results correspond well to the outdoor experience of PI-Berlin.

Table III: Hot spot formation of all modules and all shading scenarios.

	Module Type								
	a-Si			μc-Si	CIGSe			CdTe	
Scenario	1	2	3	4	5	6	7	8	9
#01	-	-	-	-	-	-	-	-	+/-
#02	-	-	-	-	-	-	-	-	+/-
#03	-	-	-	-	-	-	-	-	+
#04	-	-	-	-	-	-	-	-	+/-
#05	-	-	+/-	-	-	-	-	+/-	-
#06	+/-	-	+	-	-	+	+	-	-
#07	-	-	-	-	-	-	+/-	-	-
#08	+	+	-	+	-	-	+/-	-	-
#09	+/-	-	-	+/-	-	-	+/-	-	-
#10	-	-	-	-	-	-	+/-	-	-

Table IV: Hot spot risk of thin film pv modules

	Thin film technology			
Property	a-Si	a-Si/μc-Si	CIGSe	CdTe
Hot spot frequency	+	+	+/-	+/-
Cell damage caused by reverse bias	+	+	+/-	-
Broken glass	+/-	+/-	-	-
Power loss	+/-	+/-	+/-	+/-

The hot spots found are located at the module edge, at the cell interconnection, the junction box and the crossbars.

Furthermore the a-Si/μ-Si module investigated showed after shadowing very strong delaminations. These are visually different from TCO corrosion.

All modules investigated are certified products acc. IEC 61646. That means the prescribed hot spot test for thin film modules were successfully passed. The hot spot sensitivity of Si thin-film modules was not discovered by the tests showing that the standard is not suited to evaluate the field relevant hot spot sensitivity.

5 ACKNOWLEDGEMENTS

The authors would like to thank the whole Photovoltaik Institut Berlin team for its support. This paper was financially supported by the Federal Ministry for Economic Affairs and Energy (Contract number FKZ 03ET1312).

6 REFERENCES

[1] Lorena Alexandra Messmann-Vera, "*Einfluss der undotierten ZnOSchicht auf die elektrischen Eigenschaften von ZnO/Cu(In,Ga)Se$_2$-Solarzellen*", Schriftreihe des HZB, Institut für Technologie E-I3, Oktober 2010

[2] W. Herrmann, M.C. Alonso: "*Behaviour of Thin-Film Modules under Shading*", 19th EUPVSEC, Paris 2004

[3] Timothy J. Silverman, Michael G. Deceglie, Chris Deline, Sarah Kurtz: "*Partial Shade Stress Test for Thin-Film Photovoltaic Modules*", SPIE Optics + Photonics, San Diego, California, August 9...13, 2015

[4] Timothy J. Silverman, Michael G. Deceglie, Rebekah L. Garris, Chris Deline, Sarah Kurtz, Xingshu Sun, Muhamma d Ashraful Alam: "*Thermal and Electrical Effects of Partial Shade in Monolithic Thin - Film Photovoltaic Modules*", 42nd IEEE Photovoltaic Specialists Conference, New Orleans, Louisiana, June 14…19, 2015
[5] Steve Johnston, Ingrid Repins, Nathan Call, Rajalakshmi Sundaramoorthy, Kim M. Jones, Bobby To: "*Applications of imaging techniques to Si, Cu(In,Ga)Se₂, and CdTe and correlation to solar cell parameters*", Photovoltaic Specialists Conference (PVSC), 2010 35th IEEE
[6] A. Vasko, A. Vijh, V. G. Karpov: „*Hot spots spontaneously emerging in thin film photovoltaics*", Solar Energy, Volume 108, October 2014, Pages 264–273
[7] Philip Mack, Thomas Walter, Robert Kniese, Raymund Schäffler: „*Reverse Bias and Reverse Currents in CIGS Thin Film Solar Cells and Modules*", 23rd European Photovoltaic Solar Energy Conference, 1-5 September 2008, Valencia, Spain
[8] P-O. Westin, U. Zimmermann, L. Stolt, M. Edoff: „*Reverse Bias Damage in CIGS Modules*", 24th European Photovoltaic Solar Energy Conference, 21-25 September 2009, Hamburg, Germany
[9] S. Wendlandt et al., *Hot Spot Risk Assessment of PV Modules*, Investors day, PI Berlin, 28th September 2011, Berlin, Germany
[10] IEA PVPS Task 13: „*Performance and Reliability of Photovoltaic Systems Subtask 3.2: Review of Failures of Photovoltaic Modules*" External final report, March 2014
[11] T. Weber, J. Berghold, F. Heilmann, M. Roericht, S. Krauter, P. Grunow: „ *Test Sequence Development for Evaluation of Potential Induced Degradation on Thin-Film Modules*", 28th European Photovoltaic Solar Energy Conference, 2013, Paris, Spain

INVESTIGATION OF UV-INDUCED DEGRADATION OF DIFFERENT TYPES OF WPVS REFERENCE SOLAR CELLS

Ingo Kröger[1], Jochen Hohl-Ebinger[2], Stefan Brachmann[2], Stefan Winter[1]
[1]Physikalisch-Technische Bundesanstalt (PTB), Bundesallee 100, 38116 Braunschweig, Germany
[2]Fraunhofer Institut für Solare Energiesysteme (ISE) Heidenhofstrasse 2, 79110 Freiburg, Germany
Tel.: +49 531 592 4147, fax: +49 531 592 694147, e-mail: ingo.kroeger@ptb.de

ABSTRACT: Calibrated World Photovoltaic Scale (WPVS) reference solar cells are widely used for calibrating the irradiance of solar simulators as well as the determination of spectral mismatch correction factors. WPVS reference solar cells should provide long term stability of the short circuit current under standard test conditions < 0.3% over typical recalibration intervals of 1-2 years. The UV-irradiance of the utilized light sources (natural sunlight or solar simulators) can lead to degradation effects either in the solar cell itself or in the encapsulant. In this study different types of WPVS reference solar cells were exposed to elevated UV-irradiance in order to investigate the long term stability of the devices over a period of several hundreds of hours of permanent irradiation. Prior and after the UV-exposure high precision differential spectral responsivity (DSR) calibrations were performed in order to investigate the impact of the UV-aging on the spectral responsivity and hence the short circuit current of the individual devices. In order to identify small relative changes < 0.2% of the spectral responsivity a DSR-facility is needed with very high reproducibility. This could be realized at the DSR and Laser-DSR facilities at PTB.
Nine different types of reference solar cells were prepared by Fraunhofer ISE. They were developed to spectrally match different solar cell technologies. The BG38 and BG40 filtered devices should spectrally match a-Si solar cells, the OG590 and RG610 filtered devices should spectrally match µ-Si solar cells and the GaAs solar cell should spectrally match CdTe solar cells. The three unfiltered c-Si devices show different spectral responsivity in the IR-region. In conclusion these reference devices represent a thoroughly set covering all possible spectral responsivities needed in the PV-calibration and testing community.
Keywords: calibration, characterization, degradation, durability, encapsulation, radiation damage, reference cell, stability

1 INTRODUCTION

The comparability of different photovoltaic technologies and devices is maintained by the international standard IEC 60904 that defines the standard test condition (STC): irradiance E=1000W/m², AM1.5g spectrum and the device temperature of 25°C. If for some reason calibration or testing of photovoltaic devices of a given technology is intrinsically associated with large measurement uncertainties, generally problematic or even unreliable this might lead to a significant obstacle for a successful market entrance. Therefore, the development of measurement techniques for new photovoltaic technologies may be as important as the development of the technology itself.

For calibration and testing of photovoltaic devices solar simulators are generally used. For adjustment of the irradiance of the solar simulators to the demanded value of 1000W/m² calibrated reference devices are needed that are traceable to the SI-units.

Whenever the spectral responsivity of the device under test (DUT) differs significantly from the spectral responsivity of the reference device a spectral mismatch correction is needed for correcting the measured I_{STC} of the DUT. Generally, the uncertainty of the spectral mismatch correction factor increases with the deviation of the mismatch factor from unity. Hence a reference device that is spectrally matched to the DUT can lead to a significant reduction of measurement uncertainties.

However, the most important property of a reference device is the long term stability since recalibration intervals are often in the order of 1-2 years. If the short circuit current at standard test condition I_{STC} of the reference device changes significantly over time of use either shorter recalibration intervals are needed (that lead to extra costs) or the assigned uncertainty of the calibration value as to be increased over time until recalibration.

From experience with world photovoltaic scale (WPVS) reference devices encapsulated silicon solar cells have generally proven to be very stable. However, whenever new materials, encapsulants or solar cell technologies are incorporated into a new generation of reference devices the long term stability has to be investigated.

In this study we investigate the ultraviolet radiation (UV) induced degradation of different types of recently developed reference devices. UV radiation that occurs under real outdoor conditions but also when using solar simulators can be considered to be the most harmful. UV radiation might lead to reduced optical transmission in the encapsulants materials or defects within the semiconductor material. The degradation effect will be quantified by calibrating the reference devices prior and after enhanced UV-exposure using the differential spectral responsivity (DSR) facilities at PTB.

2 REFERENCE DEVICES

The set of reference devices was designed by Fraunhofer ISE to spectrally match the wide range of existing photovoltaic technologies (see Table 1) covering thin film technologies such as amorphous silicon, micro-crystalline silicon and CdTe as well as silicon technologies with different infrared sensitivity. Please note, that the reference device assignment A – I from

Table 1 will be used in the course of this paper when discussing the results. The spectral responsivity of these reference devices was measured by PTB and is shown in Figure 1 and clearly underlines the variety of the spectral matching.

Table 1: Reference devices

	Material	spectral match to
A	p-type c-Si + HOQ filter	c-Si/CIGS
B	p-type c-Si + cover glass	c-Si/CIGS
C	n-type c-Si + cover glass	c-Si/CIGS
D	c-Si + BG40 filter	a-Si
E	c-Si + BG38 filter	a-Si
F	c-Si + OG590 filter	μ-Si
G	c-Si + RG610 filter	μ-Si
H	GaAs	CdTe/CIS
I	GaAs + cover glass	CdTe/CIS

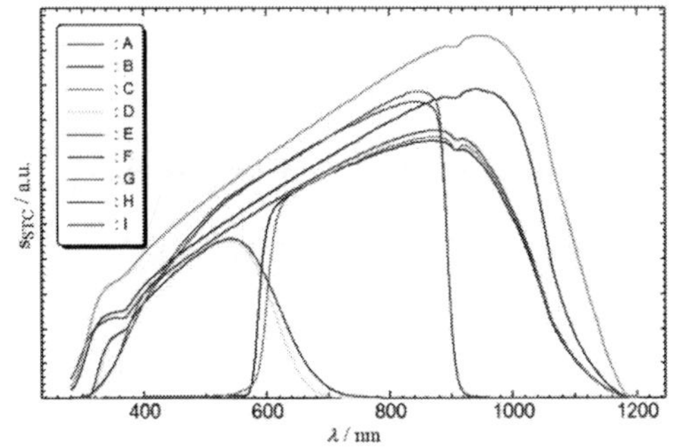

Figure 1: Spectral responsivity of the investigated reference devices. The assignment A-I can be found in Table 1.

3 EXPERIMENTAL

In this section we will describe the DSR facilities at PTB that were used for the spectral responsivity measurements, the UV-light source and finally the ageing procedure.

3.1 The DSR-facilities at PTB

The DSR facilities at PTB are primary calibration facilities that allow the calibration of reference solar cells with an expanded uncertainty of U(I_{STC}) < 0.6% for the lamp-based DSR facility and U(I_{STC}) < 0.4% for the laser-based DSR facility. At both facilities the absolute spectral irradiance responsivity at STC is measured. The I_{STC} is derived mathematically using the tabulated AM1.5g spectrum from the IEC60904-3 standard. Details about these facilities can be found elsewhere [1, 2].

However, the given uncertainties represent the absolute calibration of the I_{STC}. When it comes to ageing investigations only the relative change of the spectral responsivity and I_{STC} is discussed and statements of significance are wanted only for these changes. Therefore, all systematic uncertainty contribution that are related to the absolute value can be ignored for degradation investigations leaving only uncertainty contributions that contribute to the reproducibility of a measurement. For both facilities the reproducibility of measuring the I_{STC} of a reference device can be considered to be < 0.2%. This means that statements of significance can be made for observed ageing effect of a reference device > 0.2%. Less pronounced ageing effect cannot be stated to be significant since they also might result from the reproducibility of the measurement.

3.2 The UV-light source

For the enhanced UV-induces degradation two Labino UV-light sources were focused on the reference device. The spectral irradiance of the UV-source is shown in comparison the AM1.5g spectrum in Figure 2. It can be seen, that in the focus of the UV-sources the spectral irradiance about one order of magnitude larger that the AM1.5g spectrum. This enables enhanced UV-degradation investigations.

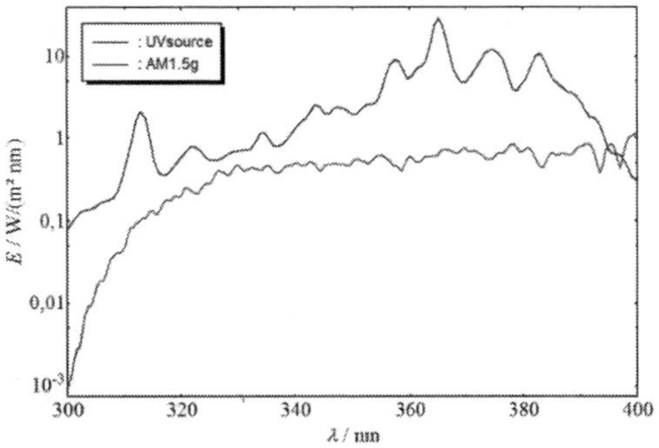

Figure 2: Comparison of the spectral irradiance of the Labino UV-light source and AM1.5g.

3.3 The degradation procedure

For investigation of the degradation the reference device is mounted on the automated sample stage of the DSR-facility. The UV-sources are mounted next to the DSR-facility optics, so that the sample stage can automatically move the reference device to the DSR measurement position as well as to the UV-exposure position in the focus of the UV-sources. It was carefully checked that there is no UV-exposure at the DSR measurement position. The degradation procedure of a reference device was performed as follows [3]:

(1) Complete absolute DSR-calibration
(2) Repeating degradation cycle
    a. Approx. 2h of UV exposure. The UV-irradiance induced current of the reference device is monitored.
    b. Measurement of the relative DSR at one given Bias-level

    This cycle is repeated up to 30 times.
(3) Complete absolute DSR-calibration

Additionally, a well characterized WPVS reference solar cell of PTB was calibrated together with the reference devices at step (1) and (3) in order to exclude drifting or other possible deviations of the DSR-facility over the several days of continuous operation. The calibration value of the PTB solar cell decreased by 0.1%. That is well inside the reproducibility of the facility.

By measuring the photocurrent during degradation, the in situ monitoring of the absolute UV-irradiance is possible. From step (1) the absolute spectral responsivity $s(\lambda)$ of the reference device is known. From (2a) the UV-induced current of the reference I_{UV} device is known at any time of the whole degradation procedure. Taking the relative spectral irradiance $E_{UV,rel}(\lambda)$, that were measured with a spectroradiometer, the absolute factor A of the spectral irradiance and hence the total UV-irradiance E_{UV} exposed to the reference device can be calculated at any moment during the degradation procedure:

$$I_{\mathrm{UV}} = \int s(\lambda) \cdot A \cdot E_{\mathrm{UV,rel}}(\lambda)\,\mathrm{d}\lambda.$$

$$\Rightarrow A = \frac{I_{\mathrm{UV}}}{\int s(\lambda) \cdot E_{\mathrm{UV,rel}}(\lambda)\,\mathrm{d}\lambda}$$

$$\Rightarrow E_{\mathrm{UV}} = A \cdot \int E_{\mathrm{UV,rel}}(\lambda)\,\mathrm{d}\lambda$$

An example of this irradiance monitoring is shown with the red curve in Figure 3. The large step at the beginning shows the switching on of the second UV-source. It can be observed, that the UV-sources are not very stable, hence in situ irradiance monitoring is a reasonable choice for irradiance determination.

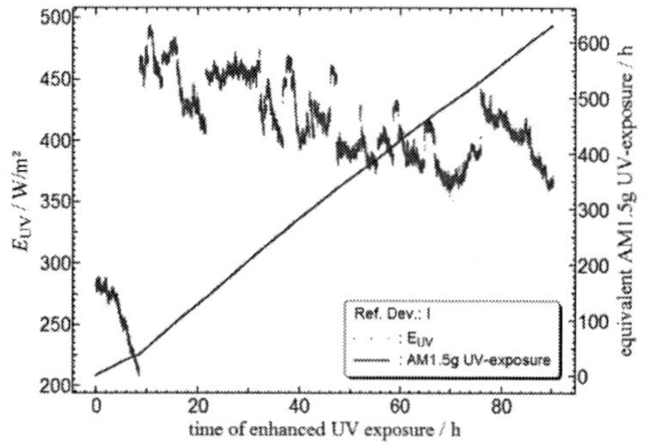

Figure 3: Monitoring the UV-irradiance during degradation by measuring the photocurrent.

The valuable information that can be derived from this data is the *equivalent AM1.5g UV-exposure*, which is an estimate of how many hours the solar cell has to be exposed to the UV-part (280nm – 410nm) of the AM1.5g spectrum in order to sustain an equivalent UV-damage comparable to the enhanced UV-degradation performed in the experiment. This value is estimated by multiplying the enhanced UV-exposure duration to the ratio of $E_{UV}/E_{AM1.5g,UV}$. This is exemplary shown as a blue graph in Figure 3. Please note, that this procedure cannot be applied to sample F and G, since these devices have negligible spectral responsivity in the UV-region. Hence an estimated value is given.

4 RESULTS AND DISCUSSION

The results of the UV-degradation experiments are summarized in Table 2 that shows the equivalent exposure time of the individual reference devices and the measured change of the I_{STC} after the UV-exposure. Please note, that the equivalent exposure time of reference device B, C and I are twice as long as the other devices. The exposure time can be considered to be rather high for a reference solar cell since reference devices are mostly illuminated for very short time periods when adjusting the irradiance level of a solar simulator.

The right hand side of Table 2 and Figure 4 denote the measured relative change of the short circuit I_{STC}. It can be observed that apart from device D and E all observed ageing effects are well within the reproducibility range of the facility. This reproducibility is underlined by the drop of 0.1% of the I_{STC} of the internal PTB reference that has not been exposed to UV-irradiation. Hence these reference devices can be considered to very stable.

Table 2: Impact of UV-degradation on short circuit current of the individual devices.

	Equivalent exposure / h	ΔI_{STC}
A	330	-0.18%
B	514	+0.13%
C	530	-0.01%
D	290	-0.35%
E	281	-0.46%
F	ca. 280	+0.11%
G	ca. 280	+0.15%
H	253	-0.15%
I	630	+0.06%
PTB	0	-0.10%

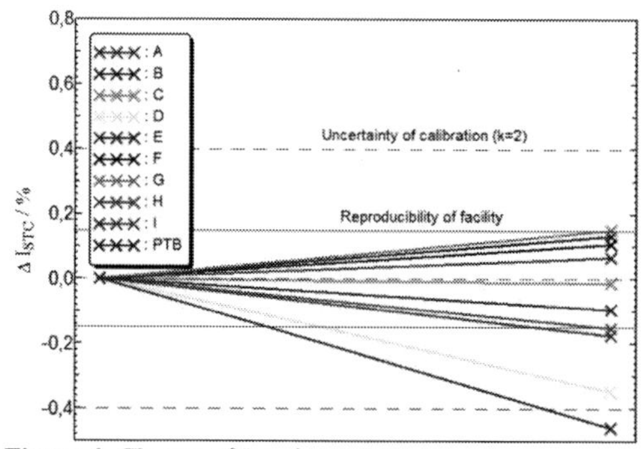

Figure 4: Change of I_{STC} due to UV-exposure

On the other the degradation of reference device D and E can be considered to be significant.

More insight in these changes is gained from the changes of the spectral responsivity that are shown in Figure 5. There it can be observed that the devices crystalline silicon and GaAs based devices B, C, H and I do not show any significant UV-degradation in the whole spectral range.

The HOQ-, BG40- and BG38 filtered devices A, D and E show a very similar and significant drop of the spectral responsivity below ≈ 500 nm. For all three devices this effect is in the same order of magnitude. Since the devices D and E are filtered in the infrared this leads to the significant drop in the I_{STC}. In contrast for the HOQ-filtered device A the impact of this change of spectral responsivity on the overall *relative* change current is of course much smaller.

The OG590 and RG610 filtered devices F and G show a very significant change in the spectral responsivity in the blocking region < 560 nm. There the spectral responsivity increases significantly after UV-exposure (please note the different percentage scale in these graphs). This indicates a degradation of the filters leading to a higher transmission and hence to a higher spectral responsivity and a higher current. However, since the degradation only occurs in the blocked spectral range the effect on the overall I_{STC} is only minor.

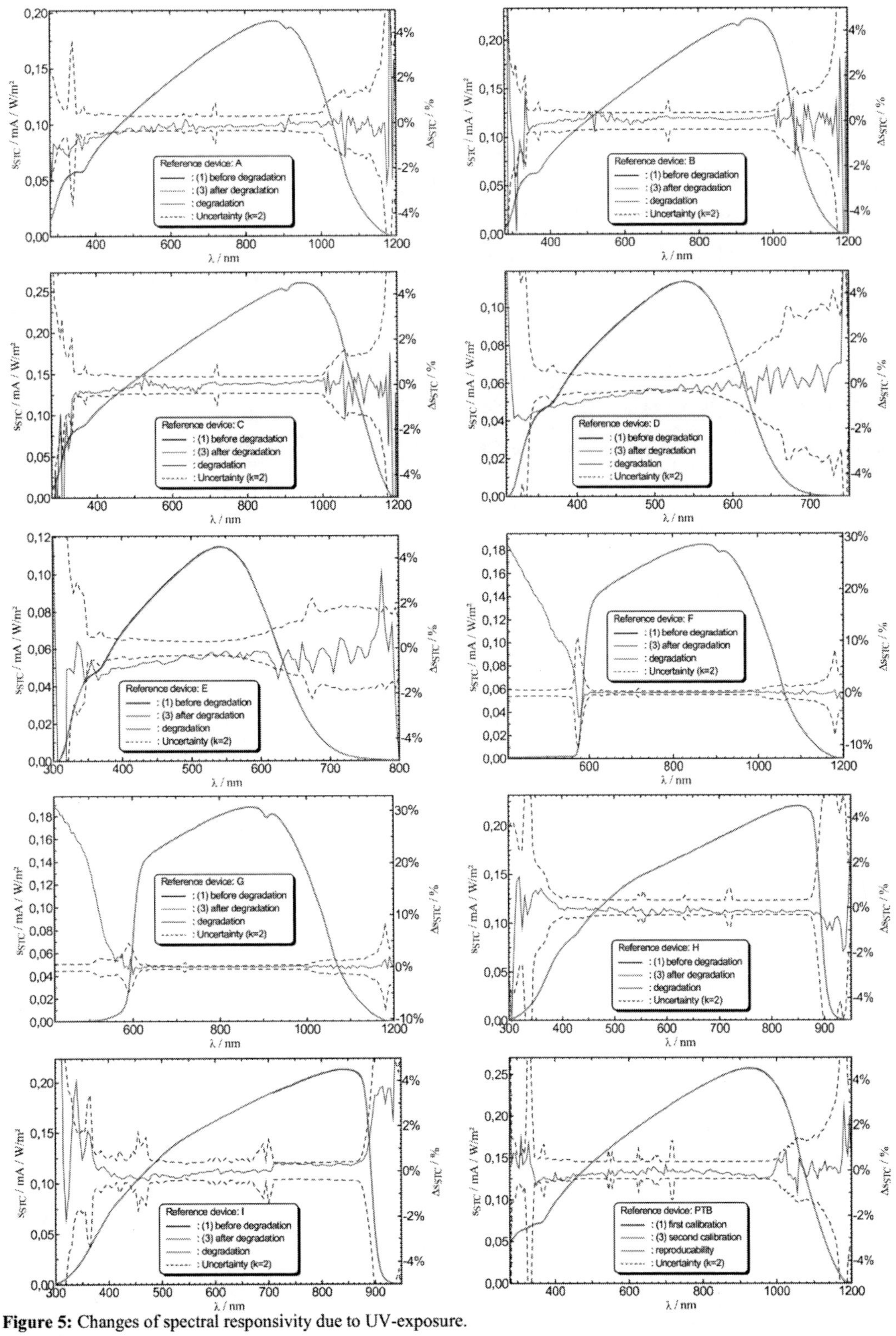

Figure 5: Changes of spectral responsivity due to UV-exposure.

5 SUMMARY

We investigated the UV-induced degradation of a variety of different (filtered) reference solar cells developed by Fraunhofer ISE. The measurements were performed by the high precision differential spectral responsivity facilities at PTB. The degradation was performed with enhanced UV irradiation with spectral irradiance between 280 nm – 410 nm. AM1.5g equivalent UV exposure durations between 250 hours up to 630 hours were achieved. Considering the spectral data, UV-degradation effects could be identified and the impact on the I_{STC} was discussed.

In summary, apart from the BG38 and BG40 filtered devices, the reference solar cells showed no significant degradation with respect to the I_{STC}.

6 ACKNOWLEDGEMENTS

The research work leading to this article was partly carried out within the BMU project FKZ 0325242 and the EMRP ENG55 project "Photoclass". The EMRP is jointly funded by the EMRP participating countries within EURAMET and the European Union.

7 REFERENCES

[1] S. Winter et al, Proceedings. 16th European Photovoltaic Solar Energy Conf. (2000), 4ff
[2] S. Winter et al, Measurement 51 (2014), 457ff
[3] S. Winter et al, Proceedings. 22nd European Photovoltaic Solar Energy Conf. (2007)

INFLUENCE OF LIGHTNING STRIKES ON PHOTOVOLTAIC MODULES PROPERTIES

I.Naxakis, V. Perraki and E. Pyrgioti
Department of Electrical and Computer Engineering, University of Patras, GREECE
Patras-Rio, Greece 26500
mail to: naxakis@ece.upatras.gr

ABSTRACT: The objective of this work is to present how modules react in case of lightning strikes and overvoltages. The experimental process included the application of impulse voltages within the limits specified by the standards as well as, beyond these limits After each impulse voltage stress, the properties of the stressed photovoltaic module were compared with those of one with identical characteristics unstressed module in order to evaluate possible modifications in the stressed module properties. The measurements were gathered with the use of a PVPM power measuring device. The conclusions were extracted after completion of an extensive series of tests. The results confirmed that, after the completion of the tests, the tested module had not undergone power degradation, had not shown any evidence of surface defects and continued to operate reliably.
Keywords: photovoltaic, module properties, Impulse voltage, regulations

1 INTRODUCTION

The widespread adoption of the photovoltaics (PVs) forming grid connected or standalone electrical applications is a dominant issue nowadays. The ecological destruction caused by the excessive use of polluting fuels in order to increase the electricity production and the exhaustion of conventional energy deposits has pushed towards this direction. However, while photovoltaic are the fastest growing power generation technology worldwide, their growth raises new questions. For example, a difficult issue is whether the usual lightning protection system of a PV installation is sufficient under the burden of current rapid developments. In general, photovoltaic facilities are more vulnerable to direct lightning strikes than conventional low voltage (LV) power distribution systems due to their inherently uncovered locations on roofs and facades of buildings and on wide unsheltered areas. Installation of earth electrodes and ground wires has been not only the primary method for the protection of the electrical systems, (such as a photovoltaic system) but also a developing technology [1], [2], [3]. The reliable protection of valuable photovoltaic installations and their electronics against overvoltages and current surges are of great importance. [4-6]. It should be mentioned that the field experience is limited and the applied protective systems do not follow a general rule. In relevant literature there is not enough information available about the PV module behavior against lightning strikes and overvoltages. In this paper the standard IEC 61730-2 [7] was used which refers to photovoltaic module safety qualifications. The IEC 61730-2 outlines in detail many electrical and mechanical tests that have been designed to ensure efficient operation and safety over time, concerning the properties and materials of photovoltaic modules. The impulse voltage test, which is a part of [7] simulates an atmospheric event such as a lightning strike and is one of the last tests that must be conducted. Extended tests have been conducted at the Electrical and Computer Engineering Department at the University of Patras in Greece, presenting research which is related to the effect of temperature [8] as well as to the effect of solar irradiation [9] on the properties of the PV modules. Moreover, with this work the response of a photovoltaic module against a potential standard lightning strike has been experimentally studied using a high impulse voltage generator (up to 400 kV).

2 IMPULSE VOLTAGE TEST

One of the final tests, included in IEC 61730-2, is the impulse voltage test which must be conducted with acceptable results. This test is performed only for modules classified as application classes A and B, and differs in terms of requirements for each class [7]. Modules which are classified as Class A are stressed with higher impulse voltages than modules which are classified as Class B. The impulse voltage is based on the maximum system voltage of the module. For example, if the maximum system voltage is 1000 V, application Class A modules have to be tested with 8 kV (impulse voltage peak) while application Class B modules with 6 kV. Table I shows the Impulse voltage in relation to the maximum system voltage, in accordance with IEC 61730-2.

Table I: Impulse voltage versus maximum system voltage according to IEC 61730-2

Maximum System Voltage	Impulse Voltage - Application Class A	Impulse Voltage - Application Class B
1000 V	8000 V	6000 V

The tests were always conducted at room temperature conditions (19.5 ^{0}C - 21 ^{0}C, 745 mmHg - 755 mmHg) and relative humidity of less than 75%. The impulse voltage generator was set to deliver a 1.2/50 μs Impulse Voltage waveform (Figure 1) according to [7]. During the test, there was no evidence of dielectric breakdown or surface tracking of the module, so the test was considered successful [7].

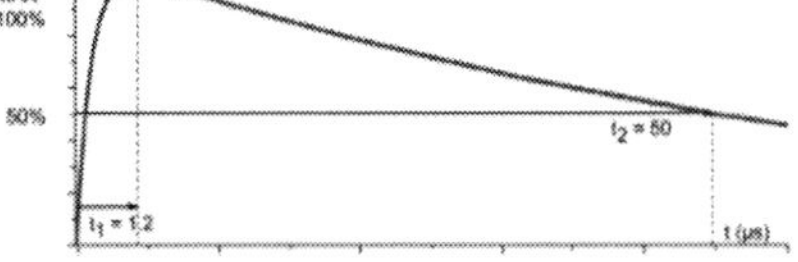

Figure 1: Impulse voltage waveform 1.2 / 50 μs. 1.2 μs is the front time and 50 μs is the time to half-value

The impulse voltage waveform was applied by the impulse voltage generator shown in Figure 2 while the front time and duration was constantly checked by an oscilloscope (Tektronix DPO 4014) which follows the IEC

1180-2 [10].

3 EXPERIMENTAL PROSEDURE

This work studies the effect of lightning on the performance of the PV modules. Different values of impulse voltages were applied on two photovoltaic modules, with the specifications shown in Table II, (Luxor LX-200M - made of monocrystalline silicon cells). The properties of a stressed module (module B) were compared with those of an unstressed reference module (module A) of the same specifications.

Table II: The specifications of the tested modules at STC provided from manufacturer

Electrical Data	LX-200M
$P_{max,STC}$ [Wp]	200
$P_{max,STC}$ Range	201.50-206.49
I_{max} [A]	5.39
V_{max} [V]	37.39
I_{SC} [A]	5.87
V_{OC} [V]	44.27
Maximum systems voltage	1000 V

Figure 2: View of the impulse-voltage generator

, **Figure 3:** Photovoltaic outdoor measuring set up

The characteristic curves were plotted with the use of the PVPM 2540C, which plots the current-voltage (I-V) and the power-voltage (P-V) curves of the photovoltaic modules. Through, a patented procedure the device calculates the module's most important parameters, such as open circuit voltage (V_{OC}), short circuit current (I_{SC}), voltage and current at the maximum power point (V_{pmax}) and (I_{pmax}), peak power (P_{max}), maximum power at standard test conditions ($P_{max,STC}$), fill factor (FF), series resistance (R_s) and parallel resistance (R_p). The collected data were transferred into a computer and were evaluated. The experimental devices for the impulse voltage tests were installed in the High Voltage Laboratory (Figure 2), whereas the devices for the evaluation of the PV properties (Figure 3) were installed on the roof at the building of the Electrical and Computer Engineering Department, at the University of Patras in Greece. In the preparatory phase, I-V and P-V characteristic curves were plotted in order to see whether the modules are identical and whether their properties comply properly with the manufacturer's specifications. Then extensive outdoor measurements were carried out and I-V and P-V curves of the tested modules were plotted corresponding to a variety of different irradiances and temperatures. Important electrical characteristics were observed. At this point it is of great importance to mention that the measurements, gathered in order to choose the appropriate ones for the comparison between the modules, lasted more than three years while the experiments on the PV modules are still under development. The results selected were the ones obtained during periods of day, with as much stable sunshine as possible and around solar noon. In that case, all the sets of measurements were conducted simultaneously for both modules, with the same PVPM and the same sensors. Therefore, the comparison of the module's photovoltaic properties was made between the plots delivered from both modules performed under almost equal irradiances and temperatures. The impulse voltage generator that was used in the next steps had to be adjusted to deliver the appropriate Impulse Voltage amplitude. "Module B" was stressed with impulse voltages into the limits of the standards as well as with voltages beyond the limits [10]. After each voltage stress, measurements were gathered and I-V and P-V curves were created in order to compare them with the ones from the unstressed "module A". At the same time, the maximum power, P_{max}, was recorded in order to evaluate possible modifications in the module B properties.

4 RESULTS

The results presented in this work, refer to commercially available monocrystalline silicon modules. The I-V and P-V characteristics extracted from measurements conducted under outdoor conditions focus on the influence of different values of impulse voltages (simulating lightning and surge conditions) on the photovoltaic parameters of the selected modules.

Figure 4(a) presents the I-V and P-V characteristics which were plotted for modules A and B in the preparatory phase. The comparison based on the formulation of these characteristics as well as on the maximum power of both modules shows a slight non-significant difference for voltages equal or higher than the V_{max} as to the current and power. These pre measures, indicating that the two modules are not entirely similar, may, nevertheless, justify future divergences between their results.

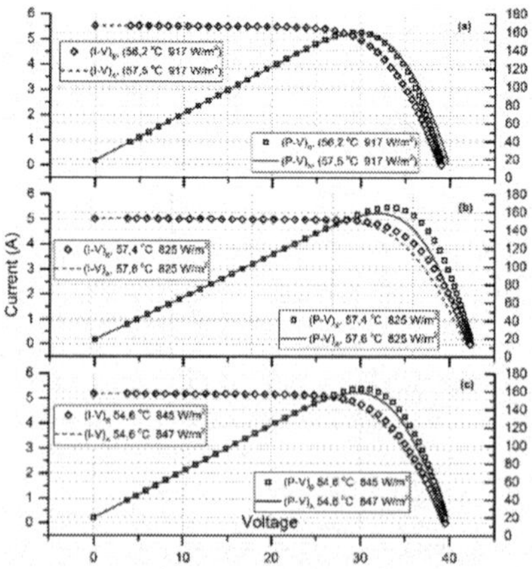

Figure 4: P-V and I-V curves. (a) Before stress. (b) After stress with 6 kV. (c) After stress with 8 kV

The I-V and P-V curves extracted from module A and the stressed module B with the impulse voltage 6 kV are shown in Figure 4(b) (Application Class B). The maximum power generated from this module was slightly higher (for a little lower temperature) than from the unstressed module, while the values of V_{OC} and I_{SC} remained almost the same.

In Figure 4(c), module B has been stressed with an impulse voltage of 8 kV (within the limits of the regulation specifications - Application class A). A non-significant difference in the values of maximum power (less than the observed in Figure 4(b)) is observed, for slightly different irradiation, while the same values of V_{OC} and I_{SC} are recorded for the two modules. Even though the regulations require the realization of tests with impulse voltage 6 kV and 8 kV, the amplitude of the developing overvoltage in a real installation can be significantly greater, exceeding the value of 30 kV [11]. This statement has been confirmed by simulation studies in our laboratory with the help of the Alternative Transients Program - Electromagnetic Transients Program (ATP - EMTP) [12]. For the above reasons, the experimental process was chosen to continue with test voltages beyond the limits set by the standards.

In Figure 5 the unstressed and stressed module I-V and P-V curves are compared for the cases where module B is stressed with 10 kV, 20 kV and 25 kV impulse voltages. Even though, the used impulse voltage far exceeded the limits dictated by the standards, the I-V and P-V curves of the two modules have the same form as in the previous cases. However, as seen in Figure 5(b), the maximum power delivered from module B (stressed with 20 kV impulse voltage) is a little higher than the power from module A, while the values of V_{OC} and I_{SC} remain almost the same. This result could be attributed to the slight difference in temperature between them. No differences are shown in the performances between the two modules in Figure 5(a) and Figure 5(c).

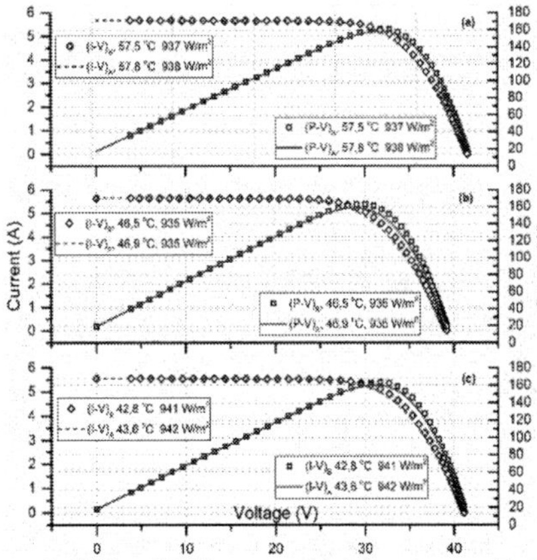

Figure 5: P-V and I-V curves. (a) After stress with 10 kV. (b) After stress with 20 kV. (c) After stress with 25 kV

As no substantial differences between the plots corresponding to module A and B (stressed up to 25 kV) are observed, it was decided to test the module with impulse voltages up to 35 kV. The P-V curves of module B after stressed with 32 kV and 35 kV impulse voltages, respectively, (Figure 6(a) and Figure 6(c)) are almost identical with those of module A. The graphs which correspond to the stress with 34 kV impulse voltage (Figure 6(b)) show that module B has a slightly higher peak power than module A, while the values of V_{OC} and I_{SC} are almost the same.

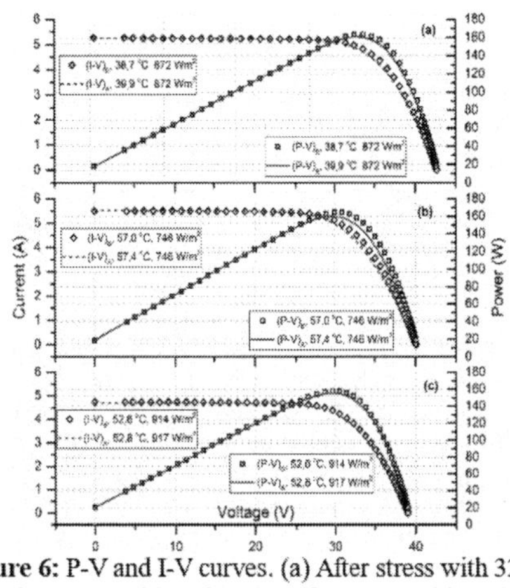

Figure 6: P-V and I-V curves. (a) After stress with 32 kV. (b) After stress with 34 kV. (c) After stress with 35 kV

During the application of the impulse voltages, in the last set of tests, however, the voltage waveform presented some interesting changes shown in Figure 7. Figure 7 depicts the waveform of the stress test with 32 kV impulse voltage. The result of the activation of the impulse voltage generator with open circuit configuration is indicated with the outer (red) line. The inner (black) line depicts the waveform after the module was connected to the impulse voltage generator.

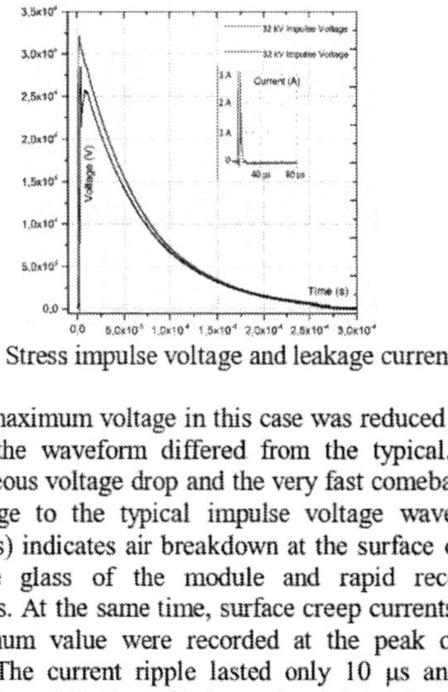

Figure 7: Stress impulse voltage and leakage currents

The maximum voltage in this case was reduced to 28 kV and the waveform differed from the typical. The instantaneous voltage drop and the very fast comeback of the voltage to the typical impulse voltage waveform (1.2/50 μs) indicates air breakdown at the surface of the protective glass of the module and rapid recovery afterwards. At the same time, surface creep currents of 3 A maximum value were recorded at the peak of the voltage. The current ripple lasted only 10 μs and the energy released (0.29 mWh) due to the air breakdown was not significant. Small surface discharges (3 - 4 mm) were visible during this test. Testing with impulse voltage higher than 32 kV shown that the electrical discharges on the air above the module's surface are multiplied. After this stress test, the module, once again, seemed to perform without any significant change in its properties.

Moreover, Figure 8 presents the effect of the application of the impulse voltage to module B onto the module's output power (Figure 8(a)) as well as the effect of irradiance onto the modules output power (Figure 8(b)). More specifically, Figure 8(a) illustrates the ratio of the measured maximum output power to the maximum output power indicated by the manufacturer for module B ($P_{B,max}/P_{man}$), as a function of the test impulse voltage (blank squares symbols). In the same figure the ratio of the maximum output power calculated under standard test conditions is also presented, (delivered by the PVPM measuring device), to the maximum output power indicated by the manufacturer, ($P_{B,STC}/P_{man}$), as a function of the test impulse voltage (blank circular symbol). A negligible slope is observed, through the linear fit of $P_{B,max}/P_{man}$ and $P_{B,STC}/P_{man}$ as a function of the test impulse voltage, which cannot, however, confirm permanent degradation of the module's B power. Figure 9(b) shows the ratio of the measured maximum output power to the maximum output power at STC for module B, ($P_{B,max}/P_{B,STC}$) as a function of the corresponding irradiance (blank square symbols) and for module A ($P_{A,max}/P_{A,STC}$) as a function of the corresponding irradiance (x cross symbols). From the linear fitting on the $P_{B,max}/P_{B,STC}$ and $P_{A,max}/P_{A,STC}$, we confirm that the reduced performances of the two modules A and B are always the same. The only difference between the linear fittings is a negligible variation in irradiations above 1000 W/m^2, which however is not enough to change the overall conclusion. Therefore, module B which has suffered multiple stresses by applying impulse voltages has not sustained degradation of its properties.

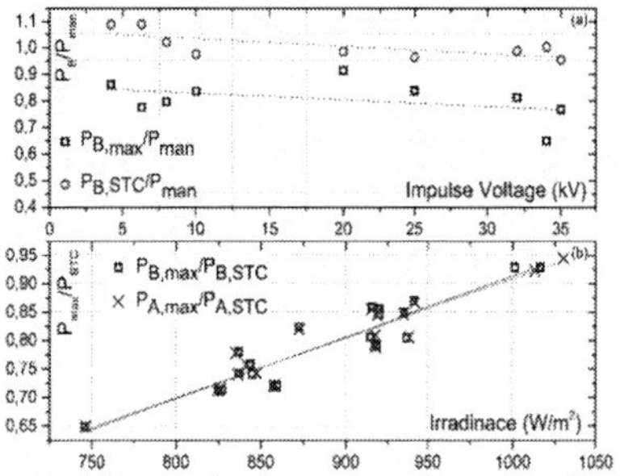

Figure 8: P-V and I-V curves. (a) $P_{B,max}/P_{man}$ and $P_{B,STC}/P_{man}$ versus impulse voltage. (b) $P_{B,max}/P_{B,STC}$ and $P_{A,max}/P_{A,STC}$ versus irradiance.

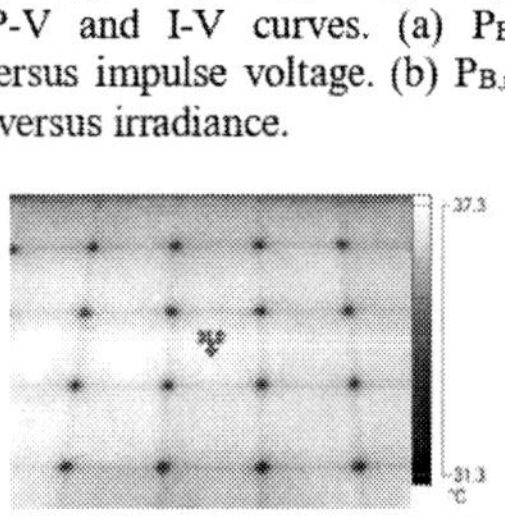

Figure 9: representative IR image for both modules

In parallel to the basic tests with the impulse voltages application on the modules a series of visual observations (Figure 9) and measurements were conducted with an infrared (IR) camera (Fluke Ti32 Thermal Imager). Thermal images of module A and B were initially taken (before the impulse voltage stress test) by using the IR camera. The thermal images captured for both the modules in every step of the procedure shown no evidence of damage. Figure 10 shows the characteristically form of the thermal image for both the modules which remained unchanged after every implementation of the voltage.

6 CONCLUSIONS

To sum up, in this work impulse voltages into the limits specified by the standards (6 kV, 8 kV), as well as beyond these limits (20 kV, 32 kV, 34 kV and 35 kV) were applied on a photovoltaic module in order to assimilate lightning and surge stress conditions on photovoltaic installations. The performance of module A with the properties of a similar unstressed module B (through the I-V, the P-V curves and the P_{max} values) was initially compared and it was noticed that the module successfully passed the tests showing that its performance was not degraded and it could reliably operate. Subsequently, we compared the properties of both modules for many different values of solar irradiation which correspond to different temperatures.

Comparison of a series of visual observations which were realized with an infrared (IR) camera has shown a smooth operation of the module both before and after the stress tests. Differences of the thermal behaviour in each photovoltaic cell were not observed either. Therefore, although module B has suffered multiple stresses, no induced degradation has occurred on its properties; thus neither proof of visual defects nor surface tracking was observed.

5 REFERENCES

[1] D. C. Carmichael and G. T. Noel, "Development of low-cost modular designs for photovoltaic array fields", IEEE Trans. Power App. Syst. 104, 5, 1005–1011 (1985).

[2] J. C. Hernandez, P.G. Vidal, and F. Jurado, "Lightning and Surge Protection in Photovoltaic Installations Power Delivery", IEEE Transactions 23, 4, 1961-1971 (2008).

[3] C. A. Charalambous, N. D. Kokkinos, and N. Christofides, "External Lightning Protection and Grounding in Large-Scale Photovoltaic Applications," IEEE transactions on electromagnetic compatibility 56, 22, 427 - 434 (2014).

[4] H. Häberlin and R. Minkner, "A simple method for lightning protection of PV-systems", Proc. 12th EPSEC, Apr. 1994, Amsterdam, Netherlands, pp. 1885–1888.

[5] H. Häberlin, "Interference voltages induced by magnetic fields of simulated lightning currents in PV modules and arrays", Proc. 17th EPSEC, Oct. 2001, Munich, Germany, pp. 2343–2346.

[6] H. Becker, W. Vaaßen, and F. Vaßen, "Lightning and overvoltage protection in PV and solar thermal systems", Proc. 16th EPSEC, May 2000, Glasgow, UK, pp. 2257–2260.

[7] IEC 61730-2:2004, PV Module Safety Qualification - Part 2: Requirements for Testing.

[8] V. Perraki and G. Tsolkas, "Temperature dependence on the photovoltaic properties of selected thin-film modules", International Journal of Renewable and Sustainable Energy 2, 4, 140-146 (2013).

[9] T. Michalacopoulos and V. Perraki, "Effect of solar radiation incidence angle on the photovoltaic module properties: modelling for Mediterranean sites", 29th EUPVSEC, Sep. 2014, Amsterdam, Netherlands, pp. 2715-2717.

[10] IEC 1180-2:1994-06, High-voltage test techniques for low-voltage equipment - Part 2: Test equipment.

[11] P. Nikolaidis, "Lightning protection of photovoltaic systems", Final year thesis, 2012, University of Patras, Greece.

[12] Alternative Transients Program (ATP-EMTP), see http://www.emtp.org/ for double exponential voltage source (last accessed January 2015).

EFFECT OF TEMPERATURE ON INSULATION RESISTANCE OF DIFFERENT PV TECHNOLOGY MODULES.

Ganesh Gowri[1], Manoj Morampudi[1], Gopal kumar[1] , Rahnuma Siddiqui[1], Rajesh Kumar[1], Sahan Raghava[1], Sneh Lata[1],Pragati Rajput1
National Institute of Solar Energy, 19th Milestone, Gwalpahari, Gurgaon-Faridabad Road, Haryana-122003, India.
Email – ganeshgowri@outlook.com

ABSTRACT: PV Technologies having same power rating with different configuration are selected for this study. Insulation Test is performed for 2 minutes at 1000 V on the modules which are kept in environmental chamber. A Temperature dosage in the increments of 10°C, ranging from 10°C to 80°C is applied on the modules. The insulation resistance, leakage current, Dielectric Absorption ratio and capacitance readings were noted down at each interval. All the modules have fulfilled the criteria of IEC 61215, IEC 61646. Theoretically from the Observations it was found that the insulation resistance decreases with increase in temperature. In addition to that a generalized model of insulation of PV module in terms of resistance and capacitance was built and simulated.
Keywords: Insulation resistance, different Technologies, Dielectric Absorption Ratio, Temperature effect, dielectric equivalent circuit, MATLAB modeling.

1 INTRODUCTION

The penetration of PV modules into the common lives of people made the things much simpler, despite the fact that power is not easily accessible by transmission lines in remote places. At the same time it is creating some problems for the consumers as well as manufacturers, like abnormal behavior of the modules with high voltage stress. The Insulation resistance is the dielectric property and safety criteria of the module defined as amount of Insulation material provided in between Solar Cells and metallic frame. Insulation resistance tests are commonly performed with a high voltage tester to appraise insulation condition of the module. Failure of electrical insulation of a PV module is the most severe fault when the modules are operating in the power plants. The modules with low insulation resistance (i.e., between frame and active cell area) working in high system voltage will induce leakage currents as it circulates in the frame and causing a severe threat to the operating personnel as well as inverters.

The Insulation resistance depends on several parameters like environmental (Irradiance, temperature, Humidity, Contamination), as shown in Fig.1. In this work the focus is mainly on the short term effect of the temperature on insulation resistance and modeling of the dielectric equivalent circuit.

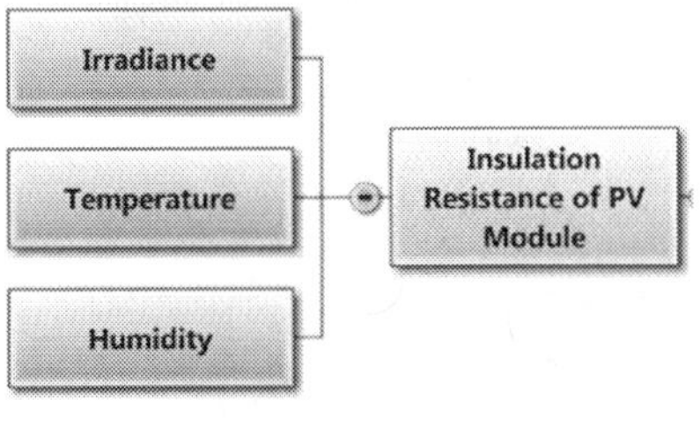

Figure 1 Factors affecting the insulation resistance.

The insulation resistance of a PV module depends on the type of insulating material and on the condition of that material. As a rule, insulation resistance is proportional to insulation thickness and inversely proportional to the surface area of the PV module. Resistance values are affected by contamination, humidity, temperature, and other factors. Low values may point to moisture ingress, dirt or damaged insulation. By periodic checking and recording of values, a good basis for judging the condition of the insulation can be built. Even if readings are higher than the minimum Suggested safe values for that PV Module, a constant downward trend should be treated as a warning of need of maintenance. Results close to suggested minimum values may be considered satisfactory, provided that these values are constant over a number of periodic checks. Care should still be taken with PV module with insulation readings nearing their minimum.

In PV installations with high operating voltages, negative charges may flow through the encapsulating material and the frame as stray current, and cause an accumulation of positive charges at the cell's surface. High temperatures and humidity further amplify the effect. Despite all research efforts by manufacturers, a significant portion of modules is still vulnerable to. In the course of a few years (or months, in some cases), they cause significant power losses. In the PV power plants now-a-days using HighVoltages to compress losses by adding more panels in series, this will also reduce the thickness (size) of the wire, perhaps the main problem is these high voltages will induce an Electric force (stress), this field is in turn applied to modules as well as hazardous to humans while cleaning and maintenance. This field will affect the module performance by means of insulation. In this field, it can occur that an inverter will protectively shut down due to a large leakage current, either in the early morning or rainy days [1]. Some authors have investigated Leakage currents in thin film modules to analyses their influence on power degradation and to estimate module lifetimes after an accelerated aging testing and simulation [2].Characterization of leakage currents and insulation resistance of PV Generators under operating conditions (meteorological variables) is explained in [3]. To analyses the behaviour a dielectric equivalent model was developed and effect of meteorological variables in field conditions shall be de delineated.

2. METHODOLOGY

The encapsulant in PV module is the dielectric in capacitor of a low capacitance value and non-defined geometry. Its electrodes are either metallic frame or metallic electrodes of the contact grid wrapped foil . Measurement capacitor ,made of conductive foils and encapsulant.

Insulation resistance is measured by applying the High voltage DC to the PV module as specified by the manufacturers system voltage according to the IEC standards[4,5] and acquiring the leakage current. In general the resistance is defined in terms of ohms Law.

$$R = \frac{V}{I(t)}$$

Equation 2 Resistance Ohms Law

Resistance and capacitance can be written as shown in the eq.3

$$\left.\begin{aligned} C &= \frac{\varepsilon_0 \varepsilon A}{d} \\ R &= \frac{\rho d}{A} \end{aligned}\right\} \Rightarrow R = \frac{\rho \varepsilon_0 \varepsilon}{C}$$

Equation 1. Combined form of Resistance and capacitance interms of physical property

V is the applied voltage and t is the time $t \leq 2$ min

R is inversely proportional to the value of capacitance and do not depend on V and size (thickness of the dielectric),

d is the distance/length between two conductors.

However, ρ is not constant, and Schottky conduction increases exponentially with $E = V/t$.

The abnormal behavior and uncertain measurements of insulation test has strived to do this study. This area required more comprehensive contemplation to assess and understand the insulation characteristics.

The methodology followed in this investigation of different PV technologies subjected to change in temperature is represented in the flowchart as indicated in Fig.2.

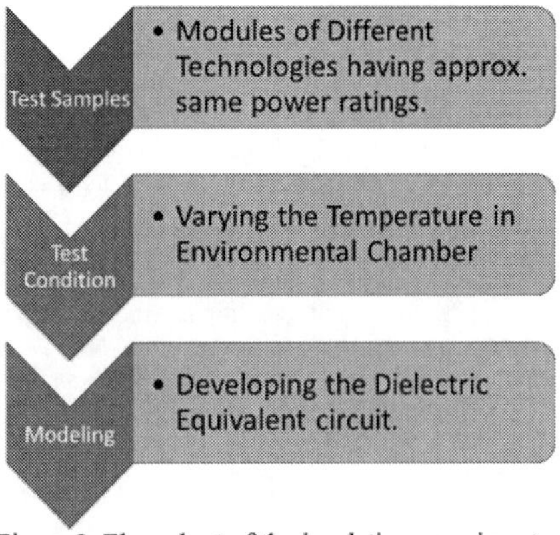

Figure 2. Flow chart of the insulation experiment

The PV modules are also categorized by structural point of view. There are approximately 4 configuration of Module assemblies such as Glass-Glass, Glass- Tedlar, Acrylic sheet, silicone-PCB mounted etc., this study deals with only two of its kind as shown in Fig.

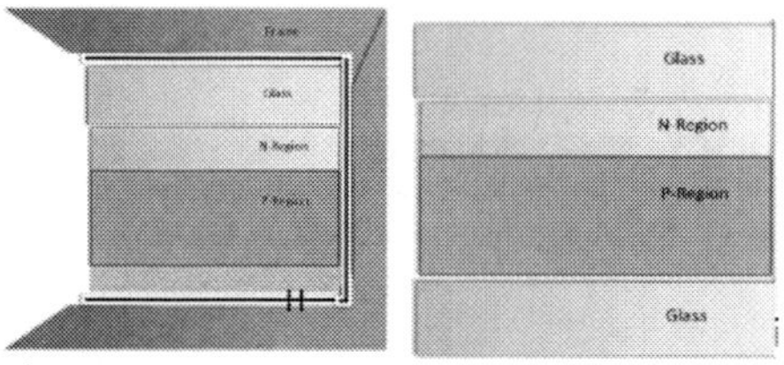

Figure 3 Cross sectional View of two major PV assemblies.

The whole layup Enclosed in the frame. The active cell area is acts like conductor and the frame acts like another conductor in between the dielectric material like EVA, Glass, Tedlar it can be represented as a capacitor.

The outline of the investigation is symbolised in lucid manner in Fig.4.

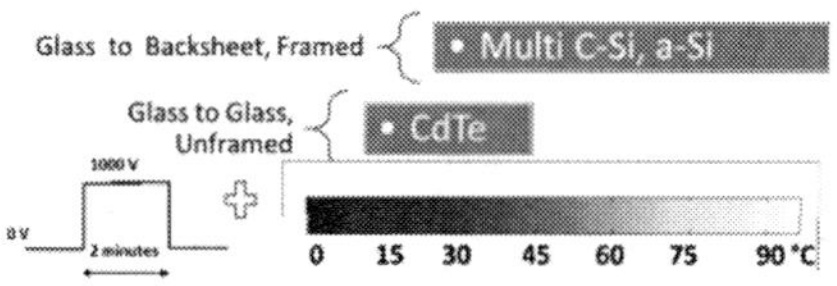

Figure 4 Graphical representation of the elements used in the effect of temperature on insulation test of Various technologies.

3 EXPERIMENTAL SETUP

This paper utilises the experimental setup at the Photovoltaic Testing Facility (PVTF) of the National Institute of Solar Energy (NISE), such as

i. Fluke Insulation Tester for applying high voltage to the PV module. Shown in Fig.5
ii. CTS Climatic Chamber for simulating the Environment conditions like temperature, humidity and Dew. Shown in Fig.6

Figure 5 Insulation Tester used for the insulation measurement

Figure 6. Environmental chamber used for varying the temperature.

The connection diagram for performing the insulation test is as shown in Fig.9 the positive and negative terminals are shorted together and brought as one common terminal, which is connected to the positive terminal of the high voltage tester and the negative terminal is connected to the module frame.

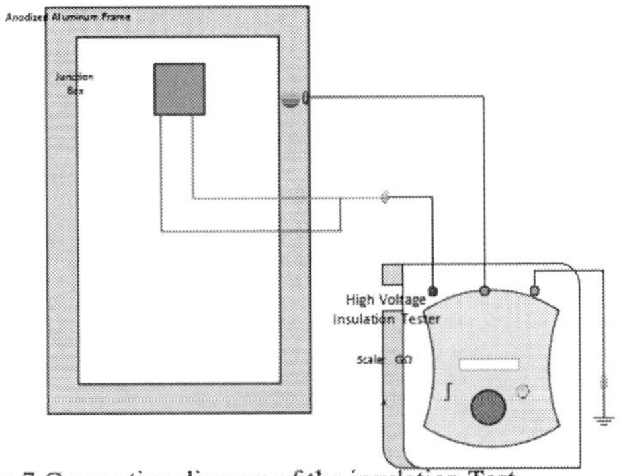

Figure 7 Connection diagram of the insulation Test

4 RESULTS AND DISCUSSION

In the insulation resistance Test, trends are of greater importance than actual values since there is no repeatability of measured values in any respect.

From this experiment it was depicted from the Fig.8 that the insulation resistance of Multi C-Si and CdTe decreases with increase in temperature whilst the a-Si slightly gains with increase in temperature.

All leakage currents are following the same pattern of Parabola. Thus a correlation may derived from the Fig. 9 that the leakage current is proportional to square of the Temperature, Also note that the leakage current of a-Si is slighty low at higher temperatures.

Due to unsymmetrical capacitances of the module the dielectric absorption rate (DAR) of the modules at different temperature is dissimilar as shown in Fig.10 . it was also illustrated that the DAR of the a-Si is higher among the other technologies in the entire temperature span .Thus giving strong evidence that the dielectric strength of the a-Si is good, whereas in case of Multi c-Si it is poor. Warning bells to the user/manufacturer.

Apart from that the simulation of Dielectric model for insulation built by using MATLAB also replicated the same leakage current as in case of tested result, multi C-si leakage current at 20°C is 24.5 Nano amps .The significance of the MATLAB modeling is to assess the leakage current characteristics and to predict the results, once we know the circuit parameters.

Table I measured insulation test data at various temperatures of different PV technology Modules

	a-Si				CdTe				Multi C-Si			
T(°C)	**R (GΩ)**	**ilc (nA)**	**DAR**	**C (µF)**	**R (GΩ)**	**ilc (nA)**	**DAR**	**C (µF)**	**R (GΩ)**	**ilc (nA)**	**DAR**	**C (µF)**
10	1.2	884	1.21	0.04	1.09	966	1.18	0.02	27.4	38.6	1.14	0.01
20	1.87	565	1.18	0.02	1.79	592	1.06	0.01	43.1	24.5	1.13	0.01
30	2.8	379	1.15	0.01	8.58	123	1.04	0.01	38.4	27.6	1.1	0.01
40	4.32	245	1.11	0.01	23.9	44.3	1.02	0.01	17.3	61.3	1.01	0.01
50	6.8	156	1.14	0.01	12.5	84.4	1.02	0.01	8.48	125	0.99	0.01
60	7.8	136	1.16	0.01	6.13	173	1.01	0.01	5.09	208	0.98	0.01
70	6.26	169	1.19	0.01	2.71	391	1	0.01	3.13	338	0.98	0.01
80	3.96	268	1.15	0.01	1.27	831	0.99	0	2.01	526	0.99	0.01

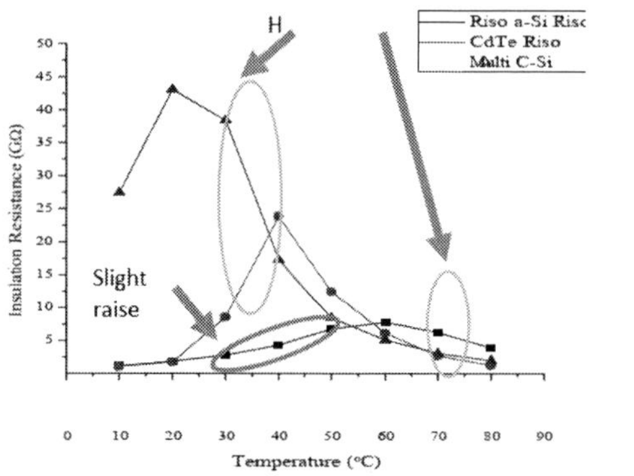

Figure 8. Insulation Resistance of Different Temp

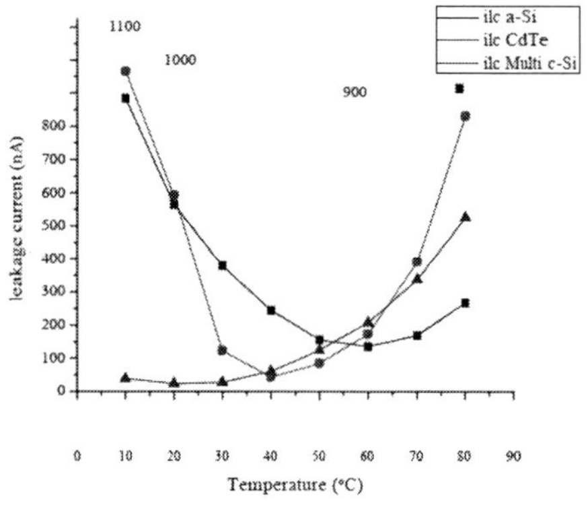

Figure 09. leakage current at different temperature

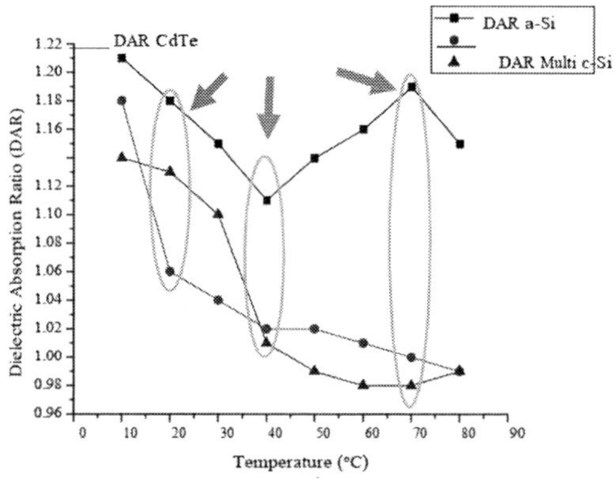

Figure 10. Dielectric Absorption ratios of various at different technology.

5. CONCLUSION

A module working under non-MPP load condition could result into raise in operating temperature hence leads to sudden fall in insulation resistance and indicates a severe threat to operating personnel. A Dielectric equivalent circuit of the PV module was obtained with the readings of Insulation test resistance, capacitance. The simulation of the circuit also replicated the same leakage current. It was observed that the results of the repeated tests were not contingent. This depicts the uncertain behavior of the leakage current paths through the module

6. ACKNOWLEDGEMENT

The authors would like to acknowledge support and help provided from National Institute of Solar Energy (NISE), Gurgaon, An autonomous institute under Minister of New and Renewable Energy (MNRE).

7. REFERENCES

[1] Guofeng Wang, Yue Jiang, Yinggang Wei, Haidan Gong, Jingbing Zhu, "Study on Insulation Resistance Test of PV Module For The 28th Pvsec 2013," in 28th European Photovoltaic Solar. Energy Conference and Exhibition .

[2] Thomas Weber, Juliane Berghold, Florian Heilmann, Margarete Roericht, Stefan Krauter and Paul h Grunow, "Test Sequence Development for Evaluation of Potential Induced Degradation on Thin-Film Modules," in 28th European Photovoltaic Solar Energy Conference and Exhibition.

[3] J.C.Hernandez, P.G.Vidal, Medina "Characterization of the insulation and leakage current of pv generator

[4] IEC61215:2005 International Standard for crystalline PV module Design Qualification and Type approval.

[5] IEC 61646 : 2005 International Standards for Thin film PV module Design Qualification and Type approvaval .

This page intentionally left blank.

DEFECT IDENTIFICATION AND CORRELATION WITH ELECTRICAL DEGRADATION OF FIELD AGED THIN FILM PHOTOVOLTAIC TECHNOLOGIES IN COMPOSITE CLIMATE

Rahul Rawat*[1], S. C. Kaushik[1], O. S. Sastry[2], Y. K. Singh[2], B. Bora[2] and Ramayan Singh[2]
[1]Centre for Energy Studies, Indian Institute of Technology Delhi, Hauz Khas, New Delhi (India)-110016
[2]National Institute of Solar Energy, Ministry of New and Renewable Energy, Gwal Pahari, Gurgaon, Haryana (India)-122003
*Corresponding Author: Tel: +91-11-26596465; Email: rawatktp@gmail.com

ABSTRACT: The thin film photovoltaic (TFPV) technologies are emerging out as an alternative to the established silicon wafer based PV technology due to competitive efficiency, and low energy payback period. However, the performance and degradation study of these technologies are important in order to ensure their stability and reliability in terms of operating lifetime. The current-voltage (I-V) characteristic of PV module are most widely used for assessing the electrical performance and degradation. In this paper, the degradation analysis of I-V characteristic of 40 CdTe and 69 micromorph TFPV modules has been carried out after long term outdoor exposure. The degradation rate of electrical parameters i.e. short circuit current; open circuit voltage; current, voltage and power at maximum power point of individual field aged module has been evaluated with respect to the name plate values. The defects of the modules have been identified using visual inspection and electro-luminescence imaging (EL) which are then correlated with the electrical degradation. The degradation of CdTe and micromorph TFPV technologies are found to be 0.55%year and 6.52% in the first year respectively with respect to the nameplate value. The power output of micromorph modules have degraded by 17% after 1 year of outdoor exposure with respect to the power output of unexposed modules due to light induced degradation.
Keywords: CdTe; Micromorph; Thin Film; PV Module; Degradation

1 INTRODUCTION

The utilization solar photovoltaic systems is growing rapidly for different power generation application i.e. grid connected, water pumping, lighting and standalone systems due to power shortage, inconsistent oil prices and environmental issues [1]. The conventional wafer based PV technology is well proven and hold major portion of PV market worldwide, however, the researchers are interested in new low cost, stable and reliable PV technologies i.e. thin film photovoltaic technology (TFPV). The TFPV technologies have thin layers of PV materials making p-n junction which are deposited and grown on a substrate in a controlled manner [2]. The TFPV technologies offers more energy yield in hot and composite climates due to low temperature coefficient with low cost as compared to the wafer based technology [3]. However, the reliability of these technologies are not yet proven for 25 years of lifetime. Although, the indoor qualification testing of TFPV modules is being done as per IEC 61646 standard in order to ensure the mechanical, thermal and electrical durability, the regress outdoor performance and degradation of study of these technologies are required in order to identify the defects and degradation rates.

In this paper the electrical degradation study and defect identifications of CdTe and micromorph TFPV technologies have been carried out after outdoor exposure of 3 and 1 years respectively in composite climate. The CdTe technology has CdS and CdTe layer in order to make p-n junction while the micromorph technology is multi junction tandem solar cell of micro-crystalline and amorphous silicon layers. Both the technologies are glass to glass in which the PV materials and other layers are sandwiched between the two toughened glasses. Jordan and Kurtz (2013) have reviewed the degradation rates of different PV technologies and found the degradation rate of CdTe modules to be 0.4%/year [4]. The micromorph TFPV technology exhibits light induced degradation during initial months of outdoor exposure and requires about 16 months to stabilize its performance with a degradation of approximately 16.5% in 12 months [5-6].

2 METHODOLOGY

The TFPV modules are installed in different grid connected PV systems at National Institute of Solar Energy, India which is considered to be in composite climate zone. The I-V characteristics of each field aged PV modules has been measured in outdoor conditions using the PVPM 1000C I-V curve tracer of Photovoltaik Engineering GmbH.

The outdoor IV characteristics of two different CdTe modules and 3 micromorph modules are shown in figure 1(a&b).

The device uses 4-wire cabling for accurate voltage, reference cell for irradiance and Pt100 RTD for module temperature measurement. The measured values and the nameplate values of the TFPV technologies have been translated to the alternate reporting conditions (ARC) which are close to the actual measurement conditions so as to reduce translation errors (Table I). The technical specifications of PV modules (name plate values) at standard test conditions (STC) and ARC has been presented in Table II.

Table I. Alternate reporting conditions

Parameter	Environmental Conditions	
Technology	CdS/CdTe	Micromorph
Irradiance	900 W/m^2	800 W/m^2
Module Temperature	60 °C	45 °C

The measured as well as name plate value of short circuit current (I_{SC}), open circuit voltage (V_{OC}), current (I_M) and voltage (V_M) at maximum power point have been translated to alternate reporting condition using eq. 1 and eq. 2 [7-9].

Table II. Technical specifications of PV modules

Parameter	**Technical Specifications**			
Technology	CdS/CdTe		Micromorph	
Number of modules	40		69	
Years of exposure	3		1	
	STC	*ARC*	*STC*	*ARC*
Power (P_M)	80 W	64.5 W	110 W	82.89 W
Voltage at MPP (V_M)	48.5 V	42.74 V	64 V	57.56 V
Current at MPP (I_M)	1.65 A	1.51 A	1.76 A	1.44 A
Open circuit voltage (V_{OC})	60.8 V	55.04 V	87 V	80.56 V
Short circuit current (I_{SC})	1.88 A	1.72 A	1.98 A	1.61 A
Temperature Coefficient of Voltage (β_V)	-0.27%/°C		–0.37%/°C	
Temperature Coefficient of Current (β_I)	0.04%/°C		0.08%/°C	

$$V_{STC} = V + \beta_V (T_{STC} - T_M) \quad (1)$$

$$I_{STC} = I\left(\frac{G_{STC}}{G_T}\right) + \beta_I (T_{STC} - T_M) \quad (2)$$

The fill factor of the PV module is the deviation of measured I-V characteristics of PV module from the I-V characteristic of the PV module at ideal condition having zero series resistance and infinite shunt resistance. It is defined as the ratio maximum power output to the theoretical power of the module as given by eq. 3.

$$FF = \frac{I_M \times V_M}{I_{SC} \times V_{OC}} \quad (3)$$

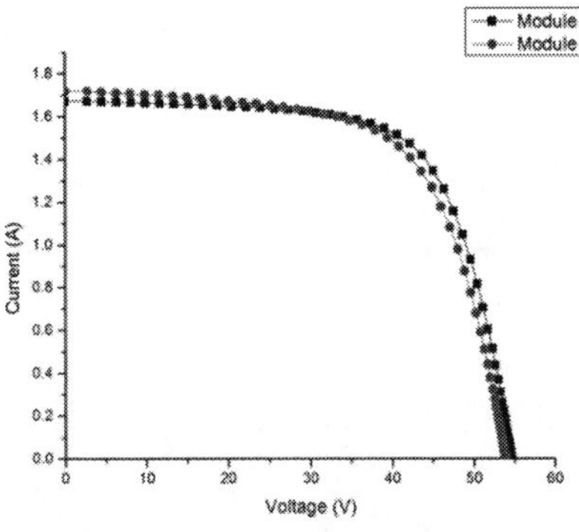

(a)

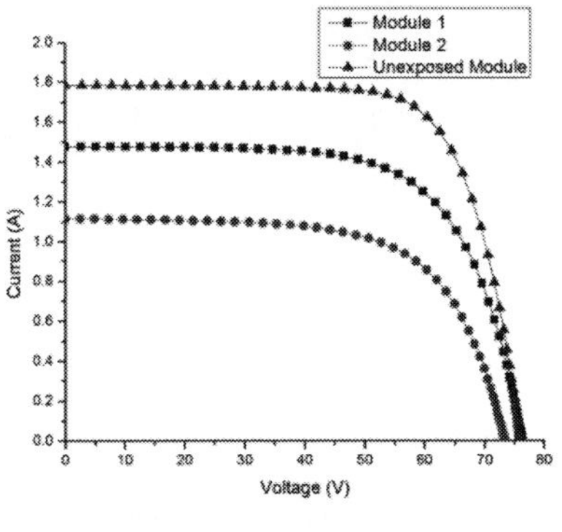

(b)

Figure 1: Outdoor measured I-V characteristics of (a) CdTe and (b) micromorph TFPV modules

The power and fill factor are the main indicators of performance degradation in a PV module. The magnitude of degradation (Y_d) of the parameter over the time (t) is given by eq. 4 where X_f is the measured parameter at STC. The degradation rate (R_d) of parameter has been determined with respect to the initial condition (X_i) i.e. name plate value as given eq. 5.

$$Y_d = X_f - X_i \quad (4)$$

$$R_d = \frac{X_f - X_i}{X_i \times t} \quad (5)$$

The variation of translated I_{SC}, V_{OC}, V_M and I_M, and measured R_S and R_{SH} with respect to the initial condition is used to assess the cause of degradation rate of power and fill factor.

The visual inspection and EL imaging of the modules have been carried out in order to identify the visual degradations such as cracks, pin holes, discoloration, rust, burn marks etc. These degradations not only affect the power output and operating lifetime of the module but also the raise the safety issues. These defects are qualitatively correlated with the measured electrical degradation.

3 RESULTS AND DISCUSSION

3.1 Degradation of CdTe TFPV modules

The percentage deviation of power, voltage and current at MPP, short circuit current, open circuit voltage fill factor of 40 CdTe TFPV modules after 3 years of outdoor exposure is shown in figure 2. The average percentage power and fill factor deviation of the modules are -1.66% and -2.22% respectively. The average power degradation rate of the modules is determined to be 0.55%/year. Out of 40 modules, the total 9, 23 and 8 modules are found to have the percentage deviation of power of below -3%, 0 to -2.99% and above 0% respectively. The short circuit and MPP current are found to have positive percentage deviation of average 2.86% and 1.71% respectively while open circuit and MPP voltage have negative deviation of average 2.39% and 3.43% respectively. It shows that the power degradation is mainly due to the MPP current degradation.

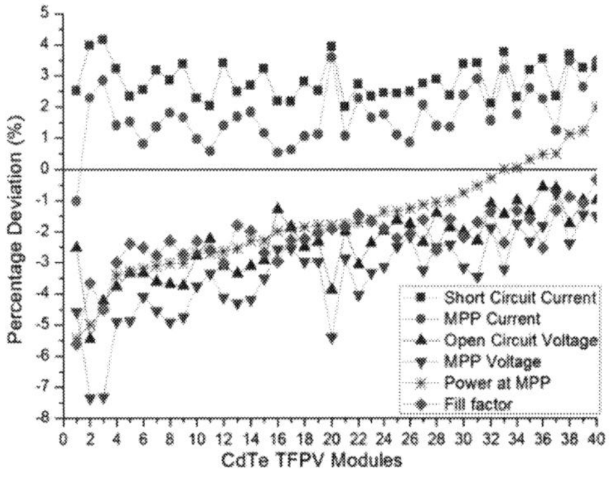

Figure 2: Percentage deviation of electrical parameters of CdTe TFPV modules after 3 years of outdoor exposure

3.2 Degradation of micromorph TFPV modules

The power of 69 micromorph modules is found to have negative deviation after 1 year of outdoor exposure as shown in figure 3 and figure 4. The unexposed modules have average power deviation of +13.1% as compared to the stabilized name plate value. The average power degradation of the technology as compared to the unexposed modules is 17.35%. The initial high degradation rate is due to the light induced degradation during stabilization of the technology, however it has an average degradation of 6.52% as compared to the stabilized nameplate value which is also on the higher side. The high degradation rate implies that the modules have not stabilized yet. The power degradation is mainly due to percentage deviation of MPP current and open circuit voltage. The total 10 modules are found to be degraded more than 10% with maximum degradation of 38.18%. The visual inspection shows the mechanical damage in 4 highest degraded modules. The total 20 modules have 5% to 10% degradation, 38 modules have 0% to 5% degradation and 1 module is found to have positive deviation.

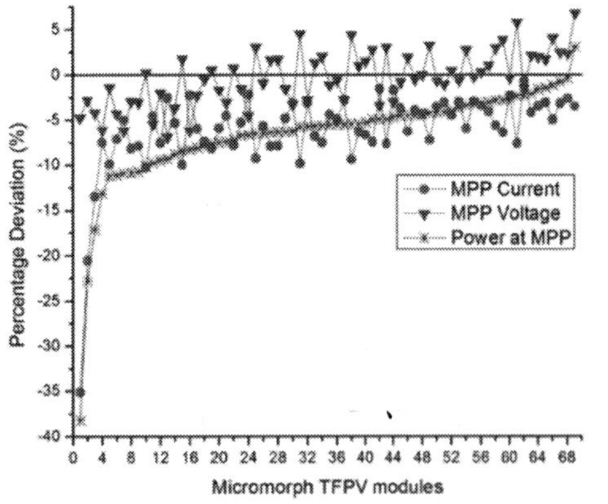

Figure 3: Percentage deviation of power, current and voltage at MPP of micromorph TFPV modules after 1 years of outdoor exposure

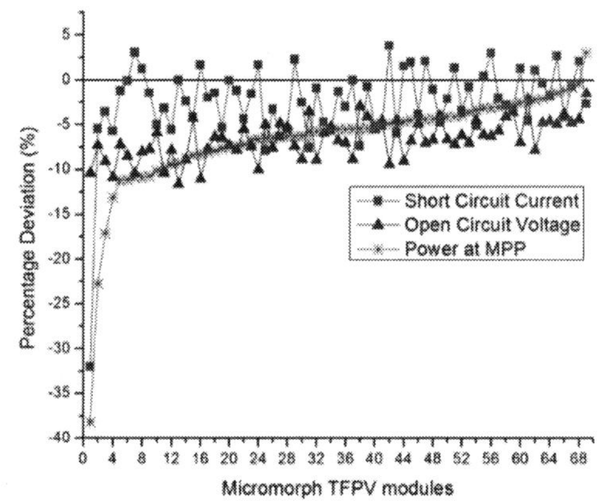

Figure 4: Percentage deviation of power, short circuit current and open circuit voltage of micromorph TFPV modules after 1 years of outdoor exposure

3.3 Correlation of the electrical degradation with electroluminescence image

The EL image of two field aged CdTe TFPV modules with high and normal degradation is shown in figure 5(a) and figure 5(b) respectively. The two CdTe modules are visually identical and doesn't show any major defects but the degradation is clearly visible in their EL images. The image of degraded module has large dark area which shows the reduced electron hole pair generation rate at the locations. The image of other module is very bright and the dark spots are clearly visible for the pin-hole defects which has also been observed in the visual inspection.

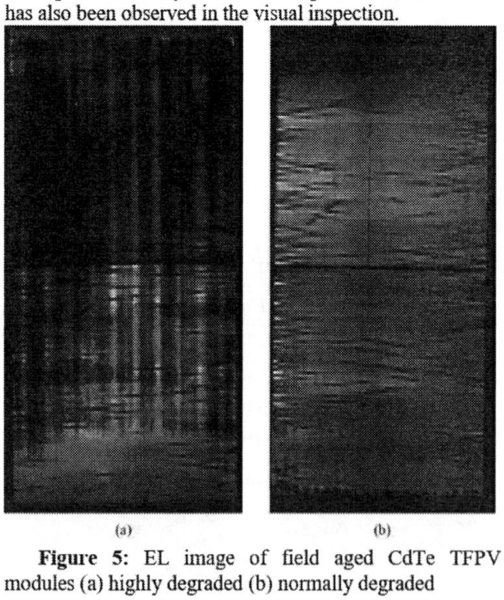

Figure 5: EL image of field aged CdTe TFPV modules (a) highly degraded (b) normally degraded

4 CONCLUSIONS

The outdoor performance degradation analysis of field aged CdTe and micromorph based TFPV technologies has been carried in the composite climate. The average degradation rates are found to be 0.55%/year in 3 years of oudoor exposure and 6.52% in 1 year of outdoor exposure for CdTe and micromorph based TFPV technologies respectively. The pin-hole defects are commonly found in the CdTe TFPV modules and cracks have been observed in few highly degraded micromorph

PV modules. However, the two TFPV modules of these technologies with high and low degradation rates cannot be differentiate by visual inspection. Among the studied two technologies, the CdTe is performing better and is more reliable for long term power generation.

ACKNOWLEDGEMENTS

The authors would like to acknowledge the financial assistance from Indian Institute of Technology Delhi and Science and Engineering Research Board (SERB), Department of Science and Technology (DST) for attending the conference and National Institute of Solar Energy (NISE) for data. Data analysis part of the work is carried out under the Solar Energy Research Institute for India and the United States (SERIIUS) project.

REFERENCES

[1] R. Rawat, S.C. Kaushik, R. Lamba, Renewable and Sustainable Energy Reviews 57 (2016) 1506–19. doi:10.1016/j.rser.2015.12.228.

[2] K.L. Chopra, P.D. Paulson, V. Dutta, Progress in Photovoltaics: Research and Applications 12 (2004) 69–92. doi:10.1002/pip.541.

[3] A.K. Tossa, Y.M. Soro, L. Thiaw, Y. Azoumah, L. Sicot, D. Yamegueu, et al. Energy 103 (2016) 261–70. doi:10.1016/j.energy.2016.02.133.

[4] D.C. Jordan, S.R. Kurtz, Progress in Photovoltaics: Research and Applications 21 (2013) 12–29. doi:10.1002/pip.1182.

[5] M.Z. Hussin, S. Shaari, A.M. Omar, Z.M. Zain, Renewable and Sustainable Energy Reviews 43 (2015) 388–402. doi:10.1016/j.rser.2014.10.093.

[6] K. Anastasios, T. Stathis, R. Evangelos, N. John, H. George, 28th European Photovoltaic Solar Energy Conference and Exhibition (2013) 3222-7.

[7] R. Rawat, S.C. Kaushik, O.S. Sastry, Y.K. Singh, B. Bora, Energy Conversion and Management 110 (2016) 42–50. doi:10.1016/j.enconman.2015.11.069.

[8] R. Rawat, S.C. Kaushik, O.S. Sastry, B. Bora, Y.K. Singh, 5th International Conference on Advances in Energy Research, (2015) 884-90.

[9] IEC 60891. Photovoltaic devices – procedures for temperature and irradiance corrections to measured I–V characteristics.

PRELIMINARY ASSESSMENT OF DEGRADATION IN FIELD-AGED MULTI-CRYSTALLINE SILICON PV MODULES INSTALLED IN HOT-HUMID CLIMATE OF MID GHANA

D. A. Quansah[1], M. S. Adaramola[1A] and G. Takyi [2A]*
[1]Renewable Energy Group, Department of Ecology and Natural Resources
Norwegian University of Life Sciences (NMBU), Ås, NORWAY
[2]Department of Mechanical Engineering
Kwame Nkrumah University of Science and Technology (KNUST), Kumasi, GHANA
**david.ato.quansah@nmbu.no (Corresponding Author), [A]muyiwa.adaramola@nmbu.no, [A]gtakyi.soe@knust.edu.gh*

ABSTRACT: A multicrystalline solar PV installation in Kumasi Ghana, a hot humid climate, was studied for degradation in key performance characteristics after 18-years of exposure. An outdoor current-voltage (I-V) characteristization of the ASE-50-PWX-D the various modules was undertaken. The data was then translated to Standard Test Conditions (STC) and compared with the primary performance characteristics of the modules. The results showed a 0.9% - 1.6% average annual decline in peak power. Short-circuit current and open circuit voltage declined annually by 0.3%-0.6% and 0.1%-0.2% respectively. Decline in the fill factor was 0.3% - 1.2% p.a. resulting from 0.7%-1.1% and 0% - 0.8% loss in current and voltage at peak power. The module had a 10-year warranty and this data shows that it has met and exceeded warranty requirements.

Keywords: Degradation, Multi-crystalline, Silicon, PV Module

1 INTRODUCTION

The Solar PV industry has grown rapidly in recent years and is set to play an even more important role in the years ahead. Between 2010 and 2014, installations grew by almost 450% - from 40 GW to 177 GW [1]. Solar PV has been projected to supply 16% of global electricity consumption by 2050 [2]. New investments in renewable energy averaged $254 billion in the 5-year period 2010-2014 [1]. As investment requirements grow, investors require assurances that modules will perform on field and last the 20-30 year periods that are typical of Solar PV project analysis.

Manufacturers seek to assure investors by granting warranties on modules sold. Current module warranties range from 20 – 25 years [3], [4] [5] and promise a minimum of 80% peak power in this period. This requires that modules loose no more than 20% of peak power by the end of the 20 – 25 year period. This implies a maximum linear degradation of 0.8 - 1% of peak power annually. There are two problems that manufacturer warranties typically encounter. The first, is, there is insufficient field data to confirm that modules will perform as predicted when operated in the different climatic conditions of the world. Qualification tests as per IEC (International Electrotechnical Commission) Standards 61215 and 61646, for crystalline Si and thin films respectively [6], [7], serve as minimum requirements for market entry and has proven quite efficient in reducing infant mortality among PV modules. These tests, however, do not determine lifetime of modules. It may require up to 20 years of waiting to verify this.

The second problem, is that, many companies hardly last the duration of the warranties they provide, making the warranties hardly enforceable. Between 2009 and 2013, over fifty (50) manufacturers closed or filed for bankruptcy with many more undergoing some kind of restructuring/merger/acquisition (see Figure 1) [8].

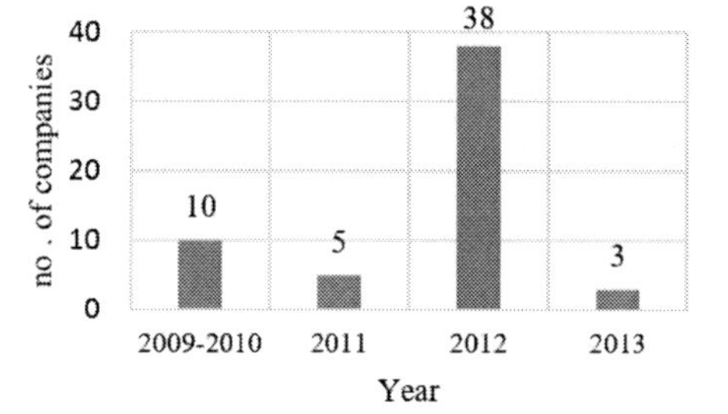

Figure 1: No. of Solar PV Companies Closing or Filing for Bankruptcy Data: GTM Research [8]

While some manufacturers are moving to offer insurance-backed warranties [5], data on long-term performance of Solar PV technologies in the different climatic conditions of the world helps to minimize investor uncertainties and also provide feedback to manufacturing companies and research institutions involved in developing solar cell materials. There is, however, little information in the open literature on the long-term performance and degradation issues on PV systems in Africa [9] [10].

In this paper, we present results of our assessment of the performance of ten (10) multicrystalline modules installed in 1998 on the roofs of the Solar Energy Applications Laboratory (SEAL) at the Kwame Nkrumah University of Science and Technology (KNUST) in Kumasi Ghana, after 18-years of exposure. Kumasi, (located at latitude 6.7°N, longitude 1.6 °W and elevation of 287 m) is a hot and humid climate with average daily temperature ranging from 24.4 °C – 27.8 °C and relative humidity of 65% - 83.5% [11].

2 METHOD

The installation under study (Fig. 2a and 2b) comprises fourteen (14) ASE-50-PWX-D modules (manufacturer - PHOTOWATT) with nominal power of 49 W (a total of 686 Wp). The modules were labelled as PWX1, PWX2,

PWX3 up to PWX 14 (Figures 1a and 1b) for the purpose of the measurements. This list excludes PWX8, PWX9 and PWX10.

Figure 2a (Left): Picture of installation (view from left)
Figure 2b (right): Picture of installation (front view)

A TRI-KA I-V (Current - Voltage) tracer (TRITEC AG, Switzerland) was used to measure the current-voltage characteristics of the modules. The measurements were conducted under two conditions – in the dusty condition, and again after cleaning. The effect of dust accumulation on the output of Solar PV modules is generally not considered as a degradation effect, since this effect is reversible [12]. Analysis of the degradation of electrical parameters of the modules was therefore based only on measurements taken after cleaning the module surface with water (Figure 3a).

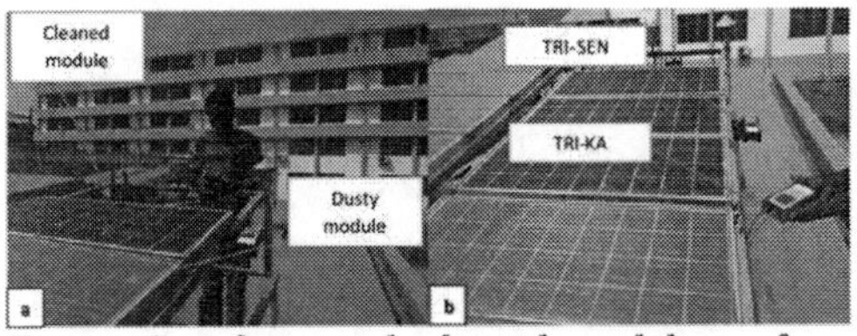

Figure 3a: dusty and cleaned module surfaces
Figure 3b: TRI-KA and TRI-SEN measurement devices

The TRI-KA I-V tracer was used together with the TRI-SEN device (both shown in Fig. 2b). The TRI-SEN device measures module temperature, irradiance and inclination, and communicates with the TRI-KA IV tracer during measurements by a wireless connection. The measurement sensitivity of the TRIKA and TRI-SEN are shown in Table 1.

Table 1: Measurement characteristics of TRI-KA and TRI-SEN

Parameter (Unit)	Measuring Range	Uncertainty
Voltage (V)	1.0 – 1000	<±1 %
Current (A)	0.1 – 15.0	<±1 %
Temperature (°C)	0 – 100	± 3 %
Irradiation (W/m²)	100 – 1200	± 5 %

Source: Product manual [13]

The I-V characteristics of the modules determined under the measured conditions of in-plane irradiance and module temperature according to IEC 61829 standard [14] were then translated to Standard Test Conditions (STC) - (1000 W/m², 25°C) - using a modified version of IEC 60891 which was developed by the Joint Research Centre of the European Commission and applied in the German 1,000 Roofs Solar Programme [15] [16] - Equations 1 – 4.

$$I_{SC,2} = I_{SC,1}[1 + \alpha(T_2 - T_1)]\frac{G_2}{G_1} \quad (1)$$

$$V_{OC,2} = V_{OC,1}\left[1 + a\ln\frac{G_2}{G_1} + b.(T_2 - T_1)\right] \quad (2)$$

$$I_2 = I_1\left(\frac{I_{SC,2}}{I_{SC,1}}\right) \quad (3)$$

$$V_2 = V_1 + (V_{OC,2} + V_{OC,1}) + R_s(I_1 - I_2) \quad (4)$$

Where:

I-current; I_{sc} - short-circuit current; G – irradiance; T - module temperature, V – voltage and V_{oc} is open-circuit voltage.

Subscripts 1 and 2 refer respectively to measured values and values at reference conditions.

And:

α is a dimensionless temperature coefficient of I_{sc} (default value is 0)

b is dimensionless temperature coefficient of Voc (default value is -0.004)

a dimensionless irradiance correction factor (default value is 0.06)

R_s: series resistance (default value is 0)

The standardized electrical characteristics (translated to STC), - Isc, Voc, FF and Pmpp -, were then compared with the original values of the manufacturer. Product documentation on the modules was not readily available from the laboratory and the module label displayed only the nominal power (49 W) under reference conditions (Fig. 4).

Figure 4: Picture of nameplate label (taken from back of module) showing nominal power (highlighted in red colour)

Other sources were therefore consulted for additional data on the module (ASE-50-PWX-D). These sources were the websites of Soltec Reimann GmbH [17] and Pearen Ventures [18], as well as the database of PVSYST 5.0 [19]. These sources host a compilation of solar module specifications spanning several years. The database of PVSYST also provided the reference curves at the various irradiance levels. Table 2 presents the primary electrical characteristics of the ASE-50-PWX-D module.

Table 2: Primary Electrical Characteristics of Module

Parameter	Value
Short-circuit current (I_{sc})	3.1 A
Open-circuit voltage (V_{oc}),	21.6 V
Nominal Power (P_{nom}), W	49 W
Current at maximum power point (I_{mpp})	2.9 A
Voltage at maximum power point (V_{mpp})	17.0 V
Fill Factor (FF)	0.73

The degradation in electrical characteristics was calculated as (Equation 5):

$$\Delta X = \frac{X_0 - X_{1,STC}}{X_0} \times 100\% \quad (5)$$

Where ΔX is the % change in electrical parameters of the module resulting from comparison of its initial value X_0 to measured value after correction to STC conditions, $X_{1,STC}$. X = I_{sc}, I_{mpp}, V_{oc}, V_{mpp}, P_{nom} and FF (defined in Table 2). The annual degradation rates, X_{ADR}, are determined as (Equation 6):

$$X_{ADR} = {\Delta X}/{y_{field}} \quad (6)$$

where y_{field} is the years of exposure, i.e. 18 years.

3 RESULTS AND CONCLUSIONS

The measured performance indicators for the modules are shown in Table 3 and shows module temperature (T_{mod}), in-plane irradiance (H_{poa}) as well as the short-curcuit current (I_{sc}), open-circuit voltage (V_{oc}), voltage and current at maximum power point (I_{mpp} and V_{mpp}), nominal power (P_{nom}) and fill factor (FF). The I-V characteristics for selected modules (irradiance levels higher than 700 W/m^2) are shown in Figure 5. Figure 6 shows the curves translated to STC using Equations 1 – 4.

Table 3: Key Performance Characteristics of modules (as measured)

Mod. ID	T_{mod} (°C)	H_{poa} (W/m²)	I_{sc} (A)	V_{oc} (V)	I_{mpp} (A)	V_{mpp} (V)	P_{nom} (W)	FF
PWX1	46.2	701	2.03	18.84	1.63	12.94	21.1	0.55
PWX2	46.7	720	2.04	18.82	1.73	14.43	25.0	0.65
PWX3	46.4	718	1.97	18.90	1.67	14.28	23.9	0.64
PWX4	46.2	671	1.96	19.00	1.67	14.17	23.7	0.64
PWX5	46.1	674	1.93	18.93	1.69	14.46	24.5	0.67
PWX6	46.3	673	1.86	18.82	1.55	14.36	22.3	0.63
PWX7	46.5	652	1.83	18.56	1.61	14.30	23.0	0.67
PWX11	45.4	742	2.14	19.24	1.74	15.06	26.2	0.64
PWX12	48.9	700	2.02	18.88	1.65	14.68	24.2	0.63
PWX13	48.1	720	2.05	18.82	1.67	14.47	24.1	0.63
PWX14	45.1	720	2.05	19.09	1.72	14.75	25.5	0.65

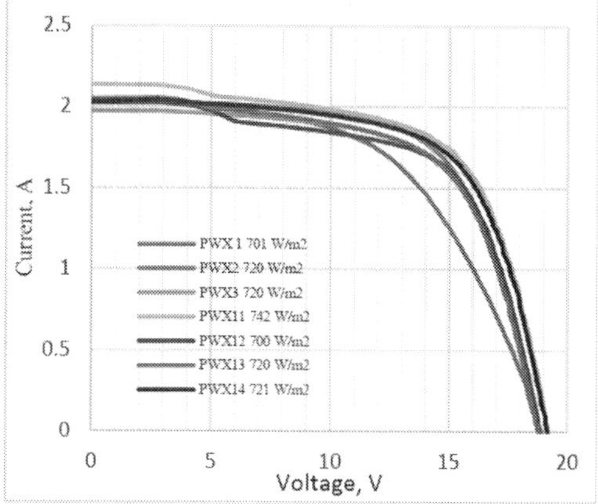

Figure 5: Measured I-V profiles of selected PV modules

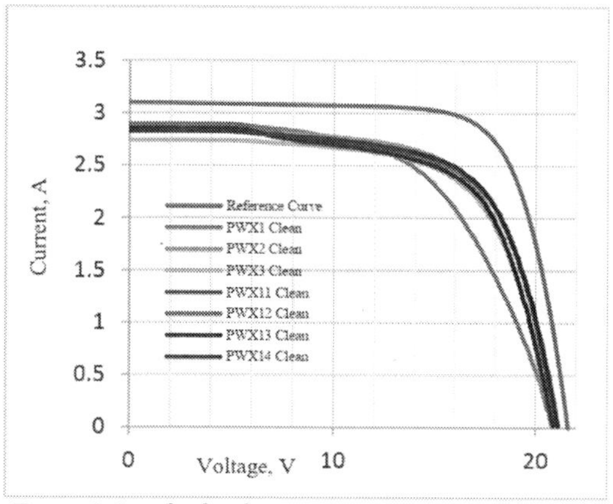

Figure 6: Standardized I-V curves of PV modules

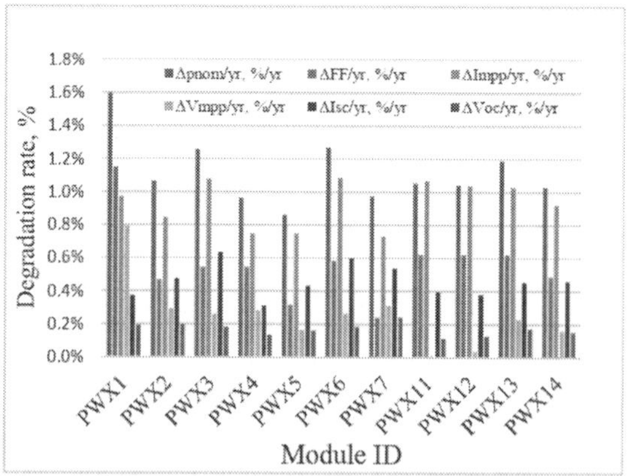

Figure 7: Degradation rates (%/year) in key parameters

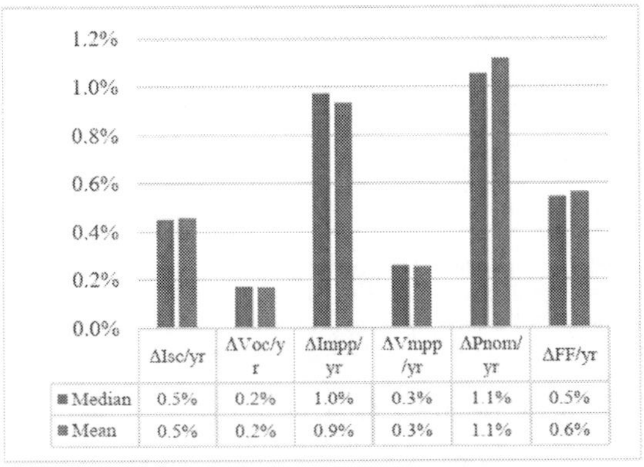

	ΔIsc/yr	ΔVoc/yr	ΔImpp/yr	ΔVmpp/yr	ΔPnom/yr	ΔFF/yr
Median	0.5%	0.2%	1.0%	0.3%	1.1%	0.5%
Mean	0.5%	0.2%	0.9%	0.3%	1.1%	0.6%

Figure 8: Mean and median of assessment parameters

Figure 7 shows, for the various modules, the average annual degradation in the key performance parameters, namely: nominal power (P_{nom}), voltage and current at maximum power point (I_{mpp} and V_{mpp}), short-curcuit current (I_{sc}), open-circuit voltage (V_{oc}) and fill factor (FF). The average degradation rates for the modules the 18-year period range from: 0.9% - 1.6% for P_{nom}, 0.3%-0.6% for I_{sc}, 0.1%-0.2% for V_{oc}, 0.7%-1.1% for I_{mpp}, 0% - 0.8% for V_{mpp} and 0.3% - 1.2% for FF on an annual basis.

Module PWX1 exhibits significantly higher degradation in power – 1.6% p.a. – as result of loss in fill factor. The nature of its I-V curve (Figure 3 and 4) suggest an increase in its series resistance [20].

The mean and median values are shown in Figure 8. For P_{nom}, I_{sc} and V_{oc}, both mean and median values are respectively: 1.1%, 0.5% and 0.2%. The median value for fill factor is 0.5% and 0.6% for mean values. Taking into account the guarantee period of 10 years and a tolerance of +/-10% [17], the module could be considered to have met and exceeded warranty expectations.

4 ACKNOWLEDGEMENT

The first author is grateful to the Quota Scholarship Programme of the Norwegian Government for sponsoring his Doctoral studies at NMBU. He is also grateful to his employer, the Kwame Nkrumah University of Science and Technology (KNUST) for granting him study leave and for equipment support. We finally would also like to acknowledge our indebtedness to Messrs. Isaac A. Edwin and Clement Appiah of the Solar Energy Applications Laboratory at KNUST for their assistance with data acquisition.

5 REFERENCES

[1] REN21, «RENEWABLES 2015 - GLOBAL STATUS REPORT,» REN21 Secretariat, Paris, 2015.

[2] IEA, «Technology Roadmap - Solar Photovoltaic Energy 2014 Edition,» IEA/OECD, 2014.

[3] CanadianSolar, «Limited Warranty Statement - Photovoltaic Module Products,» CanadianSolar, 2015.

[4] Kyocera, «Limited Warranty for Kyocera Photovoltaic Module(s),» Kyocera, 2013.

[5] Trina Solar, «25 Year Insurance Backed Warranty,» 2014.

[6] IEC, «Crystalline silicon terrestrial photovoltaic (PV) modules - Design qualification and type approval,» IEC, 2005.

[7] IEC, «Thin-film terrestrial photovoltaic (PV) modules - Design qualification and type approval,» IEC, 2008.

[8] GTM Research, «Rest in Peace: The List of Deceased Solar Companies,» GTM Research, 2013.

[9] D. C. Jordan e S. R. Kurtz, «Photovoltaic Degradation Rates—an Analytical Review,» *PROGRESS IN PHOTOVOLTAICS,* p. 12–29, 2013.

[10] A. Phinikarides, N. Kindyni, G. Makrides e G. E. Georghiou, «Review of photovoltaic degradation rate methodologies,» *Renewable and Sustainable Energy Reviews,* vol. 40, p. 143–152, 2014.

[11] RETScreen International, *RETScreen Climate Database,* 2014.

[12] IEA, «Review of Failures of Photovoltaic Modules,» IEA/OECD, 2014.

[13] TRITEC Logistics, «Operating Instructions TRI-KA,» 2010.

[14] IEC, «Photovoltaic (PV) array - On-site measurement of current-voltage characteristics,» 1995.

[15] W. Herrmann e W. Wiesner, «Current-Voltage Translation Procedure for PV Generators in the German 1,000 Roofs-Programme,» CiteSeerX, 2014.

[16] W. Zaaiman, «Solar Irradiance and Photovoltaic Measurements From Solar Radiation to PV Arrays,» AFRETEP 3rd Regional Workshop, 20th-24th 2012, Cape Town, South Africa, 2012.

[17] Soltec Reimann GmbH, «ASE SCHOTT-SOLAR,» 2016.

[18] Pearen Ventures, «PV Specifications,» 2001.

[19] PVsyst SA, «PVSyst Photovoltaic Software,» 2012.

[20] Solmetric Corporation, «Guide To Interpreting I-V Curve Measurements of PV Arrays,» Solmetric Corporation, 2010.

HOW TO REDUCE I-V MEASUREMENT DEVIATION BETWEEN RESEARCH AND PRODUCTION

Yoshikazu Takeda, Hisashi Kojima, Kazutaka Iwamoto, Yuji Fujita
Tatsuji Morishima, Kazuhiko Shibamoto, and Junichiro Abe

Kyoshin Electric Co., Ltd. (KOPEL)
Shichijo Goshonouchi Nishimachi 18
Kyoto Shimogyo-ku, Kyoto 600-8865 Japan

ABSTRACT: I-V measurement is very important for solar cell research well as for production. The probing method is often different between them, and that may create measurement errors. We have found that FF is normally measured lower in production line than in research environment, and the deviation gets bigger as the current grows and as the bus bar gets narrower. That means we need to know how the deviation is caused by the probing differences, particularly for high-efficiency cells.

We have tried various probing patterns and analyzed the cause of the measurement errors. Using the results, we will propose better probing method in order to reduce the deviation between research and production environments. In addition to the information we already published in our previous paper, this paper has added more test results and new insight.

Keywords: Electrical Properties, Evaluation, High-Efficiency

1 Introduction

In recent years, new high-efficiency cells, such as hetero-junction or PERC cells have been developed, and the conversion efficiency and Isc have been improved. At the same time, the bus bar types are diversified: in order to reduce the bus bar material cost and increase the light receiving area, very thin, partly meshed, or railroad track pattern is being used; or, instead of covering the entire back surface with metal coating, some cells (Bifacial, for instance) use bus bars on the back side also.

In research labs, cells are normally measured using the back plate electrode probing. In production lines, on the other hand, cells are often measured using the probe bars on the both surfaces (both probe bar type).

This paper identifies the issues of the probe bar probing when it is used for new types of bus bars that are developed to increase efficiency and to save cost, and makes proposals how to reduce measurement errors caused by the probing method.

2 Bare Cell Probing

Unlike research labs, production lines normally use probe bars also for backside probing. This probing difference will change measurement results between production lines and research labs.

2.1 Back Plate Electrode Type

This type uses probe bars for contact on the front side, and a metal plate on the backside.

The sample stage of this type is used in many research labs worldwide, such as SIMIT (China), ITRI (Taiwan), AIST(Japan), JET(Japan), and Fraunhofer (Germany), and KOPEL has provided our products to most of them. This type of probing has various advantages for accurate measurement: low contact resistance, and uniform voltage distribution on the backside.

The measurement data using our back plate electrode type show very strong correlation with the data measured at AIST and JET. Fig. 1 shows a picture of our sample stage of this type.

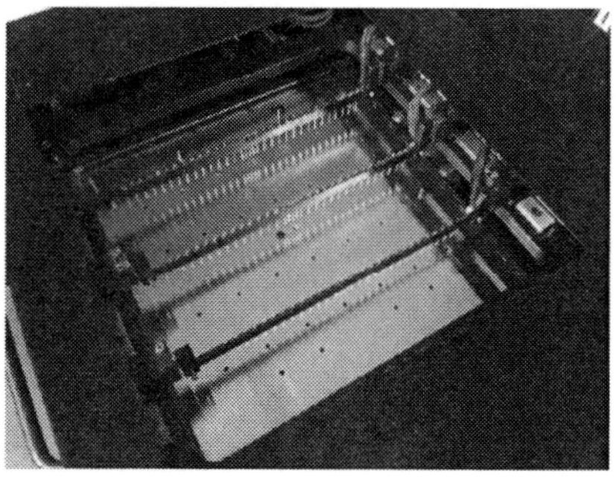

Fig. 1: Back Plate Electrode Type Sample Stage (KOPEL Product)

2.2 Both Probe Bar Type

This type uses probe bars for contact on the both sides: front side as well as backside. Compared with the back plate electrode type, the backside contact resistance is higher and this may affect FF value.

Also, the voltage distribution is different. In the case of plate, the voltage is uniform, and does not change depending on the measurement location. As to the probe bar, however, the voltage may change depending on the actual location of contact. (This applies to the front side as well.) Fig. 2 shows a picture of our sample stage of this type.

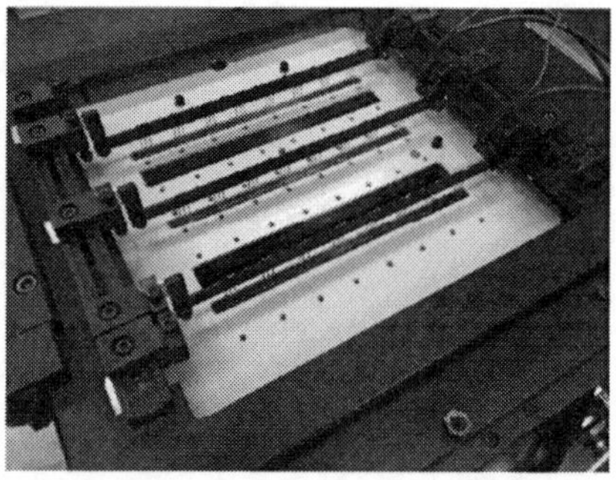

Fig. 2: Both Probe Bar Type Sample Stage (KOPEL Product)

3 Bus Bar Patterns

In recent years, bus bar types are also diversified. Complicated shape of bus bar may cause measurement errors because of its resistance and the current pattern changes. Both probe bar type probing in particular will cause a difference from the measurement in research labs.

3.1 Railroad Track, Meshed Bus Bars

Normal bus bar (Fig. 3(a)) is continuous from edge to edge of the cell, and the probe pins can contact any location on it.

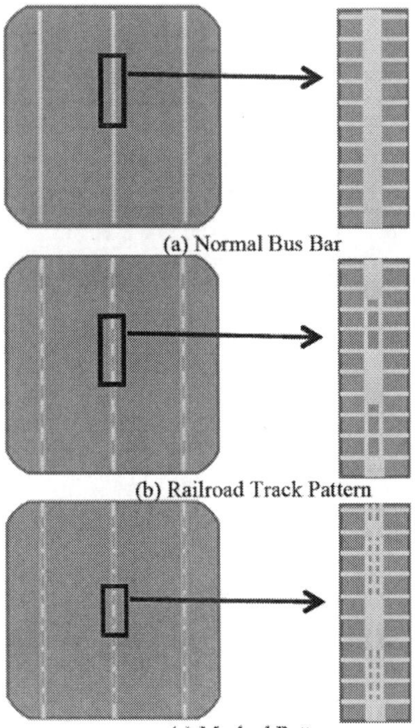

(a) Normal Bus Bar

(b) Railroad Track Pattern

(c) Meshed Pattern

Fig. 3: Cell Bus Bar Pattern

But railroad track type (Fig. 3(b)) or mesh type (Fig. 3(c)) are uncontinuous, and the contact locations and the number of pins are limited. The reduction in contact pins (I pins as well as V pins) may affect the measurement [1]. Also, the resistance changes by locations, and the voltage distribution will intricately change by the current that flows in the bus bar. This may affect the voltage measurement. [2]

3.2 Narrow Bus Bar

As the bus bar gets narrower, the resistance will increase. The higher resistance makes the voltage distribution steeper, and may cause voltage measurement error.

3.3 Backside Bus Bar

Instead of covering the entire back surface with metal coating, some cells (Bifacial cells, for instance) have bus bars on the backside also. This may affect FF measurement. [3]

The bus bar current is related to the issues described in Chapter 2 and 3 and as Isc increases, the effect on the measurement will grow.

4 Measurement Results

We have investigated the issues described in Chapter 2 and Chapter 3 using a sample stage (Both Probe Bar Type: see Fig. 2) and measured high-efficiency cells by changing probing patterns. We have used a pulse type solar simulator that is normally used in production lines.

4.1 Measurement Conditions

- Target Cells
  - Railroad track pattern (Fig. 3 (b))
  - 6in, N-type high-efficiency (with hysteresis)
  - 3 bus bar : 5 samples
  - 4 bus bar : 1 sample
  - Bifacial (Bus bar printed on the both sides. Backside has more fingers)
- Sample Stage (Fig. 2)
  - Both probe bar type
  - Probe bars: Refer to Fig. 4
  - Temperature is controlled at 25°C
- Measurement System
  - KOPEL: KSX-3000H System
    - Pulsed Solar Simulator: IEC60904-9 Class AAA, 50ms pulse width, LTI < 0.1%.
    - I-V Tester: IEC60904-1 Compliant, High-speed Bipolar Power Supply, and KOPEL Method inside.
- Measurement Method of High-efficiency PV
  - KOPEL Method
    - Dark measurement: Sweep time 20ms ×2 (Isc to Voc, Voc to Isc)
    - Dark measurement: Sweep time 600ms ×1 (Isc to Voc)
    - Lighted measurement: Sweep time 20ms ×2 (Isc to Voc, Voc to Isc)

Fig. 4: Probe Bar Configuration. 8 sets of pin groups, and each group has three pins (I, V, I). Each V pin is electrically insulated. (KOPEL Product)

We have tested each cell ten times successively without removing the contact, and averaged the ten measurements. As to 3 bus bar type, we have measured 5 cells for each pattern, and averaged the measurements. As to 4 bus bar type, we have measured one cell for each pattern ten times as described before, and averaged the measurements.

4.2 Probe Pin Patterns

By pulling out some pins of the probe bars as shown in Fig. 4, we have investigated the effects described in Chapter 2 and 3.

As to 3 bus bar cells, we have used the 6 patterns, (3BB-a) to (3BB-f), as shown in Fig. 5.

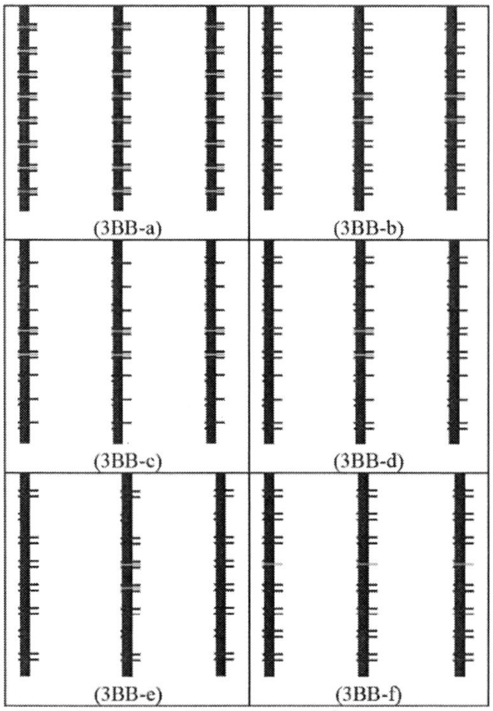

Fig. 5: Pin Patterns (3BB-a) to (3BB-f). The two probe bars on the both sides have the same pattern. Black pins are I pins, and orange are V pins.

As to 4 bus bar cells, we have used the 3 patterns, (4BB-a) to (4BB-c), as shown in Fig. 6.

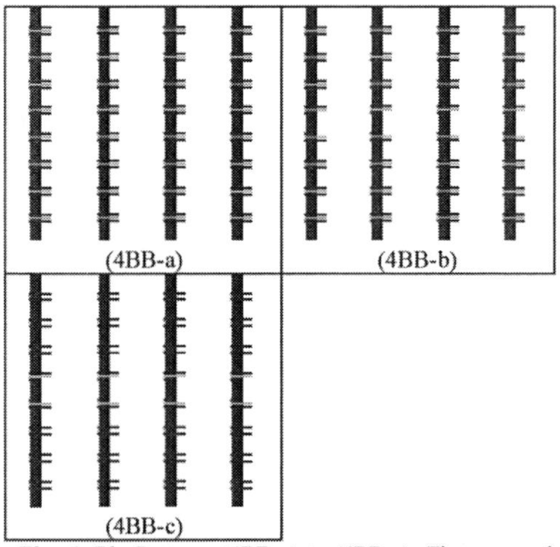

Fig. 6: Pin Patterns (4BB-a) to (4BB-c). The two probe bars on the both sides have the same pattern. Black pins are I pins, and orange are V pins.

4.3 Measurement Results

4.3.1 Measurement Results: 3 Bus Bar

Table 1 shows the test result of each pattern shown in Fig. 5. Isc and Voc show small changes (only 0.2%), but FF change depending on the pin patterns.

Table 1: Test Results on Both Probe Bar (3bus bar)

Type	Isc [A]	Voc [V]	FF [%]
(3BB-a)	9.037	0.7279	76.8
(3BB-b)	9.033	0.7280	76.8
(3BB-c)	9.041	0.7279	75.7
(3BB-d)	9.034	0.7281	76.4
(3BB-e)	9.033	0.7281	75.3
(3BB-f)	9.032	0.7279	80.4

For reference, Table 2 shows the test results using the back plate electrode type at KOPEL and JET (Japan). The data on the back plate electrode types are very consistent and show very strong correlation.

Table 2: Test Results on Back Plate Electrode (3bus bar)

	Isc [A]	Voc [V]	FF [%]
KOPEL	9.038	0.7292	78.2
JET(Japan)	9.037	0.7292	78.1
Dev.	0.0%	0.0%	0.1%

Fig. 7 shows FF test results in Table 1 and Table 2. Compared with the back plate electrode type, most of the both probe bar types are about 2% lower in FF.

Fig. 8 shows I-V curves of a cell using the both probe par types and the back plate electrode type, and Fig. 9 shows an enlarged graph around Pmpp (maximum peak power) area. Voc does not change, but the voltages around Pmpp area are shifted.

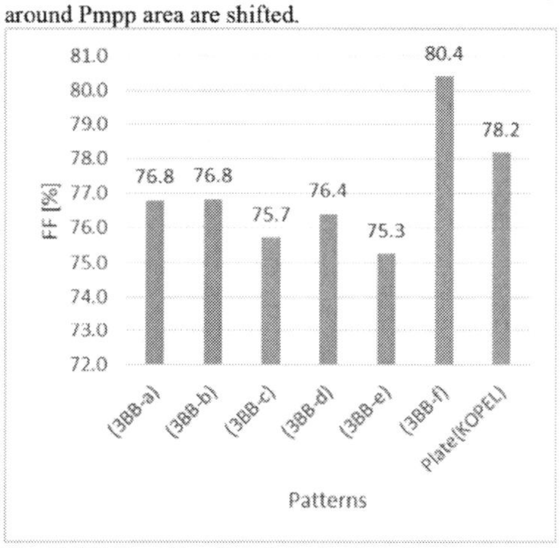

Fig. 7: FF Test Results of the both probe types and the back plate electrode type. "Plate(KOPEL)" means back plate electrode type.

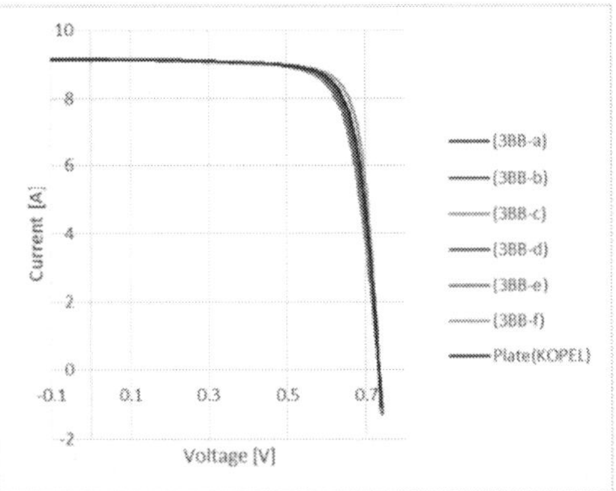

Fig. 8: I-V curves using the both probe bar types and the back plate electrode type. "Plate (KOPEL)" means back plate electrode type.

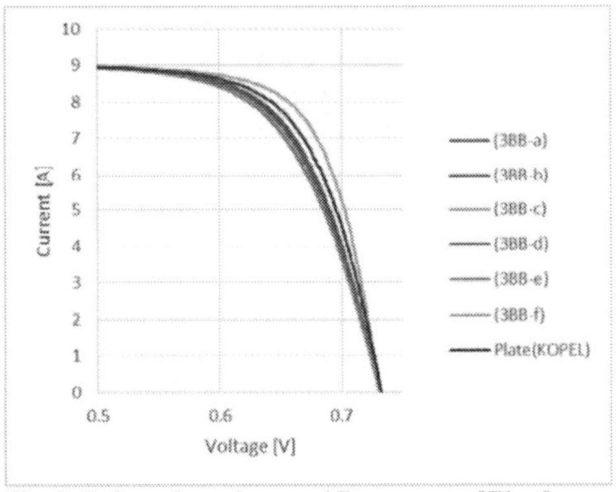

Fig. 9: Enlarged graph around Pmpp area of Fig. 8.

4.3.2 Measurement Results: 4 Bus Bar

Table 3 shows the test result of each pattern shown in Fig. 6. Isc and Voc show small changes, but FF change depending on the pin patterns. It shows a similar tendency as 3 bus bar type.

Table 3: Test Results on Both Probe Bar (4bus bar)

Type	Isc [A]	Voc [V]	FF [%]
(4BB-a)	8.969	0.7323	77.3
(4BB-b)	8.971	0.7322	77.2
(4BB-c)	8.970	0.7321	78.2

Fig. 10 shows I-V curves of a cell using the both probe bar types and the back plate electrode type, and Fig. 11 shows an enlarged graph around Pmpp area. It shows a similar tendency as 3 bus bar type.

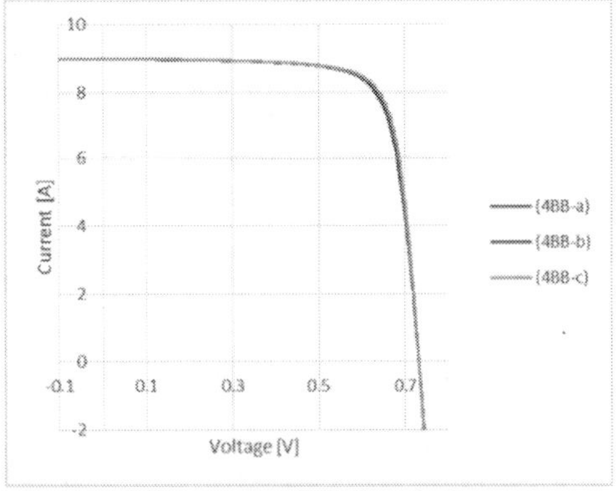

Fig. 10: I-V curves using the both probe bar types.

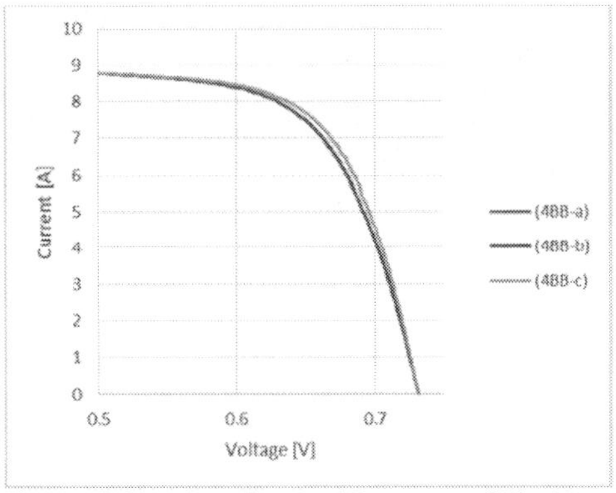

Fig. 11: Enlarged graph around Pmpp area of Fig. 10.

4.4 Considerations

4.4.1 Difference between Plate and Probe

According to Fig. 7, most of the both probe bar types show lower FF than the back plate electrode type. This is because of the increase of the contact resistance, from plate to probe bar, and from the whole surface metallization on the backside to the bus bar.

4.4.2 Effect of the number and locations of I pins

Comparing the patterns (3BB-b) and (3BB-c), the number of V pins is the same, but I pins are reduced. Fig. 7 shows FF of (3BB-c) is about 1% lower than that of (3BB-b). It is consistent with the description in the paper [3].

Then, comparing the patterns (3BB-d) and (3BB-e), the numbers of V pins and I pins are the same, but their locations are changed. Fig. 7 shows FF of (3BB-e) is about 1% lower than that of (3BB-d). This means that irregular intervals of I pins (or I pin groups) will affect FF measurement. Therefore, I pins must be arranged at regular intervals.

4.4.3 Effect of the number and locations of V pins

Comparing the patterns (3BB-a) and (3BB-b), the number of I pins and their locations are the same. Fig. 7 shows there is no big change in FF. That means there is not much change in the bus bar current, and the voltage distribution is not affected. That is the reason why FF does not change.

Comparing the pattern (4BB-a) and (4BB-b), the central I pins have been removed in (4BB-b) and the number of I pins has reduced, but FF is almost the same as (4BB-a). The current pattern in the bus bar has been changed by the I pin locations, but because of abundant V pins the voltage change caused by the current change may be averaged [2]. In fact, comparing (4BB-b) and (3BB-c), there is not much change in I pin locations, but the V pins are allocated where the currant flow may be very different (where I pins are removed). That has caused a big difference in FF measurement. To reduce the impact of contact differences, V pin and I pin (or I pin group) must be allocated in pairs.

4.4.4 Effect of the distances (between I pin and V pin, and between I pin and I pin)

Pattern (3BB-f) and (4BB-c) shows abnormal increase in FF, although the number of I pins is reduced. The cause of this abnormality is probably because of the locations of V pins. The paper [2] reports that if I pin and V pin is apart, the voltage measure at V pin is higher than the actual voltage at I pin. If the distance is bigger, the voltage distribution will spread wider. In the pattern (3BB-f) and (4BB-c), therefore, V pins measure higher voltages, and that has increased FF.

The paper [2] reports by placing a V pin by two I pins, the voltage difference between V pin and I pin can be reduced.

5 Probing Methods to Reduce Measurement Errors

From the descriptions in 4.4.1 through 4.4.4, the following conditions must be considered to reduce the measurement errors of high-efficiency cells with complex bus bars using the both probe bar type sample stage.

- Increase the number of I pins: As the number decreases, FF will drop. The number may be limited by the bus bar pattern. The paper [1] recommends the number should be 12 I pin (I pin group) or over.
- Allocate the same number of I pins (I pin group) at same intervals: The interval and number of I pins affect FF.
- Place I pin and V pin closer: As the distance grows, V pin measures higher voltage. The paper [2] recommends less than 3mm is desirable.
- Increase the number of V pins: As the number decreases, the measurement is affected by the voltage distribution caused by the bus bar current. If the voltage distribution is stable, the number of V pins can be reduced.

Based on the above conditions, Fig. 12 shows a probe bar that we can manufacture, and that can reduce measurement errors. This probe bar satisfies the above conditions, and the distance between I pin and V pin can be as small as 2mm. The number of I pins, however, depends on the bus bar pattern.

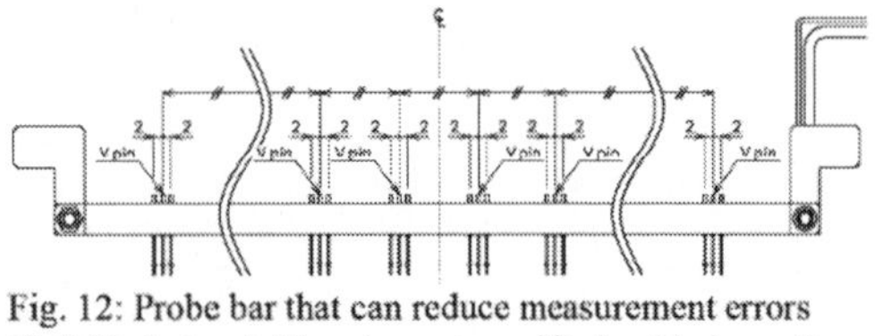

Fig. 12: Probe bar that can reduce measurement errors (Full V pin type). The pins not specified as V pin are I pins.

The high number of V pins shown in Fig. 12 may increase the cost. If the voltage distribution at the measurement in the production line can be stabilized – contact is well controlled for instance – the number of V pins can be reduced as the low cost version. The low cost version shown in Fig. 13.

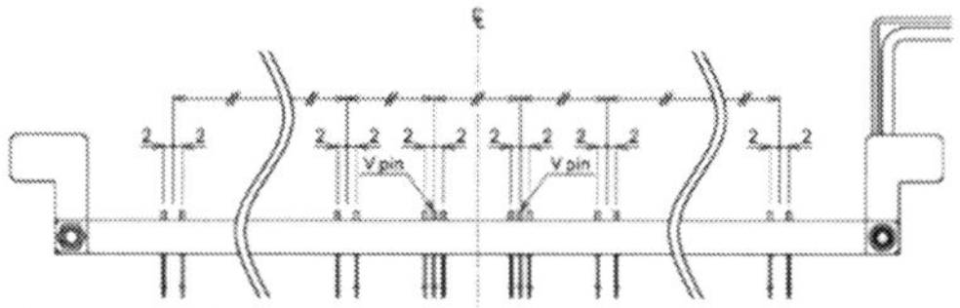

Fig. 13: Probe bar that can reduce measurement errors. (Minimum V pin type). The pins not specified as V pin are I pins.

6 Summary

We have actually tested the high-efficiency cells that have complex bus bar by the both probe bar type sample stage, and have clearly understood the issues. Increase of contact resistance – from plate electrode to probe bar, and from entire metallization to bus bar – affect FF. Reduced number of V pins does not affect FF if I pins (I pin groups) are evenly arranged and does not affect the voltage distribution. Reduced number of I pins definitely reduces FF. Not only the number, but also I pins (I pin groups) must be evenly distributed. The distance between V pin and I pin must be closer. Otherwise, V pin measures higher voltage and FF will change. By considering these issues, we have proposed probe bars that can reduce the measurement errors.

For our future study, we will investigate probing with high measurement reproducibility in production line, and probing for back-contact cells.

References

[1] Hund,B. et al, Possible Impact of Systematic Errors in IV-Measurement, 23rd EU PVSEC, Valencia Spain, 2008

[2] J. Hohl-Ebinger. et al, Contacting bare solar cells for STC measurements, 23rd EU PVSEC, Valencia Spain, 2008.

[3] Fraunhofer ISE, Solar Cell Calibration and Measurement Procedures at Fraunhofer ISE CalLab PV Cells-21. December 2012.

[4] Y. Takeda. et al, How Probing Pattern Affects on I-V Characteristic Measurements of High-efficiency Solar Cells, SNEC 2016.

NEW CROSS-LINKING ASSISTANT FOR ENCAPSULATING MATERIALS OF EVA

Yuki Kawamura1, Mabuko Yamaura1, Shuichi Asao2, Katsuhiko Shirasawa2, Hidetaka Takato2
1 NIPPON KASEI CHEMICAL CO., LTD. 2 National institute of Advanced Industrial Science and Technology
1: 34 Aza Takayama, Onahama iwaki-shi, Fukushima, Japan
Phone: +81-246-54-3115 Fax: +81-246-53-3911 e-mail: kawamura.yuuki@mw.nkchemical.co.jp
2: 2-2-9 Machiikedai, Koriyama, Fukushima, Japan
Phone: +81-24-963-1805

ABSTRACT: Potential Induced Degradation (PID) occurs at mega solar power plants and is a serious problem. The mainstream approach to PID prevention is to improve anti-reflection coating to change Si/N ratios. We, on the other hand, suggest prevention with the encapsulating material. We developed the cross-linking assistant "TENASHIELD®" which improves volume resistivity of the EVA (ethylene vinyl acetate) film. Volume resistivity of the test film is more than ten times higher than EVA film containing conventional cross-linking assistants. The power output of modules equipped with EVA film containing TENASHIELD® didn't decrease as our PID testing showed. TENASHIELD® provided superior PID prevention to EVA film. We infer that TENASHIELD® reduces the formation of voids and conductive passes when EVA film is applied in thermal cross-linking. We assume that a migration of Na ions was prevented, and PID was reduced. Conventional anti-reflection coating is effective enough for photo-voltaic cells if TENASHIELD® is used. Cost of this method may be much lower than other PID prevention methods. Thus, TENASHIELD® may contribute to the further development of mega solar power plants.
Keywords: Encapsulation, PV Materials

1 INTRODUCTION

Recently, Potential Induced Degradation (PID) occurred at mega solar power stations and becomes the serious problem. In 2012, PID test of PV modules in the market was performed by the Fraunhofer-Gesellschaft. As a result of test, the output of many modules decreased. After this report was announced, many research organization and company have been researching PID prevention.

The assumed mechanism is as follows: first, Na ions derived from glass are moved and charged on the cell surface by high voltage, next, Na ions catch electrons and electrons can't drift to the circuit [1]. The mainstream of PID prevention is to improve of anti-reflection coating that to change Si/N ratio. In case of Si in the anti-reflection coating is increased, conductivity improves and the anti-reflection coating prevents charge of the Na ions.

However, this method increases refractive indexes. High refractive indexes reduce the output because interfere with an invasion of the light. The drop of the output causes increase of the cost and obstacles to the spread of photovoltaic modules. Therefore we suggest prevention with the encapsulating material. Photovoltaic modules usually have encapsulating material between glass and cell. Encapsulating materials may prevent PID by high insulation characteristics because the Na ions pass encapsulating material and arrive at the cell surface. Ethylene vinyl acetate (EVA) is common for s encapsulating material. EVA is cross-linked by peroxide and cross-linking assistant. Volume resistivity is usually around $10^{14}\sim10^{15}$ [$\Omega \cdot$ cm].

We industrialized triallyl isocyanurate (TAIC®) for the first time in the world in 1970 and has been producing and selling it since then. In many cases, triallyl isocyanurate is used as cross-linking assistant. We developed a new cross-linking assistant we call "TENASHIELD®" relying on our expertise as a cross-linking assistant manufacturer.

2 EXPERIMENTAL APPROCH

2.1 NEW CROSS-LINKING ASSISTANT

We prepared EVA film including peroxide (PERBUTYL E, NOF CORPORATION), ultraviolet absorber (2-Hydroxy-4-n-octyloxybenzophenone, Wako), silane coupling agent (KBM-503, Shin-Etsu Chemical) and cross-linking assistant. We used four kinds of cross-linking assistant. One is the TAIC® (triallyl isocyanurate) for reference. The others are new crosslinking assistant. Three kinds of new crosslinking assistant are named TENASHIELD001®, TENASHIELD002® and TENASHIELD003®. VA content of EVA is 28%. The volume resistivity of these films was measured by SUPER MEGOHMMETER SM-8220 (HIOKI).

2.2 PV modules and PID testing

Fig.1 shows the structure of Photovoltaic module used PID test. The structure is tempered glass/EVA/solar cell/EVA/back sheet (PVF/PET/PVF). Structure of solar cell is industrial Al-BSF of 6 inch p-type CZ-Si. We used two kinds of solar cell. One was "PID free cells", the other was not PID free cell (so called "PID cells", refractive index of PECVD-SiN film : 2.0). PV modules were made using laminator.

We carried out PID test at the National Institute of Advanced Industrial Science and Technology (AIST). Test conditions are as follows. Temperature is 85°C. Humidity is 85%. Applied voltage is -1,000V. And an aluminum plate is put on the top glass of PV module (Fig.2). We compared PV power and EL images before PID testing and after PID testing respectively.

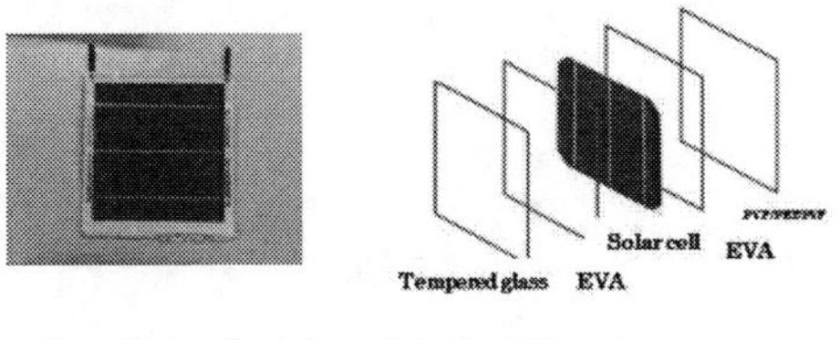

Figure1: 1cell mini module for PID test.

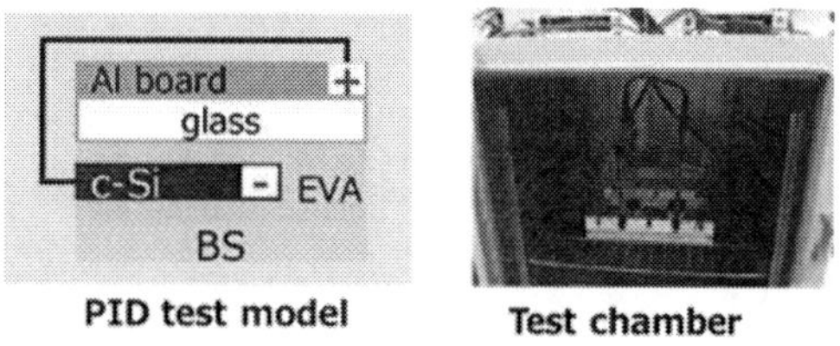

Figure2: PID test model and photograph of test chamber.

3 RESULTS AND DISCUSSION

3.1 EVA films properties

Test films containing TENASHIELD® are more than ten times higher than EVA film containing TAIC® (Table1).

Table 1: Volume resistivity of the EVA film

	Volume resistivity of the EVA film[Ω · cm]
TAIC®	2.0×10^{15}
TENASHIELD®001	9.7×10^{15}
TENASHIELD®002	1.7×10^{16}
TENASHIELD®003	2.6×10^{16}

The weatherability, the durability at high temperature of the EVA film containing TENASHIELD® were at the same level as EVA film containing TAIC® (Table 2).

Table 2: The rate of change of total light transmittance after weatherability test(2000hrs)

	The rate of change
TAIC®	2.2%
TENASHIELD®001	1.9%
TENASHIELD®002	2.4%
TENASHIELD®003	1.9%

3.2 PID test

The power output of the module comprised of EVA film containing TAIC® decreased overall (Fig.3 Ref. and Fig.4 and Fig.5). In the case of PID free cells, power output dropped to 80% after 96hrs, in the case of not PID cells, power output dropped to 27% after 96hrs. In contrast, power of the module comprised of EVA film containing TENASHIELD®001~003 didn't decrease much at all. In the case of PID free cells, power output was more than 95% after more than 200hrs (Fig.5 (a)). In the case of PID cells equipped with TENASHIELD®001, 002 power output decreased. In contrast however, applying TENASHIELD®003 those modules did not degrade (Fig.4 and Fig.5 (b)).

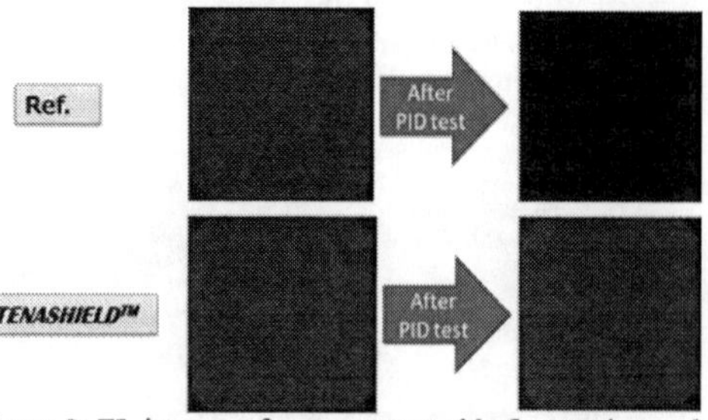

Figure 3: EL images of were compared before testing and after testing.

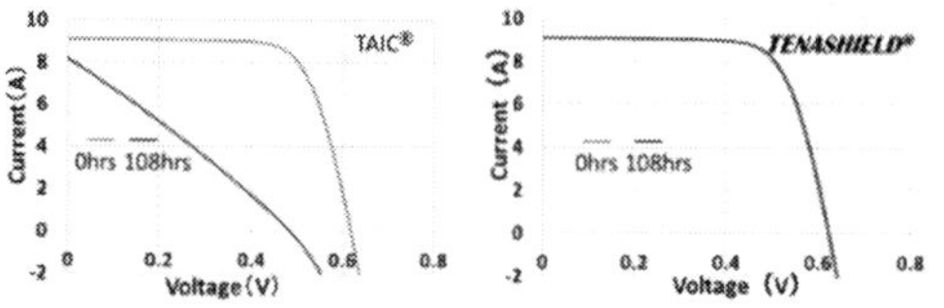

Figure 4: IV curve of PV module using TAIC® and TENASHIELD®.

3.3 Discussion

We estimate that because TENASHIELD® changed the structure of the EVA film, EVA film got PID prevention. EVA film is usually an insulation material. However, EVA film has multiple small voids. We assume that voids reduce insulation resistance because voids catch moisture and small conductive things and form conductive passes (Fig. 6). We infer that TENASHIELD® restrains the formation of voids and conductive passes when EVA film is used in thermal cross-linking. So, the volume resistivity of the EVA film was improved. No migration of ions is caused, and PID is reduced.

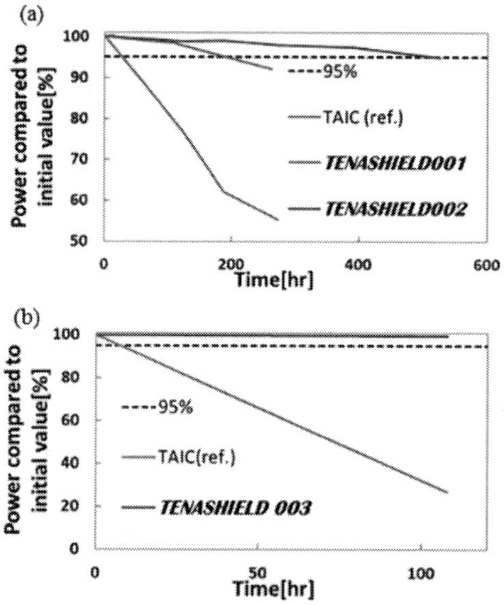

Figure 5: (a) Power compared to initial value of in the case of a PID free cells. (b) Power compared to initial value of in the case of PID cells.

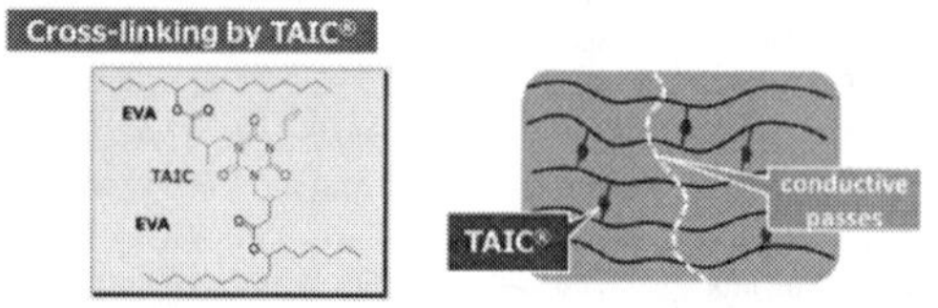

Figure 6: Model of cross-linking by TAIC® and conductive passes.

4 CONCLUSION

We have developed the new cross-linking assistant, TENASHIELD®, for improving reliability of PV modules. TENASHIELD® showed remarkable prevention effects against PID in PV modules; Volume resistivity of EVA was improved by containing TENASHIELD® and TENASHIELD® was effective for both "PID-free cells" and "PID cells". Power loss was less than 5% after test duration of 96 hours at temperature of 85°C and relative humidity of 85% with applied voltage of -1000V. Conventional anti-reflection coating is effective enough for PV cells if TENASHIELD® is used. In that case, no loss in power output caused by optical loss occurs, and cost drops. In addition, you can use TENASHIELD® to replace a conventional cross-linking assistant without any interruption of the production process. TENASHIELD® may thus contribute to the further development of mega solar power plants.

5 References

[1] Hacke, Peter, et al. "System voltage potential-induced degradation mechanisms in PV modules and methods for test." Photovoltaic Specialists Conference (PVSC), 2011 37th IEEE. IEEE, 2011.

THE IMPACT OF INDOOR AND OUTDOOR RADIOMETER CALIBRATION ON SOLAR MEASUREMENTS

Aron Habte[1], Manajit Sengupta[1], Afshin Andreas[1], Ibrahim Reda[1],
Justin Robinson[2]
[1]National Renewable Energy Laboratory, Golden, CO, 80401, USA
[2]GroundWork Renewables Inc., Logan, UT, 84321, USA

ABSTRACT: Accurate solar radiation data sets are critical to reducing the expenses associated with mitigating performance risk for solar energy conversion systems, and they help utility planners and grid system operators understand the impacts of solar resource variability. The accuracy of solar radiation measured by radiometers depends on the instrument performance specification, installation method, calibration procedure, measurement conditions, maintenance practices, location, and environmental conditions. This study addresses the effect of calibration methodologies and the resulting calibration responsivities provided by radiometric calibration service providers such as the National Renewable Energy Laboratory (NREL) and manufacturers of radiometers. Some of these radiometers are calibrated indoors, and some are calibrated outdoors. To establish or understand the differences in calibration methodology, we processed and analyzed field-measured data from these radiometers. This study investigates calibration responsivities provided by NREL's broadband outdoor radiometer calibration (BORCAL) and a few prominent manufacturers. The BORCAL method provides outdoor calibration responsivity of pyranometers and pyrheliometers at a 45° solar zenith angle and responsivity as a function of solar zenith angle determined by clear-sky comparisons to reference irradiance. The BORCAL method also employs a thermal offset correction to the calibration responsivity of single-black thermopile detectors used in pyranometers. Indoor calibrations of radiometers by their manufacturers are performed using a stable artificial light source in a side-by-side comparison of the test radiometer under calibration to a reference radiometer of the same type. In both methods, the reference radiometer calibrations are traceable to the World Radiometric Reference. These different methods of calibration demonstrated 1% to 2% differences in solar irradiance measurement. Analyzing these values will ultimately assist in determining the uncertainties of the radiometer data and will assist in developing consensus on a standard for calibration.

Keywords: Radiometer calibration, thermopile, photodiode, pyranometer, pyrheliometers, global horizontal irradiance, direct normal irradiance, diffuse global irradiance, responsivity, measurement uncertainty

1 INTRODUCTION

One of the critical components for a successful solar energy project is access to accurate solar resource data. Pyranometer and pyrheliometer radiometers are used to measure global horizontal, diffuse, plain-of-array, and direct normal solar irradiances. To acquire accurate solar resource information from these radiometers, the tools used to design, manufacture, calibrate, deploy, and maintain them must continuously improve to provide necessary assurances to solar system financiers, developers, and operators. Investors must understand the quantitative terms of the risks of their investments based on the knowledge of the solar resource inputs, and they need to be assured that system operators can operate and maintain their solar energy conversion system assets with maximum efficiency and value.

This study assesses multiple radiometer calibration methodologies and the resulting variations of irradiance measurements. Calibration is one source of uncertainty in radiometric data. Addressing the irradiance differences in data resulting from the myriad calibration methodologies helps quantify the radiometric measurement uncertainties. Further, the study serves to open a dialogue among radiometer manufacturers and calibration service providers to develop a consensus on a standard approach to calibration that will alleviate, reduce, or explain the differences in irradiance measurements due to the differing calibration methodologies.

2 METHOD

The radiometric data used in the study is traceable to the System Internationale, or SI, units through the World Radiometric Reference scale [1]. Data was collected for two months through the Measurement Instrumentation Data Center at the National Renewable Energy Laboratory (NREL). Clear-sky data were selected for the analysis based on a clearness index (Kt) greater than 0.6 (Eq.1).

$$Kt = \frac{GHI}{ETRN * \text{Cos (SZA)}} \quad (1)$$

where:
GHI = global horizontal irradiance (W/m^2)
$ETRN$ = extraterrestrial direct normal radiation (W/m^2)
SZA = solar zenith angle (degree).

The 1-minute data from the unit under test (UUT) radiometers were compared to reference GHI or direct normal irradiance (DNI) data. The reference irradiance data were obtained using a Kipp and Zonen model CHP1 pyrheliometer (serial number 140108), and reference diffuse horizontal irradiance (DHI) measurements were made by a shaded Kipp and Zonen model CM22 pyranometer (serial number 010047). These two reference radiometers provide the lowest measurement, uncertainty for GHI [2], [3], [4]. A component sum method (Eq. 2) was used to obtain reference GHI data using measured data from the above reference radiometers.

$$GHI = DNI * Cos(SZA) + DHI \quad (2)$$

The DNI UUT radiometer data were directly compared to the reference DNI. All of the radiometers included in the study were calibrated by NREL's BORCAL process and the radiometer manufacturers using indoor calibration methods. The radiometers were deployed soon after calibration by the two methods. Measurement degradation and/or sensitivity drift were

TABLE I

RESPONSIVITY VALUE CASES APPLIED IN THE STUDY. WHEN THERMAL OFFSET CORRECTION IS APPLICABLE (YES), EQUATION (3) IS USED. IF NOT APPLICABLE (NO), EQUATION (4) IS USED.

Cases	Calibration Method	Thermal Offset Correction Applicability		
		Thermopile Pyranometer	Thermopile Pyrheliometer	Silicon Photodiode Pyranometer
Case 1	BORCAL responsivity as a function of SZA	Yes	No	No
Case 2	Manufacturer calibration responsivity at manufacturer-specified SZA in degrees	N/A	N/A	N/A
Case 3	BORCAL responsivity at 45°	Yes	N/A	N/A
Case 4	BORCAL responsivity at 45°	No	N/A	N/A
Case 5	Manufacturer calibration responsivity at manufacturer-specified SZA in degrees with manufacturer-supplied measurement equation	N/A	Yes	N/A

not considered to affect the responsivity of the radiometers. Further, our analyses were well within the recommended calibration cycle of one to two years [5]. The two months of data in this study were collected from June 23, 2015, through July 19, 2015. The analyses are based on measurements under clear skies.

2.1 Background on Calibration Methodology

NREL provides the following calibration services through the BORCAL process;

1. Calibration responsivity at a 45° SZA with thermal offset correction (using Eq. 3) for thermopile pyranometers and no thermal offset correction for thermopile pyrheliometers and photodiode.
2. Calibration responsivity as a function of SZA with thermal offset correction (using Eq. 3) for thermopile radiometers and no thermal offset correction for thermopile pyrheliometers and photodiode.

$$R = \frac{(V - R_{net} * W_{net})}{GHI} \qquad (3)$$

$$GHI = DNI * Cos(SZA) + DHI$$

where:
R = the pyranometer's responsivity, in μV/ (Wm^{-2})
V = the pyranometer's sensor output voltage, in μV
DNI = the direct normal irradiance measured by a primary or standard reference pyrheliometer, measuring the beam irradiance directly from the sun's disk, in Wm^{-2}
SZA = the solar zenith angle, in degrees
DHI = the diffuse sky irradiance, without the beam irradiance from the sun's disk, measured by a shaded pyranometer, in Wm^{-2}
R_{net} = the pyranometer's net infrared responsivity, in μV/(Wm^{-2})
$Wnet$ = the effective net infrared irradiance measured by a collocated pyrgeometer, in Wm^{-2}.

Indoor calibration methods by manufacturers provide single calibration responsivity, typically at approximately normal incidence. In most cases, the indoor calibration method uses Eq. 4:

$$R_{UUT} = \frac{V_{test}}{V_{ref}} R_{ref} \qquad (4)$$

where:
R_{test} = responsivity of the radiometer under calibration, μV/(Wm^{-2})
V_{ref} and Vtest = the voltages (μV), measured using the reference and the field radiometers, respectively
R_{ref} = responsivity, μV/(Wm^{-2}) of the reference radiometer.

2.2 Measurement Analysis Method

To analyze the measured data, the radiometric data from the UUT radiometers were back-calculated to determine the raw voltage readings using Eq. 5 This equation excludes the thermal offset correction applied by NREL to the solar irradiance data used in the study. Then the different responsivity values were applied to the voltage readings to generate irradiances based on each responsivity.

The measurement equation for generating raw voltages is:

$$V_{raw} = G_m * R + R_{net} * W_{net} \qquad (5)$$

where:
V_{raw} = raw voltage readings, in μV
G_m = measured irradiance, in W/m^2
R = the responsivity, in μV/Wm^{-2} used to produce the initial GHI values (from BORCAL).
Note: G_m applies to both pyranometers and pyrheliometers UUT.

Five responsivity value cases (Table 1) were applied to the raw voltage readings. These values were obtained from the BORCAL and manufacturers' indoor calibration methods—that is, users of the radiometric data obtain calibration responsivity from one of the cases below either from NREL or manufacturers.

For thermal offset correction cases, the thermally corrected voltage reading was obtained using Eq. 6:

$$V_{cor} = V_{raw} - W_{net} * R_{net} \qquad (6)$$

where:
V_{cor} = corrected voltage, in μV

The thermal offset correction was applied using net long-wave responsivity (R_{net}) estimated or determined by blackbody characterization of the pyranometer [6] and a collocated pyrgeometer of a body type similar to the thermopile radiometer under test for determining W_{net}, using Eq.7 & 8 [7].

$$W_{net} = W_{in} - W_{out} \quad (7)$$

$$W_{out} = \sigma * T_c^{\,4} \quad (8)$$

where:
W_{in} = incoming infrared from the pyrgeometer, in W/m^2
W_{out} = outgoing infrared from the pyrgeometer, in W/m^2
T_c = case temperature of the pyrgeometer, in Kelvin (K)
Σ = 5.6704×10^{-8} $W/(m^2K^4)$ (Stephan-Boltzmann's Constant).

The measured irradiance can now be determined for each case of responsivity under consideration:

$$G_{new} = \frac{V_{cor}}{R_{case_i}} \quad \text{or} \quad G_{new} = \frac{V_{raw}}{R_{case_i}} \quad (9)$$

G_{new} is the new GHI and DNI irradiance values obtained through implementing the various calibration cases. R_{case} is the responsivity value for each case. Eq. 9 (left) was applicable for Case 1 and Case 3, and Eq. 9 (right) was used for Case 2 and Case 4; and for photodiodes and thermopile pyrheliometers, the thermal offset corrections (R_{net} and W_{net}) were not applicable.

BORCAL provides responsivity at 2° bins of SZA. For this case, irradiance values are calculated using a high-resolution responsivity and SZA. The high-resolution function is obtained by linearly interpolating the morning and afternoon averaged 2° responsivity data [8]. The authors of [9] confirmed that the responsivity as a function of SZA significantly reduces the cosine response error and ultimately the overall uncertainty of the radiometric data.

Note: The reference DNI and the DNI measured by the UUT were corrected for cosine response using the responsivity as a function of SZA method.

A percentage difference calculation was applied to assess the performance of each UUT relative to the reference instruments (Eq. 10):

$$Difference(\%) = \left(\frac{(G_{new} - G_{Ref})}{G_{Ref}}\right) * 100 \quad (10)$$

where G_{Ref} are the irradiance values for the reference instrument in W/m^2.

3 RESULTS AND DISCUSSION

Multiple radiometers were included in the study. Fig. 1 and Fig. 3–5 show the results of the comparison of the UUT to the reference GHI and DNI data. To better understand the differences due to calibration methodology, the results of the comparison were partitioned into 10° bins of SZA in which the center of the bin represented the midpoint of each 10° SZA bin. The information in the figures below is derived from 1-minute irradiance data.

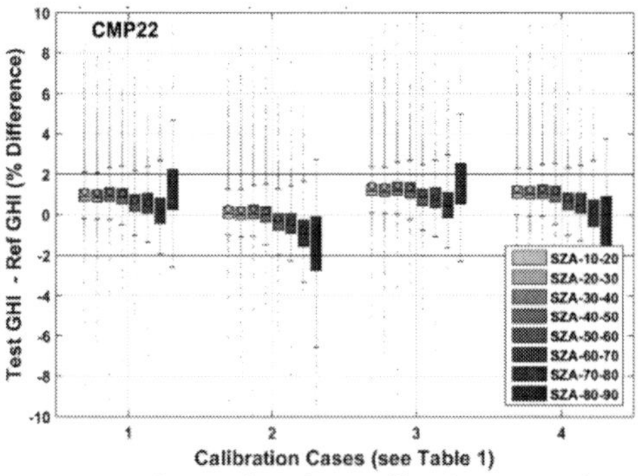

Figure 1: Indoor and outdoor calibration comparison of the reference GHI irradiance to the CMP22 test radiometer irradiance. Note: Each blue box represents a 10° bin as well as the upper quartile and the lower quartile (also called the interquartile range) of the data in each bin. The circle in each blue box is the mean, and the black line signifies the median value. Ninety-nine percent of the data set is within the whiskers; data beyond the whiskers are plotted with a symbol (dots).

The four calibration cases in Fig.1 show differences up to 2% in GHI measurement. However, the NREL responsivity as a function of SZA (Case 3) and the manufacturer calibration responsivity (Case 4) provide least differences compared to the NREL 45° SZA responsivity with and without thermal offset. One hypothetical explanation for the better agreement of the manufacturer calibration responsivity result relative to the NREL methods is that the reference DHI used in the study is from the same manufacturer and same model as the UUT (CMP22). This condition could favor the CMP22 comparison to the reference data; however, this needs further investigation.

Overall, the NREL responsivity as a function of SZA performed relatively better compared to the GHI UUT (Fig. 3 and Fig. 4). However, this method demonstrated poor performance in higher SZA (Bin 80° to 90° SZA). The reason is that the NREL 2° responsivity bins extend only up to a maximum of 80° during the BORCAL, and the same maximum responsivity number gets repeated beyond the SZA limits (circled in red in Fig. 2) on the day of calibration.

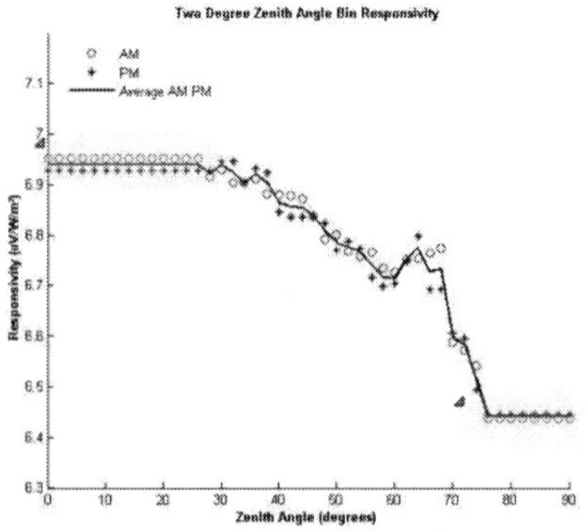

Figure 2: Example of responsivity as a function of SZA for Eppley Laboratory, Inc.'s Standard Precision Pyranometer (Model PSP) after averaging the a.m./p.m. 2° bins and interpolating among each set of bins. Responsivities for less than 26° SZA have the

same value as the 26° SZA. The same is true for responsivities greater than 76° SZA, wherein they have the same value as the 76° SZA (modified from [8]).

However, the effect is minimal for the smaller SZA bin (Bin 10° to 20° SZA, Fig. 2—red circle) for three reasons. First, the lowest SZA observed for the study location was approximately 17°; therefore, the bin range had an approximate 18.5° center point lowest SZA for this bin, which is on the higher end of Bin 10° to 20°. Second, the number of occasions when the SZA falls between 17° and 20° SZA during the study period is relatively small compared to Bin 80° to 90° SZA. Third, responsivity at a smaller SZA has much lower dependence on zenith angle compared to a high SZA.

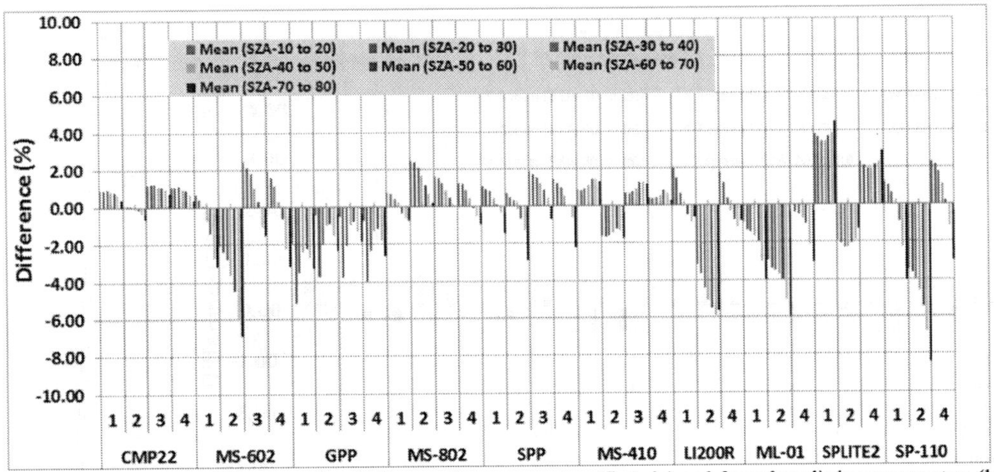

Figure 3: Mean percent differences for six thermopile pyranometers (first six) and four photodiode pyranometers (last four).

As shown in Fig. 3, some of the radiometers—such as MS-802, SPP, L200R, SP-110—in Case 1, which is NREL's BORCAL responsivity as a function of SZA, appear to perform better than they do in the rest of the cases. Overall, comparing the calibration methodologies would provide a better understanding of the performance of the radiometers in relation to SZA.

DNI UUT radiometers were also compared. Fig. 3 shows a comparison of the EKO Instruments model MS-56 to the Hukseflux model DR02.

As stated in Case 5 of Table 1, the comparisons of the MS-56 and DR02 were carried out using manufacturer-supplied equations (11 and 12, respectively).

$$G_{man_{MS-56}} = \frac{V_{raw} + k * V^2{}_{raw}}{R_{test}} \quad (11)$$

where:
G_{man}= DNI irradiance derived using manufacturer supplied equation, in W/m^2.
k = a multiplier coefficient supplied by the manufacturer, (1/ μV).

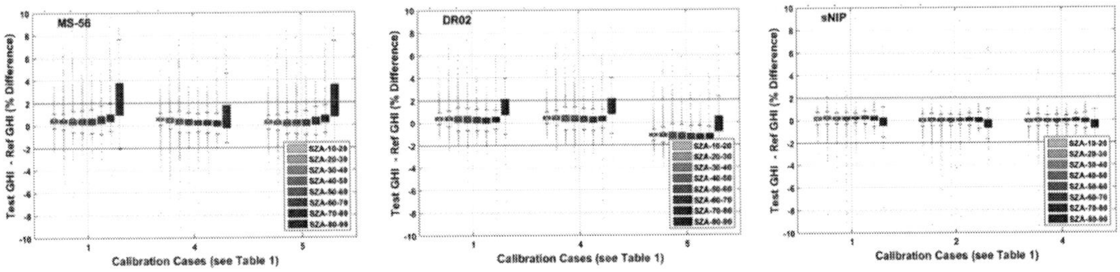

Figure 4: Mean percent differences for three thermopile pyrheliometers.

$$G_{man_{DR02}} = \frac{V_{raw}}{R_{test} * (a * Temp^2 + b * Temp + C)} \quad (12)$$

where a, b, and c are the coefficients supplied by the manufacturer, and *Temp* (degrees Celsius) is the measured ambient temperature from a collocated temperature sensor.

For MS-56 in Fig. 4, for SZA greater than 30° and lower than 60°, the NREL 45° responsivity (Case 4) method provided a better comparison to the reference data than that of the manufacturer calibration responsivity (Fig. 4); however, the manufacturer calibration responsivity performed better than the NREL responsivity as a function of SZA in most SZA bins. This could be related to no SZA dependence of pyrheliometers.

For DR02, NREL's BORCAL calibration responsivity provided a better result than that of the manufacturer-supplied responsivity; however, for the SPP, the manufacturer-supplied responsivity demonstrated relatively a better result than NREL's BORCAL methods.

4 SUMMARY AND RECOMMENDATION

The accuracy of solar radiation measurements depends on the radiometer calibration procedure. This study addressed the calibration methodology and the resulting calibration responsivity provided by the manufacturers and radiometric calibration service provider. Differences among the values from indoor manufacturers' irradiance compared to outdoor NREL's BORCAL irradiance values are observed to be on the order of 1% to 2% (Fig. 1 and Fig. 3) for pyranometers. Differences for pyrheliometers are less than 1% (Fig. 4). The study is important for quantifying the overall radiometric data uncertainty estimation and to apprise radiometric data users and calibration service providers of the variations in solar resource measurements due to calibration. The results of our study suggest a need to develop a consensus on a standard approach to calibration that will alleviate such differences in irradiance measurements due to the differing calibration methodologies. Guidance from radiometer manufacturers and calibration service providers to the users of these specialized instruments will also aid in understanding the accuracy limitations (e.g., when and where to use the calibration responsivity). For example, radiometers used indoors, such as in accelerated weathering chambers, should be calibrated indoors.

Acknowledgements

The coauthors would like to thank Dr. Andrew Clifton, manager of NREL's Sensing, Measurement, and Forecasting group, for his leadership and critical review of this work. The coauthors also gratefully acknowledge Dr. Yu Xie, memeber of NREL's Sensing, Measurement, and Forecasting group; and retired NREL employees Thomas Stoffel, Daryl Myers, and Stephen Wilcox for their guidance and review of this work. We are grateful to NREL Solar Radiation Research Laboratory employees for maintaining the outdoor solar irradiance and surface meteorological instrumentation. This work was supported by the U.S. Department of Energy under Contract No. DE-AC36-08-GO28308 with the National Renewable Energy Laboratory.

References

- Reda, I., Calibration of a Solar Absolute Cavity Radiometer with Traceability to the World Radiometric Reference NREL/TP-463-20619, 1996, 79pp. http://www.nrel.gov/docs/legosti/fy96/20619.pdf
- Habte, A.; Wilcox, S.; Stoffel, T. (2014). Evaluation of Radiometers Deployed at the National Renewable Energy Laboratory's Solar Radiation Research Laboratory. 187 pp.; NREL Report No. NREL/TP-5D00-60896. Golden, CO: National Renewable Energy Laboratory.
- Wilcox, S.M.; Myers, D. (2008). Evaluation of Radiometers in Full-Time Use at the National Renewable Energy Laboratory Solar Radiation Research Laboratory. NREL/TP-550-44627. Golden, CO: National Renewable Energy Laboratory.
- Michalsky, J.J., et al. (2011). "An Extensive Comparison of Commercial Pyrheliometers Under a Wide Range of Routine Observing Conditions." Journal of Atmospheric and Oceanic Technology (28); pp. 752–766.
- Wilcox, S.; D. Myers; N. Al-Abbadi; M. B. Mahfoodh Using Irradiance and Temperature to Determine the Need for Radiometer Calibrations NREL/CP-560-32956 2001
- Reda, I., J. Hickey, C. Long, D. Myers, T. Stoffel, S. Wilcox, J. J. Michalsky, E.G. Dutton, and D. Nelson (2005), Using a Blackbody to Calculate Net Longwave Responsivity of Shortwave Solar Pyranometers to Correct for Their Thermal Offset Error During Outdoor Calibration Using the Component Sum Method. Journal of Atmospheric and Oceanic Technology, vol. 22, pp. 1531-1540.
- Reda, I., J. Hickey, T. Stoffel, D. Myers, (2002). Pyrgeometer Calibration at the National Renewable Energy Laboratory (NREL).Journal of Atmospheric and Solar-Terrestrial Physics. Vol. 64(15), 2002; pp. 1623-1629.
- Dooraghi, M.; Habte, A.; Reda, I.; Sengupta, M.; Gotseff, P.; Andreas, A., 2014, Quantifying the Impact of Incidence-Angle Dependence on Solar Radiometric Calibration, Proceedings of the 2014 IEEE 40th Photovoltaic Specialist Conference (PVSC), 8-13 June 2014, Denver, Colorado
- Reda, I., "Method to calculate uncertainty estimate of measuring shortwave solar irradiance using thermopile and semiconductor solar radiometers," National Renewable Energy Laboratory, Golden, CO, Tech. Rep., NREL/TP-3B10-52194, 2011, 20 pp.

VALIDATION OF SATELLITE BASED SOLAR IRRADIANCE ACCORDING TO THE HELIOSAT-4-METHOD FOR GERMANY

Author for correspondence
Konstantin Ditz[1)]
Co-Authors
H. Ruf[1,4)], D. Funk[1)] Marion Schroedter-Homscheidt[2)], Carmen Köhler[3)], Gerd Heilscher[1)]

1) Hochschule Ulm, Germany
2) German Aerospace Center, Germany
3) German Weather Service, Germany
4) University of Agder, Norway

Eberhard-Finckh-Str. 11, 89075 Ulm, Phone: +49 731 50 28 350, E-mail: ditz@hs-ulm.de

ABSTRACT: The continuing growing number of volatile feed-in systems, e.g. photovoltaic (PV) systems in the low voltage grid, leads to essential changes of the grid planning and operation. For this reason the distribution grid operators will need independent switching operation in the future [1]. Therefore, the DSO is interested in knowing the actual PV feed-in power. The most detailed solution is to equip all PV systems with Smart Meters to measure the actual feed-in power. However, this solution causes additional costs for installation, service and operation. Though it's only possible to react on the volatile Feed-in power of the PV systems which could lead to high costs for control power or for the curtailment of PV systems [2]. The solar irradiance is the most important input parameter for the PV simulation. A solution without ground-based measurements is the usage of calculated solar irradiance from satellite pictures. Therefore, some providers, like Helioclim, CM SAF, OSI SAF or the new developed CAMS Radiation Service, based on the Heliosat-4-method, are available [3]. This paper focus on the validation of the solar irradiance calculated with the Heliosat-4-method provided by CAMS Radiation Service.

Keywords: Energy Meteorology, PV feed-in, Grid Planning, Solar Irradiance, CAMS Radiation Service

1 INTRODUCTION

The solar irradiance is the most important input parameter for the simulation of PV feed-in power. One opportunity is the installation of ground-based measurement devices, e.g. pyranometer, in the electric grid area. This solution leads to high installation and maintenance costs and provides only point measurements. The PV systems the error increases with an increasing distance. An alternative is the usage of calculated solar irradiance based on satellite images.

The objective of this paper is the analysis of the usability of satellite-based solar irradiance for the grid planning and the asset management of electric distribution grids. Therefore, a validation of solar irradiance calculated with the Heliosat-4-method for Germany was done. The German Weather Service (DWD) provides solar irradiation measurements from 35 weather stations at different locations as reference for this analysis. The advantage of the CAMS Radiation Service is the online availability of the solar irradiance data. Furthermore, the CAMS Radiation Service data is free of charge [4] and is an adequate source for the grid planning and the asset management.

2 APPROACH

This work validates the calculated solar irradiance of the CAMS Radiation Service for several locations in Germany. The Meteosat Second Generation satellite has a resolution of approximately 5 x 3 km² for Germany and the information has a temporal resolution of 15 minutes. In the frame of the MACC projects leading to CAMS Radiation service the irradiance is validated versus several international weather stations of the Baseline Surface Radiation Network (BSRN) [22]. The BSRN includes only one station (Lindenberg) for Germany. To improve the statement about the quality of the CAMS data for Germany irradiation measurements of 35 DWD weather stations were chosen and used as reference [5]. The regional distribution of the stations is listed in the section 3. The DWD stations are measuring the global horizontal irradiance (GHI) with a temporal resolution of one minute. However, the stored averaged data have a resolution of one hour, therefore the validation is based on hourly data. For the evaluation of the CAMS data the hourly mean values of the GHI from 2010 to 2013 were analyzed.

According to [6] the Mean (ME), the Root Mean Square (RMSE) and the correlation coefficient (CC) were calculated with the formula (1) – (3):

$$ME = \frac{1}{N}\sum_{i=1}^{N} x_s(i) - x_m(i) \qquad (1)$$

$$RMSE = \sqrt{\frac{1}{N}\sum_{i=1}^{N}\left(x_s(i) - x_m(i)\right)^2} \qquad (2)$$

$$CC = \frac{\sum_{i=1}^{N}\left(x_s(i)-\overline{x_s}\right)\cdot\left(x_m(i)-\overline{x_m}\right)}{\sqrt{\sum_{i=1}^{N}\left(x_s(i)-\overline{x_s}\right)^2\cdot\sum_{i=1}^{N}\left(x_m(i)-\overline{x_m}\right)^2}} \qquad (3)$$

with N as the whole number of measurement points, $x_s(i)$ as simulated data at the time i, $x_m(i)$ as measured value at the time i and $\overline{x_s}$, $\overline{x_m}$ as empiric mean values.

To compare the results of this work with other studies the relative values for the ME and RMSE are calculated. The correspondent relative values for the ME and RMSE (rME and rRMSE) refer to the mean value of the measured absolute value and were calculated by:

$$rME = \frac{ME}{|x_m|} \cdot 100\% \quad (4)$$

$$rRMSE = \frac{RMSE}{|x_m|} \cdot 100\% \quad (5)$$

For the calculation of the statistic errors only measured values are taken into account where the sun elevation angle is higher than 15° and the minimal measured GHI is higher than 10 W/m².

To investigate how the Heliosat-4-method works under different weather conditions the clear-sky index (kc) for the GHI values was calculated in addition. Therefore 3 categories were considered is based on [7]:

- $kc < 0.3$
- $0.3 <= kc <= 0.7$
- $kc > 0.7$

The kc is calculated as:

$$kc = \frac{GHI}{GHI_{ClearSky}} \quad (6)$$

3 DWD WEATHER STATIONS

The irradiance measurement of the DWD is based on pyranometers and so-called SCAPP (scanning pyrheliometer and pyranometer). Overall the DWD meteorological network consists of 82 automated stations and 29 manned stations as well as the BSRN station in Lindenberg. A quality evaluation and descriptions of the stations can be found in [21]. The pyranometer CM11 and CM21 manufactured by Kipp & Zonen are used, which have a measuring inaccuracy based on the daily sum less than 3 % according to [8]. The SCAPP used have a measuring inaccuracy of 10 % for daily mean values. For hourly values the deviation can be 20 % [12]. However, an investigation by the DWD shows that comparable accuracies could be achieved [13].

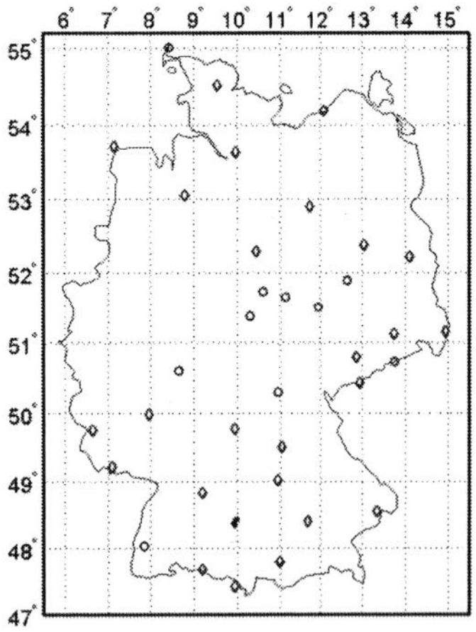

Figure 1: Overview of the distribution of the 35 DWD weather stations in Germany. Circles marks SCAPP, diamonds marks pyranometer

Figure 1 shows the distribution of the 35 DWD weather stations in Germany and the type of measurement device. In table 1 the station locations are listed as well as the coordinates and the altitude.

Table 1: Overview of the 35 selected DWD weather stations with geographical coordinates and altitude

ID	City	latitude [°]	longitude [°]	altitude [m]
1	Braunlage	51.73	10.60	607
2	Braunschweig	52.29	10.45	81
3	Bremen	53.05	8.80	4
4	Chemnitz	50.79	12.87	418
5	Coburg	50.31	10.97	345
6	Dresden	51.13	13.76	227
7	Fichtelberg	50.43	12.96	1213
8	Freiburg	48.02	7.84	236
9	Fürstenzell	48.55	13.35	476
10	Geisenheim	49.99	7.95	110
11	Giessen	50.61	8.65	203
12	Görlitz	51.16	14.95	238
13	Halle	51.51	11.95	93
14	Hamburg	53.64	9.99	11
15	Harzgerode	51.65	11.14	404
16	Hohenpreissenberg	47.80	11.01	977
17	Konstanz	47.68	9.19	443
18	Leinefelde	51.39	10.30	356
19	Lindenberg	52.21	14.12	98
20	List	55.01	8.41	26
21	Norderney	53.71	7.15	11
22	Nürnberg	49.50	11.06	312
23	Potsdam	52.38	13.06	81
24	Rostock	54.18	12.08	4
25	Saarbrücken	49.21	7.11	320
26	Schleswig	54.53	9.55	43
27	Seehausen	52.89	11.73	21
28	Stuttgart	48.83	9.20	314
29	Trier	49.75	6.66	265
30	Weihenstephan	48.40	11.70	477
31	Weissenburg	49.02	10.96	422
32	Wittenberg	51.89	12.65	105
33	Würzburg	49.77	9.96	268
34	Zinnwald	50.73	13.75	877
35	Zugspitze	47.42	9.96	2964

4 THE HELIOSAT-4-METHOD

The Heliosat-4 method is a radiative-transfer based direct method based on an important approximation: the decoupling of clear-sky irradiance calculation and a modification due to cloud properties and ground albedo [17]. This approximation is applied for the solar irradiance at ground considering clear-sky irradiances and a modification by clouds and ground albedo. Changes in clear atmosphere properties have negligible effect on the clear-sky index so that both terms can be calculated independently.

Heliosat-4 uses two different irradiance models to realize the decoupling approach mentioned before: McClear and McCloud.

McClear is a new irradiance model calculating the downwelling solar irradiances under clear-sky conditions using physical relations pre-calculated with libRadtran [19] and stored in look-up tables. This leads to calculation time savings compared to other radiative transfer models.

The McCloud model calculates the clear-sky indices for global and beam irradiance under cloudy conditions taking into account cloud properties. These required cloud properties as well as the position of clouds are also derived from satellite data provided by the AVHRR Processing scheme over Clouds, Land and Ocean (APOLLO) scheme [20].

The combination and interaction of these models as well as the input data are shown in figure 2. The APOLLO model calculates the cloud mask from multispectral satellite images. This cloud mask decides if the McCloud is applied. If clouds are detected, McCloud calculates the kc values and multiply them by the clear-sky irradiance.

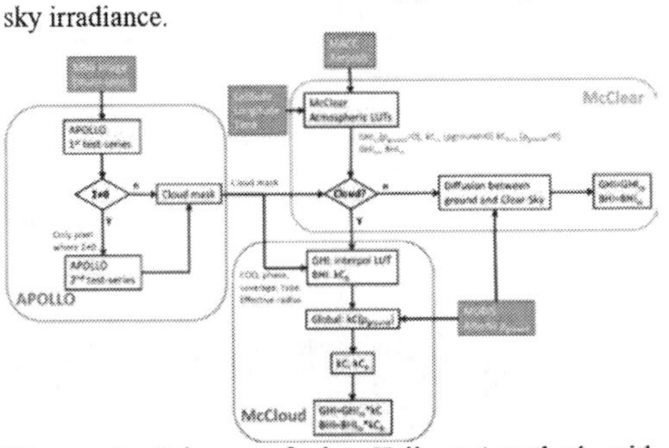

Figure 2: Scheme of the Heliosat-4-method with McClear, McCloud and APOLLO. The necessary input parameters are in green boxes.

5 STATISTICAL EVALUATION

Figure 3 shows the values for the ME. They are between -18 W/m² and 59 W/m². The ME values, except the station 21 (Norderney), are positive. This illustrates the overestimation of the solar irradiation within Germany calculated by the CAMS Radiation Service. In addition the rME values are illustrated as red crosses. These vary between -5 % and 18 %. The rME of the CAMS validation (Lindenberg) is 5.3 %. The rME values of the global validation vary between 1 – 13 % [4]. Because of that the results of this validation for Germany are plausible.

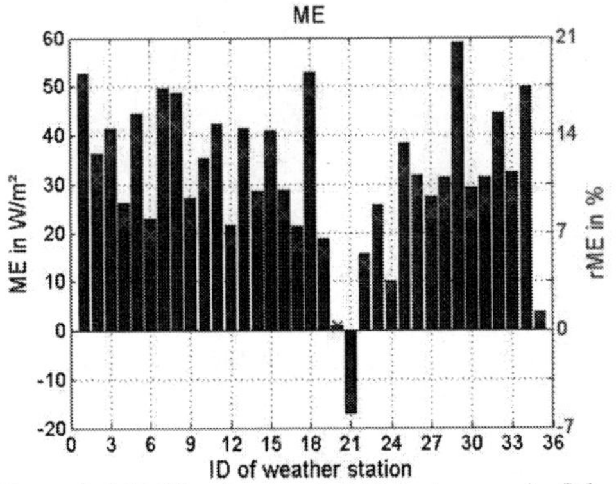

Figure 3: ME (blue bars) and rME (red crosses) of the comparison of GHI from CAMS and DWD weather stations over the period from 2010 to 2013.

The RMSE values are shown in figure 4. The RMSE vary between 75 W/m² and 178 W/m². For the station 35 (Zugspitze) a very high RMSE is visible. The reason therefore could be the geographic and climate conditions in high alpine regions and snow covering. Furthermore, the station 29 (Trier) has a high RMSE. The reason for this is unknown and further investigations will be done. The rRMSE, marked with red crosses, is between 22 % and 45 %. The rRMSE values of the validation against the BSRN stations have a variation from 11 % to 34 %. Except the two stations described before, the values in this work are plausible.

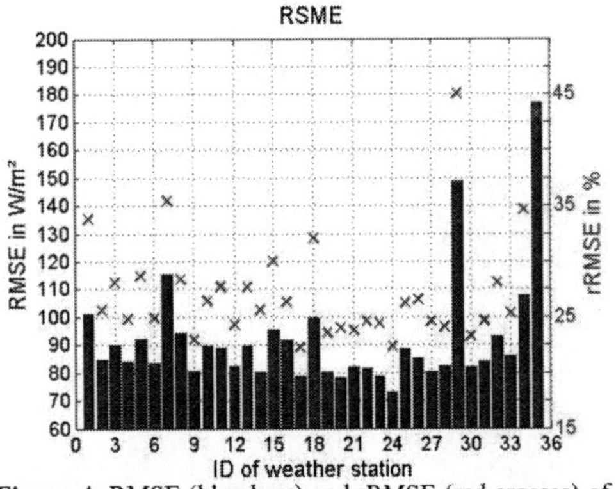

Figure 4: RMSE (blue bars) and rRMSE (red crosses) of the comparison of GHI from CAMS and DWD weather stations over the period from 2010 to 2013.

Figure 5 shows the values of the CC between the measured GHI (DWD) and the calculated GHI (CAMS). The values are between 0.89 and 0.96 except the two outliers Trier and Zugspitze. This means there is a high correlation between the ground measurements and the satellite model. The value of the CC from the CAMS validation for the station in Lindenberg is 0.95 [4]. The CC values of the worldwide validation by the CAMS Service vary between 0.91 and 0.97. The validation of this work correlates with the worldwide CAMS validation.

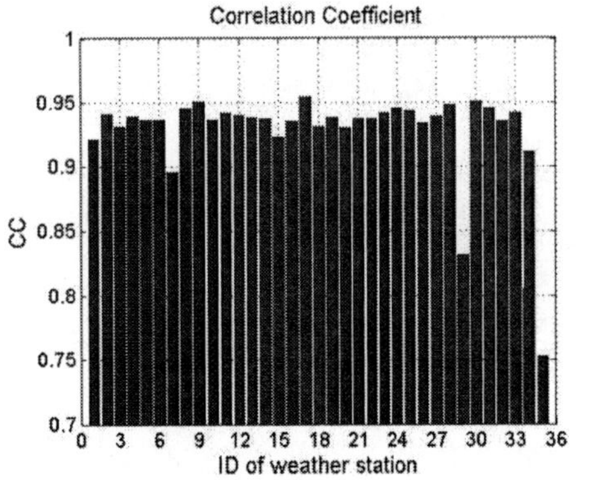

Figure 5: Correlation coefficient of the comparison between GHI of the weather stations and the calculated GHI by the CAMS Radiation Service.

6 VARIATION OF THE GHI VALUES

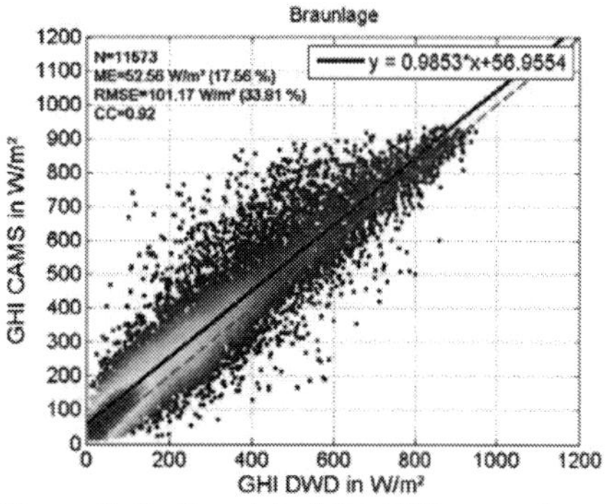

Figure 6: Scatter plot of the GHI comparison for the station in Braunlage. The color code shows the frequency of the values. Red means a high number of values in this area, blue for a low number

As example only the scatter plots for the stations in Braunlage and Lindenberg are shown. These stations are chosen because of the different variation of the GHI values. Figure 6 shows the scatter plot of the GHI values for the station in Braunlage. The regression line (black) over the whole scope shows that the CAMS Service overestimates the GHI. This is visible for 17 of the 35 stations.

Figure 7 shows the scatter plot of the GHI values from the station in Lindenberg. A look on the regression line (black) shows a small overestimation of the GHI values by the CAMS Service for the area up to approximately 500 W/m². Above this value a small underestimation is visible. The color scale shows that a high number of values are in the area of a low solar irradiation. The positive ME confirms this statement because on average there is an overestimation of the GHI. This is visible for 18 of the 35 stations. There is no connection between the variation of the results and the location of the weather stations within the results of this analysis. For the results of this analysis. As a first result of this study a similar behaviour for whole Germany will be assumed for CAMS. Further investigations has to be done to confirm this Statement.

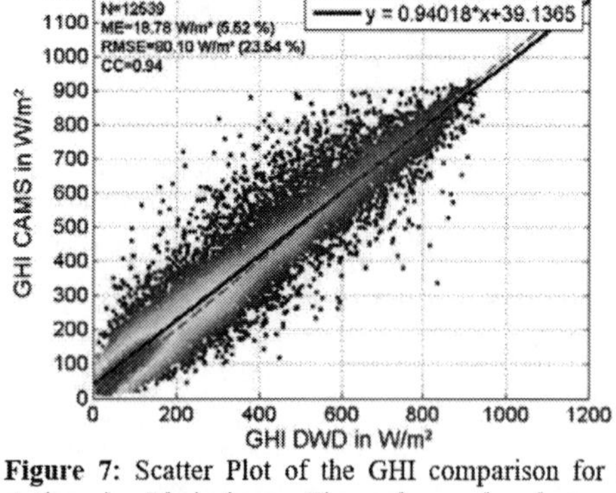

Figure 7: Scatter Plot of the GHI comparison for the station in Lindenberg. The color code shows the frequency of the values. Red means a high number of values in this area, blue for a small number

7 CONSIDERATION OF THE CLEAR SKY INDEX

The Lindenberg station was chosen to compare the results with other publications taking into account different kc values. In figure 8 the GHI scatter plot of the station in Lindenberg with kc < 0.3 (very cloudy) is shown. The graphic shows mostly higher calculated GHI values than for the measurements. This results in a ME of 63 %. The CC is 0.77.

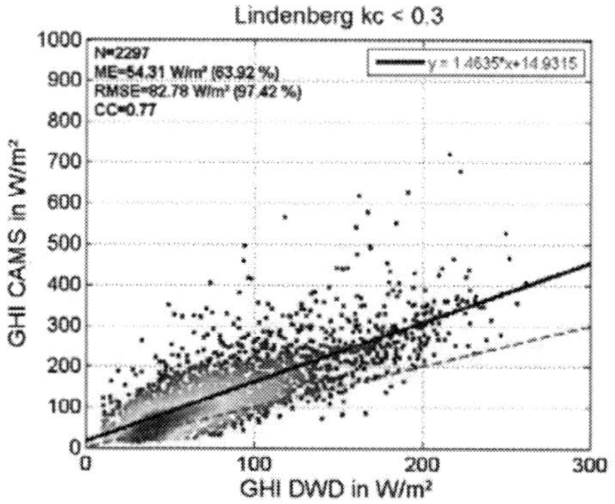

Figure 8: Scatter plot of kc < 0.3 of the station in Lindenberg

Figure 9 shows the scatter plot for the range of 0.3 ≤ kc ≤ 0.7 (cloudy and light clouds) of the station Lindenberg. The overestimation of CAMS is also visible (ME 17.2%) but lower than in the situations of kc < 0.3. The CC is 0.89.

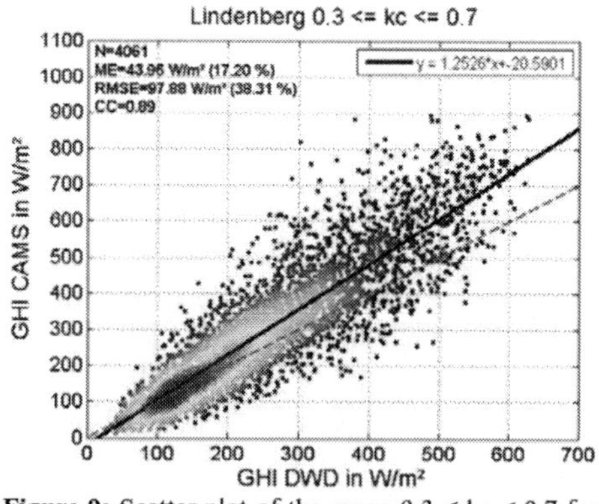

Figure 9: Scatter plot of the range 0.3 ≤ kc ≤ 0.7 for the station in Lindenberg

Finally, figure 10 shows the GHI comparison for kc > 0.7. It's visible that for cloudless days a consistent over- and underestimation of the measured GHI values exists. This reflects in the gradient of nearly 1 of the regression curve (black). The negative ME shows a very small underestimation. The analysis shows the Heliosat-4-method works well for cloudless days [26].

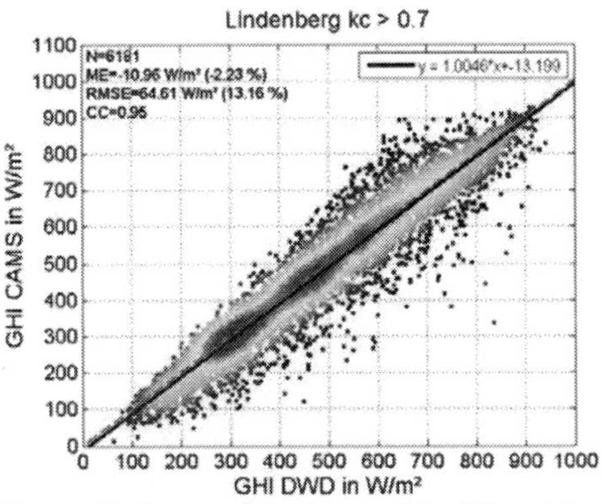

Figure 10: Scatter plot of kc > 0.7 of the station in Lindenberg

7 CONCLUSION AND OUTLOOK

The objective of this work was the validation of the CAMS Radiation Service for Germany. The analysis shows good results for the new developed Heliosat-4-method and is comparable with other studies. The results reasons for sunny days the estimation of the GHI works fine and so the basic procedure for the calculation of the ground irradiation has a good performance. A further result is the overestimation and the high deviation according to the measured GHI by the weather stations.

The stations Trier and Zugspitze shows a very high deviation compared to the other stations. While for the Zugspitze the errors could be probably explained by the snow covering for the station in Trier no reasons could be found within the actual results. Although for both stations a high correlation (CC 0.72) between the RMSE and the altitude as well as a high inverse correlation (-0.74) between CC and the altitude exist.

Furthermore, the results show the usability of the satellite-derived solar irradiance for the grid planning. For the grid planning it's important to know the voltage at the nodes and the maximum utilization of the transformers and cables. The existing deviations of the GHI from CAMS are tolerable for this application, because it's visible in the results that on cloudless days only small deviations exist. For the grid planning it is necessary to know the maximum feed-in of PV systems. The CAMS data are available over years, and can support the asset management of a DSO. It's possible to estimate how often high irradiation values in a yearly average occurs and which maximum utilization of the transformers and cables are induced and how long they occur. This information is important for the calculation of the remaining lifetime of the transformers and cables.

The CAMS Radiation Service data are available free of charge and can avoid high costs for a comprehensive irradiation measurement network. The purchase of several irradiance sensors led to high investment costs of several thousand Euros. Furthermore, yearly maintenance is necessary leading to additional costs. Although there are other providers of free irradiation data available, for example in the Climate Data Center from the DWD [23] or Heliosat [24] and from the Satellite Application Facility on Climate Monitoring (CM SAF) [25].

DISCLAIMER

The OrPHEuS project is co-financed by the European Commission under the 7th Framework Program under grant agreement and No. 608930.

The sole responsibility for the content of this publication lies with the authors. It does not necessarily reflect the opinion of the European Commission. The European Commission is not responsible for any use that may be made of the information contained therein.

SOURCES

[1] BDEW und VKU: „Praxis-Leitfaden für unterstützende Maßnahmen von Stromnetzbetreibern". Berlin 2012, Available at: www.bdew.de; latest check on 2.6.2014.

[2] Dickert, J. and Schegner, P. Prof. Dr.-Ing. (2012) "Lastgänge als Grundlage für die Analyse von Smart Grids: Anforderungen und Vergleich von Lastgangsynthesen," in Verband der Elektrotechnik Elektronik Informationstechnik e. V. (ed) VDE Kongress 2012 Smart Grid: Intelligente Energieversorgung der Zukunft, Kongressbeiträge 5. - 6. November 2012, Internationales Congresscenter Stuttgart (ICS) [Online],

[3] B. Espinar, C. Hoyer-Klick, M. Lefevre, M. Schroedter-Homscheidt, and L. Wald, "USER'S GUIDE to the MACC-RAD Services on solar energy radiation resources: v3 - July 2014 - MACC-II Deliverable 122.6," Project MACC-II, Armines, Jul. 2014.

[4] Solar Radiation Data Service, www.SoDa-pro.com

[5] Solar irradiation, provided by German Weather Service, GHI data also available on CDC (Climate Data Center) http://www.dwd.de/EN/climate_environment/cdc/cdc_node.html

[6] H. G. Beyer, J. P. Martinez, M. Suri, J. L. Torres, E. Lorenz, S. Muller, C. Hoyer-Klick, and P. Ineichen, "Report on Benchmarking of Radiation Products: Deliverable 1.1.3," Aug. 2009. [Online], http://www.mesor.org/docs/MESoR_Benchmarking_of_radiation_products.pdf

[7] V. Badescu, A. Dumitrescu; "The CMSAF hourly solar irradiance database (product CM54): Accuracy and bias corrections with illustrations for Romania; Journal of Atmospheric and Solar-Terrestrial Physics," Dezember 2012

[8] W. Riecke, „Bereitstellung von historischen Globalstrahlungsdaten für die Photovoltaik, 2. Fachtagung Energiemeteorologie," April 2011, [Online], https://www.uni-oldenburg.de/fileadmin/user_upload/physik/ag/ehf/enmet/download/proc/G01_riecke_globalstrahlungsdaten.pdf

[9] http://www.kippzonen.com/Product/13/CMP11-Pyranometer#.VlV_bXYvdD8

[10] http://www.kippzonen.com/Product/14/CMP21-Pyranometer#.VlV_v3YvdD8

[11] DWD, „Globalstrahlung - Die Energie der Sonne," [Online], http://www.rfsolar.eu/wp-content/uploads/DWD-Broschuere_Globalstrahlung-Die-Energie-der-Sonne.pdf

[12] WMO, W. M. O. (2008) "Guide to meteorological instruments and methods of observation", 7th edn, Geneva, Switzerland, World Meteorological Organization, [Online], https://www.wmo.int/pages/prog/www/IMOP/CIMO-Guide.html (Accessed 13 July 2015)

[13] K. Behrens, R. D. Grewe, DWD, "A comparison of SCAPP Radiation Data with Global,

Diffuse and Direct Solar Radiation," 2005 [Online].Available:https://www.wmo.int/pages/prog/www/IMOP/publications/IOM-82-TECO_2005/Papers/1(08)_Germany_Behrens.pdf

[14] D. Cano, J. Monget, M. Albuisson, H. Guillard, N. Regas, and L. Wald, "A method for the determination of the global solar radiation from meteorological satellite data," Solar Energy, vol. 37, no. 1, pp. 31–39, 1986.

[15] P. Ineichen, "Long term satellite hourly, daily and monthly global, beam and diffuse irradiance validation. Interannual variability analysis," University of Geneva, Geneva, Jan. 2013.

[16] Z. Qu, P. Blanc, M. Lefevre, L. Wald, P. Watts, M. Schroedter Homscheidt, G. Gesell, and L. Klueser, "Use of OCA and APOLLO in Heliosat-4 method for the assessment of surface downwelling solar irradiance," in Proceedings of the 2012 EUMETSAT Meteorological Satellite Conference, Sopot, Poland, 2012, pp. s2-03.

[17] A. Oumbe, Z. Qu, P. Blanc, M. Lefèvre, L. Wald, and S. Cros, "Decoupling the effects of clear atmosphere and clouds to simplify calculations of the broadband solar irradiance at ground level," Geosci. Model Dev, vol. 7, no. 4, pp. 1661–1669, 2014.

[18] M. Lefèvre, A. Oumbe, P. Blanc, B. Espinar, B. Gschwind, Z. Qu, L. Wald, M. Schroedter-Homscheidt, C. Hoyer-Klick, A. Arola, A. Benedetti, J. W. Kaiser, and J.-J. Morcrette, "McClear: a new model estimating downwelling solar radiation at ground level in clear-sky conditions," Atmospheric Measurement Techniques, vol. 6, no. 9, pp. 2403–2418, http://www.atmos-meas-tech.net/6/2403/2013/, 2013.

[19] B. Mayer and A. Kylling, "Technical note: The libRadtran software package for radiative transfer calculations - description and examples of use," Atmospheric Chemistry and Physics, vol. 5, no. 7, pp. 1855–1877, http://www.atmos-chem-phys.net/5/1855/2005/, 2005.

[20] K. T. Kriebel, G. Gesell, M. Kästner, and H. Mannstein, "The cloud analysis tool APOLLO: Improvements and validations," International Journal of Remote Sensing, vol. 24, no. 12, pp. 2389–2408, 2003.

[21] R. Becker and K. Behrens. Qualtiy assessment of heterogeneous surface radiation network data. Adv. Sci. Rec., 8, 93-97, 2012. doi:10.5194/asr-8-93-2012

[22] http://bsrn.awi.de/

[23] http://www.dwd.de/EN/climate_environment/cdc/cdc_node.html

[24] Müller, Richard; Pfeifroth, Uwe; Träger-Chatterjee, Christine; Cremer, Roswitha; Trentmann, Jörg; Hollmann, Rainer. (2015): "Surface Solar Radiation Data Set - Heliosat (SARAH) - Edition 1. Satellite Application Facility on Climate Monitoring," http://dx.doi.org/10.5676/EUM_SAF_CM/SARAH/V001

[25] https://wui.cmsaf.eu/safira/action/viewProduktSearch

[26] Lefevre, Mireille; Wald, Lucien: "Validation of the McClear clear-sky model in desert conditions with three stations in Israel," 2016; online available at: http://www.adv-sci-res.net/13/21/2016/asr-13-21-2016.pdf

AUTHOR INDEX

AUTHOR INDEX

AUTHOR INDEX

AUTHOR INDEX

AUTHOR INDEX

AUTHOR INDEX

AUTHOR INDEX

AUTHOR INDEX

AUTHOR INDEX

AUTHOR INDEX

AUTHOR INDEX

AUTHOR INDEX

AUTHOR INDEX

AUTHOR INDEX

AUTHOR INDEX

AUTHOR INDEX

AUTHOR INDEX

AUTHOR INDEX

AUTHOR INDEX

AUTHOR INDEX

AUTHOR INDEX

AUTHOR INDEX

AUTHOR INDEX

AUTHOR INDEX

32nd European Photovoltaic Solar Energy Conference and Exhibition (EU PVSEC 2016)

Munich, Germany
20-24 June 2016

Volume 4 of 4

Editors:

M. Topic
P. Helm
N. Taylor

ISBN: 978-1-5108-3651-8

Printed from e-media with permission by:

Curran Associates, Inc.
57 Morehouse Lane
Red Hook, NY 12571

CURRAN ASSOCIATES INC.
proceedings.com

Some format issues inherent in the e-media version may also appear in this print version.

Printed by Curran Associates, Inc. (2017)

For permission requests, please contact WIP – Renewable Energies
at the address below.

WIP – Renewable Energies
Sylvensteinstr. 2
81369 Munchen
Germany

Phone: +49 89 72012735
Fax: +49 89 72012791

wip@wip-munich.de

Additional copies of this publication are available from:

Curran Associates, Inc.
57 Morehouse Lane
Red Hook, NY 12571 USA
Phone: 845-758-0400
Fax: 845-758-2633
Email: curran@proceedings.com
Web: www.proceedings.com

32nd European Photovoltaic Solar Energy Conference and Exhibition

Proceedings of the International Conference held in Munich, Germany

20 June - 24 June 2016

Edited by:

M. TOPIČ
University of Ljubljana
Chairman of European Technology & Innovation Platform Photovoltaics
Ljubljana, Slovenia

N. TAYLOR
European Commission - DG Joint Research Centre
Ispra, Italy

P. HELM
WIP Renewable Energies
Munich, Germany

Coordination of the Technical Programme:
European Commission - DG Joint Research Centre
Via E. Fermi 1
21020 Ispra (VA)
Italy

Institutional Support:
European Commission
UNESCO – United Nations Educational, Scientific and Cultural Organization, Natural Sciences Sector
WCRE – World Council for Renewable Energy

Supporting Organisations:
ARE – Alliance for Rural Electrification
EREF – European Renewable Energies Federation
ESA – European Space Agency
EUFORES –The European Forum for Renewable Energy Sources
EUREC – The Association of European Renewable Energy Research Centres
European Photovoltaic Technology Platform
IEA PVPS – International Energy Agency Photovoltaic Power Systems Programme
REN21 – Renewable Energy Policy Network for the 21st Century

Institutional PV Industry Cooperation:
SolarPower Europe
SOLARUNITED

Supporting Associations:
AIE – European Association of Electrical Contractors
AREA – African Renewable Energy Alliance
CABA – Continental Automated Buildings Association
EPIC – European Photonics Industry Consortium
SASIA - Saudi Arabia Solar Industry Association

EU PVSEC realised by:
WIP
Sylvensteinstr. 2, 81369 München, Germany
Tel: +49 89 720 12 735, Fax: +49 89 720 12 791
Email: pv.conference@wip-munich.de
www.photovoltaic-conference.com
www.wip-munich.de

Proceedings produced and published by:
WIP
Sylvensteinstr. 2, 81369 München, Germany
Tel: +49 89 720 12 735, Fax: +49 89 720 12 791
Email: pv.conference@wip-munich.de
www.photovoltaic-conference.com
www.wip-munich.de

ISSN 2196-0992

32nd EUROPEAN PHOTOVOLTAIC SOLAR ENERGY CONFERENCE AND EXHIBITION
20 JUNE – 24 JUNE 2016
MUNICH, GERMANY

EU PVSEC COMMITTEES

INTERNATIONAL SCIENTIFIC ADVISORY COMMITTEE (ISAC)

Chair
G.F. De Santi, European Commission, DG JRC, Director of the Institute for Energy and Transport, Ispra, Italy

Secretary
N. Taylor, European Commission, DG JRC, Institute for Energy and Transport, Ispra, Italy

Committee Members
P. Frankl, Head of the Renewable Energy Division, International Energy Agency, France
P. Hofer-Noser, CTO, Meyer Burger Technology, Switzerland
P. Malbranche, CEA, Le Bourget du Lac, France
S. Nowak, Managing Director of NET Nowak Energy & Technology, Switzerland
W. Sinke, Program Development Manager, ECN Solar Energy, The Netherlands
P. Verlinden, Chief Scientific and Vice Chair, State Key Lab. TRINA Solar, China
E. Weber, Director, Fraunhofer ISE, Germany
P. Helm, WIP, Munich, Germany
A. Grassi, ETA-Florence Renewable Energies, Florence, Italy

CONFERENCE EXECUTIVE COMMITTEE

Conference General Chairman
M. Topič, University of Ljubljana, Chairman of European Technology & Innovation Platform Photovoltaics, Ljubljana, Slovenia

Technical Programme Chairman
N. Taylor, European Commission, DG JRC, Institute for Energy and Transport, Ispra, Italy

Committee Members
H. Ossenbrink, European Commission, DG JRC, Institute for Energy and Transport, Ispra, Italy
S. Nowak, Managing Director of NET Nowak Energy & Technology, St. Ursen, Switzerland
W. Sinke, Program Development Manager, ECN Solar Energy, The Netherlands
P. Helm, WIP, Munich, Germany
A. Grassi, ETA-Florence Renewable Energies, Florence, Italy

2016 SCIENTIFIC COMMITTEE
Topic Organisers and Paper Review Experts

Topic 1 – New Materials and Concepts for Solar Cells and Modules

G. Beaucarne, Dow Corning, Seneffe, Belgium
N.J. Ekins-Daukes, Imperial College London, United Kingdom
A. Martí Vega, UPM, Madrid, Spain
J. Poortmans, imec, Leuven, Belgium
M. Rusu, HZB, Berlin, Germany
J.A.M. Van Roosmalen, ECN, Petten, Netherlands

V. Bukauskas, Center for Physical Sciences and Technology, Vilnius, Lithuania
F. Falk, IPHT, Jena, Germany
J. Govaerts, imec, Leuven, Belgium
J.F. Guillemoles, CNRS, Chatou Cedex, France
M.A. Hernandez-Fenollosa, UPV, Valencia, Spain
J.C. Hummelen, University of Groningen, Netherlands
D. König, UNSW, Sydney, Australia
I. Konovalov, Ernst Abbe University of Applied Science Jena, Jena, Germany
C. Lévy-Clément, CNRS, Thiais, France
Y. Okada, University of Tokyo, Japan
B. Rech, HZB, Berlin, Germany
F. Roca, ENEA, Portici, Italy
V. Saly, Slovak University of Technology, Bratislava, Slovakia
H. Shirai, Saitama University, Japan
W. Warta, Fraunhofer ISE, Freiburg, Germany

Topic 2 – Wafer-Based Silicon Solar Cells and Materials Technology

O. Anspach, PV Crystalox Solar, Erfurt, Germany
D.L. Bätzner, Meyer Burger Research, Hauterive, Switzerland
R. Brendel, ISFH, Emmerthal, Germany
F. Ferrazza, Eni, San Donato Milanese, Italy
S.W. Glunz, Fraunhofer ISE, Freiburg, Germany
G. Hahn, University of Konstanz, Germany
D. Muñoz, CEA, Le Bourget du Lac, France
S. Riepe, Fraunhofer ISE, Freiburg, Germany
A.W. Weeber, ECN, Petten, Netherlands
P. Wohlfart, Singulus Technologies, Kahl am Main, Germany

A.G. Aberle, SERIS, Singapore
N. Azkona, UPV/EHU, Zamudio, Spain
J. Bagdahn, Fraunhofer CSP, Halle (Saale), Germany
R. Barinka, Fill Factory, Rožnov pod Radhoštm, Czech Republic
P.A. Basore, PV Specialist Services, Solana Beach, USA

D. Biro, Fraunhofer ISE, Freiburg, Germany
S. Bowden, Arizona State University, Tempe, USA
T.M. Bruton, TMB Consulting, Woking, United Kingdom
B. Dauksher, Arizona State University, Tempe, USA
M. Debucquoy, imec, Leuven, Belgium
T. Desrues, CEA, Le Bourget du Lac, France
D. Dimova-Malinovska, Bulgarian Academy of Sciences, Sofia, Bulgaria
S. Dubois, CEA, Le Bourget du Lac, France
T. Dullweber, ISFH, Emmerthal, Germany
P. Fath, RCT-Solutions, Konstanz, Germany
A.J. Galdikas, Modernios E-Technologijos, Vilnius, Lithuania
M. Hermle, Fraunhofer ISE, Freiburg, Germany
K. Hesse, Wacker Chemie, Burghausen, Germany
J. Horzel, CSEM, Neuchâtel, Switzerland
J. John, imec, Leuven, Belgium
S.D. Kim, SunPower, San Jose, USA
J. Libal, ISC Konstanz, Germany
A. Metz, h.a.l.m. elektronik, Frankfurt am Main, Germany
J.W. Müller, Hanwha Q CELLS, Bitterfeld-Wolfen, Germany
Y. Ohshita,TTI, Nagoya, Japan
S. Peters, Hanwha Q CELLS, Bitterfeld-Wolfen, Germany
K. Petter, Hanwha Q CELLS, Bitterfeld-Wolfen, Germany
R. Preu, Fraunhofer ISE, Freiburg, Germany
J. Rentsch, Fraunhofer ISE, Freiburg, Germany
E. Sauar, Brighterlite, Oslo, Norway
H. Schlemm, Meyer Burger, Hohenstein-Ernstthal, Germany
M.C. Schubert, Fraunhofer ISE, Freiburg, Germany
B. Terheiden, University of Konstanz, Germany
Y. Veschetti, CEA, Le Bourget du Lac, France
G.P. Willeke, Fraunhofer ISE, Freiburg, Germany
D. Wood, Hemlock Semiconductor, Hemlock, USA
A. Yamada, Tokyo Institute of Technology, Japan
D.L. Young, NREL, Golden, USA

Topic 3 – Thin Film Solar Cells and Modules

C.J. Brabec, University of Erlangen-Nuremberg, Germany
J. Cárabe, CIEMAT, Madrid, Spain
A. Meeder, Flisom, Dübendorf, Switzerland
J. Meier, PV Consultant, Berlingen, Switzerland
S. Siebentritt, University of Luxembourg, Belvaux, Luxembourg
M. Topic, University of Ljubljana, Slovenia
S.C. Veenstra, ECN, Eindhoven, Netherlands
W. Wischmann, ZSW, Stuttgart, Germany

C. Ballif, EPFL&CSEM, Neuchâtel, Switzerland
R. Birkmire, University of Delaware, Newark, USA
D. Cheyns, imec, Leuven, Belgium
S. Cros, CEA, Le Bourget du Lac, France
T. Dalibor, AVANCIS, Munich, Germany
P. Delli Veneri, ENEA, Portici, Italy
B. Dimmler, Manz CIGS Technology, Schwäbisch Hall, Germany
K. Durose, Stephenson Institute for Renewable Energy, Liverpool, United Kingdom
S. Gall, HZB, Berlin, Germany
I. Gordon, imec, Leuven, Belgium
S. Huang, UNSW, Sydney, Australia
S.J.C. Irvine, Swansea University, St. Asaph, United Kingdom
R. Klenk, HZB, Berlin, Germany
C.-F. Lin, NTU, Taipei, Taiwan
M.C. Lux-Steiner, HZB, Berlin, Germany
E. Mellikov, Tallinn University of Technology, Tallinn, Estonia

M. Meuris, imec/IMOMEC - ATFPV, Leuven, Belgium
C. Molpeceres, UPM, Madrid, Spain
J. Parisi, University of Oldenburg, Oldenburg, Germany
R. Po, Eni, Novara, Italy
M. Powalla, ZSW, Stuttgart, Germany
A. Romeo, University of Verona, Verona, Italy
V. Sittinger, Fraunhofer IST, Braunschweig, Germany
A. Slaoui, InESS, Strasbourg, France
A.N. Tiwari, EMPA, Dübendorf, Switzerland
A. Virtuani, SUPSI, Canobbio, Switzerland
M. Zeman, Delft University of Technology, Netherlands

Topic 4 – Concentrator and Space Applications

R. Campesato, CESI, Milan, Italy
G. Flamand, imec, Leuven, Belgium
F. Rubio Berenguel, Freiburg, Germany
C. Signorini, ESA-ESTEC, Noordwijk, Netherlands

I. Anton Hernandez, UPM, Madrid, Spain
A.W. Bett, Fraunhofer ISE, Freiburg, Germany
V. Khorenko, AZUR SPACE, Heilbronn, Germany
R. King, Arizona State University, Tempe, USA
G. Siefer, Fraunhofer ISE, Freiburg, Germany
T. Takamoto, SHARP, Nara, Japan
G. Timò, RSE, Piacenza, Italy

Topic 5 – Operations, Performance, Reliability and Sustainability of PV

J. Binder, ZSW, Stuttgart, Germany
M. de Wild-Scholten, SmartGreenScans, Groet, Netherlands
R. Gottschalg, Loughborough University, United Kingdom
U. Jahn, TÜV Rheinland Energy, Köln, Germany
N.M. Pearsall, Northumbria University, Newcastle upon Tyne, United Kingdom
M. Perrin, CEA, Le Bourget du Lac, France
C. Protogeropoulos, Phoenix Solar, Athens, Greece
K. Radouane, EDF EN, Paris La Defense, France
T. Sample, European Commission DG JRC, Ispra, Italy
W.G.J.H.M. van Sark, Utrecht University, Netherlands
K. Wambach, Wambach-Consulting, Aindling, Germany

C. Alonso-Tristán UBU, Burgos, Spain
M. Bächler, PerVorm, Ulm, Germany
T.R. Betts, Loughborough University, United Kingdom
P. Blanc, MINES ParisTech, Paris, France
R. Einhaus, Apollon Solar, Lyon, France
G. Friesen, SUPSI, Canobbio, Switzerland
V. Fthenakis, Columbia University, New York, USA
W.J. Gambogi, DuPont, Wilmington, USA
L. Gomes de Freitas, Federal University of Uberlândia, Brazil
G. Graditi, ENEA, Portici, Italy
M. Grottke, WIP - Renewable Energies, Munich, Germany
Y. Hishikawa, AIST, Tsukuba, Japan
T. Huld, European Commission DG JRC, Ispra, Italy
A.R. Lagunas, CENER, Sarriguren, Spain
O. Mayer, GE Global Research, Garching, Germany
M.A. Muñoz, UPM, Madrid, Spain
H. Nagel, Fraunhofer ISE, Freiburg, Germany
H. Nussbaumer, Zurich University of Applied Sciences, Winterthur, Switzerland
C. Nyman, Soleco, Porvoo, Finland
S. Oktik, Sisecam, Istanbul, Turkey
W. Palitzsch, Loser Chemie, Zwickau, Germany
A.H.M.E. Reinders, University of Twente, Enschede, Netherlands

J. Remund, Meteotest, Bern, Switzerland
J.N. Roy, IIT Kharagpur, Kharagpur, India
M. Schroedter-Homscheidt, DLR, Wessling, Germany
M. Sengupta, NREL, Golden, USA
R. Sinton, Sinton Instruments, Boulder, USA
J. Sutterlueti, Gantner Instruments, Schruns, Austria
S. Tselepis, CRES, Pikermi, Athens, Greece
M. Van Iseghem, EDF R&D, Moret-sur-Loing, France
E. Voroshazi, imec, Leuven, Belgium
A. Wade, First Solar, Berlin, Germany
J. Wohlgemuth, NREL, Golden, USA
M. Zehner, Rosenheim University of Applied Sciences, Germany

Topic 6 – PV Applications and Integration

F.P. Baumgartner, Zurich University of Applied Sciences, Winterthur, Switzerland
F. Bonemazzi, ENEL, Rome, Italy
F. Frontini, SUPSI, Canobbio, Switzerland
P. Malbranche, CEA, Le Bourget du Lac, France
A. Scognamiglio, ENEA, Portici, Italy
X. Vallvé, Trama TecnoAmbiental, Barcelona, Spain
I. Weiss, WIP - Renewable Energies, München, Germany

R. Burkhardt, Energy Depot, Kreuzlingen, Switzerland
S. Caneva, WIP - Renewable Energies, Munich, Germany
I. Da Silva, Strathmore University, Nairobi, Kenya
J.P. de Barros Duarte Fonseca, University of Cape Verde, Mindelo, Cape Verde
B. de Boer, TNO, Eindhoven, Netherlands
H. Fechner, University of Applied Sciences Technikum Wien, Breitenfurt, Austria
I.B. Hagemann, Architekturbüro Hagemann, Aachen, Germany
C. Helmke, Management Consultancy Research, Erfurt, Germany
E. Kaijuka-Okwenje, Makerere University, Kampala, Uganda
G. Mütter, Alternative Energy Solutions, Vienna, Austria
K. Peter, ISC Konstanz, Germany
T. Reijenga, BEAR-iD Sustainable Urban Planners + Architects, Gouda, Netherlands
S. Wittkopf, HSLU, Horw, Switzerland

Topic 7 – PV Economics, Markets and Policies

C. Breyer, Lappeenranta University of Technology, Finland
M. Getsiou, European Commission DG RTD, Brussels, Belgium
G. Masson, Becquerel Institute, Brussels, Belgium
T. Nordmann, TNC Consulting, Feldmeilen, Switzerland
S. Nowak, NET Nowak Energy & Technology, St. Ursen, Switzerland
E. Perezagua, Consultores de Energía Fotovoltaica, Madrid, Spain

I. Kaizuka, RTS, Tokyo, Japan
D. Mayer, MINES ParisTech,Paris, France
P. Mazzucchelli, EUREC,Brussels, Belgium
D. Stickelberger, Swissolar, Zürich, Switzerland
S. Szabó, European Commission DG JRC, Ispra, Italy

SUBJECT INDEX

Topic Code	**Session Type**	**Day Codes**
1 New Materials and Concepts for Solar Cells and Modules	P = Plenary Session	A = Monday, 20 June 2016
2 Wafer-Based Silicon Solar Cells and Materials Technology	O = Oral Session	B = Tuesday, 21 June 2016
3 Thin Film Solar Cells and Modules	V = Visual Session	C = Wednesday, 22 June 2016
4 Concentrator and Space Applications		D = Thursday, 23 June 2016
5 Operations, Performance, Reliability and Sustainability of PV		E = Friday, 24 June 2016
6 PV Applications and Integration		
7 PV Economics, Markets and Policies		

e.g. 3CP.1 ⇒ 3=Thin Film Solar Cells and Modules, C=Wednesday, P=Plenary Session, 1=Session 1

TABLE OF CONTENT
of EU PVSEC 2016 Proceedings Papers

New Materials and Concepts for Solar Cells and Modules

Plenary SESSION 1AP.1 New Materials and Concepts for Solar Cells and Modules

Oral PRESENTATIONS 1AO.1 Fundamental Characterisation, Theoretical and Modelling Studies

Oral PRESENTATIONS 1AO.2 Fundamental Materials Studies, their Characterization and Modelling

Oral PRESENTATIONS 1AO.3 New Materials and Concepts: Nanostructures

Oral PRESENTATIONS 1BO.12 Advanced Concepts for Modules

Visual PRESENTATIONS 1BV.6 New Materials and Concepts for Cells

Wafer-Based Silicon Solar Cells and Materials Technology

Visual PRESENTATIONS 2AV.2 Silicon Solar Cell Improvements and Innovation (II)

Visual PRESENTATIONS 2BV.7 Silicon Solar Cell Characterisation and Modelling / Manufacturing and Processing

Concentrator and Space Applications

Oral PRESENTATIONS 4CO.6 III-V-based Devices for Terrestrial and Space Applications / Concentrator and Space Systems

Visual PRESENTATIONS 5BV.1 PV Cells and Modules (I)

Visual PRESENTATIONS 5BV.2 Operation of PV Systems

Visual PRESENTATIONS 5CV.3 Solar Resource and Forecasting / Sustainability and Recycling

PV Applications and Integration

Plenary SESSION 6DP.2 PV as Part of Our Built Environment: Solutions for Integration into Building Envelops and Energy Systems

Oral PRESENTATIONS 6CO.12 Grid and Energy System Integration (III) - Technology Solutions

PV Economics, Markets and Policies

Visual PRESENTATIONS 7DV.4 PV Economics and Markets / PV Global Issues, Policies and Strategies

FOREWORD

The EU PVSEC 2016 in Munich concluded with the strong message that photovoltaic solar energy is here to stay. Research and innovation across the sector, together with its deployment and endless application option, are paving the way to putting PV at the center-stage of sustainable energy systems. 1,742 conference participants from 73 countries participated in this most inspiring platform. It provided a vital forum for information exchange in the field of PV solar energy research, demonstration and applications.

The EU PVSEC 2016 Proceedings give a comprehensive overview of the photovoltaic solar sector, its current status and future prospects in science, research, innovation, development and deployment on more than 5,000 pages. Selection for inclusion in the conference was made by paper review experts and topic organisers (see the listing on pages III-V), to whom we express our sincere gratitude for their comprehensive review work and overall contribution to the success of the conference. In addition to the 658 submitted papers, the proceedings includes 82 presentations (slides) shown during the plenary and oral presentations as well as 269 poster files of the visual presentations. In total this amounts to 1,009 publications!

The Proceedings constitute a powerful tool for targeted and quick information search and retrieval, enabling you to search by keyword, paper title and authors. We are sure that these features will help to simplify the use and exploit the full potential of this extensive source of information.

The Conference Proceedings are also published online. **A DOI code (Digital Object Identifier) has been assigned to each paper. This ensures unequivocal and permanent identification and full citability**. The papers can be viewed and downloaded from the EU PVSEC Proceedings website www.eupvsec-proceedings.com.

We are confident that these Proceedings will play an important role in providing a comprehensive overview of the current actors and activities in the global PV sector and that they will disseminate information on the state-of-the-art of technologies and applications. This can generate further research, add momentum to innovation and promote interest in PV worldwide.

We would like to cordially thank all authors and participants of the EU PVSEC 2016 for their contributions and look forward to welcoming you in 2017 at the next EU PVSEC.

The Editors

Joint Research Centre

the European Commission's in-house science service

Serving society
Stimulating innovation
Supporting legislation

EU PVSEC 2016 Highlights

Nigel Taylor
Technical Programme Chair

European Commission

First some stats ...

Participants: 1742

Presentations: 1053 (334 oral, 719 posters)

Parallel Events: 6

European Commission

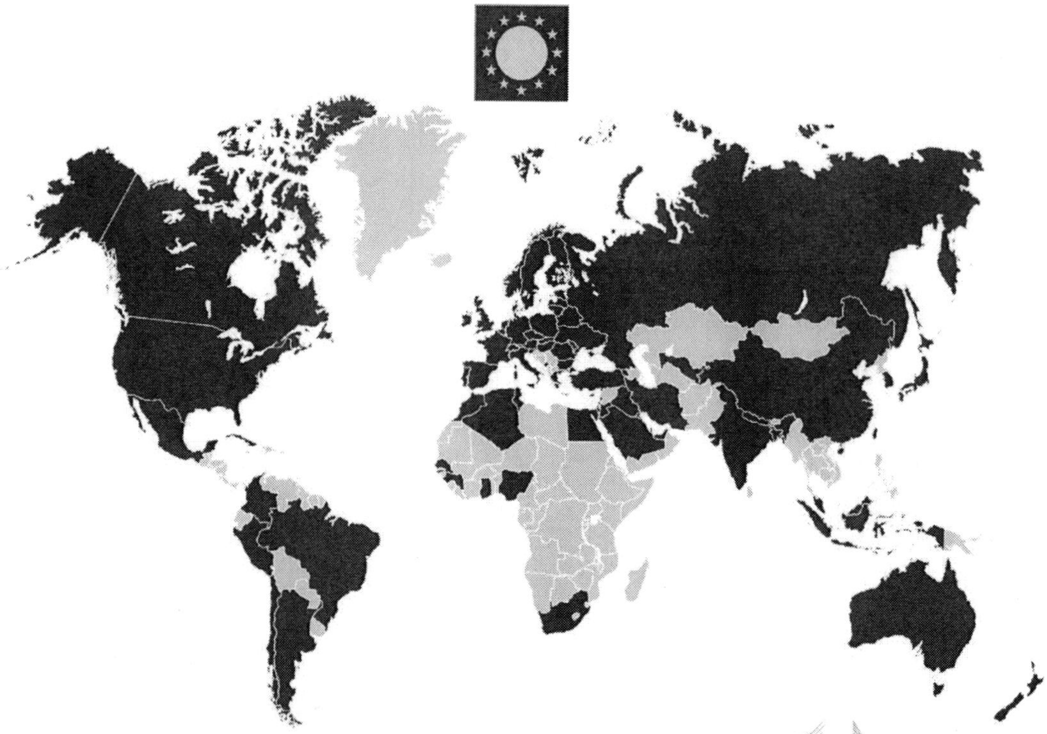

European Commission

Silicon: strength in depth

European Commission

Q.ANTUM
Champion R&D module results

Champion module **features**

- Use current production tools
- Standard soldering technique
- Same BOM as in standard modules (high quality ARC glass, back sheet)
- Standard 60 cells module size
- No additional light-trapping schemes

- Usage of champion cell technology with module optimized ARC coating
- Half cell technology and splitted junction box
- Minimized intercell distance

- η = 19.5 % (ap)*
- V_{oc} = 41.5 V
- J_{sc} = 9.30 A
- FF = 77.4 %

- P = 301 W (ta)*

Classification	Efficiency (%)	Area (cm²)	V_{oc} (V)	I_{sc} (A)	FF (%)	Test centre (date)	Description
Si (crystalline)	23.8 ± 0.5	11562 (ap)	53.4	6.32*	81.6	AIST (1/16)	Panasonic (72 cells) [11]
Si (multicrystalline)	19.5 ± 0.4	15349 (ap)	41.53	9.299*	77.4	FhG-ISE (12/15)	Hanwha Q Cells (120 cells) [12]

M. Green et al, Prog. Photovolt: Res. Appl. 2016; 24:905–913

Current Status of High-Efficiency Q.Antum Technology with New World Record Module Efficiency of 19.5%
M. Scherff, P. Kowalzik, C. Gerber, K. Duncker, M. Junghänel, C. Fahrland, S. Kunath, S. Hörnlein, M. Schütze, L. Niebergall, B. Klöter & J.W. Müller
Hanwha Q CELLS, Bitterfeld-Wolfen, Germany

European Commission

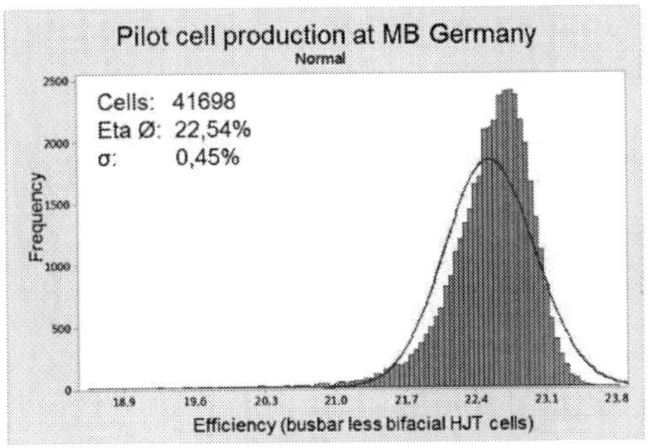

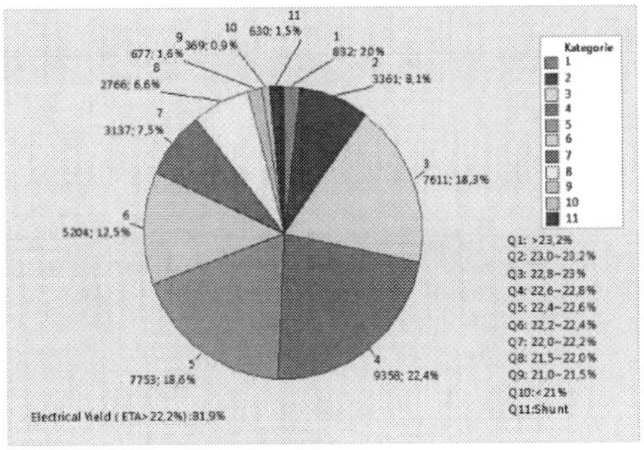

Golden Cell

Status	Cell	Area (cm²)	Eff. (%)	V_{oc} (mV)	FF (%)	J_{sc} (mA/cm²)
09/2015	Cz	239,5	23,3	742,5	81,3	38,64

- In this pilot production, total 41698 SWCT(busbar less) bifacial cells were processed
- Average ETA reached 22,54%
- 34617 cells (yield of 82%) are over 22,2%
- Record cell reached 23,3% cell efficiency, which is from its high V_{oc} and FF.

Silicon Heterojunction Solar Cells in Meyer Burger's Demo line: Results of Pilot Production on Mass Production Tools
J. Zhao et al

European Commission

Tipping point:

2BO.3 Heterojunction Solar Cell Concepts

Speaker gender balance for the first time!

European Commission

Vibrationally Induced Fluidization

1. CapEx, Full Cost, and Cash Cost will be lower than legacy processes

China: $/Kg on a Polysilicon Plant Basis[1]			
	Genesis™	H-FBR	HC-TCS/Siemens
CapEx	43	65	68
Full Cost	10.25	14.00	18.45
Cash Cost[2]	7.35	9.65	13.90

2. Advantages include
   - Only 1% yield loss to dust
   - Only 1 KwHr/Kg for V-FBR process unit – only 8 KwHr/Kg for entire plant
   - 10X higher reactor productivity
   - Self-seeding and in-situ particle size classification
   - On/off fluidization and "Ok" to temperature cycle
   - Continuous self-regulating product withdrawal

2BP.1.2 New Monosilane Fluid Bed Decomposition Technology for the Production of Solar Quality Silicon Feedstock, M. Dassel, SiTec, Augsburg, Germany

European Commission

Material Limits of Si from State-of-the-Art PLI Techniques

WAFER ELBA

Exact specification of loss mechanisms by ELBA

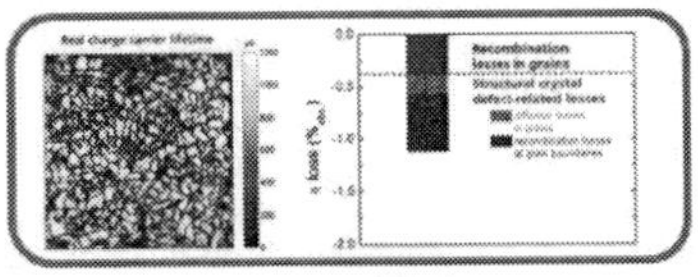

INGOT

Precise PL imaging of τ and [Fe_i] on ingot level

modulum

harmonically modulated luminescence

Fraunhofer ISE

ELBA CUSTOM APP

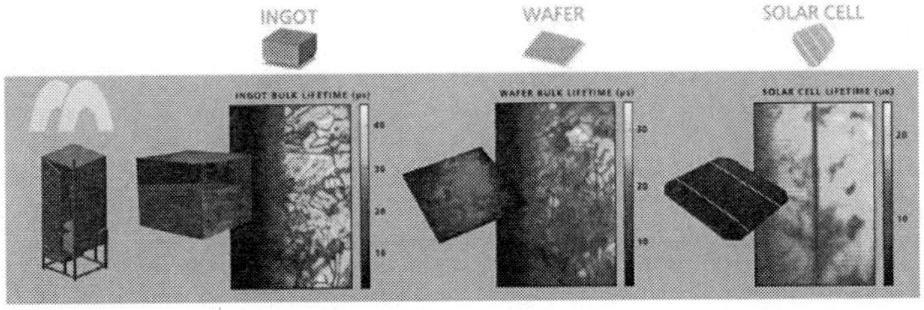

2CO.1.2 Material Limits of Silicon from State-of-the-Art Photoluminescence Imaging Techniques, F. Schindler, J. Giesecke, B. Michl, W. Warta & M.C. Schubert
Fraunhofer ISE, Freiburg, Germany

European Commission

Thin film competes on efficiency

European Commission

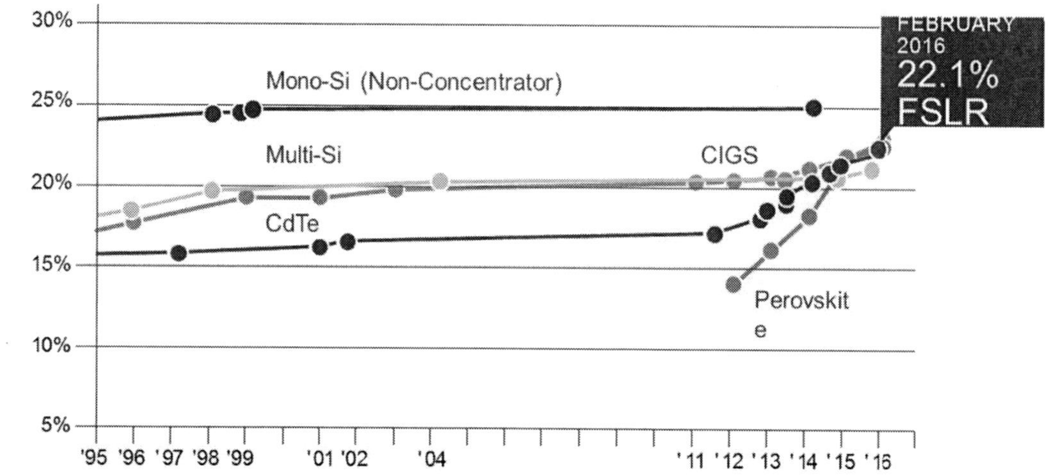

First Solar: current production module average 16.2%, top 18.2% and set to increase

European Commission

AVANCIS' Track Record of Certified Champion Efficiencies

LATEST 2016 EXTERNALLY CERTIFIED CHAMPION MODULE

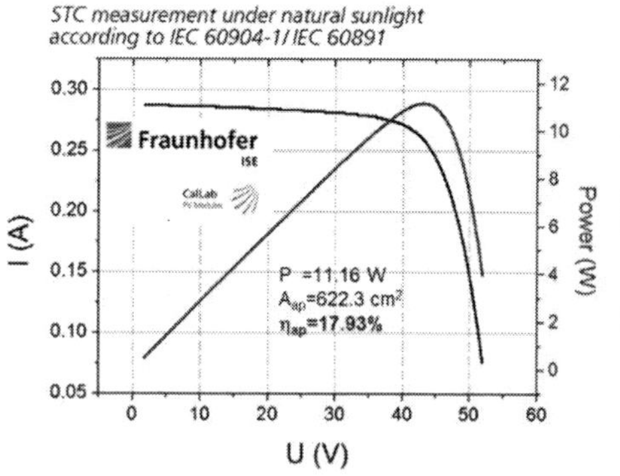

Comparison of champion modules:

	η_ap (%)	Jsc (mA/cm2)	Voc (mV)	FF (%)
2014 Champ (NREL)	16.6	38.5	621	69.4
2016 Champ (ISE)	**17.9**	39.7	627	72.2
Improvement	+7.8%	+3.1%	+1%	+4%

▶ Key improvements 2016 at a glance:

- Optimized cell width and dead area
- Improved TCO layer

▶ Major improvements on FF and Jsc

3BO.8.2 Efficiency Improvement of CIGSSe/Cd-Free Solar Module by Optimized Cell and Interconnect Design

P. Eraerds, C. Schubbert, T. Kwast, M. Grave, F. Braun, M. Algasinger, R. Lechner, T. Dalibor & J. Palm
AVANCIS, Munich, Germany

European Commission

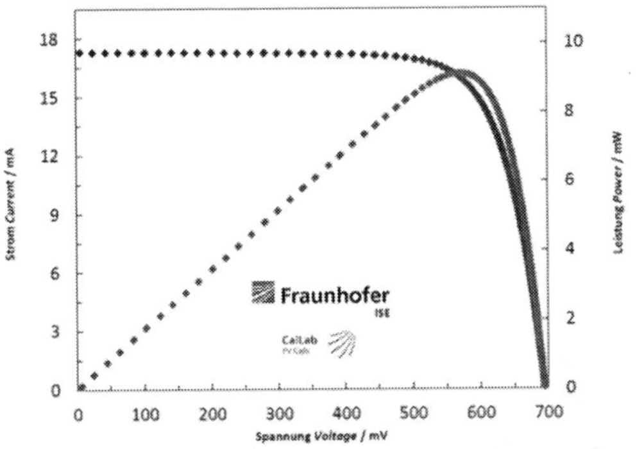

V_{OC}	= (	694.5	±	2.4	) mV
I_{SC} (Ed.2 - 2008)/3/	= (	17.28	±	0.33	) mA
J_{SC}	= (	33.58	±	0.64	) mA/cm²
I_{MPP}	=	15.94			mA
V_{MPP}	=	572.0			mV
P_{MPP}	= (	9.12	±	0.18	) mW
FF	= (	75.98	±	0.49	) %
η	= (	17.72	±	0.37	) %

- **Eff 17.72% has been approved by Fraunhofer Institute**
- **Best result for the solution coated CIGSSe cell**
  (Nano solar : Eff 17.1%)

Photo conversion efficiency of 18.7% was achieved with solution coated Cu(In,Ga)(S,Se)$_2$.

3BO.8.3 High Efficiency Solution Coated Cu(In,Ga)(Se,S)2 Thin Film Solar Cells

T. Aramoto & Y. Kawaguchi
Solar Frontier, Atsugi, Japan
Y.-C. Liao, Y. Kikuchi, T. Ohashi, H. Iida & A. Nakamura
Tokyo Ohka Kogyo, Koza-Gun, Japan

European Commission

Lab scale PSC baseline development: mini-modules

- **Perovskite module development: lab scale**
  - Glass/sputtered ITO/P1 laser scribe /e-beam TiO$_2$/spin coat/Perovskite/spin coat OMeTAD/P2 mechanical scribe/evaporate Au/P3 mechanical scribe
  - **Status Solliance**: **16 cm^2** mini-module performance of **14,2%**

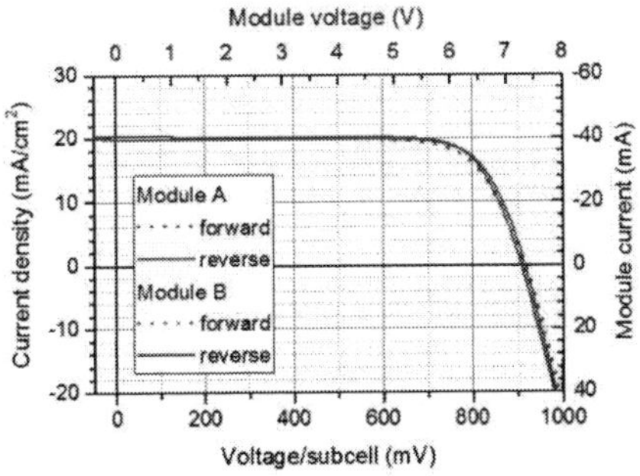

For further information:
3CO.7.5
Loss Analysis and Optimization for High Efficiency Perovskite Photovoltaic Modules
R. Gehlhaar, T. Merckx, C. Masse de la Huerta, L. Rakocevic, W. Qiu & M. Jaysankar
imec, Leuven, Belgium

Towards Roll-to-Roll Manufacturing of Perovskite Based PV Modules

R. Andriessen & P. Poodt, TNO, Eindhoven, Netherlands, T. Aernouts, imec, Leuven, Belgium. S. Veenstra, ECN, Eindhoven, Netherlands, R. Janssen & A. Createore, TU/e, Eindhoven, Netherlands, D. Vanderzande, University of Hasselt, Diepenbeek, Belgium T. Kirchartz Forschungszentrum Jülich, Germany

European Commission

Computational Screening of A_2MX_6 Compounds

Lead-free Perovskite Materials for Photovoltaic Applications

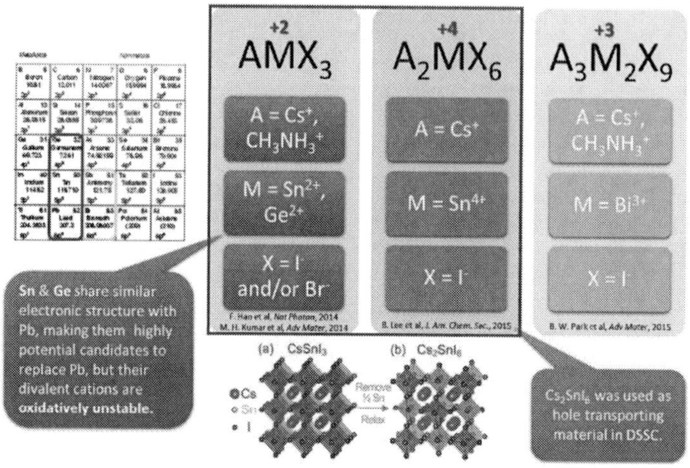

Tin(IV)-based Iodide Perovskites A2SnI6 (A = Rb+, Cs+, MA+, FA+)

Y. Chen, T. Krishnamoorthy, C. Yao, T. Baikie, S. Dharani, J.H. Fu, S.T. Chien, M. Asta, N. Mathews, L.H. Wong & S.G. Mhaisalkar
Nanyang Technological University, Singapore, Singapore & University of California, Berkeley, USA

European Commission

Concentrators:

new paradigms

European Commission

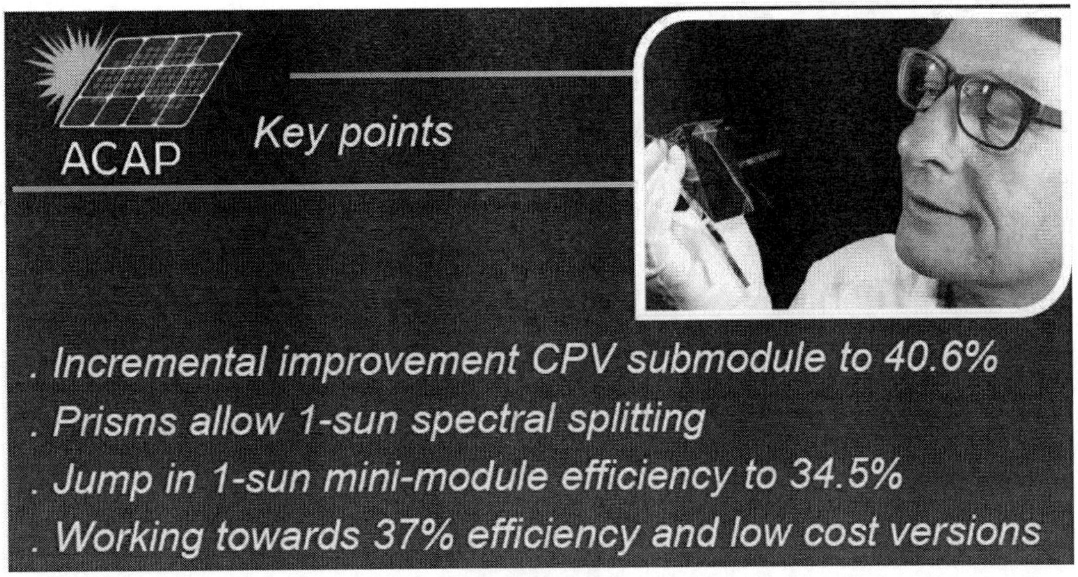

1AP.1.1 37% Efficient One-Sun Minimodule and over 40% Efficient Concentrator Submodules, M.A. Green, M.J. Keevers, B. Concha-Ramon & J. Jiang, UNSW, Sydney, Australia, P.J. Verlinden, Y. Yang & X. Zhang, Trina Solar, Changzhou, China

European Commission

4CO.6.4 Luminescent Solar Noise Barrier – Large Scale Testing and Modeling L.H. Slooff ECN, Petten, Netherlands S. Verkuilen Heijmans Wegen, Rosmalen, Netherlands M.M. de Jong & M.N. van den Donker SEAC, Eindhoven, Netherlands M. Kanellis & M.G. Debije Eindhoven University of Technology, Netherlands

European Commission

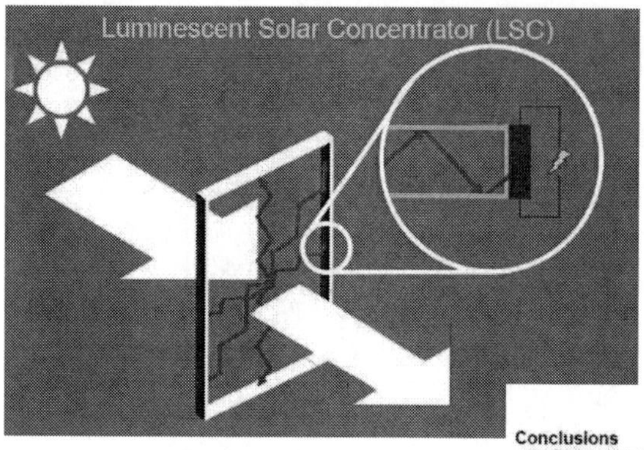

Conclusions

- Tm^{2+} doped halides discovered
    - Colorless
    - No self-absorption
    - >60% sunlight absorption
    - Expected ~15% power efficiency
- Made as both powders and thin-films
- Scattering still a challenge for BIPV

Outlook

- Not a lot known/understood (only ~4 major studies)*
    - Fundamental luminescence properties
    - Optimizing thin film growth
- Realize a lab-scale Tm^{2+} LSC prototype

1AO.2.4 Efficient Luminescent Solar Concentrators Based on Self-Absorption Free, Tm2+ Doped Halides

O.M. ten Kate, M. ten Kate, M. De Jong, O.M. ten Kate & E. van der Kolk
Delft University of Technology, Netherlands
K.W. Krämer
University of Berne, Switzerland

European Commission

Operation

European Commission

Peer-to-Peer (P2P) Forecasting:

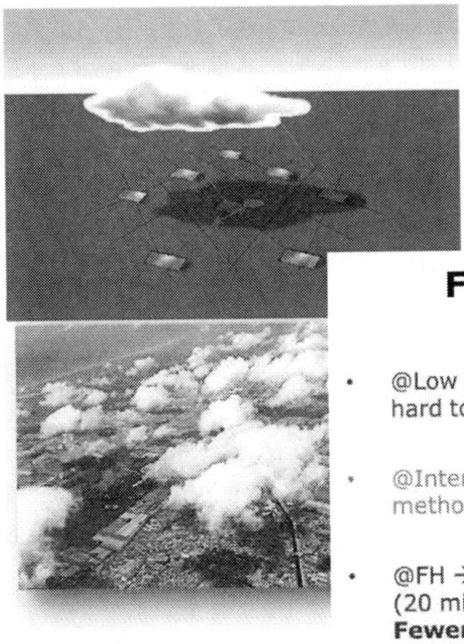

Forecast Skill vs. Forecast Horizon

- @Low FH: **persistence** hard to beat
- @Intermediate FH: P2P method shows its **value**
- @FH → T_{Obs} (20 min. in this example.): **Fewer** to **no** available periph. $k_j^*(t-\tau_{ij})$ timeseries for P2P FC

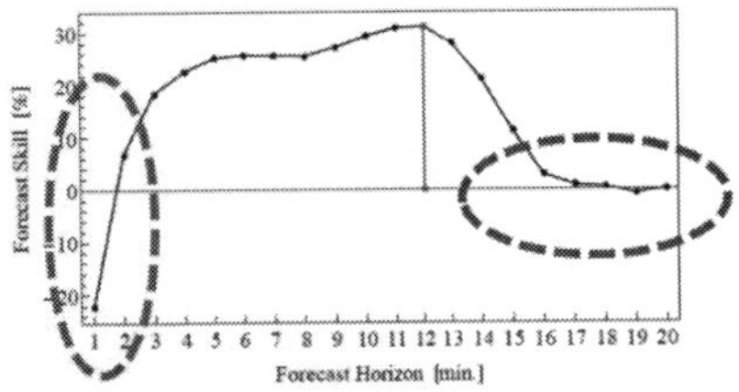

Dependence of Peer-to-Peer Solar Forecast Skill on Irradiance Variability
B. Elsinga & W.G.J.H.M. van Sark
Utrecht University, Netherlands

European Commission

Analysis of Failure Occurrence

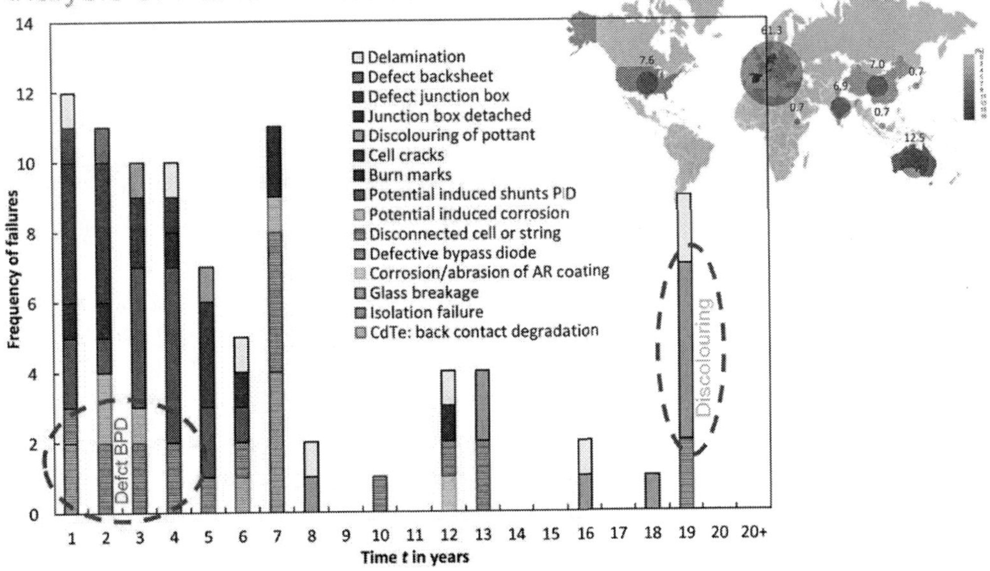

5DP.1.2 Mean Degradation Rates in PV Systems for Various Kinds of PV Module Failures
M. Köntges & S. Altmann
ISFH, Emmerthal, Germany
U. Jahn
TÜV Rheinland, Cologne, Germany

European Commission

Implementation and Monitoring

2015/16

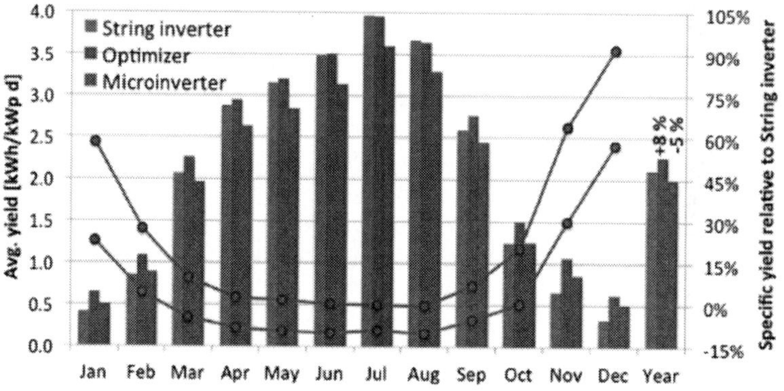

TNC *(4.6.2015 – 3.6.2016)* 12

5AO.9.4 Performance of Recent Inverter Systems under Partial Shading Conditions

R. Lingel, T. Nordmann & T. Vontobel
TNC Consulting, Feldmeilen, Switzerland

European Commission

Energy Solutions

European Commission

IS "BIPV" ALREADY "VINTAGE"...?

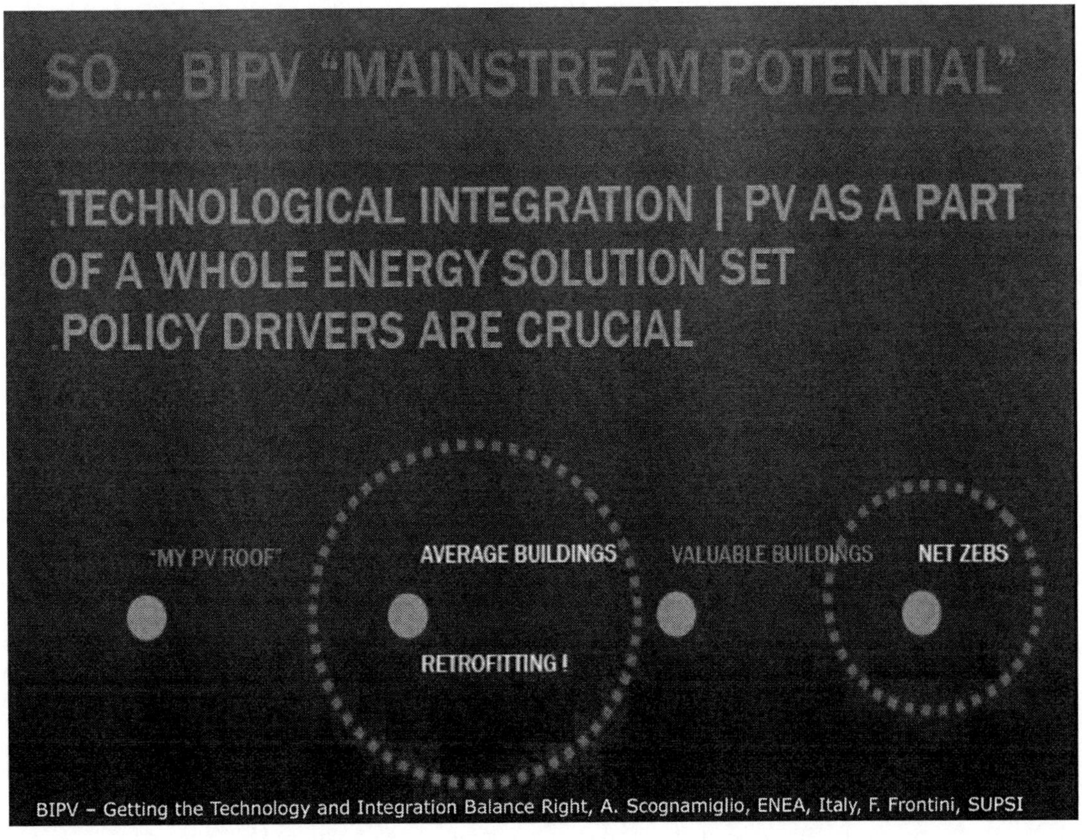
SO... BIPV "MAINSTREAM POTENTIAL"
.TECHNOLOGICAL INTEGRATION | PV AS A PART OF A WHOLE ENERGY SOLUTION SET
.POLICY DRIVERS ARE CRUCIAL
"MY PV ROOF"
AVERAGE BUILDINGS
RETROFITTING !
VALUABLE BUILDINGS
NET ZEBS
BIPV – Getting the Technology and Integration Balance Right, A. Scognamiglio, ENEA, Italy, F. Frontini, SUPSI

Kinetic envelope for heating cooling and light comfort + PV generation

Case Study

- Single Zone Office
- Zurich Weather Data
- 50 Panels in a Single Cluster
- Heating COP: 4
- Cooling COP: 3
- Lighting Load: 11.74 W/m²

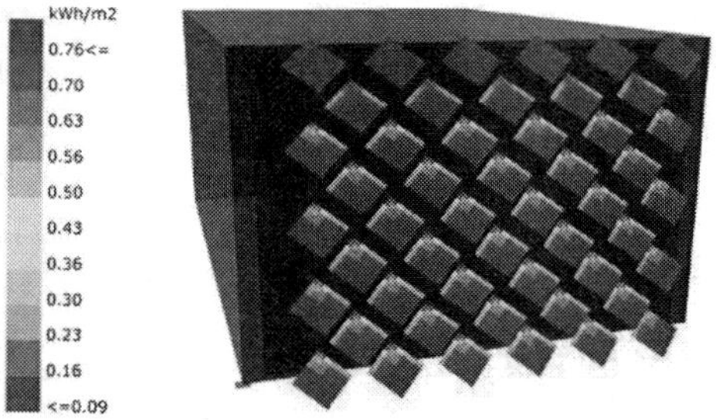

6DO.8.5 Energy Performance of PV Modules as Adaptive Building Shading Systems

J. Jayathissa, J. Schmidli, J. Hofer & A. Schlueter
ETH Zurich, Switzerland

New metrics for PV + storage performance!

Definition of the Storage Performance Index (SPI)

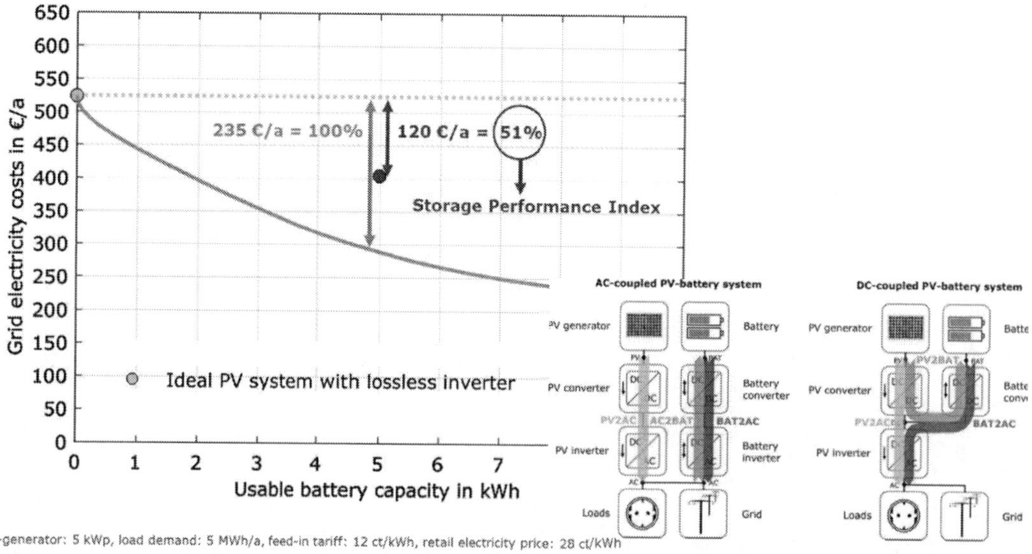

PV-generator: 5 kWp, load demand: 5 MWh/a, feed-in tariff: 12 ct/kWh, retail electricity price: 28 ct/kWh

Emerging Performance Issues of Photovoltaic Battery Systems, Johannes Weniger, Tjarko Tjaden, Joseph Bergner, Volker Quaschning, HTW Berlin - University of Applied Sciences, Germany

Thanks!

European Commission

Start working on excellent abstracts for EU PVSEC 2017!

European Commission

We thank the EU PVSEC 2016 Sponsors

SATELLITE DATA ASSIMILATION IN REGIONAL NUMERICAL WEATHER PREDICTION AS A KEY FOR BETTER CLOUD COVER FORECASTS IN TROPICAL ENVIRONMENTS

Frederik Kurzrock[1,2], Sylvain Cros[2], Fabrice Chane-Ming[3], Roland Potthast[4], Laurent Linguet[5], Gilles Lajoie[1]
[1]Université de La Réunion, UMR ESPACE-DEV, 15 avenue René Cassin, 97715 ST DENIS Cedex 9, La Réunion, France, frederik.kurzrock@univ-reunion.fr
[2]Reuniwatt SAS, 14 rue de la Guadeloupe, 97490 Sainte Clotilde, La Réunion, France
[3]Université de La Réunion, UMR LACy, 15 avenue René Cassin, 97715 ST DENIS Cedex 9, La Réunion, France
[4]University of Reading, Department of Mathematics, Whiteknights, PO Box 220, Reading RG6 6AX, United Kingdom
[5]Université de Guyane, UMR ESPACE-DEV - Centre IRD de Guyane, BP 165, 275 route de Montabo, 97323 Cayenne, France

ABSTRACT: Although the high amount of solar irradiance in the tropics is an advantage for a profitable PV production, the local meteorological conditions induce a very high variability which is problematic for a safe and gainful injection into the power grid. This issue is even more critical in non-interconnected territories where network stability is an absolute necessity and the injection of PV power has to be limited.
The basis for precise cloud evolution and subsequent irradiance forecasts are high quality atmospheric analyses for NWP. Geostationary meteorological satellites provide valuable observations of cloud properties with high spatio-temporal resolutions and allow a pertinent data assimilation.
The shortcoming is that optical and thermal channels of satellite sensors do not provide cloud properties from inside clouds. Different existing data assimilation approaches aim at deriving atmospheric analyses with most realistic cloud features, utilising geostationary satellite observations. The potential of assimilating satellite-derived cloud information in regional NWP with focus on irradiance forecasts in tropical regions has not been evaluated so far. Hence, the present work aims at evaluating the potential of geostationary satellite data assimilation in limited-area models applied to the French tropical oversea territories Reunion Island and French Guiana.

Keywords: Forecasting, Solar Radiation, Grid-Connected, Plant Control, PV System

1 INTRODUCTION

The globally installed amount of PV systems is expected to grow from 178 GW in 2014 to 540 GW by 2020 ([1]). This implies that solar power is becoming a major component of the future global energy mix and opens the way towards independence of fossil fuels and thus a reduction of global warming. Nevertheless, the exploitation of solar power is a technological challenge. As the photovoltaic power production reduces in the presence of clouds, its intermittent character constitutes an obstacle concerning grid injection. In France, the injection of intermittent renewable energies is legally capped at 30 %. This threshold has already been exceeded in Corsica (40 %) and in the French tropical oversea territories Guadeloupe (35 %), Reunion Island (39 %). In French Guiana it already reached 25 % [2].

Reliable forecasts of the solar power production facilitate a safer injection of PV power into the grids and will in the long term permit the elevation of legal injection limits. The high amount of irradiance in the tropics, combined with the fact that non-interconnected territories are obliged to be energy independent, leads to the importance of accurate irradiance forecasts in such regions if one wants to increase the injection of PV into the grids [3] [4].

Regarding the requirements for temporal (a few minutes) and spatial (a few kilometers) resolutions of day-ahead solar power forecasts, numerical weather prediction (NWP) is the only appropriate forecasting solution. Solar power production forecasts almost entirely depend upon the forecast of the Global Horizontal Irradiance (GHI), directly driven by cloud conditions. Being able to predict cloud cover and its evolution thus means being able to predict solar power production. Since many physical processes which have large impacts on cloud and radiation processes have to be parameterised in NWP models, irradiance predictions by NWP models exhibit a considerable lack of precision regarding solar energy applications [5] [6].

The temporal and spatial resolutions of global weather models (around 0.25° grid spacing and 3h temporal resolution) do not permit high-precision cloud evolution forecasts. It is therefore necessary to apply dynamical downscaling with regional NWP models in order to increase the spatio-temporal forecast resolution and permit a more realistic simulation of cloud processes.

Irradiance prediction in the tropics is particularly difficult due to pronounced convection, homogenous air masses and large temperature contrasts between land and sea [7].

2 CURRENT STATUS OF RESEARCH

Efforts are being made to improve irradiance forecasts by improving the internal model processes (e.g. WRF-Solar [8] or optimal model configuration [9]). Nevertheless, such efforts do not necessarily improve the forecast results if the initial conditions of the NWP simulation already lack precision.

2.1 Geostationary meteorological satellite data assimilation

The quality of NWP forecasts largely depends upon the initial conditions which are used to launch the NWP model. Data assimilation is the procedure which statistically combines observations and short-range NWP forecasts (first guess) in order to derive the best possible analysis of the atmospheric state on the entire NWP model grid [10]. Various meteorological observations (synoptic stations, buoys, radiosondes, aircraft measurements, satellites, radars, etc.) can potentially be used to derive the initial atmospheric state for NWP. The various types of observations have to be treated differently in the data assimilation process.

Besides the advantages of high temporal and spatial resolution and coverage, geostationary meteorological satellite observations provide pertinent information about cloud presence and evolution. Leveraging the remote sensing technology together with data assimilation techniques is a promising solution for improving solar irradiance forecasts, especially in data-sparse regions like islands.

Different existing approaches aim at deriving atmospheric analyses with most realistic cloud features, utilising satellite observations. Two approaches exist: the direct assimilation of observed radiances and the assimilation of cloud physical properties.

2.2 Direct radiance assimilation

This technique is massively performed in global NWP models by weather centers around the world.

A radiative transfer model is used to simulate a radiance observation from the model variables. Within the data assimilation procedure the model's initial state is subsequently modified by statistically combining the synthetic observations and the actual radiance observations measured by the satellite [11].

Since the optical and thermal sensors on geostationary meteorological satellites do not always capture information inside clouds, the data assimilation procedure has to adapt to different cloudiness conditions. Therefore, in many current assimilation systems, most cloud-affected radiance observations are discarded [12].

Since the direct assimilation of radiances brings along a number of critical issues to solve, the method has not yet been exploited in regional models regarding the tropics. Accordingly, an appropriate radiance data assimilation system for GHI forecasts must address the following issues:

- observation and background error estimation
- an accurate estimation of atmospheric profiles and cloud microphysical properties
- the choice of appropriate satellite channels
- optimal thinning of radiance observations
- bias correction.

2.3 Physical cloud property assimilation

Besides the direct assimilation of radiances, another approach is to assimilate geostationary satellite observations into limited-area models is the assimilation of derived cloud properties. This is an active field of research since approximately the beginning of this century [13] [14].

Recent research focusses on the assimilation of cloud-top information like cloud mask, cloud top-height, cloud-top temperature, cloud type classification, cloud liquid and ice water path and cloud albedo. In one existing approach the initial water vapour profile of the WRF model has been directly modified using cloud-top properties derived by GOES imagery [15]. Another initiative uses MSG imagery, an ensemble Kalman filter and the COSMO NWP model in order to assimilate cloud-top height and relative humidity [16]). Cloud top property assimilation has also been experimentally and successfully performed in the Rapid Update Cycle of the HIRLAM model, operated by the Royal Netherlands Meteorological Institute [17]. All recently published approaches that address cloud property assimilation show improved cloud cover forecasts or solar irradiance forecasts, respectively. This shows the general potential of such techniques to improve cloud cover and solar irradiance forecasts. Nevertheless, special attention to the tropics has not been drawn so far.

Although all approaches from recent years provide improved cloud cover forecasts or irradiance forecasts in mid-latitudes, the major limitation is that only clear-sky or completely cloud cases can be considered. Fractional clouds cause a measured signal mixing cold clouds and warmer Earth surface. As a consequence, clouds are usually considered as single-layer clouds.

3 AREAS OF STUDY

So far there has not been any research focus on the assimilation of geostationary satellite observations in regional NWP with the goal of improving cloud cover forecasts in the tropics. The aim of this work is to evaluate the benefits of this practice in the non-interconnected French oversea territories Reunion Island and French Guiana (Figure 1). Going along with the global increase of installed PV power, these zones likewise experience a rapid increase, e.g. 8.3 % increase of installed solar power on Reunion Island in 2013 [18].

While Reunion Island is surrounded by ocean for hundreds of kilometres, French Guiana has both continental and maritime influence. Reunion Island has a distinct orography with up to 3000 m high mountains which facilitate local thermal circulations. There are no major elevations in French Guiana which makes it a good site for studies of coastal clouds which are influenced by sea breeze effects.

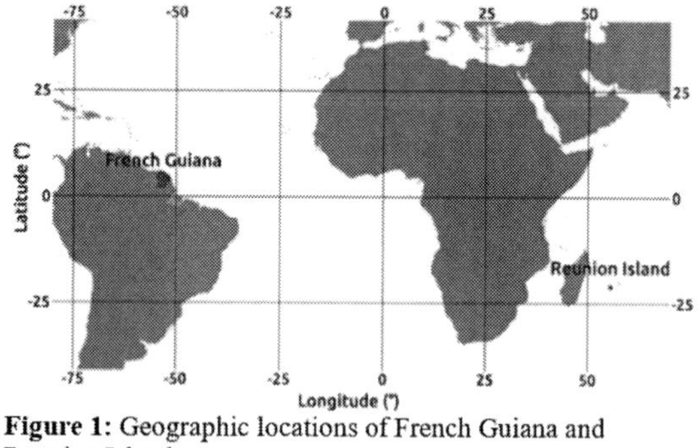

Figure 1: Geographic locations of French Guiana and Reunion Island.

4 OUTLOOK AND GOALS OF THIS WORK

For the first time, we evaluate the potential of assimilating satellite-derived cloud information in tropical regions. In a first study on Reunion Island, Meteosat-7 (IODC service) observations are assimilated in the COSMO model using its Ensemble Kalman filter (EnKF) data assimilation system [19]. This serves as an initial evaluation of remotely sensed cloud information assimilation in a state-of-the-art regional NWP modelling system, applied to the tropics. Surface irradiance measurements are used in order to validate the simulation results.

In the next step, a satellite data assimilation strategy for the WRF model is chosen. In the course of this, the results of the experiments using COSMO help approaching diverse issues: the interplay of nesting and satellite data assimilation, satellite channel selection, bias correction and observation error treatment. Applying two models to the same region allows to compare the influences of changes in the data assimilation procedure on the simulation results.

The chosen WRF satellite data assimilation strategy is then being applied and tested on Reunion Island (using Meteosat IODC data) and in French Guiana (using GOES-East data).

5 CONCLUSION

In conclusion, this work is an innovative research approach, aiming at increasing the injection of solar power into the electricity grids of non-interconnected tropical territories.

The assimilation of radiances and cloud properties into limited-area models is not yet exploited and constitutes a promising approach for improving cloud cover and solar irradiance forecasts in the tropics.

The raised solution could be a major contribution to facilitate the energy autonomy of such territories by an optimal use of the abundant solar energy yield which is largely under-exploited so far.

Reunion Island and French Guiana constitute ideal tropical, yet European experimental sites for testing solutions for solar power injection. Nevertheless, the outcome of this work will be applicable to all non-interconnected territories in tropical areas. It might permit the numerous developing and very populated countries in the tropics to increase their green economic growth by an optimal use of solar energy.

REFERENCES

[1] Rekinger, M., Thies, F., Masson, G., Sinead, O., 2015. Global Market Outlook for Solar Power 2015-2019. SolarPower Europe.

[2] Assemblée Nationale - Commission des affaires économiques (2014), rapport d'information n° 2225 sur l'adaptation du droit de l'énergie aux Outre-mer, déposé le 17 octobre 2014. http://www.assemblee-nationale.fr/14/rap-info/i2225.asp

[3] Diagne, M., David, M., Lauret, P., Boland, J., Schmutz, N., 2013. Review of solar irradiance forecasting methods and a proposition for small-scale insular grids. Renew. Sustainable Energy Rev. 27, 65–76.

[4] Nobre, A.M., Severiano Jr., C.A., Karthik, S., Kubis, M., Zhao, L., Martins, F.R., Pereira, E.B., Rüther, R., Reindl, T., 2016. PV power conversion and short-term forecasting in a tropical, densely-built environment in Singapore. Renewable Energy 94, 496–509.

[5] Kleissl, J., 2013. Solar Energy Forecasting and Resource Assessment, 1st ed. Academic Press.

[6] Sengupta, M., Habte, A., Kurtz, S., Dobos, A., Wilbert, S., Lorenz, E., Stoffel, T., Renné, D., Gueymard, C., Myers, D., Wilcox, S., Blanc, P., Perez, R., 2015. Best Practices Handbook for the Collection and Use of Solar Resource Data for Solar Energy Applications (Technical Report). National Renewable Energy Laboratory (NREL).

[7] Beucher, F., Belamari, S., & Beau, I. (2010). Météorologie Tropicale: des alizés au cyclone. Météo-France.

[8] Jimenez, P.A., Hacker, J.P., Dudhia, J., Ellen Haupt, S., Ruiz-Arias, J.A., Gueymard, C.A., Thompson, G., Eidhammer, T., Deng, A., 2015. WRF-Solar: An augmented NWP model for solar power prediction. Model description and clear sky assessment. Bull. Amer. Meteor. Soc.

[9] López-Coto, I., Bosch, J.L., Mathiesen, P., Kleissl, J., 2013. Comparison between several parameterization schemes in WRF for solar forecasting in coastal zones. Presented at the 42nd ASES Annual Conference.

[10]Kalnay, E., 2003. Atmospheric Modeling, Data Assimilation, and Predictability. Cambridge University Press.

[11]Weng, F., 2007. Advances in Radiative Transfer Modeling in Support of Satellite Data Assimilation. J. Atmos. Sci. 64, 3799–3807.

[12]Bauer, P., Ohring, G., Kummerow, C., Auligne, T., 2011. Assimilating Satellite Observations of Clouds and Precipitation into NWP Models. Bull. Amer. Meteor. Soc. 92, ES25-ES28.

[13]Bayler, G.M., Aune, R.M., Raymond, W.H., 2000. NWP Cloud Initialization Using GOES Sounder Data and Improved Modeling of Nonprecipitating Clouds. Mon. Wea. Rev. 128, 3911–3920.

[14]Yucel, I., James Shuttleworth, W., Gao, X., Sorooshian, S., 2003. Short-Term Performance of MM5 with Cloud-Cover Assimilation from Satellite Observations. Mon. Wea. Rev. 131, 1797–1810.

[15]Mathiesen, P., Collier, C., Kleissl, J., 2013. A high-resolution, cloud-assimilating numerical weather prediction model for solar irradiance forecasting. Solar Energy 92, 47–61.

[16]Schomburg, A., Schraff, C., Potthast, R., 2014. A concept for the assimilation of satellite cloud information in an Ensemble Kalman Filter: single-observation experiments. Q.J.R. Meteorol. Soc. 141, 893–908.

[17]de Haan, S., van der Veen, S.H., 2014. Cloud Initialization in the Rapid Update Cycle of HIRLAM. Wea. Forecasting 29, 1120–1133.

[18]SPL Energies Réunion, 2015. Le bilan énérgétique de la Réunion 2014. http://energies-reunion.com/publications/bilan-energetique-de-la-reunion-2/ber-2014/

[19]Schraff, C., Reich, H., Rhodin, A., Schomburg, A., Stephan, K., Periáñez, A., Potthast, R., 2016. Kilometre-scale ensemble data assimilation for the COSMO model (KENDA). Q.J.R. Meteorol. Soc. 142, 1453–1472.

GLOBAL HORIZONTAL IRRADIATION IN CASTILLA-LEÓN, SPAIN: SPATIAL ESTIMATION THROUGH INTERPOLATION METHODS

M. C. Rodríguez-Amigo[(1)], M. Díez-Mediavilla[(1)], D. González Peña[(1)], M. Dieste-Velasco[(1)], C. Alonso-Tristán[(1,*)]
[(1)]Research Group SWIFT (Solar and Wind Feasibility Technologies). Universidad de Burgos. Escuela Politécnica Superior. Avda Cantabria s/n. 09006 Burgos. Spain.
*Corresponding author: catristan@ubu.es Phone/fax number: +34 947 258925/+34 947 258910.

ABSTRACT: Four spatial interpolation methods (Inverse Distance Weighting, Spline, Kriging and Natural Neighbor) and their different variations have been employed to map Solar Global Horizontal Irradiation (GHI) in Castilla- Leon, Spain. The interpolation network consists of 71 ground meteorological stations and seven complete years of half hourly data have been used to obtain annual daily averages of GHI. The interpolation results have been tested against data from the four Spanish National Meteorological Agency (AEMET) stations available in the region using standard statistical indicators (RMSE, MBE, MAPE and MAE). An additional partial cross validation of the results has been implemented excluding five stations from the interpolation network attending different criteria to test the results of the interpolation method.

Keywords: Interpolation methods, solar radiation, Kriging, Splines, IDW, Natural Neighbour.

1 INTRODUCTION

Solar radiation prediction and forecasting are important in the generation of electricity and to provide the alternative energy. The availability of solar radiation measurements at a given location makes realistic projects and controls achievement of these activities. Resource monitoring becomes critical for resource assessment, system design, energy planning and grid management. An important objective of monitoring networks is to map and predict data of interest at unobserved locations using observed data at known locations. In the case of solar energy, Global Horizontal Irradiation (GHI) is the most common value recorded at meteorological ground stations. Several mathematical methods have been proposed to obtain values of GHI through different meteorological data such as temperature, rainfall, etc and geographical parameters but its accuracy is restricted. As the number of ground station is limited and the available data restricted, mathematical interpolation methods can provide suitable results without need to know additional meteorological parameters.

In this work, four different interpolation methods and 29 variations have been used to map annual daily average values of Global Horizontal Irradiation (GHI/MJ m^{-2} day) in Castilla-León, Spain from a network of 71 ground stations. The four AEMET (National Meteorological Agency) facilities in the region constitute the control system. Standard statistical indicators have been used to classify the interpolation methods.

2 CASTILLA-LEON: THE REGION

The region of Castilla y León, Spain covers 94226 Km^2 or 18.6% of Spain. With a very low population density and continental Mediterranean climate, the region benefits from very favorable atmospheric conditions. Solar irradiation is estimated at approximately 1450 KWh/m^2year [1]. The ambient temperature range is between 4ºC and 20ºC and there are very few cloudy days [2], which implies a highly significant potential source of energy. It is essential to estimate the exact amounts of radiation in different parts of Castilla y León for designing optimum solar facilities. Figure 1 shows the location of the region in the country and its provinces.

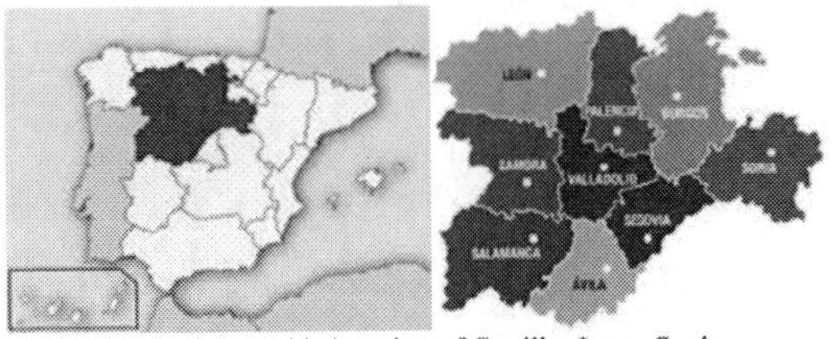

Figure 1 Geographic location of Castilla -Leon, Spain.

3 INTERPOLATION METHODS FOR CLIMATIC AND METEOROLOGICAL DATA

Interpolation methods have been traditionally applied to geospatial data. Franke [3] and Lam [4] has review and classified the interpolation techniques. Software ArcGis 10 has been used for this work. A viewable rectangular grid of pixels (raster) of the working area is built by applying the mathematical interpolation methods to the interpolation network. Several methods are implemented in the software. Main characteristic of these methods are summarized below.

3.1. Inverse Distance Weighting

The Simplest spatial interpolation method is the IDW (Inverse Distance Weighting), consisting of weighting the inverse of the distance between two points in the sample. The influence of the proximity between data can be defined in both deterministic and analytical way[5]. No maximum and minimum values are allowed as the extreme values are not included in the interpolation network [6]. The input variable is less important as higher is the distance to the calculated point. The fundamental parameter of the method is the power, assigned to calculate the influence of the known values (interpolation network) to the calculated values depending the distance between them. Power is a real positive number, assumed as 2. Higher values of power, emphasizes the influence of nearest points and makes the surface more irregular. Lower values of power, overvalue nearest points and smooth out the resulting surface. An optimization engine is available in ArcGis to fix this parameter.

3.2. Kriging method

Kriging is the other standard interpolation procedure. The interpolated values are modelled by a Gaussian process governed by a previous co-variances estimation method. This method uses a variogram model for data collection calculating the weights given to each point used in the valuation of the references. Universal Kriging assumes a structural component in the series of values. That means a local variable trend. In Ordinary Kriging local averages are not necessarily close from the population mean, so that neighboring points are hardly used for estimation. Variogram models available are lineal and cuadratic for Universal Kriging and Circular, spherical, exponential, Gaussian and linear for the Ordinary Kriging.

3.3. Natural Neighbor

Closer values to the searched point are used as input data for the interpolation. The weighting of each point is based on proportional areas. This method is known as Sibson's interpolation [7]. The natural neighbors of a point are those associated with adjacent Thiessen polygons. Voronoi's diagram with all interpolation data is built around the search point as well as the new polygons. The overlap between both areas is assumed as weighting.

3.4. Splines

Spatial interpolation by splines has frequently used map climatologic variables. A spline is a piecewise differential function defined through low degree polynomials functions which generates a surface containing all input data. Intervals are needed to avoid abrupt changes in large surfaces. Values are estimated using a mathematical function that minimizes the general curvature of the surface for obtaining a smooth surface that contains all input data. Two additional parameters are needed to control the output: weight and number of points. The weight reflects the contribution to the results of the third derivatives of the function. The number of point used for the interpolation has influence of the shape of the surface but increases the computation time.

4 NETWORK RADIATION DATA

The SIAR (*Sistema de Información Agroalimentaria para el Regadío*) [Agricultural Information System for Irrigation] radiation network was used for this study. It is a national network of weather stations managed by the Department of Agriculture that extends across the whole country. In the area under study, there are 59 stations from which data sets on GHI (Global Horizontal Irradiation), temperature, rainfall, humidity and wind speed and direction over a seven-year period were collected for this study. The location of the stations follows the directives of the WMO and AEMET [8] . The maintenance of weather stations and the calibration of different sensors is performed annually. SKYE 1110 pyranometers (Sensitivity range: 350 nm and 1100 nm; Operating temperature: -35°C to +75°C; absolute accuracy: ±5 %) and a Campbell ST1000 data logger were used to measure GHI. Pyranometer calibration followed the calibration technique specified in ISO 9847 for Solar Energy. This work is focused on daily GHI data, that means global radiation accumulated throughout the day, calculated by the trapezoids integration method [9] using the half-hourly values provided by the data logger. Daily data on GHI were filtered using the WMO directives [8]. Our work was restricted to the 44 stations where data have been of proven quality over at least 5 years and for a minimum of 340 days per year. 110080 GHI daily values were used in this work A geographical description of the network system used is shown in Table I.

Table I Geographic data of the SIAR network in Castilla-León, Spain, used in this study. Values of GHI has been calculated as annual average of daily data (2007-2013)

N^{ER}	Station	Latitude	Longitude	Height/m	GHI $MJ \cdot m^{-2} day^{-1}$
1	AV010	40.969	-4.77	921	16.980
2	BU020	42.746	-4.13	975	15.443
3	BU030	42.040	-3.77	840	16.032
4	BU040	42.346	-3.80	770	15.640
5	BU050	41.637	-3.58	870	16.145
6	BU070	42.702	-3.08	532	13.966
7	BU101	42.967	-3.24	635	12.412
8	LE010	42.565	-6.72	467	15.633
9	LE020	42.506	-5.43	791	16.391
10	LE030	42.396	-5.51	777	16.603
11	LE040	42.258	-5.74	779	16.957
12	LE050	42.218	-5.84	750	16.401
13	LE060	42.460	-5.90	835	16.691
14	LE070	42.464	-5.77	874	16.726
15	LE080	42.374	-5.02	856	16.089
16	LE090	42.441	-5.26	885	16.565
17	PA010	42.051	-4.30	868	15.986
18	PA020	41.946	-4.49	750	15.898
19	PA030	42.081	-4.72	744	16.452
20	PA040	42.271	-4.59	817	16.146
21	PA060	42.489	-4.25	821	15.455
22	PA070	42.528	-4.78	927	16.185
23	PA080	42.344	-4.28	798	16.149
24	SA010	40.590	-6.54	635	16.861
25	SA030	40.990	-5.48	815	17.349
26	SG010	41.301	-4.30	870	15.921
27	SG020	41.172	-4.48	822	17.063
28	SO010	41.461	-2.50	943	16.254
29	SO020	41.566	-3.22	855	16.562
30	SO030	41.834	-2.43	1063	16.391
31	SO101	41.734	-2.10	1043	16.480
32	VA010	42.151	-5.29	748	17.159
33	VA020	41.228	-5.21	793	17.261
34	VA030	41.310	-4.69	750	16.554
35	VA060	41,487	-5.00	658	16.483
36	VA070	41.643	-4.27	756	16.489
37	VA080	41.861	-5.07	739	17.088
38	VA101	41.711	-4.70	714	15.940
39	ZA010	42.000	-5.81	709	16.762
40	ZA020	41.482	-5.65	659	17.119
41	ZA040	41.880	-5.39	701	16.860
42	ZA050	41.782	-5.90	714	17.474
43	ZA060	41.934	-5.67	738	17.006
44	ZA080	41.508	-5.37	650	16.813

To avoid abrupt responses to the interpolation method in the border area, where the SIAR network lacks ground stations, a set of facilities in the boundary area has been used. The search has been extended to Galicia, Asturias, Cantabria, País Vasco, La Rioja, Aragón, Castilla La Mancha, Madrid, Extremadura and Portugal, looking for ground station in the surroundings areas with the same technical characteristics and recorded data over the same time period. Table II shows the geographical information (longitude, latitude, height and average daily value of GHI /MJ m^{-2} day) of these stations.

Table II Geographic location of the ground meteorological stations used to complete de interpolation network in the boundary layer. Values of GHI has been calculated as annual average of daily data (2007-2013)

N^{ER}	Station	Latitude	Longitude	Heigh/m	GHI MJ·m^{-2}day^{-1}
1	AL020	42.638	-2.52	689	13.133
2	BI020	43.171	-2.78	150	11.217
3	BI030	43.163	-3.28	300	11.800
4	ZG020	41.916	-1.75	516	15.566
5	CC020	40.298	-6.31	480	16.076
6	CC040	40.055	-6.69	273	17.258
7	CC060	40.011	-5.56	318	16.872
8	TO020	39.956	-4.70	400	17.628
9	GU010	40.927	-2.95	808	16.783
10	GU020	40.796	-1.80	1129	16.799
11	LU020	42.820	-6.92	1364	13.308
12	LU060	42.960	-6.91	910	13.126
13	OR020	42.460	-6.93	577	14.418
14	OR050	42.400	-7.30	1026	14.254
15	OR060	42.160	-7.09	851	14.763
16	CA020	43.002	-4.14	850	12.157
17	CA030	43.154	-4.26	640	9.9217
18	LO010	42.334	-3.01	1000	13.695
19	LO020	41.967	-1.97	752	14.727
20	LO030	42.125	-2.67	1235	12.801
21	MA010	40.452	-3.73	600	16.301
22	AS010	43.250	-5.78	206	11.960
23	AS020	43.335	-6.42	674	14.317
24	AS030	43.380	-5.15	50	12.153
25	PO010	41.600	-7.10	680	15.868
26	PO020	40.666	-7.10	717	16.578
27	PO030	41.100	-7.40	579	16.249

The AEMET (Agencia Estatal de Meteorología) stations located at Castilla-León have been reserved as the control system. Only four stations in the AEMET network record GHI data in Castilla-León. Table III shows geographical information (longitude, latitude, height and average daily value of GHI (MJ m^{-2} day^{-1}) of these stations used to validate the study. A partial cross-validation of the results has been also carried out: five additional controls have been used; each consisting of five stations (approximately 10% of the total amount of the SIAR network) of the interpolation network attending to different criteria.

Table III. Geographic data of the AEMET network in Castilla y León, Spain, used in this study. Values of GHI has been calculated as annual average of daily data (2007-2013)

N^{ER}	Station	Latitude	Longitude	Height/m	GHI MJ·m^{-2}day^{-1}
1	VA_AEMET	41.641	-4.75	735	16.620
2	LE_AEMET	42.583	-5.64	916	16.540
3	SA_AEMET	40.960	-5.50	790	16.868
4	SO_AEMET	41.775	-2.48	1082	16.045

5 ANALYSIS OF THE RESULTS

5.1 Data Processing

Data sets running from 2007 to 2013 (seven complete years) have been used for this study. As summary, sixty seven ground stations have been used for this study: 44 of them belong to Castilla y León and 27 are located at the boundary area to complete the available information. Figure 3 shows the distribution of all ground stations used. A total amount of 110080 daily values of GHI has been used.

Different interpolation methods implemented in ArcGis Software have been applied to the basedata: IDW, Natural Neighbor, Splin and Kriging. 29 variants of these methods have been used, changing different parameters as the number of interpolation points, weight or power.

Figure 3 Complete meteorological network used in this study. Small dots represent the interpolation network stations and large dots represent the four AEMET control stations

The four conventional statistical indicators in use: RMSE, MBE, MAE and MAPE (%) are defined as follows:

$$RMSE = \sqrt{\frac{\sum(v_r - v_i)^2}{N}} \qquad RMSE\,\% = \frac{RMSE \times 100}{Average\,(v_r)} \qquad (Eq.\ 1)$$

$$MBE = \frac{1}{N}\sum (v_r - v_i) \qquad MBE\,\% = \frac{MBE \times 100}{Average\,(v_r)} \qquad (Eq.\ 2)$$

$$MAE = \frac{1}{N}\sum |v_r - v_i| \qquad MAE\,\% = \frac{MAE \times 100}{Average\,(v_r)} \qquad (Eq.\ 3.)$$

$$MAPE = \frac{1}{N}\sum \left|\frac{v_r - v_i}{v_r}\right| \times 100 \qquad (Eq.\ 4)$$

where v_r is the experimental value and v_i is the

calculated value

5.2. Results of GHI

Maps of annual daily average GHI (MJm^{-2} day^{-1}) in Castilla-León have been calculated using IDW, Natural Neighbor, Kriging and Splines interpolation methods and their variations. A partial cross validation has been performed: five controls have been established using 5 stations in the control system (10% of the total) and 66 stations in the interpolation network. In each control (1 to 5), interpolated values have been compared to those measured at the points of the control network using the previously defined standard statistical indicators. Figure 4 shows the distribution of the stations used in each control at the region. Definitive control has been performed using the AEMET stations in the region (Control AEMET).

All the methods yielded RMSE values of below 6% and the use of different control systems (control 1 to 5) added no significant differences. Using stations belonging to the SIAR network as the control stations, Ordinary Kriging (OK) method with 35 points and quadratic semivariogram showed the highest RMSE value (5.7%). The Universal Kriging (UK) method with 35 points and quadratic semivariogram showed the lowest MAE value. Table IV shows the results of the interpolation methods and variations using AEMET stations as control system. The previously defined statistical parameters (RMSE (%), MBE (%), MAE (%) and MAPE(%)) are included. All methods present RMSE (%) values lower than 4%.

The interpolation methods undervalue GHI as the negative values of MBE show. The interpolation method that gave lowest RMSE value (Universal Kriging, 35 points, quadratic semivariogram) coincided to the lowest MBE, MAE and MAPE indicators and was considered the best method for this region and data series.

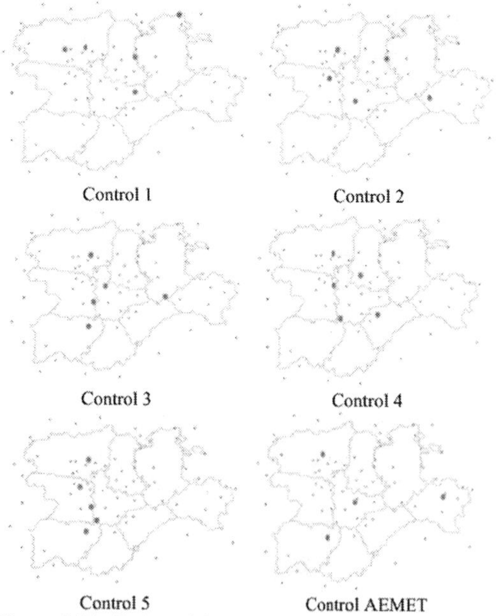

Figure 4 Distribution of the stations chosen for each control system following different criteria.

Figure 5 shows the annual daily average of GHI maps in Castilla-León calculated by different interpolation methods. Figure 5-a shows the results using UK interpolation method and lineal semivariogram (12 points) and Figure 5-b UK-lineal (35 points). That influences in grid resolution. Similar results are shown in figures 5-c and 5-d for the IDW interpolation method and in Figures 5-e and 5-f for the Regularized Spline (RS). Numerical indicators of these figures are shown in Table V. As can be seen, the resolution of the grid (number of interpolation points) is unrelated to the performance of the method: UK and RS yielded better result using 12-points interpolation rather a 35-points interpolation in opposition to IDW method, which presents better results using a-35 points interpolation.

Table IV Standard Statistical indicators (%) of the results of the different spatial interpolation methods applied to the SIAR network in Castilla y León. Control system has been implemented using the AEMET stations located at the region under study (Control AEMET)

AEMET STATIONS							
Method	Variation			RMSE	MBE	MAE	MAPE
KRIGING 12 POINTS	ORDINARY		Spherical	2.651	-0.585	2.458	2.500
			Circular	2.647	-0.583	2.454	2.445
			Exponential	2.625	-0.581	2.431	2.422
			Gaussian	3.856	-2.196	3.193	3.213
			Linear	2.641	-0.580	2.448	2.438
	UNIV ERSA L		Linear	3.362	1.694	2.641	2.650
			Quadratic	2.377	0.924	2.129	2.141
KRIGING 35 POINTS	ORDINARY		Spherical	2.625	-0.581	2.431	2.422
			Circular	2.688	-0.653	2.510	2.501
			Exponential	2.620	-0.531	2.439	2.430
			Gaussian	5.702	-2.740	4.574	4.622
			Lineat	2.596	-0.202	2.343	2.329
	UNIV ERSA L		Linear	5.103	3.191	3.670	3.698
			Quadratic	1.434	-0.082	1.260	1.252
SPLIN 12 POINTS	REGULAR IZED	WEIGTH	0.1	3.306	-1.402	2.979	2.986
			0.01	3.191	-1.218	2.943	2.947
			0.001	3.080	-1.095	2.868	2.869
	TENSION	WEIGTH	1	3.000	-1.015	2.804	2.804
			5	3.007	-0.984	2.824	2.824
			10	3.006	-0.959	2.830	2.829
SPLIN 35 POINTS	REGULAR IZED	WEIGTH	0.1	3.410	-1.340	3.124	3.133
			0.01	3.342	-1.232	3.105	3.112
			0.001	3.206	-1.112	3.006	3.010
	TENSION	WEIGTH	1	3.070	-1.0190	2.887	2.890
			5	3.070	-0.995	2.894	2.900
			10	3.066	-0.977	2.895	2.900
IDW			12 POINTS	2.377	-1.259	2.018	2.006
			35 POINTS	2.231	-1.060	1.692	1.681
NATURAL NEIGBORD			12 POINTS	2.691	-0.081	2.417	2.401

Table V Numerical results of the interpolation methods used to obtain the GHI (MJ m^{-2} day^{-1}) maps shown at figure 5.

Interpolation Method	RMSE (%)	MBE (%)	MAE (%)	MAPE (%)
UK, linear 12 pts	3.362	1.694	2.641	2.650
UK, linear 35 pts	5.103	3.191	3.669	3.698
IDW 12 pts	2.377	-1.259	2.018	2.006
IDW 35 pts	2.231	-1.060	1.692	1.681
RS, 12 pts, weight 0.01	3.191	-1.218	2.943	2.947
RS, 35 pts, weight 0.01	3.342	-1.232	3.105	3.112

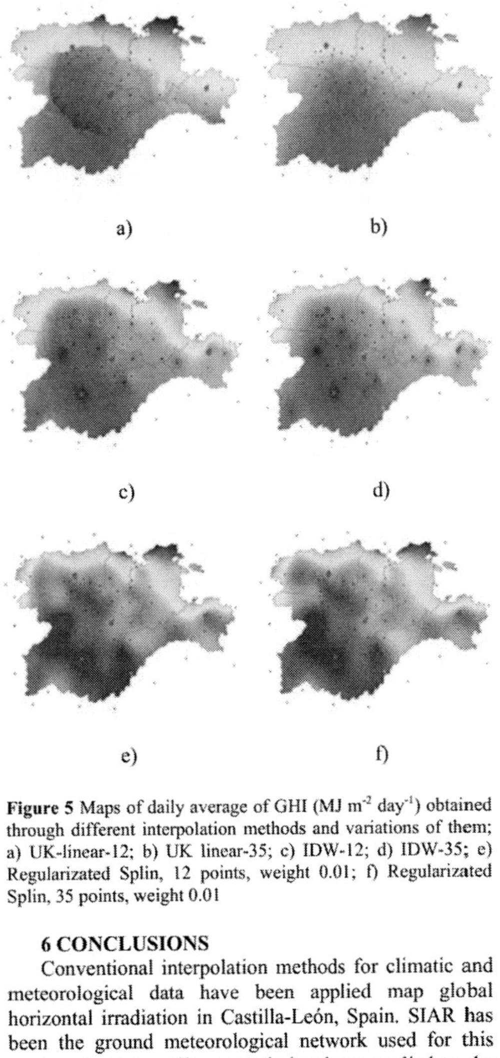

Figure 5 Maps of daily average of GHI (MJ m^{-2} day^{-1}) obtained through different interpolation methods and variations of them; a) UK-linear-12; b) UK linear-35; c) IDW-12; d) IDW-35; e) Regularizated Splin, 12 points, weight 0.01; f) Regularizated Splin, 35 points, weight 0.01

6 CONCLUSIONS

Conventional interpolation methods for climatic and meteorological data have been applied map global horizontal irradiation in Castilla-León, Spain. SIAR has been the ground meteorological network used for this work and strict quality controls has been applied to the available data. The SIAR network in the area under study has been complemented by similar ground stations in the boundary area to avoid abrupt responses of the interpolation method in the border area. As result, the interpolation network is formed by 71stations, 44 in Castilla-León and 27 in the boundary area. Seven complete years of half hourly GHI data have been used to calculate daily GHI and to obtain annual averages.

On the basis of the individual goodness of each interpolation methods, Ordinary Kriging with Gaussian semivariogram has been rejected to be applied to the area under study. This method had always obtained the highest deviation from the experimental results, regardless of the control used. Number of interpolation points used (12 or 35) had not significant influence on the results. The best interpolation method for the area under study has been UK-35 points and quadratic semi variograme taken the RMSE and MAE results into account.

To improve the results of this study, a more homogeneous distribution of the stations in the interpolation network would be advisable. As Figure 2 shows, there is an accumulation of ground stations in the centre of the region but large empty areas around the boundaries.

Acknowledgements

This research received economic support from the Spanish Government (grant ENE2011-27511 and grant ENE2014-54601-R). One of the authors, David González Peña, thanks to Junta de Castilla-León for economic support (PIRTU Program, ORDEN EDU/301/2015)

7 REFERENCES

[1] PVGIS. Available in: (http://re.jrc.ec.europa.eu/pvgis/index.htm). Last accessed: (July, 2014)

[2] Meteonorm. Available in: (http://www.meteonorm.com). Last accessed: (July, 2014)

[3] R. Franke. Scattered data interpolation: Tests of some methods. Mathematics of computation. 38 (1982) 181-200.

[4] N.S.-N. Lam. Spatial interpolation methods: a review. The American Cartographer. 10 (1983) 129-50.

[5] F.V. Gutierrez-Corea, M.A. Manso-Callejo, M.P. Moreno-Regidor, J. Velasco-Gómez. Spatial estimation of sub-hour global horizontal irradiance based on official observations and remote sensors. Sensors. 14 (2014) 6758-87.

[6] D. Watson, G. Philip. Comment on "a nonlinear empirical prescription for simultaneously interpolating and smoothing contours over an irregular grid" by F. Duggan. Computer methods in applied mechanics and engineering. 50 (1985) 195-8.

[7] R. Sibson. A brief description of natural neighbour interpolation. Interpreting multivariate data. 21 (1981) 21-36.

[8] World Meteorological Organization (WMO), Technical Regulations, Volume I: General Meteorological Standards and Recommended Practices. 2012. http://library.wmo.int/pmb_ged/wmo_49-v1-2012_en.pdf

[9] G. Peano. Applicazioni geometriche del calcolo infinitesimale, per Giuseppe Peano. fratelli Bocca 1887.

GLOBAL VERTICAL IRRADIATION IN THE FOUR CARDINAL ORIENTATIONS IN BURGOS, SPAIN

M. Díez-Mediavilla(1), M. C. Rodríguez-Amigo(1), Ana Pérez-Burgos(1,2), T. García Calderón(1), C. Alonso-Tristán(1,*)
(1)Research Group SWIFT (Solar and Wind Feasibility Technologies). Universidad de Burgos. Escuela Politécnica Superior. Avda Cantabria s/n. 09006 Burgos. Spain. (2) Dpto. Física aplicada. Facultad de Ciencias. Universidad de Valladolid. Paseo Belén, 7, 47011 Valladolid, Spain
*Corresponding author: catristan@ubu.es Phone/fax number: +34 947 258925/+34 947 258910.

ABSTRACT:
Four empirical mathematical models have been used to characterise the global vertical solar irradiation in the city of Burgos, Spain. The cardinal orientations (North, South, East and West) have been studied in order to determine the solar energy available on vertical surfaces in any orientation. The four models chosen from the literature cover all sky conditions: clear-sky (Circumsolar model), cloudly-sky (Isotropic model) and partially cover sky (Hay model and Klucher model) [1]. The empirical mathematical models allow the calculation of global vertical irradiation from experimental data of global, direct and diffuse irradiation. Results have been compared to the experimental data measured at the experimental facility located in Burgos, Spain.

Keywords: Global Vertical Irradiation, Models, Hay, Isotropic, Klucher, Circumsolar.

1 INTRODUCTION

For achieving the environmental objectives raised in the Horizon 2020, distributed electricity generation from renewable energy is essential. In this regard the potential of energy generation of BIPV facilities on actual and new buildings must be known [1, 2]. First step to achieve this goal is the knowledge of the solar resource on vertical façades in any orientation. As vertical measurement of solar irradiation is scarce, especially for different orientations to the South, mathematical procedures to estimate adequately this magnitude are necessary. Major applications of these procedures explain the existence in the literature of different and varied mathematical models for processing data of annual average of daily global horizontal radiation (GHI) to vertical (GVI) in any orientation.

Main objective of the work is the estimation of the potential of energy generation of BIPV facilities in any orientation. An experimental measurement system has been designed to measure Global Vertical Irradiation (GVI) on the four cardinal directions (North, South, West and East) and Global Beam and Diffuse irradiation on horizontal surface. Calculated and experimental values have been compared in order to determine the best model for each cardinal orientation in the area under study. The measurement campaign has extended throughout 2014. The standard statistical indicators (RMSE, MBE, MAE and MAPE) have been used to test the goodness of the empirical models.

2 THE MODELS

Models for processing GHI to GVI are many and varied. The goodness of the procedures is related to the atmospheric conditions and sky characteristics. Models are classified into parametric and non-parametric models. First ones are mathematical empiric relationship between different parameters that take into account the atmospheric conditions and sky characteristics. These models need the fit of the parametric through experimental values. No physical or geometric criteria are used to fit the non-parametric models and artificial neuronal networks, fuzzy-logic or fuzzy inference systems are used to implement the models.

Parametric models are traditionally classified as isotropic and anisotropic models. First ones assume that the diffuse component of the radiation is homogenous for the entire sky. Anisotropic models give heterogeneous behavior of the diffuse radiation in the circumsolar region or in the limit of the horizon. Liu-Jordan [3], Koronakis [4], or Tian[5] models are example of isotropic models. As anisotropic models can be cited Hay [6], Klucher [7] or Perez [8].

For this work, four models have been chosen to be applied covering all sky conditions: Circumsolar [9], Isotropic [9], Klucher [7] and Hay[6]. The four models are composition models, that means, the global radiation on a surface oriented an angle β, and tilted an angle α, $H_{\beta\gamma}$, results as sum of three components: beam $H_{b,\beta\gamma}$, diffuse $H_{d,\beta\gamma}$, and reflected $H_{r,\beta\gamma}$:

$$H_{\beta\gamma} = H_{b,\beta\gamma} + H_{d,\beta\gamma} + H_{r,\beta\gamma} \qquad (eq. 1)$$

The selected models differ in the description of the diffuse component

3 EXPERIMENTAL SECTION

The experimental data used to test the different mathematical models have been recorded at the experimental facility that the Research Group has implemented on the terrace of the Escuela Politécnica Superior, at Burgos University. This is a five floor building height and flat roof. No shadows or reflections from the surroundings can affect the measurements of the equipment located on the terrace because the horizon angle has lower elevation to 10 ° relative to the terrace.

The meteorological facility measures temperature, humidity, wind speed and direction, rainfall and atmospheric pressure. Hukseflux/SR11 Class 1 piranometers recorded global horizontal radiation, tilted, beam, and horizontal diffuse and reflected radiation (Figure 1). A specific structure allows the measurement of global vertical radiation in the directions Nord, South, East and West using PH.SCHENK 8101 Class 1 pyranometers (Figure 2). A meteorological station records data of temperature, rainfall, wind speed and direction, atmospheric pressure and global solar radiation (Figure 3). CR3000 Campbell Scientific datalogger is used to record the measurements. It is configured to measure 10 sec. data and to integrate values every ten minutes, 1 hour or 1 day, and record the results in different files. Figure 4 shows data of vertical global radiation in the four cardinal directions for a day in Burgos. In figure 4, data of global, diffuse and beam radiation can be seen

Figure 1: Experimental facility to measure global, diffuse and beam radiation.

Figure 2 Experimental facility to measure Global Vertical Irradiation located at Burgos University

Figure 3 Experimental meteorological facility located at Burgos University

All experimental data were filtered using the WMO directives [10] and recorded in files. For this work, daily values were used.

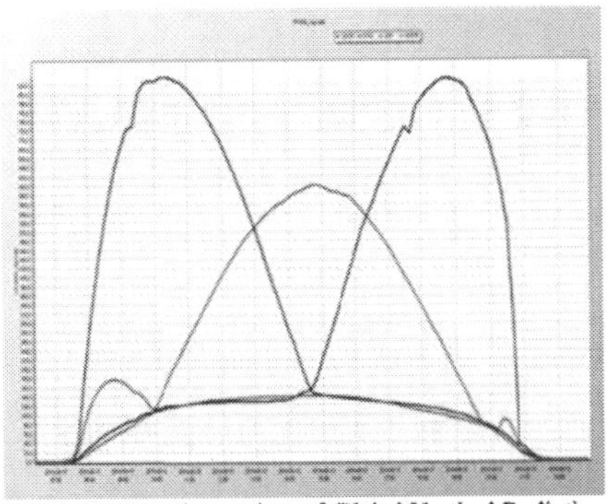

Figure 4: Ten minutes data of Global Vertical Radiation measured at the radiometric station of Burgos, Spain.

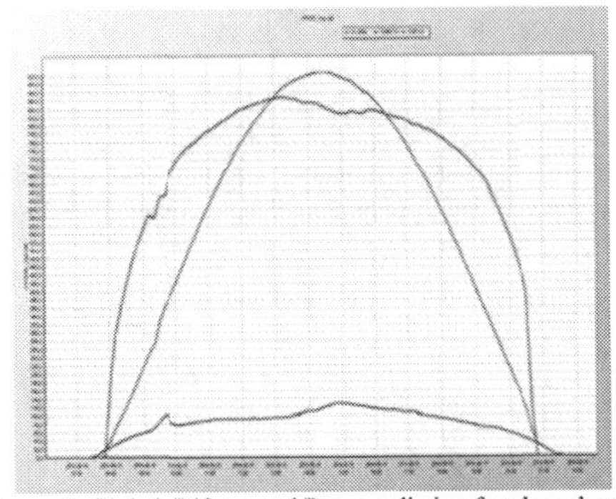

Figure 5: Global, Difusse and Beam radiation for clear sky day measured in Burgos, Spain

4 RESULTS

Daily values of Global Horizontal, Diffuse, Beam and Reflected irradiation have been calculated from the ten minutes values recorded at the experimental facility. In the same way, daily values of global vertical irradiation in the four cardinal directions have been calculated. Data of 2014 year have been used for the study. Values of Global vertical Irradiation (GVI) have been calculated from the Circumsolar (C), Isotropic (I), Klucher (K) and Hay (H) models and the results were compared to the experimental data.

Four standard statistical indicators have been used to classify the models: RMSE, MBE, MAE and MAPE (%) defined as follow:

$$RMSE = \sqrt{\frac{\sum(v_r - v_i)^2}{N}} \qquad RMSE\,\% = \frac{RMSE \times 100}{Average\,(v_r)} \qquad (Eq.\ 2)$$

$$MBE = \frac{1}{N}\sum (v_r - v_i) \qquad MBE\,\% = \frac{MBE \times 100}{Average\,(v_r)} \qquad (Eq.\ 3)$$

$$MAE = \frac{1}{N}\sum |v_r - v_i| \qquad MAE\,\% = \frac{MAE \times 100}{Average\,(v_r)} \qquad (Eq.\ 4.)$$

where v_r is the experimental value and v_i is the calculated value.

4.1. Results of GVI-North

A pyranometer north oriented only see the Sun few days a year. Then, data than the solar orbit is higher than 90º, taken the south as 0º are used for calculations in North direction (only 172 daily data). Annual averages of daily GVI (MJm^{-2} day^{-1}) have been calculated from each model and compared to the experimental values.

Table I shows the results of this comparison. As can be seen, Hay model shows the lowest values of RMSE and MAE and underestimates the annual GVI-North values. Isotropic model gives slightly higher error values and also underestimates the GVI-North calculations. Klucher model, that assumes clear sky conditions, adding cover conditions, gives lower values of the statistical parameters and Circumsolar model, defined exclusively for clear sky, overestimates largely GVI-North.

Table I Standard Statistical indicators (%) of the results

Model	Circumsolar	Isotropic	Klucher	Hay
RMSE (%)	63.80	27.66	34.80	18.31
MBE (%)	61.53	-24.49	-6.47	-13.49
MAE (%)	61.53	24.68	29.15	15.45

Results of statistical indicators of the monthly averages of daily GVI calculations (%) are shown in Figure 6. Hay model presents better behaviour for the central months of the year. Isotropic model is advisable in Mars and September and Circumsolar shows the worst results for all periods studied. This study can relate to the sky conditions in the city.

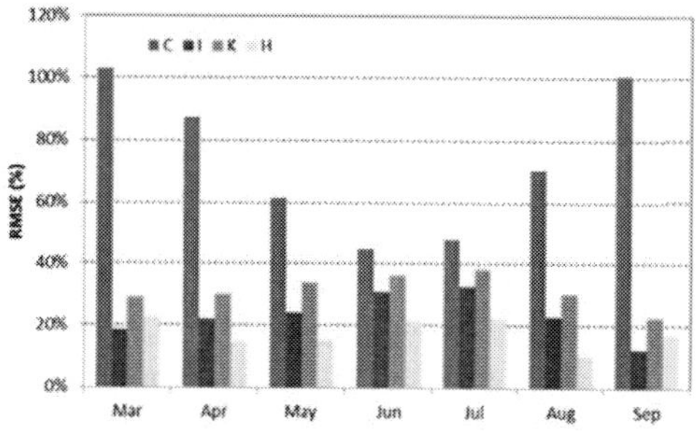

Figure 6: RMSE (%) of monthly averages of daily GVI (MJ m^{-2} day^{-1}) calculations from the four models studied.

Ranging values of daily GVI in regular intervals (Fig. 7), Circumsolar model presents the worst results for all values of GVI. Hay model fits well in all range of GVI.

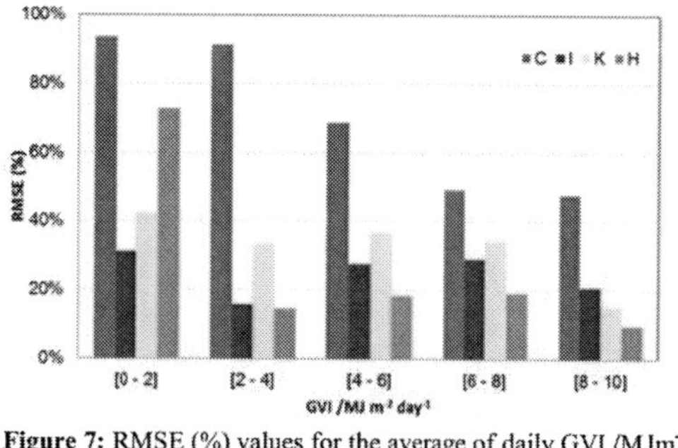

Figure 7: RMSE (%) values for the average of daily GVI /MJm^{-2}day^{-1} calculations from the four models studied, ranging daily GVI in regular intervals in the North orientation

4.2. Results of GVI-South

A vertical pyranometer South-oriented receives sunlight every day of the year. Only few days a year, in the corresponding hours of sunset and sunrise, the sensor remains in the dark. For this orientation, all models present high values of RMSE, comparable among them. When we study the monthly averages (Figure 8), MAE (%) values of all models decrease in summer time, showing the best result in August, when value of GVI is the highest of the year.

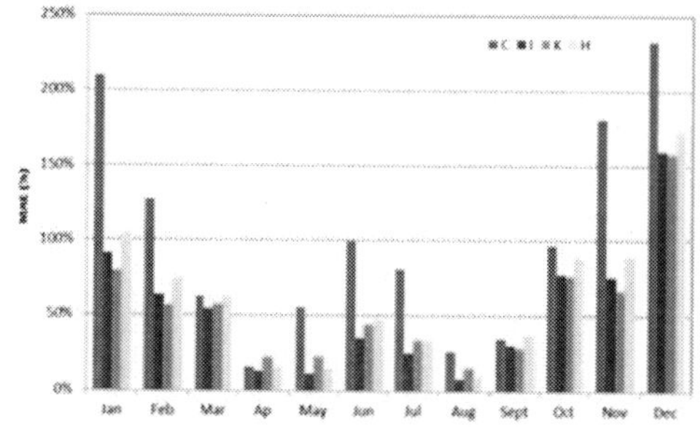

Figure 8: MAE (%) results of monthly averages of daily GVI calculations from the four models facing South.

4.3. Results of GVI-East

The azimuthal angle corresponding to this orientation is $\pi/2$, then, the sensor records sunlight only during half day, from the sunrise to midday. Values of GVI at the evening are very low and not representative. Values of RMSE (%) for the annual average of daily GVI are presented in Table 2. All models show similar results.

Table II RMSE (%) of the annual averages of daily GVI (MJ m^{-2} day $^{-1}$) calculated from the four models studied for the East orientation.

Model	Circumsolar	Isotropic	Klucher	Hay
RMSE (%)	57.13	66.13	68.22	71.51

Ranging the results into GVI intervals, values of RMSE (%) for the four models are very close for low values of GVI. When the irradiation increases, Circumsolar model shows the lowest values of RMSE and MAE, as can be seen in Figure 9.

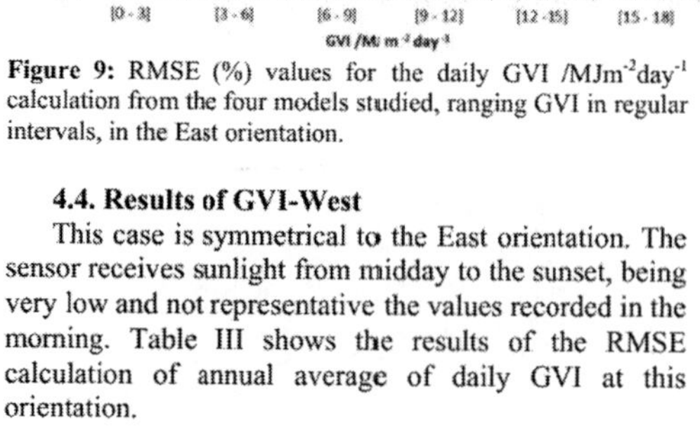

Figure 9: RMSE (%) values for the daily GVI /MJm^{-2}day^{-1} calculation from the four models studied, ranging GVI in regular intervals, in the East orientation.

4.4. Results of GVI-West

This case is symmetrical to the East orientation. The sensor receives sunlight from midday to the sunset, being very low and not representative the values recorded in the morning. Table III shows the results of the RMSE calculation of annual average of daily GVI at this orientation.

Table III RMSE (%) of the annual average of daily GVI (MJ m^{-2} day $^{-1}$) calculated from the four models studied in the West orientation.

Model	Circumsolar	Isotropic	Klucher	Hay
RMSE (%)	57.13	66.13	68.22	71.51

Ranging the results into regular intervals of daily GVI, Klucher model gives the lowest values of MAE for low values of GVI while Circumsolar model fits well when irradiation is high (clear sky situations) (Figure 10)

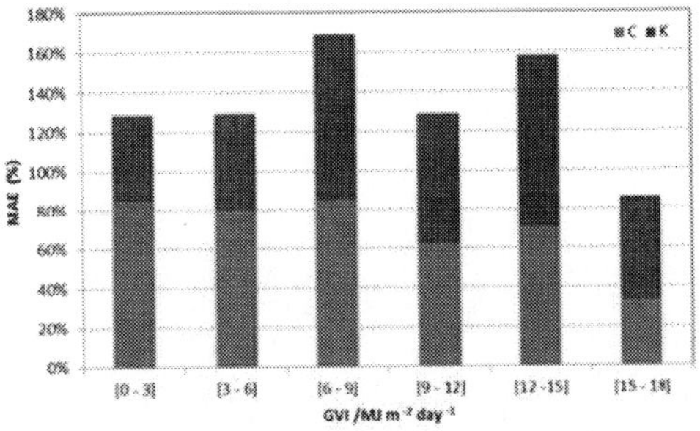

Figure 10: MAE (%) results of daily GVI (MJ m^{-2} day $^{-1}$) calculations from Klucher and Circumsolar model in the West orientation.

5 CONCLUSIONS

Annual and monthly daily average values of Global Irradiation on Vertical surfaces oriented in the four cardinal directions have been calculated through four classical empirical models using global, diffuse and direct radiation data on horizontal plane. Results have been compared to experimental data. Hay model for North orientation, Isotropic model for South oriented façades and Circumsolar model for East and West directions are the models that present the best results according to conventional statistical indicators.

Acknowledgements

This research received economic support from the Spanish Government (grant ENE2011-27511 and grant ENE2014-54601-R).

6 REFERENCES

[1] M. Utrillas, J. Martinez-Lozano, A. Casanovas. Evaluation of models for estimating solar irradiation on vertical surfaces at Valencia, Spain. Sol Energy. 47 (1991) 223-9.

[2] H.B. Tolabi, M.H. Moradi, S.B.M. Ayob. A review on classification and comparison of different models in solar radiation estimation. Int J Energy Res. 38 (2014) 689-701.

[3] B.Y.H. Liu, R.C. Jordan. The interrelationship and characteristic distribution of direct, diffuse and total solar radiation. Sol Energy. 4 (1960) 1-19.

[4] P.S. Koronakis. On the choice of the angle of tilt for south facing solar collectors in the Athens basin area. Sol Energy. 36 (1986) 217-25.

[5] Y.Q. Tian, R.J. Davies-Colley, P. Gong, B.W. Thorrold. Estimating solar radiation on slopes of arbitrary aspect. Agric For Meterol. 109 (2001) 67-74.

[6] J.E. Hay, J.A. Davies, Calculation of the solar radiation incident on an inclined surface, Proc of First Canadian Solar Radiation Data Workshop (Eds: JE Hay and TK Won), Ministry of Supply and Services Canada 1980

[7] T.M. Klucher. Evaluation of models to predict insolation on tilted surfaces. Sol Energy. 23 (1979) 111-4.

[8] R. Perez, P. Ineichen, R. Seals, J. Michalsky, R. Stewart. Modeling daylight availability and irradiance components from direct and global irradiance. Sol Energy. 44 (1990) 271-89.

[9] M. Iqbal. An introduction to solar radiation. Academic Press 1983.

[10] World Meteorological Organization (WMO), Technical Regulations, Volume I: General Meteorological Standards and Recommended Practices. 2012. http://library.wmo.int/pmb_ged/wmo_49-v1-2012_en.pdf

A NEW METHOD FOR THE BENCHMARKING OF IRRADIANCE PREDICTIONS

Antoine Guérin de Montgareuil [1], Laurent Bellemare [2], Rudy Blonbou [3], Mathieu David [4], Thierry Hedde [1], Philippe Lauret [4], Stéphane Mével [5], Jean-Philippe Morel [5], Philippe Poggi [6], Ted Soubdhan [3], Cyril Voyant [6]

[1] CEA (Atomic Energy and Renewable Energies Agency) - CEATech DPACA - Cadarache, bât. 356 - F-13108 St-Paul-lez-Durance, France - gdm@cea.fr - [2] Martinique Energy Agency - [3] Antilles-Guyane University, Large laboratory - [4] Reunion Island University, Piment laboratory - [5] Meteo France – Carpentras Radiometric Centre - [6] Corsica University

ABSTRACT: In the French overseas departments and in Corsica, insular territories, as well as in Provence, south of France, where the electric grid is weakly interconnected because of the proximity of the Mediterranean Sea and the Alps Mountains, a very high proportion of photovoltaic power will be introduced in the electric grids in the near future. In order to manage the balance between supply and demand of electricity, grid operators ask electricity producers to linearize their production and to announce in advance their energy production plans. The accuracy of these plans depends strongly on the accuracy of the prediction of the solar resource. The article proposes a new method for the benchmarking of irradiance predictions used in this context. It is shown that even for perfect energy predictions, the requirement of energy production plans brings potential costs to energy producers. The new benchmarking method compares the accuracy of the predictions with the accuracy of an ideal prediction made from the linearization of irradiance measurements and with the accuracy of the persistence prediction, which consists in using the irradiance of the current day as predicted irradiance for the next day.

Keywords: Irradiance prediction benchmarking, energy production plans, linearization of irradiance, persistence

1 INTRODUCTION

1.1 Context

The setup of high shares of photovoltaic energy in the French overseas departments and in Corsica is already limited by the ability of the grid operator to take into account the variability of the solar resource and the concomitant variability of electricity production. The problem will be the same in the near future in Provence, south of France, a sunny region where the electric grid is weakly interconnected because of the proximity of the Mediterranean Sea and the Alps Mountains.

1.2 Energy production plans and associated intrinsic financial losses

In order to limit the effects of the variability of the solar resource on the balance of supply and demand of electricity, the grid operator asks for the linearization of the photovoltaic energy production during predefined time intervals and electricity producers have to announce the linearized production for the next day.

When the electric production is lower than the planned production, penalties will be applied. When the available power exceeds the planned power, the power in excess will be lost.

Thus, since the electric power is almost proportional to irradiance and irradiance is not a linear function of time, even the best energy production plan will constantly bring excess or shortage of power and then financial losses. Furthermore, any energy production plan may be characterized by the corresponding wasted and missing energies.

Setting up energy production plans for the next day requires predictions of the solar resource for this day. Depending on the accuracy of irradiance predictions and on the algorithms used for the calculations, the energy production plans will bring more or less losses of income and penalties. Then, depending on the price of electricity, the cost of penalties and on the cost of any alternative solutions such as energy storage (which is not the topic of this article), it will possible to rate the irradiance predictions.

1.3 Present benchmarking of irradiance predictions

How "good" a prediction is depends strongly on the criteria used for the assessment of the prediction. Classical prediction methods [1, 2] forecast the average value of irradiance during predefined time intervals (typically one hour, 30 minutes, 15 minutes...). Each prediction is assigned to the end of the period for which the irradiance is predicted. For example, the prediction of irradiance from 08:00:00 AM (excluded) to 09:00:00 AM (included) is timed to 09:00:00.

The present prediction benchmarks are based on statistical quantities on the difference between the predicted and the measured averages of irradiance during the time intervals. In simple terms, there are 2 criteria for good predictions:

- An average value of the difference between predicted and measured irradiances as low as possible
- A dispersion of the difference around zero as little as possible

These benchmarks look convenient for the power exchange where there is no real interest on the way electricity is produced. This article focuses more on the benchmarking of prediction of irradiance for energy production plans.

1.4 Energy production plans and benchmarking of irradiance predictions

Despite this clear dependence of the rating of predictions on every user's context, this article proposes a simple method for the benchmarking of irradiance predictions for energy production plans, avoiding the use of complex algorithms and based on the following general assumptions:

- The production of electricity is proportional to irradiance
- Thus, t optimization of energy production plans consists in finding the optimal linearized irradiance predictions for the same time intervals
- the optimal linearized irradiance prediction must maximize the energy production, minimize the financial losses and the technical complexity at the electricity production level

1.5 Origin of data

In this article, all irradiance data come from 2 years of measurements of global horizontal irradiance by a horizontal pyranometer at Saint-Pierre, Reunion Island. Predictions have been made by the Piment Laboratory according to [3] and have been provided "as is" to illustrate the benchmarking method.

1.6 Clear sky irradiance as reference

When benchmarking predictions, as irradiance conditions vary a lot between sites and, for a same site, between time of year and seasons, it is interesting to normalize the results with the clear sky irradiance, the irradiance which would be measured with a pure atmosphere.

In this article, the clear sky irradiance has been modelled from the irradiances measured in the morning of a day with no clouds. It has been expressed as a third order polynomial function of the inverse of air-mass [4], calculated for all the days and verified for days with no clouds.

2 LINEARIZATION OF IRRADIANCE

2.1 Definitions

From one day of irradiance measurements, the available irradiation A is defined by Formula 1:

$$A = \sum_i G(t_i)\,.\,\Delta t \qquad (1)$$

where $G(t_i)$ is the average irradiance during the ith interval of Δt duration. If Δt is expressed in hours, the irradiation A is expressed in Wh/m^2.

In Figure 1, the available irradiation is represented by the zone in pink. The considered day is characteristic of the climate of St-Pierre. During the morning of this typical day, there are no clouds. Around 8:00 UT, irradiance increases and becomes higher than the clear sky irradiance because of reflection of sunlight on clouds and probably on walls surrounding the pyranometer [5]. Then, big clouds come, hide the direct irradiance and the diffuse irradiance goes down regularly until sunset.

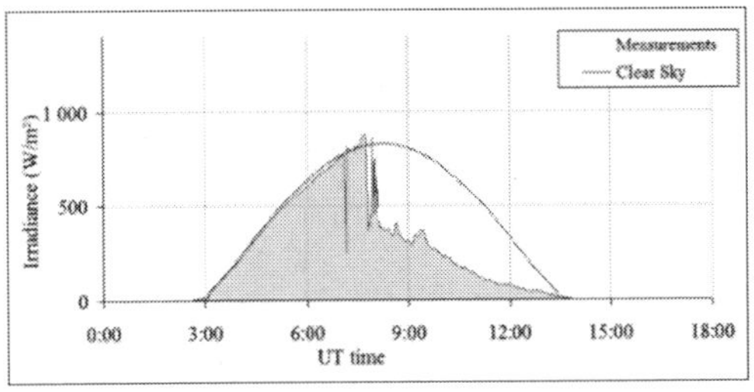

Figure 1: Horizontal irradiance measured at St Pierre, Reunion Island, 6 May 2013.

A linearization of irradiance (or shortly "linearization") is defined by the set of values of irradiance attributed to the middle of the considered time intervals (for example 00:30, 01:30, 02:30... for hourly time intervals). Values for other moments are calculated by linear interpolation of the previous set of values.

The linearized irradiation L is defined by Formula 2:

$$L = \sum_i G_{linearized}(t_i)\,.\,\Delta t \qquad (2)$$

The "wasted" irradiation W is defined by Formula 3:

$$W = \sum_i \max\left(G(t_i) - G_{linearized}(t_i), 0\right).\,\Delta t \qquad (3)$$

This quantity is calculated from the intervals where the measured irradiance exceeds the linearized irradiance (the excess of irradiance is not used to produce electricity and is "wasted"). In Figure 2 and following, the wasted irradiation is represented by the zones in green (the blue zones are described below).

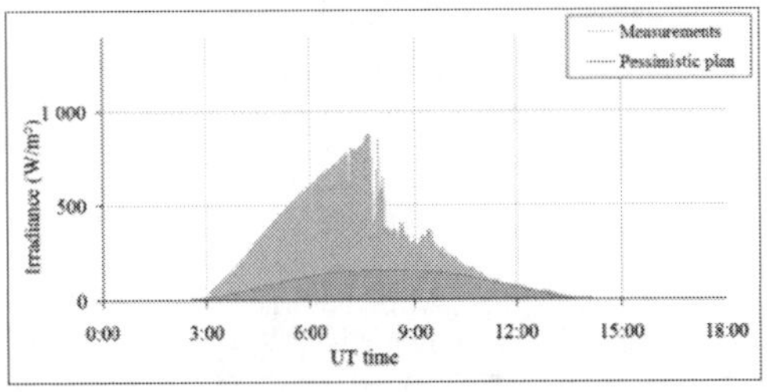

Figure 2: Insufficient use of available irradiance by a low linearization representing a theoretical overcast day

The "missing" irradiation M is defined by Formula 4:

$$M = \sum_i \max\left(G_{linearized}(t_i) - G(t_i), 0\right).\,\Delta t \qquad (4)$$

This quantity is calculated from the intervals where the linearized irradiance is greater than the measured irradiance (there is not enough irradiance to produce the announced electricity generation and the difference of irradiance is "missing"). In Figure 3 and following, the missing irradiation is represented by the zones in yellow.

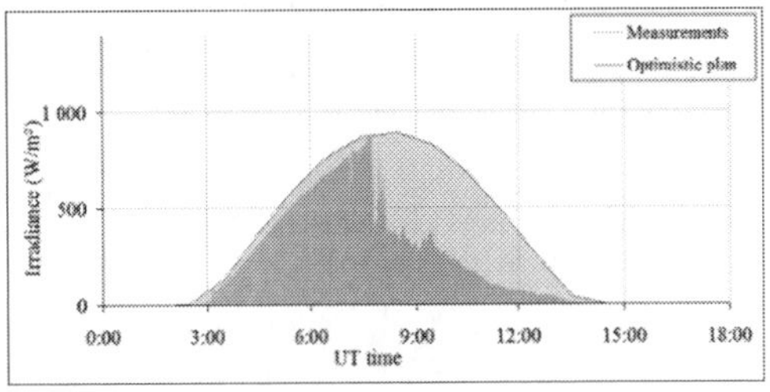

Figure 3: Linearization representing a theoretical clear sky day and constantly exceeding the existing irradiance

The "used" irradiation U is defined by Formula 5:

$$U = \sum_i \min\left(G(t_i), G_{linearized}(t_i)\right).\,\Delta t \qquad (5)$$

This quantity is calculated from the available irradiance when it is smaller than the linearized irradiance and from the linearized irradiance in the opposite case. In Figure 2 and following, the used irradiation corresponds to zones in blue.

The "bias" irradiation B is defined by Formula 6:

$$B = L - A \qquad (6)$$

The bias irradiation the difference between available and linearized irradiations. It is null if the linearized irradiance brings as much irradiation as the existing irradiance.

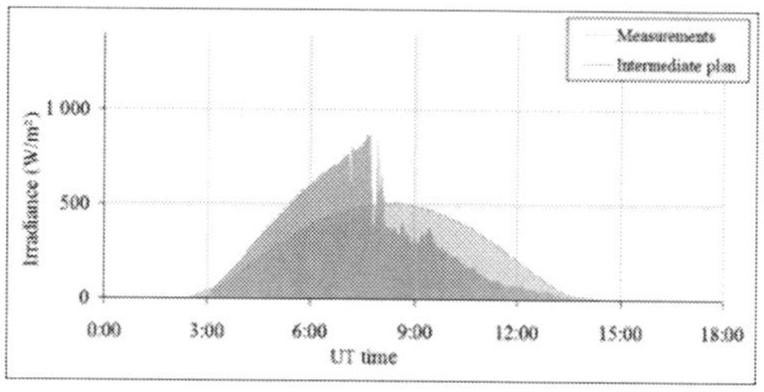

Figure 4: Linearization of irradiance with used, wasted and missing irradiation.

The linearization in Figure 4 has been designed so that the wasted and missing irradiations are equals. The bias irradiation of Figure 4 is null.

The "shifted" irradiation S is defined by Formula 7:

$$S = M + W \quad (7)$$

It represents the total of the irradiations added and subtracted to the available irradiation to get the linearized irradiation. It represents the surface between the measured irradiance and linearized irradiance curves.

Using Figure 4, it is easy to infer Formulae 8 and 9 and to express the relations between the available irradiation A, the linearized irradiation L, the used irradiation U, the wasted irradiation M (positive by definition) and the missing irradiation W (positive by definition).

$$U = A - W \quad (8)$$

The used irradiation is the available irradiation minus the wasted irradiation.

$$U = L - M \quad (9)$$

The used irradiation is also the linearized irradiation minus the missing irradiation

Finally, derived from Formulae 8 and 9, Formula 10 expresses the bias irradiation B as the difference between missing and wasted irradiations.

$$B = M - W \quad (10)$$

The definition of the bias irradiation is coherent with the usual meaning of a bias. When there is more missing irradiation than wasted irradiation, the linearized irradiation is lower than the available irradiation and the bias is negative (see Figure 2). It is positive in the opposite case (see Figure 3).

Then, mathematics (intermediate value theorem) tell us that, for any existing irradiance dataset, it is possible to define at least one linearization of irradiance with a null bias. An example can be seen in Figure 4 where the linearization has been designed so that the total surface of the green and the yellow zones are exactly the same.

All linearizations with a null bias may differ by the value of their missing and wasted irradiations, and then by their shifted irradiation. In Figure 4, the wasted irradiation is limited to the morning and the missing irradiation to the afternoon.

We feel clearly that this intermediate irradiance linearization is not as "satisfactory" as a combination of the linearization in the right part of Figure 2 and the linearization in left part of Figure 3. Such a linearization would minimize *locally* the bias between the measured irradiance and the linearized irradiances.

2.2 Benchmarking of linearizations

Given one measured irradiance dataset and one linearized irradiance dataset, the benchmarking of the linearized irradiance dataset is based on the calculation of 4 quantities:

- The wasted irradiation and the missing irradiation, as defined by Formulae 3 and 4
- the difference between these quantities (the bias irradiation defined by Formula 6)
- the sum of the wasted and missing irradiations (the shifted irradiation defined by Formula 7)

The benchmarking criteria are the following:

- If 2 linearized irradiance datasets have the same bias irradiation, the best one is the linearization with the lowest shifted irradiation
- If 2 linearized irradiance datasets have the same shifted irradiation, the best one is the linearization with the lowest bias.

The following chapter proposes a very simple algorithm to find for any measured irradiance dataset the almost optimal linearized irradiance datasets responding to these criteria.

2.3 Algorithm for the calculation of an optimal linearization

The idea of the algorithm is to keep low the difference between the irradiances and their linearization. The optimal linearization of an irradiance dataset is characterized by the set of the average values of the irradiance for every interval (these values are attributed to the middle of the intervals and the values for other moments are calculated by their linear interpolation).

Figure 5 shows the application of the algorithm to the 6th of May, 2013.

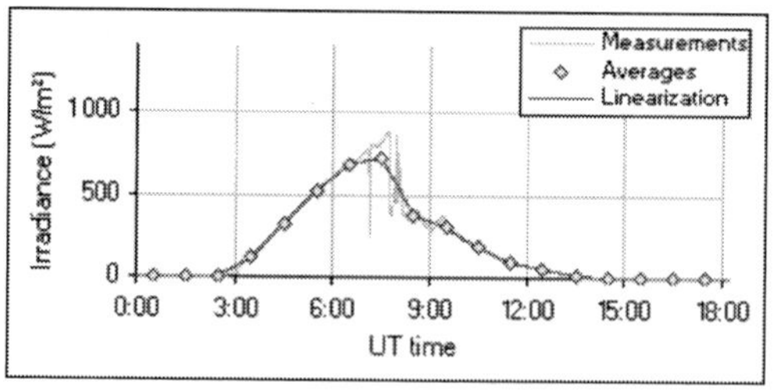

Figure 5: Algorithm for the linearization of the irradiance with null bias and minimum shifted irradiation

It can be easily demonstrated that for that algorithm, wasted irradiation equal strictly missing irradiation at the day level. So, the daily bias is null and remains low at the hourly level. It has been also verified that for most datasets, this linearization minimizes the shifted irradiance.

Used, wasted and missing irradiations are shown in Figure 6. Missing and wasted irradiations are equal and their sum, corresponding to financial losses at the production level, is almost minimum.

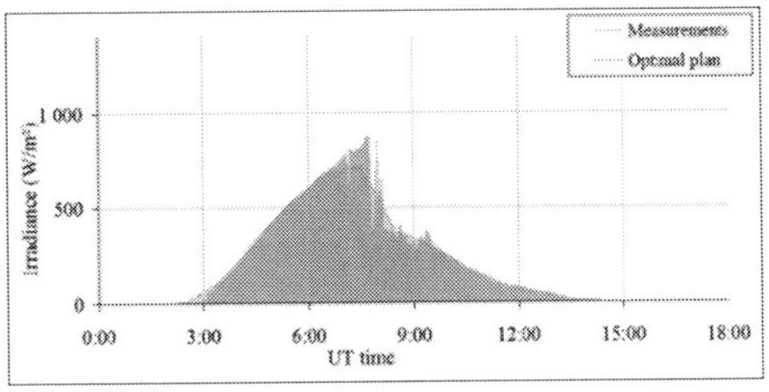

Figure 6: Used irradiation is maximum

3 BENCHMARKING GRAPH

This graph (Figure 7) allows the presentation of the missing irradiation as a function of the wasted irradiation.

It is interesting to draw the lines where the sum of these 2 quantities (the shifted irradiation) is constant (light green lines) and the lines where their difference (the bias irradiation) is constant (light blue lines).

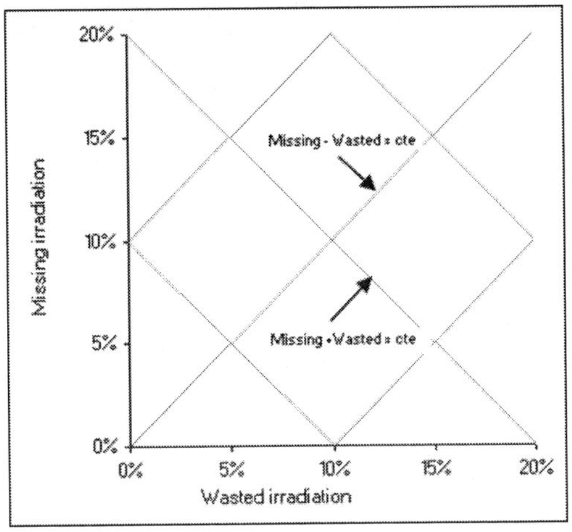

Figure 7: Graph of wasted and missing irradiations between an irradiance and a linearized irradiance

All irradiances during the 2-year period have been linearized with the simple linearization algorithm. Wasted and missing irradiations have been calculated. Figure 8 shows that for all these linearizations with a null bias, the value of the shifted irradiation varies between a maximum, around 2% and a maximum around 30%.

This value depends on the dispersion of irradiance around its linearization.

Days with many passing clouds such as the 2nd of June, 2013, present high shifted irradiations (Figure 9). For this day, the shifted irradiation represents 30 % of the irradiation of a clear sky day and 52 % of the available irradiation.

At the opposite, days without almost no clouds such as the 24th of December, 2012 (Figure 10), present a very limited difference between the irradiance and its linearization.

So, the shifted irradiation depends directly on the intrinsic dispersion of the irradiance around its linearization.

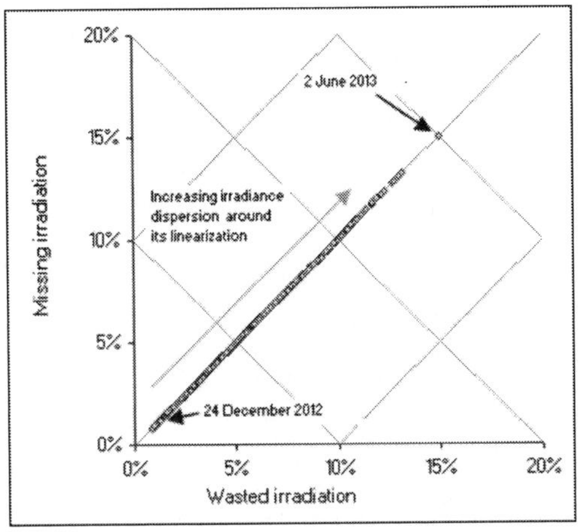

Figure 8: Daily wasted and missing irradiations for the linearization of 2 years of data

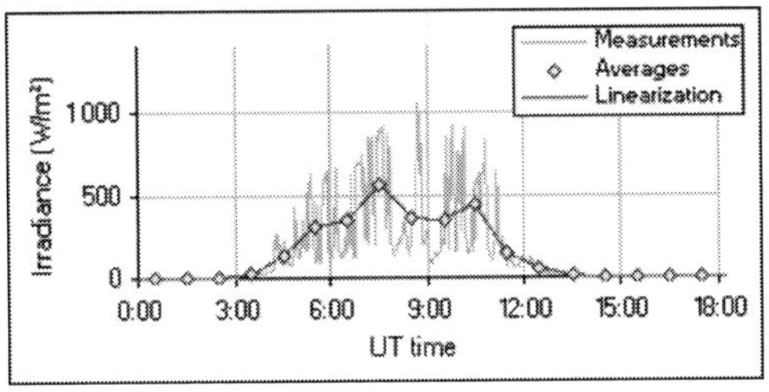

Figure 9: Very high dispersion of irradiance around its linearization, 2nd June 2012

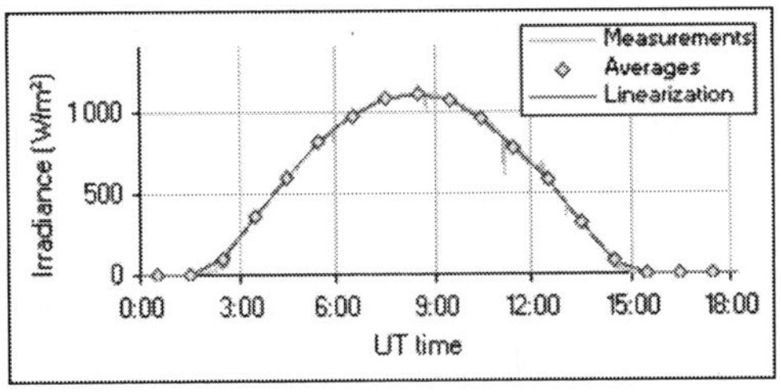

Figure 10: Very low dispersion of irradiance around its linearization, 24th December 2012

The next chapter shows the result of the application of the benchmarking of linearized irradiance to irradiance predictions.

4 BENCHMARKING OF IRRADIANCE PREDICTIONS

4.1 Algorithm for the linearization of irradiance predictions

The values attributed to the middle of the intervals are the values of the prediction of the average irradiance for every time interval.

4.2 Benchmarking of predictions based on persistence

In meteorology, the simplest prediction of weather is the persistence method which consists in saying that the weather remains identical from one day to the next day. It is well known that it is not possible to predict with accuracy the weather more than 10 days ahead with prediction methods such as observation of clouds from earth or satellites, or atmospheric calculations from physical modelling - numerical weather prediction. So, for mid and long term predictions as well as for periods where the type of weather is not changing or changing slowly, persistence predictions are meaningful.

The determination of the predicted irradiance with persistence consists in attributing to every time interval of the next day the average value of irradiance for the same time interval during the current day, corrected by the ratio between the calculated clear sky irradiations for the considered intervals.

Figure 11 shows the irradiance of a very interesting day for the benchmarking of predictions, the 8th of February, 2012, with very big clouds arriving around noon. The irradiance of the previous day (Figure 12) was less variable.

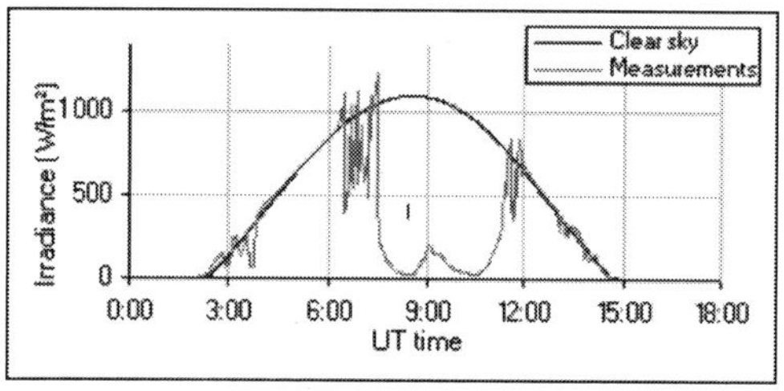

Figure 11: Horizontal irradiance measured at St Pierre, Reunion Island, 8 February 2012

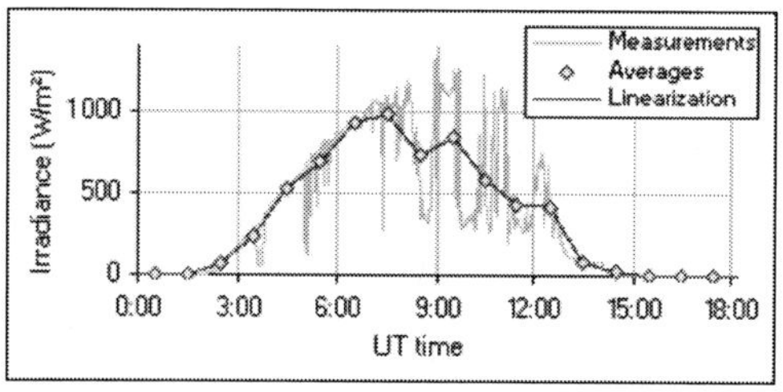

Figure 12: Horizontal irradiance measured at St Pierre, Reunion Island, 7 February 2012

Figure 13 presents in the same graph the measured irradiance, its linearization and the persistence prediction.

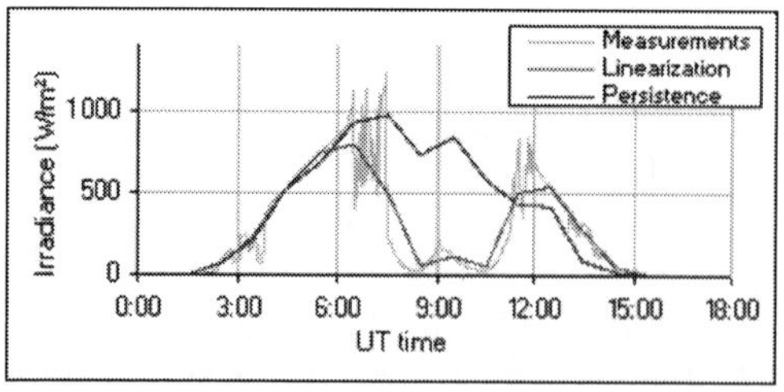

Figure 13: 8 February 2012, measured irradiance and its linearization (pink) considered as the optimal prediction, and linearized irradiance of the previous day corrected by clear sky irradiation, seen as the persistence prediction

The benchmarking of the persistence prediction for the 2 years of data is presented in Figure 14. Results vary strongly, depending simultaneously on the intrinsic dispersion of irradiance around its best possible linearization and on prediction errors.

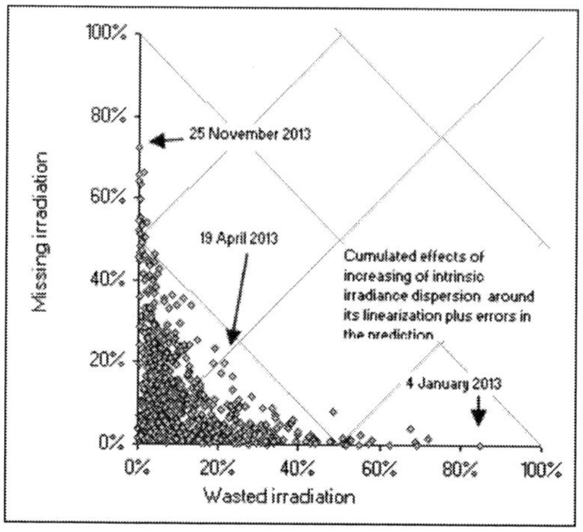

Figure 14: Daily wasted and missing irradiations for the linearization of persistence prediction for 2 years of data

Some days like the 4th of January, 2013, (Figure 15) are a lot sunnier than the previous day and persistence brings wasted irradiation.

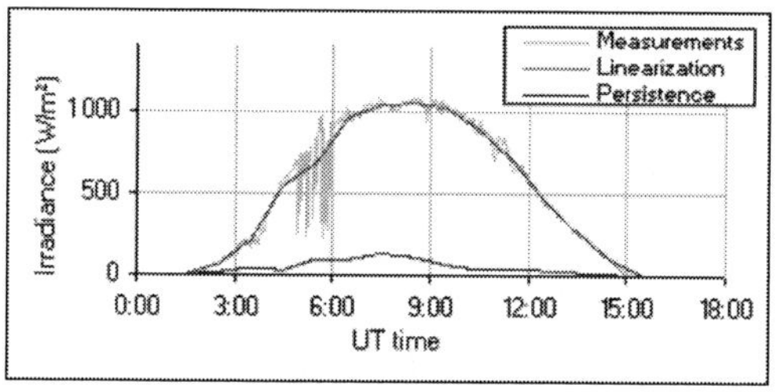

Figure 15: 4 January 2013, stable irradiance but very bad persistence prediction

Other days like the 25th of November, 2013, are in the opposite case (Figure 16). They are a lot less sunny than the previous day and a lot of missing irradiation.

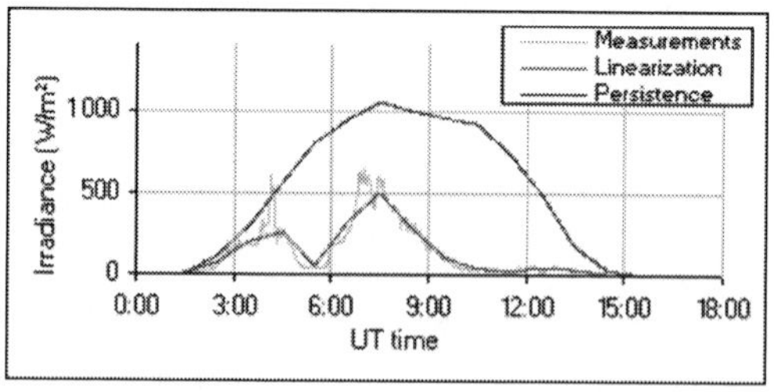

Figure 16: 25 November 2013, stable irradiance but very bad persistence prediction for the opposite reason

The 19th of April, 2013, is a good example of day with almost no difference between wasted and missing irradiations (a small bias irradiation), but with a high shifted irradiation between morning and afternoon (Figure 17).

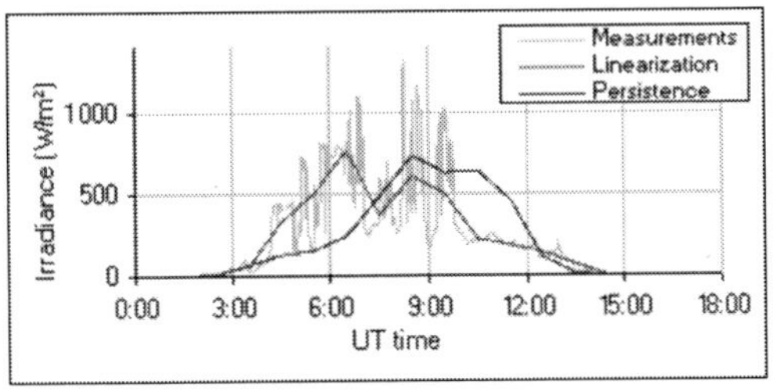

Figure 17: 19 April 2013, persistence prediction characterized by an underestimation of irradiance in the morning and an overestimation in the afternoon

4.3 Benchmarking of other predictions

Predictions are linearized as described above. Figure 18 and Figure 19 present the predictions for the 19th of April, 2013. Clearly, the 6-hour prediction overestimates the irradiance, while the 1-hour prediction is more accurate.

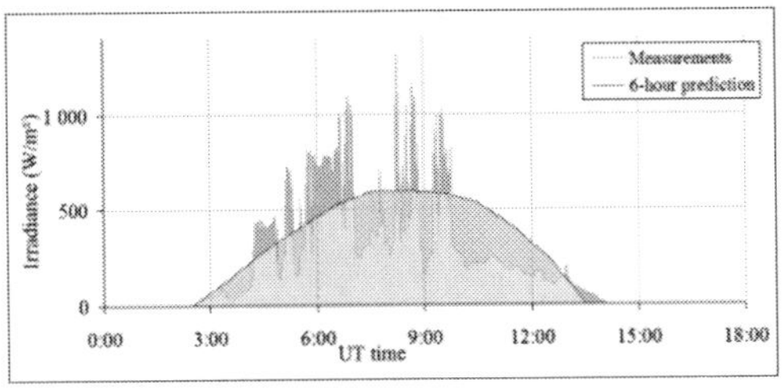

Figure 18: 19 April 2013, linearization of irradiance coming from its prediction 6 hours in advance

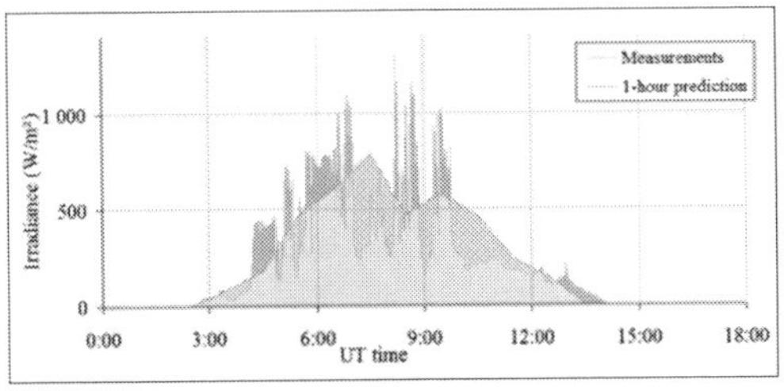

Figure 19: 19 April 2013, linearization of irradiance coming from its prediction 1 hour in advance

Figure 20 presents the results of the benchmarking of persistence and of the 6-hour and 1-hour predictions for the 19th of April, 2013, in the same graph as the benchmarking of the linearization of measured irradiance, which may be considered as the best possible prediction. The figures presented in the graph are the ratio between the wasted and missing irradiations and the clear sky irradiation.

The shifted irradiation for the linearization of the measured irradiance is around 40%, because this irradiance is highly variable. It is around 82% for persistence, because of its inaccuracy and of the intrinsic variability of irradiance. It is 68% for the 6-hour prediction and 60% for the 1-hour prediction.

The bias is null for the linearization and low for the persistence which overestimate by 2% the irradiance. The 6-hour prediction overestimates the irradiation by 8% and the 1-hour prediction is better with an overestimation of 5%. If the main criterion is the total bias of irradiation, persistence is a better prediction than the 6-hour and 1-hour predictions. None of the predictions approaches the performance of the linearization of the measurements considered as the optimal prediction.

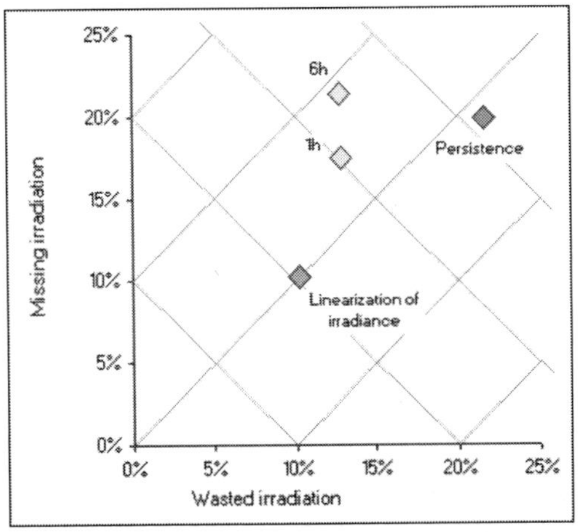

Figure 20: 19 April 2013, benchmarking of predictions and of linearized irradiance as ideal prediction.

4.4 Benchmarking of the irradiance prediction for 2 years of data

Figure 21 presents the result of the benchmarking of all the irradiance prediction for 2 years of data. The presented figures are the ratio between the total wasted and missing irradiations and the 2-year clear sky irradiation.

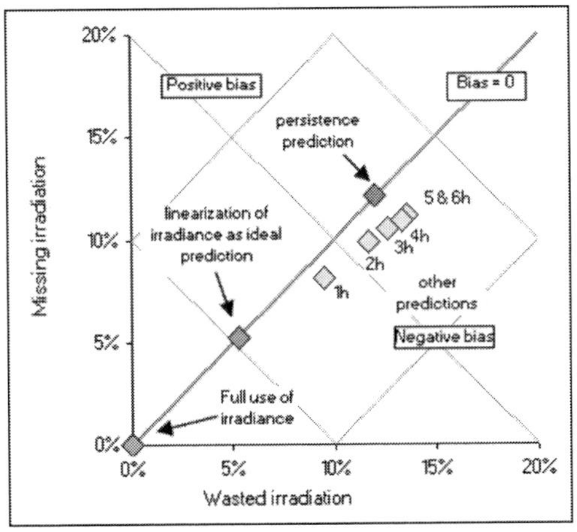

Figure 21: Benchmarking of predictions and of linearized irradiance as ideal prediction for 2 years of irradiance.

Synthetic conclusions may be drawn from this benchmarking:

- Linearization of irradiance, which is very near to the ideal prediction (null bias, minimum shifted irradiation) presents a shifted irradiation around 10.5 %, inherent to the variability of irradiance at this site. It is the intrinsic lower limit for the accuracy of predictions
- The persistence prediction has a remarkably low bias (-7‰) and a limited shifted irradiation around 24 %

- All predictions have a negative bias from -2.5% to -1.4% and underestimate the irradiance.
- The performance of the 6-hour to 3-hour predictions is very close to the performance of the persistence prediction, when considering the whole period. For long term prediction, persistence is better with its null bias
- 2-hour and 1-hour predictions are better. The 1-hour prediction has made half of the way between the persistence prediction and the ideal prediction. Once corrected of its bias error, it could be a real improvement of persistence prediction

5 CONCLUSION

In the French overseas departments and in Corsica, the grid operator asks for the linearization of production and for its announcement, which requires the best possible irradiance predictions.

This article proposes a new method for the benchmarking of the predictions. The idea of the benchmarking is to put in perspective the performance of the predictions by comparing them to an ideal prediction done with the perfect knowledge of the irradiance measurements and to the most simple and conservative prediction, the persistence prediction using the irradiance of the day as prediction for the next day.

Several new quantities have been defined for the benchmarking. They represent the overestimation and underestimation of irradiation between the real irradiance and its linearized predictions.

It can be shown that the linearization of production leads to intrinsic production losses and to potential complexity of electricity management for the electricity producers. The losses and the complexity depend partly on the intrinsic variability of the irradiance at the considered site and partly on the accuracy of the prediction methods.

The simultaneous benchmarking of the predictions with the persistence prediction and with the ideal prediction shows more clearly if improvements are necessary or if the present predictions are enough accurate.

6 REFERENCES

[1] Coimbra C. F. M., Kleissl J., Marquez R., Chapter 8 - Overview of Solar-Forecasting Methods and a Metric for Accuracy Evaluation Solar Energy Forecasting and Resource Assessment. Boston: Academic Press, 2013. p. 171-94.

[2] David M., Diagne H. M., Lauret P., Outputs and error indicators for solar forecasting models, *WREF 2012 - World Renewable Energy Forum*, Denver, United States, 2012.

[3] Lauret P., Lorenz E., David M., Solar Forecasting in a Challenging Insular Context. *Atmosphere* 2016.

[4] Guérin de Montgareuil A., Modelling of a clear sky irradiance with a third order polynomial function of the inverse of air-mass, Personal communication, 2016

[5] David M., Over irradiance explained by reflection of light on walls, Personal communication, 2016

EVALUATING A MODEL TO ESTIMATE GHI, DNI, & DHI FROM POA IRRADIANCE

Michael Gostein[1], Bill Stueve[1], Kendra Passow[2], Bill Marion[3]
[1]Atonometrics, 8900 Shoal Creek Blvd. Suite 116, Austin, TX 78757, USA
[2]First Solar, 135 Main Street, San Francisco, CA 94105, USA
[3]National Renewable Energy Laboratory, 15013 Denver West Parkway, Golden, CO 80401, USA

ABSTRACT: We evaluate the GTI-DIRINT model for estimating direct normal irradiance (DNI), diffuse horizontal irradiance (DHI), and global horizontal irradiance (GHI) from measured plane-of-array (POA) irradiance. Our motivation is to explore the potential for limiting ground-based irradiance measurements to POA to reduce instrumentation costs. We test the GTI-DIRINT model using meteorological data collected at seven First Solar PV power plants, comparing the GTI-DIRINT model, which uses POA as input, to the DIRINT model, which uses GHI as input. Furthermore, we include translation between reference module and thermopile pyranometer measurements to demonstrate the ability to use PV module POA performance data as a model input. The results indicate that the GTI-DIRINT model performs at least equivalently to the DIRINT model for fixed-tilt arrays and may have superior performance for large tilt angles encountered in single-axis tracking arrays.

Keywords: irradiance, measurement, decomposition model, system performance

1 INTRODUCTION

Solar resource assessments are essential to both pre-construction performance prediction and post-construction system validation, and may be performed with a combination of ground-based and satellite-derived weather data. Optimally, interrelating satellite and ground-based weather data with PV system performance requires knowledge of four different quantities: global horizontal irradiance (GHI), direct normal irradiance (DNI), diffuse horizontal irradiance (DHI), and plane-of-array (POA) irradiance. High-accuracy instrumentation to measure all four quantities can be deployed at meteorological stations for utility-scale projects. However, the expense may not be justified at all projects, particularly smaller power plants.

A number of models are in use to estimate one or more of the four irradiance quantities from the other quantities, allowing a reduction in the number of deployed sensors and therefore a reduction in instrumentation cost. One approach is to measure GHI only and to estimate DNI, DHI, and POA from the GHI measurements using decomposition and transposition models. However, this paper addresses an alternative approach: measuring only POA and using this to estimate the remaining three quantities. A potential advantage of deriving all quantities from POA measurement is that POA can be conveniently measured using a relatively low cost reference cell or reference module. This allows automatic spectral correction and also potential dual-purpose use for soiling measurement instrumentation which employs POA PV reference devices [1].

One of the authors, Marion, has recently developed the GTI-DIRINT model [2], [3] for estimating DNI and DHI from global tilted irradiance (GTI), which is equivalent to POA irradiance. Since GHI can be directly calculated from DNI and DHI, this model allows determination of all four quantities.

In this paper, we evaluate the GTI-DIRINT model which uses POA irradiance to determine the other three irradiance quantities, comparing its accuracy to the DIRINT model which uses GHI to determine the other three quantities. We perform this evaluation by testing both models with data collected from meteorological stations at seven First Solar PV power plants at locations in the southwestern U.S., Canada, and the Arabian Peninsula.

2 MODEL

The GTI-DIRINT model [2], [3] is a modification of the DIRINT model developed by Perez et al. which determines DNI and DHI from GHI [4]. In GTI-DIRINT, the DIRINT model is modified to use POA irradiance (or GTI, equivalently) in place of GHI [2], [3]. While the DIRINT model uses a global clearness index calculated from GHI, in GTI-DIRINT this is replaced with a global clearness index calculated from POA. Subsequently, the GTI-DIRINT model uses the same binning method and fit coefficients as DIRINT. However, GTI-DIRINT also includes an iteration procedure to ensure that the model outputs agree with the input data. In addition, the model includes special treatment for irradiance with angles of incidence on the plane-of-array which are greater than 90 degrees, a condition for which DIRINT has no provision since DIRINT uses GHI as input.

We have performed our analysis using the MATLAB version of the Sandia National Laboratories PV_LIB Toolbox, which includes an implementation of DIRINT as well as supporting functions. To implement GTI-DIRINT, we started from the DIRINT code in PV_LIB and modified it as needed per the description in [3]. However, in our implementation we also added spectral correction and inclusion of a step for incidence angle modifiers in the GTI-DIRINT iteration loop in order to use reference module data as input, as will be described further below.

3 EVALUATION DATA

For this work we have examined meteorological data collected at the seven First Solar PV power plants summarized in Table I, which include five fixed-tilt and two single-axis tracking arrays. The data from each site included secondary standard thermopile pyranometer measurements of both GHI and POA irradiance. In addition each site included pairs of First Solar CdTe reference modules used for POA soiling measurement [1]. The short-circuit current and back surface temperature of the modules are measured and used to compute irradiance. One of these modules is routinely cleaned, and we use the irradiance value from this clean module as a POA measurement. The GHI and POA thermopile pyranometers were also routinely cleaned. At the single-axis tracking sites all the POA instruments were installed

Table I: Summary of data sources

Site	Location	Array Type	Tilt (°)	Az. (°)	Years of Data
A	Southwest US	Single-Axis Tracking	-	-	2.8
B	Southwest US	Single-Axis Tracking	-	-	1.5
C	Southwest US	Fixed Tilt	25.0	180	2.7
D	Southwest US	Fixed Tilt	22.5	180	1.5
E	Southwest US	Fixed Tilt	22.5	190	1.7
F	Arabian Peninsula	Fixed Tilt	20.0	180	1.9
G	Canada	Fixed Tilt	25.0	180	3.3

on the trackers with the PV array. In addition to the above, the five fixed-tilt sites also included rotating shadowband radiometers for measurement of DNI and DHI. All sites included measurements of ambient temperature and humidity which were used as supporting inputs for the models. From each power plant we selected one instrument of each available type to use for the correlation analysis, choosing those instruments located in the lowest-numbered blocks of each power plant for consistency.

The available data from the sites spanned time periods ranging from 1.5 to 3.3 years. To perform the analysis, we aggregated the readings of each instrument into hourly averages. For each hour we used the midpoint of the hour as the timestamp, except for hours that included sunrise and sunset, for which we used the midpoint of the daylight portion of the hour as the timestamp, following [3].

4 CORRELATION EQUATIONS

We evaluated our correlation results using the mean bias deviation (MBD) and root-mean-square deviation (RMSD) statistics, expressed both in physical units and as a percent of the mean. These are calculated according to the following expressions from [3]:

$$\mathrm{MBD} = \frac{1}{n}\cdot\sum(y_i - x_i) \qquad (\mathrm{W/m^2}) \quad (1)$$

$$\mathrm{MBD} = 100\cdot\left[\frac{1}{n}\cdot\sum(y_i - x_i)\right]/\left[\frac{1}{n}\cdot\sum x_i\right] \qquad (\%) \quad (2)$$

$$\mathrm{RMSD} = \left[\frac{1}{n}\cdot\sum(y_i - x_i)^2\right]^{1/2} \qquad (\mathrm{W/m^2}) \quad (3)$$

$$\mathrm{RMSD} = 100\cdot\left[\frac{1}{n}\cdot\sum(y_i - x_i)^2\right]^{1/2}/\left[\frac{1}{n}\cdot\sum x_i\right] \qquad (\%) \quad (4)$$

Here x_i and y_i are the ith values in the x and y series that are to be correlated and n is the number of pairs of values in the series.

5. REFERENCE MODULE VS. PYRANOMETER MEASUREMENTS

Our data set includes POA irradiance measurements from both thermopile pyranometers and PV reference modules. The pyranometers measure broadband irradiance with near cosine angular response, while the reference modules have a reduced spectral and angular sensitivity matching that of the PV array. For this work we wished to explore how POA and GHI measurements could be interrelated for the combined purposes of matching historical or satellite-derived irradiance data as well as predicting PV array performance. Therefore we used the thermopile pyranometer measurements for GHI but the PV reference module measurements for POA irradiance; the POA pyranometer was additionally used as a check and a calibration reference, although it was not used as a model input. Our modeling goal is to determine how well we can predict the reference module POA data from the pyranometer GHI data (using DIRINT) and vice-versa (using GTI-DIRINT).

Before examining the performance of the models, we first established the baseline degree of correlation between the measurement instruments. Table II illustrates the RMSD of the correlation in measurements between the POA reference module, the POA pyranometer, and the GHI pyranometer. The measurements used were drawn from the two sites using single-axis tracking and only for hours near noon when the tracker tables were nearly horizontal, so that the POA reference module and pyranometer could be compared not only with each other but also with the GHI pyranometer. A spectral correction factor [6] and an incidence angle modifier factor were applied to the reference module data, using functions included in PV_LIB.

Table II shows that the two pyranometers correlate extremely well with each other, while the correlation between the reference module and either of the

Table II: Correlation of irradiance measurements when plane of array is nearly horizontal

Site	Type	GHI Pyranometer *vs.* POA Pyranometer: MBD %	MBD (W/m²)	RMSD %	RMSD (W/m²)	POA Pyranometer *vs.* POA Ref. Module: MBD %	MBD (W/m²)	RMSD %	RMSD (W/m²)	POA Ref. Module *vs.* GHI Pyranometer: MBD %	MBD (W/m²)	RMSD %	RMSD (W/m²)
A	Single-Axis Tracking	0.7	5.4	1.6	12.7	-0.8	-6.6	4.3	34.3	0.0	0.0	4.3	34.2
B	Single-Axis Tracking	1.4	11.6	1.7	14.2	-1.3	-11.3	2.9	24.5	0.0	0.0	2.5	21.3
	Avg. Abs. Value:	1.1	8.5	1.7	13.5	1.1	9.0	3.6	29.4	0.0	0.0	3.4	27.8

Table III: Modeled vs. measured results for POA and GHI

	Model/Method:	**DIRINT**				**GTI-DIRINT**			
	Input Parameter:	**GHI Pyranometer**				**POA Reference Module**			
	Output Parameter:	**POA Reference Module**				**GHI Pyranometer**			
		MBD		**RMSD**		**MBD**		**RMSD**	
Site	**Type**	**%**	**(W/m²)**	**%**	**(W/m²)**	**%**	**(W/m²)**	**%**	**(W/m²)**
A	Single-Axis Tracking	-0.9	-6.1	7.8	51.6	0.9	4.8	5.5	28.6
B	Single-Axis Tracking	4.6	30.1	13.8	90.8	-1.3	-6.8	6.3	33.8
	Average Abs. Value:	2.8	18.1	10.8	71.2	1.1	5.8	5.9	31.2
C	Fixed Tilt	0.8	3.8	10.7	49.7	0.0	0.1	7.9	33.0
D	Fixed Tilt	1.6	8.0	8.0	40.3	-1.5	-7.3	9.2	44.5
E	Fixed Tilt	0.0	-0.2	3.6	18.4	0.2	0.9	3.8	17.6
F	Fixed Tilt	-0.1	-0.5	3.0	15.4	0.5	2.4	3.6	17.3
G	Fixed Tilt	1.5	4.9	15.2	48.9	-0.2	-0.5	12.3	37.6
	Average Abs. Value:	0.8	3.5	8.1	34.5	0.5	2.2	7.4	30.0

Table IV: Modeled vs. measured results for DNI and DHI

	Model/Method:	**DIRINT**								**GTI-DIRINT**							
	Input Parameter:	**GHI Pyranometer**								**POA Reference Module**							
	Output Parameter:	**DNI**				**DHI**				**DNI**				**DHI**			
		MBD		**RMSD**		**MBD**		**RMSD**		**MBD**		**RMSD**		**MBD**		**RMSD**	
Site	**Type**	**%**	**(W/m²)**	**%**	**(W/m²)**	**%**	**(W/m²)**	**%**	**(W/m²)**	**%**	**(W/m²)**	**%**	**(W/m²)**	**%**	**(W/m²)**	**%**	**(W/m²)**
C	Fixed Tilt	-0.1	-0.3	24.4	116.2	1.7	2.1	44.7	54.6	-1.0	-4.6	29.1	138.3	16.8	20.6	61.4	74.9
D	Fixed Tilt	5.2	31.3	20.6	124.4	-7.0	-7.4	62.0	65.4	-7.0	-42.4	28.5	172.7	30.6	32.4	60.4	63.8
E	Fixed Tilt	0.4	2.7	15.0	93.5	-2.2	-2.3	39.8	40.4	-1.3	-7.8	15.9	98.8	10.3	10.4	44.9	45.6
F	Fixed Tilt	19.3	89.3	26.3	121.6	-30.3	-56.8	41.9	78.7	22.8	105.4	29.6	136.9	-31.3	-58.7	43.9	82.5
G	Fixed Tilt	11.1	29.8	38.2	102.0	-2.8	-3.7	33.2	43.7	7.6	20.4	47.8	127.7	4.4	5.8	40.4	53.2
	Average Abs. Value:	7.2	30.7	24.9	111.5	8.8	14.5	44.3	56.6	7.9	36.1	30.2	134.9	18.7	25.6	50.2	64.0

pyranometers is poorer, although still good. This residual non-correlation reflects the differences in the fundamental responses of the instruments, even though the spectral and angular corrections have been performed. We observed that when we eliminated partly cloudy days with highly variable irradiance from the dataset, the reference module vs. pyranometer correlation RMSD improved from ~30 W/m² to ~15 W/m², matching the pyranometer-pyranometer RMSD. This suggests that a possible explanation for the poorer correlations between the reference modules and the pyranometers in the full dataset is the differences in the temporal responses of these devices – the reference modules respond much more quickly than the pyranometers to changes in irradiance. Another possible explanation is poorer performance of the spectral correction factor on partly cloudy days.

In any case, for the full dataset, the correlation of the reference modules to the pyranometers shows RMSD of ~30 W/m². It should be understood that this value represents the cross-instrumental contribution to the RMSD of the correlations between model results based on either the POA reference module or the GHI pyranometer versus measurements from the other instrument, as will be discussed below.

In evaluating the models, we wished to use the MBD statistic in addition to RMSD. In order to ensure that the MBD represents the accuracy of the model results, and not any differences in instrument calibration, we matched the calibration of the reference module and the pyranometers. For the two single-axis tracking sites, we did this by setting the reference module calibration so that the spectrally corrected reference module irradiance measurements matched, on average, the GHI pyranometer during the hours when the tracker table was nearly horizontal. Note that the MBD values in the third column group of Table II are 0 as a result. For the fixed-tilt sites this method was not possible; therefore, for these sites we matched the spectrally and angularly corrected reference module irradiance measurements to the POA pyranometer. Based on the single-axis tracking sites shown in Table II, we made the assumption that at the fixed-tilt sites each POA pyranometer is also well matched (<2% MBD) with its corresponding GHI pyranometer. Although we could not directly verify this assumption, the model results indicate consistency.

6. RESULTS

Table III shows the results of the DIRINT and GTI-DIRINT models for POA and GHI. To apply the DIRINT model, we used the GHI pyranometer data as input, determined DNI and DHI from DIRINT, translated the resulting DHI to a tilted diffuse irradiance using the Perez model [5] function in PV_LIB and added a diffuse ground reflection using the ground reflection model in PV_LIB (with albedo of 0.2) to yield a total diffuse tilted irradiance (DTI). We then applied incidence angle modifier and spectral correction factors appropriate for the PV modules and combined the DNI and DTI to yield a POA irradiance value. This modeled POA was then correlated with the reference module irradiance measurements to yield the DIRINT statistics shown in Table III for the MBD and RMSD comparing the modeled and measured values. To apply GTI-DIRINT, we used the spectrally corrected POA reference module irradiance as input, determined DNI and DHI from the GTI-DIRINT model (with albedo of 0.2 and with incidence angle modifiers included as described above), and then calculated GHI by combining the DNI and DHI components. This modeled GHI was then correlated with the pyranometer measured GHI yielding the GTI-DIRINT statistics in Table III. The bottom row of the table shows the average absolute value of each column, to provide a summary value for the results.

The results show that both the model approaches perform very well, with average MBD ~1% and RMSD 7-9%. Based on both the MBD and RMSD the GTI-DIRINT performance appears at least equivalent or possibly superior to that of DIRINT, since both the MBD and RMSD of the GTI-DIRINT method were lower than those for DIRINT.

The RMSD values shown in Table III for both DIRINT and GTI-DIRINT, on average ~30 W/m^2 for the fixed-tilt sites, are somewhat larger than the equivalent values found by Marion in [3], which were only ~20 W/m^2. However, in [3], thermopile pyranometer data were used for both GHI and POA measurements, while here we use a spectrally and angularly corrected reference module for the POA measurements. In Table II and its associated discussion we already showed that by correlating the pyranometer and reference module measurements we have an additional source of variance on the order of ~30 W/m^2 due to the differences in instrument sensitivities. Therefore the final RMSD values in Table III which are themselves on the order of ~30 W/m^2 indicate that the models perform better than indicated by Table III. We confirmed this conclusion by repeating the model calculations using only the pyranometer measurements for both POA and GHI and found that for both models the correlation of modeled to measured results was significantly improved. This does not indicate superiority of the pyranometer measurements but rather that there is some variance in going from pyranometer data to PV module data and vice versa, for the reasons already discussed.

Note in particular in Table III that for the fixed-tilt sites DIRINT and GTI-DIRINT have similar performance, while for the single-axis tracking sites GTI-DIRINT performs significantly better than DIRINT, as shown by better MBD and RMSD. Figure 1 illustrates this in more detail, showing the RMSD for the model results for Site A with the data filtered to exclude tracker tilt angles above the indicated maximum values. When only low tilt angles representing mid-day hours are included, both DIRINT and GTI-DIRINT perform equivalently. But when high tilt angles representing morning and evening hours are included, the performance of GTI-DIRINT is only marginally affected while DIRINT is significantly affected. Similar results were found for Site B, explaining the unusually large RMSD for DIRINT in Table III. The superior performance of GTI-DIRINT for large tracker tilt angles is likely due to the fact that the input to

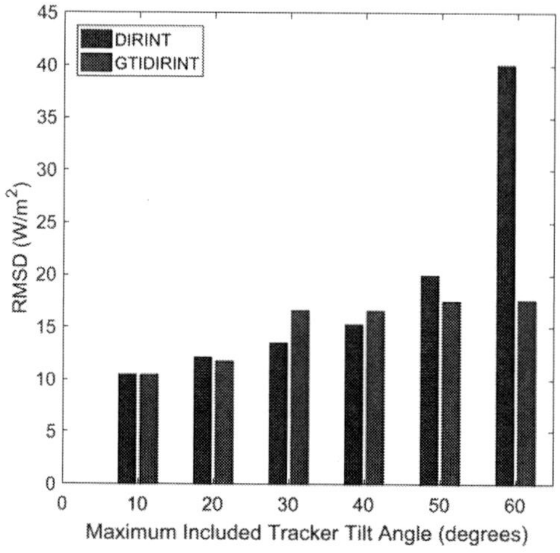

Figure 1: Variation of RMSD for modeled global irradiance translations (DIRINT: GHI to POA, GTIDIRINT: POA to GHI) versus measured values, for Site A, using only pyranometer data.

GTI-DIRINT, the POA irradiance, has a larger DNI component than the input to DIRINT, which is the GHI. Evidently the decomposition model performs better when using measurements from the plane tilted towards the sun.

In Table IV we compare the DNI and DHI results of each of the models with the DNI and DHI measured by the shadowband radiometers at the five fixed-tilt sites where these were available. The correlation statistics for DNI and DHI are not as good as those for POA and GHI, even though the modeled POA and GHI results are computed from the modeled DNI and DHI values. This is because the modeled DNI and DHI are somewhat anti-correlated; a deficit in DNI is compensated by an excess of DHI and vice versa. However, the average MBD and RMSD in Table IV are comparable to, though somewhat larger than, those in [3]. Overall the models appear to perform about equivalently, although in this case the DIRINT results are somewhat better than the GTI-DIRINT results.

7. CONCLUSIONS

We have evaluated the GTI-DIRINT model using data collected at seven sites spanning three geographic regions over periods from 1.5 to 3.3 years and found generally good performance, comparable to that obtained with the DIRINT model. We observed that the use of POA irradiance data with the GTI-DIRINT model is particularly beneficial, as compared to using GHI data with DIRINT, for the case of single-axis tracking systems. These results suggest the potential for using reference module POA measurement as the single irradiance measurement from which GHI, DNI, and DHI can be modeled, in situations where this may be advantageous. Some questions remain regarding the degree of variance in the correlation of reference module and pyranometer data, which may be explored in future work.

8. REFERENCES

[1] M. Gostein, J. R. Caron, B. Littmann, "Measuring Soiling Losses at Utility-scale PV Power Plants," in the 40th PVSC, 2014.

[2] B. Marion, "PV Derived Data for Predicting Performance," at Solar Power International, Anaheim, CA, 2015.

[3] B. Marion, "A model for deriving the direct normal and diffuse horizontal irradiance from the global tilted irradiance," Solar Energy, v. 122, pp. 1037-1046, 2015.

[4] R. Perez, P. Ineichen, E. Maxwell, R. Seals, A. Zelenka, Dynamic global-to-direct irradiance conversion models, ASHRAE Transactions—Research Series, 354-369, 1992.

[5] R. Perez, P. Ineichen, R. Seals, J. Michalsky, R. Stewart, "Modeling daylight availability and irradiance components from direct and global irradiance," Solar Energy 44 (5), 271–289, 1990.

[6] L. Nelson, M. Frichtl, A. Panchula, "Changes in Cadmium Telluride Photovoltaic System Performance due to Spectrum," IEEE Journal of Photovoltaics, v. 3, pp. 488-493, 2012.

Detailed Irradiance Statistics for the Design of PV-Systems from a Set of Ground Stations in Central Africa (Rwanda)

Hans Georg Beyer*, Fabien Habyarimana#
*Institute of Engineering, University of Agder, Grimstad, Norway
#University of Rwanda, Kigali, Rwanda

ABSTRACT: In Rwanda the capital region around Kigali city and the districts as Gasabo, Nyarugenge and Kacyiru show a good coverage by the nationwide grid whereas, in contrast more rural districts show much lower rates. For both, an increased share of PV to bulk electricity and off-grid rural electrification projects are initialized, and are in need of information on the irradiance characteristics in the country. Using data from a dedicated network of meteorological stations, that are logging the irradiance with a 15min time resolution, statistics of solar irradiance relevant for the design of both types of PV-systems are extracted and compared to the respective information for sites in Europe.
Keywords: Solar Radiation, , Sizing , Rural Electrification, Stand-alone PV Systems

1 INTRODUCTION

To support planning concerning PV application – both, on and off-grid – knowledge on the solar resource beyond the information on annual irradiance sums is needed. In countries where the respective development is still in a starting phase, the respective knowledge is not very widespread. However, besides the basic information on the solar resource given by the generally available satellite derived irradiance maps (see example in fig.1), in Rwanda (Central Africa, size ~250km*400 km) the acquisition of a new data set for solar irradiance for variety of environmental condition in the country, had been started in 2013 by the Rwanda Environment Management Authority (REMA) (see e.g. [1]). Global solar irradiance is recorded with a temporal resolution of 5min at a number of stations spread over the country (). Besides the basic reporting on the solar resource, these data can be used for the analysis of some key statistical parameters of the set, relevant for the proper sizing of the components of on- and off-grid systems. This comprises the analyses of the probability density – relevant for the proper selection of the adequate ratio of the PV STC power to the rated power of the inverter for grid connected systems. For off-grid systems, in addition the temporal correlation structure of the sets – influencing the response of the system reliability to the sizing of the storage – is of importance. In view of an in future augmented use of grid connected systems the spatial characteristics of the fluctuations of the irradiance have to be analyzed to be prepare for the respective power fluctuations in the distribution grid.

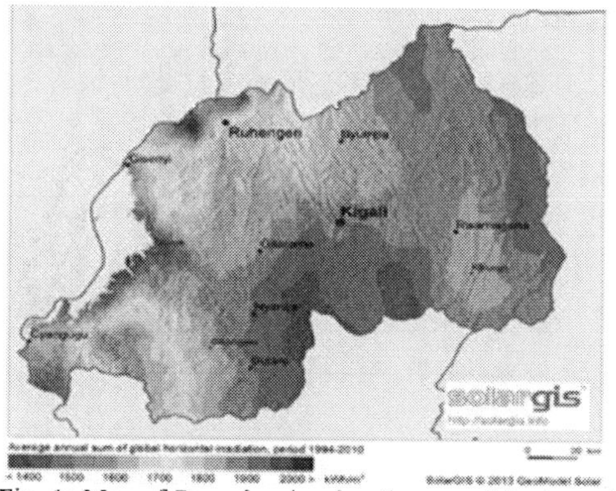

Fig. 1: Map of Rwanda, showing the annual horizontal irradiance sum (from [2]).

Fig.2: Locations of automated weather stations in Rwanda operated since 2014 by the Rwanda Environment Management Authority (REMA)

2 BASIC DATA QUALITY

The quality of the ground data set was checked by its consistency to the outcome of a clear sky model. As can be seen in Fig. 3, the measured data are well in accordance with modeled clear sky irradiance (Model [3], Linke Turbidity 3). A comparison of the station data with the satellite derived information is given in [4].

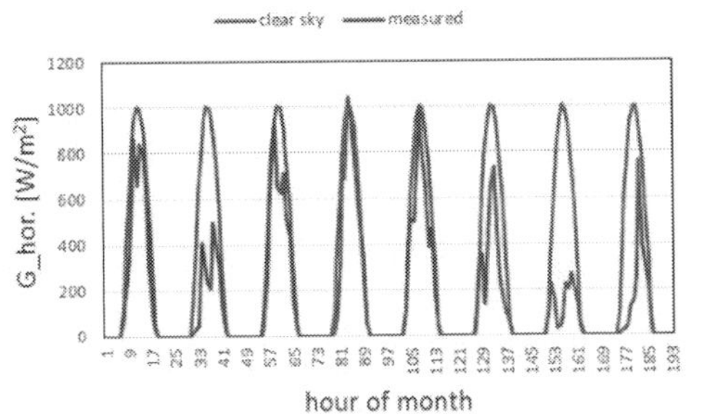

Fig.3: Data quality from the weather stations in Rwands, shown by consistency to modelled clear sky data. Section of a time series of measured data from Location Gikomera, data from July 2014.

3 STATISTICAL CHARACTERISATION

3,1 Spatial correlation

In view of a large scale integration of PV power into the nationwide electricity grid the variability of the solar input causing increased stress to the conventional generation park is of importance, Spatial inhomogeneity of the irradiance field would cause a release by levering out single point fluctuations in the lumped power output. Figs. 4 and 5 show the pattern of the clear sky index - representing the stochastic component of the irradiance variability - at 5 locations on hourly scale for examples from the months of July and December, It can be observed, that there is a remarkable similarity in the evolutions at the different sites, indicating a limited benefit of a spatial distribution of the generation capacity on the daily time scale.

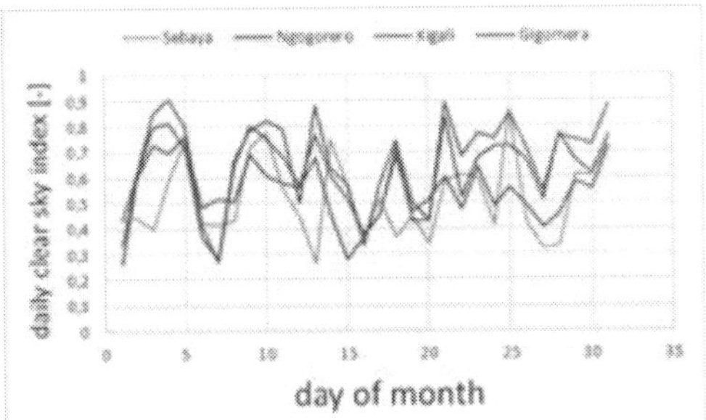

Fig.4: Time series of the daily clear sky index at 4 locations in Rwanda in July 2014.

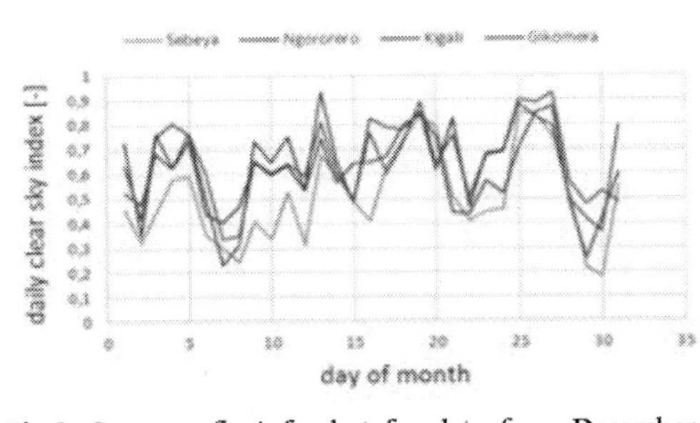

Fig.5: Same as fig.4 for but for data from December 2014.

On a shorter time scale a more distinct decorrelation of the fluctuations with increasing interstation distance can be observed. Fig.6 gives the cross correlation of the 15min clear sky indices for pairs of stations with interstation distance ranging from about 10km to about 100km for the months of July and December. The clear tendency of a decrease of correlation can be observed. These data are compared here with correlation characteristics of hourly data for pairs of sites in Germany [5], indicating a tendency for a stranger decorrelation for the conditions in Rwanda. This would cause a larger degree of smoothing of the lumped power output on the shorter time scales. The pattern for the two months inspected is quite similar.

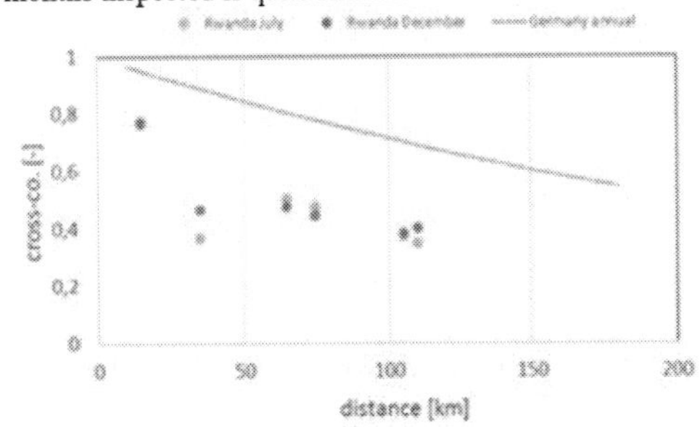

Fig.6: Cross correlation of the 15min clear sky index for pairs of sites in Rwanda with various interstation distance for the months of July and Decembercompared to the pattern of hourly data for pairs of sites in Germany (from [5]).

3,2 Temporal characteristics

For the design and performance of off-grid systems the temporal variability of the irradiance of single site data has a governing role for stress to the storage components (see e.g. [6]). As basic characterization the standard deviation of the data on various time scales can be analyzed. As example Fig.7 shows the standard deviation of clearness index on the 15min time extracted for two month (July and October) at the station Kigali, The data are presented here sorted by the average daily clear sky index. As expected, the highest standard deviation occurs for days with clearness index around 0.5 indication mixed skys. A dependence on the month can be observed with higher variability in October.

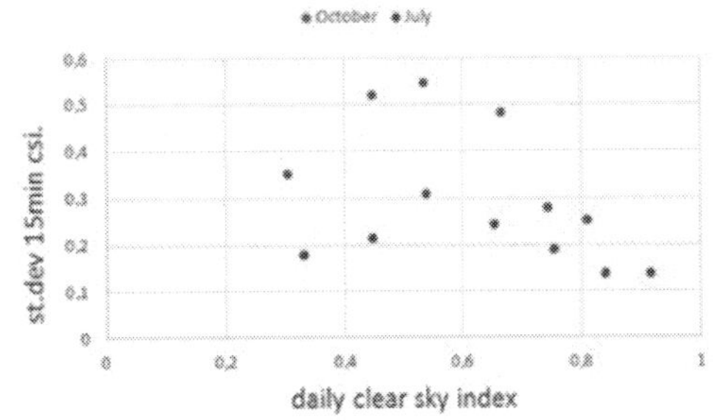

Fig.6: Standard deviation of the 15min clear sky index as function the average daily clear sky index for the months of July and October, station Kigali.

To analyze the short term temporal stability of the irradiance – relevant for e.g. the analysis of the accumulated energy flow through a storage or the cycling of back-up components in stand-alone systems – the autocorrelation of the data give first indication. Fig. 7

gives as example the autocorrelation of the 15min clear sky index for days with different daily clear sky indices. The values > 0.6 encountered here for the months July and December indicate a remarkable stability on this time scale i.e. low occurrence of high frequency cycling.

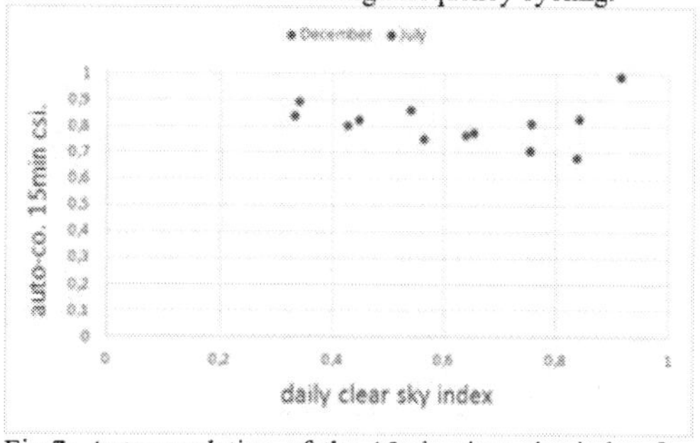

Fig.7: Autocorrelation of the 15min clear sky index for the months of July and December for the station Kigali.

4 CONCLUSIONS

The data gathered by the ground station nework in Rwanda offer a unique source for detailed irradiance statistics in central Africa. Based on the statistical characterization of these data the schemes and tools for the planning of both, grid connected and off-grid systems can by reviewed for the application in Central Africa. The first inspection presented here indicate that distinct differences as compared to the characteristics known from e.g. sites in Europe may occur.

5 REFERENCES

[1] http://www.eurekalert.org/pub_releases/2012-10/sffs-aws101212.php

[2] http://geomodelsolar.eu/home as of 15.052014

[3] Kasten, F., Parametriserung der Globalstrahlung durch Bedeckungsgrad und Trübungsfaktor, Annalen der Meteorologie, Neue Folge 20 (1983)

[4] Beyer, H.G., F. Habyarimana, Assessment of the quality of satellite derived irradiance data for Rwanda by comparison with data from a fleet of automated ground stations, ISES Solar World Congress, Daegu, South Corea (2015)

[5] Wiemken, E., et.al, Power characteristics of PV ensembles: experiences from the combined power production of 100 grid connected PV systems distributed over the area of Germany, Solar Energy, 70 (2001)

[6] Beyer, H.G., J..Luther,Performance of solar energy systems derived from synthetic radiation series - sensitivity to statistical model parameters, Solar Energy Society Proceedings Series,1, 412-416 (1990)

STOCHASTIC DOWNSCALING ALGORITHM TO GENERATE HIGH-RESOLUTION TIME-SERIES FOR IMPROVED PV YIELD SIMULATIONS

Christiane A. Duscha[1, 2], Stefan A. Buehler[2], Jorge Lezaca[1], Carl-Maria Bohny[1], Richard Meyer[1]
[1]Suntrace GmbH, Brandstwiete 46, 20457 Hamburg, Germany, +49 40 767 96 38 0,
christiane.duscha@suntrace.de, jorge.lezaca@suntrace.de, carl.bohny@suntrace.de, richard.meyer@suntrace.de
[2]University of Hamburg, Meteorological Institute, Bundesstr. 55, 20146 Hamburg
stefan.buehler@uni-hamburg.de

ABSTRACT: The output of PV plants varies strongly with variations of incoming solar radiation. As most weather data sets for PV today still are only supplied in hourly time-resolution, and as most popular yield simulation tools are only capable of handling hourly resolved input, the effect sub-hourly fluctuations on the yield of PV plants is largely neglected. However, it is assumed that especially where clouds appear frequently such variations can play an important role in estimating the yield of a PV plant. Methods to downscale coarse scale time-series data to higher resolutions have been already developed. But they still have shortcomings, generating only time series with hourly or even daily resolution and underestimating high radiation values actually observed. Depending on frequency of such 'over-radiation' situations and plant layout, this effect can lead to lower yields. In contrast to other methods, the new algorithm developed here is able to generate 1 minute resolution time series from incoherent hourly resolution time-series. Using the simulation tools SAM and PVlib for PV plants we have compared the energy yield obtained with different time-resolutions. We see that the energy yield calculation improves using the 1 min synthesized time-series compared to the 1 hour averaged.

1 RELEVANCE OF HIGH RESOLUTION TIME-SERIES FOR PV YIELD ANALYSIS

The photovoltaic effect reacts instantaneously to changes in solar radiation. Thus, the electric output of any photovoltaic device changes abruptly with changes in radiation. Solar radiation on Earth' surface changes quickly especially under partly cloudy skies. Such situations are frequent in many regions where PV plants are applied.

Due to exceedances of the maximum power limit of the PV plants inverter during partly cloudy skies, energy is lost. The strong fluctuations of solar radiation are well resolved by 1 minute resolution time series. Within 1 hour resolution time series, these exceedances are averaged out. Therefore the energy yield calculated for 1 hour resolution is often overestimated during partly cloudy skies.[1]

Today it is still widely used to calculate PV yields only in hourly time-resolution. Tools like PVsyst, which is the industry standard used today for many of the large PV simulations, handles only hourly time-resolutions

To achieve accurate energy yield predictions for solar power plants at any desired location, high quality solar radiation time-series of high temporal resolution and wide spatial and long temporal coverage are needed. Satellite derived solar radiation data provide wide spatial and long temporal coverage, but a temporal resolution of usually one hour, which is for plant simulation purposes relatively low. Therefore, an algorithm has been developed, which is capable of downscaling 1 hour resolution time series to a 1 minute resolution time-series of solar radiation. The statistical characteristics of the time-series generated by the newly implemented downscaling algorithm match the statistical characteristics of measured 1 minute resolution solar radiation time-series.

2 SCIENTIFIC INNOVATION FOR STOCHASTIC DOWNSCALING

Downscaling methods have already been invented for solar radiation time series in the past. The most common approaches use Autoregressive Gaussian Models (AGM) [2]–[4] or use Markov Transition Matrices (MTM) [5]–[8].

Hofmann et al. (2014) [6] found that with the same input data the MTM approach generates better results than the AGM approach. Especially concerning overirradiation values which are most interesting for the differences in energy yield predictions, the MTM approach performs better. Also the variability characteristics of time series generated with the MTM approach are more similar to the measurements.

The new downscaling algorithm presented here is based on the MTM approaches given by Aguiar et al. (1988) [5] and Hofmann et al. (2014) [6]. While the algorithm by Aguiar et al. (1988) [5] only generates daily resolution data from monthly resolution data, the new algorithm is able to generate 1 min resolution data from hourly resolution data. Instead of differentiating only between 3 cloud situation as introduced by Hofmann et al. (2014) [6], the new algorithm resolves more continuous cloud fractions and is able to generate a 1 minute resolution time series from an incoherent hourly resolution time series.

3 METHODOLOGY FOR STOCHASTIC DOWNSCALING

In order to synthesize 1 minute resolution time series of solar radiation, information about the transition between sequenced 1 minute time steps is needed. Therefore measured 1 minute resolution data from several stations of the Baseline Surface Radiation Network (BSRN) [9] is analysed. The information retrieved from this analysis is stored within empirically generated MTM, each one dedicated to a certain range of cloud fraction as shown in fig. 1. The rows of a MTM contain the transition probability of a given time step to every possible sequenced time step of the observed hour.

To estimate the transition probabilities between two sequenced solar radiation time steps, the solar radiation component is converted to a normalized variable: the clearness index K_t. The 1 minute resolution clearness index K_t is calculated from the measured GHI (G_{mes}) and clear sky GHI (G_{clear}) as:

$$K_t = \frac{G_{mes}}{G_{clear}} \tag{1}$$

The values of the clearness index K_t are mostly between 0 and 1. But in some cases the G_{mes} exceeds G_{clear} due to albedo or reflecting cloud effects. These over-irradiation effects often appear within high resolution time series during partly cloudy skies [10].

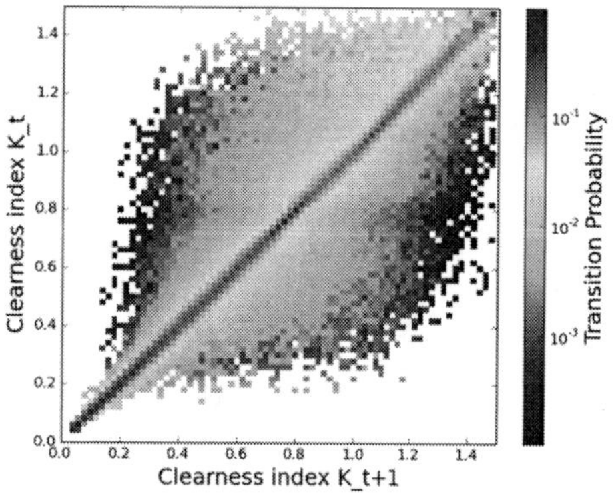

Figure 1: Cloud fraction specific MTM (hourly averaged Clearness Index between 0.7 and 0.8, means partly covered skies)

Each time step of the hour is then generated by using the cumulative transition probability function retrieved from the MTM corresponding to the observed time step (K_t) and a uniform distributed random variable x_r. Using this process the value of following 1 minute time step is found. To estimate the next time step the cumulative transition probability function corresponding to the K_{t+1} value and a new x_r, is selected as shown in fig. 2. This is repeated until 60 K_t values are found. If the mean value of these 60 K_t values exceeds the hourly input K_t by more than a set threshold (<0.03), the process is started again, until the threshold is no longer exceeded.

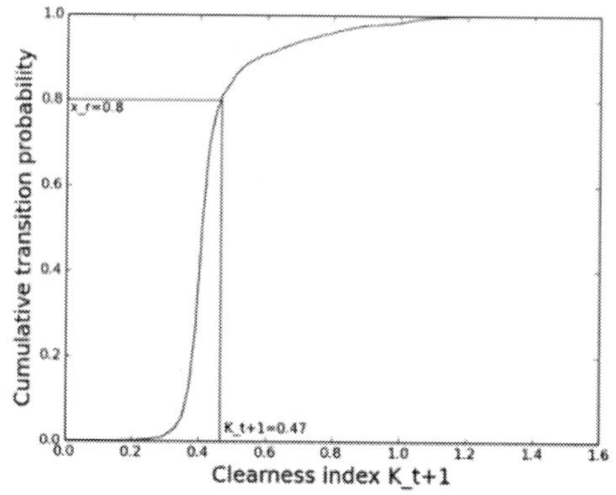

Figure 2: Cumulative transition probability corresponding to the horizontal cut on K_t =0.39 in the MTM shown in fig. 1.

Afterwards the K_t time-series is transformed to a time series of solar radiation using eq.1. The synthesized data G_{syn} is scaled with the hourly solar radiation value $G_{mes,hour}$ using :

$$G_{scal}(t) = \frac{G_{syn}(t)}{G_{syn,hour}} \cdot G_{mes,hour} \tag{2}$$

4 TRAINING OF SYNTHESIZED 1 MINUTE RESOLUTION TIME-SERIES COMPAIRED WITH 1 HOUR RESOLUTION TIME-SERIES OF SOLAR RADIATION

The downscaling algorithm is trained with 3 years of BSRN [9] measurements from 3 different stations within 3 different climate zones. The BSRN station in Lindenberg, Germany (LIN 2004, 52.21°N, 14.12°E) represents the mid latitudes (climate Zone C), while the BSRN station in Tamanrasset, Algeria (TAM 2005, 22.79°N, 5.53°E) represents the arid subtropics (climate zone B) and Ilorin, Nigeria (ILO 1996, 8.53°N, 4.57°E) represents the equatorial tropics (climate zone A) [11].

The MTM used to synthesize the 1 minute resolution data from the measured 1 hour resolution data is generally filled by the statistics using data from several stations within all observed climate zones.

The synthesized time-series are validated with the actual measured 1 minute data and compared with hourly averaged data at the 3 BSRN sites as displayed in Fig. 3,4 & 5.

The results from the BSRN site LIN are shown on fig. 3. No data from LIN was used in order to generate the MTM, which was used to synthesize the 1 minute data for that site from the averaged 1 hour measurements. By applying the Kolmorov Smirnov Integral (KSI) it is observed that the synthesized 1 minute data at the test site LIN closely approaches the actual measured 1 minute data. The 1 hour data presents bigger deviations, especially concerning the higher solar radiation values, which are most important for the energy yield estimation.

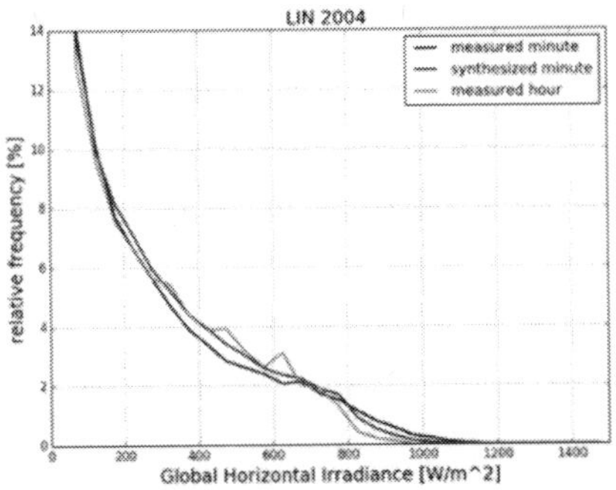

Figure 3: Measured 1 minute and 1 hour resolution GHI distribution and synthesized 1 minute resolution GHI distribution at LIN for the year 2004

The results from the BSRN site TAM are shown in fig.4. Several years of data from TAM were used to generate the MTM, which was used to synthesize the 1 minute data from the hourly averaged measurements. By applying the KSI it is observed that the synthesized 1 min data also approaches closely the measured 1 minute data. The 1 hour data presents in this case even higher deviations than the ones seen for the LIN site. In a similar way, higher solar radiation values are significantly improved by the synthesized 1 minute distribution. PV yield calculations should therfore improve using the sythesized 1 minute data. In overall the improvement obtained with the synthesized data is higher at the TAM site than at the LIN site due to its local cloud characteristics.

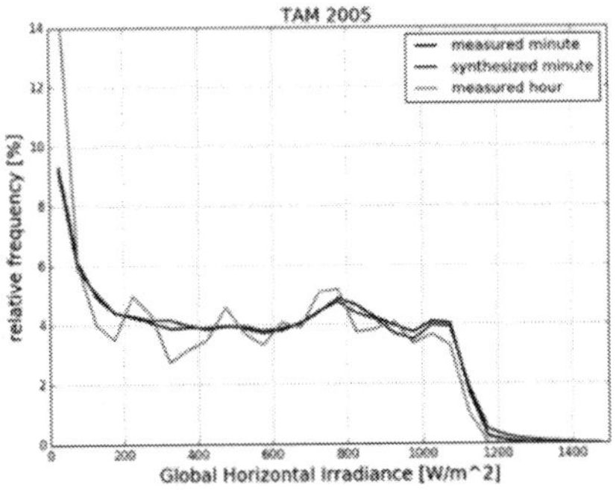

Figure 4: Measured 1 minute and 1 hour resolution GHI distribution and synthesized 1 minute resolution GHI distribution at TAM for the year 2005

Finally, the results from the BSRN site ILO are shown in fig. 5. Several years of data from ILO, but fewer than from TAM, were used to generate the MTM, which was used to synthesize the 1 minute data. By applying the KSI, similar results are obtained as for the sites LIN and TAM The improvement in high values of solar radiation is more noticeable at ILO than for the other 2 sites as displayed in fig. 5. The overall improvement for ILO is smaller than for TAM, but higher than for LIN.

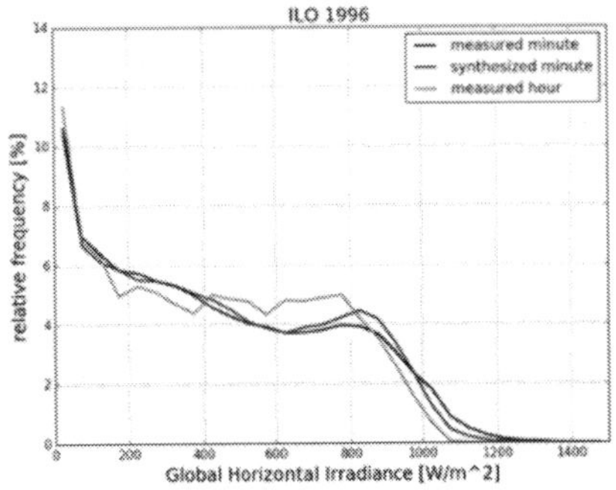

Figure 5: Measured 1 minute and 1 hour resolution GHI distribution and synthesized 1 minute resolution GHI distribution at ILO for the year 1996

From the KSI calculations it is observed that for all cases the synthesized 1 minute data deviates less from the measured 1 minute data than the 1 hour averaged data. The good match of the trained results indicates that the approach works globally. The quantitative results in form of the KSI underline that training at sites with similar cloud situations are the better choice, if sufficient number of measured years are available.

5 SPECIFICATIONS OF THE PV POWER PLANT USED FOR COMPARISON OF THE DIFFERENT TIME SERIES

Once the synthesized 1 minute solar radiation time-series have been obtained, their effect on the PV plant energy yield compared to the 1 hour averaged measured time-series must be assessed. Therefore the following plant design for simulation purposes was chosen.

In contrast to primarily technical optimization processes, the selection of modules, inverters and the system design itself focused here more on the need for comparability of data using different simulation tools than on the optimization of the energy yield in the determined site. Three software simulation tools were used (PVsyst, System Advisor Model (SAM) and the python library pvlib) to better illustrate the impact of minute to hourly data input variations. In a first step, PVsyst, as industry standard and widely applied simulation software, was used to benchmark the hourly data simulation with SAM. PVsyst itself cannot be used to simulate the effect of minute resolution since the simulation algorithms would average out these effects. In order to minimize possible side effects all detailed losses regarding for instance shading or soiling were eliminated so that only the module and inverter loss model of the respective simulation tool was taken into account including temperature effects (not using the nominal operating cell temperature (NOCT) method). This allowed reducing the difference in simulated annual energy production to a few hundred kWh (with PVsyst figures generally being a bit more conservative) for a 1.2 MWh photovoltaic power plant without any extended fine-tuning.

Even though for power plants of this size central inverters are typically not used, the setup of 4800 modules, 24 modules per string, 200 strings and 2 inverters was chosen. The choice of modules and inverters (module: SolarWorld SW250 Poly, inverter: SMA America SC500 CP-US 270 V), raised from the need of a configuration that could be reproduced in all simulation tools (with minimal deviations for base cases and standard data sets) in order to minimize side effects from deviating technical specifications. Hence, as for the plant size itself these choices are somewhat random. Same argument for simplicity and inter-comparability of the set-up applied for the mounting system and other design options (in particular for pvlib). Unlimited sheds were used and the tilt adapted according to the sites' latitudes (30° for Lindenberg, 10° for Tamanrasset and 10° for Ilorin). Although in most cases at least the relative deviation in terms of total energy production was constant, this comparison (SAM and pvlib) will required further investigation in the future. The elimination of side effects proved to be very efficient for the PVsyst-SAM benchmarking though.

6 THE EFFECT OF APPLYING SYNTHESIZED 1 MINUTE RESOLVED DATA COMPARED TO CONVENTIONAL HOURLY DATA TO ESTIMATE THE ENERGY YIELD

Energy yields are calculated with SAM and pvlib for the PV plant using 1 hour time-series and its corresponding 1 minute measured and synthesized 1 minute time-series. To validate the differences, the relative deviations (D_h, D_s) of 1 hour energy yield (E_h) as well as synthesized 1 minute energy yield (E_s) compared to the measured 1 minute energy yield (E_m) as:

$$D_h = \frac{E_h - E_m}{E_m} \cdot 100\% \quad (3)$$

$$D_s = \frac{E_s - E_m}{E_m} \cdot 100\% \quad (4)$$

From theory it is expected that the energy yield simulated for the 1 hour resolution input time-series to exceeds the 1 minute resolution time-series. Therefore it is expected that D_h>0.

In order to certify an improvement on the energy yield using the new algorithm, the absolute value of the relative deviation of synthesized 1 minute energy yield has to be smaller than the absolute value of the relative deviation of 1 hour energy yield, that is $|D_s|<|D_h|$.

The relative deviations of the energy yields obtained using 1 year time series for 3 BSRN sites is shown in fig. 6. At all stations for both simulation tools the absolute value of D_s is smaller than for D_h. Therefore in terms of yearly energy yield the synthesized 1 minute energy yield improves the simulated 1 hour energy yield. The highest relative improvement of more than 58% for both simulation tools is found at BSRN test site ILO. The relative improvement of deviation is smaller at BSRN test site TAM for both the SAM (45%) and pvlib (34%) simulation. The relative improvement of deviation is only slightly better for the pvlib simulation (47%) and the smallest for the SAM simulation (17%) at BSRN site LIN. The smaller deviations at higher latitudes occurs due to the fact that the solar radiation is in average lower and even if the partly covered skies conditions are present, the maximum specifications of the inverter are less frequently exceeded, therefore less suppression of power is seen here at the inverter level. For desert like influenced test sites as TAM partly covered skies are not as present and even if the solar radiation is high enough, 1 minute and 1 hour data will not change drastically the behaviour of the inverter regarding the suppression of power.

For all stations and both simulation tools, except for the pvlib simulation at BSRN test site LIN, the relative deviation D_h is larger than 0 as expected. Therefore at most stations the theory is valid.

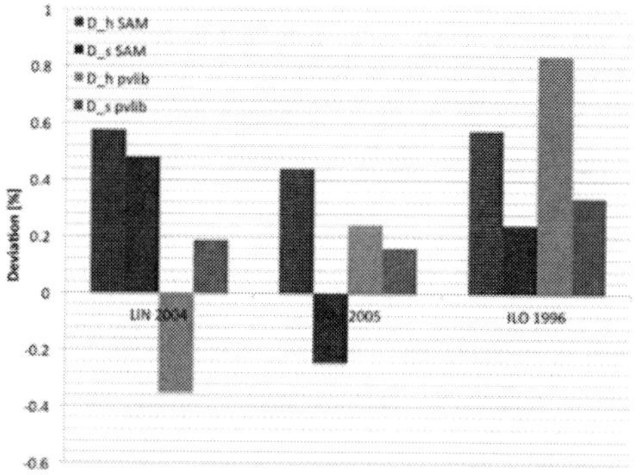

Figure 6: Deviation of hourly and synthesized 1 minute energy yield from measured 1 minute energy yield

Performance simulations for PV plants comparing the 1 hour resolved data result in higher yields compared to 1 minute data as earlier reported by Hofmann et al. (2014) [6]. The largest differences are observed for the tropical site ILO in Nigeria. Because of large presence of convective clouds, the event of partly covered skies is more probable than at the TAM site. The BSRN site ILO is therefore used in this publication for further investigations.

To achieve a better understanding of the annual cycle of energy yield simulations the monthly values of relative deviation from the 1 minute measurements at BSRN test site ILO are shown in fig. 7. The summer months clearly dominate the annual cycle of relative deviation D_h for both simulation tools. The relative deviations reach values in order between 1% and 2%. Also the relative deviations D_s are higher during the summer months. The larger amount of solar radiation and cloudiness are an explanation for this phenomenon. These results show clearly that for sites with high variability of instantaneous cloud conditions the developed algorithm provides a important tool to improve the energy yield calculations of the commonly used 1 hour time-resolution data sets.

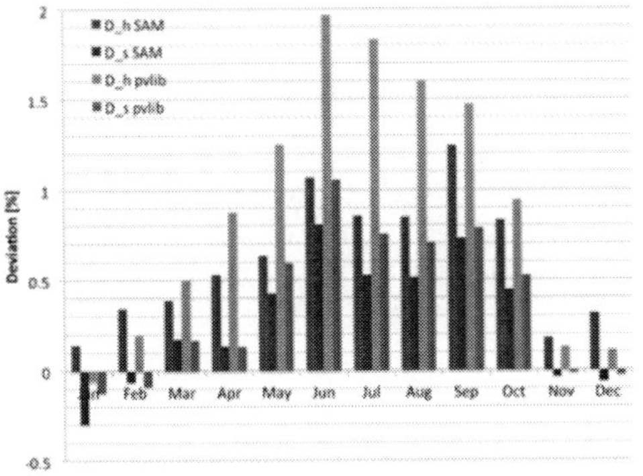

Figure 7: Deviation of hourly and synthesized 1 minute energy yield from measured 1 minute energy yield (monthly resolved) at BSRN site ILO

Still the largest improvements derived from the downscaling algorithm are during the months february, april, november and december as shown in fig. 8. In this figure it can also be observed that during the summer months this improvement is smaller. In january both 1 minute simulations of synthesized energy yield deviate more from the measured 1 minute data than the 1 hour energy yield.

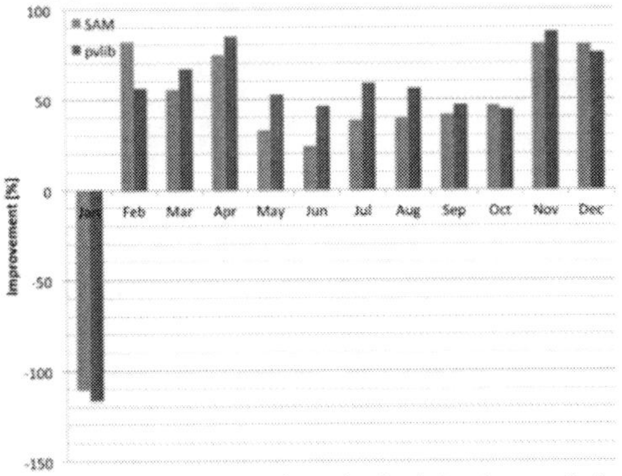

Figure 8: Improvement of synthesised 1 minute relative energy yield deviation compaired to 1 hour relative energy yield deviation disagregated by month.

7 CONCLUSIONS AND OUTLOOK

The newly developed algorithm for downscaling radiation time-series to much higher temporal resolutions works universally using actual measurements. Applying 1 minute resolved GHI time-series to calculate the yield of PV plants leads to a slight improvement of PV yields for annual sums compared to hourly data from the same source. Larger improvements are observed for shorter time periods in terms of months. Effects observed are stronger at sites and time periods where convective clouds frequently cause cloud enhanced over-radiation.

The improvement is also better at sites where larger amounts of solar radiation are present, because only if solar radiation is high enough, the maximum power limit of the PV plants inverter is exceeded. Additionally it is found that the improvement in energy yield deviation is larger, if the KSI of the synthesized solar radiation distribution is much smaller than the KSI of the 1 hour solar radiation distribution.

The results here produced by using pvlib and SAM should be further confirmed by results derived from other models like PVsol, which are also capable of working with 1 minute resolved time-series.

The procedure here was applied to global horizontal irradiance. In principle the presented algorithm is also capable of producing consistent patterns of direct and diffuse irradiance. It should be shown that this leads to more realistic results of PV yields.

The downscaling algorithm should also be applied to satellite derived solar radiation data which usually comes in a 1 hour time-resolution.

ACKNOLEGEMENTS

We acknowledge BSRN for access to the ground-measured solar radiation data at the sites LIN, TAM and ILO.

REFERENCES

[1] B. Burger and R. Rüther, "Inverter sizing of grid-connected photovoltaic systems in the light of local solar resource distribution characteristics and temperature," *Solar Energy*, vol. 80, no. 1, pp. 32–45, 2006.
[2] A. Skartveit and J. A. Olseth, "The probability density and autocorrelation of short-term global and beam irradiance," *Sol. Energy*, vol. 49, no. 6, pp. 477–487, 1992.
[3] R. Aguiar and M. Collares-Pereira, "TAG: A time-dependent, autoregressive, gaussian model for generating synthetic hourly radiation," *Sol. Energy*, vol. 49, no. 3, pp. 167–174, 1992.
[4] V. Graham, K. Hollands, and T. Unny, "A time series model for kt with application to global synthetic weather generation.," *Sol. Energy*, vol. 40, no. 2, pp. 83–92, 1988.
[5] R. Aguiar, M. Collares-Pereira, and J. Conde, "Simple procedure for generating sequences of daily radiation values using a library of markov transition matrices," *Sol. Energy*, vol. 40, no. 3, pp. 269 – 279, 1988.
[6] M. Hofmann, S. Riechelmann, C. Crisosto, R. Mubarak, and G. Seckmeyer, "Improved synthesis

of global irradiance with one-minute resolution for pv system simulations," *Int. J. Photoenergy*, pp. 1–10, 2014.

[7] A. Mellit, M. Benghanem, A. H. Arab, and A. Guessoum, "A simplified model for generating sequences of global solar radiation data for isolated sites: Using artificial neural network and a library of markov transition matrices approach.," *Sol. Energy*, vol. 79, no. 5, pp. 469 – 482, 2005.

[8] J. Bright, C. Smith, P. Taylor, and R. Crook, "Stochastic generation of synthetic minutely irradiance time series derived from mean hourly weather observation data," *Sol. Energy*, 2015.

[9] A. Ohmura, E. Dutton, B. Forgan, C. Fröhlich, H. Gilgen, H. Hegner, A. Heimo, G. König-Langlo, B. McArthur, G. Müller, R. Philipona, R. Pinker, C. H. Whitlock, and M. Wild, "Baseline surface radiation network (BSRN/WCRP) New precision radiometry for climate research," *Bull. Am. Meteorol. Soc.*, vol. 79, no. 10, pp. 215–2136, 1998.

[10] P. Emck and M. Richter, "An Upper Threshold of Enhanced Global Shortwave Irradiance in the Troposphere Derived from Field Measurements in Tropical Mountains," *J. Appl. Meteorol. Climatol.*, 2008.

[11] Kottek, M., J. Grieser, C. Beck, B. Rudolf, and F. Rubel, "World Map of the Köppen-Geiger climate classification updated." *Meteorol. Z.*, **15**, 259-263, 2006

Performance Analysis of Transposition Models Simulating Solar Radiation on Inclined Surfaces

Yu Xie and Manajit Sengupta
National Renewable Energy Laboratory
15013 Denver West Parkway, Golden, CO 80401, USA
Yu.Xie@nrel.gov
Manajit.Sengupta@nrel.gov

ABSTRACT: Transposition models are widely used in the solar energy industry to simulate solar radiation on inclined photovoltaic (PV) panels. These transposition models have been developed using various assumptions about the distribution of the diffuse radiation, and most of the parameterizations in these models have been developed using hourly measurements. Numerous studies have compared the performance of transposition models but this paper aims to understand the quantitative uncertainty in the state-of-the-art transposition models and the sources leading to the uncertainty using high-resolution ground measurements in the plane-of-array. Our results suggest that the aerosol optical depth (AOD) can affect the accuracy of isotropic models. The choice of empirical coefficients and the use of decomposition models can both result in uncertainty in the output from the transposition models. It is expected that the results of this study will ultimately lead to improvement of the parameterizations as well as the development of improved physical models.
Keywords: transposition model, solar radiation

1. INTRODUCTION

Significant growth in the number and size of solar energy projects has increased the demands for accurately measuring and modeling solar radiation. Although solar radiation data from surface measurements[1] or satellite retrievals[2] are routinely available on horizontal surfaces, photovoltaic (PV) panels are generally placed at a tilt to receive more solar radiation than horizontal surfaces. Compared to global horizontal irradiance (GHI), the partition of total solar radiation in the normal direction of inclined surfaces or plane-of-array (POA) irradiance is more closely correlated with PV system performance. As POA solar radiation measurements are generally not available and it is difficult to store POA irradiances for variable PV tilt angles, it is normally easier to compute the POA values from GHI and direct normal irradiance (DNI) using a transposition model [3-7].

Almost all transposition models compute POA irradiance from three contribution sources: direct sky radiation, diffuse sky radiation, and reflected radiation by land surface. Among the current transposition models, the contribution from diffuse sky radiation is simulated by either following empirical equations correlating it to diffuse horizontal irradiance (DHI) (hereafter referred to as the empirical model)[3, 5] or assuming the diffuse radiances are isotropic over the sky dome (hereafter referred to as the isotropic model)[6, 7]. Compared to isotropic models, empirical models account for the strong forward scattering by clouds or aerosols [8, 9]; thus, they are likely to better simulate POA irradiance on inclined PV systems. However, empirical models are developed to minimize model uncertainty from long-term observations at specific locations. The variation in time and geographic position may lead to diverse atmospheric and land surface conditions and significantly affect the accuracy of the empirical models.

Thus, accurate atmospheric and land surface information in high temporal and spatial resolutions can help reduce the uncertainties in the existing isotropic and empirical models. A reliable physics-based model that benefits from the rapid development of remote sensing technologies, e.g. remarkably improved aerosol and cloud products from the future Geostationary Operational Environmental Satellite R (GOES-R), should provide more precise estimation of PV system performance.

To understand the improvement in the future physics-based model, it is first necessary to quantitatively estimate the uncertainties in the state-of-the-art transposition models. This paper uses surface measurements at the Solar Resource Research Laboratory (SRRL) of National Renewable Energy Laboratory (NREL) to analyze the performance of the isotropic and empirical models and explore the physics behind their uncertainties[10, 11]. This study should be useful in improving the accuracy in the development of future physics-based models.

2. TRANSPOSITION MODELS

Transposition models express POA irradiance on an inclined surface by

$$POAI = POAI_d + POAI_{u,sky} + POAI_{u,ground} \tag{1}$$

where $POAI_d$, $POAI_{u,sky}$, and $POAI_{u,ground}$ represent the POA irradiances related to direct solar radiation, downwelling diffuse solar radiation, and solar radiation reflected by land surface and reaching the PV panel, respectively.

$POAI_d$ is essentially the component of direct solar radiation in the normal direction of a PV panel as follows:

$$POAI_d = DNI\cos\theta' \tag{2a}$$

where θ' is the angle between the direct solar radiation and the normal direction of the PV panel. The value of θ' can be solved by rotating a PV plane from the horizontal position to an inclined position:

$$\cos\theta' = \cos\beta\cos\theta + \sin\beta\sin\theta\cos\varphi \tag{2b}$$

where θ is solar zenith angle, φ is the relative azimuth angle, and β is the tilt angle of the PV panel.

Liu and Jordan[6] analytically derived diffuse sky radiation with isotropic radiances:

$$POAI_{u,sky} = DHI\frac{1+\cos\beta}{2} \tag{3a}$$

which has been widely used in solar energy applications. The corresponding POA irradiance from surface reflections is

$$POAI_{u,ground} = GHI\sigma\frac{1-\cos\beta}{2} \quad (3b)$$

where σ is land surface albedo.

A well-known empirical model to compute diffuse POA irradiance was developed by Perez et al. [3] (hereafter referred to as PEREZ) wherein isotropic diffuse, circumsolar and horizon brightening radiation were considered with comprehensive sets of coefficients determined from various climatic environments. Details on PEREZ are not restated here because they are described by Perez et al.[3].

3 RESULTS

We investigated 1-minute GHI, DNI, DHI, and surface albedo measurements at NREL. The isotropic model based on Eqs.(1-3) is combined with the surface measurements to simulate POA irradiance. The simulations are then compared to POA irradiance measured by a Kipp and Zonen CM 11 Pyranometer (CMP 11) and IMT Solar reference cell (IMT) on a 1-axis tracker located at NREL.

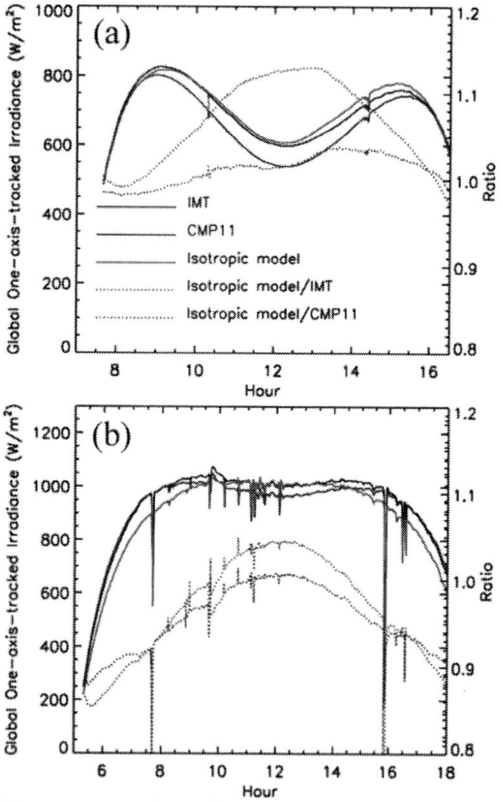

Figure 1 Comparison of POA irradiances from the model simulation to measurements at NREL (a) "winter day (b) summer day.

Figure 1 compares POA irradiances simulated by the isotropic model and measurements on 1/22/2015(a "winter" clear-sky day) and 7/30/2015 (a "summer" clear-sky day). The dashed lines represent ratios of model simulation to measurements. As shown in Fig.1a, the isotropic model agrees well with the measurements. Compared to CMP 11, the relative error of the isotropic model is within 3%. However, the POA irradiance from the IMT is systematically smaller than both the CMP 11 measurements and the model simulations, which is probably caused by angle of incidence effects and measurement biases due to spectral differences between calibration and measurement conditions. In this case the magnitude of the difference is more pronounced at noon than it is during the morning and evening. Compared to CMP 11, smaller POA irradiances observed by IMT are also shown in Fig.1b. In addition, the uncertainty from the isotropic model becomes more obvious in summer. Fig.1b shows that the simulation using the isotropic model can underestimate the POA irradiance by up to 20% when compared to both the IMT and CMP 11 measurements.

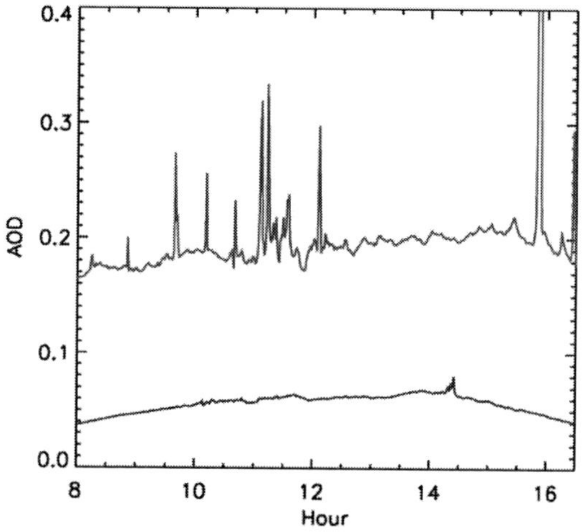

Figure 2 AODs on 1/22/2015 "winter" (blue) and 7/30/2015 "summer" (red).

The variation of the uncertainty in the isotropic model should also be related to the variation on aerosol optical depth (AOD). Figure 2 shows the AODs on 1/22/2015 and 7/30/2015 from NREL's measurements. Greater uncertainty of the isotropic model is associated with greater AOD. When the AOD is negligible, diffuse radiation in the atmosphere is mainly caused by Rayleigh scattering, wherein the spatial distribution of the scattered radiation is not sensitive to the scattering angle. For a larger AOD, using the isotropic model results in more uncertainty because the light scattering by aerosol is more significant in the forward direction compared to other directions resulting in higher anisotropic scattering.

Figure 3 compares the CMP 11 measurements on 1/22/2015 to the simulations by PEREZ. The colors in the figure represent PEREZ from different empirical coefficients. As shown, the simulated POA irradiances from PEREZ that have different empirical coefficients can be substantially different indicating that the selection of proper empirical coefficients is critical. Compared to the isotropic model shown in Fig. 1a, PEREZ does not show better performance unless the optimal coefficients are used. Thus, the isotropic model can be still used as a reference when empirical coefficients from PEREZ lead to substantial biases.

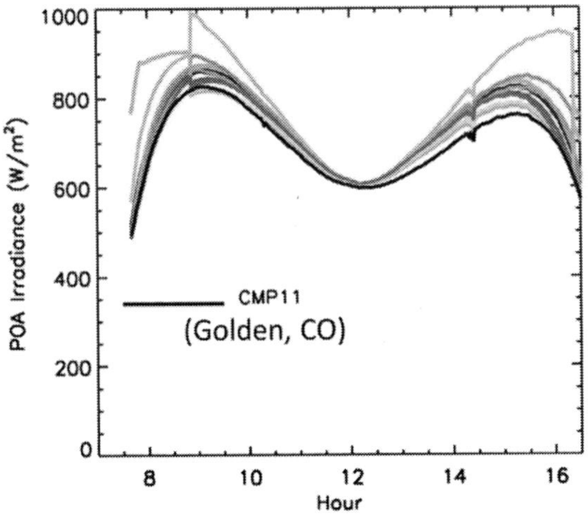

Figure 3 Comparison of POA irradiance from CMP 11 measurements on 1/22/2015 (winter) and PEREZ with different empirical coefficients.

In addition to the empirical coefficients, other input properties in transposition models also affect their accuracy. Decomposition models, e.g. DISC model, are routinely used to compute DHI and DNI, and they are used as inputs to transposition models. The simulated properties can bring additional uncertainties to POA irradiances. Figure 4 shows the difference between the PEREZ simulations with optimal coefficients and CMP 11 measurements on 1/22/2015. The solid line is associated with simulations using surface measurements of DHI, whereas the dashed line represents the simulations from computed DHI by DISC model. As shown, using the computed DHI brings additional uncertainty in the POA irradiance—by up to 80 W/m^2. This uncertainty is more obvious in the morning and evening when more diffuse radiation is available.

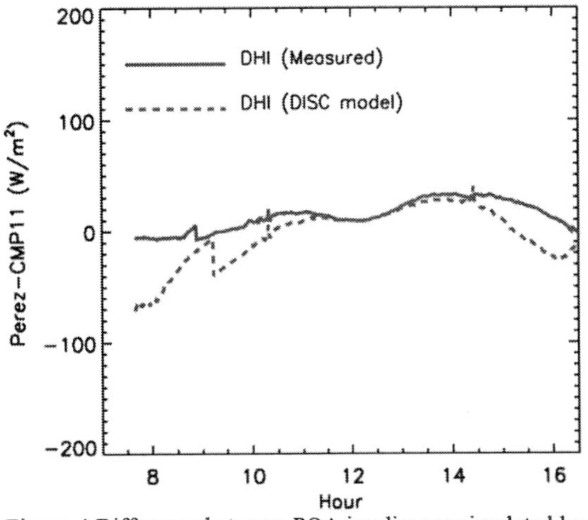

Figure 4 Difference between POA irradiances simulated by PEREZ and measured at NREL on1/22/2015. DHIs used for PEREZ are from surface measurements (solid) or computed by DISC model.

4. CONCLUSIONS

This study analyzes the performance of transposition models simulating solar radiation on inclined surfaces. Surface measurements of solar radiation on horizontal surfaces were used as inputs of an isotropic model and an empirical model, PEREZ, to compute the POA irradiances and compare them to those measured by the 1-axis CMP 11 and IMT. Our results indicate that the isotropic model is associated with larger uncertainty for larger AODs. However, the isotropic model plays an important role in transposition models because the use of empirical coefficients in the PEREZ model may lead to significantly different uncertainties when simulating the POA irradiance. The use of computed DHI and DNI from decomposition models may bring additional uncertainties to POA irradiance estimates.

4. ACKNOWLEDGEMENTS AND DISCLAIMER

This work was supported by the U.S. Department of Energy under Contract No. DE-AC36-08GO28308 with the National Renewable Energy Laboratory. Funding provided by [applicable DOE Office and Program, e.g., U.S. DOE Office of Energy Efficiency and Renewable Energy Solar Energy Technologies Program]. The U.S. Government retains and the publisher, by accepting the article for publication, acknowledges that the U.S. Government retains a nonexclusive, paid-up, irrevocable, worldwide license to publish or reproduce the published form of this work, or allow others to do so, for U.S. Government purposes.

5. REFERENCES

[1] G. M. Stokes and S. E. Schwartz, "The Atmospheric Radiation Measurement (ARM) Program: Programmatic background and design of the cloud and radiation test bed," *Bull. Amer. Meteor. Soc.*, vol. 75, pp. 1201-1221, 1994.

[2] M. Sengupta, A. Habte, P. Gotseff, A. Weekley, A. Lopez, C. Molling, and A. Heidinger, "A physics-based GOES product for use in NREL's National Solar Radiation Database," *European Photovoltaic Solar Energy Conference and Exhibition,* vol. Amsterdam, Netherlands, 2014.

[3] R. Perez, P. Ineichen, R. Seals, and J. Michalsky, "Modeling daylight availability and irradiance components from direct and global irradiance," *Sol. Energy,* vol. 44, pp. 271-289, 1990.

[4] C. Gueymard, "An anisotropic solar irradiance model for tilted surfaces and its comparison with selected engineering algorithms," *Sol. Energy,* vol. 38, pp. 367-386, 1987.

[5] D. Reindl, W. Beckman, and J. Duffie, "Evaluation of hourly tilted surface radiation models," *Sol. Energy,* vol. 45, pp. 9-17, 1990.

[6] B. Liu and R. Jordan, "Daily insolation on surfaces tilted towards the equator," *ASHRAE Journal,* vol. 3, pp. 53-59, 1961.

[7] V. Badescu, "3D isotropic approximation for solar diffuse irradiance on tilted surfaces," *Renewable Energy,* vol. 26, pp. 221-223, 2002.

[8] Y. Xie, "Study of ice cloud properties from synergetic use of satellite observations and modeling capabilities," Ph. D., Department of Atmospheric Sciences, Texas A&M University, College Station, TX, 2010.

[9] Y. Xie, P. Yang, G. W. Kattawar, P. Minnis, Y. X. Hu, and D. Wu, "Determination of ice cloud models using MODIS and MISR data," *Int. Remote Sens.,* vol. 33, pp. 4219-4253, 2012.

[10] M. Sengupta and C. Hansen, "2015 PV Solar Resource Workshop," *http://www.nrel.gov/pv/performance_reliability/pvmrw_20150227_fri.html,* vol. Golden, CO, 2015.

[11] M. Sengupta, A. Habte, S. Kurtz, A. Dobos, S. Wilbert, E. Lorenz, T. Stoffel, D. Renné, C. Gueymard, D. Myers, S. Wilcox, P. Blanc, and R. Perez, "Best practices handbook for the collection and use of solar resource data for solar energy applications," NREL, Golden, CO2015.

ALGORITHM FOR TECHNICAL AND ECONOMIC DESIGN OPTIMIZATION OF PHOTOVOLTAIC SYSTEMS

Johannes Birtel, Henrik te Heesen
Trier University of Applied Sciences, Environmental Campus Birkenfeld, Campusallee, 55768 Hoppstädten-Weiersbach, Germany, h.teheesen@umwelt-campus.de

ABSTRACT: Customers are interested in a technical and economic optimized investment to a photovoltaic system. Often, the focus of project developer is to plan and construct a PV system and maximize the own consumption rate. The economic optimization is disregarded although this is the more important factor. Here we show an algorithm, that is able to optimize a PV system made of two subsystems with different module orientation and inclination. The algorithm takes the load profile, the costs and prices, and the PV yield into account to vary the subsystem capacities and to calculate the annual energy costs. The system setup is automatically optimized by minimization of the energy costs. Thus, the algorithm is able to support customers to identify the best PV system setup for their individual energy consumption and costs.

Keywords: Modelling, Simulation, Financing

1 INTRODUCTION

The PV business has shown an extraordinary development in the past years. While the growth in Europe has declined, the newly installed capacity globally is still growing fast [1]. Today, new PV installations in particular in Germany are mainly used to increase the own consumption of the customer. Due to decreased levelized costs of electricity (LCOE) costs of PV systems and increased electricity costs a PV system can support the reduction of total energy costs. Often, the main question of design optimization of a PV system is technically driven: How can a PV system be set-up to maximize the own-consumption and autarky rate [2, 3, 4, 5]?

Our focus in this publication is to identify the technical and economic optimum of a PV system under consideration of the energy load profile of a customer. We present an algorithm which is able to calculate the best configuration of two PV subsystems to minimize the total energy costs. The PV system consists of standard PV modules and inverter. No battery is taken into account as the LCOE of batteries are still larger than the energy purchase price. In future, there may be an interest to integrate a battery to the system.

2 METHODS

Figure 1 shows the setup of the PV system for the algorithm. There are two PV subsystems with different orientation and inclination of modules. The energy of the PV system is own-consumed by the consumer. Excess energy is fed-in to the grid. If the PV yield is smaller than the energy demand, energy will be taken from the grid to fulfil the demand of the consumer.

A PV simulation is performed for each PV subsystem and the yield distribution for one year on an hourly time interval basis is stored in a databank. We use default load profiles which are provided by the energy utilities (BDEW). The type of the load profile depends on the consumption type (e. g. H0 for households, G1 for industry working from 8 h to 18 h or G3 for industry with a continuous production cycle).

The occurring costs which are considered to identify the total cost minimum are:

- Levelized costs of electricity (LCOE) of the PV system [6, 7]. P is the total capacity of the PV system

$$\mathrm{LCOE} = \mathbf{-0,63954\frac{Cent}{kWh}\cdot \ln\left(\frac{\mathit{P}}{kW_p}\right) + 14,97\frac{Cent}{kWh}} \ .$$

- Energy Purchase Price (EPP). E_a is the annual energy consumption [8]

$$\mathrm{EEP} = -1{,}468\frac{\mathrm{Cent}}{\mathrm{kWh}}\cdot \ln\left(\frac{E_a}{\mathrm{MWh}}\right) + 29{,}18\frac{\mathrm{Cent}}{\mathrm{kWh}} \ .$$

- Feed-In Tariff (FIT) and reduced EEG surcharge costs (RES) for own-consumed energy are based on the German EEG.

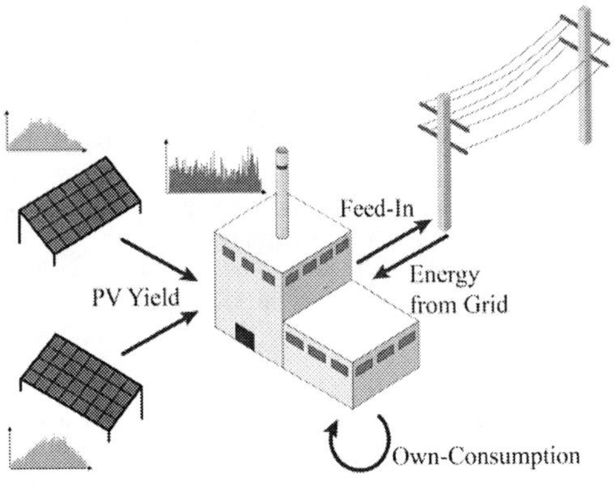

Figure 1: PV system setup. Two PV subsystems with different orientation/inclination angles feed a consumer. The PV yield is own-consumed, excess energy is fed-in to the grid. If the PV production does not fit to the energy demand, energy is taken from the grid.

3 RESULTS AND DISCUSSION

3.1 Algorithm

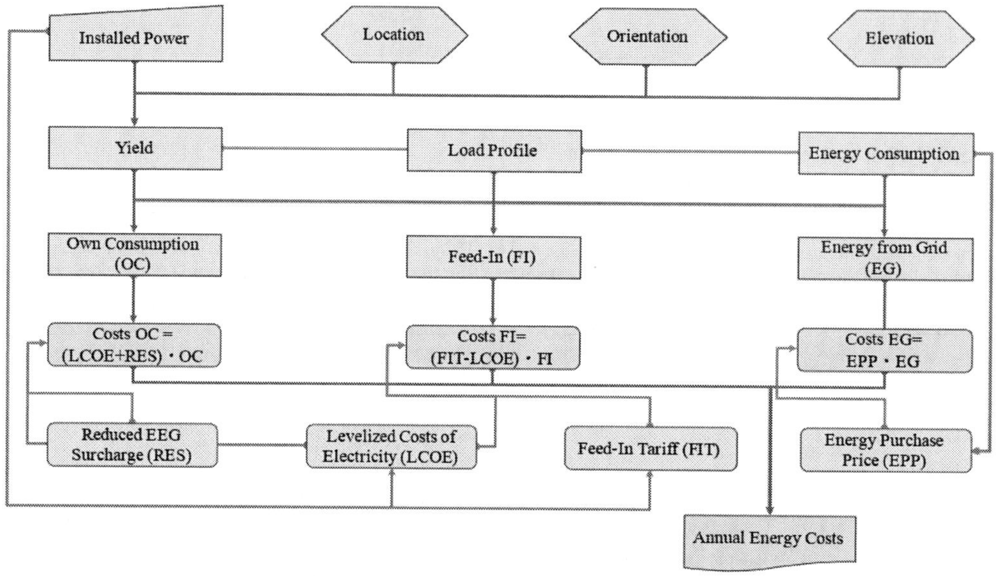

Figure 2: Algorithm for technical and economic design optimization of PV systems. Blue color represents energy, yellow are basic costs and green are derived costs. The dependencies are marked by the arrows.

The concept of our algorithm and the flow of energy and costs is shown in Figure 2. The input parameters of the algorithm are the location of the PV system, the orientation and elevation angle of the modules as well as the installed peak power. These four parameter influence the yield of the PV system. The own consumption (OC), the fed-in energy (FI) and the energy taken from the grid (EG) depend on the PV yield, the load profile and the energy consumption. The levelized costs for electricity (LCOE) and the Feed-In Tariff (FIT) can be calculated by using the installed PV peak power. The PV yield, the load profile and the energy consumption have to be considered to calculate the energy purchase price (EPP). The resulting costs for the own-consumption, the feed-in and the energy from the grid can be derived by using the corresponding energy and costs parameter. Finally, the annual energy costs can be determined.

In the first development step of the algorithm, we have fixed the location, the orientation, and the elevation of the PV modules. The peak power of the PV system can vary from 0 kWp to 500 kWp for each subsystem. We have done a PV simulation using PV*Sol premium 8.0 for a PV system located in Birkenfeld, Rhineland-Palatinate. The system is separated in two subsystems, elevation angle is 30°, module orientation is 90° (east) and 270° (west). The reference peak power is 100 kWp, the variation of the power from 0 kWp to 500 kWp is done by linear scaling of the yield.

3.2 Application of the algorithm

First, we show the results of the algorithm for an industrial consumer with G1 load profile (main usage of electric energy between 8 h and 18 h). The algorithm calculates all costs for all capacity combinations of the two subsystems with an increment of 10 kWp. The final results are the PV peak power of the first and the second PV subsystem and the minimized annual energy costs.

The annual energy costs are shown for three different energy consumption scenarios (Figure 3). Total PV peak power of 0 kWp represents the situation without an additional PV system to support the energy system of the consumer. The larger the energy consumption is, the larger the total costs get. Is the annual energy consumption 80 MWh, the optimum total capacity of the PV system is 70 kWp with an own-consumption rate (OR) of 54 % and an autarky rate (AR) of 47 %. Increasing the annual energy consumption leads to an increase of the PV capacity up to 600 kWp, if the annual energy consumption is 500 MWh. These results are of course only valid for the G1 load profile.

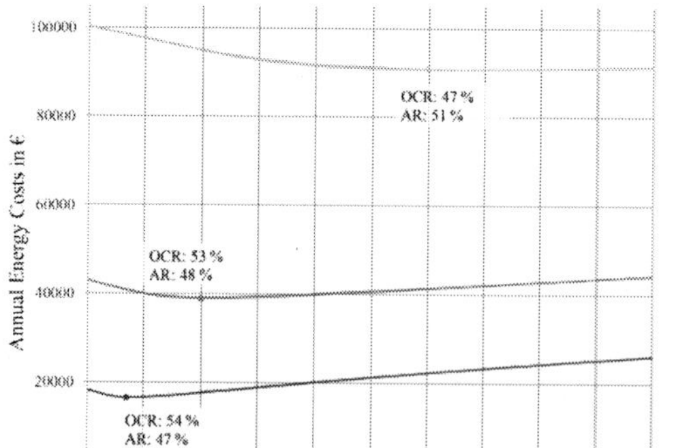

Figure 3: Annual energy costs depending on the annual energy consumption of 80 MWh (blue), 200 MWh (green) and 500 MWh (yellow) for a G1 load profile. OCR is the own-consumption rate, AR is the autarky rate.

The technical and economic optimum of the PV peak power depends on the annual energy consumption and the load profile (see Figure 4). If the energy consumption is smaller than 10 MWh, no PV system shall be installed due to increased energy costs. The maximum capacity of 1000 kWp at 700 MWh for the G1 load profile and 2000 MWh for the G3 load profile is a result of the capacity limitation.

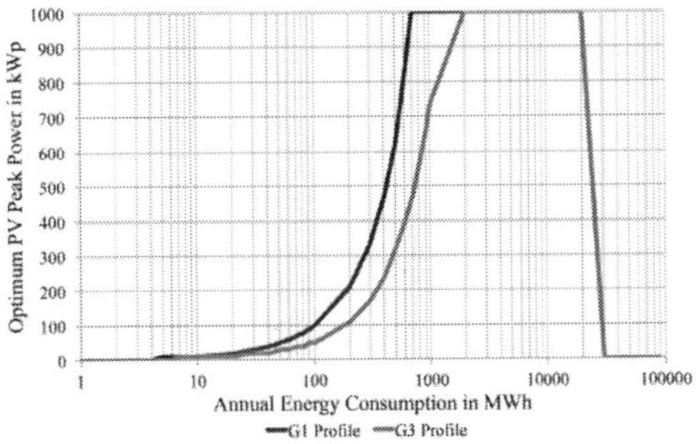

Figure 4: Optimum PV peak power in relation to the annual energy consumption. The optimum PV capacity depends on the load profile (G1 in blue, G3 in green).

The comparison of the industrial load profiles G1 and G3 shows a small difference in the optimum PV power of the total PV system (see Figure 5). The G3 profile represents industries with a constant energy load all day and night on weekdays and weekend. Thus, the capacity of the PV system is smaller than compared to an industry with a G1 load profile (largest energy load between 8 h and 18 h). The PV yield which is produced during daytime leads to a larger own-consumption rate if the main energy demand is also on daytime. If the annual energy consumption exceeds 1 GWh, the capacity of the PV system drops as the energy purchase costs decrease faster than the LCOE of the PV system (see equations above).

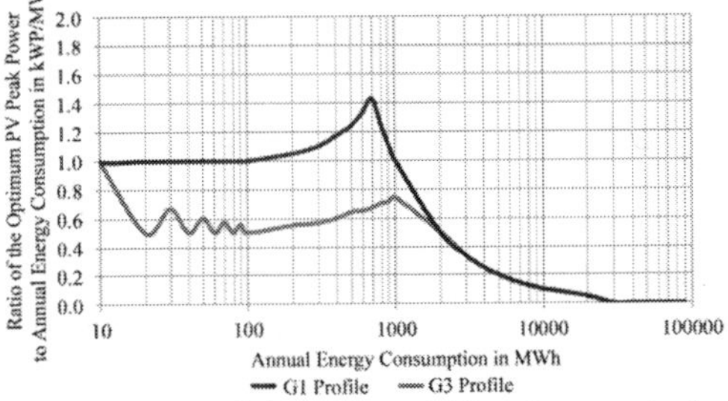

Figure 5: Ratio of the optimum PV peak power to the annual energy consumption (G1 load profile in blue, G3 in green). The vibration of the green curve is an artefact of the algorithm due to a too large step length.

Finally, we have analyzed the sensitivity of the four main input parameters – annual energy costs, energy purchase price, LCOE and the PV yield. The question has been how large is the change of the optimum PV capacity in case of parameter variation. A sensitivity of 1.0 means, that an increase of one parameter by 5 % leads to an increased annual energy costs of 5 %. Figure 6 shows for three different load profiles, that the sensitivity of the input parameter is rather small. The levelized costs of electricity are very sensitive to changes for households (H0 profile) due to a maximum load during evening hours. The accuracy of the PV yield simulation has a negligible influence on the final result.

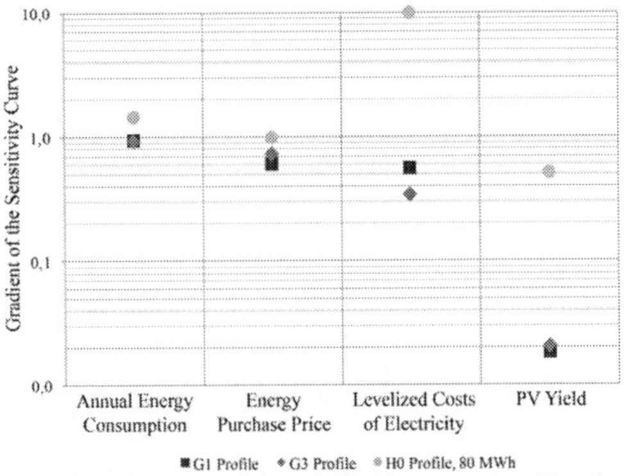

Figure 6: Gradient of the sensitivity curve regarding the four varied parameters. Positive values represent a positive slope of the curve. The annual energy consumption and the energy purchase price are the most sensitive parameters for industrial load profiles (G1 and G3). The levelized costs of electricity show the largest sensitivity for households (H0 profile).

4 CONCLUSION

We have presented an algorithm which can automatically determine the technical and economic optimum of the PV system with two subsystems regarding the energy consumption profile of the consumer. The algorithm is able to identify the best combination of two PV systems with different orientation/elevation angle of the PV modules. The results are robust against variation of the input parameter.

The algorithm will help customers to identify the best PV system setup for their individual energy consumption and costs. First, a PV simulation is performed for two different orientation/elevation angles to get the PV yield for both subsystems under consideration of the specific project parameter like the location and information about the roof. Second, the algorithm will try to figure out the best PV system configuration by minimization of the annual energy costs. If available, the load profile is included otherwise standard load profiles are used.

Next, we will integrate an automatic PV simulation. The consideration of an additional battery system is not necessary at the moment as the LCOE of batteries are still significant larger than the energy purchase price in particular for industrial customers [9].

5 REFERENCES

[1] REN21. Renewables 2016 Global Status Report. May 2016.

[2] J. Weniger, J. Bergner, T. Tjaden, V. Quaschning. Dezentrale Solarstromspeicher für die Energiewende. BWV Verlag. 2015.

[3] P. Groo, D. Heun, H. te Heesen. Anwendung eines Algorithmus zur technischen und wirtschaftlichen Auslegungsoptimierung von Photovoltaikanlagen. 31. Symposium Photovoltaische Solarenergie. 2016

[4] V. Becker, B. Schulitz, A. Lichius, H. te Heesen. Analysis for the Integration of a PV System in the Energy Concept at a German Airport. 30th European Photovoltaic Solar Energy Conference and Exhibition. P. 2754 - 2758. Sept. 2015.

[5] J. Birtel, H. te Heesen. Optimization of the Site for a Carport with a PV System Concept for Electric Vehicles. 30th European Photovoltaic Solar Energy Conference and Exhibition. P. 1717 - 1719. Sept. 2015.
[6] EuPD Research. Photovoltaik Preismonitor Deutschland. 2013. [Available online: https://www.solarwirtschaft.de/fileadmin/media/pdf/130218_EuPD_Preismonitor_q1_13.pdf]
[7] J. Kost, J. N. Mayer, J. Thomsen, N. Hartmann, C. Senkpiel, S. Philipps, S. Nold, S. Lude, T. Schlegl. Stromgestehungskosten Erneuerbare Energien. Fraunhofer Institur für Solare Energiesysteme. Nov. 2013.
[8] Bundesverband der Energie- und Wasserwirtschaft e.V. , BDEW. Strompreisanalyse März 2015.
[9] K.-P. Kairies, D. Magnor, J. Badeda, D. U. Sauer. Nachhaltige Entwicklung dezentraler Solarstromspeicher aus wissenschaftlicher Sicht. Institut für Stromrichtertechnik und elektrische Antriebe /RWTH Aachen. 2015.

MAXIMUM POWER POINT MODELING THROUGH IRRADIANCE BASED DUTY CYCLE CALCULATION

Prashant Upadhyay, Subrahmanyam Pulipaka and Rajneesh Kumar
Department of Electrical and Electronics Engineering, Birla Institute of Technology and Science, Pilani, India

ABSTRACT: This paper proposes an irradiance based modeling of duty cycle, for maximum power point operation of a panel. In existing methods for irradiance prediction, a relation was established to input maximum power duty cycle with parameters of panel. Proposed system operated with better maximum power point based on panel parameters duty cycle relation. Duty cycle of the converter is independent of feedback from output side so that the necessity of voltage sensing was eliminated and duty cycle input to converter was calculated from irradiance input on panel. Irradiance data for last 10 years was collected from NASA's website and for various level of irradiance Vmp, Imp are modeled to ensure maximum output power from converter. Duty cycle is then calculated with converter topology to trigger switch. Modeling of proposed system was done with different irradiance levels to calculate duty cycle for a year. Then obtained model was verified at predicted irradiance levels and a maximum point operation up to 98.21% was obtained. In various parameters of solar panel maximum power voltage, maximum power current and load side impedance was taken for duty cycle calculation. This proposed research work based on input effective resistance and output resistance is useful in constant voltage demand applications like battery charging or to supply constant dc load.
Keywords - MPP, Irradiance, duty cycle

1 INTRODUCTION

Solar energy has much potential to fulfill the electricity demand of world in future. As a renewable energy source various research works are done to make a pollution free efficient power generation system. Solar power generation is reached to a sufficient amount to supply more than 1% of globally installed capacity. India is highly populated and has high solar insolation mainly in desert area like shekhawati region of Rajasthan in India. With developing efficient systems for photovoltaic power generation can improve future power demand. In series of most relevant research work (Muthuramalingam & Manoharan, 2014) compared various MPPT methods experimentally with better dynamic performance. Irradiance plays important role on performance of solar panel output. Panel output power with input irradiance fluctuate due to fog, rain, dust and other weather parameters. To increase the tracking system and improve the efficiency under any whether conditions(Boukenoui, Salhi, Bradai, & Mellit, 2016) developed a fuzzy logic based MPPT to trach the identified irradiance. For irradiance prediction researchers presented artificial neural network model neural networks with exogenous inputs method performs better root mean square error was 0.243 MJ/m^2(Ahmad, Anderson, & Lie, 2015) and multilayer perception method was better with mean absolute percentage error of 5.21%(Behrang, Assareh, Ghanbarzadeh, & Noghrehabadi, 2010). Under non-uniform weather condition(Li, 2014) developed a new inductance-capacitor-diode circuit instead of DC-DC converter, analyzed with the ability of logic signal control with weather conditions.

All above discussed methods use feedback based output voltage/current sensing to achieve maximum power point operation of photovoltaic power generation systems. In proposed research a model is developed to calculate duty cycle of DC-DC converter with panel output parameters and load characteristics based on irradiance levels. Maximum power point operation is archived by calculating duty cycle for every month from the V_{MP} and I_{MP} calculated with existing irradiance prediction method. The feasibility of proposed model is tested for short interval irradiance forecasting and results obtained are discussed.

2 METHODOLOGY

The model is developed for Pilani, located at a Latitude – 28.37°N, Longitude – 75.6°E Shekhawati region of Rajasthan in India. Last 10 year's average and maximum irradiance values of this location are obtained from NASA's solar radiation data base to establish a relation on standard test conditions. From the irradiance values obtained the $V_{MP,STC}$, $I_{MP,STC}$ are calculated as given by

$$I_{MP}(G,T) = I_{MP,STC} * \frac{G}{G_{STC}} \quad (1)$$

$$V_{MP}(G,T) = V_{MP,STC} - K_2 * (T - T_{STC}) \quad (2)$$

The $V_{MP,STC}$ and $I_{MP,STC}$ are calculated from irradiance and temperature at STC.

From the boost converter relation of input output parameter, a relation is established between the input and output resistance to obtain the duty cycle.

Boost converter duty cycle relation of input voltage and output voltage is given by

$$V_o = \frac{V_{in}}{(1-D)^2} \quad (3)$$

And input output current relation as

$$I_o = I_{in}(1-D)^2 \quad (4)$$

So that the output resistance is given by

$$R_o = \frac{R_{ineff}}{(1-D)^2} \quad (5)$$

The duty cycle can be expressed as

$$D = 1 - \sqrt{\frac{R_{ineff}}{R_o}} \quad (6)$$

If the panel should operate at MPP the effective resistance is considered as R_{OMPP} at the input side and is equivalent to the ratio of V_{OMPP} and I_{OMPP}

Table 1
Operating points of panel for calculated duty cycle

Month	Average Irradiance(W/m^2)	V_{OP}(V)	I_{OP}(A)	R_{IN}(Ω)	R_O(Ω)	% Duty cycle	%MPP
Jan	363	16.69	1.19	13.16	100	63.71	99.1
Feb	455	16.91	1.51	10.48	100	67.61	99.07
Mar	588	17.52	1.88	8.24	100	71.29	97.69
Apr	711	17.54	2.33	6.82	100	73.83	98.33
May	800	18.01	2.51	6.08	100	75.33	96.07
Jun	799	17.93	2.52	6.10	100	75.28	95.92
Jul	686	17.50	2.25	7.04	100	73.44	98.37
Aug	616	17.34	2.02	7.85	100	71.96	97.2
Sep	583	17.39	1.90	8.25	100	71.26	98.67
Oct	504	17.01	1.67	9.46	100	69.24	99.1
Nov	397	16.90	1.29	12.01	100	65.33	98.79
Dec	327	16.44	1.08	14.48	100	61.94	99.24

Hence, from (4)

$$D = 1 - \sqrt{\frac{R_{OMPP}}{R_o}} \qquad (7)$$

R_{OMPP} is expressed as

$$R_{OMPP} = \frac{V_{OMPP}}{I_{OMPP}} \qquad (8)$$

From (5), (6) D is given by

$$D = 1 - \sqrt{\frac{V_{OMPP}}{I_{OPP} * R_o}} \qquad (9)$$

Duty cycle is calculated from the 3hr and 5 hr average irradiance values of 12th January, 2016 and the maximum point operation is observed by implementing PV system on a solar array simulator as shown in fig.1.

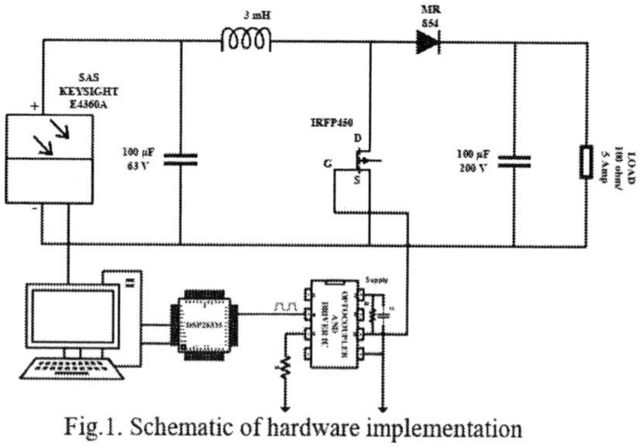

Fig.1. Schematic of hardware implementation

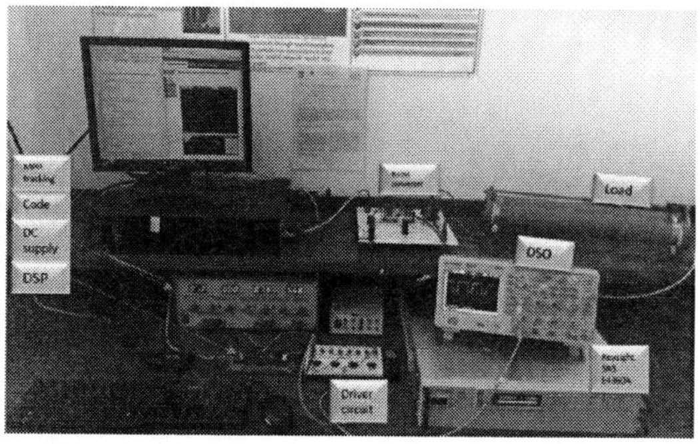

Fig.2. Experimental hardware of proposed system

A Keysight E4360A simulator is used to simulate the PV system at desired irradiance. A dsp28335 module is used to input duty cycle to the boost converter and the operating point on the curve for both the irradiance levels calculated is observed. Opto-coupler circuit is used to isolate logic circuitry and power circuitry and provided high current gate pulse to DC-DC converter using driver IR2113 IC. Solar array simulator is connected as a source to input the DC-DC converter. A resistive load of 100 ohms is connected to analyze output characteristics of converter. Results are verified and implemented with laboratory hardware prototype shown in fig. 2.

Panel output parameters and their established relation based calculation is shown in table 1. The obtained operating point of voltage and current is compared with the ideal MPP operation and the corresponding IV and PV curves are illustrated in fig. 3.

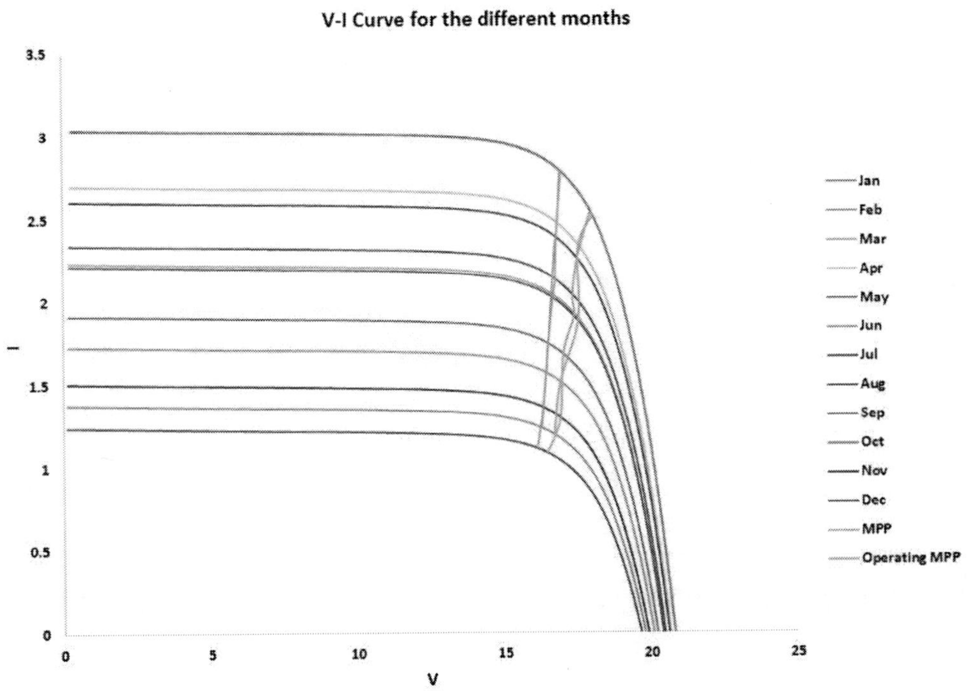

a) V-I curve

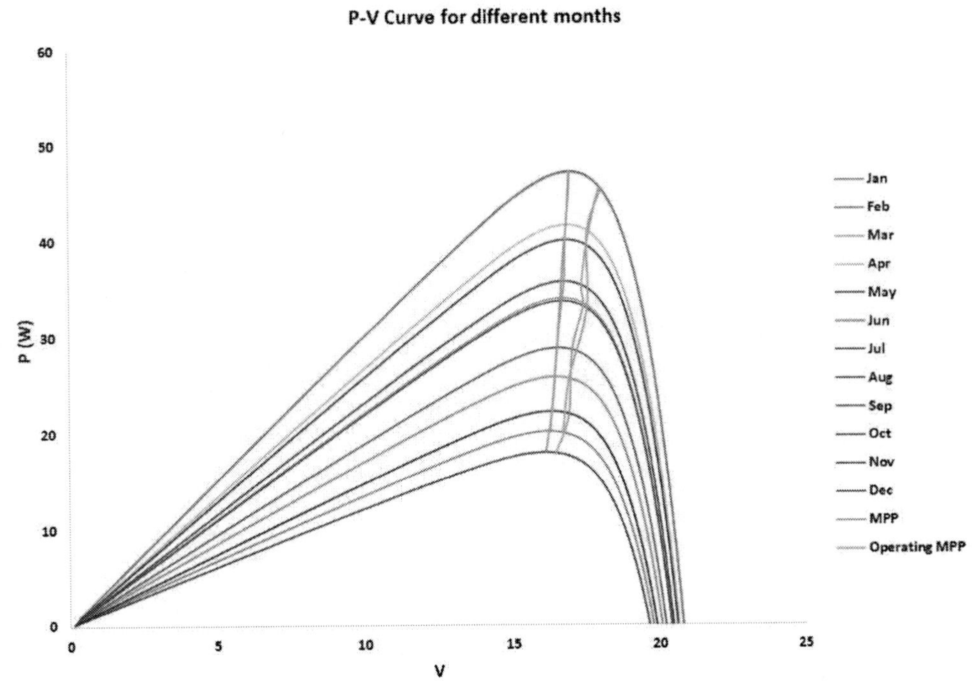

b) P-V curve

Fig. 3. V-I & P-V curves for MPP and operating point

3 RESULTS AND VALIDATION

Duty cycle is calculated for both irradiance levels and is implemented on simulator to find the operating point of panel at the calculated duty cycle. For the 3 hr. average irradiance value of 553 W/m^2, duty cycle calculated is 75.8% and at this duty cycle the operating point of the system is at 98.21% as shown in fig.4. If this duty cycle is kept constant for a day, the percentage MPP operation is obtained as illustrated in fig.5.

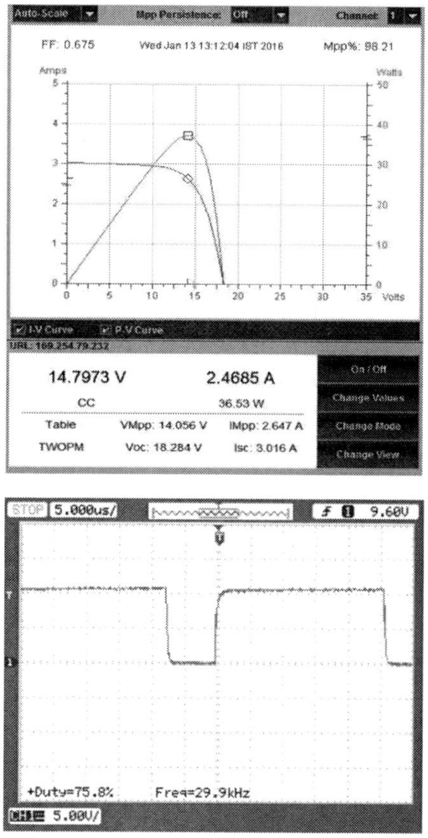

Fig. 4. Keysight SAS E4360A % MPPT and calculated duty cycle

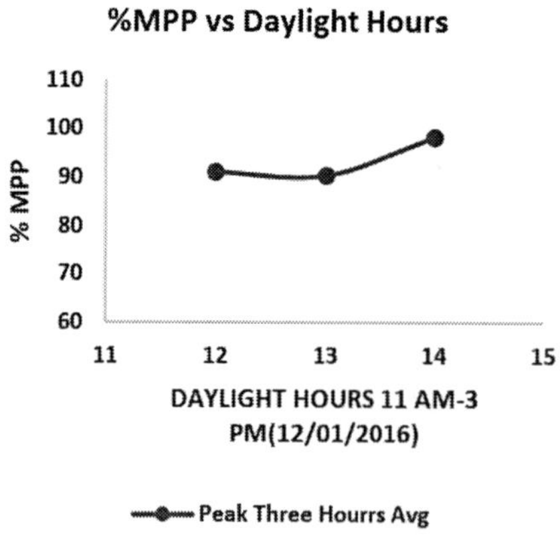

Fig. 5. % MPP throughout the Peak Three Hours Avg.

It is observed that for maximum time (6 hours) the %MPP is above 80% throughout the day.

The duty cycle is also calculated from irradiance values of daily average, 7 hour and 5 hour average as shown in table 2 and the %MPP operation is illustrated in fig.6. It is observed that the %MPP is maximum when the irradiance used to calculate duty cycle is taken for short term intervals.

Table 2
Irradiance and duty cycle calculated

Irradiance	Duty Cycle (%)	MPP (%)
Day AVG (395.11 W/m^2)	72.34	55-97.3
Peak SEVEN HRS AVG (460.2 W/m^2)	75.10	73-99.5
Peak FIVE HRS AVG (515 W/m^2)	76.03	83.45-98.5

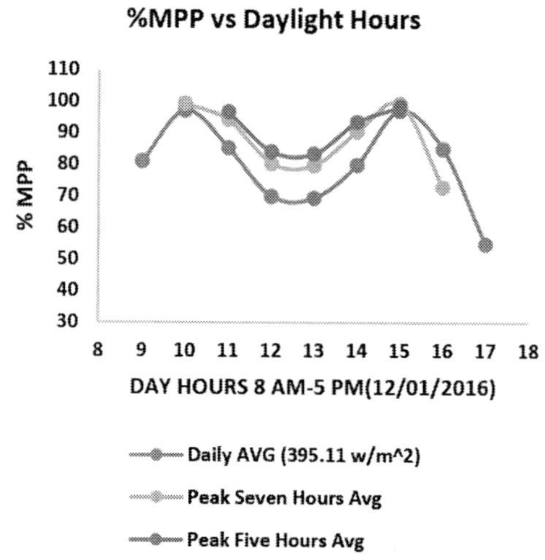

Fig. 6. %MPP at duty cycle calculated from various irradiance intervals

4 INNOVATION AND RELEVANCE

It is observed that as the time range at which irradiance obtained is decreased, the %MPP obtained is increasing. In the literature there are various models to predict irradiance on very long term basis(McCandless, Haupt, & Young, 2015) to short term span (Torregrossa, Le Boudec, & Paolone, 2016). If such models are used to predict irradiance in long term or short term, and from the obtained irradiance, if the duty cycle is calculated as proposed in this research work, the necessity of sensing the output voltage is completely eliminated. This also leads to preventing the malfunctioning of entire PV system due to errors in the sensors.

Acknowledgement

Authors would like to acknowledge the support provided by the Ministry of New and Renewable Energy, India for the project sanctioned under the grant number No. 31/9/2011-12/ST.

References

Ahmad, A., Anderson, T. N., & Lie, T. T. (2015). Hourly global solar irradiation forecasting for New Zealand. *Solar Energy*, *122*, 1398–1408. http://doi.org/10.1016/j.solener.2015.10.055

Behrang, M. a., Assareh, E., Ghanbarzadeh, a., & Noghrehabadi, a. R. (2010). The potential of different artificial neural network (ANN) techniques in daily global solar radiation modeling based on meteorological data. *Solar Energy*, *84*(8), 1468–1480. http://doi.org/10.1016/j.solener.2010.05.009

Boukenoui, R., Salhi, H., Bradai, R., & Mellit, A. (2016). A new intelligent MPPT method for stand-alone photovoltaic systems operating under fast transient variations of shading patterns. *Solar Energy*, *124*, 124–142. http://doi.org/10.1016/j.solener.2015.11.023

Li, S. (2014). A MPPT control strategy with variable weather parameter and no DC/DC converter for photovoltaic systems. *Solar Energy*, *108*(Dc), 117–125. http://doi.org/10.1016/j.solener.2014.07.002

McCandless, T. C., Haupt, S. E., & Young, G. S. (2015). A model tree approach to forecasting solar irradiance variability. *Solar Energy*, *120*, 514–524. http://doi.org/10.1016/j.solener.2015.07.020

Muthuramalingam, M., & Manoharan, P. S. (2014). Comparative analysis of distributed MPPT controllers for partially shaded stand alone photovoltaic systems. *Energy Conversion and Management*, *86*, 286–299. http://doi.org/10.1016/j.enconman.2014.05.044

Torregrossa, D., Le Boudec, J.-Y., & Paolone, M. (2016). Model-free computation of ultra-short-term prediction intervals of solar irradiance. *Solar Energy*, *124*, 57–67. http://doi.org/10.1016/j.solener.2015.11.017

COMPARATIVE LIFE CYCLE ASSESSMENT OF PV TECHNOLOGIES

Satwant Dahiya and Thomas Vogt
NEXT ENERGY - EWE Research Centre for Energy Technology at the University of Oldenburg, 26129 Oldenburg, Germany
Phone: +49 441 99906103; Fax: +49 441 99906109;
E-mail: satwant_dahiya@yahoo.in
thomas.vogt@next-energy.de

ABSTRACT: A comparative Life Cycle Assessment (LCA) of 'Silicon wafer based' (Single-Si & Multi-Si) and 'Thin-film' (CdTe & CIGS) PV types is conducted to develop a comparative assertion for Global Warming Potential (GWP), Cumulative Energy Demand (CED), Energy Payback Time (EPBT) and Energy Return on Investment (EROI). Based on these impacts, the chosen PV types are ranked for their sustainability quotient. The study followed a 'cradle-to grave' approach including the stages of raw material extraction and processing, PV cell & module manufacturing followed by use and disposal of slanted roof-mounted, grid connected, 3 kWp PV systems operating at 1700 kWh/m^2/year of solar irradiation for a lifetime of 30 years. The results indicated that Single-Si PV has the highest impacts followed by Multi-Si, CIGS and CdTe PV type. The results are also compared to the harmonized outcome from previous PV LCA studies to reflect the change in impacts with evolving technology and although the environmental ranking remains unchanged, the findings of this study shows a decrease of 7-43% for GWP, 18-31% for CED, 30-52% for EPBT & 24-75% for EROI based on PV types. An advanced contribution analysis is further carried out to determine the relative impacts of individual PV system components and to provide an insight on most impact-burdening value chain processes and input flows with high potential for subjection to ecological optimization.

Keywords: PV Module, Sustainable, Life Cycle Assessment, Environmental Impacts

1 INTRODUCTION

Declining manufacturing costs, policy impetus and consumer awareness for climate change mitigation is setting a stage for photovoltaics (PV) to play a key role in shaping the future of energy scenarios. The global cumulative installed PV capacity in 2015 reached 242 Giga Watt peak (GWp) showing annual growth rates of 41% since 2001 [1]. Although, PV is considered as one of the cleanest sources of energy, earlier LCA studies have raised questions on the sustainability of the technology. Earlier studies for instance show that PV technology is more resource and energy intensive, especially during the manufacturing and disposal phase [2]. In addition, it is estimated that PV technology extends a greater Global Warming Potential (GWP) and shows a higher energy resource intensiveness in terms of Cumulative Energy Demand (CED) when compared with other renewable technologies like wind, geothermal, hydro, etc. for per unit electricity generated [3, 4]. The increased production and use of photovoltaics at global level amplifies the seriousness of identification and evaluation of these impacts in an objective manner so that the research and development process could orient itself in tapping the ultimate potential of photovoltaics as "green" source of energy.

At the same time, the continuous evolution of technology has led to emergence of multiple PV types in the global market with a recorded improvement in conversion efficiencies. Also, the supply chains and manufacturing processes have evolved to be more economic in use of raw materials, waste production and energy use, therefore calling in for a need of an updated environmental impact analysis to reflect upon the technology as of today.

Depending on their global market share [1], four major PV types (Single-Si, Multi-Si, CIGS and CdTe) are chosen for comparison based on their GWP & CED [5]. Energy Payback Time (EPBT) [13] & Energy Return on Investment (EROI) [6] are also calculated and put side by side for different PV types to draw performance based comparisons.

Multiple PV LCA studies have been conducted in the past to analyse these impacts, but the results vary greatly with data availability, choice of system boundaries and assumptions made for individual cases [7, 8]. A "harmonization" of previous findings is attempted in multiple meta-studies [7, 8, 9] with the aim to have an absolute estimate of the impact values and to set a base for comparison, with which the outcome of current study is compared reflecting upon the change in environmental impacts of newer PV technology estimated using more accurate data and robust LCA modelling techniques.

Lastly, a 'contribution analysis' is carried out to determine the most impact intensive system components, life stages and process flows in the value chain of a PV system so that the so called "environmental hot-spots" could be identified and an ecological optimization could be aimed for during further research and development of PV technology.

2 METHODOLOGY

As defined by the Society of Environmental Toxicology and Chemistry (SETAC), Life Cycle Assessment (LCA) is a systematic approach of considering each step in the life cycle of a product, a process or a service while analyzing the associated environmental impacts [20]. ISO 14040/14044 [10] defines the framework, requirements and guidelines for conducting a LCA using 4 interlinked phases: Goal and Scope Definition, Inventory Analysis, Life Cycle Impact Assessment and Interpretation (cf. Fig.1)

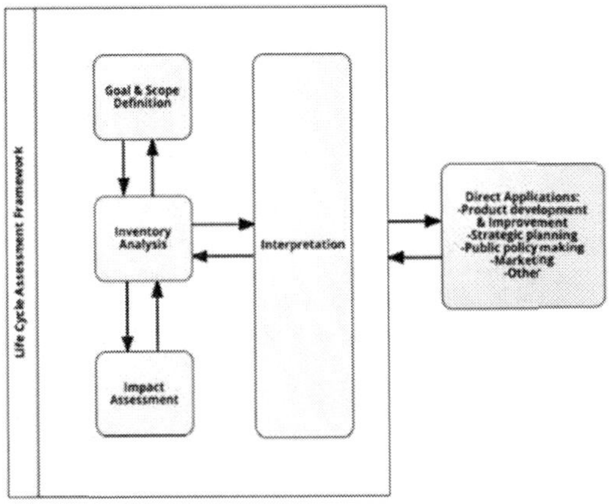

Figure 1: LCA Framework & applications as defined by ISO14040/14044 [22]

2.1 Goal and Scope

The primary goal of this study is to conduct a comparative LCA of the above listed PV product systems following a cradle to grave approach. It considers the complete PV manufacture value chain beginning from extraction of raw materials (semiconductor material) till disposal after use referred to as 'End of Life or EOL'. Recycling of modules at EOL is excluded from the assessment. The studied systems are analyzed within the geographical boundaries of Europe. The background processes used are specific to European average and are taken from Ecoinvent 3.1 database [11]. ENTSO-E electric grid (medium voltage) is used as electricity input for value chain processes and EURO 6 transport norms are assumed for the transport input processes as documented in the Ecoinvent 3.1 database [11]. The 'Life Cycle Inventory or LCI' data is collected from multiple literature sources summarized under 'Task 12" of International Energy Agency Photovoltaic Power Systems Programme (IEA PVPS) [12].

2.2 Product System & System Boundaries

A 3kWp PV system consisting of PV modules along with the Balance of System (BOS) which includes a mounting structure, 2 units of 2.5 kW inverter as well as cabling & electrical parts are collectively considered as the "product system". 1% module rejects due to manufacturing defects and 2% module repairs during installation and operational phase are assumed [13]. The PV systems are installed as slanted roof mounted systems with a performance ratio of 0.75 [14, 15] and are considered to be operational at solar insolation of 1700 kWh/m^2/year (specific to southern European location) with a lifetime of 30 years. Functional unit is considered to be 1 kWh of electricity generated (kWhp). Since the system is assumed to be grid connected, no battery is included in the assessment. The LCI is adjusted to the reference flows which are calculated for the above described product system using the average market data (conversion efficiencies and surface area dimensions) for modules available in PV market [16]. Table 1 elaborates further on module efficiencies and reference flows.

Table 1: Module efficiencies at STC (Standard Test Conditions [21]) & Reference Flows for a 3kWp module (module surface area)

PV Type	Module Efficiency (%)	Reference Flow (m^2)
Single-Si	16.80	18.39
Multi-Si	15.58	19.83
CIGS	15.97	19.34
CdTe	13.84	22.32

3 RESULTS

The LCI is modelled using 'SimaPro 8.1.0' LCA software [17] 'CED 1.09' [18] and 'IPCC 2013 100a' [19]. Life Cycle Impact Assessment (LCIA) methods are used to estimate CED and GWP respectively. The results are normalized to the functional unit. EPBT is calculated using CED and EROI from EPBT [13].

3.1 Global Warming Potential (GWP)

The GWP is calculated as 40.92, 31.27, 26.15 and 16.78 g CO_2 eq./kWhp for Single-Si, Multi-Si, CIGS and CdTe PV systems, respectively. On comparison of different PV types side by side (cf. Fig.2) it is shown that CdTe PV has the lowest GWP impacts and Single-Si PV type has the highest followed by Multi-Si and CIGS PV type with lower values. Furthermore, comparisons with previous studies have shown a decrease in impacts ranging from 7-43% depending on different PV types as seen below in Fig. 2.

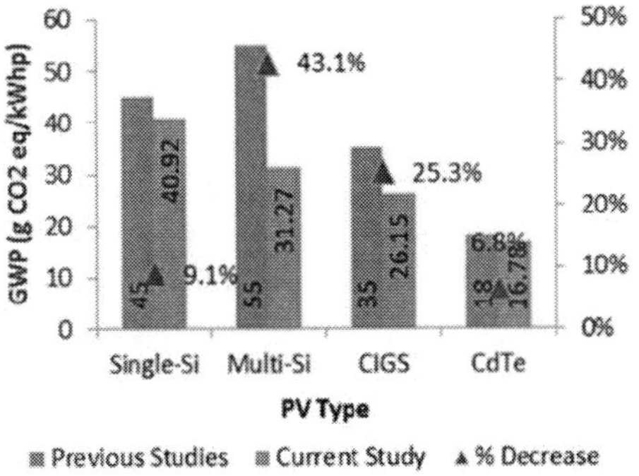

Figure 2: GWP comparison of different PV types as well as with previous (harmonized) results

3.2 Cumulative Energy Demand (CED)

CED comparison (cf. Fig. 3) follows a similar trend to the GWP with Single-Si PV having the highest (0.80 MJ/kWhp or 4554 MJ/m^2) followed by Multi-Si (0.57 MJ/kWhp or 3011 MJ/m^2), CIGS (0.46 MJ/kWhp or 2165 MJ/m^2) and CdTe (0.27 MJ/kWhp or 1410 MJ/m^2) PV. Note that in order to achieve coherence for comparison with previous studies, the CED in Fig. 3 is depicted per m^2 of module's active surface area. A decrease of 18-31% from older values can be observed, depending of different PV types.

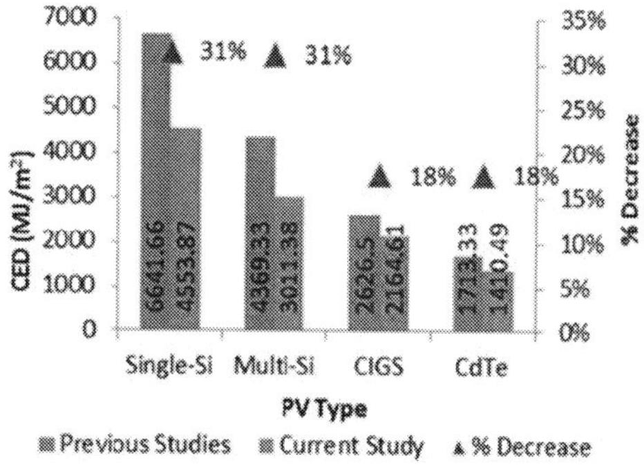

Figure 3: CED comparison of different PV types as well as with previous (harmonized) results

3.3 Energy Payback Time (EPBT)

As shown in Fig. 4 below, the decreasing order of EPBT is as follows: Single-Si (2.07 years) followed by Multi-Si PV (1.47 year), CIGS PV (1.19 years) and lastly CdTe PV (0.71 years). The comparison with older studies shows a decrease of approx. 30% for Thin Film PV and about 50% for Silicon wafer based PV.

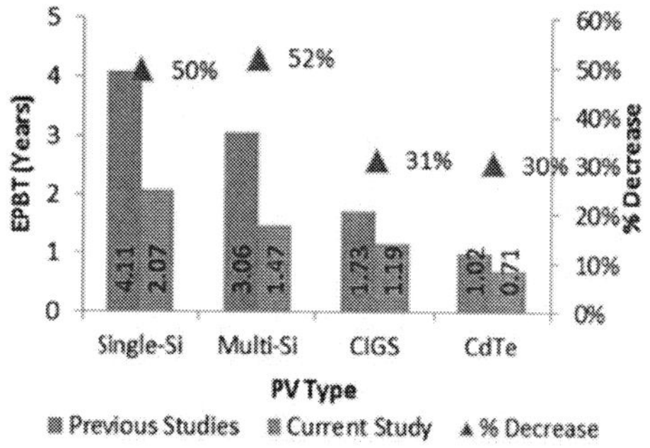

Figure 4: EPBT comparison of different PV types as well as with previous (harmonized) results.

3.4 Energy Return on Investment (EROI)

The EROI is calculated as 14.50, 20.34, 25.13 & 42.26 for Single-Si, Multi-Si, CIGS & CdTe PV type respectively (cf. Fig. 5) and the % increase in this case is 24-75% on comparison to older literature data.

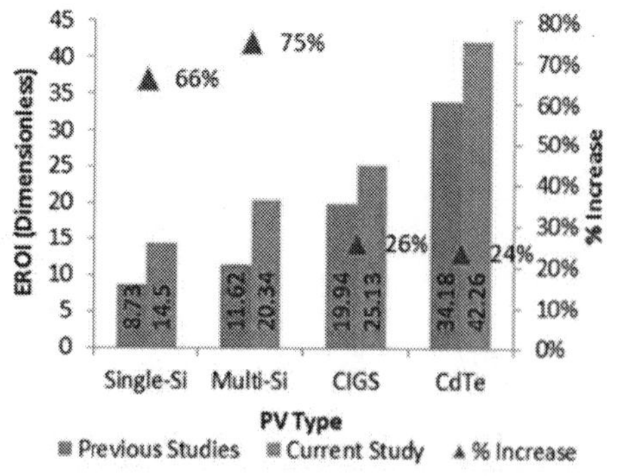

Figure 5: EROI comparison of different PV types as well as with previous (harmonized) results.

3.4 Contribution Analysis

A contribution analysis is conducted as a part of "Interpretation" phase in LCA where contribution of individual components/parts/life stages of the product system to the impact is estimated. CED is used as an example impact to assess the contribution.

A breakdown of impacts between different parts of product system shows that the PV module manufacturing has highest impacts (50 - 83%) followed by the inverters used in the lifetime (cf. Fig 6).

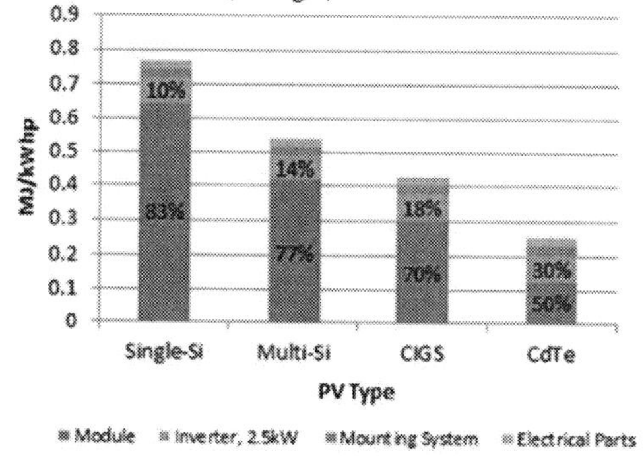

Figure 6: PV system parts contribution to CED of a 3kWp PV system

To further break down the module manufacturing chain of PV types with higher impacts (Single-Si & Multi-Si), the value chain is subjected to a subsequent contribution analysis. In Single-Si PV value chain, it is estimated that the single crystalline Silicon Ingot growing process has the maximum share in CED while the contribution of other processes is distributed more or less equally over the value chain (cf. Fig.7).

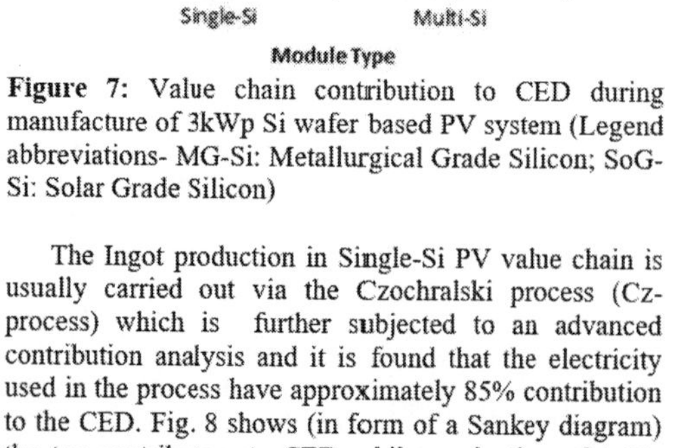

Figure 7: Value chain contribution to CED during manufacture of 3kWp Si wafer based PV system (Legend abbreviations- MG-Si: Metallurgical Grade Silicon; SoG-Si: Solar Grade Silicon)

The Ingot production in Single-Si PV value chain is usually carried out via the Czochralski process (Cz-process) which is further subjected to an advanced contribution analysis and it is found that the electricity used in the process have approximately 85% contribution to the CED. Fig. 8 shows (in form of a Sankey diagram) the top contributors to CED while production of 1 Kg single crystalline Silicon to be used as a semiconductor in Single-Si PV.

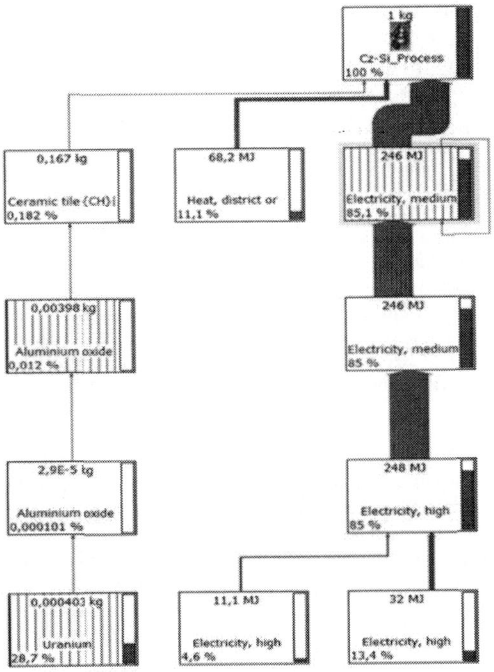

Figure 8: Sankey diagram for Czochralski-Silicon (Cz-Si) ingot growing process for Single-Si PV

The contribution analysis for GWP, EPBT and EROI also follows the same trend as of CED.

4 CONCLUSIONS

CdTe technology has shown to have the lowest CED and GWP impacts throughout its lifetime while Single-Si PV type has the highest. The difference between the two is more than two-fold for both CED and GWP. Multi-Si follows Single-Si with ~ 24% lower GWP and ~ 34% lower CED although having a difference of ~ 1% in their module efficiencies (cf. Table 1), which makes it a preferred Silicon wafer based PV technology from an environmental point of view. On basis of GWP, CED, EPBT, EROI the following sustainability ranking (high to low) is obtained for PV types in question: CdTe > CIGS > Multi-Si > Single-Si PV. The primary reason for CdTe and CIGS leading the pack is the increased performance of thin film PV (CdTe & CIGS) over the past few years in terms of conversion efficiencies and a less burdening value chain especially for the extraction and preparation of semiconductor raw material.

Also, not all life stage and processes in a PV life cycle are equal in terms of impacts. The manufacturing stage of PV is most energy intensive and has the highest GWP & CED when compared to other life stages in all four types of PV systems. In thin film PV the impacts are more or less equally divided between manufacturing of modules and BOS but in Silicon wafer based PV, the impacts are skewed more towards the production chain. The use and disposal phase CED and GWP impacts are negligible as compared to the production phase impacts. Of all four PV systems, the module production in Silicon wafer based PV has the highest contribution to both CED and GWP and upon conducting an advanced contribution analysis of module manufacturing value chain, it is shown that the 'Czochralski process' is the most energy intensive where approximately 85% of the impacts are due to the electricity used for production of single crystalline Silicon. Therefore, for ecological optimization it is recommended to use alternative sources of electricity where there is a higher share of renewables in the electricity grid or to replace the process with an alternative single silicon crystal production pathway such as Float Zone Silicon process, etc. which requires less energy in comparison to Czochralski process.

Lastly, comparison of the CED, GWP, EPBT and EROI values of this study with the pool of older studies conducted between 2000 and 2014 have shown that the GWP, CED and EPBT have decreased up to varying extents depending on PV type. There most important factors for this change are the increased efficiency of current PV types along with an increased share of renewables in the used electricity grid (ENTSO-E). Other factors are efficient production pathways, reduced production waste, use of recycled raw materials, improvement in transport efficiency, etc. Recycling of used modules and BOS components at the EOL could further decrease the impacts making PV technology competitive to other renewable energy technologies in terms of sustainability.

5 ACKNOWLEDGEMENTS

The authors would like to acknowledge and appreciate the support of Dr. Martin Vehse and Dr. Stefan Geißendörfer from the PV division of NEXT ENERGY - EWE Research Centre for Energy Technology at the University of Oldenburg, Germany. The first author would also like to thank Prof. Edward C. Moretti and Prof. Dr. Kirk W. Junker from IMES program, University of Cologne, Germany for supporting this work as a part of his master thesis.

6 REFERENCES

[1] © Fraunhofer ISE: Photovoltaics Report, updated: 6 June 2016

[2] Sherwani, A. F., and J. A. Usmani. "Life cycle assessment of solar PV based electricity generation systems: A review." Renewable and Sustainable Energy Reviews 14.1 (2010): 540-544.

[3] Pehnt, Martin. "Dynamic life cycle assessment (LCA) of renewable energy technologies." Renewable energy 31.1 (2006): 55-71.

[4] Evans, Annette, Vladimir Strezov, and Tim J. Evans. "Assessment of sustainability indicators for renewable energy technologies." Renewable and sustainable energy reviews 13.5 (2009): 1082-1088.

[5] Battisti, Riccardo, and Annalisa Corrado. "Evaluation of technical improvements of photovoltaic systems through life cycle assessment methodology." Energy 30.7 (2005): 952-967.

[6] Raugei, Marco, Pere Fullana-i-Palmer, and Vasilis Fthenakis. "The energy return on energy investment (EROI) of photovoltaics: Methodology and comparisons with fossil fuel life cycles." Energy Policy 45 (2012): 576-582.

[7] Hsu, David D., et al. "Life cycle greenhouse gas emissions of crystalline silicon photovoltaic electricity generation." Journal of Industrial Ecology 16.s1 (2012): S122-S135

[8] Kim, Hyung Chul, et al. "Life cycle greenhouse gas emissions of thin-film photovoltaic electricity generation." Journal of Industrial Ecology 16.s1 (2012): S110-S121.

[9] Bhandari, Khagendra P., et al. "Energy payback time (EPBT) and energy return on energy invested (EROI) of solar photovoltaic systems: A systematic review and meta-analysis." Renewable and Sustainable Energy Reviews 47 (2015): 133-141.

[10] Finkbeiner, M., et al. "The new international standards for life cycle assessment: ISO 14040 and ISO 14044." The international journal of life cycle assessment 11.2 (2006): 80-85.

[11] www.ecoinvent.org

[12] R. Frischknecht, R. Itten, P. Sinha, M. de Wild-Scholten, J. Zhang, V. Fthenakis, H. C.Kim, M. Raugei, M. Stucki, 2015, Life Cycle Inventories and Life Cycle Assessment of Photovoltaic Systems, International Energy Agency (IEA) PVPS Task 12, Report T12-04:2015.

[13] R. Frischknecht, G. Heath, M. Raugei, P. Sinha, M. de Wild-Scholten, V. Fthenakis, H. C. Kim, E. Alsema and M. Held, 2016, Methodology Guidelines on Life Cycle Assessment of Photovoltaic Electricity, 3rd edition, IEA PVPS Task 12, International Energy Agency Photovoltaic Power Systems Programme. Report IEA-PVPS T12-06:2016, ISBN 978-3-906042-38-1.

[14] IE Commission. "International Standard IEC 61724: Photovoltaic System Performance Monitoring Guidelines for Measurements, Data Exchange and Analysis." IEC (1998).

[15] Fthenakis, Vasilis M., Hyung Chul Kim, and Erik Alsema. "Emissions from photovoltaic life cycles." Environmental science & technology 42.6 (2008): 2168-2174.

[16]www.photon.info/photon_site_db_solarmodule_en.photon; Accessed 02.11.2015

[17] PRé Consultants (2015) SimaPro 8.1.0, Amersfoort, NL.

[18] Huijbregts, Mark AJ, et al. "Cumulative energy demand as predictor for the environmental burden of commodity production." Environmental science & technology 44.6 (2010): 2189-2196.

[19] IPCC, 2013: Climate Change 2013: The Physical Science Basis. Contribution of Working Group I to the Fifth Assessment Report of the Intergovernmental Panel on Climate Change [Stocker, T.F., D. Qin, G.-K. Plattner, M. Tignor, S.K. Allen, J. Boschung, A. Nauels, Y. Xia, V. Bex and P.M. Midgley (eds.)]. Cambridge University Press, Cambridge, United Kingdom and New York, NY, USA, 1535 pp.

[20] Fava, James A., and Society of Environmental Toxicology and Chemistry. A technical framework for life cycle assessment: workshop report; August 18-23, 1990. SETAC, 1991.

[21] Carr, A. J., and T. L. Pryor. "A comparison of the performance of different PV module types in temperate climates." Solar Energy 76.1 (2004): 285-294.

[22] „DIN EN ISO 14040: 2006-10." Umweltmanagement – Ökobilanz - Grundsätze und Rahmenbedingungen (ISO 14040: 2006).

ESTIMATING FUTURE RECYCLING QUANTITIES OF PV MODULES IN THE EUROPEAN UNION

Gerhard Kleiss
SolarWorld AG, Martin-Luther-King Str. 24, 53175 Bonn, Germany
gerhard.kleiss@solarworld.com

ABSTRACT: This work addresses an expected timeline for future quantities of to-be-recycled PV solar modules. The triggering point considered for recycling is end-of-life (breakdown, wear-out). A general adaption of the reliability engineering concept in terms of the Weibull failure model to the expected lifetime of PV modules is given and in particular a three-modal failure model (initial failures, service life failures, wear out) is reduced to two less complex failure models: a step functional model is presented, that can be applied when comprehensive knowledge for the service live failures (from a single manufacturer with precise data) is available. In case of absence of precise data a reduction to a generalizing single failure mode model is proposed and adapted. The latter concept is applicable when statements over an entire market (with a mix of products and manufactures) are done. Model calculations for the European market and the German market as an example show that the time-peak-shaved deployment of PV modules translates into a smooth and broadened distribution of the expected return time of the modules. Consequently, industry should be prepared much earlier to take back and recycle modules than commonly expected.

KEYWORDS: Reliability, Solar Modules, Recycling, Weibull Distribution

1. INTRODUCTION

With the inclusion of PV modules under the WEEE directive in Europe [1] and the finalized implementation of this directive into virtually all national legislations of member states, the question of PV recycling gains high attention. Requirements to recycling are made now by legislation and specialized recycling technologies are key to address this challenge [2], [3], [4]. First structures are operative [5] and the applied techniques are reported to deliver a considerable degree of material recovery. However, at this time the number of recycling companies offering PV recycling services seems to be limited to a small number of players. This may have several reasons, including the uncertainty about when major quantities of modules-to-be-recycled may be expected in the future. A good prediction of this triggering point is an important factor for the traditional recycling industry to estimate the best entry point into PV module recycling, as adaptions in the recycling process or the set-up of new machinery shall require invest and a robust business plan

This work adopts theoretical concepts of the theory of reliability engineering to estimate a time frame when solar PV modules may be expected for recycling. Calculations are based on historic data [6], [7] for installed capacity and projected market data. The purpose is to predict when considerable quantities of modules to be recycled may be expected. This is obviously based on historic and projected installation data of PV modules. In Industry, a single hypothesis is often adopted, namely that a PV module shall last (exactly) 25 years [8]. Such assumptions automatically lead to an expected future recycling timeline that corresponds exactly to a time-shifted (by 25 years) of the field-installation rate, whereas the 25 years are taken from the typical warranty time. However, it is obvious that this straight model may be too simplistic as PV modules reaching the end of their warranty period (e.g. 25 years) are neither expected to break down immediately after the warranty period nor to be disposed for recycling by this time [9]. On the other hand, the lifetime of modules will probably differ for various products and some will wear out before 25 years. Moreover, modules may be broken in operation during their "useful life" period due to exceptional external circumstances (e.g. catastrophic climatic impact like thunderstorms, hail) or can break down during shipment, installation, or more generally during their first fails. Ultimately, scrap from module production should be considered.

2. APPROACH

2.1. Failure model on the base of Weibull parameters and the bathtub curve.

The time scale resolution for all calculations presented here is one year. Following a general Weibull model, solar modules installed in the year i shall fail after a time x according to a parameter density function (PDF):

$$f_i(x) = \frac{\alpha}{\beta} \cdot \left(\frac{x}{\beta}\right)^{\alpha-1} \cdot e^{-\left(\frac{x}{\beta}\right)^{\alpha}} \qquad \text{(eqn. 1)}$$

where x runs from the year i onwards. The two parameters lifetime β and form factor α are assumed to be constant for all calculations presented here. The cumulated failure function CDF for modules fielded in year i is the integral of $f_i(x)$ and can be interpreted as the cumulated ratio of all defective (recyclable) modules until time x:

$$\mathrm{F}_i(x) = \int_0^x f_i(x)dx = 1 - e^{-\left(\frac{x}{\beta}\right)^{\alpha}} \qquad \text{(eqn. 2)}$$

$\mathrm{F}_i(x)$ approaches unity for large x, which simply reflects the fact that no product has an infinitive lifetime and ultimately all products shall fail:

$$\mathrm{F}_i(x \to \infty) = 1 \qquad \text{(eqn. 3)}$$

Consequently, the so-called reliability function $\mathrm{R}_i(x) = 1 - \mathrm{F}_i(x)$ is a measure for the probability of solar PV modules to survive from year i until time x (where $x = 0$ in year i as stated above). For sake of completeness, it may be noted that for $x = \beta$:

$$\mathrm{F}_i(\beta) = 1 - \mathrm{e}^{-1} = 62{,}34\% \qquad \text{(eqn. 4)}$$

The hazard function $h_i(x)$ gives the general probability for PV modules (fielded in year i) to fail at time x under condition that there was no fail until time x:

$$h_i(x) = \frac{f_i(x)}{R_i(x)} = \frac{f_i(x)}{1-F_i(x)} \qquad \text{(eqn. 5)}$$

It is the hazard function $h_i(x)$ that shows when modules failures (the triggering point for recycling modules) from the population of modules fielded in year i will occur.

$h_i(x)$ is usually plotted in the form of the well-known bathtub curve, where several sections can be considered, all of them with different form factor α :

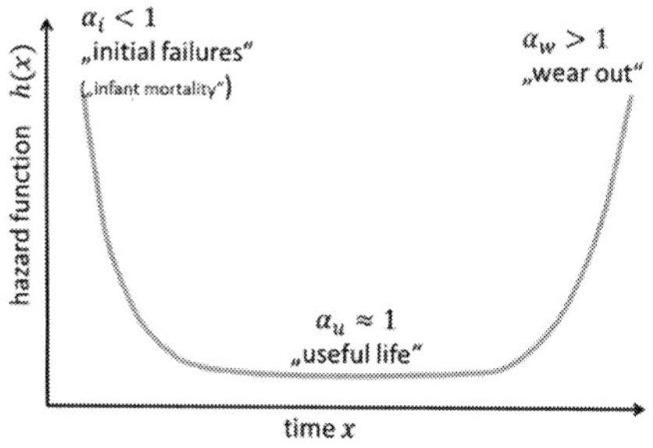

Fig. 1: Bathtub curve, schematically

When the bathtub model is applied, there should be a dedicated set of Weibull parameters for each of the 3 sections as shown in the above scheme. A simplified approach is shown later. In particular, the length of the "useful life" must be modelled, as Weibull lifetimes β may be found from data to be erratically long when wear-out is not yet visible.

2.2. Connection between Weibull statistics and service life considerations

The approach to predict service life of PV modules (e.g. defined in terms of a typical power warranty that promises a residual power of 80% after 25 years) through ALT (Accelerated "Lifetime" Testing) has been comprehensively studied recently [12]. These concepts are typically applying an Arrhenius equation to model the influence of temperature. The Arrhenius factor k is:

$$k = A \cdot e^{-\frac{E_A}{R \cdot T}} \qquad \text{(eqn. 6)}$$

The kinetic reaction model assumed is derived fom a monomolecular chemical reaction (another example would be radioactive decay): the concentration c of a substance (or a population of non-defective PV modules) at a time x can be expressed as:

$$c(x) = c(0) \cdot e^{-kx} = c(0) \cdot e^{-x \cdot A \cdot e^{-\frac{E_A}{R \cdot T}}} \qquad \text{(eqn. 7)}$$

In (eqn. 7) the Arrhenius factor k has already been inserted. $c(x)$ can be interpreted as a reliability function $R(x)$. When we choose the Weibull concept in particular as generalized failure model, we obtain:

$$R(x) = 1 - F(x) = e^{-\left(\frac{x}{\beta}\right)^{\alpha}} \qquad \text{(eqn. 8)}$$

Since in (eqn. 7) x is only used the power of 1, the Weibull-form factor parameter α in (eqn. 8) must likewise be unity ($\alpha = 1$) for $R(x) = c(x)$. This indicates that the Arrhenius model for "service live" is a description of the module's performance during the "useful life" phase (compare Fig. 1) with a lifetime parameter inverse to the Arrhenius factor:

$$\beta = \frac{1}{k} \qquad \text{(eqn. 9)}$$

2.3. Considering multiple failure modes and selection of a simulation model

When several failure modes exist, there shall be several cumulated failure functions (e.g.: F' and F'' in the case of two failure modes). Such failure modes can be completely independent from each other or competing. In the case of completely independent failure modes the resulting total failure function F may be expressed with help of the fraction q of all product that are defective due to failure function F' as:

$$F = q \cdot F' + (1-q) \cdot F'' \qquad \text{(eqn. 10)}$$

This approach (eqn. 10) is typically used to model initial failures F' that occur during the first phase "infant mortality" of the bath tub curve.

If the two failure modes are competing then the total failure function F is given by:

$$F = 1 - (1 - F') \cdot (1 - F'') \qquad \text{(eqn. 11)}$$

Consequently, a complete approach to model the bathtub curve as shown in Fig. 1 would be applying 3 failure modes ("infant mortality", "useful life" and "wear out"). As all of those are expected to follow a Weibull characteristic as given in (eqn. 2), the general formula reads [13]:

$$F(x) = 1 - q \cdot e^{-\left(\frac{x}{\beta_i}\right)^{\alpha_i} - \left(\frac{x}{\beta_u}\right)^{\alpha_u} - \left(\frac{x}{\beta_w}\right)^{\alpha_w}} - (1-q) \cdot e^{-\left(\frac{x}{\beta_u}\right)^{\alpha_u} - \left(\frac{x}{\beta_w}\right)^{\alpha_w}} \qquad \text{(eqn. 12)}$$

The subscripts i, u, w in (eqn. 12) refer to the 3 phases "(initial) infant mortality", "useful life" and "wear-out". The first term (containing the 3 exponential functions) models a possible influence of the failure mode responsible for "useful life" and "wear out" during the first "infant mortality" phase. For purposes of this investigation, a rougher approach can be selected since the exact time resolution of $F(x)$ during the initial phase is not necessary: as the time scale of this study has been selected to be one year we simplify (eqn. 12) to:

$$F(x) = 1 - q' - (1 - q') \cdot e^{-\frac{x}{\beta_u} - \left(\frac{x}{\beta_w}\right)^{\alpha_w}} \qquad \text{(eqn. 13)}$$

A second simplification has been taken in (eqn. 13), namely to assume $\alpha_u = 1$, which has been discussed earlier in section 2.2.

Furthermore, the parameter β_u can be estimated under the assumption that during the "useful life" phase this would be the dominant failure mode and failures due to "wear out" could not be seen yet. A conservative assumption backed up by data given in [14] leads to an estimation of β_u= 1000 years, corresponding to a yearly return rate of 0,1 %.

Fig. 2 gives an example for the impact of the two failure functions of "useful live" and "wear out". Contributions to the term $e^{-\frac{x}{\beta_u} - \left(\frac{x}{\beta_w}\right)^{\alpha_w}}$ from (eqn. 13) are plotted. With this parameter selection a crossover in the year 11 can be observed and after year 15, the contribution of the wear-out (red line) has clearly the largest impact. The estimation of the parameters, especially for the wear-out is a difficult task. Despite the fact that we know today that there are many failure modes [15] leading to degradation and many research is done [16] to investigate the service life of PV modules (e.g.: power loss 20% after 25 years – compare [12] and the discussion in section 2.2), there is limited data available about the real end-of-life of solar modules (i.e. recyclability). This is also due to the fact that modules have not been designed for long life times in the past [9], [17] the design life of Parameter choices in the example of Fig. 2 are based on [10] and [15]. It should be noted, that for a given known population of solar modules (e.g. from a single manufacturer using with consistent quality levels) a good judgement about the length of the "useful life" phase T may be possible. In this case another modeling approach can be used to model "useful life" and "wear out":

$$F(x) = \begin{cases} x \le \mathrm{T}:\ 1 - e^{-\frac{x}{\beta_u}} \\ x > \mathrm{T}:\ 1 - e^{-\left(\frac{x-\tau}{\beta_w-\tau}\right)^{\alpha_w}} \end{cases} \qquad \text{(eqn. 14)}$$

(eqn. 14) is a 3 parameter Weibull model and the time-shift parameter τ must be chosen to be slightly smaller than T to generate a behavior compared to Fig. 2. The clear computational advantage of (eqn. 14) in comparison to (eqn. 13) is the ease of calculating the hazard function $h(x)$ – see (eqn. 5).

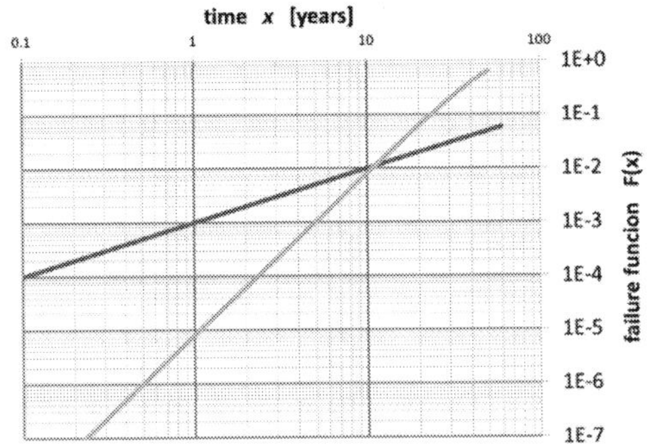

Fig.2: Failure functions F for two assumed failure mode contributions: "useful life"(β_u= 1000 years and α_u = 1, blue line) and "wear out" (β_w= 50 years and α_w =3, orange line).

3. RESULTS

3.1. General

As the "useful life" for all modules brought into the market is unknown (many different manufacturers with unknown quality) the approximation of (eqn. 14) cannot be used per se. However, within Europe, it may be generally postulated, that site dependent impacts (e.g. through temperature influence on aging [12]) is of minor importance for the sake of this model. As only the end-of-life of the modules is of interest for the question of recycling (eqn. 13) is simplified by omitting the "useful life" mechanism completely. As a second rough approximation, the losses during the early "infant mortality" phase are also discarded for the moment ($q' = 0$). The failure function becomes in this very rough approximation:

$$\mathrm{F}(x) = 1 - e^{-\left(\frac{x}{\beta_w}\right)^{\alpha_w}} \qquad \text{(eqn. 15)}$$

3.2. PV Market in Germany

Prediction calculations are based on the following scenario: Historic data show a peak of new installations during 2009 until 2013 in Germany (with highest values in 2010 through 2012). For purposes of prediction it is assumed that the level of new installations shall continue with the 2015 value.

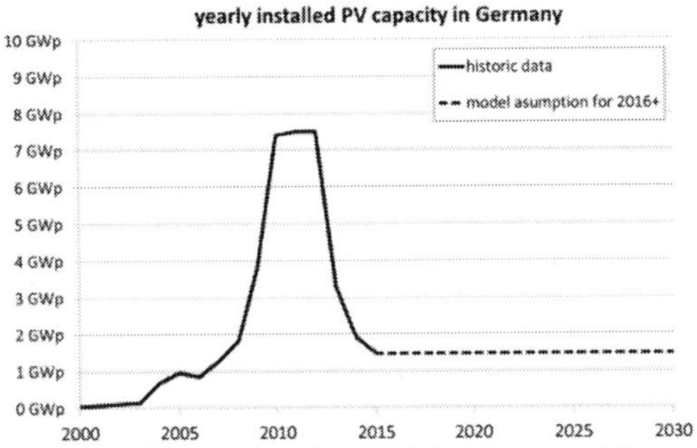

Fig. 3: historic data (until 2015) and assumed future evolution of yearly installed PV capacity in Germany.

Data shown in Fig. 3 form a set of yearly installed power $\{P_{2000}, P_{2001}, \dots, P_{2015}, \dots, P_{2030}\}$ where years before 2000 are not considered and all years after 2015 are estimated to be equal to the 2015 value.

Moreover, it is assumed that (eqn. 15) holds for all years 2000 ... 2030 and that the parameters α_w and β_w do not change over time. This is of course a rough estimation and the impact on the timeline can be tested by parameter variation.

With (eqn. 15) and the P_i data set, the Wattage of all PV modules that were fielded in year i and that are defective after x years is given by:

$$D_i(x) = \mathrm{F}(x) \cdot P_i \qquad \text{(eqn. 16)}$$

Ultimately, all modules shall fail – as (eqn. 3) postulates. $D_i(x)$ has in our model the same shape for all years and the index i denotes that the starting point ($x = 0$: fielding of the module) is in year i. This means that for a given year j there are contributions from all preceding years $i < j$ based on the prior installation rates weighed with the failure function. The model depends now again on two Weibull parameters, only: form factor α_w and life time parameter β_w.

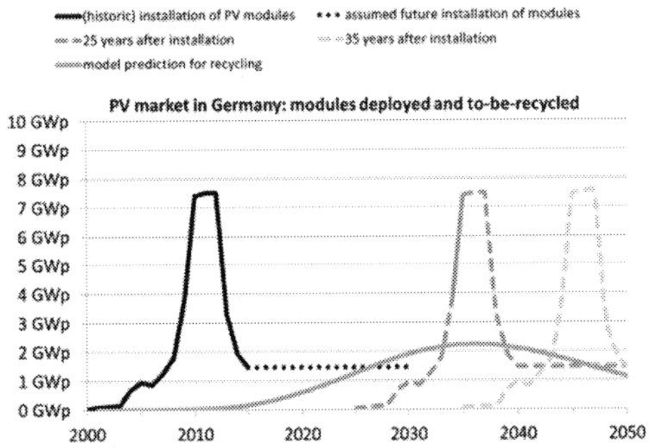

Fig. 4: modelled future wattage for recycling from installations in Germany (green line, Weibull parameters α_w=3, β_w =25 years) in comparison with the by 25 years or 35 year time-shifted curve of installation date (dashed grey lines).

Fig. 4 shows the impact of adapting the Weibull failure model on the estimation of future quantities for recycling in contrast to the simple assumption that modules would be recycled after 25 years or 35 years of service. Obviously, the two Weibull parameters have a big impact in this scenario. However, virtually all reasonable scenarios tested show that the expected quantities for recycling shall come earlier in the future compared to the time-shift scenario. The distribution is flattened and softened.

3.3. European Market

Historic Market data for prediction of future PV module recycling quantities are taken from [6] and [7]. A general annual growth rate of 15% is assumed until 2019. No statement beyond the year 2019 has been published so far.

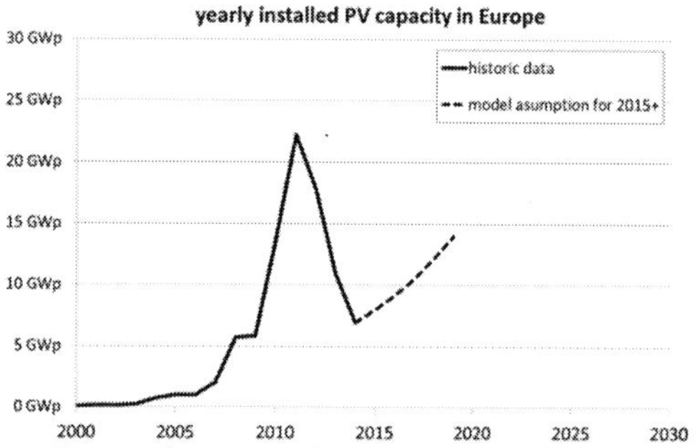

Fig. 5: historic and assumed future PV module deployment in the EU used for the recycling-model

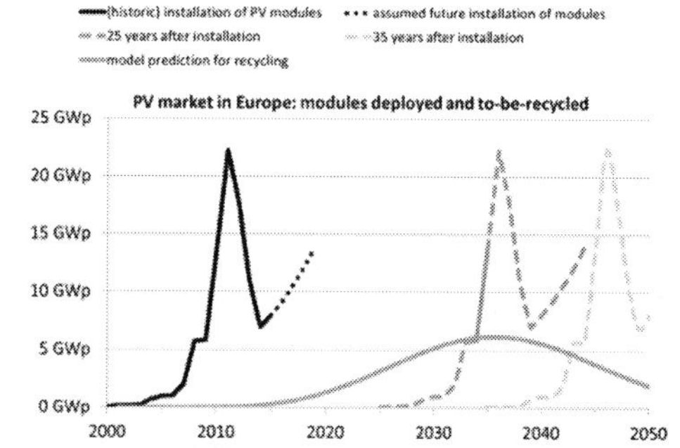

Fig. 6: modelled future wattage for recycling from installations in Europe (green line, Weibull parameters α_w=3, β_w =25 years) in comparison with the by 25 years or 35 year time-shifted curve of installation date (dashed grey lines).

It is interesting to compare values obtained on the base of this model with published recycling data. One player in the PV module recycling market reports [5] to have taken back 13'164 t of modules for recycling purposes during the years 2011 and 2015. Applying a very rough estimation of equivalence (mainly valid for c-Si product, only), namely 1 MWp ≈ 80 t, we would expect in our model 55'200 t (690 MWp) for the European market or 32'000 t (400 MWp) for the German market, only. As the total predicted quantity is higher than reported by a single market player, the general conclusion of this study ("recycling starts earlier than 25 years after deployment") is fully backed up. It is nonetheless important to continue data alignment in the future to achieve higher precision in model prediction.

4. DISCUSSION AND CONCLUSION

Based on the theory of reliability a model for the prediction of end-of-life of PV solar modules has been deducted. The bathtub curve (used for many industrial applications to model life expectancy) is adapted to predicting quantities for solar module recycling. Both the European module deployment and the German PV module deployment, show that considerable quantities are expected sooner than commonly anticipated.

5. REFERENCES

[1] The Parliament And The Council Of The European. Union, "L 197/38, pp. 38–71, 2012.

[2] A. Müller, K. Wambach, and E. Alsema, "Life Cycle Analysis of Solar Module Recycling Process," Mater. Res. Soc. Symp. Proc., vol. 895, no. February, pp. 1–6, 2006.

[3] G. Granata, F. Pagnanelli, E. Moscardini, T. Havlik, and L. Toro, "Recycling of photovoltaic panels by physical operations," Sol. Energy Mater. Sol. Cells, vol. 123, no. 2014, pp. 239–248, 2014.

[4] M. Rentoumis, I. Athanailidis, P. Koulouridakis, Y. Katsigiannis, and B. Nikolaos, "Development of a Three Dimensional Industrial Production Line Simulation for Crystalline Silicon Photovoltaic Panel Recycling," 4th Int. Symp. 26th Natl. Conf. Oper. Res. Chania - Crete, 2015.

[5] PV-CYCLE, "PV-CYCLE-Status-Report-jan-2016.pdf." 2016.

[6] EPIA, "Global Market Outlook for Photovoltaics 2014-1018," 2015.

[7] Solar Power Europe, Global Market Outlook For Solar Power / 2015 - 2019. 2015.

[8] V. M. Fthenakis, "End-of-life management and recycling of PV modules," Energy Policy, vol. 28, no. 14, pp. 1051–1058, 2000.

[9] E. D. Dunlop and D. Halton, "The performance of crystalline silicon photovoltaic solar modules after 22 years of continuous outdoor exposure," Prog. Photovoltaics Res. Appl., vol. 14, no. 1, pp. 53–64, 2006.

[10] P. D. T. O'Connor and A. Kleyner, Practical Reliability Engineering, Fith Ed. Chichester, West Sussex, UK: Wiley, 2013.

[11] M. Köntges, S. Kurtz, C. E. Packard, U. Jahn, K. Berger, K. Kato, T. Friesen, H. Liu, and M. Van Iseghem, Review of Failures of Photovoltaic Modules. 2014.

[12] M. Koehl, "Accelerated Service Life Testing of Photovoltaic Modules," in Service Life Prediction of Exterior Plastics : Vision for the future, C. C. White, Ed. Springer International Publishing, 2015, pp. 199–213.

[13] C. D. Tarum, Classification and Analysis of Weibull Mixtures, no. 724. Warrendale, PA 15096-0001 U.S.A.: The Engineering Society for Advancing Mobility Land Sea Air and Space, 1999.

[14] J. Wohlgemuth and D. Cunningham, "Using accelerated tests and field data to predict module reliability and lifetime," 23rd Eur. Photovolt. Sol. Energy Conf. Exhib., no. September, pp. 2663 – 2669, 2008.

[15] N. R. Wheeler, L. S. Bruckman, J. Ma, E. Wang, C. K. Wang, I. Chou, J. Sun, and R. H. French, "Degradation pathway models for photovoltaics module lifetime performance," Conf. Rec. IEEE Photovolt. Spec. Conf., pp. 3185–3190, 2013.

[16] J. H. Wohlgemuth and S. Kurtz, "Reliability Testing Beyond Qualification as a Key Component in Photovoltaic' s Progress Toward Grid Parity," IEEE Int. Reliab. Phys. Symp., no. February, pp. 551–556, 2010.

[17] A. Skoczek, T. Sample, and E. D. Dunlop, "The Results of Performance Measurements of Field-aged Crystalline Silicon Photovoltaic Modules," Prog. Photovolt Res. Appl., vol. 17, no. December 2008, pp. 227–240, 2009.

Interpolation Methods for Obtaining Maps on Solar Radiation for Colombian Geographical Conditions

An. Aldana[1], D. J. Rodriguez[1,2] and O. Simbaqueva[2]

[1]*Universidad Distrital Francisco José de Caldas*

Laboratorio de Investigación en Fuentes Alternas de Energía (LIFAE)

Bogotá – Colombia

[2]*Fundación Universitaria Los Libertadores*

Grupo de Investigación Energía Solar en Desarrollo de Energías Limpias (GESTEL)

Bogotá – Colombia

ABSTRACT: The study of solar radiation has many purposes not only in the field of meteorology but in the development and implementation of new technologies to contribute to a sustainable economy. For the design, implementation and management of any applications that make use of solar energy, it is necessary to know and to characterize the energy resource. One way is through solar radiation values obtained from the meteorological observation networks to reconstruct information about the solar energy potential. Climatic and atmospheric factors as well as the geographic position of Colombia, are affecting the availability of solar resources. For this reason, in this work the solar energy potential of the Colombian territory is quantified through the development of an atlas of global solar radiation, based on data collected from stations of the Institute of Hydrology, Meteorology and Environmental Studies (IDEAM) and under the time series between 2002 and 2013 on average.

Keywords: Radiation Maps, Interpolation methods, Solar energy, Colombian energy resource

1 INTRODUCTION

The study of solar radiation has many purposes not only in the field of meteorology, but also in the development and implementation of new technologies that contribute to a sustainable economy. For the design, implementation and management of any applications that make use of solar energy, it is necessary to know and characterize the energy resource. One way to do this is by solar radiation values obtained from the meteorological observation with this information solar radiation maps [1] are built.

According to the above, measurement and utilization of solar resources have taken importance globally. An example is Costa Rica, which in 2002 was used the Kriging technique for mapping monthly daily solar radiation. For this aim, they made use of the observations of 25 actinometric stations and 38 stations made from the heliophany obtained from files (IMN) National Meteorological Institute and experimental measurements in the Solar Patio of the Department of Physics of Costa Rica National University [2].

Other example, the National Renewable Energy Laboratory (NREL), whose actions are aimed at research and development of renewable energy and energy efficiency, reports through its website periodically the solar radiation maps for the energy sector and in general for the whole community interested in learning about the solar energy resource. To design these maps, satellite images were obtained from the National Administration of Aeronautics and Space (National Aeronautics and Space Administration, NASA) that shows the global solar potential per month and annual average. Equally. The model used is the UM-SBR, developed by Pinker and Lazlo at the University of Maryland, USA [3] and used in the database Data Center for Atmospheric Sciences (Atmospheric Science Data Center) of the US space agency NASA / SSE.

Climatic and atmospheric factors and the geographical position of Colombia, affecting the availability of solar resources, which is why the IDEAM has made an effort in favour of increased monitoring network of energy resources at national level [1]. For this purpose, in this work the solar energy potential of the Colombian territory is quantified through the development of an atlas of global solar radiation, based on data collected from the stations of the Institute of Hydrology, Meteorology and Environmental Studies (IDEAM) and under the time series between 2002 and 2013.

2 METHODOLOGY FOR MAPPING FOR SOLAR RADIATION: GEOSTATISTICAL WORKFLOW

The 79 selected meteorological stations are distributed throughout the country (Figure 1). Initially the calibration of radiometric measurement equipment was verified, from calibration performed at a pyranometer by comparison with a reference pyranometer. The calibration of the pyranometer allowed establishing the calibration factor and sensitivity of the instrument. The value of the standard deviation of said sensitivity radiation values are close to the average. Meanwhile, the data present in the error was less than 5%. Accordingly, it is considered that the calibration was valid.

Subsequently, an application called Analyst stations was developed, with the aim to realize the evaluation and processing of solar radiation data from terrestrial measurement points. This software has a graphical interface developed in Java, which can be connected to a MySQL database engine data to create and to store the information in a database. Thus, the database compiled from monthly averages and annual solar radiation for each station was created.

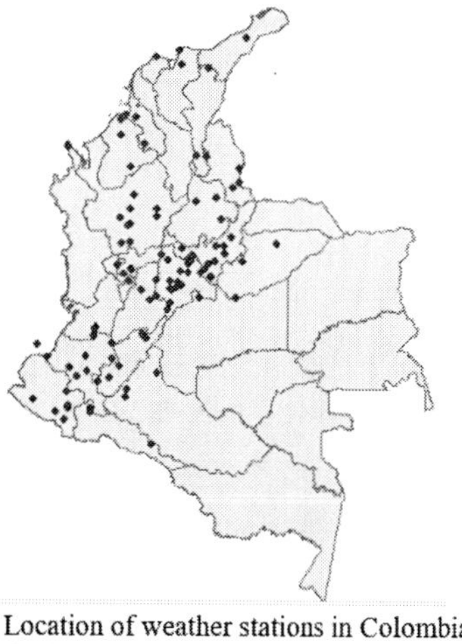

Figure 1: Location of weather stations in Colombia employed in this study

Geostatistical workflow is composed of the steps as presented below. The first is to georeference the selected study weather stations. From there, we proceed obtaining the map-containing direct monthly measurements of multi-year average radiation through of TIN surface. Followed this process, exploring statistics and spatial properties of the dataset through the analysis of statistical parameters, frequency histogram product was made. According to the values returned by the mean, median, skewness and kurtosis, and applying thereto a logarithmic transformation to ensure that the data follow an approximately normal distribution.

The generation of radiation maps begins with Ordinary Kriging. It can be developed from four types of trend: first, second, third and constant order. For this study, the trend of constant order was eliminated. Additionally, through the semivariogram map can be established whether there is an anisotropic distribution data. In this case, the data set has an isotropic behaviour. Performing a visual inspection determined that the variogram function fits a spherical type. Finally, before producing the final surface is necessary to present the results of cross-validation, which serve as a partial input basis for evaluation and selection of the best prediction model to be applied.

3 VALIDATION OF THE MODEL APPLIED

In order to evaluate the TIN surfaces of the region were used statistical and historical weather criteria.

Statistical criteria were taken according to the literature [4,5,6], these correspond to four specific criteria listed below:

Criterion 1. Average standardized near to zero,

Criterion 2. Root mean square,

Criterion 3. Average near the root mean square standard,

Criterion 4. Standardized root mean square near to one.

The analysis performed on the surface generated with Ordinary Kriging fully satisfies the proposed statistical criteria.

Climatological historical criteria of the study region were provided by the set of Global Solar Radiation Maps developed by the Mining and Energy Planning Unit (UPME) and the IDEAM in 2004 [1]. The purpose is to compare the radiation maps of 2004 and to check whether the map generated is consistent with the radiation values found until this year. From observations and analysis it is determined that the maps fully comply with these criteria. Since the developed workflow meets all the criteria, it is possible to extrapolate this scheme to every month of the year to get multiyear solar radiation maps.

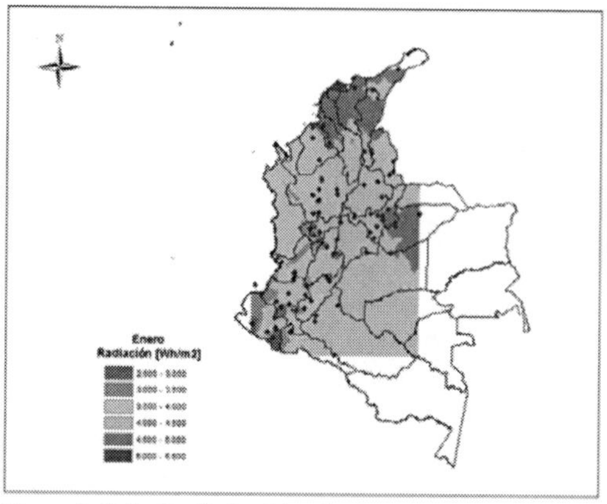

Figure 2: Solar Radiation Map for January

The results map shows the west of the country is an area which has radiation values ranging from 3500 to 4500 Wh/m^2. The northern part of the study region corresponds to higher radiation values; the trend 5000-5500 between Wh/m^2. On the map generated for this study can be seen that there is similarity for this area; it particularly presents values in this range.

Finally, it is noted that the centre of the country has average values of solar radiation ranging from 4,000 to 5,000 Wh/m^2. Similarly, it can be seen on the map generated by Ordinary Kriging, since the values for this same region are between 4000-4500 Wh/m^2.

La Guajira located at the north of the country in the Caribbean region, has historically been known as the area with the highest solar potential in the country. Based on observations to different surfaces, generated output for the year and the multi-year average it is shown that it has the highest average solar radiation for most of the year.

These results are consistent with the solar radiation maps generated for Colombia in 2004. Although current trends are not exact, it appears that quite acceptable results are presented to areas that are not covered by the radiometric network. Thus it is determined that the maps obtained fully comply with these criteria.

7 SUMMARY OF THESE NOTES

Based on the study and evaluation of changes in the Kriging method (Ordinary Simple and Universal), it was determined that the most appropriate configuration for obtaining maps of solar radiation corresponds to Ordinary Kriging.

The maps obtained by this method successfully fulfilled the criteria for statistical and climate of the study region assessment therefore represent the solar radiation behaviour in the country. Thus the maps presented in this study allowed to locate areas of high and low radiation and to establish its order of magnitude in areas where there are no radiometric stations.

Regarding the assessment of the solar resource in the region, it was observed that the months with the highest average incidence of solar radiation are January, February and December; meanwhile the months with the lowest solar radiation values are July and November. It was easy with a simple observation to determinate that La Guajira (Region located at the north of the country in the Caribbean region) is the area with the highest solar potential in the country. To have best results it is necessary to develop maps through other methods based on indirect measurements such as sunlight, humidity and temperature as a complement of this study.

8 References

[1] Atlas de Radiación Solar de Colombia. Unidad de Planeación Minero Energética (UPME), 2005 http://www.upme.gov.co/Atlas_Radiacion.htm

[2] Wright, J. Mapas de radiación solar en Costa Rica. 79-87, 2002.

[3] Pinker, R., & Laszlo, I. Modeling Surface Solar Irradiance for Satellite Applications on a Global Scale. J. Appl. Meteor, 194–211. 1992.

[4] Heine, G. W. "A Controlled Study of Some Two-Dimensional Interpolation Methods." COGS Computer Contributions 3 (no. 2): 60–72. 1986.

[5] McBratney, A. B., and R. Webster. "Choosing Functions for Semi-variograms of Soil Properties and Fitting Them to Sampling Estimates."Journal of Soil Science 37: 617–639. 1986.

[6] Oliver, M. A. "Kriging: A Method of Interpolation for Geographical Information Systems." International Journal of Geographic Information Systems 4: 313–332. 1990.

EMERGING PERFORMANCE ISSUES OF PHOTOVOLTAIC BATTERY SYSTEMS

Johannes Weniger, Tjarko Tjaden, Joseph Bergner, Volker Quaschning
HTW Berlin - University of Applied Sciences
Wilhelminenhofstraße 75A, 12459 Berlin, Germany
http://pvspeicher.htw-berlin.de

ABSTRACT: This paper sheds new light on the real energetic performance of grid-connected PV-battery systems for self-supply purposes in residential buildings. A novel simulation test for the performance assessment of residential PV-battery systems with the Storage Performance Index (SPI) is proposed. Two different systems are modelled and annual time series simulations are carried out with a temporal resolution of one second. The calculated Storage Performance Index of the AC- and DC-coupled PV-battery system under study amounts to 51% and 53%, respectively. In other words, the investigated PV-battery systems can realize about the half of the respective grid electricity cost saving potential of equivalent lossless PV-battery systems. Moreover, a detailed loss analysis reveals that the majority of the system losses are caused by the conversion losses in the power electronic components, followed by the losses due to the standby power consumption of the battery, converter and auxiliaries. The developed simulation test allows the evaluation of the energetic and economic performance of PV-battery systems of different sizes and system topologies. In this way, the comparability and transparency with regard to the performance of different products available on the market can be improved.
Keywords: PV-Battery Systems, Performance Assessment, Efficiency, Energy Losses, Loss analysis

1 INTRODUCTION

Due to the significant cost reduction of both PV systems and battery storage devices in the recent years, the market for grid-connected storage systems for residential applications has been growing rapidly. Until the end of 2015, more than 35000 small-scale battery systems have been installed in conjunction with grid-connected PV systems in Germany [1]. The main objective of such PV-battery systems in residential applications is to store surplus PV energy on-site in order to use it later to cover the electrical loads within the respective buildings [2]. In this way, the home owner's self-sufficiency is increased and the amount of electricity that has to be purchased from the grid is reduced. In other words, the battery system is operated with the aim of reducing the energy procurement from the grid by adjusting the charging and discharging power according to the resulting power flow at the point of common coupling.

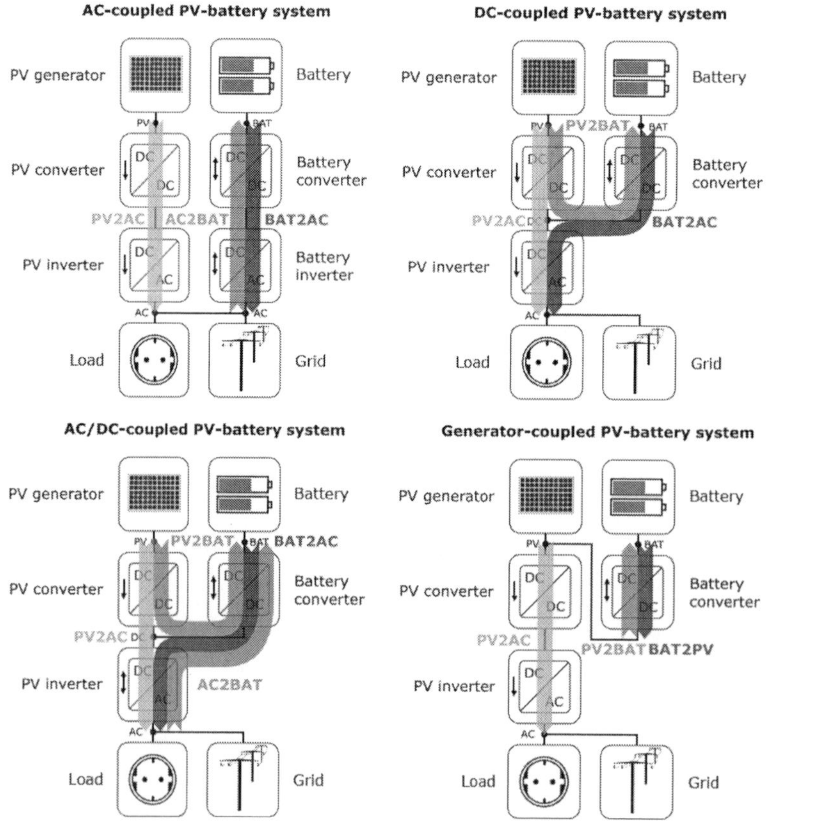

Figure 1 Schematic system configuration and relevant power flows of different topologies of PV-battery systems.

2 PV-BATTERY SYSTEM TOPOLOGIES

In general, a PV-battery system consists of a common PV generator, power electronic components, a battery device and a control unit. Today, there are numerous commercial PV-battery systems for the application in residential buildings available on the market [3]. One of the relevant distinctive characteristics between the products is the type of the connection between the battery storage and the PV system [4]. The system layouts of the most common topologies are illustrated in Figure 1. In AC-coupled systems, the PV inverter and the battery device with a bidirectional battery converter are linked to the AC-bus of the household. In contrast, the battery can also be linked to the PV inverter's DC-side. In such a DC-coupled PV-battery system, the battery can only be charged by power from the DC-bus. AC/DC-coupled systems are equipped with bidirectional AC/DC converters, so that additionally power from the household's AC-bus can be used to charge the battery. The battery unit can also be linked directly to the PV generator via a DC/DC charge regulator. Such generator-coupled battery systems supply the loads via the common PV inverter.

In order to provide a better understanding of the different system topologies, the relevant paths of energy flow are shown in Figure 1. The nomenclature of the paths is chosen with regard to the direction of the energy flows between the three points of reference on the PV generator side (PV), battery side (BAT) and AC-side (AC). The path PV2AC describes the conversion of DC-power from the PV generator to AC-power and is present in all system topologies. The charging path is described either by AC2BAT or by PV2BAT, depending on the topology. BAT2AC and BAT2PV are the discharging paths. The defined energy flow paths are required to characterize both modular system components as well as integrated systems appropriately.

3 CLASSIFICATION OF SYSTEM-RELATED LOSS MECHANISMS

Several loss mechanisms contribute to the overall energy efficiency and performance of grid-connected PV-battery systems. The different losses can be separated into four categories, which are as follows: conversion, standby, control and energy management-related losses. This section provides an overview of the various loss factors.

3.1 Conversion losses

Energy losses occur during the charging and discharging process within the battery mainly due to its internal resistance. The battery-related conversion losses can be specified with the battery round trip energy efficiency, which is calculated over one cycle that begins and ends at the same state of charge [5].

Apart from the internal battery losses, the distinct paths of energy flow depicted in Figure 1 are associated with conversion losses in the power electronic components. These converter-related losses can be separated into power-independent losses, voltage losses proportional to the power and resistive losses proportional to the square of the power [6]. At power levels of less than 20% of the nominal power, the characteristic drop in efficiency is mainly caused by the power-independent losses.

3.2 Standby losses

In addition to the losses associated with power conversion or energy storage, a standby power demand of the battery system occurs even when no power charges or discharges the battery. The standby consumption of the battery is caused by the battery management system (BMS), which observes the battery unit and fulfills safety functions. The standby consumption of the BMS can either be covered by energy stored in the battery or by energy drawn from the grid [7].

When the battery is not charging or discharging, the power electronic components remain in the idle or standby mode with respective losses. During completely discharged periods, this standby consumption has to be provided by the grid. At a fully charged state of the battery, the standby demand of the power converters is usually covered by PV generated electricity. On top of that, the additional energy consumption of the measuring devices and energy management system (EMS) has to be taken into account. It must be noted that the power demand of these components has to be supplied at all times and cannot be disregarded.

3.3 Control losses

The goal of the control unit is to balance the power flow at the point of common coupling by adjusting the charging and discharging power of the battery system. For reasons of data acquisition and signal processing, typical residential battery systems have an inherent response time of up to several seconds. Therefore, the power of the battery system cannot follow the strongly fluctuating electrical load demand and PV generation profile instantaneously. As a consequence, a mismatch between the residual power (PV output minus load) and the provided or absorbed battery power can be observed [8]. The mismatch in the power balance in buildings equipped with AC-coupled PV-battery systems is compensated by power fed into the grid or drawn from the grid. Due to the unidirectional path of DC- and generator-coupled PV-battery systems, the mismatch causes only a grid injection of such battery systems without any additional energy flows from the grid to the battery.

Further mismatch losses are related to the accuracy of the power measurements at the point of common coupling, as the control unit of a PV-battery system is usually subjected to error-prone measurements. Typical measuring devices are less precise compared to the calibrated energy meter which is relevant for billing purposes. Due to the lower measuring accuracy and the resulting control deviations, the battery power cannot be adjusted exactly according to the power flows within the building even under steady state conditions.

3.4 Energy management losses

With the introduction of the new market incentive program for small-scale battery systems in March 2016 in Germany, a limitation of the feed-in power to 50% of the rated PV power was introduced. If the PV-battery system is operated in a way that charges the battery as soon as excess PV power is available, the feed-in limitation is mainly realized by curtailing surplus PV power [9]. To avoid these unnecessary curtailment losses, the battery charging has to be scheduled based on forecasts of the PV generation and load consumption [10]. With such predictive energy management strategies, the grid feed-in can be increased due to the decline in the curtailment losses. However, the forecast errors can reduce the utilization of the battery system which results in an increase in the energy demand that has to be provided by the grid [11]. As a result, the energy management is associated with additional energy losses on top of the conversion, standby and control losses.

4 OVERVIEW ON SYSTEM PERFORMANCE TESTS

With the increasing market penetration of residential PV-battery systems, new aspects regarding the assessment of the energetic system performance are arising. Up until now, no standardized performance evaluation procedures for grid-connected PV-battery systems exist [12]. As a consequence, performance-related specifications are rarely stated in the data sheets of the products. From the end customer's point of view, the comparability of the performance between different products is rather difficult as of yet [7]. This highlights the need to develop test procedures for performance evaluation purposes of residential PV-battery systems at both the component level as well as the system level. However, when assessing the performance of the entire PV-battery system, the interaction between the system components causes a range of phenomena that are difficult to quantify if studied in isolation [5]. As an example, the common round-trip efficiency determined by the ratio of the discharged energy to the charged energy does not take all of the aforementioned loss mechanisms into account, and is thereby not adequate to assess the overall performance of grid-connected PV-battery systems [13]. Table I provides an overview of the different methods to identify the performance of grid-connected PV-battery systems. The performance evaluation procedures are described and discussed in the following subsections.

Table I Overview on different performance evaluation procedures for grid-connected PV-battery systems.

Field tests 	Black-box tests 	White-box tests 	Simulation tests 
Characteristics			
• Long-term field tests by monitoring the real operational behavior over a period of at least one year	• Short-term application tests in a laboratory environment with reference profiles or at defined operating points	• Detailed characterization tests under laboratory conditions with the aim of characterizing the efficiency of each component or path of energy flow	• Model-based simulation tests parametrized with measurements from white-box tests and based on measured load and PV output profiles
Results			
• Average operating efficiency and load distributions of distinct paths of energy flow	• Use case-specific performance indicator obtained from measurements	• Measurements of efficiency curves, standby consumption, response behavior, etc.	• Use case-specific performance indicator obtained from simulations
Advantages			
• Real operational and long-term performance can be observed	• Short time period required and good comparability of the test results	• Detailed characterization of the components and overall system	• Fast test procedure and good reproducibility of the test results
Disadvantages			
• Long time period required • Results mostly available for outdated products • Identical test conditions are hard to ensure • Limited comparability of the test results from different systems	• Measurements are already affected by the predefined profiles • Test results are only applicable to the specific use case • General validity of the reference profiles has to be demonstrated	• No single performance indicator for the end-customer can be extracted from the test results • Expensive measurement equipment is required • Poor comparability between the results of different topologies	• Detailed characterization measurements are required • Accuracy of the test results depends on the level of detail of the simulation model • Not all loss mechanisms can be modelled exactly

4.1 Field tests

The most obvious way to analyze the performance of residential PV-battery systems is to monitor the relevant energy flows in the buildings equipped with such systems. As the performance of the systems may vary during the course of the year, a measuring period of at least one year is needed to derive meaningful evaluation results. One of the most comprehensive field tests is being conducted by the RWTH Aachen University in the framework of the scientific monitoring of the German federal funding program for PV-battery systems [1]. Both average operating efficiency values and energy content distributions with regard to the charging and discharging power levels are analyzed within the project based on field data. However, one of the drawbacks of long lasting field tests is that the systems under study are mostly outdated by the time of publication of the test results. Owing to the site-specific utilization of the battery systems in distinct buildings, the results gained from different systems are not fully comparable.

4.2 Black-box tests

Another approach is to accelerate the test procedure through short-term application tests in the laboratory, also known as black-box tests. Two types of application tests can be distinguished: reference profile-based and typical operating point-based black-box tests. An application test using a reference profile sequence of four days was developed by the Fraunhofer IWES [13]. From the proposed test procedure, three different performance indicators concerning the static and dynamic efficiency as well as the degree of self-sufficiency can be derived. Referring to the EU efficiency of PV inverters, a typical operating point-based black-box test determining an EU efficiency for residential PV-battery systems was proposed by [14]. Both types of black-box tests have the disadvantage that the measurement results are already affected by the predefined test profiles, thus the test results are only applicable for the specific use case. Moreover, the general validity of the chosen reference profiles and typical operating points have not yet been fully demonstrated.

4.3 White-box tests

The aim of white-box tests is to measure the performance-relevant characteristics of the system components under laboratory conditions. Such detailed characterization tests were developed and performed by several institutions [1], [7], [13]. Apart from measuring the efficiency curves as a function of the power throughput for the different paths of energy flow, the standby consumption of the system components and the system's response behavior are measured. The experimental results obtained from such white-box tests allow the characterization of the performance-related system specifications. Despite the advantages, no single performance criterion for the end-customer can be extracted from the test results.

4.4 Simulation tests

By means of the experimental results obtained from white-box tests, detailed simulation models of PV-battery systems can be parametrized. Thus, it is possible to conduct model-based simulation tests in order to simulate the operational system behavior with realistic PV output and load profiles measured over a period of one year or longer. As a result, use case-specific performance metrics can be derived from the simulation tests. The simulation test approach allows the analysis of the relevance of various loss mechanisms separately. Other advantages of this approach are the fast test procedure and the fact that everyone can perform the tests if the underlying simulation models are freely available. However, the effort to develop and validate such universal simulation models is very high [15]. Moreover, the accuracy of the test results depends on the level of detail of the simulation models. For reasons of simplification, some loss mechanisms, such as the cell balancing of the battery unit, cannot be modelled in detail. On the other hand, once detailed performance models have been developed, the simulation results can also be used to optimize the system layout, system sizing and control algorithms.

5 PERFORMANCE EVALUATION WITH THE STORAGE PERFORMANCE INDEX (SPI)

In this section, a model-based simulation test is proposed and test results for two different PV-battery systems are presented. The goal of the simulation test is to determine the proposed Storage Performance Index (SPI).

5.1 Approach

The calculation of the Storage Performance Index aims to evaluate the performance of grid-connected PV-battery systems for self-supply purposes in residential buildings. To perform the simulation test, profiles of the PV output and electrical load demand with a 1-s resolution over a period of one year are used. In this study, the PV array's output power is calculated based on meteorological measurements provided by the University of Oldenburg, Germany [16]. The DC-power output simulations are conducted for a PV generator with a size of 5 kWp by means of empirical models [17], [18]. The load profile of one household with an annual load demand of 5009 kWh is used for the simulation test. The chosen load profile (no. 31) is taken from a database with 74 domestic load profiles [19]. In accordance with the intention to establish a replicable simulation test procedure, all used input profiles are freely available as open data.

In a second step, the PV-battery system has to be modelled according to the respective system topology, as shown in Figure 1. In this study, two different lithium-ion based PV-battery systems are modelled and analyzed. The first one is a modular AC-coupled system with 5 kWh of usable battery capacity, which comprises the following components:

- SMA Sunny Boy 5000 TL (PV inverter)
- SMA Sunny Island 3.0M (battery inverter)
- Akasol neeoQube (battery unit)
- SMA Sunny Home Manager (EMS)
- SMA Sunny Remote Control (control unit)
- SMA Energy Meter (meter)

The second one is an integrated DC-coupled system with 2 kWh of usable battery capacity, which is composed of the following components:

- SMA Sunny Boy 5000 Smart Energy (integrated system with inverter, control and battery unit)
- SMA Energy Meter (meter)

Both PV-battery systems are modelled according to publicly available data and data provided by the manufacturer. The conversion losses of the different paths of energy flow are modelled with efficiency curves as a function of the power throughput. The standby power consumption of the battery, converters and auxiliaries are set to the provided values. To incorporate the transient behavior of the battery systems in the simulation test, a first-order time delay element with a dead time is implemented in the battery control unit. It must be noted that measuring inaccuracy incurred control losses and energy management-related losses are not depicted in the simulation model.

In a further step, the identical PV-battery systems without any losses are modelled considering the same usable battery capacity. The simulation results obtained from the lossless PV-battery systems serve as the benchmark for the performance evaluation of the real PV-battery systems. Note that the conversion paths of the ideal battery systems have no power restrictions, so that the utilization of the lossless battery system depends only on the specific battery capacity and is independent from the system topology. It is thereby ensured that the results of the performance evaluation can be compared between differently sized battery systems. As the DC-power profile of the PV generator serves as an input for the system simulation, losses induced by the PV inverter are included within the proposed performance evaluation procedure. Subsequently, the performance assessment results of different systems can be compared between the four distinct system topologies presented in Figure 1.

5.2 Energetic Performance Results

This subsection focuses on the energetic results of both PV-battery systems under study. The energetic performance is evaluated by analyzing the amount of energy exchanged between the residential building equipped with the respective systems and the grid. The results obtained from the power flow simulations of both PV-battery systems are represented in Figure 2. Without a PV-battery system, the residential building's load is completely supplied by the electricity grid and the amount of energy drawn from the grid is about 5009 kWh/a. If the building is equipped with an ideal PV system with a lossless inverter rated at 5 kWp, the grid supply is reduced by about 30% to 3474 kWh/a. In addition, the energy injected into the grid amounts to 3739 kWh/a. The results obtained from the ideal PV system with a lossless inverter serve as a reference.

Both the grid supply and grid feed-in drop when adding an ideal battery system without any conversion, standby and control losses. The higher the usable battery capacity, the lower the amount of energy exchanged with the grid. The 5 kWh lossless battery system reduces the grid supply and grid feed-in to 2007 kWh/a and 2272 kWh/a, respectively. As a result, about 60% of the electrical load demand is provided by the ideal PV-battery system. However, by considering all the modelled losses of both systems, the energy drawn from the grid rises while the energy injected into the grid declines. In total, the building equipped with the AC-coupled PV-battery system draws 2308 kWh/a from the grid and feeds 2017 kWh/a into the grid. In comparison with the results obtained from simulating the ideal 5 kWh battery system, the grid supply is increased by 15% and the grid feed-in is reduced by 11%. Similarly, the system losses of the DC-coupled PV-battery system are accompanied by an increase in the energy supply from the grid of 6% and a 5% reduction in the grid injection. As such, the overall energetic performance of grid-connected PV-battery systems is interpreted as the increase in grid supply and drop in grid feed-in in comparison with the identical lossless PV-battery system.

5.3 Economic Performance Results

As the primary goal of PV-battery systems is to reduce the energy costs, it is reasonable to evaluate the operational results from the economic perspective as well. The annual costs for the procurement of electricity from the grid C_{GS} are obtained by multiplying the annual amount of energy supplied by the grid E_{GS} with the retail electricity price p_{GS}:

$$C_{GS} = E_{GS} \cdot p_{GS} \quad (1)$$

The annual revenues R_{GF} from selling the PV energy E_{GF} for a specific feed-in tariff p_{GF} are calculated via:

$$R_{GF} = E_{GF} \cdot p_{GF} \quad (2)$$

In a further step, the net grid electricity costs C associated with the energy exchanged with the grid are obtained by subtracting the annual revenues R_{GF} from the annual costs C_{GS}:

$$C = C_{GS} - R_{GF} \quad (3)$$

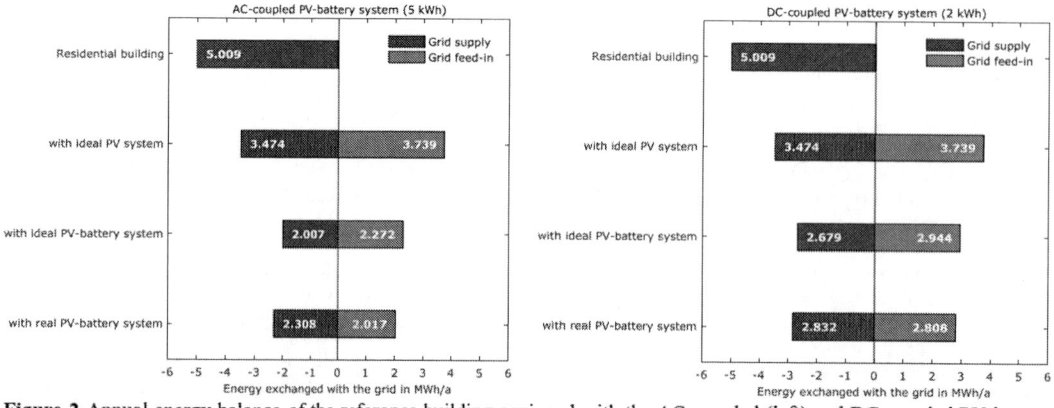

Figure 2 Annual energy balance of the reference building equipped with the AC-coupled (left) and DC-coupled PV-battery system (right).

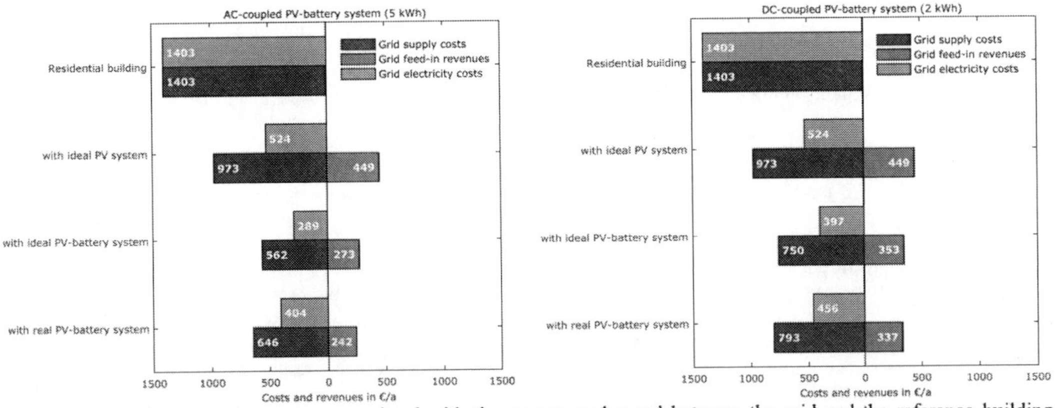

Figure 3 Annual costs and revenues associated with the energy exchanged between the grid and the reference building equipped with the AC-coupled (left) and DC-coupled PV-battery system (right).

In this way, the energy flows between the building and the grid are weighted with regard to their economic value. According to the current economic framework in Germany, the feed-in tariff is set to 0.12 €/kWh and the retail electricity price is valued at 0.28 €/kWh in this study.

The economic assessment results of both PV-battery systems are shown in Figure 3. The annual costs due to the procurement of electricity from the grid will be 1403 €/a, if neither a PV system nor a PV-battery system is installed. Owing to the revenues from the grid injection and the reduced expenses for the grid supply, the grid electricity costs are diminished to 524 €/a with the ideal PV system rated at 5 kWp. In the case in which the building is equipped with the lossless AC- or DC-coupled PV-battery system, the grid electricity costs are dropped to 289 or 397 €/a. As was to be expected, the system losses reduce the economic benefit of the battery systems and increase the grid electricity costs to 404 and 456 €/a for the AC- and DC-coupled system, respectively.

In order to enable a better comparability between the economic results of PV-battery systems of different sizes, the grid electricity costs are plotted as a function of the usable battery capacity in Figure 4 (left). The green line represents the grid electricity costs of the ideal PV-battery system, which serves as a benchmark. The yellow dot indicates the grid electricity costs of the PV system with a lossless inverter. As can be seen, the grid electricity costs of the real AC-coupled PV-battery system are increased by 115 €/a compared to the lossless system of the same usable battery capacity. For the DC-coupled PV-battery system, the losses increase the grid electricity costs by only 58 €/a. As the grid electricity cost saving potential of the ideal PV-battery system differs with the battery size, the increase in the grid electricity costs of systems with distinct usable battery capacities cannot be compared directly.

From the PV system owner's point of view, the economic benefit of the battery system consists in the reduction of the grid electricity costs compared with the grid electricity costs obtained from a PV system without a battery. Thus, the purpose of the proposed Storage Performance Index *SPI* is to determine the ratio of the grid electricity cost savings of the real PV-battery system to the grid electricity cost saving potential of the lossless PV-battery system with the same usable battery capacity:

$$SPI = \frac{C_{PV,IDEAL} - C_{PVBAT,REAL}}{C_{PV,IDEAL} - C_{PVBAT,IDEAL}} \quad (4)$$

where $C_{PV,IDEAL}$, $C_{PVBAT,IDEAL}$ and $C_{PVBAT,REAL}$ are the grid electricity costs of the ideal PV system (with a lossless inverter), ideal and real PV-battery system, respectively. For better comparability between different system topologies, the PV inverter related losses are incorporated within the economic assessment of the real PV-battery systems by taking the ideal PV system as the reference.

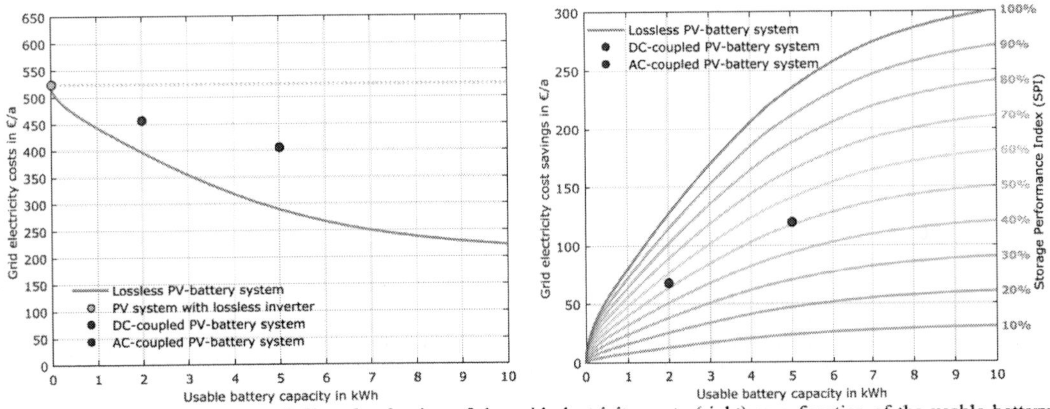

Figure 4 Grid electricity costs (left) and reduction of the grid electricity costs (right) as a function of the usable battery capacity.

In Figure 4 (right), the grid electricity cost savings of real PV-battery systems for varying Storage Performance Indices and usable battery capacities are shown. The lossless PV-battery system with a Storage Performance Index of 100% represents the upper limit. The grid electricity cost savings of the real PV-battery systems are reduced according to the specific Storage Performance Index. The results of the AC-coupled and DC-coupled PV-battery systems under study are also displayed. By comparing both systems, it can be noticed that the determined Storage Performance Indices of the AC- and DC-coupled PV-battery systems with 51% and 53% are in the same range. Taking into consideration that the layouts of both system topologies differ significantly from each other (compare Figure 1), this is a surprising result. However, due to the small battery size of the DC-coupled system and therefore low energy-related utilization of the battery conversion paths (PV2BAT and BAT2AC), the highly efficient charging path has only a small contribution to the overall system performance. This leads to the fact that the Storage Performance Index of small sized PV-battery systems is strongly affected by the PV conversion path (PV2AC). Furthermore, an identical standby power consumption has a higher negative impact on the Storage Performance Index of small-scaled PV-battery systems compared to systems with larger sized battery capacities.

For a better understanding of how the different loss mechanisms affect the Storage Performance Index, a detailed loss analysis of the systems under study is presented in Figure 5. In order to identify the contribution of each loss mechanism to the overall reduction of the Storage Performance Index, system simulations are carried out by isolating the distinct loss mechanisms from each other. The bars represent the reductions in the Storage Performance Index due to the various loss mechanisms. The first thing to note is that the biggest losses are associated with conversion in the power electronic components, which cause the Storage Performance Index to drop by approximately 26% in both systems. This is because the PV conversion path is the predominated path in both systems and the battery converter takes effect twice: during charging as well as discharging. Secondly, the reduction of the Storage Performance Index by the battery, converter and auxiliaries standby power consumption varies between 15% and 10% for the AC-coupled and DC-coupled system, respectively. Another point to note is that the energy losses occurred in the battery and the dynamic response related control losses only slightly affect the overall system performance. Because of the different sizing of the power electronics of the AC- and DC-coupled PV-battery system, the reduction in the Storage Performance Index due to the power rating of the converters varies for both systems. In this way, the contribution of each loss mechanism to the economically assessed system performance can be determined with the proposed performance evaluation procedure.

As it is a drawback of each economic assessment procedure, the test results are affected by the presumed economic input parameters. For this reason, the Storage Performance Index of both systems was calculated for a variety of feed-in tariffs and retail electricity prices. It was found that the Storage Performance Index depends only on the ratio of the feed-in tariff to the retail electricity price. Figure 6 illustrates the impact of this ratio on the Storage Performance Index of the AC- and DC-coupled PV-battery system. Without any feed-in remuneration, the Storage Performance Index is in the order of 80% and is equal to the ratio of the grid supply savings of the real PV-battery systems to those of the ideal PV-battery systems, both in comparison with the ideal PV system (see Figure 2). In such a case without a feed-in tariff, the losses in the grid feed-in do not lead to a reduction in the Storage Performance Index. For this reason, the higher the ratio of the feed-in tariff to the retail electricity price, the larger the impact of the feed-in losses on the Storage Performance Index. The results obtained for the reference case with a ratio of the feed-in tariff to the retail electricity price of 0.43 are already presented in Figure 4 (right). As soon as the feed-in tariff is larger than approximately 70% of the retail electricity price, the grid electricity costs of the real PV-battery systems are larger than those of the ideal PV system, which results in a Storage Performance Index below zero. As a consequence, the battery system cannot contribute to the reduction of the grid electricity costs. Summarizing, the calculation of the Storage Performance Index can be adopted to the economic circumstances of different countries.

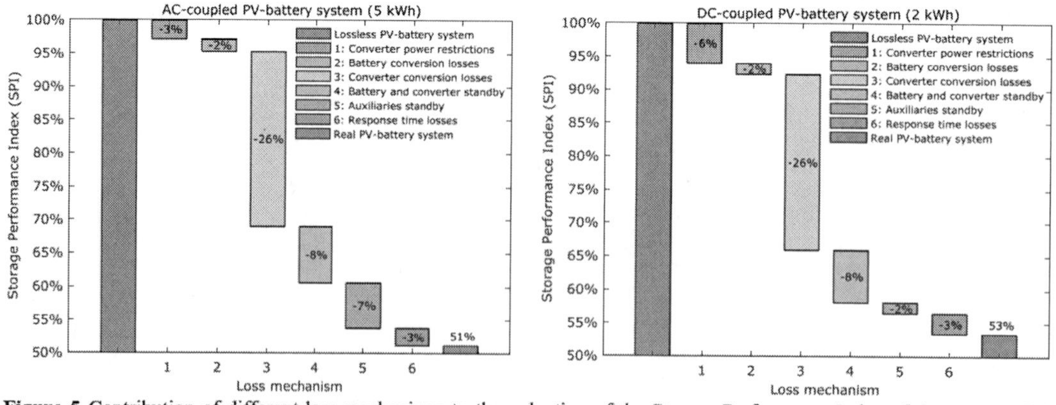

Figure 5 Contribution of different loss mechanisms to the reduction of the Storage Performance Index of the AC-coupled (left) and DC-coupled PV-battery system (right)

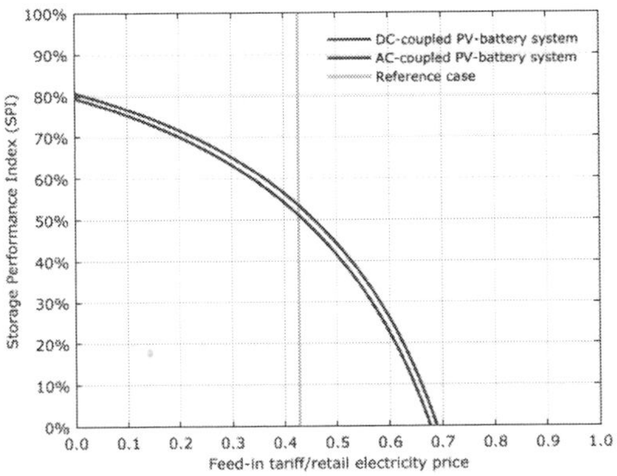

Figure 6 Variation of the Storage Performance Index of the AC-coupled and DC-coupled PV-battery system with the ratio of the feed-in tariff to the retail electricity price (reference case: feed-in tariff of 0.12 €/kWh and retail electricity price of 0.28 €/kWh).

6 CONCLUSIONS

In this paper, a novel simulation test procedure for assessing the energetic and economic performance of grid-connected PV-battery systems with the Storage Performance Index (SPI) is proposed. In a first step, the energetic behavior including the conversion, standby and control losses of two PV-battery systems are simulated and the specific annual amount of energy exchanged with the grid is determined. In a second step, the simulation results are compared with the results obtained from simulating lossless PV-battery systems of the same usable battery capacity. This comparison allows the determination of the energetic performance of the real PV-battery systems expressed by the increase in grid supply and the reduction in grid feed-in due to the system losses. In a further step, the energetic test results are assessed economically by weighting the energy exchange with the grid with the respective monetary values and calculating the grid electricity cost savings of the PV-battery systems compared with a reference PV system. The calculated Storage Performance Index of the AC- and DC-coupled PV-battery systems under study amounts to 51% and 53%, respectively. This means that the investigated PV-battery systems can realize about the half of the respective grid electricity cost saving potential. A detailed loss analysis reveals that the majority of the reduction in the Storage Performance Index is caused by the conversion losses in the power electronic components, followed by the losses due to the standby power consumption of the battery, converter and auxiliaries. The developed simulation test allows the minimization of the energy losses by improving the system layout, system sizing and control algorithms of residential PV-battery systems. Moreover, the energetic and economic performance of PV-battery systems of different sizes and system topologies can be assessed with the proposed Storage Performance Index. Further research will be carried out to evaluate the performance of more products available on the market and to improve the comparability between the products from the end customer's point of view.

ACKNOWLEDGEMENTS

The authors thank the German Federal Ministry of Economics and Technology (BMWi) and the Projektträger Jülich (PtJ) for the support of the project "LAURA" (grant agreement no. 0325716G). Furthermore, the authors thank Martin Rothert from SMA and Philipp Bartl from Akasol for providing detailed data of the modelled products. The authors are solely responsible for the content of this publication.

REFERENCES

[1] K.-P. Kairies, D. Haberschusz, J. van Ouwerkerk, J. Strebel, O. Wessels, D. Magnor, J. Badeda, and D. U. Sauer, "Wissenschaftliches Mess- und Evaluierungsprogramm Solarstromspeicher - Jahresbericht 2016," Institut für Stromrichtertechnik und Elektronische Antriebe (ISEA), RWTH Aachen, Aachen, Jahresbericht, May 2016.

[2] J. Weniger, J. Bergner, T. Tjaden, and V. Quaschning, "Economics of residential PV battery systems in the self-consumption age," in *29th European Photovoltaic Solar Energy Conference and Exhibition*, Amsterdam, 2014.

[3] M. Fuhs, "Marktübersicht Batteriespeicher," *pv magazine*, pp. 28–39, Jun-2015.

[4] J. Weniger, J. Bergner, T. Tjaden, and V. Quaschning, *Dezentrale Solarstromspeicher für die Energiewende*, 1. Berlin: Berliner Wissenschafts-Verlag, 2015.

[5] C. Patsios, B. Wu, E. Chatzinikolaou, D. J. Rogers, N. Wade, N. P. Brandon, and P. Taylor, "An integrated approach for the analysis and control of grid connected energy storage systems," *J. Energy Storage*, vol. 5, pp. 48–61, Feb. 2016.

[6] H. Schmidt and D.-U. Sauer, "Wechselrichter-Wirkungsgrade: Praxisgerechte Modellierung und Abschätzung," *Sonnenenergie*, vol. 4, pp. 43–47, 1996.

[7] C. Messner, J. Kathan, and J. Mayr, "Effizienz und Effektivität von netzgekoppelten PV-Heimspeichersystemen - Erfahrungen und Erkenntnisse aus Labortests kommerzieller Produkte," in *31. Symposium Photovoltaische Solarenergie*, Bad Staffelstein, 2016.

[8] J. Weniger, T. Tjaden, J. Bergner, and V. Quaschning, "Auswirkungen von Regelträgheiten auf die Energieflüsse in Wohngebäuden mit netzgekoppelten PV-Batteriesystemen," in *31. Symposium Photovoltaische Solarenergie*, Bad Staffelstein, 2016.

[9] J. Weniger, J. Bergner, T. Tjaden, and V. Quaschning, "Effekte der 50%-Einspeisebegrenzung des KfW-Förderprogramms für Photovoltaik-Speichersysteme," Hochschule für Technik und Wirtschaft Berlin (HTW Berlin), Berlin, Studie, Mar. 2016.

[10] J. Weniger, J. Bergner, and V. Quaschning, "Integration of PV power and load forecasts into the operation of residential PV battery systems," in *4th Solar Integration Workshop*, Berlin, 2014.
[11] J. Bergner, J. Weniger, T. Tjaden, and V. Quaschning, "Feed-in Power Limitation of Grid-Connected PV Battery Systems with Autonomous Forecast-Based Operation Strategies," in *29th European Photovoltaic Solar Energy Conference and Exhibition*, Amsterdam, 2014.
[12] K.-P. Kairies, D. Haberschusz, O. Wessels, J. Strebel, J. van Ouwerkerk, D. Magnor, and D. U. Sauer, "Real-Life Load Profiles of PV Battery Systems from Field Measurements," in *10th International Renewable Energy Storage Conference (IRES 2016)*, Düsseldorf, 2016.
[13] F. Niedermeyer, J. von Appen, T. Kneiske, M. Braun, A. Schmiegel, N. Kreutzer, M. Rothert, and A. Reischl, "Innovative Performancetests für PV-Speichersysteme zur Erhöhung der Autarkie und des Eigenverbrauchs," in *30. Symposium Photovoltaische Solarenergie*, Bad Staffelstein, 2015.
[14] A. Piepenbrink, "EU Efficiency for Home Storage Systems - A New and Simple Procedure," in *10th International Renewable Energy Storage Conference (IRES 2016)*, Düsseldorf, 2016.
[15] M. Knoop and M. Littwin, "Bewertung von PV-Speichersystemen mittels Modellbildung," presented at the 4. Treffen der Arbeitsgruppe Speicherperformance, Aachen, 22-Apr-2016.
[16] J. Kalisch, T. Schmidt, D. Heinemann, and E. Lorenz, "Continuous meteorological observations in high-resolution (1Hz) at University of Oldenburg in 2014." 10.1594/PANGAEA.847830, 2015.
[17] H. G. Beyer, G. Heilscher, and S. Bofinger, "Identification of a General Model for the MPP Performance of PV-Modules for the Application in a Procedure for the Performance Check of Grid Connected Systems," in *19th European Photovoltaic Solar Energy Conference*, Paris, 2004, pp. 3073–3076.
[18] T. M. Klucher, "Evaluation of models to predict insolation on tilted surfaces," *Sol. Energy*, vol. 23, no. 2, pp. 111–114, Jan. 1979.
[19] T. Tjaden, J. Bergner, J. Weniger, and V. Quaschning, "Representative electrical load profiles of residential buildings in Germany with a temporal resolution of one second," 2015.

EFFICIENCY AND EFFECTIVENESS OF PV BATTERY ENERGY STORAGE SYSTEMS FOR RESIDENTIAL APPLICATIONS - EXPIRIENCE FROM LABORATORY TESTS OF COMMERCIAL PRODUCTS

Christian Messner, Johannes Kathan, Christian Seitl, Stefan Hofmüller, Roland Bründlinger
Austrian Institute of Technology
Donau-City-Strasse 1
1220 Wien, Austria
Phone: +43 50550 6351
christian.messner@ait.ac.at

ABSTRACT: Decreasing feed-in tariffs, diminishing PV system costs and , simultaneously rising electricity prices provide best conditions for the use of photovoltaic battery energy storage system (PV-BESS) for residential application. System profitability depends on its ability to cover the electricity demand with the local generation provided. Apart from the sizing of the system, the performance of the power conversion system (PCS) and the control algorithm have a fundamental impact on the effectiveness of the system. Additionally missing standards for the determination of the efficiency of BESS prevent transparent and comparable evaluations for the customer. To fill this gap a number of institutes are developing specific laboratory test procedures which enable a qualitative and quantitative evaluation of the performance of residential PV-BESS. First test results with a set of commercial products in the AIT SmartEST laboratory revealed a discrepancy of overall efficiency numbers between the tests and the datasheet. Particularly at low load e.g. during the night – despite the high efficiency of the used lithium-ion batteries – low efficiency values are measured. In addition, the behaviour of the control system is strongly affected by the used metering device.

Keywords: PV Battery Home Storage System, Efficiency, Performance

1 INTRODUCTION

The implementation of battery energy storage systems (BESS) for the local consumption of locally produced PV has been increasing recently. This development is fostered by decreasing costs for PV [1], decreasing costs for BESS [2], decreasing feed-in tariffs [1], increasing electricity prices [3] and dedicated subsidy schemes [4][5][6][7] for PV-BESS.

With the growing market also a larger number of manufacturers and system providers enter the market. This has led to a significant range of available products, which is listed in [8] highlighting more than 250 products provided by different companies. The available systems exhibit a wide range of possible configurations in terms of technology (lead, lithium), connection to the grid (single phase, three phase), the topology (AC-coupled, DC-coupled), battery voltage level (low voltage, high voltage) and system size (power, capacity). All those aspects have an impact on the performance of the complete system.

In this sense performance includes the efficiency (energy conversion, battery losses) and effectiveness (ability to achieve the application aim, increase of self-consumption). As a result of this, the selection of a proper system can become a challenging task for end customers as standards for the comprehensive analysis and comparison of PV-BESS performance are still missing on a general basis.

Figure 1 shows basic components of a PV-BESS and defined power flows between individual components. Basically, electric power is either generated by the PV or provided by the grid. A certain part of the locally generated electricity is directly consumed without being stored in the battery. All excess energy is basically stored in the battery respecting the capacity and power limits. When the battery is completely charged all excess energy is fed directly to the grid. The loads are covered by the battery when PV generation is below the local electricity demand. Additionally there might be minor power flows from the grid to the PV generator or battery for either standby power provision or emergency recharge of the battery.

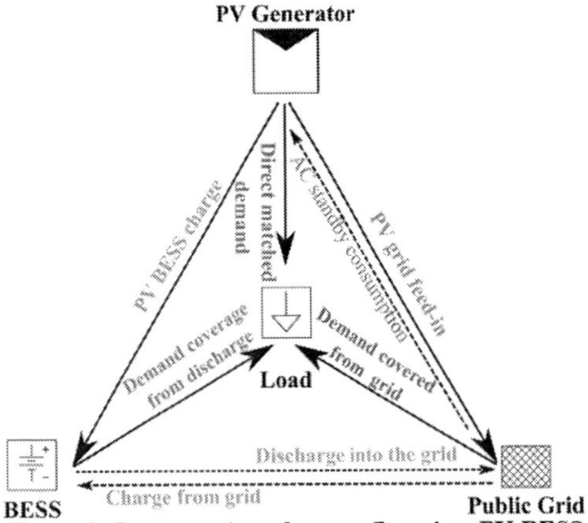

Figure 1: Components and power flows in a PV-BESS system

The evaluation of the performance of PV-BESS has to take into consideration all components and power flows as well as full and partial load operation and standby consumption of the system. A set of dedicated tests which covers a variety of system states is developed in order to analyse the performance of a PV-BESS. The test portfolio is briefly described as follows:

- Efficiency of the power conversion system (PCS) for grid feed-in, charge/discharge at full and partial load
- Round-trip efficiency (RTE) of the system for several full cycles
- Effectiveness of the control system
- Application related test for effectiveness and for efficiency evaluation

This work provides an overview of test procedures and comparative results from laboratory tests. The developed methods should provide a basis for the purpose of future standardisation.

1.1 System Configuration

As briefly discussed in the introduction there is a wide range of possible system configuration of PV-BESS available on the market. In the following chapters the relevant system aspects are introduced and discussed.

1.1.1 Storage Technology

There are basically two types of battery technologies available which dominate the market for PV-BESS. On the one hand these are lead based systems. On the other hand these are lithium based systems. The most recent numbers in Germany show that 90% of the installed systems in Q4 2015 are lithium based systems [2]. An overview of existing products [8] gives a similar picture where two thirds of the available systems are based on lithium.

1.1.2 Topology

There are AC-coupled and DC-coupled systems available for PV-BESS today. Figure 2 shows a basic representation of an AC-coupled system. This topology is widely used today. It has the advantage to be flexible in its configuration setup. Existing PV systems can be retrofitted with a battery system. Disadvantages arise from the need for a separate converter. Charging the battery from PV requires a DC/AC conversion, followed by an AC/DC conversion.

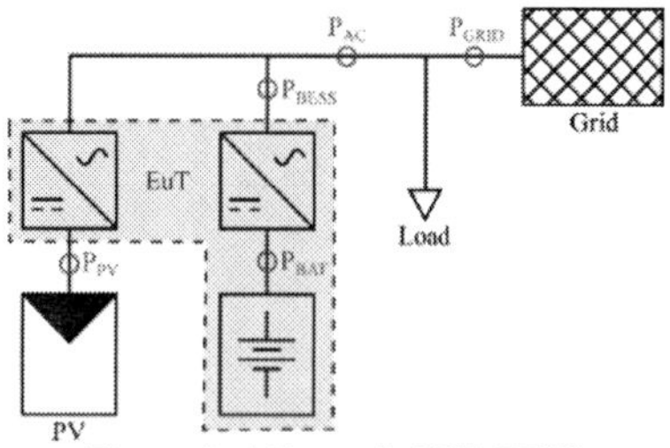

Figure 2: AC-coupled PV-BESS

Compared to the AC-coupled system, the DC coupled system requires less conversion steps for charging the battery, which in theory increases the efficiency. However, the flexibility in configuration is lower and retrofitting not always easily possible. There are three different subtypes of DC-coupled systems available today:

- **Hybrid-Inverters (Figure 3)**: A single inverter has both one terminal for PV and one for the battery. They are connected on the internal DC-bus of the system. These systems exhibit tight system integration and can be also operated bidirectional. Some hybrid inverters can be also operated AC-coupled with another PV inverter.
- **PV-generator retrofits (Figure 4)**: A charge controller for a battery is connected in parallel to an existing PV system on the DC side. This setup is often proclaimed as an economic retrofitting solution. A tight integration between PV inverter and charge controller has to be done to ensure efficient operation. A bidirectional operation is due to the topology of the PV inverter not possible.
- **Modular Systems**: This type is basically the same as a hybrid inverter but all components are separate devices. A bidirectional converter is connected on the DC-side with an MPP-Tracker and PV and a charge controller for the battery. They all share the same DC bus. Bidirectional operation is possible.

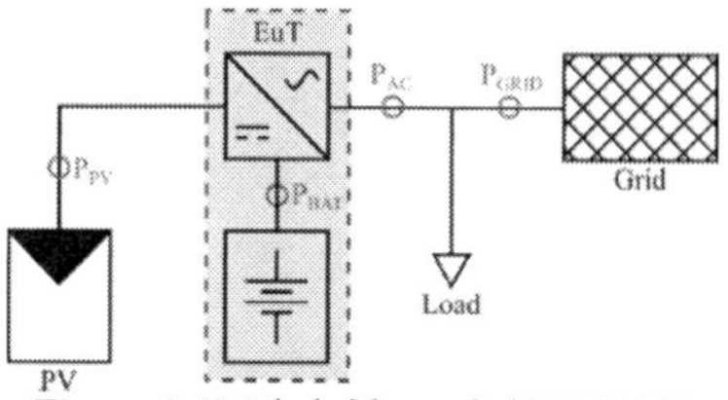

Figure 3: DC-hybrid coupled PV-BESS

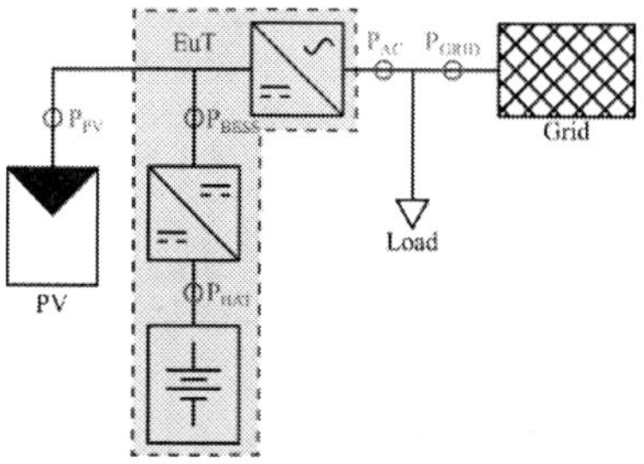

Figure 4: DC-generator coupled PV-BESS

Figure 5 gives an overview based on data from [8]. The vast majority of the available products is AC-coupled. This also represents current installations numbers in Germany where 59% of all installed PV-BESS systems are AC-coupled [2]. There are only a very few number of DC-generator retrofits (DC-gen) available. Systems shown as DC/AC can be either operated AC-coupled or DC-coupled.

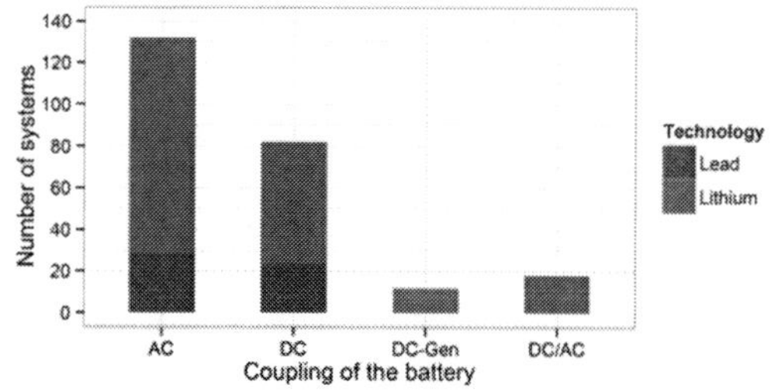

Figure 5: Available products for PV-BESS and the type of topology in use based on data from [8]

1.1.3 Electrical Connection

Figure 6 gives an overview of the available systems and their electrical connection. The majority of the systems are three phase system. Single phase systems are also widely available. Only a minor set of systems has the flexibility to be installed both single phase and three phase.

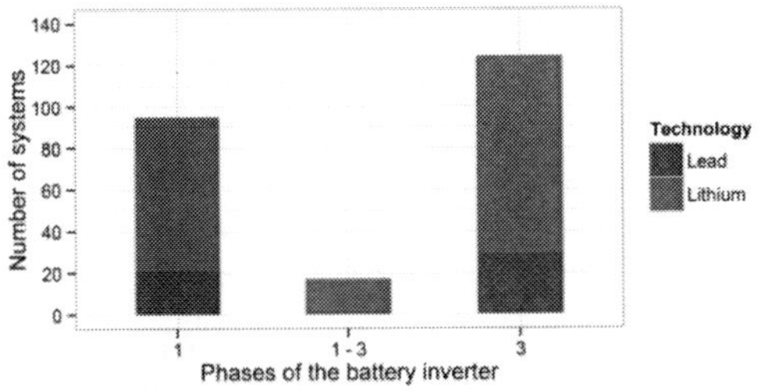

Figure 6: Number of phases for the available systems based on data from [8]

1.1.4 Battery Voltage

Figure 7 shows the nominal battery voltage in relation to the battery capacity. It can be clearly seen that nearly all systems based on lead are 48 V systems but also most of the lithium systems. There is also a visible number of systems with battery voltages over 48 V. It can be depicted that an increase of the capacity also the voltage level rises.

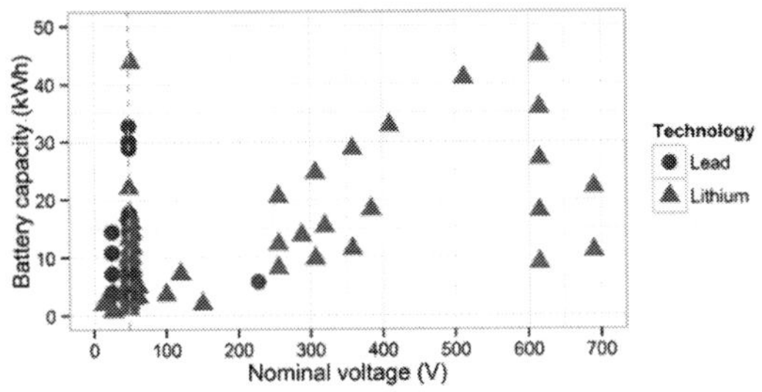

Figure 7: Nominal battery voltage and battery capacity of single available systems based on data from [8]

1.1.5 Power & Capacity

Figure 8 gives an overview of the capacity of available PV-BESS products based on data from [8]. The capacity of the available systems is between 1 kWh and 45 kWh. The most systems are available between 3 and 9 kWh. This represents also the most recent numbers from the German market [2].

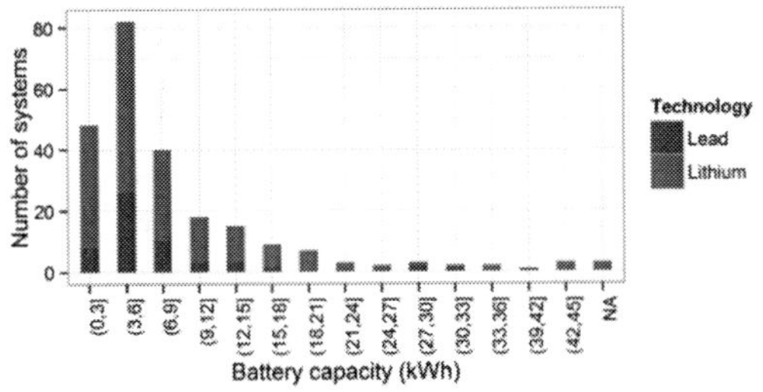

Figure 8 Distribution of the available capacity of existing storage products

1.2 Performance Evaluation of PV-BESS

The performance of a PV-BESS depends on the one hand on the system design as described in Chapter 1.1 and on the other hand on the size of the PV systems and the local demand patterns. This alone makes a comprehensive and comparable evaluation towards the system efficiency a hard task. However, there are additional aspects which complicate the analysis of PV-BESS:

- The systems have an integrated control algorithm which has to be taken into account when doing laboratory analysis. The PV-BESS is charged by activating the PV and discharged by activating the load.
- Numerous system states in the application, due to system size, as well as specific PV and load profiles are possible. Testing scenarios have to take that into account.
- The determination of the instantaneous battery efficiency is not possible in the normal operation mode of the system by use of simple power measurements.

The complexity is further increased when looking at different levels and components of the PV-BESS. In the following chapters the relevant components and their basic assessment is described.

1.2.1 Power Conversion System

The power conversion system (PCS) includes the battery converter, the PV inverter, the charge controller and MPP-Tracker. It exhibits a certain efficiency mainly depending on the power and voltage levels of this components. The second relevant aspect is the standby consumption.

1.2.2 Battery

The battery system consists of the single battery modules and a dedicated battery management system (BMS). The instantaneous efficiency of the battery cannot be determined directly when operated as intended. A cycling of the battery is required in order to determine the efficiency of the battery. As the efficiency depends on the power level such tests have to be done in multiple iterations with different power levels. Additionally, the standby consumption of the BMS has to be taken into account.

1.2.3 ICT & Controls

The control system of a PV-BESS is comprised of a so called energy management system (EMS) and a dedicated metering device installed at the point of common coupling (PCC). The control system has a significant impact on the effectiveness of the system. Response Time, settling time, steady state error, resulting energy at the POC grid, due to limited control system performance are the relevant patterns which have to be analysed.

1.3 Assessment Setup

The tests are performed in laboratory environment by use of a PV Simulator and load cascades. The PV simulator emulates characteristic I/V curves of a PV generator, as function of solar irradiance and module temperature. By the use of load cascades, the demand of electrical appliances in single family households is emulated. The test procedures are designed that charge/discharge power set points are the result of applied PV generator or/and load power. Direct control possibility of the PV-BESS, for instance over a communication protocol is mostly not available and not standardized.

Figure 9 shows the test setup. The measurement points are briefly described in Table 1. The equipment under test (EuT) is an AC or DC coupled system as described in

Section 1.1.2 (Figure 2, Figure 3, Figure 4). In case of an AC and DC generator coupled system the EuT consists of the battery inverter/converter and the PV inverter.

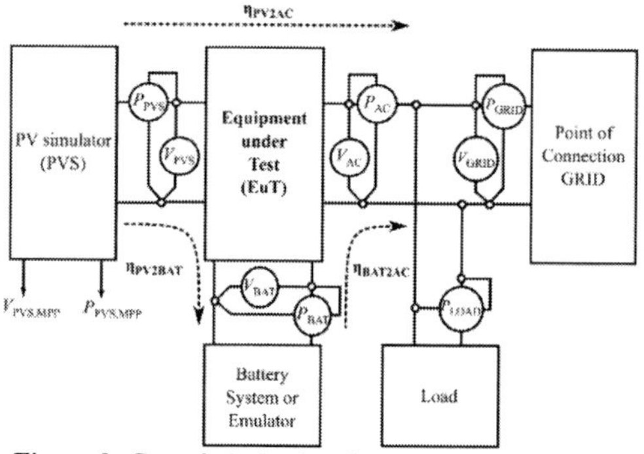

Figure 9: Generic test setup for measurements of the PCS efficiency.

Table 1: Measurement Points

Symbol	Description
P_{PVS}	DC Power output PV simulator
$P_{PVS,MPP}$	Provides MPP power
V_{PVS}	DC voltage PV simulator
$V_{PVS,MPP}$	MPP voltage
V_{BAT}	Battery voltage
P_{BAT}	Battery power (charge /discharge)
P_{AC}	Total AC power (import/export)
V_{AC}	AC voltage
P_{LOAD}	Load power
P_{GRID}	Total Grid Power
V_{GRID}	Grid Voltage

2 EFFICIENCY ASSESMENT OF THE POWER CONVERSION SYSTEM

The efficiency of the PCS is evaluated for the main operation modes of PV home storage systems [9]. The losses are determined for full and partial load operation and at different PV voltage levels. Charge/discharge efficiency is evaluated at least at the nominal battery voltage (declared by the manufacturer), within a specified tolerance band.

Table 2 describes the individual operation modes. Efficiency of each mode is defined as relation between output and input power at the system boundary of the PCS (EuT).

Table 2: Power flow from source (input) to sink (output) of the individual operation modes.

Operation Mode	P_{PV}	P_{BAT}	P_{AC}
PV grid feed-in / direct-consumption	*Input*	~ 0	*Output*
PV battery charge	*Input*	*Output*	~ 0
Battery discharge	~ 0	*Input*	*Output*
AC battery charge	~ 0	*Output*	*Input*
Standby PCS *(PCS = Sink)*	~ 0	*Input 1*	*Input 2*

2.1 Calculation Method

The efficiency is evaluable for each instant of time or time period. The instantaneous efficiency is the relation between total output and input power at the system boundary of the EuT. For reproducible results it is usually determined within a specified measurement period T_M. The conversion efficiency of the PCS $\eta_{PCS,c}$ is calculated according to Equation 1

$$\eta_{PCS,c} = \frac{\int_0^{T_M}[p_{BAT,chg} + p_{AC,exp}]dt}{\int_0^{T_M}[p_{BAT,dchg} + p_{PVS,DC} + p_{AC,imp}]dt} \quad (1)$$

where:

$p_{PVS,DC}$	PV simulator DC power
$p_{BAT,chg}$	Battery DC charge power
$p_{BAT,dchg}$	Battery DC discharge power
$p_{AC,exp}$	PV-BESS AC exported power
$p_{AC,imp}$	PV-BESS AC imported power
T_M	Measurement period

The conversion efficiency does not include maximum power point tracking (MPPT) losses for PV grid feed-in or/and PV-charge operation. The MPPT efficiency is determined according to Equation 2 [10]

$$\eta_{MPPT} = \frac{\int_0^{T_M} p_{PVS,DC}dt}{\int_0^{T_M} p_{PVS,MPP}dt} \quad (2)$$

where:

$p_{PVS,MPP}$ Provided power of the PV simulator at the maximum power point (MPP)

Considering MPPT losses, the overall efficiency (MPPT and conversion eff.) of the PCS $\eta_{PCS,t}$ is calculated to

$$\eta_{PCS,t} = \frac{\int_0^{T_M}[p_{BAT,chg} + p_{AC,exp}]dt}{P_{PVS,MPP} \cdot T_M + \int_0^{T_M}[p_{BAT,dchg} + p_{AC,imp}]dt}. \quad (3)$$

The equation is valid for steady state condition. If dynamic PV profiles are used, time synchronized measurement of provided MPP power and DC and AC measurement points are required.

2.2 Test Procedure Description

2.2.1 *PV grid feed-in / direct consumption (η_{PV2AC})*

PV power is directly consumed by local loads or fed into the utility grid (DC/AC conversion). The efficiency is determined similarly to the procedure described in the European standard EN-50530 'Overall efficiency of grid connected photovoltaic inverters' [10]. The test procedure requires the measurement of MPPT, conversion and overall grid feed-in efficiency. The test conditions are described in Table 3 (48 test points)

Table 3: Test conditions η_{PV2AC} according to EN50530

Parameter	Values
P_{PVS}	0.05, 0.1, 0.2, 0.25, 0.3, 0.5, 0.75, 1 $P_{PVS,MPP}/P_{PVS,DCr}$ (normalized to rated DC power)
V_{PVS}	V_{MPPmin}, $V_{DCrated}$ V_{MPPmax}
I/V characteristic of the PV generator	c-Si
	TF

2.2.2 *PV battery charge (η_{PV2BAT})*

PV power is entirely charged into the battery. PV charge efficiency depends on multiple system states:

- PV power P_{PVS}
- PV input voltage V_{PVS}
- Battery voltage $V_{BAT} = f(SOC, I_{BAT})$

The battery voltage is a function of state of charge (SOC) and battery current. Both is changing during a test sequence with continuous charging at different power levels.

- The voltage increases with higher SOC, whereas the current decreases during a constant power charge.
- As efficiency is evaluated for different power levels the battery voltage changes with each power step (**Figure** 7).

Efficiency characterization at multiple, constant battery voltage levels is desired. It would require the use of a battery emulator, which is mostly not possible without additional implementation effort. Modern lithium-ion based BESS are controlled by a dedicated BMS. For the purpose of safety and reliability a data exchange between PCS and BMS is required (usually based on CAN or Modbus). Information about SOC, battery current limitations etc. are communicated from the BMS to the PCS. Thus not only the emulation of a DC source/sink is required but furthermore the emulation of the entire battery system, including communication to the BMS. As no standard for the BMS → PCS communication interface exists, different manufacturers have other implementations. If a system is tested in the laboratory the emulation of the battery including the BMS requires additional time effort. A first validation and proof of concept is done in [11]. As a real battery system is used for testing, the battery voltage cannot be held at a constant level. Therefore a voltage tolerance band is defined, in which the efficiency is evaluated.

The test conditions for PV charge efficiency evaluation are described in Table 4.

Table 4: Test conditions η_{PV2BAT}

Parameter	Values
P_{PVS}	0.05, 0.1, 0.2, 0.25, 0.3, 0.5, 0.75, 1 $P_{PVS}/P_{PVS,CHGmax}$ (Normalized to max. PV charge power at the specified SOC level)
V_{PVS}	V_{MPPmin}, $V_{DCrated}$ V_{MPPmax}
I/V characteristic of the PV generator	c-Si
V_{BAT}[1]	*Option 1:* Nominal battery/inverter voltage
	Option 2: Low, nominal and high voltage levels
	Option 3: Full battery charge (V_{BATmin} to V_{BATmax}) with iterating PV power step/stair-profile [12]

2.2.3 *Battery discharge (η_{BAT2AC})*

The battery is discharged to supply the electrical demand in the household (emulated load in the laboratory). In private homes the electrical demand is usually low during the evening and night (< 400 W). Hence an appropriate efficiency characterization in this partial load area is recommended. The test conditions are described in Table 5.

Table 5: Test conditions η_{BAT2AC}

Parameter	Values
P_{LOAD}	0.025, 0.05, 0.1, 0.2, 0.25, 0.3, 0.5, 0.75, 1 $P_{LOAD}/P_{LOAD,DCHGmax}$ (Normalized to max. AC discharge power) *Additional test points:* $P_{load,add}$ = 100 W, 250 W, 500 W if $P_{load,add} < 0.025\ P_{LOAD,DCHGmax}$
V_{BAT}	*Option 1:* Nominal battery/inverter voltage
	Option 2: Low, nominal and high voltage levels
	Option 3: Full battery discharge (V_{BATmax} to V_{BATmin}) with iterating load power step/stair-profile

An example of a test sequence is shown in Figure 7. The test is performed on a DC coupled storage system (V_{BAT} > 300 V). It is seen that higher discharge power levels decrease the battery voltage more as low power levels. The battery voltage is kept at ± 2% of nominal voltage during the test. Further it is seen that especially during low power discharge the efficiency fluctuates stronger (1 s sample time of measurement). Hence appropriate measurement time and averaging is required.

[1] The test is performed with the maximum amount of battery modules (maximum power rating). If the modules are connected in series, additionally with at least the minimum amount of modules (minimum battery voltage).

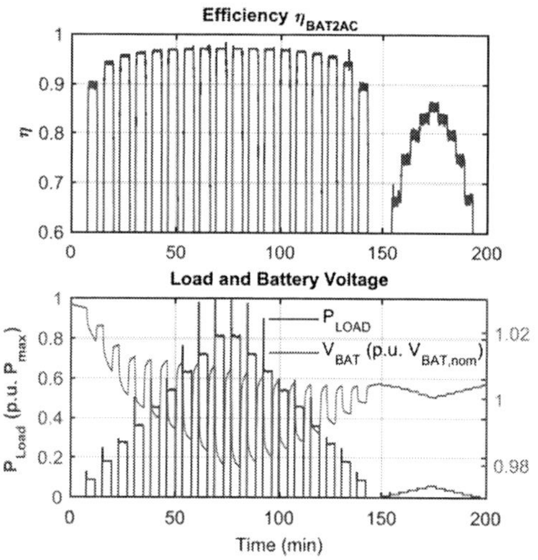

Figure 10: Example of a discharge sequence, for the demonstration of varying battery voltage during the test.

2.2.4 AC battery charge (η_{AC2BAT})

The mode 'AC battery charge' is distinguished in:

- Battery charged directly from the grid
- Battery charged from local generation (AC coupled systems)

The first operation (grid charge) is not regularly used for PV home storage systems. Grid charge may occur for battery conditioning, maintenance charging etc. and can be used for 'swarm battery' applications and ancillary grid support services. As it does not correspond to the main application of PV home storage systems, it is not considered within this work.

The latter (charge from local generation) describes the efficiency of the battery inverter for AC coupled systems. The battery is charged over a bidirectional inverter from any AC source in the home.

2.2.5 Standby consumption of the PCS

The system operates in standby mode, when no PV generation and no electrical demand is present. The energy is required to keep the PCS ready to operate. Further distinction of the standby mode is done:

- Idle Mode: The system is ready for immediate operation. A short reaction time is required, and the energy consumption is highest. In this mode the standby consumption may also vary if temporary fans are running for cooling.
- Sleep Mode: The system operates in an energy save mode, usually after a defined time with no activity. In this mode the system requires longer to 'wake up' and operate again in regularly operation.

The source for coverage of the standby consumption may vary and depend on the SOC:

- Battery is charged → Energy is withdrawn from the battery
- Battery is discharged → Energy is withdrawn from the utility grid

The manufacturer should declare the operation modes of its system. The power consumption is validated within a defined time period (e.g. 24 hrs), once with a fully charged and once with a fully discharged battery.

2.3 From single component to total PCS efficiency

The manufacturer of a battery inverter (AC coupled system) or battery converter (DC generator coupled system) usually does not provide the PV inverter for the customer. Moreover the customer is widely free in the choice which PV inverter to use. Hence the overall discharge efficiency (DC generator coupled system) or charge efficiency (AC coupled system) depends on the used PV inverter.

For efficiency evaluation the BESS manufacturer can recommend or provide a PV inverter. Otherwise the laboratory should use a standard inverter and determine a tolerance band if a better or lower performing inverter is used. This can be done in a theoretical approach:

Configuration 1: AC coupled system:

If an AC coupled system is examined, the total PV charge efficiency η_{PV2AC} depends on the used PV inverter. According to Equation 4 it is the product of PV inverter and battery inverter efficiency. The statement is true if the entire PV generated power is charged in the battery.

$$\eta_{PV2BAT}(P_{\mathrm{BESS}}) = \eta_{\mathrm{BatInv}} \cdot \eta_{\mathrm{PVInv}} \qquad (4)$$

where:

η_{BatInv}	Charge efficiency $\eta_{\mathrm{BatInv}} = f(P_{BESS}, V_{BAT})$ of the battery inverter for the AC input power P_{BESS}
η_{PVInv}	Feed-in efficiency $\eta_{\mathrm{PVInv}} = f(P_{\mathrm{BESS}}, V_{\mathrm{PVS}})$ of the PV inverter for its AC output power (input power battery inverter) P_{BESS}
$P_{\mathrm{BESS}} =$	Power at the point of common coupling PV inverter and battery inverter

For a better comparison with other systems total charge efficiency is declared as function of PV input power according to Equation 5.

$$\eta_{PV2BAT}(P_{PVS,DC}) = \eta_{PV2BAT}\left(\frac{P_{BESS}}{\eta_{\mathrm{PVInv}}}\right) \qquad (5)$$

Figure 11 shows the efficiency chart for a DC coupled three-phase system, which can be used as AC coupled system as well (bidirectional inverter). The efficiency is illustrated for the DC and AC coupled configuration. For demonstration of different PV inverter efficiencies the DC input voltage $V_{PVS,MPP}$ is varied.

The efficiency curves are illustrated separately for:

- PV inverter ($V_{MPPrated}$, V_{MPPmin})
- Battery Inverter (AC to BAT)
- Total charge efficiency (AC coupled @$V_{MPPrated}$ and V_{MPPmin})
- Total charge efficiency (DC coupled @$V_{MPPrated}$)

It is seen that the total charge efficiency decreases with lower efficiency of the PV inverter at V_{MPPmin}. It exemplifies that it is affected by the used PV inverter of the customer. For comparison the efficiency curve of the DC coupled configuration is shown.

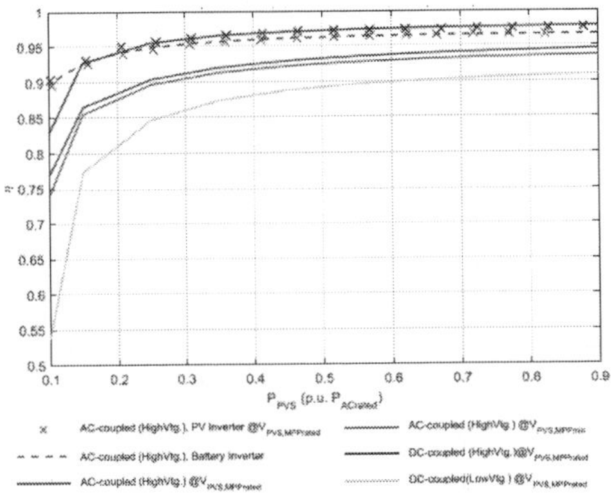

Figure 11: Efficiency chart for a PV home storage system AC and DC coupled configuration

<u>Configuration 1: DC generator coupled system:</u>

If a DC generator coupled system is examined, the total discharge efficiency depends on the used PV inverter. It is the product of PV inverter and discharge efficiency of the battery converter. The calculation is done similarly to the described procedure for the AC coupled system.

2.4 Test Results and experience from laboratory tests

This section summarizes experiences and insights gained from laboratory tests, for the evaluation of the PCS efficiency.

2.4.1 Power Rating

- Efficiency is strongly reduced at low power discharge for most systems. PV-BESS with a higher maximum AC output power rating are naturally more affected.
- Systems with lower power output profit from a higher efficiency in partial load.
- The power of an AC coupled system (battery inverter) can be sized independently from the maximum output power of the PV inverter. This can be seen as advantage for efficiency optimizations.
- Challenge for the manufacturer of a PV-BESS is to optimize the power/efficiency rating for the use in home storage application.

2.4.2 Battery Voltage

- Laboratory tests have shown that the influence of the battery voltage to the charge/discharge efficiency is marginal for high quality systems
- Hence it might be sufficient to evaluate efficiency at nominal battery / inverter DC voltage
- If the battery capacity is scalable by a series connection of modules, charge/discharge efficiency should be determined for minimum, nominal, and maximum amount of modules.
- For DC coupled systems the efficiency is decreased at lower battery voltage (DC/DC converter)
- AC coupled single-stage systems with low-frequency (LF) transformer show a stronger influence of the battery voltage to the efficiency. The efficiency is decreased with higher battery voltage.

DC input voltage V_{PVS}

- PV grid feed/in and PV charge efficiency is affected by the DC input voltage
- Recommended efficiency evaluation at three voltage levels (see EN-50530 [10])

AC vs. DC coupled system

DC coupled systems have naturally a higher charge efficiency though the conversion path is shorter. First tests have shown that there are also AC coupled system configurations (combination of PV and battery inverter) whose efficiency is higher than some of the tested DC coupled systems.

3 SYSTEM ROUND-TRIP EFFICIENCY

Instantaneous efficiency evaluation of the battery is not possible with methods of power measurements. The reason is that no input and output power is measureable at the same instant of time. Hence, the input is defined as charged energy, the output as dischargeable energy. Thus it is required to charge/discharge the battery. Best practice is to perform a number of full charge/discharge cycles. [13],[14]

The battery efficiency is the relation of the discharged energy to charged energy at the battery DC terminal

$$\eta_{BAT,RTE} = \frac{\int_0^{T_{\mathrm{dch}}} p_{\mathrm{BAT,dch}} dt}{\int_0^{T_{\mathrm{chg}}} p_{\mathrm{BAT,chg}} dt}. \qquad (6)$$

The RTE of the complete system ($\eta_{SYS,RTE}$) includes charge/discharge losses of the PCS and the battery. It is described as relation between charged PV energy and discharged AC energy resulting in

$$\eta_{SYS,RTE} = \frac{\int_0^{T_{\mathrm{dch}}} p_{\mathrm{AC}} dt}{\int_0^{T_{\mathrm{chg}}} [p_{\mathrm{PVS}} - p_{\mathrm{AC}}]\, dt}. \qquad (7)$$

Furthermore $\eta_{SYS,RTE}$ is expressed as the product of PV charge efficiency, discharge efficiency and battery RTE (Equation 8). The statement is true if the entire PV generated energy is charged only (no additional feed in, or load coverage). If the efficiency of the PCS is known (efficiency chart) a fast estimation of the system RTE is possible for different battery systems.

$$\eta_{SYS,RTE}(P_{BAT}) = \eta_{PV2BAT} \cdot \eta_{BAT,RTE} \cdot \eta_{BAT2AC} \qquad (8)$$

3.1 Proposed Test Procedure

Figure 12 illustrates one full cycle. The sequence parametrization is detailed described in Table 6.

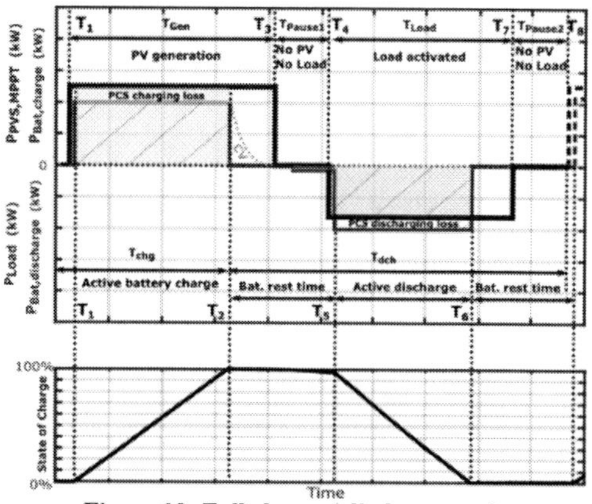

Figure 12: Full charge / discharge cycle

A full cycle requires charging the battery to the cut-off voltage. This is done because the SOC is usually best defined for a fully charged and fully discharged battery. For instance the commonly used lithium-iron-phosphate (LFP) battery cells have a very flat voltage profile over a wide SOC range. At low and high SOC states the voltage gradient is steeper, which allows the internal BMS to estimate the SOC better. The battery is considered as fully charged/discharged when the BESS terminates the charge/discharge process ($P_{BAT} \sim 0$) at the cut-off battery voltage. The test procedure requires that the SOC is identical at the beginning and end of the measurement. Otherwise not only battery losses but energy differences due to different SOC states determines the result.

Table 6: Sequence description of the full charge/discharge cycle illustrated in Figure 12

Time	Description
T1 to T3	PV generation applied for charging the battery. The test conditions are: • PV power is below maximum charge power ($P_{PVS} < P_{PVS,charge,max}$) • No additional grid feed-in • T_{GEN} allows a full battery.
T1 to T2	The battery is charged until the BMS terminates the charge process. • In this phase the system may switch to a Constant-Voltage (CV) charge, at the end of charge process. • The battery power is decreased continuously and not abrupt to zero.
T2 to T5	Option 1: No CV charge occurs • Entire PV power is fed into the grid • Battery power should be zero • Battery might start to discharge with low power during PV feed-in. Option 2: CV charge • Battery is still charged with decreasing power • Simultaneous grid feed-in and PV-charge • Efficiency is reduced due to additional PCS feed-in losses
T4 to T7	Load is activated for battery discharge The time T_{LOAD} is chosen that a full battery charge with specified charge power is possible.
T7 to T8	Rest time (Allow OCV relaxation until next cycle starts)

For maximum reproducibility, one full cycle is iterated several times (> 3) and efficiency is calculated as relation between total energy output and input. All iterations are performed with maximum discharge power and for at least two discharge power levels ($P_{LOAD,dchmax}$, $P_{LOAD,dchmax}/2$).

3.2 Test procedure according to IEC 61427-2:2015

The international standard IEC 61427-2:2015 'Secondary cells and batteries for renewable energy storage – General requirements and methods of test Part 2 On-grid applications [15] evaluates the RTE of BESS (without considerations to the PCS) for the application of "Daily PV energy storage time shift" in a seven day endurance test. The test sequence is briefly described below:

1. Full battery charge in two steps (240 and 120 min) at constant power
2. Discharge until final battery voltage, specified by the manufacturer
3. Wait 1440 min (24 h)
4. Iterate step 1 to 3 seven times.

The test sequence requires that the battery discharge power is at least 3 kW. The standard focuses primary on the battery system itself, without concerns to the system integration. An adopted standard for the whole PV storage system and the specific application of home storage systems would be useful.
Alternatively the results of the battery efficiency could serve as basis for a fast calculation of the system RTE (Equation 8). A PV-BESS was tested according to IEC 61427-2:2013 by the institute for solar technology at the technical university of Madrid. Determination of PCS and battery efficiency are described in the paper [16].

3.3 Test Results and experience from laboratory tests

This section summarizes experiences and insights gained from laboratory tests, for the evaluation of the RTE.

3.3.1 Accuracy of the SOC estimation o

The evaluation of the RTE has shown that SOC estimation of the integrated BMS is not always reproducible.

- The internal BMS does not always charge/discharge the same amount of electrical charge Q (Ampere-hours, Ah) even for identical performed cycles
- Lithium-ion based battery systems achieve coulomb efficiency ($\eta_{COL} = Q_{IN}/Q_{out}$) values close to 100 %.

- During full single charge/discharge cycles coulomb efficiency can achieve values > 100 % (more Ah discharged, than charged). Thus it discharges slightly below the declared Depth of Discharge (DOD). In order to restore the true SOC it will charge more during the next cycles.
- Hence it is necessary to determine efficiency for multiple full cycles

3.3.2 *Battery Voltage*

Efficiency of the PCS is evaluated over the entire SOC range during a full charge / discharge cycle (Average efficiency)

3.3.3 *Efficiency*

The RTE was evaluated for lithium-ion batteries. Tests have shown that:

- Lithium-ion batteries achieve high efficiencies (> 97 %) at application related C-rates < 0.7C
- Battery efficiency increases with lower charge/discharge power. Efficiency of the PCS decreases with lower charge/discharge power
- Standby consumption of the BMS of a tested high voltage battery system (10-20 W), decreases the RTE by 1 to 2 % in a multiple full charge/discharge cycle test
- System RTE consists of battery system, and PCS efficiency. **Table 7** shows a comparison of two DC coupled systems. It is seen that System A would profit from a higher battery efficiency, whereas System B would profit from a higher PCS efficiency

Table 7: System RTE of two DC coupled systems

EuT	η_{PV2BAT}	$\eta_{BAT,RTE}$	η_{BAT2AC}	$\eta_{SYS,RTE}$
System A	0.97	0.93	0.97	**0.87**
System B	0.91	0.97	0.91	**0.78**

3.3.4 *Test Sequence*

- Not only one single full cycle is sufficient for a reproducible result. Iteration of cycles is required
- To shorten the test time it is recommended to charge with nominal power and to vary only the discharge power.
- Smart charging algorithms [17] are turned off, that continuous charging is possible

4 PERFORMANCE OF THE CONTROL SYSTEM

The basic control strategy of PV home storage systems compensates grid power best possible to zero. An energy meter at the grid connection point, transmits periodically information about measurement data to the energy management system. If a surplus of PV power is fed-into the grid, it is registered and the battery starts charging. If electrical loads draw energy from the utility grid, the BESS start discharging. The battery SOC and battery power must allow required discharge.

Due to delays in the control loop, it takes a certain time until the BESS provides the requested storage power. Following situations are observed [18]:

a. Grid consumption for load coverage until requested discharge power is provided
b. Battery discharge to the grid after a reduction of load power, until discharge power is reduced
c. Grid injection of PV power until requested charge power is provided
d. Battery charge from the grid: after a reduction of PV power, until charge power is reduced

4.1 Performance evaluation of the control system

The performance is evaluated by dynamic load and/or PV profiles. For instance step/stair/ramp profiles. For appropriate testing of the control system the requested power must not exceed the maximum battery power and the SOC must be in an area where the requested power can be provided.

In control theory the settling time and steady state error is used to describe a step response:

- *Settling Time (T_s):* The settling time is the elapsed time until the storage power or grid power enters and stays within a specified error band (For instance grid power in steady state ±10 W)
- *Steady State Error (SSE):* The SSE describes the difference between the set point (zero grid power) and the actual value in a steady-state.

The characteristic indexes are explained in Figure 13. It shows the system response on a load step. The step response of the tested BESS was evaluated on the one hand with the proprietary energy meter and at the other hand with a D0-compliant smart meter (optical read out, serial communication according IEC-62056-21 [19]). It is seen that for the latter the control systems shows a second order behaviour and power is oscillating between feed-in and consumption at the grid connection point. The time T_s until the grid power stays within the error band of ±10 W is 5 s for the test with the proprietary meter. At the test with the D0-compliant meter it takes about 74 seconds. A steady state error of ±4 W is measured.

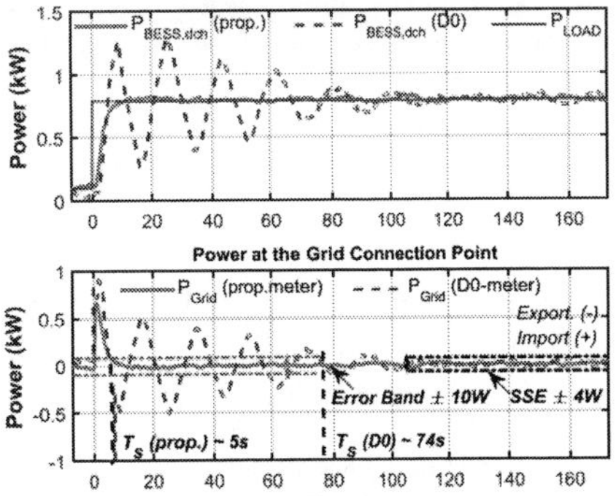

Figure 13: Step response / load Step (D0 and proprietary metering device).

An alternative approach is the evaluation of the energy-flow at the grid connection point. The test sequence consists of a dynamic profile with several load steps applied. Examples of such profiles are found in [12].

Figure 13 shows one example of such a profile, which can be used for PCS efficiency evaluation as well (see Figure 9).

It is seen that after a load step is applied, power is obtained from the grid until the discharge power fits the load power. After the load is turned off, power is discharged from the battery into the grid until the battery power is reduced to zero.

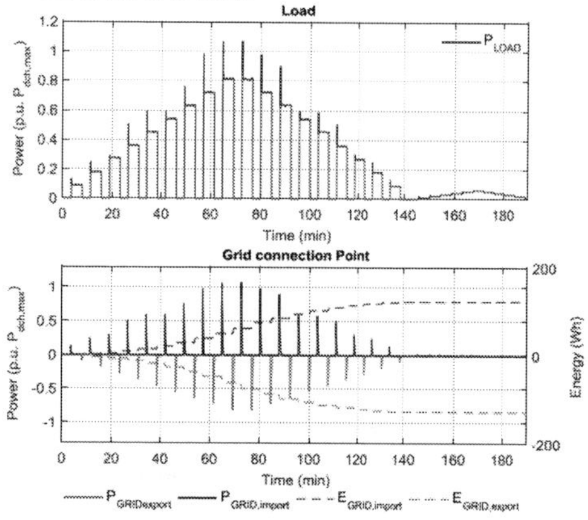

Figure 14: Load profile using several power steps

An ideal fast and accurate control system would reduce grid power immediately to zero. Hence the degree of self-coverage is 100 %. Self-coverage describes the ability of the BESS to cover the load demand with local generated- and battery discharge power. It is calculated according to Equation 9.

Self-coverage (self-sufficiency)

$$SC = \frac{\int_0^{T_M} [p_{\mathrm{LOAD}} - p_{\mathrm{GRD,IMP}}] dt}{\int_0^{T_M} p_{\mathrm{LOAD}}\, dt} \qquad (9)$$

where:

p_{LOAD} Load Power

$p_{\mathrm{GRID,imp}}$ Obtained power from the utility grid

For the load profile shown in Figure 14**Error! Reference source not found.**, a degree of 91.6 % self-coverage was achieved. The test was iterated several times and the result was reproducible.

E_{LOAD} (kWh)	$E_{\mathrm{GRID,imp}}$ (kWh)	SC (%)
1.55	0.13	91.61

4.2 Test Results and experience from laboratory tests

The time resolution and measurement accuracy of the used energy meter plays a crucial role in the performance of the control system. For instance meters based on the D0 protocol transmit the measurement data every 2-5 s to the EMS. Proprietary meters send measurement data every second or less to the BESS. Experiences and testing outcomes can generally be described as:

- Settling time may vary on the time point the load step is applied
- It results from constant meter-transmission and processing time of the system
- If the load is turned on short before data is sampled and processed the settling time is shorter.
- If the load is turned on shortly after data is sampled and processed by the EMS the settling time is longer.
- This behaviour must be considered for the design of a test-sequence. The pause between steps and the duration of steps should not be constant.
- The test can be performed also with varying PV power. A requirement is that the charge algorithm operates in zero grid-power mode (smart charge algorithms, as for feed-in limitation used must be deactivated).

5 APPLICATION RELATED EFFICIENCY

At the application related test high sampled, real irradiance and electrical load data of single family households is emulated in a multiple day test [9],[12] [20]. The test allows the determination of the efficiency and effectiveness of the system for a given profile.

- System and application related battery efficiency
- Share of direct use (self-consumption) and self-coverage (self-sufficiency).

First results have shown that such a test has considerable drawbacks:

- Temporal scalability: A relative short, multiple-day test in the laboratory might not represent effectiveness for longer periods > 1 year
- Systems size (battery capacity, PV generator peak power) is optimized for specific users (PV and load profile/annual electricity demand). Hence the test-profile(s) must cover a broad field of users
- The evaluated system efficiency depends the specific system size. For instance a PV-BESS with low battery capacity is less charged and discharged over the test time. On the one hand most generated energy is then directly fed-into the grid or used for local load coverage. As η_{PV2AC} is usually high, the system efficiency is high as well. At the other hand such a system will achieve a low self-sufficiency, as less energy is provided by the battery for local loads. Hence system efficiency and effectiveness behave inversely in this case.

Further research will be done, if such a test can be used for evaluation of an application related efficiency. Nevertheless the test is useful for verification of the appropriate system functionality during multiple days, in an application related environment.

6 SUMMARY

First results have shown that charge / discharge efficiency of the PCS is mostly, significantly decreased at low power levels. As electricity demand and power in single family households is usually low at the evening and during the night PV-BESS profit from higher discharge efficiency in this range. Figure 15 shows the efficiency of a residential PV-BESS, as function of power and voltage difference between DC link and battery. Figure 16 shows the same test result in more detail. It can be seen that efficiency decreases with a higher voltage difference (lower state of charge). Discharge efficiency up to 95% was achieved for a low-voltage AC coupled system. For an entirely transformer-less charge controller an efficiency around 98% is expected.

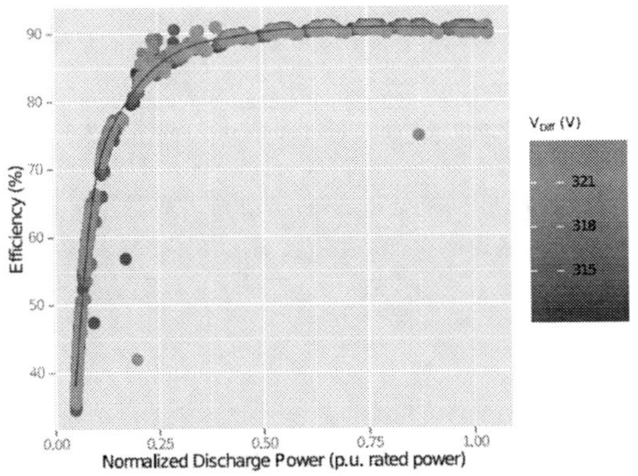

Figure 15: Discharge efficiency of a low voltage BESS as function of the difference between DC-link and battery voltage

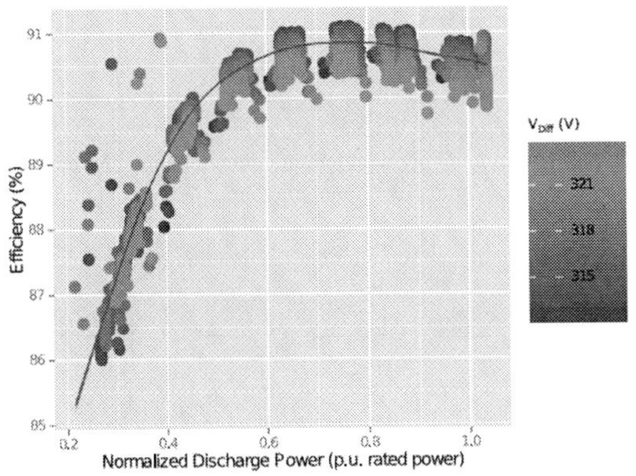

Figure 16: Discharge efficiency (zoomed-in view)

Evaluation of the round-trip efficiency consisting of full charge/discharge cycles showed that lithium based batteries can achieve a high RTE (> 95 %). The overall system RTE is decreased due to charge and discharge losses of the PCS. Depending on the charge/discharge power a system RTE below 90% is expected for most commercial available products.

The settling time until the load is covered by storage power is strongly affected by the used energy meter. The settling time varies strongly between different systems. It is expected, that due to system optimizations of the manufactures the settling time will be at least below 10 s for most future products. Recent research has shown that some manufactures declare to be below 1s settling time. Further testing will validate if their statements are true.

7 BIBLIOGRAPHY

[1] H. Wirth, "Recent facts about photovoltaics in Germany." 22-Apr-2016.

[2] K.-P. Kairies, D. Haberschusz, J. van Ouwerkerk, J. Strebel, O. Wessels, D. Magnor, J. Badeda, and D. U. Sauer, "Wissenschaftliches Mess- und Evaluierungsprogramm Solarstromspeicher - Jahresbericht 2016." 2016.

[3] Eurostat, "Electricity price statistics," 2016. [Online]. Available: http://ec.europa.eu/eurostat/statistics-explained/index.php/Electricity_price_statistics. [Accessed: 21-Jun-2016].

[4] KfW, "Merkblatt Erneuerbare Energien - KfW-Programm Erneuerbare Energien 'Speicher' - Programmnummer 275 - Finanzierung von stationären Batteriespeichersystemen in Verbindung mit einer Photovoltaikanlage."01-May-2013.

[5] Land Salzburg - Abteilung 4: Lebensgrundlagen und Energie, "Salzburg Landesförderprogramm PV-Speicher Richtlinien." 19-Jan-2014.

[6] Land Steiermark, "Anhang zur Förderrichtlinie - Technische Voraussetzungen." 2014.

[7] Amt der Oö. Landesregierung - Direktion Umwelt und Wasserwirtschaft - Abteilung Umweltschutz, "Oberösterreich Förderprogramm stationäre Solarstromspeicher." 11-Sep-2014.

[8] pv magazine, "Produktdatenbank Batteriespeichersysteme für Photovoltaikanlagen," *http://www.pv-magazine.de*, 02-Sep-2014. [Online]. Available: http://www.pv-magazine.de/marktuebersichten/batteriespeicher/produktdaten-2014/. [Accessed: 02-Sep-2014].

[9] F. Niedermayer, J. v. Appen, M. Braun, A. Schmiegel, N. Kreutzer, M. Rothert, and A. Reischl, "Innovative Performancetests für PV-Speichersysteme zur Erhöhung der Autarkie und des Eigenverbrauchs," presented at the Symposium Photovoltaische Solarenergie, Bad Staffelstein, Germany, 2015, p. 12.

[10] CENELEC, "ÖVE/ÖNORM EN 50530:2014-02-01 Gesamtwirkungsgrad von Photovoltaik Wechselrichtern (Overall efficiency of grid connected photovoltaic inverters)." ÖVE/ASI, 01-Feb-2014.

[11] Christian Seitl, Christian Messner, Hartmut Popp, and Johannes Kathan, "Emulation of a High Voltage Home Storage Battery System using a Power Hardware-in-the-Loop Approach", Proc. of IEEE IES IECON 2016 , Florence, Italy, 2016.

[12] Christian Messner, Johannes Kathan, and Johann Mayr, "Effizienz und Effektivität von netzgekoppelten PV-Heimspeichersystemen - Erfahrungen und Erkenntnisse aus Labortests kommerzieller Produkte," presented at the PV-Symposium 2016, Bad Staffelstein / Germany, 2016.

[13] SANDIA National Laboratories, "Protocol for Measuring and Expressing Performance for Energy Storage Systems," 2012.

[14] N. Martin, E. Lemaire, E. Rikos, G. Messinis, and P. Crolla, "Criteria and procedures for performance testing - Grid connected storage," 2013.

[15] IEC, "IEC 61427-2 Secondary cells and batteries for renewable energy storage -General requirements and methods of test Edition 1.0." Aug-2015.

[16] Aitor Makibar and Luis Narvante, "Characterisation and Efficiency Test of a Li-Ion energy Storage System for PV Systems," presented at the EU PVSEC, Hamburg, 2015, pp. 1652 – 1658.

[17] J. Li and M. A. Danzer, "Optimal charge control strategies for stationary photovoltaic battery

systems," *J. Power Sources*, vol. 258, pp. 365–373, Jul. 2014.
[18] Weniger Johannes, Bergner Joseph, and Quaschnig Volker, "Auswirkungen von Regelträgheiten auf die Energiefluesse in Wohngebäuden mit netztgekoppelten PV-Batteriesystemen," presented at the PV Symposium 2016, Bad Staffelstein.
[19] IEC, "IEC 62056-21:2002 Electricity metering - Data exchange for meter reading, tariff and load control - Part 21: Direct local data exchange." 2002.
[20] Hauke Loges, M.Sc., "Messtechnische Untersuchung von PV-Batteriespeichern (Jahresbericht 2015 - TU-Braunschweig - Institut für Hochspannungstechnik und Elektrische Energieanlagen - elenia)." 2015.

SHORT-TERM POWER FLUCTUATIONS IN DENSELY CLUSTERED PV-BATTERY SYTEMS

James Barry, Nina Munzke, Jorge Thomas
Karlsruhe Institute of Technology
Hermann-von-Helmholtz Platz 1, 76344 Eggenstein-Leopoldshafen, Germany

ABSTRACT: Energy storage systems coupled to solar photovoltaic arrays will become a vital pillar in the future energy landscape, both on and off the grid. The high variability of solar power can be overcome with smart technologies and rapid battery response rates. Knowledge of the prevalence (how often they occur) and persistence (how long they last) of photovoltaic power fluctuations allows storage system developers to find the optimal solution to achieve the desired level of smoothness in the resulting power output from the system. In this work power data measured at 1 Hz from the densely clustered arrays of the KIT 1 MWp solar installation are analysed using statistical techniques in order to characterise the variability of single arrays as well as that of the system as a whole. The relationship between correlation and distance is examined in detail using models from the literature, for fluctuation lengths in the range of several seconds and separation distances down to 10 metres. The smoothing that results from decorrelation can be used as an additional mitigation factor in the design of small-scale collections of solar systems, especially in the context of residential and district storage solutions.
Keywords: very short-term power fluctuations, spatial correlation, battery storage, system integration

1 INTRODUCTION

The successful integration of rapidly fluctuating renewable energy sources into the electricity grid requires storage technologies capable of fast and intelligent power control, as well as a thorough understanding of the variations themselves. Solar photovoltaic (PV) plants are particularly susceptible to large and fast changes in power output, with extreme ramp rates occurring relatively often. These fluctuations have two distinct sources: (i) the "natural" variations due to changes in the shortwave downward radiation collected by the solar modules and (ii) the "artificial" variations that come from the electronics and digital systems. The former can be divided into deterministic variations due to geographical location, time of day and time of year as well as stochastic variations due to the effects of weather, whereas the latter originate from peculiarities of individual inverter electronics and software algorithms as well as from blackouts and brownouts due to faults.

The characteristic non-Gaussian statistics of the stochastic cloud-induced fluctuations in solar irradiation and resulting PV power output have been studied extensively in the literature [1]–[3]. Whereas some authors [3]–[10] focus on the variability of the clearness index or clear-sky index, others [1], [2], [11]–[15] have analysed the power output itself. It has been shown that power ramp rates of up to 60% of peak power per second are possible [8], but that the aggregation of multiple solar plants leads to a smoothing effect. Indeed, several studies have sought to understand the spatio-temporal characteristics of solar power variability in different contexts. A simple model, based on the so-called dispersion factor, to quantify the variability reduction in the context of grid-connected systems was developed by Hoff and Perez in [11] and extended in [6]. Another model based on wavelets was developed in [8], [16]. In both cases it has been shown that the further apart the individual solar arrays are the less correlated the variations in power output become, and that this reduction in correlation is less pronounced for larger time increments. The authors of [14] apply both the dispersion factor and wavelet models to data from a number of different sites around the world, with the aim of developing a global model for spatial smoothing of PV field aggregate power output.

Apart from the recent analysis [3] using a network of densely packed radiation sensors, most prior studies have been in the context of relatively large distance and time scales, largely due to the lack of high-frequency data. In the case of smaller or off-grid systems the effects of very short-term fluctuations become important and can affect power quality, due to voltage flicker, for example. Although decorrelation between arrays can help to mitigate these effects, the integration of a battery storage system allows further control over ramp rates, thus facilitating higher PV penetration. Different battery control strategies for smoothing fluctuations have been studied before (see for example [17]–[22]), mostly in the context of large-scale PV plants.

The power variations in densely clustered PV systems such as those in residential areas could either be smoothed by small batteries in each household or by a district-scale storage solution. In this context it is useful to understand the correlation between individual array power outputs. Optimal storage system design means that one needs to take the largest fluctuation into account [20], specifically for off-grid systems. Experience with prototype storage systems has shown that in order for successful smoothing of PV fluctuations one needs control algorithms with a cycle time on the order of one fifth of a second or better. In most control systems there is a certain amount of latency, which leads to unwanted power flows. A recent simulation [23] showed that for a household storage system in Germany a dead time of 5 seconds in the control loop would lead to a monetary loss of €15 p.a., which is hardly a notable effect. However in reality these effects could be larger and should be studied in an experimental setup. In off-grid systems with 100% renewable energy supply the effects would be more dramatic, since the voltage and power quality depends on the rapid response of the storage system to variations in power input.

The very short-term fluctuations in power output from 78 solar PV arrays installed on the north campus of Karlsruhe Institute of Technology (KIT) are analysed in detail in what follows. The array separation distances range from roughly 10m to 250m and power data was recorded at 1 Hz over the course of one year. The power fluctuations of each array are characterised by their probability distributions, and the reduction in overall power variability due to decorrelation effects at small distance and time scales are shown. The application to battery storage systems of different scales is briefly discussed – a comprehensive study of this topic is left for future work.

The next section includes a description of the PV installation and data collection procedure as well as an outline of the calculations performed. The results are presented in section 3, and we summarise and conclude in section 4.

2 METHODOLOGY

2.1 Description of the PV installation

The fluctuation analysis was carried out using data from the unique "static solar-tracking configuration" 1MWp PV installation at KIT, shown in Figure 1. This system was designed and procured by Project Competence E (PCE) at KIT and includes a state-of-the-art data acquisition system (see Ref. [24] for details). There are six different PV module types (labelled 1…6) installed at a variety of inclination and orientation angles, and four inverter types (labelled A...D) connect the 102 PV arrays to the institute grid, as summarised in Table I and described in more detail in Table II. Each array has a peak power of roughly 10 kW, and the distances between the centre points of any two arrays ranges from 7.4 m to 251.6 m, which will be relevant for the analyses that follow.

Figure 1: Aerial photograph (top) of the 1MW solar installation at KIT north campus (bottom), with the weather mast 1.3km away in the background.

Table I: Description of subset of PV field with 1 Hz sampling rate (see Table II and Figure 2 for details).

Characteristic	Description
Number of arrays	92 (86)
Inverter brands	A, B, D
PV module brands	1, 2, 3, 4, 5, 6
Orientation (0° S)	-60°, -45°, -30°, -15°, 0°, 15°, 30°, 45°, 60°
Inclination	2°, 15°, 30°, 45°, 60°

Table II: Array numbers by orientation (rows), inclination (columns) and inverter type, where types A, B, C, D are green, magenta, blue and red respectively

	2°	15°	30°	45°	60°
60°E		1, 2	3, 4	5, 6	7, 8
45°E		9, 10, 11	12, 13	14, 15	16, 17
30°E		18, 19, 20, 21, 22, 23	24, 25	26, 27, 28	29, 30, 31
15°E		32, (33), 34	35, 36, (37)	38, 39, 40	41, 42, 43
00°	46, 47, 48	(44), 45	49, 50, 51, 52, 53, 54	55, 56	57, 58
15°W		59, 60, 61	62, 63, 64	65, 66, (67)	(68), 69, 70
30°W		71, 72, 73, 74, 75, 76	77, 78 79	(80), 81, 82	83, 84, 85
45°W		86, 87, 88	89, 90	91, 92	93, 94
60°W		95, 96	97, 98	99, 100	101, 102

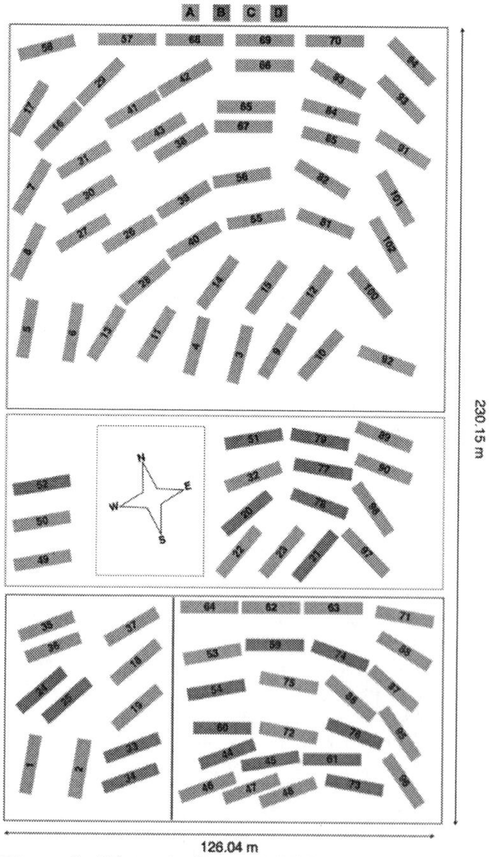

Figure 2: Schematic diagram of field layout, with array numbers and colours corresponding to those in Table II.

2.2 Data collection and filtering

The analysis was performed with DC power data collected over the 2015 calendar year from 78 PV inverters with a sampling rate of 1 Hz. The values from each inverter are sent via Ethernet to a central database, which also provides the time stamp for each time series. A brief description of the data is given in Table III. The data can be quickly accessed, visualised and downloaded via the advanced data extraction infrastructure (ADEI) system [25], developed by the Institute for Data Processing and Electronics (IPE) at KIT.

Table III: Description of the data

Quantity	Description
Start time	01/01/2015 00:00:00
End time	31/12/2015 23:59:59
Sampling rate	1 Hz
Power values per array	31 536 000
Power measurements	DC power
Precision	1 Watt
File size w/out time stamp	119 090 kB per array
File size with time stamp	702 884 kB per array

For speed of processing, specially developed software was used to access the data directly from the replicated database and filter the data for use in the MATLAB program that performed the fluctuation analysis. In order to synchronise the time series to study collective effects the original time stamp had to be modified, which introduces an error of up to 0.5 s. This did not have a significant effect on the results. Note that due to database maintenance the data from 2 July 2015 was used to replace the data from 1 July 2015 (both were clear sky days).

Several of the inverters suffered blackouts or brownouts during the course of the year, so that the data were filtered to remove these anomalous effects. The filtering algorithm was customised for the specific inverters, and consisted in setting a threshold for the relative drop in PV power per time step. In this way the "manmade" fluctuations were removed, in order to better analyse the variations from cloud-induced changes in solar radiation.

2.3 Calculation procedure

Once the data is filtered and post-processed into time series of equal lengths the fluctuations were calculated in MATLAB as follows (note that the large volume of data necessitated the use of the high-performance cluster computer bwHPC):

- Night time values were discarded using the known sunrise and sunset times for Karlsruhe. Although the time window for which the power is non-zero is slightly different for each array, these small differences are irrelevant for the fluctuation analysis (solar radiation is diffuse in the early morning and late evening).
- Each power series was normalised by its maximum value during the sample period.
- The change in power was calculated for each time series for different time intervals.
- In order to study correlations, the correlation matrix of array pair correlations was calculated for each time interval.
- The distance between each the centre of each array was calculated using coordinates provided by the solar installer.

The fluctuations for each array for the time differences $\Delta t = 1\ldots10, 30, 60, 300, 600, 1800$ seconds were calculated, along with the correlation between each pair of arrays. A fluctuation $\Delta F_i^{\Delta t}$ for array i at time t is given by

$$\Delta F_i^{\Delta t}(t) \equiv \frac{P_i(t) - P_i(t-\Delta t)}{\max(P_i)}, \qquad (2.1)$$

where $\max(P_i)$ is the largest power recorded from array i during the sampling period. The relative variability of the entire field compared to that of the individual arrays i $(i = 1..78)$ for the fluctuation length Δt is defined as

$$\mathrm{VR}_i^{\Delta t} \equiv \frac{\sigma_{\text{field}}^{\Delta t}}{\langle \sigma_i^{\Delta t} \rangle}, \qquad (2.2)$$

where $\sigma_{\text{field}}^{\Delta t}$ is the standard deviation of the fluctuations in the sum of the entire field's power, whereas $\langle \sigma_i^{\Delta t} \rangle$ is the mean standard deviation of the fluctuations in the power of a single array. The array pair correlation between two fluctuation time series $\boldsymbol{\Delta F}_i^{\Delta t}$ and $\boldsymbol{\Delta F}_j^{\Delta t}$ is calculated using the Pearson product-moment correlation coefficient

$$\rho_{ij}^{\Delta t} \equiv \frac{\mathrm{cov}(\boldsymbol{\Delta F}_i^{\Delta t}, \boldsymbol{\Delta F}_j^{\Delta t})}{\sigma_i^{\Delta t}\, \sigma_j^{\Delta t}}, \qquad (2.3)$$

that is the ratio of the covariance to the product of standard deviations. Following the analysis in [14] (see also [6]) two different models for the relationship between pair correlation and distance are fitted to the data for different fluctuation lengths Δt,

$$\rho_{ij}^{\Delta t} = \frac{1}{1+\frac{d_{ij}}{\Delta t\, v_1}} \quad \text{and} \quad \rho_{ij}^{\Delta t} = e^{\frac{d_{ij}\ln(0.2)}{1.5\,\Delta t\, v_2}}, \qquad (2.4)$$

referred to as the hyperbolic and exponential models, respectively. Here d_{ij} is the separation between array i and j, and v_1 (v_2) is a parameter representing the cloud speed.

3 RESULTS AND DISCUSSION

Of the 102 AC/DC inverters, only 82 (of types A and D) provide reliable data at the required 1 Hz sampling rate. The others either only record average values or are not equipped with sensors fast enough to measure power at 1 Hz so that there are many repeated values. A further four inverters suffered severe data blackouts and/or digital errors, so that the analysis described below was performed with a total of 78 power time series from two different inverter types: 68 from type A and 10 from type D. The 68 inverters of type A are spread out over a larger area, whereas the 10 inverters of type D are closer together (see Figure 2 and the frequency distribution in Figure 3).

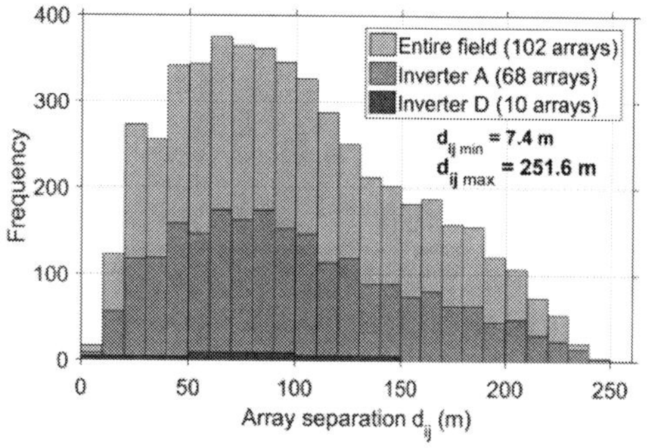

Figure 3: Histogram showing the spread of distances between PV arrays

3.1 Fluctuation probabilities

The characteristic non-Gaussian statistics of the power fluctuations can be seen in Figure 4, where extreme events manifest in the flat tails of the probability distribution. The points marked "A" and "D" are not sums but rather probability densities calculated by combining all inverters of each type into one sample set. It is clear from the plots that aggregating the PV arrays into one field reduces the frequency of extreme events – the dependence of this smoothing effect on distance and correlation will be described in the following section. Each plot shows the relative variability between the entire field of 78 arrays and the mean variation from individual arrays. The relative variability is smallest for small time increments, which correspond to low correlation coefficients (see Section 3.2).

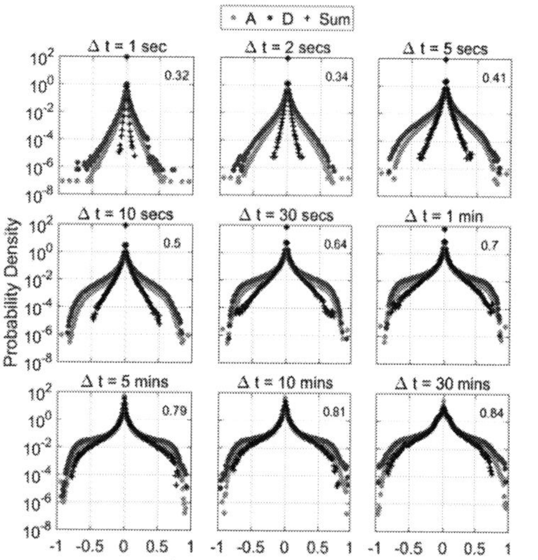

Figure 4: Probability distribution of the normalised rate of change of DC PV power from two inverter types as well as their sum, for different time increments. Relative variability is indicated in each case [cf. Equation (2.2)].

The unique nature of the PV installation also allows groupings in terms of orientation and inclination angle; however the fluctuation probability distributions did not show significant differences in this case. Indeed, by normalising the data with the expected clear sky curve one could remove effects of orientation and inclination, but this is of course model dependent. Further studies could examine the fluctuations on a seasonal level or by grouping the data into different types of days (as in [3]), so that orientation and inclination effects would come into play.

3.2 Fluctuation correlations

Correlations between power series from pairs of arrays were calculated using Equation (2.3), for the 68 inverters of type A, leading to a 68×68 symmetric matrix. Figure 5 provides a graphical representation of this matrix for different time scales, where the inverters are ordered by their number (cf. the green numbers in Table II). It is evident that correlation increases with time scale and distance, since small distances are represented by elements close to the diagonal of the matrix. The effect of shadowing can also be seen – a number of arrays show very little correlation with the others, even at large time scales.

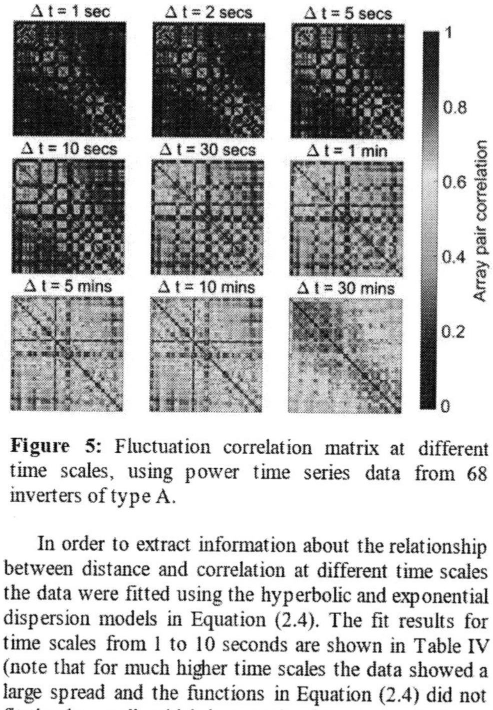

Figure 5: Fluctuation correlation matrix at different time scales, using power time series data from 68 inverters of type A.

In order to extract information about the relationship between distance and correlation at different time scales the data were fitted using the hyperbolic and exponential dispersion models in Equation (2.4). The fit results for time scales from 1 to 10 seconds are shown in Table IV (note that for much higher time scales the data showed a large spread and the functions in Equation (2.4) did not fit the data well, which is a result of the dense clustering of PV arrays). The parameters $v_{1(2)}$ represent an implied wind speed for each fluctuation length – the KIT weather mast recorded an average wind speed at 200m of 6.64m/s in 2015, which roughly corresponds to 8.86m/s at 1500km (850hPa), so that v_1 is a closer representation of the true wind speed. A more comprehensive analysis of the relationship between wind speed, correlation and distance will be left to future work [26].

The contour plot in Figure 6 shows the results of fitting the data with the exponential model for time scales from 1 to 10 seconds. Here one can read off the minimal distance required for a certain correlation level, for instance in order to reduce correlation to 0.1 for 5 second fluctuation time increments, one should place the PV arrays at least 90m apart. Note that these results are location-dependent, since the characteristic wind conditions affect fluctuation correlations. The effect of orientation and inclination has in this case not been taken into account.

Table IV: Fit parameters and R-squared goodness-of-fit for time increments from 1 to 10 seconds using correlation and distance data from 68 inverters of type A.

Δt (s)	v_1 (m/s)	v_2 (m/s)	R_1^2	R_2^2
1	8.97	26.86	0.73	0.72
2	5.44	15.43	0.74	0.76
3	4.30	11.62	0.75	0.79
4	3.74	9.70	0.76	0.80
5	3.41	8.53	0.77	0.81
6	3.18	7.72	0.78	0.82
7	3.01	7.12	0.79	0.83
8	2.88	6.65	0.80	0.83
9	2.78	6.27	0.80	0.83
10	2.69	5.95	0.81	0.83

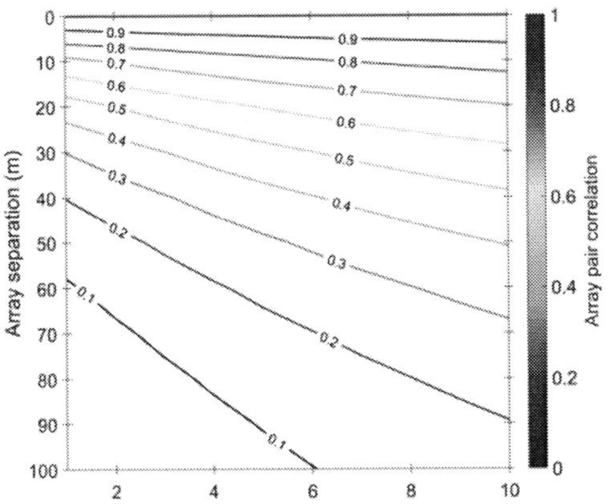

Figure 6: Contour plot of correlation as a function of array separation and time scale, using power time series data from 68 inverters of type A, with the exponential model.

3.3 Application to battery storage system control

As discussed above, battery storage systems are crucial to smooth power fluctuations resulting from cloud-induced variations in solar irradiance. The analysis of fluctuation correlations is applicable in the context of storage system design, since one can get a rough idea of the smoothing effects of aggregating several PV arrays in a mini-grid, for instance. As an example of this decorrelation effect, power curves from array 5 and 93 ($\rho_{5-93}^{\Delta t=5s}$ = -0.0005) are shown in Figure 7, along with their average, for a three minute period on 26 April 2015. Here one can see that the combined power shows a less dramatic fluctuation per installed kWp, simply because the cloud reaches the arrays at different times. The small pause in the green curve provides the storage system more time to catch up. Since most storage systems on the market have a response time of anywhere between 5 and 15 seconds this may or may not be enough to help smooth the fluctuations. It must be noted that in the grid-connected context the variations in load are more prominent and faster than those in PV generation; however in the off-grid context both should be taken into account. A thorough analysis of both load and PV fluctuations would help in understanding the required system architecture, specifically in off-grid scenarios; this will be the subject of further analysis.

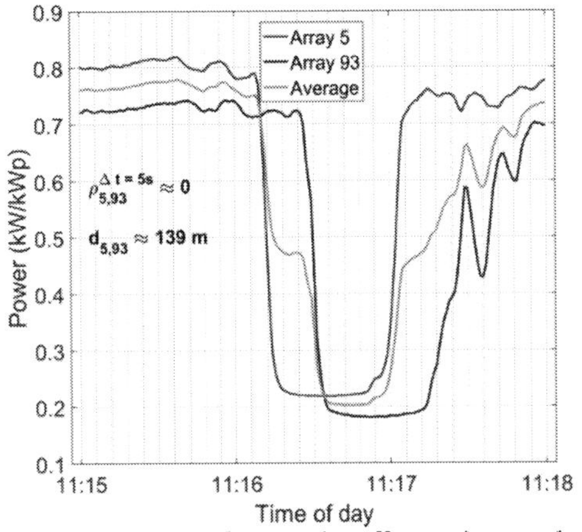

Figure 7: Example of aggregation effects using two de-correlated arrays (5 and 93).

4 SUMMARY AND CONCLUSIONS

The study of variations in PV power over short time scales provides useful information for the design of storage systems with renewable energy. Several different analytical techniques from the literature have been applied to the KIT PV power dataset, providing new insight into the behaviour of densely clustered PV arrays over an entire year. Correlation coefficients between array pairs were compared to their separation distance using well-known models, for distances between 7m and 250m, and the minimum decorrelation distances for fluctuations of 10 seconds or less were found. The preliminary results can be applied to the design of PV-battery storage systems in the residential and off-grid context, in particular when several PV arrays are installed close together. Further analyses will examine the role of orientation and inclination angles in conjunction with the fluctuation levels at different times of the year, and the study will be verified using the battery systems on site. The unique mixture of PV-battery installations at KIT, coupled with a state-of-the-art data processing system, allows for novel studies of both the characteristics of power fluctuations as well as providing a benchmark for the communication and control infrastructure required for successful PV-storage system integration.

5 ACKNOWLEGDEMENTS

The authors thank Feliks Hoxhallari for help with data processing and producing images and Petra Rohr from the IPE at KIT for database and network administration, fault-finding and more. Support by the state of Baden-Württemberg through the high-performance computer bwHPC is acknowledged, as well as the use of data from the 200 m weather mast maintained by the Institute of Meteorology and Climate Research (IMK) at KIT.

6 REFERENCES

[1] J. Marcos, L. Marroyo, E. Lorenzo, D. Alvira, and

E. Izco, "Power output fluctuations in large scale pv plants: One year observations with one second resolution and a derived analytic model," Prog. Photovoltaics Res. Appl., vol. 19, no. 2, pp. 218–227, 2011.

[2] M. Anvari, G. Lohmann, M. R. R. Tabar, M. Wächter, P. Milan, D. Heinemann, J. Peinke, and E. Lorenz, "Suppressing the non-Gaussian statistics of Renewable Power from Wind and Solar," 2015.

[3] G. M. Lohmann, A. H. Monahan, and D. Heinemann, "Local short-term variability in solar irradiance," Atmos. Chem. Phys., vol. 16, no. 10, pp. 6365–6379, May 2016.

[4] A. Woyte, R. Belmans, and J. Nijs, "Fluctuations in instantaneous clearness index: Analysis and statistics," Sol. Energy, vol. 81, no. 2, pp. 195–206, 2007.

[5] A. Woyte, R. Belmans, and J. Nijs, "Localized spectral analysis of fluctuating power generation from solar energy systems," EURASIP J. Adv. Signal Process., vol. 2007, 2007.

[6] T. E. Hoff and R. Perez, "Modeling PV fleet output variability," Sol. Energy, vol. 86, no. 8, pp. 2177–2189, 2012.

[7] R. Perez, S. Kivalov, J. Schlemmer, K. Hemker, and T. E. Hoff, "Short-term irradiance variability: Preliminary estimation of station pair correlation as a function of distance," Sol. Energy, vol. 86, no. 8, pp. 2170–2176, 2012.

[8] M. Lave, J. Kleissl, and E. Arias-Castro, "High-frequency irradiance fluctuations and geographic smoothing," Sol. Energy, vol. 86, no. 8, pp. 2190–2199, 2012.

[9] L. M. Hinkelman, "Differences between along-wind and cross-wind solar irradiance variability on small spatial scales," Sol. Energy, vol. 88, pp. 192–203, 2013.

[10] R. Calif and T. Soubdhan, "On the use of the coefficient of variation to measure spatial and temporal correlation of global solar radiation," Renew. Energy, vol. 88, pp. 192–199, 2016.

[11] T. E. Hoff and R. Perez, "Quantifying PV power Output Variability," Sol. Energy, vol. 84, no. 10, pp. 1782–1793, 2010.

[12] J. Marcos, L. Marroyo, E. Lorenzo, D. Alvira, and E. Izco, "From irradiance to output power fluctuations: the PV plant as a low pass filter," Prog. Photovoltaics Res. Appl., vol. 19, no. 5, pp. 505–510, Aug. 2011.

[13] O. Perpinan, J. Marcos, and E. Lorenzo, "Electrical power fluctuations in a network of DC/AC inverters in a large PV plant: Relationship between correlation, distance and time scale," Sol. Energy, vol. 88, pp. 227–241, 2013.

[14] J. Remund, C. Calhau, L. Perret, and D. Marcel, "Characterization of the spatio-temporal variations and ramp rates of solar radiation and PV," 2015.

[15] B. Elsinga and W. van Sark, "Spatial power fluctuation correlations in urban rooftop photovoltaic systems," Prog. Photovolt Res. Appl., vol. 23, pp. 1390–1397, 2014.

[16] M. Lave, J. Kleissl, and J. S. Stein, "A wavelet-based variability model (WVM) for solar PV power plants," IEEE Trans. Sustain. Energy, vol. 4, no. 2, pp. 501–509, 2013.

[17] J. Marcos, I. de La Parra, M. García, and L. Marroyo, "Control strategies to smooth short-term power fluctuations in large photovoltaic plants using battery storage systems," Energies, vol. 7, no. 10, pp. 6593–6619, 2014.

[18] I. de la Parra, J. Marcos, M. García, and L. Marroyo, "Control strategies to use the minimum energy storage requirement for PV power ramp-rate control," Sol. Energy, vol. 111, pp. 332–343, 2015.

[19] H. Beltran, E. Bilbao, E. Belenguer, I. Etxeberria-Otadui, and P. Rodriguez, "Evaluation of storage energy requirements for constant production in PV power plants," IEEE Trans. Ind. Electron., vol. 60, no. 3, pp. 1225–1234, 2013.

[20] J. Marcos, O. Storkël, L. Marroyo, M. Garcia, and E. Lorenzo, "Storage requirements for PV power ramp-rate control," Sol. Energy, vol. 99, pp. 28–35, 2014.

[21] I. de la Parra, J. Marcos, M. García, and L. Marroyo, "Storage requirements for PV power ramp-rate control in a PV fleet," Sol. Energy, vol. 118, pp. 426–440, 2015.

[22] J. Weniger, J. Bergner, T. Tjaden, J. Kretzer, F. Schnorr, and V. Quaschning, "Einfluss verschiedener Betriebsstrategien auf die Netzeinspeisung räumlich verteilter PV-Speichersysteme," 30. Symp. Photovoltaische Solarenergie, 2015.

[23] J. Weniger, T. Tjaden, J. Bergner, and V. Quaschning, "Auswirkungen von Regelträgheiten auf die Energieflüsse in Wohngebäuden mit netzgekoppelten PV-Batteriesystemen," in 31. Symposium Photovoltaische Solarenergie, 2016.

[24] J. Thomas and N. Munzke, "A Static Solar-Tracking Configuration for PV Power Plant to Redistribute the Daily Power Supply," in 31st European Photovoltaic Solar Energy Conference and Exhibition, 2015.

[25] S. Chilingaryan, A. Beglarian, A. Kopmann, and S. Vöcking, "Advanced data extraction infrastructure: Web based system for management of time series data," J. Phys. Conf. Ser., vol. 219, no. 4, p. 042034, 2010.

[26] J. Barry, N. Munzke, and J. Thomas, "In preparation," 2016.

ASSESSING THE POTENTIAL OF HYBRID PV–BATTERY SYSTEMS TO COVER HVAC LOADS UNDER SOUTHERN EUROPEAN CLIMATE CONDITIONS

J. Solano[a], L. Olivieri[a], E. Caamaño-Martín[a], G. Almeida Dávi[a]
[a] Instituto de Energía Solar, Universidad Politécnica de Madrid
Av. Complutense 30, 28040 Madrid, Spain. Tel.: +34 91 5495700x4323; fax: +34 91 5446341;
e-mail: juancarlos.solano@ies-def.upm.es

ABSTRACT: This paper presents theoretical and experimental works that are being carried out in a grid-connected residential building prototype available at the Solar Energy Institute. The house is provided with five rooftop photovoltaic (PV) systems distributed on different planes, a battery energy storage system (BESS) and heating, ventilation and air-conditioning system (HVAC) based on two air-to-air direct expansion reversible heat pumps. Thermal loads, HVAC consumption, PV generation and storage system are simulated using different dynamic models and validated with actual data coming from the monitoring experimental campaign, in order to determine the energy and economic suitability of using hybrid PV-battery systems in combination with reversible heat pumps and optimized control strategies to cover HVAC loads in residential buildings under southern European climate conditions. Results corresponding to the Madrid case-base show that for the building without BESS the self-sufficiency rate is about 20%, however the implementation of storage systems could duplicate this rate, depending on the BESS capacity and the PV generator nominal power.
Keywords: Hybrid, PV System, Battery Storage and Control, Building Integration, HVAC.

1 INTRODUCTION

HVAC systems represent between 40% and 60% of energy demand for buildings in Europe [1] and are also of increasing importance in buildings worldwide [2,3]. Despite the importance of thermal systems in the energy balance of buildings, a review of researches in the field of control allow concluding that specific strategies for combining local PV generation with conditioning equipment powered by electricity (especially reversible heat pumps) have not been proposed yet.

The interest of this approach has been also shown by the IEA-Solar Heating and Cooling Programme, particularly within its Task 53 [4], whose main objective is to analyze the use of solar driven systems for cooling and heating, including PV driven systems, from both technical and economic points of view.

In this work HVAC loads are analyzed under different climatic conditions. With this aim, Madrid base case has been selected in order to cover the main characteristics of the southern European climate. The selection has been made taking into account not only the Köppen climate classification [5] but also the climatic zones defined in the Spanish Technical Building Code (CTE) [6] in order to have a higher resolution about the influence of the Spanish climate peculiarities.

The main objective of this work is to assess the benefits of using hybrid PV-battery systems in combination with reversible heat pumps and optimized control strategies to cover HVAC loads in residential buildings under southern European climate conditions. This work is being developed in Spain, particularly at the premises of the Solar Energy Institute (IES-UPM) where a prototype of a Zero-Energy Residential Building, called "Magic Box", exists since 2006 (Fig. 1), which was the first European entry in the "Solar Decathlon" international competition and has been recognized internationally as outstanding urban-scale PV system integration project [7,8]. In the prototype several research initiatives have been carried out, amongst which the project "Residential electricity demand side management with PV technology" [9] is to be mentioned.

Within this project, the first theoretical and experimental evidence of the possibilities of performing an Active Demand Side Management of domestic electricity consumption was demonstrated worldwide. In particular, by using local forecasts of the expected PV generation, the so-called "deferrable" loads, for example washing machine, dryer and dishwasher, which constitute about 25% of daily load, these loads are displaced in time to make the best direct use of PV electricity by using Artificial Intelligence techniques. Together with a small local electrical storage system (7 kWh), an annual self-sufficiency of 90% can be achieved [10-12].

Figure 1: Upper and frontal (south) views of "Magic Box" prototype

In a previous work by the authors [13], it was shown that annual billing for HVAC demand in commercial buildings can be reduced up to 50% only with the use of PV without storage systems, mainly because thermal loads match reasonably well with PV energy production. In contrast, in the residential sector, most of the HVAC consumption occurs time-shifted with respect to the PV generation, therefore the addition of a BESS is required in order to decouple generation and consumption, increasing the self-consumption rate. According to this approach, the aim of this work is to propose optimized sizing strategies for BESS and PV systems, in order to cover HVAC loads in an energy and cost efficient way.

2 METHODOLOGY

2.1 Simulations

To analyze the behavior of the global system under different BESS control strategies, the following parameters are being simulated:

a) HVAC thermal loads and corresponding electricity consumption of the air-to-air direct expansion heat pumps, on the basis of annual dynamic simulations performed using *DesignBuilder* and *EnergyPlus* tools.
b) PV energy produced by "Magic Box" rooftop mounted PV systems, using data of incident irradiance on the inclined surfaces of the PV arrays and air temperature measured with a local meteorological station.
c) Performance of the storage system, based on models of batteries developed within this research.
d) Energy and cost savings that the PV-storage-HVAC system will produce in comparison with a conventional system, in which the heat pumps are powered by grid electricity.

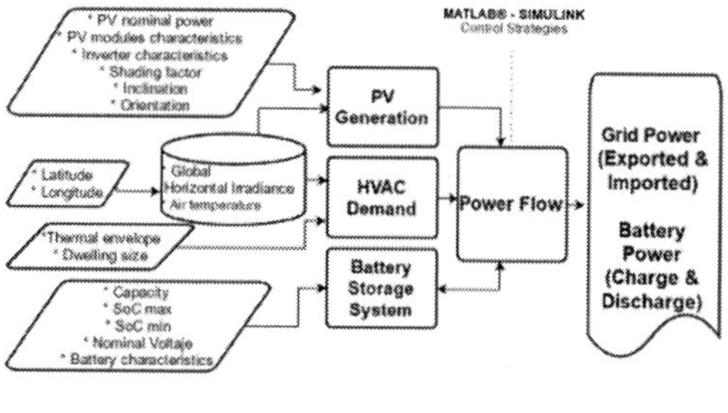

Figure 2: Simulation diagram of the system.

In Fig. 2 a diagram of the input and output variables of the whole process simulation is shown, and Fig. 3 show the electrical scheme of the system.

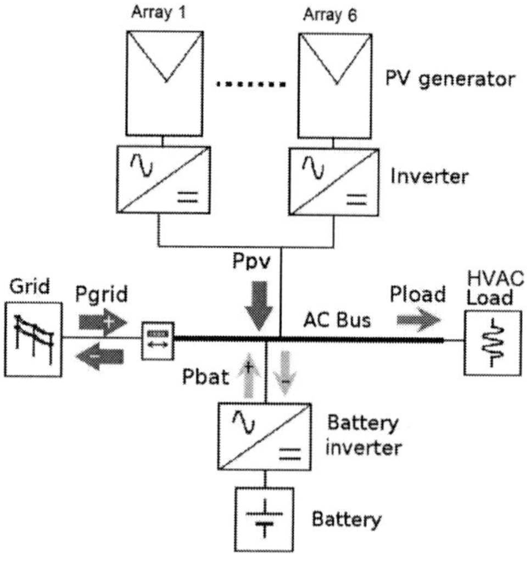

Figure 3: Electrical scheme of the entire system.

Simulations have been made with different durations (minute-based, hourly, daily, monthly, yearly, etc.) and with different configurations of PV and battery sizes. BESS can be simulated by varying storage capacity, state of charge (SoC), efficiency and general behavior according to different algorithms. These control strategies were implemented in MATLAB® - SIMULINK and they have been based on the smart grid simulator *GridSim* [14].

2.2 Experimental measurements

In order to validate the theoretical simulations, the following variables are monitored: outside temperature, outdoor relative humidity, indoor temperature (six sensors), indoor relative humidity, indoor concentration of CO_2, electricity consumption of the heat pumps and electricity generated by the PV system. All these variables are being registered every minute.

2.3 HVAC loads estimation

In order to estimate the annual thermal loads of the building and the associated electricity demand of the HVAC system, the following approach has been used.

Firstly, some of the building characteristics such as location, geometry, materials, internal gains and setup temperatures were defined in DesignBuilder. Next, the model was exported to EnegyPlus in order to accurately define HVAC systems. In particular, to define the ductless air-to-air direct expansion split systems the following EnergyPlus objects were used:

- AirLoopHVAC:UnitaryHeatPump:AirToAir:MultSpeed
- Coil:Cooling:DX:Multispeed
- Coil:Heating:DX:Multispeed
- Fan:OnOff

This configuration is the same used in the tool BEopt developed by NREL [15] to define Mini-Split Heat Pumps, i.e. the same kind of equipments installed in the experimental building. Next, the performance characteristics were tuned according to the manufacturers specification and the annual simulations were performed with a time-step of 10 minutes.

In the next step, the output of the simulation was compared with the experimentally measured data in order to have a feedback about the reliability and accuracy of the model. In particular, for some days, simulated and measured electricity consumptions of the HVAC system were compared, showing a good agreement (Fig. 4). The comparison was not performed on a annual basis because the experimental campaign was still under development when the simulations were carried out.

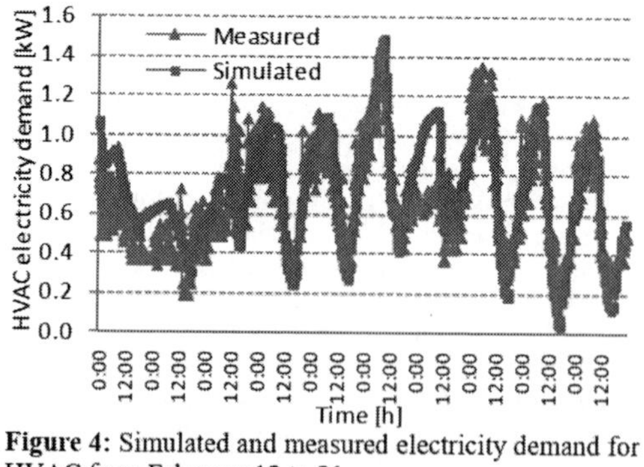

Figure 4: Simulated and measured electricity demand for HVAC from February 12 to 21.

2.4 Photovoltaic System

The PV generation system of "Magic Box" consists of six independent monocrystalline silicon PV arrays with 7 kW_p of total nominal power. This system was designed in order to exploit the different tilts of the sun along the year. To achieve this goal, the arrays are distributed in different south-oriented surfaces. Each PV array has an associated string-type inverter, therefore, the PV AC power is supplied to the AC bus. In Table I, the main features for each array are shown: tilt angle, power nominal of PV generator (P_G), maximum power inverter (P_{inv}), and the annual expected PV energy production (E_{PV}).

Table I: Characteristics of PV arrays

Arrays	1	2	3	4	5	6
Inclination	12°	12°	25°	25°	38°	90°
PG(kW)	1.45	1.44	1.34	1.34	0.80	0.61
Pinv(kW)	1.50	1.50	1.50	1.50	1.00	0.60
EPV(kWh)	1745	1683	1774	1774	1026	572

Due to the fact that the HVAC consumption is just a part of a residential load, specific PV arrays have been selected (arrays 3 & 4), so that annual generation will be comparable to the annual HVAC load (3548 kWh vs. 3230 kWh).

2.5 Battery Energy Storage System

Before installation and experimental analysis of the BEES under different control strategies, a theoretical analysis of the dynamic behavior under the expected HVAC load conditions has been done by means of simulations. This assessment is extremely important in any profitability analysis of BEES [16,17], as well as for providing insight on sizing methods for HVAC applications. In this sense, it is worth mentioning that specific powering of HVAC applications have been much less studied compared with more traditional PV applications (e.g. lighting, telecom., signaling, etc.).

According to [18], the optimal storage size for residential buildings using PV systems was about 4.5 kWh in 2013 and it will be increased significantly to 7.0 kWh in 2021 under different scenarios of electricity costs. In this work the value of 7 kWh for the case study will be used, and variations between 0 to 10 kWh for further analysis.

The house is simulated with a lead-acid battery stationary bank (it is the most used storage element in PV system [19]) coupled to the AC bus by means of a bidirectional inverter. The battery bank is divided in 4 batteries, each battery has a capacity of 145 Ah (C10) and a voltage of 12 V. Therefore, the total battery bank voltage is 48V with a capacity around 7 kWh. The battery inverter does not only implement the current conversion, but also allows for controlling the power flows in the house with high-level software battery controller that can develop different strategies [20].

2.6 Electrical Storage Control strategies

The aim of implementing control strategies in the BEES is to maximize the use of local PV production in terms of both energy (Strategy 1) and power (Strategy 2).

2.6.1 Strategy 1: Maximize the use of PV Energy (kWh balance)

In this strategy, the PV power supplies demand in self-consumption mode, excess of PV charges the battery, and if the battery is full, PV surplus is exported to the grid. Likewise, HVAC demand is supplied firstly by PV, if it demands more power, it is obtained from the battery, and finally it is imported from the grid. Under any circumstance there is power exchange between the grid and battery. Fig. 5 shows the operation of this strategy for a specific day of the year (March 3rd, 2016), where both PV generation and electricity consumption of HVAC are real measured data, whereas battery power, grid power and SoC are simulated data.

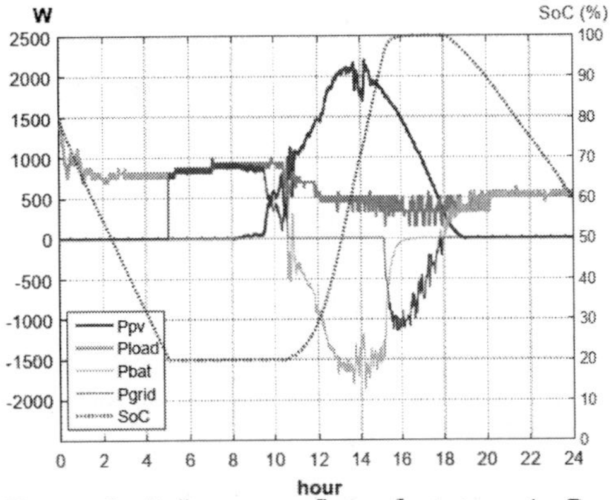

Figure 5: Daily power flow of strategy 1. P_{PV}: experimental PV power; P_{load}: experimental HVAC load; P_{bat}: simulated battery power; P_{grid}: simulated grid power; SoC: Battery state of charge.

As shown in Fig. 5, the battery supplies the HVAC demand until it reaches the minimum state of charge allowed (in this case SoC_{min} = 20%), then the power required is imported from the grid. PV generation is used first to supply the local demand, the surplus of PV charges the battery to 100%, and the excess is exported to the grid.

2.6.2 Strategy 2: Reduction of grid power (Watts balance)

In this strategy, as in strategy 1, the PV power supplies demand in self-consumption mode, excess of PV charges the battery, and if the battery is full, PV surplus is exported to the grid. The difference with strategy 1 is that the battery is discharged only to supply power peaks. HVAC demand is supplied firstly by PV and then by the grid, but only up to a limited established power. If demand exceeds this limit, the battery is discharged to supply higher powers. Under any circumstance there is power exchange between the grid and battery. Fig. 6 shows operation of strategy 2, wherein it can be seen that the battery reduces power peaks from the grid, but in terms of energy savings the outcome is minimal.

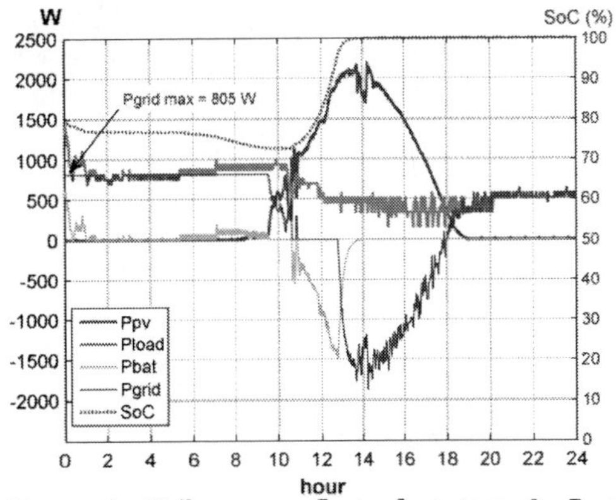

Figure 6: Daily power flow of strategy 2. P_{PV}: experimental PV power; P_{load}: experimental HVAC load; P_{bat}: simulated battery power; P_{grid}: simulated grid power; SoC: Battery state of charge.

The potential benefit of this control strategy can be explained taking into account the impact that the contracted peak power has on the electricity bill, which strongly depends on the regulatory framework. In Spain for instance the power term has a heavy impact on the electricity bill, so a BESS control strategy focused on reducing power peaks supplied by the grid is worth of study.

3 RESULTS AND DISCUSSION

Due to large number of input variables (17 according to Fig. 1), a specific case has been chosen to show the overall behavior of PV-battery hybrid system:

- Location: Madrid
- PV generator nominal power: 2.68 kW_p
- Orientation & Tilt angle: south, 25°.
- Annual PV energy generation: 3548 kWh
- BESS capacity: 7 kWh, Nominal Voltage: 48V
- Minimum SoC: 20%; Maximum SoC: 100%
- Annual HVAC electricity consumption: 3230 kWh
- Control strategies: strategy 2 in winter (November to March), strategy 1 for rest of the year.

The main results are shown below: annual power flows, self-sufficiency and self-consumption, HVAC demand duration curve, billing and economic savings.

3.1 Annual power flows

In Fig. 7 the simulation of both power exchange with the grid and power exchange with the battery are shown with a minute-based resolution over a year.

In winter months, the battery is discharged to supply power values greater than 805 W (strategy 2), and for the rest of the year (from day 90 to 304), the grid does not exceed the established maximum power, so battery uses strategy 1 to reduce energy consumption imported from the grid, as shown in Fig. 7.

The annual simulation of the entire system identifies in which periods of time (seasonal, monthly, hourly, etc.) the HVAC demand is higher. Fig. 5 and Fig. 6 show data measured on a winter day wherein electrical consumption is outside the period of PV generation. However, in the summer months, most of HVAC demand is shifted within the period of PV generation, hence self-consumption is increased and battery SoC is kept in high levels (average of daily values: 88%, being the corresponding average in winter period: 76%), as shown in Fig. 7b.

The control strategies used in batteries, allow to reduce consumption of both energy and power from the grid. Strategy 1 is useful in electricity markets whose value of energy (kWh) is high, while strategy 2 is useful in electrical markets as the Spanish one, where the contracted power (kW) has a high cost in billing (38 €/kW/year in 2015, 212% greater than in 2013). For case study, in strategy 1 (Fig. 5), the contracted power (according to the different levels of permitted power) is 1.725 kW, however, using strategy 2 in Fig. 6, the contracted power is reduced to the smallest reasonable value: 0.805 kW (47% less). The combination of both strategies, as shown in Fig. 7a, enables to limit the power from the grid in the months of peak demand (winter) and to reduce grid electricity imports in summer months.

3.2 Self-sufficiency & self-consumption

The amount of PV energy locally consumed is called *absolute self-consumption* when it is instantaneously supplying the electrical demand consumption. However, this value (expressed as a percentage) can be relative to the total generation (self-consumption), or to the total consumption (self-sufficiency) [21].

Necessary energy to supply the HVAC demand comes from two sources: PV (directly or indirectly, through the battery) and the grid. The percentage of demand supplied from PV is the self-sufficiency of the system. Likewise, the energy produced by the PV system is directed toward three destinations: HVAC, battery and the grid. Of the total PV generation, the percentage of energy directed toward both battery and HVAC is the self-consumption of the system.

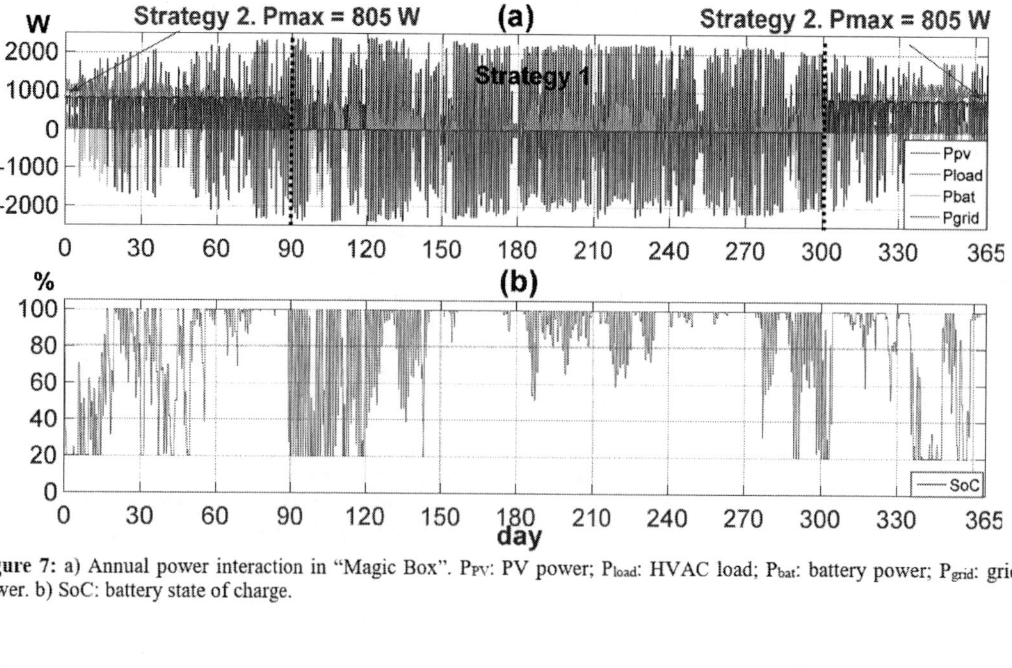

Figure 7: a) Annual power interaction in "Magic Box". P_{PV}: PV power; P_{load}: HVAC load; P_{bat}: battery power; P_{grid}: grid power. b) SoC: battery state of charge.

In Fig. 8 it shows the percentages of self-sufficiency and self-consumption for the case study.

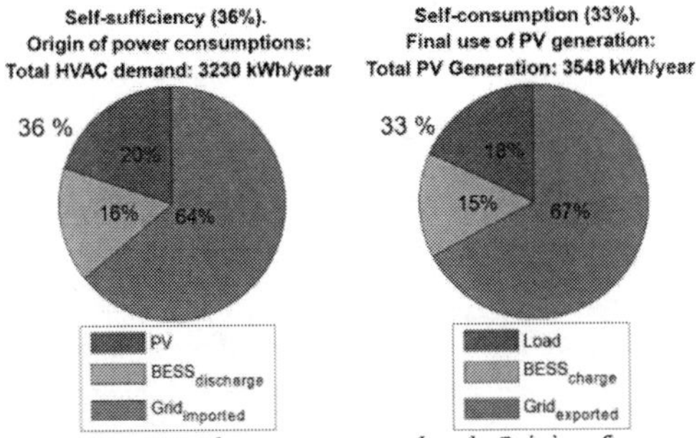

Figure 8: Annual energy accumulated. Origin of energy consumed by HVAC and self-sufficiency (left); final use of PV generation and self-consumption (right).

Using a PV system to supply directly the HVAC demand (specific case presented) only reaches 20%, i.e. 80% of energy comes from the grid (Fig. 9). When batteries are added, the energy from grid can be reduced to 64%, as observed in Fig. 8a and Fig. 9.

In Fig. 8b the final use of electricity locally generated by the PV system is shown, wherein the 18% covers instantly the HVAC demand, 15% is used to charge the batteries, and also was observed that exists 67% of PV energy that is not used directly and therefore it is exported to the grid. This energy can be used to supply other different types of demand with different load profiles that will significantly increase the self-consumption.

3.3 HVAC demand duration curve

The grid load duration curves shown in Fig. 9, which represent the grid electricity demand to supply the HVAC load, compare three scenarios: the first one (blue) is without using PV nor battery; the second curve (red) shows the scenario when using only PV (no BEES), and the third curve (yellow) shows the scenario when using PV and BEES.

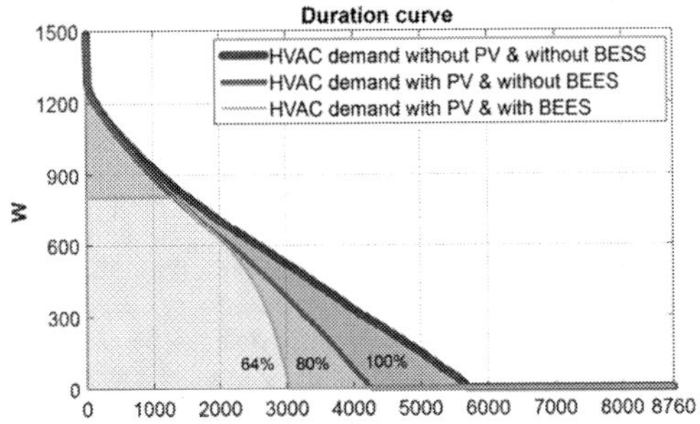

Figure 9: Duration curve of imported energy from the grid to supply the demand for HVAC under three scenarios: original demand (blue); after using PV system (red); and, after using PV - battery system (yellow).

A-long-side with the benefit for the society from increasing clean electricity penetration in the grid, for the residential user another benefit arises from the use of PV & BEES, consisting decreasing imported energy from the grid, and therefore decrease billing, as will be seen below.

3.4 Billing and economic saving

According to Spanish regulated prices currently in effect [22], specifically the 2.0 DHS tariff period (residential tariff with hourly prices), the value in euros that would be paid to the utility has been calculated over a year for the three scenarios outlined above. The invoice is divided into three values: power, energy, and taxes, as shown in Fig. 10.

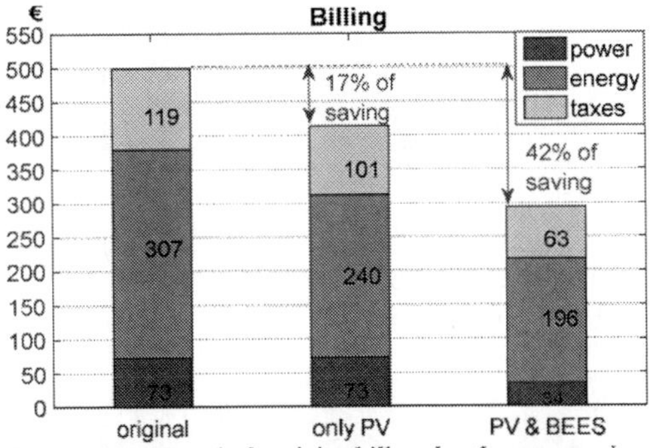

Figure 10: Annual electricity bill under three scenarios: original demand (left); after using PV system (center); and, after using hybrid PV - battery system (right).

In the specific case study, Fig. 10 shows that the use of the PV system only reduces 17% the initial billing, and adding batteries reaches 42% of economic saving.

It should be noted that annual simulations have been made with each strategy individually, but with the combination of both strategies the highest results in economic savings are obtained (5 percentage points more than strategy 1, and 10 percentage points more than strategy 2).

3.5 Variation of parameters

The variation of both PV and battery systems characteristics modify the self-sufficiency (Fig. 11), self-consumption (Fig. 12) and therefore the economic savings (Fig. 13). Thus, keeping a constant the HVAC demand, multiple simulations were performed varying the battery size between 0 to 10 kWh (viable sizes for residential buildings), and the PV system size between 0 to 7 kW_p (covering all fields of "Magic Box").

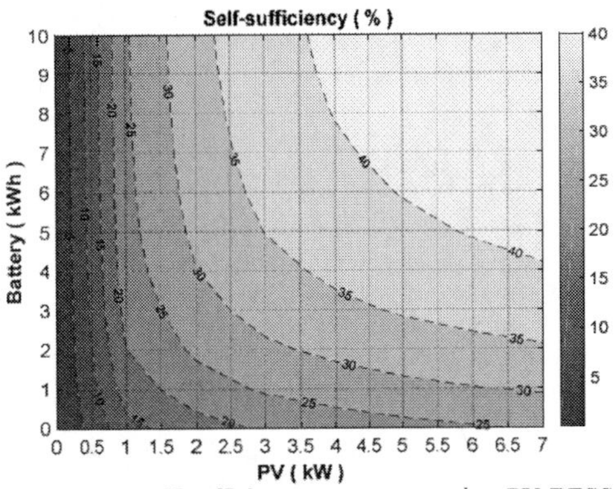

Figure 11: Self-sufficiency percentage using PV-BESS with different battery size and different PV nominal power. HVAC load is constant.

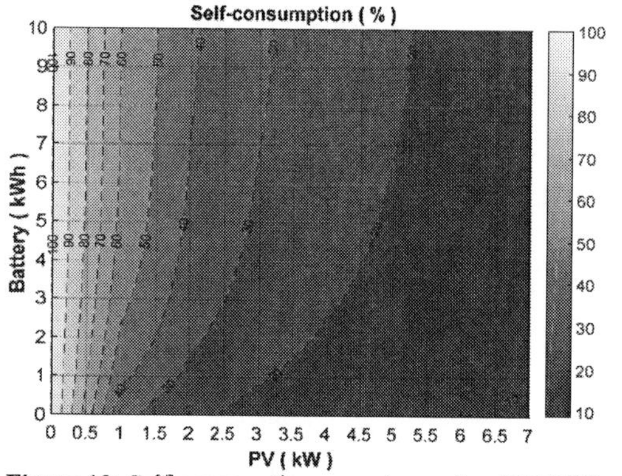

Figure 12: Self-consumption percentage using PV-BESS with different battery size and different PV nominal power. HVAC load is constant.

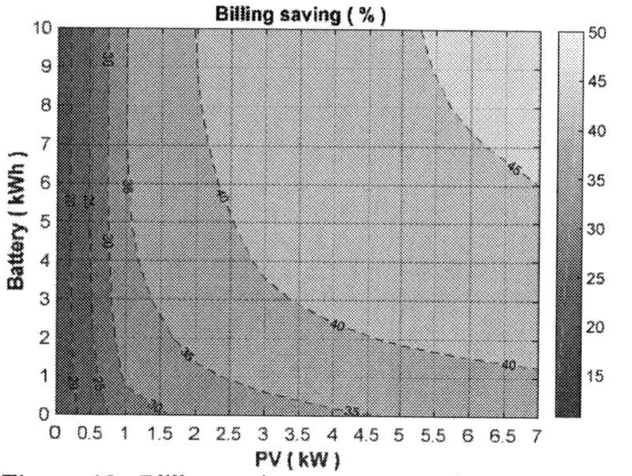

Figure 13: Billing saving percentage using PV-BESS with different battery size and different PV nominal power. HVAC load is constant.

The economic savings in billing, which is shown in Fig. 13, is the percentage that is reduced compared to the original scenario wherein the same house does not have a PV system nor a BEES.

The self-sufficiency (Fig. 11), self-consumption (Fig. 12) and billing savings (Fig. 13) rate increments are not linear with the BESS capacity, suggesting that the optimal size of the PV-battery system is a crucial factor to be established on the basis of different energy and economic approaches [23].

4 CONCLUSIONS

In this paper, theoretical and experimental works that are being carried out in a grid-connected residential building prototype available at the Solar Energy Institute, has been presented. In particular, the following conclusions have been drawn from the theoretical analysis done on the influence of the specific application (HVAC electrical supply) and the BEES control strategy on the fraction of local energy used in terms of energy (self-consumption, self-sufficiency) and economics (electricity grid bill):

a) It is important to know the annual distribution of electrical loads, to determine which season or hourly periods with higher power demand justify the existence of some kind of strategy enabling to reduce consumption from the grid. In this paper, two strategies have been proposed: the first helps to reduce energy consumption, and the second helps to reduce power. In the case study considered ("Magic Box": 2.68 kW_p coupled to a 7kWh/48V Lead-Acid battery with bidirectional inverter; annual PV generation comparable to HVAC demand), when using a combination of both strategies, it is possible to reduce both 47% of contracted power and 36% of energy consumption.
b) The implementation of local PV and electrical storage systems to power HVAC loads improves the self-sufficiency rate in varying degrees, depending not only on the PV power, but also on the HVAC load and storage control capacity. This behavior implies that the optimal sizing of PV & BEES must be carried out in each case by taking into account both energy and economic aspects.
c) It is clear that, at present, the billing savings by itself might not be enough to encourage the use of PV-battery systems, it will also strongly depend on the electricity tariff structure and energy policy in each country, in addition to PV and storage systems costs which will determine whether the investment will be profitable from the financial point of view.

5 ACKNOWLEDGEMENTS

The authors gratefully acknowledge the financial support by the National Secretary of Higher Education, Science, Technology and Innovation of Ecuador (SENESCYT) for a PhD scholarship to the first author.

The authors acknowledge support of the "Fundación Iberdrola España" by means of the "2015 Ayudas a la Investigación en Energía y Medio Ambiente" in the "Smart Grids para la eficiencia en redes eléctricas: caso práctico en la ETSIT-UPM" project. This work has been partially financed by the Spanish Ministry of Economy and Competitiveness within the framework of the project "DEMS: Sistema distribuido de gestión de energía en redes eléctricas inteligentes" (TEC2015-66126-R).

6 REFERENCES

[1] D. H. W. Li, L. Yang and J. C. Lam, "Zero energy buildings and sustainable development implications – A review," Energy, vol. 54, pp. 1-10, 6/1, (2013).
[2] L. Perez-Lombard, J. Ortiz and I. R. Maestre, "The map of energy flow in HVAC systems," Appl. Energy, vol. 88, pp. 5020-5031, 12, (2011).
[3] K. F. Fong, T. T. Chow, C. K. Lee, Z. Lin and L. S. Chan, "Comparative study of different solar cooling systems for buildings in subtropical city," Solar Energy, vol. 84, pp. 227-244, 2, (2010).
[4] IEA Solar Heating & Cooling Programme. Task 53: New Generation Solar Cooling & Heating Systems (PV or solar thermally driven systems); Available at: http://task53.iea-shc.org/, (2015).
[5] M. Kottek, J. Grieser, C. Beck, B. Rudolf and F. Rubel, "World map of the Köppen-Geiger climate classification updated," Meteorologische Zeitschrift, vol. 15, pp. 259-263, (2006).
[6] CTE. Código Técnico de la Edificación. Boletín Oficial del Estado: BOE 12/09/2013; Available at: http://www.codigotecnico.org/images/stories/pdf/ahorroEnergia/DBHE.pdf, (2013).

[7] M. Calvo-Fernández, J.E. Vega, M.A. Egido, E. Caamaño-Martín, "Spanish participation in the 2005 Solar Decathon: design and simulation of the Photovoltaic system". 20th EU Photovoltaic Conference, pp. 1958-1963, Barcelona (2005).
[8] PV database. Building integrated and urban photovoltaic solar energy projects and products. Available at: http://www.PVdatabase.org/projects_viewbest.php, (2016).
[9] Spanish Ministry of Education and Science, GEDELOS PV. Project (ENE2007-66135). National Research, Development and Technology Innovation Program 2007-2010.
[10] M. Castillo-Cagigal, E. Caamaño-Martín, E. Matallanas, D. Masa-Bote, A. Gutiérrez, F. Monasterio-Huelin and J. Jiménez-Leube, "PV self-consumption optimization with storage and Active DSM for the residential sector," Solar Energy, vol. 85, pp. 2338-2348, 9, (2011).
[11] M. Castillo-Cagigal, A. Gutiérrez, F. Monasterio-Huelin, E. Caamaño-Martín, D. Masa and J. Jiménez-Leube, "A semi-distributed electric demand-side management system with PV generation for self-consumption enhancement," Energy Conversion and Management, vol. 52, pp. 2659-2666, 7, (2011).
[12] D. Masa-Bote, M. Castillo-Cagigal, E. Matallanas, E. Caamaño-Martín, A. Gutiérrez, F. Monasterio-Huelín and J. Jiménez-Leube, "Improving photovoltaics grid integration through short time forecasting and self-consumption," Appl. Energy, vol. 125, pp. 103-113, 7/15, (2014).
[13] J. Solano, L. Olivieri, E. Caamaño-Martín and M. Egido, "Climatización eficiente mediante bombas de calor y tecnología solar fotovoltaica: Análisis de vialidad en edificios comerciales en España". XIII Congreso Ibero – Americano de Climatización y Refrigeración, (2015).
[14] ROBOLABO. GridSim Simulator manual; Available at: https://github.com/Robolabo/gridSim, (2016).
[15] BEopt™ (Building Energy Optimization). Available at: https://beopt.nrel.gov, (2016).
[16] G. Mulder, D. Six, B. Claessens, T. Broes, N. Omar and J. Van Mierlo, "The dimensioning of PV-battery systems depending on the incentive and selling price conditions," Appl. Energy, vol. 111, pp. 1126-1135, (2013).
[17] World Energy Council. E-storage: Shifting from cost to value. Wind and solar applications; Available at: https://www.worldenergy.org/e-storage-shifting-from-cost-to-value-2016/, (2016).
[18] J. Hoppmann, J. Volland, T. S. Schmidt and V. H. Hoffmann, "The economic viability of battery storage for residential solar photovoltaic systems – A review and a simulation model," Renewable and Sustainable Energy Reviews, vol. 39, pp. 1101-1118, 11, (2014).
[19] C. Salame, M. Aillerie, G. Khoury, N. Achaibou, M. Haddadi and A. Malek. Terragreen 2012: Clean energy solutions for sustainable environment (CESSE) modeling of lead acid batteries in PV systems. Energy Procedia 18pp. 538-544. (2012)
[20] M. Castillo-Cagigal, "A swarm intelligence approach based on coupled oscillators: an application in demand side management with photovoltaic distributed generation," (2014).
[21] R. Luthander, J. Widén, D. Nilsson and J. Palm, "Photovoltaic self-consumption in buildings: A review," Appl. Energy, vol. 142, pp. 80-94, 3/15, (2015).
[22] Agencia Estatal Boletín Oficial del Estado. BOE-A-2015-13782; Available at: https://www.boe.es/boe/dias/2015/12/18/pdfs/BOE-A-2015-13782.pdf, (2015).
[23] J. Weniger, T. Tjaden and V. Quaschning, "Sizing of Residential PV Battery Systems," Energy Procedia, vol. 46, pp. 78-87, (2014).

COMBINED PV SOLAR COMPRESSION COOLING AND FREEE COOLING SYSTEM

Paul Gantenbein[1], Lukas Omlin[1], Daniel Notter[1] and Andrejs Snegirjovs[2]
[1]Institute for Solar Technology SPF, HSR University of Applied Sciences Rapperswil
Oberseestrasse 10, CH-8640 Rapperswil
[2]Institute of Physical Energetics, Riga, Latvia

ABSTRACT: A photovoltaic and grid connected, electric energy driven cooling system is set up and measurements are performed. The system with a maximum cooling power of P_{cool} = 8kW can be operated in several cooling modes (predominantly) regarding comfort and energy efficiency but also can be run in a heating mode. Therefore it contains PV modules, a compression cooling machine i.e. heat pump (HP), a cold storage (cold side of the HP), an outdoor unit and cold ceilings for the cold distribution. To increase the system efficiency, while cooling is needed, the energy on the hot side of the HP is used for domestic hot water preparation. In case of favourable weather conditions the cold storage can be cooled down in the free cooling mode, the most energy efficient cooling operation mode. The outdoor unit for heat rejection is therefore positioned on a shadowy place.
A water based ethanol brine (20 vol.% ethanol) is used as the heat transfer fluid because of better thermo-physical properties than water at low temperature i.e. below 20°C?. To perform measurements a number of sensors are installed and for system operation control and data acquisition the NI LabVIEW® software is used. The data analysis is further done with MATLAB scripts. First operation and data analysis results show a good performance of the system.
Keywords: renewable energy, solar energy, solar cooling, photovoltaic

1 INTRODUCTION

The increasing comfort need in buildings and in the transportation sector will increase the cooling demand and so the growth of the cooling market can be expected. Actual observations show that air conditioning systems of a few kW cooling power in general have a short live cycle. This ends up in a large number of the yearly replaced and installed cooling systems. In Europe a rise of the share of commercial buildings equipped with cooling systems is expected to reach of at least 60% by the year 2020. The maximum potential cooling demand in Europe, if 100% of all useful space would be air-conditioned, is estimated to 1400 TWh per annum [1].

To use solar energy for cooling has the big advantage that in regions with a high cooling demand (e.g. southern Europe) a high solar irradiance is expected. The coupling of Photovoltaic (PV) modules with an electrically driven cooling machine is the basic concept of PV-based air-conditioning. For a PV electric driven cooling system all components are market available and the dynamics of the PV market leads to price reductions of almost 10% every year for PV modules. Electrically driven heating and cooling equipment such as vapour compression heat pumps, chillers or reversible heat pumps in connection with heat and/or cold storages are attractive options for the energy supply in buildings. However, except of some European regions, today only a few complete system solutions which use photovoltaics for energy supply in buildings are available in the market. Therefore, a lack of information on overall cost and performance of such systems is identified [2]. In this study an assessment of a small scale PV driven solar cooling system will be done and will be presented in a further publication.

2 SYSTEM SCHEMATIC & DATA ACQUISITION

In figure 1 the schematic shows the solar cooling system with the operation option of free cooling the cold storage tank. The cooling system is installed in a laboratory – location of Rapperswil-Jona (CH) a warm-temperate climatic zone - for thermodynamic and accelerated ageing test experiments and therefore the domestic hot water profile (Euro 2 profile low; 200liter/day) for a single family house is emulated by time controlled valves. The schematic also shows most of the measurement points used for data acquisition with the LabVIEW® software. LabVIEW® is also used for control of parts of the system, like the mixing valve position (M1) which is controlled by dew point calculation, based on the room air temperature and air humidity. Figure 2 shows photos of the PV modules, the main energy source – besides of the electric grid in which a Swiss electric energy mix can be taken as a basis – and the cold distribution sub-system realised with cold ceilings. The table 1 contains a list of the components used to set up the system. A short description is given therein and table 2 contains the thermo-physical properties of the water based brine with 20 vol.% ethanol. This brine is used in both hydraulic loops of the cooling machine (HP) – on the low temperature side and on the high temperature side of the HP. The installation of a cold storage – a thermal storage instead of an electric battery – was based on economic and cooling power reasons. Although this system is designed for cooling, predominately, a domestic hot water storage tank (DHW) is built in the system, on the hot side of the cooling machine, to increase the efficiency through heat recovery for DHW preparation.

Table I: System components.

	Type	Description
PV Modules (9 pcs.)	Sky 285	$2.65kW_p$
Inverter	TL2500	2.5kW
Cooling machine	F1155-6	8kW
Cold Storage	TPSK 500	445l
DHW Storage	SWWS 506	477l
Outdoor Unit	Heat Exchanger	Brine - Air
Cold Ceilings	Zip2	$52.5m^2$
Heat Transfer Fluid	Minoltherm	20vol.% Eth.
Meteo Station	SMA Meteo	irradiance

Table II: Thermo-physical properties of the water-ethanol mixture. Density ρ(T), heat capacity c_p(T) and kinematic viscosity ν(T) in function of the temperature.

	-5°C	0°C	5°C	10°C	20°C
ρ (g/cm³)	0.984	0.984	0.983	0.981	0.977
c_p (kJ/kg*K)	4.37	4.37	4.32	4.31	4.27
ν (mm²/s)	3.79	2.98	2.39	2.02	1.51

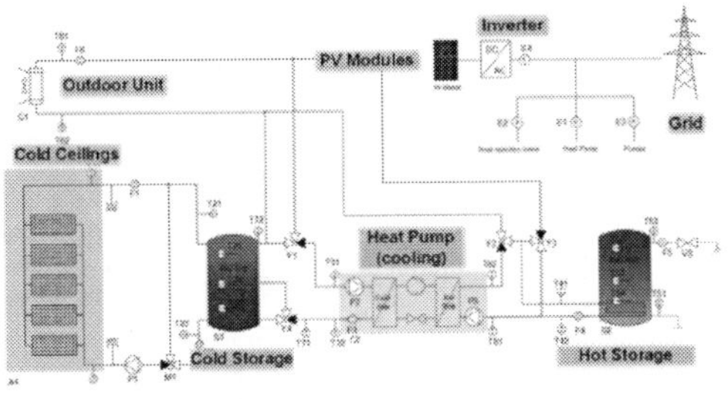

Figure 1: Schematic of the photovoltaic solar electric cooling system with public electric grid connection.

Figure 2: PV module field on the porch roof (left) and cold distribution ceilings in the laboratory (right).

3 RESULTS AND DISCUSSION

The data acquisition and control is in operation and in figures 3, 4 and 5 first measurements are shown in function of day time. During the day the PV power output follows the solar irradiation and a coincidence of their maximums can be observed. But the PV module temperature has its maximum slightly – approx. 1 h – later with a long declining time to reach its minimum short before sun rise, figure 3. During the day, the electric power supply from the PV field to the cooling system is increasing and so in the consumption peak – electricity consumption – a fifth of the needed power is supplied by the public grid, figure 4. A closer look to the power – sum of electric plus cooling and heating (DHW) – shows that in the system energy losses of approx. 600W occur, figure 5. The large heating power oscillations in the operation "batch" of 11 to 12o'clock stems from the switching of - heat recovery to the DHW - to the heat rejection to the outdoor unit. In these situations the HP control is adjusting the temperatures and the flow rates of the high and low temperature sides of the cooling machine.

The figures 6, 7 and 8 are showing measurement data in function of time of the compression cooling mode. The cooling was started at 11 o'clock and stopped at 16 o'clock. Until 13:30 the electric energy used by the cooling machine was supplied from the PV modules. For the rest of the HP running time both the PV modules and

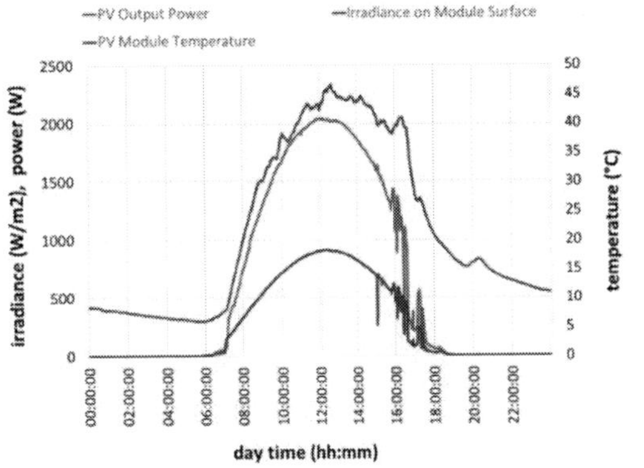

Figure 3: Solar irradiance, PV field electric power output and PV module temperature in function of day time.

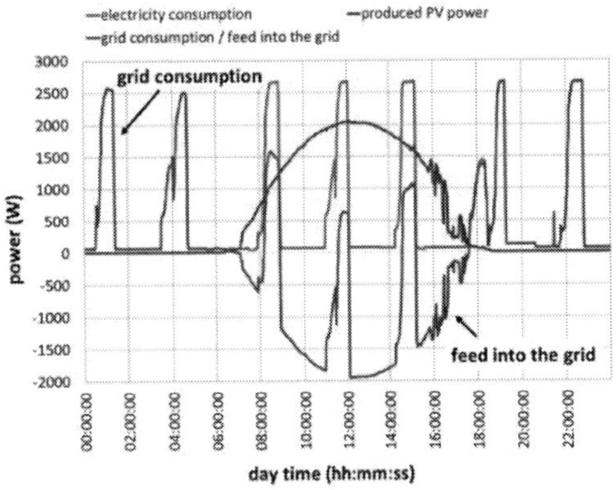

Figure 4: Compression cooling. Electric energy consumption of the system, produced PV power and grid consumption in function of day time. Positive (+) values use of electric power – negative (-) values feed into grid.

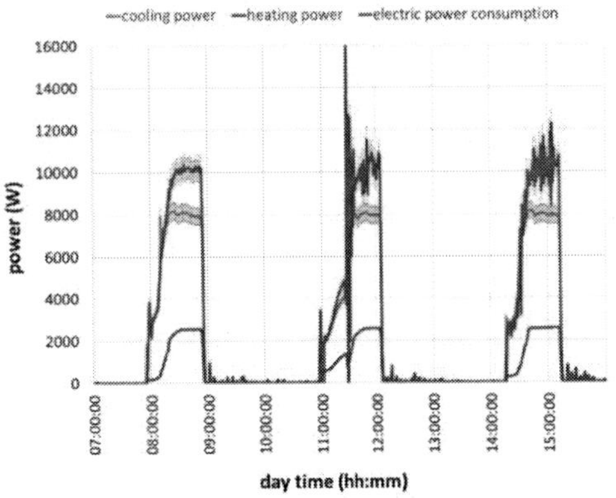

Figure 5: Compression cooling. Compression cooling machine electric power consumption, cooling power and heating power - for DHW preparation - in function of day time.

the public grid were supplying the necessary electric energy. The cooling power and the COP of the cooling system are decreasing from 7500W and 3.5 down to 4000W and 2.4, respectively. This decreasing is caused by the reduction of the cold storage temperatures, shown in figure 7, where a simultaneous reduction of the temperatures from 17°C down to -7°C (T25, T24, T23; figure 1) can be seen. Taking into account the thermos-physical properties of the water-ethanol mixture and the mentioned temperature difference, the capacity of the cold storage tank is in the range of 12kWh and thus the cold storage discharging can run until 23 o'clock with a power of

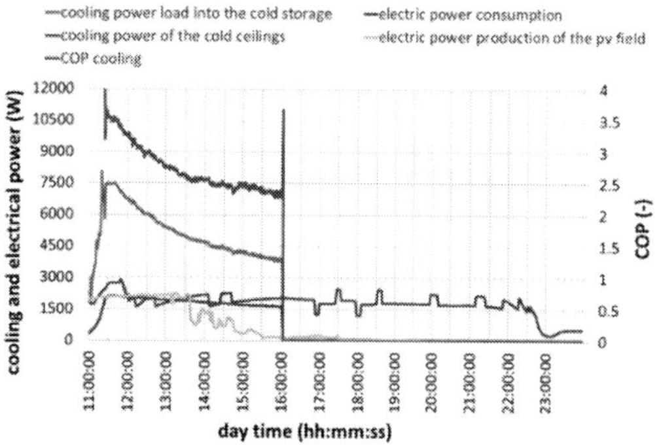

Figure 6: Compression cooling until 16 o'clock – charging the cold storage – and discharging the cold storage (room cooling) for the rest of the day.

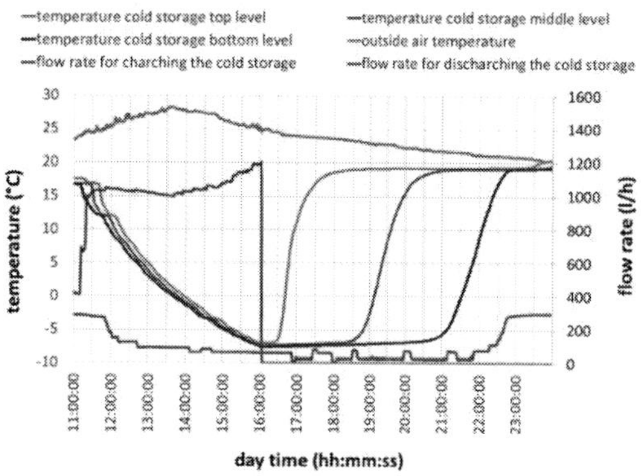

Figure 7: Compression cooling until 16 o'clock – charging the cold storage - and discharging the cold storage (cooling) for the rest of the day. A simultaneous decrease of the cold storage temperatures until 16 o'clock – no temperature stratification – and an individual cold storage temperature increase thereon (T25, T24, T23; figure 1) – indication of thermal stratification.

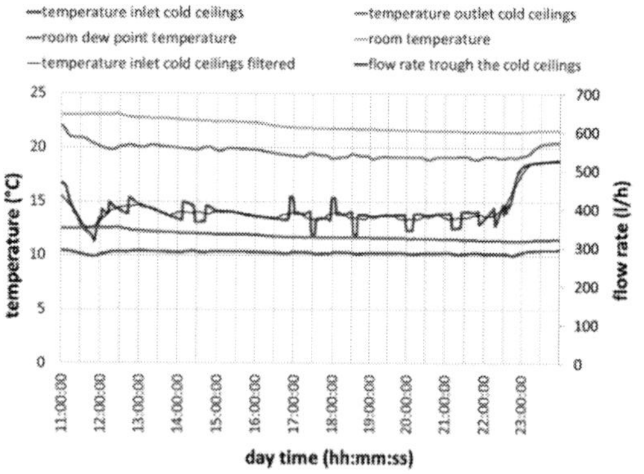

Figure 8: Compression cooling. Temperatures and flow in function of time. Room air and room air dew point temperatures and cold ceilings inlet and outlet temperatures.

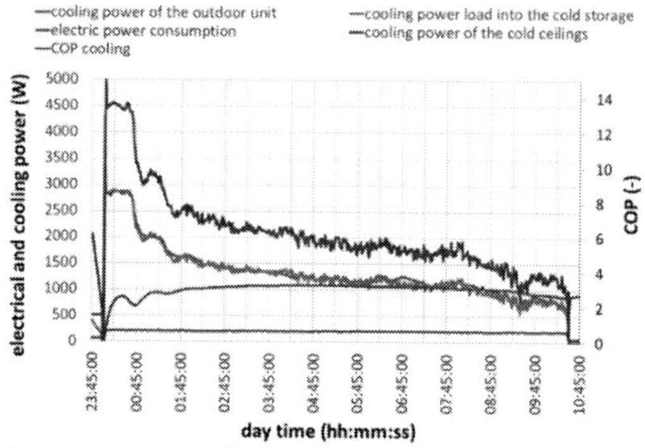

Figure 9: Free cooling. Step wise decreasing of the power and the coefficient of performance COP while the cooling power of the cold ceilings stays nearly constant at a low level.

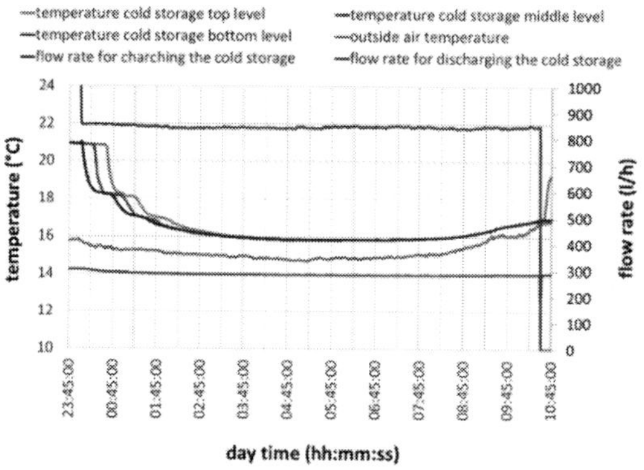

Figure 10: Free cooling. Step wise decreasing of the temperatures in the cold storage - indication of thermal stratification (T25, T24, T23; figure 1).

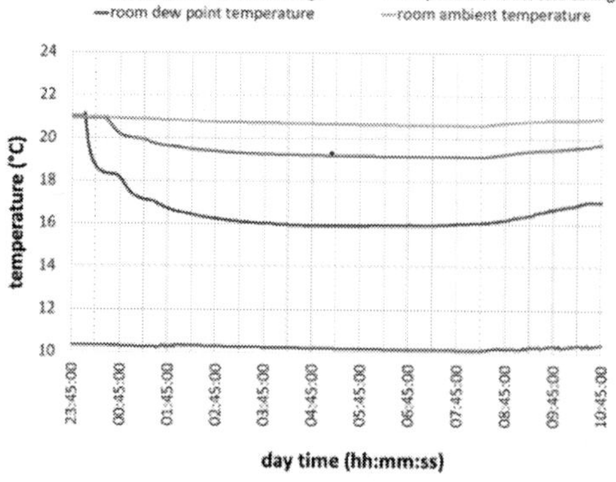

Figure 11: Free cooling. Temperatures in function of time in the cold ceilings.

1.8kW. The mentioned type of cold storage – a thermal cold storage – is in the actual economic situation more reasonable than a battery i.e. an electric energy storage.
Figure 8 shows the cooling fluid temperatures of the cold ceilings - inlet and outlet – as well as the room temperature and the dew point temperature. Using the volume flow rate (multiplied by the density and the heat capacity of the heat transfer fluid) in the cold ceilings, the cooling power is 1.8kW. The spikes in the ceiling inlet temperature (plus and minus relative to the mean i.e. filtered data) are provoked by the mixing valve operated by the dew point PID controller.
Depending of the climatic zone – the location of the building - and thus the ambient air conditions, free cooling is the most cost-efficient room air temperature treatment. A prerequisite for the function is a temperature gradient from the cold storage to the ambient, which in some regions is only the case during night time. Simple air ventilation can be applied or, like in our system, fluid circulation from the cold storage tank to the outdoor heat exchanger is an alternative. Figure 9, 10 and 11 are showing measurement results of this type of cooling mode during night time. In the first figure of the free cooling mode, figure 9, the power and the COP are decreasing step wise because of the temperature in the cold storage shows this time dependence, figure 10. As a comparison to the figure 8, in figure 11 the temperatures in the cold ceilings in function of time during the night are shown. No spikes in the fluid inlet temperature to the ceilings are occurring because the cold storage temperature is not lower than 16°C i.e. the mixing valve is not active due to no dew point controller activity. With the flow rate of 300l/h the cooling power of the ceilings is 1kW.

4 CONCLUSION & OUTLOOK

A solar electric cooling system was set up and first measurements were performed. Beside the operation with compressor cooling machine (HP), the system can also be run in a free cooling mode. For a high energy efficiency the hot side heat of the HP is used to prepare the domestic hot water (DHW). A cold storage tank opens the option of storing the solar energy for room air cooling in times of no direct solar irradiation is available. In case of cooling demand and no solar radiation the public grid acts as a backup. A water based ethanol heat transfer fluid acts as the antifreeze liquid and has the advantage of a higher heat capacity than water in the operating temperature range.
The operation will be optimized towards a higher self-supply with electric energy and measurements will be continued for analysis. A scaling tool – for a system power up to 100 kW – will be developed and will comprise a simulation template in a Polysun® environment.

5 ACKNOWLEDGMENT

Financial support by the Swiss Federal Office of Energy SFOE is gratefully acknowledged. Further we gratefully acknowledge financial support of HSR University of Applied Sciences Rapperswil and the support by the companies, who disposed the project with components.

6 REFERENCES

[1] Burkhard Sanner and Co. Common Vision for the Renewable Heating & Cooling sector in Europe, Brussels, January, 2011, 48p.
[2] Henning Hans-Martin; Motta Mario; Mugnier Daniel. Solar Cooling Handbook. Ambra; 3rd Revised & enlarged edition, September 2013. 368 p. ISBN: 3990434381.

OPTIMIZED DEMAND SIDE MANAGEMENT AND MINIMIZED BATTERY STORAGE FOR HIGH SELF-CONSUMPTION WITH PV-DRIVEN LOW-PART-LOAD HEAT PUMPS

B. Heithorst, F. Kiefer, J. Shen, A. Kastl, A. Präbst, M. Spinnler, T. Sattelmayer
Lehrstuhl für Thermodynamik, Technische Universität München
Boltzmannstraße 15, 85748 Garching

ABSTRACT: Heat pumps for domestic applications have shown to be an intelligent option for demand side management and thus enhanced self-consumption of photovoltaic power. The underlying intention is to store a surplus in energy in a cheap and long-life thermal storage instead of using batteries or burdening grid. New commercially available heat pump systems are supplied with variable frequency drives in order to reduce the energy consumption and allow part load operation. Nevertheless, they suffer from relatively high minimum part-load values and thus limited control range due to the characteristics of the used scroll compressor. In contrast to these systems, Spinnler et al. (2014) presented a novel concept for an enhanced "low-part-load" capable heat pump or compression chiller system based on swash-plate compressors typically used in mobile applications. These systems show better control characteristics and very low possible part load at a relatively high efficiency. The aim of this paper is to compare highly transient driven scroll and swash-plate based heat pumps and to develop an intelligent energy storage system design for increasing self-consumption in domestic application. A quasi-steady state computational model of the heat pump system including a transient building model was therefore implemented in Matlab. An analysis of the simulated heat pump power shows that small short-time electrical storages, combined with a thermal water storage tank could be an interesting alternative for a durable and robust Demand-Side-Management-integrated heat pump system. It could be shown that the size of the thermal storage has a significant influence on the self-consumption ratio.
Keywords: Appliances and Loads, Batteries, Demand Side, Grid Integration, Small Grid-connected PV-Systems

1 INTRODUCTION

On the way towards grid parity of photovoltaic (PV) generated electricity, self-consumption as well as load shifting have become a crucial as well as financially promising measure for grid stability in industry and private households (Büdenbender, Braun and Strauß, 2010). The general aim of Demand-Side Management systems is to shift storage capacity from grid or battery storage to a cheaper thermal storage (Arteconi et al., 2012; Thygesen and Karlsson, 2014). Latter already exists in a lot of households in form of a combined-solar stratified hot water storage tank.

The ohmic heater represents the easiest way to follow even highly transient solar resources due to the simple control and absence of moving parts. Nevertheless, it shows a poor thermodynamic efficiency when comparing to sophisticated heat pump systems. However, heat pumps are generally controlled by an on-off strategy or more advanced by a variable frequency drive (VFD). The first case essentially needs a connection to the grid or a battery storage as a buffer and performs according to the capacity of this storage.

Batteries are the state of the art energy storage (ES) technology in off-grid energy systems. Their most common representatives are Pb-acid and Li-ion batteries. For short term power storages for example electric double-layer capacitors (EDLCs) are widely used in wind energy plants and have energy densities around 0.5 to 10 Wh/kg at 10000-20000 €/kWh. They usually have a typical life time of 15 years. Problematic is the Self Discharge Rate (SDR) of 25% per 48 hours, why it only can be used for a short time period (Droege et al., 2015). Secondary batteries' energy densities vary from 30 to 190 Wh/kg. The prices ranges from 100-800 €/kWh depending on Pb-acid or Li-ion type. Their lifetime ranges from 5-20 years. The SDR ranges from 0.1 to 5% per month. (Droege et al., 2015)

Comparing the costs and lifetime, it becomes obvious why the focus is set on reducing the size of electrochemical storage systems.

For that reason, VFD driven heat pump systems including a thermal energy storage are very attractive (Arteconi et al., 2013), but they are limited by its relatively high minimum part load (Spinnler et al., 2014). Other researchers (Vrettos et al. 2013; Guo et al., 2012; Suzuki et al, 2012; Ratnam et al, 2013), investigated different operation strategies for residential buildings including thermal loads with additional battery storages to reduce electricity costs, but without explicitly addressing the challenge of PV self-consumption.

In another publication of Vrettos et al. (2013 b) the behavior of small-scale heat pump systems was studied, in order to find the smallest battery storage capacity for a high self-consumption. They found out that a combination of thermal and electrical energy leads to the best results, but did not mention an optimal ratio between them. Witzig et al. (2014) developed a simulation model using Polysun software to predict the self-consumption potential of PV driven heat pump systems. Loesch et al. (2014) also developed a model based on genetic algorithm to simulate heat pumps integrated in smart grids. They presented the possibility of an operation strategy of overheating the hot water storage tank to minimize the operation time of the heat pump in times of supply deficits.

To summarize, all analyzed papers show the high potential of thermal energy storages to increase self-consumption in PV applications of domestic households, but most of them do not investigate different heat pump types or compare different thermal storage sizes to minimize the electrical energy storage and thus reduce the costs.

In the present study, the investigated system is driven by solar energy and therefore affected by the common problem of fluctuating and time dispatched electricity generation. The PV driven heat pump system gives the opportunity to use a thermal storage as short/medium term storage (hourly, daily storage cycles). Additionally, the system can include electrical storages in order to further enhance the self-consumption and degree of autarky. The swash-plate and scroll compression chiller types are investigated to compare the influence of different minimum part loads. For the case studies one-second-sample weather data from Munich Riem, Germany, were used.

2 COMPUTATIONAL METHOD

The simulation tool is completely implemented in Matlab and optimized for fast annual calculations.

2.1 Model description

For investigating the interplay between transient power supply and heating power demand, a realistic modelling of all components is necessary. Due to the high dynamics of the heat pump compared to the thermal load of building a quasi-steady-state approach is implemented.

Building Simulation Tool

The demand side, i.e. the building, is provided by a transient building model developed at the Institute for Thermodynamics at TUM (Kastl, 2015). This building simulation tool (BST) allows a flexible modelling of complete houses including transient heat transfer through walls, windows and other building elements. Furthermore, it incorporates models for convective, long and short wave radiation heat exchange, a sun tracking, shading and weather model (cf. fig. 1). Due to an intelligent coupling of implicit calculation of heat transfer through walls with explicitly in time coupled rooms and air nodes, it allows to calculate flexible setups of complex buildings at low computational effort.

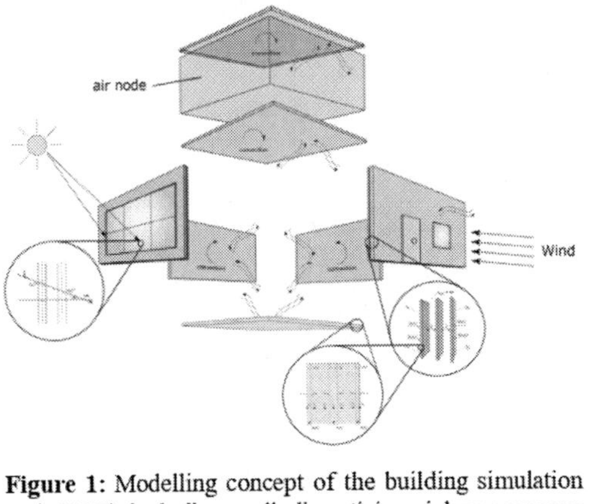

Figure 1: Modelling concept of the building simulation tool (BST) including wall discretizing, inhomogeneous wall structures and heat transfer models; adopted from (Kastl, 2015)

The model is validated against VDI 6007 (Beuken model) and the two capacity (2K) model of VDI 6020 (Rouvel and Zimmermann, 1997). The applicability of the simulation concept has been proven in different simulation studies.

For the study at hand, the simulation of the building is decoupled from the residential heating system by giving the BST a climate control strategy and thus calculating the heating demand with a black box HVAC system. Later on, as described in the following paragraphs, the heating system is simulated in a quasi-steady state approach. It is assumed, that the heating system is able to provide the heat, demanded by the black box HVAC system in the BST.

Occupancy of the building influences the system in two ways: First, a thermal load (positive or negative) by user controlled ventilation, electrical devices or the occupants themselves influences the HVAC demand. This part is neglected for the simulation. Nevertheless, secondly, the electricity consumption of the household (excluding the heat pump system) is modelled, since it significantly influences its complete electricity demand and thus the consumption of PV power. This part is modelled based on the BDEW standard load profile H0 (KommEnergie, 2011). Electricity demand for illumination is living space averaged, other appliances are taken for the complete household. The annual electricity consumption sums up to approximately 6800 kWh, whereas the simultaneity of illumination is set to 0.2 and for other consumers to 0.28.

The demand on warm water is estimated based on Jordan et al. (2001).

Energy Simulation Tool

The energy simulation tool (EST) incorporates the models for the heat pumps, thermal and electrical storages, the PV array as well as the control system. All components are calculated in a quasi-steady state approach except for the storages. The simulation is carried out with a time-step of 1 sec based on the highly resolved PV data.

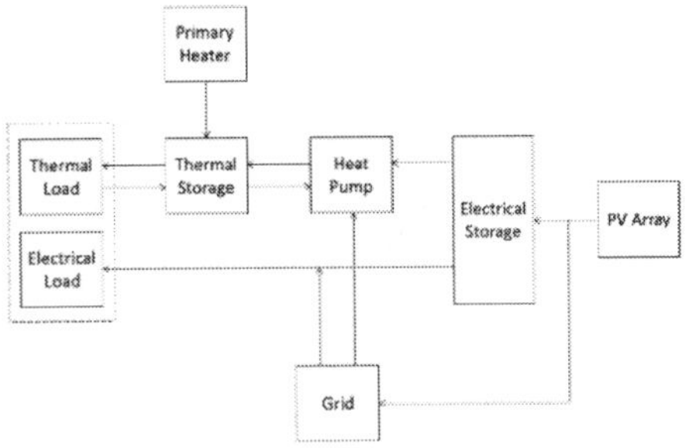

Figure 2: System configuration with all storage types and grid connection

The ***PV power*** is not simulated, but taken from measurements at Messe Riem (Munich, Germany) for a period of one year (Sept. 29, 2008 – Sept. 28, 2009), provided by Solarenergieförderverein Bayern e.V. This highly resolved PV power data allows to investigate the transient behavior of the system.

A simplified model for the ***thermal storage*** is used. It is based on an exergy balance which calculates the state of charge against ambient temperature as reference. A minimum state of charge (SOC) is calculated with a minimum useful temperature difference as

$$SOC_{min} = \frac{T_{w,use} - T_{ref}}{T_{w,HP,out} - T_{ref}}$$

with the minimum useful temperature

$$T_{w,use} = T_{w,HP,out} - \Delta T_{w,use} \, .$$

The ***electrical storage*** is modelled again in terms of its SOC. The storage capacity is defined with $E_{el,storage,max}$. Its lowest charge state is set to

$$E_{el,storage,min} = DOD \cdot E_{el,storage,max}$$

The depth of discharge (*DOD*) is a user defined value which takes into account that the storage must not be discharged below a certain voltage. Since the storage is only used for short time periods, its efficiency is set to one. If, in future simulations, a certain storage type is chosen, a more advanced model is required. For first estimations, the present approach is assumed to be sufficient.

The EST contains two different heat pump models for a heat pump driven by either a **swash-plate** or a **scroll compressor**. The heat source is ambient air and the heat sink is in both cases water.

Each model calculates the coefficient of performance (COP) for every simulated time step. It is defined as the ratio between the emitted heat in the condenser $\dot{Q}_{cond}$ and the electrical Power P_{el} to drive the process:

$$COP = \frac{\dot{Q}_{cond}}{P_{el}}$$

Therefore, the specific enthalpies of the considered refrigerant R134a are calculated after the evaporator and the condenser using the refrigerant properties (Bell et al., 2016). The following assumptions were made:

- Temperature difference between ambient air and evaporation temperature: $\Delta T_{amb,evap} = 10\ K$
- Temperature difference between condensation temperature of refrigerant and water generated by the heat pump: $\Delta T_{cond,w} = 5\ K$
- Superheating in evaporator $\Delta T_{sh} = 5\ K$
- Subcooling in condenser $\Delta T_{sc} = 2\ K$

The specific enthalpy after the compressor is calculated with the specific isentropic enthalpy. The latter is calculated with the refrigerant properties and the isentropic efficiency of the compressor. The isentropic efficiency mainly depends on the pressure ratio and the rotary speed for the scroll compressor and additionally on the cylinder stroke for the swash-plate compressor. In order to model the isentropic efficiency, experimental data were used to fit the polynomial approach of DIN EN 12900 (DIN 2013) by Spinnler et al. (2014) for both the scroll and swash-plate compressor. The COP is then calculated as

$$COP = \eta_{is} \cdot \eta_{el} \frac{h_2 - h_3}{h_{2,isen} - h_1}$$

with an assumed efficiency $\eta_{el} = 0.85$ of the compressor motor.

Due to a desired long lifetime and restrictions of the electricity provider, the heat pump remains switched off at least for 15 minutes, before it can be turned on again.

The advantage of the swash-plate compressor compared to the scroll compressor is the ability to operate at lower part loads while the response time of both compressors is nearly equal (Spinnler et al. 2014). In the present study, the parameters for maximum and minimum electrical power of the compressor are set. They have been identified from experimental results in the prior study (Spinnler et al. 2014).

Table I: Operation limits of compressors

	Scroll	Swash-plate
$P_{el,max}$ [kW]	3	3
$P_{el,min}$ [kW]	0.9	0.3

Operation Strategy

For the analysis of the system, the choice of a set of meaningful operation strategies is crucial. Three different strategies are used: A ***PV-oriented*** control strategy tries to always use available PV power and fill up the thermal storage. The heat pump follows the slope of the solar radiation as long as the minimum part load as well as the maximum compressor power are not violated. If an electrical storage is available, a surplus in electricity production can be stored. If the PV power is below the minimum part load power of the compressor and the electrical storage capacity is able to fill this lack, the compressor continues its operation at lowest part load. Additional heat demand from the building is always buffered by a supplementary heater.

The other extreme is the ***heat-oriented*** control. Here, the heat pump is operated in an on-off strategy in order to fill up the heat storage, when it reaches a predefined lower SOC.

As a combination of these two approaches, the ***PV + heat-oriented*** operation strategy is used in order to investigate a more sophisticated operation of the system. Both electrical and thermal storage capacity and their current SOC are included into the decision path (cf. fig. 3).

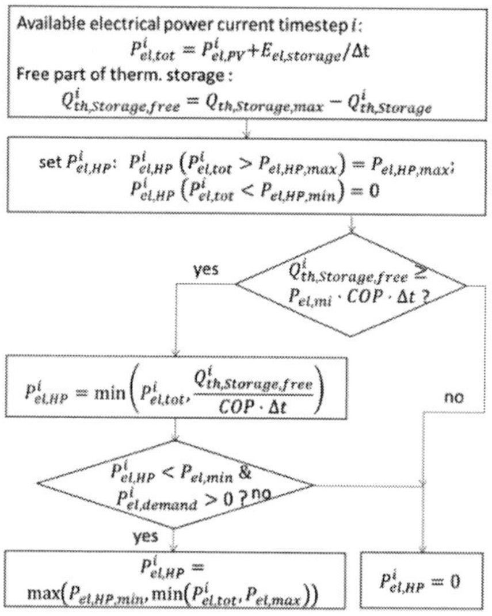

Figure 3: Flow diagraph of signal process unit for the PV + heat-oriented strategy

The strategy can be summarized as follows: The available electrical power from PV ($P^i_{el,PV}$) of the current time step i is compared to the available power from the thermal storage ($\dot{Q}^i_{th,Storage}$), that can be supplied within the current time step, as well as the available electrical power from the electrical storage ($P^i_{el,Storage}$). If the resulting available electrical power is sufficient to run the heat pump, it is started and set to the maximum power necessary to fill up the storage. A surplus in electrical power ($P^i_{el,PV,tot} - P_{el,HP,max}$) is stored into the electrical storage, if enough capacity is available. Additionally, the requirements of the heat pump system as summarized in the model description, have to be fulfilled.

Design Variation

Besides the operation strategy, the sensitivity of the system on different design parameters provides a deeper insight, especially for system optimization. From the major goals of this paper the different system configurations can be derived as follows:

(1) The system sensitivity on different thermal and electrical storage sizes shall be investigated. Table II shows the discrete thermal and electrical storage capacities.

(2) The dimensioning of the PV array has significant influence on the self-consumption and solar fraction, or vice versa, if an HP is added to an existing PV array. Therefore, different cases with 5, 10 and 25 kWp are simulated.

(3) The new concept of a swash-plate compressor based "low-part-load" HP should be investigated. As benchmark, the system is compared to a scroll compressor with a variable frequency drive.

Table II: Variations in storage capacities

Storage	Values				
E_{el} [kWh]	0	2.5	5	10	20
V_{th} [L]		100	500	1000	2000

2.2 Boundary Conditions

The simulation is carried out under the following boundary conditions:

- Ambient air temperature, pressure and humidity
- Solar irradiation
- Terrestrial and atmospheric radiation
- Wind speed and direction

The ambient air temperature and global irradiation are obtained by measurement data of Messe Riem (Munich, Germany) in 1-second time steps. For the other boundary conditions, data supplied in a test reference year (Bundesinstitut für Bauforschung, 2011) for the region of Munich, Germany in hourly resolution is used. The building is simulated with a time resolution of 90 s. The raw data of the boundary conditions are therefore averaged or interpolated, respectively.

The boundary conditions for the energy simulation tool are set up by the demand of thermal and electrical power, the PV plant power generation and the specifications of the components.

The considered building has a living space of 300 m² and is insulated with a 0.1 m layer of thermal conductivity $\lambda = 0.037 \frac{W}{m\,K}$. The windows of the building are considered as triple-glazed. The location of the building for the calculation of the sun position is set to Munich, Germany.

In order to obtain the PV power of the PV plant, the measurement data of Messe Riem is normalized to its peak power.

The DOD of the electrical storage is set to 70%, storage losses are neglected. The supply temperature of the heat pump is set to 45 °C. The return temperature is set to the storage temperature at the specific time step.

2.3 Performance Parameters

For the analysis of the results and especially for the comparison of different configurations, a set of dimensionless numbers is defined.

One optimization goal of this work is the increase of the self-consumption ratio (SCR), which is the quotient of PV energy used in the household and the total PV electricity production:

$$SCR = \frac{\int P_{el,PV,used}\, dt}{\int P_{el,PV,total}\, dt}$$

The self-supply ratio (SSR) is a measure for the autarky of a system and defined as the fraction of electrical energy supplied by the own PV plant and consumed directly in the household, divided by the total electricity consumption of the household:

$$SSR = \frac{\int P_{el,PV,self-supplied}\, dt}{\int P_{el,total\ demand}\, dt}$$

3 RESULTS AND DISCUSSION

The simulation study has been carried out for one year with the highly resolved PV data. The results from this year are analyzed in the following sections.

3.1 Annual Simulation

The analysis of a characteristic day shows the effects of the different system configurations and the influence of the operation strategy. Here, a winter day (Jan, 31. 2009) shall be analyzed exemplarily.

Figure 4 (a) shows the simulation results for both HP setups without electrical storage, a 500 liter thermal storage and a PV generator size of 10 kWp. Due to the small thermal storage size, the heat pump has to operate during night with power from the grid. Also in the morning, when the solar irradiation is too low, it might be turned on when the storage goes below its minimum SOC.

During the day, the storage is quickly recharged and the heat pump is only turned on with a fixed frequency of 15 minutes, in order to refill the storage, but not to violate the constraint of more than four on-off cycles per hour.

Figure 4 (b) shows the results for the same setup, but with a larger thermal storage of 2000 liters. It is clearly visible, why the SSR significantly increases with larger thermal storages. The heat pump operates for many hours depending on the solar power supply. Recharging during the night is not necessary. Only in the morning, power is taken from grid. In this case, a difference between the two heat pump types can be observed: The swash-plate compressor with its lower part load capability is able to operate with PV supply already in the morning. Here, the scroll compressor is not able to run exclusively with PV power. Therefore, the operation strategy fills up the necessary demand from the grid and operates the heat pump at design load and thus optimum COP.

When a small electrical storage is installed, the operation can be further optimized: Figure 4 (c) shows the profiles for the same day with the large thermal storage and an electrical storage of 2.5 kWh. This again decreases the demand from the grid significantly.

Evaluating the whole year, the following sensitivity analyses can be made. The thermal storage capacity is varied from 100 to 2000 liters (2.9 to 58.1 kWh) and the battery storage is varied from 0 to 20 kWh.

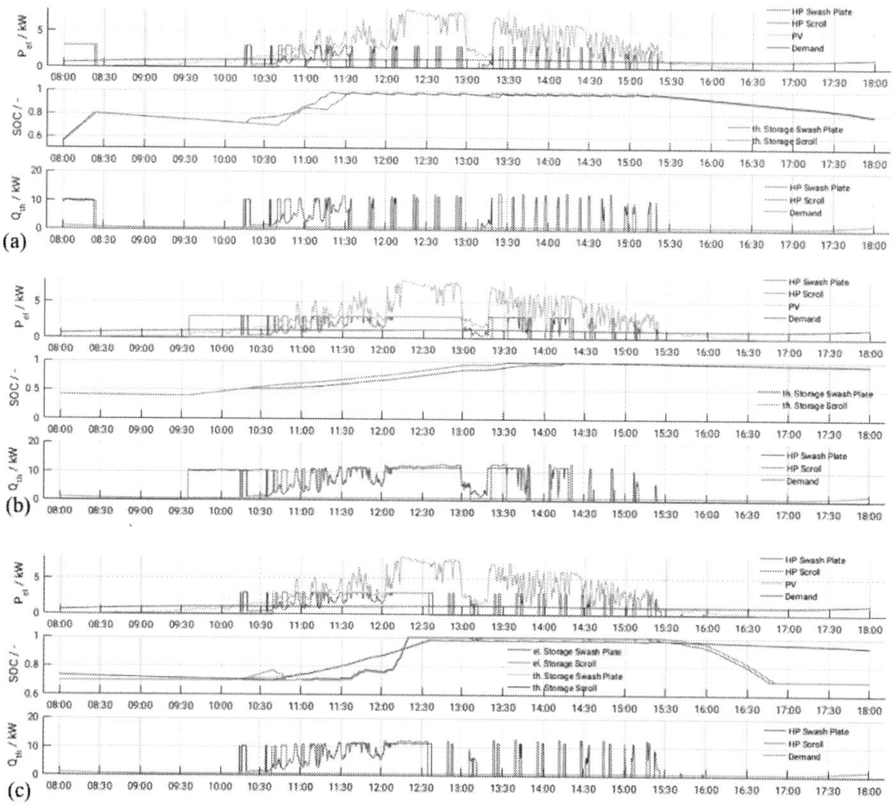

Figure 4: Plots of boundary conditions and heat pump operation for Jan, 31. 2009 with a PV array of 10 kWp: (a) $E_{storage} = 0\ kWh; V_{th} = 500\ l$; (b) $E_{storage} = 0\ kWh; V_{th} = 2000\ l$; (c) $E_{storage} = 2.5\ kWh; V_{th} = 2000\ l$

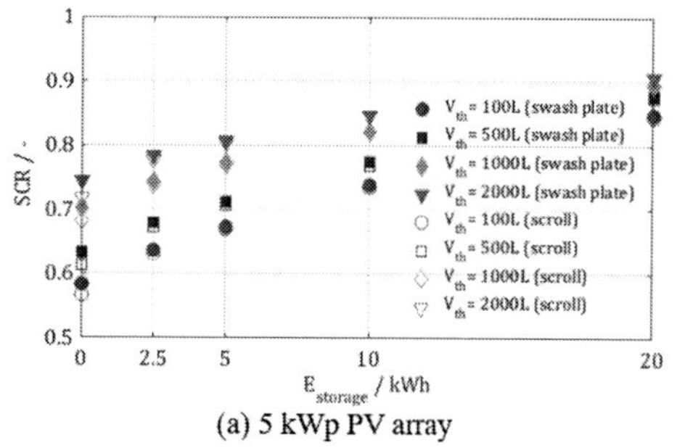

(a) 5 kWp PV array

(b) 10 kWp PV array

Figure 5 : Self consumption ratio (SCR) for different sizes of thermal and electrical storage for a PV array of 5 and 10 kWp, comparing scroll and swash-plate compressor based heat pumps

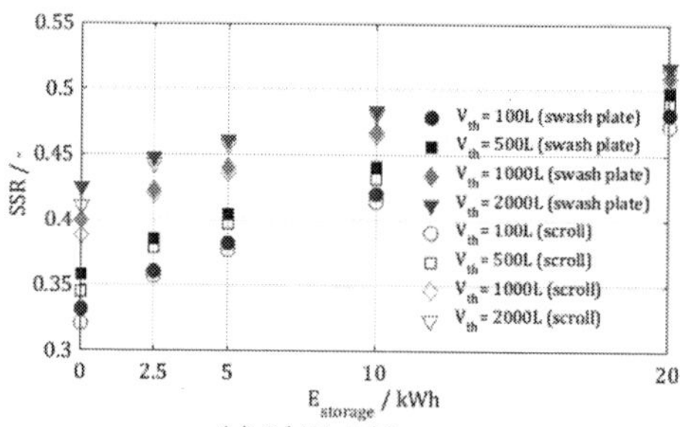

(a) 5 kWp PV array

(b) 10 kWp PV array

Figure 6: Self supply ratio (SSR) for different sizes of thermal and electrical storage for a PV array of 5 and 10 kWp, comparing scroll and swash-plate compressor based heat pumps

Figure 5 shows the SCR for a PV array of 5 (fig. 5 a) and 10 (fig. 5 b) kW_P. The results for the swash-plate compressor are compared with those of the scroll compressor. Compared to the previous study (Spinnler et al., 2014), it can be observed, that the heat demand significantly limits the obtainable SCR from 97% for unlimited heat output to values around 70% (fig. 5 a) for the smaller PV array and heat output according to the building heating demand. Thus, the building clearly is the limiting factor.

It is also obvious, that the SCR is higher for smaller installed PV peak power. For the small PV array, a significant influence of the storage sizes on SCR can be observed, whereas for a larger PV array this influence decreases. The larger PV array is clearly oversized with regard to the heat pump capacity and SCR as optimization parameter. Nevertheless, this oversizing allows a more flexible operation.

When comparing the influence of the storage ratio (SR), the benefit of increasing the thermal storage over increasing the electrical storage clearly appears, especially for small electrical storages. This behavior can also be observed for the SSR. A larger storage, thermal or electrical, of course leads to a higher degree of autarky. It is interesting that the increase from 5 to 10 kWp PV only leads to an increase in SSR of approximately 6% (cf. fig. 6).

The comparison of the swash-plate compressor based system to a scroll compressor with variable frequency drive does not show significant differences in the annual evaluation. For small electrical storages and a smaller PV array size, which have turned out to be the preferred system configuration, the highest differences between the two compressor types can be observed.

As already observed in the daily evaluation, the lower part load capability of the swash-plate compressor allows to run the heat pump in times with low PV power, where the scroll compressor is not able to operate. This is slightly advantageous for SRC and SSR.

3.2 Dynamics in Heat Pump Operation

The maximum gradient of compressor power in the real heat pump is not considered in the energy simulation tool. In order to investigate the requirements for the heat pump in terms of load variations, the gradients in heat pump power of an annual simulation without electrical storage, a thermal storage of $V_{th} = 2000$ l and a PV array of 10 kW_P are plotted in figure 7 sorted over their absolute value. It can be seen, that the main part of the gradients is lower than roughly 0.8 kW/s, while there are several outliers of about 2.1 kW/s for the scroll compressor and 2.7 kW/s for the swash-plate compressor.

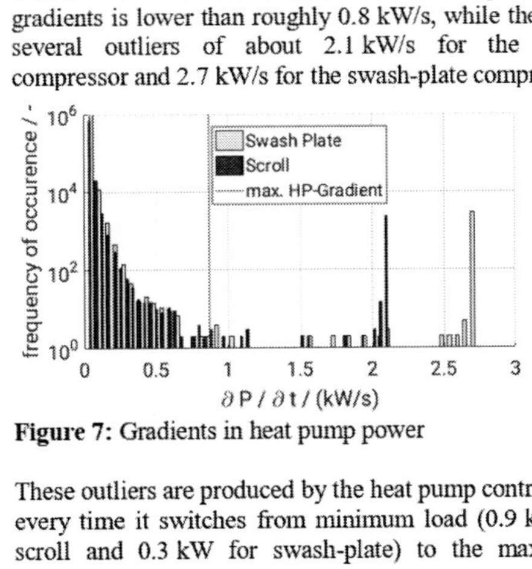

Figure 7: Gradients in heat pump power

These outliers are produced by the heat pump control unit every time it switches from minimum load (0.9 kW for scroll and 0.3 kW for swash-plate) to the maximum power of 3 kW in order to cover the heating demand. This phenomenon is also visible in figures 4 a-c.

The gradients lower than 0.8 kW/s result from fluctuations in the PV power. The heat pumps investigated in the test facility at the institute of Thermodynamics at TUM (cf. Spinnler et al., 2014) showed a maximum gradient in the compressor power of 0.87 kW/s. Thus, it is not necessary to install a low response time electrical storage such as EDLCs in order to smooth occurring PV fluctuations.

4 CONCLUSION

From the annual simulation, the following conclusions can be drawn:

The heat demand significantly limits the hours per year, in which the heat pump can be operated with PV power, for any well designed system. Only a reduction of both installed PV power and heat pump size or a significant increase of the thermal storage lead to a higher self-supply ratio.

The influence of the electrical and thermal storage on the system performance has clearly been shown. Enlarging the thermal storage has a much higher benefit on SCR and SSR, than an increase in the electrical storage. Combining this with economic considerations clearly emphasizes the supremacy of the thermal storage.

An analysis of the two compressor types (scroll and swash-plate) shows only small advantages of the swash-plate compressor, which is capable of a lower part load operation compared to the scroll compressor. This influence is rising with decreasing PV array peak power and only significant for low storage sizes (thermal and especially electrical).

From the analysis of the compressor power gradients, it can be concluded that the fluctuations in solar radiations have a lower impact on the operation than the slopes generated by on-off control. Therefore, it should be possible to follow the fluctuating PV power at all times.

Nevertheless, the influence of frequent load changes on the working life span of the compressor has to be investigated in experiments. Here, the swash-plate compressor might be especially promising, since its general application in cars already proves that it can resist frequent load changes.

Outlook

Further investigation has to be done on the optimum design for summer and winter operation. The implemented simulation model is already able to simulate the system for different PV inclination, which allows an optimization. Another interesting application is the coupled heat pump and compression chiller operation for winter and summer.

For the performance of the low-part-load heat pump, the application in bivalent operation (together with a conventional heating system for peak loads) would be interesting. Here, an especially high load range is expected.

5 ACKNOWLEDGEMENTS

The authors gratefully thank "Solarenergieförderverein Bayern e.V." (SEV) for financial support within the "PVCool" research project and both SEV and Prof. Mike Zehner for sharing the measurement data of Messe München.

6 NOMENCLATURE

Variable	Name	Unit
COP	Coefficient of performance	[-]
DOD	Depth of discharge	[-]
E	Electrical energy	[kWh]
h	Specific enthalpy	[kJ/kg]
P	Electrical power	[kW]
Q	Thermal energy	[kWh]
$\dot{Q}$	Thermal power	[kW]
SOC	State of charge	[-]
SCR	Self-consumption ratio	[-]
SSR	Self-supply ratio	[-]
T	Temperature	[°C]
t	Time	[s]
V	Volume	[L]
λ	Thermal conductivity	[W/m/K]
η	Efficiency	[-]

Index	Meaning
amb	Ambient
$cond$	Condensation
el	Electrical
$evap$	Evaporation
HP	Heat pump
i	Time step
$isen$	Isentropic
max	Maximum
min	Minimum
PV	Photovoltaic
ref	Reference
sc	Sub cooling
sh	Superheating
th	Thermal
tot	Total
use	Usable
w	Water
1	After evaporator
2	After compressor
3	After condenser

Abbreviation	Meaning
BDEW	Bundesverband der Energie- und Wasserwirtschaft (German Association for Energy and Water Sector)
BST	Building simulation tool
EDLS	Electric double layer capacitor
ES	Energy storage
EST	Energy simulation tool
HVAC	Heating, ventilation and air conditioning
PV	Photovoltaic
SDR	Self discharge rate
SR	Storage ratio
TUM	Technical University of Munich
VFD	Variable Frequency drive

7 LITERATURE

A. Arteconi, N.J. Hewitt, F. Polonara, State of the art of thermal storage for demand-side management, Applied Energy, Volume 93, May 2012, Pages 371-389

A. Arteconi, N.J. Hewitt, F. Polonara, Domestic demand-side management (DSM): Role of heat pumps and thermal energy storage (TES) systems, Applied Thermal Engineering, Volume 51, Issues 1–2, March 2013, Pages 155-165

I. Bell, J. Wronski, S. Quolin, V. Lemort, CoolProp http://www.coolprop.org/index.html (last visited: 15.06.2016)

K. Büdenbender, M. Braun, P. Strauß (2010). Technische Voraussetzungen einer Netzentlastung durch Eigenverbrauch. Presentation at 7. Fachgespräch der Clearingstelle EEG "Eigenverbrauch von Solarstrom", Berlin. URL: http://publica.fraunhofer.de/documents/N-146054.html

Bundesinstitut für Bau-, Stadt- und Raumforschung (2011). Aktualisierte und erweiterte Testreferenzjahre (TRY) von Deutschland für mittlere und extreme Witterungsverhältnisse. URL: http://www.bbsr-energieeinsparung.de/EnEVPortal/DE/Regelungen/Testreferenzjahre/Testreferenzjahre/01_start.html?nn=739044&notFirst=true&docId=743442 (last visited: 15.05.2016)

Deutsche Norm (2013). DIN EN 12900: Kältemittel-Verdichter – Nennbedingungen, Toleranzen und Darstellung von Leistungsdaten des Herstellers

P. Droege, T. Bocklisch (2015), Hybrid Energy Storage Systems for Renewable Energy Applications. 9th International Renewable Energy Storage Conference, IRES, Energy Procedia, Volume 73, 2015, Pages 103-111

Y. Guo, M. Pan, and Y. Fang (2012), Optimal power management of residential customers in the smart grid, IEEE Transactions on Parallel and Distributed Systems, Vol. 23, Issue 9, pages 1593-1606, 2012.

U. Jordan, K. Vajen (2001). Realistic Domestic Hot-Water Profiles in Different Time Scales V. 2.0, University of Marburg, Marburg, 2001

M. Loesch, D. Hofnagel, S. Steuer, T. Faßnacht, H. Schmeck (2014), Demand Side Management in Smart Buildings by Intelligent Scheduling of Heat Pumps, IEEE International Conference on Intelligent Energy and Power Systems (IEPS), 2014

A. Kastl, (2015). Entwicklung eines thermischen Modells zur Gebäudesimulation (Development of a Thermal Model for Building Simulation). Semester Thesis, Lehrstuhl für Thermodynamik, Technische Universität München. Supervisors: F. Kiefer, M. Spinnler, T. Sattelmayer. (undisclosed)

KommEnergie 2011. Standardlastprofil (SLP)
http://www.kommenergie.de/netz/lastprofilverfahren/standardlastprofile-slp/ (last visited: 15.06.2016)

E. L. Ratnam, S. R. Weller, and C. M. Kellett Fang (2013), An optimization-based approach for assessing the benefits of residential battery storage in conjunction with solar PV, 9.th IREP Symposium, Rethymon, Greece, August 2013.

L. Rouvel, F. Zimmermann (1997). Ein regelungstechnisches Modell: Teil 1 bis 2. HLH Bd. 48, S. 67-74.

M. Spinnler, L. Hörth, F. Böing, S. Wolf, T. Sattelmayer (2014). Novel Heat Pump System for Highly Transient PV Operation. PVSEC 2014, Amsterdam.

R. Suzuki, Y. Fujimoto, Y. Hayashi Fang (2012), Determination method of optimal planning and operation for residential PV system and storage battery based on weather forecast, IEEE International Conference on Power and Energy (PECon), Kota Kinabalu Sabah, Malaysia, December 2012.

R. Thygesen, B. Karlsson (2014). Simulation and Analysis of a Solar Assisted Heat Pump System with two Different Storage Types for High Levels of PV Electricity Self-Consumption. Solar Energy 103 (2014) 19-27, Elsevier.

E. Vrettos, K. Lai, F. Oldewurtel, and G. Andersson (2013), Predictive control of buildings for demand response with dynamic day-ahead and real-time prices, European Control Conference (ECC), Zürich, Switzerland, July 2013.

E. Vrettos, A. Witzig, R. Kurmann, S. Koch, G. Andersson (2013 b), Maximizing local PV Utilization using Small-scale Batteries and flexible Thermal Loads, EU PVSEC 2013; 28th European Photovoltaic Solar Energy Conference and Exhibition

A. Witzig, T. Straub, M. Hartmann, T. Sonntag (2014), Fast Simulation Algorithm for PV Self-Consumption with Heat Pump Systems: Calibration Methodology with Comprehensive Dynamic Simulation Model as Reference. 29 th European PV Solar Energy Conference, Amsterdam, 2014

SIMULATION OF LOCAL ENERGY SURPLUS USAGE IN HYBRID GRIDS WITH A HIGH PV PENETRATION RATE

Author for correspondence
David Stakic[1)]
Co-Authors
H. Ruf[1)], K. Ditz[1)], D. Funk[1)] G. Heilscher[1)], F. Meier[2)]

1) Hochschule Ulm, Germany
2) Stadtwerke Ulm/Neu-Ulm Netze GmbH, Germany

Eberhard-Finckh-Str. 11, 89075 Ulm, Phone: +49 731 50 28 353, E-mail: stakic@hs-ulm.de

ABSTRACT: The objective of the "optimizing hybrid energy grids for smart cities"- (OrPHEuS) project is the development of cooperative control strategies for smart cities and hybrid energy networks. The OrPHEuS project aims at solutions for optimizing the interaction among multiple energy grids e.g. electric power, district heating and gas, which are connected through coupling points and defined as hybrid energy grids. This paper focus on the framework developed for the situation in the two test areas close to the city of Ulm, Southern Germany. The test areas have suburban and village character with high photovoltaic penetration rate. The interaction of the key values of the heat, gas and electrical demand has been investigated. As a result the most promising configuration for the control strategies was set-up. As a part of the preprocessing for the co-simulation of the energy grids the potential storage capacity within the gas grid was calculated.

Keywords: Smart Grid, Renewable Energy, Photovoltaic, Hybrid Energy, Power to Heat, Power to Gas

1 INTRODUCTION

The objective of the "optimizing hybrid energy grids for smart cities" (OrPHEuS) project is the development of cooperative control strategies for smart cities and hybrid energy networks. The OrPHEuS project aims at solutions for optimizing the interaction among multiple energy grids, which are connected through coupling points and defined as hybrid energy grids. The project evaluates how information and communication technology (ICT) devices using control algorithms can provide strategies for energy optimization. Using the cases of existing energy system setups in two cities, Ulm (Germany) and Skellefteå (Sweden), strategies for hybrid energy grid control and cooperation will be developed and evaluated. The content of this work focusses on the Ulm test site with high photovoltaic (PV) penetration rates and investigates the different use cases for the electric network, the district heating and the gas grid based on data of the year 2014.

2 TEST SITE DESCRIPTION

The distribution system operator (DSO) at the city of Ulm provides two test areas, both with a high number of PV systems connected to the electrical grid. The grid characteristics of both test areas are a mixture of suburban and village structures.

The city of Ulm has close meshed energy grid infrastructure and high load areas nearby. The power surplus of a PV system at a single house is used close to the house and has not to be transported over a long distances to the next customers. The key figures of the different energy types are shown in Table I. For space heating and domestic hot water (DHW) are no measured values available. Therefore it was necessary to assume the energy by using housing data like usable space, year of construction of the buildings and persons. The calculated values for every household deliver the baseline for further calculations, e.g. the energy for space heating and DHW of the complete test area in Table I.

Table I: Energy key figures of the two test areas Einsingen und Hittistetten in Ulm

Key figures	Einsingen	Hittistetten
Electrical demand [GWh/a] *	1.05	1,05
Number of households [-]	186	104
PV yield [GWh/a] *	0.23	0.99
Number of PV systems [-]	21	63
PV capacity [kWp]	233	1067
Gas demand [GWh/a] *	2.26	0.57
Number of households connected to the gas grid [-]	92	6
Space heating demand [GWh/a] **	3.34	1.87
Domestic hot water demand [GWh/a] **	0.8	0.45

*Measured values
**Calculated values based on housing data

However, the seasonal variation affects the operation strategies of hybrid grids. During the summer time the PV feed-in power is high while the heat and gas demand is low and vice versa. The overview and structure of both test areas are shown in Figure 1 and 2.

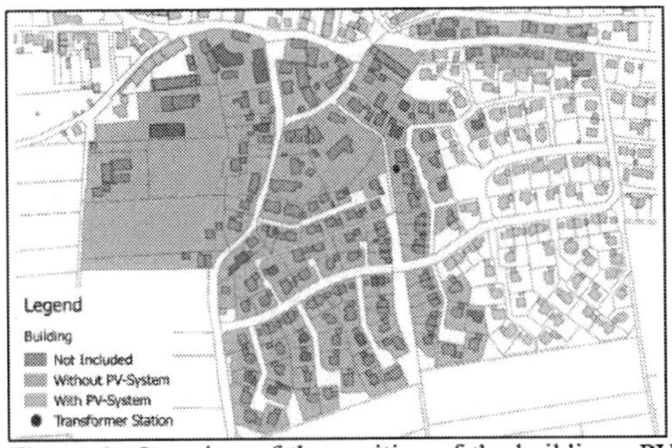

Figure 1: Overview of the position of the buildings, PV systems and the transformer station at the test area Einsingen in Ulm, Germany

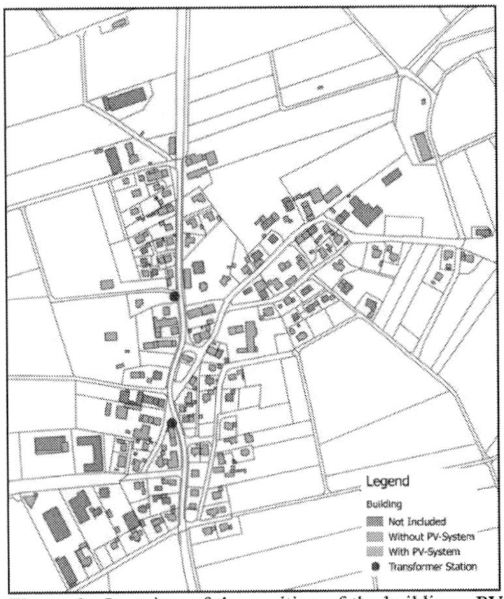

Figure 2: Overview of the position of the buildings, PV systems and the transformer stations at the test area Hittistetten in Ulm, Germany

3 SCENARIOS

All developed hybrid scenarios are compared to a reference denoted as "baseline scenario". The grid models of the baseline scenario consider constant limits of the voltage band and maximal utilization of the assets. Furthermore, it include the predicted PV yield based on a PV roof potential analysis as well as the individual ratio of PV power and local demand [2].

Nowadays, the main incentive for residential PV system installations in Germany is the self-consumption due to decreasing feed-in tariffs and increasing end-customer energy prices.

The challenges for the DSO caused by the local PV feed-in and the conventional way to solve those, delivers the technical and economic framework to evaluate the results of the developed hybrid scenarios 1 to 3. Conventional grid reinforcement means e.g. new cables to extend the hosting capacity for PV power. However, this is the most expensive solution for several cases [3]. Table II gives an overview of the maximal power values exceeding the 80% (load factor 0.8) of the rated transformer power in the three PV potential scenarios. According to the DSOs experience a load factor of 0.8 seems to be a good value to avoid stress at the medium voltage (MV) to low voltage (LV) transformer. An additional aim was to check the energy self-consumption in the grid of the test areas not only as an island, but also in context of the whole system. Electric power is most valuable energy type. Therefore the electric power should be used by other consumers surrounding the test areas instead of covering heat demand.

Table II: Overview of the PV surplus in the test area Einsingen for the different PV-scenarios and with a maximal load factor of 0.8 at the transformer.

Scenario	MWh/a	MWh/d (max)	MW (max)
50%-PV Surplus	297	3.52	0.63
75%-PV Surplus	669	6.50	1.05
100%-PV Surplus	961	8.64	1.34

The first hybrid scenario describes a local power to heat (PtH) solutions at customer/prosumer level. A comparison for all PV scenarios was done taking into account the different variations of heating rods, e-boilers and thermal storages. DHW and space heating are separately analyzed. This allows sufficient results for a recommendation of optimization strategies. A combination of 2 kW heating rod and upgraded DHW storage at each house with PV system up to 300% capacity shows the most promising results according to the co-simulation, based on the data of July 2014. However, the actual installed capacity of the DHW storage is not available. Therefore the storage capacity was assumed to cover the average head demand over the year for two days. [1]

The second hybrid scenario is almost similar to the first but assumes a district heating network (DHN). This enables a higher direct use of the PV surplus according to the increased heat demand in summer time. Furthermore the DHN can be an energy transport solution to nearby areas with a significant heat demand. In the preprocessing of this scenarios Heat pumps (HP) are analyzed. The high efficiency of the HP is used to avoid conventional heat production in the DHN. But in reasons of availability, reliability, control and cheap prizes was the heating rod preferred by the DSO.

The third hybrid scenario extends the first hybrid scenario including DHW and space heating with the gas grid. In suburban areas gas is mainly used for space heating. The efficiency from PtH is higher than for the power to gas (PtG) technology, but the PtG technology enables the gas grid as seasonal storage. The PtG systems gain the highest amount of loses within the energy conversation. Beside the seasonal thermal storages decreasing their efficiency over the time. This also has an impact on the DHN. However, PtG systems are not reasonable to avoid voltage band violations or overloaded lines because these systems are connected to a single connection point of the electric grid. Thus economic reasons it is necessary to use decentral electric power lodes to avoid grid reinforcement.

Figure 3 shows the three hybrid scenarios in one picture. The results from the co-simulations will answer the question which is the most efficient solution. The evaluating criteria for the results are:

- Avoiding voltage violation
- Avoiding grid asset stress
- Influence on gas and heat demand
- Reduction of grid reinforcement costs
- Reduction of greenhouse gas emissions

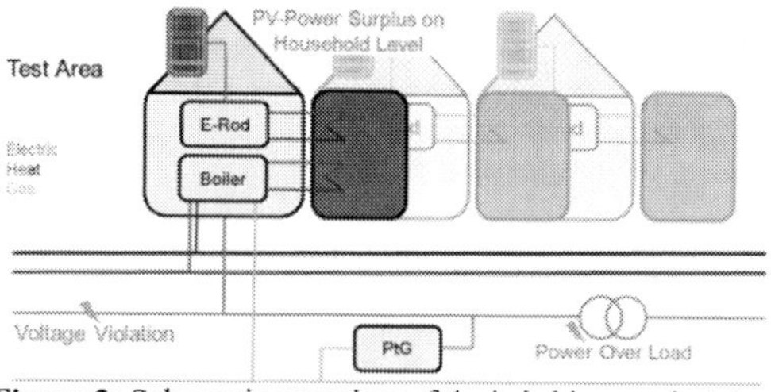

Figure 3: Schematic overview of the hybrid scenario 1 to 3 in the test area

4 CONTROL STRATEGIES

The approach of the control strategy targets are: reduction of the PV surplus due to self-consumption, surplus usage within the community, all available e-devices under central control by ICT and the usage of surplus predictions. To reach these targets four control strategies are defined, which varies the number of parameters considered by the controllers.

The first is an isolated controller observing only in-house power usage, like the heating rod, and in-house generation, e.g. PV systems.

The second is a local controller observing the electrical surplus as in control strategy one and additionally the voltage value at the point of common coupling.

The third is a centralized controller observing the major parameters of the whole grid like voltage, line loading and load flow amount and direction. The voltage is monitored at node level and involves additional heating rods in the neighborhood of the critical node. Line overloading caused by PV feed-in is solved with heating rods at both ends of the line. Additional heating rods will be activated if the load at the transformer station reach a critical level, e.g. 0.8 load factor.

The fourth control strategy considers PV feed-in forecast. The simulation will be done in 1 minute time resolution. However, due to the time needed for the simulation the investigation is done for one selected week instead of a whole year.

These variations of the control strategy are in addition to the technical scenarios.

5 SIMULATION APPROACH

The model of the electricity systems of the test areas (generation units, distribution lines and demand models) has been implemented in the simulation environment DIgSILENT PowerFactory. The implementation of the district heating system and the thermal systems (heat demand of buildings, supply and storage units) is done with the simulation environment Dymola based on Modelica Fluid library specifications and on the extension of resulting in the Modelica library “DisHeatLib” developed at Austrian Institute of Technology (AIT) [1].

The prerequisite for a wholesale model-based evaluation of hybrid system solutions is the detailed simulation of the individual subsystems including their dynamic interactions. This is done by a co-simulation approach coupling the existing domain-specific state-of-the-art tools into dynamic multi-physics simulation of the hybrid system. This approach requires an additional software layer that coordinates the execution of the individual simulation tools and the data exchange between each tool, it also enables the domain experts to use the most appropriate tools for their respective domain (Figure 4).

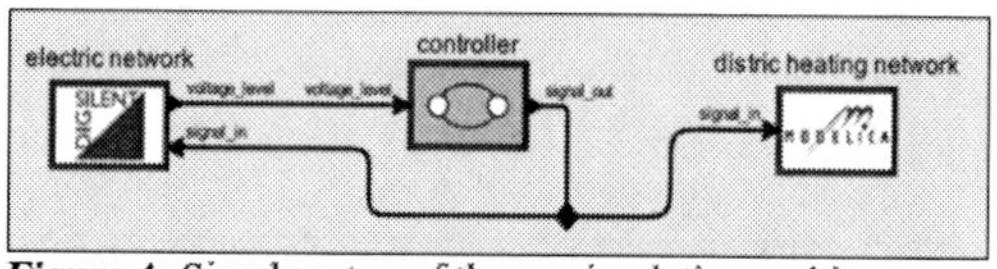

Figure 4: Simple setup of the co-simulation architecture [1]

6 RESULTS FROM THE BASELINE

To evaluate the baseline scenarios and the correlation of the energy flows in the hybrid system, the heat demand and the PV surplus in the test area Einsingen is balanced and cumulated as shown in Figure 5. For this analysis was only power and energy flows from the total test area Einsingen used, no grid and storage models or control strategies were implemented. The difference at the points of inflexion of the resulting mathematical equation (Taylor series) of the cumulated energy values allows the calculation of the seasonal PV surplus. The assumed constraint in Figure 5 is the usage of all surplus (η=1) in the test area. The seasonal PV surplus is energy that has to be transmitted into near areas, e.g. the City of Ulm, or has to be stored.

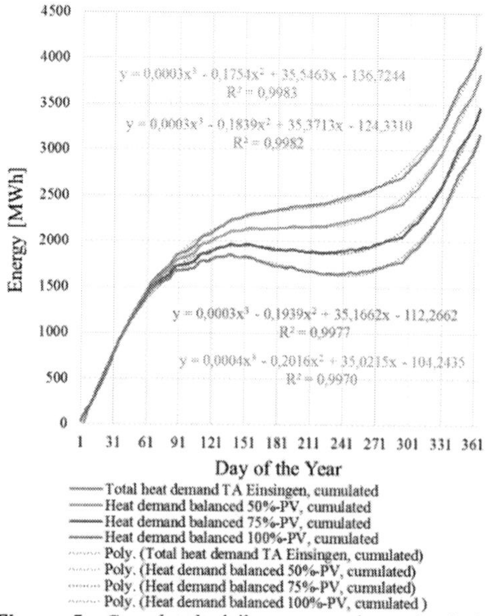

Figure 5: Cumulated daily energy demand of the electrical PV surplus and the total heat demand over on year.

The PtG approach can only balance the PV surplus with the gas consumption. The gas demand is only part of the total heat demand and is therefore lower. The lower efficiency ($\eta(H_2) \approx 0{,}75$ and $\eta(CH_4) \approx 0{,}5$) negate the effect on the sessional PV surplus not totally. Table 3 gives an overview of the result values for all three scenarios mentioned before. The additional hydrogen percentage is limited due to technical and legal (max. 5 %) specifications. This has a strong influents to the available feed-in capacity of the gas grid for hydrogen. The yearly seasonal PV surplus could cover from 7 % to 23 % of the annual heat demand depending on the different PV potential scenarios. Without the seasonal PV surplus the usage could be from 6 % to 16 %. If a seasonal PV surplus is considered the usage rate increases by 7 % of the heat demand in the 100%-PV potential scenario. This value is also the maximum for scenario 2 taking into account the DHN. However, the surplus of the heat can be used outside of the test area.

Table III: seasonal PV surplus for local power to heat, power to gas (H2) and Power to gas (CH4).

Necessary if all surplus should be used in the area	PtH	PtG (H2)	PtG (CH4)
100%-PV Surplus [MWh]	203.7	213.73	96.02
75%-PV Surplus [MWh]	86.3	126.87	43.87
50%-PV Surplus [MWh]	0	11.31	0

With the geometrical and topological information of the gas grid infrastructure the grid volume and the energy storage capacity can be calculated. The medium pressure (MP) grid is supplying a bigger area as the defined test area Einsingen. The down-scaling from the MP grid area to the test area is based on a calculation of the percental factor of the inhabitance. The MP gas grid provides a storage capacity of 207.1 kWh for methane and 3.11 kWh for hydrogen, respectively. Suburban MP gas grids, as e.g. in the test area Einsingen, cannot deliver the seasonal storage capacity which is necessary if the seasonal PV surplus will increase. The most urban areas include a high pressure (HP) distribution gas grid and can use this in combination with the PtG technology as a seasonal storage. A first estimation for the HP gas grid of the city of Ulm results in a storage capacity of 600 MWh. Combined with an average base load of 106.45 MW for summer time (June, July, August) results in power shifting a high potential. The capacity in Ulm is strong influenced of a pipe storage and therefore not so typical like the most European cities. This pipe storage is feed from the transmission grid and has about the same volume as the distribution HP gas grid of the city, but its higher pressure level results in a higher storage capacity. The HP distribution gas grid has an estimated storage capacity of 84 MWh (0.7 kWh per inhabitant of Ulm).

7 CONCLUSION AND OUTLOOK

The analysis of the test areas in case of hybrid grid infrastructure, electric power generators (PV) and demand at all give the information to develop strategies for solving the upcoming problems. These analyses have to be done individually to get the detailed potential. Decentralized controllable loads and hybrid grid coupling points are a promising technology face the challenges of decentralized power feed-in. These loads should be easy available, reliable, fast controllable and cheap. PtH systems like heating rods and e-boiler are good solutions to avoid voltage band violations, line and transformer overloading and to increase the local usage of the feed-in power. The DHN can provide additional transmitting potential. However, their storage capacity is limited. PtG systems and the urban gas grid can provide a seasonal storage and additional direct demand, but with low efficiency because of the transformation from electric to chemical energy. The described and planed co-simulation will show the combination of the technologies suitable to solve the different challenges in decentral energy supply systems by the use of available grid infrastructure.
The detailed Results will be available at the beginning of 2017 after the OrPHEuS project has finished.

DISCLAIMER

The OrPHEuS project is co-financed by the European Commission under the 7th Framework Program.

SOURCES

[1] D. Basciotti, S. Henein, E. Widl, C. Marguerite, H. Brunner, R. R. Schmidt; Austrian Institute of Technology AIT; Deliverable 4.1, WP4 System Modelling and Simulation for the Evaluation of the OrPHEuS Control Strategies; Project: OrPHEuS, OPtimising Hybrid Energy grids for smart citieS

[2] H. Ruf et al, "Active Grid Planning Based on Solar Power Roof Potential Analysis," in 27th European Photovoltaic Solar Energy Conference: Proceedings of the International Conference held in Frankfurt a. M, 24-28 September 2012, S. Nowak, A. Jäger-Waldau, and P. Helm, Eds, München: WIP, 2012, pp. 3782–3787.

[3] M. Schwarz and A. Kollmann, "Wirtschaftliche Betrachtung von Netzausbau und Smart Grid-Lösungen in der Niederspannungsebene," Elektrotech. Inftech, vol. 131, no. 3, pp. 99–104, 2014.

SOLAR PV RESOURCE FOR HIGHER PENETRATION THROUGH A COMBINED SPATIAL AGGREGATION WITH WIND

Tobias Bischof-Niemz[1], Crescent Mushwana[2], Stephen Koopman[3]

[1,2,3] Council for Scientific and Industrial Research (CSIR), Energy Centre, 627 Meiring Naude Road, Brummeria, Pretoria 0001, South Africa, Phone: +27 83 403 1108, E-Mail: TBischofNiemz@csir.co.za [2] E-Mail: CMushwana@csir.co.za [3]E-Mail: SKoopman@csir.co.za

ABSTRACT: The conventional approach to assess and integrate renewable energy (RE) resource individually has its own limitations especially in areas where different energy resources get integrated to supply power to the grid. In South Africa, the Solar PV resource potential is extremely high virtually all over the country, however, solar PV is integrated into the grid with other energy sources. The other significant RE source in South Africa is wind, therefore it is important to understand the interaction between wind and solar PV and how these would be reflected in the power system. The benefits of spatial distribution of renewables are well understood, but the impact of the combined spatial aggregation of wind and solar PV is central to the design of the future power system in a responsive manner.
A thorough understanding of the combined aggregated spatial effect of these two resources is critical in ensuring that PV is optimally rolled out in the country. The fact that solar PV has a good yield potential all over the country could lead to a situation where PV could be developed aggressively with no due regard for the spatial aggregation effects when it is combined with wind. This could result in suboptimal use of the resource and undesired grid capacity and stability problems.
Keywords: Aggregation, spatial distribution, variable renewables

1 INTRODUCTION

South Africa's excellent solar and wind resources have remained largely untapped due to the high cost of integrating variable renewables (VRE) and the perceived integration challenges due to their intermittency and variability. The high cost challenge remains in some African countries, however, in South Africa wind and solar PV have reached grid parity and are currently the cheapest low cost bulk electricity supply option. Even with these advances, the perception of negative grid impact when considering high penetration of solar and wind due to their intermittency and variability, still remains. Furthermore, the aggregated power output of solar and wind plants and their spatial distribution effect on the power system needs to be understood in order to exploit the synergies. This study highlights the benefits of aggregating solar and wind in the South African context and how this knowledge can be used to increase the penetration of solar PV into the South African grid.

2 PURPOSE OF THE WORK AND DATA SETS

2.1 Purpose

The overarching objective of this study is to increase the fact base and the understanding of aggregated wind and PV profiles in South Africa for different spatial distributions and penetration levels of wind/PV, in order to serve as input into the South African energy-planning processes (Integrated Energy Plan, Integrated Resource Plan for Electricity, strategic grid planning, Transmission Development Plan (TDP), wind and PV Strategic Environmental Assessments, etc.). This will result in higher allowable penetration levels of wind and solar PV into the South African power system, based on quantified impacts.

2.2 Data sets for the study

The existing Wind Atlas for South Africa (WASA) was used, but it only covers the southern part of the country as shown in Fig. 1. This data set was extrapolated for the entire country to create a high quality wind speed map with a data set for the entire country as shown in Fig. 2 with exceptional wind speed > 6m/s for most of the country. The solar PV map for South African is given in Fig. 3, showing an impressive > 1900 kWh/m^2 for most of the country. The source of the solar data used was from Solar radiation Data (SoDa). Solar and wind data was synchronised spatially and temporal.

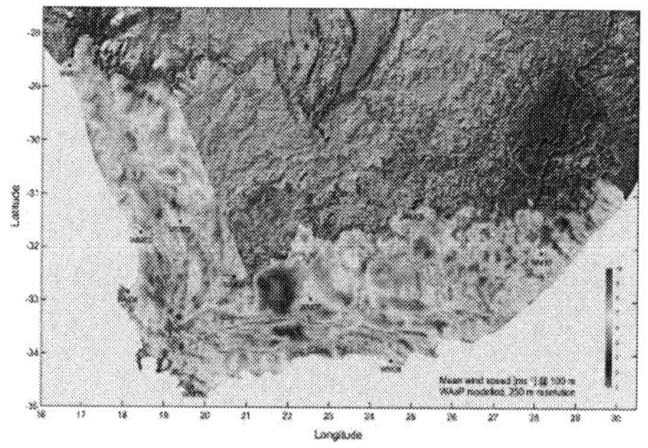

Figure 1: WASA domain

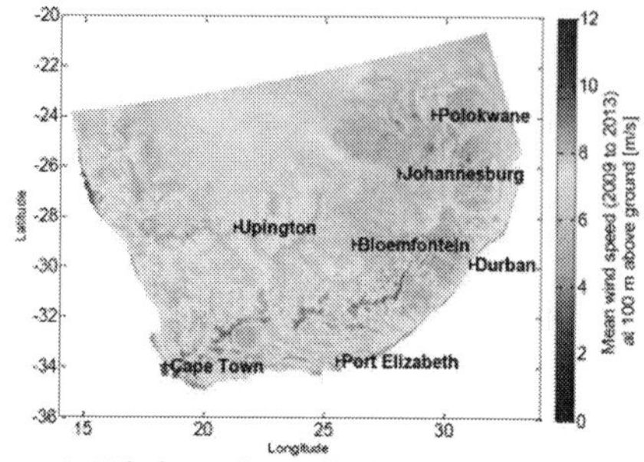

Figure 2: Wind speed map for the entire country

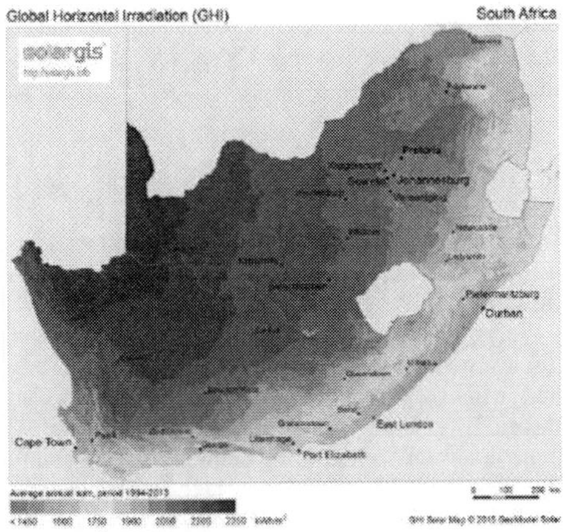

Figure 3: Solar PV resource

Table 1 gives the summary of the solar and wind and load data used in the study including the spatial and temporal resolution. The spatial resolution resulted in about 50 000 pixels (5x5 km) for the whole country.

Table I: Data sets applied in the study

Model name	WASA	SoDa	Load
Variables	Wind speed (v), temperature (T)	Solar irradiation	MW
Height	50 m, 80 m, 100 m, 150 m (v), 2 m (T)	-	-
Temporal coverage	2009 to 2013	2010 to 2012	2010 to 2012
Temporal resolution	15 min	15 min	30 min (interpolated to 15 min)
Spatial coverage	South Africa	South Africa	South Africa
Spatial resolution	5 km x 5 km	0.2° x 0.2° (similar to 5km x 5km)	All transmision substations (approx. 140) distributed countrywide.

The solar and wind data sets where aggregated and combined with load data to assess the impact on the power system and generate key metrics relevant to assess for conventional generation fleet and grid capacity assessment. In assessing the impact of variable renewables on conventional dispatchable plant the concept of residual load becomes important; where Residual load (t) = Load (t) – PV feed-in (t) – wind feed-in (t)

3 SCIENTIFIC INNOVATION AND RELEVANCE

South Africa has abundant wind and solar resources, however, in order to fully exploit their potential, it is important to understand how they should be integrated as part of the existing conventional generation fleet. The variability of wind and solar can result in limiting the perceived amount that can at the maximum be integrated into the existing generation fleet if the actual variability of wind and PV for different penetration levels and spatial distributions is not quantified. The international and national context are important aspects to consider as a higher penetration of PV is pursued.

3.1 International context

The 2015 global PV installed capacity stands at 256 GW, and almost 90% of the installed capacity was done over the last five years. In 2015, 57 GW and 63 GW of PV and wind respectively were installed as shown in Fig. 4. The drop in PV costs is one of the key drivers, and PV is expected to exceed wind in the coming years. These global trends are good indicators of where investment into the resources that drive the energy sector globally and also the projects that are being funded.

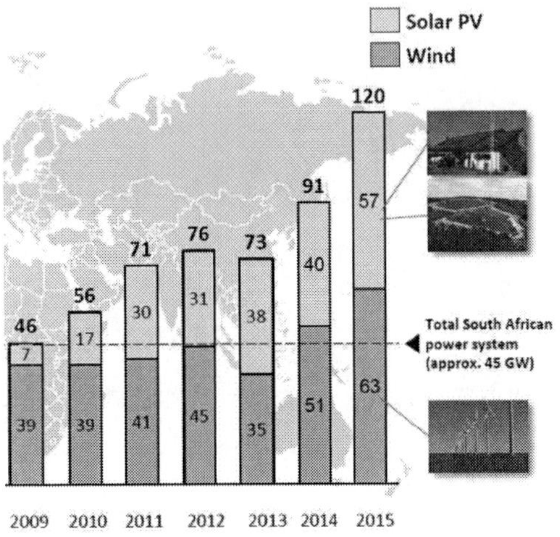

Figure 4: Global annual wind and PV installed capacities

3.2 South African context

The exploitation of the excellent solar PV and wind resources in South Africa has largely been hampered by cost and the lack of knowledge about the potential. The targets set for 2030 in the Integrated Resource Plan (IRP) are a clear indication that there was a certain level of reluctance to allocate high PV penetration in the country because the costs at the time (2009 – 2010) of the drafting of the IRP were very high in the country.

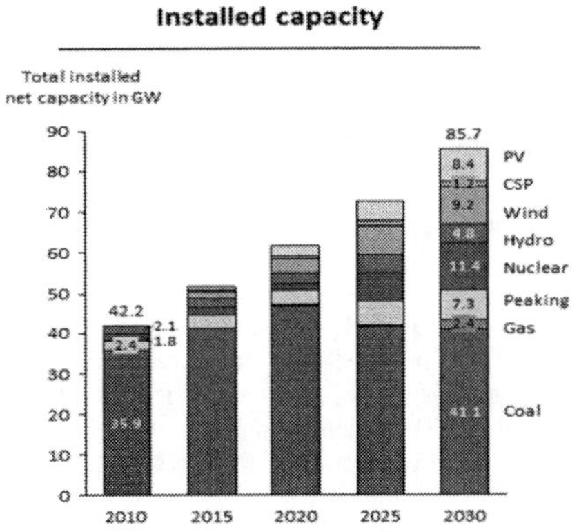

Figure 5: South Africa's installed capacity for 2030

In Fig. 5, the installed PV capacity planned for 2030 is given as 8.4 GW, which translates to only 3% of the energy mix. However, it is encouraging to see that from year 2020 to 2030, there is no significant new coal plants, but renewables, among other resources, are expected to satisfy the energy requirements.

The RE independent power producer procurement program (REIPPPP) in South Africa is implemented via a competitive bidding process by the department of energy and has resulted in significant declines in PV and wind costs as shown in Fig. 6. Therefore, since 2014 PV (≈ 6 €-ct/kWh) and wind (≈ 5 €-ct/kWh) are the cheapest new-build options based on lifetime costs per kWh; with base load coal at ≈ 5.6 – 7.7 €-ct/kWh and mid-merit coal at ≈ 9 €-ct/kWh.

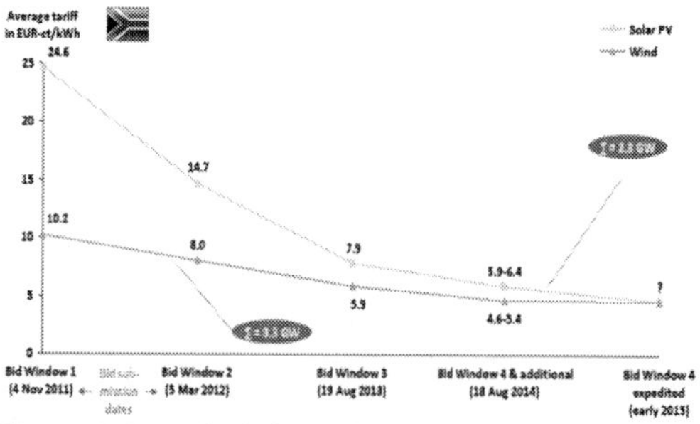

Figure 6: PV and wind costs in South Africa

3.3 Study approach

The data sets used in the study as given in Table I were key to the formulation of the study assumptions and they were central to the approach used. The approach used in the study is captured in Figure 7.

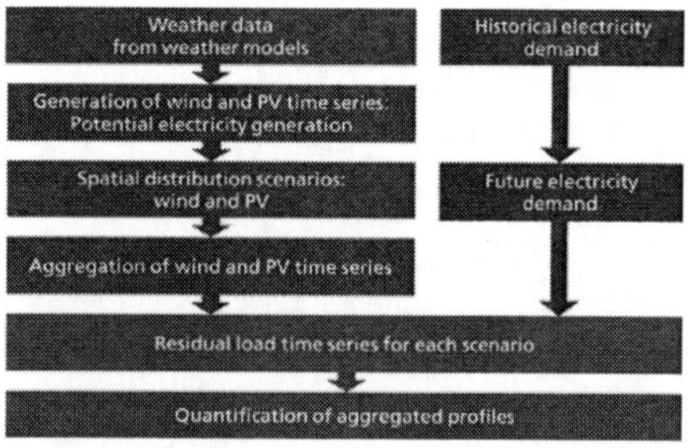

Figure 7: Study approach

The generation of the residual load (load minus PV minus wind) profile which would be a proxy for the expected power system behavior that the power system operator must respond to. The residual load profile was determined for a number of study scenarios as given in section 3.4.

One of the critical factors of aggregating wind with PV, is the selection of wind turbines to apply for harvesting wind in the diverse wind resources around the country. Five turbine types were selected to cover areas of the country from high to low wind speed areas with hub heights from 80 m to 140 m.

3.4 Study scenarios

In formulating study scenarios, it is important to assume a future system load level that the planned power system must meet. The current energy demand is around 250 TWh/yr, therefore a future power system with 500 TWh/y was assumed to be fed by a combination of PV and wind with existing conventional plant. Therefore three PV and wind combinations scenarios and penetration levels had to be defined, starting from a low energy share of 18% to a high share of 75%. In determining the combinations of wind and PV, the PV spatial profile was kept constant throughout the country because the PV profile throughout South Africa is constant. Fig. 8 shows the PV and wind combinations and penetration levels that were defined for use in the simulations.

Solar PV energy penetration (one fixed spatial distribution defined)	Wind energy penetration: 50 TWh/yr (~15 GW)	100 TWh/yr (~30 GW)	250 TWh/yr (~75 GW)
40 TWh/yr (~21 GWp)	• VRE Share: 18% • 9 Wind distribution scenarios	• VRE Share: 28% • 9 Wind distribution scenarios	• VRE Share: 58% • 9 Wind distribution scenarios
80 TWh/yr (~42 GWp)	• VRE Share: 26% • 9 Wind distribution scenarios	• VRE Share: 36% • 9 Wind distribution scenarios	• VRE Share: 65% • 9 Wind distribution scenarios
200 TWh/yr (~104 GWp)	• VRE Share: 44% • 9 Wind distribution scenarios	• VRE Share: 52% • 9 Wind distribution scenarios	• VRE Share: 75% • 9 Wind distribution scenarios

Figure 8: PV and wind combinations and shares

In addition to the wind distribution shares, the PV allocation for each penetration level was assumed as follows:

- Distributed (e.g. rooftop and small-scale): 70%
- Utility scale: 30%

Distributed PV includes all embedded PV which is mainly distributed according to population density.

The study scenarios that were simulated were based on the following three broad categories:

1. **Scientific scenarios:** these scenarios assist in stretching and testing the boundaries in terms of technical limitations and capability
2. **Defined areas scenarios:** these scenarios help to quantify and validate areas that have been selected for PV and wind development.
3. **Grid-focused scenarios:** when integrating new generation sources, the spatial extent and capacity of the grid is very important.

Table II gives all the scenarios that were simulated.

Table II: Scenarios simulated in the study

Category	Scenario	ID
1.Scientific	Uniform wind distribution	1a
	All-in-one place	1b
	High wind speeds distribution	1c
	Minimal 15-min. ramps distribution	1d
2.Defined areas	Environmental Impact Assessment (EIA) wind distribution	2a
	Renewable Energy Development Zone (REDZ) wind distribution	2b
	EIA/REDZ overlaps	2c
3.Grid-focused	Current grid wind turbine distribution	3a
	Future grid wind turbine distribution	3b

4 RESULTS AND CONCLUSION

4.1 Results

When the simulations were conducted, scenario 2c was fully represented by both 2a and 2b, and similarly, scenario 3b is represented by scenario 3a. The first simulation run showed the generation capacity that would be required from each scenario to satisfy the future demand. And Fig. 9 clearly shows that the generation capacity required to minimise ramps would be the highest of all the scenarios, however, the determining factor will be whether the ramps exhibited by the other scenarios are manageable or not. The ramps considered in the study are 15 min. ramps because these are the type of ramps that the systems operator should be able to respond to. Therefore it is critical to understand what the existing power system's 15min. ramp/MW value is, and how this value is influenced by the change in PV and wind penetration.

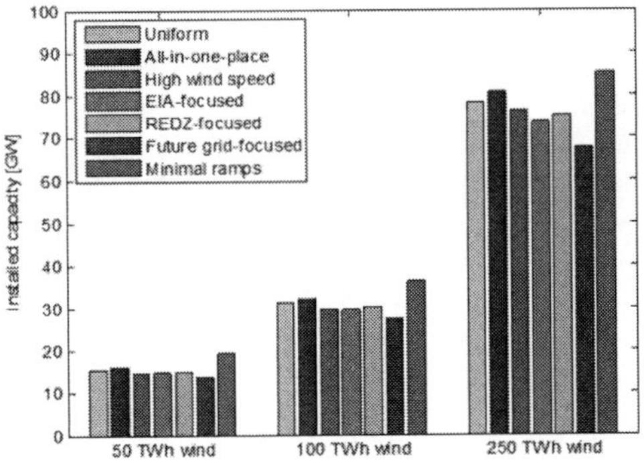

Figure 9: Installed wind capacity per scenario

The standard deviation of the 15 min. ramps for all the study scenarios was done, and the results are shown in Fig. 10. It shows the impact of different penetration levels of PV and wind on the 15 min. ramp rate that conventional power plants must deal with in the residual load (load-pv-wind). The green cross on the far left indicates the existing power system with no wind energy and has a 15 min. ramp of about 1800 MW (4% in ≈ 45 GW system). With a high penetration of wind and PV the 15 min. ramp is only about 2800 (roughly 4% in ≈75GW system – wind only). This figure shows that even with high wind and PV penetration, the 15 min. gradients are insignificant.

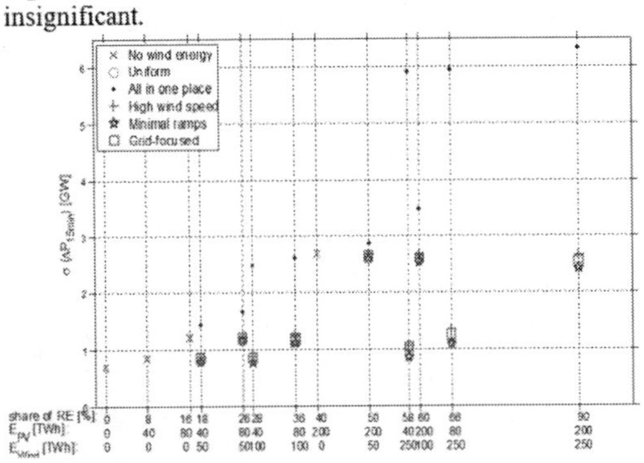

Figure 10: Short-term (15 min.) ramps for all scenarios

The results as shown in Fig. 10 clearly indicate that 15 min. ramps are not a problem at all when PV and wind plants are spatially distributed. However, the scientific all-in-one scenario (1b) exhibit a significant increase in 15 min. ramps with high penetration of both wind and PV. Even up to 36% share of variable renewables do not significantly increase the short-term ramps. The rise in 15 min. gradients due to higher PV penetration is also no cause for concern because PV is highly predictable. The study further highlights that both wind and PV have insignificant seasonal variations, thus making PV and wind very reliable.

A case study (see Fig. 11) to show the correlation between PV and wind was conducted for one area with five transmission substations where PV and wind is fed in. The results show that PV and wind are not correlated, and they complement each other with wind coinciding with the evening peak load; this pattern and behavior of wind and PV is similar throughout the country, thus allowing also for higher possible co-integration capacity of wind and PV plants because they don't reach their peak generation at the same time. It is also clear that individual plants have high variability, high ramps, and are unpredictable, however, the aggregated profiles exhibit low ramps, low volatility and thus predictability.

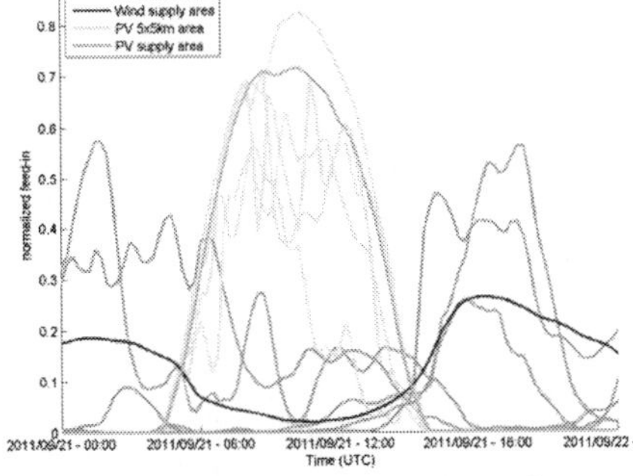

Figure 11: PV and wind diversity in a selected area

In addition, a grid capacity study was done to evaluate the grid capacity requirements for areas where PV and wind plants are integrated at one point. The study highlighted that up to 20% grid capacity can be made available at each transmission point if wind and PV are integrated, if no load is considered at the integration point, thus making the results very conservative. Fig. 12 shows the results of the grid capacity assessment. The current approach to grid capacity assessment is based on the assumption that PV and wind are highly correlated and therefore peak at the same time.

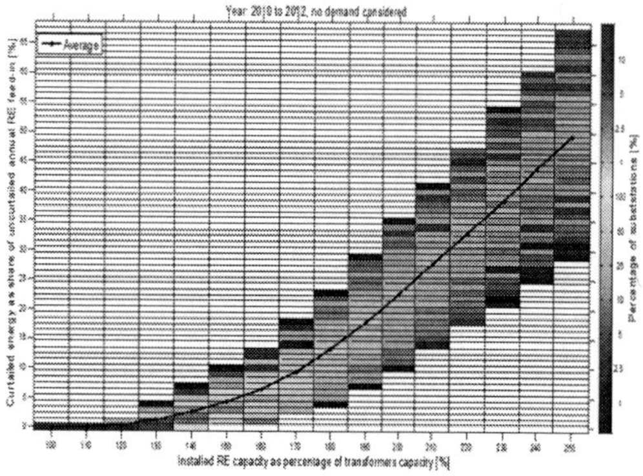

Figure 12: Grid capacity assessment

4.2 Conclusion

The study results provide the quantified fact base that will allow higher acceptable penetration levels of wind and solar PV into the South African power system. Individual solar PV and wind farms exhibit strong short-term (intra-hour) variability. However, the portfolio effect of aggregating the power output of several farms spread over a wide area leads to disappearance of intermittency and short-term fluctuations. The aggregated wind portfolio power output exhibits intra-day variability with changes over several hours. This level of variability is more manageable from a system operator's point of view.

- Solar PV and wind energy are very low-cost bulk energy providers in South Africa; resource potential far exceeds the current and future electricity demand
- Cost not a barrier anymore: new wind now costs 0.6 R/kWh (< 5 €ct/kWh) and new solar PV costs 0.8 R/kWh (6 €ct/kWh), based on actual PPA tariffs
- Short term fluctuations in the aggregated solar PV and wind power feed-in are significantly reduced by wide spatial distribution. Although PV is main cause of 15-min gradient changes, it is highly predictable.
- Solar PV and wind supply complement each other; PV high during mid-day and wind is higher in the evening to night time.
- At least 20% additional wind/PV power can be installed per substation without any curtailment of wind/PV power (40% with 5% curtailment) – no load considered, thus conservative
- Up to 30% energy share of VRE will not increase short-term ramps in the system significantly if there is a balanced combination of PV and wind in the system

5 References

[1] Wind and Solar PV Resource Aggregation study for South Africa, CSIR, March 2016 (http://www.csir.co.za/Energy_Centre/wind_solarpv.html
[2] Integrated Resource Plan 2010 -2030, South African Department of Energy
[3] F. Jones, Beyond Integration – three dynamics shaping renewable energy and the grid, a global industry report, 2015, (DNV GL)
[4] Eskom interim integrated report 2014 (http://integratedreport.eskom.co.za/Eskom_interim_integrated_report_30_Sept_2014.pdf), page 51
[5] Eskom interim financials 2014 (http://integratedreport.eskom.co.za/Eskom_condensed_interim_financials_30_Sept_2014.pdf), page 14
[6] IRP Update (http://www.doe irp.co.za/content/IRP2010_updatea.pdf), page 68 %20Methodology.zip), page 21
[7] South African Department of Energy, announcement of results of bidding windows of the Renewable Energy Independent Power Producer Procurement Programme (REIPPPP)

TECHNO-ECONOMIC OPTIMIZATION OF PHOTOVOLTAIC PLANT LAYOUT BY USING DESIGN OF EXPERIMENTS TECHNIQUES

Sascha Nadja Ringlstetter[1,2], Lukas Haack[1], Fritz Dildey[2], Lutz Sommer[1], Richard Meyer[1]

[1] Suntrace GmbH, Brandstwiete 46, 20457 Hamburg, Germany, +49 40 767 96 38 0, nadja.ringlstetter@suntrace.de, lukas.haack@suntrace.de, lutz.sommer@suntrace.de, richard.meyer@suntrace.de
[2] Hamburg University of Applied Sciences (HAW), Ulmenliet 20, 21033 Hamburg, +49 40 428 756244, f.dildey@ls.haw-hamburg.de

ABSTRACT: The conceptual design of utility scale photovoltaic power plants is commonly realized by using commercial or non-commercial simulation software tools for energy yield calculations.
Systematic techno-economic optimization by variation of several input variables with a reasonable step size can lead to a very large number of simulation runs. Also more detailed simulation algorithms increase the computational time.
Performance approximation of different plant layouts by applying Design of Experiments (DOE) methods and multiple linear regression models help to reduce required simulation cases and to identify the optimal plant layout more efficiently. Instead of step by step variation of all input parameters, only specific parameter configurations (design points) are simulated, which provides sufficient information for developing accurate regression models for approximation of the respective annual energy yield and Levelized Cost of Electricity (LCOE). The LCOE is used as measure for comparison of possible variants and identification of the optimal plant configuration.
High accuracy of the regression models could be achieved and demonstrated in a case study. With only 95 simulation runs for determining the approximation function, LCOEs of several thousands different layouts have been calculated with an uncertainty lower than 3%. Computation time has been reduced by a factor of five compared to a direct reference calculation, which corresponds to saving several hours of worktime.

Keywords: PV system, Simulation, Utilities

1 MOTIVATION

During the techno-economic optimization procedure for utility scale PV plants, several different design configurations are examined. When simulating the PV plant's energy production, by using detailed simulation software or applying high time resolutions, the calculation time increases significantly. The time required for the optimization procedure also depends on the computing power. Minimizing the number of required simulation runs and estimating all remaining possible plant configurations with satisfying accuracy can reduce the time and hardware effort. A regression model can be used to predict different output variables, such as annual energy yield, Performance Ratio (PR) or directly Levelized Cost of Electricity (LCOE). Furthermore, influence of single input parameters on the output variables e.g. over dimensioning of PV array, row distance or other factors can be analysed systematically.

This fast approximation allows analysing a PV project in conceptual design state systematically and prevents biasing the result of the techno-economic optimization by too strict limitation of the input variables.

2 METHODOLOGY

The proposed approach is based on Design of Experiments (DOE) techniques, which aim to describe a system's behaviour with as few simulation runs as possible by gaining most information for system approximation via regression models. First, a set of plant configurations (design points) is determined and simulated. Based on response results for the design point a suitable regression model is configured.

2.1 Selection of factors and responses

The first step is defining the system onto which the DOE method shall be applied. In this case, the system is a utility scale PV plant, or more accurate the model of the plant used by the PV simulation software. The selected factors and responses form the basis for DOE design. Plant parameters, which are not varied due to either software restrictions or minor importance regarding the impact on the response, are not considered in the design and are held constant. Whereas factors affecting the response significantly are considered in the design and varied according to selected design points during the simulation. A 50 MW PV power plant in sunbelt region was assessed in a case study with variation of four factors as presented in the following table I. As response values, the annual energy yield and Levelized Cost of Electricity (LCOE) have been chosen.

Table I: Factors varied in 50 MW plant case study (module materials mono-Si: monocrystalline silicon, mc-Si: multicrystalline silicon, CdTe: Cadmium Telluride; sheds width: 4 m for mono- and multicrystalline modules, 2.4 m for thin film modules)

	Type	Min	Max
Sheds tilt	continuous	10°	40°
Sheds distance (pitch)	continuous	4 m	14 m
DC/AC sizing ratio	continuous	0.8	1.6
Module technology	categorical		mono-Si/ mc-Si/ CdTe/

2.2 Selection of design type

The selection of an appropriate experimental design depends, among others, on expected response characteristic. When high order models are required for appropriate description of response characteristics, then more design points are needed to develop respective regression models. On the other hand, the maximum number of design points is limited by the affordable time and effort, also in this case of a computer experiment. The simplest experimental design is the full factorial design, in which all combinations of input parameters, each varied with a certain step size, are assessed. The resulting design points are uniformly distributed over the multi-dimensional design space formed by the design factors. Experimental designs suitable for computer simulations are random and space-filling designs, the most popular being Monte-Carlo design and Latin Square (Hypercube) design [1][2]. In this work, D-optimal and I-optimal designs have been applied due to their ability to generate designs with a particularly small number of design points, and due to literature recommendations indicating their suitability for computer experiments. The underlying principle of D-optimal designs is minimizing the determinant of $(X'X)^{-1}$ (with X being the factor matrix), which is equal to minimizing the volume of the joint confidence region on the regression coefficient vector. In I-optimal designs (also called Q-, V- or IV-optimal, standing for integrated variance), the average integrated prediction variance over the design space is minimized. [1][2].

2.3 Realization of design points

In order to obtain response values for each parameter configuration specified in the DOE design, the annual energy yield has been calculated with the commercial PV simulation software PVsyst (version 6.3.9). The time needed for each run is mainly influenced by the working principle and internal algorithms of the software, and by the plant configuration specified by the user (e.g. number of strings). In the conducted case study, each simulation run required approximately 30 seconds in batch mode.

The LCOE for each plant configuration has been determined using a company-intern cash-flow model based on standard financial assumptions.

2.4 Regression model function

After determining the results of PV generator design point simulations and financial modelling of LCOE, a multiple linear regression (MLR) is applied to determine an approximation model. In this case, least square method was used for generating polynomial model function also including interaction terms of input factors. The function represents a hyper plane in the k-dimensional design space of the k regressor variables and aims to approximate the true response of a system.

$$y = f(x_1, x_2, \ldots, x_k) + \varepsilon \quad (1)$$

Where:
$\boldsymbol{y}$ = Response
$\boldsymbol{x_i}$ = Coded variable (factor)
$\boldsymbol{\varepsilon}$ = Statistical error (deterministic experiments: bias)

The following example illustrates a second-order predictive model with $\boldsymbol{x_i x_j}$ being two-factor interaction terms of factors i and j ($i < j$) [3].

$$y = \beta_0 + \sum_{j=1}^{k} \beta_j x_j + \sum_{i=1}^{k} \sum_{j=1}^{k-1} \beta_{ij} x_i x_j + \sum_{j=1}^{k} \beta_{jj} x_j^2 + \varepsilon_i \quad (2)$$

After linearization of model terms, the model for prediction of response values $\hat{y}$ can be determined with least square regression.

$$X'Xb = X'y \Rightarrow b = (X'X)^{-1}X'y \quad (3)$$

$$\hat{y} = Xb = b_0 + \sum_{j=1}^{k} b_j x_{ij} \quad (4)$$

Where:
$\boldsymbol{X}$ = Factor matrix
$\hat{\boldsymbol{y}}$ = Predicted response
$\boldsymbol{b_i}$ = Regression coefficients

The errors between actual responses and predicted values are called residuals.

$$e = y - \hat{y} \quad (5)$$

A suitable model requires careful selection of relevant model terms and elimination of non-appropriate terms with the help of statistical measures and indicators. With regard to selection of models, backward and stepwise were proven to give better results compared to forward selection, which is in well agreement with literature recommendations [1][3]. A transformation of the response, usually a power or logarithmic transformation, may be advantageous by means of stabilizing the response variance, making its distribution closer to the normal distribution or improving the fit of the model to the design point values. They can be realized in the following form:

$$y^* = \begin{cases} y^\lambda, & \lambda \neq 0 \\ ln(\lambda), & \lambda = 0 \end{cases} \quad (5)$$

Where:
$\boldsymbol{y}$ = Response
$\boldsymbol{y^*}$ = Transformed response
$\boldsymbol{\lambda}$ = Parameter of transformation

The selection of base model order and model terms was evaluated with different selection criteria indicating the quality of the resulting predictive model. Visual tools were applied in order to identify discrepancies, such as normal probability plot, predicted vs. actual plot, residuals vs predicted and vs. factor setting plots. A box-cox plot was used as support tool for choosing an appropriate response transformation. Further measures for evaluating different combinations of model terms were

- Predicted Error Sum of Squares PRESS, an indicator for the prediction quality of the regression model,
- adjusted R^2, which is "normalized" with regard to number of design points and number of model terms,
- standard deviation of the model.

The regression function characteristics obtained with an I-optimal design based on 95 design points, meaning 95 simulations, are presented in table II. The assessment of the quality of obtained models with the help of these value was done by comparing them to other model term combinations determined with the same set of design point values or with the best fitting regression models for different numbers of design points.

Table II: Regression function characteristics for energy yield and LCOE as response values, both obtained with I-optimal design, 95 design points

	Annual energy yield	LCOE
Order of base model	quartic	quartic
Model term reduction	stepwise	backward
Response transformation	none	ln
Adjusted R2	0.99995	0.99996
PRESS	7.9E+12	9.4E-04
Standard deviation	1.6E+05	2.2E-03

2.5 Comparison with reference data

For final evaluation of quality of obtained predictive models, their response predictions were compared to a full factorial reference set covering the same design space as the predictive models that has been directly simulated and calculated. The steps and resulting number of simulations are given in table III.

Table III: Factors variation in full factorial reference data set, based on the case study (module materials mono-Si: monocrystalline silicon, mc-Si: multicrystalline silicon, CdTe: Cadmium Telluride)

	Range	Step size	No. of variations
For each technology:			
Tilt continuous	10°...40°	5°	7
Pitch	4m...14m	2m	6
DC/AC sizing	0.8...1.6	0.2	5
Sum:			210
Total sum (3 module types)			630

The predictive models obtained for different numbers of design points and with different types of designs (D- and I-optimal) were evaluated with respect to mean bias, standard deviation and maximum absolute residual of predicted responses compared to reference data.

3 RESULTS AND CONCLUSIONS

The maximum deviations between different predictive models and number of simulations are illustrated in figures 1 and 2. Both figures clearly indicate that a higher number of design points yields in smaller maximum bias. As both D- and I-optimal designs lead to comparably small deviations, both design types can be regarded as being suitable for this application with a sufficiently high number of design points. However, the quality of regression model is strongly correlated with selection of model terms as described in section 2.4.

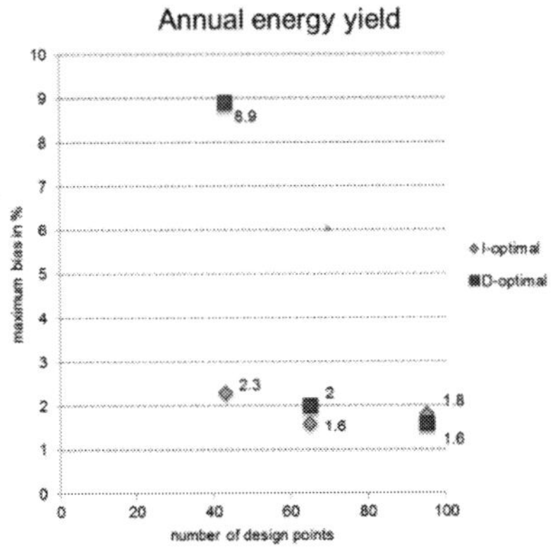

Figure 1: Maximum deviations of regression model predicted responses to reference values for annual energy yield

Figure 3 and 4 indicate that the regression models only slightly underestimate the responses for the I-optimal design based on 95 simulations. However, the model accuracy is definitely sufficient, as most absolute deviations are below 1 %.

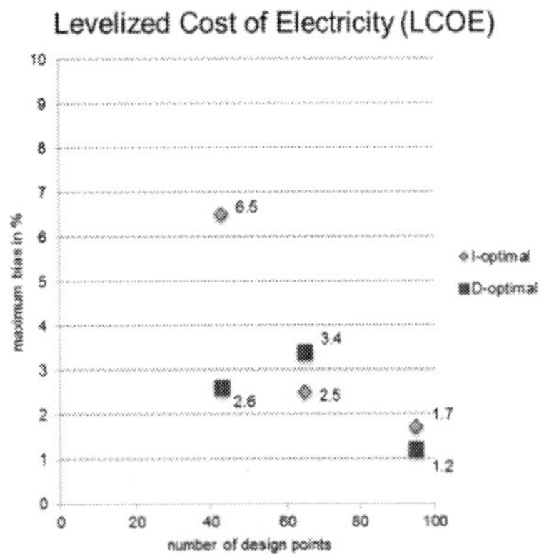

Figure 2: Maximum deviations of regression model predicted responses to reference values for LCOE

The uncertainty of the regression models can be taken into account by recalculating not only the plant layout with minimum LCOE as determined by the approximation function, but a larger number plant configurations out of the most favourable region of design space. When the top ten layouts are recalculated, it can be assumed that the real optimal layout is included in this set and thus can be determined.

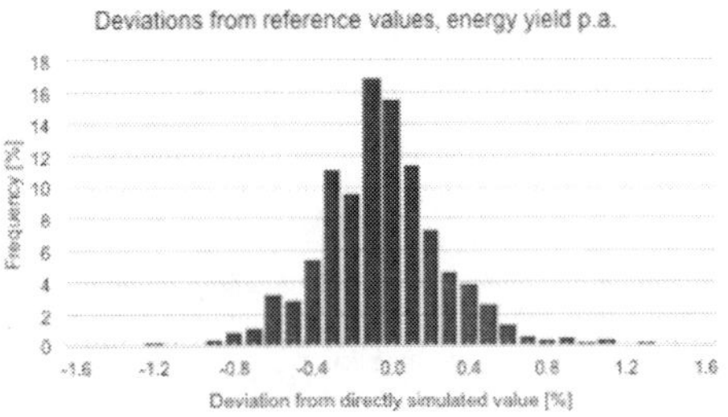

Figure 3: Distribution of deviations of predicted responses to reference values for annual energy yield, I-optimal design with 95 design points

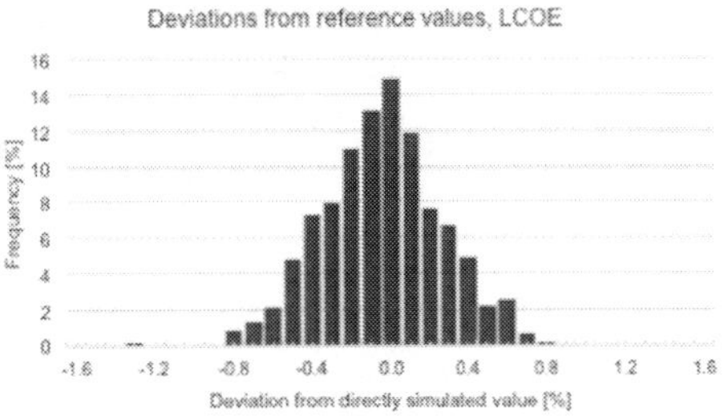

Figure 4: Distribution of deviations of predicted responses to reference values for LCOE, I-optimal design with 95 design points

Direct calculation of all 630 variants in the reference data set results in a time effort of more than 5 hours. The aim of the assessed approach was to reduce this time while maintaining a satisfying accuracy. With regard to the presented deviations, this target could be reached. The time needed for simulations was reduced by a factor of five to six. Some additional worktime is necessary for data handling and utilization of DOE software, on the other hand, a simple and fast possibility of analysing simulation results is provided by the regression function. One main advantage of obtaining a regression model function is the visualisation of predicted responses as three-dimensional surface plots and contour plots. With the help of the regression function it is possible to interpolate the response value for every plant layout within the design space. Visualisations indicating ranges of desirable or undesirable response values, which can be applied as decision support tool. Examples of such contour plots are provided by figures 5 to 8.

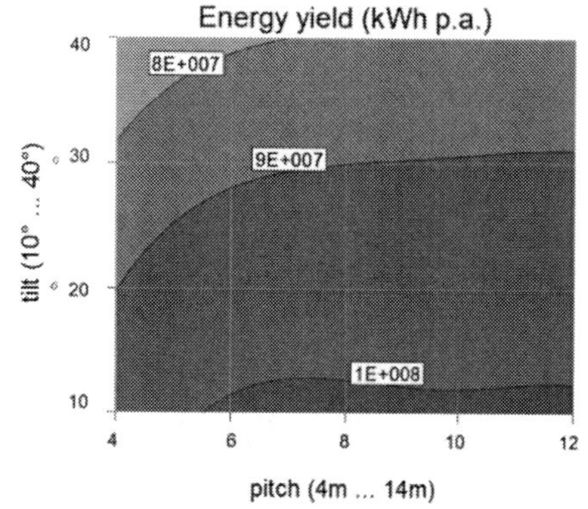

Figure 5: Interpolation of annual energy yield for different pitch and tilt values, at DC/AC sizing of 1.2 and mono-Si modules

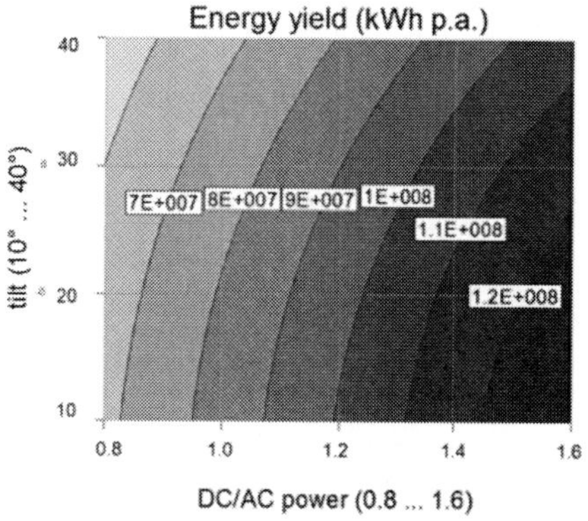

Figure 6: Interpolation of annual energy yield for different DC/AC sizing and tilt values, at pitch of 8m and mono-Si modules

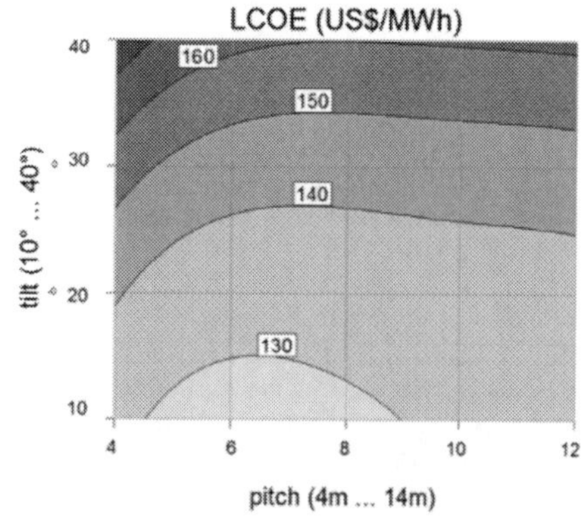

Figure 7: Interpolation of LCOE for different pitch and tilt values, at DC/AC sizing of 1.2 and mono-Si modules

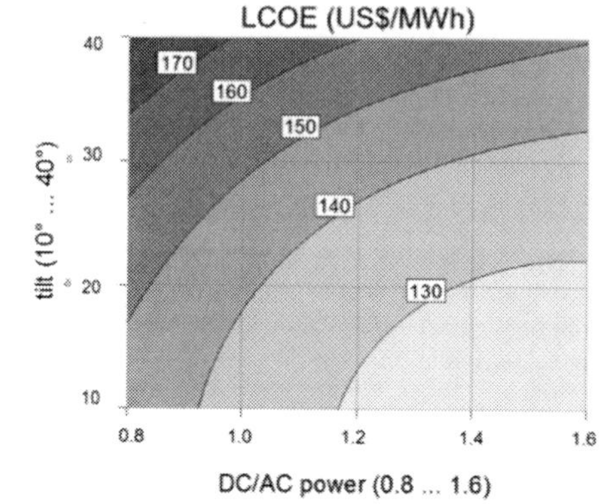

Figure 8: Interpolation of LCOE for different DC/AC sizing and tilt values, at pitch of 8m and monocrystalline modules

As illustrated by the contour plots, in this case study a small tilt angle, medium pitch and large DC/AC power sizing ratio is desirable in terms of minimized cost of electricity. All results are exemplary, as based on the applied technical and financial assumptions, such as site-specific solar resource, PV module characteristics and land costs. Especially differences between maximum energy yield and minimum LCOE are influenced by financial assumptions.

SUMMARY

The proposed method of applying DOE techniques for techno-economic layout optimization of utility scale photovoltaic plants shows high potential regarding saving of time and accuracy and reliability of results. D- and I-optimal designs with 95 design points resulted in standard deviations of 0.5% and 0.3% respectively and 1.7% and 1.2% of maximum absolute bias. An I-optimal design composed of 95 simulations and resulting in a polynomial regression function of fourth degree in terms of accuracy is most suitable for the examined case study. Time effort was reduced from more than five hours of simulation time to less then one hour, with some additional time for DOE software handling and data processing. However, the approach is based technical and financial assumptions. Knowledge of the underlying statistical methods is required for selecting the regression model terms, which significantly influence the quality of the predictive model.

REFERENCES

[1] Montgomery, D. C. (2009). *Design and Analysis of Experiments* (7th ed.). Singapore: Wiley.

[2] Siebertz, K., van Bebber, D., & Hochkirchen, T. (2010). *Statistische Versuchsplanung : Design of Experiments (DoE)*. Berlin, Germany: Springer.

[3] Myers, R. H., & Montgomery, D. C. (2002). *Response Surface Methodology : Process and Product Optimization Using Designed Experiments* (2nd ed.). New York, USA: Wiley.

NEW DESIGN CHALLENGES FOR UTILITY SCALE PV POWER PLANTS

Suzanne Santamaria[1], Esmeralda Palatnic[2], Felice Montanari[3]
Enel Green Power
[1]102 Rivonia Road, 5th Floor Block B, Sandton, Johannesburg, 2196, South Africa
[2]Av. Presidiente Riesco 5335, Piso 14, Las Condes, Santiago, Chile
[3]Via della Bufalotta, 255, 00139 Roma, Italy

ABSTRACT: In the last five years the size of utility scale PV plants has exponentially increased, while the awarded PPA prices or tariffs have exponentially decreased. Additionally, the demand for PV plants is aggressively growing in new and emerging markets and in new countries. In order to remain competitive, companies are faced with new design challenges in tough market and country conditions. The following paper examines the PV market changes, design optimization, the impact of market availability, construction planning and logistics, and the impact harsh environmental conditions has on a utility scale PV project.
Keywords: Large Grid-connected PV systems, PV Market, Design Optimization

1 INTRODUCTION

In order to remain competitive in the new and emerging markets, bidders are driven to optimize their project design, equipment selection, construction plan, and overall strategy to reduce the initial and operational costs for PV plants; while in parallel, the size of PV plants being built are growing exponentially. Even though the PV industry started from a scaled concept, the new design challenge is to consider site specific conditions and other factors imposed by the country and emerging markets.

2 EXPONENTIAL CHANGE

2.1 The Definition of Utility Scale

The demand and size of the utility scale PV plants has grown exponentially in just a few short years. For example, in 2012 the largest PV plant in operation by Enel was the 26 MWDC Stillwater Solar Plant in the United States. The largest solar power plant currently under construction in Enel's portfolio is the 254 MW Ituverava solar power plant in Brazil. Enel recently won a tender in Mexico, where the Villanueva and Villanueva 3 projects will have a combined installed capacity of 754 MWDC; these projects are expected to be operational in 2018 but the design will start now in 2016 [1].

This example shows that in just four years the plant size to be designed by Enel is already thirty times larger and that the definition of Utility Scale for PV plants is changing dramatically. This same example also shows that the location of the PV power plants is becoming more diverse and that PV is aggressively growing in the new and emerging markets.

Each market poses its own challenges based on regulations, tender rules, local content requirements, local market capability, and availability of materials or equipment. Projects are being built in different environments and harsh conditions, posing new design challenges and considerations.

2.2 The Decrease in the Price of Energy

Another exponential trend to observe is the decrease in the price of power in the PV market. When designing a utility scale PV plant the existing design for a smaller PV plant in theory can be scaled to the size desired. A contractor who can build a smaller 10 MW plant can also build a larger 100 MW plant. In markets new to renewable energy, this approach can be accepted.

However, the market quickly responds to the demand for renewables and the approved tariffs are reduced significantly. In South Africa for example, the Department of Energy manages a tender program for IPPs to bid annually to build renewable power plants. From the first bid in 2011 to the fourth bid in 2014, the awarded tariffs have dramatically decreased by 76%. This example is representative of the global PV Market trend, supported by PWC research.

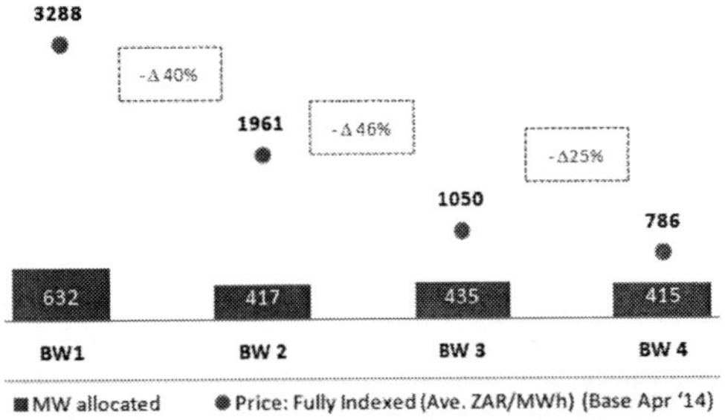

Figure 1: PV Bid Window (BW) Average Tariffs [2]

Therefore, companies building, owning, and operating utility scale PV plants must find new ways to reduce costs without impacting safety and design standards.

3 DESIGN OPTIMIZTION

3.1 Structural Design

Economies of scale also apply to the design decisions and the impact they have on the power plant cost. In a utility scale PV plant, every aspect of the design must be reconsidered.

For a fixed structure power plant designed to last 25 years or more, the best practice would be to have a bi-pole design, structural steel galvanized to 80 microns, set in concrete; however, it is not an economical design for a utility scale PV plant in competitive markets.

The structural costs can be reduced significantly when the design is changed to monopole, rammed foundation, with 25 microns of galvanization [3]. This change in design has many benefits; it increases the production rates during construction because there are less components to assemble, the use of a ramming machine allows for the work to progress faster than a design with concrete reinforced supports, removing concrete from the support structure design eliminates the need for additional concrete works on site, and reduces the amount of water required during construction on site.

Furthermore, the design should take into consideration the ease of assembly and installation of the supporting structures. The simpler a structural support is designed, the simpler and faster it will be to install. By improving the design to make the structural assembly leaner, the construction installation will also progress.

Figure 2: Example of a lean support structure design, in S. Africa

Another structural optimization is to consider the type of vegetation present at the project; in the desert, without grass or trees the height of the modules above ground can be reduced, significantly reducing the total structural steel quantities for the project [3].

Figure 3: Adams 2 PV Plant (82 MW) in S. Africa, showing that the site has been left as natural as possible, and that the panel height is above the natural grass and shrubs

In countries where local content requires the hiring of a large number of people during the operation of the plant, it becomes economical to consider seasonal tilting of the modules [3]. This allows for more production than fixed structures, and less maintenance and initial capex than tracking systems. The change to the basic monopole design from fixed to seasonal tilt is also minimal.

3.2 Electrical Optimizations

Considering the size of the new utility scale PV plants, another change is the material type for the grounding system. It is becoming more common to use aluminum in the grounding grid design because it is more cost effective than copper, and in some countries, reduces the amount of theft on site.

Small optimizations to the cable specification can have a larger impact on the material costs for the project. Choosing the proper cable size and optimizing the cable path directly impact the material procurement cost and construction costs to lay the material in trays or to bury the material in trenches.

In plants using tracking systems, there are design optimizations to consider. For example, the use of new wireless technologies like ZigBee eliminate the need for communication cables between the trackers and the conversion units. Optimizing the tracker engine location can reduce the length of cable and trenches, and the use of a larger tracker allows for more strings to be placed on a single structure. Eliminating the central axis trackers results in more trackers being in line with the north to south axis. This reduces the internal road requirements, and reduces the amount of trenches required during construction.

4 THE IMPACT OF MARKET AVAILABILITY

When designing a plant the local market availability must be considered. In many countries, bidding regulations require that a high percentage of equipment be purchased or manufactured in the country locally, which must be considered during the design. Equipment such as the medium voltage transformers, switchgear, current/voltage transformers, cable, and breakers can have limitations of availability. For example, the type and standard of medium voltage switchgear can be arc flash resistant in one country, SF6 insulated in another, or only air insulated elsewhere. Misalignment with the local market capabilities can have a large impact on the project cost when the design requires a specific product that is not available in the country, to be manufactured in that country.

The availability of equipment and materials from country to country causes companies and their engineers to re-examine best practices in design; an optimization in one country or market may not be the best solution in another. For example, the use of dry-type transformers in Italy is considered a best practice because the power plants are unmanned and the regulations require that no oil is on site to avoid possible environmental issues.

For larger plants, oil transformers are more efficient, reliable, and have a longer life, but do require containment, fire protection, and more manpower for inspections for possible oil leaks. Over the life of the plant oil type is generally considered to be more economical.

In the US and South Africa, the regulations or local content requirements, respectively, require that the plants are manned. Therefore, in the long term operational costs manpower is required regardless of the presence of an oil cooled transformer. In South Africa the labor costs are significantly less than in Italy or in the US, contributing to the oil cooled transformers competitiveness in the country.

5 CONSTRUCTION PLANNING AND LOGISTICS

As the size of PV plants being constructed is growing exponentially, there are more issues with how the plant will be built. Companies need to evaluate what equipment will be delivered, how to organize the laydown or storage area, if site camps are needed, where to place the assembly area for the structures, etc.

Figure 4: Chile Carrera Pinto PV Solar 97 MW

PV modules typically cannot be stacked, double handling of modules and other main equipment should be avoided to reduce risk of damage. Depending on where materials are stored internal plant roads may need to be improved or widened. Therefore, the design of the plant layout and the site organization plan is very important right from the start.

In order to mitigate delayed equipment deliveries on site, it is important to monitor the factory progress in designing, manufacturing, and testing of components. It is necessary to visit factories to determine if they meet company standards, if the production is aligned with the contractual or project needs, and to witness the acceptance testing to ensure the quality of the product.

5.1 Transportation

Smaller PV plants can be built near cities as they require less land. Larger utility scale PV plants require more land and are typically built in remote or rural areas. This extended distance from the manufacturing facilities has an impact on the equipment design.

When equipment needs to be transported a long distance, such as 2000 km, it can impact either the design or the transportation costs. The roads in the remote locations are commonly damaged with potholes, and portions of the site access are typically dirt roads. When equipment arrives on site, it has been subjected to additional vibrations and movement during transportation.

For example, a conversion unit skid with a precast concrete base is a combination of a flexible metallic structure being mounted to a rigid structure. Preassembled conversion unit cabins are a best practice because they reduce the construction time on site, and considering that the works are done in a factory, reduces the need for concrete work, wiring, and testing in the field. However, when the equipment arrives on site after being transported over a long distance, the flexible walls undergo a misalignment that breaks the sealed housing, allowing rainwater and dust to enter, resulting in the need for field repairs.

These challenges prompted a change in the design to remove the inverters and transformers from the conversion unit and place them outdoors [3], in order to reduce the overall size of the conversion unit being transported.

Therefore, each plant's road conditions and distance from the manufacturing facilities needs to be considered case by case to determine the best solution for the design of the equipment not only during operation, but also during the transportation.

5.2 Personnel

The construction of a utility scale PV plant is labor intensive; for a thin film PV project size of 82 MWDC such as the Adams 2 PV Plant in S. Africa, there is a peak of 800 people on site. One consideration is the capability of people to perform the works, spread out over 210 ha of land. Contractors who are more experienced and organized reduce the learning curve and production ramp up rates. If the country requires that a high percentage of the laborers be from the local communities, contractors need to have the availability of people to perform the work.

In this case, many of the people will require training by the contractor. In some countries, the laborers from the local communities cannot read and it's necessary to provide training videos.

It is important to plan the number of people needed, the types of teams required, and the number of supervisors needed to ensure that there are the right people and enough people on site to perform the works. In a larger utility scale PV plant, there is a higher number of drawings and documents to be managed by the engineering and site team, typically resulting in the need for a dedicated design documentation administrator. Due to the large site of the new PV Plants, it is also necessary to increase the number of engineers and supervisors for the project; for example, one electrical supervisor to manage the substation and switching station area, while another electrical supervisor manages the PV field. Some projects use such a large space that the site conditions vary from one side of the plant to another. This change results in different designs for the plant, and subsequently different construction methods for the installation to manage.

Furthermore, the infrastructure required to support a high number of laborers must be considered. Contractors may need to provide transportation for the laborers, personal protection equipment, work clothing, and meals.

5.3 General Optimizations

In order to reduce cost in the utility scale PV plants, a new design improvement is to leave the site as natural as possible, only removing shrubs or trees that would affect the performance of the PV modules [3]; this eliminates the need for the initial site preparation, clearing, grubbing and grading, and the need for a site finish using imported gravel. Of course in order to do this, the type of soil must be studied to consider the impact dust or soiling will have on the module performance.

6 HARSH CONDITIONS

There are new design challenges to consider as projects are being built in environments new to the PV market. These new environments are causing engineers to consider site conditions that are outside of the equipment's normal operating range, defined by the manufacturers. The same site conditions have impacts on the personnel and equipment used to construct a facility.

6.1 High Wind Speeds

Projects built in high elevations, in the desert, or in large open areas that are generally flat are all subjected to higher wind speeds. One design solution is to implement a wind fence; doing so eliminates the need to design the plant support structures to withstand the high loads. An alternate solution is to design the first row of panels to be structurally reinforced, also serving as a wind fence to the rows of panels downstream of the predominant winds.

6.2 High Elevation

Elevation directly affects the inverter efficiency, reducing performance and capacity of the DC to AC ratio. When a project is built at a higher elevation, more inverters are required as the equipment will be de rated.

In very high elevations such as 4,000m above sea level, the laborers are also medically affected. The workers feel tired, fall asleep easily, and must return to a lower elevation after a period of six to eight hours. In order to meet schedule, contractors are required to have two or three shifts of workers to meet the same production levels of a project at normal elevation.

6.3 Temperature Swings

In both high elevation and deserts, the temperature delta the equipment will experience in a single day can cause high levels of thermal expansion and contraction. If the frame or clamps supporting the PV modules are not designed properly, the modules will crack. If the supporting structure isn't designed properly, this temperature swing can cause rotation of the support structure, causing unnecessary moments on the modules, and module damage/cracking.

6.4 Desert Areas

While high irradiance levels and little rainfall or cloud coverage are desirable PV plant conditions that the desert provides, the sun and heat contribute to unfavorable working conditions for the laborers. If they are not properly hydrated, given breaks to cool off in the shade, or if they are forced to work during the hottest parts of the day, people experience heat stroke, they may faint, become ill, or suffer from other medical conditions.

For projects built in desert conditions, there is usually a lack of water, which is needed for the workers, for construction works such as concrete or dust control, and is required for the panel washing. Contractors need to consider drilling boreholes to supply water in the area, or may need to transport and store water at the site.

The main equipment is also affected by the desert conditions. Module efficiency and conversion unit performance is negatively affected by higher ambient temperatures and dust. To prevent the dust from entering the conversion unit cabin and affecting equipment efficiency, larger air filters must be installed. If the air filters are not replaced or maintained and become clogged, the temperature inside the conversion units will increase resulting in lower efficiency of the equipment and overheating.

High levels of dust, called soiling can also negatively affect the PV module performance. In Enel's Chañares (40 MW) and Finis Terrae (160 MW) PV Plant in Chile, the company experienced severe soiling due to the fine dust and sand of the desert conditions. The dust covering the panels after three months of operation had affected performance by 45%.

Figure 5: Soiling in Chile

For plant with high levels of soiling, several methods for dust control needed to be tested. At Enel's plants in Chile, modified water trucks were used to spray the modules directly, dust control measures such as planting grass or watering the roads were tried, dry cleaning the modules was tested, and even an automatic robotic system was installed. Unfortunately, dry cleaning the panels was a negative result as the panels were scratched in the process.

Figure 6: Testing the use of a robotic panel washer

The desert soil conditions can also impact the plant design. Some desert soil conditions contain high levels of iron and magnesium, which are conductive and have caused electrical arcs in the conversion units. Other desert soil types present bad conditions for conductivity, requiring a larger grounding system to be installed. Equipment must be designed to consider all negative impacts the external environment can have on the equipment during operation.

6.5 Freezing Conditions

At the Enel Sheldon Springs PV Plant in Vermont, USA (2.4 MW), the weather reached below freezing temperatures, causing the ground to freeze. The result was that several support structure stanchions and their foundations heaved above ground level and twisted, adding undue stresses to the array architecture.

Figure 7: Concrete support heaved above ground level due to freezing conditions (Vermont, USA)

This case had to be studied in order to improve the design for future projects, such as Enel's Aurora plant.

Enel is building a cluster of 16 PV Plants to total 100 MW called Aurora, in Minessota, USA, an area which is characterized by below freezing temperatures and heavy snowfall. The construction schedule took into consideration the harsh working conditions and planned for the peak labor period to occur during the non-winter months

7 SUMMARY

The way to design a utility scale PV plant has shifted to a customized approach instead of simply scaling the design. During the initial phases of a project, optimizing the structure design to reduce the total amount of materials to be procured and installed must be considered.

Materials must be suited to the application and not over designed to avoid additional costs to the project. Taking advantage of new technologies and standards can also eliminate the need for additional material procurement and installation.

Researching the types of equipment and materials that are available in a country and using the locally provided materials can also positively impact the project economics, construction schedule. Furthermore, using locally sourced products increases the response time if the equipment or materials needs to be repaired or replaces, and typically increases the availability of technical support during the start-up, testing, commissioning and operation of a PV plant.

The equipment delivery and transportation to site must also be evaluated as an input to the equipment design and selection. If equipment must be transported over long distances then it should be considered in the design in addition to the requirements of the equipment during operation.

Engineers need to evaluate the local requirements and if local labor is required, and if the plants must be manned, then the design can take advantage of the availability of people.

Lastly, site specific environmental conditions such as elevation, high wind speeds, excessive heat and dust must be evaluated and designed for project by project. While the main components and size of two plants can be similar, no two projects are ever the same.

The demand for utility scale PV will continue, and engineers will be required to continue to face new design challenges in even tougher contexts.

REFERENCES

[1] Enel Green Power, Media and Investors, press releases, accessed 5 June 2016, <http://www.enelgreenpower.com/en-GB/media_investor/press_releases/index.aspx>.

[2] Republic of South Africa Department of Energy, Renewable Energy IPP Procurement Programme, Bid Window 4 Preferred Bidders' Announcement (16 April 2015)

[3] Enel Green Power, Solar Design Unit, General Technical Specifications and Annexes (2016)

INTEGRATION OF PV TO INDUSTRIAL CONSUMERS WITH MULTIPLE GRID SUPPLY AND ENERGY MANAGEMENT IN LEBANON AND PALESTINE

Maria Anzizu[1], Guillermo Velasco[2], Hassan Harajli[2], Imad Ibrik[3] and Xavier Vallvé[1]
[1]Trama TecnoAmbiental, Av. Meridiana, 153 – E-08026 Barcelona – Spain
Tel. +34 93 446 32 34 – maria.anzizu@tta.com.es and xavier.vallve@tta.com.es;
[2] Consorci Escola Industrial de Barcelona – CEIB - guillermo.velasco@upc.edu;
[3] United Nations Development Programme-Lebanon, Beirut, Lebanon - hassan.harajli@undp.org;
[4] Energy Research Center, An-Najah University, Nablus, Palestine - iibrik@najah.edu.

ABSTRACT: Several countries of the Mediterranean Area face the increase in the energy price and the weakness of the electrical grid (limited increase in power, interruptions, etc.), which undermines security of supply in critical facilities like hospitals, schools, etc. and creates financial burdens on businesses. These electricity supply problems are common to Lebanon and Palestine because of scarce quality and regular interruptions and cuts-off. During those periods that the utility grid is absent, electricity is supplied by diesel generators (gensets). As an alternative to using, exclusively, diesel fuel generators as back-up, PV hybrid power plants, with or without storage, have been demonstrated to be a cost effective solution whilst reducing pollution and energy dependency.
The paper highlights the methodology, solution and lessons learned in Lebanon and Palestine during the development of two Projects funded by the European Union (EU): "MED-Solar [www.medsolarproject.com]" and "CEDRO IV [www.cedro-undp.org]" for the integration of PV to industrial consumers with multiple grid supply and energy management. Within the framework of the projects, 12 adapted PV hybrid demonstration plants have been constructed in Lebanon (9), Palestine (2) and Jordan (1) and are presented in the present paper.
Keywords: Design, Grid Integration, Grid Stability, PV market and Strategy.

1 INTRODUCTION

Several countries of the South Mediterranean Area face an increase in the price of electricity and weakness of the electrical grid (limited black outs, etc.), which undermines security of supply in critical facilities like hospitals, schools, etc. and creates financial burdens on businesses. These electricity supply problems are common in Lebanon and Palestine.

1.1 Lebanon electricity context

The total real installed power in the country's national grid is of 1.7 GW and the peak power demand of 2.8 GW, representing a gap of 1.1 GW. The electricity demand is estimated at about of 15 TWh, whereas the energy supplied by the grid is of 11.5 TWh (2010) [6]. The difference of approximately 23% between demand and grid production is supplied by a large number of private diesel generators. The energy delivered by the grid was primarily produced by thermal plants supplied by diesel (91.2%). Only the 6.6% is produced form RE (hydropower) and the remaining 2.2% was imported from neighbouring countries (Egypt and Syria).

Scheduled load shedding grid outages take place daily [7]. The average duration of outage is 5.2 hours, but it is higher in rural areas than in urban areas.

1.2 Palestine electricity context

The energy demand in Palestine is of 5.4 TWh, being 91.4% of this amount imported from neighbouring countries (Jordan, Israel and Egypt), whereas the remaining 8.6% is produced by a single thermal plant located in the Gaza Strip [8].

The IFC [7] reports that the number of power outages in a typical 2013 month is 8.7, and its average duration is 3.7 hours.

The above data highlight the dependence on fossil fuels for electricity generation and the dependence on other countries for energy resources supply. In this regard, the Palestinian case is particularly critical because, besides the need of importing fuel, nearly 91 % of the electricity consumed in the country has to be also imported.

1.3 Challenge

Given the high solar resource in the region, PV hybrid power plants, with or without storage, can be a cost effective alternative whilst reducing pollution and energy dependency.

Potentially, the market is of high interest. However, to achieve the market maturity, the following barriers should be overcome:

- Potential target client confidence
- Bankability of projects
- Local maturity in technology supply chain
- Specific technological challenges

The two complementary projects funded by the European Union (EU): MED-Solar and CEDRO IV have identified the market potential and are accompanying it in order to overcome the barriers by working on the implementation of demonstration projects for such grid integrated solutions.

1.4 MED-Solar Project

The MED-Solar Project [1] is linked to the multilateral Cross-Border Cooperation in the Mediterranean Sea Basin Program (CBCMED) [2]. This program is part of the European Neighbourhood Policy (ENP) [3] and its financing instrument, named European Neighbourhood and Partnership Instrument (ENPI) [4] for the 2007-2015 period. The ENPI-CBCMED program promotes the cooperation between the EU and partner country regions placed along the shores of the Mediterranean Sea.

The MED-Solar Project has the objective to promote renewable energy (RE) use and improvement of energy efficiency contributing to addressing, among others, the climate change in three target countries: Lebanon, Palestine and Jordan.

The lead partner of the MED-Solar Project has been Trama TecnoAmbiental (TTA), and the partnership is made up of the following organizations:

- Universitat Politècnica de Catalunya – (UPC), Spain.
- SOLARTYS, Spain.
- Commissariat à l'Énergie Atomique et aux Énergies Alternatives (CEA), France.
- Energy Research Center (ERC), Palestine.
- United Nations Development Programme (UNDP) - Lebanon.
- National Center for Research and development (NERC) Energy Research Program, Jordan.

In their respective countries, all the partners are forerunners in the field of solar energy.

1.5 CEDRO IV Project

The UNDP, in partnership with the Lebanese Ministry of Energy and Water, initiated in 2014 the project CEDRO (Community Energy Efficiency and Renewable Energy Demonstration Project for Lebanon) UNDP-CEDRO's fourth phase (CEDRO IV) is funded by the EU [5], managed by UNDP and TTA provides engineering and technical backstopping services.

Its objectives address climate change through green-house gas (GHG) mitigation and enhance security of supply through increased dependence on national and natural resources and diversifying the energy mix, while engaging the private sector end-user, in collaboration with the public sector. To achieve this, the project works on three levels: implementation of model end-use energy efficiency and renewable energy demonstration projects for private sector buildings and facilities; setting-up of an enabling environment for the conversion of other buildings and facilities into energy efficiency; and assisting the development of a national sustainable energy strategy and action plan.

2 SCOPE OF WORK

The market development of PV hybrid power plants in the countries was still in the pre-commercial stage and demonstration projects were required to lure in the various sectors of the economy. These projects focused on PV plants for small and medium sized private and also public facilities. A target of 12 sites (9 in Lebanon, 2 in Palestine and 1 in Jordan), each with capacity in the range of 100 and 300 kWp (STC) were to be implemented to demonstrate the technical and economic feasibility of embedded PV plants and also as a strategy to help develop local capacities by technology providers.

2.1 PV hybrid plants

Traditionally, the grid integration of PV to a public grid or to a local genset based grid combines a stochastic intermittent source and one deterministic electricity source. In this application context, there are two stochastic intermittent sources (i.e. the PV and the utility grid) and one deterministic source (i.e. the gensets). In the case of Lebanon, the utility grid has a daily scheduled intermittency (with unscheduled ones), while in the case of Palestine the intermittency is more random and less frequent.

The PV hybrid pilot plants have the following strategy of operation:

When gensets supply the loads during black outs, the PV hybrid plant offsets the use of fuel and power may be occasionally curtailed or stored according to the operating range of the gensets. When the utility grid is on, PV generation is self-consumed and any short term surplus can be fed to the grid with a net metering agreement with the utility. In three of the nine plants built in Lebanon, there is storage and dual-mode inverters that are also used as uninterruptible power supply (UPS) for selected critical loads.

In the two plants in Palestine, where the use of gensets is less frequent, the PV embedded generation reduces the electricity consumption and the peak load from the grid combined with load management.

2.2 Energy Management System – microgrid controller

Having different PV hybrid plant architectures and strategies, has required the design and testing of an Energy Management System (EMS) within the MED-Solar project.

The EMS manages and dispatches energy sources, remotely monitors the plant and ensures the proper operation of the gensets. The EMS is also able to monitor, control and manage the loads (i.e. connect or disconnect deferrable loads or gensets).

3 METHODOLOGY

The projects have had a comprehensive strategy to address and overcome the market barriers identified. Although this paper focuses more on technical aspects it is relevant to list the main activities performed:

3.1 Short listing of potential contractors

During the different phases of CEDRO (since 2007), UNDP has built up local capacity through workshops and dissemination activities and established a short list of qualified contractors that is reviewed yearly. The criteria for selection include objective technical, management and financial aspects. Contractors have been invited to bid with a limit of 2 sites per contractor which increased the local maturity in the technology supply chain.

3.2 Bankability of projects

In Lebanon, the National Energy Efficiency and Renewable Energy Action (NEEREA) is a national financing mechanism that was established in 2010. It provides low interest loans (0.6-1.0%) through commercial banks with the support of the Banque du Liban (BDL), to cover the extra costs of energy efficiency measures and/or renewable energy investments. Moreover, the EU funded projects have financed part of the solution and have accompanied the beneficiaries and contractors in obtaining particular grants for co-financing the project. The project's due diligence has been based on the fact that project engineering and supervision was qualified and independent and managed by UNDP.

3.3 Pilot projects' site selection

From an open invitation to apply, beneficiary sites for the pilot plants were selected. The main mandatory requirements were: legal registration and financial stability of the firm, space availability (non-shaded vacant space) to place the PV generator, annual electricity demand of at least 140,000 kWh and the capacity for co-investment. The scoring requisites were the performance of energy audits and implementation of energy saving measures, or the social role of the facility and the capacity to have environmental certificates, among others.

The sites selected for both projects in Lebanon were: **EMKAN**, a wholesale vegetables and fruits market in Akkar (North Lebanon), the **René Moawad Foundation - Agricultural Center of the North** (RMF-CAN) for raw material processing implemented (dairy products and oil production and fresh cutting vegetables) Zgharta (North Lebanon), the **Academy Sports Club-Medrar**, educational institution for students from nursery to grade 12 from different villages in the South of Lebanon, and the Engineering Faculty of the **American University of Beirut** in Beirut.

Within CEDRO IV project, the selected sites have been: Campus of the **Holy Spirit University of Kaslik** (USEK) in Mount Lebanon, **Libanlait**, a dairy products factory in the Bekaa, **Libanjus**, a dairy and juice factory in Beirut, and **Gemayel Freres**, a cardboard factory in Beirut

In Palestine (and as part of MED-Solar), the selected sites are the **An-Najah University Hospital** and the **An-Najah Medicine Faculty** both in Nablus.

3.4 Feasibility studies and negotiation

A complete techno-economic feasibility study of the PV hybrid plants in Lebanon was performed assessing the sensitivity of PV capacity and the related impacts, enabling the potential beneficiaries to take the final decision of go or no go and to select the most suitable size.

Once the client's confidence was gained, a memorandum of understanding (MOU) was signed and the final engineering adapted design, facing any specific technological challenges, and specifications were completed.

3.4.1 Technical Assessment approach

Inception activities to determine the physical and operational conditions of the sites were performed through site visits and data gathering of electricity consumption at the beginning. The local grids were assessed with power analysers for several days in order to confirm the load profile of the facilities and the grid stability.

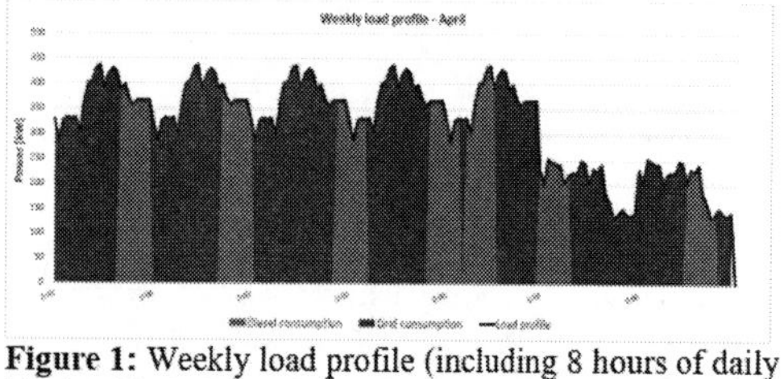

Figure 1: Weekly load profile (including 8 hours of daily blackouts) defined for a factory with 3 GWh yearly consumption

The load profile was crossed with the simulated PV electricity generation (in Figure 2, a 300 kW_{AC} PV hybrid plant) to assess:

- Solar electricity directly consumed (during normal operation of the grid and during blackouts)
- Solar electricity injected into the grid and belatedly consumed via net-metering.

Two midterm scenarios were also considered and simulated: 1. Blackouts persist and 2. National grid is upgraded and blackouts cease after 5 years.

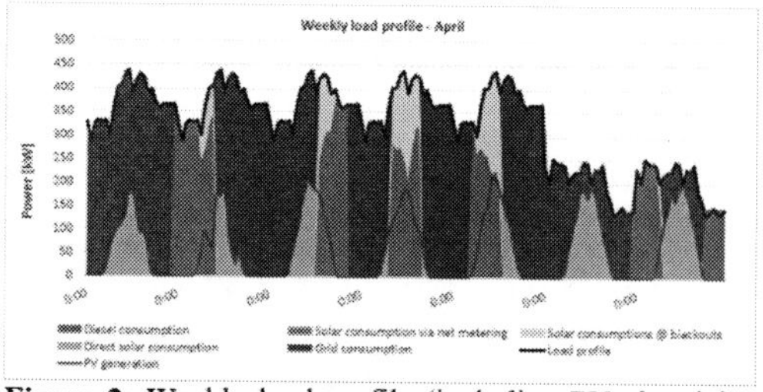

Figure 2: Weekly load profile (including PV electricity generation from a 300 kWp) defined for a factory with 3 GWh yearly consumption.

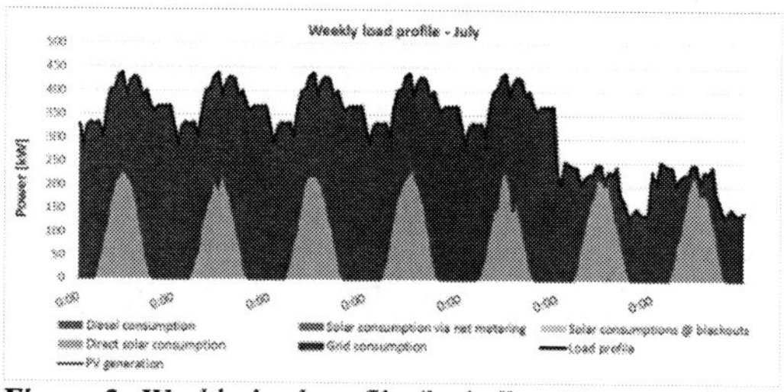

Figure 3: Weekly load profile (including PV generation from a 300 kWp) without blackouts after year 5.

3.4.2 Economic Assessment approach

Several assumptions referring to electricity prices were defined:

- The diesel and lubricant costs (i.e. diesel 0.87 USD/L and lubricant 0.018 USD/L in May 2014)
- Annual increase in the costs of diesel, lubricant and electricity from the national grid following the projected inflation of Lebanon (i.e. 3%).
- For the scenario where blackouts cease after 5 years, it has been assumed that electricity prices will increase by 30% in year 5, assuming that a 24-hour service will be offered.
- Electricity pricing from the utility grid for industrial consumers (i.e. time-of-use tariffs with seasonal variation.
- Co-investment from the final beneficiary.

The economic assessment included: the analysis of the savings, the cash flows, the levelized cost of electricity (LCOE) of the final solution compared to the LCOE at the facility without having the PV hybrid plant; the payback period (PBP); the Interest rate of return (IRR) and the net present value (NPV).

All the plants showed very interesting results (PBP close to zero, high NPV and low LCOE) given that the projects are subsidized and the project management and the engineering are covered by the MED-Solar and CEDRO IV projects. The investment from the beneficiaries has been relatively small (i.e. 10% of capital costs for public facilities and 50% for private facilities).

3.4.3 Environmental Assessment approach

The environmental impact was included in the feasibility, assessing the carbon footprint of the plant (i.e. equivalent CO_2 emissions), the equivalent CO_2 savings and the energy payback time.

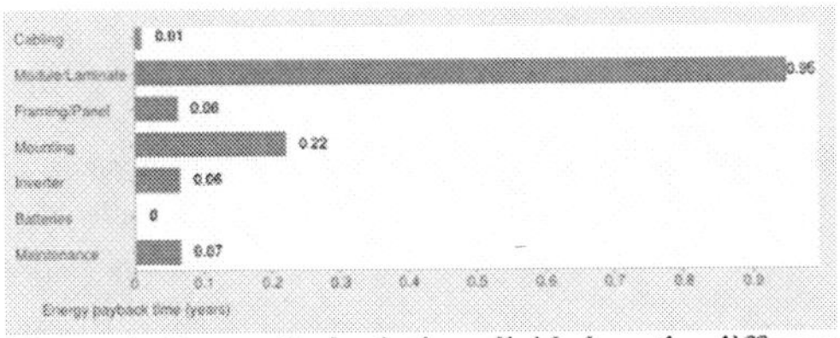

Figure 4: Energy Payback time divided on the different parts of the plant, depending on the amount of energy used to produce each part for the 300 kW_{AC} PV plant (source: LCA to go tool).

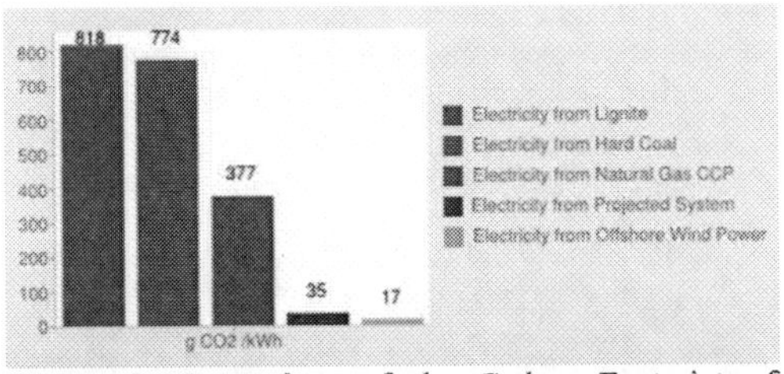

Figure 5: Comparison of the Carbon Footprint of different electricity technologies and the PV plant (300 kWp) per kWh (source: LCA to go tool, TTA).

3.4 Procurement

The technical specifications and engineering drawings for the procurement procedure were prepared and included in the invitation to bid for each site. For CEDRO, the selection of the contractor was done based on a Non-discretionary "Pass/Fail" criteria and awarded to the lowest price. For MED-Solar, a two stage procedure was followed. Once the eligibility of the proposals was confirmed, the solution with best value for money was awarded.

In the feasibility studies, a relatively high reference price, based on previous phases of the project, was considered because of the still inmmature local technology supply chain. Nevertheless, the offers that were received have shown a decrease as the project progressed with prices closer to more mature markets like the European.

3.5 Construction

Construction has been followed by Project Management of the local partners (UNDP and ERC) and the technical inspection and acceptance testing supervision by TTA. The technical inspection included the review of the installation shop drawings and the as built drawings; the technical report, the detailed planned schedule of works; a concise Environmental Statement specifying waste disposal arrangements during installation; a description of the proposed performance and acceptance testing procedure; and a letter certifying the requirements on warranties, spare parts and standards.

4 SOLUTIONS IMPLEMENTED

The adapted PV hybrid demonstration plants with or without a transient storage connected to the main grid or to the gensets have been constructed in Lebanon (9), Palestine (2) and Jordan (1 not described in this paper).

4.1 Lebanon

The PV hybrid solution was defined depending on the need of each facility:

1) In Lebanon, six facilities required PV-diesel hybrid solutions including features of genset fuel reduction and utility grid-connected (without storage). In this case the PV generator is not seen as a grid forming power source and uses only grid-dependent inverters. One of the 6 sites constructed has a PV capacity above the 60% of the nominal genset capacity, and an advance controller was specified in order to achieve adequate operation of the existing gensets (genset efficiency, minimum part load, spinning reserve, reverse current protection, etc.).

2) The other 3 pilot plants constructed are slightly different as these included short term storage in order to provide the security of supply during grid transition and integrated the EMS developed in the framework of the project. The architecture is presented in Fig. 6 where there are 2 PV generators, one one is connected to the battery inverter input node together with the non-critical loads, the grid and gensets, and the second PV generator is connected at the output node of the battery inverter along with the critical loads, creating a microgrid that can operate autonomously.

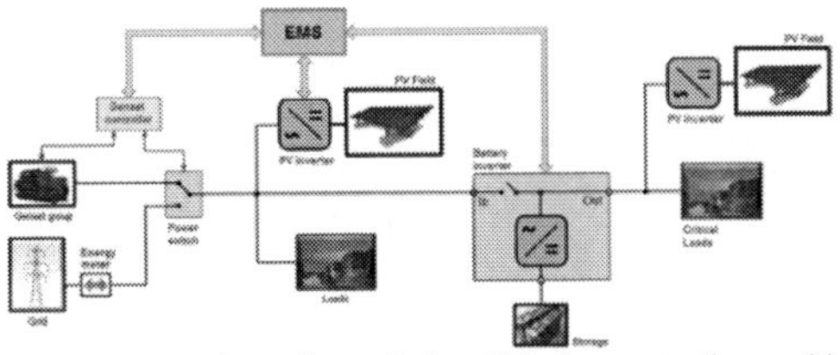

Figure 6: Schematics of the PV demonstration grid integrated plant with storage in Lebanon.

The three PV hybrid pilot plants with this architecture are:

EMKAN:

- PV capacity of 120.6 kWp
- Grid-dependent inverters:
  - Three-phase 6x20kW = 120kW
  - Single-phase (critical loads) 2x5kW = 10kW
- Battery capacity of 101,5 kWh made of two batteries, OPzS type (2.114 Ah, C_{10})
- Battery single phase inverter: 8 kVA

Grid interruption pattern is 3 times outage per day, 4 hours' typical outage duration.

Academy Sports Club-Medrar:

PV capacity of 116 kWp

- Grid-dependent inverters:
  - Three-phase 2x20kW+4x15kW = 100 kW
  - Single-phase (critical loads) 2x5kW = 10kW
- Battery capacity of 57,6 kWh made of two batteries, OPzS type (1.200 Ah, C_{10})
- Battery inverter: 8 kW

Grid interruption pattern is 3 times per day, 4 hours' typical outage duration.

RMF CAN

PV capacity of 136 kWp

- Grid-dependent inverters:
  - Three-phase 6x20kW = 120kW

  - Three-phase (critical loads): 20kW
- Battery capacity of 72 kWh made of two batteries, OPzS type (1.500 Ah, C_{10})
- Battery inverter: 3x5 kVA = 15 kVA

Grid interruption pattern is 3 hours' outage per day, 4 hours' typical outage duration

4.2 Palestine

The PV hybrid solution presented is the one constructed at An-Najah University Hospital. The use of gensets is used during outages for feeding the loads. During normal operation of the national grid, the the peak load is 170 kW.

The proposed architecture of the PV hybrid plant is shown in Fig. 7, where the connection point of the PV inverter and loads is located at the input of the battery inverter. In this architecture, the battery inverter rated power can be lower than the load peak power, and it also allows the possibility of critical load management, if required in the future.

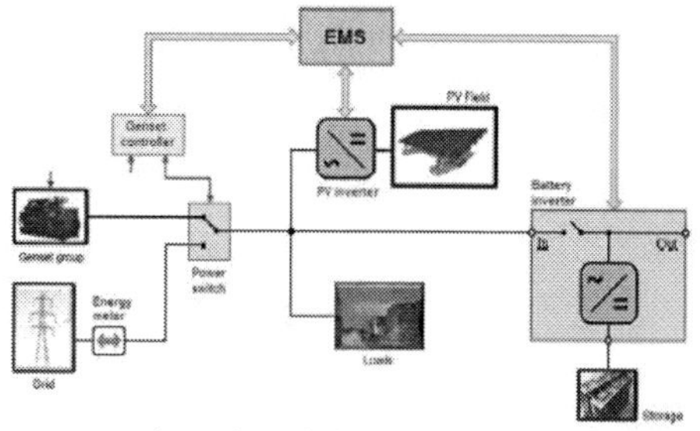

Figure 7: Schematics of the demonstration plant with storage installed in Palestine.

The PV hybrid power plant is as follows:

- PV capacity of 100 kWp
- Three-phase grid-dependent inverters: 100kW
- Battery capacity of 302,4 kWh made of two batteries, OPzV type (2x3.125 Ah, C_{10})
- Battery inverter: 6 x 8kVA = 48 kVA

Figure 8: PV hybrid power plant under construction at An-Najah University Hospital in Nablus, Palestine.

5 RESULTS AND CONCLUSIONS

Demonstration projects are key as part of a comprehensive market development strategy.

Involving also final clients and financial institutions has proven to show the way towards eventual private market deployment.

Grid integration to unreliable grids is possible but requires more site specific engineering than typical grid-tied solutions.

Twelve adapted PV hybrid demonstration plants in the range of 80kWp to 300kWp are integrated in the main grid and the gensets in Lebanon (9), Palestine (2) and Jordan (1):

- The plants have been built at industrial and public facilities, to improve their service quality and reduce their electricity expenses, as well as improve their environmental footprint, and have a better corporate image.
- Six different contractors with different brands of PV plants components have been involved and have reference projects that are being used to attrac new clients.
- Prices evolution indicate that supply chain is getting closer to maturity.
- EMSs – microgrid controllers has proven to be key to optimise the performance of this type of PV plants

Figure 9: PV hybrid power plant built at EMKAN, Akkar, north Lebanon.

6 REFERENCES

[1] http://medsolarproject.com
[2] http://www.enpicbcmed.eu
[3] http://ec.europa.eu/world/enp/index_en.htm
[4] Regulation (EC) No 1638/2006 of the European Parliament and of the Council of 24 October 2006 laying down general provisions establishing a European Neighbourhood and Partnership Instrument. Official Journal of the European Union, November 9, 2006.
[5] http://cedro-undp.org/
[6] International Energy Agency (IEA), "Lebanon: Electricity and Heat 2010". Available online at: http://www.iea.org/countries/non-membercountries/lebanon
[7] International Finance Corporation (IFC) - The World Bank. "Enterprise Surveys - What Business Experience". Data available online at: http://www.enterprisesurveys.org/Data/ExploreTopics/infrastructure
[8] Central Bureau of Statistics of Palestine (PCBS). "annual energy report for the year 2012". Data available online at: http://www.pcbs.gov.ps/.
[9] G. Velasco, F. Casellas, F. Guinjoan, H. Martínez and R. Piqué, "Grid-PV-diesel Hybrid System Management. Application to MED-Solar Project Scenarios," Environment and Electrical Engineering (EEEIC), 2015 IEEE 15th International Conference on, Rome, 2015, pp. 713-718.

PHOTOVOLTAIC AND BATTERY ENERGY STORAGE SYSTEMS IN SHOPPING MALLS: ENERGY AND COST ANALYSIS OF AN ITALIAN CASE STUDY

Grazia Barchi, Roberto Lollini, David Moser
EURAC Research - Institute for Renewable Energy Viale Druso 1, 39100 Bolzano (BZ), Italy
*Phone: +39 0471 055628; Fax: +39 0471 055699; E-mail: grazia.barchi@eurac.edu

ABSTRACT: Nowadays, due to the high diffusion and the lower cost of photovoltaic (PV) systems, there is an increasing interest in combining PV generators with battery energy storage systems (BESS) to improve load match and grid interaction, giving more flexibility to the building. The aim of this work is twofold. First, considering a specific building energy demand (i.e. a supermarket), an energy evaluation of the PV-BESS system and grid exchange is performed though dynamic simulations. Then, a cost analysis is carried out with the evaluation of the return of investment (ROI) both with and without BESS. Finally, the achievable percentage of carbon dioxide reduction with the inclusion of PV-BESS system is also estimated. The main purpose of this work is to show when the BESS becomes cost effective for the Italian market. The described analysis is part of the CommONEnergy FP7 EU project.

Keywords: *Photovoltaic, Grid Integration, Storage, Energy efficiency*

1 INTRODUCTION

Commercial buildings have large electrical and thermal loads, offering interesting opportunities to improve the energy efficiency and to reduce CO_2 emissions. Over the last few years, multiple shops and services, previously decentralized in different locations of a city, are now usually concentrated in shopping malls. Shopping malls are a very peculiar kind of buildings, because of the specific final energy uses (mainly related to lighting, ventilation, cooling and refrigeration), the complexity of the overall system, mainly driven by the need to ensure a certain sales volume, and the state of constant renovation. This means a challenging energy efficient managing.

For this reason, the CommOnEnergy EU FP7 project regards shopping malls as a great opportunity to increase energy efficiency through deep retrofitting. CommONEnergy relies on a comprehensive systemic approach involving innovative technologies and solutions as well as methods and tools to support their implementation [1]. The ambitious objectives of the CommONEnergy project are: i) to reduce energy demand by 75%, ii) to shave the power peaks and iii) to increase the use of renewable energy resources by 50%. The use of energy storage systems (ESS) and energy management systems (EMS) can effectively contribute to such objectives [1].

Typically, shopping malls have large areas (e.g. on the roof or in the parking lot) where photovoltaic (PV) systems of considerable size could be installed.

The goal of the analysis described in this paper is to understand and to explore how the combination of PV systems and BESS can contribute to meet the targets of the CommONEnergy project, when an increasing amount of renewable energy resources is available.

Similar analyses about the combination of PV generators with ESS in commercial buildings or supermarkets are reported for instance in [2] and [3]. However, while in [3] the authors deal with the method based on fuzzy logic for EMS control, the focus of this paper is mainly on the economic impact of the PV and BESS coupling in the Italian context, assuming net-metering, absence of incentives and tariff variability.

A starting point to evaluate the economic and energy benefits of introducing PV systems and ESS in shopping malls is to consider the relative overall consumption. The PV-BESS model is developed in TRNSYS, a modelling environment used to simulate complex dynamic systems, e.g. buildings and energy systems [4].

The paper is structured as follows: in Section 2 the system model of PV-BESS is described and the analyzed indices are defined in Section 3. In section 4 the simulation results in three different cases are presented. Finally conclusions and future works are reported in Section 5.

2 SYSTEM MODEL OF PV-BESS

Methodological approach foresees a parametric analysis that is performed by varying the PV nominal power (P_n) and the BESS capacity (C_{BESS}). The PV production is calculated using the meteorological and radiation data from the Weather Analytics database [5], while the consumption data were collected in a supermarket located in the city of Genoa, Italy. The simulations are performed over one year, with one-hour time resolution.

The PV-BESS system is modeled in TRNSYS using the built-in types of the electrical library [4]. The main components are: the PV system, the inverter/regulator and the battery bank.

The main input data to the PV system are the meteorological information collected in the Genoa area in 2013 and extracted from the Weather Analytics database [5], assuming to have PV crystalline modules oriented towards south with a 30° slope. The output electrical quantities are instead power, voltage and current at the maximum power point (P_{MMP}, V_{MMP}, I_{MMP}), computed as described in [4]. Once the PV parameters and the voltage of the inverter (e.g. 500 V) are set, it is possible to vary the nominal power by changing the number of strings in parallel. PV efficiency is η_{PV}=15%.

The battery model allow us to define the cell capacity and the number of cells in series and in parallel. Moreover, the total charge and discharge efficiency can be set, as wanted (i.e. 90% in our case study). The PV system, the battery bank, and the load are connected through the inverter/regulator.

The energy management system (EMS) implements different control rules according to the specific objectives. In the following, it will be shown that the EMS is able to exploit mainly the power produced by the PV system and to store the surplus into the BESS.

In the case of PV overproduction the surplus is stored into the batteries till the maximum state of charge (SOC_{max}) is reached. Afterwards, if the PV production is not enough to fulfill the energy demand, some power is drawn firstly from the battery and then from the grid. When the demand is completely supplied and the battery is fully charged (i.e. $SOC=SOC_{max}$), the PV overproduction is injected into the grid.

3 ANALYZED PARAMETERS

One of the main reasons motivating the use of PV-BESS in shopping malls is the intention to increase the exploitation of renewable energies, while decreasing the amount of power taken from the grid. In this sense, the most important two parameters to evaluate are: the self-consumption and self-production defined according to the following expressions:

$$Sc_{\%} = \frac{\sum_{i=1}^{8760}(E_{PV\_onsite} + E_{BESS})}{\sum_{i=1}^{8760}(E_{PV})} \times 100 \qquad (1)$$

$$Sp_{\%} = \frac{\sum_{i=1}^{8760}(E_{PV\_onsite} + E_{BESS})}{\sum_{i=1}^{8760}(E_{load})} \times 100 \qquad (2)$$

where E_{PV_onsite}, E_{BESS}, E_{load} and E_{PV} are the total energy over the year of PV system consumed on-site, BESS, the load consumption and PV system production respectively. Moreover, since shopping malls are usually located in city districts, it is important to take into account also the possible interaction with the grid and the role played in the future by shopping malls as flexible buildings. For this reasons, the amount of green energy produced and consumed on site, as well as the energy exchanged from or to grid over the year, is also estimated.

The sustainability of the system is quantified in terms of reduction of CO_2 emissions. Stated k_{CO2}= 0.483 as the coefficient of conversion of the electricity mix in Italy from MWh into tons of CO_2 as reported in [6], the relative reduction of CO_2 emissions is given by:

$$CO_2\ reduction_{\%} = \left(1 - \frac{CO_2\ w\ PVBESS}{CO_2\ wo\ PVBESS}\right) \times 100, \qquad (3)$$

where "*CO_2 w PVBESS*" and "*CO_2 wo PVBESS*" are the CO_2 emission with and without the PV-BESS respectively. From the economic point of view, it is interesting to compute the return of investment (ROI), taking into account the infrastructure costs (i.e. PV system, inverter, BESS) and the price of electricity. In our analysis the electricity price has been chosen on the basis of the 2013 Italian Unique National Price (PUN) available in [7] along with additional fixed costs. In particular, such fees are divided into three daily ranges, labelled as F1, F2 and F3, respectively. Moreover, in Italy the surplus of power generated by PV systems and injected into the grid is valorized through the net-billing scheme ("scambio sul posto", up to a maximum yearly amount equal to the electricity consumption) and is also included in the economic analysis. All considered costs are reported in Table 1.

Table 1 - Economic parameters

Items	Cost
PV system	2000 €/kWp
BESS	500 €/kWh
Price of electricity	PUN*+fixed costs
Net-metering scheme	100 €/MWh

4 SIMULATION RESULTS

Parametric simulations have been performed in order to analyze the energy-related and economic impact of PV-BESS systems considering three different shopping malls final energy uses:

- the total supermarket consumption (*Case A*);
- the supermarket lighting system consumption (*Case B*);
- the supermarket lighting system consumption, when this is shifted with the PV production (*Case C*).

In the following sub-sections the simulation results are presented.

4.1 Shopping mall energy consumption

Shopping malls usually comprise several shops (of various size and with different energy profiles) and at least one supermarket. Typically, a supermarket has a large area and dissipates a large amount of energy, due to the presence of constant loads, such as refrigeration cabinets, lighting, ventilation and air conditioning. The PV-BESS can match only partially the very large total demand, and this implies that a grid-connected system should be considered.

In the following, some results based on the measured consumption data in a supermarket located in Genoa, Italy, are presented. Starting from the measured data collected in 2013, the maximum and minimum values of active power demand are 257 kW and 67.6 kW, respectively. The nominal power of the PV system is 300 kWp, i.e. higher than the maximum peak consumption and achievable if all the available area is exploited. This scenario will be referred to as *Case A*.

Figure 1 shows the hourly power profiles (expressed in kW) of PV production (in red) and supermarket consumption (in blue), during a typical summer week (July in this case).

Assuming to install only the PV system or the PV system in combination with different storage capacity (C_{BESS} = {50, 100, 150, 200} kWh), the results of the simulations for the parameters described in Section 3 are summarized in Table 2.

By using the PV system only, self-production reaches a value of about 30%, and the self-consumption is near 100%, as expected. If the PV system is combined with the BESS, the self-production slightly increases.

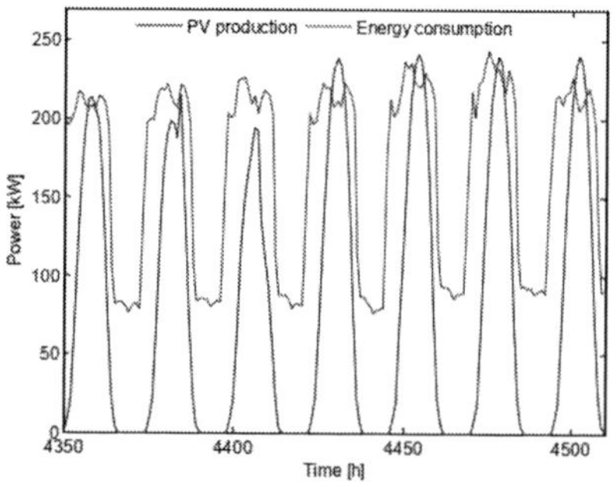

Figure 1 - Typical summer-week (July) profiles with generation (red) from a 300 kWp PV plant and supermarket consumption (blue).

Table 2 – Result values for 300 kWp of photovoltaic, different energy storage capacities and supermarket energy consumption.

Storage capacity [kWh]	**0**	**50**	**100**	**150**	**200**
Sp [%]	28.4	29	29.2	29.4	29.4
Sc [%]	93.0	93.5	94.5	94.8	95.0
Green-on-site [%]	28.40	28.80	29.04	29.16	29.22
To grid [%]	0.52	0.54	0.40	0.28	0.22
From Grid [%]	71.08	70.66	70.56	70.56	70.56
CO_2 reduction [%]	28.96	28.96	28.96	28.96	28.96
ROI [years]	7	7.52	8.18	8.55	8.92

Therefore, the largest part of the energy (around 70%) is drawn from the grid.
Since the fraction of energy taken from the grid is approximately the same in all the considered cases, the reduction of CO_2 emissions given by expression in (3) is on the same level (28.96%).
The last row of Table 2, shows the trend of the return of investment (ROI), which grows by about two years (from 7 to 8.92 years), when a 200 kWh BESS is installed.
Finally, in *Case A* the benefits of PV system in terms of self-production and CO_2 reduction are evident. On the contrary, simultaneous generation and demand as well as the net-billing scheme do not play in favor of storage systems.

4.2 Lighting energy consumption

The results in *Case A* show the advantages, but also the limitations of PV-BESS in a shopping mall, when the large energy demand of a supermarket is considered.
However, the shopping mall comprises several different loads, as mentioned in Section 4.1. One of the most significant is the lighting demand, which represents almost 40% of the total energy consumption.
In the rest of this section, the lighting system demand for the Genoa supermarket is considered to be simultaneous (*Case B*) and non-simultaneous (*Case C*) to PV production. The lighting profiles are simulated maximizing the use of natural lighting and dimmering the electrical lights, as described in [8]. The nominal power of the PV system in these cases is fixed to 100 kWp, since the maximum consumption is equal to 98.5 kW. A weekly example of *Case B* and *Case C* profiles are shown in Figure 1 and Figure 2, respectively.
Similar analyses are performed by varying the BESS capacity C_{BESS} = {25, 50, 75, 100} kWh. The results are reported in Table 3 and Table 4 for *Case B* and *Case C*, respectively.
As far as self-consumption is concerned, the benefit of the BESS is more evident than in the *Case A*. Indeed the *Sc* grows from 59% to 74% in *Case B* and from 31% to 45% in *Case C*. The reduction values of CO_2 emissions follow the same trend being strictly related to the on-site energy.
It is interesting to highlight the different, but consistent behavior of the renewable energy generated on-site and the amount of energy exchanged with the grid. In *Case B*, since the production and consumption are simultaneous, the increase of *Sc* corresponds to an increment of on-site renewable energy with a consequent decrease of the energy exchanged with the grid. On the contrary, *Case C* is favored by a higher interaction with the grid, because of the mismatch between the time when PV generation is active and the instant when the demand is satisfied. In this case, the deployment of BESS is more favorable. These results impact also on the ROI, which in *Case C* ranges from 11.7 to 13.7 years, while in *Case B* ranges from 9.3 to 11.3, as shown in the last rows of Table 4 and Table 3, respectively.
In practice, due to the presence of net-billing in Italy, the price of the storage is still too high, for having a positive effect on the payback time. Thus, the main question that arises is whether and when the cost of BESS becomes competitive.
Assuming that in the future the difference of the price between grid-in and grid-out (kWh produced by PV, virtually stored by the grid, used in a second moment) will become smaller, from Figure 4 and Figure 5 it is possible to have an idea (for *Case B* and *Case C* respectively) about how much the cost of the battery have

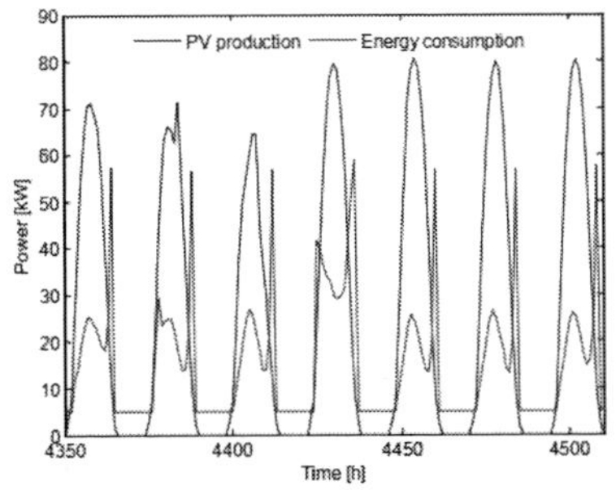

Figure 1 -Typical summer-week (July) profiles of generation from a 100 kWp PV system (red) and simulated lighting consumption (blue).

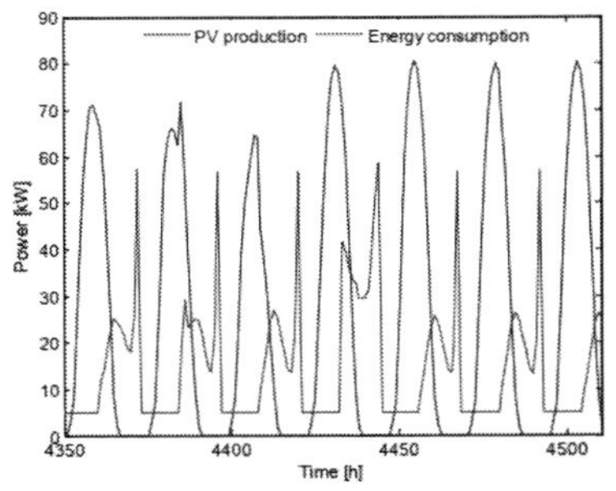

Figure 2 -Typical summer-week (July) profiles of generation from a 100 kWp PV plant (red) and simulated lighting consumption, not simultaneous with the PV production (blue).

Table 3 – Result values for 100 kWp of photovoltaic, different energy storage capacities and lighting energy consumption.

Storage capacity [kWh]	25	50	75	100
Sp [%]	44.0	46.2	49.3	55.2
Sp [%]	59.0	61.9	66.1	74.0
Green-on-site [%]	34.36	36.74	40.28	47.84
To grid [%]	21.90	20.47	18.24	13.38
From Grid [%]	43.74	42.80	41.48	38.78
CO_2 reduction [%]	44.00	46.19	49.27	55.23
ROI [years]	9.3	10.1	10.7	11.3

Table 4 – Result values for 100 kWp of photovoltaic, different energy storage capacities and lighting energy consumption.

Storage capacity [kWh]	25	50	75	100
Sc [%]	23.2	27.1	30.1	33.9
Sp [%]	31.0	36.3	40.3	45.5
Green-on-site [%]	15.45	18.64	21.17	24.67
To grid [%]	51.23	50.11	49.23	48.01
From Grid [%]	33.33	31.25	29.60	27.32
CO_2 reduction [%]	50.02	54.55	57.96	62.42
ROI [years]	11.7	12.2	12.8	13.7

to decrease. The trend of the ROI is given by changing the values of net-billing from 100 €/MWh to 0 €/MWh (blue lines), as the valorization of PV hourly surplus (up to a maximum PV electricity production equal to the consumption on an annual basis), when only the PV system is considered. Then, assuming to have for example 50 kWh of BESS, we calculated the relative suitable cost to have the same ROIs (green lines).

5 CONCLUSIONS

In conclusion, this paper presents some results about the possible use of PV-BESS in an Italian shopping mall. Three different scenarios, based on three different consumption profile are reported. The main consideration is that the PV-BESS system improves, accordingly to the analyzed cases, the self-consumption and reduce the energy exchanged with the grid. The ROI slightly increases from the scenario with only PV system and with significantly BESS capacity. However, currently the price of the BESS is still high and some consideration about the suitable cost, in case of simultaneous and non-simultaneous cost are also provided.

At the end the main consideration is that the PV-BESS system is and will be interesting also in Italy for commercial (e.g. shopping mall) buildings. This will be more advantageous if the PV-BESS will be associated to smaller load (e.g. EV-mobility) or if the cost of the BESS will decrease and/or the value of net-billing.

ACKNOWLEDGEMENT

The research leading to these results has received funding from the European Community Seventh Framework Programme (FP7/2007-2013) under grant agreement n. 608678.

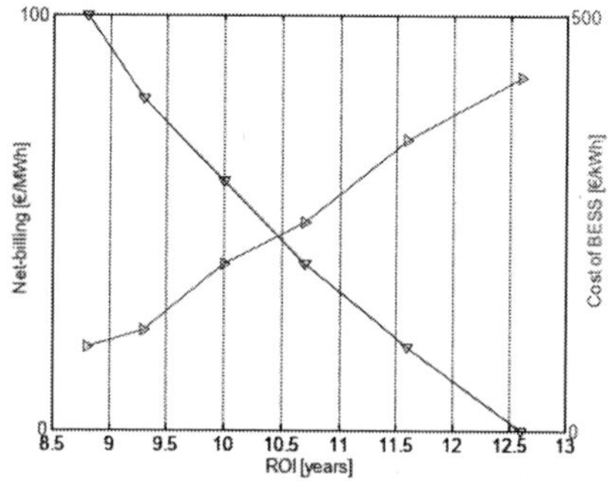

Figure 4 - Trend of the price of the BESS and net-billing over the ROI for 50 kWh of BESS, when the production is simultaneous with the consumption.

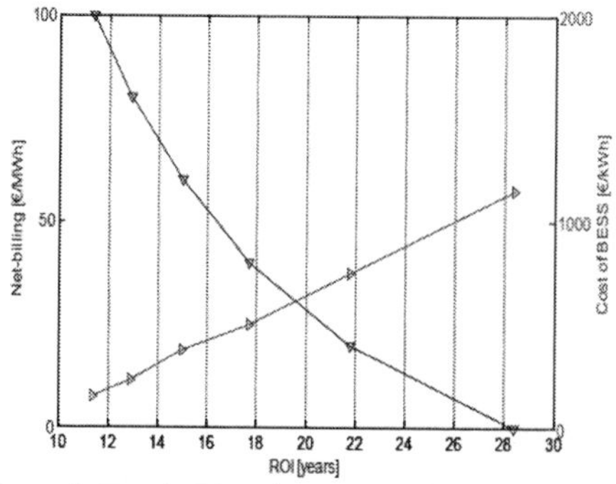

Figure 5 - Trend of the price of the BESS and net-billing over the ROI for 50 kWh of BESS, when the production is simultaneous with the consumption.

REFERENCES

[1] http://www.commonenergyproject.eu/

[2] Stadler, Michael. (2009). Effect of Heat and Electricity Storage and Reliability on Microgrid Viability: A Study of Commercial Buildings in California and New York States. Lawrence Berkeley National Laboratory.

[3] He Zang, Arnaud Davigny, Jonathan Sprooten, Benoit Robyns, Frederic Colas, et al.. "Energy Management Strategy for Commercial Buildings Integrating PV and Storage Systems". 2012.

[4] http://sel.me.wisc.edu/trnsys/

[5] Weather Analytics.

[6] Covenant of Mayors, "How to develop a sustainable Energy Action Plan (SEAP)", Bruxelles 2010.

[7] http://www.mercatoelettrico.org/it/

[8] G. De Michele, U. Filippi Oberegger, L. Baglivo, "Coupling dynamic energy and daylighting simulations for complex fenestration systems", BSA 2015, Feb. 2015, Bolzano.

ANALYSIS OF STATIONARY ELECTRICAL STORAGE SOLUTIONS FOR RESIDENTIAL DISTRICTS WITH HIGH PHOTOVOLTAIC PENETRATION

R. Völker, T. Kilper, F. Schuldt, K. von Maydell, C. Agert
NEXT ENERGY, EWE Research Centre for Energy Technology at the University of Oldenburg,
Carl-von-Ossietzky-Str. 15, 26129 Oldenburg, Germany
Corresponding author: R. Völker, rasmus.voelker@next-energy.de

ABSTRACT: Residential districts with high photovoltaic (PV) penetration can be equipped either with decentralized inhouse storage systems or with centralized district storage systems in order to increase the nearby self-consumed PV electricity. The objective of this work was to develop and to provide the methodical instruments to evaluate electrical storage systems for residential districts. As part of this work a case study based on a simulation model compared centralized and distributed storage solutions. It was found that the centralized storage solution exhibits several benefits like a significantly reduced storage capacity for reaching equal self-sufficiency rates and throughput (in this case study -18 % and -12 %, respectively). Moreover a conducted market analysis showed that larger battery storage systems also offer decreased specific system costs. Furthermore due to the boundary conditions given by the seasonal varying supply of solar irradiation and the specific load characteristics in residential buildings, the simulation series illustrated an economic optimum at 20 kWh of usable district storage capacity.
Keywords: distributed and centralized storages, modelling, PV system, electrical reference profiles

1 INTRODUCTION

The capital costs for photovoltaic systems have significantly and continuously decreased over the last decade. As a function of the particular local conditions coevally the height of the levelized cost of electricity (LCOE) of PV systems has also considerably decreased. Today PV has already passed the grid parity in many countries worldwide, in Germany the grid parity has been passed for rooftops PV systems in 2012 [1]. Parallel to the decreased PV LCOE also the German EEG feed-in-tariff for PV electricity has been steadily lowered. Due to this process many PV electricity producer experienced a paradigm shift. They stopped feeding the entire PV yield into the grid. Instead, it was now more attractive to consume as much as possible PV electricity by oneself.

Households with rooftop PV systems ordinarily consume 20-40 % of their self-produced PV electricity. Electrical storage solution can significantly increase the onsite self-consumption rate up to 100 % in dependence of the storage capacity and the installed PV nominal power. Nowadays economic reasonable dimensioning of PV storages can raise the self-sufficiency rate from 15-35 % (without storage integration) up to approximately 70-80 % (with storage integration) [2]. This enables residentials with an installed PV system on the roof top to reach a certain rate of self-sufficiency reducing the dependency of increasing grid electricity tariffs.

In order to achieve both high PV electricity self-consumption rates and high self-sufficiency rates in residential districts, there are two possible alternative solutions:

a) The storage integration takes place in a decentralized way on residential building level.

b) The storage integration takes place on residential district level with a common centralized mass storage for all residential buildings.

In our contribution we investigate and compare these two alternative storage solutions for residential districts with high PV penetration in order to address the question if one of these two approaches has a fundamental benefit concerning the occurring overall costs. Our investigations base on a market analysis and on modeling and simulating the PV storage operation in a residential district environment.

2 METHODOLOGY

2.1 Model Description

Basis for the assessment of distributed and centralized storage approaches in residential districts with high PV penetration was a simulation model consisting of 10 one family houses (see Figure 1). Each one family house has a PV rooftop installation in the range of 3-10 kWp (see Table I). Moreover, in one type of the model all one family houses are coevally equipped with distributed storage solutions exhibiting a usable storage capacity of 2-10 kWh. The second model type contains a centralized district storage system with a varied usable storage capacity instead of the installation of distributed storages in each one family house.

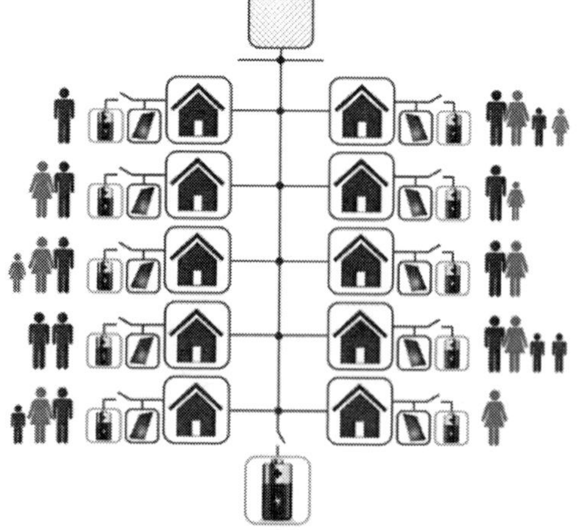

Figure 1: Conception of the modeled residential district.

In order to get representative results statistical data were taken to characterize the simulation model.

Therefore the varying number of residents in each household represents the statistical number of residents in a German district. These data were collected by the German Federal Statistical Office [3]. It has a significant effect on the consumer load profile characteristics because the amount of energy consumed by the households is mainly dependent on the number of residents. For reflecting the typical annual German household electricity demand, data collections of the German Federal Ministry for the Environment, Nature Conservation, Building and Nuclear Safety were taken into account [4] (see Table I). Based on these data the electricity demand and load profiles were modeled for each family house of the residential district. Annual electricity consumptions in the range 1500-4000 kWh were defined depending on the number of household inhabitants.

The dimensioning of PV generator size and residential storage capacity of each one family house (Table I) is based on the corresponding annual electricity consumption. Target of dimensioning was to realize both high PV self-consumption and high self-sufficiency rates. Furthermore the PV generator size is oriented on the typical installed PV system size in Germany in the year of 2015 [5]. The key data of the distributed residential storage systems like capacity, power (c-rate) and efficiency vary and depend on the selected manufacturer and type. Only residential storage systems were considered that are funded by the KfW Group in 2015 [6].

Table I: Key simulation data of the residential district model.

Number of inhabitants	Annual electricity demand [MWh]	Installed PV power [kWp]	Usable electrical storage capacity [kWh]
1	1.5	4	2.0
2	2.1	5	4.0
3	2.7	7	4.0
2	3.0	6	4.6
3	3.5	9	4.7
4	3.0	8	8.0
2	2.7	6	5.8
2	2.5	5	2.0
4	4.0	10	10.0
1	2.2	3	2.0
24	27.2	63	47.1

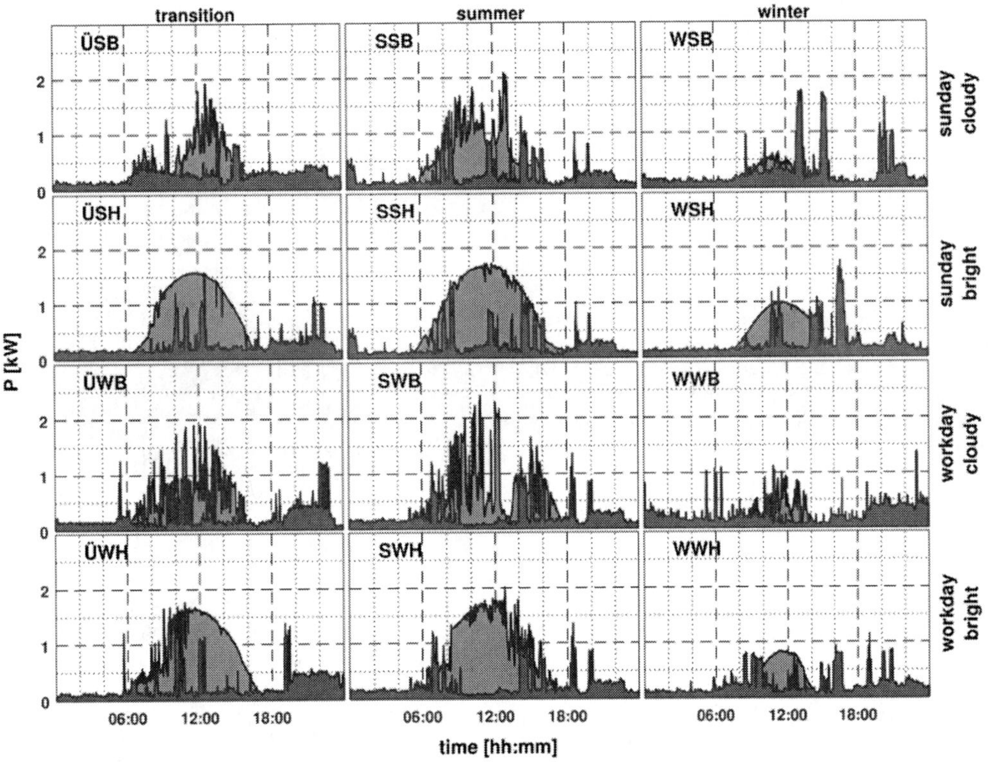

Figure 2: Electrical load profiles according to VDI 4655 [7] (blue line and area) and PV generation profiles [8] (orange line and area) of reference days. The profiles relate to a one family house with 2.2 MWh annual electricity demand and with a 3 kWp PV installation exhibiting a performance ratio of 80 %.

2.2 High-resolution load and PV generation profiles

Highly resolved load and PV generation profiles are a key issue for a realistic simulation of PV storage systems. Real load profiles that were available had rarely no representative character. Therefore, guideline VDI 4655 standard load profiles were taken from [7]. One objective of this guideline is to provide reference electrical load profiles in a one minute resolution for simulation and design purposes. The methodology of the VDI 4655 provides the possibility to create a typical test reference year based on representative days. A typical day represents the electrical load profile for each day category comprising the electricity demand that can be calculated in dependency of the number of residents living in a household or by a given yearly electricity demand. One calendar year is categorized by summer (S), winter (W) and transition (Ü). All profiles during one week are subdivided into working days (W) and Sundays (S) partitioned into bright (H) and cloudy (B) sky. Moreover, VDI 4655 guideline divides Germany into different climate zones taking variable weather conditions into account. However, VDI 4655 guideline possesses no standardized PV generation profiles. Hence we applied data from [8] that provides PV generation profiles in the manner of the VDI 4655.

Applying the mentioned key simulation data (Table I) and load and PV generation profiles, a test reference year with the climate conditions of Oldenburg (Northern Germany) was generated along the guideline methodology. By this procedure an entire reference year was disposable with one minute data resolution ensuring a reliable simulation basis. Figure 2 shows exemplarily the load and PV generation profiles of the one family house with an electricity demand of 2.2 MWh. The corresponding installed PV capacity was 3 kWp, as performance ratio a value of 80 % was defined.

3 RESULTS AND DISCUSSION

3.1 Market analysis

First a market analysis of more than 100 residential and district storage types of 26 different manufacturers has been conducted for the German market in 2015. For this analysis the overall system costs were taken. Thus the costs include the battery, power electronics, housing and assembly.

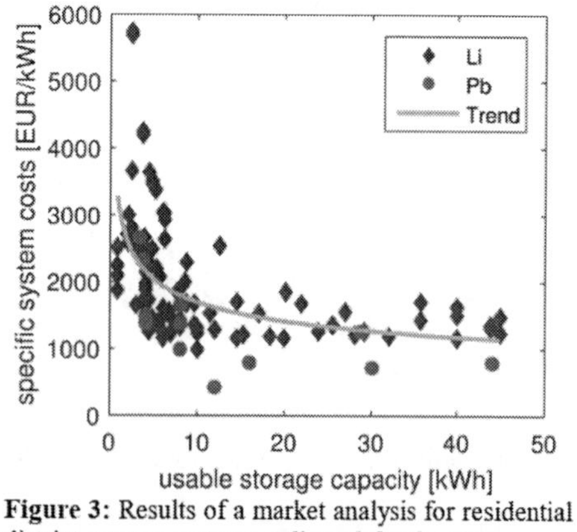

Figure 3: Results of a market analysis for residential and district storage systems collected for the German market in 2015 showing the correlation between usable storage capacity and specific system costs.

Figure 3 shows the results of this analysis that exposes the tendency for storage systems to drop in specific costs (€/kWh) by increasing storage capacity. This can be attributed to the effect that large district storage systems basically requires less capital costs due to less specific costs per usable kWh storage capacity. Likewise, the specific costs for power electronics decrease with increasing nominal power.

3.2 Self-sufficiency rate

The electricity supply of the modelled residential district was simulated for one calendar year including a couple of scenarios:

a) First the simulation was performed without storage integration. The obtained value served as reference in order to study the self-sufficiency rate gain due to storage integration.
b) Afterwards the distributed storage scenario of Table I was simulated with a usable cumulative storage capacity of 47.1 kWh.
c) The scenario of centralized district storage was totally simulated 14 times varying the usable storage capacity from 5-70 kWh in 5 kWh steps. The c-rate of the district storage was varied in relationship to the capacity.

The generated simulation results offer a time-resolved examination of all load-flows in the residential district model. First the self-sufficiency rate for the district was analyzed and the obtained simulation results are shown for each calendar week in Figure 4. Seasonal differences were taken into account by the reference year indicating lower self-sufficiency rates in the winter and transition period compared to the summer period. The reference case without storage integration and the scenario with distributed residential storages are also presented.

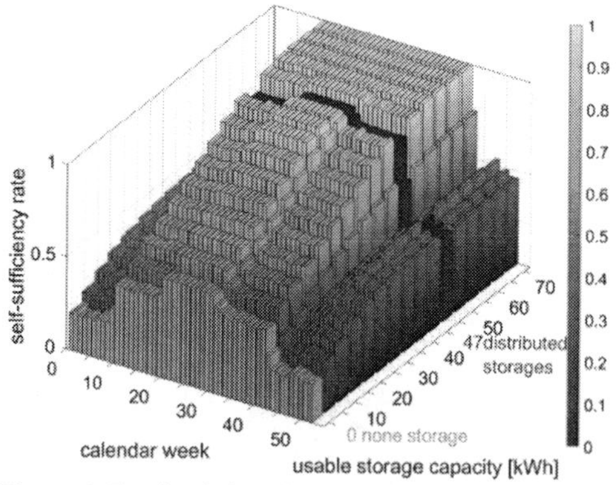

Figure 4: By simulation obtained self-sufficiency rate of a modeled PV equipped residential district within a calendar year as function of the usable storage capacity of a centralized district storage. Data in orange indicate the self-sufficiency rate without storage integration (reference). As comparison the red filled bars show the simulation results for the case of the distributed storages with a usable total capacity of 47.1 kWh.

An overview of the annual self-sufficiency rate as function of the usable storage capacity is shown in Figure 5. Without storage integration, the total PV capacity of 63 kWp leads to a self-sufficiency rate of 37 %. By installing the centralized district storage the self-sufficiency rate increases up to 80 % at the maximum usable storage capacity of 70 kWh. For the scenario of distributed residential storages with a usable total capacity of 47.1 kWh, a self-sufficiency rate of 74 % was determined. For realizing the same self-sufficiency rate with a centralized district storage solution, a significantly less capacity of 38.7 kWh is required (-18 %).

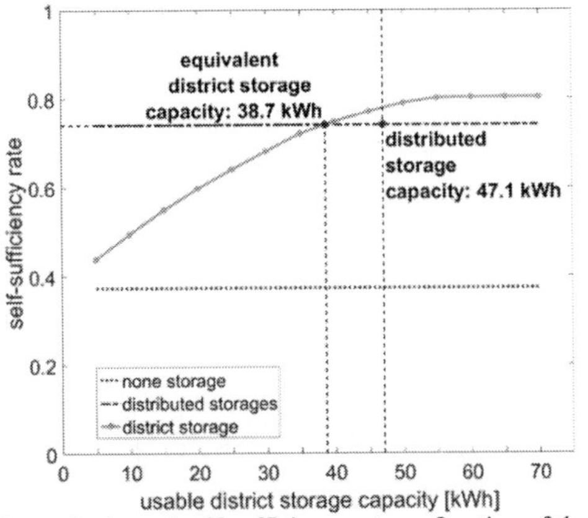

Figure 5: Annual self-sufficiency rate as function of the district storage capacity. The self-sufficiency rate for the distributed storage scenario is shown as constant (blue dash-dotted line). The red markers represent the required storage capacity resulting in the same self-sufficiency rate for both cases (central vs. distributed storage approach).

3.3 Throughput

One important criterion for the assessment of the cost effectiveness of a storage solution is the scored throughput as measure for the utilization level. Both considered storage integration scenarios for the residential district (centralized vs. distributed) were compared referring to this and the results are shown in Figure 6.

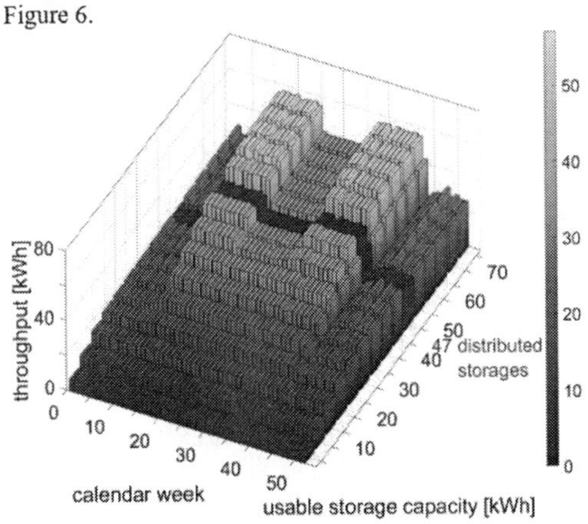

Figure 6: Throughput vs. calendar week for each varied district storage capacity (5-70 kWh). The red filled bars mark the simulation result for the scenario of the distributed storage solution with a usable cumulative capacity of 47.1 kWh.

Like in Figure 4 all week results of a calendar year are illustrated for both the scenario of a distributed storage solution (usable total capacity 47.1 kWh) and the scenario of a centralized storage solution (usable capacity 5-70 kWh). With increasing centralized storage capacity a considerable difference occurs concerning the throughput in the winter months and the remaining part of the year. The lower throughput during the winter months can be observed at usable storage capacities of ≥ 20 kWh. Moreover, for usable storage capacities ≥ 40 kWh the throughput values increase variably so that the highest throughput values are achieved during the transition time.

The overall annual throughput for each simulated usable storage capacity is shown in Figure 7. The distributed storage solution with a usable total storage capacity of 47.1 kWh achieves an annual throughput of 13.3 MWh. For scoring the same annual throughput, the centralized storage solution requires merely 41.4 kWh usable storage capacity (-12 %).

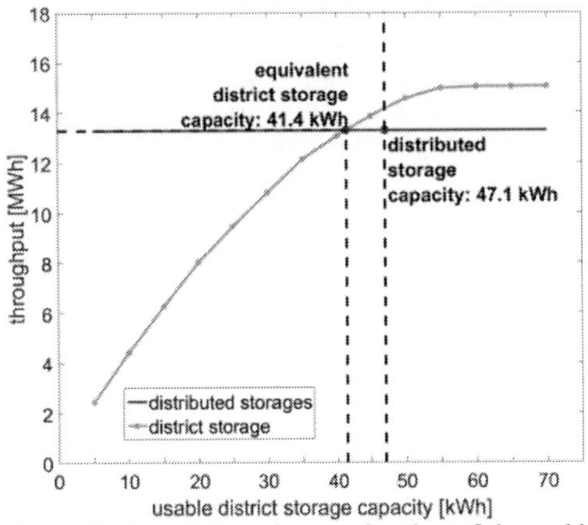

Figure 7: Annual throughput as function of the usable district storage capacity. The orange line is displaying the results for the centralized district solution whereas the constant blue line is representing the results for the distributed residential storages. The intersections representing the equality of throughput of both solutions are marked with red dots.

3.4 Further aspects

Existing buildings in residential districts usually offer a lot of roof space for the installation of PV systems. Providing space for on-site electrical storage systems, however, was not state of the art when the buildings were constructed. Thus the installation of such storage systems in existing buildings could be limited. Even in modern buildings it could be difficult to find enough space for an on-site PV storage since these buildings are often equipped with huge ventilation systems. Therefore in many cases there is no acceptance to provide space for more technical installations like electrical storages.

Finally it has to be pointed out that in addition to the realization of high PV self-consumption rates and high self-sufficiency rates the establishment and dissemination of electricity storage systems basically has an essential meaning on macroscopic scale in matters of the transformation of the global energy system toward a full supply with renewable energies. The approach to buffer electricity produced by volatile renewable energy sources nearby the consumer instead of a long distance transmitting may lead to the benefit of significantly reduced investment costs concerning the expansion of electrical transportation grids.

4 CONCLUSION

In this case study we compared centralized and distributed storage solutions for residential districts with high PV penetration. It was found that the centralized storage solution exhibits several benefits like a significant reduced storage capacity for reaching equal self-sufficiency rates and throughput values (in this case study -18 % and -12 %, respectively). Moreover, larger battery storage systems also offer decreased specific system costs.

The case study also demonstrated that the self-sufficiency rate of PV equipped residential districts can be distinctly increased by storage integration. Due to the boundary conditions given by the seasonal varying supply of solar irradiation and the specific load characteristics in residential buildings, the simulation series in 5 kWh storage capacity steps illustrated an economic optimum at 20 kWh usable storage capacity. An increased storage capacity leads to nearly no further self-sufficiency gain in the winter period due to the limited supply of solar irradiation. If the storage capacity is further increased, also for the summer period a saturation effect arises.

5 REFERENCES

[1] H. Wirth, "Recent Facts about Photovoltaics in Germany", Fraunhofer ISE, 2015 www.ise.fraunhofer.de/en/publications/veroeffentlichungen-pdf-dateien-en/studien-und-konzeptpapiere/recent-facts-about-photovoltaics-in-germany.pdf, [accessed 26th Dec. 2015]
[2] V. Quaschning, "Solare Unabhängigkeitserklärung", Photovoltaik, p.50, 2012
[3] German Federal Statistical Office, "Wirtschaftsrechnungen", 2013 www.destatis.de/DE/Publikationen/Thematisch/EinkommenKonsumLebensbedingungen/EinkommenVerbrauch/EVS_HausGrundbesitzWohnverhaeltnisHaushalte.html, [accessed 3rd Nov. 2015]
[4] Federal Ministry for the Environment, Nature Conversation, Building and Nuclear Safety, "Stromspiegel für Deutschland", 2014 http://www.diestromsparinitiative.de/fileadmin/bilder/Stromspiegel/Brosch%C3%BCre/Stromspiegel2014web_final.pdf, [accessed 3rd Nov. 2015]
[5] Federal Network Agency, "PV-Datenmeldungen und EEG-Vergütungssätze", 2015 http://www.bundesnetzagentur.de/DE/Sachgebiete/ElektrizitaetundGas/Unternehmen_Institutionen/ErneuerbareEnergien/Photovoltaik/DatenMeldgn_EEG-VergSaetze/DatenMeldgn_EEGVergSaetze_node.html, [accessed 5th Nov. 2015]
[6] B. Schwartz, "Marktüberblick über ausgewählte Stromspeicher", housewarming2015, Hamburg, 2015, [online], available www.elbcampus.de/web/adb/output/asset/3117
[7] VDI 4655 2008-05, "Reference load profiles of single family and multi-family houses for the use of CHP systems", 2008
[8] P. Klement, T. Thomsen, F. Lammers, M. Zobel, B. Hanke, K. v. Maydell, Proc. of the 29th European Photovoltaic Solar Energy Conference (EUPVSEC), p.3576, 2014

ADVANCED SIMULATION PLATFORM FOR THE INTEGRATION OF PHOTOVOLTAICS INTO POWER SYSTEMS: SPIDER

Franck BOURRY[1], Benjamin GUINOT[2], Cyril BOURASSEAU[2], Sébastien REVOL[3], Saadia DHOUIB[3]
[1]Univ. Grenoble Alpes, INES, F-73375 Le Bourget du Lac, France
[2] CEA, LITEN, F-38054 Grenoble, France
[3] CEA, LIST, F-38054 Grenoble, France
Phone: +33 (0)4 79 79 21 57 - Fax: +33 (0)4 79 68 80 49 franck.bourry@cea.fr

ABSTRACT: The fast development of PV systems, and their associated large scale integration into power systems, lead to the need for new simulation methods and tools. More precisely, modeling and simulation tools are necessary as a test-bench, where energy management algorithms are first developed and tested before being deployed to real equipments. Another objective of simulation is to carry on technic and economic analyses, for sizing purposes for example.
In this context, the CEA developed an advanced simulation platform, which addresses various PV applications such as self-consumption systems, microgrids or utility scaled PV systems. The software is called SPIDER as Simulation Platform for the Integration of Distribution Energy Resources. SPIDER is a standardized platform based on a generic open-source modeling environment (Papyrus). SPIDER relies on the "model based designed" approach where models representing the physical system are associated to models representing the multi-level system control.
SPIDER is operational and is used as a support for Energy Management Systems (EMS) development.
Keywords: Simulation, Modeling, System, Grid Integration

1 INTRODUCTION

For various PV applications, ranging from self consumption systems or microgrids to utility-scale PV systems, a solution to increase the share of photovoltaics into power systems consists in improving the system control, so that it becomes able to cope with the PV variability and uncertainty.

The resulting system control, also called energy management system, then consists in complex algorithms, where optimal planning modules taking into account PV production forecasts are combined with low level equipment controls. In order to shorten the development time for these complex control systems, a modeling and simulation platform is necessary: the different energy management algorithms can then be developed and validated using a simulated environment before being deployed to real equipments.

Also, the required simulation platform must be modular in order to consider complex physical systems, with various electric architectures. The electricity grid must be taken into account when necessary. Consequently, the platform must be based on standardized model libraries which enable the user to build its system for his case study. Different levels of model precision can be handled; for example, an electricity storage system can be modeled as either as a single component taking into account the system charge and discharge power and its losses, or as a subsystem including converter, battery and auxiliaries models. The choice of the model precision for a given case study then depends on the objective of the study, or on the granularity of the control algorithms to be developed.

A deep analysis of the state-of-the-art regarding modeling and simulation tools revealed that such modular simulation platforms, able to combine models of the physical systems with multi-level control algorithms, where missing. CEA then developed SPIDER (Simulation Platform for the Integration of Distribution Energy Resources), which can be seen as an integration and simulation platform, based on existing standardized model libraries.

The paper first presents the model driven engineering method, which has been applied for the development of SPIDER. The architecture of the simulation platform, as well as the adopted standard for model integration, are also presented in this first section.

A next section details the current status, and the on-going work about the development of the SPIDER platform. Final concluding remarks are presented.

2 MODEL DRIVEN ENGINEERING APPLIED TO ENERGY MANAGEMENT SYSTEM

2.1 Methodology

SPIDER has been developed following the "Model Driven Engineering" (MDE) methodology. MDE uses the UML, standing for Unified Modeling Language, which is a graphical modeling language, including a set of diagrams such as class diagram or activity diagram. SysML, standing for System Modeling Language, is an extension of the UML, focusing on system modeling. Then, the SPIDERml metamodel, has been defined as an extension of the UML/SysM, dedicated to energy system modeling.

The activity diagram related to SPIDER led to the definition of the several concepts associated to modeling and simulation for Energy Management Systems (EMS), such as "component" or "control" concepts. Based on this activity diagrams, class diagrams of the main concepts have been designed. The class diagram informs about concepts (system, component, control) and the links between these concepts, such as inheritance, inclusion. Figure 1 shows an example of class diagrams defining the concepts "simulator", "component", "control", "system", "parameter", "port" and "link", as well as the interactions between these concepts.

Both activity and class diagrams have been build using the Eclipse Papyrus software, which is an open-source and recognized modeling software, developed by CEA List.

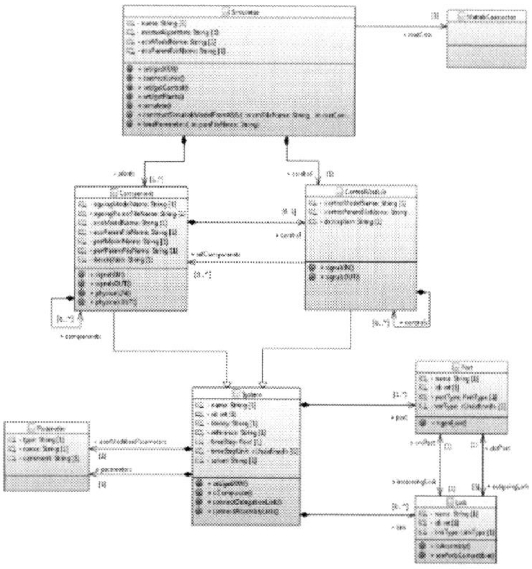

Figure 1: Class diagram defining the concepts "simulator", "component", "control", "system", "parameter", "port" and "link", as well as the interactions between these concepts

2.2 Model architecture

Regarding the "control" concept, a generic multi-level architecture for Energy Management System (EMS) has been proposed. Such architecture defines the EMS as a combination between planning controls and one operation control.

- The planning control aims at computing system set points for a given look ahead horizon. It may be based on generation or consumption power forecasts. It may also include optimization methods and associated models. Generation unit commitment based on optimization methods such as dynamic programming or Mixed Integer Linear Programming are examples of planning controls. The EMS may include from zero to several planning controls.
- The operation control applies the optimal trajectories provided by the planning modules to compute the physical system component set points, also using measures from the plant model.

Such combination between the planning and operation controls is illustrated in figure 2

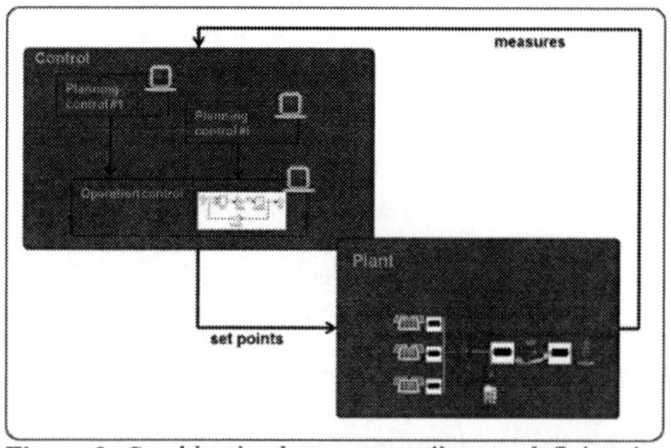

Figure 2: Combination between ass diagram defining the concepts simulator", "component", "control", "system", "parameter", "port" and "link", as well as the interactions between these concepts

2.3 Model standard

SPIDER has been developed with the objective to integrate models provided by different modeling tools. Consequently, it is compliant with the Functional Mockup Interface (FMI) standard which defines a model format particularly adapted to model exchange. More precisely, the role of the FMI standard is for model interoperability is to make a model (component, control, sensor, etc.) independent from the environment/tool (Simulink, Modelica/Dymola) in which it has been developed. The FMI for co-simulation defines a model as a zip file including a dll file, which is compiled C code containing solver of the native environment (Simulink, Dymola, etc.), and an xml file containing information on parameters types, names and values, inputs and outputs of the model. The figure 3 illustrates the FMI standard.

Also, the Common Information Model (CIM) standard is to be considered for the description of the electrical architecture.

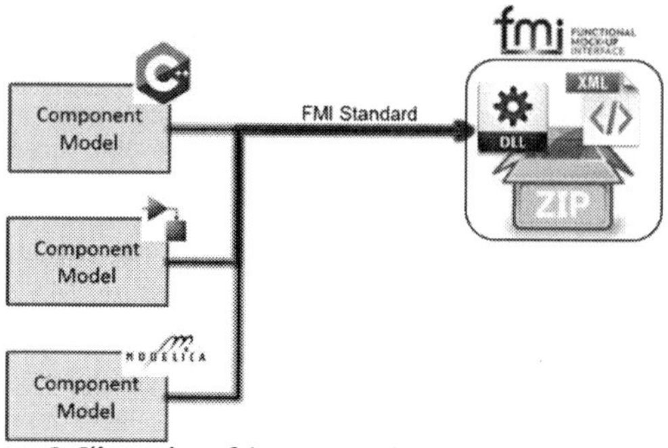

Figure 3: Illustration of the FMI standard

3 SIMULATION PLAFORM STATUS AND PERSPECTIVES

3.1 Current status

The SPIDER consists in a set of 3 main parts:

1. Component and model libraries. The models are developed by technology experts. They can be developed based on Modelica modeling language, or using Mathworks products, or even directly in C++. Efforts are currently done to develop methods to generate documentation automatically from the model, including model unitary tests.

2. A system modeler: this modeler is based on the Eclipse Papyrus software. It allows the construction and parameter setting of the simulator based on component and control models taken from the libraries.

3. A set of C++ libraries, which include optimization and sensitivity analysis modules.

These 3 main parts are represented in Figure 4.

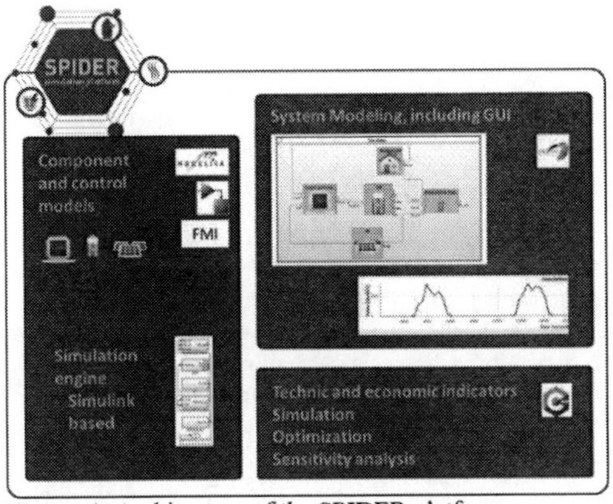

Figure 4: Architecture of the SPIDER platform

The Papyrus system modeler offers the user a Graphical User Interface (GUI). This GUI gives the user the possibility to:

- Import control and component models;
- Build a simulator by drag-and-drop models from the imported libraries, and connecting their ports;
- Set the model parameters;
- Define technic or economic indicators;
- Run a single simulation;
- Run a sensitivity analysis on a selection of system parameters;
- Run optimization to determine the optimal parameters according an indicator which has to be minimized or maximized.

An overview of the SPIDER Graphical User Interface is given in figure 5.

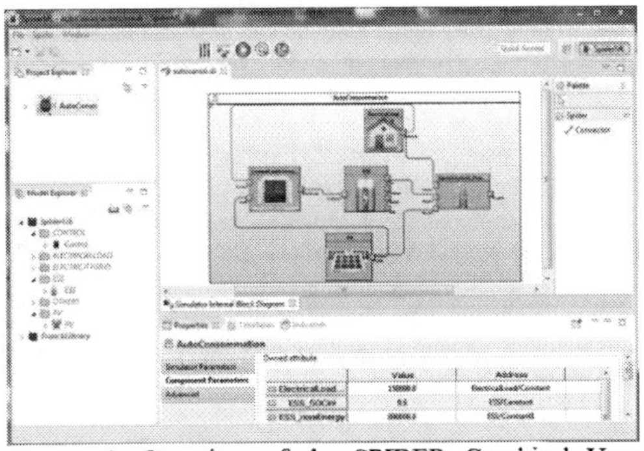

Figure 4: Overview of the SPIDER Graphical User Interface

3.2 On-going work

Regarding FMI integration and simulation engine, a simulation master is under development. This entity is an algorithm which coordinates the simulation between the different FMI models included in the simulator. The actual operational master for the SPIDER current version relies on MATLAB-Simulink. The under-development master is developed in Papyrus, making SPIDER independent from any additional simulation software. The master algorithm is developed using activity diagram. It consists in the development of a discrete event timer, and an interface between Java (Papyrus) and FMU based on Java-Native Access.

Also, efforts are made for the development of a language for the definition of energy management strategies. Three approaches are proposed for the SPIDER user to define its operational control using the Graphical User Interface:

1. Strategy representation using activity diagram;
2. Strategy definition using Papyrus scripting language ALF, which is directly related to activity diagrams;
3. Strategy representation using state machines;

These three approaches are derived from SPIDER capacity to handle UML/SYSML.

7 CONCLUSIONS

The conclusions can be summed up in the following points:

- Having a framework for EMS development and benchmarking is a current industrial need; also, Model Based Design approach is an adequate solution;
- Going towards standards for model interoperability is a necessity; the FMI standard brings an interesting solution;
- What makes SPIDER a unique tool is the fact that it relies on standards, and not on any commercial products, and that it will allow the user to define its own energy management strategy;
- The SPIDER simulation platform is now operational as a test bench for EMS development programs at INES; however, a single « monolithic » tool is not the solution; SPIDER must be considered as a method, combined with component and control libraries, and with generic modules.

PROBABILISTIC EVALUATION OF UK DOMESTIC SOLAR PHOTOVOLTAIC SYSTEMS: AN INTEGRATED GEOGRAPHICAL INFORMATION SYSTEM PV ESTIMATION TOOL

Philip A. Leicester*, Nick Doylend, Paul Rowley
Centre for Renewable Energy Systems Technology (CREST), School of Mechanical, Electrical and Manufacturing Engineering, Loughborough University, Loughborough, LE11 3TU, United Kingdom.
* Tel.: +44 (0)1509 635343, Email: p.a.leicester@lboro.ac.uk

ABSTRACT: It is shown how key predictor parameters for the spatial estimation of PV yield, self-consumption and thereby economic and social indicators can be extracted from a GIS system and introduced into a Bayesian Network model. This model endogenises the uncertainties and incorporates spatial variability inherent in these parameters. Empirical monthly and annual yield measurements obtained from over 600 PV installations have been obtained and compared with estimated yields obtained by two key solar tools used for performance estimation in the UK – these are PVGIS and the UK Government's Standard Assessment Procedure (SAP) for domestic buildings. Mean bias estimates and root mean square error estimations were obtained for each tool and the results used to construct an uncertainty distribution in PV yield prediction given key input parameters such as system rating, orientation and tilt. This uncertainty was used to furnish a probabilistic graphical model with a prior distribution for PV yield estimation. This was integrated into a Geographical Information (GIS) system furnished with roof and building stock parameters including roof attributes obtained from lidar data. Elements held in a vector layer of the GIS system can be selected and the resultant distributions of input parameters automatically fed to the model to yield a posterior distribution of the PV yield. The model is able to propagate the yield uncertainty to other probabilistic models, including ones which predict the internal rate of return and self-consumption. The latter is in turn predicted by empirical marginal distributions of domestic electricity consumption. Thus with a given posterior distributions of PV yield, new posterior distributions for the internal rate of return, self-consumption and carbon emission savings are automatically calculated. By integration with GIS this novel approach allows the spatial analysis of the uncertainty pertaining to representative risk factors for PV adoption in the UK, and facilitate the estimation by installers, investors, and local authorities in a manner which endogenises uncertainty.
Keywords:

1. INTRODUCTION

At a whole system level photovoltaic (PV) technologies are seen as a valuable means to deliver both (i) an effective net energy gain to support multiple consumption patterns of society, and (ii) a decreasing reliance on non-renewable primary energy sources resulting in long term energy sustainability [1]. To support the emergence of a viable innovation system, in which PV can overcome market barriers and contribute to the energy mix, many nations have provided subventions or other market interventions to enable it to compete with incumbent technologies [2]. In the UK the domestic adoption of PV has been rapidly accelerated using the FiT mechanism, introduced in April 2010, which has driven the installation of almost 4GW on UK domestic roofs by April 2016 [3].

The cumulative cost of this subsidy, over the 6 years of FiT operation, commensurate, with a rapid reduction in total system cost, has led to intense political pressure to reduce the real and perceived burden on the electricity bill payer by lowering the FiT rate. The UK's Department of Energy and Climate Change proposed that a target hurdle rate for investors of 4% was desirable and suggested it should only be realised in the most favourable locations [4]. Following consultation the generation FiT has been significantly reduced, though not as drastically as first proposed [5]. However, adopters and investors are exposed to greater risk of not making an economic return, and a slow-down in the rate of installation in the UK has been observed [3].

This has given greater impetus to understanding the uncertainty in the financial return. Research has been carried out to investigate the sensitivity of PV yield to a number of technical system parameters [6]. Taylor et al. have recently calculated a mean yearly integrated performance ratio of 83%, with a standard deviation of 7% (a standard error of 8%) using empirical data for over 7000 mainly domestic systems [7]. This error depends, of course, on many predictor variables, each with their own uncertainties. In particular monthly global horizontal irradiation, estimated by interpolating between ground station measurements, had an RMSE of 4.5%; further errors result from decomposition into direct and diffuse components and transposition to the plane of array [8].

However knowledge of uncertainties in the performance of PV is not enough to understand resultant uncertainties in economic performance. Under the UK's FiT scheme, value is created for the generation, export and the displacement of imported electricity [9]. Thus Leicester et al. have investigated the uncertainty in self-consumption and characterised a distribution with a mean value of 34% and a standard error of circa 60%, depending on the housing stock studied [10]. This and other parameters have been used to furnish a probabilistic model with the required distributions for a probabilistic discounted cash flow analysis to yield a measure of the uncertainty in a net present value calculation for domestic PV [11]. This research has pioneered the employment of probabilistic graphical models (PGM) in order to endogenise the uncertainties in renewable energy generation and allow uncertain or variable inputs to furnish investors or policy makers with probabilistic outputs [12].

Specifically a PGM approach using Bayesian Networks (BN) has been developed in order to qualitatively define the parameter space, and to quantify the conditional independency relationships, between them using a directed acyclic graph and underlying conditional probability tables. In recent years, the BN community has employed graphical information systems (GIS) to furnish BN with the requisite marginal distributions for key parameters [13]. This allows the model to deliver spatially disaggregated probabilistic outputs.

The objective of this paper is to show how GIS/BN integration can be developed using publically available datasets so as to deliver a spatial probabilistic PV evaluation tool. This can yield spatially disaggregated probabilistic outputs for key indicators allowing community planners to evaluate social, environmental impacts of PV. The paper presents the material as follows. In the Section 2 the use of GIS to furnish BN with probabilistic inputs is presented. In Section 3, GIS data to deliver building stock attributes, and PV yield, household incomes are explored. In Section 4 the use of a GIS interface to select evidence for the probabilistic model is demonstrated. Finally we present a discussion of the utility of this approach, followed by conclusions and suggestions for further work.

2. PROBABILISTIC EVIDENCE FOR BAYESIAN NETWORKS

An entire knowledge domain defined by a set $\boldsymbol{V} = \{V_1, V_2, \cdots, V_n\}$ of random variables can be represented by a joint probability distribution (JPD), $P(V_1, V_2, \cdots, V_n)$. A jpd with many variables, each with a number of potential values (states) can be mathematically intractable due to an inordinate total number of states. In contrast, a Bayesian network (BN) is a significantly more compact representation of the knowledge domain. A BN is a couple $(\boldsymbol{G}, \boldsymbol{P})$, where $\boldsymbol{G} = (\boldsymbol{N}, \boldsymbol{E})$ is a directed acyclic graph with a set of nodes $\boldsymbol{N}$, each representing a variable in $\boldsymbol{V}$, and $\boldsymbol{E}$ is a set of directed edges, which represent conditional dependencies between them. The latter are encapsulated as conditional probability tables (CPT). The joint probability distribution for a BN can be factorised using the chain rule, represented by Equation 1, where $\boldsymbol{\pi}_{V_i}$ is the set of parent nodes of node V_i. Thus each term in the product is a CPT, or, in the case where the set of parent nodes is empty, its marginal probability distribution. The BN renders the algorithmic calculation of the prior distribution of each variable mathematically tractable. For a brief introduction in to BN theory see reference [11] and references therein.

$$(V_1, V_2, \cdots, V_n) = \prod_{i=1}^{n} P(V_i|\boldsymbol{\pi}_{V_i}) \qquad \text{Equation 1}$$

The crucial concept to grasp for the application of BN to a problem domain is the concept of evidence in order to make prognostic and diagnostic inferences. Evidence means that the model user has new knowledge about the probability distribution applied to one or more variables, and applies these to the model. This results in a new calculation of the JPD and furnishes all the remaining variables with a new posterior distribution. Evidence is of two principal types. Firstly, a variable may be fixed to a specific state i.e. the variable is fixed to one specific value of its potential states. For example, consider a distribution of the discretised states for the PV system rating. The prior distribution of system ratings in a dataset might be as in Table 1. The user could then update the model by applying evidence such that the system rating were set to 4 kWp. This is known as hard evidence since the variable has been fixed to this value. In Table 1 the hard evidence is set to 4kWp. A second type of evidence might be one where the user applies a new distribution which updates the prior distribution. This is called probabilistic evidence since the evidence represents a new local probability distribution applied to the variable in the model [14].

Table 1 Example of a prior distribution of a parameter modified with hard evidence and probabilistic evidence.

	System Rating (kWp)				
	1	2	3	4	5
Distribution	Probability (%)				
Prior Distribution	5	20	40	30	5
Hard evidence	0	0	0	100	0
Probabilistic Evidence	0	15	40	35	10

3. PROBABILISTIC EVIDENCE FURNISHED FROM A GIS SYSTEM

For this work several BN have been developed to understand the relationship between building stock parameters, building energy consumption, PV yield and self-consumption. These have been linked together to deliver an object oriented BN which integrates several knowledge domains to deliver a whole system model. The key elements of this model are reproduced in Figure 1. The individual BN models can be seen in Reference [11].

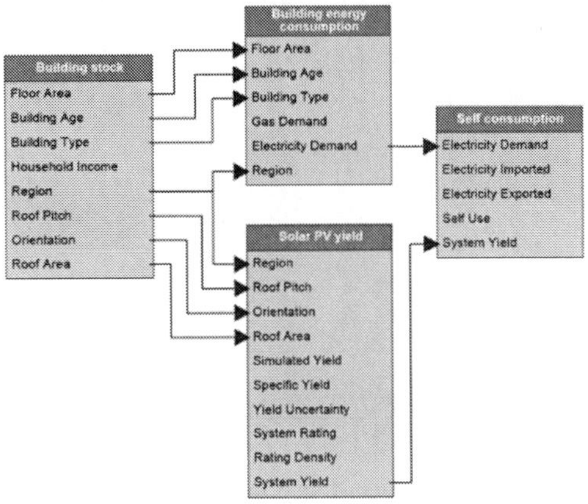

Figure 1. Object oriented BN model for domestic PV

The building stock BN model is key since it delivers probabilistic distributions for its parameters to the building energy consumption and PV yield models. These in turn predict the distribution of self-consumption [10]. The challenge is to extract distributions from GIS vector layers and to furnish the BN model with these distributions as probabilistic evidence. The distributions are determined from the features selected by the GIS user.

There are two approaches which decision and policy makers might employ to achieve this. The first is to select specific geographic areas based on boundary polygons (for example a local authority or community). All the features within the boundary are then used to generate distributions for the required parameters. The second approach is to use the GIS interface to select specific features, for example a row of houses, or those matching particular search criteria, such as all south-facing properties which have a roof size of 25m^2 or greater. Here we demonstrate the approach using UK census areas known as lower super output areas (LSOA). Figure 2 shows a vector map of one such census area in the South West of England.

The selection of a census area can be used to trigger a software script in order to calculate the distributions required as probabilistic evidence in the BN model. This involves a frequency count of discrete variables, such as building type or, for a continuous variable the allocation to an appropriate interval in a discretised distribution. Some parameters are held in vector layer attribute tables which are linked to the selected features. Others may be calculated using attribute parameters from one or more layers. In the following subsections the parameter sources and processing are described.

Figure 2 Census area LSOA Kerrier 008B in Camborne, Cornwall, England

3.1. DWELLING FLOOR AREA

Building footprints are available in the UK's OS MasterMap® Topography Layer which provides streets and building vector layers [15]. For the correct estimation of energy consumption, using the Building Energy Consumption BN, footprints need to be converted to floor areas. Thus the footprint area is multiplied by the number of floors, the latter estimated using building height data. This has recently been included for many conurbations in the UK MasterMap Building Heights geospatial data [16]. Building footprints in the MasterMap datasets do not always represent single dwelling units. Using the OS AddressBase® File [17], which provides a point vector layer for postal addresses, the number of dwelling units within the building footprint can be estimated using spatial queries. Figure 3 shows an example row of dwellings evaluated using this method. It can be seen that the building footprint on the far left encompasses four address points and the next one encompasses two. The remaining building footprints have only one.

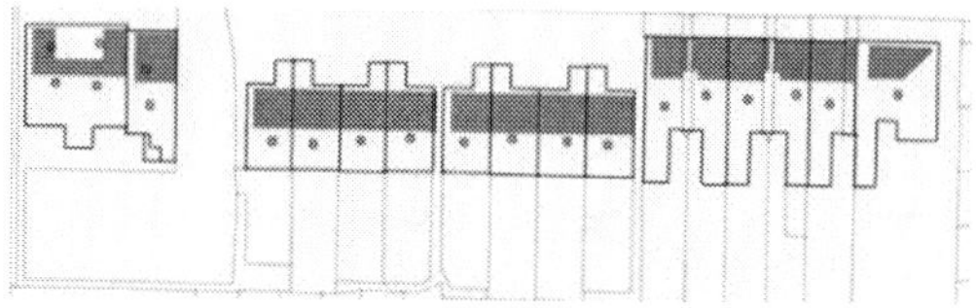

Figure 3 Building footprints with address points

It is important that the floor area is calculated correctly since, as previous work has shown [10], it is an important predictor of building energy demand by a domestic unit.

3.2. BUILDING TYPE AND AGE

Two other attributes, which have been shown to influence domestic energy consumption, are the Building type and the Building Age. These are available in commercially available GIS datasets from the GeoInformation Group [18]. Figure 4 shows the distributions of Building Age determined for four different LSOA census areas in England.

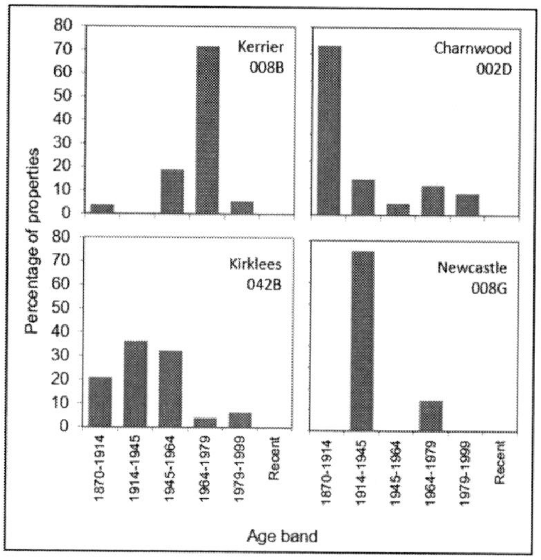

Figure 4 Distributions of Building Age

3.3. ROOF AREA AND ORIENTATION

Publically available lidar data, in combination with building footprints discussed above to serve as cookie-cutters in the GIS data layer, can be used to estimate roof orientations and tilts using methods reported by Palmer et al [19]. Roof areas can also be calculated using lidar data. Whilst this can yield azimuth and tilt distributions for most building stock in the UK, visual inspection does show that there is a high degree of potential error. For example stock which exhibits a high density of complex roof structures common in the UK, such as dormer windows, skylights, hip-roofs and intersecting roofs can overestimate the PV potential of the housing stock. Table 2 shows the percentage of buildings afflicted by problems which are difficult to estimate using automated algorithmic approaches. These were assessed using aerial photography in combination with GIS data. Only two-thirds of dwellings had roofs unaffected by structural constraints or shading. This included significant number of apartment dwellings which had no dedicated roof [12].

Table 2 roof assessments in four census areas (% suitable)

Assessment	Kerrier 008B	Kirklees 042B	Charnwood 002D	Newcastle 008G	TOTAL
Suitable	71.0	66.8	57.2	73.3	66.7
Affected by shading	9.7	11.1	9.4	8.2	9.6
Apartment (No roof)	14.0	11.1	26.4	15.3	16.9
Structural Constraints	5.2	11.0	7.1	3.2	6.8

Nevertheless these techniques enable the creation of GIS datasets of large areas of building stock from which parameter distributions to serve as probabilistic evidence can be determined.

3.4.SOCIO-ECONOMICS

A key objective was to be able to evaluate the socio-economic impacts of PV. For example, there is a need to understand the impact of low carbon technologies on fuel spending and the overall household income. To this end household income distributions for census areas were calculated using the iterative proportional fitting technique (IPF) [20]. This technique uses a reference dataset of anonymised census data with income linked to other social parameters to calculate the distribution of income in a target dataset for which only the social parameters are known. This showed that, typically, equivalised household incomes are variable, exhibiting a coefficient of variation of over 50%. This delivers a marginal distribution of income for the census area (Figure 5). This data can then be further linked to building stock using the IPF technique. As a reference dataset the UK's English Housing Survey [21] was used to estimate income for the housing stock in candidate census areas.

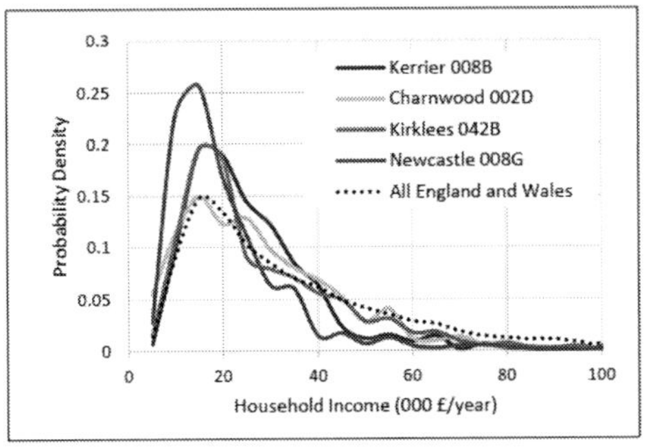

Figure 5 Household income distributions for census areas

Earlier studies have shown that domestic energy consumption, both electricity and gas, are directly and indirectly influenced by household income; the indirect influence is via the dwelling floor area [22] [23]. The result of the IPF technique applied to the building stock data is shown in Figure 6. This shows that the latter relationship has been effectively modelled in the simulated dataset. This probabilistic interpolation of income into the building stock model and the relationships modelled between building parameters, building energy consumption, and self-consumption of PV generated electricity ensure that the GIS layers deliver useful probabilistic evidence to the BN model in Figure 1.

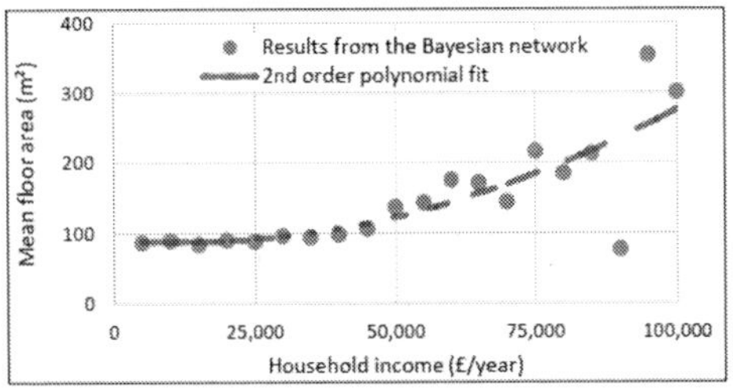

Figure 6 Result of modelled relationship between floor area and household income using iterative proportional fitting

3.5.PV YIELD

Whilst above the characterisation of building stock and occupancy factors which can then introduce probabilistic distributions in to the Building Stock BN model, it is also pertinent to note the sources of data used to estimate the PV Yield. Two PV yield estimation tools have been evaluated for the degree of uncertainty they introduce in to the model, PVGIS [24] and the UK SAP. Estimates from these tools where compared to the empirical annual yields for over 600 UK installations obtained from the Sheffield Microgeneration Database [25]. Figure 7 shows the result for PVGIS using the CMSAF database. Mean bias estimates and root mean square error estimations were obtained for each tool and the results used to construct an uncertainty distribution in PV yield prediction given key input parameters such as system rating, orientation and tilt. This analysis furnishes the model with an uncertainty distribution for the yield estimation by PVGIS or the SAP. In the application developed in this work, PVGIS was used to furnish the model with estimations.

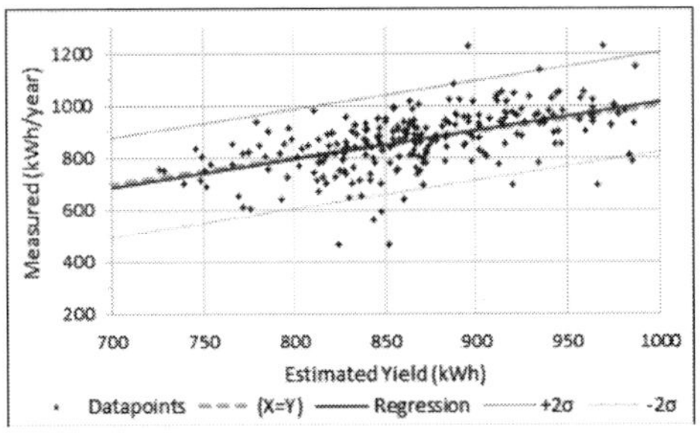

Figure 7 Measured versus PVGIS/CMSAF estimated yield for 600 UK systems

In this section a number of parameters have been presented which can be extracted from GIS layers in order to then be summarised as probability distributions. These are summarised in Table 3.

Table 3 Summary of parameters obtained from GIS vector layers and source data

Parameter	GIS Source	Comments
Floor area	Mastermap building footprints	Calculated using address point data and building heights
Building Type	Geoinformation group	Commercially available data; can be ignored if total energy not required.
Building Age		
Roof Slope	Lidar data	Calculated and added to attribute table
Roof areas for PV		
Orientation		
Household income	Census data	Iterative proportional fitting to produce simulated data added to attribute table
PV Yield	PVGIS or SAP	Uncertainty must be incorporated using empirical data.

4. APPLICATION OF PROBABILISTIC EVIDENCE TO THE BAYESIAN NETWORK

When selecting GIS features, either by selecting a polygon, or a large number of features using QGIS, it is a simple task to iterate through the selection set to obtain attribute values with which to construct probability distributions for the required parameters in the BN model. This was implemented as a QGIS plugin using a routine written in R [26]. This calculated the required distributions using an R-GIS implementation which were then provided to an RNetica software layer [27]. The latter uses the Netica-C API to establish the distributions in the target Netica BN model.

The application of probabilistic evidence is not straightforward in Netica, or indeed, in most commercial BN software [14], since it requires the application of evidence using a dummy node which is a child node of the one to which the evidence is being applied [28]. It is even more complicated if the new evidence is a set of multiple probabilistic evidences to be applied to a number of target nodes – this is predominantly the case when deriving a set of new distributions from a GIS system. In theory is a straightforward application of Bayes' rule to each node in turn. However the process is noncommutative since the order in which multiple CPTs are adjusted to set target nodes to their desired distribution is significant. In other words, the application of Bayes' rule to set the desired distribution to the second node influences the first and so on. In order to solve this, an iterative proportional fitting procedure as proposed by Pan et al [29] has been applied using a software algorithm written in R. This cycles round each target node, adjusting the dummy nodes' CPTs in turn. It was found that the CPTs rapidly converge such that the correct distribution representing the probabilistic evidence is applied to all the target nodes.

This algorithm, once all the probabilistic evidence from a GIS selection has been applied to the BN model, returns the new posterior distributions to the GIS interface for the user to evaluate.

5. DISCUSSION

In this section some key results obtainable from a GIS/BN integrated model are presented. These results are the probability distributions for any variable in the integrated BN model. As such the results are not unlike those we have previously published [10] [11]. The significance of this approach is discussed with regard to spatial decision making for the deployment of low carbon technologies. The further development of this approach is discussed in the context of the low carbon transition, new value propositions for decentralised energy business models and the management of risk.

Figure 8 shows the model's posterior distributions when a specific census area is selected.

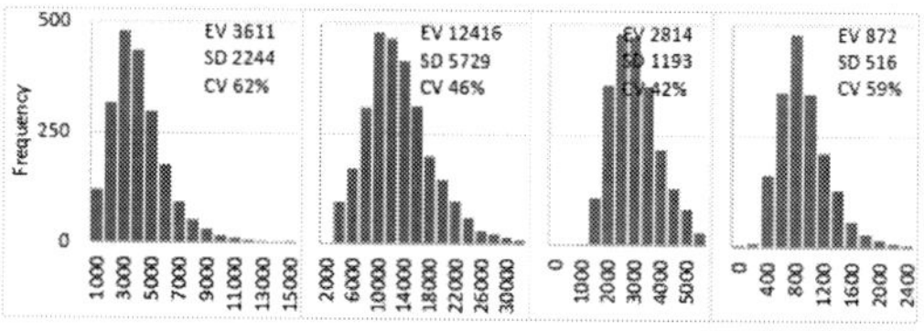

Figure 8 Posterior distributions of key parameters following selection of a census area: (from left to right) Electricity consumption, Gas consumption, PV yield, Self-consumption

As reported in [11] these output disitrbutions can be used to furnish probabilistic discounted cash flow analysis and energy affordability indicators such as for example the percentage of household income spent on fuel following the installation of PV. The fact that these distributions can be obtained for any selection of properties in a GIS model equipped with appropriate vector layers and attribute data provides a powerful evaluation tool for decision and policy makers.

One of the benefits of this technique is that is provides a means of accessing the power of probabilistic graphical models while retaining a GIS interface that may be more familiar or more intuitive to stakeholders in the decision-making process. This paper is therefore presents the development of an integrated GIS/BN model as a logical development of a probabilistic analysis of a low carbon intervention such as PV.

The approach is very extendable. It can be easily adapted to yield a number of other impact indicators which will benefit from a spatially disaggregated analysis. For example the evaluation of low voltage grid impacts is a candidate application since there are a number of studies which predict grid impacts such as voltage and thermal issues as a function of the penetration of PV [30]. This requires both an assessment of demand and generation which the model can assess using the selection of building stock.

A third area is the evaluation of community wide cash flows for the evaluation of virtual power flows and the influence of energy storage on the value proposition which can be realised. It is clear that a probabilistic approach is required for an accurate evaluation of risk.

6. CONCLUSION

A Bayesian network which probabilistically estimates generation, demand and direct self-consumption for

domestic PV has been successfully integrated with a GIS system. The latter has been furnished with a number of commercial and publically available datasets which allow the extraction of probability distributions based on a selection of GIS features.
The problem of supplying these multiple probabilistic evidences to the BN has been solved using an iterative algorithm. The model then furnishes a user with posterior distributions of all variables based on the selection of GIS features. This has potential for further development to incorporate other low carbon technologies and evaluate new business models such as virtual power stations and integrate energy storage using s method which endogenises uncertainties and helps evaluate risk.

7. ACKNOWLEDGMENTS

The authors wish to acknowledge the financial support of the Engineering and Physical Sciences Research Council (EPSRC) of the United Kingdom through EPSRC grants EP/K022229/1 (WISE PV - Whole System Impacts and Socio-economics of wide scale PV integration) and EP/K02227X/1 (PV2025 - Potential Costs and Benefits of Photovoltaics for UK-Infrastructure and Society).

8. REFERENCES

1 M. Raugei and E. Leccisi, "A comprehensive assessment of the energy performance of the full range of electricity generation technologies deployed in the United Kingdom", Energy Policy, vol. 90, pp. 46-59, 2016.
2 A. Campoccia, L. Dusonchet, E. Telaretti and G. Zizzo, "An analysis of feed'in tariffs for solar PV in six representative countries of the European Union", Solar Energy, vol. 107, pp. 530-542, 2014.
3 Department of Energy and Climate Change, "Solar photovoltaics deployment", gov.uk, 2016. [Online]. Available: https://www.gov.uk/government/statistics/solar-photovoltaics-deployment. [Accessed: 31- May-2016].
4 Department of Energy and Climate Change, "Consultation on a review of the Feed-in Tariffs scheme", 2015. Available: https://www.gov.uk/government/uploads/system/uploads/attachment_data/file/469476/Consultation_on_a_Review_of_feed-in_tariff_scheme.pdf. [Accessed: 31- August- 2015].
5 Department of Energy and Climate Change, "Review of the Feed-in Tariffs Scheme - Government Response", 2015. Available: https://www.gov.uk/government/uploads/system/uploads/attachment_data/file/487300/FITs_Review_Govt__response_Final.pdf. [Accessed: 06-January-2016].
6 B. Goss, R. Gottschalg and T. Betts, "Uncertainty Analysis of Photovoltaic Energy Yield Prediction", in 8th Photovoltaic Science Application and Technology (PVSAT-8) Conference and Exhibition, Newcastle, England, 2012, pp. 157-160.
7 J. Taylor, J. Leloux, L. Hall, A. Everard, J. Briggs and A. Buckley, "Performance of Distributed PV in the UK: a Statistical Analysis of over 7000 Systems", in EU PVSEC, Hamburg, 2015, pp. 2271-2276.
8 G. Saluja and T. Muneer, "An Anisotropic Model for Inclined Surface Solar Irradiation", Proceedings of the Institution of Mechanical Engineers, Part C: Journal of Mechanical Engineering Science, vol. 201, no. 1, pp. 11-20, 1987.
9 R. Cherrington, V. Goodship, A. Longfield and K. Kirwan, "The feed-in tariff in the UK: A case study focus on domestic photovoltaic systems", Renewable Energy, vol. 50, pp. 421-426, 2013.
10 P. Leicester, C. Goodier and P. Rowley, "Probabilistic analysis of solar photovoltaic self-consumption using Bayesian network models", IET Renewable Power Generation, vol. 10, no. 4, pp. 448-455, 2016.
11 P. Leicester, C. Goodier and P. Rowley, "Probabilistic evaluation of solar photovoltaic systems using Bayesian networks: a discounted cash flow assessment", Prog. Photovolt: Res. Appl., p. n/a-n/a, 2016.
12 P. Leicester, "The Development Of Object Oriented Bayesian Networks To Evaluate The Social, Economic And Environmental Impacts Of Solar PV", PhD Thesis, Loughborough University, 2015.
13 D. Landuyt, S. Broekx, R. D'hondt, G. Engelen, J. Aertsens and P. Goethals, "A review of Bayesian belief networks in ecosystem service modelling", Environmental Modelling & Software, vol. 46, pp. 1-11, 2013.
14 A. Mrad, V. Delcroix, S. Piechowiak, P. Leicester and M. Abid, "An explication of uncertain evidence in Bayesian networks: likelihood evidence and probabilistic evidence", Applied Intelligence, vol. 43, no. 4, pp. 802-824, 2015.
15 OS MasterMap Topography Layer [GML geospatial data], Coverage: UK, Updated Jan 2016, Ordnance Survey, GB. Using: EDINA Digimap Ordnance Survey Service, <http://edina.ac.uk/digimap>, Downloaded: January 2016
16 OS MasterMap Building Heights [CSV geospatial data], Coverage: UK Various, Updated Dec 2014, Ordnance Survey, GB. Using: EDINA Digimap Ordnance Survey Service, <http://edina.ac.uk/digimap>, Downloaded: January 2016
17 OS AddressBase [CSV Geospatial Data] Coverage: UK, Updated Dec 2013, Available from Ordnance Survey, Adanac Drive, Southampton SO16 0AS
18 Geoinformation group datasets
19 D. Palmer, I. Cole, R. Gottschalg and T. Betts, "Assessment of potential for photovoltaic roof installations by extraction of roof tilt from light detection and ranging data and aggregation to census geography", IET Renewable Power Generation, vol. 10, no. 4, pp. 467-473, 2016.
20 B. Anderson, "Estimating small-area income deprivation: an iterative proportional fitting approach". In: Tanton, R., Edwards, K. (Eds.), Spatial Microsimulation: A Reference Guide for Users, Understanding Population Trends and Processes, vol. 6. Springer, Netherlands, pp. 49–67, Chapter 4, 2013.
21 EHS data reference
22 K. Steemers and G. Yun, "Household energy consumption: a study of the role of occupants", Building Research & Information, vol. 37, no. 5-6, pp. 625-637, 2009.
23 S. Kelly, "Do homes that are more energy efficient consume less energy?: A structural equation model of the English residential sector", Energy, vol. 36, no. 9, pp. 5610-5620, 2011.

24 ECJRC (European Commission Joint Research Centre), "Photovoltaic Geographical Information System (PVGIS) Version 4", [online computer program] Available from: http://re.jrc.ec.europa.eu/pvgis/ [Accessed 12-Jan-2016].
25 G. Colantuono, A. Everard, L. Hall and A. Buckley, "Monitoring nationwide ensembles of PV generators: Limitations and uncertainties. The case of the UK", Solar Energy, vol. 108, pp. 252-263, 2014.
26 R Core Team, "R: A language and environment for statistical computing", R Foundation for Statistical Computing, Vienna, Austria, [online computer program] Available from https://www.R-project.org/.
27 R. Almond "RNetica: R interface to Netica(R) Bayesian Network Engine", R package version 0.4-4, 2015, [online computer program] Available from http://pluto.coe.fsu.edu/RNetica
28 J. Pearl. "Jeffery's Rule, Passage of Experience, and Neo-Bayesianism". In H.E. et al. Kyburg, Jr., editor, Knowledge Representation and Defeasible Reasoning, 245-265. Kluwer Academic Publishers. 1990.
29 R. Pan, Y. Peng, Z. Ding, "Belief update in Bayesian networks using uncertain evidence. In: Proceedings of the 18th IEEE International Conference on Tools with Artificial Intelligence (ICTAI'06), pp 441–444, 2006.
30 A. Navarro and L.F. Ochoa, "Probabilistic impact assessment of low carbon technologies in LV distribution systems," IEEE Transactions on Power Systems, vol 31, no 3, p 2192-2203, 2016.

HIKARI: A POSITIVE ENERGY BUILDING WITH AN ARCHITECTURALLY INTEGRATED PV FACADE AND A PV ROOFTOP SYSTEM (190 KWP)

Gaiddon, B.[1], Valentin, M.[2], Alfonsi, L.[3], Laquerriere, M.L.[4], Gouranton, G.[5], Corgier, D.[6]
[1]Hespul, 14 place Jules Ferry, 69006 Lyon, France
email: bruno.gaiddon@hespul.org
[2]SPL Lyon-Conluence, 73 rue Smith, 69002 Lyon, France
email: mvalentin@lyon-confluence.fr
[3]Bouygues Immobilier, 186 Avenue Thiers, 69006 Lyon, France
email: l.alfonsi@bouygues-immobilier.com
[4] Tecsol, 20, Boulevard Eugène Deruelle, 69003 Lyon, France
email: mll@tecsol.fr
[5]Terre Ciel Energies, 2, allée du Parc Montaury, 64600 Anglet, France
email: germain.gouranton@terrecielenergies.com
[6]Manaslu, Savoie Technolac, 73370 Le Bourget-du-Lac, France
email: david.corgier@cmdl.pro

ABSTRACT: In the center of the Lyon-Confluence district in Lyon, France, the public company in charge of the redevelopment of the area, SPL Lyon-Confluence, launched in 2011 an international design competition for the selection of a real estate developer to build a positive energy building on a 3.400 m2 piece of land called P plot.
The team Bouygues Immobilier/SLC in association with Kengo Kuma from Japan has been selected by the jury for the design of a group of 3 buildings called HAKARI with dwellings, offices places and shops for a total floor area of 12.300 m2. In order to build a positive energy building to respect the requirements set by SPL Lyon-Confluence, the selected team decided to equip this building with 2 kinds of PV systems: a see-through PV facade on the MINAMI building (21 kWp) and a rooftop PV system on the roof of each of the 3 buildings (168 kWp). The official opening ceremony of this building took place on Thursday 17 September 2015.
Keywords: Solar Architecture, Building Integrated PV (BIPV), Building Integration, Façade

1 INTRODUCTION

Lyon-Confluence is the name of an urban project started in 2003 that aims at redeveloping a former industrial area of 150 ha in the city center of Lyon and building 1.000.000 m2 of new dwellings, offices and shops by 2030.

This project has also very ambitious targets in terms of sustainability that are described in a sustainable action plan design with WWF within the One Planet Living initiative [2]. One of the most ambitious goals set in this action plan is the zero carbon objective defined as follow: the 1 million m2 of floor area of new buildings to be built in the area by 2030 should not lead to any increase of the CO2 emissions of the area.

In order to reach this goal, SPL Lyon-Confluence, the public company in charge of the redevelopment of the area, has undertaken several concrete measures to:

- Drastically reduce the consumption of new buildings in the area,
- Refurbish the existing buildings of the area located in the north of the district,
- Massively install, on almost each new block, renewable energy systems such as wood chip boilers, solar thermal systems and PV systems.

This article provides information on the most recent building that has been delivered in this area: a positive energy building of 12.300 m2.

2 FEASABILITY STUDY

In 2009, SPL Lyon-Confluence selected the team Herzog & de Meuron (architects/urban planners) from Switzerland and MDP - Michel Desvigne (landscaper) from France to design the second phase of the Lyon-Confluence project. This team released in December 2009 the masterplan for the redevelopment of the former wholesale market and, in September 2010, the feasibility study of a new block called P Block.

The P Block is a central element of the Lyon-Confluence project, even if the area of this piece of land is smaller than other blocks with around 3 380 m² (91m x 37m). It is surrounded by the Ice-Rink to the north, the water place and quai Antoine Riboud) on the south to the east by the Cours Charlemagne, the main street of Lyon-Confluence with the tram line, just in front of the Leisure Centre and the Rhône-Alpes region Headquarters (see Figure 1).

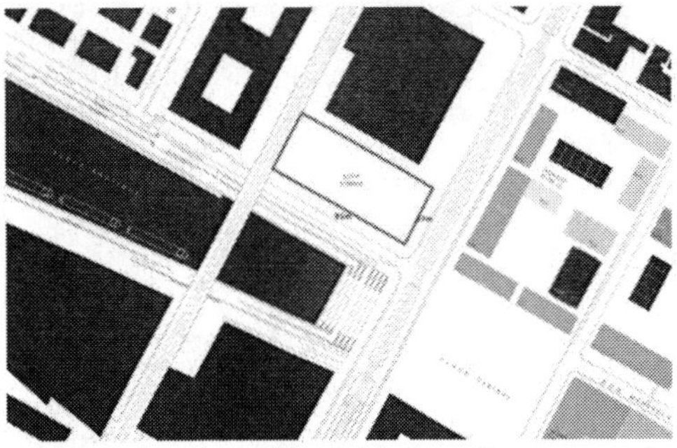

Figure 1: location of the P block at the corner water place / tram line

Herzog & de Meuron proposed that this block, as most blocks of the area, should have multiple functions (offices, dwellings and shops) and gave an estimation of the floor area potential of the building to be built on this piece of land:

- 7.000 to 8.000 m² of offices,
- 3.000 to 4.000 m² of dwellings,

- 1.000 to 1.500 m² of shops.

Herzog & de Meuron then proposed several possible building shapes with the same expected floor area per usage that can be foreseen on this plot (see Figure 2).

Figure 2: example of possible shape proposed by Herzog & de Meuron (offices in purple, dwellings in yellow and shops in orange)

3 INTERNATIONAL DESIGN COMPETITION

SPL Lyon-Confluence was in charge of organising an international competition to select the developer that will be in charge of the design and construction of the P Block building within the Lyon-Confluence area.

Based on the feasibility study done by Herzog & de Meuron for this piece of land, SPL Lyon-Confluence and its assistants, the Environmental engineering office TRIBU and Hespul, set up the environmental and energy targets of this building. Thanks to the financial support expected within the EC funded project NEXT-Buildings related to very low energy buildings [3], the energy performance requirements of this building were more ambitious than what was achieved for recent buildings of this area and was set at a very strict level: make the P Block building a positive energy building.

More precisely, this objective was defined as the following: the positive energy balance of the plot over the year shall consider the whole energy consumption (and not only the heating consumption) and shall be calculated with the yearly primary energy consumption and the yearly primary energy produced by the building (see Figure 3)

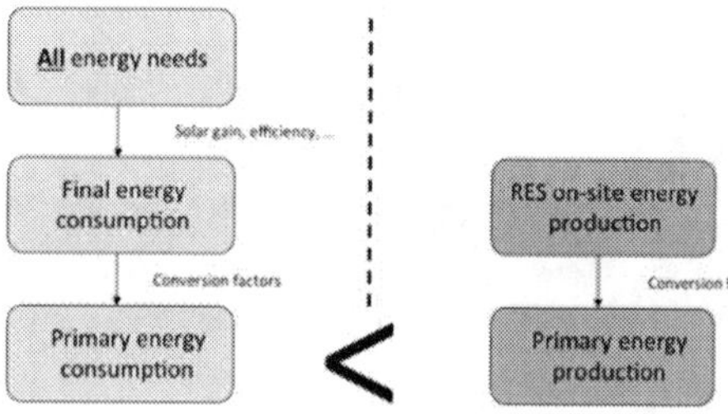

Figure 3: Definition of the positive energy target

This energy performance target and the primary energy factors to be used (3.2 for electricity and 0.2 for biomass) have been included in the environmental guidelines used for the international competition to select a developer for the design and the construction of this building.

The first round of this international competition was lunched on January 2011 by SPL Lyon-Confluence and was quite successful since 17 teams composed of developers, architects, engineering offices and environmental consultants applied to this bid. It took then 3 months for SPL Lyon-Confluence to analyze the applications and to short-list 4 teams of developers for the second round of the competition:

- ALTAREA-COGEDIM (Developer) with Bernard Reichen, Barbosa & Guimaraes (Architects),
- BOUYGUES Immobilier/SLC (Developer) with Kengo Kuma/CBR Architectes (Architects),
- ICADE (Developer) with Shigeru Ban/Rue Royale Architectes (Architects)
- VINCI (Developer) with Dominique Perrault/Mateo Arquitectura (Architects)

This time, each team had to pre-design a building and to send a comprehensive report on the architecture features and the energy performance assessment of the building.

SPL Lyon-Confluence received from each selected team building proposals with high quality architecture and very low energy needs so that it was not easy for the jury to select one of them. After a full audit of the application report and an oral presentation of each project, the jury decided, in September 2011, to award the proposal lead by Bouygues Immobilier/SLC in association with Kengo Kuma.

4 DESIGN PHASE

The awarded project proposed by Bouygues Immobilier/SLC is called HIKARI, which means natural light in Japanese. It is composed of 3 buildings for a total floor area of approx. 12.300 m2:

- HIGASHI (East), an office building with shops on the ground floor,
- MINAMI (South), in the middle of the plot, an apartment building with 36 dwellings and shops on the ground floor,
- NISHI (West), 3060 m2 with 5 storeys of offices and 4 top-end penthouse apartments on the 2 upper storeys and shops on the ground floor (see Figure 4 and Figure 5).

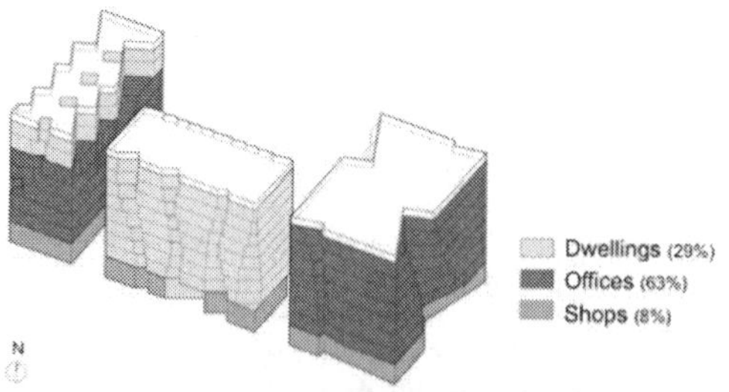

Figure 4: The 3 buildings of the HIKARI project: HIGASHI on the right, MINAMI in the middle and NISHI on the left

The design team was composed of the following companies:

- Project developers: Bouygues Immobilier and SLC Pitance,
- Architect: Kuma Associates Europe,
- Contractor: SETEC Bâtiment,

- PV engineering: Tecsol,
- Energy consultant: Manaslu Ing.

Figure 5: Virtual image of the Southern façade of the HIKARI building

4 PV SYSTEM DESIGN

4.1 Overall energy strategy

The design of a positive energy building requires to strongly focus on energy systems in order to produce on site and with renewable energy sources the energy necessary to operate the building. Thus, in order to provide electricity, heat and cold to the building, the design team decided to use 3 kinds of renewable energy sources available on site, which are solar, biomass and ground water, with 3 main energy systems which are solar PV modules, a cogeneration system (CHP) and an absorption chiller (see Figure 6).

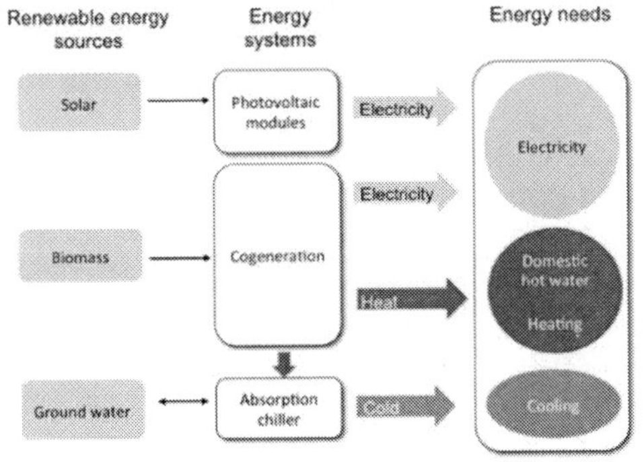

Figure 6: Overview of the energy system design

The HIKARI building is in fact equipped with 2 kinds of PV systems:

- A see-through PV façade on the MINAMI building,
- A rooftop PV system on the roof of each of the 3 buildings.

The technical details of these PV systems, which have been designed in partnership with Toshiba as a supplier, are given below.

4.2 See-through PV façade

The see-through PV façade has a power of 21 kWp and is used as balcony for the dwellings of the MINAMI building and is highly visible from the street and from the inside of the flats. Consequently, the design team paid special attention to its design and to safety issues.

As Kengo Kuma, the architect of this building, decided to design an organic building with cuttings in the façade in order to improve natural light inside the building, he also tried to avoid the use of regular PV modules and designed 9 different layouts of PV modules that give the impression of random positioning of PV cells (see figure 7).

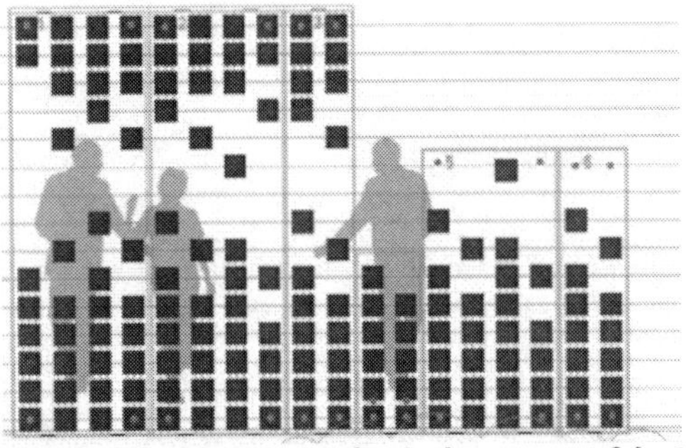

Figure 7: Drawing of 6 PV façade elements out of the 9 available layouts

The quality of the backside of PV modules is also an important issue since, in case of see-through PV façade, this becomes one element of the dwelling inside design. Thus, the design team selected high quality mounting technics based on a bolted glazing façade with custom-made metal supports and no visible DC wires (see figure 8).

Figure 8: artist impression of the back side of the PV façade from the inside of a dwelling

As this see-through PV façade is also used as balcony, an approval of the mounting technique of the PV façade by an accredited and independent body was necessary.

First, a prototype of the see-through PV façade was installed for one flat of the MINAMI building during summer 2014.

Then, a comprehensive set of tests done by a testing laboratory and among them the M50/900J test that consists in smashing the PV façade with a 50 kg sand bag from a height of 1,2 m in order to create a chock of 900 Joules according to the NFP 01-012 standard with P 08-302 testing protocol (see Figure 9). All these tests have been successful so the PV façade was approved for safety on September 2014.

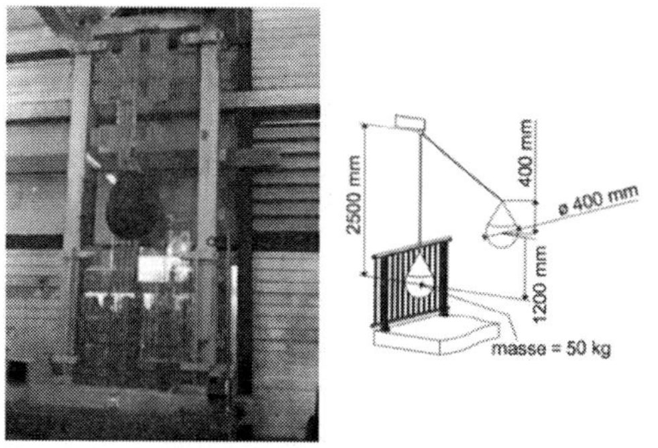

Figure 9: M50/900J chock test on the PV module

Detailed features of the see-through PV façade are:

- Cells: 1120 x Neo Solar Power NP6 multi crystalline cells (5" – 3,8 W)
- PV modules: AGC SunEwat / thickness of 21,6 mm / 9 different layouts
- Mounting technic: Bolted glazing façade with custom-made metal supports
- Inverters: 3 x SMA Tripower – 18,5 kVA in total
- Expected yield: 15 MWh/year

4.3 Rooftop PV system

The rooftop PV system has a power of 168 kWp and covers almost all the roof area of the 3 buildings (HIGASHI, MINAMI and NISHI) and has also an aesthetical role as it has been designed to give a homogenous appearance to the roof surface, as it is visible from surrounding hills.

Detailed features of the rooftop PV system are:

- PV modules: 700 x Panasonic HIT-N240 (240 Wp)
- Mounting technique: custom-made metal frames - tilt angle: 10° - east-west
- Inverters: 13 x SMA Tripower – 143 kVA in total
- Expected yield: 182 MWh/year

Figure 10: Rooftop PV system of HIKARI

5 ACHIVEMENTS

HIKARI is now completed and the official opening ceremony took place on Thursday 17 September 2015.

The current PV market in France concentrates on low cost and/or rural systems. High quality and customized systems, such as the see through PV modules, in metropolitan areas are an exception.

With its central location in Lyon, the high visibility of the PV façade and its excellent design quality, HIKARI is a new BIPV show-case. All technical details of this building are available in the deliverable D5.1 of the NEXT-Buildings project "Report on the P Plot overall construction process" [4]. A comprehensive monitoring campaign is now about to start. First results of the operation of this building and its PV system will be available at the end of 2016.

Figure 11: HIKARI building in La Confluence area in Lyon (September 2015)

Figure 12: Detail view of the see-through PV façade

6 ACKNOWLEDGEMENTS

Authors wish to thank the European Commission as this project received funding from the European Community's Seventh Framework Programme (FP7/2007-2013) under the grant agreement N° ENER/FP7/284533 (NEXT-BUILDINGS).

Authors also thank the New Energy and Industrial Technology Development Organization from Japan (NEDO) for the financial support provided to Toshiba for the construction of this building within the Lyon Smart Community project.

8 REFERENCES

[1] www.lyon-confluence.fr

[2] http://wwf.panda.org/

[3] www.next-buildings.com/

[4] Gaiddon, B., *'Report on the P Plot overall construction process'* Deliverable D5.1 of the NEXT-Building project, Lyon, 2015

VISUAL DESIGN OF PV-MODULES – A CRUCIAL FACTOR FOR FAÇADE APPLICATION ACCEPTANCE

Geissler, A.[1)], Fornaro, P.[2)], Bianco, A.[2)]
[1)]University of Applied Sciences and Arts Northwestern Switzerland, [2)]University of Basel
[1)]Institute Energy in Building, St. Jakob-Strasse 84, CH-4132 Muttenz; [2)]Digital Humanities Lab, Bernoullistr. 32, 4056 Basel

ABSTRACT: Increasing the acceptance of façade-integrated PV (photo voltaic) is of great interest. Architects will often be opposed to integrate PV in their design based on visual aspects of available PV products. The project described in this paper addresses this topic by bringing together a known technology - sandblasting of glazing surfaces – and standard, off the shelf PV-modules. The final goal of the project is to provide architects with a catalogue of possible glass-surface visual design elements. Each design element will be given with approximate reduction values for radiation transmission. Thus, ideally a design that is desired can be estimated in regard to PV-performance-values to be expected. A first set of design elements is shown and corresponding radiation transmission reduction values are given. The design elements are differently sandblasted and coloured glass surfaces with various grades of opacity. The reduction in transmission as compared to a standard low iron glass is found to be in the range of 10 to 80 %, mainly depending on the glass colour. Some examples for possible designs are shown. The relative loss in performance compared to an untreated PV-module of the same type is found to be of the same order of magnitude as the accepted spread in module power according to manufacturer data. This is viewed to be an encouraging result.

Keywords: Building Integrated PV (BIPV), Façade, Surface Modification

1 INTRODUCTION

Replacing fossil fuels by renewable energy sources is paramount in regard to a sustainable energy supply. The application of photovoltaic modules (PV) on buildings is to date mostly restricted to roofs. Using building façades would greatly increase the available area for harvesting solar power or can even be necessary for larger buildings to reach NZEB status (see e.g. [1], [2]). Often, though, architects dislike the visual appearance and/or limited visual design options of standard PV modules. There are several solutions to this dilemma available on the market to date (e.g. [3], [4], [5]), which basically focus on introducing the possibility to give colour to the cover glass. Each of these solutions has its merits, however, additional design possibilities should help increase acceptance of façade integrated PV with architects and thus pave the path to a wider application of BIPV in façades. A "sounding board" held with various local key players of the building market gave the following results:

- Overall costs of façade BIPV should not exceed that of similar, non-PV façades. Overall meaning of course that PV yield shall be taken into account.
- A peak yield is not necessarily crucial. If the design appeals, it was felt that many architects and clients would accept efficiency losses of up to 30 %. Some present even allowed for up to 50 % loss.
- Imitations (e.g. wood surface, rough plaster or similar) were not deemed acceptable, in general. Those present agreed that surface treatments must do justice to the material "glass" or have an artistic aspect.
- Older buildings (Wilhelminian style) or heritage buildings should not be considered for façade BIPV, as this would be an unacceptable encroachment on the building character.
- Façade BIPV was generally deemed acceptable for buildings dating from approximately 1950 onwards which more often than not feature simple, unstructured façades.

The method considered in this paper for the change of visual appearance is the sandblasting with optional addition of colour. The treatment is applied to standard PV modules. This approach has several advantages. Not only can this technique be regarded as proven, it is also applicable to any type of available PV modules. Also, the "glass like" surface remaining with above-mentioned other available solutions is mitigated through sandblasting the glass. In the following, possible visual effects are introduced and a first indication of the expected loss of PV performance due to the surface treatment is given.

2 VISUAL EFFECTS BY SANDBLASTING GLASS SURFACES

Figure 1 shows examples of sandblasted glass surfaces. It is also possible to add colour to the sandblasted surfaces in the process [6]. The visual effect generated depends mainly on grain type, grain size, blast density and duration and blasting power. The grain size used is typically in the range of 50 – 400 μm. Figure 2 shows a detailed view of two sandblasted glass surfaces. Finding the "right" combination of the parameters mentioned in order to achieve a desired result requires extended experimentation and long experience, however. The technique has been used for several decades and there are many examples on buildings world wide available. The sandblasted surface is typically sealed to protect the "chipped" surface structure and has been shown to be durable. Figure 3 shows examples of a sandblasted standard CI(G)S PV modules.

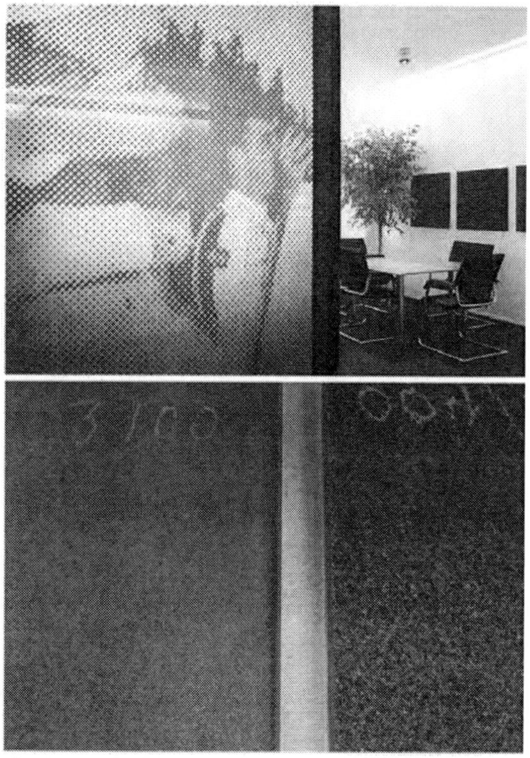

Figure 1: Sandblasted glass. A picture on a glass compartment wall (top) and samples with additional colour (bottom) [6].

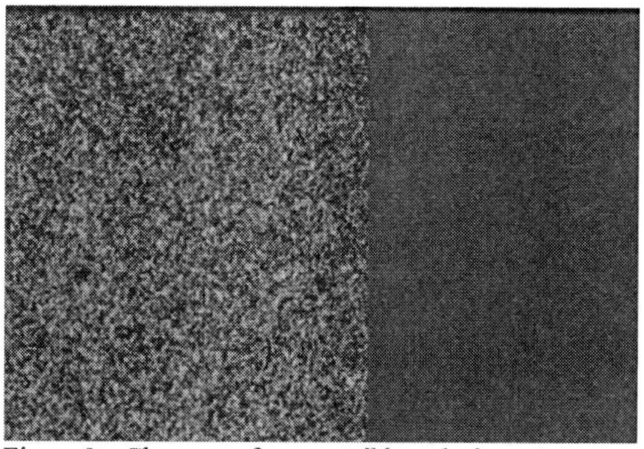

Figure 2: Close-up of two sandblasted glass specimens with different grain-sizes blasted for laboratory measurement.

Figure 3: Standard CI(G)S PV modules after treatment of glass cover with sandblasting techniques (top) and sandblasting & colouring techniques (centre, bottom) [6], [7].

3 EXPECTED PV PERFORMANCE

3.1 Laboratory Measurements

3.1.1 Measurement Setup

The measurement set-up is a combination of a flash-light with a colour spectrum similar to daylight and with a colour temperature of 6'400 K (see Figure 4). The flashlight triggers the spectral measurement device Sekonic C-700R that is able to measure the following parameters:

- Total incoming luminous exposure H [lx s]
- The spectral distribution of the incoming light in the extended visible range from 380nm – 780nm.

The measurements are done in a black painted lab on an optical table to reduce errors due to light scattering and errors from vibration or mechanical instabilities.

After calibrating the setup, the measurements are done by the following procedure:

- The light transmission of the PV-glasses is measured for three incoming light angles (0°, 45° and 65°, where 0° is equivalent to the glass surface normal direction) and each measurement is performed 5 times to average intrinsic errors of the components used (5 % for the spectral measurement device, < 0.3 EV for flash).
- For the 0° measurement setups for direct and diffuse light are used.
- The averaged measured data is set in relation to the reference measurement (untreated low iron glass). The ratio thereof is stored as transmission and loss information for each type of glass treatment.

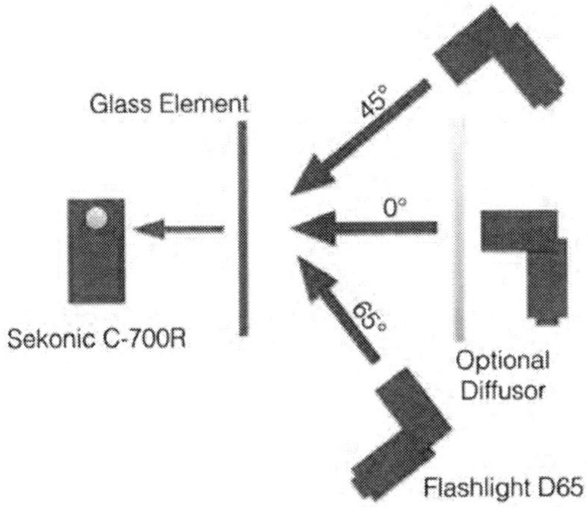

Figure 4: Laboratory Measurement setup consisting of spectral measurement device, flash light and test specimen.

3.1.2 Measurement specimens

The laboratory measurements are to give initial guidelines in regard to the effect of considered surface treatments on glass light transmission and thus likely PV performance degradation. The set of test glasses consists of:

- four blasting grades / particle size ranges (S1 – S4) with two blasting densities ("open", "closed") each,
- magenta-, blue-, yellow- and green-coated glasses with one and two coating layers and
- three different sealing types for protection (A, B, C).

The test glasses are based on standard four-millimetre thick low iron glass. Figure 5 shows some examples of the test specimens. Overall, 56 individual glass specimens were tested to date.

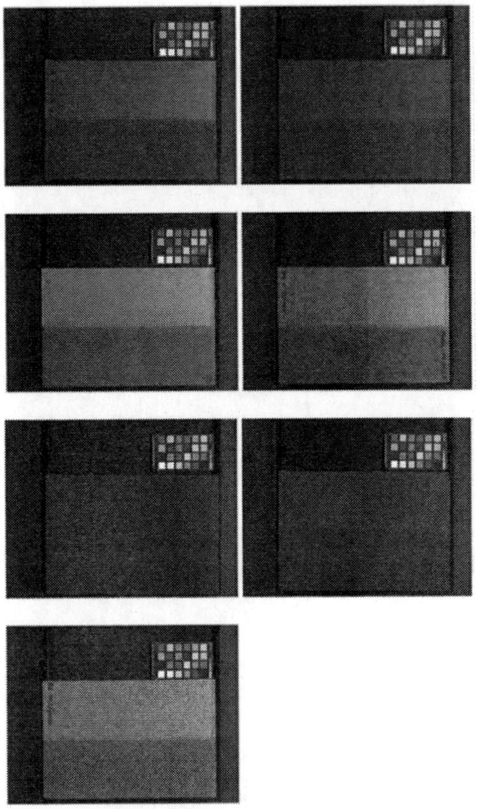

Figure 5: Sandblasted and coloured glass specimens. The pictures show a selection of the test specimens with black and white back-layers.

3.1.3 Laboratory Results

The measurement results show that the surface modification due to sandblasting and the choice of coating can influence the transmission of the PV-glass elements significantly. It is straight forward that colour coatings absorb incoming light relative to the density of the colorants or pigments applied. Figure 6 shows the change in spectral transmittance of two coloured test specimens compared to the reference glass. The values for the red and green glass examples are given as the ratio between the transmittance of coloured and the non-coloured glass. The surface modification due to sandblasting leads to surface scattering, subsurface-scattering, absorption and reflection. In the measurement procedure, the ratio between those effects of surface modification has not been studied in a first step, because the change in overall transmittance is of major interest in regard to PV performance loss. Table I summarizes the results obtained in the form of loss factors compared to an untreated cover glass.

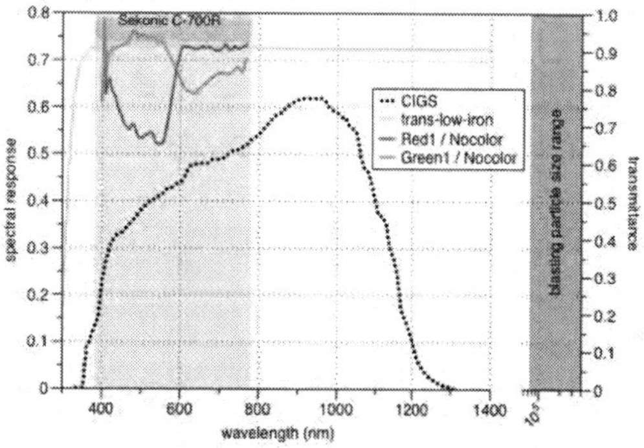

Figure 6: Spectral response of a typical CI(G)S cell, transmittance of a standard low iron glass pane and exemplary relative spectral transmittance values from the laboratory measurements for red and green coatings. The relative transmittance values use the scaling "transmittance" on the right.

Table I: Relative loss of light transmission of sandblasted and sandblasted & coloured glass specimens compared to an untreated but otherwise identical glass pane (4 mm low iron glass). The uncertainty shows the standard deviation of the mean of five measurements with each specimen. S1 to S4 are different blasting grain size intervals, "open" means a less dens blasting structure, "closed" a more dens structure.

Glass	Sandblasted	Sandblasted + Blue	Sandblasted + Red	Sandblasted + Yellow	Sandblasted + Green
0°, no diffusor					
S1 Open	0.15 ± 0.01	0.24 ± 0.01	0.30 ± 0.01	0.12 ± 0.02	0.08 ± 0.02
S1 Closed	0.23 ± 0.01	0.32 ± 0.01	0.43 ± 0.00	0.14 ± 0.01	0.19 ± 0.01
S2 Open	0.27 ± 0.03	0.44 ± 0.02	0.46 ± 0.00	0.18 ± 0.01	0.26 ± 0.00
S2 Closed	0.20 ± 0.01	0.49 ± 0.01	0.58 ± 0.00	0.19 ± 0.01	0.35 ± 0.01
S3 Open	0.12 ± 0.01	0.35 ± 0.01	0.43 ± 0.01	0.13 ± 0.01	0.20 ± 0.01
S3 Closed	0.20 ± 0.01	0.50 ± 0.00	0.56 ± 0.00	0.23 ± 0.01	0.33 ± 0.00
S4 Open	0.09 ± 0.02	0.31 ± 0.02	0.35 ± 0.01	0.14 ± 0.02	0.19 ± 0.01
S4 Closed	0.20 ± 0.01	0.46 ± 0.01	0.57 ± 0.00	0.24 ± 0.01	0.35 ± 0.00
65°, no diffusor					
S1 Open	0.17 ± 0.02	0.31 ± 0.01	0.32 ± 0.01	0.18 ± 0.01	0.13 ± 0.01
S1 Closed	0.36 ± 0.00	0.57 ± 0.01	0.65 ± 0.00	0.34 ± 0.01	0.41 ± 0.01
S2 Open	0.37 ± 0.01	0.62 ± 0.02	0.69 ± 0.00	0.42 ± 0.00	0.49 ± 0.01
S2 Closed	0.43 ± 0.01	0.70 ± 0.01	0.75 ± 0.00	0.48 ± 0.01	0.60 ± 0.00
S3 Open	0.31 ± 0.00	0.45 ± 0.01	0.57 ± 0.01	0.31 ± 0.01	0.42 ± 0.01
S3 Closed	0.42 ± 0.00	0.72 ± 0.00	0.73 ± 0.00	0.49 ± 0.01	0.60 ± 0.00
S4 Open	0.21 ± 0.01	0.39 ± 0.02	0.45 ± 0.00	0.26 ± 0.01	0.30 ± 0.01
S4 Closed	0.41 ± 0.00	0.70 ± 0.01	0.77 ± 0.00	0.46 ± 0.00	0.59 ± 0.00
0°, with diffusor					
S1 Open	0.04 ± 0.01	0.17 ± 0.01	0.26 ± 0.00	0.04 ± 0.03	0.07 ± 0.02
S1 Closed	0.11 ± 0.02	0.26 ± 0.01	0.41 ± 0.01	0.07 ± 0.02	0.18 ± 0.03
S2 Open	0.14 ± 0.01	0.40 ± 0.02	0.44 ± 0.01	0.11 ± 0.01	0.24 ± 0.01
S2 Closed	0.17 ± 0.01	0.42 ± 0.01	0.56 ± 0.01	0.14 ± 0.01	0.30 ± 0.00
S3 Open	0.14 ± 0.01	0.27 ± 0.01	0.41 ± 0.01	0.07 ± 0.01	0.14 ± 0.01
S3 Closed	0.20 ± 0.01	0.47 ± 0.00	0.53 ± 0.01	0.15 ± 0.01	0.30 ± 0.01
S4 Open	0.12 ± 0.01	0.21 ± 0.02	0.29 ± 0.01	0.05 ± 0.02	0.12 ± 0.02
S4 Closed	0.20 ± 0.01	0.41 ± 0.01	0.53 ± 0.01	0.15 ± 0.01	0.30 ± 0.01

3.1.4 Expected overall performance reduction

The PV module cover glass will often not be sandblasted / treated on the full surface. Also, currently it is being looked into the possibility to reduce the actually affected area in sandblasted regions by using printing screens. The overall power loss of a PV module will thus depend on treated area and affected area in the treated area. This can be expressed e.g. by eqn. (1).

$$p_t = p_0\left(1 - \prod r_k\right) \qquad (1)$$

Where

p_t expected module power / yield with treatment,

p_0 module power / yield without treatment,

r_k loss factors / cover ratios.

For example, if 50 % of the module surface has a design-treatment and the loss (the reduction of transmission as compared to an untreated cover glass) due to the treatment is 20 % (e.g. 0° w/o diffusor, S2, closed, no colour), overall a 10 % power / yield reduction is expected. If the actually treated area can be reduced to e.g. 85 % without changing the visual effect, then an overall power / yield reduction of 8.5 % would be expected (see section 3.3). Of course, actual solar radiation situations will be a mix of direct and diffuse radiation where the direct radiation will have varying angles of incidence.

3.2 Full scale comparison measurements

At the time of writing of this paper, two full-scale measurement campaigns have been initiated. The PV modules are mounted in a vertical position facing south in both cases. The measurement site is not shaded. The data is acquired via the optimizer for each separate module and consists of hourly or daily yield.

A) A selected set of surface designs applied to PV-modules is submitted to a full scale measurement campaign in order to "ballpark" the actual performance which can be expected (see Figure 3 for the designs considered, here).

B) A series of basic design elements are applied to PV-modules of different types and also measured in full scale in comparison to untreated, otherwise identical PV-modules. Figure 7 shows the surface treatments used, here.

Results for A) are shown in Table II. Results for B) are not available at this stage. The initial results from A) are quite encouraging. Performance loss due to the surface treatments according to Figure 3 is in the order of magnitude of the variation between the PV-modules themselves.

Prior to surface treatment for B), the PV-modules used are measured without treatment for several days in order to ascertain the spread of their yield under identical boundary conditions (see Table III for results). It can be seen that especially the selection of CI(G)S modules have a quite large spread in yield. The largest spread observed here is found for the "CIS" modules considered and lies just inside the manufacturer data given. Of course, this spread must be taken into account in the evaluation of the treated modules.

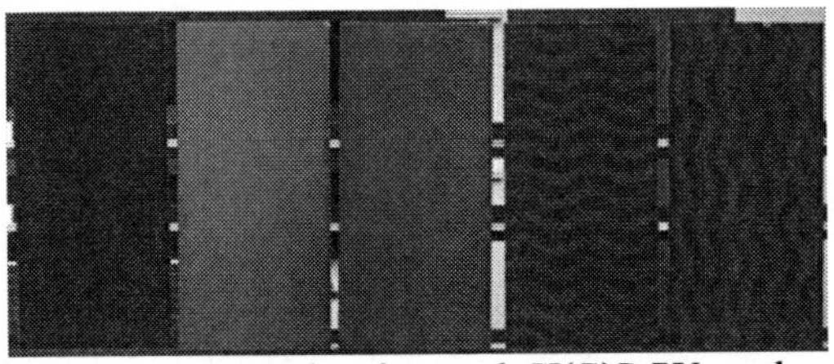

Figure 7: Untreated and treated CI(G)S PV modules. From left to right: Untreated reference module, "closed", "open", 50 % coverage with "arbitrary" designs (horizontal and vertical waves). These measurements are on-going at the time of writing of this paper.

Table II: Measurements w/ designs according to Figure 3. PV-module vertical, orientation south, location Basel (Switzerland), the measurement period is April 16th through May 3rd 2016 (18 days). The leftmost column (typeset bold) refers to an untreated module of the same type as the treated modules.

	Un-treated module	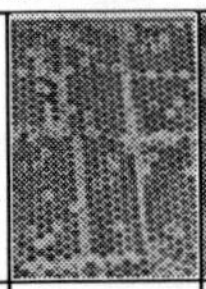			
yield	**6000**	5290	5870	5560	Wh
rel. yield	**100.0**	88.2	97.8	92.7	%
yield diff	-	-11.8	-2.2	-7.3	%

Table III: Reference measurements, PV modules vertical, orientation south, location Basel (Switzerland), measurement period March 22nd through April 6th, 2016 (16 days).

	Ref. module	Module 1	Module 2	Module 3	Module 4
	m-Si				
yield (Wh):	**8436**	8512	8502	8758	8528
rel. yield (%)	**100.0**	100.9	100.8	103.8	101.1
yield diff. (%)	-	0.9	0.8	3.8	1.1
	c-Si				
yield (Wh):	**7648**	7850	7687	7480	7637
rel. yield (%)	**100.0**	102.6	100.5	97.8	99.9
yield diff. (%)	-	2.6	0.5	-2.2	-0.1
	CIS				
yield (Wh):	**4937**	5190	5449	4839	5211
rel. yield (%)	**100.0**	105.1	110.4	98.0	105.6
yield diff. (%)	-	5.1	10.4	-2.0	5.6
	CIGS				
yield (Wh):	**4029**	4131	3973	4026	4211
rel. yield (%)	**100.0**	102.6	98.6	99.9	104.5
yield diff. (%)	-	2.6	-1.4	-0.1	4.5

3.3 Optimization procedure

As described in section 1.1.1, the sandblasting technique most appropriate for the visual design of PV modules is still in development. Figure 8 (centre and right) shows initial approaches to the above-mentioned reduction of covered area by using screening techniques. The potential of such optimisations is not yet fully tapped. However, it is expected that it should be possible to avoid more than 10 % actual coverage without significant impact on the visual effect.

The other area of improvement identified was which colour type to use. Currently, colours with embedded reflecting particles are being evaluated for suitability.

Figure 8: Detail examples (bottom row) of the sandblasted surface structures based on the PV modules shown in Figure 3. "Closed" structure with very fine grain on the left, more open structures in the middle and on the right with larger and again fine grain, respectively.

4 CONCLUSIONS & OUTLOOK

New design possibilities are an important component for mass application of Façade-BIPV. The sandblasting technique is a very promising option in this regard. A wide variety of surface designs can be applied to standard PV modules by a proven technique. The loss in efficiency depends on structure density and colour choice. The optimization of the blasting grain sizes, structure density and type of colour used is on-going. Initial results obtained are quite promising.

Architects will be provided with a wide range of possible design elements. Apart from the continued optimizations mentioned, further areas of interest are e.g. questions of product warranty and loss thereof due to "tampering" with the product "PV module". These and other practical aspects of the solution proposed will be considered in the further scope of the project of which the results described herein are a part. Implementation of a PV façade cladding featuring modules treated by sandblasting and colouring is scheduled for a building in Basel near the end of the current year.

Acknowledgements

This work is being supported by the Swiss Federal Office of Energy (SFOE) through the research project "Development of new designs of photovoltaic glass panels to improve acceptance in society", contract number SI/501274-01. The core project team "solarglaslabor" (see [8]) consists of Markus Bloch, raumweg GmbH; Martin Gruber, Gewerbeverband Basel; Dominic Müller, Solvatec and Marc Weidmann, Atelier Weidmann. The authors would like to thank the project team for the collaboration opportunity.

References

[1] M. Hall and A. Geissler, "Möglichkeiten und Grenzen von grossen Nullenergiegebäuden," Schlussbericht BFE SI/500950-01, www.fhnw.ch/habg/iebau, 2015.

[2] A. Scognamiglio and F. Garde, "Photovoltaics' architectural and landscape design options for Net Zero Energy Buildings, towards Net Zero Energy Communities: spatial features and outdoor thermal comfort related considerations," in *Progress in Photovoltaics*, Amsterdam, The Netherlands, 2014.

[3] M. Osborne, "Can white PV panels do the same as blue LEDs?," *PV-Tech*, 28-Oct-2014. [Online]. Available: http://www.pv-tech.org/news/36977. [Accessed: 01-Feb-2016].

[4] J. Wellstein, "Gebäudehülle: Solarstrom inklusive," *Haustech*, no. Nr. 5, pp. 42–44, May-2015.

[5] S. Pelisset, M. Joly, V. Chapuis, A. Schüler, S. Mertin, V. Hody-Le Caër, and C. Ballif, "EFFICIENCY OF SILICON THIN-FILM PHOTOVOLTAIC MODULES WITH A FRONT COLOURED GLASS," in *Proceedings CISBAT 2011*, Lausanne, Switzerland, 2011, pp. 37–42.

[6] M. Weidmann, "Atelier Weidmann," *Atelier Weidmann GmbH*. [Online]. Available: http://www.atelier-weidmann.ch/index.php?page=de/home.html. [Accessed: 01-Feb-2016].

[7] M. Gruber, "Neues Design für Solarfassaden," *KMU Magazin, Gewerbeverband Basel-Stadt*, no. 2, p. 35, Sep-2015.

[8] M. Bloch, "ARGE solarglaslabor," *www.solarglaslabor.ch*. [Online]. Available: http://www.solarglaslabor.ch. [Accessed: 20-Jun-2016].

INTEGRATION OF TRACKLESS HOLOGRAPHIC CPV MODULES IN BUILDINGS AND URBAN FURNITURE

Rodríguez San Segundo HJ*, Pérez López FJ, Villamarín Villegas AM and Calo López A
Instituto Holográfico Terrasun, S.L.
P.T. Tecnobahía - Ctra. Sanlúcar, km. 7, Edificio ReTSE
11500 El Puerto de Santa María (Cádiz, Spain)
Tel. +34 956 549 354, Fax +34 956 549 416
*Author for correspondence. E-mail: hrodriguez@holotech.es

ABSTRACT: Concentration photovoltaic (CPV) systems are usually destined for large field applications, because the conventional optics applied require mandatory tracking, which implies larger size systems, larger land use and a high quantity of additional hardware. Thus, the use of CPV is very restricted when applied to small size systems, let alone integrated applications. However, the very compact Low CPV (LCPV) module named SoHo3X has shown that it can avoid tracking by using holograms, and yet provide up to 50% more capacity factor than a conventional, fixed flat module, at very competitive costs. This work presents the performance measurements for the first time and uses them to calculate performance at BIPV representing positions, tilts, orientations and latitudes, in order to show how it can be used both in BIPV and also as integrated photovoltaics in urban furniture. The advantage versus integrating conventional modules is not only a higher production, but also a much lighter, more aesthetic product, with less need for solar cell area (concentration factor is around 3X). Conclusion is that SoHo3X shows a much better performance in most cases, with a few exceptions for vertical position at high latitudes. Mechanical load simulations have also been carried out, showing that SoHo3X complies with the strictest mechanical load regulations, fulfilling class I mechanical stability for wind loads, in opposition to conventional flat PV modules.
Keywords: Building Integration, Hologram, Concentrators, Performance

1 INTRODUCTION

It is a well-known fact that big urban environments are the main energy consumers. The difference in consumption from urban to rural areas will increase with the ever-increasing number of electric gadgets such as smart phones, tablets, and even with the expected increase in electric vehicles [1]. However, electricity generation is generally very far from these high consumption areas. Even those utilities assuring a high percentage of green origin of their electricity, like the Munich Electric Utility SWM, import their power even from distant countries across Europe and the world [2].

Transport along these thousands of kilometers does not only imply higher costs, but also higher carbon footprint, as well as higher energy losses along the way.

Building and urban integration of electricity generation is therefore desirable in order for green energy to become even greener and more efficient. Traditionally, however, Building (and Urban) Integrated Photovoltaics (BIPV) has the disadvantage of less energy production due to forced tilts and orientations, and also of higher costs due to the use of specifically designed, more expensive solar modules and the higher Balance-of-System costs (BoS) implied by the integration itself.

For more than 20 years, hologram use as efficient optics for solar concentrators [3-5] has had low success in the right design for efficient, compact and inexpensive devices. The company IHT has succeeded with a previously presented novel concept of Low Concentrated Photovoltaic (LCPV) device, named SoHo3X [6-9]. This device shows *modularity* from 20 watts-peak (W_p) and integrates a *self-supporting structure*, both aspects making it a very good candidate for integration into virtually any building or urban furniture. In addition, costs are similar to those of conventional flat PV modules, and mounting on urban furniture is straightforward, so that no additional BoS is expected.

SoHo3X was designed for optimum operation at horizontal position, facing South and tilted at latitude. However, the very good results obtained, and its ability for urban integration, lead the authors to investigate on other positions, orientations, tilts and for various latitudes. Such is the object of the present work.

2 EXPERIMENTAL MEASUREMENTS

2.1 Indoor measurements

As shown in Figure 1 in percentage of power gain over conventional, fixed flat silicon PV modules, SoHo3X prototypes with the same silicon cell type were submitted to power measuring by a sun simulator at several angles of incidence, simulating the sun movement along the year ("winter-summer") and along the day ("morning-evening"). The SoHo3X prototypes submitted to all tests presented here incorporated very early versions of holograms, which only diffracted 150 nm of the incoming light around the green region of the spectrum [7]. Therefore, as expected, the "winter-summer" measurements show decay in performance even below flat modules for angles of incidence over 30° from normal (see Figure 1b). Improvements for higher bandwidth holograms are expected in future work.

In opposition, performance along the day showed very steep power gains for the early morning and late evening hours, even reaching over 400% more power than conventional fixed flat PV modules (Figure 1c).

2.2 Outdoor measurements

These measurements along the day have been confirmed by outdoor measurements, as shown in Figure 2, which represents normalized power versus global incident irradiance (GII). SoHo3X power measurements were taken along several days, confirming higher performance than fixed flat conventional PV module. The latter, as known, adapt to the cosine curve of GII.

Calculations derived from these experimental data for

the city of Seville, taking sun position data from the sunearthtools.com database [10] and irradiation on a surface oriented South and tilted to latitude from the PVGIS database [11], showed an impressive potential for SoHo3X, of 50% more electricity production along one whole year than a conventional PV module.

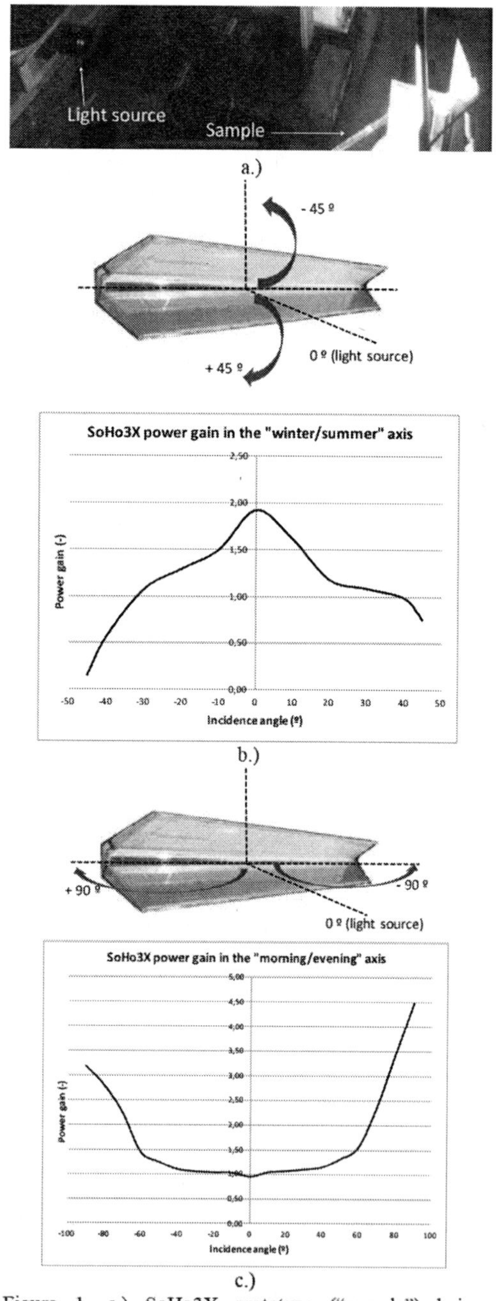

Figure 1. a.) SoHo3X prototype ("sample") being submitted to indoor power measuring under sun simulator ("light source"); b.) Mode of measurement (above) and results (below) of SoHo3X performance as power gain over flat PV module, in the "winter-summer" simulation; c.) Mode of measurement (above) and results (below) of SoHo3X performance as power gain, in the "morning-evening" simulation.

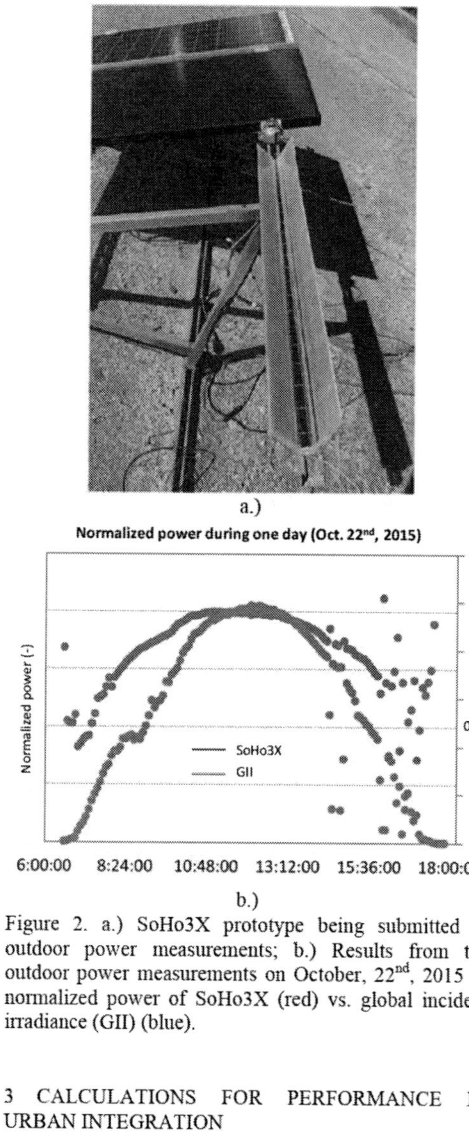

Figure 2. a.) SoHo3X prototype being submitted to outdoor power measurements; b.) Results from the outdoor power measurements on October, 22nd, 2015 as normalized power of SoHo3X (red) vs. global incident irradiance (GII) (blue).

3 CALCULATIONS FOR PERFORMANCE IN URBAN INTEGRATION

The above-mentioned encouraging results, together with the attractiveness of the product, which is colorful and changes color along the day, as well as its modularity from 20 W_p power and the easy integration in buildings and urban furniture thanks to its self-supported structure, make this photovoltaic device ideal for urban integration.

The authors have thus investigated the performance of SoHo3X for the many possible situations a photovoltaic module may have when integrated in cities. As an example of this, Figure 3 shows the picture of a streetlight with integrated SoHo3X in vertical position with three simultaneous orientations: East (E), South (S) and West (W), while keeping the harmony of the circular shape of the streetlight pole.

For this work, variations in position, tilt and orientation have been simulated for several cities located across Europe, as well as for other three big Asian capitals located at lower latitudes.

Figure 3. Streetlight with integrated SoHo3X device oriented East, South and West simultaneously.

Figure 4 shows the simulated orientations, tilts and positions, where:

- Three orientations have been chosen: East (E), South (S) and West (W).
- Three tilts have been chosen: solar cell plane at 0°, solar cell plane at 45°, and solar cell plane at 90° vs. horizontal plane.
- Two different positions of SoHo3X have been simulated: horizontal, i.e. with the longitudinal axis following the horizontal plane, and vertical, i.e. with the longitudinal axis following the vertical plane. For this latter position, only 90° tilt is simulated, as other tilts make no sense, as they can be obtained by the horizontal position. The vertical position does include the special case of the streetlight with simultaneous E-S-W orientations (see Figure 3).

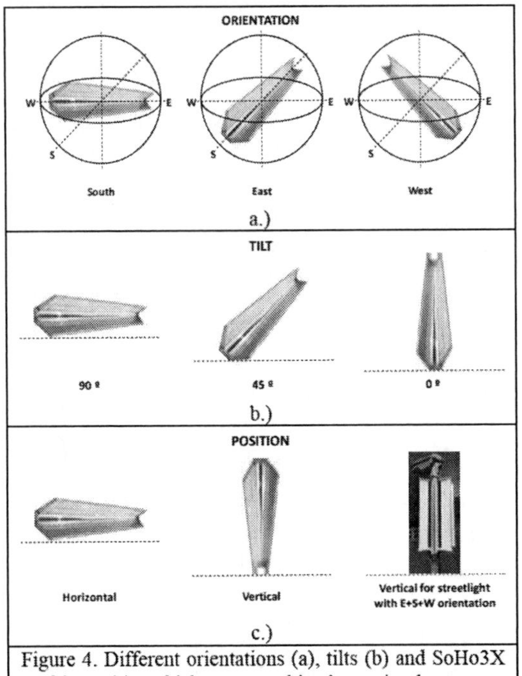

Figure 4. Different orientations (a), tilts (b) and SoHo3X positions (c), which are combined to simulate many possible situations of the SoHo3X device when integrated in buildings or urban furniture.

Results from the simulations for all situations and latitudes are summarized in Table I, represented as percentage of yearly energy production gain of SoHo3X along a full year versus a fixed flat module with same tilt and orientation, except for the case of solar streetlight, where the three simultaneous orientations of SoHo3X are compared with a standard fixed flat module oriented South. Data for the simulation have also been taken from the databases sunearthtools.com [10] for solar position and PVGIS [11] for daily irradiation.

Table I. Percentage of yearly energy production gain of SoHo3X, along one full year, versus a fixed flat module with same tilt and orientation. (*) For streetlight (simultaneous E-S-W orientation), comparison is made with standard module at 90° tilt and oriented South.

		HORIZONTAL POSITION									VERTICAL POSITION			
		EAST			SOUTH			WEST			TILT 90º			
LATITUDE	CITY	0º	45º	90º	0º	45º	90º	0º	45º	90º	E	S	W	E-S-W(*)
41.4N	BARCELONA	22	65	10	11	42	6	-21	34	35	-4	-1	-7	-22
52.5N	BERLIN	21	68	19	7	35	9	-24	21	24	-13	-10	-16	-28
44.4N	BUCAREST	18	52	24	10	41	16	-24	10	29	-4	-3	-8	-26
47.5N	BUDAPEST	23	72	17	12	43	12	-21	27	31	-9	-4	-11	-23
50.9N	COLOGNE	20	78	23	11	42	16	-21	24	33	-12	-8	-14	-27
55.7N	COPENHAGEN	22	63	16	7	31	7	-25	20	29	-13	-12	-17	-30
53.4N	DUBLIN	22	62	21	4	28	9	-27	23	43	-14	-12	-15	-30
53.6N	HAMBURG	21	62	17	4	32	7	-27	18	19	-13	-12	-17	-29
38.7N	LISBON	18	66	8	17	44	11	-17	38	44	1	4	-4	-15
51.5N	LONDON	22	70	21	7	35	9	-24	20	22	-13	-10	-18	-17
40.4N	MADRID	20	54	3	10	41	3	-23	32	23	-3	1	0	-15
45.5N	MILAN	21	56	7	5	36	2	-26	25	16	-6	-4	-10	-16
48.1N	MUNICH	23	62	14	7	36	7	-25	21	16	-9	-6	-13	-26
48.9N	PARIS	21	69	16	11	41	12	-21	27	32	-9	-7	-12	-25
41.9N	ROME	21	51	7	10	40	3	-23	30	23	-4	-1	-6	-20
48.2N	VIENNA	22	68	14	8	40	8	-23	26	25	-9	-5	-12	-25
52.2N	WARSAW	22	56	14	6	35	8	-26	20	13	-11	-9	-15	-27
13.8N	BANGKOK	-3	28	-14	30	69	19	-15	32	13	20	17	31	23
24.7N	RIYADH	7	46	-11	18	58	10	-9	45	38	22	12	17	9
1.4N	SINGAPORE	-14	8	-20	45	83	28	-14	31	15	25	29	32	57

The research summarized in Table I allows to withdraw the following conclusions:

a.) Most horizontal positions of SoHo3X imply a gain in yearly electricity generation as compared to a conventional fixed flat silicon PV module.

b.) An exception are those modules oriented west and tilted 0° (i.e. laid flat on the ground). This latter case is due to the asymmetry of the SoHo3X "wings". In this situation, the west-wing casts shadow over the module most of the time. It only needs to be turned 180° with respect to the longitudinal axis, and then the "East-0°" gain will be accomplished.

c.) Another exception are low latitudes (below 25°), for which also the east-0° and the east-90° situations are clearly unfavorable. This is due to the fact that, for these latitudes, the sun is most of the time very high, and the gains at the steep angles of morning and evening shown in Figure 1c are less evident.

d.) However, this same fact favors the high gains obtained in vertical positions in these low latitudes. Clearly the low sun positions for higher latitudes make vertical position not recommendable from a strict point of view of productivity (see below how streetlights can overcome this obstacle). Therefore, vertical positioning would also be recommendable at low latitudes.

e.) For high latitudes, the highest gain is obtained for those modules oriented east and tilted 45°, with over 50% more electricity production than conventional modules. This is due to the fact that the highest gain is obtained at very steep morning angles, which happen in these latitudes quite often in the summer months (Figure 1c).

f.) For low latitudes, the highest gain is for modules

oriented south and tilted 45°, fully in accordance with the sun's very high elevation angle most of the time.

These results will allow to estimate immediate performances for the many possibilities of urban integration along the many cities from the Equator up to over 50° latitude.

4 THE SPECIAL CASE OF SOLAR STREETLIGHT

Even though Table I shows that vertical positioning of SoHo3X is unfavorable for European latitudes, including the simultaneous E-S-W orientation of solar streetlights (very favorable for latitudes below 25°, though), there are several reasons that encourage the authors to examine the possibilities that this application provides. This reasons are:

a.) Unlike conventional solar modules, SoHo3X allows real integration *around* the solar streetlight pole.

b.) Plug-in kit with clamps makes installation easy and very cost-effective, as well as very fast.

c.) Unlike conventional solar modules, no special further requirements are needed, for instance reinforce the streetlight foundations, or adapt special structures. Mechanical stability and resistance to winds has been calculated for the streetlight with coupled SoHo3X as represented in Figure 3 (see below). Class I resistance (120 km/h winds in flat areas) is achieved, which means that resistance criteria are met for all cities around the world.

Mechanical stability has been calculated according to the international standards for mechanical resistance of streetlights, taking the strictest class (class I) into account. This class implies a terrain category I, i.e., the installation of streetlights along open seashores, lakes with exposed wind areas below 5 km in extension, or flat and open grounds with no obstacles. Wind speed is set at 120 km/h. Calculations have been made for the worst case scenario, which is normal incidence of wind on the modules (see Figure 5). Comparison between a conventional flat silicon module of 60 W_p, and a SoHo3X kit of three SoHo3X devices of 20 W_p each has been used for this simulation, as the one in Figure 3. This kit is named SoHo3X SL60 (which stands for "StreetLight 60 W_p").

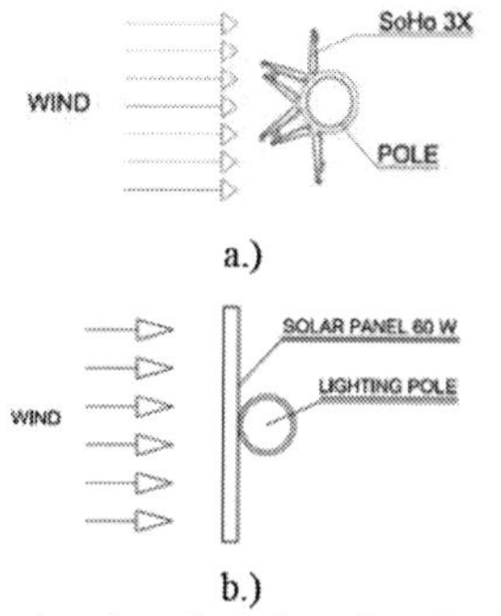

Figure 5. Premises for calculation of mechanical load on streetlight with kit SoHo3X SL60 (a) and conventional 60 W_p silicon module (b). Wind incidence normal to module plane; wind speed 120 km/h; terrain class I.

Commercial streetlights with conventional dimensions, ranging from 4 m height to 12 m height have been used in these calculations as holders for the solar devices. Table II clearly shows that the SoHo3X SL60 kit withstands mechanical stability for all kinds of streetlights, whereas a conventional 60 W_p flat module would need extra foundations in order to pass the test. In plain English: SoHo3X can be directly installed around any existing streetlight in any city without extra works, against the complexity of doing so with conventional modules. This reduces costs significantly.

Table II. Compliance with international mechanical stability standards for SoHo3X SL60 kit and for a conventional 60W_p flat silicon module, for several commercially available streetlights.

Commercial streetlight	Height (m)	Bottom diameter (mm)	Top diameter (mm)	Cylinder thickness (mm)	SoHo3X SL60 compliance	60 Wp flat module compliance
PRODUCT A	4	102	60	4	YES	NO
PRODUCT B	5	113	60	4	YES	NO
PRODUCT C	6	124	60	4	YES	NO
PRODUCT B	7	35	60	4	YES	NO
PRODUCT D	9	157	60	4	YES	NO
PRODUCT E	11,7	186	60	4	YES	NO
PRODUCT F	4	124	76	3	YES	YES
PRODUCT G	5	120	60	3	YES	NO
PRODUCT H	6	132	60	3	YES	NO
PRODUCT I	7	144	60	3	YES	NO
PRODUCT J	9	184	76	4	YES	NO
PRODUCT K	12	220	76	4	YES	YES
PRODUCT L	12	246	102	4	YES	YES

Given these results, it is interesting to study the number of SoHo3X SL60 kits installed around already existing streetlights, that would be needed to provide right-on-spot electricity to, for instance, cover the annual charging of all the smartphones in a certain city. This has been calculated based on the results obtained in Table I for the vertical simultaneous E-S-W orientation. Table III summarizes the results found for several European cities. The number of necessary kits is extremely low, far below 1% of the total number of streetlights in a city. This kit is then a very interesting application to be promoted by local authorities at low cost, thus not only charging all the city's smartphones, but also enhancing the consciousness for green energies among the population. In the future, given the number of streetlights still available for use, it can be extended for the charging of the ever increasing number of electrical gadgets, including electric vehicles.

Table III. Yearly consumption of electricity in charging all city smartphones for several European cities; number of SoHo3X SL60 kits that cover this consumption needs; and annual emission of CO_2 avoided.

City	Number of smartphones (milliards)	Annual energy consumption (*) (kWh)	SoHo3X SL60 annual production (kWh/unit)	Total number of SoHo3X SL60 kits (units)	CO2 avoided (kg/year)
Barcelona	1.136	4.146	52,10	80	2.281
Berlin	2.100	7.665	32,26	238	4.216
Bucharest	950	3.468	39,22	88	1.907
Budapest	720	2.628	37,11	71	1.445
Cologne	660	2.409	31,46	77	1.325
Copenhagen	444	1.621	33,46	48	891
Dublin	325	1.186	33,53	35	652
Hamburg	1.080	3.942	32,16	123	2.168
Lisbon	300	1.095	52,87	21	602
London	5.780	21.097	39,34	536	11.603
Madrid	2.201	8.034	56,70	142	4.419
Milan	780	2.847	47,46	60	1.566
Munich	900	3.285	34,56	95	1.807
Paris	1.078	3.935	35,03	112	2.164
Rome	1.740	6.351	49,52	128	3.493
Vienna	864	3.154	36,23	87	1.734
Warsaw	697	2.544	31,76	80	1.399
TOTAL	21.755	79.406		2021	43.673

(*) Average battery capacity 10Wh considered (2.600mAh, 3,85V). One charge daily

5 CONCLUSIONS

SoHo3X is a novel low CPV solar device that incorporates the combination of holograms with conventional optics as a concentrator. Power has been measured for different incident angles in both longitudinal ("morning-evening") and transversal ("winter-summer") directions, resulting in a potential for yearly electricity production increase over 50% with respect to a fixed, flat conventional silicon PV module with the same orientation (South) and tilt (to latitude). This fact, together with the design features that make SoHo3X prone to integration in buildings and urban furniture, have encouraged the authors to calculate performance for a number of different tilts, orientations and positions at several latitudes ranging from the Equator to over 50° North.

Results show that gain in electricity production is general for most situations for a horizontal position of SoHo3X, while vertical position is clearly also favorable at lower latitudes. Not only does the device produce more than a conventional module: it is also visually attractive, easily installed and also price competitive. It is concluded that SoHo3X is the perfect match for urban integration. Several examples of how this integration would work are represented in Figure 6.

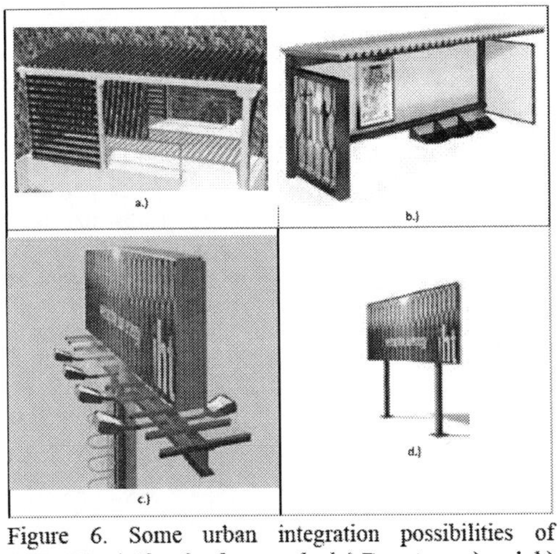

Figure 6. Some urban integration possibilities of SoHo3X. a.) Sketch of a pergola. b.) Bus stop. c.) and d.) Street signs.

Regarding vertical positioning, an interesting application is the use of SoHo3X simultaneously oriented east-south-west, that can be installed *around* the pole of any existing streetlight (see Figure 3). Mechanical stability calculations show an enormous advantage versus conventional flat modules, because the latter do not comply with regulations for most of the existing streetlights, while SoHo3X does. Therefore, the unfavorable vertical position for production at high latitudes is compensated by the fact that installation is straightforward, fast, and does not need extra work like more foundations. The low investment and the low number of devices needed for, for instance, covering the whole year of charging the smartphones of the whole population make this option attractive for local governments.

6 ACKNOWLEDGEMENTS

The authors wish to acknowledge the Spanish Ministry of Economy and Competitivity by funding through projects PTQ-12-05437 and EMP-TU-2014-3347.

7 LITERATURE

[1] Paiano, A., Lagioia, G. and Cataldo, A., *Energia, Ambiente e Innovazione*, 3/2011, pp. 75-82
[2] https://www.swm.de/dam/jcr:6b51fbab-df16-493b-92e4-5c4385e50dfd/map-renewable-energies.pdf
[3] Stojanoff C, Proc. SPIE, vol. 7957, pp. 0L-1-0L-14, 2011
[4] Imenes AG and Mills DR, Solar Energy Materials and Solar Cells, 84, pp. 19-69, 2004
[5] Loicq, J., *et al.*, in Proc. of SPIE, Vol. 9140, pp. 91400T-1—91400T-12, 2014
[6] Rodríguez San Segundo HJ, Villamarín Villegas AM and Calo López A, in Proc. of the 28th European PVSEC, pp 484-488, 2013
[7] A.M. Villamarín, Fco. J. Pérez, Antonio Calo L., Hugo-J. Rodríguez. S.S., in Proc. of SPIE, Vol. 9140, pp. 91400R-1–91400R-13, 2014
[8] Rodríguez San Segundo, HJ, Villamarín Villegas, AM, Pérez López, FJ and Calo López A, in Proc. of the 29th EU PVSEC, pp. 3815-3820, 2014
[9] Rodríguez San Segundo, HJ, Villamarín Villegas, AM, Pérez López, FJ and Calo López A, in Proc. of the 30th EU PVSEC, pp. 1399-1402, 2015
[10]http://www.sunearthtools.com/dp/tools/pos_sun.php?lang=es
[11]http://re.jrc.ec.europa.eu/pvgis/apps4/pvest.php?lang=en&map=europe

ELECTRICAL DESIGN AND LAYOUT OPTIMIZATION OF FLEXIBLE THIN-FILM PHOTOVOLTAIC MODULES

Johannes Hofer, Zoltan Nagy, Arno Schlueter
Architecture & Building Systems, Institute of Technology in Architecture, ETH Zurich
John-von-Neumann Weg 9, 8093 Zürich, Switzerland

ABSTRACT: Due to the low weight, thinness and the possibility to adapt to non-standard shapes, flexible thin-film photovoltaic (FPV) modules offer new opportunities for building integrated photovoltaics (BIPV). Even though very important for BIPV applications, the performance of FPV modules in partial shading and under curvature has been rarely investigated. In addition, conventional modeling tools can simulate module electrical performance only for planar surfaces. In this contribution, we present a new modeling environment for the calculation of the electricity yield of FPV modules in partial shading and under curvature. We investigate how FPV module design, in terms of cell dimension, interconnection and bypass diode integration, influences module performance in partial shading using experimental measurements and simulation. The methodology is applied in a case study to optimize the design and layout of FPV modules on a doubly curved roof shell structure.
Keywords: Thin Film PV Module, Flexible Substrate, Modelling, Building Integrated PV (BIPV), System Performance

1 INTRODUCTION

Buildings are responsible for more than one third of total greenhouse gas emissions in Europe [1]. Refurbishment of old buildings, improving energy efficiency and reducing material use over the whole life-cycle are key strategies for the reduction of building related energy use and emissions. In addition, the EU aims for all newly constructed buildings to have close to zero net energy consumption by 2020 [2]. Integrating photovoltaics (PV) in buildings is indispensable for meeting the net zero energy target [3]. In recent years, there has been substantial progress in improving the efficiency and reducing the cost of thin-film PV modules. For example, copper indium gallium selenide (CIGS) modules reach an efficiency of more than 16% today [4,5]. Due to the low weight, its thinness and the possibility to adapt to non-standard shapes, flexible thin-film PV (FPV) modules offer new opportunities for building integrated photovoltaics (BIPV). Even though very important for BIPV applications, the performance of FPV modules in partial shading and under curvature has been rarely investigated [6]. Typically, partial shading and curvature lead to efficiency degradation due to non-uniform irradiance and electrical mismatch of (a) cells within a module and (b) modules within the PV array.

Conventional modeling tools can simulate module irradiance and electrical performance only for planar surfaces. In this contribution, we present a new simulation environment for the calculation of the electricity yield of FPV modules. The developed simulation tool couples irradiation analysis and electrical modeling of FPV modules with high spatial resolution. We investigate how FPV module design, in terms of cell geometrical structure, cell interconnection, and bypass diode integration, influences module performance in partial shading using experimental measurements and simulation. We apply the methodology to a case study of FPV module layout on a double curved roof shell. The developed method can be used to optimize FPV module design and interconnection, including the use of distributed power electronic converters, for the case study building.

First, the procedure to simulate module solar irradiance and electrical energy yield for different module designs is explained. Then results on FPV module performance in partial shading and under curvature are presented. Finally, the the application of the developed methodology in a case study is demonstrated.

2 METHODOLOGY

The geometry of curved FPV modules is simulated within the Rhinoceros 3D/Grasshopper environment [7,8] using a geometric triangulation method developed for that purpose [9]. It allows us to simulate the layout of FPV modules on arbitrarily curved surfaces and to assess module curvature, twist, and other structural integration parameters. The method is fully parametric so that variations in parameters such as module dimension and orientation can be quickly analyzed.

Irradiation on the FPV modules is calculated based on the cumulative sky approach [10,11] as implemented in Ladybug [12] and Radiance [13] utilizing the Perez All-Weather luminance distribution model and EnergyPlus weather files with hourly resolution. In the default case, the sky is divided using the Tregenza scheme. A higher resolution can be achieved with the Reinhart sky patch subdivision [10]. The irradiation data used in this work was exported from the Meteonorm software and is based on long-term measurements of a MeteoSwiss weather station in Zurich, Switzerland.

The electric PV model used in this work is suitable to assess the influence of module partial shading and curvature on FPV module current-voltage (I-V) characteristics and follows the approach described in [14]. As shown in Fig. 1, monolithic thin-film cells are discretized into sub-cells to assess the influence of non-uniform irradiance. The sub-cell I-V curve is calculated based on sub-cell irradiance and temperature using the standard equivalent circuit model [15]. Solar radiation analysis and electrical modeling are performed using the same sub-cell level resolution. In case of partial shading, some of the cells are in reverse bias and act as a load. In order to account for this effect, the reverse characteristic of cells is modeled. The module I-V curve is calculated by summing the current of parallel connected sub-cells and subsequently the voltage of series connected cells. Bypass diodes are either integrated between cells in the module or in the module junction box. The array performance is calculated based on the I-V curves of modules and their interconnection.

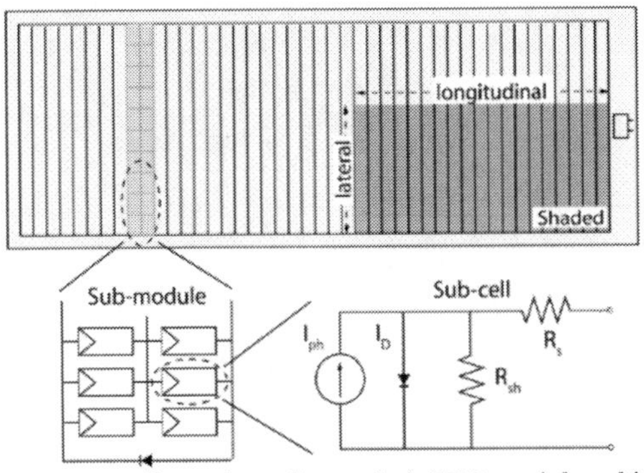

Figure 1. Illustration of a typical FPV module with monolithically interconnected cells. In order to achieve higher resolution in the model, cells are disaggregated into sub-cells. Each sub-cell is modeled using the standard equivalent circuit with a single diode, series and shunt resistance.

The analysis was performed for two commercial thin-film CIGS (copper indium gallium selenide) modules. The differences in module design are illustrated in Fig. 2. Module 1 consists of 44 cells in series with one integrated bypass diode for every two cells. The cells span the full width of the module. Module 2 consists of 36 cells in series, however, one cell only spans half of the module width and the series connection of cells extends in the lateral module direction. A bypass diode is integrated for every cell.

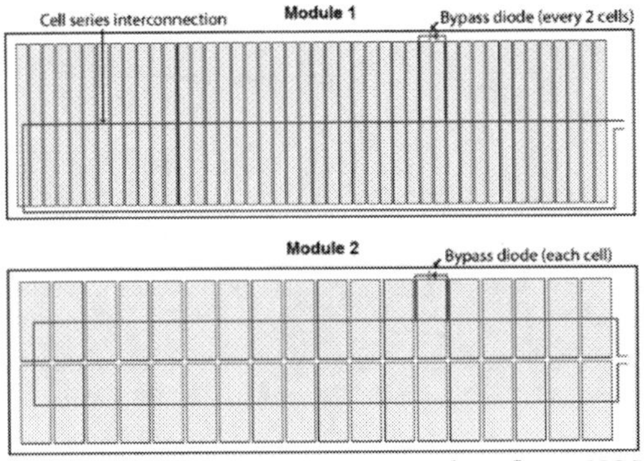

Figure 2. Cell design and interconnection of two CIGS modules analyzed in this work.

3 RESULTS

3.1 Performance in partial shading

In order to assess the influence of module design on the electricity yield in partial shading, the I-V curves of different FPV modules were measured in controlled shading experiments using a PVPM2540C measurement device from PV Engineering. The irradiance in the unshaded region was measured with an external reference cell and module temperature with a thermocouple sensor. The modules were mounted on a building façade and a non-transparent object was moved along the modules to incrementally adjust the shaded area with a step size of one cell in longitudinal and 10% in lateral direction (see Fig. 3). For an initial assessment presented in this paper, shaded regions were fully shielded from solar irradiation.

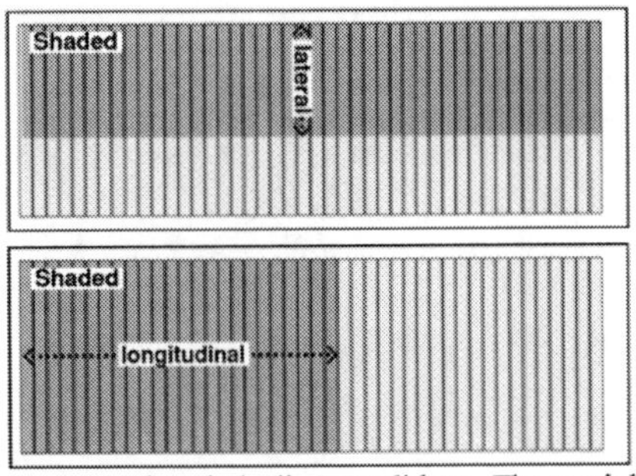

Figure 3. Analyzed shading conditions. The module is shaded in distinct steps in lateral and longitudinal direction.

For each shading condition, the MPP was inferred from the power-voltage curve of the modules. Fig. 4 compares the measured power loss relative to the unshaded condition as a function of the shaded area for both modules. As shown in Fig. 4a, the power output of module 1 decreases by the same amount as the shaded area in lateral direction increases. For the same type of shading pattern, the power loss of module 2 is higher due to the series connection of cells in lateral direction. If for example 40% of a module is shaded in lateral direction, the decrease in power is ca. 40% for module 1 and 80% for module 2. In the opposite case, with an incremental increase of longitudinal shading, module 1 performs better again due to lower power losses in bypass diodes. The step profile in the power loss of module 1 can be explained by the placement of bypass diodes for every second cell. Fig. 5 shows, that the power loss of module 1 as calculated with the model described above, is in good agreement with the measurement results.

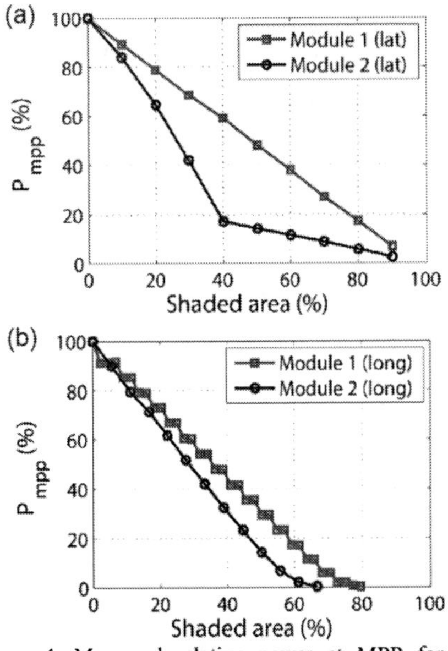

Figure 4. Measured relative power at MPP for two different flexible CIGS modules as a function of the shaded area for shading in (a) lateral and (b) longitudinal module direction.

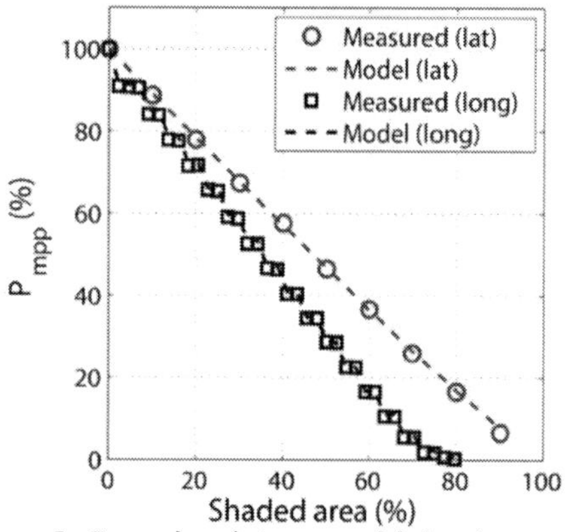

Figure 5. Comparison between modeled and measured power loss in partial shading for module design 1.

3.2 Performance of curved modules

To assess the performance of curved FPV modules, module irradiance and efficiency were been simulated for different types of curvature and weather conditions. As an example, Fig. 6 shows the simulated irradiance on a curved FPV module oriented from east to west for one specific hour. To simulate the corresponding IV characteristic and power output of module 1, the resolution in longitudinal direction corresponds to the number of cells of module design 1 (cf. Fig. 2). Module irradiance was coupled with the electrical model described in section 2. In addition to the curved module, a flat and horizontally mounted module was simulated for comparison.

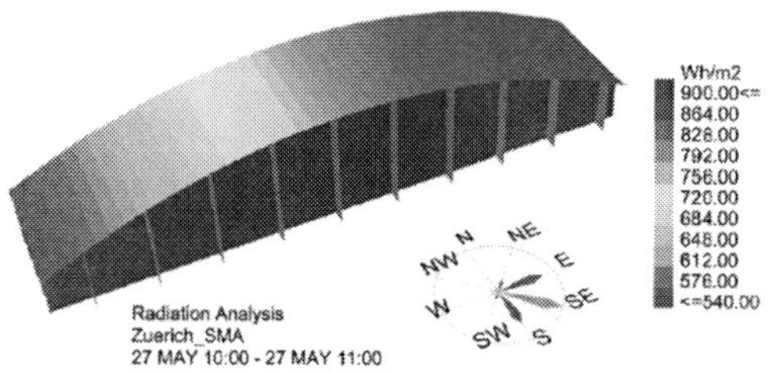

Figure 6. Simulated irradiance on a curved FPV module during one sunny hour in May for the used weather file of Zurich.

The resulting module irradiance and electrical power at MPP during one clear day in May are shown in Fig. 7. While irradiance is very similar for the flat horizontal and curved module, the electrical power output of the curved module is significantly lower in the morning and afternoon.

The dependence of power output can be also analyzed in terms of module efficiency as shown in the upper graph of Fig. 8. For flat horizontally oriented modules, the efficiency decreases slightly at solar noon due to module heating. On the other hand, the efficiency of the curved module decreases significantly in the morning and afternoon. The variation of cell irradiance over the day is shown in the bottom graph of Fig. 8. Electrical mismatch between cells due to non-uniform irradiance within the module is the main reason for the observed efficiency loss. Balanced cell irradiance at solar noon coincides with the highest module efficiency.

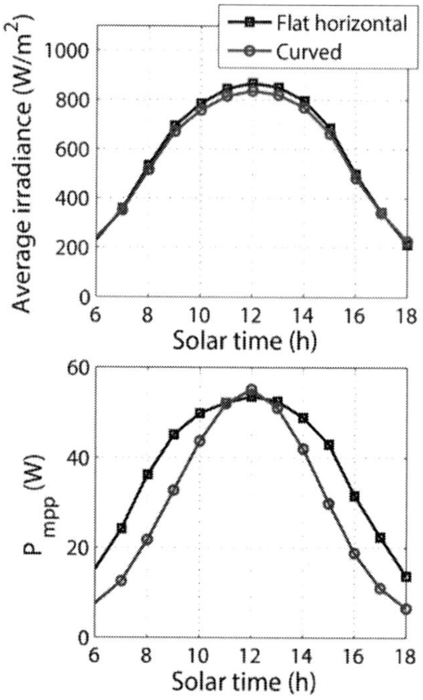

Figure 7. Simulated module irradiance and electrical power at MPP for module design 1 during one clear day in May. Results are shown for the same module in a curved and flat horizontal position.

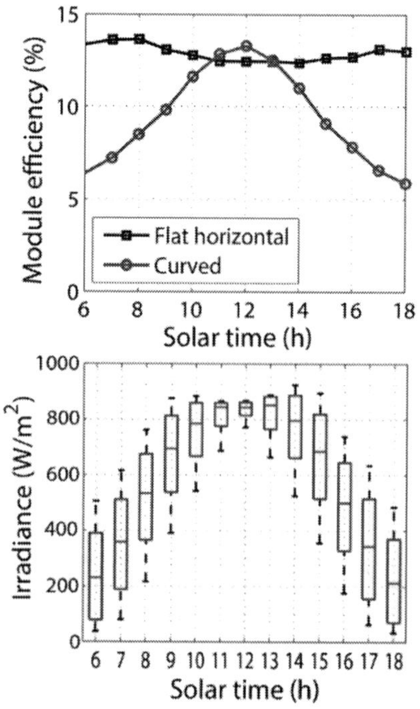

Figure 8. Top: module efficiency corresponding to the results shown in Fig. 7. Bottom: cell irradiance distribution for the curved module over the day.

The simulation results have been compared to experimental measurements. For this purpose, a prototype of a substructure with the same geometry of the top surface as the simulated test module has been built (Fig. 9). The I-V curves of the curved module and of a reference module in a flat horizontal position were continuously measured. Simultaneously, five pyranometers measured incident irradiance along the module to infer module efficiency.

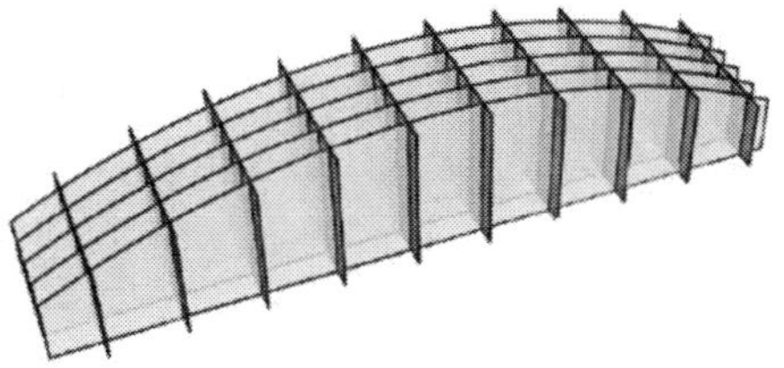

Figure 9. Illustration of the module holder used to measure the I-V characteristic of an FPV module in a predefined curved shape.

Initial measurements were performed during a clear day in May and good agreement between simulated and experimentally measured efficiency has been observed, very similar to the results shown in Fig. 8.

3.3 Building integration

The work presented in this paper is applied in the context of the NEST HiLo building [16]. HiLo features several innovations, including a vaulted, ultra-slim roof shell based on a geometrically optimized doubly curved form that integrates solar gains, thermal insulation, and conditioning of the indoor space. The methodology presented in this work is used for the design and layout of FPV modules on the HiLo roof. The FPV modules will be arranged in narrow strips to follow the curved roof geometry.

Fig. 10a shows one possible module layout with a panel size of 2m length and 0.6m width, and an orientation of 20° relative to the horizontal. Also shown is solar irradiance on the modules from 11:00 to 12:00 for a clear day in May. The distribution of irradiance between the cells of those modules is shown for module 1-12 in Fig. 10b. Generally, the strong variation of irradiance between cells within a module can lead to electrical mismatch and efficiency loss as shown in the previous section. Different FPV module designs will be investigated that minimize these effects. The approach described in section 3.2, will allow us to build a prototype setup based on the actual roof geometry and to experimentally test FPV module performance. The results can be compared with simulated data to further calibrate the model and optimize module design.

The electrical mismatch between different modules within a string can be reduced by using distributed power electronic converters such as DC power optimizers. Knowledge of the FPV module operational performance will inform the selection of those components.

Finally, the influence of strip orientation on the number of modules, average irradiance, efficiency and electricity yield has been analyzed. Initial results show that power output is mainly influenced by the number of modules that can be placed on the roof and that average module irradiance is independent of strip orientation. The orientation of 20° as shown in Fig. 10a, provides the highest energy yield and an acceptable visual appearance.

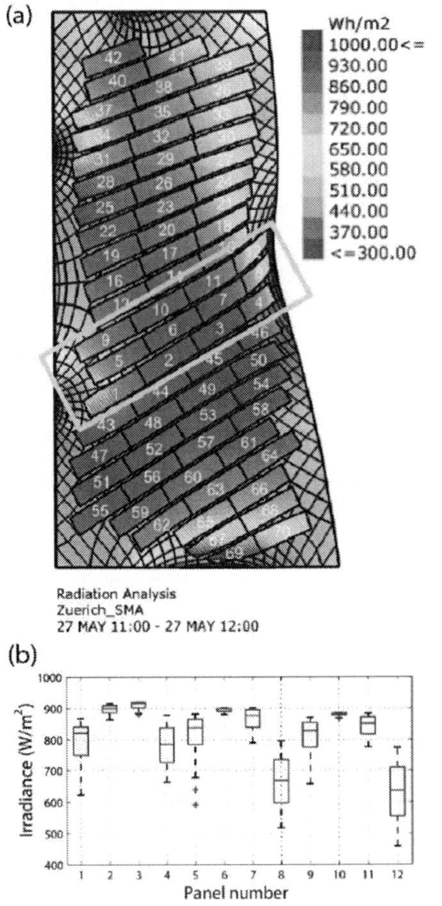

Figure 10. (a) Possible PV module layout on the NEST HiLo building and hourly solar insolation. (b) Distribution of irradiance between modules 1-12 and between the cells of each module.

4 CONCLUSIONS

In this work, we presented a novel simulation methodology for the calculation of electricity yield of FPV modules in partial shading and under curvature. We showed that differences in the cell dimension and bypass diode integration of currently available FPV modules have a significant impact on their performance in partial shading. We applied the methodology to a case study of FPV module integration on a doubly curved roof shell and showed how design parameters such as the orientation of FPV module strips influence overall energy yield. The methods developed in this work will inform the design of FPV modules and their layout on the NEST HiLo building.

ACKNOWLEDGEMENT

This research has been financially supported by CTI within the SCCER FEEB&D (CTI.2014.0119) and by the Building Technologies Accelerator program of Climate-KIC.

REFERENCES

[1] https://ec.europa.eu/energy/en/topics/energy-efficiency/buildings, accessed 04.02.2016
[2] Energy Performance of Buildings Directive (2010/31/EU)
[3] Scognamiglio A, Garde F. Photovoltaics' architectural and landscape design options for Net Zero Energy Buildings, towards Net Zero Energy Communities: spatial features and outdoor thermal comfort related considerations. Progress in Photovoltaics: Research and Applications, 2014.
[4] Nam J, Kang Y, Lee D, Yang J, Kim YS, Mo CB, Park S, Kim D. Achievement of 17.9% efficiency in 30×30 cm2 Cu (In, Ga)(Se, S) 2 solar cell sub-module by sulfurization after selenization with Cd-free buffer. Progress in Photovoltaics: Research and Applications, 2015.
[5] http://miasole.com/wp-content/uploads/2015/08/FLEX-02N_Datasheet_1.pdf
[6] Sharma P, Duttagupta SP, Agarwal V. A novel approach for maximum power tracking from curved thin-film solar photovoltaic arrays under changing environmental conditions. IEEE Transactions on Industry Applications, 50(6):4142-51, 2014.
[7] Rhinoceros 3D v5. https://www.rhino3d.com/
[8] Grasshopper - algorithmic modeling for rhino. http://www.grasshopper3d.com/
[9] Groenewolt A, Bakker J, Hofer J, Nagy Z, Schlueter A. Methods for modelling, analysis and optimisation of bendable photovoltaic modules on irregularly curved surfaces (in review)
[10] Ibarra D, Reinhart CF. Solar availability: a comparison study of six irradiation distribution methods. In Proceedings of the 12th Conference of International Building Performance Simulation Association, Sydney, Australia (Vol. 1416), 2011.
[11] Robinson D, Stone A. Irradiation modelling made simple: the cumulative sky approach and its applications. In PLEA Conference, 2004.
[12] Roudsari MS, Pak M, Smith A, Gill G. Ladybug: a parametric environmental plugin for grasshopper to help designers create an environmentally conscious design. Proc. of the 13th International IBPSA Conference, Lyon, France, 2013.
[13] Ward GJ. The RADIANCE lighting simulation and rendering system. Proc. of the 21st annual conference on Computer graphics and interactive techniques, pp. 459-472. ACM, 1994.
[14] Hofer J, Groenewolt A, Jayathissa P, Nagy Z, Schlueter A. Parametric analysis and systems design of dynamic photovoltaic shading modules. Energy Science & Engineering, 4(2), 134-152, 2016.
[15] Mermoud A, Lejeune T. Performance assessment of a simulation model for PV modules of any available technology. Proc. of the 25th European Photovoltaic Solar Energy Conference, 2010.
[16] http://hilo.arch.ethz.ch/

ZIGZAG STRUCTURE IN FAÇADE OPTIMIZES PV YIELD WHILE AESTHETICS ARE PRESERVED

R.M.E. Valckenborg[1]*, W. van der Wall[4], W. Folkerts[1], J.L.M. Hensen[2] and A. de Vries[3]
[1] Solar Energy Application Centre (SEAC), High Tech Campus 21, NL-5656 AE Eindhoven, The Netherlands
[2] TU/e, PO Box 513, Vertigo 6.18, NL-5600 MB Eindhoven, The Netherlands
[3] Holland Solar, Korte Elisabethstraat 6, NL-3511 JG Utrecht, The Netherlands
[4] Wallvision, Deken de Bruijnplein 3, NL-5591 CZ Heeze, The Netherlands

*Corresponding author: valckenborg@seac.cc; Tel: +31 652803729

ABSTRACT: In the Netherlands, the possibilities for façade PV applications draw more attention lately due to a combination of government incentives, declining PV-module prices and upcoming legislation from the European NZEB-directive (2010/31/EU). In this context, start-up company Wallvision has introduced a ZigZag structured façade in which the PV-modules are hidden from ground level, and the decorative panels are customized to the aesthetical preferences of the architect. Understanding the PV-yield as a function of the geometry and other system components is crucial. For this purpose a customized geometrical model has been developed that can handle the varying solar path and the specular and diffuse reflection of the decorative panels. The experimental validation of the model is done with a newly built test façade (of 3 x 3 ZigZag cassettes, area ~ 20m^2) in the SolarBEAT outdoor research facility of SEAC. Measurements include the irradiance at various points and kWh-yield per minute time interval. The aesthetic aspect of the ZigZag façade is as important as the PV-performance part of the product, and is therefore included in this paper.
Keywords: Façade, PV System, Modelling

1 INTRODUCTION

Façades are a significant part of the potential BIPV market area. Up till now, facades did not get the same amount of attention as flat and pitched roofs in the Netherlands. This is logical, as one recalls that rooftop PV (whether it is BAPV or BIPV) gives more yield per invested kWp than a vertical façade PV system. And this rooftop part of the total PV-market is still not completely cashed in the Netherlands. However, things are changing rapidly due to a combination of incentives from the government and declining prices for PV-modules and other system components. The incentives are: net-metering scheme for households and kWh-subsidy for larger (>15 kWp) systems. On top of that, the NZEB-directive for Europe (2010/31/EU) is being rolled out in the Netherlands. This made people aware of the fact that for new-built houses since 1 January 2015, PV is an effective option for meeting the criteria for energy performance. Moreover it is expected that from the end of 2018 for government buildings and from the end of 2020 for all utility buildings the criteria for near zero energy building have to be met. A PV generating part of the building skin could be crucial to meet these new energy criteria. Therefore several start-up companies are exploring the opportunities of PV façades. Among them, the company Wallvision has introduced a ZigZag structured façade in which the PV-modules stay out of sight for pedestrians and other traffic in the streets at ground level (see Figure 1).

Figure 1: Sketch of the ZigZag structured façade principle.

Only the decorative panels are visible and give the façade a customized aesthetics (see Figure 2). Customization includes various materials, colors and prints, like e.g. a company logo. The assembly of one PV-module and one decorative panel with the same width is called a ZigZag-cassette. Optionally the decorative panel can also be chosen in such a way that it has a maximum light reflection onto the PV-module below, to maximizethe PV-yield.

Figure 2: Picture of first commercial ZigZagSolar façade, taken from ground level. Decorative panels in this façade are colored from dark gray at the lowest row till light gray at the uppermost row.

This working principle may sound quite easy. But understanding the electrical performance of the whole façade is fascinatingly complex. First of all, the trajectory of the sun follows a certain path depending on the exact location of the installation. Even in a small country like the Netherlands this gives rise to variations influencing electrical performance. As one can imagine, the incoming total sun light (as a sum of a direct and a diffuse contribution) is reflected -both specular and diffuse- by the decorative panel and will partly hit the underlying PV-module. Albedo effects of the surrounding built environment have a different effect on the PV yield opposed to 'standard' vertically installed PV-modules. The tilt angles of the product itself are adjustable. Wallvision, SEAC and Technical University Eindhoven are a.o. participating in the Dutch ZonneGEVEL-project aiming to conduct an extensive research on the performance of this ZigZag-structured façade.

2 HOW CAN THE ELECTRICAL PERFORMANCE BE UNDERSTOOD

The major research question of the ZonneGEVEL-project is to fully understand the electrical performance of a ZigZag-structured façade including the following aspects:

- Location of installation (latitude, longitude).
- Orientation towards the sun (azimuth: from East via South towards West, even installations towards NE and NW are investigated). Please note that the renovation market of the Dutch building sector is much larger than the new-built market. And the façade in a renovation project is generally not adjustable anymore.
- Geometry of the ZigZag-cassette (length & width PV-module + length & width reflector + tilt angle β of the module & angle α_{ref} between decorative panel and vertical).
- Albedo of ground reflection, ρ_{ground}.
- Decorative material (with a reflection decomposed towards direct & diffuse light).

In order to answer this question, it is needed to have a clear understanding of the irradiance falling on the PV-module for any moment of the day, any day of the year, and with respect to all above mentioned possible variations of the ZigZag-cassette installed at any location on Earth were potential customers like to utilize their façade. Therefore the research is clearly split in a two-step approach. In the first part of the project, focus is set on the irradiance research. Thereafter (starting around mid-2016), the measured and modelled irradiance will be used as input to predict AC-kWh yield, which will be validated with electrical measurements. Clearly this paper focusses on the irradiance research, but also some preliminary AC yield measurements are presented.

3 EXPERIMENTAL SETUP

A vertical test façade of width 6.6 m and height 4.5 m is built in the SolarBEAT facility. This outdoor research infrastructure for innovation on BIPV(T) has been presented to the solar community in the previous two PVSEC conferences [1][2]. The upper part of the test façade gives room for 3 rows of 3 ZigZag-cassettes with β=32° and α_{ref}=28°. The lower part of the test façade gives room for a grid of 2 rows with 5 CIGS-modules incorporated in a innovative façade mounting system; see Figure 3.

Figure 3: Picture of the ZonneGEVEL test façade. The upper part holds 3 rows of 3 ZigZag-cassettes. The lower part holds 10 CIGS-modules. Pyranometer and photodiodes are used for measuring irradiance (yellow circles).

4 IRRADIANCE MODELING

4.1 Two dimensional geometrical modeling

Because of the complexity of the PV-system, it was decided to start modelling the irradiance in the cross-sectional cut of the ZigZag-cassette (see Figure 4).

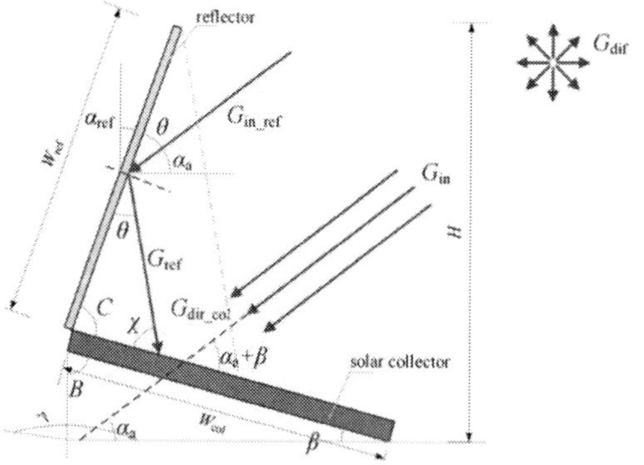

Figure 4: Cross-sectional cut of the ZigZag-cassette as starting point of first geometrical modelling.

The total irradiance falling on the PV-module is now described by:

$$G_{tot\_col} = G_{in} + G_{dif} + G_{ref} \qquad (1)$$

where: G_{tot_col} is the total irradiance on the PV-module, G_{in} is the beam irradiance that is not obstructed, G_{dif} is the (isotropic) diffuse irradiance, and G_{ref} is the reflected beam irradiance. For time resolution, it was decided to start with a one hour step interval, because of two reasons: a) it keeps simulation times for a full-year analysis (8760 hours) acceptable, b) TMY-data is generally available in a one hour time resolution. This

2D-model is used on various locations with façade orientation of E, S, SW and W, and also for many variations of the geometry of the ZigZag-cassette and the material of the decorative panel. We present the results for a South façade installed in Helsinki, with a decorative panel that is reflecting quite efficiently. The cassette geometry has been varied in the maximum possible range of module tilt angle β, and decorative panel angle α_{ref}: see Figure 5.

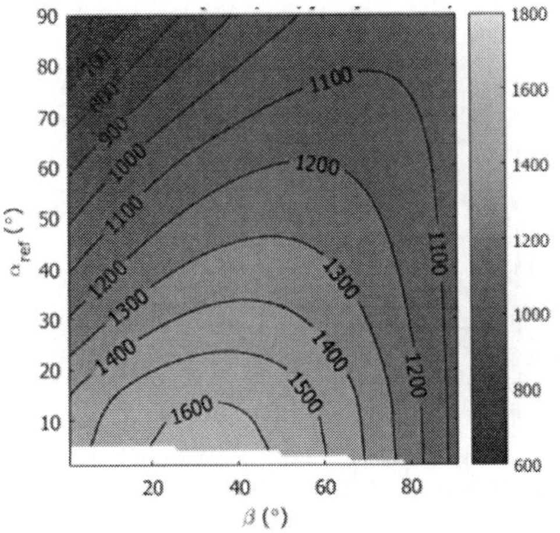

Figure 5: Total irradiance in kWh/m²/year as function of module tilt β, and angle between decorative panel and vertical α_{ref}.

Now we can search for optimal angles at which the irradiance onto the PV-panel is maximum. The color plot of Figure 5 shows this maximum irradiance for low α_{ref} and β around 35°. It should be noted that these are results for one ZigZag-cassette. Low α_{ref} implies that fewer cassettes will fit in the total façade. That is not an issue for very large façades. However, when one would like to maximize the total yield of the façade, then this gives an additional constraint. From the -more time consuming- calculations of multiple cassettes we observed that the optimum is shifting towards less steep decorative panels.

4.2 Three dimensional geometrical modeling including the HDKR-diffuse model.

Although the model gave many insights into a ZigZag structured façade, it was recognized by experts that the diffuse light modelling is insufficiently taken into account. Therefore a major upgrade of the model has been made. The total irradiance is now described by:

$$G_{tot\_col} = G_{in} + G_{dif\_iso} + G_{dif\_cs} + G_{dif\_hz} + G_{ref\_ground} + G_{ref\_obj\_sp} + G_{ref\_obj\_dif} \quad (2)$$

with a newly decomposition into three diffuse terms and three reflected terms. It goes beyond the scope of this paper to explain each term in full detail. In short the diffuse terms with subscripts iso, cs, and hz respectively, refer to isotropic diffuse, circumsolar diffuse, horizon brightening diffuse. The reflection terms with subscripts ground, obj_sp, and obj_dif respectively, refer to ground reflected, object specular reflected, and object diffuse reflected. As for diffuse irradiance, there are several models with a different treatment of the three components. Comparison and evaluation have been made continuously [3][4][5]. Here the model developed by Reindl et al. [5] is used because of its superior accuracy [6], which was an improvement to the works of Hay and Davies [7] and Klucher [8], therefore abbreviated as the HDKR model. As for reflected irradiance, the presented terms are written down from text book material [9] following the route via a view factor. An extensive paper about the model is soon to be published in a scientific journal.

For this more accurate model, it is important to consider the precise location of the receiving spot of the irradiance. Therefore a grid has been put on the PV-module. For each grid point the irradiance is calculated with formula 2 as a function of time. The coding is done in Matlab®. The user has multiple options to visualize the results; see e.g. Figure 6.

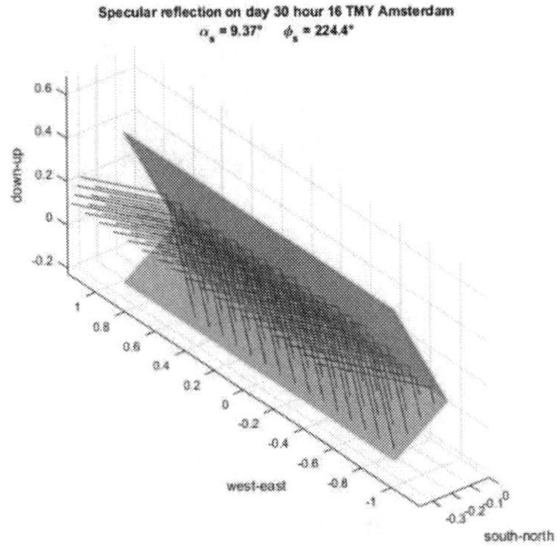

Figure 6: The intensity and direction of the specular reflection component only, for an arbitrary moment. Animated movies for full days are also available.

For the remainder of this paper, we will show the irradiance at exactly the middle of the PV-module, as this is also the location of the pyranometer in the experimental setup.

5 MEASUREMENT RESULT (IRRADIANCE & ELECTRICAL)

5.1 Irradiance

At SolarBEAT a central solar measurement station measures every minute the DHI, DNI, and GHI [1][2]. Moreover on the pitched roof of the nearby test cabin (just visible in Figure 3) the G_{POA} is measured under 35°. These irradiance measurements are a kind of reference. Now, in the ZigZag-façade the global vertical irradiance (GVI) is measured, and the G_{POA} in the middle of the cassette, as seen in Figure 3. These are all plotted in Figure 7 for the 11th May 2016. We deliberately picked this day because it shows many steep variations in irradiance, which is the most difficult situation to model. If the model succeeds on this kind of day, then it will also succeed on completely sunny and overcast days.

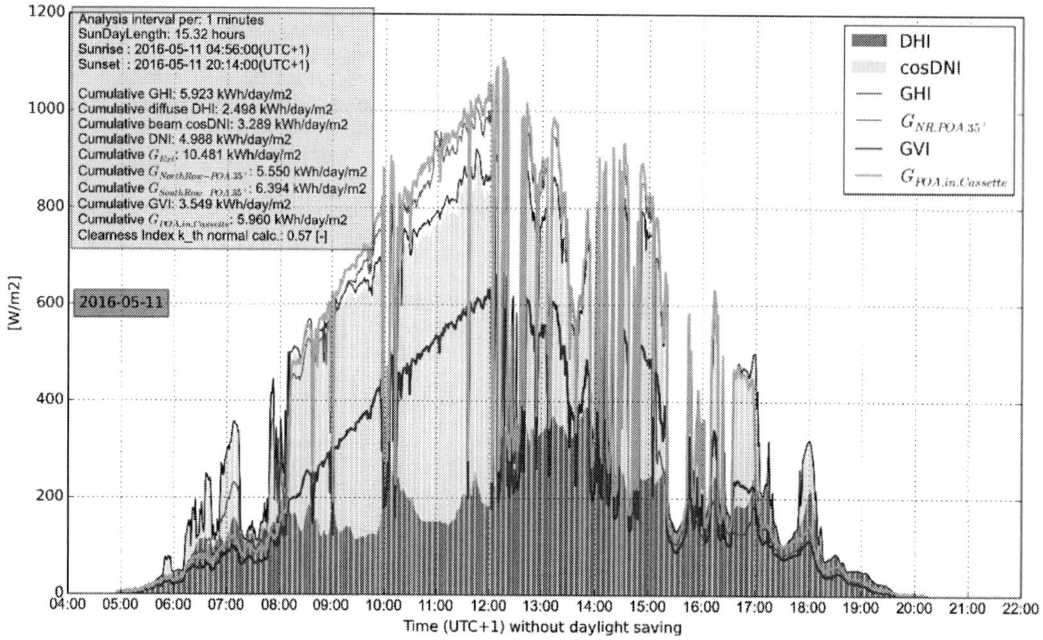

Figure 7: GHI (blue), cosDNI is the beam perpendicular to horizontal (yellow), DHI (grey), GVI (red), G_{POA} of nearby pitched 35° roof (magenta), and G_{POA} 'inside the cassette' (orange) as a function of time for the irradiance fluctuating 11th May 2016. In the green text box, the daily cumulative values are given.

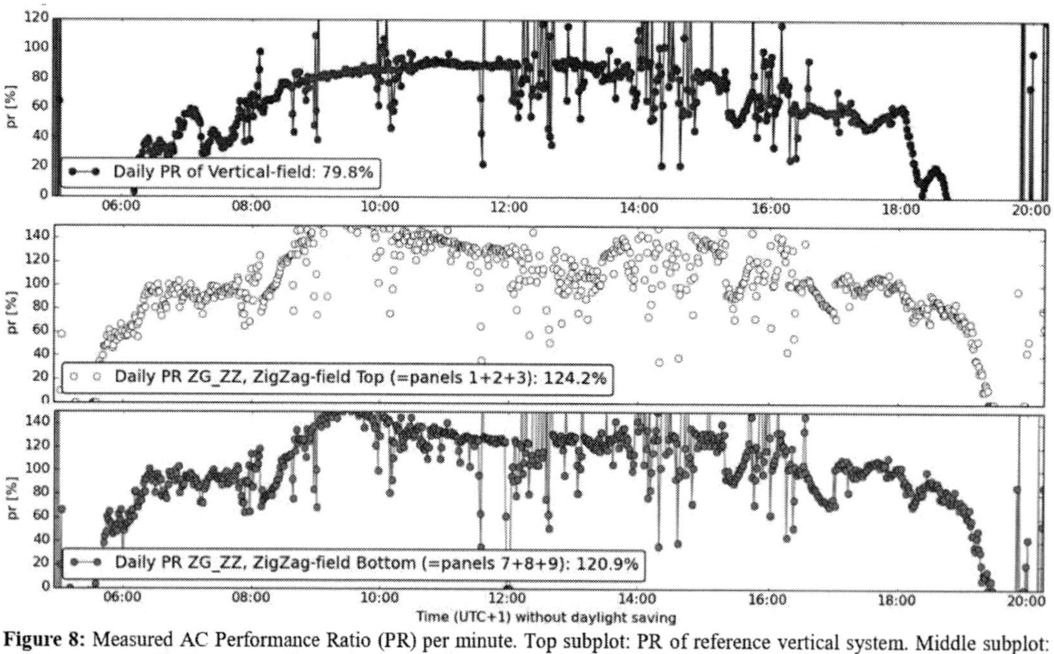

Figure 8: Measured AC Performance Ratio (PR) per minute. Top subplot: PR of reference vertical system. Middle subplot: PR of Top row of ZigZag-cassette with white reflector. Bottom subplot: PR of Bottom row of ZigZag-cassette with grey reflector. Same pyranometer (GVI) has been taken for calculating all 3 subplots.

It should be noted from Figure 7, that the pyranometer inside the cassette (orange line) receives the same amount of irradiance as the pyranometer on the nearby pitched roof (magenta line). On the one hand inside the ZigZag-cassette, the hemisphere is blocked more, giving lower reception of diffuse light. On the other hand inside the ZigZag-cassette, an extra amount of reflected light is received. Apparently for this example day, both effects are of the same order of magnitude and the net result is the observation that the PV-module in the ZigZag cassette receives as much irradiance as a pitched roof PV-module under the same angle. It remains to be seen (and doubtful) if this will hold for all the days of the year. The upcoming one full year of outdoor measurement will give an answer to this interesting question.

5.2 AC Electrical yield

As explained earlier in this paper, we do not have yet the precise relation between AC electrical yield and irradiance. But we assume a linear relationship between yield (e.g. kWh total row per minute) and irradiance integrated over the area in the cassette, e.g. [kWh/m^2] * (area PV-modules in row) per minute. An assumption which is questionable because we suspect that the irradiance is quite non-uniform distributed over the PV-module. But it would be very unsatisfactory to ignore the first electrical yield results from the new experimental setup. Therefore we present the AC performance ratio for the same day as the irradiance plot; see Figure 8.

First of all it can be seen that the vertical system has a PR=80% which is good. As all PRs are calculated with the same vertical pyranometer GVI, it is possible to compare the subplots with each other. So, in the middle subplot, we observe a different pattern which is caused by the continuously varying effect of reflection into the cassette. At the end of the day, the PR=124%. This is much more than 100%, because the specific yield (expected yield) is calculated based on the vertical pyranometer. One should note that the ZigZag-cassette construction prevents the façade to be fully covered with PV-modules. Hence when taking this geometry effect into account, the PR drops well below 100%, even below 80% for cassette geometries needed for optimal yield in North-West Europe. Finally in the bottom subplot, the pattern is very similar to the middle subplot, from which it can be concluded that the different decorative panels are not influencing the yield that much. At the end of the day, the PR=121%, which is 'just' 3% lower than the result of the white decorative panel.

5.3 Validation of Irradiance modelling

The installation of the pyranometer in the ZigZag-cassette enables us to validate the irradiance model. The advanced 3D model has been run with a time step of one minute. Real measured GHI, DNI, and DHI from the central measurement station have been used as input (opposed to the default TMY-data that one would take when modeling a virtual installation in an arbitrary city).

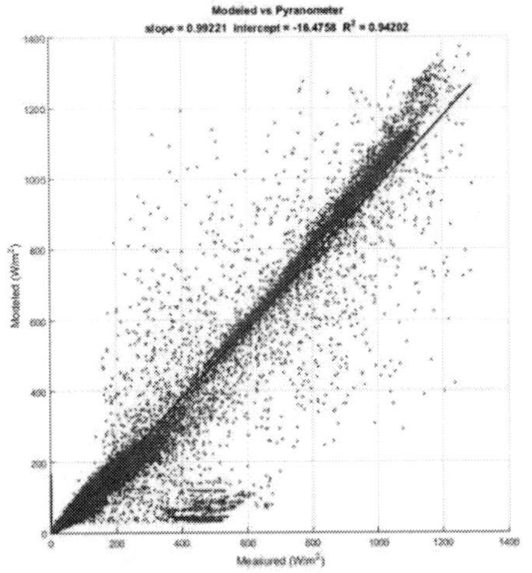

Figure 9: Modelled irradiance versus measured irradiance for the pyranometer in the middle of the central ZigZag-cassette. Every minute during daytime of May 2016 is incorporated without any filtering.

In the scatter plot of Figure 9 it can be seen that the modelled irradiance correlates quite well (R^2=0.94) with measured irradiance. One should note that May 2016 has more than 25000 minutes during daytime. Points far away of the central part are expected to be resulting from moments at which the irradiance fluctuates quite heavily. One cloud of points is clearly pointing towards a remaining issue: during sunrise and sunset, the model is severely underestimating irradiance. The reflection on nearby objects and the hemisphere blockage at the low solar angles are not modelled correctly, because these nearby objects are not taken into account. This is clearly a point of attention for future modelling work.

6 AESTHETICS

Aesthetics are very subjective and difficult to quantify. Each customer will have his own preferences. That is not a problem because the ZigZag-cassette can be tuned to specific customer wishes in terms of chosen geometry and decorative panel. In Figure 8, we saw the difference between a white and a grey decorative panel for one day only. Now summing up for all days in May 2016 - which have been measured with a time resolution of about 15 seconds without any system failure (i.e. about 100.000 measurements for each system) – the monthly result is:

Decorative panel	**Monthly AC-yield**	**Monthly PR**
Grey	28.6 kWh	115%
White	30.7 kWh	123%

If striving for maximum yield, the customer will choose for a white decorative panel. However, if the grey colour is more appealing to him (or e.g. a company colour), then the result in the table shows that the effect on system PR is not dramatic: 'just' 8% for this May 2016. Of course yearly PR can deviate from this result as the relative contribution of reflection becomes more important for winter months as we have learned from previous modelling.

In principle, any decorative panel can be chosen by the customer. Real measurements of electrical performance for any decorative panel is very time consuming. But as long as the specular and diffusive (Lambertian) reflection coefficients of the preferred decorative panel are known, we can use the model to predict the yearly irradiance inside the cassette. And assuming that AC yield will closely follow irradiance, this gives predicted kWh per year for a specific designed ZigZag-structure. In this sense we conclude that aesthetics are truly preserved, or even optimized(!), with a very acceptable compromise towards maximum yield that would come out of a ZigZag-cassette with a white decorative panel.

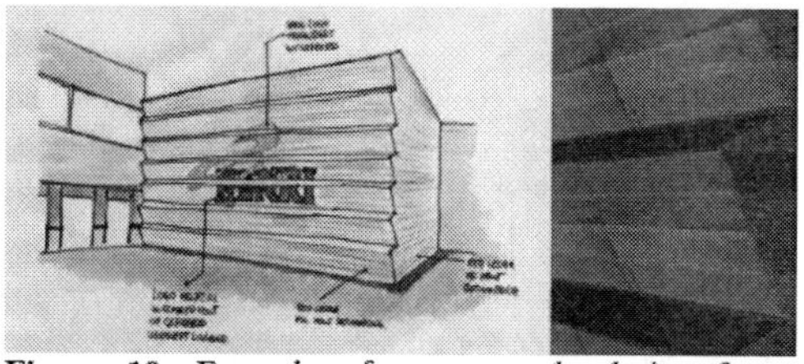

Figure 10: Example of custom-made design for a company that has two preferences: wood-finish appearance of the façade and the company logo printed on the decorative panels.

7 CONCLUSIONS

An improved irradiance model for the ZigZag-structured façade is using the HDKR-model for the diffuse components, and a threefold model for the reflected light based on view factor. For a precise implementation, it has been found necessary to perform this irradiance model on multiple locations at the PV-module, preferably in a grid with a spacing of about 10 cm. The validation of this model with measurements of all minutes of May 2016 looks very promising. Only during low solar altitude, the model has difficulty with nearby objects giving rise to albedo and blockage.

The experimental test façade has been realized in the first quarter of 2016. Analyzing the AC yield of the vertical system, that serves as a kind of reference, shows acceptable daily PRs between 75% and 80%. The PR of the ZigZag-cassettes are understood to be well above 100% because the same vertically installed pyranometer has been used as with the reference system. One should note, that in a ZigZag-structured façade, the complete façade area cannot be cladded by PV-modules only. Therefore, when comparing the total system yield of the vertical system and the ZigZag-system, the presented PR of the ZigZag-system should be divided by a factor calculated from the chosen angles in the cassette.

Finally, regarding the AC yield, it can be observed that the grey reflector produces less than the white reflector. The difference of 8% has only validity for this specific experimental setup in May 2016. The remaining months of the one full year outdoor testing will be studied closely, to be able to draw a more general conclusion.

As suggested by the title of this paper, the customer has the final say in the aesthetics of the façade. In principle he can pick any color and any decorative material (that could come from normal façade panel manufacturers). With the help of the model shown in this paper, an annual yield of a custom-made façade can be predicted and communicated to the customer in order to make the best compromise between maximum yield and aesthetics.

8 REFERENCES

[1] R.M.E. Valckenborg, A. de Vries, W. Folkerts, G.P.J. Verbong, 'The BIPV research facility SolarBEAT in the Netherlands', 6AV4.25, PVSEC 2014

[2] R.M.E. Valckenborg, J.L.M. Hensen, W. Folkerts, A. de Vries, 'Characterization of BIPV(T) Applications in Research Facility SolarBEAT', 6AV5.26, PVSEC 2015

[3] P.G. Loutzenhiser, H. Manz, C. Felsmann, P.a. Strachan, T. Frank, G.M. Maxwell, 'Empirical validation of models to compute solar irradiance on inclined surfaces for building energy simulation'. Sol.Energy 81 (2007), 254–267. doi:10.1016/j.solener.2006.03.009

[4] A.M. Noorian, I. Moradi, G.A. Kamali, 'Evaluation of 12 models to estimate hourly diffuse irradiation on inclined surfaces', Renew. Energy 33 (2008), 1406–1412. doi:10.1016/j.renene.2007.06.027

[5] D.T. Reindl, W.A., Beckman, J.A. Duffie, 'Evaluation of hourly tilted surface radiation models', Sol. Energy 45 (1990), 9–17. doi:10.1016/0038-092X(90)90061-G

[6] C.A. Gueymard, 'Direct and indirect uncertainties in the prediction of tilted irradiance for solar engineering applications', Sol. Energy 83 (2009), 432–444. doi:10.1016/j.solener.2008.11.004

[7] J.E. Hay, J.A. Davies, 'Calculation of the solar radiation incident on an inclined surface', in: Proc. of First Canadian Solar Radiation Data Workshop (Eds: JE Hay and TK Won), Ministry of Supply and Services Canada (1980).

[8] T.M. Klucher, 'Evaluation of models to predict insolation on tilted surfaces', Sol. Energy 23 (1979), 111–114. doi:10.1016/0038-092X(79)90110-5

[9] J.A. Duffie, W.A. Beckman, 'Solar Engineering of Thermal Processes', 4th edition (2013). Wiley.

9 ACKNOWLEDGEMENT

This work is supported by 'Netherlands Enterprise Agency' (RVO) and the Dutch Topteam Energy via the project: 'ZonneGEVEL' with grant number TEZG114003.

BIFACIAL PV INTEGRATED ON BUILDING BALCONIES

S. Freitas[1], M.C. Brito[1]
[1]Instituto Dom Luiz – Faculdade de Ciências, Universidade de Lisboa
Faculdade de Ciências da Universidade de Lisboa, Ed. C8 (8.3.33), Campo Grande, 1749-016 Lisboa, Portugal.
Tel.: +351 962 497 716, e-mails: srefreitas@fc.ul.pt, mcbrito@fc.ul.pt

ABSTRACT: In this work we explore the potential improvement of PV yield by modules with bifacial solar cells integrated into balconies with glass/mirror doors and side walls reflecting radiation to the back surface of the modules. PV potential production was estimated considering reference weather data for 3 different locations, with a wide range of latitudes: Oslo (Norway, 59.95°N 10.75°E), Lisbon (Portugal, 38.71°N 9.14°W) and Abu Dhabi (United Arab Emirates, 24.47°N 54.37°E). Optimized balcony geometries and materials were found for each case, maximizing the PV generation density according to the façade orientation and local solar radiation time series. The results show that the use of bifacial PV modules on balconies with proper reflective materials and optimized geometries can lead to an increment in PV yields of around 20% (and up to almost 30%), with the greater impact observed in eastern/southern façade orientations. However, while lower latitudes with high levels of irradiation might be promising for bifacial PV (even for north facing facades), higher latitudes with low direct sunlight are not attractive to the application of bifacial PV except for the south facing facade. Generally southern and northern facades should be wider and with smaller offsets, while the others vary from intermediate width to large depth as an attempt to gather most of the diffuse reflected radiation.
Keywords: Bifacial PV, building facades, DIVA-for-Grasshopper, optimization

1 INTRODUCTION

The urban environment is becoming the new playfield for photovoltaics, namely when building applied/integrated photovoltaics (BAPV/BIPV) are considered. Along with the continuous decrease of PV system costs and the increase in popularity of solar generated electricity, exciting research is being conducted in order to create new concepts and technologies that facilitate the adoption of PV as part of the architectonical composition of the building while replacing or performing as a construction element. Roof-mounted conventional c-Si modules are still the prime approach in most projects, but façade PV systems are also becoming popular, mainly due to 3 key factors: 1 - it has been shown that facades may double the PV potential of buildings in the urban context [1], 2 - new emerging solar cell technologies suited for vertical elements are getting into the PV market and becoming more affordable [2]; and 3 – the aesthetical side is meeting optimized performance as design professionals are more aware of software tools to model and simulate building integrated energy systems (the most common are Energy Plus-based [3]).

Typically building facades are a mix of wall and window surfaces, so a balcony represents not just a way of making additional quality space in a flat but also an interesting opportunity for the integration of PV modules for electricity generation, since these spaces are highly exposed to daylight/solar radiation, which can be enhanced by the use of highly reflective surrounding materials. In fact, many solar cell technologies, such as building integrated concentrating modules, transparent solar cells, semi-transparent modules (i.e. modules with interval between cells or solar cell material with a degree of transmissivity) or bifacial modules, can be applied to a balcony. However, if some technologies may permit more radiation into the balcony space, they may also decrease the electricity output due to lower efficiency.

Due to the growing interest in bifacial PV and its potential to harvest reflected solar radiation [4-8], in this paper we focus on bifacial PV integrated on balconies and on how their performance varies under different locations and balcony geometries and materials. Optimized solutions are suggested for each orientation.

2 METHODOLOGY

The estimation of solar irradiance in complex structures and environments is not trivial, and three components are required for a detailed study: the solar radiation model, the digital representation of the surfaces and context, and results visualization and processing [9]. Thus, the present study is based on a 3-step methodology. First, a case-study design of a building with one balcony in each one of its façades was modelled (Fig.1) in the 3D modelling software Rhinoceros [10] using the integrated graphical algorithm editor called Grasshopper [11]. This tool allows for straightforward parametrical studies on the influence of different balconies' depth, width and offset in the solar radiation harvested by bifacial PV modules.

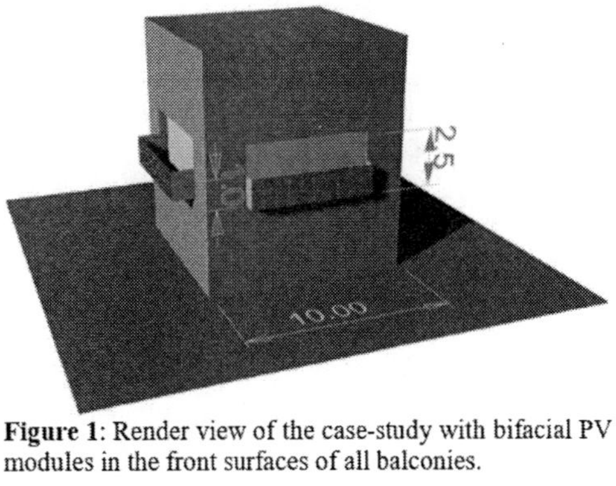

Figure 1: Render view of the case-study with bifacial PV modules in the front surfaces of all balconies.

Second, the solar radiation data series for three locations with dissimilar climate conditions were considered: Oslo (Norway, 59.95°N 10.75°E), Lisbon (Portugal, 38.71°N 9.14°W) and Abu Dhabi (United Arab Emirates, 24.47°N 54.37°E). Then, employing the plug-in DIVA [12] (a Radiance-based tool able to include the effects of a wide variety of material types, sky and lighting conditions, and physical phenomena related to light propagation of a specific scene), we assigned different material types to each of the balconies' surfaces, including a PV module material to the front and rear surfaces [13]. After this, we computed the yearly solar irradiation in these surfaces with a 0.1m² point grid resolution. The solar radiation model behind DIVA's estimations is that of Perez [14] with Tregenza sky patch model [15] and a cumulative sky approach [16] to reduce computation time. Average electricity production was estimated considering front and rear efficiencies of 16% and 14% [17] respectively, and an 80% and 70% performance ratio to approximate the observations reported in [4-7]. For simplicity, impacts of non-homogeneous irradiation and detailed electrical behaviour of the series interconnections between solar cells were not considered.

Finally, each one of the balcony designs underwent a single-objective optimization process using Galapagos [18], an optimization module inside Grasshopper. The evolutionary solver was selected, with default options, aiming for maximum PV generation per module unit area. In the total of 6 optimizations, it took around 15 seconds to compute each genome with low quality Radiance parameters. The final 6 optimized solutions were simulated a second time using high quality Radiance parameters in order to produce more accurate estimates.

3 RESULTS

3.1 Optimized balcony geometry

The door and balcony height were kept constant (2.5m and 1m, respectively). The depth and offset of the balcony were allowed to change between 0.1m and 2m, and the width could vary from 1m to 7m in the course of the optimization process.

Figures 1, 2 and 3 summarize the depth, width and offset values of the designs that maximize PV generation density (illustrated in detail in A. Supplementary material), and Table 1 shows the different materials assigned to the different surfaces.

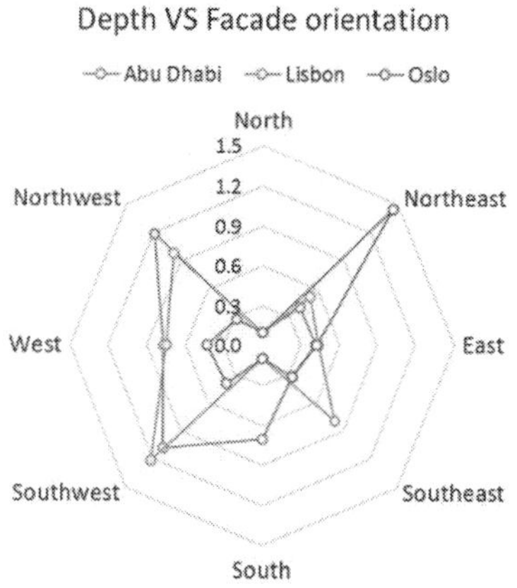

Figure 2: Balcony depths [m] obtained in the geometry optimization process.

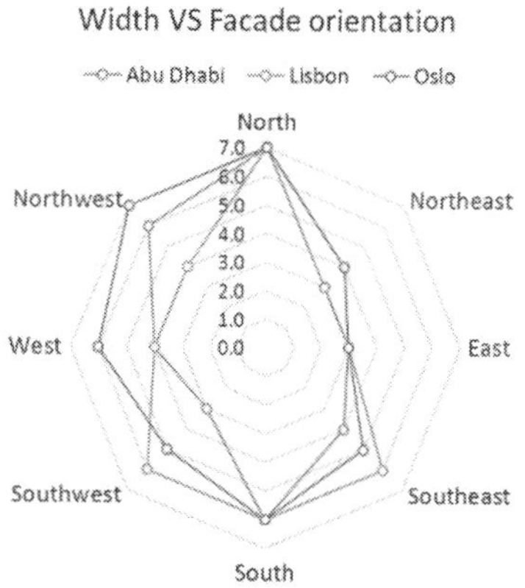

Figure 3: Balcony widths [m] obtained in the geometry optimization process.

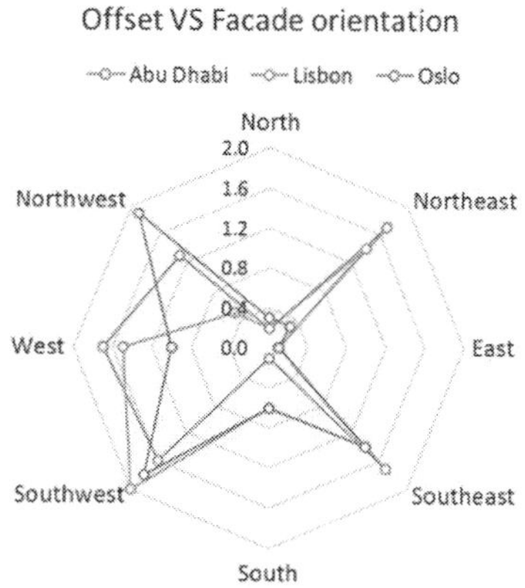

Figure 4: Balcony offsets [m] obtained in the geometry optimization process.

It can be noted that for Abu Dhabi and Lisbon there is a clear preference for deeper balconies on the South-West facing facades, while for Oslo they should be less deep, except for those facing the Northeast (Fig. 2). As for the width of the balconies, all locations feature large widths in the North-West-South orientations, with Oslo featuring the widest ones (Fig. 3). The offsets also show a clear trend to small values for the South, East and North orientations, but higher values for the remaining ones (Fig. 4).

As far as the surface materials are concerned, as expected high reflective materials such as white walls and mirrors are more favourable. However, the Southwest, Northwest and Northeast were attributed slightly less reflective materials. This would require more careful inspection, since there is no trivial way to find if the interaction of light with surfaces really caused this preference, or if it has to do with methodological limitations, such as the fact that neither the low quality ray-tracing or the annual cumulative sky approach are ideal .

Table 1: Materials assigned to the optimized balcony designs (generic wall with 50% reflectance, white wall with 70% reflectance, sheet mirror, double glazing pane).

	ABU DHABI		LISBON		OSLO	
	Walls	Door	Walls	Door	Walls	Door
East	white	mirror	white	mirror	white	mirror
Southeast	white	mirror	white	mirror	generic	mirror
South	white	mirror	white	mirror	white	mirror
Southwest	generic	glazing	generic	glazing	generic	mirror
West	white	mirror	white	glazing	generic	mirror
Northwest	generic	mirror	generic	glazing	white	glazing
North	white	mirror	white	mirror	white	mirror
Northeast	white	glazing	white	glazing	generic	glazing

From the PV generation density results for the optimized configurations (Fig. 5) we can conclude that the most interesting facades for the integration of bifacial PV are all but those North-facing in Abu Dhabi and West-North orientation in Lisbon, while only the South facades are appealing in Oslo.

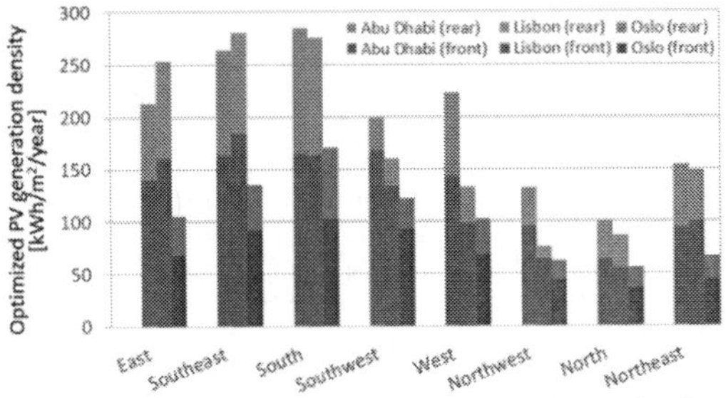

Figure 5: Estimated PV generation density [kWh/m^2/year] for each orientation and location – optimized dimensions and materials.

Although the increment in total incident radiation is generally very high even in orientations that would be discarded at a first glance, it can be noted that some still do not justify the application of bifacial PV panels since the level of annual electricity produced is too low. This is the case of most north facing balconies that feature small offsets but large widths.
In Abu Dhabi, balconies achieve levels of PV generation close to or above 1000 kWh/year, but a deeper study should be conducted in order to account for the effect of high ambient temperatures on the PV generation.
On the other hand, and as expected, Oslo designs feature very low levels of PV generation with wide north facing balconies as an attempt to gather most of the radiation. However, only the south facing balcony is able to yield sufficient electricity to be considered as a worth investment at today's market conditions.
In what concerns Lisbon optimized designs, the conclusions are similar to Abu Dhabi's scenario, with a preference for small East and West facing balconies and wide South and North facing facades.

3.2 Impact of poor reflective surfaces

In order to understand the importance of the door, floor and wall materials to the main contributions to reflected radiation, we assigned poor reflecting materials to the surfaces representing the doors, floors and side walls of each optimized design and ran the solar radiation simulations again in high quality mode. The comparison of the PV yields in this new scenario with the optimized geometries from 3.1 can be found in Fig. 6.

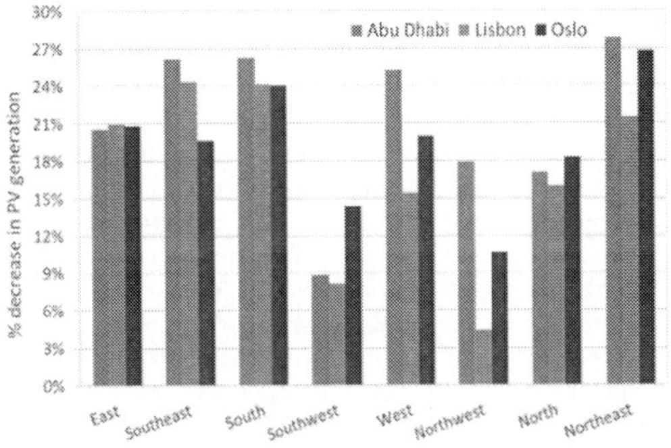

Figure 6: Decrease in PV generation from the optimized dimensions and material to the scenario with poor reflective materials.

As expected, the orientations that feature higher increments due to the use of bifacial PV modules are more sensitive to the use of materials with lower reflectivity. Three important effects can be pointed out:
1 - orientations such as East-South are highly direct sunlit, so a small change in reflectivity of surrounding materials has a great impact;
2 - orientations that feature low levels of direct irradiation mainly rely on the reflected diffuse irradiation, so poor reflective surfaces will also have a strong impact;
3 – materials with different physical properties ray-trace differently, thus probably there are better geometries for this new set of balcony materials.
It should be underlined that lower reflectivity of the balcony door would typically be associated to higher transmissivity and therefore better natural illumination of the indoor space thus, as always, there ought to be a design compromise between electricity generation, other energy requirements of the building and aesthetics.

3.3 Impact of partial-shading

The presence of a person in the balcony area may also cause the radiation to interact differently with the balcony materials and thus decrease the PV production of bifacial modules that rely on the reflected radiance.
We attempted to study such impact by including a human-like figure near each balcony's door. A low reflective material "dark blue jeans" [19] was assign to it and the radiation simulations re-ran, this time only for the optimized south facing balcony in Lisbon.

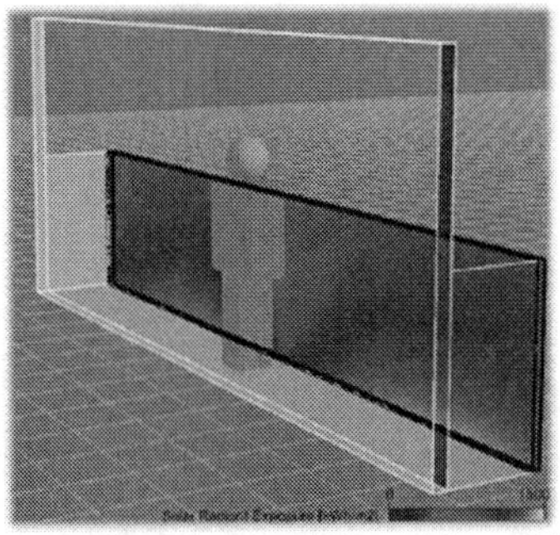

Figure 7: Detail of the simulation with partial-shadow cast by a person standing in a south facing balcony, in Lisbon.

Although Fig. 7 depicts the cumulative annual irradiation in the rear surface of the bifacial module, and it would be more realistic to assume permanent furniture and not a person in the balcony), the simulation allows us to understand the significant impact that this new element would have on the inter-reflections between the balcony surfaces. The decrease in PV density generation with the optimized materials is around 5%. Such reduction points out the importance of the bypass technology on bifacial modules and/or their connection to micro-inverters, since these are able to lessen the dramatic impacts of partial-shading in the operation of PV modules. In order to explore this phenomenon at a deeper level, sub-hourly simulations would be preferable, as well as a detailed sketch of the solar cells and their series-connections.

4 DISCUSSION

The potential for bifacial PV modules integrated on façade balconies was presented in this paper. The nature of light interaction with different surface material was taken into account in the optimization of the depth, width and offset of balconies facing different orientations. Three locations at different latitudes and with dissimilar solar radiation profiles were considered in this study: Abu Dhabi, Lisbon and Oslo.

It was observed that the optimization of balcony geometry and surface materials leads to an average increment of around 20% (from 4% and up to almost 30%) in comparison with a low reflective scenario, with the greater impact being observed in eastern/southern façade orientations. Clearly, lower latitudes that feature high levels of irradiation might be promising for bifacial PV (even north facing facades if the module prices drop further). On the contrary, higher latitudes with low direct sunlight are not attractive to the application of bifacial PV except for the south facing facade. In general, we can take that southern and northern facades should be wider and with smaller offsets, while the others could vary from intermediate width to large depth as an attempt to gather most of the diffuse reflected radiation.

Comparing the bifacial PV yield with a hypothetical monofacial yield in a flat PV wall (i.e. PV area equal to the balcony door area), in average the bifacial generation would be only -15% from the optimized scenario, ranging from +8% to -32% in Abu Dhabi, +7% to -49% in Lisbon and +6% to -39% in Oslo, in the whole set of orientations. Still, a deeper analysis would provide more insight into the impacts of partial-shading, that are expected to happen due to the nature of a space such as a balcony.

With these findings we conclude that there is high potential for bifacial PV in several locations around the world, both in terms of electricity generation and aesthetical solution for a better integration into the building envelope.

5 ACKNOWLEDGMENTS

Part of this work was supported by MIT Portugal Program for Sustainable Energy Systems, FCT grant SFRH/BD/52363/2013,Projects MITP-TB/CS/0026/2013 (SUSCITY), PTDC/EMS-ENE/4525/2014 (PVCITY) and UID/GEO/50019/2013 - Instituto Dom Luiz, and Fundação Calouste Gulbenkian.

6 REFERENCES

[1] Redweik, P., Catita, C., Brito, M., 2013. Solar energy potential on roofs and facades in an urban landscape. Solar Energy 97, 332 – 341

[2] Bonomo, P., Chatzipanagi, A., Frontini, F., 2015. Overview and analysis of current BIPV products: new criteria for supporting the technological transfer in the building sector. International journal of Architecture Technology and Sustainability, Volume 0, 67 – 85

[3] Energy Plus: Third-Party Graphical User Interfaces, https://energyplus.net/interfaces

[4] Soria, B., Gerritsen, E., Lefillastre, P., 2013. Vertical façade integration of bifacial PV modules: outdoor testing & optical modeling, 28th EU PVSEC

[5] Lindslay, A., Chiodetti, M., Dupeyrat, P., Binesti, D., Lutun, E., Radouane, K., 2015. Key elements in the design of bifacial PV power plants, 31st EU PVSEC

[6] Reise, C., Schimd, A., 2015. Realistic yield expectations for bifacial PV systems – an assessment of announced, predicted and observed benefits, 31st EU PVSEC

[7] Comparotto, C., Noebels, M., Wefringhaus, E.,Ferretii, N., Mancini, G., Berghold, J., Einhaus, R., Madon, F., 2015. Energy yield estimation of monofacial and bifacial solar modules, 31st EU PVSEC

[8] Fertig, F., Nold, S., Wohrle, N., Greulich, J., Hadrich, I., Kraub, K., Mittag, M., Biro, D., Rein, S., Preu, R., 2016. Economic feasibility of bifacial silicon solar cells. Prog. Photovolt: Res. Appl. 24, 800-817

[9] Freitas, S., Catita, C., Redweik, P., Brito, M.C., 2015. Modelling solar potential in the urban environment: State-of-the-art review. Renewable and Sustainable Energy Reviews 41, 915–931

[10] Rhinoceros 3D (2016), https://www.rhino3d.com/

[11] Grasshopper (2016), http://www.grasshopper3d.com/

[12] Diva For Rhino 4. (2016), http://diva4rhino.com/

[13] Lighting Materials for Simulation: Sunpower E20 Solar Panel Cell, http://www.lighting-materials.com/materials/665

[14] Perez, R., Ineichen, P., Seals, R., Michalsky, J., Stewart, R., 1990, 'Modeling Daylight Availability and Irradiance Components From Direct and Global Irradiance', Solar Energy, vol. 44, issue 5, pp. 271-289

[15] Tregenza, P.R., 1987, Subdivision of the sky hemisphere for luminance measurements. Lighting Research and Technology, 19:13–4

[16] Robinson, D., and Stone, A., 2004, Irradiation modeling made simple: the cumulative sky approach and its applications, PLEA2004 - The 21st Conference on Passive and Low Energy Architecture. Eindhoven, The Netherlands, 19 - 22 September 2004

[17] Primsolar (2016), BIFACIAL MODULE MODEL Bi60-343BSTC, http://www.prismsolar.com/pdf/bi60specs.pdf

[18] Galapagos (2016), http://www.grasshopper3d.com/group/galapagos

[19] Lighting Materials for Simulation: Dark Blue Jeans, http://www.lighting-materials.com/materials/998

A. Supplementary material

Figures A1 to A3 present the optimized balcony geometries and solar irradiation levels in bifacial PV modules for the 3 locations analysed.

- Abu Dhabi

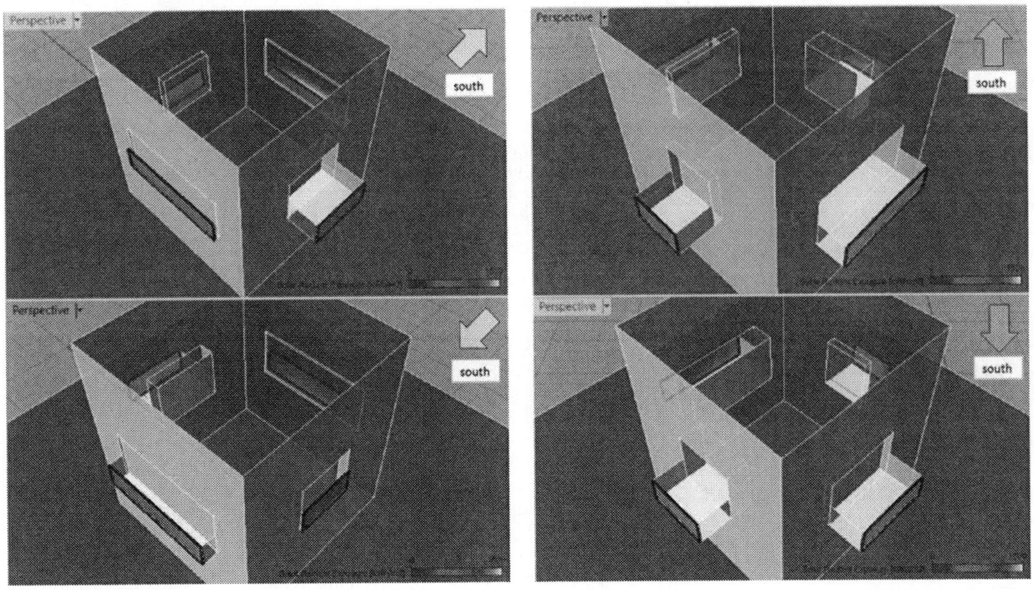

- Lisbon

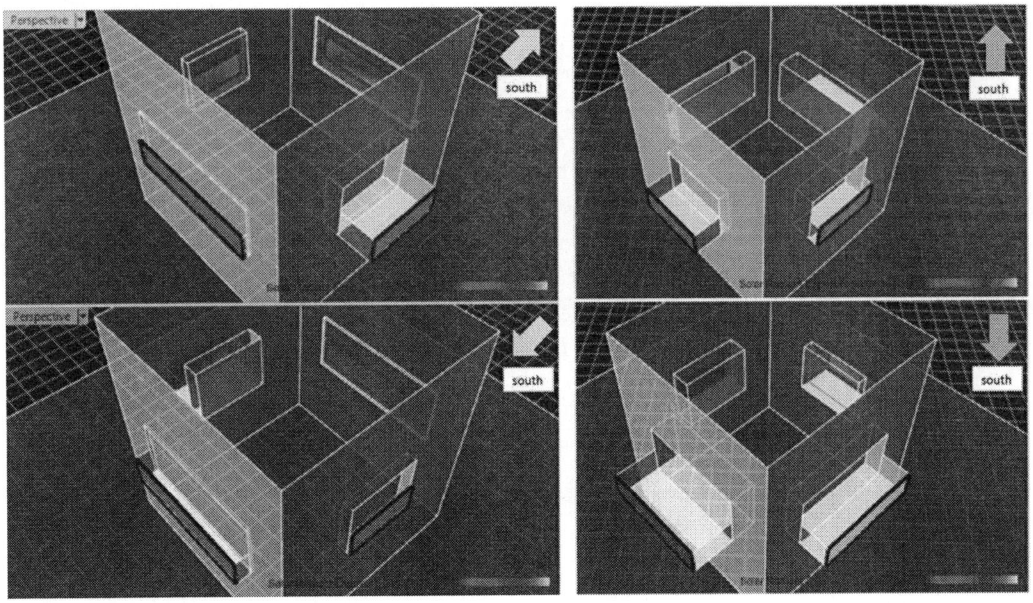

- Oslo

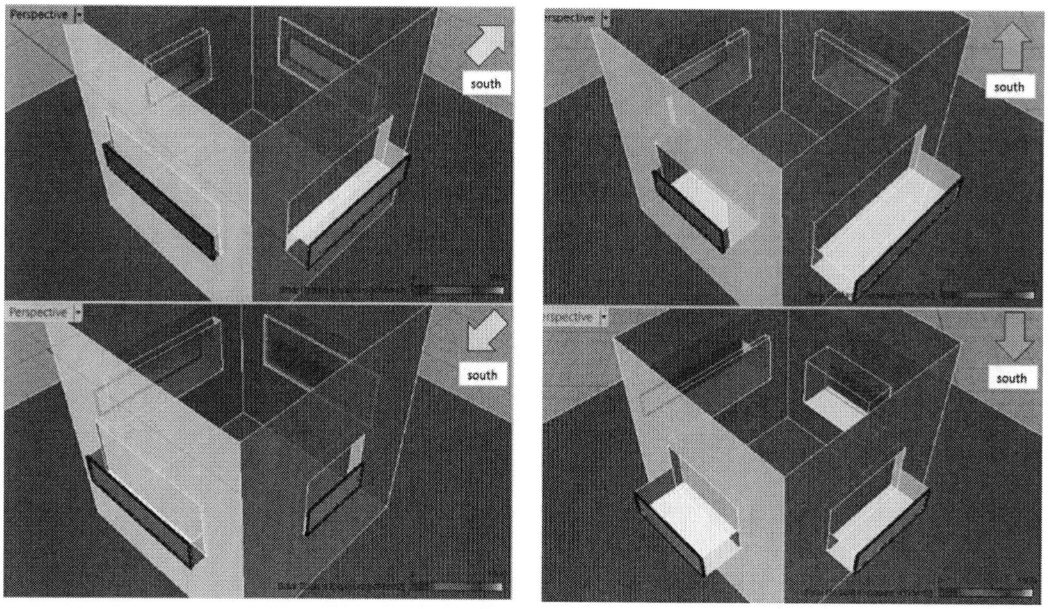

INDOOR AND OUTDOOR CHARACTERIZATION OF INNOVATIVE COLORED BIPV MODULES FOR FAÇADE APPLICATION

F. Frontini[1], P. Bonomo[1], E. Saretta[1*], T. Weber[2], J. Berghold[2]

[1] SUPSI-ISAAC, Campus Trevano, CH 6952, Canobbio, Switzerland
[2] PI Photovoltaik-Institut Berlin AG, Wrangelstr. 100, D-10997 Berlin, Germany
* mail to: erika.saretta@supsi.ch; phone +41 052 666 6320, fax +41 052 666 6349

ABSTRACT: The integration of photovoltaic technologies in buildings is considered as promising way to profit by renewable energies in the built environment, with many advantages. Nevertheless, dealing with PV in architecture is still a challenge for architects and engineers both because of aesthetic and technical issues. There are at least two ways to integrate photovoltaic technologies in the building skin design concept: emphasizing the use of solar energy (*high-tech approach*) or trying to hide them with the use of a *camouflage* technology, thanks to the use of colour filters or painting techniques.

This paper focus on the investigation of the performance behaviour of different coloured modules developed within the European Project SMART-FLeX. Within the project a special digital technique has been developed to print the front glass of BIPV modules, offering to the architect many possibility of design. It is clear, that paintings on front glass will shadow the solar cells, influencing in that way the performance of modules. In order to assess the impact of this technology on module performance extensive indoor measurements have been done by the authors. Therefore, prototype modules of different colour and printing degrees were assembled and investigated under standard and weak-light conditions, electroluminescence tests, spectral responsivity and temperature coefficient measurements. Outdoor assessment in two different location (Lugano and Berlin) will also provide important information on modules behaviour in real condition.

Keywords: BIPV, building integration, performance, colours

1 INTRODUCTION

Several aspects have to be synergistically evaluated when photovoltaics (PV) is integrated in the building envelope, such as technical, technological and aesthetical aspects among others.

With regard to aesthetical aspects, the transfer of new technologies in the building skin always opens the design approach to a basic question: make it visible or invisible? The semantics of the "new" technology becomes one of the most challenging aspect in the technological innovation process in architecture. In the transfer process of PV in architecture, namely BIPV, a lot of design approaches and aesthetical principles have been adopted in recent years. On one hand, BIPV projects highlighted the "pure" architectural language of PV, communicating visibly the aesthetics of solar cells. On the other hand, architects and engineers have attempted to use a more "traditional" design in BIPV language with the principle to "filter" or to "hide" the expression of PV technology in the building skin, in accordance to a "camouflage" approach. For this reason, the lexicon of PV modules within a façade becomes part of the building and urban quality, entering in a direct relation with the surrounding environment, the streets and other common point of views. In this architectural debate, involving PV as a new alternative for building cladding, a growing interest is linked to the glass treatment as a possibility to customize BIPV modules.

In the framework of the European Project SMART-FleX [1], a novel digital printing technique has been developed for the front of BIPV glass modules, offering to the architects the possibility to "*camouflage*" the PV cells and, at the same time, design such BIPV module with their own drawings and sketches, as well as with the shapes that better fit in the building envelope. Indeed, one of the goals of the project is to provide and demonstrate the flexibility and the adaptability of manufacturing of customized BIPV glass modules capable to be integrated harmoniously in the building skin. Currently, the project is running in the demonstration phase and such printed and tailored BIPV modules are integrated in a facade on an office building [2].

However, the use of a digital – coloured – printing on the front glass of BIPV modules causes the shadowing of the PV cells and the affection of the module performances in terms of power output and energy yield. Therefore, this paper focus on the investigation of the performances of different coloured SMART-FleX prototypes in order to characterize such innovative products and to provide preliminary results on the influence of the printings for offering to the architect effective design possibilities.

2 PURPOSE AND METHODOLOGY

SMART-FLeX project is proposing an integrated solution for building design with PV installation, through the introduction of automated production lines. This enables the manufacturing of BIPV products customizable in shapes and colours for being fitted perfectly into the building envelope.

Together with the possibilities to produce customized BIPV modules, the projects partners developed an innovative solution to colour the glass elements hiding the PV cell (*camouflage* technology). Such solution is made by means of a digital ceramic-based printing. Moreover, the digital printer is able to print high definition pictures (720 dpi) on the glass, enabling architects for limitless design of façade (either with or without solar cells).

Nevertheless, it is clear that paints covering solar cells will influence performance of modules, but the actual amount of power loss is not known and will be determined in this work. For this reason special prototypes have been produced.

2.1 Prototypes

All the prototypes are glass/glass modules made of multi c-Si cells laminated within two thermal toughened glass panes. The front glass is painted with coloured dots that can be printed in the outer side (position 1) or in the inner side (position 2). The prototypes have dots in three different colours (white, black and green) and with different printing degree (ranging from 0 % colour to 100 %). The prototypes are listed in Table I and a picture of the prototypes can be seen in Fig. 1.

Table I: SMART-FleX prototypes. If not specified, dots are printed in the inner side (position 2)

	Module	*Dim.* *[m]*	*n° of strings / n° of cells per string*	*Printing degree* *[%]*
A	Large – white dots	2x1.7	12 / 10	0; 30; 60; 70; 80; 100
B	Large – black dots	2x1.7	12 / 10	0; 30; 60; 70; 80; 100
C	Large – black dots – *position 1*	2x1.7	12 / 10	0; 30; 60; 70; 80; 100
D	White dots	1x1.7	10 / 6	10; 20; 30; 40; 50; 60; 70; 80; 90; 100
E	Green dots	1x1.7	10 / 6	10; 20; 30; 40; 50; 60; 70; 80; 90; 100
F_a	Transparent, a	1x1.7	1 x 60	30%
F_b	Transparent, b	1x1.7	1 x 60	30%
G	White dots	1x1.7	1 x 60	30%
H_a	Black dots, a	1x1.7	1 x 60	30%
H_b	Black dots, b	1x1.7	1 x 60	30%
I	Green dots	1x1.7	1 x 60	30%

Figure 1: SMART-FleX prototypes. Left picture: "Large - black dots (B)" - prototype with 12 different strings in 6 different printing degrees. Right picture: "Green dots (I)" - prototype with 60 cells and with 30% green colour

Some of the prototypes were also equipped with a further result of the SMART-FleX project: an innovative power tracker [2]. This power tracker has 3 input channels enabling to take three strings from a single BIPV module and 1 voltage output, e.g. to a micro inverter. These three channels work independently of one another and can be controlled by MPP tracking method. When one or more solar cells are shaded in a single string, the two non-affected strings will generate power. In addition, when solar cells are shaded in two strings, only one string will perform. The modules equipped with SMART-FleX trackers were compared to modules with conventional junction boxes and are categorized as "_b".

2.2 Methodology

To evaluate to what extent the digital printing influence the energy performance of the module, both indoor and outdoor tests have been performed.

In detail, the colour and the printing degree of the prototypes have been defined as the *influencing parameters* – the factors that affect the energy performances of the painted module compared to a transparent module used as reference. Therefore, the test results are evaluated with regard to such aspects.

2.2.1 Indoor characterization

Indoor characterization tests have been performed both in the laboratory of ISAAC-SUPSI and in the laboratory of PI-Berlin in order to evaluate the influence of the colour and printing degree to the power parameters at STC and at different irradiance values (weak-light conditions). Electroluminescence (EL) were performed additionally to assess the quality of the modules/cells and to assess if EL is still possible when modules are printed with colors. Then, both spectral response and temperature coefficient measurements were performed as well at SUPSI labs. Indoor tests have been performed to the modules A to I.

2.2.2 Outdoor characterization

Outdoor characterization tests have been performed both in the test facilities of ISAAC-SUPSI and at PI-Berlin to evaluate the energy behaviour of the prototypes in real conditions. The prototypes with homogeneous colour printing have been installed in Lugano (Switzerland) and in Berlin (Germany) for several weeks and the resulting data have been analysed and compared. The following module types were installed in Lugano: F_a, G, H_a and I and in Berlin: F_b, G, H_a and H_b. Specifically, two modules have been equipped with the Smart-FleX trackers in Berlin.

Figure 2: SMART-FleX prototypes installed at ISAAC-SUPSI outdoor test facility in Lugano (Switzerland)

To evaluate the energy behavior a monitoring system was installed. At SUPSI the four modules were equipped with Maximum Power Point Trackers (MPPT 3000) developed by SUPSI and adapted to their voltages and current ranges in order to optimize the measurement accuracy. As well as recording the voltage and current of the maximum power point, the full I-V curve was measured every five minutes.

A pyranometer was installed on the stand to monitor the irradiance at 45°. Meteorological data, such as global and diffuse horizontal irradiance, wind speed and ambient temperature, as well as the module performance data, were recorded simultaneously with a resolution of one minute. In order to achieve a high level of reliability in the inter-comparison, improved data quality control and an advanced data analysis procedure were used [3].

Figure 3: SMART-FleX prototypes installed at PI-Berlin outdoor test facility in Berlin (Germany)

At PI-Berlin the monitoring was realized together with the power conversion from direct to alternating current (DC → AC). Therefore, commercial available micro inverters from Enphase (ME250) were installed. A system related web interface enabled the authors to monitor each module (micro-inverter) with energy and power data time resolved (5 min steps). In parallel the module temperatures on the module back side (30sec steps) and the prevailing weather condition were monitored (5 sec steps). The pyranometer was installed to monitor the irradiance in module plane at 30°.

3 INDOOR CHARACTERIZATION

The investigated prototypes are the modules type A to I As a preliminary step, electroluminescence tests have been performed on the prototypes to the evaluate the presence of damaged cells that can affect the efficiency. From such test, no damages arise, hence the prototypes have been subjected to the following indoor tests.

The module types A to C, which are made of 12 strings with 6 different area degrees (ranging from 0 to 100%) and the modules types F to I have been subjected to the power output tests both at STC and at weak-light conditions.

The modules D and E, which are made of 10 strings with 10 different area degrees (ranging from 10% to 100%), have been subjected to the power output test at STC (SUPSI).

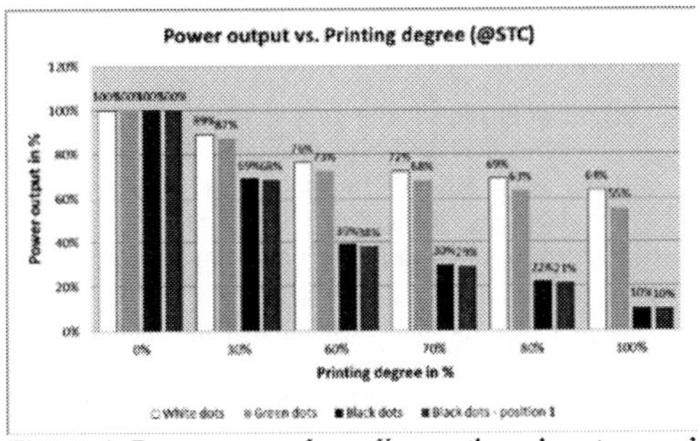

Figure 4: Power output depending on the colour type and on the printing degree – calculation based on a single cell

Thanks to the data recorded at STC and to interpolation methods, the power output of the prototypes as function of colour type and printing degree has been obtained (Figure 4).

It is worth to note that the printing of the black dots in position 1 or in position 2 does not significantly affect the power output in this conditions. Moreover, when the printing degree increases, the reduction of the power output is almost linear. Such reduction trends can be observed also when prototypes are subjected to power measurement tests at weak-light conditions. For instance, Figure 5 shows the power output of the module type A (white) depending on the printing degree for different irradiance levels.

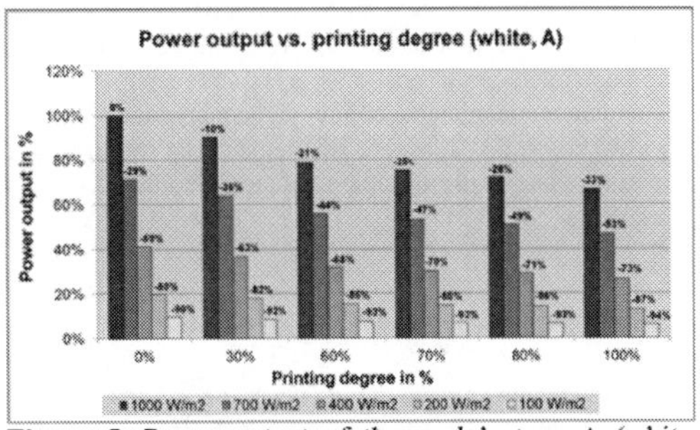

Figure 5: Power output of the module type A (white coloured) depending on the printing degree for different irradiation levels

Since the real energy outputs of PV modules operating outside depends on several factors (e.g. incident irradiance, spectral irradiance distribution, module temperature, …) further indoor tests have been performed such as the spectral responsivity test and the temperature coefficient measurements.

In order to better understand the effects of the colour types on the power output of the module, the spectral responsivity (SR) has been investigated at ISAAC-SUPSI labs for the module type F to I [4]. From such measurements, it arises that the black dot paint absorbs the wavelengths corresponding both to the visible light and to the infrared light, and the green dots absorb the wavelengths corresponding to the blue/green colour (see Fig. 6).

Therefore, such results have been used to correct the power output of the modules investigated indoor at STC.

The corrected power outputs do not have significant differences with the non-corrected power outputs as can be seen in Table II.

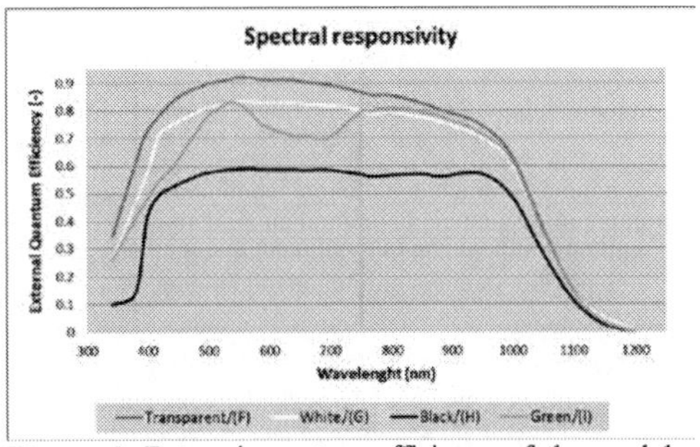

Figure 6: External quantum efficiency of the modules type F to I

Table II: Power output of the modules type F to I corrected with the mismatching factor (MMF) of the SR test

	Module /(F_a)	*Module /(G)*	*Module /(H_a)*	*Module /(I)*
$P_{NOM}(W)$	234.0	212.4	156.8	203.6
MMF	1.0	1.0	0.99	0.99
$P_{NOM\text{-}SR}$ (W)	234.0	212.4	156.8	203.6

Moreover, the temperature coefficients of current, voltage and maximum power of the modules type F and G have been assessed. From such test, it arises that the colour paint on the front glass has almost no impact on the temperature coefficients of the modules at laboratory and controlled conditions.

However, further tests have been performed outdoor to investigate the influence of the printing coloured dots on the energy performance of the modules in real conditions.

4 OUTDOOR CHARACTERIZATION

The outdoor tests have been performed on the modules type F to I (60 cells modules). They are mainly aimed for investigating the influence of the printing dots on the energy behaviour of the module when exposed outdoor at real conditions. Furthermore, the Smart-FleX trackers have been installed on some prototypes in order to evaluate their influence on the energy output of the modules. Results on power tracker performance will be presented in another publication and will be not discussed here.

In detail, the modules type F_a, G, H_a and I have been equipped with conventional junction box and installed at the ISAAC-SUPSI outdoor test facility in Lugano (Switzerland).

The modules of type F_b and H_b have been equipped with Smart-FleX trackers. Such modules have been installed at the PI-Berlin outdoor test facility together with the modules of type H_a and G.

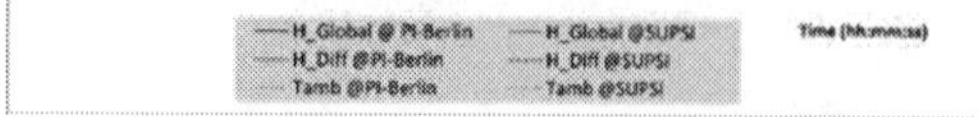

Figure 7: Comparison between the local conditions at the outdoor test facility in Lugano (21.5.2016 – very clear sky) and at the outdoor test facility in Berlin (6.6.2016 – very clear sky)

Figure 7 shows the comparison between the local conditions (irradiance and ambient temperature) for a very clear sky day at the ISAAC-SUPSI outdoor test facility in Lugano and at the outdoor test facility in Berlin.

It is worth to note that the irradiance is almost the same but the ambient temperature is higher in Lugano than in Berlin in the average amount of 3 degrees. Such ambient air temperature differences involve a variation in the modules temperatures in Lugano and Berlin (Figure 8 and 9).

The daily power output and temperature of the modules installed in Lugano are shown in Figure 8, whereas the ones of the modules installed in Berlin are shown in Figure 9.

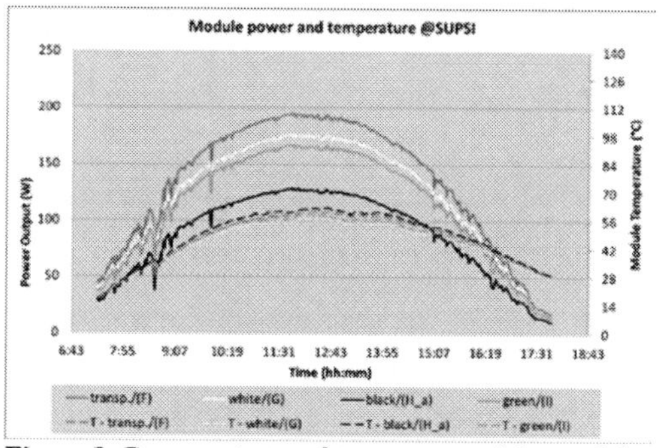

Figure 8: Power output and temperature of the modules type F to I as a function of the colour type (data for Lugano on 21.5.2016, very clear sky)

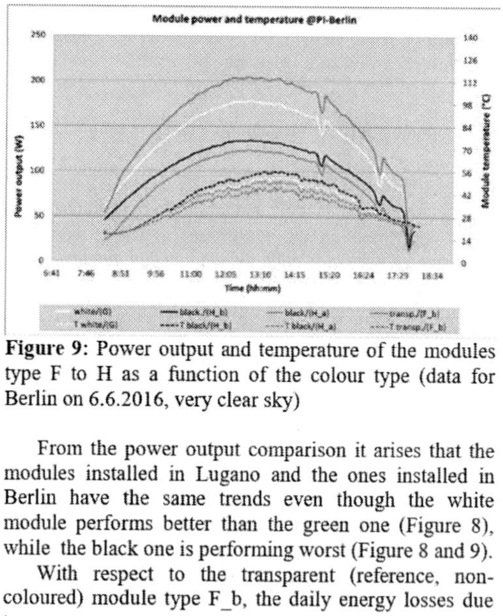

Figure 9: Power output and temperature of the modules type F to H as a function of the colour type (data for Berlin on 6.6.2016, very clear sky)

From the power output comparison it arises that the modules installed in Lugano and the ones installed in Berlin have the same trends even though the white module performs better than the green one (Figure 8), while the black one is performing worst (Figure 8 and 9).

With respect to the transparent (reference, non-coloured) module type F_b, the daily energy losses due to:

- white dots (30% printing degree) range from 10% (@ISAAC-SUPSI) to 12% (@PI-Berlin);
- green dots (30% printing degree) is about 15% (@ISAAC-SUPSI);
- black dots (30% printing degree) range from 34% (@PI-Berlin) to 35% (@ISAAC-SUPSI);

for the modules equipped with conventional junction box.

Forthermore, the performance ratio (PR) of the modules installed in Lugano are shown in Figure 10. It is important to note that it is difficult to assess the influence of the colour types on the PR values due to both the uncertainty in the indoor and in the outdoor measurements.

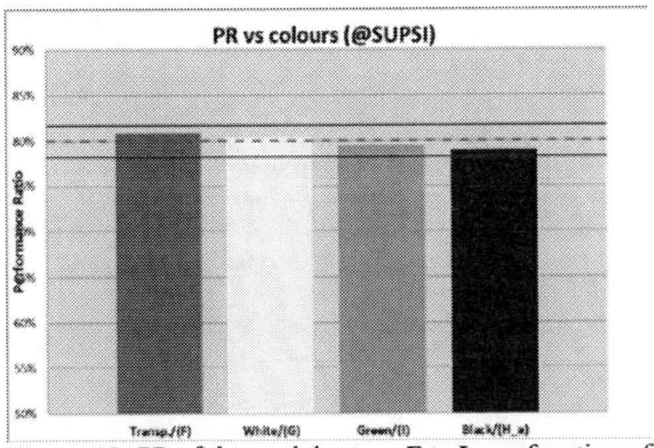

Figure 10: PR of the modules type F to I as a function of the colour type in Lugano (21.5.2016, very clear sky). Red lines represent the uncertainty measurement values

From the evaluation of the data recorded in Lugano during the 4th week of May 2016, it is possible to depict the modules temperature behaviour for the different coloured modules in comparison with the non-coloured module (type F) when the irradiance increases (Figure 11). It is worth to note that the trend line for the black module has the highest slope. This is due to the fact that the black dots absorbs more light (Figure 6) than other coloured dots (as shown in the spectral responsivity test) and it converts the absorbed light into heat, which is transferred to the module.

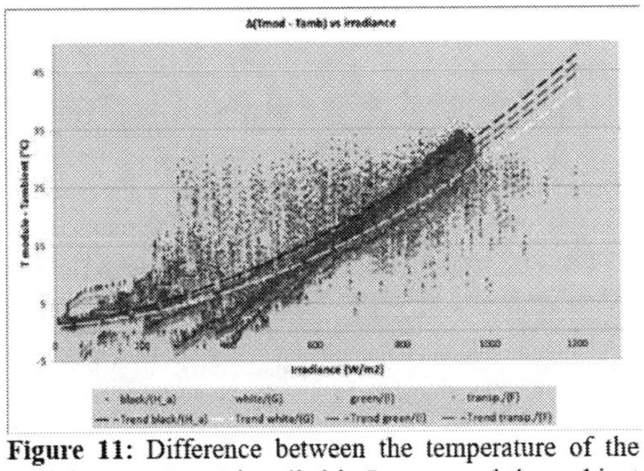

Figure 11: Difference between the temperature of the modules type F to I installed in Lugano and the ambient temperature over a week as a function of the irradiance

5 CONCLUSIONS AND FUTURE CHALLENGES

From the indoor and outdoor characterization, one can conclude that the choice of bright colours (green and white), for the printed dots, allows the module to perform better, also when installed in real outdoor conditions, with daily energy losses not more than 15%. In detail, the modules with black dots have the worst energy behaviour.

Furthermore, the positioning of the printing on the inner side or on the outer side of the front glass does not affect significantly the energy output when measured in controlled environment (solar simulator). However, the durability aspects for the modules with printing on the outer side should be evaluated with specific tests, such as abrasion test like cleaning or a standard abrasion tests [5] and climatic test to assess the life time of the painting.

Considering such preliminary results, further investigation should be performed in order to provide more detailed information about the influence of the colour types and the printing degrees/patterns especially for non-uniform coloured modules. However, such paper demonstrates that the printing of coloured dots on the front glass of BIPV modules can be an innovative and energy feasible solution for the design of active painted facade.

ACKNOWLEDGEMENTS

The work has been supported by the European Commission in FP7 through the funding of the project SMART-FLeX (Grant Agreement n°ENER/FP7/322449). Authors would like to thank the industrial partners Glassbel (especially Mr. Tomas Lenkimas) and Viasolis (mainly Mr. Rimvydas Karoblis and Mr. Modestas Pikutis) for the manufacturing of the modules and their valuable support.

REFERENCES

[1] EU Project SMART-FleX. http://www.smartflex-solarfacades.eu/

[2] Ulbikas J., Galdikas A., Stonkus A., "SMART-FLeX – demonstration of integrated approach for design of glass BIPV façade", 30th EUPVSEC 2015, Hamburg, September 2015.

[3] Dittmann S. et al., "Energy yield measurements at SUPSI – importance of data quality control and its influence on kWh/Wp inter-comparison", 26th EUPVSEC, Hamburg, September 2011.

[4] M. Pravettoni, "Characterization of the electrical performance of photovoltaic cells and modules: a review", in *Energy Science and Technology*, Vol. 6: Solar Engineering II (Photovoltaics & Solar Cells), pp. 1-36, edited by Umesh Chandra Sharma, Ram Prasad and Sri Sivakumar, Studium Press LLC, USA (2015)

[5] N. Ferretti et al., "Investigation on the impact of module cleaning on the antireflection coating", 32nd EUPVSEC 2016, Munich, June 2016

QUANTIFICATION OF GLARE FROM SUNLIGHT REFLECTED ON SOLAR INSTALLATIONS

Mattia Battaglia[1], Florian Ruesch[1], Andreas Bohren[1] and Stefan Brunold[1]
[1]*Institut für Solartechnik SPF,*
Oberseestrasse 10, CH-8640 Rapperswil, Switzerland

ABSTRACT: Several methods to assess glare in an indoor and outdoor environment are reviewed. Furthermore, measurements of the reflection properties of different glass types and other materials used in the building envelope are presented. The bidirectional reflectance distribution function (BRDF) is used to characterize the different materials and to evaluate their potential of creating discomfort glare in combination with reflected sunlight according to different thresholds. The results are used to discuss the applicability of the different thresholds for developing a reliable framework for the classification of reflections. In addition, recommendations for which glass type performs best with respect to glare are given for different situations.
Keywords: Optical Properties, Safety, Building Integration

1 INTRODUCTION

Glare from solar installations has led to various neighborhood disputes, which sometimes even ended in court. There is no widely accepted method for the quantification of this effect and the legal practice is non-uniform and sometimes even contradictory. Existing approaches for the identification of inadmissible glare from sunlight reflected at solar installations are based on the duration of the appearance of glare (acceptable if shorter than 30 min for less than 30 days in the year [1]).

For the determination of this time period a threshold is needed that identifies glare from given luminance and illuminance values. Due to the non-availability of such a threshold, current glare evaluations usually assume mirror-like (i.e. specular) reflection in which glare is present whenever the reflected image of the sun can be seen by an observer under this simplified conditions. However, most glasses used for solar pv-modules and thermal collectors have a structured and therefore scattering surface. Also scattered light can be at high levels of luminosity, which makes it important to consider the scattering of surfaces in order to achieve a detailed image of a given situation with potentially disturbing glare. There is an increasing number of methods to quantify glare discomfort or glare probability, that are mainly aimed to be used in the field of natural and artificial indoor illumination [2], [3] or outdoor illumination at night or dawn [6]. In this work, these different ways to identify glare are evaluated in the context of solar installations. Based on the results, their suitability to act as an assessment criterion for future cases is discussed. In addition, possibilities to classify glasses with different scattering properties according to their glare potential are debated.

2 METHOD

2.1 Measurement

An imaging sphere from Radiant Zemax was used to characterize the bidirectional reflection distribution function (BRDF) of different glass types frequently used in solar installations (thermal collectors and pv-modules). The BRDF-Function has the following form:

$$BRDF(\theta_i, \phi_i; \theta_r, \phi_r) = \frac{dL_r(\theta_i, \phi_i; \theta_r, \phi_r; E_i)}{dE_i(\theta_i, \phi_i)} \quad (1)$$

L_r represents the reflected Luminance while E_i is the incident illuminance. An illustration of the different involved angles is shown in Figure 1. In addition to solar glasses, other materials frequently used in the building envelope (tiles, metals for facade and roof, paints, flagstones) were recorded, too. In imaging sphere measurements, the distribution of the reflected radiation in the whole semisphere for a given incident angle is given. The incident polar angle can be varied along a circular track. For the database of this work 10° steps were used for the incident polar angle. The complete BRDF can then be used to quantify the illuminance and the luminance of a reflection depending on the local situation.

By assuming discrete parallel light from the sun for the incident angle, the reflected luminance takes the following form, where L_{sol} and ω_{sol} are the luminance and the apparent size of the sun respectively.

$$L_r(\theta_r, \phi_r) = BRDF(\theta_{sol}, \phi_{sol}; \theta_r, \phi_r) L_{sol} \omega_{sol} \cos\theta_{sol} \quad (2)$$

The reflected illuminance for an observer at position x that sees the surface of reflection under a solid angle ω_x can be computed by taking the sum over all discrete BRDF values associated with reflection angles that are element of the given solid angle.

$$E_x = \sum_{\theta_r, \phi_r \in \omega_x} L_r(\theta_r, \phi_r) \sin\theta_r \cos\theta_r \,\Delta\theta_r \Delta\phi_r$$

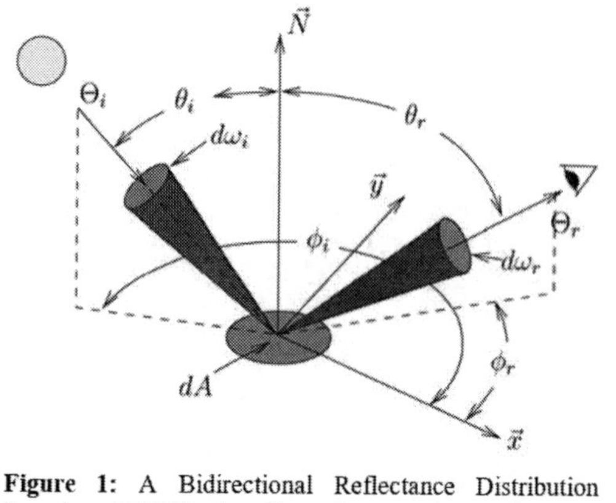

Figure 1: A Bidirectional Reflectance Distribution Function (BRDF) describes the complete reflectance properties of a surface.

2.2 Data analysis

Reflections in 3-dimensional space constitute a multidimensional problem. The surfaces orientation, the suns position and the observers position relative to the surface lead to various situations in which glare can possibly occur. To reduce the dimensionality of the problem, two well defined example cases are defined. The considered reflection geometry is illustrated in Figure 2. The reflecting surface is assumed to be quadratic with side length L. The observer sees the surface from a distance D under an angle θ. The evaluation was done for different values of θ since the reflection properties are strongly dependent on the reflection angle. As a measure for the apparent size of the surface the ratio D/L was used. This ratio defines a dimensionless distance. As the value of D/L increases the apparent size of the surface for the observer decreases.

The data was evaluated according to the 30-minute-criterion [1]. To study in detail the effect of reflection angles, the two example cases were defined. They differ in the path of the reflected image in the perspective of the observer. In the first case, the sun moves along the axis A_1 that is indicated in Figure 2. In this case, the time that is needed for the center of the reflection to move through the surface remains constant for all values of θ. The other case describes the situation in which the sun moves along the perpendicular axis A_2. In this case, the (visual) path of the sun in the surface is significantly shortened as θ approaches 90°. By using the BRDF for each value of D/L, the reflected luminance distribution and the reflected illuminance is determined for the measured 10° steps of θ for both cases. The incident illuminance of the sun was set to $10^6\ lux$

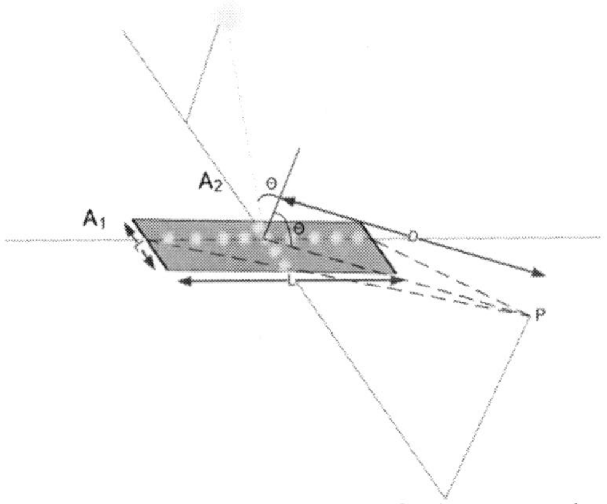

Figure 2: The ratio D/L is a measure for the apparent size of the reflection surface for an observer.

2.3 Definitions of glare

Visual comfort of a human observer depends on multiple different parameters. Considered as most important are the total amount of light, the distribution of light in the visual field as well as the adaptation state of the observer's eye.

Maximal Luminance

A simple criterion is provided by the maximal luminance to which the human eye can adapt. This threshold is in a range between $10^4 - 1.6 * 10^5\ cd/m^2$. In this work outdoor situations with bright background illumination are considered. For this reason the upper limit of $1.6 * 10^5\ cd/m^2$ is used for farther evaluations. In order to quantify the degree of visual comfort or discomfort for given lighting situations, various different indices have emerged based on experiments with humans and animals. An extensive review can be found in [2]. However, most of the indices address to describe the visual comfort in the presence of artificial and natural lighting coming from different sources in an architectural setting. Those indices usually rely on a defined line of sight and weight the sources according to their position in the visual field.

Daylight Glare Probability

A widely used example of such a visual comfort index is the Daylight Glare Probability [4].

$$DGP = 5.8 \times 10^{-5} E_v + 9.2 \times 10^{-2} \log\left(1 + \sum_i \frac{L_i^2 \omega_i}{E_v^{1.9} P_i^2}\right) + 0.16 \quad (3)$$

The used parameters are: The vertical illuminance E_v at eye-level (lux).The luminance $L_{s,i}$ of the i-th glare source (cd/m^2). The solid angle $\omega_{s,i}$ subtended by the i-th glare source (sr). The Guth's position index P_i of the i-th source in the visual field.

The resulting Daylight Glare Probability value (DGP) ranges from 0 to 1 indicating the percentage of people who experience glare in the given situation. A value of 0.45 is usually considered to be the threshold for inadmissible glare [5]. While this index is designed to deal with multiple uniform glare sources, the requirement for this work is to include the effect of a large light source with a highly non uniform luminance such as solar installations with structured glass surfaces. For this purpose, each solid angle element of the measured BRDF was considered to be a single object in the sum in the last term of Equation (1), while the effect of the Guth's position effect was neglected assuming that the observer has the light emitting surface in the center of the visual field.

De Boer rating by ASSIST

The alliance for solid-state illumination systems and technologies ASSIST recommended a different index to quantify the glare of a light emitting source [6]. This discomfort glare index is designed to deal with artificial light sources in a dark environment and is not validated for the present situation of reflected sunlight at daytime. This index accounts for the distribution of the source illuminance by including both the total illuminance E_L(lux) of the source as well as the maximum luminance L_L(cd/m2). It is defined by

$$DG = \log(E_L + E_s) + 0.6 \log\left(\frac{E_L}{E_s}\right) - 0.5 \log(E_A) \quad (4)$$

$$DB = 6.6 - 6.4 \log DG + 1.4 \log\left(\frac{50'000}{L_L}\right) \quad (5)$$

The additional parameters are the illuminance coming from the surrounding of the source E_s (lux) and the ambient illuminance E_A (lux). The resulting Index value DB has the following interpretations:

1	Unbearable
3	Disturbing
5	Just permissible
7	Satisfactory
9	Just noticeable

In this work a De Boer rating of 4 was considered to be the threshold for inadmissible glare.

After image potential

If a high amount of irradiance is absorbed by the retina in the eye, the observer experiences a temporary after image potential on the affected area of the visual field or, in the worst case, a permanent eye damage can occur. Detailed experiments have been done with rabbits in order to determine the maximal intensity that is still safe for an eye [7]. This approach was adapted for applications with concentrating solar installations [8]. In order to develop a safety measure for illuminance to the eye, the retinal irradiance E_r is computed from the irradiance at the entrance of the eye E_c.

$$E_r = E_c\left(\frac{{d_p}^2}{f^2\omega^2}\right)\tau \qquad (6)$$

The parameters describing the physical properties of the eye are the pupil diameter d_p, the focal length of the eye ω and the blink response time τ. For this retinal irradiance there exist different safety measures. Permanent eye damage is not possible with normal blink response times of ~0.5 s and without concentrating optics. Therefore, the important measure for this work is the threshold for after image potential which can be expressed as follows in W/cm^2:

$$E_{r,flash} = \frac{3.59 \times 10^{-5}}{\omega^{1.77}} \qquad (7)$$

3 RESULTS

In general, the duration of glare is the shorter, the smaller the apparent size of the surface. The computed reflected luminance and reflected illuminance values for different values of D/L and different angles θ can be used to determine the minimal dimensionless distance D/L for which the duration of glare is shorter than 30 minutes. In the following, this minimal distance will be referred to as 'safe distance'. The evaluation was done for the above described criteria for glare where for the maximal luminance threshold the upper limit $1.6 * 10^5\ cd/m^2$ was used. The results for three solar glasses with different scattering properties are shown in Figure 3 (DGP), Figure 4 (maximal luminance), Figure 5 (De Boer rating) and Figure 6 (after image potential). The values for white plaster and a zink-coated steel sheet are given as a comparison.

The theoretical values of perfectly flat float glass are added for all glare ratings. This was necessary since the sharp, in theory Dirac like peak of a specular surface cannot be resolved by a practical measurement with the imaging sphere. The different glare criteria lead to different values for the safe distance. However, the Daylight Glare Probability, the Maximal Luminance criterion and the De Boer rating all lead to a comparable classification of the different materials in terms of their potential glare range. The largest range is marked by the safe distance of the structured glass followed by the measured float glass. The white plaster exhibits the smallest range of potential glare. For observation angles θ steeper than 50° the prismatic glass has a safe distance that is comparable to the white plaster. In case of higher angles, the prismatic structure loses its scattering effect, which leads to an increased safe distance. The glare range of the steel sheet lies well above the corresponding values of the prismatic glass and the plaster.

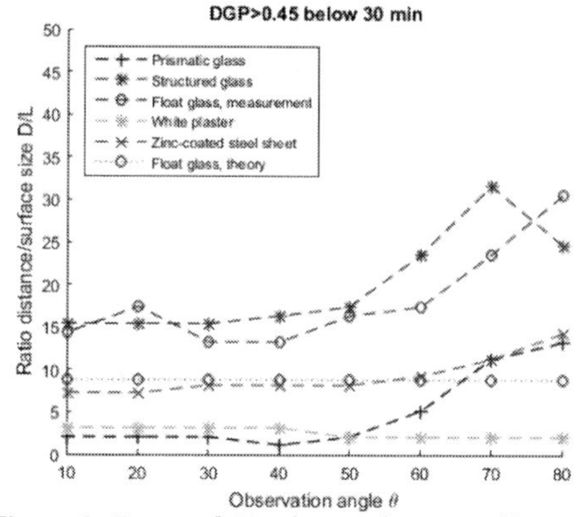

Figure 3: Range of 30 minutes glare according to the Dalight Glare Probability Rating (Case 1).

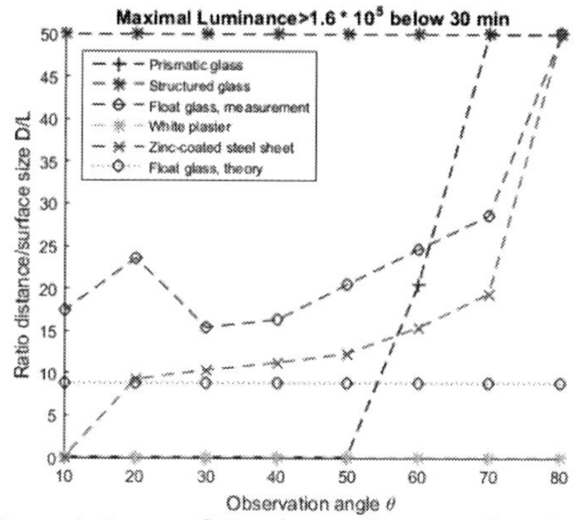

Figure 4: Range of 30 minutes glare according the the maximal luminance criterion (Case 1).

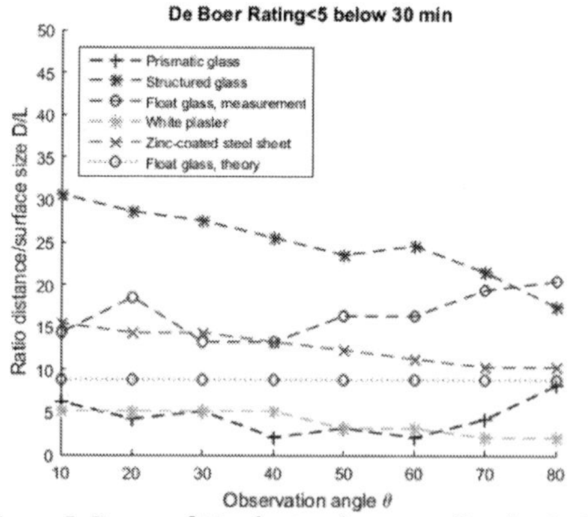

Figure 5: Range of 30 minutes glare according the the De Boer Rating (Case 1).

The results of the after image potential criterion exhibit a different characteristic. According to this evaluation method, the float glass has the longest reaching glare impact. This can be explained by the high threshold of this criterion. The reflected illuminance of the scattering surfaces drops below this high threshold at very short distances for which the float glass still shows a reflection longer than 30 minutes.

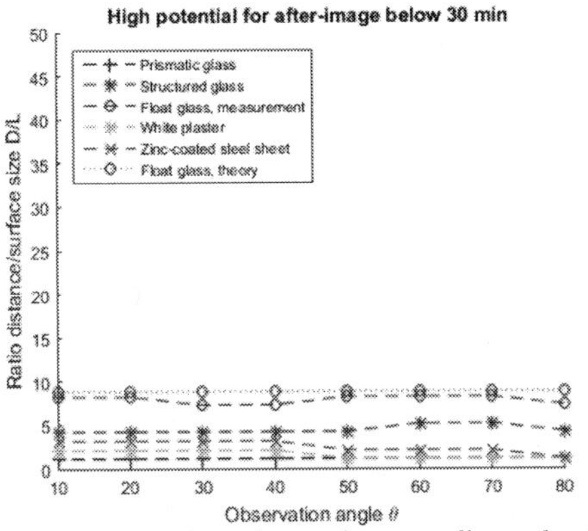

Figure 6: Range of 30 minutes glare according to the after image potential criterion (Case 1)

In the defined case 2 the image of the sun moves along the direction that appears shortened. Figures 7-10 show the results for the measured materials analogously to the results of case 1. For case 2 the safe distance of flat glass strongly drops at high observation angles θ as a consequence of the shorter appearance of the reflected sun path. On the other hand, scattering materials lead to similar results in both cases.

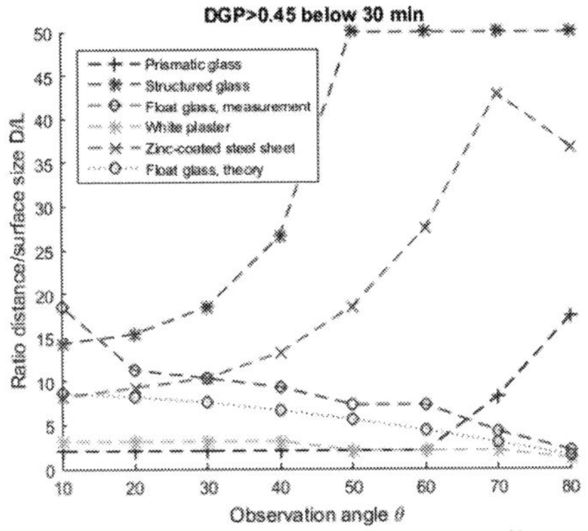

Figure 7: Range of 30 minutes glare according to the Daylight Glare Probability Rating (case 2).

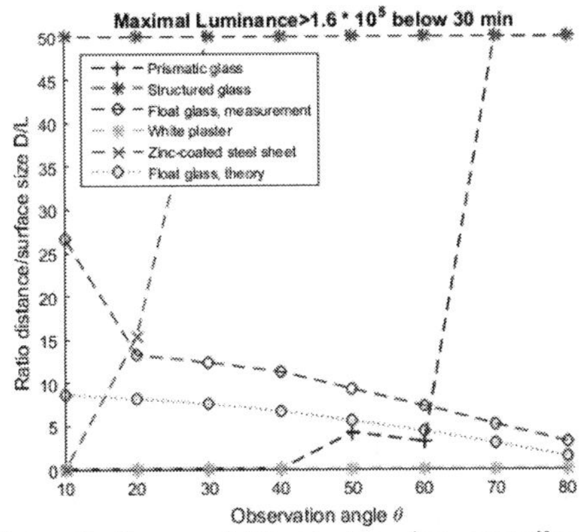

Figure 8: Range of 30 minutes glare according the maximal luminance criterion (case 2).

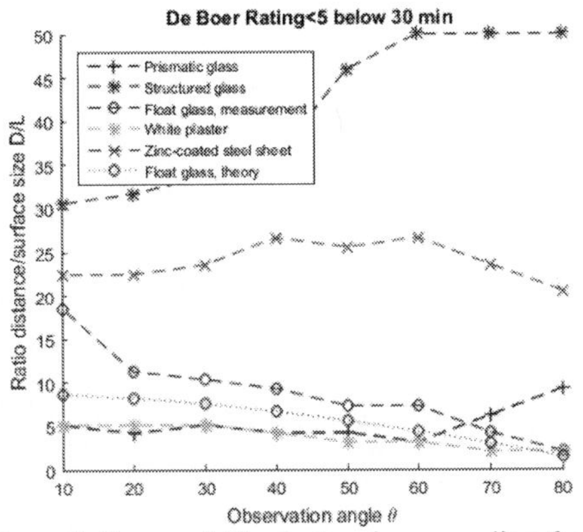

Figure 9: Range of 30 minutes glare according the De Boer Rating (case 2).

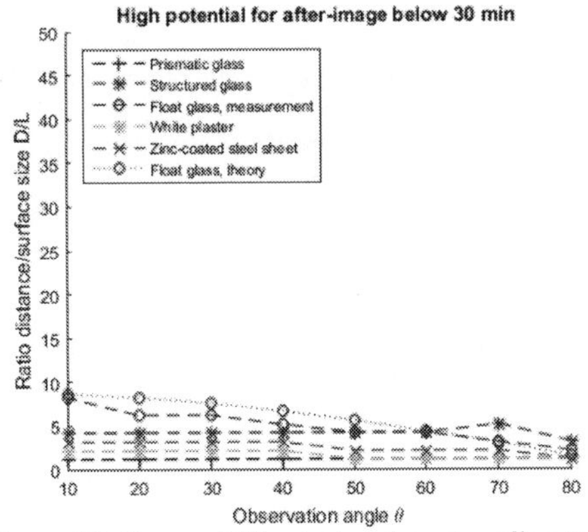

Figure 10: Range of 30 minutes glare according to the after image potential criterion (case 1)

4 DISCUSSION AND CONCLUSION

4.1 Applicability of the used assessment methods and glare thresholds

The used 30 minutes criterion is widely used for glare evaluations in the German speaking countries. However, it was first applied for the periodic shadow cast of wind power plants based on a psychological study where the time limit of 30 minutes was arbitrarily chosen. It is questionable if the impact of a reflective glare source that is only locally present in the field of view develops the same impact on a subject as periodic change in ambient illumination caused by rotating wind turbines. Therefore, the establishment of the 30 minutes criterion would need some further experimental validation.

The Daylight Glare Probability and the De Boer rating were not designed to deal with bright daylight outdoor glare sources. Their application in the setting of reflected solar radiation in broad daylight exceeds their experimental validation. The Daylight Glare Probability can be used for situation in which reflected daylight is visible through a window. The after image potential criterion was originally designed to determine flash blindness of pilots after the detonation of a nuclear bomb. Therefore, it sets the threshold at a much higher value than the discomfort glare ratings, leading to a very short range of potential glare. This criterion is useful in situation were security issues are dominant such as solar plants near airports or highways. However, for these situation the acceptance glare during less than 30 minutes is meaningless. The maximal luminance threshold is based on physical properties of the eye. By its definition it completely neglects the total illuminance of the reflection as well as the apparent size of the source of glare. This apparent size of the source has an important effect on discomfort glare according to all other indices, which suggests that it has to be taken into account. In conclusion, all available glare thresholds are not universally applicable as a criterion to determine discomfort glare from reflecting surfaces in bright daylight situations. However, they include the main physical properties and relationships between those properties that are important for glare. The indices can be used for some suitable settings and in order to compare the different materials among each other.

4.2 Glare from solar installations

While further work has to be done to establish a reliable criterion for disturbing glare from reflected sunlight in outdoor conditions, it is already possible to point out some conclusions based on the elaborated results:

- Glare caused by reflected sunlight is not a problem that is restricted to solar installation as other materials of the building envelope can act as a source of glare too.
- The widely used, slightly structured (matt) glass can have a large range of potential glare compared to float glass and highly structured prismatic glass.
- Depending on the setting, different glare thresholds are applicable for the evaluation. When temporary loss of the ability to see is most important (e.g near airports or highways), the after image potential criterion can be applied. If the reflection is seen through a window of an office building, the Daylight Glare Probability metric is appropriate.
- In settings in which an observer can see a reflection in near to normal angles θ, the use of prismatic glasses is favorable since in this range the maximal luminance is very efficiently decreased. This kind of situation can be encountered in the presence of solar façades. Also during approach for a landing near an airport near perpendicular reflections can occur.
- Roof integrated solar installations usually lead to reflection geometries with very flat angles θ. In this regime the float glass performs best, since the small apparent size of the reflecting surface often leads to short durations during which the image of the sun can be observed.

6 References

[1] "Hinweise zur Messung, Beurteilung und Minderung von Lichtimmissionen." Bund/Länder-Arbeitsgemeinschaft für Immissionsschutz (LAI), 11-Mar-2015.

[2] S. Carlucci, F. Causone, F. De Rosa, and L. Pagliano, "A review of indices for assessing visual comfort with a view to their use in optimization processes to support building integrated design," *Renew. Sustain. Energy Rev.*, vol. 47, pp. 1016–1033, Jul. 2015.

[3] "Blendung durch natürliche und neue künstliche Lichtquellen und ihre Gefahren." [Online]. Available: http://www.ssk.de/SharedDocs/Beratungsergebnisse_PDF/2006/Blendung_Lichtquellen.pdf?__blob=publicationFile. [Accessed: 16-Jun-2016].

[4] J. Wienold and J. Christoffersen, "Evaluation methods and development of a new glare prediction model for daylight environments with the use of CCD cameras," *Energy Build.*, vol. 38, no. 7, pp. 743–757, Jul. 2006.

[5] H. Moshammer and M. Kundi, "Medizinische Beurteilungsgrundlagen der Passiven Blendung." Dec-2013.

[6] J. D. Bullough, K. S. Hickcox, and N. Narendran, "A Method for Estimating Discomfort Glare from Exterior Lighting Systems," *ASSIST Recomm.*, vol. 9, no. 1, Apr. 2011.

[7] D. H. Sliney and B. C. Freasier, "Evaluation of optical radiation hazards," *Appl. Opt.*, vol. 12, no. 1, pp. 1–24, Jan. 1973.

[8] C. K. Ho, C. M. Ghanbari, and R. B. Diver, "Hazard Analyses of Glint and Glare From Concentrating Solar Power Plants," in *Proceedings of SolarPACES*, Berlin, Germany, 2009.

6 Acknowledgments

The financial support of the Swiss Federal Office of Energy is gratefully appreciated.

INTEGRATION OF PV MODULES IN ENERGY YIELD OPTIMIZED CARBON CONCRETE COMPOSITE FACADES

Sebastian Schindler*1), Adrian Heller2), Michaela Gorges3), Carsten Rudolf 4), Linda Dämmig5), Jens Schneider 1)
1) Fraunhofer-Center for Silicon Photovoltaics CSP, Otto-Eißfeldt-Str. 12, 06120 Halle (Saale), Germany
2) Leipzig University of Applied Sciences, Architecture Institute, Leipzig, Germany
3) TU Dresden, Institute of Construction Materials, Dresden, Germany
4) Solar Valley GmbH, Juri-Gagarin-Ring 152, 99084 Erfurt, Germany
5) SGB Steuerungstechnik GmbH, Marschnerstraße 13, 04109 Leipzig, Germany
* Corresponding author: +49 345 5589 5523, sebastian.schindler@csp.fraunhofer.de

ABSTRACT: Carbon reinforcement of concrete façades allows for lighter weight and more flexible shapes in modern buildings. The German C3 (carbon concrete composite) project aims at developing this technology further. Part of the project is integration of photovoltaics into the façade. We investigate both (i) optimization of energy yield from façade elements and (ii) technical realization of applying photovoltaics modules to C3 elements. Building façades are optimized in shape in order to maximize annual energy yield. The compatibility of solar modules and concrete are investigated with three approaches: (1) Vacuum lamination of solar modules onto prefab façade elements, (2) integration of prefab PV modules into the concrete placing process, (3) hindsight attachment of prefab PV modules onto prefab concrete façade elements. We report on the successful development of all three approaches, show initial mock-ups and share our vision of future integration into modern and existing buildings.

Keywords: Building Integrated PV (BIPV), Module, Module Integration, Module Manufacturing

1 MOTIVATION

It is an important and popular fact that most of the economic, cultural, and societal activities in modern societies take place in an urban environment [1]. For instance, in Germany around 75% of the population live in cities [2]. These conditions clearly bear a wide range of challenges and potentials. Particularly of interest to our modern society is the question of how houses should be constructed to save energy and resources and to avoid greenhouse gas emissions. The need for new and innovative solutions in this sector has been emphasized even further since the Paris climate summit in November 2015 [3].

Germany has pioneered the wide-spread distribution and use of renewable energies in the past and now has aimed to achieve a climate-neutral energy supply of all buildings until 2050. New buildings should be designed and old buildings should be renovated so that their energy demand can be covered without any energy contributions from fossil fuels. This ambitious target can only be reached if both more renewable energies are applied in the building sector and less energy is consumed throughout the full life cycle of a building (construction, operation, and deconstruction). Moreover, this restructuring of urban environments needs to be not only accepted but actively supported by communities. Thus, comfort and aesthetics have to be met by future solutions.

Carbon fibre reinforced concrete allows for a significant decrease in wall thicknesses and for a reduction of the amount of concrete needed for the production of building elements. This directly reduces the amount of CO_2 produced during the production of concrete. Due to the excellent mechanical properties and the inert nature of carbon, much less concrete is needed to achieve the same degree of stability and durability with carbon reinforcement of concrete than with steel reinforcement of concrete. Even more, it allows the design of new types of lightweight and spherically-shaped construction elements. Thus, architects may get access to an unprecedented freedom in forming the envelopes of buildings. This freedom could increase the structural attractiveness of buildings for both inhabitants and investors. Finally, from the point of interconnection technology, the carbon fibre structure opens possibilities for integrating additional electrical component like PV modules, electric energy storage or light emitting diodes into the building envelope.

This paper presents the first results from a project funded by the German ministry of research and education investigating solutions for the integration of photovoltaics into carbon concrete elements.

2 FAÇADE SURFACE SIMULATION

Façades are typically non-optimally oriented for energy yield. Façades are not necessarily south-facing and are typically vertical, assuming planar façades. Furthermore, they suffer from shading of neighbouring buildings, trees etc. C3 allows façades to assume shapes with different faces in three dimensions. Therefore, it is possible to develop building envelopes with fully optimized solar yields.

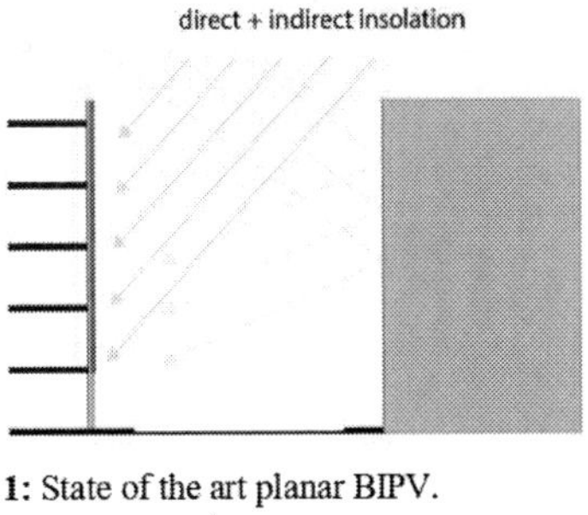

Figure 1: State of the art planar BIPV.

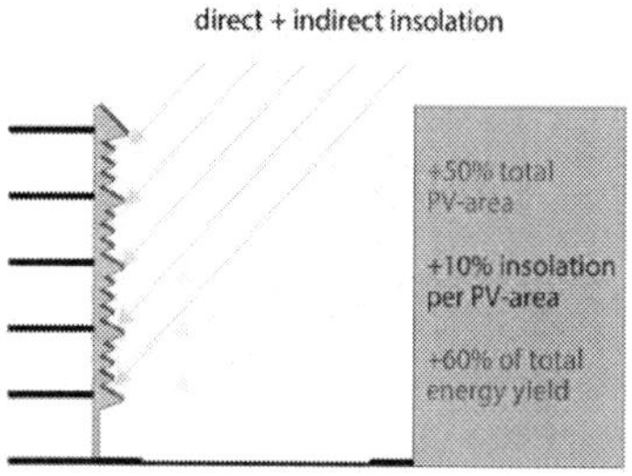

Figure 2: Possible gains of 3D optimized BIPV.

We use computer-aided parametric design processes to take a full approach of integral, functional and optimised solar photovoltaics. First, simulations of different "handmade" test façades appeared to have a remarkable influence on the overall solar performance as compared to a planar building envelope. The following examples show different approaches to work with three-dimensional façade geometries.

Figure 3: Planar Facade.

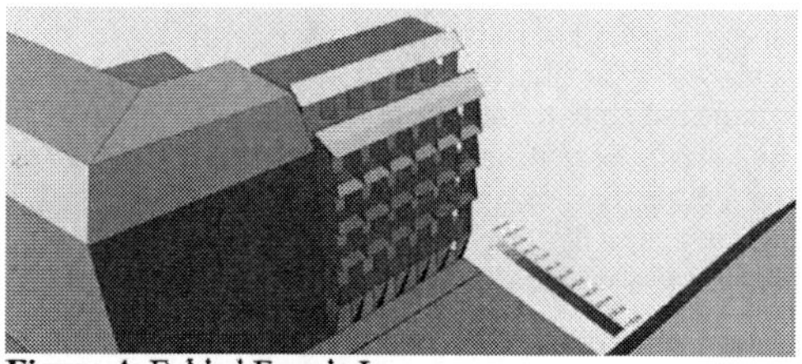

Figure 4: Folded Facade I.

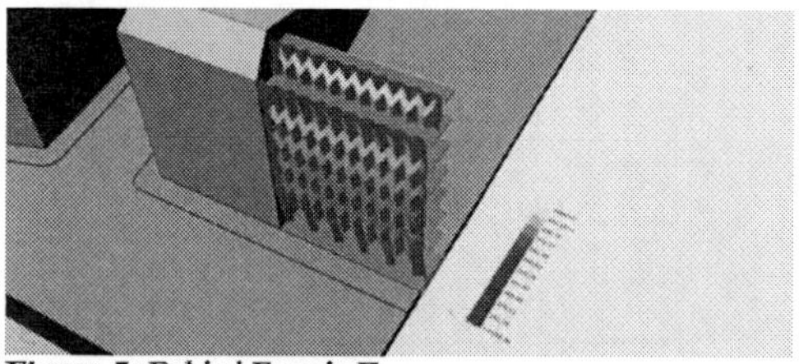

Figure 5: Folded Facade II.

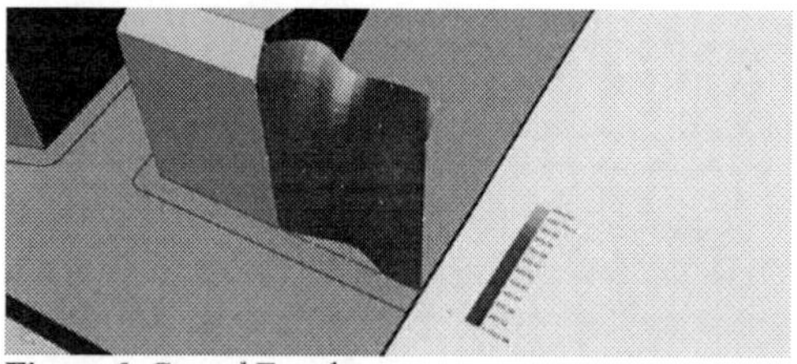

Figure 6: Curved Facade.

The simulation results showed that for example the folded geometry (figure 4) increases the useable PV-area up to 50%, the average insolation per PV-area rises to at least 10% and the total energy yield reaches a plus of 60%, compared to the planar reference. Next, we implement different generic and parametric façade models to which we then apply metaheuristic algorithms in order to generate several optimised façade designs.

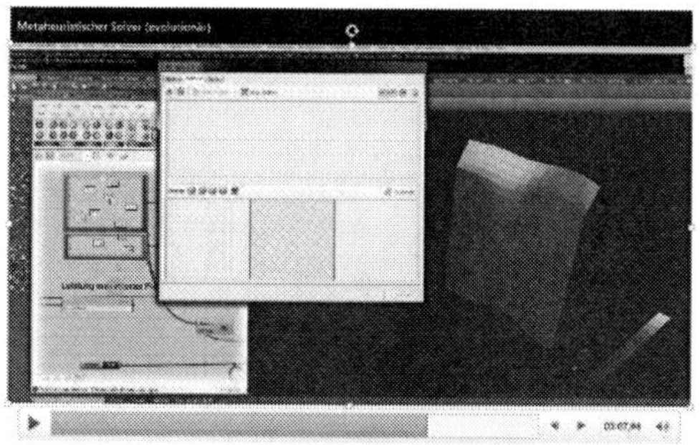

Figure 7: Screenshot from Generic solver.

In order to optimize flexibility in the 3D design process it is necessary to get PVs in smaller units which can be recombined in to clusters in the concrete elements. Realization of such smaller PV module units and their integration on or in the concrete façade elements is addressed within this paper.

3 TECHNOLOGY OPTIONS

Fundamentally, there are three technological options for attaching PV laminates to concrete façade elements: (1) embedding of prefabricated PV modules into the concrete, (2) direct lamination of the PV module stack onto a prefabricated concrete element, (3) separate fabrication of both concrete element and PV laminate and subsequent attachment of both.

In all façade variations tested the same carbon reinforcement (TUDALIT-BZT1-TUDATEX) and a flowable, fine-grained concrete mixture is used for the production of the façade elements.

3.1 Embedded PV Modules in concrete element

The embedment of PV modules in the concrete element leads to a plane, homogenous surface of the façade without gaps or offsets between the concrete and PV-module surface. After the production of the façade element, the PV module is completely integrated and cannot be removed without damaging the concrete or the PV module. All electrical connections can be integrated in the reinforcement-layer of the concrete element and, therefore, hidden inside the structure. The element could provide a box for the connection among the façade elements. If a PV module exceeds its lifetime or must be exchanged, the whole façade element must be replaced.

3.2 PV module lamination on concrete element

PV Modules can be laminated directly onto concrete elements. Here the concrete can replace the backsheet of the module. This approach promises low material cost, but poses challenges on module reliability. High sodium concentration in the concrete can reduce the lifetime of cells and modules comparable to effects know for potential induced degradation (PID) [4, 5]. The alkaline nature of concrete and a large variety of chemical components in the ppm range, both from deliberate additives as well as residual contaminants, have a high potential of causing degradation effects in combination with higher temperature that occur during lamination and insolation. Such effects have not been fully investigated during this project, but occurrence of unintended consequences is expected. Some risks can be mitigated by using diffusion barriers like backsheets between solar cells and concrete elements. On the other hand, modules with even thinner front glasses can be established as the whole module with be mechanically supported by the concrete back side. Furthermore additional design aspects and aesthetics (cell gaps, coloration, etc.) can be realized which are now not possible with standard backsheets.

One drawback of laminating PV modules directly onto the concrete is the higher energy consumption during heating of the stack since the concrete has a high thermal capacity.

In a first attempt, lamination of a single cell in a ETFE/EVA/cell/EVA/Concrete stack was laminated with an extended preheating time at 80 °C for 30 min, a reduced pressure of 400 mbar and an extended crosslinking time of 60 min at 155 °C. The result is shown in Fig. 1. The sample shows little delamination and some bubbles at the perimeter. Such issues are acceptable for first experiments. Further experiments and tests are required to explore this approach further.

Figure 8: Standard solar cell laminated on a fiber-reinforced concrete element.

3.3 Revisable PV module on concrete façade

In this approach both the concrete element and the PV laminate are prefabricated and later attached to one another. Rawlplugs in concrete allow for screwing the PV laminate with holes to the concrete element. We selected a solution with connections on the concrete element and the PV module for an easily detachable solution.

The synergies of such a solution are somewhat limited, but on the other hand, both the concrete element and the PV laminate can be fabricated using standard processes.

A large benefit of such attachable modules is that they allow for revision of the original design. While concrete facades have lifetimes of up to 50 years, solar modules are expected to last 20 to 25 years. Furthermore, in the case of damage to one module, replacement is possible without the removal of the entire façade.

4 PRODUCT DEVELOPMENTS

4.1 Façade element with PV integration

To embed the PV module into the concrete element, the cover glass of a complete, fully-functioning PV module is fixed in the element mold. The backsheet, depending on its preparation, can be used as a spacer towards the element reinforcement. Two different types of backsheets (A & B) with different preparation were used (see Table I and Figure 9).

Table I: Types of backsheets for embedment in concrete

	A	B
Plain	X	X
Strips glued with Silicone	X	X
Folded strips	X	

Figure 9: Backsheet with folded strips (left) and glued strips (right).

The strips of the backsheets were cut to fit into the mesh of the carbon reinforcement, which was placed 2-3 mm above the PV module (Figure 10). In between the PV modules concrete spacers were used to keep the distance between reinforcement and mold (Figure 11).

Figure 10: Backsheet strips cut to fit the carbon mesh.

Figure 11: Form, PV modules and reinforcement prepared for casting.

A flowable, fine-grained concrete mixture was prepared and poured into the element form. The connection ribbons of the PV modules were let out of the element for functionality tests post- production. The casted element was covered with a moist cloth and plastic foil for about 24h. Afterwards the element was demolded and stored in a climate room at 20°C and 65% r.H. The finished element can be seen in Figure 12. The functionality test under laboratory conditions showed no loss of efficiency.

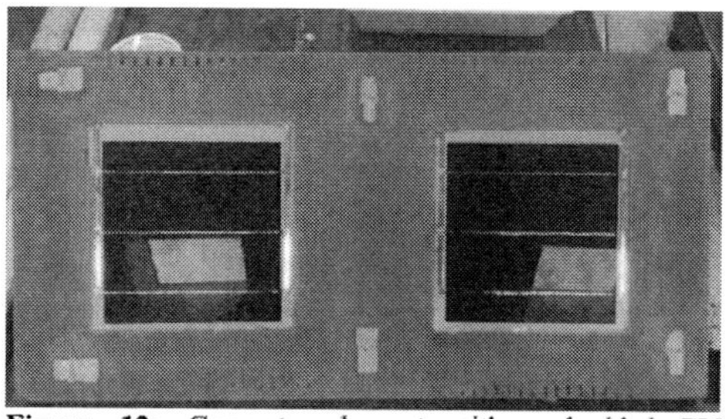

Figure 12: Concrete element with embedded PV modules.

In a pullout test, a stamp was glued to the glass surface. While the element was fixed in a testing machine (ZWICK 1445), the PV module was pulled from the concrete. The results of the test show that the applied strips result in larger forces necessary to pull the PV module out of the concrete element with backsheet B. No improvement could be measured with backsheet A. In both cases, the pull-out forces were much higher than a possible maximum wind force of 1.44 kN/m² (Table II)

Table II: Pulled test results.

Material	Type of backsheet preparation	Pullout force [kN/m²]
A	Plain	94.82
	Strips glued with Silicone	74.07
	Folded strips	88.54
B	Plain	74.15
	Strips glued with Silicone	126.45

4.2 Revisable PV Module on façade element

For the revisable module set-up, standard screw and anchors have been used to attach buttons to the concrete façade surface (see Figure 13).

Figure 13: Concrete element with attached buttons.

PV modules were manufactured with button counterparts attached to the backsheet of a mini module (see Figure 14). In the future, the revisable module's holes will infiltrate the entire module, including the front cover of glass.

Figure 14: Revisable module on concrete element.

4.3 Outlook on further developments and challenges

The initial test of attaching PV modules to concrete elements shows that both directly embedding PV modules in façade elements as well as the detachable modules are possible. Reducing module size is key to satisfying architectural needs for the effective integration of photovoltaics into urban environments.

Now, we look into the challenge of integrating PV modules into façades with non-planar optimized surfaces. PV modules in facades with optimized non- planar surfaces. Two main designs will be followed. Flexible PV modules are required in order to adapt to curved façade surfaces (see Figure 15). Using very small cells allows PVs to follow such a curved façade. It might be beneficial to interconnect such cells with wire interconnection technology, like that developed by Meyer Burger or Schmid [6, 7]. It is the goal of this project to process such mock-ups within this year.

Figure 15: Design study on curved façade element with smaller PV cells attached.

The other non- planar design is a facetted design with 9 separate faces. Small mini modules including individual 5" back contact cells like those produced by Sunpower [8] will be used and interconnected inside the concrete element.

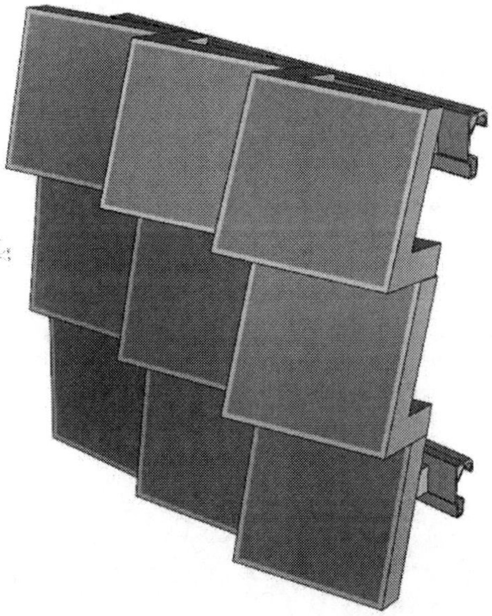

Figure 16: Design study of façade element with facetted faces.

5 CONCLUSION

In this paper we present an optimization of façade surfaces in order to increase energy yield in PV façade systems in urban environments. From these optimized surfaces, requirements for solar module design and technologies for attaching PV-modules to façades are derived. Smaller modules with more flexible designs are required for successful integration and adaption to new design. Embedding PV modules in the concrete produces a new type of concrete solar module that is interesting and aesthetically appealing. Detachable PV modules, attached to the concrete by screws, buttons, or anchors, allow for design revision, offer flexibility, and permit easy replacement of PV modules in the case of defects.

Currently, only mock-ups are developed. Interaction of the components must be investigated in order to understand and avoid any potential defects or degradation in the new module composed of carbon-reinforced concrete in conjunction with PV modules. Mechanical and thermomechanical calculations must still confirm that these modules can pass wind load requirements in different building envelopes. Eventually such modules must pass building regulation requirements in order to be implemented in façade elements. There is a lot of work ahead; however, this work is a starting point for decreasing our reliance on non-renewable energy sources in infrastructure.

6 ACKNOWLEDGEMENT

This work has been funded by the federal ministry of education and research within the project "Zwanzig20 – Carbon Concrete Composite C3" (03ZZ0305B)

7 REFERENCES

[1] Survey "World Development Indicators", World Bank, 2015

[2] Text of the Paris Agreement 2015, available at http://unfccc.int/paris_agreement/items/9485.php

[3] S. Kohler, "Energiewende in Deutschland – Roadmap bis 2020/2015", presented at „Umweltkonferenz Natur &Gesundheit", Salzburg, 9.3.2013

[4] Berghold J, Frank O, Hoehne H, Pingel S, Richardson B, Winkler M. Potential Induced Degradation of solar cells and panels. Proc. 25th European Photovoltaic Solar Energy Conf., Valencia, Spain, 2010, pp. 3753-3759.

[5] Naumann, Volker, et al. "Micro structural root cause analysis of potential induced degradation in c-Si solar cells." Energy Procedia 27 (2012): 1-6.

[6] Braun, S., Micard, G. & Hahn, G.2012, "Solar cell improvement by using a multi busbar design as front electrode", Energy Procedia, Vol. 27, pp. 227–233.

[7] Meyer Burger: Meyer Burger's revolutionary SmartWire Connection Technology significantly lowers photovoltaic production costs and boosts PV system energy yield by 10%, Meyer Burger Press Release, 11.02.2013

[8] Mulligan, William P., et al. "Manufacture of solar cells with 21% efficiency." Proc. 19th EPVSEC 387 (2004).

ENERGY PERFORMANCE OF PV MODULES AS ADAPTIVE BUILDING SHADING SYSTEMS

P. Jayathissa*, J. Schmidli, J. Hofer, A. Schlueter
Architecture and Building Systems, Institute of Technology in Architecture, Department of Architecture, ETH Zurich
John-von-Neumann Weg 9, Zurich, Switzerland
jayathissa@arch.ethz.ch

Shading systems improve building energy performance by controlling solar gains and natural lighting. Integrating photovoltaics opens new opportunities for building integrated photovoltaics by combining the benefits of adaptive shading with facade integrated solar tracking. Furthermore, it reduces the building energy demand and simultaneously generates electricity onsite. This paper presents a methodology for simulating the photovoltaic electricity production of a dynamic facade mounted PV system in combination with the energy consumption of a building through shading. The simulation is conducted within the parametric Rhino / Grasshopper environment using a high resolution radiation analysis for the calculation of PV electricity generation. Building energy analysis is conducted through DIVA / EnergyPlus. From this simulation we can determine the optimum hourly position and orientation of the PV panels, not only for optimal energy harvest, but also for the overall balance of the room.
Keywords: Building Integrated PV (BIPV), Modelling, Tracking

1 INTRODUCTION

Buildings are at the heart of society and currently account for 32% of global final energy consumption and 19% of energy related greenhouse gas emissions [1]. Nevertheless, the building sector has a 50-90% emission reduction potential using existing technologies [1]. Within this strategy, building integrated photovoltaics (BIPV) have the potential of providing a substantial segment of a building's energy needs [2]. Even the photovoltaic (PV) industry has identified BIPV as one of the four key factors for the future success of PV [3].

Dynamic building envelopes have gained interest in recent years because they can save energy by controlling direct and indirect radiation into the building, while still responding to the desires of the user [4]. This mediation of solar insolation offers a reduction in heating / cooling loads and an improvement of daylight distribution [5]. Interestingly, the mechanics that actuate dynamic envelopes couples seamlessly with the mechanics required for facade integrated PV solar tracking.

Previous BIPV research analyses electricity production and building energy demand for static BIPV shading systems [6] [7] [8] [9]. This paper expands on this work by analysing dynamic PV shading systems, while also taking into account mutual shading amongst modules and its effect on PV electricity generation. The approach allows us to reduce efficiency degradation due to partial shading of PV modules [10].

The work presented in this paper is applied in the context of the Adaptive Solar Facade (ASF) project [11]. The ASF is a lightweight PV shading system composed of CIGS panels, that can be easily installed on any surface of new or existing buildings. An example an ASF can be seen in Figure 1.

This paper will present a methodology of simulating an ASF while simultaneously calculating the energy demand of the office space behind the facade.

Figure 1: An example of an ASF constructed at the House of Natural Resources [11]

2 METHODOLOGY

To study the electricity generation and building energy consumption, a 3D geometry of the room and solar facade is built using the Rhinoceros software [12], and its parametric modelling plugin Grasshopper [13]. The solar facade consists of 400mm CIGS square panels that can rotate in two degrees of freedom. On the horizontal axis, the panels can move from 0° (closed) to 90° (open) position in steps of 15°. In the vertical axis, it can move from 45° to -45° in 15° steps. Existing ASF systems [11] have independently actuated panels and a continuous range of actuation, however for simplicity, we group all panels into one cluster that moves in unison. This leaves us with 49 possible dynamic configurations of the facade system. A corresponding workflow can be seen in Figure 2.

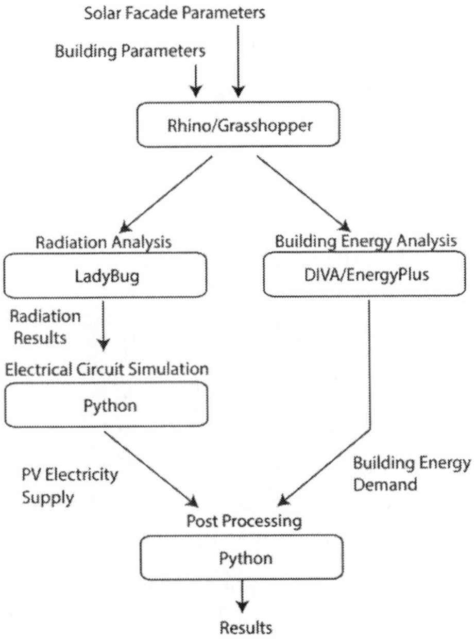

Figure 2: Simulation Workflow

2.1 Building Energy Demand

The building energy simulation is conducted using EnergyPlus [14] through the DIVA [15] interface. A summary of the energy simulation parameters can be found in Table 1. The geometric solar facade is interpreted in EnergyPlus as an external shading system. A simulation of each possible dynamic configuration of the facade is run for each hourly timestep of the year using a weather file for Zurich, Switzerland.

Table I: Summary of main assumptions for the calculation of operational emissions

Building Settings	
Office Envelope	Roof: Adiabatic Floor: Adiabatic Walls: Adiabatic Window: Double Glazed (e=0.2) 3mm/13mm air
Thermal Set Points	Heating 22°C Cooling: 26C°
Building System	Hydronic Heating: COP=4 Hydronic Cooling: COP=3 Lighting Load: 11.8W/m^2 Lighting Control: 300lx
Occupancy	Office: Weekdays from 8:00 – 18:00 People Set Point: 0.1persons/m^2 Ventilation: 0.0094m^3/s/person
Local Assumptions	
Weather File	Zurich-Kloten, Switzerland

2.2 PV Electricity Supply

A solar radiance simulation is run in parallel using Ladybug [16], which uses Radiance [17] to determine the incident insolation on the solar facade. The approach enables us to calculate solar irradiance on the modules with high spatial resolution including the effect of module mutual shading as seen in Figure 3. The results

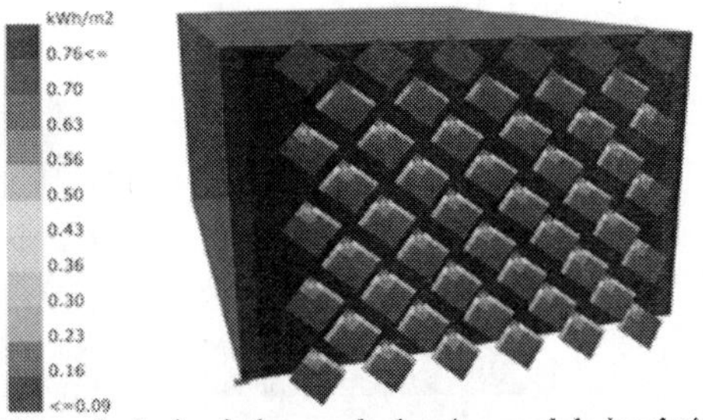

Figure 3: A simulation result showing module insolation from 11:00-12:00 on the 11 August for the used weather file and a specific module orientation.

are coupled to an electrical circuit simulation of thin-film PV modules with sub-cell level representation [18]. This model uses the standard equivalent circuit model to calculate sub-cell I-V curves with a single diode, one series resistor, and one shunt resistor [19]. In addition to the irradiation dependency, the PV simulation includes temperature dependency. The variation in efficiency due to temperature is calculated through a linear dependence of cell temperature to irradiance, where thermal imaging was used to determine the correlation factor.

2.3 Optimisation

Results of the PV Electricity Supply, and Building Energy Demand for each possible configuration of the ASF is post processed in Python [20]. The optimum configuration that minimises building energy consumption, and maximises PV electricity production is calculated for each hourly timestep.

3 RESULTS

3.1 Determinations of Optimum Configurations

The optimal configurations of the ASF can be visualised using carpet-plots. Figure 4 and 5 details the optimum angles of the panels to maximise PV generation, and minimise heating, cooling and lighting demands independently. For the simplicity of visualisation we have split the representation of the data to altitude and azimuth angles. We can see in Figure 4(a) how open configurations (light coloured) are chosen to minimise the building heating demands during the winter months and early mornings of spring and autumn. Likewise closed configurations (dark colours) are the preferred solutions to minimise the cooling demand during the summer months.

The PV optimisation, best seen in Figure 5(d), follows a path similar to a classic solar tracking model. The variation from a solar tracking model is due to the effects of self shading. This disparity will be further analysed in Section 3.3.

When the four optimisation cases are combined to achieve the configurations for total energy minimisation we get some interesting results seen in Figure 4(e). We can see that during the twilight hours of the morning and evening the open positions are preferred to minimise the use of artificial lighting. During the summer afternoons, closed positions to minimise cooling demands are preferred, however enough light is allowed to pass to sufficiently illuminate the space. When we also include the PV electricty optimisation we notice a strong

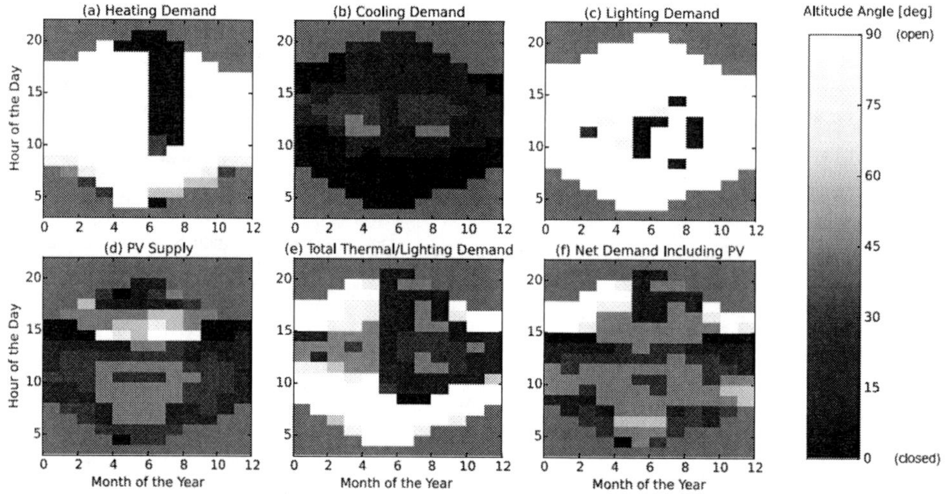

Figure 4: Carpet plots detailing the optimal altitude angles to minimise the (a) heating demand, (b) cooling demand, (c) lighting demand, and (d) maximise PV electricity production. Figure (e) details the combinations for optimum building thermal management without PV production, (f) also includes the PV production. Small angles correspond to closed positions, whereas large angles represent open positions. The corresponding azimuth angles for each hour can be seen in the following Figure (5).

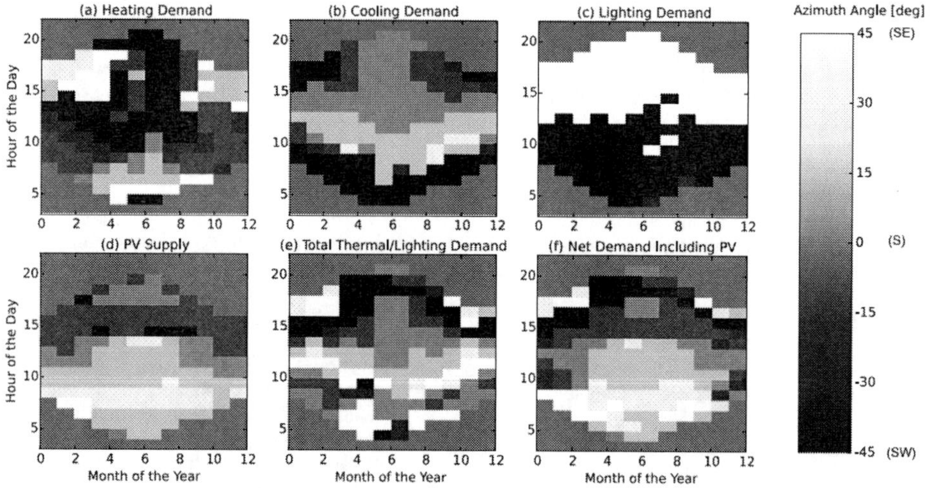

Figure 5: Carpet plots detailing the optimal azimuth angles to minimise the (a) heating demand, (b) cooling demand, (c) lighting demand, and (d) maximise PV electricity production. Figure (e) details the combinations for optimum building thermal management without PV production, (f) also includes the PV production. Negative angles correspond to the panels facing west, whereas positive angles represent east-facing panels. The corresponding altitude angles for each hour can be seen in the previous Figure (4).

tendency of the ASF to follow an optimal PV production pattern. This, however changes if the building system becomes more inefficient. Less efficient heating, for example, would result in configurations optimised for heating overpowering those of PV electricity generation.

3.2 Net Energy Consumption

Figure 6 shows the net energy use at the optimal angles described in Section 3.1. Red colours detail the energy consumption intensity, and blue colours detail the PV electricity supply. It is interesting to see in Figure 6(f) how the combination of electricity generation and adaptive shading can compensate for the energy consumption of the building during most sunlit hours. Overall the PV electricity compensates for 41% of the energy demand of the office behind the facade during the course of the year.

3.3 Solar Tracking vs. Optimal PV Harvest

Our model which optimises the photovoltaic electricity harvest was compared to an optimal solar tracking model. Figure 7(a) shows radiation incident on the PV panels for the two models including the effects of self shading. The

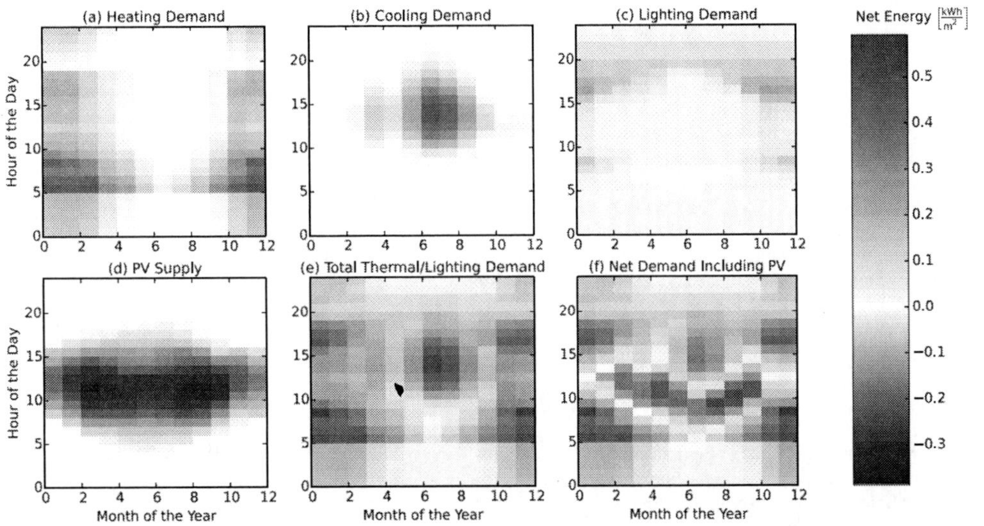

Figure 6: Carpet plots detailing the net energy consumption. Each square represents the total energy consumption for that specific hour of the entire month. Red colours detail the energy demand, while blue colours detail the energy supply.

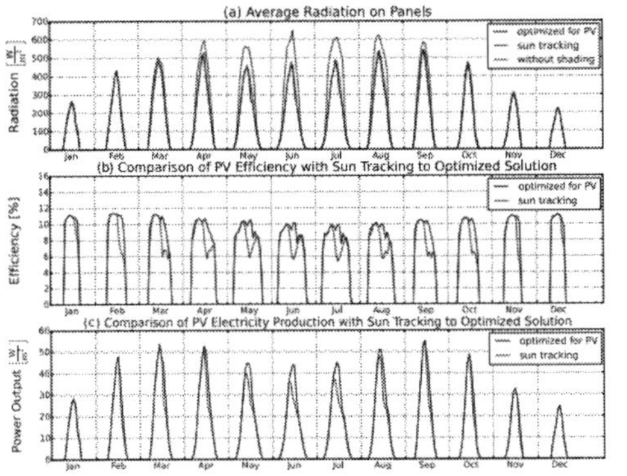

Figure 7: Variation in energy production between optimal solar tracking, and optimal electricity harvest

red line represents the solar tracking model without self shading. Even though the radiation incident on the PV panels for both models are similar, the effect of self shading reduces the efficiency of the PV panels as seen in Figure 7(b). As a result, the optimum angles for dynamic BIPV modules does not follow a solar tracking model, unless the panel spacing is large enough to prevent self shading.

4 DISCUSSION AND CONCLUSION

In this paper, we present a simulation methodology to evaluate a dynamic photovoltaic shading system, combining both electricity generation, and the energy demand of the building. This is achieved through the use of DIVA/EnergyPlus for the building energy calculation, LadyBug/Radiance for the radiation simulation, and Python for the PV electrical simulation. A post processing algorithm then determines the optimum system configuration for control, and delivers a final energy performance assessment of the system. The methodology can be applied to evaluate different PV system geometries, actuation control strategies, building systems, building typologies and climates.

Our utilisation of this model to evaluate the Adaptive Solar Facade (AFS) details the advantages of an adaptive system to a static system. The ASF is able to orientate itself to the most energy efficient position, thus finding the optimum balance between PV generation, and daylight control to minimise heating, cooling and lighting loads. The optimum orientation however, strongly depends on the general efficiency of the building. Decreasing the efficiency of the heating, cooling or lighting systems will give higher preference for configurations optimised for building thermal management through adaptive shading, than for PV electricity production.

When the model was run, purely to optimise photovoltaic electricity production, we see that the optimal angles do not follow a classic solar tracking model. This is because an optimal solar tracking model results in high module self shading which reduces the overall efficiency of the PV panels. This methodology, presented in this paper, can therefore also be applied to dynamic photovoltaic systems in general where self shading is present.

This work ultimately presents a methodology for the planning and optimisation of sophisticated adaptive BIPV systems. Future work will use this methodology to determine the environments and building typologies that could benefit from adaptive BIPV systems.

REFERENCES

[1] Fifth assessment report, mitigation of cliamte change, Intergovernmental Panel on Climate Change (2014) 674–738.
[2] P. Defaix, W. van Sark, E. Worrell, E. de Visser, Technical potential for photovoltaics on buildings in the eu-27, Solar Energy 86 (9) (2012) 2644–2653.
[3] M. Raugei, P. Frankl, Life cycle impacts and costs of photovoltaic systems: current state of the art and future outlooks, Energy 34 (3) (2009) 392–399.
[4] R. Loonen, M. Trˇcka, D. Co´stola, J. Hensen, Climate adaptive building shells: State-of-the-art and future challenges, Renewable and Sustainable Energy Reviews 25 (2013) 483–493.
[5] D. Rossi, Z. Nagy, A. Schlueter, Adaptive distributed robotics for environmental performance, occupant comfort and architectural expression, International Journal of Architectural Computing 10 (3) (2012) 341–360.
[6] M. Mandalaki, K. Zervas, T. Tsoutsos, A. Vazakas, Assessment of fixed shading devices with integrated pv for efficient energy use, Solar Energy 86 (9) (2012) 2561–2575.
[7] S.-H. Yoo, H. Manz, Available remodeling simulation for a bipv as a shading device, Solar Energy Materials and Solar Cells 95 (1) (2011) 394–397.
[8] F. Sara, et al, Maximizing the solar photovoltaic yield in different building facade layouts, EU PVSEC (2015).
[9] P. Jayathissa, Z. Nagy, N. Offedu, A. Schlueter, Numerical simulation of energy performance and construction of the adaptive solar facade, Proceedings of the Advanced Building Skins 2 (2015) 52–62.
[10] J. Hofer, A. Groenewolt, P. Jayathissa, Z. Nagy, A. Schlueter, Parametric analysis and systems design of dynamic photovoltaic shading modules, EU PVSEC 2015 conference, Hamburg, Germany (paper in review).
[11] Z. Nagy, S. Bratislav, P. Jayathissa, M. Begle, J. Hofer, G. Lydon, A. Willmann, A. Schlueter, The adaptive solar facade: From concept to prototypes, under review.
[12] Rhinoceros v5 (2015).
URL https://www.rhino3d.com/
[13] Grasshopper - algorithmic modeling for rhino (2015). URL http://www.grasshopper3d.com/
[14] D. B. Crawley, L. K. Lawrie, C. O. Pedersen, F. C. Winkelmann, Energy plus: energy simulation program, ASHRAE journal 42 (4) (2000) 49–56.
[15] Diva for rhino. URL http://diva4rhino.com/
[16] M. Roudsari, M. Pak, A. Smith, Ladybug: A parametric environmental plugin for grasshopper to help designers create an environmentally-conscious design.
[17] G. J. Ward, The radiance lighting simulation and rendering system, in: Proceedings of the 21st annual conference on Computer graphics and interactive techniques, ACM, 1994, pp. 459–472.
[18] H. Johannes, A. Groenewolt, P. Jayathissa, Z. Nagy, A. Schlueter, Parametric analysis and systems design of dynamic photovoltaic shading modules, Energy Science and Engineering.
[19] A. Mermoud, T. Lejeune, Performance assessment of a simulation model for PV modules of any available technology, Proceedings of the 25th European Photovoltaic Solar Energy
Conference, WIP, Munich, 2010, iD: unige:38547.
[20] Python. URL https://www.python.org/

Technical Evaluation of BIPV power generation potential in EU-28

Adel El Gammal *– Becquerel Institute - a.elgammal@becquerelinstitute.org - mob +32475676805*
David Mueller *– Merck KGaA - david.mueller@merckgroup.com*
Hannah Buerkstuemmer *– Merck KGaA- hannah.buerckstuemmer@merckgroup.com*
Renaud Vignal *– ArcelorMittal – renaud.vignal@arcelormittal.com*
Philippe Macé *– Becquerel Institute – p.mace@becquerelinstitute.org*

ABSTRACT: The present study evaluates the potential PV capacity that could be installed on roof and façade surfaces of buildings in the EU-28. The paper then evaluates the potential power generation and the theoretical PV penetration in key countries and in Europe as a whole. Drawing from methodological approaches previously developed in earlier papers, the paper presents a modular approach that provides a base study from where more focused research can be derived as needed. The paper assesses in detail each of the underlying assumptions and makes use of most recent information available on the EU building stocks. The power generation potential is calculated using the current state of the art in building applied and building integrated PV technologies (BAPV / BIPV). Finally, a sensitivity analysis is carried out on the major assumptions and the results are discussed and compared to those provided in earlier reference studies.

Keywords: economic analysis, photovoltaic production, BIPV, BAPV, distributed generation, roofs, façades.

1 INTRODUCTION

1.1 Context of the Research

The unprecedented decrease of PV technology costs has also driven down the price for PV generated kWh much faster than expected. Standard applications of PV are becoming increasingly competitive in Europe and it is expected that under the regulatory pressure on the energy performance of buildings, advanced integration of PV in the built environment will become an important focus in both the building and the PV industries.

This is especially the case in Europe, as a consequence of high population density and therefore, the increasing scarcity of affordable land surface for large-scale ground mounted PV plants. It is therefore of major interest to assess the technical potential of BIPV, i.e. the total net surface offered by roofs and facades in the built environment that is suitable for PV.

Although various studies have discussed the subject, the figures most often cited in the literature refer to the original study of S. Nowak & al. performed in the framework of IEA PVPS in 2002[1]. The present study attempts to provide an updated view on this estimation, also using the most recent information available on building stocks as well as more precise solar irradiation calculations.

1.2 Building Applied vs. Building Integrated PV

An important distinction exists between Building Applied PV (BAPV) and Building Integrated PV (BIPV).

BAPV typically refers to situations where a solar installation (PV modules) is applied on (and in addition to) the envelope of a building, be it on facades or more frequently on roofs. In such case, the PV functionality is simply added to an existing building, such as modules fixed onto the existing roof or façade. BAPV represent the vast majority of existing PV installations on buildings.

In contrast, BIPV describes a situation where the PV module makes an integral part of the building envelope and contributes to the envelope functionalities, such as (but not limited to) water tightness, mechanical stability, noise reflection, thermal insulation, heat reflection, shading, etc... BIPV should therefore rather be considered as a building material to which PV functionality was added. BIPV that typically, until recently, meant "additional costs", has been mostly limited up to now to more emblematic high-end architectural projects. However several major building materials producers have announced to work on highly standardized, innovative and affordable new BIPV products thus paving the way towards mass adoption.

1.3 Purpose of this study

The present study aims only at evaluating the roof and façade available surfaces that are suitable for PV generation and consequently, the power generation potential of buildings in Europe, notwithstanding the fact PV could either be effectively integrated (BIPV) or applied (BAPV) to it. There is no attempt in the present study to discuss the distribution of BIPV and BAPV, even if the former is expected to grow in the future, driven both by improved aesthetics and increasingly attractive economics.

For the sake of ease in this study, "BIPV" will be used indistinctively when referring to "BIPV" or "BAPV".

2 Methodology

2.1 Overview

Working on the complex manifold of building data obviously requires the use of statistical approaches and a number of assumptions on how to model the building's population.

For instance, in Germany only, it is estimated that about 38 million buildings exist[2]. Considering only residential buildings in Germany, about 22 different generic building typologies[3] can be distinguished.

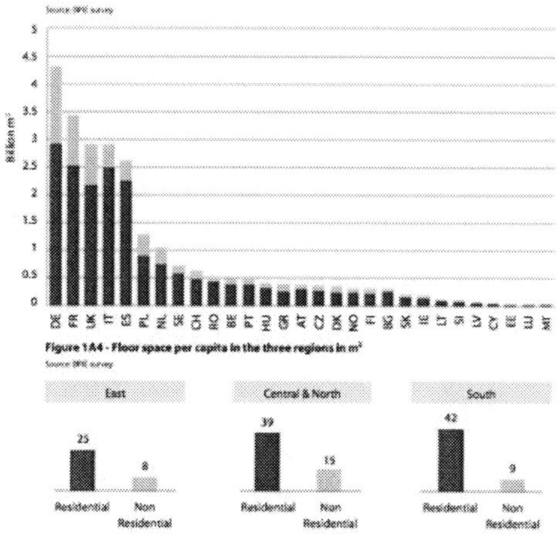

Figure 1: Residential / Non residential floor area per capita in EU

Figure 1 illustrates the vast diversity of situations encountered in Europe and therefore the need for using carefully chosen sets of assumption and statistical analysis.

In the wake of the EU Energy Performance of Buildings Directive[4], numerous studies exist now that describe the EU stock of buildings. Unfortunately, in most cases, building information has been collected in the view of socio-demographic and energy conservation analysis and always refer to gross or net floor surfaces. Only very few information exist on the height of buildings (e.g. their number of floors), a variable that is essential to e.g. obtain façade surfaces from floor surfaces.

Two main studies have been performed after the baseline of S. Nowak & al. (ref 1). The study of R. Defaix & al.[5] provides additional insights on the evaluation of the number of floors per building segment. The study of H. Lehmann & al.[6] uses an alternative approach based on the correlation found between population density and facades and roof areas per capita. However based on our extensive analysis of existing building data we're considering this approach as less robust than the previous ones.

2.2 Methodological Approach

The methodological approach used in this paper is inspired from the original approach of S. Nowak & al. (ref 1) but has been developed independently and modified to integrate more precise building data and PV yield information that are now available.

Also, all base assumptions have been reconsidered and reassessed independently from the original study.

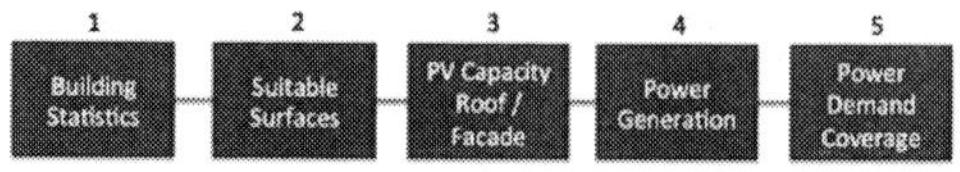

Figure 2: Methodological Overview

Figure 2 provides a synthetic overview of the methodological process used in the present paper for deriving potential BIPV penetration from raw building statistics.

2.2.1 – Building Statistics

The EU Climate and Energy Framework[7] and the EPBD Directive[8] have resulted in an increased focus on the EU building stocks, leading to a number of detailed EU funded research projects on the subject.

The paper draws essentially from these sources, which can be generally found on the website of the Building Performance Institute Europe[9]. In particular, the paper has used data from the Eurostat database[10], the Buildings Database[11] the Episcope Platform[12], the Entranze Data Mapping Tool[13] as well as the Tabula Web Tool[14].

Additional country specific studies and databases have been used when possible.

In this paper, we have analysed data for 10 EU representative countries, representing about 80% of EU-28 population and Switzerland: Austria, Belgium, France, Italy, Germany, Netherlands, Poland, Romania, Spain, Switzerland, and United Kingdom.

These countries therefore can be considered as highly representative of EU-28 as a whole, in particular for deriving average power demand coverage potential.

2.2.2 – Suitable Surfaces

Building statistics provide quite detailed information on gross and net floor areas, buildings segmentation as well as some information of generic building typologies for each country.

In order to derive roofs and façade surfaces from floor surfaces, assumptions have been made to obtain a generic number of floors per buildings segment. Our detailed analysis of buildings typologies and distribution across Europe resulted in a EU-28 average of 2,25 floors (ground floor = 1floor) per building in the residential segment and of 2 floors in the non-residential sector (a similar figure can be found in US EIA[15] regarding non-residential buildings on the densely populated areas of US East Coast). This parameter is essential as it conditions the translation of floor surface into ground surface, from where roof and façade surfaces can be derived. As such, these assumptions have been carefully considered based on building typologies available in the Tabula Web Tool[16] and on analysis provided in the paper from Defaix & al (ref 5). A sensitivity analysis was carried out considering a variation of +/- 10% of this parameter, in the residential and non residential segments.

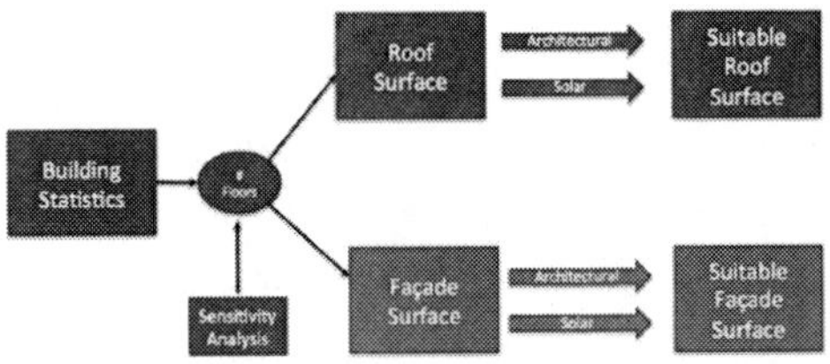

Figure 3: Determination of Suitable Surfaces

Roof surface can be derived from the ground surface, which is calculated from the total floor surface and number of floors. However, roofs can have various slopes, varying from flat (0°) to sharp slope (60°). Building statistics show however that in average about 1/4 of total roof surfaces are flat and 2/3 of tilted roofs surfaces have slopes lower or equal to 30°.
Considering this slopes distribution, and the fact tilted roofs present a relatively greater surface than flat roofs relative to the same ground area, the calculated average ratio is 1,2 m2 of roof surface for every 1m2 of ground surface.

The next step is to define what portion of that roof is suitable for PV use. Two factors will be analysed:

- The architectural suitability. It reflects the portion of roof surface that can be used for PV when removing all surfaces that are not available for urbanistic reasons, geometric constraints, presence of other roof constructions, such as HVAC or chimneys, and shadowing created by other roof superstructure.
- The solar suitability. We will arbitrarily consider only surfaces that are able to provide at least 80% of the maximum yield provided at the same location for a module with optimal orientation. Any surface leading to an inferior yield will be considered as unsuitable for PV.

Considering a selection of representative roof typologies, it has been assumed that 50% of the roof surface is architecturally suitable both in the residential and non-residential segments.
Regarding solar suitability, we must consider all combinations of tilt angle (slope of the roof) and azimuth angle (deviation from meridian direction) that will provide at least the pre-defined acceptable yield.
Considering an average EU latitude, computations have been done for a typical distribution of roof slopes in both the residential and non-residential segments, and assuming a random azimuth distribution. For both segments calculations show that on average, 63% of roof surfaces satisfy the conditions of solar suitability.

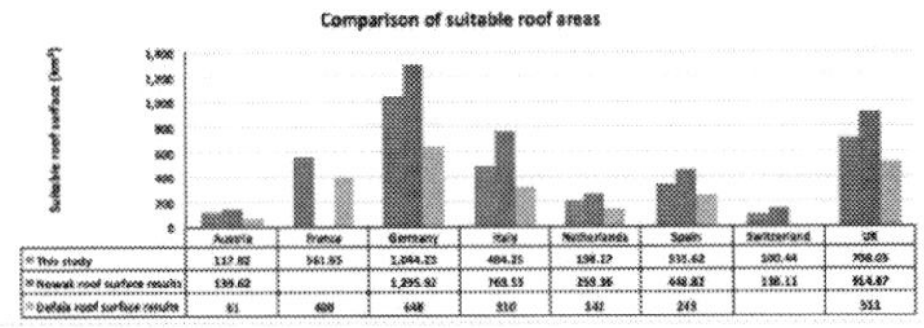

Figure 4: Suitable Roof Surfaces

Figure 4 provides the calculated roof surface (blue bar) and is compared with results of the baseline study of Nowak & al. (ref 1) and with the results of the study of P.R. Defaix & al. (ref 5).
The current study, based on most recent data consistently provides results within the range defined by the 2 previous studies.

Façade surface will be derived again from estimated ground surfaces and number of floors.
As façade surfaces vary linearly with the ground perimeter and not ground surface, average ground surfaces have been calculated separately for the residential and non-residential segments, which typically feature very different ground surfaces. Calculations have yielded an average of 2 m2 and 1,2 m2 façade area per 1 m2 ground surface, respectively in the residential and non-residential segments.

	Residential		Non-Residential	
	Roofs	Facades	Roofs	Facades
1m2 Ground Floor	1,2 m2	2 m2	1,2m2	1,2 m2
Architectural Suitability	50%	15%	50%	15%
Solar Suitability	63%	41%	63%	41%
PV Suitability	38%	12%	38%	7,4%
PV suitable surface	0,38 m2	0,12 m2	0,38 m2	0,07m2

Figure 5: Key surface parameters

Facades can be very diverse considering the various typologies and orientations. They can range from plain facades with no opening to window facades, and include all intermediate situations.
Considering openings are usually found on well-exposed facades and the need to ensure a visually pleasing PV organisation, we have taken the conservative assumption that only 15% of façade surfaces will be architecturally suitable for PV.
Regarding solar suitability, it is assumed that all facades have a 90° tilt angle, and a random azimuth distribution.
Considering the same average EU latitude, we have calculated that only 41% of façade surfaces would be satisfying the solar suitability criteria.

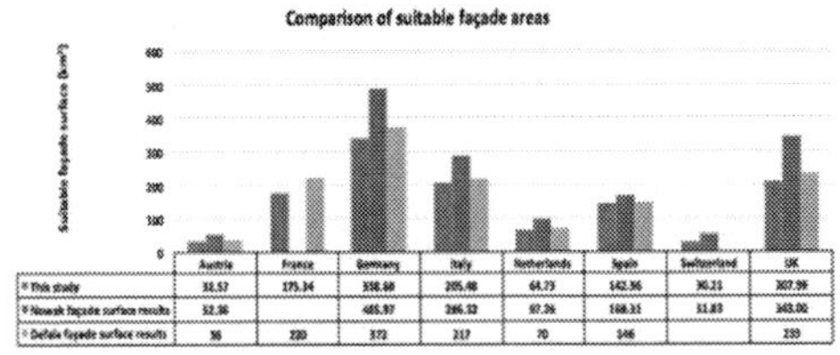

Figure 6: Suitable Facade Surfaces

Figure 6 provides the calculated facade surface (blue bar) and is compared with results of the baseline study of Nowak & al. (ref 1) and with the results of the study of P.R. Defaix & al. (ref 5).
The current study, again consistently provides results within the range defined by the two previous studies.
It is interesting to note, that suitable façade area is generally less than one third of roof suitable surface.

2.2.3 – PV capacity on roofs and facades
In this paper, we have considered mainstream commercial opaque BAPV or BIPV products, typically consisting of c-Si or CI(G)S technology.
It must be emphasized that novel BIPV products (e.g. based on new technologies such as dye or organic cells) are expected to come on-line in the short to medium term and have not been considered in our studies yet. These new technologies promise numerous benefits such as flexible form factors, semi-transparency, ease of integration and better performance under difficult light conditions (e.g. shadows and diffuse light). They ultimately should offer additional PV capabilities to buildings so that the estimation provided in this paper should be considered as a minimum PV penetration theoretically achievable.
We have considered an average solar efficiency of 15%, meaning every 1 m2 of PV would represent a capacity of 150Wp.
This capacity should obviously be calculated on a country-by-country basis, given the variance in solar irradiation, and separately for roofs and façades, considering both provide different yields at any given location.

2.2.4 – Power Generation
Power generation is calculated for each country separately for roof and façade surfaces.
The paper has used PVGIS[17] tool from the JRC. In the case of countries that span over a significant range of latitudes, a weighted average of solar yields has been calculated on different location points of the country.
The more recent SAF-Climate PVGIS irradiation database has been used, with the rather conservative assumption that all installations would be integrated (BIPV), thus with a lower yield due to higher assumed operating temperatures.

2 EU BIPV TECHNICAL POTENTIAL

Based on the preceding, the technical potential of 10 EU countries and Switzerland has been calculated.
A sensitivity analysis has been performed on the estimated average number of floors, for a +/- 10% variance of this parameter on both residential and non-residential segments.
The analysis shows that the estimated power generation exhibits a lower variance, which confirms the robustness of the methodological approach, and the reliability of the derived figures.

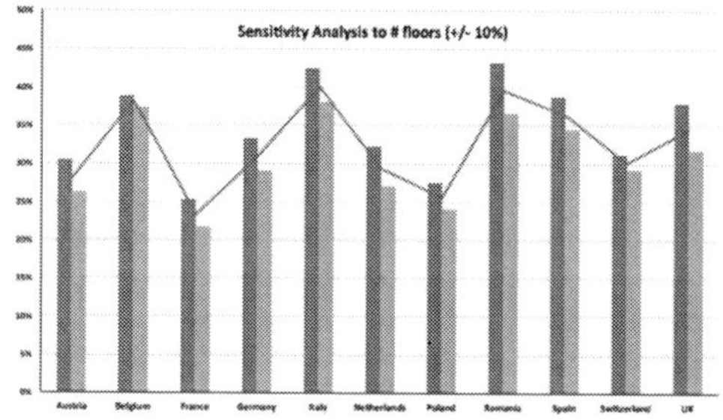

Figure 7: PV coverage of power demand by country

The BIPV coverage of the demand varies from country to country within a range of 24% to 40%.
As expected the coverage rate generally varies and depends on the irradiation level (latitude of the country), the general structure and typology of the built environment as well as on the very country specific overall power intensity. Not surprisingly, France, which has high power intensity due to e.g. widely spread electric domestic heating, has the lowest demand coverage, while Spain and Italy achieve much a higher coverage of about 40%.

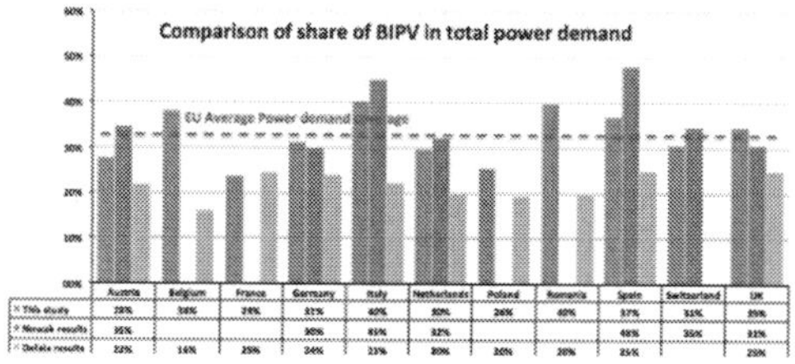

Figure 8: Comparison between main studies

Figure 8 shows the comparison of results calculated in the present paper (blue bar) with the results provided by the baseline study of Nowak & al. (ref. 1) and the study of P.R. Defaix (ref. 5).
The comparison confirms once again the good alignment of this study, which generally provides demand coverage ratios comprised in the range of the 2 other studies and confirms the excellent estimation that was performed 15 years ago by Nowak & al. (ref.1).
The weighted average maximum demand coverage of BIPV only is about 32% at the EU level, confirming and the huge potential of buildings for distributed power generation

3 FURTHER ANALYSIS

The present study provides a refined estimation of the global technical generation potential of buildings in Europe, assuming all suitable surfaces of the envelope carry PV generating capabilities.
Buildings Segmentation
Although special care have been brought to ensure choices of parameters that satisfactorily represent the distribution of buildings, a finer segmentation of buildings would allow to better represent the huge diversity of buildings typologies (e.g. office buildings, sport facilities, hotels, restaurants, education, retail buildings, agricultural and industrial warehouses that all

belong to the same non-residential segment) and of the attractiveness they represent in terms of power generation potential.
In addition, such a finer segmentation would allow further assumptions on self-consumptions ratios that are expected to increasingly affect the distributed generation economics.

In depth Country Analysis

As evoked earlier, EU countries present a significant level of diversity when it comes to the distribution and typology of the built environment. The present study was based on an extensive set of EU-wide studies and research projects, providing country specific information, collected through similar methodologies and presented under comparable formats. More specific country analysis could be performed by using, when available, country specific data and statistics on its building stock, that in some cases allow a much finer segmentation and analysis than that performed uniformly across all EU-countries as in the present study.

Novel PV technologies and applications

As mentioned in § 2.2.3, the present paper assumes only mainstream mass market commercial technologies are used. It is however anticipated that, more sophisticated front-end treatments on classical cells, or the use of novel technologies such as dye or organic cells will provide additional BIPV solutions in the future, notably due to more flexible module formats, mechanical flexibility, light weight, and useable transparency, potentially allowing other applications and/or additional areas of the envelope (e.g. windows) to be used.

5 CONCLUSION

The present study unequivocally confirms the huge theoretical potential that buildings represent in terms of distributed power generation, with roof surfaces being the main contributors to power generation, However, a significant number of technical, economic and non economic obstacles need to be overcome and streamlined before such potential could, even partially, be exploited through massive implementation of BAPV / BIPV applications.
While major building material manufacturers are working on the integration of the PV functionality into the building material, one can expect an increasing number of cost attractive PV enabled materials to come on-line.
The continuous downward pressure on PV costs, and the increasing focus of policymakers on implementing self-consumption schemes should further drive the interest for BAPV/BIPV.
Finally, deep renovation and new constructions constitute an ideal moment to consider an economically attractive implementation of PV-enabled envelope components. While there is a pressure to triple the annual renovation rate of the EU building stock from the current 1% to 3% by 2020, policy makers should carefully establish the proper framework conditions that will make the use of PV-enabled envelope components a standard practice across EU in all deep renovations or new construction.

[1] Technical Report IEA - PVPS T7 – 4, S. Nowak & al.,

[2] Estimating the number of buildings in Germany , Martin Behnish and A. Ultsch

[3] Episcope Tool; Tabula building Typology for Residential buildings, Germany

[4] Energy Performance Directive and Energy Efficiency Directive

[5] Technical potential for photovoltaics on buildings in the EU-27 P.R. Defaix a,1,2, W.G.J.H.M. van Sark a,□, E. Worrell b, E. de Visser c

[6] Asessment Of Roof & Façade Potentials For Solar Use In Europe, Harry Lehmann and Stefan Peter

[7] 2020 Climate & Energy Package and 2030 Climate and Energy Framework

[8] Energy Performance Directive and Energy Efficiency Directive

[9] Building Performance Institute Europe.

[10] Eurostat Database

[11] Buildings Database

[12] The Episcope Platform

[13] Entranze Data Mapping

[14] Tabula Web Tool

[15] EIA: US Energy Information Administration Commercial Buildings Energy Consumption Survey CBECS) - 2012

[16] Ibid 12

[17] PVGIS tool from the EU Joint Research Center

DESIGNING HIGH EFFICIENT SOLAR POWERED OLED LIGHTING SYSTEMS

Rasmus Overgaard Ploug[1], Peter Behrensdorff Poulsen[1,*], Sune Thorsteinsson[1], Anders Thorseth[1], Dennis Dan Corell[1], Jesper Wolff[1], Carsten Dam-Hansen[1], Arnold Knott[2]
[1]Department of Photonics Engineering, Technical University of Denmark, Frederiksborgvej 399, Building 130, DK-4000 Roskilde, Denmark
[2] Department of Electrical Engineering, Technical University of Denmark, Richard Petersens Plads 325, DK-2800 Kgs. Lyngby, Denmark
*Phone +45 21325110, E-mail: ppou@fotonik.dtu.dk

ABSTRACT: OLEDs used in solar powered lighting applications is a market of the future. This paper reports the development of electronic Three-Port-Converters for PV OLED product integration in the low-power area respectively for 1-10 Wp and 10-50 Wp with a peak efficiency of 97% at 1.8 W of PV power for the 10 Wp version. Furthermore, we present measurements of state-of-the-art commercial available OLED with regards to the luminous flux, luminous efficacy, luminance homogeneity, temperature dependency and IV characteristic of the OLED panels. In addition, solar powered OLED product concepts are proposed.
Keywords: Stand-alone PV Systems, High-Efficiency, Design, OLEDs.

1 INTRODUCTION

Organic Light-Emitting Diodes (OLEDs) are a promising new lighting technology, as they are very thin uniform light sources which are pleasing to look directly at without the need to encapsulate it in heavy optical components to hide the light source. By employing optically transparent components, transparent organic light emitting diodes (TOLEDs) have also been realized. When operating, TOLEDs emit light from both their bottom and top surfaces, while in non-operational mode they look like ordinary transparent glass. TOLEDs can be applied to a variety of window-like displays, bi-directional lighting facilities and eyewear displays [1]. It also makes them suited for combinations with photovoltaics, since the same area can be used to harvest solar radiation energy and irradiate light. Furthermore, OLEDs brings new possible form factors into the lighting sectors which makes new design possibilities available for innovative lighting products. Their ability to be deposited on a flexible substrate making flexible light sources available to the market gives unique opportunities in differentiations in the lighting market with similar arguments as flexible solar panels of thin-film, organic and polymer photovoltaics. OLEDs are still very expensive compared to other lighting technologies but in the display market they are competitive to LCD and have the later years gained ground in the display market e.g. in mobile phones, TVs and laptops.

Light-2-Light (L2L) stand-alone self-powered products (as sketched in figure 1) based on photovoltaics, OLEDs and batteries has not yet been seen on the market to the author's knowledge. Since OLEDs are very sensitive to e.g. humidity, oxygen and solar radiation it requires extra care in the encapsulation of the OLEDs when used outdoor. Many scientific problems are yet to be solved and successful solar powered OLED products might first reach the market some years in the future.

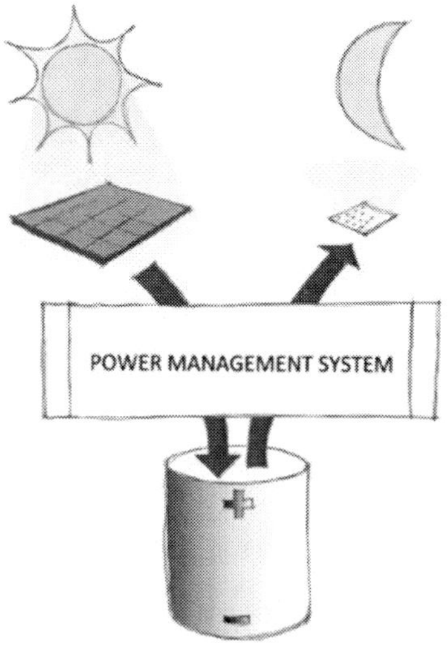

Figure 1: Principle sketch of a Light-2-Light system.

OLED technology is interesting for L2L due to the electric characteristics, which resembles conventional LED, and the potentially very low production cost of OLEDs when roll-to-roll techniques can be used for mass production [2]. In this paper we present measurements of state-of-the-art commercial available OLED products. The investigation will focus on the performance with regards to light output vs. electrical input. The characteristics studied are: the luminous flux and luminous efficacy as a function of power and luminance homogeneity and IV characteristics of the OLED panels.

Furthermore, in this work it is sought to lower the barrier towards design and implementation of solar powered OLED products by proposing a topology to a high efficient converter system for use in L2L OLED systems and investigates new concepts that could potentially exploit the uniqueness of the OLED and photovoltaics in combination.

2 OLED LIGHTING

The OLEDs have been characterized in the nearfield goniophotometer at DTU Department of Photonics Engineering, which for the 10×10 cm samples can measure the intensity distribution, luminance and spectral radiant intensity in the visible range. This paper presents measurements of luminous flux, luminous efficacy, luminance homogeneity, IV characteristics and the result of a temperature dependency investigation.

In figure 2 the measurement of luminous flux and luminous efficacy is shown as a function of the power consumption from low currents up to the rated maximum. It can be seen that the efficacy decreases with increasing power consumption with the decrease falling off rapidly in the low power range.

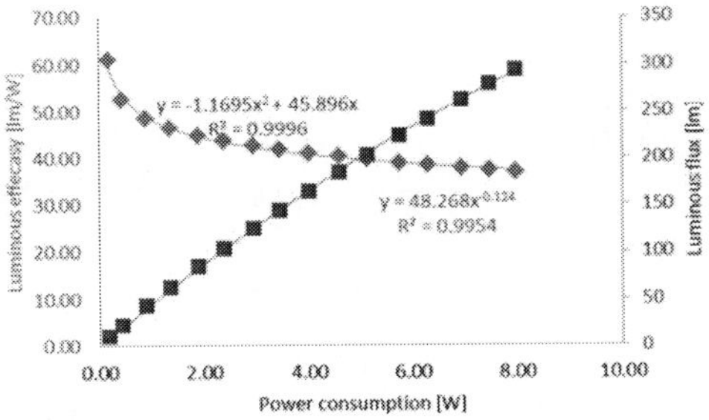

Figure 2: Luminous flux and luminous efficacy as a function of the consumed power.

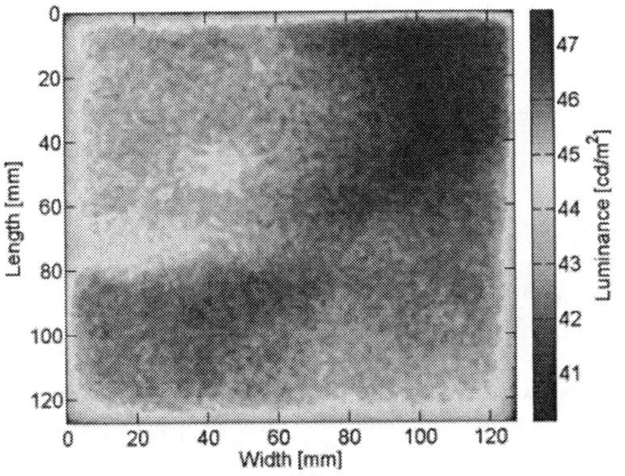

(a)

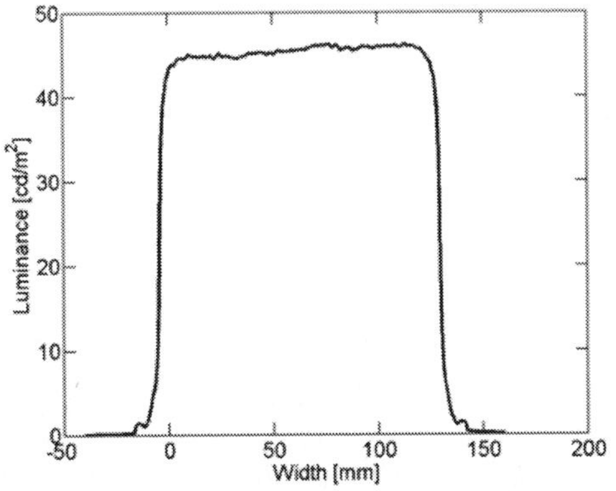

(b)

Figure 3: Luminance map of an OLED with the colours used as intensity scale i.e. not colour distribution (a) and a horizontal cross-section taken in the centre of the map (b).

The OLEDs have also been measured using an imaging luminance meter, which shows the uniformity of the appearance. The results are shown in figure 3. The luminance of an OLED tile like this, with a 4% gradient in the cross section might in some cases seem disturbing to a human viewer.

The forward IV characteristics of an OLED was measured and plotted in figure 4. The OLED shows a diode characteristic with a threshold voltage around 7.5 V.

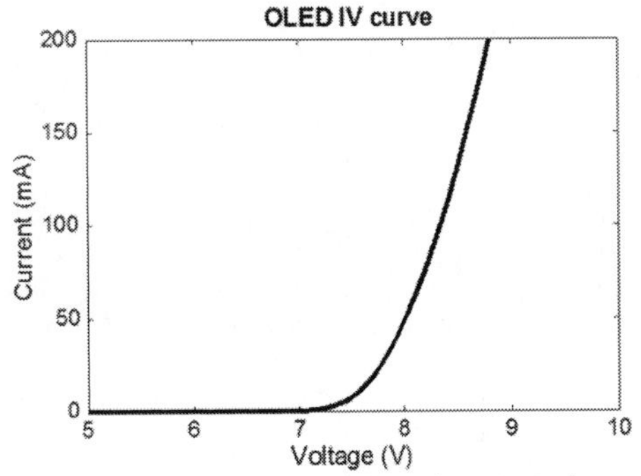

Figure 4: OLED IV characteristics. The OLED shows a diode characteristic with a threshold voltage around 7.5 V.

Lastly the temperature dependence of selected OLED was investigated. The temperature dependency is a critical parameter in outdoor applications due to the potentially large variations in weather conditions. For one OLED panel we measured a very small reduction in light output of 0.04% per °C in the temperature range 33-50 °C, with the temperature measured on the mounting plate of the OLED. Another OLED showed higher temperature dependence with a reduction of 0.14 % per °C.

3 ELECTRONIC CONVERTER

In order to design an efficient electronic converter for an outdoor L2L system two operating modes for the system are identified as

1. Energy harvesting during the day: Conversion of energy from the PV panel to the energy storage with high efficiency maintaining Maximum Power Point Tracking (MPPT) to harvest as much energy as possible
2. Energy usage during the night: Conversion of energy from the energy storage to satisfy the electrical operating point of the OLED panel.

To fulfil these two operating modes a switch-mode Three-Port-Converter (TPC), as shown in figure 5, has been developed.

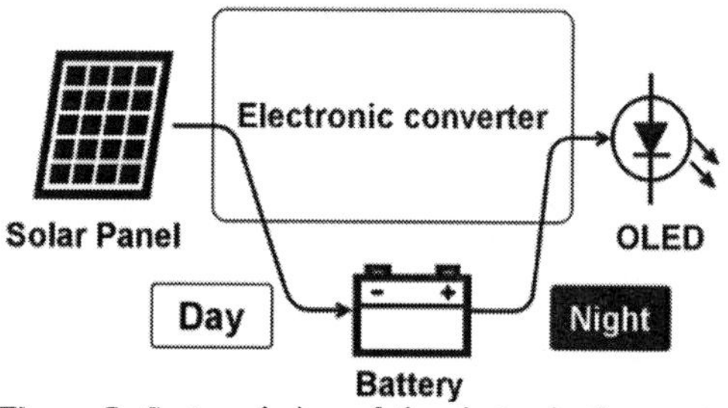

Figure 5: System design of the electronic three-port-converter for the L2L application.

3.1 Topology

For this study the voltage restrains for the PV panel, battery and OLED was chosen such that the PV panel voltage and the OLED operating voltage must be larger than the maximum battery voltage. Hence, in the daytime, the PV voltage has to be stepped down to the battery voltage (buck-type conversion) and during the night the battery voltage has to be stepped up to meet the OLED requirements (boost-type conversion). While one could design two separate converters – one for each of the operating modes – a conceptual topology has been suggested to meet both demands in one topology in order to reuse components between each operating mode and thereby save on total cost. This topology is shown in figure 6. It can be seen that the inductor L1 and switch Q2 are reused between each operating mode while Q4 and Q5 serve as power control switches (either fully on or fully off) preventing unwanted currents to run in the system. The electrical design and properties of this topology have been documented in [3] and a slightly different version with a tapped-boost step-up converter in [4]. In order to cover a full range of applications covering panels from 1-50 Wp, a low power range of 1-10 Wp and a high power range optimized for 10-50 Wp is developed. This project documents the 1-10 Wp prototype.

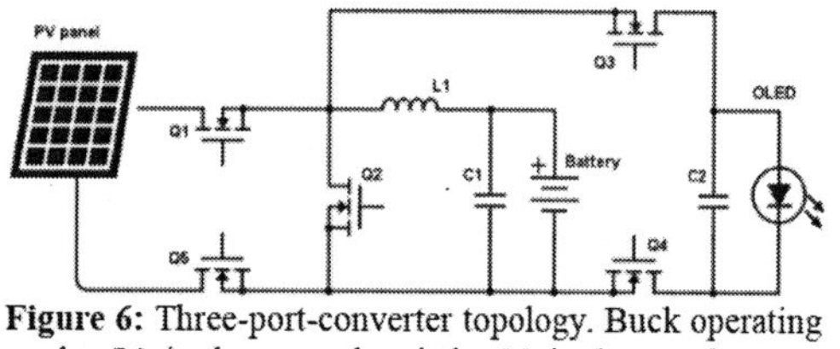

Figure 6: Three-port-converter topology. Buck operating mode: Q1 is the control switch, Q2 is the synchronous rectifier, Q5 is on, Q3 and Q4 is off. Boost operating mode: Q2 is the control switch, Q3 is the synchronous rectifier, Q4 is on, Q1 and Q5 is off.

3.2 Implementation and results

Apart from seeking a high efficiency energy conversion by proper component selection and converter switching frequency, it is furthermore sought to optimize the converter switching behaviour for having highest efficiency in the lower part of its operating range. It is desired to have a high performance in a L2L system also on days with low solar irradiation. Efficiency wise this means that the system should be optimized for having its highest efficiency in its low power operating area as this would be the 'bottleneck' in securing high reliable performance. Furthermore, L2L products are often placed in areas with a lot of shadows on the solar panels e.g. in parks or in the urban environments, making peak power operation rare. An optimization for the electronic power converter for this low power area was described in [3] and the resulting energy conversion efficiency graphs can be seen in figure 6. The converters peak efficiency is 97 % when operating at 1.8 W with PV panel voltage of 9 V. The conversion from battery to light is of similar efficiency.

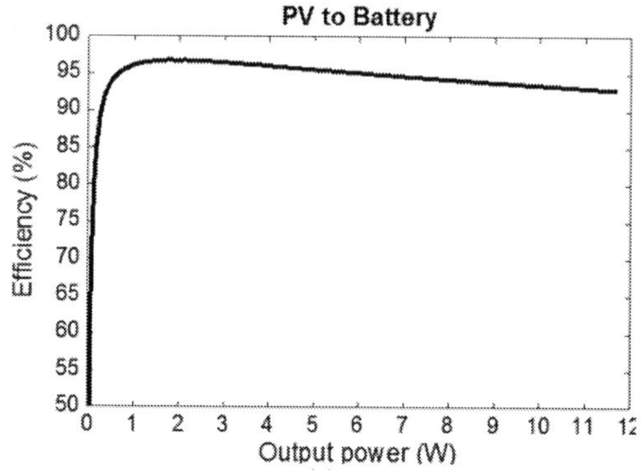

(a)

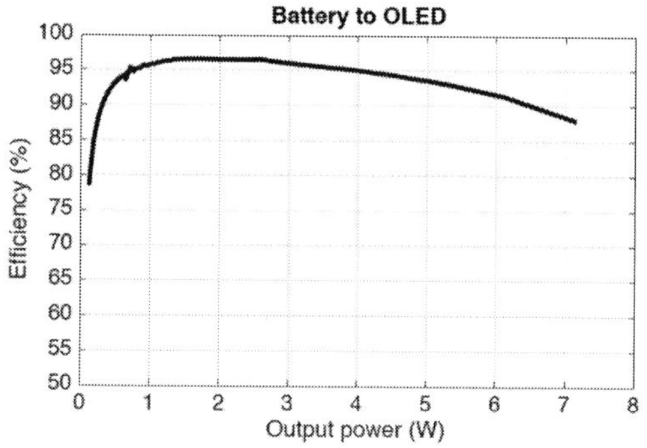

(b)

Figure 7: (a) Measured efficiency curve of the 10 W converter when harvesting energy from PV panel to the battery. It can be seen that the converter excels in having its highest efficiency at the lowest power obtaining 94 % at 0.5 W and peaking at 97% at 1.8 W. (b) Efficiency curve of the converter operating in boost mode. The converter reaches 96 % efficiency at the 1.275 W corresponding to an OLED operating point of 8.5 V at 150 mA.

A more in depth view of the efficiency during energy harvesting of the electronic converter can be seen in figure 8.a. Here the efficiency has been measured in the entire converter operating range in order to assess the energy harvesting behaviour for all possible solar irradiation conditions. This was done in an automated setup using a solar array simulator as the PV and a source meter unit as the battery. Also the tracking efficiency of the MPPT algorithm was measured as well using panel characteristics with 1 maximum power point (figure 8.b). These two measurements were then combined in order to estimate the total energy harvesting yield (figure 8.c). Note that the converter efficiency measurements differ slightly from figure 7.a since, in the latter measurements are made with a 4.0 V fixed voltage source-meter instead of a battery with a voltage floating in the range 3.2 V to 4.2 V.

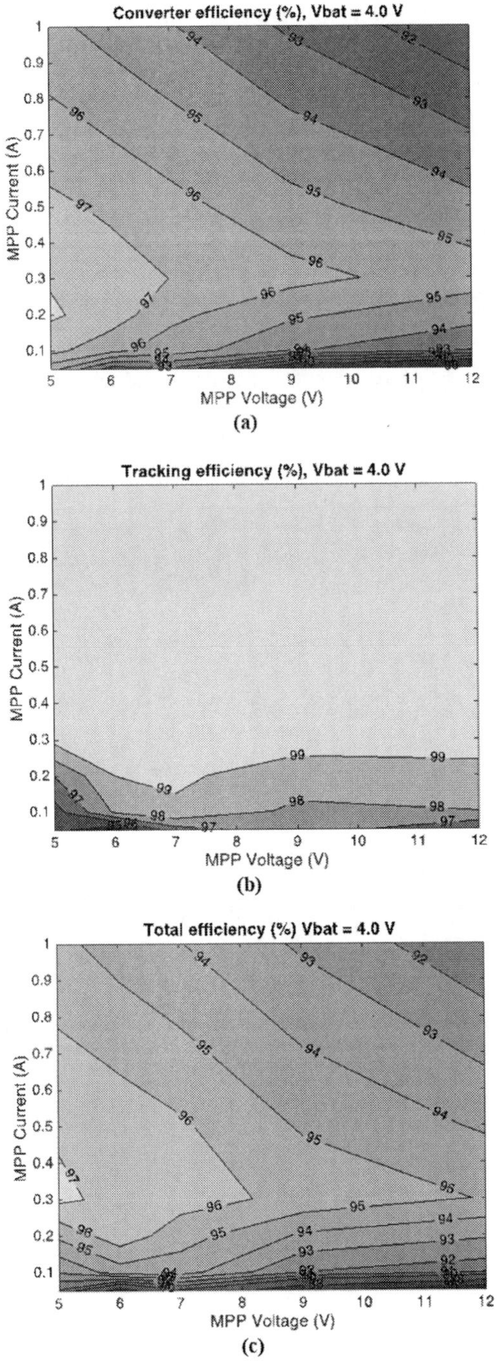

Figure 8: Converter energy harvesting efficiency measured in its entire operating range. (a) shows converter efficiency, (b) shows the MPP tracking efficiency and (c) shows the converter and tracking combined efficiency.

4 SOLAR POWERED OLED LIGHTING CONCEPTS

The OLED technology brings many new unique design possibilities into the market of stand-alone self-powered lighting. New designs should utilize the surface light concept in new products with new possibilities in mating the very thin and flat OLED structure with PV panels which also are flat structures. Transparent OLED (TOLED) technology could also lead to new 'sandwich' structures simply by placing the TOLED on top of a PV-panel making small devices that are able to both illuminate and harvest energy from the same surface. Also flexible OLEDs paired with flexible PV-panels can potentially introduce new unique design possibilities.

As an example of a product idea it is suggested to use the OLED lighting structure to give the possibility of making self illuminating signs which are perfect examples of utilizing self-powered L2L systems in remote locations where grid-connection is not an opportunity. A possible design concept for a L2L OLED sign is included in figure 9 for inspiration purposes.

Figure 9: OLED based sign with PV integrated in the mast.

5 CONCLUSIONS

OLEDs used in solar powered lighting applications are a market of the future. To break down the barriers towards exploiting the unique potentials of PV and OLEDs in combination we have reported the development of a Three-Port-Converter for PV OLED product integration for the 1-10 Wp range with a peak efficiency of 97% at 1.8 W of PV power. Furthermore, we have presented measurements of state-of-the-art commercial available OLED with regards to the luminous flux, luminous efficacy, luminance homogeneity and IV-characteristic select OLED panels. In addition, solar powered OLED products are elaborated and a design suggestion of a solar powered OLED sign is presented.

REFERENCES

[1] C. W. Joo, J. W. Shin, J. Moon, J. W. Huh, D. H. Cho, J. Lee, S. K. Park, N. S. Cho, J. H. Han, H. Y. Chu, and J. I. Lee, "Highly efficient white transparent organic light emitting diodes with nano-structured substrate," *Org. Electron. physics, Mater. Appl.*, vol. 29, no. 2016, pp. 72–78, 2016.
[2] R. R. Søndergaard, M. Hösel, and F. C. Krebs, "Roll-to-Roll fabrication of large area functional organic materials," *J. Polym. Sci. Part B Polym. Phys.*, vol. 51, no. 1, pp. 16–34, Jan. 2013.
[3] R. O. Ploug, A. Knott, "Photovoltaic OLED Driver for Low-Power Stand-Alone Light-to-Light Systems", *Elektronika Ir Elektrotechnika*, vol 22, no. 3, 2016. Available: http://dx.doi.org/10.5755/j01.eie.22.3.15314
[4] M. C. Mira, A. Knott, M. A. E. Andersen, "Three-port converter for low-power stand-alone light-to-light systems (submitted for publication)", *IEEE Energy Conversion Congress & Expo (ECCE) 2016.*

SELECTION OF WEATHER PROFILE(S) FOR TESTING PERFORMANCE OF SPV PUMPS IN INDIAN CLIMATE

Kamlesh Yadav[AB], O.S. Sastry[A], B.Bora[A], M.kumar[A], R.Singh[A], Atul Kumar[B], B.Prasad[B]
[A]National Institute of Solar Energy (NISE), Gurgaon, India
[B]The Energy and Resources Institute, New Delhi, India
*sastry284@gmail.com, Tel: +91-0124-2579213, Telefax No. : +91-124-2579207

ABSTRACT: Indian climate is broadly divided into 4 climatic zones, Hot & dry, Hot & humid, Composite and cold & Cloudy. Due to escalated targets for SPV water pumping systems installation in the country there is a large demand for testing and certification of SPV pump samples. Indian test labs involved in testing and certifications are developing a generalized radiation and temperature profiles to cover the above 4 climatic zones such that the certified pump provide required performance in these climates. For testing samples quickly and yet ensure covering all climatic zones is with minimum number of profiles is an involved and challenging problem. This paper presents two profiles of radiation and temperature, representing summer and winter profiles, which is able to cover best and least performance of pump during a year. Further, using these two profiles performance data of two sample pump systems evaluated in-terms of PV array power output, quantity of water supplied (liters/ watt/ day), MPPT, Motor, pump and overall system efficiency.
Keywords: SPV water Pumps, MPPT and Pump efficiency, Radiation and temperature profile, water throughput.

1 INTRODUCTION

Pumping of water for irrigation and drinking purpose are very important applications of Solar Photovoltaic (SPV) technology. Ministry of New and Renewable Energy (MNRE) the nodal ministry of government of India has set a huge target of more than o.1 million pump systems for the year 2015-16 [1], consequently the state governments also announced SPV pump targets matching the federal government. To ensure quality of installed systems, MNRE also issued set of performance guidelines and specifications of the pumping systems. To test, evaluate and certify samples. National Institute of Solar Energy (NISE) developed test set-ups, methodology, parameters to be monitored, report format and accredited test laboratories in the country.

The programme of SPV water pumping system is expected to generate huge demand for testing and certification of pump samples. The certification testing should be comprehensive, quick and should cover all requirements. Keeping in view the requirements NISE has designed typical profiles for radiation and temperature to be used with commercial PV array simulators available in the market. This method is easier, quicker and accurately provides the maximum and minimum performance of the SPV pump under summer (worst case) and winter (best case) radiation conditions, covering most of the sites in the country.

2 EXPERIMENTAL PROCEDURE

The SPV pumping test facility established at NISE mainly comprises of a sump well (4 m x 2.5 m x 2.5 m) and overhead platform of (3.5 m x 2 m x 2 m) which can make suction head up to 7 m and a total static head up to 75m including the friction head [2]. For irradiance and temperature data is logged with data logger make: Campbell CR-1000, pyrometer Make: EKO and RTD-type thermocouples are used to record SPV module temperature. Solar array simulator make: Terra SAS (ETS600X), 15 KVA capacities is used for irradiance and temperature profile simulation and for power measurements, a power analyzer make: Zimmer (LMG-450) is used with the pumps. The PV array simulator provides a means of comparing the performance by simulating exactly the same SPV characteristics during experimentation. The TerraSAS software also records the data of PV power, voltage, current and time-stamping. Continuous monitoring of the flow of water and Total dynamic head (TDH) is maintained through flow meters and pressure gauges. To benchmark the results during the experiment, the day's radiation profile is maintained using the array simulator. All the experimental results are recorded using PV simulator software and power analyzer [3].

3 RESULT AND ANALYSIS

SPV water pumping systems performance is mainly dependent on PV module, Pump controller, motor and pump. For efficient operation of the system all the three components should be perfectly match with each other. Performances of Photovoltaic modules depend upon radiation, air-mass, temperature, humidity and wind speed; which vary from place to place. In all of these affecting causes module temperature which is site specific is another major contributing factor affecting the performance of PV Technology. Ambient temperature and wind speed affects temperature of the Module. With rise in module temperature, voltage and power decreases, current will have little gain but overall power decreases with increase in module temperature [4-5]. SPV technologies are offered with different temperature coefficient for power, voltage and current by the manufacturers: which makes their performance location and site specific. Keeping the varying nature of voltage and power, MPPT of the pump controller should be designed accordingly for complete range of operation, and further duty point design of pump should be taken care because the power availability from the module vary whole day.

So the complete characterization of the SPV water pumping system it should be tested all the available environment conditions. Thomas Huld et al. [6] proposed Low-Irradiance-Low Temperature, Medium-Irradiance-Medium- Temperature, Medium-Irradiance-High-Temperature, High-Irradiance-Low-Temperature, High-Irradiance-High-Temperature and Normal Irradiance-Cool-Environment conditions for the characterization of the PV modules

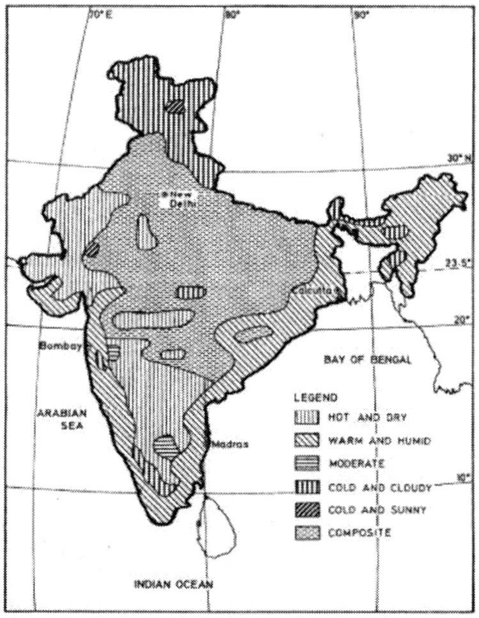

Figure 1: Climatic zones of India [7]

Further from the figure 1. India is divided in Hot and Dry, Warm and humid, Moderate, Cold and cloudy, Cold and Sunny and Composite climatic zones. Most of the area is covered by composite climate, warm and humid, hot and dry climate.

The temperature and relative humidity variation is given in following table.

Table 1 Climatic zones of India [7]

Climatic Zone	Mean monthely temperature (◦C)	Relative Humidity (%)
Hot and Dry	>30	<55
Warm and humid	>30	>55
Moderate	25-30	<75
Cold and cloudy	<25	>55
Cold and Sunny	<25	<55
Composite	This applies, when six months or more do not fall within any of the above categories	

In recent period SPV pumps are tested and evaluated in the real outdoor field condition by the test labs. Further evaluation is done based on the liter/Watt/day, PV energy to water output efficiency and total water output on a clear sunny day with three times tracking of SPV panel, under the "Average Daily Solar Radiation" condition of 7.15 KWh/ m^2 on the surface of PV array [8]. But irradiation and temperature vary with season and site to site due to this variation the PV module generated power [9], MPPT & conversion efficiency [10] of the pump controller and pump and motor performance varies. So testing on any random day can't fulfil comprehensive characterization of the SPV pumps and as well as the condition for qualification for all the manufacture are not same without doing temperature correction and water output extrapolation for 7.15 kWh/m^2 irradiation. While doing temperature correction and water output extrapolation the error probability in results and complexity increases. So for quick, error free and complete characterization of SPV pumps this paper proposes two days profile one for cold day and one for hot day with same irradiation of 7.15 kWh/m^2, which can be simulated on the solar array simulators

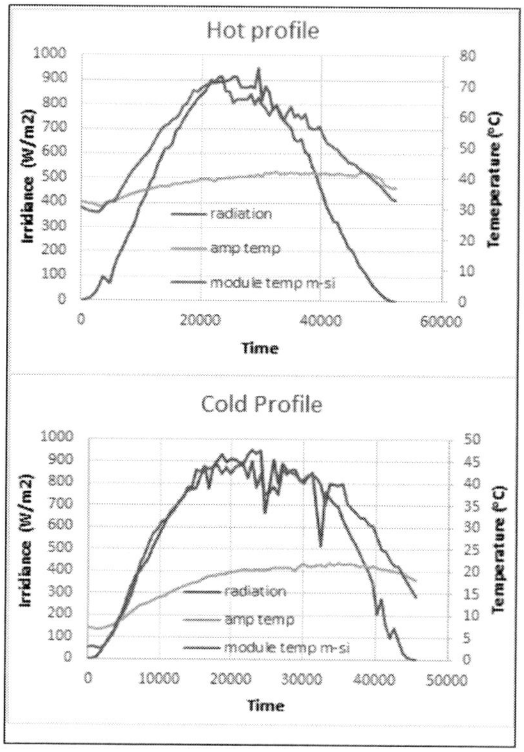

Figure 2: Cold and Hot day Profile of same energy =7.15 kWh/m^2

After the analysis of the whole year data of 2010 for NISE it is found that the peak module temperature is 43 °C at 950 W/m^2 radiations on 18th January and peak module temperature is 73 °C at 892 W/m^2 radiations on 15 may. In above profiles irradiance energy is purposefully kept at 7.15 kWh/m^2 for both the hot and cold day by adding the extra data points and following the same patterns of those days. So irradiation is same for both day and there is only temperature variation.

So the proposed profiles quickly and accurately provides the maximum and minimum performance of the SPV pump under summer (worst case) and winter (best case) radiation conditions, covering most of the sites in the country. Only regions like Leh and Ladakh is not covered by this profile which are negative temperature region. This can be simulated separately on demand of implantation agencies.

Further the study also compares the performance of the two SPV pump systems based on the PV array power output, quantity of water supplied (Total water output and liters/ watt/ day), MPPT, Motor, pump and overall system efficiency. First pump tested is for drinking water purpose of 1 hp capacity with 1000 Wp module at 60 meter duty head and second pump is 5 hp with 5000 Wp module at 20 meter duty head for irrigation purpose. From the below table the overall efficiency of 1st pump is 6.10 % and 6.76 % correspondingly for hot and cold profile. In case of 2nd pump is 5.95 % and 7.0 % correspondingly for hot and cold profile.

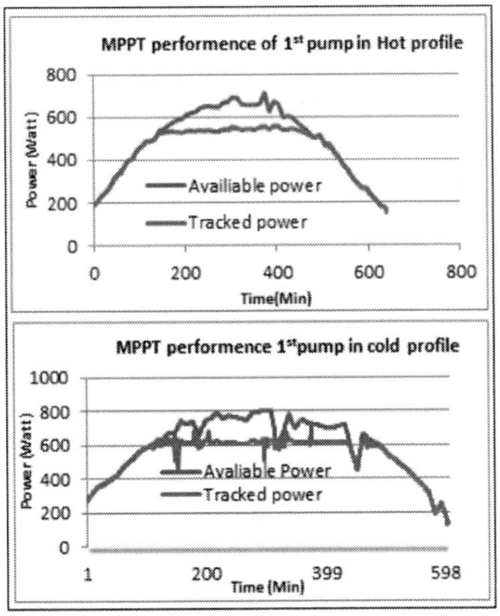

Figure 3: MPPT performance of 1st Pump in Hot and cold days

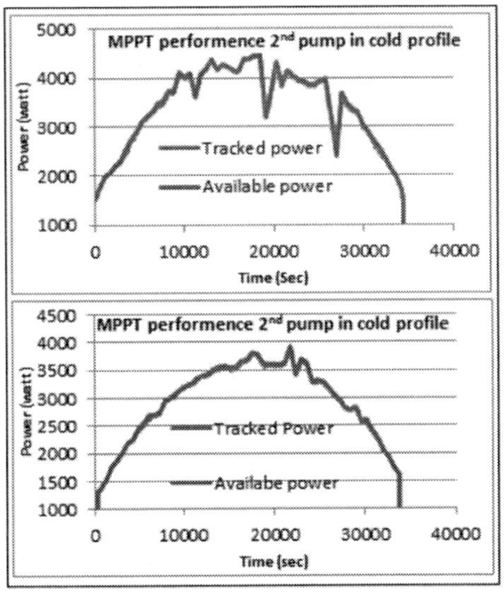

Figure 4: MPPT performance of 2nd Pump in Hot and cold days

With data below it is clear In case of the second pump motor and pump efficiency is less but is consistent so there is loss only because of PV module. As the PV module output decreases with temperature. But in case of pump 1 st pump, controller is not well optimized. Possible reasons may be MPPT efficiency and peak power limitation [3].

Table 2 Performance parameters for 1st and 2nd pump

	Pump sample	Tracked Power kWh	Available Power kWh	Total output delivered in Kilo Liter	L/W/D	Module eff %	MPPT %	Motor Pump eff %	Overall system eff radiation to water
1.	Hot Profile	4.81	5.30	17.817	17.81	11.11	90.76	60.50	6.10
	Hot Profile	4.81	5.30	17.817	17.81	11.11	90.76	60.50	6.10
2.	Hot Profile	4.81	5.30	17.817	17.81	11.11	90.76	60.50	6.10
	Hot Profile	4.81	5.30	17.817	17.81	11.11	90.76	60.50	6.10

4 CONCLUSION

So the study clearly establish the importance of hot and cold day profile which quickly covers and yet ensure the best and least performance of pump during a year. With using same irradiation energy it also ensures the comprehensive characterization of the SPV pumps. Further, using these two profiles performance of the pump in-terms of PV array power output, quantity of water supplied (liters/ watt/ day), MPPT, Motor, pump and overall system efficiency can be obtained accurately.

ACKNOWLEDGEMENT

This research is based upon work supported in part by the Solar Energy Research Institute for India and the U.S. (SERIIUS) funded jointly by the U.S. Department of Energy subcontract DE AC36-08G028308 (Office of Science, Office of Basic Energy Sciences, and Energy Efficiency and Renewable Energy, Solar Energy Technology Program, with support from the Office of International Affairs) and the Government of India subcontract IUSSTF/JCERDC-SERIIUS/2012 dated 22nd Nov. 2012

REFERENCES

[1] http://www.mnre.gov.in/

[2] O.S.Sastry, S. Agarwal, D. Adiraju, R. Kumar and B. Bandopadhyay, "Performance evaluation of solar photovoltaic lighting and water pumping systems in India," 26th European Photovoltaic Solar Energy Conference and Exhibition 25 (2013) pp. 351–370

[3] Kamlesh Yadav, O.S.Sastry, R. Wandhare, N. Sheth, M. Kumar, B. Bora, Rashmi Singh , Renu and A. Kumar "Performance Comparison of Controllers for Solar PV Water Pumping Applications" Solar Energy 119 (2015) 195-202

[4] Effect of module temperature and seasonal variation analysis of the outdoor performance of CdTe at composite climate of India, Birinchi Bora, O.S. Sastry, Yogesh Kumar Singh, Ramayan Singh, Kamlesh Yadav, Vikrant Sharma, Manander Bangar, Mithilesh Kumar, Invertis Journal of Renewable Energy, Vol. 5, No. 3, 2015

[5] Effect of module temperature and seasonal variation analysis of the outdoor performance of CdTe at composite climate of India, Birinchi Bora, O.S. Sastry, Yogesh Kumar Singh, Ramayan Singh, Kamlesh Yadav, Vikrant Sharma, Manander Bangar, Mithilesh Kumar, Invertis Journal of Renewable Energy, Vol. 5, No. 3, 2015

[6] Data sets for energy rating of photovoltaic modules, Thomas Huld, Ewan Dunlop, Hans Georg Beyer, Ralph Gottschalg, Solar Energy 93 (2013) 267–279

[7] http://mnre.gov.in/solar-energy/ch2.pdf

[8] http://mnre.gov.in/file-manager/offgrid-solar-schemes/NABARD-operational-Guidelines-of-Solar-Water-Pumping-Programme.pdf

[9] Effect of module temperature and seasonal variation analysis of the outdoor performance of CdTe at composite climate of India, Birinchi Bora, O.S. Sastry, Yogesh Kumar Singh, Ramayan Singh, Kamlesh Yadav, Vikrant Sharma, Manander Bangar, Mithilesh Kumar, Invertis Journal of Renewable Energy, Vol. 5, No. 3, 2015

[10] Comparison of Photovoltaic Array Maximum Power Point Tracking Techniques, Trishan Esram, Student Member, IEEE, and Patrick L. Chapman, Senior Member, IEEE, IEEE transactions on energy conversion, vol. 22, no. 2, june 2007

MODEL-BASED DESIGN AND SIMULATION OF CONTROL STRATEGIES TO MAXIMIZE THE PV HOSTING CAPACITY IN ISOLATED DIESEL NETWORKS

USING SOLAR SHORT-TERM FORECASTS FOR PREDICTIVE CONTROL OF DIESEL GENERATION

D. Peters[1], R. Völker[1], T. Kilper[1], M. Calais[2], T. Schmidt[3], C. Carter[2], K. v. Maydell[1], C. Agert[1]
[1]NEXT ENERGY, EWE-Research Centre for Energy Technology at the University of Oldenburg
Carl-von-Ossietzky-Straße 15, 26129 Oldenburg, Germany
[2]School of Engineering and Information Technology, Murdoch University, Murdoch WA 6150, Australia
[3]Department of Energy and Semiconductor Research, Carl von Ossietzky University Oldenburg, Germany
Corresponding author: D. Peters, dorothee.peters@next-energy.de

ABSTRACT: Increasing Photovoltaic (PV) system penetration in small, isolated networks faces a number of technical, regulatory and economic challenges but also opportunities. The replacement of costly and polluting diesel generation is highly desirable, however, a key technical issue is system stability, particularly on intermittent cloudy days. In this context, the step load response capability and spinning reserve provided by diesel generation are critical to cover for the sudden loss of PV power. This work develops and tests the application of short-term forecasts using 'sky imagers' in the control of generation capacity. The benefits for system stability shall be examined using an example isolated network modeled in PowerFactory that simulates frequency and voltage at the generator busbar. To reduce forecast errors, ensemble forecasts for all PV systems in the network are used. The results show a favourable low impact of PV power on voltage and frequency, if realistic ramp rates for dispersed PV systems are used. The envisaged fuel savings from reducing spinning reserve requirements are achieved but stay below 1 %. With the inclusion of short-term storage capacity, further benefits are expected. In addition, the high dependency of possible PV penetration on generator dispatch and loading is shown.
Keywords: Hybrid PV, Diesel mini-grids, Grid Stability

1 INTRODUCTION

In Western and Northern Australia many small isolated networks exist, which are powered by diesel generators. Ecological and economic reasons, with Levelised Cost Of Electricity (LCOE) from PV becoming distinctly cheaper than the LCOE from diesel generation, provide the motivation to replace diesel generation with photovoltaics. However, the increasing number of installed PV systems imposes difficulties for the network operators. As a response, some of them have prohibited or restricted the connection of additional PV systems.

Case studies carried out in different grids in Western Australia and the Northern Territory identified current common problems and fears of stakeholders concerning further PV installations [1,2]. The step load response capability and spinning reserve of conventional generation are especially critical for system stability to cover for the sudden loss of PV power due to the intermittent nature of the solar resource on cloudy days. Introducing sky imager forecasting, into the dispatch control of diesel generators, shall lead to higher PV hosting capacity and more efficient generator operation. Sky imager forecasts were chosen due to their short term forecast horizon of approx. 0-900 s, which corresponds to the time frame in which regulation reserves are the primary means of stabilizing the network.

2 MODEL DESCRIPTION

The developed network model in PowerFactory[1] includes key components which are diesel generators, cumulated PV systems and loads. Load profiles, as well as the capacity of diesel generation and PV systems, were sized according to data given for the Carnarvon grid [1].

The result is an example isolated network with five diesel generators (320 kW each), 15 equally sized PV systems (110 kW in total) and peak load of 1,100 kW.

2.1 Control model

The diesel generator control consists of a tuned standard governor and voltage controller, secondary and dispatch controller (see [3] for controller parameters). Dispatch control connects and disconnects generators according to load requirements. For additional deployment, a maximum loading criterion is used (80 % of rated capacity). As a consequence, at least 15 % of online capacity is available as spinning reserve for load fluctuations, because reliable operation is possible up to 95 % of rated output, taking high temperature derating into account. The minimum allowed loading is 30 % of rated capacity. To operate at high loading, which is the most efficient, a generator is switched off when remaining generators will be loaded below the allowed maximum loading. An additional criterion is that sufficient spinning reserve is maintained to cover for PV power fluctuations. The total spinning reserve equals the greater of the required SR for load fluctuations and the required SR for PV power fluctuations. Delay times for starts and stops are defined to stabilize the control.

2.2 Irradiance data

Sky imager forecasts and corresponding real data were collected by a pyranometer network within the "High Definition of Clouds and Precipitation for advancing Climate Prediction HD(CP)^2" measurement campaign HOPE and was kindly provided by the Energy Meteorology Group at Carl von Ossietzky University Oldenburg [4]. The irradiance level is scaled using the clear sky model by Haurwitz [5] and validated in terms of cumulative ramp distribution with irradiance data for Australia obtained from the Australian Bureau of Meteorology [6].

For the example network, irradiance and forecast data of 15 sensors in an area of 3 x 3 km^2 are used.

[1] Power system analysis software by DIgSILENT

2.3 Predictive Control

The predictive control algorithm sets spinning reserve requirements according to the irradiance forecast. An area ensemble forecast is used, as single point forecasts can have root mean square errors (RMSE) of up to 200 W/m² for heterogeneous cloud conditions [4]. The area ensemble forecast combines average, minimum and maximum values of single point forecasts inside the respective area (see Fig. 2). The RMSE between average forecast and measured irradiance is reduced to 90-120 W/m² depending on the forecast horizon. Outliers lying below the minimum forecast are still occurring with 25 % probability and have a mean deviation of approx. 50 W/m² from the minimum forecast. In a further processing step, for all time steps '*t*', the minimum of the available forecasts for each forecast horizon is used for predictive control. Below minimum outliers can be reduced to 2 % and their mean deviation to below 10 W/m².

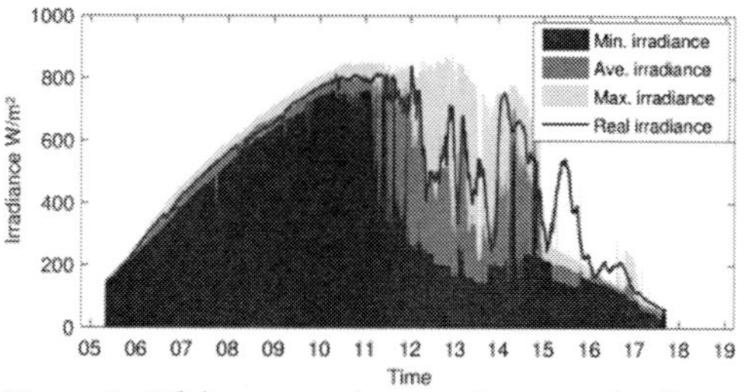

Figure 2: Minimum, maximum and average irradiance forecasts for 10 minute forecast horizon for 15 sensor locations on 06.05.13. The locations are spread over an area of 3 x 3 km². Real irradiance (red line) is shown for comparison.

3 SIMULATION & RESULTS

For comparison between normal and forecast-aided dispatch, three cases with different spinning reserve (SR) criteria are established. In the base case, SR equals day-time PV capacity within the network. The second case provides no additional SR to cover for PV power fluctuations. The third case uses predictive control where SR equals the fluctuating PV power derived from the sky imager ensemble forecast for long-term simulation (3.3), estimated for maximum PV penetration (3.1, 50 % of rated capacity) and equal to the ramp rate for stability analysis (3.2).

3.1 Maximum PV penetration

The maximum PV penetration is calculated according to the apparent load using the operational limits of diesel generators being a 30 % minimum loading and an 80 % maximum loading. Fig. 3 displays the results for the base case. The colours denote the number of generators dispatched, while no operation is possible in the white area (top graph). The amount of possible PV penetration strongly depends on the load.

For each colour region, the diagonal border represents the increasing load which can be covered due to PV power. The horizontal border sets in when spinning reserve requirements, according to PV capacity, have grown too large for the respective number of operating generators and so additional generation capacity, due to the PV supplying much of the load, has to be made available as spinning reserve. This is not apparent if no SR or less SR in predictive control is applied (see Fig. 4).

The curve in the upper plot of Fig. 3 shows a summer peak load time series [1] and summer clear sky irradiance at a PV penetration of 60 % of peak load to illustrate the day-time dependence. The rated capacity of 60 % is not achieved in January, because the PV systems are modeled with fixed tilt, which is optimized for the whole year.

This penetration level is only theoretically achievable for averaged high loads (> 50 %) and irradiance profiles.

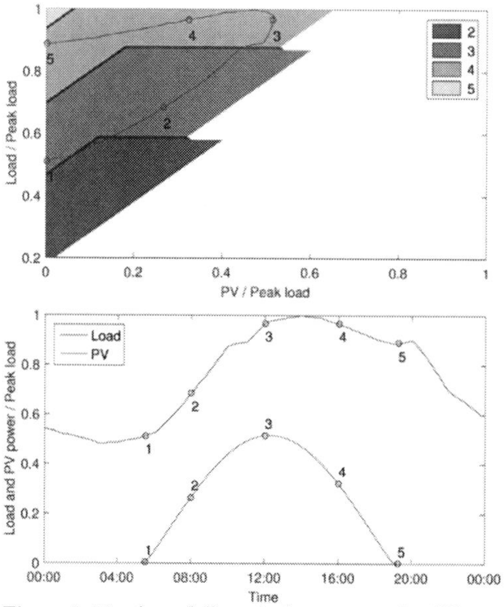

Figure 3: Number of dispatched generators for different PV and load cases (top) and daily time series (top - red line, bottom) for summer peak load and peak PV at 30 % PV penetration. The labelled data points are at sunrise (1), 8 am (2), noon (3), 4 pm (4) and sunset (5).

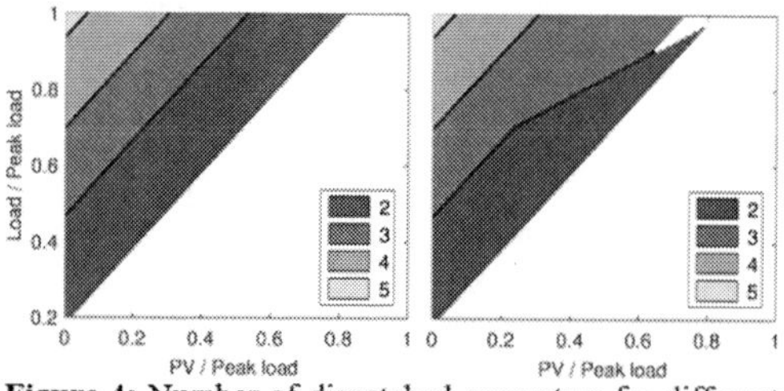

Figure 4: Number of dispatched generators for different PV and load cases for no SR (left) and predictive control (right).

3.2 Stability analysis

For short term simulations of a few minutes, the worst ramps that occur in the HOPE dataset were extracted for time intervals of 15 s, 1 minute and 10 minutes (see Table I). The RMS-Simulation is carried out for 250 s with 15 second PV power ramp for load cases of 50 %, 75 % and 100 % of peak load and PV penetration of 10-50 %. Increasing PV penetration and decreasing load increase the deviations in voltage magnitude and frequency. Voltage and frequency deviations are still below 0.3 % for all cases. The generators' step load capability is sufficient, as long as they are not overloaded, which is the case for 40-50 %

PV penetration and 50 % load. In the case of generator loading, using low SR leads to higher loading for high PV penetrations of 40-50 %, while there is little difference for penetrations up to 30 %.

Table I: PV power ramps derived from HOPE data and expected ramp found in network stability tests for comparison (*).

Time interval in s	Relative Power drop in %	Change rate in %/ s
15	23	1.53
60	27	0.45
600	47	0.08
0.01(*)	50	5000

3.3 Long-term analysis

Long-term simulations are conducted with scaled load time series 'summer peak' and 'winter average' from [1]. The rated capacity of PV systems is set to 30% of peak load, as this penetration is the highest possible with a winter average day-time load of 50 %. Four days from the HOPE data set are chosen according to a classification of 15-minute mean and standard deviation of clear sky index k_i with criteria displayed in Tab. II.

Table II: Classification of irradiance conditions.

Irradiation conditions Date (#days)	Statistical Criteria for 15-min mean and standard deviation
Clear 04.05.13 (5)	$\bar{k}_i > 0.9$ for 75 % $\sigma(k_i) < 0.1$ for 50%
Highly fluctuating 18.04.13 (9)	$\sigma(k_i) > 0.1$ for 50%
Overcast 06.04.13 (14)	$\bar{k}_i < 0.5$ for 75 % $\sigma(k_i) > 0.1$ for 50%
Mixed conditions 06.05.13 (24)	not in previous classes $\sigma(k_i) < 0.1$ for 50%

Fig. 5 shows the resulting spinning reserve for the base case and the predictive control (PC) case, which has significantly lower spinning reserve conditions. The number of generators dispatched, their start counts and loading are also assessed.

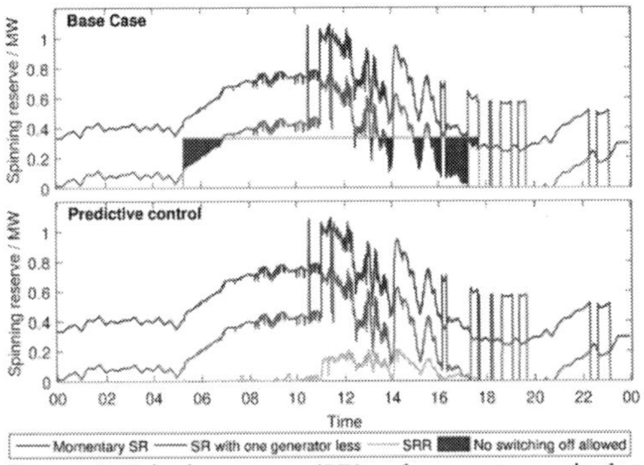

Figure 5: Spinning reserve (SR) and momentary spinning reserve for base case (top) and predictive control (bottom) (6.5.13, summer). The red area marks the times when no further generator can be switched off due to SR requirements (SRR).

Benefit from PC in terms of reduced generator run hours is determined for overcast irradiance with summer peak profile (1.5 hours less) and for all irradiance conditions with winter average profile (2.7 to 4.6 hours less). The start counts are high for both spinning reserve policies (8 - 15 start counts for summer and 2 - 10 for the winter load profile) and show mixed behaviour throughout the scenarios. Unlike expected, the highest start counts are found in overcast irradiance conditions. The greatest benefit from using PC is achieved in the scenario: winter average load and intermittent conditions (5 less start counts).

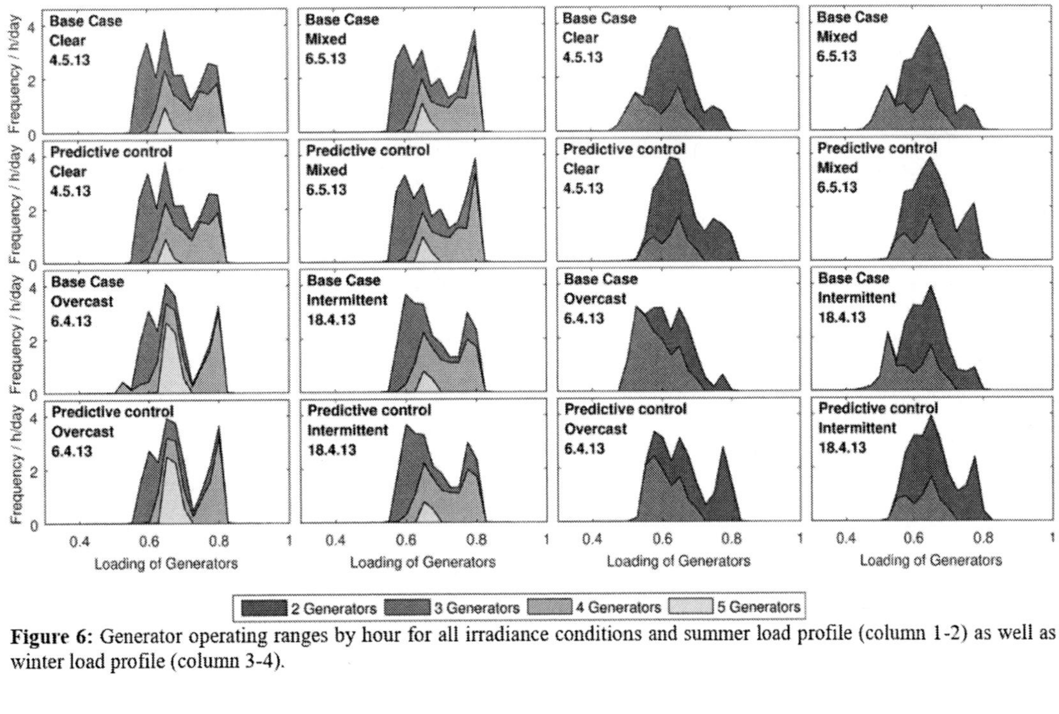

Figure 6: Generator operating ranges by hour for all irradiance conditions and summer load profile (column 1-2) as well as winter load profile (column 3-4).

Generator loading is nearly identical for the base case and the predictive control case in summer and ranges between 50 and 80 % (see Fig. 6, columns 1-2). For the winter average load profile, the base case show more time with low generator loading in the range of 40 to 50 % (see Fig. 6, columns 3-4), when 3 generators are dispatched, while with predictive control higher generator loading in the range of 70 to 80 % can be seen when only 2 generators are dispatched.

Fuel consumption is derived from the simulated loading and interpolated efficiencies given in [7]. For the summer peak load profile, the mean consumptions for the base case and the predictive control case are very similar, with differences below 0.2 % for all irradiance scenarios (see Tab. III). Using the winter average profile, the fuel consumption is reduced using predictive control (approx. 0.7 %) for clear, mixed and intermittent irradiance conditions, while in overcast conditions, normal dispatch consumes less fuel. The overall fuel consumption in winter is significantly less, due to the reduced load compared to the summer peak profile. Given the simulated fuel consumption without PV, the fuel saving with PV is approx. 8 %.

Table III: Estimated fuel consumption.

Load profile	Dispatch method	Mean fuel consumption kl/ day
Summer peak load	Base case	5.15 ± 0.22
	Predictive control	5.14 ± 0.22
	Without PV	5.63
Winter average load	Base case	3.15 ± 0.12
	Predictive control	3.12 ± 0.12
	Without PV	3.41

4 DISCUSSION

The frequency and voltage deviations seem low in view of the restrictive grid connection rules introduced by some network operators. An explanation can be the coarse resolution of load time series (15 minutes), which does not include step changes. High resolution load time series will likely reveal higher deviations in voltage and frequency. The stability of controls is tested in terms of irradiance fluctuations and they allow for 30 % PV capacity penetration, if the minimum load is 50 %.

The profiles used for PV systems are modeled with the assumption of similar dispersed systems, therefore ramp rates and maximum hosting capacity cannot be transferred to utility scale plants. However ramp rates found for these systems are low compared with ramps found in network stability tests (50 % power reduction in 0.01 s, see Tab. I).

The high number of start counts can be attributed to the thresholds used in the dispatch controller. For operation with high loading and therefore high efficiency, generators are switched off, when the remaining generators will be just under their maximum allowed loading. Hence, the demand can only vary in a low range before another switching operation is triggered. Under these circumstances generators can be switched off by predictive control, if irradiance conditions are stable for the forecast time horizon. If long minimum run times for generators are introduced, benefits of PC could shrink. Therefore generators that allow a high number of starts and stops are preferred, when using predictive control.

Looking at the prevalent irradiance conditions in BOM data from the Geraldton Airport, 42 % of days are clear, 27 % of days are mixed and intermittent, and only 4 % of days had overcast irradiance conditions. Preferably, predictive control shall enable operation with reduced number of generators in stable irradiance conditions (clear, overcast) and reliably provide enough spinning reserve for intermittent and mixed conditions.

As the SR requirement for load is often higher than for PV, the benefits of predictive control are rather low for the simulated PV penetration of 30 % of peak load but are expected to grow with higher PV penetration, which requires more spinning reserve in base case.

5 CONCLUSION & OUTLOOK

Based on a generic grid model, the presented simulations involve integration of distributed PV up to 30 % of peak load with 8 % fuel savings. The implementation of predictive control led to reduced spinning reserve requirements and a reduction in fuel consumption by up to 0.7 %, depending on the scenario. Fuel savings are expected to increase with even higher PV penetration. The analysis of maximum PV penetration for this generator set-up shows possible hosting capacities up to 60 % for averaged load profiles, by accounting for the time dependency of load and PV power. Short-time energy storage is needed to reduce voltage and frequency fluctuations and other adverse effects on the distribution system e.g. voltage rise in feeders, due to PV power injection.

An assessment of including short-term irradiance forecasts in the smart operation of such energy storage systems appears to be the next step. Prospectively, energy storage capacity and rated power can be optimized. The resulting benefits are estimated to be greater than for the PV and diesel only case.

6 REFERENCES

[1] APVA/ CEEM. Carnarvon: A Case Study of Increasing Levels of PV Penetration in an Isolated Electricity Supply System. A report by the UNSW Centre for Energy and Environmental Markets for the Australian PV Association, 2012.

[2] APVA/ CEEM. Alice Springs: A Case Study of Increasing Levels of PV Penetration in an Electricity Supply System. A report by the UNSW Centre for Energy and Environmental Markets for the Australian PV Association, 2011.

[3] D. Peters. Master's thesis, University of Oldenburg, Germany, 2016.

[4] T. Schmidt, J. Kalisch, E. Lorenz, and D. Heinemann. Evaluating the spatio-temporal performance of sky imager based solar irradiance analysis and forecasts. Atmos. Chem. Phys., 16(5):3399-3412, 2016.

[5] M. Reno, C. Hansen, and J. Stein. Global horizontal irradiance clear sky models: Implementation and analysis. Technical report, Sandia, 2012.

[6] Australian Bureau of Meteorology. One minute solar data. http://reg.bom.gov.au/climate/reg/oneminsolar.

[7] Diesel Service & Supply. Approximate diesel fuel consumption chart. http://www.dieselservice andsupply.com/ Diesel_Fuel_Consumption.aspx, accessed 09.05.2016., n.d.

INDUSTRIAL HYBRID SYSTEMS WITH HIGH PV PENETRATION – PERFORMANCE ANALYSIS AND KEY SUCCESS FACTORS

Prof. Dr.-Ing. Antonio Notholt, Mohamed Mostafa, Volker Wachenfeld
Reutlingen University
Alteburgstraße 150, 72762 Reutlingen, Germany, +49 7121 271-7031, Antonio.Notholt@reutlingen-university.de
SMA Solar Technology AG
Sonnenallee 1, 34266 Niestetal, Germany, +49 561 9522-421056, Volker.Wachenfeld@sma.de

ABSTRACT: Since the first industrial-scale hybrid system was installed by SMA in 2012, information about the performance of several hybrid systems around the world has been monitored. This paper analyses the performance of SMA's largest PV-Diesel hybrid system in the industrial-scale installed in Bolivia in 2014 and summarizes the lessons learned by managing this system with large-scale energy storage. The paper finally concludes with an outlook for future hybrid systems.

Keywords: Hybrid, Battery Storage and Control, Plant Control

1 INTRODUCTION

In December 2014, SMA together with Isotron and Guaracachi-ENDE commissioned the first PV-Diesel-Hybrid System in the Megawatt range with large-scale storage at a city with 60,000 inhabitants in Bolivia. This unique system uses power-optimized energy storage to provide ancillary services as well as help in the stabilization of the system, whilst allowing for a high PV penetration. Since it commissioning, the system has been operating continuously, providing cost effective solar power to the inhabitants of the city of Cobija [1].

The system is equipped with 5 MWp of PV modules, feeding into a local diesel grid via 6 central inverters, each with a rated power of 880 kVA. The Diesel grid is set up by 9 CAT Diesel gensets with power in the range of 1000kVA to 1600kVA each. The genset system is governed by a state-of-the-art controller. The battery system is Li-Ion based and has a nominal power of 2 500 kVA with a nominal capacity of 1 MWh

2 SYSTEM CONTROLLER

To guarantee the overall system stability, a system controller developed by SMA,"the Fuel Save Controller", was installed as part of the PV-Diesel hybrid system. The system controller analyses the current operation parameters of the hybrid system, which are load behavior, available diesel power, current genset operating status and available MPP PV power.

The load is computed from high speed genset measurements and dedicated measurement units at the PV and storage plant. The genset parameters such as state and availability are directly provided by the genset controller via a Modbus interface. Out of these parameters, the controller unit calculates a maximum limit for PV injection and a dispatch setpoint for the storage system, taking into account:

- The current genset status and the available spinning reserve of the rotating generators.
- The power delivery capability of the battery system
- The minimum load demand of the Diesel gensets needed to operate in an efficient mode
- The minimum runtime of the Diesel gensets
- The maximum expected loss of PV generation in the upcoming time period
- The currently available PV resource provided by a weather station

With this set of information, the PV generation can be limited to a safe maximum value whenever irradiation is high or load demand is not sufficient. During operating hours with significant load demand and PV generation, a curtailment of the PV will typically not be necessary.

The controller also takes over the dispatch functions of the battery system. Although the battery inverter itself reacts to frequency events through an autonomous P(f) characteristic curve, the controller gives setpoints to adjust for PV losses due to clouds or loss of other generation facilities through a smooth transition with a maximum ramp rate control.

In addition, the controller is taking care of the reverse power protection of the gensets. If there is a sudden loss of dominating loads, the system controller automatically reduces the PV generation and thus prevents the integrated genset protection mechanisms from tripping. This solution is marketed by SMA as the Fuel Save Solution however, the main responsibilities of the system controller remain the same across solution providers.

3 PERFORMANCE INDICATORS AND RESULTS

This section presents the Key Performance Indicators (KPI) of the system. Additional to the traditional KPIs used in PV system, a set of KPIs that can be useful for evaluating other operational issues have been calculated and will be discussed.

3.1 Results

Table I shows the reporting sheet for the studied hybrid system performance.

Although the performance ratio is lower than expected for grid connected systems, it is well within the expected range of typical hybrid systems. The performance ratio can be improved if the availability of the system can be increased. The production factor, which does not take into account the downtime of inverters nor the time in which these operate in curtailed mode shows a very good performance of the system.

Table I: Traditional Key Performance Indicators of the System studied

KPI	Actual	Remarks
Irradiation	1700 kWh/m²·a	
Performance ratio [2]	0.6652	0.6-0.85
Production Factor	0.8028	0.8-0.9
Availability (daytime)	93.5 %	> 98 %
PV Energy produced	5 655 MWh	
Load energy consumption	45 580 MWh	
Solar Energy Yield	1 131 kWh/kWp	
Reporting period from 10th of June 2015 to 9th of June 2016.		

Table II shows some other performance indicators which give valuable information about potential O&M improvements. The most relevant of these indicators is the amount of time in which the inverters are running on MPP mode (not being curtailed), with 88.5% of the time being a very good figure. Other important indicator is the average reserve power during the day. A typical system without PV would try to have a reserve power of around 10%-20% of the load. This value is not being changed by the inclusion of the PV+Battery system and is a positive effect of having the energy storage provide the necessary reserve power.

Table II: Extended Key Performance Indicators

KPI	Actual	Remarks
Estimated daytime SF*	25.6 %	
Estimated Fuel Savings**	1 516 m³	
Storage energy throughput	190.6 MWh	(0.53 C/day)
Unrestricted operation	88.5 %	of time
Reserve power factor (day)	82%	(18% of load)
Genset average loading (day)	59.8%	
Genset average loading (night)	63.3%	

*Daytime is assumed to have an irradiation > 100W/m²
**Estimated fuel savings are based on benchmarked best conversion efficiencies of the machinery used (0.268 L/kWh)

The average genset loading remains little affected (compare average loading day and night) by the PV injecting large amounts of power. This has again to do with the fact that almost no additional reserve power is required to compensate for large gradients in PV production, since these are taken care by the battery.

Even with the services provided by the battery it can be seen that the battery system is operating well within their limits (manufacturers specify average energy throughputs of less than one full cycle a day).

Since the main improvement of this system, against state-of-the art plant controllers is the dispatching of the battery system, the operation of the energy storage is described in further detail.

The battery system covers two important functions: The first one firming of renewables in order to guarantee a ramp rate in the PV production which is compatible with the capabilities of the generator system. This can be seen in Figure 1. This figure shows the comparison of the PV-production gradients throughout the year with and without battery. It is possible to see the dramatic decrease of the probability of the system seeing a large rate of change (from 0.4% to 0.06%) from the complementary system (PV+Battery). While rate of changes of 25%-50% of the nominal power (1250-2500 kW/min) would be typical for this system (~35h/year), the battery system almost smooths them out completely.

Figure 2 shows a detail of the positive effect of the use of energy storage in the rate of change of power of the PV system. This figure also depicts a typical ramp of PV in the studied system and the resulting loading on the generator with battery support.

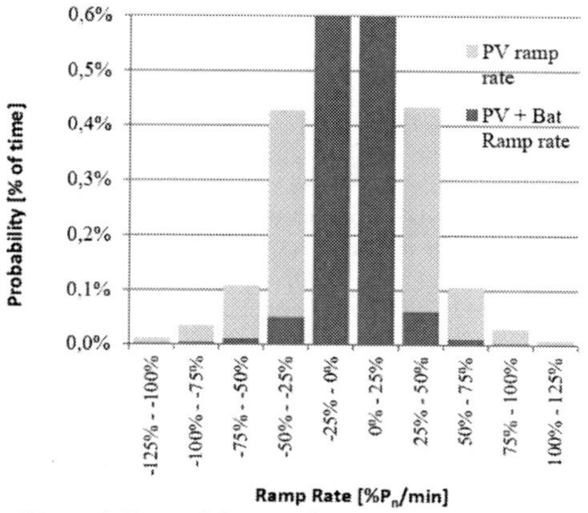

Figure 1: Rate of change of PV power output in the system

3.2 Discussion

Although the performance ratio is quite useful to evaluate the performance of grid connected systems, it is not able to describe completely the operation of a hybrid system; moreover, it is unsuitable for describing the target of a fuel saving application.

Figure 2: Example of Ramp rate control (14-06-2015)

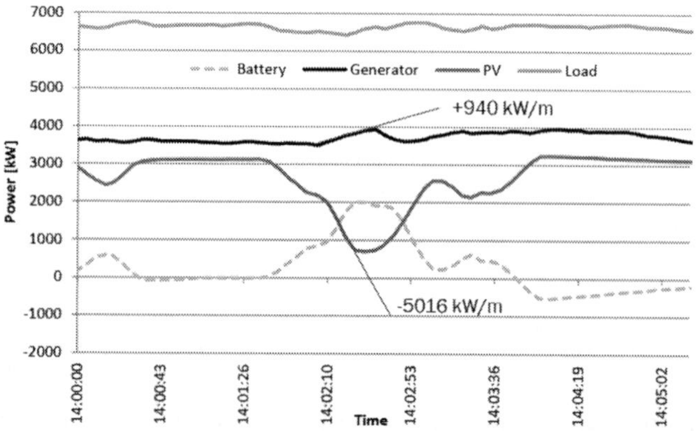

Literature such as [3] proposes simple methods in which the performance of a hybrid system may be assessed. Of special attention is the KPI called Hybrid

System Fuel Consumption (HSFC), which is in fact the ratio between the energy consumed and the fuel consumed. Higher energy throughputs per liter fuel could describe the "efficiency" of the combiner PV/Diesel system. This paper also uses other modified KPI in order to evaluate different operational aspects of the hybrid system. The Production Factor for instance, evaluates the design of the PV system as such as it takes into account only the part of the time where PV power can completely be injected to the grid by the inverter.

4 LESSONS LEARNED AND OUTLOOK

Since 2012, SMA has commissioned more than 70 "Fuel Save Systems" throughout the world in various industries and with different configurations. The successful fuel savings and the accuracy of the design tools have been proven and validated.

It has also been verified that the accuracy of the assumptions regarding load profile and energy consumption considerably impact the final performance and consequently the profitability of the system. The efforts being made by VDE to standardize methods for the load profiling of hybrid systems [4] can be seen as a response to this very important factor in the design and development of these projects.

Regarding monitoring and operation, a remote connection to the system has proven helpful and worth investment since the service fleet is able to solve a majority of the system requests for service remotely, thus saving time and money.

The system's interaction with the operator shall be improved. The operators shall be able to understand why the system is reacting and shall allow the crew to find ways of optimizing the system (for example through rescheduling of loads, etc.) The added value of the PV should be visible. In this project, the solar resource (calculated MPP power) was also added in order to give the operator the possibility to decide if some gensets may be stopped.

Typical performance indicators for PV systems may not entirely represent the actual performance of the system and/or help in the analysis of long term faults. Indicators such as the amount of energy produced by the system per unit fuel can describe a system more accurately.

PV-Diesel hybrid systems, with storage are able to operate more efficiently, as the required reserve power is delivered from the battery system thus allowing for less generator sets running and thus increasing their average load.

4.1 Outlook on new projects

SMA and its subsidiary SMA Sunbelt Energy GmbH have commissioned a PV Hybrid System on the Caribbean island of St. Eustatius, Caribbean Netherlands, to completely electrify the island with utility scale PV, battery storage and diesel generators. Up until March 2016 the island's power supply was fully based on diesel gensets. By integrating 1,9 MWp of PV power and 1 MW of batteries with the SMA Fuel Save Controller 2.0, the local utility company is now able to reduce its fossil fuel consumption significantly by up to 30 percent and save more than 800.000 liters of diesel fuel and 2,200 tons of CO_2 per year.[5]

6 CONCLUSIONS

This article has described the typical characteristics of a system controller used in PV-Diesel-Hybrid-Systems with Fuel Save Applications. Based on the first large-scale reference of the company, the performance of such a system was compared and verified with design assumptions. It also underlined the importance of the use of more adequate KPIs for assessing the performance of such systems. The energy throughput per unit fuel (kWh/L) described in the literature was proposed.

The 70 currently operating fuel save systems, report successfully fuel savings in their operation. From these a set of recommendations have been derived.

The next-generation hybrid systems for fuel-save applications are characterized by the integration of high power, low energy storage as well as a tighter integration between the system controller and the genset's power management. Additionally, energy management services like shifting of PV energy through the battery are expected to furtherly improve the performance of this type of systems.

7 REFERENCES

[1] SMA Solar Technology, Press Release on the World's largest PV-Diesel Power Plant, August 14th, 2014.
[2] B. Marion, Performance Parameters for Grid-Connected PV Systems, 31st IEEE Photovoltaics Specialists Conference and Exhibition, 2005.
[3] A. Swingler, A pragmatic performance reporting approach for describing PV Hybrid systems within Minigrids, Proceedings of the 5th PV-Hybrid and Mini-Grid Conference, Tarragona, 2010.
[4] Punzalan, Keith. The VDE Test specification: An overview. Presentation. Event: Bridgring Hybrid Power Projects and finance, June 2016.
[5] SMA Solar Technology AG, Press Release on the Sustainable power supply for carribean island St. Eustatius, 2016-05-24.

Energy Forecast for mobile Photovoltaic Systems with Focus on Trucks for cooling Applications

M. Kühnel, B. Hanke, S.Geißendörfer, K.v.Maydell, C.Agert
NEXT ENERGY - EWE Research Centre for Energy Technology at the University of Oldenburg
Carl-von-Ossietzky-Str. 15, 26129 Oldenburg, Germany

For a viable and commercial attractive integration of vehicle integrated photovoltaic applications (ViPV) energy forecasting is required as a foundation for business case calculation. The developed algorithm facilitates the forecast of annual power distribution and solar energy yield along any given track tested on a German use case. A high temporal and spatial resolution of the meteorological database providing ambient temperature, wind speed and global horizontal irradiance was determined as a necessity to preserve the irradiance distribution. For ViPV module temperature, respectively module efficiency benefits strongly from head wind. As a consequence, the performance of ViPV under motion was identified as superior to non-mobile pv installation. The potential annual energy yield of the 3 showcasing commercial semi-trailer trucks operating in Germany is identified to be 3-7MWh depending on the used cell technology. The potential energy gain due to head-wind cooling is estimated to 20kWh-75kWh per truck and year.
Keywords: refrigeration, modeling, ViPV, energy forecast

1. INTRODUCTION

The idea of PV supported refrigeration on the transport of perishable goods was first published by Bahaj [1], Bahaj and James [2]. At nearly the same time Sandia National Laboratories published a feasibility study on solar powered refrigeration for transport applications [3].

Almost 15 years later vehicle integrated pv (ViPV) remains a niche product. The International Energy Agency (IEA) stated 2012 a problematic situation within the logistic sector; an increasing demand on transported goods but only minor adaptation to renewable energies [4]. Instead of the aspired reduction of CO_2 emissions road logistics experienced an increase of 13% CO_2 emissions between 1995 and 2013 [5]. The stringent directives for the reduction of CO_2 emissions and the use of fossil fuel in total, as well as political instabilities in oil and gas producing countries brought ViPV back on screen. Green logistics [6] is nowadays a growing sector on the way to serve this increasing demand of a growing sustainable acting community and ViPV or pv supported cooling applications in transportation are again in the focus [7].

For a feasible business plan and the calculation of the systems payback time a profound knowledge of power distribution and potential energy yield is necessary. The aim of this study is to close this gap and develop a suitable algorithm for the potential energy yield along any given track of roof mounted and back insulated pv system for mobile applications.

2. MOTIVATION

Modern pv system software like INSEL[1] and PVSyst[2] offers a wide range of functionality but lack capabilities to determine the yield of a moving pv installation. Especially the reduction of the cell temperature caused by additional head-wind and its influence on the module efficiency needs to be elaborated. Further effects are a potential additional heat load on the vehicle caused by the increased absorption of pv module material compared to conventional rooftops. The aim of this study is to close the gap and develop a suitable algorithm for a mobile energy forecast of a roof mounted and back insulated pv system for cooling applications.

Prior work on the topic from Kronthaler et al. [8, 9] emphasizes the benefit of the head wind on solar cell performance of ViPV. Limitations of the Kronthaler model are the assumed constant head-wind for three different scenarios (highway-scenario: 25m/s) and the stationary meteorological weather profile. Whereas the implemented algorithm of this study operates with real commercial lorry track profiles and high resolution weather data along any given track.

3. METHODOLOGY

The energy yield for a given track is calculated in a two step-simulation process. An overview of the simulation scheme is shown in Fig. 1. The first block deals with the generation of a meteorological dataset along a given track including wind speed (FF), ambient temperature (T_a) and global horizontal irradiance (G). Within the second block the potential production of two module types is calculated, as well as the additional heat load transferred into the lorries and the benefit of a reduced cell temperature caused by the head wind.

[1] INSEL: www.insel.eu
[2] PVSyst: www.pvsyst.com

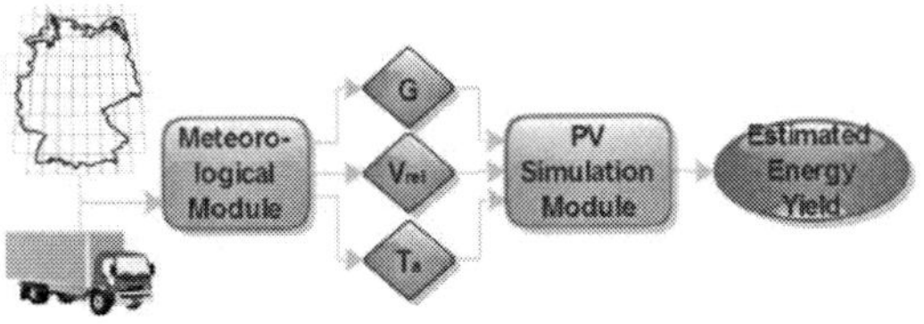

Figure 1: Scheme of the two stepped energy forecast for mobile applications

3.1. Generation of the meteorological profile

A climatological data grid with 5km spatial and 1min temporal resolution covering the area of Germany is generated. Every grid knot contains climatological data of a typical meteorological year generated by meteonorm[3] including ambient temperature, wind speed and global horizontal irradiance.

The data is organized in 1° x 1° containers based on WGS84 reference ellipsoid in the HDF5[4] data format to provide a fast and efficient access. Each geographical track position given in latitude, longitude and minute of year is matched within the meteorological grid. Two different matching methods are stressed against each other, the nearest neighbor method and a three point barycentric interpolation [9].

As module cooling caused by wind has a significant influence on the pv module temperature [10] a special focus is spent on the benefit of the combination of ambient wind with vehicle motion induced head wind.

3.2. Cell Temperature

The temperature evolution of PV cells, insulated at the back, and additional heat load on the roof of a lorry caused by this cells is investigated with a temperature experiment using the solar climatic test board SC 600 (CC)[5]. Within this CC the ambient temperature and humidity can be controlled. In this conditioned environment a white metal plate as reference, a commercial thin film module and a solar thermal acceptance plate where exposed to an G=1010±29.2W/m 2 by a metal halide lamp producing a dense multi linear spectrum similar to the suns spectrum with different ambient temperatures. T_a has been set between 0°C and 50°C with a step size of 5K. T_a and T_m are measured with thermocouples and recorded with myPCLab data acquisition tool. The u-value of Styrofoam with a thickness of 5cm is given in literature as $u_{box} \approx 0.85 W/m^2 K$.

3.3. PV simulation and heat load

The pv-simulation is based on the open source python library PVLIB developed by *Sandia National Laboratories*[6] and is exemplary computed for two different module types, a flexible thin-film module based on the CIGS (copper, indium, gallium and selenite) technology and a mono-crystalline silicon module (c-Si) for building integrated pv (BiPV).

[3] www.meteonorm.com/en
[4] www.hdfgroup.org/HDF5/
[5] ATLA Material and testing solutions
[6] http://www.sandia.gov

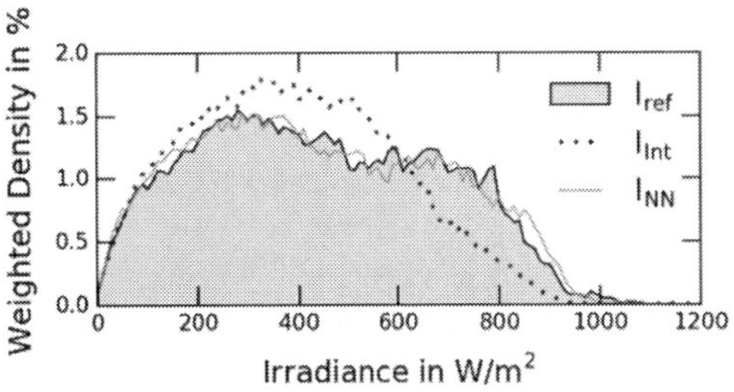

Figure 2: Annual irradiance distribution at a station (ref) and along a track calculated with two different matching methods, nearest neighbor (NN) and barycentric interpolated (int)

For the calculation of cell (T_c) and module temperature (T_m) meteorological parameters (irradiance, ambient temperature, wind) and mounting and technology depending empirical parameters (a, b, ΔT) are needed:

$$T_m = G \cdot e^{a+b \cdot FF} \cdot T_a \qquad (1)$$

$$T_c = T_m + \frac{G}{G_{ref}} \cdot \Delta T \qquad (2)$$

The measured data of the white reference plate is fitted to Eq.1 to determine the normal temperature evolution on a semi-trailer roof without a PV installation under varying irradiance, airstream and temperature conditions.

For the power calculation the 5-parameter model of de Soto et al. [11] is used. The five input parameters (ideality factor, diode reverse saturation current, photocurrent, series and shunt resistance at STC[7]) are mostly provided by the manufacturer's datasheet.

4. RESULTS

4.1 Matching methods

The effect of spatial interpolation on the annual distribution of irradiance is investigated on the basis of a gps-track profile of a lorry working in a single urban region. Figure 2 displays the comparison of the irradiance distribution of a stationary profile with the two simulated profiles. Spatial interpolation causes through averaging a distinct underestimation of irradiance values greater 600W/m 2 and a corresponding overestimation of medium values. The nearest neighbor methods results in a distinct higher conformity with the stationary distribution of the same area. Such an underestimation of high irradiation occurrences will have an serious impact on the calculated power and energy yield and should be avoided as it might cause a false inverter dimensioning. The computed curve of a thin film module (dotted line) based on the empirical temperature equation by pvlib (cf. Eq. 2) displays a good match with the measured values of the swiss SiS Tech module. The solar thermal acceptance plate on the other hand reveals the same behavior as the roof mounted crystalline (dashed line) simulated with pvlib.

[7] Standard test conditions: G=1000W/m², T_c =25°C, AM1.5g

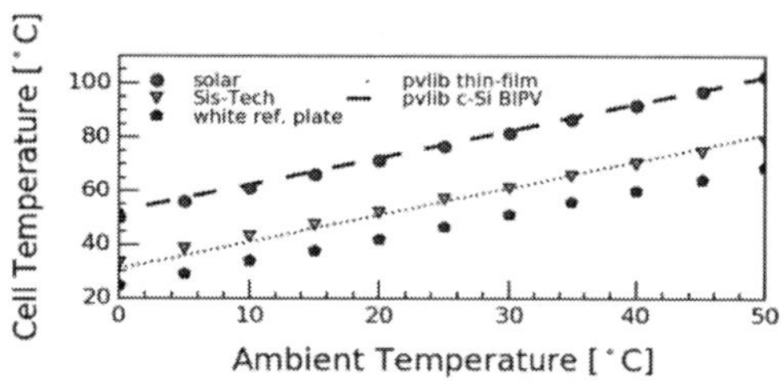

Figure 3: Temperature evolution under an irradiance of 1010W/m² at the backside of a white metal plate, a commercial thin film module and a solar thermal plate. The numerical calculation for a building integrated pv (BiPV) is shown as a dashed line and for thin-film module as a dotted Line.

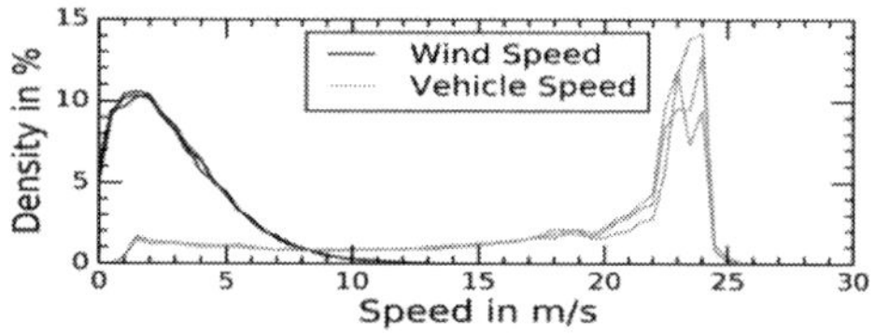

Figure 4: Annual distribution of ambient wind and vehicle speed of 3 lorries.

4.2 Module Temperature

The measured temperature evolutions of the three tested materials are displayed in Fig. 3. All three plates exhibit a linear behavior as a function of the ambient temperature within the investigated temperature range. An insolation of 1010W/m² causes for the solar thermal plate and the building integrated temperature model of pvlib a temperature increase of over 50°C compared to the ambient temperature.

4.3. Head wind chill

A comparison of the annual ambient wind speed and vehicle speed, respectively the head-wind distribution of three far-distance lorries is shown in Fig. 4. The bulk of ambient wind speed is below 5m/s, whereas the most frequent head-wind is 22m/s. This is only valid for far-distance transport, the annual head-wind profile of urban operating lorries is equally distributed over the whole speed range. The effect of varied airstream on the cell temperature and resulting efficiency (η) is display in Fig. 5. For an ambient temperature of T_a = 5°C and an irradiance of G=1000W/m 2 the module efficiency of the investigated thin film module improves from 10%-10.5% (considering ambient wind only) to 11.1%-11.25% and for the mono crystalline pv-panel from 19.5%-20.2% to 21.7%-22.4%.

During drive times, merely 20% of the year, the improvement increases to more than 2.5% for CIGS installation and nearly 4% for a c-Si installation. Different cell temperatures at the same solar irradiance and different material surfaces can explain the difference in performance enhancements.

4.4. Additional heat load

The additional heat-load on the cooling system of a semi-trailer is strongly depending on the insulation level, the covered area and used cell technology. Figure 6 displays the additional heat load during one year depending on the insulation level u_1 = 0.4, u_2 = 0.144 and u_3 = 0.072W/m²K. The energy necessary for the compensation of the additional heat-load is calculated with an coefficient of performance of 0.6 for the cooling system. The higher heat impact of a crystalline installation on a bad insulated truck can counter its higher power output. The additional heat load is ΔE_{CIGS} = 0.8 % - 4.6 % and $\Delta E_{c\text{-}Si}$ = 1.6 % - 9 %.

4.5 Net Benefit

The energy content of one liter diesel equates 9.86kWh for the higher heating value. With a combustion efficiency of around 35% the estimated diesel savings for the CIGS technology are around 850l per year. With the c-Si module installation the diesel savings are approximately 2000l per year. The summed results of the crystalline and thin-film pv installation along the three annual far-distance lorry profiles are listed in Tab. 1. With the European standard insulation (0.144W/m²K) the annual CO_2 saving per lorry is 2.8t for the CIGS installation and more than 5.7t for the c-Si installation. The summed annual reduction of CO_2 emissions of the more than 50000 freezer trucks operating in Germany equipped with crystalline solar cells would be more than 320kt.

5. CONCLUSION

Remarkable is the positive effect of head wind on module temperature, forcing a gain in efficiency of $\Delta\eta_{CIGS} \approx$ 0.5-1% and $\Delta\eta_{c\text{-}Si} \approx$ 1.5-2.9%.

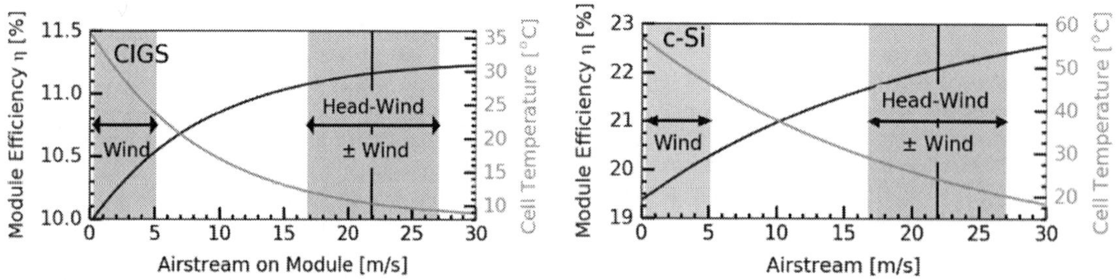

Figure 5: Evolution of cell temperature and efficiency of two different cell technologies as a function of the airstream (G=1000W/m², T_a =5° C), mean wind speed and the mean head wind plus/minus ambient wind are indicated in light grey.

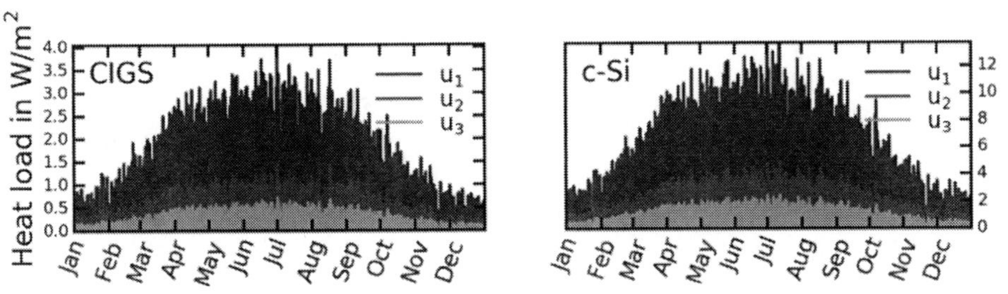

Figure 6: Daily maximum heat load transferred into the lorry by a CIGS and c-Si installation, depending on insulation level (u_1 =0.4W/m² K, u_2 =0.144W/m²K and u_3 =0.072W/m²K

Table I: Results and identified profits along three annual far-distance truck tracks (A-C) for c-Si and CIGS cell technology. The calculations are based on an insulation of u_2= 0.144W/m²K.

	A	B	C	
E_{sun} [MWh/m²a]	1	1	1	
$E_{density}$ [kWh/m²a]	107	107	107	CIGS
$E_{heat\ load}$ [kWh/a]	31	30	30	CIGS
E_{net} [MWh/a]	3	3	3	CIGS
Diesel [1] (μ=35%)	871	865	869	CIGS
CO_2 [t]	2.9	2.9	2.9	CIGS
$E_{density}$ [kWh/m²a]	210	209	210	c-Si
$E_{heat\ load}$ [kWh/a]	136	134	135	c-Si
E_{net} [MWh/a]	6.79	6.75	6.78	c-Si
Diesel [1] (μ=35%)	1969	1957	1965	c-Si
CO_2 [t]	6.4	6.3	6.4	c-Si

The increased heat transfer into the lorry lowers significantly the net energy yield. This emphasizes the necessity of a good thermal insulation of the lorry. The potential annual electrical energy yield simulation results in 7MWh per truck with an energy density of around 200kWh/m²a for c-Si installation and 3MWh with a density of 100kWh/m²a for the CIGS installation. With an annual potential CO_2 saving of around 3-6t per truck.

However, up to now there is no software package published for the accurate calculation of the energy yield potential along a track. Recent studies on the energy gain for mobile pv applications included neither non-static weather time-series nor operational drive profiles.

6. ACKNOWLEDGEMENT

I would like to thank *Uhlhorn Logistics* for supporting this work with telemetric data of their lorries and the working group of energy meteorology of the University of Oldenburg, PhD. Th. Schmidt in particular for providing reference irradiance data.

REFERENCES

[1] A.S. Bahaj. Photovoltaic power for refrigeration of transported perishable goods. In Conference Record of the Twenty-Eighth IEEE, volume 28, pages 1563–1566. IEEE, IEEE, Sep 2000. doi: 10.1109/PVSC.2000.916195.

[2] A.S. Bahaj and P.A.B. James. Economics of solar powered refrigeration transport applications. In IEEE, editor, Conference Record of the Twenty-Ninth IEEE, volume 29, pages 1561– 1564. IEEE, IEEE, 2002. doi: 10.1109/PVSC. 2002.1190911.

[3] David Bergeron. Solar powered refrigeration for transport application – a feasiblility study. Sandia report, Sandia National Laboratories, Dec 2001. URL http://prod.sandia.gov/techlib/access-control.cgi/2001/013753.pdf.

[4] IEA. Energy technology perspectives 2012, pathway to a clean energy system, 2012. URL https://www.iea.org/publications/freepublications/publication/ETP2012_free.pdf

[5] BMUB. The german governments climate action program 2020. Cabinet Decision, December 2014. URL http://www.bmub.bund.de/en/topics/climate-Energy/climate/national-climate-policy/Climate-action-programme/.

[6] Fraunhofer-Institut fuer Materialfluss und logistik IML. Green logistic. http://www.iml.fraunhofer.de. URL http://www.iml.fraunhofer.de/de/themengebiete/umwelt_ressourcenlogistik/umwelt_ressourcen/Green_Logistics.html.

[7] Guido Bröer. Solar-Kühltransporter entwickelt. Solarthemen, June 2015. URL http://www.solarthemen.de/index.php/2015/06/18/solar-kuehltransporter-entwickelt/.

[8] L. Kronthaler, L. Alberti, and D. Moser. The Effect of Head-Wind Cooling on the Electrical and Thermal Behaviour of Vehicle Integrated Photovoltaic (VIPV) System. In 29th European Photovoltaic Solar Energy Conference and Exhibition, pages 3541 – 3544, 2014. doi: 10.4229/EUPVSEC20142014-6BO.9.6.

[9] L. Kronthaler, L. Maturi, D. Moser, and L. Alberti. Vehicle-integrated Photovoltaic (ViPV) Systems: Energy Production, Diesel Equivalent, Payback Time; an Assessment Screening for Trucks and Busses. In Ninth International Conference on Ecological Vehicles and Renewable Energies (EVER), 2014. doi: 978-1-4799-3787-5/14/$31.00l'2014IEEE.

[10] C. Schwingshackl, M. Petitta, J.E. Wagner, G. Belluardo, D. Moser, M. Castelli, M. Zebisch, and A. Tetzlaff. Wind effect on pv module temperature: Analysis of different techniques for an accurate estimation. Energy Procedia, 40:77–86, 2013. doi: 10.1016/j.egypro.2013.08.010.

[11] W De Soto, S. A. Klein, and W.A. Beckman. Improvement and validation of a model for photovoltaic array performance. Solar Energy, 80: 78–88, 2006. doi: 10.1016/j.solener.2005.06.010.

Integration of reverse osmosis seawater desalination in the power sector, based on PV and wind energy, for the Kingdom of Saudi Arabia

Upeksha Caldera, Dmitrii Bogdanov, Svetlana Afanasyeva, Christian Breyer
Lappeenranta University of Technology, Skinnarilankatu 34, 53850 Lappeenranta, Finland,
E-mail: upeksha.caldera@lut.fi, christian.breyer@lut.fi

Abstract

Saudi Arabia is in the midst of redefining the vision for the country's future and creating an economy that is not dependent on fossil fuels. This work presents a pathway for Saudi Arabia to transition from the 2015 power structure to a 100% renewable energy based system by 2050 and analyse the benefits of integrating the power sector with the growing desalination sector.

It is found that Saudi Arabia can transition to a 100% renewable energy power system by 2040 whilst meeting the growing water demand through seawater reverse osmosis (SWRO) desalination plants. The dominating renewable energy sources are PV single-axis tracking and wind power plants with 210 GW and 133 GW, respectively. The levelised cost of electricity (LCOE) of the 2040 system is 48 €/MWh. By 2050, PV single-axis tracking dominates the power sector due to the further reduction in the capital costs alongside cost reductions in supporting battery technology. This results in 80% share of solar PV in the total electricity generation. Battery storage is required to meet the total electricity demand and by 2050, accounts for 48% of the total electricity demand. The LCOE is estimated at 38 €/MWh, required capacity of PV single-axis tracking is 369 GW and wind power plants 75 GW. In the integrated scenario, due to flexibility provided by the SWRO plants, there is a reduced demand for battery storage and power-to-gas (PtG) plants. In addition, the ratio of the energy curtailed to the total energy generated is lower in all time periods from 2020 to 2050, in the integrated scenario. As a result, the annual levelised costs of the integrated scenario is found to be 2% - 4% less than the non-integrated scenario.

Keywords: *100% renewable energy, Saudi Arabia, energy transition, desalination, solar PV, wind energy, sector integration*

1 Introduction

The Kingdom of Saudi Arabia (KSA) is the largest country on the Arabian Peninsula and is the 13th largest in the world. As of 2015, Saudi Arabia had a population of approximately 31,540,000 – the highest amongst the Gulf Cooperation Council (GCC) countries [1,2]. Between 2011 - 2013, the electricity energy consumption per capita was estimated at 8.41 MWh/capita, the 5th largest in the GCC region [3]. At the current annual increase of 6%, electricity demand is expected to exceed 630 TWh equal to a peak capacity of 120 GW by 2032 [4, 5].

Saudi Arabia has 1/5th of the world's proven oil reserves and is the largest producer and exporter of petroleum products [6]. In addition KSA is the 9th largest exporter of natural gas [6]. Saudi Arabia's economy is based on oil revenues and during the time period 2010 – 2013, oil revenues contributed up to 91% of the country's national income [7]. This is reflected in the recent low oil prices and the resulting slower growth of the Saudi economy. Al Bassam [7] explains that although the government has had plans to diversify the economy and decouple economic growth from oil, there has in fact been an increase in dependence on oil revenues over the last few years. In addition to the low oil prices, growing concern is the share of local fossil fuel reserves, driven by increasing income and population, being used to meet the country's energy demand. According to the King Abdullah City for Atomic and Renewable Energy (KACARE) [8], the country's demand for fossil fuels is expected to grow from 3.8 million barrels of oil equivalent in 2010 to 8.3 million barrels of oil equivalent in 2028. Moser et al. [9] suggests that at the current rate of consumption, Saudi Arabia could become an oil importer by 2030. By 2015, Saudi Arabia's power plants capacities were almost equally separated into oil based thermal power plants and gas based power plants [10].

Economic reforms have been discussed in the past but it is only recently that reforms have been discussed with haste. A recent report in the Guardian [11] suggests that swift actions were triggered by the 100 billion € deficit the Saudi government incurred in 2015. Consequently, in April 2016, King Salman bin Abdulaziz presented details of the Saudi vision 2030 plan and heralded a Saudi future without oil [11]. A key component of the new vision is the growth of Saudi Arabia's renewable energy sector. It is planned to have 9.5 GW of renewable capacity by 2023, as an initial stage to meet the country's increasing energy demands [12]. In spite of the country's wealth of solar and wind resources, Saudi Arabia had only 83 MW of renewable power plants, all solar photovoltaic installations, by 2015 [10]. Figure 1 presents the total installed capacities, by the beginning of 2015, in Saudi Arabia and illustrates the almost complete reliance on oil and gas in the current power sector [10]. The country lies within the sunbelt region of the world and the average population weighted global horizontal irradiation is estimated to be 2158 kWh/(m^2·a) [13, 14]. In the adjacent country of United Arab Emirates (UAE), with similar solar resources, recent bids have been made for utility scale PV plants with an electricity production cost as low as 2.99 USD cents/kWh [15].

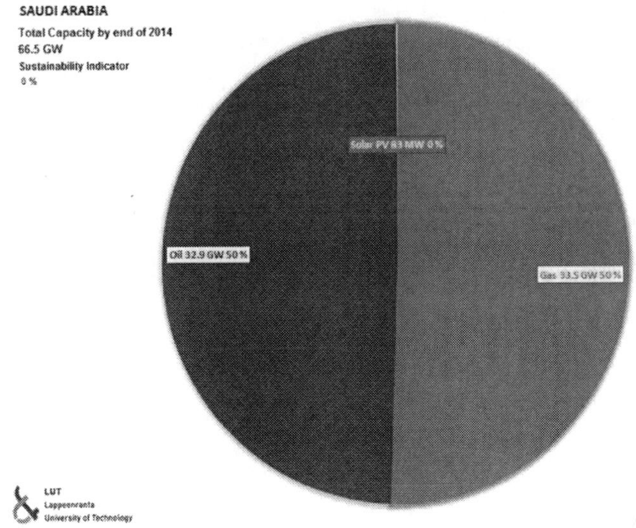

Figure 1 Total Installed Capacities by the beginning of 2015 in Saudi Arabia [10]

The Saudi Vision 2030 plan also outlines the need for the protection and better management of renewable water resources in the country [16]. Saudi Arabia is one of largest arid countries without any permanent rivers or lakes. Whilst the global average renewable water resource per capita per year is 6000 m^3, Saudi Arabia has only 84.8 m^3/(capita·a) [17]. However, in spite of the water scarcity, Saudi Arabia has the third highest water consumption per capita at 250 liters/(capita·d). This is only behind the United States and Canada. In 2010, 58% of the country's total water demand was met through non-renewable ground water resources, 33.5% by surface water and renewable ground water, 6% by desalinated water and 2.2% by waste water reuse [18]. The country's water demand is expected to increase by 56% by 2035. Meanwhile, at the current rate of water withdrawal, ground water aquifers are expected to provide potable water only for the next 10 – 30 years [19].

To augment the fresh water resources, Saudi Arabia relies on seawater desalination, particularly to meet the municipal and industrial water demands. By the end of 2015, Saudi Arabia accounted for 15% of the global installed desalination capacity [20]. With the diminishing of fresh water resources, seawater desalination is expected to play a pivotal role in meeting Saudi Arabia's future water demands. However, desalination is an energy intensive process compared to traditional water treatment methods. Saudi Arabia is reported to use 25% of the domestic oil and gas production in desalination plants and the share is expected to increase to 50% by 2030 [21]. By 2015, reverse osmosis (RO) desalination plants accounted for 49.6% of the total installed capacity and thermal desalination technologies accounted for 48%.

Table 1 illustrates the energy consumption of the most prevalent desalination technologies in Saudi Arabia [22]. In the past, Saudi Arabia has relied on thermal desalination technologies. However, due to the lower energy consumption and improvement in technology, reverse osmosis is expected to dominate the Saudi market in the future [18].

Table 1 Average energy consumption of prevalent seawater desalination technologies in the KSA [22]

	Electrical energy consumption (kWh_e/m^3)	Thermal energy consumption (kWh_{th}/m^3)
Seawater Reverse Osmosis (SWRO)	4.0	-
Multi Stage Flash (MSF)	2.5 -5.0	85
Multiple Effect Distillation (MED)	2.0 - 2.5	65

In Caldera et al. [23], it was shown that the global water demand of 2030 can be met by seawater reverse osmosis (SWRO) plants powered by 100% hybrid renewable energy power plants at a cost level competitive with that of fossil powered SWRO plants today. This system eliminates the reliance of SWRO desalination plants on non-renewable fossil fuels and concerns about greenhouse gas emissions. Meanwhile, reflecting the Saudi government's vision of high renewable energy capacities, there is recent literature that discusses the 100% renewable energy transition of different countries and regions [24- 31].

In this paper, motivated by the growing demand for seawater desalination in Saudi Arabia and transition towards 100% renewable energy power systems, we pose the following research question: What are the impacts on Saudi Arabia's electricity and water production costs when the seawater desalination demand is integrated into the country's 100% renewable energy (RE) based power sector? The energy transition pathway, for Saudi Arabia, from the current fossil based power system to a 100% RE based system is first found. Then, the seawater desalination demand is integrated into the power system, to understand the impacts on the cost of the power and water sectors. Thus, the results will illustrate an optimal transition path for Saudi Arabia to meet the country's future electricity and water demands through a 100% RE system.

2 Methodology

2.1 Overview

The key objective of our work is to understand the energy transition pathway to a 100% renewable energy power system for Saudi Arabia and the impacts of integrating the large seawater desalination demand into the energy system.

The approach that was taken to answer this research question can be summarized as follows:

1. The energy transition from the current (as of the beginning of 2015) fossil based power system in KSA to a 100% renewable energy based power system by 2050, in 5 year time steps, is found. After 2015, there are no new fossil powered thermal plants allowed. The existing fossil based power systems are phased out based on their lifetimes. The increase in total electricity demand and population over the years is accounted for. The optimal mix of renewable power systems to replace the phased out fossil power systems are found and the resulting system LCOE is found for every time step.
2. Seawater desalination demand in KSA from 2015 up to 2050 is determined and the corresponding desalination capacities integrated into the system. After 2015, only seawater reverse osmosis plants are allowed to be installed due to the dominance of the technology in the Saudi Arabian desalination market. Multi stage flash desalination capacities that were online up to 2015 are included and phased out based on the lifetimes.
3. The energy transition for KSA from the 2015 fossil based power system to the 100% renewable energy based power system by 2050 is found with seawater desalination integrated. Ultimately, the total cost of the power and water sectors are compared to determine the impacts of integrated seawater desalination in the energy system.

For the design and analysis of the energy transition, the energy model, used and further developed at LUT, is used. The energy model is based on the linear optimization method with interior point optimization and designed in an hourly temporal and 0.45° x 0.45° spatial resolution. The energy model allows for the design of local, national, regional or global energy systems. It is composed of all relevant power generation and storage technologies, respective installed capacities and different operation modes of these technologies. A key feature of the model is its flexibility and expandability besides the hourly resolution for a full real year.

The sections that follow present an overview of the technical and financial assumptions of the energy model, the power plants and seawater desalination plants. The data assumed for the time period from 2015 – 2050 is presented.

2.2 Model Overview

A key requirement of the LUT energy systems model is to match the power generation and demand for every hour of the applied year. The hourly modeling enables a more accurate system description including synergy effects of different system components for power system balance [28]. The aim of the system optimization is to minimize the total annual cost of the installed capacities of the different technologies, cost of energy generation and generation ramping. The full description of the model, its input data including RE resources and technical assumptions can be found in Bogdanov and Breyer [28] and Barbosa et al [32].

The three main components of the energy model can be listed as below:

- Technologies for the conversion of RE into electricity
- Energy storage technologies
- Energy sector bridging technologies to provide more flexibility to the complete energy system

Aghahosseini et al. [33] describes the various components of the LUT energy model and how the model determines the optimal 100% renewable energy mix for Iran for the year 2030.

For Saudi Arabia, the model described by Aghahosseini et al. [33] was modified with the addition of thermal desalination technologies as another desalination technology and a subsequent energy transition, i.e. the system setup found for one period had been adjusted by the capacities being beyond their technical lifetime as a starting value for incremental optimization for the new period. Figure 2 illustrates the modified energy model. For the energy transition, the model determines the optimal combination of the components that meets the electricity demand of every hour for the time period from 2015 to 2050, in 5 year time steps.

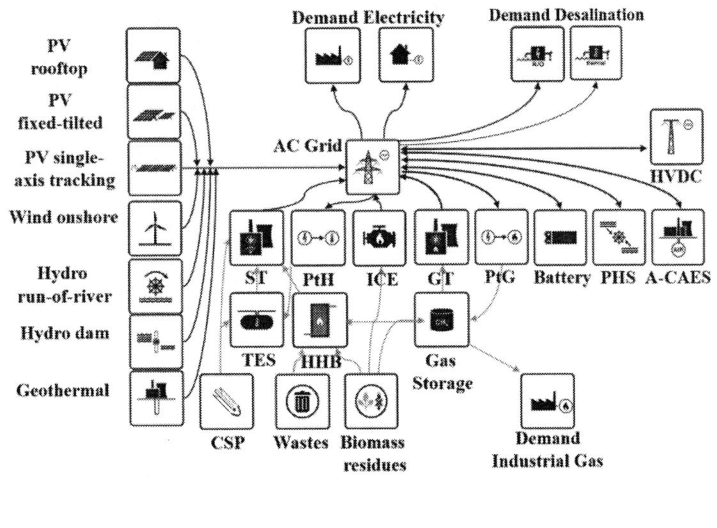

Figure 2 Block diagram of the LUT energy systems model used for Saudi Arabia

2.3 Power Plant Capacities - Technical and Financial Assumptions

The average rate of increase in electricity demand in KSA is 6% and as of 2015, for a total population of 32 540 000, the total energy consumption of the country was 285 TWh. To determine the future energy system, the forecasted total energy demand numbers are required. Al-Saleh et al. [34] presents the total energy demand in Saudi Arabia up to 2050, based on the current rate of increase.

Table 2 presents these electricity consumption assumptions. Figure 3 illustrates the hourly load profile for Saudi Arabia assumed for 2015. As expected, during the summer period of May to September, the highest loads occur. For the subsequent time periods, the load profile presented is varied depending on the total electricity consumption for the period.

Table 2 Variation in the electrical energy consumption of Saudi Arabia from 2015 up to 2050

	Total Electricity Consumption (TWh)
2015	285
2020	350
2025	420
2030	510
2035	580
2040	650
2045	740
2050	840

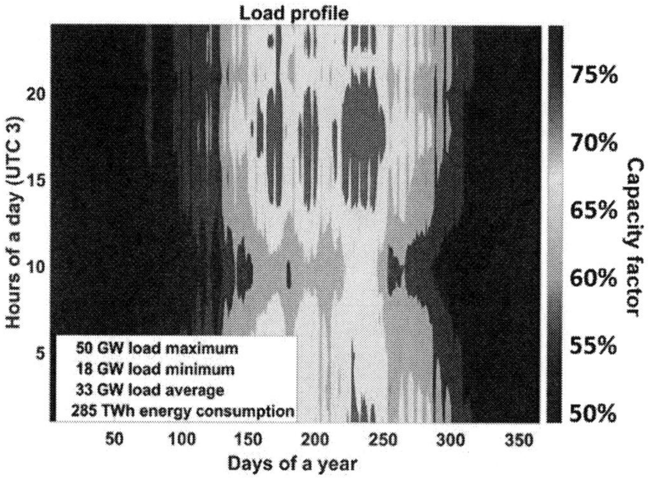

Figure 3 Aggregated load curve for the year 2015 for Saudi Arabia

The model takes the 2015 installed power plant capacities, presented in Figure 1, corresponding lifetimes, total electrical energy demand, and optimizes the mix of renewable energy plants needed to be installed to achieve a 100% renewable energy power system by 2050. The optimization for each time period is carried out on basis of assumed costs and technological status of the renewable energy technologies.

The capital expenditures (capex), operational expenditures (opex) and efficiency variations of all the power sector components, from 2015 up to 2050, assumed in the model is presented in the Appendix (Table 1). Figure 4 is a visual representation of the capex numbers. A learning curve approach based on several references, is used to identify the variation in capex for the different components. The references for the different components are presented in the Appendix (Table 1).

The capex and opex numbers generally refer to a kW of electrical power. For the case of water electrolysis the capex and opex numbers refer to a kW of hydrogen thermal combustion energy, and for CO_2 scrubbing, methanation and gas storage to a kW of methane thermal combustion energy. The financial assumptions for storage systems refer to a kWh of electricity, and gas storage refers to a thermal kWh of methane at the lower heating value. Weighted average cost of capital (WACC) is set to 7% for all years. The technical assumptions concerning power to energy ratios for storage technologies, and efficiency numbers for generation and storage technologies are provided in the Appendix (Table1 and Table 2).

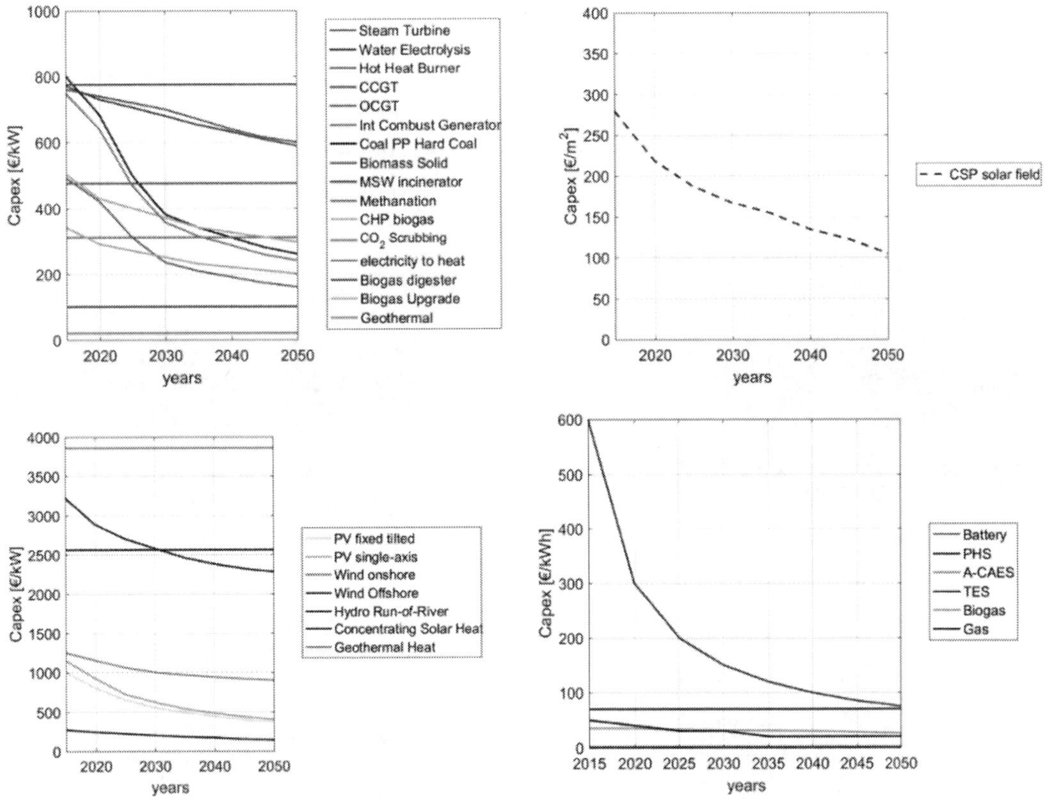

Figure 4 Top Left : Variation in capex of power system components assumed in the model
Top Right : Variation in capex of the CSP solar field assumed in the model
Bottom Left : Variation in the capex of the RE components assumed in the model
Bottom Right : Variation in the capex of the storage components assumed in the model

To determine the capacities of the different renewable energy power plant components, it is necessary to understand the local resources available. The approach discussed by Aghahosseini et al. [33] is used to identify the biomass and geothermal potential available in KSA.

Table 3 presents the biomass and geothermal potential results for KSA and it is assumed that these are the amounts available for every time period from 2015 to 2050.

Table 3 Estimation of biomass potentials and geothermal energy potential for Saudi Arabia from 2015 – 2050

Solid waste [TWh_{th}/a]	Solid biomass [TWh_{th}/a]	Biogas sources [TWh_{th}/a]	Geothermal Potential [TWh_{th}/a]
1.92	4.14	1.97	57.53

Feed-in full load hours (FLH) for Saudi Arabia is also computed, as described in Bogdanov and Breyer [28], on the basis of the 0.45° × 0.45° spatially resolved data. The average FLH y for CSP, PV optimally tilted, PV single-axis tracking and wind power plants in Saudi Arabia are presented in Table 4. Wind energy has the highest FLH, followed by PV single-axis tracking, CSP and PV optimally tilted. Figure 5 presents the aggregated profiles of CSP solar field, wind energy power generation, solar PV generation (optimally tilted and single-axis tracking) normalized to maximum capacity and country-average.

Table 4 Average full load hours (FLH) for CSP, PV optimally tilted, PV single-axis tracking and wind onshore power plants in Saudi Arabia

PV single-axis	PV optimally tilted	CSP	Wind onshore
FLH	FLH	FLH	FLH
2443	1830	2382	2538

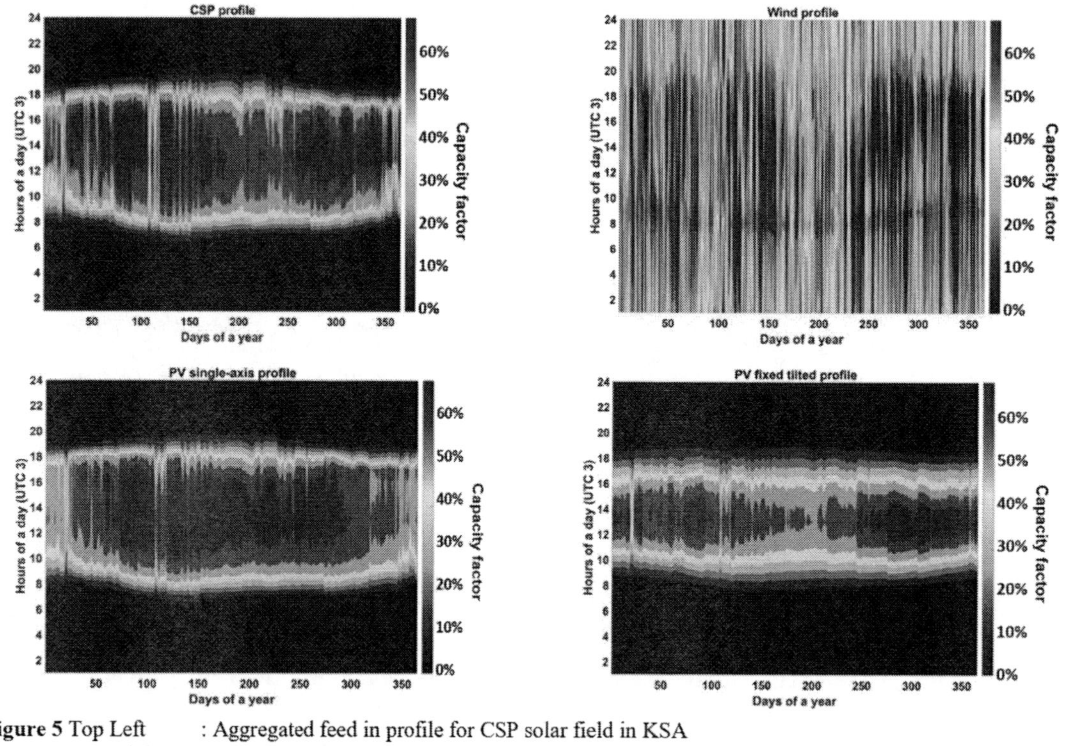

Figure 5 Top Left : Aggregated feed in profile for CSP solar field in KSA
Top Right : Aggregated feed in profile for wind power plant in KSA
Bottom Left : Aggregated feed in profile for single-axis tracking PV in KSA
Bottom Right : Aggregated feed in profile for optimally tilted PV in KSA

The model specifies the upper limits and lower limits for the renewable energy power plant capacities. Lower limits are the installed capacities of renewable energy plants by 2015. For Saudi Arabia this is only 81.6 MW of solar PV plants as listed in Table 5. In periods after 2015 the lower limit is set to the capacity of the previous period adjusted by the capacities reached the end of their technical lifetime. The upper limits are calculated according to Bogdanov and Breyer [28] and the relevant results for KSA are listed in Table 6. However, for solid biomass residues, biogas and waste-to-energy plants it is assumed, due to energy efficiency reasons, that the available and specified amount of the fuel is used during the year.

Table 5 Lower limits of installed capacities [MW] considered in this study for Saudi Arabia

	Solar PV
KSA	81.6

Table 6 Upper limits on installable capacities in Saudi Arabia in units of GW_{th} for CSP and GW_{el} for all other technologies.

	Area [1000 km²]	Solar CSP	Solar PV	Wind
KSA	446.5	19347	9673	722

2.4 Seawater Desalination Capacities - Technical and Financial Assumptions

Table 7 represents the total online capacities of seawater desalination in Saudi Arabia by 2015 [20].

Table 7 Online capacities of seawater desalination in Saudi Arabia by 2015

Desalination technology	Total online capacity (m^3/day)
MSF	5,564,469
SWRO	3,166,680
MED	1,268,805
Other technologies, including those that use brackish water	220,324

The desalination capacity required for the time period from 2015 – 2050 is determined using the methodology described in Caldera et al. [23]. The approach is based on the water stress and the water demand values in Saudi Arabia for an optimistic future scenario. The desalination capacities are calculated assuming a 100% utilization of the available renewable water resource. The National Water Strategy (NWS) for Saudi Arabia outlines specific amounts of non-renewable water to be consumed from 2015 up to 2040, such that by 2040 there is 5 billion m^3 of non-renewable water resource being withdrawn annually [35]. Therefore, the actual desalination demand also should take into account the usage of non-renewable water resources. For the model, the non-renewable water resource being withdrawn is deducted from the modelled desalination capacity to find the actual desalination capacity required. However, the non-renewable water resource numbers found in the NWS are based on a percentage of renewable water resource used. In the model, the desalination capacities are based on optimal use of the available renewable freshwater resource. As a result, the actual non-renewable water resource withdrawal decreases. This is ultimately used to find the actual desalination capacity required for the Kingdom of Saudi Arabia from 2015 – 2050. Table 8 represents the numbers and the final desalination capacities needed. From 2015 onwards, only the installation of SWRO desalination plants are allowed. Through the use of the Desal database, the MSF plants are separated out into cogeneration and stand-alone MSF plants. Stand-alone MSF plants have a dedicated steam source to provide heating for seawater to produce potable water. In contrast, cogeneration MSF plants use the steam source to produce both potable water and electricity [36]. Al Mutaz and Al Namlah [36] explains the technical and economical advantages of cogeneration MSF plants over the stand alone plants. It has to be mentioned other desalination technologies such as MED and brackish water reverse osmosis (BWRO) are excluded from the present study.

Table 8 Desalination capacities required to meet KSA's increasing total water demand

		2015	2020	2025	2030	2035	2040	2045	2050
Population	mill	31.50	34.40	36.85	39.13	41.24	43.14	44.76	46.06
Total water demand	mill m^3/day	65.4	68.5	75.6	81.0	88.6	96.7	105.0	112.6
Desalination demand based on [23]	mill m^3/day	19.1	21.8	27.5	32.0	38.4	45.2	53.8	58.7

Final non-renewable water resource used	mill m^3/day	8.9	7.9	5.8	3.9	6.2	8.5.	0	0
Actual desalination demand	mill m^3/day	10.2	13.9	21.7	28.2	32.3	36.7	53.8	58.7
Installed capacities									
SWRO	mill m^3/day	5.9	11.1	19.9	26.7	31.4	36.8	53.8	58.7
MSF Stand Alone	mill m^3/day	1.52	0.6	0.2	0.1	0.01	0	0	0
MSF Cogeneration	mill m^3/day	2.73	2.20	1.6	1.3	0.8	0	0	0

The technical and financial parameters of the desalination technologies from 2015 – 2050 are presented in Table 9. Similarly, the capex, opex and energy consumption of the water transportation and water storage are presented. There is no variation in the latter numbers from the time period 2015-2050.
The future capex of SWRO desalination plants and the energy consumption was estimated as follows:

1. Using the Global Water Intelligence (GWI) Desal Database the average capex of plants of the capacity range of 10,000 m^3/day up to 50,000 m^3/day, contracted in all years from 1980 – 2015, was found [20]. It is assumed that all the plants to be built in Saudi Arabia in the future, would have a capacity of 50,000 m^3/day or higher. Then the methodology discussed by Loutatidou et al. [37] to determine the EPC costs for SWRO desalination plants up to 2030 was utilized. In our study, this methodology was used to estimate the capex costs for SWRO plants of capacity 50,000 m^3/day from 2020 up to 2050. The average capex value for 2015 was estimated from the GWI Desal Database.
2. Al Zahrani et al. [38] illustrates the energy consumption of SWRO plants with hydraulic turbines for seawater of different salinity levels. For 35 parts per million (ppm) salinity level, the energy consumption is approximately 3.1 kWh/m^3. Assuming that the other processes such as seawater intake, pretreatment, post-treatment and brine discharge take 1 kWh/m^3, the total energy consumption is 4.1 kWh/m^3. This is in line with a presentation [39] on the desalination plants in Masdar, UAE, where energy consumption of SWRO plants is estimated at 4.5 kWh/m^3. However, test plants built recently have an energy consumption of 3.3 kWh/m^3. These low energy consumption trends are also observed in recently built large SWRO plants such as Ghlalilah, in the United Arab Emirates [40]. The total energy consumption of this plant is estimated to be just under 3 kWh/m^3 [40]. However, the larger scale plants built recently use a pressure exchanger as an energy recovery device (ERD). For the purpose of our work, it is assumed that after 2015, all SWRO plants will utilize pressure exchangers. According to Al Zahrani [38], the energy consumption with a pressure exchanger is about 2.6 kWh/m^3. Thus by 2020, it is assumed that the total energy consumption of the plants will drop to 3.6 kWh/m^3. This is feasible as there are currently large scale plants with similar or lower energy consumption values. Meanwhile, Elimlech et al. [41] presents the optimal energy consumption of the SWRO plants from 1970 – 2008. It is shown that the decrease in energy consumption between 2004 and 2008 is 5%. From 2020 onwards, there is assumed to be about a 5% decrease in energy consumption every 5 years. Elimelech et al. [41] explains that the practical minimum energy consumption of an SWRO plant with a 50% recovery rate and seawater salinity of 35 ppm, is 1.56 kWh/m^3. Assuming 1 kWh/m^3 for the other processes, the total minimum energy consumption is about 2.56 kWh/m^3. Thus by 2050, an energy consumption of 2.6 kWh/m^3 is assumed.

Table 9 also presents the parameters for the MSF co-generation and stand-alone plants utilized in the model sourced from El-Nashar [42]It is assumed that all co-generation plants in KSA are a combination of a MSF desalination plant and OCGT – unfired with Heat Recovery Generator System (HRGS). The capex numbers are obtained as an average of all MSF plants from 1980 to 2015. After 2015, it is assumed that there are no more MSF plants built. Thus, there are no technical and financial parameters listed for MSF plants after 2015. The gain output ratio (GOR) indicates the amount of water produced in kg per kg of steam used in the MSF plants. The general value of the GOR is about 8 for MSF plants located in Saudi Arabia, based on the GWI Desal Database. The power to water ratio (PWR) is specific for cogeneration plants and indicate the amount of power produced per m^3 of water produced a day. For MSF plants with an OCGT power plant and HRGS, this is about 2.25 kW/ (m^3·d).

Table 9 Technical and financial parameters of the seawater desalination technologies from 2015 – 2050

			2015	2020	2025	2030	2035	2040	2045	2050
Sea Water Reverse Osmosis	**Capex**	€/(m³·day)	1150	960	835	725	630	550	480	415
	Opex fix	€/(m³·day)	46	38	33	29	25	22	19	17
	Energy consumption	kWh/m³	4.1	3.6	3.35	3.15	3	2.85	2.7	2.6
	Lifetime	years	25	25	30	30	30	30	30	30
Multi Stage Flash for cogeneration Gain Output Ratio: 8 Power to Water: 2.25 kW/(m³·day)	**Capex**	€/(m³·day)	2000	2000	2000	2000	2000	2000	2000	2000
	Opex fix	€/(m³·day)	100	100	100	100	100	100	100	100
	Thermal energy consumption (Total gas input required for water and electricity)	kWh_{th}/m^3	202.5	202.5	202.5	202.5	202.5	202.5	202.5	202.5
	Electrical energy consumption	kWh_{el}/m^3	2.5	2.5	2.5	2.5	2.5	2.5	2.5	2.5
	Lifetime	years	25	25	25	25	25	25	25	25
Multi Stage Flash for stand alone Gain Output Ratio: 8	**Capex**	€/(m³·day)	2000	2000	2000	2000	2000	2000	2000	2000
	Opex fix	€/(m³·day)	100	100	100	100	100	100	100	100
	Thermal energy consumption	kWh_{th}/m^3	85	85	85	85	85	85	85	85
	Electrical energy consumption	kWh_{el}/m^3	2.5	2.5	2.5	2.5	2.5	2.5	2.5	2.5
	Lifetime	years	25	25	25	25	25	25	25	25
Water Transportation										
Piping	**Capex**	€/(m³·a·km)	0.053	0.053	0.053	0.053	0.053	0.053	0.053	0.053
	Fixed Opex	€/(m³·a·100 km)	0.023	0.023	0.023	0.023	0.023	0.023	0.023	0.023
	Lifetime	years	30	30	30	30	30	30	30	30
Vertical Pumping	**Capex**	€/(m³·h·m)	15.4	15.4	15.4	15.4	15.4	15.4	15.4	15.4
	Fixed Opex	€/(m³·h·m)	0.3	0.3	0.3	0.3	0.3	0.3	0.3	0.3
	Energy consumption	kWh/(m³·h·100 m)	0.36	0.36	0.36	0.36	0.36	0.36	0.36	0.36
	Lifetime	years	30	30	30	30	30	30	30	30
Horizontal Pumping	**Capex**	€/(m³·h·km)	19.26	19.26	19.26	19.26	19.26	19.26	19.26	19.26
	Fixed Opex	€/(m³·h·km)	0.4	0.4	0.4	0.4	0.4	0.4	0.4	0.4
	Energy consumption	kWh/(m³·h·100km)	0.04	0.04	0.04	0.04	0.04	0.04	0.04	0.04
	Lifetime	years	30	30	30	30	30	30	30	30
Water Storage	**Capex**	€/m³	65	65	65	65	65	65	65	65
	Fixed Opex	€/m³	1.3	1.3	1.3	1.3	1.3	1.3	1.3	1.3
	Lifetime	years	30	30	30	30	30	30	30	30

2.5 Results

The following two scenarios are simulated by the model:

1. Non-integrated scenario
   In this scenario, the model determines the optimal transition pathway for Saudi Arabia's power sector, discussed in Section 2.4, and separately the power plant capacities required for seawater desalination demand, discussed in section 2.5. Thus, the sum of the power plant capacities required for both sectors present the total power capacities required to meet KSA's power and water demands in the energy transition.
2. Integrated scenario
   In this scenario, the seawater desalination demand is integrated into Saudi Arabia's power sector, throughout the transition from the fossil based power system in 2015 up to a 100% renewable energy based system in 2050.
   The desalination plants, integrated into the power system, will allow for the optimal use of the hourly energy produced by the renewable energy power plants. The excess energy produced by the renewable energy power plants can be stored as desalinated water and at times of low energy production, the stored water can be used. Thus, the desalination plants offer additional flexibility to the energy system.
   The model determines the power capacities required for the integrated scenario and the two scenarios are compared to understand the impacts of the integrated desalination plants.

In the following sections, the technical and financial results of the energy transition for the integrated scenario are presented. The financial results are ultimately compared with that of the non-integrated scenario to learn the impacts of integrated seawater desalination plants.

2.5.1 Integrated Scenario - Power Sector

Figure 6 (left) presents the installed capacities of the different types of power plants required for the energy transition from 2015 up to 2050. In 2015, the total power capacity was 69.3 GW, out of which 33 GW were internal combustion engines and 36.2 GW were gas turbines. The model is restricted from installing more fossil fuels thermal power plants after 2015 and can only install renewable technologies to meet the growing electricity demand, however power-to-gas based gas turbines are not restricted. Internal combustion generators are eliminated from the system by 2035 after exceeding their technical lifetime. Thermal power plants running on natural gas are eliminated from the system by 2040. Power-to-gas based electrolyser and methanation plants are used to produce synthetic natural gas (SNG) that is then used as fuel for existing open cycle gas turbines (OCGT) or combine cycle gas turbines (CCGT).

The total capacity of all plants required by 2050 is about 480 GW. PV single-axis tracking accounts for 368 GW and wind power plants account for 75 GW of the 480 GW required by 2050. Figure 6 (right) presents the electricity production of the different power plants. After 2030, electricity production from wind power plants remain constant at 321 TWh. The installed capacities of PV single-axis tracking continue to grow and start to dominate the Saudi power sector. By 2050, PV single-axis tacking contributes 914 TWh out of the total annual electricity generation of 1108 TWh. Figure 6 (right) also illustrates the diminishing use of fossil oil and gas in the energy system. The power capacities required for the energy transition can be found in the Appendix (Table 3).

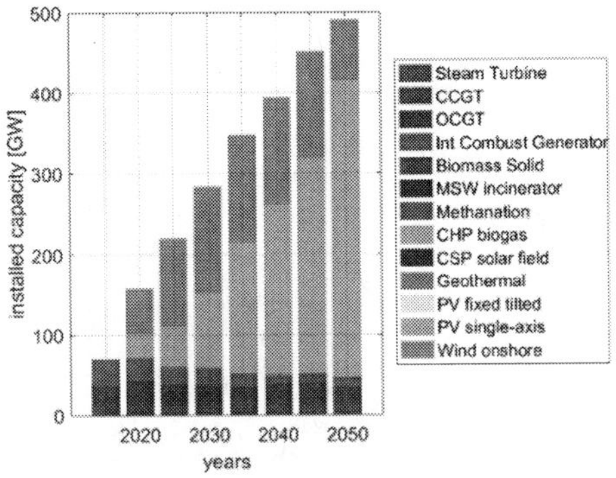

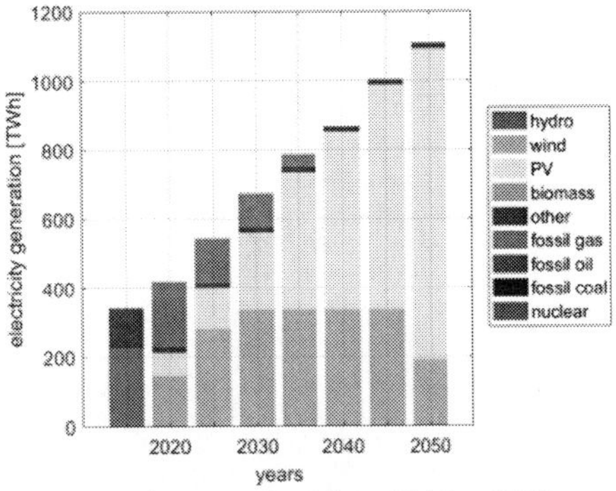

Figure 6 Left : Installed capacities of the different power plants required from 2015 to 2050
Right : Electricity production of the different power plant categories from 2015 to 2050

Figure 7 (left) illustrates the full load hours (FLH) of the different power plants utilized by the model. The FLH of PV single-axis tracking, PV fixed tilted and wind onshore plants are constant at 2443, 1830 and 2538 hours. The model determines the optimal FLH and the capacities of the power plants. The FLH of the internal combustion generator rapidly diminishes from 3500 to approximately 43 FLH by 2035. The high cost of oil renders the existing oil powered thermal power plants too expensive to run. In addition, the FLH of the CCGT reduces from 8322 FLH to 1540 FLH in 2050. The lower cost of the natural gas, in contrast to oil, still enables natural gas powered thermal plants to be cost effective. However, after 2035, as shown in Figure 6 (right), there is no more contribution from natural gas powered thermal plants. The same trend is observed by the OCGT plants where the FLH of the plants reduce from 3500 to 357 FLH. The remaining gas plants are used with SNG to store and produce electricity as required in the Saudi power system.

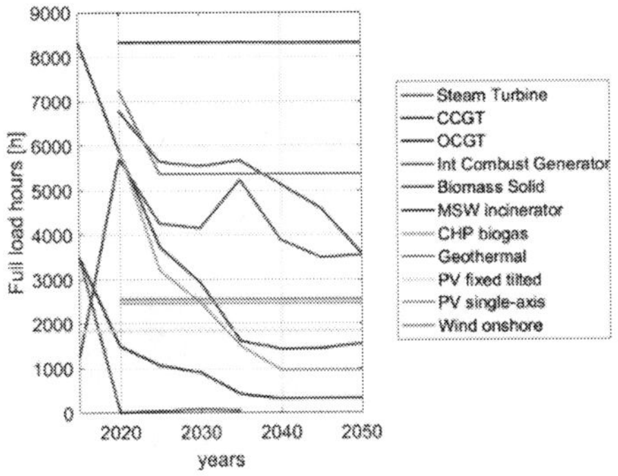

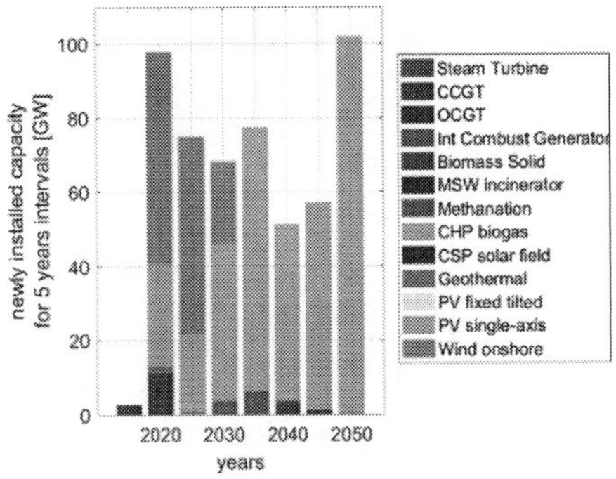

Figure 7 Left : FLH variation of the different power plants used in model
Right : Additional capacities of different power plants required from 2015 to 2050

Figure 7 (right) represents the additional capacities of different power plants necessary to meet the growing electricity demand and to replace the fossil based thermal power plants that are being phased out. Initially, there are more wind power plant installations than PV power plants. However, after 2030, the installation of wind power plants cease and PV single-axis tracking start to dominate the power sector. Methanation units are added for the first time in 2030, with additional capacities added until 2045, to meet the gas demands of the energy system through SNG.

Figure 8 (top left) represents the additionally installed storage capacities for every 5 years period. The large capacity of gas storage is required in 2040 as it is created for the first time, and thereafter, less additional storage is required. Figure 8 (top right) represents the ratio of the total storage output for every time period. By 2025, electricity storage systems come into effect as it becomes a more cost effective option to balance the power system than the use of fossil powered thermal plants. By 2050, batteries provide a total output of about 401 TWh_{el} which accounts for about 48% of the total electricity demand and about 75% of the direct electricity supply.

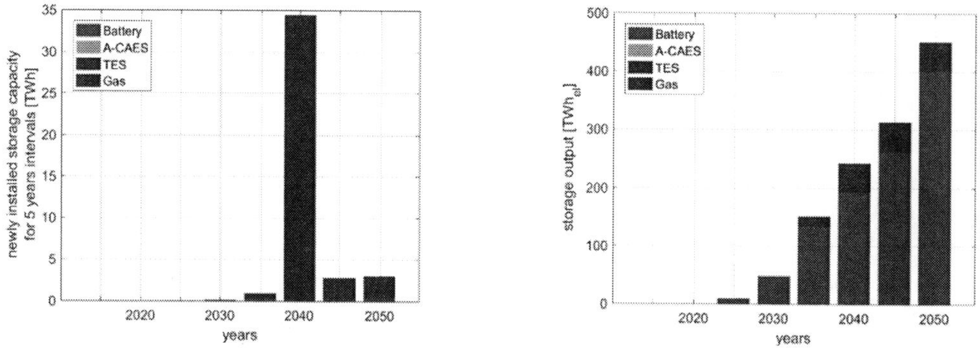

Figure 8 Top Left : Additional storage capacity required from 2015 to 2050
Top Right : Storage output required from 2015 to 2050

The total levelised cost of electricity of the system is the sum of the levelised cost of primary electricty generation (LCOE primary), levelised cost of storage (LCOS), levelised cost of curtailment (LCOC), fuel cost and the carbondioxide emission cost (CO_2 cost). Figure 9 (top left) represents the contribution of the different components to the total LCOE from 2015 to 2050. In 2015, the fuel cost accounts for the largest contribution due to the fuel prices of 52.50 €/MWh$_{th}$ for oil and 21.8 €/MWh$_{th}$ for gas. As the FLH of the fossil powered thermal plants decrease, as illustrated in Figure 7 (left), the fossil fuel consumption and consequently the contribution to the LCOE decreases.

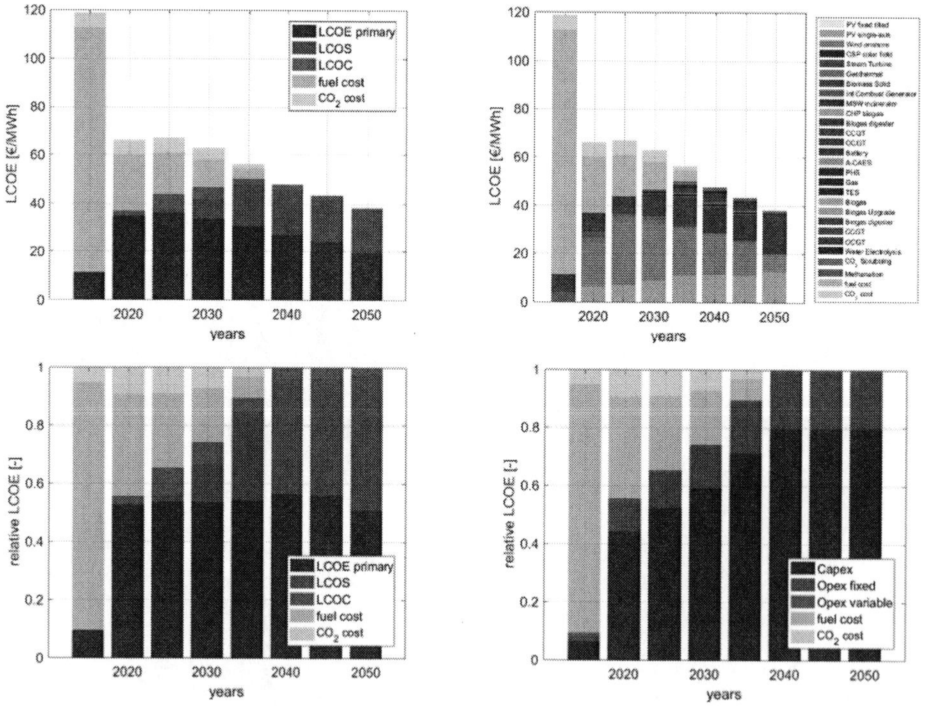

Figure 9 Top Left : Contribution of different components to the total LCOE from 2015 to 2050
Top Right : Detailed contribution of components to the total LCOE from 2015 to 2050
Bottom Left : Relative contribution of different components to the total LCOE from 2015 to 2050
Bottom Right : Relative contribution of financial components to the total LCOE from 2015 to 2050

The LCOC is as a result of curtailment of the excess energy produced by the renewable energy power plants. The remaining thermal power plants contribute to the CO_2 emissions cost. The batteries and gas storage contribute to the LCOS of the power system. The figure illustrates that with the transition to renewable power plants and the phasing out of the thermal powered plants, the total LCOE of the system decreses. By 2050, the LCOE is estimated to be about 38 €/MWh as opposed to 119 €/MWh in 2015. Figure 9 (top right) separates the LCOE primary and LCOS to the individual components and presents the specific contributions towars the LCOE. By 2050, the largest contributors to the cost will be the PV single-axis tracking and battery plants. The contribution of wind decreases after 2030, due to no more wind plant installations and a highly competitive combination of hybrid PV-battery plants, as already found by Afanasyeva et al. for the case of Morocco [43]. Figure 9 (bottom right) presents the relative contribution of the different components to the LCOE. It can be seen that for every time period, the LCOE primary accounts for almost 50% of the final LCOE. The relative contribution of the fossil fuel cost ,and therefore also the CO_2 emission costs, disappear. The relative contribution of the storage increases and ultimately by 2050, the LCOS and LCOE primary have almost equal contributions to the LCOE. This is also reflected in Figure 9 (top right). Figure 9 (bottom left) presents the relative contribution of the financial components to the LCOE. In 2015, the fossil fuel costs determine the LCOE. As the fossil fuel usage diminishes, the total capex of the power system is the largest relative contributor to the LCOE. From 2040 onwards, when there is a 100% renewable energy system, the capex of the system contributes upto 80% of the LCOE. The fixed opex of the energy system componets accounts for the remainder of the LCOE.

Figure 10 illustrates the annual CO_2 emissions during the energy transition illustrated in the previous figures. The annual CO_2 emissions are reduced from about 200 Mton/a to 0 by 2040. The red line diagram represents the ratio of CO_2 emitted for every kWh of electricity produced. In 2015, this value is at about 800g of CO_2 per kWh and drops to 0 by 2040. The CO_2 emission calculations are explained in [43]

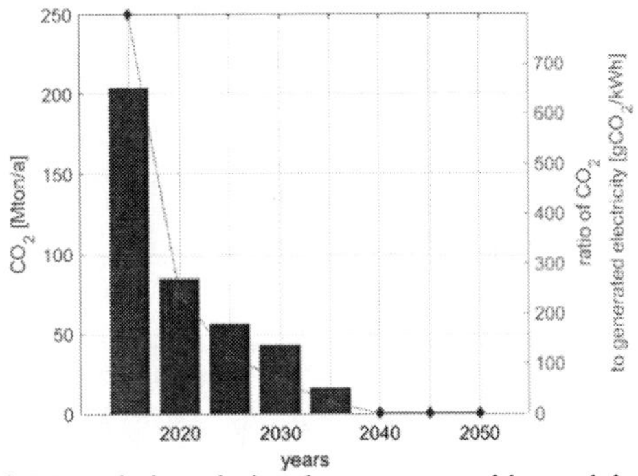

Figure 10 Annual CO_2 emissions during the energy transition and the ratio (blue bars) CO_2 emissions per kWh of electricity produced (red line)

2.5.2 Integrated Scenario – Desalination Sector

Figure 11 illustrates the growth of the Saudi Arabian desalination sector assumed in the model. The difference between the total water demand and the installed desalination capacity is met by renewable water sources and non-renewable groundwater sources. As explained in section 2.4, the desalination demand increases to meet the growing total water demand and is dominated by SWRO. The MSF plants are phased out based on the lifetime of the plants. By 2050, it is expected that there is a total water demand of 112 million m^3/day and SWRO plants meet 52 % of the total water demand. In addition, water storage is only found to play a considerable role after 2030. Figure 12 illustrates the variation in water storage for the year 2015. It can be seen that water storage acts as a weekly or seasonal storage, similar to A-CAES or PtG storage. With the increase in SWRO desalination capacity there is an increase in water storage capacity. This behavior of the water storage over time has to be further studied.

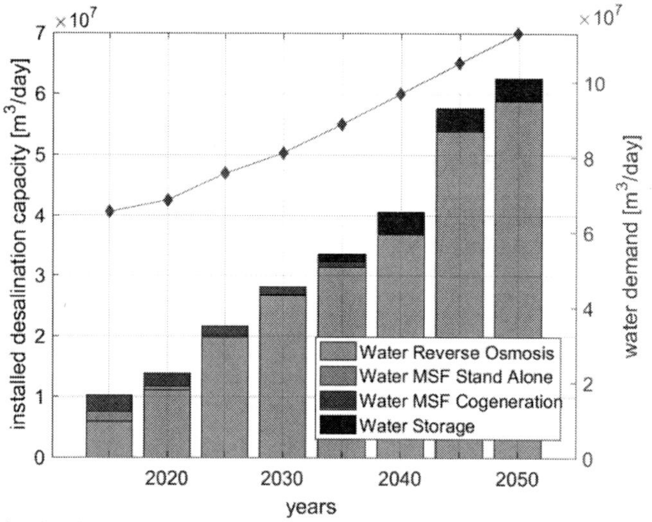

Figure 11 Water desalination capacities required to meet KSA's total water demand from 2015 to 2050

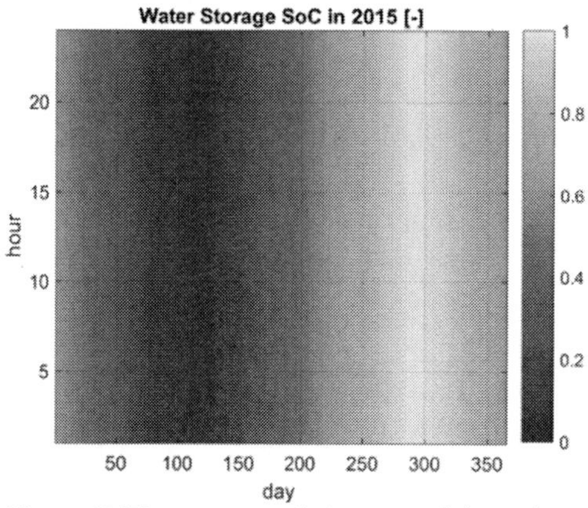

Figure 12 Water storage relative state of charge in 2015

Figure 13 (top left) presents the capex breakdown of the desalination plant capacities installed for the 5 year intervals. The contribution of the vertical transportation increases as the installed desalination capacities increase. By 2050, the vertical transportation infrastructure is the second largest contributor to the capex. The total capex required for the period from 2015 up to 2050 for the desalination sector is 910 billion €.

Figure 13 (top right) presents the variation in the annual fixed opex of the desalination capacities. With the increase in desalination capacities, the fixed opex increases. The fixed opex excludes the electricity consumption of the desalination plants and water transportation infrastructure. Figure 13 (bottom left) represents the variation in the annual variable opex. The annual variable opex accounts for both the gas and electricity consumption of the desalination plants. The MSF plants require both thermal and electrical energy. As a result, when the MSF plants are phased out and only SWRO plants exist, the annual variable opex is lower. The variation in the electricity cost maybe explained by the considerable increase in installed SWRO capacities with time and the variation in energy efficiency of the plants.

Finally, Figure 13 (bottom right) illustrates the contribution of the different aspects of the desalination system to the final LCOW for every time period. The LCOW of the final system decreases from 2.7 €/m^3 in 2015, assuming non subsidized oil and gas prices, to 0.60 €/m^3 in 2050. The costs reduction is mainly due to the phasing out of gas consumption, decrease in electricity costs but also in an increase in the efficiency of RO desalination plants in the decades to come.

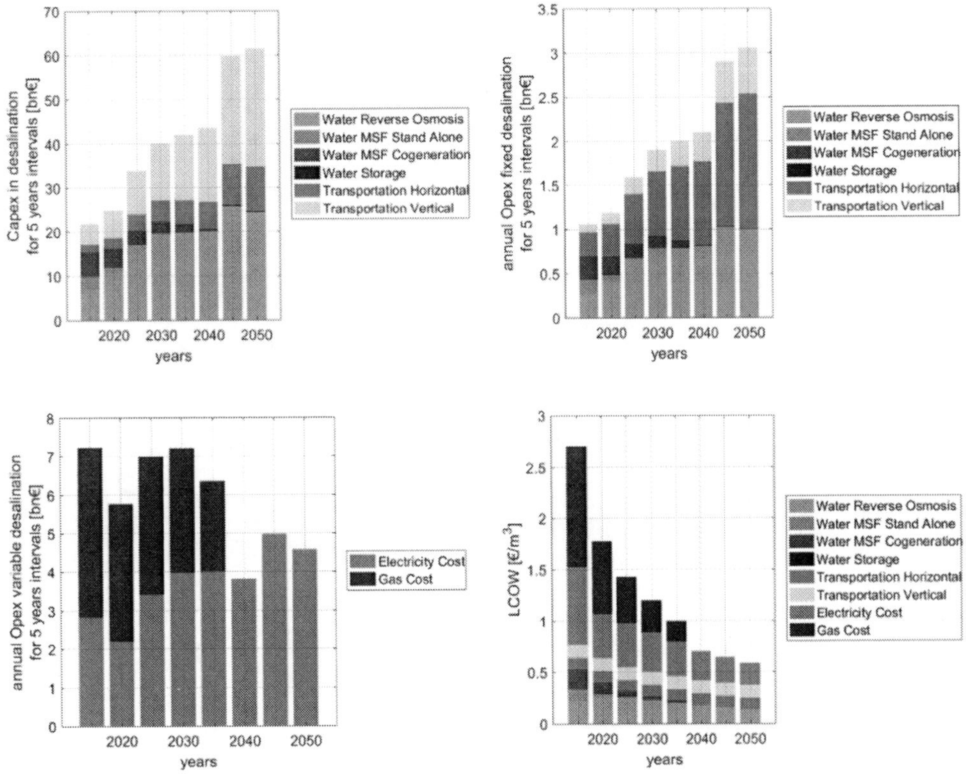

Figure 13 Top Left : Capex of all desalination sector components from 2015 to 2050
Top Right : Annual fixed opex for desalination sector components from 2015 to 2050
Bottom Left : Annual variable opex for desalination sector components from 2015 to 2030
Bottom Right : LCOW variation from 2015 to 2030

2.5.3 Comparison of the Integrated and Non-Integrated Scenarios

As discussed in Section 2.5, the model was run for a non-integrated scenario where the power capacities to meet the electricity demand and the desalinated water sector are found separately. To compare the integrated and non-integrated scenarios, the annual levelised cost for both scenarios from 2020 to 2050 was found. This is represented in Figure 14 (top). The annual levelised cost for the non-integrated scenario is higher than that for the integrated scenario. It is found that the approximate percentage benefit of the integrated scenario is between 2.5% to 4%, as a consequence of increased flexibility to the integrated energy system.

Figure 14 (bottom) illustrates the curtailment of electricity for the two scenarios, for the time period from 2015 to 2050. The bars represent the value of the energy curtailed and the dashed lines represent the ratio of the curtailed energy to the total generated electricity. It can be seen that there is more curtailed energy in the integrated scenario. However, the ratio of the curtailed energy to the total electricity generated is lower in the integrated scenario. This can be attributed to the increased flexibility provided by the desalination plants and therefore better utilization of the hourly produced renewable energy, hence the integrated energy system can be run in a more efficient way than the separated systems.

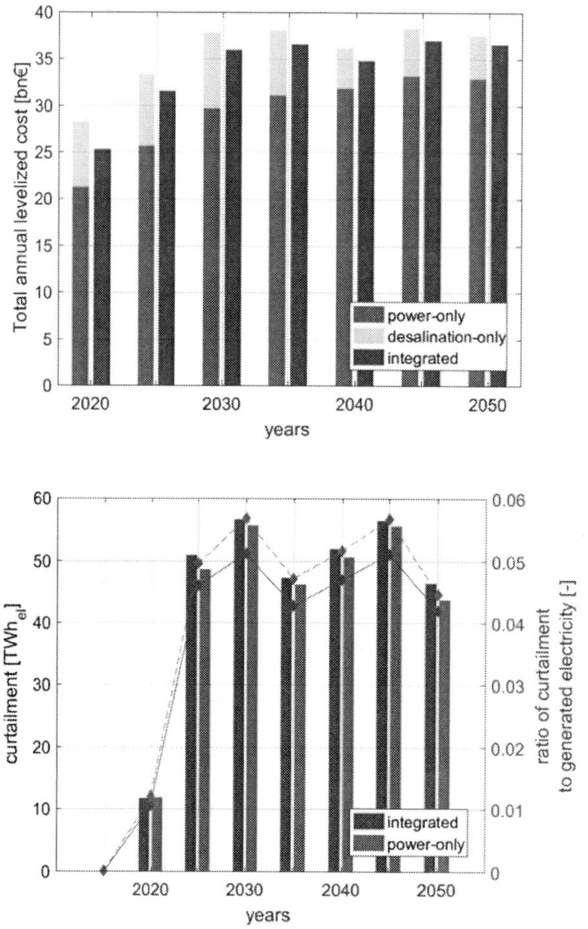

Figure 14 Top : Comparison of total annual levelised cost for the integrated and non-integrated scenarios
Bottom: Comparison of the curtailed electricity for the integrated and non-integrated scenarios

It has to be mentioned that in the integrated scenario, for the total time period of 2015 to 2050, there was a reduction in required battery storage and power-to-gas (PtG) electrolyser capacity by 2% and 9%, respectively, compared to the separated scenario. For the integrated scenario the battery storage and the PtG electrolyser capacity was estimated to be 1266 GWh and 9.85 GW_e respectively. For the non-integrated scenario, the corresponding values were estimated to be 1290 GWh and 10.81 GW_e.

3 Discussion

The work presented prove that it is in fact possible for the Kingdom of Saudi Arabia to achieve a 100% renewable energy sector before the proposed year 2050. Figure 6 (right) illustrates that this milestone is feasible for Saudi Arabia by 2040. By 2040, PV single-axis tracking and wind power plants will dominate the Saudi power sector with 210 GW of the former and 133 GW of the latter. However, by 2050, due to the steeper decrease in the PV single-axis tracking and battery capex than the wind onshore plants, as presented in Figure 4 (bottom left), it is more economical to have PV single-axis tracking as the dominant energy source in the power sector of Saudi Arabia. By 2050, about 369 GW of PV single-axis tracking and 76 GW of wind power plants are required. This result documents the outstanding impact of low cost solar PV supported by low cost battery storage which leads to a solar PV electricity generation share of 80%, which is significantly higher than the average of about 40% found in the global average for year 2030 assumptions [24], but also as the 48% solar PV share for the MENA region [44]. However, comparable results had been found already earlier for the case of Israel [29], where the solar PV share had be found for cost optimized systems to about 90% of the total electricity supply, however for less good wind conditions as in Saudi Arabia, but for year 2030 assumptions.

With the phasing out of fossil fueled thermal power plants, battery storage comes into play by 2025. By this time period, about 24% of the energy generation share is accounted for by fossil fueled thermal power plants. As the electricity demand increases and the share of renewables in the system increases, the battery storage required continues to increase. By 2050, batteries provide a total output of about 401 TWh_{el} accounting for approximately 48% of the total electricity demand. For a 100% RE scenario by 2040, the required battery storage output is 193 TWh_{el} accounting for approximately 30% of the total electricity demand. Between, 2030 and 2050 the battery full charge cycles required are between 300 and 310 cycles a year, indicating the almost daily usage of the battery capacity.

PtG plants come into effect in 2030 with 3.5 GW_{el} of input capacity and increases to 9.85 GW_{el} by 2050. Similarly, the FLH of the PtG plants increase from about 5490 hrs in 2030 to 6650 hrs by 2050. However, the gas storage output contributes about 7% towards the total electricity demand by 2050, significantly less than the 48% of batteries.

The LCOE of the energy system is dominated by the high fuel cost in 2015. However, with the phasing out of the fossil powered thermal plants and the decreased consumption of fossil fuel, the LCOE of the energy system reduces from 119 €/MWh_{el} in 2015 to 48 €/MWh_{el} in 2040 and further to 38 €/MWh_{el} in 2050.
Figure 9 (bottom left) illustrates that by 2020 onwards, the dominating contributor to the LCOE is the $LCOE_{primary}$. With the increase in storage capacities, both the $LCOE_{primary}$ and LCOS are the largest contributors to the system LCOE. Figure 9 (bottom right) further shows that by 2050, the largest contributors are PV single-axis tracking and battery storage. However, by 2040, there is still a significant contribution by the existing wind power plants to the LCOE. This is in spite of the fact that new installations of wind power plants cease after 2030, due to the arising competitiveness of hybrid PV-battery plants.

In addition to the increasing electricity demand, there is also an increasing desalination demand in the Kingdom of Saudi Arabia. As can be seen in Figure 11, the capacities of SWRO desalination increases to meet the total water demand of KSA. The difference is to be met through renewable water resources and some non-renewable ground water resources. The existing MSF cogeneration and stand-alone plants are phased out by 2035. In an integrated scenario, the desalination capacities are used to better utilize the hourly renewable energy production of the power system. Figure 13 presents the variation in the LCOW from 2.7 €/m^3 in 2015 to 0.6 €/m^3 in 2050. As the MSF plants are phased out, there is no gas cost to the final LCOW. In addition to producing potable water, the MSF cogeneration plants also produce electricity. However, cogeneration plants are usually designed with a low power to water ratio, as water is the main product.
The reduction in LCOE of the system, contributes to the rapid decrease of the LCOW. The electricity costs for the desalination sector includes the energy consumption for water production at the desalination plant and transportation of water (vertical and horizontal transportation) to the demand site. By 2050, there is almost equal contributions from the SWRO desalination plant capital costs and the electricity costs to the final LCOW.

To understand the impacts of integrating the desalination and power sectors in KSA, the LUT energy systems model was used to analyse both an integrated and non-integrated scenario. Thus, for the non-integrated scenario, the same desalination capacities are required, but they are not used to increase the flexibility of the power system. Figure 14 (top) illustrates that the annualized levelised cost of the non-integrated scenario is higher than that of the integrated scenario for all time periods from 2020 to 2050. In fact it is found that between 2040 and 2050, this reduction is between 2% – 4%, representing an annual cost decrease between 0.9 – 1.3 bn€. The lower cost is attributed to the reduced amounts of electricity storage, in the form of batteries and power-to-gas, required in the integrated scenario. Figure 14 (bottom) highlights the lower curtailed energy ratio in the integrated scenario and the better utilization of the hourly renewable energy production.

Therefore, the above work presents a pathway for Saudi Arabia to transition to a 100% renewable energy power sector and shows that the integration of the country's power and desalination sectors provides the least cost solution for both sectors.

It may be argued that the current electricity production costs in Saudi Arabia are lower than what our work presents for 2015. However, the artificially low fossil fuel prices in Saudi Arabia are a result of the high subsidies which according to Al-Iriani et al. [45]

accounts for 10% of the country's GDP. As the energy consumption increases, due to the heavily subsidized fuel prices, in form of opportunity costs since the export value would be much higher than the subsidized domestic consumption, the economic price the country has to pay increases. With the decrease in oil prices and resulting revenue, as well as concerns about availability of fossil fuel resources to meet the country's growing energy demand, Saudi Arabia has started to transition to less or no subsidies. Thus, our work offers a solution for Saudi Arabia to meet its growing energy demands through renewable energy without hindering the future economic growth of the country.

This research also addresses how Saudi Arabia can meet the country's growing water demand through the use of 100% renewable energy powered SWRO desalination plants. The average LCOW of KSA is found to be 0.6 €/m^3 by 2050. This includes the cost for water desalination, water transportation to the demand site and water storage. Fthenakis et al. [46] estimated the current LCOW of a 190,000 m^3/day SWRO plant, located in Al-Khajfi, on the east coast of KSA, powered by fixed-tilted PV to be 0.70 €/m^3. Current water production costs of fossil powered SWRO plants in KSA lie between 0.65 €/m^3 – 1.90 €/m^3. However, this excludes the cost of transporting the desalinated water to the demand site and water storage.

4 Conclusion

There is growing interest in the energy transition towards a 100% renewable energy (RE) based power system in many countries and regions. Meanwhile, as the global renewable freshwater resource diminishes, seawater reverse osmosis (SWRO) desalination is expected to play a key role in securing future water supplies.
Saudi Arabia is the world's largest producer of crude oil and the 11th largest consumer. Economic growth of the country is tied closely to oil prices. To secure the country's economy, the government recently heralded a future without oil. The new vision calls for increase in renewable energy capacities and at least 9 GW of installed RE capacity by 2023. Saudi Arabia is also the largest producer of desalinated water in the world and desalination will remain vital to the country's future water supply.

The purpose of our work is to analyse the impacts of integrating the large seawater reverse osmosis desalination demand into the country's 100% RE based power system by 2050. It is found that the integration of the power sector and the desalinated water sector, allows for a reduction of 2% - 4% of the levelised annual costs, in comparison to the non-integrated scenario. The LCOE and LCOW for Saudi Arabia by 2050, in the integrated scenario, is estimated to be 38 €/MWh and 0.60 €/m^3, respectively. By 2050, PV single-axis tracking dominates the power sector in Saudi Arabia with about 80% of total generated electricity due to the further reduction in capex of solar PV and supporting battery technology over wind power plants.

Thus, the results present a least cost transition path for Saudi Arabia to meet the country's future electricity and water demands through a 100% RE system.

However, there are gaps in the research methodology and data for Saudi Arabia that could be summarised as below:

1. The desalination sector should also account for the multiple effect distillation (MED) capacities in Saudi Arabia.
   MED has a lower thermal energy consumption and electrical energy consumption.
2. The industrial gas demand of Saudi Arabia should be accounted for. This would result in an increase in synthetic natural gas production and excess heat in the system, which could be utilized for lowering the thermal energy demand of desalination technologies like MED.
3. Better understanding of the impacts of water storage in the integrated scenario. In the Vision 2030 document for Saudi Arabia, it is mentioned that one of the government's policies is to boost water storage and have three strategic days of water storage by 2020.
4. No well-defined learning curve for SWRO desalination plants. This makes it difficult to project the future SWRO costs.

By filling in the gaps in our research, we can further refine our results and provide the optimal transition pathway for Saudi Arabia to finally start the country's future without oil. By capitalising on the country's excellent solar and wind resources through the implementation of renewable energy, Saudi Arabia can indeed meet the country's future electricity and water demands in a lucrative manner.

Acknowledgements

The authors gratefully acknowledge the scholarship offered by the Reiner Lemoine Foundation and public financing of Tekes, the Finnish Funding Agency for Innovation, for the 'Neo-Carbon Energy' project under the number 40101/14.

Nomenclature

a	annum
A-CAES	adiabatic compressed air storage
Capex	capital expenditures
CCGT	combined cycle gas turbine
CSP	concentrating solar power
crf	capital recovery factor
LCOE	levelised cost of electricity
LCOW	levelised cost of water
MSF	multi stage flash
MED	multi effect distillation
Opex	operational expenditures
PtG	power-to-gas
PV	photovoltaic
RE	renewable energy
SEC	specific energy consumption
SNG	synthetic natural gas
SWRO	seawater reverse osmosis
WACC	weighted average cost of capital

5 References

[1] United Nations, 2015, Probabilistic Population Projections based on the World Population Prospects: The 2015 Revision. Population Division, DESA, New York City, NY, http://esa.un.org/unpd/ppp/

[2] Salam A A, Elsegaey I, Khraif R, Al-Mutairi A, 2014, Population distribution and household conditions in Saudi Arabia: reflections from the 2010 census, *SpringerPlus*, 3:530, 1-13

[3] The World Bank, 2016, World Development Indicators : Electric power consumption (kWh per capita), Washington, USA, http://data.worldbank.org/indicator/EG.USE.ELEC.KH.PC

[4] Nachet S and Aoun M-C, 2015, The Saudi electricity sector: pressing issues and challenges, Institut Francais des Relations Internationales Centre Energie, Paris, France, www.ifri.org/sites/default/files/atoms/files/note_arabie_saoudite_vf.pdf

[5] King Abdullah City for Atomic and Renewable Energy, Saudi Arabia's renewable energy strategy and solar energy deployment roadmap, KACARE, Riyadh, www.irena.org/DocumentDownloads/masdar/Abdulrahman%20Al%20Ghabban%20Presentation.pdf

[6] World Energy Council, 2014, Global Energy Transitions : A comparative analysis of key countries and implications for the international energy debate, Berlin, Germany, www.atkearney.com/documents/10192/5293225/Global+Energy+Transitions.pdf/220e6818-3a0a-4baa-af32-8bfbb64f4a6b

[7] Al Bassam A B, 2015, Economic diversification in Saudi Arabia: Myth or reality?, *Resources Policy*, 44, 112-117

[8] King Abdullah City for Atomic and Renewable Energy, 2016, The Vision : Energy Sustainability for Future Generations, Riyadh, Saudi Arabia, www.kacare.gov.sa/en/FutureEnergy/Pages/vision.aspx

[9] Moser S, Swain M, Alkhabbaz H M, 2015, King Abdullah Economic City : Engineering Saudi Arabia's post-oil future, *Cities*, 45, 71-80

[10] Farfan J and Breyer Ch, 2016, Structural changes of global power generation capacity towards sustainability and the risk of stranded investments, submitted

[11] Black I, 2016, Saudi Arabia approves ambitious plan to move economy beyond oil, The Guardian, London, United Kingdom , www.theguardian.com/world/2016/apr/25/saudi-arabia-approves-ambitious-plan-to-move-economy-beyond-oil

[12] Apricum The Cleantech Advisory, 2016, Saudi Arabia unveils new plan including 9.5 GW of renewable energy, Berlin, Germany, www.apricum-group.com/saudi-arabia-announces-9-5-gw-renewable-energy-target-new-king-salman-renewable-energy-initiative/

[13] Almasoud A H, Gandayh H H, 2014, Future of solar energy in Saudi Arabia, *Journal of King Saud University – Engineering Sciences*, 27, 153-157

[14] Breyer Ch and Schmid J, 2010, Population density and area weighted solar irradiation: global overview on solar resource conditions for fixed-tilted , 1-axis and 2-axis PV systems, 25th EUPVSEC/5th World Conference on Photovoltaic Energy Conversion, 6-10 September, Valencia

[15] PV Magazine, 2016, Masdar selected by DEWA to lead consortium on 800 MW solar PV project in Dubai, http://www.pv-magazine.com/news/details/beitrag/masdar-selected-by-dewa-to-lead-consortium-on-800-mw-solar-pv-project-in-dubai_100025158/?utm_source=twitterfeed&utm_medium=facebook#axzz4ClgJ6D3I

[16] Kingdom of Saudi Arabia, 2016, Vision 2030, Riyadh, http://vision2030.gov.sa/en

[17] DeNicola E, Aburizaiza S O, Siddque A, Khwaja H, Carpenter O D, 2015, Climate Change and Water Scarcity: The Case of Saudi Arabia, *Annals of Global Health*, 81-3, 342-352
[18] Global Water Intelligence Desal Data, 2016, Country Profile Saudi Arabia, https://www.desaldata.com/countries/31
[19] Drewes J E, Garduno R P C, Amy G, 2012, Water reuse in the Kingdom of Saudi Arabia – status, prospects and research needs, Water Science and Technology : Water Supply, 12, 926-936
[20] Global Water Intelligence Desal Data, 2016, Plant Database, www.desaldata.com/projects
[21] United Nations World Water Assessment Programme, 2014, The United Nations World Water Development Report 2014: Water and Energy, Paris, UNESCO, http://unesdoc.unesco.org/images/0022/002257/225741E.pdf
[22] Al-Kharaghouli A and Kazmerski L L, 2013, Energy consumption and water production cost of renewable energy powered desalination processes, *Renewable and Sustainable Energy Reviews*, 24, 343-356
[23] Caldera U, Bogdanov D, Breyer Ch, 2016, "Local cost of seawater RO desalination based on solar PV and wind energy: A global estimate", *Desalination*, 385, 207-216
[24] Breyer Ch, Bogdanov D, Gulagi A, Aghahosseini A, Barbosa L S N S, Koskinen O, Barasa M, Caldera U, Afanasyeva S, Child M, Farfan J, Vainikka P, 2016, On the Role of Solar Photovoltaics in Global Energy Transition Scenarios, 32nd EU PVSEC, Munich, June 20-24
[25] Breyer Ch, Koskinen O, Bogdanov D, 2016, Europa 100% erneuerbar, 14. Nationale Photovoltaik-Tagung, Bern, February 22-23
[26] Bogdanov D and Breyer Ch, 2015, Eurasian Super Grid for 100% Renewable Energy power supply: Generation and storage technologies in the cost optimal mix, ISES Solar World Congress 2015, Daegu, Korea, November 8-12
[27] Gulagi A, Bogdanov D, Breyer Ch, 2016, Solar Photovoltaics – A driving force towards a 100% renewable energy system for India and the SAARC region, EUPVSEC, 20th -24th June, Munich
[28] Bogdanov D and Breyer Ch, 2016, North-East Asian Super Grid for 100% Renewable Energy supply: Optimal mix of energy technologies for electricity, gas and heat supply options, *Energy Conversion and Management*, 112, 176-190
[29] Bogdanov D and Breyer Ch, 2015, The Role of Solar Energy towards 100% Renewable Power Supply for Israel: Integrating Solar PV, Wind Energy, CSP and Storages, Proceedings of the 19th Sede Boqer Symposium on Solar Electricity Production, Sede Boqer, February 23-25
[30] Connolly D and Mathiesen B V, 2014, A technical and economic analysis of one potential pathway to a 100% renewable energy systems, *International Journal of Sustainable Energy Planning and Management*, 1, 7-28
[31] Center for Alternative Technology, 2015, Who is getting ready for zero? A report on the state of play of zero carbon modelling, Powys, Wales, http://zerocarbonbritain.org/en/ready-for-zero
[32] Barbosa L S N S, Bogdanov D, Vainikka P, Breyer Ch, 2016, Hydro, wind and solar power as a base for a 100% Renewable Energy supply for South and Central America. submitted.
[33] Aghahosseini A, Bogdanov D, Ghorbani N, Breyer Ch, 2016, The role of a 100% renewable energy system for the future of Iran: Integrating solar PV, wind energy, hydropower and storage, 11th International Energy Conference, Tehran, May 30-31
[34] Al-Saleh Y, Upham P, Malik K, 2008, Renewable Energy Scenarios for the Kingdom of Saudi Arabia, Tyndall Climate Change Research Center, United Kingdom, www.tyndall.ac.uk/sites/default/files/wp125.pdf
[35] Ministry of Water and Electricity, 2013, National Water Strategy: The roadmap for sustainability, efficiency and security of the water future in the KSA National Water Strategy, presentation for Water Arabia 2013, www.sawea.org/pdf/waterarabia2013/Tue_Panel_Discussion/National_Water_Strategy.pdf
[36] Al-Mutaz I S and Al-Namlah M A, 2004, Characteristics of dual-purpose MSF desalination plants, *Desalination*, 166, 287-294
[37] Loutatidou S, Chalermthai B, Marpu P, R, Arafat H, 2014. Capital cost estimation of RO plants: GCC countries versus southern Europe, *Desalination*, 347, 103-111
[38] Al-Zahrani A, Orfi J, Al-Suhaibani Z, Salim B, Al-Ansary H, 2012. Thermodynamic Analysis of a Reverse Osmosis Desalination Unit with Energy Recovery System, *Proceedia Engineering*, 33, 404–414
[39] M+W Group, 2016, Solar Seawater Desalination. Visit of the MASDAR test plants in Ghantoot, UAE, M+W Central Europe GmbH Global Business Unit Energy, Germany
[40] Water Online, 2016, Ghalilah Desalination Plant Honored With Distinction Award For Desalination Plant Of The Year At 2016 Global Water Awards, www.wateronline.com/doc/ghalilah-desalination-plant-honored-plant-of-the-year-at-global-water-awards-0001
[41] Elimelech M and Phillip A W, 2011, The Future of Seawater Desalination: Energy, Technology and the Environment, *Science*, 333, 712-717
[42] El-Nashar A M, 2001, Cogeneration for power and desalination – state of the art review, *Desalination*, 134, 7-28
[43] Afanasyeva S, Breyer Ch, Engelhard M, 2016, The Impact of Cost Dynamics of Lithium-Ion Batteries on the Economics of Hybrid PV-Battery-GT Plants and the Consequences for Competitiveness of Coal and Natural Gas-Fired

Power Plants, 10th International Renewable Energy Storage Conference (IRES 2016), Düsseldorf, March 15-17
[44] Aghahosseini A., Bogdanov D., Breyer Ch., 2016. The MENA Super Grid towards 100% Renewable Energy Power Supply by 2030, 11th International Energy Conference, Tehran, May 30-31
[45] Al-Iriani A M, Trabelsi M, 2016, The economic impact of phasing out energy consumption subsidies in GCC countries, *Journal of Economics and Business*, article in press
[46] Fthenakis V, Atia A A, Morin O, Bkayrat R, Sinha P, 2016, New prospects for PV powered water desalination plants: case studies in Saudi Arabia, Progress in Photovoltaics : Research and Applications, 24, 543-550
[47] Pleßmann G, Erdmann M, Hlusiak M, Breyer Ch, 2014, Global Energy Storage Demand for a 100% Renewable Electricity Supply, Energy Procedia, 46, 22-31
[48] European Commission, 2014. ETRI 2014 - Energy technology reference indicator projections for 2010-2050. EC Joint Research Centre Institute for Energy and Transport, Petten, Netherlands
[49] Vartiainen E, Masson G, Breyer Ch, 2015, PV LCOE in Europe 2015-2050, 31st EU PVSEC, Hamburg, September 14-18
[50] Fraunhofer ISE, 2015, Current and Future Cost of Photovoltaics. Long-term Scenarios for Market Development, System Prices and LCOE of Utility-Scale PV Systems, study on behalf of Agora Energiewende, Freiburg and Berlin, www.agora-energiewende.de/fileadmin/Projekte/2014/Kosten-Photovoltaik-2050/AgoraEnergiewende_Current_and_Future_Cost_of_PV_Feb2015_web.pdf
[51] Neij L, 2008, Cost development of future technologies for power generation – A study based on experience curves and complementary and bottom-up assessments, *Energy Policy*, 36, 2200-2211
[52] Haysom J E, Jafarieh O, Anis H, Hinzer K, Wright D, 2014, Learning curve analysis of concentrated photovoltaic systems, *Progress in Photovoltaics*, 23, 1678-1686
[53] Kutscher C, Mehos M, Turchi C, Glatzmaier G, Moss T, 2010, Line-Focus Solar Power Plant Cost Reduction Plant, National Renewable Energy Laboratory, Colorado, United States
[54] Sigfusson B, Uihelein A, 2015, Joint Research Center Geothermal Energy Status Report, European Commission Joint Research Center Institute for Energy and Transport,
[55] Joint Research Center Science and Policy Reports, 2014, Energy Technology Reference Indicator projections for 2010-2050, Joint Research Center Institute for Energy and Transport
[56] Agora Energiewende, 2014, Stromspeicher in der Energiewende, Berlin, September, www.agora-energiewende.de/fileadmin/Projekte/2013/speicher-in-der-energiewende/Agora_Speicherstudie_Web.pdf
[57] Breyer Ch, Tsupari E, Tikka V, Vainikka P, 2015, Power-to-Gas as an emerging profitable business through creating an integrated value chain, *Energy Procedia*, 73, 182-189
[58] International Energy Agency, 2003, World Energy Investment Outlook, International Energy Agency, Paris, France
[59] McDonald A, Schrattenholzer L, 2001, Learning rates for energy technologies, *Energy Policy*, 29, 255-261
[60] Urban W, Lohmann H, Girod K, 2009, Abschlussbericht für das BMBF-Verbundprojekt Biogaseinspeisung, Fraunhofer UMSICHT
[61] Hoffmann W, 2014, Importance and evidence for cost effective electricity storage, 29th EU PVSEC, Amsterdam, September 22-26

Appendix

Table 1: Technical and Financial Assumptions of all energy system components used in the energy transition from 2015 to 2050 for KSA. Most assumptions are taken from Pleßmann et al. [47] and European Commission [48] and further references are individually mentioned.

Name of component			2015	2020	2025	2030	2035	2040	2045	2050	Reference
PV optimally tilted	Capex	€/kWp	1000	800	650	550	490	440	400	370	[49] [50]
	Opex fix	€/(kWp a)	15	12.0	10	8	7	7	6	6	
	Opex var	€/(kWh)	0	0	0	0	0	0	0	0	
	Efficiency	%	16	18	20	22	24	26	28	30	
	Lifetime	years	30	30	35	35	35	40	40	40	
PV single axis tracking	Capex	€/kWp	1150	920	720	620	535	480	435	400	[49] [50]
	Opex fix	€/(kWp a)	17.3	13.8	10.8	9.3	8.025	7.2	6.525	6	
	Opex var	€/(kWh)	0	0	0	0	0	0	0	0	
	Efficiency	%	16	18	20	22	24	26	28	30	
	Lifetime	years	30	30	35	35	35	40	40	40	
Wind onshore	Capex	€/kW	1250	1150	1060	1000	965	940	915	900	[51]
	Opex fix	€/(kW a)	25	23	21	20	19	19	18	18	
	Opex var	€/(kWh)	0	0	0	0	0	0	0	0	
	hub height	m	120	150	150	150	150	150	150	150	
	Lifetime	years	25	25	25	25	25	25	25	25	
CSP (solar field, parabolic trough)	Capex	€/m^2	270	240	220	200	180	170	150	140	[52] [53]
	Opex fix	%	2.3	2.3	2.3	2.3	2.3	2.3	2.3	2.3	
	Opex var	-	0	0	0	0	0	0	0	0	
	Lifetime	years	25	25	25	25	30	30	30	30	
Geothermal	Capex	€/kW	5250	4970	4720	4470	4245	4020	3815	3610	[54] [55]
	Opex fix	€/(kW a)	80.0	80.0	80.0	80.0	80.0	80.0	80.0	80.0	
	Opex var	€/(kWh)	0	0	0	0	0	0	0	0	
	Lifetime	years	40	40	40	40	40	40	40	40	
Water electrolysis	Capex	€/kW	800	685	500	380	340	310	280	260	[56] [57]
	Opex fix	€/(kW a)	32	27	20	15	14	12	11	10	

	Opex var	€/(kWh)	0.0012	0.0012	0.0012	0.0012	0.0012	0.0012	0.0012	0.0012	
	Lifetime	years	30	30	30	30	30	30	30	30	
Methanation	Capex	€/kW	492	421	310	234	208	190	172	160	[56] [57]
	Opex fix	€/(kW a)	10	8	6	5	4	4	3	3	
	Opex var	€/(kWh)	0.0015	0.0015	0.0015	0.0015	0.0015	0.0015	0.0015	0.0015	
	Lifetime	years	30	30	30	30	30	30	30	30	
CO2 scrubbing	Capex	€/kW	749	641	470	356	314	286	258	240	
	Opex fix	€/(kW a)	29.9	25.6	18.8	14.2	12.6	11.4	10.3	9.6	
	Opex var	€/(kWh)	0.0013	0.0013	0.0013	0.0013	0.0013	0.0013	0.0013	0.0013	
	Lifetime	years	30	30	30	30	30	30	30	30	
CCGT	Capex	€/(kW_{el})	775	775	775	775	775	775	775	775	[58]
	Opex fix	€/(kW_{el} a)	19.375	19.375	19.375	19.375	19.375	19.375	19.375	19.375	
	Opex var	€/(kWh)	0	0	0	0	0	0	0	0	
	Efficiency		58	58	58	58	59	60	60	60	
	Lifetime	years	35	35	35	35	35	35	35	35	
OCGT	Capex	€/(kW_{el})	475	475	475	475	475	475	475	475	[58]
	Opex fix	€/(kW_{el} a)	14.25	14.25	14.25	14.25	14.25	14.25	14.25	14.25	
	Opex var	€/(kWh)	0	0	0	0	0	0	0	0	
	Efficiency		43	43	43	43	43	43	43	43	
	Lifetime	years	35	35	35	35	35	35	35	35	
Steam turbine (CSP)	Capex	€/(kW_{el})	760	740	720	700	670	640	615	600	
	Opex fix	€/(kW_{el} a)	15.2	14.8	14.4	14	13.4	12.8	12.3	12	
	Opex var	€/(kWh)	0	0	0	0	0	0	0	0	
	Efficiency	%	42	42	42	43	44	44	45	45	
	Lifetime	years	25	25	25	25	30	30	30	30	
Steam turbine (coal-fired PP)	Capex	€/(kW_{el})	1500	1500	1500	1500	1500	1500	1500	1500	[58] [59]
	Opex fix	€/(kW_{el} a)	20	20	20	20	20	20	20	20	
	Opex var		0	0	0	0	0	0	0	0	
	Efficiency	%	45	45	45	45	46	46	47	47	

	Lifetime	years	40	40	40	40	40	40	40	40	
Biomass CHP	Capex	€/kW	3400	2900	2700	2500	2300	2200	2100	2000	
	Opex fix	€/(kW a)	238	203	189	175	161	154	147	140	
	Opex var	€/(kWh)	0.001	0.001	0.001	0.001	0.001	0.001	0.001	0.001	
	Efficiency	%	36	37	40	43	45	47	47.5	48	
	Lifetime	years	30	30	30	30	30	30	30	30	
Biogas CHP	Capex	€/kW	503	429	400	370	340	326	311	296	
	Opex fix	€/(kW a)	20.1	17.2	16.0	14.8	13.6	13.0	12.4	11.8	
	Opex var	€/(kWh)	0.001	0.001	0.001	0.001	0.001	0.001	0.001	0.001	
	Efficiency	%	35	36	39	42	44	46	46	47	
	Lifetime	years	30	30	30	30	30	30	30	30	
Waste incinerator	Capex	€/kW	5940	5630	5440	5240	5030	4870	4690	4540	
	Opex fix	€/(kW a)	267.3	253.35	244.8	235.8	226.35	219.15	211.05	204.3	
	Opex var	€/(kWh)	0.0069	0.0069	0.0069	0.0069	0.0069	0.0069	0.0069	0.0069	
	Efficiency	%	27	31	32.5	34	35.5	37	29.5	42	
	Lifetime	years	30	30	30	30	30	30	30	30	
Biogas digester	Capex	€/kW	771	731	706	680	653	632	609	589	
	Opex fix	€/(kW a)	30.8	29.2	28.2	27.2	26.1	25.3	24.3	23.6	
	Opex var	€/(kWh)	0	0	0	0	0	0	0	0	
	Efficiency	%	100	100	100	100	100	100	100	100	
	Lifetime	years	20	20	20	20	25	25	25	25	
Biogas upgrade	Capex	€/kW	340	290	270	250	230	220	210	200	[60]
	Opex fix	€/(kW a)	27.2	23.2	21.6	20	18.4	17.6	16.8	16	
	Opex var	€/(kWh)	0	0	0	0	0	0	0	0	
	Efficiency	%	98	98	98	98	98	98	98	98	
	Lifetime	years	20	20	20	20	25	25	25	25	
Battery, Li-ion	Capex	€/(kWh_{el})	600	300	200	150	120	100	85	75	[61]
	Opex fix	€/(kWh_{el} a)	24	12	8	6	4.8	4	3.4	3	
	Opex var	€/($kWh_{throughput}$)	0.0002	0.0002	0.0002	0.0002	0.0002	0.0002	0.0002	0.0002	

	Efficiency	%	90	91	92	93	94	95	95	95	
	Lifetime	years	15	20	20	20	20	20	20	20	
Adiabatic compressed air energy storage (A-CAES)	Capex	€/kWh	35.0	35.0	33.0	31.1	30.4	29.8	28.0	26.3	
	Opex fix	€/(kWh a)	0.46	0.46	0.43	0.40	0.40	0.39	0.36	0.34	
	Opex var	€/(kWh)	0.0012	0.0012	0.0012	0.0012	0.0012	0.0012	0.0012	0.0012	
	Efficiency	%	54	59	65	70	70	70	70	70	
	Lifetime	years	40	55	55	55	55	55	55	55	
Gas storage	Capex	€/kWh_{th}	0.05	0.05	0.05	0.05	0.05	0.05	0.05	0.05	
	Opex fix	€/(kWh a)	0.001	0.001	0.001	0.001	0.001	0.001	0.001	0.001	
	Opex var	€/(kWh)	0	0	0	0	0	0	0	0	
	Lifetime	years	50	50	50	50	50	50	50	50	

Table 2: Energy to power ratio of the storage technologies

Technology	Energy /Power Ratio (hrs)	Self-Discharge
Battery	6	0
A-CAES	100	0.001
Gas Storage	80*24	0

Table 3: Key power capacities required for the energy transition pathway for KSA from 2015 to 2050

		2015	2020	2025	2030	2035	2040	2045	2050
PV single-axis tracking	GWp	0	27.7	48.7	91.2	162.4	210.0	266.1	368.0
PV optimally tilted	MWp	81.6	81.6	81.6	81.6	81.6	72.5	72.5	0
Wind power plants	GW	0	57.2	110.3	132.5	132.5	132.4	132.4	75.3
Geothermal	GW	0	1.8	2.6	2.6	2.6	2.5	2.5	2.5
Battery storage output	TWh	0	0	9.7	46.9	133.3	192.9	262.3	401.1
PtG electrolyser input	GW_e	0	0	0	3.5	9.6	9.8	9.8	9.8

A Cost Optimal Resolution for Sub-Saharan Africa powered by 100% Renewables for Year 2030 Assumptions

Maulidi Barasa, Dmitrii Bogdanov, Ayobami Solomon Oyewo, and Christian Breyer
Lappeenranta University of Technology, Skinnarilankatu 34, 53850 Lappeenranta, Finland
E-mail: christian.breyer@lut.fi

ABSTRACT:
This paper determines a least cost electricity solution for Sub-Saharan Africa (SSA). The power system discussed in this study is hourly resolved and based on 100% Renewable Energy (RE) technologies. Sub-Saharan Africa was subdivided into 16 sub-regions. Four different scenarios were considered according to the setup in high voltage direct current (HVDC) transmission grid. One integrated scenario that considers water desalination and industrial gas production were also analysed. This study uncovers that RE is sufficient to cover 866.4 TWh estimated electricity demand for 2030 and additional electricity needed to fulfil 319 million m³ of water desalination and 268 TWh_{LHV} of synthetic natural gas demand. Existing hydro dams can be used as virtual batteries for solar PV and wind electricity storage, diminishing the role of storage technologies. The results for total levelised cost of electricity (LCOE) decreases from 57.8 €/MWh for a highly decentralized to 54.7 €/MWh for a more centralized grid scenario. For the integrated scenario, including water desalination and synthetic natural gas demand, the levelised cost of gas and the levelised cost of water are 113.7 €/MWh_{LHV} and 1.39 €/m³, respectively. A reduction of 6% in total cost and 19% in electricity generation was realized as a result of integrating desalination and power-to-gas sectors into the system.

Keywords:
100% Renewable Energy, Sub-Saharan Africa, hourly resolution, energy system model, grid integration, sector integration, economics.

1. Motivation

The need for revolution in the energy sector globally is vital, in view of tackling the problems of global warming, pollution, climatic changes, and energy security. The usage and further dependence on fossil fuel in pursuit of energy generation will engender increase of greenhouse gas emissions and more extreme climate impacts, the energy sector accounts for roughly two-third of all anthropogenic greenhouse gas emissions, there is imminent need for a paradigm shift from the conventional to renewable energy resources which will be consistent, clean and cost effective (Greenpeace, 2015; IPCC, 2014; Jacobson and Delucchi, 2011). Investment in sustainable energy infrastructure is the crucial link between economic growth, development, and climatic action. Renewable electricity generation will ensure the reduction of carbon dioxide emissions in electricity generation and to meet climatic targets. Renewable energy optimization solution will reduce the dependency on fossil fuel as predominant energy sources in the power system and address socio-economic development needs and vulnerability to environmental change (UNFCCC, 2015; CLIMATEACTION, 2015).

The Sub-Saharan Africa region continues to face significant energy crises; despite the unique potential of energy sources in the region, severe energy shortage is yet to be conquered and access to electricity eludes millions of people in the region (IEA, 2014b). Billions of dollars are spent annually on inefficient and often dangerous alternatives such as kerosene lamps, candles, flashlight, or other fossil-fueled powered stopgap technologies, in 2012 almost 16 TWh amount of electricity demand was served by backup generators for service and industrial activities in SSA (IEA, 2014b; BNEF, 2016). The current energy challenge in Africa requires a rapid increase in energy supply (growth and development of energy) for the continent due to growing population, unprecedented economic progress and need for reliable modern energy access is expected to require at least double energy supply by 2030 and might even triple for electricity (IRENA, 2015). Africa's electricity generation varies significantly between African countries, North Africa (more than 99% electricity access) dominates the continent in terms of electricity generation from regional perspective (KPMG, 2015), while Sub-Saharan Africa is starved for electricity, only seven countries, Cameroon, Cote d'Ivoire, Gabon, Ghana, Namibia, Senegal and South Africa have electricity access rates exceeding 50%. The average consumption in Sub-Saharan Africa (except South Africa) is only about 150 kWh/capita (Castellano et al., 2015). Electricity demand in Africa was 385 TWh and 621 TWh in 2000 and 2012, respectively, and estimated to increase to about 1,258 TWh and 1,869 TWh by 2030 and 2040, respectively. The Sub-Saharan African demand will be about 812 TWh and 1,297 TWh by 2030 and 2040, respectively (IEA, 2014b). According to Greenpeace (2015), by 2050 95% of the electricity produced in Africa will come from renewable energy sources, whereas wind, solar photovoltaics (PV), concentrating solar thermal power (CSP), and geothermal energy will contribute 80% to the total electricity generation. Already by 2020 the share of renewable electricity production will be 31% and 65% by 2030, the installed capacity of renewables will reach about 380 GW in 2030 and 1,390 GW by 2050.

A brief summary of various studies on the trend of renewable energy share in African energy systems is presented in Table 1. Greenpeace (2015) reveals the possibility of achieving 100% renewable energy in Africa by 2050 and WWF (2015) affirms this possibility. The 2030 renewable energy map for Africa by IRENA (2015) reports that the renewable share in the generation mix in the power sector can grow by 50% while the new energy policies scenario of the IEA (2014b) for 2040 reveals that hydropower account for 26% of electricity generation and other renewable will reach 12 GW (15%). Energy demand keeps growing as population increases in the world (Asif and Muneer, 2007). Africa is expected to contribute more than half of the global population growth between now and 2050, i.e. 1.3 billion people will be added in Africa of the 2.4 billion projected increase in global population between 2015 and 2050 and even more for the following decades (UN, 2015). Most developing countries and Africa need to plan a secure and sustainable means of meeting the current energy shortage problem and future energy demand, due to the fast growing and projected increase. In particular the lack of access to affordable electricity in the Sub-Saharan Africa region requires a massive expansion of access to electricity (IRENA, 2015; Castellano, 2015). Over 620 million people in Sub-Saharan Africa are without access to electricity and nearly 730 million rely on the traditional use of solid biomass for cooking (IEA, 2014b). Sub-Saharan Africa has a huge untapped renewable energy resources potential, widely spread across the region which could provide affordable, sufficient and secure supplies of energy for the continent need.

Therefore, investments in the energy system in Sub-Saharan Africa with a focus on increasing energy accessibility and affordability, will improve the quality of life, life expectancy and economic growth (UN, 2015; IRENA, 2013).

Table 1: Studies on trends of renewable energy shares in the African energy system

Reference	Scope	Findings
Greenpeace (2015)	Global	The energy revolution scenario 2015 reveals a progressive share of renewables in African energy system till year 2050: 31%, 65% and 100% by 2020, 2030 and 2050, respectively, in electricity generation. Increase in installed capacity from 380 GW (2030) to 1,390 GW (2050).
IRENA (2015)	Africa	Reduction of about 310 Mt of CO_{2eq} emission by 2030 with 50% share of RE in the power sector. Solar capacity could reach 90 GW, hydropower and wind 100 GW each.
IEA (2014b)	Sub-Saharan Africa	Electricity generation according to the 2040 new policies scenario is 1,540 TWh; hydro accounts for 26% of the total generation and other renewable which includes solar, wind, bioenergy, geothermal and CSP will reach 12GW (15%) by 2040
WWF (2015)	Uganda	Transition to 100% RE in Uganda requires, adoption of on- and off-grid solar energy and increasing use of other RE. With a target of 60% RE system in 2030 and close to 100% by 2050.

Africa's energy sector is vital to its development, therefore effective energy planning, optimal design, wise utilization of all available renewable energy (RE) resources and maximum synergy between various resources and different regions (regional electricity networking due to disperse energy resource) of Africa will positively impact on energy systems of the continent (IRENA, 2015). In view of designing and optimizing a renewable based energy system several research have been carried out aiming at increasing utilization of renewables in renewable energy rich regions (such as Northeast Asia), by utilizing high voltage direct current grids (HVDC) for interconnections within the regions (Bogdanov and Breyer, 2016).

Access to modern energy services can be secured by RE, especially solar PV (IEA, 2014). Globally, solar PV installations have shown high growth rates and one of the fastest growing renewable technology with continuous decreasing generation cost (Nygaard et al., 2014b; Breyer and Gerlach, 2013). PV in Africa has transcended the government-donor niche to a commercial based market development (Hansen et al., 2014). Solar PV technology currently plays a dominant role in the off-grid market, 4 million solar products were sold globally in the last half of 2015, and Sub-Saharan Africa sale accounts for 2.2 million (54.3%) (GOGLA and Lightning Africa, 2016). PV will play a crucial role towards a 100% renewable energy system in Africa and requires the adoption of on- and off-grid solar energy and increasing use of other RE (WWF, 2015). Mini-grids and other off-grid technologies account for 26 TWh and 12 TWh power generation, respectively, in Sub-Saharan Africa in the new policies scenario 2040, solar PV contributes 37% and 47% of the technology mix (IRENA, 2015). Solar home systems and mini-grids provide technically feasible and economical solutions to energy challenges, where the cost of grid extension is not economical (Niez, 2010). In Africa, solar has the largest RE resource potential and the high quality resource is widely available and thefor the PV and CSP technical power could be as high as 6,567 TWh and 4,719 TWh, respectively. The rapid cost reduction for PV, has led to increasing annual capacity additions and enhancing off-grid access (IRENA, 2012). The Sub-Saharan Africa energy sector is vital to its development, therefore effective energy planning, optimal design, and wise utilization of the untapped, abundant renewable energy resources across Sub-Saharan Africa will positively impact on energy systems of the region.

The idea of a super grid has recently attracted more attention by Gobitec and the Northeast Asian Super Grid initiative, influenced by EU-MENA Desertec (Mano et al., 2014; Dii, 2012). The super grid approach enables connections within a region or continent to attain synergy effects by the harnessing of RE resources and help to realize a 100% RE electricity supply system which can set the path for a 100% RE supply (Bogdanov and Breyer, 2016). The main focus of this paper is to design an optimal energy system in Sub-Saharan Africa, which is cost efficient and competitive, a 100% RE-based system via optimal design and wise utilization of all available RE resources and supportive storage and grid technologies.

2. Methodology

In this study, the power system model was developed based on linear optimization of energy system parameters under applied constraints and is composed of a set of RE generation and storage technologies, as well as water desalination and synthetic natural gas (SNG) generation sectors, which operates as flexible demand. A fully integrated scenario that considers furthermore heat and mobility demand have to be modeled, to understand the whole energy system. However, the heat and mobility sectors were not considered in this study. As the applied energy system model has already been described in Bogdanov and Breyer (2016), the following sections do not include a detailed description of the model, its input data, and the applied technologies. However, it presents a comprehensive definition of all additional information that has been assumed for the model in the present study. Further technical and financial assumptions can be found in the Supplementary Material in the appendix of this paper.

2.1. Model Overview

The energy system model used in this study was developed with a perfect foresight of RE power generation and power demand. The use of multi-node approach allows for the description of any desired configuration. The main study constraint for the optimization is to guarantee that for every hour of the year the total electric generation within a sub-region covers the local demand from all considered sectors and enables a precise system description including synergy effects of different system components for the power system balance.

The outcome for the system optimization is the minimization of the total annual energy system cost, calculated as the sum of the annual costs of installed capacities of the different technologies, costs of energy generation and generation ramping. The model also includes distributed generation and self-consumption of residential, commercial, and industrial electricity consumers (prosumers) by installing respective capacities of rooftop PV systems and batteries. For these prosumers the target function is the minimal cost of consumed energy calculated as the sum of self-generation, annual cost, and cost of electricity consumed from the grid, less benefits from selling excess energy. The model flow diagram that has all the considered input data, system models and model output data is presented in Figure 1.

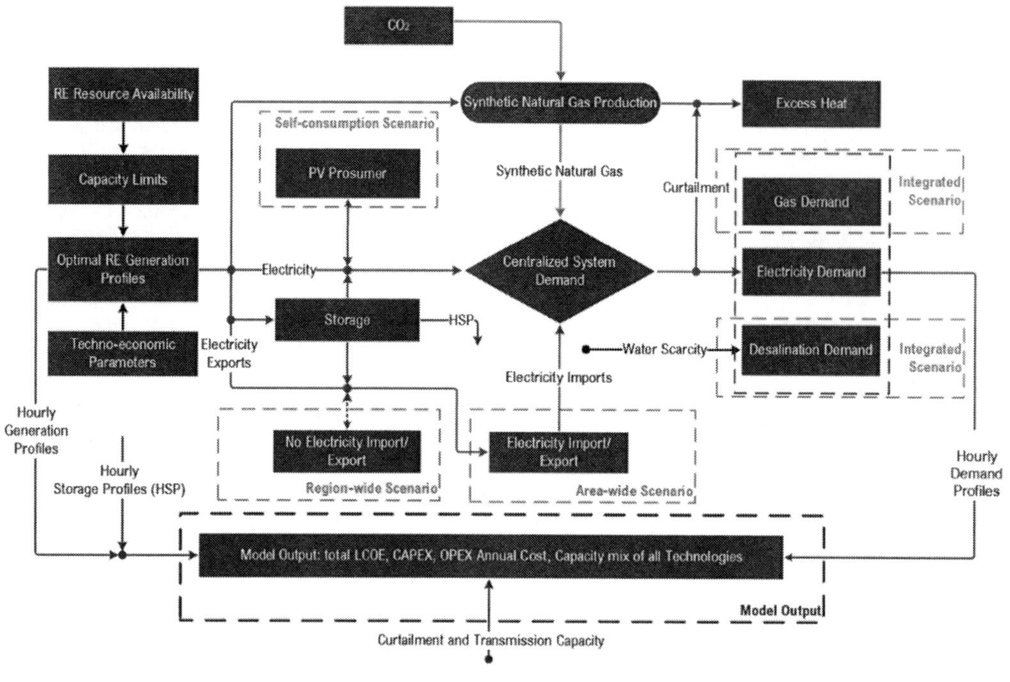

Fig. 1. Model framework diagram for the study scenarios.

2.2. Input data

Additional information to Bogdanov and Breyer (2016) about geothermal data, water desalination and industrial gas demand are defined in this section. The geothermal potential is evaluated based on available information on the surface heat flow rate (IASPEI, 2015; AAPG, 2015) and surface ambient temperature for the year 2005 globally. For areas where surface heat flow data are not available, an extrapolation of existing heat flow data was performed. Based on that, temperature levels and available heat at the middle depth point of each 1 km thick plate, between depths of 1 km and 10 km (Chamorro et al., 2014a, 2014b; Huenges, 2012) globally with 0.45°x0.45° spatial resolution, are derived.

Water desalination demand is projected based on water consumption levels and future water stress (Luck et al., 2015). Water stress occurs when the water demand exceeds renewable water availability during a given period. It is assumed that water stress, greater than 50% shall be covered by seawater reverse osmosis (SWRO) desalination. Transportation costs are also taken into account, as described by Caldera et al. (2016). The energy consumption of horizontal and vertical pumping are 0.04 kWh/ ($m^3 \cdot h \cdot$100km) and 0.36 kWh/ ($m^3 \cdot h \cdot$100m), respectively (Caldera et al., 2016).

Present industrial gas consumption is based on natural gas demand data from the International Energy Agency statistics (IEA, 2015) and natural gas consumption projections for the year 2030 were calculated considering industrial annual growth projections based on the World Energy Outlook (IEA, 2014b).

2.3. Applied Technologies

The technologies applied in the cost optimization for Sub-Saharan Africa can be classified into four main categories: conversion of RE resources into electricity, energy storage, energy sector bridging (for definition, see later), and electricity transmission. The technologies for converting RE resources into electricity applied in the model are ground-mounted (optimally tilted and single-axis north-south oriented horizontal continuous tracking) and rooftop solar PV systems, concentrating solar thermal power (CSP), wind onshore turbines, hydro power (run-of-river and dams), biomass plants (solid biomass and biogas), waste-to-energy power plants and geothermal power plants. The energy storage technologies used in this study are battery storage, pumped hydro storage (PHS), adiabatic compressed air energy storage (A-CAES), thermal energy storage (TES) and power-to-gas (PtG) technology. PtG includes synthetic natural gas (SNG) synthesis technologies: water electrolysis, methanation, CO_2 scrubbing from air, gas storage, and both combined and open cycle gas turbines (CCGT, OCGT). SNG synthesis process technologies have to be operated in synchronization because of hydrogen and CO_2 storage absence. Additionally, there is a 48-hour biogas buffer storage, and a part of the biogas can be upgraded to biomethane and injected into the gas storage.

The energy sector bridging technologies give more flexibility to the entire energy system, thus reducing the overall cost. One bridging technology available in the model is PtG technology for the case that the produced gas is consumed in the industrial sector and not as a storage option for the electricity sector. The second bridging technology is seawater reverse osmosis (SWRO) desalination, which couples the water sector to the electricity sector. The technologies are represented on two levels: power distribution and transmission within the sub-regions are assumed to be based on standard alternating current (AC) grids which are not part of the model and inter-regional transmission grids modeled by applying high voltage direct current (HVDC) technology.

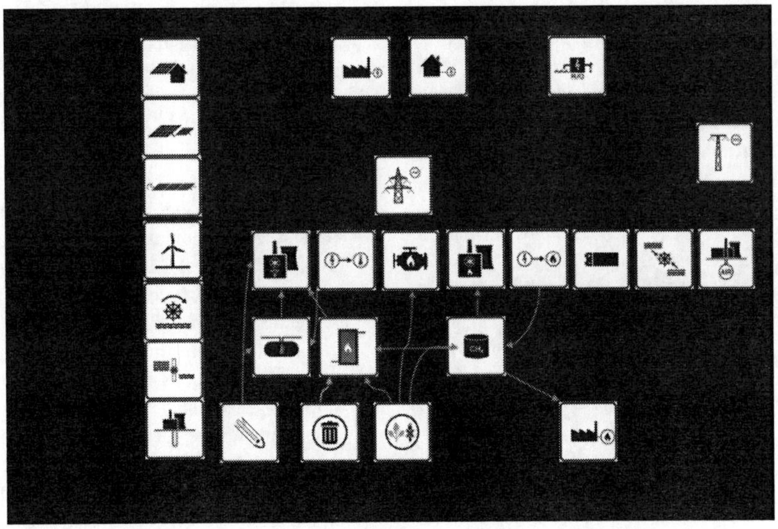

Fig. 2. Block diagram of the energy system model for Sub-Saharan Africa.

Power losses in the HVDC grids consist of two main elements: length dependent electricity losses of the power lines and losses in the converter stations at the interconnection with the AC grid. An energy system mainly based on RE and in particular intermittent solar PV and wind power requires different types of flexibility for an overall balanced and cost optimized energy mix. The four broad categories are generation management (e.g. hydro dams or biomass plants), demand side management (e.g. PtG, SWRO desalination), storage of energy at one location and energy shifted in time (e.g. batteries), and transmission grids connecting different areas and energy shifted to the site (e.g. HVDC transmission). The full model block diagram is presented in Figure 2.

3. Scenario assumptions

3.1. Regions subdivision and grid structure

This study considered 51 countries that are merged or subdivided into 16 sub-regions of the Sub-Saharan Africa super regional network based on area, population, and national grid connections. The 16 interconnected sub-regions include: West-West (Senegal, Gambia, Cape Verde Islands, Guinea Bissau, Guinea, Sierra Leone, Liberia, Mali, Mauritania and Western Sahara), West-South (Ghana, Cote d'Ivoire, Benin, Burkina Faso and Togo), West-North (Niger and Chad), South-Nigeria (South Nigeria), North-Nigeria (North Nigeria), Sudan-Eritrea (Sudan and Eritrea), Ethiopia (Ethiopia), Somalia (Djibouti and Somalia), Kenya-Uganda (Kenya and Uganda), Tanzania (Rwanda, Burundi, Tanzania), Central (Central African Republic, Cameroon, Equatorial Guinea, Sao Tome and Principe, Democratic Republic of the Congo [Kinshasa], Republic of the Congo [Brazzaville] and Gabon), South-West (Angola, Namibia and Botswana), South Africa (the Republic of South Africa and Lesotho), South-East (Malawi, Mozambique, Zambia, Zimbabwe and Swaziland) and Indian Ocean (Comoros Islands, Mauritius, Mayotte, Madagascar and Seychelles).

In this paper four scenarios for energy system development options are discussed:

1. Regional-wide, which assumes that the sub-regions operate independently (with no HVDC grid interconnections), and the electricity demand has to be covered by the respective region's supply;
2. Country-wide energy system, which assumes that the sub-regions are interconnected via

HVDC lines but within the borders of nations;

3. Area-wide energy system, which assumes that all sub-regions are interconnected via HVDC lines;

4. Integrated scenario, which takes into account additional demand sectors (SWRO desalination and industrial gas demand) to the area-wide energy system scenario. In this scenario, RE sources combined with PtG technology are used not only for electricity generation and storage options within the system but also as energy sector bridging technologies to cover water desalination and industrial gas demand, increasing the flexibility of the system.

Fig. 3. Sub-Saharan African sub-regions and HVDC transmission lines configuration.

The Sub-Saharan African regions' subdivision and grid configuration are presented in Figure 3. Existing HVDC interconnections of Sub-Saharan African regions are shown by dashed lines. The structure of the HVDC grid for the scenarios (solid lines) is based on the existing configuration of Sub-Saharan Africa Power Pools.

3.2. Financial and technical assumptions

The model optimization is undertaken based on cost assumptions derived from technological status for the year 2030 and the overnight building approach, which is also applicable to nuclear energy (Carvalho and Sauer, 2009). The financial assumptions for capital expenditures (CAPEX), operational expenditures (OPEX) and lifetimes of all components are provided in the Supplementary Material (in Table I). CAPEX and OPEX are quantified estimates by weighting the hourly LCOE per technology (generation, transmission, and storage) balanced with hourly demand for 2030. Weighted average cost of capital (WACC) is set at 7% for all scenarios, but for residential PV self-consumption, WACC is set to 4%, due to lower financial return requirements. The technical assumptions of power to energy ratios for storage technologies, efficiency numbers for generation and storage technologies, and power losses in HVDC power lines and converters are provided in the Supplementary Material (Tables II, III, and IV, respectively). The electricity prices for residential, commercial, and industrial consumers in most of the Sub-Saharan Africa countries for the year 2030 are obtained from Gerlach et al. (2014b) and aggregated based on the populations weighted estimates as presented in the Appendix (in Table V). Prosumers will use PV to supply a portion of their electricity needs. Thus, prosumers will cut the slice of electricity demand and get remunerated for any surplus generation, which does not top their annual consumption. The operation cost for prosumers depend on market conditions for the year 2030 and covers the cost of new rooftop PV systems and batteries. Excess generation, which cannot be self-consumed by the solar PV prosumers, is assumed to be fed into the grid for a transfer selling price of 2 €cents/kWh.

3.3. Feed-in for solar and wind energy

The feed-in profiles for solar CSP, PV optimally tilted and single-axis tracking, and wind energy were calculated according to Bogdanov and Breyer (2016). The aggregated profiles of solar PV generation (optimally tilted and single-axis tracking), wind energy power generation, and CSP solar field, normalized to the maximum capacity averaged for Sub-Saharan Africa are presented in Figure 4. The computed average full load hours (FLH) for optimally tilted, single-axis tracking PV systems, wind power plants, and CSP are provided in the Supplementary Material (Table VI). The feed-in values for hydro power are computed based on the monthly resolved precipitation data for the year 2005 as a normalized sum of precipitation in the regions. Such an estimate leads to a good approximation of the annual generation of hydro power plants.

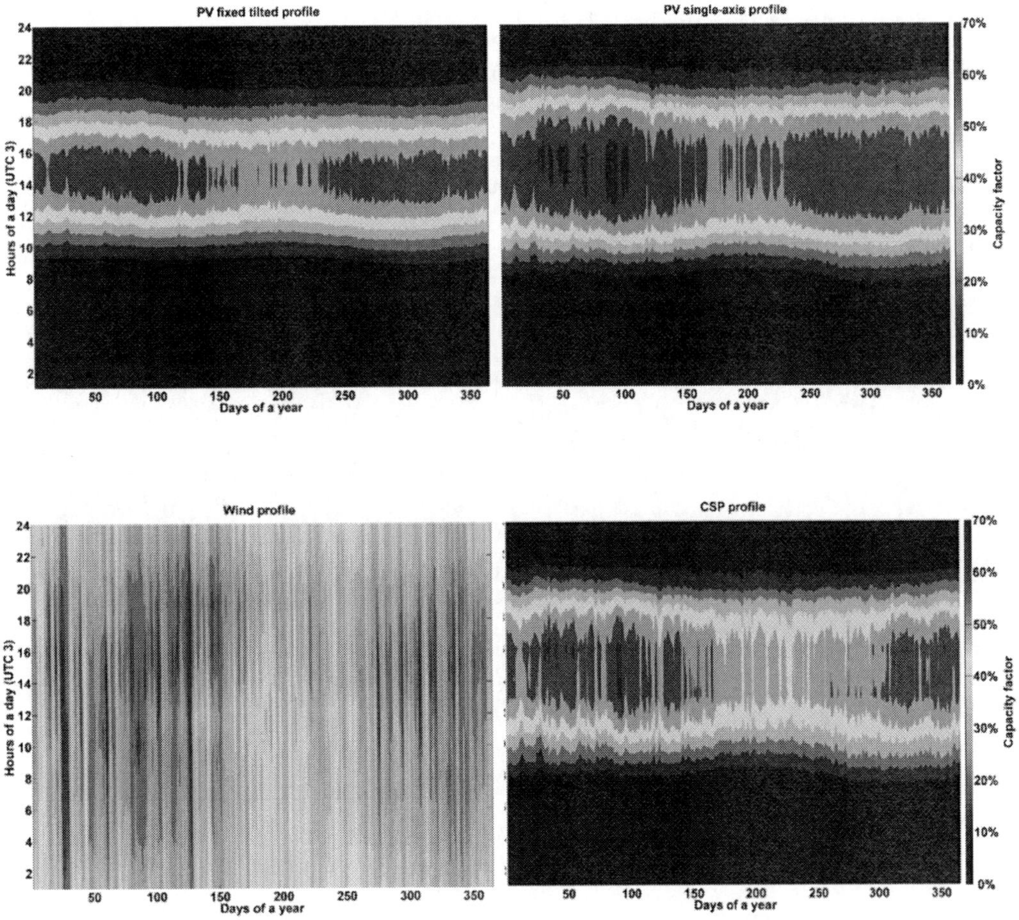

Fig. 4. Aggregated feed-in profiles for PV optimally tilted (top left), PV single-axis tracking (top right), 3 MW 150 m hub height wind turbine (bottom left) and CSP solar field (bottom right).

3.4. Biomass and geothermal heat potentials

Biomass and waste resource potentials are mainly taken from DBFZ (2009) and classified as described in Bogdanov and Breyer (2016). Costs for biomass are calculated using data from the International Energy Agency (IEA, 2012) and Intergovernmental Panel on Climate Change (IPCC, 2011). For solid wastes a 100 €/ton gate fee for incineration is assumed. Calculated solid biomass, biogas, solid waste, and geothermal heat potentials are provided in the Supplementary Material (Table VII). Prices for biomass fuels are provided in the Supplementary Material (Table VIII), and price differences between countries are explained by various waste and residue component shares. Heating values are based on lower heating values (LHV). Regional geothermal heat potentials are calculated based on spatial data for available heat, temperature, and geothermal plants for depths of 1 km to 10 km. For each 0.45°x0.45° area and depth, geothermal LCOE is calculated and optimal well depth is determined. It is assumed that only 25% of available heat will be used as an upper resource limit. The total available heat in the region is calculated using the same weighed average formula as for solar and wind feed-in explained in Bogdanov and Breyer (2016), except for the fact that areas with geothermal LCOE exceeding 100 €/MWh are excluded.

3.5. Upper and lower limitations on installed capacities

Lower limits are taken from Farfan and Breyer (2016) and upper limits were estimated based on Bogdanov and Breyer (2016). Lower limits on already installed capacities in Sub-Saharan Africa are provided in the Supplementary Material (Table IX) and all upper limits of installable capacities are summarized in Supplementary Material (Table X). For all other technologies, upper limits are not specified. However, for biomass residues, biogas, and waste-to-energy plants it is assumed, due to energy efficiency reasons, that the available and specified amount of the fuel is used during the year, except for solid biomass which can be used up to the full resource potential.

3.6. Load

The synthetic load data are generated based on hourly load data on a national level and takes into account local phenomena such as gross domestic product, population, temperature, and power plant structure. Figure 5 represents the area-aggregated demand of all sub-regions in Sub-Saharan Africa. Electricity demand increase by the year 2030 is estimated using IEA data (IEA, 2014b) and local data. Solar PV self-consumption prosumers have a significant impact on the extra load on the energy system as illustrated in Figure 5 (right). The total electricity demand and the peak load are reduced by 11.2% and 5.2%, respectively.

Industrial gas demand values (expressed as gas demand excluding electricity generation and residential sectors) and desalinated water demand for Sub-Saharan Africa sub-regions are presented in the Supplementary Material (Table XI), gas demand values are based on the (IEA, 2015). Desalination demand numbers are based on water stress and water consumption projections obtained from (Caldera et al., 2016).

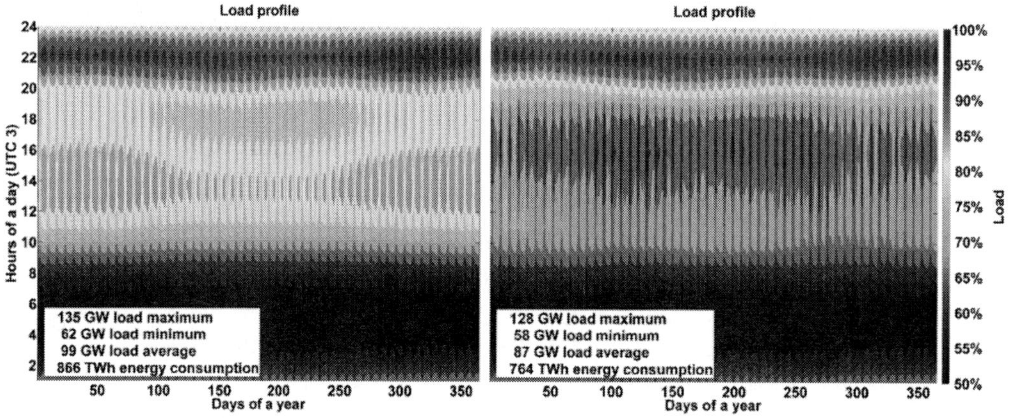

Fig. 5. Aggregated load curve (left) and system load curve with prosumers influence (right) for the year 2030.

4. Results

To describe the cost structure of the different scenarios, a set of central parameters are computed according to the methodology described in Bogdanov and Breyer (2016).

4.1. Key findings on the optimized energy system structure and costs

The section summarizes the findings from the optimized cost modeling approach for 100% RE in Sub-Saharan Africa. The cost minimized electric power system configurations are derived for all scenarios, the given constraints and characterized by optimized installed capacities of RE electricity generation, storage and transmission for every modeled technology, leading to respective hourly electricity generation, storage charging and discharging, electricity export, import, and curtailment. The average financial results for the study scenarios are expressed as levelised cost of electricity (LCOE), levelised cost of electricity for primary generation (LCOE primary), levelised cost of curtailment (LCOC), levelised cost of storage (LCOS), levelised cost of transmission (LCOT), total annualized cost, total capital expenditures, total renewables capacity and total primary generation, as presented in Table 2.

Table 2 displays the financial results for all four analyzed scenarios. The role of HVDC transmission is important in achieving the least cost solution for Sub-Saharan Africa for the year 2030. It can be observed that the installation of HVDC transmission lines leads to a significant reduction of electricity cost of the entire system; a comparison of the region-wide and area-wide scenario shows a 5.4% and 5.8% decrease in LCOE and the annual expenses for the system, respectively. The same conclusion can be drawn regarding grid utilization which shows a reduction in installed capacities by 5.3% and in total electricity generated by 2.1%. Grid utilization results in a considerable reduction of storage utilization (Table 3), whereas the cost of transmission is relatively small in comparison to the decrease in the primary generation and storage costs. Storage costs decrease more significantly than curtailment costs in the case of broader grid utilization, leading to a reduction of 18.8% in the area-wide scenario compared to the region-wide scenarios, respectively; however, the impact of excess energy on total cost is rather low.

The integration of water desalination and industrial gas sectors results in a further reduction in LCOE by 23.4% compared to the area-wide open trade scenario. The cost reduction benefits are a result of a massive reduction cost of storages, and a minor decrease in the cost of curtailment. Storage cost reduces by 48.3% since industrial gas and desalination sectors decrease the need for long-term storage utilization, giving additional flexibility to the system. The utilization of low-cost wind and solar electricity results to increased system flexibility as can be seen in Table 3, this results to an 8.7% decrease in primary electricity generation cost. A comparison of the area-wide and area integrated scenario emphasizes the increased flexibility in the system by an additional PtG capacity of 47.0 GW ; this occurs despite the availability of low cost biogas, since the biogas resource potential cannot cover the full gas demand in the energy system as tabled in the Supplementary Material (Tables VII and XI). The numeric values for LCOE components in all sub-regions and scenarios are summarized in the Supplementary Material (Table XIV).

Regarding the RE installed capacities, all the RE technologies present a reduction of total installed capacity with an increase of grid utilization (Table 3); in all the scenario, solar PV technologies have the largest overall share compared to alternative technologies. The share of PV dominates as a result of interconnectivity reaching 55.5%, as the share of wind drops to 30.8%. The rest of the RE technologies take the remaining, with hydro, biomass and geothermal taking a slight share that ranges from 4.7% to 6.6%, 4.5% to 7.2%, and 1.0% to 1.3%, respectively. The share for hydro and geothermal show some minor changes as a result of interconnection and integration of gas and desalination sectors. Hydro dams also act a virtual storage to the system and provide required flexibility to the system. Interconnection and integration have considerable impacts on storage capacities.

The cost of most RE technologies has declined and additional expected technical advances would result in further cost reductions. PV self-consumption overview is given in the Supplementary Material (Table XII). Self-generation plays a crucial role in Sub-Saharan Africa owing to high electricity prices throughout in the region and low self-consumption LCOE. Self-consumption covers 84.7% of residential prosumers' demand, 88.0%, and 89.4% of generated PV electricity for commercial and industrial prosumers.

Despite the fact that an upper limit 50% higher than the current capacity was considered for hydro dams and hydro RoR plants, the total hydropower plants' installed capacity practically did not change considering all the studied scenarios. It is not necessary to utilize all the available hydro power capacities considering PV and wind seemed to be more cost competitive options for Sub-Saharan Africa.

As seen results for the area-wide scenario, the interconnection of sub-regions in Sub-Saharan Africa has considerable impacts on storage capacities. Gas storage dominates the proportions for storage capacities in all the scenarios followed by A-CEAS. However, when the storage throughputs are considered, there is a complete shift of dominance to battery storage which is utilized especially during the night hours, in the absence of solar PV. The gas storage has a moderately low throughput of 45.9 TWh compared to batteries storage with the highest throughput of 77.5 TWh. Besides gas, batteries, other storage solutions utilized include, PHS which has the least throughput of 0.83 TWh considering it is only utilized in the South African sub-region and A-CAES which is still surpassed by batteries and gas storage of 10.5 TWh. The storage capacities of batteries, PHS, A-CAES, PtG and gas turbines decrease as a result of interconnection and integration of gas and desalination sectors. However, gas storage increases from 18.9 TWh to 24.3 TWh as a consequence of the integration of gas and desalination sectors. PtG electrolysers have a rather low installed capacity in the area-wide and area-wide scenario, since PtG is not needed much for seasonal storage. Hydro dams have a key role as virtual batteries for solar and wind long-term balancing, reducing interregional electricity trade, electricity transmission costs and reducing the seasonal storage contribution of PtG. The total LCOE decline from the region-wide to the integrated scenario by 18.8% as results of geographic integration and sector coupling.

Table 2: Financial results for the four scenarios applied in Sub-Saharan Africa.

Scenario	Total LCOE [€/MWh]	LCOE primary [€/MWh]	LCOC [€/MWh]	LCOS [€/MWh]	LCOT [€/MWh]	Total ann. cost [b€]	Total CAPEX [b€]	RE capacities [GW]	Generated electricity [TWh]
Region-wide	58	38	2	18	0	50	429	404	981
Country-wide	58	38	2	18	0	50	428	403	980
Area-wide	55	36	2	15	2	47	416	382	960
Integrated	47	34	2	9	2	62	563	544	1327

Table 3: Overview on installed RE technologies and storage capacities for the four scenarios.

RE Technology	Units	Region-wide	Country-wide	Area-wide	Integrated
PV self-consumption	[GW]	61.3	61.3	61.3	61.3
PV optimally tilted	[GW]	1.4	1.4	1.4	1.4
PV single-axis tracking	[GW]	134.6	133.9	111.5	239.1
PV total	[GW]	197.3	196.7	174.2	301.9
CSP	[GW]	0.0	0.0	0.0	0.0
Wind energy	[GW]	125.9	126.1	138.7	167.6
Biogas power plants	[GW]	4.5	4.2	1.6	1.3
Biomass power plants	[GW]	1.7	1.7	1.7	1.7
MSW incinerator	[GW]	23.1	23.2	21.5	21.5
Geothermal energy	[GW]	5.1	4.9	4.6	5.4
Hydro Run-of-River	[GW]	5.2	5.2	4.7	5.4
Hydro dams	[GW]	20.4	20.4	20.4	20.4
Battery PV self-consumption	[GWh]	72.2	72.2	72.2	72.2
Battery total	[GWh]	262.6	268.6	240.3	218.4
PHS	[GWh]	3.2	3.2	3.2	3.2
A-CAES	[GWh]	623.2	579.7	432.5	335.0
Gas storage	[GWh_{th}]	25,754	25,948	18,936	24,304
PtG electrolyzers	[GW_{el}]	9.1	9.1	6.5	53.9
CCGT	[GW]	16.9	16.7	13.6	9.8
OCGT	[GW]	13.2	13.0	10.3	15.3
Steam Turbine	[GW]	30.1	29.7	23.9	25.1

An overview of the electricity generation curves for the area-wide scenario can be seen in Figure 6. All 8760 hours of the year are sorted according to the net hourly supply (generation minus load), which is represented by the black line. A higher electricity generation than demand can be observed for 4400 hours of the year, which is used for charging storage. This is caused by a high electricity generation from inflexible energy sources, due to high shares of solar PV and wind energy in Sub-Saharan Africa energy mix and a higher solar irradiation and wind speed in the region during these hours of the year. As a consequence, flexible electricity generation options (such as hydro dams, biomass, and biogas) and discharge of storage plants are needed. The inflexible electricity generation reduces significantly in comparison to the decrease in electricity demand for the remaining hours of the year, thus increasing the need for flexible electricity generation, energy storage discharge and grid utilization. The storage plants are deployed for about 4400 hours of the year in charging mode and about 4200 hours in discharging mode. Electricity curtailment is significant for about 1000 hours in the year and not applied during almost the entire period since the existence of storage and HVDC transmission lines enables storing electricity or that sub-regions with best RE resources to export electricity to the ones with a shortage in RE resources.

If a sub-region analysis is considered, as presented in Figure 7, noticeable differences can be observed between the scenarios, especially between the area-wide and the integrated scenarios. More demand in the case of a RE-based energy system can change the entire system structure because of shifting optimal cost structure parameters and areas being confronted with their upper resource limits. For region-wide and area-wide scenarios, the share of PV dominates in almost all the sub-regions considered; apart from Kenya-Uganda, Somalia-Djibouti, and West-North where wind energy plays a key role due to favorable wind resources. Wind plays a key role in the integrated scenario, in which an additional electricity demand for additional water desalination and SNG production was included, especially the sub-regions that have excellent wind conditions and, therefore, low cost wind energy.

4.2. Main findings on the optimized energy system structure in a sub-region analysis

The region-wide scenario neglects the broad cost saving opportunities available as a result of HVDC interconnection. The integration of water desalination and industrial gas sectors to the power system increases the electricity demand but also the energy system flexibility. The total installed capacities reduce as a result of HVDC transmission grid interconnection as seen in Figure 7 and Table 3. This decrease occurs mainly because of a reduction in capacity by 17.2% for PV single-axis in Sub-Saharan Africa, which is seen when region-wide to area-

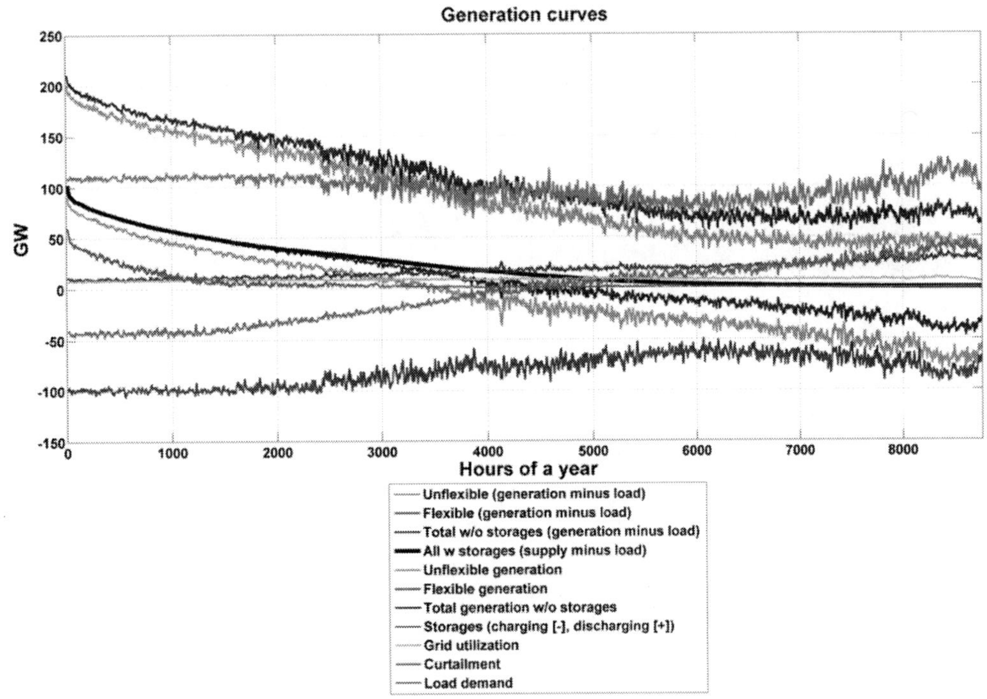

Fig. 6. Electricity generation duration curves for the area-wide open trade scenario for Sub-Saharan Africa.

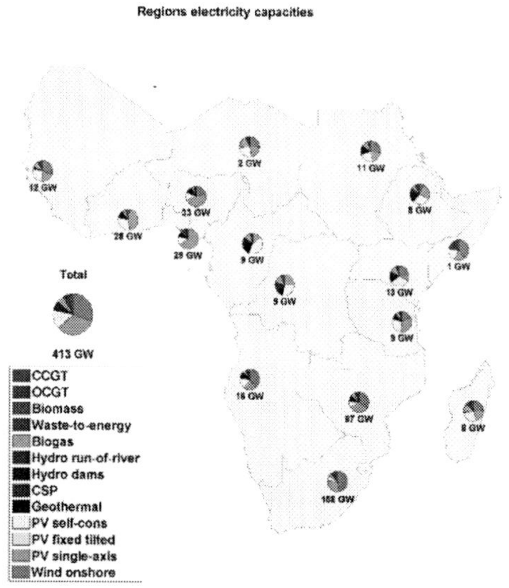
Regions electricity capacities
Total
413 GW
CCGT
OCGT
Biomass
Waste-to-energy
Biogas
Hydro run-of-river
Hydro dams
CSP
Geothermal
PV self-cons
PV fixed tilted
PV single-axis
Wind onshore

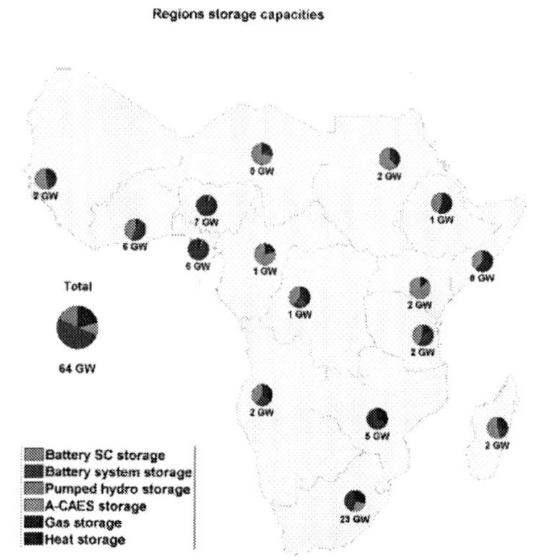
Regions storage capacities
Total
64 GW
Battery SC storage
Battery system storage
Pumped hydro storage
A-CAES storage
Gas storage
Heat storage

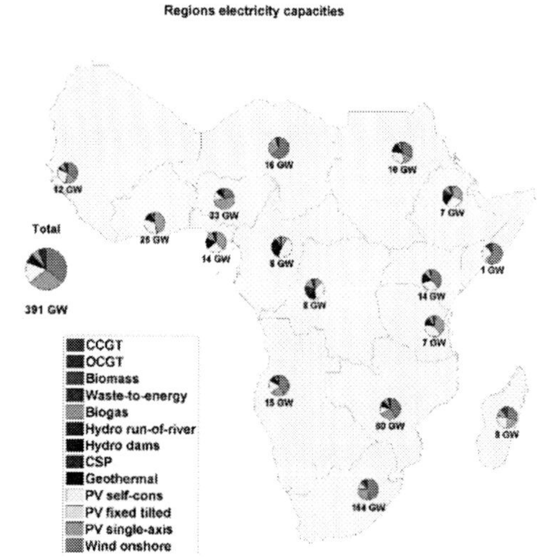
Regions electricity capacities
Total
391 GW
CCGT
OCGT
Biomass
Waste-to-energy
Biogas
Hydro run-of-river
Hydro dams
CSP
Geothermal
PV self-cons
PV fixed tilted
PV single-axis
Wind onshore

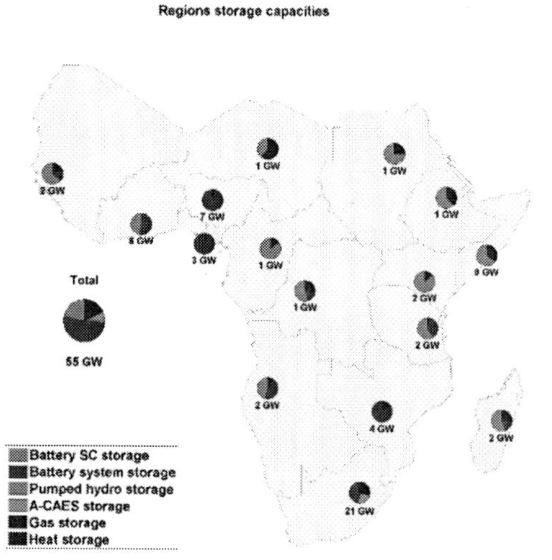
Regions storage capacities
Total
55 GW
Battery SC storage
Battery system storage
Pumped hydro storage
A-CAES storage
Gas storage
Heat storage

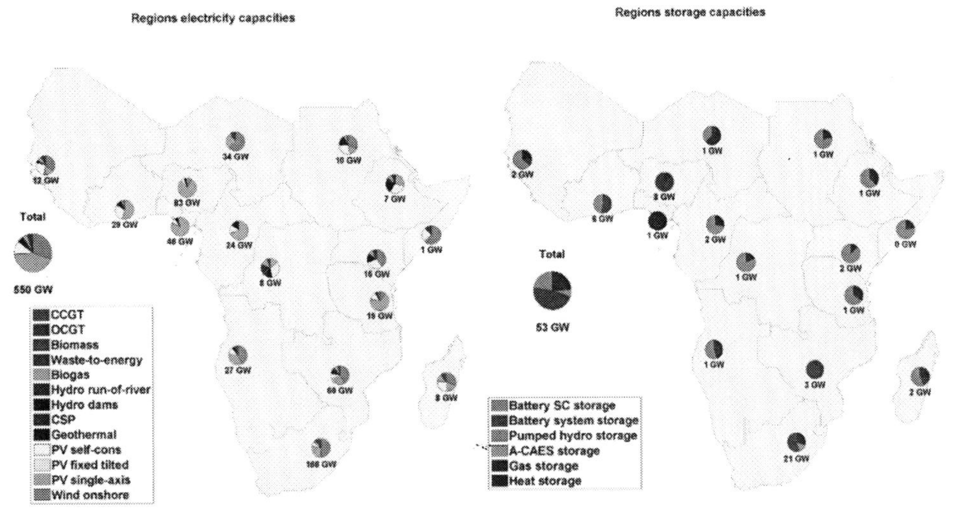

Fig. 7. Installed capacities of RE generation (left) and storages (right) for region-wide (top), area-wide (center) integrated (bottom) scenarios.

wide scenarios are compared; with an exception of North-Nigeria and West-North that showed an increase in total RE installed capacities by 9.9% and 898%, respectively. The significant reduction in PV shares is countered by an increase in wind capacities by 10.2% for a mainly in the West-North sub-region which shows a 20-fold increase from 0.49 GW to 10.4 GW. This results highlight the role played by wind as more sectors are integrated into the system. The structure of HVDC power lines and utilized RE resources strongly influences the total storage capacity needed. In this context, the already installed hydro dams are an important RE source that can act as virtual batteries for long-term storage and seasonal balancing. The regional capacities also vary between sub- regions. Data of storage systems' discharge capacities, annual energy throughput, and full load cycles per year are summarized in the Supplementary Material (Table XIII). The generation capacities of storage technologies decrease with the geographic integration by applying an HVDC transmission lines. State-of-charge profiles for the area-wide scenario for battery, PHS, A-CAES and gas storages and hydro dams are provided in the Supplementary Material (Fig. 4 and Fig. 5). The state-of-charge diagrams show the system optimized operation mode of the different storage technologies: mainly daily (battery, PHS), mainly weekly (A-CAES) and mainly seasonal (gas, hydro dams). Additionally, the absolute and percentage hourly grid capacity profiles in the integrated scenario for the different times of a day and days of a year are as presented in the supplementary Figure 6).

4.3. Electricity import/export

In the case of the region-wide open trade scenario, all sub-regions of Sub-Saharan Africa need to match their demand using only their RE resources. Nonetheless, in the case of the country-wide and area-wide open trade scenarios, a division of sub-regions into net exporters and net importers with interregional electricity flows can be observed. An annual import and export diagram for area-wide open trade is presented in Figure 8. Differences in generation and demand are mainly due to export and import, but in a minor quantity also due to storage losses, and for the area-wide integrated scenario due to energy consumption for SNG production. Figure 8 also gives a good overview of regions' RE resources; net exporters are sub-regions with the best renewable resources and net importers are sub-regions with moderate ones.

The results in supplemental Table XIV summarizes the import/export shares in all regions and scenarios. The share of export is defined as the ratio of net exported electricity to the generated primary electricity of a sub-region and the share of import is defined as the ratio of imported electricity to the electricity demand. The area average is composed of sub-regions' values weighted by the electricity demand. West-North emerges as the major net exporter in Sub-Saharan Africa, located in the Sahara Desert, the region depicts a surprisingly favorable wind and solar generation with a low electricity demand. West-North exports 36 TWh of electricity to South-Nigeria the major net importer in Sub-Saharan Africa, with a higher population. The rest of the net importing regions include Tanzania, West-West, Congo, and South-East.

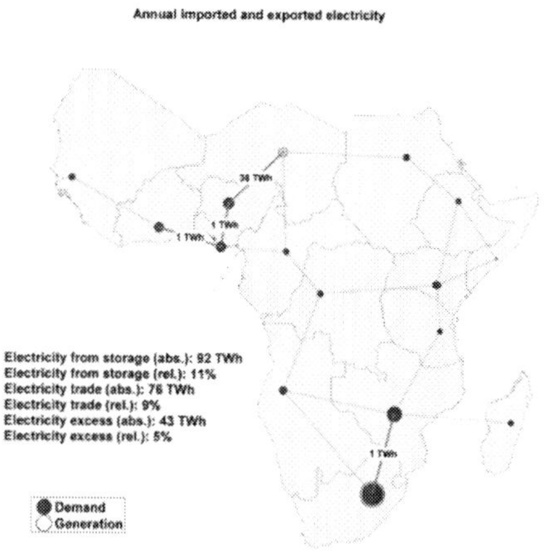

Fig. 8. Annual generation and demand diagram for the area-wide open trade scenario.

The remaining sub-regions are classified as balancing sub-regions since electricity is both imported and exported in the long term (hourly) and in the long term (annually). Hourly resolved profiles for a regional generation in importer sub-region (South-Nigeria), balancing sub-region (South Africa), and exporting region (West-North) are presented in the Supplementary Material (Fig. 1, 2 and 3, respectively). For the integrated scenario a drastically increased electricity demand for the desalination and SNG producing sectors changes the picture dramatically; SNG producing regions tend to increase the intra-regional electricity generation to fulfill the growing demand.

The Sankey flow diagram reveals more about the energy flows, all the way from generation to end use and the efficiency of each energy conversion. The Sankey flow diagram in Figure 9, summarizes the integrated scenario and is comprised of the primary RE resource generators, the energy storage technologies, HVDC transmission grids, the total demand for each sector and system losses. The potentially usable heat and ultimate system losses includes the difference between the primary power generation and final electricity demand. Both are comprised of curtailed electricity; the heat produced by biomass, biogas, and waste-to-energy power plants; the heat of transforming power-to-hydrogen in the electrolyzers, hydrogen-to-methane in methanation and methane-to-power in the gas turbines; and the efficiency losses in A-CAES, PHS, battery storage, as well as by the HVDC transmission grid. At the same time the end use comprises of the electricity, gas, and desalination sectors. Additional sectors in the integrated scenario cause additional capacity for generation from renewables, mainly wind energy and solar PV. An overview of the power transmission lines, their key parameters, and the percentage of grid utilization can be found in the Supplementary Material (Table XV and Fig. 6). The energy flow for the region-wide and area-wide scenarios are presented in the Supplementary Material (Fig. 7 and 8).

5. Discussion

5.1 Cost implications for large-scale HVDC interconnections

From the presented results for Sub-Saharan Africa, it can be confirmed that different grid configurations as witnessed in the different scenarios will lead to several cost ramifications. Appropriate approaches for institutional, technical, and economic design and implementation are crucial to carrying out grid expansion programmes in Sub-Saharan Africa.

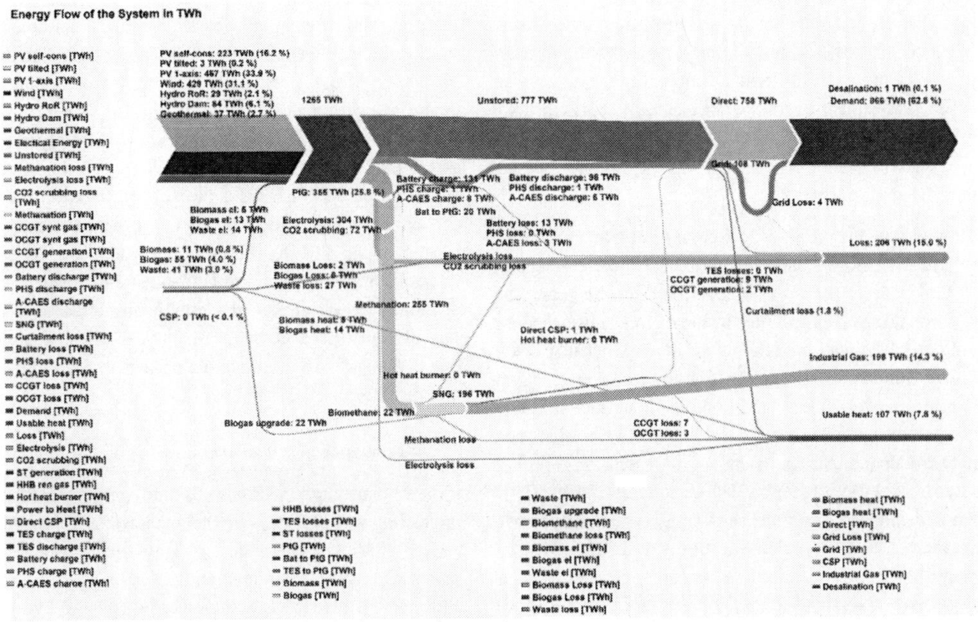

Fig. 9. Energy flow of the system for the integrated scenario.

The electricity consumption in most Sub-Saharan countries is low due to a variety of reasons amidst a poorly developed grid infrastructure (World Bank, 2008). Thus, for the region to engage in a large-scale energy projects, it calls for additional investment in HVDC power lines to facilitate cross-border transmission. The majority of Sub-Saharan countries have recurrent power shortages, almost 40% of the grid was built or refurbished decades ago, and it is estimated that 18 bUSD is required for the renovation of the existing grid (Szabo et al., 2013). In an effort to tackle the inefficiencies associated with the Sub-Saharan Africa power supply, joining national markets can provide the economy of scale to overcome this. Regional power pools need to be established to promote cross-border trade in electricity and incentivize capacity investment. HVDC interconnectors that could enhance renewable electricity trade between power pools with diverse energy mixes (Rosnes, 2012). There are discussions about interconnection across the continent and facilitating the creation of a Pan African power market (Hammons and Naidoo, 2010). The power market will comprise a consolidation of regional power pools namely, Southern African power pool (SAPP), West African Power Pool (WAPP), Central African Power Pool (PEAC) and the East African Power Pool (EAPP). This study has however subdivided the power pools further and highlights 16 sub-regions as discussed in Section 3.1. The linkage of these sub-regions via HVDC transmission lines can address the challenges of intermittency in RE technologies. The installation of HVDC transmission lines between sub-regions leads to a significant cutback of overall cost, storage sizes, and RE installed capacities in the 100% RE-based system and this reduces the total system cost. A comparison of the region-wide to country-wide scenarios, shows negligible difference regarding cost since only Nigeria (with an enormous population) was subdivided. A slight drop in the total levelized cost of electricity from 57.8 €/MWh to 57.7 €/MWh is noticeable. The same argument is feasible when the annualized cost is also compared. The annualized cost shows a slight decrease from 50.3 b€/a to 50.2 b€/a. However, when each the country-wide or the region-wide scenario are compared to area-scenarios, a sizeable reduction in LCOE by 5.5% from about 57.8 €/MWh to 54.6 €/MWh and annualized cost from around 50.2 b€/a to 47.3 b€/a. In parallel, the capex requirements are also reduced in a similar comparison by 3.0% from 429.7 b€ to from 416.8 b€, which is less as the annualized cost, since the biomass fuel cost remain stable. Additional costs of HVDC transmission lines (23.1 b€ annual cost for the area-wide scenario) are compensated by a substantial decrease in generation and storage capacities enabled by lower losses and costs of energy transmission compared to energy storage, access to low-cost electricity generation in other regions and finally a more efficient use of energy system components.

Transmission and distribution (T&D) investments are key for any developing power grid system, particularly for systems that are expanding to meet universal access goals (Bazilian et al., 2011). Eberhard and Shkaratan (2012), estimates about 37% of the investment cost is used for new transmission and distribution networks. Unfortunately, the use of HVDC transmission lines interconnections may have limited impact in improving electricity access especially in rural localities in Sub-Saharan Africa because of the complexity of rural settlement (geographical constraints), poorly developed, and cost of extending grid infrastructure. Sub-Saharan Africa has generally lagged behind other regions of the world in terms of infrastructure and power sector investment and performance (Eberhard and Shkaratan, 2012).

5.2 Role of off-grid PV

The traditional approaches to electrify Sub-Saharan Africa by grid extensions have not contributed to the eradication of rural poverty (Szabo et al., 2013). Over the years, the substantial subsidies provided by the region's utilities leave millions of African households in the dark because a too limited number of households are connected to the grid (World Bank, 2008). This means that much more investment would be required to achieve full electrification in Sub-Saharan Africa through grid extension options. Based on the World Bank (2006) a total of 280 bUSD is required for 200 million new connections to provide electricity to 100% electrification by 2030. Off-grid renewable energy technologies systems, especially solar PV based technologies (solar home system and mini-grid) provides technically feasible and economical solutions to energy challenges, where the cost of grid extension is not economical (Niez, 2010; Breyer et al., 2011). Furthermore, the choice of technologies for rural electrification should consider the economic viability, geographical potential, and compatibility of (renewable) technology options, spatial and techno-economic analysis helps in identifying least cost electrification option for remote areas (Szabo et al., 2011; Szabo et al., 2013). Access to modern energy services can be enhanced by RE, especially solar PV. Access to clean and reliable energy constitutes an important prerequisite for fundamental determinants of human development, contributing, among other things, to economic activity, income generation, poverty alleviation, health, education, and gender equality. Due to their decentralized nature, solar PV can play an important role in fostering rural electrification (IPCC, 2011).

The solar PV technologies predominantly dominate the off-grid market, a recent market research conducted in the last half of 2015, on sales of pico-PV and solar home systems reveals that about 4 million products have been sold globally, and Sub-Saharan Africa accounts for about 2.2 million (54.3%) (Lightning Africa, 2016). Mini-grid and off-grid technology account for 26 TWh and 12 TWh power generation respectively in Sub-Saharan Africa in the new policies scenario 2040, solar PV contributes 37% and 47% of the technology mix (IEA, 2014b). Mini-grids and off-grid PV capacity are predicted to reach 100 GW, representing about 5% of total capacity by 2030 (2% by 2050), which would be a significant increase from current trends (IEA, 2014a). For instance, in Uganda, on-grid electricity will become an increasingly important supplier of energy, though will still account for less than 25% of the overall energy supply by 2050 (WWF, 2015). This means that 75% of Ugandans will not rely on grid connectivity for the provision of energy. Off-grid based RE solutions like mini-grids, and in particular solar home systems (SHS) provide a least cost solution for realizing the full electrification for the rural people in Sub-Saharan Africa (Cader et al., 2016a; Bertheau et al., 2014 Breyer et al., 2011; Breyer et al., 2012; WWF, 2015). PV-based mini-grids can be established for initial electrification but also for upgrading existing and costly off-grid diesel-grids (Cader et al., 2016b). Furthermore, SHS can act as an interim solution until a grid is planned and constructed or economic power makes the establishment of mini-grids feasible, since the is no financial drawback for the operators due to the very low amortization periods of SHS of 6 to 18 months (Breyer et al., 2012; 2011)

5.3 Relevance of PV prosumers

PV self-consumption influences the energy sector in an interesting way. To measure the impacts of PV prosumers, region-wide, country-wide, and area-wide open trade scenarios are also calculated without PV self-consumption and the total demand is assumed to be covered by a more centralized system. A comparison of self-consumption-scenarios to base scenarios shows the following additions in the annualized costs; 48.6 b€ compared to 50.3 b€ for region-wide scenarios (+3.4%), 48.5 b€ compared to 50.2 b€ for country-wide scenarios (+3.4%), 45.6 b€ compared to 47.3 b€ for area-wide scenarios (+3.6%). PV self-consumption induces additional costs because of a different target function of prosumers. However, prosumers reach their minimum annual electricity cost. The LCOE for PV self-consumption then must be lower than the grid electricity selling price, but can be higher than the total system LCOE. The cost of PV installations is projected to drop because of technological innovation, product optimization, economies of scale, among other reasons. This trend will prompt a further cut of LCOE for PV self-consumption. PV self-consumption can also reduce the stress on local distribution and transmission grids in particular in combination with batteries, as also documented by a reduction in the Sub-Saharan African peak-load by 5.2%. Due to the high electricity prices in most of the Sub-Saharan African countries, the value of the savings brought about by PV self-consumption expands and improves the return on investment (Rickerson et al., 2014), which is further accelerated by broadly use of diesel generation even used occasionally as a substitute for base load power plants. So, the prolonged or avoided distribution and transmission capacity may result in a fast growing base of PV prosumers which could set up a component of competition to monopoly utilities in Sub-Saharan Africa. A comparison between self-consumption and base scenarios also depicts additional battery storage demand because of PV prosumers. Electricity generation from prosumers causes some positive and negative distortion in the system demand profile (Fig. 5), i.e. the system reacts by installing more flexibility granting capacities, such as low-cost RE or further storage capacities, which increase the system costs as well. However, due to some slight inflexibility caused constrained PV use mainly during the daytime, impact of prosumers is constrained to a reduction of daytime peaks as opposed to night peaks. Thus, the most expensive peak hours throughout the year are substantially reduced by 5.2% by PV self-consumption, which exhibits a substantial economic value. The electricity consumption in the centralized system is higher from the morning hours till the evenings compared to a system with PV prosumers influence. For the region-wide scenario, a comparable low-cost increase due to the decentralized generation can be explained by the fact that additional disturbance cost in the system (provoked by prosumers) is compensated by access to low-cost residential electricity (for residential consumers WACC is assumed to be 4%). Finally, PV self-consumption is of particular benefit in region-wide and country-wide scenarios, particularly in areas with zero impact on rooftops. PV use for local electricity generation reduces the need to develop long-distance transmission to transmit remote RE sources to load centers. Some countries may have to carry out PV self-consumptions as a policy measure for the creation of local value chains and less supply risk due to higher electricity imports.

5.4 Demand for Desalination and Industrial Gas

The integration of water desalination and industrial gas sectors results in additional electricity generation that covers projected natural gas demand and renewable water demand by SNG generation and SWRO desalination, respectively. In parallel with supplying demand, such an integration gives the additional system flexibility, especially for seasonal fluctuation compensation. The availability of RE in Sub-Saharan Africa is sufficient to cover additional electricity demand for producing 268 TWh_{LHV} of SNG and 320 million m^3 of renewable water. The cost of renewable water seems to be quite affordable at 1.4 €/m^3. Adding 357 TWh_{el} for gas synthesis and SWRO desalination induces an additional installation of RE generation capacities of 128 GW of PV and about 48 GW of wind energy. As well, former long-term gas storage is partly substituted by short-term battery storage. Additional benefits as a results of integration of water desalination and industrial gas sectors includes an 8.4 TWh_{el} (-17.2%) reduction of efficiency losses of storages turbines; a 15.6 b€ increase in annual system cost; 357 TWh increase in the electricity demand and a 10.6 TWh increase in curtailed electricity. Despite the extra cost for the generation, transmission, integrating water desalination and gas sectors the combined benefits as a result of integration results in a 37.2% reduction in LCOS; likewise, the total LCOE decreases by 7.7 €/MWh (-14.2%). This decrease in total LCOE can be confirmed by the substantial reduction of 37.1% in the cost of storages due to the more flexibility provided to the system by industrial gas synthesis and the desalination sector. This results are reaffirmed by Castellano et al. (2015), who determine that significantly increasing regional integration could save more than 40 bUSD in capital spending, and save the African consumer nearly 10 bUSD per year by 2040, as the levelized cost of energy falls from 70 USD/MWh to 64 USD/MWh. However, the cost impacts of the surplus of energy on the total cost are minimal.

5.5 Overall relevance of solar PV

Solar PV technologies emerge as the dominant technologies in Sub-Saharan Africa for the year 2030. IRENA (2012) estimates that Sub-Saharan Africa has a technical potential of 9261 TWh for both CSP and PV with around plus or minus 50% certainty. This demonstrates that single-handedly, solar technology is sufficient enough to power about 10 times the electricity demand for Sub-Saharan Africa (866 TWh) based on our estimates. As seen in the area-wide scenario, PV technologies generate 364.9 TWh (38%). Castellano et al. (2015) conclude that higher levels of integration would result in larger regional gas options being favored over some of the smaller in-country solar and wind additions, leading to an increase in carbon emissions. However, the results of this study show that solar PV still dominates with the integration of gas and desalination sectors for the 100% RE case. As seen in the area-wide integrated scenario, PV technologies generate 635.6 TWh (47.8%). The high share of PV can be explained with the fact that PV is the least cost RE technology in most of the Sub-Saharan Africa and decreasing cost of battery storage further pushes this trend. The large and desert countries, with the top five Chad, Democratic Republic of the Congo [Kinshasa], Mali, Niger and Sudan holding about 40 percent of the solar PV potential capacity (Castellano et al., 2015). However, winds prevail in Kenya-Uganda, Somalia-Djibouti, and West-North because of the excellent wind resources. Solar PV is followed by wind onshore which generates 138.7 TWh (36.3%). The rest of the RE technologies (including hydro, geothermal and biomass) take up the remaining 18.1% of the energy supply. Based on Breyer and Gerlach (2013) and Vartiainen et al. (2015), solar PV LCOE are projected to decline by further 30-40% from 2020 to 2040 as a consequence of progressive efforts to reduce materials use, improved efficiency and to the development of the manufacturing processes. It is also expected that the end results of a steady decline in PV cost will be the further dominance of PV technologies in the market beyond the year 2030 (Castellano et al., 2015; Breyer et al., 2016). These revelations are in line with Greenpeace (2015) advanced energy [r]evolution scenario, which projects that the hydro power generation capacity will be outstripped by solar energy after the year 2020 and that hydro power will have a smaller proportion of 41 GW (11%) compared to solar energy at 177 GW (47%) by 2030.

All the three wind energy dominated sub-regions are net-exporters. Compared to the net-importers like South-

Nigeria, the net-exporters have a more favorable renewable resource mix. Cross-border energy supply often also provides greatly enhanced diversification of energy source, a key component of energy security (Hammons and Naidoo, 2010). This results shows that the gains from electricity trade differ between sub-regions depending on resource availability and the resource mix. However, electricity trade in Sub-Saharan Africa requires significant cooperation and coordination among countries. Hence, achieving optimal power trade will require a range of important political, legal, and economic commitments Eberhard and Shkaratan (2012). As the penetration levels for variable renewables connected to the grid rises, it also becomes important to consider the impact on grid stability and reliability. Africa needs a strategic approach to improve its economy based on energy security-development linkage policies. However, diversification of renewable resources, spreading their location over different climate zones and grid interconnections with neighboring countries can all play an important role in ensuring grid stability and energy security (IRENA, 2012). In order to facilitate electricity trade exporting and importing sub-regions, the national utilities will need to engage into long-term bilateral contracts for the sourcing and consumption of electrical energy.

5.6 Large-scale hydropower in Sub-Saharan Africa

The Grand Inga project on the Congo River envisages the installation of 40 GW of hydro generating capacity, which would make it the largest hydro facility in the world (IRENA, 2015). IRENA, (2015), estimates that around 92% of technically feasible potential has not yet been developed. Given that large hydro projects often have outputs far more than the country's electricity demand, it is necessary to develop these as regional projects. The construction and successful operation of the major hydro dams in Sub-Saharan Africa usually requires cross-border co-operation between countries that may be suffering from political instability and even civil disorder. In such circumstances, achieving the necessary water access agreements across national boundaries may prove challenging. However, it can be assumed that the creation of regional power pools will boost cross-border transmission in Sub-Saharan Africa. The expansion of hydropower capacity is expected to further progress by reaching 93 GW in 2040, with several major projects (such as Inga III and the Grand Renaissance Dam) coming online incrementally over the projection period (IEA, 2014b).

In this respect, the water footprint, and human displacements as results of for large-scale hydro electricity generation, the social costs have to be taken into account, since many parts of Africa are already experiencing water shortage with about one-third of Africa's productive area already classified as dry-land. Thus, stakeholders have to consider the impact of hydropower development on local populations, their impacts on water use and rights, as well as issues over the biodiversity impacts of large-scale hydropower developments. Most refurbishment projects focus on the electro-mechanical equipment, but can involve repairs or redesigns of intakes, penstocks, and tail races. For large hydropower plants, economic lifetimes are at least 40 years, with 80 years lifetimes as upper limit meanwhile for small-scale hydropower plants, the typical lifetime can be 40 years or less. In this case refurbishment of hydro projects fall into two categories namely life extension and upgrades (IRENA, 2012). The costs of life extension and upgrades for old hydro plants are have estimated that life extensions cost around 60% of electro-mechanical costs and upgrades anywhere up to 90% depending on their extent (IRENA, 2012). Considering, hydropower has a particular cost advantage compared to other renewable sources that have not yet been amortized, an additional 500 €/kW capex should be considered for old hydropower plants (>50 years). Considering solar PV and wind energy are cheaper alternatives compared to hydro power, it may be not necessary to build large hydro dams to utilize all the available capacity in Sub-Saharan Africa.

Further review of the energy mix, shows contradictions of the results of this research to the IEA (2014b) Africa Energy Outlook, which reports that the total renewable share will grow to 44% of Sub-Saharan Africa, being in contradiction to the findings of this research that 100% RE are already low cost for 2030 financial assumption. According to IEA's findings, the power generation capacity in 2040 will include a dominating hydro power capacity mainly based on large-scale hydro power generation providing a 26% supply share. IEA (2014b) projects that the coal demand in Sub-Saharan Africa is expected to grow for power generation and coal-to-liquid production. However, the immense societal and environmental burden of coal consists of the impacts of lethal chemicals and heavy metals emissions that cause severe health problems and respective high health costs aside from mortality related to air pollutants discharged through coal combustion that is nevertheless not captured. Epstein et al. (2011) and the International Monetary Fund (IMF, 2015) have already pointed out that these high cost represent subsidies which places coal electricity in a not competitive position to solar PV and wind energy due to its very high societal cost. The most direct risks and hazards presented by coal sludge, coal slurry, and coal waste impoundments; the total contributions of nitrogen deposition to eutrophication of fresh and coastal sea water and the extensive appraisal of impacts owing to an increasingly unstable climate (Monks et al., 2009).

This study considers the hourly system balance between, generation, demand, grid transmission, charging and discharging storage units. The storage units get charged when power generation surpasses demand, thus PV use results to reliance in battery storage, notably during night hours. This makes solar PV a key attribute to a substantial proportion of battery use. However, solar PV compliments well with a wind power generation especially during night hours to cut down on battery reliance. Additional generation from flexible biomass and hydro dam plants, backup discharge from large-scale storage, facilitates grid utilization and thus significantly reducing the storage needs. In the integrated scenario, installed capacities for PV and wind energy increase by 40.9% and 54.9% contrasted to the area-wide scenario, correspondingly, because of the higher demand for electricity, the low cost of PV and wind energy and enhanced system flexibility. Based on the demand projections for 2030 by IRENA (2015), forecasts have a slight dissimilarity from our results. IRENA predicts that the RE generation for entire Africa (including North Africa) will surpass 1000 TWh attaining 50% of the total power generation. Around one-third of total power generation will come from hydro power plants with the remainder coming from solar PV, CSP, wind energy, and geothermal. The results of this study clearly indicate that a much higher RE generation could be achieved mainly driven by low cost solar PV and wind energy.

5.7 Sustainable biomass use

The biomass resources available for electricity production for all the study scenarios in consideration amounts to 355 TWh_{th}/a which is inadequate when compared to 866 TWh_{th}/a needed to cover the total electricity demand for both the region-wide and area-wide or to 1224 TWh_{th}/a required for the integrated scenario. However, renewable energy technologies could still play a vital role in cooking and water heating. Globally, 100% renewable energy would need bioenergy from an additional 250 million ha of crops and tree plantations by 2050 plus 4.5 billion m^3 of wood from multiple sources. However, the global or a regional electricity demand can be covered by an optimal renewable mix (IRENA, 2015; Castellano, 2015; WWF, 2015 Greenpeace, 2015; Bogdanov and Breyer, 2016). As seen in the Greenpeace (2015) energy [r]evolution scenario, renewables can provide 68% of Africa's total heat demand in 2030 and 91% in 2050. The future models will include additional demand sectors for heating, cooking, and mobility. Further review of the energy mix shows a complete contradiction to IEA (2014b) Africa Energy Outlook, which reports that bioenergy (biomass) use for cooking dominates the primary energy demand with 80% of household consumption. Based on IEA (2014b), biomass use will reach 5699 TWh_{th} (40% of primary demand) and continue to dictate the energy mix in 2030. The measure at which wood fuel is renewable energy depends on how much is produced or consumed (IEA, 2014). Current biomass assessments and models do not take these land use efficiency measures into account (Batidzirai et al., 2012). IRENA (2015), considers 70% of thermal energy demand comes from coal, oil, and natural gas. However, with a high portion of residue use (30% is a mixture of biomass and waste products).

Traditional biomass use, characterized by low efficiency and adverse impacts on human health and living conditions from the high concentrations of particulate matter and carbon monoxide, among other pollutants (IPCC, 2012). In Sub-Saharan Africa, almost three quarters of those people using traditional biomass do not have access to energy efficient cooking stoves. The result is wasteful fuel use and serious health effects from wood smoke, which along with coal smoke kills almost 2 million people a year (WWF, 2011b). In this context, non-combustion-based RE power generation technologies have the potential to reduce local and regional air pollution significantly and lower associated health impacts compared to fossil-based energy production. Thus, an additional generation from other sources is needed. The results show that wind energy and solar PV are more cost competitive in Sub-Saharan Africa compared to biomass. A closer look at the area-wide scenario reveals that the LCOE for the solid biomass, waste-to-energy, and biogas power plants are 60.6 €/MWh, 86.0 €/MWh and 78.2 €/MWh, respectively compared to solar PV and wind energy LCOE which are 26.1 €/MWh and 36.1 €/MWh. Solar PV and wind are more cost competitive in comparison to biomass when it comes to electricity production. As a consequence, the biomass usage in the model is still low at 34.3% of the available sustainable resource potential with solid biomass, waste, and biogas producing 58.5 TWh_{th}, about 4% of the total generation by renewables. The synthetic natural gas and the heat generated by the system as a by-product of biogas and biomass CHP plants, waste-to-energy incinerators, gas turbines, electrolyzers and methanation units has the potential to replace some biomass used for heat purposes. Based on IEA (2014b), the growth in gas use is insignificant and is focused on cooking and water heating in gas-rich countries, mainly Nigeria, Mozambique, and Tanzania. As illustrated in the energy flow diagrams (Fig. 9, Supplementary Material Figs. 7 and 8), the usable heat increases from 74 TWh_{th} per annum for the area-wide scenario up to 106 TWh_{th} for the region-wide scenario. The higher efficiency as a results of interconnectivity plays a key role in the low amount of heat generated in area-wide scenario. According to Greenpeace (2015) advanced energy [r]evolution scenario higher efficiency gains can be achieved in the heating sector compared to the electricity sector. About 1000 TWh can be saved as a result of the introduction of high energy standards and highly efficient technologies, e.g. for industrial and commercial process heat, cooking and air conditioning. The waste heat from biomass and gas power plants is evenly distributed over the year. Cooling demand is mainly included in electricity demand numbers and therefore does not generate an additional demand. For the integrated scenario the amount of usable heat is slightly smaller than for the area-wide scenario, at 122 TWh_{th}, due to increased efficiency in gas turbines, which covers the heat losses in methanation and electrolysis. Synthetic fuels can also have an alternative use in mobility. Based on the Greenpeace (2015) advanced energy [r]evolution scenario, the generation of synthetic fuels will substitute fossil fuels in Africa by the end of 2050. Greenpeace (2015) estimates that 560 TWh will be utilized for hydrogen generation and 790 TWh for synthetic liquid fuel generation intended for the transport sector. Fasihi et al. (2015, 2016) conclude that renewable electricity based synthetic fuels are a real option for decarbonizing the energy system for the period of the year 2030 and beyond. The findings for the Sub-Saharan Africa 100% renewable resource-based energy system clearly show the potential of the region for RE generation and for a global climate change mitigation strategy. The results of a fairly low LCOE for the year 2030 (in all the considered scenarios) added to the already existing RE policies and low carbon development plans can boost the development of a renewable power system in Sub-Saharan Africa in the coming decades.

5.8 Competitiveness of 100% RE

Renewable energy technologies are cost competitive compared to the high cost alternatives; for instance, the low carbon based energy systems and non-renewable options, such as nuclear energy, natural gas and coal carbon capture and storage (CCS) as highlighted by Agora Energiewende (2014b). The LCOE of the alternatives are as follows (Agora Energiewende, 2014): 112 €/MWh for new nuclear (assumed for 2023 in the UK and Czech Republic), 112 €/MWh for gas CCS (assumed for 2019 in the UK) and 126 €/MWh for coal CCS (assumed for 2019 in the UK). However, a report published by the European Commission (2014b) concludes that CCS technology is not likely to be commercially available before the year 2030. The findings for Europe are assumed to be also valid for Sub-Saharan Africa in the mid-term. Research from climate change modeling, such as Griffin et al. (2014) conclude no CCS and no nuclear energy would be too high in cost and therefore not possible, however in Luderer et al. (2014) and its supplementary it is stated that the 2050 capex assumptions for solar PV and wind energy in Griffin et al. (2014) are 1200-1400 USD/kW and 1000 USD/kW, which is fine for wind energy but fully wrong for solar PV since we have already today capex which are 10-20% lower and the learning curve is continuing (Vartiainen et al., 2015; Fraunhofer ISE, 2015; Breyer and Gerlach, 2013). More realistic capex assumptions would be 300 – 400 €/kW in 2050 according to Vartiainen et al. (2015). Such limited solar PV industrial insights are a key reason why in many energy scenarios a misleading conclusion is drawn, that fossil-CCS and/or nuclear fission would be without any alternative. The 100% renewable resource-based energy system options for Sub-Saharan Africa presented in this work seem to be considerably lower in cost (about 48 - 57%) than the other alternatives, which have still further disadvantages. Globally, the end use combustion of carbon fuels is the principal source of potential emissions from the production of reserves, accounting for about 90% of total estimated emissions, on average (Heede, 2016). Climate change mitigation is one of the key driving forces behind a growing demand for RE technologies. The price of emissions could influence electricity prices, so as to reduce fossil fuel consumption and to promote cleaner technologies with benefits on air quality and climate change (Garcia-Martos et al., 2013). Castellano et al. (2015) conclude that, if Sub-Saharan Africa aggressively promotes renewables, it could obtain a 27% reduction in CO_2 emissions; this would result in a 35% higher installed capacity base and 31% higher capital spending, or an additional 153 bUSD. Uncertain future value of CO_2 emissions abatement will not affect the renewable energy drive in Sub-Saharan Africa. Bazilian et al. (2012) foresees a decrease, in relative terms, of carbon-intensive resources in Africa in the coming two decades, even without an explicit focus on climate change mitigation. In addition to reducing GHG emissions, renewable energy technologies can also offer benefits with respect to air pollution and health compared to fossil fuels. These include nuclear melt-down risk, nuclear terrorism risk, unsolved nuclear waste disposal, remaining CO_2 emissions of power plants with CCS technology, a diminishing conventional energy resource base, and high health cost due to heavy metal emissions of coal-fired power plants. Nuclear energy produces both operational and decommissioning wastes. The radioactive waste possesses a threat to the environment and is unsafe for humans. The Chernobyl and Fukushima nuclear melt-downs have shown the disastrous effects of nuclear radiation on humans. Nuclear energy also calls for massive subsidies required for development and operation and loan guarantees. Dittmar (2012) also emphasizes the mentioned limitations on nuclear fission, but also points out that the financial and human research and development resources spent for nuclear fusion are no help either for the energy problems in the world and even worse these resources are not available for research of pathways towards a low cost energy future.

The findings for the Sub-Saharan Africa 100% renewable resource-based energy system illustrate the capability of the region for RE generation and a global climate change mitigation scheme. The results of a rather low LCOE for the year 2030 (in all the considered scenarios) included to the existing RE policies, and low-carbon development programs can boost the development of a renewable power system in the Sub-Saharan Africa in the future years.

6. Conclusion

This research work establishes that a 100% renewable resources-based energy system is a technically and economically practical solution for Sub-Saharan Africa. RE technologies can generate sufficient power to provide for all electricity demand in Sub-Saharan Africa for the year 2030 at a low overall cost of 47-58 €/MWh$_{el}$, and this depends on the intensity of geographic integration and energy sector coupling. The power required to cover PtG technology, and SWRO desalination demand can be produced by RE sources, providing the region 100% renewable synthetic natural gas and clean water delivery. However, the synthetic gas price is perhaps high than current, and government regulation and subsidies are still needed to ensure the commercial viability of this synthetic fuel.

In the recent past, the crucial issue of dependency on biomass energy has widely contributed to deforestation in various countries in Sub-Saharan Africa. This problem can be dealt with by shifting to a least cost energy mix with additional generation from solar PV and wind energy. The HVDC interconnection of the sub-regions and the integration of gas and desalination sectors significantly influences the mix of renewable technologies and thus the storage requirements for the energy system. This results in a reduction of the required generation and storage capacities, thus reducing the total LCOE. Meanwhile, PV self-consumption causes a moderate increase in total electricity costs of about 3-4%, which may be outweighed by other benefit positive effects.

Industrial SNG generation as seen from the integrated scenario increases the demand side flexibility and can also be used for long-term storage as a by-product service as a consequence reducing reliance on battery storage. The system further can curtail heat losses through the SNG production, this results in flexibility and reduces the system costs.

A fully integrated renewable energy system has to be simulated and deeply studied to understand the findings for Sub-Saharan Africa better. However, this research work indicates that a 100% renewable resources-based energy system is a real low cost option for a recent future.

Acknowledgements

The authors gratefully recognize the governmental financing of Tekes (Finnish Funding Agency for Innovation) for the 'Neo-Carbon Energy' project under the number 40101/14. The authors would like to thank Javier Farfan, Upeksha Caldera, Otto Koskinen, and Michael Child for valuable support.

Supplementary Material

Supplementary data associated with this article can be found in the appendix of this paper.

References

AAPG, 2015. Global Heat Flow Database. American Association of Petroleum Geologists, Tulsa, USA, www.datapages.com/gis-map-publishing-program/gis-open-files/global-framework/global-heat-flow-database

Agora Energiewende, 2014. Comparing the cost of low-carbon technologies: what is the cheapest option? report by Prognos AG on behalf of Agora Energiewende, p. 10-13. www.prognos.com/fileadmin/pdf/publikationsdatenbank/140417_Prognos_Agora_Analysis_Decarbonisationtechnologies_EN.pdf

Asif M. and Muneer T., 2007. Energy supply, its demand and security issues for developed and emerging economies, Renew. Sustain. Energy Rev., vol. 11, no. 7, pp. 1388–1413.

Batidzirai B., Smeets E.M.W., Faaij A.P.C., 2012. Harmonising bioenergy resource potentials - Methodological lessons from review of state of the art bioenergy potential assessments. Renewable and Sustainable Energy Reviews, 16, 6598–6630.

Bazilian M., Nussbaumer P., Rogner H.H., Brew-Hammond A., Foster V., Pachauri S., Kammen D.M., 2012. Energy access scenarios to 2030 for the power sector in sub-Saharan Africa. Utilities Policy, 20, 1–16.

Bertheau P., Müller H., Blechinger P., Seguin R., Breyer Ch., 2014. Energy Storage Potential for Solar Based Hybridization of Off-Grid Diesel Power Plants in Tanzania, Energy Procedia, 46, 287-293.

[BNEF] - Bloomberg New Energy Finance, 2016. Off-Grid Solar Market Trends Report 2016. London. www.gogla.org/wp-content/uploads/2016/03/Off-Grid-Solar-Market-Trends-Report-2016.pdf.

Bogdanov D. and Breyer Ch., 2015. North-East Asian Super grid for 100% Renewable Energy Power Supply: Distributed Small-scale and Centralized Large-scale Solar PV as a Major Energy Source, 31st European Photovoltaic Solar Energy Conference. Hamburg, September 14-18.

Bogdanov D. and Breyer Ch., 2015. Eurasian Super Grid for 100% Renewable Energy power supply: Generation and storage technologies in the cost optimal mix, ISES Solar World Congress. Daegu.

Bogdanov D. and Breyer Ch., 2016. North-East Asian Super Grid for 100% Renewable Energy supply: Optimal mix of energy technologies for electricity, gas and heat supply options. Energy Conversion & Management, 112, 176-190.

Bossio D. and Kim G., 2008. Conserving Land, Protecting Water, Reading UK: International Water Management Institute, Colombo, Sri Lanka. www.cabi.org/cabebooks/ebook/20083237767.

Breyer Ch., Werner C., Rolland S., Adelmann P., 2011. Off-Grid Photovoltaic Applications in Regions of Low Electrification: High Demand, Fast Financial Amortization and Large Market Potential, 26th EU PVSEC, Hamburg, September 5-9

Breyer Ch., Gaudchau E., Gerlach A., Hlusiak M., Cader C., Bertheau P., and Wasgindt V., 2012. PV-based mini-grids for electrification in developing countries: An overview on market potential and business models, study of the Reiner Lemoine Institut gGmbH on behalf of the cdw Stiftungsverbund gGmbH, December 11. www.researchgate.net/publication/261345306_PV-based_Mini-Grids_for_Electrification_in_Developing_Countries_-_An_overview_on_market_potentials_and_business_models.

Breyer Ch. and Gerlach A., 2013. Global Overview on Grid-Parity, Progress in Photovoltaics: Research and Applications 21, 121-136.

Breyer Ch., Bogdanov D., Gulagi A., Aghahosseini A., Barbosa L.S.N.S., Koskinen O., Barasa M., Caldera U., Afanasyeva S., Child M., Farfan J., Vainikka P., 2016. On the Role of Solar Photovoltaics in Global Energy Transition Scenarios, 32nd EU PVSEC, Munich, June 20-24

Cader C., Barasa M., Gulagi A., Breyer Ch., 2016a. Solar Home Systems, Mini-Grids and Grid Extensions in Kenya: Policy implications for electrification approaches building on high resolution population data and satellite night light imagery. submitted

Cader C., Bertheau P., Blechinger P., Huyskens H., Breyer Ch., 2016b. Global cost advantages of autonomous solar-battery-diesel systems compared to diesel-only systems, Energy for Sustainable Development, 31, 14-23

Caldera U., Bogdanov D., Breyer C., 2016. Local cost of seawater RO desalination based on solar PV and wind energy: A global estimate. Desalination, 385, 207–216.

Carvalho J.F., Sauer I.L., 2009. Does Brazil need new nuclear power plants?. Energy Po- see supplememlicy 37, 1580-1584.

Castellano A., Kendall A., Nikomarov M., and Swemmer T., 2015. Brighter Africa: The growth potential of the Sub-Saharan electricity sector. McKinsey Co. New York, United States. www.mckinsey.com/industries/electric-power-and-natural-gas/our-insights/powering-africa

Chamorro C.R., García-Cuesta J.L., Mondéjar M.E., Pérez-Madrazo A., 2014a. An estimation of the enhanced geothermal systems potential for the Iberian Peninsula. Renewable Energy 66, 1-14

Chamorro C.R., García-Cuesta J.L., Mondéjar M.E., Pérez-Madrazo A., 2014b. Enhanced geothermal systems in Europe: An estimation and comparison of the technical and sustainable potentials. Energy 65, 250-263

CLIMATEACTION, 2015. Produced for COP21 United Nation Climatic Change Conference Paris. London. www.climateactionprogramme.org/bookstore/book_2015

DBFZ, 2010. Deutsches BiomasseForschungsZentrum Endbericht Globale und regionale räumliche Verteilung von Biomassepotenzialen. (D. Thrän, K. Bunzel, U. Seyfert, V. Zeller, K. Müller, B. Matzdorf, H. Weimar, Eds.). D-04347 Leipzig. http://literatur.ti.bund.de/digbib_extern/dn047244.pdf

Dii, 2012. 2050 Desert Power: Perspectives on a Sustainable Power System for EUMENA. Munich. www.trec-uk.org.uk/reports/D /DPP_2050_Study.pdf

Dittmar M., 2012. Nuclear energy: Status and future limitations, Energy, 37, 35-40

Epstein P.R., Buonocore J.J., Eckerle K., Hendryx M., Stout B.M., Heinberg R., Clapp R.W., May B., Reinhart N.L., Ahern M.M., Doshi S.K., Glustrom L., 2011. Full cost accounting for the life cycle of coal, Annals of the New York Academy of Sciences, 1219, 73-98.

Eberhard A. and Shkaratan M., 2012. Powering Africa: Meeting the financing and reform challenges. Energy Policy, 42, 9–18.

European Commission, 2014a. ETRI 2014 - Energy technology reference indicator projections for 2010-2050. EC Joint Research Centre Institute for Energy and Transport, Petten, Netherlands

European Commission, 2014b. Integration of renewable energy in Europe. study prepared by KEMA Consulting, DNV GL – Energy, Imperial College and NERA Economic Consulting on behalf of DG Energy, Brussels, p. 7-8 and 54.

Farfan J. and Breyer Ch., 2016. Structural changes of global power generation capacity towards sustainability and the risk of stranded investments, submitted.

Fasihi M., Bogdanov D., Breyer Ch., 2016. Techno-economic Assessment of Power-to-Liquids (PtL) Fuels Production and Global Trading Based on Hybrid PV-Wind Power Plants, 10th International Renewable Energy Storage Conference (IRES 2016), Düsseldorf, March 15-17

Fasihi M., Bogdanov D., Breyer Ch., 2015. Economics of global LNG trading based on hybrid PV-Wind power plants, 31st EU PVSEC, Hamburg, September 14-18

Fraunhofer ISE, 2015. Current and Future Cost of Photovoltaics. Long-term Scenarios for Market Development, System Prices and LCOE of Utility-Scale PV Systems, study on behalf of Agora Energiewende, Freiburg, Berlin, www.agora-energiewende.de/fileadmin/Projekte/2014/Kosten-Photovoltaik-2050/AgoraEnergiewende_Current_and_Future_Cost_of_PV_Feb2015_web.pdf

Garcia-Martos C., Rodriguez J.,Sanchez M.J., 2013. Modelling and forecasting fossil fuels, CO_2 and electricity prices and their volatilities. Applied Energy, 101, 63–375.

Gerlach A., Werner Ch., Breyer Ch., 2014. Impact of Financing Cost on Global Grid-Parity Dynamics till 2030. 29th EU PVSEC, Amsterdam, September 22-26, DOI: 10.4229/29thEUPVSEC2014-7DO.15.4. Available online at: www.researchgate.net/publication/266558306_Impact_of_Financing_Cost_on_Global_Grid-Parity_Dynamics_till_2030

Global Economy and Development at Brookings, 2015. COP21 at Paris: What to expect. Washington, 2015. www.brookings.edu/~/media/Research/Files/Reports/2015/11/16-paris-climate-talks/COP21atParis.pdf?la=en

GOGLA and Lightning Africa, 2016. Global solar off-grid semi-annual market report, London. www.gogla.org/sites/www.gogla.org/files/recource_docs/glb_solar_og_market_report_h22015_public_web.pdf

Greenpeace, 2015. Energy [r]evolution – A sustainable world energy outlook 2015. Amsterdam, Netherlands .www.greenpeace.org/international/Global/international/publications/climate/2015/Energy-Revolution-2015-Full.pdf

Griffin B., Buisson P., Criqui P., Mima S., 2014. White Knights: will wind and solar come to the rescue of a looming capacity gap from nuclear phase-out or slow CCS start-up?, Climatic Change, 123, 623-635

Hammons T.J. and Naidoo P., 2010. Africa-integrated gas and electricity transmission planning in power generation and HVDC engineering in harnessing large-scale hydroelectric sites for interconnected regional power systems. Energy Systems, 1, 79-112

Hansen U. E., Pedersen M. B., and Nygaard, I. 2014. Review of Solar PV Market Development in East Africa. UNEP Risø Centre, Technical University of Denmark. (UNEP Risø Centre Working Paper Series; No. 12).

Hoffmann W., 2014. Importance and evidence for cost effective electricity storage. 29th EU PVSEC, Amsterdam, September 22-26.

Heede R. and Oreskes N., 2016. Potential emissions of CO_2 and methane from proved reserves of fossil fuels: An alternative analysis. Global Environmental Change, 36, 12–20.

IASPEI, 2015. International Association of Seismology and Physics of the Earth's Interior, The International Heat Flow Commission, Colorado, USA. IHFC database. Available online at: www.heatflow.und.edu/index2.html

[IEA] – International Energy Agency, 2012. Technology Roadmap – Bioenergy for Heat and Power. IEA Publishing, Paris.

[IEA] – International Energy Agency, 2014a. Technology Roadmap – Solar Photovoltaic Energy. IEA Publishing, Paris.

[IEA] – International Energy Agency, 2014b. World Energy Outlook 2014. IEA Publishing, Paris.

[IEA] – International Energy Agency, 2015. IEA Statistics. Paris, Available online at: www.iea.org/statistics/statisticssearch/

[IMF] - International Monetary Fund, 2015. How Large Are Global Energy Subsidies?, IMF Working Paper, WP/15/105, IMF. Washington. www.imf.org/external/pubs/ft/wp/2015/wp15105.pdf

[IPCC] – Intergovernmental Panel on Climate Change, 2011. Special report on Renewable Energy sources and Climate Change mitigation. Intergovernmental Panel on Climate Change, Geneva. www.ipcc.ch/pdf/special-reports/srren/SRREN_FD_SPM_final.pdf

[IPCC] – Intergovernmental Panel on Climate Change, 2014. Synthesis Report, Contribution of Working Groups I, II and III to the Fifth Assessment Report of the Intergovernmental Panel on Climate Change [Core Writing Team, R.K. Pachauri and L.A. Meyer (eds.)]. IPCC, Geneva. www.ipcc.ch/pdf/assessment-report/ar5/syr/SYR_AR5_FINAL_full.pdf

[IRENA] – International Renewable Energy Agency, 2012. Prospects for the African power sector. Scenarios and Strategies for Africa Project, Abu Dhabi. www.irena.org/DocumentDownloads/Publications/Prospects_for_the_African_PowerSector.pdf

[IRENA] – International Renewable Energy Agency, 2012. Renewable Energy Technologies: Cost Analysis Series. Hydropower, Volume 1: Power Sector, Issue 3/5, Abu Dhabi. www.irena.org/documentdownloads/publications/re_technologies_cost_analysis-hydropower.pdf

[IRENA] – International Renewable Energy Agency, 2013. Africa renewable future the path to sustainable growth. Abu Dhabi. www.irena.org/DocumentDownloads/Publications/Africa_renewable_future.pdf

[IRENA] – International Renewable Energy Agency, 2015. Africa 2030: Roadmap for a Renewable Energy Future. REmap 2030 Programme, Abu Dhabi. www.irena.org/DocumentDownloads/Publications/IRENA_Africa_2030_REmap_2015_low-res.pdf

[ITRPV] – International Technology Roadmap for Photovoltaic, 2016. International Technology Roadmap for Photovoltaic (ITRPV) – 2015 Results, 7th ed., study by VDMA Photovoltaic Equipment and SEMI, Frankfurt a.M., http://www.itrpv.net/Reports/Downloads/

Jacobson M.Z. and Delucchi M.A., 2011. Providing all global energy with wind, water, and solar power, Part I: Technologies, energy resources, quantities and areas of infrastructure and materials, Energy Policy, 39, 1154–1169.

Komoto K., Ito M., Van der Vleuten P., Faiman D., Kurokawa K. (eds.), 2009. Energy from the desert – very large scale photovoltaic systems: socio-economic, financial, technical and environmental aspects. Earthscan, London.

KPMG, 2015. Power in Africa. Amsterdam. www.kpmg.com/Africa/en/IssuesAndInsights/Articles-Publications/General IndustriesPublications/Documents/Power%20in%20Africa%20%20sector%20report%202015.pdf

Lightning Africa, 2016. Global solar off-grid semi-annual market report, London. www.gogla.org/sites/www.gogla.org/files/recource_docs/glb_solar_og_market_report_h22015_public_web.pdf

Luderer G., Krey V., Calvin K., Merrick J., Mima S., Pietzcker R., Van Vliet J., Wada K., 2014. The role of renewable energy in climate stabilization: results from the EMF27 scenarios, Climatic Change, 123, 427-441

Mano S, Ovgor B, Samadov Z, Pudlik M, Jülich V, Sokolov D, et al., 2014. Gobitec and Asian super grid for renewable energies in Northeast Asia, report prepared by Energy Charter Secretariat, Energy Economics Institute of the Republic of Korea, Energy Systems Institute of the Russian Federation, Ministry of Energy of Mongolia, Japan Renewable Energy Foundation. www.energycharter.org/fileadmin/DocumentsMedia/Thematic/Gobitec_and_the_Asian_Supergrid_2014_en.pdf

Monks P.S., Granier C., Fuzzi S., Stohl A., Williams M.L., Akimoto H., Glasow von R., 2009. Atmospheric composition change - global and regional air quality. Atmospheric Environment, 43, 5268–5350.

Niez A., 2010. Comparative Study on Rural Electrification Policies in Emerging Economies: Keys to successful policies", IEA Energy Papers, No. 2010/03, OECD Publishing, Paris. DOI: http://dx.doi.org/10.1787/5kmh3nj5rzs4-en

Niez A., 2010. Comparative Study on Rural Electrification Policies in Emerging Economies: Keys to successful policies", IEA Energy Papers, No. 2010/03, OECD Publishing, Paris. DOI: http://dx.doi.org/10.1787/5kmh3nj5rzs4-en

Nygaard I. Hansen U. E., Pedersen M. B., and Mackenzie G. A, 2014. Measures for the diffusion of solar PV are aligned in technology action plans for six countries in Africa, Proc. from 1st Africa Photovolt. Sol. Energy Conf. Exhib. March 27-29 2014, Durban.

Pleßmann G., Erdmann M., Hlusiak M., Breyer Ch., 2014. Global energy storage demand for a 100% renewable electricity supply. Energy Procedia 46, 22-31.

Rosnes O. and Vennemo H., 2012. The cost of providing electricity to Africa. Energy Economics, 34, 1318–

1328.
Szabo S., Bodis K., Huld T., Moner-Girona M., 2011. Energy solutions in rural Africa and mapping electrification costs of distributed solar and diesel, Environmental Research Letters, 6.
Szabo S., Bodis K., Huld T., Moner-Girona M., 2013. Sustainable energy planning: leapfrogging the energy poverty gap in Africa, Renewable and Sustainable Energy Reviews, 28, 500–509.
Urban W., Lohmann H., Girod K., 2009. Abschlussbericht für das BMBF-Verbundprojekt Biogaseinspeisung. Fraunhofer UMSICHT.
[UN] – United Nations, 2015. Population Division. World Population Prospects: The 2015 Revision. New York. http://esa.un.org/unpd/wpp/Publications/Files/Key_Findings_WPP_2015.pdf
[UNFCCC] – United Nations Framework Convention on Climate Change, 2015 Adoption of the Paris Agreement – Proposal by the President. UNFCCC, Paris. http://unfccc.int/resource/docs/2015/cop21/eng/l09.pdf
Vartiainen E., Masson G., Breyer Ch., 2015. PV LCOE in Europe 2015-2050. 31th EU PVSEC, Hamburg, September 14-18; 10.4229/31stEUPVSEC2015-7DO.15.1. Available online at: www.researchgate.net/publication/281939918_PV_LCOE_in_Europe_2015-2050
[WEC] – World Energy Council, 2015. Energy Trilemma Index: Benchmarking the sustainability of national energy systems. London. www.worldenergy.org/wp-content/uploads/2015/11/20151030-Index-report-PDF.pdf
[WWF] – World Wildlife Fund, 2011a. Chapter 1: Forests for a living planet. WWF Living Forests Report, Gland, Switzerland. www.wwf.se/source.php/1359561/Living Forests Report_Chapter 1_2011.pdf.
[WWF] – World Wildlife Fund, 2011b. Chapter 2: Forests and energy. WWF Living Forests Report, Gland, Switzerland. http://wwf.panda.org/about_our_earth/deforestation/forest_publications_news_and_reports/living_forests_report/.
[WWF] – World Wildlife Fund, 2011c. New Generation Plantations Bioenergy and Carbon Report 2011, Gland, Switzerland. . http://wwf.panda.org/about_our_earth/deforestation/forest_publications_news_and_reports/living_forests_report/.
[WWF] – World Wildlife Fund, 2015. Energy report for Uganda: A 100% renewable energy future by 2050. Kampala. https://d2ouvy59p0dg6k.cloudfront.net/downloads/energy_report_for_uganda_2015_1.pdf
[WWF] – World Wildlife Fund, 2016. 100% Renewable energy by 2050. Gland, 2011. https://www.wwf.or.jp/activities/lib/pdf_climate/green energy/WWF_EnergyVisionReport.pdf
World Bank, 2006. An investment framework for clean energy and development - Progress Report; DC2006-0012, Washington, D.C. http://siteresources.worldbank.org/DEVCOMMINT/Documentation/21046509/DC2006-0012%28E%29-CleanEnergy.pdf.
World Bank 2008. Underpowered: The State of the Power Sector in Sub-Saharan Africa. World Bank, June, 2008. Washington, D.C. www.eu-africa-infrastructure-tf.net/attachments/library/aicd-background-paper-6-power-sect-summary-en.pdf

Nomenclature Table

A-CAES	adiabatic compressed air energy storage
Capex	capital expenditures
CCGT	combined cycle gas turbine
CCS	carbon capture and storage
CHP	combined heat and power
CSP	concentrating solar thermal power
GDP	gross domestic product
GT	gas turbine
HVDC	high voltage direct current
LCOC	levelized cost of curtailment
LCOE	levelized cost of electricity
LCOS	levelized cost of storage
LCOT	levelized cost of transmission
OCGT	open cycle gas turbine
Opex	operational expenditures
PHS	pumped hydro storage
PtG	power-to-gas
PV	photovoltaic
RE	renewable energy
RoR	run-of-river
SNG	synthetic natural gas
ST	steam turbine
SWRO	seawater reverse osmosis
TES	thermal energy storage
WACC	weighted average cost of capital
Subscript	
el	electricity
fix	fixed
th	thermal
HHV	base on higher heating value of fuel
LHV	base on lower heating value of fuel

Supplementary Material

Table I: Financial assumptions for energy system components. (Pleßmann et al., 2014; Komoto et al., 2009; Hoffmann, 2014; Agora Energiewende, 2015; Urban et al., 2009; European Commission, 2014a; Vartiainen et al., 2015; Caldera, 2016)

Technology	Capex [€/kW]	Opex fix [€/(kW.a)]	Opex var [€/(kWh)]	Lifetime [a]
PV optimally tilted	550	8	0	35
PV single-axis tracking	620	9	0	35
PV rooftop	813	12	0	35
Wind onshore	1000	20	0	25
CSP (solar field)	528	11	0	25
Geothermal	4860	87	0	30
Hydro run-of-river	2560	115.2	0.005	60
Hydro dam	1650	66	0.003	60
Water electrolysis	380	13	0.0012	30
Methanation	234	5	0.0015	30
CO_2 scrubbing	356	14	0.0013	30
CCGT	775	19.4	0.001	30
OCGT	475	14.25	0.001	30
Steam turbine	600	12	0	30
Hot heat burner	100	2	0	30
Heating rod	20	0.4	0.001	30
Biomass CHP	2500	175	0.001	30
Biogas CHP	370	14.8	0.001	30
Waste incinerator	5240	235.8	0.007	20
Biogas digester	680	27.2	0	20
Biogas upgrade	250	20	0	20
	Capex [€/(m³·a)]	Opex fix [€/(m³·a)]	Opex var [€/m³]	Lifetime [a]
Water desalination	2.23	0.09	0	30
	Capex [€/(kWh)]	Opex fix [€/(kWh.a)]	Opex var [€/kWh]	Lifetime [a]
Battery	150	10	0.0002	10
PHS	70	11	0.0002	50
A-CAES	31	0.4	0.0012	40

TES	24	2	0	20
Gas storage	0.05	0.001	0	50
	Capex [€/(m³)]	Opex fix [€/(m³·h·a)]	Opex var [€/m³]	Lifetime [a]
Water storage	65	1.3	0	30

	Capex [€/(kW$_{NTC}$·km)]	Opex fix [€/(kW$_{NTC}$·km·a)]	Opex var [€/kWh$_{NTC}$]	Lifetime [a]
HVDC line on ground	0.612	0.0075	0	50
HVDC line submarine	0.992	0.0010	0	50
	Capex [€/(m³·h·km)]	Opex fix [€/(m³·h·km·a)]	Opex var [€/m³·h·km]	Lifetime [a]
Horizontal pumping and pipes	19.3	0.39	0	30
Vertical pumping and pipes	15.5	0.31	0	30

Table II: Efficiencies and energy to power ratio of storage technologies. Assumptions are mainly taken from Pleßmann et al. (2014).

Technology	Efficiency [%]	Energy/Power Ratio [h]	Self-Discharge [%/h]
Battery	90	6	0
TES	90	8	0.002
PHS	85	8	0
A-CAES	70	100	0.001
Gas storage	100	80*24	0

Table III: Efficiency assumptions for energy system components for the 2030 reference years. Assumptions are mainly taken from Pleßmann et al. (2014), and from Hoffmann (2014).

	η_{el} [%]	η_{th} [%]
CSP (solar field)		51
Steam turbine	42	
Hot heat burner		95
Heating rod		99
Water electrolysis		84

Methanation		77
CO_2 scrubbing		78
CCGT	58	
OCGT	43	
Geothermal	24	
Biomass CHP	40	45
Biogas CHP	42	43
Waste incinerator	34	
Biogas upgrade		98

Table IV: Efficiency assumptions for HVDC transmission (Dii, 2012).

	Power losses
HVDC line	1.6%/1000 km
HVDC converter pair	1.4%

Table V: SSA end-user grid electricity costs for year 2030. Assumptions for most of the countries were taken from Gerlach et al. (2014).

Region	Electricity costs [€/MWh]		
	Residential	Commercial	Industrial
Total area	140	137	129
West-West	189	189	196
West-South	175	197	215
West-North	189	173	148
South Nigeria	46	81	117
North Nigeria	46	81	117
Sudan Eretria	177	138	81
Ethiopia	191	162	107
Somalia	198	168	117
Kenya Uganda	234	186	132
Tanzania	157	166	174
Central	166	177	183
Congo	73	123	164
South-West	116	99	80
South Africa	71	57	43
South-East	61	58	53
Indian Ocean	143	141	136

Table VI: Average full load hours and LCOE for CSP, optimally tilted and single-axis tracking PV systems, and wind power plants in Sub-Saharan Africa regions. Abbreviation: full load hour, *FLH*.

Region	Pop. [mil. Pop]	Electr. demand [TWh]	CSP FLH	PV fixed tilted FLH	PV single-axis FLH	Wind FLH	CSP LCOE [€/MWh]	PV fixed tilted LCOE [€/MWh]	PV single-axis LCOE [€/MWh]	Wind LCOE [€/MWh]
Total area	1348	1813	2105	1713	2258	3239	56	29	25	33
West-West	87	23	1811	1615	2099	1484	65	31	27	71
West-South	121	56	2456	1870	2520	3540	48	27	23	30
West-North	43	4	1312	1469	1832	743	89	34	31	142
South Nigeria	127	72	1922	1697	2197	2064	61	30	26	51
North Nigeria	132	74	2259	1811	2363	3481	52	28	24	30
Sudan Eretria	55	26	2132	1771	2370	3190	55	28	24	33
Ethiopia	149	20	2133	1788	2442	4753	55	28	23	22
Somalia	20	2	2136	1735	2408	4004	55	29	24	26
Kenya Uganda	115	34	2165	1755	2409	2724	54	29	24	39
Tanzania	111	19	1660	1589	1992	1162	71	32	29	91
Central	52	24	1569	1554	1917	968	75	32	30	109
Congo	109	27	2330	1815	2413	2796	50	28	24	38
South-West	34	35	2487	1822	2456	3294	47	28	23	32
South Africa	51	301	2107	1730	2286	2658	56	29	25	40
South-East	106	131	2105	1713	2258	3239	56	29	25	33
Indian Ocean	36	17	1811	1615	2099	1484	65	31	27	71

Table VII: Regional biomass (DBFZ, 2009) and geothermal energy potentials.

Region	Biomass potential [TWh_{LHV}/a]			Geothermal
	Solid waste	Solid biomass	Biogas sources	Potentials [TWh_{th}/a]
Total area	41.5	258.5	55.6	358.0
West-West	1.4	9.4	2.2	0.0
West-South	2.1	8.4	2.7	0.0
West-North	1.2	11.7	1.3	2.6
South Nigeria	3.2	32.0	3.8	72.5
North Nigeria	2.7	27.1	3.3	18.8
Sudan Eretria	1.1	12.8	4.8	0.0
Ethiopia	5.0	43.1	5.0	0.0

Somalia	0.0	4.3	0.0	0.0
Kenya Uganda	2.4	4.4	3.7	4.4
Tanzania	1.1	0.0	1.1	0.0
Central	1.7	9.3	2.2	252.1
Congo	3.7	31.3	3.3	6.3
South-West	1.6	8.4	1.1	1.3
South Africa	7.9	35.6	9.1	0.0
South-East	5.1	11.8	8.0	0.0
Indian Ocean	1.3	9.1	3.7	0.0

Table VIII: Regional biomass costs, calculated based on biomass sources mix in the region. Solid wastes cost are based on assumption of 100 €/ton gate fee paid to the MSW incinerator.

Region	Biomass costs [€/MWh_{th}]		
	Solid waste	Solid biomass	Biogas sources
Total area	-20.3	7.7	3.3
West-West	-20.3	5.3	2.7
West-South	-20.3	5.5	1.0
West-North	-20.3	5.3	2.0
South Nigeria	-20.3	10.5	0.6
North Nigeria	-20.3	10.5	0.6
Sudan Eretria	-20.3	5.3	6.0
Ethiopia	-20.3	14.0	1.6
Somalia	-20.3	9.7	3.4
Kenya Uganda	-20.3	5.3	5.2
Tanzania	-20.3	5.3	1.0
Central	-20.3	5.3	4.1
Congo	-20.3	15.9	1.9
South-West	-20.3	5.3	1.3
South Africa	-20.2	10.0	9.0
South-East	-20.3	5.4	5.1
Indian Ocean	-20.3	5.4	8.0

Table IX: Lower limits of installed capacities in SSA regions. Data were taken from Farfan and Breyer (2016).

Region	Installed capacity [MW]

	Solar PV	Wind	Hydro RoR	Hydro dams	PHS
Total area	1424.9	805.3	5541.9	15749.2	1580.0
West-West	109.7	26.0	29.4	347.8	0.0
West-South	45.4	0.0	269.3	2009.7	0.0
West-North	4.7	0.0	0	0.0	0.0
South Nigeria	22.3	6.0	0	0.0	0.0
North Nigeria	23.3	6.2	434.6	1536.8	0.0
Sudan-Eretria	8.4	0.0	425.5	2722.2	0.0
Ethiopia	15.0	171.2	423.2	1464.4	0.0
Somalia Djibouti	2.9	0.0	0	0.0	0.0
Kenya-Uganda	44.8	19.3	176.9	1093.3	0.0
Tanzania	31.7	0.0	320.2	374.5	0.0
Central	13.4	0.0	445.1	629.9	0.0
Congo	3.8	0.0	1402.3	566.3	0.0
South-West	31.4	0.2	274.2	949.0	0.0
South Africa	1035.1	570.2	69	612.3	1580.0
South-East	25.8	0.3	1238.4	3420.3	0.0
Indian Ocean	7.3	6.0	33.8	22.7	0.0

Table X: Upper limits on installable capacities in Sub-Saharan Africa regions in units of GW_{th} for CSP and GW_{el} for all other technologies.

Region	area [1000 km^2]	Limits [GW] Solar CSP	Solar PV	Wind	Hydro RoR	Hydro dams	PHS
Total area	25304.9	230137.9	115068.9	8591.8	8.3	23.6	2.4
West-West	2947.3	28919.8	14459.9	1079.7	0.0	0.5	0.0
West-South	1005.5	9049.7	4524.9	337.9	0.3	3.0	0.0
West-North	2207.8	19869.9	9935.0	741.8	0.0	0.0	0.0
South Nigeria	853.3	7679.5	3839.6	286.7	0.0	0.0	0.0
North Nigeria	938.5	8446.9	4223.4	315.3	0.7	2.3	0.0
Sudan-Eretria	2627.1	23644.2	11822.1	882.7	0.6	4.2	0.0
Ethiopia	1127.1	10144.2	5072.1	378.7	0.6	2.3	0.0
Somalia Djibouti	660.7	5945.9	2973.0	222.0	0.0	0.0	0.0
Kenya-Uganda	818.7	7368.2	3684.1	275.1	0.3	1.6	0.0
Tanzania	999.3	8993.3	4496.6	335.7	0.5	0.6	0.0
Central	1737.1	15634.3	7817.2	583.7	0.7	0.9	0.0
Congo	2345.4	21108.7	10554.3	788.1	2.1	0.8	0.0
South-West	3114.1	28026.6	14013.3	1046.3	0.4	1.4	0.0
South Africa	1250.3	11252.4	5626.2	420.1	0.1	0.9	2.4
South-East	2080.6	18725.6	9362.8	699.1	1.9	5.1	0.0
Indian Ocean	592.1	5328.7	2664.4	198.9	0.1	0.0	0.0

Table XI: Annual industrial gas (IEA, 2014; IEA, 2015) and water (Caldera et al., 2016) demand for year 2030.

Region	Annual gas demand TWh_{th}	Annual electricity demand for gas synthesis TWh_{el}	Annual water desalination demand 10^6 m^3	Annual electricity demand for water desalination TWh_{el}
Total area	354.8	197.1	319.1	2.2
West-West	7.8	5.5	0.0	0.0
West-South	0.0	0.0	0.0	0.0
West-North	100.0	54.4	0.0	0.0
South Nigeria	103.6	56.0	60.9	0.0
North Nigeria	0.0	0.0	1.1	0.3
Sudan-Eretria	0.0	0.0	23.9	0.0
Ethiopia	0.0	0.0	0.0	0.1
Somalia Djibouti	0.0	0.0	0.0	0.0
Kenya-Uganda	32.7	17.4	0.0	0.0

Tanzania	41.8	23.3	0.0	0.0
Central	0.0	0.0	0.0	0.0
Congo	28.1	15.3	233.1	0.0
South-West	40.2	23.3	0.0	1.8
South Africa	0.4	2.2	0.0	0.0
South-East	0.0	0.0	0.0	0.0
Indian Ocean	0.2	0.3	0.0	0.0

Table XII: Overview on prosumers electricity costs installed capacities and energy utilization for Sub-Saharan Africa.

Prosumers parameters	Residential	Commercial	Industrial
Electricity price [€/kWh]	0.0963	0.0959	0.0930
PV LCOE [€/kWh]	0.0255	0.0341	0.0339
Self-consumption PV LCOE [€/kWh]	0.0302	0.0387	0.0380
Self-consumption PV and Battery LCOE [€/kWh]	0.0569	0.0625	0.0590
Self-consumption LCOE [€/kWh]	0.0566	0.0622	0.0588
Benefit [€/kWh]	0.0397	0.0337	0.0342
Installed capacities	Residential	Commercial	Industrial
PV [GW]	23.5	13.8	24.0
Battery storage [GWh]	31.8	15.7	24.7
Generation	Residential	Commercial	Industrial
PV [TWh]	40.2	23.6	41.3
Battery storage [TWh]	10.3	5.1	7.9
Excess [TWh]	5.0	2.3	3.5
Utilization	Residential	Commercial	Industrial
Self-consumption of generated PV electricity [%]	84.7	88.0	89.4
Self-coverage market segment [%]	11.9	10.5	9.6
Self-coverage operators [%]	59.6	52.5	48.2

Table XIII: Overview on storage capacities, throughput, and full cycles per year for the four scenarios for Sub-Saharan Africa.

			Region-wide	Country-wide	Area-wide	Integrated
	Battery SC	[GWh_{el}]	72	72	72	72
Storage capacities	Battery system	[GWh_{el}]	190.4	196.4	168.1	146.2
	PHS	[GWh_{el}]	3.2	3.2	3.2	3.2
	A-CAES	[GWh_{el}]	623.2	579.7	432.5	335.0
	Gas	[GWh_{th}]	25753.7	25947.9	18935.5	24303.7
	Battery SC	[TWh_{el}]	322.9	322.9	322.9	322.9
Throughput of storages	Battery system	[TWh_{el}]	317.6	317.9	322.5	319.5
	PHS	[TWh_{el}]	319.1	319.3	322.6	320.6
	A-CAES	[TWh_{el}]	264.0	264.0	262.8	241.7
	Gas	[TWh_{th}]	77.3	100.7	76.4	66.8
	Battery SC	[-]	190.4	196.4	168.1	146.2
Full cycles per year	Battery system	[-]	72.2	72.2	72.2	72.2
	PHS	[-]	3.2	3.2	3.2	3.2

	A-CAES	[-]	623.2	579.7	432.5	335.0
	Gas	[-]	25753.7	25947.9	18935.5	24303.7

Table XIV: Total LCOE components in all sub-regions.

Region-wide	LCOE primary [€/MWh]	LCOC [€/MWh]	LCOS [€/MWh]	LCOT [€/MWh]	LCOE total [€/MWh]	
Area average	37.5	2.0	18.2	0.0	57.7	
West-West	35.5	1.4	19.8	0	56.7	0.0
West-South	42.3	1.2	22.2	0	65.7	0.0
West-North	32.6	2.4	13.4	0	48.4	0.0
South Nigeria	50.8	0.6	17.7	0	69.1	0.0
North Nigeria	41.7	0.7	18.8	0	61.2	0.0
Sudan-Eretria	26.1	2.6	9.7	0	38.4	0.0
Ethiopia	44.7	3.6	13.1	0	61.4	0.0
Somalia Djibouti	24.5	3	27.3	0	54.8	0.0
Kenya-Uganda	36.6	0.9	10.5	0	48.0	0.0
Tanzania	38.9	1.4	25.2	0	65.5	0.0
Central	50.3	0.9	12.4	0	63.6	0.0
Congo	25.8	0	10.7	0	36.5	0.0
South-West	39.1	1.3	13.8	0	54.2	0.0
South Africa	31.8	3.1	23.5	0	58.4	0.0
South-East	40.8	2.7	10.5	0	54.0	0.0
Indian Ocean	41.5	1.2	16.5	0	59.2	0.0
Country-wide	LCOE primary [€/MWh]	LCOC [€/MWh]	LCOS [€/MWh]	LCOT [€/MWh]	LCOE total [€/MWh]	
Area average	37.5	2.0	18.1	0.1	57.7	
West-West	35.5	1.4	19.8	0.0	56.7	0.0
West-South	42.3	1.2	22.2	0.0	65.7	0.0
West-North	32.6	2.4	13.4	0.0	48.4	0.0
South Nigeria	50.6	0.2	18.1	0.3	69.2	0.0
North Nigeria	40.2	0.6	19.1	0.3	60.2	0.0
Sudan-Eretria	26.1	2.6	9.7	0.0	38.4	0.0
Ethiopia	44.7	3.6	13.1	0.0	61.4	0.0
Somalia Djibouti	24.5	3.0	27.3	0.0	54.8	0.0
Kenya-Uganda	36.6	0.9	10.5	0.0	48.0	0.0

Tanzania	38.9	1.4	25.2	0.0	65.5	0.0
Central	50.3	0.9	12.4	0.0	63.6	0.0
Congo	25.8	0.0	10.7	0.0	36.5	0.0
South-West	39.1	1.3	13.8	0.0	54.2	0.0
South Africa	31.8	3.1	23.5	0.0	58.4	0.0
South-East	40.8	2.7	10.5	0.0	54	0.0
Indian Ocean	41.4	1.3	16.5	0.0	59.2	0.0
Area-wide	LCOE primary [€/MWh]	LCOC [€/MWh]	LCOS [€/MWh]	LCOT [€/MWh]	LCOE total [€/MWh]	export (-)/ import (+) [%]
Area average	36.4	1.7	14.8	1.7	54.6	
West-West	35.4	1.4	16.4	0.9	54.1	-2.9
West-South	40.4	0.3	21.2	1.0	62.9	2.0
West-North	30.6	2.1	5.4	8.2	46.3	-80.3
South Nigeria	47.9	0.0	19.6	10.5	78.0	50.8
North Nigeria	38.2	0.5	18.7	0.9	58.3	-1.2
Sudan-Eretria	24.4	1.0	7.6	1.3	34.3	-1.9
Ethiopia	45.2	4.4	7.3	1.0	57.9	1.7
Somalia Djibouti	24.8	5.0	12.6	2.5	44.9	-3.9
Kenya-Uganda	35.3	0.5	8.4	1.3	45.5	-12.4
Tanzania	40.1	0.4	20.7	2.8	64.0	19.7
Central	50.6	0.1	10.4	1.1	62.2	-1.6
Congo	25.2	0.0	7.2	1.1	33.5	5.2
South-West	38.9	0.9	8.3	1.1	49.2	-3.2
South Africa	32	3.0	20.9	0.6	56.5	-0.4
South-East	42.2	1.7	5.2	1.4	50.5	0.8
Indian Ocean	38.9	1.6	15.7	0.9	57.1	-0.2
Integrated scenario	LCOE primary [€/MWh]	LCOC [€/MWh]	LCOS [€/MWh]	LCOT [€/MWh]	LCOE total [€/MWh]	export (-)/ import (+) [%]
Area average	34.4	1.5	9.3	1.7	46.9	
West-West	35.9	1.6	14.7	0.9	53.1	-4.3
West-South	38.4	0.7	18.1	0.7	57.9	2.2
West-North	30.5	3.1	4.0	7.7	45.3	-84.7
South Nigeria	37.4	0.3	0.2	7.1	45.0	45.1
North Nigeria	31	0.8	7.7	0.3	39.8	-1.6
Sudan-Eretria	24.7	1.0	7.6	1.2	34.5	-0.3

Ethiopia	45.9	4.4	7.2	0.9	58.4	1.5
Somalia Djibouti	24.8	5.8	10.6	2.2	43.4	-3.6
Kenya-Uganda	35.5	1.1	8.0	1.6	46.2	-15.6
Tanzania	31.4	0.7	4.4	1.7	38.2	13.7
Central	40.2	0.3	3.7	0.9	45.1	5.7
Congo	27.2	0.1	4.6	1.1	33.0	-1.0
South-West	35.6	0.9	2.8	0.4	39.7	-2.0
South Africa	31.9	2.4	16.9	0.5	51.7	-1.1
South-East	42.3	1.5	4.8	1.4	50.0	1.4
Indian Ocean	38.7	1.9	15.1	1.0	56.7	-0.2

Table XV: Overview on electricity transmission lines parameters for the area-wide open trade scenario.

	Region 1	Region 2	Length [km]	Capacity [GW]	Utilisation [%]
1	West West	West South	2581.50	0.35	77
2	West South	South Nigeria	451.37	1.08	55
3	West North	North Nigeria	990.68	4.63	90
4	West North	Sudan Eritrea	2125.10	0.68	69
5	West North	Central	1101.80	0.43	75
6	South Nigeria	North Nigeria	580.07	1.03	71
7	South Nigeria	Central	1244.90	0.21	84
8	Sudan Eritrea	Ethiopia	1149.30	0.32	60
9	Ethiopia	Somalia Djibouti	1166.40	0.09	60
10	Ethiopia	Kenya Uganda	1357.50	0.17	67
11	Kenya Uganda	Tanzania	789.55	0.74	73
12	Kenya Uganda	Congo	2805.60	0.26	85
13	Tanzania	South East	2058.10	0.39	78
14	Central	Congo	1152.90	0.31	67
15	Congo	South West	591.38	0.59	68
16	South West	South Africa	3237.90	0.00	45
17	South West	South East	2418.10	0.27	82
18	South Africa	South East	1077.30	3.32	69
19	South East	Indian Ocean	2351.70	0.26	71

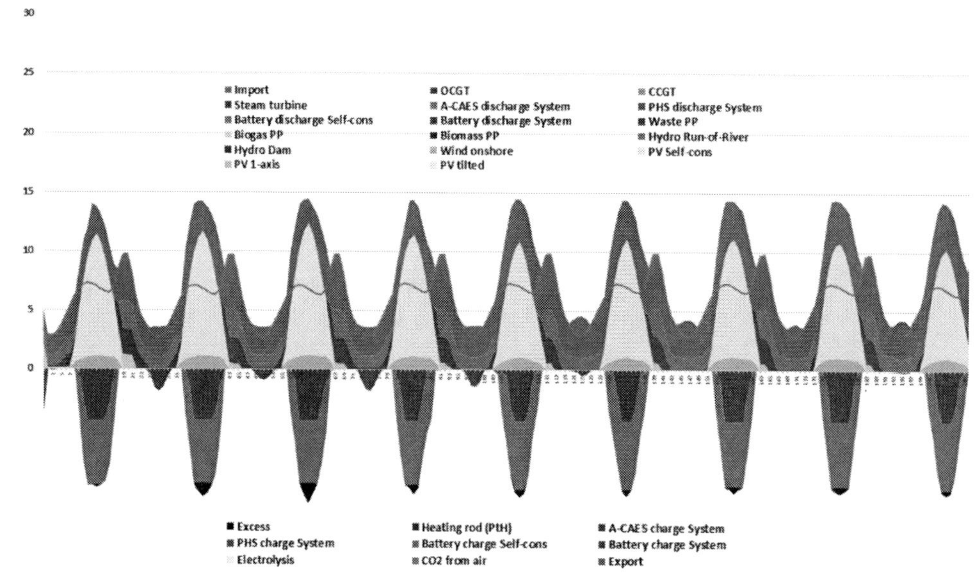

Fig. 1. Hourly generation profile for South-Nigeria example for an importing region.

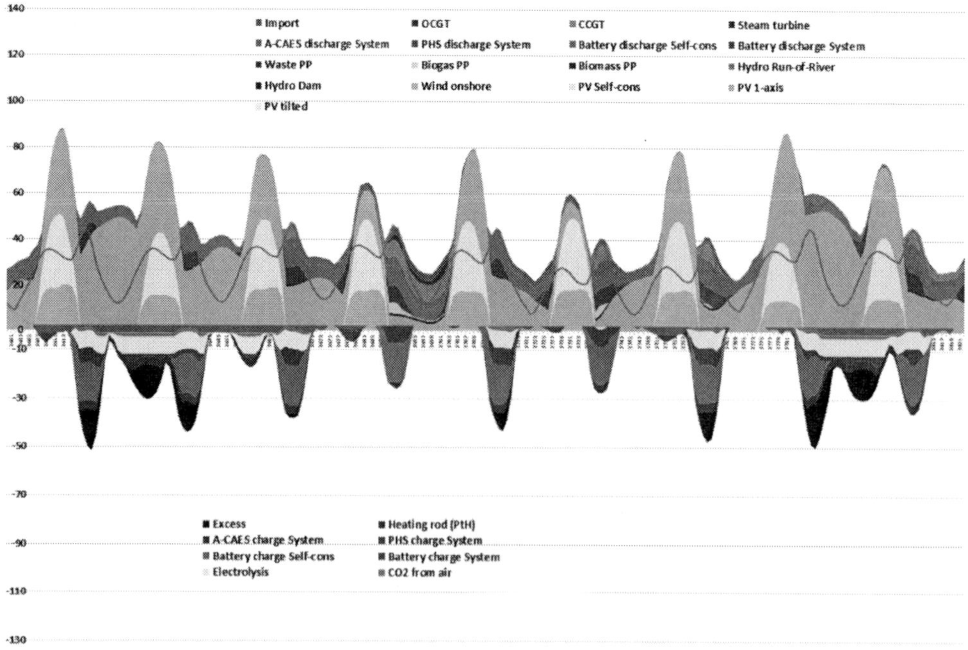

Fig. 2. Hourly generation profile for South Africa, example for a balancing region.

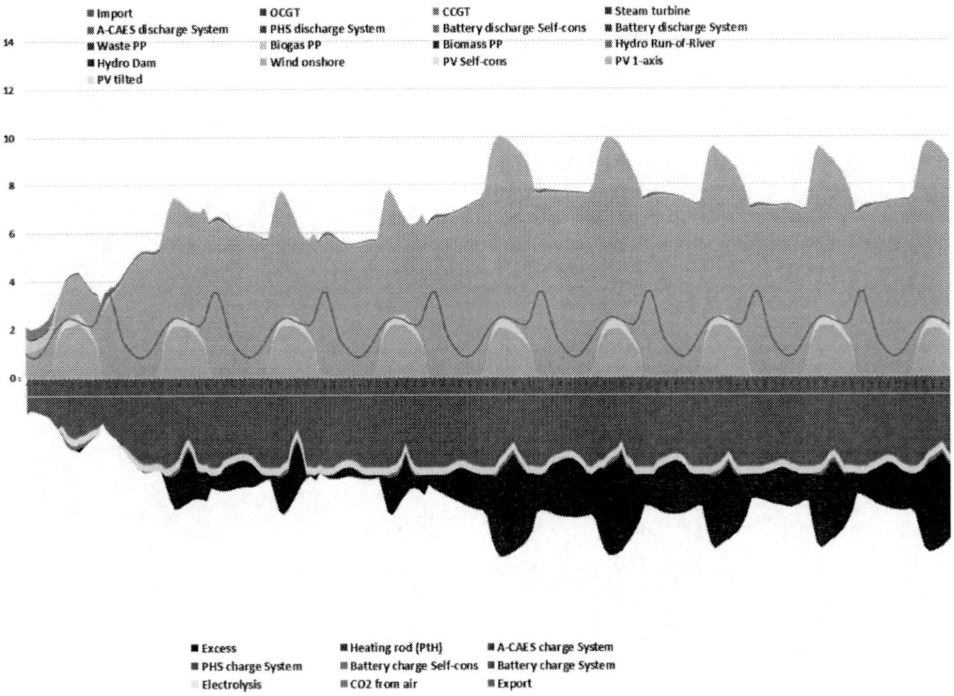

Fig. 3. Hourly generation profile for West North, example for an exporting region.

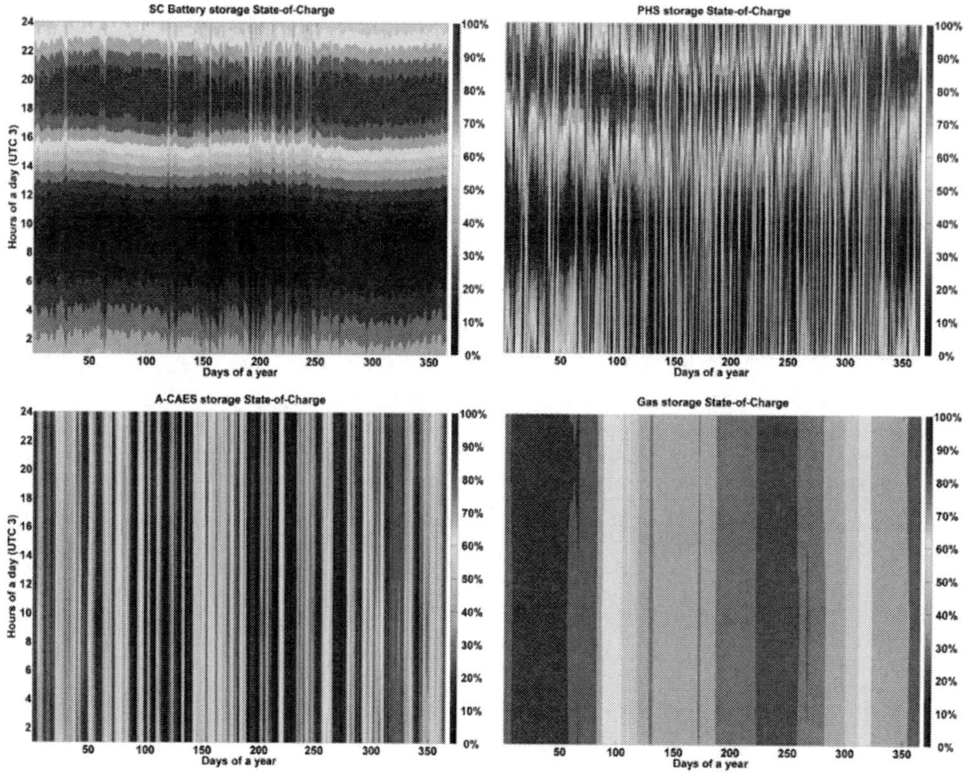

Fig. 4. Aggregated state-of-charge for the storages in the integrated scenario: battery (top left), PHS (top right), A-CAES (bottom left) and gas storage (bottom right).

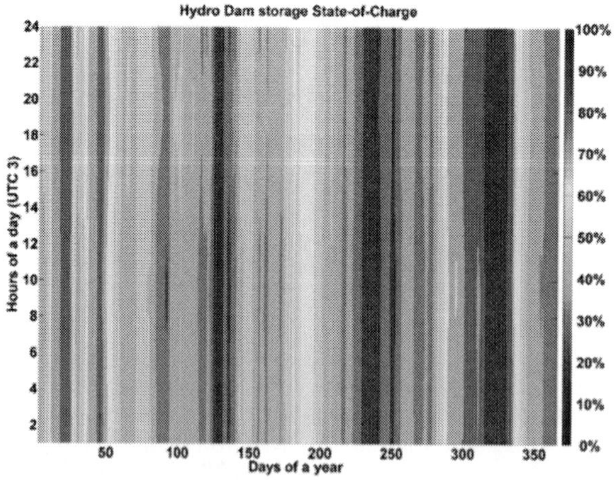

Fig. 5. State-of-charge for hydro dams in the integrated scenario.

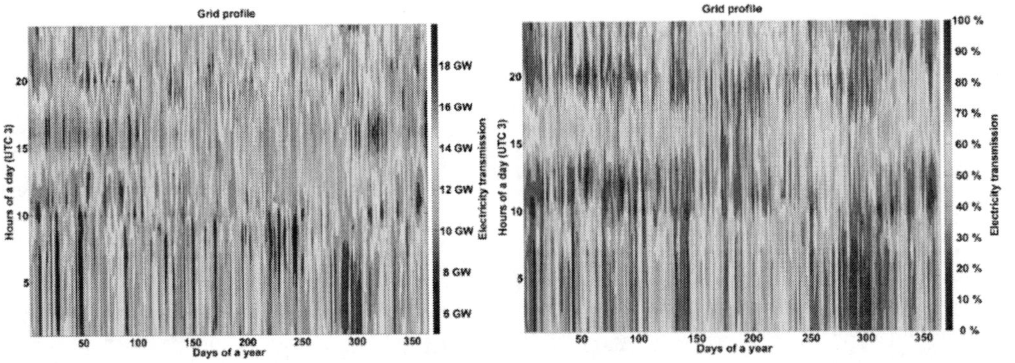

Fig. 6. Grid capacity profiles, absolute (left) percentage (right) for the integrated scenario.

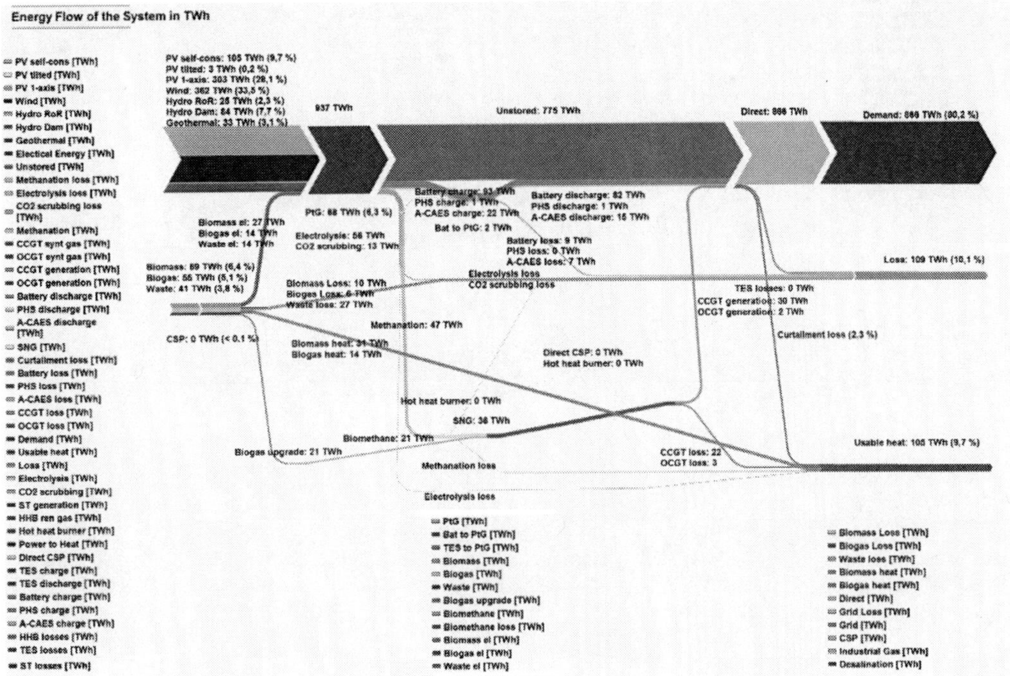

Fig. 7. Energy flow of the system for the region-wide open trade scenarios.

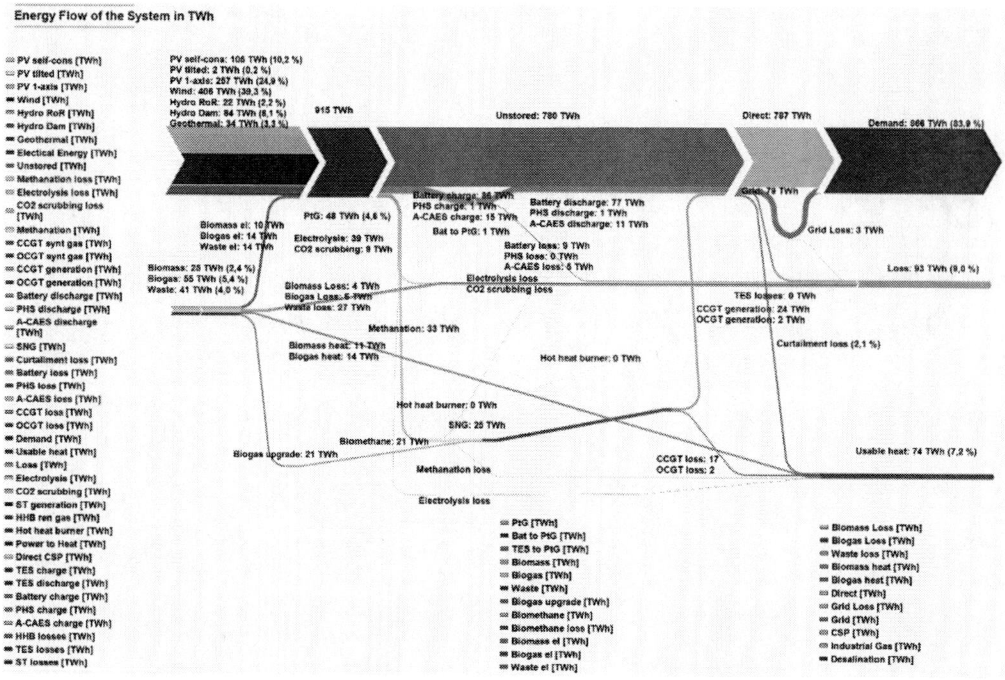

Fig. 8. Energy flow of the system for the area-wide open trade scenarios.

SOLAR PHOTOVOLTAICS – A DRIVING FORCE TOWARDS 100% RENEWABLE ENERGY FOR INDIA AND SAARC

Ashish Gulagi*, Dmitrii Bogdanov and Christian Breyer
Lappeenranta University of Technology, Skinnarilankatu 34, 53850 Lappeenranta, Finland
E-mail: Ashish.Gulagi@lut.fi, Christian.Breyer@lut.fi

ABSTRACT:
The developing region of SAARC (South Asian Association for Regional Cooperation) is home to a large number of people living below the poverty line. In future, providing affordable, access to all, reliable, low to zero carbon electricity in this region will be the main aim of electricity generation. A cost optimal 100% renewable energy based system is simulated for this region for the year 2030 on an hourly resolved basis for an entire year. The region was divided into 16 sub-regions and three different scenarios were set up based on the level of high voltage direct current (HVDC) grid connections. The results obtained for a total system levelised cost of electricity (LCOE) showed a decrease from 71.6 €/MWh in a decentralized to 67.2 €/MWh for a centralized grid connected scenario. An additional scenario was simulated to show the benefits of integrating industrial gas production and seawater reverse osmosis desalination demand which was reflected as the system cost decreased by 5% and the total electricity generation decreased by 1%. The results show that a 100% renewable energy based system could be a reality in the SAARC region with the cost assumptions used in this research and it may be more cost competitive than the nuclear and fossil carbon capture and storage (CCS) alternatives.

Keywords: 100% Renewable Energy, India, energy system optimization

1 Introduction

Energy is critical, directly or indirectly, to the entire process of evolution, growth and survival of all living beings and it plays a vital role in the socio-economic development and human welfare of a country and any uncertainty in its supply can threaten the functioning of the economy, particularly in developing countries [1]. The region of interest for this research is the developing region of South Asia which is made up of the following countries Afghanistan, Bangladesh, Bhutan, India, Maldives, Nepal, Pakistan, and Sri Lanka also called as SAARC (South Asian Association for Regional Cooperation). This region is home to 23% of the total world population, large number of them living in extreme poverty and lacking access to clean and reliable energy services [2]. The future of electricity generation for this region will be based on providing affordable, access to all, reliable, low to zero carbon electricity [3]. A report published by WWF lists ten recommendations for a 100% renewable energy future: Developing new and existing renewable energy sources to provide clean energy, sharing and exchange of clean energy through grids, making the use of sustainable resources in different areas [4] being the top two recommendations. A least cost energy system needs to be developed without compromising the above mentioned objectives.

The ways to achieve sustainable growth is to generate electricity from the various renewable sources, of which solar energy is the most promising [5]. According to IPCC, solar has the greatest resource potential among all the available renewable energy sources and only a small fraction of this energy would be sufficient to meet our current energy demands [6]. Despite the technical potential and the exponential growth of the PV market in recent years, contribution of solar energy to the global energy supply mix is still on a small level [7].

The largest country in the SAARC region in terms of population and gross domestic product (GDP) is India with 1.2 billion inhabitants which accounts for 17% of world population [8] and one in five of the population – 240 million people – still lack access to electricity [9]. India is the fourth-largest consumer of energy, accounting for 4.9% of global consumption, which is dominated by coal and imported oil [10]. With a high rate of population growth and GDP expected to grow at 8% per annum till 2030 and 6% beyond that [11], it is evident that the demand for electricity is expected to grow in the future [12]. To satisfy the growing electricity demand in a responsible and sustainable way, India needs to shift away from the fossil fuels to renewable energy sources. The National Action Plan on Climate Change (NAPCC) in June 2008 identified the development of solar energy technologies in the country to be pursued as national mission [13]. According to WBGU [14] and IPCC [15], India is most vulnerable to climate change, due to flooding, change in monsoon cycle and water scarcity.

Solar energy, if harnessed properly can play a big part in reducing dependence on coal and other fossil fuels, as India lies in the sunniest regions of the world with 250-300 clear sunny days [16]. The average solar radiation incident over India varies from 4 kWh/day to 7 kWh/day with sunshine ranging between 2300 and 3200 h in most parts of India [17]. The solar radiation received over the Indian land area is estimated to be 5 million TWh/year which is far more than its annual consumption [18]. The full load hours for the SAARC regions is given in Fig. 1.

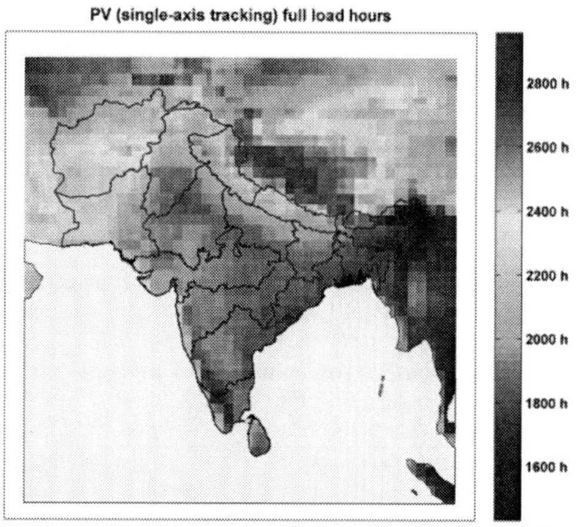

Fig. 1: Full load hours for PV (single-axis tracking) for the SAARC region.

The steps taken by the government of India are in the right direction in terms of utilizing the solar potential to provide electricity for all in a sustainable way and reduce the effects of climate change. Solar power in India is poised to grow significantly with 'The National Solar Mission' programme as a major initiative of the Government of India. A scheme for development of Solar Parks, Ultra Mega Solar Power Projects, canal top solar projects and solar pumps for farmers is at different stages of implementation. The government's goal of 'Electricity for All' is sought to be achieved by the above programmes that would require huge investments, infusion of new technology and international support [19]. The main driver is 'The National Solar Mission' which aims to promote the development and use of solar energy for power generation with an ultimate aim of making solar cost competitive with fossil based energy options through long term policy, large scale deployment goals, aggressive R&D and domestic production of critical raw materials [20].

India is set to become the world's second largest market for solar PV [21] and solar power will find a breakeven to the price of imported coal in 2015 and domestic coal in 2019 [22]. According to the energy minister of India, a new coal power plant will give costlier power than a solar plant which is due to the rapid decrease in the solar prices in recent years [23]. India has ambitious plans to expand the deployment of solar and wind power. The targeted levels of deployment for renewables is 175 GW by 2022 (of which 100 GW is solar and 60 GW is wind) are a powerful statement of intent from the Government of India [24]. In projections, India achieves its climate pledge that 40% of the installed power generation capacity in 2030 is planned to be based on non-fossil fuel [25]. According to KPMG [22], by 2024/25, 166 GW of solar capacity will be added. During COP 21, India launched the International Solar Alliance (ISA), of all countries located in between Tropic of Cancer and Tropic of Capricorn [26]. ISA is conceived as a coalition of solar resource rich countries to address their special energy needs and will provide a platform to collaborate on addressing the identified gaps through a common, agreed approach

The paper is organised as follows: Section 2 gives the methodology, the input data and the technologies used for the simulations. This is followed by the assumptions and the description of the different scenarios used for the simulations. Section 4 explains the results of the simulations. This is followed by the discussion of the results. Finally, conclusions are drawn in section 6.

2 METHODOLOGY

The model used in this study is based on linear optimization of the energy system parameters under previously defined constraints which are applied to the system and the assumptions for the future RE power generation and demand, required storage technologies including additional water desalination and synthetic natural gas generation which are the flexible demands in the model. One of the key constraint for the system optimization is matching of power generation and demand on an hourly basis for a particular year. The main aim of the system optimization is the minimization of the total annual energy system costs which is a sum of the installed capacities of the different technologies, costs of energy generation and generation ramping. Also, the system consists of prosumers for residential, commercial and industrial sectors. The prosumers install the respective capacities of rooftop PV systems and batteries. Minimizing the cost of consumed energy is the main aim of the target function for the prosumers, which is calculated as a sum of self-generation, annual cost and cost of electricity consumed from the grid. However the extra-benefits from selling excess electricity are not included in the study. The flowchart of the model is presented in Fig. 2.

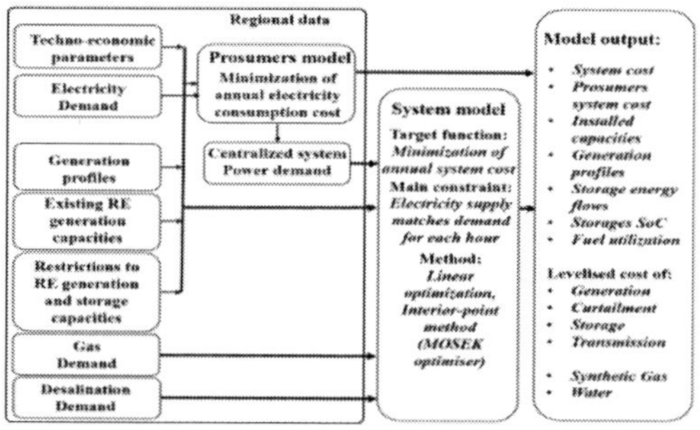

Fig. 2. Model flow diagram with the input data, system model optimization and output data [27]

The detailed description of the model can be found in Bogdanov and Breyer [28] and the next sections does not repeat the same information related to the model, input data related to it and the applied technologies. The technical and financial assumptions related to the model is found in the supplementary material. However all the additional assumptions considered for the model in this study is discussed in detail.

2.1 Input Data for the model

Detailed information of the input data used for the model is given in Bogdanov and Breyer [28] and additional calculations related to geothermal, desalination water demand and industrial gas demand data is described here.

- The potential for geothermal for the sub-regions is calculated on the available information related to heat flow rate and ambient temperature of the surface [29,30] for the year 2005. For the sub-regions where the heat flow data was not available, extrapolation was performed to get the required data. Based on the available data, different temperature levels and available heat at the mid-point of a 1 km thick deep layer and for between the depths of 1 km to 10 km [31,32,33] globally with 0.45 ° x 0.45 ° spatial resolution, required potential is derived.

- For every sub-regions, projected water desalination demand is calculated on projections for water consumption and stress level [34]. It is assumed, water stress more than 50% will be covered by seawater desalination. Detailed calculations for the technical constraints and financial cost of seawater reverse osmosis desalination are described Caldera et.al. [35].
- Natural gas consumption was derived in the non-energy sector for every sub-region. Industrial gas consumption data are based on IEA statistics for non-energy sector demand [36].

2.2 Applied Technologies

For the SAARC region, technologies applied for energy system optimization can be divided into four main categories.

- *Technologies for converting renewable energy sources into electricity*

The technologies used for converting RE sources into electricity are two different types of ground mounted PV systems (optimally tilted and single-axis north-south oriented horizontal continuous tracking) and rooftop PV, concentrating solar thermal power (CSP), onshore wind turbines, hydro power divided into run-of-river and dams, biomass (biogas and solid biomass), waste-to-energy power plants and geothermal power plants.

- *Energy storage*

The energy storage technologies used in the model are battery storage, pumped hydro storage (PHS), adiabatic compressed air energy storage (A-CAES), thermal energy storage (TES) and power-to-gas (PtG) technology. Technologies such as water electrolysis, methanation, CO2 scrubbing from air, gas storage, and both combined and open cycle gas turbines (CCGT, OCGT) are part of the synthesis of SNG and its reconversion to electricity. The PtG technologies have to be operated in synchronization because of the absence of hydrogen and CO2 storage. There is a 48-hour biogas buffer storage and part of the biogas can be upgraded to biomethane and introduced to the gas storage.

- *Energy bridging technologies*

The bridging technologies used in this model provide the required flexibility to the energy system and helps in reducing the overall cost. For example, gas produced from PtG can be used for industrial gas demand rather than storage for electricity sector. Similarly, seawater reverse osmosis (SWRO) desalination couples with the water sector than electricity sector.

- *Electricity transmission technologies*

The power transmission within the sub-regions is assumed to be based on alternating current (AC) grids which is not included in the model and between the sub-regions on high voltage direct current (HVDC). Loss of electricity due to the length of the transmission lines and in converter stations at the interconnection with the AC grid form a major component of the power losses in HVDC grids. The full block model diagram is presented in Fig. 3.

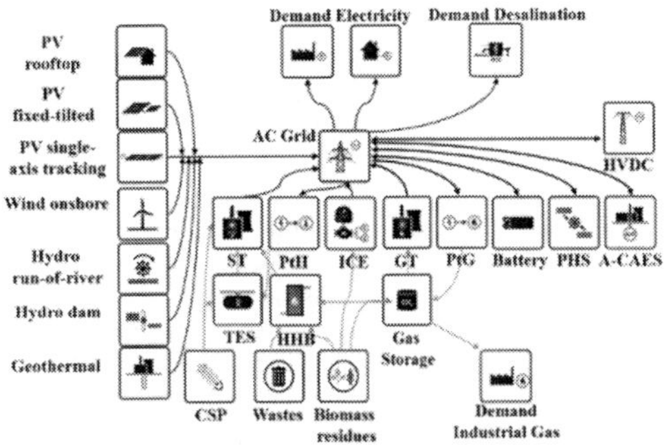

Fig. 3. Block diagram of the all the energy technologies applied in the model for the SAARC region [27]

3 SCENARIO ASSUMPTIONS FOR THE SAARC REGION

3.1 Subdivision of the region and grid structure

The SAARC region is subdivided into 16 sub-regions, based on the population distribution, electricity consumption and countries grid structure. The different sub-regions can be seen from the Fig. 3. India is subdivided into 10 different regions, Pakistan into two regions and the other remaining countries are treated as individual regions. The grid connection between the regions is shown in Fig. 4 which includes interconnections within the countries shown by dark line and between the countries shown by dotted lines.

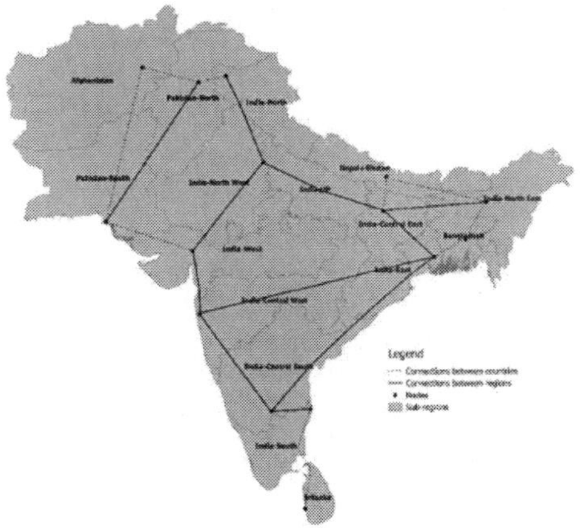

Fig. 4. The different SAARC regions and HVDC grid configuration

3.2 Applied scenarios

The different scenarios taken into consideration in this paper for the analysis of the energy system of SAARC region are:

- Region-wide scenario, in which the regions are independent of each other and have no interconnections so the demand for electricity is covered by the respective region's own generation capacity.
- Country-wide scenario, in which regions of the same country are interconnected via HVDC lines
- Area-wide scenario, in which the energy systems of the countries are interconnected

- Integrated scenario: area-wide scenario plus SWRO desalination and industrial gas demand, where PtG technology is used not only as a storage option but also covers industrial gas demand. This increases the flexibility of the system.

3.3 Financial and technical assumptions

The optimization of the model is carried out on an assumed cost basis and the state of technology for the year 2030. The financial assumptions for all the energy system components, HVDC transmission lines and converter stations which are given as net transmission capacity (NTC) for the 2030 reference year are tabulated in the Supplementary Material (Table I). Weighted average cost of capital (WACC) is set to 7% for all scenarios, but for residential PV prosumers WACC is set to 4% due to lower financial return requirements. The technical assumptions related to energy to power ratios for storage technologies, efficiency numbers for generation and storage technologies and power losses in HVDC power lines [37] and converters are presented in the Supplementary Material (Tables II - IV). Electricity prices for residential, commercial and industrial consumers for all the countries is taken from Gerlach et al. [38]. The electricity prices for Nepal and Bhutan are assumed to be similar to India. The electricity prices for 2030 are calculated according to the assumptions from Werner et al. [39]. The electricity prices for all the regions is provided in the Supplementary Material (Table V). The excess electricity generated by the prosumers is assumed to be fed into the grid for a transfer selling price of 2 €cents/kWh. The prosumers have to satisfy their own annual demand before they can sell electricity to the grid. The renewable energy investments' financial incentives, such as Renewable Energy Certificate mechanism, Perform Achieve Trade (PAT), are not considered in the assumptions, since a cost-based model is applied.

3.4 Feed-in for solar and wind energy

The feed-in profiles for solar CSP, optimally tilted and single-axis tracking PV, and wind energy were calculated according to Bogdanov and Breyer [28]. The aggregated profiles of solar PV generation single-axis tracking and wind energy power generation normalized to maximum capacity averaged for the SAARC region are presented in Fig.5. The computed average full load hours (FLH) for optimally tilted, single-axis tracking PV systems, wind power plants and CSP are provided in the Supplementary Material (Table VIII).

For the hydro power, feed-in values are computed based on the monthly resolved precipitation data for the year 2005 as a normalized sum of precipitation in the regions. Such an estimate leads to a good approximation of the annual generation of hydro power plants.

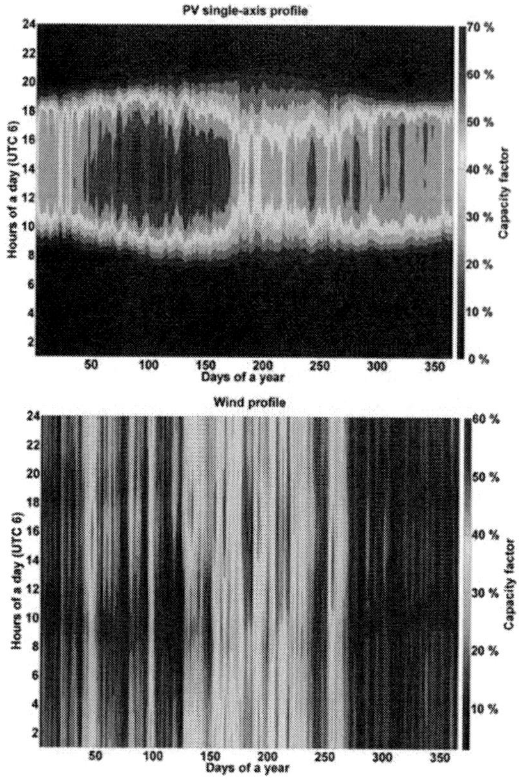

Fig. 5: Yearly profile for PV single-axis tracking (top) and wind onshore (bottom).

3.5 Biomass and geothermal potential

The potential for biomass and waste resources are taken from DBFZ [40] and divided into different categories as described in Bogdanov and Breyer [28]. The cost associated with all the biomass resources is calculated using data from International Energy Agency [41] and Intergovernmental Panel on Climate Change [42]. For solid fuels a 75 €/ton fee for the waste incineration is assumed. The heating values are based on lower heating values (LHV).

The geothermal heat potential for all the regions were calculated based on the spatial data for available heat, temperature and geothermal plants for depths from 1 km to 10 km. For each 0.45 ° x 0.45 ° area and depth, LCOE for geothermal is calculated and optimal depth is determined. The assumption for available geothermal heat is that only 10% of it will be utilized as an upper resource limit. The total available heat for all the regions was calculated using the same weighted average formula as for solar and wind feed-in as described in Bogdanov and Breyer [28], with an exception of areas with geothermal LCOE exceeding 100 €/MWh, which are excluded. The calculated potentials for solid biomass, biogas, solid waste and respective costs, and geothermal heat potentials are provided in the Supplementary Material (Tables VI and VII).

3.6 Upper and lower limitations on installed capacities

The upper limits for all the RE technologies (optimally tilted PV, wind turbines, hydro power) and for pumped hydro storage were calculated according to Bogdanov and Breyer [28] and summarized in

Supplementary Material (Table X). The upper limits for all other technologies are not specified. The data for the lower limit for all the above RE technologies is taken from Farfan and Breyer [43] and summarized in Supplementary Material (Table IX). For biomass residues, biogas and waste to energy plants it is assumed, due to energy efficiency reasons, that the available and specified amount of the fuel is utilized during the year.

3.7 Load

The load profile for each region is calculated as a fraction of the total demand in that particular country based on synthetic load data weighted by the region's population. Fig. 6 represents the area aggregated demand profile for all the regions considered in SAARC. Solar PV prosumers have a significant impact on the residual load as observed from overall energy demand and maximum load which are reduced by 9% and 0.2% respectively (Fig. 6). The demand for gas in industries and desalination water demand for the SAARC region is given in the Supplementary Material (Table XI).

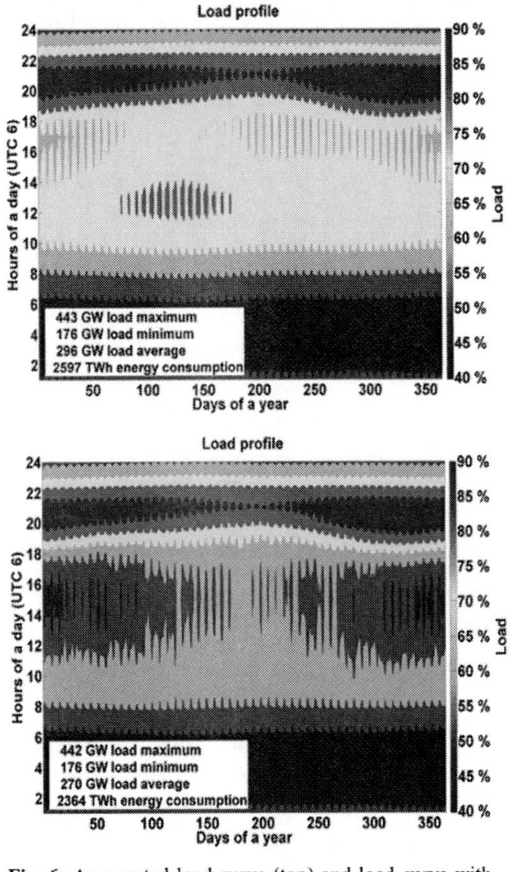

Fig. 6: Aggregated load curve (top) and load curve with prosumers influence (bottom) for SAARC region for the year 2030

4 RESULTS

4.1 Structure and cost of an optimized energy system

The cost structure of the different scenarios were analyzed with the set of parameters calculated as given in Bogdanov and Breyer [28] plus the model extension described in section 2.1. The computed key financial numbers for the different scenarios for the SAARC region is presented in Table II. The key financial results are given as the total system (LCOE) levelised cost of electricity (including PV self-consumption and the centralized system), levelised cost of electricity for primary generation (LCOE primary), levelised cost of curtailment (LCOC), levelised cost of storage technologies (LCOS), levelised cost of transmission (LCOT), total annualized cost, total capital expenditures and other results which are included are the total renewable capacity and total primary generation

The benefit due to interconnection of the sub-regions via HVDC power lines has a positive impact on the LCOE and total annual cost of the system, but this impact is only marginal due to limited grid utilization. LCOE and annual cost of the system are decreased by 6.1% and 6.9% respectively from region-wide to area-wide scenario. Also, grid utilization decreases installed capacities and total electricity generated from RE sources by 12.1% and 4.4%, respectively. The cost of transmitting electricity with HVDC power lines is small in comparison to the cost of storage therefore total system cost and LCOE are reduced. The transmission lines decrease the need for storage technologies, since energy shifted in time (storage) can be partly cost effectively substituted by energy shift in the location. Also, cost of curtailment is reduced when electricity is transmitted to other sub-regions as seen in Table I. The components that make up the LCOE in region-wide, country-wide, area-wide and integrated scenario is presented in the Supplementary Material (Table XII).

The installed capacities of all the renewable energy technologies show a decrease with increase in installation of HVDC lines as shown in Table II. The total installed capacities of PV decrease by 16.7% from region-wide to area-wide scenario due to efficient use of solar resources available in the region. The installed capacity for wind shows a slight increase in the area-wide scenario due to low solar irradiation in the monsoon months. The installed capacities of PV and wind in the integrated scenario increase due to the additional demand of seawater desalination and industrial gas. In the SAARC region, PV is the least cost RE source followed by wind energy. The share of PV single-axis tracking and PV self-consumption of the total solar PV installed capacity for the area-wide scenario is 82.6% and 15.3%, respectively.

In the integrated scenario, additional flexible demand from the desalination and industrial gas sectors leads to an increase in installed capacities of the low cost solar by 38.3% and wind resources by 64.7% with respect to the area-wide scenario. There is a slight increase in the installed capacities of the biomass resources. The capacity of the hydro dams do not change as it provides the system with required flexibility. Despite an upper

Table I: Financial results for the four scenarios applied for the SAARC region

2030 Scenarios	Total LCOE	LCOE primary	LCOC	LCOS	LCOT	Total ann. cost	Total CAPEX	RE capa-cities	Gener-ated electri-city
	[€/MWh]	[€/MWh]	[€/MWh]	[€/MWh]	[€/MWh]	[b€]	[b€]	[GW]	[TWh]
Region-wide	71.6	42.3	1.5	27.8	0.0	187	1539	1377	2948
Country-wide	69.6	41.9	1.1	25.5	1.1	181	1468	1294	2865
Area-wide	67.2	41.4	0.7	22.7	2.3	174	1421	1210	2818
Integrated scenario	67.9	40.8	1.4	22.6	3.1	299	2562	2213	4988

Table II: Installed RE technologies and storage capacities for the four scenarios for SAARC region.

		Region-wide	Country-wide	Area-wide	Integrated scenario
PV self-consumption	[GW]	145	145	145	145
PV optimally tilted	[GW]	21	23	3	3
PV single-axis tracking	[GW]	782	721	640	1131
PV total	[GW]	947	889	789	1280
CSP	[GW]	0	0	0	0
Wind energy	[GW]	242	229	245	694
Biomass power plants	[GW]	64	64	61	65
MSW incinerator	[GW]	3	3	3	3
Biogas power plants	[GW]	21	16	22	14
Geothermal power	[GW]	2	6	8	8
Hydro Run-of-River	[GW]	22	21	21	21
Hydro dams	[GW]	35	35	35	35
Battery PV self-consumption	[GWh]	4	4	4	4
Battery System	[GWh]	1389	1522	1450	1682
Battery total	[GWh]	1393	1526	1454	1686
PHS	[GWh]	44	44	44	44
A-CAES	[GWh]	2416	551	3	3187
Heat storage	[GWh]	0	0	0	0
PtG electrolyzers	[GW_{el}]	43	30	20	115
CCGT	[GW]	50	43	41	99
OCGT	[GW]	3	2	2	0
Steam Turbine	[GW]	0	0	0	0

limit of 50% higher than the current capacity which was considered for hydro dams and hydro run-of-river plants, installed capacities do not increase in the integrated scenario as PV and wind are least cost technologies in the SAARC region.

The generation curves for electricity can be represented for the whole year divided into 8760 hours and sorted according to the generation minus the load, which is represented by a black line as shown in Supplementary Material (Fig. 6) for the area-wide scenario. All the storage technologies used in the system are charged for about 3500 hours in a year which is due to higher electricity generation than demand. As the solar and wind are highly inflexible generation sources in these particular hours in SAARC region therefore there is a high electricity generation. The other flexible generation options such as hydro dams, biomass, biogas and discharge of storage technologies are required to balance the high inflexibility. The inflexible electricity generation options reduce significantly in the other hours of the year as the electricity demand decreases and there is a need for flexible electricity generation options, discharge of storage technologies and utilization of the grid. Curtailment of electricity is for only some hundred hours of the year since for all the other hours the HVDC lines enable the export of the electricity from the best RE producing sub-regions to other sub-regions of remaining demand.

4.2 Sub-regional analysis on an optimized energy system

The sub-region's distribution of system optimized RE resources can be observed from Fig. 7 and circles represent demand and generation with some sub-regions with the best renewable resources are net exporters and the others are net importers. The share of export is defined as the ratio of net exported electricity to the generated primary electricity of a sub-region and the share of import is defined as the ratio of imported electricity to the electricity demand. The area average is electricity demand. The sub-regions with best renewable energy resources help balance out sub-regions where the availability of renewable resources is scarce. For region-wide scenario, an individual SAARC sub-regions need to match their own demand using their own available RE resources. When the sub-regions are interconnected, as in case of country-wide, area-wide and integrated scenario, sub-regions act as net exporters and importers of electricity. Fig. 7 points out the net exporters and importers of electricity in an area-wide and integrated scenario. The differences observed between the demand and generation are mainly due to import and export and also due to losses related to storage. The net exporter regions for SAARC are: Afghanistan, Sri Lanka, India North and India Northeast due to excess of very good RE resources. The net importer regions are: Pakistan North, India Northwest, Bangladesh and India South. Due to a high electricity demand for additional desalination and SNG production, the integrated scenario tends to increase the electricity generation between the regions to fulfill the increased demand. The hourly resolved profiles for the net exporting regions Afghanistan and Sri Lanka and net importing region Pakistan North are presented in the Supplementary Material (Fig. 3, 5 and 4, respectively). The import/export shares in all regions and scenarios are summarized in the Supplementary Material (Table XI).

The grid utilization profile for the SAARC region can be found in the Supplementary Material (Fig. 8). The electricity trade increases during the night and morning hours almost throughout the year and more so in the monsoon period. This can be explained by the high share of solar PV generation in the region where electricity is imported by the sub-regions with high inflexible generation. In the monsoon period, due to less solar PV electricity generation, more trading of wind energy takes place within the sub-regions. The capacity of the power lines and grid utilization between the sub-regions for the area-wide open trade scenario is shown in the Supplementary Material (Fig. 8) and (Table XVII).

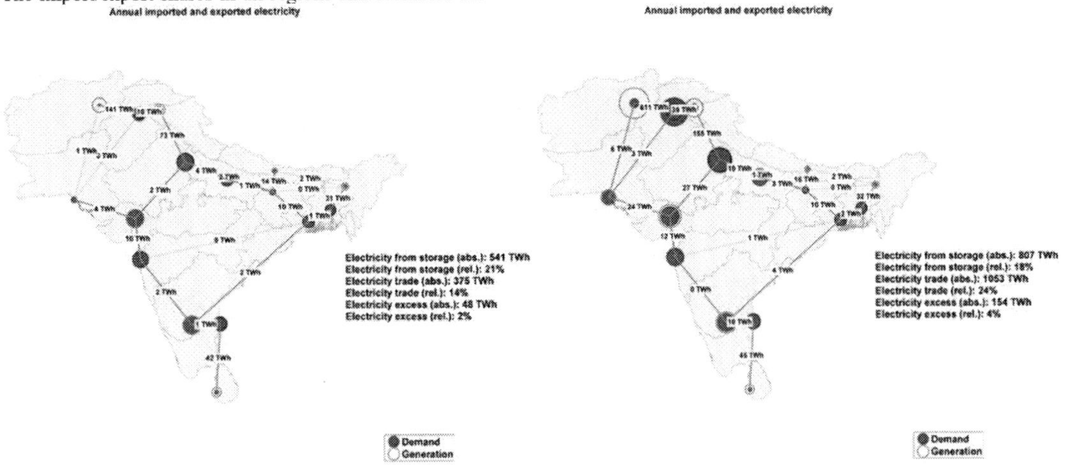

Fig. 7: Sub-region's annual import and export of electricity for area-wide (left) and integrated scenario (right).

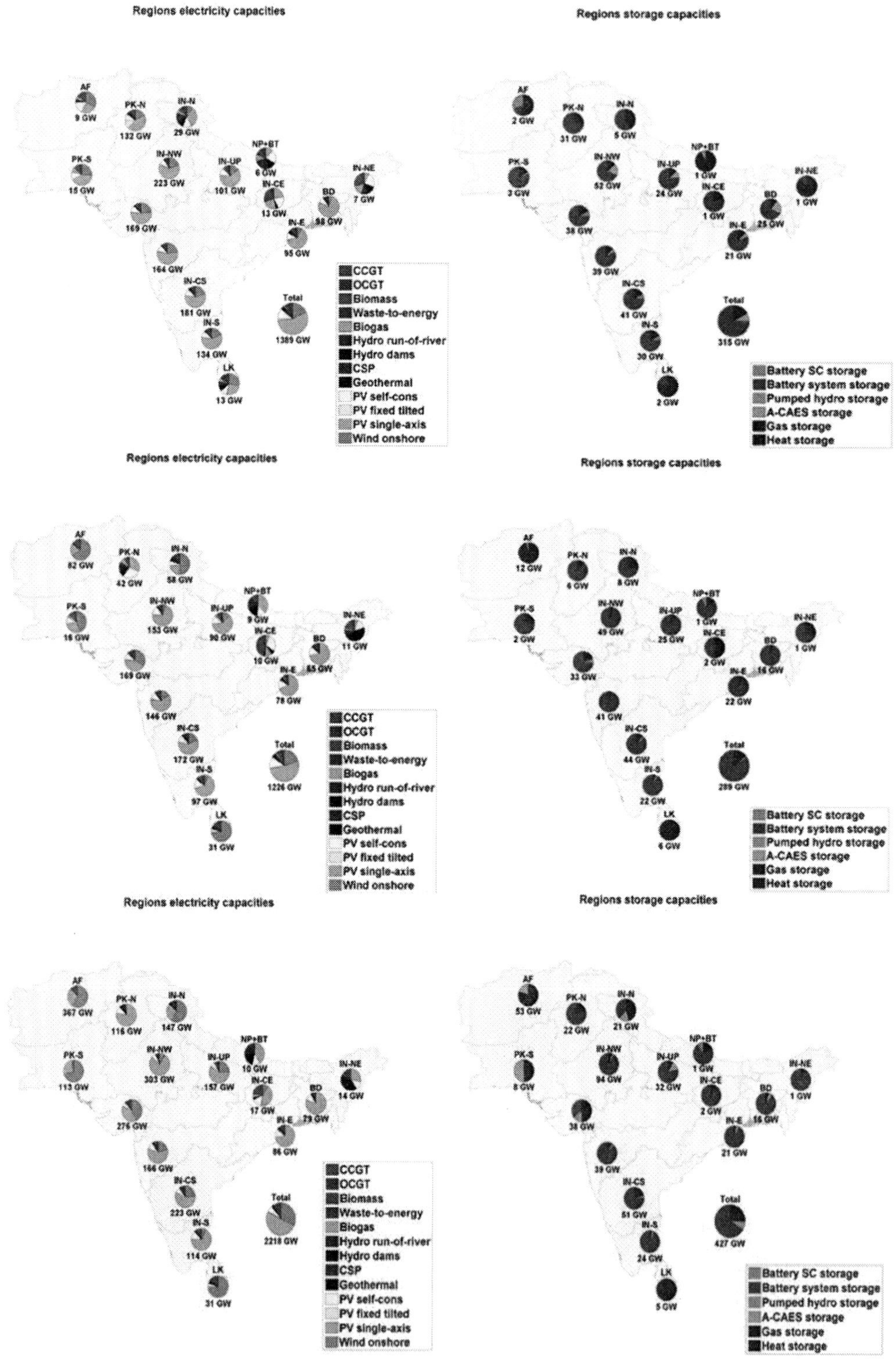

Fig. 8: Installed capacities RE generation (left) and storage capacities (right) for the SAARC sub-regions for region-wide (top), area-wide (center) and integrated (bottom) scenarios.

The installed capacities for RE generation and storage technologies for all sub-regions in region-wide, area-wide and integrated scenarios is shown in Fig. 8. The installed solar PV capacities exceed 50% of the total RE installed capacities in the Western and Southern part of India despite full load hours (FLH) of wind being comparable or exceeding FLH of solar PV. The installed capacities of PV was 947 GW (region-wide), 889 GW (country-wide) 789 GW (area-wide) and 1280 GW (integrated). The decrease in the installed capacities of PV from region-wide to area-wide scenario is observed due to proper utilization of PV resources due to transmission of electricity and decrease in curtailment. The highest installed capacity for PV was found in the Northwest region of India which has one of the best solar resources. For the total electricity generation mix, PV contributes to 60%, 58%, 53% and 49% for region-wide, country-wide, area-wide and integrated scenario, respectively. This is the highest solar PV share we have found in our global studies in any region in the world [44]. Prosumers play a vital role in providing flexibility to the system and installing low cost RE options and more storage capacities. PV electricity generation for prosumers is 234 TWh and battery storage contributes to 1.3 TWh.

It is observed in the sub-regions that have excellent wind conditions, low cost wind energy is the next preferred technology after solar PV, which is lowest in cost. In Sri Lanka and Afghanistan, installed capacities of wind energy is 71% and 65% of the total RE installed capacity in the area-wide scenario, as these countries have one of the best wind resources in the region and export of wind energy takes place from Sri Lanka to high demand centers in Southern India.

The total storage capacity required is greatly influenced by the connection of the sub-regions via HVDC transmission lines and RE generation and demand in a particular sub-region. Also, the mix of storage technologies for a particular sub-region depends on the above mentioned factors. The high share of PV in the system goes together with a high share of short term storage. Batteries play a vital role in shifting PV based electricity in the afternoon and at night. The throughput of batteries contribute 64%, 74%, 77% and 86% of the total storage annual generation for region-wide, country-wide, area-wide and integrated scenarios. The throughput of A-CAES, and gas storage technologies decreases by 99.8% and 30.5% respectively, from the region-wide to the area-wide scenario. The discharge capacities, annual throughput of storage technologies and full load cycles per year are provided in the Supplementary Material (Table XIII). State of charge profile diagrams for the area-wide scenario for battery, PHS, A-CAES and gas storage are given in the Supplementary Material (Fig. 7).

PV self-consumption does not influence the system in a big way in the SAARC region due to low electricity prices. PV self-generation covers 38.8%, 36.8% and 37.3% of residential, commercial and industrial prosumers demand, respectively. An overview on the PV self-consumption is provided in the Supplementary Material (Table XIII).

4.3 Energy flow for the optimized power systems for SAARC

The energy flow of the system from generation to demand for the integrated scenario is presented in Fig. 9. The energy flow diagram is made up of RE resources, storage technologies for the generated energy and the transmission of this energy via HVDC grids. The end use of electricity for the integrated scenario consists of electricity, desalination and industrial gas demand. The potentially usable heat generated and the losses incurred are comprised of curtailed electricity, heat produced by biomass, biogas and waste-to-energy power plants, heat generated from electrolyzers for transforming power-to-hydrogen, in the methanation process transforming hydrogen-to-methane, and methane-to-power in gas turbines. Efficiency losses incurred in A-CAES, PHS, battery storage and HVDC transmission grid losses form part of the overall losses. The energy flow diagrams for the region-wide and area-wide scenario are presented in the Supplementary Material (Fig. 9 and 10, respectively).

4.4 Effect of monsoon on the energy system

Interesting observations were made in the simulations regarding the effect of monsoon on the energy system. Monsoon in India normally lasts from June to September and the starting period depends on a particular region. In India solar is a constant resource for energy but in the period of monsoon, there is a reduced solar activity due to cloudy and rainy days. But in the monsoon period it was observed that there is a substantial increase in wind resources as shown in Fig. 5. For the Western region of India a particular week in the summer months and in the monsoon period shows the flexible operation mode of the RE system, as depicted in the Supplementary Material (Fig. 1 and 2). In the period of low solar radiation, wind and to some extent hydro balance the system to provide electricity and to keep the energy system running without power failure.

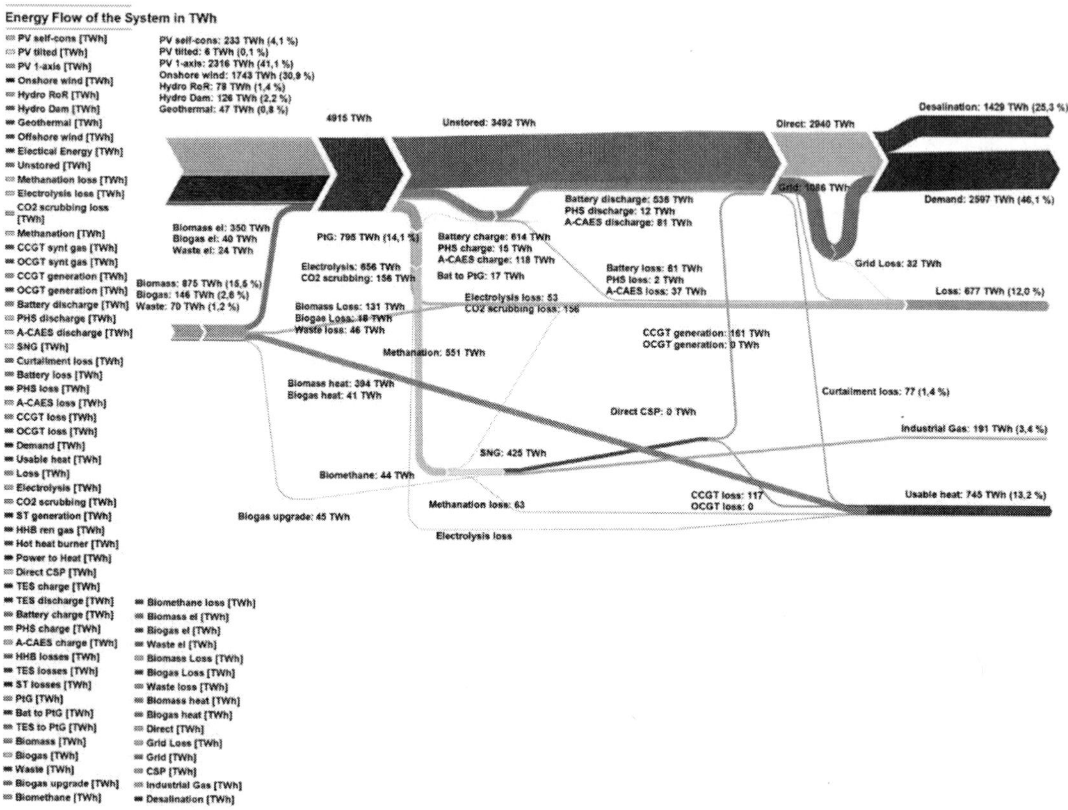

Fig. 9. Energy flow of the system in an integrated scenario

5 DISCUSSION

The installation of a HVDC transmission grid between sub-regions enables a significant decrease in the cost of electricity and in installed capacities of RE technologies in a 100% RE based system. The benefit due to grid integration varies for different regions of the world [28,44]For the SAARC region benefit due to grid integration seems to be marginal as local storage options seems to be more cost effective than transmission of the electricity. As there is a highly significant influence of solar in the system and almost stable solar conditions all around year, batteries store this energy to help balance the night time demand.

The total levelised cost of electricity in the SAARC region decreased from 71.6 €/MWh for the region-wide open trade scenario to 69.6 €/MWh for the country-wide open trade scenario and 67.2 €/MWh for the area-wide open trade scenario. The total annualized cost of the system decreased from 187 b€ to 174 b€ for region-wide to area-wide scenario. The capital expenditures for the system decreased from 1539 b€ to 1421 b€ for the region-wide to the area-wide open trade scenario, respectively. For the country-wide and the area-wide open trade scenario, the cost incurred in installations of HVDC transmission lines is compensated by a decrease in installed capacities of electricity generation sources and storage capacities, which enable lower efficiency losses and import of low cost electricity from other regions. The installation of HVDC lines may not cover the non-electrified people in the rural areas of the SAARC region, which in fact is home to largest non-electrified population in the world. The best way to bring electricity to these people is to install RE-based mini-grid and solar home systems depending on the population density. Also, grid extension can be a solution for the people living near the grid [45,46,47].

PV technologies play a vital role in the SAARC region as it has the highest share in the installed capacities for a 100% RE energy system. The SAARC region is blessed with plenty of sunshine and this is reflected in high FLH all over the sub-regions and PV is the least cost source of electricity in all the sub-regions. The installation of distributed small-scale PV and more centralized utility-scale PV already achieved grid-parity and respective profitability in most parts of the world [48,49].

The self-consumption of the generated electricity from PV plays a vital role in the power sector and has a noticeable impact on the system parameters. The comparison between the decentralized system and centralized system gives vital insights into the total annual costs of the system. The total annualized cost for a more centralized 100% RE system is 1.1%, 1.2% and 3.1% lower than decentralized system for region-wide, country-wide and area-wide scenarios, respectively. However, potential positive effects at the distribution grid level and a lower risk level of power cuts has been not taken into account in the modelling. The target function used for prosumers is different than for a centralized generation this gives an additional costs to the system. Prosumers tend to reach minimum annual cost of electricity consumption. To get most out of PV self-consumption, its LCOE must be lower than the grid electricity purchase price but it can be higher than the total system LCOE. In addition to prosumers' higher electricity generation cost, there is a tendency to increase the cost of the system by installing more flexible options like low cost RE or more storage capacities, which induce a disturbance in the system demand profile. However, the peak demand of the entire system is reduced marginally (Fig. 6), particularly the noon time demand is reduced by PV prosumers. PV self-consumption can be particularly valuable in area constraint regions of SAARC, since rooftop area can be utilized for local electricity generation, which in turn reduces losses incurred due to electricity transmission.

The storage requirements for the SAARC region is mainly based on batteries due a high influence of solar PV on the system. The batteries provide 74% - 91% of the total stored electricity. The other storage technology which plays a big role in the region-wide scenario is the A-CAES storage which acts as a mid-term storage for storing wind energy and discharging at the time of low solar radiation [50]. The impact of A-CAES on the system decreases as the level of grid integration increases due to transferring of electricity via grids in a larger area being more economical than mid-term storage.

The integrated scenario presents a possibility to cover the projected natural gas demand in the industrial sector (except for demand in power generation and residential use) by flexible generation of synthetic natural gas (SNG), and providing clean water in water stressed areas by SWRO desalination. Providing clean water in the SAARC region is most important in the future because lot of regions are under severe water stress. The flexibility provided by the integrated scenario to the system is most useful in compensating seasonal fluctuations. The abundance of solar and wind resources in all the sub-regions of SAARC is sufficient to cover additional demand for electricity required for producing 190.7 TWh_{LHV} of SNG and 298.4 billion m^3 of renewable water. The total electricity required for gas synthesis and SWRO desalination water is 1619.3 TWh_{el} which leads to additional installations of 491 GW of PV and 449 GW of wind energy. The additional demand for gas synthesis leads to a substantial increase of electrolyzer units of about 95 GW (+83%) compared to the area-wide scenario. The cost of desalinated water obtained is 1.2 €/m^3 and cost of SNG is 141.8 €/MWh.

The different processes used for converting RE sources to electricity give rise to heat as a by-product. The heat generated from biogas and biomass CHP plants, waste-to-energy incinerators, gas turbines, electrolyzers and methanation plants can be used for the heating demand in the industrial sector which is still not integrated into the system. Also, curtailed electricity can be converted to heat and the excess heat can be stored in heat storage and used when required by the heat sector. The area-wide open trade scenario generates usable heat of 595 TWh_{th} per year, for the region-wide scenario it is 533 TWh_{th} per year, and for the integrated scenario it is 745 TWh_{th} per year. The higher usable heat in the integrated scenario is due to a higher curtailment of electricity and more SNG production. The heat generated as a byproduct of biomass and biogas plants is evenly distributed over the year. The demand due to the cooling sector is included in the electricity demand and therefore there is no additional demand for cooling is considered.

Coal is India's primary source of energy and the country is the world's third largest coal producer after China and the United States [51] which is followed by oil and gas. According to International Monetary Fund India has huge subsidies on coal and other fossil fuels [52]. India's post tax subsidies on coal for the year 2015 were at 196 bUSD. Also, the government supports coal mining through research and development and several tax benefits for coal transportation [53]. Further, coal-fired power plants are associated with high health costs and heavy metal emissions [54,55,56] which are actually not yet taken into account in India in optimizing the societal cost of energy supply. The total subsidies for the coal-based and gas-based electricity generation can be estimated for the year 2010 to 84 €/MWh_{el} and 15 €/MWh_{el}, based on estimated total subsidies of 8.7 USD/GJ (coal) and 2.2 USD/GJ (gas) according to the IMF (2014), primary energy demand for electricity of 2338 TWh_{th} (coal) and 302 TWh_{th} (gas), an electricity generation of 653 TWh_{el} (coal) and 118 TWh_{el} (gas) according to the IEA [57] and long-term USD/€ exchange rate of 1.3. Taking away the subsidies from the fossil fuels and investing in RE would enable India to achieve the climate change mitigation goals and zero carbon emissions.

The IEA recently published its 'India Energy Outlook 2015' [9], which projects installed capacity of 182 - 221 GW of solar PV and 134–160 GW for other renewables in India for the year 2040, which is in drastic contrast to the findings of this study. The cost assumptions used in the IEA study for India seems to be questionable. The PV LCOE is assumed to be 82 USD/MWh (61.5 €/MWh for 1.33 USD/€) which seems to be too high for the year 2040. For European solar PV power plants the capital expenditures are 850 €/kWp already achieved in 2015 and expected to further decrease to 470 €/kWp in 2030 [49]. According to KPMG [22], the installed capacity for PV will be 166 GW in 2024/25 compared to 2033/34 as predicted by IEA. KMPG assumes 550 – 670 USD/kWp capex for utility-scale PV plants for the year 2025which is even lower than assumed in this study. The IEA (2015) report does not reflect the already achieved cost reductions of solar PV nor the future cost reduction potential, which is in drastic contrast to the more market related report by KPMG [22] and Vartiainen et al. [49]. Moreover the IEA [9] report is also in drastic violation to the set renewables and in particular solar energy targets of the Government of India [24,58].

The results obtained show a low LCOE for the year 2030 in all the scenarios considered in this study. The growth of RE policies in this region has been remarkable in recent years particular so in India with ambitious projects from the Ministry of New and Renewable Energy [58,59] and the formation of the International Solar Alliance at COP 21 [26]. These initiatives will support the development of a 100% RE based system for the future in this region. The obtained results from this study when compared with recent alternatives for non-renewable low carbon technology options in Europe such as nuclear energy, natural gas and coal carbon capture and storage [60], which can partly comply with the climate change mitigation policy for a low carbon based energy system. According to Agora Energiewende [60], the LCOE of the alternatives are 112 €/MWh for a new nuclear plant (assumed for 2023 in the UK and Czech Republic), 112 €/MWh for gas CCS (assumed for 2019 in the UK, and 126 €/MWh for coal CCS (assumed for 2019 in the UK). A report by the European Commission [61], indicates that

CCS technology will not be available till the year 2030, and a report by Citigroup questions whether it will be ever profitable at all [62]. The results obtained for a 100% renewable energy based system shows the available least cost RE electricity generation options which would help achieve the goal of net zero GHG emissions set at COP 21 [63]. The results of this paper indicate various scenarios where a 100% RE-based system is possible and lower in cost than the high risk options which have disadvantages related to proliferation risk, nuclear melt down, unsolved nuclear waste disposal, CO_2 emissions from power plants with CCS technology, health risk due to heavy metal emissions from coal fired power plants and diminishing fossil fuel reserves. Also, nuclear fission has limitations same as mentioned above and also the financial and human research and development resources spent will not solve the energy problems in the world [64]. The criteria for low cost fully sustainable are not satisfied by the above mentioned alternative options.

6 CONCLUSION

In the SAARC region, a 100% RE-based system is achievable and the real policy option. Renewable Energy sources can cover the electricity demand for 2030 by sectors such as power, SWRO desalination and synthetic natural gas demand by the industrial sector using PtG technology. The proposed energy system configuration can handle the hurdle due to the monsoon season pretty effectively. The LCOE obtained was from 71.6 – 67.2 €/MWh depending on the geographical and sectorial integration. The obtained price range for electricity is lower than for non-renewable energy resources matching climate change targets. The heating demand in the industries and residential sectors may be partly covered by the excess heat generated as a by-product of synthetic natural gas generation and conversion of curtailed electricity to heat. In all the scenarios, solar PV plays a vital role in power generation followed by wind energy in all the regions except Sri Lanka, Afghanistan and Pakistan South where wind is the least cost energy due to very high FLH. For the scenarios, the storage requirements are mainly based on batteries, which provide 74% – 91% of the total stored electricity. The role of other storage technologies, especially A-CAES, which has a vital role in the region-wide scenario as a mid-term storage between batteries and PtG particularly in areas of high wind share and high seasonal variation. The HVDC transmission grid plays a vital role in the transmission of low cost electricity from Afghanistan and India North to Pakistan North, where trading is more cost competitive than local storage technologies available and also due to high demand of electricity. As the level of grid integration increases, economic benefit due to A-CAES is reduced but other storage technologies such as batteries and PtG are still required. A slight increase of 1% - 3% in the total cost of electricity because of PV self-consumption is due to the utilization of solar electricity and in particular respective batteries for self-consumption at a higher cost level. The most important advantage of PV self-consumption is the reduction of noon and afternoon peak hours in the year. In the integrated scenario, seasonal SNG storage is substituted by the industrial SNG generation for the electricity sector. The system restricts synthetic natural gas production in the case of energy deficit as a major source of flexibility. A 100% renewable energy system for India and SAARC seems to be highly attractive, in particular due to the fact that it cost less than only the subsidies for a coal-based energy system.

7 ACKNOWLEDGEMENTS

The authors gratefully acknowledge the public financing of Tekes, the Finnish Funding Agency for Innovation, for the 'Neo-Carbon Energy' project under the number 40101/14. The first author would like to thank Fortum Foundation for the valuable scholarship. The authors would like to thank Michael Child for proofreading.

8 APPENDIX A. SUPPLEMENTARY MATERIAL

9 REFERENCES

[1] Chakrabarti S., Rathore G.S., Kumar S., Hada A.S., Bajaj R., 2013. Energy Statistics, Central Statistics Office, Ministry of Statistics and Programme Implementation, Government of India, New Delhi. http://mospi.nic.in/mospi_new/upload/Energy_Statistics_2013.pdf

[2] [SAARC] - South Asian Association for Regional Cooperation Group on Statistics, 2014. SAARC Statistics, Kathmandu. http://saarcstat.org/content/saarc-figures

[3] Cochran J., et al., 2013. Market evolution: Wholesale electricity market design for 21st century power systems, Technical report for NREL, Colorado.

[4] Singer S., 2010. The energy report: 100% renewable energy by 2050, report by Ecofys, Utrecht.

[5] Kapoor K., Pandey K. K., Jain, A. K., Nandan A. 2014. Evolution of solar energy in India: A review. Renewable and Sustainable Energy Reviews, 40, 475-487.

[6] [IPCC] – Intergovernmental Panel On Climate Change, 2012. Renewable Energy Sources and Climate Change Mitigation. Cambridge University Press, NewYork:

[7] [REN21] – Renewable Energy Policy Network for the 21st Century, 2016. Renewables 2016 – Global Status Report, Paris, www.ren21.net/wp-content/uploads/2016/06/GSR_2016_Full_Report_REN21.pdf

[8] [UN] - United Nations, Department of Economic and Social Affairs, Population Division, 2015. World Population Prospects: The 2015 Revision. New York.

[9] [IEA] – International Energy Agency, 2015. India Energy Outlook, IEA publications, Paris.

[10] British Petroleum, 2015. BP Statistical Review of World Energy, London.

[11] TERI and WWF, 2011. The Energy Report India: 100% Renewable Energy by 2050, New Delhi. http://awsassets.wwfindia.org/downloads/the_energy_report_india.pdf.

[12] Garg P., 2012. Energy Scenario and Vision 2020 in India, Journal of Sustainable Energy & Environment vol. 3, pp 7-17.

[13] Sharma N. K., Tiwari P. K., Sood Y. R., 2012. Solar energy in India: Strategies, policies, perspectives and future potential. Renewable and Sustainable Energy Reviews, 16(1), 933-941

[14] [WBGU] - German Advisory Council on Global Change, 2007. Climate Change as a Security Risk, Berlin, Earthscan, London www.wbgu.de/fileadmin/templates/dateien/veroeffentlichungen/hauptgutachten/jg2007/wbgu_jg2007_engl.pdf

[15] [IPCC] - Intergovernmental Panel on Climate Change, 2014. Special report on Impacts, Adaption and Vulnerability. Intergovernmental panel on climate change; Geneva.

[16] Arvizu D., Balaya P., Cabeza L., Hollands T., Jäger-Waldau A., Kondo M., Konseibo C., Meleshko V., Stein W., Tamaura Y., Xu H., Zilles R., 2011: Direct Solar Energy. In IPCC Special Report on Renewable Energy Sources and Climate Change Mitigation [O. Edenhofer, R. Pichs-Madruga, Y. Sokona, K. Seyboth, P. Matschoss, S. Kadner, T. Zwickel, P. Eickemeier, G. Hansen, S. Schlömer, C. von Stechow (eds)], Cambridge University Press, Cambridge, United Kingdom and New York, NY, USA.

[17] Prakash, R., & Bhat, I. K. 2009. Energy, economics and environmental impacts of renewable energy systems. Renewable and sustainable energy reviews, 13(9), 2716-2721.

[18] Kumar A. and Kumar K., 2010 Renewable energy in India: current status & future potentials.Renew Sustain Energy Rev; vol.14: 2434 – 42.

[19] [UNFCCC] – United Nations Framework Convention on Climate Change, 2015b. India's intended nationally determined contribution: working towards climate justice, Technical report, Bonn.

[20] Pratap A., Ram M., Pathanjali A. P., 2013. Powering Ahead with Renewables - Leaders and Laggards, report by Greenpeace India, Bangalore. http://www.greenpeace.org/india/Global/india/report/2013/powering-ahead-with-renewables.pdf

[21] Trivedi P., 2015. Is India caught between a coal and a solar future?, Thomson Reuters Foundation News, London, December 9. http://news.trust.org//item/20151209120231-vj9ds/?source=hpMostPopularBlogs

[22] KPMG, 2015. The Rising Sun - Disruption on the horizon, Report published in ENRich – KPMG Energy Conference, India. https://www.kpmg.com/IN/en/IssuesAndInsights/ArticlesPublications/Documents/ENRich2015.pdf

[23] King E., 2016. Solar is now cheaper than coal, says India energy minister, ClimateChangeNews, London, April 18. http://www.climatechangenews.com/2016/04/18/solar-is-now-cheaper-than-coal-says-india-energy-minister/

[24] Buckley T. and Sharda J., 2015. India's Electricity-Sector Transformation, Institute of Energy Economics and Financial Analysis, Ohio, http://ieefa.org/wp-content/uploads/2015/08/IEEFA-Indian-Electricity-Sector-Transformation-August-2015.pdf

[25] Ross K. and Damassa T., 2015. Assessing the Post-2020 Clean Energy Landscape, Technical note by World Resources Institute (WRI), Washington D.C. http://www.wri.org/sites/default/files/WRI-OCN_Assessing-Post-2020-Clean-Energy-Landscape.pdf.

[26] [ISA] - International Solar Alliance, 2015. Working Paper on International Solar Alliance, Ministry of New and Renewable Energy, Govt. of India, Gurgoan. http://isolaralliance.com/pdf/ISA-Working-Paper.pdf.

[27] Barbosa L.S.N.S., Bogdanov D., Vainikka P., Breyer Ch., 2016. Hydro, wind and solar power as a base for a 100% Renewable Energy supply for South and Central America, submitted.

[28] Bogdanov D. and Breyer Ch., 2016. North-East Asian Super Grid for 100% Renewable Energy supply: Optimal mix of energy technologies for electricity, gas and heat supply options. Energy Conversion and Management, 112, 176-190.

[29] IASPEI, 2015. International Association of Seismology and Physics of the Earth's Interior, The International Heat Flow Commission, Colorado, USA. IHFC database. www.heatflow.und.edu/index2.html

[30] AAPG, 2015. Global Heat Flow Database. American Association of Petroleum Geologists, Tulsa, USA, www.datapages.com/gis-map-publishing-program/gis-open-files/global-framework/global-heat-flow-database

[31] Chamorro C.R., García-Cuesta J.L., Mondéjar M.E., Pérez-Madrazo A., 2014a. An estimation of the enhanced geothermal systems potential for the Iberian Peninsula. Renewable Energy 66, 1-14.

[32] Chamorro C.R., García-Cuesta J.L., Mondéjar M.E., Pérez-Madrazo A., 2014b. Enhanced geothermal systems in Europe: An estimation and comparison of the technical and sustainable potentials. Energy 65, 250-263.Citi Research Equities, 2015. Paris Climate Change Agreement, Sydney, December 15.

[33] Huenges E., 2012. Geothermal Energy Systems, Exploration, Development and Utilization, 2nd edition, Weinheim, Germany.

[34] Luck M., Landis M., Gassert F., 2015. Aqueduct Water Stress Projections: Decadal projections of water supply and demand using CMIP5 GCMs, Washington DC, World Resources Institute.

[35] Caldera U., Bogdanov D., Breyer Ch., 2016. Local cost of seawater RO desalination based on solar PV and wind energy: A global estimate, Desalination, 385, 207-216.

[36] [IEA] – International Energy Agency, 2013. Energy Balances of Non-OECD Countries, IEA publications, Paris.

[37] Dii, 2012. 2050 Desert power – perspectives on a sustainable power system for EUMENA. Dii, Munich.

[38] Gerlach A., Werner Ch., Breyer Ch., 2014. Impact of Financing Cost on Global Grid-Parity Dynamics till 2030. 29th EU PVSEC, Amsterdam.

[39] Werner C., Breyer C., Gerlach A., and Beckel O., 2012. Photovoltaic With Energy Storage: An Overview On Economics, System Design And Politics, 27th EU PVSEC, Frankfurt, September 24-28.

[40] DBFZ, 2009. Regionale und globale räumliche Verteilung von Biomassepotenzialen. German Biomass Research Centre. [in German]

[41] [IEA] - International Energy Agency, 2012. Technology Roadmap – Bioenergy for Heat and Power. IEA Publishing, Paris.

[42] IPCC] – Intergovernmental Panel on Climate Change, 2011. Special report on Renewable Energy sources and Climate Change mitigation. Intergovernmental Panel on Climate Change, Geneva.

[43] Farfan J. and Breyer Ch., 2016. Structural Changes of the global power generation capacity towards sustainability and the risk of stranded investments, submitted.

[44] Breyer Ch., Bogdanov D., Gulagi A., Aghahosseini A., Barbosa L.S.N.S., Koskinen O., Barasa M., Caldera U., Afanasyeva S., Child M., Farfan J., Vainikka P., 2016. On the Role of Solar Photovoltaics in Global Energy Transition Scenarios, 32nd EU PVSEC, Munich, June 20-24.

[45] Bertheau P., Blechinger P., Cader C., Breyer Ch., 2012. Business Models for the Implementation of Renewable Energy Systems for upgrading Diesel Generators in Developing Countries for the Examples of Peru, Tanzania and Cambodia. 5th International Conference on Integration of Renewable Energy and Distributed Energy Resources (IRED), Berlin, December 4-6.

[46] Breyer Ch, Werner C., Rolland S., Adelmann P., 2011. Off-Grid Photovoltaic Applications in Regions of Low Electrification: High Demand, Fast Financial Amortization and Large Market Potential. 26th EU PVSEC, Hamburg, September 5-9.

[47] Breyer Ch., Gaudchau E., Gerlach A.-K., Hlusiak M, Cader C., Bertheau P., Wasgindt V., 2012. PV-based Mini-Grids for Electrification in Developing Countries - An overview on market potentials and business models. A study of the Reiner Lemoine Institut gGmbH on behalf of the cdw Stiftungsverbund gGmbH, December 11.

[48] Breyer Ch. and Gerlach A., 2013. Global Overview on Grid-Parity, Progress in Photovoltaics: Research and Applications 21, 121-136.

[49] Vartiainen E., Masson G., Breyer Ch., 2015. PV LCOE in Europe 2015-2050. 31th EU PVSEC, Hamburg, September 14-18.

[50] Gulagi A., Aghahosseini A., Bogdanov D. and Breyer Ch., 2016.Comparison of the potential role of adiabatic compressed air energy storage (A-CAES) for a fully sustainable energy system in a region of significant and low seasonal variations, 10th International Renewable Energy Storage Conference, March 15-17.

[51] Enerdata, 2014. Global Energy Statistical Yearbook 2014, Grenoble. (https://yearbook.enerdata.net/)

[52] Coady D., Parry I., Sears L., Shang B., 2015. How large are global energy subsidies?, International Monetary Fund, Washington.

[53] Ministry of Finance, 2015. Indian Public Finance Statistics, Government of India, New Delhi (http://finmin.nic.in/reports/IPFStat201415.pdf)

[54] Duguid B. (ed.), 2010. Ontario's Long-Term Energy Plan, Ontario Ministry of Energy, Toronto.

[55] Epstein P. and Buonocore J., 2011. Full cost accounting for the life cycle of coal, Annals of the New York Academy of Sciences, 1219, 73–98.

[56] Buonocore J. J., Luckow P., Norris G., Spengler J. D., Biewald B., Fisher J., and Levy J. I., 2016. Health and climate benefits of different energy-efficiency and renewable energy choices, Nature Climate Change, 6, 100–105.

[57] [IEA] – International Energy Agency, 2012. World Energy Outlook 2012, IEA publications, Paris

[58] [JNNSM] - Jawaharlal Nehru National Solar Mission, 2011. Initiative by Gov. of India, New Delhi, http://mnre.gov.in/filemanager/UserFiles/jnnsm_gridconnected_24082011.pdf

[59] [GOI] – Government of India, 2015. Report on India's Renewable Electricity Roadmap 2030: Towards Accelerated Renewable Electricity Deployment. NITI Aayog, Govt. of India, New Delhi.

[60] Agora Energiewende, 2014. Comparing the cost of low-carbon technologies: what is the cheapest option? report by Prognos AG on behalf of Agora Energiewende, Berlin. p. 10-13. www.prognos.com/fileadmin/pdf/publikationsdatenbank/140417_Prognos_Agora_Analysis_Decarbonisationtechnologies_EN.pdf

[61] European Commission, 2014. Integration of renewable energy in Europe, study prepared by KEMA Consulting, DNV GL – Energy, Imperial College and NERA Economic Consulting on behalf of DG Energy, Brussels, p. 7-8 and 54.

[62] Citi Research Equities, Paris Climate Change Agreement, Sydney, December 15, 2015.

[63] [UNFCCC] – United Nations Framework Convention on Climate Change, 2015a. Adoption of the Paris Agreement – Proposal by the President, UNFCCC, Paris. http://unfccc.int/resource/docs/2015/cop21/eng/l09.pdf

[64] Dittmar M., 2012. Nuclear energy: Status and future limitations, Energy, 37, 35-40.

[65] Pleßmann G., Erdmann M., Hlusiak M., Breyer Ch., 2014. Global energy storage demand for a 100% renewable electricity supply. Energy Procedia 46, 22-31.

[66] Komoto K., Ito M., Van der Vleuten P., Faiman D., Kurokawa K. (eds.), 2009. Energy from the desert – very large scale photovoltaic systems: socio-economic, financial, technical and environmental aspects. Earthscan, London.

[67] Hoffmann W., 2014. Importance and evidence for cost effective electricity storage. 29th EU PVSEC, Amsterdam, September 22-26.

[68] Agora Energiewende, 2015. Current and Future Cost of Photovoltaics – Long-term Scenarios for Market Development System Prices and LCOE of Utility-Scale PV Systems. Report by Fraunhofer Institute for Solar Energy Systems on behalf of Agora Energiewende, Berlin. www.fvee.de/fileadmin/publikationen/weitere_publikationen/15_AgoraEnergiewende-ISE_Current_and_Future_Cost_of_PV.pdf

[69] Urban W., Lohmann H., Girod K., 2009. Abschlussbericht für das BMBF-Verbundprojekt Biogaseinspeisung. Fraunhofer UMSICHT, Liepzig. [in German].

[70] European Commission, 2014. ETRI 2014 - Energy technology reference indicator projections for 2010-2050. EC Joint Research Centre Institute for Energy and Transport, Petten, Netherlands.

Supplementary Material:

Table I: Financial assumptions for energy system components [35,49,65-70]

Technology	Capex [€/kW]	Opex fix [€/kW]	Opex var [€/kWh]	Lifetime [a]
PV optimally tilted	550	8	0	35
PV single-axis tracking	620	9	0	35
PV rooftop	813	12	0	35
Wind onshore	1000	20	0	25
CSP (solar field)	528	11	0	25
Hydro run-of-river	2560	115.2	0.005	60
Hydro dam	1650	66	0.003	60
Geothermal energy	4860	87	0	30
Water electrolysis	380	13	0.0012	30
Methanation	234	5	0.0015	30
CO_2 scrubbing	356	14	0.0013	30
CCGT	775	19.4	0.001	30
OCGT	475	14.25	0.001	30
Steam turbine	600	12	0	30
Hot heat burner	100	2	0	30
Heating rod	20	0.4	0.001	30
Biomass CHP	2500	175	0.001	30
Biogas CHP	370	14.8	0.001	30
Waste incinerator	5240	235.8	0.007	20
Biogas digester	680	27.2	0	20
Biogas upgrade	250	20	0	20
	Capex [€/kWh]	Opex fix [€/kWh]	Opex var [€/kWh]	Lifetime [a]
Battery	150	10	0.0002	10
PHS	70	11	0.0002	50
A-CAES	31	0.4	0.0012	40
TES	24	2	0	20
Gas storage	0.05	0.001	0	50
	Capex [€/(kW$_{NTC}$*km)]	Opex fix [€/(kW$_{NTC}$*km)]	Opex var [€/kWh$_{NTC}$]	Lifetime [a]
HVDC line on ground	0.612	0.0075	0	50
HVDC line submarine	0.992	0.0010	0	50
	Capex [€/kW$_{NTC}$]	Opex fix [€/kW$_{NTC}$]	Opex var [€/kWh$_{NTC}$]	Lifetime [a]
HVDC converter pair	180	1.8	0	50
	Capex [€/(m^3·a)]	Opex fix [€/(m^3 a)]	Opex var [€/m^3]	Lifetime [a]
Water desalination	2.23	0.09	0	30
	Capex [€/(m^3·h·km)]	Opex fix [€/(m^3·h·km·a)]	Opex var [€/m^3·h·km]	Lifetime [a]
Horizontal pumping and pipes	19.3	0.39	0	30
Vertical pumping and pipes	15.5	0.31	0	30

Table II: Efficiencies and energy to power ratio of storage technologies. Assumptions are mainly taken from Pleßmann et al. [65].

Technology	Efficiency [%]	Energy/Power Ratio [h]	Self-Discharge [%/h]
Battery	90	6	0
PHS	85	8	0
A-CAES	70	100	0.001
TES	90	8	0.002
Gas storage	100	80*24	0

Table III: Efficiency assumptions for energy system components for the 2020 and 2030 reference years. Assumptions are mainly taken from Pleßmann et al. [65], and from Hoffmann [67].

	η_{el} [%]	η_{th} [%]
CSP (solar field)		51
Steam turbine	42	
Hot heat burner		95
Heating rod		99
Water electrolysis		84
Methanation		77
CO_2 scrubbing		78
CCGT	58	
OCGT	43	
Geothermal	24	
Biomass CHP	40	45
Biogas CHP	42	43
Waste incinerator	34	
Biogas upgrade		98

Table IV: Efficiency assumptions for HVDC transmission according to Dii [37].

	Power losses
HVDC line	1.6 % / 1000 km
HVDC converter pair	1.4%

Table V: Regional grid electricity costs acccording to Gerlach et al. [38].

Region	Electricity costs [€/MWh]		
	Residential	Commercial	Industrial
Total area	83	91	97
India East	81	92	100
India Centraleast	81	92	100
India West	81	92	100
India Centralwest	81	92	100
India North	81	92	100
India Northwest	81	92	100
India Uttar Pradesh	81	92	100
India South	81	92	100
India Centralsouth	81	92	100
India Northeast	81	92	100
Bangladesh	100	78	55
Nepal + Bhutan	110	110	100
Pakistan South	55	73	90
Pakistan North	55	73	90
Afghanistan	150	140	130
Sri Lanka	55	69	82

Table VI: Regional biomass potentials [40] and geothermal energy potentials

Region	Biomass potential [TWh_{th}/a]			Geothermal Potential
	Solid waste	Solid biomass	Biogas sources	[TWh_{th}/a]
Total area	69.6	874.9	145.7	198.4
India East	5.8	74.6	14.6	6.8
India Centraleast	5.6	66.3	14.1	3.8
India West	5.6	94.4	14.3	1.1
India Centralwest	5.8	90.2	14.6	0.0
India North	1.2	38.6	3.1	0.0
India Northwest	5.8	90.4	14.7	0.0
India Uttar Pradesh	8.3	97.7	21.1	0.0
India South	6.0	79.4	15.1	0.0
India Centralsouth	4.6	72.0	11.7	0.0
India Northeast	1.6	35.8	4.0	122.7
Bangladesh	7.2	53.0	6.9	33.5
Nepal + Bhutan	1.3	13.5	1.8	29.6
Pakistan South	3.2	21.0	2.5	0.0
Pakistan North	6.0	39.6	5.2	0.0
Afghanistan	0.0	0.0	0.0	0.8
Sri Lanka	1.6	8.3	1.2	0.0

Table VII: Regional biomass costs.

Region	Biomass costs [€/MWh]		
	Solid waste	Solid biomass	Biogas sources
Total area	-9.5	8.0	6.7
India East	-10.2	8.6	8.9
India Centraleast	-10.2	7.9	8.9
India West	-10.2	10.4	8.9
India Centralwest	-10.2	10.0	8.9
India North	-10.2	13.3	8.9
India Northwest	-10.2	9.9	8.9
India Uttar Pradesh	-10.2	7.7	8.9
India South	-10.2	8.8	8.9
India Centralsouth	-10.2	9.9	8.9
India Northeast	-10.2	12.0	8.9
Bangladesh	-10.2	5.3	3.6
Nepal + Bhutan	-10.2	5.3	5.5
Pakistan South	-10.2	5.3	3.1
Pakistan North	-10.2	5.3	3.1
Afghanistan	0.0	0.0	0.0
Sri Lanka	-10.2	8.2	3.0

Table VIII: Average full load hours and LCOE for PV single-axis tracking, PV optimally tilted, solar CSP and wind power plants in SAARC sub-regions.

Region	Pop. [mio. Pop]	Electr. demand [TWh]	PV-single-axis FLH	PV optimally tilted FLH	CSP FLH	Wind FLH	PV single-axis LCOE [€/MWh]	PV optimally tilted LCOE [€/MWh]	CSP LCOE [€/MWh]	Wind LCOE [€/MWh]
Total area	1922	2597	2071	1645	1922	1934	28	31	62	61
India East	159	181	1870	1535	1683	1263	30	33	70	84
India Centraleast	154	47	1995	1595	1910	1608	29	32	61	66
India West	156	313	1995	1624	1860	2394	29	31	63	44
India Centralwest	159	303	1950	1579	1785	2298	29	32	66	46
India North	34	77	2515	1878	2277	2440	23	27	51	43
India Northwest	161	375	2011	1633	1968	1626	28	31	60	65
India Uttar Pradesh	230	200	2100	1639	2165	1748	27	31	54	61
India South	166	244	1946	1536	1597	2174	29	33	73	49
India Centralsouth	128	331	1937	1559	1653	2431	29	32	71	44
India Northeast	44	27	1808	1525	1587	998	31	33	74	106
Bangladesh	186	162	1843	1516	1681	1151	31	33	70	92
Nepal + Bhutan	34	19	2334	1789	2193	1579	24	28	53	67
Pakistan South	85	34	2236	1775	2091	2333	25	28	56	45
Pakistan North	160	241	2187	1697	2153	1522	26	30	54	70
Afghanistan	44	15	2340	1802	2230	3449	24	28	53	31
Sri Lanka	22	28	1870	1535	1683	1263	30	33	70	84

Table IX: Lower limits of installed capacities in the SAARC sub-regions. Data taken from Farfan and Breyer [43]

Region	Installed capacity [MW]			
	Solar PV	Wind	Hydro RoR and dams	PHS
Total area	3461.7	23883.5	44120.7	22032.0
India East	35.7	0.5	3782.8	32.0
India Centraleast	18.0	0.0	372.0	0.0
India West	1304.4	3339.1	4342.1	9600.0
India Centralwest	289.9	3929.7	3212.3	2000.0
India North	6.8	0.0	5863.6	0.0
India Northwest	1281.5	7935.5	2712.5	0.0
India Uttar Pradesh	36.2	0.0	533.6	0.0
India South	74.2	5621.3	3933.5	3200.0
India Centralsouth	203.7	2809.9	6627.7	7200.0
India Northeast	0.0	0.0	1316.8	0.0
Bangladesh	140.0	1.0	232.0	0.0
Nepal + Bhutan	6.0	0.0	2104.0	0.0
Pakistan South	0.0	105.0	26.6	0.0
Pakistan North	45.5	49.6	7176.3	0.0
Afghanistan	5.0	0.0	315.0	0.0
Sri Lanka	15.0	92.0	1570.0	0.0

Table X: Upper limits on installable capacities in SAARC sub-regions in units of GW_{th} for CSP and GW_{el} for all other technologies.

Region	area [1000 km^2]	Limits [GW] Solar CSP	Solar PV	Wind	Hydro RoR	Hydro dams	PHS
Total area	5212	79749	23456	1751	32	35	44
India East	252	3849	1132	85	3	3	0
India Centraleast	174	2660	782	58	0	0	0
India West	509	7781	2289	171	2	4	19
India Centralwest	443	6776	1993	149	2	2	4
India North	331	5070	1491	111	5	3	0
India Northwest	438	6708	1973	147	2	2	0
India Uttar Pradesh	241	3686	1084	81	0	0	0
India South	284	4349	1279	96	3	3	6
India Centralsouth	352	5386	1584	118	5	5	14
India Northeast	255	3904	1148	86	1	1	0
Bangladesh	148	2258	664	50	0	0	0
Nepal + Bhutan	186	2839	835	62	2	1	0
Pakistan South	488	7469	2197	164	0	0	0
Pakistan North	394	6032	1774	132	3	8	0
Afghanistan	652	9979	2935	219	0	0	0
Sri Lanka	66	1004	295	22	1	1	0

Table XI: Annual industrial gas demand [36] and water demand [35] for year 2030 in the SAARC sub-regions

	Annual gas demand		Annual electricity demand for gas synthesis	Annual water desalination demand	Annual electricity demand for water desalination
	[Bcm]	[TWh_{th}]	[TWh_{el}]	[mil. m^3]	[TWh_{el}]
Total	18.0	190.7	298.7	298437	1428.6
India East	0.9	9.9	10.9	597	1.9
India Centraleast	0.9	9.5	9.8	456	1.8
India West	0.9	9.7	15.8	29030	119.4
India Centralwest	0.9	9.9	13.3	2666	13.3
India North	0.2	2.1	3.3	226	4.1
India Northwest	0.9	10.0	11.9	74945	333.4
India Uttar Pradesh	1.3	14.3	17.2	19325	91.7
India South	1.0	10.3	10.6	7419	29.7
India Centralsouth	0.7	7.9	13.1	9242	47.8
India Northeast	0.3	2.7	2.8	0	0.0
Bangladesh	1.7	17.6	29.5	0	0.0
Nepal + Bhutan	0.0	0.0	0.0	0	0.0
Pakistan South	2.8	30.1	55.8	28910	119.9
Pakistan North	5.3	56.6	104.7	115432	568.7
Afghanistan	0.0	0.0	0.0	10190	96.9
Sri Lanka	0.0	0.0	0.0	0	0.0

Table XII: Total LCOE components in all sub-regions of SAARC

Region-wide	LCOE primary [€/MWh]	LCOC [€/MWh]	LCOS [€/MWh]	LCOT [€/MWh]	LCOE total [€/MWh]	
Area average	42.3	1.5	27.8	0.0	71.6	
India East	44.9	1.4	26.2	0.0	72.5	
India Centraleast	45.8	0.1	4.3	0.0	50.3	
India West	41.8	1.2	26.9	0.0	70.0	
India Centralwest	42.6	1.1	28.9	0.0	72.5	
India North	46.1	0.1	15.7	0.0	62.0	
India Northwest	42.3	1.8	32.8	0.0	76.9	
India Uttar Pradesh	41.2	2.6	26.5	0.0	70.3	
India South	44.2	1.6	27.7	0.0	73.4	
India Centralsouth	40.8	1.1	29.5	0.0	71.4	
India Northeast	50.1	1.5	3.5	0.0	55.1	
Bangladesh	40.9	2.7	36.9	0.0	80.4	
Nepal + Bhutan	48.3	4.4	10.1	0.0	62.9	
Pakistan South	43.0	0.5	18.7	0.0	62.3	
Pakistan North	39.6	1.5	28.6	0.0	69.7	
Afghanistan	34.6	2.8	32.3	0.0	69.7	
Sri Lanka	43.6	1.4	15.8	0.0	60.9	
Country-wide	**LCOE primary [€/MWh]**	**LCOC [€/MWh]**	**LCOS [€/MWh]**	**LCOT [€/MWh]**	**LCOE total [€/MWh]**	**export (-)/ import (+) [%]**
Area average	41.9	1.1	25.5	1.1	69.6	6.4
India East	41.8	0.2	22.7	1.4	66.2	7.8
India Centraleast	51.2	0.1	6.0	3.9	61.3	11.9
India West	43.8	1.4	22.3	1.5	68.9	-5.1
India Centralwest	41.7	0.8	26.9	0.9	70.3	3.6
India North	46.4	1.5	13.0	2.4	63.2	-20.4
India Northwest	38.2	0.6	32.5	1.5	72.7	8.4
India Uttar Pradesh	41.4	1.3	21.3	1.1	65.1	2.4
India South	43.9	1.2	25.2	0.8	71.1	3.5
India Centralsouth	40.5	0.8	29.3	0.7	71.3	-3.1
India Northeast	51.3	0.1	0.6	5.2	57.2	-51.0
Bangladesh	40.9	2.7	36.9	0.0	80.4	0.0
Nepal + Bhutan	48.3	4.4	10.1	0.0	62.9	0.0
Pakistan South	44.2	1.0	15.4	1.1	61.7	-5.0
Pakistan North	39.8	1.4	28.2	0.2	69.5	0.7
Afghanistan	34.6	2.8	32.3	0.0	69.7	0.0
Sri Lanka	43.6	1.4	15.8	0.0	60.9	0.0
Area-wide	**LCOE primary [€/MWh]**	**LCOC [€/MWh]**	**LCOS [€/MWh]**	**LCOT [€/MWh]**	**LCOE total [€/MWh]**	**export (-)/ import (+) [%]**
Area average	41.4	0.7	22.7	2.3	67.2	15.6
India East	41.8	0.3	24.1	1.1	67.3	6.1
India Centraleast	49.1	0.2	7.9	2.8	60.0	9.9
India West	43.1	1.1	22.4	0.9	67.4	-2.5
India Centralwest	41.5	0.8	27.3	0.8	70.5	3.5
India North	45.0	0.9	12.5	3.6	62.0	-42.3
India Northwest	38.1	0.2	30.7	2.0	71.1	17.9
India Uttar Pradesh	40.7	1.3	23.4	0.9	66.3	2.2
India South	43.9	0.5	20.9	2.0	67.3	16.8
India Centralsouth	40.8	0.5	27.1	0.7	69.1	-1.5
India Northeast	55.9	0.5	3.1	5.0	64.6	-51.0
Bangladesh	41.9	0.4	23.5	2.4	68.1	18.9
Nepal + Bhutan	46.3	0.3	3.6	3.7	53.9	-45.7
Pakistan South	44.8	1.4	13.0	1.3	60.5	-7.1
Pakistan North	42.9	0.4	10.3	9.3	63.0	53.0
Afghanistan	32.1	1.9	23.4	6.9	64.3	-68.8
Sri Lanka	34.2	1.3	24.3	5.9	65.7	-45.2

Integrated scenario	LCOE primary [€/MWh]	LCOC [€/MWh]	LCOS [€/MWh]	LCOT [€/MWh]	LCOE total [€/MWh]	export (-)/ import (+) [%]
Area average	40.8	1.4	22.6	3.1	67.9	25.4
India East	45.2	0.4	23.4	1.0	70.0	6.6
India Centraleast	56.3	0.5	6.3	1.7	64.8	5.3
India West	45.0	1.8	22.9	1.1	70.8	-2.9
India Centralwest	44.3	0.9	25.1	0.8	71.1	2.9
India North	44.3	2.3	23.2	4.5	74.3	-57.6
India Northwest	37.8	0.5	34.5	2.1	74.9	22.6
India Uttar Pradesh	44.1	2.1	23.6	1.0	70.9	4.0
India South	46.9	0.7	19.8	1.8	69.2	18.4
India Centralsouth	41.6	0.8	28.5	0.6	71.4	-3.0
India Northeast	60.2	0.9	3.9	3.9	68.8	-49.2
Bangladesh	40.4	0.6	19.8	1.6	62.4	17.1
Nepal + Bhutan	48.2	0.9	2.8	3.1	55.1	-46.6
Pakistan South	44.2	3.1	4.9	1.4	53.6	-9.0
Pakistan North	33.5	0.7	13.2	15.0	62.3	70.0
Afghanistan	31.5	2.2	22.7	5.2	61.6	-65.4
Sri Lanka	34.8	1.5	21.6	4.7	62.7	-47.6

Table XIII: Prosumers electricity costs, installed capacities and electricity utilization for SAARC.

Prosumers parameters	Residential	Commercial	Industrial
Electricity price [€/kWh]	0.080	0.089	0.096
PV LCOE [€/kWh]	0.027	0.036	0.036
Self-consumption PV LCOE [€/kWh]	0.033	0.043	0.043
Self-consumption PV and Battery LCOE [€/kWh]	0.034	0.043	0.044
Self-consumption LCOE [€/kWh]	0.034	0.043	0.043
Benefit [€/kWh]	0.046	0.046	0.053
Installed capacities	Residential	Commercial	Industrial
PV [GW]	30	36	78
Battery storage [GWh]	1.8	0.5	1.5
Generation	Residential	Commercial	Industrial
PV [TWh]	49	59	126
Battery storage [TWh]	0.6	0.2	0.5
Excess [TWh]	8.8	8.7	20.5
Utilization	Residential	Commercial	Industrial
Self-consumption of generated PV electricity [%]	81.7	85.0	83.6
Self-coverage market segment [%]	7.8	7.4	7.5
Self-coverage operators [%]	38.8	36.8	37.3

Table XIV: Storage capacities, throughput, full cycles and utilization of A-CAES potential per year for the four scenarios.

			Region- wide	Country-wide	Area-wide	Integrated
Storage capacities	Battery SC	[TWh_{el}]	0.0	0.0	0.0	0.0
	Battery system	[TWh_{el}]	1.4	1.5	1.4	1.7
	PHS	[TWh_{el}]	0.0	0.0	0.0	0.0
	A-CAES	[TWh_{el}]	2.5	0.6	0.0	3.2
	Gas	[TWh_{th}]	100.2	85.4	79.1	208.0
Throughput of storage technologies	Battery SC	[TWh_{el}]	1.2	1.2	1.2	1.2
	Battery system	[TWh_{el}]	444.8	483.9	456.9	551.3
	PHS	[TWh_{el}]	11.2	11.5	11.1	12.5
	A-CAES	[TWh_{el}]	53.4	11.8	0.1	0.0
	Gas	[TWh_{th}]	180.8	145.0	125.8	81.4
Full cycles per year	Battery SC	[-]	320.6	320.6	320.6	320.6
	Battery system	[-]	320.1	318.0	315.1	327.7
	PHS	[-]	255.0	259.0	251.3	283.3
	A-CAES	[-]	21.7	21.5	26.6	25.5
	Gas	[-]	1.8	1.7	1.6	1.3
Utilization of A-CAES potential	Storage used	[TWh]	2.5	0.6	3.2	3.2
	Storage available	[TWh]	70210.6	70210.6	70210.6	70210.6

Table 17: Electricity transmission lines parameters for the area-wide scenario for SAARC.

Region 1	Region 2	Length [km]	Capacity [GW]	Utilization [%]
India East	India Centraleast	508.0	3.3	67
India East	India Centralwest	1810.6	0.2	77
India East	India Centralsouth	1718.3	2.1	73
India East	Bangladesh	264.6	1.7	62
India Centraleast	India Northwest	944.1	1.5	73
India Centraleast	India Uttar Pradesh	486.5	2.1	58
India Centraleast	Nepal + Bhutan	259.4	2.1	76
India West	India Centralwest	485.5	4.8	64
India West	India Northwest	848.3	3.9	72
India West	Pakistan South	669.9	1.1	70
India Centralwest	India Centralsouth	924.9	2.6	75
India North	India Northwest	711.1	13.5	69
India North	Pakistan North	192.8	9.7	63
India Northwest	India Uttar Pradesh	461.7	3.3	63
India South	India Centralsouth	306.6	5.1	53
India South	Sri Lanka	815.5	9.1	68
India Northeast	Bangladesh	321.3	5.7	74
India Northeast	Nepal + Bhutan	727.1	0.6	69
Pakistan South	Afghanistan	1210.5	0.5	76
Pakistan North	Afghanistan	391.2	23.0	70

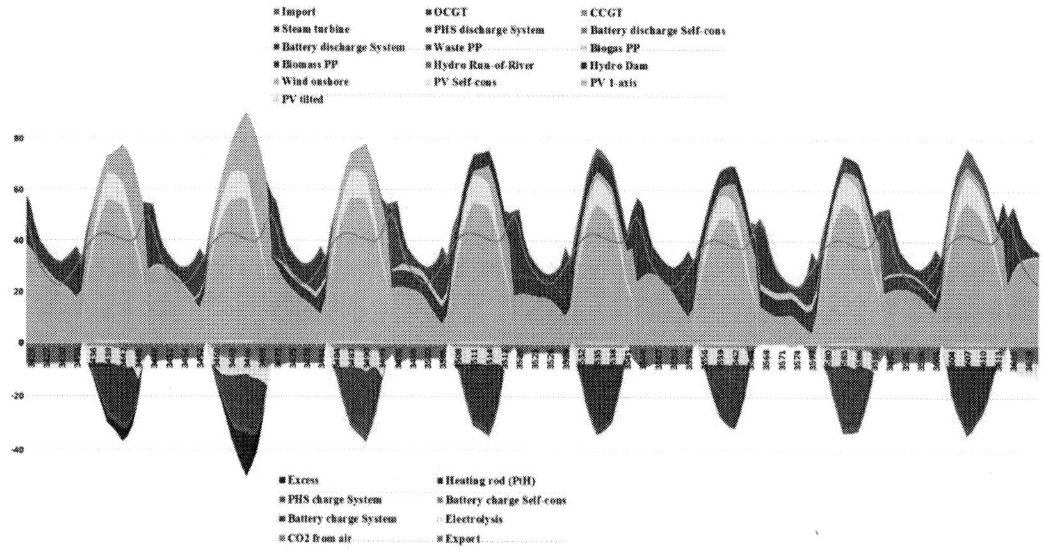

Fig. 1: Hourly generation profile for a representative week in summer month for India West.

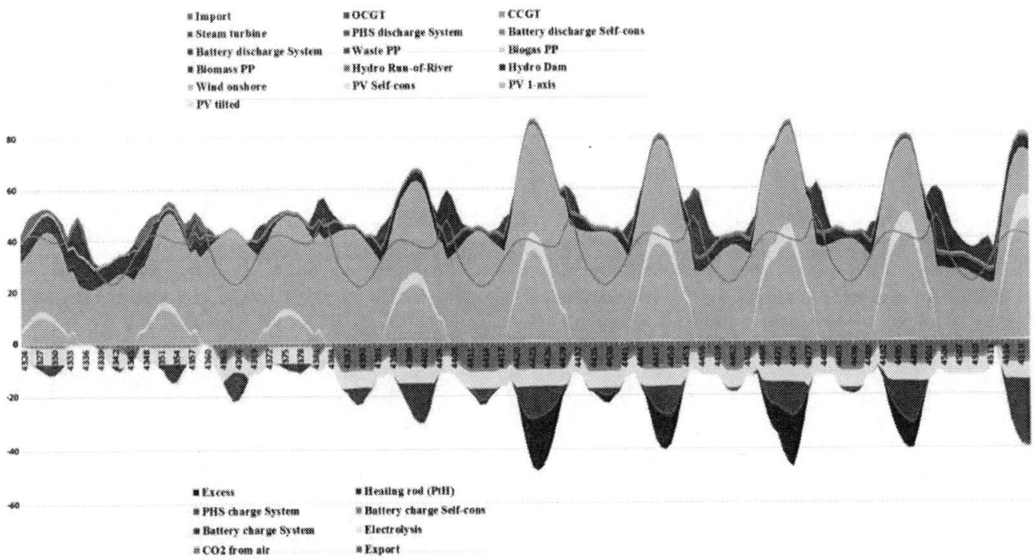

Fig. 2: Hourly generation profile for a representative week in monsoon month for India West .

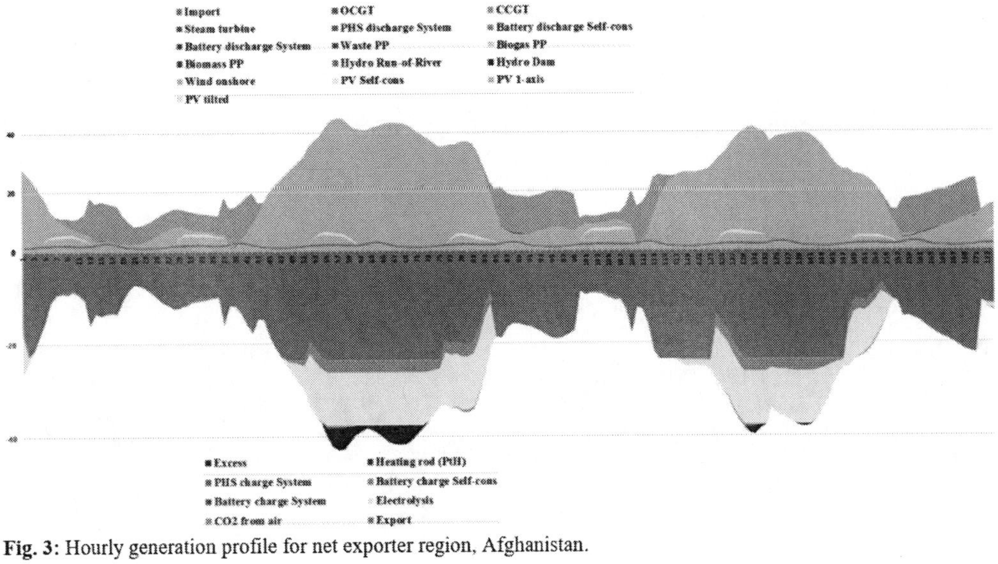

Fig. 3: Hourly generation profile for net exporter region, Afghanistan.

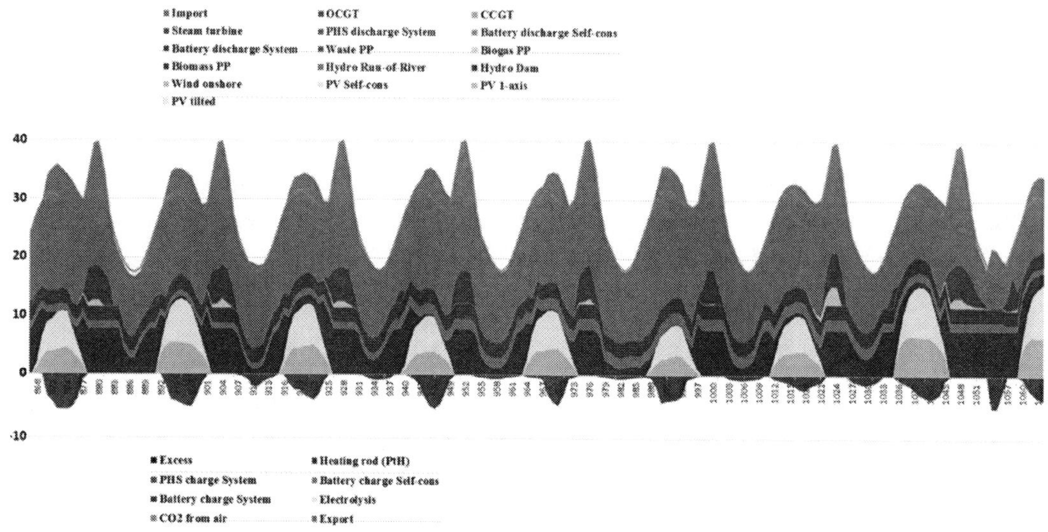

Fig. 4: Hourly generation profile for a net importer region, Pakistan North.

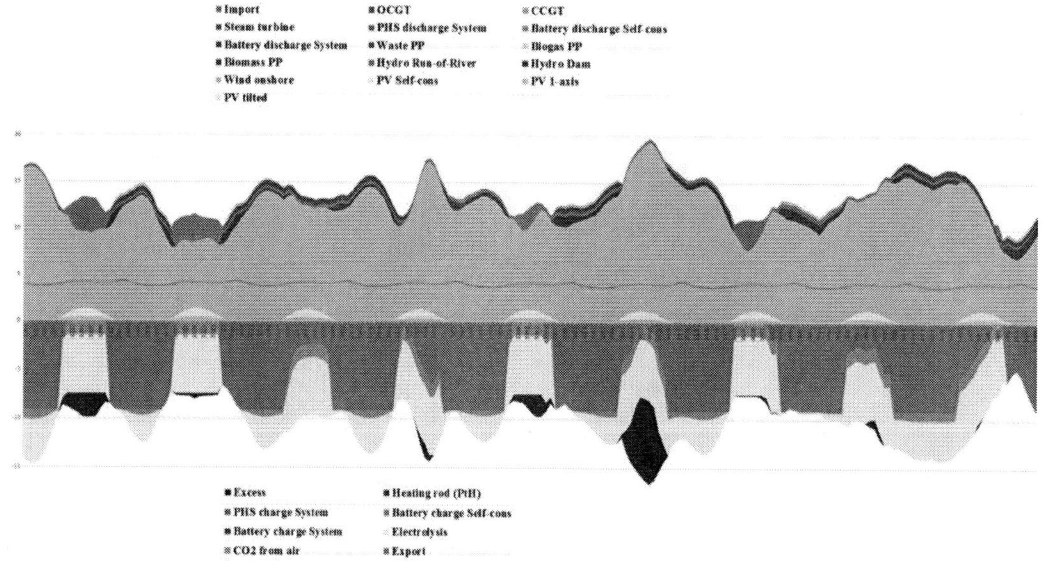

Fig. 5: Hourly generation profile for Sri Lanka.

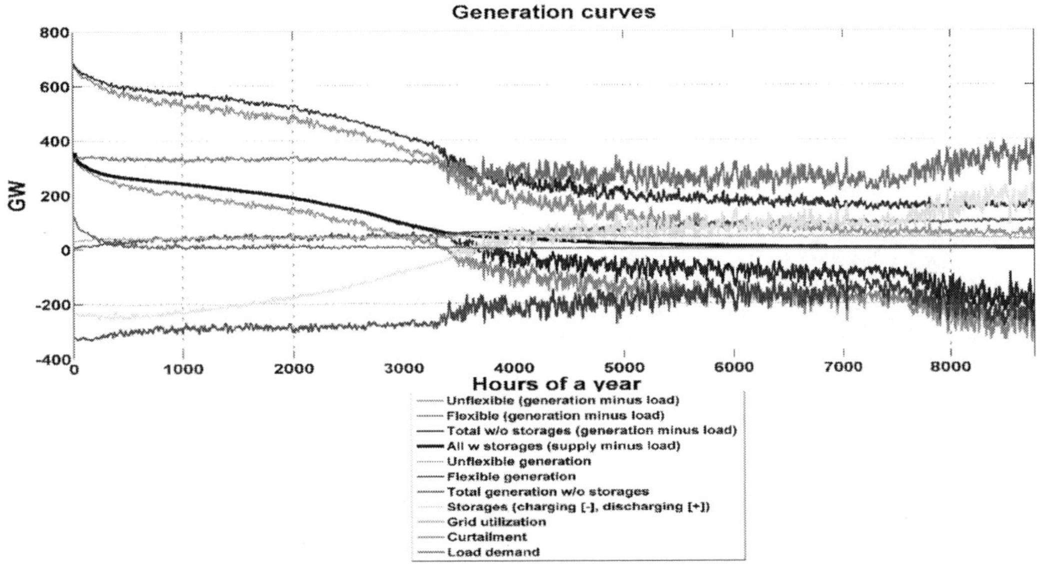

Fig. 6. Electricity generation curves for a whole year for area-wide open trade scenario for the SAARC region.

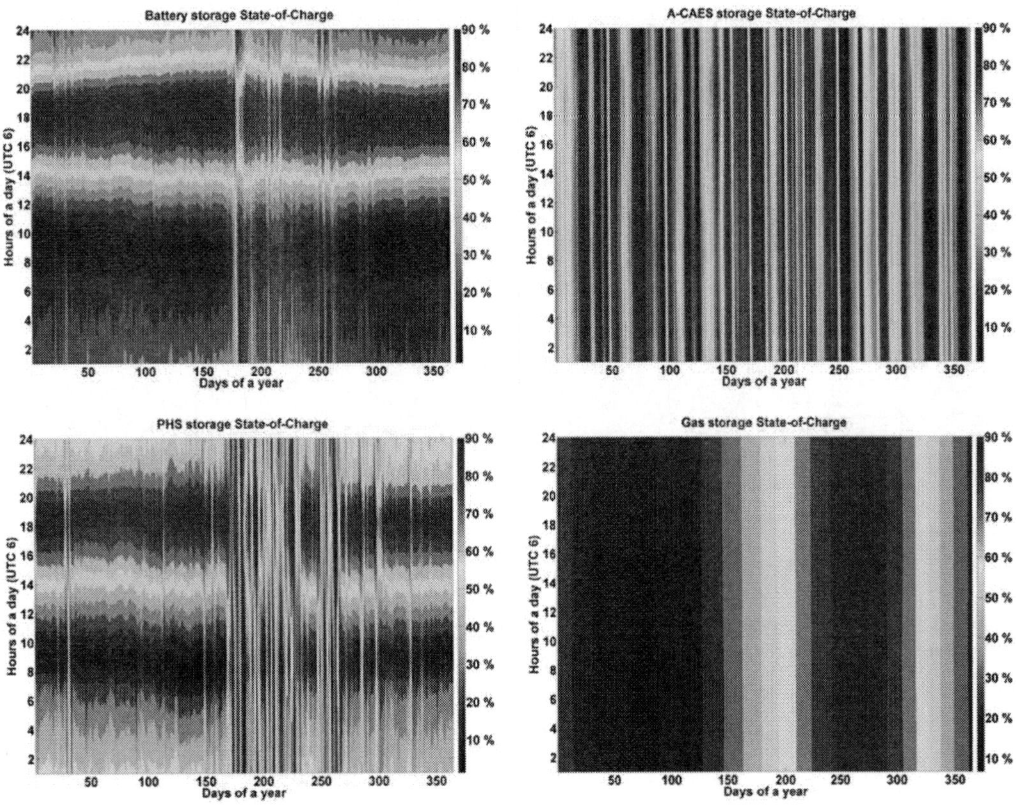

Fig. 7: Aggregated yearly state-of-charge for storage technologies, battery (top left), A-CAES (top right), PHS (bottom left), gas storage (bottom right).

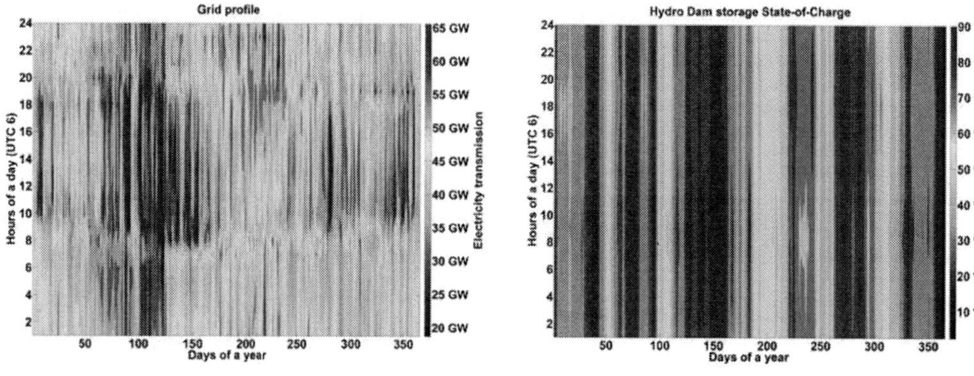

Fig. 8. Profile for interregional electricity trade between regions for area-wide open trade scenario (left) and hydro dam storage (right).

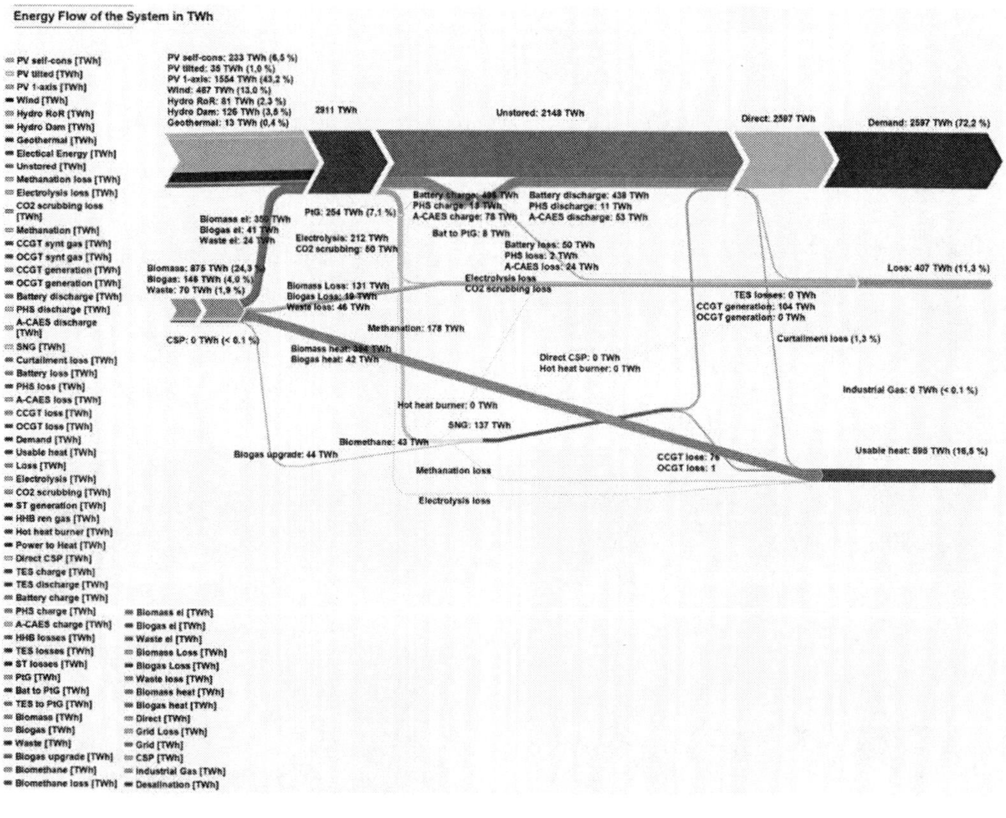

Fig. 9: Energy flow of the system for the region-wide open trade scenario for 2030.

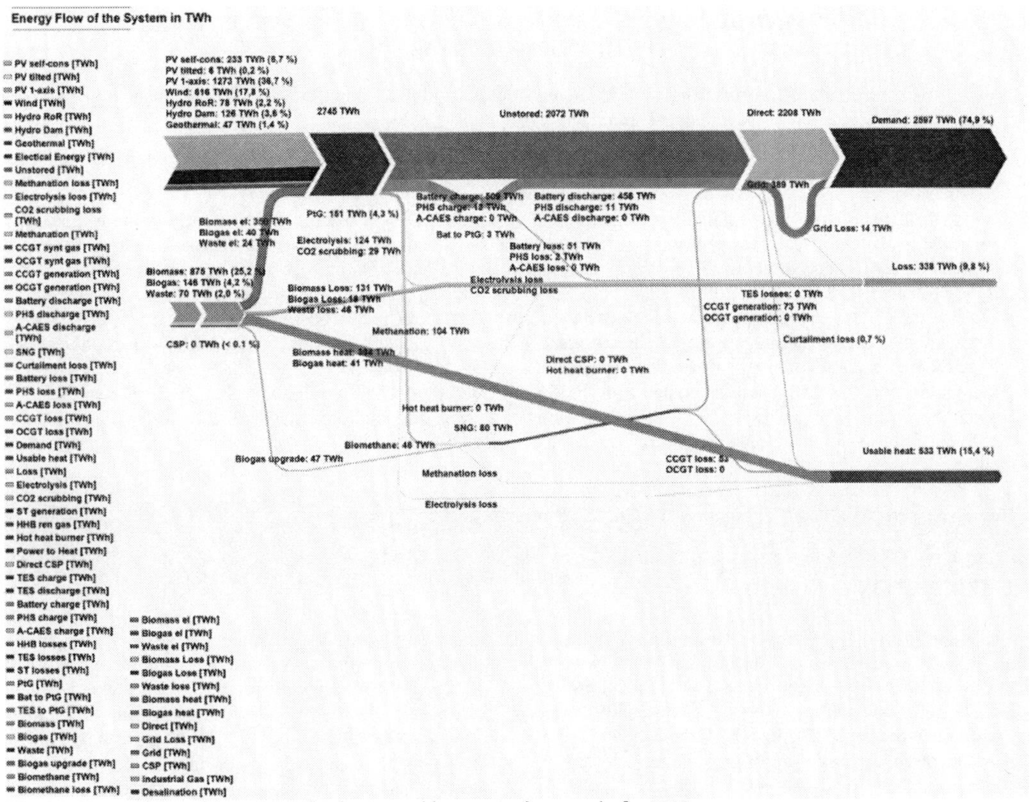

Fig. 10: Energy flow of the system for the area-wide open trade scenario for 2030

NATIONWIDE PHOTOVOLTAIC HOSTING CAPACITY IN THE FINNISH ELECTRICITY DISTRIBUTION SYSTEM

Jukka Lassila, Ville Tikka, Jouni Haapaniemi, Michael Child, Christian Breyer, Jarmo Partanen
Lappeenranta University of Technology
Skinnarilankatu 34, 53850 Lappeenranta, Finland, E-mail: jukka.lassila@lut.fi

ABSTRACT: In a large-scale integration of distributed generation (DG), a hosting capacity reflects the technical limit for the adaption. Hosting capacity depends strongly on the existing electricity infrastructure and a profile of electricity demand of end-customers. Hosting capacity and economic feasibility of large-scale integration of PV relate strongly to each other. The target of the study is to illustrate what are the limits and possibilities for the large-scale solar PV integration in Finland from the electricity distribution infrastructure perspective. The study utilizes a nationwide database of buildings and roof surface areas, a case specific network data from actual distribution areas and nationwide statistics of distribution system operators. Hourly-based load and generation data are used considering geographical location and point of compass of individual buildings. The study shows that there is significant residential rooftop solar PV capacity available according to the building and electricity infrastructure information. The main outcome is that implementing this amount of solar PV in residential level does not bring significant challenges for electricity distribution system.

Keywords: solar PV, Finland, grid integration, distribution grid, hosting capacity

1 INTRODUCTION

The role of distributed energy resources (DER) in the energy system is emphasized. The last ten years have shown a dramatic increase in solar PV installations. Prices of installations (€/kW) have decreased steadily by 20% when the worldwide capacity of PV has doubled [1]. This has increased interest for PV installations also in low solar radiation intensive areas such as Nordic environment (Fig 1). A summary of the current status of solar PV in Finland is provided by IEA PVPS [2].

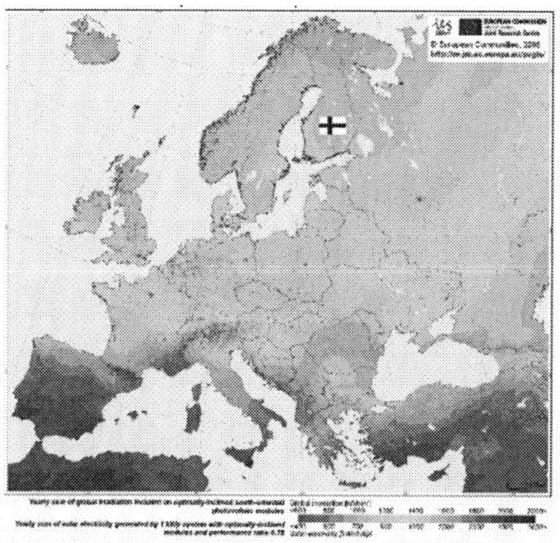

Figure 1: Solar electricity potential [3].

The role of the existing electricity distribution infrastructure in PV studies is crucial. There are numerous examples where technical challenges have been simulated and reported after wide scale installation of PVs. The challenges have related to for instance voltage quality, thermal stress and system harmonics. By the modernized network planning and operation principles, technical challenges can be avoided or minimized. At the same time, economic feasibility of the integration of distributed generation is improved.

1.1 Background

Nordic electricity distribution infrastructure differs significantly from for instance central or southern European electricity infrastructure. This is due to the high share of electricity based heating in households. Electricity heating (direct or with water heating tank) is in 50 % of residential houses in Finland [4]. In new houses, heating solutions are based mostly on heat pump technology (air-to-air, air-to-water). High role of electricity heating has been a main reason, why electricity distribution networks are strong. Typical network connection for the end-customer is three-phase with 25 A fuses in voltage level of 400 V (main voltage). This allows at maximum 17 kW continuous peak load consumption (or generation) for the end-customer. Background information of Finland and Finnish electricity distribution is presented in Table I.

Table I: Background information of Finland and electricity distribution utilities (2014 and 2015). [5], [6]

Main information, Finland	
Citizens	5.5 million
Surface area	338 441 km²
Density	18 citizens/km²
Households	2.6 million
Power system and distribution utilities	
Peak power of the system	15 100 MW
Annual delivered energy in whole power system	81.1 TWh
Distributed energy (low voltage customers) Distributed energy (medium voltage customers)	35.4 TWh 11.5 TWh
Total number of DSOs	81
Electricity end-users (from which residental %)	3.47 million (28%)
Medium voltage (MV) networks, 10 or 20 kV	141 000 km
Low voltage (LV) networks, 0.4 kV	240 000 km
Distribution substations (MV/LV)	139 000
Price of electricity (energy taxes + VAT + distribution/transmission + energy), year 2014	15.4 cent/kWh (EU-28: 20.6)

Consumption and generation curves in the Finnish electricity system are presented in Fig 2. Strong dependence of consumption on wintertime low temperatures, electric heating can be seen.

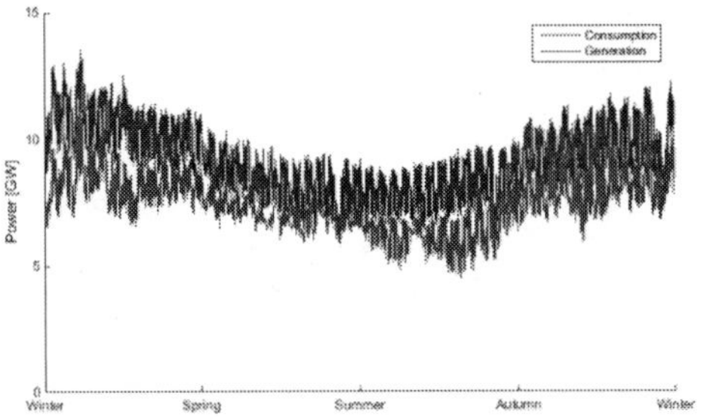

Figure 2: Total consumption and generation in the Finnish electricity system in 2015. [6]

1.2 Main steps of the study

The target of the study is to illustrate the limits and possibilities for the large-scale solar PV integration [7],[8] in Finland from the electricity distribution infrastructure perspective. The study utilizes a nationwide database of buildings and roof surface areas, a case specific network data from actual distribution areas and nationwide statistics of distribution system operators.

The main steps of the study are illustrated in Fig. 3. In the first phase of the study nationwide rooftop area analysis is done using public data related to the buildings. The target of the analysis is to define maximum amount of PV panels (m^2 and kWp) which could be installed on rooftops. This phase does not indicate what are the limits for installations from distribution network perspective.

In the second phase PV generation for buildings is defined from peak power perspective. Generation estimation process takes into account geographical location of building, point of compass and tilt angle (as fixed value). Assumption is that all the buildings have ridge type of roof and half of the building's roof top area is projected to the east-south-west direction. Different obstacles and restrictions on the roof (chimneys, ventilation pipes, ladders, etc.) have been taken into account so that PV panels utilize only 70% of the roof half area. As a result, a maximum hourly power for each building are defined.

In the third phase, hosting capacity for distribution system is defined. This takes into account actual capacity of distribution system as well as actual measured hourly based load of low-voltage end-customers. As a result, this phase gives an estimation of the hosting capacity for solar PV generation in a distribution system.

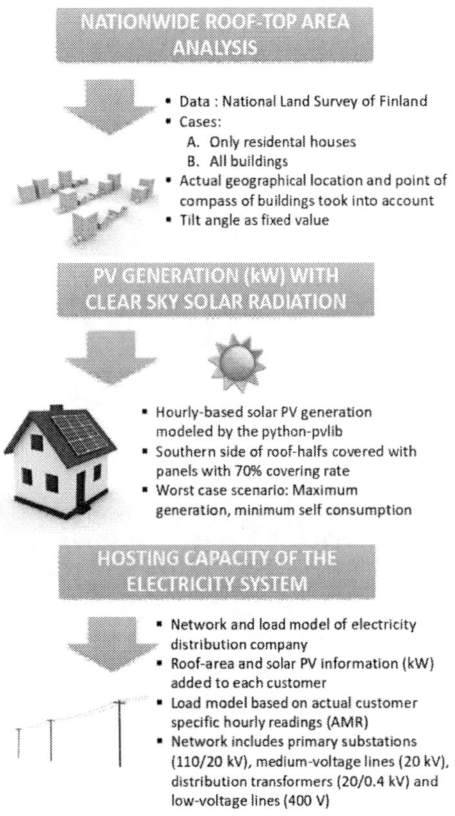

Figure 3: Main steps of the hosting capacity study.

2 DEFINING ROOF AREA CAPACITY AND SOLAR PV GENERATION

In this chapter, methodology by which building data is used in rooftop area capacity analysis is presented. Also principles of defining PV generation potential is described.

2.1 Rooftop area capacity

In this paper, hosting capacity is limited to the building installations and therefore the summary of roof surface areas defines maximum amount of PV installations. Ground –based PV solar installations are not taken into account. Data used in the study is publicly downloadable from database of National Land Survey of Finland (NLS) [9]. Building related information is included in the Topographic database. Raw data is in shapefiles, which include polygons of the buildings. Polygons include the coordinates of every corner of the buildings, building type information and number of floors in the building classified in three floor categories. These categories are one or two floors, more than three floors and unknown number of floors. The buildings are classified in the database as residential, office/public, leisure, industrial, ecclesiastical or other buildings. Fig. 4 shows the buildings in Finland.

Figure 4: Buildings in Finland map. Red marks are presenting urban or city area buildings, green ones rural area buildings.

In Fig. 4 it can be seen that most of the buildings are located in the southern Finland. There are noticeably less buildings in the north. The Helsinki metropolitan area clearly stands out in the middle of the southern coast of Finland. In this study, the focus is on residential as well as additional buildings located close to them (for instance warehouses and garages). It is assumed that roofs of buildings are pitched roof type. This is because the type of the roof structure is not included in public database. This assumption means that half of the rooftop area is projected to the east-south-west direction.

In Table II the main information related to the number of buildings and rooftop-areas is presented. The total number of buildings in Finland is a little bit over 5 million from which residential buildings is 1.3 million (about 242 million m^2 of rooftop surface area). In the same table, maximum nominal capacity [MVA] of PV panels is presented. This number illustrates the roof area capacity for installations, but it does not answer to the question how electricity distribution network can host the generated power. This is discussed later in chapter 4. Numbers presented in Table II are based on the assumption that half of the roof top area is projected to the east-south-west direction and 70% of this is utilized to PV panels. In other words, from the total rooftop area of the building 35% is covered by panels (1/2 x 70% = 35%). With these assumptions, Finnish building infrastructure offer room together for 34 400 MVA of solar PV installations.

Table II: Main information related to the rooftops.

Building type	Number	Rooftop surfaces, km^2	Avg. size, m^2/roof	Rooftop potential, S_n, MVA
Residental	1 319 444	242.2	184	12 109
Public	71 685	62.2	868	3 110
Leisure	488 763	36.9	75	1 844
Industry	30 259	47.5	1 570	2 375
Other	3 127 670	299.3	96	14 966
Total	**5 037 821**	**688**	**137**	**34 400**

For residential houses, this estimation provides totally about 12 000 MVA PV solar nominal capacity. However, this figure does not illustrate real peak power generation, which could get during sunny day because buildings are located in different geographical locations and their point of compasses are differing from each other's. In Fig. 5 a distribution of rooftop's point of compass is presented. It can be seen that buildings are rather evenly distributed in Finland.

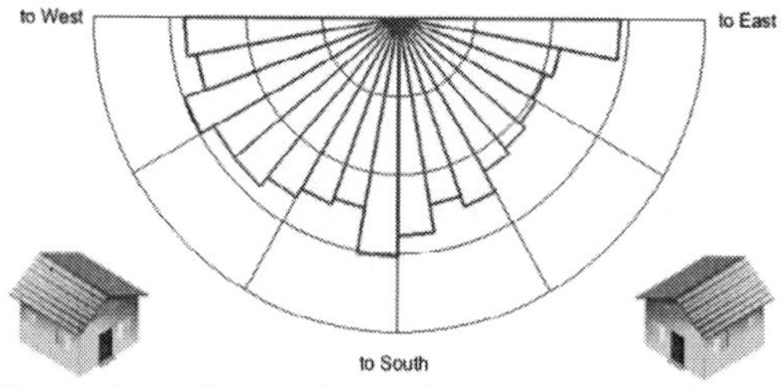

Figure 5: Rooftop distribution (point of compass) for residential buildings in Finland.

2.2 PV generation estimations

Nordic countries belong to the mooderate solar radiation intensive areas. Because of the cold wintertime and high share of electric heating, the maximum PV generation exists in the moment, when the load levels are relatively low in the distribution networks. In hosting capacity studies, peak powers (kW) are in key role instead of annual energy production (kWh). In this study, the maximum PV generation is evaluated by estimating clear sky irradiation to the panel surfaces taking into account geographical location of each rooftop, point of compass and tilt angle (as fixed value). Available rooftop area of each building is reduced by multiplier discussed earlier (70%). Hourly-based solar PV generation is modeled by the python-pvlib [10] introduced by Andrews et.al. [11]. Modeled irradiation values are used to estimate maximum production of the rooftop PVs' for every hour of the year. Average sizes of PV power plants and results of maximum PV power generation for different building types are presented in Table III. It can be seen that if all the roofs of Finnish buildings would be covered with solar PV (34 400 MVA), generated maximum peak power would be about 20 800 MW during a sunny day. Residential buildings (about 12 000 MVA) would generate approximately 7 350 MW maximum peak. In this study, the main interest is in hosting possibilities of this power.

Table III: Peak powers of PV generation if all the rooftops of buildings would be covered with PV panels.

Building type	Rooftop potential, S_n, MVA	Avg. size kVA/roof	PV peak, P_{max}, MW
Residental	12 109	9	7 350
Public	3 110	43	1 880
Leisure	1 844	4	1 110
Industry	2 375	78	1 450
Other	14 966	5	9 033
Total	**34 400**	**7**	**20 800**

Table III illustrates nationwide values for the whole Finland. For the case distribution company the rooftop data is limited to the residential houses and buildings, which are located nearby (max 200 m) to the network connection point of a low voltage customer. In the analyses, all the buildings inside this range were linked to the customer connection point (electricity distribution network database). This way all the generated power was allocated to the right node of low voltage network. Buildings which distance from the closest electricity connection point was over 200 m were not taken into account (Fig. 6).

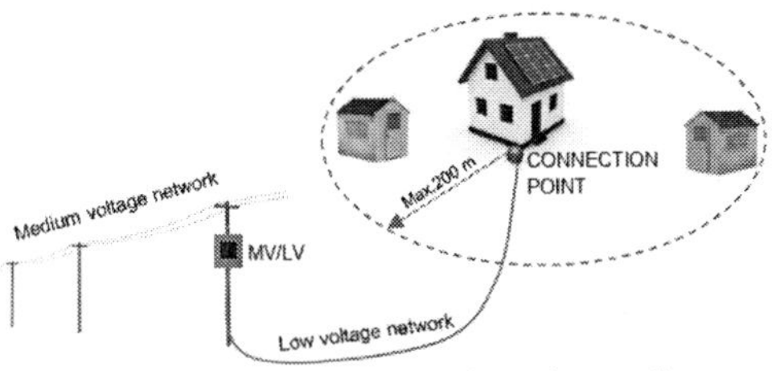

Figure 6: Electricity connection point and range for including additional buildings into same customer point.

3 DEFINING HOSTING CAPACITY IN ELECTRICITY DISTRIBUTION NETWORK

This chapter presents main defining principles of hosting capacity in electricity distribution networks in Finland. The study is based on a network and load model from actual Finnish electricity distribution company. The developed network model includes voltage levels from primary substation (110/20 kV) to the medium voltage networks (20 kV), distribution transformers (20/0.4 kV) and finally to the low-voltage networks (400 V) and actual end-customers (Fig. 7). Load information contains customer specific hourly AMR (Automatic Meter Reading) measurements during a two year period. In Finland, 100% of all electricity end-users have AMR-measurements and hourly data is used daily for network management and electricity market activities. Information of PV generation (kW-size and generation profile) capability as discussed in previous chapter is implemented for each low-voltage customer reflecting actual possibilities (rooftop areas) to generate electricity on the customer's premises. In other words, actual network data, actual load measurements and simulated individual PV generation meet each other in actual customer connection point.

3.1 Background of the network load

PV hosting capacity depends on the two main facts: present capacity of distribution infrastructure and use of electricity in the point where PV is connected to the grid. As discussed earlier, customer-end connections are relatively strong in Nordic countries due to high share of electric heating. Typical network connection is three-phase with 25 A fuses in voltage level of 400 V (main voltage). This allows at maximum 17 kW continuous peak load consumption (or generation) for the end-customer. In Fig. 8 annual load curve for the group of residental customers is presented. In Fig 8b the same curve is presented for a few summer days. The powers presented in figures respresent a typical primary substation (110/20 kV) locating in a rural area

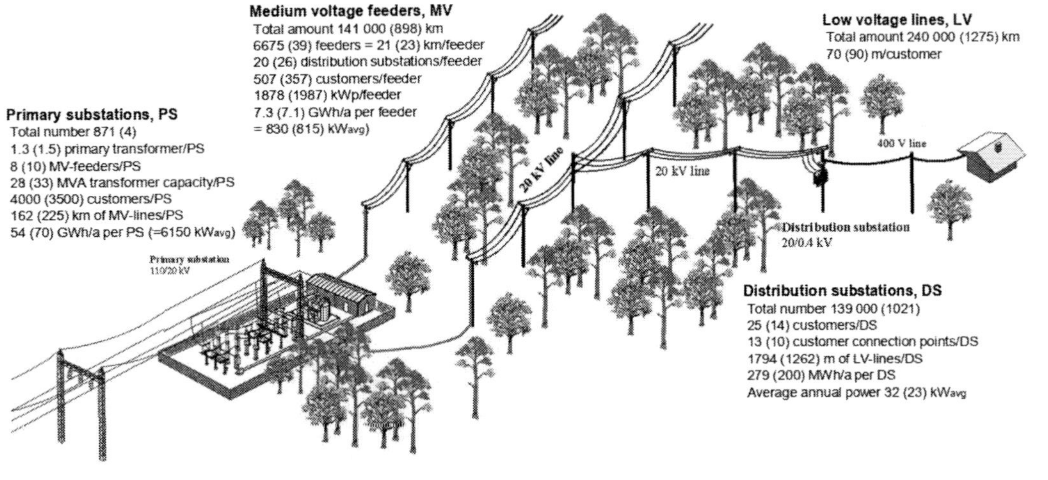

Figure 7: Key numbers of electricity distribution in Finland and in the case company (in brackets).

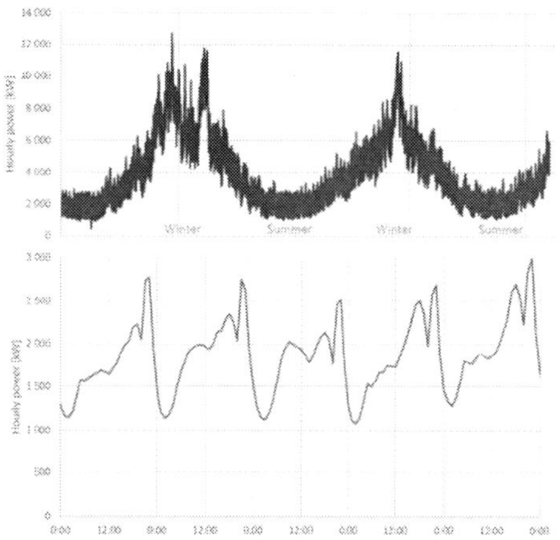

Figure 8: Typical a) load curve of electricity end-users (households, total 4600 customers) of whole year and b) summer time load curve of five days.

It can be seen that the system loads in the wintertime are several times higher than summertime loads. Within a day, peak load hours are usually in the early nighttime. This is the case when electricity is used in households to heat up water. In wintertime, nighttime peak is even higher when also heating of house is based on electricity (water boilers). Tariff-structure of electricity sets incentives to adjust heating loads (automatically) on nighttime hours.

Load curves play a role in hosting capacity analyses. Based on the curves presented in Fig. 8, there seems to be possibilities for self-consumption of PV generation also in summertime. This is the case in a primary substation and a medium-voltage level. When studying the situation from the end-customer perspective, possibilities for self-consumption are significantly lower. Fig. 9 presents a typical load curve of an individual residential customer. A strong seasonal dependency and the effect of electric sauna (random peaks) can be seen clearly.

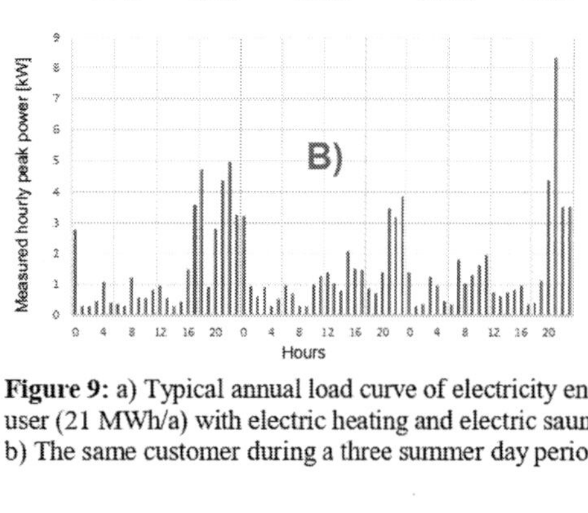

Figure 9: a) Typical annual load curve of electricity end-user (21 MWh/a) with electric heating and electric sauna. b) The same customer during a three summer day period.

It can be seen from Fig. 9b that possibilities for self-consumption are relatively low for a single end-customer. Daytime load varies significantly and there are moments, when most of the generated PV electricity would be supplied back to distribution grid. The highest electricity consumption is during evening hours due to cooking, sauna and water heating. This decreases possibilities to benefit of customer own load (self-consumption) in hosting capacity analyses. The following results illustrate the situation with and without customer load.

3.2 Power flow and hosting capacity

In this paper, the main target has been to analyze and define the hosting capacity for solar PV in the distribution networks. The study takes into account thermal and voltage restrictions coming from the present electricity infrastructure and possibilities coming from the present use of electricity (self-consumption of generated electricity). On a residential customer level, the hosting capacity is primarily limited by the connection point capacity. In most of the cases, an agreement between a low-voltage residential customer and a local electricity distribution company (DSO) allows 3 x 25 A continuous demand to be transmitted through the connection point. The peak powers of different types of electricity consumers do not usually occur simultaneously. Load curves are intersecting, and the peak power of the sum loads of different customers is usually lower than the sum of peak powers of the individual customers. With solar PV, simultaneous generation peaks during the sunny days create a risk to overload feeder lines and transformers even the generation in a single customer connection point would not exceed the present connection agreement (fuse size) with local DSO.

Loading of distribution system are estimated by traditional power-flow simulations. Customer loads are implemented into the model as measured hourly powers from AMR. These customer-specific hourly loads are summarized in low-voltage feeders and distribution transformers (20/0.4 kV) and from there through medium voltage lines (20 kV) up to primary substation transformers (110/20 kV). This way the load behavior in each node of the distribution network is known in hourly resolution. Voltage levels used in the model correspond to the actual voltage levels used in the case company. For instance, the primary substation busbar voltage is 20.7 kV. The algorithm uses this value for the start nodes of the medium voltage feeders and voltage levels after that depend on losses based on line (MV and LV) and transformer impedances, distribution transformer voltage rates (usually 20/0.4 kV) and load currents (hourly AMR loads). Thermal limits of each network components are included into analyses.

Fig 10. illustrates a simulated solar PV generation for the whole case company in the situation, where residential houses are included in the studies. The annual peak load of the low-voltage customers in the company is usually between 40–50 MW. The simulated solar PV generation in the beginning and end of the year (January-February and November-December) is not limited. In reality, generation possibilities are minimal due to low irradiation level and snow bed covering ground and roofs. Thus, solar PV generation does not effect to the wintertime peak loads. Instead, in summer time loading of primary transformers is strongly increased - now to the negative direction (supplying from MV and LV feeders to the HV network). In Fig. 10b situation is presented during summer time.

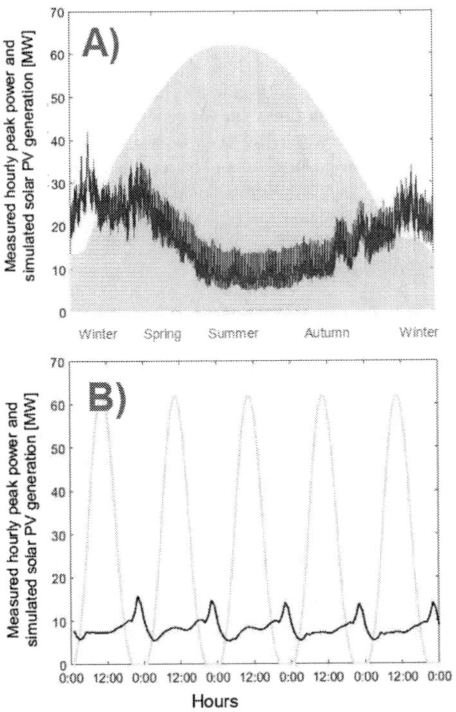

Figure 10: Solar PV generation curve and load curve of whole case company of a) whole year and b) few days when only residential houses are included in study.

Customer-end situation is illustrated in Fig. 11.

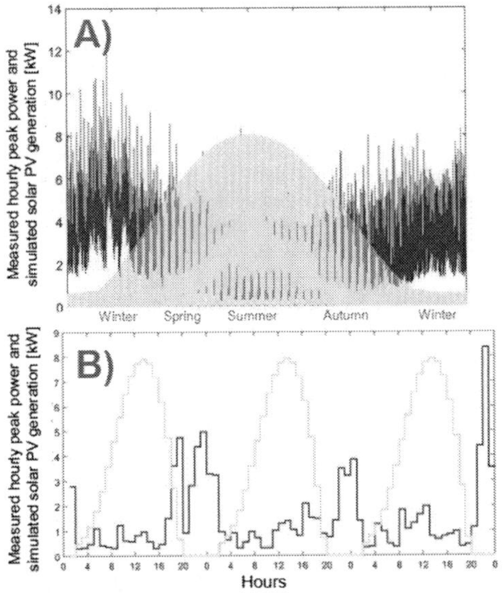

Figure 11: Solar PV generation curve and typical load curve of a single electricity end-users of a) whole year and b) few days.

As in earlier case, it can be seen that solar PV generation does not effect on wintertime peak loads. During summertime, powers vary strongly from peak loads (electric sauna/water heater) to daytime peak generation. In this particular single customer case, size of residential house roof limits the power plant size to the 9 kWp. Generated hourly peak is approx. 8 kW. As in many cases, most of the generated electricity is supplied back to the network due to low level customer loads during daytimes.

3.2 Case area and scalability

The case company and network with load reflects relatively well typical Finnish electricity distribution company operating both in rural and urban area. This helps to scale results of hosting capacity nationwide. In Fig. 12 annual delivered energies in different utilities are presented.

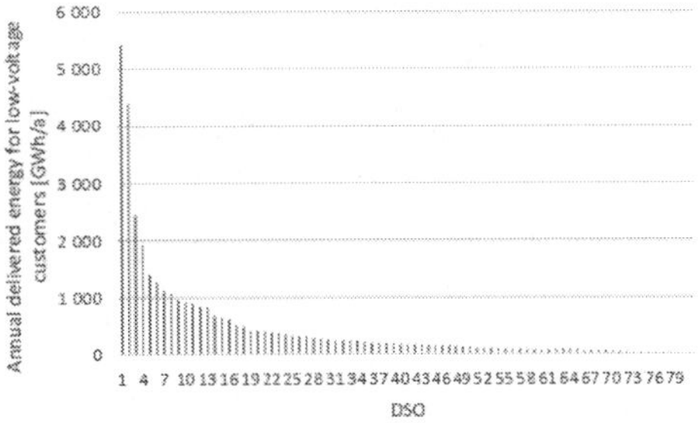

Figure 12: Annual delivered energies for low-voltage customers in different distribution utilities (total 81 DSOs) [4]

In Fig. 13 annual average consumption of low-voltage customer in different utilities are presented. On the average low-voltage customer uses from 7 MWh to 19 MWh per year depending on utility. In the case company average annual consumption is 14 MWh/a for end-customer.

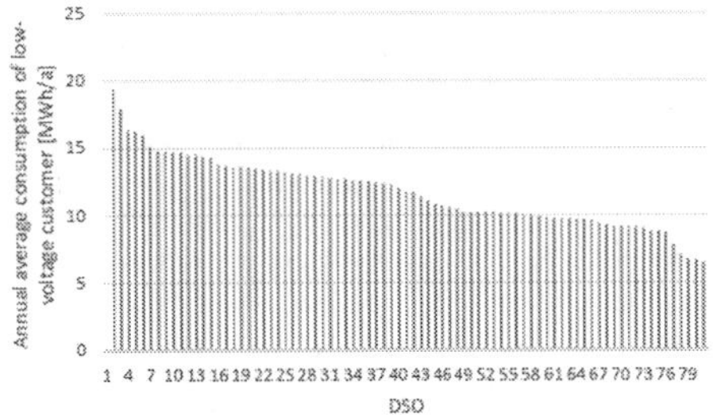

Figure 13: Annual average consumption of electricity of low voltage customer. [4]

In Fig. 14 present system peak powers and annual delivered energies in different DSOs are presented. Strong correlation with these factors reflect similar peak operating time in different companies in low voltage networks. For the case company peak power is 78 MW and annual delivered energy for low voltage customers about 200 GWh. If considering only low voltage customers, their summary peak power is 49 MW.

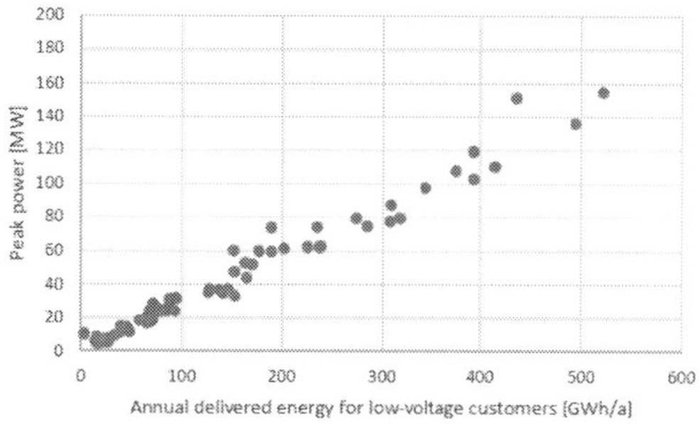

Figure 14: Present system peak powers in different DSO networks and annual delivered energies (limited to 600 GWh/a and 200 MW at maximum, total 65 DSOs)

When considering network structure, case company reflects nationwide averages. In Finland, there are totally about 6700 medium-voltage feeders which forms 141 000 km of medium voltage lines. On the average, this makes 21 km/feeder. For the case company number is 23 km/feeder. With number of distribution transformers nationwide average is 20 transformers/feeder and for the case company 26 transformers/feeder. In Fig. 15 DSO specific average values for number of customers in distribution substation and average annual load of the distribution substation are presented.

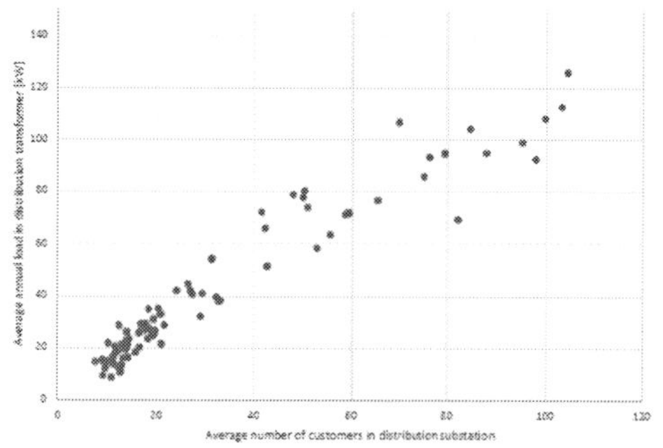

Figure 15: DSO specific average values for number of customers in distribution substation and average annual load of the distribution substation.

Even the size of the case company represents only a small share of the whole Finnish electricity distribution system, the case network with typical low voltage customer loads reflects still very well actual urban and rural area electricity distribution.

4 RESULTS

In this chapter, the main results of the study are presented. The case network plays a significant role in the analyses due to the possibility to use detailed information of distribution network and loads together with building data. The results reflect a situation during the most challenging hours from the hosting capacity perspective. Results are presented for each level of the distribution system: from the end-customer to the primary substation level. In the hosting capacity both thermal tolerance and voltage rise perspectives are considered.

4.1 Main results

For the case area (about 14 000 customers) the main background numbers and results are following:

- Measured maximum hourly peak load: 78 MW, from which 49 MW in low voltage network
- Simulated rooftop solar PV generation peak in low voltage network: 160 MW, from which
  - residential house rooftops: 63 MW
  - other rooftops nearby: 97 MW
- Distribution network hosting capacity: 60–70 MW of generation peak power without significant challenges

The main result is that a generated PV peak power can be higher than measured actual peak load in low voltage network. This is the case even PV generation is not significantly intersecting such as traditional load power curves are. It is good to notice that the values presented above are peak powers for generation. Actual installation sizes (kWp) are significantly higher, for instance for residential house rooftops installation capacity is about 100 MVA which can generate a 63 MW peak.

It can be seen that the generated powers differ significantly depending on the installation scale: are the installations limited to the residential buildings only (generated peak 63 MW) or are additional buildings (generated 97 MW) close to the low voltage customer (max. 200 m range) taken into study allowing them to be connected to the same connection point with a residential house. Especially in a rural area, rooftop areas grow significantly when nearby buildings (garages, storage buildings, and sheds, etc.) are taken into analyses.

Present wintertime peak load for low-voltage customers (49 MW) is close to the amount of peak power, which could be generated by residential rooftop PV panels (63 MW). This indicates already, that hosting capacity should not set significant challenges in the present infrastructure if PV installations are implemented only to the residential rooftops.

Due to seasonal difference in timing of the present peak load and simulated peak PV generation, the role of self-consumption in hosting capacity study is not significant. For the case company, demand for electricity is approx. 12 MW in a low voltage network. This number represents the situation only during two-year hourly measurements and thus, from statistical and risk management perspective, it is reasonable to study also the case, when there is no load at all or load level is very small. However, self-consumption due to electric vehicles (EV) and stationary battery energy storage systems (BESS) can be part of the solution to improve the hosting capacity even as solar PV grow in number.

4.2 Detailed results

In following figures (16-21), the results of hosting capacity analyses from the thermal tolerance perspective are presented for the case company. In Fig. 16 PV generation peak power against customer fuse size in low voltage networks is presented. It can be seen that a customer connection point does not set limits for the PV installations. This is especially the case, when only residential house rooftops are included into study. When additional buildings inside a rage of maximum 200 m from customer connection point are covered with solar

PV and connected to the same network connection point, about 20% of the low-voltage customers should restrict generation power during peak moments or update their connection size agreement from 25 A for instance to 35 A with local DSO.

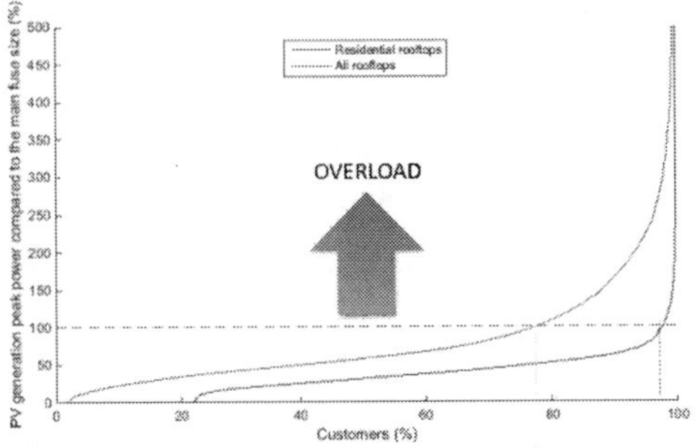

Figure 16: PV generation peak power against customer fuse size in low voltage networks.

In Fig. 17 load rate distribution of low voltage lines are presented during the PV peak generation. Dimensioning of the present low voltage system is relatively high due to high wintertime heating loads. Thus, in both scenarios, only few percent of all lines are thermally overloaded.

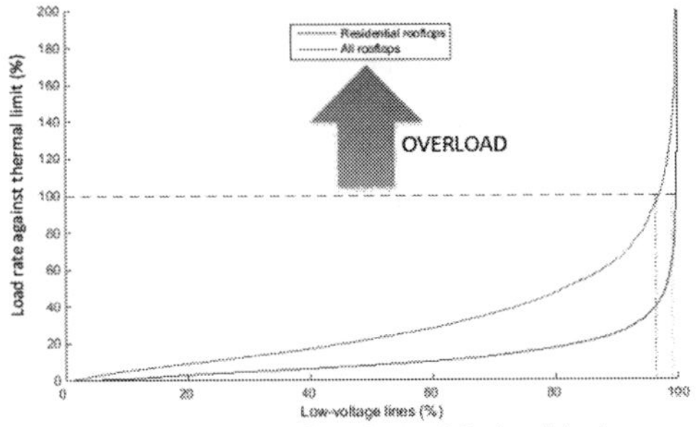

Figure 17: Load rate against thermal limit of the low voltage lines during PV peak generation.

In Fig. 18 load rate distribution of distribution transformers (20/0.4 kV) is presented during the PV peak generation. Compared to the earlier figures, the difference in overloading of transformer is significant depending on the installation scale, whether solar PV installations are limited to the residential buildings only or additional buildings close to the low voltage customer are taken into study. This can be explained by the fact that traditional dimensioning of a transformer takes into account of intersecting of powers. The peak power demands of different types of electricity consumers do not usually occur simultaneously, but the power curves are intersecting, and the peak power of the sum loads of different customers is lower than the sum of peak powers of individual customers. With solar PV, simultaneous generation peaks during sunny days lead overloading. In the case company, residential house PVs do not affect challenges for distribution transformers, but taking all the buildings nearby into account leads to the situation where about 60% of the transformers are overloaded. In Nordic operation environment overloading of transformers is accepted commonly due to the fact that peak loads are mostly in the time when cooling conditions are the best (wintertime). With solar PV, overloading is not recommended due to weak cooling conditions during peal generation (summertime peak).

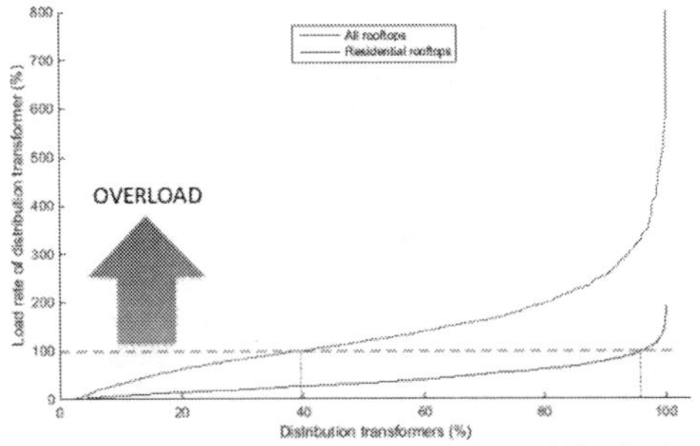

Figure 18: Load rate against thermal limit of distribution transformers during PV peak generation.

In Fig. 19 load rates of primary substation transformers (110/20 kV) are presented during the PV peak generation. The result corresponds with the earlier figure where distribution transformers were discussed. When taking into account the actual sizes of each primary transformer (MVA), transformer capacity should be increased on the average by 20–30% in 'All rooftops'–scenario to avoid overloading.

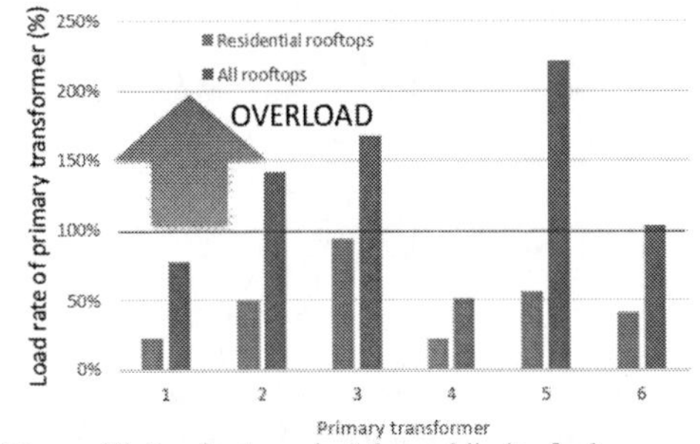

Figure 19: Load rate against thermal limit of primary substation transformers during PV peak generation.

In addition to thermal limit analyses, also possibility to voltage rising [12], [13] during peak generation and low self-consumption has to be studied. As in the thermal limit studies, Nordic way of dimensioning distribution networks against high wintertime heating loads enables solar PV penetration from voltage control perspective.

Fig. 20 illustrates voltages in the low voltage networks in the case where rooftops of residential buildings are covered with PV. From DSO perspective, possibility of over voltages in low voltage networks during peak generation can be partly avoided by controlling busbar voltage in primary substations. The figure shows difference of scenarios where DSO operates the network with 20.7 kV or 20 kV busbar voltage. It can be seen that the voltages do not exceed 440 V (+10%) widely if only residential building roofs are included into study.

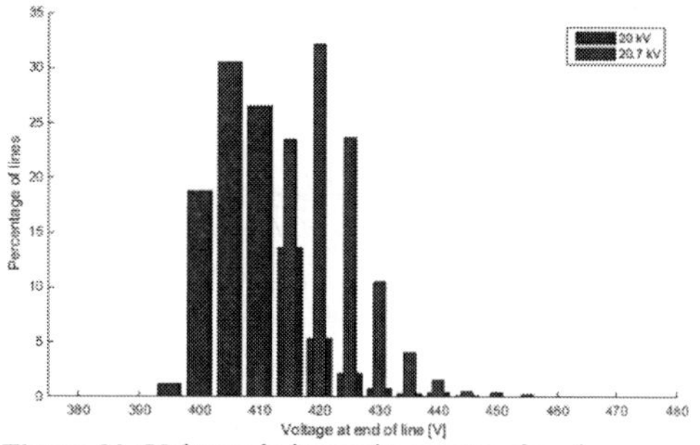

Figure 20: Voltages in low voltage networks when busbar voltage is 20.7 kV or 20 kV. Only rooftops of residential houses included.

Fig. 21 shows the results when all the rooftops of different buildings are covered with PV. It can be seen that voltages would rise widely over 440 V.

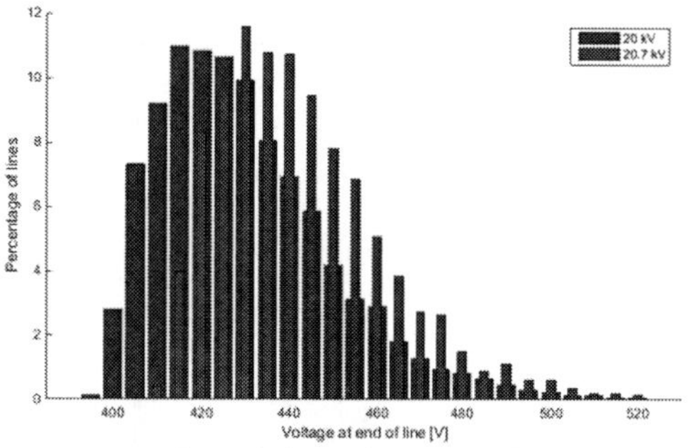

Figure 21: Voltages in low voltage networks when busbar voltage is 20.7 kV or 20 kV. All rooftops nearby residential houses included.

Fig. 20 and 21 indicate that voltage level on primary substation busbar has a high impact on voltage levels in the low voltage networks. In Finland medium voltage networks are usually operated on 20.5–21 kV voltage level to prevent a too low voltage drop during winter peak loads. If busbar voltage control is put in practice for solar PV management, one have to take care that voltages stay in allowed range during power changes in all parts of distribution system. In the hosting capacity analyses it is assumed that all the low-voltage customers in distribution substations participate on PV generation as long there are roofs available to be covered with PV. That way the similar needs for voltage control concern all customers (the same direction of voltage control).

Risk of over voltages can be managed also with control of generation. In residential rooftop PV scenario it was estimated that voltages in customer connection points would stay under U_n + 10% if only 2.5% of total generation peak was reduced. On all buildings scenario 30% of total generation should be decreased to avoid too high voltage rise.

5 CONCLUSIONS

The study shows that there is significant residential rooftop solar PV capacity available according to the building and electricity infrastructure information. There are more than 5 million buildings in Finland, which form about 690 million m^2 of roof surface area. Residential buildings (about 1.3 million buildings) constitute about 240 million roof-m^2. The results show that if residential building roof-areas would be fully utilized for solar PV generation, the level of PV installations would be about 12 000 MWp. These installations would provide approximately 7 350 MW peak power on sunny day when geographical locations and direct of compass of buildings are taken into account. This would be close to the spring and summer time peak loads (9 000–10 000 MW) in the present Finnish power system. The study shows that implementing this amount of solar PV in residential level does not bring significant challenges for electricity distribution system. PV generation peaks can be higher than wintertime peak loads. Main reason for the promising situation is the present electricity infrastructure – dimensioning of the network is based on wintertime peak loads (electric heating and electric sauna). In the most of the cases electricity network connection size of residential buildings is 3x25 A or higher which allows 17 kW powers to be transmitted from or into the distribution network. For distribution transformers simultaneous behavior of solar PV generation peaks have to be taken into account to avoid overloading. This is the case, if installation of PV panels is not limited to the residential house roofs but also nearby buildings are taken into account. However, before massive penetration of PV power plants, smart grid solutions with demand response, electric vehicles and battery energy storages have to become more widely spread allowing even higher hosting capacity than discussed in the study.

The results of the paper help different stakeholders to define the PV hosting capacity in electricity distribution networks and understand possibilities in large-scale PV integration in the Nordic environment.

ACKNOWLEDGEMENTS

The authors gratefully acknowledge the public financing of Tekes (Finnish Funding Agency for Innovation) for the 'Neo-Carbon Energy' project under the number 40101/14.

REFERENCES

[1] IRENA International Renewable Energy Agency, Renewable Energy Technologies: Cost Analysis Series. Volume 1: Power Sector Issue 4/5. June 2012. [Online]. Available: www.irena.org/DocumentDownloads/Publications/RE_Technologies_Cost_Analysis-SOLAR_PV.pdf

[2] National Survey Report of PV Power Applications in FINLAND 2014, The IEA Photovoltaic Power Systems Programme (IEA-PVPS), http://iea-pvps.org/index.php?id=146

[3] Šúri M., Huld T.A., Dunlop E.D. Ossenbrink H.A., 2007. Potential of solar electricity generation in the European Union member states and candidate countries. Solar Energy, 81, 1295–1305, http://re.jrc.ec.europa.eu/pvgis/

[4] Motiva, Sähkölämmityksen tehostamisohjelma Elvari 2008-2015 loppuraportti, in Finnish (Electric Heating efficiency program's final report), 2015. http://www.slideshare.net/MotivaOy/elvari-loppuraportti

[5] EMA, Energy Market Authority, official statistics of electricity distribution companies 2014. www.energiavirasto.fi

[6] Fingrid, load and generation statistics 2015 www.fingrid.fi/EN/ELECTRICITY-MARKET/LOAD-AND-GENERATION/Pages/default.aspx

[7] Kosonen A., Ahola J., Breyer C. Large Scale Solar Power Plant in Nordic Conditions. 16 th EU Conference on Power Electronics and Applications (EPE 14), August 2014, Lappeenranta, Finland https://www.researchgate.net/publication/265141779_Large_Scale_Solar_Power_Plant_in_Nordic_Conditions

[8] Child M., Haukkala T., Breyer C. The Role of Solar Photovoltaics and Energy Solutions in a 100% Renewable Energy System for Finland in 2050. 31 st European Photovoltaic Solar Energy Conference, September, 2015, Hamburg, Germany https://www.researchgate.net/publication/281859358_The_Role_of_Solar_Photovoltaics_and_Energy_Solutions_in_a_100_Renewable_Energy_System_for_Finland_in_2050

[9] National Land Survey of Finland (Maanmittauslaitos), "List of the NLS open digital datasets". [Online]. Available: www.maanmittauslaitos.fi/en/professionals/digital-products/datasets-free-charge/list-nls-open-digital-datasets

[10] pvlib-python documentation. [Online]. Available: http: pvlib-python.readthedocs.org

[11] Andrews R., Stein J. S., Hansen C., and Riley D.,"Introduction to the open source pv lib for python photovoltaic system modelling package," in 40th IEEE Photovoltaic Specialist Conference, 2014.

[12] Finnish Standards association, SFS-EN 50160 Voltage characteristics of electricity supplied by public electricity networks, 2010.

[13] Finnish Standards association, SFS-EN 50438 Requirements for micro-generating plants to be connected in parallel with public low-voltage distribution networks, 2015.

PERFORMANCE ANALYSIS AND YIELD ASSESSMENT OF SEVERAL UNCOVERED PHOTOVOLTAIC-THERMAL COLLECTORS: RESULTS OF FIELD MEASUREMENTS AND SYSTEM SIMULATIONS

Corry de Keizer*[1], Minne de Jong[1], Munish Katiyar[2], Wiep Folkerts[1], Camilo Rindt[2], Herbert Zondag[2,3]
[1] Solar Energy Application Centre (SEAC), High Tech Campus 21, 5656 AE Eindhoven
[2]Technical University Eindhoven, Eindhoven, the Netherlands
[3]ECN, Petten, the Netherlands
*Corresponding author: dekeizer@seac.cc

ABSTRACT: A PVT collector combines a PV module with a solar thermal absorber and produces electricity and heat. Interest in PVT systems is growing, since these potentially generate more energy per m^2 than PV-only systems. Furthermore, a large share of the residential energy use consists of heat. Within the Dutch project WenSDak, a consortium of eight companies and three knowledge institutes developed five different PVT concepts and evaluated their thermal and electrical performance. In this paper, we focus on three types of uncovered PVT collectors. There is a large difference between the measured thermal performance of the three different collectors. These differences can be explained by the PVT collector design. The absorption of all PVT modules is quite good (0.9 – 0.94). A good heat conduction from the PV cells to the absorber leads to a high peak collector efficiency, while insulation at the back of the PVT module improves the performance at higher temperatures. Performance of some collectors can be improved by e.g. heat conducting paste, however, this is also more expensive. System simulations with PVT as part of a solar heating system are carried out in TRNSYS and will yield valuable information on the annual yield for a specific system.
Keywords: Building Integrated PV (BIPV), Thermal Performance, PVT Systems

1 INTRODUCTION

Hybrid PV-thermal (PVT) systems have been around for several decades [1]. Currently, there is a renewed interest in the development and application of PVT systems. In the Netherlands, this mainly holds for uncovered PVT systems, often in combination with a (ground source) heat pump. The renewed interest for PV-T collectors is, besides declining PV prices, related to European regulations aiming towards (near) zero-energy buildings in the near future. Several full-roof (BI)PV solutions have been developed for such buildings, limiting the roof space available for solar thermal. Since a large share of the energy demand in the built environment in continental Europe consists of heat, PVT systems may result in higher energy and exergy yields and a more aesthetic uniform appearance of the roof. Furthermore, they may result in higher combined efficiencies, with relatively little additional costs in comparison to standard PV.

The research presented in this paper is part of the Dutch WenSDak project in which 8 industrial partners and 3 research institutes develop and analyse 5 different concepts for the combination of solar heat and solar electricity in one roof. This paper focuses on the electrical and thermal performance of the three uncovered PVT systems in the project that were developed by different project partners. Uncovered PVT modules do not have an additional air or gas layer with a glass cover and therefore, show lower PV temperatures than covered PVT systems and lower thermal performance at higher fluid temperature.

The thermal and electrical performance of three different types of PVT collectors was analysed in a field test at SEAC's field test facility 'SolarBEAT'. The performance data of the different collectors can be used as input for annual simulations of a typical system.

2 METHOD

2.1 Definition of the PVT collectors

The different types of PVT collectors are (see also Figure 1):

- A: CIGS panel with clamped absorber and insulation produced by Solartech (Energiedak MEP panels)
- B: c-Si PV with uninsulated absorber clamped to the back of the module, produced by GEO Holland
- C: building integrated c-Si PV with in-roof absorber and insulation produced by Dimark Solar

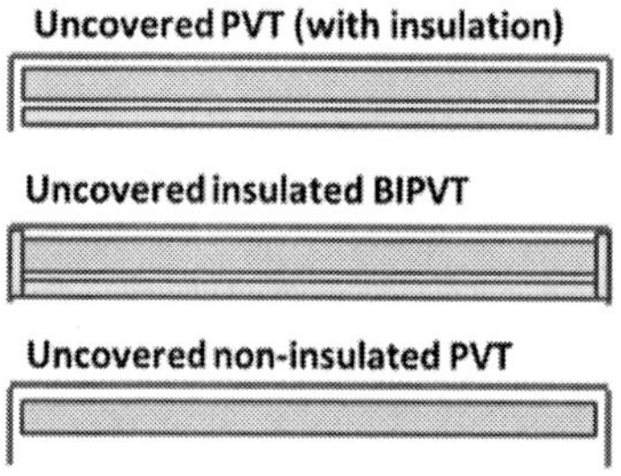

Figure 1: Three different PVT concepts (bright blue line: PV module, blue block: thermal absorber with fluid channels, yellow: insulation)

2.2 Field test design and operation

The three uncovered PVT systems were installed on the experimental outdoor field test facility of SEAC as shown in Figure 2. The following setup was installed:

- System A: 4 PVT collectors and 4 PV modules; results are shown for 4 PVT collectors, gross area of 4.4 m^2, flow rate 24 l/m^2h
- System B: 5 PV collectors and 3 thermal absorbers; results are shown for 2 PVT collectors, gross area of 3.3 m^2, fluid flow rate 74 l/m^2h
- System C: 3 PVT collectors and 3 thermal collectors; analysis of 2 PVT collectors, gross area of 3.5 m^2, flow rate 18 l/m^2h

The outdoor facility includes a solar measurement station with which direct, diffuse and global horizontal irradiance are measured, as well as wind speed and direction and ambient temperature.

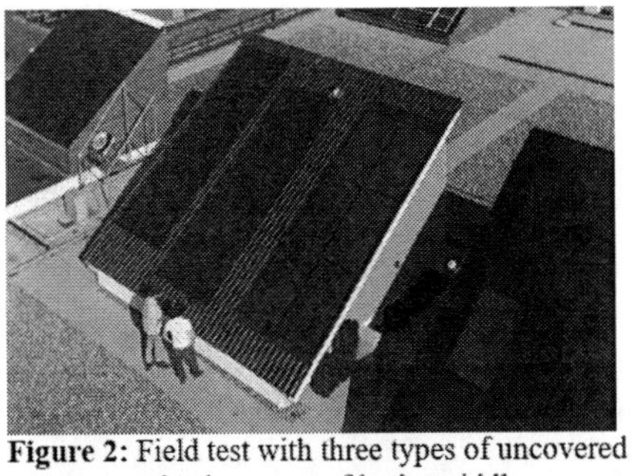

Figure 2: Field test with three types of uncovered PVT systems on the dummy roof in the middle

Furthermore, a thermal loop was designed and installed to allow for measurements on solar thermal collectors and PVT systems. Excess heat produced by the thermal systems is dumped in the university's aquifer. A combination of heaters, valves, chillers and control technology make sure the liquid is preconditioned to the specified temperature. The input temperature for the systems can be set between 7 and 80°C. A 25 % glycol solution is used in the PVT loop.

The collectors of each system are thermally connected in series. For each system, different flow rates are defined. However, the inlet temperature of the first collector of each system is the same. The flow rates were decided upon by the supplier of the system, to match the flows that are used in real systems. Each PVT panel is electrically connected to a DC/DC SolarEdge power optimizer. The power optimizers are in series connected to the SolarEdge AC/DC inverter. Therefore, the electrical and thermal performance of the PVT modules is measured at maximum power point.

The following measurement equipment is installed:

- Meteorological measurements: Global tilted irradiance (secondary standard pyranometer), pyrgeometer, in-plane wind speed and direction, ambient temperature.
- PV performance measurements: Measurements are done at MPP (maximum power point). DC voltage and DC current (via a shunt) are measured for each PVT collector separately.
- Thermal performance measurements: input and output temperature of each collector (Pt100, 1/3B) and flow rate (Electromagnetic sensor, one per series of collectors).
- Datalogging: All sensors are connected to a Yokogawa MW100 datalogger. Data is recorded every minute and uploaded every night to a database.

The field test has been running for a full year, from June 2015 to May 2016.

2.3 Evaluation of thermal performance

For analyzing the thermal efficiency, we followed the steady state analysis for unglazed collectors as described in the ISO 9806 norm [2], though there are some differences , like building integration of the PVT modules in our test site and the flow rate. The thermal efficiency (η_{th}) is calculated by Equation 1. The effective irradiance is calculated by equation 2, with the in-plane irradiance as an input. Furthermore, the pyrgeometer is used to calculate the long-wave irradiance (EL). The coefficients η_{th}, b_u, b_1 and b_2 are fitted by using a least squares method and equation 3.

Equation 1

$$\eta_{th} = \frac{\dot{Q}}{A_G \cdot G''} = \frac{\rho \cdot c_p \cdot \dot{V} \cdot (T_{out} - T_{in})}{A_G \cdot G''}$$

Equation 2

$$G'' = G_{POA} + \frac{\varepsilon}{\alpha}(E_L - \sigma T_a^4) \text{ with } \varepsilon/\alpha = 0.98$$

Equation 3

$$\eta_{th} = \eta_{0,th}(1 - b_u u) - (b_1 + b_2 u)\frac{(T_m - T_a)}{G''}$$

With:

A_G	Gross collector area (m^2)
b_1	heat loss coefficient (W/m^2K)
b_2	wind dependence of the heat loss coefficient (J/m^3K)
b_u	collector efficiency coefficient (wind dependence) (s/m)
c_p	Specific heat capacity (J/kgK)
G_{POA}	Global irradiance in plane-of-array (W/m^2)
G''/G_{eff}	Net irradiance (W/m^2)
T_a	Ambient temperature (K)
T_m	Mean collector temperature (Tin+Tout)/2 (K)
T_{in}	Input temperature (K)
T_{out}	Output temperature (K)
$\dot{Q}_{th}$	Heat flow (W)
u	Wind speed (m/s)
$\dot{V}$	Volume flow (l/s)
α	Solar absorptance (%)
ε	hemispherical emittance (%)
$\eta_{0,th}$	Peak collector efficiency (η_{th} at $T_m = T_a$)
η_{th}	Thermal collector efficiency, with reference to T_m
ρ	Density (kg/l)
σ	Stefan-Boltzmann constant (W/m^2K^4)

3 FIELD TEST RESULTS

3.1 Introduction

PVT collectors produce both heat and power. The energy yield depends on different factors, of which the most important factors are: Irradiance, ambient temperature, average fluid temperature and wind speed. Figures 3a and 3b show the produced power (heat and electricity) in W/m^2 for the three collector types. We chose two sunny days with a low (7°C) and a high (35°C) inlet temperature in summer 2015.

Irradiance in the plane-of-array is shown in green. The thermal heat is depicted by continuous lines, while in the dotted lines the electrical power is added. System B (red) does not have any insulation at the back and therefore acts as a heat exchanger at night and produces heat in Figure 3b, when the ambient temperature is higher than the collector temperature. While the PV yield is in a similar range on the two days, the thermal yield depends largely on the inlet temperature of the water.

Please note, that the average collector temperature is very different for the different collectors. System C operates at higher temperatures and therefore, produces more useful heat. System A and C perform better at higher temperatures due to the insulation at the back.

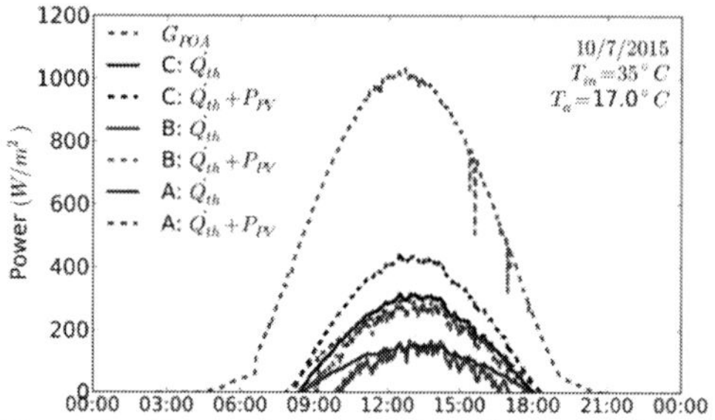

Figure 3a: Thermal (continuous lines, $\dot{Q}_{th}$) and additional electrical power output (dashed lines, $\dot{Q}_{th}+P_{PV}$) per m^2 for a day with a fluid input temperature of 35°C. In-plane irradiance (G_{POA}) is shown in green.

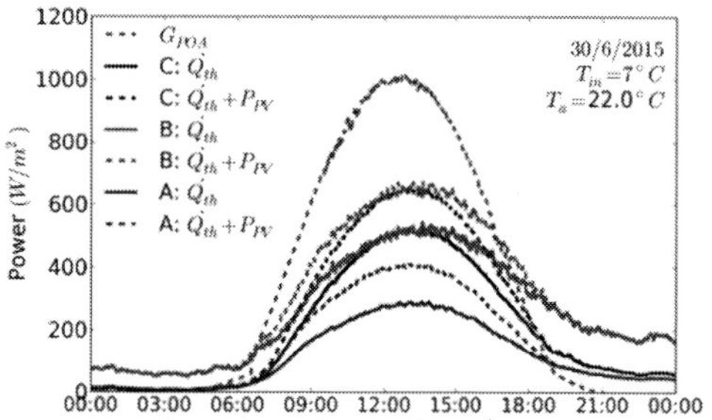

Figure 3b: Thermal (continuous lines, $\dot{Q}_{th}$) and additional electrical power output (dashed lines, $\dot{Q}_{th}+P_{PV}$) per m^2 for a day with a fluid input temperature of 7°C. In-plane irradiance (G_{POA}) is shown in green.

3.2 Thermal performance

The thermal efficiency is calculated based on measured data for a one year period from June 2015 to May 2016. The absorption of the different PVT panels was measured by ECN with an integrating sphere and was between 0.90 and 0.94 for the different panels. The emission is approximately 0.9.

The thermal collector efficiency curves for collector A, B and C are shown in Figure 4 for a wind speed of 0 and 3 m/s, with the PV in MPP. The PV efficiency (12-14%) is additional. The collectors show very different performance features. Collector C performs the best of the three measured collectors. The other collectors can also perform well in system configurations that have a low demand temperature. E.g. systems that are connected to a ground-source heat pump, often operate below ambient temperature.

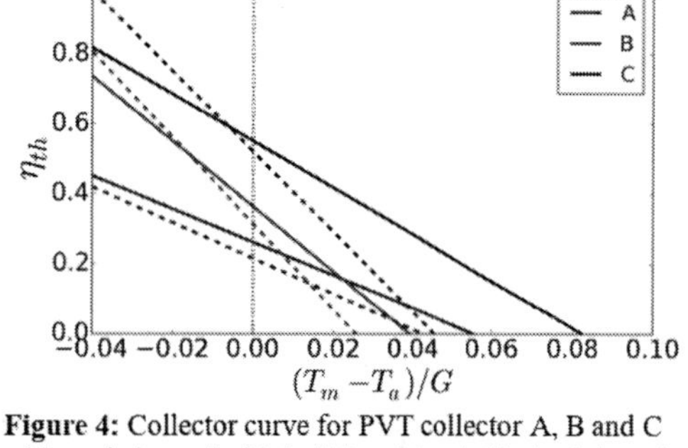

Figure 4: Collector curve for PVT collector A, B and C with a wind speed of 0 (solid) and 3 m/s (dashed), based on measured data from June 2015 to May 2016, with PV operational and measured in MPP

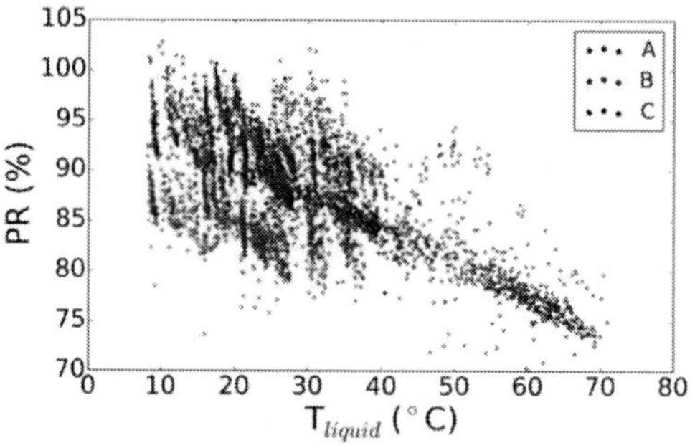

Figure 5: Correlation between the DC performance ratio and the average fluid temperature.

3.2 Overall performance

The electric performance of PVT modules is influenced by the thermal part. Figure 5 shows the correlation of the DC PV performance ratio (PR) on the average fluid temperature of the PVT collectors. It shows that the heat conduction between the PV cells and the thermal absorber of PVT collector C is good, the PR reduces with higher liquid temperatures. The mean fluid temperature is one of the main drivers of the PR for collector C. It also shows that the heat conduction between the PV cells and the thermal absorber for systems A and B is not very good.

The coefficients and the average electrical efficiency for all collectors are shown in Table I.

Table I: Performance indicators

Col-lector	η_{th}	b_u (s/m)	b_1 (W/m^2 K)	b_2 (J/m^{3}K)	η_{el} (DC)
A	26 %	0.06	4.7	0.1	12.2 %
B	37 %	0.04	9.3	1.0	13.8 %
C	55 %	0.02	6.7	1.6	12.7 %

The electrical average efficiency of the three uncovered PVT systems is between 12.2 and 13.8 %. The difference is partially caused by a different peak power per square meter. The thermal efficiency parameters are for a large part caused by the thermal conduction between the PV and the thermal collector together with the insulation on the back of the system. A better thermal contact leads to a higher η_0, but also to higher heat loss (b_1) and wind dependency of the heat loss (b_2) parameters, also if the back of the panel is insulated or integrated in the roof.

4 CONCLUSIONs

There is a large difference between the measured thermal performance of the three different collectors. These differences can be explained by the PVT collector design:

- The absorption of all PVT modules is quite good (0.9 – 0.94).
- A good heat conduction from the PV cells to the absorber leads to a high peak collector efficiency

- Insulation at the back of the PVT module improves the performance at higher temperatures.

PVT collector C performs the best, since it has a good heat conduction and insulation at the back. However, for systems that need low temperature heat, the other systems also can supply heat. Furthermore, the performance of some collectors can be improved by e.g. heat conducting paste, however, this may not lead to a techno-financial optimal situation.

System simulations with PVT as part of a solar heating system are carried out in TRNSYS and will yield valuable information on the annual yield for a specific system.

REFERENCES

[1] Zondag, H.A., Flat-plate PV-thermal collectors and systems: a review, Renewable and Sustainable Energy Reviews 12 (4) (2008) 891–959

[2] EN ISO 9806, Solar energy - Solar thermal collectors - Test methods (ISO 9806:2013)

ACKNOWLEDGEMENTS

This work is supported by the ‘Netherlands Enterprise Agency’ (Rijksdienst voor Ondernemend Nederland, RVO) and the Dutch Topteam Energy via the project: WenSDak with grant number TEZG113008.

SIMULATION OF THE LOAD FLOW AT THE TRANSFORMER IN LOW VOLTAGE DISTRIBUTION GRIDS WITH A SIGNIFICANT NUMBER OF PV SYSTEMS USING SATELLITE-DERIVED SOLAR IRRADIANCE

Holger Ruf[1,3], Marion Schroedter-Homscheidt[2], Hans Georg Beyer[3], Gerd Heilscher[1]

[1]*Ulm University of Applied Science, Ulm, Germany, ruf@hs-ulm.de, heilscher@hs-ulm.de*
[2]*Deutsches Zentrum für Luft- und Raumfahrt e.V., Oberpfaffenhofen, Weßling, Germany, marion.schroedter-homscheidt@dlr.de*
[3]*University of Agder, Grimstad, Norway, hans-georg.beyer@uia.no*

ABSTRACT: A scheme using satellite-derived irradiance measurements to model the load flow of a medium-to-low voltage transformer in a low voltage distribution grid area is described. This scheme is validated with measurements at the transformer from a test site with 21 residential PV systems in the city of Ulm, Germany. The irradiance data are provided by the SOLEMI, the Helioclim-3 and the CAMS Radiation Service. The load flow simulation model is based on a PV model taking into account nominal power and location of the PV systems as provided by the distribution system operator and orientation angles taken in high resolution from aerial laser-scan data. The load data are assumed using load profile time series from standard load profiles and measured profiles. Modell quality is assessed by a comparison of modelled and measured power. Average mean errors of power are found in the range of 14.0 % to 16.6 % and the RMSE is between 41.1 % and 42.3 % for the different input data sources. Overall, the study indicates the usefulness and applicability of satellite-derived irradiance data for utility applications like load flow monitoring of transformers in areas with a high amount of residential PV systems.
Keywords: Grid Integration, PV System, Distributed System Operators (DSO), CAMS Radiation Service

1 INTRODUCTION

The successful implementation of a feed-in tariff in Germany and many other countries has led to a growing number of small photovoltaic (PV) systems with less than 30 kWp. Recently, more than 1.5 million PV systems are connected to the electric grid and over 96 % are installed at the low voltage level in Germany [1] and do not provide feed-in time series information to the distribution system operators (DSO).

The responsibility for grid stability and power quality at this level rests on the DSO. They have to plan, operate and maintain the grid in an economic way ensuring that no voltage band violations and overloading of grid elements do occur. The instant voltage within a grid is strongly dependent on the load flow. The calculation of the load flow allows the simulation of voltage values and is a pre-requisite to monitor or model any voltage band violation risks. DSOs are currently reconsidering their traditional approaches and need more information about the influence of the volatile decentralized renewable energy to the voltage in the grids. This includes e.g. feed-in power of decentralized generators, their volatility and possible ramp rates, distribution of voltage values and probabilities for voltage band violations.

While the interconnections between the high and medium voltage grid are monitored widely in real-time and remote controllable actuator exist to influence the load flow, there are normally no measurement devices providing time series in the low voltage grid. Currently, only meters in the households and power drag-indicators in transformer stations exist. This does not solve the need for monitoring the grid state because an annual energy feed-in or demand value is not sufficient for planning and operating low voltage grids with volatile feed-in power characteristics in an economic way avoiding any unnecessary high costs as a result of grid asset oversizing.

The efficiency of new established technologies such as controllable transformers, reactive feed-in power by inverters and the related control strategies requires more time-dependent information about the behavior of low voltage grids with large amounts of PV feed-in power. PV feed-in power variability as well as power and voltage ramp rates is needed.

One option contributing to the closure of this information gap is using data on irradiance conditions taken from remote sensing technologies as suggested in [2]. With 15 minute resolved irradiance data from the Meteosat-Second-Generation (MSG) meteorological satellite a near real-time estimation of the state of a low voltage grid gets possible. Also, an assessment of historical irradiance time series at a location under the assumption of nowadays or future PV shares is possible for planning purposes, as such data exists since 1991 for any location in Europe, Africa and the Middle East.

The usage of satellite-based irradiance data has a long history and is applied in broad field of disciplines, e.g. climate, material sciences, agriculture, and most relevant in our case for solar energy siting and system planning [3]. Irradiance data from satellites is also the state-of-the-art in the management of large scale transmission systems with high a significant amount of PV penetration in Europe and U.S. [4], [5].

However, the application of such data on distribution grid level is still a topic in research and development. Several studies have been carried out to investigate the influence of PV systems to distribution grids considering meteorological parameters: Ground measurements of irradiance provide high accuracy and are often used in PV feed-in power studies, e.g. [6]. Nevertheless, such ground measurements are point measurements only and do not fully represent the natural spatial and temporal variability of a distributed PV fleet in an area as e.g. a regional utility grid. In contrast, satellite-based irradiances provide the spatial information, but with restrictions in the available spatial and temporal resolution with their km sized pixels and their 5 to 15 minute temporally resolved observation scheme. Within each pixel they provide an average value, which is known to be more representative for the situation as soon as the area of interest is larger than 20x20 km² [7], [8]. It is currently not fully known how the trade-off with respect to DSO management between spatial averaging observations as providing by the satellite and the pin point, but very accurate measurements of the pyra-nometer looks like.

In [9] the impact of PV systems on the load flows in the distribution grid on both low and medium voltage levels for two extreme cases – clear sky and extremely fluctuating conditions is calculated. The load flow simulation on medium voltage level of a PV fleet distributed over an area of 11 km x 15 km using irradiance derived from satellite and assuming a singular orientation for all systems is performed for 20 days in [10]. The influence of PV feed-in power to the voltage level in an island grid also taking into account satellite irradiance data is analysed in [11]. It is shown that the variability of voltage variations in the island grid is well described on the basis of using satellite observations. In [12] satellite irradiance data is used to derive statistic values for a given location to generate synthetic high resolution PV feed-in power by Monte-Carlo simulations. These synthetic feed-in power pro-files are used for the calculation of the hosting capacity for distributed PV in model grids, taking into account various load profiles.

However, those studies focus on larger areas and higher voltage levels. A larger area benefits from a sufficient number of both PV systems and consumers that allows the usage of statistical equivalent characteristics and lumped PV models, representing the average response of the PV fleet.

In this paper, we continue these applications of satellite data for the management of electricity grids, but on scales of interest being typical in the low voltage grid in an area served by a single transformer station. Satellite-based irradiance time series are used as input for the simulation of the load flow at a low voltage grid transformer.

The irradiance at ground level is provided by irradiance services using the Heliosat-2 and Heliosat-4 method based on satellite images. The PV feed-in power is computed with a non-linear PV model while the load is assumed by load profiles scaled to the annual consumption. The load flow at the transformer is computed as 15-minute averages and validated against measurements at the transformer.

Section 2 describes the data used for the simulation and validation. The simulation model for the load flow at the transformer is introduced in section 3. In section 4 the method is validated against measurement with respect to different satellite-based input data sources. Finally, conclusions and future directions are given in Section 5.

2 DATA SOURCE

2.1 Test site

The investigated test site is a suburb in the city of Ulm, southern Germany (Figure 1). At the test site (470 x 615 m²) 133 houses are connected via eight feeder lines to a 630 kVA medium-to-low voltage transformer

There are 21 roof-mounted residential PV systems installed with an overall nominal power Pnom = 233 kWp, ranging from 2.2 kWp to 47.84 kWp. The average nominal power per roof of 11.07 kWp is close to the average value for southern Germany [13]. The PV systems are distributed over the whole test site. APLC-based measurement device logs the voltages and load flows per second for each single feeder at the transformer with an accuracy of 4%.

Usually, the DSO only knows the location and the nominal power of the PV system. The tilt and azimuth angles were extracted by laser-scan data provided from an extended roof potential analysis for electric grid planning based on [14]. The error of these angles is assumed with less than 10°.

Figure 1: Aerial image of the test site. The border is marked with the blue polygon. Installed PV-Systems and the transformer (highlighted by the image of the trans-former tower) are marked.

2.2 Meteorological Data

Irradiance data are taken from the SOLEMI service [15] with an hourly time resolution and the HelioClim-3 data base (HC3v4) [17] and Copernicus Atmosphere Monitoring Service (CAMS) Radiation Service [16] with a 15-minute time resolution. The MSG satellite images used have a temporal resolution of 15 minutes and a spatial pixel resolution of 3 km² at the sub-satellite point. The spatial resolution at the test site is approximately 5 x 3 km². Originally, the Heliosat method was introduced by Cano et al [18] and further improved, e.g. [19]-[22]. Both HC3v4 and SOLEMI are based on the Heliosat-2 method [17] while CAMS Radiation Service uses the new Heliosat-4 approach [16].

Further meteorological data required for the simulation are ambient temperature and wind speed which are taken from the roof-mounted meteorological station operated by Ulm University of Applied Science (latitude 48.42°N, longitude 10.00°E, height above sea level 550 m). Temperature is measured with a Thies 'hygro-thermo transmitter compact'. Wind speed and direction are measured with a Thies 'windsensor classic' which is a cup-anemometer in combination with wind direction sensors. The distance to the test site is 11 km.

The distance in the measurements leads to deviations due to local effects on the air temperature as urban heat islands or roof temperature as well as wind effects based on the orography and building structures. Their influence is unknown, however according to [23] an error of 2% in air temperature leads to an error in the annual PV yield of 0.5% and 50% in wind speed at module to 1.5% in the annual PV yield, respectively. The authors assume that the approach is realistic and justified according to the general data availability practice in DSO even if the risk of inaccuracy is existing.

The investigation period for transformer simulation is 1st January to 31th December 2012 and takes SOLEMI, HC3v4 and CAMS Radiation Service into account.

The study of the PV model accuracy considering the CAMS Radiation Service is described in [24].

2.3 Load Data

For the private households in the test area standard load profiles (SLP) as well as an average profile based on 145 measured load profiles (NoSLP) [28] are assumed. Both SLP and NoSLP have a temporal resolution of 15 minutes. The NoSLP profiles are from randomly selected households in Ulm, monitored from May 2009 to April 2010. Their average is re-sorted with taking into account to the weekdays and weekends in 2012. The influence of weather conditions on the household day to day variability is negligible.

SLPs as communicated by the energy industry [25] are usually normalized to an annual consumption of 1,000 MWh and scaled to the real measured annual consumption. Due to the statistical nature, the usage of these profiles is appropriate for clusters with at least 150 households [26]. It is assumed that deviations from the German SLP will be small at the test site with the 133 houses.

The annual consumption in the area was 1,051 MWh in 2012, with a distribution of 85 % related to private households, 4 % to agricultural and 11 % to commercial activities. For the connected commercial and agricultural consumers the respective standard load profiles as given in [27] are used.

3 TRANSFORMER LOAD FLOW MODEL

The load flow at the transformer P_{Trafo} is given as residuum of the power demand P_{Load} and the PV feed-in power P_{PV}. P_{Load} is approximated using the SLP and NoSLP time series for residential customers taking into account the annual consumption and the corresponding SLP is used for both commercial and agricultural customers, respectively. The resulting load profile time series is multiplied by the annual consumption of the area which is given by the sum of the consumer energy meter values provided by the DSO [27].

The power generation by PV systems P_{PV} can be described with the knowledge of irradiance and temperature – and the knowledge of the orientation and inclination of the single PV systems, the module types and local shading effects (Figure 1). These orientation values and the local shading can be derived from a solar roof potential analysis [14]. The calculation of the 15 minute averaged feed-in power for each PV system is performed with a non-linear PV model based on the PVLib [29]. The input parameters are diffuse (DHI) and direct (BHI) irradiances on the horizontal planes as provided by the satellite products SOLEMI, HC3v4 and CAMS Radiation Service as well as the ambient temperature to calculate the cell temperature reducing the efficiency of the PV array. The detailed PV model description as well as its validation is given in [24].

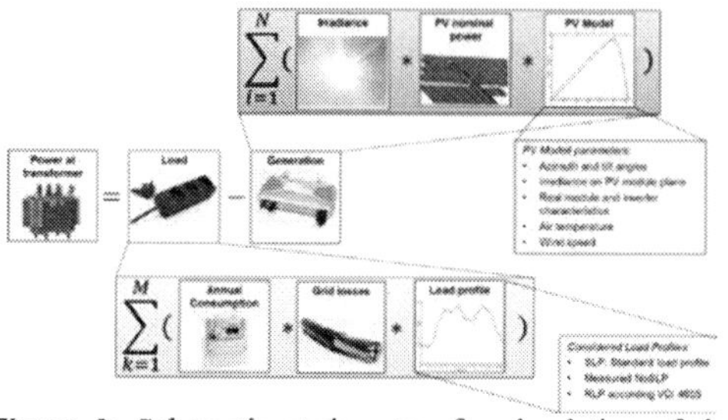

Figure 2: Schematic study setup for simulation of the medium-to-low voltage transformer load flow

The losses within the grid are estimated by a grid simulation without PV as 2.17 % of the overall annual consumption. Variation of the loss value by PV feed-in power is neglected.

4 RESULTS

4.1 Metrics

The metrics used to validate the results are mean error (ME), root-mean-squared error (RMSE) and correlation coefficient (CC) as:

$$ME = \frac{1}{N}\sum_{i=1}^{N} x_s(i) - x_m(i)$$

$$RMSE = \sqrt{\frac{1}{N}\sum_{i=1}^{N}\left(x_s(i) - x_m(i)\right)^2}$$

$$CC = \frac{\sum_{i=1}^{N}\left(x_s(i) - \overline{x_s}\right)\cdot\left(x_m(i) - \overline{x_m}\right)}{\sqrt{\sum_{i=1}^{N}\left(x_s(i) - \overline{x_s}\right)^2 \cdot \sum_{i=1}^{N}\left(x_m(i) - \overline{x_m}\right)^2}}$$

with N as the total number of measurement data points, $x_s(i)$ as simulated data at time i and $x_m(i)$ as measured data at time i. The relative values of ME and RMSE are rated to the absolute load flow value at the transformer.

4.2 Transformer load flow

As example, the result for 3rd August 2012 is shown in Figure 3. There is no deviation between measurement and calculation between 0:00 to 2:00 UTC and 21:00 to 24:00 UTC. In the morning hours after 2:00 UTC the calculated active power increases following the load profile while the measurement is almost constant at 50kW. At approximately 7:00 UTC the difference between the assumed load profile and the simulation increases because of the increasing impact of the PV feed-in power. At the same time the measured active power decreases as well. The measurement and the calculation reach the zero line at around 8:30 UTC for the first time. The following negative active power values represent a so-called reverse load flow. The PV systems feed-in more power than consumed in the test site. This state changes several times during the day depending on the amount of solar irradiance. The PV feed-in power decreases again in the afternoon following the diurnal cycle. Nevertheless, the standard load profile remains too high. There is a difference between simulation and measurement probably because the true consumption is lower than assumed. The calculated active power and assumed load profile converge at around 19:30 UTC.

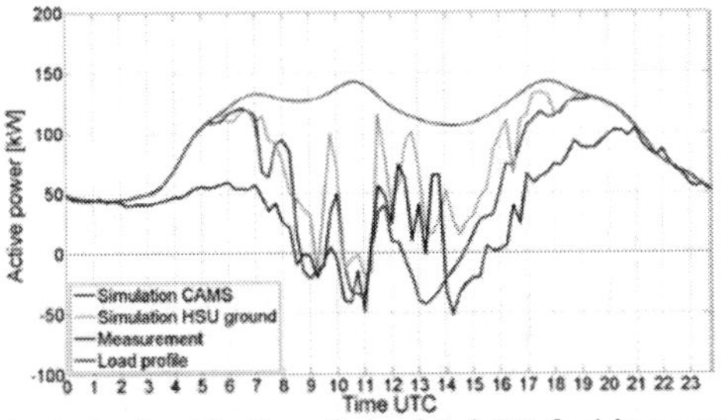

Figure 3: Combination of simulated PV feed-in power and assumed standard load profile on 3rd August 2012 compared with measurements at the transformer. The blue line is the simulated active power at the transformer using CAMS Radiation Service, the green line is the simulation using the ground measured irradiance and the black line is the measurement. The red line indicates the assumed standard load profile for this day.

The average of each 15-minute interval of the simulated year 2012 is illustrated in Figure 4. The proposed model underestimates the consumption during night and overestimates the consumption during day. Higher deviations occur during morning and evening hours. However, during these hours the PV feed-in power is low because of the low GHI values. Therefore, these deviations are caused by the used load profiles. Both SLP and NoSLP results show the similar demand overestimation in the evening while the hours before noon are much closer to the measured load profile. The use of NoSLP from measurements already in the Ulm area obviously is not solving the issue fully. Dealing with this mismatch between the general and statistical averaging nature of standard load profiles and the real conditions in our selected residential area is an important research question, but beyond the scope of this study.

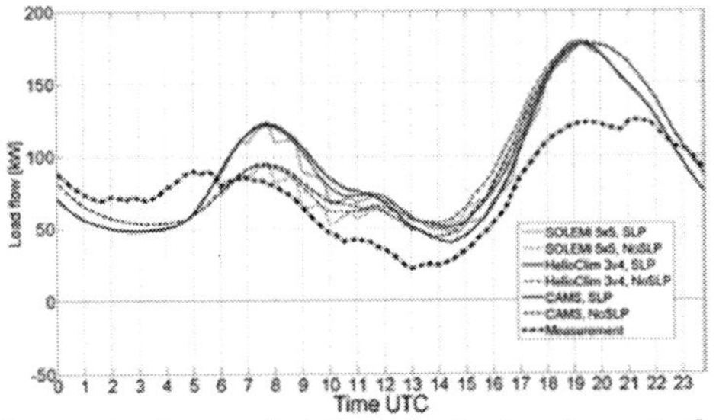

Figure 4: Averaged 15-minute load values at the transformer of the year with simulated (coloured) and measured (chain dotted line) values

The relative errors of the transformer load flow rME and rRMSE are related to the absolute measured loading of the transformer (Table I and Table II). The load flow scheme shows a positive bias in the range of 14.0 % to 16.6 % using both SLP and NoSLP considering the overall time period. The bias increases to 44.3 % to 51.9% for SLP and 35.1% to 42.6% for NoSLP, respectively for daylight hours (sun elevation angle larger than 0°). If the PV feed-in is not taken into account the rME is 63.8 %, rRMSE 153.9 % and CC of 0.35 for the overall day and to rME of 143.1 %, rRMSE 164.7 % and CC of 0.30 for the period with sun elevation angles above 0°, respectively.

Table I: Statistical measures for simulation considering all sun elevation angles (day and night)

All sun elevation angles	SLP	NoSLP
SOLEMI		
rME [%]	14.0	14.0
rRMSE [%]	41.1	40.7
CC	0.97	0.97
HC3v4		
rME [%]	16.6	16.6
rRMSE [%]	41.4	41.3
CC	0.97	0.97
CAMS Radiation Service		
rME [%]	14.9	14.9
rRMSE [%]	42.3	41.0
CC	0.97	0.97

Table II: Statistical measures for simulation considering only sun elevation angles >0° (daylight hours)

All sun elevation angles	SLP	NoSLP
SOLEMI		
rME [%]	44.3	35.1
rRMSE [%]	66.6	62.2
CC	0.87	0.83
HC3v4		
rME [%]	51.9	42.6
rRMSE [%]	67.3	63.6
CC	0.88	0.84
CAMS Radiation Service		
rME [%]	46.9	37.6
rRMSE [%]	69.8	62.8
CC	0.86	0.84

The variation of the statistical measures versus the time is shown in Figure 5.

The irradiance measurement sources do not cause large deviations. The oscillation effect of SOLEMI due to the hourly time resolution is visible in the ME. CAMS Radiation Service shows a tendency to lower ME values during the afternoon hours. These ME curves are the combination of the diurnal cycle of the PV feed-in power and the mismatch between the load profiles and the measurement. The ME of SLP and NoSLP is below 0 kW during the night, that is, an underestimation of the load flow. The SLP profile shows a higher ME before noon than NoSLP. However, the NoSLP profile has a higher ME during the afternoon than SLP.

There are minor deviations in the RMSE curves because of the variation of the irradiance measurement sources. The RMSE considering CAMS Radiation Service is slightly higher in the morning than when considering SOLEMI and HC3v4. The SOLEMI RMSE curve also shows the oscillation effect of the interpolation. In the afternoon the RMSE of HC3v4 is higher than that of SOLEMI and CAMS Radiation Service. The RMSE of SLP and NoSLP increases during the morning hours and falls into a trough. This trough is caused by the change of the ME from negative to positive while the standard deviation is still low. Before noon the SLP has higher values of RMSE than NoSLP. Around noon all profiles shows similar amplitude. In the afternoon the RMSE of the SLP drops to a local minimum while the RMSE of the NoSLP increases. All profiles have an additional peak in the evening. These RMSE curves are the combination of the diurnal cycle of the PV feed-in power and the mismatch between the load profiles and the measurement. The morning and evening peaks are not a result of the PV feed-in power because of its low impact related to assumed consumption by the load profiles at these times.

Basically, the CC also follows the diurnal cycle. During the day the CC is in a range of 0.6 to 0.8 and during the night in the range of -0.2 to 0.4 for both SLP and NoSLP. The effects of the different irradiance measurement sources lead to small effects during the day. In terms of CC the best performance has HC3v4 followed by SOLEMI and CAMS Radiation Service.

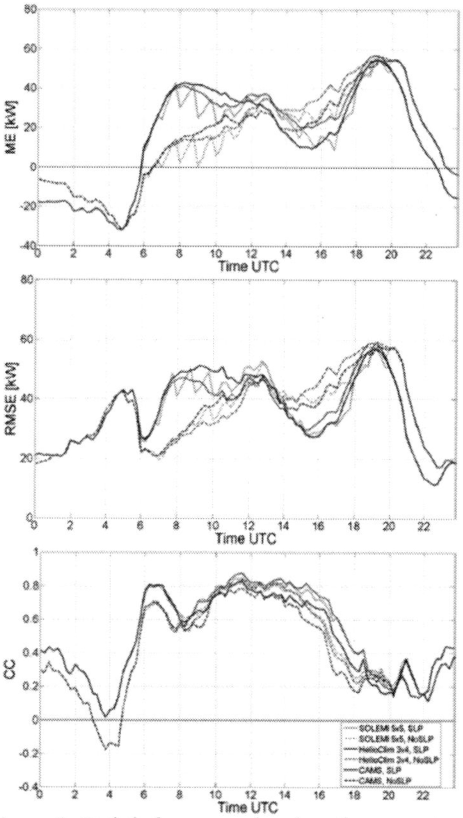

Figure 5: Statistical measures based on the comparison of calculated annualized average load flow time series against the annualized average measurement. The irradiance data sources are colour-coded (SOLEMI is green, HC3v4 is red, CAMS is blue). The applied load profiles are marked with the different line types (SLP is solid, NoSLP is dashed).

5 CONCLUSION AND OUTLOOK

The usefulness of satellite derived irradiance data for transformer load flow calculations of a low voltage transformer in a small-scale distribution grid is investigated.

The calculation scheme proposed uses a PV model considering the nominal PV power, the location and the orientation angles as site-specific input parameters. As the orientation angels are normally not available from the DSO and have to be provided by external data sources as e.g. airborne laser-scan data as done in this study. The neglect the PV system orientation will lead to higher coincidence factors of the PV feed-in power. The time series of the load is assumed with standard load profiles and measured profiles scaled to the annual consumption at the test site.

The assumed load profiles have a strong influence to the results. Especially, the deviation during morning and evening hours, with less PV influence, leads to high error values. The SLP shows higher errors before noon. Dealing with this mismatch between the general and statistical averaging nature of standard load profiles and the real conditions in our selected residential area is an important research question, but not in the scope of this study.

The approach introduced will be extended to a load flow calculation of the connected grid by using detailed grid structure information. This allows the calculation of voltage values.

In view of predictive future grid operations, the scheme can work with satellite-based irradiance data in the planning phase. Additionally, in online monitoring applications satellite-based nowcasting by using cloud-motion-vectors from satellite scenes can be applied, before calculating the irradiance.

These methods will support grid operators in the strategic planning of their infrastructure re-enforcement in areas with already increased PV shares. Based on multi-annual historic irradiance databases it will allow simulating larger PV shares being expected in the future. Hence, priority setting in grid enforcement decisions can be based on the detailed basis of realistic and both spatially and temporally resolved input data. Additionally, the use of near real-time satellite information will allow DSOs the monitoring of their assets in near real-time and for short-term predictions in daily grid operations.

6 ACKNOWLEDGEMENT

The authors would like to thank Florian Meier at Stadtwerke Ulm/Neu-Ulm Netze GmbH for providing smart meter measurements and electric grid-related data as well as discussions about the DSO interests.

The research leading to these results has received funding from the European Unions´s Seventh Framework Programme (FP7/2007-2013) under grant agreement No. 608930 (project OrPHEuS), and from the German Federal Ministry of Education and Research support code 03FH030I3 (project NATHAN-PV).

7 REFERENCES

[1] DGS, EEG Anlagenregister: Data 03/2015. [Online] Available: http://www.energymap.info/ download.html (2015, Apr. 24)
[2] www.endorse-fp7.eu
[3] M. Lefevre, P. Blanc, B. Espinar, B. Gschwind, L. Menard, T. Ranchin, L. Wald, L. Saboret, C. Thomas, and E. Wey, "The HelioClim-1 Database of Daily Solar Radiation at Earth Surface: An Example

of the Benefits of GEOSS Data-CORE," Selected Topics in Applied Earth Observations and Remote Sensing, IEEE Journal of, vol. 7, no. 5, pp. 1745–1753, 2014.

[4] J. Kühnert, E. Lorenz, and D. Heinemann, "Satellite-Based Irradiance and Power Forecasting for the German Energy Market," in Solar energy forecasting and resource assesments, J. Kleissl, Ed, Oxford, Waltham: Academic Press, Elsevier, 2013, pp. 267–297.

[5] D. Renné, "Emerging Meteorological Requirements to Support High Penetrations of Variable Renewable Energy Sources: Solar Energy," in Weather matters for energy, A. Troccoli, L. Dubus, and S. E. Haupt, Eds, 2014, pp. 257–273.

[6] Y. Ueda, K. Kurokawa, K. Kitamura, M. Yokota, K. Akanuma, and H. Sugihara, "Performance analysis of various system configurations on grid-connected residential PV systems," Solar Energy Materials and Solar Cells, vol. 93, no. 6–7, pp. 945–949, http://www.sciencedirect.com/science/article/pii/S0927024808004212, 2009.

[7] Perez R., Seals R., Zelenka A., 1997. Comparing satellite remote sensing and ground network measurements for the production of site/time specific irradiance data. Solar Energy, 60, 89–96

[8] Zelenka A., Perez R., Seals R., Renne, D., 1999. Effective accuracy of satellite-derived hourly irradiances. Theoretical and Applied Climatology, 62, 199–207

[9] R. Pardatscher, R. Witzmann, G. Wirth, G. Becker, M. Garhamer, and J. Brantl, "Untersuchung zu den Auswirkungen von Photovoltaikeinspeisung auf das Nieder- und Mittelspannungsnetz" Internationaler ETG-Kongress 2011: Proceedings.

[10] L. Grossi, G. Wirth, E. Lorenz, A. Spring, and G. Becker, "Simulation of the Feed-in Power of Distributed PV Systems," in 29th European PV Solar Energy Conference and Exhibition: Proceedings, Amsterdam, The Netherlands,2014, pp. 2916–2920.

[11] E. Rikos, S. Tselepis, C. Hoyer-Klick, and M. Schroedter-Homscheidt, "Stability and Power Quality Issues in Microgrids Under Weather Disturbances," Selected Topics in Applied Earth Observations and Remote Sensing, IEEE Journal of, vol. 1, no. 3, pp. 170–179, 2008.

[12] C. Bucher, J. Betcke, G. Andersson, B. Bletterie, and L. Küng, "Simulation of Distribution Grids with Photovoltaics by Means of Stochastic Load Profiles and Irradiance Data," in 27th European Photovoltaic Solar Energy Conference: Proceedings, Frankfurt a. M, Germany, 2012, pp. 3795–3800.

[13] G. Wirth, A. Spring, G. Becker, R. Pardatscher, M. Lödl, R. Witzmann, J. Brantl, M. Garhamer, and H. Wagenhäuser, "Felduntersuchung der Netzauslastung und Spannungsanhebung durch PV-Anlagen," in Proceedings of the 26. Symposium Photovoltaische Solarenergie, Kloster Banz, Bad Staffelstein, Regensburg, 2011.

[14] H. Ruf, D. Funk, F. Meier, and G. Heilscher, "Bestimmung von Ausrichtungs- und Neigungswinkeln von Bestandsanlagen mittels LIDAR Daten," in Proceedings of the 30. Symposium Photovoltaische Solarenergie, Kloster Banz, Bad Staffelstein, Regensburg, 2015.

[15] https://wdc.dlr.de/data_products/SERVICES/SOLARENERGY/description.php

[16] Z. Qu et al, "Fast radiative transfer parameterisation for assessing the surface solar irradiance: The Heliosat-4 method," Submitted to Meteorologische Zeitschrift, 2016.

[17] C. Rigollier, M. Lefèvre, and L. Wald, "The method Heliosat-2 for deriving shortwave solar radiation from satellite images," Solar Energy, vol. 77, no. 2, pp. 159–169, 2004.

[18] D. Cano, J. Monget, M. Albuisson, H. Guillard, N. Regas, and L. Wald, "A method for the determination of the global solar radiation from meteorological satellite data," Solar Energy, vol. 37, no. 1, pp. 31–39, 1986.

[19] H. G. Beyer, C. Costanzo, and D. Heinemann, "Modifications of the Heliosat procedure for irradiance estimates from satellite images," Solar Energy, vol. 56, no. 3, pp. 207–212, 1996.

[20] L. F. Zarzalejo, J. Polo, L. Martín, L. Ramírez, and B. Espinar, "A new statistical approach for deriving global solar radiation from satellite images," Solar Energy, vol. 83, no. 4, pp. 480–484, 2009.

[21] R. Mueller, K. Dagestad, P. Ineichen, M. Schroedter-Homscheidt, S. Cros, D. Dumortier, J. Olseth, G. Piernavieja, C. Reise, L. Wald, and D. Heinemann, "Rethinking satellite-based solar irradiance modeling - The SOLIS clear-sky module," Remote Sensing of Environment, vol. 91, no. 2, pp. 160–174, 2004.

[22] A. Oumbe, Z. Qu, P. Blanc, M. Lefèvre, L. Wald, and S. Cros, "Decoupling the effects of clear atmosphere and clouds to simplify calculations of the broadband solar irradiance at ground level," Geoscientific Model Development Discussions, vol. 7, no. 2, pp. 2007–2032, 2014.

[23] S. Krauter, P. Grunow, A. Preiss, S. Rindert, and N. Ferretti, "Inaccuracies of input data relevant for PV yield prediction," in 2008 33rd IEEE Photovolatic Specialists Conference (PVSC), pp. 1–5.

[24] H. Ruf, M. Schroedter-Homscheidt, H. G. Beyer, and G. Heilscher, "Quantifying residential PV feed-in power in low voltage grids based on satellite-derived irradiance data with application to power flow calculations," Submitted to Solar Energy, 2016.

[25] www.bdew.de

[26] K. Engels, Probabilistische Bewertung der Span-nungsqualitat in Verteilungsnetzen, 1st ed. Aachen: Klinkenberg Verlag, 2000.

[27] www.ulm-netze.de

[28] G. Heilscher, M. Gmehlin, J. Conrad, and A. Ehm, "Concept for Large-Scale-Deployment of Grid Connected PV in the Private Sector at the Edge of Grid Parity," in 25th World Conference on Photovoltaic Energy Conversion: Proceedings, Valencia, Spain, 2010, pp. 5305–5308.

[29] J. S. Stein, "The photovoltaic Performance Modeling Collaborative (PVPMC)," in Photovoltaic Specialists Conference (PVSC), 2012 38th IEEE: Date, 3-8 June 2012, Piscataway, NJ: IEEE, 2012, pp. 003048–003052.

STUDY ON OPTIMAL INSTALLED CAPACITY OF PHOTOVOLTAIC AND BATTERY TO MINIMALIZE TOTAL COST IN FACTORY

Yoshiki MINAMISHIMA, Satoshi TAKAYAMA, Atsushi ISHIGAME
Osaka Prefecture University
1-1 Gakuen-cho, Naka-ku, Sakai, Osaka 599-8531, Japan
Phone: +81 72 252 1161, e-mail: swb01164@edu.osakafu-u.ac.jp
Masayasu TAKEUCHI
NISSHIN ELECTRIC CO. LTD.
47, Umezu-Takase-cho, Ukyo-ku, Kyoto 615-8686, Japan

ABSTRACT: In this paper, a method of scheduling scheme for the electric supply of Factory Energy Management System (FEMS) is presented. Additionally, optimal installed capacity of photovoltaic (PV) cells and battery to minimalize total cost is calculated. PV output including uncertainty is forecasted by using Just-In-Time Modeling (JIT Modeling), and scheme for the electric supply is scheduled based on the forecasted data. Furthermore, simulation for scheduling scheme is carried out with proposed method in various capacity of PV cells and battery, and optimal capacity to minimalize total cost including each initial cost is calculated.
Keywords: Battery Storage and Control, Photovoltaic, Cost reduction

1 INTRODUCTION

Recently, renewable energy (RE) such as PV is installed greatly in Japan promoting a new buying and selling rule. This is because, combination operations of distributed generation (DG) held by user and demand load is necessary. One of the purpose of the optimal management is to produce the minimum energy cost. Previous research has been somewhat focused in user satisfaction [1] and minimum CO2 emissions [2].

However, most research has focused on Home Energy Management System (HEMS) and Building Energy Management System (BEMS) rather than on FEMS. FEMS has features that more capacity of PV cells and battery can be installed than in HEMS and BEMS, and the basic charge of electricity has a big impact on total cost. The basic charge of electricity is calculated by annual peak electric power demand. This rate system is applied in many countries [3]. However, electric power supply facilities are equipped based on peak electric power demand. As a result, increase of power consumption causes increase of annual basic charge of electricity under 30-Minute Balancing Rule. Therefore, in order to reduce total cost, it is necessary that maximum demand power defined as mean power in 30 minutes is reduced by using DG in facilities.

In this paper, it is assumed that FEMS has the four types of suppliers which are power grid, PV, battery, and cogeneration system (CGS) using the liquefied natural gas (LNG). The scheme for electric supply is generally scheduled in the previous day. However, PV output depends on the weather of the day. Therefore, the day-ahead scheduling must be based on PV forecast output. In this paper, we propose the method of the day-ahead scheduling by using predicted solar irradiance with JIT Modeling [4].

On the other hand, this method has risk that if real output of PV is less than forecast output. As a result, there is a possibility that the maximum demand power exceeds that of scheduled in the previous day. In order to solve this problem, we propose the method with robustness against prediction error of PV output by estimating probability density function (PDF) of solar irradiance [5]. Specifically, probabilistic restriction can enable scheduling which PV real output is more than plan output with high probability. In comparison with the previous method, this method has the merit which the maximum demand power in the day-ahead scheduling is hardly exceed on the day. However, there is the demerit which the target value of the maximum demand power becomes higher if probabilistic restriction is too strictly. From these points, this paper compare the total cost between the day-ahead scheduling by using predicted solar irradiance and estimating PDF of solar irradiation.

In this paper, proposed method are evaluated through the computational simulation in various capacity of PV cells and battery, and optimal capacity to minimalize total cost including each initial cost is calculated.

2 SYSTEM MODEL

2.1 System model

System model in this paper is shown in Fig.1. Power resources in the factory are the CGS, PV, and battery. These three resources and power grid supply the electricity to the loads in the factory. It is assumed that the reverse power flow is not sold. Accordingly, if the PV output is larger than demand, surplus power must be absorbed by the battery. If battery cannot absorb surplus power, electric supply and demand are adjusted by the PV output curtail under 30-Minute Balancing Rule.

2.2 Prediction of PV output

In this research, we proposed FEMS control using the PV output forecast, and used JIT Modeling as forecast method. MSM-GPV [6] is used as the input data for JIT Modeling. MSM-GPV data is meteorological numerical

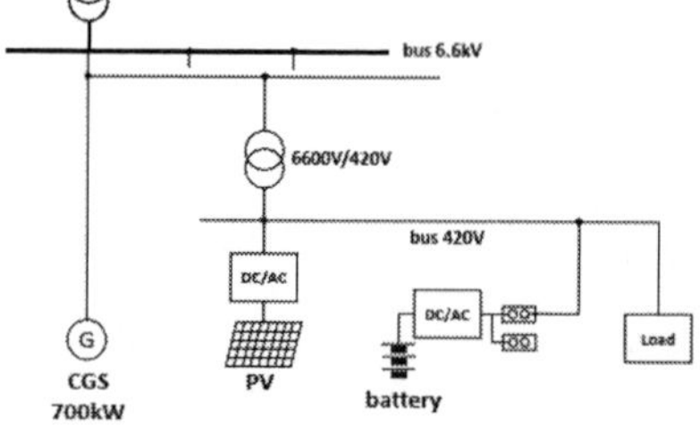

Fig.1. System model for FEMS

forecast model based on a physical model. It is calculated in intervals of 3 hours, and consists of hourly physical predicted values for 33 hours in advance at each node of a 5-km grid throughout Japan. However, in this research, day-ahead scheduling is built under 30-Minute Balancing Rule, the average of sequential two data complement the data of half hour. MSM-GPV is consisted temperature, low cloud amount, middle cloud amount, upper cloud amount, and sun altitude. Actual solar irradiance data is in accordance with the data of Maebashi Local Meteorological Observatory in Japan.

In predicting solar irradiance with JIT, the degree of similarity between the predicted data of target time and the past data in the database is evaluated by the Euclidean distance. Here, the number of neighbors is set as K = 20. We refer to [4], and the method of calculating the output can be written as:

$$y^q = \left(\sum_{j=1}^{K} w_j y_j\right) \Big/ \left(\sum_{j=1}^{K} w_j\right) \tag{1}$$

$$w_j = \left(1 - \left(\frac{d_j}{d_K}\right)^3\right) \tag{2}$$

where y^q is the forecasted PV output for the predicted data of target time, y^j is the output in the neighborhood, d_j is a distance of the predicted data of target tune and the neiborhood, and d_K is the distance of the the data of predicted hour and the neiborhood which is the farthest.

PV output by calculating solar irradiance is:

$$s_{PV} = C_{PV} \times s_{solar} \times \eta_{season} \times \eta_{PCS} \tag{3}$$

where s_{PV} is the PV output, C_{PV} is the capacity of PV, s_{solar} is the solar irradiance, η_{season} is the seasonable efficiency, and η_{PCS} is the PCS efficiency.

2.3 Estimating PDF of PV output

We refer to [5], and the method of estimating PDF of PV output can be written as:

$$\hat{p}(y) = \frac{1}{K}\sum_{j=1}^{K} w_j g(y - y_j) \tag{4}$$

where $\hat{p}(y)$ is the estimated PDF and $g(x)$ is the kernel function. Setting of kernel function and band width is referred to the reference [5].

PDF of PV output can be represented by:

$$P_{PV} = P_{solar} \times C_{PV} \times \eta_{season} \times \eta_{PCS} \tag{5}$$

where P_{PV} is the estimated PDF of PV output and P_{solar} is the estimated PDF of solar irradiance.

3 APPROACH

The day-ahead scheduling is decided based on load demand prediction and PV output prediction of the next day by the optimized calculation in order to minimalize the power generation cost of the day. However, because the load of this factory remains almost unchanged on weekdays, this paper simulates on the presupposition that load is already known.

The scheme for battery charge at night and discharge at peak hours shall not change from the day-ahead scheduling. However, if surplus power of PV is generated, battery charges surplus power without regard to the day-ahead scheduling. Moreover, electricity from power grid and CGS output shall be able to change in accordance with the load.

By the way, a monthly basic charge of electricity is decided based on the maximum demand power. Accordingly, the cost including basic charge of electricity cannot be minimalized by only minimalizing the power generation cost. In order to solve this problem, a constraint condition is provided.

3.1 The optimization problem

(a) Formulated objective function

According to 30-Minute Balancing Rule, we set the number of time section as T=48, and the power generation cost is minimalized of the day. Therefore, the objective function can be formulated as:

$$\min \sum_{t=1}^{T} \left\{ c(t) s_{grid}(t) + c_g s_{CGS}(t) + \frac{c_{min}}{\eta_{in}\eta_{out}} s_{bat}^{out}(t) \right\} \tag{6}$$

where $c(t)$ is the electric rate at t, $s_{grid}(t)$ is the electricity from power grid at t, c_g is the cost of gas per unit energy, $s_{CGS}(t)$ is the CGS output at t, c_{min} is the minimum cost per unit energy, η_{in} is the charging efficiency of battery, η_{out} is the discharging efficiency of battery, and $s_{bat}^{out}(t)$ is the amount of battery discharging at t.

(b) 30-Minute Balancing Rule

Demand and supply must be balanced in half hour. This rule is written as:

$$s_{grid}(t) + s_{CGS}(t) + s_{bat}^{out}(t) + s_{PV}^{f}(t) = d(t) + s_{bat}^{in}(t) \tag{4}$$

where $s_{PV}^{f}(t)$ is the PV forecast output at t, $d(t)$ is the demand at t, and $s_{bat}^{in}(t)$ is the amount of battery charging at t.

(c) CGS restriction of activation stop times

In CGS, if waste heat is utilized, the cost of gas per unit energy is generally smaller than the electricity from power grid. However, CGS must be performed periodic inspection. Therefore, in this paper, if equation (8) is satisfied, CGS is activated.

$$d(t) - s_{PV}^{f}(t) \geq s_{grid}^{max} \tag{8}$$

where s_{grid}^{max} is the maximum demand power so far. If the net load which is subtracting PV output from the load is larger than the maximum demand so far, CGS is activated.

(d) Restriction of battery charging and discharging

In order to reduce the basic charge of electricity, peak power demand must be reducing. Restriction of battery charging and discharging is described by:

$$s_{bat}^{out}(t) + s_{CGS}^{max} + s_{PV}^{f}(t) \geq d_{goal} \tag{9}$$

where s_{grid}^{max} is the CGS maximum output, and d_{goal} is the target value which is established by the capacity of battery and the naximum demand power so far. d_{goal} is the maximum value which is satisfied with:

$$\sum_{t=1}^{T} \max\{0, d(t) - s_{grid}^{max} - s_{CGS}^{max} - s_{PV}^{f}(t)\} \leq C_{bat}(SOC_{max} - SOC_{min}) \quad (10)$$

where C_{bat} is the capacity of battery, SOC_{max} is the maximum SOC limit, and SOC_{min} is the minimum SOC limit. That is to say, if CGS is operated with maximum output but the maximum demand is larger than so far, battery discharging complement insufficient power. However, if battery discharging cannot complement all insufficient power due to the restriction of battery capacity, battery use for peak load leveling within the capacity region.

(e) Battery charging

Battery is charged at night as N=18 in accordance with the amount of battery discharging in optimal solution. Therefore, less charge-discharge cycle can be realized. Specifically, the amount of battery discharging can be written as:

$$s_{bat}^{in}(t) = \left(\sum_{t=1}^{T} s_{bat}^{out}(t)\right) / N \quad (11)$$

$$s_{grid}^{charge}(t) = \frac{s_{bat}^{in}(t)}{\eta_{in}} \quad (12)$$

where $s_{grid}^{charge}(t)$ is the surplus the electrical power from power grid at t.

Moreover, if the PV output is larger than the load, surplus power is charged battery based on:

$$s_{bat}^{in}(t) = \left(s_{PV}^{real}(t) - d(t)\right) * \eta_{in} \quad (13)$$

(e) Transition of SOC

Transition of SOC of the battery can be written as:

$$SOC(t+1) = SOC(t) + \frac{\eta_{in} s_{bat}^{in}(t) - s_{bat}^{out}(t)}{C_{bat}} \quad (14)$$

where $SOC(t)$ is the SOC of the battery at t.

(f) Restriction of capacity

Restrictions of capacity in regard to CGS output and SOC of the battery are defined as:

$$0 \leq s_{CGS}(t) \leq s_{CGS}^{max} \quad (15)$$

$$SOC_{min} \leq SOC(t) \leq SOC_{max} \quad (16)$$

3.2 Probabilistic restriction

In case of using estimated PDF of PV output, equation (8) and (9) replace equation (17) and (18).

$$d(t) - P_{PV}^{f}(t) \geq s_{grid}^{max} \quad (17)$$

$$s_{bat}^{out}(t) + s_{CGS}^{max} + P_{PV}^{f}(t) \geq d_{goal} \quad (18)$$

where $P_{PV}^{f}(t)$ is the PV output as randam variable at t. The probability distribution function of PV output is defined as:

$$F_{PV}^{f}(t)\left(d_{goal} - s_{bat}^{out}(t) - s_{CGS}^{max}\right) = P\left(d_{goal} - s_{bat}^{out}(t) - s_{CGS}^{max} \leq P_{PV}^{f}(t)\right) \quad (19)$$

Then the restriction (18) can be represented as the probabilistic restriction:

$$F_{PV}^{f}(t)\left(d_{goal} - s_{bat}^{out}(t) - s_{CGS}^{max}\right) \geq \alpha_{ks} \quad (20)$$

In other words, equation (20) is the restriction which the amount of battery discharging must be higher than being able to reduce the peak power more than target value over probability α_{ks}. Also, equation (17) can be represented as the probabilistic restriction.

3.3 Change the scheduling

In regard to the electrical power from power grid and CGS output, day-ahead scheduling is changed based on Fig.2 in accordance with prediction error of PV output. s_{PV}^{dif} is difference between PV real output and PV plan output. Specifically, if PV real output is larger than plan output, the electrical power from power grid is reduced by the amount of error. If PV real output is smaller than plan output, we confirm whether or not electrical power from power grid can be operated as scheduled by changing CGS output from day-ahead scheduling. If CGS cannot complement all amount of the error, CGS output change to maximum output, and electrical power from power grid provide necessary supplies. In this case, there is a risk that we buy electricity beyond target value.

Moreover, if the battery cannot be charged because of restriction (16), PV is shut off and 30-Minute Balancing Rule is satisfied.

4 SIMULATION

4.1 Simulation conditions

In various capacity of PV cells (0 to 2250kW by a step of 15kW) and battery (0 to 1500kWh by a step of 10kWh), are simulated by using proposed method. Moreover, in each case, total cost is caluclated, and we investigate optimal capacity to minimalize total cost. Then the simulation period is one year. PV equipment cost is refered to [7], and life span is 18 years. Battery cost is 50,000 JPY per kWh, and life span is 12 years.

Basic charge of electricity is decided by the annual maximum demand, and is defined as:

$$\left(c_{base} * s_{grid}^{max}\right) * 12 \quad (19)$$

where c_{base} is basic charge of electricity per kW.

Table.1. shows the simulation condition. The load using in the simulation is based on the real load of the factory in 2014. In JIT Modeling, we built a database based on measured data from 2010/1 to 2013/12 in Maebashi, and predicted PV output of 2014/1 to 2014/12 at the same point. In this simulation, day-ahead scheduling is built by using PV forecast output in 2014, and real time operation is simulated based on PV real output in 2014.

4.2 Simulation Result

First, the minimum total cost and the optimal capacity of PV cells and battery to give the minumum total cost are

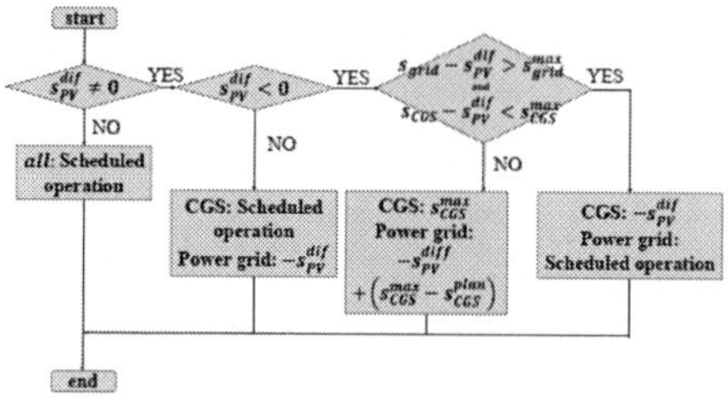

Fig.2. Control rule for change of plan

Table.1 System parameter and energy cost

Power grid	c=17.13 JPY/kWh, c_{base}=1782 JPY/kW
CGS	15.18 JPY/kWh s_{CGS}^{max}=700kW
PV	η_{season}=0.85(winter), 0.8(spring) 0.85(summer),0.9(autumn), α=0.95
Battery	η_{in}=η_{out}=0.85, SOC_{min}=0.1, SOC_{max}=0.9

shown in Table.2. Moreover, the total costs without PV cell and battery are also shown in Table.2. Proposed method II is the case which day-ahead scheduling is built by using fomula (1). Proposed method II is the case which day-ahead scheduling is built by using fomula (4). In method II, probavility α_{ks} is 0.2. Therefore, PV real output is larger than forecast output over 80%. The total cost with method I and with method II are almost same, and total cost is reduced about 4 million JPY using proposed method in comparison with not introducing PV cells and battery. Moreover, the total cost is almost same as shown in Table.2. However, the optimal capacity of PV cells and battery with method II is larger than with method I. In order to examine the reason, we show Fig.3, Fig.4, and Fig.5. Fig.3 and Fig.4 show measured value and planned value of one day. The planned value of Fig.3 is calculated with method I. The planned value of Fig.4 is calculated with method II. Method I has less error between measured value and planned value. On the other hand, time interval when planned value exceed measured value is long in Method I. This is because, if same target value is set, method I can schedule with less capacity of battery than method II. In addition, Fig.5 shows the difference in total cost between method I and method II in each capacity of PV cells and battery. Plus means that total cost with method II is smaller than with method I. From Fig.5, in case of large capacity of PV cells and battery, method II with robustness against error is effective. In case of large capacity of PV, if planned value exceeds measured value, deviation quantity from a target value is larger. This is because, if the capacity of battery is enough, method II which has lower probability of exceeding target value can reduce maximum demand.

One day simulation result in the optimal capacity with method II is shown in Fig.6. The negative output of battery mean that battery is charged. Fig. 6 shows that battery is charged at night in accordance with the amount of battery discharging on the next day. This is because, this method can realize once a day charge-discharge cycle. However, in t=17 (number of time section), surplus power of PV is generated but battery cannot be absorbed due to SOC restriction. Accordingly, PV must be shut off in this time.

Finally, the total cost in various capacity of PV cells and battery with method II is shown in Fig.7. When we

Table.2 Comparison of total cost

	PV (kWh)	Battery (kW)	Total cost (million JPY)
Not installed PV and battery	0	0	109.36
Proposed method I	1050	510	105.56
Proposed method II	1185	570	105.75

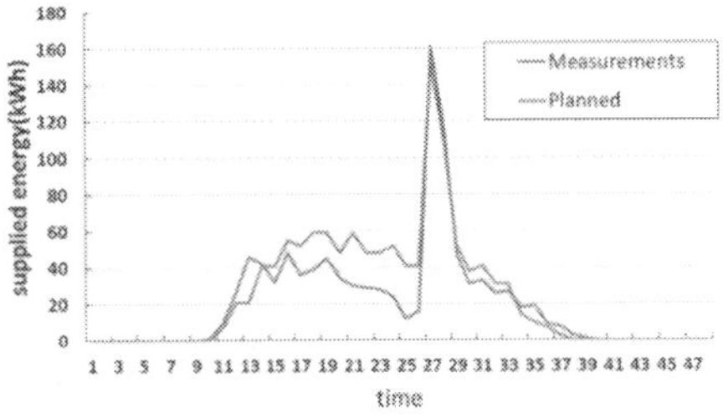

Fig.3. Prediction of PV output from (1)

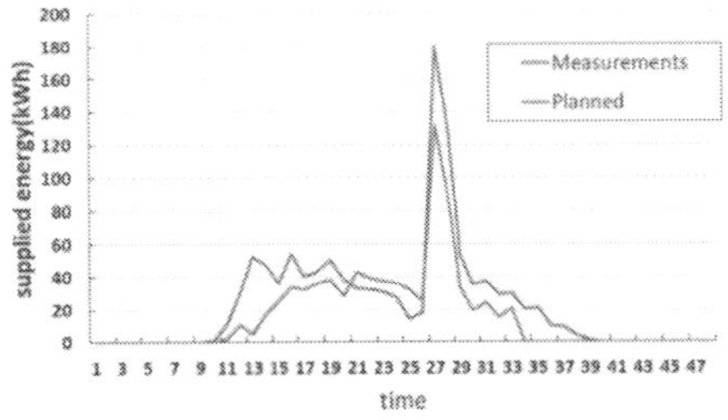

Fig.4. Prediction of PV output from (2)

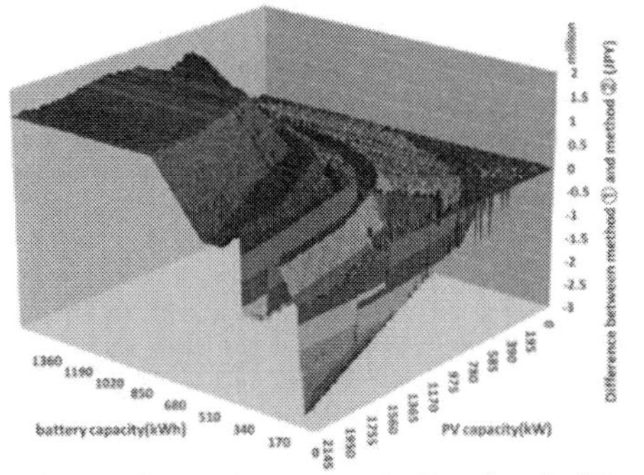

Fig.5. Difference between methodI and methodII

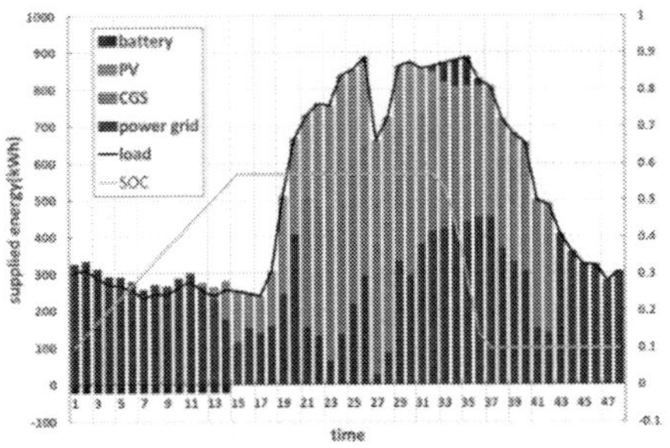

Fig.6. One day simulation result

focus on total cost change of PV capacity, we find that the shape of the graph is secondary function type with a valley. One of the reasons is the case of large capacity of PV cells. Because surplus power beyond battery charging is generated in holidays with lower load, the merit of large capacity of PV cells may be insignificant. The other reason is that, as Fig.6 suggests, in case of large capacity of PV cells, because maximum demand hour shift from noon to sunset, the merit of reducing maximum demand by generating PV may be insignificant. Then, when we focus

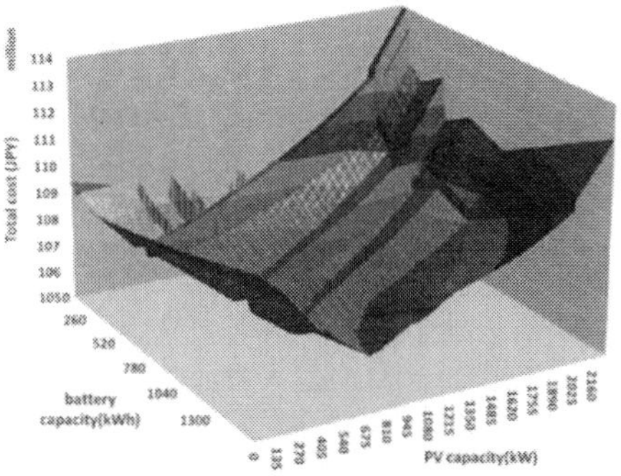

Fig.7. Total cost in each PV and battery capacity

on total cost change of PV capacity, we find that there is little merit of increasing battery capacity. However, in case of large capacity of PV cells, total cost reduction by introducing battery becomes higher. In this result, it seems advisable that battery is used with PV.

5 CONCLUSION

In this paper, a method of scheduling scheme for the electric supply of FEMS is proposed, and optimal installed capacity of PV cells and battery to minimalize total cost is calculated. From the simulation result, in case of small capacity of PV cells and battery, it is more effective that day-ahead scheduling is built by using PV forecast output. In case of large capacity of PV cells and battery, it is more effective that day-ahead scheduling is built by using PDF of PV output which has robustness against error. Moreover, it seems advisable that battery is used with PV.

Future work is consideration of a rate system by time zone and examination of the parameters α_{ks}. Moreover, demand control by analyzing composition of load is simulated.

6 REFERENCES

(1) Sungjin Lee, Beom Kwon, and Sanghoon Lee: "Joint Energy management System of Electric Supply and Demand in Houses and Buildings", IEEE TRANSACTIONS ON POWER SYSTEMS, VOL.29, NO.6, NOVEMBER 2014, pp.2804-2812 (2014)

(2) Kentaro Shimomachi, Toshiyuki Iwami, Ryoichi Hara, Hiroyuki Kita, Masatoshi Noritake, Hideyuki Hoshi, Keiichi Hirose: "Development of EMS for Microgrid for Office Building", IEEJ Transactions on Power and Energy, Vol.136, No.4, pp.373-381 (2016)

(3) Mitsubishi Research Institute, inc.: URL:http://www.meti.go.jp/meti_lib/report/2012fy/E002274.pdf

(4) Takanobu Suzuki, Yusuke Goto, Takahiro Terazono, Shinji Wakano, Takashi Oozeki: "Forecast Method of Solar Irradiance with Just-In-Time Modeling", IEEJ Transactions on Power and Energy, Vol.131, No.11, pp.912-919 (2011)

(5) Tomohide Yamazaki, Hayato Homma, Shinji Wakano, Yu Fujimoto, Yasuhiro Hayashi: "Estimation Method of Prediction Interval of Solar Irradiance Based on Just-In-Time Modeling for Photovoltaic Output Prediction", IEEJ Transactions on Power and Energy, Vol.135, No.3, pp.160-167 (2014)

(6) Research Institute for Sustainable Humanosphere: URL:http://database.rish.kyoto-u.ac.jp/arch/jmadata/gpv-original.html

(7) Japan Renewable Energy Foundation: URL:http://www.renewable-ei.org/images/pdf/20150226/PV_Cost_Report_2014.pdf

PV INTEGRATION AND PRICE-BASED DEMAND SIDE MANAGEMENT: OPTIMUM TIME-OF-USE TARIFFS

Nikolas Philippou, Venizelos Venizelou, George Makrides, Maria Hadjipanayi, Venizelos Efthymiou and George E. Georghiou
FOSS Research Centre for Sustainability Energy, Photovoltaic Technology Laboratory
Department of Electrical and Computer Engineering, University of Cyprus, Nicosia, Cyprus

ABSTRACT: Growing electricity demand along with environmental concerns promote the use of renewable energy sources (RES). In the Mediterranean region, the high solar energy potential renders the use of photovoltaic (PV) technologies as one of the most favourable green sources. However, high penetration levels of distributed grid-connected PV systems aggravate the intermittency and grid stability issues especially in cases where the grid is not interconnected and there is a weak correlation between the electricity demand and PV generation. To this extent, a price-based demand side management (DSM) tool has been developed in order to emulate the effectiveness of Time of Use (ToU) tariffs. Through the application of such tariffs, demand response (DR) actions can be facilitated aiming to partly mitigate PV operational issues and alleviate congestion problems. For this purpose, three hundred households in Cyprus with grid-connected PV systems installed at their rooftops were equipped with smart meters (SMs) in order to acquire consumption and production profile details of typical Cypriot prosumers. A sensitivity analysis was performed where the Load Factor (LF) was compared under different peak clipping and load shifting scenarios.
Keywords: Demand-Side, Grid Management, Photovoltaic, Small Grid-connected PV Systems.

1 INTRODUCTION

The integration of renewable energy sources (RES) and particularly grid-connected photovoltaic (PV) systems offers decentralized and distributed electricity generation however, their high penetration may lead to grid stability and reliability issues. To this effect, demand side management (DSM) actions have been introduced worldwide aiming to control the demand and allow higher penetration of RES [1]–[6]. The main DSM schemes include applications of automatic control systems, energy storage systems (ESS) and price signals. Price-based DSM programs offer an alternative to the traditional flat tariffs and comprise of Critical Peak Pricing (CPP), Real-Time Pricing (RTP) and Time-of-Use (ToU) tariffs [4], [7]. Amongst the different schemes, ToU tariffs are commonly preferred because the price of energy consumption is fixed for different periods of the day in contrast to other price-based DSM programs where the price fluctuates following the real time cost of electricity [4], [7].

The development of an efficient DSM scheme offers the advantage of generating cost reductions for grid utilities and the increase of operational efficiency. However, in order to achieve a balanced DSM scheme, the existing flat tariffs need to be transformed to ToU tariffs providing the necessary monetary incentives for domestic consumers to flatten their load profile.

Even though ToU tariff schemes offer the advantage of price certainty, the effectiveness of such schemes must be verified prior to implementation because of the risk of a new peak appearing through load shifts at cheaper price periods, posing negative effects on the optimal operation of the system [3], [8]–[10].

In this work, a price-based DSM tool has been developed and implemented based on real data from households in Cyprus [11]. The tool derives an optimum Time of Use (ToU) tariff on a seasonal basis aiming to motivate residential customers to shift load from peak to valley periods. These actions will have an impact both on the demand side and on the grid as the customer will benefit from a lower electricity bill and at the same time peak demands are mitigated. In this scope, smart meters (SMs) have been deployed at three hundred households with grid-connected PV systems installed at their rooftops, in order to acquire consumption and production profile details of typical Cypriot prosumers. The selected pilot sites that already have a 3 kWp grid-connected PV system, have been offered a ToU tariff allowing rates and charges to vary based on the time and date of consumption, i.e. day or night and seasonality.

2 PROSUMER FEATURES

In support of this work, three hundred prosumers in Cyprus have been selected through the implementation of the SmartPV project (http://www.smartpvproject.eu/), in order to collect real measurements from a representative sample for the Cypriot consumer. Smart meters (SMs) have been installed to acquire consumption and production profile details. The selected pilot sites are equipped with a 3 kWp grid-connected PV system as is the case with the existing net metering scheme for households in Cyprus.

In addition, the prosumers are divided into three groups each receiving a distinct method of monitoring their energy habits: one group has In-House Displays (IHDs), the other one has access to their production and consumption through a web application, and the third one receives information about their energy habits through the traditional bi-monthly mail bill. This will facilitate an in-depth examination of the prosumers' response based on the information sent to them [12].

3 METHODOLOGY

3.1 Methodology to find the optimum ToU blocks

The dynamic ToU tariff tool aims to derive the optimum ToU blocks and tariffs given the load profile. This will enable demand response (DR) since higher tariffs will occur in peak periods while lower tariffs in off-peak periods. The algorithm takes as input the total electricity demand and based on statistical analysis it derives the ToU blocks [11]. An optimization process is then performed on the derived blocks. The objective function of the optimization procedure minimizes the RMSE as described by Eq.1.

$$\text{RMSE} = \sqrt{\frac{1}{n}\sum_{k=1}^{n}(ToUb_k - P_k)^2} \qquad (1)$$

where $ToUb_k$ is the derived block period, P_k is the load profile and n is the total sampling interval. Based on this equation, $ToUb_k$ is the variable to be optimized and it changes according to the desired levels. To achieve this, the developed optimization tool uses a hybrid optimization function. In particular, simulated annealing [13] and pattern search [14] are the two solvers used to derive the optimum result.

3.2 Methodology to find the optimum ToU tariffs

Beyond the identification of the ToU blocks the dynamic ToU tariff tool derives the optimum ToU tariffs. The optimized ToU tariffs are calculated using the well-known fmincon solver [15], which estimates a constrained minimum of a scalar function of several variables. The tariffs are calculated on the participants' average load profile on a seasonal basis with the constraint of maintaining the electricity bill the same as in the case of flat tariff charge if the load profile remains unchanged. For this purpose, the existing flat tariff, the fixed charge and the fuel price adjustment are required to calculate the average electricity bill of the participants and set the off-peak price for the ToU tariff. The optimum ToU tariffs are calculated using the following equality and inequality criteria:

$$hp - mp = mp - lp \qquad (2)$$
$$lp = (flat - fc + fa) * 0.65 \qquad (3)$$
$$hp > flat\_tariff \qquad (4)$$
$$lp < flat\_tariff \qquad (5)$$

where,
$hp = high\, price$
$mp = medium\, price$
$lp = low\, price$
$$fc = \frac{fixed\ cost}{bimonthly\ energy\ cosnumption}$$
$fa = fuel\ adjustment$

The objective function minimizes the absolute error (AE) between the electricity bill using ToU and the flat tariffs.

$$\text{AE} = |ToUprice - flat\_tariff| \qquad (6)$$

Therefore, the values for hp and mp vary within the equality and inequality criteria until the minimum feasible AE is achieved.

4 RESULTS

4.1 ToU tariffs and the tool

Table I presents the derived ToU periods and tariffs for each category for every season based on one year of collected datasets from the participants.

Table I. Derived ToU blocks and tariffs for the summer and autumn season.

Block	Price (€cents /kWh)	Period Winter (Dec – Mar)	Summer (Jun – Sep)	Middle (Apr, May, Oct, Nov)
Peak	18,85	16:00 – 21:59	09:00 – 18:59	08:00 – 20:59
Shoudler	14,85	06:00 – 15:59 22:00 – 23:59	07:00 – 08:59 19:00 – 00:59	06:00 – 07:59 21:00 – 23:59
Off-peak	10,85	00:00 – 05:59	01:00 – 06:59	00:00 – 05:59

Figure 1 presents the dynamic ToU tariff tool as this is derived based on the simulation performed using datasets received for the summer period of 2015. The first graph in the tool (in the top right corner) exhibits the total electricity demand for the summer period. The ToU blocks are derived from the optimization process finding the minimum RMSE between the total electricity demand and the ToU blocks. These ToU blocks are applied to the extracted load profile from the participating customers in order to derive the optimum ToU tariffs for the summer season.

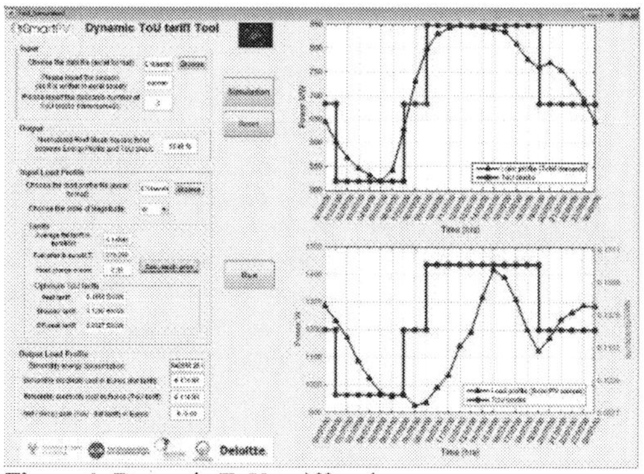

Figure 1. Dynamic ToU tariff tool.

Figure 2 (a), (b) and (c) show that for the summer, middle and winter season respectively, the ToU tariffs and the load profile of the participants are split into three categories, A for off-peak period, B for shoulder period and C for peak period.

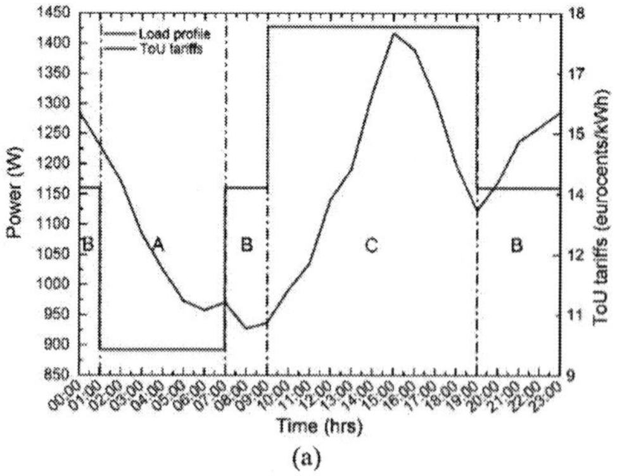

(a)

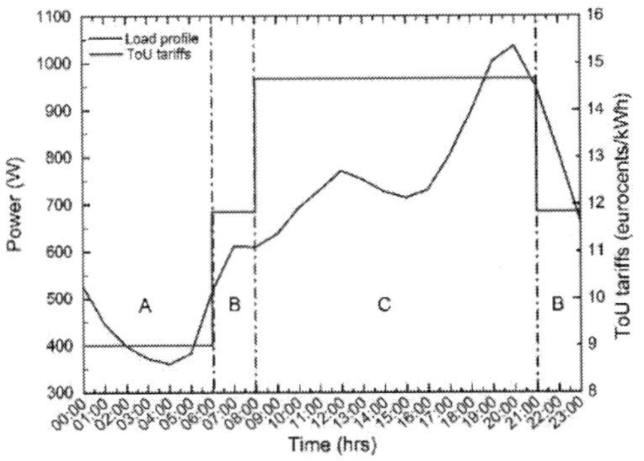

(b)

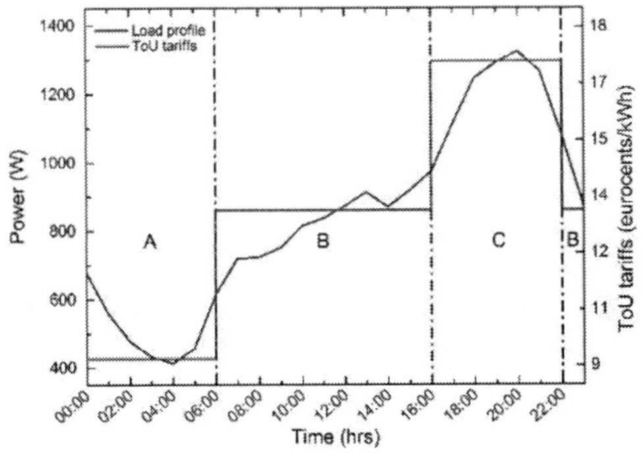

(c)

Figure 2. Derived ToU tariffs and participant average load profiles for: (a) summer, (b) middle and (c) winter season.

4.2 Peak clipping scenarios

Analysis of one year of data from three hundred prosumers in Cyprus has been undertaken in order to investigate the ToU tariffs, DR scenarios and associated financial implications. Sensitivity analysis has also been performed to examine the effect of different peak clipping scenarios on the residential load factor (LF) and the economic impact for the summer and autumn season. The LF is defined as the average load divided by the peak load in a specified time period (7):

$$\text{Load factor } (LF) = \frac{average\ load}{maximum\ load\ in\ given\ time\ period} \tag{7}$$

Preliminary results for the analysis of peak clipping variations are given in Table II. The sensitivity analysis considers a peak clipping variation from 5 - 20 % in steps of 5 %. This scenario was initially examined as ToU tariffs are more likely to affect customers by shaving peak demand rather than shifting demand from peak to off-peak periods [7]. The maximum residential switchable load was found to be up to 20 % therefore this was set as the cap for this analysis [8], [9]. The LF for the summer period can be increased from 42.50 % to 52.61 % in the case of 20 % peak clipping. On the other hand, during the autumn period the LF can be increased from 33.50 % to 41.35 % in the case of 20 % peak clipping as can be seen in Table II.

Table II. Residential load factor for different scenarios.

	LF - summer (%)	LF - middle (%)	LF – winter (%)
Average profile	40.65	32.94	32.48
Case 1: 5 % peak clipping	42.69	34.57	34.09
Case 2: 10 % peak clipping	44.95	36.37	35.87
Case 3: 15 % peak clipping	47.48	38.40	37.86
Case 4: 20 % peak clipping	50.32	40.67	40.11

4.3 Load shifting scenarios

To further investigate the impacts of DR on the load profile, the Load Shifting (LS) technique was applied. LS is also one of the classic forms of load management taken by the utilities, which involves shifting loads from peak to off-peak periods. To evaluate the performance of this method, a DR tool has been developed. The tool replicates the load shifting technique using an optimization function in order to maximize the LF.

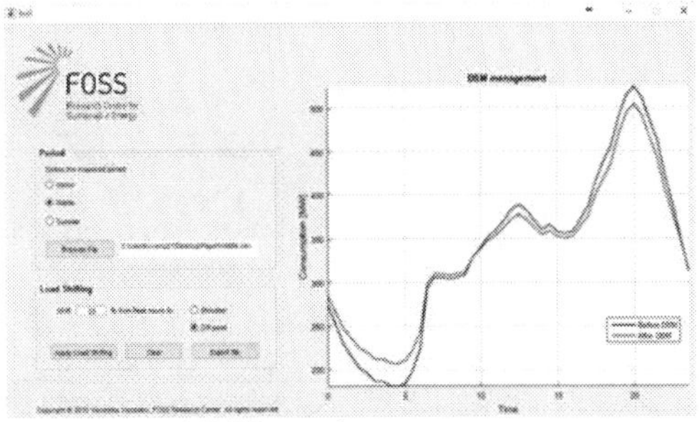

Figure 3. Load Shifting tool.

The algorithm implemented for the optimization function is summarized in the following steps:

Table III. Optimization Algorithm steps

1. REPEAT
2. AT each time slot DO
3. Find $Load_{new}$ → max (Load_Factor)
4. s.t.
5. Selected time_block
6. $Energy_{new} = Energy_{old}$
7. $Peak_{new} < Peak_{old}$
8. END

The optimization function is applied on the participants' average load profile on a seasonal basis, where the largest percentage of the shifted load is gathered either on the shoulder or off-peak periods, depending on the Distribution System Operator's (DSO's) preference. The following Figures illustrate the seasonal load profiles for both scenarios, a) shoulder and b) off-peak period. Each scenario is divided into four cases in which the shifted load percentage is between 5 - 20% in steps of 5%.

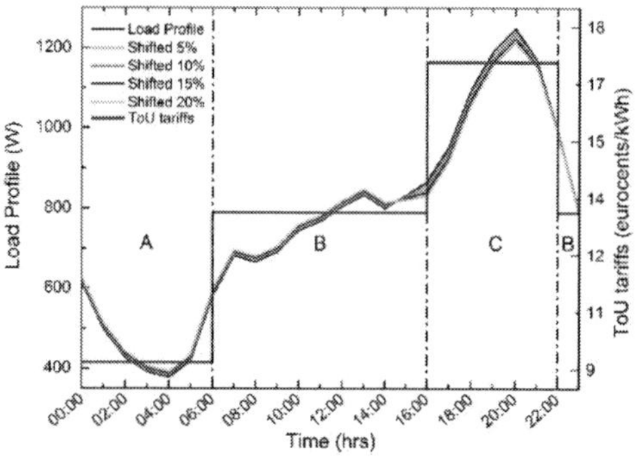

Figure 4. LS technique for the winter season (shoulder period scenario).

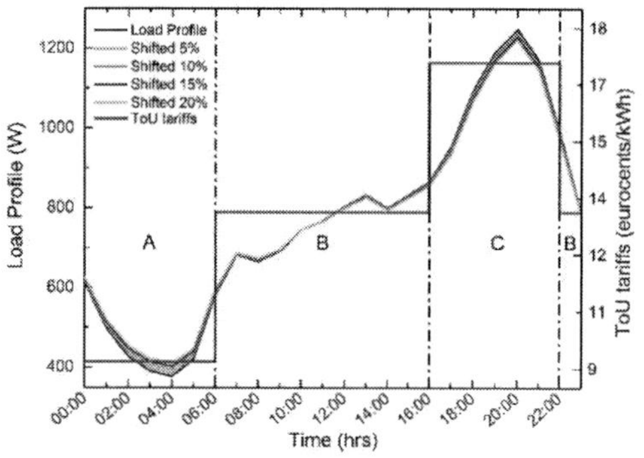

Figure 5. LS technique for the winter season (off-peak period scenario).

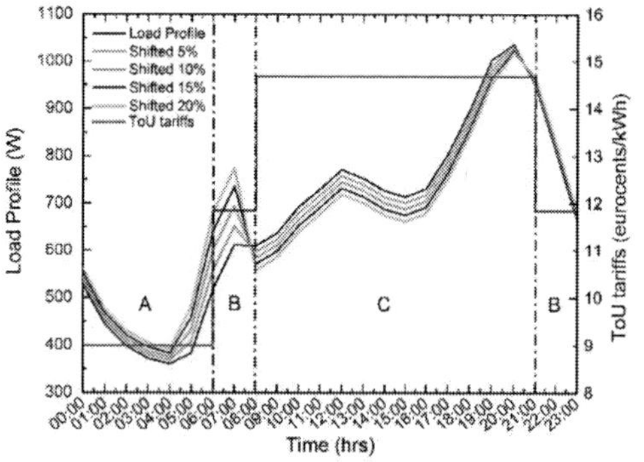

Figure 6. LS technique for the middle season (shoulder period scenario).

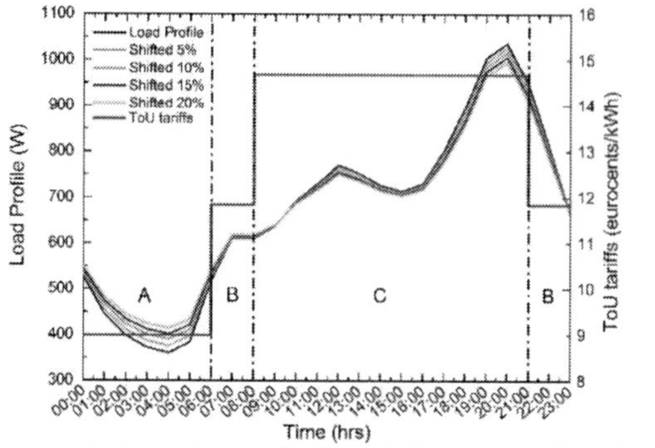

Figure 7. LS technique for the middle season (off-peak period scenario).

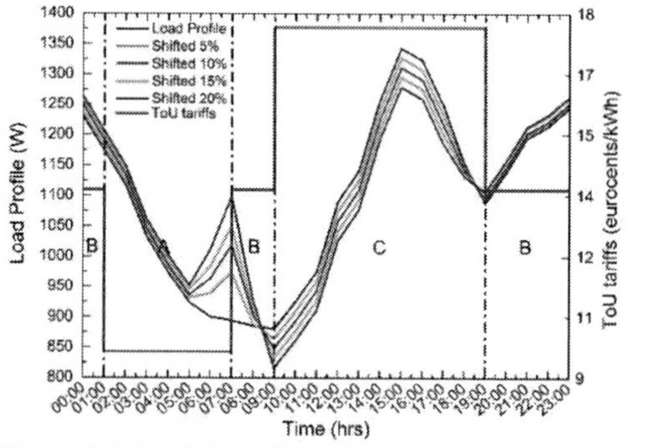

Figure 8. LS technique for the summer season (shoulder period scenario).

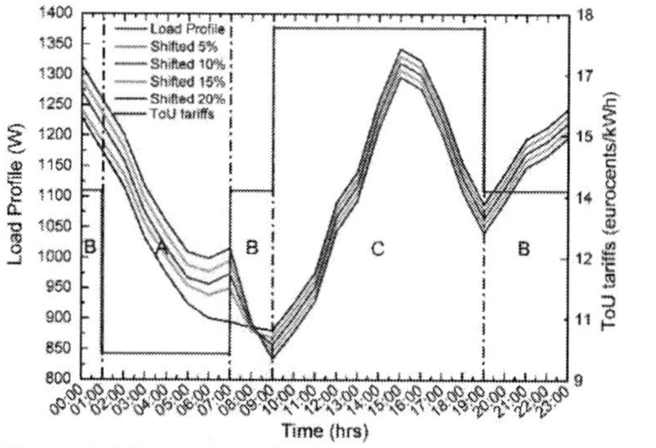

Figure 9. LS technique for the summer season (off-peak period scenario).

It is obvious that a high shifted load percentage, gathered on the shoulder period, offers a relatively high possibility of creating a second peak demand. However, this depends on the load profile before the LS technique is applied.

Table IV. Seasonal shifted Load percentage allocation base on the period.

	Summer (%)	Middle (%)	Winter (%)	Period selected
Case 1: 5 % LS	3.54	3.33	4.05	shoulder
	3.74	2.79	2.47	off-peak
Case 2: 10 % LS	7.11	6.76	8.16	shoulder
	7.04	5.55	4.92	off-peak
Case 3: 15 % LS	10.67	10.04	12.26	shoulder
	10.58	7.97	7.36	off-peak
Case 4: 20 % LS	14.22	13.30	16.36	shoulder
	14.1	10.5	9.7	off-peak

Table IV shows the seasonal shifted load percentage based on the period selected by the user. The percentage on the selected period should be higher in all seasons, however that seems not to be the case for the off-peak period of the winter season. This is due to the fact that the duration of the shoulder time block is twice the size of the off-peak.

In order to evaluate the impact of the applied technique on the load profile, the LF for each one of the cases was identified. The two following Tables show the LF for each one of the examined cases as well as before the load shifting is applied.

Table V. Residential load factor for the load shifting technique (shoulder period scenario).

	LF - summer (%)	LF - middle (%)	LF - winter (%)
Average profile	40.65	32.94	32.48
Case 1: 5 % LS	41.29	33.34	32.69
Case 2: 10 % LS	41.79	33.48	32.90
Case 3: 15 % LS	42.30	33.36	33.12
Case 4: 20 % LS	42.83	33.21	33.33

Table VI. Residential load factor for the load shifting technique (off-peak period scenario).

	LF - summer (%)	LF - middle (%)	LF - winter (%)
Average profile	40.65	32.94	32.48
Case 1: 5 % LS	41.15	33.25	32.64
Case 2: 10 % LS	41.51	33.56	32.80
Case 3: 15 % LS	41.88	33.88	32.97
Case 4: 20 % LS	41.34	34.21	33.13

The results indicate that the load profile can benefit from the specific DR technique as the LF is fairly increased in all cases.

5 CONCLUSIONS

The development of a dynamic Time-of-Use (ToU) tariff tool is presented in order to promote effective Demand Side Management (DSM) practices in the electricity network of Cyprus. Different peak clipping and load shifting (LS) scenarios were demonstrated using the seasonal load profiles derived from the collected datasets. The results derived from peak clipping showed that the load factor (LF) for the summer period can be increased from 40.65 % to 50.32 % in the case of 20 % peak clipping. For the middle period an improvement from 32.94 % to 40.67 % can be achieved and lastly for the winter period the LF can be increased from 32.48 % to 40.11 %. On the other hand, when LS techniques are implemented the maximum LF depends on where the load is shifted to (e.g. from peak to shoulder etc.). For the summer and winter season, the maximum LF when load is shifted to the shoulder period was 42.83 and 33.33 % respectively. Finally, the LF for the middle season is maximized (34.21 %) when LS is performed to off-peak periods.

6 ACKNOWLEDGEMENT

The SmartPV project is co-financed by the program LIFE (LIFE+ Environment Policy and Governance) of the European Union under the grant agreement number LIFE 12/ENV/CY/000276.

7 REFERENCES

[1] S. Funke and M. Speckmann, "Demand Side Management in Private Households – Actual Potential, Future Potential, Restrictions," in *Energy-Efficient Computing and Networking*, Springer Berlin Heidelberg, 2011, pp. 29–36.

[2] L. Gelazanskas and K. A. A. Gamage, "Demand side management in smart grid: A review and proposals for future direction," *Sustain. Cities Soc.*, vol. 11, no. 2014, pp. 22–30, Feb. 2014.

[3] S. Gottwalt, W. Ketter, C. Block, J. Collins, and C. Weinhardt, "Demand side management—A simulation of household behavior under variable prices," *Energy Policy*, vol. 39, no. 12, pp. 8163–8174, Dec. 2011.

[4] B. Severin, M. Jaske, and A. Rosenfeld,

"Dynamic Pricing, Advanced Metering, and Demand Response in Electricity Markets." Center for the Study of Energy Markets, p. 103, 2002.

[5] E. Commission, "European SmartGrids Technology Platform: Vision and strategy for European electricity networks for the future," Luxembourg, 2006.

[6] M. A. López, S. de la Torre, S. Martín, and J. A. Aguado, "Demand-side management in smart grid operation considering electric vehicles load shifting and vehicle-to-grid support," *Int. J. Electr. Power Energy Syst.*, vol. 64, pp. 689–698, Jan. 2015.

[7] M. H. Albadi and E. F. El-Saadany, "A summary of demand response in electricity markets," *Electr. Power Syst. Res.*, vol. 78, no. 11, pp. 1989–1996, Nov. 2008.

[8] J. Torriti, "Price-based demand side management: Assessing the impacts of time-of-use tariffs on residential electricity demand and peak shifting in Northern Italy," *Energy*, vol. 44, no. 1, pp. 576–583, Aug. 2012.

[9] J. Thumim, "Investigating the potential impacts of Time of Use (TOU) tariffs on domestic electricity customers Report to Ofgem," 2014.

[10] C. M. Flath, "An optimization approach for the design of time-of-use rates," in *IECON - 39th Annual Conference of the IEEE Industrial Electronics Society*, 2013, pp. 4727–4732.

[11] N. Philippou, M. Hadjipanayi, G. Makrides, V. Efthymiou, and G. E. Georghiou, "Effective dynamic tariffs for price-based Demand Side Management with grid-connected PV systems," in *PowerTech, 2015 IEEE Eindhoven*, 2015, p. 5.

[12] B. Davito, H. Tai, and R. Uhlaner, "The smart grid and the promise of demand-side management," 2010.

[13] L. Ingber, "Adaptive simulated annealing (ASA): Lessons learned," *Invit. Pap. to a Spec. issue Polish J. Control Cybern. "Simulated Annealing Appl. to Comb. Optim.*, 1995.

[14] T. G. Kolda, R. M. Lewis, and V. Torczon, "A generating set direct search augmented Lagrangian algorithm for optimization with a combination of general and linear constraints," 2006.

[15] R. H. Byrd, J. C. Gilbert, and J. Nocedal, "A trust region method based on interior point techniques for nonlinear programming," *Math. Program.*, vol. 89, pp. 149–185, 2000.

LARGE-SCALE INTEGRATION OF RENEWABLE ENERGY SOURCES: TECHNICAL AND ECONOMICAL ANALYSIS FOR THE ITALIAN CASE

Matteo Giacomo Prina[1,2,*], Giulia Garegnani[1], David Kleinhans[3], Giampaolo Manzolini[2], Roberto Vaccaro[1], Stefan Weitemeyer[4], David Moser[1]
[1]Institute for Renewable Energy, EURAC Research, Viale Druso 1, 39100 Bolzano (BZ), Italy
[2]Dipartimento di energia, Politecnico di Milano, Via Lambruschini, 4, 20156 Milano (MI), Italy
[3]NEXT ENERGY, EWE Research Centre for Energy Technology at the University of Oldenburg, 26129 Oldenburg, Germany
[4]University of Oldenburg, 26111 Oldenburg, Germany
*Phone: +39 0471 055587; Fax: +39 0471 055699;
E-mail: MatteoGiacomo.Prina@eurac.edu
Giulia.Garegnani@eurac.edu
david.kleinhans@next-energy.de
giampaolo.manzolini@polimi.it
roberto.vaccaro@eurac.edu
stefan.weitemeyer@uni-oldenburg.de
david.moser@eurac.edu

ABSTRACT: The integration of high shares of variable renewable energy sources is an aspect that is acquiring importance as it can lead to integration problems. The three main options to achieve this task are the enlargement and reinforcement of the grid, the use of storage devices, and demand side management. A modelling approach is applied in this study to evaluate the maximum level of renewable energy integration for the Italian power system when adopting electric storage. Firstly, the integration trend has been investigated starting from the production profile of variable renewable energy sources. The results show that the optimal combination of wind and solar energy to maximize the integration of renewables is 55% and 45%, respectively. Afterward, the impact of storage systems has been evaluated in the case of high renewable energy penetration. An optimization approach is then applied to assess the best mix of solar and wind energy while minimizing the costs. Results show that the best share of renewable energies stabilizes on 60 - 70% of wind production and 40 - 30% of solar production due to higher costs of rooftop photovoltaic systems.
Keywords: PV System, Grid Integration, Storage.

1 INTRODUCTION

A future power system with high share of renewable energy production present the problem of integration of non-dispatchable power production into the system. An available solution to increase renewable energy integration is to link the electricity sector with the heat or transport sector [1-3]. For example, Nastasi *et al.* have analyzed how to integrate heat and electricity by means of hydrogen synthesis [4]. Concentrating only on the electricity sector, the three main options are (i) the reinforcement of the grid, (ii) the use of storage devices, and (iii) the demand side management. Steinke *et al.* have compared the installation of storage system capacity with the expansion of the electric grid [5].
Weitemeyer *et al.* have developed a modelling approach to evaluate the impact of storage devices and their contribution to go towards a 100% renewable energy system, and applied the model both to the German [6] and to the entire European electricity system [7]. Alhamwi *et al.* have followed this approach to estimate the optimal renewable power generation mix for Morocco [8].
This paper investigates the large-scale integration of variable renewable energy sources (VRES) from a technical and economical point of view. Using long-term power generation and load data with hourly resolution, the methodology developed by Weitemeyer *et al.* [6, 7] is applied to the Italian case.
The novelty of the present paper is given by the expansion of the existing model, studying the renewable energy integration not only from a technical but also from an economical point of view. For this purpose a multi objectives optimization approach needs to be developed to optimize the renewables mix by total annual costs minimization and the maximization of renewable energy integration function.

2 METHODOLOGY

The analysis starts from a detailed data set that is composed by six years data series with hourly resolution (i.e. the considered period is 2009 to 2014). In this way it is possible to catch the renewable energy sources variability from year to year and take it into account for the future scenarios. Limitations and constraints of the national transmission grid are not considered. The two VRES taken into account are on-shore wind energy and photovoltaics (PV) solar energy.
As in [6, 7] the production from VRES is matched with the demand load time series $L(t)$, which is available from ENTSOE [9]. The remaining electricity not covered by VRES is produced by fully flexible power plants, that can increase their production from 0 to 100% of the load in one hour. In this first step, the electricity generation which cannot be fed into the system is curtailed. The production profile and the absolute annual production for Italy of on-shore wind and solar, are taken from the STRATEGO project outputs [10] and from the GSE reports (2009 to 2014) [11], respectively. These two different data were used to create the two curves of renewable production, on-shore wind $W(t)$ and solar $S(t)$.
In practice, a normalization with respect to the average electricity demand was carried out as in Eq. 1 and 2 (where N is the number of hours in the considered time span):

$$w(t) = W(t) \cdot \frac{\sum_{i=1}^{N} L(t_i)}{\sum_{i=1}^{N} W(t_i)} \tag{1}$$

$$s(t) = S(t) \cdot \frac{\sum_{i=1}^{N} L(t_i)}{\sum_{i=1}^{N} S(t_i)} \tag{2}$$

The Eq. 3 gives the mismatch between generation and demand [6]:

$$\Delta_{\alpha,\gamma}(t) = \gamma \cdot (\alpha \cdot w(t) + (1-\alpha) \cdot s(t)) - L(t) \tag{3}$$

The parameters α and γ express the share of wind and solar power generation of the overall electricity demand. In particular, the parameter α expresses which percentage of the electricity generation from RES is covered by wind and, thus, the factor (1-α) expresses the part covered by solar power. The parameter γ is the average renewable energy power generation factor. γ = 1 value means that the electricity demand can be entirely covered by RES on the considered time span (i.e. without taking into account concurrence of production and demand). In order to further clarify the concept, it is possible to say that when γ is equal to 1, Eq. 4 is always verified:

$$\sum_{i=1}^{N} (\alpha \cdot w(t_i) + (1-\alpha) \cdot s(t_i)) = \sum_{i=1}^{N} L(t_i) \tag{4}$$

The mismatch function, Eq. (3), is now analyzed with respect to the resulting overall integration of renewables in the system. When $\Delta_{\alpha,\gamma}(t)$ is negative, its absolute value is stored and weighted by the average load, which indicates the required share of backup energy as a function of α and γ. Afterwards, a renewable integration function RE(α, γ) is defined as 1 minus the required share of backup energy. RE(α, γ) is by definition between 0 and 1 and measures the share of renewable integration. For more details on the approach it can be referred to [6].

Once the mathematical problem is defined, the best mix of variable renewable energy sources can be investigate to understand the compatibilities between the production profile of wind and solar energy. Secondly, the impact of storage systems has been evaluated to study the potentialities of these types of system in the increasing of the renewable integration into the grid. The impact of different size of storage capacities H^{max} (in units of L(t)) have been analyzed. The storage system is modelled through the following parameters: η is the round-trip efficiency of the fully flexible storage system and H^{max} is the size of the storage. For the analysis, sizes of the storage are assumed equal to 0, 2, 4, 6, 8, 10, 12, 14, 24, 36, 48, 72, 168, 360, 720, 1440 average load hours (av.l.h.). For Italy 1 av.l.h. is equal to 36.9 GWh. The initial charging of the storage has been set to 50% of the H^{max}. The storage time series $H^{\eta}_{\alpha,\gamma}(t)$, for a fixed share of α and γ, describes the available electricity to the grid and is given by Eq. 5:

$$H^{\eta}_{\alpha,\gamma}(t) = \begin{cases} if\ \Delta_{\alpha,\gamma}(t) \geq 0 \\ \min[H^{max}, H^{\eta}_{\alpha,\gamma}(t-1) + \eta \cdot \Delta_{\alpha,\gamma}(t)] \\ if\ \Delta_{\alpha,\gamma}(t) < 0 \\ \max[0, H^{\eta}_{\alpha,\gamma}(t-1) + \Delta_{\alpha,\gamma}(t)] \end{cases} \tag{5}$$

The quantity of electricity that it is lost due to curtailment, efficiency losses, and storage fully charged is calculated through the overproduction function and is expressed by Eq. 6:

$$O^{\eta}_{H}(\alpha,\gamma) = \frac{\sum_{i=1}^{N} max[0, \Delta_{\alpha,\gamma}(t_i) - (H^{\eta}_{\alpha,\gamma}(t_i) - H^{\eta}_{\alpha,\gamma}(t_i - 1))]}{\sum_{i=1}^{N} L(t_i)} \tag{6}$$

Consequently, the renewable integration function RE(α, γ) is obtained through the Eq. 7:

$$RE^{\eta}_{H}(\alpha,\gamma) = \gamma - \max[\gamma - 1, O^{\eta}_{H}(\alpha,\gamma)] \tag{7}$$

After this technical analysis based on [6], the methodology is further developed and an optimization analysis, evaluating also economic aspects, is presented. The objectives of the analysis are the minimization of total annual costs and of the share of demand covered by fully flexible power plants or (1 - RE(α, γ)). The best value of α for each γ has been assessed not considering storage systems. The costs of the different technologies are shown in Table 1 where the PV one is referred to the rooftop Italian situation. The costs of the remaining back-up energy are not taken into consideration.

Table I: Parameters for the economic analysis [12].

Technology	*C_{Inv} = Investment cost [€/kW]*	*$C_{O\&M}$ = O&M cost [%]*	*n = Lifetime [years]*
Wind	1300	3	20
Rooftop PV system	2000	2.1	20

The investment costs are converted into annual costs by means of the Eq. 8, where i is the discount rate that is assumed equal to 3 %:

$$A_{inv} = Capacity_{VRES} \cdot C_{inv} \cdot \frac{i}{[1-(1+i)^{-n}]} \tag{8}$$

The total annual costs are calculated through the sum of the capital costs (namely the investment costs converted in annual values) and operation and maintenance costs that are evaluated through the Eq. 9:

$$A_{O\&M} = Capacity_{VRES} \cdot C_{O\&M} \cdot C_{inv} \tag{9}$$

The chosen optimization model is a Multi Objective Evolutionary Algorithm (MOEA) specifically designed for multi objective problems [13, 14]. It starts with a number of random solutions to initialize the population. Each solution or individual is evaluated running the simulation model. The outputs are the total annual costs and the value of RE(α, γ), renewable integration function. Each individual of the population is evaluated on these two objectives. A new population is created through different operators like parent selection, crossover and mutation. The flow diagram of this optimization model is shown in Figure 1.

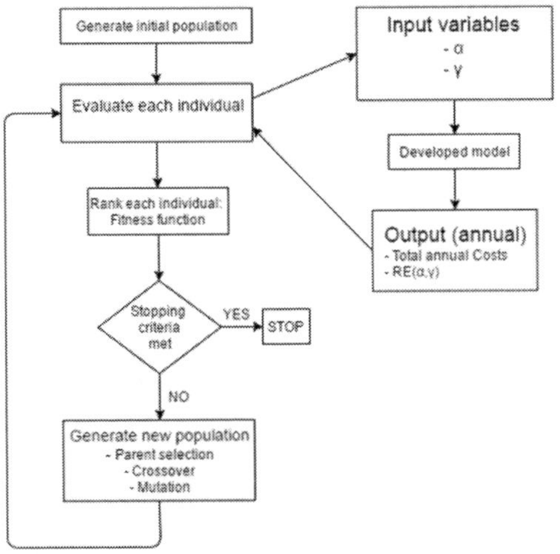

Figure 1: Flow diagram of the Multi Objective Evolutionary Algorithm (MOEA) connected to the model. For a similar approach see [15].

3 RESULTS

The resulting renewable integration function RE(α, γ) as function of the parameter γ for Italy is shown in Figure 2. In this first case storage systems are not considered. Until γ=0.3 there is a small overproduction and the curves rise linearly like the perfect integration one. At higher values the curves bend down due to overproduction (implying curtailment) that occurs more frequently during the considered time span.

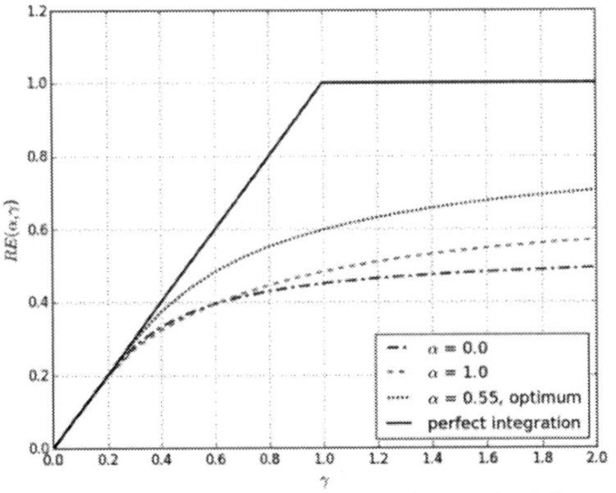

Figure 2: Renewable integration function RE(α, γ) for different value of α, not considering the role of storage.

The worst case with respect to the integration is the curve with α=0, which is the only solar power case. It results a RE(α, γ) value of 0.45 at γ=1. The other extreme, the only wind power case, α=1, produces at γ=1 a value of RE(α, γ) equal to 0.48. In absence of storage, the optimum α that allows for the best integration of renewables is α=0.55. The indicator used in the optimization to find the optimum α was the integral of the curves between γ=0 and γ=2. This value is rather low in comparison to results obtained for Germany (cf. [6]).

Figure 3 shows the PV and wind production curve compared to the load curve in the case in which γ is equal to 1 and α is the optimum, equal to 0.55.

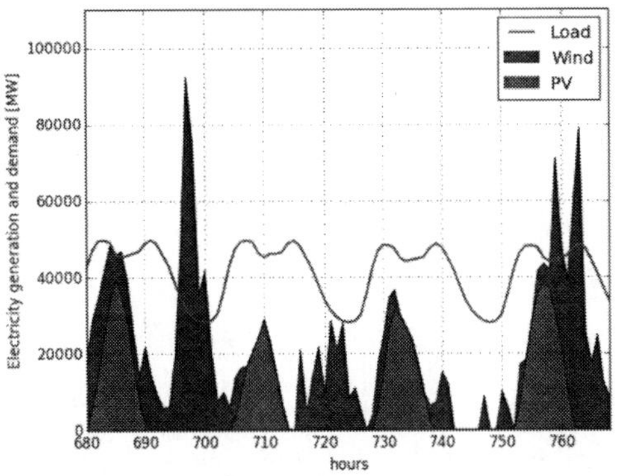

Figure 3: PV, s(t), and wind, w(t), production curve setting γ=1 and α=0.55.

From now on, the optimum wind share of α=0.55 will be used for the further calculations. Storages in the system can directly affect the mismatch function, Eq. (3), and by this means improve the integration of renewables. Figure 4 shows the renewable integration function at the increasing of the storage size for η=0.8, which would correspond to pumped-hydro or battery storage system. It is important to underline that the highest value of the storage capacity produces an almost linear increase of RE(α, γ) with γ, albeit at a lower slope compared to the perfect integration one due to storage losses.

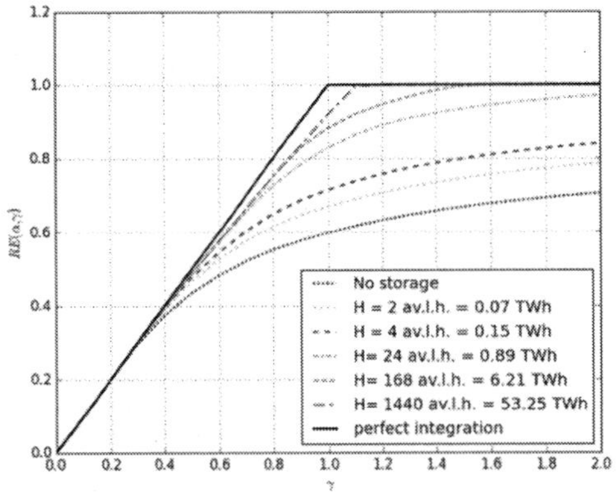

Figure 4: Renewable integration function RE(α=0.55, γ) for different capacity of the storage with η=0.8, utilizing the optimum α, equal to 0.55.

Figure 5 shows the renewable integration function for two different storage classes: highly efficient storage (a pumped-hydro or battery storage system) with low capacity (H = 4 av.l.h. = 0.15 TWh, η=0.8) and a large storage with low efficiency (H= 168 av.l.h. = 6.21 TWh, η=0.3) that could describe a seasonal storage based on synthetic hydrogen. Until γ=0.4 overproduction is negligible and thus storage system is not needed. After this threshold, for a share of α equal to 0.55, a small but efficient storage can increase the integration of RES up to γ=1. For higher values of γ a seasonal large and low efficient storage leads to higher values of RE(α, γ) compared to a small but efficient storage, which is consistent with the results obtained for Germany as described in [6].

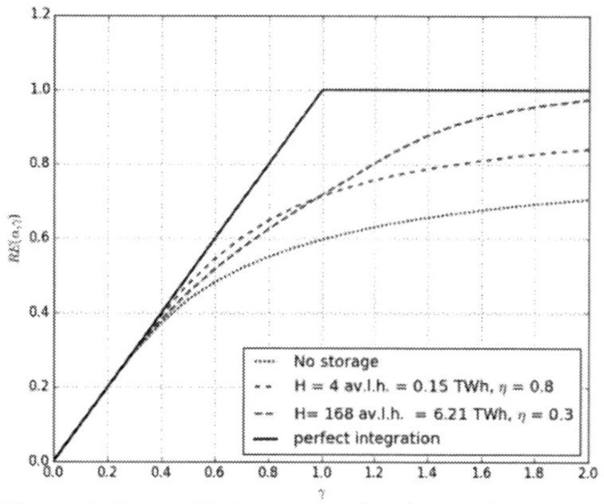

Figure 5: Renewable integration function RE(α=0.55, γ) for different types and capacity of storage, utilizing the optimum α, equal to 0.55.

Figure 6 shows the results of the optimization analysis and the Pareto front of the best solutions. The objective of the optimization analysis are the minimization of the total annual costs and the maximization of the renewable energy integration. For this reason, this part consider also an economic analysis in addition to the technical one. In order to reach a RES integration of almost 40% the solution with 100% wind is selected being the cheapest technology. No overproduction occurs in this first part with only wind power and the PV capacity remains equal to zero due to the higher costs. In order to reach higher values of RES integration, a combination of production from wind and PV is needed. The best share of wind and PV, in order to maximize the RES integration and minimize the total annual costs, stabilizes between 60% and 70% of wind production and 40-30% of PV production. The required capacity to reach 70% of RES integration without storage would be 250 GW of wind power and 200 GW of PV. The PV and wind power capacity is obtained considering an average value of the full-load hours over the considered six years equal to 1248 for PV and 1714 for wind power. It is important to underline that the current values are based on theoretical calculation and no territorial constraints or potential installed values of the Italian energy system are taken into account.

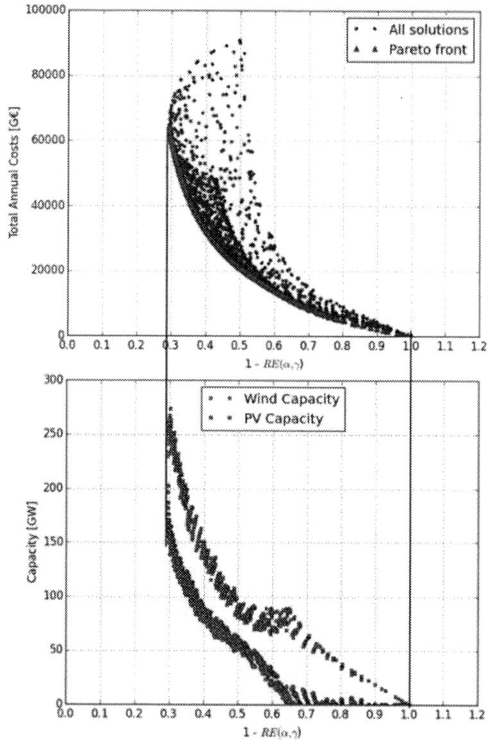

Figure 6: The graph on the top shows the results of the optimization analysis. On the bottom, the graph shows the trend of the PV and wind capacity (depending on the inputs of the model: α and γ) of the simulated solutions on the Pareto front.

4 CONCLUSIONS

In order to reach a future power system fully based on renewable generation, high penetration of generation from wind and solar sources is foreseen and, thus, issues related to the non-dispatchable power production must be faced. A theoretical model has been applied to the Italian case study in order to assess from a technical and economical point of view the level of renewable energy integration into the electric grid. This model, indeed, consider only the electricity sector and does not take into account possible interactions with heat and transport sector that could further improve the integration of renewables. Firstly, the best mix of electricity generation by VRES has been inspected studying the production profile of solar and wind energy, evaluated through time-series of 6 years with hourly resolution. The results have shown that the optimum electricity generation covered by wind (α) maximizing the renewables is equal to 0.55 and the remaining 45% by PV. Using an optimal mix between wind and photovoltaic power generation and assuming fully flexible power plants, a share of almost 40% of the electricity demand can be provided by VRES without the need of storage.

Secondly, the impact of storage systems has been studied. The results have shown that small but highly efficient storage devices should be favored over large but less efficient (seasonal) ones to reach a share of about 70% of the electricity demand being met by RES. In order to increase renewable energy integration, a seasonal large and low efficient storage is favored from a technical point of view.

To summarize, an optimization analysis has been implemented in order to find the best mix of renewable generation from a technical and economical point of view. A multi objective optimization algorithm has been applied to the problem. The results have shown that in order to reach a RES integration of almost 30% the solution with 100% wind is selected because is the cheapest technology. No overproduction occurs in this first phase with only wind power and PV capacity remains equal to zero due to the higher costs. In order to reach higher values of RES integration at low costs a combination of production from wind and PV is preferable. The best share of wind and PV production stabilizes between 60% and 70% of wind production and 40-30% of PV production. These values allow for a renewable energy integration of almost 70%.

The optimization analysis hence provides additional information compared to the technical analysis. It permits to explore the variation of the best share of wind and PV production at the increase of the renewable energy integration. The objective of costs minimization produces a difference in the results especially in the first phase with $RE(\alpha, \gamma) \leq 0.3$. Here, wind power production is preferred because of the higher costs of rooftop PV systems. At the decreasing of these costs this share would increase. The extreme ideal case is the one in which PV and wind have the same cost per unit of produced electricity and thus the optimum share is the one found through the technical analysis.

5 ACKNOWLEDGEMENTS

The authors wish to acknowledge the use of EURAC internal funding for the project "Regional Energy Modeling".

6 REFERENCES

[1] B. V. Mathiesen, H. Lund, D. Connolly, H. Wenzel, P. A. Østergaard, B. Möller, S. Nielsen, I. Ridjan, P. Karnøe, K. Sperling, and F. K. Hvelplund, "Smart Energy Systems for coherent 100% renewable energy and transport solutions," Appl. Energy, vol. 145, pp. 139–154, May 2015.

[2] K. Hedegaard, B. V. Mathiesen, H. Lund, and P. Heiselberg, "Wind power integration using individual heat pumps – Analysis of different heat storage options," Energy, vol. 47, no. 1, pp. 284–293, Nov. 2012.

[3] P. A. Østergaard, "Wind power integration in Aalborg Municipality using compression heat pumps and geothermal absorption heat pumps," Energy, vol. 49, pp. 502–508, Jan. 2013.

[4] B. Nastasi and G. Lo Basso, "Hydrogen to link heat and electricity in the transition towards future Smart Energy Systems," Energy.

[5] F. Steinke, P. Wolfrum, and C. Hoffmann, "Grid vs. storage in a 100% renewable Europe," Renew. Energy, vol. 50, pp. 826–832, Feb. 2013.

[6] S. Weitemeyer, D. Kleinhans, T. Vogt, and C. Agert, "Integration of Renewable Energy Sources in future power systems: The role of storage," Renew. Energy, vol. 75, pp. 14–20, Mar. 2015.

[7] S. Weitemeyer, D. Kleinhans, L. Wienholt, T. Vogt, and C. Agert, "A European Perspective: Potential of Grid and Storage for Balancing Renewable Power Systems". Energy Technology, vol 4, pp. 114–122, Jan. 2016.

[8] A. Alhamwi, D. Kleinhans, S. Weitemeyer, and T. Vogt, "Moroccan National Energy Strategy reviewed from a meteorological perspective," Energy Strategy Rev., vol. 6, pp. 39–47, Jan. 2015.

[9] ENTSOE. < https://www.entsoe.eu/Pages/default.aspx>

[10] STRATEGO project. <http://stratego-project.eu/>

[11] GSE. <http://www.gse.it/it/Statistiche/RapportiStatistici/Pagine/default.aspx>

[12] EnergyPLAN costs database. <http://www.energyplan.eu/useful_resources/costdatabase/>

[13] X. Yu and M. Gen, Introduction to Evolutionary Algorithms, vol. 0. London: Springer London, 2010.

[14] K. Deb and D. Kalyanmoy, Multi-Objective Optimization Using Evolutionary Algorithms. New York, NY, USA: John Wiley & Sons, Inc., 2001.

[15] M. S. Mahbub, M. Cozzini, P. A. Østergaard, and F. Alberti, "Combining multi-objective evolutionary algorithms and descriptive analytical modelling in energy scenario design," Appl. Energy, vol. 164, pp. 140–151, Feb. 2016.

CONDITIONS IN WHICH A PHOTOVOLTAIC SYSTEM IS MORE VIABLE THAN A LOW-TEMPERATURE SOLAR THERMAL SYSTEM

Isidoro Lillo[1], Manuel Silva[1], Sara Moreno[1] and Elena Pérez[2]
1 Department of Energy Engineering. University of Seville.
2 Andalusian Association for Research and Industrial Cooperation. Group of Thermodynamics and Renewable Energy.
Camino de los Descubrimientos s/n. 41092, Seville, Spain. Phone/Fax number: (+34)954487236/(+34)954487233.
E-mail: isidorolillo@etsi.us.es

ABSTRACT: Photovoltaic (PV) and low-temperature solar thermal (LT-STH) are the most widely used technologies in the building sector. This study determines, depending on the most influential parameters, the conditions in which a PV system is more beneficial than a LT-STH system from an economical, energetic and environmental point of view. The parameter used for economically evaluated both technologies is the levelized cost of energy (LCOE). Moreover the Greenhouse Gas Emissions factor is employed for an environmental evaluation. The main results shown on this study reveal that in most cases PV is economically and environmentally more viable than LT-STH, although it would be necessary to analyze the particular conditions of each site.
Keywords: Economic analysis, viability, CO_2 emissions, zero-energy building.

1 INTRODUCTION

Renewable energies have experienced a noteworthy surge during the last few years [1,2]. Years ago it was rare to see solar collectors or windmills around our cities, but nowadays they have become usual elements [3,4]. With a particular focus in the building sector, it is safe to say that the technologies that have achieved greater growth are PV and LT-STH [5,6].

The function of the PV technology analyzed in this paper is to generate electrical energy for residential buildings. This energy is allocated to conventional uses as lighting, appliance or for the generation of thermal energy. On the other hand it is considered that LT-STH technology exclusively generates thermal energy to obtain hot water through the increase of temperature by the effect of solar radiation. This means that PV technology presents higher exergetic efficiency than LT-STH technology. Against this, LT-STH technology is mainly characterized by higher energetic efficiency than PV technology [7].

From an economic point of view it can be highlighted that PV technology has reduced significantly its costs and new laws in different countries as Spain or Germany have been developed to allow the self-consumption. PV technology also contributes significantly to achieve the European law target about nearly zero-energy building [8,9]. For this reason and because the available free area in buildings is limited, the following question arises: For which technology would it be the use of the free area more economically and environmentally viable? The aim of this study is to provide a useful tool to decide in function of market conditions.

2 CURRENT STATE

To achieve the objective set out above is necessary to establish the relation existing among the most influential parameters. The selected ratios are listed below:

* Investment cost rate: $C_{I\_PV}/C_{I\_LT\text{-}STH}$
* Energetic efficiency rate: $\eta_{PV}/\eta_{LT\text{-}STH}$
* Exergetic efficiency rate: $\eta_{x,PV}/\eta_{x,LT\text{-}STH}$
* Electrical and thermal energy price rate: P_{EE}/P_{THE}
* Environmental conversion factor rate: CF_{EE}/CF_{THE}

The analysis of the current market situation shows that the investment cost of a PV system is between 150€/m^2 and 600€/m^2 depending on their characteristics. In the case of LT-STH the lowest limit of this range is 200€/m^2 and the highest 600€/m^2 [10-12]. Considering this information, the investment cost rate ($C_{I\_PV}/C_{I\_LT\text{-}STH}$) ranges from 0.25 to 3.

As noted earlier another important relationships between PV and LT-STH technologies are the energetic and exergetic efficiencies. According to the current state of technology the energetic efficiency rate ($\eta_{PV}/\eta_{LT\text{-}STH}$) changes from 0.2 to 0.8 while the exergetic efficiency rate ($\eta_{x,PV}/\eta_{x,LT\text{-}STH}$) varies from 2 to 7 [12].

To provide an economic evaluation it is necessary to know the price of energy replaced by solar technologies, in this case electrical energy from the grid and thermal energy obtained by any of the conventional energy sources. The market conditions and the country under analysis affect significantly the value acquired by the ratio P_{EE}/P_{THE}. Average European values indicate that this relationship evolves from 1 to 3 [13].

Finally it is analyzed the relationship between the electrical and thermal environmental conversion factor, CF_{EE}/CF_{THE} that varies from 1 to 4 [14].

3 HYPOTHESES

It is also important to summarize the hypotheses considered to frame the conditions in which this analysis is performed:

* Annual operation and maintenance costs: 1% of investment cost for PV technology and 5% of investment cost for LT-STH technology.
* Replacement costs: 30% of investment cost in year 15 for LT-STH technology, this cost is not considered for PV technology.
* Consumer price index: 3%.
* Energy price index: 1.5%.
* Lifetime: 30 years for both technologies.
* Energetic storage is not considered in this study.

4 METHODOLOGY

As mentioned previously the objective of this study is to provide a useful tool to decide which technology, PV or LT-STH, is more viable. The tool comprises a set of graphs in which the evolution of main economical and environmental parameters can be analyzed in function of rates specified in paragraph 2.

The first part of this analysis is based on the variability of LCOE and savings obtained using each kind of technology. In the second part, from an environmental perspective, it is estimated the impact avoided by the use of each technology.

4.1 Levelized cost of energy

LCOE can be calculated by the following equation in which C_t represents life cycle costs and E_t represents the electrical or thermal energy generated. Furthermore the economical parameters r and s represent the consumer and energy price index respectively [15].

$$LCOE = \left[\sum_{t=0}^{n} \frac{C_t}{(1+r)^t}\right] \Big/ \left[\sum_{t=0}^{n} \frac{E_t}{(1+s)^t}\right]$$

Life cycle costs of this expresion are calculated as the sum of the annual values of investment costs (C_I), operation and maintenance cost ($C_{O\&M}$) and replacement costs (C_R):

$$C_t = C_I + C_{O\&M} + C_R$$

Thermal and electrical energy are obtained using area (A), incident solar radiation (I) and the energetic efficiency of photovoltaic or thermal technology (η_{PV} or $\eta_{LT\text{-}STH}$) as follow:

$$E_E = A \cdot I\, \eta_{PV} \qquad E_{THE} = A \cdot I\, \eta_{LT-STH}$$

The relation between the LCOE of PV and LT-STH technologies is shown below:

$$\frac{LCOE_{PV}}{LCOE_{LT-STH}} = \frac{\left[\sum_{t=0}^{n} \frac{C_t}{(1+r)^t}\right]\Bigg|_{PV} \cdot \left[\sum_{t=0}^{n} \frac{E_t}{(1+s)^t}\right]\Bigg|_{LT-STH}}{\left[\sum_{t=0}^{n} \frac{C_t}{(1+r)^t}\right]\Bigg|_{LT-STH} \cdot \left[\sum_{t=0}^{n} \frac{E_t}{(1+s)^t}\right]\Bigg|_{PV}}$$

Taking into account the especified hypotheses in third paragraph, the LCOE resulting relationship expressed in terms of energetic efficiency and investment cost of PV and LT-STH technologies is:

$$\frac{LCOE_{PV}}{LCOE_{LT-STH}} = \frac{[C_I + 0.196 C_I]\big|_{PV}}{[C_I + 0.980 C_I + 0.193 C_I]\big|_{LT-STH}} \cdot \frac{A \cdot I\, \eta_{LT-STH}}{A \cdot I\, \eta_{PV}}$$

The simplification of the expression above is:

$$\frac{LCOE_{PV}}{LCOE_{LT-STH}} = \frac{1.196}{2.173} \frac{C_I\big|_{PV}}{C_I\big|_{LT-STH}} \cdot \frac{\eta_{LT-STH}}{\eta_{PV}}$$

4.2 Savings

Another fundamental issue related to the suitability of installing a PV or a LT-STH system in a building is the saving generated by replacing a conventional source of energy. This saving is obtained as the product of the generated energy and the cost of the conventional replaced energy. It is considered that in the case of a PV system the generated energy replaces the electricity from the grid, while in the case of a LT-STH system the generated energy replace the thermal energy produced from conventional sources as oil or natural gas. The following equation shows this relation:

$$\frac{S\big|_{PV}}{S\big|_{LT-STH}} = \frac{P_{EE}}{P_{THE}} \cdot \frac{E_E}{E_{THE}} = \frac{P_{EE}}{P_{THE}} \cdot \frac{A \cdot I\, \eta_{PV}}{A \cdot I\, \eta_{LT-STH}}$$

Analyzing from an energetic point of view this equation results:

$$\frac{S\big|_{PV}}{S\big|_{LT-STH}} = \frac{P_{EE}}{P_{THE}} \cdot \frac{\eta_{PV}}{\eta_{LT-STH}}$$

The exergetic efficiencies of a PV and a LT-STH system are shown below:

$$\eta_{x,PV} = \frac{I \cdot A \eta_{PV}}{I \cdot A \left(1 - \frac{298}{5000}\right)}$$

$$\eta_{x,LT-STH} = \frac{I \cdot A\, \eta_{LT-STH} \left(1 - \frac{298}{333}\right)}{I \cdot A \left(1 - \frac{298}{5000}\right)}$$

If the previous savings relationship is analyzed from an exergetic perspective the result is:

$$\frac{S\big|_{PV}}{S\big|_{LT-STH}} = \left(1 - \frac{298}{333}\right) \cdot \frac{\eta_{x,PV}}{\eta_{x,LT-STH}} \cdot \frac{P_{EE}}{P_{THE}}$$

4.3 Environmental impact

The relationship of the environmental impact avoided by the use of PV and LT-STH technologies instead of conventional energies is measured in terms of CO_2 emissions avoided ($GHG_{PV}/GHG_{LT\text{-}STH}$):

$$\frac{GHG\big|_{PV}}{GHG\big|_{LT-STH}} = \frac{CF_{EE}}{CF_{THE}} \cdot \frac{E_E}{E_{THE}} = \frac{CF_{EE}}{CF_{THE}} \cdot \frac{A \cdot I\, \eta_{PV}}{A \cdot I\, \eta_{LT-STH}}$$

This relationship expressed in terms of the parameters defined in paragraph 2 is:

$$\frac{GHG\big|_{PV}}{GHG\big|_{LT-STH}} = \frac{CF_{EE}}{CF_{THE}} \cdot \frac{\eta_{PV}}{\eta_{LT-STH}}$$

5 RESULTS

This paragraph presents the most outstanding results of this study.

5.1 Levelized cost of energy

Figure 1 shows the evolution of the ratio $LCOE_{PV}/LCOE_{LT\text{-}STH}$ depending on the investment costs and the energetic efficiencies relationships. The area of the graph above the $LCOE_{PV}/LCOE_{LT\text{-}STH}$ equal to one indicates that it is more expensive to generate electrical energy with a PV system than thermal energy with a LT-STH system, as the graph shows this is the situation in most cases.

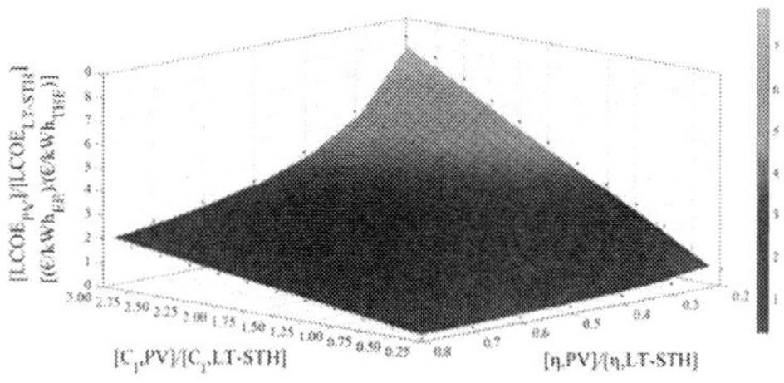

Figure 1: Levelized costs of energy of PV and LT-STH systems

The possibilities of use offered by electrical energy are much greater than the ones offered by low temperature thermal energy. The energy quality is measured by the exergetic variable. All this means that it is necessary to extend this analysis.

5.2 Savings obtained with PV and LT-STH systems

The second graph of this study shows the relationship of the savings generated by replacing the consumption of electrical energy by a PV system and the savings generated by replacing the thermal energy from a conventional source by a LT-STH system depending on prices and energetic efficiency rates.

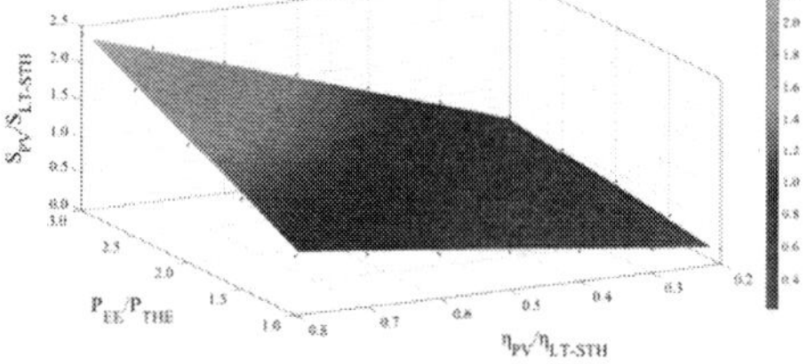

Figure 2: Savings obtained with PV and LT-STH systems depending on prices and energetic efficiency rates.

The surface area between the blue line ($S_{PV}/S_{LT\text{-}STH}$=1) and the yellow line ($S_{PV}/S_{LT\text{-}STH}$=2.5) in Figure 2 indicates the conditions in which the PV system provides greater savings than those generated by the LT-STH system.

Figure 3 shows the evolution of savings of PV and LT-STH technologies ratio depending on prices and exergetic efficiency rates.

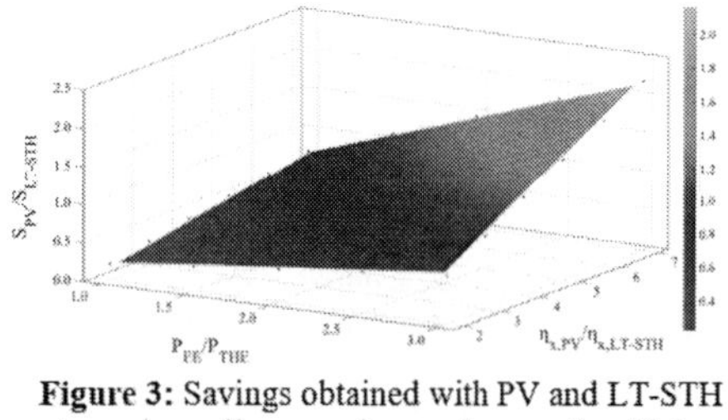

Figure 3: Savings obtained with PV and LT-STH systems depending on prices and exergetic efficiency rates.

The next figure shows the influence of the energetic efficiency ($\eta_{PV}/\eta_{LT\text{-}STH}$), the investment costs ($C_{I\_PV}/C_{I\_LT\text{-}STH}$) and the electrical and thermal energy prices (P_{EE}/P_{THE}) rates in a new parameter defined as savings over initial investment for each technology.

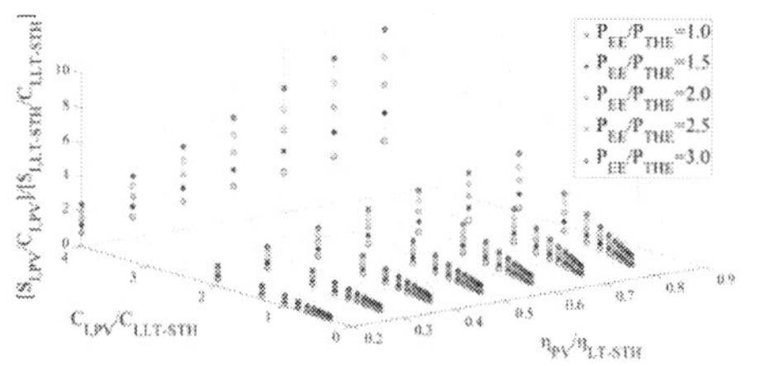

Figure 4: Evolution of savings over initial investment.

5.3 Environmental impact

The fifth and last graph included in this section of results shows the evolution of the $GHG_{PV}/GHG_{LT\text{-}STH}$ ratio depending on the energetic efficiency and the environmental conversion factor rate.

As can be noted from the graph provided when the CF_{EE}/CF_{THE} takes the value 1 the environmental impact avoided with a PV system is lower than the avoided with a LT-STH system. At present and for most countries this factor is usually greater than one, so the opposite extreme is analyzed. When the CF_{EE}/CF_{THE} coefficient takes the value 4 and from an energetic efficiency of 0.3, the environmental impact avoided by using a PV system is higher than the avoided when a LT-STH is employed.

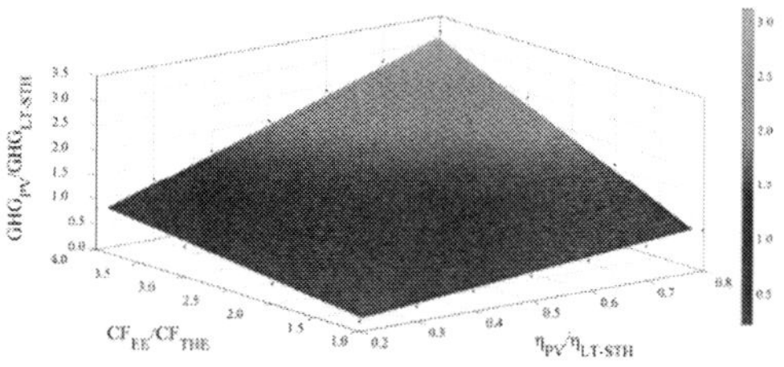

Figure 5: Environmental impact avoided by PV and LT-THS technologies.

5 CONCLUSIONS

As results show, nowadays PV technology is, in most cases, economically and environmentally more viable than LT-STH technology.

The thermal storage linked to LT-STH technology is more economical and reliable than the electrical storage associated to the PV technology, mainly chemical storage batteries. If costs and efficiencies related to energy storage would have been taken into account in this study, LT-STH systems would be more viable than PV systems.

6 REFERENCES

[1] G. Wetstone, K. Thornton, R. Hinrichs-rahlwes, S. Sawyer, M. Sander, R. Taylor, D. Rodgers, M. Alers, H. Lehmann, M. Eckhart, and D. Hales, "Renewables 2016 Global Status Report," 2016.

[2] N. Kannan; D. Vakeesan, "Solar energy for future world: A review," Renew. Sustain. Energy Rev., vol. 62, pp. 1092–1105, 2016.

[3] M. amado; F. Poggi; A. R. Amado, "Energy efficient city: A model for urban planning,"

Sustain. Cities Soc., 2016.
[4] J. Kanters; M. Wall, "A planning process map for solar buildings in urban environments," Renew. Sustain. Energy Rev., vol. 57, pp. 173–185, 2016.
[5] M. Tripathy; P.K. Sadhu; S.K. Panda, "A critical review on building integrated photovoltaic products and their applications," Renew. Sustain. Energy Rev., vol. 61, pp. 451–465, 2016.
[6] A.K. Pandey; V.V. Tyagi; J.A.L. Selvaraj; N.A. Rahim; S.K. Tyagi, "Recent advances in solar photovoltaic systems for emerging trends and advances applications," Renew. Sustain. Energy Rev., vol. 53, pp. 859–884, 2016.
[7] A. G. H. C. Good; I. Andresen, "Solar energy for net zero energy buildings - A comparison between solar thermal, PV and photovoltaic-thermal (PV-T) systems," Sol. Energy, vol. 122, pp. 986–996, 2015.
[8] W. Pan; K. Li, "Clusters and exemplars of buildings towards zero carbon," Build. Environ., vol. 104, pp. 92–101, 2016.
[9] A. Hermelink; S. Schimschar; T. Boermans; L. Pagliano; P. Zangheri; R. Armani; K. Voss; E. Musall, "Towards nearly zero-energy buildings. Definition of common principles under the EPBD," 2013.
[10] J. Rawlins and M. Ashcroft, "Small-scale Concentrated Solar Power - A review of current activity and potential to accelerate deployment," 2013.
[11] International Renewable Energy Agency, "Solar Heat for Industrial Processes",2015.
[12] C. Parrado; A. Girard; F. Simon; E. Fuentealba, "2050 LCOE (Levelized Cost of Energy) projection for a hybrid PV (photovoltaic)-CSP (concentrated solar power) plant in the Atacama Desert, Chile," Energy, vol. 94, pp. 422–430, 2016.
[13] R.M. Lazzarin, "Solar cooling: PV or thermal? A thermodynamic and economical analysis," Int. J. Refrig., vol. 39, pp. 38–47, 2014.
[14] E. Commission, "Eurostat." [Online]. Available: http://ec.europa.eu/eurostat. [Accessed: 09-Jun-2016].
[15] U.S. Department of Energy, "U.S. Energy Information Administration." [Online]. Available: https://www.eia.gov. [Accessed: 09-Jun-2016].
[16] P. Konstantin, Praxisbuch Energiewirtschaft: Energieumwandlung, - transport und *-beschaffung im liberalisierten Markt*, Springer. Berlin, 2009.

PHOTOVOLTAIC PLANT ORIENTATION STRATEGIES TO MINIMIZE GRID EXCHANGE IN FREE FIELD AND BUILDING INTEGRATED SETUPS

Philip Ingenhoven, Grazia Barchi, Marco Lovati, David Moser
Institute for Renewable Energy, EURAC Research, Viale Druso 1, 39100 Bolzano (BZ), Italy
*Phone: +39 0471 055634; Fax: +39 0471 055699; E-mail: philip.ingenhoven@eurac.edu

ABSTRACT: High penetration of photovoltaic generation might lead to problems in the distribution grid such as instability, frequency deviations and reverse power flow. Parts of these problems can be mitigated by minimizing the exchange from and to the grid. A possible strategy to minimize the power injected into the grid and to maximize the power used to supply the load is to design a PV system in order to optimize the matching with the load. The classical strategy used in PV installation is to maximize the production. However, to address load matching, power maximization is not anymore the suitable approach. For this purpose, the aim of this work is to present two different strategies to optimize the orientation of photovoltaic modules. The first considers every possible orientations and no restriction on the surface area as possible in a free field installation. The second considers the surface of a building and tries to optimize the module orientations taking into account typical constraints from the built environment (e.g. limited area, shading from nearby building, etc).

Keywords: photovoltaic orientation optimization, building integrated PV (BIPV), grid integration, load matching, self-consumption maximization.

1 INTRODUCTION

The cumulative worldwide photovoltaic (PV) capacity is growing and in Europe alone has reached about 90 GW [1]. The most part of the installed PV system are grid-connected, this means that the PV generators are part of the distribution feeder and can cause different problem due to their intermittent nature [2]. Since the PV system's production is strictly dependent on the availability of the solar resource, this implies that its production cannot be sure all the time. In terms of distribution grids, this situation creates a miss correlation between the time when the power from PV flows into the grid and the instant when the power is required by the loads. Critical situations happen when, for example during the morning hours, the PV production is very high and the energy demand is lower. The low correlation between the PV generators and the energy demand is cause of over voltage and power flow problems [3]. In the distribution grid the utility distributor has to assure the voltage requirements defined for the European Country in [4].

However, the large voltage oscillation is not the only possible problem caused by grid-connected PV but the increase of reverse flows and the interaction of the harmonics mainly introduced by the inverter should affect the feeder at low distribution level [5].

The most common practice to install a PV system is to place south oriented modules with a tilt optimized in order to maximize the production over the year. The high power coming from PV has to be regulated before flows into the grid are allowed to prevent overvoltage as reported in [7], or part of energy stored in an energy storage system [6]. However, even if the energy storage is acquiring a growing importance in several technologies (i.e. hydrogen, electrochemical...), the cost are still high and their application is not always feasible.

For this reasons, an optimization of load matching by analyzing various combinations of PV orientation and inclination is a possible strategy that can support even higher penetration rates without investments in the current grid infrastructure.

Building integrated photovoltaic (BIPV) has been demonstrated to be a suitable solution for load matching as reported in [7]. This strategy will change if other objectives are considered i.e. module degradation, load matching, soiling, etc.

In [8] is mentioned that the load matching can be performed following two ways. The first is the modulation of the demand to the generation through the Demand Side Management (DSM), while the second is to tune the generation to the building energy needed.

The purpose of this work is to analyze the generation profiles of various PV configurations and see how these match the load profile. The two methods presented in this paper can help to change the current paradigm from yield maximization to grid compliance optimization. With the rising degree of PV penetration, the importance of grid servicing installation is getting obvious. As the price of storage solutions is still not financially viable, we propose strategies which may relieve the strain on the grid without the financial impact of storage systems. The algorithms and strategies developed in this work will help PV installers, grid planners as well as architects that are interested in BIPV solutions.

2 METHODS AND RESULTS

In this section methods for different situation of PV integration are presented. If there are no special restrictions a simple optimization of azimuth and tilt finding the minimal grid exchange can be done. The situation changes if a building integrated solution is to be found. Here the geometry of the building has to be considered carefully. In the following two section the two situation are described in detail.

2.1 Method for free-field load matching

The first approach presented in this section, namely *method A,* consider to achieve the load-matching without limited area or restrictions.

A year of meteonorm™ data of radiation and temperature was used to compute the PV production. The in-plane irradiation is reconstructed using the direct, diffuse and global horizontal irradiance. An array of possible tilt and azimuth angles from α=-90°(east) to 90°(west) of south for the azimuth and from β = 5° to 90° tilt angle (both in steps of 5°) is built in order to perform the optimization input. To compute the in-plane irradiance the following isotropic equation is used [11]:

$$G_\theta = G_D \cdot cos\theta + G_d \cdot \frac{(1 + cos\beta)}{2} + \rho \cdot G_H \cdot \frac{(1 - cos\beta)}{2} \quad (1)$$

where θ is the angle of incidence, which itself is a function of time and is computed every 15 minutes. The angle β denotes the tilt of the module. G_D is the direct, G_d is the diffuse, and G_H is the global horizontal irradiance. The production is then computed for a polycrystalline module using the equations presented in [11]. The various parameters defined in [11] and used in this work were measured from a poly-silicon PV module in indoor measurements. To predict the temperature behavior correctly we used measured data from [12], where the Ross factor for a building integrated system using poly-silicon PV was found to be about K=0.03 °C m^2/ W. The power in each tilt/azimuth combination was normalized to an installed power of P_{Wp}=1kWp.

For each combination of tilt (β) and azimuth (α) the power vector $P_{\beta,\alpha}$ (in time with ΔT = 15 min) was found. As load profile, the residential consumption provided by the German electricity provider Westnetz [i] were used. The load profile is normalized to 1*MWh* of consumption which is a good approximation of the average residential electricity consumption per capita [13]. The optimization function is given by computing the difference between production and daytime consumption *C* every 15 minutes, like in the following equation:

$$\int dt \left| \sum_{\beta,\alpha} a_{\beta,\alpha} P_{\beta,\alpha} - C \right| = Minimum \quad (2)$$

where the coefficients $a_{\beta,\alpha}$ indicates the power installed for each tilt and azimuth combination. The resulting non-zero coefficients are given in Table 1:

Table 1 – Values of coefficients result ($a_{\beta,\alpha}$) from the optimization

		Azimuth α				
		-90°	-30°	-25°	70°	90°
Tilt β	90°	82.9	62.4	97.5	33.9	110.9

This result is remarkable that is favors purely façade installation and discards all other more efficient inclinations, in terms of total production. In total 387.7 *Wp* have to be installed per person, which covers 64.6 % of the day time load and 36% of the total load. The same amount of PV installed following a yield maximization strategy with tilt β=40° and azimuth α= 20° west would produce the amount of energy over the year, which would net cover 104.8% of the daytime load.

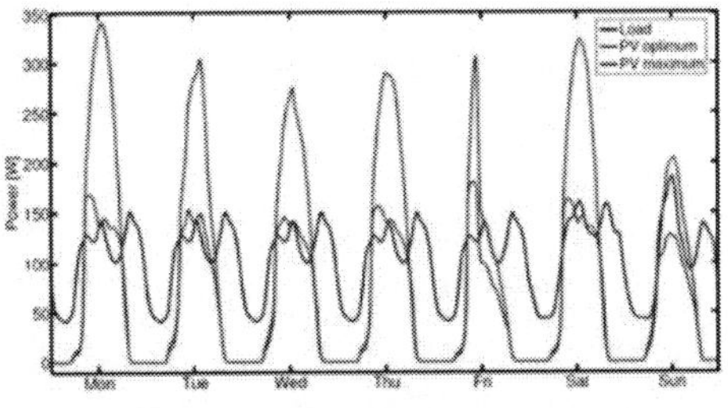

Figure 1- Difference between a maximal (red) and the optimal PV installation (green) in comparison with the load profile (blue) for a week in May.

In terms of energy exchange with the grid the found optimum has minimal grid exchange, whereas the maximally producing configuration shows a large over production especially in summer, which might cause problems to the grid. In quantitative terms we find that in the case of maximal production the over production during the day amounts to 37.6% of the daytime load whereas the demand from the grid amounts to 32.8%. In the minimal grid exchange case we find that the over production is massively reduced to only 5.4%, however the energy taken from the grid increases only slightly to 40.8%. Figure 1 shows the difference between a maximal (red) and the optimal PV installation (green) in comparison with the load profile (blue) for a week in May.

2.2 Method for building integration

The second approach, called *method B,* considers the restriction due to a building integrated installation. The computation is implemented in Daysim based on Radiance [14] combined with a genetic optimization algorithm performed in Octopus [15]. Given a geometry for the building, the degrees of freedom consist in the possibility of varying tilt angle (β) and percentage of coverage (κ) over each of the building's surfaces. For each combination (tilt east, tilt south, tilt west, quantity of installed PV in percentage for east, south, west) the power output of every HOY (Hour Of a Year) is calculated as

$$P_{out}(\mathrm{HOY}) = \eta \cdot PR \cdot \textstyle\sum_{n=1}^{n_c} H_n \cdot A_n, \quad (3)$$

where P_{out}(HOY) is the instantaneous electricity production of the system, η is the module efficiency, PR is the performance ratio of the system assumed 0.8. n is the number of surfaces in the system, H_n is the n-th irradiation retrieved from RADIANCE, A_n is the n-th area of a PV module. Note that n depends on the three percentage of coverage chosen for each configuration, and that these percentages influence the time of the day in which the n-th module receives irradiation. The configuration is evaluated through the value:

$$f = \textstyle\sum_{HOY} \left| P_{out}(HOY) - Ab \cdot L_{person} \right|, \quad (4)$$

where *f* is the function to be minimized, P_{out} is the power output of the system in a specific hour, L_{person} is the average consumption for a single person described above and *Ab* is the number of inhabitants in the building. The value *f* is the power that has to be exchanged via the grid, either as consumption or as over production. Minimizing *f* will give the PV configuration with the best load matching, just as in *method A*. The main difference between the two analyses is the use of RADIANCE [16] which uses ray tracing approach instead of the simple geometric approach of *method A*. This way local shading and the building geometry can be taken into account in detail. The irradiance data as well as the consumption profile used is the same as in *method A*. The main result is that the optimal configuration depends on the ratio of the area available for PV installation and the load of the specific building, see Figure 2-4. For a low load the optimal result does not differ much from the free field solution. However when the load is increased more and more southerly installed PV modules are necessary to meet the demand.

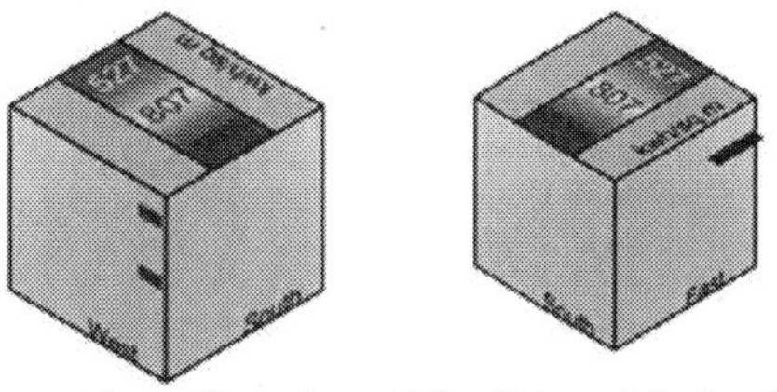

Figure 2 -configuration of the PV module for the building occupied by 1 inhabitant. (ca. 0.8 kW_{peak})

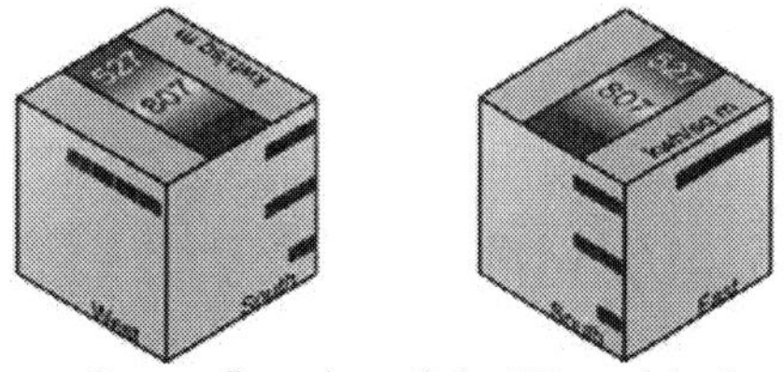

Figure 3 - configuration of the PV module for the building occupied by 5 inhabitant (ca. 2.5 kW_{peak}).

Figure 4 -configuration of the PV module for the building occupied by 10 inhabitant (ca. 5.6 kW_{peak}).

Table 2 - Results from the optimization (*method B*)

	west	south	east
1 inhabitants	Tilt 90° % 10	Tilt N.A. % 0	Tilt 0° % 10
5 inhabitants	Tilt 80° % 20	Tilt 90° % 30	Tilt 80° % 20
10 inhabitants	Tilt 70° % 40	Tilt 90° % 60	Tilt 90° % 60

3 CONCLUSIONS

In this paper the optimal orientation of PV modules for load-matching is presented with two methods. The complementarity of these methods is

The results for the free field analysis (*method A*) show that the optimal PV orientation is given by 90° tilted modules mainly installed facing east and west with smaller number of modules installed facing in southern directions. However this configuration is not feasible in an in urban context, where the surfaces are usually limited and the effects of shading by nearby buildings are not negligible. For that reasons a second analysis (*method B*) investigation introducing the restrains mentioned above was performed. The main results is that the optimal configuration depends on the ratio of the area available for PV installation and the load of the specific building. For a low load the optimal result does not differ much from the free field solution. However when the load is increased more and more southerly installed PV modules are necessary to meet the demand. In the future it is planned to implement this method to existing buildings and extent the analysis of the grid exchange and include information about the power flow in more detail. Possible extension of the work will include an economic analysis comparing the cost for not fully utilizing the PV potential by not using a maximally productive orientation with the cost of storage devices.

4 AKNOWLDEGMENT

The work was possible thanks to the project Flexi BIPV 5-1a-232, which has received funding from the ERDF managed by the Province of Bolzano.

REFERENCES

[1] EPIA, "Global market outlook for photovoltaics 2015-2019," Jun. 2015.

[2] J. V. Paatero and P. D. Lund, "Effects of large-scale photovoltaic power integration on electricity distribution networks," Renewable Energy, vol. 32, no. 2, pp. 216–234, Feb. 2007.

[3] S. Conti, S. Raiti, G. Tina and U. Vagliasindi, "Study of the impact of PV generation on voltage profile in LV distribution networks," *Power Tech Proceedings, 2001 IEEE Porto*, Porto, 2001, pp. 6 pp. vol.4-.

[4] CENELEC European Committee for Electrotechnical Standardization, EN 50160, Voltage characteristics in public distribution systems.

[5] J. H. R. Enslin et al., "Harmonic interaction between large numbers of photovoltaic inverters and the distribution network," Power Tech Conference Proceedings, 2003 IEEE Bologna, 2003, pp. 6 pp. Vol.3-.

[6] A. Zahedi, Maximizing solar PV energy penetration using energy storage technology, Renewable and Sustainable Energy Reviews, Volume 15, Issue 1, January 2011, Pages 866-870

[7] R. Tonkoski, L. A. C. Lopes and T. H. M. El-Fouly, "Coordinated Active Power Curtailment of Grid Connected PV Inverters for Overvoltage Prevention," in *IEEE Transactions on Sustainable Energy*, vol. 2, no. 2, pp. 139-147, April 2011.

[8] Voss, Karsten [et. al], "Load Matching and Grid Interaction of Net Zero Energy Buildings", in Proceedings of EUROSUN 2010 International Conference on Solar Heating, Cooling and Buildings, Graz, Austria, 28 Sept.- 1 Oct., 2010, 8p.

[9] W. Sprenger, Electricity yield simulation of complex BIPV systems, H. F.ISE, Ed. Freiburg/Brsg.: Fraunhofer Verlag, 2013.

[10] A. Padovan and D. Del Col, "Measurement and modeling of solar irradiance components on horizontal and tilted planes," Solar Energy, vol. 84, no. 12, pp. 2068–2084, Dec. 2010.

[11] Yordanov, G.H., "Relative efficiency revealed: Equations for k1-k6 of the PVGIS model," in *Photovoltaic Specialist Conference (PVSC), 2014 IEEE 40th* , vol., no., pp.1393-1398, 8-13 June 2014

[12] L. Maturi, G. Belluardo, D. Moser, and M. Del Buono, "BiPV System Performance and Efficiency Drops: Overview on PV Module Temperature Conditions of Different Module Types," *Energy Procedia*, vol. 48, pp. 1311–1319, 2014.

[13] http://shrinkthatfootprint.com/average-household-electricity-consumption Home Electricity Use Per Person 1157 kWh/year for Italy source: enerdata
[14] REINHART, Christoph; BRETON, Pierre-Felix. Experimental validation of Autodesk® 3ds Max® Design 2009 and DAYSIM 3.0. Leukos, 2009, 6.1: 7-35.
[15] http://www.food4rhino.com/project/octopus?etx
[16] WARD, G. RADIANCE reference manual. IEA SHC Task, 1996, 12.

[i] *westnetz.de*

PV PLANT REPOWERING: INCREASING THE INSTALLED PV POWER; A CHALLENGING APPLICATION FOR AN INTELLIGENT ACTIVE POWER CURTAILMENT

Rodrigo Estrella[1], Jonas Hagemann[1], Kristin Albers[2], Manfred Kammerer[2]
[1] Skytron Energy GmbH, Ernst-Augustin-Straße 12, D-12489 Berlin
[2]Parabel GmbH, Holländerstraße 34, 13407 Berlin

ABSTRACT: The repowering of the existing PV plant Krempendorf I having an apparent power of 22.48 MVA (DC peak power of 24.3 MWp), was carried out through the Krempendorf II extension with an apparent power of 7.84 MVA (DC peak power of 10.0 MWp). The following specific requirements had to be fulfilled: Since the grid connection point was not updated, the apparent power of both plants had to be limited to 22.48 MVA. Due to different feed-in tariffs, the old plant with a higher tariff had to be given priority while the plant extension had to deliver power that whenever possible, did not exceed the apparent power limit. The implemented control concept had to avoid tripping the overcurrent protection and, at the same time, maximize the production of the plant extension (Krempendorf II). In order to fulfil these requirements a new control feature had to be developed. In this paper the methodology for developing this feature is presented, together with the results including real measurements.

Keywords: Plant Control, Grid Integration, Energy Performance

1 INTRODUCTION

The "70 percent" regulation in the EEG 2012 was implemented by the German government in order to improve grid stability owing to the high penetration of PV power into the public grid. This regulation results in a reduction of a plant's annual energy yield by three to eight percent for small PV installations [1][2]. Applying this oversizing factor to Krempendorf I, the installed nominal power had to be increased to 34.3 megawatt peak (a factor of 1.43). The required area and the grid infrastructure were available at the Krempendorf I site. A yield analysis performed by Parabel GmbH resulted in an average reduction in annual energy production of four percent (losses caused by curtailment of the PV plant Krempendorf II). Taking both plants' yields together, the reduction was found to be around 1.2 percent.

Nevertheless, due to different feed-in tariffs, the power delivery of the Krempendorf I plant had to take priority over that of the plant extension Krempendorf II.

Under the requirements defined in the business case, skytron energy was in charge of developing the control concept. Part of the challenge was to avoid the overcurrent protection being tripped by rapid cloud movement, yet avoiding further investment in additional faster metering or communication systems within the PV plant.

In section 2 the repowering concept proposed by Parabel is described together with the control architecture proposed by skytron energy. The methodology for the development of the overshoot protection feature is presented in section 3. Section 4 shows real measurements from the repowered PV plant together with a data analysis. Finally, conclusions and insights are presented in section 5.

2 REPOWERING CONCEPT

The repowering concept proposed by Parabel is illustrated in Figure 1. The Krempendorf I plant must run without any limits as long the grid operator does not send a curtailment command. The remaining power must be provided by the plant Krempendorf II as long as the limit on current is not exceeded.

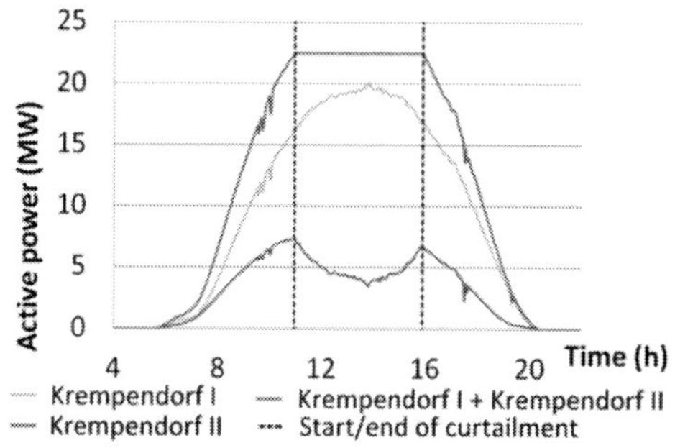

Figure 1: Active power profile of the proposed repowering concept (clear sky)

Figure 2 shows the control architecture that was implemented. Each PV plant has been equipped with its own controller. Krempendorf I (green) is configured with a nominal power of 22.48 MW and Krempendorf II with 7.84 MW. The controller of plant Krempendorf II receives the sum of both plant's power output (red line) as its feedback input.

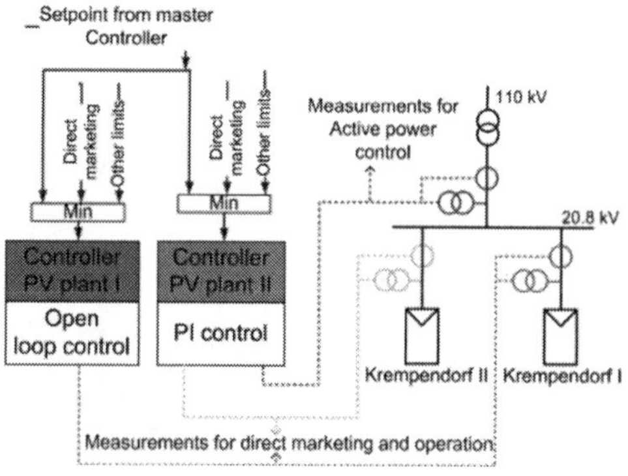

Figure 2: Proposed control architecture

2.1 Setting for the overcurrent protection relay

In order to protect the point of common coupling (PCC) of both plants (Krempendorf I and Krempendorf II) against overload, an overcurrent protection relay was installed. Given a nominal apparent power of 22.48 MVA at 20.8 kV, the maximal allowed current is 624 A. The protection relay was configured to trip at a current of 690A with a time delay of eight seconds, meaning that the relay will trip when a current 10.6 percent above the nominal value persists for more than eight seconds. The time counter is only reset when the current drops below the trip limit again.

2.2 The problem of moving clouds

With the proposed control concept shown in Figure 2, cloud movement will cause rapid power variations in the Krempendorf I plant; these will be seen as disturbances by the PI controller of the plant Krempendorf II (Figure 3).

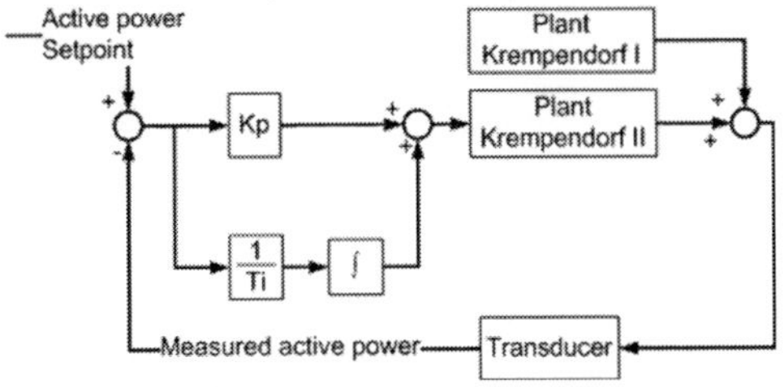

Figure 3: Control diagram of the plant Krempendorf II

When a cloud temporarily covers the PV panels, the power of PV plant I will decrease, causing PV plant II to deliver its full available power. When the cloud moves away the power of PV plant I will recover within a short time. Due to this fast increase there is the risk of a power overshoot that could cause the protection relay to trip.

In case of a tripping of the protection relay the PV plant II is disconnected, the reconnection would have to be carried out by hand. Since no technician is located permanently at the plant, the energy losses due to a relay tripping can be significant.

Therefore, an intelligent control feature had to be developed which was able to predict risky situations and react accordingly in order to reduce overshoots and avoid tripping the overcurrent protection.

3 THE OVERSHOOT PROTECTION FEATURE

3.1 Simulation model

In order to develop the overshoot protection feature, a simulation model was implemented using Matlab Simulink. The model includes the power plant controller (PPC) from skytron energy, generic inverter models and an electrical model. The electrical part of the model includes transformers and connection lines (Figure 4).

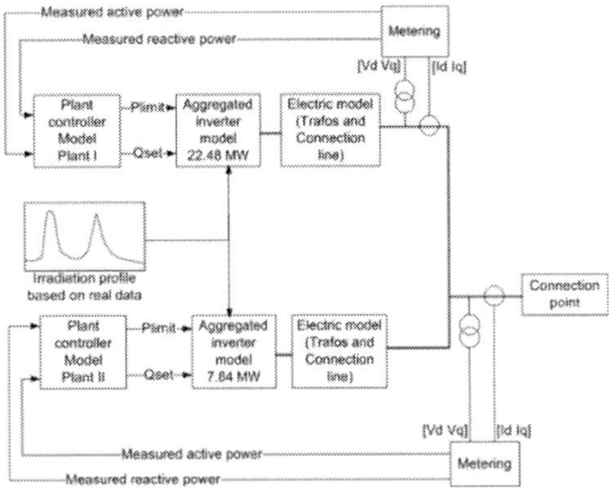

Figure 4: Simulation model in Matlab Simulink

The time constants of the inverter models were parametrized to match the real reaction times of the real power plants. The PI controller was also tuned. The detailed procedure of tuning is out of the scope of this paper.

Additionally, the PPC of PV plant II was configured with a constraint on its active power gradient. This feature limits the rate of change of the active-power set point based on the actual power. The gradient limit was configured to a maximum of ten percent of the nominal power per minute. The gradient limitation is available as a standard function in skytron energy's PPC.

3.2 Simulation of a moving cloud event

Active power measurements at the plant Krempendorf I on 26th May 2015 were used to extrapolate an irradiation profile. The behavior of the active power at the PCC was simulated (Figure 5).

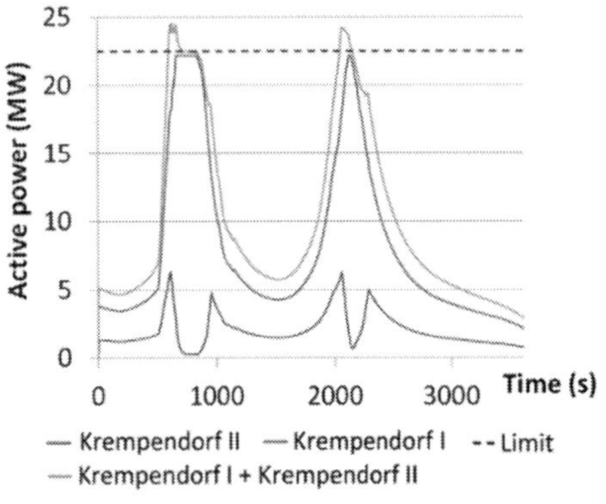

Figure 5: Simulation of a moving clouding event

It can be noticed that after a cloud moves away, there is a rapid increase in the power output of Krempendorf I (blue line), causing an overshoot in the power at the point of common coupling (green line). This overshoot occurs because the PI controller of the Krempendorf II plant requires time to compensate for the disturbance. A detailed view of the controller signals of Krempendorf II can be seen in Figure 6.

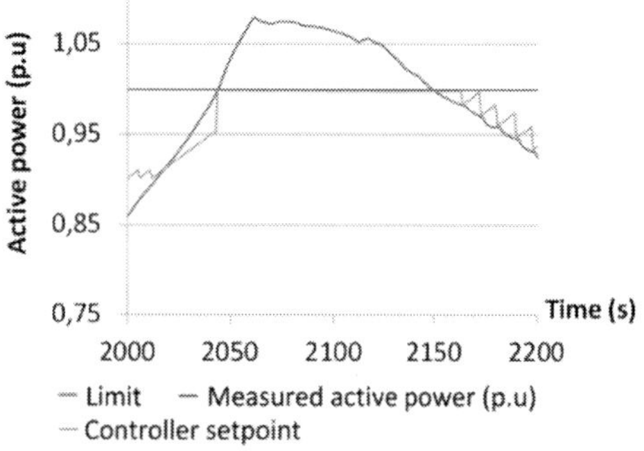

Figure 6: Active power controller signals during an overshoot event

The controller set point (green line) has a limit on the active power gradient. The overshoot of the measured value (red line) is more than five percent. The objective was to find a method to reduce this overshoot without major hardware modifications (i.e. without additional communication lines or a faster controller).

3.3 Implementation of the overshoot protection feature

References can be found to methods for improving PI controller performance by disturbance events. One mechanism would be the addition of a feed-forward path, however this has the disadvantage of requiring additional inputs for the PI controller [3]. In the current case, additional, fast measurement data would have been required for the Krempendorf I plant, which was not feasible under the time and budget constraints for the project.

Another method would have been to use two independent control loops for both plants, then to implement an override strategy at the Krempendorf II plant, using the total power of both plants as a constraint as indicated in [4]. But this solution presents the same problems as the solution mentioned above.

An alternative solution was found which uses a predictive procedure to trigger the overriding of the controller output. The trigger is activated when a risky situation is detected. It considers just the total power delivered by both plants at the point of connection. This feature was called "overshoot protection".

The operation of the PI controller with overshoot protection can be described by three states that are controlled by means of a trigger (Figure 7).

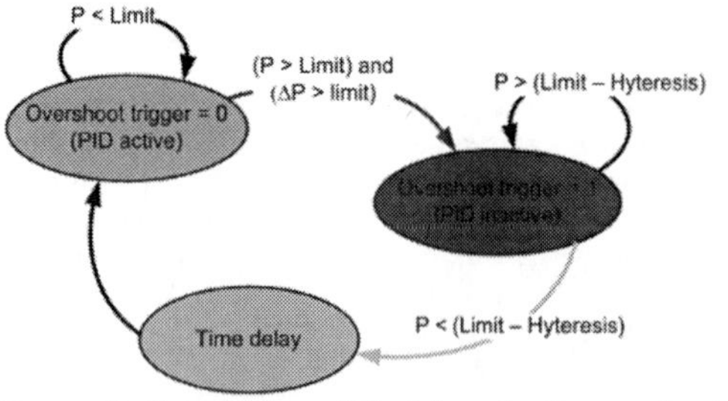

Figure 7: State diagram of the trigger for the overshoot protection

The states of the overshoot protection are:

- Overshoot protection-inactive (green), when the power is below the critical limit.
- Overshoot protection-active (red), when the power is above the critical limit and also the power gradient.
- Transition (yellow), when the power returns to an acceptable level. The trigger is still active but can be deactivated after a configurable time delay.

In the case of the Krempendorf plant, two identical trigger blocks were implemented (Figure 8). When the power is below P1, there is no output limitation (green zone). The yellow region of the graphic represents the values where there is a risk of the protection relay being tripped. When the power lies between the values P1 and P2 and the power gradient is greater than ΔPmax1, the first trigger is activated. When the power is greater than the limit P2 and the power gradient is greater than ΔPmax2, the second trigger is activated. The override of the PI controller occurs when either of the triggers is active.

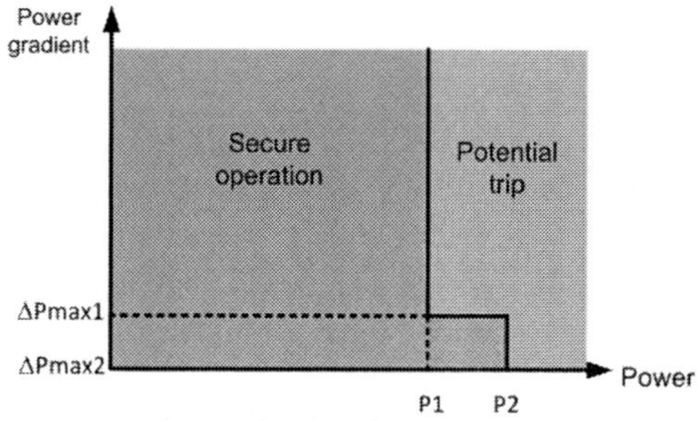

Figure 8: Diagram showing the gradient limits for trigger activation

The trigger produces three actions in the PI controller: override the controller error state to zero, reset the integrator and override the control signal to have a configurable safe value (Figure 9).

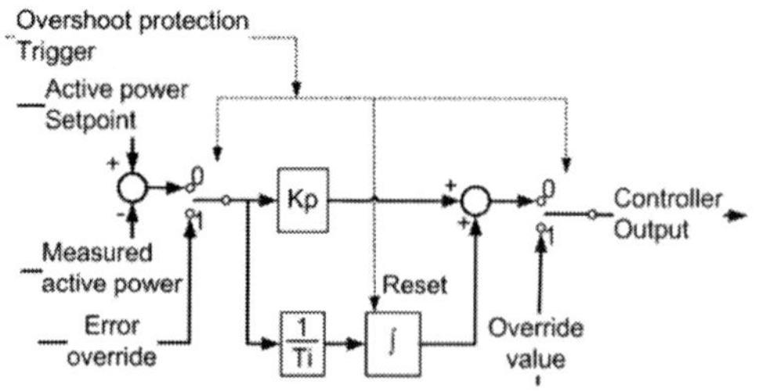

Figure 9: PI controller with override to prevent overshoot conditions

3.4 Simulation results using overshoot protection

After developing the model for the overshoot protection, the feature was tested for the same scenario as previously; the results are shown below (Figure 10). A significant reduction in the amount of overshoot (red line) can be observed.

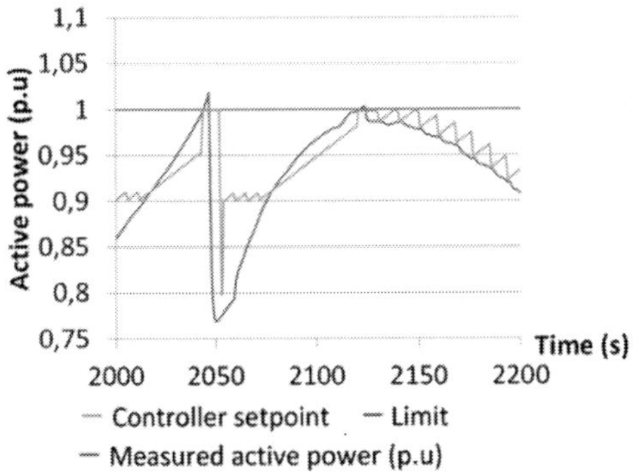

Figure 10: Active power controller signals using the overshoot protection

The overshoot is significantly reduced (red line). Figure 11 shows the triggering of the overshoot protection against the PI controller output. The figure shows how the output of the controller (red line) is reset to zero when the overshoot protection is triggered (blue line). The results correspond to the expected behavior.

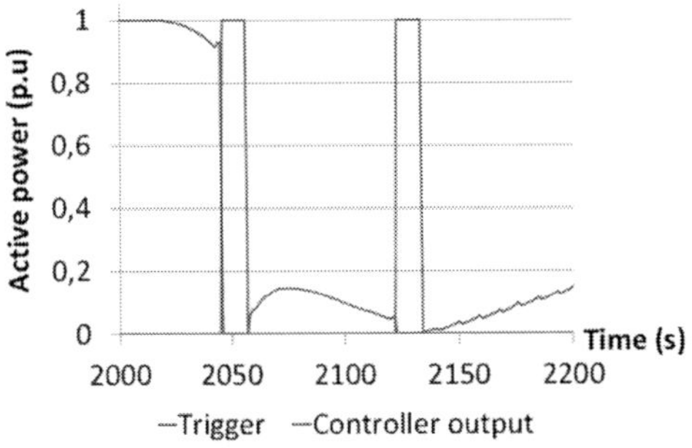

Figure 11: Trigger signal and controller output during overshoot protection events

4 FIELD RESULTS

In this section, real data from the monitoring and the control system are presented and analyzed.

4.1 Controller response during an overshoot event

In order to show the results in the field, a day with intermittent moving clouds was chosen. The day chosen was the 4th May 2016. The data for this day was analyzed at a one second resolution (Figure 12).

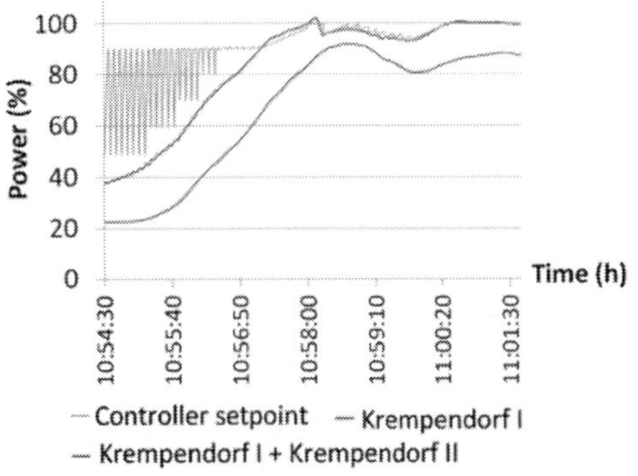

Figure 12: Power measurements during an overshoot event

For the event shown, the resulting overshoot (red line) did not trip the protection relay. In addition, according to the event logger in the PPC, the overshoot protection was triggered at the time 10:58:07 am.

Figure 13 shows the signals in the PI controller of Krempendorf II. The overshoot protection causes a reset in the integral part of the PI controller (green line) and the controller output is reset to zero (red line).

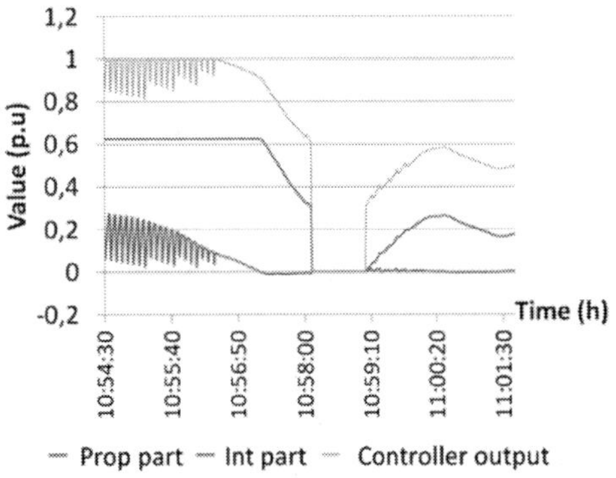

Figure 13: Signals of the active power controller

4.2 Active power injection during a cloudy day

The daily performance was also analyzed for the 4th May using data at a one minute resolution (Figure 14).

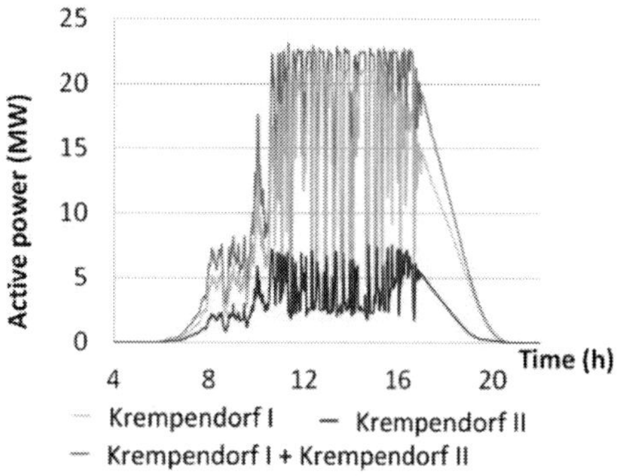

Figure 14: Active power measurements on a day with moving clouds

It can be seen that the combining output of both plants using the proposed control concept also reduces the power fluctuations at the point of common coupling (PCC).

4.3 Energy yield analysis

The energy production of the expanded Krempendorf plant was analyzed for the period August 2015 to April 2016. The data used has a resolution of 15 minutes, with each measurement point averaging the individual one-minute values. Using these measurements, the energy yield has been calculated for each month (Figure 15).

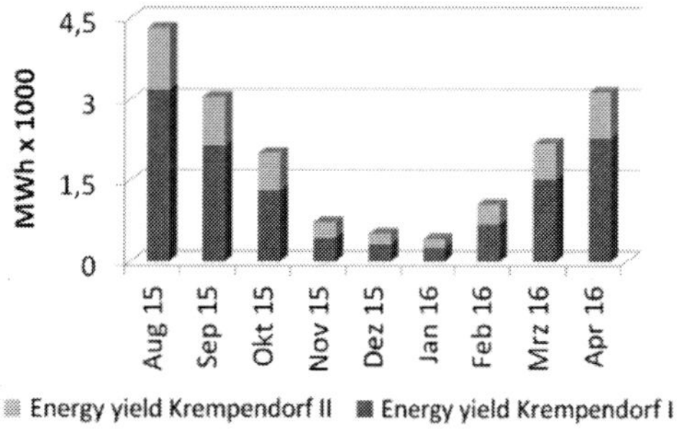

Figure 15: Energy yield from August 2015 to April 2016

Below in Table 1 the main results from the energy yield analysis are presented.

Table 1: Results of energy analysis for the period from August 2015 to April 2016

	Value
Energy production measured at PV plant Krempendorf I	12,084.7 MWh
Energy production measured at PV plant Krempendorf II	5,271.9 MWh
Energy production measured at the point of common coupling (PCC)	16,830.8 MWh
Cable and substation losses	3.03 %
Increase of energy production	43.6 %

5 CONCLUSIONS

Based on the results presented here, plant expansion without increasing grid connection capacity appears to make sense since one can avoid additional investment costs (ten percent or more) in the grid infrastructure.

Further, the losses caused by curtailment of the retrofitted plant are around just four percent. These losses can be compensated by improving the performance ratio; this was the case for Krempendorf II.

The repowering concept presented here can be implemented for all existing PV plants with no tracking system and an installed power of 100 kWp and more. Repowering would not be not economically attractive for plants with a lower output.

The control concept presented here fulfilled the behavior required by Parabel's business case.

The overshoot protection feature has been proved under real field conditions.

The overshoot protection features could also be used for other applications, such as PV plants with other generators or loads connected to the same bus which have strict power injection constraints. Examples would be island systems fed with diesel generators, systems connected to weak grids or virtual power plants.

6 REFERENCES

[1] „Erneuerbare-Energien-Gesetz (EEG) in der ab 1.1.2012 geltenden Fassung, beschlossen am 30.6.2011 vom Deutschen Bundestag". .

[2] B. Engel, M. Rothert, M. Breede, T. Degner, G. Arnold, T. Reimann, und P. Strauss, „Netzintegration von Photovoltaikanlagen: Anforderungen an Stromrichter bei hohem PV-Anteil im Insel- und Verbundnetz", gehalten auf der 16. Kasseler symposium energie-systemtechnik, Kassel, Okt-2011.

[3] H. Lutz und W. Wendt, *Taschenbuch der regelungstechnik*. Europa lehrmittel, 2014.

[4] A.H. Glattfelder und W. Schaufelberger, „PI control with output constraints", in *Control Systems with Input and Output Constraints*, London: Springer, S. 101–103.

EVAPORATING PURE RAINWATER TO INCREASE THE YIELD OF COMMERCIAL-SIZE PV ARRAYS

Nicolas Cristi, André Macq and Laura Martin-Carron (*)
SUNiBRAIN S.A.S.,
52 Blv. Gabriel Koenigs, F-31300 Toulouse, France
(*) corresponding author e-mail: laura.martin@sunibrain.com

ABSTRACT: This paper summarizes 5 years of R&D efforts towards for the conception of PV module cooling equipment that would be economically sound for retrofitting existing PV plants in temperate and tropical climates. It is well established that cooling Si-based PV modules does increase their power output by ~0.45%/°K. The initial objective of the research was to identify the most effective method to cool large arrays of modules: the outcome was to maximize the evaporation of droplets of pure rainwater entering into contact with the solar glass surface. The next objective was to devise a method that would optimize use of the water resources, (i.e. of rainwater collected in the winter) and maximize the overall power output gains across the year. This resulted in connected equipment driven by industrial algorithms, which continuously calculate 2 main factors: the "cooling opportunity" and the "cooling cost", thereby ensuring at any instant the financial viability of cooling the PV modules. The paper shows the relationship between solar irradiation and power output and detailed comparative results obtained from the all-year-long cooling of existing commercial-size PV plants in Europe.
Keywords: Water Cooling, Silicon modules, Photovoltaic Panels

1 INTRODUCTION

The performance of a PV system depends on several parameters and there are numerous issues that can reduce the expected efficiency of commercial PV plants; among them, one is high operational cell temperature and another is the dirt and dust particles accumulated on the front surface of the PV panel. It is well known that operational temperature can decrease the performance by up to 18%-29% compared to panels held at standard test temperature [1-3], whereas dust loss can amount to 30%.

The heat issue has been thoroughly researched and different cooling strategies have been developed to reduce module temperature and increase efficiency [4-8], although water cooling has clearly proved to be one of the most effective techniques [9,10]. In spite of the many systems proposed for water-based cooling, none of them demonstrated economic feasibility.

In this paper we present a summary of 5 years of R&D efforts towards for the conception of PV module cooling equipment that is economically sound for retrofitting existing PV plants in temperate and tropical climates [11,12].

Company SUNiBRAIN of France developed a commercial system that cools, cleans and performs a monitoring of the PV modules, optimizing the resources used. The initial objective of the research was to identify the most effective method to cool large arrays of modules: the outcome was to maximize the evaporation of droplets of pure rainwater entering into contact with solar glass surface [13]. The next objective was to devise a method that would optimize use of the water resources, (i.e. rainwater collected in the winter) and maximize power output gains across the entire year.

2 SYSTEM DESCRIPTION

The performance validation was based on the use of a twin solar system (specially built as real-size test equipment offering the possibility of real-time comparison) during periods of autonomous production and operation of the cooling/cleaning system without human intervention. The first complete calendar year of fully autonomous operation was 2013.

The installation consists of a steel roof with mounting systems and PV modules with two absolutely identical solar generators of which one only is equipped with the cooling/cleaning system. In order to prevent temperature contagion between the two generators, they are separated by a buffer zone (Fig. 1 top).

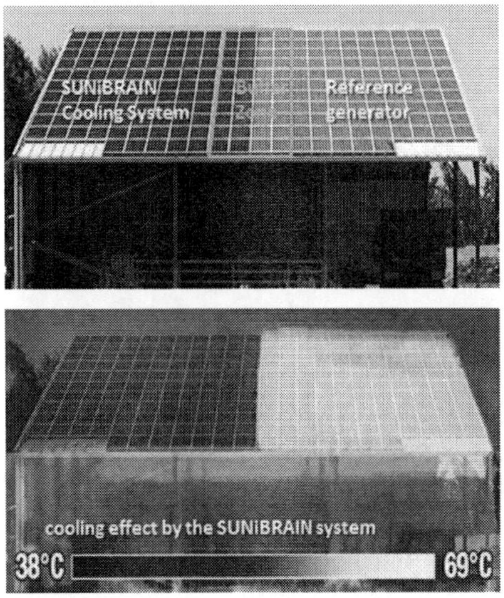

Figure 1: (Top) Test System consisting in 2 twin generators separated by a buffer zone. (Bottom) Thermographic image of the roof showing the cooling effect by the SUNiBRAIN system.

Each solar generator is instrumented with various measurement instruments (separate electric meters, multiple sensors for t°, wetting, irradiance, weather parameters, etc.) to allow various analyses including:

- Continuous monitoring of production gains
- Analysis of t ° and PR curves
- Daily study of strings current and voltage as a function of t° differentials
- Frequency distribution of operating powers over time

- Calculation of production gains and of changes in the performance ratio

This system is driven by industrial algorithms, which continuously calculate 2 main factors: the "cooling opportunity" and the "cooling cost", thereby ensuring the financial viability of cooling PV modules at any instant (Fig. 2).

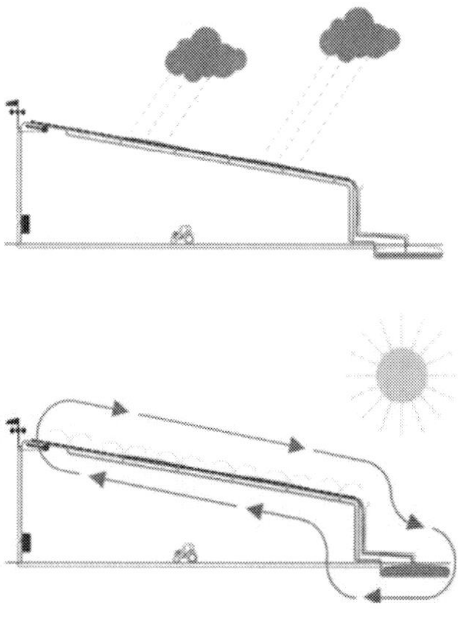

Figure 2: Function diagram of the system developed to cool and clean on-roof PV installations.

3 RESULTS

3.1 Performance

In Figure 3, each bar represents the kWh hourly output of the inverters of twin PV installations over a day: SUNiBRAIN-equipped generator in blue, reference generator in red.

The difference between the red and the blue bars represents the output differential of the two inverters, i.e. the SUNiBRAIN gain.

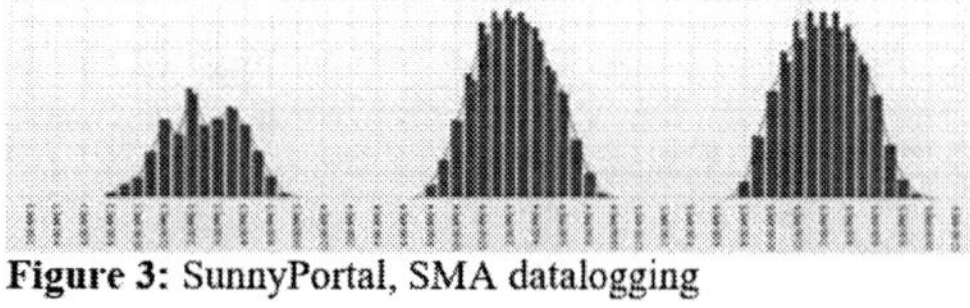

Figure 3: SunnyPortal, SMA datalogging

Figure 4 shows the relationship between solar irradiation and power output. The power output gain (red curve, right scale) and the temperature drop (blue curve, left scale) obtained from real-time comparison of the modules of each of the twin solar generators is presented. The power output differential (red) is not a perfect inverse of the t° differential (blue), but it is close. This very strong correlation demonstrates how the SUNiBRAIN system is capable of anticipating the cooling opportunities, i.e. the moments when it is economically useful to cool the solar field.

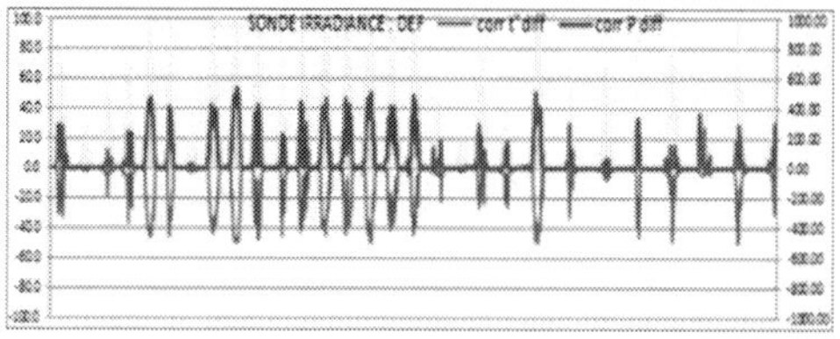

Figure 4: 30-days of t° differential (blue, in °K, left axis), power differential (red, in %, left axis) and solar irradiation (yellow, right axis)

The monthly performance gains shown (Fig.5) represents the differential output of the two PV fields, namely:

$$Gain(\%) = \frac{\left(PowerOutput_{SIBGenerator} - PowerOutput_{ref.Generator}\right)}{PowerOutput_{ref.Generator}}$$

The monthly distribution of gains varies quite widely from one year to another, depending on monthly weather conditions (wind, precipitations and cloud cover). However, the total annual gain is statistically much more stable from one year to another.

The total power output differential in 2013 – i.e. the average annual gain – was 6.24%

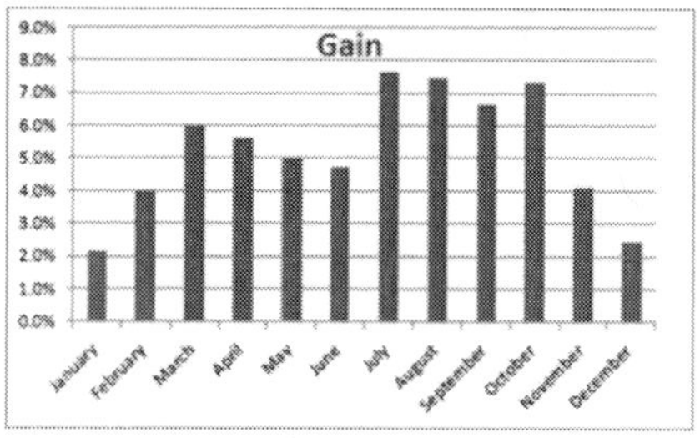

Figure 5: Monthly performance: Power output gain from cooling the panels.

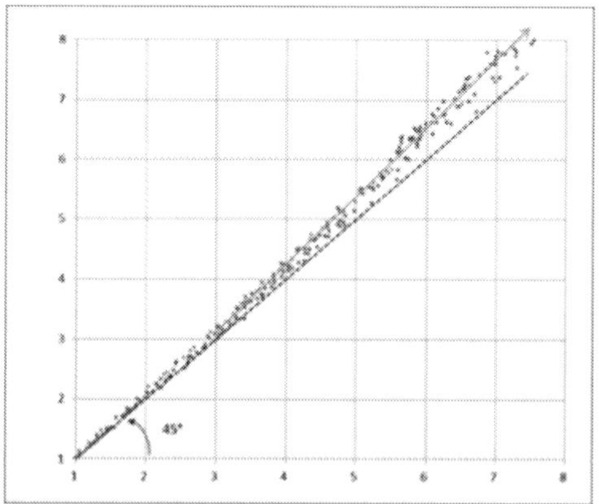

Figure 6: Full-year comparison of daily power output (in kWh) from non-cooled (H axis) vs. cooled- PV (V axis).

Graph in Figure 6 shows the power output over a full day of production. The position of each dot depends on the total daily output of the reference generator (X) and total daily output of the generator equipped with SUNiBRAIN (Y). Each dot above the red line represents a day when the SUNiBRAIN generator produced more output than the reference generator. Dots on the 45% line correspond to days when the SUNiBRAIN cooling system is not working at all (e.g. freezing conditions, excessive wind, rain, etc..) This happens especially in the bottom of the graph, i.e. by days of rather weak sunshine.

3.2 Predictability of performance and PR validation

Whenver data is available to compute or assess the frequency distribution of the power output of a solar generator (on the above graph the blue bars, left scale) and expressed for each percentile group of power output (here on the x-axis, expressed as a %age of the inverters nameplate capacity), and knowing the efficiency of the inverter at each power level, as well as the average gain generated by SUNiBRAIN at each power level (green curve, left scale), it is possible to interpolate correctly, in terms of volume and frequency, the gains for each power level (blue curve, right scale). Using integral calculus, this blue curve provides an estimate the average gains of equipping an existing PV generator with SUNiBRAIN.

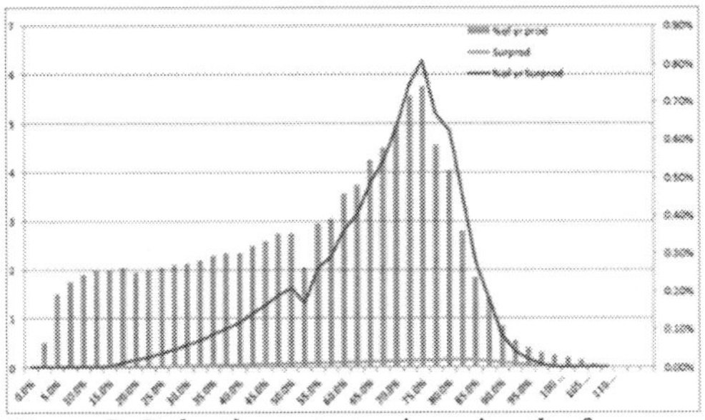

Figure 7: Estimating output gains using the frequency distribution method. Interpolation based on statistical data obtained from inverter manufacturer SMA, Inc. for a solar generator near the French city of Colmar.

In a similar way, whenver irradiance data is available, simple methods make it possible to validate, in terms of performance ratio, the benefits from installing SUNiBRAIN on a PV generator.

5 CONCLUSIONS

The cooling system is able to accumulate power generation gains not only in summer, but under very many meteorological conditions throughout the year.
Several years of developments allowed us to improve our models of the cooling mechanism of solar arrays.

The key progress factors were the integration of factors, such as such as hygrometry and air density, which influence the rate of water evaporation. Beyond instrumentation, the cooling equipment's firmware was reprogrammed to make better use of such environmental data, and to derive in real time more realistic financial estimates of the "cooling opportunity" and the "cooling cost", thereby ensuring at any instant the financial viability of using spending rainwater to cool the PV modules. The result is industrial equipment (incl. rainwater recycling and storage) whose cost can be amortized over a period of only a few years even in central European climates.

4 REFERENCES

[1] "On the temperature dependence of photovoltaic module electrical performance: A review of efficiency/power correlations" E Skoplaki, JA Palyvos, Solar Energy 83 (2009) 614–624

[2]H. Youichi, I. Tutomu, and T. Tatsuo, "Output variation of photovoltaic modules with environmental factors- II: Seasonal variation," J. Solar Energy, vol. 63, 1998, pp. 185-189.

[3] Radziemska, E. and Klugmann, E. "Thermally affected parameters of the current-voltage characteristics of silicon photocell", Energy Conversion and Management, Vol.43, p.p. 1889-1900, 2002

[4] Odeh, S., Behnia, M., "Improving photovoltaic module efficiency using water cooling", Heat Transfer Eng., Vol.30 (6), pp. 499–505, 2009

[5] Krauter, S. "Increased electrical yield via water flow over the front of photovoltaic panels." Solar Energy Materials & Solar Cells, Vol. 82, pp. 131–137, 2004

[6] Kristijan Brecl, Marko Topič "Performance of PV Modules with regard to the mounting situation", 29th European Photovoltaic Solar Energy Conference and Exhibition

[7] Valeh-E-Sheyda P, Rahimi M, Parsamodghadam A, Masahi MM. "Using a wind driven ventilator to enhance a photovoltaic cell power generation", Energy Build, Vol.73, pp.115–9. 2014

[8] Y. Tripanagnostopoulos1, P. Themelis1 and J.K. Tonui2 "Natural flow building integrated Air-cooled Photovoltaics", 23rd European Photovoltaic Solar Energy Conference and Exhibition

[9] S. Nižetić, D. Čoko, A. Yadav, F. Grubišić-Čabo, "Water spray cooling technique applied on a photovoltaic panel: The performance response", Energy Conversion and Management, Vol. 108, pp. 287–296, 2016

[10] Yuzuru UEDA, Kosuke KUROKAWA, Makoto KONAGAI, Sadayuki TAKAHASHI, Akito TERAZAWA, Hiroyuki AYAKI, "Five years demonstration results of floating PV Systems with water spray cooling", 27th European Photovoltaic Solar Energy Conference and Exhibition

[11] André Macq, Loïc Mercier des Rochettes, Laura Martin-Carron, Nicolas Cristi, Marie-Pierre Gleizes, Carole Bernon and Pierre Glize, "A use of artificial intelligence for improving pv array performance (empirical approach)", Proceedings 32nd European Photovoltaic Solar Energy Conference and Exhibition (to be published), 2016

[12] Laura Martin-Carron, Roland Becker, Didier Graebling, Robert Luce, Daniel Ugarte, André Macq and Nicolás Cristi, "Air cooling of photovoltaic panels: a numerical approach", Proceedings 32nd European Photovoltaic Solar Energy Conference and Exhibition (to be published), 2016

[13] K. S. Birdi, D. T. Vu, A. Winter, "A study of the evaporation rates of small water drops placed on a solid surface", J. Phys. Chem., 1989, 93 (9), pp 3702–3703, May 1989

OPTIMISING THE SELF-CONSUMPTION OF SOLAR POWERED SMART MICRO GRIDS

A. Mahran, A. Minde, K. Peter, J. Glatz-Reichenbach
International Solar Energy Research Center - ISC Konstanz, Rudolf-Diesel-Str. 15, D-78467 Konstanz, Germany
Phone: +49-7531-3618 351, e-mail: joachim.glatz-reichenbach@isc-konstanz.de

ABSTRACT: Due to the continuous rise of retail electricity price and the falling remuneration of photovoltaic (PV) electricity feed-in tariffs (FIT), consumers who own PV installations are encouraged to increase their self-consumption. Furthermore, the intermittency of renewable energy resources (RES) hinders the exclusive reliance on RES as a sustainable energy resource. Additionally, batteries still reveal several technical and economic disadvantages when used to increase self-consumption. This research presents an approach to model a smart neighbourhood by using a multi-agent system (MAS). The model's consensus is to increase the self-consumption as well as PV-based self-generation of an overall smart grid community. Furthermore, updated PV yield forecasts in addition to recorded load consumption profiles are used as input to the MAS to invoke demand-side management and urge participants to exchange their excess energy. In order to minimise the misfit between forecasted and measured PV yields, a recursive optimisation algorithm is applied to improve future estimations, based on past prediction errors. Results show that the self-consumption of single-users as well as the entire community – without using batteries – can exceed 60% with a potential to achieve full grid independency when supported by electrical storage. Moreover, community consumption can utilise all the electricity generated by PV, which leads to less grid congestion by feed-ins from renewable energy sources.

Keywords: Information and Communication Technologies (ICT), Smart Grid, Multi-Agent System (MAS), Agent-based Modelling, PV prediction, Demand Side Management

1 INTRODUCTION

The EU funded project CoSSMic [1] (Collaborating Smart Solar-powered Microgrids) aims by incorporating ICT as an intelligent approach to model and optimise the operation of smart grids. The objective of this project is to demonstrate an intelligent approach towards increasing the self-consumption of consumers, which are collaborating in a smart neighbourhood through peer-to-peer (P2P) communication [2-8]. Information about weather forecast, PV yield, load consumption profiles, instrumentation signals and readings, for example, temperatures and battery state of charge (SoC), etc. is exchanged across different stakeholders in the system to plan for instant energy exchange. Subsequently, increasing the self-sufficiency of various participants in the system leading to less required energy storage units (i.e. rather expensive batteries). The underlying innovation applied in such a system is an agent-based modelling in which all system components are modelled in a multi-agent environment. Representative agents are collaborating and communicating autonomously to achieve their objectives according to certain rules based on requirements, preferences and restrictions [2-7].

In order to maximise the utilisation of the generated energy by means of PV within a regarded community, the expectations of power generation and consumption profiles must be as close as possible. By a detailed knowledge of the load profiles of the integrated consuming devices, which can be reached by experimental records, the only free parameter so far is the expected PV yield. A prognosis of the future supplied amount of energy from the installed PV systems is essentially required. A reliable weather forecast is therefore the inevitable tool of choice to modulate the expected PV yield.

The overall concept of this project is currently tested in field trials within the City of Konstanz, Germany and the Province of Caserta, Italy. Furthermore, however, the objective of this paper is to present results obtained from the simulation phase using input data from the actual participants of the project. These data cover updated weather predictions provided by German Weather Service (DWD), PV installation details and a continuous record of energy consumption of various loads using smart plugs and meters. Some selected days are used in the simulation to evaluate the performance of the concept at different operating conditions in terms of load consumption, solar irradiance and ambient conditions. Such data are obtained from onsite surveys and used as input for the model, as explained in the following sections.

2 SCIENTIFIC INNOVATION AND RELEVANCE

Many studies which discuss the operation of PV in smart grids only cover battery integration with smart energy management devices in single-user residential applications as an approach towards increasing their self-consumption [9-14]. This is achieved by offsetting part of the peak generation to partly cover the load consumption in the early morning or evening hours. Nonetheless, fewer studies focus on instantaneous energy sharing and utilisation through the application of MAS in smart communities. Once implemented, tested and set into operation, this concept would revolutionise the collaborative operation of smart micro grids based on ICT and the Internet of Things (IoT), and promote for consumer integrity, energy efficiency and the reduced need of using expensive batteries.

In addition, Smart Python multi-Agent Development Environment (SPADE) is well suited to model, for example, the operation of smart neighbourhoods, as a multi-agent modelling environment, since less work has been presented using it [15, 16]. Additionally, SPADE shows a much faster response time when compared with Java Agent Development Framework (JADE), which is already applied in many applications [17].

The smart neighbourhood model consists of different users, which are categorised based on their role in the system, i.e. as consumer, producer or prosumer. A prosumer in this case is defined as a consumer with an additional PV installation, and does both; consumes and produces energy. The working principle of the model is to allow for a prioritised mechanism of energy exchange in which the primary preference is for the PV system at the prosumers' site to fulfil the demands of its consumers, then to supply other load demands in the community and finally to feed the remaining capacity into the grid. Such working rules are defined in each supplier's agent representing a PV installation. Similarly, consumer agents follow certain rules in order to initiate the communication procedures within the operation of the program by sending energy supply requests from various supplier agents. The community operating preferences and workflow diagram of the agent environment are shown in Figures 1 and 2, respectively.

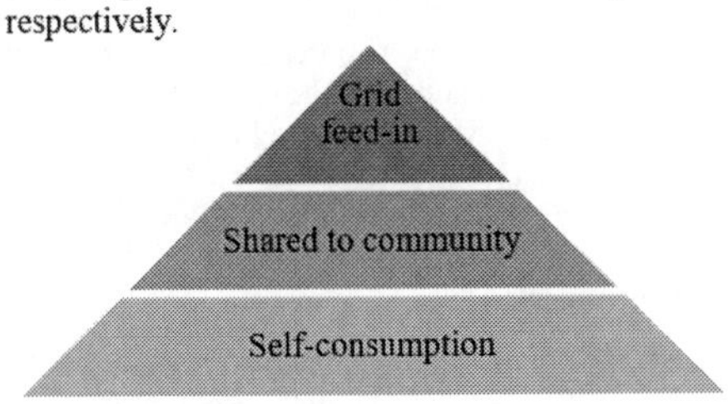

Figure 1: Operating preferences of the community model for excess PV generation, which is allocated to supply first the load consumption of its participant, secondly to supply other neighbours' energy demand and finally feed into the public grid.

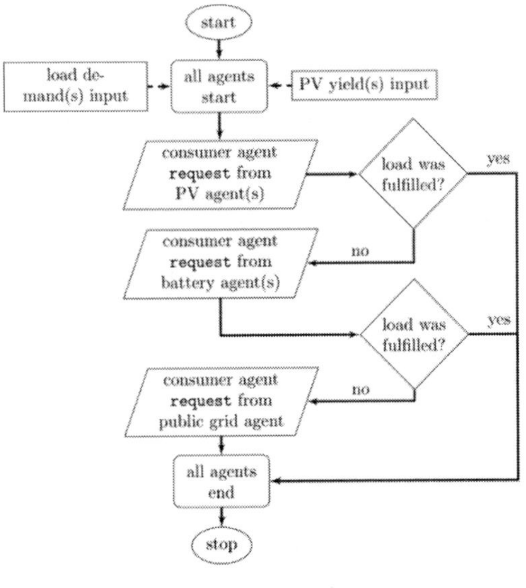

Figure 2: Workflow diagram of the multi-agent system representing the operation of a single-user model with a battery unit. The communication is always invoked by the consumer agent, with the primary objective to fulfil its load demand from the available energy resources inside the system. In this system, the available resources are a PV plant, a battery and the public grid.

In many research works which discuss the integration of batteries for smart grid operation, the definition of self-consumption alone is ambiguously specifying the overall efficiency of the system in terms of how self-sufficient the system operates independently from the grid (i.e. the focus lies on maximum self-consumption) with a minimum grid feed-in from PV yield [11-13]. Subsequently, two definitions are claimed in order to evaluate the operation of the model.

- **Self-consumption**, or consumption autarky, is defined as the percentage of the load demand that is covered from the local PV unit(s) in the community.

The self-consumption ε_c of a consumer, who owns a PV installation is expressed by Equation 1.

$$\varepsilon_c = \frac{E_{load,PV} + E_{load,comm}}{E_{load}} \qquad (1)$$

Where $E_{load,PV}$ is the load demand of the consumer which is covered by its own PV yield, $E_{load,comm}$ is the additional load demand which is covered by the PV yields through energy exchange from the community and E_{load} is the total load demand of the consumer.

- **Self-generation**, or production autarky, is the share of the total PV yield that is utilised by the local load demand of the same consumer as well as what can be supplied in the community (i.e. that is not fed into the grid).

The self-generation ε_p of the same consumer, who owns a PV installation is expressed by Equation 2.

$$\varepsilon_p = \frac{E_{PV,load} + E_{PV,comm}}{E_{PV}} \qquad (2)$$

Where $E_{PV,load}$ is the consumer PV yield which is allocated to cover its own load demand, $E_{PV,comm}$ is the additional PV yield which is exchanged in the community to cover other consumers' load demands and E_{PV} is the total yield of the PV installation.

3 APPROACH

The community model consists of different user-applications (i.e. residential, commercial and industrial) which are located in the City of Konstanz, Germany. In addition, it has a single PV generator whose role is to feed the required load demands of the community. The presented investigation is done for two specific days to compare the behaviour of the system during two different case studies representing a cloudy day (October 14th) and a sunny day (August 5th) of the year 2015. Using the definitions of consumption and production autarkies, the performance of the model is evaluated for both days. As depicted in Figure 3, the model consists of three households – two of them with PV rooftop installations (i.e. prosumers), one office building, a small industrial complex, a PV generator and the public grid. Tables 1 - 3 summarise the load demands during both evaluation days as well as the installed PV capacities of the participating users respectively.

Table 1: Daily average load demands in kWh for each participant in the community during a cloudy day (October 14th).

User ID	C1	N1	R1	R2	R3	Total
Load demand (kWh)	48	1626	23	4	18	1719

Table 2: Daily average load demands in kWh for each participant in the community during a sunny day (August 5th).

User ID	C1	N1	R1	R2	R3	Total
Load demand (kWh)	48	1720	10	3.5	10.7	1792

Table 3: The installed PV capacity in kW_p at each participant site in the community of together 99.5 kWp.

User ID	G1	C1	N1	R1	R2	R3
Installed PV capacity (kWp)	60	12.4	6.9	10	0	10.2

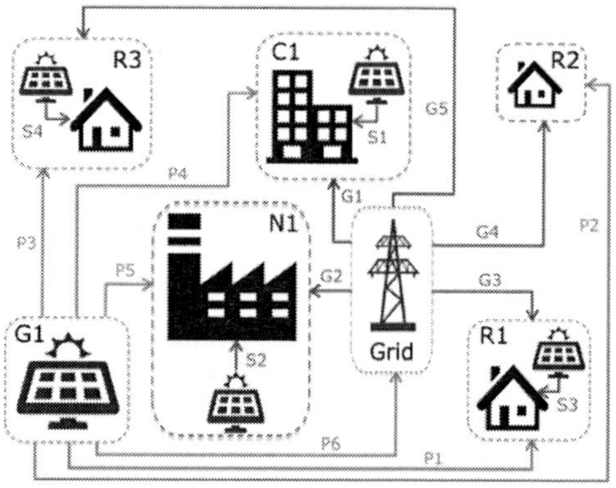

Figure 3: Schematic diagram of the community model applied to simulate the real operation of multi-user smart grids. The model consists of four prosumers (two residential, one commercial and one industrial), one residential consumer, one PV generator and the public grid. The arrows represent the supply schemes of the model, whereas the tags represent the supply preferences.

The operating scenario of the model involves certain working preferences that are followed in order to fulfil the load demand of each consumer with a priority of supplying it from its own PV generator. Following, further PV generation of the prosumers C1, R1, and R3 is directed to supply the industrial load since it shows the main energy requisition in the system due to its significant load demand. Furthermore, the generator G1 is responsible for supplying the remaining load demands of R1, R2, R3, C1 and N1. In case of a PV surplus, for example, due to a sunny day, it is fed into the grid. In contrast, the grid is committed to take care of covering any supply deficit that occurs after all PV supply. These preferences are set as internal rules for each respective agent in order to spur them to follow such rules.

In order to improve the autonomous responsiveness of the model, the PV generation and load consumption of the model elements are forecasted and recorded, respectively. This is useful to ease the anticipation of the operating behaviour of the model. As for the load consumption, smart plugs and meters are used to record details about home appliances and specific equipments. The acquired load profiles get processed and logged with a modified version of the open-source web application emoncms[1], running on a single-board computer, and provide an estimation for future load profiles by curve fitting a set of past profiles [8]. Weather forecasts from DWD are available on a FTP[2]-server and can be accessed on each site. Therefore, using an open-source PV yield prediction algorithm, PV production can be predicted with an acceptable tolerance [8]. This algorithm relies only on straightforward accessible data and hardware. Consequently, a system without further required knowledge to setup or unnecessary cost can be utilised for further improvements.

A first attempt to improve the performance of the prediction by integrating a simple linear optimisation, adverts high potential and serves for further investigations. This attempt shows that the incorporation of historical measurement data to estimate the actual PV systems efficiency could prove great potential, since many influences are challenging to simulate without the implementation of additional sensors or more complex forecast models. Some environmental factors, like the soiling of the PV modules' glass and the reduced energy yield, as well as its cleaning again during rainy periods, are slow processes, which could be adjusted by a dynamic optimization. With a separately optimised efficiency value for each forecasted hour of the day, shading issues, stemming, for example, from trees or chimneys, can be addressed, as their importance increase for certain solar elevations during specific seasons of the year.

Furthermore, reduced efficiencies caused by elevated cell temperature or non-optimal solar inverter loads, depending mainly on the suns' incidence angle towards the PV module, occur more often at specific times of the day and seasonal periods.

Despite the applied dynamically optimised simulation results incorporating not negligible uncertainties, the major uncertainty is still the radiation forecast itself. During stable or closely forecasted weather conditions though, e.g. during high atmospheric pressure periods over several days in the summer, the prediction error may tend quickly to converge to zero.

Figure 4 shows the error distribution for a residential yield prediction with a 12 kW_p PV installation for each hour of the day for a period of 122 fully measured and valid days between 15th Oct 2015 and 31st Mar 2016. The prediction is calculated with a first static efficiency estimation of 92% [8]. The graph displays the median prediction error, including upper (light blue) and lower

[1] http://emoncms.org/

[2] File transfer protocol

(dark blue) quartiles of the error distribution. Positive values indicate an overestimation, relative to the measured PV-yield values in percentage points, whereas negative values indicate an underestimation.

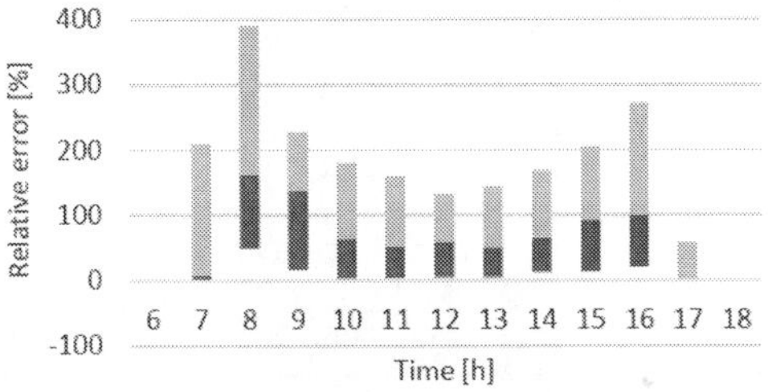

Figure 4: Prediction error distribution of 122 consecutive days from 15th October, 2015 to 31st March, 2016.

With a median error of 34%, the error distribution for the analysed roof installation shows a clear tendency for the early morning, and late afternoon hours of overestimation, which could be explained by the fact that the modules are shadowing themselves. Figure 5 shows the error distribution for the same time period and days, with the hourly efficiency values being recursively optimized by a standard basic least squares filter. The reduced median error to 0.2% and overall error distribution make the potential for this dynamic approach apparent.

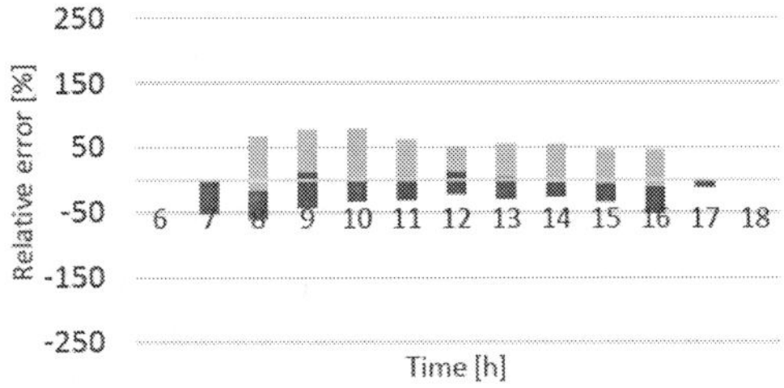

Figure 5: Prediction error distribution of 122 consecutive days from 15th Oct 2015 to 31st Mar 2016 for recursively optimised hourly efficiency values.

4 RESULTS

This section presents the outcome from two study cases. First case is the operation of the community model on both, a cloudy day as well as a sunny day, with a comparison between the values of self-consumption and self-generation before and after energy exchange using MAS is evaluated. The second case introduces an additional PV generator in the energy exchange of the smart community instead of feeding the electricity into the public grid. Such approach is used in order to optimise the model operation. Figure 6 represents an example of the load demand and PV generation of the industrial prosumer denoted as "N1". More details about this prosumer are expressed in Tables 1 - 3. It is obvious that the load demand of this participant in more significant than its PV production. Therefore, it represents a noteworthy potential while optimising the operation of the model.

- Case one: cloudy vs. sunny

Figure 7 depicts the energy flows from locally produced PV electricity in the community after supplying part of the energy demands of the consumers using MAS on a cloudy day. The remaining energy demand is supplied by the public grid. In comparison with the overall supply from public grid, PV units already supplied 10% of the total energy demands of the community on that day as compared to 2.5% before energy sharing. Further, during the sunny day, PV units covered 38% as compared to 5.8% before energy sharing. Due to the significant power demand of the industrial block within the community, neither PV electricity is stored in batteries, nor fed into the public grid. Subsequently, all electricity generated from PV in the community is used to cover the community demands and therefore community self-consumption is increased on both days. Such increases represent tremendous potential of the concept to maximise the reliance on PV electricity in smart communities. Detailed values of both self-consumption and self-generation before and after energy sharing during cloudy and sunny days are summarized in Tables 4 - 7, respectively.

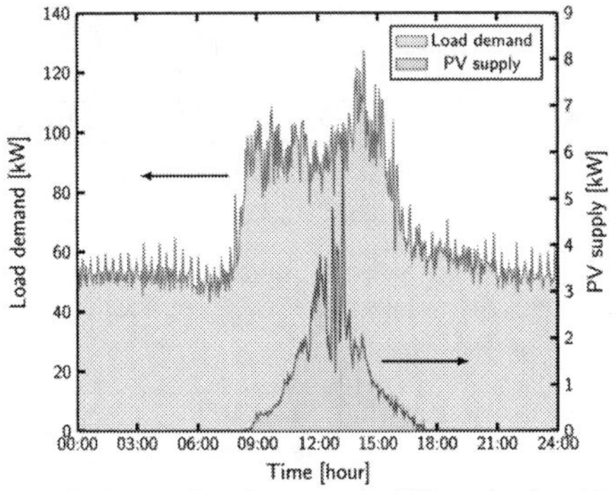

Figure 6: Comparison between the PV production (right axis) and load demand (left axis) in kW on a cloudy day (2015, Oct 14th) for an industrial prosumer "N1" which is located in Konstanz. Notice that the load demand and the PV generation occur significantly simultaneously.

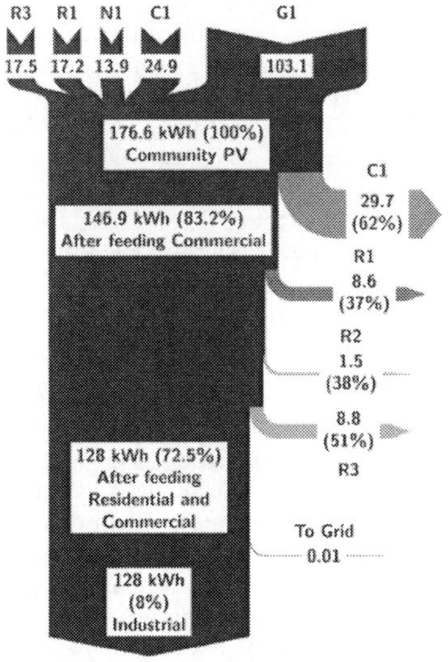

Figure 7: Energy flow diagram in kWh from local PV units within the community after exchanging excess energy using MAS on a cloudy day (2015, October 14th) in Konstanz. Notice that the percentages of the call-outs represent the self-consumption of the load demand, which is fulfilled from local PV supply.

Table 4: Self-consumption and self-generation values of the community members before energy sharing on a cloudy day.

User ID	G1	C1	N1	R1	R2	R3	Overall
Self-cons. [%]	–	44	1	18	0	36	2.6
Self-gen. [%]	0	85	100	24	–	36	25.7

Table 5: Self-consumption and self-generation values of the community members after energy sharing using MAS on a cloudy day.

User ID	G1	C1	N1	R1	R2	R3	Overall
Self-cons. [%]	–	62	8	37	38	51	10
Self-gen. [%]	100	100	100	100	–	100	100

Table 6: Self-consumption and self-generation values of the community members before energy sharing on a sunny day.

User ID	G1	C1	N1	R1	R2	R3	Overall
Self-cons. [%]	–	80	2.8	67	0	82	5.8
Self-gen. [%]	0	48	100	10	–	13	15

Table 7: Self-consumption and self-generation values of the community members after energy sharing using MAS on a sunny day.

User ID	G1	C1	N1	R1	R2	R3	Overall
Self-cons. [%]	–	82	36	74	72	83	38
Self-gen. [%]	100	100	100	100	–	100	100

- **Case two: additional PV generator**

In this case an additional PV plant is added to seek for the optimum operating point of the model. This plant is located in Konstanz and has a capacity of 140 kWp with inclination and orientation angles of 14° and 140°, respectively. Figure 8 depicts the energy flow diagram of the locally generated PV electricity including the additional PV plant G2. Community self-consumption and self-generation increased both by a factor of about ten from 5.8% to 60% and from 6.3% to 65%. Detailed values of both self-consumption and self-generation before and after energy exchange on a sunny day are compiled in Tables 8 and 9, respectively.

As a final optimisation step, the capacity of the added PV plant is varied, while keeping the remaining parameters, to study the effect on both, the self-consumption and self-generation of the community, and the outcome is visualised in Figure 9. The results show that the self-consumption of the community increases from 38% to 63% as the additional PV generator installed capacity is increased. Nonetheless, the community self-generation is decreasing due to additional capacity, which is fed into the public grid. The latter effect is occurring to the additional generators' self-generation as well. This trade-off between increasing the community self-consumption while reducing its self-generation by adding additional PV generation is an expected drawback that could be overcome, for example, by introducing electrical storage.

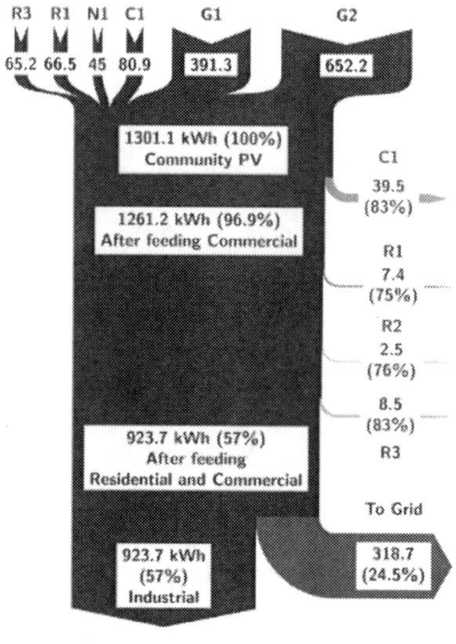

Figure 8: Energy flow diagram in kWh from local PV units plus the additional PV generator G2 within the community after exchanging excess energy using MAS on a cloudy day (2015, October 14th) in Konstanz. Notice that the percentages of the call-outs represent the self-consumption of the load demand, which is fulfilled from local PV supply.

Table 8: Self-consumption and self-generation values of the community members including the additional PV generator G2 before energy sharing on a sunny day.

User ID	G1	G2	C1	N1	R1	R2	R3	Overall
Self-cons. [%]	–	–	80	3	67	0	82	5.8
Self-gen. [%]	0	0	47.6	100	10	–	13	6.3

Table 9: Self-consumption and self-generation values of the community members including the additional PV generator G2 after energy sharing using MAS on a sunny day.

User ID	G1	G2	C1	N1	R1	R2	R3	Overall
Self-cons. [%]	–	–	83	59	75	76	83	60
Self-gen. [%]	40	100	100	100	100	–	100	65

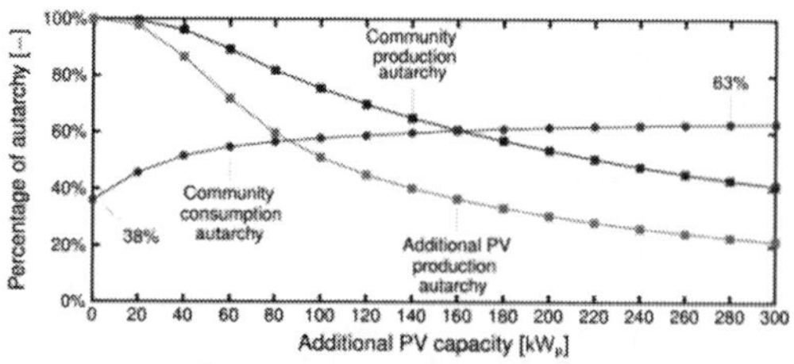

Figure 9: Variation of the self-consumption (consumption autarky) and self-generation (production autarky) values for the community as well as the production autarky of G2 versus the additional installed PV capacity of G2 in kW_p.

5 SUMMARY

This work presents a summary of system simulations within the EU supported project CoSSMic, in which the primary target is to exchange excess electrical energy across different neighbours inside a smart community, without installed electrical storage and to minimise public grid feed-in. Two exemplary days (e.g. a sunny and cloudy one) are used to compare the operation of the community under different operating conditions. In addition, two different performance indicators are defined and calculated, in order to evaluate the operation of the model; self-consumption and self-generation. Most reliable PV yield prediction results are achieved, by recursively optimising the prediction probability of each hour of the day, based on prior measurements. As a result, the median error of the yield prediction is reduced down to 0.2% within the analysed time period of 122 days during an autumn and winter season. A Smart Python multi-Agent Development Environment – SPADE – is used to model the community behaviour. Results show, that the community self-consumption can exceed 60% while most of the locally generated PV electricity is utilised. Additionally, when energy storage is used, the potential of approaching full grid independency could be achieved. Yet, a significant storage capacity would be necessary to cover the entire load demand of the community and to reduce the expected increase of PV feed-in into the public grid. Therefore, a trade-off is presented to obtain the optimum operating point.

6 ACKNOWLEDGMENT

This work is financially supported by the EU Project CoSSMic under FP7 Grant agreement no. 608806.

7 REFERENCES

[1] http://cossmic.eu

[2] A. Amato, B. Di Martino, M. Scialdone, S. Venticinque, S. Hallsteinsen and S. Jiang, A Distributed System for Smart Energy Negotiation, Springer International Publishing, ISBN 978-3-319-11691-4, pp. 422-434, 2014

[3] A. Amato, B. Di Martino, M. Scialdone and S. Venticinque, Towards a (SLA) for Collaborating Smart Solar-Powered Micro-Grids, 2014 International Conference on Intelligent Networking and Collaborative Systems, Salerno, Italy, September 10-12, 2014, pp. 592-597, 2014

[4] A. Amato, B. Di Martino, M. Scialdone and S. Venticinque, An Agent-based Approach for Smart Energy Grids, (ICAART) 2014 - Proceedings of the 6th International Conference on Agents and Artificial Intelligence, Volume 2, ESEO, Angers, Loire Valley, France, 6-8 March, 2014, pp. 164-171, 2014

[5] A. Amato, B. Di Martino, M. Scialdone and S. Venticinque, Multi-agent Negotiation of Decentralized Energy Production in Smart Micro-grid, Intelligent Distributed Computing (VIII) - Proceedings of the 8th International Symposium on Intelligent Distributed Computing, IDC, Madrid, Spain, September 3-5, 2014, pp. 155-160, 2014

[6] L. Tasquier, M. Scialdone, R. Aversa, S. Venticinque (2014). Agent based negotiation of decentralized energy production. In: Proceedings of 8th International Symposium on Intelligent Distributed Computing (IDC-2014). STUDIES IN COMPUTATIONAL INTELLIGENCE, p. 59-67.

[7] S. Jiang, S. Venticinque, G. Horn, S. Hallsteinsen, M. Noebels, A distributed agent-based system for coordinating smart solar-powerd microgrids, accepted at: IEEE Technically Sponsored SAI Computing Conference 2016, 13-15 July 2016 | London, UK

[8] M. Noebels, J. Glatz-Reichenbach, A. Mahran, A. Minde, K. Peter, Developing and investigating a smart solar powered energy system for increased PV self-consumption, Proceedings 31st EUPVSEC, p.2676-9, ISBN: 3-936338-39-6, 2015

[9] For examples of self-consumption optimisation, see: http://files.sma.de/dl/2485/SI_EIGENVERBR-AEN130610W.pdf

http://en.sma-sunny.com/2014/07/07/sunny-boy-smart-energy-a-practical-test/

[10] SMA: SMA Smart Home. http://www.sma.de/fileadmin/intersolar/SMART_HOME-KEN1421W.pdf

[11] M. Castillo-Cagigal, E. Caamañno-Martin E. Matallanas, D. Masa-Bote, A. Gutierrez, F. Monasterio-Huelin, J. Jimenez-Leube: {PV} self-consumption optimization with storage and Active {DSM} for the residential sector. In: Solar Energy 85 (2011), Nr. 9, 2338 - 2348

[12] M. Castillo-Cagigal, A. Gutierrez, F. Monasterio-Huelin, E. Caamaño-Martin. D. Masa, J. Jimenez-Leube, A semi-distributed electric demand side management system with PV generation for self-consumption enhancement. In: Energy Conversion and Management 52 (2011), Nr. 7, 2659 - 2666

[13] E. Caamaño-Martin, D. Masa-Bote, A. Gutierrez Martin, M. F. Monasterio-Huelin, M. Castillo-Cagigal, F. J. Jimenez-Leube, J. Porro, Optimizing PV use through active demand side management (2009)

[14] M. J. van d. Kam, W. G. J. H. M. van Sark, Increasing Self-consumption of Photovoltaic Electricity by Storing Energy in Electric Vehicle using Smart Grid Technology in the Residential Sector, A Model for Simulating Different Smart Grid Programs (2014)

[15] A. Kantamneni, L. E. Brown, G. Parker, W. W. Weaver, Survey of multi-agent systems for microgrid control. In: Engineering Applications of Artificial Intelligence 45 (2015), 192 -203

[16] R. Leszczyna, Evaluation of Agents Platforms. In: European Commission Joint Research Centre scientific and technical reports (2008). http://publications.jrc.ec.europa.eu/repository/bitstream/JRC47224/reqno jrc47224.pdf

[17] A. Amato, B. Di Martino, M. Scialdone, S. Venticinque, Design and evaluation of P2P overlays for energy negotiation in smart micro-grid. In: Computer Standards and Interfaces (2015)

EVALUATION OF LOAD MATCHING AND GRID INTERACTION INDEXES OF A NET PLUS-ENERGY HOUSE IN BRAZIL WITH A HYBRID PV SYSTEM AND DEMAND-SIDE MANAGEMENT

G. Almeida Dávi[a], M. Castillo-Cagigal[a,b], E. Caamaño-Martín[a], J. Solano[a]
[a] Instituto de Energía Solar, Universidad Politécnica de Madrid
[b] ETSI Telecomunicación, Universidad Politécnica de Madrid
Av. Complutense 30, 28040 Madrid, Spain (Tel.: +34 914 533 556/Fax: +34 915 446 341); email:
giovani.almeida@ies-def.upm.es

ABSTRACT: This paper analyzes four scenarios of PV system configuration and proposes an optimized hybrid PV system size in order to improve Load Matching and Grid Interaction (LMGI) indicators, describing the yearly and daily variations of the indexes. A net plus-energy house is modeled in EnergyPlus under humid subtropical and dry tropical climates and electricity flows in the house and with the power grid are based on a high level battery controller simulated with different PV-battery sizes. An economic analysis is presented through two supporting schemes in order to perform analyses of Levelized Cost of Electricity (LCOE), grid parity and system payback time. The results demonstrated the combination of load-shifting demand-side management and small-scale PV-battery capacities provide improvements on building self-consumption and on the effects on the grid. The economic study indicates the building reaches better profitability under current Brazilian net-metering scheme in scenarios without batteries, reducing the payback time and LCOE sensibility analysis indicate the importance in reducing PV costs and to promote financial incentives.
Keywords: Battery storage and control, Load Matching and Grid Interaction (LMGI), Economic analysis

1 INTRODUCTION

In recent years, the ways with which new buildings are being designed have undergone changes, insomuch that introduce new applications of hybrid energy systems. Net plus-energy building (NPEB) is defined as a house that generate at least as much primary energy as the building uses over the course of the year. In this context, the net plus balance means that the annual electrical energy surplus fed-in to the grid is greater than the annual electrical energy imported from the grid. Net-plus energy buildings can reach their full potential in terms of energy conservation and global reduction of greenhouse gas emissions when Load Matching and Grid Interaction (LMGI) issues are taken into account [1]. However, with rapid development of energy technology, LMGI aspects need to be investigated more precisely when involves hybrid PV systems. Load Matching (LM) refers to how the local energy generation compares with the building load by describing the degree of the utilization of on-site PV electricity generation related to the local demand [2] (e.g. the percentage of load covered by on-site generation over a period of time). Grid Interaction (GI) refers to the energy exchange between the building and an energy infrastructure, typically the power grid. (e.g. peak powers delivered to the electricity distribution grid) [2].

In countries where feed-in tariffs (FIT) are applicable, the PV systems owners receive a fixed rate for each kWh fed into the power grid. However, FITs are being phased out in most countries, and PV promotion is switching to self-consumption schemes from a building integration perspective [3]. Recently in Brazil, the Brazilian National Electrical Energy Agency (ANEEL) approved Normative Resolution nº 687/2015 [4], which established an energy compensation system (net metering), whereby the active power injected into the grid by a consumer unit with distributed micro or mini generation offsets the active power consumption. This resolution also assigns the installed PV capacity must be limited to the grid connection capacity.

In this study we analyze the possibilities for the energy provision of a sustainable house using new generation of PV system combined with small-scale energy storage system and Demand-Side Management (DSM), in order to optimize the self-consumption of generated electricity. To this aim, we propose the study of a net plus-energy house prototype developed to the international competition Solar Decathlon Europe 2012, named Ekó House [5], adapted to Brazilian weather conditions through thermal-energetic simulations. Simulations were carried out in different scenarios by combining different storage capacities and the use or not of DSM. The response of different LMGI parameters as well as economic prospects have been analyzed.

2 METHODOLOGY

2.1 Building simulations

Energy modeling of the house was done in EnergyPlus environment (version 8.1). The Energyplus models were based on typical electricity consumption of Brazilian domestic consumers. The analyses were performed for two cities with different solar resource: São Paulo (SP) (Lat. 23.5°S) and Brasília (Lat. 15.7°S). São Paulo has a humid subtropical climate and Brasília is characterized by tropical climate with dry season in the winter and hot summer [5]. For the studied locations the weather data used for the simulations studies were taken from SWERA database [6]. The HVAC system model consists of a Variable-Air-Volume (VAV) heat pump [5]. In order to reduce HVAC load, the indoor air temperature was kept between 22.5 °C and 25.5 °C in summer and from 21 °C to 23.5 °C in winter, based on ASHRAE comfort zone [7]. By profiting the high degree of on-site generation allied with low house energy consumption an electrical boiler of 2 kW nominal operating capacity was modeled with nominal thermal efficiency of 90%.

2.2 The demand and its management

Ekó House electricity consumption simulations include typical electrical appliances of a highly electrified house provided with on-site generation: lighting system, cooking appliances, entertainment appliances, HVAC system and electrical boiler. Tab. I resumes annual electricity of the electrical loads considered for the case study. Deferrable loads are appliances that can be displaced along the day and the user or a domotic system can program the time the appliance works, e.g., programmable washing machine. Non-deferrable loads are not controllable appliances that represent instantaneous consumption like lights, fridge and computers [8]. Cooling space and cooking appliances appear to be the principal electricity sources based on energy usage in each location and in São Paulo the consumption of electrical boiler is high due to lower temperatures in the southeast of Brazil.

Demand-side management involves load shifting strategy in order to control the time-of-use of the deferrable loads, preferably to solar periods where the load is compensated by PV electricity and to off-peak periods where the retail electricity price is lower.

Table I: Annual electricity consumption for the test case study.

Load	Consumption (kWh/y)		Share of total (%)	
	Brasília	SP	Brasília	SP
Deferrable				
Clotheswasher	122	122	2.6	2.32
Clothesdryer	150	150	3.2	2.85
Dishwasher	253	253	5.3	4.8
Non-deferrable				
Lights	690	690	14.6	13.1
Cook	1131	1131	23.88	21.45
Fridge	637	637	13.44	12
Boiler/Pump	353	883	8.05	19
Enter. appliances	392	392	8.3	7.45
HVAC	979	898	20.63	17.03
Total	4707	5156	100	100

2.3 Hybrid photovoltaic system and Battery controller

The entire PV array was simulated considering m-Si PV modules and 18.5% efficiency. The PV modules were oriented to the north and the tilt angles were approached to the latitude of each location to maximize annual PV output. The simulated PV generator is expressed by Sandia model [9-11]. Fig. 1 summarizes an electrical diagram of the system with an overview of relevant nomenclature: P_{PV} is the PV generation electricity, P_L is the load consumption electricity, P_{Imp} is the imported electricity from the power grid, P_{Exp} is the exported electricity to the grid (-) and P_{Bat} is the electricity exchanged with the batteries in charge (P_{Char} (+)) or discharge (P_{Disc} (-)).

The installation of a storage system was investigated to optimize the self-consumption and the degree of self-sufficiency of the house. Lead-acid batteries coupled with PV generators have the main advantage to provide higher levels of self-consumption, to reduce the net exported energy and the current issues with PV-grid interaction.

On the second step of the study, simulations were carried out in a high level battery controller. The controller actuates to inject electricity from the storage system to the load when the PV generation is lower than the load consumption or when the Sun goes down, as well as allows to supply the demand during the peak pricing period. As a strategy, the batteries just can be charged with PV energy and the priority is to use energy from the PV system to supply the load as well as the battery State of Charge (SoC) is respected to preserve the battery lifetime. For midcharge (20% ≤ SoC ≤ 95%) the controller performs the self-consumption maximization by no exchanging energy with the grid [8].

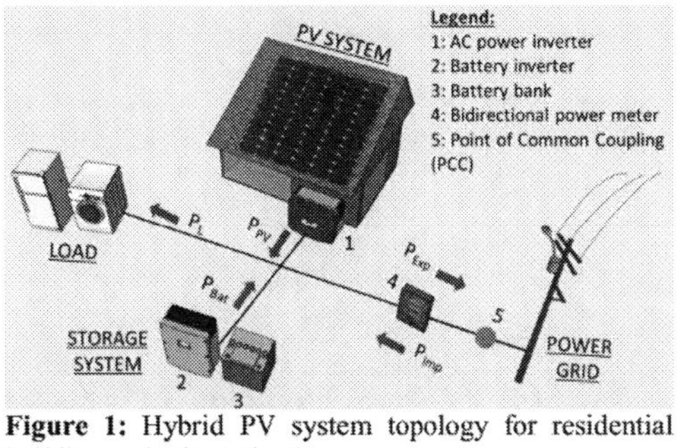

Figure 1: Hybrid PV system topology for residential building under investigation.

2.4 Evaluation equations

The Load Cover Factor, ξ_L, defined by [8], represents the percentage of the electrical demand covered by the PV system. A ξ_L closer or equal to one the local generation matches the local consumption. The Supply Cover Factor, ξ_G, described in [13], is related to the PV-generated electricity that supplied the loads with respect to the total generated electricity. A system operating with ξ_G values closer to unit is a system that does not export large amounts of energy to the grid.

$$\xi_L = \frac{E_{PV\rightarrow L} + E_{Disc}}{E_L} \quad (1) \; ; \; \xi_G = \frac{E_{PV\rightarrow L} + E_{Disc}}{E_L} \quad (2)$$

Where $P_{PV\rightarrow L}$ is the electricity produced by the PV system and directly consumed by the load. The Loss of Load Probability (LOLP) represents the fraction of time that the local generation does not cover the building demand [14]. With LOLP analysis it is possible to access the building energy autonomy (A_b) as the fraction of time when 100% of the load can be matched by PV-origin [1].

$$LOLP = \frac{\int_{\tau 1}^{\tau 2} f(t)}{T} \begin{cases} f(t) = 1, \; if \; P_{net} < 0 \\ f(t) = 0, \; if \; P_{net} \geq 0 \end{cases} \quad (3)$$

$$A_b = 1 - LOLP \quad (4)$$

Where P_{net} is net exported power, T is the evaluation period and τ is the time interval. Grid interaction parameters indicate the magnitude of the power injected or purchased by the building. The capacity factor shows the total energy exchange with the grid divided by the exchange that would have occurred at nominal connection capacity [14]. The capacity factor of imported electricity (CF_{Imp}) and capacity factor of exported electricity (CF_{Exp}) measure the degree of imported and exported electricity with the grid, respectively, normalized with respect to the grid connection capacity (P_{GCC}).

$$CF_{Imp} = \frac{|E_{Imp}|}{P_{GCC} \cdot T} \quad (5) \; ; \; CF_{Exp} = \frac{|E_{Exp}|}{P_{GCC} \cdot T} \quad (6)$$

The Dimensioning Rate is defined as the peak power exchange over the connection capacity and in the assumption of "self-consumption" connection mode, we defined DR in terms of imported power (DR_{Imp}) and exported power (DR_{Exp}).

$$DR_{Imp} = \frac{\max(P_{Imp})}{P_{GCC}} \quad (7) \; ; \; DR_{Exp} = \frac{\min(P_{Exp})}{P_{GCC}} \quad (8)$$

The grid interaction indexes describe the mean grid stress using the standard deviation of the grid interaction over a given period [15]. The grid interaction index of exported electricity ($f_{Grid,Exp}$) gives the ratio of the exported electricity into the grid compared to the maximum value over a year. The grid interaction index of imported electricity ($f_{Grid,Imp}$) indicates the variability of the imported electricity from the grid within a year normalized on the maximum absolute value within an annual cycle. High values of $f_{Grid,Exp}$ means large amount of surplus electricity not profited by the building and high values of $f_{Grid,Imp}$ means the building is strictly depend of the power grid performance.

$$f_{Grid,Imp} = STD\left(\frac{P_{Imp}}{max(|P_{Imp}|)}\right) \quad (9)$$

$$f_{Grid,Exp} = STD\left(\frac{P_{Exp}}{max(|P_{Exp}|)}\right) \quad (10)$$

The Levelized Cost of Electricity (LCOE) is the cost of generating electricity delivered to the load or power grid during hybrid PV system lifetime (n) and covers all system investments and operation and maintenance costs (C) in the kth year. The LCOE is given by Eq. 11, where ρ is the annual discount rate. The investment cost (I) is given by Eq. 12, where I_0 is the initial investment, $C_{inverter}$ and $C_{battery}$ are, respectively, the replacement costs of inverter and batteries.

$$LCOE = \frac{I + \sum_{k=1}^{n} \frac{C_k}{(1+\rho)^k}}{\sum_{t=1}^{T} \frac{(E_{PV} - E_{Exp})_k}{(1+\rho)^k}} \quad (11)$$

$$I = I_o + C_{inverter} + C_{battery} \quad (12)$$

3 RESULTS

The work analysis considers four scenarios as given by Tab. II. The scenarios S1 and S2 do not consider storage system and the scenarios S3 and S4 were simulated with batteries through GridSim [12]. The grid connection capacity (P_{GCC} = 6 kW), not inferior to the PV installed capacity power (P_0), according to Brazilian regulations, was designed according to the maximum annual building demand. It was investigated different scenarios of installed PV output (from 1 to 10 kWac) and battery nominal capacity (C_{Bat}) from 1.92 to 13.44 kWh (40-280 Ah) in order to evaluate an optimized system capacity. The depth of discharge considered in the simulations is 80%.

Table II: Building electrical installation scenarios considered in the study.

Scenario	Configuration
S1	PV system (without DSM and storage)
S2	PV system and DSM
S3	PV system and storage
S4	PV system, DSM and storage

A combination of annual simulations results for GI indexes as a function of installed PV output and C_{Bat} variations to São Paulo is summarized in Fig. 2. The results for $f_{Grid,Imp}$ using standard deviation shows that the index declines with increasing PV and storage capacities due to the fact that the building self-sufficiency grows with DSM and storage system strategies. It is important to emphasize that this index describes the fluctuation of the energy exchange with the grid, not the amount of the exchanged electricity [15]. Even in the scenarios of minimum hybrid PV sizes, the maximum imported power fluctuation is just 14% by virtue of the effect of storage in supplying the load leading $f_{Grid,Imp}$ to low and safety margins. On the contrary, the variability of the exported power over the grid grows by increasing the PV size but decreases by increasing the storage capacity. As long as a high PV capacity is combined with a small battery bank, the batteries do not have enough capacity to store sunny days PV surplus over local demand, so that $f_{Grid,Exp}$ can reach up to about 25%. In this sense, smaller PV sizes are more technically effective in the terms of excess of electricity and grid impact, being in the best case $f_{Grid,Exp}$ just 4%.

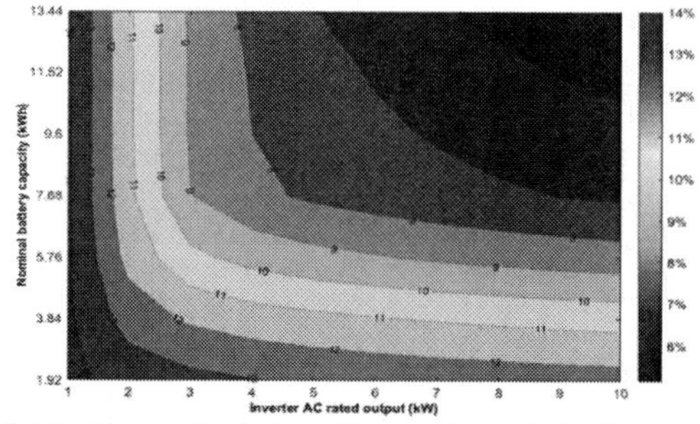

(a) Grid interaction index of imported electricity ($f_{Grid,Imp}$)

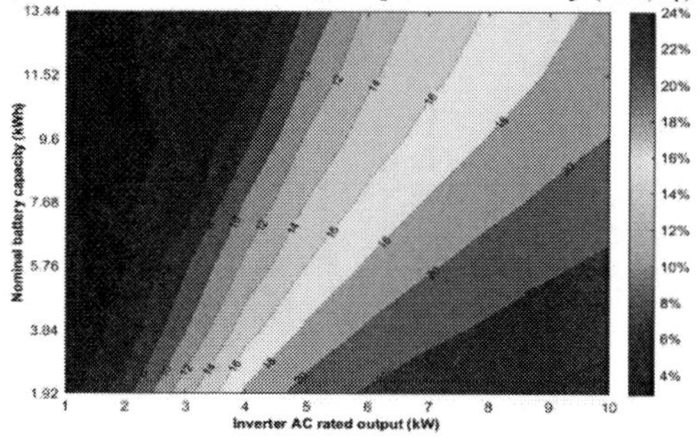

(b) Grid interaction index of exported electricity ($f_{Grid,Exp}$)

Figure 2: Annual mean of grid interaction indexes as function of the inverter size and installed battery capacity in S4.

Figures 3 presents the annual mean ξ_G with inverter AC output deviations. The supply cover factor decreases when the PV size grows, because the P_{PV}-P_L correlation diverges when the PV production grows under a fix consumption pattern. The PV electricity directly consumed by the loads is improved in the scenario S4 with the displacement of deferrable loads to the solar periods and in addition the surplus electricity is profited to match the night-time loads. The optimal PV system size for self-consumption is around 3 kW prior to curve saturation and ξ_G grows from each scenario: 0.27 in S1, 0.36 in S2, 0.55 in S3 and 0.7 in S4. However in the scenarios with storage (S3 and S4) the curves declination is more pronounced and the saturation will occur with highest PV capacities.

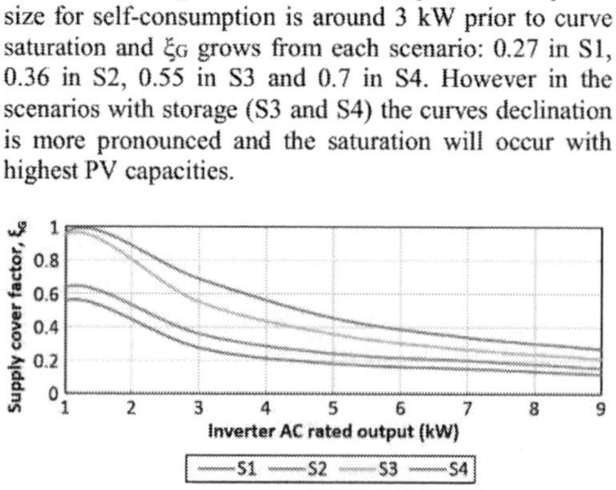

Figure 3: Annual supply cover factor (ξ_G) under different PV system topologies. C_{Bat}: 5.76 kWh.

In order to find the optimal battery bank size that is

more efficient for self-consumption Fig. 4 presents variations of C_{Bat} as function of ξ_L. Small-scale battery sizing up to 5.76 kWh (0.5 days of functional autonomy) contributes more for the growth of ξ_L as can be observed in straight line tendency. In the transition between 0.5 and 1 day of autonomy lies a maximum ξ_L and a C_{Bat} higher than 1 day of autonomy should be avoided since contributes less to the growth of ξ_L. By reviewing the previous analyzes, the optimized hybrid PV system corresponding to 3.22 kW_p and 5.76 kWh battery capacity results in a ξ_L of 55%, a ξ_G of 70% and an A_b of around 40%. In this case, the house has a large potential to consume the generated local electricity but it loses in energy autonomy when compared to the larger installed capacities.

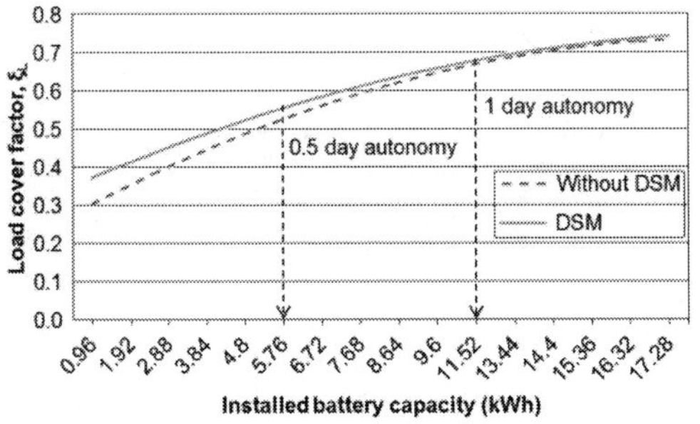

Figure 4: Sizing of battery bank as function of load cover factor in S3 and S4. P_0: 3.22 kW_p.

Figure 5 indicates results of daily electricity flows in the house operating in São Paulo. Without DSM the deferrable loads are allocated when the house occupants get home after work (18:00 h). The PV system charges the battery up to 100% SoC and the battery supplies the load until it reaches the minimum SoC allowed (SoC_{min} = 20%), after which the house extracts electricity from the grid. It is possible to note the convenience to displace deferrable loads to solar periods because the consumption is very low in the afternoon when the PV generation is available. In this sense, from S3 to S4, ξ_G increases 30% while ξ_L increases 20%.

Figure 6 emphasizes monthly variations of ξ_L and ξ_G for the two investigated locations. The monthly daily mean ξ_L ranges from 0.2 to 0.3 in S1 and from 0.25 to 0.35 in S2. With batteries, it was reported major improvements and thus ξ_L ranges from 0.45 to 0.6 in S3 and from 0.47 to 0.67 in the optimized scenario S4. In addition, the application of DSM increases ξ_L and it depends on the amount of "deferrable" loads that are displaced, where the bigger the amount of "deferrable" loads are the higher the parameter is reached [17]. The seasonal variations are more pronounced in winter, where the electricity consumption profile changes due to different temperatures in the country. From May to August less PV energy is produced, the cooling load decreases, the boiler demands electricity before sunset and there are around two hours less of Sun, thus decreasing ξ_L. In Brasília, this effect is not pronounced because it presents better potential of solar energy and even the consumption for DHW is lower than in São Paulo. Otherwise, PV production increases in summer, battery charging potential increases and ξ_L reaches the highest values, up to about 0.7 in São Paulo and 0.6 in Brasília. In São Paulo, ξ_G grows significantly in winter months because the on-site generation gets smaller and the P_{PV}-P_L correlation is optimized, achieving around 0.8 in S4. The same happens in Brasília with the increasing of cooling space consumption in summer, making ξ_G superior than remainder of the year.

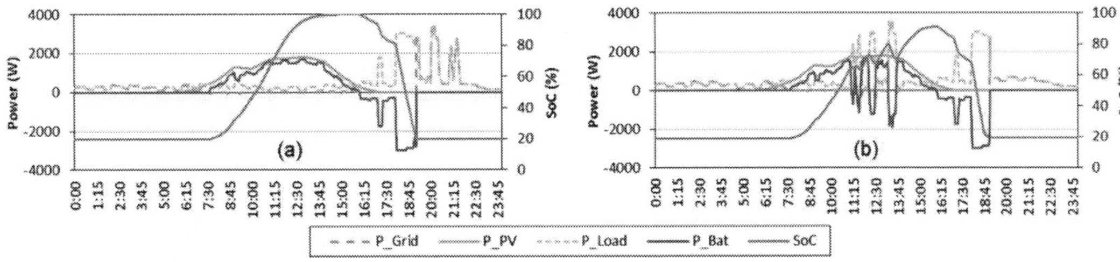

Figure 5: Electricity flows and battery SoC in typical day: (a) Scenario without DSM (S3): ξ_L: 0.54, ξ_G: 0.62, A_b: 0.4. (b) Scenario with DSM (S4): ξ_L: 0.70, ξ_G: 0.90, A_b: 0.37. P_0: 3.22 kW_p. C_{Bat}: 5.76 kWh.

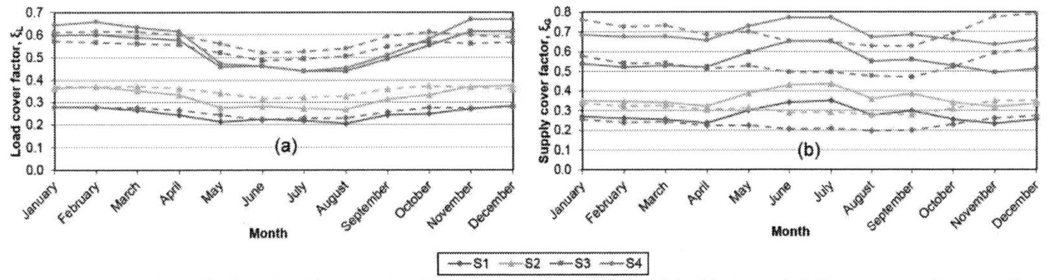

Figure 6: Annual LM indicators. (a) Annual daily mean load cover factor (ξ_L). (b) Annual daily mean supply cover factor (ξ_G). Continuous lines: São Paulo. Dashed lines: Brasília.

The duration curve given by Fig. 7 represents the time duration the building imports or exports electricity, indicating the power levels that electricity exchanges with the grid occur over the year. Positive values mean that electricity is fed into the grid while negative values mean electricity imports from the grid.

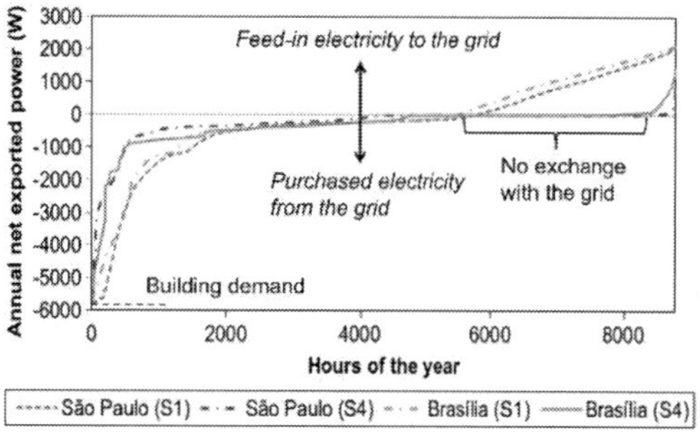

Figure 7: Duration curve for electricity exchanges with the power grid.

The duration of purchased electricity and imported power peaks are reduced with storage in both locations. In a general context, it can be seen the power reduction from 6 kW up to 2 kW implies a lower grid request time if compared to less than 1 kW where the house demands electricity during a larger number of hours. In fact, this power range represents the night-time demand the storage system is not able to supply. As indicated by Fig. 8, there is no exchanges between the building and the grid for a large time interval, representing 35% of the annual period. Similarly, the annual duration of exported electricity is very short in S4 due to the fact the surplus energy not profited by the load is expected to charge the batteries.

Table III highlights the annual results of LMGI indicators calculated for the two studied locations in the scenarios S1 and S4.

Table III: Annual results of LMGI indicators for the studied locations. C_{Bat} of 5.76 kWh (fully charged to discharge, 1h: 4.6 kW), P_0 of 3.22 kW$_p$ and P_{GCC} of 6 kW.

	São Paulo		Brasília	
Scenario	S1	S4	S1	S4
Load Matching indicators				
ξ_L	0.24	0.55	0.25	0.60
ξ_G	0.27	0.68	0.22	0.70
LOLP	0.65	0.60	0.60	0.50
A_b	0.36	0.40	0.40	0.50
Grid interaction indicators				
DR_{Imp}	1.0	0.93	0.95	0.91
DR_{Exp}	0.40	0.30	0.46	0.31
$f_{Grid,Exp}$	25.4%	7.80%	25.5%	7.39%
$f_{Grid,Imp}$	13%	10.23%	13.6%	12.82%
$P_{net} \approx 0$	0.52%	35.42%	0.40%	37.48%
CF_{Imp}	7.60%	5.75%	6.65%	5.60%
CF_{Exp}	6.45%	0.52%	7.80%	0.50%

The results for ξ_L show an increase of 30% from the scenarios S1 to S4 and up to 60% of the load requirements can be met by local generation with respect to the designed system capacity. The supply cover factor (ξ_G) can be increased even more, as it can be deduced when comparing S4 with S1 (40% in São Paulo and 50% in Brasília), showing that up to 70% of the PV electricity is directly profited by the loads. The high annual values of ξ_G for S4 evidence that load and generation profiles are well correlated and the building profits efficiently from locally generated electricity, thus improving the self-consumption capability.

The loss of load probability (LOLP) indicates that around 60% of the time the load is not covered by PV and thus the building must purchase power from the grid. The smallest LOLP occurs for the case S4 in Brasília and then the energy autonomy (A_b) corresponds to 50%, increasing 10% with respect to the scenario without battery. The building energy autonomy constraint is associated to the sizing of the small-scale storage system and the limitation of the PV capacity in order to improve the load matching potential. However, the results evidence that the growth of the hybrid system size improves the energy autonomy.

The results for DR_{Imp} show that the P_{GCC} designed according to the maximum annual electrical demand ensures a safety fraction of an equivalent nominal connection capacity and the grid connection does not exceed the allowed limit. The dimensioned DR_{Exp} in S4 is smaller than the PV system without storage by reason of the designed PV capacity of 3.22 kW$_p$ is far from the designed connection capacity of 6 kW and, in addition, the surplus electricity is used to charge the batteries. Thus, DR_{Exp} is reduced around 10% while the exported power over the grid connection is reduced. The $f_{Grid,Exp}$ indicator is reduced from 25% in S1 to around 8% in S4 mainly due to the capability of the system to reduce grid exchange fluctuations carried out by the battery controller in order to complete daily SoC. The low percentage of the ratio of the imported power metering normalized on the maximum absolute value ($f_{Grid,Imp}$) means an almost constant import instead of high fluctuations, according to energy simulations, and the building does not depend considerably on the power grid performance.

In the scenario S1, the building night-time consumption is strictly dependent of the grid and the purchased electricity can be reduced when the PV generates power. In this sense, the probability of the building acting autonomously from the grid ($P_{net} \approx 0$) is near to zero in the scenario S1 but it increases to 40% due to the use of storage.

The capacity factor suggests about the total energy exchanged with the grid over the year. Through the comparison of normalized imported electricity (CF_{Imp}) it can be concluded that the effect of storing energy decreases the amount of purchased energy and only around 6% of the equivalent energy use of the grid connection is used. The results show the normalized exported electricity (CF_{Exp}) is near zero when applying energy storage by virtue of the efficiency in charging the batteries with PV surplus electricity, and with the displacement of deferrable loads the PV supplies more efficiently the loads, decreasing in effect the amount of feed-in electricity into the grid.

3.5 Levelized Cost of Electricity and payback time

The investigations for the optimized system were performed in scenarios S1 and S4 considering the cost of PV modules, inverter and battery bank obtained from typical Brazilian PV solar system installed at the end of 2015 [18]. Considering added taxes and national taxes the investment cost is 7.24 R$/W$_p$ in S1 (equivalent to 1.80 €/W$_p$) and 9.92 R$/W$_p$ (equivalent to 2.55 €/W$_p$) in S4. The economic analysis were performed considering two supporting schemes: (1) actual Brazilian net-metering in S1, admitting the consumer unit gets credits in amount of active energy for the PV surplus fed into the grid. The credits can be used to compensate the purchased electricity from the grid in a period up to 60 months; (2) no financial compensation for PV excess in S4. It was considered batteries replacement each 10 years and PV generator lifetime of 30 years.

The annual savings obtained by using the energy credits is directly related to the power utility tariffs and the financial incentives. Fig. 8 shows the payback time is shorter for the lowest values of PV cost and discount rate and the higher the credits are, the higher the annual savings will be, leading to shorter return of the investment. Considering the actual investment cost, the payback time is reached from near to 13 years in S1 and from 21 years in S4. By assuming an optimistic financial scenario (i.e. ρ = 4% and PV cost = 6 R\$/W_p), the payback time can be achieved in 15 years in S1 and similarly in 21 years in S4. The investment is more profitable under Brazilian net-metering scheme (S1), due to the fact the building exports larger amount of electricity into the grid than operating in S4 mode and gets credits from the PV excess, that compensate the energy usage and reduce electricity costs. In S4, for discount rates above 7% the payback approaches or is above the PV generator lifetime, not profitable to invest, according to Fig. 8 (b). The main restriction of this scenario is that the initial investment is higher due to the cost of the storage system, the consumer is not compensated by the night-time purchased electricity that the small-scale storage system is not able to supply and the absence of credits makes the investment less attractive.

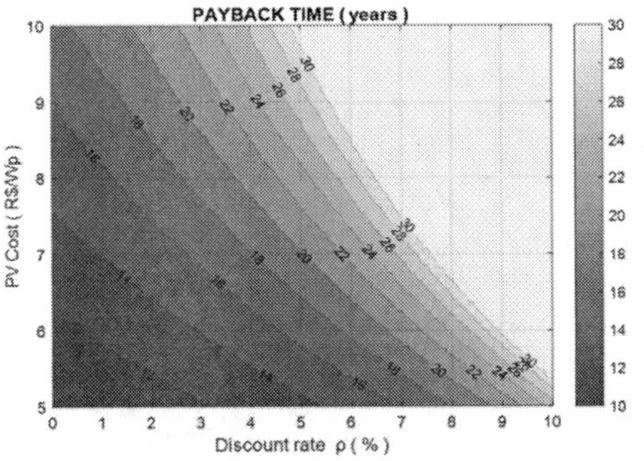

(a) Scenario S1 without storage system.

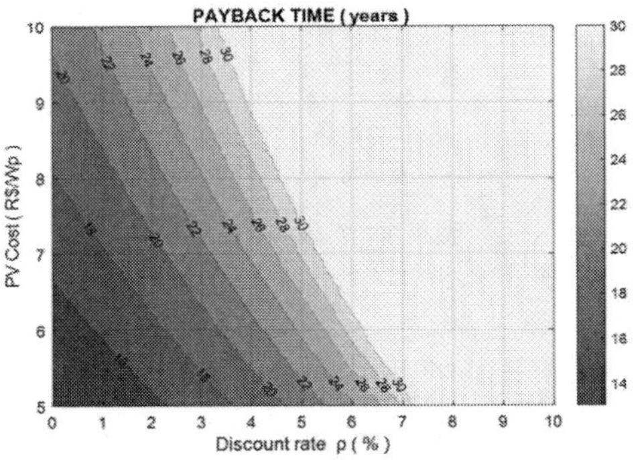

(b) Scenario S4 with storage system.

Figure 8: Payback time perspective as function of discount rate and PV cost deviations for São Paulo.

Figure 9 presents results of LCOE in function of discount rate variations in S4, without compensation for PV excess. As expected, the LCOE is commonly higher in scenarios with battery by virtue of high initial investment when considering costs of battery system and batteries replacement. In consequence, low annual discount rates are requested to the investment to be profitable. By virtue of the low electricity tariff in São Paulo (0.5 R\$/kWh), the discount rate must be around 2% for the PV system achieves grid parity. However, a grid-connected PV system without batteries, supported by the Brazilian energy compensation system (net-metering), can achieve grid parity up to 6% discount rate [5], closest to a mature market in Brazil. Despite relatively high irradiation levels, PV LCOE is higher in Brazil than in other countries mainly due to higher installation prices caused by customs duties levied on PV equipment and by the immaturity of the PV market [19].

While in São Paulo PV technology is still far from being competitive against grid tariff, in other Brazilian locations where the electricity tariffs and irradiation levels are relatively higher (i.e. Brasília and Belo Horiz.) grid parity and payback time could be reached sooner [5].

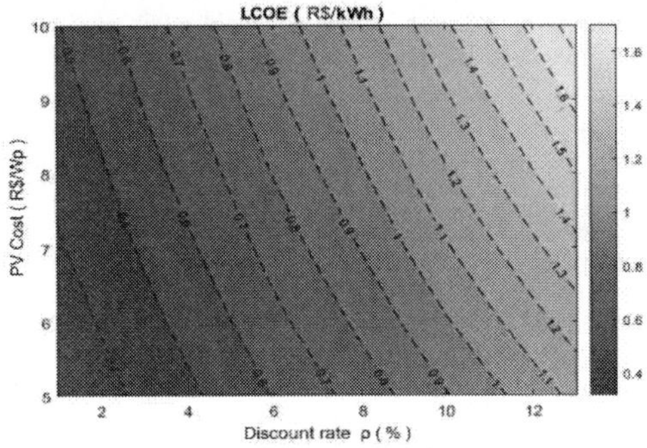

Figure 9: LCOE as function of discount rate and PV cost deviations for São Paulo in the scenario S4.

4. FINAL REMARKS AND CONCLUSIONS

In this paper, simulations results have been established that LMGI indicators present different responses with respect to four scenarios of PV system configuration and the selection of suitable LMGI indicators leads for an important task by reason of strengths and limitations of each one. Furthermore, the study analyzed an optimized PV-battery system, of special interest of building designers and operators. In general, by growing PV-battery capacity, load matching indexes tend to be improved due to the best potential in covering load consumption. However, the degree of self-consumption (ξ_G) decreases when the PV size grows, because the P_{PV}-P_L correlation diverges when growing the PV production. The straight line tendency with respect to battery capacity variations showed that the smaller sizes contribute more to the growth of ξ_L.

In GI terms, graphical representations in carpet plots showed that the ratios of imported-exported electricity into the grid are reduced by increasing storage capacity and the variability of exported power over the grid indicated that small PV sizes are more technically effective when considering PV surplus. It has been shown that curtailment of imported electricity effected with DSM and battery reduces $f_{Grid,Imp}$ to low and safety margins (6-14%), and for the designed system, according to Tab. III, $f_{Grid,Exp}$ is satisfactorily reduced from 25% in

S1 to 8% in S4.

Monthly calculations of ξ_L and ξ_G indicated there are no general rules for load matching by virtue of it depends of decoupled factors (i.e. electricity consumption, PV and ambient temperature). The results showed ξ_L has more potential in Brasília most of the year due to the better solar conditions. However, the house potential in profiting on-site generation is mostly higher in São Paulo winter due to higher consumption combined with lower PV production, which means less excess of PV electricity.

The duration curves in Fig. 8 provided useful information's about PV-grid interaction, indicating the duration of purchased electricity and power peaks over the grid connection are reduced in both locations when applying energy storage. The duration of exported electricity demonstrated to be very short in S4, leading to the reduction of voltage rise associated with reverse power flows, contributing in a positive manner for the grid operation when comparing with larger system capacities.

The application of DSM and storage evidenced increases in the building self-sufficiency and improvements on the grid interaction. The annual results of LMGI indicators showed the LM indexes (ξ_L, ξ_G, LOLP, A_b) presented great improvement from S1 to S4: 30% for ξ_L, 40% for ξ_G and notably 10% for A_b in Brasília. The proposed GI indicators ($f_{Grid,Imp}$, CF_{Imp}, DR_{Imp}) showed slight differences between the two test case scenarios. The results of CF_{Imp} index assigned only around 6% of equivalent energy use of the grid connection was used and this information can be useful for both target groups, building and grid designers. Some indexes were reduced greatly due to the effects of DSM and storage, as expressed by DR_{Exp}, $F_{Grid,Exp}$ and CF_{Exp}. Notably, the normalized exported electricity (CF_{Exp}) decreased near to zero due to system design improvements, and for high penetration situations of renewable energy sources into the grid this condition can be taken into account for a grid design perspective. In terms of the probability that the building acting autonomously of the grid ($P_{net} \approx 0$) it was notified a great increasing of 40%, suitable to guide design process of the next generation of self-sufficient houses.

The study determined under Brazilian net-metering scheme the building reaches better economy by using energy credits generated by PV excess, reducing substantially the electricity costs and the return of investment, but it was denoted the importance of grid tariffs and solar resource to achieve greater profitability. The paper demonstrated the NPEB was set to achieve grid parity with very low discount rates in S4, far from a real market, and highlighted the needs to reduce PV-battery costs and to promote financial incentives in order to achieve shorter payback time.

In conclusion, the parameters proposed in this paper are suitable for analyzing the complexity of NPEB-grid interaction and the implication of hybrid PV on load supplying, aspects of great interest for the development of smart grids and sustainable cities.

5. ACKNOWLEDGEMENTS

G. Dávi wishes to acknowledge the Brazilian National Research Council (CNPQ) for financial support to carry out the studies which led to this article. The authors acknowledge support of the Spanish Ministry of Economy and Competitiveness within the framework of the project "Sistema distribuido de gestión de energía en redes eléctricas inteligentes" (TEC2015-66126-R).

REFERENCES

[1] J. Salom, A.J. Marszal, J. Widén, J. Candanedo, K.B. Lindberg, Analysis of load match and grid interaction indicators in net zero energy buildings with simulated and monitored data, Appl. Energy. 136 (2014) 119-131.

[2] J. Salom, J. Widén, J. Candanedo, I. Sartori, K. Voss, A. Marszal, Understanding net zero energy buildings: Evaluation of load matching and grid interaction indicators, in: 12th Conf. of the Building Performance Simulation Association, Sydney (2011).

[3] IEA PVPS Task 1, Trends 2015 in photovoltaic applications, Tech. Rep. 27, International Energy Agency (IEA) (2015).

[4] Agência Nacional de Energia Elétrica (ANEEL), Resolução Aneel nº 687/2015 (2015).

[5] G.A. Dávi, E. Caamaño-Martín, R. Rüther, J. Solano, Energy performance evaluation of a net plus-energy residential building with grid-connected photovoltaic system in Brazil, Energy Buildings. 120 (2016) 19-29.

[6] SWERA. Solar and Wind Energy Resource Assessment, <http://en.openei.org/apps/swera> (Acessed 2015-10-27).

[7] ASHRAE Standard 55, Thermal Environmental Conditions for Human Occupancy (2013).

[8] M. Castillo-Cagigal, E. Caamaño-Martín, E. Matallanas, D. Masa-Bote, A. Gutiérrez, F. Monasterio-Huelin, J. Jiménez-Leube, PV self-consumption optimization with storage and Active DSM for the residential sector, Solar Energy. 85 (2011) 2338-2348.

[9] D.L.King, W.E. Boyson, J.A. Kratochvil, Photovoltaic Array Performance Model, Sandia National Laboratories, Albuquerque, NM 87185 (2003).

[10] G. Barker, Predicting Long-Term Performance of Photovoltaic Arrays. Submitted as deliverable for NREL subcontract LAX-1-30480-02 (2003).

[11] EnergyPlus Engineering Reference, Lawrence Berkeley National Lab. Berkely: LBL.

[12] M. Castillo-Cagigal, E. Caamaño-Martín et al, Self-consumption of PV electricity with active demand side management: the GEDELOS-PV system, 25th EU PVSEC Conference Valencia, Spain (2010).

[13] E. Matallanas, J. Solórzano, M. Castillo-Cagigal, et al, Electrical energy balance contest in Solar Decathlon Europe 2012, Energy Buildings. 83 (2014) 36-43.

[14] B. Verbruggen, R. De Coninck, et al, Grid impact indicators for active building simulation, Innovative Smart Grid Technologies (ISGT), IEEE PES (2011) 1-6.

[15] K. Voss, I. Sartori, A. Napolitano, S. Geier, H. Gonzalves, M. Hall, P. Heiselberg, J. Widén, J. Candanedo, E. Musall, B. Karlsson, P. Torcellini, Load matching and grid interaction of net zero energy buildings, Proceedings of EuroSun, Graz, AT (2010).

[16] C.M. Colson, M.H. Nehrir, A review of challenges to real-time power management of microgrids, Power & Energy Society General Meeting. IEEE. 09. (2009) 1-8.

[17] E. Matallanas, M. Castillo-Cagigal et al, Analysis of the self-consumption possibilities in small grid-connected PV systems in Spain. Proceedings of the 26th European Photovoltaic Solar Energy Conference. Munich, Germany. (2011) 4619-4624.

[18] Photon, Photon international GmbH, Germany. November (2015).

6.1 GRID AND ENERGY SYSTEM INTEGRATION - Submission for poster presentation

ADVANCED PV INVERTER FUNCTIONS: VERIFICATION TEST PROCEDURE

Joachim Freis, Marijo Cosic and Bengt Jaeckel

UL International GmbH, Admiral-Rosendahl-Strasse 9, 63263 Neu-Isenburg (Zeppelinheim), Germany, Bengt.Jaeckel@gmail.com, Tel. +49-162-2621901

ABSTRACT:
In recent years the installed PV and wind capacity increased significantly. Due to the higher penetration of renewable fluctuating energy sources in the local grid new challenges rose up and needed to be implemented by grid suppliers and inverter manufacturers to maintain a stable grid. To avoid large over voltages and frequency fluctuations that may cause total grid black-outs new advanced requirements were set for all inverter type system that feed power into the grid. Those new requirements for the US are designed to regulate, operate and manage the local grid to keep it stable and safe for all grid connected power consumers (e.g. avoid overvoltage that may damage electronic appliances).
This will be important because of further increased penetration of renewable energies in certain areas in combination with limited grid resources. Grid expansion is typically expensive and will take years and therefore is one of the last options. With the help of smart and advanced features build into inverters more green energy can be feed-in while allowing grid operators to more easily manage the grid.
Advanced inverter functions are defined (or are being developed) in international and national standards and will be mandatory in the near future, if not already. The scope of those functions is to provide more flexibility for the grid to enable more renewable energy sources to be connected to it while keeping the grid reliable. The new requirements based on different local requirements and standards can be easily evaluated using the presented approach with an automated test setup in combination of reduced time and cost.
The upcoming revision of UL 1741 includes new requirements, where new tests and functions are described to stabilize and serve the public electrical grid and allow tailoring the function of the inverter depending on the need of the public grid. A protocol like Sunspec allows the parameterization remotely and helps to speed the testing. The possibility to manage and control inverters remotely will be of general importance. Most manufactures already include some of those functions in their products. There reliability and interoperability has still to be proven and our presented approach is designed to do so.

Keywords: Photovoltaic, Inverter, Sunspec, Grid, Automation

1. INTRODUCTION

In recent years the installed renewable energy like PV, but also Wind, capacity increased significantly and will increase further in the future. This more and more distributed energy production leads to new challenges for the electric grid as e.g. local voltage increase (see schematic in Figure 1). Grid operators struggle with grid stability while PV and wind are pushing conventional power plants out of the grid [1].
To maintain grid stability new installed power sources will have to have smart and advanced functions including remote control to interact and react not only passive like to today mostly.
Those new requirements, functions and features are currently under development and are getting standardized. The presented approach will help to test, verify and compare different hardware and software and quantifies the outcome based on set grid parameters. The standardization for these protocols among all manufactures is important as this can lead hopefully in the future to develop grid requirements faster. This standardized grid requirements can then be internationally standardized and allow each utility still to be flexible for local needs better then today. On the other hand this protocols and standardizations are helping manufacturers to reduce cost by maintaining less grid compliances for different markets. Further this helps testing laboratories to develop and maintain one program that can test for multiple grid standards faster and more reliable. The Sunspec protocol is a major step forward to this and used in the following approach

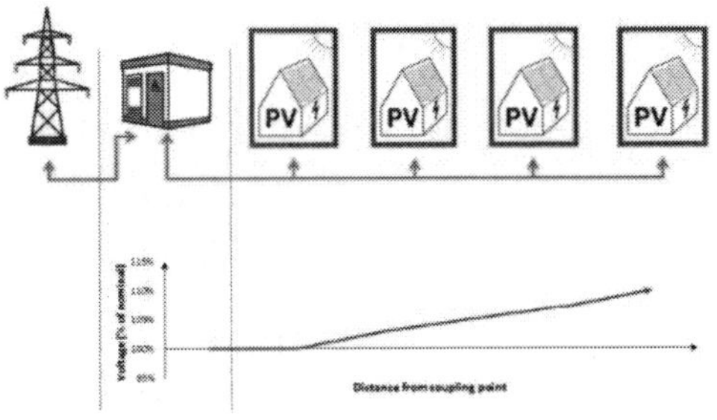

Figure 1: Schematic of a local grid with high penetration of PV. Going away from the controller station/coupling point the voltage will increase if PV generators feed into the grid line. If a certain limit is reached the grid-operator lowers the output of the renewable energy source. This will lead to no further voltage increase and a more or less stable grid, but it also lowers the income for the e.g. PV system owner. Smart and advanced functions can increase the amount of energy that can locally be feed into the grid. This will be beneficial for both sides: The grid and the system owner.

2. SCOPE and APPROACH

Previous requirements in the US led to certain cut-off scenarios where no further power feed-in was allowed. This is not in the interest of the power producer. With the new regulations and by the use of new advanced inverter

features in the US like California Rule 21 which is integreted in the upcoming UL 1741SA will allow actually more renewable power that can be feed into the grid. Voltage and frequency stabilization are increased by allowing the inverters to supply reactive power into the grid further the possibility in the new rule to allow reducing active power as a function of frequency and voltage. In combination with remote access and smart metering new business cases for PV and other renewable suppliers/parties can be created in the future.
The local/ national requirements still differ by countries. To be able to deploy the same type of inverter in different countries the advanced functions must be easily, if needed remotely, adjustable to the local requirements. This can lead to more flexibility and interaction for grid operators with a standardized protocol/standards like Sunspec.
The aim of this paper is to show similarities and differences in the grid standards/approach for the upcoming standards in US (UL1741SA [3]) to in this example Germany (VDE 4105 [4]) and the benefits in automating test setups with the ability of modern protocols which are integrated in new inverters to speed up the testing and certification process in general.
The presented approach is to test and simulate the interaction of inverters to different grid scenarios. This is of great importance since local and regional requirements are developed currently and with this approach compliance can be easily studied and verified.
Chapter 3 gives a comparison in the upcoming UL 1741SA standard to the grid requirements versus the Germany VDE AR-N 4105. After this the following chapter will show the approach that we have taken to build up a software system to interact with the measurement devices and the interaction with the inverters by using advanced inverter protocols.

3. UL 1741 vs VDE AR-N 4105

Based on different Grid architecture and topology the requirements vary by country and region. This paragraph highlights the most important ones in the upcoming UL 1741SA to the VDE ARN 4105.

a. US SMART-GRID-SERVICES (UL 1741SA):

UL 1741 is the North American Inverter safety standard. Currently it is updated with a supplement to include advanced inverter functions for grid support. This supplement defines the evaluation criteria for utility interactive inverters which are not covered in IEEE 1547 [6], yet.
The following advanced inverter functions are implemented/currently under revision in the UL 1741SA:

- Specific power factor / Reactive power feed-in
- Low- / High voltage ride through
- Low- / High frequency ride through
- Normal and soft-start ramp rate
- Volt-Var Mode (VV) – Q(V) Function
- Volt-Watt (VW) - P(V)Frequency-Watt (FW) (optional)

As an example the allowed frequency field is shown in Figure 2.

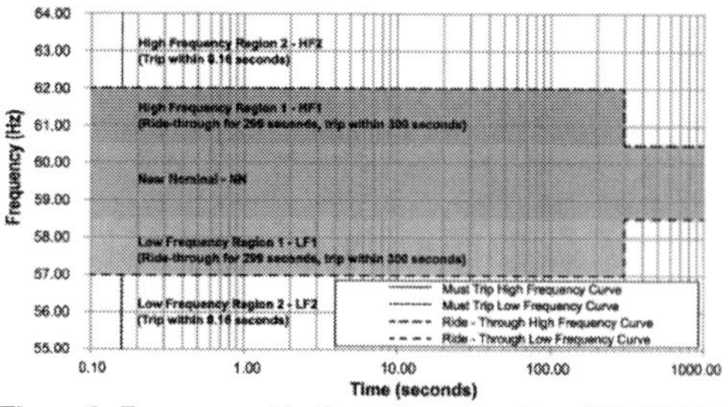

Figure 2: Frequency ride through regions from UL 1741SA [2].

b. GERMAN APPLICATION GUIDE (VDE AR-N 4105)

This application guide describes the requirements for power generation systems that are connected to a network operator's low-voltage network and operated in parallel with this network. The aim of this is to maintain the reliability and safety of the network operation.
Besides grid supporting functions more requirements such as system reactions (e.g. Harmonics, inter-harmonics, Flicker, Commutation notches, voltage unbalance, higher frequencies) are implemented.
The advanced inverter functions are described in the part "Behavior of the power generation system at the network" and contains following functions:

- Low- / High frequency ride through
  - Active power feed in at overfrequency
  - No disconnetting at underfrequency
- Reactive power feed-in
  - Displacement factor / active power characteristic curve cos φ (P)
  - Fixed power factor cos φ

The part „Construction of the power generation system / network and system protection (NS Protection)“ describes the functions below:

- Voltage drop protection (U<)
- Rise-in-voltage protections (U>;U>>)
- Frequency decrease protection (f<)
- Frequency increase protection (f>)

The principle for the network support is that the power generation systems shall be able to contribute to the static voltage- and frequency stability in the low-voltage network. Thereby the slow voltage changes are maintained within limits in the distribution network.

c. COMPARISON BETWEEN UL1741SA & VDE AR-N 4105:

The above mentioned grid functions differ due to the different frequencies and grid-voltages.
The essential advanced inverter functions are in both standards the same, but UL1741SA describes different testing procedures. The VDE V 0124-100 [5] standard describes the testing procedures for the VDE ARN 4105. Additional to the advanced grid functions other requirements for system reactions like Flicker, Harmonics, etc. are included in this application guide, UL1741SA refers to IEEE1547 which contains these.
With the upcoming update UL1741SA is getting a little closer to the German VDE ARN 4105 requirements. Still there are differences which in the laboratory (testing) needs to be considered and requires different settings.
The next chapter introduces the SUNSPEC alliance which helps testing laboratories and was used in our

approach in creating modular testing software to minimize testing time.

4. SUNSPEC

The Sunspec alliance is a global federation of solar developers, manufactures, service providers and financiers that establish information standards for the distributed energy industry.
The Sunspec alliance specified standard information models, data formats, communication protocols and system interfaces that enable renewable energy plants to interoperate with other system components, software applications, financial systems and smart grid [2].
With this open communication protocols and industry-standard device it is possible to configure, control and monitor inverter's parameters.
This approach allows setting any characteristics of the inverter via remote functions. By that it should be possible to test different grid-codes compliances inverters very fast.
For more details about the specification it is referred to www.sunspec.org.

5. TESTING SOFTWARE/ENVIRONMENT

Based on a variable input (PV simulator) that feeds a variable power signal into the inverter with advanced functions the interactions are investigated. For the investigation the following setup and software was build: The equipment under test (EUT) is connected to a Grid simulator (programmable AC-power supply) and a PV-Simulator (DC-power-supply). A scope (High-speed-Data-acquisition system) is used for measurement of AC-voltages and currents. The power-analyzer measures and analyzes AC voltages, currents, power-factor, active power, reactive power, apparent power and the DC-voltages and currents. The inverter is connected by an communications device and tested as an assembly which is connected to a computer for experimental setup, control and data storage. The setup is schematically shown in Figure 3.

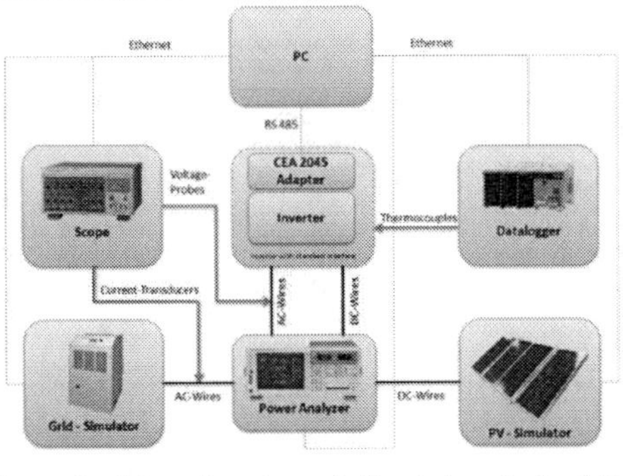

Figure 3: Schematic setup of the test and simulation software.

Due to the complex test-setup and especially test-sequences for new smart functions UL is using a programmable setup and uses highly automated self-developed test software which utilizes also the described Sunspec (see chapter 4) function to control the EUT.
Hundreds of measurements can be performed in hours and a pass/fail criteria as well as automated test-reports can be delivered. Different tests and grid-codes can be easily added and tested due to the modular approach of the software.

Each instrument is put into a separated module and utilizes an event bus system to communicate with other modules to visualize data or perform tests. So each module is loosely coupled and can be easily replaced by more efficient modules or with control modules for specific testing requirements.
The following diagram (see Figure 4) shows the Basic architecture of the software. Like described each device or function is written in a separated module and communicating over an event bus, which can transport data (measurement points) or switch, control commands to modules.

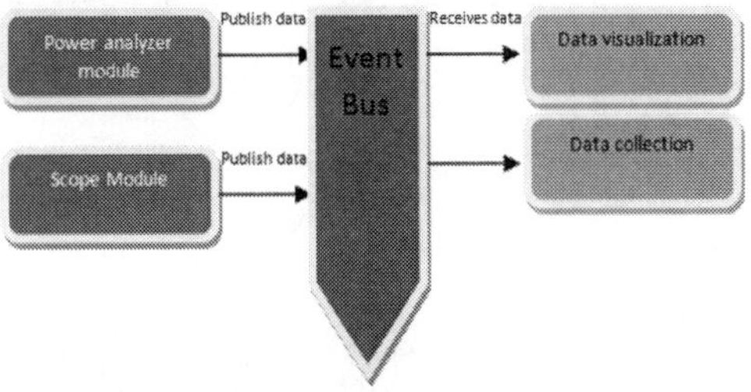

Figure 4: Basic modular architecture. Modules are sending data or commands over the event bus. This helps to decouple the testing software and allows the software to be easily adjusted and modified by just programming separated modules which can be plugged into the environment over the event bus system.

For a system architecture the modular approach helps to reduce dependencies and a small team can work independently without interfering each other.
The first prototype was able to be established with a small team of 2 people in 3 weeks with the basic function. Figure 5 shows the testing software interface with allows the user to see the measurement instrument and EUT live with values. The user can interact via a graphical interface with the instruments and see the results live. Figure 5is a separated module which uses the base architecture (see Figure 4) without interfering with the data sampling as the module only listens on the measurement data over the event bus and visualize them.

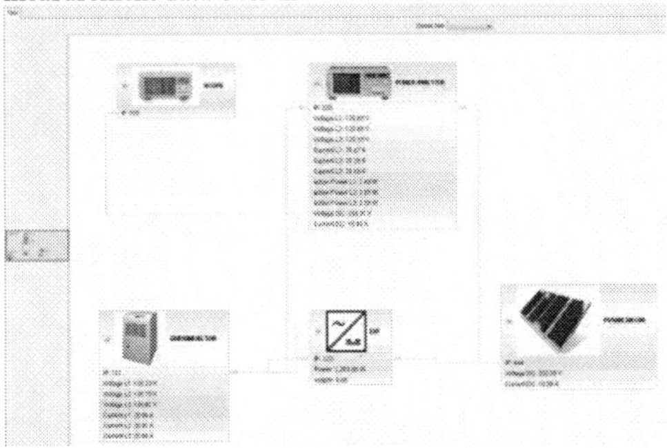

Figure 5: Graphical user interface to interact with the instruments and to visualize in real time the current collected data and the connection to the devices.

6. DISCUSSION and CONCLUSION

The software is still in an early stage and is further developed. We were able by utilizing the modular decoupled developing approach to build up an user interface and modules to speak to the measurement devices. The Sunspec library has allowed us to be able to interact with the inverters. Unfortunately not all functions were supported by the inverters. The following shows some of the characteristic (Figure 6 and Figure 7) which were measured during the setup and development phase of the software and the test setup for the new advanced function for the upcoming UL 1741SA revision.

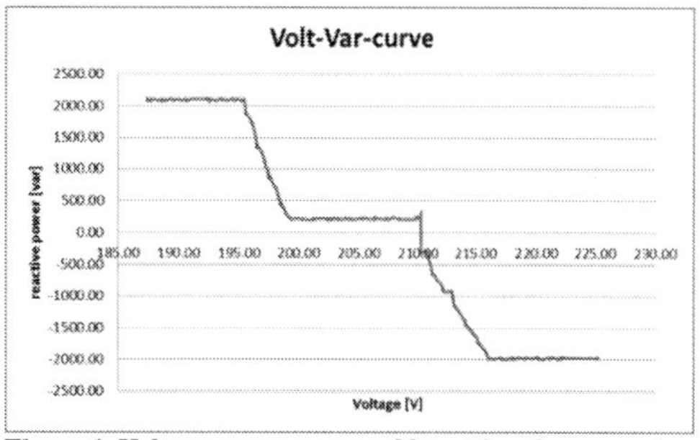

Figure 6: Volt var curve measured by setting the parameters with the Sunspec protocol.

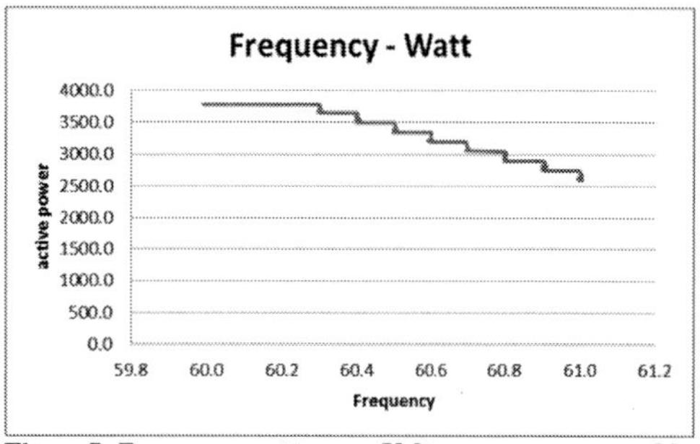

Figure 7: Frequency watt curve. Volt var curve measured by setting the parameters with the Sunspec protocol.

Advanced inverter functions are defined and can be evaluated using the presented approach. The scope of those functions is to provide more flexibility for the grid to enable more renewable energy sources to be connected to it while keeping the grid reliable.

The upcoming revision of UL 1741SA includes new requirements. The Sunspec protocol allows new possibilities for the local grid suppliers to monitor and maintain their grid and testing laboratories to control inverters remotely like in our presented approach.

Most manufactures already include some of those functions in their products. There reliability and interoperability has still to be proven and our presented approach is designed to do so automatically in the future.

7. REFERENCES

[1] http://www.dnhk.org/fileadmin/ahk_niederlande/Bilder/Dienstleistungen/2_connect_Business/W1-A_Smart_Grids_Thomas_Wenner.pdf

[2] http://sunspec.org/wp-content/uploads/2015/10/SunSpecESNAOpenMeetingOct13.pdf

[3] UL 1741SA

[4] VDE AR-N 4105

[5] VDE V 0124-100

[6] IEEE 1547.1

DEFINITION OF A DESALINATION-REFRIGERATION UNIT POWERED BY A SOLAR PHOTOVOLTAIC THERMAL COLLECTORS PVT: A CASE STUDY FOR DAKHLA MOROCCO

Mouhsine Ibrahimi[1] *, El Mehdi Elkhattabi[2], Abdelaziz Arbaoui[1], Youssef Aoura[1]
[1]ENSAM, BP 40 24, UMI, Meknes, MAROC
[2]USMBA, Fez, Morocco
Fax: (+212)05 35467163/64 Email: ibrahimi.mouhsine@hotmail.fr

ABSTRACT: The use of solar energy especially for desalination of seawater and refrigeration is well adapted for the Saharan region where the climate is extremely hot, the fresh water is scarce and solar energy potential is high. This paper presents an innovative mechanism reflected in a combination of mechanical vapor compression (MVC) desalination unit and absorption refrigeration machine (ARM) powered by solar Photovoltaic Thermal (PVT) collectors. The idea consists to recover the heat generated by the condenser of cold production machine in order to heat part of the intake seawater and improve the coefficient of performance (COP) of the (ARM). The PVT system provides the electrical energy required by the compressor and thermal energy for cold production and/or to heat the intake seawater. The proposed approach is illustrated using a case study for a typical summer day in the Dakhla city sited at the south of Morocco by using the global model, which includes SOLAR-MVC-ARM models. The influence of the combination on the power consumption and COP of MVC-ARM system powered by solar energy is studied. The results demonstrate that the new system reduces the electric consumption of the compressor and improve the COP of the ARM. Moreover, it is possible to operate the cycle at a low generator temperature. This study testifies that the combining system is efficient compared to each separate systems.
Keywords: Refrigeration, Desalination, Solar Radiation, South of Morocco.

1 INTRODUCTION

Nowadays, The most three critical problems that face human life are energy crisis, water crisis and pollution [1]. Fresh water supply and sustainable energy sources are of the most important topics on the international environment and development plans. Furthermore exploiting inexpensive energy sources like solar becomes important choice, especially for the systems whose, demand and the availability of solar irradiation are positively correlated in time and intensity. Moreover, using the combined systems is one of the ways to improve the total energy conversion efficiency of this system. On the other hand, as freshwater scarcity continues to increase with the increasing continuous of population, desalinating seawater is becoming a necessity for meeting freshwater needs especially in the North Africa and other regions. Water scarcity is a threat to over 40% of the global population [2].

Numerous papers report studies on the impudence of temperature levels and components irreversibility on COP, using either first or first and second law analysis [3-8] The first major project on an all solar absorption refrigeration system was undertaken by Trombe and Foex (1964) [9]. Ammonia-water solution is allowed to flow from a cold reservoir through a pipe placed at the focal line of a cylinder-parabolic reflector. Heated ammonia-water vaporized in the boiler is subsequently condensed in a cooling; Karagiorgas et al. [10] investigated the application of renewable technologies in the European tourism industry and identified a large number of solar thermal systems but only a few solar cooling systems.

However, the two major types of technologies that are used around the world for desalination can be broadly classified as either membrane such as reverse osmosis (RO) or thermal including single effect evaporation (SEE), multiple effect evaporation (MEE), multi-stage flash (MSF), thermal vapor compression (TVC) and mechanical vapor compression (MVC) [11,12]. Both technologies need energy to operate and produce fresh water.

The inception of commercial mechanical vapor compression (MVC) units dates back to the early 1970s [13,14]. MVC units have been evolved to become a mature technology over past decades. However, initial costs, system design and energy consumption remain challenging problems. Efficient use of energy in such energy- intensive operations is crucial to reduce the net energy consumption and to compete with other desalination technology. The MVC process remains to be attractive and competitive for production capacities less than 5000 m3/d [15]. An early report by Matz and Fisher [14] in 1981 showed that either the RO or MVC system has a definitive edge regarding total production cost. However, expansion of the MVC process remained limited. In 1994, only 200 MVC units with very small unit capacity are reported by Zimmerman [16].

Several water desalination plants driven by solar energy are in operation. In Abu Dhabi, a solar desalination plant operated since 1984 [18]. It consists of a multi-effect stack evaporator with an average yearly production of 85 m3/d. In Almeria (Spain), a Multi Effect Distillation (MED) solar seawater desalination plant of 72 m3/d was tested and a Wind/PV/RO of around 3 m3/d for seawater desalination in Lavrio, Greece [17].

The detailed SOLAR-MVC-ARM model is quite complex to solve, since it contains many coupled variables and many constraints corresponding to complex physics phenomena

In this study, we describe a thermodynamic model of a combination between single-effect mechanical vapor compression (MVC) desalination unit, absorption refrigeration system and Photovoltaic Thermal PVT solar system. The model is solved on 21 June, Dakhla-Morocco city (23° 41' 5 N -15° 57' 28 O) which is poor in water resources, rich in solar sources and Characterized by a hot climate.

2 SYSTEM DESCRIPTION

The schematic of global system is shown Figure 1. the absorption machines uses the working pair NH_3/H_2O, the technology used for desalination is based on mechanical vapor compression technology and the global system is powered by photovoltaic thermal collectors.

The MVC design model is based on the design step in the product lifecycle. More detailed of this approach is presented in previous work [21].

2.1 MVC System

Figure 1 (Black dashed line) shows the sequence of operations in an MVC system. Electricity is needed for compressor, circulation pumps, water treatment processes and for other component. The main components of the MVC unit are the mechanical vapor compressor, evaporator/condenser heat exchanger, and a circulation pumps. The demister serves to filter the water vapor before joining the compressor. The feed preheater is a plate type heat exchanger, which recovers part of the sensible heat found in the distillate and brine stream to heat the intake seawater. The thermal energy recovered is then sprayed over the evaporator tubes of the evaporator/condenser. The feed temperature is further increased to the brine boiling temperature and subsequently evaporation commences. The produced steam is drawn through the demister to the compressor, which is powered using the PV/T collectors in order to add an additional amount of superheat, increases the pressure and temperature of the steam by compression. The superheated vapor that has passed through the condenser heat exchanger tube releases its latent heat, which can be the heat source of the evaporator. At the same time, the vapor is condensed into purified water. The condensed distillate product and the brine stream flow through the preheaters where it exchanges heat with the intake stream. Moreover, distilled water is treated before being transported to the consumer.

2.2 ARM System

Figure 1 (Red dashed line) illustrates the main components of the absorption refrigeration cycle. The Absorption refrigeration machine consists of an evaporator, a condenser, a generator, an absorber, an internal heat exchangers and a solution pump. The generator takes the heat from PVT collectors and generates vapor. The vapor consists of a high mass fraction of refrigerant with a low mass fraction of absorbent. The generated vapor enters the condenser, which liquefies the refrigerant vapor. The liquid refrigerant flows into the evaporator where the cooling capacity occurs. The refrigerant leaves the evaporator as saturated vapor. The weak refrigerant-absorbent solution, consisting of a low mass fraction of refrigerant with a high mass fraction of absorbent, which is coming from the generator, flows to the solution heat exchanger. There it is cooled down by transferring heat to the strong solution from the absorber. The cooled weak solution enters the absorber precooler where it is cooled down even further. The weak solution absorbs the pure refrigerant vapor inside the absorber. Finally, the resulting strong solution flows into the generator and the absorption refrigeration cycle reconvenes.

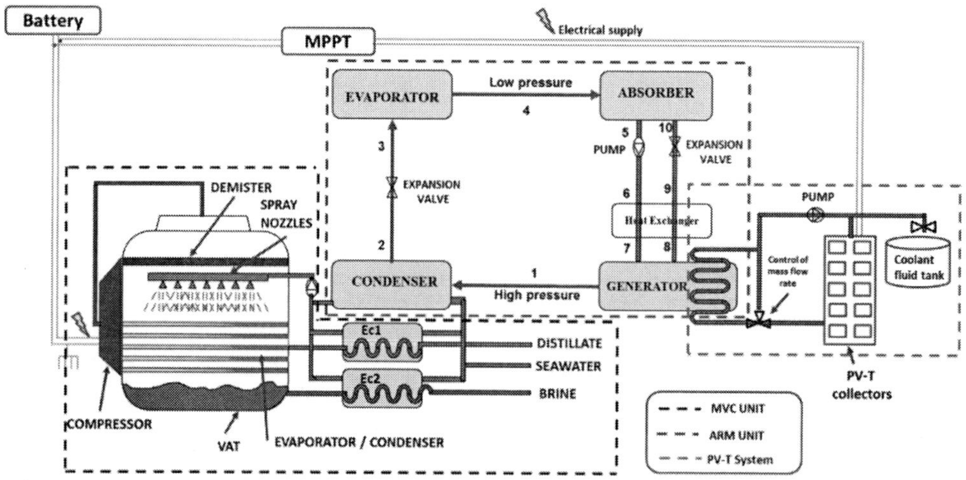

Figure 1: The global configuration of the coupled desalination-refrigeration system

2.3 The global coupled system.

A coolant fluid crosses the PV/T field and transports the cumulated heat from PV/T panels to the generator where pumped strong solution from the absorber will be heated. Electrical efficiency is improved by varying the fluid flow rate [19]. Moreover, the strong solution can also be heated electrically through an electrical resistance in the case where the desalination unit is out of service. The produced electrical energy is essentially consumed by vapor compressor in the MVC unit. A maximum power point tracker (MPPT) which is an electronic unit forces PV cell to provide maximum power [20]. As the coupled system is autonomous, the maximum power point tracker (MPPT) is connected to batteries so as to guarantee system supply in case of weather disturb. Hence, cold and water are produced only when there is available solar energy. The coupling of the units is based on the recovery of heat release by the condenser of cold production machines to heat part of the intake seawater in order to improve the COP of the (ARM) system and the efficiency of the MVC unit.

3 PROCESS MODELING

The system consists of the following subsystems: the MVC, ARM and the solar models.

3.1 Estimation of solar energy absorption

The solar constant received according to day (j) of the year can be calculated with the following formula:

$$\mathrm{E_{Sol}} = 1367\times\left(1+0.0334\times\frac{\cos(j-2.7206)}{365.25}\right) \quad (1)$$

The incidence coefficient (CI) is defined in part by the height and azimuth of the sun and on the other part by the direction (o) and the inclination (i) of the receiving solar collector:

$$\mathrm{CI} = \sin(\mathrm{i})\times\cos(\mathrm{h})\times\cos(\mathrm{o}-\mathrm{a})+\cos(\mathrm{i})\times\sin(\mathrm{h}) \quad (2)$$

The Direct solar irradiance on a receiver plane (o, i) is given by:

$$\mathrm{G} = \mathrm{CI}\times\mathrm{E_{Sol}}\times e^{(-E_R\times m\times T_L)} \quad (3)$$

m is the atmospheric masse, ER is the integral Rayleigh optical thickness and TL is the Linke turbidity factor.

The produced electrical energy to powered the compressor can be expressed by:

$$P_{Elec} = G\times\eta_{El}\times A_{PVT} \quad (4)$$

The thermal energy transferred to the working fluid in the generator can be written as:

$$P_{th} = G\times\eta_{th}\times A_{PVT} \quad (5)$$

3.2 ARM modeling

The flow rates of the strong and weak solutions in the ARM can be determined by using equations (6) and (7):

$$m_7 = m_1 + m_8 \quad (6)$$

$$m_7 X_7 = m_1 + m_8 X_8 \quad (7)$$

The energy increase by pumping is

$$W_{pump} = m_6(h_5 - m_6) \quad (8)$$

Energy balances for the absorber, condenser, generator and evaporator yield

$$Q_a = m_4 h_4 + m_{10} h_{10} - m_5 h_5 \quad (9)$$

$$Q_C = m_1(h_1 - h_2) \quad (10)$$

$$Q_g = m_1 h_1 + m_8 h_8 - m_7 h_7 \quad (11)$$

$$Q_e = m_1(h_4 - h_3) \quad (12)$$

The system performance is measured by the coefficient of performance (COP):

$$COP = \frac{Q_e}{Q_g - W_{pump}} \quad (13)$$

The ranges of pressure,, concentration and enthalpy can be calculated based on the thermodynamic properties of ammonia–water [22]

3.3 MVC unit modeling

The mass flow rates of seawater and brine are obtained from the following system of equations:

$$M_f = M_d + M_b \quad (8)$$

$$M_b = \frac{X_f}{X_b} M_f \quad (9)$$

The amount of heat given by the distillate and the brine in the heat exchanger is given by equations (10) and (11):

$$Q_{E1} = M_d\ C_{pd}\ (T_d - T_o) \quad (10)$$

$$Q_{E2} = M_b\ C_{pb}\ (T_b - T_o) \quad (11)$$

The following equation translates the interaction between the evaporator and the condenser

$$Q_{ev} = Q_{con} = A_e U_e (T_d - T_b) \quad (14)$$

The specific power consumption of the compressor can be expressed by:

$$W = (\mathrm{H}_s - \mathrm{H}_d)(1000/3600) \quad (20)$$

The specific heat transfer area of the evaporator is calculated from the following relationship:

$$A_e = \frac{M_d\lambda_d + M_d C_{pv}(\mathrm{T}_s - \mathrm{T}_d)}{U_e(\mathrm{T}_d - \mathrm{T}_b)} \quad (21)$$

The temperature of the seawater in the condenser outlet is given by equation:

$$T_{ocw} = \frac{Q_C}{M_f Cp_f} + T_{cw} \quad (20)$$

4 RESULTS AND DISCUSSION

A systematic approach has been carried out to study the effect of the combination on the performance of the considered desalination-refrigeration system. The results are given for a case study in the Dakhla city sited at the south of Morocco. Predictions of the mathematical model are shown in Figures. 2–4.

The evolution of specific power consumption of the compressor (W) as a function of the boiling temperature (Tb) is shown in Figure 2. With the increase of Tb, W decreases, In addition it is expected that the combined system under these conditions, reduce the compressor consumption and resulted in savings approximately 0.45 KWh/m^3. this is because of sea water heating in the condenser of the ARM unit.

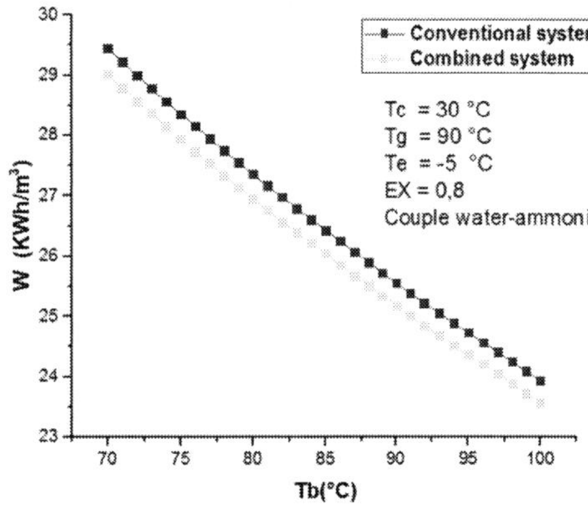

Figure 2: Influence of Tb on the Specific power consumption

The comparison of COP values vs generator temperatures for absorption cycles is shown in Figure 3. As seen, the combined system allowed improving coefficient of performance (COP). Therefore, it is possible to operate the cycle at a low generator temperature. The recovery of the heat generated by the capacitor through the exchanger with seawater, contributed to permission the heat transfer and improved COP

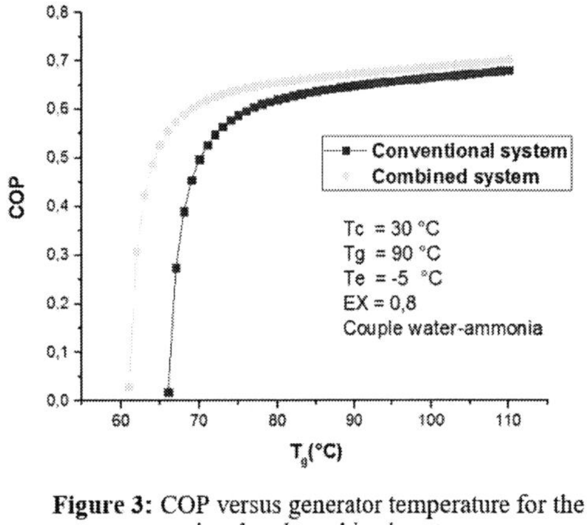

Figure 3: COP versus generator temperature for the conventional and combined system

Solar irradiation received by a sensor during a given day varies according to the incidence coefficient, and weather condition. The electric and thermal power produced by PVT panels varies throughout the sunshine duration.

Figure 4 gives the comparison of COP values versus the daily solar irradiation for combined and conventional system in 21 June for Dakhla city and the influence on COP for a boiling temperature T_b=90°C and temperature difference of saturated compressed vapor and boiling brine ΔT=5°C

Figure 4 shows also, the comparison of COP values vs generator temperatures for NH_3-H_20 absorption cycle. The COP values for both cases increase with generator temperatures. There exists a low generator temperature limit for the cycle. The cycle cannot be operated at generator temperatures lower than its limit. For the coupled cycle, a lower generator temperature can be used than for the conventional one. This is an important point for utilizing coupled system with solar energy since fluid temperatures for PVT solar collectors are generally below 100°C. It is shown that, for daily solar irradiation higher than 700W/m², the COP does not suffer major change. On the other hand, the combined cycle gives the best performance. However, the differences among them are not very remarkable.

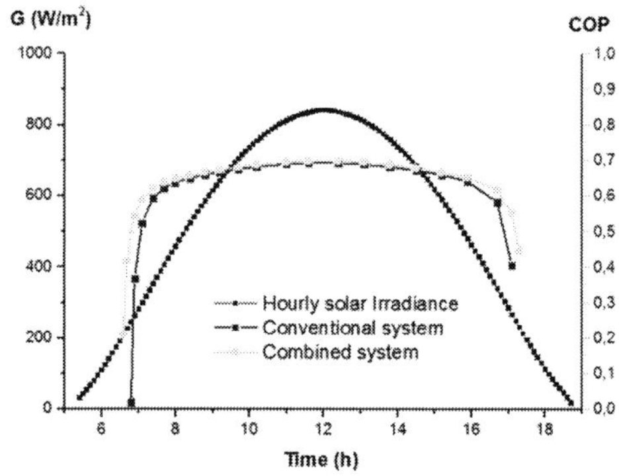

Figure 4: Evolution of COP during 21 June.

4 CONCLUSION

In the present work, a program able to examine the performance of a combination of desalination-refrigeration unit powered by solar PVT collectors in Dakhla region situated in the south of Morocco has been developed. Based on the obtained results, the new system proposed an important possible application for needs agriculture reflected in the use of desalinate seawater for irrigation and the refrigeration of perishable products (vegetables, fruits...) directly in the field after harvesting. This process remains to be attractive, especially for the Saharan region where the climate is extremely hot, the fresh water is scarce and solar energy potential is high. For these reasons, we proposed this prototype

REFERENCES

[1] Ahmed A.A. Attia, 2012. Thermal analysis for system uses solar energy as a pressure source for reverse osmosis (RO) water desalination, Solar Energy 86, 2486–2493.

[2] International Water Management Institute. In Water for food, water for life: a comprehensive assessment of water management in agriculture, Earthscan, 2007; 57.

[3] W.F. Stocker, L.D. Reed, Effect of operating temperatures on the coefficient of performance of aqua-ammonia refrigerating systems, ASHRAE Transactions 77 (1971) 163_170.

[4] C. Keizer, Absorption refrigeration machines, Ph.D. Thesis Delft University of Technology, Delft, Holland, 1982.

[5] O.E. Ataer, Energy analysis of an aqua-ammonia absorption refrigeration system, Progress in the Science and Technology of Refrigeration in Food Engineering, International Institute of Refrigeration,

Commissions B2, C2, D1, D2/3, Dresden, Germany, 1990, pp. 387_392.
[6] O È.E. Ataer, Y. GÈguÈs, Comparative study of irreversibility in an aqua-ammonia absorption refrigeration system, International Journal of Refrigeration 14 (1991) 86_92.
[7] A. Ming, Optimization of ammonia absorption refrigeration process: effects of seasonal ambient temperature fluctuations, International Journal of Refrigeration 14 (1991) 341_344.
[8] R. Kahn, G. Alefeld, S. Hammerer, R. Pfeifer, M.L. Tomasek, An ammonia_water absorption cycle with high temperature lift, in: Proceedings of the International Absorption Heat Pump Conference, AES-Vol. 31, 1993, pp. 93_100.
[9] Trombe, F., and Foex, M., 1964. Economic Balance Sheet of Ice Manufacture with an Absorption Machine Utilizing the Sun as the Heat Source, New Courses of Energy, Vol. 4, U.N. Publication Sales No. 63.1.38, pages 56-59
[10] Karagiorgas M, Tsoutsos T, Drosou V, Pouffary S, Pagano T, Lara GM, et al. HOTRES: renewable energies in the hotels. An extensive technical tool for the hotel industry. Renew Sustain Energy Rev; in press, doi:10.1016/j.rser.2004.09.012.
[11] O. Lefebvre, R. Moletta, Treatment of organic pollution in industrial saline wastewater: a literature review, Water Res. 40 (2006) 3671–3682.
[12] D.F. Zhao, J.L. Xue, S. Li, H. Sun, Q.D. Zhang, Theoretical analyses of thermal and economical aspects of multi-effect distillation desalination dealing with high salinity wastewater, Desalination 273 (2011) 292–298.
[13] Matz R, Fisher U. A comparison of the relative economics of sea water desalination by vapor compression and reverse osmosis for small to medium capacity plants. Desalination 1981;36:137–51.
[14] Matz R, Zimerman Z. Low-temperature vapour compression and multi-effect distillation of seawater. Effects of design on operation and economics. Desalination 1985;52:201–16.
[15] Al-Sahali M, Ettouney H. Developments in thermal desalination processes:design, energy, and costing aspects. Desalination 2007;214:227–40.
[16] Z. Zimerman, Development of large capacity high efficiency mechanical vapor compression (MVC) units, Desalination, 96 (1994) 51–58.
[17] Zejli, D., Ouammi, A., Sacile, R., Dagdougui, H., & Elmidaoui, A. (2011). An optimization model for a mechanical vapor compression desalination plant driven by a wind/PV hybrid system. Applied Energy, 88(11), 4042-4054.
[18] El-Nashar, Ali M. "Performance of the solar desalination plant at Abu Dhabi."Desalination 72.3 (1989): 405-424.
[19] Fudholi Ahmad, Sopian Kamaruzzaman, Yazdi Mohammad H, Ruslan Mohd Hafidz, Ibrahim Adnan, Kazem Hussein A. Performance analysis of photovoltaic thermal (PVT) water collectors. Energy Convers Manage 2014;78:641–51.
[20] Bhatnagar Pallavee, Nema RK. Maximum power point tracking control techniques: state-of-the-art in photovoltaic applications. Renew Sustain Energy Rev 2013;23:224–41
[21] M. Ibrahimi, A. Arbaoui, Y. Aoura, L. Ouzizi. (2015). Embodiment design of a mechanical vapor compression desalination unit, presented at the IEEE IRSEC15 Conference, Marrakech, Morocco
[22] ASHRAE, ASHRAE Handbook, Fundamentals, Chapter 17, p. 17.45 & p. 17.81. ASHRAE, 1791 Tullie Circle, N.E., Atlanta, GA 30329, 19

DEMAND SIDE POWER MANAGEMENT OF A GRID CONNECTED SOLAR PV SYSTEM WITH VANADIUM REDOX FLOW BATTERY STORAGE

Ankur Bhattacharjee[1], Hiranmay Saha[2]

[1] Centre of Excellence for Green Energy and Sensor Systems,IIESTShibpur,Howrah,India, ankurbhattacharjeejpg@gmail.com
[2] Centre of Excellence for Green Energy and Sensor Systems, IIEST Shibpur, Howrah, India, sahahiran@gmail.com

ABSTRACT: This paper presents an optimized energy management scheme for grid connected storage enabled solar PV power system dedicated towards demand side satisfaction under intermittent solar insolation. Vanadium Redox Flow Battery (VRFB) claims to be a promising energy storage technology to increase the supply reliability of large-scale PV applications due to its scalability, independent power and energy capacity and longer life cycle as compared to other conventional battery energy storages. During PV Power generation deficiency, the battery will be in discharging mode to supply the required load power and thus shaving the peak load in that period. The two way power flow is executed by placing a suitable bi-directional DC-DC converter between the point of common coupling (PCC) and the VRFB. A very low available state of charge (SOC) [1] less than 10% of VRFB storage and inadequate power generation from solar PV source together causes peak load demand dissatisfaction [2] which can be satisfied by importing power from the AC grid. Considering a DC load a suitable rectifier unit along with necessary filter is placed between the PCC and the grid. The efficiency of the solar PV source is enhanced by an efficient Maximum Power Point Tracking Technique (MPPT). The entire power management scheme [3] has been designed with the real life data of 1kW 6hr VRFB storage and the battery SOC [4-6] prediction algorithm is also implemented in MATLAB/Simulink environment to lengthen storage lifetime and the demand side reliability also. The presented research work has dealt with the simulation of optimized penetration of solar PV, VRFB and AC grid to smoothen the peak load demand for two practical case studies and the results are very useful for field validation.

Keywords: Solar PV, MPPT, VRFB storage, SOC

1 PURPOSE OF THE WORK AND APPROACH

The main emphasis of this research work is on the load demand shaving with optimized penetration of solar PV, VRFB and Grid. The power management algorithm will decide the charging and discharging of the VRFB by online monitoring of battery SOC under variable solar insolation and fixed load demand. The overall block diagram representation of the proposed power management scheme is shown in figure1,

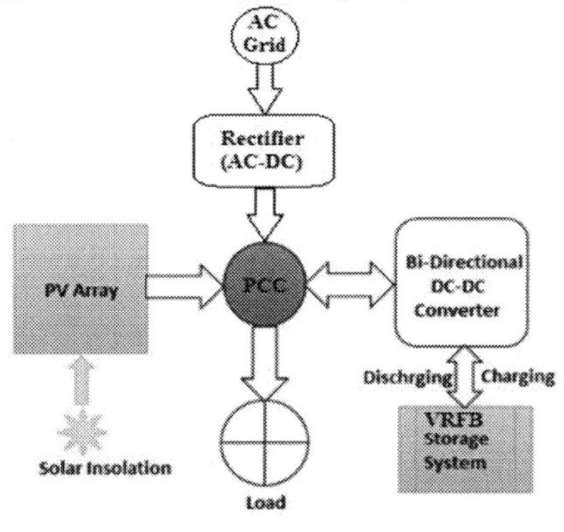

Figure 1 Block diagram representation of the proposed power management scheme

2 SCIENTIFIC INNOVATION AND RELEVANCE

A generic mathematical model of a 1.2 kW solar PV source is designed in MATLAB/Simulink environment and the power output is enhanced by implementing Perturb & Observe MPPT algorithm and a DC-DC Buck converter. The electrochemical model of a 16 cell, 1kW 6hr VRFB storage system is designed and the battery SOC is monitored by a prediction algorithm. The repetitive deep discharge of VRFB is avoided by the power management algorithm thus lengthening the VRFB storage lifetime and enhancing the demand side power supply reliability even under variable solar insolation. The modelling of the whole grid connected solar PV power system with Vanadium Redox Flow Battery storage and the algorithm for optimized penetration of these three sources to meet the required load demand draws the novelty of this research work.

3 VRFB ELECTROCHEMICAL AND ELECTRICAL EQUIVALENT MODEL

In the electrochemical model [3] of VRFB the relations between the stack voltage V_Stack as a function of the operating conditions such as the concentration of four oxidation states of Vanadium; $c(V^{2+})$, $c(V^{3+})$, $c(V^{4+})$ and $c(V^{5+})$, the proton concentration $c(H^{+})$, and the electrolyte flow rate Q are established.

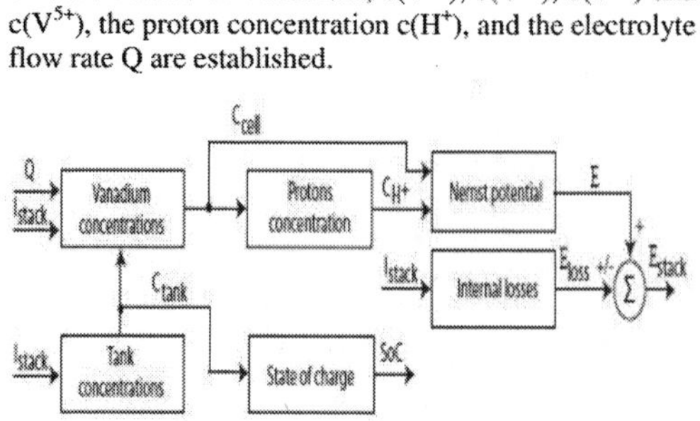

Figure 2 Electrochemical SOC estimation model of VRFB

The electrical equivalent model for SOC estimation is done in MATLAB/Simulink and to validate this model a practical VRFB having 16 cells, 6kWh has been chosen. The relation between battery stack voltage [4] and SOC is established in this model based on Nernst equation and relation between different electrical parameters given by equation 1 -3,

$$V_{cell} = V_{equilibrium} + \frac{2RT}{F} \ln\left(\frac{SOC}{1-SOC}\right) \quad (1)$$

$$V_{stack} = n \times V_{cell} \quad (2)$$

$$V_{battery} = V_{stack} - I_{stack}\left(R_{Reaction} + R_{Resistive}\right) - \eta \quad (3)$$

Where,

E_{Cell} = Average voltage of a single cell VRFB (Volts)

E_{Stack} = Stack Voltage of VRFB (Volts)

n = No. of cells of VRFB

$E_{Battery}$ = Terminal voltage of VRFB (Volts)

I_{Stack} = Stack current of VRFB (Ampere)

$R_{Reaction}$ = Equivalent resistance of reaction inside the VRFB stack (Ω)

$R_{Resistive}$ = Equivalent resistance of parasitic losses (Ω)

The proposed VRFB energy storage is modeled by considering a real life technical specifications as shown in table 1,

Parameters	Specifications
Cell configuration	16 Cell stack
Power Rating	1kW
Energy capacity	6kWh
Voltage Range	20 – 27.2 Volts
Rated Current	60 Amperes
Temperature limit	0 – 45°C
Electrolyte flow pressure	0 – 60 kPa
$R_{Reaction}$	0.045 Ω
$R_{Resistive}$	0.03 Ω
R_{Fixed}	13.889 Ω
$C_{Electrodes}$	0.15 F

Table1 Technical specification of VRFB storage system

The MATLAB/SIMULINK model of 1kW 6hr VRFB storage is shown in figure 3,

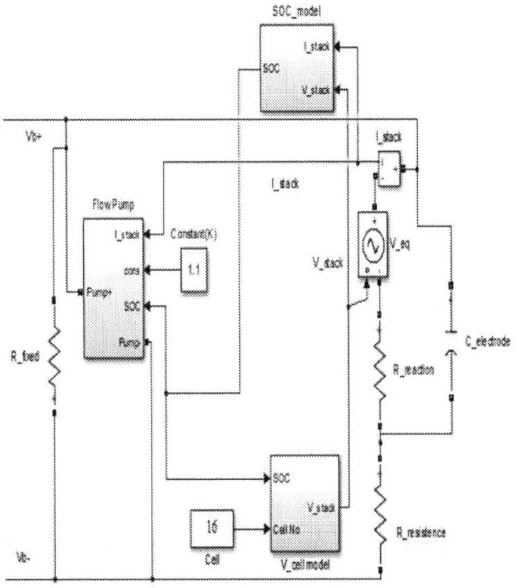

Figure 3 MATLAB/SIMULINK model of VRFB for SOC estimation

4 SIMULATION RESULTS

The entire model of the proposed Grid connected VRFB storage integrated Solar PV power system has been simulated under a particular solar insolation (W/m^2) data for a period of time and the results are shown in figure 4 to figure 13 respectively, for two different case studies. The VRFB power output curves are observed as positive and negative while charging and discharging respectively, depending on the PV power output and load demand.

4.1 Practical Case Study 1

The simulation results for the first case study are shown here,

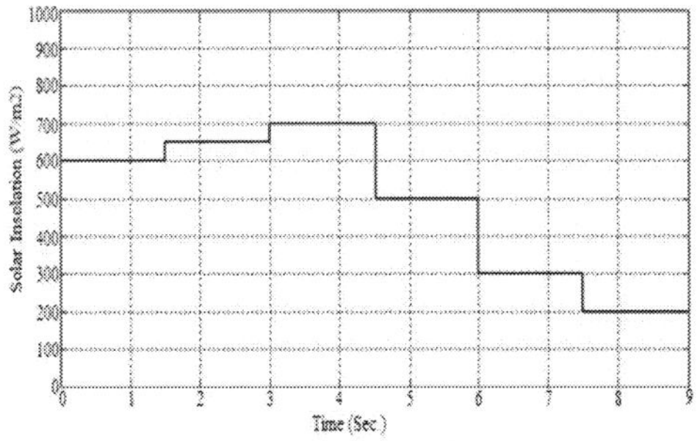

Figure 4 Solar Insolation (W/m^2) pattern for case study 1

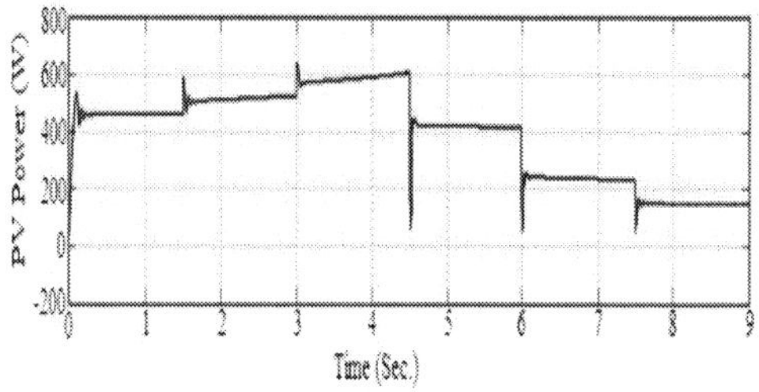

Figure 5 Solar PV array output power under varying insolation

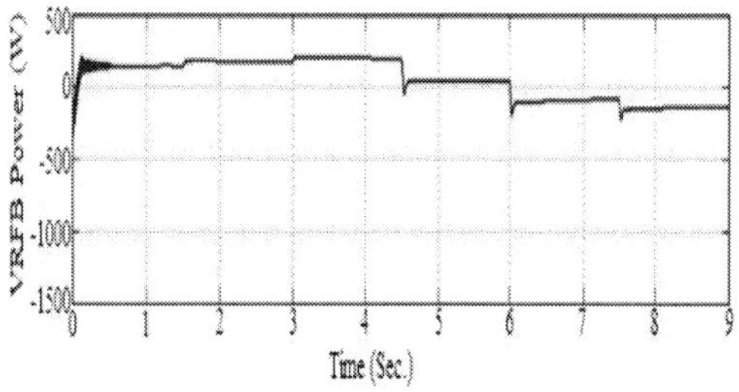

Figure 6 VRFB Storage power (+ve for charging, -ve for discharging)

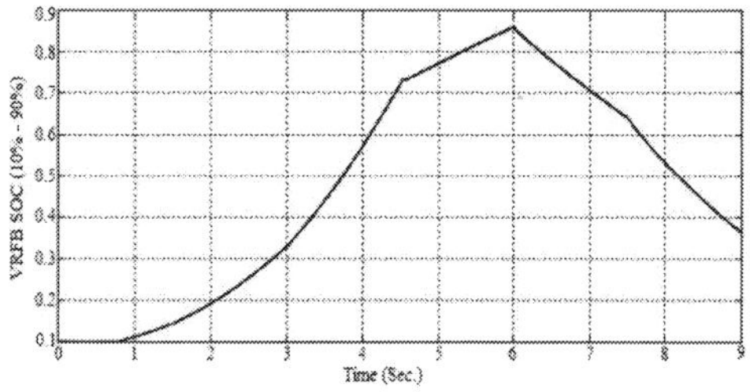

Figure 7 VRFB SOC (10% - 90%)

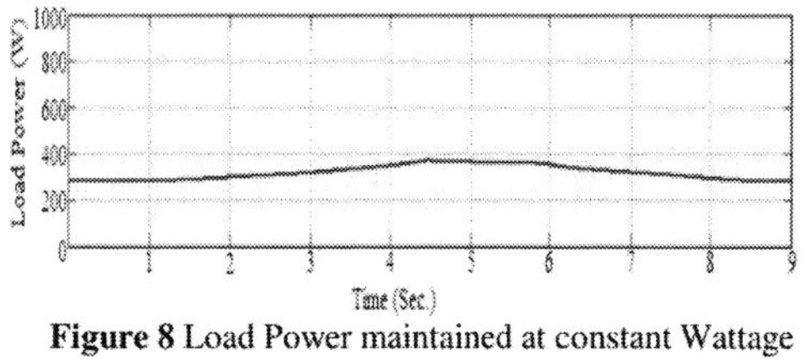

Figure 8 Load Power maintained at constant Wattage

4.2 Practical Case Study 2

The simulation results for the first case study are shown here,

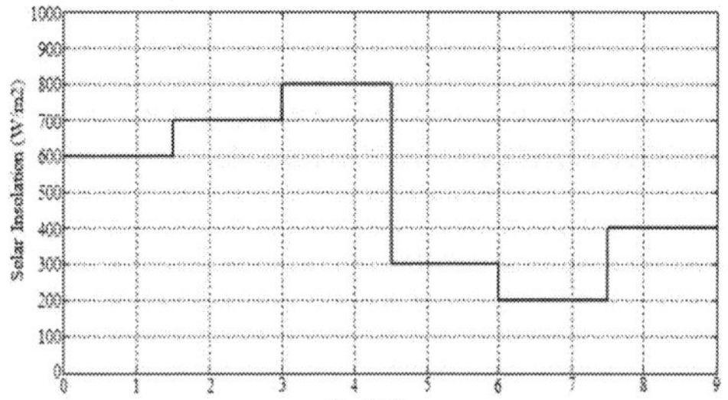

Figure 9 Solar Insolation (W/m^2) pattern for case study 2

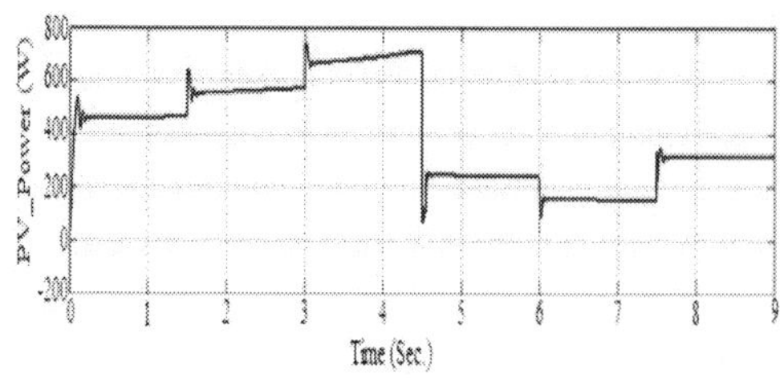

Figure 10 Solar PV array output power under varying insolation

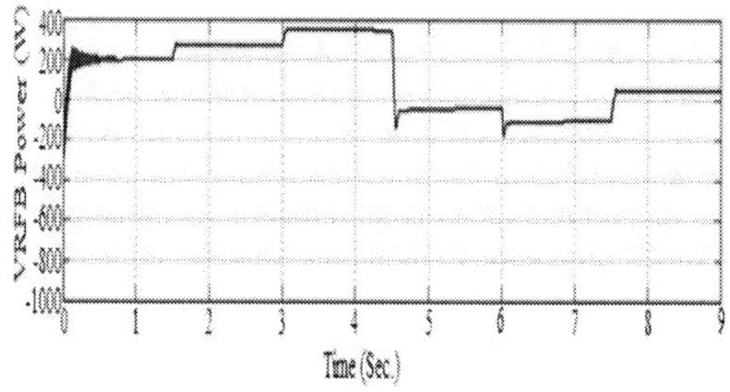

Figure 11 VRFB Storage power (+ve for charging, -ve for discharging)

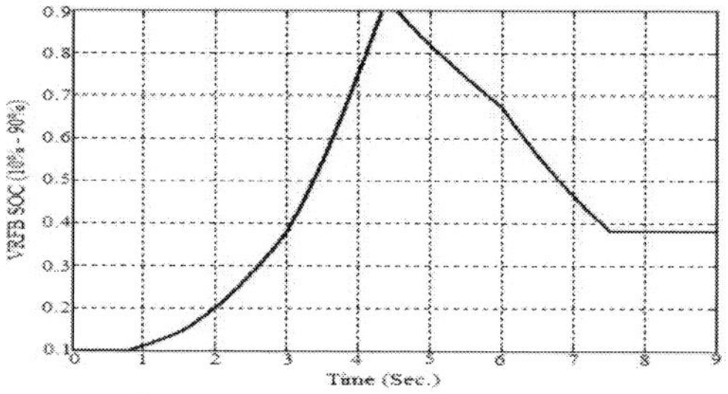

Figure 12 VRFB SOC (10% - 90%)

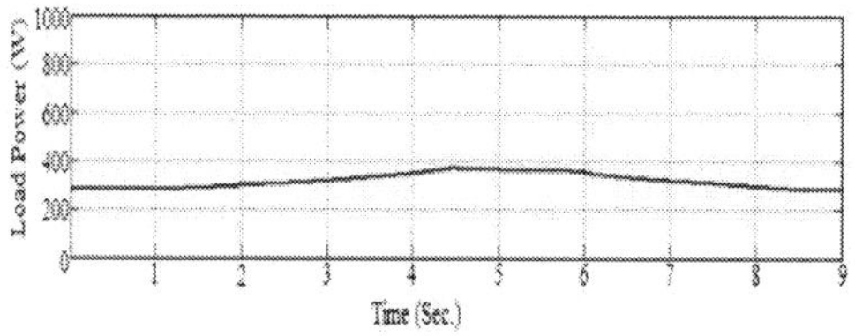

Figure 13 Load Power maintained at constant Wattage

5 CONCLUSION

The simulation results show the online SOC prediction of VRFB storage and the fixed load demand of 300 Watt is managed under two case studies with variable solar insolation data. In both the cases the proposed power management algorithm helps the VRFB storage neither to be overcharged nor deeply discharged even after repetitive charging and discharging in order to match the required load demand. The penetration of utility grid is executed only in the worst situation of poor solar PV power output and very low SOC of VRFB storage thus making the overall PV power system cost effective and efficient. The simulation results help in validation of the proposed power management scheme in practical field.

REFERENCES

[1] A.Bhattacharjee, H.Saha, "A Generic Electro-Thermal Characteristic Model of Vanadium Redox Flow Battery under Different Charging Conditions", *5th International Conference on Advances in Energy Research*, 15-17th Dec. 2015, pp. 240.

[2] G. Wang, M. Ciobotaru, and V. G. Agelidis, "Power smoothing of large solar PV plant using hybrid energy storage", *IEEE Trans. Sustain. Energy*, vol. 5, no. 3, Jul. 2014, pp. 834–842.

[3] L. Baumann, J. Grahn, E. Boggasch, and L. Khl, "Integration of a vanadium-redox-flow-battery into a hybrid energy system", *7th International Renewable Energy Storage Conference, Berlin, Germany*, 2012.

[4] C. Blanc, "Modeling of vanadium redox flow battery electricity storage system", *Ph.D. dissertation, Echole Polytechnique Federale DeLausanne,* 2009.

[5] M. -H. Li, T. Funaki, and T. Hikihara, "A Study on the Modeling of Output Terminal Voltage for Redox Flow Battery Based on Charge and Discharge Experiment", *Industry Applications Society, IEE Japan*, 2006, pp. 585-586.

[6] M. Skyllas-Kazacos, D. Kasherman, D. Hong, and M. Kazacos, "Characteristics and performance of 1 kW UNSW vanadium redox battery," *Journal of Power Sources*, vol. 35, no. 4, 1991, pp. 399-404.

TOWARDS A NOVEL PROPOSAL OF A SOLAR POLYGENERATION SYSTEM FOR MOROCCO'S PUBLIC HOSPITALS

Souad Lakjiri[1], Mohammed Ouassaid[2], Mohammed Cherkaoui[3]
Department of Electrical Engineering, Mohammadia School of Engineers, Mohammed V University in Rabat, Morocco
Email: [1] s.lakjiri.al@gmail.com, [2]ouassaid@emi.ac.ma, [3]cherkaoui@emi.ac.ma

ABSTRACT: The idea of integrating polygeneration system in hospitals came from the characteristic that healthcare facilities require simultaneously different forms of energy (heat, steam, electricity and cooling) with the necessity of permanently guaranteed availability of the energy throughout the year. Therefore, since Morocco expects to benefit largely from solar energy, a solar polygeneration system applied for hospitals seems the adequate way to respond to the National Energy Efficiency strategy launched by the Morocco's Ministry of Health. The purpose of this paper is to illustrate and examine energetic problematic faced by most hospitals in Morocco presented through a case study of a 250-bed hospital, and to establish a configuration of a solar polygeneration system that will remedy energetic concerns. The proposed system, consisting of both Concentrating Photovoltaic Thermal Systems (CPV/T) ,and Combined Cooling, Heating and Power plant(CCHP), recovers the energy rejected by the overall CPV, incinerator and boiler, in order to predict an optimal size of both CPV/T and CCHP plant which still require a considerable investment. In addition, it will achieve greater energy efficiency, provide more comfort to patients and caregivers, and reduce greenhouse gas emissions.
Keywords: Polygeneration system, energy efficiency in healthcare facilities, Combined cooling Heating and Power, Incinerator; Moroccan hospitals, Concentration Photovoltaic Thermal systems.

1 INTRODUCTION

The energy of today makes awake the entire world, according to the World Energy Council [1].This is because of the uncertainty of energy prices, climate risks, electric storage, and so on. Also; it makes leaders of energy busy about renewable energy and the energy efficiency according to the same council. Indeed, the world is, faced to the challenge to provide everyone the energy, while the forecasting unveils energetic deficit in the near future. In particular, a modern world demands access to electrical energy for everyone, such human rights requires.

The report of World Energy Council [2], estimates that 70% of electricity is produced basing on fossil fuels, especially in countries in way of development, which causes a high gas emissions. The strategies of energy mix and energy efficiency were, so, driven in these countries, including Morocco that is concerned about the energetic state and global climatic.

Morocco has, since 2006, initiated programs, to contribute to green energetic economic, to reduce the energetic dependence ([3] reports that it imports almost 91% of its energy), and to remedy the infrastructures that do not take into account the energy efficiency .Several programs and projects in partnership with institutions and international technical and financial partners have so been launched, particularly in building sector which consumes a significant portion of the energy, in comparison with other sectors, and whose the public hospitals sector is part.

Indeed, hospitals are considered among the largest energy consumers in a country. Moreover, this type of building, has the specificity that it requires not only enough installations that should meet the high hospitals demands of energy, but also the necessity of a guaranteed availability of the energy along the hours of a day, the days of a week, the weekends and overall the year.

Few previous studies have proposed, to respond to the simultaneous demand of different forms of energy in hospitals, plants with trigeneration engine, trigeneration integrating the existing boilers or polygeneration plant based on renewable energy [4,5,6]. In this paper, the present basic study doesn't include details about the requested dimensions of the proposed polygeneration plant and therefore the energy saving, however, this approach undertakes an optimal exploitation of both renewable energy, and energy released by hospitals' consumers, other than boilers, such the incinerator; to predict an optimized size of polygeneration plant.

2 PURPOSES OF THE WORK AND METHODOLOGY

The purpose of this paper is to give a proposal for a solar polygeneration system, based on the analysis of the current energetic situation within Moroccos public hospitals. It reports the databases collected from the Mohamed V hospital of Tangier in Morocco, as part of the strategy of the national energy efficiency, launched by the Ministry of Health and which aims to rationalize the use of energy in all its forms including electricity, drinking water consumption, medicinal gases consumption, heating and air conditioning.

This paper is in two parts. The first part illustrates and examines energetic concerns in Moroccos' public hospitals given through the Mohamed V hospital with 250 beds capacity. This hospital presents approximately the same barriers and difficulties faced by the most of other hospitals in Morocco.

The second part determines the consumers which can be beneficed by the proposed system of polygeneration, and the essential components of this system that will increase the energy efficiency in hospitals and reduce the size of CCHP and CPV/T systems.

2.1 Energy concerns faced by most Moroccan healthcare facilities

Case study: The Mohamed V 250-beds regional hospital of Tangier is considered as medium capacity in Morocco classification, and supports the entire population of the region with the specialties: Surgery, maternity, pediatrics, dialysis and emergency. It was built on the surface of 7500 m^2 and consists of two floors, a ground floor and a basement. The Table I presents some global indicators of Mohamed V hospital.

Actually, it is equipped by solar panels for the production of domestic hot water (20 m^2 of solar collectors are installed); but the solar water heating is usually not sufficient to meet the overall need of hot water, and requires a supplemental heating. Also, it is equipped by a fuel boiler for heating and to produce steam for laundries and sterilization. Nevertheless, the boiler is nonoperational by the staff assuming that it means important energy losses whilst the current use of fuel in such this installation will be more expensive than simply considering electricity for the consumers of energy they can supply.

All this affects the comfort of patients and caregivers which is shown with the electric consumption ratio (2555kWh/bed/year compared to 9700 kWh/bed/year as European average according to Hespul association). Given that the energy consumption in Moroccos' public hospitals ratio is already largely dominated by electric consumption compared with the international ratio, this lower ratio highlights that it does not mean a high efficiency applied, but a lack of comfort.

Concerning the electrical installations, a transformer with 400KVA feeds the electrical demands of the hospital facilities:

- Artificial lighting that is considered one of the biggest electricity consumers in a hospital in general, consumes in Mohamed V hospital 22% of the total consumption according to the current data of the hospital.
- The central sterilization uses direct electrical heating containing two autoclaves, with a capacity of 60 kg. It consumes 25% of overall electric consumption of the hospital. Sometimes, the two autoclaves drive simultaneously, which causes a power demand of112 kWh amounting to 64% of overall power demand.

2.2 Some energy efficiency features required to deal with some weaknesses of the current situation and the insufficiency of some ones already taken

The manual of the " Energy 2000"established in Switzerland, as an extension of the "Energy Management in Hospital "coordinated by the International Energy Agency, [7],is still an important reference in hospitals' installations. It explains methods, with the objective to save energy costs without impairing the everyday operation in the hospital. A among it, avoiding the coincidence peak loads. In fact, when hospitals operate in permanent time, the technician should drive the equipment with intensive power consumption in low tariff times "off peak hours"(sterilization,..), switch off temporally some machines(refrigeration, ventilation,..) and avoid simultaneous running of large consumers equipments.

The energy saved, and consequently, the cost saving, with the forms of control [8], calculated to save heating cooling and ventilation energy demand; such presence detection and timer, are valuable for the lighting system. Even though they are already used in some Moroccan hospitals, like the important reduction of incandescent lamps, use of electronic ballasts, etc.; the uncertainty of prediction of the occupancy, and the permanent circulation in some halls like in the emergency service,

Table I: Global indicators for the hospital of Mohamed v

In-service date	1992
Surface built	7500 m²
Beds capacity	250 beds
emergencies admissions	134069/Year(On average)
External Consultations	15578/Year(On average)
Admissions	24197/year(On average)
Admissions (Hospital Day)	3886/An (On average)
Days of hospitalization	75066/Year (On average)
Laboratory Exams	111687/Year (On average)
Surgical Interventions	4790/Year
Births	9975/Year
Average beds rotation	96,78%
Occupancy rate	82,26%
Number of admissions per day	76,93%

make this methods unable to achieve significant savings, and create however turbulences.

The Fig.1 resumes the weakness of the current situation and the insufficiency of some energy efficiency features already taken

2.3 Heat energy in healthcare facilities

Apart from the direct electrical heating, most of the water consumed in healthcare facilities (domestic hot water, water for sterilization...) requires a system heating, either with hot water, superheated water or steam networks. The Fig.2 shows the temperatures that requires each system in a hospital according to [7].The heating generation, reports the same reference, covers theses temperatures that vary depending on the season and on and on the hours of the day.

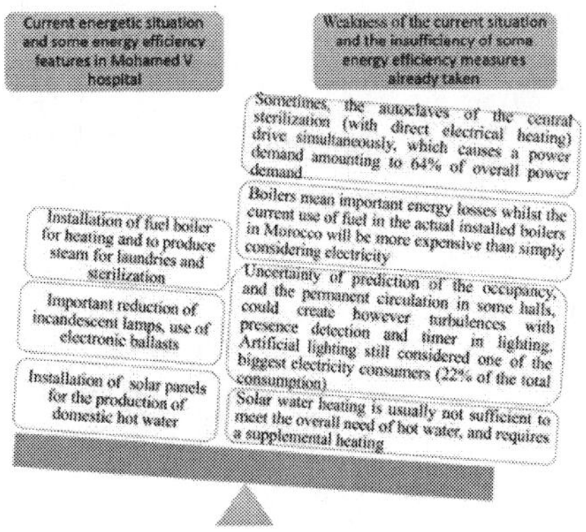

Fig.1: Weakness of the current situation and the insufficiency of some energy efficiency mesures already taken

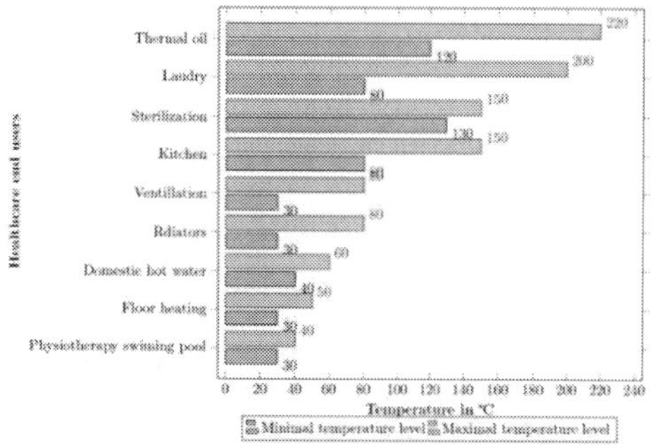

Fig. 2: Minimal and maximal Levels of temperature required for usual energy end users in a hospital.

3 RESULTS AND DISCUSSION

3.1 Proposed solar plygeneration plant for Moroccan public hospitals

Despite some energy efficiency measures already taken in Moroccan hospitals, there is a need to provide more comfort to patients and caregivers, reduce energy costs and reduce greenhouse gas emissions. This will be allowed by integrating CPV/T and by recovering, for a CCHP plant, the important losses of energy, especially in boiler and incinerators. The configuration shown in the Fig.3 schematizes global components of the renewable polygeneration plant, and consumers that will be beneficiaries. It includes:

- CPV/T system: Consisting of the receiver, optics and sun tracking system. The receiver includes photovoltaic cells and the cooling mechanism (by mean of water-glycol solution) that preserves the cell electric efficiency and provides the thermal energy once the refrigerant fluid reached the temperature required. The proposed CPV/T system can change the size according to the number of the cells and to the concentration factor;
- CCHP plant consitsts of:
  - □ A Heat Recovery Steam Generator (HRSG) wherein the saturated steam is produced by recovering the energy rejected by the flue gas emitted by both the incinerator, and the boiler. A cleaning system of the flue gas is also considered;
  - □ A Micro gas turbine and a generator;
  - □ An Absorption Chiller (LiBr – H2O) coupled to a cooling tower.
- Auxiliary boiler;
- Incinerator: the incinerator consists of a waste loading system, rotary combustion chamber (rotary Kiln where waste is decomposed into solid and gaseous products during a thermal treatment) and the afterburner chamber.

3.2 Profitability of installing a solar polygeneration plant in Moroccan public hospital

In previous research [4], D. Boschiero studied the efficiency of a proposed trigeneration system, established in a Brazilian university of 403 beds and constructed in area of 65.000 m². He concludes a Primary Energy Saving (PES) of 26,6%, allowing simultaneous production of steam, electricity, hot water and chilled water with an exergy efficiency of 41,21%, and an Energy Utilization Factor(EUF) of 85,52.

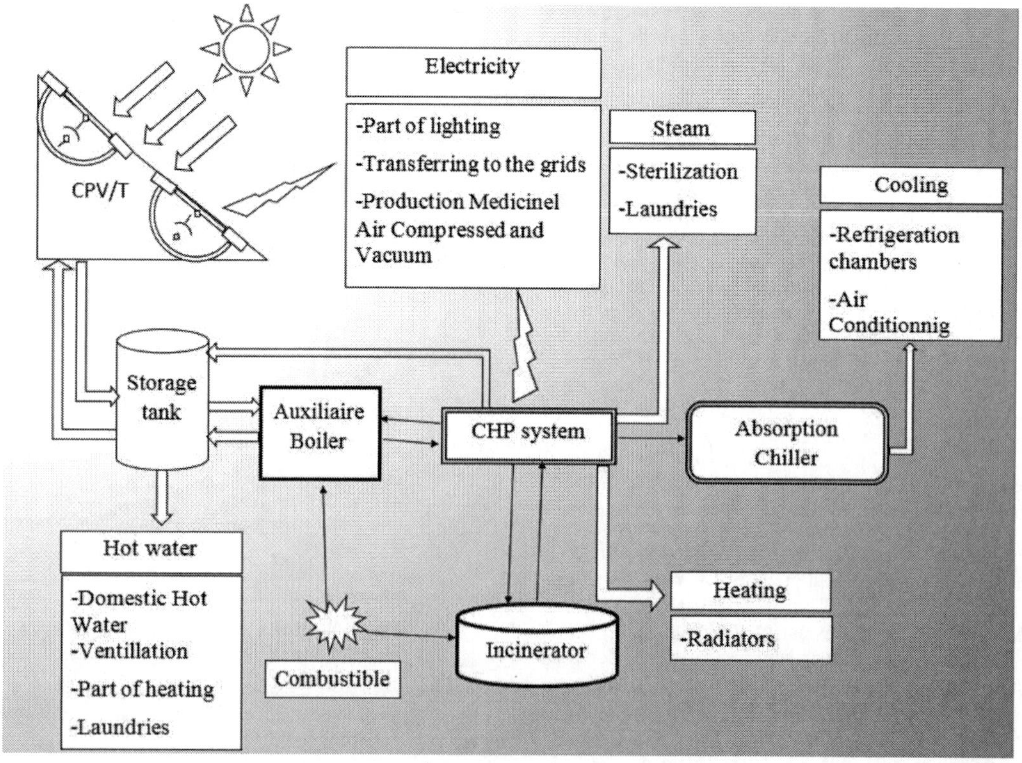

Fig.3: Proposed polygeneration scheme for Moroccan public hospitals.

The study shows that a trigeneration system can compete intensively centralized thermal plants.

The system proposed in this paper, includes, in addition to the trigenation system that reuses the energy rejected from the boiler:

- Concentrating Photovoltaic Thermal Systems (CPV/T).Indeed, [9] shows that CPV/T can be exploited efficiently in hospital buildings since they require electrical and thermal energy simultaneously;
- Recover technology of the energy emitted by the incineration. The study [10] explains how energy in the forms of hot water and electricity can be produced, in an incineration plant, with a heat recovery system consisted of Heat Recovery Steam Generator (HSGS), and the micro turbine with a generator. Indeed, for the incineration of medical waste, the temperature of the flue gas can reach above 1100°C.

The suggested polygeneration plant, consisting of CPV/T and Combined Cooling, Heating and Power (CCHP) combining boilers and incinerators installed with common recovery technologies, will cover efficiently the Mohamed V hospital's needs of sufficient hot water and steams, and reduce energy costs for lighting. Moreover, with the additional chilled water produced, this plant will be also a significant source of energy for air-conditioning.

The Fig.4 shows the benefits of the suggested solar polygeneration plant in addition to reducing the carbon dioxide emissions, and providing a secure and guaranteed energy for hospitals.

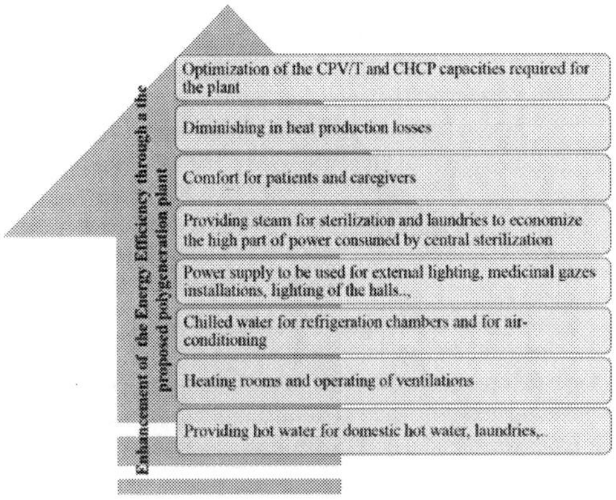

Fig.3:Benefits of the suggested solar polygeneration plant for Moroccan public hospitals.

4 CONCLUSION AND FURTHER WORK

The idea of a polygenration plant in public Moroccan hospitals was justified by examining energetic consumption and energetic installations in the case of Mohamed V hospital. Indeed, despite some actions already taken for energy efficiency in Moroccan hospitals, there is need for a supply of energy produced. This will be allowed by recovering the important losses of energy, especially in boilers and incinerators to reduce high energy costs, provide patients and staff with all the necessary comfort and facilities, and reduce carbon dioxide emissions. The global components of the proposed renewable polygeneration plant, and consumers that will be beneficiaries were determined as well, exploiting largely the solar energy. The proposed plant will allow for healthcare facilities, not only for Moroccos' ones, greater efficiency in providing electricity, heat, cooling, hot water and steams.

The additional benefit of producing the energy by recovery systems, other satisfying, as needed, energy demand, and preventing high loss of energy in energetic installations; is eventual transferring of exceeding electricity to the grids and thus, hospital buildings will become a productive environment , not just consumers as are normally considered.

As a result, this renewable polygeneration system seems the adequate solution for high energy efficiency in hospitals, to be demonstrated through a computational simulation program and detailed by auxiliary equipments, investments required and the correspondent economic payback time. Furthermore, the optimal exploitation of CPV/T, boilers and the incinerators, and the rationalization of the energy and water consumption ; will reduce the size of the technology of the Combined Heating and Power (CHP) system which still requires a considerable investment.

REFERENCES

[1] World Energy Council, World Energy Issues Monitor.World Energy Council, 2015.

[2] World Energy Council, Global Electricity Initiative 2014 Report. World Energy Council, 2014.

[3] "maroc2014sum," OECD/IEA, Tech. Rep., 2014.

[4] Annamaria Buonomano, Francesco Calise, Gabriele Ferruzzi, and Laura Vanoli, "A novel renewable polygeneration system for hospital buildings: Design, simulation and thermo-economic optimization," Applied Thermal Engineering, vol. 67, no. 12, pp. 43 – 60, 2014.

[5] Denilson Boschiero, "An energy and exergy analysis of a high-efficiency engine trigeneration system for a hospital: A case study methodology based on annual energy demand profiles," Energy and Buildings, vol. 76, pp. 185 – 198, 2014.

[6] G. Pagliarini, C. Corradi, and S. Rainieri, "Hospital {CHCP} system optimization assisted by {TRNSYS} building energy simulation tool," Applied Thermal Engineering, vol. 44, pp. 150 – 158, 2012.

[7] Pierre Chuard, Franois Vuille, and Pierre Jaboyedoff, Manuel d'exploitation des installations dans les hôpitaux. Notions pratiques de gestion énergétique.Federal Department of Transport, Communications and Energy, 1993.

[8] Velimir ongradac, Bogdan Prebiraevi, and Neboja Petrovaki, "Methods for assessing energy savings in hospitals using various control techniques," Energy and Buildings, vol. 69, pp. 85 – 92, 2014.

[9] Jee Joe Michael, Iniyan S, and Ranko Goic, "Flat plate solar photovoltaic thermal (PV/T) systems: A reference guide," Renewable and Sustainable Energy Reviews, vol. 51, pp. 62 – 88, 2015.

[10] Janusz Wojciech Bujak, "Production of waste energy and heat in hospital facilities," Energy, vol. 91, pp. 350 – 362, 2015.

ESS MANAGEMENT IN GRID CONNECTED PV SYSTEMS: FROM SIMULATION TO EXPERIMENT ON FIELD

Francesco De Lia, Salvatore Castello, Riccardo Schioppo, Roberto LoPresti, Massimo Izzi, Mario Tucci
ENEA C.R. Casaccia DTE-FSN-TEF, Via Anguillarese 301 – 00123 – Rome, Italy

ABSTRACT. In this work are reported the analysis as obtained from an advanced simulator implemented to model the dynamic behavior of grid connected systems composed of photovoltaic plant, electric storage systems of different technologies and electric loads. By means of a specially developed modeling, the behavior of the systems from the electric and energy point of view has been analyzed in detail. The simulator allows the calculation for the entire system life time and for fixed size of photovoltaic generator and storage system capacity, the energy flows among the individual systems and all electrical quantities of interest sampling at one second.
The behavior of the energy management system (EMS) has been also modeled. Optimized management logic have been developed in the simulator and then uploaded in the controller of the demonstration plant that has been installed close to ENEA Centre at Lampedusa island.
The economic benefits introduced by electric storage systems in grid-connected photovoltaic plants has been evaluated with the aid of the simulator for different types of users.
Keywords: PV system, Storage, Simulation, Economic Analysis

1. INTRODUCION

In recent years, the photovoltaic plants in Italy have experienced an increasing interest as demonstrated by the 18.9 GW connected to the public grid that actually covers about 8.2% of the national energy requirements [1].
The uncertainty of daily power generation and the limited ability to perform predictions of energy supply in short and medium-term, requires to smartly control the energy flows and power distribution in regional areas by means of appropriate Energy Management System (EMS). In this scenario the energy storage system (EES) can become a great relevance option to improve the efficient use of energy and further promote the renewable energy sources in the energy demand. Indeed, in addition to optimal use of energy resources from the user's perspective, ESS also plays a strategic role as decoupling time between production and consumption. Moreover, the storage of the energy produced by grid connected photovoltaic plant represents a smart way to reduce the impact on the electric network grid and to enhance the economic benefits for the end user.
The Italian Ministry of Economic Development gave to ENEA a public fund to evaluate the profitability of introducing ESS in grid connected PV system for end Users.
ENEA has produced a study in which has been developed an advanced simulator and an experimental Plant built in the Lampedusa island.

2. THE SIMULATION MODEL

This study suggests an appropriate model to smart manage the energy flows among photovoltaic, batteries, electrical distribution grid and end-users. In particular the study focuses on the economic benefits of PV systems with storage in residential and commercial applications from the user's perspective. To validate the proposed model, able to finely describe the behaviour of each component of the entire system, it has been loaded into the EMS of a prototype of PV and Storage plant installed at Lampedusa island, in south Italy (figure 1). In this plant two different electrical storage systems (Lithium Polymer and VRLA lead acid) have been selected in order to evaluate their performances in real operating conditions.
To address the topic this study suggests a model of the PV system with storage taking into account different types of customer load profiles for domestic and commercial applications, different storage technologies and their costs, the benefits related to electric Grid services, the benefits achievable by the end user and the evolution of the electricity costs and charges related to Grid use.

Figure 1 – The PV array on Lampedusa island

Then a mathematical model of the entire system has been developed on Matlab-Simulink software platform in which PV, Storage, User load, EMS and Utility Grid models are specifically designed as depicted in Fig. 2.

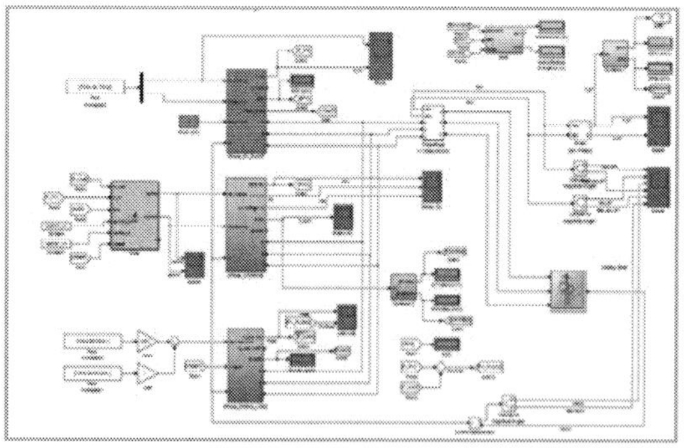

Figure 2 - The Simulator developed in Matlab-Simulink platform.

This model simulates and manages the energy flows from PV to Storage or to the grid according to the boundary conditions as defined by the charge of batteries and the user needs. In particular the Simulator finely describes the behaviour of all components of the system as: Storage, PV, user-load, EMS and Grid in terms of Voltage, Current, Power and Energy. Since economical models are implemented too, the model can predict the benefits for the end users as in presence of Italian net metering service (i.e. Scambio Sul Posto) or when

simplified purchase and resale arrangements (i.e. Ritiro Dedicato) is adopted. Of course, each component used within the model has been validating through a demonstration plant installed at Lampedusa Island in Italy. The simulated behaviour of a PV system with storage at a sampling rate of 1 sec allows an assessment of the real energy self-sufficiency and self-consumption parameters per year for the user once is known its load profile, the size of PV, the technology and capacity of the ESS. In particular, the model allows to evaluate the sensitivity of input parameters as well as the trend of ESS cost. The effects of smart management of ESS is performed taking into account user load profiles, PV generation and the storage technologies selected as the most suitable for specific application. This approach also allows to optimize the battery capacity for specific application.
Different charge battery control strategies can be developed ranging from the classical charging for maximizing self-consumption (SC) to the forecast charging strategies.
The model has been used to evaluate the effect of storage and ESS on residential and commercial users representative of the North, Central and Southern Italy, and the economic benefits for the user depending on the Storage technology. Concerning the interface with utility Grid, the model has been used to evaluate management strategies of the energy flows among the photovoltaic system, the storage system, the user loads and the utility Grid taking into account different electricity tariffs related to the various types of user.

3. MODELLING RESULTS

The model able to describe the energy flows between PV plant, storage, electrical network grid and user has been detailed designed in Matlab Simulink software ambient as reported in Fig.1. It allows a fine monitor of different parameters, at a sampling rate of 1sec, of the entire plant and its subsystems (PV, ESS, User-load and utility Grid) as obtained through related models, as partially reported in the plots of Fig. 2. The model permits the setting up of different strategies for managing the energy flows to optimize the behaviour of the entire Plant. Indeed it is possible to perform a sensitivity analysis regarding most relevant technical and economic parameters in order to identify the optimal sizes of the various components as obtained through focused simulations.
In this work we have addressed several case studies for residential and commercial users, with different PV and ESS systems. As far as the ESS we have modelled both Lithium than Lead Batteries.
The following figures illustrate some results obtained by the simulator for a case of study on Commercial User with an annual consumption of 147,000 kWh connected to a 150 kWp PV System with Li-ESS (300 kWh – 110 Ah). Figure 3 shows the electrical values as obtained by simulations in January for the commercial user. As evident, during sunny days the batteries don't completely discharge during the night due to the particular user profile, reported in Figure 4, that is quite synchronous with solar radiation. This situation doesn't occur during cloudy days in which batteries are completely discharged during the night.
The models account for bidirectional converter efficiency curve and battery efficiency, therefore the AC Round Trip efficiency can be calculated, as shown in Figure 5. The different values reported in the figure are due to different magnitude of DC current that flows in the batteries and different daily profile of SOC.

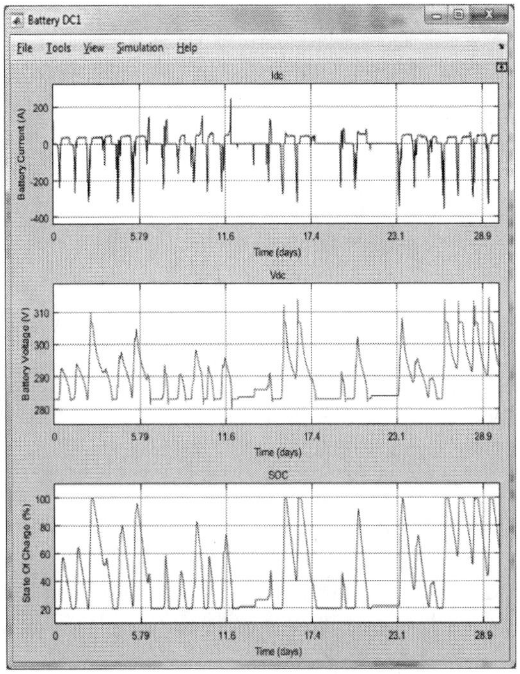

Figure 3 - Battery pack values occurred in January referred to the Commercial User.

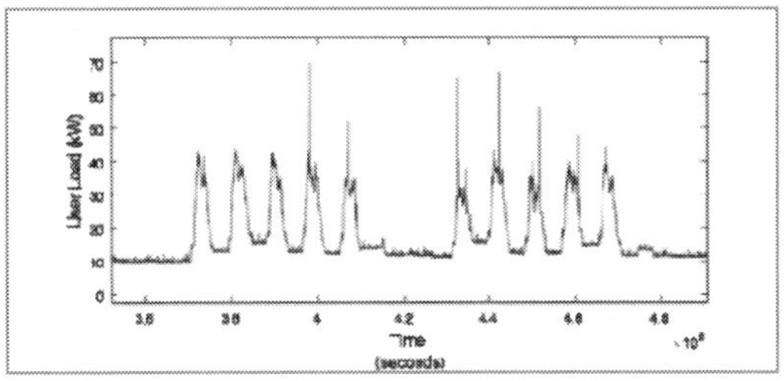

Figure 4 - Two tipical weekly user profile for the Commercial User.

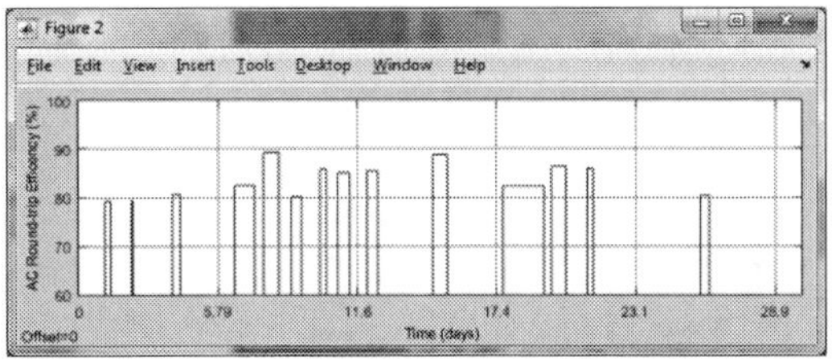

Figure 5 – AC Round Trip Efficiency of lithium ESS occurred in January referred to the Commercial User.

The AC Round Trip Efficiency parameter is also depending on the charge control strategy adopted. In this work we have considered the common Maximizing Self-Consumption strategy, nevertheless smarter control strategies [2], [3] are going to be implemented in the next months, firstly in the Simulator and then in the plant.

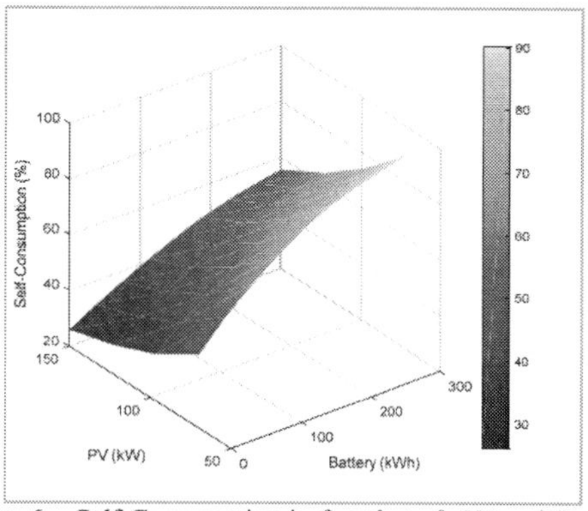

Figure 6 – Self-Consumption in function of PV and Storage sizes (Commercial User case study).

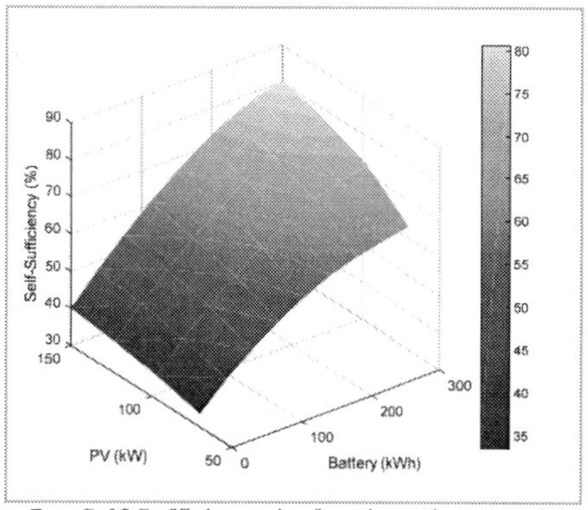

Figure 7 – Self-Sufficiency in function of PV and Storage sizes (Commercial User case study).

4. EXPERIMENT ON FIELD

We have performed preliminary test on the battery pack in order to check the accuracy of some models adopted.
The figure 8 shows the comparison of electrical values of DC lithium battery pack obtained with the simulations (continuous curves) with those measured in the Lampedusa Plant (dotted curves) during Discharge and Charge phases at constant power.

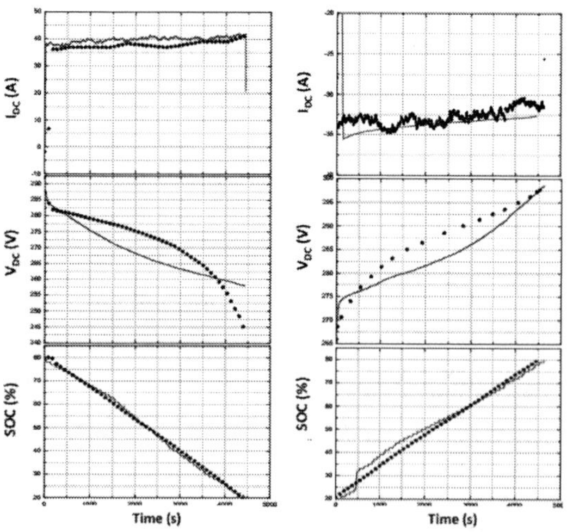

Figure 8 - Discharging and Charging of Battery pack at constant power (P=10 kW).

Since we are in intermediate phase of the project, the validating phase is still not completed. The models need to be tuned and this can be done during the experimental phase that will start in the next months.

5. ECONOMIC EVALUATIONS

In the economic evaluation we have not taken into account incentives currently available in Italy (tax reduction on 50% of the capital cost for residential users). The evaluation of the economic benefits due to the introduction of electric storage systems (ESS) in the grid-connected photovoltaic plants has been achieved on the entire system life time through economic simulations for different types of users at different size of photovoltaic generator and storage system capacity.
The economic analysis has been quantified taking into account the following quantities:

- the amount of electricity taken from the grid and the related expenditure in accordance with current tariff structure in Italy
- the amount of electricity fed into the grid and the related remuneration both in the case of net billing option so-called "Scambio Sul Posto" (SSP) and electricity sale named "Ritiro Dedicato" (RID)
- the Self-Consumption quotas and the annual Self-Sufficiency Energy
- the overall economic benefit in the case of SPP and in the case of RID

The analysis has been conducted for four typical users:

- medium residential users,
- residential users with high consumption (in this case, the simulations have been performed both with Li-ion, and Pb accumulation systems),
- a laboratory at ENEA Research Centre,
- commercial user (typically an office),

The maximum size of the PV generators have been chosen so that they are able, at the irradiation conditions taken into consideration, to provide the energy required by the loads. The maximum capacities of storage systems have been chosen so that they are able to accumulate the energy produced by the minimum size of PV for a period of 4 hours. The sizes of PV generator and of the storage system capacity adopted to carry out the simulations are reported in Table 1 for each type of user. In Table 2 are reported the costs of the PV plant, bidirectional converter (BC) and batteries used to perform the economic analysis

Table 1 – Hypothesized sizes of PV generator and ESS.

User Type	Consumpt. MWh/y	Size of PV generator (kW)			ESS capacity (kWh)		
		From	to	step	from	to	step
Redidential	3.100	1.5	3	0.5	0	6	0.75
Redidential	6.000	2.5	4	0.5	0	10	1.25
ENEA Lab	32.000	15	30	5	0	60	7.5
Commercial	174.000	70	150	26.6	0	300	37.5

Table 2 – Costs of PV plant, bidirectional converter and battery used in the simulations.

	Costs							
PV plant (€/kW)	2,000							
BC (€/kW)	500				50			
Li battery (€/kWh)	1,000	500	100		1,000	500	100	
Pb battery (€/kWh)		300	100	50		300	100	50

Simulations relating to the different types of users have been performed on the basis of described conditions. The results (for a typical Italian residential user of 3100 kWh/year) are summarized in Figures below that show the annual economic benefit for different sizes of the PV generator and of the storage system.
The figure 9 shows the annual economic benefit for the different type of users. In particular, the highest values of economic benefit is obtained for domestic users with high consumption (with both batteries to Li that Pb). These benefits increase by increasing the storage system capacity. Discrete annual economic benefit values are also reported for commercial users. In this case, the economic benefit decreases as the storage system increases because of the relative bill structure.

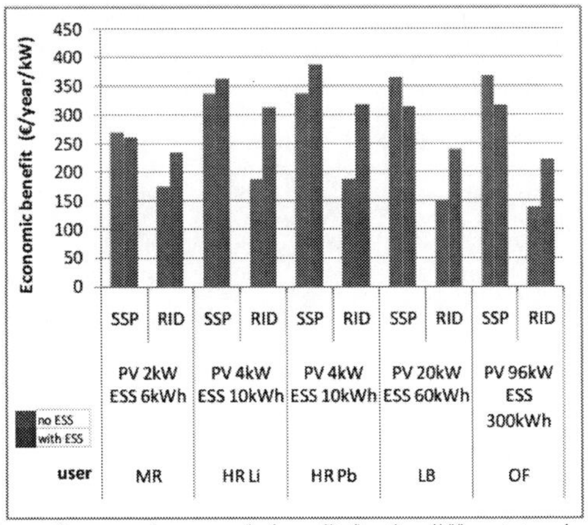

Figure 9 - annual economic benefit for the different typology of users.

Concerning the payback period calculation these conditions have been assumed:

- a life time of approximately 12 years for the Li batteries and of about 6 years for Pb.
- an operation and maintenance cost that affects with an annual fee of 3% of the initial cost of the system.

The following figures show the values of payback period for the different sizes and costs of the PV generator and the storage system, both in the case of SSP and RID.

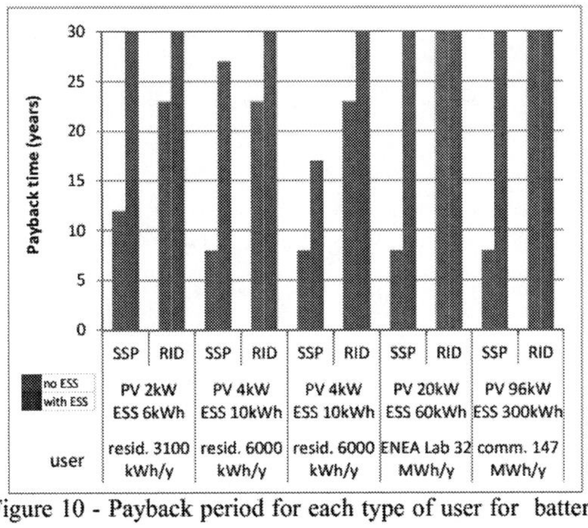

Figure 10 - Payback period for each type of user for battery cost of 500 € / kWh for Li to 300 € / kWh for Pb.

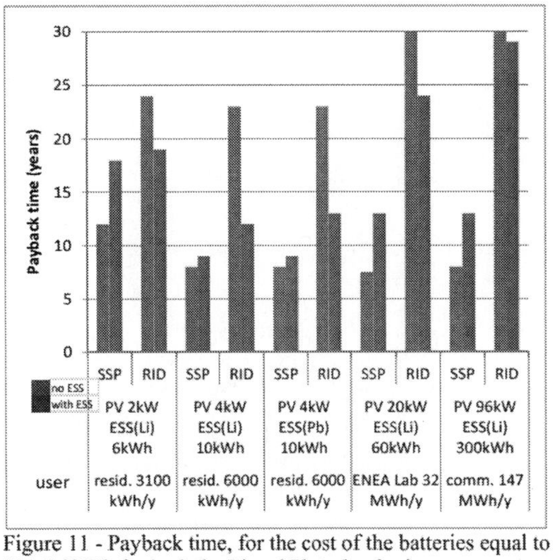

Figure 11 - Payback time, for the cost of the batteries equal to 100 € / kWh for both the Li and Pb technologies.

Figure 10 shows the payback period for each type of user for battery cost of 500 € / kWh for Li to 300 € / kWh for Pb. Figure 11 reports the payback time, for the cost of the batteries equal to 100 € / kWh for both the Li and the Pb.
These figures show that for cost of the storage system of 500 €/kWh, acceptable values of the payback time (about 10 years) are obtained with only modest capacity of the storage system and for users who operate in SSP scheme. For almost all users, increasing the capacity of the storage system, and therefore the cost of the plant, increases the payback period to over the lifetime of the plant.
Nevertheless at very low cost of the storage system (100 €/kWh) is obtained a reduction of the payback time also at the highest values of the storage system capacity. Generally in the SSP option there is an increase of the payback time at the increase of storage system capacity. However, in typical situations of energy-intensive user, the payback time does not vary appreciably with the presence or absence of the storage system. Therefore, in such situations becomes relevant the added value provided by the storage system in terms of network services and continuity of supply in emergency situations. In case of RID option, being less remunerated the kWh fed into the grid compared to the SSP option. Indeed there is a reduction in the payback time in the presence of the storage system. In the RID regime, however, the payback times are generally higher than in the SSP option.

In conclusion The ESS begin to be advantageous, in terms of reduction of the PBT, in case of domestic users characterized by a high consumption (> 6000 kWh) and operating in RID scheme if the battery (Lithium) cost reduces below 250 € / kWh, as shown in Figure 12.

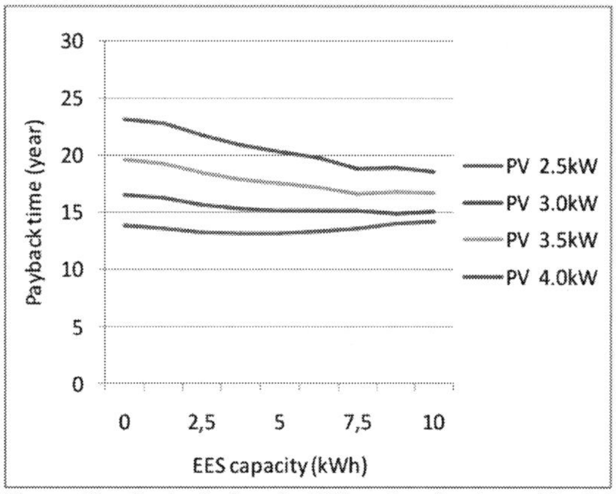

Figure 12 - Payback time in RID option for the residential user (6000 kWh/y) with a foreseeing Lithium batteries costs of 250 €/kWh.

6. CONCLUSIONS

In this work a study and an advanced simulator with appropriate models to smart manage the energy flows among photovoltaic, batteries, electrical distribution grid and end-users, is presented. The simulator, developed on Matlab-Simulink platform, permits to setting up different strategies for managing the energy flows in order to optimize the behaviour of the entire Plant. The optimized strategies can be loaded into the EMS that is an important component of the experimental Plant built in Lampedusa Island. The models adopted in the simulator are going to be evaluate on the field in the next months and relevant informations about the real behaviour of the batteries (Lithium and Lead acid) in real applications will be obtained.
The simulator is able to evaluate not only all the electrical quantities in a generic PV + ESS grid connected plant but also can evaluate the real annual economic benefit for residential or commercial users.
In Italy, since the end of the national incentives, the options for end users to add value to the electricity produced by their PV and ESS systems are SSP (Net metering), RID (simplified purchase & resale arrangements) and Self-consumption that is always present in these systems.
In Italy, a residential user opts for SSP option because it ensures a greater economic benefit with respect to RID option. In SSP presence, the ESS are not economic convenient because the Grid is still a near zero cost Electrical Storage System. However under the hypothesis of RID option the use of storage systems become interesting if the costs of the Lithiun batteries will drop from the actual 700 - 1,000 € / kWh to below 250 € / kWh.
Besides, there are other benefits that ESS is able to provide at the end user. In fact, thanks to "peak shaving", the ESS helps to reduce the contractual power and consequently to reduce the fixed costs of the bill, while from the grid point of view the ESS allows to limit the power fed into the grid from the PV. In this perspective it will be necessary to define energy policies so that the implementation of large-scale PV systems with storage does not result in imprudent investments, negative effects on the networks as well as an unfair distribution of costs.
To achieve these objectives should be appropriate to create a regulatory framework in order to:

- promote strategies aimed at avoiding spikes or network withdrawals through the introduction of appropriate incentives;
- manage critical aspects of the transmission and distribution network arising from possible expansions at national level of the photovoltaic;
- improve the reliability and security of networks.

REFERENCES

[1] S. Castello, S. Guastella, F. Tilli: "National Report of PV Power Application in Italy", IEA PVPS report, Oct. 2015.
[2] Jiahao Li, Michael A. Danzer: "Optimal charge control strategies for stationary photovoltaic battery systems", Journal of Power Sources 258 (2014) pagg. 365 – 373.
[3] J. Weniger, J. Bergner, V. Quaschning: "Integration of PV power and load forecasts into the operation of residential PV battery systems" 4th Solar Integration Workshop 10-11 November 2014 at ISE.

PV MICRO-INVERTER SYSTEM USING AN IN-HOME DISPLAY AND MOBILE ELECTRONIC DEVICES FOR DISPLAYING DIAGNOSTIC AND OPERATING PARAMETERS

Rito Mijarez[1], Fernando Martínez[1], Ángel Gomez[1], Javier Antunez[1] and David Pascacio[1], Gustavo Vázquez[2]
[1]Instituto de Investigaciones Eléctricas
62490, Cuernavaca, Morelos, México
[2]Diseño e Ingeniería Sustentable
Hermosillo, Sonora, México

ABSTRACT: Self-contained PV systems consist of a commercial PV module and a small inverter or micro-inverter that is mounted typically under the PV module, and is interconnected to either 120 or 240 AC voltage grids. Generally, PV modules are guaranteed to perform at specified levels of output for 20 or 25 years, which is longer than the 5 to 15 years inverters guarantee. Hence, it is important the development of cost-effective systems to monitor the performance of these inverters and the whole PV system. This work presents a PV micro-inverter system that has access to the inverters and PV module by using power-line carrier communications (PLC) connected to an in-home display located inside the consumer residence. The in-home display is a standalone user interface that permits the user not only display diagnostic and operating parameters of the inverter, but also input parameters such as reactive and active power adjustment required by the grid utility, and enhance safety and installation. Alternatively, the in-home display can be connected via Bluetooth to a mobile electronic device without the need of a network communications transmission to a remote Web server.
Keywords: Photovoltaic; micro-inverter; monitoring

1 INTRODUCTION

The development self-contained power generation photovoltaic (PV) systems for consumers and power utilities is a reality that offers a safe energy option for reducing greenhouse gas emissions, cut energy costs and lessen the bottleneck caused by distribution lines. PV systems are basically composed of commercial PV modules and inverters. Conventional PV systems employ a combination of PV panels connected in series or in parallel to obtain a desired DC voltage link and output power. In order to extract the maximum power and maximum energy produced from solar panels, several maximum power point tracking (MPPT) techniques are used [1]. Nonetheless, PV panels are subjected to different conditions of solar irradiation and temperature that produce an effect denominated partial shading, which lead to system inefficiencies, since MPPT techniques involve the overall number of PV panels. To overcome this drawback a power electronic inverter is installed to every PV module. This inverter system is commonly referred as micro-inverter or AC modules [2]. The micro-inverters are designed to work in grid-tied mode and they cease operating in the event of grid supply failure. Generally micro-inverters are installed along with every PV panel and are connected in parallel to feed power into the grid as depicted in Figure 1.

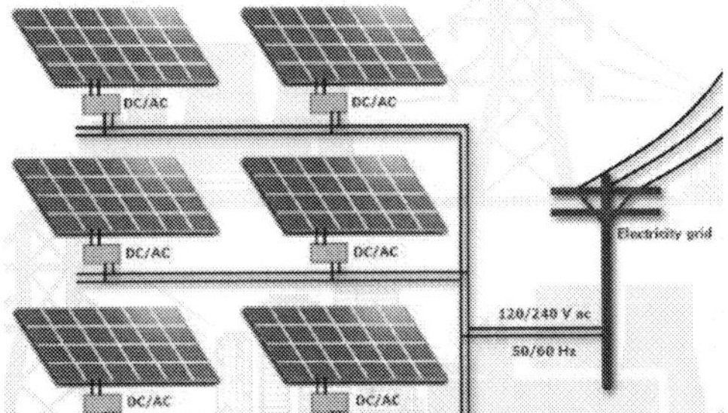

Figure 1. Grid connected operation of PV panels and integrated micro-inverters.

However, micro-inverters can last less time than the PV modules, yielding unattractive return of the user inversion. Moreover, the micro-inverter market worldwide is encountering sceneries where prices continue to fall and the request for more services is increasing, and expansion into new markets with fragile grids demand for more flexible micro-inverters architectures [3].

The approach taken in this work was to integrate communication technologies and protocols and digital signal processing capabilities to the micro-inverter design, as well as integrating sensors such as accelerometers to the PV modules that can also be monitored during installation. The inverter is connected via PLC to a low-cost in-home display with input/output capabilities. Alternatively the in-home display can be connected to a mobile electronic device such a mobile phone via Bluetooth. Performance of the whole system can be monitored, but also diagnostic parameters of inverter critical elements can be displayed.

2 MICRO-INVERTERS MAIN GOAL DESIGN

Micro-inverters are typically installed on a roof under the PV panel. They are difficult to reach for servicing and possibly hazardous owing to the risk of falling. Due to the location where are mounted, micro-inverters need to be able to endure harsh environmental conditions and provide fault free operation normally for 25 years. However, in a PV system the prevailing failure element that has been identified is the micro-inverter, which has to be replaced at least once during the lifetime of a PV system. Designers have been trying to eliminate components susceptible to heat and other forms of stress. However, a single micro-inverter component, i.e. the electrolytic capacitor, can reduce the lifetime and Mean Time between Failures (MTBF) of the device. Figure 2 depicts the failure rate and probability distribution function of micro-inverters, where it can be observed that the dominant failure mode of the electrolytic capacitor is the deterioration resulting from long use [4].

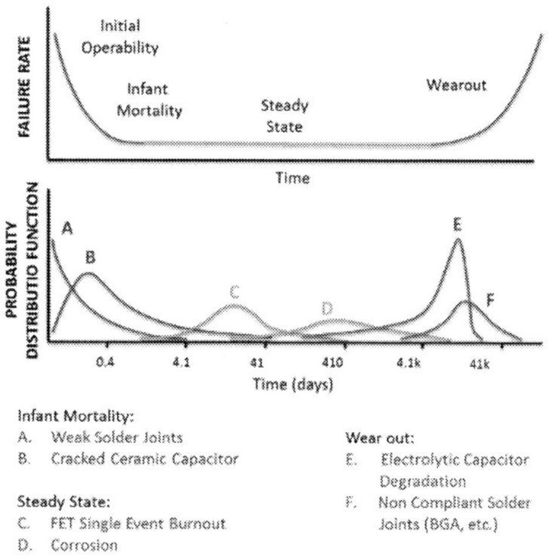

Figure 2. Failure rate and probability distribution function of micro-inverters.

Designers around the work are looking for alternatives for replacing the electrolytic capacitor. New capacitors which can withstand higher temperatures and wear-out have already been considered for the micro-inverter industry. Companies already are using thin film capacitors that are known to be temperature tolerant and considering the use of costly Silicon Carbide (SiC) semiconductors that possess high endurance and operate at extended temperature ranges suitable for the oil and aerospace industries.

As a result, reliability is the most important goal aiming to match the life of PV panels. Hence, micro-inverters performance monitoring is of particular relevance.

3 MICRO-INVERTER ARCHITECTURE

The PV system general scheme presented in this work is depicted in Figure 3.

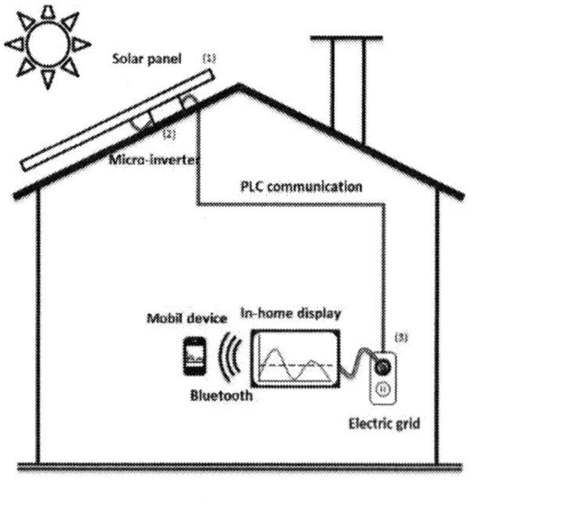

Figure 3. PV system general scheme using an in-home display and a mobile electronic device.

The proposed PV system architecture employs a two-stages topology with two high gain, high efficiency Flyback DC/DC converters, which includes a MPPT algorithm and a DC/CA single phase converter, controlled by a Texas Instruments Digital Signal Controller (DSC) TMS320F28035. Figure 4 shows the block diagram of the system.

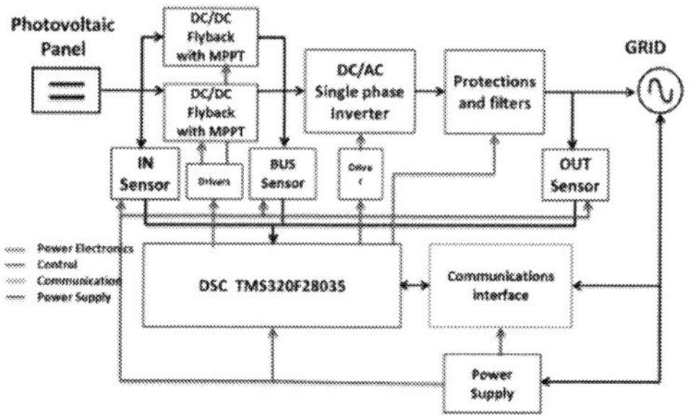

Figure 4. Block diagram of the PV system architecture

The dual DC/DC topology provides a voltage boost of 350– 400 Volts for PV panels outputs between 20 to 45 Volts. The following stage is a standard H-Bridge DC/AC topology. EMI issues are alleviated by using output filters and protections that also provide the required grid interconnection. A communication interface is included, which performs wireless data exchange protocols using PLC to communicate with micro-inverters and Bluetooth to communicate with mobile devices such as a smart phone or tablet. Figure 5 shows the micro-inverter diagnostic software application executed in a smart phone.

Figure 5. Micro-inverter diagnostic software application executed in a smart phone.

DSC combines high-performance DSPs operation and microcontroller functions. The DSC support high speed mathematical calculations for real time control algorithms and integrate control peripherals, such as analog to digital converters (ADCs) and pulse-width-modulated outputs (PWMs), on-chip flash memory and communication ports. These capacities offer a suitable solution for micro-inverters, since it can perform real-time control of the DC/AC converter bridge, MPPT algorithms, protection circuitry, directly sense inputs, control power FETs and carry out communication protocols [5].

Figure 6 shows the single-phase schematic diagram and equivalent circuit of the DC/DC converter operating under unity power factor in steady state considering a total output power of 300 W. The output voltage, V_o, the output current, I_o, the input current, I_i, and the magnetizing current, I_L, are given by equation (1)

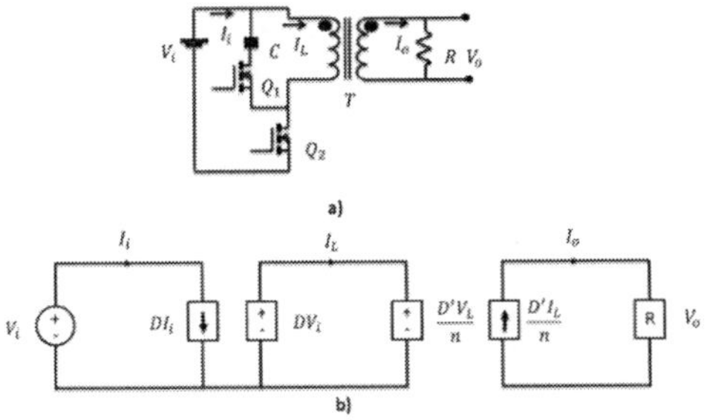

Figure 6. Single-phase schematic diagram and equivalent circuit of the DC/DC converter operating under unity power factor in steady state.

$$V_o = \frac{D}{D'} n V_i \qquad I_o = \frac{D' I_L}{n} \qquad I_L = \frac{I_i}{D} \tag{1}$$

Where D is the duty cycle, $D' = 1\text{-}D$ and n is the transformation ratio. Figure 7 shows the voltage doubler and the DC/AC inverter schematic diagram.

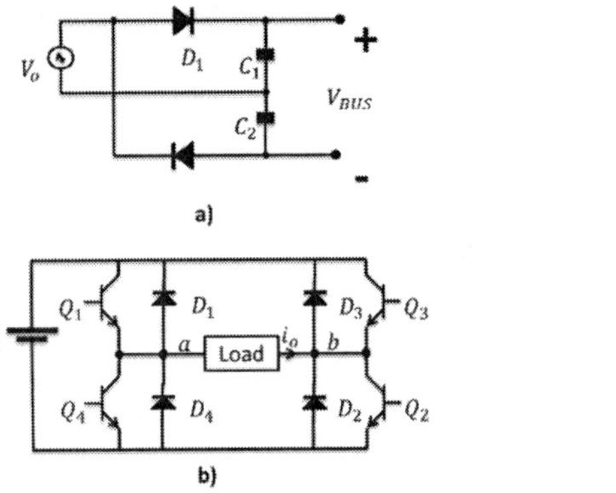

Figure 7. Voltage doubler and the DC/AC inverter schematic diagram.

The output voltage, V_o, the output current, I_o, considering RL load is given by equation (2)

$$v_o = \sum_{k=1,3,5,\dots}^{\infty} \frac{4V_s}{k\pi} sen(k\omega t) \qquad i_o = \sum_{k=1,3,5,\dots}^{\infty} \frac{4V_s}{k\pi\sqrt{R^2 + (k\omega L)^2}} sen(k\omega t - \theta_k) \tag{2}$$

Where the phase angle of the k component and the THD are given by equation (3)

$$\theta_k = tan^{-1}\left(k\omega \frac{R}{L}\right) \qquad THD = \frac{1}{V_1}\left(\sum_{k=2,3,\dots}^{\infty} V_k^{\,2}\right)^{1/2} \tag{3}$$

Where f is the frequency of the duty cycle, $\omega=2\pi f$ and k is the Fourier series component.

4 HARDWARE IMPLEMENTATION AND RESULTS

The 3D hardware design and its actual implementation of the presented PV system is shown in Figure 8.

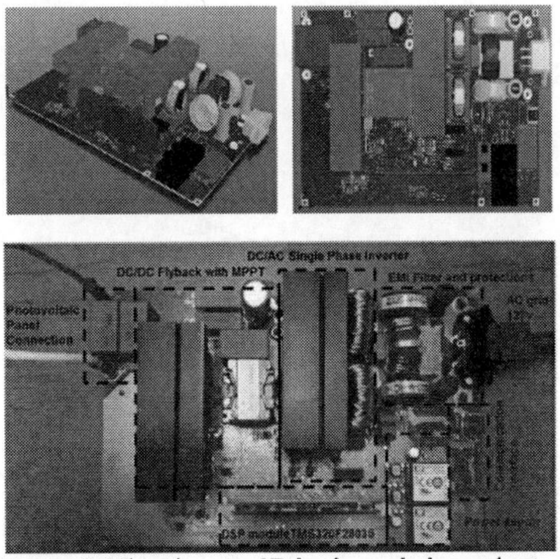

Figure 8. Micro-inverter 3D hardware design and actual implementation.

The micro-inverter was tested for different operation conditions. First a DC power supply model 6274B, which has a maximum capacity of 60V@15A, was used to emulate the PV panels, limiting its output to 20 – 45 Volts at 8.5 Amperes to match the micro-inverter input. Table I shows the attained results.

Table I: Micro-inverter results using DC power supply model 6274B to emulate the PV panels.

V_{PV}	I_{PV}	P_{IN}	Bus (V)	V_O	I_O	P_O	Efficiency
25.54	5.78	147.62	274.61	127.92	1.05	134.32	0.91
25.08	6.01	150.73	299.69	120.29	1.10	132.32	0.88
26.09	6.21	162.02	274.57	119.73	1.21	144.87	0.89
28.90	5.94	171.67	274.65	117.84	1.25	147.30	0.86
28.88	6.35	183.39	299.97	121.73	1.37	166.77	0.91
25.65	7.42	190.32	299.51	118.71	1.41	167.38	0.88
35.99	7.49	269.57	299.70	119.67	1.96	234.55	0.87
42.01	7.48	314.23	299.73	120.15	2.22	266.73	0.85

Posteriorly, the micro-inverter was fed by a 300 W PV panel. Table II show these results.

Table II: Micro-inverter results using PV panel.

V_{PV}	I_{PV}	P_{IN}	Bus (V)	V_O	I_O	P_O	Efficiency
23.53	8.50	200	299.50	120.00	1.42	170.00	0.85
28.82	8.50	245	299.50	120.00	1.76	210.70	0.86
31.76	8.50	270	299.50	120.00	1.89	226.80	0.84

Figure 9a depicts the voltage grid and the output current of the micro-inverter CD/CA. Figure 9b and 9c shows input and output power and its efficiency, respectively, of the micro-inverter for 31 measurements.

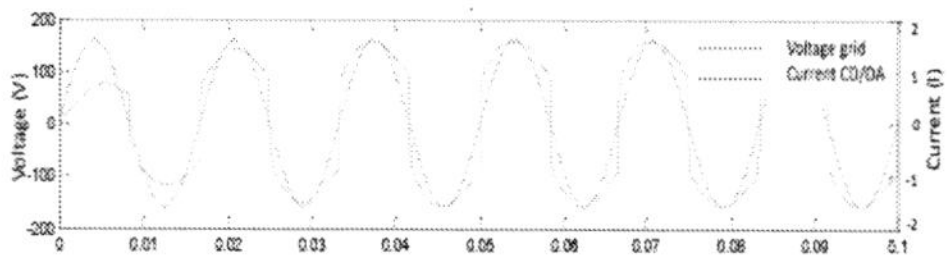

a)

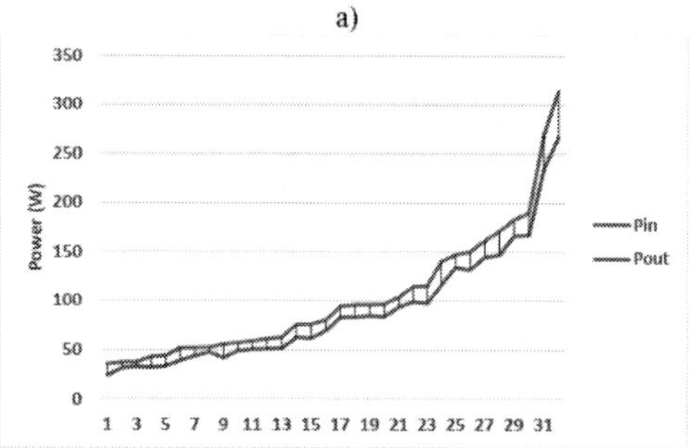

b)

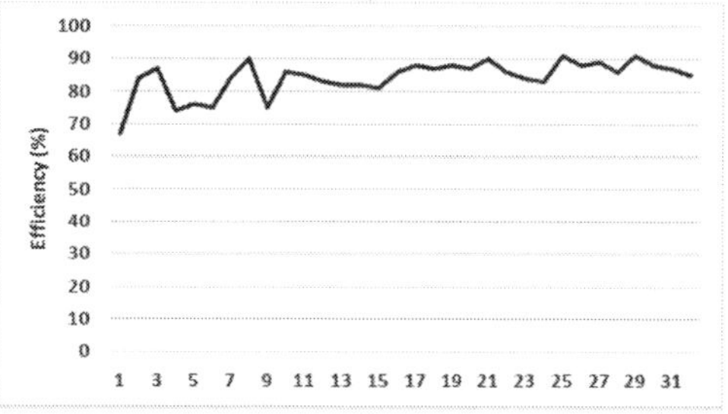

c)

Figure 9. Voltage grid and the outputs current of the micro-inverter CD/CA; b) micro-inverter input and output power; c) micro-inverter efficiency.

The PV micro-inverter shows a good efficiency, but it is still behind the string inverters which can reach up to 98%, due to the use of transformer-less topology that increment their efficiency. However, generally PV inverters do not have an interactive user interface to receive control parameter adjustment coming from either the user or the grid utilities, and display diagnostic and operating parameters via a low cost in-home display without the need of remote web server, which limits its application.

5 CONCLUSIONS

This paper has presented a flexible PV micro-inverter architecture that uses a low cost in-home display and mobile devices. The PV micro-inverter system is capable of displaying, via PLC or Bluetooth communication interfaces, key operating parameters such as DC bus voltages related to the performance of critical capacitors. The micro-inverter also possesses the versatility of receiving commands to perform tunable power factor, required for electricity companies to attain grid stability. Although the legislation of many countries have not approved reactive power feeding coming from the PV micro-inverters, Advance Metering Infrastructures (AMI) systems have been implanted in many places around the world and it has been anticipated that PV micro-inverters with communication interfaces and power factor adjustment resourcefulness could advantageously be integrated to AMI systems.

REFERENCES

[1] R. Kale, S. Thale and V. Agarwal, Design and Implementation of a solar PV panel integrated inverter with multi-mode operation capability, IEEE 39th Photovoltaic Specialists Conference (PVSC), 2013.

[2] J. Bzura, The AC module: an overview and update on self-contained modular PV systems, IEEE, 2010.

[3] S. Moskowitz, The Global PV Inverter Landscape 2015: Technologies, Markets and Prices. Solar Research, 2015.

[4] S. Sinapis and W. Folkerts, MLPM Benchmark Report, SEAC 2013.

[5] E. Sambuis, Digital Signal Controllers Improve Efficiency for Solar Power Inverters, Proceedings of the 2006 IEEE International Symposium on Electronics & the Environment, pp. 77 - 82, 2006.

LEAF ROOF – DESIGNING LUMINESCENT SOLAR PV ROOF TILES

Angèle Reinders[1], Guillaume Doudart de la Grée[2], Argyrios Papadopoulos[2], Alexander Rosemann[2], Michael G. Debije[3], Mark Cox[4], Zachar Krumer[1]

a.h.m.e.reinders@utwente.nl
1)University of Twente, ARISE, Design, Production & Management, Enschede, 7500 AE, The Netherlands
2)Eindhoven University of Technology, Building Physics & Services (BPS), Eindhoven, 5600 MB, The Netherlands
3)Eindhoven University of Technology, Chemical Engineering and Chemistry, Eindhoven, 5600 MB The Netherlands
4)Eindhoven University of Technology, Innovation Lab, Eindhoven, 5600 MB, The Netherlands

ABSTRACT: The Leaf Roof project on the design features of PV roof tiles using Luminescent Solar Concentrator (LSC) technology [1] has resulted in a functional prototype . The results are presented in the context of industrial product design with a focus on the aesthetic aspects of LSCs [2]. This paper outlines the design of Leaf Roof tiles under consideration of simulation results, experimental measurements on dyes' absorption spectra, and the energy performance of Leaf Roof elements in the context of their geometry and colors.

Keywords: building integrated PV (BIPV), PV module, ray tracing,

1 INTRODUCTION

The 3TU Lighthouse Leaf Roof project focuses on creating a roof design inspired by the natural shape of leaves. Each of the leaf-shaped shingles incorporates a Luminescent Solar Concentrator (LSC) [1]. An LSC is a technology for harvesting solar energy that is comprised of a transparent, thin, shaped plate acting as a lightguide with a large surface area. This lightguide consists of a material with a refractive index higher than air containing luminescent pigments, usually called a 'dye'. Photons originating from solar irradiance that enter the LSC are absorbed by dye molecules and are subsequently re-emitted at longer wavelengths. A large fraction of the re-emitted radiation is trapped in the lightguide by total internal reflection (TIR) at the material's surfaces. Because the re-emitted radiation has longer wavelengths, the probability of a secondary interaction such as re-absorption by the dye is reduced. TIR is interrupted at the interface between the lightguide and an edge attached solar cell, where the photons get converted into electricity

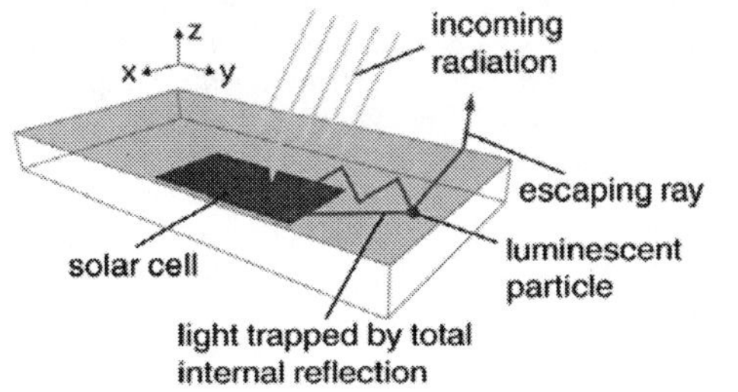

(see Figure 1).

Figure 1: Working principle of an LSC taken from [1].

The use of LSCs could allow more freedom of building orientation and roof inclination compared to conventional PV systems as they are capable of harvesting both direct and indirect illumination [3]. This project has successfully demonstrated a "Leaf Roof" concept [1] including improved aesthetic design considerations [2], see Figure 2.

The Leaf Roof project comprises a collaboration by Eindhoven University of Technology and University of Twente in the framework of the Dutch Lighthouse program of 3TU.Bouw which is focused on speculative research related to the theme 'Energy and the Built Environment'.

In the following sections we will introduce the reader to optical simulations of various configurations of Leaf Roof tiles (Section 2), design features (Section 3) and the presentation of the first prototypes of the Leaf Roof concept. The paper will be completed by Section 4 with a discussion and conclusions.

Figure 2: Visualisation of the Leaf Roof concept.

2 SIMULATIONS

In order to identify the most suitable features of an LSC for use as a roof tile, several numerical simulations have been carried out at University of Twente. Three-dimensional models of waveguides with defined optical properties of the materials involved have been simulated using commercial ray-tracing softwares PVtrace and LightTools, which enable performance evaluation of an optical systems without the physical realization of the system itself. For both software packages, the absorption and emission spectra of dyes (see next section) can be used as an input. The simulated evaluation of the influence on LSC harvesting performance considered parameters such as:

Table I: Simulation results for variation of lightguide shapes for a four cells' configuration

Configuration	4 units fused	Saturn	Rhombic	Barrel	Barrel 80mm	Rectangle
Layout	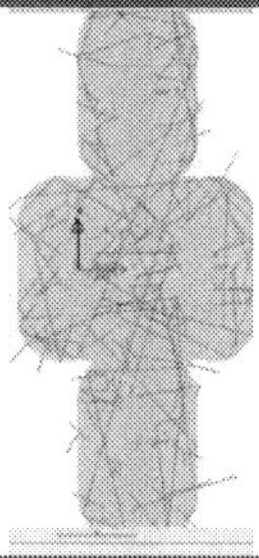	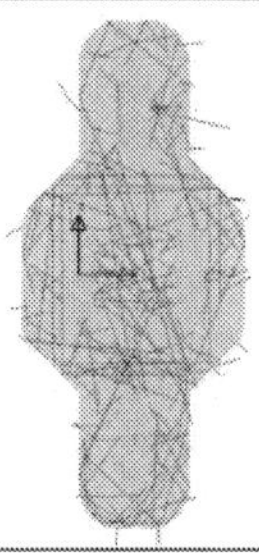	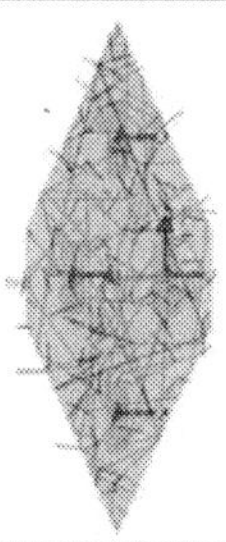	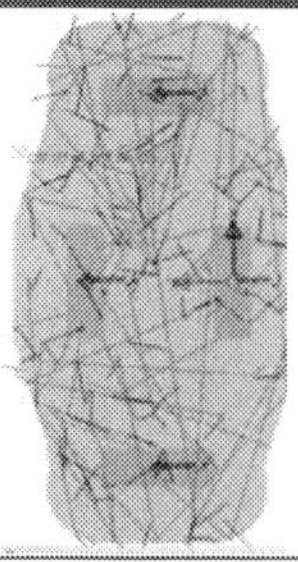	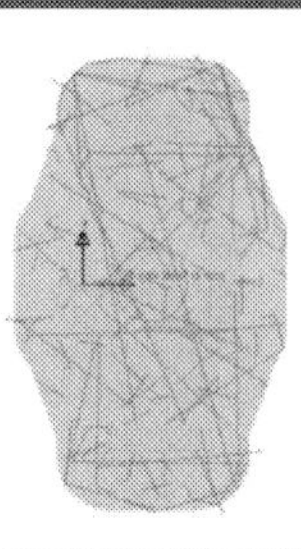	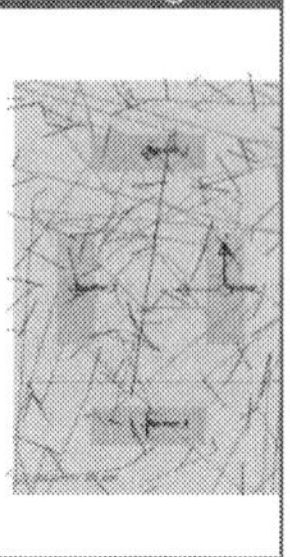
# Harvested rays	18809	19562	20661	20305	18697	14882
Area (cm²)	1290	1510	1380	1560	2110	1400
Photons per cm²	14.6	13.0	15.0	13.0	8.7	10.6

Table II: Simulation results of six cells' configuration

Configuration	Veins parallel	Eye	Skewed parallel	Snowflake	Parallel
Layout	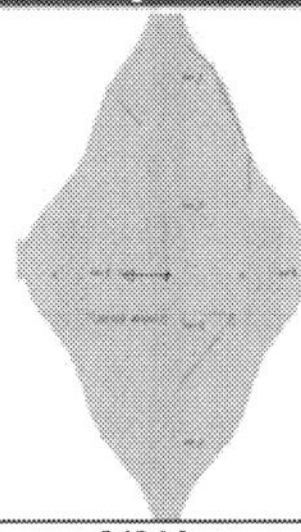	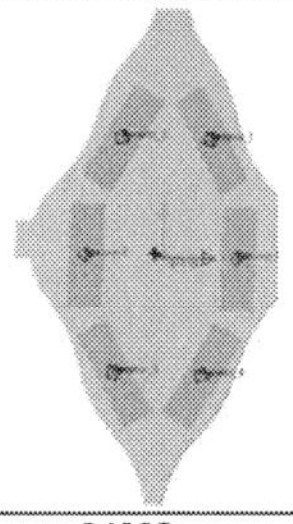	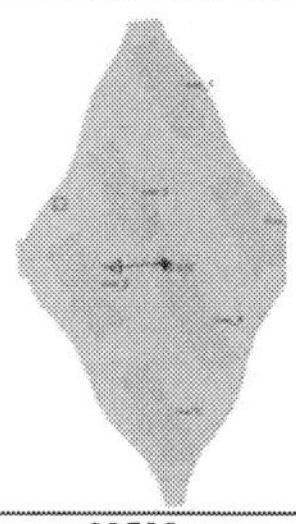	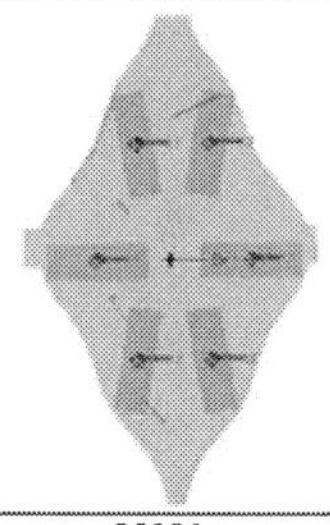	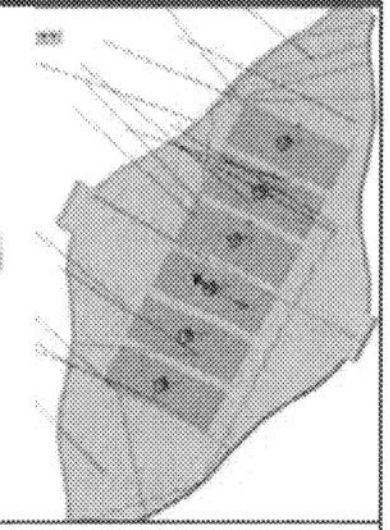
# Harvested rays	24246	24088	23708	23381	21222
Rays per cell	4041	4015	3951	3897	3537

- Shape of a Leaf Roof tile,
- Tile size and tile thickness,
- Dye content, absorption and emission characteristics,
- Position and size of the PV cells.

This led to many results which have been partially reported in [1]. In this paper, two new key simulation results will be addressed in more detail: the overall shape of the tile and the configuration of the PV cells.

We defined a waveguides with the shape as shown in Table I that could be potentially used as a roof tile. Surfaces were chosen to be transparent. The luminescence properties were included by defining a phosphor that models the properties of the commonly used dye for LSCs Lumogen Red 305 with the corresponding absorption and emission spectra, see Figure 3. The luminescence quantum efficiency is assumed to be 95% [4]. The refractive index of the waveguide was chosen to be constant at 1.49, which corresponds to that of PMMA. The solar cells were defined as perfect absorbers in optical contact with the waveguide from one side. Their refractive index is chosen to be 4.08, which corresponds to that of silicon [5]. The wave guide material was considered to be a perfect transmitter unless the radiation was absorbed by the phosphor. Unfortunately LightTools has no built in model for the Stokes-shift upon re-emission. Therefore it was modelled by cutting of a part of the emission spectrum that is shorter than the wavelength of the previously absorbed photon. In this way it is ensured that the photons do not gain energy upon re-emission while preserving the distribution of wavelengths as defined by the emission spectrum. Also the top surface of the solar cells and the bottom surface of the waveguide were put into optical contact. The dimensions of solar cells in the simulations were 125 x 42 x 4 mm³. Each simulation involved 50000 rays at 550 nm to be traced. The amount of rays absorbed by the solar cell is the quantity that determines the performance of the luminescent solar concentrator configurations.

Table I provides an overview of the simulation results which clearly point out an advantage of a rhombic (leaf-like) shape over other possible, selected shapes with respect to photon density at the rear-mounted PV elements. Subsequent simulations, shown in Table II, revealed that within the rhombic shape, optimum arrangements for the PV cells are those that do not concentrate all PV cells at the center of the tile. Moreover, an orientation of the solar cells parallel to the long axis of the tile led to the best performance.

Adding to this, to understand the influence of the number of corners in the shape of a waveguide, we varied the shape while keeping it a regular polygon with a fixed apothem of 200 mm. The number of angles was varied between four and eleven with the additional shapes with 20, 40 corners and with a circle as the infinite limit. The thickness of 4 mm was not altered. Results are shown in Figure 3. If applying a correction factor for the total area of the light guides it can be concluded that the absorption of photons does not depend on the area.

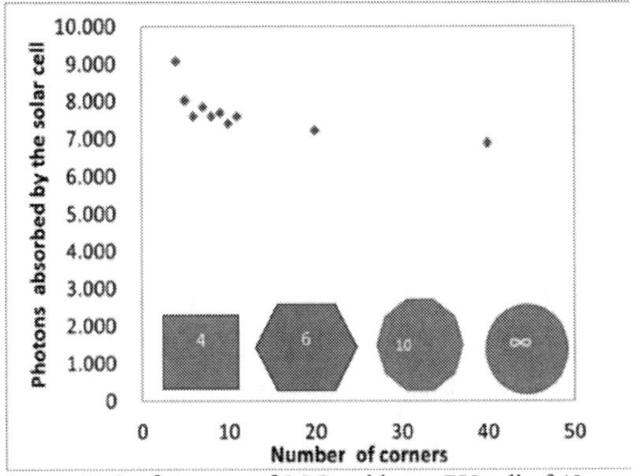

Figure 3: Performance of LSCs with one PV cell of 42 x 125 mm in the shape of regular polygon. Size of square shape 400 x 400 mm and radius of circle 200 mm.

3 DESIGN OF THE PROTOTYPE

The first pre-concepts of the roof tiles were simple rectangles and evolved via a rhombus to a leaf shape. The first leaf-shaped tile was developed resembling the leaf of a European birch, a common tree in the Netherlands. In order to benefit, in terms of aesthetics, from the chromatic characteristics of dyes, and to avoid a flat, monochromatic impression of the roof, it has been decided that different dye colors should be used in one roof.

The pattern was composed of different colors, among them three different green variations, orange and red. Three guiding criteria are considered during design of the leaf shapes:

- degree of physical panel overlap,
- photon travel distance, and
- positioning of the attached solar cells.

Overlapping of tiles must be reduced to enhance the energy performance (maximize the exposed surface, hence light harvesting) and reduce costs (minimizing the overlap, hence material demand). The travelling distance of photons in the lightguide must not be longer than ten centimeters in order to minimize extinction (according to the exponential Beer-Lambert law for absorption of photons in matter) and re-absorption losses due to the dye [6]. In order to maximize the utilization of light reflected from edges via reflective foils, the orientation of the solar cell relative to the LSC system is of importance. Different colors for dyes of the Lumogen series of BASF have been chosen to realize the LSC leaf tiles.

Absorbance spectra for the various plates containing the fluorescent dyes were measured on a Shimadzu UV-3102 PC spectrophotometer, and the emission from one edge of smaller, square plates were measured by an SLMS 1050 integrating sphere (Labsphere) equipped with a diode array detector (RPS900, International Light). The LSC samples were placed in a custom-made sample holder with a black background and exposed to a collimated light source from a 300 W solar simulator with filters to approximate the 1.5 AM (global) solar spectrum (Lot-Oriel), located at a distance of about 15 cm. The normalized results are shown in Figure 4.

The electrical performance of the LSC device plates depends on both the choice of fluorescent dye as well as the attached photovoltaic cell. For example, given similar fluorescent quantum yields (the yields of the Lumogen dye series in a PMMA host are generally >95% [7]), the Red305 will both absorb (due to the broader absorption range) and emit more incident solar photons than Yellow 083. Likewise, the spectral matching of the applied photovoltaic with the emission spectra of the dye will influence the total conversion to electricity as not all cells process all wavelength of incident light the same [8]. Thus, selecting different cells for different dye-doped LSC plates could aid in achieving higher efficiencies. However, in the Leaf Roof project, electrical performance is not the only consideration, and perhaps not even the most important. The aesthetics of the roof plays a vital role; the appearance requires the inclusion of multiple colors, despite potentially diminished performance. With this in mind, concepts such as color blending could be very important: that is, the mixing of multiple dyes into a single LSC 'leaf' to achieve a specific visual color. Multi-dye combinations have been shown to maintain high emissive fluorescent quantum yields [7], and allow for a much broader spectrum of possible visual effects for the leaves.

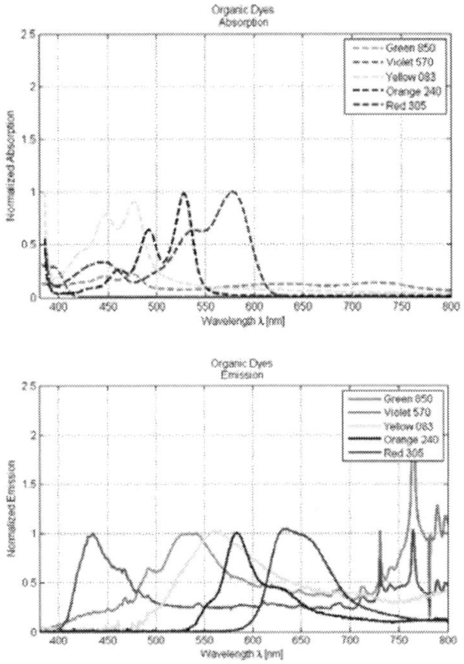

Figure 4: Normalized spectral absorbance (above) and emittance (below) of four different dyes of the Lumogen series of BASF.

For the prototype design, a tile configuration consisting of 6 PV cells with a customized size of 52 cm^2 (resulting from cutting complete cells in 3 slices) has been chosen. The total area covered by PV cells is therefore 312 cm^2. The solar cells used are Sunpower C60 solar cells (GEN C BIN J) with a 22.5% efficiency under Standard Text Conditions (STC). Table III shows the effect of cutting these cells in three slices based on information from the datasheets of the manufacturer.

Table III: Specifications of PV cells used in Leaf Roof*.

Sort of PV cell	P_{mpp} (Wp)	η_{STC} (%)	V_{mpp} (V)	I_{mpp} (A)	V_{oc} (V)	I_{sc} (A)
Complete cell	2,42	22,5	0,582	5,93	0,687	6,28
Cut in 3**	0,81	22,5	0,582	1,39	0,687	1,46

*) based on spec sheets of Sunpower C60 cells
**) by laser cutting

Various cell configurations for the Leaf Roof elements have been evaluated with regards their efficiency by numerical simulations (see previous section). However during lab tests of prototypes it has been observed that dark elements such as PV cells concentrate irradiation, resulting in the heating of the local surroundings, leading to a temperature gradient in the plastic (PMMA) waveguide. As a result, this created a small bend along the Y-axis of the prototype. The resulting mechanical stresses might lead to the breakage of the interface between cell and waveguide. To overcome this, a configuration was chosen that allows a greater distance between cells, as shown below in Figure 5, instead of the simply basing positioning solely on optical simulations, labeled Veins Parallel and Eye, shown in Table II.

Figure 5: Photo of a prototype of a red Leaf Roof tile.

The present prototype of a Leaf Roof tile, shown in Figure 5, consists of a 5 mm thick LSC waveguide of PMMA containing 80 ppm of Lumogen® F Red 305 dye. It is equipped with SunPower™ C60 solar cells, cell-wiring and a metallic back reflector made from copper. The total area of the front surface of the Leaf Roof element is 0.11 m^2, leading to a theoretical efficiency of 4,4% with an Impp of 1,39 A and a Vmpp of 3,49 V, without encapsulation with the PMMA waveguide.

Figure 6: A Leaf Roof configuration in various colors at the Dutch Design Week in 2015.

The real efficiency is expected to be slightly higher, because of the beneficial effect of the waveguide on photon capturing. For instance, a preliminary measurement of this prototype by the 'Eternal Sun' unit at the TU/e campus has shown an energy yield which is 9 % higher compared to a reference prototype with a transparent PMMA waveguide (without dyes). The various colours of the Leaf Roof are shown in Figure 6.

4 DISCUSSION AND CONCLUSIONS

At the moment flash tests are being executed and outdoor monitoring of several prototypes with different dyes will happen in summer of 2016. These data will add to our growing knowledge about building integrated PV and designing with LSC photovoltaic solar technologies.

There are a number of additional considerations that should be addressed for the Leaf Roof to become a reality. One will be the eventual wiring plan for a complete roof demonstration unit, see Figure 7. Current balancing will be important, for, given similarly size leaf tiles, the different color plates will produce different output currents. This could be adjusted materially, but adjusting dye concentrations to achieve similar output currents, for example, but this puts too many constraints on the appearance of the roof as a whole. Rather, the corrections should be done electrically.

Also, it should be noted that the Leaf Roof PV modules need not be contained to a flat architecture as a traditional silicon-based solar panel. Expanding into the third dimension by curved PV elements could have a number of benefits: aesthetically - allowing more design freedom for appearance -, electrically – since curved surfaces have been shown to be capable of as good or better performance to flat [9-10] - , and mechanically - additional shape could allow for enhanced performance of other activities for the roof, such as sloughing off of snow and/or rain. Summarizing, Leaf Roof PV modules could potentially solve many existing issues with regular flat plate silicon PV modules, for which reason it would be worthwhile to further explore LSC PV modules for building integrated PV applications in the nearby future.

Figure 7: Example of a Leaf Roof configuration with various colors applied in a rendered modern architecture building.

7 REFERENCES

[1] G.C.H. Doudart de la Gree, A. Papadopoulos, M.G. Debije, M.G.D.M. Cox, Z. Krumer, A.H.M.E. Reinders and A. Rosemann. "A new design for luminescent solar concentrating PV roof tiles", in 42nd IEEE Photovoltaic Specialists Conference, 2015

[2] A.H.M.E. Reinders, J.C. Diehl and J.C. Brezet, The Power of Design, London, Wiley and Sons, 2012.

[3] M.G. Debije and V.A. Rajkumar, "Direct versus indirect illumination of a prototype luminescent solar concentrator", Solar Energy, Vol. 122, pp. 334-340, 2015.
[4] L.H. Slooff, E.E. Bende, A.R. Burgers, T. Budel, M. Pravettoni, R.P., Kenny, E.D. Dunlop, and A.C. Büchtemann, "A Luminescent Solar Concentrator with 7.1% power conversion efficiency" , Physica Status Solidi - Rapid Research Letters, Vol. 2(6), pp. 257-259, 2008.
[5] C. Corrado, S.W. Leow, M. Osborn, E. Chan, B. Balaban and S.A. Carter, "Optimization of gain and energy conversion efficiency using front-facing photovoltaic cell luminescent solar concentrator design", Solar Energy Materials and Solar Cells, Vol. 111, pp. 74-81, 2013.
[6] M.G. Debije, J.P. Teunissen, M.J. Kastelijn, P.P.C. Verbunt, P.P.C. and C.W.M. Bastiaansen, "The effect of a scattering layer on the edge output of a luminescent solar concentrator", Solar Energy Materials and Solar Cells, Vol. 93, pp. 1345-1350, 2009.
[7] L. R. Wilson and B. S. Richards, "Measurement method for photoluminescent quantum yields of fluorescent organic dyes in polymethyl methacrylate for luminescent solar concentrators", Appl. Optics, Vol. 48, pp. 212-220, 2009.
[8] H. Field, in 26th IEEE Photovoltaic Specialists Conference, 1997.
[9] B. Vishwanathan, A.H.M.E. Reinders, D.K.G. de Boer, L. Desmet, A.J.M. Ras, F. H. Zahn and M.G. Debije, "A comparison of performance of flat and bent photovoltaic luminescent solar concentrators", Solar Energy, Vol. 112, pp. 120-127, 2015.
[10] K.R. McIntosh, N. Yamada and B.S. Richards, "Theoretical comparison of cylindrical and square-planar luminescent solar concentrators", Appl. Phys. B, Vol. 88, pp. 285-290, 2007.

TUNABLE SHADE WINDOWS WITH INTEGRATED LUMINESCENT SOLAR CONCENTRATORS AND HIGH EFFICIENCY LIGHTING

Paolo Bernardoni, Michele Tonezzer, Donato Vincenzi, Silvio Fugattini, Vincenzo Guidi
University of Ferrara - Department of Physics and Earth Sciences - Sensor and Semiconductor Laboratory
Via Giuseppe Saragat 1 - Building C, 44122 Ferrara (FE), Italy
Corr.Auth. E-Mail: paolo.bernardoni@unife.it Tel: (+39) 0532 974271

ABSTRACT: In this work we present the prototype of a smart window that combines multiple features as energy production, energy storage, lighting and shading control. The energy production is provided by a Luminescent Solar Concentrator (LSC) coupled with silicon solar cells on its sides, the energy is stored in a Lithium-ions battery and is used to provide either illumination through an array of LEDs or shading control through a Liquid Crystal Display (LCD) or Polymer Dispersed Liquid Crystal (PDLC) panel. All these devices are connected as a stand-alone system hosted inside a standard window frame.
Keywords: Building Integrated PV (BIPV), Concentrators, Stand-alone PV Systems

1 Our proposal

"Smart" windows are designed to integrate a number of innovative features not present in the traditional building fixtures, these features can range from the power generation to the shading control or integrated illumination and are typically coupled with good thermal insulation characteristics, which are essential for their employment in sustainable [1] and "smart" architecture designs.
Our proposed application consists in a smart window whose energy source is represented by a transparent Luminescent Solar Concentrator (LSC) [2] coupled with solar cells, the energy is stored in Li-ion batteries integrated into the window frame and can be used to power either a PDLC (Polymer Dispersed Liquid Crystal) [3] or LCD (Liquid Crystal Display) panel in order to control the window transparency plus a RGB LED (RED-Green-Blue Light Emitting Diodes) array for high efficiency illumination or light effects.
All these features are embedded into a traditional window frame that can be used as a drop-in replacement for existing fixtures.

2 Power generation

An LSC photovoltaic panel generates the energy consumed by the other devices integrated into the window.
This device consists in a slab made of a plastic material (PolyMethylMethAcrylate PMMA) doped with a fluorescent dye (BASF Lumogen F Violet 570) that absorbs light in the UV spectrum with a fluorescence peaked at 430nm, the emitted light is trapped inside the slab by total internal reflection and is guided towards the sides.
The concentrated light coming out of the sides is collected and converted by an array of monocrystalline silicon solar cells arranged on one side of the LSC, the other three sides are covered with a high efficiency multilayer reflective film which purpose is to reduce the amount of cells needed for a given power output.
In fact the reflective film redirects back on the fourth side (which hosts the cells) the light concentrated by the LSC on the other three sides effectively increasing the concentration factor of the system at cost of the optic losses caused by the absorption of the film and increased path length of the light inside the concentrator.
The optimal optical coupling between the cells and the concentrator is obtained using an optical UV-curing glue to attach the cells, this glue ensures optimal coupling having a refractive index (n_{glue}=1.506) close to that of PMMA (n_{PMMA}=1.489), good thermal stability and a fast assembly time since it hardens in a few seconds (t_{min}=15s) upon exposure to UV light at 365nm.

3 Energy storage and lighting

The energy produced by the LSC panel is stored in lithium-ions batteries, the charging process is managed by the main electronic board (see Fig. 1) that tracks the maximum power point of the panel and regulates the charging current of the battery, safety is assured by an overvoltage-undervoltage protection and a protection from overheating.

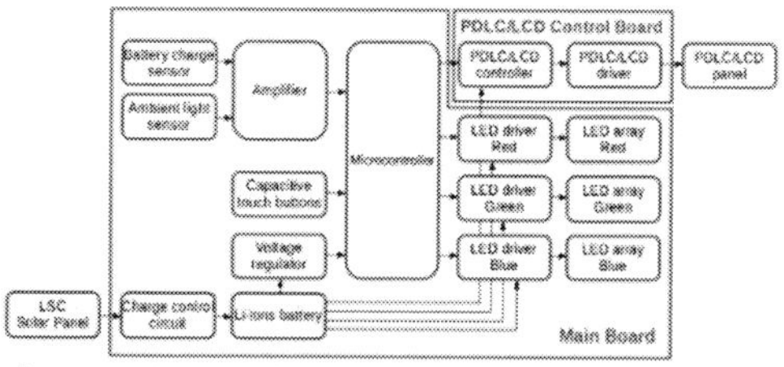

Fig. 1: Scheme of the electronics installed into the window frame: most functions are integrated on the main board while the PDLC/LCD controller and driver are on a different board to ensure a better modularity of the design since the two technologies have different driving specifications

The lighting is provided by an RGB LED array which is also integrated on one side of the control board, the use of these devices permits to provide illumination with a wide range of colors including white light obtained powering all the LEDs with the same current.
The LEDs are powered by three high efficiency boost LED drivers, one for each colour, that raise the voltage of the battery from the 2.7-4.2V provided by the lithium cells to the 11-23V needed by the array.
The color and intensity of the lighting (and shading) are controlled by four capacitive touch buttons that are part of the main board and can be operated from one side of the window frame.

4 Tunable shading

The transparency of the window can be changed by mean of devices based on two different technologies: a PDLC (see Fig. 2) or an LCD panel.

Fig. 2: The PDLC panel is transparent when completely activated (left) and has a milky appearance when switched off (right)

A PDLC panel consists of a layer of liquid crystals dispersed in a viscous polymer matrix: these liquid crystals, if no external electric field is supplied, have a random orientation and hence scatter the incoming light giving to the panel a milky appearance
If an external electric field is provided then the liquid crystals orientate themselves along the field with minimal effects on light transmission making the panel effectively transparent.
In order to ensure this polarization of the liquid crystals a high voltage must be applied to the panel, typically between 50 and 70V to achieve the maximum transparency, at lower voltages the liquid crystals are not completely polarized so the transparency can be tuned by changing the voltage applied to the panel.
Applying such high voltages to the panel can result in a permanent polarization so to ensure the reversibility of this effect the voltage must be continuously alternated with a frequency high enough to avoid a visible flickering typically above 50Hz.
The second option available to control the window transparency is an LCD panel (see Fig. 3), in this device the liquid crystals act as a polarizing filter whose orientation with respect to a static filter allows to control the amount of light passing through the window from a maximum of 50% when the crystals are not polarized to virtually the 0% (see Fig. 3).

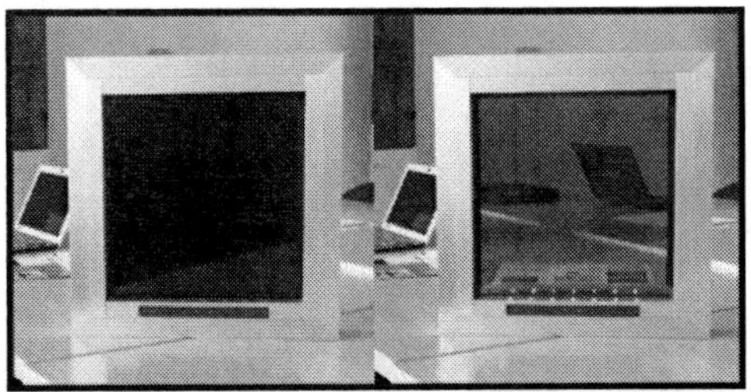

Fig. 3: The LCD panel is dark when activated (left) and becomes transparent once switched off (right)

LCD panels driving don't need the high voltage required by the PDLC (~5V instead of >50V) so the driving circuit requires only one third of the power consumed by the PDLC driver.

5 Frame assembly

The LSC and the PDLC/LCD are assembled as a double glazed panel: this structure, which provides a good thermal insulation, is installed in a standard aluminium window frame with an internal thermal break that ensures the appropriate insulation.
The battery, PDLC/LCD driver and main board (which also hosts the LED array and the capacitive buttons see Fig. 4) are designed to be fitted inside the frame leaving visible only the LEDs, the light sensor and the touch controls.

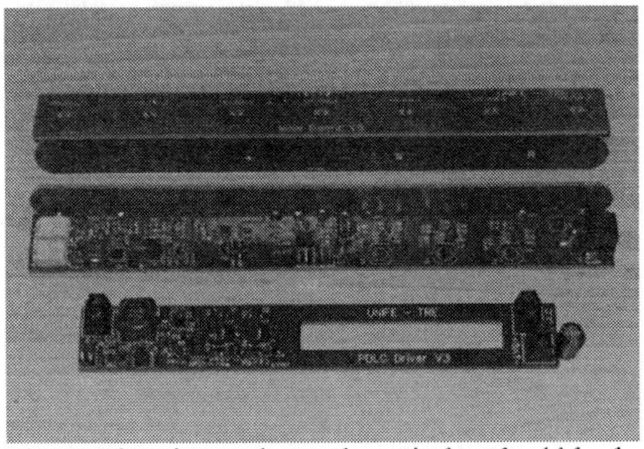

Fig. 4: The picture shows the main board which also hosts the capacitive buttons plus the LED array and the PDLC driver that has a slot to allow for the installation of the battery

This results in a stand-alone system which offers advanced features packed in a traditional window frame that can be effectively employed in building renovation projects without any need for additional connections.

6 References

[1] Sekhar, S. C., and Kenneth Lim Cher Toon. "On the study of energy performance and life cycle cost of smart window." Energy and Buildings 28.3 (1998): 307-316.

[2] Hermann, A. M. "Luminescent solar concentrators - a review." Solar Energy 29.4 (1982): 323-329.

[3] Spruce, G., and R. D. Pringle. "Polymer dispersed liquid crystal (PDLC) films." Electronics & communication engineering journal 4.2 (1992): 91-100.

[4] Wu, Bao-Gang, John H. Erdmann, and J. William Doane. "Response times and voltages for PDLC light shutters." Liquid Crystals 5.5 (1989): 1453-1465

SELF-SHADING IN BIFACIAL PHOTOVOLTAIC NOISE BARRIERS

M.M. de Jong[1*], M.N. van den Donker[1], S. Verkuilen[2], W. Folkerts[1]
[1]Solar Energy Application Centre, High Tech Campus 21, Eindhoven, The Netherlands
[2]Heijmans Wegen BV, Rosmalen, The Netherlands
*email: dejong@seac.cc

ABSTRACT: There is a large market potential for Bifacial Photovoltaic Noise Barriers (PVNB) due to their high energy output for any road orientation. A crucial parameter affecting their energy output is the so-called 'self-shading'. In this research we evaluate the effect of self-shading on the energy production of both an east and west facing bifacial PVNB. We followed an approach consisting of 1) Experimental investigation by accurately monitoring the power output of two bifacial PVNB prototypes, 2) Shading simulation using in-house developed simulation methods. Results show that in particular noise barriers on north-south aligned roads suffer from self-shading because the support structure of the noise barrier itself casts a shadow on the PV modules around noon. A strong correlation is found between clearness index and energy yield loss. On sunny days the energy loss can be as high as 5%. The year average for the loss is about 3%. Clear design guidelines for minimizing the impact have been derived. We conclude that our measurement method, simulation tools and derived design guidelines provide a way to assess and minimize the effect of self-shading on the energy output of bifacial photovoltaic noise barriers, leading to better energy yields and more market success for this type of application.
Keywords: BIPV, Bifacial, Field Testing, Performance

1 Introduction

This paper evaluates the performance of a bifacial photovoltaic noise (PVNB) barrier and more in particular the effect of self-shading on the performance of a bifacial PVNB along north-south aligned roads.
There is a growing interest in photovoltaics in the built environment and placing solar modules on or in noise barriers is an attractive option. Whereas most photovoltaic installations are oriented southward, the orientation of a PVNB cannot be chosen freely, but is governed by the orientation of the adjacent road. For roads running from north to south using bifacial PV modules for PV noise barriers will result in a large increase in electrical output. A system facing east and west will show a minimum in electricity output at solar noon, as no direct light can reach the solar modules. Around this time, the carrying structure of such a system will cast a shadow on the solar system itself, which is called self-shading. How much this has a negative effect on the electricity output depends on the design of the system; The size of the supporting structure, the placement of the cells within the module and the string layout will influence the power output of the modules. The negative effects of self-shading can be mitigated by smart solar cell placement within the modules, a smart design of the string layout and an optimal design of the supporting noise barrier structure. Nevertheless, optimal cell placement and stringing might limit the cell density (or cell area coverage ratio) within the module. Furthermore, in general the supporting structures for noise barriers are rather large. Because of the noise blocking functionality of the noise barrier, the barrier will need a certain mass. Furthermore, the nature of the noise barrier structure will induce high wind loads on the structure. For these reasons, noise barriers will need larger support structures than would be ideal for the solar modules.
Although the use of bifacial solar cells in solar noise barriers has been demonstrated before [1], the role of self-shading has not been studied in detail. In this paper we evaluate the effect of self-shading on the electrical output of a bifacial solar noise barrier pilot system built in 's Hertogenbosch, the Netherlands.

Figure 1: (left) The SONOB Living Lab test site, showing two differently oriented noise barriers. The third row from the top holds the bifacial solar cells. (Right) Self-shading by the carrying structure on the rear side of the bifacial cells.

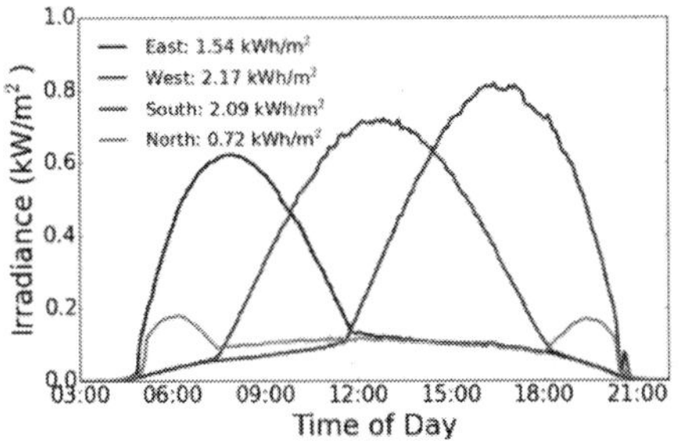

Figure 2: Irradiance in-plane for both sides of both barriers on 30 June 2015, an almost completely clear day.

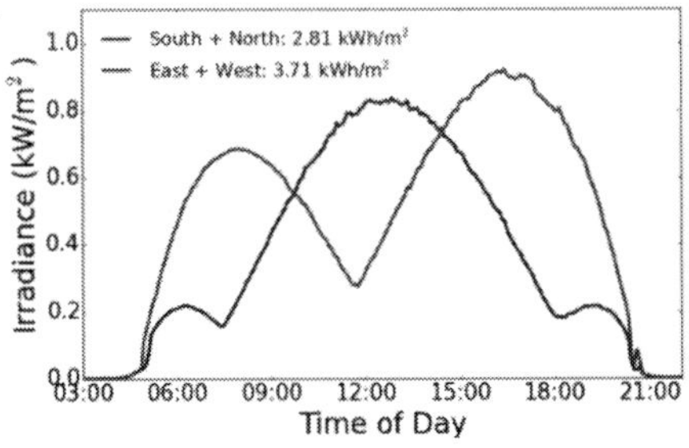

Figure 3: Added in-plane irradiances for both barriers. The legends also show the total amount of irradiance received.

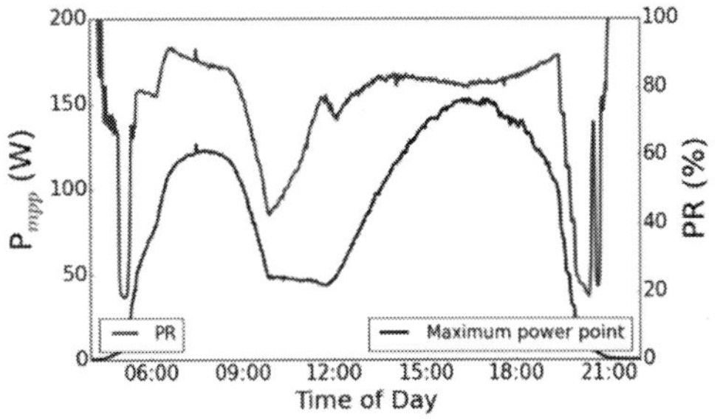

Figure 4: Measured maximum power point and performance ratio of the east-west facing barrier.

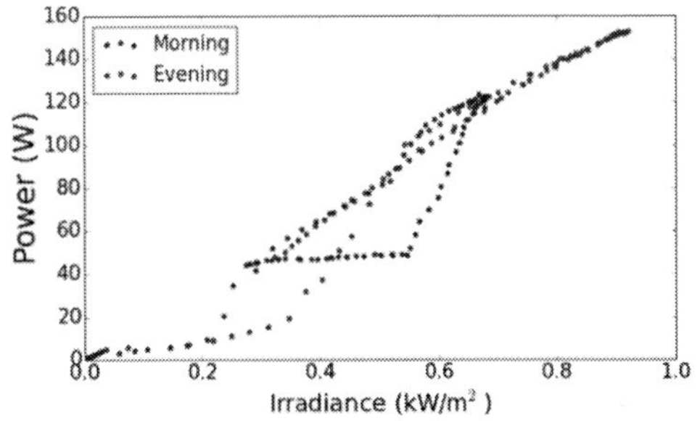

Figure 5: Measured maximum power point plotted against the irradiance. The self-shading-periods in the morning (blue) and evening (red) hours are easily identified.

2 Experimental Setup

For the SONOB project [2,3] a 'Living Lab' for testing a diversity of solar cell techniques in a solar noise barrier was set up by a consortium of companies and knowledge institutes consisting of Heijmans, Van Campen industries, Airbus Defence and Space, Scheuten, TU Eindhoven, ECN and SEAC. The pilot setup consists of two noise barriers of 5 meters wide and 4.5 meters high, one facing south and north, and one facing east and west. Both barriers are inclined 15° backwards, to the north and east, respectively. While in the project we investigate the performance of a number of different solar cell technologies and orientations, this paper will focus on the behavior of bifacial modules in the barrier facing east and west. The row holding bifacial solar cells consists of two modules that both contain 48 NSP NS6MN '1960' bifacial solar cells, produced by Scheuten glass. The cells are placed approximately 20 cm away from the sides and top of the modules to mitigate the effect of self-shading. Flash tests at Scheuten showed a front side STC rated power of 207 Wp and a rear side rated power of 190 Wp. The setup is presented in figure 1 (left), showing both barriers and the different PVNB solar concepts. The third row from the top contains the bifacial solar modules. Figure 1 (right) shows an instance of self-shading by the carrying structure on the rear side of the bifacial solar cells, in this case the east facing side of the PVNB facing east and west.

On top of both barriers, two EKO MS-802 secondary standard pyranometers are placed in-plane to measure the irradiance in the four planes of the barriers. A separate pyranometer is used to measure the global horizontal irradiance, to be able to determine the clearness index. An on-site weather station measures ambient temperature and wind speed. Every two minutes an IV-curve is measured using an EKO MP-160 IV-tracer, while in between the measurements the modules were in open-circuit.

3 Results

As a case study, we analysed the performance of the east and west facing PVNB on a sunny summer day, showing an almost unclouded sky during the complete day. Figure 2 shows the irradiation measured by the in-plane pyranometers on June 30, 2015. The figure shows the irradiation in all planes of the two differently oriented barriers. The irradiation on the south facing side shows a peak around noon. Because in summer the sun rises in the northeast and sets in the northwest, the irradiance on the north facing barrier shows two peaks: in the early morning and in the late evening. The east and west facing side receive a peak in the morning and evening, respectively. For all faces, there is a significant diffuse contribution to the total irradiance throughout the day.

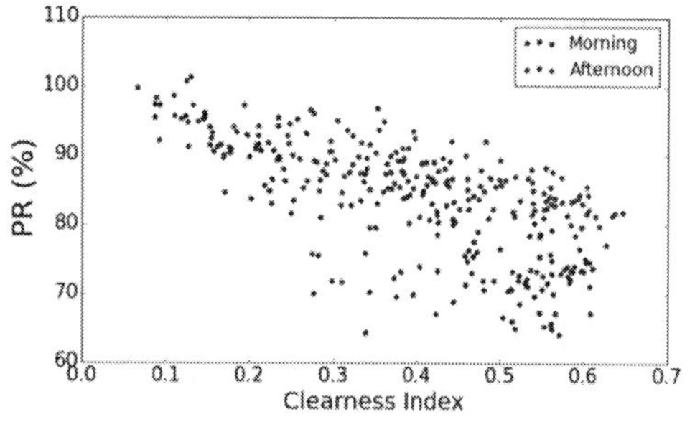

Figure 6: There is a correlation between PR and clearness index that is stronger in the morning (self-shading) than in the afternoon.

Figure 3 shows the added in-plane irradiances as 'seen' by the bifacial modules. The bifacial irradiation profile for the east and west facing barrier shows a double peak structure, showing a peak in the morning and a peak in the afternoon. Around midday, the irradiation is parallel to the barrier and no direct light is received by the modules, but there is still a substantial contribution by diffuse light. The asymmetry in morning and afternoon illuminations is caused by the tilted angle of the barriers. The south and north facing barrier shows a more conventional irradiation profile.

Figure 4 shows the maximum power points (P_{mpp}) of the east and west facing barrier together with the corresponding DC performance ratio (PR). The power output profile also shows a double peak profile, but at 9:30 the P_{mpp} reaches a minimum and is stable to 12:00, whereas the irradiation profile does not show this 'flat zone'. If we look at the PR, we see a significant dip during this period. This dip in PR is caused by shading of the support structures. In this particular case, the self-shading by the side support and the top support occur simultaneously. Another large dip in PR is observed around 5:30, which is caused by the support structure on the north side, as in summer the sun rises in the northeast. A third dip in PR at 20:00 is caused by an external shading source.

In figure 5 we plotted the P_{mpp} plotted against the bifacial irradiance of the east and west facing module. In this way we can easily identify the periods in which the installation suffers from self-shading, where the points deviate from a linear dependence.
The effect of self-shading is most pronounced in sunny conditions. In overcast conditions, the diffuse nature of the irradiation will mitigate the negative effects. To make an assessment of the year-round effects of this self-shading, we divided the measured performance data in two parts: morning performance, when self-shading occurs and afternoon performance, when there is no self-shading. The resulting PRs as a function of the clearness index are presented in figure 6. We clearly see a negative correlation between PR and clearness index. This can partly attributed to the temperature effect of crystalline silicon; in sunny conditions, the cells heat up. The fact that this correlation is stronger in the morning can be attributed to the presence of self-shading in the morning. Throughout the year the average morning PR was 72.9%, whereas the average PR in the afternoon was 86.3%.

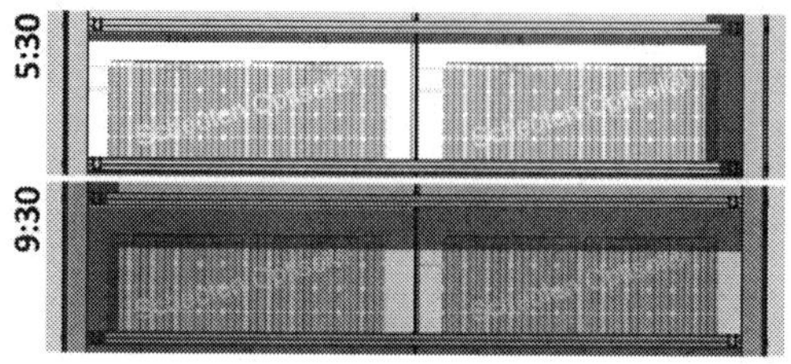

Figure 7: Images from a Sketchup model showing rear-side self-shading pattern at 5:30 and 9:30 of the east and west facing barrier.

Taking into account the lower rear side efficiency of the bifacial solar cells, the total irradiation on the front and back side and the lower morning PR, we estimate the loss of energy yield due to self-shading to be 3.0% in this configuration.

To visualize the self-shading, a model was created in Sketchup. By using the shadow feature of the program we can see when self-shading becomes a problem. Figure 7 shows the modelled self-shading of the east and west facing barrier on June 30 at 5:30 a.m., about an hour after sunrise and at 9:30 a.m., when the south support structure casts a shade on the rear side of the bifacial modules. We see that both the vertical and the horizontal support structures cause self-shading. Using this visualization we can derive design guidelines for bifacial installation in the built environment or infrastructure in any orientation.

This work has been supported by the Netherlands Enterprise Agency (RVO) under TEZG114009.

References

[1] Th. Nordmann et al., 15 Years Of Practical Experience In Development And Improvement Of Bifacial Photovoltaic Noise Barriers Along Highways And Railway Lines In Switzerland, 27th EUPVSEC proceedings, 2012
[2] http://www.seac.cc/projects/solar-noise-barriers-sonob/
[3] http://www.heijmans.nl/en/news/unique-field-test-started-solar-noise-barriers/

THERMAL MODEL OF BUILDING INTEGRATED AIR TYPE PHOTOVOLTAIC-THERMAL SYSTEM UNDER VARYING CONDITIONS

Andri Jagomägi
Tallinn University of Technology
Ehitajate tee 5, Tallinn 19086, Estonia
+372 5666 4834, andri.jagomagi@gmail.com

ABSTRACT: We present the dynamic thermal model of air type photovoltaic-thermal systems. Our model divides such a system into finite slabs with the spatial independent variable z in the direction of the moving air. The set of equations consists of six differential equations that are solved using the Runge-Kutta numerical solving method. The accuracy of the model is tested via measurement setup that consist of a sixty cell monocrystalline photovoltaic module with the air channel behind. The comparison of the measurements with the results of the simulations show that the model predicts well the backside temperature of the module and the exit temperature of the flowing air. The model responds well to the rapid changes in the heat fluxes and can be used to study the feasibility or optimize the performance of air type photovoltaic-thermal applications. The model is more accurate at the end of the channel and at air flow rates higher than 6 l/s.

Keywords: Building Integrated PV (BIPV), Modelling, Roofing Systems, Thermal Performance, Ventilation

1 INTRODUCTION

Building Integrated Photovoltaics (BIPV) is one of the fastest growing markets in the energy sector [1]. In addition to the electrical energy production, the BIPV modules replace the conventional building materials, thus, they have additional roles in the building environment. The BIPV products alter the aesthetics of the building, regulate the indoor climate, change the cost of the construction and influence the thermal balance of the building. From the economic side, the BIPV modules reduce the bill of the construction materials. In order to decrease the payback time of the investment, the BIPV systems may coproduce the electrical and the thermal energy simultaneously. One possibility to manage the waste heat that is produced by the BIPV modules is to use the air as the thermal energy exchange medium between the modules and the building. These air-type photovoltaic-thermal (APVT) systems can be used for the direct heating of buildings, pre-heating of the ventilation air, increasing the COP value of the heat pump or drying of agricultural products. In order to study the economic and technical possibilities of these solutions a thermal model of the APVT system must be used. Several steady state models of APVT systems are presented in Refs. [2] - [18]. A dynamic thermal model of the ordinary photovoltaic modules under varying conditions is presented in Ref. [19].

In the present paper we present the dynamic thermal model of the APVT system that can be used to model the thermal and the electrical performance of such a system under rapidly varying environmental conditions, which is very common in the Northern European climate.

2 EXPERIMENTAL

In order to validate the model we developed the outdoor test bench that is depicted on the Fig. 1. The experimental setup consists of 60 cell PV module that is embedded into the insulation material. The width of the insulation layer at the back of the module is 70 mm. Between the module and the insulation there is rectangular 30 mm airgap for the forced flow. The driving pressure is generated by the controllable speed fan shat is installed at the end of 2 m straight tube with the diameter of 200 mm. Just in front of the fan there is a filter box that helps to minimize the effects of the ambient pressure drop in the air channel that is generated by the wind. The air flow speed in the duct is measured with thermal dispersion flowmeter that allows accurate readings at low flowrates down to 0.1 m/s. The temperatures of the backside of the module, the front side of the insulation layer, inlet air, and the outlet air are measured with low thermal capacity thermocouples. The cross section of the experimental setup is depicted on the Fig. 2.

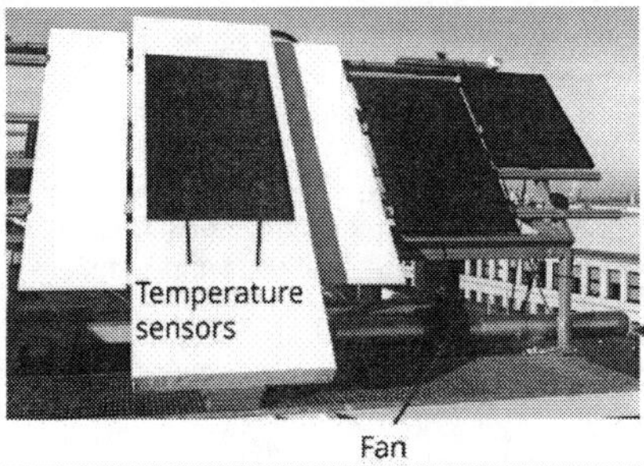

Figure 1: Outdoor air type photovoltaic-thermal test bench. In order to simulate the roof integrated conditions a 60 cell photovoltaic module is placed into the insulating material. Between the backside of the PV module and the insulation there is a 30 mm airgap for the airflow. The hot air outlet is in the top part of the backside of the structure. In order to reduce the front surface cooling by the wind that is coming from the back, the sides of the module are protected by plywood sheets. The tilt angle of the module is 48 degrees from the horizontal. The experimental setup is located on the roof of Tallinn University of Technology building no 6.

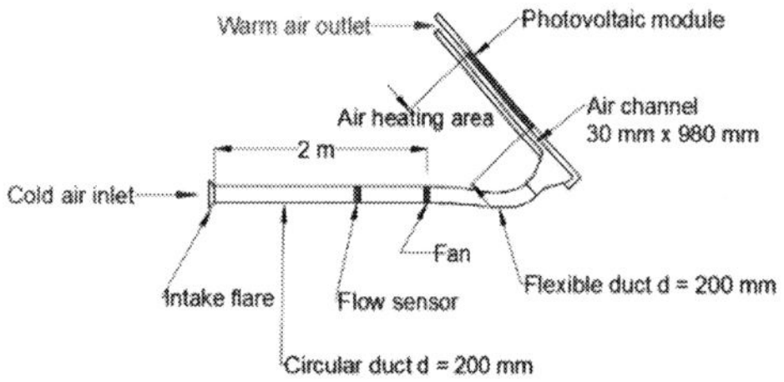

Figure 2: The schematic drawing of the experimental setup. The airflow is controlled by variable speed fan that is installed 2 m upstream from the inlet. The thermal dispersion flow sensor is 2/3 from the total straight distance of the tube from the outlet.

In addition, environmental parameters like in-plane irradiation, wind speed, wind direction, ambient air pressure, ambient air relative humidity, and ambient air temperature were measured. The irradiation was measured using the Kipp&Zonen CMP11 pyranometer and the weather parameters with the Vaisala Weather Transmitter WXT510. All the parameters were recorder with the period of 10 s and stored in the database for the further analysis.

3 MATHEMATICAL MODEL

3.1 Heat balance of the APVT

The temperature of the PV module inside the APVT system is affected by the heat exchange of the front surface with the environment and the heat exchange of the back surface with the insulation layer. The convectional heat exchange between ambient air and the front surface is mainly influenced by the ambient temperature and the wind speed. In addition, the front surface gains heat via solar irradiation and loses the heat via infrared (IR) heat exchange with the sky. The back surface of the PV module changes the heat with flowing air in the channel via convection and with insulation via IR heat exchange. The back side of the insulation is mainly influenced by the convection of the ambient air. Because of the low heat conduction of the insulation layer, we assumed in our simulations that the temperature of the backside of the insulation layer is equal to the ambient temperature.

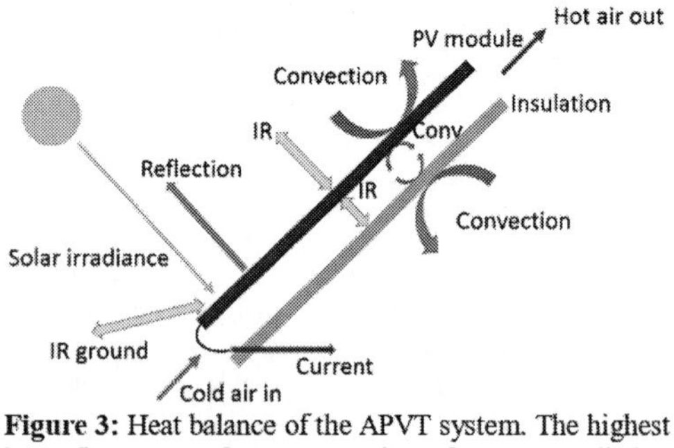

Figure 3: Heat balance of the APVT system. The highest heat fluxes are front convection, front IR radiation exchange, solar irradiance, photovoltaic current, and the convection in the channel.

3.2 Solar irradiance absorption

The majority of the solar irradiance is absorbed by the PV cells in the module. The solar irradiance absorption rate is determined by the glass transmittance τ_{gl} and PV cell absorption α_{pv}:

$$\alpha_{\mathrm{pv}}\tau_{\mathrm{gl}}G \quad (1)$$

In our simulations we assumed that the glass transmittance coefficient is 0.93 and the PV cell absorption coefficient is 0.97.

3.3 Infrared heat exchange of PV module front glass with the sky and the ground

The glass surface changes the IR heat with the cold sky according to the Stefan-Boltzmann law. The IR radiation emitted by the glass is:

$$IR_{gl} = \sigma * \epsilon_{gl} * T_{gl}^4 \quad (2)$$

where σ is the Stefan-Boltzmann constant, ϵ_{gl} is the glass emissivity, which we took 0.84, and T_{gl} is the glass absolute temperature.

The IR radiation that the module receives from the sky is:

$$IR_{sky} = \sigma * VF_{sky} * T_{sky}^4 \quad (3)$$

where VF_{sky} is the sky viewfactor and

$$T_{sky} = 0.05532 * T_{amb}^{1.5} \quad (4)$$

is the sky temperature, and T_{amb} is the ambient air temperature. The Eq.(4) is the approximation of sky temperature over all different cloud conditions. We could have incorporated more accurate three cloud level sky temperature models or made the sky temperature measurements, however, that would not have been practical because this kind of data is rarely available for other locations. Thus, we decided to work with the sky temperature approximation.

The IR radiation that the module receives from the ground is:

$$IR_{gr} = \sigma * \epsilon_{gr} * VF_{gr} * T_{gr}^4 \quad (5)$$

where ϵ_{gr} is the ground emissivity, VF_{gr} ground viewfactor and T_{gr} is the ground temperature. All the temperatures in the IR equations are in absolute units.

3.4 Infrared heat exchange of PV module back sheet with insulation

The back sheet of the PV module changes the IR radiation with the insulation layer through the 30 mm airgap. The equation that covers this phenomenon is the heat exchange between two infinite flat plates [20]:

$$IR_{i1} = \frac{\sigma * (T_{bs}+T_{i1}) * (T_{bs}^2+T_{i1}^2) * (T_{bs}-T_{i1})}{1/\epsilon_{bs} + 1/\epsilon_i - 1} \quad (6)$$

where ϵ_{bs} is PV module back sheet emissivity (white plastic 0.84), ϵ_i is insulation surface emissivity (black paint 0.95), T_{bs} is the PV module back surface absolute temperature, and T_{i1} is the absolute temperature of the surface of the insulation.

3.5 Convectional heat exchange of the module and the ambient air

The front surface of the PV module loses the heat to the environment also via convection. According to the

Newton law of cooling, the heat loss rate depends on the temperature difference between the glass surface T_{gl} and the ambient air T_{amb}:

$$CV_{gl} = h_w * (T_{gl} - T_{amb}) \quad (7)$$

where the heat transfer coefficient h_w depends on the wind speed ws:

$$h_w = 5.8 + 2.1 * ws \quad (8)$$

From the literature it is possible to find way too many different least square fittings for these linear parameters. In addition, it is rather difficult to choose which one of them is the best. Thus, the parameters in Eg.8 were derived from the best fit of our measurements of the module temperature and the wind speed while keeping the air flow in the channel zero.

3.6 Convectional heat exchange between the air and the channel walls

Similar to the front surface convection, the heat exchange between the channel walls and the forced air flow is governed by the Newton law of cooling. The convection at the back surface of the PV module is:

$$CV_{bs} = hconv * (T_{bs} - T_{air}) \quad (9)$$

and at the front surface of the insulation:

$$CV_{i1} = hconv * (T_{i1} - T_{air}) \quad (10)$$

where T_{air} is the air temperature in the channel. The heat exchange rate $hconv$ in the rectangular tube is defined as:

$$hconv = \frac{k_m}{D_h} * Nu \quad (11)$$

where k_m is the thermal conductivity of the air, which consists in the reality of the mixture of the dry air and the water vapor, Nu is the Nusselt number that gives the ratio of the convective to the conductive heat transfer, and $D_h = 2(Wd)/(W+d)$ is the hydraulic diameter of the rectangular channel with the width of W and the thickness of d.

The parameter that greatly influences the accuracy of our model is the Nusselt number Nu. The Nusselt number tells us how many times the heat transfer of the moving air increases compared to the air that stands still. The answer to this question is complex and has been studied over a century. The Nusselt number depends on the flow type (laminar or turbulent). In order to detect the nature of the hydrodynamic flow we have to bring in the Reynold's number, which is the ratio of the flow medium inertial forces to the viscous forces:

$$R_e = \rho_m * v_{\text{air}} * D_h / \mu_m \quad (12)$$

where ρ_m is the density of the air in the channel, μ_m is the dynamic viscosity of the air and v_{air} is the flow speed. The rule of thumb is that below the $R_e = 3000$ the flow is laminar and above this value it is turbulent. In our specific case this limit corresponds to the air flow speed in the tube around 1.05 m/s.

For the turbulent flow $R_e > 3000$ we used the Nusselt number [21]:

$$Nu_t = 0.0158 R_e^{0.8} + (0.00181 R_e + 2.92) e^{-0.03795 \frac{z}{D_h}} \quad (13)$$

where z is the distance from the entrance. Accordingly, the heat transfer rate is faster at the beginning of the channel and decreases towards the end of the channel (see Fig. 3).

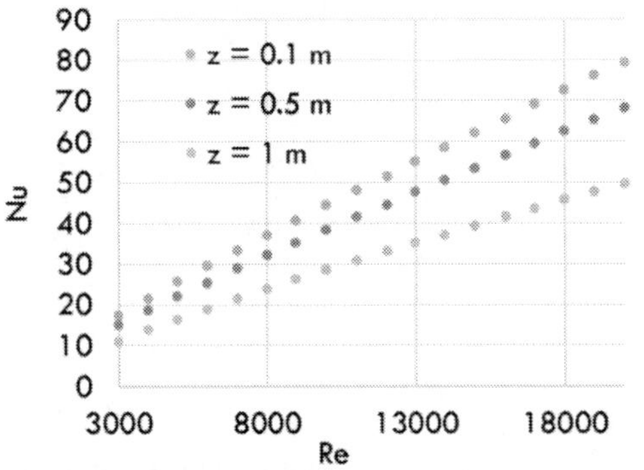

Figure 3: The dependence of the Nusselt number on the Reynold's number and the distance from the entrance. The heat transfer rate increases with the increasing flow speed and decreases towards the end of the air channel with the increasing distance z from the channel entrance.

In the case of laminar flow at low flow rates that we assume to exist when $R_e < 3000$, we also have to take into account the effect of the free convection in the channel. Thus, the Nusselt number during the laminar flow is the sum of the free and forced Nusselt numbers:

$$Nu_{lam} = Nu_{free} + Nu_{forced} \quad (14)$$

The Nusselt number for the laminar forced flow in the rectangular ducts has been analyzed in Ref.[22]:

$$Nu_{forced} = 7.55 + \frac{0.024 x^{-1.14} (0.0179 Pr^{0.17} x^{-0.64} - 0.14)}{(1 + 0.0358 Pr^{0.17} x^{-0.64})^2} \quad (15)$$

where $x = z/(2 Pr R_e d)$, Pr is the Prandtl number, and d is the thickness of the air channel. The Prandtl number is the ratio of the viscous diffusion rate to the thermal diffusion rate and can be calculates as $Pr = c_m \mu_m / k_m$, where c_m, μ_m, and k_m are the specific heat, dynamic viscosity and the thermal conductivity of the air in the channel.

For the modeling of the free convection part of the Nusselt number we used the approximation of the vertical isothermal channel, where the average Nusselt number can be expressed as [23]:

$$Nu_{free} = \left(\frac{576}{(Ra * \frac{d}{L})^2} + \frac{2.873}{(Ra * \frac{d}{L})^{0.5}} \right)^{-0.5} \quad (16)$$

where L is the length of the channel and the Ra is the Rayleigh number $Ra = Gr Pr$. The Grashof number Gr is the ratio of the buoyancy forces to the viscous forces and can be calculated as:

$$Gr = \frac{g \beta (T_{bs} - T_{air}) * d^3}{v_m} \quad (17)$$

where $\beta = 1/T_{air}$ is the air thermal expansion coefficient, $v_m = \mu_m/\rho_m$ is the air kinematic viscosity, and μ_m is the air dynamic viscosity.

3.7 Definition of the geometry and the set of equations

In order to model the heat transfer through the different layers of the structure finite slab method was used. The Fig.4 shows the structure of the layers as well as the temperature nodes. With one exemption, each component of the structure forms one layer in our mathematical model. To allow some nonlinear temperature profile in the insulation layer, we divided it into two slabs. For the simplicity we assumed that the backside of the insulation is at the temperature of the ambient air. The assumption is based on the low thermal conductance and relatively high thickness (70 mm) of the insulation layer. The CD defines the thermal conductance thought the layers and can be calculated via general form:

$$CD = \frac{k(T_2 - T_1)}{x} \quad (18)$$

where T_1 and T_2 are the temperature nodes, x is the slab thickness, and k is the material thermal conductivity.

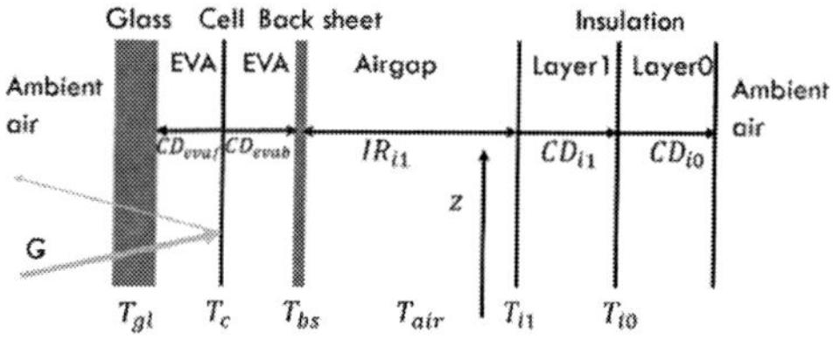

Figure 4: The division of the test structure into the modeling slabs and temperature nodes. The labels CD with arrows denote the conductances with the positive heat flow direction.

First, we simplified the governing thermal equations by assuming that the cells, glass, and the back sheet have uniform temperature over their cross section. Second, the layer of the cells consist of the cells and half of the EVA from each side. And third, we assumed that the temperature distribution over the PV module sidewise is uniform. The spatial temperature change is allowed through the cross section of the structure and along the z axis with the direction of the air flow.

As a result we get six differential equations that describe the heat balance of each slab.

Airgap:

$$\frac{\partial T_{air}}{\partial z} = \frac{hconv}{v_{air} d \rho_m c_m}(T_{bs} + T_{i1} - 2T_{air}) \quad (19)$$

Insulation layer 0:

$$\frac{m_{i0}c_i}{2}\frac{\partial T_{i0}}{\partial t} = CD_{i1} - CD_{i0} \quad (20)$$

Insulation layer 1:

$$\frac{m_{i1}c_i}{2}\left(\frac{\partial T_{i0}}{\partial t} + \frac{\partial T_{i1}}{\partial t}\right) = -CD_{i0} + IR_{i1} - CV_{i1} \quad (21)$$

Back sheet:

$$\frac{m_{bs}c_{bs} + m_{evab}c_{eva}}{2}\frac{\partial T_{bs}}{\partial t} = -IR_{i1} + CD_{evab} - CV_{bs} \quad (22)$$

Cell:

$$\frac{2m_c c_c + m_{evab}c_{eva} + m_{evaf}c_{eva}}{2}\frac{\partial T_c}{\partial t} = -CD_{evab} + \alpha_{pv}\tau_{gl}G - PV + CD_{evaf} \quad (23)$$

The symbol PV denotes here the power that is being drawn away from the system by the photovoltaic current.

Front glass:

$$\frac{m_{gl}c_{gl} + m_{evaf}c_{eva}}{2}\frac{\partial T_{gl}}{\partial t} = -CD_{evaf} + \alpha_{gl}G - CV_{gl} - IR_{gl} + IR_{sky} + IR_{gr} \quad (24)$$

In addition, the initial conditions are defined as:

$T_{i0}^{t=0}$=C1, $T_{i1}^{t=0}$=C2, $T_{bs}^{t=0}$=C3, $T_{air}^{t=0}$=C4, $T_{tc}^{t=0}$=C5, $T_{gl}^{t=0}$=C6 (25)

where C1-C6 are constants. The Table I shows the values of the physical constants that we used in the simulations.

Table I: Definition of the physical constants that we used in the simulations. All the parameters were taken from the manufacturers' datasheets or the material databases.

Constant	Meaning	Value
α_{pv}	absorption of cells	0.97
τ_{gl}	transmittance of glass	0.93
α_{gl}	absorption of glass	0.01
ε_{gl}	emissivity of glass	0.84
ε_i	emissivity of insulation	0.95
ε_{bs}	emissivity of back sheet	0.84
	thermal conductivity ($W \cdot m^{-1} \cdot K^{-1}$)	
k_i	insulation	1500
k_{bs}	back sheet	1223
k_{eva}	EVA	2400
k_c	cell	710
k_{gl}	glass	767
	material density (kg/m^3)	
ρ_i	insulation	30
ρ_{bs}	back sheet	1420
ρ_{eva}	EVA	950
ρ_c	cell	2330
ρ_{gl}	glass	2450
	specific heat capacity (J·Kg/K)	
c_i	insulation	1500
c_{bs}	back sheet	1223
c_{eva}	EVA	2400
c_c	cell	710
c_{gl}	glass	767
	layer thickness (mm)	
x_{i0}	insulation layer 0	35
x_{i1}	insulation layer 1	35
x_c	cell	0.18
x_{evaf}	EVA front	0.45
x_{evab}	EVA back	0.45
x_{bs}	back sheet	0.30
x_{gl}	glass	3.2

In addition to the physical constants in the Table I, we had to take into account the real parameters of the air, which consists of the mixture of the dry air and the water vapor. Thus, the parameters of the air depend on the atmospheric pressure, relative humidity, and the temperature. These environmental parameters are constantly changing. Accordingly, the physical parameters of the air in our model like density, dynamic viscosity, thermal conductance, thermal capacity, and kinematic viscosity are also constantly changing. The air parameters are not only changing in time, but also along the spatial axis z as the air heats up in the channel. The air physical parameter estimation in this study was done according to the Ref. [24].

3.8 Solving the problem

The set of differential equations (19-25) were solved numerically using the Runge-Kutta method in the simulation software Mathematica.

4 RESULTS AND DISCUSSION

Below we present the measurement and the simulation results of the experiment that was made on the 25th of February 2015. As can be seen from the Fig.5, the sky conditions during the experiment were partially cloudy. The ambient temperature was steady ranging from +4°C to +7°C. Also, the wind speed was rather steady ranging from 1 to 4 m/s.

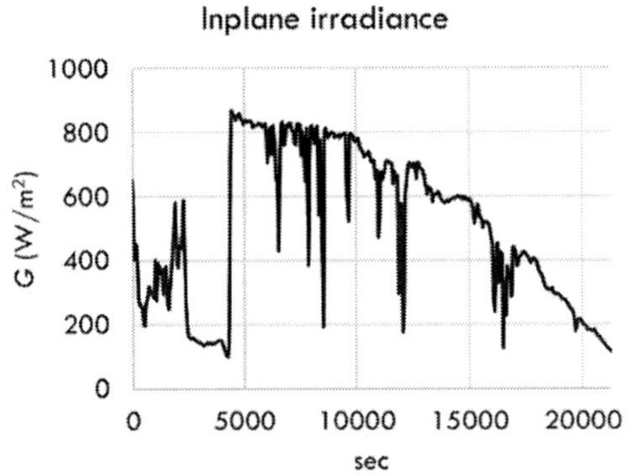

Figure 5: The in-plane irradiance level during the experiment. The experiment was started after the irradiance had reached over 800 W/m^2.

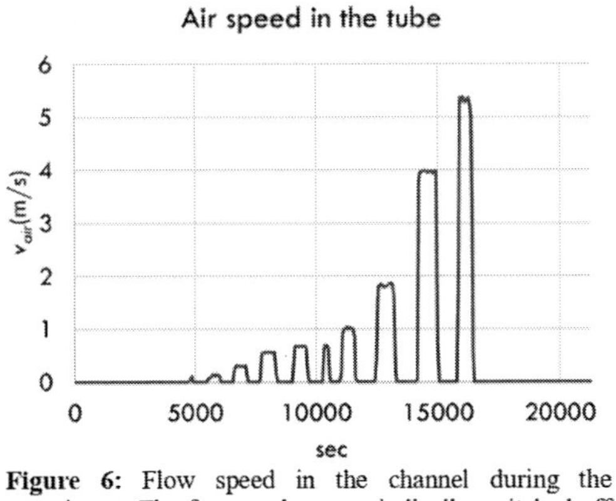

Figure 6: Flow speed in the channel during the experiment. The fan speed was periodically switched off and on with increasing intensity.

In order to introduce some dynamics to the experiment, we periodically changed the air flow speed in the channel. First we switched on the air flow with the speed of 0.1 m/s for approximately 10 minutes. Then we switched off the fan in order to let the system gain some energy from the sun. The resting period was also about 10 minutes. These cycles were repeated periodically with the increasing fan speed. The detailed air flow profile can be seen on the Fig 6.

The measurement results of the module temperature along with the simulations are seen on the Fig.7 and Fig.8. On the Fig.7 the measured data points represent the average temperature of the two temperature sensors that were attached to the lower part of the module and on the Fig.8 they represent the average of the two temperature sensors that were attached to the upper part of the back side of the module. The detailed positions of the sensors are shown on the Fig.1.

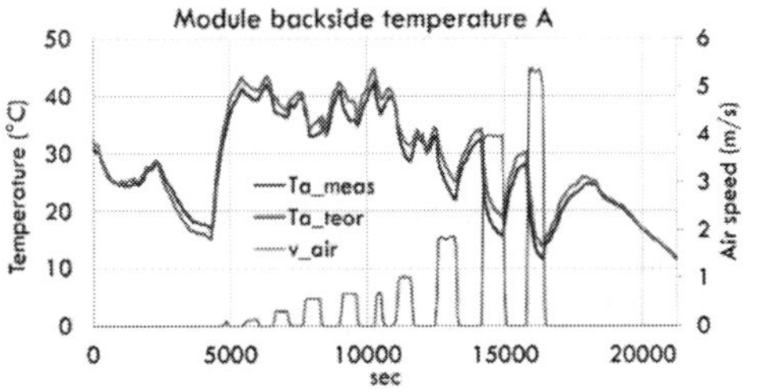

Figure 7: The measurement results of the temperature of the **lower part** of the PV module compared with the simulated values. The red line represents the measured and the green theoretical values. The blue line represents the air speed in the channel.

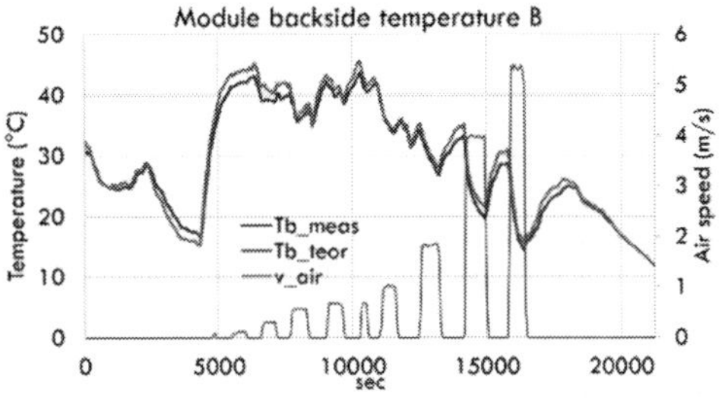

Figure 8: The measurement results of the temperature of the **upper part** of the PV module compared with the simulated values. The red line represents the measured and the green theoretical values. The blue line represents the air speed in the channel.

As it can be seen from the figures above, the model predicts the module temperature well. However, the model tends to be more accurate in the upper part of the module than in the lower part. This can be explained with the fact that the nature of the air flow in the beginning of the channel is not fully known to us. Thus, at high flow speeds, the model underestimates the heat exchange rate at the entrance of the channel. Nevertheless, as we move upstream the model gets more accurate and the temperature of the module at the end of the channel can be predicted with higher accuracy. When it comes to the applications the air channels on top of the roof will be most likely longer than one module, thus, we believe that in real applications the entrance effect can be neglected.

The Fig.9 shows the accuracy of the outlet air temperature prediction of our model.

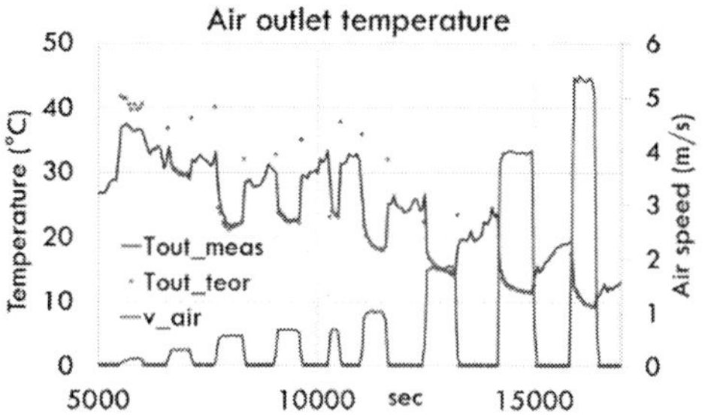

Figure 9: The measurement and the simulation results of the air temperature at the channel outlet. The red line represents the measured values, the green dots represent the simulated points and the blue line shows the air speed in the tube.

As it can be seen, the model works well when the fan is on and the air speed in the system is well defined. In case the fan is in the acceleration or deceleration phase and the flow regime in the channel is not fully developed, it is rather difficult to predict what the air temperature at the exit is. We believe that the outliers are caused by the fact that we cannot measure accurately the speed of undeveloped flow. In addition, the model makes biggest mistakes when the air speed in the tube is less than 0.2 m/s. At this level, the natural convection starts to play bigger role in the air movement in the channel. Also, it becomes more difficult to perform accurate flow speed measurements.

5 CONCLUSIONS

We have created a reliable model for predicting the thermal properties of air type photovoltaic-thermal systems. The model responds well to the rapid changes of the air flow in the heat exchange channel. The model is more accurate at the end of the air channel than in the beginning. Also, it works best at the air flow speed that exceed 0.2 m/s which corresponds approximately to 6 l/s. In our experiment the length of the channel was relatively short 1.6 m. However, the model predicted well the temperature of the module and the air at the exit of the channel. Thus, we can be confident that this model can be used for validating and optimizing the possible applications of air type photovoltaic-thermal systems. The future work involves testing of the model at larger scale and performing the feasibility studies of real air type photovoltaic-thermal applications.

6 ACKNOWLEDGEMENTS

This work was supported by institutional research funding IUT (IUT19-28) of the Estonian Ministry of Education and Research, and by the European Union through the European Regional Development Fund, projects TK141 and AR12044.

7 REFERENCES

[1] Building Integrated Photovoltaic Technologies and Markets 2015-2022, n-tech Research 2015

[2] Adel A. Hegazy, Energy Conversion & Management 41 (2000) 861-881

[3] Arvind Tiwari et al., Solar Energy Materials & Solar Cells 90 (2006) 175–189

[4] Arvind Tiwari et al., Solar Energy Materials & Solar Cells 91 (2007) 17–28

[5] Anand S. Joshi et al., Renewable Energy 32 (2007) 2223–2241

[6] S.C. Solanki et al., Applied Energy 86 (2009) 2421–2428

[7] Swapnil Dubey et al., Applied Energy 86 (2009) 697–705

[8] Luis Miguel Candanedo Ibarra et al., Proceedings International High Performance Buildings Conference (2010)

[9] Teo Han Guan, B.Sc.Eng. thesis "Photovoltaic Thermal (PV/T) System: Effect of Active Cooling" (2010)

[10] F. Sarhaddi et al., Energy and Buildings 42 (2010) 2184–2199

[11] Sanjay Agrawal et al., International Journal of Engineering, Science and Technology Vol. 3, No. 1, 2011, pp. 244-254

[12] Rakesh Kumar et al., Applied Thermal Engineering 31 (2011) 1402-1410

[13] R. K. Koech et al., International Journal of Scientific & Engineering Research, Volume 3, Issue 10, October-20

[14] Mohsen Mahdavi Adeli et al., Journal of Mechanical Engineering 58(2012)5, 309-318

[15] A. B. Kasaeian et al., IJE TRANSACTIONS B: Application Vol. 26, No. 8, (August 2013) 913-926

[16] Jin-Hee Kim et al., Energy Procedia 48 (2014) 888 – 894

[17] K.E. Amori, M.A. Abd-AlRaheem, Renewable Energy 63 (2014) 402-414

[18] S.A. Brideau, M.R. Collins, Solar Energy 102 (2014) 234–246

[19] G. Notton et al., Applied Thermal Engineering 25 (2005) 2854–2877

[20] S.Goodman, Journal of Research of the National Bureau of Standards 58, (1957) 2732

[21] H.M.Tan et al., Solar Energy 12 (1969) 513-516

[22] Shah, R.K. and Bhatti, Laminar Convective Heat Transfer in Ducts, Handbook of Single-Phase Convective Heat Transfer, Kakaç, S., Yüncü H. and Hijikata K. (eds.), John Wiley & Sons, USA, 1987.

[23] Deborah A. Kaminski, Michael K. Jensen, Introduction to Thermal and Fluids Engineering (Updated Edition), Wiley 2011

[24] P.T. Tsilingiris, Energy Conversion and Management 49 (2008) 1098–1110

THERMAL ANALYSIS OF A BIPV/T PROTOTYPE FOR FODDER DRYING

Ya Brigitte ASSOA[1], Simon BODDAERT[2], Philippe MALBRANCHE[1]
[1] CEA, LITEN, F-73375 Le Bourget du Lac, France.
50, avenue du Lac Léman 73377 LE BOURGET DU LAC
[2] CSTB / Renewable Energy Division / PV Innovation, Sophia Antipolis, France
290 route des lucioles BP 209 – F 06904 Sophia Antipolis CEDEX

ABSTRACT: The thermal analysis of a building integrated photovoltaic/thermal prototype (BIPV/T) suitable for fodder drying is presented. This component consists in photovoltaic modules inserted in a metal absorber.
Similar test models were installed at INES site at Le Bourget du Lac and at CSTB site at Sophia Antipolis and compared. The experimental setup in both sites was composed of two test benches differentiated by the insulation of the backside of one of the absorbers. A PV field was used as reference. The insulation of the test bench leads to the heating of PV modules of nearly 10°C in days considered.
The overall thermal conductance of each test bench was determined using a linear equation representing the impact of solar radiation on the heating of PV modules. This parameter decreases with the level of PV modules cooling. Also, this linear equation could be used for a first prediction of PV modules temperature in various building integration configurations. As further studies, the BIPV/T component will be analyzed on longer periods on the two sites. Then, the prototype will be integrated into a 35m² test bench of fodder drying installation at INES site in order to optimize its design (air flow rate, ducts size…).
Keywords: Building Integrated PV (BIPV), Rooftop, Thermal Performance

1 INTRODUCTION

This study is part of a research project called "Rezospring" consisting in developing a prototype of solar Photovoltaic/Thermal (PV/T) hybrid air collector suitable for drying of fodder and industrial products and for heating [1]. This component is designed to solve some solar PV/T collectors [2] [3] [4] weak points such as the unstuck of PV modules from their metal support structure. Thus, it consists in photovoltaic modules inserted and not stuck in a metal panel comprising an insulated air layer at the underside. Identical test benches were installed at INES site at Le Bourget du Lac and at CSTB site at Sophia Antipolis in order to take into account different climates. In this paper, the thermal results of these experimental studies are presented and analyzed. The impact of the insulation of the backside of the metal absorber is evaluated. Finally, a simplified linear correlation between PV modules temperature, exterior ambient air temperature and solar incident radiation is determined.

For each test bench, the value of overall thermal conductance for each test bench is determined and analyzed.

2 THE EXPERIMENTAL SETUP

2.1 Description of the solar PV/T collector

Since the thermal production of solar PV/T collectors is not always concomitant with energy needs (in building for example), the applications of this kind of systems are still limited. The advantage of the fodder drying is that this application is realized during summertime, at the moment when the PV modules most need to be cooled by extracting heat losses.

The prototype of solar PV/T hybrid air collector studied is composed of photovoltaic modules inserted in a ribbed metal absorber. (see. Fig. 1)

2.2 The monitoring system

At INES site (Le Bourget du Lac), the experimental setup comprised a PV field used as reference and two test benches differentiated by the presence of an insulation layer at the backside of one of the absorbers. The test benches were grid connected. A pyranometer and an anemometer were mounted in order to measure values of solar radiation in the collector plan and of wind velocity and direction. The ambient temperature was obtained with a sheltered PT100 sensor. Temperature sensors permitted to measure in different points the temperature at the backside of PV modules, of the metal absorber and of the insulation layer. Also, electrical transducers were used to measure electrical data (AC and DC). A comparative experimental campaign was performed with a 1 min time step using a data logger.

Figure 1: Test benches at INES site (Le Bourget du Lac)

At CSTB site (Sophia Antipolis), two similar tests benches were mounted. Bench 2 (see square on the left side on Fig. 2) represents the solar component studied mounted on part of a real roof composed of a wooden frame and of a double insulation layer. Bench 3 (see square on the right side on Fig. 2) is an equivalent structure without insulation. Figure 3 presents the Bench 1 which is used as reference.

Figure 2: Bench 2 (insulated) and Bench 3 (non-insulated) at CSTB site (Sophia Antipolis)

Figure 3: Bench 1 on a mounting structure at CSTB site

Figure 4: Monitoring system at CSTB site

Each bench is equipped with an independent weather station, composed of a pyranometer in the bench plane, an ultrasonic anemometer for wind velocity and direction and an ambient temperature measurement sensor. Each photovoltaic module is equipped with at least one temperature sensor (see Fig. 4). One module of each array is equipped of six temperature sensors in order to validate the mean temperature of each PV module. A data control unit permitted collecting data and realizing first information treatment. Specific current and voltage sensors permitted monitoring AC and DC production.

3 ANALYSIS OF RESULTS

An analysis of the experimental results was performed in order to evaluate first the impact of the insulation of the backside of the solar component studied. Then, the impact of the solar radiation on PV modules heating was estimated by determining the overall thermal conductance.

3.1 Experimental parametric study

The thermal behavior of the three test benches at INES site were analyzed on various days. Figures 5 to 7 presents the results obtained on a sunny day in summer. The maximal solar radiation was of nearly 1050W/m^2 with wind velocities lower than 1m/s.

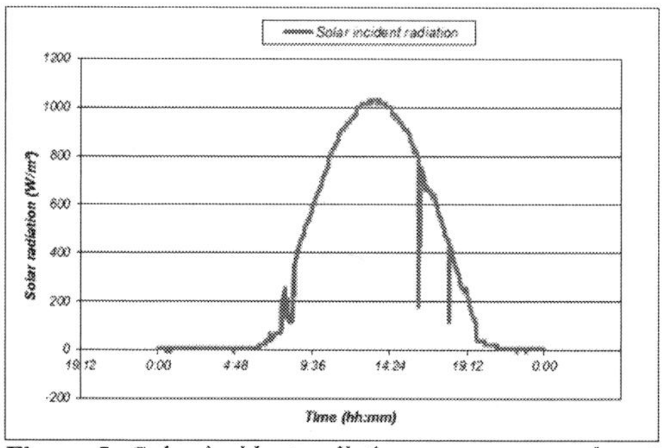

Figure 5: Solar incident radiation on a summer day at INES site

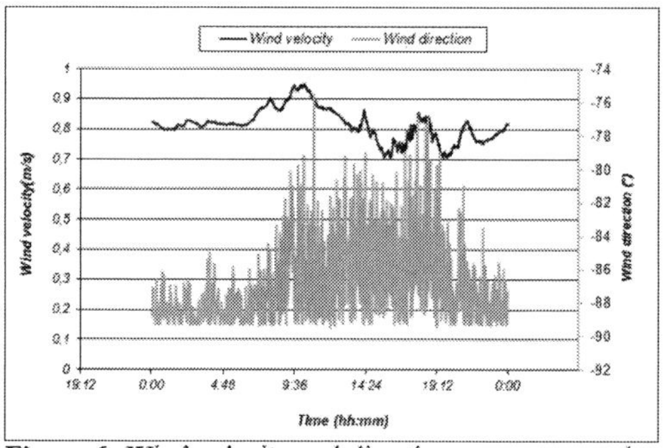

Figure 6: Wind velocity and direction on a summer day at INES site

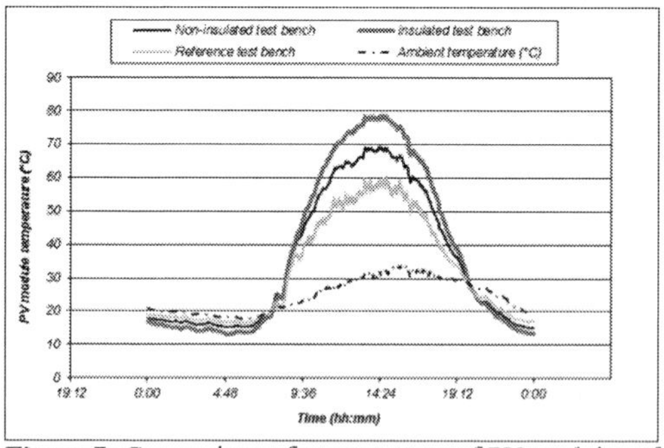

Figure 7: Comparison of temperatures of PV modules of the non-insulated test bench, of the insulated test bench and of the reference PV field on a summer day at INES site

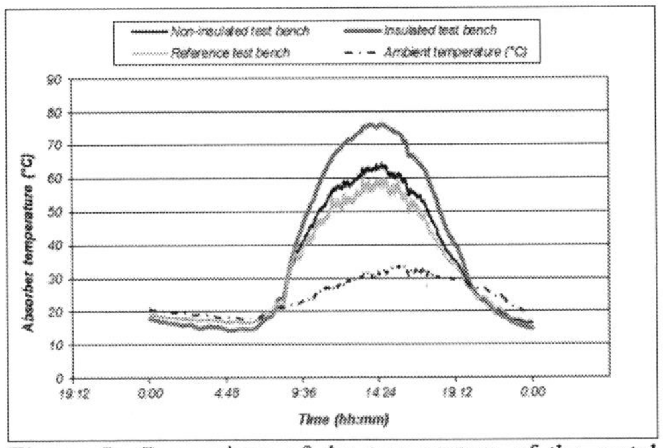

Figure 8: Comparison of the temperature of the metal absorber of the non-insulated test bench, of the insulated test bench and of the reference PV modules during on one day in summer at INES site

The comparison of the thermal behavior of the three test benches shows that the insulation of the backside of the metal absorber leads to an increase of PV modules temperature of nearly 10°C on this sunny day. (See Fig. 7)

The observation of Figure 7 and Figure 8 shows that the curves related to the metal absorber and to PV modules are very close in the two configurations (insulated and non-insulated). The differences could be due to the existence of some air layers between the metal absorber and PV modules.

At CSTB site, a similar comparison was performed on a day in cold period on benches 1 to 3. Figure 9 and Figure 10 confirm the results obtained at INES site on the summer day. Thus, the insulation of Bench 3 leads to the heating of PV modules of nearly 10°C.

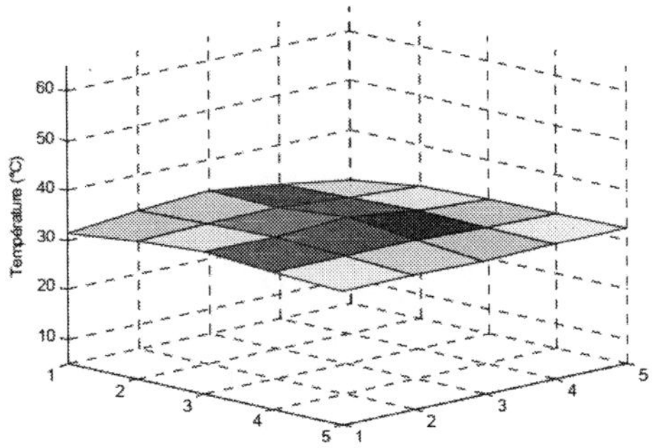

Figure 9: Thermal mapping of the Bench 3 (non-insulated) at CSTB site

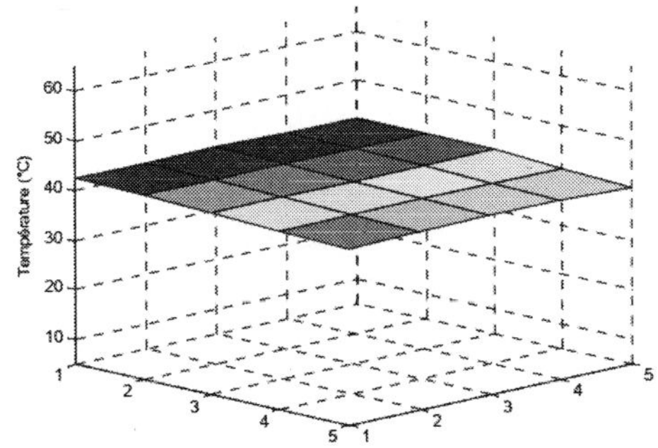

Figure 10: Thermal mapping of Bench 2 (insulated) at CSTB site

3.2 Analysis of PV modules heating

For the three test benches, the impact of solar radiation on the difference of temperatures between PV modules and ambient temperature is evaluated by determining the overall thermal conductance noted k (°C.m²/W) and given by Equ. 1. (a =0 in [5])

$$T_{PV} - T_{amb} = k \cdot G + a \tag{1}$$

With T_{PV} the PV module temperature (°C), T_{amb} the ambient temperature (°C) and a a parameter considering among all, the wind effect.

At INES site, the value of the overall thermal conductance was determined on the summer day, by neglecting parameter a (see Fig. 11).

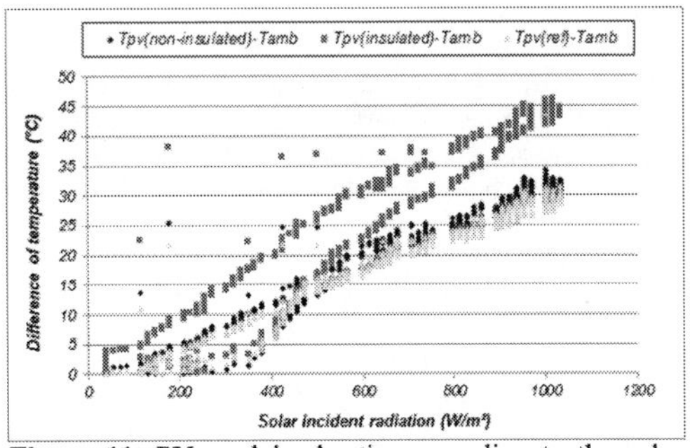

Figure 11: PV modules heating according to the solar incident radiation during a summer day at INES site

At Sophia Antipolis, the analysis was realized during a cold week considering the parameter a. (see Fig. 12)

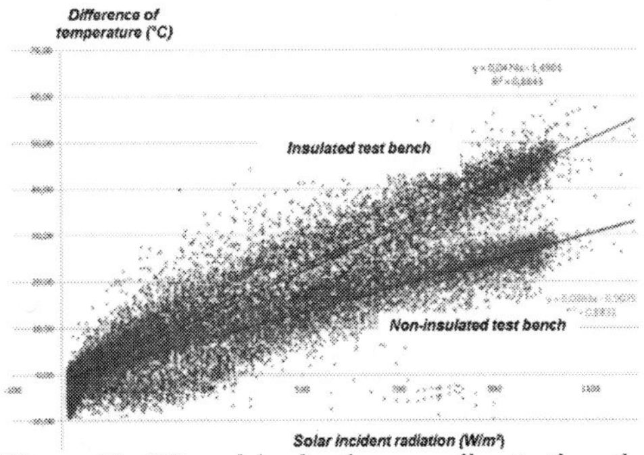

Figure 12: PV modules heating according to the solar incident radiation during a cold week at CSTB site

Table I presents the values of overall thermal conductance obtained on the both site.

Table I: Overall thermal conductance k (C.m²/W) (1: at Le Bourget du Lac: (a=0) on a day; 2: at Sophia Antipolis (a estimated) on a week period)

Configuration	1	2
Reference test bench	0.0287	-
Non-insulated test bench	0.0305	0.0283(a=-1.4901)
Insulated test bench	0.0433	0.0474(a=-0.9075)

The comparison of k values in Table I shows that the insulation of the solar component has an impact on the overall thermal conductance which increases, as expected. Thus, this parameter decreases with the level of PV modules cooling.

The overall thermal conductance k values obtained are coherent with the range presented by Skoplaki et al. in 2008 (between 0.02 and 0.04°C. m^2/W) [5].

The linear equation (see Equ. 1) could be used for a first prediction of PV modules temperature according to solar radiation and ambient temperature in various building integration configurations and to estimate the level of ventilation of a PV system (cooling).

4 CONCLUSION

A thermal analysis of two configurations of a solar Photovoltaic/Thermal (PV/T) hybrid air collector suitable for drying applications was performed at Le Bourget du Lac and at Sophia Antipolis. The test benches were differentiated by the insulation or not of the backside of the metal absorber and were compared to a rack-mounted PV field used as reference. On both sites, on a day or a week period, the results show that the insulation of the test bench leads to the rise of PV modules temperature of nearly 10°C.

Moreover, the overall thermal conductance for the three test benches was determined on a sunny day using a simplified linear correlation relating the difference between the PV modules temperature and the ambient air temperature and the solar incident radiation. The analysis of the value obtained has shown that the insulation of the metal absorber has an impact on the overall thermal conductance which increases, as expected. This linear equation could be used for a first prediction of PV modules thermal behaviour in various building integration configurations. It also permits to evaluate the level of ventilation of PV modules based on the overall thermal conductance.

As further studies, the prototype will be analyzed after integration into a 35m² -test bench of fodder drying installation at INES site in order to optimize its design (air flow rate, ducts size…). Moreover, the thermal analysis of the prototype will be performed on longer periods.

5 ACKNOWLEGMENT

Authors are grateful to SOLAREZO Company, CEA and CSTB teams which have work on the Rezospring project, namely, F. Chervet, F. Sauzedde and B. Boillot.

6 REFERENCES

[1] Y. B. Assoa, F. Sauzedde, B. Boillot, S. Boddaert, F. Chervet, 28th European Photovoltaic Solar Energy Conference (2013) 3986 - 3988
[2] J. C. Hollick, Renewable Energy 15 (1998) 195-200.
[3] G. Vokas, N. Christandonis, F. Skittides, Solar Energy 80 (2006) 607-615.
[4] A. Ibrahim, M. Y. Othman, M. H. Ruslan et al, Renewable and Sustainable Energy Reviews 15 (2011) 352–365.
[5] E. Skoplaki, A. G. Boudouvis, J. A. Palyvos, NTUA, (2008).

EXPERIMENTAL ANALYSIS OF THE PERFORMANCE OF FAÇADE-INTEGRATED BIPV IN DIFFERENT CONFIGURATIONS

Giel Van den Broeck, Wout Parys, Hans Goverde, Stijn Van der Putten, Jef Poortmans, Johan Driesen, Kris Baert, Dirk Saelens
EnergyVille
Thor Park 8300
3600 Genk, Belgium

ABSTRACT: The purpose of this paper is to quantitatively compare different façade-integrated BIPV configurations that relate to curtain walls and comprise both cold, ventilated façades and warm, unventilated façades and a different number of glazing sheets to improve aesthetics. Therefore, this paper introduces an experimental set-up and analyses the experimental results that were obtained over one year between May 2015-2016. Both temperatures and IV characteristics were continuously logged every 15 seconds. The different module configurations experience increased operating temperatures, optical losses and partial shading due to building-integration and are compared to a reference PV module over the observed time period in order to quantify the contribution of the different loss factors. The results show energy-yield losses, ranging from -11% to -27% as compared to the reference PV module.

1 INTRODUCTION

Building Integrated PV (BIPV) systems consist of solar photovoltaic cells and modules that are integrated in the building envelope as part of the building structure, thus replacing conventional building materials [1]. BIPV is now being proposed as an economically viable solution to the increasing demand for renewable electricity generation, since the added cost of photovoltaic cells to the overall building component's cost results in conceivable payback times [2]. This evolution represents a significant paradigm shift, as the integration of PV in a building component requires a broader assessment of the resulting component than was needed before for either a PV panel or a building component separately.

BIPV are classified into two main categories, being *roofing BIPV* (r-BIPV) and *façade-integrated BIPV* (f-BIPV) [2], [3]. Although r-BIPV currently represents the highest share of commercially available BIPV products [3], [4], studies highlight the economic potential of f-BIPV because of increased space availability for instance in office buildings with a significant façade relative to roof surface and where part of the roof surface is occupied by HVAC systems [5]. Further investment cost reductions are expected to offset energy-yield reductions due to sub-optimal f-BIPV inclination and orientation. Furthermore, differently oriented f-BIPV systems result in a more constant or load matching electricity generation profile throughout the day [6]. That techno-economic potential consequently motivates the further investigation of f-BIPV.

Many buildings nowadays, especially non-residential buildings, are constructed as a concrete frame structure applying non-load bearing lightweight walls, so-called curtain walls. These typically consist of aluminum frame floor-to-floor modular components containing both a transparent – window – part and an opaque part, covering the building's structure. These façade components are sometimes configured as a *shadow box*, where a glass panel covers the entire component and a cavity is left behind the external glazing and the insulation panel constituting the opaque part. In this way, the outside appearance of the building is more uniform and aesthetically pleasing. Curtain walls offer promising advantages for high-volume production of BIPV components, since their dimensions are partially standardized.

The purpose of this work is to quantitatively compare different f-BIPV configurations that relate to curtain walls and comprise both cold, ventilated façades and warm, unventilated façades and a different number of glazing sheets. The comparison is based upon an experimental measurement set-up, deployed in Leuven, Belgium where IV characteristics and temperatures of the BIPV components are acquired every 15 seconds to study both the electrical and building physical aspects. Although the energy-yield reduction due to elevated temperatures and non-uniform irradiation is widely understood, the authors have identified a need for a *quantitative* assessment of those phenomena in outdoor conditions.

This paper is structured as follows. The first section describes the experimental set-up, including the different BIPV component configurations and the measurement equipment, followed by the second and third section displaying and interpreting the measurement results.

2 DESCRIPTION OF THE EXPERIMENTAL SET-UP

An experimental BIPV test set-up has been configured at the south-west facing façade of a test building, depicted in Figure 1 [7].

Figure 1 Test building with a detailed view of the BIPV set-up on the south-west façade

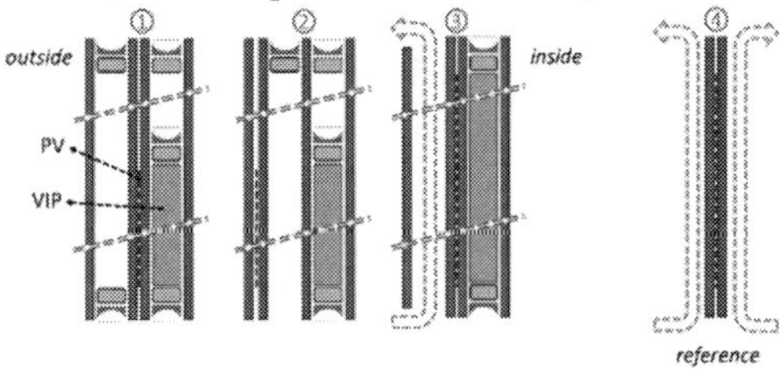

Figure 2 Three f-BIPV triple glazing configurations and a building-applied reference module

The set-up compares 3 f-BIPV configurations (configuration 1 to 3 in Figure 2) and a glass-glass PV module serving as a reference (configuration 4 in Figure 1). All BIPV components are scaled down mock-ups of triple glazing curtain wall façade elements.

Configuration 1 is built up to function as a shadow box, where the PV cells are placed in the middle glass pane in order to improve aesthetics in front of a low thermally conductive vacuum insulation panel (VIP). That is the least favorable position, as the absorbed heat cannot easily be removed. In configuration 2, the PV cells are placed in the front pane, losing the shadow box effect but increasing the irradiance as less light is reflected and absorbed by intermediate glass panes and allowing front natural ventilation to cool down the PV cells. Finally, configuration 3 is a variant of configuration 1, with a naturally ventilated front cavity in order to reduce cell temperatures. The glazing panes have a thickness of 4 mm and thermal conductivity of 1 W/mK.

The experimental set-up contains two modules (A and B) of each configuration intended to provide verifiable results. Each module contains 6 PV cells connected in series, except for module 3 that consists out of 12 cells. The cell parameters are provided in Table 1. All modules are mounted in the vertical plane, including one conventional vertical PV module that serves as a reference.

Thermal and electrical measurements are performed in the experimental set-up to study both thermal aspects from a building-physics point-of-view and the IV curves from an electrical point-of-view. Temperature and heat flux sensors are applied as depicted in Figure 3 to evaluate thermal heat flows and the module's insulation characteristics. Furthermore, the electrical measurement set-up, acquires the IV characteristics of the different modules every 15 seconds [8]. Acquiring IV curves instead of solely output power measurements enables to investigate the electrical characteristics like temperature-related open-circuit voltage and irradiance-related short-circuit current more in detail.

Table 1 Solar cell parameters

Quantity	Symbol	Value
Ideality factor	n	0.840
Short-circuit current	$I_{sc,0}$	8.655 A
Open-circuit voltage	$V_{oc,0}$	0.626 V
Maximum power	$P_{mpp,0}$	4.349 W
Voltage temperature coefficient	β_{Voc}	-2.040 mV/°C

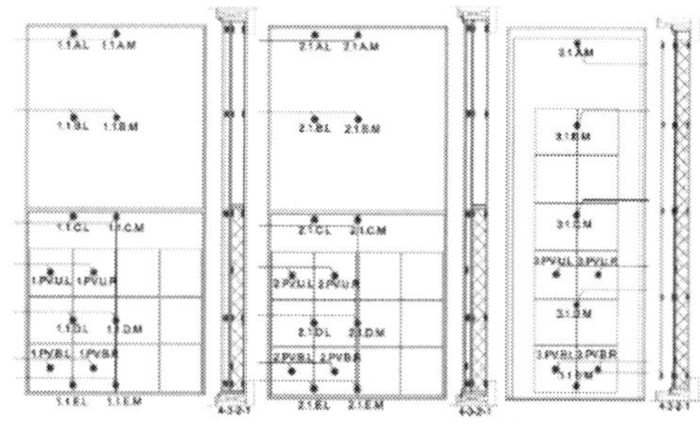

Figure 3 Location (red dots) of the thermocouples in module 1, 2 and 3

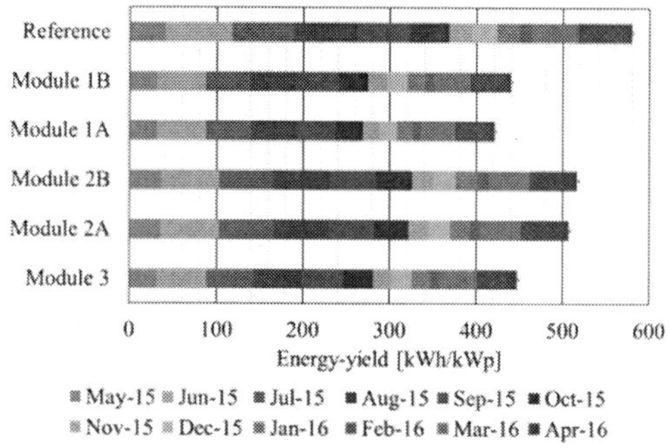

Figure 4 E-yield of the different BIPV modules*

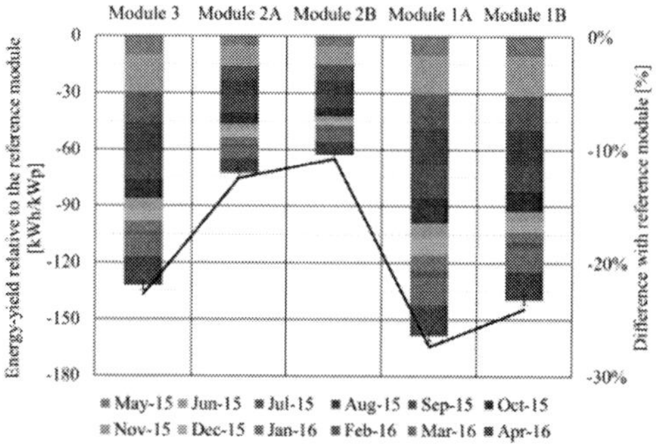

Figure 5 E-yield relative to the reference module*

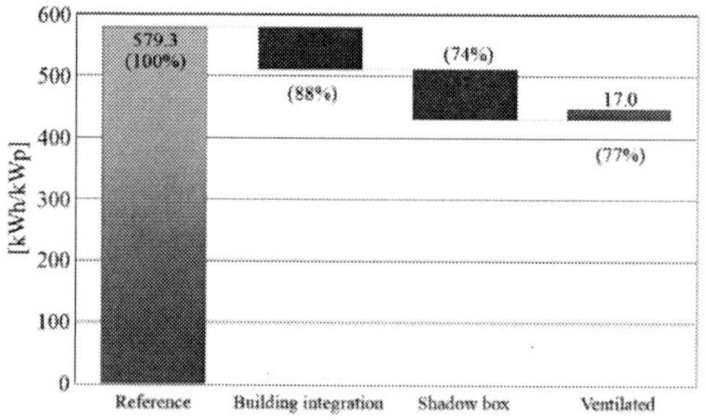

Figure 6 Comparison of the loss contributing factors for May 2015-2016*

3 ENERGY-YIELD COMPARISON

Figure 4 depicts the energy-yield measured for the period May 2015-2016 for different modules and the differences with respect to the reference module are enlarged in Figure 5. As a result of sub-optimal operating conditions, being increased operating temperatures due to less heat removal and decreased irradiance due to additional glazing panes in front of the PV cells, all configurations generate 11% to 27% less electricity as compared to the reference module.

By comparing the different module configurations, the phenomena affecting the energy-yield can be subdivided in four categories which are discussed subsequently: the effect of building-integration, the effect of the shadow box, the effect of ventilation and the effect of partial shading. The first three effects are depicted in Figure 6, showing a loss breakdown analysis. The influence of partial shading will be discussed separately.

* Results from May-15 and Feb-16 are incomplete due to data loss

3.1 Effect of building integration:
configuration 2 versus 4

As a result of the integration of PV cells in the building envelope, the PV cells experience less cooling and as a consequence higher operating temperatures. That effect is quantified by comparing configuration 2 to the reference module 4. As depicted in Figure 5, the energy-yield of module 2A is reduced by 13% as compared to the reference module.

3.2 Effect of the shadow box: configuration 1 versus 2

As compared to configuration 2, the PV cells in configuration 1 are integrated behind the shadow box (Figure 2) that reduces the energy-yield by 19%, as the irradiance is reduced and the heat removal is impeded by the VIP.

Based upon the energy-yield measurements, the negative contribution of both effects cannot be exactly distinguished, but an attempt is made to quantify the contribution of each effect. The transmittance (in the visible spectrum) of configuration 1 and 3 is 90.5% relative to configuration 2 and 4, corresponding to 9.5% optical losses and accordingly, 60% of the 81.3 kWh/kWp difference between configuration 1 and 2 (with and without shadow box) can be attributed to optical losses. The remaining 40% (32.7 kWh/kWp) can be attributed to reduced electrical conversion efficiency due to elevated cell temperatures. Accordingly, optical losses (predominantly) and efficiency reduction both significantly reduce the energy-yield after applying a shadow box.

3.3 Effect of ventilation: configuration 1 versus 3

To improve the heat removal, the cavity is ventilated in configuration 3, thereby improving the energy-yield by 3.9% as compared to configuration 1 when the shadow box is not ventilated.

When estimating the optical losses using the transmittance coefficient from the previous paragraph (90.5%) and comparing it to configuration 2, 76% of the 64.3 kWh/kWp energy-yield difference as compared to configuration 2 can be attributed to optical losses (the optical losses for configuration 1 and 3 are thus assumed to be the same). The remaining 24% (15.7 kWh/kWp) is again due to electrical conversion efficiency reduction. Accordingly, when comparing the thermal losses between configuration 3 (15.7 kWh/kWp) and 1 (32.7 kWh/kWp), a difference of 17.0 kWh/kWp is observed. Accordingly, the calculations indicate that the thermal losses are halved by enabling ventilation.

3.4 Effect of partial shading: module A versus B

The energy yield was also affected by partial shading caused by the frame, as can be observed by comparing the energy yield of modules A and B from the same configurations in Figure 7. Since all the solar cells in the PV modules are series-interconnected, the current in the whole module is reduced when partial shading occurs, thereby reducing the energy-yield. As depicted in Figure 7, the reduced energy-yield is observed in a systematic manner throughout the observed period. As compared to the reference module, module 1A and 1B yielded 27% and 24% less energy respectively and module 2A and 2B yielded 13% and 10% less energy respectively. The difference is less pronounced in configuration 2 as the cells were less shaded by the edges of the supporting frame as compared to configuration 1 with the shadow box. The partial shading effect is also clearly noticeable in Figure 8a, where the difference in short-circuit current for both modules is highlighted. The effect of partial shading is hence not negligible and should be considered within the design process of a BIPV installation.

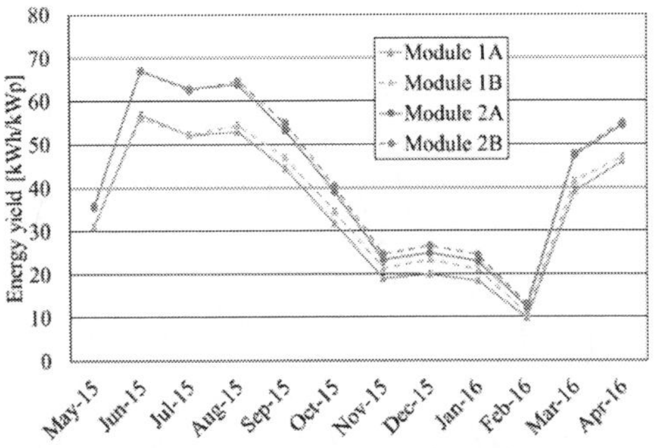

Figure 7 Energy-yield of module 1 and module 2*

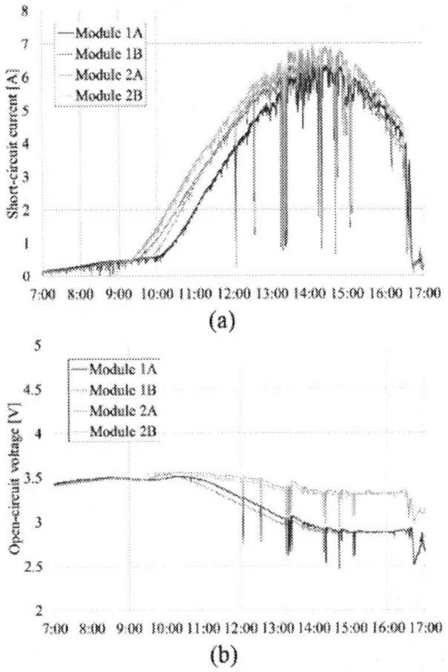

Figure 8 Short-circuit current (a) and open-circuit voltage (b) of the PV module (September 29, 2015)

4 OPERATING TEMPERATURES

Apart from electrical measurements, the spatial and temporal temperature distribution has been measured as well. High temperatures are undesirable as they reduce the solar cell efficiency on the one hand and will have an impact on the durability of the building components. In addition, the heat flux to the inside that is introduced will increase the cooling load.

The temperature distribution for the different modules is depicted in Figure 9. The figure shows temperature readings from a temperature sensor mounted behind the glazing sheet containing the solar cells. The legend entries 'U', 'B', 'R', 'L' refer to the upper, lower, right and left temperature sensors respectively.

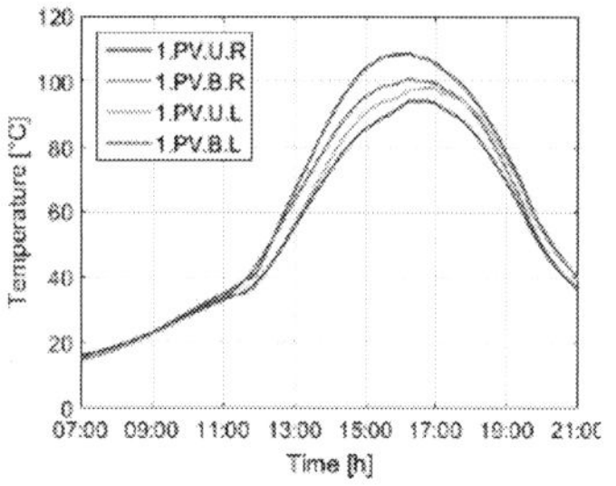

(a) Module 1

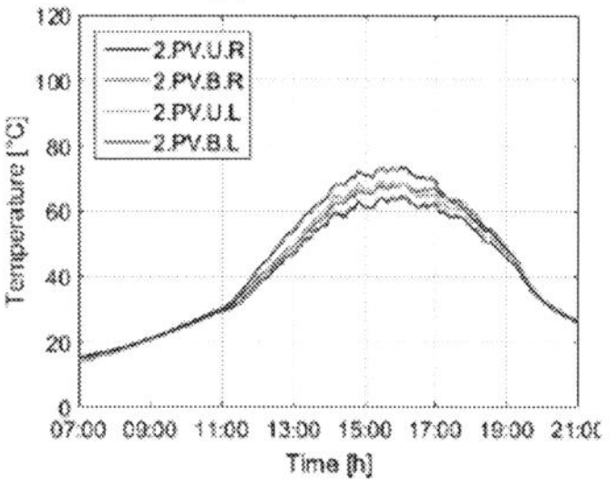

(b) Module 2

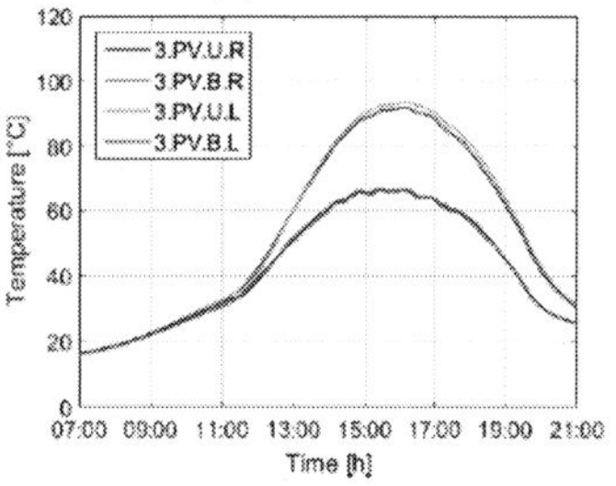

(c) Module 3

Figure 9 Measured module temperatures (August 22, 2016)

The upper temperature sensors clearly indicate higher operating temperatures than the lower ones, due to the vertical temperature distribution. Especially the third module experiences a 30°C temperature difference between the upper and lower temperature sensors, since natural ventilation is enabled. As the incoming air temperature is rapidly raised by the entering air flow, the air temperature will saturate and impede heat removal from the upper solar cells.

Furthermore, the measured operating temperatures in configuration 2 are 20-40°C lower as compared to configuration 1 and 3 where the solar cells are beside the VIP. Enabling natural ventilation reduces the maximum operating temperature by 20°C in the investigated set-up, but still temperatures near 100°C do occur that have an impact on the functional requirements of the building element.

5 CONCLUSIONS

This paper presented an experimental analysis of the electrical and thermal behavior of three distinct façade-integrated BIPV configurations over the period May 2015-2016. The configurations differ with respect to the number of glazing panes that improve aesthetics, the presence of vacuum insulation panels that restrict heat removal from the PV cells and with respect to the presence of natural ventilation. The three different configurations have been compared to a reference PV module.

Four different effects that affect the energy-yield were quantitatively assessed, being (1) building-integration, (2) addition of a shadow box in front of the PV cells, (3) ventilation and (4) partial shading by the frame edges. The results have been summarized in Figure 6, indicating that optical losses (due to reflection and absorption by the glazing panes) and efficiency reduction (due to elevated cell temperatures) reduce the energy-yield appreciably. Calculations indicated that optical losses prevail, although more detailed simulations should confirm that conclusion.

With respect to the reference PV module, the energy-yield of the BIPV modules is reduced by 11-27%. 10 to 13% reduction is observed for a module where cells are covered by a single glazing pane, while 24 to 27% reduction is observed for cells mounted behind a shadow box. The results indicated that the energy-yield improves by 3.9% when natural ventilation is enabled.

Observed partial shading by the frame edges on similar modules resulted into observed energy-yield disparities of 3-5%pt, relative to the reference module.

In the configuration where the solar cells are installed next to a VIP, temperatures were observed that well exceed 100°C. Enabling ventilation reduced the temperature by at least 20°C.

ACKNOWLEDGEMENT

Research funded by EFRO, Province of Limburg, Agentschap Innoveren & Ondernemen (Belgium) and a PhD grant of the Research Foundation Flanders (FWO).

REFERENCES

[1] K. Sinapis and M. van den Donker, "BIPV REPORT 2013 - State of the art in Building Integrated Photovoltaics," 2013.

[2] G. Verberne, P. Bonomo, and F. Frontini, "BIPV Products for Facades and Roofs: A Market Analysis," in *European Photovoltaic Solar Energy Conference and Exhibition*, 2014.

[3] I. Cerón, E. Caamaño-Martín, and F. J. Neila, "'State-of-the-art' of building integrated photovoltaic products," *Renew. Energy*, vol. 58, pp. 127–133, Oct. 2013.

[4] SUSPI and SEAC, "BIPV Product Overview for Solar Facades and Roofs," 2015.

[5] K. Fath, J. Stengel, W. Sprenger, H. R. Wilson, F. Schultmann, and T. E. Kuhn, "A method for predicting the economic potential of (building-integrated) photovoltaics in urban areas based on hourly Radiance simulations," *Sol. Energy*, vol. 116, pp. 357–370, Jun. 2015.

[6] F. Frontini, T. Friesen, C. von Ballmoos, and S. Di Gregorio, "Palazzo Positivo: renovation of a residential building in Switzerland with BIPV facades," in *European Photovoltaic Solar Energy Conference and Exhibition*, 2014.

[7] J. Langmans and S. Roels, "Experimental analysis of cavity ventilation behind rainscreen cladding systems: A comparison of four measuring techniques," *Build. Environ.*, vol. 87, pp. 177–192, May 2015.

[8] G. Van den Broeck, B. Herteleer, H. Goverde, F. Catthoor, K. Baert, and J. Driesen, "Substring-level Energy-yield Assessment of Photovoltaic Modules subject to Partial Shading Conditions," in *European Photovoltaic Solar Energy Conference and Exhibition*, 2015, pp. 1–5.

A MULTI CRITERIA OPTIMIZATION TOOL FOR BIPV OVERHANGS

Marco Lovati*, Jennifer Adami, Giuseppe De Michele, Laura Maturi, David Moser.
Institute for Renewable Energy, EURAC Research, Via G. Di Vittorio 16, 39100 Bolzano (BZ), Italy
***Phone: +39 0471 055667; Fax: +39 0471 055699; *E-mail: marco.lovati@eurac.edu**

ABSTRACT:
BiPV (Building Integrated Photovoltaics) are complex architectural systems based on the concept of multifunctionality[1]. Once a specific system is chosen, they need a fine tuning of its parameters to be a desirable option. Effects such as energy yield, cost analysis, impact on the energy demand of the building and on the internal comfort should be optimized. Despite a growing number of building performance analysis tools, the performance optimization tools are still in short supply[2]. In this paper a new method is shown for the optimization of fixed BiPV overhangs taking into account the electricity production and impact on the building energy demand for heating and cooling. The case study for the optimization is a block of of three high rise residential buildings in a contemporary urban environment.
Keywords: Building Integrated PV (BIPV), Simulation, Software.

1 INTRODUCTION

The value of BIPV and its benefits have been recognized by the industry and by the scientific community since more than a decade [3]. Despite the general consent, the built examples are still a strict minority of the new buildings, and the main type of PV integration is, by far, the application on the roof. This is due, at least in part, to the difficulty of integration in the complex system of a building [1]. Often the poor market performance of BIPV is believed to come from the lack of proper products [4][5], nevertheless there should be at least a set of possibilities to use what the industry produces most, and what is cheap nowadays. In this paper, we show a type of integration that does not require any special product, indeed the PV technology is integrated in the whole energy concept of the building (see also [6]). In the architectural practice building features are often described in a conceptual statement [7] [8]. In many countries, especially in Europe, Asia and the Middle East, designers look at historical buildings as examples of clever and sound environmental design, and these buildings are analyzed in a conceptual way. The energy design by conceptual statements has been a source of numerous clever solutions in vernacular architecture.
The study is based on the use of BiPV window overhang (i.e. shading devices to improve solar gain over the year) on a district scale.
A methodology to optimize the configuration of a BiPV system was developed and tested on a district case study.
Contrary to most of the performance analysis tool [2], where the performance of a PV system is evaluated given a chosen area, tilt angle and orientation, the developed methodology allow the planners to decide the strategy of integration and the software simply makes sure that it works by optimizing its geometrical features. The main innovation of the method consists in the input, which is parametric and based on a "architectural system"(i.e.). This approach is especially relevant for BiPV because they are always part of an architectural system. Architectural systems are meant in this paper as specific structures within the building with a particular function (e.g. complex fenestration systems, ventilated facades, overhang, Trombe walls and so on) and they are associated with a specific model and physics (natural ventilation in a ventilated façade, solar heat gain in overhangs, etc). From this conceptual operation the simulation engine produces a model to study the system specific behavior and optimize its parameters. As a case study three high rise residential buildings [Figure 1] in Bolzano were studied.

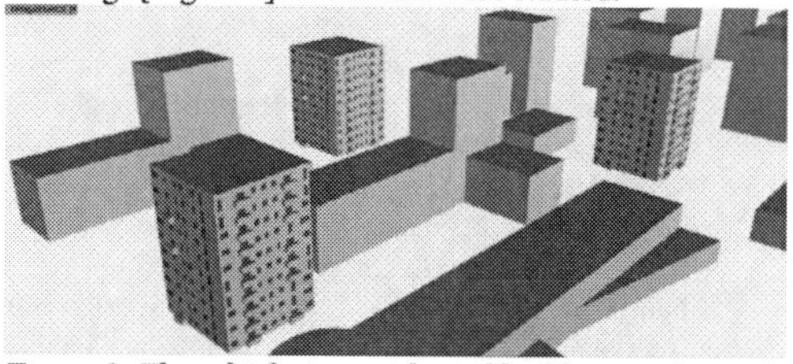

Figure 1: Three high rise residential buildings under study

2 METHOD

The method consist in two distinct parts, which are optimization phase and evaluation phase. The optimization phase finds a desirable way to configure the BiPV overhangs, the evaluation phase access their performance and produces results such as electricity production and energy savings.

2.1 optimization

The optimization of the system is obtained by creating a pool of different possibilities and testing them through simulation to access their performance. The possibilities represent different combination of parameters in a parametric description of the geometry. The whole system can be described by seven parameters (φ_{south}, φ_{east}, φ_{west}, L_{south}, L_{east}, L_{west}, κ) [Figure 1].

gene	units	boundaries	sketch
φ south φ east φ west	Deg °	0°;90°	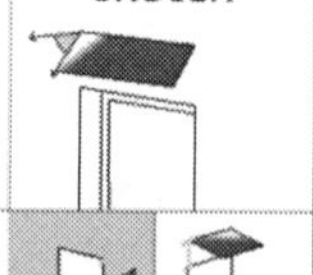
L south L east L west	m	0; max	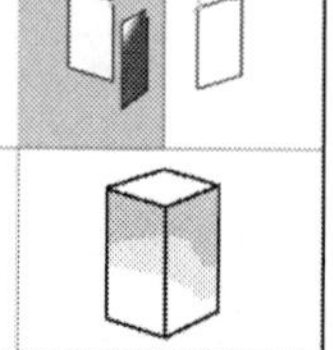
κ	%	0;100	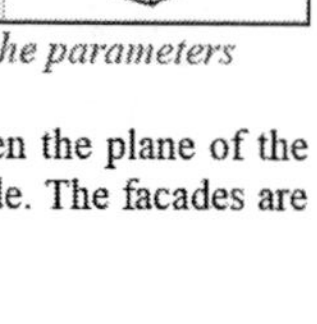

Figure 2: schematic representation of the parameters

$\Phi_{cardinal}$ represents the angle between the plane of the PV module and the normal to the façade. The facades are grouped by their direction, the angle should be $0 \leq \Phi_{cardinal} \leq 90°$.

$L_{cardinal}$ represents the distance from the window to the edge of the PV module, it is $0 \leq L_{cardinal} \leq 1$ and represent the proportion between the distance of the module in the combination and the maximum distance allowed.

κ represents the percentage of PV modules that are placed in the system, therefore the system capacity, it is $0.1 \leq \kappa \leq 1$ and refers to a capacity of $P_{max} \cdot \kappa$ ($P_{max} \approx 180$ [kW_{peak}]). All these seven parameters, in order to create a finite number of combinations, are made discrete (i.e. divided into steps). The number of combinations depends on the number of steps of each parameters, in this example each parameter has the same number of steps therefore can be calculated as

$$n_c = n_s^{n_p},$$

Equation 1: number of possible combinations for the parameters in the BiPV system.

Where: n_c is the number of combinations, n_s is the number of steps for each parameter and n_p is the number of parameters. Having 7 parameters of 6 steps each in the system, the number of possible combinations counts to 279'936.

Part of these combinations has been evaluated through the use of 3 PI (performance indicators): the annual energy consumption savings [kWh/m^2 y], the annual cumulative production of the system [kWh/y] and the average specific cumulative irradiation of the system [kWh/m^2 y], which is proportional to the yield [kWh/kWp]. The annual energy consumption savings are accessed by calculating the sum of the ideal h&c (heating and cooling) loads for each combination of parameters. The best performing combinations have lower h&c loads compared to the baseline (absence of BiPV), therefore they produce energy savings. The model for the evaluation of the heating and cooling demand of the building consists of 30 thermal zones (one for each storey of each building). For the estimation of the internal loads a report from ENEA referring to 40 dwellings in northern Italy was used [9]. Both the annual cumulative production and the average specific cumulative irradiation of the system are calculated starting from the annual cumulative irradiation values retrieved from the Energy+ software [10], these values are calculated on a specific set of surfaces corresponding to the PV overhangs. The surfaces analyzed are generated from every combination of parameters analyzed. Knowing the geometry of the system and the cumulative irradiation associated to each surface, the annual electricity production of the system can be calculated as

$$E_c = \eta \cdot PR \cdot \sum_{n=1}^{n_c} H_n \cdot A_n,$$

Equation 2:annual cumulative electricity production of the system

where: **E_c** is the annual electricity production of the system, **η** is the module efficiency, **PR** is the performance ratio of the system assumed 0.8. **n** is the number of surfaces in the system, **H_n** is the n-th irradiation retrieved from Energy+, **A_n** is the n-th area of the surface and is multiple of a single module's area. The E_c variates with each configuration because n_c depends on the parameter κ and the whole array of H_ns is influenced by $\Phi_{cardinal}$ (and perhaps by $L_{cardinal}$ near strong gradients of irradiation due to near shading such as other buildings). The yield is simply

$$EY_c = \frac{E_c}{P_c},$$

Equation 3:electric yield

Where: EA_c is the yield and P_c is the capacity of the system in the configuration. P_c is in turn calculated as

$$P_c = 1 \cdot \eta \cdot \sum_{n=1}^{n_c} A_n,$$

Equation 4: installed power capacity of the configuration

where 1 is the number of kW of irradiation in S.T.C (standard test condition). In building performance simulation the number of combinations is often imposing [2], and it is not practical to launch the simulations needed to retrieve the 3 PI for each combination. Because of this a search algorithm based on GA (genetic algorithms)[11] was used. Because of the evaluation of 3 PI the GA used was multi criteria. In a single criterion GA every individual (combination of values for each parameter) is evaluated through a single value (fitness value). If the criteria for evaluation are multiple, it can happen that two individuals represent the best solution for two different criteria. In this case the solution is represented by a Pareto front, or the collection of points whose performance cannot be further improved. No individual among the simulated ones had surpassed all the performance values of any point in the Pareto front. All the individuals in the Pareto front are better in at least one PI than their neighbors (i.e. the individuals with a similar performance).

2.2 evaluation

In this study there is no increase of complexity of the model between simulation phase and optimization phase, therefore the main difference is in the type of output and in the presence of additional calculations. Simulation was performed only on a sample of combinations, which were chosen from the Pareto front of the optimization output. For the chosen configurations hourly and monthly output were examined, and calculations considering the NPV (net present value) were performed with

$$NPV = \left(\sum_{t=1}^{life} E_c \cdot deg(t) \cdot P_E(t) - OPEX_c \right) - CAPEX_c,$$

Equation 5: Net Preent Value for any configuration of BiPV

Where: **NPV** is the net present value considering a time frame of **life** years, **deg(t)** is a coefficient of degradation of the performance assumed equal to 0.5 [12], **PE(t)** is the price of the electricity assumed stable and equal to 0.234 [€/kWh] [13]. **$OPEX_c$** is the maintainance associated with the configuration assumed to be 40 [€/kWp year], it is directly proportional to the capacity installed. **$CAPEX_c$** is the initial investment to purchase the system assumed 2000 [€/kWp]. The actualization of the cost of money was not considered in this calculation.

The formula for NPV was derived from [14].

3 RESULTS

As explained in the 2.1 the solution of the GA consists in a group of combinations that represent the best in the three performance indicators. Two solutions were evaluated [Figure 3].

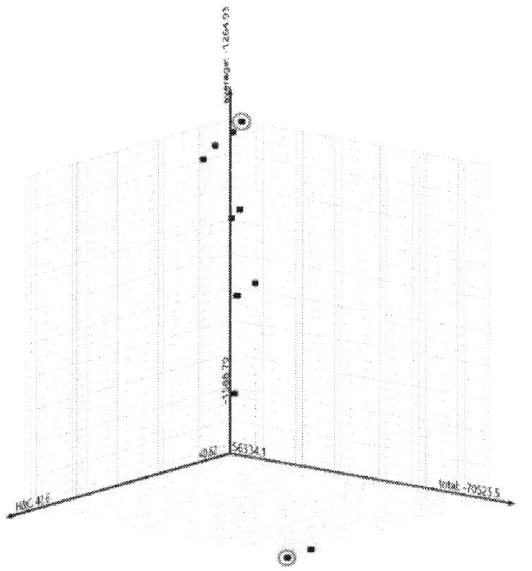

Figure 3: scheme of the Pareto front: the three axes represent the three performance indicators ("total" is the annual electricity production, "average" is proportional to the yield of the system, "H&C" is the consumption for heating and cooling), the origin would represent the ideal solution, the two circles represent the two solution evaluated.

The two solution chosen for evaluation were the one with best yield and the one that most reduces heating and cooling needs [Figure 4].

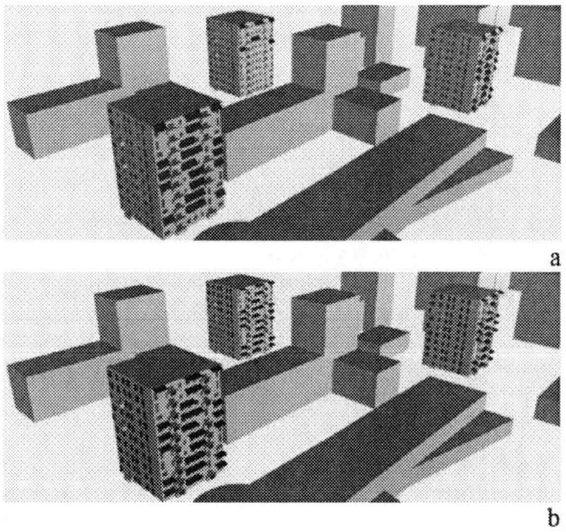

Figure 4: solutions chosen for evaluation: a) best electric yield, b) best reduction in heating and cooling needs.

The annual cumulative ideal heating and cooling demand is shown in [Table 1]

Individual	Heat	Cool	H+C	% Base
Baseline	18.9	24.6	43.5	1.00
Best Yield	20.4	22.3	42.7	0.98
Best Energy Savings	19.0	19.9	39.0	0.90

Table 1:ideal annual cumulative needs of the three buildings in [kWh/m2 year].

The installation of the BiPV overhangs results in a slight reduction of the annual cumulative energy demand

In [Figure 5] the energy demand values are converted into electric energy by applying a COP equal to 3. The total electricity consumption of the "best savings" scenario shows a moderate annual decrease (10%) from the baseline. the energy demand is in large part compensated by BiPV production in both the configurations.

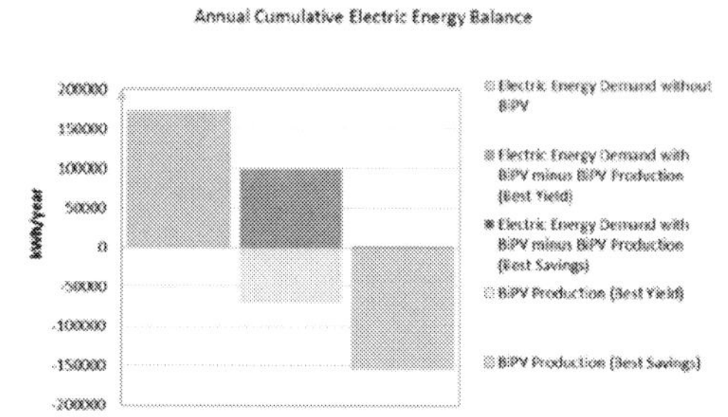

Figure 5: annual cumulative electric energy balance (demand for heating and cooling and electricity production)

[Figure 6] Refers to the best energy savings scenario. Shows that the overhangs has negative impact in the heating season. Nevertheless the savings in the cooling season cause a net positive impact during the year. Considering electric generation, the net energy balance of the BiPV system is positive in every month.

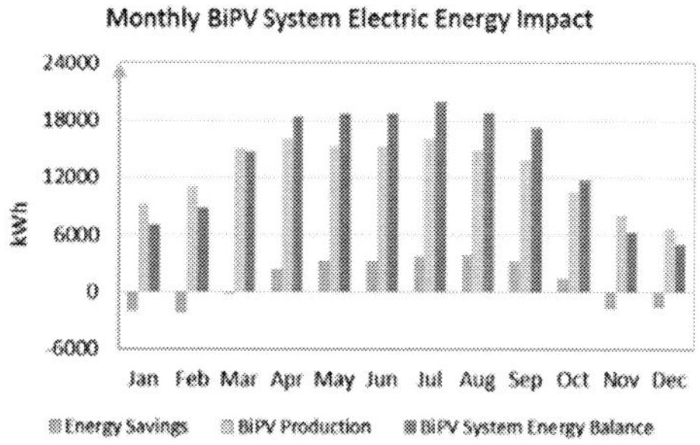

Figure 6: monthly energy impact of the BiPV system.

[Figure 7] Shows the monthly electric energy demand for heating and cooling of the building (COP = 3), The demand as modified by the presence of the overhangs, and The energy balance considering electicity production from BiPV minus demand from the building. Again the chart refers to the best energy savings scenario.

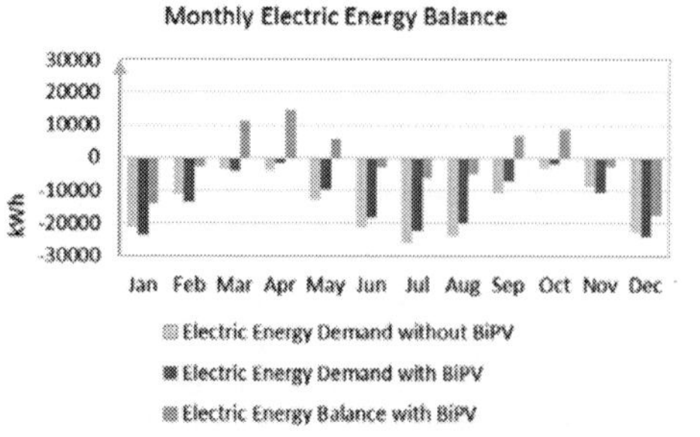

Figure 7: Monthly electric energy balance. in yellow the monthly cumulative electricity production minus the electric energy demand for heating and cooling.

The result of the NPV calculations are shown [Figure 8]. As expected, the best yield scenario has a faster payback time. The best saving scenario should not be seen as an investment but indeed as a component that improves the quality of the building.

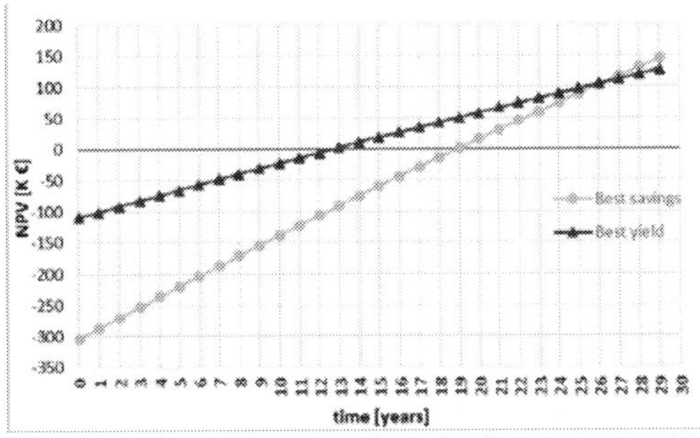

Figure 8: NPV of the two evluated solutions over a 30 years time span.

4 DISCUSSION AND CONCLUSIONS

The experience is satisfactory for some aspects but it is unfortunately still far from being a useful tool in the practice. The results show that the impact of the configurations in the Pareto front is considerable in the energy economy of the building block, in fact installing the PV overhangs we can have a reduction of around 10% of the energy consumption for heating and cooling. This savings are, from the photovoltaic point of view, a completely free positive side effect, The PV modules are therefore truly multi-functional in their usage. There are tough issues that cannot be ignored, the two main ones are the lack of consideration for the daylight and the excessive simulation time. The lack of daylight calculations leads to consider acceptable solutions that are not, the need for reduction of solar gain in peculiar moments of the day could lead to the almost complete shutting of some windows. This is not a viable strategy therefore at least a basic form of daylight consideration should be implemented. Considering the complexity of an average daylight simulation, the impact on the simulation time would be absurd, there is the need for a simplified approach able to guarantee a minimum standard at the city block level. Another issue is the simulation time: in the present experience, to access the impact of the overhangs for each individual of the genetic pool, a simulation with energy plus was performed. The evaluation time for each individual could be reduced either by simplification of the model (from openings to a mere glazing percentage), or by simplification of the procedure. A good strategy could be to launch an energy plus simulation for the baseline scenario, and simply post process the output by considering variations in the hourly solar gain.

An interesting experiment would be to apply a similar optimization method on a building in a much earlier phase of the design, this would make the optimization much faster by acting on a simplified (massing) model, and provide a better integration into the design workflow. The approach used in this example, tough situated in a late stage of the design, could indeed turn out useful in case of building retrofit.

5 AKNOWLEDGEMENTS

The authors thanks Flexi BIPV 5-1a-232, founded from the ERDF and managed by the Province of Bolzano.

6 REFERENCES

[1] R. J. Yang, "Overcoming technical barriers and risks in the application of building integrated photovoltaics (BIPV): hardware and software strategies," *Autom. Constr.*, vol. 51, pp. 92–102, 2015.

[2] S. Attia, M. Hamdy, W. O'Brien, and S. Carlucci, "Assessing gaps and needs for integrating building performance optimization tools in net zero energy buildings design," *Energy Build.*, vol. 60, pp. 110–124, May 2013.

[3] P. Heinstein, C. Ballif, and L.-E. Perret-Aebi, "Building integrated photovoltaics (BIPV): review, potentials, barriers and myths.," *Green*, pp. 125–156, 2013.

[4] P. Bonomo, A. Chatzipanagi, and F. Frontini, "Overview and analysis of current BIPV products: new criteria for supporting the technological transfer in the building sector," *Vitr.-Int. J. Archit. Technol. Sustain.*, no. 1, pp. 67–85, 2015.

[5] L.-E. Perret-Aebi, J. Escarré, H.-Y. Li, L. Sansonnens, F. Galliano, G. Cattaneo, P. Heinstein, S. Nicolay, J. Bailat, S. Eberhard, and others, "When PV modules are becoming real building elements: White solar module, a revolution for BIPV (Presentation Recording)," in *SPIE Optics+ Photonics for Sustainable Energy*, 2015, p. 95630Q–95630Q.

[6] J. Adami, "A support tool for the early design phase of BiPV towards nearly zero energy districts.," Universita di Venezia (IUAV), 2016.

[7] N. Lechner, *Heating, cooling, lighting: Sustainable design methods for architects*. John wiley & sons, 2014.

[8] H. Fathy, "Natural energy and vernacular architecture," *Princ. Ex. Referances Hot Clim. Univercity Chic. Press Ed Chic.*, 1986.

[9] F. Cervarolo, B. Di Pietra, F. Margotta, and G. Puglisi, "ODESSE: simulazione dinamica del sistema edificio-impianti per la climatizzazione estiva," ENEA, 2010.

[10] P. Ellis, R. Liesen, and C. Pedersen, "Energyplus Experimental Data Validation Work: Development and Validation of the Unvented Trombe Wall Model and Other Heat Balance Components," *CERL Final Rep. DACA42-01--0004 Task*, vol. 3, 2003.

[11] L. Davis, "Handbook of genetic algorithms," 1991.

[12] G. Belluardo, P. Ingenhoven, W. Sparber, J. Wagner, P. Weihs, and D. Moser, "Novel method for the improvement in the evaluation of outdoor performance loss rate in different PV technologies and comparison with two other methods," *Sol. Energy*, vol. 117, pp. 139–152, Jul. 2015.

[13] "Eurostat: Energy price statistics." [Online]. Available: http://epp.eurostat.ec.europa.eu/statistics_explained/index.php/Energy_price_statistics. [Accessed: 28-Oct-2014].

[14] Eero Vartiainen, Gaëtan Masson, and Christian Breyer, "PV LCOE in Europe 2014-30," 2015.

OBSTRUCTION SURVEYING METHODS FOR PV APPLICATION IN URBAN ENVIRONMENTS

S. Freitas[1], A. Cristóvão[2], R. Silva[1], M. C. Brito[1,2]

[1] Instituto Dom Luiz - Faculdade de Ciências, Universidade de Lisboa.
[2] Departamento de Engenharia Geográfica, Geofísica e Energia (DEGGE).
Faculdade de Ciências da Universidade de Lisboa, Campo Grande, 1749-016 Lisboa, Portugal.
e-mails: srefreitas@fc.ul.pt; andree_cristovao@hotmail.com; rasilva@fc.ul.pt; mcbrito@fc.ul.pt.

ABSTRACT: Solar photovoltaics (PV) is becoming trendy as a means to achieve energetic sustainability in cities, where a significant amount of electricity is consumed. In order to estimate the solar potential in all points of a building, district or the whole metropolitan region, the elements that might obstruct the sky must be taken into account for both direct and diffuse components of solar radiation. In this work, we focus on 3 state-of-the-art methodologies for Sky View Factor (SVF) estimation in urban environments: 1) fisheye photographic method (the standard reference method), 2) GPS receiver accounting for satellite visibility and signal intensity, and 3) airborne Light Detection And Ranging (LiDAR) and a ray-tracing method with different sky dome divisions. Our findings point out that in comparison with the fisheye photographic both GPS and the ray-tracing/LiDAR methods slightly overestimates SVF when the level of obstruction is low, but tends to overestimate in more closed spaces such as canyons with tall buildings and courtyards and in vertical facades. Still, the GPS can be considered a good method for obstruction surveying when there is a trade-off between accuracy, execution time and economy of resources.
Keywords: Urban PV, Sky View Factor, Fisheye, LiDAR, GPS

1 INTRODUCTION

The constant increase of population in urban areas and the technological development that is required to meet its needs is leading to an increasing electricity demand in metropolitan areas. Thus, it is crucial to consider and explore new ways of producing electricity closer to where it is needed. In this sense, PV urban applications appear not only as promising but also as logical solutions, since the solar resource is fairly high in most densely populated cities. To successfully disseminate this vision, more detailed approaches have to be taken in terms of the study of urban photovoltaic potential, since the cities are complex, featuring a huge variability in building morphology and streets profiles.

The study of the Sky View Factor (SVF) parameter gains particular importance to correctly describe the open sky access and the degree of obstruction at certain places. Although SVF has always had special relevance as one of the most important parameters in the study of urban heat island phenomena [1, 2], the study carried out in this paper is focused on the capacity of SVF to give insight into obstructions to solar radiation at places that might be suited for the installation of PV systems. The SVF can also be a measure of the amount of diffuse solar radiation that reaches a certain point when isotropic distribution is considered [3, 4].

Several methods exist to estimate SVF and they can be classified into four main categories: analytical, photographic, GPS and software methods [5]. The photographic method is a standard procedure for the evaluation of sky obstruction, however other methodologies might be much more cost-competitive and less time consuming than this standard procedure when medium to large urban areas are under analysis. In this study the photographic, GPS and ray-tracing with LiDAR data approaches will be explored.

2 METHODOLOGY

2.1 Fisheye photographic method

In what concerns accuracy, this method is considered to be the best for the assessment of SVF in small areas. In this study, an Olympus SP-350 photographic camera was used with a Nikon FC-E8 fisheye converter, a fisheye lens commonly used to capture the entire surroundings of a shooting place [2, 6], and known for its 180 degree hemispheric view, depicting with great quality the full horizon skyline. The shots were taken at 1.20m height using a tripod along different places at the Faculty of Science of the University of Lisbon (FCUL) campus, including some vertical facade points (Fig. 1).

After collecting all the photos, the SkyViewFactor Calculator software [7] was used to estimate the respective SVF values. This tool allows the upload of single fisheye photographs and estimates the SVF employing two different methods: the more recent pixel-based method introduced by [8], and the annulus method presented in [9].

A weakness recognised to the photographic method relates to the ideal meteorological conditions in which the surveys has to be carried out. Overcast sky are preferable, since the sky looks homogenous, thus avoiding any excess of direct radiation or even different types of clouds that could jeopardize the images processing [10]. Still, the SkyViewFactor Calculator allows small adjustments on photos in order to draw the proper skyline delineation in case the software misses any part of it or does improper recognition of non-sky pixels.

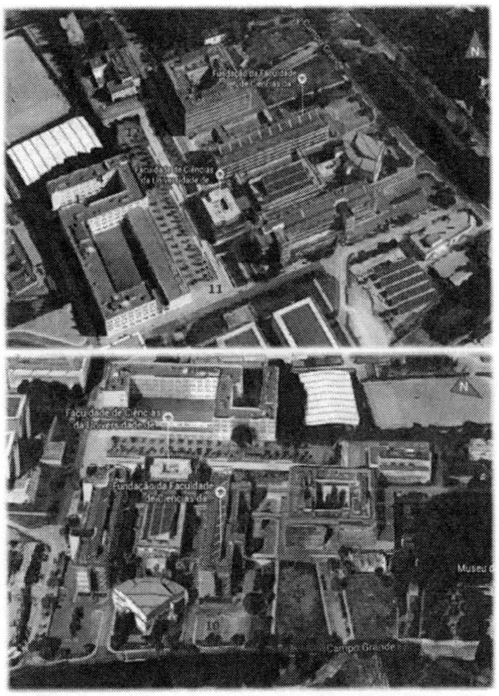

Figure 1: Location of the 13 shooting places in FCUL campus. Points 1, 3 and 4 belong to a vertical façade.

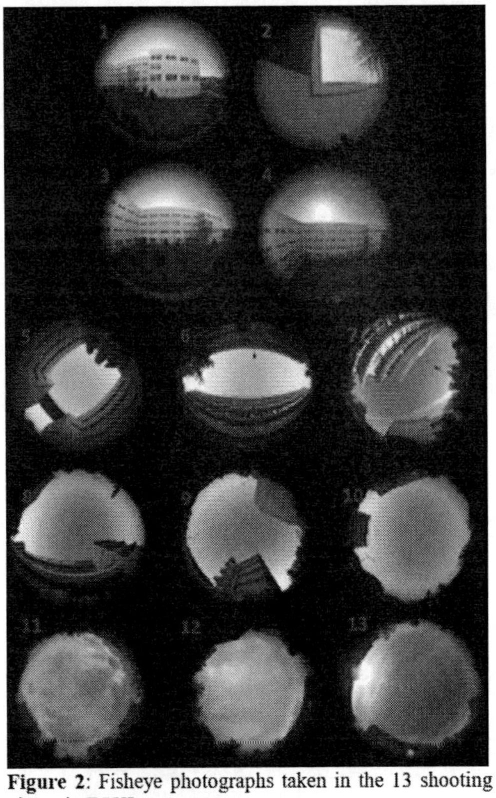

Figure 2: Fisheye photographs taken in the 13 shooting places in FCUL campus.

2.2 GPS method

The use of GPS to assess the SVF of a location is an uncommon but promising alternative due to its low cost and tolerance to different meteorological conditions [10]. It also allows for the portability that lacks to many others SVF surveying methods. The equipment used in this approach was a Garmin eTrex 30, a receiver typically used in outdoor activities, such as Geocaching, which served the purpose of displaying the number of satellites (N) and their satellite status index (SSI), which reflects the signal to noise ratio, in a 1 (weak signal) to 10 scale [11]. It is assumed that a higher signal intensity is associated to a low level of obstruction, and therefore a higher SVF value. Before the actual data acquisition, a prediction of the satellites location was checked in order to see how would be the availability in terms of satellites along the all hemisphere [12]. A wide distribution of satellites in the sky means that the receiver is going to collect information from all directions, and therefore identify possible obstructions from any point on the horizon. To guarantee this assumption, data acquisition was made during a 10 hour period, one time per hour, for each place already photographed with the fisheye lens, thus ending with 10 surveys per local. After achieving the stabilization of the signal, which takes around 40 seconds for nearly open sky areas and 70-80 seconds for high obstruction level places, a screenshot was taken to save the information on the receiver. This procedure was made for each survey hour of each place. On the screenshots taken, the green colour corresponds to GPS while the blue represents the GLONASS satellites, as seen in Fig. 3.

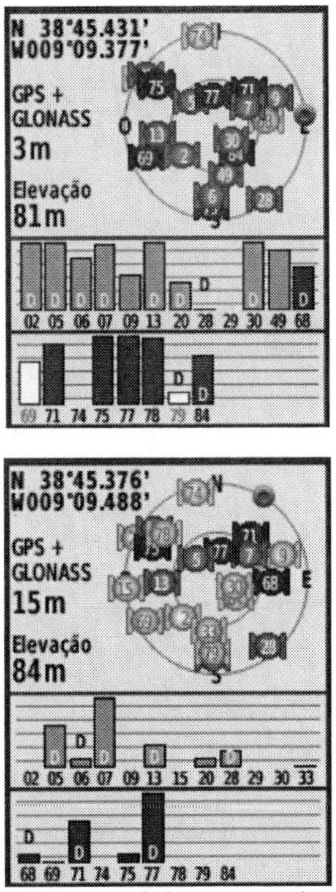

Figure 3: Screenshots of the GPS receiver display n places with: top) – low obstruction, bottom) – high obstruction.

On Fig. 3 - left) an example of a low obstruction area is represented, where the number of visible satellites is high, as well as their signal intensities (green and blue bars). On the other hand, Fig. 3 - right) shows a pattern of a place where obstructions have a higher impact, as the visible satellite are fewer, and SSI values are lower. To find the SSI value for each satellite, the bars were given values according to the line they were reaching in the screenshot display, except the white bars, which represent the signal instability and therefore were not counted.
This way we were able to obtain an index that correlates with the SVF for each of the shooting places, eq. 1:

$$SVF \propto \frac{\sum SSI}{N} \qquad [1]$$

where $\sum SSI$ is the sum of all SSI values and N is the number of satellites contributing for the signal.

2.3 Ray-tracing/LiDAR method

A digital surface model from the area that this study is focused on was already built from LiDAR data (Fig. 4), and the respective SVF map achieved through a ray-tracing method, done by the SOL algorithm presented with further detail in [13].
To ensure that the value taken from the SOL SVF maps would be the closest location to the place where the fisheye measurements were actually made, a relative location process was performed. For each of those sites, using the SVF map previously generated, the cell coordinates were recorded and then a mean with the eight surrounding cell values was made, in order to minimize the error of an eventual dislocated cell choice.

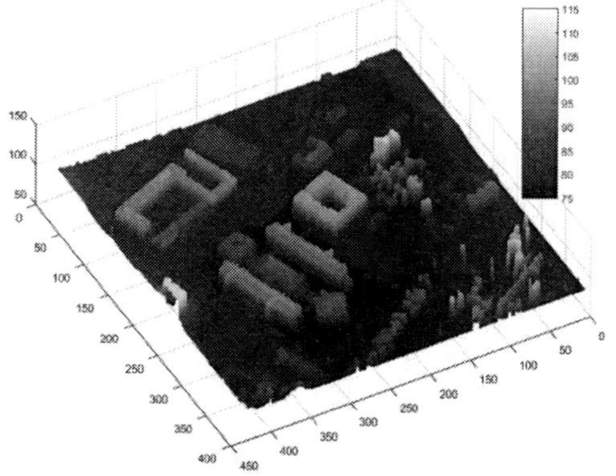

Figure 4: Digital Surface Model (DSM) of FCUL campus.

Other tests were carried out by changing the number of sky vault divisions in SOL's algorithm (Fig. 5). The source code considers 1081 divisions, which were changed to 400, 290 and 145 (as the standard reference from the International Commission on Illumination (CIE). The use of these coarser sky vaults was intended to test if there would be any change in terms of accuracy regarding the actual fisheye values, as well as the possibility to lower the computation time required by the 1081 divisions. The SVF acquiring process used the same relative location system as described before.

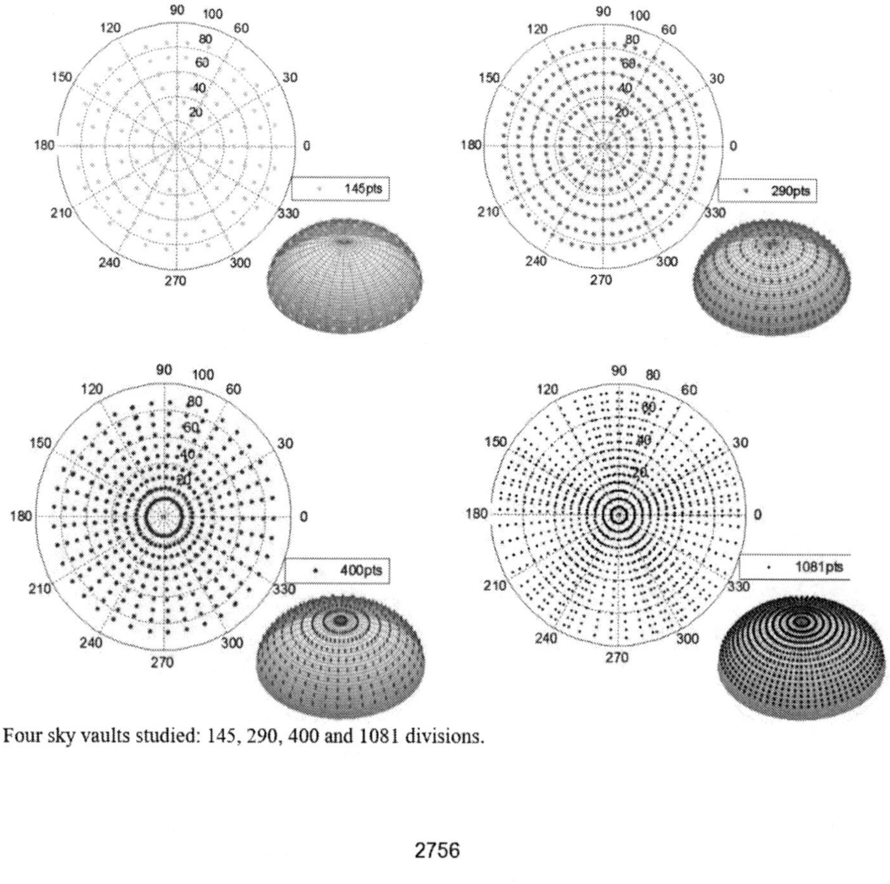

Figure 5: Four sky vaults studied: 145, 290, 400 and 1081 divisions.

3 RESULTS

3.1 LiDAR vs Fisheye

As described above, a backward ray-tracing algorithm was employed using a resampled LiDAR point cloud (1 point per square meter) to estimate sky view factor on the 13 points that are closer to the real shooting place locations. This procedure requires a careful inspection on the most appropriate sky vault for ray-tracing (i.e. how many sky divisions and how they are distributed). The comparison between the SVF obtained with the 4 sky vaults (Fig. 5) and the fisheye photographic method are shown in Fig. 6.
It can be noted that the SVFs in the lower and intermediate floors of the building facade (places 1 and 3) and in the ground-level corner (place 2) are clearly overestimated by the LiDAR method. On the contrary, open field areas (places 11 – 13) and wide streets (places 8 – 10) reveal very close results. However, the method seems to fail when buildings and vegetation start to significantly fil the near horizon (places 2, 6 and 7).

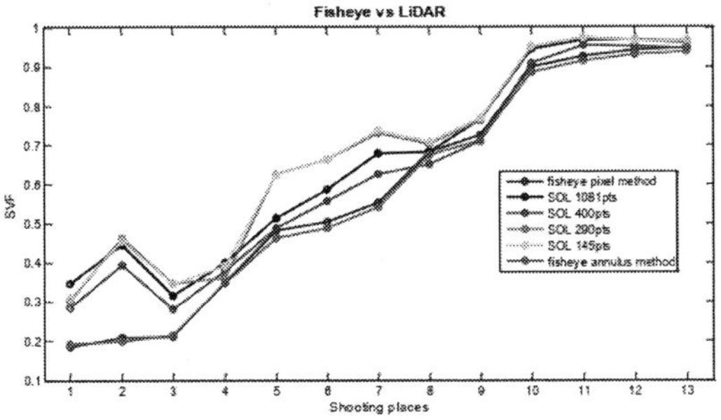

Figure 6: SVF values obtained though the LiDAR and ray-tracing method with 4 different sky vaults in comparison with the fisheye photographs processed with annulus and pixel methods.

Places 4 and 5 show an unexpected accuracy that deserves some comments. Since place 4 is on a vertical façade an overestimation of the same order that was verified for places 1 and 3 was expected. However, the fact that it is on the last floor of the building might tell us that the DSM might either be giving a wrong building height or there are some features that are not being correctly represented. On the contrary, the SVF in place 5 should be underestimated by the LiDAR method, as the vertical façade points are derived from the terrain height and the building height, the hole that exists in the building that appears in photo 5 (Fig. 2) is closed in the DSM representation, and thus less sky would be visible.
Given the lack of plausible explanation, the comparison of SVF obtained with the different sky vaults may enlighten us. It is interesting to note that the results from the four different sky vaults are actually very close, except for places 2, 5 and 6. Sky models with 145 and 290 do not show significant differences between them, which was expected as one is just twice the number of divisions of the other. The main difference between the 145 and 290 to the 400 and 1081 sky models is that the number of divisions per zenith band is kept the same in the last two, which means that, in places such as 6 and 7, the 145 and 290 sky models will always overestimate the SVF relatively to the fisheye photographic results. This is because the algorithm within the SkyViewFactor Calculator employs a distribution of pixels/divisions also with the constant number of divisions per zenith band.
We can conclude that overall the sky with 400 divisions seems to be the most appropriate for ray-tracing/LiDAR methodologies, but it should be kept in mind that for some points it is impossible to get exact values since a DSM derived from LiDAR involves an uncertainty in representation as the grid resolution becomes lower (for instance, $1m^2$ is still low for the correct representation of corners such as place 2).

3.2 GPS vs Fisheye

In Fig. 7, the fisheye photographs taken in places 6 and 13 are shown with the superimposed hourly distribution of satellites.

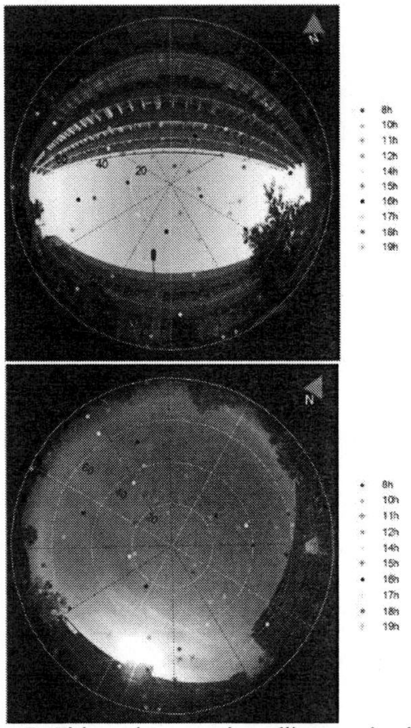

Figure 7: Fisheye image and satellite populated sky graph of: shooting place 6 (tall building canyon with fisheye SVF=0.50 and SSI/N=6.61), and shooting place 13 (open field area with fisheye SVF=0.94 and SSI/N=8.80).

After compiling this new SSI/N index for all the 10 hours for all places considered, the variability of the signal occurred during this period is represented in the boxplots of Fig. 8 - top, where the SVF was derived taking into account eq. 1. In order for both variables to share the same range of values, for visualization purposes, the SSI/N measurements were normalized respective to its maximum possible value (i.e. 10):

$$\frac{SSI}{N}_{norm} = \frac{\frac{SSI}{N}}{max.possible\ value} = \frac{\frac{SSI}{N}}{10} \qquad [2]$$

A certain degree of variability is present in all GPS recordings, being more significant for locations that are subject to either strong or intermediate levels of obstruction (i.e. SVF lower than 0.7). Additionally, a positive bias is observed, predominantly in more obstructed locations (i.e. SVF up to 0.6). On the contrary, for more open places a negative bias is observed.
Discarding the influence of the weather on the signal that reaches the GPS receiver (since it has no impact on the signal strength [14]), two possible explanations for both obstructed and open locations could be: 1) the sampling of satellite signals obtained in a specific instant may not be sufficiently representative of the place in study, but be dependent on factor such as the number of points, spatial distribution, etc.; 2) the North quadrant of the sky dome is not covered with satellites due to the geostationary trajectory "blind spot". An additional justification for the significant positive bias on obstructed locations would be that the satellite signal suffers reflection from obstacles, which may block signals to the GPS.
The addition of the GLONASS constellation signals, in average, did not result in significant changes in the SSI/N median value but led to a lower variability for almost every location (Fig. 8 - bottom). This means that adding information from more satellites is not relevant when several measurements are taken for the same location, as it already implies information from a bigger number of wide spread different satellites. Nonetheless, it helps reducing the sensibility of individual measurements when information is still scarce, and therefore it provides a better signal spatial coverage, even if not filtered for low solar height values.
After having the set of SSI/N measurements taken at different times of the day, the final estimation would result from averaging all values, aggregating information from both GPS and GLONASS satellite constellations.

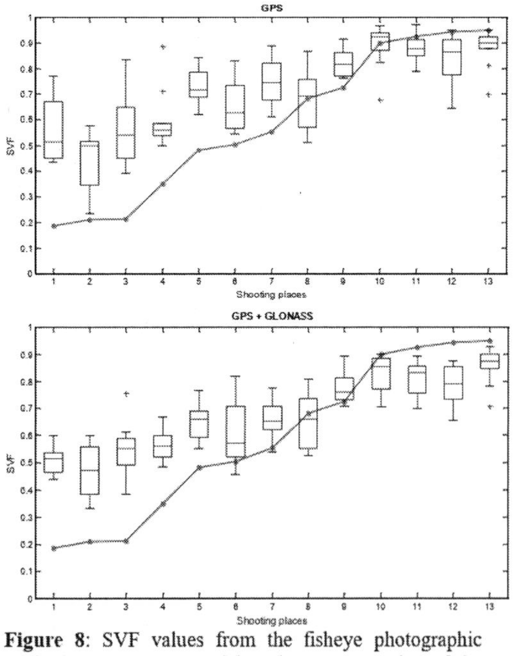

Figure 8: SVF values from the fisheye photographic method (green curve) and boxplot representation of the hourly SSI/N index measured over a 10-hour period: top) GPS satellites only, bottom) GPS plus GLONASS.

4 CONCLUSIONS

The LiDAR method seems to overestimate the SVF when the level of horizon obstruction is high (i.e. in places such as canyons with tall buildings and courtyards) which is associated to the limitation of LiDAR in the representation of vegetation and complex shaped features and/or use of aged LiDAR data
The signal intensity/number of visible satellites (SSI/N) index given by the GPS method is shown to be correlated to the SVF but features a relatively high level of variation throughout the day - geostationary satellites movement, which results in an ever changing quantity of information sources and their position
Fisheye photographic method is a standard procedure but LiDAR and GPS can be cost-competitive alternatives, with GPS being a decent SVF estimation method when there is a trade-off between accuracy, execution time and economy of resources, which will certainly facilitate the deployment of PV in urban environments

5 ACKNOWLEDGMENTS

Part of this work was supported by MIT Portugal Program for Sustainable Energy Systems, FCT grants SFRH/BD/52363/2013 and PD/BD/106007/2014, Projects MITP-TB/CS/0026/2013 (SUSCITY), PTDC/EMS-ENE/4525/2014 (PVCITY) and UID/GEO/50019/2013 - Instituto Dom Luiz, and Fundação Calouste Gulbenkian.

6 REFERENCES

[1] T. R. Oke, 1981. Canyon geometry and the nocturnal urban heat island: comparison of scale model and field observations. Journal of Climatology, Vol. 1,237-254
[2] C. S. B. Grimmond, S. K. Potter, H. N. Zutter, C. Souch, 2001. Rapid methods to estimate sky-view factors applied to urban areas, Journal of Climatology, Vol. 21, 903-913
[3] K. Yoon, G, Yun, J, Jeon, K. S. Kim, (2014). Evaluation of hourly solar radiation on inclined surfaces at Seoul by Photographical Method, Solar Energy, Vol. 100, pp 203-216
[4] Joze Rakovec, Klemen Zaksek, 2012. On the proper analytical expression for the sky-view factor and the diffuse irradiation of a slope for an isotropic sky, Renewable Energy, Vol. 37, pp. 440-444
[5] L. Chen, E. Ng, X. An, C. Ren, M. Lee, U. Wang, Z. He, 2012. Sky view factor analysis of street canyons and its implications for daytime intra-urban air temperature differentials in high-rise, high-density urban areas of Hong Kong: a GIS-based simulation approach, International Journal of Climatology, Vol. 32, 121–136
[6] J. Unger, 2009. Connection between urban heat island and sky view factor approximated by a software tool on a 3D urban database, Int. J. Environment and Pollution, Vol. 36, 59-80.
[7] Lindberg, F., Holmer, B., 2010. Sky View Factor Calculator User Manual – Version 1.1. University of Gothenburg.
[8] Holmer, B., Postgard, U. and Eriksson, M., 2001. Sky view factors in forest canopies calculated with IDRISI, Theoretical and Applied Climatology, Vol. 68, pp.33–40.

[9] Johnson, G.T. and Watson, J.D., 1984. The determination of view-factors in urban canyons', Journal of Climate and Applied Meteorology, Vol. 23, pp.329–335.
[10] Chapman, L., Thornes, J.E. and Bradley, A.V., 2002. Sky-view factor approximation using GPS receivers, International Journal of Climatology, Vol. 22, pp.615–621.
[11] Chapman, L., & Thornes, J., 2004. Real-Time Sky-View Factor Calculation and Approximation. J. Atmos. Oceanic. Technol. 21, 730-741.
[12] Map of Live Satellite Positions, 2016: https://in-the-sky.org/satmap_radar.php
[13] Redweik, P., Catita, C., Brito, M.C., 2013. Solar energy potential on roofs and facades in an urban landscape, Solar Energy, Vol 97, 332-341.
[14] Haase, J., Ge, M., Vedel, H., Calais, E., 2003. Accuracy and Variability of GPS Tropospheric Delay Measurements of Water Vapor in the Western Mediterranean, Journal of Applied Meteorology, Vol. 42, 1547 - 1568

SIMULATION OF MISMATCH LOSSES FOR PARALLEL CONNECTION OF CIGS MODULE STRINGS WITH DIFFERENT ORIENTATIONS IN BIPV SYSTEMS

Rolf Wächter, Andreas Jenninger, Tobias Repmann
Manz CIGS Technology GmbH
Alfred-Leikam-Str. 25, 74523 Schwäbisch Hall, Germany
Tel. +49 791 95665-326, fax -111, e-mail rwaechter@manz.com

ABSTRACT: Parallel "polystring" connection of CIGS module strings of four different orientations (south, west, north, east) to a common single-MPP inverter in a vertical façade PV system has been simulated, taking into account the influence of irradiance and module temperature on MPP voltage. The MPP voltage increases with irradiance but decreases with the increasing temperature that follows from higher irradiance. It was found that these two effects compensate each other very well over a wide irradiance range. Accordingly, the overall annual DC energy yield loss caused by voltage mismatch is only around 1 % compared to the standard setup where different orientations are connected to separate MPP trackers or inverters. Hence the system configuration can be designed with a reduced number of inverters and at reduced costs. Since the four sides of such a PV system never see sunlight all at the same time, the polystring configuration has the additional advantage that the inverter(s) can be chosen much smaller; the simulation shows that only about half as much inverter power is required as for the conventional separate-inverters-per-orientation setup. All in all, parallel connection of different orientations is an advantageous option that offers economic benefits in façade and other systems.
Keywords: Building Integrated PV (BIPV), Polystring, CIGS, Simulation, Cost reduction

1 INTRODUCTION

There is growing interest in building-integrated or building-attached PV systems (BIPV), and particularly in the use of CIGS modules as optically attractive façade cladding for buildings. Such systems often comprise modules with two or more different orientations. To a certain extent this applies even to some roof systems, e.g. east/west roofs. The situation is, however, more common in façade systems, where modules oriented in more than two different directions are frequently found. The reasons for this can be manifold, for example uniform building design, usage of a maximum surface area for energy production, or levelling the photovoltaic energy production during daytime. The conventional approach to such a situation is to connect each orientation to a separate inverter or at least a separate maximum-power-point (MPP) tracker, requiring a large number of inverters. Alternatively, differently oriented but otherwise electrically identical module strings can be connected in parallel to one common MPP tracker ("polystring" configuration) if some reduction in annual energy yield is accepted. Up to now, polystring configurations have been suggested mainly for east-west roof systems [1, 2]. The discrepancies in irradiance and module temperature between differently oriented module strings are much more pronounced in vertical BIPV systems though, so it needs to be investigated whether the polystring concept is suitable for these systems at all. In addition, the higher number of different orientations often found in such systems makes the situation more complex.

Due to the fact that differently oriented modules reach their peak power production at different times, less total inverter power is required in polystring configuration than for the standard case of individual inverters per orientation. On the other hand, differently oriented modules will in general have different MPP voltages due to their individual irradiance/temperature situation, so a parallel connection of such strings will result in mismatch losses. The MPP voltage increases with increasing irradiance, but decreases with increasing module temperature. Since the module temperature increases with irradiance, these two main effects are expected to counterbalance each other partially. This work aims to quantify the mismatch losses resulting from these differences in MPP voltages by simulation and to show the benefits of extending the polystring concept to BIPV systems.

2 SIMULATION APPROACH

2.1 Description of BIPV system

The exemplary building-integrated PV system simulated in this work has four vertical façades oriented in the four cardinal directions (south, west, north, east) and is located in Stuttgart, Germany. Each of the four façades is equipped with a rear-ventilated PV generator with a nominal power of 1 kWp, made from standard frameless Manz CIGS modules. The modules of each façade are series-connected to form strings of identical length, which are then connected in parallel, using string diodes to prevent back currents.

The number of modules per string or in total does not influence the results (all the strings need to have the same length, however) and is therefore not specified. Likewise, the size of 1 kWp per façade is only chosen for convenience; the simulation results can be scaled to any PV generator size. The critical parameters are the orientation of the façades and their nominal PV power ratio.

The example system chosen here is a fairly straightforward but yet typical system for BIPV on multi-storey buildings. Nevertheless, the polystring concept as well as the simulation approach presented here can easily be applied to more complex BIPV systems.

Two configurations of this BIPV system are compared in this work: For the "polystring configuration" (Fig. 1 (a)), all module strings of the four façades are connected in parallel to a common inverter with a single MPP tracker. All strings therefore need to run at a common voltage, determined by the common MPP of the whole system. In contrast, each façade in the "conventional configuration" (Fig. 1 (b)) is connected to a separate inverter and thus can run at its individual MPP voltage.

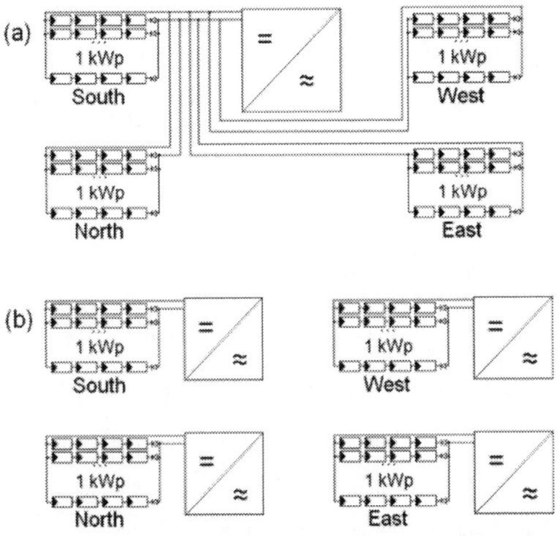

Figure 1: Circuit diagrams of PV system configurations simulated in this work. (a) Polystring configuration – strings of all four façades are connected in parallel to a single inverter; (b) conventional configuration – each façade has its own inverter.

2.2 Simulation of irradiance and module temperature

The commercial PV system design software "PV*SOL premium" [3] was employed to simulate the BIPV system described above. The extensive functions of PV*SOL for calculating DC and AC energy yields were not used as the software is not designed for simulation of polystring configurations. Instead, the software was utilized only to simulate the in-plane irradiance and module temperature for each of the four façades over a full reference year. A time resolution of one minute was chosen for these data.

2.3 Simulation of I-V curves

A set of I-V curves of an actual Manz CIGS module (Fig. 2), measured on a solar simulator at irradiances of 1000, 700, 200, 100, and 0 W/m² and corrected to 25 °C, was taken as the basis for calculation. Individual I-V curves as a function of irradiance and module temperature were obtained from this set by first interpolating to the desired irradiance and then applying a temperature correction to the desired temperature using

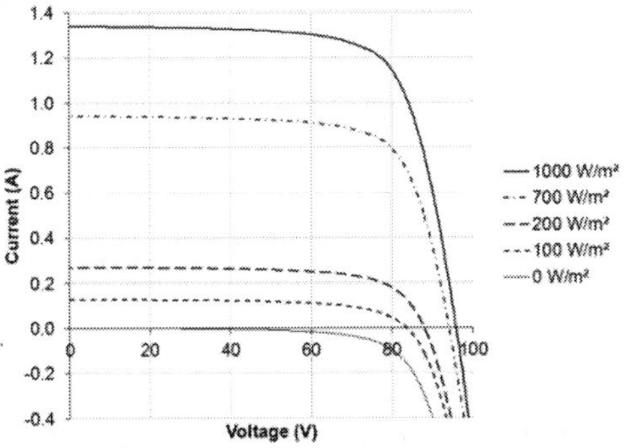

Figure 2: I-V curves of a Manz CIGS module at different irradiances, used as the basis of I-V curve simulation.

temperature coefficients for voltage and current of -0.29 %/K and +0.05 %/K, respectively. Only voltage differences, not current differences, play a role for the subsequent mismatch calculation of strings connected in parallel, so the temperature coefficient for voltage is the essential one. This calculation was performed for each minute of the year and for each of the four façades, using the irradiance and module temperature data from the PV*SOL simulation.

2.4 Calculation of DC energy yields

To simulate the polystring configuration, i.e., the parallel connection of modules from all four orientations, the I-V curves of the four façades were added (adding the respective currents) for each simulated minute. The output power of the polystring configuration was determined from the maximum power point of the resulting I-V curve, scaled to the nominal power of 1 kWp per orientation. Numerical integration of this power (i.e., addition of the power values for every simulated minute, multiplied with the time difference of one minute) yielded the annual DC energy yield for the polystring configuration.

For comparison, the same calculation was performed for each of the four orientations separately, using the individual MPPs of the four façades. The sum of these four energy yields is the annual DC energy yield of the PV system in conventional configuration.

3 RESULTS AND DISCUSSION

3.1 MPP voltage mismatch

Fig. 3 shows the simulated I-V curves for modules on each of the four façades at 9:00 in the morning on a sunny summer day as an example. According to the PV*SOL simulation, the east façade receives relatively strong irradiance with almost 700 W/m²; the south façade, less than half of that value. Consequently the modules on the east façade are significantly warmer (33.9 °C for east vs. 23.2 °C for south). Both west and north façade receive only indirect light at this moment.

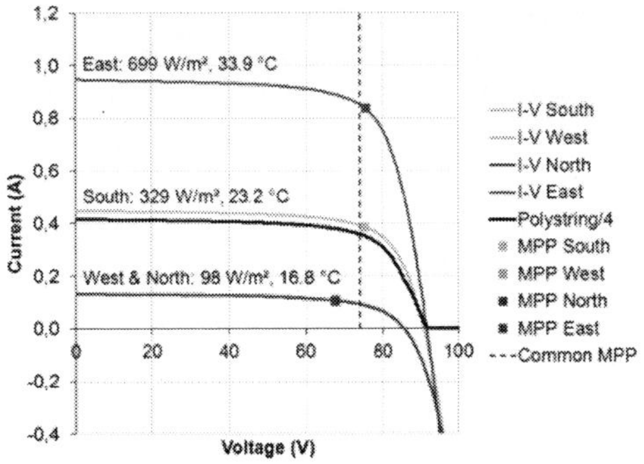

Figure 3: Simulated I-V curves for the four orientations (coloured) and summation curve for their parallel connection (black, scaled to the same nominal power) at 9:00 on a sunny summer day. The squares show the individual maximum power points of the four orientations; the vertical dashed line marks the common MPP voltage at which the modules operate in polystring configuration.

The individual MPP voltages of the east and the south façade, marked by the squares in Fig. 3, differ by less than 0.5 %. The temperature difference alone would cause the voltage for east-facing modules to be about 3 % lower than for south-facing modules; the irradiance difference alone would cause a voltage difference of more than 3 % in the opposite direction. It can be seen that these two linked effects on the voltage compensate each other remarkably well for the Manz CIGS modules considered here. Only at low irradiance, as on the west and north façade at this moment, the MPP voltage differs more strongly.

The black I-V curve in Fig. 3 is the average of the four individual I-V curves (currents added, then divided by 4 for easier comparison) and thus represents the polystring configuration. The MPP voltage of this I-V curve is marked by the dashed vertical line; this is the voltage that the modules of all four façades operate at when connected in parallel. It is determined mainly by the façades with the highest irradiance and accordingly is close to their individual MPP voltages, so the mismatch power loss for these façades is small. The individual MPP voltages of the weakly-irradiated west and north façade are about 9 % below the common MPP voltage, but as neither of these façades contributes significant power, the total power loss due to voltage mismatch at the moment shown here is only 0.85 %.

Fig. 4 shows the MPP voltages resulting from this simulation over the course of a complete sunny summer day, along with the irradiance values for each façade. It can be seen that the MPP voltages for different orientations are practically identical as long as the corresponding irradiance is above 200 W/m². This shows that there is good compensation of temperature and irradiance effects on the voltage over a wide irradiance range. The common MPP voltage in polystring configuration is always close to and slightly below the individual MPP voltage of the façades with highest irradiance. The slight overall voltage decrease towards the afternoon, caused by rising ambient temperature, affects all orientations and therefore does not influence the mismatch significantly.

For overcast days, exemplified in Fig. 5, the irradiance values on the four façades are very close to each other, and so are the MPP voltages. Hence the mismatch loss caused by polystring configuration is negligible on such days.

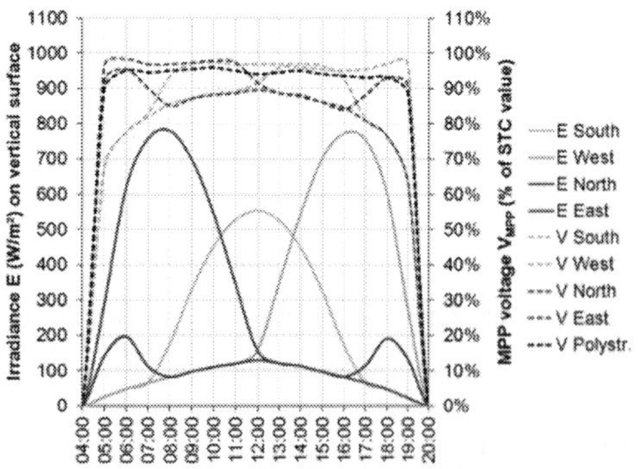

Figure 4: Simulated irradiance (solid lines) and MPP voltage (dashed lines) for the four façades on a sunny summer day, hourly values. Also shown is the resulting common MPP voltage for polystring configuration (black dashed line).

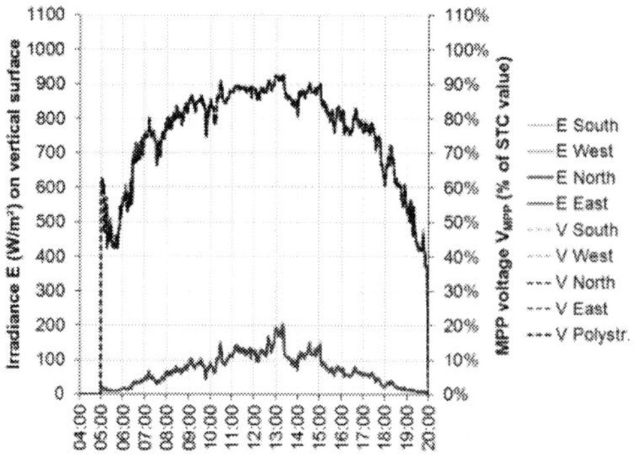

Figure 5: Simulated irradiance (solid lines) and MPP voltage (dashed lines) for the four façades on an overcast summer day. Also shown is the resulting common MPP voltage for polystring configuration (black dashed line). The curves for different orientations are nearly identical and therefore cannot be distinguished.

3.2 Annual energy yields and mismatch loss

Table I presents the annual DC energy yields calculated from the simulation results for each façade in conventional configuration as well as for the whole system in conventional and in polystring configuration. The loss in annual DC energy yield for the polystring configuration, which is caused by the voltage mismatch between different façades, is only 1.1 % compared to the conventional configuration.

Table I: Annual DC energy yields for individual façades and for the complete BIPV system in conventional and polystring configuration.

Orientation	Annual DC energy yield (kWh)
Conventional configuration:	
South (1 kWp)	827
West (1 kWp)	609
North (1 kWp)	295
East (1 kWp)	636
Total (4 kWp)	**2367**
Polystring configuration:	
Total (4 kWp)	**2341**

3.3 Inverter sizing

The inverter of a PV system needs to match the power delivered by the modules. In reality as well as in simulations with sufficient time resolution, however, a few short but high power peaks occur, in which the modules may even deliver more than their nominal power. This can happen, for example, if the sun comes out between clouds, which provide extra reflection, while the modules are still cool. Cutting off such power peaks by choosing a smaller inverter reduces energy yield only marginally and thus is the more economical option than using an inverter big enough even for the highest power peaks.

The trade-off between inverter size and energy yield loss can be determined from the simulation results, too. Fig. 6 shows the required inverter power as a function of cut-off loss for each orientation in conventional configuration (four individual, differently-sized inverters) as well as for the total system in polystring configuration (one common inverter).

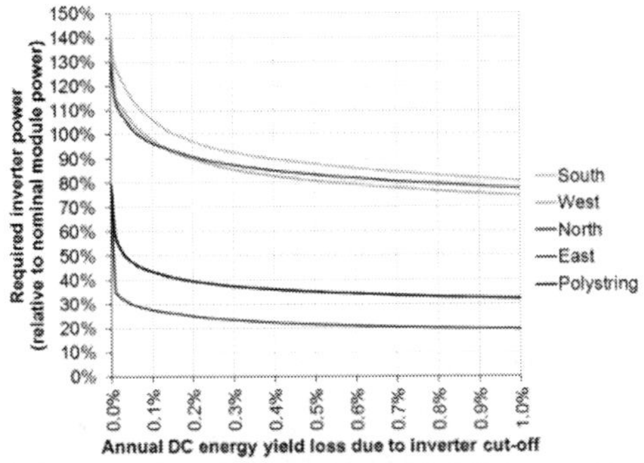

Figure 6: Required inverter power (relative to nominal module power) for each of the four façades in conventional configuration (coloured) and for the complete PV system in polystring configuration (black) as a function of annual DC energy yield loss due to inverter cut-off.

The exact economic optimum of inverter size and corresponding cut-off loss depends on many factors. So for further discussion, the acceptable cut-off loss is set to 0.5% for both configurations here. Table II lists the required inverter sizes for this value. As can be seen, the total required inverter power for conventional configuration is 68.3 % of the nominal PV module power, split up into four inverters of very different sizes. The polystring configuration, on the other hand, requires only one inverter with a power of 35.1 % of the nominal PV module power. So applying the polystring configuration to such a PV system not only reduces the number of inverters from four to one, but at the same time halves the required total inverter power. Choosing a different cut-off loss value does of course alter the absolute amounts of required inverter power, but the inverter power ratio for polystring/ conventional configuration of roughly ½ is valid over a reasonable range of cut-off loss values.

Table II: Required inverter power for the four façades in conventional configuration and for the complete BIPV system in polystring configuration, both calculated for a cut-off loss of 0.5 % annual DC energy yield.

Orientation	Required inverter power (% of nominal module power)
Conventional configuration:	
South	87.7
West	80.8
North	21.4
East	83.4
Total	**68.3**
Polystring configuration:	
Total	**35.1**

4 CONCLUSIONS

The simulation presented here shows that the polystring configuration is suitable even for BIPV systems with PV modules facing in many different directions. The influences of irradiance and temperature on module voltage compensate each other very well for irradiance values above 200 W/m². Hence the energy yield losses caused by voltage mismatch are small and amount to only 1.1 % over a full year for the PV system presented here. Compared to the conventional, individual-inverters-per-orientation configuration, the polystring configuration not only requires fewer inverters and thus simplifies the system configuration, but it also halves the total required inverter power by making use of the fact that the irradiance maxima of the four sides occur at different times. These economic advantages far outweigh the minimal losses in energy yield. The polystring configuration should therefore seriously be considered when planning BIPV systems.

REFERENCES

[1] D. Staudacher, 25th European Photovoltaic Solar Energy Conference and Exhibition / 5th World Conference on Photovoltaic Energy Conversion, 2010, Valencia, Spain
[2] T. Straub et al., 27. Symposium Photovoltaische Solarenergie, 2012, Bad Staffelstein, Germany
[3] Software "PV*SOL premium", version 7.5, http://www.valentin-software.com/

APPRECIATING PERFORMANCE OF A BIPV LAB IN BANGALORE (INDIA)

MONTO MANI, GAYATHRI AADITYA, BALAJI, N.C.
Indian Institute of Science, Bangalore: 560012, INDIA
Email: monto_mani@hotmail.com

ABSTRACT: A BIPV (Building Integrated Photovoltaic) is essentially PV integrated as a building envelope (roof, wall, façade, and fenestration) regulating indoor thermal comfort and generating on-site energy. Tropical regions are solar rich (>4 kWH/m2/day) and generally warmer (20~30°C) and impose a specific need for preventing indoor heat build-up and increased natural ventilation. Roofs have been found to be the single largest contributor of indoor heat gain, and with PV integration this gain is amplified due to low thermal mass and higher radiation. This results in a stratified temperature gradient between the (PV) roof and the floor below in conjunction with a higher natural buoyancy of warm air below the PV panel.

In tropical conditions, PV performance is characterized by higher a higher operating temperature that is generally detrimental to its efficiency and aggravates indoor heat buildup when integrated as the roof. The interplay between indoor thermal comfort and PV performance needs to be carefully investigated particularly the influence of stratified indoor temperature on thermal comfort and the natural buoyancy induced ventilation on PV performance. The current study is an experimental investigation into a 5.25 kWp roof integrated BIPV lab (see Figure 1) located at the campus of the Indian Institute of Science, Bangalore (India).
Keywords: Building Integrated PV (BIPV), Thermal Performance, Energy Performance, Ventilation

1 INTRODUCTION

BIPV is an integration of photovoltaics and buildings where PV also plays the role of a building envelope besides generating electricity. This integration of PV in buildings needs to be carefully discerned based on both the thermal (comfort) performance of the building and PV output. The current paper primarily presents salient observations based on the performance of a 5.25 kW_p BIPV system installed at the Centre for Sustainable Technologies, Indian Institute of Science (IISc), Bangalore, India. While buildings are evaluated by their thermal performance, PV is evaluated based on its output. With PV as the building envelope it is necessary to understand the influence of PV performance on buildings and vice versa (Figure 1). For successful BIPV integration, both performances (thermal and electrical) need to be evaluated and provided due consideration.

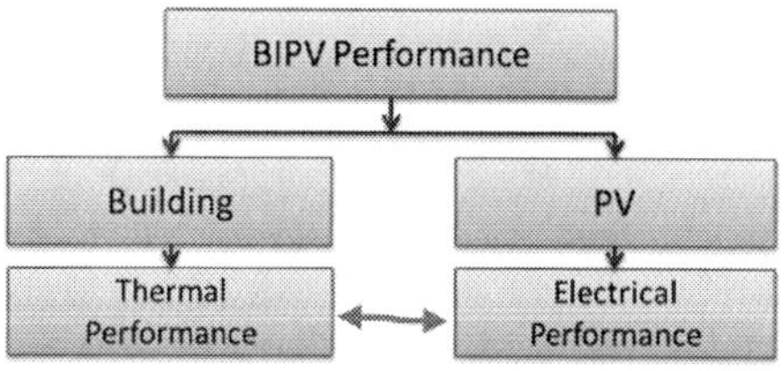

Figure 1: BIPV performance

For BIPV as compared to ground-mounted, the operating ventilation regimes vary with these being more confined with PV integrated in buildings. Building orientation and solar geometry are fundamental to both building thermal performance and PV performance. Just as indoor thermal comfort (temperature) determines the performance of a building, the PV cell temperature also determines its output. Both thermal comfort and cell-temperature can be directly or indirectly traced to the building (and PV) design-exposure to the sun. The temperature of a solar cell in operation increases phenomenally (especially true in tropical regions), resulting in decrease in the output.

The cell temperature depends on the incident solar radiation, ambient air temperature, air-flow, radiant surfaces, and cooling mechanisms and temperature distribution [1]. Temperatures of PV panel can reach as high as 95°C [2] in which case wind movement above and below the panel can aid in lowering the panel temperature and thereby improve system out. An earlier study [3] indicated that wind speeds above 5 m/s can bring down cell operating temperatures by 10°C, thereby enhancing the cell conversion efficiency and the power output of the system. The operating wind-regimes and its effectiveness in (natural) thermal management of PV cell temperature varies from a ground mounted system and a BIPV. The wind-regimes are relatively confined in a BIPV and less effective in naturally regulating (lowering) PV cell temperature (see Figure 2). Ventilated air-cavities behind PV panels in a BIPV (roof and/or façade) can have a significant bearing on lowering cell temperature.

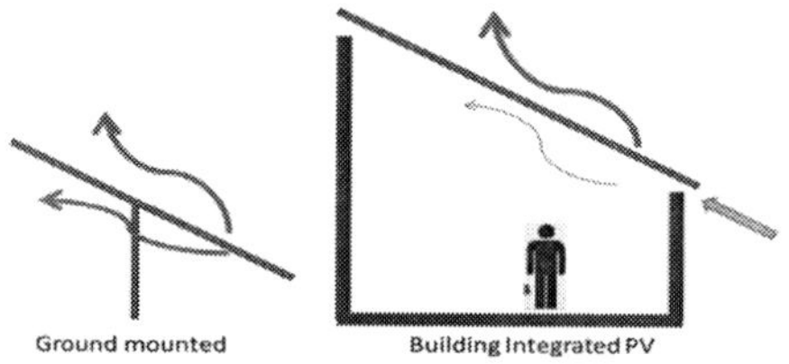

Figure 2: Wind (ventilation) regimes between a ground-mounted PV vs BIPV

Façade PV systems without gap (for air movement/ventilation) have been found to have a 10 % loss of performance, with 5 cm gap a 5% loss and for a 15 cm the loss was found to be negligible [4]. "*The individual frames of each BIPV module can intrude into the air flow path. In order to reduce internal pressure drops in the cavity, the rear of the BIPV needs to be smooth, and the cross-section of the duct large. However, a smooth surface minimizes the area for heat transfer, and a large cross-section reduces the air velocity over the surface*" [5]. The current paper provides a preliminary insight into BIPV performance based on a real-time study on a 5.25 kW_p BIPV facility. Salient observations made are in terms of:

- Stratified room temperatures below PV roof
- Ventilation under PV panels impacts its operating temperature
- Heat Gain/Loss through the PV roof crucial in determining indoor thermal comfort
- Low thermal-mass of PV greatly influences building heat gain/loss

2 METHODOLOGY AND STUDY OVERVIEW

The methodology includes, firstly discerning peculiarities of the building following PV (roof) integration, followed by an appraisal of PV performance. Studies into the thermal performance of the building and the PV roof have been based on real-time measurements. These included air and surface temperature measurement, heat-flux measurements and PV performance measurements based on an integrated Grid-tie Inverter. Following these real-time measurements, specific study was conducted on the role and performance of the masonry walls and PV as a building envelope based on a BIPV case-study. This included studies into transmittance, admittance, time-lag and decrement factor. Relevant codes were referred to.

3 BIPV CASE-STUDY

A dedicated 5.25 kW_p BIPV facility at IISc Bangalore has been studied real-time for both building thermal response and PV output. Bangalore lies Latitude 12.58° E with moderate climate. The PV panels are mono-crystalline 2006 manufactured. The roof is installed south-facing with a 15 Deg slope and ventilators installed at the top and bottom edge of the roof. This is to lower the temperature of the PV panels with natural ventilation drafts below the panel. The building comprises a load-bearing masonry structure of Stabilized Mud Blocks (SMB) with a PV integrated roof (see Figure 3) with intermittent glass panels for natural lighting.

Figure 3: BIPV lab at the Indian Institute of Science

4 BUILDING THERMAL PERFORMACE

Studies on the thermal performance of the building included temperature measurements at various levels from the floor. Characteristic temperature stratification was observed below the PV roof. The stratification was more uniform with the ventilators closed. The PV panel temperatures often exceed 80°C resulting in consistently high air-temperatures below the panel (see Figures 4 and 5).

It can be observed from the figures that near (38.1mm in Figure 4 and 0.5m in Figure 5) the PV panel temperatures are high in both cases of ventilators opened and closed conditions. Thereafter, the temperatures are showing the decreasing trend in ventilators opened conditions, than in closed conditions. Ventilation however does not significantly change the indoor thermal comfort conditions (discussed subsequently).

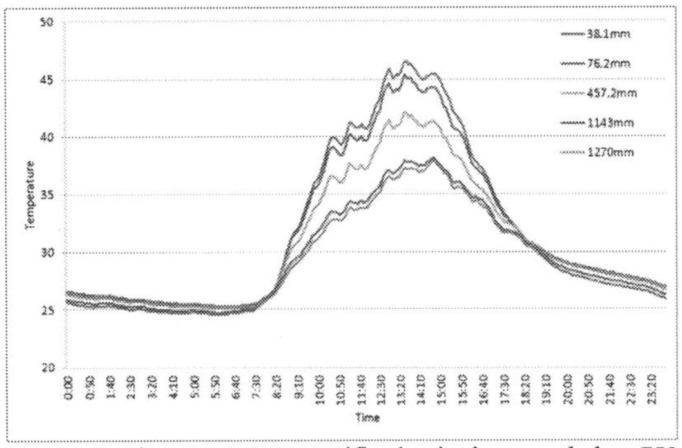

Figure 4: Air-temperature stratification in the room below PV panel with ventilators closed (1 July 2011)

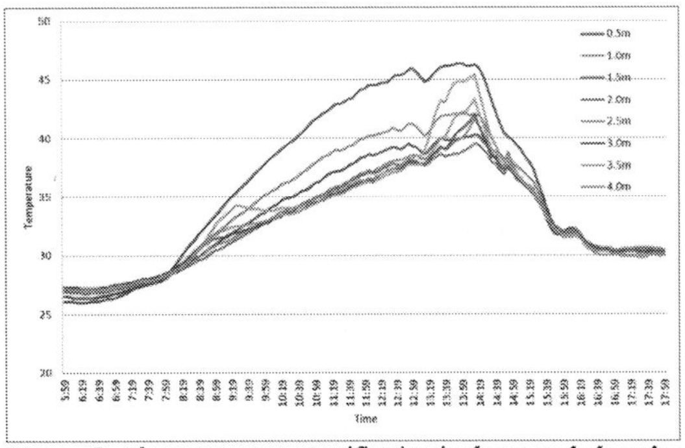

Figure 5: Air-temperature stratification in the room below the PV panel with ventilators open (17 April 2012)

4.1 HEAT-FLUX STUDIES

A real-time study was carried out using Portable Heat Flow Meter (KEM HFM-201) to measure heat flow through all the four-walls and underneath the PV panels. Heat-flux was recorded for two days one each in November 2010 and June 2011 at an interval of one hour (see Figures 7 and 8). The transmittance across the PV roof was found to be the single largest contributor of heat gain and loss in the building (see Figure 6). During mid-day the PV roof was responsible for 54% of heat-flux within the building, while at mid-night its share was nearly 43%. It thus becomes inevitable to further study the individual thermal performances of the building envelope.

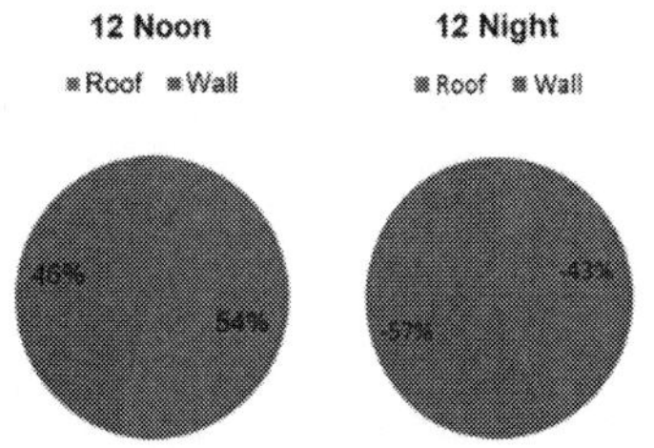

Figure 6: Heat-flux share of the PV roof

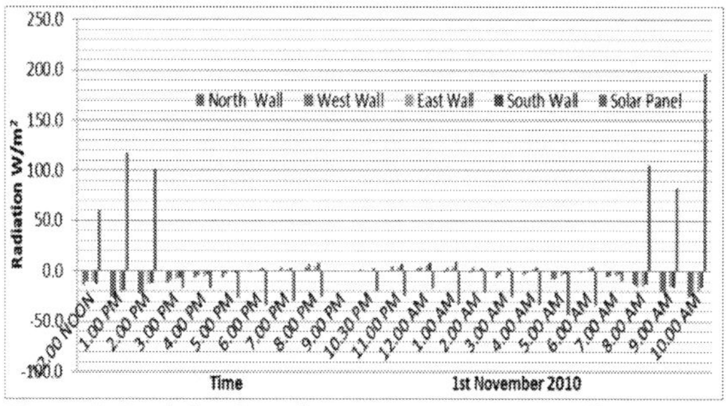

Figure 7: Real-time heat flux measurements across five building elements (four SMB masonry walls and one PV roof) in November 2011

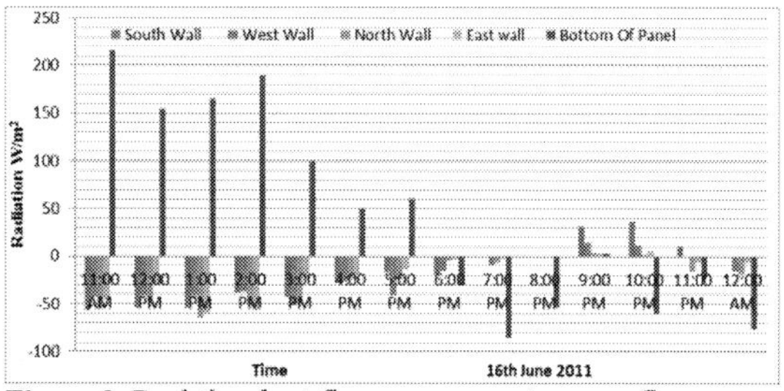

Figure 8: Real-time heat flux measurements across five building elements (four SMB masonry walls and one PV roof) in June 2011

4.2 DETAILED THERMAL STUDY

It is appropriate to understand the properties of PV as a building material that has an effect on the thermal performance of the BIPV system in response to external climatic condition. Based on the real-time observations, theoretical analysis was performed to understand the influence of the wall and PV panel roof on inside air temperatures. The influence of SMB masonry wall on indoor air temperature was studied by dividing the room into two thermal zones (Zone 1 and Zone 2) (see Figure 9). This was justified based on the observed air-temperature stratification observed below the PV roof (see Figures 4 and 5). Further, the influence of PV roof on the indoor thermal environment was studied under two conditions, viz., ventilator open and closed conditions. The study is based on the measured air velocity in the particular zones and conditions of the room. Air velocity is a basic parameter required for calculating material heat transfer coefficients, which in turn influences the thermal performance of the wall and/or roof elements. Since the ventilation regimes (near the PV) and air temperatures varied across the height of the room, the observed air-velocities also differed. ISO method was used to study dynamic thermal parameters such as thermal admittance, time lag, decrement factor and thermal transmittance (U-value), as influenced by internal convective heat transfer coefficients (hi) due to varying air velocities.

The surface heat transfer coefficient is influenced by the wind velocity and direction, surface shape and roughness and, temperature difference between surface and air [6]. For actual outside building-surface, Ito et al. [7] obtained h_o as a function of wind speed (v_f) and direction as:

$$h_o = 18.63\ V^{0.605} \text{ in W/m}^2\text{ K} \quad (1)$$

where, for windward surface, velocity can calculated as per Equations 2a and 2b

$$V = 0.25\ v \text{ for } v_f > 2 \text{ m/s} \quad (2a)$$

$$V = 0.50\ v \text{ for } v_f < 2 \text{ m/s} \quad (2b)$$

For the leeward surface, velocity is calculated as

$$V = 0.3 + 0.05\ v_f \quad (3)$$

where, v_f is the air velocity in m/sec

In this study, the outside surface heat transfer coefficient (h_o) 18.63 W/m^2 K was considered constant. Inside surface heat transfer coefficient (h_i) are computed for the corresponding thermal zones and air velocities (see Table 1).

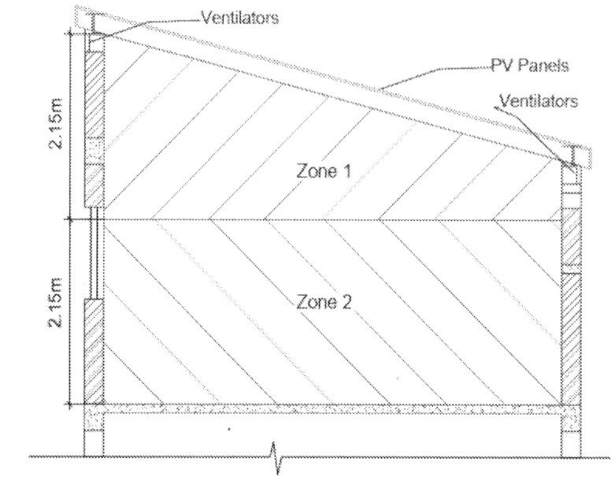

Figure 9: BIPV Building study thermal zones (1 & 2)

From measured data, the average air velocity in Zone 1 and Zone 2 was found to be 1.8 and 0.2 m/s respectively. The respective inside surface heat transfer coefficients were computed based on Equation 1. The outside heat transfer coefficient is taken as constant of 19.86 W/m^2 K based on Indian Standard IS 3792-1978 [8].

Table 1: Surface heat transfer coefficients

Zones (for wall)	Outside surface heat transfer coefficients (h_o) (W/m^2 K)	Inside surface heat transfer coefficients (h_i) (W/m^2 K) (air velocity)
Zone 1	19.86	10.54 (1.8m/s)*
Zone 2	19.86	9.17 (0.2m/s)*
Conditions (for PV Panel roof)		
Ventilator opened	19.86	10.54 (1.8m/s)*
Ventilator closed	19.86	9.70 (0.8m/s)*

( )* - air velocity

In the case of the SMB wall, thermal performance parameters are shown in Table 2.

Table 2: Dynamic thermal properties of the CSSB wall envelope

Zone	Inside surface heat transfer coefficient (hi) (W/m^2 K)	Admittance (W/m^2 K)	Decrement factor	Time lag (Hours)	U value
Zone 1	10.54	6.345	0.592	5.278	3.416
Zone 2	9.17	5.857	0.572	5.389	3.260

For the PV roof, thermal performance parameters were studied for two cases, i.e. with ventilators below are roof in open and closed conditions. The air velocities at the surface of the PV panel varied between 1.8 ~ 0.8 m/sec for open and closed conditions respectively. The thermal performance parameters of the PV roof are shown in Table 3.

Table 3: Dynamic thermal properties of the PV panel roof

Vent	Inside surface heat transfer coefficient (h_i) (W/m^2 K)	Admittance (W/m^2 K)	Decrement factor	Time lag (Hours)	U value
open	10.54 (1.8m/s)	6.782	1.000	0.043	6.781
close	9.70 (0.8m/s)*	6.433	1.000	0.045	6.432

* IS 3792-1978 [8] specified h_i value

The inside surface heat transfer coefficient (h_i) under ventilator closed condition is 9.70 W/m^2K which is nearly that of still air. IS 3792-1978 [8] recommends inside film coefficient at still air to be 9.36 W/m^2 K

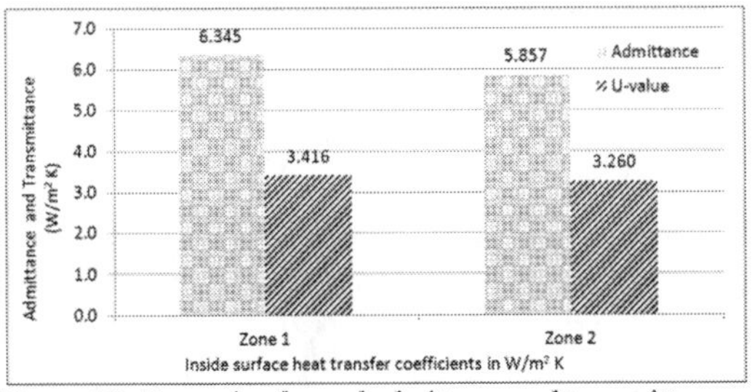

Figure 10: Zone-wise thermal admittance and transmittance of SMB masonry wall

As the inside surface heat transfer increases from Zone 2 to Zone 1, the admittance values decreases and the corresponding U-value also decreases (see Figure 10). Heat flow through the walls in Zone 1 is comparable to Zone 2, due to high thermal transmittance values. Further the value of time lag and decrement factor also decreases (see Figure 11). There is however not much variation in time lag and decrement factor between the two zones. Balaji et. al [9] earlier also reported the influence of surface heat transfer coefficient to be negligible on time lag and decrement factors.

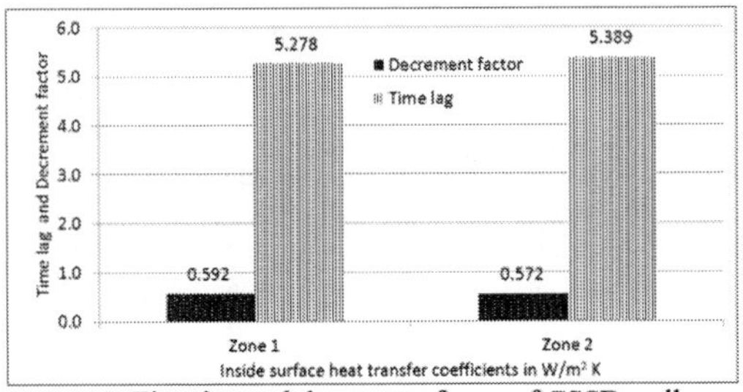

Figure 11: Time lag and decrement factor of CSSB wall at Zone 1 and Zone 2

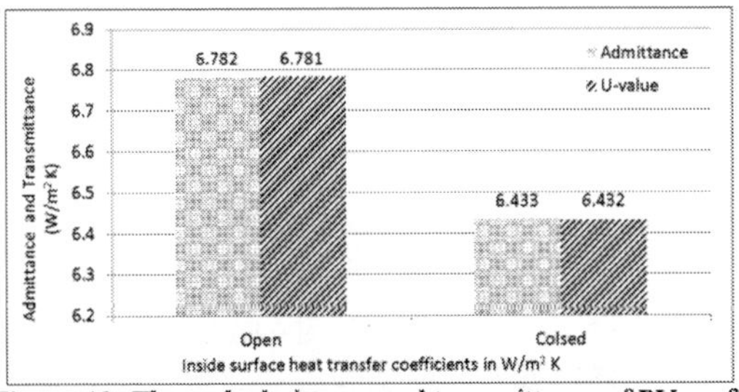

Figure 12: Thermal admittance and transmittance of PV roof under ventilator open and closed condition

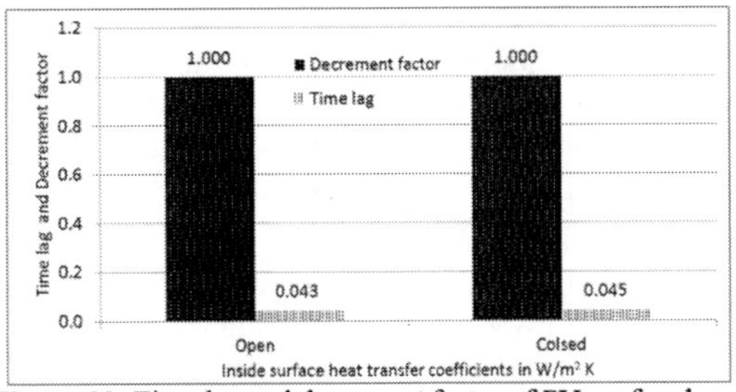

Figure 13: Time lag and decrement factor of PV roof under ventilator open and closed condition

For the PV panel roof, between open and closed ventilator condition, the admittance and transmittance values tend to reduce (see Figure 12) while the time lag and decrement factor remain comparable (see Figure 13). The noticeably higher admittance and transmittance values under open conditions may be attributed to the increased air-speed below the PV panel, which increases the surface heat transfer coefficient by increasing the heat flow across the panel. Another reason may also be the low thermal mass of 0.0083 J/K m^2 for PV when compared to SMB masonry wall with higher thermal mass of 0.370 J/K m^2 (see Table 4). The heat storage capacity of PV is thus much lower than SMB, resulting in much higher heat transfer through PV panels.

Table 4: Thermal mass of few building materials

Bldg part	Material	Specification				Thermal mass
		Thick (mm)	Conductivity (W/m-K)	Specific Heat (J/kg- K)	Density (kg/m^3)	(ρ Cp) (x 10^6) (J/K m^2)
Wall	Brick# wall	230	0.811	880	1820	**0.368**
Wall	Cellular# concrete	100	0.188	1050	704	**0.74**
Wall	Dense# concrete	100	1.74	880	2410	**0.21**
Roof	RCC roof#	100	1.58	880	2288	**0.20**
Wall	SMB	230	1.56	900	1800	**0.37**
PV Roof	PV (5 layers configurations)*					**0.0083**
	Glass	3	1.8	500	3000	
	EVA	500×10^{-9}	0.35	2090	960	
	PV Cell	225×10^{-9}	148	677	2330	
	EVA	500×10^{-9}	0.35	2090	960	
	Tedlar	0.1	0.2	1250	1200	

(IS 3792 - 1978 2004) [8]
*(Armstrong & Hurley 2010) [10]

Further, from Tables 2 and 3 it is evident that the time lag and decrement factor of a building envelope (masonry wall and PV) depends on the thermal mass, which plays a major role in regulating the indoor thermal performance. The thermal mass (heat storing capacity) of the PV can be understood through its time lag and decrement factor [11]. Overlooking PV as an energy generating system, its low time-lag and decrement factor makes it unsuitable for building roofs in tropical conditions. However, appropriate design including ventilated double roofs could alleviate problems associated with high heat gain/loss.

4.3 PV PERFORMANCE AND THERMAL COMFORT

Thermal comfort regimes under stratified conditions have been studied in conjunction with PV performance and its operating temperatures. Indoor air temperatures at different levels at room during typical winter and summer periods during mid-day and night have been measured. Figures 14 and 15 illustrate temperature stratification in relation to average human seating (1.1 m) and standing (1.8 m) height from the floor. During the day there is a range of ~7°C and ~3.2°C between November and June, while at night the range is much lower between ~0.8°C and ~0.9°C between November and June.

The PV back-sheet temperature and the air temperature at different levels where measured, from roof to ground at an interval of 0.5m. It was also observed that due to PV's low thermal mass, the ceiling temperature follows the solar radiation trend, i.e. the cell (panel) temperature peaks with a peak in the solar radiation (around solar noon), and reaches in excess of 60°C.

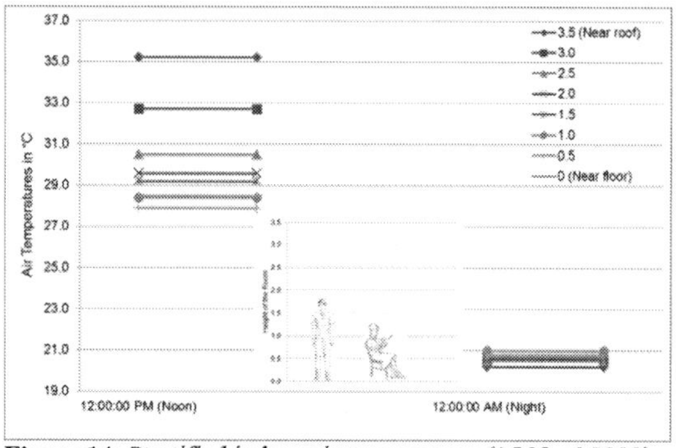

Figure 14: Stratified indoor air temperatures (15 Nov' 2013)

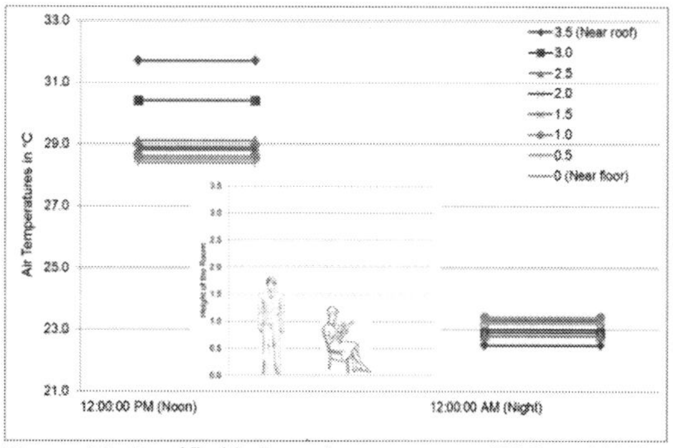

Figure 15: Stratified indoor air temperatures (15 June 2013)

In April the energy generated from the PV roof is around 17.75 kWh (units) instead of 22.8 units due to high operating temperatures. Similarly in September, around 17.63 units were generated as against 22.73 units due to temperature related losses. Figures 16 and 17 illustrates various PV and thermal comfort parameters for typical April and September days. Futher, due to high operating cell temperature and thermal transmittance the indoor temperatures are higher than ambient outdoor themperature predominantly through the year. To mitigate discomfort nearly 1.2 units of electricity would be required daily to operate (100W) fans.

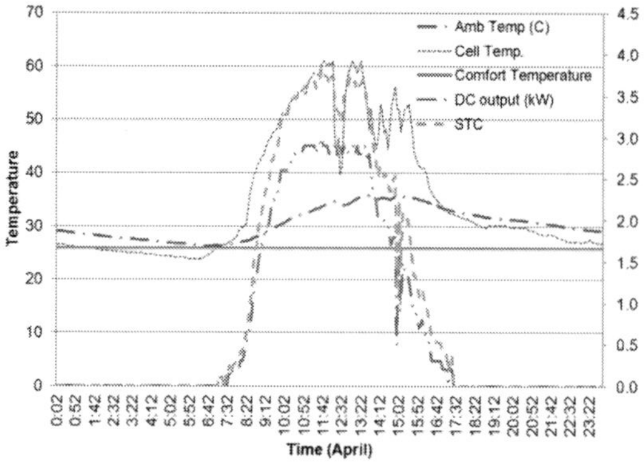

Figure 16: Typical daily PV performance and indoor thermal comfort (April)

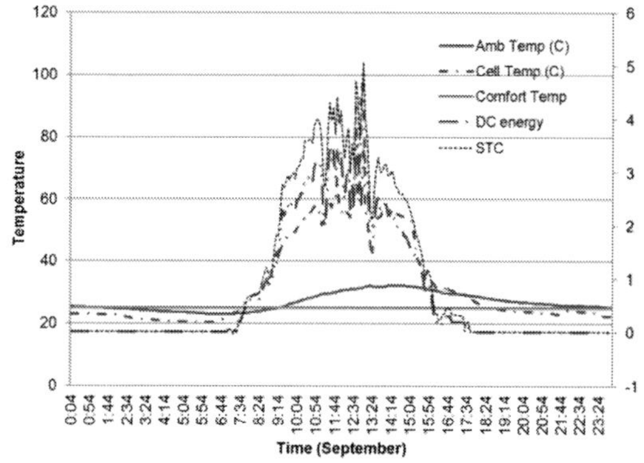

Figure 17: Typical daily PV performance and indoor thermal comfort (September)

To further understand the influence of cell-temperature under open and closed ventilator condition on PV performance a comparison was made between performance of actual (observed) vs. theoretic PV output. The trends during early and late hours of the day show a better performance of the open system over the closed. This could be attributed to improved ventilation below the panel (further possibly aided by external wind movement). However mid-day performance seems comparable when the cell-temperatures are high for both open and closed ventilator systems. However, these observations need further study and substantiation.

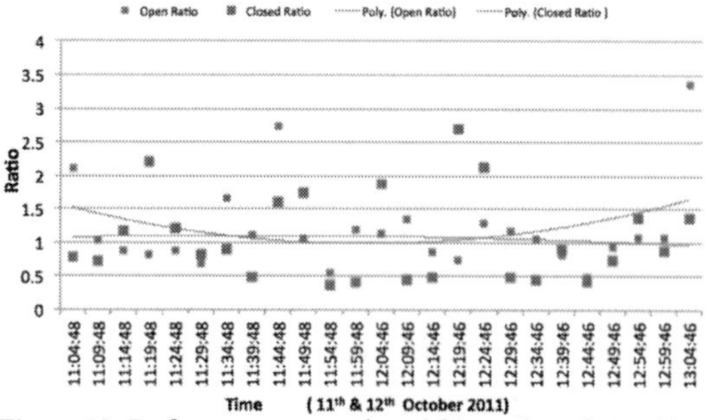

Figure 18: Performance comparison (observed vs. theoretic) between open and closed PV output

5 CONCLUSIONS

BIPV performance appraisal requires interconnected investigation involving building thermal performance and PV performance. Both systems fundamentally operate in response to the sun. The current study involves real-time study into a roof integrated BIPV in Bangalore, India. With PV as the roof, high cell temperatures (average 70°C) and low thermal mass of PV 0.0083 J/K m^2 have a significant bearing on indoor thermal comfort and PV performance. Both thermal comfort and building thermal performance requires consideration of stratified air-temperatures occurring below the roof. It was also observed that time lag and decrement factor of PV panel roof are low and can be attributed to low thermal mass of PV. Ventilation plays major role in both electrical and thermal performance of BIPV.

6 ACKNOWLEDGMENT

This work is partially supported by the Robert Bosch Centre for Cyber Physical Systems (RBCCPS) at the Indian Institute of Science, Bangalore. Further in part under the US–India Partnership to Advance Clean Energy-Research (PACE-R) for the Solar Energy Research Institute for India and the United States (SERIIUS), IUSSTF/JCERDC-SERIIUS/ 2012 dated 22nd November 2012.

The authors sincerely thank the management of the Electronics Division—Bharat Heavy Electricals Limited, Bangalore (India) and the Indian Institute of Science, Bangalore for facilitating the BIPV experimental facility at the Center for Sustainable Technologies.

7 REFERENCES

[1] Brinkworth, B.J., Marshall, R.H. & Ibarahim, Z., 2000. A Validated Model of Naturally Ventilated PV Cladding. Solar Energy, 69(1), pp.67–81.

[2] Schams, B. & TamizhMani, G., 2011. BAPV modules with different air gaps: Effect of temperature on relative energy yield and lifetime. In 37th IEEE Photovoltaic Specialists Conference, Washington, USA, pp. 3213–3217.

[3] Rao, A. & Mani, M., 2013. Evaluating the Nature and Significance of Ambient Wind Regimes on Solar Photovoltaic System Performance. In M. Baratieri and F. Patuzzi (eds.). Building Simulation Applications BSA 2013, Bozen-Bolzano University Press, pp. 395–405.

[4] Prasad, D. & Snow, M., 2005. Designing with Solar Power: A Source Book for Building Integrated Photovoltaics, Earthscan.

[5] Norton, B. et al., 2011. Enhancing the performance of building integrated photovoltaics. Solar Energy, 85(8), pp. 1629–1664.

[6] Cole, R.J. & Sturrock, N.S., 1977. The convective heat exchange at the external surface of buildings. Building and Environment, 12(4), pp. 207–214.

[7] Ito, N., 1972. Field experiment study on the convective heat transfer coefficient on exterior surface of a building. Journal Name: ASHRAE Trans.; (United States); Journal Volume: 78, pp. 184–191.

[8] IS 3792 - 1978, 2004. Guide for heat insulation of non-industrial buildings, New Delhi.

[9] Balaji N.C., Monto Mani & Venkatarama Reddy, B.V., 2013. Thermal performance of the building walls. In M. Baratieri and F. Patuzzi (eds.), Building Simulation Application 2013, Bozen-Bolzano University Press, pp. 151–160.

[10] Armstrong, S. & Hurley, W.G., 2010. A thermal model for photovoltaic panels under varying atmospheric conditions. Applied Thermal Engineering, 30(11-12), pp.1488–1495.

[11] Balaras, C.A., 1996. The role of thermal mass on the cooling load of buildings: An overview of computational methods. Energy and Buildings, 24 (1996), pp. 1—10.

OUTDOOR CHARACTERIZATION OF INNOVATIVE BIPV MODULES FOR ROOF APPLICATION

C. Polo Lopez[1], F. Frontini[1], P. Bonomo[1], F. Cais[2], D. Salvador[2]
[1]SUPSI, Institute for Applied Sustainability to the Built Environment (ISAAC), Lugano, Switzerland
[2]Tegola Canadese, Via dell'industria, 21 - 31029 Vittorio Veneto (TV) Italy
Mail to: francesco.frontini@supsi.ch; phone +41 (0)58 666 63 20 fax +41 (0)58 666 63 49

ABSTRACT: In the framework of the European project Construct PV, two different real scale roof mock-up have been installed, one at SUPSI campus in Lugano (Switzerland) and one at Tegola Canadese Headquarter in Vittorio Veneto (Italy). Purpose of this paper is the investigation of the BIPV solar shingle developed within the project that consists of a durable laminated PV glass (containing crystalline solar cells) integrated onto a waterproofing back membrane bitumen-based with the aim of ensuring an easy-mounting and on-site customization. The test-facilities are aimed to demonstrate and to show the Construct PV technology (visits, professional development and student's training, etc.) as well as to monitor the main energy performance of the new BIPV shingle in comparison to other BIPV roof systems already on the market. Electrical BIPV performances (energy production, performance ratio, temperature of modules, etc.) and some building performances (watertightness, temperatures, ventilation, etc.) are analyzed and discussed, showing strengths and weaknesses of the technology in a multicriteria perspective involving both energy and constructive requirements. This paper presents the preliminary achievements of the monitoring campaign carried out during the first months of installation.

Keywords: BIPV, Building Integration, Performance, Roof integration

1 INTRODUCTION

Renewable energy systems, such as Photovoltaics (PV), play an important role in the scenario identified by the European Directive on Net Zero Energy Buildings (ED 2010/31/EU) [1], and are mandatory in order to reduce the energy demand of buildings and to contribute to a more rational use of the energy resources.

Within the European project Construct-PV (http://www.constructpv.eu/) new solutions have been developed with the purpose of achieving a cost-effective integration of photovoltaic (PV) technologies in the building envelope, both in roof and façade. Construct-PV research project (FP7-ENERGY-2011-2) has the goal to develop and demonstrate customizable, efficient and low cost BIPV for opaque surfaces of buildings [2-4]. The consortium involves selected partners, connecting all the actors of the value chain, including manufacturers, architects, research centers and contractors working on PV and building sectors.

In a first part of the project a detailed analysis of the BIPV market (WP1) has been performed in order to evaluate the attractiveness of innovative BIPV solutions and to understand their evolving opportunities and threats [5]. Furthermore, other aspects as the Life Cycle Analysis (LCA) and Life-cycle cost analysis (LCCA) have been investigated to determine the most cost-effective options among different competing alternatives [6]. All these analyses contributed to identify the essential requirements and features which have been translated into product technical specifications during conceptual and detailed design of Construct-PV solutions. The process involves the manufacturing phase (WP2 and WP3), the development and demonstration of a radical evolution of BIPV opaque products, from BIPV-modules to multifunctional building systems with added functionalities (WP4), besides considering these new components within an integrated approach of the whole building process (WP5). For this reason, a set of demonstrative installations have been realized within the project, including two small scale test-facilities (that will be discussed in this paper) for roof and a real-scale mock-up, prepared by the coordinator of the project Züblin AG for the façade systems, that will have a round-trip around European events in next years [7]. Moreover, two real scale buildings will also be used as large-scale demonstrators in Athens and Stuttgart and they will be completed in the next months to respectively show the roof and façade BIPV technologies.

An important part of the process of product development is undoubtedly establishing their technical, technological and constructive requirements to meet the recent new standards for BIPV products [8], such as the essential building requirements as specified in the European Construction Product Regulation CPR 305/2011, and the applicable electro-technical requirements as stated in the Low Voltage Directive 2006/95/EC / or CENELEC standards. For this reason it has been necessary to perform specific tests for the electrical and constructive characterization of the products also by defining proper standard references and testing procedures for Construct-PV products.

The test facilities are aimed at both demonstrating and disseminating the Construct PV technology (visits for roof industries and associations, professionals and student's training) as well as monitoring the performances of the new Construct-PV solutions and to compare them with existing systems on the market.

2 SMALL-SCALE TEST-FACILITIES FOR DEMONSTRATING THE ROOF BIPV SHINGLE

As mentioned before, a couple of small scale facilities for testing and demonstrating the Construct-PV roof shingles have been installed both at of Tegola Canadese S.p.a. headquarter in Vittorio Veneto, Italy and at SUPSI Campus, Lugano, Switzerland. These outdoor facilities consist of two roof test-stands representing different solutions to be monitored in parallel to compare the ConstructPV shingle with another BIPV roof solution already commercially available.

In Vittorio Veneto, the ConstructPV product (Figure 1, Stand 1) has been compared with another BIPV solar shingle already developed by Tegola Canadese S.p.a (large size solar modules TEGOSOLAR 136®) and available on the market (Figure 1, Stand 2). Tegosolar is a photovoltaic flexible membrane that uses thin-film triple-junction amorphous silicon technology. Its flexibility and lightness make Tegosolar an ideal solution for any type of roof, with the possibility of installation on pitch roof from 5° to 60°, offering high aesthetic and production standards. In addition to that, Tegosolar offers waterproofing of the roof as the supporting part represents the bituminous membrane for roof covering.

Figure 1: The two test-stands of the outdoor test facility: demonstration site at Tegola Headquarters (Vittorio Veneto, Italy).

At SUPSI test-facility, the ConstructPV solar tile (Figure 2, Stand 1) has been compared to the BIPV in-roof system developed by Meyer Burger AG, also partner of the project (large size solar tile, MegaSlate®), commercially available (Figure 2, Stand 2). The MegaSlate® solar roof system uses high-efficiency monocrystalline cell technology being an architecturally appealing product, especially suited to new buildings and roof refurbishments. This system offers the possibility to replace conventional roof covering with all the benefits of a conventional roof and it is fitted to roofs with pitches of 20° or more. Even lower pitches are possible if a sub-roof for heavy-duty applications with sealed seams is used.

Figure 2: Outdoor test facilities: demonstration site at SUPSI CAMPUS (Lugano, Switzerland).

The object of the investigation is the BIPV solar shingle developed within Construct PV consortium that is mainly conceived as a roof active tile for an easy-mounting and on-site customization. It consists of a durable laminated PV glass (containing crystalline solar cells provided by the partner Meyer Burger AG) integrated and glued into a rear waterproofing back-sheet layer made of a bitumen-based membrane.

Modules with different dimensions can be produced as follows (Figure 3), but not only.

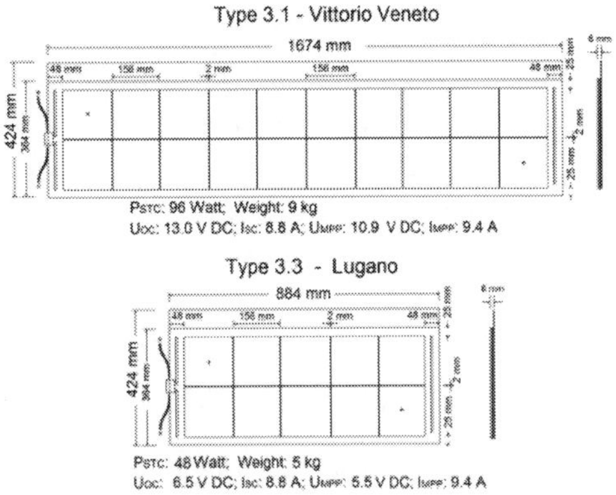

Figure 3: Construct-PV prototypes installed in the two test stand (Type 3.1, 20 solar cells prototype installed in Vitorio Veneto and Type 3.3, 10 solar cells in Lugano).

The main goals of these installations are:

- Demonstrate the technology;
- Show innovation system and identify similarities / differences;
- Detect and prevent manufacturing defects that may result in new product development;
- Benchmark the technology with other BIPV solution;
- Evaluate the construction and electrical performance of the system;
- Evaluate the temperature effects in the inner layers of the building envelope.

3 OUTDOOR TEST-STANDS AT SUPSI CAMPUS

The two test stands (Figure 2) consist in two metallic framed structures of anodized aluminum supporting an orientable roof (1,8 x 3,0 m), including a complete construction package, where BIPV solar systems are integrated as the tiling external layer.

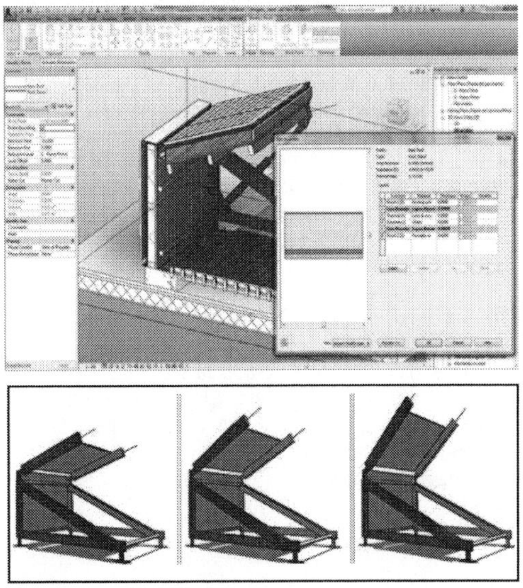

Figure 4: Planning and design of the test facility developed at SUPSI campus, Switzerland.

The test facilities have been designed on the specific purpose of testing other BIPV complete system solutions (roof and façade) as well. SUPSI has truly experience to investigate all aspects concerning BIPV photovoltaic systems in the specific conditions of installation [10-13] and other different testing and monitoring activities are usually performed contributing to better development of new products and providing support to industry. The small demonstrators allow changing the tilt angle of the roof to perform different test under real operation and weather conditions (Figure 4).

3.1 Set up and installation, assembly process.

The complete roof structure has been installed to simulate the real behaviour of the systems under BIPV tiling. The real building layering that includes substructures, supporting layers, thermal insulation, and ventilation air gap have been built onto the load-bearing metal structure of the stands in order to simulate the real building envelope conditions in terms of constructive, architectural and technological aspects.

The new Construct-PV solar shingles are designed as a replacement of the conventional roof tiling with watertightness. They are installed with a constructive procedure defined by Tegola Canadese mainly based on the use of simple and safe operations in the construction site such as bonding, nailing and sealing, not significantly different from the installation of the conventional roof tiles. Since these solar modules include waterproofing membrane (BIPV back-sheet layer made of a bitumen-based membrane), they are fixed onto a continuous OSB rear wooden board with a similar overlapping to conventional bitumen tiles, without additional auxiliary frames normally used as a channel for water drainage and retention hooks, as it happens in the roof solution with MegaSlate.

The solar tiles are arranged so that they are double-layer overlapped; the vertical and lateral overlapping allows the rain to run off properly with good watertightness. The installation has been carried out by accredited roofers and skilled tradesmen of Tegola together with the non-specialized staff of SUPSI showing that they can be easily and quickly installed also because the bitumen membrane allows to be cut directly on site to better adjust the installation tolerances.

The outdoor test-stands have provided valuable information to define the best process of installing the new product and to detect and solve potential problems during product assembly. The modular design ensures a simple, speedy and therefore cost-effective installation as well as a functional and aesthetically appealing roof-integrated solution.

Six solar tiles BIPV modules have been installed in each test facility:

- Construct-PV prototypes: small solar tile Tegola with 10 SWCT monocrystalline cells (Figure 5, Stand 1);
- MegaSlate®, BIPV in-roof system: large solar tile, 30 monocrystalline cells with three strips busbar (Figure 5, Stand 2).

Figure 5: Assembly sequence of the demonstration site at SUPSI (Lugano, Switzerland).

3.2 Monitoring equipment and test procedures.

Each module was equipped with a Maximum Power Point Tracker (MPPT 3000) developed by SUPSI aimed at measuring the electrical performances every five minutes on the purpose of this task (Pm, Voc, Isc and IV curves). The MPPT were adapted to the modules voltage and current range to optimize measurement accuracy. This aspect motivates that in the case of Stand 1, the MPPT has to measure the power output of a couple of Tegola BIPV modules installed by sets of two panels connected in series to each MPPT (as similar to a large solar tile of 20 cells, Type 3.1 Figure 3), while the MB MegaSlate are monitored individually. Module temperature was measured with a PT100 fixed on the back of the module. In the case of Construct-PV Tegola prototypes this sensor has been inserted between glass and bitumen before manufacturing. Additionally, other temperature sensors (PT100) have been positioned in the different layers of the whole roofs' packages and also behind the conventional tiles, as well as in different positions at the top and bottom of the roof system (near ridge and eaves) to check differences.

Furthermore, air flux meters for the evaluation of the air speed velocity in the cavity of the roof were placed in the middle of the ventilation chamber under the BIPV modules and other water sensors were installed to survey the tightness of the roof (even though a significant natural ventilation was not registered since the limited dimension of the roof pitches).

Meteorological data as global and diffuse horizontal irradiance, wind speed and ambient temperature as well as the module performance data were recorded simultaneously with a resolution of one minute. Two pyranometers were installed on the stand to monitor the irradiance at roof planes. In order to achieve a high level of reliability in the inter-comparison, improved data quality control and advanced data analysis were used.

Main aspects analyzed:

- Outdoor characterization of BIPV modules electrical performances (Pm, Voc, Isc and IV curves);
- Energy production (Final Yield FY, performance ratio PR, temperature of modules Tbom, etc.);
- Performances of the whole roof system installed (tightness, in-layers temperatures, ventilation, etc.).

In the following table (Table I) the main specifications of the BIPV modules monitored are described: Construct-PV Tegola prototype couple of modules in-roof (C5) and MegaSlate BIPV module (B3) as seen in Figure 6. Another couple of ventilated free-standing Construct-PV Tegola prototype module (D1) has been located on the top of the Stand 1(Figure 6a) to be used as reference (with free back ventilation).

Table I: BIPV modules monitored in SUPSI outdoor test demo site have been measured in SUPSI accredited lab.

SUPSI Demo stands	BIPV modules monitored	Rated Power P_{STC} [W_p]	Module Area A [m^2]	Temperature Coefficient $\gamma(P_{MPP})$ [%/K]
STAND 1	Construct-PV C5 / D1 (Tegola)	96.00	0.65	-0.39
STAND 2	MegaSlate ® B3 (Meyer Burger)	135.00	0.84	-0.42

a)

b)

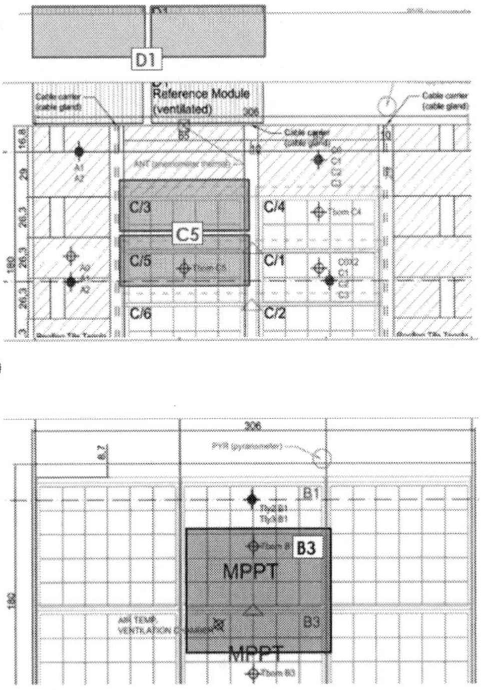

Figure 6: Technical drawings of the outdoor test stands at SUPSI: a) Stand 1, BIPV roof system Tegola developed in ConstructPV project; b) Stand 2, BIPV roof system MegaSlate®, developed and commercialized by Meyer Burger AG.

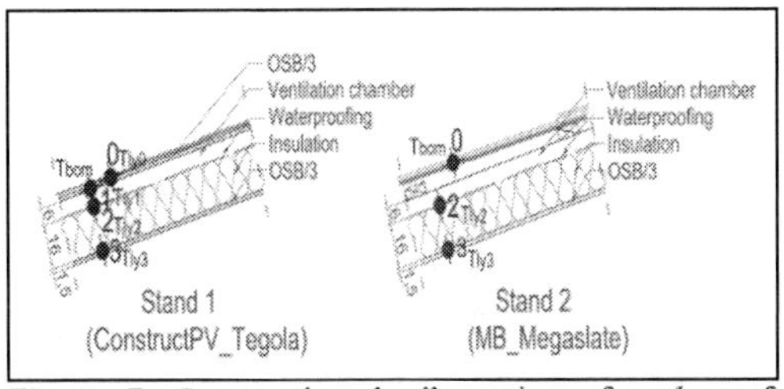

Figure 7: Constructive detail, section of each roof package with the position of the in-layers temperature sensors.

3.3 First results of the monitoring campaign (SUPSI).

In the following graphics (Figure 8) the outdoor energy yield measurements, Final Yield, Yf [kWh/kWp], the Performance ratio, PR [%] and the Energy Production per square meter, E_{PV} [kWh/m^2] are shown. The monitoring period here presented consider the last four months, from February to May 2016.

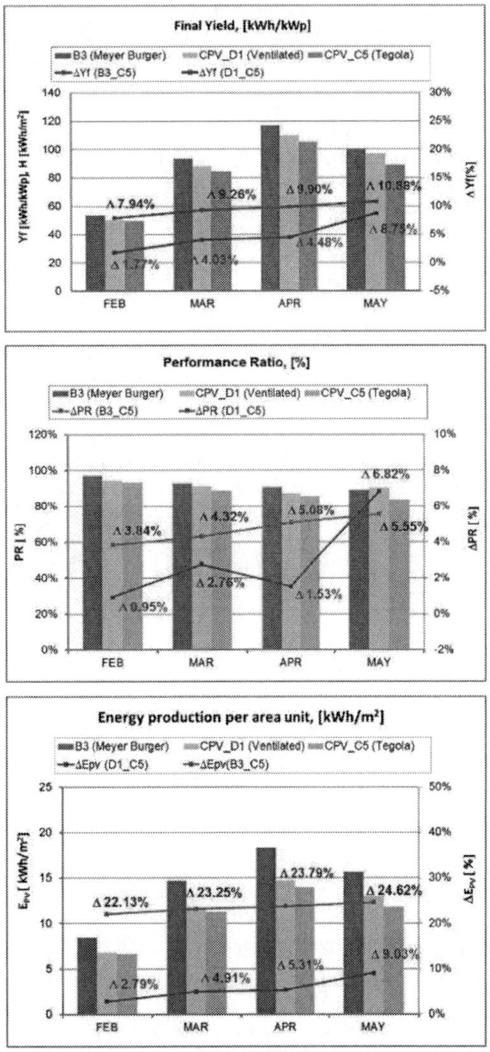

Figure 8: Results of the first months of the monitoring campaing from the beginning of February to May 2016.

The final PV system yield Yf [kWh/kWp] is the net energy output E_{PV} divided by the real d.c power Pmax $_{G.STC}$ (STC – Standard Test Condition 1000 W/m^2, 25°C, AM 1.5), measured in SUPSI laboratory. Results have shown differences in Final Yield between the big MegaSlate tile and the small tile Construct-PV Tegola in a range of 7.9% to 10.9% (red line) while differences between the same type of module full ventilated (D1) or in-roof installed varies from almost zero to 8.7% in May (blue line).

The Performance Ratio, PR [%] is essentially the final yield divided by the reference yield (Yf/Yref). As value independent from the irradiation is therefore useful to compare systems. Findings from this first period of monitoring have shown that the comparison between MegaSlate and Construct-PV new prototype have been up to 5.5% maximum while differences between the two Tegola modules, in-roof and free-standing, depending on the different installation conditions, have reached the 6.8% in May which can be attributed to higher temperature of the unventilated module.

Nevertheless, the Energy Production per square meter E_{PV} [kWh/m^2], that is the energy delivered by the PV modules considering the real module area (including also not active parts), as expected, is higher in the MegaSlate big tile than the small tile developed in the research project reaching values from 22.1% to 24.6% of increase. As in the following table (Table II), differences are reduced by almost 7% if the larger Construct-PV solar tile prototype of 20 cells is considered (Type 3.1, Figure 3) instead of the small solar tile prototype of 10 cells (Type 3.3, Figure 3) as seen in Table III. In this case, a greater impact on energy production per square meter is observed due to as result of fewer cables ducts.

Table II: Differences in Energy production per area unit performances [kWh/m^2] between MegaSlate® BIPV module and Construct-PV prototype of 10 cells (Type 3.3) installed at Lugano, SUPSI Campus.

E_{PV} [kWh/m^2]	Construct-PV (mc-Si SWCT)	MegaSlate® (mc-Si)	Δ E_{PV} [%]
FEB	6.58	8.45	**-22.13%**
MAR	11.26	14.67	**-23.25%**
APRIL	13.98	18.34	**-23.79%**
MAY	11.83	15.69	**-24.62%**
TOTAL	**43.65**	**57.16**	**-23.63 %**

Table III: Differences in Energy production performances [kWh/m^2] between MegaSlate® PV module and Construct-PV prototype of 10 cells (Type 3.3) and prototype of 20 solar cells (Type 3.1).

	Construct-PV Type 3.3 C5 (10 cells)	Construct-PV Type 3.1 C5 (20 cells)	Δ E_{PV} [%]
FEB	22.13%	14.87%	**-7.26%**
MAR	23.25%	16.10%	**-7.15%**
APRIL	23.79%	16.88%	**-7.10%**
MAY	24.62%	17.59%	**-7.03%**
TOTAL	**23.63%**	**16.51%**	**-7.12%**

It has been found also small Energy production differences registered between D1 (free-standing module) and C5 Construct-PV Tegola (in-roof module), that vary from 2.8% to 9.0% (Figure 8).
These differences can be attributed to the different operating temperatures of the modules, some mismatch effects caused by series-connected of each couple of modules measured by the MPPT tracker device and by local shadows caused by the cable ducts at the edge of the modules (Figure 9), aspect that needs to be further investigated and that will be optimize later on.

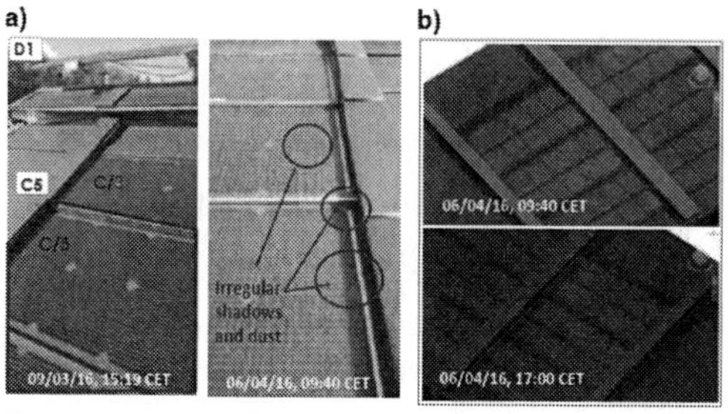

Figure 9: a) Detail pictures of the test Stand 1 (C5 and D1 prototypes) in a clear sky day of March and April; b) 3D simulation, shadowing analysis performed with Revit software at different times of day.

Focusing on a short period of monitoring in clear sky days (Figure 10), it is possible to observe the clear differences in the back of module temperature of the BIPV modules under discussion. As noticed, in Figure 11, when the ambient temperatures are higher the difference in the back of module temperature increased.

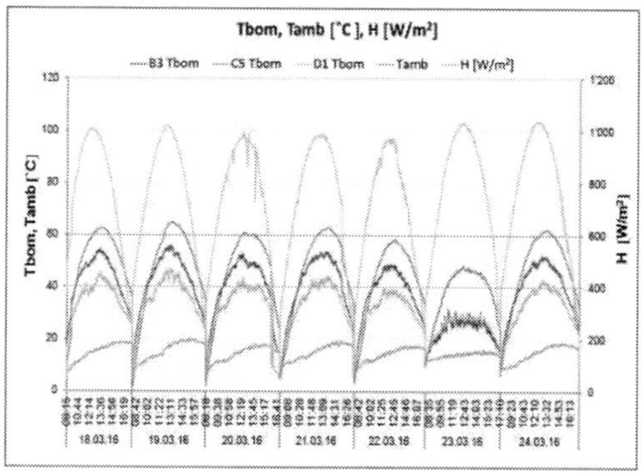

Figure 10: Back of the module temperatures (T_{bom}) monitored in a week period of March (18/03/16-25/03/16) with clear sky conditions.

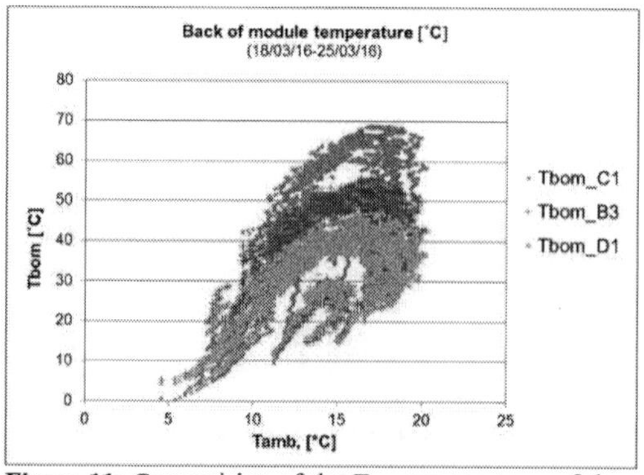

Figure 11: Comparision of the T_{bom} temperatures of the BIPV modules regarding the outdoor ambient temperature.

The period under investigation showed large

differences on temperatures of the back of the module, Tbom [°C] (maximum differences more than 12 °C between Stand 1 and 2). Regarding the temperatures of the inner layers, of the roof packages (referred to Figure 7 temperature sensors) for the period reported here (18/03/16 - 25/03/16) display great differences especially at the level of the layer 2 under the ventilation chamber.

It has been noticed that in the case of Stand 2 with MegaSlate BIPV modules the temperatures at this level decreases up to even 11.3 °C (Figure 12). This condition is mainly influenced by the general weak air flow in the air gaps of the two stands and also by a customized constructive detail of the roof in the Stand 1 at the ridge joint, that further obstacles natural ventilation.

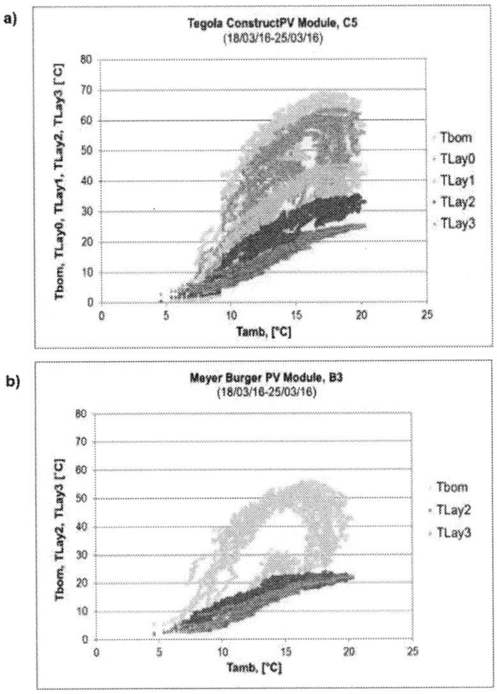

Figure 12: a) Inner layer temperatures of the Construct-PV solar tile prototype (T_{bom}, T_{Lay0}, T_{Lay1}, T_{Lay2}, T_{Lay3}); b) Inner layer temperatures of the MegaSlate BIPV module (T_{bom}, T_{Lay2}, T_{Lay3}) monitored during the same period.

4 OUTDOOR TESTS-STANDS AT TEGOLA HEADQUARTERS

On summer 2015 two small-scale demonstrators (Figure 13) have been realized and installed on the roof over the research center of Tegola Canadese headquarters, in Vittorio Veneto (Italy).

The roof is regularly used as a testing center, in which several products of TEGOLA production and R&D have been installed to analyze behavior of the building components with ageing. This roof is a challenging demosite to analyse photovoltaics, since it is a south east exposed roof, with 18% of slope.

First step of installation has been the roofing products removal and the preparation of a flat waterproof substrate, in order to reproduce a typical roof of southern Europe. The test roof consists of two separate stands of BIPV products, parallel installed by torch on procedure on the roof. No retro ventilation has been provided to BIPV that has been installed in direct contact with bituminous waterproofing substrate.

The roof covered with BIPV was a normal concrete roof that has been covered with 3 layers of waterproof membrane (two old layers, one the brand new installed on purpose for this demo site). Second step of installation has been the positioning of the BIPV roofing products on the test stand roof demo site:

- Construct-PV prototypes: large solar tile Tegola with 20 SWCT monocrystalline cells (Figure 13, Stand 1);
- Tegosolar 136®, a thin film technology PV shingle (triple-junction amorphous silicon, a-Si) produced by Tegola Canadese (Figure 13, Stand 2).

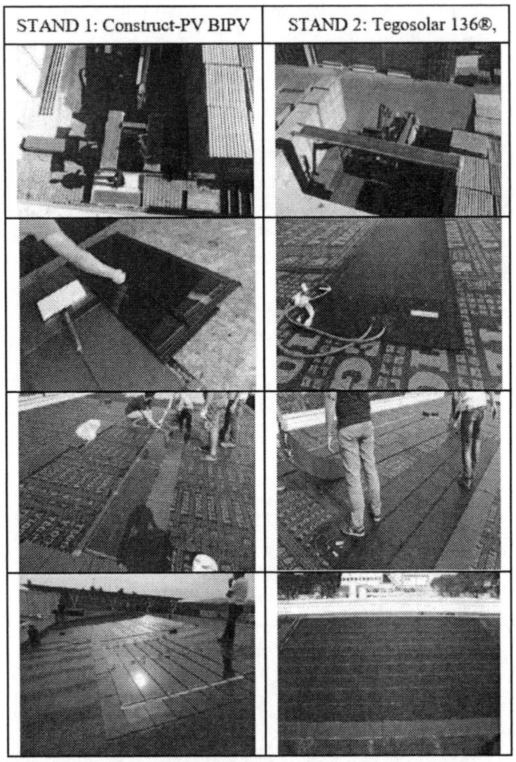

Figure 13: Assembly sequence of the demonstration site at TEGOLA Headquarter (Vitorio Veneto, Italy).

4.1 Monitoring equipment and PV module specifications

In the following table (Table VI) are described the main specifications of the BIPV modules monitored are described: Construct-PV Tegola prototype (Stand 1) and Tegosolar 136 (Stand 2).

Table VI: BIPV modules monitored in TEGOLA outdoor test demo site.

TEGOLA Demo stands	BIPV modules monitored	Rated Power P_{STC} [kW_p]	Module Area A [m^2]	Number of BIPV modules n. [ud.]
STAND 1	ConstructPV (Tegola Canadese)	1.34	8.53	14
STAND 2	Tegosolar 136 (Tegola Canadese)	1.36	21.22	10

The stands have been equipped with a monitoring system that allows a continuative analysis by measurement different parameters, with installations of two inverters of 1.5 kWp and a cluster controller, supplied by SMA (Figure 14).

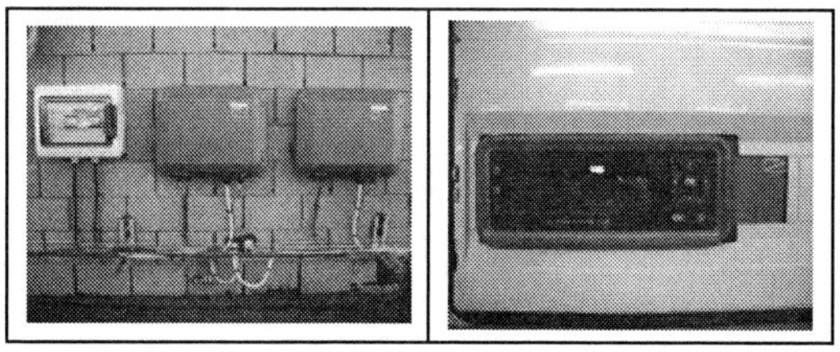

Figure 14: Assembly sequence of the demonstration site at TEGOLA.

4.2 First results of the monitoring campaign (TEGOLA HEADQUARTERS).

Preliminary results obtained by monitoring activity are compared, showing strengths and weaknesses of the technology; each stand was equipped with an Inverter and a cluster controller that both allow Tegola Research operators to investigate performances. The roof temperature was measured with a PT100 fixed on the close Tegola 200 kWp PV plant, and a pyranometer as well gives an estimation of the irradiation.

In the following tables and pictures (Figures 15 to 17) blue columns represent performances of Tegosolar BIPV modules; red columns refer to Construct-PV Tegola prototype. In clear sky days Tegosolar performs better than ConstructPV prototype (Figure 15a).

a)

TEGOLA Demo stands	BIPV modules monitored	Energy Production E_{PV} [kWh] (22/05/16)	Total Yield Yf [kWh/kWp] April 2016
STAND 1	Construct-PV	7.74	5.49
STAND 2	Tegosolar 136	8.57	6.30

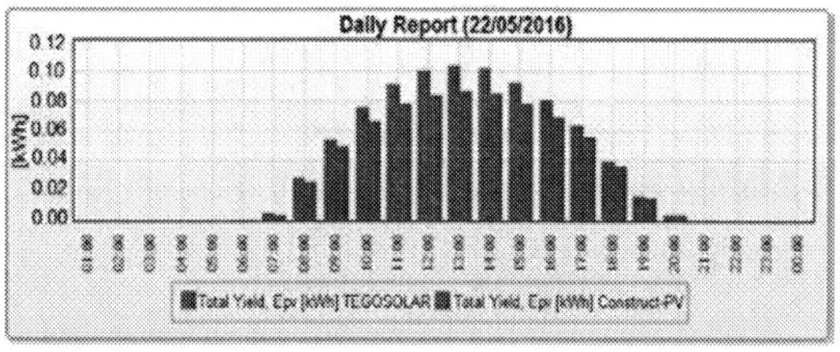

b)

TEGOLA Demo stands	BIPV modules monitored	Energy Production E_{PV} [kWh] (11/05/16)	Total Yield Yf [kWh/kWp] May 2016
STAND 1	Construct-PV	0.81	0.60
STAND 2	Tegosolar 136	0.77	0.56

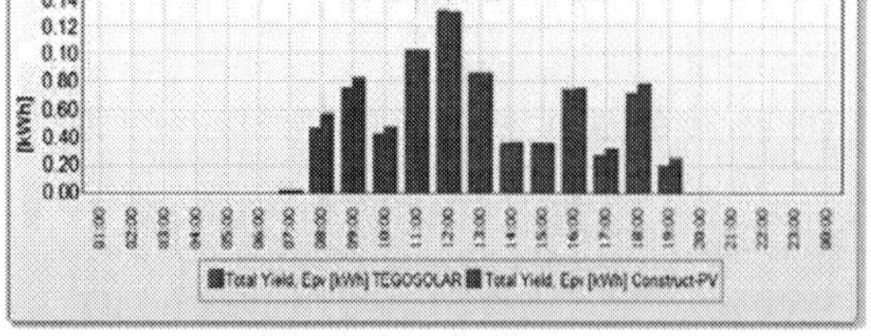

Figure 15: Daily Energy production, E_{PV} values [kWh] and Final Yield, Yf [kWh/kWp] measured in a clear sky day and a cloudy day of May 2016.

Analyzing in detail the monitoring results during the day, Tegosolar performs better especially in sunny day mid hours, while in the morning or in the afternoon Construct-PV prototype is equal or even better (Figure 15a). Besides in cloudy days Construct PV prototype is more performant than Tegosolar (Figure 15b).

The following tables and pictures show the electrical performance results of the monitoring campaign since February to May 2016. Figure 16 shows the total monthly Energy production, E_{PV} values [kWh] and the Final Yield, Yf [kWh/kWp] measured in February (16.a) and March (16.b) 2016. Figure 17 shows the same values (total monthly Energy production, E_{PV} and tFinal Yield, Yf) registered in the months of April (17.a) and May (17.b). Considering the entire period of analysis, Tegosolar BIPV roof system performs better (Figure 18) in terms of energy produced versus installed power. Results have shown differences in Final Yield, Yf [kWh/kWp] between the Tegosolar system and the Construct-PV Tegola new prototype in a range of 0.32% in February to 9.34% in May (black line in Figure 18).

a) FEBRUARY 2016

TEGOLA Demo stands	BIPV modules monitored	Energy Production E_{PV} [kWh] FEBRUARY	Total Yield Yf [kWh/kWp] FEBRUARY 2016
STAND 1	Construct-PV	38.850	28.91
STAND 2	Tegosolar 136	39.440	29.00

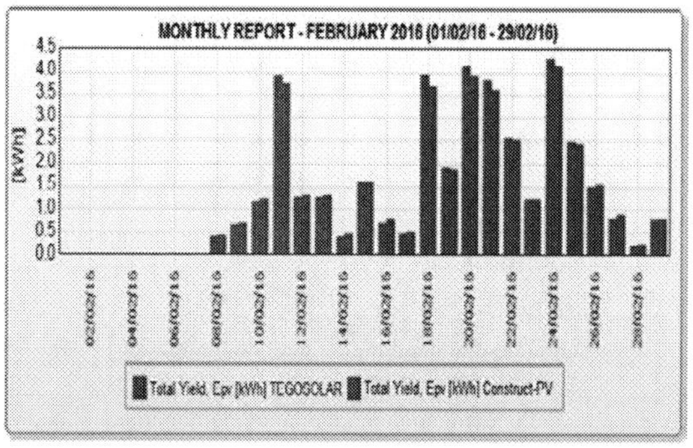

b) MARCH 2016

TEGOLA Demo stands	BIPV modules monitored	Energy Production E_{PV} [kWh] MARCH	Total Yield Yf [kWh/kWp] MARCH 2016
STAND 1	Construct-PV	108.91	81.04
STAND 2	Tegosolar 136	114.97	84.54

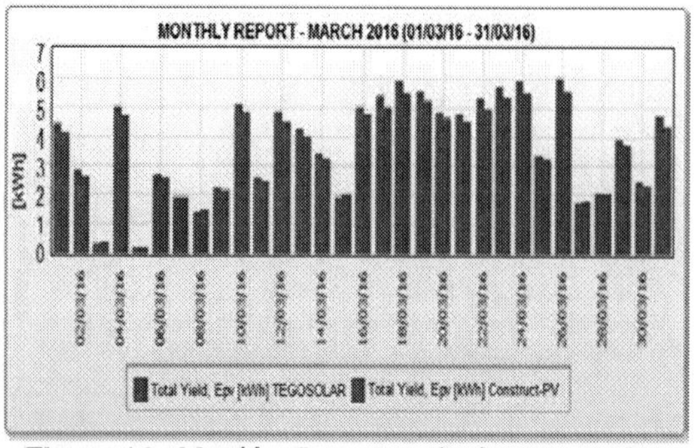

Figure 16: Monthly Energy production, E_{PV} values [kWh] and Final Yield, Yf [kWh/kWp] measured in February and March 2016.

a) APRIL 2016

TEGOLA Demo stands	BIPV modules monitored	Energy Production E_{PV} [kWh] APRIL	Total Yield Yf [kWh/kWp] APRIL 2016
STAND 1	Construct-PV	135.76	101.02
STAND 2	Tegosolar 136	146.74	107.90

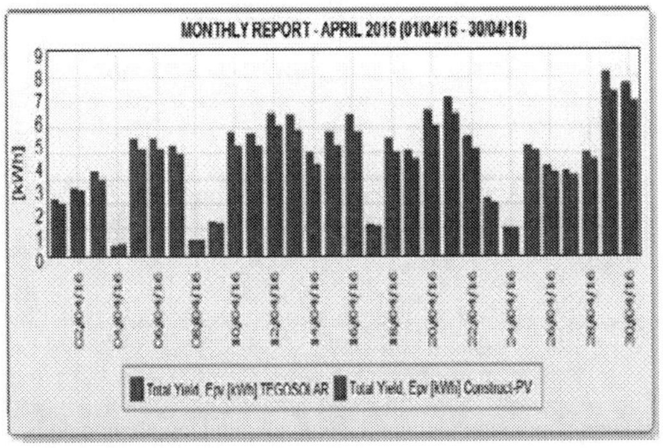

b) MAY 2016

TEGOLA Demo stands	BIPV modules monitored	Energy Production E_{PV} [kWh] MAY	Total Yield Yf [kWh/kWp] MAY 2016
STAND 1	Construct-PV	153.92	114.53
STAND 2	Tegosolar 136	171.80	126.33

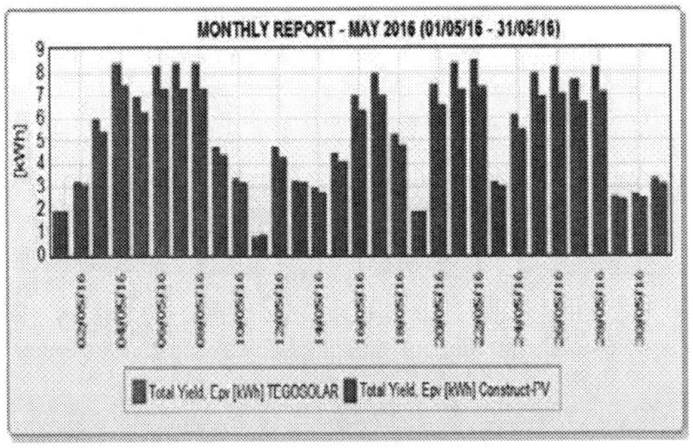

Figure 17: Monthly Energy production, E_{PV} values [kWh] and Final Yield, Yf [kWh/kWp] measured in April and May 2016.

Construct-PV prototype, installed on the top of non-ventilated roof, over a waterproofing membrane by torch on method, is producing a power output of only 10.40% less, in the worst case (May), referred to the benchmark Tegosolar installed in the same conditions (Figures 16 and 17).

In addition, the Performance Ratio, PR [%] differences in this first period of monitoring comparison between Tegosolar system and new Construct-PV BIPV roof have shown reaching values from 0.17% in February (cold month) to 6.63% in May (mild climate). Nevertheless, Construct-PV has a higher efficiency per square meter, since area covered by the prototype is mostly 50% of the installation area of the benchmark. As seen in Table V the Energy Production per square meter E_{PV} [kWh/m^2], is significantly higher in the case of Tegola Construct-PV new prototype (mc-Si SWCT technology) than the Tegosolar BIPV module (a-Si technology) reaching values from 55.1% to 59.1% of increase.

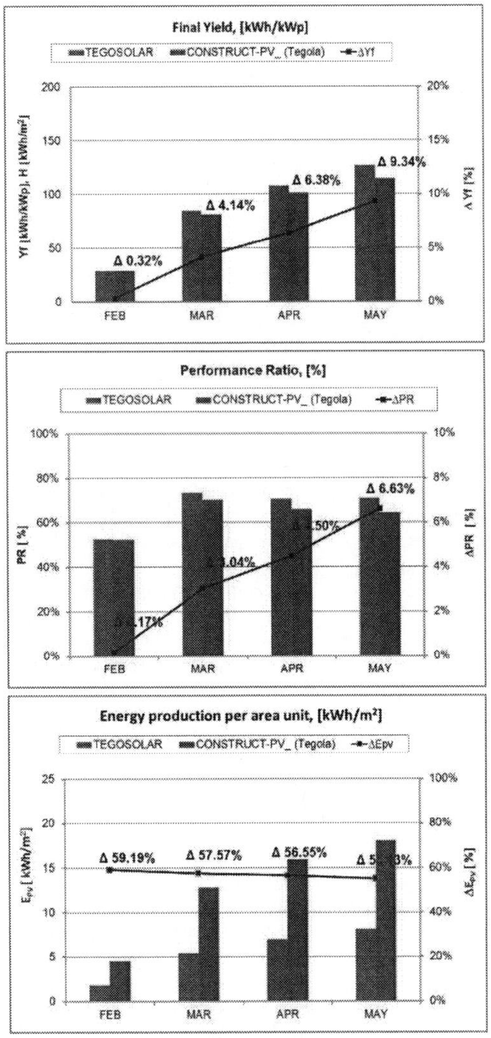

Figure 18: Results of the first months of the monitoring campaing from the beginning of February to May 2016.

Table V: Differences in Energy production performances [kWh/m^2] between Tegosolar 136® PV module and Construct-PV prototype of 20 cells (Type 3.1) installed at Vittorio Veneto, Tegola Headquarters.

E_{PV} [kWh/m^2]	Construct-PV (mc-Si SWCT)	Tegosolar 136® (a-Si)	**Δ E_{PV} [%]**
FEB	4.55	1.86	**+59.19%**
MAR	12.76	5.42	**+57.57%**
APRIL	15.92	6.91	**+56.55%**
MAY	18.04	8.10	**+55.13%**
TOTAL	**51.28**	**22.29**	**+57.54%**

5 CONCLUSIONS

The paper presented two demonstration sites where two new PV shingles, developed within the Construct PV European project, have been installed and monitored for different months. Comparison with commercially available BIPV systems have been presented. The performances of the BIPV Construct-PV solar tiles prototypes result influenced mainly by both the type of installation (in-roof) and the dimension of the solar shingle (small solar tile of 10 cells or large solar tile of 20 cells).

The small-scale installation does not allow relevant natural back-ventilation, like in a real roof. This has an impact on the operative temperature of the BIPV prototypes. Moreover, for the monitoring period studied so far, it has been found that the new BIPV Tegola system compared to the conventional Tegola roof tile (not PV) imply, an increase in the roof temperature of about 10 °C at the back of the tiling layer (Layer1) and of 4 °C within the air gap (Layer 2). Since the test-stands have a very small pitch (1.8 x 3.0 m) this does not allow proper natural ventilation and air flow, so that the temperature conditions can be considered worse than in the real behavior. The ventilated chamber used in such a roof installation, in fact, represents in real conditions one of the most effective technical solutions for natural ventilation in sloped roof so that lower temperatures are expected in the air gap as well as possible consequent improvement of PV performance (and of course of the roof in summer).

Another point that could be remarked is that the option of a large solar tile BIPV prototype is better than the small one in terms of energy production per square meter, E_{PV} values, as the impact of the cable ducts shadowing less affected module surface. Anyway, the monitoring campaign still continues and will be important to further analyze the following aspects:

1) Evaluate the soiling/shadowing and temperature effects as function of the tilt angle;
2) Evaluate of possible effects of back ventilation;
3) Assess the impact of the mounting structure (metal channel) and local shadowing that could be caused by cable ducts.
4) Evaluate the behavior of the different installations in the summertime, to see if the difference between power outputs of Construct PV prototype towards Tegosolar is increasing or remains stable.
5) Evaluate the real temperature coefficient of the Construct PV prototype by those measurements in the jobsite.

In conclusion, even though the systems developed seem to have a bit lower performance in terms of energy performance as compared to other high efficient products available on the market, considering the strategy of a low final cost of the installation that the producer aims at achieving the cost-effectiveness of the product in its life-cycle could be demonstrated with further investigation.

Finally, it is important to highlight that the new system enables an easy and fast mounting having other advantages as for example, can be nailed or welded, could be used in roofs even with small slopes, as 10°, the dimension can be adjusted in site, etc., all aspects that allow a cost-effective installation as well as a functional and aesthetically appealing roof-integrated solution with regard to other BIPV solutions for roof.

AKNOWLEDGMENTS

Construct-PV: The research leading to these results received funding from the European Community's Seventh Framework programme (FP7/2007-2013) under grant agreement no 295981 [www.constructpv.eu]. The authors would like to thank Mr. Christof Erban from Meyer Burger AG for his collaboration in this research and for providing the PV modules. We also would like to thank the colleague of SUPSI, namely Domenico Chianese, Gabi Friesen, Sebastian Dittmann, Enrico Burà and Boris Margna for their valuable contribution and support.

REFERENCES

[1] Directive 2010/31/EU of the European Parliament and of the Council of 19 May 2010 on the Energy performance of buildings. Official Journal of the European Union.
[2] B. Van Berkel, T. Minderhoud, A. Piber, G. Gijzen. (2014). Title: Design Innovation from PV-Module to Building Envelope: Architectural Layering and Non Apparent Repetition. Proceedings of the 29th European Photovoltaic Solar Energy Conference and Exhibition. Amsterdam, The Netherlands. Pages 3606-3612.
[3] F. Frontini, A. Scognamiglio, G. Graditi, C. Polo Lopez, M. Pellegrino (2013). Title: From BIPV to Multifunctional Building Component. Session Reference: 5CO.8.1. Conference Proceedings at European Photovoltaic Solar Energy Conference and Exhibition. Paris, France. Pages 3757-3761.
[4] J. Bezler, K. Fath, R. Hecker. Title: Customised photovoltaic modules for building envelopes – Construct PV, Research, Developments, Innovation 2014/15 – Züblin-Strabag
[5] E. Delponte, F. Marchi, F. Frontini, C. Polo-Lόpez, K. Fath, M. Batey (2015). Title: BIPV in EU28, from Niche to Mass Market: An Assessment of Current Projects and the Potential for Growth through Product Innovation. Proceedings of the 31st European Photovoltaic Solar Energy Conference and Exhibition 14-18 Sept. 2015 Hamburg, Germany. Pages 3046 - 3050.
[6] R. Manariti, G. Urbano, F. Marchi, E. Delponte. Title: Life Cycle Analysis and Life Cycle Costing Analysis of Building Integrated PhotoVoltaics, Convegno dell'Associazione Rete Italiana LCA 2016, Ravenna, Italy.
[7] 10th Energy Forum on Advanced Building Skins. Special Session BIPV 6 - From Design Concepts to Real Buildings: How Stakeholders Envision BIPV. Bern, Switzerland, 03-04 November 2015.
[8] EN 50583-1:2016. Photovoltaics in buildings. BIPV modules
[9] F. Frontini, G. Friesen, T. Friesen; C. S. Polo López (2013). Title: Experience on the behaviour of different BiPV solution: electrical and thermal performance of roof and façade mounted micromorph modules. Conference Proceedings at 28th European Photovoltaic Solar Energy Conference and Exhibition. Paris, France. Pages 4357-4360.
[11] F. Frontini, G. Friesen, T. Friesen, C. S. Polo López Title: Experimental testing under real conditions of different solar building skins when using multifunctional BIPV systems. Energy Procedia V48, December 2014, Pages 1412-1418.

ARCHITECTURAL SOLUTION FOR USING AREA OF SIDE STREETS AND ALLEYS TO UTILIZE SOLAR PANELS

Abbas Rahmani
PhD researcher of KIT, Faculty of Architecture, Karlsruhe, Germany
Phone: +49 (0)157 31419403, Fax: +98 (0)87 33236750, Email: abbas.rahmani@kit.edu

ABSTRACT: Alleys and streets have occupied significant portion of the whole area of a city. Based on current urban patterns of European countries more than half of a city's area is allocated to streets and communication routes. On the other hand, spreading public transportation, bicycle network and sidewalks could decrease co2 emission and would provide a more healthy urban life. How could we use this considerable area to generate electricity by utilizing solar panels? In this work a multipurpose solar town has been prepared with innovative architectural aspects of main streets, alleys, metro and its connections, bike routes and sidewalks. The side streets and alleys are covered with a transparent layer holding more than 20000 separate solar panels. This solar system has occupied 15% of the whole area of the concept and would be able to work as a solar power plant complex. Practical capacity of generating electricity of this solar power station could be more than 2.5 MWP. It could prepare an off-grid solar town that would be independent from the state electrical grid. The whole solar town could be supposed as a zero-energy concept. The purposed concept would use considerable area of the alleys for generating electricity and would decrease the costs of equipment, installation and maintenance of the solar system compared to traditional pattern. The proposed solar system would smoothly switch the need of the concept from fossil fuels into green energy resources.

Keywords: BIPV, PV modules, solar architecture, energy efficiency, aesthetic characteristic, side street, zero-energy, sustainable architecture, urban pattern, physical activities

1 PURPOSE OF THE WORK

To switch from fossil fuels into green energies in building sector, we have to replace new forms of architecture with the pattern of our current buildings [1]. The new concepts must be aware of upcoming technological products and the tradition of the client tastes [2]. In this concept the solar panel installed on the transparent layer is supposed to have the current technology but they are utilized on the top of the alley's surface not on the top of the roof of buildings. The purposed concept would be able to use the considerable area of the alleys for generating electricity, so the combination of buildings with alleys would propose a semi zero-energy solar town. In this method the buildings are roof-free and the possibility of having diverse usages specially sport activities on these flat roofs is available. This concept is trying to prepare a more dynamic solar town in its daily activities by authorizing the users to have their direct access only while they are walking, jogging, cycling, etc. So the tradition of having close access to the flats or commercial zones by car is replaced with jogging or walking toward them. For using this concept more amount of physical activities would be done on a regular basis. More healthy transportation based on the more physical activities is supposed to be prepared by giving the inhabitants more facilities for cycling, walking and jogging in short distances. For having a double-emphasis on cycling and walking in short distances, the designed sidewalks and bike routes lead the users only into the metro stations. The soft slope of the curved ramps in the entrance of metro stations encourages the users for a comfortable cycling into metro stations. Also the sidewalks and bike routes have good views with different sceneries of the solar town. Parallel lines of greenbelt with short-root trees accompany the users in their short trips and prepare them a high quality route that is much more refreshing compared to the car routes. The master route of metro is located at the lower level of the main highway. Along the route there are metro stations in coordination with the upper highway.

2 APPROACH

2.1 Location

Location of the solar town has been considered in Stuttgart, Germany that has appropriate urban scale. As a result the represented views are based on annual altitudes and azimuths of sun in Stuttgart. Total area of the selected site is 90 hectares containing highways, metro routes and greenbelts. It is divided into three separate complexes that each of them is 30 hectares. Two complexes are in the west of the highway and one of them is in the east of it. The main access is supposed to be a north-south highway. It is a two-level route that is divided into metro route in lower level and car route in upper level. At the east and west end of the solar town there are two other highways for a better traffic circulation. The city center is supposed to be in the south of the solar town. Each complex has a two-way street for vehicle access in the south and four separate pedestrians and bicycle routes that connect the buildings to metro station and sidewalk of the main highway (Pic 1).

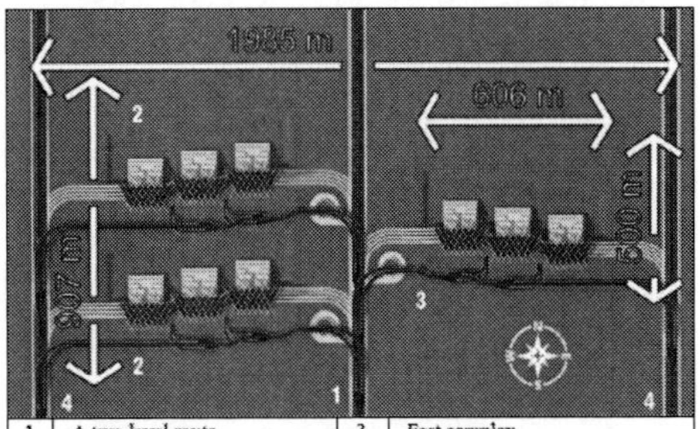

1	A two-level route	3	East complex
2	West complexes	4	East and west highways

Picture 1: Site plan - main accesses and dimensions

2.2 The whole complex

The whole concept is divided to three complexes. A complex has its particular metro station, car access, four bike routes and four sidewalks. Each complex has three units. Each unit is a combination of a southern structure and a northern building. A southern structure is in composition with the main structure of the relative building and its main duty is preparing a transparent layer that solar panels are installed on it. A northern building is a six-storey building that has different usages in its levels. To create visual difference between the complexes, color of transparent layers differs from one complex to the other one. The installed solar panels in each complex have the same technology. 40% of each structure would be remained with transparent material that could have different color and 60% of it would be covered with solar panels as the opaque portion of the structure. This obvious difference in the color could make the mentioned complexes more recognizable from each other and easier to be found by their users (Pic 2).

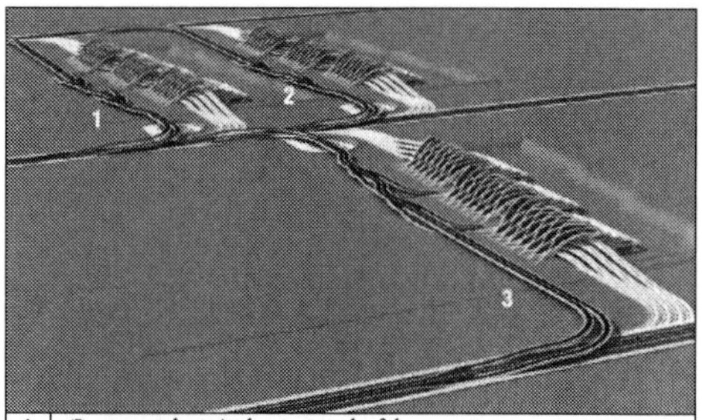

1	Green complex - At the west-south of the concept
2	Blue complex - At the west-north of the concept
3	Red complex - At the east of the concept

Picture 2: Three complexes and its connections

2.3 Two-level highway

The main two-level highway consists of multi functions to prepare facilities for city passengers with different tastes; the people who wish to do their intra city travels with public transportation system like metro lines, or who wish to use their private cars, as well as who wish to cycle or walk in a more healthy transportation system. Sidewalks and bike routes have been considered along the main highway. Different profiles of greenbelt have been allocated to divide the vehicle routes, bike routes and sidewalks from each other. Level of sidewalks and bike routes are at lower level compared to the highway for more security and less noises of vehicles. Bike routes are 10 cm upper than sidewalks, so it would be safer for pedestrians and more convenient for cyclists. The parallel sidewalks and bike routes of the main highway connect the complexes to each other. So the users could take these routes to go from one complex to the other one. Vehicles could take the exit ramps to go to the car entrances of each complex via two-way streets. In reverse direction similar ramps are available to connect the two-way streets to the main highway (Pic 3).

1	Main two-level highway	4	Bike route
2	Metro lines	5	Greenbelt
3	Sidewalk	6	Exit ramp

Picture 3: Section - accesses to a complex - connection of metro station with bike routes, sidewalks and alleys

2.4 Complex

Main elements of each complex are;

2.4.1 A metro station + bike parking zone

Width of each metro station is 19.00 m that could be suitable for the whole capacity of a complex. In the center point there is a bike parking zone for the users that prefer to walk from metro wagons into the ground floor then pick their bikes from parking zone and take one alley while they are cycling into their destination.

2.4.2 Four separate alleys

Width of each alley is 9.60 m. It is divided into a bike route, a greenbelt, a sidewalk and a second greenbelt.

2.4.3 A two-way street + two car entrances

Width of each two-way street is 18.84 m. A narrow greenbelt and two narrow sidewalks have been considered in each side. Along the two-way street there are two car entrances to enter into parking zone of complex by car.

2.4.4 Three units

Each unit is combination of; a southern structure holding PV modules and a northern building.

2.4.5 A common open space

Area of each common open space is 25000 m^2. It could be suitable for lots of other activities in the front of shopping and commercial centers in the first floor of each building (Pic 4).

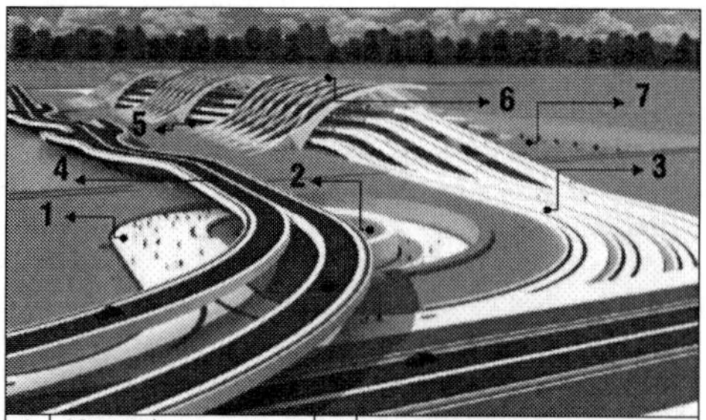

1	Metro station	4	Two-way street
2	Bike parking zone	5	Car entrance
3	Alley	6	PV modules
		7	Common open space

Picture 4: A complex and its main routes in connection with metro station

2.5 Four main alleys

These four separate alleys are as the healthy routs of each complex because of the great sport activities done on them. These daily activities are along a small portion of the whole daily distance of the user trips between the complex and other commercial or residential centers of the city. So the concept proposes a regular- short trip that its distance would be less than one kilometer per day. These healthy routes are divided to four separate levels and each level would take the users directly to a specific level of the complex. The angle of each route with horizon is 9°, so it could be convenient for cycling and walking toward the complex. Along the routs B, C and D the width of greenbelt would increase to 7.20 m. These wide greenbelts would accompany the users along the different entrances of the buildings in a more attractive path (Pic 5).

1	Route A	3	Route C
2	Route B	4	Route D

Picture 5: Four bike routes and sidewalks toward a complex

2.6 Car entrance

Two car entrances are provided in each complex. Each entrance is located in the gap of between two neighbor units. After exiting from the two-way street the cars could enter the complex and turn right or left to the purposed car parking. The car parking zones are in the 3rd floor of each building and its level is exactly at the level of the car entrance, so the traffic of cars between entrance and relative parking zones would be done easily. Alternatively before going to the car parking zones, the cars could drop their passengers off; straight ahead a sloped ramp is available as a main sidewalk for the passengers to walk smoothly toward the common open space. These car entrances act as opening for the sidewalks and bike routes. Natural airflow with wide view to south side is available for the cyclists and walkers of the route "A", "B", "C" and "D" through these openings (Pic 6).

1	Two-way street	4	Route C
2	Route A	5	Route D
3	Route B	6	PV modules

Picture 6: Four bike routes and sidewalks

2.7 Connection of alleys with buildings

In the junction of; route "A" with 4th floor, route "B" with 5th floor and route "C" with 6th floor there are entrances with appropriative bike parking zones. Route "D" has access to the 7th floor as the roof of the building with comprehensive view toward the north and the common open-space (Pic 7).

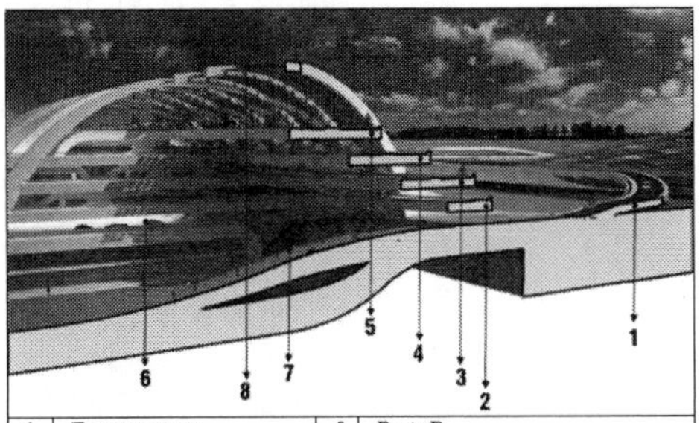

1	Two-way street	5	Route D
2	Route A	6	Car parking zone
3	Route B	7	Sidewalk
4	Route C	8	PV modules

Picture 7: Section of a complex passing from car entrance - sidewalk toward common open-space

2.8 Priority of sidewalks and bike routes

Pedestrians and cyclists are only allowed to access to the shopping centers that are located in common open-space. The intermediate ramp connects car parking to the open-space. Inhabitants of 4th, 5th and 6th floor could directly access to the shopping centers via interior elevators. Similar elevators are available for the cyclist and walker in 7th floor. Different views of the concept have obvious emphasis on encouraging the inhabitants to cycle or walk rather than using private cars. In the intersection of neighbor units there are lots of special routs that are allocated for pedestrians. The condition is the same for the cyclists but private cars are only allowed to access directly to the 3rd floor, leave their car and walk toward their destination. This purposeful strategy in the pattern of the architectural aspects of the concept could encourage the users to have more physical activities on a regular basis (Pic 8).

1	Intermediate ramp	5	Route D
2	Route A	6	4th floor
3	Route B	7	5th floor
4	Route C	8	PV modules

Picture 8: Section of a complex passing from car entrance - sidewalk toward common open-space

2.9 Unit

Each complex is divided to three units in a west-east direction. A unit is mainly divided into a structure on south and a building on north. On south, the main structure is a woven composition of fourteen curved elements. These intersecting elements hold a transparent semi-glass layer that 2581 solar panels are installed on it. This structure could also be useful to maintain the system like cleaning the surfaces, repairing the panels, etc. Each solar panel is a squad of 1.20 m × 1.20 m. The transparent layer is only due to hold the solar panel modules, so they could be colorful. As a result the color of each unit differs from one complex into the other one. The complexes are known as blue, green and red complex respectively. In technical aspect, this curved structure prepares different tilts for the solar panels [3]. As the altitude of sun differs in each season, the different tilts of solar panels would prepare the best seasonal efficiency for one piece of the unit to generate electricity. In structural aspect, these elements could be used to transfer the lateral forces of the unit into the ground in a bracing system (Pic 9).

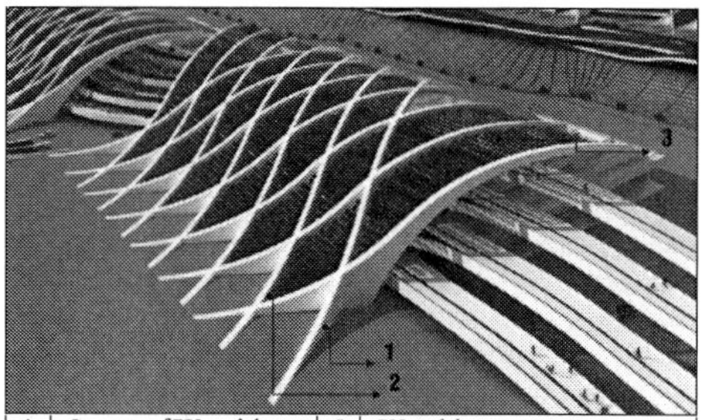

1	Structure of PV modules	3	PV modules
2	Transparent layer		

Picture 9: South side of a unit

2.10 Building of a unit

On north side of a unit, a seven-storey building is located with diverse usages in each level. The first and second floors are mainly allocated to commercial and business zones like shopping centers, malls, etc. The third floor is allocated to car parking and is exactly in the same level of the car entrance. The fourth, fifth and sixth floors are mainly allocated to residential zones. Appropriative openings have been included in these floors, so these residential flats could get enough light from east, north and west. Additionally extensive terraces have been considered for these levels that could be allocated for any kind of open-space activities like roller skating ground, children's playground or some kinds of ball games like badminton, volleyball, basketball, etc. Thickness of the floors is 1.80 m, so planting of short root trees is possible in each floor. The seventh floor as the final roof of the building has the best view to the whole complex and is allocated to open-space activities like jogging, cycling, etc. The routes "A","B","C" and "D" are connected to the floors "4","5","6" and "7" respectively. The connection of routes with building creates the main entrances of the building in each level. Before each entrance, bike parking zones are available for cyclists and pedestrians (Pic 10).

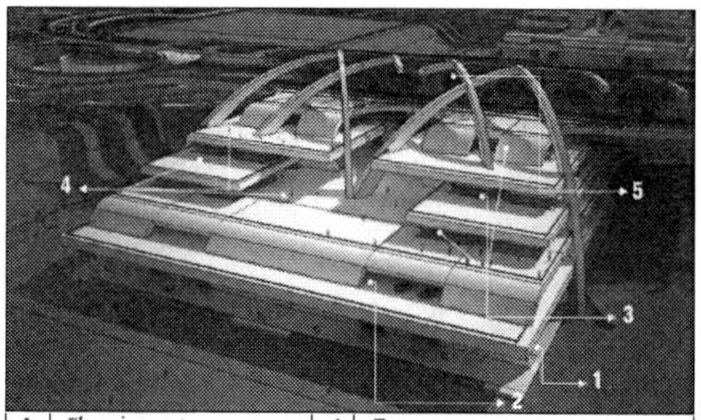

1	Shopping center	4	Terraces
2	Car parking	5	PV modules
3	Residential zones		

Picture 10: A building and its main zones

2.11 Levels of a building

First and 2nd floors are allocated to commercial centers mostly shopping centers. By allocating a shopping center at the first floor of each building, the inhabitants of residential zones in upper levels are supposed to limit their shopping trips to their own building as far as possible. They could use internal elevators that their required electricity would directly come from the solar panel system of the relative unit. So this composition of the units could strongly highlight the contribution of architectural concepts in "controlling co2 emissions strategies" in three different aspects; using bicycles instead of private cars in short distances, eliminating intra city travels & generating the electrical energy of vertical trips by PV modules. 3rd floor is a semi-open space and is allocated to car parking. Additionally, this semi- open space could be used for public activities in each building like meetings, speeches, group celebrations, etc. 4th and 5th floors are allocated to residential flats. The required light for interior subdivisions is available from east, north and west sides. These flats could have diverse designs and different areas. 6th floor is allocated to trade and administrative offices. This level has a comprehensive view to the complex and serves its inhabitants with luxury services. The terraces are for private usages. 7th floor is allocated to public activities in a semi-open space. It has wide view to both sides of the complex; to the common open-space and to the alleys and structure of solar panels. It is supposed that the users would be more familiar with the whole system of the complex while they are visiting through the 7th floor. The roof of the 6th floor acts as appropriative areas for 7th floor that would be mainly used for entertainment activities (Pic 11).

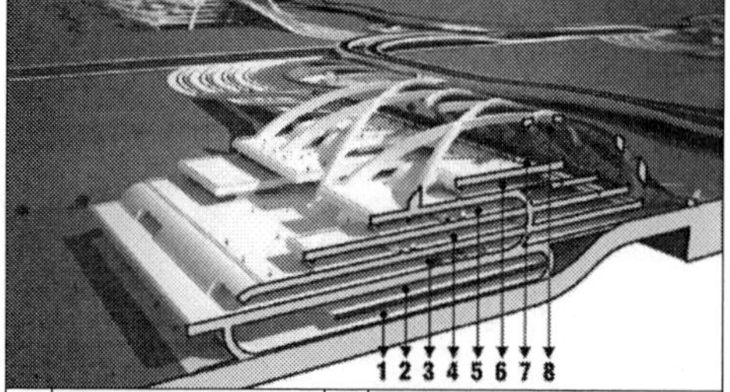

1	Shopping center	5	Residential / Trade offices
2	Second floor- shopping	6	Administrative / Sport centers
3	Car parking zone	7	Semi-open space.
4	Residential	8	PV modules

Picture 11: Section of a complex passing from building - main floors

2.12 Electrical capacity

Based on current technology theoretical capacity for a solar panel with the area of 1.36 m2 is 200 WP. This capacity would be prepared in ideal situation in terms of tilt and azimuth of the panels. With regard to the curved shape of the structure that the solar panels are installed on it, each solar panel would have different amount of tilt. So the average of final capacity in practical would definitely be less than the theoretical one. A ratio of 0.60 has been multiplied to calculate the practical capacities. Practical capacity of generating electricity of the solar town is more than 2.5 MWP. It could prepare an off-grid solar town that would be independent from the state electrical grid. The whole solar town could be supposed as a zero-energy concept by adding passive thermal strategies (Pic 12).

	Module	Unit	Complex	Solar town
Total number of modules	1	2581	7743	23229
Total area of modules (m²)	1.36	3510	10530	31591
Theoretical capacity (kWp)	0.20	516.20	1548.60	4645.60
Practical capacity (kWp)	0.12	309.00	929.00	2787.00

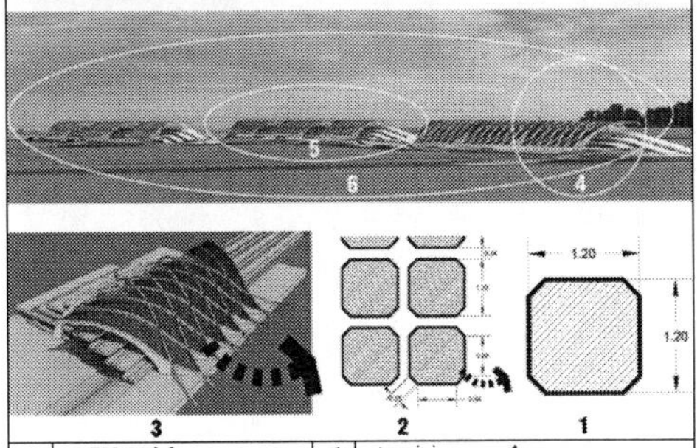

1	A PV module	4	A unit in a complex
2	Array of modules	5	A complex
3	An individual unit	6	Whole concept (solar town)

Picture 12: Solar town and subdivisions

3 SCIENTIFIC INNOVATION AND RELEVANCE

In addition to the mentioned advantages of the suggested concept compared to the traditional pattern that are; providing more physical activity, contribution in decreasing co2 emissions, decreasing intra city travels and providing a semi zero-energy complex (with little dependence to electrical grid), there are some additional advantages as below;

- More portion of absorption for generating electricity
- More compact equipment, less additional costs [5]
- More aesthetic values [6]
- Less traffic junctions
- Visibility of solar system
- Using southern side of buildings to utilize PV modules
- Ideal shadow as well as canopy
- No sloped roof - Diverse functionality on flat roof
- Coherent additional component – Cheaper utilization
- Natural ventilation to cool the modules, better efficiency in hot months
- Packing the modules
- Separate spaces for equipment and batteries
- Southern structure as the main component without modifying the whole concept
- More saving time + more physical activities + more sustainability
- More greenbelts

4 RESULTS AND CONCLUSIONS

- Each complex has three units with specific area that three separate structures have prepared for utilization of solar panel. These structures are covering four parallel alleys and are occupying 15% of the whole area of the complex. The two-way street in the south side of each complex is parallel to three units with appropriative distance. The distance prepares opportunity for additional installation of solar panel on the top of street by adding similar structures holding solar panels. In this situation the portion of area for utilization of solar panel would increase to 33% of the whole area of the complex.

- In traditional situation the only available application in residential sector is the roof of buildings and would seldom exceed from 25% of the whole complex area. In the purposed concept the portion of absorption could easily reach into 33% of the whole area of complex.

- In traditional pattern in building sector the most available application is roof, so each building needs separate equipment. In the suggested pattern total costs of equipment, installation and maintenance is much less than the traditional system. Each unit has common equipment like batteries, invertors, etc.

- In the purposed concept a common architecture has been applied for the whole structures holding solar panels. This integrated system would prepare more aesthetic values in urban patterns compared to the tradition system that each building uses its roof as component in an inharmonic urban pattern. The form of each structure holding solar panels fulfills diverse duties in structural, architectural and electrical fields. The transparent layer could have different colors [4].

- The units of each complex are accessible via four alleys. In Alley B, C and D wide greenbelts have been considered in parallel to the sidewalks and bike routes. Total area of these greenbelts would be much more than the area of limited greenbelt in traditional patterns. Three upper alleys have two separate greenbelts. These greenbelts are covered with grass and short-root trees and would provide a more attractive atmosphere for cycling, walking, jogging, roller skating, etc.

- Each complex has a two-street way for accessing private cars to the complex that are completely separated from access of cyclists and walkers. This system would definitely provide a smoother traffic system and lots of junctions of pedestrians and cyclists would be eliminated.

- Each complex is multi-function with a different combination of three units. The floors of units have different usages like; residential flats, commercial zones, administrative departments and shopping centers. As a result lots of daily activities will be done in the concept and it would have positive impression into the traffic of relative city.

• Atmosphere of each alley is a combination of diverse activities like; group and individual walking, cycling, jogging, roller skating and other physical activities. Also routes of "B", "C" and "D" have greenbelts in the whole path. Natural airflow also is available through numerous openings. The roofs are semi-transparent and a smooth light would pass toward the alleys in diverse colors.

• The transparent layer of each unit holds the solar panels. This semi-glass material has different colors in each complex. Visibility of the whole solar system is achieved through this transparency so the alleys could get appropriative amounts of light during the day. Also the users would be able to be in contact with the installed solar system in a closer connection. Contribution of the whole concept to decrease co2 emissions would be understood by the local inhabitants and other users both in interior floors and in far view to the complexes.

• A central void allows the natural daylight to pass and reach into the other floors. Diverse openings provide opportunity for solar panels to be seen in each floor. This is an important point in several ways; firstly different floors would have natural day-lighting and secondly the solar system would be understandable in the whole concept. So the users will be reminded that they are using a zero-energy complex.

5 REFERENCES

[1] R. Krippner, T. Herzog, (2011), Architectural aspects of solar techniques – Studies on the integration of solar energy systems, in Proceedings Eurosun 2000, 3rd ISES-Europe Solar Congress, Cophenagen, Denmark: : International Solar Energy Society.

[2] T.E. Kuhn, S. Herkel, H.M. Henning, (2009), Active Solar Facades (PV and Solar Thermal).

[3] Arbitrary Orientation and Tilt, Retrieved from http://pveducation.org/pvcdrom/properties-of-sunlight/arbitrary-orientation-and-tilt

[4] S. Mende, F. Frontini, J. Wienold, (2011), Comfort and building performance analysis of transparent building integrated silicon photovoltaics, 12th Conference of International Building Performance Simulation Association, Sydney, 14-16 November.

[5] IEA SHC Task 41: Solar Energy and Architecture, Report T.41.A.1, Building integration of Solar, Thermal and Photovoltaics: barriers, needs, strategies

[6] T. OTIS, (2011), Is Solar Design a Straitjacket for Architecture? Conference Proceedings of the 27th International conference on Passive and Low Energy Architecture (PLEA), Louvain-la-Neuve, Belgium.

SMART-FLEX SOLUTION WAY FORWARD FOR COST COMPETITIVE BIPV PRODUCTION?

Juras Ulbikas, Algirdas Galdikas, Andrius Stonkus
Applied Research Institute for Prospective Technologies
Vismaliuku str. 34, LT-10243, Vilnius, Lithuania
Phone: +370 5 2500610; E-mail: juras.ulbikas@protechnology.lt

ABSTRACT: Implementing Nearly Zero Energy Building concepts BIPV solutions can play significant role only if they are competitive in terms of costs, performance and visual appearance comparing with other RES technologies. SMART-FLeX project is targeting these issues proposing integrated solution for glass façade buildings design, flexible manufacturing production and cost effective installation of tailored BIPV products according architect's vision. Project results are demonstrating competitiveness of BIPV products in terms of flexibility of building design and production as well as cost effective integration into the building. Further steps leading to competitive prices for products developed using SMART-FLeX solution comparing to other construction products will be discussed to ensure EU wide multiplication of the proposed approach in other than glass façade buildings and wide market deployment of SMART-FLeX based BIPV products.
Keywords: BIPV Solution, Glass Façade, Cost Effective

1 PURPOSE OF THE WORK

The purpose of the project SMART-FLeX is to demonstrate feasibility and cost effectiveness of integrated approach in BIPV products implementation for particular BIPV application in construction business – namely for glass façade buildings. Integrated approach proposed by the SMART-FLeX is linking building design process and BIPV product implementation by means of automated production line of increased flexibility when automated production of PV products with tailored according architect's vision visual appearance becomes feasible. Further cost reduction is achieved in building construction step by proposed plug-and-play solution for BIPV products integration into building envelope.

2 APPROACH

Proposed twofold approach targeting at demonstration of significant cost reduction of BIPV elements on product implementation and on integration into the building steps, is thought to open this way wider acceptance of BIPV products by construction market. Designed in SMART-FLeX project approach is focusing on steps which to our understanding are of main importance for cost effectiveness and market acceptance of the BIPV product, namely a) attractive to architects and designers building design tools implementation, b) automated production of tailored BIPV elements and c) cost effective integration of BIPV elements into the building and building electricity grid.

3 SCIENTIFIC INNOVATION AND RELEVANCE

Innovative solutions introduced in the project were targeting at eliminating barriers for BIPV market acceptance along value chain for construction product starting from building design, moving onto production process and ending at construction process itself. For the building design process SMART-FLeX is introducing design tool providing architect with freedom for building design with PV integration. On production level project is proposing flexible automated manufacturing concept for tailored to architect's vision BIPV product. For the construction process project is implementing plug-and-play solution for BIPV product integration into the building. Project is also proposing innovative approach for power production by PV plant monitoring and maximizing the power production (MPPT) from each string of cells under different weather and shadowing conditions.

4 RESULTS AND CONCLUSIONS

Project is in the final stage of implementation and main achieved project results will be presented and discussed during the presentation. Final months of the project are devoted to the testing and economical evaluation of the project results on both product production and PV plant operation levels. Project demonstration site of area ~ 600 m2, with integrated 180 BIPV façade elements of tailored design and different sizes (maximum size up to 1.7 x 3.5 m) with power trackers of special design, micro inverters and mounting system was implemented and now is under thorough monitoring.

Cost efficiency of the proposed by SMART-FLeX approach was evaluated taking into account final product cost estimations and savings achieved on the each production step. Limitations for product flexibility and visual appearance comparing to market requirements are identified and measures to improve product acceptability are thought. To have attractive visual appearance instead of the use of colored solar cells foreseen in the project, glass painting technology was utilized, this way providing architect with freedom to put any desired picture on the façade. Solutions eliminating the need to put smart-box on the surface of the PV module to further improve visual appearance of the BIPV product were proposed.

Despite successful demonstration of BIPV solution for glass façade buildings further steps are needed to transfer proposed solution to other building types with even further cost reduction of the BIPV products to become competitive with convenient construction products for windows, walls or roofs.

5 ACKNOWLEDGEMENT

The SMART-FLeX project has received funding from the European Union's Seventh Framework Programme, managed by the European Commission under Grant Agreement No. ENER/FP7/322449/SMART-FLeX.

OPTIMIZATION OF STAND-ALONE PV POWER SYSTEMS WITH HYBRID ENERGY STORAGES BASED ON ULTRA CAPACITORS

Sergey M. Karabanov[1], Dmitry V. Suvorov[1], Evgeny V. Slivkin[1], Andrey S. Karabanov[2], Dmitry Yu. Tarabrin[1]
1- Ryazan State Radio Engineering University
59/1 Gagarina St., Ryazan 390005, Russia
2- Energy Ltd.
51 B Novaya St., Ryazan 390027, Russia
e-mail: pvs.solar@gmail.com

ABSTRACT: The improvement of stand-alone PV power systems characteristics, including increase of their service life is the actual direction of ground solar power development. Now one of the prospective directions is the use of ultra capacitors in stand-alone PV power systems. Their application as an additional energy storage decreases considerably the load on the basic energy storage – accumulator, and increases its lifetime. The majority of stand-alone PV power systems uses unattended gel accumulators the service life of which does not exceed 3-7 years. The use of ultra capacitors in hybrid PV energy systems allows to increase essentially the accumulator lifetime. But the problem of optimum relationship of basic (accumulator) and additional (ultra capacitor) energy storage capacities and study of factors affecting this relationship remains not sufficiently investigated.
Keywords: stand-alone PV power system; ultra capacitor; battery service life, hybrid energy storage

1 INTRODUCTION

One of the principal elements of a PV power system is the energy storage device. Its service life is rather short and is defined by electrical load intensity. The majority of stand-alone PV power systems uses unattended gel accumulators the service life of which does not exceed 3-7 years. Due to this reason the use of ultra capacitors (electric double layer capacitors) in stand-alone PV power systems is the advance development tendency.

The application of ultra capacitors as energy storage components in PV systems allows to solve the following problems [1, 2, 3]:

- operation optimization of MPPT devices;
- smoothing of PV generator power at partial shadowing [4];
- increase of output peak power of the energy system.

The key advantages of ultra capacitors are:

- possibility of deep charge/discharge without reducing service life;
- long service life (several tens of years);
- possibility of operation at low temperatures;
- relative simplicity of charge/discharge cycles.

The purpose of the work is optimization of stand-alone PV power systems with hybrid energy storages based on ultra capacitors as far as the regularities of optimum relationship of basic (accumulator) and additional (ultra capacitor) energy storage capacities and study of factors affecting this relationship concerns.

2 STRUCTURE OF STAND-ALONE PV POWER SYSTEMS USING HYBRID ENERGY STORAGE DEVICES

Figure 1 shows the structure of stand-alone PV power systems with the use of hybrid electric power accumulators based on ultra capacitors.

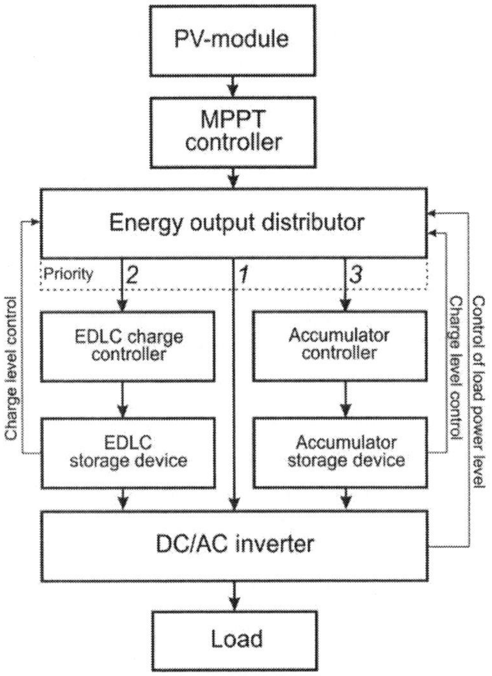

Figure 1: Structure of stand-alone PV power systems using hybrid electric energy storage devices based on ultra capacitors

A stand-alone PV power system with the use of hybrid electric power accumulators based on ultra capacitors includes the following components:

- PV modules;
- MPPT device for maximum power takeoff;
- energy output distributor;
- ultra capacitors charge controller;
- accumulator charge controller;
- basic accumulator energy storage device;
- ultra capacitor energy storage device;
- DC/AC inverter.

Electric energy generated by PV modules from MPPT

output means of output energy distributor is distributed between load (in current consumption), ultra capacitor energy storage device and basic accumulator energy storage device. Ultra capacitor and gel accumulator batteries are joined with an inverter connected directly to the current consumable load. Besides, the inverter is connected with the MPPT controller output for direct energy transfer without accumulation (buffering in the storage battery). The charge algorithm is the following: if the power generated by modules exceeds current consumption, then power is supplied, first of all, for the ultra capacitor charging and after its complete charge – to the additional charge of the storage battery. The algorithm of application of accumulated energy is similar to the previous one: if power consumed by the load at the current moment exceeds the power generated by PV modules, then first of all the energy stored in the ultra capacitor battery is used. Soon after this energy is used, the accumulator energy will be used.

The mathematical model of stand-alone PV power systems with the use of hybrid electric energy storage devices based on ultra capacitors includes the system of integral equations and logical conditions determining the algorithm of the system operation. The mathematical model algorithm is described in detail in [1].

2. INITIAL CONDITIONS OF THE NUMERICAL EXPERIMENT

At performing calculations the following initial conditions of the numerical experiment were used:

The power dynamics of insolation [5] normalized for some value of installed maximum peak power P_{PV_max} at the MPPT controller output was used as $P_{PV}(t)$ dependence. For calculations the value of installed peak power of 2.5 kW was used. As initial point of time the beginning of a calendar day equal to 00:00, the final point of time – the end of the day, 24:00, were used.

The limiting depth of discharge of a storage battery W_{bat_min} was set at the level of 25% of W_{bat_max}.

The maximum storage capacities are the following:

- storage battery– 9600 Wh (48 V, 200 Ah);
- the capacity of the ultra capacitor battery changed within the limits of 48-960 Wh.

Initial conditions for energy storage devices:

- the storage battery charge is at the level of 50% of W_{bat_max};
- zero initial charge of the ultra capacitor battery.

The controllers efficiency was set at the level:

- inverter efficiency $\eta_{DC/AC}$– 90%;
- efficiency of the ultra capacitor charge controller – 90%;
- efficiency of storage battery charge controller η_{ELDC}– 90%.

The model of the time dependence of the energy consumption P_{load} was presented by the function including 3 components:

- some level of continuous round-the-clock consumption P_{load_1} (refrigerator, wireless phone, domestic appliances in a sleep mode, TV antenna, etc.) at the level of 200 W;
- periodic average consumption level in the morning (08:00 – 10:00) and in the evening (18:00 – 22:00) (illumination, TV set, computer equipment, etc) at the level of 400 W and average consumption level P_{load}_2 in the morning (08:00 – 10:00) at the level of 200 W;
- periodic peak consumption level P_{load_3} of about 2500 W, 5 peaks per 5 minutes in the morning (08:00 – 10:00), in the daytime (10:00– 18:00) and in the evening (18:00 – 22:00) (electric tea kettle, electric stove, electric iron, washing machine, microwave oven, etc).

As a whole the above conditions are typical for stand-alone PV power systems.

3. THERESULTSOF NUMERIC EVALUATION OF STAND-ALONE PV POWER SYSTEMS WITH THE USE OF ULTA CAPACITORS. THE ANALYSIS OF THE OBTAINED DATA

The numerical experiment the results of which are given further was carried out with the use of the developed mathematical model of a stand-alone PV power system based on hybrid electric energy storage devices on the basis of ultra capacitors for the above conditions.

Figure 2 shows time dependence of the PV module output power and model dependence of the load power consumption. One can see that only the energy reserved in the storage battery is used at night. In the morning and in the evening during the most intensive electricity consumption the energy both generated by a PV module and energy reserved in the energy storage device are used. In the daytime mainly PV module energy is used.

It should be noted that the considerable part of morning and evening time and completely during the whole day period the current power consumption level is lower than that of the generated power.

However, during connection of consumers with high power consumption, the used power significantly exceeds the power generated in the day time at maximum insolation. In traditional stand-alone PV power systems it results in short-term pulsed discharge of the storage battery.

Figures 3, 4 show time dependences of discharge-charge power of the storage battery with the capacity of 9600 Wh without ultra capacitor energy storage device and at various capacities of an ultra capacitor (48, 96, 192 Wh).

One can see that use of one energy storage in the battery leads to its multiple discharge-charge cycles that results in decrease of the storage battery service life. At additional energy storage (ultra-capacitor) the number of cycles of discharge-charge is reduced with increase of its capacity.

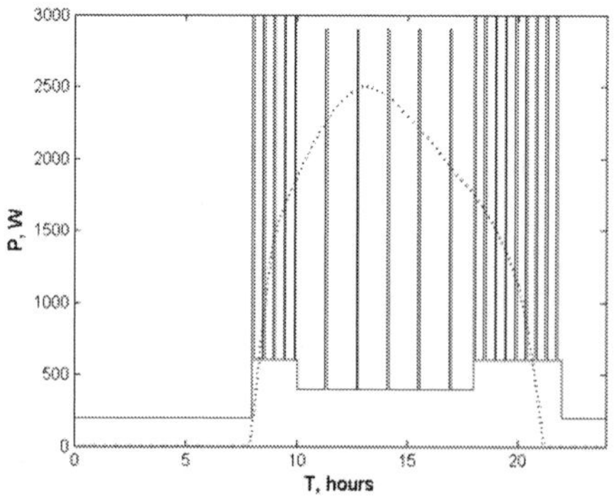

Figure 2: Time dependences of the PV module output power and load power consumption

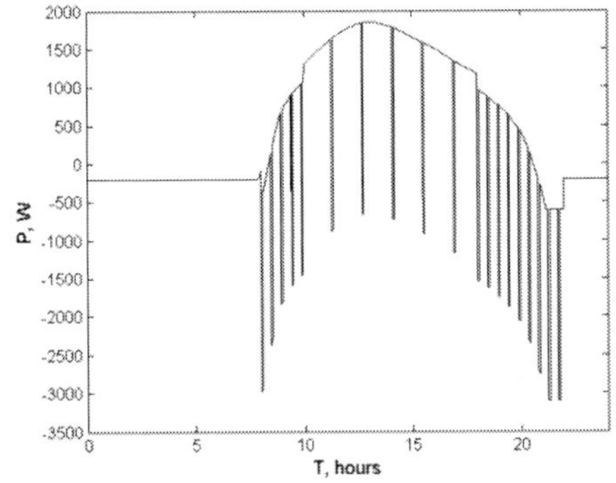

Figure 3: Time dependence of discharge-charge power of 9600 Wh storage battery without ultra capacitor

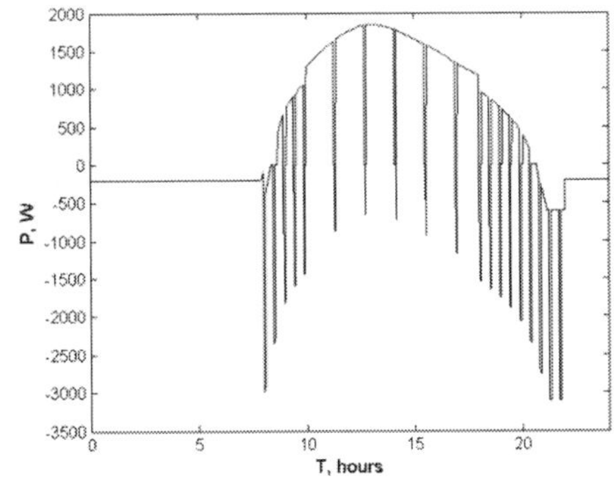

a

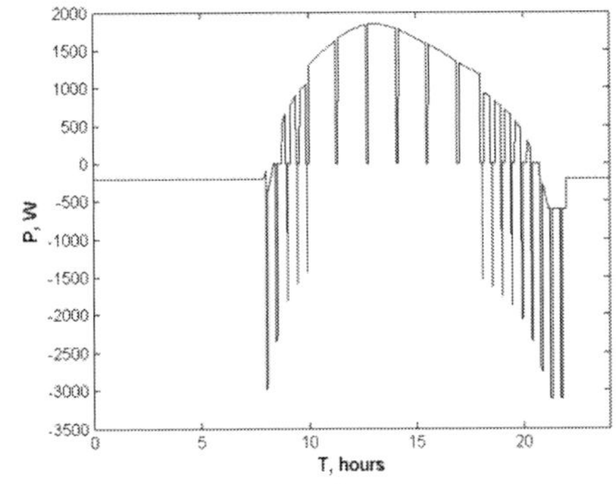

b

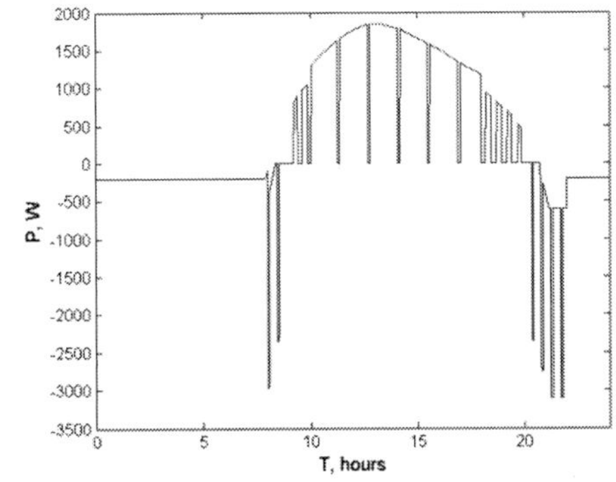

c

Figure 4: Time dependence of the discharge-charge power of 9600 Wh storage battery with 48Wh (a), 96Wh (b) and 192Wh (c) ultra capacitor capacities.

Comparing Figure 3 and Figure 4 it can be seen that the use of hybrid energy storage devices on the basis of ultra capacitors will allow to decrease significantly the load on the main storage device – the storage battery; and number of discharge-charge cycles. It is provided due to the fact that the ultra capacitors energy storage device carries out the function of the power buffer, the stored energy of which is used for ensuring power supply of short-term high-power consumers.

Figure 5 shows the time dependence of energy consumed by the load W_{load}, energy from the storage battery W_{bat} and ultra capacitor W_{ELDC} at various capacities of an ultra capacitor (48 Wh and 480 Wh).

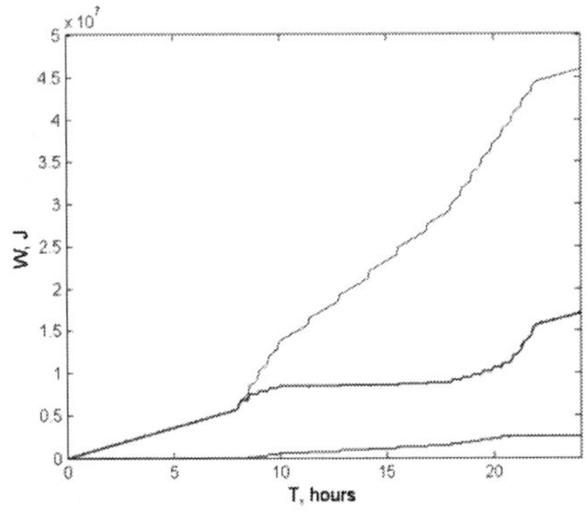

a

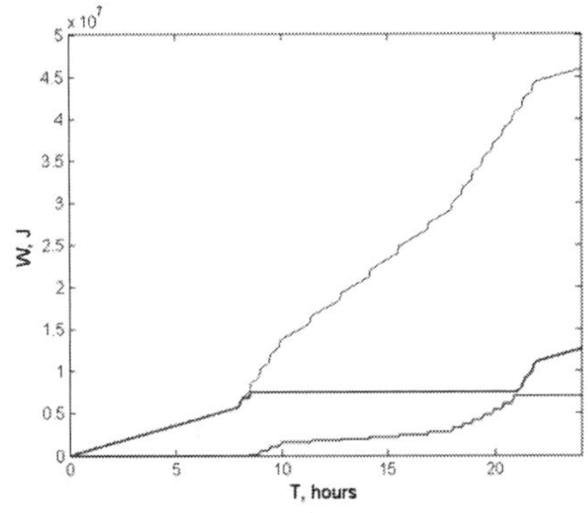

b

Figure 5: Time dependence of energy consumed by the load W_{load}, energy from the storage battery W_{bat} and ultra capacitor W_{ELDC} at various capacities of an ultra capacitor: 48 Wh (a) and 480 Wh (b).

The analysis of time dependence of energy consumed by load W_{load}, energy used by load from a storage battery W_{bat} and ultra capacitor W_{ELDC} shows that under such calculation conditions at about 10-45% of the total energy consumed by the load is carried out according to the cycle "ultra capacitor battery - load", and about 55-90% of consumption – according to the cycle "storage battery - load" (Fig. 5). The remaining part of energy is consumed by load from a PV module directly.

Figure 6 shows the dependences of energy consumed by the load (for 24 hours) from the battery and energy consumed from the ultra capacitor battery on the capacity of the ultra capacitor battery at direct storage battery capacity. It can be seen that in the area of small ultra capacitor battery capacity (up to 200 Wh) the share of energy consumed according to the cycle " ultra capacitor - load" increases sharply and then the growth is sharply slowing down. Thus it can be concluded that there is an optimal ratio of capacities of the storage battery and power consumed from the ultra capacitor battery. For the numerical experiment conditions it is equal to 200 Wh, or 2.08% of the capacity of the main storage battery.

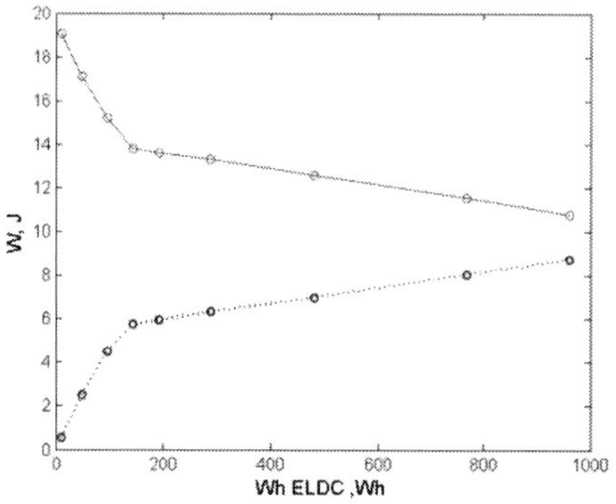

Figure 6: Dependencies of energy consumed by the load (24 hours) from the storage battery and energy consumed from the ultra capacitor battery on the capacity of ultra capacitor battery at a direct storage battery capacity.

So, the presented calculations show that the application of ultra capacitors in stand-alone PV power systems allows to increase considerably the storage battery service life.

4. CONCLUSION

It is shown that the use of hybrid energy storage devices on the basis of ultra capacitors will allow to decrease significantly the storage battery load and to increase its service life due to the fact that ultra capacitors perform the function of the power buffer, the stored energy of which is used for ensuring power supply of short-term high-power consumers.

It has been established that the ratio of energy capacities of ultra capacitors and storage battery of 2.08% is the optimum one for a stand-alone PV power system under numerical modeling conditions. It is obvious to suggest that the optimum ratio for other typical conditions will be close to this value.

Moreover, the application of ultra capacitors in stand-alone PV power systems is reasonable:

- for the purpose of maximum peak power increase;
- for the purpose of elimination of negative influence of temporary shadowing owing to cloudiness.

5 REFERENCES

[1] S.M. Karabanov, D.V. Suvorov, E.V. Slivkin, Y.V. Kukhmistrov, G.P. Gololobov, The Efficiency Analysis of Stand-Alone PV Power Systems with the Use of Ultra Capacitors, Proceedings 29th European Photovoltaic Solar Energy Conference, pp. 3756 – 3760.

[2] S.M. Karabanov, D.V. Suvorov, Y.V. Kukhmistrov, E. V. Slivkin, The study of PV modules with electric double layer capacitors integrated in their structure, Proceedings of 28th European Photovoltaic Solar Energy Conference, pp. 511 – 516.

[3] S. Karabanov, Y. Kukhmistrov. D. Ponomarev, Electric Double Layer Capacitors on the Base of Aprotonic Solvents, the Equivalent Circuit and Modes of Operation, 1994, pp. 32-34.

[4] S. Karabanov, Y. Kukhmistrov, The Method of the Electric Double Layer Capacitor Production, Patent 2139587 RU, MKI H01G 9/00, November 10, 1999

[5] Handbook of Photovoltaic Science and Engineering, Second Edition, 2011, John Wiley & Sons, Ltd.

[6]www.mpoweruk.com/life.htm.

A POWER MANAGING UNIT FOR MOBILE, STANDALONE AC SOLAR PV INSTALLATIONS

Driss Oulad-abbou*, Said Doubabi*, Ahmed RACHID#
*Department of Applied Physics, Faculty of sciences and technology, Cadi Ayyad University, Marrakech, Morocco
#Laboratoire des Technologies Innovantes, University of Picardie Jules Verne, Amiens, France
driss.ouladabbou@gmail.com, s.doubabi@uca.ma, rachid@u-picardie.fr

ABSTRACT: Battery bank is an important element in photovoltaic (PV) installations, since it permits to supply energy when there is a lack of power. In addition, battery bank is the most sensitive element of PV installations. Hence, their use requires more attention in order to increase their lifetime. In the light of this vision, the main idea of this paper is to reduce the stress on the battery bank, while keeping the use of one DC/DC converter in a standalone PV installation. To this purpose, a managing unit is used to manage the flow of energy between the solar PV installation components. These components are: PV generator, battery bank, and the AC load. The control algorithm is explained. Simulation results are also analyzed.
Keywords: energy management, Battery storage, standalone photovoltaic installation.

1 INTRODUCTION

Solar energy has become one of the most challenging energy sources. For this reason, many domestic, industrial and commercial applications are based on solar energy.

In Morocco, especially Marrakesh, solar energy is widely available, as shown in figure 1. Hence, it is preferable to use it as energy source, which is clean and renewable.

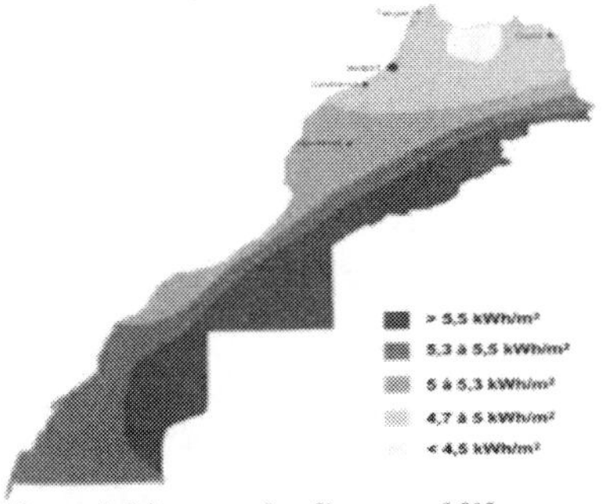

Figure 1: Moroccan irradiance card [1]

The concept of developing and using solar energy has become more popular in our country. This is reflected by the increased number of project based on solar energy. **Noor** station project is a concrete example. In addition, the rise in using standalone AC photovoltaic installations in Marrakesh, leads us to propose some low cost solutions, based on components available in the local market.

The basic components of a standalone AC photovoltaic system are: photovoltaic generator, charge regulator, which include or not the MPPT technique, batteries and DC/AC converter.

The architecture that has been widely used is shown in figure 2 (a) [2], it consists to directly connect the battery bank to the DC side of the DC/AC converter. The maximum of power is extracted from PV panels and supplied to the battery bank using a DC/DC converter. However, this structure leads to fast charging/discharging and ageing of batteries [3].

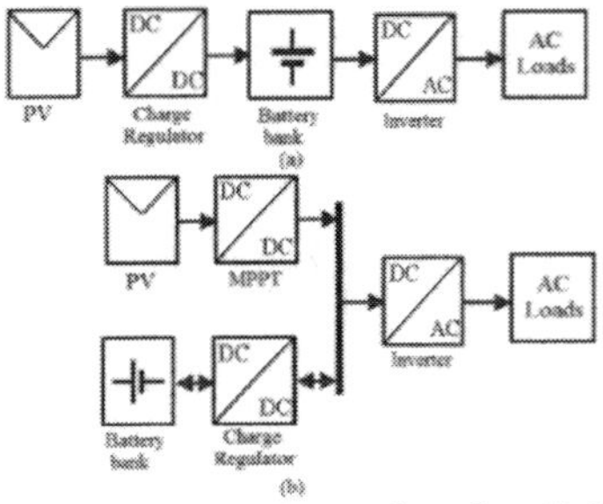

Figure 2: (a) direct connecting configuration, (b) DC coupled structure

In order to achieve a good use of batteries, a DC coupled structure has been used (figure 2 (b)) [4][5][6]. The battery bank is connected to the DC bus through bidirectional DC/DC converter which permits to achieve optimized charge/discharge operations [7]. This structure is widely used for battery charging applications, especially for electric vehicle and hybrid energy sources [8][9].

By considering the bidirectional DC/DC converter losses, in addition to its increased cost, our purpose is to reduce dissipating components number in a standalone photovoltaic AC installation. For this, a managing unit is proposed in order to reduce the stress on batteries. This is ensured while maintaining the good operation of the whole installation (figure 3). This study was done by considering an AC charger of an electric bike (e-bike) as AC load.

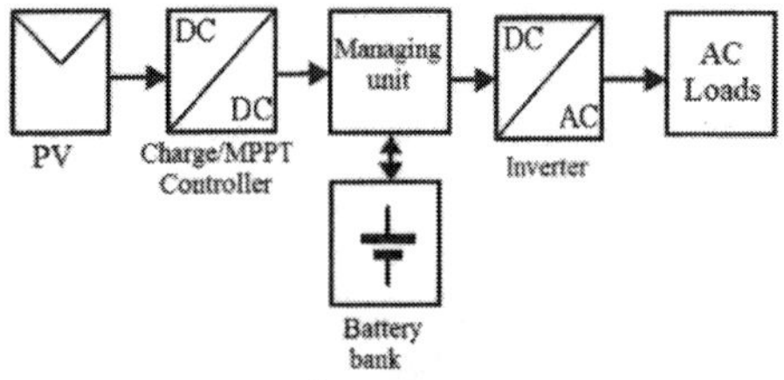

Figure 3: The proposed structure

This work was supported by the institute of research on solar energy and new energies Morocco: IRESEN-project VERES

The rest of paper is organized as follows. Section II, gives the configuration the proposed structure. Then, the control algorithm of the managing unit is given in Section III. Simulation results and discussions are presented in Section IV, followed by the conclusion and perspectives in section V.

2 CONFIGURATION OF THE PROPOSED STRUCTURE

The system scheme is shown in figure 4. The Managing unit is composed of two parts; the fisrt one is the power part that contains the three power switchs, which are used to control flows of energy, from PV to battery bank, from PV to the inverter end finaly from battery bank to the inverter. The second part is the controlling part. Basing on values of input paramers this controlling part control the three power switchs. The algorithm describing the functioning of our managing unit is described in the next section.

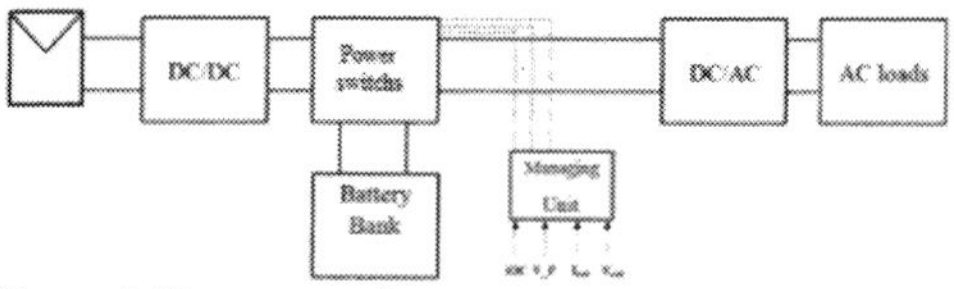

Figure 4: The system scheme.

3 CONTROL ALGORITHM

The flowchart of the control algorithm is given in figure 5. Solar PV panels supplies power to the load. Battery bank is used as extra source. This latter supply power when solar energy is not sufficient to satisfy the load demands.

The control algorithm consists to detect the presence of an AC loads. The presence of load is detected using a presence sensor V_P, if no charge is connected, then the battery is charged using constant current-constant voltage (CC-CV) algorithm. In the first phase, the maximum of power is extracted and delivered to the battery bank using a constant voltage algorithm ensured with a P&O algorithm. Then, in the CV phase a constant voltage charging phase is applied using the same PI controller used before, so one voltage controller is used to ensure MPPT and CV charging.

If the load is connected to the charging station, the first step is to calculate its nominal power, to this end the first step consist to supply energy from the battery bank, and then the nominal power is calculated. After the maximum of power is extracted from PV generator using constant voltage algorithm, the power is supplied to the AC load and the rest is used to charge the battery bank respecting the two phases indicated above, if the energy demand of AC load is not satisfied, all the PV energy is supplied to this loads. If the energy is still lower than the charge demands, the rest of energy is extracted from the battery bank if its state of chare permits this operation of course.

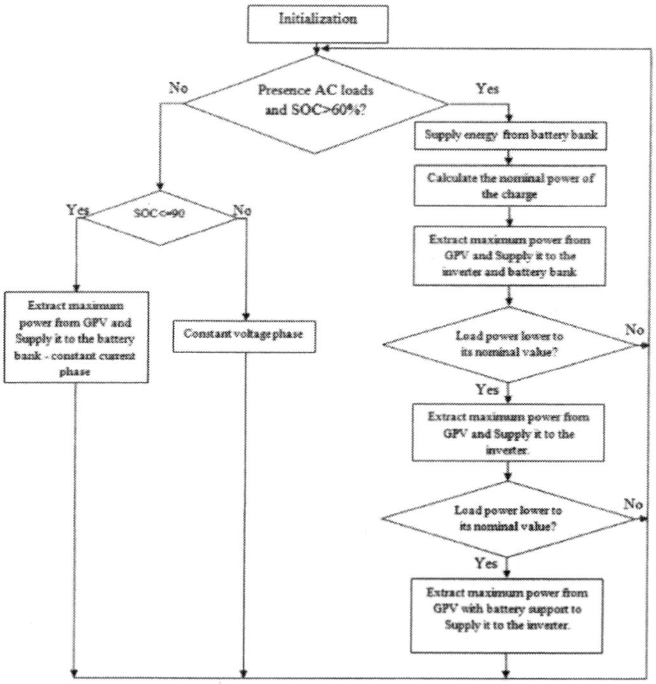

Figure 5: The control algorithm

6 SIMULATION RESULTS

The simulation scheme is shown in figure 6. The PV panels used in simulations deliver a power of 400Wc. These latter were modeled as described in [10].

Simulations were done using an irradiance of 700W/m2, which is the average value of irradiance in Marrakech.

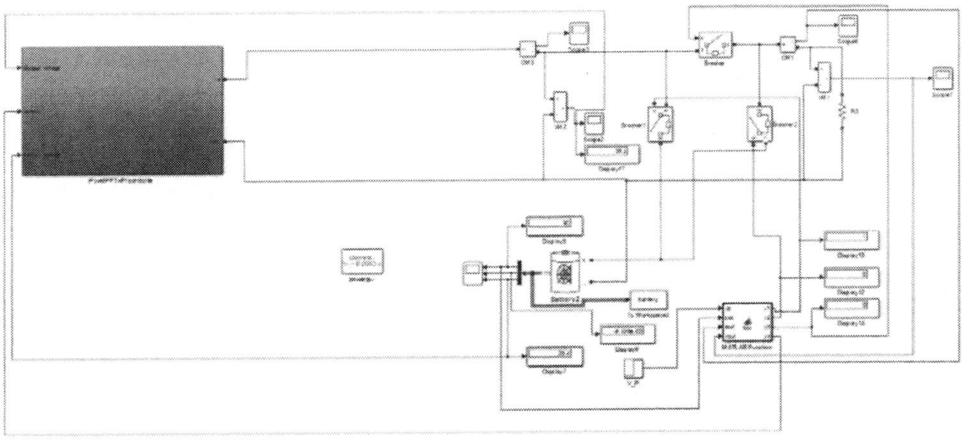

Figure 6: Simulation scheme

In order to simplify this study, the inverter and the output load are modeled as resistive load.

The simulation results are showen in figures 7, 8, 9, 10 and 11.

Figure 7, shows the SOC, battery voltage and battery current evolutions when the maximum of power extracted by the MPPT is used to charge the battery. The charging current is 10A. This phase is the constant current phase. The constant voltage operation is ensured using a PI controller, which consist to apply a constant charging voltage when the SOC reach 90%.

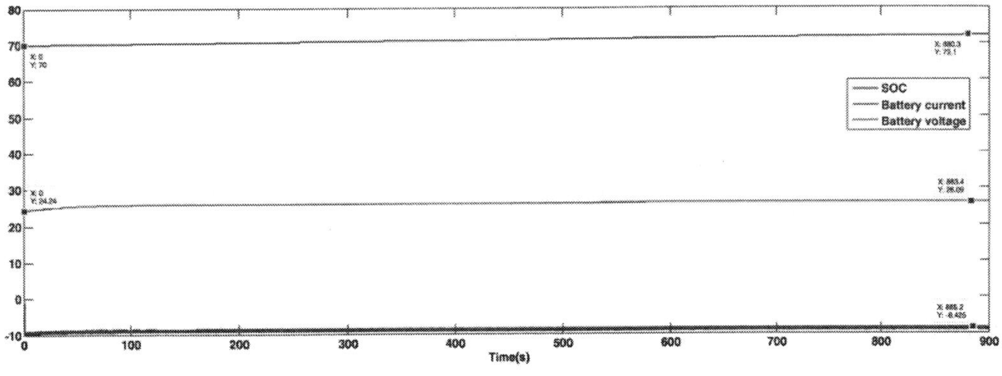

Figure 7: Evolutions of battery SOC (%), battery Current (A) and battery voltage (V) during charging

In figure 8, the evolution of battery parameters, output voltage and current, and PV current are illistrated. It is clear that the first step is supplying energy from battery bank (if it is enough charged) to the load in order to calculate its nominal power, then the PV generator supply the energy to the load, if this energy is not enough the battery supply the rest, hence, the load power is already equal to its nominal value as shown in figure 8.

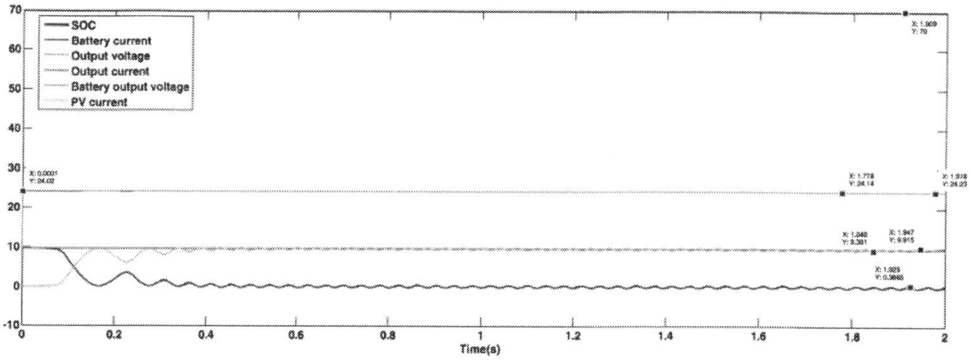

Figure 8: Evolutions of battery SOC (%), voltage (V) and current (A), PV current (A), output current (A) and voltage (V); the load is connected to the station.

The functioning was tested during the transition from no vehicle present to presence of vehicle, in figure 9 the vehicle was absent, all the PV energy is used to charge the battery respecting the charging phases. On t=2s a vehicle is connected to the station the same procedure described in figure 8 is applied in order to supply the nominal power to the load. The evolution of battery voltage, current and SOC, PV current, output current and vehicle presence in this case are shown in figure 10.

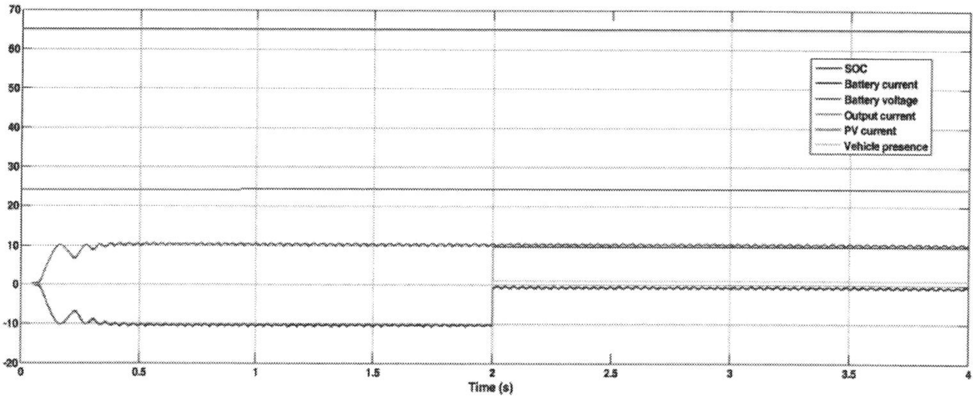

Figure 9: Evolutions of battery SOC (%), voltage (V) and current (A), PV current (A), output current (A) and Vehicle presence, when the load is connected on t=2s.

6 CONCLUSIONS AND FUTURE WORKS

This paper has presented a managing unit of PV installation and a solar charging station in particular case. The simulation results show the usefulness of the proposed managing unit, it can be seen also, that the managing unit ensures good performance either the vehicle is present or not. The battery stress is reduced by given the priority for supplying energy from PV generator. Moreover, this combination of this unit and the standard PV topology permits proves a low cost and simple design which makes the design and sizing easy.

In addition, the control algorithm could be implemented on a low cost microcontroller. The components used are low cost and widely available in the local market.

Our priority for further work is to prove the validity of this work through experimental tests, and then to test the ability of this system, to guarantee good performances in practical cases.

4 REFERENCES

[1] Moroccan Agency for Solar Energie "www.masen.org.ma", Home >MOROCCAN SOLAR PLAN >Sunshine map. 02 June 2015.
[2] Nayar, C. V., Ashari, M., & Keerthipala, W. W. L. (2000). A grid-interactive photovoltaic uninterruptible power supply system using battery storage and a back up diesel generator. IEEE Transactions on Energy Conversion, 15(3), 348–353. http://doi.org/10.1109/60.875502
[3] Fakham, H., Lu, D., & Francois, B. (2011). Power control design of a battery charger in a hybrid active PV generator for load-following applications. IEEE Transactions on Industrial Electronics, 58(1), 85–94. http://doi.org/10.1109/TIE.2010.2062475
[4] Lu, D., Fakham, H., Zhou, T., & François, B. (2010). Application of Petri nets for the energy management of a photovoltaic based power station including storage units. Renewable Energy, 35(6), 1117–1124. http://doi.org/10.1016/j.renene.2009.12.017
[5] Guo, F., Member, S., Inoa, E., Choi, W., & Wang, J. (2012). Study on Global Optimization and Control Strategy Development for a PHEV Charging Facility. IEEE Transactions on Vehicular Technology, 61(6), 2431–2441. http://doi.org/10.1109/TVT.2012.2195787
[6] Capasso, C., & Veneri, O. (2015). Experimental study of a DC charging station for full electric and plug in hybrid vehicles. Applied Energy, 152, 131–142. http://doi.org/10.1016/j.apenergy.2015.04.040
[7] Lo, Y. K., Chiu, H. J., Lee, T. P., Purnama, I., & Wang, J. M. (2009). Analysis and design of a photovoltaic system DC connected to the utility with a power factor corrector. IEEE Transactions on Industrial Electronics, 56(11), 4354–4362. http://doi.org/10.1109/TIE.2009.2030216
[8] Preetham, G., Member, S., Shireen, W., & Member, S. (2011). Photovoltaic Charging Station for Plug-In Hybrid Electric Vehicles in a Smart Grid Environment.
[9] De Freige, M., Ross, M., Joos, G., & Dubois, M. (2011). Power & energy ratings optimization in a fast-charging station for PHEV batteries. 2011 IEEE International Electric Machines and Drives Conference, IEMDC 2011, 486–489. http://doi.org/10.1109/IEMDC.2011.5994646
[10] Nemes, C., Munteanu, F. An analysis of a photovoltaic panel model: Comparison between measurements and analytical models. 2012 International Conference and Exposition on Electrical and Power Engineering (EPE), 939-944.http://dx.doi.org/10.1109/ICEPE.2012.-6463811

DESIGN AND COST OPTIMIZATION OF SMALL-SCALE PV-POWERED REVERSE OSMOSIS DESALINATION (CASE STUDY)

Safaa Hajji[1], Anas Bentamy[2], Ndiaga Mbodji[3], Ali Hajji[4]
[1,2]Al Akhawayn University, Ifrane, Morocco
[3,4] Process Engineering and Environment Research Unit, Institut Agronomique et Vétérinaire Hassan II, Rabat, Morocco
safaa-hajji@hotmail.com m.ndiaga@yahoo.fr ahajji61@hotmail.com

ABSTRACT: Photovoltaic-powered reverse osmosis (PVRO) desalination can potentially be the solution to two major global issues: water and energy. This technique not only increases water supply but also decreases the dependency on fossil fuels.

The objective of the present paper deals with the design of a small-scale (20 m^3/day) PVRO desalination unit, given the daily water needs, in addition to conducting an economic analysis to determine both the total investment and the specific desalination cost (MAD/m^3). For this case study in Dakhla (South of Morocco) which supplies drinking water for a coastal hotel, the system configuration that has the lowest investment and specific water costs is the one with an energy recovery system and separate components. The specific energy consumption of such system was found to be 2.92 kWh/m^3. The total costs over a 10-year period for this system is 1,056,024 MAD, producing water at a cost of 15.47 MAD/m^3.

An effective methodology is used for determining the optimal design configuration for a PVRO desalination station. For our case study, the results show two optimal solutions, both of which operate with one machine only. The first one, which yields the lowest initial investment (828,000 MAD), consists of installing one RO machine with a nominal capacity of 36 m^3/day, 23 kWp of PV power and 50.5 kWh of battery storage. The second optimal combination, which generates the lowest cost of water during 10 years (14.28 MAD/m^3), consists of installing one bigger RO machine having a capacity of 72 m^3/day, 32 kWp of PV power and 38.5 kWh of battery storage.
Keywords: Desalination, Reverse osmosis, Photovoltaic, Morocco, System design, Cost analysis, Optimization, Stand-alone PV Systems

1 INTRODUCTION

Although about 97% of the Earth's surface is covered by water, only 3% of this water is actually fresh water and hence, many regions of the world are facing enormous water shortages. According to the World Health Organization, in 2005, half a billion people lived in water-scarce and water-stressed countries; many of them are forced to drink contaminated water, which often causes disease and can lead to their early death [1]. Population growth and the pollution of natural sources of freshwater are causing a significant and on-going reduction in the worldwide per capita availability of clean drinking water. Additionally, increasing water demands for agriculture, industry and tourism lead to further water shortages around the globe.

Similarly to water, energy is another important global issue. The growth in the world population, which has reached over 7 billion, has led to the significant increase in energy demand in different sectors. This increase in energy consumption causes the depletion of energy sources, or fossil fuels to be more precise. Besides the decline in their reserves, fossil fuels, especially coal and oil, have two other main drawbacks: unstable prices and negative effects on the environment. This is why alternative renewable energy sources must be increasingly included in the energy mix of any country. However, despite the enormous growth rates of some renewable like wind and solar in the past decades, their contributions in absolute terms must still grow considerably.

In Morocco as well, water and energy are critical issues. It is a water-scarce country confronted with declining groundwater reserves, due to their overexploitation, and a strong dependence on rain-fed agriculture. According to Morocco's Ministry of Energy, Mines, Water and Environment (MEMWE), the country's renewable water resources are limited and the amount that can actually be used has been estimated at no more than 22 billion m^3/year [2]. The average amount is estimated at 630 m^3 per inhabitant per year, which is very low compared to the world's average water footprint, which is 1,243 m^3/capita/year [2]. Additionally, the lack of a functioning sanitation network and wastewater treatment system causes already scarce water resources to become contaminated and unsuitable for multipurpose use.

Concerning the Moroccan energy sector, it is characterized first by a high dependency on imports. In fact, the country's energy demand depends to about 95% on imports and is thus depending on the instability of prices for oil, gas and coal. According to the Ministry of Economy and Finance (MEF), Morocco's gross energy bill for 2014 was 65.5 billion MAD, of which 57.8 billion MAD was spent on crude oil and petroleum products [3].

Desalination of sea and brackish water already plays a major role in the provision of freshwater for domestic, agricultural and industrial uses in many parts of the world and is a thriving industry that has doubled its capacity to almost 70 million m^3/day in the last ten years or so [4]. In this field, reverse osmosis and various thermal methods such as multi-stage flash and multi-effect distillation dominate the industry. Whilst it is highly successful in delivering freshwater from saline water, desalination is a very energy-intensive process, consuming vast quantities of fossil fuels worldwide. As a result, desalination is a major contributor to CO_2 emissions and will undeniably play a part in global warming.

The objective of the present work is to present a design methodology of a reverse osmosis system and optimize the cost of a small unit of reverse osmosis.

Section 2 below briefly presents the various existing desalination processes and the operating principle of photovoltaic (PV) powered reverse osmosis (RO) desalination system, PVRO for short, and its singularities, whereas section 3 describes the steps of the design methodology as applied to the case of a coastal hotel in Dakhla, south of Morocco. Section 4, on the other hand optimizes the system configuration that has the lowest investment and specific water costs and finally section 5 presents the main conclusions.

2 DESALINATION PROCESSES AND OPERATING PRINCIPLE OF PVRO

Desalination, as the name clearly states, is the process by which salt (and other minerals) is removed from saline water. This salty water which can either be seawater or brackish water, is treated using one of the numerous existing separation technologies to give potable water. This desalinated water is then used for human consumption, and in some cases, for irrigation as well, whereas the salts that have been removed are "concentrated in a stream of water called the brine reject, disposed either to the sea or to a saline aquifer or in evaporation ponds" [5]. Desalination, also known as desalting, is composed of three major steps which are the pre-treatment of the salty water, the actual separation and the final post-treatment of the potable water. Additionally, desalination requires some form of energy to operate and provide water of a certain quality which satisfies international water norms.

There exists numerous types of desalination technologies and they can be classified in two different ways; either according to the form of energy involved or according to the phenomena involved as detailed in Figure 1.

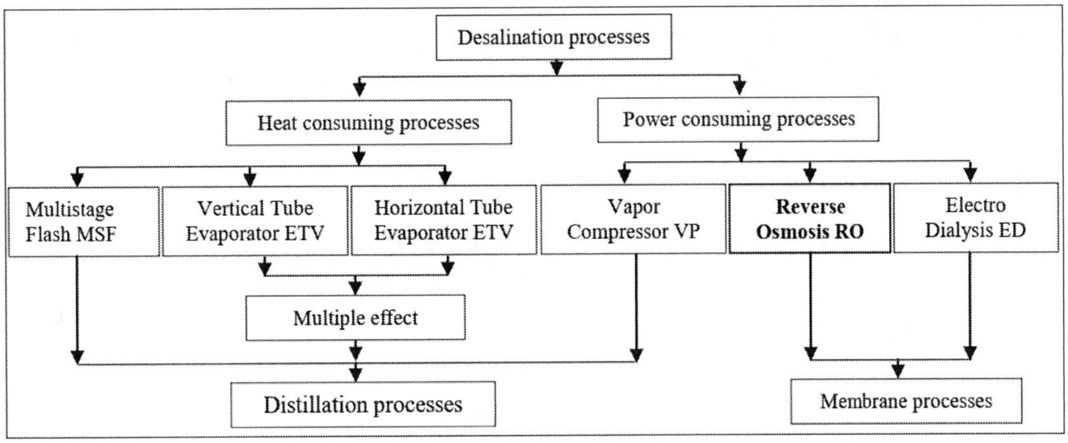

Figure 1: Principal desalination technologies

Reverse osmosis (RO) desalination is a membrane technology that has been developing over the past 40 years or so but it was not until the 21st century that it started growing at a very high pace, especially compared to thermal or distillation techniques. In fact, RO is currently the leading desalination technology, holding 60% of the total installed capacity in the world [6]. This is mainly due to the advances in the energy recovery technology as well as other developments that have improved its reliability, costs as well as the energy consumption, which are all important aspects in desalination. Reverse osmosis desalination uses the RO process to remove the salts from brackish or seawater in order to provide fresh, potable water. During this process, pressure is applied on the saline water, consuming energy, and producing fresh water and a highly concentrated salt byproduct called brine. The amount of pressure, and hence of energy, required depends on the salinity of the feed water; the more concentrated the feed water is, the higher the pressure is. The operation pressure for brackish water systems varies from 15 to 25 bars whereas it ranges from 54 to 80 bars for seawater systems [7]. This process allows the removal of 95% to 99% of the dissolved salts, which are discharged through the brine. This brine, also called concentrate, is rejected at high pressure, and in some cases, this energy is recovered through the use of energy recovery systems (ERS).

All RO desalination plants or systems are composed mainly of:

Feed pump: conveys the saline, feed water from its source to the pretreatment system and the HP pump;

Pretreatment: reverse osmosis units require intensive, proper pretreatment in order to keep the membranes clean and therefore increase their efficiency, lifespan and decrease future operating costs;

Pressurization: after the pretreatment process, the saline water needs to be pressurized before entering the RO membranes. This process, called pressurization, is accomplished by one or more high pressure (HP) pumps, which represent the heart of a RO desalination system;

Membrane separation: after the pretreated feed water is pressurized to the adequate operating pressure, it then enters the RO membrane, which removes the salts from the brackish or seawater;

Post-treatment: The last step in the desalination process is the post-treatment of permeate or the product water before transferring it to the distribution system;

Energy recovery system: recovers the energy from the high-pressure brine in order to reuse it to increase the pressure of the low-pressure feed water.

Brine disposal: whether the RO unit includes an ERS or not, brine is always present. Brine, the main byproduct of desalination, is nothing but a stream of water with a very high salt concentration and contains other chemicals as well which have been used mainly in the pre and post-treatment stages.

As the name clearly states, a photovoltaic (PV) powered reverse osmosis (RO) desalination system, PVRO for short, is nothing but a desalination system that uses the RO technology to remove salts from either brackish or seawater only it uses electricity generated from PV panels to power the system instead of using the conventional utility grid. PVRO desalination offers a practicable solution to the water scarcity problem and is in fact considered to be one of the promising technologies when it comes to producing fresh, potable water especially in remote, sunny areas where solar energy is abundant and/or connection to the electric grid is absent. Additionally, PVRO systems have proved to be very efficient in particular for small-scale usage. The performance of such systems depends on numerous parameters, especially meteorological conditions. The PVRO desalination system consists of the following components: an RO desalination unit, a PV generator, a charge regulator, an inverter, and storage batteries, as shown in Figure 2.

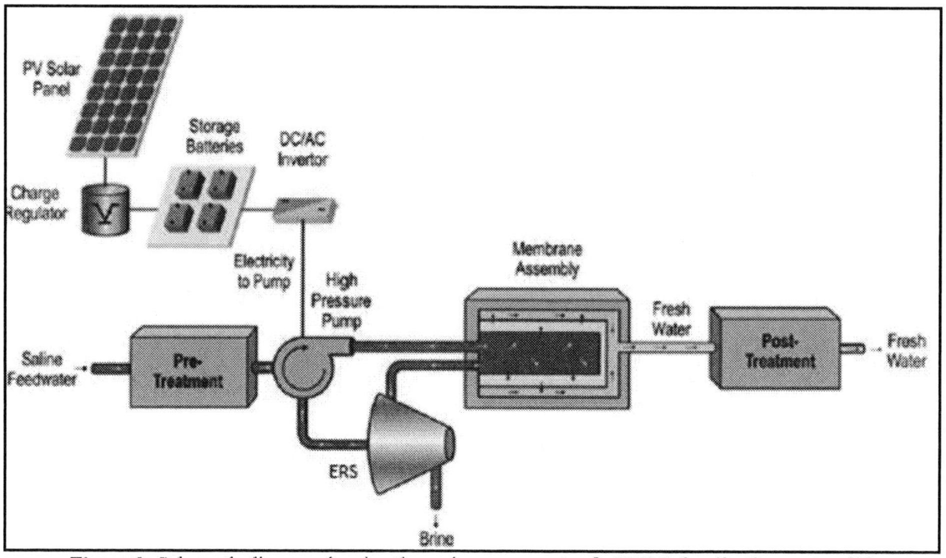

Figure 2: Schematic diagram showing the main components of a PVRO desalination system [8]

3. DESIGN METHODOLOGY

The methodology used in the design of such PVRO system consists of 6 steps depicted in Figure 3 and explained below.

Step 1: Identify the location
↓
Step 2: Define the daily water needs and operation time
↓
Step 3: Identify feed water and product water
↓
Step 4: Size the equipment
↓
Step 5: Refine the configuration
↓
Step 6: Perform the financial analysis

Figure 3: Steps of the methodology

Step 1: Identify the location

The project consists of the design of a stand-alone reverse osmosis desalination plant for a small isolated hotel that will be built not too far away from the city of Dakhla, in the southern part of Morocco.

Step 2: Define the daily water needs and operation time

The daily capacity required by the client for this project is 20 m³, which he believes is sufficient to cover the hotel water needs. Seawater will be used as feed water for the station and the operating time is 7 hours.

Step 3: Identify feed water and product water characteristics

The salt concentration, the temperature and the recovery rate of the feed water are respectively 38.2 g/L, 25 °C and 43% whereas the concentration of the product water is 0.5 g/L.

Step 4: Size the equipment

In this step, we size the different system components and the results are in Table I.

Step 5: Refine the configuration

There are different RO system configurations or with different technologies where the desalination unit include only an HP pump, or by integrating the HP pump and ERS in one component, or by integrating both the pressure exchanger (PX) and the booster pump in one device, or using separate components as summarizes in Table II. Table III provides the power, the specific energy and the PV capacity for the different system configurations, and we conclude that installing an ERS reduces the energy consumption by 35 to 45%, which is important. After comparing the energy consumption of all possible configurations, both with and without ERS, we conclude that the system with the least energy consumption is the one operating with an ERS and with separate components, instead of combining the HP pump and the ERS in one component. The power required in this case is equal to 8.35 kW and the specific energy consumption is 2.92 kWh/m3. Cases 2 and 3 have the same power requirements.

Step 6: Perform the financial analysis

Table IV gives the cost components of all four system configurations and computes the cost of water in MAD/m³ computed over a 10-year period. From the table below, we conclude that the system configuration that has the lowest costs and that produces the water with the least possible costs is the one with an energy recovery system and separate components (HP pump, pressure exchanger and booster pump). The total costs for this system is 1,056,024 MAD, whereas the water cost is 15.47 MAD/m³. The RO desalination system with an ERS and separate components is not only the most energy-efficient choice but is also the most economical one, which explains our final choice.

Table I: Results of the sizing

Cases	Value	Unit
Feed pump	1.15	kW
Osmotic pressure of feed water	32.40	bar
Osmotic pressure of product water	0.43	bar
Surface area of the membrane	231	m²
Number of modules	32.00	
Pressure vessels	8.00	
HP pump without ERS	13.90	kW
Chemical dosing pump	0.05	kW
Cleaning pump	0.10	kW
HP pump with ERS	6.14	kW
Pressure exchanger	5.56	kW
Booster pump	0.86	kW
Integrated HP pump and ERS	8.69	kW
PV Capacity	38.50	kWp

Table II: Reverse osmosis system configurations used in the Analysis

Cases	Description	HP Pump	ERS	Booster	Energy storage
Base case	No ERS	Yes	No	No	No
Case 1	Int. HP/ERS	Yes	Integrated with HP Pump	No	No
Case 2	Int. PX/booster	Yes	Yes	Integrated with PX	No
Case 3	Sep. components	Yes	Yes	Yes	No

Table III: Summary of the power, specific energy, and PV capacity for the three system configurations

Power and energy	Without ERS	Integrated HP/ERS	Separate components
Installed power (kW)	15.25	10.05	8.35
Specific energy (kWh/m³)	5.34	3.51	2.92
PV Capacity (kWp)	38.50	25.40	21.20
% change	0%	- 35%	- 45%

Table IV: Summary of the CAPEX and OPEX as well as the cost of water for the four system configurations

Component	Without ERS	Integrated HP/ERS	Integrated PX/Booster	Separate components
Pre and post treatment	52,216	52,216	52,216	52,216
RO unit	351,316	427,216	423,066	375,466
PV system	480,256	351,530	258,798	258,798
Others	44,600	44,600	44,600	44,600
CAPEX	831,572	778,746	681,864	634,264
% change	0.0%	-6.4%	-18.0%	-23.7%
OPEX (over 10 years)	491,533	482,289	430,090	421,76
Total costs (CAPEX & OPEX)	1,323,105	1,261,035	1,111,954	1,056,024
% change	0.0%	-4.7%	-16.0%	-20.2%
Water cost (MAD/m^3)	19.12	18.27	16.23	15.47

4. OPTIMIZATION

This section consists of the design and implementation of an optimization model for stand-alone PVRO desalination units. The goal is to determine the effect of changing certain system parameters on the investment cost as well as on the water cost over a long period of time (10 to 20 years), in order to find the optimal solution, and apply it to our case study.
Figure 4 shows the average hourly electrical power production from a solar PV system as well as the consumption of a 10 kW reverse osmosis desalination unit.

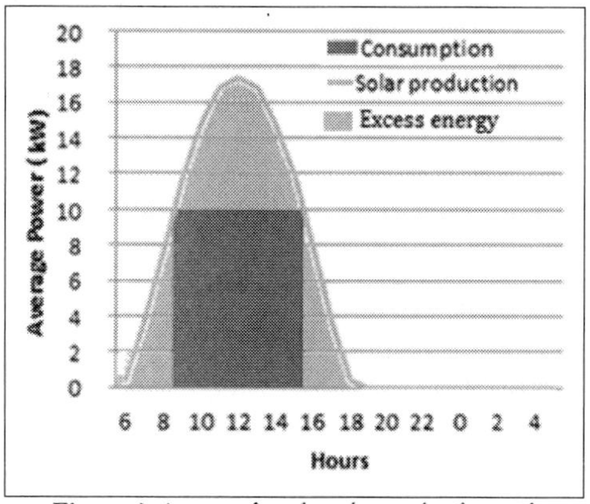

Figure 4: Average hourly solar production and consumption of a 10 kW RO system, without energy storage

From the graph, we notice that a significant amount of electrical energy produced by the PV system (shown in green) is wasted, whereas it could be used to produce more water. This energy can be used in at least two ways:

- It can be stored in batteries and later used to operate the desalination unit for longer hours.
- It can be used to run a smaller machine, in hours where the PV system is still producing electrical energy enough for a small machine to operate but not enough for larger machines.

Or it can be used as a combination of both; part of the excess energy is used to run smaller machines whereas another part is stored in batteries for later use, as shown in Figure 5.

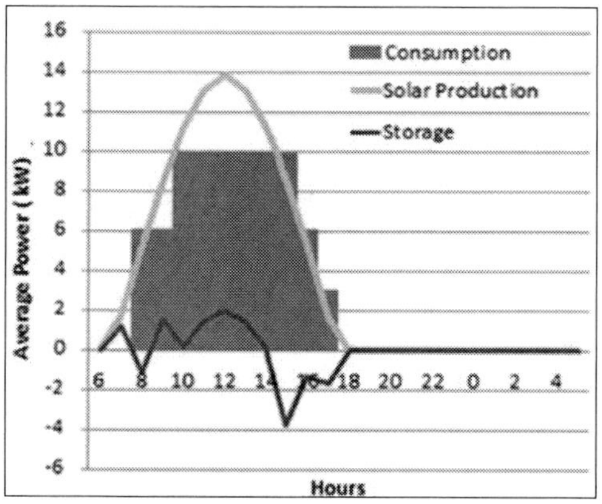

Figure 5: Average hourly solar production and consumption of a 10 kW RO system (3 RO machines) with energy storage

From the graph above, we notice that there are three machines with different capacities and each one operates depending on the amount of electrical energy available. We can also see how the storage varies with time; the graph line is increasing when there is extra energy and the batteries are charging, while it is decreasing when batteries are discharging.
Whether the user decides to install batteries or to add smaller RO machines to exploit the excess energy, he needs to be aware of the additional costs that will be involved, both in the initial investment and other costs such as replacement costs. The question then becomes whether this increase in costs will be offset by the gain in water production. This is where we notice the need for an optimization model, especially since there are too many combinations and configurations to analyze.

The following typical methodology was followed to determine the optimal design and operating configuration:

- Modeling the PVRO desalination system;
- Developing a computer tool to solve the governing equations;
- Conduct a parametric study to determine the best combination of the varying parameters (RO machines, PV and batteries) that yield the optimum cost.

Once the optimization opportunity has been identified, the next step was to develop a conceptual model for the PV-powered RO system. The model is divided into two main components, which are the technical model and the financial model, as shown in Figure 6. In order to solve the model described above, we needed to design a calculation tool on Excel once again and implement a number of macros in Visual Basic (VBA). The tool analyzes each of the main components of a PVRO desalination unit: PV system, batteries and RO unit as well as a detailed cost analysis of up to 20 years.

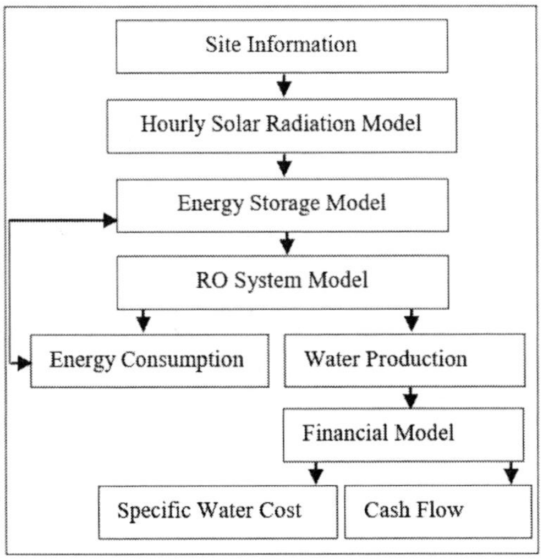

Figure 6: Technical and financial component of the PVRO desalination conceptual model

Table V below presents the three types of machines used.

Table V: RO machines used

Machine	Capacity (m^3/day)	Flow rate (m^3/h)	Power (kW)	Price (MAD)
Type-1	72	3.00	10.05	470,000
Type-2	36	1.50	5.15	270,000
Type-3	20	0.83	3.07	220,000

In order to find the optimal solution, the routine "Optimization" had to be run for each of the following five cases:

Case 1: one Type-1 machine;

Case 2: two Type-2 machines;

Case 3: one Type-2 machine and two Type-3 machines;

Case 4: one Type-2 machine;

Case 5: one Type-3 machine.

To have a clear and visual graphical representation of the results provided by our Excel tool, we used a 3D software to represent all five cases. The initial investment and cost of water over 10 years are the two parameters chosen. The results for case 1 are shown in Figures 7 and 8, and similar patterns were recorded for the remaining cases.

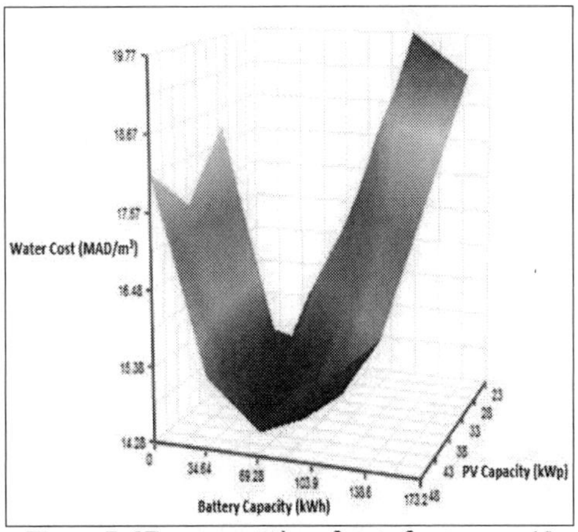

Figure 7: 3D representation of cost of water over 10 years as a function of storage capacity and PV capacity

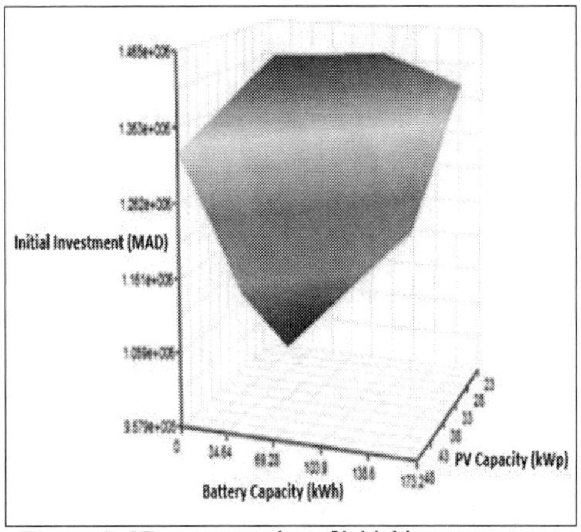

Figure 8: 3D representation of initial investment as a function of storage capacity and PV capacity

The client imposed the following constraints, an annual water production of 7,300 m^3 (20 m^3 per day) and a budget of 1.5 million MAD.

After gathering all the output tables and graphs for the five cases, we deduce a table that summarizes the two best scenarios for each case: the lowest cost of water over 10 years and the lowest initial investment, shown in Table VI. We extracted the following optimum solutions:

- Minimum initial investment: case 4, combination 4.2 is the optimum solution with a total initial investment of 827,944 MAD (see Table VI);
- Minimum cost of water over a 10-year period: case 1, combination 1.1 is the optimum solution with a water cost equal to 14.28 MAD/m^3 (see also Table VI).

The final optimum solution is determined by the client himself, depending on his goals and priorities; if he is looking for a long-term solution, the best would be to go for option 1.1 (Table VI) which yields water at the lowest cost, whereas if his goal is short-term and is rather looking to minimize the capital costs, then the best option is 4.2 as indicated in Table VI.

Table VI: Summary of the optimal solutions for the five cases

Cases	PV capacity (kWp)	Battery capacity (kWh)	Water production (m^3/year)	Investment (MAD)	Cost of water (MAD/m^3)
Case 1.1	32	38.5	11,709	1,148,778	14.28
Case 1.2	23	13.3	9,333	957,944	14.76
Case 2.1	32	49.5	12,623	1,238,778	14.31
Case 2.2	23	0	7,820	1,002,944	17.68
Case 3.1	32	42	11,785	1,393,778	16.54
Case 3.2	23	0	7,857	1,172,944	19.97
Case 4.1	23	50.5	7,731	827,944	17.59
Case 4.2	23	50.5	7,731	827,944	17.59
Case 5.1	40	122.3	7,320	1,191,000	18.71
Case 5.2	40	122.3	7,320	1,191,000	18.71

5. CONCLUSIONS

The computer tool was developed then used on a specific case study for a PVRO station for a hotel in the southern city of Dakhla, having a daily capacity of 20 m^3. Different scenarios were analyzed to determine the best scenario in terms of energy efficiency and costs. The results show that the most energy-efficient and cost-effective system configuration comprises an energy recovery system and separate components (HP pump, pressure exchanger and booster pump). The energy consumption of such system is of 2.92 kWh/m^3, whereas the total costs over a 10-year period for this system is 1,056,024 MAD, producing water at a cost of 15.47 MAD/m^3.

After analyzing the energy graphs provided by the tool, we noticed that a significant amount of electrical energy produced by the PV system is wasted, whereas it could be used in different ways to produce more water. This energy can be either be stored in batteries and later used to operate the desalination unit for longer hours or it can be used to run a smaller machine, in hours where the PV system is still producing electrical energy enough for a small machine to operate but not enough for larger machines. This proves that there is an opportunity for optimization: determining the optimal design configuration for a PVRO desalination station. Using a computer tool on Excel to find the best combination of RO machines, PV capacity and energy storage capacity that maximizes the water production and energy usage from the PV system while minimizing the total costs. The tool was then used on the same case study and two optimal solutions have been found, both of which operate with one machine only. The first one, which yields the lowest initial investment (828,000 MAD), consists of installing one RO machine with a nominal capacity of 36 m^3/day, 23 kWp of PV power and 50.5 kWh of battery storage. The second optimal combination, which generates the lowest cost of water during 10 years (14.28 MAD/m^3), consists of installing one bigger RO machine having a capacity of 72 m^3/day, 32 kWp of PV power and 38.5 kWh of battery storage. The optimal water cost found is more interesting than the lowest cost of water found in the literature for similar PVRO desalination systems, which was of about 15.14 MAD/m^3 (\$1.85/m^3), for a brackish water PVRO unit in Nicosia, Cyprus [9].

References

[1] World Health Organization, Desalination for safe water supply, Geneva, Switzerland, (2007).

[2] Ministry of Energy, Mines, Water and the Environment, "Chiffres clés" 2013, Rabat, Morocco, (2014).

[3] Ministry of Economy and Finance, Financial Bill – Annual Rapport, (2015).

[4] Poseidon Resources Corporation, Desalination worldwide: worldwide seawater desalination capabilities, (2010). Seawater desalination, http://hbfreshwater.com/desalination-101/desalination-worldwide, (2013).

[5] Nicos X. Tsiourtis, Desalination and the environment, Desalination, pp. 223-236, (2001).

[6] J. E. Miller, Review of water resources and desalination technologies, Albuquerque: Sandia National Laboratories, (2009).

[7] K. S. Tree, Water industry report: desalination, San Diego: World Trade Center, (2013).

[8] A. A. Al-Karaghouli, L. L. Kazmerski, Renewable energy opportunities in water desalination, Desalination: Trends and Technologies, Colorado: InTech, (2011).

[9] E. S. Mohamed, D. Manolakos, S. Kyritsis, P. Soldatos, Design and cost analysis of an autonomous brackish water reverse osmosis (BWRO) desalination plant powered by photovoltaic cells and battery bank: case study in Nicosia, Cyprus, Desalination, pp. 317-322, (2008).

RURAL WATER SUPPLY IN ETHIOPIA WITH PV PUMPS

Christer Nyman, Soleco Ltd ,Londbölentie 27, FIN-06100 Porvoo, Finland email: info@soleco.fi
Tameru Beshah, BISIT_Group, Erikaweg 9, D-50169 Kerpen, Germany, e-mail: tameru.beshah@bisit-group.com
Tesfaye Bayou Woldekirkos, Solatec, P.O.Box 30792 Addis Ababa Ethiopia email: solatec.tesfaye@gmail.com

ABSTRACT: Over half of the rural population in Ethiopia still lack access to water. To improve access, the use of solar and wind energy for water pumping is an attractive option in the rural areas since over 83% of the population has no access to the electric grid to power mechanical water pumps. The purpose of this project is therefore to promote and pilot the use of solar and wind energy for water pumping in rural areas of Ethiopia, and so initiate development of a long term investment in these technologies under the Government's Universal Access Programme (UAP) where they are appropriate and the most suitable. The Pilot project has become necessary to address the design, implementation and support issues of solar & wind pump systems, and to plan for future expansion phases of this component of the UAP. The challenge for the Pilot project will be to go beyond infrastructure (hardware) provision, and is also to develop capacities and mature the market for these technologies to be sustainable.
Keywords: see enclosed list of keywords

1 Purpose of project

The purpose of this project is to promote and pilot the use of solar and wind energy for water pumping in rural areas of Ethiopia, and so initiate development of a long term investment in these technologies under the Government's Universal Access Programme (UAP) where they are appropriate and the most suitable. This will help ensure an increased and sustained supply of water at lower financial, economic, environmental and social costs than with fossil fuel powered pumping systems. The purpose of this pilot program is: [i]

- to bring about an increased and sustainable supply of clean water for the rural areas through the use of indigenous renewable energy resources (solar & wind) as an alternative to fossil fuel driven pumps at lower financial, economic, environmental and social costs;
- through training of the rural population to bring about technology transfer of solar and wind energy systems that are the future choice of alternative energy resources for scattered rural settlements;
- to bring about sense of empowerment to the rural communities on technology selection, operation and maintenance of systems installed in their areas;
- to accelerate the employment of solar and wind energy systems as an environmentally friendly energy resource on the everyday life of the rural people;
- to stimulate private sector involvement in the procurement, installation, O&M and funding of systems;
- Approximately 70 water supply schemes utilizing solar and wind are effectively functioning and supplying clean potable water to 130,000 people in the project areas.
- Improved capacity at community level to operate and maintain solar or wind powered water supply systems, and at National, Regional and District levels to implement future schemes.
- Increased involvement of private sector in providing equipment and services.
- Awareness created amongst government, end users and other sectors of society on the benefits of solar/wind technologies, thereby triggering an increased demand

2. Implementation

Many of the UAP's new schemes require bigger and automatic pumps. However, since access to electric services is only about 17% in the country, most rural areas do not have the opportunity to use grid electricity to drive the pumps, and have instead relied on diesel driven generators to provide power. Diesel/petrol pumps have many drawbacks such as high running and maintenance costs, unreliable supply of fuel, and poor availability of spare parts.

The conditions for solar based pumping offer several advantages such as low running costs and minimal environmental impact. While the upfront investment on solar energy based water supply schemes is higher than for conventional systems, the financial costs on a life cycle basis are often more favorable since the much lower running costs due to the almost free supply of energy offsets the incremental initial investment cost. Maintenance costs of solar systems are also low, since the photovoltaic cells have been proven to be very reliable in practice. As such the operation and maintenance costs can easily be covered by a community.

In order to succeed in supplying potable water in the rural areas of the country within a relatively short period of time, a fast and successful dissemination of solar and wind pumping systems will be decisive. In turn, for a fast and sustainable dissemination of these systems, the Woredas will have to play a key role in all activities, from planning, sizing of the systems, installation up to managing the systems through the local communities. For the Woredas to play this crucial role, above all the Regional states, supported by Federal ministry, have to take different measures and approaches.

3. Lessons learned from the Pilot project

As the solar and wind pumping systems are stand-alone systems, their sustainability depends very much on the competence and enthusiasm of the Woredas and particularly the local communities. The activities of the Pilot project showed this clearly. In places where the Woredas are relatively competent and enthusiastic, much more self-initiative have been observed. The following main isuues are importnat for furhter implementaion:

- Systematic building of the capacities of the Woreda's Water & Energy with solar and wind pump technologies
- Organize dependable Operation & Maintenance including supply chain that includes the private sector
- Implement dependable technologies with quality equipment
- Enter a strategic co-operation with dependable solar and/or wind manufacturers
- allocate budget and mobilize funds in order to enable the Regions at different level deliver potable water supply to the rural population of the country

4. Installations.

Forty-nine PV pumps of different sizes have been installed in the following regions, Tigray, Amhara, Oromya and SNNPR:

Total power of the installed PV is 240 000 W, with minimum size of 750 W, maximum size of 12000 W. The minimum head is 35 m, maximum head 172 m , minimum water quantity Q per day is 6 m^3 maximum 47 m^3/day, total water of all 49 sites is 1430 m^3/day. The maximum hydraulic load equivalent HEQ is 6674 m^4 and minimum 287 m^4. The theoretical design of the system is presented in figure 1. For final design different software from manufacturer has been used.

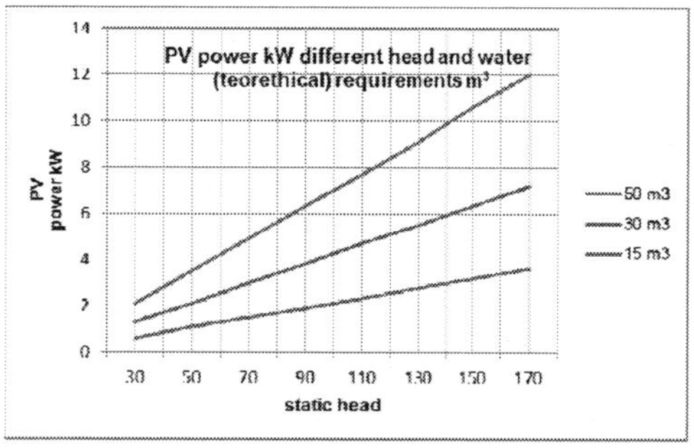

Figure 1. Required theoretical PV power for different heads and amount of water pumped 15, 30, 50 m3.

Site model 1 small system

The pictures 1 to 4 are examples of the PV pumping sites, with sizes of 750 W, 2000 W, 4000 W and 12000 W.

Picture 1. 750 W Pump system

Picture 2. 2000 W Pump system

Picture 3 4000 W pump system

Picture 4 12000 W pump system

5. System costs

A cost model has been calculated from the investment costs, and lifecycle costs of the water pump. This first model is based on the amount of water available and a portion of that is paid for (100%). The more customers, the more income is generated. The higher water price also leads to higher prices for the beneficiaries.
Depending on the daily hydraulic energy equivalent (daily water demand x total effective head) for each site, sites are classified as Small system; Medium system and large system.
A large system with pumping head of 160 m, daily water volume 37 m^3 is used for the economical calculations.

These figures are the average numbers of the 11 big systems installed.

During lifetime (20 years) the following costs for PV and diesels have been assumed. The figure 2 shows the lifecycle costs of a PV pump compared to a diesel pump. Operating and maintenance costs include the salaries for guards, technician and accountant for operating the pump systems.

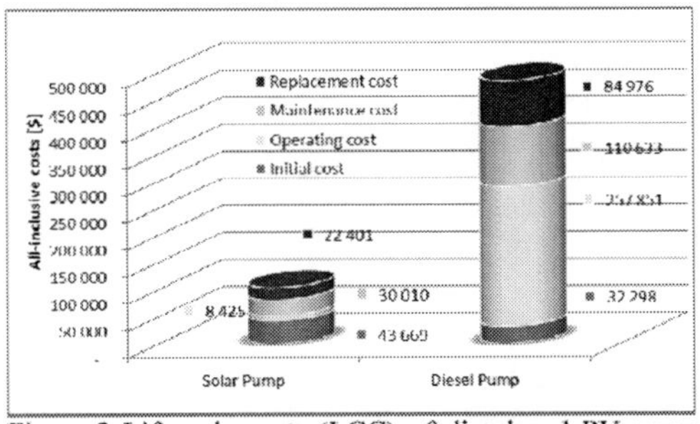

Figure 2 Lifecycle costs (LCC) of diesel and PV pump system.

6. Breakeven with diesel pump and PV

The breakeven of PV pump compared to diesel pump is 1,4 year, assuming that costs of diesel pumping is for installing a new diesel system. Figure 3.

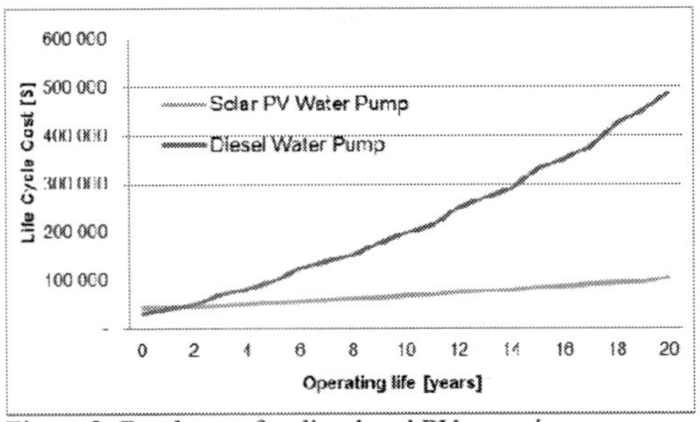

Figure 3. Breakeven for diesel and PV pumping

7. Water costs

The calculation of payback of for the PV pump system, including all LCC, is used with a water fee of 1 USD/m3. This correspond to a fee of 0,40 BIRR/20 l (jerry can what is commonly paid for water). The payback time for this system is 8 years. The payback time as function of other water fees is seen in the figure 4. In figure 5 the increasing water cost is reflecting the household costs per month.

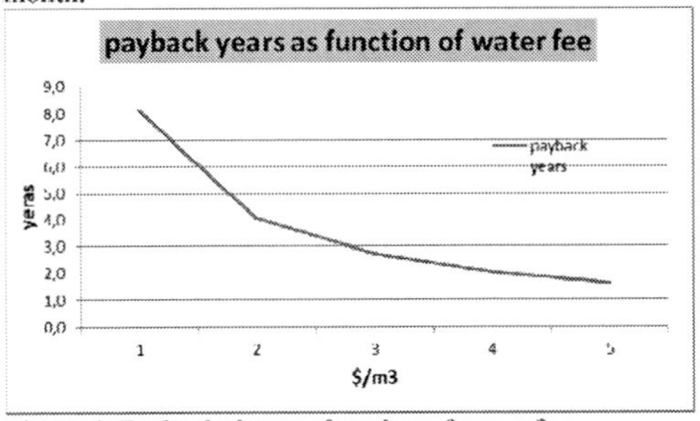

Figure 4. Payback time as function of water fee.

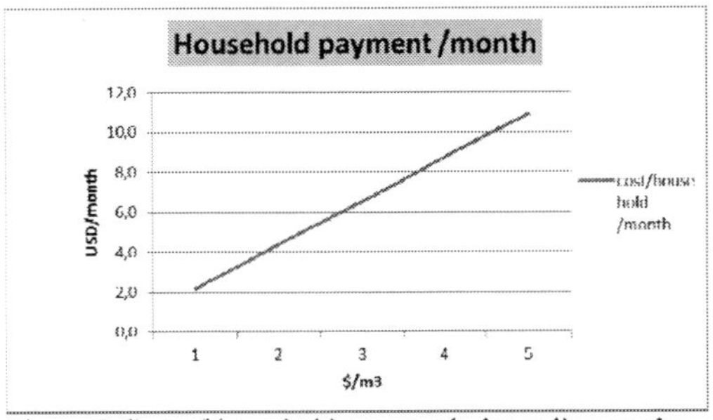

Figure 5 Cost of household per month depending on the water fee cost $/m3.

8. Conclusions

To improve the access of clean water supply to the rural areas, the Government of Ethiopia has been planning to employ alternative energy resources like solar and wind for water pumping in the rural areas. In the past, although there were some sporadic efforts by some stakeholders to use solar and wind energy resources for water pumping, to draw lessons for scaling up, there were no sufficient projects under implementation that comprehensively addresses the use of these energy resources for water supply on sustainable basis. In order to help catalyze their widespread use in its UAP, the Government of Ethiopia approached the AWF to finance a Pilot program on the utilization of solar and wind energy for water supply schemes.

The Pilot project has become necessary to address sites selection, design, implementation and support issues of solar & wind pump systems, and to plan for future expansion phases of this component of the UAP. Although the primary target had been to benefit 130,000 people with 70 pumps (40 solar + 30 wind), the Pilot project had to settle with 67 pumps (49 solar + 18 wind) and was able to cover the needs of over 102,000 people in the 4 Regions (Amhara, Oromia, SNNPR and Tigray).

References

[i] UTILISATION OF SOLAR AND WIND ENERGY FOR RURAL WATER SUPPLY IN ETHIOPIA, Appraisal report, African water facility 2008

SIZING OF PV ARRAY FOR WATER PUMPING APPLICATION

A. F. Almarshoud
Dept. of Electrical Engineering, Qassim University, Buraydah, Saudi Arabia
dr_almarshoud@qu.edu.sa

ABSTRACT: Utilization of solar energy in water pumping is one of the most economically viable photovoltaic (PV) applications especially in remote areas. The accurate sizing of PV array is essential, as under sizing will make the pumping system unable to satisfy the water demand. On the other hand, over sizing of PV array leads to extra cost. This paper presents a novel approach for sizing the PV array of water pumping systems. This approach depends on the solar irradiance data of the shortest sunny day over the year to increase the reliability of pumping system. This approach depends on determining the proper starting value of the solar irradiance on the global solar radiation curve for tilted or moving surfaces. A design algorithm has been proposed for applying this approach including the converting process of solar radiation values from horizontal surface to tilted surface. A case study from Buraydah city in the middle of Saudi Arabia has been selected for applying the sizing approach.
Keywords: PV pumping; PV array sizing; water-pumping

1 INTRODUCTION

Several papers were published last decade about utilizing solar energy in water pumping applications either for residential water supply or crops irrigation. All these papers use the same method of pump sizing which mainly depends on the flow rate of water and the dynamic water head, but they differ in the methodology of sizing the PV array which mainly depends on the efficiency of PV modules and the quality and quantity of solar irradiance available in the location of water source. Most of published researches use the monthly average of daily Global Horizontal Insolation GHI (kWh/m2/day) as a base for sizing the PV array. This methodology is not always adequate, especially in case of using large size of pumps, which wouldn't start unless there is enough amount of solar radiation. Also, in this method the determination of operation hours is difficult as it depends on the sunshine during the day. To overcome this problem, a battery bank for at least one day of autonomy must be added which will increase the reliability of pumping system, but this will increase the capital and maintenance costs at the same time.

In this paper a simple and accurate approach has been proposed for sizing the PV array, this approach depends on accurate measurements of solar radiation with resolution of five minutes at the most. The instantaneous values of daily global horizontal irradiance GHI (W/m2) of the shortest sunny day in the year has been used as a basis for sizing the horizontal PV array. Also, a transformation model has been used for converting the solar radiation data on horizontal surface to tilted surface for sizing the tilted PV array. The shortest day in the year usually be in December in the northern hemisphere, also the selected day should be sunny because, the cloudy day will lead to overestimation of PV array. Figure 1 shows the difference in solar radiation between sunny and cloudy day. In addition to that, the paper investigates the effect of using a sun-tracking system aiming at increasing the reliability and decreasing the capital cost. The proposed approach has been applied using a real solar radiation data measured in Buraydah City in the middle of Saudi Arabia for the last three years.

2 LITERATURE REVIEW

Several papers had been published dealing with PV water pumping systems. They presented various methods for sizing the PV array. Some authors used a previously sized PV array that matching the nominal rated power of motor–pump system [1-4]. Others changed the configuration of the previously sized PV array based on hourly solar radiation [5-7].

One simple sizing method for PV array of pumping systems utilizes a rule-of-thumb which attempts to capture the integrated effects of subsystem efficiency and flow losses. The rule-of-thumb relates the daily hydraulic energy (product of daily water flow by head length, m4/day) to the required rated power of PV array [8-11]. The value of hydraulic energy was estimated to be 0.6 m4/day/Wp [10]. This value, also, has been roughly confirmed by other authors [12, 13]. Using a rule-of-thumb, like this method greatly simplifies the sizing of PV array as it gives the required array power rating as a function of the required daily water flow and head.

Another method simplifies the sizing of PV array by using the monthly average of daily solar insolation data to calculate the associated peak sun hours, this design method is common in the literature [14,15]. Noble [16] and Burkhartzmeyer [17] summarized practical methods for PV array sizing based on the idea of peak sun hours. Also Cuadros et. al. [18], used the monthly average of daily global solar radiation and the monthly average of daily sunshine hours to size the PV array for each month alone, then finally select the biggest size to increase the reliability in low radiation months.

Some authors used the monthly average of solar radiation on tilted surface of the worst month of the year to size the PV array required to operate the water pump. [19], while, Azzedine et. al. [20], used the worst monthly average of solar radiation in addition to the maximum monthly of water demand to size the PV array. Similarly, Campana et. al. [21], used the worst daily collectable solar energy in addition to the system efficiency to size the PV array. Kaldellis [22], developed a simple algorithm for sizing the PV array based on the monthly average of global horizontal irradiance only, while, Zvonimir and Margeta [23], developed more complicated approach using dynamic programming to size the PV array based on the average of solar energy available per day, taking into account all relevant elements of PV pumping system.

Some authors gave the reliability of water pumping system more attention by using the concept of loss of power supply probability (LPSP) as a criteria for sizing both, the PV array and the water storage tank [24,25].

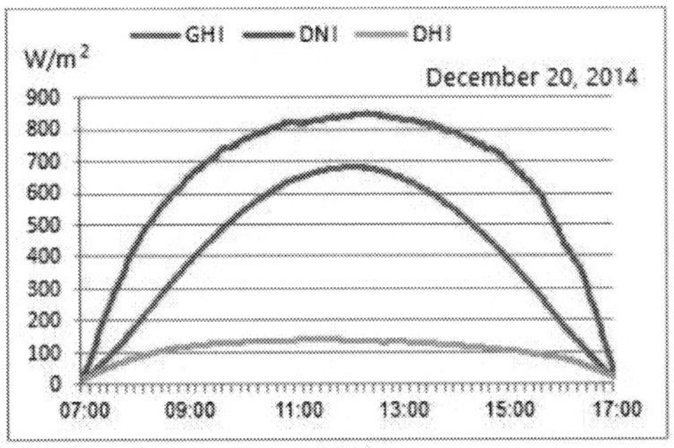

(a)

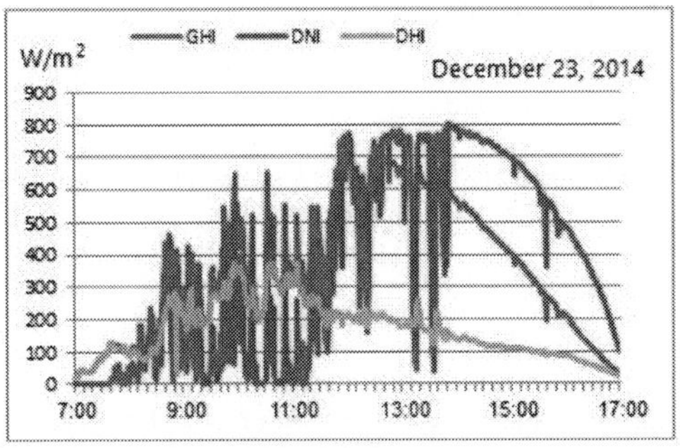

(b)

Fig. 1: Solar radiation curves of sunny day (a) and cloudy day (b) for Buraydah City on December 2014

3 THE PROPOSED METHOD

It is clear from the previous review that there are many methods followed for sizing the PV array. These methods differ according to the available data of solar radiation and sunshine hours. Some of these methods are very simple [1-11], and some are complicated [23-25]. Others are in between [12-22]. Usually, the electric motors draw high current at starting. So, the PV array must be able to provide the required current at starting. All the previous methods didn't take the starting current problem into account. As a result, the starting of the pump will be delayed until the solar radiation be enough for providing the required starting current. The accurate sizing of PV array is essential because the under sizing will make the pumping system unable to satisfy the water demand which means fail the project. On the other hand, the over sizing of PV array leads to extra cost where the PV modules represent about 60% of the capital cost which finally leads to bad economic feasibility for the project. The present approach depends on selecting a proper value on the daily curve of global tilted solar radiation of the shortest sunny day over the year which guarantees pumping the daily demand and motor starting. The motor will start as a sufficient amount of solar radiation will exists.

3.1 Converting from global horizontal to global tilted irradiance

The solar radiation transposition between the horizontal and tilted planes was investigated using various methods in the literature, here the commonly transposition models used in solar energy engineering practice: Hay [26], Klucher [27], Skartveit [28], Perez [29], Reindl [30], Duffie and Beckman [31], Muneer [32], ASHRAE [33] and Gueymard [34]. The investigation of properties and differences between these models is beyond the scope of this study. This study will adopt the simple isotropic model of diffuse radiation presented by Liu and Jordan [35] for its simplicity and accuracy specially in clear sky case as follows:

$$G_t = G_{dn} \cos\theta + G_d R_d + \mu G R_r \quad (1)$$

Where G_t is the global irradiance on tilted plane whose its tilt angle is β degrees from horizontal, G_{dn} is the direct normal irradiance (DNI), G_d is the diffuse horizontal irradiance (DHI), G is the global horizontal irradiance (GHI), θ is the angle of incidence of the sun rays on the tilted plane, R_d is the diffuse transposition factor, μ is the foreground's albedo, and R_r is the transposition factor for ground reflection. R_d and R_r could be calculated from the relations 2 and 3 given below.

$$R_d = (1 + \cos\beta)/2 \quad (2)$$
$$R_r = (1 - \cos\beta)/2 \quad (3)$$

The fraction of shortwave solar energy reflected from the earth, back into space depends mainly on the earth foreground. The foreground's albedo, μ could be either measured, modeled or estimated as given below.

$$\mu = \begin{cases} 0.2: & \text{ordinary ground or vegetation} \\ 0.8: & \text{Snow cover} \\ 0.15: & \text{Gravel roof} \end{cases}$$

The incidence angle θ of the sun rays is the angle between the sun rays and the normal of tilted plane, it may be calculated for a south-facing, tilted surface in the northern hemisphere as follows [36]:

$$\cos\theta = \sin(L - \beta)\sin\delta + \cos(L + \beta)\cos\delta\cos h \quad (4)$$

In case of using full tracking system [36]:

$$\cos\theta = 1 \quad (5)$$

Where L is the local latitude, h is the hour angle and δ is the declination angle and may be calculated as follows [36]:

$$h = \begin{cases} +0.25M: & \text{for afternoon} \\ -0.25M: & \text{for morning} \end{cases} \quad (6)$$

where M is number of minutes from local solar noon

$$\delta = 23.45 \sin\left[\frac{360}{365}(284 + N)\right] \quad (7)$$

where N is the day number within the year

In case of using full tracking system, the tilt angle β varies with varying the sun, so it could be calculated as follows [36]:

$$\cos\beta = \sin L \sin\delta + \cos L \cos\delta \cosh \quad (8)$$

So, this variable value of β, will lead to variable values for R_d & R_r based on the value of sun hour (h).

3.2 Sizing of Motor-Pump system

The sizing of motor-pump system could be estimated on the basis of instantaneous water flow and the Total Dynamic Head (TDH) which is the sum of static head of water in the well, discharge head, drawdown head, discharge pressure and friction losses in pipeline. The following formula has been used for sizing the motor-pump system:

$$P_{pump}(W) = \frac{\rho g Q H}{3.6\, \eta_m\, \eta_p} \quad (9)$$

Where ρ is the density of water (1.0 kg/L), g is the gravity acceleration (m/sec^2), Q is discharge rate (m^3/h), H is the total dynamic head (m), η_m is the motor efficiency and η_p is the pump efficiency.

3.3 Sizing of the PV array

The PV-array sizing could be done by matching the instantaneous power needed by the motor-pump system with the output instantaneous power of PV array which mainly depends on the incident global solar radiation on the tilted surface of PV array.

The instantaneous output power of PV array is given by:

$$P_{pv} = A_{pv}\, G_t\, \eta_{pv}\, (1 - \lambda_m)(1 - \lambda_c) \quad (10)$$

Where G_t is the instantaneous global solar radiation (w/m^2) on the tilted surface of PV array, A_{pv} is the area of the PV array (m^2), η_{pv} is the efficiency of PV array under operating condition, λ_m and λ_c are miscellaneous losses of PV array and power conditioning losses respectively.

By connecting motor-pump system to PV array through inverter, Eqns. 9 and 10 could be equalized as follows:

$$P_{pv} = \frac{P_{pump}}{\eta_{inv}} \quad (11)$$

$$A_{pv}\, G_t\, \eta_{pv}\, (1 - \lambda_m)(1 - \lambda_c) = \frac{\rho g Q H}{3.6\, \eta_m\, \eta_p\, \eta_{inv}} \quad (12)$$

Then the size of PV array needed for water pumping system could be obtained as follows:

$$A_{pv} = \frac{\rho g Q H}{3.6\, G_t\, \eta_{pv}\, \eta_m\, \eta_p\, \eta_{inv} (1-\lambda_m)(1-\lambda_c)} \quad (13)$$

Where η_{inv} is efficiency of inverter and η_{pv} could be obtained as follows:

$$\eta_{Pv} = \eta_r\, [1 - \alpha_p\, (T_c - T_r)] \quad (14)$$

Where η_r is the PV module efficiency at reference temperature (T_r = 25°C), and α_p is the temperature coefficient for module efficiency. T_c is cell temperature and related to the mean ambient temperature T_a as follows:

$$T_c - T_a = \frac{219 + 832 K_t}{800} (NOCT - 20) \quad (15)$$

where NOCT is the Nominal Operating Cell Temperature and K_t is the clearness index.

3.4 Sizing algorithm of PV array

The following algorithm has been proposed for sizing the PV array of water pumping system:

1- Choose the shortest sunny day (Usually in December for northern hemisphere) for using its solar radiation data (GHI, DNI and DHI) as a design input.
2- Convert the daily radiation data (GHI, DNI and DHI) to tilted global irradiance G_t at chosen tilt angle using Eqn. 1.
3- Calculate the operation period of pump using the given data of (daily demand of water and rate of charge of selected pump)
4- Determine the starting value of global tilted irradiance; G_t on the new curve of step 2 by projecting the operation period on the global tilted irradiance curve.
5- Calculate the total area of PV array required for driving the pump using Eqns. 13-15
6- Calculate the number of PV modules by dividing the total area of PV array by the area of the selected PV module

4 CASE STUDY

To prove the validity of the proposed method, a case study for Buraydah city in the middle of Saudi Arabia has been chosen for applying the design algorithm of the proposed method. Tables 1, 2 and 3 illustrate the required input data of water pumping system, components of electrical system and metrological data of chosen site (26.34° N, 43.76° E) respectively, while Figure 2. Shows the instantaneous solar radiation of GHI, DNI and DHI during one sunny day (20 December 2014) of chosen site.

Table 1: Data of water pumping system

Daily demand	200 m^3	TDH	176 m
Pump type	Multi-stage, submersible	Motor efficiency	85 %
Pump model	Grandfous (150S400-18)	Motor output power	30 kW
Pump efficiency	70.6 %	Rated Voltage	3x380 V
Pump rated head	195 m	Rated current	66.5 A
Pump rated flow	36 m^3/hour	Starting current	300 A
Pump speed	3450 rpm	Frequency	50 Hz
Motor model	MS6000QFT40	Power factor	0.87

Table 2: Electrical system specifications

PV module	SunPower (SPR-210-BLK)	PV module type	Mono-Crystalline
Peak power	210 W	Temp. Coefficient	0.38 %
Rated Voltage	40.0 V	Open Circuit Voltage	47.7 V
Rated Current	5.25 A	Short Circuit Current	5.75 A
Efficiency	16.9 %	Dimensions	156 x 80 cm
Misc. losses	2.0 %	conditioning losses	1.5 %
NOCT	45 °C	Inverter Efficiency	97.5 %
* Electrical data of PV module measured at STC (GHI: 1000 w/m^2, air mass: 1.5 g, Cell Temp.: 25 °C)			

Table 3: Monthly average of mean daily weather data

Month	Ambient Temp.	GHI	Atm. Pressure	Wind speed	Clearness Index (K)
	°C	kWh/m^2/d	kPa	m/s	
January	14.2	4.30	94.2	2.4	0.66
February	17.0	5.43	93.9	2.6	0.65
March	21.9	6.07	93.7	2.5	0.58
April	28.2	5.95	93.7	2.9	0.61
May	31.9	7.03	93.5	2.7	0.64
June	34.3	8.18	93.1	2.7	0.72
July	36.2	8.16	92.9	2.6	0.75
August	36.5	7.54	93.0	2.5	0.71
September	33.7	6.70	93.4	2.4	0.69
October	28.0	5.67	93.8	2.7	0.66
November	20.2	4.46	94.0	2.5	0.62
December	15.8	4.03	94.3	2.4	0.60
Annual	26.5	6.13	93.6	2.6	0.66

Applying the design algorithm:

Step 1: The chosen sunny day is (20 December 2014), and its solar radiation data are shown in Figure 2.

Step 2: The tilted global irradiance G_t has been calculated using Eqns. 1-8 for 2 cases:

a) A fixed tilt angle = 41.3° (location latitude +15°), and south facing.

b) A full sun-tracking system.

Figure 3 shows the calculated tilted global irradiance curves G_t for the two cases.

Step 3: from the rate of charge of selected pump and the daily demand of water (Table: 1), the operation time has been calculated (5.55 hours).

Step 4: The starting values of global irradiance G_t required for pumping the daily demand for both cases is 655 W/m^2 for case (a) and 810 W/m^2 for case (b) as shown in Figure 4.

Step 5: The total area of PV array required for driving the pump has been calculated using Eqns. 13-15, found to be 290.7 m^2 for case (a), and 235 m^2 for case (b).

Step 6: Number of PV modules has been calculated by dividing the total area of PV array by the area of selected PV module (SPR-210-BLK), and they are 227 for case (a) and 184 for case (b).

These are the minimum number of selected PV module (SPR-210-BLK) that guarantee the required power for pumping the daily demand of water, based on available solar radiation in the region. Table 4 summarizes the sizing results for both cases.

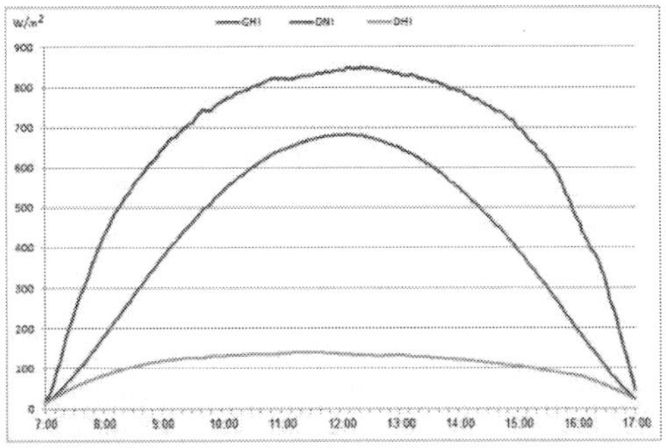

Fig. 2: GHI, DNI and DHI of one sunny for Buraydah City on day December 20, 2014

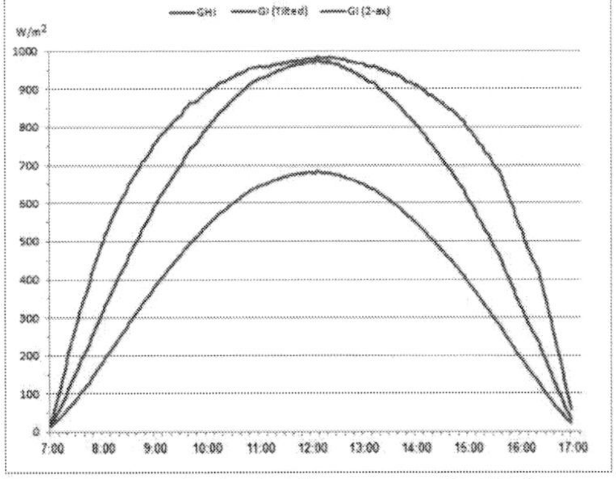

Fig. 3: Converted solar irradiance curves for tilted surface and dual axis tracking

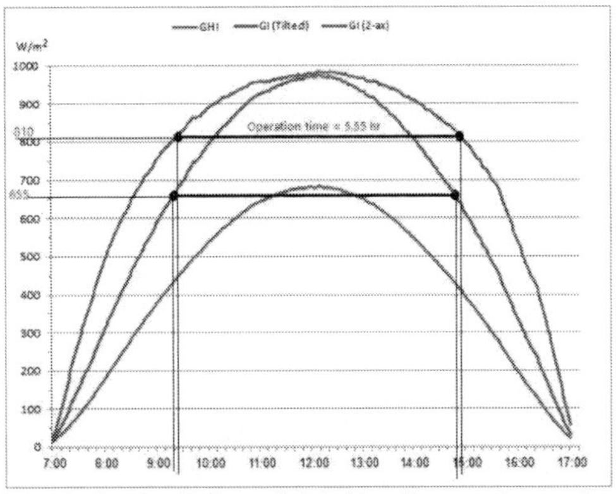

Fig. 4: Starting Values of global irradiance and operation time of pump for both tilted and full tracking surfaces

Table 4: Summery of sizing results

	case (a)	case (b)
Charge rate	36 m^3/hour	36 m^3/hour
Operation time	5.55 hours	5.55 hours
Value of G_t at starting	655 w/m^2	810 w/m^2
Area of PV array	290.7 m^2	235 m^2
No. of PV modules	227	184

The configuration of PV array (i.e. connection of modules in parallel and series) depends on the requirements of charger or inverter and the value of starting current of motor and method of starting used.

It is noticed that there is potential solar energy not harvested yet as indicated by regions A, B and C in Figure 5, this potential solar energy may be harvested by adding battery bank to the system, the stored energy may be used in the next day for contribution in the starting current of motor or may be used to extend the operation period if needed or may be used during cloudy periods.

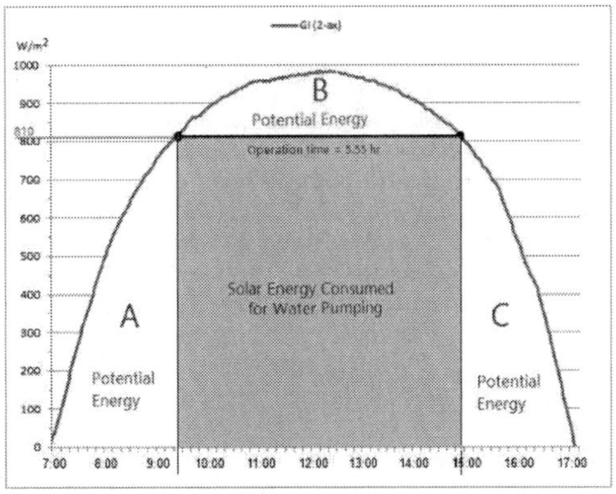

Fig. 5: The consumed and potential energy in full tracking case

5 CONCLUSION

A novel approach for sizing the PV array of water pumping systems has been proposed. This approach uses the solar radiation data (DNI, GHI, DHI) of the shortest sunny day over the year as a base for sizing the PV array. The approach has been applied on a real case from Buraydah City in the middle of Saudi Arabia through following a simple algorithm. The results showed that a 30 kW pump required to deliver a 200 m^3 of water daily from a well of a 176 m depth , may be driven by a PV array consists of 227 modules of type (SPR-210-BLK) if the PV array is fixed and tilted with (41.3°) and south facing, or by a PV array consists of 184 modules of the same type in case of using dual-axis tracking system. The application on the case study showed the ease of applying the proposed method while the needed solar radiation data are available.

ACKNOWLEDGMENT

The author would like to express the sincere appreciation to the staff of King Abdullah City for Atomic and Renewable Energy (K.A.CARE) for Facilitating access to the required data and their continuing support.

REFERENCES

[1] Meah K, Fletcher S, Ula S. Solar photovoltaic water pumping for remote locations. Renew Sustain Energy Rev 2008;12:472–87.

[2] Wagdy R, Nourb MA. Optimum design of a photovoltaic powered pumping system. J Power Sources 1994;50:l–9.

[3] Mokeddem A, Midoun A, Kadri D, Said H, Iftikhar RA. Performance of a directly-coupled PV water pumping system. Energy Convers Manag 2011;52:3089–95.

[4] Ben Ghanem Belgacem. Performance of submersible PV water pumping systems in Tunisia. Energy for Sustainable Development 2012;16:415-420

[5] Firatoglu Z, Yesilata B. New approaches on the optimization of directly coupled PV pumping systems. Sol Energy 2004;77:81–93.

[6] Benghanem M, Daffallah KO, Joraid A, Alamri and Jaber A. Performances of solar water pumping system using helical pump for a deep well: a case study for Madinah, Saudi Arabia. Energy Convers Manag 2013;6:50–6.

[7] Sotirios B. Skretas, Demetrios P. Papadopoulos. Systematic procedure for efficient design of electric water pumping systems fed by PV or/and WECS: Application using measured meteorological data for the city of Xanthi/Thrace, Greece. Energy Convers Manag 2008;49:596–607

[8] Badescu V. Dynamic model of a complex system including PV cells, electric battery, electrical motor and water pump. Energy 2003;28(12):1165–81.

[9] Benghanem M, Daffallah KO, Alamri SN, Joraid AA. Effect of pumping head on solar water pumping system. Energy Convers Manag 2014;77:334–9.

[10] Omer, A.M. Solar water pumping clean water for Sudan rural areas. Renewable Energy. 2001; 24: 245–258.

[11] Stassen, G., Olivier, A.C. A market overview of commercial renewable energy technologies in SADC. J. Energy S. Afr. 2000;11 (3):244–249.

[12] Diarra, D.C., Akuffo, F.O. Solar photovoltaic in Mali: potential and constraints. Energy Convers. Manage. 2002;43:151–163

[13] Vilela, O.C., Fraidenraich, N., Tiba, C.. Photovoltaic pumping systems driven by tracking collectors. Exp. Simul. Solar. Energy. 2003;74 (1): 45–52.

[14] Kenna, J., Gillett, B. Solar Water Pumping. A Handbook. Intermediate Technology Publications, London, 1985.

[15] Hankins, M. Stand-Alone Solar Electric Systems. The Earthscan Expert Handbook for Planning, Design, and Installation, 2010.

[16] Noble, N. Solar (Photovoltaic) Water Pumping, Technical Brief, Practical Action, May 2012. Accessed online at <http://www.practicalaction.org/solar-photovoltaic-waterpumping-1> accessed April 2016..

[17] Burkhartzmeyer, B. Solar Water Pumping Project, Ghana. Joint Project of World Vision and Always On Solar Project, October 2008. Available on-line at <http://www.sunepi.org/SunEPI/Ghana_files/Ghana_Solar_Pumping_Project1.pdf > accessed April 2016..

[18] Cuadros F, Rodnguez FL, Marcos A, Coello J. A procedure to size solar powered irrigation (photo irrigation) schemes. Sol Energy 2004;76:465–73.

[19] Munzer S.Y. Ebaid, Hasan Qandil, Mahmoud Hammad. A unified approach for designing a photovoltaic solar system for the underground water pumping well-34 at Disi aquifer. Energy Convers. Manage. 2013;75: 780–795

[20] Z. Abidin Firatoglu, Bulent Yesilata. New approaches on the optimization of directly coupled PV pumping systems. Solar Energy. 2004;77: 81–93

[21] Pietro Elia Campana, Hailong Li, Jinyue Yan. Dynamic modelling of a PV pumping system with special consideration on water demand. Applied Energy. 2013;112: 635–645

[22] Kaldellis JK, Spyropoulos GC, Kavadias KA, Koronaki IP. Experimental validation of autonomous PV-based water pumping system optimum sizing. Renew Energy 2009;34:1106–13.

[23] Zvonimir G, Margeta J. A model for optimal sizing of photovoltaic irrigation water pumping systems. Sol Energy 2007;81:904–16.

[24] Yahia B, Arab AH, Azoui B. Optimal sizing of photovoltaic pumping system with water tank storage using LPSP concept. Solar Energy 2011;85:288–94.

[25] Belkacem Bouzidi. New sizing method of PV water pumping systems. Sustainable Energy Technologies and Assessments. 2013;4:1–10

[26] Hay J.E., Calculation of monthly mean solar radiation or horizontal and inclined surfaces. Solar Energy. 1979;23: 301-307.

[27] Klucher T.M., Evaluation of models to predict insolation on tilted surfaces. Solar Energy. 1979;23: 111-114.

[28] Skartveit A. and Olseth J.A., Modelling slope irradiance at high latitudes. Solar Energy. 1986; 36: 333-344.

[29] Perez R., et al., Modeling daylight availability and irradiance components from direct and global irradiance. Solar Energy. 1990; 44: 271-289.

[30] Reindl D.T., et al., Evaluation of hourly tilted surface radiation models. Solar Energy. 1990; 45: 9-17.

[31] Duffie J.A. and Beckman W.A., Solar Engineering of Thermal Processes, 2nd edition, Wiley, 1991.

[32] Muneer T., ed. Solar radiation and daylight models, 2nd edition, Elsevier, 2004.

[33] ASHRAE, Handbook of Fundamentals, SI Edition, American Society of Heating, Refrigerating and Air Conditioning Engineers, Atlanta, GA, 2005.
[34] C.A. Gueymard, From global horizontal to global tilted irradiance: How accurate are solar energy engineering predictions in practice? Solar 2008 Conf., San Diego, CA, American Solar Energy Soc. 2008.
[35] Liu BYH and Jordan RC. The long term average performance of flat-plate solar energy collectors. Solar Energy 1963; 7: 53-74.
[36] Soteris A. Kalogirou, "Solar Energy Engineering-Processes and Systems",1st edition, Elsevier Inc, 2009.

ABBREVIATIONS AND SYMBOLS:

DHI	*Diffuse horizontal irradiance*
DNI	*Direct normal irradiance*
GHI	*Global horizontal irradiance*
NOCT	*Nominal Operating Cell Temperature*
TDH	*Total Dynamic Head*
A_{pv}	*The area of the PV array*
g	*The gravity acceleration*
G	*The global horizontal irradiance (GHI),*
G_d	*The diffuse horizontal irradiance (DHI),*
G_{dn}	*The direct normal irradiance (DNI)*
G_t	*The global irradiance on tilted plane*
h	*The hour angle*
H	*Total Dynamic Head (TDH)*
K_t	*The clearness index*
L	*The local latitude*
Q	*The discharge rate*
R_d	*The diffuse transposition factor*
R_r	*The transposition factor for ground reflection*
T_a	*The mean ambient temperature*
T_c	*The cell temperature*
T_r	*The reference temperature (25°C)*
β	*Plane tilt angle from horizontal*
δ	*The declination angle*
η_{inv}	*The efficiency of inverter*
η_m	*The motor efficiency*
η_p	*The pump efficiency*
η_{pv}	*The efficiency of PV array under operating condition*
η_r	*The PV module efficiency at reference temperature*
θ	*The angle between the sun rays and the normal of tilted plane*
λ_c	*power conditioning losses*
λ_m	*miscellaneous losses of PV array*
μ	*The foreground's albedo*
ρ	*The density of water*

PERFORMANCE OF SPV WATER PUMPING SYSTEM AT LOWER IRRADIANCE CONDITION

Manander Banger, Birinchi Bora, Renu, O S Sastry*, Rashmi Singh, Supriya Rai
National Institute of Solar Energy 122003, Gurgaon, India
*sastry284@gmail.com, Tel: +91-0124-2579213, Telefax No. : +91-124-2579207

ABSTRACT: India has six different climatic zones and some climatic zones have higher occurrence of lower irradiance range through- out the year. However, there are possibilities of deployment of solar water pump for irrigation and drinking water purpose. It has been observed that for cloudy regions of India maximum occurrence in terms of energy falls within irradiance range 600-850 W/m2, ambient temperature range 25-350C. Usually performances of the SPV water pump system are designed and analyzed at high radiation level but it needs to be studied at low radiation also. This will help for the selection of suitable pump and optimized sizing of the array size in case of cloudy and subtropical climatic zones.
Keywords: Solar photovoltaic water pump, low radiation, climatic zone, Optimized sizing

1 INTRODUCTION

India is an agriculture based country and agriculture plays the main role for GDP growth. Due to lack of reliable conventional power supply in rural areas of India SPV water pump system plays essential part to fulfill the farmer's water demand for crops. Under the National Solar Mission (NSM), Govt. of India has proposed to deploy at least 0.1 Million pumps [1, 2]. There are several parameters which influence the performance of SPV pump system but one of the main parameter is solar radiation. According to Bansal and Manke, India has six climatic zones and there is a lot variation in radiation level all over the country in different climatic zones [3, 4, and 5]. Dry and Hot climatic zone have high radiation level whereas cloudy and subtropical climatic zone have more diffuse radiation. In case of region with monsoon and cloudy area the irradiance level are very low. The performance of a SPV system will not be same in different climatic conditions as the Irradiance varies over every zone. Usually performances of the SPV water pump system are designed and analyzed at high radiation level but it needs to be studied at low radiation also. This will help for the selection of suitable pump and optimized sizing of the array size in case of cloudy and subtropical climatic zones.

2 METHODOLOGY

National Institute of Solar Energy (NISE) is an autonomous institute of Ministry of New and Renewable Energy (MNRE) working as a research and development, testing and certification body for SPV water pumping system in India. It has SPV water Pump testing facility for analysis and performance evaluation of capacity ranging from 1 HP to 10 HP with an array capacity ranging from 200 Wp to 10 kWp both for surface and submersible DC as well as AC pumps as per MNRE specifications 2015-16 and IEC standards. NISE has a sump well (4m x 2.5m x 2.5m), with static head of 4 m, and an overhead platform (3.5 m x 2m x 2m) with static head of 7 m, to create suction/ delivery head for different heads pumps varying from 10 m up to 70 m, controlled by pressure gauge and throttle valve and standard meters. Pyranometer in plane with the surface of the module are used to measure the radiation level. A 3 HP and a 5 HP AC Submersible SPV water pumping system has been used for this study with multi-crystalline PV array technology and tested at 30 meter and 70 meter head of operation. The performance of the SPV pump can be characterized by using system and subsystem efficiency. This can be defined as:

$$\textit{Sub system efficiency} = \frac{\rho Q H}{P_{spv}}$$

$$\textit{System efficiency} = \frac{\rho g Q H}{G A}$$

Where, ρ= density of water, g= acceleration due to gravity, Q= flow rate, H= head, in meter, Ps =Input power, G= Irradiance W/m^2, A= Area of module, m^2

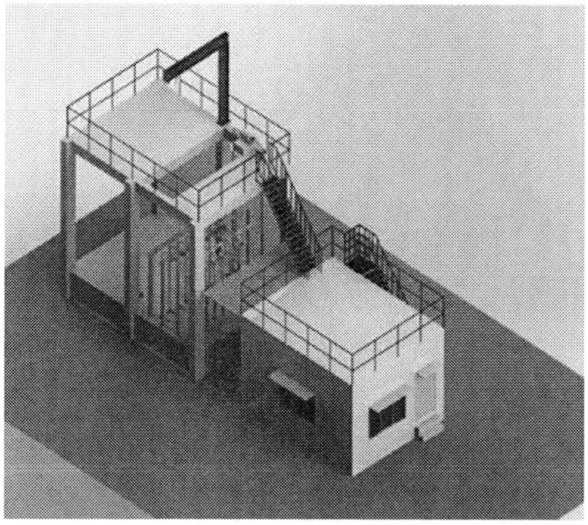

Figure: 1 Solar Water Pump Testing Setup

4 RESULTS

SPV pump systems are designed for the high input power of operation. But for the performance analysis it is always better to quantify the operating condition of the pump before the deployment. For this, contour plot analysis is done to find out the frequently occurred irradiation range for cloudy region of India. The figure 2 shows the global irradiation as a function of irradiance and ambient temperature. It has been observed that maximum occurrence in terms of energy falls within irradiance range 600-850 W/m2, ambient temperature range 25-350C for cloudy regions. However the frequency of occurrence for lower irradiance range is also high in cloudy region of India.

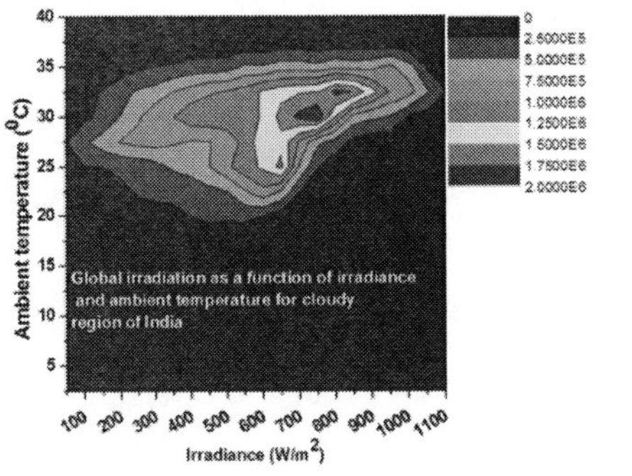

Figure 2 Irradiation of a cloudy region (Tezpur, India)

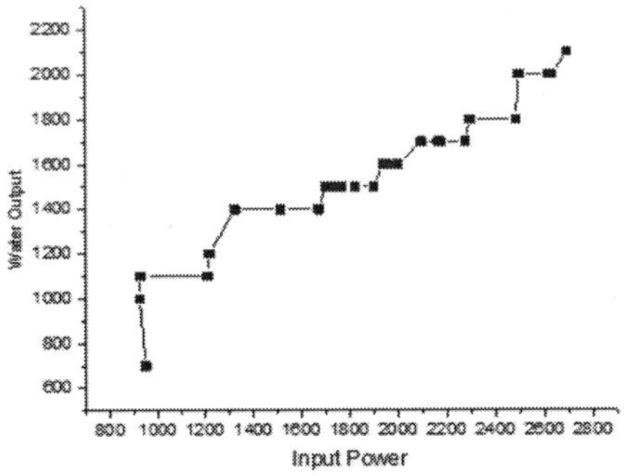

Figure 3 Input Power Vs Water Output of the SPV Pump

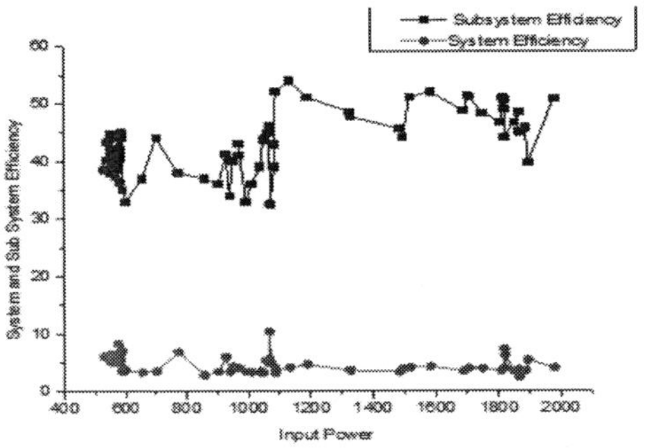

Figure 4 Input Power Vs System Subsystem Efficiency of SPV pump

The performance of a SPV water pumping system depends mainly on the power output of module technology at specified irradiation and temperature, which varies in different climatic conditions in different seasons. The performance of the solar photovoltaic pump is plotted in the figure 3 & figure 4. It has been observed that the water output behaves almost linearly with the input power. The system efficiency ranges from 3 to 7% and subsystem varies from 40 to 50 %. Comparative table No. 1 shows the performance of a solar water pump system at low and high radiation. From the table it has been observed that the water output is almost half for low irradiation condition as compared to high irradiation condition. Usually the water output is linearly dependent on the input power. So for increasing the water output for low irradiance condition it is advisable to use large size of PV array. The designing of the PV pump should be optimized in terms of input power. For designing of SPV water pump for low irradiance condition following points can be incorporated:

- Sizing the PV array by using lower irradiance condition
- Optimized tilt of the PV array
- Load curve based on the time of application
- Use of storage tank

Table 1: Performance of a solar water pump system at low and high radiation

	System at low Radiation	System at High Radiation
Average System Efficiency (%)	4.46	6.07
Average Sub System Efficiency (%)	43.8	48.2
Average Water Output (litre per day)	60847	131900
Average irradiation (kWh/m^2)	3.5	5.4

A.Sizing the PV array by using lower irradiance condition

The sizing of the PV module should be based on the frequently occurred condition of the particular site. The number of module in the series should be such that the threshold voltage should be maintained at very low radiation level of the site. The optimum sizing of the module in parallel should be such that the operating point of the pump should lies most of the time where the optimum efficiency attain by the pump most of the time.

B.Optimized tilt of the PV array

As the radition level is low, to use the whole availabe radiation level it is very much necessary to track the PV array with respect to sun. However for lowering the cost of tracking, it can be done on monthly basis. The monthly optimum tilt can be calculated by using the geometrical calculation. If the system is in fixed tilt then it can be fixed on the basis of time of load of operation.

C.Load curve based on the time of application

The analysis about the load curve is very much necessary before the designing and sizing the pump. The time of use may be different and based on that the array size and orientation of series-parallel connection needs to be improved. The analysis of loss of load probability for the particular site based on the time of operation and user specifications also is very much important.

D.Use of storage tank

The storage tank is one of a good solution for low irradiance conditions because in case of cloudy regions the variation in the irradiance is very high. So when the high radiation is available it is advisable to use the SPV pump in full swing and can store the water for the required use.

5. CONCLUSION

From this study, it is observed that the water output of SPV water pumping system depends mainly on the power output and irradiance. To optimize the performance of SPV water pump system need to change the size and orientation, and configuration of series & parallel connection of SPV array capacity for low radiation climatic zone.

ACKNOWLEDGEMENT

This research is based upon work supported in part by the Solar Energy Research Institute for India and the U.S. (SERIIUS) funded jointly by the U.S. Department of Energy subcontract DE AC36-08G028308 (Office of Science, Office of Basic Energy Sciences, and Energy Efficiency and Renewable Energy, Solar Energy Technology Program, with support from the Office of International Affairs) and the Government of India subcontract IUSSTF/JCERDC-SERIIUS/2012 dated 22nd Nov. 2012.

REFERENCE

[1] Bansal N.K. and G. Minke,"Climatic zones and rural housing in India," Kernforschungsanlage, Juelich, Germany, 1988.
[2] K. Yadav, O.S. Sastry, R. Wandhare, M. Kumar, B. Bora, Y. K. Singh, R. Singh, R.Rani, Renu, M. Bangar, R. Singh and A. Kumar "Performance Comparison of Pump Controllers for Solar PV Water Pumping Applications, Solar Energy 119 (2015) 195–202.
[3] Nagae S, Toda M, Minemoto T, Takakura H, Hamakawa Y. Evaluation the impact of solar spectrum and temperature variations on output power of Silicon based photovoltaic modules. Solar Energy Materials & Solar Cells. 2006; 90: 3568 – 75.
[4] www.mnre.gov.in

OPTIMUM ARRAY SIZING OF SOLAR PHOTOVOLTAIC WATER PUMPING SYSTEM

Renu[1, 2], Birinchi Bora[1, 2], Manander Bangar[1], O.S. Sastry[1]*, B. Prasad[2]
[1]National Institute of Solar Energy (NISE), Gurgaon, Haryana, 122003, India
[2]The Energy and Resources Institute, Delhi, India
*sastry284@gmail.com, Tel: +91-0124-2579213, Telefax No. : +91-124-2579207

ABSTRACT: Modeling for optimum sizing of the array for operating a Solar pump is done for the composite climatic zone of India based on the most favorable sub system efficiency and flow rate for a particular Solar pump. The sizing of the array should be such that the solar pump operates in its most favorable region of performance for most of the time for a particular site. The performance of a particular SPV array technology can be characterized by its performance ratio at that particular site.
Keywords: Water pump, SPV, Optimum, Array Sizing

1 INTRODUCTION

India is a vast country having different climatic zones based on irradiance, temperature, humidity, precipitation, wind and sky conditions. The major climatic zones are Warm and Humid, Hot and dry, Composite, Cold and Cloudy, Cold and Sunny and moderate [1]. SPV water pumping system plays a vital role for irrigation in rural and remote areas of country where grid connectivity is erratic or not available. High initial costs, necessitates the optimum designing of the system considering above concerns. However, due to support in the form of subsidy, the pumps are available to farmers in a reasonable cost. Under the Jawaharlal Nehru National Solar Mission (JNNSM) [2] the proposal is to deploy at least 0.1 Million pumps. Variation in the performance of the pumps from in different climatic zones is a big concern.

Varieties of pumps are promoted depending on the water requirement, and depth of water table. Both DC and AC operation based pumps are used for irrigation, drinking water applications and industrial applications from 10 m up to 100 m having capacities ranging from 0.5 hp to 10 hp depending on the flow rate of water desired, head of operation and availability of source. VFD driven submersible AC pumps are mostly used above 20 meters, up to 100 meters in most of the state's due to easily availability and less cost. A pump's performance can be determined using subsystem efficiency and flow rate desired for the application.

Subsystem efficiency (Esub) or wire to water efficiency is defined as the total hydraulic energy output divided by total PV power input.

$$E_{sub} = \frac{\rho . g . h . m}{V . I}$$

Flow rate (Q) is defined as the water output of the pump & motor per unit time.

$$Q = \frac{Volume\ of\ Water}{time}$$

Performance Ratio (%) can be defined as the ratio between final yield and reference yield of a particular technology.

$$PR = \frac{Final\ Yield}{Reference\ yield}$$

$$= \frac{DC\ power\ generation / Total\ Capacity\ of\ the\ array}{Total\ Irradiation / 1000\ W/m^2}$$

2 APPROACH

National Institute of Solar Energy (NISE) is an autonomous institute of Ministry of New and Renewable Energy (MNRE) working as a research and development, testing and certification body for SPV water pumping system in India. It has a SPV water Pump testing facility for analysis and performance evaluation of capacity ranging from 0.2 hp to 10 hp with an array capacity ranging from 200Wp to 10 kWp both for surface and submersible DC as well as AC pumps as per MNRE specifications 2015-16 and IEC standards [3,4,5]. NISE has a sump well (4m x 2.5m x 2.5m), with static head of 4 m, and an overhead platform (3.5 m x 2m x 2m) with head of static 7 m, to create suction/ delivery head for different heads pumps varying from 10 m up to 100 m, controlled by pressure gauge and throttle valve and standard multi-meters, Pyranometer at the horizontal and at the surface of the module are used to measure the radiation level [6]. The Solar PV modules are tested as per IEC standards [7, 8] for their reliability. A 3 hp AC surface SPV water pumping system has been used for this study having a multi-crystalline array technology with capacity of 2878 Wp (STC) tested at 20 m head of operation.

3 RESULTS AND DISCUSSION

The data under discussion is from a 3 hp AC pump installed at the test rig of NISE. It can be observed from Fig. 1, the input power fed by the array varies with radiation, and this affects the flow rate of the pump which can be seen in Fig. 2 and Fig, 3. In different seasons and at different places of operation, the intensity and ambient temperature varies widely which affects the performance of SPV water pump.

Irradiation at NISE has been plotted as a function of in-plane irradiance and ambient temperature for a particular year on seasonal basis. It is found that in for summer season the irradiation is found maximum in the range of 600 to 900 W/m^2 in-plane irradiance and in the ambient temperature range of 30 to 40 °C. For monsoon season the irradiation is found maximum in the range 600 to 900 W/m^2 in-plane irradiance and in the ambient temperature range of 25 to 30 °C; post monsoon season the irradiation is found maximum in the range of 500 to 900 W/m^2 in-plane irradiance and in the ambient temperature range of 20 to 30 °C. PR lies mostly in the range of 0.85 to 0.95 for winter season, 0.8 to 0.95 for Monsoon, 0.85 to 0.95 for Post Monsoon 0.8 to 0.85 for summer. It has been observed from the above figure that, the multicrystalline technology is performing best in winter season whereas its performance is worst in summer season. So the PV array size should be designed according to summer season if the pump is designed for whole year as stated above.

The power output from the PV module can be estimated by

$$\boldsymbol{Power\ output} = \boldsymbol{Irradiance} \times \boldsymbol{Performance\ Ratio}$$

We have calculated PR from performance data measured on 10 minute basis of a particular year for multicrystalline technology on seasonal basis, from which the seasonal array output of the multicrystalline technology for the year 2011 on seasonal basis by using the equation (1). The result is compared with the actual measured value of the same year 2011 for multicrystalline technology for all the four seasons of the year for this site. The error in estimation of the array output is highest in winter season i.e., 25 % whereas it is lowest in the month of monsoon, about 1%. The error for post monsoon season is plotted in the figure. It is observed that the pattern of error in estimation for the other seasons follow the same trend. It is shown in the figure that, for higher ambient temperature and higher Irradiation range the error is minimum whereas for the irradiation and ambient temperature range 700 W/m2, 20 °C and 1000W/m^2, 30 °C it is maximum.

So, for avoiding oversized design of PV array in summer season, it is necessary to use an optimum array capacity taking into account the operating temperature at time of operation for a specific application. But if we want to operate the pump for any application for the whole year, than the array sizing must be designed taking into account the loss of power due to high module temperature range in summer season.

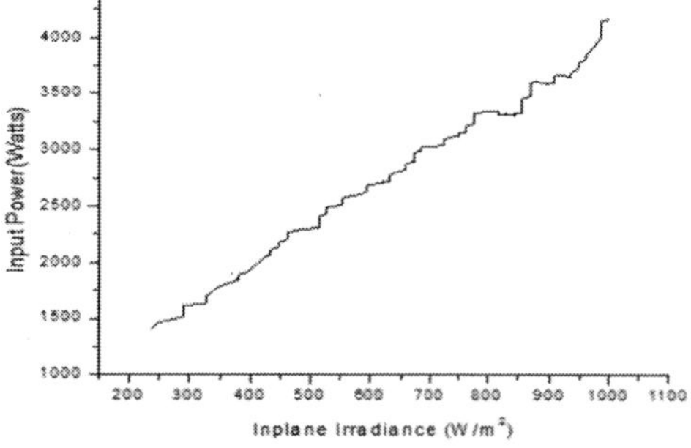

Figure 1: In-plane irradiance Vs Input power

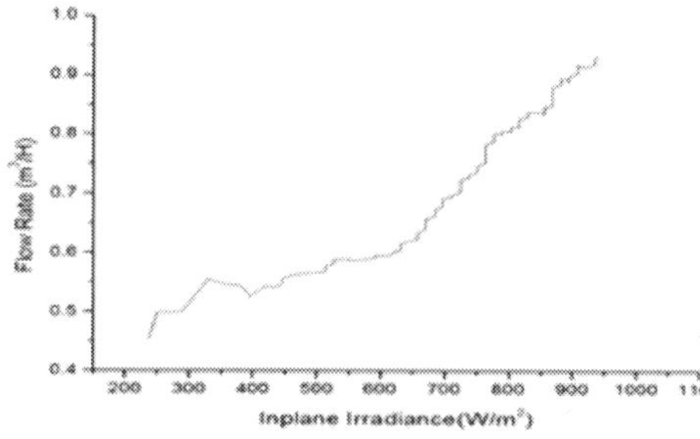

Figure 2: In-plane irradiance Vs flow rate

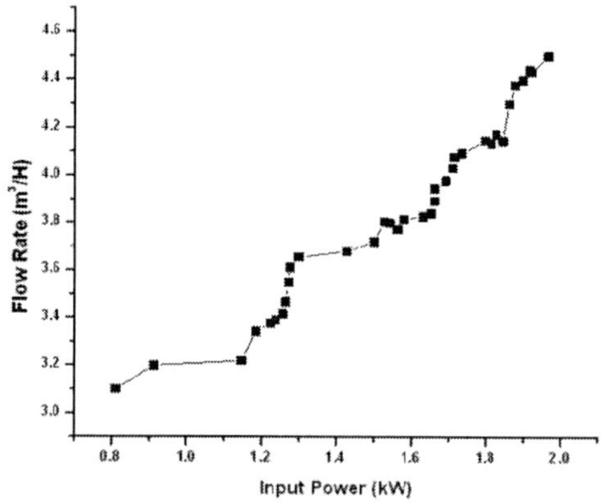

Figure 3: Variation of Flow rate vs. Input Power

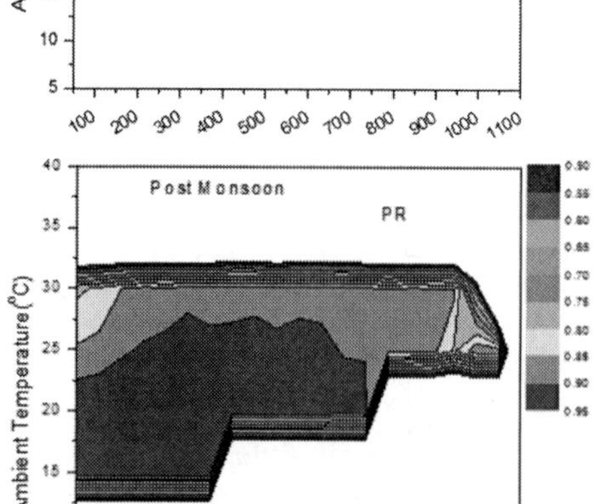

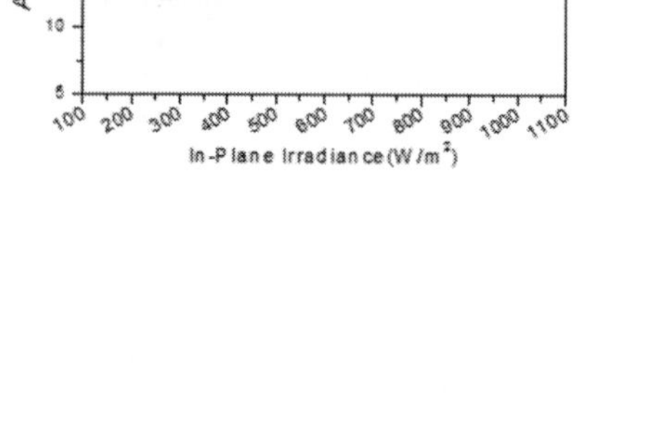

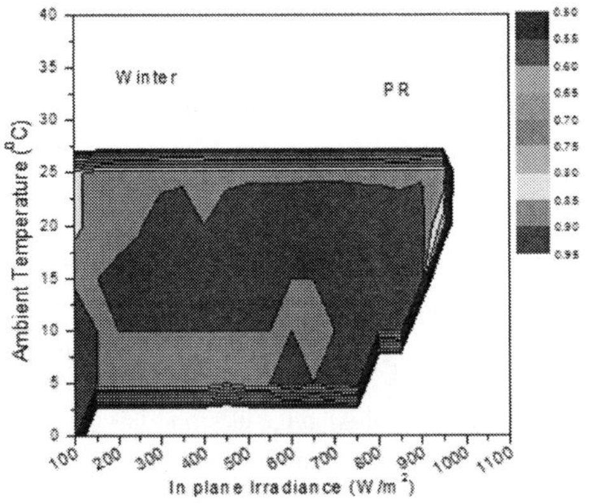

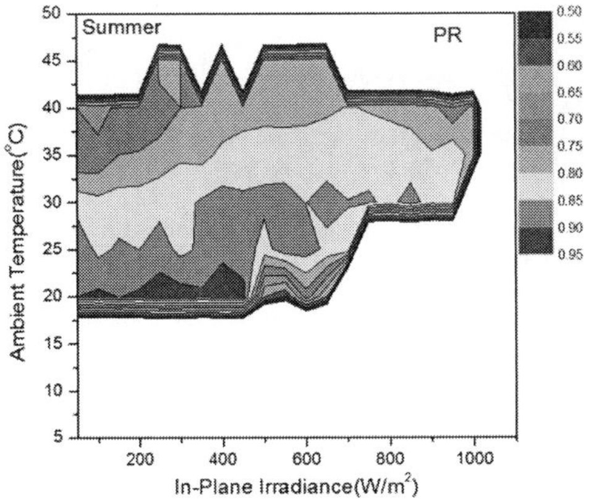

Figure 4: Performance Ratio (PR) of multicrystalline technology in different seasons as a function of In-plane irradiance (W/m^2) and ambient temperature at NISE

From this study, it is observed that the performance of a SPV water pumping system depends mainly on the power output of module technology at specified irradiation and temperature, which varies in different climatic conditions in different seasons. So, a SPV water pumping system should be selected based on its performance vs. input power from the PV module technology. The sizing of its PV array should be done based on its subsystem efficiency and flow rate after selection of its favorable region of operation.

4 ACKNOWLEDGEMENT

This research is based upon work supported in part by the Solar Energy Research Institute for India and the U.S. (SERIIUS) funded jointly by the U.S. Department of Energy subcontract DE AC36-08G028308 (Office of Science, Office of Basic Energy Sciences, and Energy Efficiency and Renewable Energy, Solar Energy Technology Program, with support from the Office of International Affairs) and the Government of India subcontract IUSSTF/JCERDC-SERIIUS/2012 dated 22nd Nov. 2012.

5 REFERENCES

[1] Bansal N.K. and G. Minke, Climatic zones and rural housing in India, Kernforschungsanlage, Juelich, Germany, 1988.
[2] www.mnre.gov.in
[3] MNRE Specifications 2013-14, Ministry of New & Renewable Energy, Government of India.
[4] IEC 62253
[5] IEC 61702
[6] O.S. Sastry, S. Agarwal, D. Adiraju, Rajesh Kumar and Bibek Bandopadhyay, Performance evaluation of solar photovoltaic lighting and water pumping systems in India, 26th European Photovoltaic Solar Energy Conference and Exhibition.
[7] IEC 61215
[8] IEC 61730 part1 and 2
[9] K. Yadav, O.S. Sastry, R. Wandhare, M. Kumar, B. Bora, Y. K. Singh, R. Singh, R.Rani, Renu, M. Bangar, R. Singh and A. Kumar "Performance Comparison of Pump Controllers for Solar PV Water Pumping Applications".

PRODUCT INTEGRATED PV: THE FUTURE IS DESIGN AND STYLING

Wouter Eggink & Angèle Reinders
University of Twente, Department of Design, Production and Management, Faculty of Engineering Technology
PO Box 217, 7500 AE Enschede, The Netherlands.
w.eggink@utwente.nl
a.h.m.e.reinders@utwente.nl

ABSTRACT: In this paper we explore how PV powered products have been designed in the past. For this purpose we have drawn a historical time line of the design features of PV powered products in the context of main stream design and styling. Our time frame is 1970 till 2016, focusing in first instance on consumer products and PV powered lamps. We will discuss how we learn from these experiences for the design and styling of future PV powered products. Because design and styling is considered one of the key features to a well-designed product, this study is relevant for the acceptability of PV products in nowadays consumer markets. Despite many studies on the technical functioning and energy performance of PV powered products, little attention has been paid to a constructive analysis of design and styling. Therefore with this study we make a start with identifying which design and styling factors can contribute to a higher appreciation for PV powered products.
Keywords: Design, Sustainable, Sociological, Stand-alone PV Systems, Module Integration

1 INTRODUCTION

The success of a product in nowadays consumer markets is strongly dominated by its design and styling, which apart from technical features, usability and production costs is considered to be one of the key factors to a well-designed product. Knowing this, it's surprising to notice that only a small share of PV powered products [1] [2] has met aesthetic requirements, as can be seen in Figure 1.

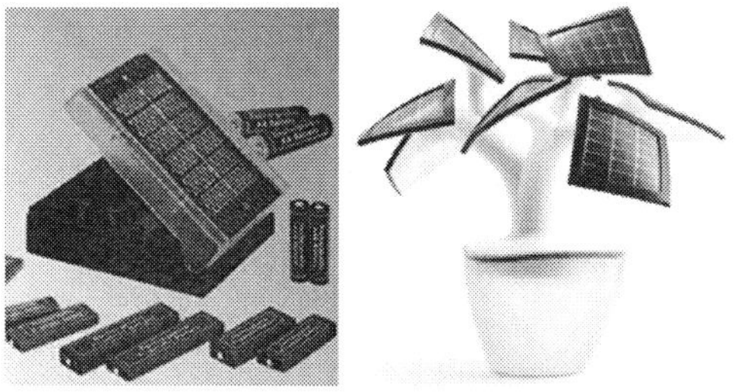

Figure 1: Chargers available on the market in 2015: left, a model without attention to design and styling, and right, a charger with great aesthetic appeal for those who like organic design by Vivien Muller

This lack of attention for aesthetic requirements has dramatic effects on the interest of consumers to purchase PV powered products. For instance recently executed research [3] with 100 users shows that just 32% of the respondents felt that the looks of a PV product are acceptable; 26% and 32% thought that the looks were resp. a bit and not acceptable at all. As such about 80% of the respondents wouldn't buy a PV powered product, moreover because they find the products' performance disappointing and their use complicated. Also initial research with 15 student groups, designing future PV concepts, showed that it yields better results when design and styling is specifically addressed in the development process [4].

Despite many studies on the technical functioning and energy performance of PV powered products, little attention has been paid to a constructive analysis of design and styling. As such we fill in a gap in existing knowledge regarding PV products. It may therefore be time to change this situation and explore in detail how the design and styling of PV powered products can be improved in the near-by future. To evaluate this to our best understanding, our research questions are: how have PV powered products been designed in the past and how are these PV-based designs related to mainstream design and styling of regular products (without PV) in the past and present? Finally we would like to know what can be learned from these experiences for the design and styling of future PV powered products.

Hence this study will provide relevant findings and yield significant guidelines for a higher appreciation of PV powered products.

2 APPROACH REGARDING INNOVATION AND DESIGN

In the following, we will explore our research questions by means of an historical time line from 1970 till 2015 of the design features of PV powered products in the context of main stream design and styling of regular products (without PV). From there, we will derive design and styling factors which can positively affect the appreciation of PV powered products and some strategies for a more meaningful design of contemporary PV integrated products.

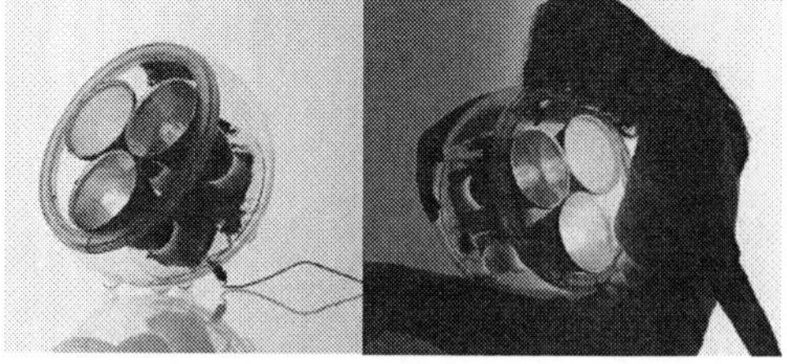

Figure 2: Yang Metamorfosi lighting system by Carlotta de Bevilacqua (2003).

Roberto Verganti [5] has shown that product innovation can be derived not only from technological developments, however also from design. In what he calls Design Driven Innovation, he shows that innovation is derived by radically changing the meaning that products incorporate for their users, either symbolic or emotionally. A canonical example is the living room lamp with adjustable lighting colors that was designed in 2003 by Carlotta de Bevilacqua (Figure 2). This changed the meaning of the product from "just a lamp" into a "system to influence the atmosphere of the room".

In the same manner, you can imagine the transformation from a simple "battery charger" into a product to "affirm your sustainable, natural lifestyle". Respectively represented by the two objects in Figure 1.

To our experience, innovative products are best accepted when they are innovative in both technology and design. In innovation, however the technology is mostly addressed first, because in first instance a product should be able to technically function. Therefore, from a technology perspective and according to evolutionary design theories we may for instance expect a transition of the application of photovoltaic cells on hard casings towards full integration in products made of flexible materials and soft fabric, such as the lamp shown in Figure 1 (left) and clothing and a tent in Figure 3. However, the latter are mostly just a normal sweater and a normal tent with PV technology applied on top of them. With our study we would like to facilitate the making of the transformation from "products with photovoltaic cells applied" into integrated and desirable "objects for a sustainable contemporary lifestyle".

Figure 3: Photovoltaic solar cells integrated in soft materials such as, left, a jumper by Pauline van Dongen (2015), and right, a refugee tent by University of Twente (2016).

3 RESULTS

Our comparison of the history of PV powered products and main stream design is represented by the timeline in Figure 5. From this overview, we can observe that for about two decades, PV powered calculators were the most dominant PV powered product on the market.

They were received with great enthusiasm in the 70-ies. This enthusiasm was mainly based on technical features which were enabled by PV solar cells and the effect that it made the owner feel modern. In fact this was Rogers' [6] diffusion theory in practice; only innovators and early-adopters purchased a PV powered calculator mainly also because of environmental concerns. For instance, the journal New Scientist of 20th July 1978 reported "The concerned environmentalist can now calculate the downfall of society without eating into the world's resources in the process. A new calculator that is coming on to the market ....{..}...doesn't need an on/off switch because the power comes from a small panel of solar cells". However if looking at existing pocket calculators without solar cells at that time, it seems that the design of the solar powered calculator is lagging about 10 years behind. It may therefore be concluded that till 2010, PV products were largely in the functionality/performance phase of evolutionary product design [7], while other products with similar functions had already reached a phase which contained more meaning. This changed after 2010; then we see new designs of solar chargers that resemble trees or flowers, indicating a meaning which represents sustainability.

In the design world it is already since the 70-ies common to experiment with meaning and emotion in the design of products [8], as is shown in the lower part of Figure 3. Sustainability is on the one hand incorporated by the choice of materials – like the cork of the TV of Starck, the solar cells in the Solar Chandelier and the recycled fridges of the Endless Chair. In other occasions the sustainability is represented by the organic shaping of elements of a product – like the Bone Chair by Laarman or the 3d printed lamp by Grossman. These approaches can be translated to PV products to make them more attractive. Moreover, integrated approaches like in the design by Howard are reflected in solar powered lamps that are also chargers at the same time. This decreases the former gadget image of PV powered products and makes them aesthetically pleasing as well.

4 APPLICATION

We will illustrate the switch from products with PV technology attached to an integral PV product design with some conceptual designs by our students in the studies of Industrial Design Engineering. The projects were executed by couples of two students who followed the course Sources of Innovation at the Master level. The students applied our integral design approach, called Innovative Design & Styling [4] in their projects. This approach is especially targeted at the integration of functional and aesthetic design in the development of technology based innovations.

The first example is the "Tulip" charging station for electric vehicles shown in Figure 4.

Figure 4: Meaningful integration of PV technology in product designs (1) "Tulip" charging system for electric vehicles by University of Twente (2014).

In the concept, the (concentrated) PV technology is an important part of the appearance of the product. It also provides largely for the association with nature and sustainability that constructs the meaning of the device.

The charging station also incorporates shading for both the vehicle and the users, and a little bench. To make a pleasant experience during the charging time.

The second example is the "Solar Parasol" pictured in

Figure 5: Historical time line of the design features of PV powered products compared to main stream designs.

Figure 6. Here the concentrated PV technology is presented as a flower that will harvest solar energy during the day, and will radiate light during the night. The transformation between harvesting and radiating is made visible by rotating the parabolic shapes upside down. During the day the parabolic shapes will track the sun like sun-flowers, communicating the harvesting process to the users. During the night the shapes represent parasols, that are a familiar sight at terraces and other urban public spaces. In this way, the technology is again fully integrated in the product design, and not only in a functional manner, but also on the level of the symbolic meaning of the device.

Figure 6: Meaningful integration of PV technology in product designs (2) "Solar Parasol" by University of Twente (2014).

4 DISCUSSION AND CONCLUSION

Because design and styling is considered one of the key features to a well-designed product, this study is relevant for the acceptability of PV products in nowadays consumer markets. Despite many studies on the technical functioning and energy performance of PV powered products, little attention has been paid to a constructive analysis of design and styling. Therefore with this study we make a start with identifying which design factors can contribute to a higher appreciation for PV powered products. These factors are:

1. During the design process, please try to perceive "products with photovoltaic cells applied" as integrated and desirable "objects for a sustainable contemporary lifestyle".
2. Use the features of PV cells (color, structure and geometries) as materials to design with instead of perceiving PV cells as materials with difficult design features (rigid, dark color, fragile) that have to be applied on top of an existing product.
3. Exploit the environmentally sound image of PV cells in each product designed with PV technologies by emphasizing sustainable design and the use of sustainable materials.
4. Connect to contemporary design styles as applied in regular products (without PV) to make PV powered products a desirable purchase with a high cool- or cute-factor.
5. Besides a well-functioning technical design please consider the societal impact of the autonomous use of a PV powered product and match the design of these products with this.

This list of factors is not extensive and of course we will need to study these in more depth and for a longer period. However, eventually we could develop guidelines for the design of successful and well-appreciated PV-powered products. In our paper we have shown that these factors above mentioned can be successfully applied in design projects, as has been done in the project about CPV in the built environment at University of Twente.

5 REFERENCES

[1] Apostolou G., Reinders A.H.M.E., "Overview of design issues in product-integrated photovoltaics", Energy Technology, Wiley-VCH Verlag GmbH, Vol. 2, pp. 229 – 242, 2014

[2] Reinders A.H.M.E., Sark W.G.J.H.M. van, "Product-Integrated Photovoltaics", in Sayigh A. (Ed.) Comprehensive Renewable Energy, Elsevier, Oxford, Vol. 1, 709-732. 2012

[3] Apostolou G., Reinders A.H.M.E., How do users interact with PV-powered products? Investigating 100 lead users and 6 PV products, Journal of Design Research, 2016

[4] Eggink, W., Reinders, A.H.M.E.,"The Design and Styling of Technology-based Innovations", Proceedings of the 5th World Conference on Design Research (IASDR); Chibaura Institute of Technology; Tokyo; August 26 - 30, 2013

[5] Verganti, R., "Design-Driven Innovation, Changing the Rules of Competition by Radically Innovating What Things Mean." Boston, Massachusets, Harvard Business Press, 2009

[6] Rogers, E.M. "Diffusion of Innovations", Simon & Schuster, 1995 (1962)

[7] Eger, A. O. and Drukker JW., "Phases of Product Development: A Qualitative Complement to the Product Life Cycle." Design Issues, Vol. 26(2): pp. 47-58, 2010

[8] Eggink, W., "A Chair to Look to the Moon: What We Can Learn from Irrational Design History for Contemporary Design Practice." Design Principles and Practices, Vol. 3(1): pp. 103-114, 2009

A NEW PHOTOVOLTAIC CHARGING TOPOLOGY AND REGENERATIVE BRAKING ANALYSIS FOR SOLAR TRICYCLE

Mohamed DAHBI[1], Issam SALHI[1], Said DOUBABI[1] & Ahmed RACHID[2]
[1]Laboratory of electrical systems and telecommunications, Cadi Ayyad University, Marrakech, Morocco
[2]Laboratoire des technologies Innovantes, Picardies Jules Verne University, Amiens, French
[1]Tel : +212-640-27-68-64
dahbi.mohamed2@gmail.com , isalhi@yahoo.fr , s.doubabi@uca.ma , rachid@u-picardie.fr

ABSTRACT: Marrakesh city has known an increasing of based-fuel tricycles. This is due to the difficult access to some narrow districts and its compacted size. The mean problem in this case is the pollution and noise. In this paper, an electric tricycle was equipped with solar panels to show their impact on the battery state of charge (SOC). Using regenerative braking was also analyzed. Based on different resistive forces on the vehicle, simulations of different parameters are presented including the battery state of charge. A proposed topology to rely different components is also discussed to get more range and more robustness during the operating periods. Two daily paths of a tricycle in Marrakech city was used in simulations, one for touristic uses, while the other is for goods transportation. Results has shown that this vehicle range has increased when using solar panels as additional source of power and the regenerative braking when the path is filled with braking phases.
KEYWORDS: Electric tricycle, solar tricycle, maximum power point tracker, PWM controls mode, regenerative braking, solar panels, Brushless DC motor.

1 INTRODUCTION

Marrakesh city is the fourth largest city in Morocco. It is a famous city since it's a worldwide touristic destination. However, the majority of its alleys are narrow, especially in the old city. This fact causes serious problems, such as the access of transportation vehicles to such areas.

The tricycle has become a very common means of transportation in the streets of Marrakesh thanks to many advantages. Some of these latter are: (i) the tricycle reduced size allowing it to circulate through streets with more flexibility; (ii) its low price; (iii) it does not need a license to drive … Nowadays; the use of such tricycles has significantly increased in various fields. It's largely used for goods transportation, and even adjusted for people transport. Some tricycle tours are even proposed for tourists to discover different areas of the city (figure1). Further, Moroccan government also encouraged the spread of the tricycles thanks to the initiative of National Human Development. This initiative has allowed different categories of people, which are among the disabled, to integrate into the social environment by giving them license to start this business.

Figure 1: Different tricycle uses

However, till now, all the used tricycles are based on fuel. The Increased number of tricycles has vicious impact on the environment especially as most of districts in Marrakesh and tourist places are narrow and with limited aeration. This fact increases the pollution degree and noise.

Consequently, using electric transportation based on renewable energy sources might be a solution to reduce gas emissions and the cost of the electrical energy. Renewable energy sources such as solar and cell fuel have become widely used for many reasons such as: (i) lack of natural resources, (ii) oil price instability, (iii) global warming. In the light of this vision, and because Marrakesh is also known by its large solar potential, the use of solar electric tricycles might become an alternate to replace fuel-based ones.

Many studies were done in order to commercialize electric and solar electric vehicles [1]-[3] [6]. Still, the vehicle autonomy is the main problem. These studies have treated different kind of transportation, like bicycles and tricycles. In [1], a realization and testing of photovoltaic charging system, suitable for the management of an electrical powered assisted tricycle, is explained. An electric system design for electric reverse tricycle for the Elderly is suggested in [2]. While [3] has presented a solar battery charging station that can power modified rickshaws that are torque sensor based and electrically assisted. However, these papers results treat only pedal assisted vehicles.

In this paper, we propose a photovoltaic charging topology for solar electric tricycle. Solar panels will be added on the top of the tricycle to analyze their impact on the vehicle autonomy. For example, we can distinguish moments of the day when the tricycle could be waiting, on the parking, for other mission. During such moments, the tricycle could take profits from sun to recharge the batteries.

On the other hand, when rolling, the tricycle might encounter many braking phases, especially in traffic lights. The back-up energy during regenerative braking could help to increase the tricycle autonomy. This energy is usually lost by using the mechanical braking. In this paper, we studied, in addition to the proposed charging topology, the use of regenerative braking on the solar tricycle.

This paper is divided in six sections. Section II presents the implemented topology to charge the battery using solar panels. Section III presents the solar panel characteristics and their model. In section IV, brushless DC motor (BLDCM) model and resistive forces equations are detailed. The simulation results are discussed in section V, followed by a conclusion in section VI.

2 THE IMPLEMENTED TOPOLOGYT

The global studied system synoptic is shown in figure 2. To design a reliable relay system we proposed a new charging topology. There will be two modules. Each one is composed by two solar panels, which are connected in parallel with a DC/DC converter. This latter is used to charge the 60V/80Ah battery. Figure 3 presents one of the two modules topology of the proposed charging system. The simulations of the overall scheme require modeling of each part.

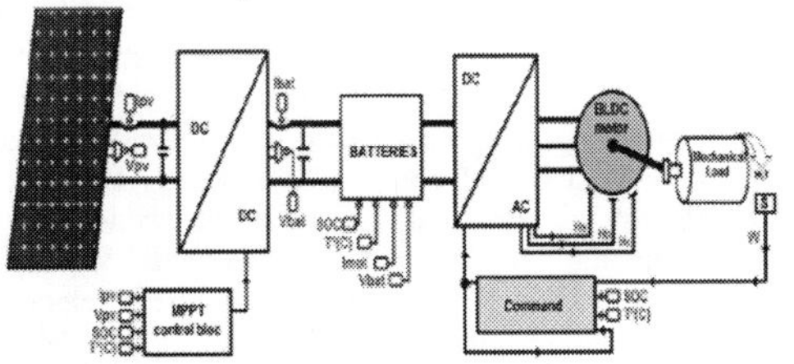

Figure 2: Global studied system synoptic

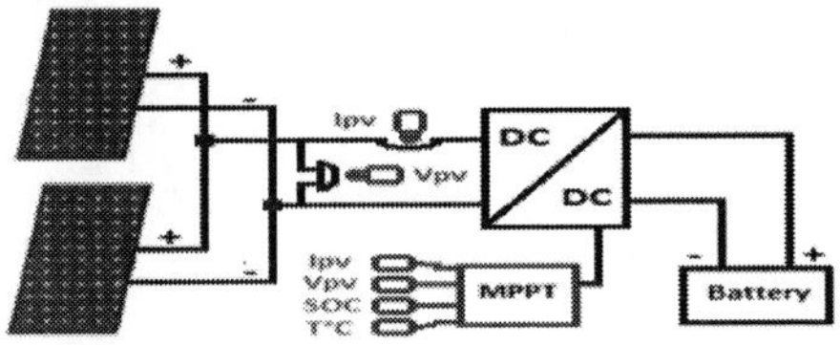

Figure 3: Charging system topology

3 SOLAR PANEL MODELING

A solar cell has a non-linear (I-V) and (P-V) characteristics and is represented as a current source [4] [5]. For our study, we used the equation based model presented in [5]. Equation (1) presents the current versus voltage equation of a solar panel.

$$I = N_p I_L - N_p I_0 * \left(e^{\frac{q*(U+I(N_s/N_p)*R_s}{N_s*A*K*T}} - 1 \right) - \frac{U + I\left(N_S/N_P\right)*R_S}{\left(N_s/N_p\right)*R_P} \quad (1)$$

Where :

- I_L the generated current under a given insolation
- I_0 Saturation current
- q charge of an electron : 1.602 e^{-19} C
- K Boltzmann constant : 1.38046 e^{-22} $J.K^{-1}$
- T temperature of PV module K
- A diode quality factor
- R_S series resistance
- R_P parallel resistance
- I photovoltaic cell output current
- U photovoltaic cell output voltage
- N_S number of cells connected in series
- N_p number of cells connected in parallel

The generated and the saturation currents of the solar cell depend on the temperature and the irradiation values. The equations linking between these parameters are presented in (2) and (3):

$$I_L = \left(I_{L,STC} + K_I + \Delta T\right) * \frac{G}{G_{STC}} \quad (2)$$

$$I_0 = I_{0,STC} * \left(\frac{T}{T_{stc}}\right)^2 * e\left[\frac{q \cdot E_g}{K.A} * \left(\frac{1}{T_{STC}} - \frac{1}{T}\right)\right] \quad (3)$$

Where :

- $I_{L,STC}$ the generated current under Standart Test Condition (STC)
- $I_{o,STC}$ the saturation current under Standard Test Condition (STC)
- K_I coefficient of current
- G_{STC} Irradiance level of 1000W/m²
- E_g band gap energy (1.124 ev)
- T_{STC} Temperature of 25°C (289.15K)

In the other hand, the system performance depends also on the implemented system area. The tricycle size has limited the number of solar panels that can be implemented, 4 solar panels are used. The solar panels power varies depending on different parameters such as temperature and irradiation; figure 4 shows the irradiation effect on the solar panels current and power with a constant temperature, while figure 5 presents simulation results of the solar panels performance with a constant temperature but with different irradiation values. Solar panel characteristics are presented in table 1.

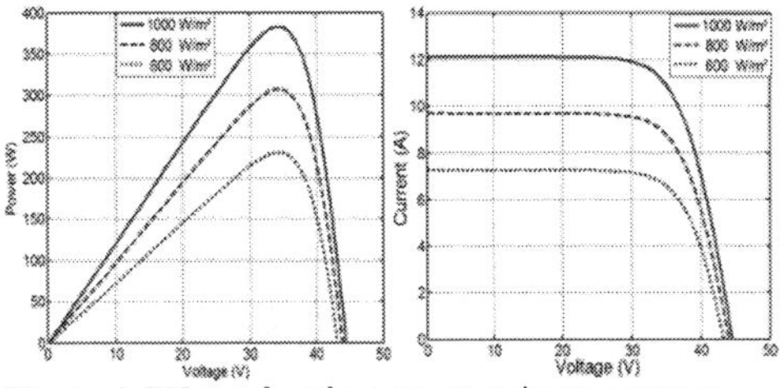

Figure 4: PV panels string current and power versus voltage for variable irradiation value in a constant temperature (25°C)

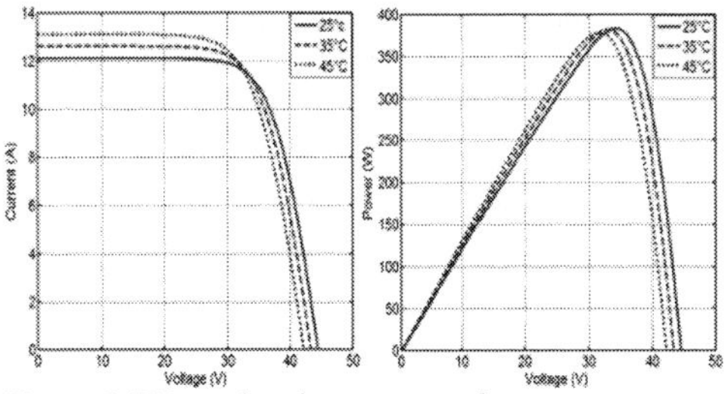

Figure 5: PV panels string current and power versus voltage for variable temperature value in a constant irradiation (1000W/m²)

Because of irradiation and temperature parameters are changing when the tricycle is exposed to sunlight, using maximum power point tracker (MPPT) is necessary. Different MPPT methods are proposed in different studies [7] [8] [9].For our use, the P&O method is implemented because of its ease of programming and does not require complicated necessities [7].

Table 1: Used solar panel parameters

Maximum Power (W)	200W
Optimum Power Voltage (V_{mp})	36.3 V
Optimum Operating current (I_{mp})	5.51 A
Open Circuit Voltage (V_{oc})	43.70 V
Short Circuit Current (I_{sc})	6.06 A
Module Efficiency (%)	17.30%

Using DC/DC converters is necessary to track the maximum power point. Boost or buck-boost converters are generally the most used to extract the maximum power from the solar panel. For our use, the boost converter is chosen because of its better performances for the proposed topology. One part of the global schematic consisting of the solar panel model, the DC/DC converter with its MPPT controlling system and the battery as energy source is presented in figure 6 using MATLAB software.

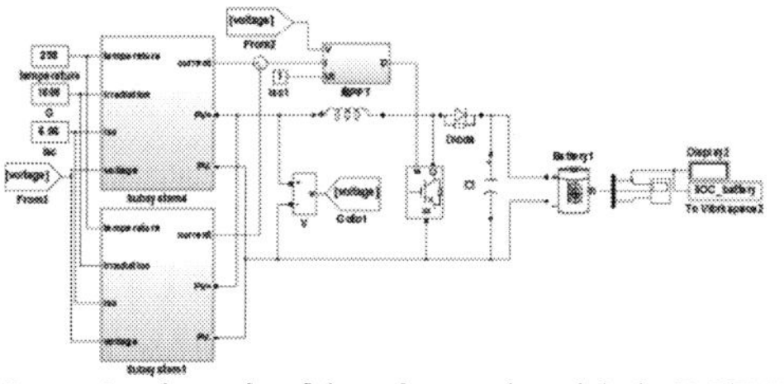

Figure 6: schematic of the solar panel model, the DC/DC converter with its MPPT controlling system and the battery

This part of the controlled system treats only the charging part of the battery, on the other hand, the battery is connected to the motor as a load charge which consumes energy and reduces the autonomy of the battery

4 BLDC MOTOR MODEL AND RESISTIVE FORCES EQUATIONS

The BLDCM is widely chosen to drive electric vehicles specially scooters and tricycles. This is due to its high efficiency, high power/weight ratio, high torque/current ratio and low cost of maintenance [10].The BLDCM is a 3 phase synchronous motor controlled by an inverter. Unlike DC motor, where the commutation is done using brushes, the BLDCM is controlled electronically by a driver circuit as presented in figure 7 [11-14].To simulate the operation of the studied BLDCM, we used Matlab/simulink environment.

This circuit is composed by a battery as power source, three bridges of two MOSFET's and the motor modeled as a resistance connected in series to an inductor and the back-EMF.

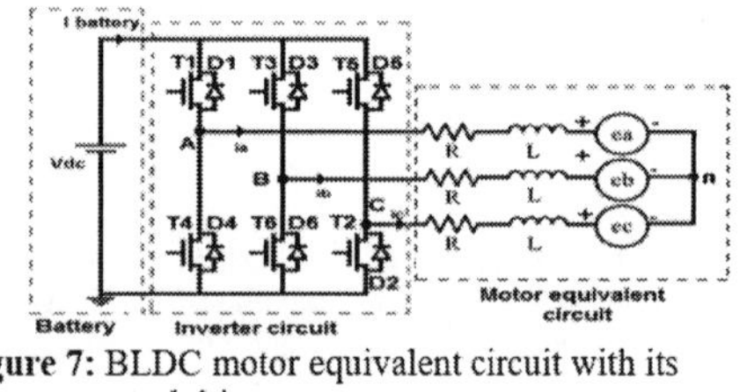

Figure 7: BLDC motor equivalent circuit with its inverter control driver

The commutation of the MOSFET's inverter is done depending on the rotor position. This latter is known by using usually three Hall Effect sensors [15]. The combinations of these three sensors are analyzed to control after the right MOSFET to conduct. Two MOSFETs are conducting at the same time to insure the right function of the motor. Table 2 presents the sequence control to rotate the motor in clockwise direction.

Table 2: Sequence control to rotate the motor in clockwise direction

Sector	Hall A	Hall B	Hall C	Active switches
1	0	1	1	T1-T4
2	0	0	1	T1-T6
3	1	0	1	T3-T6
4	1	0	0	T3-T2
5	1	1	0	T5-T2
6	0	1	0	T5-T4

Controlling the speed of the motor is done by using the pulse with modulation (PWM) control mode. The duty ratio is set by an analog to digital conversion and is applied on the gate of the MOSFETs. This duty ratio depends on the set point delivered from the accelerator. For regenerative braking, MOSFETs are controlled to make the engine as a function generator and supplies current back to the battery. The detailed BLDC motor modeling presented as follow.

Modeling the BLDC motor can be done using the abc phase variable model, the transforming to d-q axis or modeling the entire inverter and the motor in one bloc presentation. The model equations are presented as follow choosing the abc representation for its preference:

$$\begin{bmatrix} V_{an} \\ V_{bn} \\ V_{cn} \end{bmatrix} = \begin{bmatrix} r_a & 0 & 0 \\ 0 & r_b & 0 \\ 0 & 0 & r_c \end{bmatrix} \begin{bmatrix} i_a \\ i_b \\ i_c \end{bmatrix} + \begin{bmatrix} L_a & 0 & 0 \\ 0 & L_b & 0 \\ 0 & 0 & L_c \end{bmatrix} \frac{d}{dt} \begin{bmatrix} i_a \\ i_b \\ i_c \end{bmatrix} + \begin{bmatrix} E_a \\ E_b \\ E_c \end{bmatrix} \quad (4)$$

Where:

- V_{an}, V_{bn} and V_{cn} are the phase voltages
- r_a, r_b and r_c are the phases resistances
- L_a, L_b and L_c are the phase inductance
- i_a, i_b and i_c are the phases currents
- E_a, E_b and E_c are the back-EMFs

The electromagnetic torque is given by (5):

$$T_e = \frac{E_a i_a + E_b i_b + E_c i_c}{W} \quad (5)$$

Where:

- W is the mechanical speed of the motor

In addition to electric equation, the mechanical one is presented as follow:

$$\frac{dW}{dt} = \frac{1}{J}(T_e - T_l - BW) \quad (6)$$

Where:

- B is the viscous friction coefficient.
- J is the moment of inertia.
- T_L is the load torque.

Further, different resistive forces that might occur into the tricycle movement are taken in consideration in our simulations. These resistive forces will present a load torque applied on the motor as presented in equation (7). This load torque is represented by the acceleration force, the aerodynamic resistive force, the rolling resistance force, and the hill climbing force [16] [17].

$$F_{resist} = F_{acc} + F_{aero} + F_{rol} + F_{c\lim p} = m\frac{dV}{dt} + \frac{1}{2}C_d.\rho.A.V_r^2 + m.g.C_r + m.g.\sin(\alpha) \quad (7)$$

Where:

- m mass of the vehicle (Kg)
- V vehicle speed (m/s)
- C_d drag coefficient
- A Frontal area (m²)
- ρ air density (Kg/m³)
- V_r relative speed (m/s)
- C_r the rolling coefficient
- g gravitational force (m/s²)
- α road gradient (rad)

Using MATLAB software, the vehicle performance in term of power and torque are presented for two cases: when the tricycle is empty (weight of 300Kgincluding the driver), and when filled (weight of 600Kg including the driver).We may consider a 5% hill climb resistive force. Simulation results are presented in figure 8.

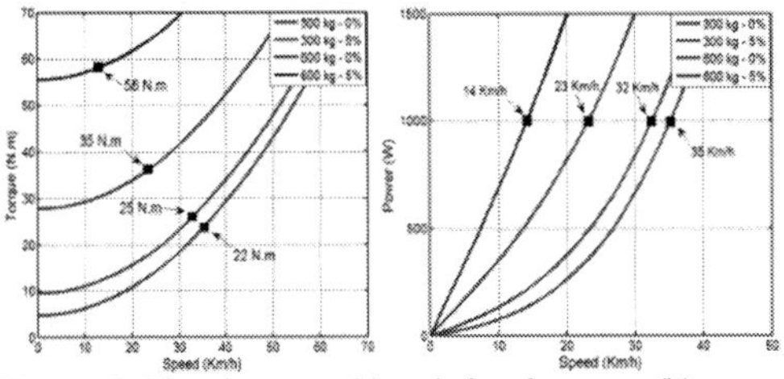

Figure 8: Electric power (a) and electric torque (b) versus vehicle speed (C_d =1.2, A=1.2, C_r = 0.007, g=9.81)

The simulation results show that the use of a 1000W power system, the speed of the vehicle can reach 35Km/h. Depending on the tricycle weight and the degree of slope, the speed varies and can be reduced to 14Km/h for the case of 600Kg and 5% as degree of slope. The torque value in this case is almost 60N.m.

The studied tricycle would be equipped with a 1000W BLDC motor, a 60V/80Ah battery, a DC/AC converter, plus the controlling system. Simulation of the global system is presented in the next section.

5 SIMULATION RESULTS

This work presents simulations of two daily paths of a solar tricycle. The first path is used for goods transportation is a 7.8 km, while the second one is for a common touristic tour in Marrakesh (10 km), as it's shown by figure 9.

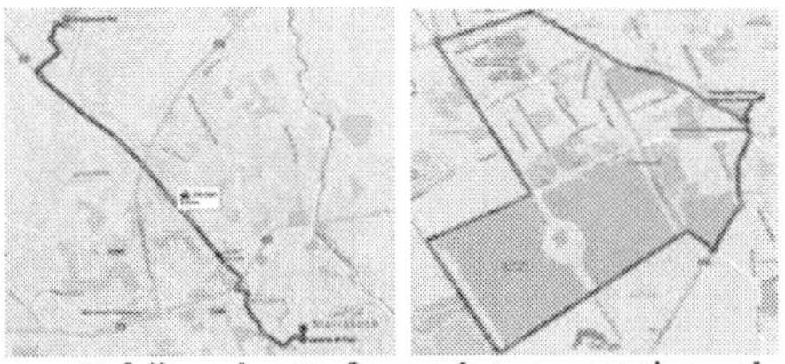

Figure 9: followed route for goods transportation and touristic use

At first, the simulation of the BLDC motor connected with only the battery as energy source is made with and without activation of the regenerative braking to observe the impact of this latter and most of this, is this command is needed or not. Figures 10 to 13 show the speed waveform, the load torque applied on the motor in the followed path, the torque motor response, the battery SOC and the phase current of the motor respectively for both with and without regenerative braking mode for both paths.

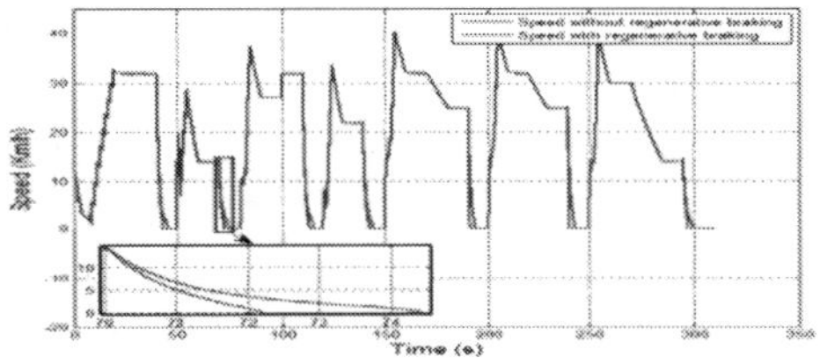

Figure 10: speed waveform of the motor response using and without using regenerative braking

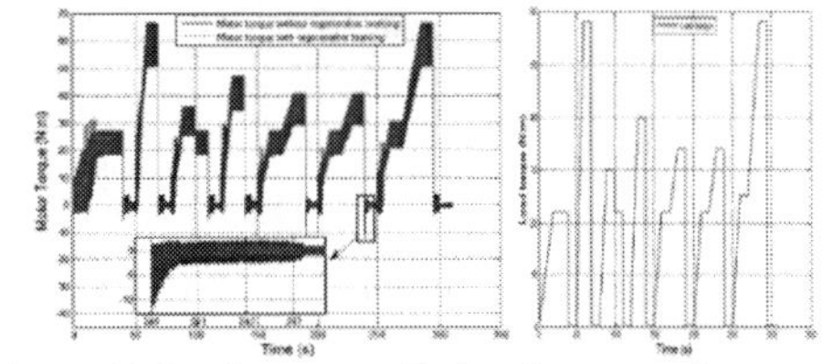

Figure 11: Load torque applied on the motor, the torque motor response using and without using regenerative braking

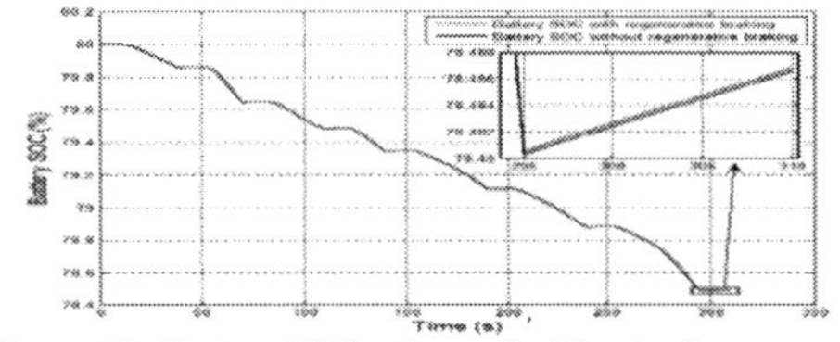

Figure 12: Battery SOC using and without using regenerative braking

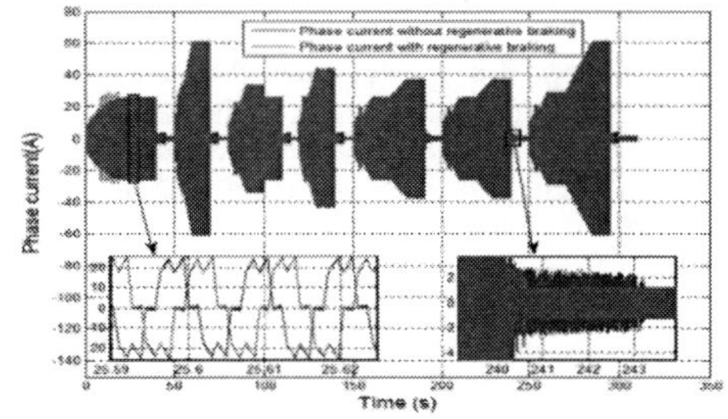

Figure 13: Phase current waveform

Simulation results have shown that using solar panels as additional energy source has increased the battery state of charge as presented in figure 14. This means increasing of the tricycle operating range. Figure 15shows the solar panel and the battery charging currents.

It can be noticed in some parts of the followed path motor consumption is less than power provided from solar panels system which can be exploit to control the motor directly by the solar panel power.

This controlling mode may have a large current spike which needs a smooth change from the solar panel power to the battery power. Also the real need of power by the motor may not be equal to the solar panel power which requires alternating between different sources of energies to detect the performed one. This controlling strategy may increase the battery live but needs more components.

6 CONCLUSION

An electric tricycle can represent an alternate means for fuel-based transportation. The vehicle autonomy represents an important issue. In order to increase the operating range of the electric tricycle, this latter was equipped with solar panels with a new topology linking between different components. The analysis of the regenerative braking was also done. Marrakesh city was chosen because of its high solar potential and the spread of tricycles for different uses. Simulation results were presented treating two daily paths. They have shown that the battery autonomy has increased using the regenerative braking control when the follow route is filled with braking phases. Adding photovoltaic panels as additional power source has widely increased the battery autonomy.

7 REFERENCES

[1] MM. Caruso, V. Cecconi, G. Cipriani, V.Di Dio, A.O.Di Tommaso, F. Genduso, D.LaCascia, R. Lauricella, R. Liga, R. Miceli, G. Prestigiacomo, G. Ricco Galluzzo, M. Trapanese, "A photovoltaic charging system of an electrically assisted tricycle for touristic purposes,"2013.

[2] Y. Choi, Y. Son, and U. Joe, "Electric System Design for Electric Reverse Tricycle for the Elderly Steering Structure Design Driving Power Performance Analysis for Selecting Motor," 2013.

[3] T. Faraz and a. Azad,"Solar battery charging station and torque sensor based electrically assisted tricycle," Proc. - 2012 IEEE Glob. Humanit. Technol. Conf. GHTC 2012, pp. 18–22, 2012.

[4] M. Datta, T. Senjyu, A. Yona, and T. Funabashi, "A fuzzy based method for leveling output power fluctuations of photovoltaic-diesel hybrid power system," Renew. Energy, vol. 36, no. 6, pp. 1693–1703, 2011.

[5] C. Nemes and F. Munteanu, "An Analysis of a Photovoltaic Panel Model Comparison Between Measurements and Analytical Models," 2012 Int. Conf. Expo. Electr. Power Eng., no. Epe, pp. 939–944, 2012.

[6] S. a K. Mozaffari Niapour, S. Danyali, M. B. B. Sharifian, and M. R. Feyzi, "Brushless DC motor drives supplied by PV power system based on Z-source inverter and FL-IC MPPT controller," Energy Convers. Manag., vol. 52, no. 8–9, pp. 3043–3059, 2011.

[7] M. Azzouzi, "Comparaison between MPPT P&O and MPPT Fuzzy Controls in Optimizing the Photovoltaic Generator," Int. J. Adv. Comput. Sci. Appl., vol. 3, no. 12, pp. 57–62, 2012.

[8] M. A. Aredes, B. W. França, and M. Aredes, "Fuzzy Adaptive P &O Control for MPPT of a Photovoltaic Module," no. April, pp. 120–129, 2014.

[9] M. A. Sahnoun, H. M. R. Ugalde, J.-C. Carmona, and J. Gomand, "Maximum Power point Tracking Using P&O Control Optimized by a Neural Network Approach: A Good Compromise between Accuracy and Complexity," Energy Procedia, vol. 42, pp. 650–659, 2013.

[10] R. Shanmugasundram, K. M. Zakaraiah, and N. Yadaiah, "Modeling, simulation and analysis of controllers for brushless direct current motor drives," J. Vib. Control, no. May 2012, 2012.

[11] L.Varghese and P. G. Student, "Modeling and Design of Cost Efficient Novel Digital Controller for Brushless DC Motor Drive," 2014.

[12] P. Pillay and R. Krishnan, "Modeling, Simulation, and Analysis of Permanent-Magnet Motor Drives, Part 11: The Brushless DC Motor Drive," Ind. Appl. IEEE Trans., vol. 25, no. 2, pp. 274–279, 1989.

[13] Zhaojun Meng, Rui Chen, Changzhi Sun, and Yuejun An, "The mathematical simulation model of brushless dc motor system," 2010 Int. Conf. Comput. Appl. Syst. Model. (ICCASM 2010), vol. 12, no. Iccasm, pp. V12–625–V12–629, 2010.

[14] Y. Li and J. Rong, "The study of new modeling method for permanent magnet brushless DC motor," 2011 Int. Conf. Electron. Mech. Eng. Inf. Technol., vol. 9, pp. 4713–4716, 2011.

[15] N. B. Bahari, A. Jidin, A. Rahim, N. Othman, and M. Manap, "Controller for Brushless DC Motor Drives," pp. 152–155, 2012.

[16] P. Prochazka, I. Pazdera, P. Vorel, and D. Cervinka, "Design of Small Electric Car," pp. 359–364, 2012.

[17] A. Adnan and D. Ishak, "Finite Element Modeling and Analysis of External Rotor Brushless DC Motor for Electric Bicycle," no. SCOReD, pp. 16–18, 2009.

DESIGN, CHARACTERIZATION AND MODELLING OF HIGH EFFICIENT SOLAR POWERED LIGHTING SYSTEMS

Peter Nymann[a], Frederik Svane[a], Peter Behrensdorff Poulsen[a,*], Sune Thorsteinsson[a], Johannes Lindén[a], Rasmus Overgaard Ploug[a], Maria C. Mira[b], Arnold Knott[b], Ib Mogensen[c], Kris Retoft[c]
[a]Department of Photonics Engineering, Technical University of Denmark, Frederiksborgvej 399, Building 130, 4000 Roskilde, Denmark
[b]Department of Electrical Engineering, Technical University of Denmark, Richard Petersens Plads 325, DK-2800 Kgs. Lyngby, Denmark,
[c]Out-sider, Kigkurren 8A, Islands Brygge, 2300 København S, Denmark
[*]Phone: +45 21325110, ppou@fotonik.dtu.dk

ABSTRACT: This paper discusses some of the major challenges in the development of L2L (Light-2-Light) products. It's the lack of efficient converter electronics, modelling tools for dimensioning and furthermore, characterization facilities to support the successful development of the products. We report the development of 2 Three-Port-Converters respectively for 1-10Wp with a peak efficiency of 99.1% at 1.5 W output power at PV to battery and almost similar characteristics for a 10-50 Wp. Furthermore, a modelling tool for L2L products has been developed and a laboratory for feeding in component data not available in the datasheets to the model is described. A living lab facility is realized to field test prototypes of L2L lighting products in their development state to validate the modelling tool and tweak the parameters in the system for optimized performance the product. Finally, the model was validated against a field test from the living lab over a four-day period with an offset of 5.3 %.
Keywords: Stand-alone PV Systems, High-Efficiency, Design, Modelling, Simulation

1 INTRODUCTION

In the last few years significant progress has been made in research and development of renewable energy systems. Solar energy is one of the most reliable, daily available and environment friendly renewable energy sources, being unlimited, clean and free [1], [2]. Furthermore, due to the ability of generating off-grid electricity, sustainable energy systems have attracted research interest in the last decades [3]. Due to the rapidly increasing efficiency of light emitted diodes (LEDs) stand-alone combinations of PV modules, battery storage and LED luminaires are becoming more attractive for a variety of lighting applications [4]. The systems operate entirely on DC and are often called Light-to-light (L2L) systems, see figure 1.

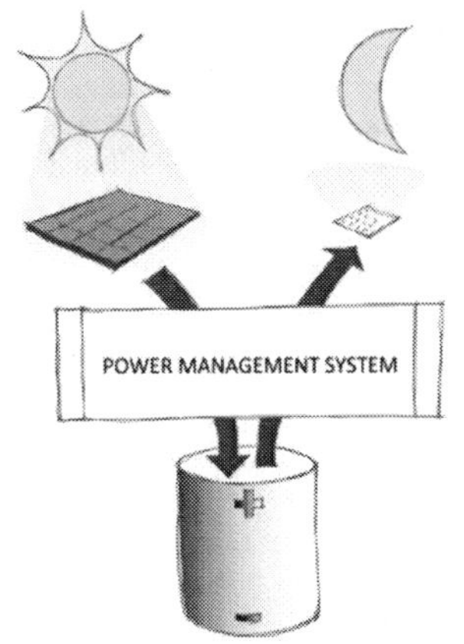

Figure 1: A principle sketch of a Light-2-Light system

Park lights and bollards are examples of L2L systems and these systems offer lighting solutions, for places where lighting is not feasible due to very high cabling costs of up to e.g. 700 €/m in Copenhagen and up to 2000€/m in Berlin [5]. Some of the major obstacles still remaining for successful L2L systems are 1) the lack of efficient converter electronics for small PV systems, (PV powers of 1-50 Wp) 2) the lack of a dimensioning and assessment tool for L2L systems and 3) the knowledge centers and laboratory facilities to support the complicated process of designing successful L2L products. The latter being an interdisciplinary technical task involving at least knowledge, tools, and test equipment for the technology fields of lighting, photovoltaics and power electronics. A series of Danish research projects has addressed these issues and the results are reported here.

2 ELECTRONIC CONVERTER

An electronic converter for outdoor light-to-light purposes should be able to work in two modes:

1. In the daytime: Convert energy from the PV panel while maintaining Maximum Power Point Tracking (MPPT) in order to harvest as much energy to the battery from the available solar energy as possible.
2. In the night: Convert stored energy from the battery to the operating point of the LED panel.

Hence this application calls for a Three-Port-Converter (TPC) with two operating modes.

To cover the full range of 1-50 Wp of solar panels with high efficiency, 2 versions of converters was developed. A low range for PV powers 1-10 Wp and a high power range of 10-50 Wp, giving the versatility of choosing solar panels in the whole span. The efficiency curve for the 10 Wp version is shown in figure 2 and figure 3. It is sought to achieve the peak conversion efficiency in the lower part of its operating range since L2L products often are placed in areas with a lot of shadows on the solar panels e.g. in parks or in the urban environments, making peak power operation rare.

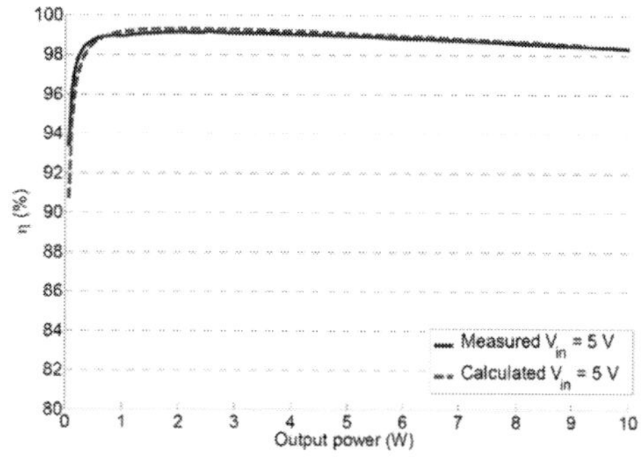

Figure 2: Efficiency curve of the 10 W converter when harvesting energy from PV panel to the battery.

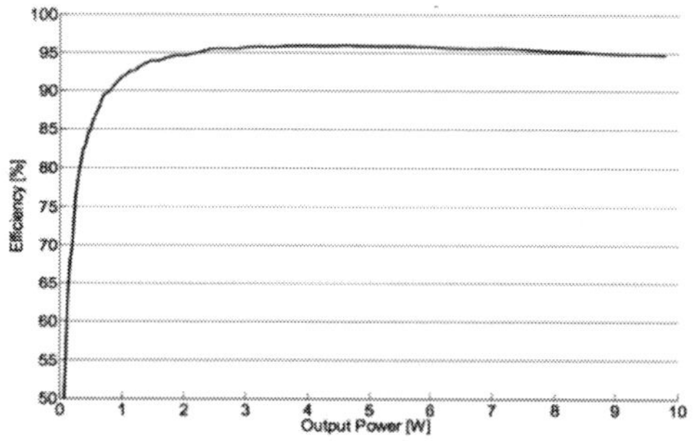

Figure 3: Efficiency curve of the LED driver stage peaking at 3W output power with 97.3%

The converters peak efficiency is 99.1 % when operating at 1.5 W. The output power and the LED driver stage present a peak efficiency of 97.3% at 3W output power. The electronics converter work is documented in [6] and [7].

3 SOFTWARE FOR DIMENSIONING AND SIMULATION OF PV LED APPLICATIONS

The authors have investigated several software tools on the market and none was found suited for the task of dimensioning and simulating performance of solar powered L2L products. A customized tool is therefore developed in MATLAB as a starting point with input from several specialized tools. The block diagram, figure 4, shows the design of the simulation tool, which essentially is a computer model of a basic PV application, including PV panel, electronics, battery and power consumption in the form of an LED luminaire. The L2L dimensioning tool differs since it uses highly time resolved weather data in 1-minute resolution. This is important when modelling solar products operating most of its time in the shadows from buildings and other artefacts. The efficiency curve, figure 2 of the developed converter is rather flat in the whole irradiation interval from 0-1 sun. The full electronic controller unit, however, consists of both converter and controller with a microcontroller operating in several states, running algorithms, and controlling other external devices as a battery management system, motions sensors etc. The real energy conversion functions are therefore usually far from linear and depend heavily on irradiation level on the PV panel, temperature, battery voltage and other parameters which need to be modelled for precision. The energy flow is basically made up of:

1. Energy flow from the sun to the PV panel
2. Energy flow from the panel – through the charger – and into the battery
3. Energy flow from the battery – through the discharger – into the LEDs
4. Stand-by consumption of the electronics controller due the intelligent disposition of the energy for lighting (the lighting scheme) and the loss in its external components (if any).

The datasheets of small solar panels, PV/LED controllers and batteries are rarely having data for this type of modelling to be done.

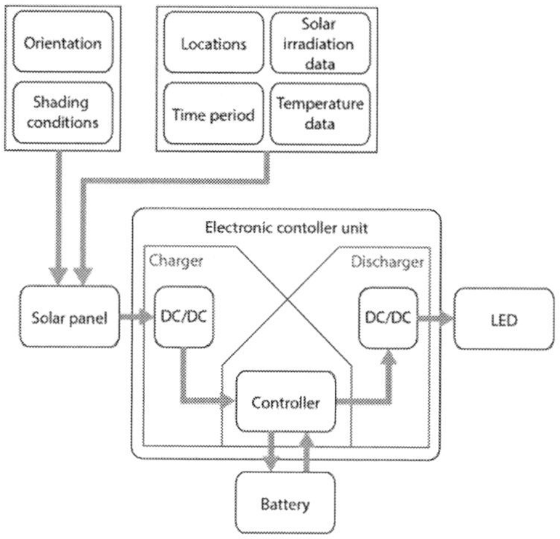

Figure 4: Block diagram of L2L modelling tool

4 LABORATORIES FOR CHARACTERIZATION AND MONITORING OF L2L PRODUCTS

There has been installed two laboratories, a specialized lab to feed the mathematical simulation tool with accurate and sufficient data for modelling of L2L products and support the development of successful solar powered LED products, and a living lab for installation and monitoring the energy flow inside the L2L products. This will be expanded in the coming years at the university campus of the Technical University of Denmark around the building of Department of Photonics Engineering in Roskilde. The modelling tool can thereby be validated against periods with known weather data where data logging has been implemented in the products. The differences between simulated performance of a product and the logged performance is due to inaccuracies in the model and is a valuable tool to optimize it. Figure 5 shows the data feed-back loop.

The specialized lab is composed of the following facilities:

Component characterization

- Solar cell electrical (I, V, Irradiation 0-1 sun AM 1.5)

- Solar panel electrical (I, V, Irradiation 0-1 sun AM 1.5)
- Battery (I, V, Charge, Temp)
- LED electrical and photometrical (I, V, lumens)
- Electronics controller
    - Charging efficiency (Emulation of the PV panel and the battery, allowing mapping of the full operating range for the charger)
    - Discharge efficiency (Vbat, I_{LED}, V_{LED})
- Shade measurement/simulation

Luminaire characterization

- Spectroradiometry
- Photometry
- Goniospectroradiometry
- Goniophotometry

The measurements are tailored to fit into the L2L modelling tool and are more or less automated to characterize the individual components in the many different operating states needed. The luminaire characterization is not part of the model but a part of the process of creating highly efficient high end L2L products. The light rays need to illuminate the desired object nicely without the light being lost in reflectors and diffusers, giving rise to glare or other waste mechanisms. Here good optical engineering can reduce the total amount of light needed, and thus improving the efficiency. The relevant metric is useful lux or useful lumens which can be modelled by use of e.g. DIALux and RELUX which in turn can be used iteratively with the L2L model.

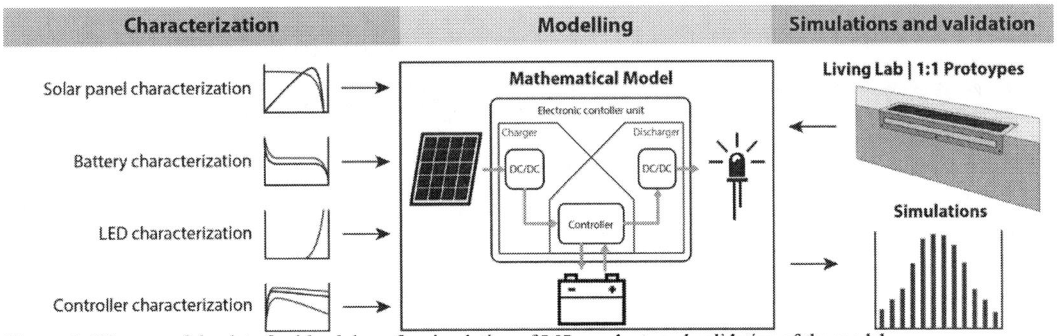

Figure 5: Diagram of the data feed-back loop for simulation of L2L products and validation of the model

5 MODEL VALIDATION

The model has been validated using the components of the "Lighttube" developed by Out-sider, which consists of a 4 W mono crystalline silicon cell, a NiMH 3.6V 4Ah battery, the control unit discussed and eight 0.3 W LEDs (Samsung LM561B Plus). For this validation weather data from the weather station at DTU RISØ Campus has been used for the simulation, alongside a logging device connected to similar components placed in the Living lab at DTU RISØ. The shading profile used in the simulation was created with a Solmetric SunEye, which is basically a fish eye camera that captures a picture of the hemisphere. It is then adjusted for the model and can be seen in figure 6.

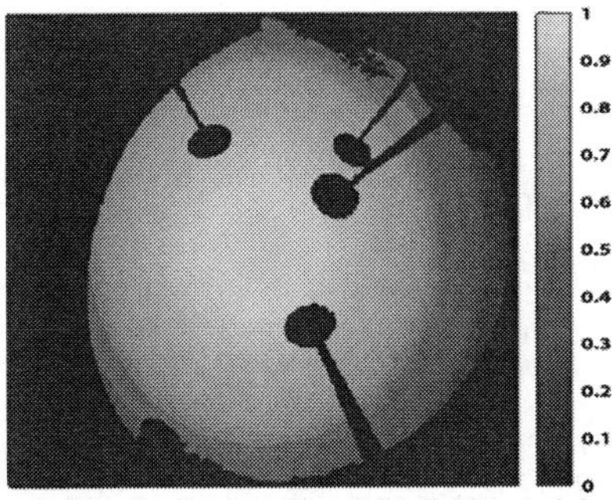

Figure 6: The shading profile of the field tested device at the Living Lab @ DTU RISØ Campus.

Since standalone systems are straight forward products they usually are not equipped with Sun tracking technology, therefore the Sun's position for all times of the year has been calculated from the SunEye's compass and the belonging software. This shade profile and the Sun's position are used to calculate the direct irradiation and the fraction of open sky, used to determine the contribution of the diffuse light, assuming that the diffuse light is isotropically distributed. The contribution of the diffuse light is calculated to be 81 % of the total diffuse light, with a fraction of open sky at 70 %, due to the angle dependency of light from normal incident. Using the shadow profile created by the fish-eye picture from the SunEye the weighting, which can be seen in figure 6 as the scale, is calculated by giving each pixel a coordinate corresponding to the distance from normal incident, i.e. the center of the picture, from which the corrected contribution of diffuse light is calculated. This could be have been done more exact by projecting the corrected contribution down onto the plane of the panel, but was not done, because of SunEye's high pixel resolution.

The models assumes that the shadows are binary, meaning that they either cover the entire panel completely or not at all, which is assumed to be an acceptable approximation for small PV panels. This approximation will not be valid for larger panels with bypass diodes. Data from four days in May and June have been used to compare the model to the device in the lab. The comparison has been made between the effects of the different components of the device, which can be seen in figure 7 with 1 minute resolution.

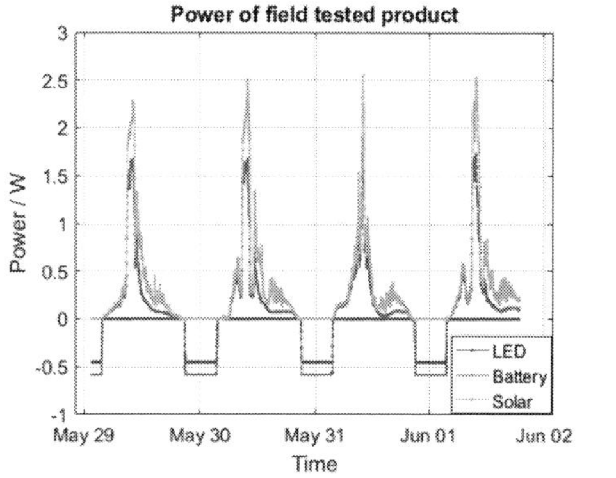

Figure 7: The power of the field tested device at the living lab @ DTU RISØ Campus

The differences in power from the PV panel to the battery and from the battery to the LED are the losses in the system, e.g. during conversion and standby. On the logged plot it can be seen that the power increases slowly when the PV panel is exposed to increasing irradiation in the early hours, where as it can be seen from the simulated plot in figure 8 that the power increases abruptly.

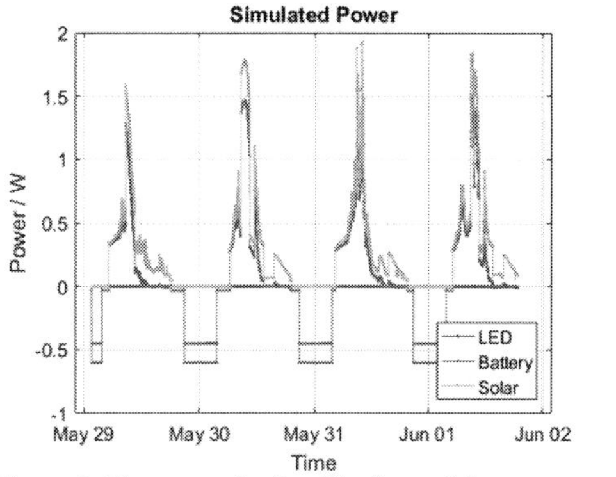

Figure 8: The power simulated by the model

This is due to the PV panel not being characterized for low levels of irradiance, which results in the current over the PV panel cutting off until the irradiance value has surpassed a certain value. This however is not of great importance since the lowest levels of irradiation do not produce very significant amounts of energy.

From figure 9, it can be seen that around midday the power on the battery drops relatively compared to the power on the solar cell, and the power on the solar cell begins to fluctuate, which is due to the battery being nearly fully charged. This consequence is not as eminent on the plot of the simulated power, which is a combination of the battery not being completely characterized, and the fact that the simulated battery is fully charged earlier than the logged battery. The characterization of the battery during charge and discharge has not been done for varying currents, resulting in the assumption; the voltage on the battery is equal to the open circuit voltage on the battery which in turn results in deviations from the field tested device when the current is either high or low relatively. This however does not play an important role in the total amount of energy accumulated. The fact that the simulated battery is fully charged earlier than the logged battery, can be explained by looking at the morning hours on figure 9. Here it can be seen that after the abrupt increase in power for the simulated solar panel, the effect is much larger than for the field tested device, i.e. the simulated battery is charged more during the early hours. This also explains why the peaks of the simulated and field tested powers are so different from 9 to 10 o'clock, since the current drops when the battery is almost fully charged. The field tested battery thus has a higher power than the simulated in the time interval from 9 to 16 o'clock, where the simulated battery has a higher power from 5 to 9 o'clock. This fallout is also believed to be caused by lacking characterization of the PV panel at low levels of irradiance, and is of much higher significance than the first missing hour by the abrupt start caused by the same lack of characterization.

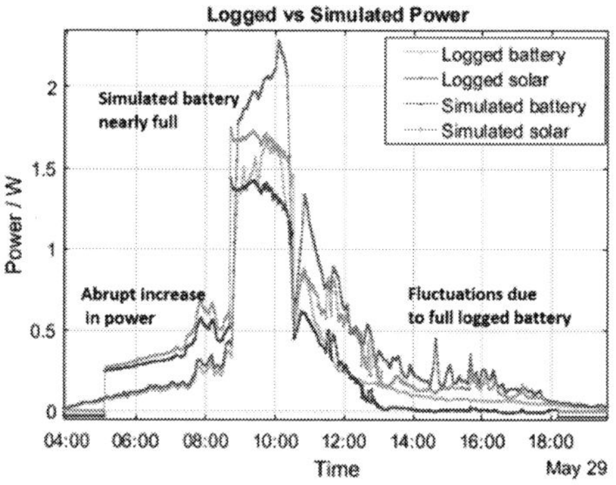

Figure 9: Comparison of the logged and simulated power on the 29.th of May

When integrating the power of the different components over the entire time span, the total change in energy is found. The relative differences in energy flow can be seen in table 1. These indicate that over a period of 4 days, the simulated values of energy harvested and consumed are very close to the logged values. However, as can be seen from figure 9, the solar power is not harvested simultaneous in the simulation and for the logged device.

Table I: Energy flow over the entire time period.

	Solar	Battery	LED
Logged	29.4 Wh	3.1 Wh	12.0 Wh
Simulated	30.2 Wh	3.3 Wh	12.6 Wh
Relative deviation	2.9 %	4.5 %	5.3 %

6 CONCLUSION

Three major challenges in the development of L2L products is the lack of efficient converter electronics, modelling tools for dimensioning and furthermore, characterization facilities to support the successful development of the products. Above we report the development of 2 Three-Port-Converters respectively for 1-10Wp and 10-50 Wp PV panels with a peak efficiency of 99.1% at 1.5 W of output power for the 10 Wp version PV to battery conversion state. Furthermore, a modelling tool for L2L products has been described and validated, but still allow for small improvements through a better characterization of the battery and of the PV panel. A laboratory for feeding in component data not available in the datasheets to the model is described. A living lab for field test of L2L product to validate the modelling tool and assisting in the development process of L2L products is also under further development.

7 REFERENCES

[1] G. Spagnuolo, G. Petrone, S. Araujo, C. C. Araújo, E. Friis-Madsen, E. Gubía, D. Hissel, M. Jasinski, W. Knapp, M. Liserre, P. Rodriguez, R. Teodorescu, and P. Zacharias, "Renewable energy operation and conversion schemes: A summary of discussions during the seminar on renewable energy systems," *IEEE Ind. Electron. Mag.*, vol. 4, no. 1, pp. 38–51, 2010.

[2] A. Al Nabulsi and R. Dhaouadi, "Efficiency optimization of a dsp-based standalone PV system using fuzzy logic and dual-MPPT control," *IEEE Trans. Ind. Informatics*, vol. 8, pp. 573–584, 2012.

[3] S. S. Thale, R. G. Wandhare, and V. Agarwal, "A Novel Reconfigurable Microgrid Architecture With Renewable Energy Sources and Storage," *IEEE Trans. Ind. Appl.*, vol. 51, no. 2, pp. 1805–1816, 2015.

[4] D. L. Meier, L. Harvest, and S. Marys, "Stand-Alone Solar Generator with LED Floodlights for Outdoor Sign Illumination," pp. 2748–2750, 2014.

[5] P. B. et al. Poulsen, "Udvikling af CO2 neutralt byrumsarmatur, Final report," 2013.

[6] M. C. Mira, A. Knott, and M. A. E. Andersen, "Three-Port Converter for Low-Power Stand-Alone Light-to-Light Systems."

[7] R. O. Ploug and A. Knott, "Photovoltaic OLED Driver for Low-Power Stand-Alone Light-to-Light Systems," no. V, 2016.

HOW SOLAR ENERGY CONNECTED TO DEVELOPMENT IN RURAL INDIA

Arkadeep Kumar
Georgia Institute of Technology
Atlanta, Georgia, USA
arkadeepkumar@gatech.edu

ABSTRACT: Solar energy is being in various ways worldwide to provide access to electricity in off-grid, especially rural areas. The current works aims to study the impact of community-led solar energy in low-resource settings in remote villages in Sunderbans, on the Bay of Bengal coast of West Bengal in India. The study analyzes how solar lanterns which were deployed to replace kerosene lamps for education became helpful in more ways than the intended outcome. The model of community-led and maintained solar power helped in finding new livelihood and income sources by solar entrepreneurs, in charge of maintaining and renting out lanterns after charging them daily. Further these lanterns helped in small businesses in the area to operate with a safer light source than fire lamp. Thus the intervention became an example of holistic approach to development of the low-resource community. Studying the results of this project might help to inform other such projects being planned in similar low-resource contexts. Further, the metrics of how solar energy intervention is connected to development can help policy makers and government institute suitable programmes.
Keywords: Developing countries, Dissemination, Education and Training, Rural Electrification, Solar lanterns

1 INTRODUCTION

Solar energy has been one of the important ways to provide access to energy in areas off the grid. More often these areas overlap with those of low-resource environments, to communities with limited financial resources. Hence due to the reason of being remote and limited finances, these communities have limited access to energy. In providing access to energy in such remote rural contexts, photovoltaic solar power has evolved to be a helpful technological intervention, which also aid development. This paper would look into how solar energy has been deployed in low-resource communities in rural India for a variety of applications, and how that has improved lives, by creating opportunities for education, better living conditions, or aided in their livelihood.

There has been some prior work on using solar technologies in developing countries and resource-constrained situations. Some work related was also found for off-grid and rural electrification. The status of off grid rural electrification in South Asia and the best practices was discussed by Palit and Chaurey [1]. Moreover, the use of specifically solar energy in rural electrification in South Asia has been reported [2]. Chaurey and Kandpal's work studied the viability of the central charging model in using solar energy for rural electrification in India [3]. Prior work focusing on solar powered LED lanterns as affordable and eco-friendly lighting system in the developing contexts was done by Ghosh et al. [4]. The introduction of these solar lighting systems could also help with the lighting for eco-tourism in the remote islands off Bay of Bengal (Sunderbans), as was discussed by Ghosh et al. [5]. Apart from lighting as an application in these low-resource contexts, other varied applications of solar power include enabling female farmers in rural South India in the state of Kerala with solar pumps [6].

With the lowering of costs over the past few years and increasing affordability of solar powered devices, there is opportunity of using solar power to provide access to energy in remote areas, and even in low-resource environments. Apart from the literature in research domain cited, there were multiple reports on use of solar power in resource constrained settings in online news sharing websites such as The Better India [7]. Some of the other initiatives which are bringing solar energy in low-resource contexts, similar to those we discuss include that of d.light [8] from the Stanford design school. Recent efforts by Menda foundation along with SELCO India [9] has aimed to enable educational aids with solar powered classroom projectors. A related initiative, which is based on a slightly different model, is Empower generation in Nepal [10]. Thus there have been a number of projects related to education and empowerment of people. Other that education and livelihood, solar power also enabled the people in these low resource contexts with updated information about weather, particularly for agriculture and fishing, when the deployment was in rural coastal areas. In the following sections we study such a project deployed in remote villages in Sunderbans in West Bengal, India.

2 BACKGROUND

Although technological developments to access electricity have made huge strides world-wide, villages in remote Sunderbans region, in coastal West Bengal off Bay of Bengal have limited access to electricity. The main occupation of the community living there is agriculture, being on the fertile delta of the river Ganges, or fishing in the seas. Owing to the remote location, infrastructure has not developed for power supply from the grid. Thus solar power makes a viable source of electricity in these areas. The community living in these areas has had to depend mostly on fossil fuels or bio-mass for their energy needs. One of the factors proving to be a hindrance to education of children was lack of light in the evenings (after 4.30 or 5pm). Lanterns using kerosene as fuel were being used. Kerosene is expensive for the socio-economic condition of the community. Moreover the burning of kerosene produced gases like CO_2 (carbon dioxide), with the danger of CO (Carbon monoxide) being formed. The light of the kerosene lanterns are often low in intensity and not adequate for children to study, straining their eyesight. Thus the lanterns posed a risk to deteriorate the health of the users. The particular NGO working in a number of villages in this area started their solar light programme as a few years ago, with the intent of providing an alternative to kerosene lamps. It was named Green Energy for Environment and Nature (GREEN). While the primary intent was to help with the education of the children, there were some other positive outcomes of the project which impacted community members like parents of these children in their household work, and small businesses like local shopkeepers and tailors. The model of renting out the lamps by local villagers who were responsible for charging and maintaining the devices gave rise to new income methods and 'solar entrepreneurs". The introduction of the solar lanterns thus provided holistic development for the community. Moreover the introduction of solar powered systems was a step towards environmental sustainability.

The initial plan started with lamps being provided in group study areas for the students (locations were after-school tutoring centers, which are called coaching centers). The NGO enabled some of the community members (7 villagers) with solar power systems – who charged the lanterns and rented them out every evening, only to get them back for charging again in the morning. The system consisted of solar panels or modules installed outside, with charging stations installed within the house. The other villagers obtained cheap and affordable solar power from them. In this paper, we discuss how the community led model helped in adoption of the technology and also the outcomes in the next few sections.

3 APPROACH

The author has been a volunteer with Asha for Education, a global fundraising organization with numerous chapters in universities worldwide. The particular location where deployment of solar lanterns took place is one such project of a NGO supported by Asha for education [11]. The author accessed project reports, and corresponded with the NGO coordinators and volunteers for the study. The goal was to study the deployment of solar lanterns, and the aspect of community-led engagement of local community. The motivation of the study was to find key take-aways for similar deployment of solar energy in low resource contexts.

4 FINDINGS

4.1 Solar power system

Each solar power system consisted of six solar panels. From the panels, the power came to the charging stations, each panel being connected to one station. Each station could provide power for ten solar lanterns. Thus each solar power system, catered to sixty solar LED lamps. These lamps were maintained by the people, called "solar entrepreneurs", who rented them out and collected a fee of INR 2/day/lamp. The lamps were rented in the late afternoon or evening and returned to the charging station for next morning where it was charged during 6am - 12 noon. The solar entrepreneurs were given training on proper handling of the lanterns. The users were educated about the technique of proper use of the lantern by the same entrepreneurs. The concerns about the upkeep and maintenance were addressed by assigning unique number to each lamp – which was recorded against the name of the person renting out the device for the day. This helped in following up with the user in case the lantern was returned in bad condition.

4.2 Stakeholders

The author found a number of stakeholders associated with the deployment, use, and sustainability of the project.

4.2.1 Students, Teachers and Households

Students who studied in school and colleges, and their parents reported that they rented lights from these 'solar entrepreneurs". The solar lamps helped the children to study in the evenings, which was earlier dependent on the kerosene lamps. The teachers who taught in the after-school coaching centers could teach better in groups, and with the better lighting intensity. The parents also reported that they were able to do their household work and be more productive in the evening with the solar lamps. Figure 1 shows one such after-school tutoring center using solar lanterns in the evening.

Figure 1. Solar lamps in an after-school coaching center helps students study, fighting erratic power supply in rural Sunderbans near the Bay of Bengal coast in West Bengal, India [Pictures from project documents in [11] http://www.ashanet.org/yale/].

4.2.2 Solar entrepreneurs

The introduction of this this technology created a new income source for some of the community members, named solar entrepreneurs. These members were given the solar panels by the NGO, which was setup on their house (rooftops or outside). They gave space for the solar charging stations and the solar lanterns inside their house. Thus the technology intervention also enabled some self-employment and generation of income for these community members. To note, there was almost equal participation of men and women of the community members in being a solar entrepreneur. Figure 2 shows solar panels outside the house [12]. Figure 3 shows the charging station with lamps.

Figure 2. Solar panels installed outside the house of the solar entrepreneur [Mukti YouTube page].

Figure 3. Solar lamps being charged through a central facility or charging station [Picture from project documents of Mukti Asha for Education page].

4.2.3 NGO coordinators

The field coordinators of the NGO reported that the solar powered lanterns had been helping the students to study at night and thus helping in the aspect of education, one of the primary goals of the NGO working in the region. The other members of the NGO explained how the area being so remote was cut off from the grid in terms of electricity from the state government, and the solar lantern has solved part of the problem about access to electricity and safer light source at night without hazards of fire or health concerns.

4.3 Costs and sustainability

The comparative cost of solar power (INR 2 per recharge) led to INR 60 per month, much lower than the reported INR 280-300 the each household spent on kerosene, when compared to their monthly income around INR 700-800.

Budget details for the project:

i. Set up cost of 1 solar charging model with 60 lanterns- INR 1,50,000/ (For this particular project 2/3 money was given from TERI or The Energy and Resource Institute of India, and 33% from Asha for Education)

ii. Project implementation and administrative charges INR 3000/-

iii. Set up maintenance charges INR 2000/-

The batteries for the solar lamps were expected to last 2-3 years, which had a high cost of replacement (approximately INR 700-800). With the model of minimal cost and providing the solar entrepreneurs as the people to maintain the lamps, the project made sure that the technology was sustainable. Further instead of owning the lamps, the model of renting with minimal fee for charging made sure that the lamps were affordable for all. Along with helping with a light source, the model also helped create income sources for some community members (solar entrepreneurs), sometimes involving women in the house, which was one of the unexpected positive outcomes.

4.4 Performance

The solar lamps were found to function even in monsoon time. During normal days with adequate sunlight, it got fully charged during 6am-12noon/1pm time and gave light for 7-8 hours. The performance dropped to 4 hours of light during the monsoon days.

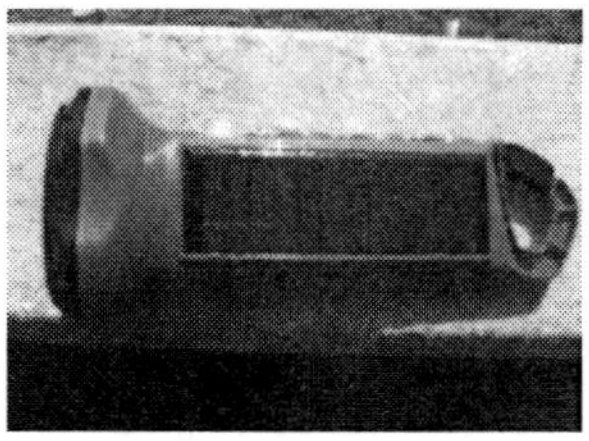

Figure 4 Recent developments include torches with integrated solar panels embedded on the device, thus decentralizing solar energy. [muktiweb.org]

4 RESULTS AND DISCUSSION

The solar lanterns have reduced the burning of kerosene to a large extent, avoiding the harmful effect of burning the fuel, both on the health of the users (children and household), as well as the environment. The light was reported to be better than the kerosene light in terms of brightness, which has reduced the concern of eyesight being strained due to studying at low light conditions. The study suggests how the introduction of the solar lanterns led to holistic development of the community. In addition to providing them with power, which aided daily chores, including education of the children, it also taught the lessons of sustainability and renewable energy. In a community with such low-resources, the concept of sharing and using the technology instead of owning them helped in the technology being affordable, to drive adoption. Also the system gave rise to a new form of livelihood among the community members who called themselves "solar entrepreneurs", who were responsible for upkeep of the whole system including the solar lanterns. Thus we find there were affordances of the solar lanterns- or use in other forms apart from intended use, with this example of deployment.

The development projects in rural India sometimes depend upon access to electricity. For example, interventions using mobile based agricultural advice [13], or maternal health services [14], and other information and communication technologies for development (ICTD) depend on energy supply to charge mobile phones. Solar power delivered to such off-grid remote areas could enable such ICTD deployment.

5 CONCLUSIONS

Solar lanterns were introduced in remote areas of Sundarbans in West Bengal with a goal of providing safer lighting systems for students to study in the evening. The model included a community-led effort of maintaining the solar power system and recharging the lamps daily. Apart from the intended outcome of helping with the education in such low –resource contexts, the solar lanterns also gave rise to better livelihood of the users, in doing their household work and small businesses operating in the evenings. The model of having community members who maintained and rented out the devices gave rise to a new form of income for some community members who called themselves solar entrepreneurs. The model became sustainable and replicated in neighboring villages Thus, there was a holistic development model being brought to these areas. This project provides insight into affordances of a technological intervention in low resource contexts. The solar lantern project here provides insights into future deployment of similar systems in low resource contexts.

5 ACKNOWLEDGMENTS

The author is grateful to Asha for Education, and team Mukti for support of this work through the documentation of this project and the images.

6 REFERENCES

[1] D. Palit and A. Chaurey, "Off-grid rural electrification experiences from South Asia: Status and best practices," *Energy for Sustainable Development*, vol. 15, pp. 266-276, 9// 2011.

[2] D. Palit, "Solar energy programs for rural electrification: Experiences and lessons from South Asia," *Energy for Sustainable Development*, vol. 17, pp. 270-279, 6// 2013.

[3] A. Chaurey and T. C. Kandpal, "Solar lanterns for domestic lighting in India: Viability of central charging station model," *Energy Policy*, vol. 37, pp. 4910-4918, 11// 2009.

[4] B. Ghosh, K. Ghosh, and N. Cowern, "Solar Powered LED Lantern: An Eco-Friendly Affordable Lighting System for the Developing World," *25th European Photovoltaic Solar Energy Conference and Exhibition / 5th World Conference on Photovoltaic Energy Conversion, 6-10 September 2010, Valencia, Spain,* pp. 4939 - 4941, 2010.

[5] B. Ghosh, S. Bardhan, and K. Hazra, "Solar PV Electricity : The Input for Ecotourism," *26th European Photovoltaic Solar Energy Conference and Exhibition*, pp. 4061 - 4063, 2011.

[6] E. Schüpbach, U. Muntwyler, A. Vezzini, A. Müller, and D. Urena, "Introducing Solar Water Pumps to Female Farmers in India," *29th European Photovoltaic Solar Energy Conference and Exhibition, Amsterdam, Netherlands,* pp. 4106 - 4108, 2014.

[7] T. B. India, "http://www.thebetterindia.com," *The Better India*.

[8] "www.dlight.com," *D.light*.

[9] "http://www.mendafoundation.com," *Menda Foundation and SELCO*.

[10] "http://empowergeneration.org/," *Empower Generation*.

[11] "Mukti page for Asha for education https://www.ashanet.org/projects/project-view.php?p=671, http://www.ashanet.org/projects/project-view.php?p=1099," *Asha for Education*.

[12] Mukti, "muktiweb.org."

[13] R. Gandhi, R. Veeraraghavan, K. Toyama, and V. Ramprasad, "Digital green: Participatory video for agricultural extension," in *Information and Communication Technologies and Development, 2007. ICTD 2007. International Conference on*, 2007, pp. 1-10.

[14] N. Kumar, T. Perrier, M. Desmond, K. Israel-Ballard, V. Kumar, S. Mahapatra, *et al.*, "Projecting health: Community-led video education for maternal health," in *Proceedings of the Seventh International Conference on Information and Communication Technologies and Development*, 2015, p. 17.

TRUE COMPETITIVENESS OF SOLAR PV – A EUROPEAN CASE STUDY

Eero Vartiainen
Fortum Power and Heat Oy
POB 1, FI-00048 Fortum, Finland, e-mail: eero.vartiainen@fortum.com

Gaëtan Masson
Becquerel Institute
Rue d'Arlon 69-71, Brussels 1040, Belgium, e-mail: g.masson@becquerelinstitute.org

Christian Breyer
Lappeenranta University of Technology
Skinnarilankatu 34, FI-53850 Lappeenranta, Finland, e-mail: christian.breyer@lut.fi

ABSTRACT: Photovoltaics (PV) has become one of the cheapest forms of electricity production during the past few years. The Levelised Cost of Electricity (LCOE) of PV has already reached grid parity with retail electricity price in many markets. However, the concept of grid parity is not always well defined. In this paper, the retail electricity price includes only the variable (per kWh) components in the customer bill since the fixed monthly or annual fees cannot be saved by the prosumer. Furthermore, the ratio of self-consumption of own production has to be taken into consideration. Self-consumption is valued for the variable electricity retail price, whereas the surplus production can often be valued only for the wholesale electricity price or part of it. True grid parity or competitiveness has already been reached in many European countries like Germany and Italy with all realistic Weighted Average Costs of Capital (WACC) and self-consumption ratios. In moderate irradiation and low electricity price markets like Finland and Sweden, true grid parity in rooftop segments will be reached most probably within 3-10 years without any subsidies. The work for this paper has been carried out under the framework of the EU PV Technology and Innovation Platform.
Keywords: LCOE, electricity price, economic analysis, prosumer, PV self-consumption, grid parity

1 INTRODUCTION

Photovoltaics (PV) has become one of the cheapest forms of electricity production during the past few years. Cost of PV generation has already reached grid parity with retail electricity price in many markets. However, the concept of grid parity is not always well defined. The purpose of this paper is to investigate what is the true competitiveness of PV with retail electricity price, i.e., comparing PV cost with only the variable part of the retail electricity price in the customer bill. The fixed annual or monthly fees the customer has to pay cannot usually be saved by self-consumption, and must therefore be excluded from the comparison.

Moreover, the aim is to analyse how the ratio of self-consumption of PV production affects the true grid parity. Self-consumption is valued for the variable retail electricity price, but surplus production can only be valued for the price which is paid to the customer when electricity is fed into the grid. This price is often the electricity wholesale price minus a small administrative fee. As the end result of the comparison, the time of true grid parity is determined for three market segments in ten European countries.

In order to make fair comparisons with electricity prices, the concept of Levelised Cost of Electricity (LCOE) is often used. The LCOE here is defined as the average generation cost, i.e., including all the costs involved in supplying PV at the point of connection to the grid. Possible grid integration cost is excluded here, but it can be argued whether it is fair to burden such costs solely on PV. After all, the old inflexible baseload generation technologies like coal and nuclear power do not have to pay grid integration cost either. On the other hand, this study do not take into account the various societal and environmental benefits of PV.

The PV LCOE here includes all the costs and profit margins of the whole value chain including manufacturing, installation, project development, operation and maintenance (O&M), inverter replacement, dismantling, etc. Residual value of the PV system after dismantling has not been considered since there is no agreed price for the second hand modules. PV LCOE also includes the cost of financing but excludes the profit margin of electricity sales and thus represents the generation cost, not the electricity selling price which can vary depending on the market situation.

The PV LCOE, expressed in €/kWh in real money, can be defined by equation (1):

$$LCOE = (CAPEX + \Sigma\,[OPEX(t)\,/\,(1+WACC_{nom})^t])\,/\,\Sigma\,[Util_0 \cdot (1-Degr)^t\,/\,(1+WACC_{real})^t] \quad (1)$$

where t is year number ranging from 1 to the economic lifetime of the system; $CAPEX$ is total investment expenditure of the system, made at $t = 0$ in €/kWp; $OPEX(t)$ is operation and maintenance expenditure in year t in €/kWp; $WACC_{nom}$ is nominal weighted average cost of capital per annum; $WACC_{real}$ is real weighted average cost of capital per annum; $Util_0$ is initial annual utilisation in year 0 in hours without degradation; $Degr$ is annual degradation of the nominal power of the system per annum; and

$$WACC_{real} = [(1 + WACC_{nom})\,/\,(1 + Infl)] - 1 \quad (2)$$

where $Infl$ is the annual inflation rate.

Average value of PV electricity P_{ave} in this paper is defined by the equation:

$$P_{ave} = SC * P_{retail} + (1 - SC) * P_{surplus} \quad (3)$$

where SC is the ratio of self-consumption of the PV production; P_{retail} is variable retail electricity price; and $P_{surplus}$ is the value of surplus electricity fed into the grid. P_{retail} excludes any fixed monthly or annual fees in the customer bill which cannot be saved by self-consumption. In the absence of electricity storage, self-consumption is the amount of PV generation which can directly be used by the consumer to reduce the electricity taken from the grid.

2 PV LCOE INPUT PARAMETERS

2.1 WACC and inflation

The input data for the PV LCOE calculation follows the example of previous papers by the authors [1, 2]. In those, it was concluded that apart from the location, Weighted Average Cost of Capital (WACC) is the most important parameter affecting the PV LCOE. Since there is no fuel cost related to PV, capital expenditure (CAPEX) has usually relatively bigger influence than operational expenditure (OPEX) on PV LCOE.

In this paper, all results are given in real 2015 money. As we are using nominal WACC rates, inflation has to be taken into account in order to arrive at real values. For example, a 4% nominal WACC with 2% inflation rate corresponds to a 2% real WACC. Because the WACC rates are highly subjective and depend among other things on the country, market segment, investor type and risk appetite, we have included a set of 4 different WACC rates in our analysis. For residential PV systems (5 kWp), 0%, 2%, 4% and 6% nominal WACC is used. For commercial (50 kWp) and industrial systems (1 MWp), 2%, 4%, 7% and 10% nominal WACC is used. Inflation rate is set at 2% which is the historical average inflation of Euro area.

2.2 Capital expenditure

The capital investment of a PV system can be divided into two components: the PV modules and the Balance of System (BoS). The BoS includes e.g. mounting structures, cabling, inverters, transformers and other electrical components, grid connection, infrastructure, installation work, planning, documentation and other work. CAPEX in this paper is the all-inclusive turnkey PV system price that needs to be paid up front. It is assumed here that the CAPEX is paid in full during the year of installation of the system and the system starts producing electricity from the next year.

PV module price here is assumed to follow the learning curve which has been observed for many decades. Each time the global cumulatively produced volume has doubled, the average price has been reduced by at least 20% [3-5]. During the last 8 years, the price has decreased significantly faster [6], but a conservative 20% learning rate is assumed here.

In order to establish the future price for PV modules according to the learning curve, a projection for global cumulative installation volumes is needed. For 2016-20, the medium scenario of SolarPower Europe's Global Market Outlook 2016-20 [7] is used, giving a compound annual growth rate (CAGR) of about 15%. From 2020 to 2030, 10% CAGR is assumed for the annual market size. After 2030, a CAGR of 2% is assumed which corresponds to about the average growth of global electricity consumption by IEA [8] in their 4DS scenario. It can be noted that a comparison of this base scenario to the rather extreme slow (CAGR 0%) and high (CAGR 15%) growth scenarios from 2020 to 2030 would lead to a maximum difference of only about one doubling in the cumulative volume in 2030, meaning that the module price uncertainty from the volume is within +/-20% [1].

The price development of solar inverters is assumed to follow the same learning curve as the PV modules [9]. For the price development of other Balance of System (BoS) components, the methodology introduced by Fraunhofer ISE [4] is used. About half of the BoS cost components are proportional to the total area of the modules, and thus, inversely proportional to the module efficiency [2]. In addition, there is other cost reduction potential driven by, e.g., standardisation and modularisation, increase in DC voltage and more efficient installation processes.

The average module efficiency was about 16% in 2015 and the development is assumed to follow the average 0.4 percentage point annual increase of the past decade [10]. Fraunhofer ISE report [4] gives the 2015 average BoS price for a 1 MWp industrial system in Germany. It is true that the BoS price is currently not the same for all countries, due to local conditions related to e.g. grid connection, labour cost or higher profit margins. However, it is assumed that the prices will converge over time in Europe and therefore, only one price for each market segment is used here. A sensitivity analysis with a +/-20% CAPEX variation is presented in Section 5.

For the residential and commercial rooftop market segments, the BoS price is naturally higher. Based on the data gathered by German Solar Energy Association (BSW-Solar) [10,11] and current PV module prices in Europe [12], it can be estimated that the BoS price for a commercial (50 kWp) rooftop system is about 50% and for a residential (5 kWp) system about 120% higher than for an industrial (1 MWp) system. Fig. 1 shows the CAPEX price development for the various market segments used in this study.

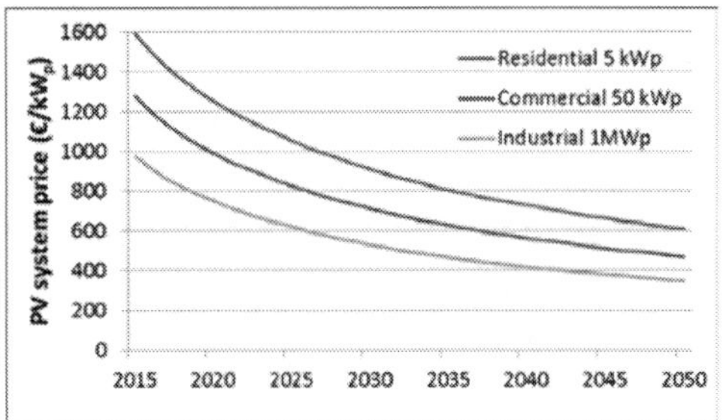

Figure 1: PV system CAPEX development 2015-50 for the three market segments. Prices are in real 2015 money and without VAT.

2.3 Operational expenditure

The operational expenditure (OPEX) of PV systems consists mainly of the operation and maintenance (O&M) cost because there is no fuel cost related to PV electricity generation. Historical OPEX data from different countries varies greatly and it is difficult to find a consensus opinion. In the past, many European countries had a very high feed-in tariff (FiT) which allowed high margins in both the system CAPEX and OPEX. Recently, the increased competition has clearly reduced the price of OPEX [13,14]. For example in Germany, the O&M price for large systems is already below 10 €/kWp/year [15]. Total OPEX price of 20 €/kWp/year is used here for 2015 for all three market segments [2]. It is assumed that 50% of the OPEX is area-dependent, and thus reducing with the efficiency improvement of the modules. By 2050, this will lead to a 30% reduction of the OPEX. It is also assumed that standardisation, more efficient processes and competition will result in a further 30% reduction of the OPEX by 2050 compared with 2015. Fig. 2 shows the OPEX price development for 2015-50.

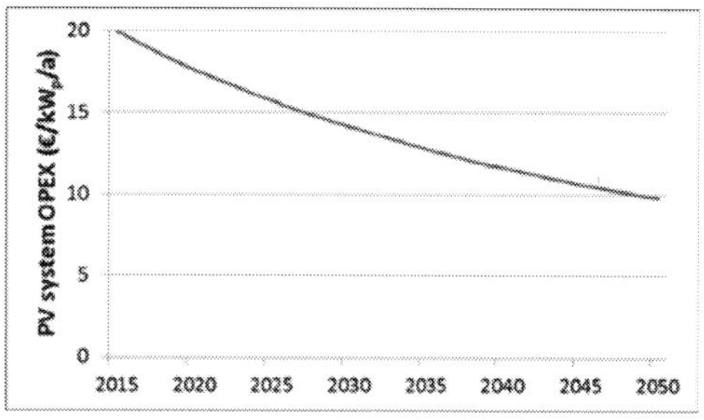

Figure 2: PV OPEX price development for 2015-50. Prices are in real 2015 money.

2.4 Yield, degradation and system lifetime

The annual yield of a PV system depends on the local irradiation and performance ratio (PR). Ten European locations were selected for this study: Helsinki, Stockholm, London, Berlin, Amsterdam, Brussels, Paris, Rome, Istanbul and Madrid, representing the most populous cities in each country.

The irradiation values are given according to Solargis database averages for 20 years [16]. The average PR in 2015 was assumed to be 0.775 in the warmest countries (Spain, Turkey and Italy) and 0.8 in the other locations. The main reason for the difference is the negative temperature coefficient of the majority of PV modules, which means that the average operating efficiency of modules is lower in warmer climates.

The average PR of PV systems has been increasing over the years. A study by Fraunhofer ISE [17] showed that the median PR of the monitored German systems increased by more than 5 percentage points during the first decade of this century. This trend is expected to continue because of, e.g., more efficient inverters, less ohmic losses with higher voltage modules, better temperature coefficients and better low light response of the modules. It is assumed here that the PR of all systems will increase by 0.5 percentage points per year from 2015 to 2030 and then will remain the same from 2030 to 2050. The annual irradiation is assumed to be stable. Table I shows the annual yields for the chosen locations with GHI and irradiation for a surface tilted 30^0 towards South, which gives almost the maximum annual yield for all locations. The annual yields are calculated for the tilted surface with the given performance ratios.

Table I: Annual GHI, irradiation on surface tilted 30° to South, and yield for different locations in 2015 and 2030; source for irradiation data: Solargis, 2016

	Irradiation (kWh/m^2)		Yield (kWh/kWp/a)	
	GHI	30°S	2015	2030
Stockholm	950	1160	930	1020
Helsinki	950	1160	930	1020
Amsterdam	1030	1200	960	1050
Paris	1130	1310	1050	1150
Brussels	1030	1200	960	1050
Istanbul	1500	1680	1300	1430
London	1000	1160	930	1020
Berlin	1030	1200	960	1050
Madrid	1710	2000	1550	1700
Rome	1580	1830	1420	1560

Note that there are several locations with similar yields, but the electricity price in these countries may be different, and therefore, the competitiveness of PV varies. The annual yields in Table I are initial values, i.e., without any degradation. The power guarantees (80% of nominal power after 25 years) of most PV module manufacturers would mean maximum average degradation of 0.9% per year. In reality, most systems in Europe degrade far less and, e.g., an average degradation of 0.2% per year has been reported for German rooftop systems [18]. A conservative value of 0.5% per year is used here for properly installed PV systems, based on the findings of IEA PVPS Task 13 recent report [19].

Clearly, system lifetime is related to the degradation of the system. A system lifetime of 30 years was recommended by IEA PVPS Task 12 for life cycle assessment studies [19] and reflects the quality of current PV systems, even though it is expected that the technical lifetime will increase in the future and give added financial, environmental and social benefits.

3 VALUE OF PV ELECTRICITY

3.1 Value of self-consumption

PV LCOE is compared with the average value of generated PV electricity for the three market segments in ten countries. The self-consumed PV generation is valued for the variable retail electricity price the customer has to pay for each consumed kWh in the electricity bill. Eurostat [20] gathers the electricity price data from all European countries and publishes the annual averages in several consumption bands. The consumption bands selected for the price comparison with different sizes of PV systems are listed in Table II.

Table II: PV system sizes and consumption bands where PV LCOE is compared with the average electricity value

PV system size	Annual consumption
Residential 5 kWp	5-15 MWh
Commercial 50 kWp	20-500 MWh
Industrial 1 MWp	2-20 GWh

In the residential segment, annual 10 MWh consumption would give an average 1.1 kW power load in the house. Considering that many household activities do not happen during the midday hours when the PV generation is the highest, it is clear that significant surplus generation will accumulate. In this study, we will consider a range from 25% to 75% self-consumption for the residential PV systems. With proper sizing of the PV system, 50% self-consumption can often be reached, while higher self-consumption ratios may require shifting part of the consumption to the hours of PV generation, or storing the surplus electricity.

In the commercial segment, most of the consumption happens during daytime and the best PV generation hours. Annual 250 MWh consumption would give an average 29 kW load which is close to the maximum generation power of the PV system. It is likely that most of the PV generation can be self-consumed and even all at the higher end of the consumption band. A range from 50% to 100% self-consumption is considered here. Because of the higher return expectations in the commercial segment, it is not likely that PV systems with lower than 50% self-consumption will be installed.

In the industrial segment, 10 GWh annual consumption would give an average 1.1 MW power load. This means that e.g. in the case of continuous manufacturing industry, all PV generation could be self-consumed in practice. However, at the lower end of the consumption band, some surplus generation will accumulate. We consider 50% to 100% self-consumption also in the industrial segment.

Eurostat divides the retail electricity price into three different components: energy, grid cost, and taxes and other fees. In most cases, the energy price and taxes are almost entirely variable, i.e., the price is charged per kWh. However, this is not the case with the grid cost. There is often a fixed annual, monthly or daily fee and also a per kW charge which depends on the maximum power or current that can be withdrawn from the grid. These fixed and power charges the customer cannot avoid with own PV generation, they have to be paid anyway. Therefore, the fixed charges have to be deducted from the grid cost component when comparing the retail electricity prices with PV LCOE. In some cases, it might be possible to shave off peak loads with PV or electricity storage, and thus reduce the power charge, but that is not considered here.

Unfortunately, Eurostat does not give separately the variable and fixed costs in the annual statistics. There was a recent study ordered by the European Commission [21], which has tables on the fixed and variable share of the distribution tariff in various countries. However, it does not cover all the focus countries of this study and the consumption bands selected do not match the ones used here. Therefore, a survey of the distribution tariffs of the biggest grid companies was made for each country and the results are used here. In some markets, there are big differences between companies and it is very difficult to establish an average. Another problem is that the tariff structure can vary with different maximum power consumption, fuse size or even by contract. In this study, 10 kW maximum power consumption was used for the residential, 100 kW for the commercial and 2 MW for the industrial segment.

It must be noted that the division to different market segments varies by country or sometimes by region or even by customer. Therefore, it is not always obvious which consumption band and which distribution tariff should be used. For that reason, the average share of the fixed component in the grid cost, given by Table III, must be taken as only indicative. It is also clear that the balance between variable and fixed or power-related grid costs is changing because of the increasing amount of grid-connected distributed renewable power systems which reduce the revenue for the distribution companies. There is a trend to increase the fixed or power-related component, even towards 100% of the grid cost. How far this can go is still an open question, as the consequences to energy efficiency measures should be carefully considered.

It is interesting to note that the share of fixed grid cost in residential segment is already 100% in the Netherlands, whereas it is very small in the commercial and industrial segments. Apart from the Netherlands, the same average share is used for all segments in each country, although differences do exist as explained. The effect of a different fixed share can be seen in the sensitivity analysis.

Table III: The share of fixed and power-related components in the grid cost in each country

Country	Fixed share
Sweden	60%
Finland	30%
Netherlands	10% / 100% (residential)
France	35%
Belgium	30%
Turkey	0%
UK	10%
Germany	10%
Spain	60%
Italy	40%

Fig. 3 shows the average residential variable electricity retail price in each country. Note that the estimated fixed component according to Table III has been deducted from the grid cost, as is the VAT on the fixed cost from the taxes and fees. It is interesting to see that grid cost is, apart from Belgium, relatively small, typically 20% or less of the total variable retail price. This means that even though the fixed share might increase in the future, it does not significantly reduce the competitiveness of PV since the PV LCOE is continuing to decrease.

In Italy, there still exist a progressive tariff, which means that the more electricity you consume, more you pay. However, this is going to change in 2018, which will reduce the competitiveness of small PV systems with low electricity consumption. Another notable observation is that the variable retail price in Germany is more than double of that in Sweden and Finland. This has a profound effect on the PV competitiveness since the level of annual yield is quite similar in these countries.

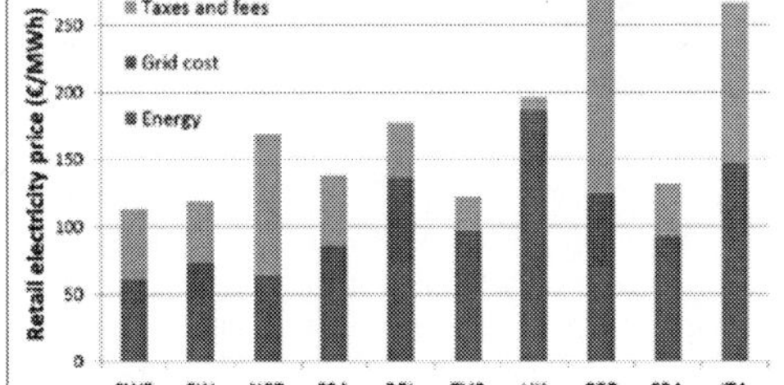

Figure 3: Average residential variable retail electricity price in Europe 2015 (source: Eurostat). Fixed/power-related component has been deducted from the grid cost and taxes. Self-consumption tax of 44.5 €/MWh has been deducted from the energy price in Spain.

Fig. 4 shows the average commercial variable electricity retail price in each country. Notable exceptions are the self-consumption taxes in Spain and Italy which have been deducted from the Eurostat energy price.

Moreover, 40% of the EEG charge has been deducted from the German taxes and fees column.

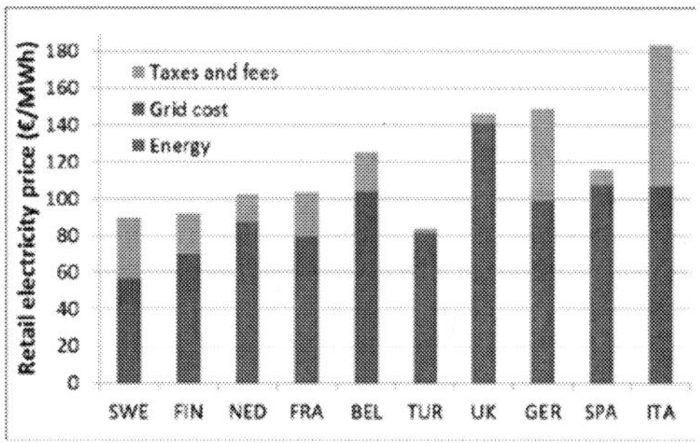

Figure 4: Average commercial variable retail electricity price in Europe 2014 (source: Eurostat). Fixed/power-related component has been deducted from the grid cost. Self-consumption tax in Spain (22 €/MWh) and Italy (5%) has been deducted from the energy price. 40% of the EEG charge in Germany has been deducted from the taxes and fees.

Fig. 5 shows the average industrial variable electricity retail price in each country. The same exceptions apply as in the commercial segment.

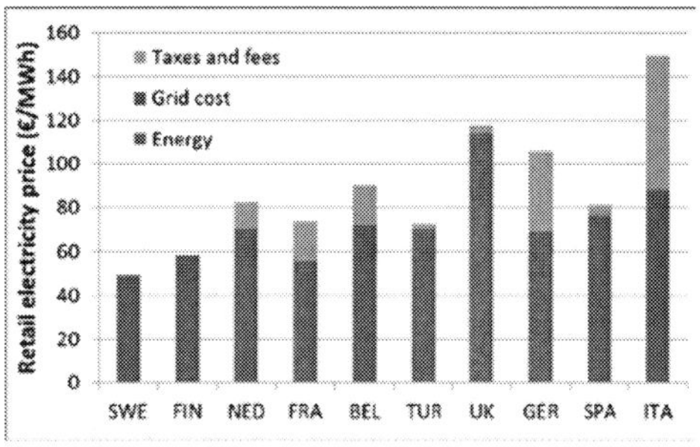

Figure 5: Average industrial variable retail electricity price in Europe 2014 (source: Eurostat). Fixed/power-related component has been deducted from the grid cost. Self-consumption tax in Spain (13 €/MWh) and Italy (5%) has been deducted from the energy price. 40% of the EEG charge in Germany has been deducted from the taxes and fees.

3.2 Value of surplus electricity

The value of surplus PV generation which cannot be self-consumed immediately depends on many things. In many countries, there still exist feed-in tariffs, or similar schemes as green certificates, which usually guarantee a relatively good remuneration for the surplus PV generation. However, the feed-in tariff era is coming to an end as PV generation is already competitive in many markets and subsidies are not needed anymore. Moreover, the European Union has issued directives pushing to replace progressively feed-in tariffs by market-based instruments, at least for the larger systems. For this reason, feed-in tariffs are not taken into account here. There are also other forms of incentives, such as net metering and billing, which essentially means a feed-in tariff at the retail electricity price, investment grants, income tax deductions or green electricity certificates. Since these incentives can also be regarded as temporary solutions, they are neither included in this analysis.

In general, the surplus PV generation should be sold in the wholesale market where the value of electricity changes over time. The most common market place is the electricity day-ahead and intraday market where the price changes hourly or even with shorter time period like every 15 minutes in Germany. Since most electricity consumers are not market players themselves, they usually have to pay a small administrative fee to a market player which can be the electricity retail company or an aggregator for selling the electricity to the market. There are sometimes also grid fees related to the fed-in electricity. For example in Finland, the distribution companies can charge a fee up to 0.7 €/MWh from customers with systems below 1 MW. On the other hand, there is a grid benefit payment in Sweden where the customer receives about 5 €/MWh for the electricity fed in to the grid.

In this analysis, the fed-in surplus PV generation is valued at the annual average spot market price minus a 10% administrative fee. The average spot market electricity prices in 2015 are presented in Fig. 6. In recent years, these prices have been going down and are already at a level which will not alone encourage investments to any new generation capacity. It is possible that the prices continue to fall, especially during daytime when a lot of surplus PV generation is available. However, the average volume-weighted price that PV generation gets is still above the annual average prices in most markets. It can also be questioned, what would be the fair remuneration for PV if its obvious societal and environmental benefits were also taken into account. For example, Sinha et al. [22] concluded that the cost of non-use of renewable electricity, mainly solar PV, would be in the range of 49-390 €/MWh, comprising grid stabilising, financial fuel hedging, value of avoided capacity etc.

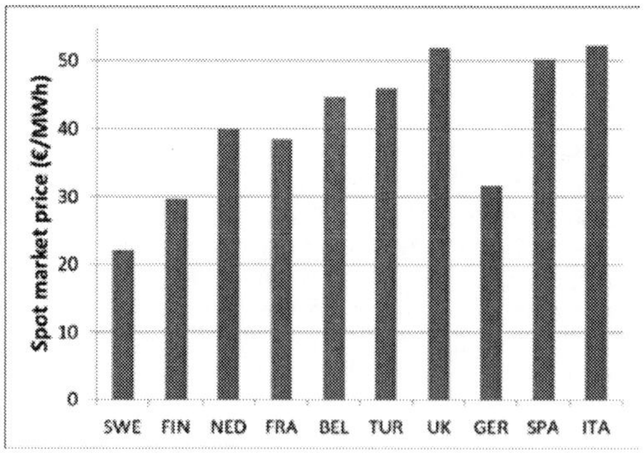

Figure 6: Average spot market electricity price in 2015 in ten European countries.

There is currently one notable exception regarding the value of the surplus PV generation. In Spain, grid feed-in is not remunerated at all for smaller than 100 kWp systems. In this analysis, it means that the value of surplus PV generation for residential and commercial market segments in Spain is zero. In the industrial segment, the 6% electricity tax has to be deducted from the wholesale market price in Spain.

4 RESULTS

4.1 Residential 5 kWp PV system

Fig. 7 shows the PV competiveness for a 5 kWp residential system in Sweden. PV LCOE with 0% nominal WACC (blue+red columns) was about 85 €/MWh in 2015. Average value of PV electricity P_{ave} with 50% self-consumption (black solid line) is about 70 €/MWh in real 2015 money, which means that true grid parity in this case was not yet reached in 2015. However, with 75% self-consumption (black dashed line), P_{ave} is about 90 €/MWh, which means that true grid parity was reached already in 2015 with 0% nominal WACC. With 2% nominal WACC (green column added), true grid parity with 75% self-consumption is reached by 2018 when PV LCOE falls to 90 €/MWh. With 4% nominal WACC (brown column added) this would happen in 2023 and with 6% (yellow column added) in 2028. With 25% self-consumption (black dotted line), P_{ave} is less than 50 €/MWh which means that true grid parity is not reached before 2030 even with 0% WACC. This shows that with low spot market prices and in the absence of feed-in tariffs, proper sizing of the PV system is imperative in order not to generate too much surplus.

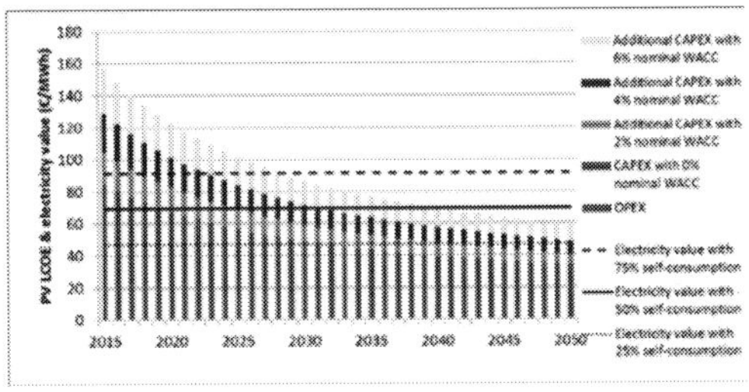

Figure 7: Comparison of PV LCOE with average residential PV electricity value for a 5 kWp rooftop PV system in Stockholm, Sweden. Prices are in 2015 real money.

Figures 8-16 show the PV competiveness for a 5 kWp residential system in Helsinki, Amsterdam, Paris, Brussels, Istanbul, London, Berlin, Madrid and Rome, respectively. It is evident that a country like Italy with good irradiation and high retail prices is already in true grid parity with all realistic interest rates and self-consumption rates. Germany is also in true grid parity apart from the 25% self-consumption rate with higher interest rates. The same applies to UK which has a lower electricity retail price but higher spot market price than Germany.

Turkey has excellent solar resource but the retail electricity price is very low, which means that true grid parity with low self-consumption and high interest rates is a few years away. In Spain, a relatively high self-consumption rate is a must because of the zero compensation for the surplus electricity. Belgium and the Netherlands have relatively high retail prices but moderate irradiation. France has slightly lower retail price but higher irradiation. Finland is similar to Sweden apart from the slightly higher spot market price.

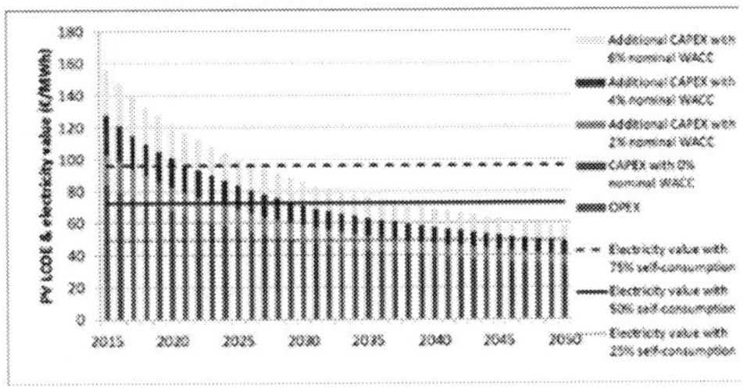

Figure 8: Comparison of PV LCOE with average residential PV electricity value for a 5 kWp rooftop PV system in Helsinki, Finland. Prices are in 2015 real money.

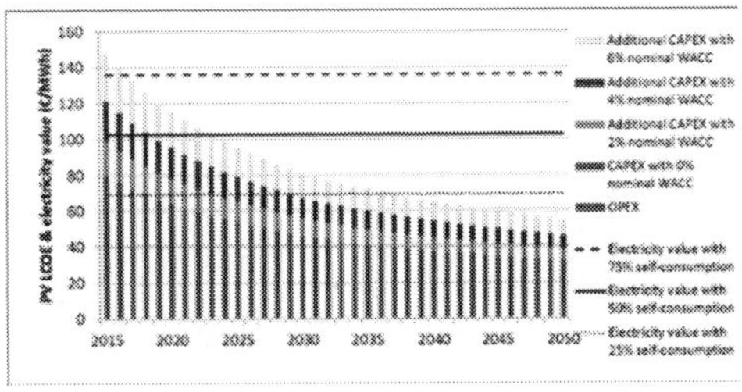

Figure 9: Comparison of PV LCOE with average residential PV electricity value for a 5 kWp rooftop PV system in Amsterdam, Netherlands. Prices are in 2015 real money.

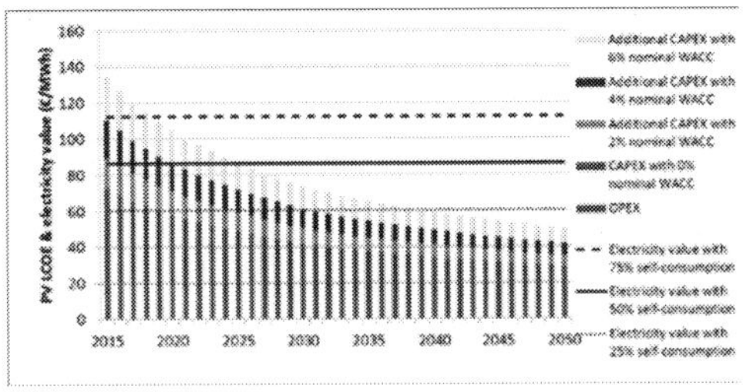

Figure 10: Comparison of PV LCOE with average residential PV electricity value for a 5 kWp rooftop PV system in Paris, France. Prices are in 2015 real money.

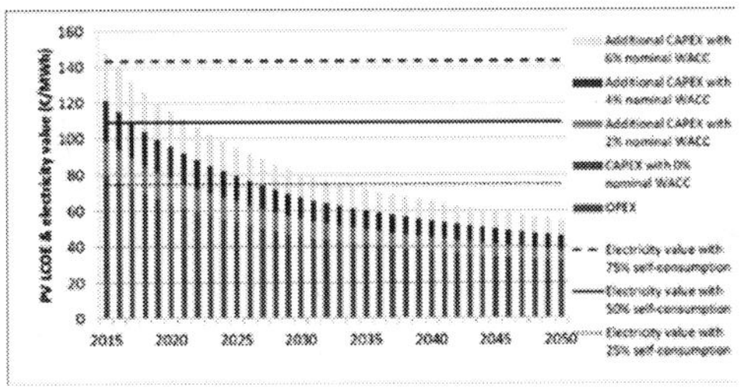

Figure 11: Comparison of PV LCOE with average residential PV electricity value for a 5 kWp rooftop PV system in Brussels, Belgium. Prices are in 2015 real money.

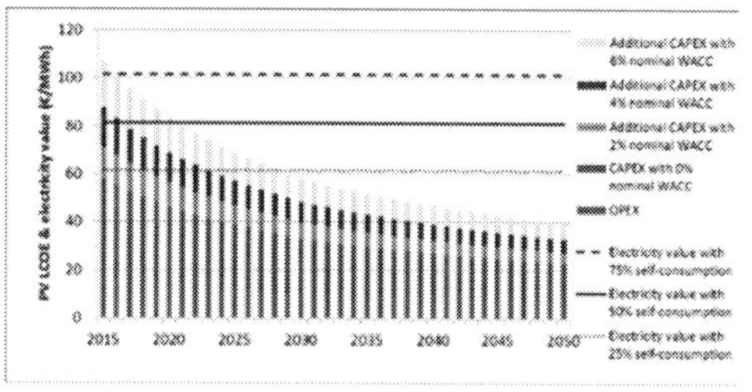

Figure 12: Comparison of PV LCOE with average residential PV electricity value for a 5 kWp rooftop PV system in Istanbul, Turkey. Prices are in 2015 real money.

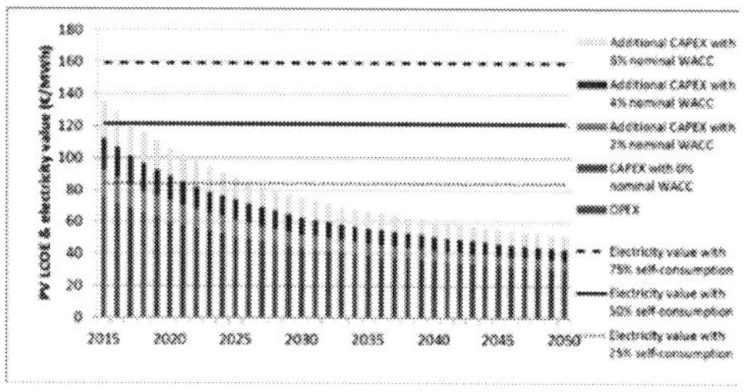

Figure 13: Comparison of PV LCOE with average residential PV electricity value for a 5 kWp rooftop PV system in London, UK. Prices are in 2015 real money.

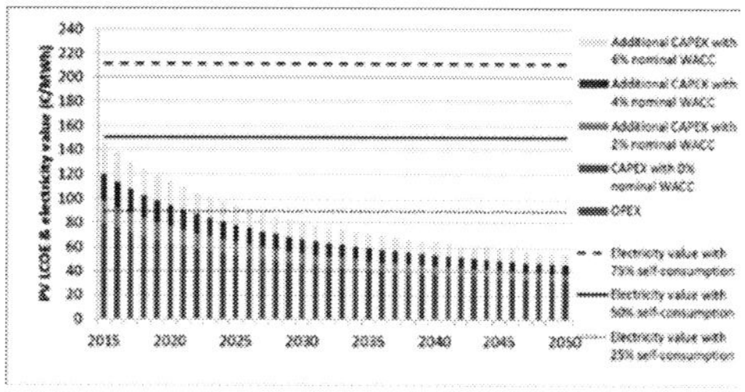

Figure 14: Comparison of PV LCOE with average residential PV electricity value for a 5 kWp rooftop PV system in Berlin, Germany. Prices are in 2015 real money.

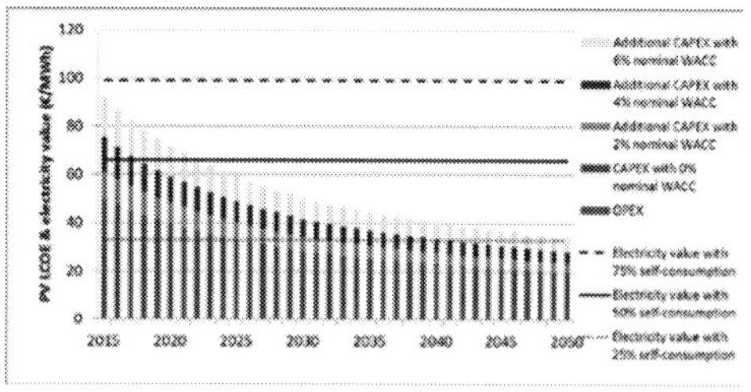

Figure 15: Comparison of PV LCOE with average residential PV electricity value for a 5 kWp rooftop PV system in Madrid, Spain. Prices are in 2015 real money.

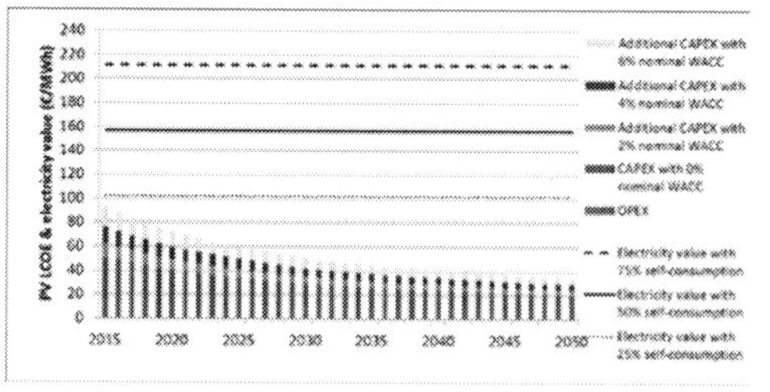

Figure 16: Comparison of PV LCOE with average residential PV electricity value for a 5 kWp rooftop PV system in Rome, Italy. Prices are in 2015 real money.

Table IV summarises the years when true grid parity is reached in the residential segment with 50% self-consumption in ten European countries. With 2% nominal WACC, all countries except for Sweden and Finland are in true grid parity by 2016. With 4% nominal WACC, the same is true within a few years. With 6% nominal WACC, only Italy, Germany and the UK are in true grid parity during this decade.

Table IV: Summary of when true grid parity is reached in the residential segment with 50% self-consumption with 4 different nominal WACCs. Parity = true grid parity has been reached by 2015.

	Nominal WACC			
	0%	2%	4%	6%
Stockholm	2020	2025	2032	2040
Helsinki	2019	2024	2030	2038
Amsterdam	Parity	Parity	2019	2023
Paris	Parity	2016	2021	2026
Brussels	Parity	Parity	2018	2022
Istanbul	Parity	Parity	2017	2021
London	Parity	Parity	Parity	2018
Berlin	Parity	Parity	Parity	Parity
Madrid	Parity	Parity	2018	2022
Rome	Parity	Parity	Parity	Parity

It is likely that these results are rather conservative because the wholesale electricity prices and energy component in the retail price is quite low at the moment. We have assumed that they stay at the current real level. Moreover, the self-consumption rates are bound to get higher with the increase of battery systems. This issue will be discussed in section 6.

4.2 Commercial 50 kWp PV system

As an example of the PV competitiveness in the commercial segment, Fig. 17 shows the comparison of PV LCOE with the average value of PV electricity in Finland. A self-consumption range from 50% to 100% is shown since the electricity use in commercial buildings concentrates more on daytime compared with residential buildings. On the other hand, higher interest rates are used since companies usually expect faster returns than households.

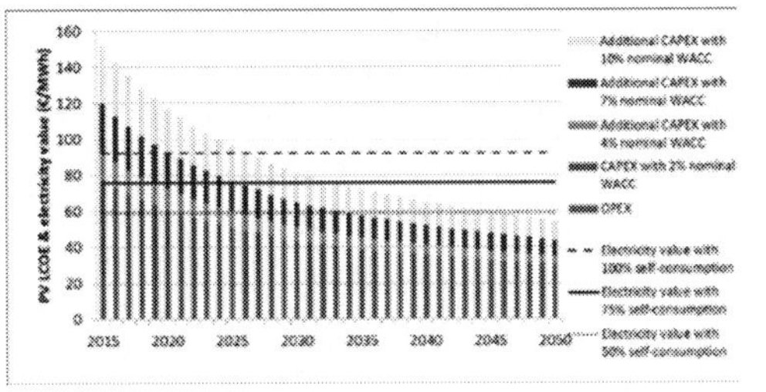

Figure 17: Comparison of PV LCOE with average commercial PV electricity value for a 50 kWp rooftop PV system in Helsinki, Finland. Prices are in 2015 real money.

As can be seen, true grid parity with 75% self-consumption and 2% nominal WACC is reached in 2016. With 4% nominal WACC, it is reached in a few years. With 100% self-consumption, true grid parity already exists with lower than 4% WACC, and with 7% nominal WACC is reached in 5 years and with 10% nominal WACC in about 10 years. It must be emphasised that 10% nominal WACC is very high. For example, with 70/30 debt to equity ratio and 4% interest on debt, 10% nominal WACC would mean 24% return on equity.

For other countries, the results with 75% self-consumption in commercial segment are quite similar to the ones with 50% self-consumption in residential segment, as can be seen in the summary of Table V. Probably the biggest difference is in Spain where PV in commercial segment is more competitive than the residential one. This is because the average electricity price that the consumer can save with own PV generation is relatively higher because of the lower self-consumption tax. For this reason, true grid parity in Spain is reached by 2016 with all interest rates like in Italy, while it has been reached in Germany and the UK with 7% or lower nominal WACC.

Table V: Summary of when true grid parity is reached in the commercial segment with 75% self-consumption with 4 different nominal WACCs. Parity = true grid parity has been reached by 2015.

	Nominal WACC			
	2%	4%	7%	10%
Stockholm	2016	2020	2027	2034
Helsinki	2016	2019	2026	2033
Amsterdam	Parity	2016	2022	2028
Paris	Parity	Parity	2019	2025
Brussels	Parity	Parity	2017	2023
Istanbul	Parity	Parity	2018	2024
London	Parity	Parity	Parity	2020
Berlin	Parity	Parity	Parity	2019
Madrid	Parity	Parity	Parity	2016
Rome	Parity	Parity	Parity	Parity

4.3 Industrial 1 MWp system

Competitiveness in the industrial segment with 100% self-consumption is very similar to the one in the commercial segment with 75% self-consumption. In manufacturing industries, 100% self-consumption is rather easy to achieve with a proper sizing of the PV system relative to the electricity consumption. On the other hand, self-consumption rate is not as important in industrial as in the other segments because the retail electricity price is closer to the spot market price. Fig. 18 shows the comparison of PV LCOE with the average value of PV electricity in Finland.

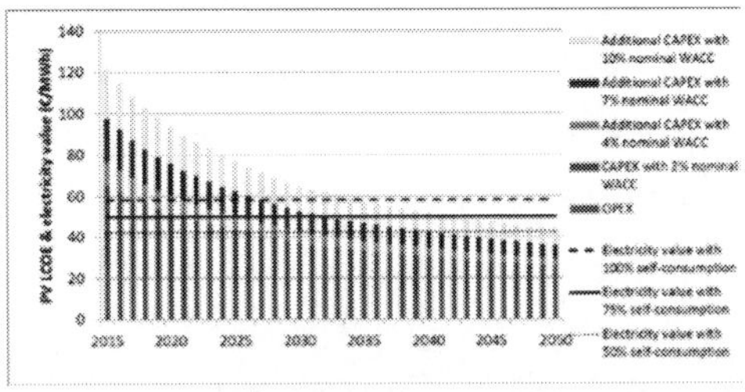

Figure 18: Comparison of PV LCOE with average industrial PV electricity value for a 1 MWp PV system in Helsinki, Finland. Prices are in 2015 real money.

Table VI summarises the years when true grid parity is reached in the industrial segment with 100% self-consumption in the ten countries. Apart from Sweden and Finland, all countries achieve true grid parity slightly earlier than the commercial segment with 75% self-consumption. Italy, Spain, Germany and the UK are all in true grid parity by 2017 with all interest rates. It must be noted that this study applies mainly to the small industries, heavy industries usually get their electricity very close to the wholesale price which is lower than the retail price.

Table VI: Summary of when true grid parity is reached in the industrial segment with 100% self-consumption with 4 different nominal WACCs. Parity = true grid parity has been reached by 2015.

	Nominal WACC			
	2%	4%	7%	10%
Stockholm	2022	2026	2033	2043
Helsinki	2018	2021	2027	2035
Amsterdam	Parity	Parity	2018	2023
Paris	Parity	Parity	2018	2023
Brussels	Parity	Parity	2016	2021
Istanbul	Parity	Parity	Parity	2019
London	Parity	Parity	Parity	2016
Berlin	Parity	Parity	Parity	2017
Madrid	Parity	Parity	Parity	Parity
Rome	Parity	Parity	Parity	Parity

5 SENSITIVITY ANALYSIS

In order to get an understanding of the various parameters affecting the PV competitiveness, a thorough sensitivity analysis was made. As a base case, a residential 5 kW_p system in Finland with 50% self-consumption, 2% nominal WACC, 26 €/MWh value for surplus generation, 2015 average variable retail electricity price, 930 kWh/kW_p annual yield, 80% performance ratio, 1.6 €/W_p (+ VAT) CAPEX and 20 €/kW_p/a OPEX in 2015, 30% fixed share of the grid cost, 30 years system lifetime, 0.5% annual degradation and 20% learning rate was used. Fig. 19 shows the sensitivity on the various parameters compared with the base case as a difference in years when true grid parity is reached.

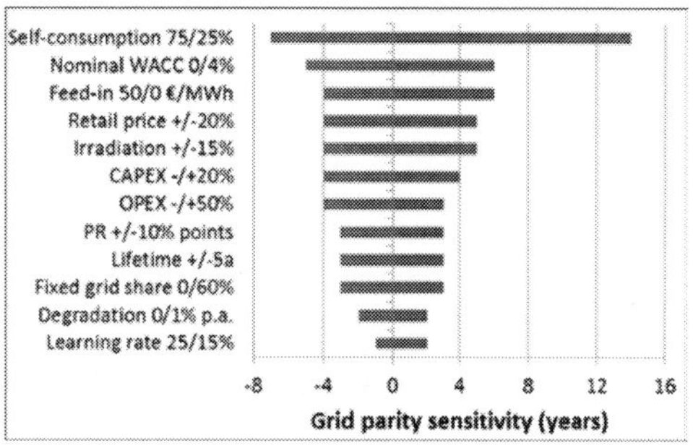

Figure 19: Sensitivity of true grid parity year on various input parameters compared with a residential 5 kW_p system in Finland with 50% self-consumption, 2% nominal WACC, 26 €/MWh value for surplus fed-in electricity, 2015 average variable retail electricity price, 930 kWh/kW_p annual yield, 80% performance ratio, 1.6 $€/W_p$ (+ VAT) CAPEX and 20 $€/kW_p/a$ OPEX in 2015, 30% fixed share of the grid cost, 30 years system lifetime, 0.5% annual degradation and 20% learning rate.

As is evident from Figs. 7-18, the level of self-consumption plays a major role in the PV competitiveness. The difference between 75% and 25% self-consumption ratio compared with 50% is from -7 to +14 years. This is because the value of self-consumed PV generation, i.e. the retail electricity price, is currently much higher than the value of surplus generation, which is often the spot market price. This means that in the absence of lucrative feed-in tariffs, it does not pay off anymore to generate much surplus by oversizing the PV system in relation to the electricity consumption.

The second most important parameter is the cost of capital or WACC. A difference of -/+2% in nominal WACC means -5 to +6 years in the true grid parity year. The value of surplus generation is also very important as it can range from zero to about 50 €/MWh in the ten countries in this study. Retail electricity price is almost as significant and it does not vary only by country but has a wide range between different companies within one country. In this study, it was assumed that the retail electricity price will stay at its current real value over time. But even assuming a 2% real annual increase, equaling 4% nominal annual increase, would mean about 10% real increase over 5 years, which would bring true grid parity only 2 years earlier.

A more profound change can be foreseen in the structure of the distribution grid tariff which in many countries has shifted emphasis from variable or per kWh price to the fixed or power-related fee during the recent years. However, as this sensitivity analysis shows, the effect of increasing the fixed share of the grid component does not have a major impact on the PV competitiveness, as the grid component is relatively small compared with the energy or taxes and fees component, typically 20% or less of the total retail electricity price. It is likely that more significant threats to PV competitiveness are the various legal, tax and regulatory changes which happen at alarming frequency in some countries and make it very difficult to create a confident environment for the PV investors.

Yield or irradiation and performance ratio obviously have some effect on PV competitiveness, as many European countries cover a geographically diverse area. And even at the same location, there might be different PV systems having varying yields because of different orientation, inclination or shadowing from the nearby buildings or trees. However, the effect of yield is not as significant as self-consumption or interest rate. The same can be said about CAPEX or OPEX which also have significant variation both between countries and installations in the same area. This is a very important conclusion, as it means that PV competitiveness does not solely rely on continuing decrease of investment cost and more efficient system operation. It can also be noted that increasing the learning rate of PV modules and inverters from 20% to 25% would only influence true grid parity by 1-2 years. PV system lifetime has some effect with low interest rates. Other parameters like PV module degradation have only a minor effect on PV competitiveness.

For comparison, Fig. 20 shows the sensitivity in the commercial segment in Finland with 7% nominal WACC and 75% self-consumption ratio. The main difference to the residential segment is that the self-consumption ratio is not as significant because the average retail electricity price in the commercial segment is closer to the wholesale price. On the other hand, the effect of interest rate is greater since the WACC rates are higher. Wholesale electricity value has smaller effect since the amount of feed-in is lower. PV system lifetime is also less significant because of the higher WACC rates.

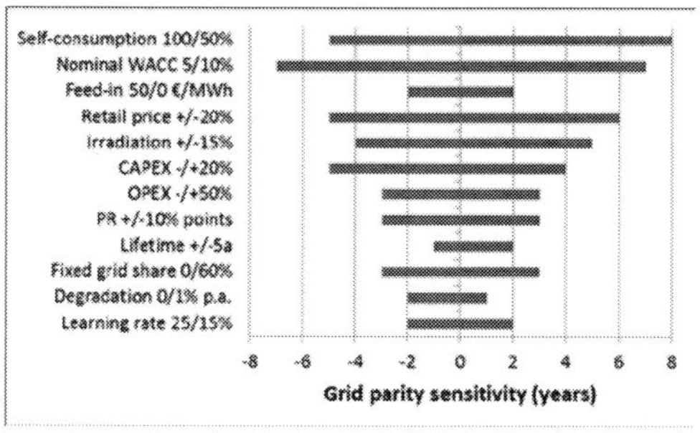

Figure 20: Sensitivity of true grid parity year on various input parameters compared with a commercial 50 kW_p system in Finland with 75% self-consumption, 7% nominal WACC, 26 €/MWh value for surplus fed-in electricity, 2014 average variable retail electricity price, 930 kWh/kW_p annual yield, 80% performance ratio, 1.3 $€/W_p$ CAPEX and 20 $€/kW_p/a$ OPEX in 2015, 30% fixed share of the grid cost, 30 years system lifetime, 0.5% annual degradation and 20% learning rate.

It is obvious that the various subsidy mechanisms still have a role to play in the countries and market segments where true grid parity has not yet been reached. For example in Sweden, subsidised PV competitiveness already exists with low interest rates because of tax deductions, investment grants, and green electricity certificates in the residential and commercial segments. In Finland, household income tax deduction is not quite enough for the residential segment, but a typical 25% investment grant in the commercial segment makes PV already viable with a 4% nominal WACC.

6 DISCUSSION

The analysis in this paper clearly indicated that for the financial attractiveness of PV prosumer systems the

PV self-consumption is the key parameter to be increased. This is expected to be easier for commercial and industrial PV systems, since a good match of the load profile to the generation and the ratio of total generation on available roofs to the total demand can be better achieved. However, for residential PV prosumers the mismatch of PV generation and the load profile is rather inconvenient, since social activities are often at other places during daytime and for the times of high electricity demand in the buildings, which is typically in the evening hours, the PV generation is typically low or zero.

Batteries are an excellent component to overcome the mismatch of generation and demand for residential PV prosumers. Batteries start to become financially attractive in cases when the difference of the variable electricity price to the wholesale price is the same or larger than the cost of storing a respective kWh in a battery system. For example, in Germany this leads to the following requirement: variable electricity price is 270 €/MWh and wholesale price 31 €/MWh which equals a value for a stored MWh of 239 €. Assuming for the battery a technical lifetime of 15 years, full charge and discharge availability, nominal WACC of 2% and annual OPEX of 2% of CAPEX, roundtrip efficiency of 90% and 200 full charging and discharging cycles a year, leads to a breakeven CAPEX for the battery of about 500 €/kWh of storage capacity.

The battery systems in the German market currently cost about 1350 €/kWh per usable storage capacity on average, with a range from about 750 €/kWh up to 2500 €/kWh including all the leading battery system providers in Germany [23]. There are about 10 different offers from 4 different companies, including the top 3 battery system suppliers, for less than 1000 €/kWh available in the market. The cost of battery systems currently decline by about 18% p.a. [24], which will lead to an accelerated market growth in coming years, and a breakeven price within 3 years for the aforementioned 3 suppliers.

Battery system prices are expected to decline substantially in the years to come, since the learning rate of batteries is about 15-20% [25], which is comparable to PV modules. The least cost market prices for comparable automotive Li-ion batteries are 145 USD/kWh for delivery in the years 2016 to 2018 [26]. The PV self-consumption can be increased for a country such as Germany up to 70% for economic system designs with a most cost-efficient PV system size of 0.8 kWp and usable battery capacity of 1.1 kWh per MWh of annual electricity consumption [27].

In most European countries a heating system is required from autumn to spring. In particular in the autumn and in the spring months the solar PV system can provide the major part of the energy for heat pumps, which can be used in a most efficient way. However, in the winter time the PV system can only contribute partly to the heat demand. Depending on the location, the energy efficiency of the building and the type and size of the building, a family may need about 2500-5000 kWh of electricity for operating the heat pump. The direct hot water demand can be covered from spring to autumn almost fully with PV electricity. Having some hundred liters of water as a storage reservoir available for the heating system optimizes the utilization of PV electricity and maximizes the self-consumption. In most countries a sustainable electricity supply could be wind energy through the winter months, which shows best production then.

Electric vehicles are a further option to increase the self-consumption. Assuming a specific energy demand of 20 kWh/100 km and a usage of 10000 km per year is equivalent to a further self-consumption potential of up to 2000 kWh per year. At least a part of it may be realized, however this is strongly dependent whether the electric vehicle is grid-connected to the residential PV system during daytime.

Summing up, when net metering or net billing schemes are not present, batteries become increasingly attractive to increase the self-consumption for residential PV prosumers to values of 70% depending on the PV system sizing, the electricity demand and the load profile. Electric heat pumps and electric vehicles can further improve the self-consumption and may become increasingly attractive for PV prosumers in the years to come. This is also a further example how the formerly separated energy sectors power, heat and mobility converge towards electricity in the future. A detailed analysis would require an hourly calculation to match best the PV generation with the respective load profiles, the flexibility options and storage capacities.

However, it needs to be highlighted that zero impact areas, and in particular the roofs should be fully covered by PV to enable a maximum PV generation, which will be finally needed in all European countries. PV self-consumption as only income option for the PV system operator is not appropriate enough to ensure the maximum PV electricity generation. From a societal perspective it is not efficient not to use the whole available zero impact rooftop area for PV generation. Therefore further policy tools will be needed finally which guarantees a maximum PV generation to reach the COP21 targets for a net zero CO_2 emission society by the mid-21st century and still create an attractive business case for investors.

7 CONCLUSIONS

This paper has assessed whether PV systems in key European markets could have reached or could reach in the coming years a reasonable level of competitiveness. This has been assessed for PV prosumers, under current self-consumption regulations, with existing and forecasted PV system prices, and under the constraining tax policies existing for prosumers in several European countries. The conclusion is rather simple: if PV systems could be financed with a cost of money corresponding to the low level of risk associated with PV systems, they would be competitive already now in most European countries and for most market segments. A higher cost of capital would delay the moment where PV could reach competitiveness under the current conditions, especially in the Nordic countries. With reasonable cost of capital, self-consumption schemes without incentives and market-based remunerations for excess PV electricity injected into the grid, PV will be competitive in almost all cases within the coming ten years.

Current PV system prices in Europe already allow a good level of competitiveness for PV installations in most segments. Further cost decline of battery storage and integration of PV systems with electric heat pumps and electric vehicles will make residential PV self-consumption increasingly attractive in more and more countries in Europe.

8 ACKNOWLEDGEMENTS

The work for this paper has been carried out under the framework of the EU PV Technology and Innovation Platform (ETIP PV). The authors would like to thank ETIP PV Steering Committee and Expert Pool Members for their valuable contributions and comments during the work. The OPEX data provided by Bloomberg New Energy Finance and solar irradiation data obtained from SolarGIS database by GeoModel Solar is also gratefully acknowledged.

9 REFERENCES

[1] E. Vartiainen, G. Masson, Ch. Breyer, PV LCOE in Europe 2014-30, European PV Technology Platform (2015).
[2] E. Vartiainen, G. Masson, Ch. Breyer, PV LCOE in Europe 2015-50, 31st European PV Solar Energy Conference and Exhibition (2015).
[3] IEA, Technology Roadmap – Solar Photovoltaic Energy, 2014 edition (2014).
[4] Fraunhofer ISE, Current and Future Cost of Photovoltaics. Long-term Scenarios for Market Development, System Prices and LCOE of Utility-Scale PV Systems. Study on behalf of Agora Energiewende (2015).
[5] ITRPV, International Technology Roadmap for Photovoltaic, Seventh Edition, 2015 Results (2016).
[6] I. Theologitis & G. Masson, Potential for cost reduction of PV technology – Impact of CHEETAH research innovations, 31st European PV Solar Energy Conference and Exhibition (2015).
[7] SolarPower Europe, Global Market Outlook for Solar Power 2016-2020 (2016).
[8] IEA, Energy Technology Perspectives (2016).
[9] Hoffmann, Winfried, The Economic Competitiveness of Renewable Energy (2014).
[10]Fraunhofer ISE, Photovoltaics Report, updated 17 November 2015.
[11] BSW-Solar, Statistical data on the German Solar power industry, factsheet, April 2014.
[12] pvXchange price index, pvxchange.com (2016).
[13] Bloomberg New Energy Finance, Maintaining Returns: PV O&M Opportunities & Prices (2013).
[14] Bloomberg New Energy Finance, PV O&M Index 2015: emphasis on boosting returns, September 2015.
[15] M. Bächler, Technical Risk Management during O&M of PV plants, Solar Bankability project presentation (2016).
[16] Solargis database (2016).
[17] N.H. Reich, B. Müller, A. Armbruster, K.Kiefer, W.G.J.H.M. van Sark & Ch. Reise, Performance ratio revisited: Are PR>90% realistic?, Proceedings of the 26th European Photovoltaic Solar Energy Conference and Exhibition (2011).
[18] P. Fritze, F. Flade & G. Becker, 15 Years of degradation: Modules of the 1 MW PV-system at Munich Trade Fair Centre under review. Proceedings of the 28th European Photovoltaic Solar Energy Conference and Exhibition (2013).
[19] IEA PVPS, Analytical Monitoring of Grid Connected Photovoltaic Systems. Report 13-03:2014.
[20] Eurostat database, ec.europe.eu/eurostat/data (2016).
[21] European Commission, Study on tariff design for distribution systems, Final Report (2015).
[22] P. Sinha, M. de Wild-Schloten, A. Wade & Ch. Breyer, Total Cost Electricity Pricing of Photovoltaics, 28th European PV Solar Energy Conference and Exhibition (2013).
[23] C. Werner, Private communication on current market overview of battery prices in Germany based on real offers, Chris Werner Energy Consulting, June 2, 2016.
[24] Germany Trade & Invest, Batteries for Stationary Energy Storage in Germany: Market Status & Outlook, Berlin, March 2016.
[25] W. Hoffmann, Importance and evidence for cost effective electricity storage, Proceedings of the 29th European Photovoltaic Solar Energy Conference and Exhibition (2014).
[26] InsideEVs, LG Chem "Ticked Off" With GM For Disclosing $145/kWh Battery Cell Pricing, October 23, 2015, insideevs.com/lgchemtickedgmdisclosing 145kwhbatterycellpricingvideo/
[27] J. Weniger, T. Tjaden, V. Quaschning, Sizing of residential PV battery systems, Energy Procedia, 46, 78-87 (2014).

On the Role of Solar Photovoltaics in Global Energy Transition Scenarios

Christian Breyer[1], Dmitrii Bogdanov[1], Ashish Gulagi[1], Arman Aghahosseini[1], Larissa S.N.S. Barbosa[2,3], Otto Koskinen[1], Maulidi Barasa[1], Upeksha Caldera[1], Svetlana Afanasyeva[1], Michael Child[1], Javier Farfan[1], Pasi Vainikka[2]

[1] Lappeenranta University of Technology, Skinnarilankatu 34, 53850 Lappeenranta, Finland, E-mail: christian.breyer@lut.fi
[2] VTT Technical Research Centre of Finland Ltd., P.O. Box 20, 53851 Lappeenranta, Finland
[3] Luiz de Queiroz College of Agriculture, University of São Paulo, Av. Pádua Dias, 11, Caixa Postal 9, 13418-900, Piracicaba, Brazil

ABSTRACT
The global energy system has to be transformed towards high levels of sustainability for executing the COP21 agreement. Solar PV offers excellent characteristics to play a major role for this energy transition. Key objective of this work is to investigate the role of PV for the global energy transition based on respective scenarios and a newly introduced energy transition model developed by the authors at the Lappeenranta University of Technology (LUT). The available energy transition scenarios have no consensus view on the future role of PV, but a progressive group of scenarios present results of a fast growth of installed PV capacities and a high energy supply share of solar energy to the total primary energy demand in the world in the decades to come. These progressive energy transition scenarios can be confirmed by the LUT Energy system model. The model derives total installed solar PV capacity requirements of 7.1 – 9.1 TWp for today's electricity sector and 27.4 TWp for the entire energy system in the mid-term (year 2030 assumptions set as reference). The long-term capacity is expected to be 42 TWp and due to the ongoing cost reduction of PV and battery technologies, this value is found to be the lower limit for the installed capacities. The cost reductions are taken into account for the year 2030, but are expected to further proceed beyond this reference year. Solar PV electricity is expected to be the largest, least cost and most relevant source of energy in the mid- to long-term for the global energy supply.

Keywords
PV demand, energy transition scenario, hourly resolution, 100% renewable energy

1. Motivation

Further development of human welfare is at a crossroads. For several decades humankind has needed the capacity of more than one planet Earth [1], and 50% of this need is due to the resource exploitation and emissions of the energy system. Access to (diminishing) energy resources [2] have caused in the past and will cause in the future dramatic economic, social, political and military conflicts. Poverty in the world needs to be tackled [3,4] for a still growing world population [5]. The only pathway to manage all these different major problems is a transition [6] towards a fully sustainable energy system, one that is able to cover accelerated growth in demand for energy.

The two key resources for very large scale renewable energy (RE) harvesting are the wind resource and the direct solar resource. The two major solar technologies are solar photovoltaics (PV) and concentrating solar thermal power (CSP), although the future cost competitiveness of CSP is more and more questioned [7]. Solar PV is the fastest growing energy technology in the world [8] and reached a level of 50 GW of new capacities added annually. Financial renewable energy experts expect installations to grow to annually 80 GW (2020), 143 GW (2030) and 206 GW (2040) [7]. The International Energy Agency (IEA) lag behind by projecting annual installations of 37.1 GW (2013-2020), 32.6 GW (2020-2025), 33.6 GW (2025-2030), 34.4 GW (2030-2035) and 33.2 GW (2035-2040) [9,10], which is not only in contradiction to the already achieved level of installations of 50 GW in 2015 [11] and the expectation of 60 GW in 2016 and 70 GW in 2017 [12], but also to the fact that solar PV becomes the least cost source of electricity in a fast growing number of regions in the world [13-16]. IEA scenarios have a more than 20-years track record of being fully incompatible with real world PV installations [17] and it remains unclear why it is so complicated for the IEA to catch up with the level of insights of other institutions.

The two observations, that solar PV is the fastest growing energy technology in the world and the highly blurry view of international institutions on the future of solar PV, led us to the following research question discussed in this paper: What is the mid- to long-term solar PV demand in a world based on sustainable energy resources?

For answering this question a two-step approach as explained in section 2 is applied. The role of solar PV in the major energy transition scenarios is reviewed in section

3. Based on modeling results of Lappeenranta University of Technology (LUT) the role of solar PV for the financial assumptions of the year 2030 is investigated and summarized in full hourly resolution on a global level for the very first time in section 4. The LUT results are discussed in comparison to the major energy transition scenarios and an outlook is provided in section 5. The conclusions on the research question are summarized in section 6.

2. Methodology

This section is subdivided in the methodology for reviewing the global energy transition scenarios (section 2.1) and introducing the LUT Energy system model (section 2.2).

2.1 Review of global energy transition scenarios

An overview on the major global energy scenarios for the years 2030, 2050 and 2100 is compiled and the role of PV is extracted from these benchmarking publications. As far as possible the total installed capacity numbers are extracted, as well as the solar PV supply share of the total primary energy demand for the power sector and the total energy system. The scenarios are taken from Bloomberg New Energy Finance (BNEF) [7], Greenpeace [18], International Energy Agency (IEA) [9], Photovoltaic Power Systems Programme of the IEA (IEA-PVPS) [19], International Renewable Energy Agency (IRENA) [20], World Wildlife Fund (WWF) [21], International Institute for Applied Systems Analysis (IIASA) Global Energy Assessment (GEA) [22], Intergovernmental Panel on Climate Change (IPCC) [23], German Advisory Council on Global Change (WBGU) [24], Jacobson and Delucchi [25] and Shell [26]. Massachusetts Institute of Technology (MIT) [27]. BP [28] and Exxon Mobil Corp. [29] had to be excluded due to no disclosed PV numbers alongside with very small supply shares for renewables.

2.2 LUT Energy system model

The LUT Energy system model is based on linear optimization of the energy system parameters under previously defined constraints applied to the system. Assumptions for the future RE power generation and demand and required storage technologies are also considered. Additional water desalination and synthetic natural gas (SNG) generation are the flexible demands in the model, and, therefore, can act as storage technologies as well. One of the key constraint for the system optimization is matching of power generation and demand on an hourly basis for a particular year. The main aim of the system optimization is the minimization of the total annual energy system costs which is a sum of the installed capacities of the different technologies, costs of energy generation and generation ramping. Also, the system consists of prosumers for residential, commercial and industrial sectors. The prosumers install the respective capacities of rooftop PV systems and batteries. Minimizing the cost of consumed energy is the main aim of the target function for the prosumers. The flowchart of the model is presented in Figure. 1.

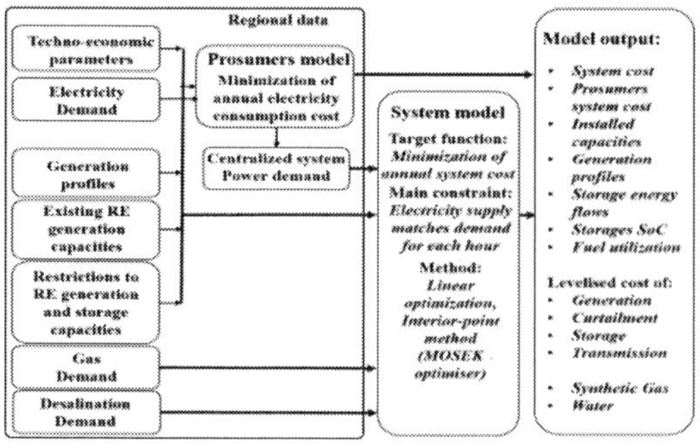

Fig. 1. Model flow diagram with the input data, system model optimization and output data

The detailed description of the model can be found in Bogdanov and Breyer [30]. Detailed information of the input data used for the model is given in Bogdanov and Breyer [30] and additional calculations related to geothermal, desalination water demand [31] and industrial gas demand data is described here [32].

The technologies used for converting RE sources into electricity are two different types of ground mounted PV systems (optimally tilted and single-axis north-south oriented horizontal continuous tracking) and rooftop PV, concentrating solar thermal power (CSP), onshore wind turbines, hydro power (run-of-river and dams), biomass (biogas and solid biomass), waste-to-energy power plants and geothermal power plants.

The energy storage technologies used in the model are residential and system level battery storage, pumped hydro storage (PHS), adiabatic compressed air energy storage (A-CAES), thermal energy storage (TES) and power-to-gas (PtG) technology. Technologies such as water electrolysis, methanation, CO_2 scrubbing from air, gas storage, and both combined and open cycle gas turbines (CCGT, OCGT) are part of the synthesis of SNG and its reconversion to electricity. The PtG technologies have to be operated in synchronization because of the absence of hydrogen and CO_2 storage. There is a 48-hour biogas buffer storage and part of the biogas can be upgraded to biomethane and introduced to the gas storage.

The bridging technologies used in this model provide the required flexibility to the energy system and help in reducing the overall cost. For example, gas produced from PtG can be used for industrial gas demand rather than storage for electricity sector. Similarly, seawater reverse osmosis (SWRO) desalination couples with the water sector than electricity sector.

The power transmission within the sub-regions is assumed to be based on alternating current (AC) grids which is not included in the model and between the sub-regions on high voltage direct current (HVDC). Loss of electricity due to the length of the transmission lines and in converter stations at the interconnection with the AC grid form a major component of the power losses in HVDC grids.
The full block model diagram is presented in Figure 2.

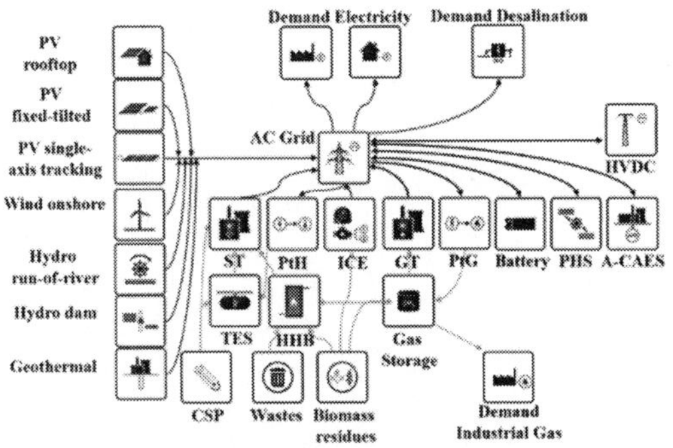

Fig. 2. Block diagram of the all the energy technologies applied in the LUT Energy system model.

The feed-in profiles for solar technologies were calculated based on NASA data [33,34] for direct and diffuse solar irradiation, wind speed and temperature for the year 2005 reprocessed by the German Aerospace Center [35].

The results of the energy system analyses are based on 100% renewable energy (RE) and are integrated on a global scale for the very first time. The key advantage of the LUT results is the hourly modelling of the energy system for an entire year based on RE resource data for solar, wind and hydro on a high spatial resolution of 0.45° × 0.45° or higher and consequent solving of a least cost target function. None of the benchmarking energy scenarios is performed on an hourly resolution for an entire year [36]. However, this is of high importance since an energy system mainly based on PV and wind energy is characterised by a high degree of intermittency. The results are based on three scenario set-ups, which reflect geographic integration and sectorial integration: region-wide (rather distributed; sectors: electricity), area-wide (rather centralised; sectors: electricity) and integrated (rather centralised; sectors: electricity, industrial gas and desalination).

The world had been subdivided into 145 sub-regions, which are used as building blocks for different setups of modeling. The main focus of research had been so far to aggregate the 145 sub-regions into 9 major regions, which form the main body of the results of this paper. Other research aggregates the sub-regions, so that the world can be represented by 23 regions [37], or an integrated analysis for Europe-Eurasia-MENA [38] or the East Asian Super Grid [39], all in full hourly resolution and interconnected. The 9 major regions are: Europe [40], Eurasia [41], MENA [42], Sub-Saharan Africa [43], India/ SAARC [32], Northeast Asia [30], Southeast Asia and Pacific [44,45], North America [46] and South America [47,48]. Solar PV is represented in the model by ground-mounted optimally tilted and single-axis tracking PV power plants and prosumer rooftop systems, enhanced by batteries in the cases of financial attractiveness for the prosumers.

3. Review of global energy transition scenarios

The results for the energy transition scenarios are summarized in Table 1. The global PV capacities range from 728 - 3725 GWp (2030), 1066 – 6678 GW (2040), 6745 – 32,700 GWp (2050) and 32,700 – 133,000 GWp (2100). The large ranges clearly indicate that there is no consensus on the expected capacities. The range in solar PV electricity contribution to the total electricity demand is 2.9% - 15.9% (2030), 3.9% - 18.2% (2040) and 19.9% - 29.0% (2050), whereas the findings for 2050 cannot be compared to 2030 and 2040 since only two progressive scenarios provided respective numbers for 2050. Notably, the World Energy Outlook (WEO) of the IEA sets in all cases the minimum expectation, which had been already criticised in the past [10,17,49-52]. It remains still unclear why the IEA cannot close its gap in PV insights compared to all other scenario makers, since the IEA should have excellent access to all energy-related information globally. The share of solar PV in the global electricity supply in the year 2050 is expected by Greenpeace for about 20%, which is surprisingly low and by WWF for about 30%, which remains also moderate given the fact that solar PV is already today the least cost source of electricity in a fast growing number of regions around the planet.

Table 1: Benchmarking global energy transition scenarios and the respective role of solar PV. Abbreviations: total primary energy demand, *TPED*, and electricity, *elec*. The colour code indicates the role for solar PV as follows: little (red), moderate (orange) and substantial (green).

		PV capacity				share PV on TPED-elec			share solar total on TPED-all			
		2030	2040	2050	2100	2030	2040	2050	2030	2040	2050	2100
	[Ref]	[GWp]	[GWp]	[GWp]	[GWp]	[%]	[%]	[%]	[%]	[%]	[%]	[%]
BNEF	[7]	1799	3687			7,7 %	14,3 %					
Greenpeace - ER	[18]	2839	4988	6745		11,4 %	16,5 %	19,9 %	6,1 %	14,1 %	22,2 %	
Greenpeace - Adv ER	[18]	3725	6678	9295		13,7 %	18,2 %	20,2 %	7,7 %	18,4 %	29,1 %	
IEA - NPS	[9]	728	1066			2,9 %	3,9 %		0,6 %	0,9 %		
IEA - 450	[9]	938	1519			4,4 %	6,6 %		0,9 %	1,8 %		
IEA-PVPS	[19]	1570	3930	11010	133000				0,7 %	1,4 %	3,3 %	25,3 %
IRENA REmap - mix	[20]	1760				15,9 %		16,7 %	17,0 %		22,0 %	
IRENA REmap - doubling	[20]	2520						21,7 %			29,0 %	
WWF	[21]					7,6 %	16,3 %	29,0 %	7,1 %	17,8 %	30,6 %	
IIASA-GEA - Efficiency	[22]								6,9 %		23,7 %	
IIASA-GEA - Mix	[22]								7,7 %		20,6 %	
IIASA-GEA - Supply	[22]								9,0 %		17,0 %	
IPCC - 5th AR WGIII - MESSAGE	[23]								6,4 %	12,9 %	15,3 %	42,8 %
IPCC - 5th AR WGIII - REMIND	[23]								0,4 %	2,3 %	6,6 %	40,3 %
IPCC - 5th AR WGIII - GCAM	[23]								0,4 %	1,1 %	1,3 %	2,7 %
WBGU	[24]								4,0 %	10,6 %	27,7 %	66,9 %
Jacobson and Delucchi	[25]	32700	32700	32700	32700						40,0 %	40,0 %
Shell - Mountains	[26]								1,5 %	2,1 %	3,6 %	37,7 %
Shell - Oceans	[26]	1800		20000					3,2 %	8,2 %	13,6 %	
min		730	1070	6750	32700	2,9 %	3,9 %	19,9 %	0,4 %	0,9 %	1,3 %	2,7 %
max		3730	6680	32700	133000	15,9 %	18,2 %	29,0 %	9,0 %	18,4 %	40,0 %	66,9 %
average		1960	3640	15950	82850	9,1 %	12,6 %	21,5 %	7,0 %	7,7 %	17,7 %	36,5 %

More interesting is the discussion of the findings on the share of solar energy in global energy supply. It should be discussed first why this is the key metric for the overall assessment of the scenarios. First of all, most selected scenarios provide information for the share of solar energy in the total energy supply and second the shift to power megatrend moves the electricity generation in the center of the energy system. Finally nearly all energy sectors will be based on electricity, such as electricity-based heating (e.g. electric heat pumps), electricity-based mobility (e.g. electric vehicles, electricity based synthetic fuels [53,54]) and electricity-based chemicals [55]. Remaining non-electric energy sources are mainly geothermal energy and biomass, which is limited and has to fulfil sustainability criteria. Fossil fuels are not compatible to the COP21 targets and carbon capture and storage (CCS) may be never profitable [56]. Nuclear energy violates all sustainability criteria [18,21,24] and thus needs to be phased-out. The solar energy technologies may be finally represented by solar PV, since solar thermal collectors for heating are higher in cost than the solar PV plus heat pump alternative [57]. Furthermore, concentrating solar thermal power (CSP) plants cannot catch up anymore with solar PV power plants as indicated by extremely low market expectations of the market research analysts of BNEF [7], but also by recent research [58,59]. CSP could have still a chance, since the low-cost thermal energy storage and the secured capacity of the steam turbine are very valuable, but very high growth rates would be required to compensate the comparable low learning rate of 10-12% for CSP [60] compared to 20-23% for PV modules [61-63] and about 16% for total PV systems [63] to reduce the cost gap to solar PV [60]. Challenging for CSP are hybrid PV-battery-GT plants which are expected to be financially attractive in the 2020s [68], since these new hybrid PV plants are technically fully comparable to CSP and depending on the cost development of battery storage they may become highly attractive, which is rather likely due to the high learning rates of batteries which are around 20% [64,65] and the very high growth rates, which reduces the cost currently on a 10-20% rate per year [65]. As a consequence the probability is high that in the year 2050 the solar energy technologies are dominated by solar PV.

The solar energy share in TPED range in the regarded energy transition scenarios is 0.4% - 9.0% in 2030, 0.9% - 18.4% in 2040, 1.3% - 40.0% in 2050 and 2.7% - 66.9% in 2100. Two of the three major energy scenarios for the IPCC 5th assessment report belong to the most conservative scenarios which is surprising, since solar PV would offer an excellent opportunity to fight climate change, however an analysis on the cost assumptions in the scenarios on solar PV reveals fully outdated cost assumptions for 2050 with 1000 – 2100 USD/kWp and 1250 USD/kWp for REMIND and GCAM, respectively [66]. The capital expenditure for today is around 700 €/kWh for India and China, and in Germany around 850 €/kWp, which is mainly explained by the anti-dumping and anti-subsidy duties on imports of PV modules from China in Europe. Why two of the leading scenarios for the IPCC energy system modeling are based on fully outdated cost assumptions remains unclear. The consequence is that compared to industry-related cost insights for 2050 of 320 – 400 €/kWp [67] a difference of factors of about 1.9 – 4.9 need to have a dramatic impact on the role of solar PV in the IPCC energy system scenarios. Keeping in mind that the IPCC has been recently requested to conduct an assessment for the 1.5°C scenario [73], the authors of this paper strongly recommend to correct the outdated PV cost assumptions in the IPCC energy system scenarios as soon as possible. However, the MESSAGE model also used for IPCC energy system modeling showed much better results and consequently it had been grouped in the moderate group. The IEA scenarios belongs again to the most conservative scenarios for the years 2030 and 2040.

The share of solar energy supply on the TPED in the progressive scenarios reach about 18% for 2040, typical values of 20-30% for 2050, except Jacobson and Delucchi [25] who expect around 40%. Most interesting is the expected share of solar PV for the year 2100, which reaches values of 40% and higher for the two IPCC models (MESSAGE and REMIND), Jacobson and Delucchi and remarkably 67% for the WBGU [24] scenario from the year 2003, even the Shell Mountains scenario is close to 40%. The authors of this paper find the first indications that the optimistic findings for the year 2100 of the progressive scenarios may be more realistic in the mid of the 21st century, due to the very fast and steep cost decline of solar PV and supporting technologies, which will be discussed in more detail in the next section.

Summing up, the energy transition scenarios can be grouped into rather progressive ones and rather conservative ones. The progressive studies are: BNEF [7], Greenpeace [18], WBGU [24], IEA-PVPS [19], IRENA [20], WWF [21], IIASA GEA [22] and Jacobson and Delucchi [25]. The conservative ones are: IEA WEO [9], IPCC [23] and Shell [26]. It needs to be mentioned again that Massachusetts Institute of Technology (MIT) [27], BP [28] and Exxon Mobil Corp. [29] had to be excluded due to no disclosed PV numbers alongside with very small supply shares for renewables. Shares of solar PV in the power sector are up to about 14% (2030), 18% (2040) and 29% (2050) and the shares of solar energy in the energy system are expected to be up to about 9% (2030), 18% (2040), 40% (2050) and 67% (2100).

4. 100% Renewable Energy system modeling by LUT

The results of the LUT energy system model simulations for 100% RE for the weather year 2005 and the cost and demand year 2030 assumptions and the integrated scenario are depicted for a global overview in Figure 3. More detailed results are shown for all 145 sub-regions globally aggregated to the nine major regions for Northeast Asia, Southeast Asia and India/SAARC (Fig. 4), Europe and Eurasia (Fig. 5), MENA and Sub-Saharan Africa (Fig. 6) and North America and South America (Fig. 7). Detailed information on all 145 sub-regions can be found in the respective publications [30,32,40-48].

The mentioned assumptions lead to a global cumulated PV demand of 9.1 TWp (region-wide), 7.1 TWp (area-wide) and 12.0 TWp (integrated) for 2030 data (Table 2, Figs. 3 - 7). The share of the prosumer driven PV capacity is found to be between 31-52% on a global average, depending on the applied scenarios, reaching 3.7 TWp in total. The share of PV fixed tilted plants on the total ground mounted utility-scale PV power plants are found to be 28% (region-wide), 6% (area-wide) and 2% (integrated) on a global average.

The energy contribution share of PV to the total energy demand is found to be between 36-42% on a global average, reaching 14,900 TWh/a (region-wide), 11,950 TWp (area-wide) and 21,670 TWp (integrated). The area-wide scenario is characterized by a full interconnection of all sub-regions within a major region. This leads to a more efficient utilization of all system components in general, but also to a better integration of the more volatile wind resources, both leading to a reduction in PV capacities and generation. The global average of the solar PV electricity share is 42% (region-wide), 36% (area-wide) and 41% (integrated). The three regions with the highest fraction of solar PV electricity in the supply for the integrated scenario are Northeast Asia, Southeast Asia and India/SAARC, all between 48-51% and representing slightly more than half of global population. The lowest contribution share is given in Eurasia, since prosumer PV obtains the lowest share due to highly subsidized electricity prices and excellent wind resource availability. In general it can be stated that solar PV is complemented quite well by wind energy, as already found earlier by Gerlach et al. [68], but also by hydro power, in particular hydro dams, since they act as virtual batteries balancing solar PV and wind generation, shown best for the case of Brazil [69], but also more flexible biomass power plants balance the resource fluctuations of solar PV and wind energy.

Table 2: Global solar PV demand subdivided by the nine major regions in the world. Data are based on [30,32,40-48] and visualized in more detail in Figs. 3-7, with updated results for Northeast Asia based on latest assumptions for all major regions.

	Population 2030	Electricity demand 2030	Electricity demand 2030	PV prosumer	PV plants	PV total	PV electricity	PV share
integrated		electricity	integrated					
	[mil]	[TWh]	[TWh]	[GW_p]	[GW_p]	[GW_p]	[TWh]	[%]
Northeast Asia	1546	9878	13496	1509	2806	4315	6986	48 %
Southeast Asia	646	1630	2635	150	609	758	1425	51 %
India/ SAARC	1922	2597	3376	145	815	960	1880	50 %
Eurasia	244	1450	2550	92	171	263	388	15 %
Europe	675	4183	5127	608	353	991	1384	27 %
MENA	529	1813	7917	85	1668	1755	4098	49 %
Sub-Saharan Africa	1384	866	1223	61	241	302	636	48 %
North America	558	6059	10304	812	1038	1850	3452	32 %
South America	445	1813	2780	268	496	764	1419	48 %
World	7949	30289	49408	3730	8197	11958	21668	41 %
				31 %	69 %			

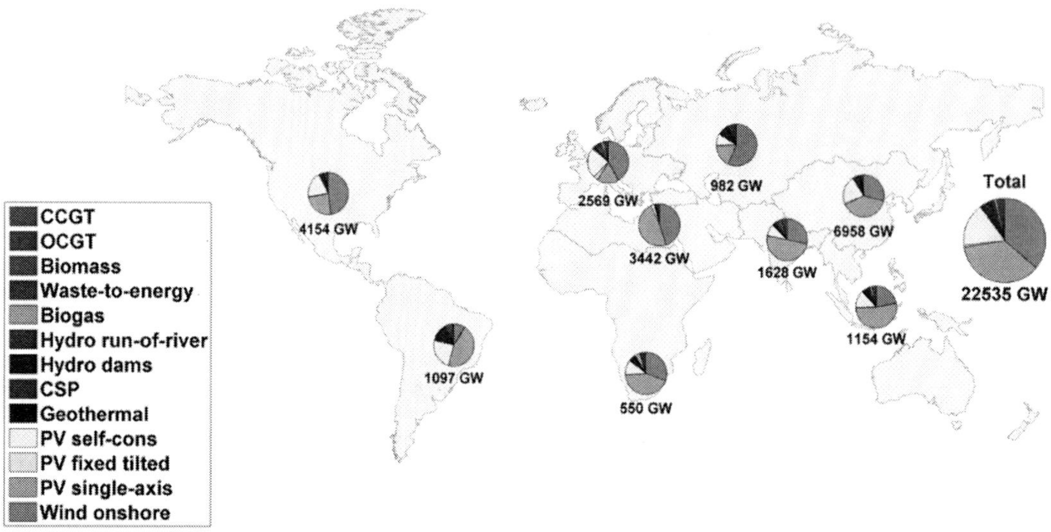

Fig. 3. Installed capacities for the integrated scenario in a global overview structured in the 9 major regions: Northeast Asia, Southeast Asia, India/SAARC, Eurasia, Europe, MENA, Sub-Saharan Africa, North America and South America. Data are taken from [30,32,40-48] which are visualized in more detail in Figs. 4-7.

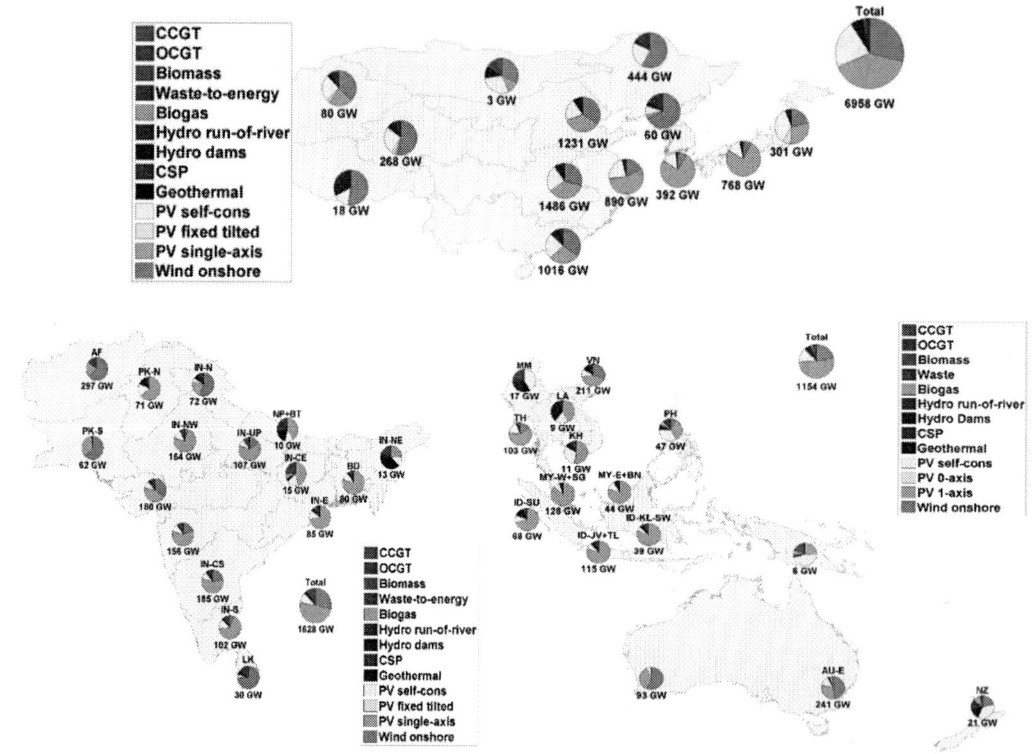

Fig. 4. Installed capacities for the integrated scenario for Northeast Asia (top) [30], India/SAARC (bottom left) [32] and Southeast Asia (bottom right) [44,45].

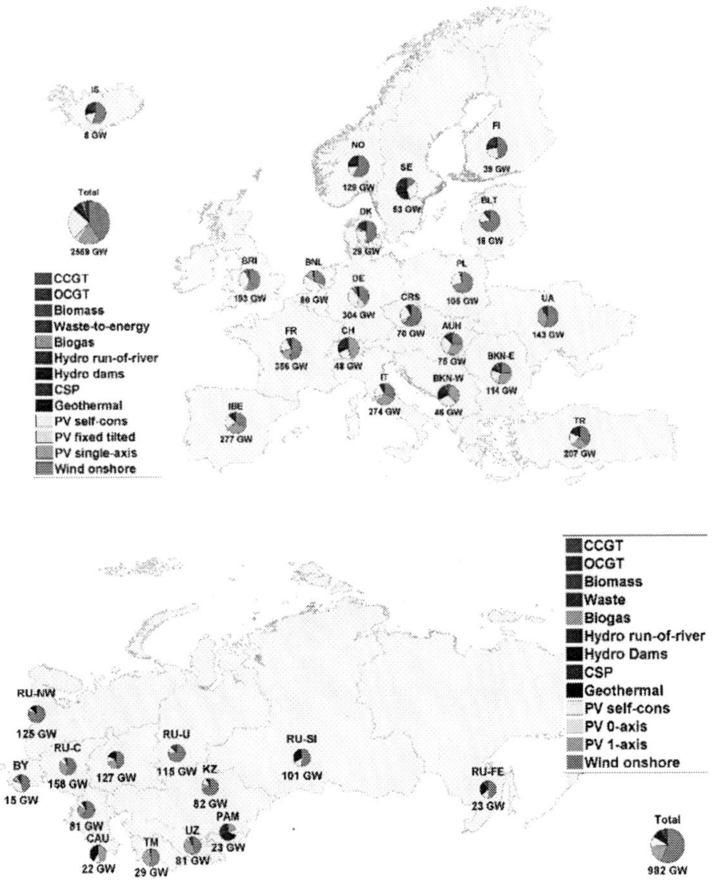

Fig. 5. Installed capacities for the integrated scenario for Europe (top) [40] and Eurasia (bottom) [41].

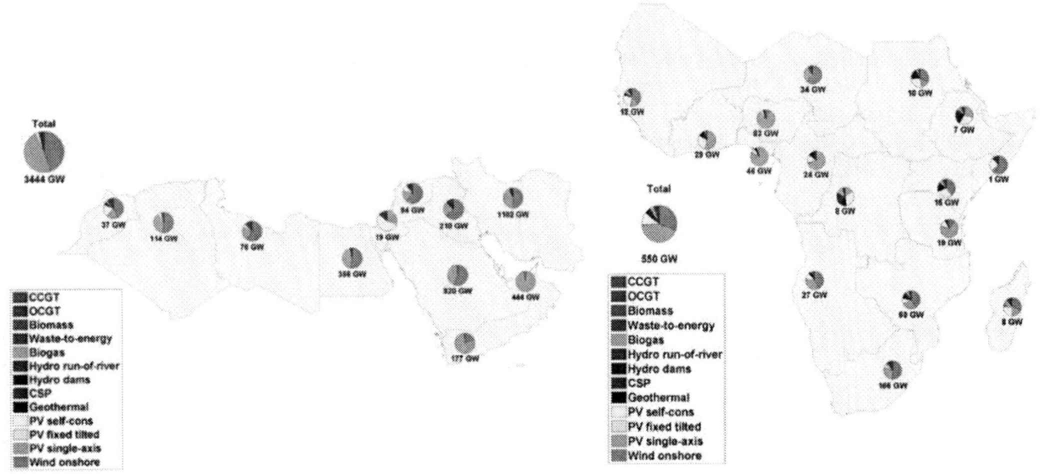

Fig. 6. Installed capacities for the integrated scenario for MENA (left) [42] and Sub-Saharan Africa (right) [43].

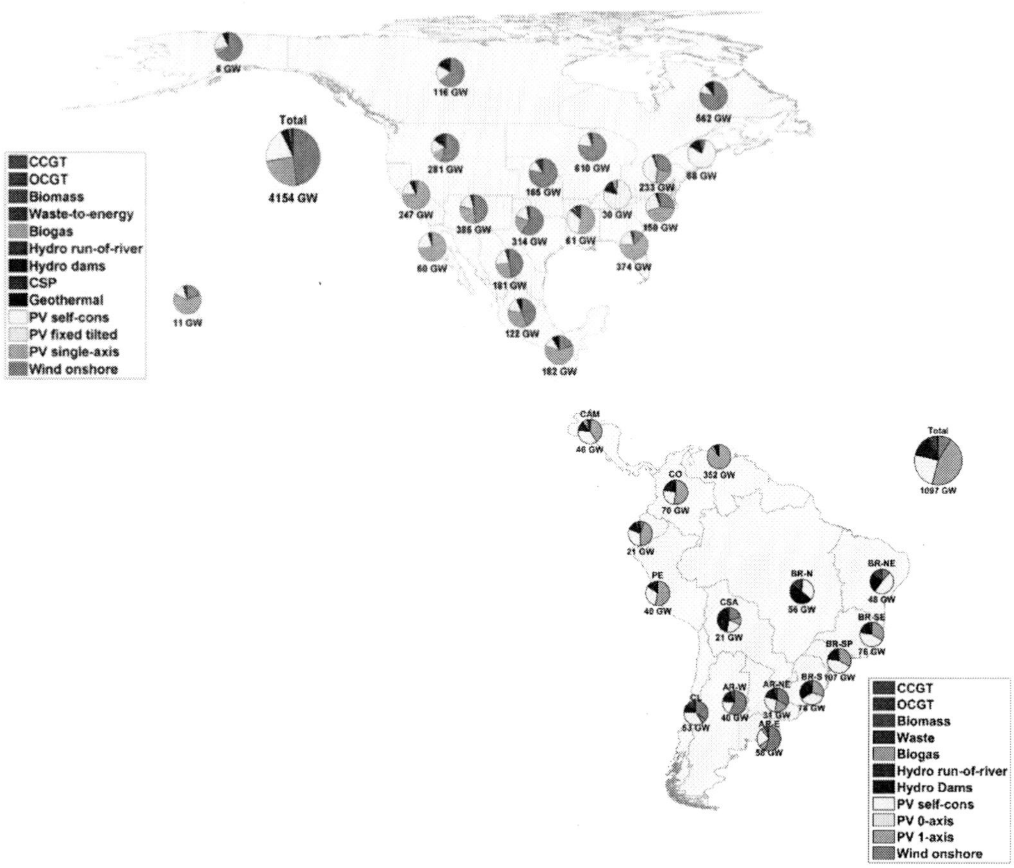

Fig. 7. Installed capacities for the integrated scenario for North America (top) [46] and South America (bottom) [47,48].

Fluctuations which cannot be balanced by a flexible generation and by the geographic integration function of grids are balanced by storage. The most relevant storage technology are batteries, since they provide 63% (region-wide), 69% (area-wide) and 75% (integrated) of all stored electricity. The remaining storage demand is covered by pumped hydro storage, which is operated in a very similar manner as batteries, adiabatic compressed air energy storage (A-CAES) and seasonal power-to-gas (PtG) storage. A-CAES shows its economic optimized value add in weekly storage and corresponds, due to weekly resource fluctuations, best to wind energy. It had been observed that geographic integration (area-wide) substitutes A-CAES almost completely, since within continental grids the wind resource fluctuations occur not in the total continental area and therefore can be balanced by continental grids. This effect is discussed in more detail in Gulagi et al. [70]. For seasonal balance hydro dams are well suited as well as PtG storage. The global PtG capacity demand for seasonal storage is found to 450 GW_{el} for the region-wide rather decentralised scenario approach and this demand shrinks drastically to 242 GW_{el} for the area-wide more centralized scenario approach. The PtG capacities in the integrated scenario increase substantially to 2093 GW_{el}, which is a consequence of industrial gas demand, since it is assumed that the industrial gas demand cannot be supplied anymore by fossil natural gas in 100% RE system, but by biomethane and RE-based PtG. Seasonal storage for the power system is then a by-product of the large industrial gas capacities, as qualitatively already expected by Agora Energiewende [71]. The overall storage demand, measured in electricity provided by storage in relation to the electricity demand is found to be 19% (region-wide), 14% (area-wide) and 10% (integrated). This means that most of the generated electricity is used directly without storing it and the geographic distribution of large interconnected grids reduces the storage demand by about 25% (relative), however the sector integration (here with desalination and industrial gas demand) further reduces the storage demand by slightly less than 30% (relative). Continental grids provide flexibility for the energy system in the geographic dimension, whereas a sector integration provides flexibility in the temporal dimension.

The total battery storage capacities are 11.4 TWh_{el} (region-wide), 9.1 TWh_{el} (area-wide) and 9.9 TWh_{el} (integrated).

The respective shares of prosumer PV battery to the total battery capacities are 44% (region-wide), 54% (area-wide) and 50% (integrated), i.e. around half of the global stationary battery demand is driven by prosumer PV end-users.

The LUT Energy system model is further developed in its functionality, but for the results presented in this paper about 45% of the total primary energy demand is covered in the integrated scenario. On this basis, the results can be roughly extrapolated to estimate the total PV capacity demand, which is found for the year 2030 assumptions to be 27.4 TWp, since the integrated scenario covers roughly 45% of TPED (not yet included are the sectors: heating, mobility, industry and others). Already now, these findings can be further extrapolated to estimate the long-term PV demand (for the year 2100) based on 10 billion people on planet Earth and a per capita demand equal to current levels in Europe, which leads to about 42.3 TWp, and represents a solar PV supply share of roughly 45% of TPED.

A further key result of the LUT Energy system model are the costs of the energy system. It is not presented in detail here, since the energy transition scenarios do not present the cost in much detail, neither the input assumptions of the technology cost nor on the output cost of the system. Some scenarios give some insights, but detailed information are missing for all scenarios. Nevertheless, the LUT Energy system model delivers detailed cost results, which are presented in much detail in the respective publications [30,32,40-48]. One of the most interesting results of the 100% RE system model with 2030 assumptions are the low cost of the energy systems around the world. Part of the cost calculation are the levelised cost for generation, storage, curtailment and HVDC power lines summarised in the total levelised cost, based on capital expenditures, operational expenditures, lifetimes, fuel cost for biomass, technical efficiencies and cost of capital. The calculation of the levelised cost is presented in much detail in Bogdanov and Breyer [30]. The global demand weighted average cost are for the region-wide scenario 65 €/MWh (range of 58 – 72 €/MWh), for the area-wide scenario 58 €/MWh (range of 53 – 67 €/MWh) and the integrated scenario 49 €/MWh (range of 42 – 64 €/MWh). The cost degression from region-wide to area-wide is due to more efficient use of all system components due to geographic flexibility provided by the grid interconnection of sub-regions and the further cost degression from the area-wide to the integrated scenario is due to the provided flexibility of the sector coupling, in particular since the SNG production delivers seasonal storage more or less as a by-product of the industrial gas supply. The electricity trading among the sub-regions is in the global average about 15% of the total demand in the integrated scenario (and similar in the area-wide scenario), which implies that the 100% RE system solution shows a strong decentral distributed structure. Solar PV electricity is typically not traded among the sub-regions, since the solar resource is quite evenly distributed and roughly accessible in the same hours and transmission over many thousands of kilometres has been found to be not economical attractive, mainly due to the cost competitiveness of local storage [44,45,30,72].

5. Discussion

The key results of the LUT Energy system modelling are a PV capacity demand of 7.1 – 9.1 TWp for power sector only consideration and 12.0 TWp for energy sectors representing about 45% of TPED for year 2030 assumptions. The share of solar PV electricity of the total generation is about 41%. The extrapolation to the full energy system leads for 2030 assumptions to 27.4 TWp and for an expected demand in the year 2100 to 42.3 TWp.

This can be compared to the findings of the major energy transition scenarios. In capacity units for the year 2040 and 2050. Comparable numbers for 2040 are only presented by the Greenpeace Energy [R]evolution [18] scenarios with 5.0 and 6.7 TWp. IEA WEO [9] projections with 1.1 and 1.5 TWp are only a fraction of what could be possible. Jacobson and Delucchi [25] also describe a 100% RE system, but for all energy sectors and find a demand of 32.7 TWp, which is close to the extrapolation of the 27.4 TWp found with the LUT Energy system model. Findings for the year 2050, in which the energy system should be close to a net zero system to reach the COP21 targets [73] indicated significantly more progressive expectations of other energy transition scenarios, such as 6.7 – 9.3 TWp for Greenpeace Energy [R]evolution, 11.0 TWp for the PV Power Systems Programme of the IEA, 20.0 TWp for the Shell Oceans scenario and 32.7 TWp derived by Jacobson and Delucchi. Summing up, Jacobson and Delucchi find similar capacities for their 100% RE system, however they derive it not on an hourly basis and therefore do not have battery storage in their energy system, which seems not realistic given the fast cost decline of batteries, but it is a consequence of their chosen equilibrium approach. For all other energy transition scenarios for the 2040 assumptions only the Greenpeace Energy [R]evolution scenario finds comparable PV capacities and for the year 2050 comparable findings are derived by Greenpeace, IEA-PVPS and Shell Oceans.

The share of solar PV in the electricity sector is assumed by Greenpeace to be at around 20% and by WWF at 29%, which are both closest to the 41% found with the LUT Energy system model, but both are significantly lower than the LUT Energy system model results. Taking into account all solar energy technologies then the picture is found to be different, since several energy transition scenarios find substantial shares for the year 2050, such as 22% and 29% by Greenpeace, 30% by WWF, 17-24% by IIASA GEA, 28% by the WBGU and 40% by Jacobson and Delucchi.

Noteworthy, the IPCC scenarios are all below 20% and two are between 1% and 7%, clearly indicating the outdated cost assumptions.

Latest results on energy system transition modelling for the case of Saudi Arabia [74] indicates that the solar PV share for the energy supply will substantially grow beyond the year 2030, since the solar PV share grew up to 80% for Saudi Arabia for the year 2050 as a consequence of ongoing learning curve progress of solar PV and batteries..

The results of the LUT Energy system model for 2030 assumptions cannot be compared to 2100 results of other scenarios, since the ongoing cost decline of solar PV and battery technologies will further increase the PV share, but as a consequence the already found 41% energy supply share can be expected as the lower limit of the long-term solar PV share. Comparable shares are expected by two IPCC models indicating again their outdated cost assumptions, which are not so relevant in the long-term. Jacobson and Delucchi find with 40% a very comparable PV share, as well as the long-term share of the Shell Mountains scenario. Outstanding is the expectation of the WBGU energy transition scenario with 67% solar energy supply share in the year 2100.

The biophysical limits of the global energy system have to be taken strictly into account since humankind is starting to transform planetary systems which led to the begin of a new geological age, the Anthropocene [75]. Bardi [76] added recently a biophysical interpretation emphasising the coming age of solar energy, which is fully supported by the findings of this paper. The substantial capacities of solar PV required for the 100% RE scenario presented in this research can be checked for the area needed to establish that energy harvesting infrastructure. The key results of this study indicate PV capacities of 12.0 TWp, 27.4 TWp and 42.3 TWp. The prosumer PV share is 3.7 TWp, which is allocated on rooftops, which are zero impact areas [77], hence do not trigger additional area. The remaining capacities for land demand are 8.3 TWp, 23.7 TWp and 38.6 TWp. The specific capacity density is derived in the LUT Energy system model to 75 MW/km^2 [30], hence an area of 110,670 km^2, 316,000 km^2 and 514,670 km^2 is needed, representing 0.07%, 0.21% and 0.35% of the global land mass. This very small fraction of global land mass demand seems to be acceptable taken in to account that the current energy system is to more than 50% responsible for the ecological footprint of humankind which equals 1.5 times the capacity of planet Earth [1].

Some of the results of the LUT Energy system modelling are noteworthy for a further discussion. The LUT Energy system model is the only one which calculates on a full hourly resolution on a global level 100% RE with a high geographical resolution of 145 sub-regions (which are composed by resource data for solar and wind in a 0.45° × 0.45° resolution). None of the global energy transition scenarios show this high level of temporal and spatial resolution. This leads to several implications, such as no quantifiable storage capacities and a weak description of the complementarity of solar PV, wind energy, hydro dams and biomass plants. In addition, the most important balancing function of power grids can be not considered in the energy transition scenarios, at all, as well as the cost reducing and efficiency increasing function of a power-based sector coupling is also beyond the methodological features of those models, since they cannot model the flexibility on an hourly scale.

The highly competitive levelised cost results for the LUT Energy system model scenarios with 65 €/MWh (region-wide), 58 €/MWh (area-wide) and 49 €/MWh (integrated), needs to be further highlighted, since this means that 100% RE systems are substantially lower in cost as the cost assumptions for fossil carbon capture and storage (CCS) and new nuclear plants in Europe, which are expected to be around 112 €/MWh for new nuclear (assumed for 2023 in the UK and Czech Republic), 112 €/MWh for gas CCS (assumed for 2019 in the UK) and 126 €/MWh for coal CCS (assumed for 2019 in the UK) [78]. However, a report published by the European Commission [79] concludes that CCS technology is not likely to be commercially available before the year 2030.

The LUT Energy system model is currently further developed to reach a full energy sector description including the sectors heating, mobility and industrial demand. Furthermore, the first energy transition simulations can be done on a single sub-region basis [74] and the coverage of the full global energy system is in preparation.

There had been estimates of the long-term solar PV demand carried out recently leading to slightly more than 90 TWp [80,81] based on a 40% energy supply share of PV. This is in substantial deviation by a factor of two to the found 42.3 TWp in this paper. The preliminary hypothesis for this deviation is that the enormous efficiency improvement of an energy system based on renewable electricity is underestimated in the recent estimates, since today's TPED per capita of Europe is extrapolated for a world population on the same wealth level. However the current energy system is very inefficient, which can be illustrated best by the low average efficiency of the installed base of thermal power plants of only 35% [9], the low efficiency of current combustion cars of less than 20% or the limited efficiency of heating systems to slightly less than 100% - compared to close to 100% utilisation of renewable electricity plants, 70-80% efficient electric vehicles or 300-400% efficient electric heat pumps. The shift to power megatrend may have the potential to increase the total energy system efficiency dramatically. This hypothesis can be more investigated by energy system transition scenarios taking into account very high shares of renewable electricity for the total energy supply.

6. Conclusions

The research question of this paper is about the mid- to long-term solar PV demand in a world based on sustainable energy resources.

The installed solar PV capacity hit 237 GWp by end of 2015 and it is a small fraction of the capacity in the decades to come. The net zero greenhouse gas emission target of the COP21 agreement in Paris [73] may lead to a global energy system based on very high shares of RE. A 100% RE system modeling on the basis of high temporal and spatial resolution based on the LUT Energy system model leads to an energy supply share of solar PV of slightly more than 40% and solar PV capacities of about 27 and 42 TWp from the mid to the end of the 21st century. This outcome is lower than recent long-term PV demand estimates of about 90 TWp [80,81] based on a comparable PV supply share despite an identical assumed energy services demand, since the efficiency gains of the shift to power megatrend may be higher than previously anticipated (e.g. heat pumps and electric vehicles are 3-4 times more efficient and power-to-gas/liquids may be comparably efficient to biofuels and the conventional fossil fuel value chain [82,83]) and system benefits from sector integration. The most progressive global benchmarking scenarios can be confirmed by the key findings of this study, however, now based on a more sophisticated fundamental methodology. Furthermore, clear result of 100% RE system analysis based on LUT Energy system modeling are the low cost between 49 – 65 €/MWh for the given scenario assumptions, which is substantially more competitive than low carbon fossil-CCS and nuclear alternatives.

Acknowledgements:

The authors gratefully acknowledge the public financing of Tekes (Finnish Funding Agency for Innovation) for the 'Neo-Carbon Energy' project under the number 40101/14. Ashish Gulagi and Upeksha Caldera would like to thank Fortum Foundation and Reiner Lemoine-Stiftung, respectively, for the valuable scholarship.

References:

[1] [WWF] – World Wild Fund for Nature International, 2014. Living Planet Report 2014: Species and spaces people and places, WWF, Zoological Society of London, Global Footprint Network, Water Footprint Network, Gland
[2] [EWG] - Energy Watch Group, 2013. Fossil and Nuclear Fuels – the Supply Outlook, EWG, Berlin
[3] Rogelj J., McCollum D.L., Riahi K., 2013. The UN's 'Sustainable Energy for All' initiative is compatible with a warming limit of 2 °C, nature climate change, 6, 545-551
[4] Oxfam Australia, 2015. Powering up against Poverty – Why Renewable Energy is the Future, Melbourne, www.oxfam.org.au/what-we-do/food-and-climate/this-is-climate-in-action/coal-and-poverty/
[5] [UN] - United Nations, 2013. World Population Prospects – The 2012 Revision, UN, New York, http://esa.un.org/wpp/
[6] Child M. and Breyer Ch., 2016. Transition and transformation: a review of the concept of change in the progress towards future sustainable energy systems, submitted
[7] [BNEF] - Bloomberg New Energy Finance, 2015. New Energy Outlook 2015, BNEF, London, June 23
[8] [REN21] – Renewable Energy Policy Network for the 21st Century, 2016. Renewables 2016 – Global Status Report, Paris, www.ren21.net/wp-content/uploads/2016/06/GSR_2016_Full_Report_REN21.pdf
[9] [IEA] – International Energy Agency, 2015. World Energy Outlook 2015, IEA, Paris, www.worldenergyoutlook.org/
[10] [EWG] – Energy Watch Group, 2015. IEA creates misleading future scenarios for solar power generation, Berlin, November 27, http://energywatchgroup.org/iea-creates-misleading-future-scenarios-solar-power-generation/
[11] [IEA-PVPS] - International Energy Agency Photovoltaic Power Systems Programme, 2016. Snapshot of Global PV Markets 2015, St. Ursen, www.iea-pvps.org/fileadmin/dam/public/report/statistics/IEA-PVPS_-__A_Snapshot_of_Global_PV_-_1992-2015_-_Final.pdf
[12] PV Market Alliance, 2016. The PV Market Alliance forecasts a growing global PV market exceeding 60 GW in 2016 and 70 GW in 2017, Hong Kong, Brussels, Madrid, Tokyo, San José, June 16, http://pvmarketalliance.biz/
[13] [SEPA] – Solar Electric Power Association, 2015. Utility Solar Market Snapshot – Sustained Growth in 2014, Washington, May, www.solarelectricpower.org/media/322918/solar-market-snapshot-2014.pdf
[14] pv-magazine, 2016. Mexico awards more than 1 GW of solar at US$40.50/MWh, Berlin, March 30, www.pv-magazine.com/news/details/beitrag/mexico-awards-more-than-1-gw-of-solar-at-us4050-mwh_100023944/#axzz4BlYC8nq7
[15] pv-magazine, 2016. Peru awards 185 MW of solar PV at US$48/MWh, Berlin, February 17, www.pv-magazine.com/news/details/beitrag/peru-awards-185-mw-of-solar-pv-at-us48-mwh_100023273/#axzz4BlYC8nq7
[16] pv-magazine, 2016. Third phase of Dubai's DEWA solar project attracts record low bid of US 2.99 cents/kWh, Berlin, May 2, www.pv-magazine.com/news/details/beitrag/third-phase-of-dubais-dewa-solar-project-attracts-record-low-bid-of-us-299-cents-kwh_100024383/#axzz4BlYC8nq7
[17] Metayer M., Breyer Ch., Fell H.-J., 2015. The projections for the future and quality in the past of the World Energy Outlook for solar PV and other

Renewable Energy technologies, 31st EU PVSEC, Hamburg, September 14-18
[18] Greenpeace International, 2015. energy [r]evolution – A sustainable world energy outlook 2015, Amsterdam, a report commonly published with GWEC and SPE
[19] Komoto K., Breyer Ch., Cunow E., Megherbi K., Faiman D., Vleuten van der P., (Eds.), 2013. Energy from the Desert: Very Large Scale PV Power – State of the Art and Into the Future, Earthscan, London, publication of IEA-PVPS Task8
[20] [IRENA] – International Renewable Energy Agency, 2016. REmap: Roadmap for a Renewable Future, 2016 ed., IRENA, Abu Dhabi, www.irena.org/remap
[21] [WWF] - World Wild Fund for Nature International, 2011. The Energy Report: 100% Renewable Energy by 2050, WWF, Ecofys and Office for Metropolitan Architecture, Gland
[22] [IIASA] - International Institute for Applied Systems Analysis, 2012. Global Energy Assessment (GEA), IIASA, Laxenburg, www.globalenergyassessment.org
[23] [IPCC] - Intergovernal Panel on Climate Change, 2014. IPCC 5th Assessment Report: Working Group III – Mitigation of Climate Change, IPCC, Geneva
[24] [WBGU] – German Advisory Council on Global Change, 2003. World in Transition – Towards Sustainable Energy Systems, Berlin
[25] Jacobson M.Z. and Delucchi M.A., 2011. Providing all global energy with wind, water, and solar power, Part I: Technologies, energy resources, quantities and areas of infrastructure, and materials, Energy Policy, 39, 1154-1169
[26] Royal Dutch Shell, 2013. New Lens Scenarios: A Shift in Perspective for a World in Transition, The Hague
[27] [MIT] – Massachussetts Insitute of Technology, 2015 Energy and Climate Outlook, MIT, Massachusetts
[28] BP, 2015. BP Energy Outlook 2035, London
[29] Exxon Mobil Corp., 2015. Outlook for Energy: A View to 2040, Irving
[30] Bogdanov D. and Breyer Ch., 2016. North-East Asian Super Grid for 100% Renewable Energy supply: Optimal mix of energy technologies for electricity, gas and heat supply options, Energy Conversion and Management, 112, 176-190
[31] Caldera U., Bogdanov D., Breyer Ch., 2016. Local cost of seawater RO desalination based on solar PV and wind energy - A global estimate, Desalination, 385, 207-216
[32] Gulagi A., Bogdanov D., Breyer Ch., 2016. Solar Photovoltaics – A driving force towards a 100% renewable energy system for India and the SAARC region, 32nd EU PVSEC, Munich, June 20-24
[33] Stackhouse P.W. and Whitlock C.H., (eds.), 2008. Surface meteorology and solar energy (SSE) release 6.0, NASA SSE 6.0, earth science enterprise program. National Aeronautic and Space Administration (NASA), Langley; 2008. http://eosweb.larc.nasa.gov/sse/
[34] Stackhouse P.W. and Whitlock C.H., (eds), 2009. Surface meteorology and solar energy (SSE) release 6.0 Methodology, NASA SSE 6.0, earth science enterprise program, National Aeronautic and Space Administration (NASA), Langley, http://eosweb.larc.nasa.gov/sse/documents/SSE6Methodology.pdf
[35] Stetter D., 2012. Enhancement of the REMix energy system model: global renewable energy potentials optimized power plant siting and scenario validation. Dissertation, Faculty of Energy-, Process and Bio-Engineering, University of Stuttgart
[36] Koskinen O. and Breyer Ch., 2016. Energy Storage in Global and Transcontinental Energy Scenarios: A Critical Review, 10th International Renewable Energy Storage Conference (IRES 2016), Düsseldorf, March 15-17
[37] Bogdanov D. and Breyer Ch., 2016. Global grid for 100% renewable energy supply, Energy Systems Conference 2016: 21st Century Challenges, London, June 14-15
[38] Bogdanov D., Koskinen O., Aghahosseini A., Breyer Ch., 2016. Integrated renewable energy based power system for Europe, Eurasia and MENA regions, 5th International Energy and Sustainability, Cologne, June 30 – July 1
[39] Gulagi A., Bogdanov D., Fasihi M., Breyer Ch., 2016. East Asian Super Grid: Can Australia become an electricity source for Asia?, 6th Solar Integration Workshop, Vienna, November 14-15, submitted
[40] Breyer Ch., Koskinen O., Bogdanov D., 2016. Europa 100% erneuerbar, 14. Nationale Photovoltaik-Tagung, Bern, February 22-23
[41] Bogdanov D. and Breyer Ch., 2015. Eurasian Super Grid for 100% Renewable Energy power supply: Generation and storage technologies in the cost optimal mix, ISES Solar World Congress 2015, Daegu, Korea, November 8-12
[42] Aghahosseini A., Bogdanov D., Breyer Ch., 2016. The MENA Super Grid towards 100% Renewable Energy Power Supply by 2030, 11th International Energy Conference, Tehran, May 30-31
[43] Barasa M., Bogdanov D., Oyewo S., Breyer Ch., 2016. A Cost Optimal Resolution for Sub-Saharan Africa powered by 100 Percent of Renewables by the Year 2030, 32nd EU PVSEC, Munich, June 20-24
[44] Breyer Ch., Gulagi A., Bogdanov D., 2015. South-East Asia and the Pacific Rim Super Grid for 100% RE power supply, IEA-PVPS Task 1 Meeting – GÜNDER Workshop, Istanbul, October 27
[45] Gulagi A., Bogdanov D., Breyer Ch., 2016. Southeast Asia and the Pacific Rim Super Grid for 100% Renewable Energy Power Supply, submitted
[46] Aghahosseini A., Bogdanov D., Breyer Ch., 2016. 100% Renewable Energy in North America and the Role of Solar Photovoltaics, 32nd EU PVSEC, Munich, June 20-24
[47] Barbosa L.S.N.S. Bogdanov D., Vainikka P., Breyer Ch., 2015. Complementarity of hydro, wind and solar power as a base for a 100% RE energy supply: South America as an example, Rio 15 – World Climate & Energy Event, Rio de Janeiro, September 4
[48] Barbosa L.S.N.S, Bogdanov D., Vainikka P., Breyer Ch., 2016. Hydro, wind and solar power as a base for a 100% Renewable Energy supply for South and Central America, submitted

[49] Gredler C., 2008. Das Wachstumspotenzial der Photovoltaik und der Windkraft – divergierende Wahrnehmungen zentraler Akteure, Diploma thesis, University Salzburg
[50] Norwegian Climate Foundation, 2014. IEA and Solar PV: Two Worlds Apart, Bergen, report 1/2014
[51] Carbon Tracker Initiative, 2015. Lost in Transition: How the energy sector is missing potential demand destruction, London October, www.carbontracker.org/report/lost_in_transition/
[52] [BNEF] – Bloomberg New Energy Finance, 2015. Future of Energy Summit, Shanghai, November 4
[53] Fasihi M., Bogdanov D., Breyer Ch., 2015. Economics of global LNG trading based on hybrid PV-Wind power plants, 31st EU PVSEC, Hamburg, September 14-18
[54] Fasihi M., Bogdanov D., Breyer Ch., 2016. Techno-economic Assessment of Power-to-Liquids (PtL) Fuels Production and Global Trading Based on Hybrid PV-Wind Power Plants, 10th International Renewable Energy Storage Conference (IRES 2016), Düsseldorf, March 15-17
[55] Palm E., Nilsson L.J., Åhman M., 2016. Electricity-based plastics and their potential demand for electricity and carbon dioxide, Journal of Cleaner Production, 129, 548-555
[56] Citi Research, 2015. Paris Climate Change Agreement, Sydney, December 14
[57] Tjaden T., Krien U., Breyer Ch., 2013. Simulation und techno-ökonomischer Vergleich von solarthermischen Heizungskonzepten und Photovoltaik-Wärmepumpen-Kombinationen im Wohnungssektor, 23. Symposium Thermische Solarenergie, Bad Staffelstein, April 24-26
[58] Afanasyeva S., Breyer Ch., Engelhard M., 2016. The Impact of Cost Dynamics of Lithium-Ion Batteries on the Economics of Hybrid PV-Battery-GT Plants and the Consequences for Competitiveness of Coal and Natural Gas-Fired Power Plants, 10th International Renewable Energy Storage Conference (IRES 2016), Düsseldorf, March 15-17
[59] Breyer Ch. and Afanasyeva S., 2016. Hybridization of PV Systems for a better Match of Demand Requirements and a faster Diffusion of solar PV, Global Photovoltaic Conference, Daegu, April 6
[60] Breyer Ch., Afanasyeva S., Brakemeier D., Engelhard M., Giuliano S., Puppe M., Schenk H., Hirsch T., Moser M., 2016. Assessment of Mid-Term Growth Assumptions and Learning Rates for comparative Studies of CSP and Hybrid PV-Battery Power Plants, SolarPACES, Abu Dhabi, October 11-14, accepted
[61] Fraunhofer ISE, 2015. Current and Future Cost of Photovoltaics. Long-term Scenarios for Market Development, System Prices and LCOE of Utility-Scale PV Systems, study on behalf of Agora Energiewende, Freiburg, Berlin, www.agora-energiewende.de/fileadmin/Projekte/2014/Kosten-Photovoltaik-2050/AgoraEnergiewende_Current_and_Future_Cost_of_PV_Feb2015_web.pdf
[62] [ITRPV] – International Technology Roadmap for Photovoltaic, 2016. International Technology Roadmap for Photovoltaic (ITRPV) – 2015 Results, 7th ed., study by VDMA Photovoltaic Equipment and SEMI, Frankfurt a.M., http://www.itrpv.net/Reports/Downloads/
[63] Kersten F., Doll R., Kux A., Huljić D.M., Görig M.A., Breyer Ch., Müller J., Wawer P., 2011. PV-Learning Curves: Past and Future Drivers of Cost Reduction, 26th EU PVSEC, Hamburg, September 5-9
[64] Hoffmann, W., 2014. Importance and evidence for cost effective electricity storage, 29th EU PVSEC, Amsterdam, September 22-26
[65] Sandalow D., McGormick C., Rowlands-Rees T., Izadi-Najafabadi A., Orlandi I., 2015. Distributed Solar and Storage - ICEF Roadmap 1.0, Innovation for Cool Earth Forum (ICEF) and Bloomberg New Energy Finance (BNEF), London, www.bbhub.io/bnef/sites/4/2015/12/distributed_solar_and_storage-icef_roadmap.pdf
[66] Luderer G., Krey V., Calvin K., Merrick J., Mima S., Pietzcker R., Van Vliet J., Wada K., 2014. The role of renewable energy in climate stabilization: results from the EMF27 scenarios, Climatic Change, 123, 427-441. plus Supplementary Material
[67] Vartiainen E., Masson G., Breyer Ch., 2015. PV LCOE in Europe 2015-2050, 31st EU PVSEC, Hamburg, September 14-18
[68] Gerlach A.-K., Stetter D., Schmid J., Breyer Ch., 2011. PV and Wind Power – Complementary Technologies, 26th EU PVSEC, Hamburg, September 5-9
[69] Barbosa L.S.N.S., Farfan Orozco J., Bogdanov D., Vainikka P., Breyer Ch., 2016. Hydropower and Power-to-Gas Storage Options: The Brazilian Energy System Case, 10th International Renewable Energy Storage Conference (IRES 2016), Düsseldorf, March 15-17
[70] Gulagi A., Aghahosseini A., Bogdanov D., Breyer Ch., 2016. Comparison of the Potential Role of Adiabatic Compressed Air Energy Storage (A-CAES) for a Fully Sustainable Energy System in a Region of Significant and Low Seasonal Variations, 10th International Renewable Energy Storage Conference (IRES 2016), Düsseldorf, March 15-17
[71] Agora Energiewende, 2014. Stromspeicher in der Energiewende, Berlin, September, www.agora-energiewende.de/fileadmin/Projekte/2013/speicher-in-der-energiewende/Agora_Speicherstudie_Web.pdf
[72] Bogdanov D. and Breyer Ch., 2016. Influence of Storage Technologies' Cost Development on the Economics of Power Transmission Grids, 10th International Renewable Energy Storage Conference (IRES 2016), Düsseldorf, March 15-17
[73] [UNFCCC] – United Nations Framework Convention on Climate Change, 2015. Adoption of the Paris Agreement – Proposal by the President, UNFCCC, Paris, December 12, http://unfccc.int/resource/docs/2015/cop21/eng/l09.pdf
[74] Caldera U., Bogdanov D., Afanasyeva S., Breyer Ch., 2016. Integration of reverse osmosis seawater desalination in the power sector, based on PV and wind energy, for the Kingdom of Saudi Arabia, 32nd EU PVSEC, Munich, June 20-24
[75] Waters C.N., Zalasiewicz J., Summerhayes C., Barnosky A.D., Poirier C., Gałuszka A., et al., 2016. The Anthropocene is functionally and

stratigraphically distinct from the Holocene, Science, 351, 6269
[76] Bardi, 2016. What Future for the Anthropocene? A Biophysical Interpretation, BioPhysical Economics and Resource Quality, 1, 2
[77] Denholm P. and Margolis R.M., 2008. Land-use requirements and the per-capita solar footprint for photovoltaic generation in the United States, Energy Policy, 36, 3531-3543
[78] Agora Energiewende, 2014. Comparing the cost of low-carbon technologies: what is the cheapest option?, report by Prognos AG on behalf of Agora Energiewende, p. 10-13. www.prognos.com/fileadmin/pdf/publikationsdatenbank/140417_Prognos_Agora_Analysis_Decarbonisationtechnologies_EN.pdf
[79] European Commission, 2014. Integration of renewable energy in Europe, study prepared by KEMA Consulting, DNV GL – Energy, Imperial College and NERA Economic Consulting on behalf of DG Energy, Brussels, p. 7-8 and 54.
[80] Gerlach A., Breyer Ch., Fischer M., Werner Ch., 2015. Forecast of Long-Term PV Installations – Discussion of Scenarios ranging from IEA to the Solar Economy, 31st EU PVSEC, Hamburg, September 14-18
[81] [FVV] – Forschungsvereinigung Verbrennungskraftmaschinen, 2013. Kraftstoffstudie - Zukünftige Kraftstoffe für Verbrennungsmotoren und Gasturbinen, study performed by Ludwig-Bölkow-Systemtechnik (lbst) on behalf of FVV, FVV, Frankfurt, November
[82] Hall C.A.S., Lambert J.G. Balogh S.B., 2014. EROI of different fuels and the implications for society, Energy Policy, 64, 141-152
[83] Breyer Ch., Heinonen S., Ruotsalainen J., 2016. New Consciousness: A societal and energetic vision for rebalancing humankind within the limits of planet Earth, Technological Forecasting and Social Change, accepted

THE RELEVANCE OF PV IN THE OPTIMISATION OF SYNERGIES AMONG HYBRID ENERGY GRIDS IN SMART CITIES – THE ORPHEUS PROJECT

TOPIC 7: PV ECONOMICS, MARKETS AND POLICIES
SUB-TOPIC 7.2: PV GLOBAL ISSUES, POLICIES AND STRATEGIES

Author for correspondence
Silvia Caneva, WIP – Renewable Energies, Sylvensteinstr. 2, 81369 Munich, Germany
Phone: +49 89 720 12 733, E-mail: silvia.caneva@wip-munich.de

Co-Authors
I. Weiss[1)], S. Betz[1)], S. Challet[1)], G. Heilscher[2)], H. Ruf[2)], D. Stakic[2)], K. Ditz[2)], D. Funk[2)], F. Meier[3)]
A. Schuelke[4)], T. G. Noh[4)], A. Papageorgiou[4)], S. Nicolas[4)]

[1)] WIP – Renewable Energies, Germany
[2)] Hochschule Ulm, Germany
[3)] SWU Netze GmbH, Germany
[4)] NEC Europe Ltd., NEC Laboratories Europe, United Kingdom

ABSTRACT

At present, energy grids (electricity, heating/cooling, gas) still mainly operate independently and do not make use of synergies between them. Although interactions and synergies are increasingly apparent, they neither have been comprehensively analysed nor implemented in practise. The optimisation of such synergies is under investigation in the OrPHEuS project, which has the aim of improving the cooperation among energy grids enabling, at the same time, high renewable resources penetration and customer satisfaction. Overall, the technology innovation of the OrPHEuS project is related to the provisioning of a cross-domain multi-utility energy management system. The system targets an increase of efficiency for renewable energy generation and minimizes the overall energy demand and CO_2 emissions in smart cities by exploiting a higher level of Information and Communication Technologies (ICT) in all energy systems.

The OrPHEuS project investigates the implementation of control strategies for the optimisation of synergies among energy grids on specific use cases scenario in two demonstration sites located in the city of Skellefteå in Sweden and in the city of Ulm in Germany. While the use cases scenario to be implemented in the city of Skellefteå is mainly focused on the minimisation of oil usage for heat production, the use cases scenario for the city of Ulm focuses on strategies to optimise the usage of PV surplus from decentralised systems and reduce the stress on the low voltage distribution grid.

This work presents the key role of PV in the optimisation of synergies among energy grids and the benefits from the economic, social and technical points of view. In Germany, more and more households are feeding electricity into the grid. In a suburban area near the city of Ulm, under investigation during the OrPHEuS project, already today about 24% of the electricity in the local grid has been generated by PV panels. Further grid penetration by solar energy risks to overload the grid and to jeopardize public acceptance of the PV technology. One solution could be the renewal of the power grid, but this investment is usually the most expensive solution for network operators. Therefore, the European project OrPHEuS has been developing and simulating innovative control strategies to deal with the future problems faced in Ulm. The researchers of the OrPHEuS consortium have been investigating how energy grids like electricity, gas and district heating can be combined in an intelligent way and use the surplus of solar power to cover part of the thermal energy demand.

The OrPHEuS project is a European project, co-financed by the European Commission under the 7th Framework Programme. The project started in September 2013 and will end in December 2016. It involves nine partners representing the relevant stakeholders in the field of control strategies for hybrid energy grids:

- Research organisations: Hochschule Ulm, Technische Universität Wien, Austrian Institute of Technology, Deutsches Zentrum für Luft und Raumfahrt (DLR) and Luleå University of Technology
- Energy actors: the distributor system operator in Ulm, Stadwerke Ulm/Neu-Ulm Netze, and the utility Skelleftea Kraft in Skelleftea
- ICT providers: NEC

The project is coordinated by WIP – Renewable Energies.

The partners involved in the OrPHEuS project activities described in this paper are mentioned as Co-Authors.

Detailed information on the OrPHEuS project is available at the project website: www.orpheus-project.eu

Keywords: Distribution System Operators (DSO), Photovoltaic, Utilities, Grid Control, Grid Integration, Hybrid Energy Grids

1. PURPOSE OF THE WORK AND APPROACH

The overall aim of the OrPHEuS project is the elaboration of a hybrid energy network control system for smart cities implementing novel cooperative control strategies for optimal interactions between multiple hybrid energy grids. The optimization of synergies among hybrid energy grids will lead to a better integration of renewable energy sources by enabling, at the same time, energy savings and reduction of CO_2 emissions in smart cities. The OrPHEuS control strategies are implemented in two demonstration sites: Ulm in Germany and Skellefteå in Sweden. The energy grids under investigation are hybrid energy grids. In ***hybrid energy grids***, the transformation of energy from one system into another and vice versa is necessary. The transformation can be done through physical systems connecting different energy grids, which are called ***coupling points***. These systems transform input power into another type of power, for example electrical power into heat power. The transformation process could be unidirectional or bidirectional and transforms one energy type into one or more others.

The following coupling points in the demonstration sites of Ulm and Skellefteå have been identified:

- For the connection between the electrical and the district heating grid: e-boiler, heat-pump and CHP (Combined Heat and Power)
- For the connection between the gas and the district heating grid: CHP and gas boiler
- For the connection between the gas and electrical grid: CHP and Power-to gas
- For the optimisation of the control strategies: thermal storage and electric storage

The coupling points are one of the ***key elements*** of the OrPHEuS project since they identify and allow the cooperation among hybrid energy grids.

The OrPHEuS researchers have identified use cases for the implementation of the control strategies to be validated with ***real-life data:***

- Reduction of peak oil boiler usage for district heating in Skellefteå, Sweden
- Storing surplus photovoltaic power as heat in households in Ulm, Germany

Once the use cases and the coupling points were defined, system modelling and the development of a simulation tool could start.

2. SCIENTIFIC INNOVATION AND RELEVANCE

It is important to highlight that the component and system models included in the simulations for the demo sites of Ulm and Skellefteå have been developed on the basis of real-life data and manufacturer data provided respectively by the distribution system operator Stadtwerke Ulm/Neu-Ulm Netze GmbH and the utility Skellefteå Kraft, both partners of the OrPHEuS project. This is one of the ***added values*** of the OrPHEuS project.

An investigation method has been designed to develop and evaluate cooperative hybrid energy network control strategies for each one of the two use cases. This is of high relevance for future National and European projects about hybrid grids.

Indeed, due to its high novelty aspect, at the time being, ***no other projects dealing with the optimisation of the cooperation among hybrid energy grids exist***. Investigation of this hybrid evolution path is still in its infancy. A thorough investigation, which includes major stakeholders that are modelled from ***a real-world city***, will open more convincing ***new possibilities for energy network evolution.***

3. RESULTS AND CONCLUSIONS

The demonstration side in Ulm covers the area of one low voltage grid transformer in the village of Einsingen. The test site contains a substation with a 630 kVA transformer (10 kV/0.4 kV). 133 houses are connected via eight feeder lines to this transformer. The houses are single family houses or multi-family houses as shown in Figure 1.

Figure 1 – Test Area Einsingen in Ulm, Germany, with the position of the PV systems and transformer

21 PV systems are installed and connected to the electrical grid in the test area Einsingen. The corresponding PV capacity is 233 kWp. The electrical demand in the test area corresponds to 1.05 GWh/a, while the PV yield is 0.23 GWh/a. Therefore 24 % of the electrical demand is covered by PV generation. 92 households are connected to the gas grid. The gas demand amounts to 2.26 GWh/a. The space heating demand corresponds to 3.06 GWh/a, while the domestic hot water demand corresponds to 0.27 GWh/a. The gas demand, electrical demand and PV yield are measured values from the meters. The space heating demand and domestic hot water demand are calculated values based on the national standard VDI 4655.

The first results obtained within the OrPHEuS project were related to the definition of the hybrid scenarios to be investigated. Three hybrid scenarios have been proposed for the test area in the City of Ulm:

- Power to Heat (PtH) scenario al local level for Domestic Hot Water (DHW) – ***Scenario 1***
- Power to Heat (PtH) scenario at district level for the District Heating Network (DHN) – ***Scenario 2***
- Power to Gas (PtG) scenario – ***Scenario 3***

The investigation method consists of a two-step subsequent holistic investigation with technical simulation and economic-model based analysis.

The investigated scenarios are compared to a baseline scenario in order to define the social, economic and technical benefits that they would provide.

This paper focuses on the results obtained until now in regards to the ***Power to Heat (P2H)*** scenario 1 for the use case ***"Storing surplus of photovoltaic power as heat in households for Domestic Hot Water (DHW) demand".***

A graphical representation of the selected use case is provided in Figure 2.

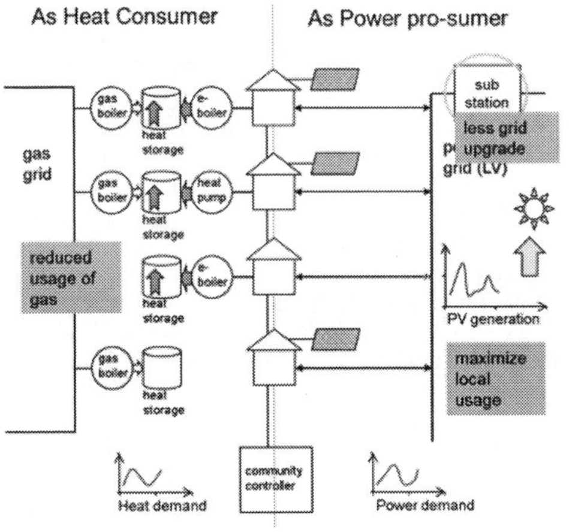

Figure 2 – Graphical representation of the use case "Storing surplus of photovoltaic power as heat in households"

The related cooperative control strategies for this ***Power to Heat (P2H)*** scenario investigate the possibility of ***storing PV surplus power as heat for Domestic Hot Water (DHW) demand*** under the following conditions:

- Surplus reduction by self-consumption - ***Control 1***
- Local reactive: surplus usage in community - ***Control 2***
- Global reactive: all e-devices under central control - ***Control 3***
- Global reactive with surplus predictions - ***Control 4***

Figure 3 shows details of the results from the simulations carried out for Control strategy 2: "Local reactive: surplus usage in community"

Scenario	Heat from all fuels	
January Baseline	**66780 kWh**	
50% PV, January with control 2	44604 kWh	- 33 %
75% PV, January with control 2	34605 kWh	- 48 %
July Baseline	**68846 kWh**	
50% PV, July with control 2	13051 kWh	- 81 %
75% PV, July with control 2	9040 kWh	- 87 %

Figure 3 - Results from the simulations for the Power to Heat (P2H) Scenario 1, Local reactive: surplus usage in community Control 2, for the use case "Storing surplus photovoltaic power as heat in households".

The simulations show that PV surplus power, which has previously not only been wasted but also overloaded LV (Low Voltage) network, can successfully be captured and stored in the form of heat. The investigation also shows the needed number of new hybrid devices, and required controls for them, to accomodate various levels of future (higher) PV penetration expected at the demo site. It shows that grid element upgrade due to PV surplus can be avoided via hybridization.

From the results summarized in Figure 3, it can be seen that during the winter time, having January as baseline scenario, a PV penetration of 50% could lead to savings of 33% on Domestic Hot Water (DHW) heating fuels. With a PV penetration of 75% savings on DHW heating fuels can reach 48%. The simulations carried out for the summer period, having July as baseline scenario, show that the savings on DHW heating fuels can be up to 81% with a PV penetration of 50% and up to 87% with a PV penetration of 75%.

4. ACKNOWLEDGEMENT

The OrPHEuS project is a European project, co-financed by the European Commission under the 7th Framework Programme. The project started in September 2013 and will end in December 2016. It involves nine partners representing the relevant stakeholders in the field of control strategies for hybrid energy grids. The project is coordinated by WIP – Renewable Energies. The logo of each OrPHEuS partners is shown below.

The partners involved in the OrPHEuS project activities described in this paper are mentioned as Co-Authors.

4. DISCLAIMER

The sole responsibility for the content of this publication lies with the authors. It does not necessarily reflect the opinion of the European Commission. The European Commission is not responsible for any use that may be made of the information contained therein.

INTERACTIVE WEB-SERVICE FOR ENVIRONMENTAL MULTI-CRITERIA LIFE CYCLE ASSESSMENT OF PHOTOVOLTAIC SYSTEMS WORLDWIDE

Paula PEREZ-LOPEZ[1], Isabelle BLANC[1], Benoît GSCHWIND[1], Philippe BLANC[1], Lionel MENARD[1], Rolf FRISCHKNECHT[2], Philippe STOLZ[2], Yvonnick DURAND[3], Garvin HEATH[3]
[1]MINES ParisTech, [2]treeze ltd, [3]ADEME, [4]NREL
[1]Sophia Antipolis BP 207, F-06904, Cedex France, tel.: +33 (0)4 97 15 7055;

ABSTRACT: Photovoltaic (PV) technologies constitute a leading renewable energy source with a worldwide installed capacity of 135 GW in 2013 that may increase to nearly 4700 GW in 2050. To achieve this production level while minimizing environmental impacts, decision makers must rely at national level on relevant technological, economic and planning aspects which are highly geographically dependent. The access to performance data is a critical issue in the decision-making process and determines the successful development of efficient PV systems. For this reason, a new interactive tool is proposed here to provide the users with easy-to-use data and maps for the solar irradiation and screening level environmental results of representative PV technologies. The calculation procedures account for the geographic location and the PV system layout (installation, orientation and inclination angles). The tool has a worldwide coverage with a multi-criteria scope, both in terms of the numerous technological scenarios and of the wide range of environmental indicators. Moreover, the user is given the possibility to compare the PV environmental performance to the corresponding country electricity mix environmental footprint.

Keywords: Photovoltaic, Web-Service, environmental performance, solar irradiation data, IEA PVPS Task 12, OGC standards.

1 INTRODUCTION

Photovoltaic (PV) technologies constitute a leading renewable energy source in the context of energy transition [1,2]. The worldwide installed capacity is expected to increase from 135 GW in 2013 to nearly 4700 GW in 2050 [2]. Although PV systems rely on sunlight, which is an extensive renewable source, and cause nearly no emissions during the use (power generation) phase, other processes over the production chain corresponding to materials extraction, components' manufacturing, transport and end-of-life management should be computed to quantify the environmental costs over the whole life cycle of the panels [3,4]. To achieve the worldwide production target while maintaining a low level of environmental impacts in the upstream and downstream stages, techno-economic and planning criteria affecting the theoretical generation potential of PV systems need to be included. Geographically-dependent factors that may alter the PV power production and, thus, the environmental performance should also be taken into account [5,6].

The access to detailed data that support the decision-making process is essential to enhance the development of more efficient PV systems [6]. With the aim of contributing to the supply of accessible information, we propose an interactive online tool to provide the users with easy-to-use data and maps for the solar irradiation and environmental performance results at screening level of representative PV technologies, according to a Life Cycle Assessment (LCA) approach. This tool was developed in the framework of Task 12 of the Photovoltaic Power Systems Program (PVPS) established by the International Energy Agency (IEA). The calculation procedures account for the geographic location and the PV system layout, namely installation, orientation and inclination angles. The tool has a worldwide coverage with a muti-criteria scope, both in terms of the numerous technological scenarios [7,8] and the wide range of environmental indicators [9-11]. Moreover, the user is given the possibility to compare the PV environmental performance to the corresponding country electricity mix, inventoried according to treeze Ltd. [12].

2 PURPOSE OF THE WORK

The aim of this work is to propose an easy-to-use interactive tool to assess the environmental footprint of current and prospective representative PV systems at screening level accounting for its geographic location. To enlarge the audience and to ease their access, the tool is available online [http://viewer.webservice-energy.org /project_iea/]. The geo-localized environmental performances provided by our tool are compared with the environmental footprint of the national electricity mix on the same methodological background to allow decision makers to position the environmental interest of PV systems.

3 APPROACH

Although renewable sources, and particularly PV systems, exhibit remarkably lower emissions and better environmental performances than fossil sources during their utilization, their impact during manufacturing and installation as well as end-of life management phases should be quantified to assess their sustainability [1-4].

Life cycle assessment (LCA) is one of the main tools recommended for the environmental evaluation of emerging technologies such as PV systems [3,4,13]. LCA is a standardized methodology that aims at assessing the potential environmental impacts of processes over their whole life cycle from the extraction and processing of raw materials to the use and end-of life stages. The environmental impacts are quantified by classifying the emitted substances into specific impact categories and converting their values into reference units through sets of characterization factors [14].

Environmental performances for energy systems are expressed as the ratio of their environmental impacts issued from LCA, over the produced energy along their life time (Eq. 1). For PV systems, the environmental impacts are calculated according to the methodological guidelines, published in the framework of the IEA PVPS Task 12 [9]. The quantification of multi-annual mean of surface solar global irradiations enables determining the total energy production at a specific location over a given life time.

$$\text{Environ perform} = \frac{\text{Total impact over lifetime}}{\text{Generated electricity}} \qquad (1)$$

The developed tool allows obtaining the environmental performances of PV systems in the form of a Web-client invoking a Web-Service that combines algorithms for the calculation of PV life cycle impacts and annual solar irradiation [6]. The Web-Service is based on representative life cycle inventories developed within IEA PVPS program [7,8]. The expected electric yearly production of the PV systems has been computed from the freely available solar irradiation NASA SSE database (https://eosweb.larc.nasa.gov/sse). More precisely, the 1°x1° worldwide maps of multiannual (22 years) average of monthly global and diffuse irradiations have been used to compute the expected yearly global irradiations on the tilted plane corresponding to the PV systems applying the same algorithm used for the NASA SSE database itself which corresponds to the RETScreen method [15].

After obtaining the annual irradiation for a given location, our tool computes the total electricity generated over the whole life cycle of the PV system, as well as the environmental performance, and provides the results to the user.

Regarding the range of results provided by the tool, the Web-client has been developed to propose an interactive user interface matching the foreseen request from the users. The tool offers the user the possibility to obtain multi-criteria LCA results including diverse impact categories of widespread impact assessment methodologies such as ILCD 2011 Midpoint+, Cumulative Energy Demand, IMPACT 2002+, IPCC 2013 and USETox among others [9-11].

3 SCIENTIFIC INNOVATION AND RELEVANCE

This Web-Service is the first tool to propose a worldwide coverage of environmental performance of PV systems. It is developed using a state of the art interoperable and open standards Web-Service framework from the Open Geospatial Consortium (OGC), namely Web Processing Service (WPS). It combines the latest updated life cycle inventories, published in 2015 by the IEA Task 12, and the solar irradiation database compiled from the worldwide NASA SSE database for a wide range of inclinations and orientations. A new Web-client based on a previously developed version [6] is proposed for the user to compare the PV performance to the national LCA electricity mix from the WPS. The implemented LCA categories cover a wide range of environmental issues.

4 RESULTS

To obtain the environmental performance for a specific impact category, the user needs to make a choice for the following parameters:

(1) the PV technology,
(2) the type of system (power output and mounting),
(3) the techno-economic context,
(4) the performance ratio,
(5) the orientation and inclination,
(6) the PV lifetime and
(7) the environmental impact assessment method and impact category.

Figure 1 shows the cascading menu developed to facilitate the selection of the parameters.

a)

b)

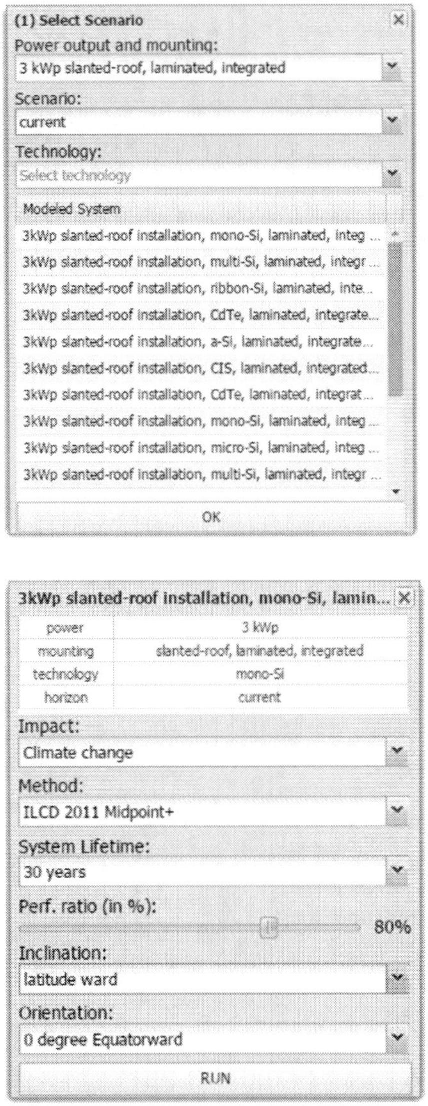

Figure 1: Web-Service cascade menu for the selection of: a) PV technology, mounting and techno-economic context and b) impact assessment method, category, and energy production conditions including the performance ratio, the lifetime, the inclination and the orientation.

The interactive Web-client allows the user to investigate many different technological scenarios for representative 3kWp or 570kWp systems (Figure 2) combined with different options of installations, orientation and inclination. The techno-economic context is taken into account by selecting either "current" or "prospective" scenarios, following the scenarios defined by IEA PVPS Task 12 [7-8].

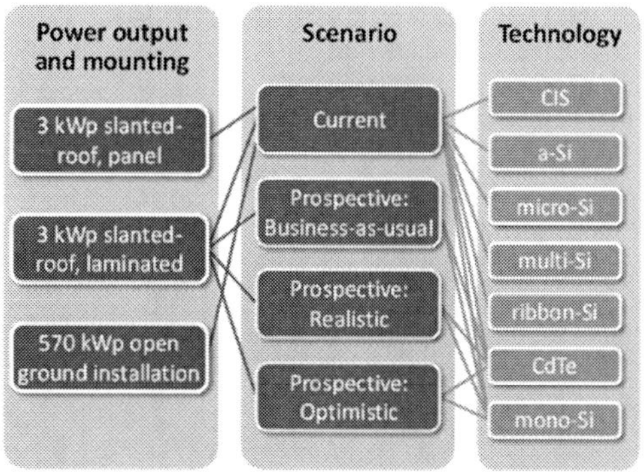

Figure 2: Available production scenarios provided by the Web-Service represented by links between boxes.

After the user's selection of the requested parameters, the Web-Service provides worldwide maps of LCA results at screening level for the selected impact category. By clicking on a specific location on the map the user get for this location: the latitude, the longitude, the mean annual solar irradiation, the environmental impact over the whole life cycle, the performance as impact per kWh, the environmental footprint of the corresponding country electricity mix for an equivalent level of energy production and the relative difference between the local PV environmental performance minus the performance of the country electricity mix. The user can export those data for further processing with the "export" button.

An illustration of the environmental performance for a specific representative PV is given in Figure 3, obtained by operating the Web-Service http://viewer.webservice-energy.org/project_iea/.

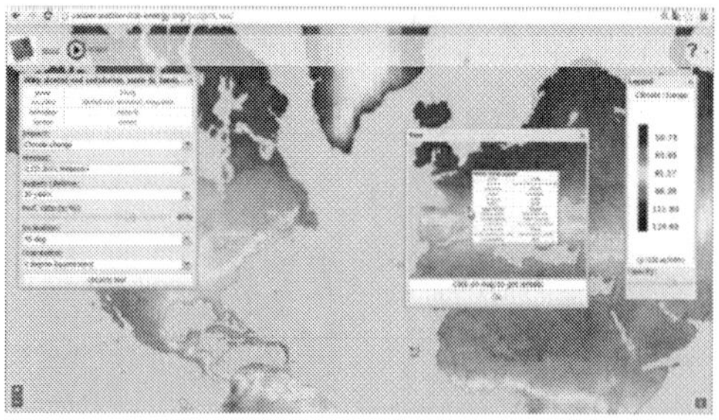

Figure 3: Worldwide map of climate change (ILCD 2011 Midpoint + method) obtained from the Web-Service for 3 kWp slanted-roof laminates of current CdTe technology with a 30-year lifetime at 45° inclination and south orientation.

LCA practitioners can process the environmental results provided by the Web-Service to carry out three types of comparative environmental analyses:

1. A given technology A vs another technology B: Figure 4 shows the comparison of the existing laminate technologies for the impact categories of climate change (equivalent to carbon footprint), non-renewable cumulative energy demand and human toxicity (cancer effects) according to ILCD 2011 Midpoint+ method.

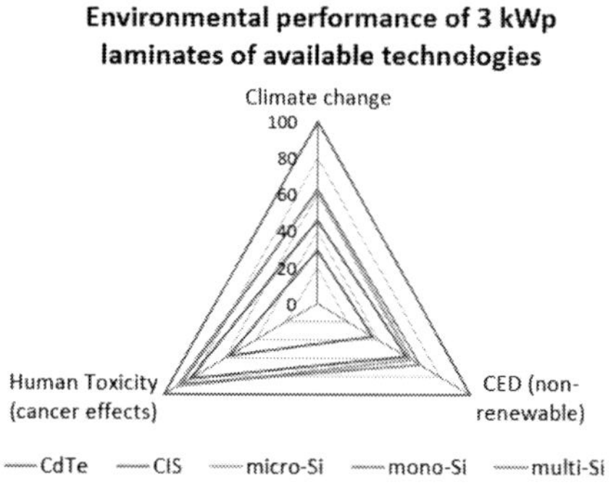

Figure 4: Comparative environmental performance of 3 kWp laminates of current PV technologies available in the Web-Service, operating at 45° inclination and south orientation

2. The current vs prospective techno-economic context for a given technology: Three prospective techno-economic contexts are provided for mono-Si and CdTe laminates. The business-as-usual scenario estimates the future context by considering the continuation with current policies and no additional goals. The realistic scenario corresponds to a situation in which energy policies would give a relatively high priority to the empowering of renewable energy policies. The optimistic scenario considers the implementation of ambitious energy policies and high development of efficient renewable energy technologies.

The environmental benefits of mono-Si prospective scenarios in the three selected categories are depicted in Figure 5. According this figure, the business-as-usual scenario may lead to limited impact reductions ranging between 10% and 20%, whereas the realistic scenario involves significant improvements from 35% to nearly 50% and the optimistic scenario may allow higher reductions, between 40% and 60%.

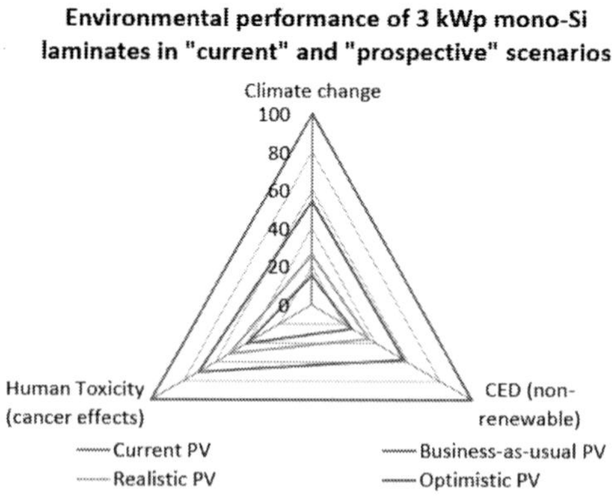

Figure 5: Comparative environmental performance of current and prospective 3 kWp mono-Si laminates operating at 45° inclination and south orientation.

3. A given technology vs the current country electricity mix: In Figure 6, we propose the comparison of PV and country mix environmental performances in different countries and geographic locations. The locations were selected to represent the extreme environmental performances within each country. As shown in Figure 6, the comparative results of the climate change for the mono-Si laminates and the corresponding country electricity mix are highly dependent on the geographical location and the specific characteristics of each national grid. Thus, the differences are rather limited in the case of France due to the low carbon emissions associated with the high percentage of nuclear energy and low production from fossil fuels. On the contrary, countries with a high percentage of electricity coming from fossil sources, such as Germany or US, exhibit a remarkably lower impact on climate change of the PV systems than that of the country mix.

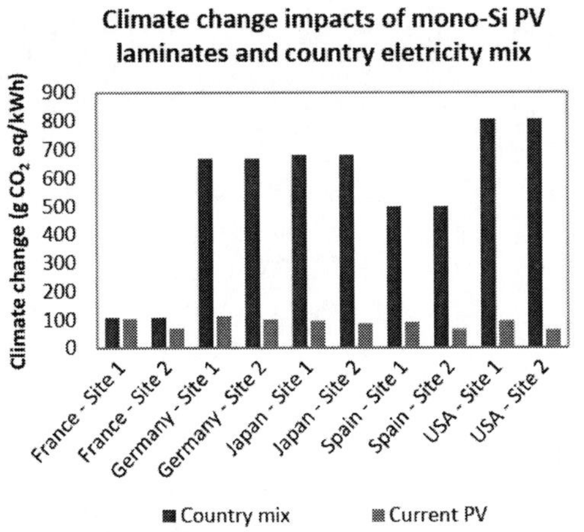

Figure 6: Comparative environmental performance of current 3 kWp mono-Si laminates operating at 45° inclination and south orientation with respect to country electricity mix.

5 CONCLUSIONS

We developed an easy-to-use interactive Web-Service based on current OGC standards to provide environmental data on existing PV technologies. The tool offers accessible and adapted information to support decision-making processes.

Analyzing the differences of the PV performance of the available technologies and providing comparisons with respect to the performance of the corresponding national electricity mix open the way to understand the positioning of PV mix relying on a multi-criteria assessment. Easy and open access to this key information is essential to feed the debate on energy transition.

6 ACKNOWLEDGMENTS

This research has been undertaken with the framework and financial support of the International Energy Agency in the framework of Task 12 of the Photovoltaic Power Systems Program (PVPS), as well as with the financial support of ADEME (France).

7 REFERENCES

[1] Hong, S., Chung, Y., Woo, C., 2015. Scenario analysis for estimating the learning rate of photovoltaic power generation based on learning curve theory in South Korea. Energy 79:80-89.

[2] International Energy Agency (IEA), 2014. Technology roadmap: Solar photovoltaic energy. Technical Report.

[3] Raugei M, Frankl P. Life cycle impacts and costs of photovoltaic systems: Current state of the art and future outlooks. Energy 2009; 34(3): 392-399.

[4] Beylot A, Payet J, Puech C, Adra N, Jacquin P, Blanc I, Beloin-Saint-Pierre D. Environmental impacts of large-scale grid-connected ground-mounted PV installations. Renewable Energy 2014; 61: 2-6.

[5] Fthenakis V, Mason JE, Zweibel K. The technical, geographical and economic feasibility for solar energy to supply the energy needs of the US. Energy Policy 2009; 37(2):387-399.

[6] Ménard, L., Blanc, I., Beloin-Saint-Pierre, D., Gschwind, B., Wald, L., Blanc, P., Ranchin, T., Hischier, R., Grianfranceschi, S., Smolders, S., Gilles, M., Grassin, C., 2012. Benefit of GEOSS interoperability in assessment of environmental impacts illustrated by the case of photovoltaic systems. IEEE Journal of Selected Topics in Applied Earth Observations and Remote Sensing 5:1722-1728.

[7] Frischknecht, R., Itten, R., Wyss, F., Blanc, I., Heath, G., Raugei, M., Sinha, P., Wade, A., 2014. Life cycle assessment of future photovoltaic electricity production from residential-scale systems operated in Europe. International Energy Agency (IEA). PVPS Task 12, Report T12-05-2015.

[8] Frischknecht, R., Itten, R., Sinha, P., de Wild-Scholten, M., Zhang, J., Fthenakis, V., Kim, H.C., Raugei, M., Stucki, M., 2015. Life cycle inventories and life cycle assessment of photovoltaic systems. International Energy Agency (IEA). PVPS Task 12, Report T12-04:2015.

[9] Frischknecht, R., Heath, G., Raugei, M., Sinha, P., de Wild-Scholten, M., Fthenakis, V., Kim, H.C., Alsema, E., Held, M., 2016. Methodology guidelines on Life Cycle Assessment of photovoltaic electricity, 3rd edition, IEA PVPS Task 12, International Energy Agency Photovoltaic Power systems Programme. Report IEA-PVPS T12-06:2016.

[10] Frischknecht, R., Jungbluth, N., Althaus, H.J., Bauer, C., Doka, G., Dones, R., Hischier, R., Hellweg, S., Humbert, S., Köllner, T., Loerincik, Y., Margini, M., and Nemecek, T., 2007. Implementation of life cycle impact assessment methods. Ecoinvent report No. 3, v2.0. Swiss Centre for Life Cycle Inventories, Dübendorf (Switzerland).

[11] IPCC, 2013. Climate Change 2013: The physical science basis. Contribution of working group I to the fifth assessment report of the Intergovernmental Panel on Climate Change. Stocker, T.F., Qin D., Plattner, G.K., Tignor, M., Allen, S.K., Boschung, J., Nauels, A., Xia, Y., Bex, V., Midgley, P.M. (eds.). Cambridge University Press, Cambridge (United Kingdom) and New York (NY, USA), pp. 1535.

[12] Itten, R., Frischknecht, R., Stucki, M., 2012. Life cycle inventories of electricity mixes and grid. Version 1.3. treeze, Uster (Switzerland).

[13] Wender, B.A., Foley, R.W., Prado-Lopez, V., Ravikumar, D., Eisenberg, D.A., Hottle, T.A., Sadowski, J., Flanagan, W.P., Fisher, A., Lauring, L., Bates, M.E., Linkov, L., Seager, T.P., Fraser, M.P., Guston, D.H., 2014. Illustrating anticipatory Life Cycle Assessment for emerging photovoltaic technologies. Environmental Science & Technology 48:10531-10538.
[14] ISO 14040:2006. Environmental management – Life cycle assessment – Principles and framework.
[15] Stackhouse, P.W., Whitlock, C.H. (eds.), 2014. Surface meteorology and Solar Energy (SSE) Release 6.0 Methodology, version 3.1.2, NASA, pp. 62 (available: https://eosweb.larc.nasa.gov/sse/documents/SSE6Methodology.pdf last access: 2016/01/20).

7DO.14.1 A SNAPSHOT OF GLOBAL PV MARKETS - THE LATEST SURVEY RESULTS ON PV MARKETS AND POLICIES FROM THE IEA PVPS PROGRAMME 2016

Ir. Gaëtan Masson, IEA PVPS, Belgium g.masson@iea-pvps.org
Pius Hüsser, Nova Energie, Switzerland Pius.huesser@novaenergie.ch
Izumi Kaizuka, RTS Corporation , 4F 3-19-2 Hatchobori, Chuo-ku, Tokyo 104-0032, Japan, kaizuka@rts-pv.com
Johan Lindahl, Spokesperson, Svensk Solenergi, Holländargatan 17 111 60, Sweden, johan@svensksolenergi.se
Arnulf Jaeger-Waldau, EC-JEC, Italy, Arnulf.JAEGER-WALDAU@ec.europa.eu
Angelo Baggini, Copper Alliance, Italy, angelo.baggini@unibg.it
Francesca Tilli, Italy, Gestore Servizi Energetici, francesca.tilli@gse.it
Guastella Salvatore, Italy, RSE, Salvatore.Guastella@rse-web.it
Peter Ahm, PA Energy, Denmark, ahm@paenergy.dk
Grégory Neubourg, APERE, Belgium, gneubourg@apere.org
José Donoso, UNEF, Spain, j.donoso@unef.es
Chinho Park, Yeungnam University, Korea, chpark@ynu.ac.kr
Vicente Salas, UC3M, Spain, vicente.salas@uc3m.es
Jero Ahola, LUT, Finland, jero.ahola@lut.fi
(on behalf of Task 1 of the IEA PVPS Programme)

ABSTRACT: This snapshot report gives information on the development of PV power applications in the PVPS member and non-member countries and is largely based on the information provided by the 24 IEA PVPS countries plus the European Union through its European Commission. The report includes information on national market developments and comments about the most important changes and trends in PV support policies at the end of the former year, in this case 2015. The International Energy Agency – Photovoltaic Power System Programme (IEA PVPS)'s Task 1 is responsible for strategy and outreach within the IEA PVPS program. This includes market and industry analysis. A key deliverable of Task 1 is the annual ***Snapshot of Global PV Markets*** publication, together with the TRENDS IN PV APPLICATIONS.

This paper presents the latest survey results for the calendar year 2015 concerning PV markets and policies, as well as other key issues. An increasing number of national markets experienced notable growth in 2015 with impacts on policy development. While the final figures for 2015 will continue to be refined in the coming months, preliminary figures show that close to 51 GW of PV systems have been installed in the world last year. Some important trends observed are as follows:

- The **global PV market grew to at least 50.7 GW** in 2015, compared to around 39.7 GW in 2014 (revised figure).
- **Asia** scored the very first place again in 2015 with more than 57 % of the global PV market.
- The market in **Europe** grew to 8 GW for the first time in years. It has decreased significantly from 22.5 GW in 2011 to 17.5 GW in 2012, 11 GW in 2013 and 7 GW in 2014. But in 2015 it grew for the first time in years at 8 GW.
- The **Asia Pacific** region installing close to 31 GW in 2015 has experienced the fastest market development in 2013 as well as 2014. China took the first place (with 15.2 GW installed), ahead of Japan (10.8 GW) and the USA (7.3 GW). The first European country ranked fourth, with the 3.8 GW installed in United Kingdom.
- In the top 10 countries, there are 3 European countries (UK, Germany and France), 5 Asia-Pacific countries (China, Japan, Korea, Australia, India), no country in Africa (South Africa installed relatively small amounts of PV in 2015) and two countries in the Americas region (USA and Canada).
- **Germany, Italy and Greece** have now enough PV capacity to produce respectively **7.1 %, 8.0 %** and **7.4 %** of their annual electricity demand with PV. 15 countries have enough PV to produce at least 1.3 % of their electricity demand by PV.
- PV represents **3.8** % of the electricity demand in Europe and **7%** of the peak electricity demand.
- **22 countries had at least 1 GW** of cumulative PV systems capacity at the end of 2015 and 7 countries installed at least 1 GW in 2015.
-

1 INTRODUCTION

1.1. Purpose of The Work

The objective of the series of annual Snapshot and Trends reports – which have been published since 1992 (Trends) and 2013 (Snapshot) – is to present and interpret developments in both the PV systems and components being used in the PV power systems market and the changing applications for these products within that market. These trends are analyzed in the context of the business, policy and non-technical environment in the reporting countries.

These reports are prepared to assist those responsible for developing the strategies of businesses and public authorities, and to aid the development of medium term plans for electricity utilities and other providers of energy services. It also provides guidance to government officials responsible for setting energy policy and preparing national energy plans. Finally, it represents the most comprehensive and complete PV market analysis on a global level.

1.2. Approach

Key data for this publication were drawn mostly from national survey reports (up to 2014) and information summaries, which were supplied by representatives from each of the reporting member countries of IEA PVPS (for 2015 data). These national survey reports and the annual Snapshots can be found on the website www.iea-pvps.org. Information from the countries outside IEA PVPS are drawn from a variety of sources and, with every attempt made to ensure their accuracy.

The 24 countries currently participating in the IEA PVPS programme are Australia (AUS), Austria (AUT), Belgium (BEL), Canada (CAN), China (CHN), Denmark (DNK), Finland (FI), France (FRA), Germany (DEU), Israel (ISR), Italy (ITA), Japan (JPN), Korea (KOR), Malaysia (MYS), Mexico (MEX), the Netherlands (NLD), Norway (NOR), Portugal (PRT), Spain (ESP), Sweden (SWE), Switzerland (CHE), Thailand (THA) Turkey (TUR) and the United States of America (USA). The European Commission (EC), the Solar Power Europe association (SPE), the US Solar Electric Power Association (SEPA) and the US Solar Energy Industries Association (SEIA) and Copper Alliance are also members.

1.3. Results And Conclusion

Trends in photovoltaic applications is currently being compiled and the following results are based on the snapshot of global PV data collected in 2015, together with data published in National Survey Reports.

2 MARKET DEVELOPMENT

2.1. Cumulative Global Installed Capacity

The 24 IEA PVPS countries represented around 197 GW of cumulative PV installations altogether, mostly grid-connected, at the end of 2015. The other 39 countries that have been considered and are not part of the IEA PVPS programme represented 30 additional GW, mostly in Europe: UK with close to 9.1 GW, Greece with 2.6 GW, Czech Republic with 2.1 GW installed, Romania with 1.3 GW and Bulgaria with 1.0 GW and below the GW mark Slovakia. Outside of Europe, the major non-IEA PVPS countries that accounted for the highest cumulative installations in 2015 were India with more than 5.1 GW, South Africa with 1 GW, Taiwan with 1 GW and in Chile with 0.8 GW. Other countries installed significant amounts of PV but their tracking is not easy since they are not reporting official numbers. Pakistan for instance has installed around 800 MW cumulative, based on PV modules import numbers but this remains to be verified. Ukraine has also installed more than 800 MW so far, despites the losses of Crimean installations to Russia.

Meanwhile a difference of 1 or 2 GW out of 227 GW represents an error margin less than 2% that can be considered as scientifically acceptable. It must be noted that several countries are reporting officially AC numbers. These AC numbers refer to the theoretical output of PV systems connected to the grid, with a methodology that can differ according to the country and the segment. The correction factor to transform these AC numbers into DC numbers (that are compiled here) differs according to the country and the market segment. While for residential installations, the ratio between DC power and AC power is relatively small (and sometimes close to 1 or even below 1 in case of an inverter slightly larger than the PV installation itself, for ground-mounted installations, the ratio DC/AC can go up to 1.4 for utility-scale applications. While such a number seems high, ratios around 1.3 are mentioned by First Solar in 2016 (Annual report 2015). In that respect, AC reporting countries are mentioned below with their AC market size. Japan that reports in AC, reports also the equivalent DC number according to the Japanese grid connection regulations: 9797.4 MW AC correspond in Japan to 10811.4 MW DC.

The reporting in AC amounts to at least 40 GW (34 GW in Japan, 5 GW in Spain, plus Canada and Austria for instance). Assuming an average correction factor of 10% instead of 15%, these 40 GW could generate a difference of 2 additional GW. These uncertainties correspond today to less than 1% of the total installed capacity. Meanwhile, it is important to consider AC and DC numbers since most tenders (for instance in Chile or UAE) consider AC number. The 200 MW_{AC} tender in Dubai that got the 58.5 USD/MWh PPA corresponds in fact to a 250 MW_{DC} PV plant.

Finally the question of the non-reporting countries is at the core of the debate on the final numbers. In this paper, we build on data produced by a consortium of researchers (Ch. Werner & All) that estimated the installations in non-reporting countries from this paper to 228 GW.

The figure 1 below illustrates the evolution of cumulative installed PV power in the world.

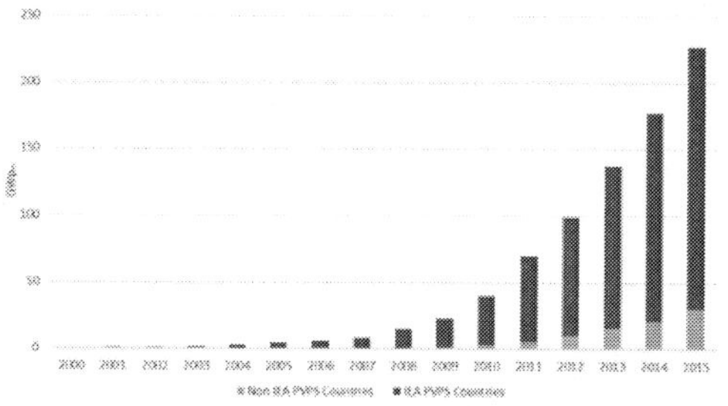

Figure 1: Cumulative installed PV power in the World (GW-DC)

2.2. Annual Installed Capacity in 2015

The IEA PVPS countries have installed 40 GW of PV. While they are difficult to track with a high level of certainty, installations in non-IEA PVPS countries were pushing the installed capacity above 50 GW in 2015.

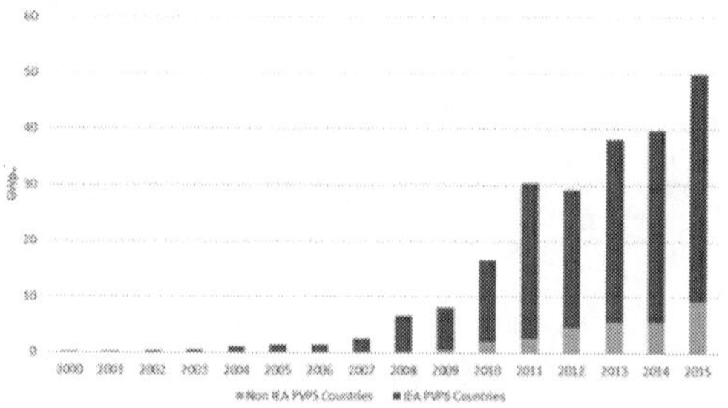

Figure 2: Annual installed PV power in the World (GW-DC)

With around 50.7 GW, the market grew in 2015 by around 25%, again the highest installation ever for PV. China installed 15.2 GW in 2015, according to the National Energy Administration, a record level significantly higher than the 10.6 GW and 10.95 GW from the last two years (2014, 2013) that placed the country in first place with regard to all time PV installations in 2013. This is perfectly in line with their political will to develop renewable sources and in particular PV in the short to medium term. Meanwhile more was initially planned for and the lower level reached can be explained by local difficulties to ramp up the distributed PV market and probably the beginning of rather high curtailment rates in some provinces far from consumption centers.

The second place goes to Japan for the second year in a row, with 10.8 GW installed in the country in 2015, up from 2014 numbers. The USA installed 7.3 GW, up from 6.2 GW of PV systems in 2014, with a growing share of large utility-scale PV compared to rooftop installations.

The UK grew significantly again in 2015, becoming the first country for PV installations in Europe with 3.8 GW. The market has been powered by the rather unique set of support schemes in Europe still favoring ground-mounted installations. More than half of all 2015 installations occurred anyway in the first quarter of the year, which predicts a market decline in 2016.

India installed 2.1 GW in 2015, while a growing number of installations are reported, amidst huge expectations powered by the policy decision to install up to 100 GW before 2022.

Together, these 5 countries represent 77% of all installations recorded in 2015, roughly the same percentage as in 2014 while the top 5 of cumulative capacities reached 70% in terms of installed capacity (China, Germany, Japan, USA, Italy).

Germany installed 1.46 GW, a major decline after the three years with PV installations around 7.5 GW, followed by 2013 at 3.3 GW and 2014 at 1.9 GW. The total installed PV capacity in Germany is now more than 39.6 GW but the country was for the first time passed by China that became number one with more than 43 GW of cumulative PV capacity at the end of 2015.

The following four places go to the France (0.9 GW), Korea (1.0 GW), Australia (0.9 GW) and Canada (0.6 GW). Together these top 10 countries cover 88% of the 2015 world market, a percentage that is decreasing slowly.

Amongst countries that in the past installed significant levels of PV, Italy installed only 300 MW, down from the 9.3 GW in 2011, 3.6 GW in 2012 and 1.7 GW in 2013. It has reached a capacity of 18.9 GW.

Several other countries where the PV market used to develop in the last years have performed in various ways: Belgium installed only 95 MW and has now reached more than 3.2 GW. Some countries that grew dramatically over recent years have now stalled or experienced limited additions: Spain (56 MW) now totals 5.4 GWDC of PV systems followed by Czech Republic (2 MW) at 2.1 GW.

In Europe, net-metering systems allowed the market to grow quickly in Denmark but the transition to self-consumption pushed the installations down despites a significant market improvement in 2015 (180 MW) thanks to utility-scale plants. In the Netherlands (500 MW estimated), 2015 saw significant additions while the market stabilized in Switzerland (300 MW) and in Austria (150 MW).

Malaysia installed 63 MW for the fourth year of its Feed-in Tariff (FiT) system. Taiwan installed around 400 MW in a growing market. In Latin America, official data for Chile shows the installation of 800 MW in two years, a first step towards PV deployment in the region. Several additional GW of PV plants have been validated in Chile, while projects are popping up in Brazil and Honduras (400 MW). The real PV development of grid-connected PV plants has finally started in the region but much more is expected in 2016, especially with the competitive tenders that have been won in Mexico, Peru or Brazil.

In the Middle East, Israel continued in a similar way (200 MW), while the PV installations in Turkey have finally started but slower than expected with around 208 MW installed in 2015. Many new projects have been announced, especially in the UAE and in Egypt and other countries are rapidly building PV plants, such as Jordan or Algeria (268 MW in 2015).

PV installations have been reported in dozens of countries in the world, with 2.3 GW installed in non-reporting countries (source: Ch. Werner and all, op. cit.). Since this number comes mainly from imports of modules in these countries, it should be considered as the maximum maximorum. To the 40.7 GW in IEA-PVPS countries, 7.7 GW can be added in other countries where PV installations are followed properly (those mentioned in this document with installation numbers) and 2.3 GW additional installations, pushing the 2015 installations numbers up to 50.7 GW.

2.3. Shift to Asia

While Europe represented a major part of all installations globally in 2011, the share of Asia and America started to grow rapidly from 2012, with Asia taking the lead. This evolution is quite visible from 2011 to 2014, with the share of the Asia Pacific region growing from 18% to close to 60% in 2014 and 2015, whereas the European share of the PV market went down from 74% to 15% in five years. This trend shows that the development of PV globally is no longer in the hands of European countries but rather has become global, powered by the demand for electricity and the population per region rather than only policy decisions.

Finally, the share of the PV market in the Middle East and in Africa remains relatively small compared to other regions of the world, despite the growth of the South African market and the numerous projects in UAE or Egypt.

2.4. Utility-scale PV in progress

The evolution of grid-connected PV towards a balanced segmentation between centralized and decentralized PV has reversed course in 2013 and continued its trend in 2014 and 2015: centralized PV has evolved faster and most of the major PV developments in emerging PV markets are coming from utility-scale PV. Globally, centralized PV represented more than 60% of the market in 2015, mainly driven by China, the USA, and emerging PV markets.

In 2014, utility-scale represented more than 55% of the market and 64% in 2015 (32.6 GW out of 50.7 GW). Prospects for the coming years are going in the same direction with many countries targeting large-scale ground-mounted PV (Brazil, Chile, South-Africa, Indonesia…). Europe was in 2015 the only region where the share of distributed PV was higher than centralized in 2015 and higher than 50%. Moreover distributed installations have stagnated in 2015 compared to 2014 in absolute terms (18 GW compared to 19 GW) and declined significantly in relative terms.

3 POLICY FRAMEWORK

PV development over the last ten years has been powered by the deployment of support policies, aimed at reducing the gap between PV's cost of electricity and the price of conventional electricity sources. These support schemes took various forms depending on the local situations and evolved to cope with unexpected market evolution or policy changes. The figure 3 below shows that only 0.2% of the world PV market has been driven by pure self-consumption or the sole competitiveness of PV installations in 2015. Tenders granting PPAs have represented less than 6% of the new PV capacities in 2015. Self-consumption represented, together with similar schemes such as net-billing or net-metering close to 15% of the new installations. It also means 78.3% of the global PV market depends on financial support schemes. In 2014 and 2015 a large part of the market remained dominated by FiT schemes (more than 59%) outside of tendering processes. Subsidies aiming at reducing the upfront investment (or tax breaks) represent around 16% of the incentives. Incentivized self-consumption including net-billing and net-metering was the main incentive in 2015 for 16% of the world market. Various forms of incentivized self-consumption schemes exist, such as in Italy with the *Scambio Sul Posto*, Israel, or Germany.

The emergence of calls for tenders has been confirmed again in 2015, with new countries using this legal tool to attribute remunerations to PV projects under certain conditions. Germany, France, Peru, Brazil, Mexico, Dubai (UAE), Jordan and many others have joined the list of countries using calls for tenders to grant PPAs for PV plants. The result of these calls for tenders is a guaranteed payment for PV electricity, or in other words, a FiT. Such tenders represented around 5.6% of the world market in 2015 and is increasing.

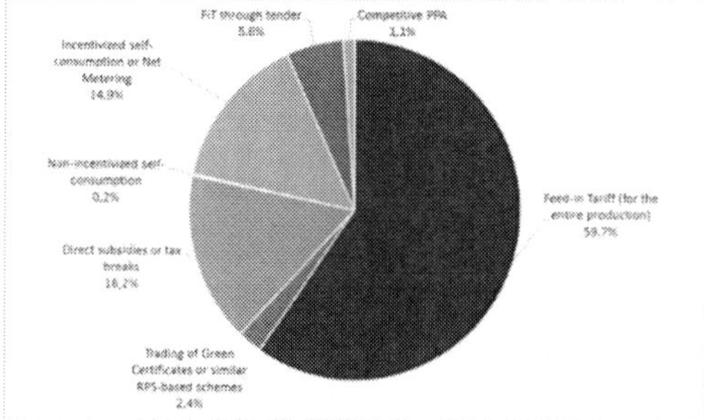

Figure 3: Market Incentives and Enablers in 2015

Historically the dominance of feed-in tariffs and direct subsidies is similar but even more visible in Figure 6 that compiles installations until the end of 2014. Incentives can be granted by a wide variety of authorities or sometimes by utilities themselves. They can be unique or add to each other. Their lifetime is in general quite short, with frequent policy changes, at least to adapt the financial parameters. Next to central governments, regional states or provinces can propose either the main incentive or some additional ones. Municipalities are more and more involved in renewable energy development and can offer additional advantages. In some cases, utilities are proposing specific deployment schemes to their own customers, in general in the absence of national or local incentives.

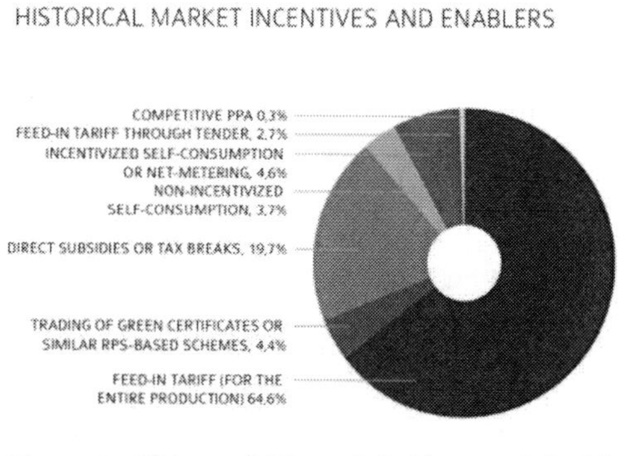

Figure 4: History of PV market drivers and Enablers until 2014 (IEA-PVPS Trends report 2015)

4 PV IN THE POWER SECTOR

Figure 5 shows how PV theoretically contributes to the electricity demand in the IEA PVPS countries, based on the PV base at end 2015. Italy remains the number one country with 8% of its electricity that will come from PV in 2016 based on 2015 total level of installations. This number can be translated into 15 to 16% of the peak electricity demand.

In Germany, with more than 6.7%, the 39.7 GW installed in the country produce up to 50% of the instantaneous power demand on some days, and around 13% of the electricity during the peak periods. Several countries outside the IEA PVPS network have the ability to produce significantly more than 3% of their electricity demand: Honduras (not featured in the figure below) with 12% of its electricity demand, Greece (around 7.6% based on the 2014 installed capacity), Bulgaria and the Czech Republic. Spain remains below the 4% mark as well as Belgium, which is producing 3.6% of its electricity thanks to PV. Romania, Japan, Australia, Switzerland, Slovenia and Israel are above the 2% mark. Denmark and the UK are approaching the 2% mark, while Austria, France, Portugal and Chile are still below the 1.5 % mark. Thailand and the Netherlands passed the 1% of the electricity demand mark in 2015. Many other countries have lower production numbers.

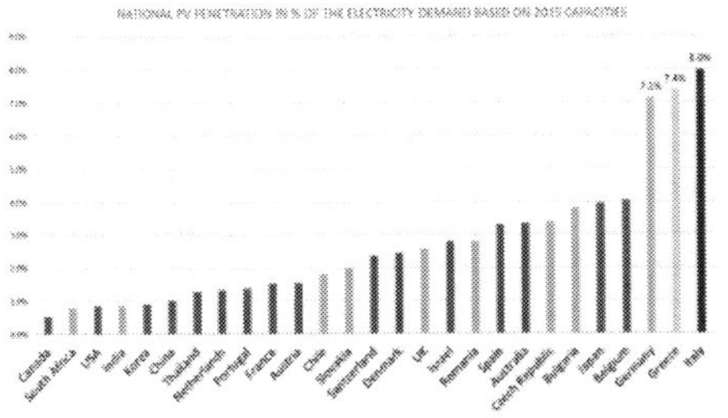

Figure 5: Theoretical PV production share by country based on 2015 cumulative installed capacity.

5 CONCLUSION

The year 2015 experienced a renewed growth of the PV market and confirmed the Asian leadership on the PV market and industry. PV is entering rapidly into a new era where the PV market will be concentrated in countries with energy needs. Two of the top three markets in 2015 were located in Asia (China and Japan), followed by Europe as a whole and the US PV market. This trend should be confirmed again in 2016, with Asia consolidating the core of the PV market, followed by the Americas and Europe, while India is going to modify these results in the coming years.

The challenges are still numerous before PV can become a major source of electricity in the world. The way how distribution grids could cope with high shares of PV electricity, generation adequacy and balancing challenges in systems with high shares of variable renewables, the evolution of energy storages technologies and the cost of transforming existing grids will be at the cornerstone of PV deployment in the coming years. Moreover, the ability to successfully transform electricity markets to integrate PV electricity in a fair and sustainable way will have to be scrutinized.

The road to PV competitiveness is open but remains complex and linked to political decisions. Nevertheless, the assets of PV are numerous and as seen in this edition of the IEA PVPS Snapshot report, the appetite for PV electricity grows all over the world. The road will be long before PV will represent a major source of electricity in most countries, but as some European countries have shown in recent years, PV has the ability to continue progressing fast.

6 REFERENCES

1. 2014 National Survey reports of IEA-PVPS countries, on www.iea-pvps.org
2. 2015 IEA-PVPS Trends in PV Applications
3. IEA-PVPS Snapshot of Global PV Markets 2015
4. GLOBAL PHOTOVOLTAICS in 2015, analysis of current solar energy markets ANd Hidden growth regions, Ch. Werner, A. Gerlach, Ch. Breyer, G. Masson, 2016, 32nd EU-PVSEC

GLOBAL PHOTOVOLTAICS IN 2015
ANALYSIS OF CURRENT SOLAR ENERGY MARKETS AND HIDDEN GROWTH REGIONS

Chris Werner[1], Alexander Gerlach[2], Christian Breyer[3], Gaëtan Masson[4]

[1] Chris Werner Energy Consulting, 06846 Dessau, Germany
Phone: +49 (0) 176 64093993, e-mail chris.werner@cw-energyconsulting.de
[2] Hanwha Q.CELLS GmbH, Sonnenallee 17-21, 06766 Bitterfeld-Wolfen, Germany,
[3] Lappeenranta University of Technology, Skinnarilankatu 34, 53850 Lappeenranta, Finland
[4] Becquerel Institute, 69-71 Rue D'Arlon, 1040 Brussels, Belgium

ABSTRACT
The installed capacity of photovoltaics (PV) is rising steadily. Most PV is installed in highly electrified countries as on-grid systems. Further, there is an increasing number of emerging markets that show high growth rates and a relevant installed PV base. However, reliable installation rates for PV are available only for a small number of countries. Not only in emerging markets, even for some developed countries exact numbers of cumulative PV installation are rarely available. For the end of 2015 IRENA reports 222,360 MWp of globally installed PV capacity, giving clear national specific data for 117 countries. IEA PVPS provides a number of 227,100 MWp installed by providing detailed data for 36 countries.
This paper gives an overview of installed PV capacities for all countries in the world and incl based on the examination of publically accessible data. Furthermore, an analysis of the development of cumulative PV installations in recent years is given. Resulting from this evaluation, PV installations are localized in 196 countries, representing 230,046 MWp.

Keywords
PV Markets, Installed Capacity, PV Installation, Market Growth

1 MOTIVATION

The global photovoltaic (PV) market is the fastest growing of all renewable energy markets [1]. However, many successful countries, as measured by their renewable energy investments per gross domestic product, are typically not known. Public reports do not usually show any data for installed capacities in small PV markets. These markets are usually named as Rest of World (RoW). However, they often show a constant growth of installed PV capacity. Furthermore, in a significant number of those small markets, a rapidly rising growth rate is identifiable. With rising photovoltaic capacities and growing economies in these markets, they have a potential to become important photovoltaic markets in the future. Thus, RoW markets have to be better understood.

As a result, an analysis of cumulative installed PV capacity for all countries in the world is strongly required. This paper gives an analysis of photovoltaic capacities and growth rates ranging from 2009 to today for every market. Further data starting in 2001 are available, but would exceed the scope of this paper. On this basis, an estimation of future growth rates and PV capacities expected to be installed in all markets can be performed.

2 EXAMINATION METHOD

For numerous PV markets no reliable estimation of installed PV is available due to a lack in public authority monitoring. This is especially valid for emerging markets, but a number of developed countries as well. To enhance a reliable valuation of the photovoltaic markets of all countries, the examination of several sources is necessary to ensure a credible estimation.

A basic source is the international customs database, monitored by the 'Market Analysis and Research' section of the International Trade Centre, which is an agency of UN's World Trade Organization. This database contains the customs data of all countries worldwide since the year 2000, receiving the data from the particular national customs authorities. The database provides the opportunity to evaluate the total value of imported and exported products for each country per year. Furthermore, a possibility of evaluating bilateral trades for a specific product is given. Products are classified in several specific product numbers, so-called HS Codes, obligatory for more than 200 countries, customs and economic unions, representing more than 98% of world's trade [2]. HS Code group 854140 represents 'photosensitive semiconductors devices, photovoltaic cells and light emitting diodes'. More specific descriptions with detailed determination of codes between light emitting diodes and PV are available for approximately 95% of the data. Otherwise, an experienced ratio of 80% PV and 20% light emitting diodes is used. Surprisingly, a large number of developing countries' customs agencies have reported detailed data. It is valid that: *'imports A from B'* = *'exports B to A'*. Data have been assigned for 190 countries in a period of 2000 until 2014. The basis of the calculation of PV capacities per country is the following assumption: *'import A'* – *'exports A'* + *'production A'* = *'market A'*. The production volumes refer to the publications of the magazine 'Photon' on the topic of production output [3]. From this, attention is paid to the fact that approximately 25% of annual production will be installed in the following year.

Since data include the monetary value of the products, a conversion to PV capacities is necessary. For this purpose a reliable estimation of PV prices per Wp for every country is needed. These price data refer to a worldwide annual average PV module price per Wp [4-6]. Since prices depend on countries' locations and market sizes, there is a significant ratio of divergence. Therefore, all countries are classified in five main groups with one of them having two subgroups, each representing a group of countries that are nearly comparable in market size and economic conditions. The world's average price is multiplied by a factor quoting the ratio of countries' prices for each of these groups. The ratio is based on experience and statements of several countries' experts that have been contacted [7-11]. This has been validated in previous publications [12-16]. Since the world's given average annual PV prices are selling prices of modules' producers, a factor of 10% is added considering distributor margins, shipping, etc. Therewith, prices for the year 2015 vary between 0.66 USD/Wp for advanced PV market countries and 1.62 USD/Wp for small PV applications in African and Asian low-income countries as can be seen in Table 1.

To lower the probability of errors, further sources with installation rates estimated by IRENA, International Energy Agency Photovoltaic Power Systems Programme (IEA PVPS), journal du photovoltaique, the European Commission, Hanwha Q CELLS market research and public resources are used [1, 17-23]. Some countries report PV installation data by giving DC capacities while others refer to their installed AC capacities. Thus a small probability of mismatch between report data and those number calculated by using the DC capacity according to the customs database is possible. However numbers referring to DC capacities are favored.

Table 1: PV module price for conversion of customs values into PV [5, 6]. Abbreviations: North America (NA) and Latin America (LA).

[USD/Wp]	year	2000	2001	2002	2003	2004	2005	2006	2007	2008	2009	2010	2011	2012	2013	2014	2015
average world price:	*100%*	5.17	4.96	4.36	4.13	4.24	4.55	5.07	4.57	4.38	2.53	2.05	1.65	1.21	0.89	0.85	0.74
countries >100MW/y	*90%*	4.66	4.47	3.92	3.72	3.82	4.09	4.56	4.11	3.94	2.28	1.85	1.49	1.09	0.80	0.76	0.66
countries >10MW/y	*110%*	5.69	5.46	4.79	4.55	4.67	5.00	5.57	5.02	4.82	2.78	2.26	1.82	1.33	0.98	0.93	0.81
countries >5MW/y	*150%*	7.76	7.44	6.54	6.20	6.36	6.82	7.60	6.85	6.57	3.79	3.08	2.48	1.82	1.34	1.27	1.11
countries <5MW/y	*180%*	9.31	8.93	7.84	7.44	7.63	8.19	9.12	8.22	7.89	4.55	3.70	2.97	2.18	1.60	1.52	1.33
countries <1MW/y	*200%*	10.35	9.93	8.71	8.27	8.48	9.10	10.13	9.14	8.76	5.06	4.11	3.30	2.42	1.78	1.69	1.47
Africa & Asia	*220%*	11.38	10.92	9.59	9.10	9.33	10.01	11.14	10.05	9.64	5.57	4.52	3.63	2.66	1.96	1.86	1.62
Europe & NA & LA	*190%*	9.83	9.43	8.28	7.86	8.06	8.64	9.62	8.68	8.33	4.81	3.90	3.14	2.30	1.69	1.61	1.40

Further, all sources are weighted in a certain order. For countries that have their own cell or module production, the arithmetic mean of all sources except customs data is taken for estimation of installation rates, due to production margins and local PV value chain structures influencing the reliability of the customs database. In case of countries having no production, the average of all sources is used.

Due to statistical impacts of additional databases, it is possible in individual cases that estimated installed PV capacities at the end of 2015 can be less than those estimated at the end of previous years.

3 OVERVIEW ON PV INSTALLATIONS 2015

Within the use of this examination, the world's total installed PV capacity is estimated to be up to 230,046 MWp at the end of 2015. It is expected that there is a further 487 MWp that had been installed before the year 2000. PV installations are found in almost all countries. Appendix Table 1 includes the results for PV installation rates for all countries in the world at end of 2015.

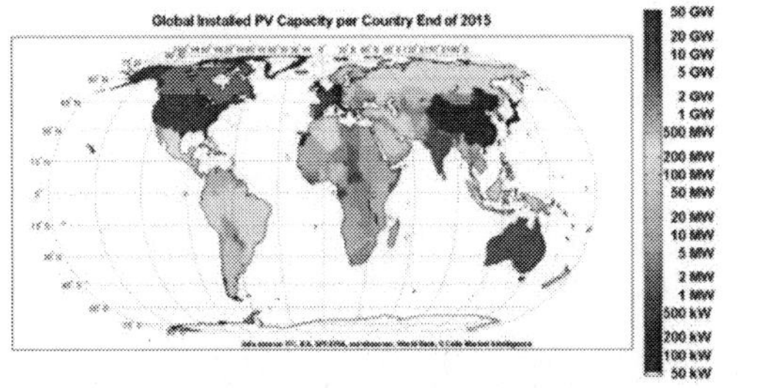

Figure 1: Total installed PV capacity per country by end of 2015.

As can be seen in Figure 1, China, with a share of 19% of total global installed PV, is now the country with leading PV installations having replaced Germany that used to be the country with highest PV installed for more than 10 years. Further, four markets show more than 10 GWp of PV capacity. Another 16 markets claim more than 1 GWp of PV capacity by the end of 2015. Those 21 largest markets are China (44,420 MWp), Germany (39,700 MWp), Japan (33,870 MWp), the United States (25,990 MWp), Italy (18,910 MWp), United Kingdom (9,080 MWp), France (6,540 MWp), India (5,180 MWp), Spain (5,180 MWp), Australia (5,020 MWp), Republic of Korea (3,360 MWp), Belgium (3,220 MWp), Greece (2,610 MWp), Canada (2,530 MWp), Czech Republic (2,170 MWp), Thailand (1,690 MWp), the Netherlands (1,440 MWp), Switzerland (1,360 MWp), Romania (1,350 MWp), South Africa (1,260 MWp) and Bulgaria (1,030 MWp) which in total amount to 94% of global PV installations. Furthermore, there are 29 additional countries claiming more than 100 MWp per country (Taiwan, Austria, Israel, Ukraine, Chile, Denmark, Pakistan, Slovakia, Portugal, Honduras, Turkey, Algeria, Slovenia, Russia, Mexico, Malaysia, Luxembourg, United Arab Emirates, Iran, Philippines, Bangladesh, Peru, Reunion, Brazil, Indonesia, Saudi Arabia, Sweden, Jordan, Puerto Rico). Some of them (Taiwan, Austria, Ukraine) are close to stepping over the 1,000 MWp threshold. Remarkable is the highly increasing share of emerging markets among those countries. Further other emerging and developing countries show a significant installed PV base, in particular Guatemala (83 MWp), Dominican Republic (75 MWp) or Ecuador (74 MWp). Further, even countries that belong the least income countries show respective numbers of PV like Ghana (73 MWp), Nigeria (68 MWp), Kenya (54 MWp) or Senegal (38 MWp). In Bangladesh claiming 198 MWp, high ratios of the installed PV are realized as off-grid systems, i.e. 5 MWp solar home systems with an average size of 50 Wp represent a solar power solution for 100,000 families [24].

Europe, by far, used to lead in installed PV, whereas in 2015 Europe and the Asia-Pacific region show comparable cumulative PV installations, with 98,193 MWp and 96,445 MWp, respectively. Thus within the three years from 2012 to 2015, the European domination disappeared with a decrease from 70% down to 42% of any PV installed. North America claims 12% of world PV with 28,774 MWp installed. The regions of Latin America (2,459 MWp), Africa (2,441 MWp) and the Middle East (1,733 MWp) represent only 3% of global PV, even if they are the ones with highest solar irradiations [25].

Approximately 67% of global installed PV capacity (156,280 MWp) is installed in countries that have a gross domestic product (GDP) per capita higher than 30,000 USD per year. Low income countries (GDP/capita <10,000 USD) claim 61,715 MWp in 145 countries. Low income countries have a more than hundred-fold increase in their PV capacity since 2009, which is ten times more than high income countries that increased their installed PV capacity only by a factor of eight during the same period of time [12]. Countries with a GDP per capita between 10,000 USD and 30,000 USD represent only 12,050 MWp, thus less than least income countries do.

Correlating to last year, the majority of 65% of total PV at end of 2015 is located in countries with medium irradiation (1,500-2,000 kWh/(m²·a)) [12-15, 25]. In 2015 photovoltaic capacity grew especially in these medium regions by 36% on average. Countries having less than 1,500 kWh/(m²·a), which are mainly highly developed countries in Europe and North America, represent 29% of total installed PV and further showed much lower growth rates of only 14%. Those countries predestined to use PV due to their very high insolation (>2,000 kWh/(m²·a)) claim only 4.9% of global PV being installed there. However, with an average increase of installed PV by 101%, they show by far the highest growth ratios.

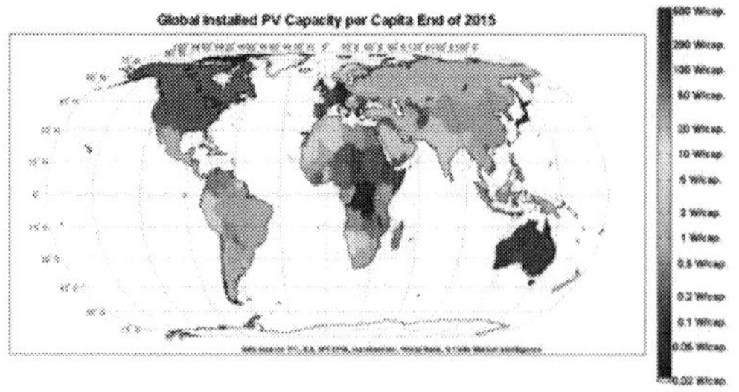

Figure 2: Installed PV capacity per capita by the end of 2015. Population data have been taken from the United Nations [26].

Figure 2 shows the PV installations per capita per country. Germany (479 Wp/cap.), Luxembourg (458 Wp/cap.), Italy (310 Wp/cap.), Belgium (289 Wp/cap.), Japan (266 Wp/cap.), Greece (234 Wp/cap.), Australia (215 Wp/cap.), Reunion (205 Wp/cap.), Czech Republic (202 Wp/cap.), Switzerland (168 Wp/cap.), Guam (155 Wp/cap.), Martinique (154 Wp/cap) are the ones with more than 150 Wp/capita. Another 15 countries show more than 100 Wp/capita (Guadeloupe, United Kingdom, Bulgaria, Denmark, Slovenia, French Guiana, Malta, Israel, Austria, Spain, Slovakia, France). Especially, highly developed countries in Europe, North America, East Asia or Australia, with respectively high electricity consumption per capita, show significant rates. Furthermore, island states show significantly high PV installations per capita. In total, 15 of those are able to be found in the top 50 ranking in PV per capita. This might be due to PV being a clear cost effective energy source that enables those remote island countries to be saved from high prices for conventional electricity generation by diesel power plants [27, 28]. As a world average, every human claims 32.2 Wp PV in the statistical view.

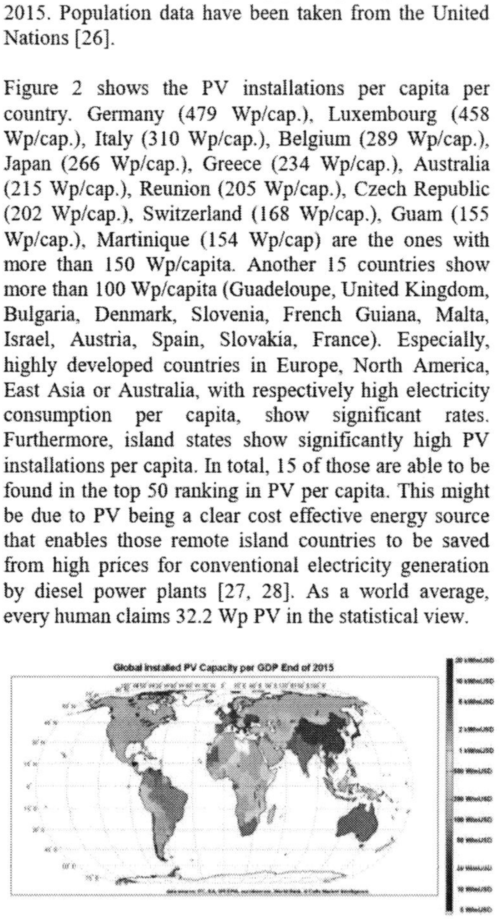

Figure 3: Global overview on cumulative PV installations per million USD of GDP by the end of 2015. Data for GDP have been taken from the World Bank [29].

An indicator of how much capital is invested relatively in PV is recognizable in installed PV capacities per countries' GDP (Figure 3). With the use of this comparative parameter, markets are able to be described as overcoming different financial and economic possibilities. Some countries have reached significant ratios of PV installations per one million USD of GDP. Honduras (23.4 kWp/mUSD) and Bulgaria (19.0 kWp/mUSD) are by far the country that relatively invested most of its economic power into PV. Followed by Germany (11.0 kWp/mUSD), the island state of Kiribati (10.0 kWp/mUSD) and the Czech Republic (10.0 kWp/mUSD). Furthermore, small countries that often are rarely recognized due to having installed only a few MWp can be distinguished. Thus, countries like Micronesia (6.1 kWp/mUSD), Malta (6.1 kWp/mUSD), Cape Verde (5.5 kWp/mUSD), Mauretania (4.3 kWp/mUSD), Palau (4.2 kWp/mUSD) or the Solomon Islands (3.4 kWp/mUSD) invest more of their economic power in PV than the bulk of highly developed countries. In comparison to this, highly developed countries such as Ireland (0.043 kWp/mUSD), Brazil (0.070 kWp/mUSD) or the countries of the Arab peninsula or Scandinavia have a vast potential of investing more capital in PV. Especially, emerging markets, like the African ones or island states, have realized PV as being able to improve significantly people's living conditions and save enormous financial resources in the mid- to long-term. Due to a fast amortization of PV systems in rural areas of developing countries [24, 30, 34], these markets have no need for subsidies and make off-grid PV countries foremost in proportion of PV expenditures to GDP.

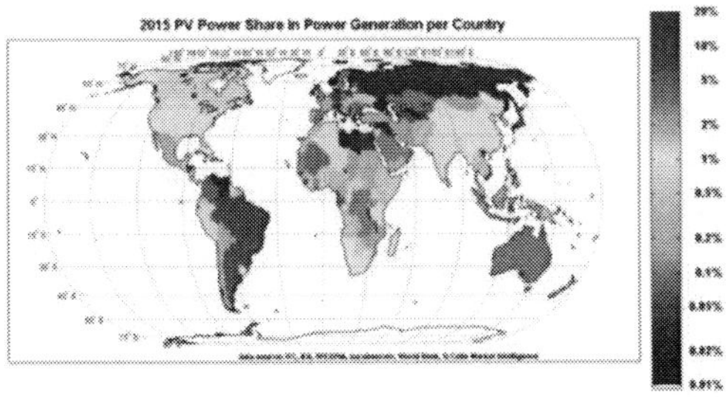

Figure 4: Global overview of PV share in the respective countries total power generation [31].

As a consequence of high ratios of PV to GDP, various countries have already reached a measurable share of PV in their total power generation (Figure 4). Thus, 12.5% of Honduras' power generation capacity is complemented by PV. Kiribati, with 12.2%, as well claims more than 10% of its power generated by solar PV. Further Italy (8.5%), the Atlantic island of St. Helena (8%), Germany (7.1%), Cape Verde (6.7%), Greece (6.3%), Guinea-Bissau (6.1%), Solomon Island (6%), Equatorial Guinea (5.4%), Sierra Leone (5.3%), Comoros (5%). In total 69 countries show more than 1% of PV share in their power generation with a global average of 1.48% of all power generated by solar PV. Especially, European, Western African, Asian countries and island states show high ratios of PV in their power generation. While regarding the African and insular countries that represent high ratios of PV in their power generation, it is obvious that they are the ones using a bulk of diesel power plants for generation of electricity. Countries leading in usage of diesel power, due to their remote location usually lead in shares of PV in their total power generation. This might be an indicator that PV is not only used as small off-grid PV solar home systems for sustainable electrification of rural areas, but also as a sustainable alternative for larger solutions as applied in hospitals, schools, supply of commericial appliances and in more and more cases as an upgrade for existing diesel/ oil fired power plants. This is especially prevailing in the environment of high diesel prices and remote areas to be supplied with affordable electricity [30, 32, 33].

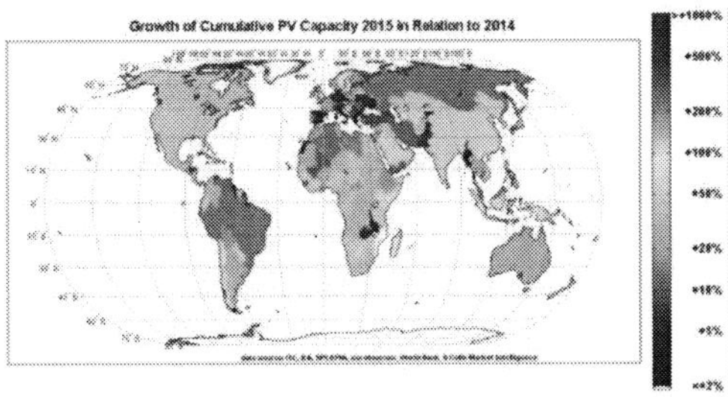

Figure 5: Ratio of added PV capacity in 2015 to cumulative PV installations in 2014, representing growth in PV markets' cumulative capacity.

Figure 5 gives an overview of PV market growth in 2015. Again, a few countries have reached huge growth rates in 2015. In Azerbaijan, the cumulative installed PV capacity has grown by more than 878% from 1.1 MWp to 10.1 MWp. Further, other emerging countries have shown high growth rates in 2015. PV capacity in Antigua and Barbuda at the end of 2015 was more than 606% higher than the end of 2014, reaching 3.6 MWp installed. Pakistan added some 659 MWp in 2015 reaching 780 MWp of cumulative capacity, equivalent to a growth rate of 545%. This indicates largely growing PV markets especially in emerging countries all over the world. In Europe, finally after years of being about to take-off, Turkey increased its PV capacity by 415% last year and is on the way to becoming the next big European market. In addition Hungary (336%), Russia (223%), Poland (196%), Ireland (143%), Kazakhstan (117%) and Armenia (105%) have also more than doubled their installed PV. The previously dominant German market reached a growth rate of only 4%, adding 1,450 MWp, and thus ranked among the 20 least growing markets in relative terms. Nearly half of the least growing markets with less than 10% growth are European. Czech Republic and Greece showed the lowest growth rates ever recorded in this serial of publication with a growth of below 1% each. In contrast, several African markets more than doubled the installed capacity per country. Enhanced by electricity shortage, South Africa added some 345 MWp equivalent to a growth of 38% in 2015. Further Algeria shows high growth rates of 265% reaching 285 MWp installed. Chinese PV installations remain at huge growth by about 56% after 55% in 2014 and 155% in 2013 and thus has been the first time ever market to surpassed the 15-GWp-threshold of new installations. Apart from Southern and Eastern Asia, especially Central Asia shows some PV markets starting to increase their PV investments. The MENA region shows a growth of 54%, led by Iran and Jordan with growth in their PV capacity of 283% and 269%, respectively. Further in absolute terms Israel remains to keep the Middle East lead with 205 MWp of new PV installations. Growth rates in the Americas are quite different, however. While markets such as Honduras, Uruguay (196%), Guatemala (163%), the Dominican Republic (143%) or Panama (143%) have grown at a high rate, the Brazilian (13%) and Argentinean (18%) markets are still far behind their great potentials [34-41]. US and Canadian PV capacity grew by about 41% and 32%, respectively.

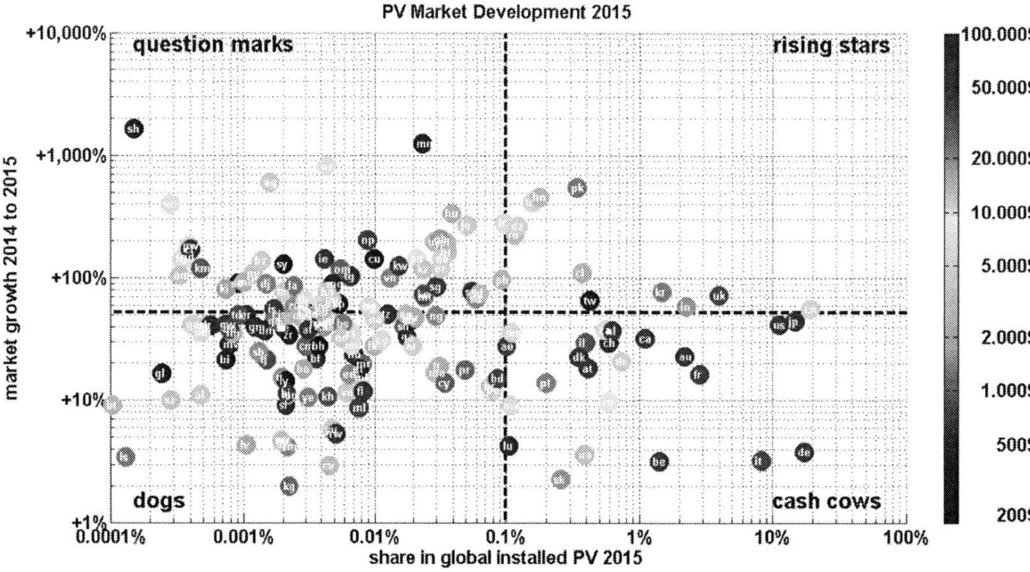

Figure 6: Cumulative installed PV by end of 2015 (x-axis) and PV capacities growth rate in 2015 (y-axis). Every bubble represents one country. The color code of the bubbles refers to the GDP/capita (left axis). The code inside of the bubble represents the internet-domain of the relevant country.

Figure 6 shows a segmentation of PV markets. Further, an attractiveness of PV markets can be represented. For this, a portfolio analysis chart is applied to PV markets with the use of a logarithmic order. 'Cash cows' are mature markets that claim significant installed PV capacities (more than 0.1% of global installed PV at the end of 2015) with relatively low growth rates, but still remaining the most important driver of the global PV market. Germany, Italy, Belgium, France or Australia are typical representatives of this PV market segment. Markets like Denmark, Slovakia or Slovenia are sorted to this segment as well. This segment claims 69% of total installed PV capacity. Even with very low growing markets, such as Germany (4% in 2015) or Italy (3%), standing for annual installation of a few 100 MWp due to having an installed PV capacity base of few GWp, this is the most important PV market segment. Furthermore, 'cash cows' mostly are countries with high incomes. In total 28 countries belong to this group, whereas 2014 it have been 24.

'Rising stars' are probably the most interesting market segment, as respective installed PV capacities (>0.1% of global PV in 2015) and growth rates are above the world's long term average (>48%) and led to significant installations in the relevant year. China, the United Kingdom, India, Korea, Pakistan, Chile, Taiwan, Honduras, Turkey, Algeria and Russia are those that represent this market segment in 2015. Therefore, this segment stands for 28% of total global PV installations. An interesting question is whether these markets will stay in this segment. In the short- to mid-term, 'rising stars' will become 'cash cows', as such high growth rates will not be able to keep up with increasing cumulative PV capacities in the years to come. Ten countries that in 2014 had been sorted as 'rising stars', are now in the 'cash cow' group (United States, Japan, the Netherlands, South Africa, Thailand, Israel, Mexico, United Arab Emirates, Malaysia and Luxembourg).

Markets that show high growth rates, but relatively low installed PV can be named as 'question marks'. They account for 1.6% of global PV installations. Markets such as the Philippines, Indonesia, Sweden, Hungary, Poland or Saudi Arabia show high growth rates, but because of their low installed cumulative PV capacity, they are marginal in global PV. However, if these markets keep their high growth rates for several years, becoming a 'rising star' can be a valid opportunity. Most interesting in this segment are markets close to the threshold of becoming a 'rising star'. Jordan, Hungary and the Philippines are among those that are expected to become new 'rising stars' in the years to come.

At first view 'dogs' seem to be markets of less interest since only 1.1% of total PV is installed in those countries. But especially these are non-feed-in tariff driven markets which often claim a high share of off-grid PV applications, such as Tanzania, Kenya or Zambia.

The development of PV in emerging markets could be rather different than in the markets where PV developed until now. PV starts to develop in several emerging countries based on calls for tenders for large-scale PV systems, rather than rooftop PV. This development will be even stronger in places where neither the population, nor companies will have the ability to finance PV systems for self-consumption. This trend is already visible in 2014 with the share of large-scale PV increasing in all regions except in Europe. In parallel, small off-grid applications are starting to develop for electrification of households with very low electricity consumption, as in the case of Bangladesh for instance. This development implies the possible cohabitation in emerging countries between Pico PV and solar home systems and utility-scale plants, leaving the rooftop market aside.

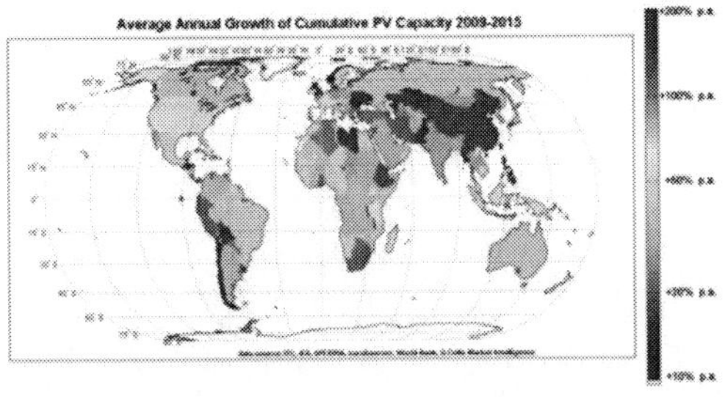

Figure 7: Average annual growth rate of cumulative PV capacity between 2009 and 2015 [12-15].

Figure 7 shows the average annual growth rate in the period from 2009 to 2015. The highest average annual growth rates within this period are able to be identified in Eastern Europe, Central, Southern and Eastern Asia. Azerbaijan (217%), Romania (207% p.a.), Honduras (185% p.a.) and Kyrgyzstan (183% p.a.) grew annually by more than 180% on average since 2009, mainly caused by the high growth in PV capacities in one particular year. For Europe, Lithuania (167% p.a.), the United Kingdom (159% p.a.) and Ukraine (155% p.a.) show high average growth rates, as well. In Africa, Ghana leads average growth rates from 2009 to 2015 with 125% p.a. The Americas claim several high growth regions, while others show less growth.

However, high average growth rates are no indicator of a sustainable and constant growth. Deviations in PV markets' annual growth rates are significant, due to markets often being driven by politically driven feed-in tariffs. Thus, average annual growth rates have to be regarded in the context of annual growth rate standard deviation (Figure 8).

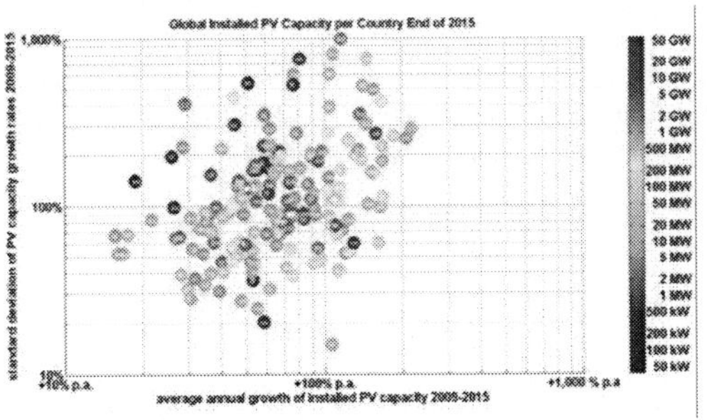

Figure 8: Average annual growth rate (x-axis) and growth rate standard deviation (y-axis) in the period from 2009 to 2015. Every bubble represents one country. The color code of the bubbles refers to the installed PV capacity at the end of 2015 (left axis). The code inside of the bubble gives the internet-domain of the respective country.

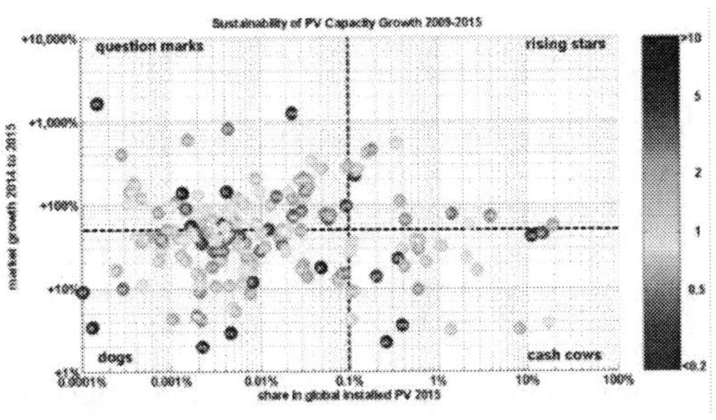

Figure 9: Ratio of standard deviation to average growth of PV capacity in the period from 2009 to 2015. The blue color refers to very unstable PV markets, while red refers to relatively constant growing photovoltaic capacities.

Sustainable growth is a significant but constant growth. Very constant growing PV markets can be identified by the ratio of standard deviation to average growth.

As can be seen in Figure 8 and Figure 9, such significant but constant growth in recent years can be identified especially in the United States, Tanzania, Portugal, Haiti, Japan, Ivory Coast, Saudi Arabia and a variety of further African states. All of them show a ratio of standard deviation to average growth of less than 0.40 (refer to Figure 9). Portugal represents an average annual growth rate of 33% in the period from 2009 to 2015. In 2010 Tanzania's PV capacity grew by 81%, in 2011 by 86%, in 2012 by 78%, in 2013 by 50%, in 2014 by 33% and in 2015 by 51% to 28.8 MWp at the end of the year. This is an example of a very constant growth in PV and can definitely be regarded as sustainable. If this constant and high growth continues, Tanzania could reach 46 MWp by the end of 2016. As another example is for the United States, the PV capacity grew by 53% in 2010, 65% in 2011, 78% in 2012, 65% in 2013, 51% in 2014 and 41% in 2015, which leads to an average annual growth rate of 58% with a relatively low standard deviation of only 12%. If this constant growth goes on steadily, 40,000 MWp might be cumulatively installed by the end of 2016 in the US. Markets such as Denmark, Ukraine, Slovakia or even the United Kingdom show high annual growth rates, but high standard deviations, due to variable growth with boom years being followed by such with very little growth [12, 13, 14]. Thus, they show ratios of standard deviation to average growth of more than two. By following this approach, the most variable markets are among others Cape Verde (ratio of 17.1), Kyrgyzstan (ratio of 11.8), Belize, Botswana, Slovakia, Zimbabwe and Denmark showing ratios of more than five.

With the use of this evaluation of average annual growth, the constancy of growth rates, the help of other publications [1, 2, 16-23] and the knowledge of five years of analysis of cumulative PV capacities [12-15], a rough forecast of cumulative PV capacities for the years 2016 and 2017 is possible. In 2016, the global cumulative installed PV capacity is likely to increase by 26% to 289,800 MWp by the end of 2016 (new installations of about 59,800 MWp in 2016) and by the end of 2017, 353,100 MWp (new installations of about 63,200 MWp in 2017) of cumulative installed PV capacity can be expected.

4 CONCLUSIONS

In total, 230,046 MWp can be allocated to 196 countries by end of 2015. 52,477 MWp have been newly installed. The global trade volume of PV modules reached 31.5 billion USD in 2015.

The most significant amounts of PV are still installed in mature markets in Europe and some other highly developed countries. However, a relevant paradigm shift to developing and emerging markets as the new growth regions can be seen. Some low and middle income countries already show remarkable numbers of newly installed PV.

In the years to come, new growth markets in Central and South America (Guatemala, Caribbean, Ecuador), Asia (the Philippines, Pakistan, Iran, Jordan, Indonesia) and Africa (Algeria, Ghana, Nigeria, South Africa, Namibia) will increasingly play a major role within the next years of global PV markets development. In Europe, Turkey, Poland and Russia show new growth perspectives, as well as the Caucasus states do.

One is able to assess that every world region might represent a great PV market potential. Enormous growth rates in specific regions of the world show that PV is on the way to becoming an important part of global electricity supply [34]. Some developing countries are on the road to solar electrification, others use PV as an opportunity to decrease suffering from having no access to electricity in remote locations. Lower energy costs enabled by reduced dependence on energy imports will contribute to leading these countries to sustainable development and stable growth.

REFERENCES

[1] [IRENA] – International Renewable Energy Agency, 2016. Renewable Capacity Statistics 2016, Dubai, http://www.irena.org/DocumentDownloads/Publications/IRENA_RE_Capacity_Statistics_2016.pdf

[2] International Trade Centre Market Analysis Tool, 2015. Product Map, Market Analysis and Research Section, International Trade Centre, Geneva, www.trademap.org/

[3] Photon International, 2012. Zellproduktion 2011, Photon Magazin, 2012(4), [plus respective data for the years 2000-2011 either of Photon Magazin or Photon International]

[4] Breyer Ch., Birkner C., Meiss J., Goldschmidt J.C., Riede M., 2013. A top-down analysis: determining photovoltaics R&D investments from patent analysis and R&D headcount, Energy Policy, 62, 1570-1580

[5] Photon, 2016. Factory Gate c-Si Module Prices, Photon Consulting, Aachen

[6] Hanwha Q CELLS Market Intelligence, 2011-2016. Intern Price Snapshot, Hanwha Q CELLS, Bitterfeld-Wolfen

[7] Werner Ch., Gerlach A., Adelmann P., Breyer Ch., 2011. Global Overview on Cumulative Installed Photovoltaic Power, Proceedings 26th European Photovoltaic Solar Energy Conference, Hamburg

[8] Szewzcyk W., 2016. private communication, PV insights for Poland, May 31

[9] Zschuckelt D., 2016. private communication, PV insights for Israel and Turkmenistan, June 10

[10] Linde J., 2015. private communication, PV insights for South Africa, March 30

[11] Vartiainen E., Masson G., Breyer Ch., 2015. PV LCOE Europe 2014-30, European Photovoltaic Technology Platform, Brussels

[12] Werner Ch., Gerlach A., Adelmann P., Breyer Ch., 2011. Global Overview on Cumulative Installed Photovoltaic Power, 2nd Symposium Small PV-Applications, Ulm

[13] Werner Ch., Gerlach A., Adelmann P., Breyer Ch., 2011. Global Overview on Cumulative Installed Photovoltaic Power, Proceedings 26th European Photovoltaic Solar Energy Conference, Hamburg

[14] Werner Ch., Gerlach A., Adelmann P., Breyer Ch., 2013. Analysis and Forecast of Global Installed Photovoltaic Capacity and Identification of Hidden Growth Markets, Proceedings 28th European Photovoltaic Solar Energy Conference, Paris

[15] Werner Ch., Gerlach A., Breyer Ch., Masson G., 2014. Global Installed Photovoltaic Capacity and Identification of Hidden Growth Markets, Proceedings 29th European Photovoltaic Solar Energy Conference, Amsterdam

[16] Werner Ch., Gerlach A., Breyer Ch., Orlandi S., Masson G., 2015. Latest Developments in Global Installed Photovoltaic Capacity and Identification of Hidden Growth Markets, Proceedings 31st European Photovoltaic Solar Energy Conference, Hamburg

[17] [PVMA] – PV Market Alliance, 2016. The PVMA Global PV Market Report, Brussels

[18] Tsagas I., 2016. Cyprus: Construction begins on five PV plants; storage to emerge, PV magazine, Berlin, 2016, April 22

[19] Lopez D., 2016. Panama reached 43 MW of installed solar PV in 2015, PV magazine, Berlin, 2016, March 18

[20] [IEA PVPS] - International Energy Agency Photovoltaic Power Systems Programme, 2016. Snapshot of Global PV Markets 2015, Paris

[21] Journal des Énergies Renouvelables, 2016, Photovoltaic Barometer, EurObserv'er, 2016 (April), Paris

[22] [REN21], 2016. Renewables 2016 – Global Status Report, REN21, Paris

[23] Hanwha Q CELLS Market Intelligence, 2011-2015. Market Research, Hanwha Q CELLS, Bitterfeld-Wolfen

[24] Breyer Ch., Werner Ch., Rolland S., Adelmann P., 2011. Off-Grid Photovoltaic Applications in Regions of Low Electrification: High Demand, Fast Financial Amortization and Large Market Potential, Proceedings 26th European Photovoltaic Solar Energy Conference, Hamburg

[25] Breyer Ch. and Schmid J., 2010. Population Density and Area weighted Solar Irradiation: Global Overview on Solar Resource Conditions for Fixed Tilted, 1-Axis and 2-Axis PV Systems, 25th EU PVSEC, Valencia, DOI: 10.4229/25thEUPVSEC2010-4BV.1.91

[26] [UN] - United Nations, 2013. World Population Prospects: The 2012 Revision, UN Department of Economic and Social Affairs, Population Division, Working Paper No. ESA/P/WP.228, New York
[27] Breyer Ch., Gerlach A., Schäfer D., Schmid J., Fuel-Parity: New Very Large and Sustainable Market Segments for PV Systems, IEEE EnergyCon, Manama, 2010, December 18–22
[28] Blechinger P., Cader C., Bertheau P., Huyskens H., Seguin R., Breyer Ch., 2016. Global Analysis of the Techno-Economic Potential of Renewable Energy Hybrid Systems on Small Islands, Energy Policy, published online 7.4.2016
[29] World Bank, 2013. Gross Domestic Production until 2012, The World Bank Group, World Development Indicators database, Washington, http://data.worldbank.org/indicator/NY.GDP.MKTP.CD
[30] Cader C., Bertheau P., Blechinger P., Huyskens H., Breyer Ch., 2015. Global cost advantages of autonomous solar-battery-diesel systems compared to diesel-only systems, Energy for Sustainable Development, 31, 14-23
[31] [CIA], 2016. The World Factbook. Central Intelligence Agency, Washington
[32] Werner Ch. and Breyer Ch., 2012. Analysis of Mini-grid Installations: An Overview on System Configurations, Proceedings 27th European Photovoltaic Solar Energy Conference, Frankfurt, DOI: 10.4229/27thEUPVSEC2012-5DO.9.6
[33] Bertheau P., Cader C., Huyskens H., Blechinger P., 2015. The Influence of Diesel Fuel Subsidies and Taxes on the Potential for Solar-Powered Hybrid Systems in Africa, Resources, 4, 673-691
[34] Breyer Ch., Bogdanov D., Gulagi A., Aghahosseini A., Barbosa L.S.N.S., Koskinen O., Barasa M., Caldera U., Afanasyeva S., Child M., Farfan J., Vainikka P., 2016. On the Role of Solar Photovoltaics in Global Energy Transition Scenarios, 32nd EU PVSEC, Munich, June 20-24
[35] Bogdanov D. and Breyer Ch., 2016. North-East Asian Super Grid for 100% Renewable Energy supply: Optimal mix of energy technologies for electricity, gas and heat supply options, Energy Conversion and Management, 112, 176-190
[36] Gulagi A., Bogdanov D., Breyer Ch., 2016. Southeast Asia and the Pacific Rim Super Grid for 100% Renewable Energy Power Supply, submitted
[36] Gulagi A., Bogdanov D., Breyer Ch., 2016. Solar Photovoltaics – A driving force towards a 100% renewable energy system for India and the SAARC region, 32nd EU PVSEC, Munich, June 20-24
[37] Bogdanov D. and Breyer Ch., 2015. Eurasian Super Grid for 100% Renewable Energy power supply: Generation and storage technologies in the cost optimal mix, ISES Solar World Congress 2015, Daegu, Korea, November 8-12
[38] Aghahosseini A., Bogdanov D., Breyer Ch., 2016. The MENA Super Grid towards 100% Renewable Energy Power Supply by 2030, 11th International Energy Conference, Tehran, May 30-31
[39] Barasa M., Bogdanov D., Oyewo S., Breyer Ch., 2016. A Cost Optimal Resolution for Sub-Saharan Africa powered by 100 Percent of Renewables by the Year 2030, 32nd EU PVSEC, Munich, June 20-24
[40] Aghahosseini A., Bogdanov D., Breyer Ch., 2016. 100% Renewable Energy in North America and the Role of Solar Photovoltaics, 32nd EU PVSEC, Munich, June 20-24
[41] Barbosa L.S.N.S, Bogdanov D., Vainikka P., Breyer Ch., 2016. Hydro, wind and solar power as a base for a 100% Renewable Energy supply for South and Central America, submitted

Appendix Table 1: Alphabetical list of all countries in the world and their cumulative installed PV capacity at the end of 2015. Data are compiled as documented in the paper and based on IRENA, IEA PVPS and further publications plus ITC data [1-23].

	PV capacity at the end of 2015	PV capacity at the end of 2014 [16]	PV capacity at the end of 2013 [15]	PV capacity at the end of 2012 [14]	PV capacity at the end of 2011	PV capacity at the end of 2010 [13]	PV/capita at the end of 2015	PV/GDP at the end of 2015	PV share in power generation 2015
	[MWp]	[MWp]	[MWp]	[MWp]	[MWp]	[MWp]	[Wp/cap]	[Wp/mUSD]	[-]
Afghanistan	7.30	5.30	4.60	3.10	1.60	1.30	0.24	404.8	0.32%
Albania	1.10	0.99	0.30	0.14	0.10	0.03	0.35	84.9	0.02%
Algeria	285.00	78.00	17.60	17.60	15.20	7.10	7.27	1,435.5	1.06%
Andorra	0.14	0.12	0.07	0.05	0.05	0.05	1.77	-1.0	0.00%
Angola	13.00	9.80	9.50	6.70	3.60	2.90	0.61	124.9	0.45%
Antigua&Barbuda	3.60	0.51	0.29	0.17	0.12	0.09	40.00	3,201.2	0.00%
Argentina	45.00	35.20	32.70	29.90	14.90	11.70	1.09	100.9	0.06%
Armenia	0.76	0.37	0.26	0.20	0.14	0.07	0.26	75.0	0.02%
Australia	5,020.00	4,080.00	3,200.00	2,380.00	1,380.00	532.00	215.05	3,626.8	3.45%
Austria	940.00	794.00	639.00	420.00	184.00	101.00	110.65	2,250.7	1.51%
Azerbaijan	10.10	1.10	0.52	0.24	0.19	0.18	1.07	159.3	0.08%
Bahamas	13.40	9.40	7.90	6.90	0.56	0.21	35.54	1,702.1	1.37%
Bahrain	8.70	6.80	5.70	4.90	0.39	0.05	6.53	-1.0	0.00%
Bangladesh	198.00	172.00	140.00	83.60	58.50	34.70	1.26	1,769.3	0.73%
Barbados	9.20	6.00	3.90	2.20	0.38	0.14	32.28	2,496.6	0.00%
Belarus	20.80	13.10	7.60	6.60	2.50	2.10	2.22	323.6	0.06%
Belgium	3,220.00	3,120.00	2,990.00	2,620.00	2,030.00	796.00	289.99	6,266.3	3.78%
Belize	3.10	1.30	1.10	0.97	0.03	0.03	9.34	2,141.6	0.74%
Benin	4.90	4.40	1.60	1.10	0.76	0.55	0.47	671.7	0.82%
Bermuda	3.30	2.40	1.40	0.97	0.33	0.22	50.77	593.8	0.00%
Bhutan	0.24	0.22	0.20	0.19	0.17	0.02	0.32	130.9	0.02%
Bolivia	6.60	5.60	4.50	4.10	3.20	3.10	0.62	275.6	0.17%
Bosnia-Herzegov.	10.10	5.70	4.00	1.90	0.54	0.13	2.64	553.7	0.10%
Botswana	9.10	5.70	4.90	4.40	3.30	2.70	4.50	567.3	0.52%
Brazil	174.00	154.00	109.00	78.90	39.90	26.60	0.87	70.3	0.05%
Brunei	1.80	1.30	0.95	0.14	0.01	0.01	4.31	110.0	0.08%
Bulgaria	1,030.00	1,020.00	1,000.00	920.00	149.00	17.40	142.60	19,236.3	4.64%
Burkina Faso	8.40	6.90	5.80	4.90	3.50	1.80	0.50	808.0	1.48%
Burma	53.90	4.00	23.00	11.40	3.70	0.71	1.01	-1.0	1.08%
Burundi	1.70	1.40	1.20	0.97	0.33	0.19	0.17	721.7	0.87%
Cambodia	10.30	9.30	7.70	5.60	3.80	2.10	0.68	802.8	0.45%
Cameroon	6.90	5.40	3.30	2.50	1.40	0.98	0.31	272.5	0.19%
Canada	2,530.00	1,910.00	1,320.00	877.00	628.00	240.00	71.91	1,423.1	0.60%
Cape Verde	10.50	10.20	9.90	9.20	8.50	7.90	21.04	5,523.0	6.72%
Central Afr. Rep.	0.92	0.34	0.28	0.25	0.19	0.15	0.20	424.8	0.89%
Chad	2.10	1.40	1.00	1.00	0.68	0.54	0.16	198.5	1.96%
Chile	847.00	404.00	134.00	13.00	4.60	3.70	48.07	3,371.9	2.27%
China	44,420.00	28,470.00	18,310.00	6,890.00	3,250.00	861.00	32.06	6,066.7	1.05%
Colombia	26.20	20.00	17.60	12.10	7.20	6.90	0.54	77.9	0.07%
Comoros	1.10	0.50	0.25	0.19	0.17	0.16	1.50	1,802.2	5.03%
Congo, DR	5.10	3.80	3.10	1.90	1.40	0.99	0.08	325.8	0.10%
Congo, Rep.	6.10	4.10	3.60	2.30	2.00	1.70	1.37	422.9	1.08%

Costa Rica	15.60	11.50	8.50	6.10	1.20	1.00	3.20	380.2	0.24%
Croatia	47.00	31.80	23.60	12.30	11.90	11.30	10.96	760.7	0.35%
Cuba	22.80	9.40	8.00	3.20	2.80	2.10	2.02	-1.0	0.22%
Cyprus	79.10	69.40	34.00	23.70	11.30	6.20	89.38	3,165.0	3.31%
Czech Republic	2,170.00	2,170.00	2,140.00	2,060.00	1,960.00	1,950.00	202.77	10,045.8	3.59%
Denmark	801.00	654.00	573.00	376.00	17.80	6.00	142.55	2,401.0	2.58%
Djibouti	3.40	1.80	1.80	1.80	1.60	1.40	3.89	2,743.8	2.02%
Dominica	1.10	0.81	0.13	0.08	0.07	0.04	15.28	2,308.6	2.34%
Dom. Rep.	75.30	31.00	21.50	11.50	4.50	1.90	7.24	1,352.6	1.01%
Ecuador	74.60	34.00	14.50	10.20	4.50	1.20	4.74	960.1	0.52%
Egypt	40.00	26.60	23.20	11.00	6.00	5.50	0.49	169.5	0.05%
El Salvador	6.60	4.40	3.10	1.10	0.41	0.32	1.04	285.8	0.21%
Equatorial Guinea	3.90	2.70	2.40	2.10	0.68	0.54	5.15	232.0	5.48%
Eritrea	2.40	1.60	1.40	0.63	0.46	0.41	0.38	920.3	1.50%
Estonia	14.90	12.80	8.50	4.80	3.40	3.00	11.58	672.5	0.19%
Ethiopia	17.70	15.00	13.40	12.30	9.00	6.90	0.19	558.2	0.60%
F.S. Micronesia	1.90	1.40	0.51	0.15	0.15	0.15	18.27	6,123.4	1.72%
Fiji	4.60	3.20	2.20	1.30	0.65	0.47	5.22	1,204.8	0.94%
Finland	19.00	17.00	12.60	38.20	34.50	26.10	3.50	72.2	0.02%
France	6,540.00	5,620.00	4,690.00	3,990.00	2,850.00	1,050.00	101.72	2,352.8	1.67%
Gabon	0.95	0.67	0.47	0.40	0.35	0.17	0.57	50.6	0.08%
Gambia	2.80	2.00	1.40	0.94	0.75	0.69	1.51	3,117.1	2.18%
Georgia	0.36	0.36	0.36	0.36	0.36	0.36	0.08	24.9	0.01%
Germany	39,700.00	38,250.00	36,350.00	32,360.00	24,760.00	17,210.00	479.89	11,025.2	7.19%
Ghana	73.20	23.80	15.70	5.00	1.80	0.84	2.83	1,868.4	1.03%
Greece	2,610.00	2,590.00	2,570.00	1,530.00	624.00	206.00	234.54	9,011.6	6.34%
Greenland	0.56	0.48	0.21	0.06	0.06	0.06	9.82	-1.0	0.00%
Grenada	0.87	0.35	0.24	0.10	0.10	0.10	8.21	1,114.9	0.91%
Guam	25.60	0.13	0.13	0.13	0.13	0.13	155.15	-1.0	2.89%
Guatemala	82.70	31.40	3.60	2.00	1.80	1.80	5.35	1,760.3	1.62%
Guinea	7.70	5.10	3.20	2.20	1.80	1.20	0.66	1,511.0	1.40%
Guinea - Bissau	1.70	1.20	0.99	0.82	0.62	0.47	1.00	1,756.6	6.11%
Guyana	4.50	4.30	4.00	2.90	2.70	1.50	5.63	1,746.4	1.15%
Haiti	8.30	5.80	4.10	2.50	1.40	0.86	0.80	1,129.8	3.15%
Honduras	409.00	73.60	5.00	2.80	1.70	1.10	50.51	23,473.7	12.55%
Hong Kong	2.50	2.50	2.50	0.90	0.33	0.14	0.35	10.1	0.01%
Hungary	88.90	20.40	14.40	3.90	3.40	1.80	8.93	641.0	0.28%
Iceland	3.90	2.50	1.50	0.90	0.53	0.33	11.82	277.1	0.00%
India	5,180.00	3,270.00	2,610.00	1,460.00	470.00	177.00	4.14	2,765.8	0.97%
Indonesia	145.00	83.00	46.70	34.70	13.30	9.20	0.58	171.3	0.13%
Iran	223.00	58.30	9.10	9.10	9.10	4.30	2.88	433.8	0.19%
Iraq	10.40	7.20	4.50	4.00	3.10	2.20	0.31	57.6	0.03%
Ireland	9.70	4.00	2.00	10.90	7.00	5.00	2.10	43.9	0.03%
Israel	895.00	690.00	455.00	243.00	158.00	40.20	115.74	3,684.2	2.69%
Italy	18,910.00	18,320.00	17,780.00	16,300.00	12,810.00	3,490.00	310.05	8,625.4	8.58%
Ivory Coast	4.20	2.80	2.30	1.60	1.20	0.93	0.21	174.5	0.13%
Jamaica	11.00	6.60	2.70	0.43	0.43	0.43	3.95	762.5	0.62%
Japan	33,870.00	23,420.00	13,500.00	7,230.00	4,910.00	3,610.00	266.39	5,743.8	4.64%
Jordan	117.00	31.70	4.80	1.10	0.79	0.79	16.08	4,056.8	1.35%

Kazakhstan	54.80	25.20	12.80	4.70	1.30	0.29	3.33	291.4	0.09%
Kenya	54.90	31.60	25.40	17.80	12.30	8.70	1.24	1,632.9	1.41%
Kiribati	1.70	0.94	0.51	0.34	0.20	0.15	16.67	10,198.9	12.25%
Korea, DPR	12.40	7.70	5.30	3.30	2.40	2.20	0.50	-1.0	0.12%
Korea, Rep.	3,360.00	1,900.00	1,350.00	974.00	786.00	655.00	68.21	3,014.9	0.99%
Kuwait	35.50	15.70	9.30	6.00	3.50	1.80	10.54	201.0	0.12%
Kyrgyzstan	5.10	5.00	4.10	1.20	0.02	0.02	0.92	822.9	0.08%
Laos	5.60	3.00	2.70	1.80	0.68	0.68	0.83	680.7	0.29%
Latvia	2.40	2.30	1.50	1.20	0.68	0.11	1.17	84.3	0.04%
Lebanon	6.80	4.10	2.10	0.74	0.74	0.71	1.41	169.6	0.09%
Lesotho	0.30	0.29	0.26	0.18	0.12	0.01	0.14	118.8	0.00%
Liberia	1.30	0.92	0.87	0.67	0.49	0.36	0.30	841.2	0.67%
Libya	4.80	4.20	3.90	2.80	2.50	2.20	0.77	-1.0	0.03%
Lithuania	72.10	60.70	37.00	8.80	0.74	0.62	23.90	1,681.7	0.76%
Luxembourg	243.00	233.00	104.00	50.90	33.30	30.10	458.49	4,104.7	0.00%
Macedonia	18.10	15.70	11.20	6.50	2.90	0.38	8.59	1,733.9	0.36%
Madagascar	11.20	5.90	5.70	3.50	2.30	1.50	0.49	1,130.0	0.99%
Malawi	1.80	1.40	1.30	1.00	0.55	0.37	0.11	320.2	0.15%
Malaysia	250.00	229.00	73.70	61.90	13.50	11.30	8.41	868.3	0.30%
Maldives	4.40	4.20	2.30	1.80	1.20	0.23	12.75	2,043.1	0.00%
Mali	17.60	16.20	15.50	13.80	10.60	3.00	1.15	1,651.6	3.49%
Malta	55.60	32.80	21.60	14.60	9.60	1.10	129.60	6,075.3	4.72%
Mauritania	18.70	15.60	12.60	7.80	1.70	0.91	4.81	4,376.1	3.42%
Mauritius	15.50	12.10	11.00	4.10	1.60	1.30	12.46	1,378.8	1.05%
Mexico	254.00	187.00	125.00	51.70	35.30	30.30	2.08	219.3	0.19%
Moldova	2.30	1.20	0.91	0.63	0.20	0.16	0.66	327.9	0.05%
Mongolia	5.00	4.50	3.50	3.50	3.30	3.10	1.76	570.7	0.18%
Morocco	67.10	57.40	36.90	23.70	18.00	16.80	2.03	676.3	0.43%
Mozambique	11.90	11.80	9.60	3.50	2.50	1.20	0.46	946.8	0.17%
Namibia	23.20	15.90	8.50	3.20	2.80	2.30	10.07	1,859.3	1.03%
Nepal	20.40	6.70	6.20	3.80	3.70	3.20	0.73	1,075.0	1.10%
Netherlands	1,440.00	1,050.00	695.00	262.00	139.00	86.10	85.92	1,722.3	1.22%
New Caledonia	6.80	4.30	3.80	3.40	3.20	2.80	26.56	-1.0	0.43%
New Zealand	43.70	30.10	23.50	14.50	7.50	5.50	9.70	-1.0	0.14%
Nicaragua	10.90	7.50	4.70	3.60	2.30	2.20	1.79	1,131.2	0.49%
Niger	4.60	3.30	2.00	1.80	1.80	0.80	0.26	764.5	0.94%
Nigeria	68.10	45.60	37.80	28.70	23.30	11.60	0.39	279.1	0.43%
Norfolk Island	0.24	0.24	0.24	0.24	0.17	0.07	-1.00	-1.0	#DIV/0!
Norway	16.10	13.00	11.10	8.40	7.80	7.30	3.19	32.8	0.01%
Oman	12.80	5.90	4.70	2.20	1.60	0.70	4.86	178.3	0.11%
Pakistan	780.00	121.00	45.50	18.20	10.80	4.00	4.28	3,701.2	1.66%
Panama	48.30	19.90	8.90	6.00	1.50	0.78	12.50	1,542.4	0.93%
Papua New	9.10	6.40	4.20	3.00	1.20	1.00	1.24	734.2	0.43%
Paraguay	2.70	1.30	0.74	0.37	0.18	0.06	0.40	103.8	0.05%
Peru	183.00	163.00	100.00	49.20	14.80	10.20	6.02	1,034.3	0.82%
Philippines	215.00	109.00	21.00	16.20	14.30	12.30	2.19	956.5	0.52%
Poland	81.60	27.60	7.90	5.80	2.90	1.80	2.14	158.2	0.06%
Portugal	461.00	405.00	283.00	231.00	166.00	126.00	43.46	1,940.5	1.51%
Puerto Rico	113.00	96.00	96.00	1.90	1.90	1.90	30.64	1,144.2	1.04%

Qatar	41.00	30.80	13.10	5.00	2.40	1.20	18.90	237.0	0.22%
Rep. of Palau	0.90	0.32	0.19	0.14	0.14	0.14	42.86	4,227.3	#DIV/0!
Romania	1,350.00	1,230.00	1,050.00	69.60	7.70	5.50	62.21	7,113.7	3.26%
Russia	268.00	83.00	13.00	114.00	58.20	35.70	1.88	141.1	0.03%
Rwanda	11.80	11.20	2.60	2.10	1.20	0.75	1.00	1,857.1	4.73%
Sao Tome&Princ.	0.18	0.14	0.08	0.02	0.01	0.01	0.93	725.0	0.00%
Saudi Arabia	137.00	81.60	55.20	33.40	21.30	9.10	4.75	237.5	0.11%
Senegal	38.70	27.60	19.10	15.60	10.10	4.70	2.74	2,678.5	2.55%
Seychelles	1.20	0.85	0.53	0.16	0.10	0.05	12.90	1,132.5	0.71%
Serbia	11.00	10.40	5.70	3.50	1.30	0.48	1.16	254.1	0.05%
Sierra Leone	4.80	4.40	4.00	2.80	0.40	0.30	0.79	1,633.3	5.30%
Singapore	68.40	37.00	14.30	6.40	4.70	3.00	12.64	279.2	0.00%
Slovakia	589.00	576.00	559.00	531.00	470.00	145.00	108.07	6,131.3	2.11%
Slovenia	285.00	283.00	270.00	208.00	87.20	36.30	137.55	5,668.3	2.60%
Solomon Island	3.00	2.40	2.20	1.30	0.55	0.32	5.35	3,461.5	5.93%
Somalia	2.10	1.10	0.32	0.16	0.11	0.08	0.20	-1.0	1.20%
South Africa	1,260.00	915.00	252.00	118.00	68.00	39.50	23.87	3,135.9	1.03%
Spain	5,180.00	5,180.00	5,020.00	4,790.00	4,420.00	3,840.00	110.38	3,507.4	3.21%
Sri Lanka	23.00	18.00	15.30	10.10	8.30	6.70	1.08	388.6	0.33%
St. Helena	0.35	0.02	0.01	0.01	0.01	0.01	87.50	-1.0	8.09%
St. Lucia	0.80	0.33	0.07	0.07	0.07	0.07	4.40	660.6	0.00%
St. Vincent&Gre.	0.65	0.13	0.06	0.06	0.06	0.06	5.96	940.5	0.93%
Sudan	9.50	6.80	5.80	4.20	2.70	2.10	0.25	148.3	0.30%
Suriname	4.60	3.30	1.10	0.79	0.56	0.48	8.53	1,068.7	0.44%
Swaziland	0.66	0.60	0.54	0.50	0.46	0.34	0.53	166.3	0.08%
Sweden	127.00	71.50	40.40	21.90	14.60	11.20	13.27	235.5	0.09%
Switzerland	1,360.00	1,050.00	748.00	414.00	214.00	90.70	168.36	2,062.8	2.75%
Syria	4.60	2.00	1.70	1.50	1.20	0.84	0.21	-1.0	0.03%
Taiwan	975.00	590.00	342.00	207.00	66.00	23.00	41.79	-1.0	0.57%
Tajikistan	3.40	2.80	2.50	1.70	0.70	0.33	0.41	521.3	0.04%
Tanzania	28.80	19.10	14.40	9.60	5.40	2.90	0.58	1,206.3	1.04%
Thailand	1,690.00	1,400.00	926.00	439.00	213.00	46.50	25.22	4,889.0	1.65%
Togo	15.20	7.50	3.30	2.00	1.40	0.93	2.23	4,124.0	2.41%
Trinidad&Tobago	4.50	3.90	3.30	1.80	0.38	0.14	3.36	191.5	0.09%
Tunisia	41.30	28.30	25.00	15.90	9.80	3.20	3.76	889.4	0.48%
Turkey	364.00	70.70	48.70	33.40	13.90	6.20	4.86	469.8	0.27%
Turkmenistan	4.60	2.60	1.70	0.85	0.50	0.13	0.88	163.9	0.06%
Uganda	20.90	13.20	9.40	6.90	5.00	4.90	0.56	1,242.4	1.17%
Ukraine	886.00	855.00	692.00	416.00	193.00	8.70	19.58	5,421.5	0.62%
UAE	233.00	182.00	118.00	118.00	44.00	15.50	24.93	646.8	0.45%
United Kingdom	9,080.00	5,260.00	3,370.00	1,810.00	955.00	72.00	143.82	3,713.9	2.57%
United States	25,990.00	18,370.00	12,200.00	7,400.00	4,160.00	2,520.00	81.21	1,733.7	0.97%
Uruguay	66.40	22.40	6.00	4.50	2.40	2.00	19.49	1,430.0	1.03%
Uzbekistan	5.60	3.50	2.70	1.40	0.33	0.23	0.19	123.6	0.02%
Venezuela	14.40	12.90	10.10	6.60	4.40	2.60	0.47	45.5	0.02%
Vietnam	30.00	15.00	11.80	11.80	9.70	7.20	0.33	242.6	0.04%
Western Sahara	0.25	0.25	0.25	0.25	0.25	0.25	0.44	-1.0	0.53%
Yemen	7.30	6.60	1.30	0.73	0.50	0.29	0.30	230.1	0.35%
Zambia	5.00	4.80	4.00	2.70	1.80	1.30	0.34	260.4	0.11%

Zimbabwe	6.7	4.30	3.90	3.30	3.00	3.00	0.47	693.9	0.17%

Appendix Table 2: Alphabetical list of all countries in the world with data for reasons of further analyses and comparing.

	Population [26]	GDP [29]	GDP/capita	Total power plant capacity in operation [31]	Population weighted irradiation [25]
	[1000pop]	[bnUSD]	[USD]	[MW]	[kWh/(m²·a)]
Afghanistan	30,552	18.03	590	431.3	2,164
Albania	3,173	12.96	4,084	1,693.0	1,923
Algeria	39,208	198.54	5,064	9,271.1	1,993
Andorra	79			46.6	
Angola	21,472	104.12	4,849	1,112.0	2,084
Antigua&Barbuda	90	1.12	12,495	68.9	
Argentina	41,446	446.04	10,762	29,935.9	1,962
Armenia	2,977	10.14	3,405	3,216.9	1,830
Australia	23,343	1,384.15	59,296	54,765.6	1,914
Austria	8,495	417.66	49,165	18,632.7	1,389
Azerbaijan	9,413	63.40	6,736	7,053.8	1,685
Bahamas	377	7.87	20,882	562.7	2,198
Bahrain	1,332			5,113.1	
Bangladesh	156,595	111.91	715	6,216.2	1,908
Barbados	285	3.69	12,930	262.1	
Belarus	9,357	64.27	6,869	7,998.1	1,264
Belgium	11,104	513.86	46,277	17,065.8	1,203
Belize	332	1.45	4,360	130.9	1,807
Benin	10,323	7.29	707	112.0	1,895
Bermuda	65	5.56	85,494	213.9	
Bhutan	754	1.83	2,432	1,501.5	1,943
Bolivia	10,671	23.95	2,244	1,268.7	2,037
Bosnia-Herzegovina	3,829	18.24	4,764	3,598.4	1,548
Botswana	2,021	16.04	7,936	155.6	2,302
Brazil	200,362	2,476.65	12,361	103,198.2	1,883
Brunei	418	16.36	39,138	959.3	1,915
Bulgaria	7,223	53.54	7,413	10,764.6	1,606
Burkina Faso	16,935	10.40	614	275.1	2,164
Burma	53,259			1,788.2	1,939
Burundi	10,163	2.36	232	40.9	1,803
Cambodia	15,135	12.83	848	276.0	1,937
Cameroon	22,254	25.32	1,138	898.9	1,875
Canada	35,182	1,777.79	50,531	126,081.0	1,554
Cape Verde	499	1.90	3,810	81.8	2,283
Central Afr. Rep.	4,616	2.17	469	26.2	2,031
Chad	12,825	10.58	825	187.0	2,222
Chile	17,620	251.19	14,256	12,946.5	2,124
China	1,385,567	7,321.94	5,284	704,649.7	1,631
Colombia	48,321	336.35	6,961	14,287.8	1,732
Comoros	735	0.61	830	14.2	2,287
Congo, DR	67,514	15.65	232	1,469.8	1,848
Congo, Rep.	4,448	14.43	3,243	189.0	1,638

Costa Rica	4,872	41.03	8,422	2,290.7	1,735
Croatia	4,290	61.79	14,403	3,976.9	1,586
Cuba	11,266			4,221.9	1,987
Cyprus	885	24.99	28,240	1,560.2	2,244
Czech Republic	10,702	216.01	20,184	15,921.1	1,251
Denmark	5,619	333.62	59,373	10,596.3	1,287
Djibouti	873	1.24	1,419	99.5	2,318
Dominica	72	0.48	6,618	23.9	2,386
Dom. Rep.	10,404	55.67	5,351	3,910.1	1,995
Ecuador	15,738	77.70	4,937	4,086.8	1,660
Egypt	82,056	236.00	2,876	22,719.0	2,242
El Salvador	6,340	23.10	3,643	1,341.0	2,206
Equatorial Guinea	757	16.81	22,202	52.8	1,632
Eritrea	6,333	2.61	412	144.0	2,215
Estonia	1,287	22.15	17,214	3,150.5	1,298
Ethiopia	94,191	31.71	337	832.0	2,205
F.S. Micronesia	104	0.31	2,984	31.8	2,016
Fiji	881	3.82	4,334	221.2	1,976
Finland	5,426	263.18	48,503	16,538.8	1,181
France	64,291	2,779.72	43,237	115,630.7	1,441
Gabon	1,672	18.79	11,239	378.9	1,667
Gambia	1,849	0.90	486	62.3	2,129
Georgia	4,341	14.43	3,325	3,454.2	1,678
Germany	82,727	3,600.83	43,527	132,257.1	1,222
Ghana	25,905	39.18	1,512	1,951.7	1,852
Greece	11,128	289.63	26,027	14,866.3	1,753
Greenland	57			141.5	
Grenada	106	0.78	7,362	51.2	2,317
Guam	165			563.1	2,212
Guatemala	15,468	46.98	3,037	2,098.4	1,997
Guinea	11,745	5.10	434	381.1	2,046
Guinea - Bissau	1,704	0.97	568	15.3	2,089
Guyana	800	2.58	3,221	171.3	1,784
Haiti	10,317	7.35	712	244.8	2,146
Honduras	8,098	17.42	2,152	1,595.4	1,932
Hong Kong	7,204	248.73	34,526	0.0	1,506
Hungary	9,955	138.70	13,932	9,108.2	1,445
Iceland	330	14.07	42,651	2,388.7	
India	1,252,140	1,872.85	1,496	157,537.3	2,032
Indonesia	249,866	846.34	3,387	37,324.1	1,809
Iran	77,447	514.06	6,638	55,211.9	2,041
Iraq	33,765	180.61	5,349	11,956.8	2,073
Ireland	4,627	220.82	47,725	7,263.6	1,055
Israel	7,733	242.93	31,415	12,590.8	2,247
Italy	60,990	2,192.36	35,946	104,897.3	1,720
Ivory Coast	20,316	24.07	1,185	1,293.1	1,818
Jamaica	2,784	14.43	5,182	1,116.4	2,132
Japan	127,144	5,896.79	46,379	269,284.0	1,578
Jordan	7,274	28.84	3,965	2,294.2	2,103

Kazakhstan	16,441	188.05	11,438	19,077.0	1,709
Kenya	44,354	33.62	758	1,467.4	2,124
Kiribati	102	0.17	1,634	9.4	2,178
Korea, DPR	24,895			9,285.0	1,874
Korea, Rep.	49,263	1,114.47	22,623	77,082.3	1,770
Kuwait	3,369	176.59	52,416	10,967.0	2,134
Kyrgyzstan	5,548	6.20	1,117	3,762.7	1,840
Laos	6,770	8.23	1,215	681.3	1,829
Latvia	2,050	28.48	13,893	2,479.6	1,307
Lebanon	4,820	40.09	8,318	2,377.6	2,159
Lesotho	2,074	2.52	1,217	78.5	
Liberia	4,294	1.55	360	15.9	1,781
Libya	6,202			7,956.5	2,110
Lithuania	3,017	42.87	14,210	4,755.1	1,277
Luxembourg	530	59.20	111,700	1,706.0	
Macedonia	2,107	10.44	4,954	1,673.8	1,718
Madagascar	22,925	9.91	432	396.4	2,091
Malawi	16,363	5.62	344	310.9	2,088
Malaysia	29,717	287.93	9,689	25,591.4	1,766
Maldives	345	2.15	6,242	101.4	
Mali	15,302	10.66	696	588.6	2,185
Malta	429	9.15	21,333	588.8	2,188
Mauritania	3,890	4.27	1,099	212.9	2,202
Mauritius	1,244	11.24	9,037	810.2	2,244
Mexico	122,332	1,158.15	9,467	59,542.8	2,136
Moldova	3,487	7.02	2,012	2,999.6	1,492
Mongolia	2,839	8.76	3,086	1,028.7	1,910
Morocco	33,008	99.21	3,006	5,658.6	2,153
Mozambique	25,834	12.57	487	2,321.2	2,026
Namibia	2,303	12.48	5,418	385.6	2,352
Nepal	27,759	18.98	684	722.3	2,176
Netherlands	16,759	836.07	49,888	21,529.8	1,242
New Caledonia	256			437.0	2,129
New Zealand	4,506			9,681.4	1,644
Nicaragua	6,080	9.64	1,585	728.5	2,016
Niger	17,831	6.02	337	135.9	2,382
Nigeria	173,615	243.99	1,405	11,101.1	1,978
Norfolk Island				0.0	
Norway	5,043	491.06	97,376	30,092.5	1,103
Oman	2,632	71.78	27,273	5,750.5	2,239
Pakistan	182,143	210.74	1,157	21,071.4	2,135
Panama	3,864	31.32	8,105	1,378.2	1,728
Papua New Guinea	7,321	12.39	1,693	630.6	1,825
Paraguay	6,802	26.01	3,824	257.6	1,898
Peru	30,376	176.93	5,825	6,141.1	2,006
Philippines	98,394	224.77	2,284	17,393.9	1,842
Poland	38,217	515.67	13,493	36,481.1	1,235
Portugal	10,608	237.57	22,396	15,437.4	1,891
Puerto Rico	3,688	98.76	26,778	0.0	2,142

Qatar	2,169	172.98	79,752	5,440.2	2,065
Rep. of Palau	21	0.21	10,138	26.8	2,020
Romania	21,699	189.78	8,746	21,228.9	1,500
Russia	142,834	1,899.09	13,296	228,525.8	1,403
Rwanda	11,777	6.35	540	53.0	1,831
Sao Tome&Princi|	193	0.25	1,286	11.0	
Saudi Arabia	28,829	576.82	20,008	48,349.1	2296
Senegal	14,133	14.45	1,022	719.4	2,126
Seychelles	93	1.06	11,394	140.7	2,168
Serbia	9,511	43.29	4,552	10,108.5	1,573
Sierra Leone	6,092	2.94	482	76.8	1,861
Singapore	5,412	245.02	45,274	11,053.6	
Slovakia	5,450	96.06	17,626	7,461.8	1,286
Slovenia	2,072	50.28	24,266	3,244.3	1,483
Solomon Island	561	0.87	1,545	38.1	1,953
Somalia	10,496			15.6	2,100
South Africa	52,776	401.80	7,613	44,063.8	2,166
Spain	46,927	1,476.88	31,472	85,938.6	1,886
Sri Lanka	21,273	59.19	2,782	2,850.0	1,813
St. Helena	4			3.0	2,149
St. Lucia	182	1.21	6,654	74.4	
St. Vincent&Gre.	109	0.69	6,341	45.9	2,277
Sudan	37,964	64.05	1,687	1,586.3	2,271
Suriname	539	4.30	7,986	516.5	1,872
Swaziland	1,250	3.97	3,175	124.4	1,982
Sweden	9,571	539.28	56,345	33,635.8	1,218
Switzerland	8,078	659.31	81,618	16,572.6	1,467
Syria	21,898			7,705.3	2,026
Taiwan	23,330			49,160.3	1,632
Tajikistan	8,208	6.52	795	4,909.0	1,996
Tanzania	49,253	23.87	485	1,186.6	2,043
Thailand	67,011	345.67	5,158	34,706.4	1,903
Togo	6,817	3.69	541	208.7	1,931
Trinidad&Tobago	1,341	23.50	17,522	1,672.4	2,136
Tunisia	10,997	46.43	4,222	3,589.8	1,916
Turkey	74,933	774.78	10,340	40,541.9	1,839
Turkmenistan	5,240	28.06	5,355	3,661.3	1,894
Uganda	37,579	16.82	448	640.5	1,980
Ukraine	45,239	163.42	3,612	54,084.4	1,398
UAE	9,346	360.25	38,545	25,191.4	2,261
United Kingdom	63,136	2,444.88	38,724	88,434.3	1,128
United States	320,051	14,991.30	46,840	1058,995.2	1,796
Uruguay	3,407	46.43	13,629	1,139.9	1,853
Uzbekistan	28,934	45.32	1,566	13,121.2	1,976
Venezuela	30,405	316.48	10,409	26,904.1	1,934
Vietnam	91,680	123.68	1,349	15,984.4	1,665
Western Sahara	567			235.7	2,209
Yemen	24,407	31.72	1,300	1,203.1	2,295
Zambia	14,539	19.20	1,321	1,881.4	2,201

Zimbabwe	14,150	9.66	682	2,005.8	2,221

TECHNICAL PARAMETERS USED IN PV FINANCIAL MODELS: REVIEW AND ANALYSIS

Jan Vedde[1], Mauricio Richter[2], Caroline Tjendgdrawira[2], Bert Herteleer[3,6], Magnus Herz[4], U. Jahn[4], B. Stridh[5], Lyndon Frearson[6] and Mike Green[7]

[1]SiCon, J N Vinthersvej 5, DK-3460 Birkerød, Denmark, Jan.Vedde@mail.dk,
[2]3E sa, Quai à la Chaux 6, 1000, Brussels, Belgium,
[3]KU Leuven, Oude Markt 13 - bus 5005, 3000 Leuven, Belgium
[4]TÜV Rheinland Energie und Umwelt GmbH, D-51101 Cologne, Germany
[5]ABB Corporate Research, Sweden
[6]CAT Projects, Alice Springs NT, Australia
[7]M.G. Lightning ltd, Israel

ABSTRACT: When photovoltaic (PV) projects are developed and constructed for investment purposes, financing from professional investors and banks are required, and financial models are developed that describe the expected cash-flow generated by the PV plant over the economic lifetime of the project. The calculation is based on models that require both technical and financial inputs. In this paper we will review and analyse the technical assumptions in these models and discuss how the technical assumptions are typically used and to what extent the uncertainty related to the input parameters are appropriately addressed. Finally, a calculation and visualisation method is presented that highlights the nature and effect of this uncertainty.
Keywords: PV financial models, Risk assessment, Uncertainty,

1 INTRODUCTION

PV financial models are used by project developers, banks and asset managers to evaluate the profitability of a PV project. The task is to predict the discounted cash flow as accurate as possible, to assess if the project represents an attractive investment opportunity. The most important key performance indicator (KPI) is the Internal Rate of Return (IRR) or Return on Investment (ROI) on the invested capital, but as investors seldom like surprises, also the uncertainty related to IRR is of interest. One method to present this perspective to the investors is by stating the minimum energy production that is expected with either a 90% or 50% probability of exceedance (P90 and P50 values respectively). To calculate energy yield expectations demonstrate a willingness to address the risk profile of the project, this approach however may conceal the impact of other sources of uncertainty on the projects' achievable IRR.

In this paper we will discuss how technical parameters are used in PV investment models and how uncertainty issues can be addressed. This paper presents the preliminary results of the IEA PVPS Task13 Subtask 1 that are working on *Economics of PV System Performance and Reliability*.

In a previous paper [1], a screening of 84 project presentations covering 9 different countries, several technologies and different business concepts where performed to assess the technical parameters used as well as the described strategies on how to deal with the uncertainty related to these parameters. In a discussion of these observations, we found that those parameters that directly influence the energy yield calculation were typically described in detail whereas those technical issues that influence the CAPEX and OPEX values were not addressed at all.

In this paper we'll present an update of the analysis and present an alternative strategy to deal with such uncertainties.

2 DATA COLLECTION AND ANALYSIS

The data that were used for this analysis was collected by the authors of this paper guided by a questionnaire that organised the observations by:

- General project data
- Irradiation & Key Performance Indicators
- Technical data & system design data
- System cost
- Business model
- Operation & Maintenance & Monitoring

2.1 General project data

Under this topic data has been collected on the background, purpose, ownership, development history, location, module-plane orientation and size (power and area) of the project. Although not all project presentations covered all specified questions, these data are in general quite detailed and descriptive. They represent those facts that are often made public and do not reflect technical issues that introduce ambiguity to the calculations. A great number of technical details are provided, but except for the PV module tolerance none of these parameters have direct impact on the calculated energy yield or financial performance uncertainty.

2.2 Irradiation & Key Performance Indicators

Under this topic data has been collected on the solar resource at the site, and the character and origin of the irradiance data used to estimate this resource potential.

As a general observation, the irradiation data are often collected, analysed and presented with a great deal of professionalism. It suggests that these data and conclusions – including uncertainty estimates – can be considered as highly trustworthy. However, the data used reflects historical data. There is an uncertainty in future irradiation over such a long period as the lifetime of a PV system, since weather effects can affect the irradiation. In addition there is variation between different years that can be up to ±10%.

TECHNICAL PARAMETERS USED IN PV FINANCIAL MODELS: REVIEW AND ANALYSIS

Jan Vedde[1], Mauricio Richter[2], Caroline Tjendgdrawira[2], Bert Herteleer[3,6], Magnus Herz[4], U. Jahn[4], B. Stridh[5], Lyndon Frearson[6] and Mike Green[7]

[1]SiCon, J N Vinthersvej 5, DK-3460 Birkerød, Denmark, Jan.Vedde@mail.dk,
[2]3E sa, Quai à la Chaux 6, 1000, Brussels, Belgium,
[3]KU Leuven, Oude Markt 13 - bus 5005, 3000 Leuven, Belgium
[4]TÜV Rheinland Energie und Umwelt GmbH, D-51101 Cologne, Germany
[5]ABB Corporate Research, Sweden
[6]CAT Projects, Alice Springs NT, Australia
[7]M.G. Lightning ltd, Israel

ABSTRACT: When photovoltaic (PV) projects are developed and constructed for investment purposes, financing from professional investors and banks are required, and financial models are developed that describe the expected cash-flow generated by the PV plant over the economic lifetime of the project. The calculation is based on models that require both technical and financial inputs. In this paper we will review and analyse the technical assumptions in these models and discuss how the technical assumptions are typically used and to what extent the uncertainty related to the input parameters are appropriately addressed. Finally, a calculation and visualisation method is presented that highlights the nature and effect of this uncertainty.
Keywords: PV financial models, Risk assessment, Uncertainty,

1 INTRODUCTION

PV financial models are used by project developers, banks and asset managers to evaluate the profitability of a PV project. The task is to predict the discounted cash flow as accurate as possible, to assess if the project represents an attractive investment opportunity. The most important key performance indicator (KPI) is the Internal Rate of Return (IRR) or Return on Investment (ROI) on the invested capital, but as investors seldom like surprises, also the uncertainty related to IRR is of interest. One method to present this perspective to the investors is by stating the minimum energy production that is expected with either a 90% or 50% probability of exceedance (P90 and P50 values respectively). To calculate energy yield expectations demonstrate a willingness to address the risk profile of the project, this approach however may conceal the impact of other sources of uncertainty on the projects' achievable IRR.

In this paper we will discuss how technical parameters are used in PV investment models and how uncertainty issues can be addressed. This paper presents the preliminary results of the IEA PVPS Task13 Subtask 1 that are working on *Economics of PV System Performance and Reliability*.

In a previous paper [1], a screening of 84 project presentations covering 9 different countries, several technologies and different business concepts where performed to assess the technical parameters used as well as the described strategies on how to deal with the uncertainty related to these parameters. In a discussion of these observations, we found that those parameters that directly influence the energy yield calculation were typically described in detail whereas those technical issues that influence the CAPEX and OPEX values were not addressed at all.

In this paper we'll present an update of the analysis and present an alternative strategy to deal with such uncertainties.

2 DATA COLLECTION AND ANALYSIS

The data that were used for this analysis was collected by the authors of this paper guided by a questionnaire that organised the observations by:

- General project data
- Irradiation & Key Performance Indicators
- Technical data & system design data
- System cost
- Business model
- Operation & Maintenance & Monitoring

2.1 General project data

Under this topic data has been collected on the background, purpose, ownership, development history, location, module-plane orientation and size (power and area) of the project. Although not all project presentations covered all specified questions, these data are in general quite detailed and descriptive. They represent those facts that are often made public and do not reflect technical issues that introduce ambiguity to the calculations. A great number of technical details are provided, but except for the PV module tolerance none of these parameters have direct impact on the calculated energy yield or financial performance uncertainty.

2.2 Irradiation & Key Performance Indicators

Under this topic data has been collected on the solar resource at the site, and the character and origin of the irradiance data used to estimate this resource potential.

As a general observation, the irradiation data are often collected, analysed and presented with a great deal of professionalism. It suggests that these data and conclusions – including uncertainty estimates – can be considered as highly trustworthy. However, the data used reflects historical data. There is an uncertainty in future irradiation over such a long period as the lifetime of a PV system, since weather effects can affect the irradiation. In addition there is variation between different years that can be up to ±10%.

2.3 Technical data & system design data

Under this topic technical data has been collected on the components and engineering concepts used in the projects. Most of the data provided represent technical facts that are presented with many details including estimates of uncertainties. The general impression is that the values have been calculated by engineers for the specific project and has been "topped off" with an uncertainty value selected from experience or judgement by the installer.

2.4 System cost

Under this topic we collect information about the cost or price of the project. The costs are supposed to include not only the direct engineering-procurement-construction (EPC) cost but also the project development cost. The ambition was to analyse the cost-structure with a breakdown into categories of: construction (site preparation cost, civil works (installation of DC, AC, fence, safety & security components and waste removal), mounting structures, PV modules, Balance of System (BOS) (i.e. cables, monitoring system, fence, etc.), insurance and grid connection cost. In general, this kind of information was not available in the screened project presentations, most likely due to the sensitive commercial nature of these figures.

2.5 Business model

Under this topic we collect information on the nature of revenue generation as well as information about the financial structure, taxation and other financially related aspects of the project. At a minimum it should be possible to understand if the power is to be sold under market conditions, subsidised by a feed-in-tariff (FIT), green certificates or tax credit or whether the power will be self-consumed in whole or in part, or sold via a private/public grid to a third party according to a Power Purchase Agreement (PPA).

Information of this kind was either not available or not transferred to the questionnaire. Some screened projects included a reference to the FIT scheme but otherwise this kind of information has not been collected.

2.6 Operation & Maintenance and Monitoring

Under this topic we collect information about the land/roof lease conditions, insurance conditions, organisation and content/extent of Operation and Maintenance (O&M) activities planned for the plant.

As the scope of this work concern technical assumptions, the uncertainty in general financial parameters like interest rate and equity share etc. is not addressed.

Although some of the screened projects do provide information about the lease and insurance costs, in general, information of this kind has not been transferred from the screened presentations to the questionnaire.

3 DISCUSSION

The ambition of this paper is to investigate how technical assumptions are used in PV financial models. From the data collection process we observe, that the presentations in general contain a lot of technical details and specific parameter values. It is also clear that the technical topics are mostly covered in the section that describe the expected energy yield, whereas the sections describing the CAPEX and OPEX values are treated as if they do not depend on technical assumptions at all.

This distinction may reflect the fact that energy yield calculations - in theory - can be performed with a high degree of accuracy by use of dedicated software packages like PVsyst, PV*SOL or similar models. In contrast, the assumptions used in other parts of the financial model are generally defined in legal and economic terms and may by tradition not be presented with an associated uncertainty specified with a one-sigma standard deviation as is typical done for a normal distribution.

3.1 Technical assumptions used in energy yield calculations

The generated income of the project is given by the amount of energy that can be delivered to the user/grid and the price that is paid for this energy. As mentioned above the energy yield can be calculated by a cascade of interlinked sub-models that each solves a particular question about energy conversion but still the character and the input assumptions to be used for these sub-models varies considerably [2].

One example is the sub-models that are used to calculate the DC-cable loss. The physical principles involved are well known as they rely on Ohm's law. The input parameters to this model also seem straightforward and should be available very exact, as the cable resistivity, length and cross-section all are parameters that are well specified in the engineering documents. The only remaining unknown in the model is the current which is supposed to be calculated in a previous sub-model dealing with the DC-output of the PV module. However several additional issues like current loops flowing among modules in the string and between strings on a combiner box due to voltage mismatch are not known with the same level of precision as the cable characteristics and the overall quality of the energy calculation will therefore include both well identified (e.g. cable lengths) and more "guestimated" (e.g. module mismatch) elements that both contributes to the overall uncertainty.

Other sub-models may have similar characteristics, and contain a core based on an exact physical principle, but where the actual implementation requires user input of estimates and assumptions that cannot be provided by direct observation or measurement, and then will reflect experience (or lack thereof) or personal bias of the person that operates the model. The importance of the individual preferences of the model operator is a generally known problem as also reported by Cameron et al. following a PV modelling workshop at Sandia Labs [3].

3.2 Technical assumptions used in OPEX calculations

The data collected in the questionnaire does not indicate that the operational expenditures reflect any technical assumptions at all. Based on experience however, this is known to be incorrect. For example, the observation that the expected inverter lifetime is normally shorter than the project economic lifetime indicates that technical knowledge is required to prepare a correct budget for operational expenditures. For budget purposes it is of importance to know the technical parameters of Mean Time Between Failures (MTBF) and life expectancy profiles of the equipment to prepare a good O&M budget [4].

As the monitoring of the plant will focus on the achieved Performance Ratio (PR), it is of importance to ascertain if a PR guarantee can be obtained from the O&M provider and how this may be linked - during the first years - to the EPC contract. The legal and financial tenets at stake will in this case depend entirely on the details of such technical definitions included in the O&M contract.

It is also not uncommon to link the O&M fee to the level of Technical Availability (TA) that the O&M provider can ensure [5]. In a more advanced model one would then see that a reduction of the TA would have two interlinked impacts; the energy sale would go down but also the O&M fee would be reduced.

3.3 Technical assumptions used in CAPEX calculations

In the financial model there is normally only room for one CAPEX value that can be either the construction cost for a project under development or the sale price. In both cases the many technical assumptions that have been considered before the "correct" CAPEX value can be chosen, are not very clear.

The construction cost is supposed to reflect the quality of components and workmanship used in the project which then should be reflected in the EPC contract. It may be difficult though, to extract such considerations about performance, quality control and inspection criteria of the PV modules, inverters, sub-structures, and BOS components from the project presentations. Also on a system level the test procedure and acceptance criteria for commissioning of the installation is often quite detailed and based on technical considerations that are reflected in the CAPEX.

Large PV installations are considered investment opportunities and will be traded among investors at a price that reflect the generated IRR. As with the EPC-contract, the sales-price could be adjusted according to the realised production or PR during the first two years of operation based on formulas specified in the contract.

Table 1. Input parameters for a hypothetical 1 MWp project used as an example. The values used are inspired by the data collected in the questionnaire but does not reflect any specific project. With Best Case/Worst Case estimates around the Best estimate, a triangular function is used in this example to describe the uncertainty of the parameter in question.

1. Project	**Unit**	Best estimate	Worst case	Best case
Module Power	[W_p]	250	250	255
Number of modules	[pcs]	4.000	Not applicable	
2. Site selection				
Horizontal global irradiation	[$kWh/m^2/yr$]	1000	±3%	
Irradiation increase due to tilt angle	[%]	15%	Not addressed in this example	
3. System design				
Irradiation loss due to soiling	[%]	2,0%	±1%	
Irradiation loss due to shading	[%]	1,0%	±0.5%	
Module energy loss due to incidence effect (IAM)	[%]	3,0%	±0.5%	
Module energy loss due to temperature	[%]	1,0%	±0.5%	
Array mismatch loss	[%]	1,0%	±0.5%	
Ohmic wiring dc-loss	[%]	1,0%	±0.2%	
DC/AC-sizing factor	[W_p/W]	120%	Not addressed in this example	
Inverter loss during operation (conversion efficiency)	[%]	2,5%	±1%	
Transformer loss	[%]	1,5%	±0.8%	
4. Project and system components costs				
Turnkey unit price	[€/Wp]	1,0	1,2	1,0
5. Power production & sales				
Start year	[year]	2016	Not addressed in this example	
Technical lifetime of installation	[year]	25	Not addressed in this example	
Technical availability	[%]	98%	95%	99%
PV system [module] degradation	[%/a]	0,30%	0,50%	0,20%
Feed-In-Tariff for first 20 year of operation	[€/kWh]	0,10	Not applicable	
Market price for electricity sales (to be used from year 21)	[€/kWh]	0,05	0,03	0,07
6. Land lease, O&M and guarantees				
O&M - fixed yearly fee	[€/kW_p]	20	15	25
Inverter replacement (according to expected lifetime)	[year]	15	10	16
7. Financing				
Economical operational lifetime	[year]	25	Not in scope of this paper	
Real discount rate	[%]	5%	Not in scope of this paper	
Debt ratio	[%]	90%	Not in scope of this paper	
Equity	[%]	10%	Not in scope of this paper	
Bank Interest – loan	[%]	3%	Not in scope of this paper	
Amortisation and depreciation	[year]	15/25	Not in scope of this paper	

Likewise, all other possible deviations of the realised project from the marketing prospectus must be accounted for in the CAPEX value. The adjustment procedure is based on Technical Due Diligence (TDD) reports prepared by Technical Advisors (TA). A negative deviation registered in this report may result in a reduction in the sales price according to a predetermined formula or a hold-back in the pay-out until the problem has been solved by the seller. These technical problems could reflect issues with substructure statics and protection, inverter power-curve adequateness to the general PV system, an observation of microcracks in the PV modules that may negatively impact the realised power at some time in the future or many other technical issues that could be identified during the TDD process.

In any case, the conception that the CAPEX is given by one single number in the investment model is not always correct and it may be of advantage to present this uncertainty in the CAPEX value in the financial model by adding a tolerance, interval or "error-bar" to the number, when it is referenced in the presentation text or calculations.

4 EXAMPLE: CALCULATION WITH UNCERTAINTY

From the analysis of data from the 84 screened PV projects, it seems clear that the uncertainty in most technical parameters are only addressed when a reference can be made to scientific literature or standard practices used among the technical advisors. The uncertainty is always reported as a symmetric standard deviation around the expectation value and whereas this practice may be generally accepted in the yield assessment community due to simplification, convention and tradition, we find that it does not adequately address the true character of this issue, as it's also expected that some of uncertainties may be single sided/biased/binned or otherwise following a non-normal distribution.

However, even with non-normal input variable distributions, it's fairly easy to perform project calculations where this variation is taken into account. In spreadsheets such as Excel, Monte Carlo simulations can be performed such that all possible combinations of inputs variables are identified with an assigned probability and the financial calculation will then result in output parameter distributions that have the character of a distribution rather than a single number[1].

As an example we have calculated the distribution of expected variations in income and NPV for a reference 1 MW_p PV project based on various distributions of the input variables given in Table 1. In order to illustrate how handling of non-normal distributions can be done, the input distributions for these calculations have been chosen as triangular functions with the target value as the top-point and with the Minimum/Maximum values defining the two foot points as illustrated in Figure 1.

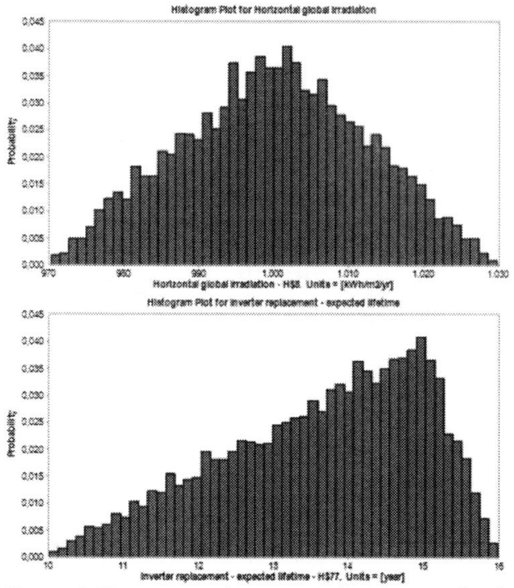

Figure 1 Histograms of two input distributions for the financial calculation: Horizontal global Irradiation (top) and Inverter replacement lifetime (bottom).

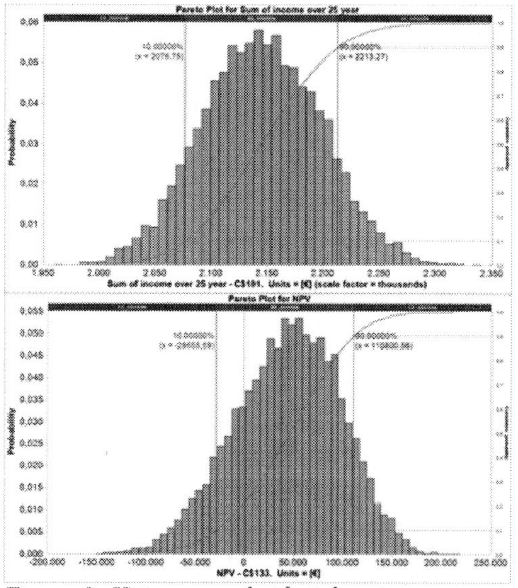

Figure 2 Histograms of selected output parameter distributions from the financial model: Sum of generated income over the project lifetime (top) and Net Present Value at 5% discount rate of cash flow (bottom).

5 CONCLUSIONS

We have screened 84 PV project presentations and analysed how uncertainty in the technical assumptions has been addressed. We find that those parameters that are used in calculating the energy yield are often well specified and referenced to fundamental physical models and also are presented with uncertainty estimates. We also find that within the overall yield calculation methodology, sub-models are included that more reflect a generally accepted industry consensus on loss factors

[1] In this paper we have used ModelRisk 5 from VOSE software with 10.000 iterations.

than a fundamental physical understanding and scientific collection of model parameters.

Whereas this energy yield calculation strategy has proven to be effective in providing investor confidence into the financial models used, it is surprising how limited information the presentations provide on the dependencies of CAPEX and OPEX values on technical issues. The project presentations may include detailed descriptions on module technology, local planning condition; grid connection agreements, O&M contractor etc. although the link between these qualitative descriptions and the input parameters used in the financial model are generally not presented.

As the uncertainty of input parameters cannot always be assumed to follow normal distributions, it's necessary to introduce non-analytical methods for calculation of the overall uncertainty of the financial outputs. In this paper we propose to use numerical methods such as Monte Carlo calculations that give the modeller full freedom to select any type of probability distribution for the input variables and also offer graphical presentations of the output with the expected uncertainties that may give investors a simple and intuitive illustration of the financial risk profile of the PV project.

ACKNOWLEDGEMENTS

The contribution from 3E s.a. to this work has received funding from the European Union's Horizon 2020 research and innovation programme under the grant agreement No 649997 (Solar Bankability).
Jan Vedde wishes to thank the Danish Energy Agency for support according to the EUDP programme with project number 64014-0519.

REFERENCES

[1] Vedde, Jan et al. 2016, Technical assumptions used in PV financial models. Submitted to *Progress in Photovoltaics*.
[2] Jordan, D.C. and Kurtz, S.R. 2013. Photovoltaic degradation rates – an analytical review. Progress in photovoltaics: Research and Applications. 21, 1 (2013), 12–29.
[3] Cameron, C.P., Stein, J.S. and Tasca, C.A. 2011. PV performance modelling workshop summary report. Technical Report #SAND2011-3419. Sandia National Laboratories.
[4] Köntges, M., Kurtz, S., Packard, C., Jahn, U., Berger, K.A., Kato, K., Friesen, T., Liu, H. and Iseghem, M.V. 2014. Review of Failures of Photovoltaic Modules.
[5] Roessler, T. and Sauer, K.J. 2012. Modeling of PV Module Power Degradation to Evaluate Performance Warranty Risks. 27th European Photovoltaic Solar Energy Conference and Exhibition (2012), 3142–3147.

IMPACT OF ENERGY STORAGE IN CONJUCTION WITH SOLAR PV ON WHOLESALE ELECTRICITY PRICES

Adel El GAMMAL (Becquerel Institute, 69-71 Rue d'Arlon-1040 Brussels, a.elgammal@becquerelinstitute.org)
Frederico SANCHES (Green Giraffe, 30 Crown Place London EC2A 4EB, f.sanches@green-giraffe.eu)
Gaëtan MASSON (Becquerel Institute, 69-71 Rue d'Arlon-1040 Brussels, g.masson@becquerelinstitute.org)
Henri GOUZERH (Green Giraffe, 33 rue du Louvre 75002 Paris, h.gouzerh@green-giraffe.eu)
Nicolas GOURVITCH (Green Giraffe, 33 rue du Louvre 75002 Paris, n.gourvitch@green-giraffe.eu)
Thi Mai Nhan NGO (Becquerel Institute, 69-71 Rue d'Arlon-1040 Brussels, t.ngo@becquerelinstitute.org)

ABSTRACT: Previous studies have shown that the total electricity demand retreated by PV production is well correlated to the prices, following an exponential curve (R^2 around 69%) and the Merit Order Price is quite dependent on the particular electricity demand profile. Since 2011, the penetration rate of PV is also a significant explanatory variable, suggesting that PV could be replacing base load capacity. Ultimately it has been observed that the negative pressure of PV production on wholesale prices is well reflected in Germany, Austria, France, Switzerland and Italy (GAFSI). The objective of this study is to quantify the impact on the wholesale electricity prices at GAFSI level that would be achieved by adding de-centralized storage capability to the PV capacity and shifting the use of centralized storage (e.g. pump-hydro storage) to benefit from the merit order effect. This study has three main findings: firstly, without storage, the cumulative gain of having PV capacity over the past 10 years is around EUR 31 bn for the whole GAFSI region and the average wholesale prices would have been 4 % higher had there not been any PV; secondly, combining PV with decentralized storage system does not currently contribute positively on the aggregate metrics, though in 2014 and 2015 small increases in the merit order effect were observed suggesting that some optimization may be possible; thirdly, where the centralized storage is used in conjunction with PV, the aggregate gain increases to EUR 33 bn (with a storage capacity of 25 GW).

Keywords: economic analysis, photovoltaic production, electricity prices, merit order effect, energy mix.

1 INTRODUCTION

1.1 General statements on renewable energies

The EU member states have committed to a drastic increase of the share of renewable energy ("RE") sources in their energy mix. PV, like most renewables behaves very differently from conventional generation sources and therefore requires special considerations from a policy, from an electrical system and from an electricity market perspectives:

- Intermittency: PV production is not dispatchable and can only be forecasted with a certain level of uncertainty. As electricity is difficult to store, this leads to require spare peak production capacities to be available and in some cases grids to be upgraded.
- Specialization: PV can be centralized in the case of large ground mounted systems, but is typically decentralized in the case of distributed rooftop applications, which may require grid upgrade works
- Support to transition: Although the cost of photovoltaic (PV) modules has dramatically decreased over the past few years (from approx. 4 €/Wp to approximately 0.5 €/Wp), PV full competitiveness has not yet been reached in all market segments and in all EU Member States, and PV producers will continue temporarily to need monetary support (revenue schemes such as FiT, CfD, ROCs, GC, tax incentives, direct subsidies,..) as well as non-financial support measures (e.g. priority access to the grid, self-consumption provisions,..).

1.2 Observed impacts of RE on electricity markets

The effects of RE generation on electricity markets are still unclear and certainly dependent on each country's energy mix.

A 2013 quarterly report on European electricity markets [1] illustrates this conundrum: *"Intermittent power generation sources, such as wind and solar, played an increasingly important role in the power mixes of many European countries during the second quarter of 2013. In Central Western and Central Eastern Europe, high levels of renewables generation contributed to the lowest wholesale power prices observed in the last few years. Frequent occurrences of negative prices in many European markets signal the need for better integration of renewables into the power grid. On a Sunday afternoon in mid-June, wind and solar assured more than 60% of power generation in Germany, resulting in negative hourly prices in the whole Central Western Europe region."*

1.3 Purpose of this study

The purpose of the study is not to discuss the performance of various policy frameworks in supporting the deployment of RE, but rather to shed light on the gain generated by the downward pressure on market spot prices when renewables produce electricity through the Merit Order Effect, and to quantify the corresponding monetary benefit. This benefit will have to be put in perspective with the costs associated to the deployment of RE, among which the cost of support policies, the impacts on the energy mix (including the possible need for capacity reserves), the grid upgrade costs, and negative prices appearing when overproduction occur as a result of RE generation.

The purpose of this study assesses the impact of adding a storage to the installed PV capacity in terms of cost savings at a national level. To be specific, the impact on the wholesale electricity prices at country level that would be achieved by adding decentralized storage capability to the PV capacity and shifting the use of pump-hydro storage to benefit from the Merit Order Effect.

Additionally, the study measures the benefit generated by the PV production over the past 10 years in Germany, Austria, France, Switzerland and Italy ("GAFSI"). Such analysis will be performed based on historical and statistical methods. The behavior of that benefit with respect to the penetration rate of PV within GAFSI's energy mix and with the correlation between PV production and electricity demand will also be assessed. This study builds on previous work by the authors where similar calculations were carried out for Italy as well as for GAFSI up until 2013 [2][3].

2 THE MERIT ORDER EFFECT

2.1 Preliminary definitions and scope of work

- Cost of electricity
  a. The levelised cost of electricity ("LCOE") is the average cost of a megawatt hour ("MWh") produced by a given plant, including the fuels cost required to produce a MWh, but also the operating expenses (maintenance, taxes, ...) as well as the amortization of the investment. The LCOE of PV plants was divided by more than five over the past years,
  b. The marginal cost of electricity represents at a given time, the cost to generate an additional MWh of electricity, e.g. only the cost of gas for a gas fired power plant. By definition, the marginal cost for PV amounts to zero (no fuel cost).
- Price of electricity
  c. Retail prices paid by end consumers are set in long term fixed price contracts and include all costs of generating, transporting and distributing electricity, the margins of the various operators plus applicable taxes;
  d. Wholesale prices (also referred to as wholesale spot market prices) are traded on electricity markets daily and will be the focus of this study, i.e. the authors will not tackle prices observed on other markets (such as the futures market for long term trades, the intraday market or the capacity reserve market),
  e. Price paid to RE electricity producers, which typically includes the revenue of support schemes for RE producers. This study does not focus on this metric.
- Net value for society
  f. It includes a monetary part (i.e. the total energy bill of the country), but also
  g. All sorts of non-monetary externalities (e.g. intermittency of renewable energy, fossil fuels depletion risks, energy security, dependency on unstable foreign countries, pollution, public health, climate change...)
  h. This study only focuses on monetary terms, and attempts to scrutinize if the injection of PV generation capacity into the grid lowers or increases the wholesale prices of electricity at the level of the GAFSI countries.
- Merit order effect ("MOE")
  i. The MOE is the downward pressure on prices exercised by RE sources when they feed electricity into the grid (detailed in section 2)
  j. In order to quantify that phenomenon, one must compare the historical payments for energy of the country/countries over a certain period to what such payments would have been had there been no RE production. This requires to simulate what the prices of electricity would have been then (see section 3)
  k. This study proposes a method based on the fact that for a given country, in a sufficiently short time frame (typically a year), there is a direct relationship between instantaneous demand and electricity price.

2.2 Electricity markets

An electricity market is an exchange platform matching supply and demand for electricity.

- Demand for electricity is a short-term phenomenon and was considered inelastic to price in the study because most consumers are supplied on long-term contracts.
- Supply: Typical energy mix can be split in three categories of energy sources:
  l. Base load, such as nuclear and coal fired power plants, to sustain a constant level of production. It is unable to adapt to short-term variations in electricity demand. It typically has high fixed costs and low marginal costs.
  m. Peak load, such as gas fired turbines, to adapt to high sudden demand. Units are usually smaller with low fixed costs and high marginal costs. These production facilities have a lower utilization rate but charge high prices because of the instantaneous shortage in supply.
  n. Mid load are intermediate generation units, with slower ramp up capabilities than Peak load generators but also lower marginal costs.
- Market coupling: power markets in Europe are currently not integrated with other countries' electricity markets. Their organization at national level can have an impact on the electricity prices since renewables can affect the demand profile generating downward pressure on prices when they feed electricity into the grid. Thanks to a market coupling the energy produced by PV could be counterbalanced between countries in order to be able to manage the electricity prices according to the different country demand profiles. As the study is made at GAFSI level rather than on a per country basis, the authors have opted to focus on the weighted average price ("WAP") of the market prices observed in each country by their respective overall electricity consumption (see section 2.3).

As electricity cannot be easily stockpiled (except to a certain extent when hydro installations are available) there needs to be a perfect clearance at each time between demand for electricity and power injected into the grid. Each supplier estimates its demand profile and purchases electricity accordingly: therefore power plants with the lowest marginal cost will be tapped in first. The system operator is ultimately responsible to guarantee security and adequacy of supply but will settle mismatches between supply and demand on a bilateral basis with balancing costs.

In market environments, prices at a given time are thus determined by the most expensive power producers able to satisfy the demand (i.e. with the highest marginal costs) and are imposed on all other producers (since in a purely competitive market, equilibrium between supply and demand is met when price equals marginal cost). This "uniform pricing" principle is the case in most markets, but there are indeed some markets with "pay-as-bid" clearance.

In a perfectly competitive and transparent market, it is then possible to build the relationship between the electricity demand at a given time and the associated price by sorting energy sources in growing order of marginal cost. This step function is called the merit order curve ("MOC"). The width of each step represents the supply capacity of an energy source while its height is its marginal cost (see figure 1 below).

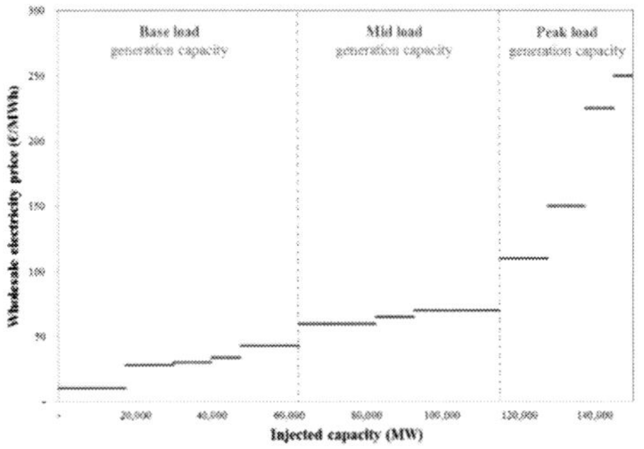

Figure 1: MOC of a fictive electricity market (source: authors)

Therefore, the electricity wholesale prices will be determined as the intersection of the instantaneous demand and the MOC, representing the marginal cost for a given production.

In the MOC context, the case of RE is unusual since they do not behave as a base, mid or peak power plant. As previously mentioned, power suppliers always purchase the renewable power injected in the network (due to the priority of RE production in the grid and because renewables always align to the lowest price) and since electricity demand is inelastic in the short-term, any RE production will decrease demand for other power sources. Since the MOC has a positive slope, this translates into lower wholesale electricity prices. This is defined as the MOE and displayed in figure 2 below.

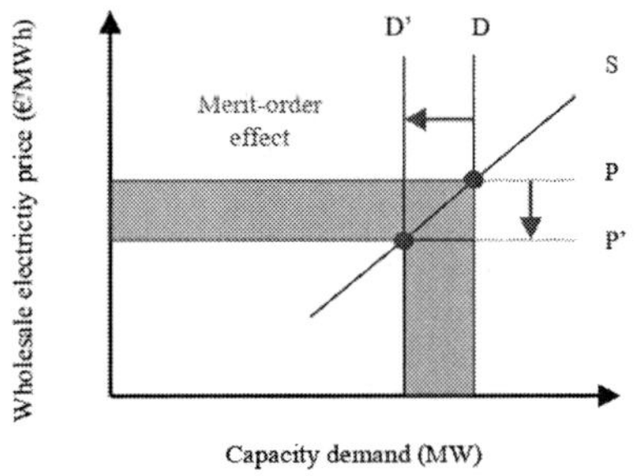

Figure 2: MOE of PV generation (source: authors)

This paper attempts to build the MOC and quantify the MOE for GAFSI between 2006 and 2015.

2.3 Set of assumptions and protocol of the study

The authors based their study on the GAFSI because they estimated it is the set of European countries that fits best with the following necessary criteria for the study.

GAFSI has experienced a massive growth of installed PV capacity over the past years. It is thus possible to measure the MOE behavior with respect to the PV penetration rate. Still, at the GAFSI scale the latter remains moderate (about 5% in 2014), so peak PV production does not lead to important market distortions on a recurrent basis, like negative prices as are observed e.g. in Germany. Intuitively, this underlines the fact that PV production does not replace base load production but rather mid or peak load production (again at the GAFSI level). The results provided in this study are not robust to negative prices (see Section 3) but as stated, fortunately less than 0.5% of negative WAP were observed at the GAFSI level between 2006 and 2014 (less than 1% in Germany alone, and none in Italy).

Up until recently there was little incentive for self-consumption in most GAFSI countries (especially in Germany) and most electricity imports and exports occur within the GAFSI countries: in 2013, exports to and imports (respectively net export) from GAFSI were insignificant compared to the overall internal electricity production, about 6% (respectively 1% in [4]). Taken individually, the countries composing GAFSI show significantly higher shares of imports and exports to production: France alone is the biggest electricity exporter in Europe, mainly to Switzerland (30%) and Italy (21%), and the vast majority of its imports are from Germany (47%) and Switzerland (23%) [7] hence the rationale for studying the GAFSI system and internalize the vast majority of inter-country electricity trades.

The authors thus deemed reasonable to assume that the internal electricity consumption in GAFSI satisfactorily matches the total electricity demand for GAFSI's electricity (and consequently electricity production).

Only a portion of the total electricity production (and thus consumption) is traded on electricity markets: the relation between electricity spot prices is de facto not obvious. Since only market prices were available the authors had to assume that all the electricity consumed is traded on the day-ahead market. This is a heavy assumption but traded volumes in European day-ahead markets have increased significantly over the last years, fluctuating above 40% since 2010 and reaching 52% of total electricity consumption in Q1 2013 [1]. Such assumption can further be justified by the fact that the mechanics of the day-ahead market are, in the long run, internalised in all the other contracts (long term purchase agreements, futures market...).

Thanks to the conjunction of these assumptions it appears reasonable to assume that all electricity (and therefore solar electricity) produced is fed into the grid and consumed (i.e., there is no self-consumption, no export or import). The GAFSI electricity market is considered as efficient (overall prices reflected through electricity market mechanisms) and shows a diversified energy mix, which leads to a MOC that is easily extractable (contrary to e.g. France alone with a vast majority of nuclear and hydro power) and most importantly invariant. Any injection of PV production in the grid should, under such observations, be reflected into a right shift of the MOC, as displayed in figure 3 below, leading to an overall decrease in prices.

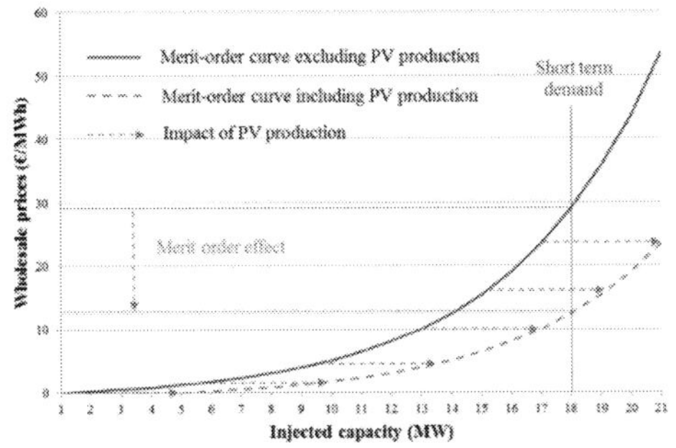

Figure 3: Impact of PV production on the MOC (source: authors)

As mentioned earlier (see definition of the MOE in section 2.1), in order to properly assess the impact of renewable electricity on market prices, the authors simulated what the wholesale price profile over the study period would have been had there been no PV generation. This implied intuiting certain properties of the MOC, i.e. of the relationship between wholesale prices and electricity consumption:

a. To analyse the rightward shift that RE production leads on the MOC, the authors worked on the total consumption of electricity (equal to demand) retreated by the PV production in order to work in a referential with a fixed MOC.
b. If the time period is short enough, the MOC of a country does not vary materially. This assumes that gas and coal prices are not too volatile, and that the country's energy mix does not vary too much (new facilities built or old ones shut down) which is among the set assumptions made above for GAFSI,
c. It is intuitive (and will be further tested) that the MOC has an exponential shape.

3 HISTORICAL ANALYSIS

3.1 MOC and MOE computation methodology

Hourly time series for PV production were reconstructed, between 2006 and 2015, from real solar irradiation data coupled with temperature data as follows:

- The solar irradiation profile was provided by GeoModel solar [6] for 60 regions/cities within GAFSI (regional main cities), using data from the Meteosat Second Generation ("MSG") satellite in the original 15-minute or hourly step time series format for the period from 2006 – 2013.
- In order to achieve a harmonized data set for the whole study period, temperature (with its original time step of 1h/3h) was also resampled to a 15 minute or hourly step time series, and
- The irradiation data set was then transformed into a normalized production of PV plants using a specific performance ratio varying with temperature, and integrated into hourly values (based on the 15 minute or hourly time step profiles).
- The solar irradiation profile for period from 2014-2015 was provided by PV GIS. The basis for the calculation of PV power output is the hourly solar radiation data estimated from satellite images. The method for deriving solar radiation at ground level from satellite data has been described in Mueller et al., 2009 and further validated in Huld et al., 2012. The resulting data consist of hourly maps of global and direct solar irradiance with a spatial resolution of about 4km. The solar radiation data has been supplemented with data on temperature and wind speed from the ECMWF Operational Forecast product (www.ecmwf.int). The temporal resolution is 3-hourly, interpolated to hourly values, while the spatial resolution of 7.5 arc-minutes. These data have been combined with models for inclined-plane irradiance and PV module performance to produce hourly maps of PV power output for each kWp of PV capacity installed. The models take into account the effects of shallow-angle reflectivity, module temperature and low irradiance as well as the cooling effect of wind, see Huld and Gracia Amillo, 2015. The calculation was performed for crystalline silicon modules mounted south-facing with a 20degree inclination. From the maps the average hourly PV production per region was found by area-weighted averaging over each pixel within a given region.

The final GAFSI PV production time series were obtained based on the hourly normalized PV production weighted by the hourly PV capacities in the GAFSI regions.

The GAFSI regions were grouped in 60 zones:

- Germany: 16 regions
- Austria: considered as one region
- France: 22 regions (including Corsica)
- Switzerland: considered as one region
- Italy: 20 regions.

The hourly PV production series obtained are the closest achievable estimation of the real PV production.

The hourly time series for the total electricity consumption [8] and wholesale electricity prices [9] were extracted from public databases.

As mentioned in Section 2 above, it was assumed that the MOC is invariant, and equivalently, that any PV production reduces demand for peak load and mid load generation sources. Based on the sets of values obtained previously, a MOC is obtained through a linear regression of the logarithm of wholesale prices (outliers - negative wholesale prices – are insignificant and have been disregarded) on the total electricity consumption net of PV production for a given period of time. Should the estimates thus derived be statistically significant, the wholesale prices that would have been observed, during such period of time, for a theoretical electricity market without any PV generation capacity, can be simulated with the following formula (with a and b the results of the regression):

$$MOC{:}: electricity\ consumption\ (c) \rightarrow wholesale\ price\ (p)$$
$$(c) \rightarrow e^{a+b*c}$$

The MOE over a certain period of time is derived from the corresponding MOC as the difference, on the considered period of time, between the total electricity spending of a theoretical electricity market without PV power plants (using the MOC to simulate the theoretical wholesale electricity prices) and the actual spending for electricity consumed. The MOE is the additional amount that would have been spent for the same consumption profile but without PV generated electricity, or equivalently, the MOE represents the monetary gain induced by PV production over a period of time.

Normalized by the total electricity generated by PV power plants, it provides an order of magnitude of a bonus price that can remunerate the PV asset operators on top of the wholesale price. Such quantity is thereafter defined as the merit-order price (the "MOP").

$$MOE = \sum (p_{no\,pv,simulated} - p_{pv,observed}) * c_{observed}$$

$$MOP = MOE \ / \ Total\ PV\ production$$

3.2 PV production computation methodology

3.2.1 PV production curve without storage system

The hourly PV production series is computed as the result of multiplying hourly irradiation profile with country cumulative PV installation.

$$PV\ production = Irradiation * PV\ installation\ capacities$$

3.2.2 PV production curve with decentralized storage system

When PV panels are attached together with a decentralized storage system. The PV production is calculated taken into accounted pre-set assumptions regarding storage size and the charging and releasing time.

First of all, to simplify the calculation process, it is assumed that the electricity produced by PV system will first be used to charge the de-centralized storage until it is full. Secondly, the moment when the sun sets and PV panel stops to produce electricity, the storage then starts injecting into the grid with equal amount per hour so that at the end of the day the storage is fully discharged and ready for a new round in the following day. It is believed that the assumption is in line with reality since during night time the demand for electricity is higher and production is limited due to reduction in several energy generation sources.

Thirdly, the size of the storage systems chosen for this study are 10GW and 25 GW for the extreme case per nation.

Finally, the study considers 10% as the energy loss ratio for the round trip electricity makes through storage and back to the grid.

3.2.3 PV production curve with pump-hydro storage

In this case, the process of calculating PV prodution curve remains the same as in the case of PV system with decentralized storage.

Due to the difference in the way de-centralized storage and pump-hydro system operate, it is the load curve that is affected in this case. To be clear, using the same assumption that the electricity generated from PV will firstly be stored, during this time of the day, the load curve is increased by the same amount of electricity that is generated by PV for storing while during night time, this accumulated elctricity is injected back to the grid causing decline in the load curve. Therefore, the load curve is decreased linearly from sunset until the end of the day.

Similar to the previous case, the study considers 25% as the energy loss ratio for the round trip electricity makes through storage and back to the grid with pump-hydro system. As for the previous case the size of the storage facilities are 10 GW and 25 GW.

3.3 Results for the historical analysis in simple case where no storage system is added (i.e. PV only).

The MOCs have first been estimated on a country per country basis but the correlation factors proved to be quite poor: Germany alone showed about 57%. Adding Austria, France, Switzerland and Italy, the biggest trading partners of Germany allowed to increase the correlation factor to about 66% as shown below, comforting the assumption that external trades do have a non-negligible impact on the MOE estimation.

The MOCs that were calculated for each year between 2007 and 2015 (dotted curves in Figure 4 and results in table I below) as well as for the whole 2007-2015 period (plain curve in the Figure 4 and results in table II below) display fairly significant correlations between wholesale prices and electricity consumption (R^2 factor above 66%), which validates the set of assumptions listed in section 2.3.

Overall, the above results tend to validate the assumption that the MOC has an exponential shape and does not vary much over time. For each MOC, a MOE and MOP were calculated (see tables below).

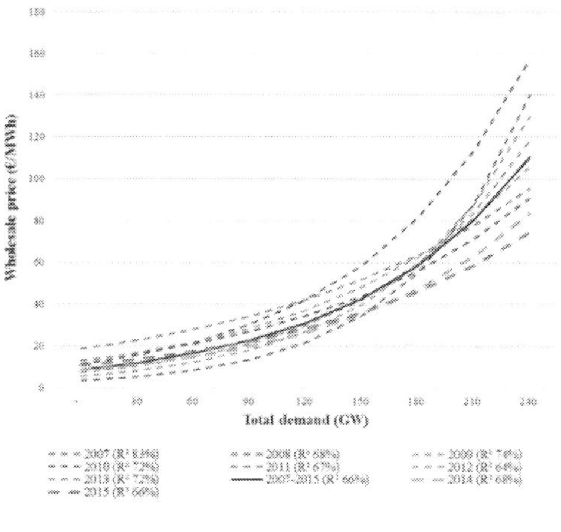

Figure 4: MOC for years from 2007 to 2015 and aggregated for the 2007-2015 period (source: authors' calculation)

Table I: MOC, MOE and MOP results for each year between 2007 and 2015

	MOC			MOE	MOP
Years	a	b	R^2	(€)	(€/MWh)
2007	1.19	1.56 10^{-5}	83%	0.6 10^9	137
2008	2.42	1.10 10^{-5}	68%	0.9 10^9	148
2009	1.71	1.32 10^{-5}	74%	1.0 10^9	108
2010	2.56	8.12 10^{-6}	72%	1.2 10^9	73
2011	2.92	6.84 10^{-6}	67%	2.6 10^9	67
2012	2.56	8.77 10^{-6}	64%	4.8 10^9	83
2013	2.00	1.15 10^{-5}	72%	6.2 10^9	97
2014	2.12	9.62 10^{-6}	68%	4.2 10^9	64
2015	2.37	8.09 10^{-6}	66%	4.1 10^9	57
2007-15	2.17	1.06 10^{-5}	66%	30.7 10^9	91

Table II: MOC, MOE and MOP results with the 2007-2015 MOC applied to all other years (source: authors' calculation)

	Yearly MOCs		2007-2015 MOC	
	MOE	MOP	MOE	MOP
Years	(€)	(€/MWh)	(€)	(€/MWh)
2007	$0.6\ 10^9$	137	$0.4\ 10^9$	100
2008	$0.9\ 10^9$	148	$0.6\ 10^9$	102
2009	$1.0\ 10^9$	108	$0.8\ 10^9$	88
2010	$1.2\ 10^9$	73	$1.6\ 10^9$	98
2011	$2.6\ 10^9$	67	$3.5\ 10^9$	92
2012	$4.8\ 10^9$	83	$5.3\ 10^9$	92
2013	$6.2\ 10^9$	97	$5.7\ 10^9$	89
2014	$4.2\ 10^9$	64	$5.7\ 10^9$	87
2015	$4.1\ 10^9$	57	$6.7\ 10^9$	92
2007-15	$30.7\ 10^9$	91	$30.7\ 10^9$	91

It is to be noted that if the MOP were to be paid to PV plants as a bonus on top of the market price for every MWh of PV produced, the total tariff received would be close to 140 €/MWh, as the average WAP is close to 50 €/MWh, which is above the range of the current feed-in-tariffs (and other support mechanisms) offered in EU countries (though in line with amounts offered only 2-3 years ago).

Equivalently, the average market price would have been 4.0% higher (2.0 €/MWh) had there been no PV generation between 2007 and 2015, as shown in table III below.

Table III: Average market price reduction between 2007 and 2015 (based on the 2007-15 MOC)

	Average WAP		Price reduction	
	Excl. PV	Incl.PV		
Years	(€/MWh)	(€/MWh)	(€/MWh)	(%)
2007	50.5	50.2	0.3	0.5%
2008	50.8	50.4	0.4	0.8%
2009	47.1	46.6	0.6	1.2%
2010	51.6	50.6	1.0	2.0%
2011	49.5	47.2	2.3	4.7%
2012	49.4	46.0	3.4	7.0%
2013	49.0	45.2	3.7	7.6%
2014	48.8	45.1	3.8	7.7%
2015	50.9	46.6	4.3	8.4%
2007-15	49.8	47.8	2.0	4.0%

Those results are in line with what other research such as Frank Sensfuß, Mario Ragwitz and Massimo Genoese (2008) [10].

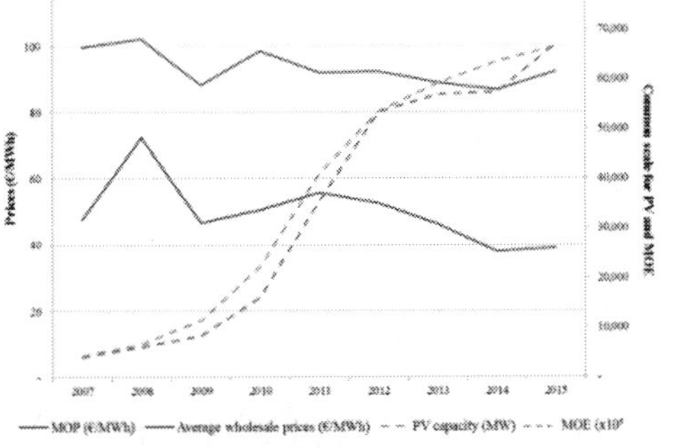

Figure 5: MOP (based on the 2007-15 MOC), average wholesale prices, MOE (based on the 2007-15 MOC) and PV capacities installed between 2007 and 2015 (source: authors' calculation)

The MOPs calculated above shows a downward trend with quite some variability. An intuitive interpretation could be that the MOE is less efficient as the PV penetration increases. In reality, only 10 years of history is rather short to validate any theory on which driver may affect the MOE. Such drivers could be the correlation between the solar irradiation and consumption profiles or the PV penetration rate.

In order to have a more reliable answer, the authors adopted a statistical analysis to observe the MOE variations with a large number of irradiation profiles, with different electricity demand profiles and different shares of installed PV capacity.

For the sake of consistency, and to allow an adequate comparison between the various metrics detailed section 4, the remainder of the study assumed that the MOC is the one obtained for the 2007 to 2015 period.

3.4 Results for the historical analysis in the case where de-centralized storage is added to the PV capacity

The addition of de-centralized storage does not significantly impact the MOC compared to the MOCs shown in Figure 4, and are thus not included in this paper.

The introduction of de-centralized storage (i.e. storage installed together with PV capacity) appears to reduce the MOE of PV. This can be seen by the decrease in calculated aggregate MOE values compared to the scenario described in Section 3.3 (PV without any additional storage) in the tables below. Similarly, the average electricity market price reduction is lower than if no storage had been added.

Table IV: MOC, MOE and MOP results with the 2007-2015 MOC applied to all other years including 10GW of de-centralized (storage source: authors' calculation).

	Yearly MOCs		2007-2015 MOC	
	MOE	MOP	MOE	MOP
Years	(€)	(€/MWh)	(€)	(€/MWh)
2007	$0.4\ 10^9$	111	$0.3\ 10^9$	85
2008	$0.7\ 10^9$	131	$0.5\ 10^9$	91
2009	$0.9\ 10^9$	98	$0.7\ 10^9$	82
2010	$1.1\ 10^9$	69	$1.5\ 10^9$	94
2011	$2.5\ 10^9$	65	$3.4\ 10^9$	91
2012	$4.7\ 10^9$	82	$5.2\ 10^9$	92
2013	$6.1\ 10^9$	96	$5.6\ 10^9$	89
2014	$4.3\ 10^9$	66	$5.8\ 10^9$	88
2015	$4.2\ 10^9$	59	$6.7\ 10^9$	94
2007-15	29.910^9	90	$29.9\ 10^9$	90

Table V: Average market price reduction between 2007 and 2015 (based on the 2007-15 MOC) including 10GW of de-centralized storage.

	Average WAP		**Price reduction**	
	Excl. PV	**Incl.PV**		
Years	(€/MWh)	(€/MWh)	(€/MWh)	(%)
2007	50.5	50.3	0.2	0.4%
2008	50.8	50.5	0.3	0.7%
2009	47.1	46.6	0.5	1.1%
2010	51.6	50.7	1.0	1.9%
2011	49.5	47.3	2.3	4.6%
2012	49.4	46.1	3.4	6.9%
2013	49.0	45.3	3.7	7.5%
2014	48.8	45.1	3.8	7.7%
2015	50.9	46.6	4.3	8.5%
2007-15	49.8	47.8	2.0	4.0%

The negative impact observed on the aggregate MOE is even more pronounced when going from 10 GW to 25 GW of storage (approximately EUR 8 bn decrease in MOE compared to the no additional storage scenario). Similarly, the calculated market price reduction as a result of the combination of PV and storage would be lower with the higher storage capacity. This would suggest that the addition of storage means the MOE is less efficient.

It may be that the approximations made for the calculation of the storage capacity for the purposes of this calculation are perhaps overly simplified. It is, for example possible that not all electricity produced would be fed into the storage units as a priority. In reality the decision on whether to inject electricity into the grid or into storage will be driven by market prices/demand, which the calculations do explicitly account for. Similarly, the re-injection of the electricity into the grid is likely to be linked to demand and price factors and will thus most likely not occur linearly between sunset and midnight. So it would be conceivable, for example, that it is more favorable to inject electricity from storage into the grid during the evening peak demand period rather than in the hour before midnight where market spot prices are likely to be low anyway.

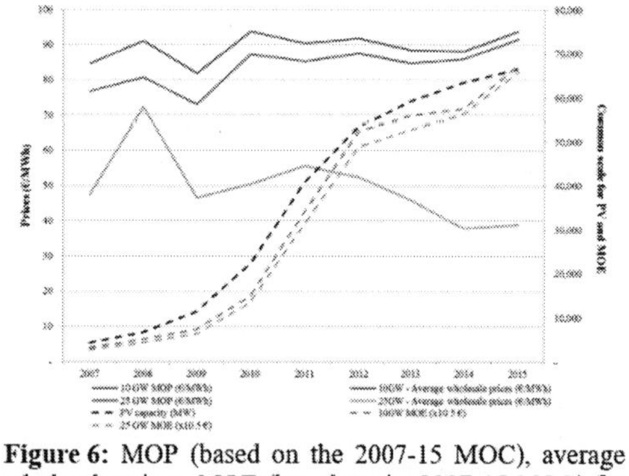

Figure 6: MOP (based on the 2007-15 MOC), average wholesale prices, MOE (based on the 2007-15 MOC) for both cases of de-centralized storage (i.e. 10GW and 25GW installed capacity) and PV capacities installed between 2007 and 2015 (source: authors' calculation)

That being said, the MOE for 2014 and 2015 do actually increase compared to the case without any storage, an effect which is more pronounced when 25GW of storage is included. Therefore, it may be that adding 10/25GW of storage was simply too much for early years were the installed PV capacity was of the order of and sometimes even lower than the storage capacity. This could mean that if all PV electricity generated in the day-time is stored the benefit of re-injecting it during night-time is outweighed by an increase in price during day time (when PV would otherwise produce electricity and thus lower the market spot prices).

Table VI: MOC, MOE and MOP results with the 2007-2015 MOC applied to all other years including 25 GW of de-centralized (storage source: authors' calculation).

	Yearly MOCs		**2007-2015 MOC**	
	MOE	**MOP**	**MOE**	**MOP**
Years	(€)	(€/MWh)	(€)	(€/MWh)
2007	$0.4\ 10^9$	99	$0.3\ 10^9$	77
2008	$0.6\ 10^9$	116	$0.4\ 10^9$	81
2009	$0.8\ 10^9$	87	$0.6\ 10^9$	73
2010	$1.0\ 10^9$	65	$1.3\ 10^9$	87
2011	$2.3\ 10^9$	62	$3.1\ 10^9$	85
2012	$4.3\ 10^9$	77	$4.9\ 10^9$	88
2013	$5.6\ 10^9$	90	$5.3\ 10^9$	85
2014	$4.2\ 10^9$	65	$5.6\ 10^9$	86
2015	$4.2\ 10^9$	55	$5.6\ 10^9$	92
2007-15	$28.4\ 10^9$	87	$28.4\ 10^9$	87

Table VII: Average market price reduction between 2007 and 2015 (based on the 2007-15 MOC) including 25GW of de-centralized storage.

	Average WAP		**Price reduction**	
	Excl. PV	**Incl.PV**		
Years	(€/MWh)	(€/MWh)	(€/MWh)	(%)
2007	50.3	50.1	0.2	0.4%
2008	50.6	50.3	0.3	0.6%
2009	47.0	46.6	0.5	1.0%
2010	51.4	50.5	0.9	1.8%
2011	49.4	47.3	2.1	4.3%
2012	49.3	46.1	3.2	6.5%
2013	48.8	45.3	3.5	7.2%
2014	48.7	45.0	3.7	7.7%
2015	50.8	46.5	4.3	8.4%
2007-15	49.6	47.7	1.9	3.8%

3.5 Results for the historical analysis in the case where centralized storage (e.g. pump-hydro) is added to the PV capacity.

The addition of centralized storage does not significantly impact the MOC compared to the MOCs shown in Figure 4, and are thus not included in this paper (as was the case for de-centralized storage). Contrary to the observation made for the de-centralized storage, the MOE increases with the addition of centralized storage. This is reflected in an increase of the MOE (of around EUR 1 bn) value over the 10 years studied as well as in higher observed MOPs and a larger reduction in WAP. In the case with 10 GW additional storage capacity the total MOE increases by around EUR 1 bn.

As was the case without the addition of storage capacity the calculated MOPs seem to be trending downwards, though with significant variability. The calculated MOPs broadly follow the trends observed for average wholesale prices over the same time period.

Figure 6 shows that increasing the storage capacity from 10 to 25 GW has a negligible impact on the average market price. Conversely, it is clear that the increase in storage capacity seems to lead to an increase in MOP, especially for the later years. It appear that this effect is amplified with increasing installed solar capacity. In section 4 some statistical analysis is carried out to test this potential correlation.

Table VIII: MOC, MOE and MOP results with the 2007-2015 MOC applied to all other years including 10GW of centralized (storage source: authors' calculation).

	Yearly MOCs		**2007-2015 MOC**	
Years	**MOE** (€)	**MOP** (€/MWh)	**MOE** (€)	**MOP** (€/MWh)
2007	0.6 10^9	139	0.4 10^9	101
2008	0.9 10^9	149	0.6 10^9	104
2009	1.0 10^9	110	0.8 10^9	90
2010	1.2 10^9	74	1.7 10^9	101
2011	2.6 10^9	68	3.7 10^9	95
2012	4.9 10^9	85	5.5 10^9	95
2013	6.4 10^9	100	5.9 10^9	92
2014	4.5 10^9	67	5.9 10^9	90
2015	4.4 10^9	57	6.9 10^9	96
2007-15	31.7 10^9	94	31.7 10^9	94

Table IX: Average market price reduction between 2007 and 2015 (based on the 2007-15 MOC) including 10 GW of centralized storage.

	Average WAP		**Price reduction**	
Years	**Excl. PV** (€/MWh)	**Incl.PV** (€/MWh)	(€/MWh)	(%)
2007	50.4	50.2	0.3	0.5%
2008	50.8	50.4	0.4	0.8%
2009	47.1	46.6	0.6	1.2%
2010	51.7	50.6	1.1	2.0%
2011	49.6	47.3	2.4	4.8%
2012	49.5	46.0	3.5	7.1%
2013	49.1	45.3	3.8	7.7%
2014	48.9	45.1	3.9	7.9%
2015	51.0	46.6	4.4	8.6%
2007-15	49.8	47.8	2.0	4.1%

This effect is further amplified when introducing more storage capacity, as seen from the case with 25GW per country, as evidenced by a further increase (with respect to the 10 GW centralized storage case) of EUR 1.2 bn in the calculated MOE. The MOP available for PV producers would be around 98EUR/MWh, making the total tariff payable close to 150EUR/MWh (average market price only decreased minimally), a 10EUR/MWh increase from the PV only case.

The decrease in average market price is more modest (in the range of 0.1%). It is worth noting that, considering the rather simple approximations taken in order to simulate the acquisition and re-injection of electricity by centralized storage facilities such significant impacts were measured. It is, therefore, conceivable that by using a more sophisticated approach (e.g. one that is linked to peak demand and or spot prices) could further improve the MOE as a result of PV installation. Furthermore, it appears to validate the assumptions made for these calculations.

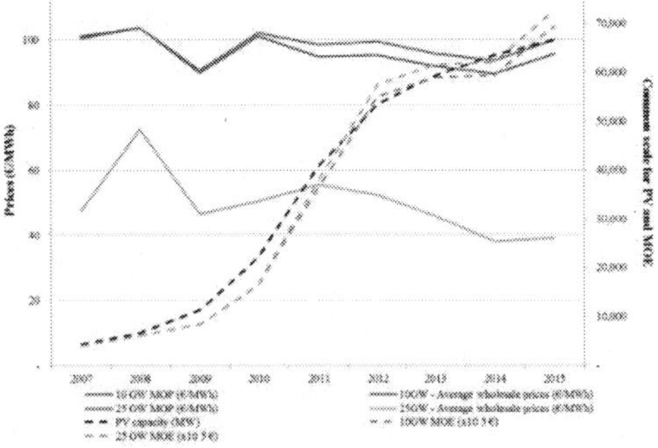

Figure 7: MOP (based on the 2007-15 MOC), average wholesale prices, MOE (based on the 2007-15 MOC) for both cases of centralized storage (i.e. 10GW and 25GW installed capacity) and PV capacities installed between 2007 and 2015 (source: authors' calculation)

Table X: MOC, MOE and MOP results with the 2007-2015 MOC applied to all other years including 25 GW of centralized (storage source: authors' calculation).

	Yearly MOCs		**2007-2015 MOC**	
Years	**MOE** (€)	**MOP** (€/MWh)	**MOE** (€)	**MOP** (€/MWh)
2007	0.6 10^9	140	0.4 10^9	100
2008	0.9 10^9	151	0.6 10^9	104
2009	1.0 10^9	112	0.8 10^9	91
2010	1.2 10^9	75	1.7 10^9	102
2011	2.7 10^9	70	3.8 10^9	98
2012	5.0 10^9	86	5.7 10^9	99
2013	6.6 10^9	102	6.1 10^9	96
2014	4.7 10^9	70	6.2 10^9	94
2015	4.6 10^9	63	7.3 10^9	100
2007-15	32.9 10^9	98	32.9 10^9	98

Table XI: Average market price reduction between 2007 and 2015 (based on the 2007-15 MOC) including 25GW of centralized storage.

	Average WAP		**Price reduction**	
Years	**Excl. PV** (€/MWh)	**Incl.PV** (€/MWh)	(€/MWh)	(%)
2007	50.3	50.0	0.3	0.5%
2008	50.7	50.3	0.4	0.8%
2009	47.1	46.5	0.6	1.2%
2010	51.6	50.5	1.1	2.1%
2011	49.7	47.3	2.4	4.8%
2012	49.7	46.1	3.6	7.2%
2013	49.3	45.4	3.9	7.9%
2014	49.0	45.1	3.9	8.0%
2015	51.2	46.7	4.5	8.8%
2007-15	49.8	47.7	2.1	4.2%

4 STATISTICAL ANALYSIS

4.1 Statistical average merit-order effect computation where PV without storage

Ten years of historical data is rather too short to assess with a high level of reliability the MOE at the scale of a country. Policy makers and economic agents may have an interest in predicting its value for the years to come as it will impact the forward prices and may influence their decisions. The MOE is by definition dependant on the PV production profile and how it correlates with the consumptions profile. A statistical approach is thus proposed, in order to observe how the MOE behaves with respect to different PV production profiles.

The data used for this section are the hourly PV production profiles (MWh) and the hourly installed PV capacity (MWp) between 2006 and 2015, to obtain hourly profiles (i.e. the ratio of the PV production by the installed PV capacity (MWh/MWp)). There is one additional year (2006) compared to the previous section, as the authors did not have access to the wholesale prices for that year to run the historical calculations.

A Monte Carlo simulation was run on a central hourly yield profile over a year. Each hourly yield for each of the main 60 regions of GAFSI (see section 3.1) was modelled as an independent random variable following a normal law (there are therefore 8760 Gaussian random yield variable). Each hourly yield has 10 samples (for 2006 to 2015), from which the authors have calculated an average and a standard deviation.

The Monte Carlo simulation was run 100 times (on each of the 8760 random yields just defined) for each of the 60 main GAFSI cities, leading to 6000 random yearly yield profiles. The authors then calculated 100 PV production profiles for each of the 10 years of historical PV installation capacities (2006-2015) around the 60 main GAFSI cities (hence a total of 6000 PV production profiles). The authors could also calculate 100 profiles of demand in GAFSI retreated by PV production for each of the 10 years of historical demand profiles, leading to 81 combinations.

For each of these 100 combinations, 100 values of the MOE and MOP were calculated, using the MOC calculated between 2007 and 2015. The authors used the same MOC throughout the combinations in order to have a common basis for comparison. A statistically meaningful average of the MOE and the MOP were then calculated for each combination.

4.2 Results

The results tables should be read as follows:

- Along columns the amount of installed PV capacities varies (from 2006 to 2015)
- Along rows the profile of total electricity demand varies as observed from 2006 to 2015.

Diagonal values show a more reliable (statistically meaningful) estimate of the MOE and MOP for a given year (note that the MOP are almost identical to the ones on the last column of table II, with a 3% tolerance).

Table XII: MOP and MOE varying with the PV penetration rate (installed capacity, in columns) and with electricity demand profiles (i.e. with the PV irradiation profiles, in rows)

MONTE-CARLO (100 draws)			PV installed capacities									
			2006	2007	2008	2009	2010	2011	2012	2013	2014	2015
Consumption profiles	2006	MOP (€/MWh)	99	98	99	99	97	94	92	90	89	89
		MOE (m€)	259	376	571	920	1,660	3,403	5,223	6,181	6,752	7,113
	2007	MOP (€/MWh)	102	102	101	101	100	97	94	92	91	90
		MOE (m€)	265	386	586	944	1,713	3,511	5,352	6,310	6,877	7,245
	2008	MOP (€/MWh)	104	103	103	103	101	98	95	94	93	92
		MOE (m€)	270	392	598	957	1,730	3,549	5,423	6,406	7,001	7,370
	2009	MOP (€/MWh)	90	90	89	89	88	85	83	82	81	80
		MOE (m€)	234	340	517	832	1,509	3,090	4,725	5,577	6,091	6,418
	2010	MOP (€/MWh)	101	101	101	101	99	96	94	92	91	90
		MOE (m€)	263	383	583	941	1,702	3,491	5,337	6,300	6,883	7,247
	2011	MOP (€/MWh)	98	97	97	97	95	92	90	89	88	87
		MOE (m€)	254	370	562	905	1,634	3,338	5,134	6,070	6,627	6,984
	2012	MOP (€/MWh)	100	99	99	99	97	94	92	90	89	89
		MOE (m€)	259	377	574	922	1,664	3,409	5,228	6,183	6,750	7,109
	2013	MOP (€/MWh)	98	98	98	98	96	92	90	89	88	87
		MOE (m€)	255	371	563	908	1,639	3,351	5,144	6,035	6,646	7,003
	2014	MOP (€/MWh)	98	97	97	97	95	92	90	88	87	87
		MOE (m€)	254	369	561	903	1,631	3,337	5,117	6,046	6,602	6,955
	2015	MOP (€/MWh)	105	105	104	104	102	99	96	95	94	93
		MOE (m€)	273	398	602	972	1,752	3,580	5,499	6,502	7,110	7,491

Table XIII: relative variations of the MOP compared to the 2006 value (99 €/MWh) with the PV penetration rate (installed capacity, in columns) and with electricity demand profiles (i.e. with the PV irradiation profiles, in rows)

Relative variations to 2006 MOP		PV installed capacity										Maximum variation
		2006	2007	2008	2009	2010	2011	2012	2013	2014	2015	
Consumption profiles	2006	-	-0.3%	-0.6%	-0.6%	-2.5%	-5.6%	-7.8%	-9.2%	-10.1%	-10.7%	10.7%
	2007	2.6%	2.4%	2.1%	2.1%	0.6%	-2.6%	-5.6%	-7.3%	-8.4%	-9.0%	11.6%
	2008	4.4%	4.1%	3.8%	3.6%	1.6%	-1.7%	-4.3%	-5.8%	-6.8%	-7.5%	11.8%
	2009	-9.5%	-9.7%	-10.0%	-10.0%	-11.4%	-14.3%	-16.7%	-17.9%	-18.9%	-19.4%	9.9%
	2010	1.9%	1.7%	1.5%	1.7%	0.0%	-3.2%	-5.9%	-7.3%	-8.4%	-9.0%	10.9%
	2011	-1.7%	-1.9%	-2.2%	-2.2%	-4.0%	-7.2%	-9.5%	-10.8%	-11.8%	-12.4%	10.6%
	2012	0.2%	0.0%	-0.3%	-0.3%	-2.3%	-5.5%	-7.8%	-9.1%	-10.1%	-10.7%	10.9%
	2013	-1.3%	-1.6%	-1.9%	-1.9%	-3.8%	-7.0%	-9.2%	-10.5%	-11.5%	-12.1%	10.8%
	2014	-1.8%	-2.0%	-2.3%	-2.3%	-4.2%	-7.4%	-9.8%	-11.1%	-12.0%	-12.7%	10.9%
	2015	5.6%	5.4%	4.9%	4.9%	2.8%	-0.7%	-3.0%	-4.4%	-5.4%	-6.0%	11.6%
Maximum variation		15.1%	15.1%	14.9%	14.9%	14.2%	13.6%	13.6%	13.6%	13.5%	13.4%	

As displayed in the tables above, the MOP varies more along columns than along rows. This indicates that the MOP does not depend much on the penetration rate of PV in the country's energy mix (total PV capacity installed), but rather varies significantly with the electricity demand profile, or to be precise, with the correlation between demand and PV production during the year.

This effect seems to ease gradually from 2011 onwards. The variance of the MOP with PV penetration rate (years in the x-axis corresponds to a given capacity profile, i.e. columns in the Table XII) is shown in Figure 8. As more capacity is installed, PV production is ensured to supply a larger portion of the electricity demand from customers. This is in line with what had been observed in the authors' previous study in 2014. Figure 8 below shows the continuation of the trend observed for 2011-2013 into the final two years. Nevertheless, the observed effect seems to have plateaued since circa 2011, perhaps suggesting that a 'critical mass' in terms of PV capacity was reached meaning that further increases do not impact the variance of MOPs as significantly.

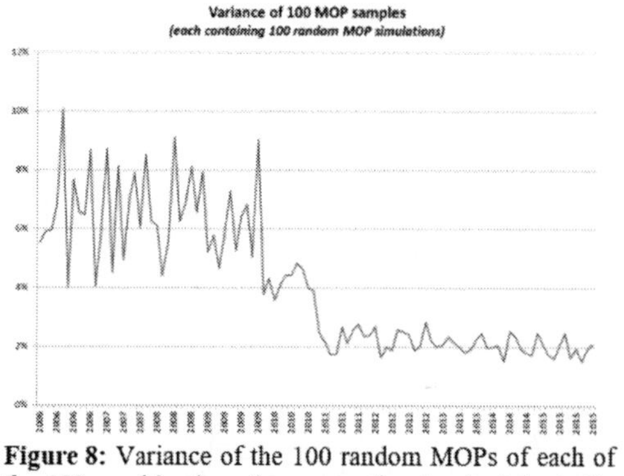

Figure 8: Variance of the 100 random MOPs of each of the 100 combinations (source: authors' calculation). On the x-axis the years are ordered between 2006 and 2015 and each year has 10 combinations (one per electricity

demand profile).

4.2.1 Statistical average merit-order effect computation in the case where de-centralized storage is added to the PV capacity.

The results are to be read as for Section 4.2. As for the PV only case, the diagonals should indicate a more reliable and statistically meaningful estimate of the MOE/MOP for each year.

Table XIV: MOP and MOE varying with the PV penetration rate (installed capacity, in columns) and with electricity demand profiles (i.e. with the PV irradiation profiles, in rows) for the case with 10 GW of de-centralized storage.

MONTE-CARLO (100 draws)		PV installed capacities									
Consumption profiles		2006	2007	2008	2009	2010	2011	2012	2013	2014	2015
2006	MOP (€/MWh)	94	93	93	93	91	87	86	85	84	83
	MOE (m€)	251	364	552	885	1,594	3,242	4,999	5,923	6,474	6,802
2007	MOP (€/MWh)	96	96	95	95	94	90	88	86	86	85
	MOE (m€)	257	374	567	908	1,642	3,341	5,117	6,028	6,588	6,920
2008	MOP (€/MWh)	98	98	98	97	95	91	90	89	88	87
	MOE (m€)	263	382	580	930	1,669	3,392	5,214	6,168	6,749	7,092
2009	MOP (€/MWh)	85	85	85	85	83	80	78	77	77	76
	MOE (m€)	229	332	504	808	1,458	2,967	4,558	5,398	5,895	6,201
2010	MOP (€/MWh)	97	96	96	96	94	91	88	88	87	87
	MOE (m€)	259	376	570	917	1,660	3,381	5,191	6,148	6,715	7,071
2011	MOP (€/MWh)	93	93	92	92	90	87	85	85	84	83
	MOE (m€)	249	362	548	880	1,585	3,233	4,978	5,898	6,450	6,777
2012	MOP (€/MWh)	94	94	93	93	91	87	86	85	84	84
	MOE (m€)	252	366	554	888	1,595	3,244	5,004	5,930	6,484	6,801
2013	MOP (€/MWh)	93	92	92	92	90	86	84	83	83	82
	MOE (m€)	248	360	545	874	1,571	3,188	4,917	5,822	6,364	6,689
2014	MOP (€/MWh)	95	94	94	94	92	88	87	86	85	84
	MOE (m€)	254	368	558	894	1,607	3,275	5,037	5,965	6,528	6,855
2015	MOP (€/MWh)	102	101	101	100	98	95	93	92	91	91
	MOE (m€)	272	396	599	958	1,726	3,511	5,414	6,414	7,012	7,373

However, in this case the MOP values are not as well aligned with some discrepancies of over 10% with respect to the historical MOP calculations described in Section 3. This is not entirely surprising as changing the PV production profile is similar to changing the energy mix of a country/region. This could create higher variance leading to a poorer match between statistical and historical data.

Table XV: relative variations of the MOP compared to the 2006 value (94 €/MWh) with the PV penetration rate (installed capacity, in columns) and with electricity demand profiles (i.e. with the PV irradiation profiles, in rows) for the case with 10 GW of de-centralized storage.

Relative variations to 2006 MOP	PV installed capacity 2006	2007	2008	2009	2010	2011	2012	2013	2014	2015	Maximum variation
Consumption profiles 2006	-	-0.4%	-0.9%	-1.2%	-3.2%	-6.9%	-8.6%	-9.4%	-10.3%	-11.0%	11.0%
2007	2.5%	2.2%	1.8%	1.4%	-0.1%	-4.0%	-6.4%	-7.8%	-8.8%	-9.4%	11.9%
2008	5.0%	4.6%	4.0%	3.8%	1.5%	-2.5%	-4.6%	-5.6%	-6.5%	-7.2%	12.2%
2009	-8.9%	-9.2%	-9.7%	-9.8%	-11.5%	-14.8%	-16.3%	-17.5%	-18.3%	-18.9%	10.0%
2010	3.1%	2.8%	2.5%	2.5%	0.7%	-2.9%	-4.9%	-6.0%	-7.0%	-7.5%	10.6%
2011	-0.8%	-1.0%	-1.6%	-1.7%	-3.7%	-7.1%	-8.9%	-9.8%	-10.7%	-11.3%	10.5%
2012	0.4%	0.0%	-0.6%	-0.8%	-3.0%	-6.8%	-8.4%	-9.4%	-10.3%	-10.9%	11.3%
2013	-1.2%	-1.8%	-2.1%	-2.4%	-4.5%	-8.3%	-10.0%	-10.9%	-11.7%	-12.3%	11.1%
2014	0.8%	0.5%	-0.1%	-0.3%	-2.2%	-5.8%	-7.7%	-8.8%	-9.6%	-10.2%	11.1%
2015	8.4%	8.0%	7.4%	7.0%	4.9%	1.0%	-0.9%	-1.9%	-2.9%	-3.5%	11.9%
Maximum variation	17.3%	17.3%	17.1%	16.8%	16.4%	15.7%	15.6%	15.6%	15.5%	15.4%	

As was the case for the PV only case, the MOP varies more strongly as a function of consumption profile (i.e. along columns) than PV penetration (along rows). Although this observation is slightly less pronounced for the more recent years (as for the PV only case). The reasoning must be different, even if numerically, the maximum variations along columns are much higher for both of the de-centralized storage cases compared to the PV only case.

Table XVI: MOP and MOE varying with the PV penetration rate (installed capacity, in columns) and with electricity demand profiles (i.e. with the PV irradiation profiles, in rows) for the case with 25 GW of de-centralized storage.

MONTE-CARLO (100 draws)		PV installed capacities									
Consumption profiles		2006	2007	2008	2009	2010	2011	2012	2013	2014	2015
2006	MOP (€/MWh)	88	87	87	87	85	82	81	80	79	78
	MOE (m€)	233	338	512	821	1,482	3,020	4,651	5,502	6,025	6,337
2007	MOP (€/MWh)	90	90	89	89	88	85	82	81	80	80
	MOE (m€)	239	348	526	842	1,530	3,110	4,763	5,609	6,125	6,440
2008	MOP (€/MWh)	92	92	91	91	89	86	84	83	82	82
	MOE (m€)	245	355	538	862	1,554	3,185	4,859	5,760	6,286	6,615
2009	MOP (€/MWh)	80	80	80	80	78	76	74	73	72	71
	MOE (m€)	212	309	468	753	1,360	2,773	4,257	5,035	5,499	5,797
2010	MOP (€/MWh)	91	91	90	91	89	86	84	83	82	82
	MOE (m€)	241	350	531	856	1,551	3,164	4,862	5,758	6,291	6,621
2011	MOP (€/MWh)	87	87	87	87	85	82	81	80	79	78
	MOE (m€)	232	337	511	821	1,483	3,024	4,655	5,514	6,021	6,340
2012	MOP (€/MWh)	88	88	87	87	85	82	81	80	79	78
	MOE (m€)	233	339	515	823	1,484	3,025	4,659	5,517	6,032	6,346
2013	MOP (€/MWh)	86	86	86	86	84	81	79	78	78	77
	MOE (m€)	230	333	504	812	1,457	2,970	4,580	5,419	5,923	6,234
2014	MOP (€/MWh)	90	89	89	89	87	84	82	81	80	80
	MOE (m€)	237	345	523	837	1,515	3,094	4,745	5,606	6,133	6,454
2015	MOP (€/MWh)	96	96	95	95	93	90	88	87	86	86
	MOE (m€)	255	370	560	899	1,623	3,310	5,092	6,021	6,584	6,627

Table XVII: relative variations of the MOP compared to the 2006 value (88 €/MWh) with the PV penetration rate (installed capacity, in columns) and with electricity demand profiles (i.e. with the PV irradiation profiles, in rows) for the case with 10 GW of de-centralized storage.

Relative variations to 2006 MOP	PV installed capacity 2006	2007	2008	2009	2010	2011	2012	2013	2014	2015	Maximum variation
Consumption profiles 2006	-	-0.4%	-0.8%	-0.8%	-3.0%	-6.1%	-8.0%	-9.1%	-10.0%	-10.7%	10.7%
2007	2.5%	2.2%	1.9%	1.7%	0.1%	-3.3%	-6.0%	-7.4%	-8.5%	-9.3%	11.8%
2008	5.1%	4.7%	4.3%	4.0%	1.9%	-1.6%	-3.9%	-5.1%	-6.2%	-6.9%	12.0%
2009	-8.5%	-8.8%	-9.1%	-9.2%	-10.8%	-13.8%	-15.7%	-16.8%	-17.8%	-18.4%	9.9%
2010	3.8%	3.5%	3.2%	3.3%	1.7%	-1.6%	-3.8%	-5.1%	-6.1%	-6.9%	10.7%
2011	-0.2%	-0.5%	-0.8%	-0.9%	-2.9%	-5.9%	-8.0%	-9.0%	-10.0%	-10.6%	10.5%
2012	0.4%	0.0%	-0.4%	-0.5%	-2.6%	-6.0%	-7.9%	-8.9%	-10.0%	-10.6%	11.1%
2013	-1.3%	-1.7%	-2.1%	-2.1%	-4.3%	-7.6%	-9.5%	-10.5%	-11.5%	-12.2%	10.9%
2014	2.2%	1.8%	1.5%	1.2%	-0.7%	-3.9%	-6.2%	-7.4%	-8.4%	-9.2%	11.4%
2015	9.7%	9.3%	8.8%	8.6%	6.5%	2.9%	0.7%	-0.5%	-1.7%	-2.4%	12.1%
Maximum variation	18.2%	18.1%	17.9%	17.8%	17.3%	16.7%	16.4%	16.3%	16.1%	16.0%	

4.2.2 Statistical average merit-order effect computation in the case where centralized storage (e.g. pump-hydro) is added to the PV capacity.

The results are to be read as for Section 4.2. Similar to the PV only case, the diagonal values show a more reliable (statistically significant) estimate of the MOE and MOP for a given year. For the 10 GW storage case, note that the MOP values in the diagonal are almost identical to those in the last column of table IX (within a 3% tolerance).

Table XVIII: MOP and MOE varying with the PV penetration rate (installed capacity, in columns) and with electricity demand profiles (i.e. with the PV irradiation profiles, in rows) for the case with 10 GW of centralized storage.

MONTE-CARLO (100 draws)		PV installed capacities									
Consumption profiles		2006	2007	2008	2009	2010	2011	2012	2013	2014	2015
2006	MOP (€/MWh)	100	100	100	100	98	95	93	91	90	90
	MOE (m€)	251	380	578	930	1,679	3,440	5,281	6,247	6,820	7,188
2007	MOP (€/MWh)	103	103	103	103	101	98	95	93	92	92
	MOE (m€)	269	391	594	956	1,736	3,555	5,421	6,385	6,973	7,342
2008	MOP (€/MWh)	106	105	105	105	103	99	97	95	94	93
	MOE (m€)	275	399	607	975	1,762	3,605	5,514	6,514	7,115	7,491
2009	MOP (€/MWh)	92	91	91	91	90	87	84	83	82	82
	MOE (m€)	238	347	527	847	1,535	3,145	4,812	5,683	6,203	6,554
2010	MOP (€/MWh)	104	104	103	104	102	99	96	94	93	93
	MOE (m€)	270	393	598	964	1,747	3,583	5,472	6,467	7,058	7,435
2011	MOP (€/MWh)	101	101	100	100	98	95	93	91	90	90
	MOE (m€)	262	381	580	933	1,686	3,454	5,298	6,259	6,843	7,203
2012	MOP (€/MWh)	103	103	102	102	100	97	95	93	92	92
	MOE (m€)	268	389	591	951	1,718	3,521	5,389	6,392	6,981	7,351
2013	MOP (€/MWh)	98	98	97	97	95	92	90	89	88	87
	MOE (m€)	254	371	563	906	1,635	3,350	5,138	6,080	6,636	6,991
2014	MOP (€/MWh)	101	100	100	100	98	95	92	91	90	89
	MOE (m€)	262	380	578	930	1,681	3,440	5,274	6,229	6,807	7,173
2015	MOP (€/MWh)	108	108	108	108	105	102	99	98	97	96
	MOE (m€)	282	408	622	#####	1,806	3,695	5,677	6,713	7,334	7,734

Table XIX: relative variations of the MOP compared to the 2006 value (100 €/MWh) with the PV penetration rate (installed capacity, in columns) and with electricity demand profiles (i.e. with the PV irradiation profiles, in rows) for the case with 10 GW of centralized storage.

Relative variations to 2006 MOP	PV installed capacity 2006	2007	2008	2009	2010	2011	2012	2013	2014	2015	Maximum variation
Consumption profiles 2006	-	-0.2%	-0.5%	-0.5%	-2.5%	-5.6%	-7.8%	-9.1%	-10.1%	-10.7%	10.7%
2007	2.8%	2.7%	2.4%	2.3%	0.6%	-2.4%	-5.5%	-7.1%	-8.2%	-8.8%	11.7%
2008	5.1%	4.9%	4.5%	4.3%	2.3%	-1.1%	-3.7%	-5.3%	-6.3%	-7.0%	12.1%
2009	-8.9%	-9.1%	-9.3%	-9.3%	-10.8%	-13.7%	-16.0%	-17.4%	-18.3%	-18.8%	9.9%
2010	3.4%	3.2%	3.0%	3.2%	1.6%	-1.7%	-4.4%	-5.9%	-7.0%	-7.7%	11.1%
2011	0.3%	0.1%	-0.2%	-0.2%	-2.1%	-5.2%	-7.6%	-8.9%	-9.9%	-10.5%	10.9%
2012	2.3%	2.1%	1.9%	1.8%	-0.2%	-3.4%	-5.7%	-7.1%	-8.0%	-8.7%	11.0%
2013	-2.6%	-2.7%	-3.0%	-3.1%	-5.0%	-8.1%	-10.3%	-11.6%	-12.6%	-13.1%	10.6%
2014	0.1%	-0.2%	-0.5%	-0.5%	-2.4%	-5.6%	-8.0%	-9.4%	-10.3%	-11.0%	11.1%
2015	7.8%	7.8%	7.2%	7.1%	4.9%	1.4%	-1.0%	-2.3%	-3.4%	-4.0%	11.8%
Maximum variation	16.7%	16.6%	16.5%	16.3%	15.7%	15.1%	15.1%	15.1%	14.8%	14.8%	

This observation is no longer valid for the 25 GW storage case. One reason for this is that with 10 GW, the changes made to the consumption/load curve were not significant enough to impact the variance of the Monte Carlo calculations. Conversely, once we introduce 25 GW it has a much more significant impact.

Table XX: MOP and MOE varying with the PV penetration rate (installed capacity, in columns) and with electricity demand profiles (i.e. with the PV irradiation profiles, in rows) for the case with 25 GW of centralized storage.

MONTE-CARLO (100 draws)			PV installed capacities									
			2006	2007	2008	2009	2010	2011	2012	2013	2014	2015
Consumption profiles	2006	MOP (€/MWh)	99	99	99	99	97	94	92	90	102	102
		MOE (€)	259	376	571	920	1,660	3,403	5,223	6,181	265	386
	2007	MOP (€/MWh)	102	101	100	97	94	92	104	103	103	103
		MOE (€)	587	944	1,715	3,512	5,350	6,302	270	392	596	958
	2008	MOP (€/MWh)	101	98	86	94	90	90	89	89	88	85
		MOE (€)	1,732	3,542	5,429	6,408	234	340	517	832	1,508	3,087
	2009	MOP (€/MWh)	83	82	101	101	101	101	99	96	94	92
		MOE (€)	4,724	5,579	264	304	582	939	1,704	3,488	5,336	6,301
	2010	MOP (€/MWh)	98	97	97	97	95	92	90	89	100	99
		MOE (€)	254	370	562	905	1,633	3,349	5,132	6,087	259	377
	2011	MOP (€/MWh)	99	99	97	94	92	90	98	98	97	98
		MOE (€)	575	923	1,664	3,408	5,230	6,136	256	371	564	908
	2012	MOP (€/MWh)	96	92	90	89	98	97	97	97	95	92
		MOE (€)	1,637	3,353	5,145	6,087	254	369	561	903	1,630	3,339
	2013	MOP (€/MWh)	90	88	103	103	104	104	102	98	96	95
		MOE (€)	5,112	6,049	907	1,054	603	970	1,751	3,578	5,501	6,506
	2014	MOP (€/MWh)	94	94	97	97	95	92	90	88	87	87
		MOE (€)	6,423	6,779	359	901	1,626	3,340	5,114	6,046	6,823	6,942
	2015	MOP (€/MWh)	105	105	104	104	102	99	96	95	94	93
		MOE (€)	274	397	606	969	1,750	3,581	5,487	6,498	7,109	7,422

Table XXI: relative variations of the MOP compared to the 2006 value (99 €/MWh) with the PV penetration rate (installed capacity, in columns) and with electricity demand profiles (i.e. with the PV irradiation profiles, in rows) for the case with 25 GW of centralized storage.

Relative variations to 2006 MOP		PV installed capacity										Maximum variation
		2006	2007	2008	2009	2010	2011	2012	2013	2014	2015	
Consumption profiles	2006	-	-0.3%	-0.6%	-0.5%	-2.5%	-5.8%	-7.9%	-9.1%	2.8%	2.4%	*11.6%*
	2007	2.2%	2.0%	0.7%	-2.6%	-5.6%	-7.2%	4.4%	4.1%	3.8%	3.6%	*11.7%*
	2008	1.7%	-1.7%	-4.3%	-5.8%	-9.5%	-9.7%	-10.0%	-10.0%	-11.4%	-14.4%	*16.0%*
	2009	-16.6%	-17.9%	1.9%	1.7%	1.5%	1.6%	0.0%	-3.2%	-5.9%	-7.3%	*19.8%*
	2010	-1.7%	-1.9%	-2.2%	-2.2%	-4.1%	-7.1%	-9.5%	-10.8%	0.2%	0.0%	*11.0%*
	2011	-0.3%	-0.3%	-2.3%	-5.5%	-7.8%	-9.1%	-1.3%	-1.5%	-1.9%	-1.8%	*8.8%*
	2012	-3.8%	-7.0%	-9.3%	-10.5%	-1.8%	-2.0%	-2.3%	-2.3%	-4.3%	-7.3%	*8.7%*
	2013	-9.8%	-11.1%	3.9%	3.8%	5.0%	4.9%	2.8%	-0.7%	-3.0%	-4.3%	*16.0%*
	2014	-5.0%	-5.6%	-2.3%	-2.4%	-4.2%	-7.4%	-9.2%	-11.1%	-12.0%	-12.6%	*10.3%*
	2015	5.6%	5.4%	4.9%	4.8%	2.8%	-0.6%	-3.1%	-4.4%	-5.3%	-6.1%	*11.2%*
Maximum variation		*22.3%*	*23.3%*	*14.2%*	*15.4%*	*14.5%*	*14.7%*	*14.4%*	*15.2%*	*15.8%*	*18.0%*	

In the 10 GW storage case the variation along columns is larger than the variation along rows, in line with all the previous analysis in this study. However, when looking at the variations as a function of PV penetration vs consumption profile for the 25 GW storage capacity case this is clearly no longer applicable. There are some years where the variation as a function of PV penetration is larger than for the consumption profile. This could be because some of the GAFSI countries might have reached large enough amounts of PV production so that the full amount of storage is used over a period of only a few hours (e.g. PV capacity from one of the later years with the consumption of one of the earlier years). If 26 GW represents a large portion of the total consumption during that time period (or even if that is similar to the total consumption) it could create some unusual scenarios. This would be reflected in the in higher variability of MOPs.

5 SOCIAL BENEFITS: ARE WHOLESALE PRICES ACTUALLY DECREASING?

It has been estimated in sections 3 and 4 that (i) the MOP is close to 90 €/MWh and could be paid to PV plants as a bonus on top of the market price for every MWh of PV produced, and (ii) that the average market price would have been 4.0% higher (2.0 €/MWh) had there been no PV generation between 2007 and 2015.

In order to estimate the net social gains for society, several factors must be also taken into account such as the costs of the various support schemes to PV (tax incentives, feed-in tariffs), additional reserve capacity, grid upgrades on the one hand, and, indirect gains through taxes, job creations and reduction of negative externalities (pollution, energy dependent on Russian gas), hedging against fuel price volatility on the other hand. As stated before, this is not the purpose of this paper.

6 CONCLUSION

This study aimed at quantifying the savings incurred in Germany, Austria, France, Switzerland and Italy as a whole over the past 10 years as a result of the decrease in electricity spot market prices observed when PV plants feed electricity into the grid. It assessed the effect in two additional cases where PV has a decentralized storage system as well as a centralised storage system (e.g. pump-hydro).

For the case where there is no storage involved, the authors showed that the total electricity demand retreated by PV production is well correlated to the prices, following an exponential curve (R^2 around 66%). That good correlation proves that the strong set of assumptions was acceptable (low share of electricity export or import to production, no self-consumption, all electricity demand traded on the spot market, stable energy mix and price behaviour). The MOE, i.e. the aggregated energy bill saving in a year, has of course increased as more PV plants were installed, reaching a circa € 31 Bn cumulative amount by end 2015. If such benefit could be monetised by the public authorities, it could be invested in the infrastructure necessary as a result of the introduction of renewables into the energy mix such as grid reinforcement works or to allow spare peak producers to remain profitable. Over the 10 year period, the MOP (MOE expressed per MWh of PV production) is close to 91 €/MWh, which could be paid as a bonus to PV producers on top of their sale on the spot market. Also, market prices would have been 4.0% higher (2.0 €/MWh) had there been no PV production.

In order to test what the impact could be on the MOE (and related metrics) if significant amounts of energy storage facilities were installed, two types of storage were identified: centralized and de-centralized. In the former case it was found that the MOE amount decreased by EUR 800 M after the introduction of 10 GW of storage and a by a further EUR 1.5 bn when considering 25 GW of storage. The MOP decreased and the average market price decrease due to PV was lower than in the PV only case. Though the impact on those is relatively minor (in the order of 3-4 EUR/MWh for the MOP and around 0.1 EUR/MWh for the average market price decrease). Nevertheless, the MOP/MOEs increased year-on-year for 2014 and 2015 compared to the PV only case. This could be an indication that the amount of storage introduced were too high in the early years. To confirm these results/hypotheses additional calculations would be needed amongst others, to test the impact of gradually scaling up the storage systems over the years along with PV capacity; test the impact of more sophisticated re-injection methodologies (to better correlate the injection to the demand curve) and simply use smaller amounts of storage capacity, particularly in the early years when PV capacities are still relatively low.

For the centralised storage the results are quite different. In this case the MOE amount increases by EUR 1 bn for 10 GW of centralized storage and by an additional EUR 1.2 bn for 25 GW of storage. This leads to an increase of MOP by up to 8 EUR/MWh (25 GWh of storage). This would mean that PV plants could be paid almost 150 EUR/MWh including the WAP (around 50 EUR/MWh).

A statistical approach enabled the authors to simulate a large number of PV production profiles, and calculate the MOP with 10 different electricity demand profiles (those observed between 2006 - 2015) and 10 penetration rates (PV capacity installed between 2006-2015). The statistical analysis revealed that the MOP is generally quite dependant on the particular electricity demand profile, and therefore on how well it is correlated to the PV production profile. Introducing a storage element allowed the authors to test if this would be an efficient way of optimising the MOE for PV (by improving the correlation between injection of electricity produced by PV and demand profile). With a relatively simple approach it was possible to significantly improve the MOE and MOP using a form of centralized storage. This suggests that there is room for further optimisation and, thus, improvement of the MOE. Since 2011, the penetration rate of PV is also a significant explanatory variable, suggesting that PV could be increasingly replacing mid load capacity.

Ultimately it has been observed that the negative pressure of PV production on wholesale prices is well reflected in Germany, Austria, France and Switzerland.

Another contribution of this analysis is the quantitative assessment of the market coupling effect on PV impact on the electricity markets (through WAP). The same methodology could be applied to wind energy. In fact, as PV, it also represents a good share of the power generation of the countries analysed here. Further studies could sign the pathway to an integrated electricity market which will not only exploit the negative pressure of PV on electricity market but also what are at the moment known as weaknesses of the RE (i.e. intermittency).

This paper opens the route for other improvements:

- take account of export/import and self-consumption;
- run the Monte Carlo on irradiation (which is expected to follow more closely a Gaussian behaviour) rather than yield.
- Analyse the combined effect of Wind and PV electricity fed into the grid on wholesale prices.
- Use more sophisticated approaches in the simulation of storage capacities to improve the correlation between electricity injection from PV production and electricity demand.

7 REFERENCES

[1] Quarterly report on European Electricity Markets, Market Observatory for Energy, DG Energy, Volume 6, issue 2, second quarter 2013.

[2] H. Gouzerh, M. Carton, N. Gourvitch and G. Masson Quantitative analysis of the merit order effect from photovoltaic production in Italy

[3] H. Gouzerh, S. Orlandi, N. Gourvitch and G.Masson Impact of PV production on electricity prices

[4] Quarterly reports on European electricity markets (http://ec.europa.eu/energy/observatory/electricity/electricity_en.htm)

[5] RTE annual report (http://www.rte-france.com/uploads/Mediatheque_docs/vie_systeme/annuelles/Bilan_electrique/bilan_electrique_2013.PDF)

[6] Solar irradiation data provided by GeoModel solar

[7] Monthly regional PV capacities provided by GSE

[8] Consumption data: https://www.entsoe.eu/

[9] Wholesale prices: http://www.mercatoelettrico.org

[10] Frank Sensfuß, Mario Ragwitz and Massimo Genoese, "The merit-order effect: a detailed analysis of the price effect of renewable electricity generation on spot market prices in Germany" (2008)

[11] Francesco de Mango, presentation on "Market Integration of PV: the impact of photovoltaic on electricity markets" during the 28th EUPVSEC (4 October 2013) in Paris.

[12] Solar irradiation data provided by PV GIS for period 2014 – 2015

[13] Mueller R.W., Matsoukas C., Gratzki A., Behr H.D., Hollmann R. The CM-SAF operational scheme for the satellite based retrieval of solar surface irradiance - A LUT based eigenvector hybrid approach. Remote Sensing of Environment, 113, 1012-1024 (2009)

[13] Huld, T., Mueller R., Gambardella A. A new solar radiation database for estimating PV performance in Europe and Africa. Solar Energy, 86, 1803-1815 (2012)

[14] Huld T. and Gracia Amillo A. Estimating PV Module Performance over Large Geographical Regions: The Role of Irradiance, Air Temperature, Wind Speed and Solar Spectrum. Energies, 8, 5159-5181 (2015)

ELECTRIC VEHICLES POWERED WITH PV ELECTRICITY AS A NEW DRIVER FOR PHOTOVOLTAICS

Urs Muntwyler, Eva Schuepbach
Berne University of Applied Sciences (BFH), Engineering and Information Technology
Photovoltaic Laboratory (PV-Lab), Jlcoweg 1, CH-3400 Burgdorf, Switzerland
phone: +41 (0)34 426 68 37, fax: +41 (0)34 426 68 63, urs.muntwyler@bfh.ch

ABSTRACT: Marketing PV electricity increasingly becomes a challenge as the feed-in tariff schemes are blocked, either by overbooking or by political actions against a higher contribution of electricity from PV. Self-consumption of PV electricity is, therefore, imperative. Moreover, it is economically attractive as current electricity costs from PV are about 0.25€/ kWh (or even below). However, most private households are relatively small consumers as compared to the production power of their PV installation. This is especially the case from spring to autumn, when surplus electricity is produced from the PV installations. To optimize the own consumption during this "high production mode", it makes sense to install a battery with a 10-20 kWh storage capacity. When additional, strong electricity consumers like heating and hot water systems with heat pumps are connected, the leverage effect is increased and burning of fossil fuel can be avoided. Yet, heating is mostly needed in the cold season, when PV installations have less overproduction. The most efficient all-year strong electricity consumer is an electric vehicle (EV). Having evolved from pioneering solar cars in the mid-1980s, EVs today offer an astonishing high economic benefit as PV electricity consumers. Even more so in the near future, when EV batteries (40kWh and larger) will serve to handle the electric load management and internal combustion engines will, by 2050, entirely be replaced by EVs (International Energy Agency scenario).

Keywords: Solar Cars, Electric Vehicles, Photovoltaic, Consumption, Economics

Figure 1: Mercedes/ alpha real, winner of the first solar car race in the world ("Tour de Sol") in 1985 across Switzerland [1]. Photo: Urs Muntwyler.

Figure 2: Opel Ampera charging at the solar carport in front of the Photovoltaic Laboratory (PV LAB) at Bern University of Applied Sciences BFH, Switzerland, in 2015. Photo: Urs Muntwyler.

Figure 3: Some of the 58 participants driving their solar cars during the "Tour de Sol" 85. Photo: Urs Muntwyler.

1. PURPOSE

The PV market needs to look out for new opportunities, one being the coupling of PV with electric vehicles (EVs). Both, the technology and organization require attention with this avenue. The technical aspect is the coupling of the PV production with the need of

EVs to be able to charge in a defined time slot. The organizational aspect concentrates on the fact that EVs are often parked at greater distance away from PV systems over the day.

2 EVs CAN RUN ON MUCH CHEAPER ENERGY THAN IEC VEHICLES - WITH PV

Electric Vehicles (EVs) have an electric consumption between 12-20kWh / 100km (BMW i3 - Tesla S). A typical car drives 12'000km - 15'000km per year in Switzerland and thus consumes 1'440 - 3'000kWh / year. A car with an internal combustion engine (IEC) has a consumption of 6 l - 10 l / 100km or 720 l - 1'500l / year.

Electricity costs from PV are about 0.25€ / kWh (or even below), which corresponds to 360 – 750€ per year for a "solar powered" EV. For a car with an IEC and a gasoline price of 1.5€ / liter, between 1'080€ - 2'250€ per year need to be spent. Assuming a lifetime of 8 years for a car (or over 100'000km), the difference of 5'760 € and 12'000€ represents the profit of an EV-driver feeding the EV with PV electricity (money that may be spent on a higher price of the EV).

If the gasoline price increases, the difference is even bigger and an EV linked to a PV installation is even more advantageous. The EV may have further advantages such as lower service and maintenance costs, lower taxes and sometimes lower insurance costs. On the other hand, extra infrastructure like a charging station is required, or certain risks associated to the lifetime of the battery are encountered (unless the producer handles it).

3 THE ENERGY CONSUMPTION OF A HOUSE IN SWITZERLAND

The vision for the future is that all single houses can produce more energy with an own PV installation than their residents consume. Based on the new legal regulations [see Fig. 4], a house in Switzerland, where considerable heating is needed in winter, has the energy needs as shown in Table 1.

Consumer	Unit	Consumption	Energy
Heat	100 $m^{2(1)}$	30 $kWh/m^{2\,(2)/(8)}$ / heat pump	1'000 kWh
Hot water	4 x 50 l /person	4x50 l à 43K $^{(3)(9)}$ / heat pump	1'950 kWh
Electricity	household	4 x 1'000kWh $^{(4)}$	4'000 kWh
MIV (E-mobile)	15'000 km /year $^{(5)}$	16 kWh /100km $^{(6)}$	2'400 kWh
Grey energy house	100 $m^{2(1)}$	20.55 $kWh/m^{2\,(7)}$	2'055 kWh
Total energy			**11'405 kWh**

Table 1: Energy consumption of a house in Switzerland.

(1) Surface for living (BGF - Bruttogeschossfläche)
(2) Consumption of 3 l/m^2 BGF (MuKen 2014 limit, see [3]. Some organizations recommend 50% of the legal limits for a "Plusenergiehaus" (September 2013).
(3) Consumption of hot water per year = 4 x 50 l / 100 l (430K/8,60K) x 365 days + 2 kWh/day heat losses (own experience with 400-500 l tanks used in Switzerland).
(4) Electricity consumption per person: 1'000kWh in "Plushouses" (experiences of architect Werner Setz, 2013).
(5) Annual mileage CH.
(6) Consumption of an electric car like Renault ZOE.
(7) Grey energy of a massive house with good insulation over 50 years (ETHZ "Pinpoint Fakten der Bauphysik, Keller und Rutz, p. 117).
(8) Assumption: HP-efficiency factor COP3 (1'000kWh electricity + 2'000 kWh ambient heat).
(9) Assumption: HP-efficiency COP3 or solar hot water-installation factor 3 (which is difficult to achieve!)

As compared with official energy standards, considering the individual mobility (with an EV) and the grey energy of the house is new [3b]. Integration of the individual mobility and grey energy is, however, necessary for individuals who want to meet the electricity needs for most of their activities with PV electricity production. The life style beyond EV is not taken into account.

4 FIRST CALCULATION PV INSTALLATION

In order to cover an annual consumption of about 11'400kWh, a PV-installation of ca. 11-12kWp power is necessary. The associated PV module surface is 60-80m^2 and the costs are in the range of 22'000-36'000€ (with 2-3€/Wp). The surface of 60-80m^2 is quite big but is no problem for a new house with a well-planned roof surface. 60m^2 can even be placed on most retrofitted houses. If the retrofit process includes the entire house, turning the roof to achieve an east-west orientation is the most efficient solution (Fig. 5).

Figure 4: Transformation of a house (built in the 17th century) near Zürich (Switzerland) into a so-called "plus house"; PV installation is oriented towards the south.

Figure 5: Multi-family retrofit house (built in 1947) in Thun (Switzerland) producing >180% of the households' energy consumption. East-west oriented PV roof of 35kWp (the orientation of the roof was originally south-north) provides enough energy to cover individual mobility and the grey energy as well.

5 OWN CONSUMPTION OF PV ELECTRICITY

Increasingly, electricity companies pay a very low tariff for PV electricity that is fed into the grid. Today, this is about 6 cents € per kWh. If the consumer needs to buy electricity from the same companies, it costs about 25 cents € per kWh. Assume a single house produces about 25% of the consumed electricity on its own. If a private household has an installed PV of 10kWp and an electric consumption of 8'000 kWh per year, only 25% or 2'000 kWh will be consumed (calculations are based on [5]) and 75% are fed into the grid. The cost calculation (see below) provides a net gain.

- 2'000 kWh at 0.25€ = 500€
- 6'000 kWh at 0.06€ / kWh = 360€

However, the rental costs for an electric meter in Switzerland is almost as high as the financial returns from the feed-in tariff. Many house owners hence seek to store the surplus electricity (that they produce with their own PV installation) for use at some later time during the day. When installing a battery with 10 kWh storage capacity, the own consumption can be increased to 54%; with a battery of 20 kWh to 62%. Tuning big electricity-consuming machines (dish washer or washing machine) to the production pattern of the PV installation only offers a small efficiency increase. Rather, it makes sense to install a battery with a 10 - 20kWh storage capacity.

6 HOME STORAGE POSSIBILITIES FOR PV ELECTRICITY

Surplus electricity produced from an own PV installation is mainly an issue during "high production mode" (i.e., from spring to fall). Assume two storage systems in a private household, being a hot water system (water tank) and an electric vehicle. To optimize the own consumption, a storage capacity of 10-20kWh per day would be ideal. Heating, in most cases, cannot be taken into account and would require very big seasonal storage. This can be done, in theory, but is more suited for special projects above the cost optimum.

Now let us look at some of the specifics of home storage: The hot water consumption depends on the number of people living in the household (residents). With four residents, we calculate an annual consumption of 1'950kWh with a heat pump boiler (electric boiler will be forbidden in Switzerland in the future). The consumption might be reduced to 1'400kW per year in a household with two residents and about 4-6kWh/ day can hence be stored. Water could be heated (and stored) during the day. This is different from the current practice of Swiss households, which heat water overnight (due to low electricity costs at night, which triggers the use of the band production from nuclear power plants in Switzerland). But the Swiss Governmental "Energy Strategy 2050" (Fig. 6) plans on phasing out nuclear power and new modes of water heating and hot water storage are thus necessary.

The biggest single electricity consumer is the electric car and, therefore, the electric car is the focus of our further reflections. Today, storage can be achieved with a stationary storage system. But in the very near future, EVs will have 40kWh batteries and larger [7] and can serve to control the electric load management.

Average driving distance of a car in Switzerland is about 40 km / day, which corresponds, depending on the size of the EV, to an electric consumption of 5-8kWh/ day. This results in a total storage capacity for the PV surplus electricity production of about 9-14kWh at our "model household" and "model home". With this self-consumption of about 60%, our calculation in Section 5 improves to:

- 4'800 kWh at 0.25€ = 1'200€
- 3'200 kWh at 0.06€/ kWh = 192€

If an electric car is used as a storage for (surplus) electricity produced from a PV installation, the car must be parked near this installation. This could be solved with a "roaming system", similarly to the telecom industry. The current market is, however, still closed for small users up to 100'000 kWh.

7 SWISS "ENERGIESTRATEGIE 2050": PV IS THE MAIN NEW SOURCE

Following the nuclear accident at Fukushima in 2011, the Swiss Government decided to phase out the existing 5 five nuclear power plants in Switzerland ("Swiss Energy Strategy 2050"). Their production has to be replaced by mainly new renewable energies (Fig. 6). The plans are to install 12TWh PV by 2050 (with a total expected consumption and production of about 60TWh). This is equal to an installed power of about 12GWp (Fig. 7).

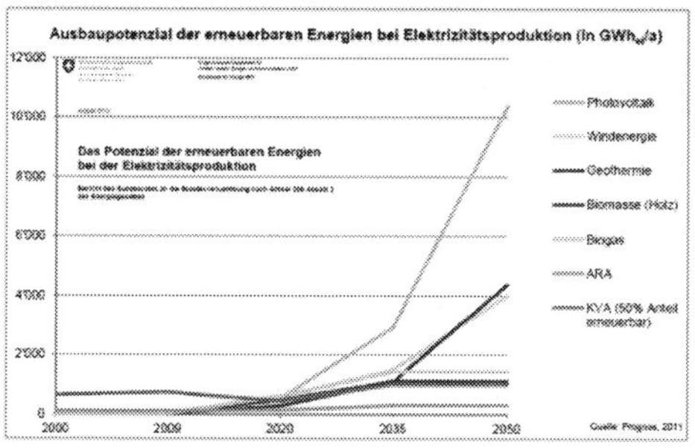

Figure 6: Energy strategy of the Swiss Government for 2050. The main new electricity source is PV. As domestic wind energy and geothermal energy are not very realistic, even more PV electricity will be needed

8 PV ELECTRICITY FOR EVs – A BUSINESS CASE FOR STAKEHOLDERS

There are about 5.9 million cars registered in Switzerland, among them are 4.5 million private cars [8]. If 80% of these are electric vehicles (EVs) in 2050, Switzerland has an electric car fleet of about 4,7 million EVs with an electric consumption of 12 kWh / 100 km (similar to BMW i3).

The electricity needed for the motorized individual traffic will be: 4.7 million x 1'800 kWh = 8,46TWh, reflecting an additional PV power of ca. 9GWp installed in Switzerland. With additional 9GWp PV consumed by the EVs, the electric vehicles should be charged in the daily peak time.

In this example, the electricity production sources only include water (hydro) and the sun (PV). In the future, additional minor sources as wind, biomass with waste-heat-cogeneration etc. can balance the production mix.

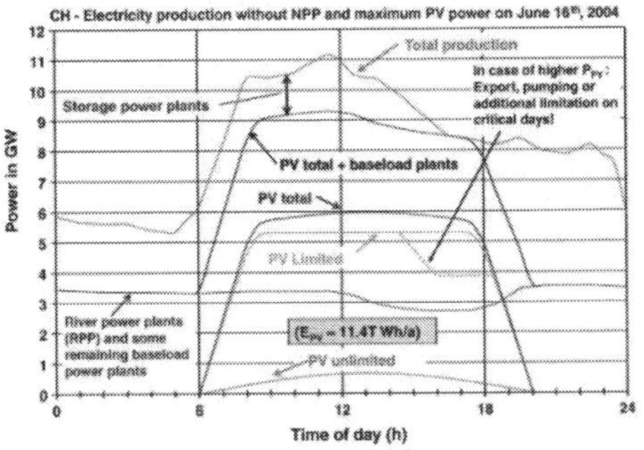

Figure 7: Forecast of the daily electricity production pattern in Switzerland with 12GWp PV. Hydro power is the only additional source (total annual consumption is 60TWh) [9].

9 THE PV MARKET IS GROWING QUICKER THAN THE EV MARKET

As shown in Fig. 8, there are still not many EVs on sale worldwide. Calculated with the figures shown in Section 8, we need only 1.08GWp PV power to cover the electricity needs of the more than 500'000 new EVs from 2015. Compared with the over 50GWp, there is still a lot of electricity for other purposes.

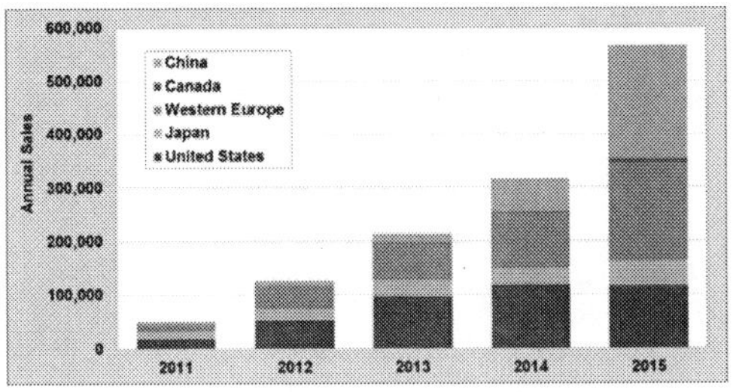

Figure 8: Small but steadily growing market: EV sales worldwide. Source: Argonne National Laboratory, USA (2016).

The worldwide sales forecasts of about 100 million EVs and PHVs cars for 2050 [10] will create a demand of additional 200TWh electricity production per year. If half of the production came from PV (and the other half from wind), this would create an additional PV market of about 100GWp per year. In the longer perspective, the market would go down to about 30GWp per year as the lifetime of PV installations is 25-40 years but a car only lasts for about 12 years. But this is still relevant for the worldwide PV market.

The IEA PV roadmap from 2010 [11] estimates an annual PV market of 117GWp for grid-connected installations and a total of about 4'000TWh. The additional PV power of 30-50GWp for EVs would be a relevant number for the total PV market. These market forecasts must be viewed with caution, however, as forecast of the International Energy Agency (IEA) in the field of renewable energies normally lag behind the reality [12].

If the estimated 2 billion cars in 2050 (the 1 billion barrier was passed in 2012) were electrified with 50% coming from PV, about 2'000TWh were needed. This represents a production rate of 1'000 kWh / kWp and a total of 2'000GWp installed. With a production figure of 1'500 kWh / kWp, still 1'300GWp PV power is needed.

10 CO_2 BALANCE OF AN ELECTRIC VEHICLE

Critical voices with regard to electric vehicles claim that EVs are "dirty" because of the electricity mix. Here, we like to address this criticism as follows (see Table 2):

The CO_2 - balance of an electric vehicle depends on the production mix of the electricity. Electric vehicles running with electricity produced from brown and stone coal are worse than gasoline vehicles. This explains why Toyota in Japan forced the development of hybrid gasoline vehicles in the 1990es.

As it is cheaper to produce the electricity with PV than buying gasoline for a gasoline car (if gasoline is not subsidized [13]), EVs should directly be linked with a PV installation.

Increasingly, electric utility companies and car dealers (for example BMW in Lausanne, Switzerland [14]) offer PV- and hydro-electricity for EVs. Other electricity production mixes (with wind, hydro- and PV mixes) could also be considered.

11 REPLACING GASOLINE CARS BRINGS BENEFIT FOR MOST COUNTRIES

Unfortunately, the huge difference in costs for running an ICE and an EV is neither completely understood by the general public nor by politicians. This is very unfortunate as most countries import fossil fuel, whereas renewable energies could be produced locally by these countries.

In the case of Switzerland, 4.9 million IECs could be replaced by EVs. To propel these EVs, we need about 8.46TWh electricity (produced locally from PV). This extra consumption of electricity could easily be covered with an additional installation of 9GWp.

In 2014, Switzerland paid about 2.2 billion Swiss francs for gasoline per year [15]. PV would cost 1,033 billion Swiss francs annually [16]. By replacing gasoline cars with EVs running on electricity from local PV installations, Switzerland could save 1 billion per year (prices in 2016). Shifting towards locally-produced PV electricity would hence offer the country and its citizens a huge financial benefit. Additional positive effects such as likely reduced air and noise pollution, resulting decreased health costs and costs associated to less damage experienced by other systems (vegetation) have not been included in these reflections.

12 MANAGING THE PEAK POWER PRODUCTION OF THE PV

Compared to the planned 12TWh PV electricity in the year 2050 ("Energiestrategie 2050" by the Swiss Government), the 9GWp is an additional interesting market for PV. With this additional PV power, the PV peak power will exceed the maximum consumption in Switzerland [9]. This peak has to be managed in different ways (see Table 3).

Model	CO_2-emissions	TWIKE	BMW i3	Renault ZOE	Opel Ampera	Tesla	Renault Clio 1.2	BMW X5 31
Electric consumption per 100 km [kWh]		5	12	14	15	20	gasoline	gasoline
Energy source	[kg/kWh]	[g/km]	[g/km]	[g/km]	[g/km]	[g/km]	[g/km]	[g/km]
Brown coal (lignite)	1.23	61.5	147.6	172.2	184.5	246		
Stone coal (anthracite)	1.08	54	129.6	151.2	162	216		
Gas	0.64	32	76.8	89.6	96	128		
PV	**0.03**	**1.5**	**3.6**	**4.2**	**4.5**	**6**		
Wind	0.01	0.5	1.2	1.4	1.5	2		
CO_2 EU-Mix (only emissions)	0.46	23	55.2	64.4	69	92		
CO_2 (GE-Mix)	0.48	24	57.6	67.2	72	96		
CO_2 EU- LCA Mix (with plant, etc.)	0.578	28.9	69.36	80.92	86.7	115.6		
CO_2 CH-Mix certified (Frischknecht 2012)	0.014	0.7	1.68	1.96	2.1	2.8		
Petrol/gasoline	---	0	0	0	0	0	**137**	**197**

Table 2: Legislation for fleets EU CO_2-emissions: 95 g/ km from 2020 onwards.

No.	Action	+ PV power	Costs of the action
1	Limit 30% of the PV power	+42%	3%
2	Regulated consumers	+20%	< 5%
3	**Consumers as EVs**, heat pumps with storage etc.	+20%	0-50%
4	Limit 50% of the PV power	+100%	18%
5	East-west PV installations	+50%	ca. 50%
	Local batteries	+50%	50% (goes down)
7	Pump storage plants	+50%	50%
8	Grid reinforcement (high voltage)	about 20%	about 25%

Table 3: Peak power management.

Further optimization of the PV power could be reached by optimizing charging regimes in relation to PV production, the production forecast, the needed charging energy, etc. These measures all aim at a better use of PV electricity for EVs and a lower impact on the electric grid network in Switzerland. Research avenues in this direction are currently being underway at our University of Applied Sciences BFH, see for example [17].

13 CONCLUSIONS AND OUTLOOK

Replacing inefficient IECs by EVs is attractive for countries with a clean electricity production mix and also interesting for individuals with EVs. Both, countries and individuals can save a lot of money in a time then EVs are not more expensive than IEC vehicles. We predict that this will happen in several years, in around 2020. At this time, the policy makers will have to lower the tax difference by creating new tax regimes to cover the costs of road infrastructure. Actually, the bulk of these costs is today packed into the gasoline price.

The economic advantage of EVs is already reality by, e.g., using certain EVs (like a Nissan Leaf) propelled with PV. The dreams of the solar car pioneer in the 1980es have already become true and are, today, a reality for everyone!

Acknowledgements
We appreciate funding from the Swiss Centre of Competence in Energy Research "SCCER FURIES", the Swiss Federal Office of Energy (SFOE) and Bern University of Applied Sciences BFH.

References

[1] U. Muntwyler et al., Solarfahrzeuge, Ringier Verlag Zürich, Switzerland (1985).
[2] MuKen (2014). www.endk.ch/de/energiepolitik-der-kantone/muken
[3a] U. Muntwyler, Photovoltaik-Wahlmodul (Module offered in the Swiss Master of Science in Engineering Degree), BFH, Switzerland (2016).
[3b] E. Schuepbach, U. Muntwyler, Sustainability in Engineering Sciences (Module offered in the Bachelor of Science Degree), Department of Engineering and Information Technology, BFH, Switzerland (2016).
[4] M. Wackernagel, B. Beyers, Der Ecological Footprint - Die Welt neu vermessen, Europäische Verlagsanstalt EVA (2010).
[5] HTW Berlin, PV Unabhängigkeitsrecher (2016). Link: www.pvspeicher.htw-berlin.de
[6] N. Heynen, Internal Report, Bern University of Applied Sciences BFH, PV LAB (2012).
[7] Opel Ampera-e, Tesla S and X, Tesla 3, etc.
[8] www.auto-schweiz.ch (2015).
[9] H. Häberlin, Photovoltaik, VDE/electrosuisse Verlag, p. 373 (2010).
[10] IEA, HEV Roadmap (2009), www.iea.com
[11] IEA, PV Roadmap (2010), www.iea.com
[12] H.-J. Fell (MP, Germany) and R. Rechsteiner (MP, Switzerland), personal communication.
[13] Fossile energies are subsidized with about 500 billion US$ (IEA World Energy Report 2015)
[14] Rochat Solaire SA, personal communication, May 2016.
[15] Swiss Federal Office of Energy, Gesamtenergiestatistik, p. 32, p. 45, p. 49 (2014)
[16] Costs per Wp: 2 CHF / 3% interest / 25 years lifetime of a PV installation
[17] P. Amiet, M. Briggeler, Internal Report, Bern University of Applied Sciences BFH, Department of Engineering and Information Technology, (2016).

GAMIFYING THE ENERGY TRANSITION

Benedict O'Donnell
Heliocentric Solutions, 56 Vincent Square, SW1P 2NE, London, UK
+44 (0)207 828 2820, benedict.odonnell@heocs.com

Dennis Pfahl & Josef Mehling
Prometeruse, Motzstrasse 63, 10777 Berlin, Germany

ABSTRACT: Sustained public support will prove vital in harnessing the opportunities offered by competitive photovoltaics. Rather than appeal to consumers directly, our project opens the dialogue with their children. Prometeruse takes the form of a citizen science experiment in which teenagers report the carbon footprint of their school and race to reduce it using means that include rooftop photovoltaics. A network of academics active in the energy and climate sciences provide the teachers in our participating schools with course material that draws the link between the technologies tested on campus, the macroeconomic policies enacted by governments and the lifestyle choices made by households. This year, the pilot phase of our project has brought six schools together across France and Germany. Its participants are releasing rare energy consumption data into the public domain, demonstrably improving their grasp of sustainable development and eagerly adopting technologies such as smart meters for their personal use. If such interest can be extended to schools across all 28 Member States, tomorrow's consumers and taxpayers could make the difference in navigating Europe through its 2050 Energy Roadmap.

Keywords: Implementation, Marketing, Rooftop, Sociological, Energy Funding and Incentives

1 INTRODUCTION

The rapid expansion of photovoltaics in today's conventional energy market gives rise to challenges as well as opportunities. One priority in sustaining the current growth rate in global installed capacity (Figure 1) is to shed the sector's reliance on government incentives. This objective has become realistic over the past few years since the levelised cost of energy produced by rooftop photovoltaic systems has dropped beneath electricity retail rates in a growing number of markets [1]. Unsubsidised rooftop photovoltaic systems now offer an investment opportunity that is safe from swings in government policy, and an overall market that could dwarf that of subsidised photovoltaics today.

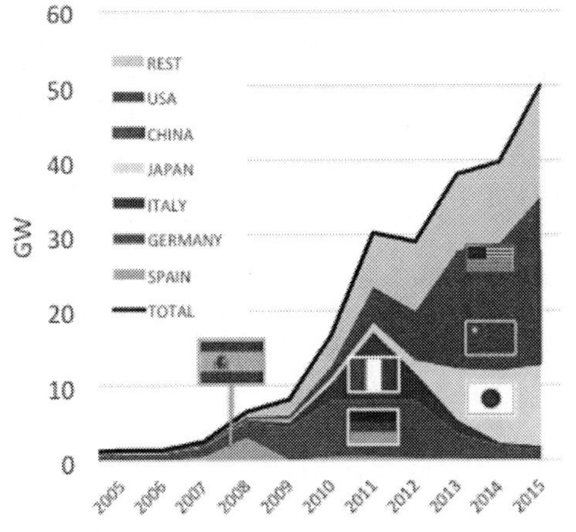

Figure 1: Annual installed photovoltaic capacity by country [2].

The catch is that while electricity markets including that of Germany, Italy and Spain, have reached socket parity years ago [3], consumers have not yet installed significant numbers of unsubsidised rooftop systems in them (Figure 1). This paper presents a consumer engagement project that could help tip the transition into motion. By offering consumers first-hand evidence of the economic and environmental benefits of solar energy, the Prometeruse project aims to unlock the vast potential of rooftop markets and invite consumers to take a more active role in the global transition towards sustainable energy sources.

2 GAMIFICATION

Public outreach initiatives have been key in driving the photovoltaic industry to its current state of maturity. Efforts have historically focused on informing consumers, typically through unilateral axiomatic teaching, of the undesirable consequences of climate change. Reports from government agencies, academic institutions and private activists have, over the past decades, galvanised the first international climate treaties and public subsidy schemes for renewable energy [4]. They have however fallen short of convincing consumers to adopt photovoltaic technology independently. A more pervasive approach may be needed for the next step of the energy transition in order to communicate the business case of rooftop photovoltaic systems and allow non-specialists to gauge their economic and environmental benefits. One practice that is gaining ground in education and civic outreach is gamification [5]. In this paper, we present a brief introduction to the use of game elements in communication and outline how they might be suited to gathering support for the global energy transition.

2.1 A new method of communication

Games offer a convenient medium to interest audiences over periods of time long enough for them to acquire new skills and understand complex issues. In the past decade, language schools [6], military outfits [7] and government agencies [5] have made notable advances in exploiting game elements to broadcast information or improve teaching methods.

The practice of gamification remains at an early stage in its development. Researchers and game designers are still investigating the psychological triggers that motivate players [7]. Current understanding is that successful games share core mechanics such as tailoring difficulty levels [8] and timing rewards [9] to satisfy an innate sense of competition in players. Many also fix reliable boundary rules within which players can explore open-ended solutions using their imagination [10]. These guidelines may help develop games that are entertaining enough to capture the attention of broad audiences.

2.2 Current barriers to gamifying the energy transition

Regardless of the quality of individual games, one challenge to gamifying outreach campaigns is that all people are not keen on playing games. Senior energy consumers, in particular, are less likely to relate to games as a medium than youngsters, and are often unfamiliar with the technologies used to play them. Even when addressing gamers, outreach efforts must compete for their attention with other leisure activities, including other games that have larger budgets and are exclusively designed for entertainment. Unlike leisure games, the objective of outreach activities is not just to convince consumers to play their game but also to change their daily habits. Given limited real-world rewards that game developers can offer, it is challenging to achieve the timing and rapid feedback necessary to sustain player motivation in real life. Even savings on electricity bills will only be experienced at the end of the month, long after the game is over.

3 PROMETERUSE

The Prometeruse project is exploring a way around these barriers. Rather than approach energy consumers directly, the project opens the dialogue with 15-17 year old secondary school students. Teenagers are particularly efficient vectors of technology adoption within their communities. They pick up consumer-products and IT services faster than their elders and help them understand how they work. Teenagers are also versatile learners, as they are still being taught the skills necessary to grasp the technical aspects of climate change and the energy transition. They are forming their opinions and tend to be more receptive to new ideas on these issues. Also, during the period of the day in which teenagers are in school, serious games need not compete with other sources of entertainment for their attention. The content of outreach games geared towards secondary school students need only ensure they are as educational and engaging as the classes that they fit into.

3.1 Project description

In 2015, the pilot run of the Prometeruse project launched activities in 6 secondary schools across two cities in Europe (Figure 2). In most schools, the project worked with one classroom. Most classrooms were composed of around 30 students. In each coming year, the project aims to boost these figures.

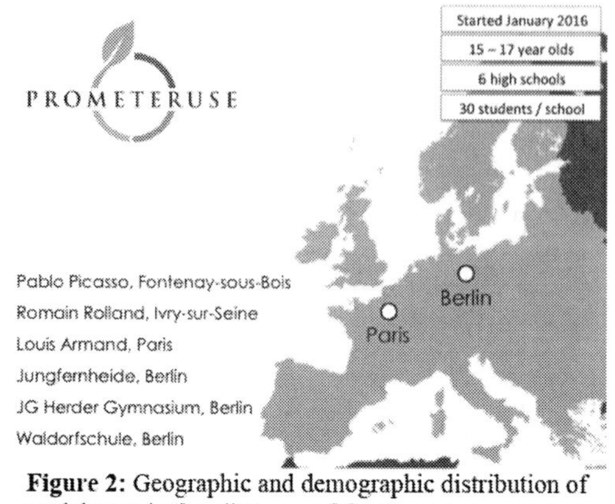

Figure 2: Geographic and demographic distribution of participants in the pilot year of the Prometeruse project.

3.2 Competition

The core objective of the Prometeruse project is to set up a yearly competition in which students from different countries race to reduce the carbon footprint of their buildings. To gauge their progress, the project is setting up means of tracking over time the electricity and gas consumption in participating schools (Figure 3).

Time-resolved energy consumption data offers consumers an unbiased view of the financial and environmental cost of their lifestyle choices. As the project unfolds, schools take part in an international citizen science experiment that is progressively uploading electricity load profiles from single buildings into the public domain. These datasets are key to demonstrating the benefits of sustainable energy solutions (including, for instance, rooftop photovoltaics) and training students to quantify the environmental impact of different measures towards sustainability. The measurements also provide a solid framework and tangible goal to create a stimulating game with. Charting the energy consumption profile for each school online and offering a prize at the end of the year touches drivers of institutional pride and competition among students that would be hard to replicate among adults.

At present, the data is being collected and shared with teachers and students using energy probes and an internet platform developed by the clean technology firm NorthQ. By 2020, similar data could be recovered in buildings across the EU using the smart meters being installed in them. In order for broader audiences to derive the same gain from the data as our project participants, they must be capable of accessing it, translating it from kilowatt hours into wasted euros and carbon emissions, understanding the meaning of this waste and implementing solutions to it. All these tasks could be facilitated by the trusted climate consultants that we are training within the Prometeruse project as teenagers relay the experience acquired at school to their parents at home.

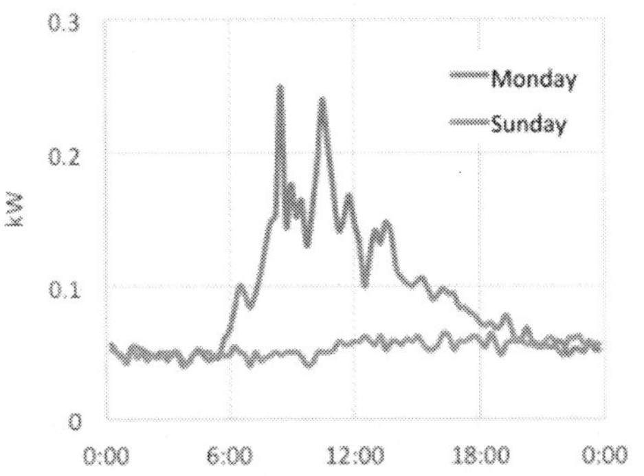

Figure 3: Electricity consumption in a Paris secondary school on Sunday 19 June and Monday 20 June 2016.

3.3 Teaching science

To support the learning outcomes of the carbon footprint competition, the Prometeruse project has rallied a network of scientists to write and deliver science lessons on climate change. The objective of these lessons is to equip students with the knowledge necessary to quantify, from primary source data, the impact of lifestyle choices and energy technologies on their carbon footprint. To link these daily activities to macroeconomic developments in the energy sector and changes in the Earth's climate system, the lessons break down information in reports by the International Panel on Climate Change into words that teenagers can relate to [12].

The first draft of these lessons was delivered between January and June 2016 and is now being peer-reviewed by climate scientists and energy experts for next year. In future installments of the project, lesson plans and teaching material will be provided to science and geography teachers in each participating school. The project will try to ensure that invited speakers (typically climate scientists or an experts active in the energy sector) deliver at least one of the lessons. The personal connection reinforces the feeling of involvement among our audience and makes the project more stimulating.

3.4 Language exchanges

Learning about climate change need not be confined to the science class. One popular activity tested this year was to organise language exchanges between students in Paris and in Berlin. English teachers in both countries swapped climate change essays and established an international poetry circle on the topic. As the competition matures, this kind of partnership may be exploited to pool the efforts of secondary schools in teams so that they can swap experience and advice on bringing down the carbon footprint of their schools together. Cross-border language exchanges also capture the international nature of the climate change challenge and the collaborative dimension of ongoing efforts to deal with it.

3.5 Role-playing game

In December 2015, the Prometeruse project supplied its participating schools with a board game based on the COP21 as a warm up to the activities of the project [13]. Its objective was to sensitise students to international environmental efforts unfolding around them and help them draw the link with the content of their science lessons. Each student was given a nation to represent and the task for each class was to agree on a treaty that would curb greenhouse gas emissions while meeting the national objectives of each player before their counterparts in Paris did. A variant of the game is now under development for students to play at home before the COP22 in Marrakech. In doing so, Prometeruse intends to engage with the parents and broader community of its student participants.

5 CONCLUSIONS

Gamification offers a promising tool to gather public support for the global energy transition. The solar energy community must still learn to wield it to its full potential. The Prometeruse project is investigating how best to target games to an appropriate audience, support them with entertaining activities and develop synergies with established institutions (such as public education). The approach is designed to unlock sustainable, and underexploited, markets for unsubsidised rooftop photovoltaics, as well as to equip future voters with the understanding needed to make informed decisions about the ongoing transformation of their energy system.

6 REFERENCES

[1] M. Bazilian, I. Onyeji, M. Liebreich, I. MacGill, J. Chase, J. Shah, D. Gielen, D. Arent, D. Landfear, and Z. Shi, "Re-considering the Economics of Photovoltaic Power Morgan Bazilian," *Renew. Energy*, vol. 53, pp. 329–338, 2013.

[2] IEA PVPS, "Snapshot of Global PV 1992-2015. Photovoltaic Power Systems Programme," *Int. Energy Agency*, 2015.

[3] B. O'Donnell, "Tailoring Rooftop Photovoltaics for Self-Consumption," in *29th EU-PVSEC*, 2014, pp. 4062 – 4064.

[4] S. C. Moser, "Communicating climate change: history, challenges, process and future directions," *Wiley Interdiscip. Rev. Clim. Chang.*, vol. 1, no. 1, pp. 31–53, Jan. 2010.

[5] T. Susi, M. Johannesson, and P. Backlund, "Serious games: An overview," 2007.

[6] W. L. Johnson, H. H. Vilhjálmsson, and S. Marsella, "Serious games for language learning: How much game, how much AI?," in *AIED*, 2005, vol. 125, pp. 306–313.

[7] G. Richter, D. R. Raban, and S. Rafaeli, "Studying gamification: the effect of rewards and incentives on motivation," in *Gamification in education and business*, Springer, 2015, pp. 21–46.

[8] A. M. Glenberg and D. A. Robertson, "Indexical understanding of instructions," *Discourse Process.*, vol. 28, no. 1, pp. 1–26, 1999.

[9] J. L. Elman, *Incremental learning, or the importance of starting small*. University of California, San Diego, 1991.

[10] A. L. Brown, "The advancement of learning," *Educ. Res.*, vol. 23, no. 8, pp. 4–12, 1994.

[11] S. E. Beatty and S. Talpade, "Adolescent influence in family decision making: a replication with extension," *J. Consum. Res.*, vol. 21, no. 2, pp. 332–341, 1994.

[12] B. O'Donnell and P. Jouy, "Teaching climate and energy science with the Prometeruse project," *EDULEARN15 Proc.*, pp. 3800–3804, 2015.
[13] B. O'Donnell, D. Pfahl, J. Mehling, and F. Gabriel, "Walk the Climate Talk," in *European Conference on Games Based Learning*, 2016, p. 749.

PHOTOVOLTAIC DEVELOPMENT STANDARDIZING BASED ON ROADMAPS AND TECHNOLOGY READINESS LEVELS

P. Baliozian[2], S. Mourad[1], D. Morse[3], S. Kim[1], L. Friedrich[1] and R. Preu[1]
[1]Fraunhofer Institute for Solar Energy, Heidenhofstraße 2, 79110 Freiburg, Germany
[2]Reinhold Schneider Straße 37, 79117 Freiburg, Germany
[3]Rathausstraße 17, 79232 March, Germany
Contact email: psb02@mail.aub.edu

ABSTRACT: A technology roadmap serves as a basis for strategic planning and decision-making by providing an in-depth analysis of the progress and milestones of technologies and changing market trends and by identifying actions and time frames to achieve higher performance, greater market penetration, and overall technology improvement. Another beneficial tool for a better technology assessment is the technology readiness level (TRL) system indicating the maturity of a technology in its developing stages. Our approach was to analyze the mentioned PV roadmaps' contents and introduce the existing TRLs. After analyzing the roadmaps, we emphasize on the International technology roadmap for photovoltaics (ITRPV) as our roadmap of choice, since 93% of it focused on technology projections. We propose a way to list technology roadmap topics, research and development fields and PV technology by using the ITRPV structure and content. The idea of combining the TRL concept to PV technologies is introduced supporting the introduction of a standardized "keyword". This keyword takes the form *"{Topic}-{R&D Field}-{PV technology}-TRL{XX}"*. In this paper a method is suggested to propose independent parameters affecting TRL for each of the materials, processes and products. In turn these parameters have threshold values for each TRL, thus allowing a quantitative way to assign TRLs. The effort to have a structured common keyword system for all entities and persons in photovoltaics research and development aims to contribute in the innovation management of PV.
Keywords: Photovoltaics, Roadmaps, Technology Readiness Level (TRL), innovation management

1 INTRODUCTION

Innovation management (IM) in photovoltaics (PV) enables efficient and effective technology research and development using qualitative and quantitative means. The main aim of the PV industry is to reduce costs, which in turn leads to further penetration of PV into the energy market. Roadmaps that take technology as a main criterion are convenient projection tools for the PV research and industry community, while TRLs give a maturity scale for technologies. The conjunction of roadmaps and TRLs contributes positively as an innovation management tool by planning the individual maturity of the development and relating it to the projections of the future. This paper aims to portray roadmaps and technology readiness levels as support tools in managing PV technology development.

The first section of the paper includes a brief overview of the roadmap concept in the PV industry, the method used to analyse the different roadmaps as well as an explanation of the technology readiness level concept. The second section includes the results of our PV roadmaps' content categories analysis, Fraunhofer ISE - TRL stages for PV cell production, and a new keyword proposal for standardizing PV technologies. The third and last section provides a modelling approach of how TRLs could be assigned by having reference values to independent parameters affecting TRL advancement. In other words, the paper gives an idea of how each PV technology present in roadmaps could be associated with a technology readiness level and contribute to the innovation management of upcoming technologies.

2 ROADMAPS AND TECHNOLOGY READINESS LEVELS IN PHOTOVOLTAICS

2.1 Photovoltaic roadmap overview:

The objective of a technology roadmap from an industry vantage point is to support in strategy development and R&D planning. Various PV roadmaps are already in existence, among those the International technology roadmap for PV (ITRPV), the International Energy Agency PV roadmap, SunShot initiative of the US Department of Energy, and the Chinese 12th five-year plan for the solar PV industry. Their focus might vary depending on country target, the principle is the same, provide a way to identify, evaluate, and research the appropriate path to achieve desired scientific and technological objectives. Roadmaps are widely used by PV manufacturers, suppliers and research institutes, making it a popular strategy planning tool aiming for comprehensive solutions. Some R&D groups establish internal PV roadmaps to structure their innovation goals and paths. For instance, Trina Solar has structured a roadmap for its solar cell production line starting from their simple p-n junction screen printed solar cell until aiming to reach the production of Tandem interdigitated back contact solar cells in the upcoming years [15]. Internal roadmaps act as guides for companies to picture their own development by setting their own goals.

In order to study the content of PV roadmaps mentioned above, four main focus categories were taken into consideration. These categories are technology, economy, climate, and legislation. Our definition of these categories is as follows:

- Technology: analysis of the technical development trajectories; example: silicon wafer thickness projections.
- Economy: financial analysis including cost calculations and financial projections; example: learning curves and cost estimates.
- Climate: environmental analysis; example: CO_2 equivalents of PV electricity production.

- Legislation: legal and policy analysis; example: renewable energy incentives, subsidies...

These categories were chosen due to their importance in the scope of PV development, as well as their availability in the roadmaps that were analysed. After identifying these categories in the roadmaps, the approach is to quantify these categories within each roadmap in the form of relative content. To get the results, table of contents and outlines of the roadmaps were checked. The number of pages encompassing the categories was divided by the total pages of the roadmap (excluding introduction and summary pages). This way can provide an approximate relative category content in each roadmap. For a more exact content analysis other methods could be considered.

2.2 TRL overview

According to DW. Engel, technology readiness shows the "degree of maturity of a technology, the extent to which a technology is appropriate for deployment in a real operational environment" [14]. Technology Readiness Level (TRL) was introduced by the NASA Goddard Space Flight center in the 1970's as a method to assess the readiness and risk of space technology [1]. It is a method that is based on a levelling approach to indicate a readiness levels of a technology in development. The current NASA TRL descriptions are defined into nine levels, from concept observations until flight mission readiness.

In addition to that, the European Commission in turn has established within the 'Horizon 2020' program TRL explanations tailored to assess 9-level TRLs and 10-level manufacturing readiness levels (MRLs) for technology development from basic principle observations to competitive manufacturing [4]. (Table I: lists the European Commission TRL levels). EU defined TRLs are seen as a beneficial tool for characterizing the maturity of technologies such as photovoltaics. By doing so, the comparison of different technology maturities is possible, in turn supporting R&D planning and guiding investors' decision making.

Table I: TRL levels as defined by the European Commission [4]

TRL	Description
1	Basic principles observed
2	Technology concept formulated
3	Experimental proof of concept
4	Technology validated in lab
5	Technology validated in relevant environment (industrially relevant environment in the case of key enabling technologies)
6	Technology demonstrated in relevant environment (industrially relevant environment in the case of key enabling technologies)
7	System prototype demonstration in operational environment
8	System complete and qualified
9	Actual system proven in operational environment (competitive manufacturing in the case of key enabling technologies; or in space)

3 RESULTS

3.1 Photovoltaic Roadmap concept

After studying the roadmaps mentioned in this report, we were able to approximate the relative content showing the points of focus of the different roadmaps. (Figure: 1). This quantitative approximation of the content in each roadmap aims to provide an easy guideline to target groups wishing to access PV roadmaps based on content focus discussed above. An indicative table (table: II) of the roadmaps' focuses was also created.

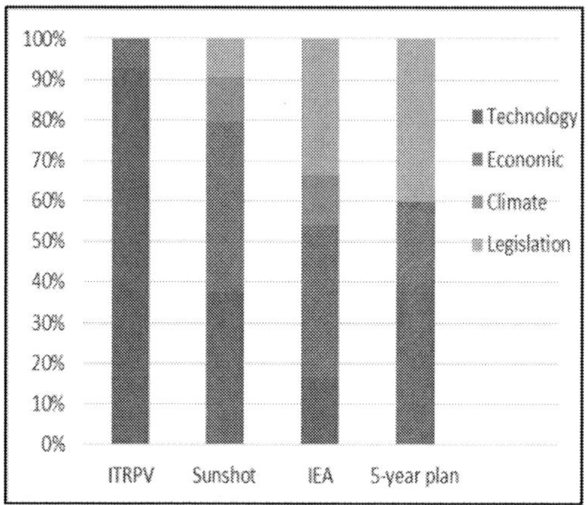

Figure 1: Relative content of discussed categories in different existing roadmaps

The results show that ITRPV is by far the main roadmap focusing on PV technology (~ 93% of content), with only ~7% focusing on cost reduction and learning curves. The rest of the roadmaps had a mosaic of different focus points as well as a difficulty to distinguish between content. However, the ITRPV had a consistency in its discussion about technological projections which makes it favourable in our case. A technological projection example from ITRPV is the decrease of Silver use per cell from 100 mg/cell in 2015 to 40 mg/cell in 2025. Another example is the increase of the module's anti-reflection coating lifetime from 12 years in 2015 to 30 years in 2025 [6].

Table II: Qualitative indication of discussed parameters in different PV roadmaps

	Parameters Discussed			
Roadmap	Technology	Economic	Climate (CO_2equ)	Policy /Legal
ITRPV	☼☼	☼	---	---
IEA	☼	☼	☼	☼
SunShot	☼	☼	☼	☼
5-yr plan	☼	☼	---	☼

Key: ☼☼=extensive discussion (>50%); ☼= brief discussion (>1%); ---: not discussed (0%)

Based on the ITRPV roadmap, a framework for structuring PV technology roadmaps is introduced. There are three sections covering the PV value adding chain (materials, processes, and products). The table (Table: III on the next page) gives an example of how technology roadmaps contents could be structured by ITRPV topics. [6] The content is not fix, other R&D fields and PV technologies could be added making the table dynamic and prone to modifications.

3.2 Photovoltaic TRL concept combined with the roadmap concept

In relation to the PV technology, the ITRPV color marking is used for selected parameters in order to describe the maturity of a technology: Green indicates that the technology is in use. Yellow means that an industrial solution is known but is not yet used in mass production. Orange means that an interim solution exists, but it is too expensive. Red indicates that there is no known industrial solution available. The approach of this maturity assessment is performed throughout the value chain (see Table IV). [6]

Table IV: ITRPV Color marking to depict the maturity of technologies. [6]

Green	Industrial solution exists and is being optimized in production
Yellow	Industrial solution is known but not yet used in mass production
Orange	Interim solution is known, but too expensive or not suitable for production
Red	Industrial solution is not known

For the purpose of analysis of photovoltaic technologies, Fraunhofer ISE introduced technology readiness levels that consists of ten different levels. The approach is in internal use since 2013 and has been slightly revised in the meantime to improve and accommodate for changes in the technology. In comparison to the other description by NASA or European Commission, it is more quantitative and uses the selected PV cell technology specialized parameters. The table below (table V) shows the Fraunhofer ISE TRLs.

Table V: Fraunhofer ISE - TRL stages for PV cell production

Process TRL	Description process TRL
TRL 0	Process simulation, functionality and economical potential demonstrated
TRL 1	Laboratory process, functioning and reproducible on small areas
TRL 2	Laboratory process, efficiency potential demonstrated potential on small areas
TRL 3	Laboratory process, functional and reproducible on industrial standard large and full areas*
TRL 4	Laboratory process, efficiency potential demonstrated on industrial standard large and full areas*
TRL 5	Process demonstrated as pilot process in environment suitable for production (throughput)
TRL 6	Process used as pilot process on tools suitable for production, > 1000 cells manufactured, sample modules manufactured and certified
TRL 7	Process used in pilot production with 6 hours/7 days on production tools
TRL 8	Process used in mass production with 24 hours/7 days on production tools
TRL 9	Process used in mass production with 24 hours / 7 days on production tools, production experience > 2 years

* industrial standard large area = (125x125 mm² or 156x156 mm²)

To have scientists and research centers informed about the status of a certain technology readiness level a keyword that contains information about a material, process or product along with the TRL level is a crucial tool. We propose to add it as a property of each published work. By this it is easier to track publications in certain R&D fields with a link to the technology readiness level. For that, certain codes for each research or technology have to be introduced and put in conjunction with the TRL. Figure 2 shows how the keyword might look like. A suggested sample code for p-type multi-crystalline cell production at pilot stage with more than 5000 cells produced could be "Prdct-Wafertyp-PmcSi-TRL06". To make the process objective, we encourage the formation of a committee of experts that approves and verifies whether the right TRL is given to technologies.

{Topic}-{R&D Field}-{PV technology}-TRL{XX}

Topic: Material, Process, or Product;
R&D Field: technology research field eg: wafer type;
PV technology: specific technology eg: p-type multicrystalline;
TRL: from 0 to 9

Figure 2: Keyword format

With these determined TRLs, the uncertainty in regards to performance, schedule, budget of a project of a technology can be reduced [13]. Therefore, managers can use the information to make decisions for the funding level needed, direction, size, timeline of the project with minimal risk.

Table III: Technology roadmap content listing structure based on ITRPV content.

Topic	R&D Fields	PV technology
Materials	Feedstock Technology	Fluidized bed reactor
		Upgraded metallurgical silicon
	Wafer Sawing Technology	Diamond wire sawing
		Other technologies eg.kerf less
	Wafer Thickness	Multi wafer approaches 150 µm
		Mono thickness reduction follows module technology (120µm)
	Silver (Ag) per Cell	reduction per cell
		Copper as substitute
		Lead free pastes
	Module Glass	polymer-based front cover
	AR-Glass Lifetime	Long term stability of Antireflection coating
	Cell Interconnection	Lead-free materials
		wire-based
		compatible with thinner wafers
		Low temperature approaches
	Module Encapsulants	Polyolefin
	Module Backsheets	Glass
		APA polyamide
Processes	Ingot Mass	mc–Si
		monoSi
		monolike-Si
		HPmc-Si
	Crystallization & Wafering	Crystal growth throughput
		Wafer sawing throughput
	Throughput Front-end / Back-end	tool throughput
	Throughput lamination / stringing	stringing time
		lamination encapsulants & lamination time
	J0 Values (p-type)	improved diffusion
		new pastes
		ramping of rear side passivation
	rear-side passivation	Al2O3 PECVD w/ capping layer
		Al2O3 ALD w/ capping
		SiOx PECVD w/ capping
	Emitter Sheet Resistance for n-type	Homogeneous diffusion
		Selective emitter techniques
	P- doping	Homogeneous phase diffusion
		Selective etch back
		Homogeneous ion implantation
		selective ion implantation
		selective laser doping
		selective printed dopant
	B diffusion	BBr3 thermal diffusion
		Ion implantation
		APCVD
	Finger width	Front side grid finger width below 30µm
	Metallization	Screen printing
		direct plating
		stencil printing
		plating of seed layer
		inkjet/aerosol/dispensing
	Busbar technologies	3 busbar technology
		4 busbar technology
		5 busbar technology
		busbarless
	Cell to Module Power Ratio	light management
		interconnection technologies and encapsulants
Products	Wafer Types	p-type mc-Si
		p-type mono-Si
		p-type HPmc-Si
		n-type mono-Si
		p-type monolike-Si
	Mono Crystallization Technology	Cz
		Continuous growth Czochralski (CCz)
	Cell Technologies	BSF
		PERC
		PERT
		Si-Heterojunction (SHJ)
		Back contact
		Si-based tandem
		monofacial c-Si
		bifacial c-Si
	Full Cells vs. Half Cells	full cell
		half cell
	Module Sizes	60 cells
		72 cells
		80 cells
	Junction-Box Functionality	Standard J-Box
		DC/DC converter
		Microinverter (DC/AC)
		Integrated circuit switch

Abbreviations:

AR- Glass: Anti-reflection glass
APA: polyamide
mc-Si: multicrystalline silicon
mono-Si: monocrystalline silicon
HPmc-Si: High Performance multi-crystalline
Al2O3: Aluminum oxide
PECVD: Plasma Enhanced Chemical Vapor Deposition
ALD: Atomic Layer Deposition
SiOx: Silicon oxides
BBr3: Boron Tribromide
APCVD: Atmospheric Pressure Chemical Vapor Deposition
Cz: Czochralski
BSF: Back Surface Field
PERC: Passivated Emitter Rear Contact
PERT: Passivated Emitter Rear Totally-diffused
DC: Direct current
AC: Alternating current

3.3 Modelling Approach

In order to assign a TRL, a quantitative model approach is proposed to have a better understanding of the status of a PV technology maturity. Independent parameters are assigned covering the whole TRL affecting aspects of the PV technology being studied. For the three topics 'materials', 'processes' and 'products' independent parameters are assigned. One can restrict the material influencing factors into three independent parameters namely: Availability, Quantity needed and Quality. The process parameters are: Throughput, Yield and Cost of Ownership. Having a satisfactory readiness level in the material and process of a specific technology one can restrict the parameter space of the product into four independent parameters namely: Area of the module, Production scale, Efficiency of energy conversion and Cost of material. Table VI lists the parameters aforementioned topics.

Table VI.: Table of parameters comprising materials, processes and products

Materials	Availability	Quantity needed	Quality	
Processes	Throughput	Yield	Cost of Ownership	
Products	Area of the module	Production scale	Efficiency	Cost of material

To further explain the modelling method, the four independent parameters of the products are used (the same could be done to parameters of the material and process). The product parameter space can be mapped to each TRL with a certain reference value for every level as shown in figure 3 with R being the quantitative reference threshold. This reference value can be determined by industry experts or via back logging to the history of technology to establish valid experience with what could be the threshold of a certain TRL. The superscript of R indicates the parameter and the subscript indicates the technology readiness level. It is important to mention that an advancement in one parameter doesn't mean an increase in the TRL unless all parameters pass their respective thresholds (referring to formula 1). The TRL of a product P having an Area P^A, production scale P^P, Efficiency P^E and Cost of material P^C can reach a TRL x if the parameters satisfy formula 1 stated below.

$$R^A \rightarrow \{0;1;2;...9\} : trl\left(P^A, P^P, P^E, P^C\right) = x \quad if \; P^i \geq R^i_x$$

$$\text{and } P^i < R^i_{x+1} \quad if \quad x < 9 \forall i \in \{A;P;E;C\} \forall R^i_x \in R$$

Formula 1: TRL mapping

$$\begin{pmatrix} R_0^A & R_0^P & R_0^E & R_0^C \\ R_1^A & R_1^P & R_1^E & R_1^C \\ R_2^A & R_2^P & R_2^E & R_2^C \\ R_3^A & R_3^P & R_3^E & R_3^C \\ R_4^A & R_4^P & R_4^E & R_4^C \\ R_5^A & R_5^P & R_5^E & R_5^C \\ R_6^A & R_6^P & R_6^E & R_6^C \\ R_7^A & R_7^P & R_7^E & R_7^C \\ R_8^A & R_8^P & R_8^E & R_8^C \\ R_9^A & R_9^P & R_9^E & R_9^C \end{pmatrix} \Rightarrow \begin{pmatrix} TRL0 \\ TRL1 \\ TRL2 \\ TRL3 \\ TRL4 \\ TRL5 \\ TRL6 \\ TRL7 \\ TRL8 \\ TRL9 \end{pmatrix}$$

$R^{parameter}_{TRL}$: *reference threshold for parameters A, P, E, &C at every TRL*

A: Area of Module

P: Production Scale

E: Efficiency

C: Cost of Material

Figure 3: Mapping TRL to reference thresholds and production parameters.

As mentioned before, the TRL review can be processed by a committee of experts. The keyword assignment can take part of the common review process applied on submitted papers for publication. Figure 4 sketches a control process that can be familiarised to research papers for assigning the TRL keyword introduced above (Figure 4).

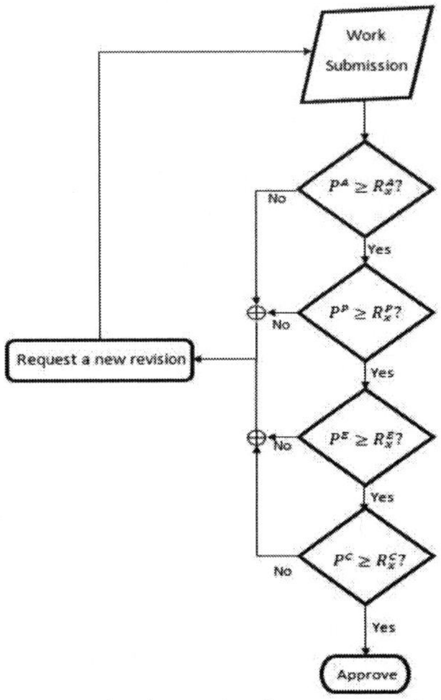

Figure 4: Product keyword assignment Control Process for publications.

4 CONCLUSION AND OUTLOOK

Innovation management is necessary for an efficient and effective innovation process in photovoltaics. Roadmaps and TRLs should remain essential tools used for strategy planning in photovoltaics. In this paper, a way of structuring technology roadmaps was proposed, derived from the ITRPV structure. Fraunhofer ISE TRL for PV cell production was introduced. In addition to that, a keyword is suggested to include the TRL level as a standard for publications. Finally, a modelling approach is proposed to link threshold references to production parameters for assigning TRLs.

After combining the notions of roadmaps and TRLs, future studies are requiring further analysis with empirical data taken from real photovoltaics R&D examples. The PV community has to keep on collaborating for the formation of standards. With the quantified TRLs, one can assign the status of a technology to all value-adding steps of the PV value chain. This in turn aids in establishing a learning curve of the market and support in evaluating the time needed for the transition between technology readiness levels. Hence, the time needed for the maturity of a certain technology. Consequently, one can draw a trajectory for a successful development of a project. Through empirical data research, sensitivities of each parameter on the total production cost can be obtained in the future. Relating TRL to invested money is benevolent for managing resources. Having the TRL defined levels as an assessment framework for current technologies helps in having better future projections, avoiding budget slips and minimizing risks.

5 ACKNOWLEDGEMENTS

The authors would like to acknowledge the contributions of all co-workers at Fraunhofer ISE, especially Stefan Rein, Jochen Rentsch and Sebastian Nold who co-initiated the Fraunhofer ISE TRL for crystalline silicon solar cell technology.

6 REFERENCES

[1] Azizian, N. Sarkani, S., Mazzuchi, T. A Comprehensive Review and Analysis of Maturity Assessment Approaches for Improved Decision Support to Achieve Efficient Defense Acquisition, paper presented at World Congress on Engineering and Computer Science, San Francisco, USA, (2009)

[2] Chinese 12th five-year plan for the solar PV industry, China's National Energy Administration (2012)

[3] EU Strategic Energy Technology Plan (SET Plan), EU Commission (2007)

[4] European Commision, Appendix G of Horizon 2020 – Work Program 2014-2015, (2014)

[5] Garcia, M.L. and Bray, O.H. Fundamentals of Technology Roadmapping. Strategic Business Development Department Sandia National Laboratories (1997).

[6] International Technology Roadmap for Photovoltaics, 6th Edition, Revision 1 (2015)

[7] SunShot Vision Study, U.S Department of Energy, National Renewable Energy Laboratory (2012)

[9] Technology Roadmap Solar Photovoltaic Energy, International Energy Agency (2014)

[10] T. Fleischer et al. / Technological Forecasting & Social Change 72 (2005) 1112-1121

[11] U.S Department of Energy. Technology Readiness Assessment Guide (DOE G 413.3-4A). Washington, DC: U.S. Government Printing Office. (2011)

[12] Kelly, P. and Kranzburg M. Technological Innovation: A Critical Review of Current Knowledge.San Francisco (1978)

[13] Mankins, J. Technology readiness and risk assessments: A new approach. Artemis Innovation Management Solutions LLC, Ashburn, VA, USA (2009)

[14] DW. Engel et al. / Development of Technology Readiness Level (TRL) Metrics and Risk Measures. US Department of Energy. (2012)

[15] P. Verlinden TRINA Solar, Challenges and Opportunities of High-performance Solar Cells and PV Modules in Large Volume Production (2015)

ROOFTOP PV POTENTIAL ESTIMATIONS: AUTOMATED ORTHOGRAPHIC SATELLITE IMAGE RECOGNITION BASED ON PUBLICLY AVAILABLE DATA

Kai Mainzer, David Schlund, Sven Killinger, Russell McKenna, Wolf Fichtner
Chair of Energy Economics, Karlsruhe Institute of Technology (KIT)
Hertzstr. 16, 76187 Karlsruhe, Germany
kai.mainzer@kit.edu
Tel.: +49 721 608-44589
Fax: +49 721 608-44682

ABSTRACT: This paper presents a new approach for the estimation of rooftop photovoltaic potentials. It extracts building footprints from OpenStreetMap and combines these with orthographic satellite images from Bing Maps. These images are analyzed using image recognition techniques in order to identify the ridge line as well as structures like chimneys and windows on each building's roof. In combination with statistical assumptions about the roof's inclination angle, the exact shape, size and orientation of the partial roof areas can be calculated for each building. For the resulting areas, an irradiance simulation is conducted and combined with a PV electricity yield model in order to calculate electricity generation profiles with a high spatial and temporal resolution. The main advantages of this approach over methods which typically employ 3D models are, that no resources are required for data acquisition and it can be applied worldwide. The method is applied to German cities and an evaluation indicates a success rate of over 70%.
Keywords: Rooftop, Potential, Modelling

1 INTRODUCTION AND RELATED WORK

For the future development of photovoltaics (PV) and as a quantitative foundation for energy political discussions and decisions, the PV potential in a given area is usually a critical piece of information. Especially for cities or other regions which develop climate protection or energy concepts, the determination of these potentials represents an indispensable first step.

Because of this, a large number of methods for the estimation of PV potentials have been presented thus far; [1] and [2] give an overview. PV potential studies based on readily available statistical and meteorological data (e.g. [3]) can be conducted for large areas. Their geographical resolution, however, is usually limited to the city level. Other methods that rely on Light Detection And Ranging (LiDAR) data (e.g. [4]), stereo image photogrammetry (e.g. [5]) or 3D city models (e.g. [6]) allow for higher resolutions and can thus deliver detailed results down to the single-building-level. The data requirements for these methods, however, are usually quite extensive, involving for example surveying flights in order to obtain aerial images. This data acquisition can be quite expensive and time consuming and is usually not carried out for larger areas. Especially in the case of smaller municipalities, this kind of data is often not available.

2 PURPOSE OF THE WORK AND APPROACH

This paper presents a new approach for the estimation of PV potentials, which aims to provide high resolution results, i.e. down to an individual building level, but without the need for extensive data gathering. It employs solely publicly available data and can thus be used worldwide and without additional cost. **Figure 1** illustrates the presented approach and the steps 1-8 are subsequently explained in detail.

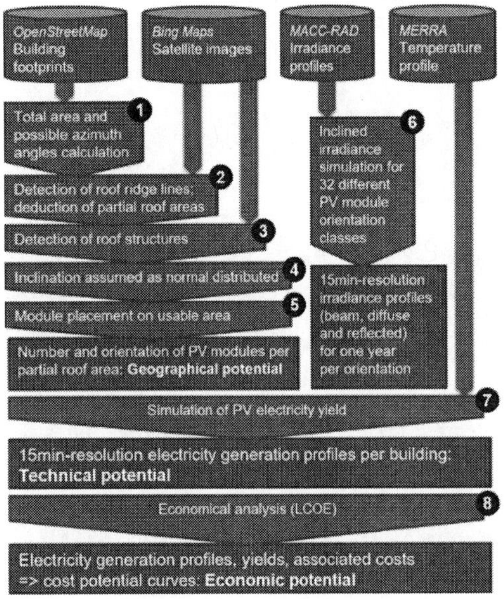

Figure 1: Overview of the presented approach

1. Based on building data extracted from OpenStreetMap [7], the size and position of all building footprints in the area of interest are determined. These are used to clip the satellite images to the correct shape. Additionally, the azimuth angles of the building outlines (i.e. the building walls) are determined as a basis for the angles of possible roof ridge lines, as these are usually parallel to the building walls. Very large (e.g. industrial) buildings are not further analyzed and assumed to have flat roofs.
2. For each building, the satellite image (source: Bing Maps [8]) covering the buildings' area is analyzed using image recognition techniques, yielding the size and shape of each partial roof area. This is done by first applying some pre-processing (which entails a bilateral filter for edge-preserving noise reduction,

a color filter which creates a black & white image by weighted adding of the red, green and blue color components, and histogram equalization, which enhances the overall contrast of the image). Next, the Canny Edge Detection algorithm and a Hough Line Transformation are applied. These functions are provided by the open computer vision library OpenCV [9], algorithmic descriptions can be found, for example, in [10]. As a result, these algorithms yield a number of prominent lines on the satellite image. These lines are further analyzed using logical conditions in order to determine which line (if any) represents the roofs' ridge line. For example, a ridge line usually has the same angle as the building walls, runs through the middle of the building and separates areas of different brightness in the image. If the ridge line is found with a sufficient degree of certainty, it can be used to deduce the partial roof areas (which face in different azimuth directions) of the building. An example for this process can be seen in **Figure 2**. In cases when the ridge line cannot be found, e.g. when the satellite image shows a flat roof building, is outdated, too small or has a weak contrast, an empirical distribution function is applied to classify the building. For slanted roofs, it is then assumed that the longest wall of a building is parallel to the roofs ridge, and the roof is split into partial areas based on that assumption.

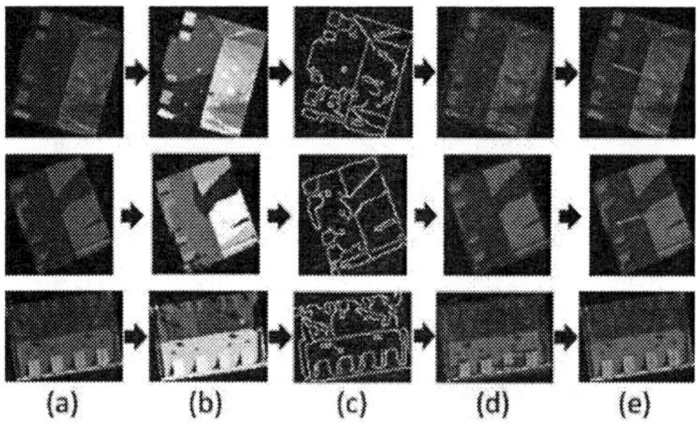

Figure 2: Process of roof ridge line detection: original satellite image (a), after pre-processing (b), after Canny Edge Detection (c), after Hough Line Transformation (d), final determined azimuth angles (e). Source: own depiction with map data from OpenStreetMap, image data from Bing Maps.

3. In most cases, only part of the roof area can be used for PV applications, since most roofs contain structures like chimneys, windows, etc. In previous PV potential studies, this fact has usually been accounted for by subtracting a fixed share of the roof area. The method presented here uses the satellite image to identify these roof structures. To achieve this, methods for contour detection and polygon approximation from OpenCV are employed in order to identify objects on the partial roof areas. All objects that fulfil certain criteria (based on size and shape) are subtracted from the usable area. An example of the roof structure detection can be seen in **Figure 3**.

Figure 3: Examples for roof structure detection. Source: own depiction with image data from Bing Maps.

4. For each building, the roof tilt angle is estimated based on a normally distributed random variable about a mean of 35°, with a standard deviation of 15° (assumption based on empirical building data). As a result, all partial roof areas with their respective tilt and azimuth angles are given.
5. In the next step, an algorithm runs incrementally over the usable area and attempts to fit as many PV modules as possible within each partial roof area. The result of such a module placement can be seen in **Figure 4**.

Figure 4: Result of PV module placement in an urban area. Source: own depiction with image data from Bing Maps.

6. The incoming irradiance on the modules' surfaces is simulated for 32 possible module orientations (8 azimuth and 4 tilt classes). Hence, the sun position at the location of interest is calculated for each 15 minute timestep for one year. This position is then combined with irradiance data (direct and diffuse irradiance on a horizontal plane, provided by [11]) in order to simulate the beam, diffuse and reflected irradiance components (calculation based on [12] and [13]) on the tilted plane for each orientation class. Shading (e.g. by other buildings) cannot be considered here, because no data on the height of buildings is available.
7. The placed modules are then combined with their respective irradiance profiles in a PV yield model (adapted from [14], based on [15], [16] and [17]) which also considers module efficiency (crystalline silicon technology is assumed), temperature [18], inverter efficiency based on workload, as well as other factors to determine the electricity generation potential for each roof plane and time step.
8. In the last step, an economic analysis is conducted. Using investment, operating costs, lifetime and yields, the levelized costs of electricity (LCOE) are calculated by dividing discounted lifetime costs by discounted lifetime generation for all potential PV installations (cost factors based on [19]). By aggregating the possible yearly electricity generation and sorting by ascending LCOE, a cost-potential curve can be generated from these calculations. An example can be seen in **Figure 5**. The economic potential could now be derived by defining a maximum LCOE and selecting only those PV installations with lower costs. However, it should be considered that the economic viability of PV

installations is usually also dependent on further individual factors, e.g. the share of own consumption.

3 RESULTS AND DISCUSSION

The described method is implemented in Java and highly automated, and can thus be used to automatically analyze a large number of buildings for any given area. The method has been applied to identify the PV potential for various municipalities.

The figures below illustrate an exemplary result for the city of Karlsruhe: the derived usable roof areas (**Figure 5**), the cost potential curve (**Figure 6**) and the electricity generation profile, if the complete technical potential were to be exploited (**Figure 7**).

Figure 5: Usable roof areas in Karlsruhe.

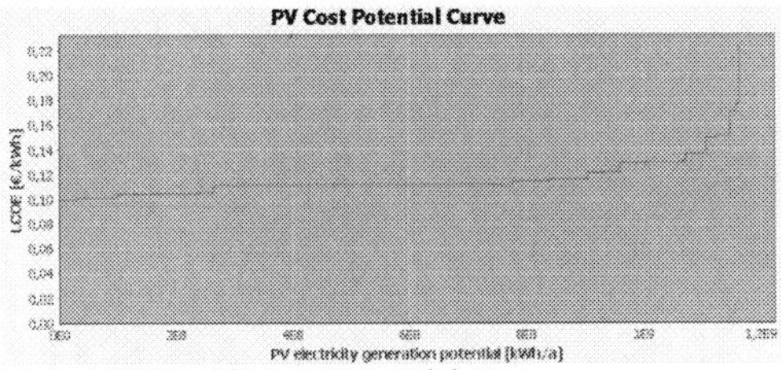

Figure 6: Resulting cost potential curve.

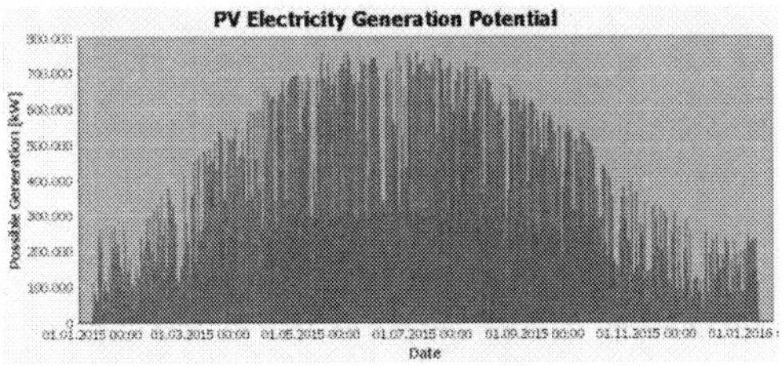

Figure 7: Resulting electricity generation profile.

Unsurprisingly, we can observe that the total potential depends mainly on the number of buildings in a region, whilst the cost range depends on the local irradiance.

The method has been validated for two different regions. First, a manual validation has been carried out by visually rating the resulting roof ridge lines found by the algorithm for 1,000 randomly selected buildings in Karlsruhe, Germany. A second validation (after applying some improvements, e.g. pre-processing of images and parameter optimizations) has been carried out by comparing the partial roof areas derived by this methodology for the German municipality Miesbach with data from a 3D city model. This model has been created using LiDAR data and was kindly provided by the Bavarian State Office for Survey and Geoinformation [20]. In this case, 589 buildings could be compared. The results from both validations are listed in Table I.

Table I: Results from the validations

Region	Karlsruhe	Miesbach
No ridge line	277 (28%)	186 (32%)
Correct ridge line	436 (44%)	335 (57%)
Wrong ridge line	287 (29%)	68 (12%)
Total	1.000	589

The second validation suffers from a generally worse image quality (typical for rural areas), but benefits from the above mentioned methodological improvements that have been conducted after the first validation.

The method still fails in the case of poor image quality (e.g. outdated imagery, low resolution images or weak contrasts) and when structures on the roof (e.g. windows or existing PV modules) are falsely identified as the roof's ridge line. Currently, only four-sided buildings are processed, and only saddle roofs are detected.

From the validations we conclude that the method presented here has a failure rate (wrong ridge line) of 12 to 29% which is significantly better than the ~50% which would be achieved by pure guessing based solely on the building outline.

The novelty of this approach is the automated extraction of building ridge lines as well as the detection of roof structures, using image recognition techniques on orthographic satellite images (see steps 2 and 3 above). A similar method has been described in [21], with some differences in the algorithms used for roof detection and without the use of building footprints. Additionally, that study relies on "very-high-resolution aerial images" which are not publicly available. To the authors' knowledge, a comparable method has never been applied for the estimation of PV potentials. Thus we conclude that the presented methodology could improve the quality and extent of PV potential estimations for urban areas in the absence of extensive data.

With the high detail of the results this method offers, it can be used to identify the PV potential as an input for energy system models that work with high geographic and temporal resolutions. This way, these models can determine the optimal investments under given restrictions and objectives, e.g. they could select the optimal PV installations that should be built in order to match the local demand and reduce the strain on the power grid or to replace as much fossil power production as possible. An example for such a model is described in [22].

4 CONCLUSION AND OUTLOOK

The validation has shown that the developed method is able to automatically determine and economically assess the technical potential for roof-mounted PV on buildings, in any area of the world for which satellite images, building outlines and climate data are available, with a good degree of accuracy.

Future work will focus on further validating the algorithm and improving its parameters by comparing against larger sets of 3D model buildings and iterating over different parameter combinations. A further extension of the method could be achieved by automatically identifying existing PV modules based on the satellite images.

5 REFERENCES

[1] Schallenberg-Rodríguez J. Photovoltaic techno-economical potential on roofs in regions and islands: The case of the Canary Islands. Methodological review and methodology proposal: Renewable and Sustainable Energy Reviews 2013;20:219–39.

[2] Martín-Chivelet N. Photovoltaic potential and land-use estimation methodology: Energy 2016;94:233–42.

[3] Mainzer K, Fath K, McKenna R, Stengel J, Fichtner W, Schultmann F. A high-resolution determination of the technical potential for residential-roof-mounted photovoltaic systems in Germany: Solar Energy 2014;105:715–31.

[4] Jakubiec JA, Reinhart CF. A method for predicting city-wide electricity gains from photovoltaic panels based on LiDAR and GIS data combined with hourly Daysim simulations; 2013, accessed 13 May 2013.

[5] Wittmann H, Bajons P, Doneus M, Friesinger H. Identification of roof areas suited for solar energy conversion systems: Renewable Energy 1997;11(1):25–36.

[6] Strzalka A, Alam N, Duminil E, Coors V, Eicker U. Large scale integration of photovoltaics in cities: Applied Energy 2012;93:413–21.

[7] OpenStreetMap-Contributors. OpenStreetMap; 2016, http://www.openstreetmap.org/, accessed 20 June 2016.

[8] Microsoft. Bing Maps; 2016, maps.bing.com.

[9] Bradski G. The OpenCV Library: Dr. Dobb's Journal of Software Tools 2000.

[10] Burger W, Burge MJ. Digital Image Processing: An Algorithmic Introduction Using Java. Springer London, London; 2008.

[11] MINES ParisTech / Transvalor S.A. MACC-RAD: Time series of solar radiation data; 2015, http://www.soda-pro.com/web-services/radiation/macc-rad.

[12] Perez R, Ineichen P, Seals R, Michalsky J, Stewart R. Modeling daylight availability and irradiance components from direct and global irradiance: Solar Energy 1990;44(5):271–89.

[13] Quaschning V. Regenerative Energiesysteme: Technologie - Berechnung - Simulation. 8th ed. Hanser, München; 2013.

[14] Schubert G. Modellierung der stündlichen Photovoltaik- und Windstromeinspeisung in Europa. 12th ed., Graz, Austria; 2012.

[15] Drews A, Keizer AC de, Beyer HG, Lorenz E, Betcke J, van Sark W, Heydenreich W, Wiemken E, Stettler S, Toggweiler P, Bofinger S, Schneider M, Heilscher G, Heinemann D. Monitoring and remote failure detection of grid-connected PV systems based on satellite observations: Solar Energy 2007;81(4):548–64.

[16] Macêdo WN, Zilles R. Operational results of grid-connected photovoltaic system with different inverter's sizing factors (ISF): Prog. Photovolt: Res. Appl. 2007;15(4):337–52.

[17] Huld T, Gottschalg R, Beyer HG, Topič M. Mapping the performance of PV modules, effects of module type and data averaging: Solar Energy 2010;84(2):324–38.

[18] MINES ParisTech / Transvalor S.A. MERRA re-analysis Web service; 2015, http://www.soda-pro.com/web-services/meteo-data/merra.

[19] Wirth H. Recent Facts about Photovoltaics in Germany, Freiburg; 2016.

[20] Bavarian State Office for Survey and Geoinformation. 3D-Gebäudemodell; 2015, http://www.ldbv.bayern.de/produkte/3dprodukte/3d.html, accessed 17 June 2016.

[21] Hazelhoff L, With P de. Localization of buildings with a gable roof in very-high-resolution aerial images: Visual Information Processing and Communication II 2011(7882 788208-1).

[22] Killinger S, Mainzer K, McKenna R, Kreifels N, Fichtner W. A regional optimisation of renewable energy supply from wind and photovoltaics with respect to three key energy-political objectives: Energy 2015.

PRO-PV LOCAL BUILDING POLICY – STATE OF PROGRESS OF THE LYON-CONFLUENCE SOLAR CITY PROJECT

Gaiddon, B.[1], De l'Epine, M.[1], Valentin, M.[2], Vignali, E. [2], Lapray, K.[3], Zanni, O. [3],
[1]Hespul, 14 place Jules Ferry, 69006 Lyon, France
email: bruno.gaiddon@hespul.org , melodie.delepine@hespul.org
[2]SPL Lyon-Confluence, 73 rue Smith, 69002 Lyon, France
email: mvalentin@lyon-confluence.fr , evignali@lyon-confluence.fr
[3]TRIBU Lyon, 103 avenue de Saxe, 69003 Lyon, France
email: klapray@tribu-concevoirdurable.fr , lyon@tribu-concevoirdurable.fr

ABSTRACT: Following the conclusions of the successful task on Urban scale PV systems (Task 10) of the Photovoltaic Power System Program (PVPS) of the International Energy Agency (IEA), a pro-PV local building policy has been set-up for the purpose of the Lyon-Confluence urban project, a 150 ha redevelopment in the city center of Lyon, France, in order to make real-estate developers and architects design and build low energy buildings equipped with PV systems. This pro-PV local building policy was first tested in 2004 within the EC funded CONCERTO initiative by SPL Lyon-Confluence, the public company in charge of this redevelopment, for the construction of the first dwellings and office spaces of the area (A, B and C blocks). Now, with 10 blocks of the Lyon-Confluence area already equipped with one or several PV systems for a total power of 1.1 MW, this pro-PV local building policy has been improved and the use of PV will increase with more and larger systems. Thus, 6 other buildings will soon be equipped with PV systems for an additional power of more 1,2 MWp that will make the Lyon-Confluence area a significant solar city showcase.
Keywords: PV Market, Solar Architecture, Building Integrated PV (BIPV), R&D and Demonstration Programmes

1 THE SOLAR CITY CONCEPT

The concept of solar city was addressed in detail from 2004 to 2009 by experts from many countries involved in Task 10 of the Photovoltaic Power System Program (PVPS) of the International Energy Agency (IEA) related to Urban Scale PV applications [1].

Conclusions of the IEA PVPS Task 10 have been summarized in a book published in 2009 [2] with a detailed analysis of success factors and potential problems of urban-scale PV systems, based on the feedback from urban areas where the installation of significant amounts of PV had been completed or was planned. In all of the areas reviewed in this book the quantity of PV is significant, impacting substantially on the area where it is located. This is the case for instance of two major urban projects in the Netherlands both of which are considered to be worldwide best practices:

- The Waterkwartier district of the Nieuwland expansion area of the city of Amersfoort completed in 1999 (1 MW),
- The 'City of the Sun' in Heerhugowaard, Alkmaar and Langedijk completed in 2009 (5 MW).

Figure 1: The 'City of the Sun' in Heerhugowaard, in the Netherlands built in 2009 (5 MW)

Such world best practices have been used as background for the specification of the Lyon-Confluence urban project.

2 THE LYON-CONFLUENCE URBAN PROJECT

The goal of the Lyon-Confluence project is to redevelop a former industrial area of 150 ha in the city center of Lyon and to build 1.000.000 m2 of new dwellings, offices and shops by 2030 [3]. The district is divided into 3 different areas:

- An area of 74 ha with 600.000 m2 of existing buildings called "Sainte-Blandine" (green area below),
- 400.000 m2 of new buildings built from 2003 to 2016 on 41 ha in the western part, near the Saone river (ZAC 1 - red area below),
- 400.000 m2 of new buildings to be built from 2016 to 2030 on 35 ha in the eastern part near the Rhône river (ZAC 2 - blue area below).

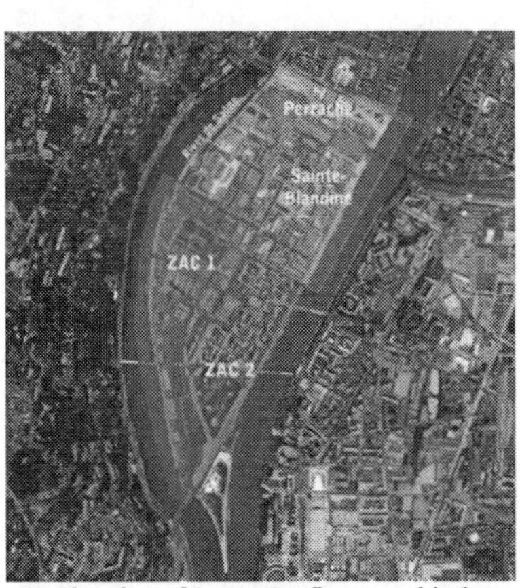

Figure 2: Site plan of Lyon-Confluence with the existing area (in green) and the 2 new areas (in red and blue).

On top of this global objective, this project has very ambitious targets in terms of sustainability that are described in a Sustainable Action Plan design with WWF[4]. One of the most ambitious goals set in this action plan is the Zero Carbon objective defined as follow: the 1 million m2 of floor area of new buildings to be built in the area by 2030 should not lead to any increase of the CO2 emissions of the area.

In order to reach this goal, SPL Lyon-Confluence, the public company in charge of the redevelopment of the area, has undertaken several concrete measures to:

- Drastically reduce the consumption of new buildings in the area,
- Refurbish the existing buildings of the Perrache/Sainte-Blandine district located in the north of the area,
- Massively install, on almost each new block, renewable energy systems such as wood chip boilers, solar thermal systems and PV systems.

This article provides information on the tasks undertaken by SPL Lyon-Confluence to make PV systems installed in this area by third parties such as real estate developers.

2 FIRST PV SYSTEMS IN THE AREA

In 2004, SPL Lyon-Confluence launched an international competition for the construction of the first 3 blocks of the area for a total of approx. 80.000 m2 composed of dwellings, offices and shops (A, B and C blocks). For the first time in France, this international design competition included a section on energy efficiency and renewable energy systems. The environmental engineering office TRIBU defined the environmental and energy targets of this group of buildings:

- Consumption for heating and domestic hot water under 60 kWh/m2/year,
- 80% of the consumption provided by renewable energy systems.

At that time this was less than 50% of the energy limit than written in the national thermal regulations and was made possible with the financial support of the European Commission within the CONCERTO initiative (€4 million of funding provided) [5].

Three real-estate developers associated with world-class architects such as Massimiliano Fuksas, MVRDV and Tania Concko were selected to build these first blocks. These real-estate developers had to comply with the local building regulations set out by SPL Lyon-Confluence and build efficient buildings powered mainly by renewable energy systems: wood chip boilers, solar thermal systems and also PV systems (see table I and figure 3).

Table I: features of renewable energy systems installed on the first building blocks of the area

	A Block	B Block	C Block
Wood boilers (kW)	540	2x540	540
Solar thermal systems (m2)	-	240	250
PV systems (kWp)	**79**	**121**	**65**

Figure 3: Architecturally integrated PV systems of A block (79 kWp)

This demonstration project had a significant effect on other new buildings built in the area since several real-estate developers choose to install PV systems on there buildings despite not being obliged to and not receiving specific subsidies.

3 PV POTENTIAL OF THE AREA

In order to understand how much PV could be installed on existing buildings of the area, SPL Lyon-Confluence undertook, in 2009, a PV potential study, a first of its kind in France. The main deliverable of this PV potential study was a solar cadaster of the area designed in partnership with other experts from Europe within the EC funded POLIS project [6].

Figure 4: Map of the area with PV potential in kWp for the existing building stock (colour code: the darker the more share of the roof can be used)

The PV potential study concluded that most of the housing blocks where not suited for the installation of PV systems due to the complex decision making process of jointly-owned buildings and the small size of each PV system. However the study concluded that 1,2 MW of PV could be easily installed on 8 existing public buildings such as primary schools, high schools and sports halls. The PV potential study lead to the fostering of several PV projects, such as the one on the gymnasium currently under design (206 kWp).

4 POSITIVE ENERGY BUILDINGS

In 2011, SPL Lyon-Confluence decided to go a step further and to launch an international design competition for the construction of a 12.300 m2 positive energy building on a 3.400 m2 piece of land called P plot. The team Bouygues Immobilier/SLC (real-estate developer) in association with Kengo Kuma from Japan (architect) was selected by the jury for the construction of this group of 3 buildings called HAKARI, with dwellings, offices places and shops. In order to build a positive energy building to respect the requirements set out by SPL Lyon-Confluence, the selected team decided to equip this building with 2 kinds of PV systems: a see-through PV facade on the building located in the middle of the block (21 kWp) and a rooftop PV system on the roof of each of the 3 buildings (168 kWp) [7]. The official opening ceremony of this building took place on Thursday 17 September 2015.

Figure 5: HIKARI, a positive energy building with a PV façade (21 kWp) and a rooftop PV system (168 kWp)

This demonstration project was also judged successful and the positive energy requirements set by SPL Lyon-Confluence for this block has been replicated to future blocks of the area.

4 PV SYSTEMS UNDER DESIGN

At the time of writing, 6 new PV systems for a total of a 1,2 MWp are under design. Two of these should be installed on existing buildings as a direct result of the PV potential study undertaken in 2009:

- A 206 kWp PV system on the flat roof a gymnasium (see figure 6),
- A 250 kWp PV system on the sloped roof of an office building that will be dedicated to French Tech companies. For this project, a specific study is underway within the new Horizon 2020 Smart Cities and Communities program [8] to check if self-consumption of the energy produced could be a new business model for the operation of PV systems.

Also, there are in this area 2 new blocks under construction with the same energy performance requirements as the HIKARI buildings, i.e. positive energy buildings, on which architecturally integrated PV systems are planned:

- A 427 kWp system for the A3 block (28.000 m2) called – Ynfluence and built by ICADE,
- A 250 kWp system for the B2 block (12.000 m2) built by OGIC (see figure 7).

Figure 6: future PV system of the gymnasium (206 kWp)

Figure 7: artist impression of future PV system of the B2 block (250 kWp)

5 SUMMARY OF THE STATE OF PROGRESS AND PERSPECTIVE

Today, as a direct outcome of the pro-PV local building policy set up by SPL Lyon-Confluence, 10 blocks of the Lyon-Confluence area already equipped with one or several PV systems for a total power of 1.1 MW. Also, thanks to the improvement of this local policy that now requires real-estate developers to design and build positive energy buildings, and building on other local initiatives, 6 additional PV systems for a total of 1,2 MWp are currently under design. Thus, as the realistic PV potential of the area is assessed at 6,8 MW, the total PV power in this area will reach 2,2 MW once these PV systems are installed, one third of the total potential will be reached (see figure 8)

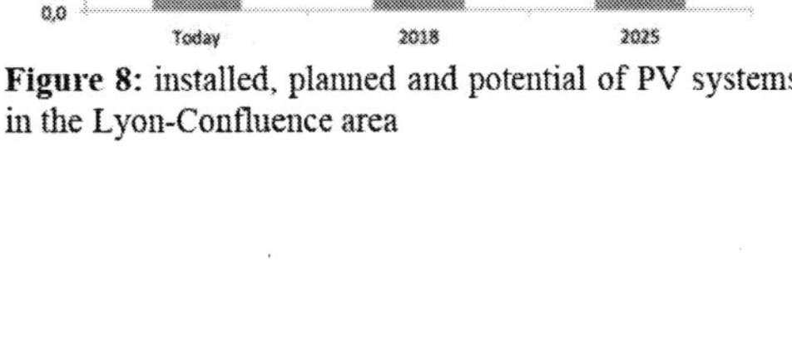

Figure 8: installed, planned and potential of PV systems in the Lyon-Confluence area

Lyon-Confluence become a significant solar city showcase, in the same way as the government district in Berlin, Germany, the Nieuwland project and the 'City of the Sun' project in the Netherlands.

Figure 8: Site plan of the Lyon-Confluence area with existing PV systems and PV systems under design

7 ACKNOWLEDGEMENTS

Authors wish to thank all experts involved in Task 10 of the Photovoltaic Power System Program (PVPS) of the International Energy Agency (IEA) related to Urban Scale PV applications.

Authors also wish to thank the European Commission for the funding received to demonstrate first of its kind buildings or processes in this area through several Framework Programmes such as FP6 (CONCERTO - Renaissance project), FP7 (Very low energy buildings - NEXT-Buildings project) and Horizon 2020 (Smart Cities and Communities – SMARTER TOGETHER project).

8 REFERENCES

[1] www.iea-pvps.org
[2] Gaiddon, B., Kaan, H., Munro, D., '*Photovoltaics in the Urban Environment, Lessons Learnt from Large-scale Projects*', ISBN 978-1-84407-771-7, Earthscan, London, 2009
[3] www.lyon-confluence.fr
[4] http://wwf.panda.org
[5] http://www.renaissance-project.eu
[6] http://www.polis-solar.eu
[7] Gaiddon, B., '*Report on the P Plot overall construction process*' Deliverable D5.1 of the NEXT-Building project, Lyon, 2015
[7] www.smartertogether.eu

PROGRESS OF SOLAR PHOTOVOLTAIC SYSTEMS IN INDIA

Saravanan Vasudevan[1], Arumugam Murugesan[2], Venkatesan Radha[3]
[1,2]Department of Electrical & Electronics Engineering,
Arunai Engineering College, Tiruvannamalai - 606 603, Tamil Nadu, India.
[3]Infineon Technologies AG, Neubiberg -85579, Germany.
[1,2]Phone: 04175 255101, +91 9994062645, +91 9443237011, [1,2]Fax: 04175 237789
[3]Phone: +49 89 234 89104, +4917688262365, [3]Fax: +49 89 234 9551929
Email: [1]vsaranaec@yahoo.co.in, [2]drmarumugam@yahoo.com, [3]Venkatesan.Radha@infineon.com

ABSTRACT: Energy independence is the biggest freedom a country should have. India can achieve this in the power sector by giving importance to the renewable energy. Especially the recent development is found to be much more promising for the solar photovoltaic means of power generation through various initiatives by government policies, public-private-partnerships and investments. This paper presents the perspectives and insights of solar photovoltaic systems in India by analyzing information reported in various literatures, government and International agencies.

Keywords: India, Photovoltaic, National Solar Mission, National Clean Energy Fund, Initiatives

1 INTRODUCTION

Renewable energies can substantially meet the increasing energy demands and mitigate greenhouse gas emissions. They will also provide direct and indirect social benefits such as reducing the health and environmental impacts associated with the use of fossil and nuclear fuels, thereby improving standard of living, educational opportunities, creating jobs, reducing poverty, increasing gender equality and ensuring energy security. Solar photovoltaic system is a reliable technology option to realize the shift to a decarbonised energy supply with a significant potential for long term growth nearly in all world regions, due to geographical advantage and availability of sun. This paper presents the perspectives and insights of solar photovoltaic systems in India through various initiatives by government policies, public-private-partnerships and investments in support for the growth of solar photo voltaic systems.

The Government of India have recently announced a massive renewable power production target of 175 GW by 2022; this comprises generation of 100 GW from solar, 60 GW from wind, 10 GW from biomass, and 5 GW from small hydro power projects [1]. However, the estimated solar power potential in India is about 748.98 GWp [2]. The breakup of year wise solar power target is listed in Table: I

Table I: Revised target (Year wise) to achieve Jawaharlal Nehru National Solar Mission

Year	Rooftop (in MW)	Ground Mounted Solar Power Projects (in MW)	Total (in MW)
2015-16	200	1,800	2,000
2016-17	4,800	7,200	12,000
2017-18	5,000	10,000	15,000
2018-19	6,000	10,000	16,000
2019-20	7,000	10,000	17,000
2020-21	8,000	9,500	17,500
2021-22	9,000	8,500	17,500
Total	40,000	57,000	93,237 *

*6762.853 MW commissioned up to 31.03.2016
Source: Ministry of New and Renewable Energy, Government of India, 31.03.2016

Almost all the state governments in India have released their state renewable energy policy and the stipulated targets. Also electricity regulators have issued regulatory and tariff orders. Such proactive action at the apex level has resulted in a commendable penetration of renewable energy in the overall Indian Energy Mix [3]. Various policy measures supporting capacity addition of solar projects in India are National Action Plan of Climate Change, National Solar Mission, State policies, Renewable Purchase Obligations (RPO) and Renewable Energy/Green Certificates. Activities like solar radiation resource assessment/ forecasting etc. are carried out for promoting solar energy in India. Many States have come up with their own solar policy and RPO's in compliance with their priorities and targets. [4]

Global investment in renewables rose from USD 55 billion in 2004 to more than USD 260 billion in 2014, with renewable power growing by 85% ,to constitute 30% of installed capacity. In India, already renewable investments are made approximately upto 5.5 billion USD and it is expected to make additional investments of 80 billion USD by 2022.[5]

Solar power comes under various programmes such as Roof top PV & Small Solar Power Generation Programme (RPSSGP) by Indian Renewable Development Agency, various batches and phases of JNNSM, NTPC Vidyut Vyapar Nigam Limited and Solar Energy Corporation of India (SECI) through schemes like state specific bundling scheme, Viability Gap Funding (VGF) and Generation Based Incentive (GBI) and so on from National Clean Energy Fund. The break-up of the 100 GW solar power target set by the Government of India for 2022 have to be made through the framework as given in Fig.1.

2 PV POLICIES IN INDIA

Energy policy is a strategy in which government decides to address the issues of energy development along with the development of the energy industry to sustain its growth; including energy production, distribution and consumption. [6] IREDA had released the compendium of state government policies on renewable energy sector in India.

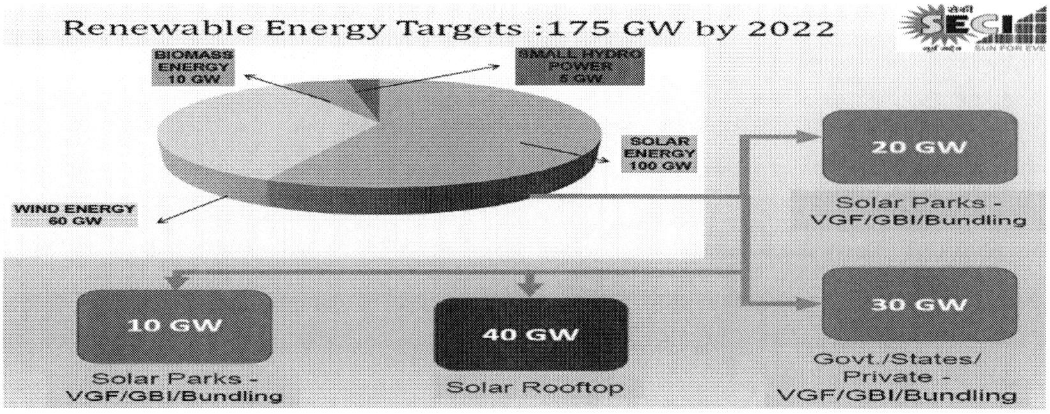

Figure 1: Break-up of the 100 GW solar power target and the framework set by the Government of India for 2022

It contains a comprehensive summary of all States/Union Territories policies on various renewable sectors, covering the aspects related to land acquisition, power evacuation facilities, frame work for grid connectivity and other financial aspects. A detailed report by IREDA contains the combined summary of CERC & SERCs Regulations / Tariff Orders of various states for solar PV and solar thermal power projects. It gives the details of Tariff/Control period, Financial/Normative parameters, CDM benefits, subsidy/incentives, taxes & duties, evacuation facilities, and transmission/wheeling/banking charges.[7] Shreemat Pandey [8] detailed the various determinants viz solar policy, responsibilities of state nodal agency and effective environmental governance of the state of Rajasthan, India. Anjali Bhide et.al [9] discussed the National Electricity Policy 2005 and Electricity Act 2003 and issues related to renewable based rural electrification systems in India. Gireesh Shrimali et.al [10] had discussed about various perspectives such as project financing, certification, metering, bench marking of grid connected and off grid solar policies. Fig. 2 describes the weighted Average price (Rs./kWh) of solar power generated through Feed in tariff (FiT), tendering and Viability Gap Funding/Generation based incentive scheme.[11].

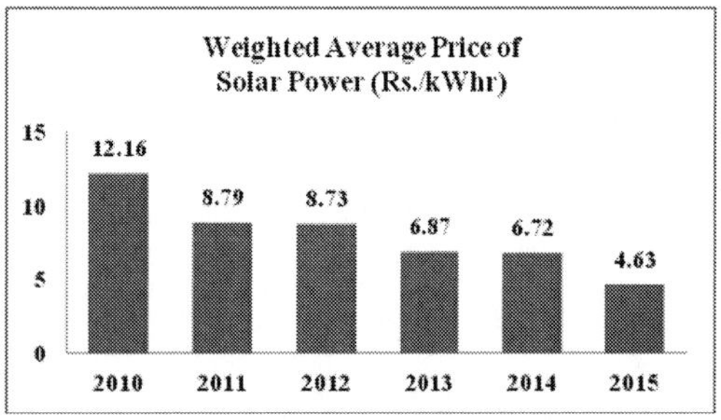

Figure 2: Price of solar power reduction to achieve grid parity

3 SOLAR PV – STATUS IN INDIA

India's National Tariff Policy was amended in January 2016 to prescribe long term growth trajectory of solar specific renewable purchase obligations (RPO) set as 8 % by 2022 for all the states and union territory in India to safeguard the environment and to promote renewable energy and energy security. The information regarding tentative state-wise break-up of solar power target/RPO requirement to be achieved by the year 2022 and current grid connected PV capacity in various states as on 31, March 2016 is given in Table II and the off-grid PV capacity is about 307.31MW. It is expected to commission 3220 MW and 570 MW under state government and Government of India policies respectively in 2015 -2016. It is proposed to commission 5464 MW and 6697 MW under state government and Government of India policies respectively in 2016-2017 through tendering. [12]

4 SOLAR PV – INITIATIVES

Ministry of New and Renewable Energy, Government of India laid the foundation stone for the International Solar Alliance Headquarters and Inaugurated the Interim Secretariat at Gurgaon, Haryana, India on 25th January 2016. Solar Energy Corporation of India (SECI), an implementing agency of MNRE is set to meet National Solar mission objectives and carry out the implementation work of solar PV projects [13]. The Govt. of India has provided a range of measures for implementing rooftop solar PV installations in the country for the successful implementation of the National Solar Mission [14] which are listed below:

1. Off-Grid & Decentralized Solar Applications – Grid connected rooftop and small solar power plants programme – Achievement –Linked Incentives and Awards.
2.Installation of grid-connected rooftop solar PV power plants through Multi Government Agencies (MGAs), State Nodal Agencies (SNAs) from National Clean Energy Fund (NCEF)
4.Installation of grid connected rooftop solar PV power plants in warehouses in various states across the country through NCEF.
5.Setting up of grid connected & Off grid solar PV power projects by defence establishments under Ministry of Defence and Para Military Forces with Viability Gap Funding.
6. Development of Solar Cities programme.

5 CONCLUSIONS AND RECOMMENDATIONS

India had made substantial progress in establishing policies to promote investment in renewable energy especially solar photovoltaic systems through investor meets, MoU with various countries and by establishing International Solar Alliance. Besides it needs to develop its grid infrastructure, energy institutions and markets in order to support the integration of renewable into national grid systems. By doing so, India can ensure affordable, reliable and sustainable solar voltaic power.

ACKNOWLEDGEMENTS

This work is part of the Central Power Research Institute Project, A Government of India Society, Ministry of Power, Bangalore. The support of this organization is gratefully acknowledged to carry out the work. The authors thank the management of Arunai Engineering College, Tiruvannamalai for providing the opportunity and facilities to do this work.

Table II: Commissioning Status of Grid connected solar power projects under JNNSM and State wise Solar RPO Requirement by 2022

Sl. No	State/Union Territory	Total Commissioned Capacity till 31.03.2016 (MW)	Solar RPO Required (2021 – 22) in MW	State wise estimated solar power potential GWp
1	Andhra Pradesh	572.966	5357	38.44
2	Arunachal Pradesh	0.265	42	8.65
3	Assam	-	801	13.76
4	Bihar	5.100	2969	11.20
5	Chhattisgarh	93.580	2171	18.27
6	Delhi	14.28	1870	2.05
7	Goa	-	369	0.88
8	Gujarat	1119.173	9403	35.77
9	Haryana	15.387	4030	4.56
10	Himachal Pradesh	0.201	521	33.84
11	Jammu & Kashmir	1	1319	111.05
12	Jharkhand	16.186	500	18.18
13	Karnataka	145.462	5961	24.70
14	Kerala	13.045	1429	6.11
15	Madya Pradesh	776.370	5636	61.66
16	Maharashtra	385.756	12611	64.32
17	Manipur	-	118	10.63
18	Mizoram	0.100	76	9.09
19	Meghalaya	-	182	5.86
20	Nagaland	-	60	7.29
21	Odisha	66.920	2200	25.78
22	Punjab	405.063	3917	2.81
23	Rajasthan	1269.932	6556	142.31
24	Sikkim	-	5	4.94
25	Tamilnadu	1061.820	8903	17.67
26	Telangana	527.843	4457	20.41
27	Tripura	5.0	123	2.08
28	Uttar Pradesh	143.495	13248	22.83
29	Uttarakhand	41.145	561	16.80
30	West Bengal	7.772	4603	6.26
31	Andaman & Nicobar	5.100	-	0.79
32	Lakshadweep	0.750	-	
33	Puducherry	0.025	229	
34	Chandigarh	6.806	50	
35	Daman & Diu	4.0	189	
36	Dadar & Nagar Haveli	-	502	
37	Others (PSU/Channel Partner) under Rooftop	58.311	-	
TOTAL		6762.853	100000	748.98

REFERENCES

[1] Scaling up of Grid Connected Solar Power Projects from 20,000 MW by the year 2021-22 to 1, 00,000 MW by the year 2021-22 under National Solar Mission. Ministry of New and Renewable Energy, Government of India. 2015http://mnre.gov.in/information/current-notices/ [Accessed on July 06, 2015]
[2] State wise Estimated Solar Power Potential in the Country, Solar R&D Division, Ministry of New & Renewable Energy, Government of India (Accessed on Nov 24, 2014)
[3]http://www.ireda.gov.in/writereaddata/AtlasPotentialLandRE/Program.htm [Accessed on October 27, 2015]
[4]Business Model for Solar Park http://www.mnre.gov.in/file-manager/.../Business-models-for-Solar-Park_PwC.pdf [Accessed on Oct 21, 2014]
[5]www.irena.org [Accessed on 06.11.2015]
[6]K.H. Solangi, M.R. Islam, R. Saidur, N.A. Rahim, H. Fayaz.A review on global solar energy policy. Renewable and Sustainable Energy Reviews 15 (2011) 2149–2163
[7]Compendium of regulations & tariff orders issued by regulatory commissions for renewable energy sources in India,http://www.ireda.gov.in/writereaddata/CompendiumRegulationTariffOrders/Program.ht[Accessed on October 27, 2015]
[8]Shreemat Pandey, Vijai Shanker Singh, Naresh Pal Gangwar, M.M. Vijayvergi,Chandra Prakash, Deep Narayan Pandey. Determinants of success for promoting solar energy in Rajasthan, India.Renewable and Sustainable Energy Reviews 16 (2012) 3593– 3598.
[9]Anjali Bhide, Carlos Rodríguez Monroy. Energy poverty: A special focus on energy poverty in India and renewable energy technologies.Renewable and Sustainable Energy Reviews 15 (2011) 1057–1066.
[10]Gireesh Shrimali, Sunali Rohra. India's solar mission: A review. Renewable and Sustainable Energy Reviews 16 (2012) 6317–6332.
[11]http://www.ujwalbharat.gov.in/ (Accessed on 12.11.2015)
[12]http://mnre.gov.in/information/current-notices/ [Accessed on 05.04.2016]
[13]http://www.seci.gov.in/content/innerpage/intro-solar-park.php (Accessed on October 27, 2015)
[14]Technical Manual for Banks & FIs on Grid-Connected Rooftop Solar Power, The Energy and Resources Institute 2015 (Accessed on 28.01.2016)

"PV Economics, Markets and Policies"
Building a level playing field for distributed solar photovoltaic in Australia

Institute for Sustainable Futures /University of Technology Sydney
Level 11, Building 10, 235 Jones Street Ultimo, Sydney, NSW 2007 (PO Box 123), Australia
sven.teske@uts.edu / +61 415072557 / http://www.uts.edu.au

Dr Sven Teske, Jay Rutovitz, Lawrence McIntosh

The University of Technology, Sydney (UTS) - Institute for Sustainable Futures (ISF), Australian grid operators, retailers and local councils across Australia developed a new and innovative business concept for solar PV potentially in combination with wind, geothermal gas cogeneration or (de-) centralized storage technologies. This project combines 'virtual net metering' and 'local electricity trading' to enable councils and – in the future – household customers to use PV as economic power supply within a fully liberalized market without governmental support mechanisms. The aim of this project was to create a sustainable business model for the PV industry as well as for grid operators and retailers. Integral parts of this project are 5 trials of specific projects under real market conditions. Besides, the involved grid operators and retailers seeking to develop products and specific technical requirements for billing on the basis of this research.

Keywords: Business models, market integration, Australia

Building a level playing field for distributed solar photovoltaic with local network credits and virtual net metering

Getting the right market structures to enable local energy could both encourage the growth of distributed solar photovoltaic generation and provide new business for network operators and retailers. The introduction of local network charges, calculated to reflect the costs of using only a small portion of the network, and the ability to net off local generation at other sites (virtual net metering) could provide this structure.

The electricity market in Australia is undergoing an unprecedented transformation. Electricity consumption from the grid is falling, and customers' interaction with the energy system evolving. As load defection increases, enabled by solar PV and battery storage systems, it could precipitate a difficult situation for network operators, where rising prices push customers to reduce consumption or even disconnect from the grid. Remaining customers face higher and higher prices as they continue to pay for network infrastructure.

Reducing network charges for local energy and introducing virtual net metering is a proactive approach to managing the transition to an electricity market with high contributions from distributed generators, and keeping networks competitive by reducing load defection. This could benefit electricity consumers (prices remain lower), local generators (networks can provide regulation and back-up services very effectively), and network operators (the revenue base is maintained, and the need for augmentation reduced).

Local network credits and virtual net metering – what are they and why do we need them?

The concept of local network charges and virtual net metering has been developed under an ARENA (Australian Renewable Energy Agency) funded project to undertake five trials around the National Electricity Market and support a rule change to introduce local network charges. The project team includes researchers, project proponents, network businesses, and electricity retailers. The objective is to develop better market structures for local generators and consumers.

Several **network service providers** took part in trials for a Local Network Credit. Network businesses are at the centre of the transformation in our electricity system, and sharply aware of the pressures from falling consumption from the grid, combined with the continued need to deliver safe and reliable grid services. Grid operators put their perspective on the potential benefits and challenges associated with offering tariffs that facilitate local energy.

Retailers are dealing with the requirement to change business models or see their business disappear. Several retailers took part in trials, to see what it offers to customers, see what the requirements are in terms of billing and metering systems, and to see how enabling local energy can fit in with new business models.

Local Electricity Trading (LET)

LET is an arrangement whereby generation at one site is "netted off" at another site on a time-of-use basis, so that Site 1 can 'sell' or transfer generation to nearby Site 2. The exported electricity is sold or assigned to another site for billing purposes. LET can be applied in a number of different ways:

- A single generator-customer can transfer generation to another meter(s) owned by the same entity (e.g. a Council has space for generator at one site and demand for renewable energy at a nearby facility);
- A generator-customer can transfer or sell exported generation to another nearby site;
- Community-owned renewable energy generators can transfer generation to local community member shareholders; and

Community retailers can aggregate exported electricity generation from generator-customers within a local area and resell it to local customers

Local Network Charges (LNC)

Local network charges are reduced network tariffs for electricity generation used within a defined local network charge according to the calculated long-term benefit to the network. The rationale for a local network charge is to address some aspects of inequitable network charges levied on a generator/consumer pair; dis-incentivise duplication of infrastructure (private wires) set up to avoid network charges altogether; and maintain use of the electricity network area. This recognises that the generator is using only part of the electricity network and may reduce the network.

Interaction of local network charges and local electricity trading

Local Network Charges and LET are independent but complementary concepts with different effects on a consumer's energy bills. In most cases, the Local Network Charge will reduce the network charge portion of electricity bills, while Local Electricity Trading may reduce the combined energy and retail portion of bills for local generation.

About the project and trials

The objective of the project is to create a level playing field for local energy, by facilitating the introduction of Local Network Charges and Local Electricity Trading. The key outputs are:

a. Improved stakeholder understanding of the concepts of local network charges and Local Electricity Trading;

b. Economic modelling of the benefits and impacts of local network charges and Local Electricity Trading;

c. A recommended methodology for calculating local network charges;

d. An assessment of the metering requirements and indicative costs for the introduction of Local Electricity Trading, and consideration of whether a second rule change proposal is required to facilitate its introduction; and

e. Support for the rule change proposal for the introduction of a Local Generation Network Credit submitted by the City of Sydney, the Total Environment Centre, and the Property Council of Australia.

f. Five 'virtual trials' of local network charges and Local Electricity Trading in New South Wales, Victoria, and Queensland.

The virtual trials aim to test the impact of Local Network Charges and Local Electricity Trading on local distributed energy projects, particularly the economic impacts, and to assess the real-world requirements for the measures to operate.

Methodology

An excel business case model was constructed to compare local generation projects under the current market conditions with the same generator installed with the two measures under investigation in the trials, namely Local Electricity Trading (LET) and a Local Network Credit (LNC) using two methodologies. The measures are considered together and separately. In order to see the effect of these measures, eight different scenarios were defined. The model calculates the changes in costs for the proponent sites as a result of the new generation, including the local generation site (LG site) and whatever trading sites are included in the trial (called the LET sites). The model also calculates the financial impact on the network business and the retailer (this does not include implementation costs).

The projects were generally at various stages of development, but all the installations are under serious consideration by the proponents, and it was expected that the trial would assist with decisions on whether to go ahead, as well as with project sizing.

The Model

In the excel business case model, all input data for the local generation side (LG) was arranged in one sheet, so specific parameters such as payback time or interest rate could be changed easily to test the influence on trial results.

Both the generation profile(s) and all demand profiles – from the local generation site (LG) as well as the LET "netting off" sites were uploaded in hourly steps. The netting off step includes a cascade, which can include up to 10 different demand profiles.

The third step of the calculation involved detailed input of consumption tariffs and the Local Network Credit (LNC) tariff. The LNC tariffs were calculated from each network partner's data, using the methodology developed for this project. The consumption tariffs include times for shoulder, peak and off peak, and the energy and network charges, including capacity, volume, and fixed charges where applicable.

Due to "time-of-use" dependent tariffs and LNCs, the shape of generation and demand profiles have a significant impact on the trial results and whether or not a project is profitable.

Steps four and five processed all inputs of LG and LET sites in sub calculations, which are summarized in a comprehensive result overview for each scenario.

Finally, a standardised report sheet provides an overview to key results in the form of tables, texts and figures.

The scenarios

The trial compares the business case for the new generation in current conditions, and with and without the new measures. Costs are calculated for the generation site and any netting off sites included in the trial in all scenarios. All scenarios except BAU (no 1) include the new local generation.

The different scenarios are:

1. **BAU**: business as usual – current electricity and network charges, without any new generation.

2. **Current market**: installation of new generation, with the market as it is now. (exported electricity is valued according to the retailer buy-back rate).

3. **LET only**: Local Electricity Trading in place for the exported electricity, but no LNC paid. Exports from the generation site are netted off at whatever LET sites are included, and any remaining residual exports are valued according to the retailer buy-back rate.

4. **LNC (M1)**: includes new generation, with payment of a Local Network Credit using methodology 1 (volumetric only).

5. **LNC (M2)**: includes new generation, with payment of a Local Network Credit using methodology 2 (combined volumetric and capacity payment)

6. **LET and LNC (M1)**: new generation with both measures in place, using the LNC methodology 1.

7. **LET and LNC (M2)**: new generation with both measures in place, using the LNC methodology 2

8. **Private wire**: some of the project sites could be connected via a private wire, so that all generation would be 'behind-the-meter' on a single metering point.

The Local Network Credit methodology was developed as part of this project. All calculations were performed using the excel model.

Briefly, the calculation of the LNC has two parts:

1. Value setting (the base value of the LNC). We used the same value setting methodology that network businesses use for regular tariffs i.e. the Long Run Marginal Cost (LRMC) of the network.

2. Tariff setting (the application of a tariff structure to the base LRMC value). We applied two different tariffs:
   - Volumetric tariff (methodology 1)
   - Combined volumetric and capacity tariff (methodology 2)

Winton – North West Queensland

Table 1 Trial description

Proponent	Winton Shire Council
Network service provider	Ergon Energy
Electricity retailer	Ergon Energy
Generator	310kW new Geothermal planned
Location	New geothermal plant (generation site) and 29 Winton Council sites (netting off sites)
Generation/customer model	Single entity, 1-to-1 transfer between a new generator and 29 Winton Council sites.
Project status at time of trial	Winton Council going out to tender for geothermal plant and private wire, but would prefer to use existing distribution infrastructure if suitable arrangements can be made.

Table 2 Key financial and market inputs

Technology		Geothermal
Electrical capacity	kW	310
Generator cost	\$/kW	1,900,000
Generator cost/ kW	\$	6,129
Generator O+M Cost (fixed)	\$/a	50,000
Interest rate	%/a	5.1%
Discount rate	%/a	5.0%
Inflation rate	%/a	2.43%
Private wire capital costs	\$	890,000
Private wire OPEX	\$/a	8,900
CO2 equivalent - replaced power	$kgCO^2/kWh$	0.93
Gas emission factor	ct/kWh	n/a
Other charges (AEMO, RET, SRES, NSW EES, VEET)	\$/MWh	n/a
Large Scale Generation Certificates (LGCs)	Year	50
LGC's credited until	ct/kWh	2030
Retailer buy back rate	ct/kWh	4.5
Retailer margin	%	1.2
Network connection level		HV

Figure 1 Winton Shire Council: combined site portfolio annual energy cost by scenario

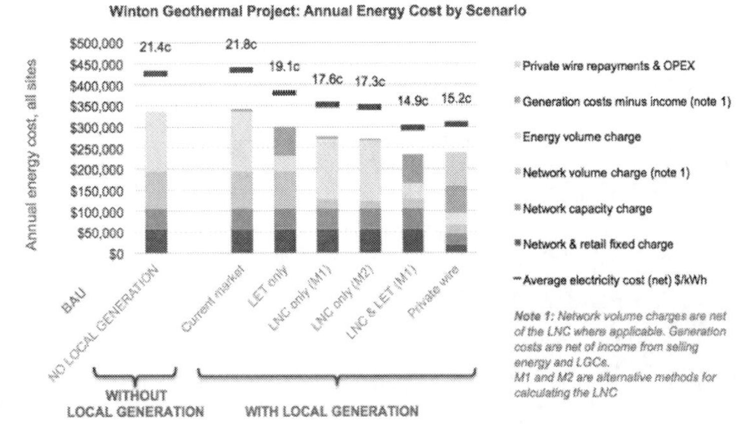

Note that costs are modelled, and may be different from actual project outcomes.

Table 3 gives the annual savings, the lifetime benefit, and the Internal Rate of Return for the project in each scenario. The private wire results in the greatest benefit, with estimated annual saving of $15,400. The next most advantageous is the scenario with both the new measures, the Local Network Charge and Local Electricity Trading.

Network charges are the most significantly affected in the private wire case, with a loss of $251,200. From the network's point of view, this is 1.8 times worse than the next best scenario of a loss of $139,800 in the LNC case.

The current market scenario still appears to have a small lifetime benefit, despite showing a loss in the annual savings. This is because annual savings are for the first year, and include the capital repayment on the geothermal system. The lifetime benefit includes the effect of inflation, and over time the capital repayments remain the same while energy and network costs increase as a result of inflation.

Table 3 Summary effect on Winton Shire energy costs by scenario

	Current market	LET only	LNC only (M1)	LNC only (M2)	LNC and LET [1]	Private wire
Annual savings	-$5,500	$36,900	$60,300	$64,600	$104,800	$96,900
Lifetime benefit	-$442,000	$586,000	$1,156,000	$1,261,000	$2,237,000	$2,201,000
IRR	4.0%	8.0%	9.9%	10.3%	13.2%	10.5%

Willoughby – North Sydney/New South Wales/Australia

Table 4 Trial description

Proponent	Willoughby Council
Network service provider	Ausgrid
Electricity retailer	Energy Australia
Generator	173kW cogeneration installed, operated to supply 85% of heat demand
Location	Willoughby Leisure Centre (generation site) and the Willoughby Council Administration Centre (netting off site)
Generation/customer model	Single entity, 1-to-1 transfer between two Willoughby Council sites, the Leisure Centre and the Administration Centre
Project status at time of trial	The business case is calculated for a new cogeneration plant, assumed to match the Leisure Centre heat load. An existing 173kW cogeneration is currently operated under a connection agreement with a minimum import of 15kW. However, for consistency between trials, results are presented for a new cogeneration plant, including capital cost. The results for a changed operational regime for the existing plant are also presented.

Table 5 Key financial and market inputs

Technology		Co-generation
Electrical capacity	kW	173
Generator cost/ kW	$/kW	4,335
Generator cost (total)	$	750,000
Generator O&M Cost (variable)	ct/kWh	1.89
Generator O&M Cost (fixed)	$/a	3,600
Interest rate	%/a	5%
Discount rate	%/a	5%
Inflation rate	%/a	2.43%
CO2 equivalent - replaced power	kgCO2/kWh	0.97
Gas emission factor	kg/GJ	51.3
Other charges (AEMO, RET, SRES, NSW EES)	ct/kWh	1.20
Large Scale Generation Certificates (LGCs)	$/MWh	50
LGC's credited until	Year	2030
Retailer buy back rate	c/kWh	3.5[1]
Retailer margin	%	2.0%
Network connection level		2 (Distribution Substation)

Note 1: Estimated by ISF

Trial results

The total cost shown in the graph is the energy cost, net of costs and income, for the two Willoughby Council sites. The table shows the results by stakeholder. The project has a positive payback for the Council even though there is a loss in the first year under the current market conditions (shown in the annual savings). This is because of inflation, which means that the capital repayments decline relative to energy costs in future years.

Figure 2 Scenario Summary Results

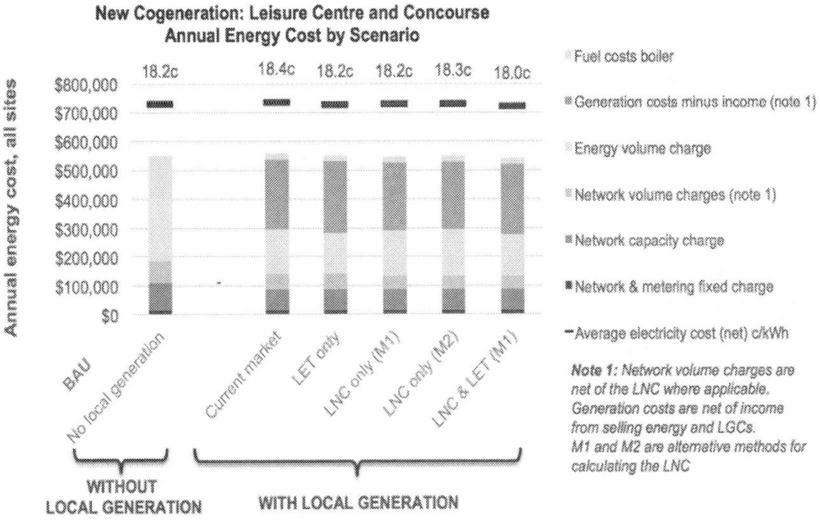

Table 6 Results by stakeholder

	Current market	LET only	LNC only (M1)	LNC & LET (M1)
Annual savings compared to BAU	-$6,000	$400	-$100	$6,300
Lifetime benefit	$302,000	$464,000	$452,000	$613,000
Effect on network charges (annual)	-$43,900	-$43,900	-$49,700	-$49,700
Effect on retailer income (annual)	-$21,200	-$26,300	-$21,200	-$26,300
Greenhouse emission reduction (all scenarios with new local generation)				871 tons/yr

Conclusion

All scenarios that include LET have a positive result to the council from year one. Current market conditions and the 'LNC only' scenarios initially do not provide a positive cash flow to the council, however inflationary effects over time mean that later years do lead to an overall benefit in even these scenarios. Unsurprisingly, the best outcome for the Council is the introduction of both the Local Network Charge and Local Electricity Trading mechanisms. The next most advantageous is the scenario with Local Electricity Trading only.

Byron Shire Council /New South Wales/Australia

Table 7 Trial description

Proponent	Byron Shire Council
Network service provider	Essential Energy
Electricity retailer	Origin Energy
Generator	150kW new Solar PV
Location	Cavanbah sports centre (generation site) and the West Byron Sewage Treatment Plant (netting off site)
Generation/customer model	Single entity, 1-to-1 transfer between two Byron Shire Council sites. The Cavanbah sports centre with low consumption and good roof space transfers energy to the nearby sewage treatment works which has high consumption but little space.
Project status at time of trial	25 kW installed, with very small amount of export. Council is investigating adding 150kW at the Sports Centre, with a significant proportion of the generation exported to supply the West Byron Sewage Treatment Plant.

Table 8 Key financial and market inputs

Technology		Solar PV
Electrical capacity	kW	150
Generator cost/ kW	$/kW	1,888
Generator cost (total)	$	283,161
Generator O&M Cost (fixed)	$/a	1,500
Interest rate	%/a	6.0%
Discount rate	%/a	6.0%
Inflation rate	%/a	2.43%
Private wire capital costs	$	200,000
Private wire O&M cost	$/a	2000
CO2 equivalent - replaced power	kgCO2/kWh	0.97
Other charges (AEMO, RET, SRES, NSW EES)	c/kWh	1.20
Large Scale Generation Certificates (LGCs)	$/MWh	50
LGC's credited until	Year	2030
Retailer buy back rate	c/kWh	calculated from Pool price [1]
Retailer margin	%	2.0%
Network connection level		1 (Low Voltage network)

Note 1: The retailer buyback rate was calculated on a daily basis as the average daily price accessed from AEMO's historical record less $5/MWh as an administration fee identified by Origin Energy.
Note 2: The retailer margin is an assumed figure, and is not based on information from AGL

Trial results

The net energy cost for the two Byron Shire Council sites is shown in Figure 3 for each scenario. This includes the energy and network charges, capital repayments on any new infrastructure, such as the solar panels and the private wire, and any income the generator may receive. Income includes renewable energy credits, the new LNC, and any buy back income from electricity, which is exported and not used at the netting off site. All scenarios except the current market conditions result in a saving compared to business as usual, so the project would have a cost benefit with either of the new measures in place, or with a private wire, with the assumptions used.

Figure 3 **Byron Shire Council: Cavanbah Sports Centre & West Byron STP annual energy cost by scenario**

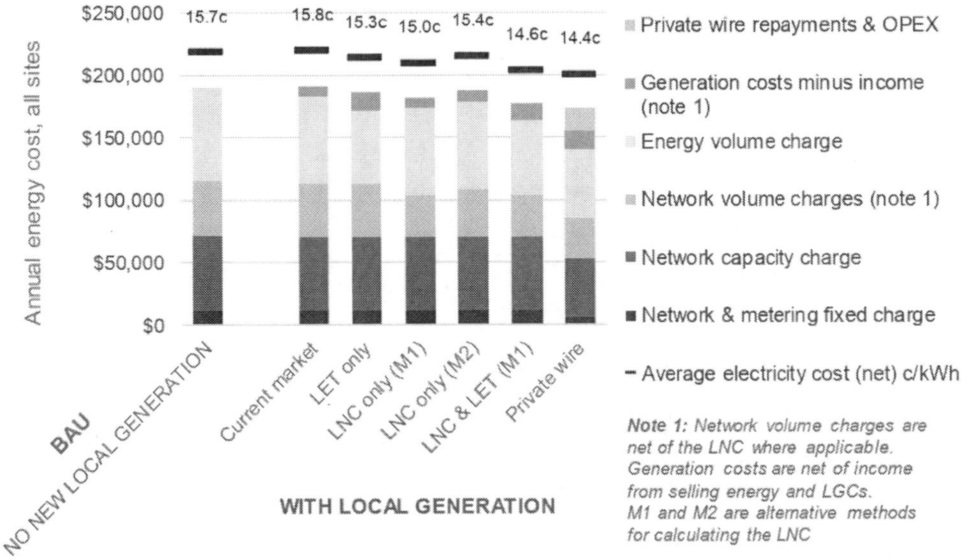

Table 9: Summary effect on Byron Shire energy costs by scenario (both sites combined)

	Current market	LET only	LNC only (M1)	LNC only (M2)	LNC and LET [1]	Private wire
Annual savings (1ST year)	-$1,200	$3,800	$7,400	$2,700	$10,000	$15,400
Lifetime benefit	$12,000	$139,000	$230,000	$110,000	$297,500	$578,000
IRR	6.5%	9.3%	11.1%	8.7%	12.4%	12.7%

Conclusion

All scenarios except the current market conditions result in a saving compared to business as usual, so the project has a cost benefit with the assumptions used if either LET or LNC is available, or with a private wire. The private wire scenario results in the greatest benefit, with estimated annual saving of $15,400. The next most advantageous is the scenario with both the new measures, the Local Network Credit and Local Electricity Trading. Network charges are the most significantly affected in the private wire case, with a loss of $29,400..

THE PV MARKET DEVELOPMENTS IN GREECE, FEED-IN-PREMIUM SCHEME

S. Tselepis
CRES, Center for Renewable Energy Sources and Saving
19th km Marathonos Av., 19009 Pikermi, Athens, Greece
Tel: +30-210-6603370, Fax +30-210-6603318
e-mail: stselep@cres.gr

Abstract: This paper presents the current state of the PV market in Greece, the achievements so far and the projections of the PV market for next years. The PV market environment in terms of new PV development policies and deployment, such as the new feed-in-premium scheme and the virtual net-metering are presented.
Keywords: Greece, PV Market, Feed-in-Premium

1. OVERVIEW OF THE GREEK PV MARKET

In the following paper the PV market achievements so far, the current environment in terms of PV development policies such as the feed-in-premium scheme are going to be presented. During the period 2008-2013, a total amount of around € 5 billion were invested in Greece in photovoltaics and the cumulative installed capacity is currently standing at 2600 MWp. For 2013, despite the economic crisis, the momentum due to the attractive feed-in tariff contracts that certain developers were still holding boosted the annual installed capacity to 1042.5 MWp. In the year 2013 the energy produced by photovoltaic systems in Greece reached 6.7% of the electricity consumed and for the year 2014 it was roughly 7%. The development of the annual and cumulative PV system capacity in Greece is presented in Figure 1. The installed capacity in 2014 and 2015 collapsed to 16.95 and 10.3 MWp respectively. Furthermore, the increasing penetration of PV systems should also push the authorities to study, plan and adopt technical and regulatory measures in order to allow higher penetration of RES in the future. In this direction a new law, 4203/2013 was passed by the parliament that allows net-metering for PV systems on buildings. The PV market environment in terms of new PV development policies and deployment, such as virtual net-metering and the new feed-in-premium scheme are going to be presented. The net-metering did not regenerate the PV market in Greece as the economy is still frozen due to uncertainty and the capital controls measures applied since June 2015.

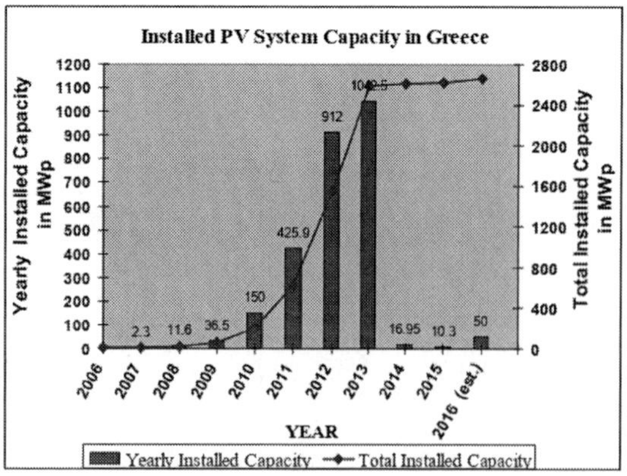

Figure 1: Historic PV market data, yearly and cumulative introduction of PV systems in Greece (CRES estimate for 2016).

Since 2010 the feed-tariffs (FiT) for new PV systems were reduced four times, the first time in 2010, then twice during 2012, and finally in May 2013. Now the current feed-in tariffs for PV systems under 100 kWp are standing at 1.2 times the System Marginal Price (SMP) of electricity of the previous year and for systems over 100 kWp, it is 1.1 times the System Marginal Price (SMP). For all PV systems in the non-interconnected system (islands) the feed-in tariff stands at 1.1 times the System Marginal Price (SMP). Given that the average SMP for 2015 was 50.4 Euro/MWh, it is calculated that for the year 2016 the FiTs in the Interconnected Electrical System for PV Systems under 100 kW stands at 60.48 Euro/MWh, while the for PV Systems over 100 kW it stands at 55.44 Euro/MWh. Regarding the feed-in-tariffs for new PV systems under 10 kWp on buildings, for the year 2016, the FiT is standing at 110 €/MWh, digressing to 80 €/MWh in the year 2019 [1].

In Table 1, the PV electricity generation and power consumption in Greece for the year 2045 is presented. Practically all the PV installed capacity is connected to the low and medium voltage grid and up to now the PV generated electricity is experienced by the System Operator as a negative load. This can be tolerated without any negative effects when the instant power penetration level is relatively low and does not create any operation and security related problems. Nevertheless, there are already locations in the distribution grid where the local penetration has increased significantly and measures are taken at the moment to limit new capacity in these grid segments by not allowing any new PV installations. In 2013 the PV electricity consumed in Greece accounted 6.7% of the annual consumption in the interconnected power system, while for 2014 and 2015 it was estimated that it covered more about 7.1% of the total annual electricity consumption in Greece.

Table 1: Statistics on PV electricity generation and power consumption in Greece for the year 2014.

Statistic	Value
Installed PV capacity nationwide, 12/2014	2.596 GWp
Installed PV capacity (interconnected system), 12/2014	2.436 GWp
Peak load (interconnected system, 2014)	9.263 GW
Total energy consumption in 2014 (interconnected system)	50.742 GWh
Total energy generated by PV in 2014 (interconnected system)	3558 GWh
Share of generated PV electricity to energy consumption (interconnected system), 2014	7,01 %
Share of installed PV capacity connected to HV level, 2014	2,2%
Share of installed PV capacity connected to MV level, 2014	45,6%
Share of installed PV capacity connected to LV level, 2014	52,2%

2. NEW ENERGY POLICY DEVELOPMENTS

As the Energy Union strategy [2] places consumers at the core of the EU energy policy, encouraging them to get involved in the energy transition, to benefit from new technologies to reduce their bills and participate actively in the market, while ensuring protection for vulnerable consumers. At the same time, the achievement of the Energy Union requires a fundamental transformation of Europe's energy system. Renewable energy is essential for this transformation to take place as it contributes to all of the Energy Union objectives:

- the delivery of security of supply,
- a transition to a sustainable energy system with reduced greenhouse gas emissions,
- industrial development leading to growth and jobs and lower energy costs for the EU economy.

Thanks to technology development and innovation driven by EU and national policies, over the last few years we have seen the realization of attractive renewable energy technologies, for both large and small-scale use, alongside considerable cost reductions such as the PV modules decreasing in cost by 80% between 2008 and 2012. As a result, businesses and households can increasingly produce and consume, some or all, of their own electricity, either instantaneously, by synchronizing their consumption, or in a deferred manner through decentralized storage, behind the connection point with the grid (i.e. the electricity meter). Through the process of 'self-consumption', passive consumers are therefore becoming active 'prosumers', therefore producers and consumers of renewable energy.

3. FEED-IN-PREMIUM SCHEME IN GREECE

A proposal for a new support scheme for RES was submitted in February 2016 for consultation by the Hellenic Ministry for Energy and Environment regarding financial support and operation of the renewable energy and combined heat and power (CHP) systems. The goal of the proposal is to establish a new support structure for renewable energy projects in Greece following the EU Guidelines on State aid for environmental protection and energy 2014–2020 ("EEAG 2014-2020"), which requires the gradual phasing-out of subsidies and a move to a market-based system from 2017 and on. The public consultation period ended on March 2016. Once the new support scheme for renewable energy projects in Greece is finalized then it must be approved by the European Commission prior to any legislation being proposed to the Greek Parliament.

In this context, for the development of renewable energy technologies an important parameter in the energy system design is the establishment of a secure investment climate with concrete implementation schedules and support for the energy generated by these technologies and the implementation of a support scheme through market-based mechanisms achieving an optimal relationship of cost to benefit for the society.

Given the ongoing process to develop a new market model for electricity in Greece, which is expected to be completed by the end of 2017, the national strategic objective is to protect consumers from a higher than required support level regarding the renewable energy generation. The objective is the new market model to allow both new investments that will exploit the significant domestic renewable energy (RE) resources for electricity generation and the progressively greater involvement and commitment of these new RE systems at the electricity market level, as well as monitoring the average electricity production costs of renewables in order to avoid unnecessary aid in their operation.

Moreover, from January 1, 2017, the Greek State is planning to adopt a support scheme through competitive submission tenders in order to identify the optimal level of support that generated electricity from RES plants requires in order to meet the national energy objectives.

It is noted, that the new annual installed capacity of RES for the following years it expected to range between 400 and 500MW.

So far in Greece, the support scheme was organized in the context of a compensation scheme for electricity with a fixed tariff (Feed-in-Tariffs). The new proposed support scheme for RES and CHP energy will be based on the development of a mechanism supporting the operation of these units (operating aid), providing a premium, on top of the price of the defined wholesale electricity market. This price premium will be guaranteed for the period of validity of the feed-in premium contract of each electricity RE system, depending on the RES technology. The financial support is provided in the form of a supplementary price (Feed-in-Premium = FiP), taking into account the income from the participation in the electric power market.

From now on, all the new photovoltaic projects with an installed capacity more than 500kWp will be supported exclusively through competitive bidding procedures. These procedures will start in the second half of 2016 and within the year 2016 it is expected to have at least two different bidding rounds with available power significantly greater than that defined as a minimum for the pilot procedures of about 5% of new renewable power installed in the period 2015-2016. Therefore, it is expected that the power capacity to be launched for competitive bidding will be of the order of 200 MW per year.

The PV market is generally divide in 3 segments, see Table 2, in the segments above and under 500 kWp and the PV systems on buildings under 10 kWp. For the category of photovoltaic systems up to 500kW it was decided to further analyze the prices resulting from the application of competitive processes that will take place within 2016 and not to fix at this point in time the tariff. Therefore, the feed-in tariff (FiT) for this PV system category of projects (<500 kWp), will be maintained for the time being at the level of existing support. The existing methodology for determining the feed-in tariff value is calculated each year by taking the average of last years' Marginal System Price and multiplying by a factor of 1.1. This results in a FiT of about 55.5 Euro/MWh, a compensation that is not considered viable by the PV market actors. The new procedure for competitive bidding that are going to be launched in 2016, will reenergize the national PV market and representative PV system costs will be available to decide on the new FiT support, or reference tariff (RT), as will be called from now on, for the under 500 kWp PV segment. In each case, the systems of this PV market segment will be also eligible for inclusion in the self generation and consumption scheme with electricity price compensation that is usually more attractive than the FiT or the expected RT, while it is also planned to have the option for participation in the competitive bidding process, even for this PV system segment.

Table 2: Proposed reference tariffs for various PV system segments.

PV system category	Reference Tariff Euro/MWh	Project IRR
PV sys. > 500 kWp	Competitive bidding process	-
PV sys. ≤ 500 kWp	2015 tariff (60-70 Euro/MWh) *	-
PV sys. on Buildings < 10 kWp	110	10%

* New tariff by the end of 2016

For the 32 independent electrical power systems of the islands that are not connected to the mainland power system the daily electricity markets is not going to be established now. These island electricity markets will start operating after the development of the necessary infrastructure, which is planned to be completed in stages over the next three years. Consequently, for the non-interconnected islands all new electricity generation projects from renewable energy sources and CHP will fall into the support scheme based on feed-in tariffs (FiT) until the completion of the necessary infrastructure and the gradual entry into full operation of the daily electricity market by the end of 2019.

From now on, all new RES and CHP plants will be subject to the new mandatory support scheme in the form of additional supplementary revenues (Feed-in Premium) received from participating in the wholesale electricity market, even if their contract was put into effect before the transition to the new electricity market model. These new plants will have the obligation to undertake all the balancing services foreseen for these units when the complete market model will be put into operation.

4. CONCLUSIONS

Regarding the sustainability of the PV market in Greece it is noted that as the FiT for 2016 will be 1.1 to 1.2 times the average System Marginal Price in the previous year (2015), which was about 50.4 €/MWh. Therefore, FiTs of 55.5 to 60.5 €/MWh are very low tariffs for the new investments under current cost of money and the difficult economic environment in Greece. Therefore, the FiTs should be reconsidered in order to render the investments in PV systems attractive. This is expected to happen at the second half of the year 2016 after the launch of the competitive bidding procedures when representative PV system costs and economic conditions will be known to decide on the new FiT support level. In comparison, it is mentioned that the FiTs for wind and other mature RES technologies range between 82 and 105 €/MWh in the mainland, depending on installed system capacity and whether there is a subsidy on initial investment or not.
On the other hand, the new Net-Metering PV program, as proved by the study cases [3], is considered attractive and given the right economic environment it could become a significant market segment in the following years. The existing experience and know how of member states with on-line information platforms and application procedures for small scale RES projects should be investigated.

The Net-metering program is under amendment allowing legal entities of public or private law with activities of public interest, as well as entities registered in the Register of Farmers and Farm exploitations of Law 3874/2010, exclusively for farming activities, to apply the virtual net-metering scheme which permits compensation of energy produced by the photovoltaic power system with the power consumed at the auto-producer plants, of which at least one is not in the same or adjacent area with the solar system, and can be supplied from different grid connection points. Virtual Net-metering concerns photovoltaic systems installed in the same prefecture of the electricity consumption facilities to which they are assigned and linked to the same Interconnected Network.
In the non-interconnected islands, the PV systems will be installed in the same prefecture and at the same electrical system with consumption facilities to which they are assigned for virtual Net-metering.
As consumers have an interest in maximising the rate of energy self-consumption in order to increase their energy savings and reduce their exposure to electricity prices, preference for self-consumption motivated over simple net-metering schemes should be widely incentivized in the future. In the context of a smart grid environment, self-consumption has the potential to drive consumers' uptake of flexibility measures, while at the same time help facilitating the system integration of variable renewable energy.
Finally, attention and a special FiT should be given to the developing sector of Building Integrated PV systems (BIPV). The integration gives added value to the PV system and promotes the involvement of architects, thus introducing BIPV in the mainstream of building design. In any case, action should be taken to modify the building code in Greece in order to allow PV integration on buildings.

5. REFERENCES

[1] S. Tselepis, "THE PV MARKET DEVELOPMENTS IN GREECE", 29th EUPVSEC Amsterdam 22 - 26 Sept 2014, 7AV 6 22, Conference proceedings.

[2] Framework Strategy for a Resilient Energy Union with a Forward-Looking Climate Change Policy, COM/2015/080 final.
[3] S. Tselepis, "The PV Market Developments in Greece, Net-Metering Study Cases", 31st EUPVSEC September 2015, Hamburg, Germany. 7DV.4.26

INNOVATIVE SERVICE AND BUSINESS MODEL FOR PHOTOVOLTAIC POWER PLANTS ON MULTIPLE DWELLINGS IN AUSTRIA FOR SELF-CONSUMPTION

Susanne Woess-Gallasch[a], Walter Aichinger[b], Martin Auer[c], Gerhard Korpitsch[c], Heidemarie Rest-Hinterseer[d], Daniel Steiner[a]
[a]JOANNEUM RESEARCH Forschungsgesellschaft mbH
Leonhardstraße 59, A-8010 Graz, Austria
[b]Ebner Aichinger Guggenberger Rechtsanwälte GmbH
Sterneckstraße 35, A-5020 Salzburg, Austria
[c]KW Solartechnik GmbH,
Triester Straße 478, A- 8055 Graz, Austria
[d]Arbeitsgemeinschaft Erneuerbare Energie Salzburg,
Auerspergstr. 20, A-5020 Salzburg, Austria

In Austria multiple dwellings have hardly been used for the installation of solar photovoltaic (PV) power plants. The aim of the project PV4residents (call "Stadt der Zukunft" of the Austrian Ministry for Transport, Innovation and Technology, project no. 850126) is to enable a higher application of them on such buildings by addressing a prosumer approach. The produced electricity will be consumed to a large share by the residents and not fully fed into the grid, unburdening the grids and bringing possible economic benefits to the residents. This requires a new business model. The legal form of a cooperative society – already applied in Germany - will be examined also for Austria with a different legal framework. Home building companies in Graz and Salzburg, residents of two multiple dwellings, an electricity provider, and other stakeholders participate in the project. Technical options, as well as legal, administrative and user related barriers will be investigated and an economic assessment will be carried out. On the basis of this information the service and business model is currently under development. The project started in November 2015 and will be finalized by the end of this year.
Keywords: PV systems, prosumer, economic analysis, business model

1 INTRODUCTION

Austrian electricity prices to be paid by private households have increased very slowly over the last years from about 19.09 ct/kWh in 2009 to 19.96 ct/kWh in 2015 (mean values with taxes included in Austria). PV feed-in tariffs for new contracts (>20 kWp) decreased significantly from 39,98 ct/kWh in 2009 to 11,5 ct/kWh in 2015 and are lying now significantly under above mentioned electricity prices for households (see Figure 1). Prices for PV modules from manufacturers in Austria (weighted average) diminished significantly from 1,958 € / kWpeak in 2010 to 595 € / kWp in 2014, the prices for a 10 kWp entire PV system with grid connection (turnkey plant) from 3,233 € / kWp in 2010 to 1,471 € / kWp in 2014 [1].

These developments make self-consumption of PV power for residents economically more attractive. Whereas household owners of single-family houses are already quite active in installation of PV plants on their roofs for their own use and for feed-in to the grid, this does not happen for multiple dwellings in Austria [5].

One important reason for that is the Austrian legal energy framework such as the Austrian General Act on Electricity (Elektrizitätswirtschafts- und Organisation-sgesetz 2010 ElWOG 2010), which doesn't allow direct sale of PV electricity to residents. This law is currently under amendment.

The aim of this project is to develop a special business model that enables a stronger penetration of PV plants on multiple dwellings. It is planned to find options for the residents of the multiple dwelling to consume a large share of the produced electricity on-site and surplus electricity will be distributed to the municipality. Residents will have also the opportunity to participate in financing the PV power plant.

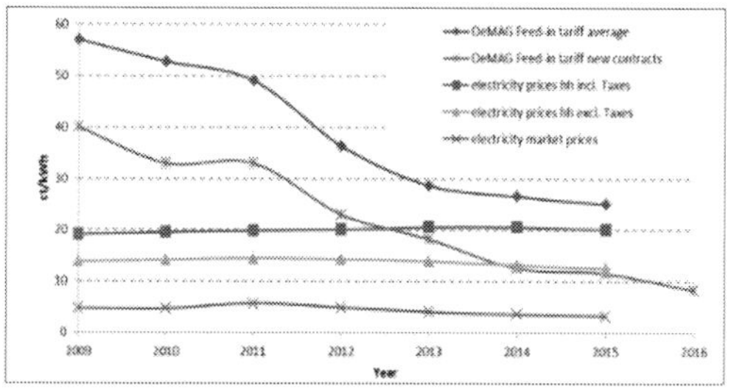

Figure1: Feed-in tariffs and electricity prices for households (with and without taxes) and electricity market prices for Austria (sources: [2], [3], [4]).

2 APPROACH

The project PV4residents is currently under way. Two representative multiple dwellings in Graz and Salzburg have been chosen. This ensures that the proposed business model can be examined in real-life conditions.

In the course of this study barriers for implementing such a business model are under evaluation. Technical and legal as well as administrative issues are currently assessed (e.g. the administrative modality of a PV-cooperative society), and user behavior patterns and load profiles will be examined. All the important stakeholders are involved in this process. The study comprises essentially 7 steps.

2.1 Outline of the new Business Model (Step 1)

Based on already existing experiences [5] [6] [7] [8] [9] it is planned in a first step to outline the new business model with special needs for multiple dwellings and the use of produced electricity mainly by the residents.

Planned participants in the business model are documented in Figure 2.

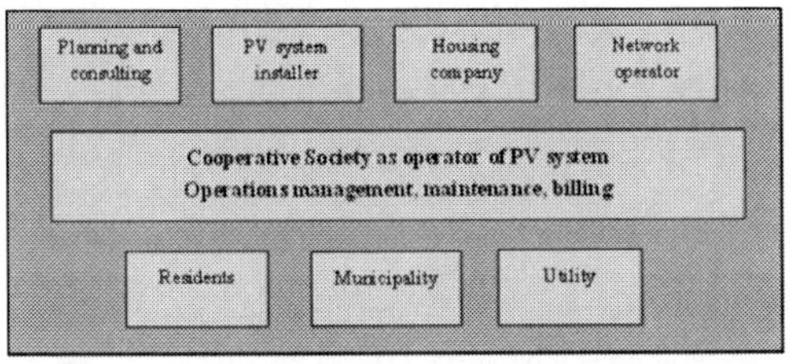

Figure 2: Planned participants in the business model to be developed

Two housing companies in Graz and Salzburg are involved, with their help two model houses have been selected. Representatives of the municipalities in Graz and Salzburg participate for a check if surplus electricity can be used by them for their near-by energy needs. If the roof of the dwelling is large enough, then the PV plant could also be extended for further municipal needs. Also representatives of the two regional state governments and the Austrian Association of Cities will participate as stakeholders of the project.

2.2 Technical aspects (Step 2)

One technical concept of a PV installation on a multiple dwelling under evaluation is outlined in Figure 3. The aim of this concept is to find the best technical requirement for a multiple dwelling to use the self-generated solar electricity primarily where it can be consumed. Two cases of PV plants are considered: roof-mounted PV and building-integrated PV (BIPV).

By applying storage solutions and load management the produced electricity should be used as far as possible for electricity needs of the residents and for community needs (e.g. walkway lightning, common areas) of the multiple dwelling. Remaining electricity can be used by near-by energy needs of the municipality.

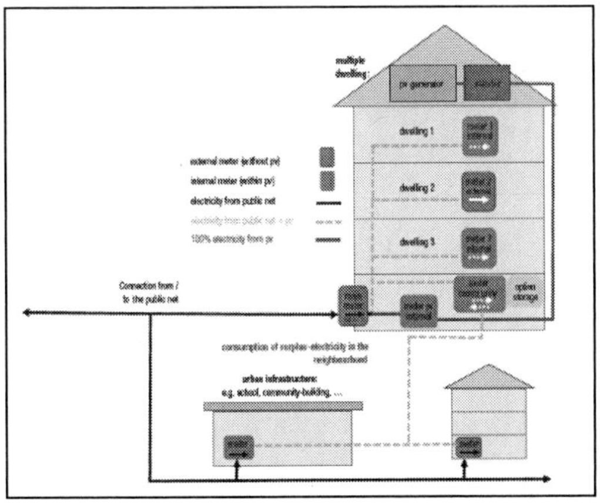

Figure 3: Possible technical concept of a PV installation on a multiple dwelling and on-line consumption by residents and by near-by energy needs of the municipality

In case that the municipality is interested in more electricity from PV, it can be evaluated if it is possible to enlarge the PV installation for their needs (depending on available areas of the dwelling).

2.3 Legal analysis (Step 3)

Step 3 deals with clarification of legal questions concerning non-profit housing development companies with common eco-power production by PV plants on multiple dwellings. This includes an analysis of Austrian

- energy framework
- housing legislation
- clarification of questions of liabilities and insurance
- treaties and agreements on taxation law
- Participatory concepts under consideration of finance market regulations

Individuals and multiple dwellings cooperatives have increasingly shown interest in taking part in the generation of electricity from PV. For the operation and exploitation of a PV plant the legal form of a cooperative society – already applied in Germany [10] - is under examination for Austria.

2.4 Administrative aspects (Step 4)

Step 4 addresses administrative aspects such as clarification of efforts for administration and models of reimbursement of costs, and the evaluation of costs and consequences for operating PV plants.

2.5 User-specific aspects (Step 5)

The adaptation of habits of users to reduce the consumption of electricity and to shift power use to times of solar power production is vital for any effort to energy saving measurements. Results of an Austrian study show [11] that energy saving and the use of interactive technologies for better implementation of solar power into everyday life are for consumers of high interest.

Residents of the chosen multiple dwellings are involved as stakeholders in the project. A survey will be carried out. They will receive information about demand-side management and possible changes in habits and how to reduce and shift electricity consumption. Smart meter user interface will enable them to actively maximize the consumption of their own produced electricity

2.6 Economic analysis (Step 6)

In the economic analysis chosen PV plants – roof-mounted PV and BIPV – are under evaluation. Net Present Value method over the entire lifetime of a PV plant of 25 years is used. Levelized Costs of Electricity (LCOE) will be calculated and economic benefits for residents evaluated. Investment schemes with possibility of residents′ participation are under development.

2.7 Final definition of the business model (Step 7)

The results from step 1 to 6 will be summarized, evaluated and are the basis for the final definition of the new business model. The results will be presented and discussed with stakeholders. The practicability of the developed business model will be evaluated by stakeholders and external experts and resulting final adjustments included.

3 RESULTS AND CONCLUSIONS

The new service and business model for PV plants on multiple dwellings focuses specifically on how the produced electricity can be consumed in an economically feasible way by the residents themselves, surplus electricity from PV can also be used by other local energy needs of the municipality.

Energy savings and emission reduction potential based on the adaption of residents behavioral patterns and on matching the use of electronic devices with the production profile of the PV plant will also be outlined. The following results are planned to be covered in the project:

- Development of a specific and robust finance-, service- and business model with participation of all necessary partners for the promotion of electricity self-production from a cooperative PV plant on a multiple dwelling:
- Clarification assistance to adapt the Austrian legal framework;
- Information how residents of multiple dwellings can participate directly in self-consumption of PV-based electricity; guidance to energy saving behavior;
- Clarification how and if electricity surpluses can be delivered to municipality infrastructure;
- Strategy of using generated know-how of this exploration project for real implementation.

The effective diversification of energy supply and the increase of the use of renewable energy require a strong commitment of the relevant stakeholders and citizens. Based on the practical experience gathered from the application of the shareholder models to the application of renewable energy an innovative participatory concept attracting the private involvement and private capital for leveraging and financing concrete investments will be elaborated.

3.1 References

[1] P. Biermayr et al.: Innovative Energietechnologien in Österreich Marktentwicklung 2014. Bundesministerium für Verkehr, Innovation und Technologie, Berichte aus Energie- und Umweltforschung 11/2015.

[2] E-Control: http://www.e-control.at/de/statistik/oeko-energie/oekostrommengen

[3] Statistic Austria: http://www.statistik-austria.at/web_de/statistiken/energie_und_umwelt/energie/preise_steuern/index.html

[4] Eurostat: http://appsso.eurostat.ec.europa.eu/nui/submitViewTableAction.do;jsessionid=9ea7d07d30def1a6bd842cdd415eb87faa8130110c4b.e34OaN8PchaTby0Lc3aNchuNa3qLe0

[5] K. Giselbrecht et al.: Marktmodelle für GIVP-Mehrparteien-Immobilien im intelligenten, dezentralen Energiesystem. Neue Energien 2020, Endbericht, Wien, Dezember 2011.

[6] G. Brandstetter, R. Haslinger: Sonnenstrom in Bürgerhand. Spezialausgabe. Sonnenstrom Ausgabe 2/12.

[7] V. Engelbert: StrombBIZ Geschäftsmodelle dezentrale Stromerzeugung und Distribution. Besonderheiten der rechtlichen Feasibility. Themenworkshop "Der Strom von Nebenan", Vienna, 17.2.2016 http://www.hausderzukunft.at/results.html/id8333.

[8] K. Seufert: Virtuelle Zählpunkte. Messkonzepte für Solarstromlieferungen in Mietsgebäuden. DGS Landesverband Franken, Sonnenenergie 3, 2014, 32.

[9] BSW-Solar 2016: PVFINANCING. EU Horizon 2020 Project, 01 2015 - 06 2017. Downloadable: http://www.pv-financing.eu/

[10] N. Ferchl, Heidelberger Energiegenossenschaft eG: Neue Geschäftsmodelle für PV-Genossenschaften. Möglichkeiten und Grenzen. Presentation, Berlin 2015.

[11] M. Stutz et al.: Smart Grids Modellregion Salzburg – Häuser als interaktive TeilnehmerInnen im Smart Grid. Begleitforschung. FFG-Project-No. 829996, Vienna, 2015.

THE STUDY OF THE EFFICIENCY OF PHOTOVOLTAICS AND WIND POWER ENERGY

Pavel P. Bezrukikh[1], Sergey M. Karabanov[2], Pavel P. Bezrukikh (Jr.)[3]
1- G.M. Krzhizhanovsky Power Engineering Institute
19, Leninsky Prospekt, Moscow 119071, Russia
2- Ryazan State Radio Engineering University
59/1 Gagarina St., Ryazan 390005, Russia
3- LUCOIL JSC
11, Sretensky Blvd, Moscow 101000, Russia
e-mail: pvs.solar@gmail.com

ABSTRACT: The paper presents the research results of energy, ecological, operational, economic and social efficiency of renewable energy (photovoltaic (PV) and wind power) and comparison results of renewable and traditional energy efficiency. It is established that wind and PV power stations are more effective, than thermal stations; their ecological efficiency is considerably higher, than that of thermal stations; the use of PV and wind power stations increases social stability.
Keywords: renewable energy; indicator, energy efficiency; ecological efficiency

1 INTRODUCTION

Efficiency has a complex character and represents a set of a few efficiencies: economic efficiency, energy efficiency, operational efficiency, ecological and social efficiency.

Economic efficiency is estimated according to the levelized cost of electricity (LCOE) value [9-13].

The energy efficiency indicator is the period during which any power station generates such energy which has been consumed for production of materials, equipment, transportation charges, installation, operation and utilization [1-4].

Operational efficiency (the term is introduced by the authors) is estimated by a few indicators: the volume of power reservation for wind power and photovoltaic stations, the possibility of electric power generation forecasting, the possibility of stability of power system operation at randomly changing generated power of renewable energy stations, the possibility of power and voltage regulation at the point of connection of a renewable energy source to a power system, water consumption at operation [5-8].

The ecological efficiency indicator is estimated by emission of CO_2 and other harmful and greenhouse gases appearing in the course of production of materials, equipment, transportation, installation, service, operation and utilization of a power plant [1-4].

Social efficiency is evaluated according to a few factors: ecological safety, quantity of created workplaces and regional development.

The paper presents complex research results of renewable energy efficiency indicators and the results of its comparison with traditional power stations efficiency. The paper uses the data of works presented by different institutions and the results of the authors' own calculations.

2 MAIN RESULTS AND DISCUSSION

The present paper used mathematical calculations and data bases from different sources for the analysis of different renewable energy efficiencies and their comparison with the efficiency data of traditional power stations [1-14].

2.1 Economic Efficiency [9-13]

There is an opinion about high cost of the electricity from renewable energy power stations and their low competitiveness. However this opinion is wrong. According to the position of leading experts of the World Bank the traditional financial analysis is not able to consider adequately future risks connected with fuel prices. It also completely ignores environmental and expenses and health care connected with emissions at power plants burning fossil fuel. If we consider costs of a complete technical cycle, some renewable sources can even now compete with traditional energy sources.

However, the potential of renewable energy technologies is not realized completely because of various barriers of the market, such as state support of traditional fuels [9]. According to the World Bank the annual state financing of gas industry in Russia amounts to USD 25 bln, power generating industry – USD 15 bln.

For evaluation of economic efficiency the data obtained in works [11] and [12] was used. Table 1 shows the summarized results for 17 types of power plants. Levelized cost of electricity (LCOE) was used for economic efficiency evaluation.

Obviously, by 2020 at PV stations (USD 1253/MWh) taking into account rapid unit cost reduction and efficiency increase of solar cells, levelized cost of electricity close to the cost of advanced coal and gas-turbine power plants will be achieved. In other words, the last barrier to broad-scale development of photovoltaics will be removed.

2.2 Energy Efficiency [1-4, 6-8]

Energy efficiency of renewable energy has tremendous advantage over thermal power.

The energy efficiency is estimated according to the formula:

$$T_{epb} = E_{con} / E_y$$

where T_{epb} is the energy payback time, E_{con} is energy consumed for the production of a station, materials, equipment, operation, servicing; E_y – the electric energy annual production of a station.

Table 2 shows the aggregated data on power consumption in the wind turbine life cycle [1].

The energy production depends on wind

characteristics at the place of installation of a wind power plant and is determined by the capacity utilization factor (CUF).

Table I: Estimation of levelized cost of electricity of power plants in 2010 and power plants put into operation in 2018

Station type	Levelized cost of electricity (LCOE), USD/MWh					
	Minimum		Average		Maximum	
	2010	2018	2010	2018	2010	2018
Normal coal	90.1	89.5	99.6	100.1	116.3	118.3
Advanced coal	103.9	11.6	112.2	123.0	126.1	137.9
Advanced coal with CCS technology	129.6	123.9	140.7	135.5	162.4	152.7
Natural gas (standard)						
Normal with combined cycle	61.8	62.5	68.6	67.1	88.1	78.2
Advanced with combined cycle	58.9	60.0	65.5	65.6	76.1	76.1
Advanced with combined cycle with CCS technology	82.8	87.4	92.8	93.4	107.5	107.5
Normal gas-turbine	94.6	104.0	132.0	130.3	149.8	149.8
Advanced gas-turbine	80.4	90.3	105.3	104.6	119.0	119.0
Advanced nuclear	108.4	104.4	112.7	108.4	120.1	115.3
Geothermal	85.0	81.4	99.6	89.6	113.9	100.3
Biomass	101.5	98.0	120.2	111.0	142.8	130.8
Wind	78.2	73.5	96.8	86.6	114.1	99.8
Wind, offshore	307.3	183.0	330.6	221.5	350.4	294.7
Photovoltaic	122.2	112.5	156.9	144.3	245.6	224.4
PV thermal	182.7	190.2	251.0	261.5	400.7	417.6
Hydro	58.6	54.4	89.9	90.3	149.7	149.2

Table II: Energy consumption lines in 2 MW Nordex wind turbine life cycle [1]

		Without utilization (disposal)		With utilization (recycling)	
			share		share
Materials	J	$1.7594 \cdot 10^{13}$	0.898	$1.7594 \cdot 10^{13}$	1.41
	kWh	$4.89 \cdot 10^{6}$		$4.89 \cdot 10^{6}$	
Production	J	$1.3593 \cdot 10^{12}$	0.069	$1.3593 \cdot 10^{12}$	0.11
	kWh	$3.77 \cdot 10^{5}$		$3.77 \cdot 10^{5}$	
Transport	J	$2.4336 \cdot 10^{11}$	0.012	$2.4336 \cdot 10^{11}$	0.02
	kWh	$6{,}76 \cdot 10^{4}$		$6.76 \cdot 10^{4}$	
Operation	J	$1.6778 \cdot 10^{11}$	0.009	$1.6778 \cdot 10^{11}$	0.01
	kWh	$4.66 \cdot 10^{4}$		$4.66 \cdot 10^{4}$	
Decommissioning	J	$2.1826 \cdot 10^{11}$	0.011	$- 6.8512 \cdot 10^{12}$	- 0.55
	kWh	$6.06 \cdot 10^{4}$		$- 1.903 \cdot 10^{4}$	
Total	J	$1.9583 \cdot 10^{13}$	1.0	$1.2513 \cdot 10^{13}$	1.0
	kWh	$5.439 \cdot 10^{6}$		$3.476 \cdot 10^{6}$	

Table III: Energy payback time of 2MW Nordex Wind turbine, CUF = 0.4, life time – 25 years [1]

	Without utilization (disposal)	With utilization
Energy consumed for a wind power plant construction, J	$1.95 \cdot 10^{13}$	$1.25 \cdot 10^{13}$
Energy consumed for a wind power plant construction, kWh	$5.41 \cdot 10^{6}$	$3.47 \cdot 10^{6}$
Average annual energy production by a wind power plant at CUF – 40%, kWh/year	$7 \cdot 10^{6}$	$7 \cdot 10^{6}$
Electric energy production by a wind power plant during service life of 25 years, kWh	$175 \cdot 10^{6}$	$175 \cdot 10^{6}$
Ratio of the energy produced by a wind turbine during service life to energy consumed for its construction	32.32	50.43
Energy payback time, months	9.27	5.94

According to the calculations for Nordex 2 MW wind power plant at CUF of 0.4 (very good wind conditions) the energy payback time is 0.76 year, or 9.27 months without utilization and 5.94 months with utilization [1]. During service life of 25 years a wind power plant generates electric power of 32.2 times more than it was used during its life cycle without utilization and 50.4 times more than with utilization (Table 3).

The calculation of T_{epb} for a ground wind power station with the total capacity of 300 MW containing 182 Vestas V82 wind turbines taking into account the joining equipment at CUF of 0.4 and life time of 20 years shows that the energy payback time is 7.2 months.

The above data confirmed by numerous calculations made in different countries show that for a wind power station the energy payback time is less than one year (6 - 10 months).

Slightly different situation (which changes quickly) is in photovoltaics. In particular, in Portugal for 62 MW PV station at solar radiation of 1900 kWh/cm^2 per year the energy payback time is 5.9 years [9]. The main contribution to the bound energy value is made by technological processes of silicon and solar cells production. This study has been made in 2005 for the station built in 2001-2003 where it is mentioned that the advanced silicon production technologies will allow to reduce the energy payback time to 1.2–1.8 years [8].

In the research made by International Energy Agency in 2011 for PV plants on roofs built in 2005-2006 in Southern Europe where solar radiation is 1700 kWh/m^2 per year, the energy payback time for the plants of monocrystalline silicon with the efficiency of 14% is 2.2 years, polycrystalline silicon with efficiency of 13.2% – 2 years, thin film silicon with efficiency of 12% – 1.7 years, CdTe modules with efficiency of 10.9% – 0.75 years [8].

So, wind power and photovoltaics have high energy efficiency.

The picture is quite different for electric and thermal energy production at thermal and nuclear power plants. In terms of energy they are never repaid as during service life they constantly consume energy in the form of fuel and transform it to electric and/or thermal energy with the coefficient much less than unity. This is the difference of our point of view from many others according to which the fuel consumption during operation is not taken into account.

Therefore, using renewable energy sources the mankind can switch to effective electricity generation at wind power, PV and fast neutrons nuclear power stations. As for hydroelectric power stations similar research was not carried out. This is the immediate task of research. However there is no doubt in their energy efficiency.

2.3 Operational Efficiency [5-8]

The term "operational efficiency" has not become conventional yet. This is the authors' suggestion. The operational efficiency indicators cover the following range of questopns: a) whether 100 % power reservation of wind and solar stations operating in a power supply system is required; b) whether it is possible to forecast the electricity generation of these power plants for the short, medium and long-term periods; c) whether a power supply system can work persistently at arbitrarily changing power from wind and PV stations; d) whether it is possible to regulate the power and voltage in the connection point of wind and PV stations to a power supply system, e) water consumption.

Power generation forecasting: Operating experience of wind and PV stations confirms the possibility to forecast power generation for various periods: year, month, days and even hours, with probability to 95%. First, it has become possible thanks to accumulation of data on the annual wind speed and solar radiation at wind and PV stations during operation, to dispersal of wind and PV stations on the country territory. Secondly, use of the wind speed forecast according to NASA database. On the website of Red Eléctrica de España it is possible to observe the share of power generation at all types of stations, including the forecast and actual power generation at any time.

Power reservation: On the basis of rather exact forecasting it is possible to solve the main problem – the necessity to reserve power of a wind station. According to the conclusion of Utility Wind Group (the organization combining 55 USA electric companies and having wind power stations in its power supply systems), the outdated and nonprofessional opinion which is often expressed in power industry is that wind stations need reservation of all power transferred by them. Now it is clear that at a moderate percentage of wind power is necessary to have rather low, than one to one, and often close to zero, additional generating power for compensation of wind power stations instability [9].

In the research carried out in the USA in 2004 for Minnesota Board of Trade it has been shown that inclusion of 1500 MW wind power stations in the state power supply system will require additional 8 MW (0.5%) station on traditional fuel [15].

In the power supply system of Russia the installed capacity utilization factor, e.g., in 2014, was 0.50. In principle, it means that there is always power reserve. The problem is load peak which happens in certain days in December – January. This circumstance has to be taken into account at the choice of the location for construction of wind power and PV stations. Thus, the problem of power reservation of wind and PV stations in Russia is not an obstacle for development of wind energy, at least, till that time, when 10% share of wind and PV power in total amount of electricity generation will be achieved.

Power regulation: Of course, the opportunity to increase power of wind stations is limited by the wind power density and in this context it is possible to say that such opportunity is unavailable. Regulation of actual power towards reduction of wind station power is unlimited: it is possible to disconnect a single wind power plant or to regulate the blade angle of rotation.

It should be specially noted that wind power stations and new design of PV stations containing an inverter or a frequency converter have the controlled reactive power range equal to the station power. It means that modern wind and PV stations can sustain voltage in the point of connection to a power supply system at the set level. This property of a wind power station is not yet completely understood by experts.

Water consumption: Water requirement belongs to operational efficiency of wind and PV stations. Specific loss of water at a wind power station is 475 times less than that of nuclear power plant; 400 times less than at coal power stations and 275 times less than at gas-fired power plants. Specific loss of water at PV stations is 20 times less than at a nuclear power plant, 17 times less than at coal power stations and 14 times less than at gas-fired power plants (Table 4) [6,8].

Table IV: Non-recoverable losses of water at the USA power plants

Power plant type	NPP	Coal power station	Oil power plant	Gas power plant	Wind power plant	PV station
Water specific loss, l/kWh	2.3	1.9	1.6	0.95	0.004	0.11

And it is apart from water requirement for a steam-water cycle of thermal and nuclear power plants and cooling water of turbine condensers, whereas at wind and PV stations the permanent water consumption is intended only for cleaning wind turbine blades and PV modules.

2.4 Ecological Efficiency [1-4, 6-8]

It is generally agreed that wind, photovoltaic, hydraulic and geothermal power stations do not produce CO_2, they are environmentally friendly. However, they consist of components at which production of which CO_2 and other noxious emissions occur. The paper provides the assessment of this effect.

The name of the research in this field is life cycle emission study.

Photovoltaic stations: Greenhouse gases emission recounted in equivalent CO_2 emission in 2009 was 29, 28 and 18 g of CO_2 /kWh for monocrystalline, polycrystalline silicon and CdTe solar cells, respectively.

NO_x emission for the above solar cells is equal to 0.067, 0.058 and 0.03 g/kWh, respectively. Heavy metal emission for CdTe solar cells in the form of cadmium emission is rather low: 0.02 g/kWh. For comparison: in the USA the emissions of a coal-fired power plant for 1 kWh are 1 kg of CO_2/kWh, 8 g of SO_2, 3 g of NO_x and 0.4 g of flying particles.

The noxious emission from PV stations is less than from coal-fired power plants for CO_2 – 35–55 times, NO_x – 44–100 times.

Wind power stations: According to the research of World Steel Association in 2008: for 160 MW offshore wind power plant (eighty 2MW wind turbines)in Denmark in North Sea CO_2 emission was 7.6 g/kWh, according to [12] - 6.6 g/kWh, according to [3] - 8 g/kWh. So, CO_2 emission at wind power plants construction is two orders less than that of thermal stations.

Tables 5 and 6 show summarized data of noxious emissions from thermal stations in the USA and Russia (according to the Energy Strategy Institute, Russia). This information confirms the difference in degree of ecological impact or renewable and thermal energy.

Table V: Average readings of gas specific emission by USA fuel power industry

Gas	Specific values at fuel utilization, g/kWh			
	Coal	Natural gas	Oil	Average
Carbon dioxide (CO_2)	967	468	708	690
Sulphur dioxide (SO_2)	6.1	0.0032	3.1	3.63
Nitrogen oxides (NO_x)	3.45	0.82	0.95	2.22

Table VI: Average specific values of noxious emission from Russian thermal power plants

Contaminator type	Specific values at fuel utilization, g/kWh		
	Oil with 1% sulphur	Coal with 1% sulphur	Natural gas
Carbon dioxide (CO_2)	738	1142	547
Sulphur dioxide (SO_2)	4.7	1.4	–
Nitrogen oxides (NO_x)	2.3	4.0	1.8
Ash and slag	–	52	–
Fly ash	–	0.33	–

2.5 Social Efficiency

Besides decrease of noxious emissions influencing negatively the population health, the significant social effect of renewable energy development is job creation for equipment production and operation of renewable energy facilities. Herewith workplaces directly relating to renewable energy (direct) and the ones in linked industries (indirect) are taken into account.

According to estimation of REN-21, 2011 the total number of workplaces in renewable energy sector in the world in 2010 was 3.5 million. The greatest contribution is made by biofuel production processes – 1.5 million workplaces [14].

However for Russia social efficiency of renewable energy is not settled by the quantity of workplaces.

Up to 20 million people in Russia live in regions with unstable power supply, including remote places where power supply is carried out from outmoded diesel power plants. Often the population in these areas can use electricity two times a day for 2-3 hours in the morning and in the evening. Diesel fuel delivery to these regions became more complex. The diesel fuel cost increases manifold due to delivery difficulties. Therefore the electricity cost exceeds the tariffs of the centralized power supply zones by 5–10 times and the government needs to subsidize the population for electricity payment.

The use of renewable energy sources for power supply allows to curtail drastically the electricity cost and to improve the population living conditions and development conditions of local production.

3 CONCLUSIONS

In terms of energy wind, PV and hydro-electric

power stations have global advantage over thermal stations as they compensate the energy consumed for their creation (wind power stations – in about one year, PV stations – in 2-5 years), but thermal stations do not compensate it ever.

Even taking into account CO_2 emission at materials and equipment production and also construction and erection work at wind power and PV stations their ecological cleanliness is rather higher, than that of thermal stations.

Social efficiency of renewable energy is characterized by availability of direct and indirect workplaces with high stability degree as this branch is less subjected to crisis impacts. Lack of issue of CO_2 and other polluting emissions improves population living conditions. Small power plants affect significantly local production development and living comfort of people not connected to centralized electricity network.

According to foreign practice special energy reservation of wind power and PV stations is not required in case of their optimum combination with other power plants types. As a rule, existing reserves in power supply systems are enough for compensation of production instability at wind power and PV stations. Vast experience of wind power and PV stations operation together with modern methods allows to forecast power of these stations for the settling periods with high probability (95% and more).

By 2018 according to minimum and average normalized cost of electric power, wind stations will be more effective than all types of coal power plants, gas power plants with a combined advanced cycle and CO_2 burial, gas-turbine stations, nuclear power plants.

Hydroelectric stations will be more effective than other renewable energy plants. PV stations will approach improved coal power plants.

So, even now we can speak about competiveness of renewable energy.

4 REFERENCES

[1] Ch. Ghenai, Life cycle analysis of wind turbine, Ocean and Mechanical Engineering, 2012, Florida Atlantic University, www.intechopen.com.

[2] N. D'Souza, E. Ghegbaje-Das and P. Shonfield, Life cycle assessment of electricity production from a Vestas V 112, Final Report, vol. 112, Turbine Wind Plant, February 2011, PE North West Europe ApS.

[3] V. Fthenakis, H.C. Kim, R. Frichknecht, M. Rongei, P. Sinha and M. Stucki,. Photovoltaic power systems programme. Life cycle inventories and life cycle assessment of photovoltaic systems, IEA, October, 2011.

[4] K. Suomalainen, Environmental life cycle assessment of a large-scale grid connected PV plant: case study of Moura MW PV Power Plant, Goteborg, Chalmers University of Technology, 2006.

[5] J.Ch. Smith, M.R. Milligan, E.A. DeMeo and B. Parsons, Utility wind integration and operating impact state of art, IEEE Transactions on Power Systems, vol. 22, no. 3, August 2007.

[6] P.P. Bezrukikh and P.P. Bezrukikh (Jr.), On indicators of energy status and renewable power efficiency under the conditions of economic crisis, Economic Questions, 2014, no. 8, pp. 92-105.

[7] P.P. Bezrukikh and P.P. Bezrukikh (Jr.), World wind energy. Status and development forecast, Elektricheskie Stantsii (Power Plants), no. 12, 2015, pp. 27-29.

[8] P.P. Bezrukikh, S.M. Karabanov and P.P. Bezrukikh (Jr.), The efficiency of the world renewable energy, Energy, no. 12, 2015.

[9] A. Cabraal, S. Aqarwal and M. Takahashi, Rising to the challenge – the whys and whens of renewable energy, Renewable Energy World, 2007, vol. 10, no. 4, pp. 117–129.

[10] Energy subsidies in the contemporary world. G20 countries, under the editorship of L.M. Grigogiev and A.N. Kurdin, Moscow, Asmin-Print Ltd. 2014.

[11] Levelized cost of new generation resources in the annual energy outlook, 2011. Report of the US Energy Information Administration (EIA) of the US Department of Energy (DOE). January 23, 2012.

[12] US Energy Information Administration, Annual Energy Outlook 2015, April 2015. DOE/EIA-0383.

[13] S.M. Karabanov, P.P. Bezrukikh and T.A. Shushkanova, Energy of the future. Photovoltaic conversion of solar energy. Thin film solar cells and modukes (technology, economy, market analysis, development prospects, Moscow, Energy, 2014.

[14] REN 21 Renewables. Global Status Report, 2012, 2013, 2014, 2015

[15] R. Zavadil, J. King, L. Xiadong, M. Ahlstrom, B. Lee, D. Moon, C. Finley, L. Alnes, L. Jones, F. Hudry, M. Monsteaim, S. Lai and J. Smith, Wind Integration Study, Final Report, 2004

100% RENEWABLE ENERGY IN NORTH AMERICA AND THE ROLE OF SOLAR PHOTOVOLTAICS

Arman Aghahosseini, Dmitrii Bogdanov and Christian Breyer
Lappeenranta University of Technology, Skinnarilankatu 34, 53850 Lappeenranta, Finland
E-mail: Arman.Aghahosseini@lut.fi, Christian.Breyer@lut.fi

ABSTRACT:
Renewable energy (RE) has been already viewed as a minor contributor in the final energy mix of North America due to cost and intermittency constraints. However, recent dramatic cost reductions and new initiatives using RE, particularly solar PV and wind energy, as a main energy source for the future energy mix of the world pave the way for enabling this source of energy to become cost competitive and beneficial in comparison to fossil fuels. Other alternatives such as nuclear energy and coal-fired power plants with carbon capture and storage (CCS) cannot play an important role in the future of energy system, mainly due to safety and economic constraints for these technologies. Phasing out nuclear and fossil fuels is still under discussion, however the 'net zero' greenhouse gas emissions agreed at COP21 in Paris clearly guides the pathway towards sustainability. Consequently, RE would be the only trustable energy source towards a clean and sustainable world. In this study, an hourly resolved model has been developed based on linear optimization of energy system parameters under given constraints with a bright perspective of RE power generation and demand for North America. The geographical, technical and economic potential of different types of RE resources in North America, including wind energy, solar PV, hydro, geothermal and biomass energy sources enable the option to build a Super Grid connection between different North American regions' energy resources to achieve synergy effects and make a 100% RE supply possible. The North American region, including the US, Canada and Mexico in this paper, is divided into 20 sub-regions based on their population, demand, area and electricity grid structure. These sub-regions are interconnected by high voltage direct current (HVDC) power lines. The main objective of this paper is to assume a 100% RE-based system for North America in 2030 and to evaluate its results from different perspectives. Four scenarios have been evaluated according to different HVDC transmission grid development levels, including a region-wide, country-wide, area-wide and integrated scenario. The levelized cost of electricity (LCOE) is found to be 63 €/MWh$_{el}$ in a decentralized scenario. However, it is observed that this amount decreases to 53 €/MWh$_{el}$ in a more centralized HVDC grid connected scenario. In the integrated scenario, which consists of industrial gas production and reverse osmosis water desalination demand, integration of new sectors provides the system with required flexibility and increases the efficiency of the usage of storage technologies. Therefore, the LCOE declines to 42 €/MWh$_{el}$ and the total electricity generation is decreased by around 6.6% in the energy system compared to the non-integrated sectors due to higher system efficiency enabled by more flexibility. The results clearly show that a 100% RE-based system is feasible and a real policy option.

Keywords: 100% renewable energy, North America, energy system optimization, grid integration, solar PV, wind energy, storage.

1 INTRODUCTION

Over the past few decades, the global population has increased and living standards have advanced dramatically in many parts of the world. Consequently, the demand for energy is increasing, particularly in Organization for Economic Cooperation and Development (OECD) countries mainly because of their lifestyles and high economic development. In fact, the US consist of less than 5% of the world's population [1] and yet consume as much as 25% of total primary energy demand [2]. As the world's population continues to grow, the current situation in energy supply and demand, climate change impacts and depletion of fossil resources will continue to worsen if these issues are not addressed. Phasing out nuclear and fossil fuels is still under discussion, however the 'net zero' greenhouse gas emissions target around mid-21st century agreed at COP21 in Paris clearly guides the pathway towards sustainability [3].

The geographical, technical and economic potential of different types of renewable energy (RE) resources in North America[1], including solar photovoltaic (PV), wind energy, hydro power, geothermal, waste-to-energy and biomass energy resources enable a successful project of building a Super Grid connection between different North American regions' energy resources to achieve synergy effects and make a 100% RE supply possible [4]-[7]. As the cost of RE resources begins to match conventional technologies, transition to such variable RE sources has become more realistic. North America has experienced significant solar photovoltaics (PV) and wind market growth since the past few years [8]. US is the main leader of solar potential in North America with total installed capacity of around 25.99 GW by the end of 2015 [9], also driven by the fact that solar PV is in more and more states the least cost power source [10]. At the end of 2015, the PV market had surpassed the limit of 28 GW installed capacity in North America [9],[11].

In the coming years, it is expected that the cost of solar PV and wind energy will further decrease, while the installed capacity will grow significantly. However, reduction of cost and increase the installed capacity mainly depends on the defined target and goal. International Energy Agency (IEA) anticipated in its World Energy Outlook report [12] that the installed capacity of PV in North America will arise to around 100 GW by 2030. In the New Policies Scenario, solar PV is expected to be a key low-carbon technology in many regions of the world, exceeding 1000 GW of installed capacity globally by 2040. This is one of the most

[1] North America in this study refers to the US, Canada and Mexico.

conservative projections [13] While Fraunhofer institute for solar energy system (ISE) [14] in the Average Value Scenario predicted that the installed capacity of PV will reach 2016 GW globally by 2030. In this report, the levelized cost of electricity (LCOE) predicted for solar PV is around 45-70 €/MWh by 2030. However, the latest report of International Renewable Energy Agency (IRENA) [15] has shown that LCOE for solar PV will fall by 59% in comparison to the year 2015 and become 44[2] €/MWh by 2025. However, already in the year 2016 several PV projects had been contracted for prices well below 37 €/MWh, for instance in Mexico and Peru [16],[17] As a result, different institutes and agencies forecast the future of RE resources according to their visions and reality is progressing very fast, typically faster than the expectations and forecasts. The financial analysts of Lazard mentioned utility-scale PV projects in 2015 being possible at a cost level of 37-43 €/MWh and expecting a further decline to 32-34 €/MWh in 2017 [18]

Renewable policy has been largely driven in the US by supply security concerns on the federal level, and economic activity and greenhouse gas mitigation concerns on the state level [19]. Recently, US federal and some State Governments have supported extending many forms of renewable power generation. On the federal level, reducing the dependence on oil imports is crucial. Renewables can help to meet this independence as part of the strategy. The national energy strategy of the US has been classified an "all-of-the-above strategy" [19]. The key goals being highlighted in this strategy are generating more electricity from RE resources, consuming less oil while holding electricity consumption constant, achieving significant economic and energy security benefits, reducing carbon emissions in the energy sector and as a consequence tackle the challenge posed by climate change [19].

Canada is a world leader in the production and consumption of RE. In the electricity sector, hydropower is the largest RE source in Canada, accounting for around 60% of electricity generation in the country. Other sources of RE, such as biomass, wind, tidal and solar contribute to rising the share of renewable to over 63% [21]. Therefore, over 63% of Canada's electricity generation does not emit greenhouse gases. The rest of the required power come from nuclear with almost 14% and fossil fuels with 22% of total share [21]. The growth in the use of RE for electricity generation is a result of a number of targeted, complementary policy initiatives undertaken by federal and provincial governments. All provinces across Canada have been increasing the use of RE through a number of initiatives, including offset programs procurement through requests for proposals, standard offer and feed-in tariff programs, and legislated renewable portfolio standards. For example, requests for proposals have been used in all provinces with the exception of Alberta, which has an offset program in place. Renewable portfolio standards are legislated in Nova Scotia, New Brunswick and Prince Edward Island, while standard offer and feed-in tariff programs are used in Ontario, Nova Scotia, Prince Edward Island and British Columbia [21].

Canada has a cross-border electricity trade with the United States and renewables play a key role in this trade. Provinces that have significant hydropower resources, such as Quebec, Newfoundland & Labrador, Manitoba, British Columbia and Yukon are the major net exporters of electricity. In recent years, Manitoba has exported more than a quarter of its electricity production into the US Midwest [21]. In the Midwest, it will help balance wind that is being added to the grid in states like North Dakota.

Mexico has overall targets of 24.9% electricity from clean energy sources by 2018, 35% by 2024, 40% by 2035 and 50% by 2050 [22]. Clean energy is defined to include renewables, cogeneration, nuclear energy, fossil fuels with carbon capture and storage (CCS) and other low-carbon technologies. However, nuclear energy and fossil fuels with CCS have not proven to be a kind of clean energy. Since coal-based CCS would still emit heavy metals causing severe health problems [20] and the CO_2 extraction with less than 90% efficiency is too low for a net zero world, and nuclear energy violates all sustainability criteria. The 2014 Special Programme for the use of RE (PEAER) set a target of 24,345 MW of RE capacity by 2018 disaggregated as follows: hydropower (13,030 MW), wind (8922 MW); geothermal (1018 MW), bioenergy (784 MW) and solar (627 MW) [22]. Solar PV has proven to be a least cost source of electricity according to results of tenders in 2016 with 40.5 USD/MWh for more than one GW of capacities [16]. The recent evolution of RE policy in Mexico was driven by the Law for the Use of Renewable Energies and Financing the Energy Transition, which was published on 28 November 2008. Its goal was to encourage the use of RE and clean technologies for electricity generation.

Different scenarios of energy systems based on an entirely sustainable energy system or high shares of RE had been already discussed for several countries and regions. Integrated Canada-US power sector modelling has already done by Martinez et al. [23] and the result provides the groundwork for these improvements.

This study aims at designing an optimal and cost competitive 100% RE power system for North America considering a potential evolution of the generation mix that takes into account:

- the actual electricity trade and transmission infrastructure of different sub-regions of North America;
- an optimal system design and wise utilization of considered available RE resources;
- synergy between various resources and different regions that increase the efficiency of the power sector.

2 METHODOLOGY

An hourly resolved model has been developed based on linear optimization of energy system parameters under given constraints with a bright perspective of RE power generation and demand. The main features of the model are its flexibility and expandability. The model is composed of several power generation and storage technologies, respective installed capacities and different operation modes of these technologies, which are used to supply the electricity demand of power, water desalination and synthetic natural gas (SNG) generation sectors [24],[25]. The energy system model can be separated into market and regulatory models. Regulatory

[2] US Dollar to Euro foreign exchange rate is assumed 1.35 in this paper.

models only take the long-term energy system structure into account and short-term market mechanisms are not considered. This model has already been applied to several regions up to now [24]-[30], and a detailed description can be found in those studies. Further technical and financial assumptions can be found in the Supplementary Material in the appendix of this paper.

2.1. Model summary

A multi-node approach enables a definition of a grid configuration in scenario assumptions for power transmission interconnections among the sub-regions. The main constraint for the optimization is the matching of the power generation and demand for every hour of the applied year. The computation time takes longer when the hourly resolution is used in the model. Nevertheless, it guarantees that for every hour of the year that the total generation of a sub-region and electricity import cover the local electricity demand from all considered sectors and enables a precise system description including synergy effects of different system components for the system balance. The main target of the system optimization is the minimization of the total annual energy system cost, calculated as the sum of the annual costs of installed capacities of the different technologies, costs of energy generation and generation ramping. The system also includes distributed generation and self-consumption of residential, commercial and industrial electricity consumers (prosumers) by installing respective capacities of rooftop PV systems and batteries. For these prosumers the target function is minimal cost of consumed energy calculated as the sum of self-generation, annual cost and cost of electricity consumed from the grid, minus benefits from the selling of excess energy. The model flow diagram that contains all the considered input data, system models and model output data is presented in Figure 1.

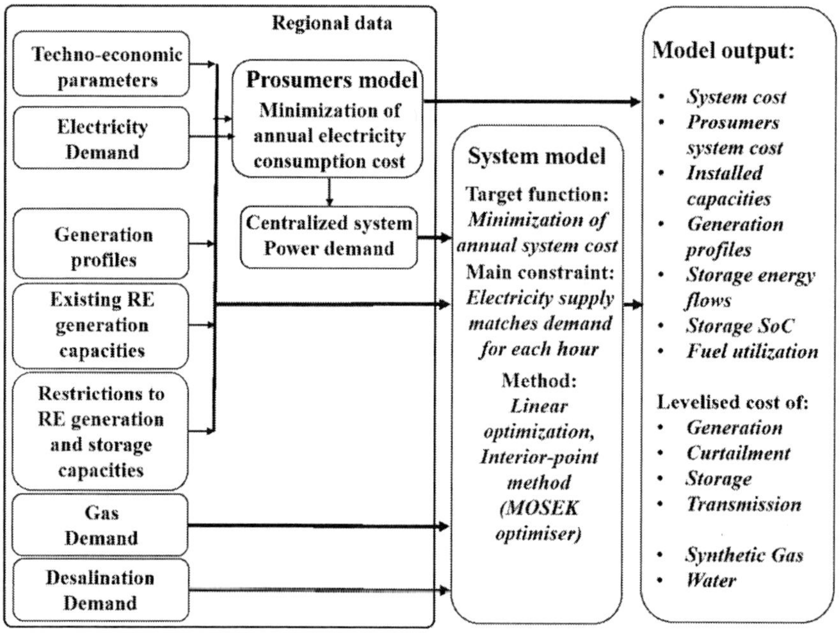

Figure 1: Model flow diagram that is considered in this paper.

2.2. Input data

The model uses several types of input data and constraints which are listed below:

- historical weather data for direct and diffuse solar irradiation, wind speed and precipitation amounts;
- synthetic load data for every sub-region based on the available hourly load data at a national level. In addition, local data of each sub-region are also taken into consideration, such as population, temperature, gross domestic product and power plant structure;
- synthetic geothermal energy potential;
- non-energy natural gas consumption sector for every sub-region;
- projected water desalination demand for every sub-region;
- technical characteristics of energy generation, storage and transmission technologies, such as power yield, energy conversion efficiency, power losses in transmission lines and storage round trip efficiency;
- technical characteristics of seawater reverse osmosis (SWRO) desalination;
- capital expenditures, operational expenditures and ramping costs for all technologies;
- electricity costs for residential, commercial and industrial consumers,
- limits for minimum and maximum installed capacity for all energy technologies,
- configuration of regions and interconnections.

The datasets for wind speed, solar irradiation components and precipitation are collected from NASA databases [31],[32] and partly reprocessed by the German Aerospace Center [33]. The spatial resolution of the data is 0.45° × 0.45°. The time resolution is hourly for wind speed and solar irradiation, and monthly for precipitation. IEA data [12] are used to estimate natural gas consumption in the industrial sector for the year 2030 (gas demand for electricity generation, residential and transportation sectors are not considered in this study). Water demand is calculated based on water consumption projections and future water stress [34]. It is assumed that water stress greater than 50% shall be covered by seawater desalination. Water transportation costs are also taken into account, which is described in Caldera et al. [35]. Annual industrial gas demand and water demand are presented in the Supplementary Material (Table 11).

Geothermal data are evaluated based on existing information on the surface heat flow rate [36],[37] and surface ambient temperature for the year 2005 globally. For areas where surface heat flow data are not available, the extrapolation of existing heat flow data was performed. Based on that, temperature levels and available heat of the middle depth point of each 1 km thick layer, between depths of 1 km and 10 km [38]-[40] globally with 0.45° × 0.45° spatial resolution, are derived. The calculated geothermal potentials are presented in the Supplementary Material (Table 6).

The potentials for biomass and waste resources are taken from German Biomass Research Center [41]. All bio-based by-products and waste are divided into three different components: solid waste, solid biomass and biogas. Solid wastes consists of municipal and industrial used wood; solid biomass includes straw, wood and coconut residues; biogas is comprised of excrement, municipal biowaste and bagasse. The costs for biomass are calculated using data from [42] and [43]. For solid fuels a 50 €/ton gate fee for the waste incineration is assumed. The results for regional biomass potentials and costs are provided in the Supplementary Material (Table 6 and 7, respectively).

All the input data can be found in the Supplementary Material.

2.3. Applied technologies

Four types of technologies are applied in North America energy system optimization, which consist of conversion of RE resources into electricity, energy storage, energy sector bridging and electricity transmission. The full model block diagram is shown in Figure 2.

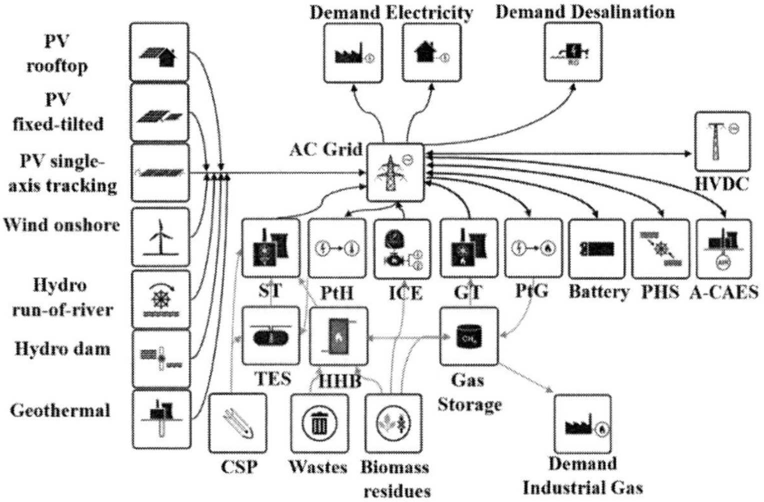

Figure 2: Block diagram of the energy system model for North America.

The technologies for converting RE resources into electricity that are used in this model are solar PV (distributed rooftop systems, ground-mounted and single-axis tracking plants), concentrating solar thermal power (CSP), onshore wind turbines, biomass plants (solid biomass and biogas), waste-to-energy power plants, hydropower (run-of-river and dam) and geothermal energy.

In addition to the mentioned energy sources, the energy storage technologies used in this model are batteries, pumped hydro storage (PHS), thermal energy storage (TES), adiabatic compressed air energy storage (A-CAES) and power-to-gas (PtG) technology. PtG includes synthetic natural gas (SNG) with the SNG synthesis technologies: water electrolysis, methanation, CO_2 scrubbing from air, gas storage, and both combined and open cycle gas turbines (CCGT, OCGT). SNG synthesis process technologies have to be operated in synchronization because of hydrogen and CO_2 storage absence. Additionally, there is a 48-hour biogas buffer storage and a part of the biogas can be upgraded to biomethane and injected into the gas storage.

The energy sector bridging technologies provide more flexibility to the entire energy system, thus reducing the overall cost. One bridging technology available in the model is PtG technology for the case that the produced gas is consumed in the industrial sector and not as a storage option for the electricity sector. The second bridging technology is SWRO desalination, which couples the water sector to the electricity sector.

The electricity transmission technologies are represented on two levels: power distribution and transmission within the sub-regions are assumed to be based on standard alternating current (AC) grids which are not part of the model and inter-regional transmission grids modeled by applying high voltage direct current (HVDC) technology. Power losses in the HVDC grids consist of two major components: length dependent electricity losses of the power lines and losses in the converter stations at the interconnection with the AC grid.

An energy system mainly based on RE and in particular intermittent solar PV and wind energy requires different types of flexibility for an overall balanced and cost optimized energy mix. The four major categories are generation management (e.g. hydro dams or biomass plants), demand side management (e.g. PtG, SWRO desalination), storage of energy at one location and energy shifted in time (e.g. batteries), and transmission grids connecting different locations and energy shifted in location (e.g. HVDC transmission).

3 SCENARIO ASSUMPTIONS

3.1. Regions subdivision and grid structure

The three North American countries, including the US, Canada and Mexico, were clustered into 20 sub-regions. The subdivision and grid configuration of North America's regions are presented in Figure 3.

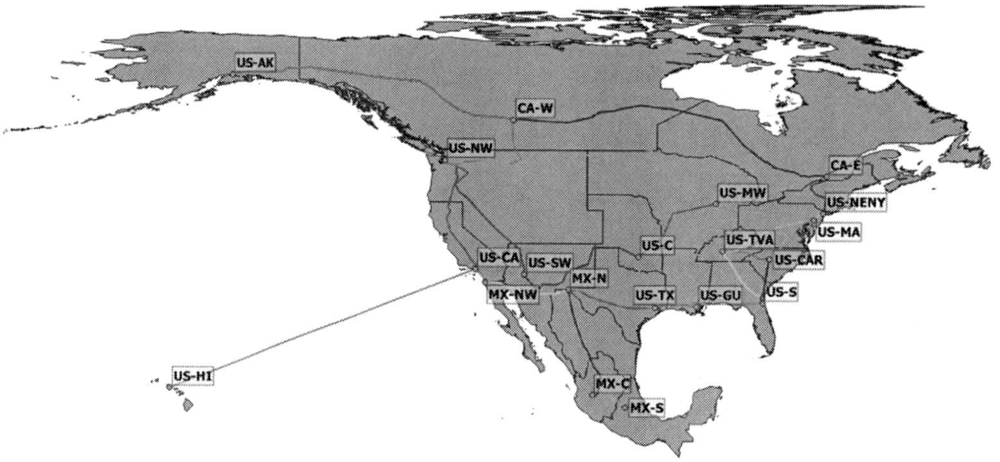

Figure 3: The figure represents North American sub-regions, including the US, Canada and Mexico, and HVDC transmission lines configuration. This region is divided into 20 sub-regions: US-New England & New York (US-NENY), US-Mid-Atlantics (US-MA), US-Carolinas (US-CAR), US-Southern (US-S), US-Tennessee Valley Authority (US-TVA), US-Midwest (US-MW), US-Central (US-C), US-Texas (US-TX), US-Southwest (US-SW), US-Northwest (US-NW), US-California (US-CA), US-Alaska (US-AK), US-Hawaii (US-HI), US-Gulf (US-GU), West Canada (CA-W), East Canada (CA-E), Northwest Mexico (MX-NW), North Mexico (MX-N), Central Mexico (MX-C) and South Mexico (MX-S). Behrmann Cylindrical Projection is used for all the maps in order to compress the northern part of Canada which is not the main discussion in this study.

In this paper four scenarios for energy system development options are discussed:

- Region-wide energy systems, in which all the regions are independent (no HVDC grid interconnections) and the electricity demand has to be covered by the respective region's own generation;
- Country-wide energy system, in which the regional energy systems are interconnected by HVDC grids within the borders of nations;
- Area-wide energy system, in which the region-wide energy systems are interconnected;
- Integrated scenario, the area-wide energy system scenario with SWRO desalination and industrial gas demand. In this scenario, RE combined with PtG technology are used not only as electricity generation and storage options within the system, but also as energy sector bridging technologies to cover water desalination and industrial gas demand, increasing the flexibility of the system.

3.2. Financial and technical assumptions

The model optimization is carried out on an assumed cost basis and technological status for the year 2030 and the overnight building approach. The financial assumptions for capital expenditures (capex), operational expenditures (opex) and lifetimes of all components are provided in the Supplementary Material (Table 1). Weighted average cost of capital (WACC) is set to 7% for all scenarios, but for residential PV self-consumption WACC is set to 4%, due to lower financial return requirements. The technical assumptions concerning power to energy ratios for storage technologies, efficiency numbers for generation and storage technologies, and power losses in HVDC power lines and converters are provided in the Supplementary Material (Tables 2, 3 and 4, respectively). Electricity prices in residential, commercial and industrial sectors for the US's regions are taken from [44]-[49], for Canada's regions are taken from [50]-[56] and for Mexico's regions are taken from [57]-[59]. Prices are provided in the Supplementary Material (Table 5). The electricity price is on a country basis. However, for the sub-regions electricity prices are the weighted average of a sub-region. Then, electricity prices are extrapolated in order to calculate electricity prices for the year 2030 using the approach of Gerlach et al. [60]. The excess electricity generation, which cannot be self-consumed by the solar PV prosumers, is assumed to be fed into the grid for a transfer selling price of 2 €cents/kWh. Prosumers cannot sell to the grid more power than their own annual consumption.

3.3. Feed-in for solar and wind energy

The feed-in profiles for solar CSP, optimally tilted and single-axis tracking PV, and wind energy were calculated according to Bogdanov and Breyer [24],[25]. The aggregated profiles of solar PV generation (optimally tilted and single-axis tracking), wind energy power generation and CSP solar field, normalized to maximum capacity averaged for North America are presented in Figure 4. The computed average full load hours (FLH) for optimally tilted, single-axis tracking PV systems, wind power plants and CSP are provided in the Supplementary Material (Table 8).

The feed-in values for hydropower are computed based on the historical weather data for precipitation for the year 2005 as a normalized sum of precipitation in the regions. Such an estimate leads to a good approximation of the annual generation of hydropower plants.

3.4. Upper and lower limitations on installed capacities

Lower limits of installed capacities are taken from Farfan and Breyer [11] and upper limits are calculated according to Bogdanov and Breyer [25]. Lower limits on already installed capacities in North America sub-regions are provided in the Supplementary Material (Table 9) and all upper limits of installable capacities in North America sub-regions are summarized in the Supplementary Material (Table 10). For all other technologies, upper limits are not specified. However, for solid biomass residues, biogas and waste-to-energy plants it is assumed, due to energy efficiency reasons, that the available and specified amount of the fuel is used during the year.

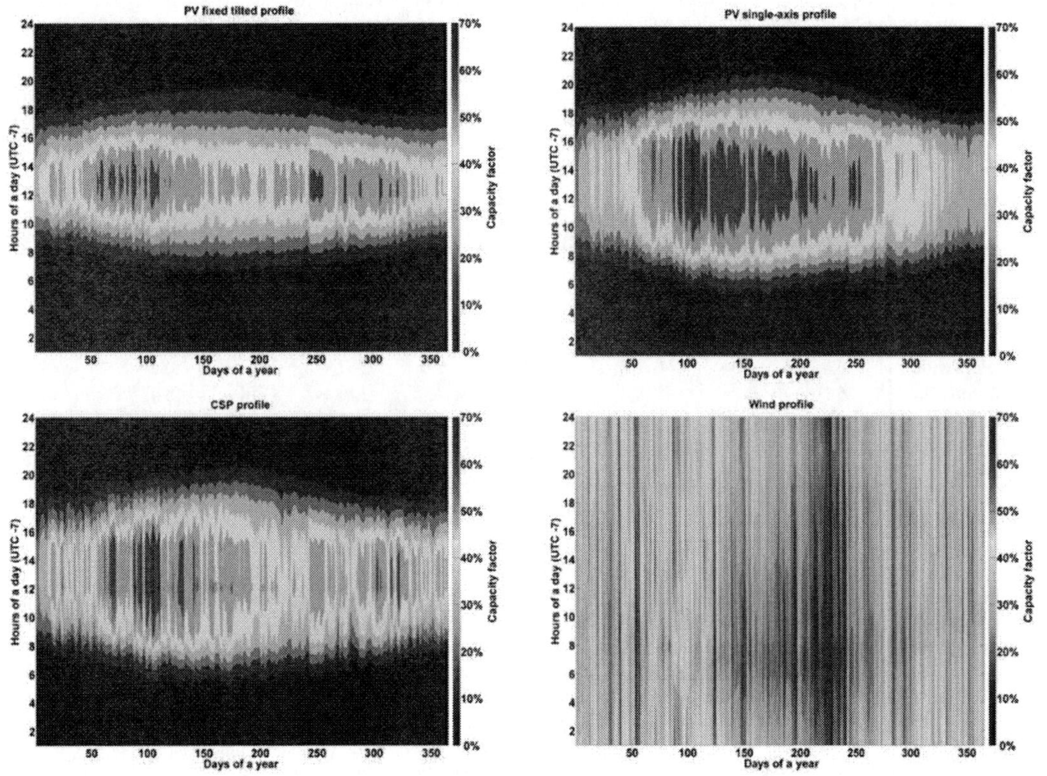

Figure 4: Aggregated feed-in profiles for optimally tilted PV (top left), single-axis tracking PV (top right), CSP solar field (bottom left) and wind power plant (bottom right) in North America.

3.5. Load profile

The demand profiles for sub-regions are computed as the total demand of regions based on synthetic load data. Figure 5 represents the area-aggregated demand of all sub-regions in North America. Current electricity demand is collected from local sources [46]-[48], and the electricity demand growth rate by the year 2030 is estimated using IEA data [12] According to the IEA [12], the electricity demand is set to continue growing by 0.9%

p.a. in North America by 2030. Solar PV self-consumption prosumers have a substantial impact on the residual load demand in the energy system as shown in Figure 5 (right). The overall electricity demand and the peak load are dropped by 19.7% and 8.5%, respectively.

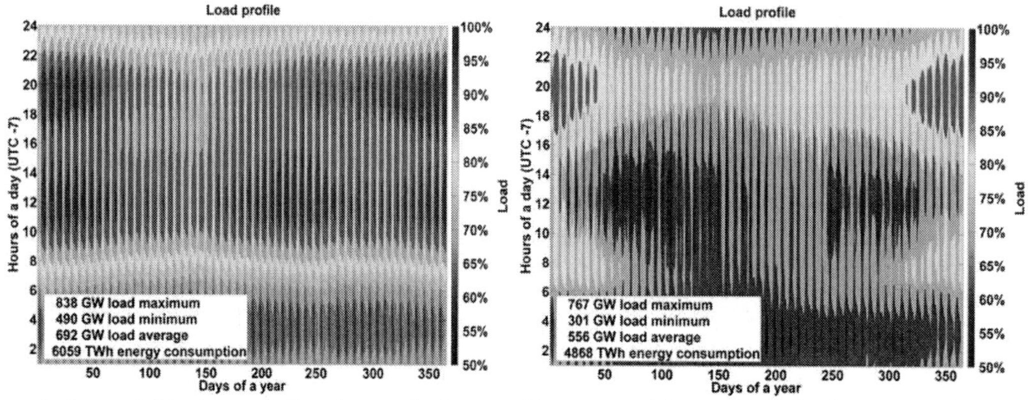

Figure 5: Aggregated load curve (left) and system load curve with prosumers influence (right) for the year 2030 in North America.

4 RESULTS

For each scenario, a cost minimized electrical energy system configuration is applied for the given constraints and characterized by optimized installed capacities of RE electricity generation, storage and transmission for every technology used in the model. Therefore, the final results include hourly electricity generation, storage charging and discharging, import and export of electricity between sub-regions, and curtailment. The key financial results of the different scenarios for the total energy system are presented in Table 1. The main financial parameters give data on levelized cost of electricity (LCOE) (including PV self-consumption and the centralized system), levelized cost of electricity for primary generation (LCOE primary), levelized cost of curtailment (LCOC), levelized cost of storage (LCOS), levelized cost of transmission (LCOT), total annualized cost, total capital expenditures, total RE capacity and total primary generation.

Table 1: Key financial results for the four analyzed scenarios applied for North America.

	Total LCOE	LCOE primary	LCOC	LCOS	LCOT	Total ann. cost	Total CAPEX	RE capacities	Generated electricity
	[€/MWh]	[€/MWh]	[€/MWh]	[€/MWh]	[€/MWh]	[b€]	[b€]	[GW]	[TWh]
Region-wide	63	38	2.8	23	0.0	387.6	3502	3354	7207
Country-wide	56	36	2.5	15	3.1	340.6	3114	2941	7062
Area-wide	53	35	2.1	12	3.9	320.6	2945	2760	6861
Integrated scenario	42	33	1.2	5	3.4	498.2	4668	4152	10975

From Table 1, it can be clearly observed that the integration benefit is significant in both electricity cost and annual expenditures due to connection of different regions via HVDC transmission lines. Electricity cost of the entire system in the area-wide open trade scenario plunged by about 16% compared to the region-wide scenario. Grid utilization decreases the primary energy conversion capacities and generation by 18% in terms of installed capacities and by 5% in terms of generated electricity in reference to the region-wide scenario. Grid utilization leads to a considerable reduction of storage utilization (Table 2), especially A-CAES, whereas cost of transmission is relatively small in comparison to the decrease in primary generation and storage costs. Curtailment costs decrease by half in comparison to storage costs in the case of broader grid utilization, leading to a reduction of about 25% in the area-wide scenario compared to the region-wide scenario. However, the impact of excess energy on total cost is rather low. The power line capacities for the electricity trade between the sub-regions for the area-wide open trade scenario is shown in the Supplementary Material (Figure 7 and Table 15).

A further decrease in LCOE of 16% can be reached by the integration of water desalination and industrial gas sectors compared to the area-wide open trade scenario. This cost reduction can be explained by a decrease in storage cost by 58% since industrial gas and desalination sectors decrease the need for long-term storage utilization, giving additional flexibility to the system. Primary electricity generation cost declined by 6% in the integrated scenario compared to the area-wide scenario and the reason could be an increase in the flexibility of the system and the utilization of low-cost wind and solar electricity as can be seen in Table 2. For the case of biogas, a fraction of 32% of the biogas used in biogas power plants in the area wide-open trade scenario is reallocated in the integrated scenario from the electricity sector to the industrial gas demand for efficiency reasons.

The LCOE components and the import/export share in the region-wide, country-wide, area-wide and integrated scenario are presented in the Supplementary Material (Table 12). The share of export is defined as the ratio of net exported electricity to the generated primary electricity of a sub-region and the share of import is defined as the ratio of imported electricity to the electricity demand. The area average is composed of sub-regions' values weighted by the electricity demand.

Concerning the total installed capacities, for PV total the installed capacities decrease by 32% from the region-wide to area-wide scenario while for wind energy the installed capacities increase by 9% from the region-wide to area-wide scenario. The reason for increasing the installed capacity of wind is that the LCOE for wind energy in some sub-regions is lower than solar PV. Therefore, the system prefers to install wind rather than PV single-axis tracking and exchange this electricity with neighboring regions via power lines. In addition, the wind resource is at a really excellent level in several parts of North America, particularly in the US-Central, US-Midwest, US-Texas, East Canada, West Canada, the west part of US-Alaska and the east part of US-Northwest. In the integrated scenario, installed capacities of PV and wind increase due to the additional demand of seawater desalination and industrial gas demand. The high share of solar PV and wind energy can be examined by the fact that these are the least cost RE resources in the region. It is noticeable to mention that a fast cost reduction of solar PV technology and battery storage is assumed in the next 15 years. The share of PV single-axis tracking and PV self-consumption of the total solar PV installed capacity for the region-wide scenario are 56% and 43%, respectively. However, the installed capacity of PV single-axis tracking in the area-wide scenario decrease to 35%, while the share of PV self-consumption increase to 63% of the total installed capacity of PV. Compared to all the available RE resources in the region, PV and wind seems to be financially more attractive technologies due to the excellent availability of the respective resources.

The grid interconnection decreases the need of energy storage. Installed capacities of batteries, A-CAES, PtG, PHS and gas turbines decrease with interconnection of the regions as summarized in Table 2, whereas the share of heat storage increases by 325% in the integrated scenario due to storage of additional usable heat generated in the system, however on a very low level.

A division of regions into net importers and exporters can be observed for the area-wide scenario and the integrated scenario, which are presented in Figure 6, and where sub-regions with the best renewable resources are net exporters and the others are net importers. In the region-wide scenario, all the individual sub-regions of North America need to match their own demand using their own RE. The regions can import or export electricity depending on their demand and generation. The differences between the demand and production are mainly due to import, export and storage losses. For the integrated scenario, the difference is due to the energy consumption for SNG production (Figure 7). The net importer regions in North America are US-Mid-Atlantics, US-Tennessee Valley Authority, US-New England & New York, US-Gulf, US-Carolinas, US-Southern, US-California, US-Southwest, Northwest Mexico, Central Mexico and North Mexico. The net exporter regions are East Canada, US-Midwest, US-Central, US-Northwest, US-Texas and West Canada. In comparing net importer and exporter regions, US-Alaska and South Mexico have a balancing system of import and export. Due to a high electricity demand for additional desalination and SNG production, the integrated scenario tends to increase the electricity generation between the regions to fulfill the increased demand. Hourly resolved profiles for the net exporting region, East Canada, the net importing region, the US-Mid-Atlantics, and the balancing region, South Mexico, are presented in the Supplementary Material (Figure 1, 2 and 3, respectively).

Table 2: Results on the installed RE technologies and storage capacities for the four scenarios.

		Region-wide scenario	Country-wide scenario	Area-wide scenario	Integrated scenario
PV self-consumption	[GW]	811.9	811.9	811.9	811.9
PV optimally tilted	[GW]	20.6	20.6	20.6	20.3
PV single-axis tracking	[GW]	1062.3	566.0	453.8	1018.2
PV total	[GW]	1894.8	1398.5	1286.3	1850.4
CSP	[GW]	0.0	0.0	0.0	0.0
Wind energy	[GW]	1004.9	1134.2	1096.6	1979.6
Biomass power plants	[GW]	22.1	14.1	8.2	4.2
MSW incinerator	[GW]	3.9	3.8	3.8	3.8
Biogas power plants	[GW]	51.9	68.0	67.6	45.8
Geothermal power	[GW]	19.4	5.2	5.2	4.6
Hydro Run-of-River	[GW]	29.3	29.2	29.2	26.2
Hydro dams	[GW]	158.3	167.5	178.3	178.2
Battery PV self-consumption	[GWh]	1410.9	1410.9	1410.9	1410.9
Battery System	[GWh]	1234.4	592.9	483.3	402.4
Battery total	[GWh]	2645.3	2003.8	1894.2	1813.3
PHS	[GWh]	34.9	34.9	34.9	24.5
A-CAES	[GWh]	5821.6	185.9	1.3	4.8
Heat storage	[GWh]	0.4	1.0	0.0	1.7
PtG electrolyzers	[GW_{el}]	113.6	83.7	53.6	442.3
CCGT	[GW]	159.4	132.7	102.8	8.8
OCGT	[GW]	64.5	76.9	42.3	54.6
Steam Turbine	[GW]	0.0	0.0	0.0	0.1

Figure 7 gives an overview on the installed capacities for RE generation and storage for all sub-regions in the region-wide, area-wide and integrated scenarios. In the region-wide and area-wide scenarios, the share of solar PV dominates the system for most of the sub-regions, except for northern sub-regions such as Eastern Canada, Western Canada, US-Alaska, US-Midwest and US-Central where excellent FLH of wind resource leads to an increase of the share of wind energy among other RE resources. Solar PV comprises of 56% of total RE resources in the region-wide scenario. Among all sub-regions, Northwestern Mexico and US-Southern have the highest amount with 74% and 70% of total share, respectively. In area-wide open trade scenario the share of solar PV still remains the highest, while it decrease by 10% in comparison to the region-wide scenario. It is noteworthy that the share of wind energy increases by 10% in the area-wide scenario. In the integrated scenario, solar PV stayed the main energy in most of the regions, while the share of wind energy increases significantly. A 48% of RE resources comes from wind energy in the integrated scenario, followed by solar PV by 45%, hydropower by 5% and the rest of resources by 2%.

Despite the fact that hydropower is one of the main energy resources in Canada and accounts for 32% and 36% for Eastern Canada and Western Canada in the region-wide open trade scenario, respectively, it does not have a significant role in other sub-regions. It is noticeable to mention that hydropower has the highest share by 36% among all RE resources in Western Canada, following by solar PV and wind energy in the region-wide and area-wide scenarios. However, the share of hydro declines in the integrated scenario to 17%. It can be explained by the fact that when the demand for water desalination and industrial gas added to the integrated scenario, wind energy is the most cost-competitive option compared to other resources. Therefore, the model prefers to use more wind energy to the system. On the other hand, US-Tennessee Valley Authority has only 4% share of hydropower in the region-wide scenario, while the rate increases in the area-wide and integrated scenario by 10% and 14%, respectively. A state of charge profile for hydro dams is shown in the Supplementary Material (Figure 5). As can be seen, the highest filling state of hydro reservoirs is in the summer months of a year.

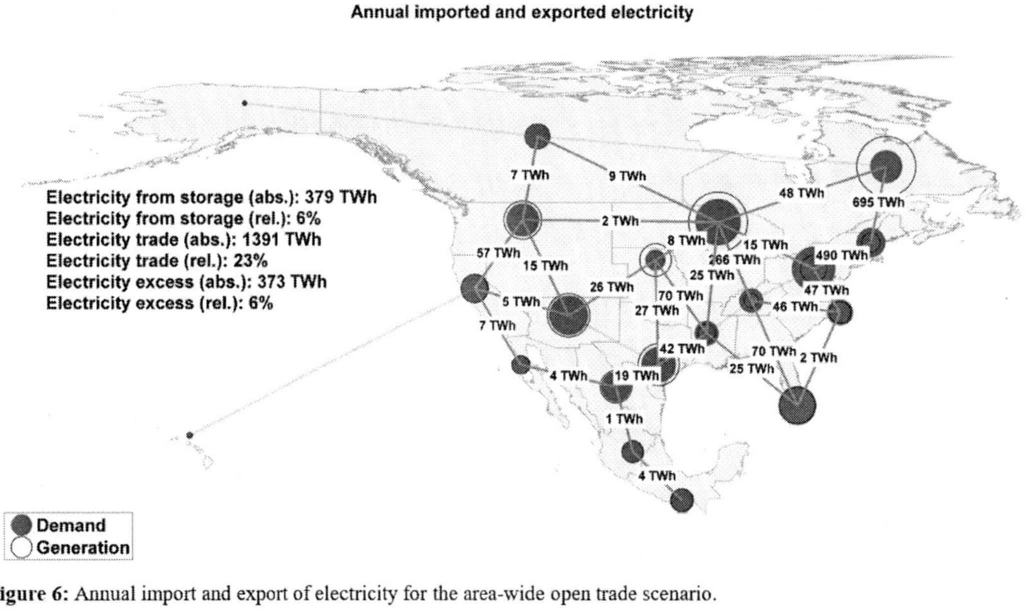

Figure 6: Annual import and export of electricity for the area-wide open trade scenario.

The interconnection of the sub-regions via HVDC transmission lines decreases the installed capacities of PV by 32%, as shown in Figure 7 and Table 2, from the region-wide to the area-wide scenario. In the case of the integrated scenario, installed capacities for PV and wind increase significantly by 44% and 81% compared to the area-wide scenario, respectively, due to the higher demand for electricity and increased system flexibility. PV self-consumption plays a key role in North America due to high enough electricity prices. PV self-generation covers 74.6%, 79.7% and 81.9% of residential, commercial and industrial prosumer demand, respectively. An overview of PV self-consumption is provided in the Supplementary Material (Table 13).

The interconnected HVDC transmission grid, together with the RE generation and demand, significantly influence the requirement of total storage capacity, but also change with the need of different storage technologies for the energy system in the whole region. The throughputs of the battery system, A-CAES and gas storage decrease by 57.8%, 99.9% and 31.8%, respectively, from the region-wide to the area-wide scenario. Therefore, A-CAES storage technology experienced the highest decline in comparison to other storage technologies, which had been already observed in other regions in the world [27] while the need for PHS technology increased from 7.9 TWh_{el} to 8.4 TWh_{el} in the area-wide scenario. The installation of HVDC lines leads to a reduction of the storage technology utilization as the transmitted electricity is lower in cost in many cases than storage options. Installed capacities of batteries, PHS, A-CAES, heat storage, PtG and gas turbines decrease due to the expansion of the grid. The storage discharge capacities, annual throughput of storage and full load cycles per year are provided in the Supplementary Material (Table 14). State of charge profile diagrams for the area-wide scenario for battery, PHS, gas storage and A-CAES is given in the Supplementary Material (Figure 4).

Various electricity generation curves for the area-wide scenario are illustrated in the Supplementary Material (Figure 6). A full year divided into 8760 hours sorted according to the generation minus the load, which is shown by a black line. Electricity generation is more than demand for about 5000 hours of the year, which is used for charging storage. The reason for high electricity generation can be attributed to the inflexibility of energy sources, due to the high share of solar and wind energy and a higher solar irradiation and wind speed during these hours of the year in North America. Consequently, other flexible electricity generation options such as hydro dams, biomass, biogas, geothermal energy and discharge of storage plants are required. As observed for the other hours of the year, the inflexible electricity generation options reduces significantly as the electricity demand decreases and there is a need for flexible electricity generation options, discharge of storage technologies and utilization of the grid. There is curtailment in about 1000 hours of the year, but for all the other hours the HVDC lines enable the export of the electricity from the best RE producing sub-regions to other sub-regions of remaining demand or due to charging of storage.

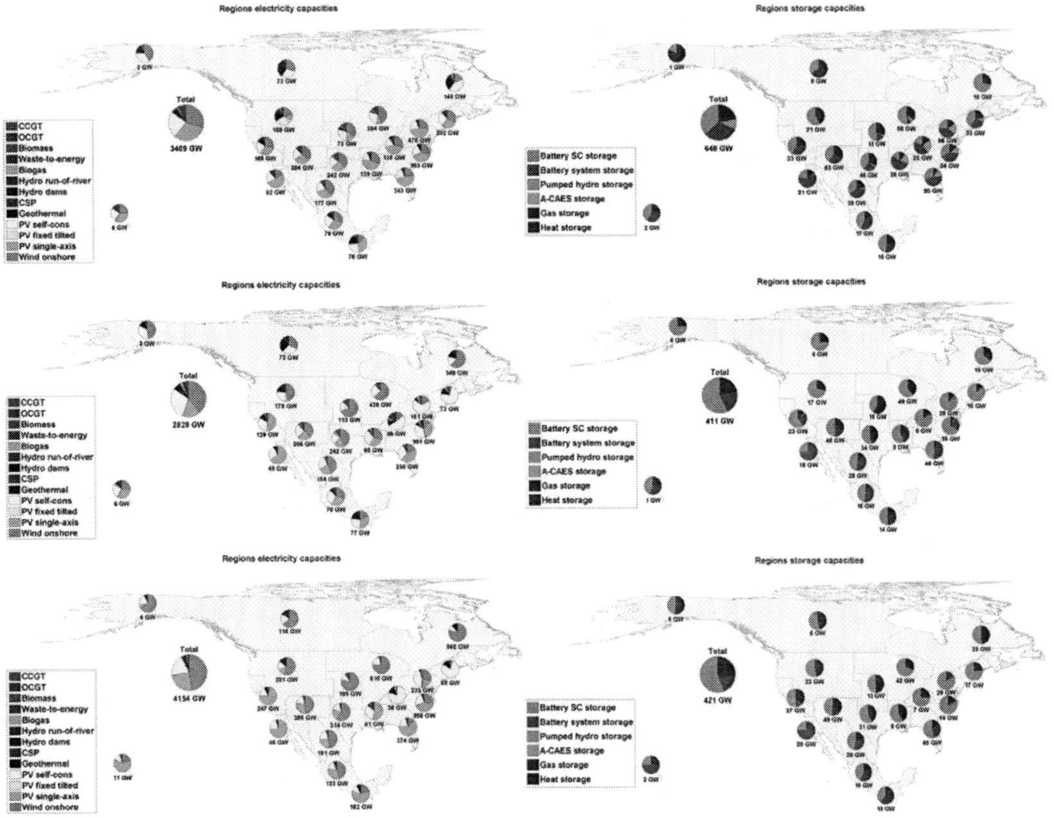

Figure 7: Installed capacities RE generation (left) and storage capacities (right) for region-wide (top), area-wide (center) and integrated (bottom) scenarios for North Americas.

The grid utilization profile for North America can be found in the Supplementary Material (Figure 7). It can be observed that the grid utilization is distributed between different hours of a day. However, it is mostly utilized in the morning and evening hours. The lowest electricity transmission is between 10 am to 15 pm, especially in the first half of the year. This can be explained by seasonal variation in the region, where most trading of the electricity takes place. In order to satisfy the morning and evening demand, transmission of electricity takes place between the regions. In hours of very good solar conditions, grid utilization is at its minimum. The capacities and utilization of the transmission lines between different regions is provided in the Supplementary Material (Table 15). The largest electricity transmission capacity is from Eastern Canada to US-New England & New York with 90.2 GW and also from US-New England & New York to US-Mid-Atlantics with 64.9 GW.

The key findings of the energy system from generation to demand in the integrated scenario can be depicted by an energy flow diagram (Figure 8). The flow diagram is comprised of the primary RE resources, the energy storage technologies, HVDC transmission grids, total demand of each sector and system losses. The sector use of electricity for the integrated scenario consists of electricity, desalination and industrial gas demand. The usable heat generated and ultimate system losses can be explained as a difference between primary electricity generation and final electricity demand, which includes curtailed electricity, heat produced by biomass, biogas and waste-to-energy power plants, heat generated from electrolysers for transforming power-to-hydrogen, in methanation process transforming hydrogen-to-methane and methane-to-power in gas turbines. Efficiency losses lead to A-CAES, PHS, battery storage and HVDC transmission grid losses. The energy flow diagrams for the region-wide and area-wide scenario are presented in the Supplementary Material (Figure 8).

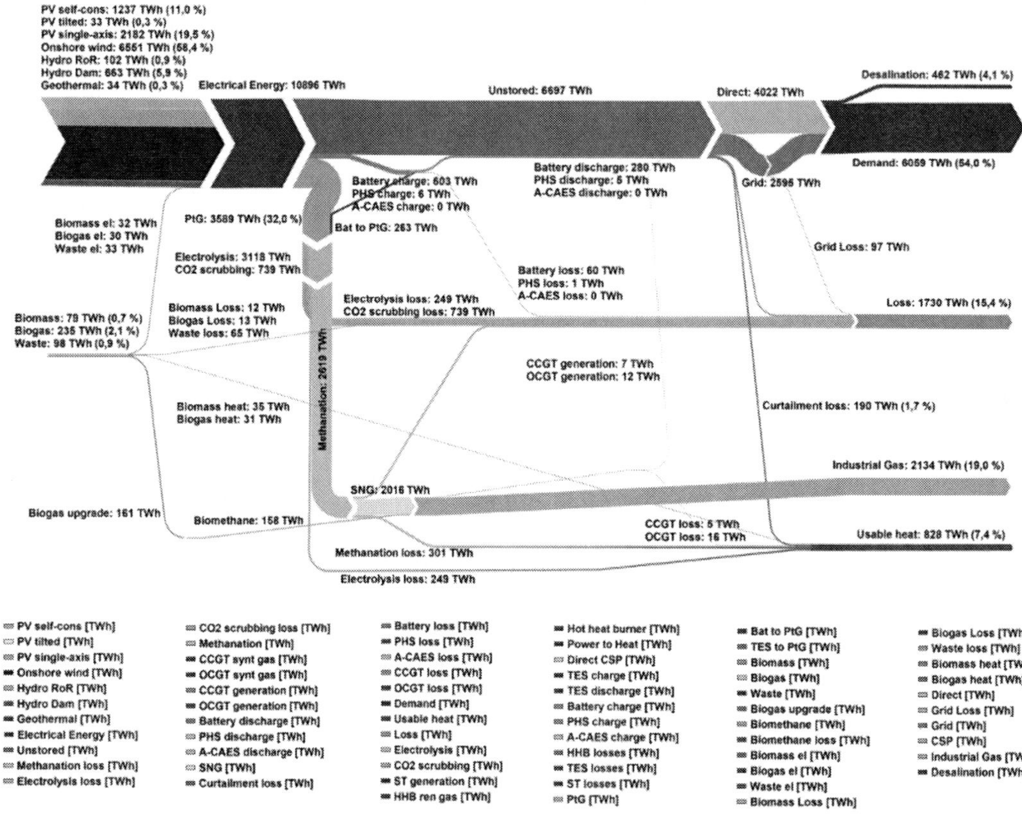

Figure 8: System energy flow for the integrated scenario.

5 DISCUSSION

The electricity cost is decreased remarkably in the RE-based system due to the installation of a HVDC transmission grid. The reduction of electricity cost by connection of HVDC grid has been also observed in other studies carried out with the same criteria [24]-[30]. The total levelized cost of electricity in the region decreased from 63 €/MWh for the region-wide open trade scenario to 53 €/MWh for the area-wide open trade scenario and to 42 €/MWh for the integrated scenario. By comparing with the region-wide scenario, it can be highlighted that regions such as US-Southern, US-Midwest and US-California experienced the highest reduction of total LCOE in the area-wide scenario, by around 17%. In contrast, the total LCOE rose in some regions such as US-Tennessee Valley Authority, US-New England & New York and US-Mid-Atlantics from region-wide to area-wide scenario, by 16%, 12% and 8% respectively. This can be justified on the basis that in these three regions the electricity cost is lower when they are independent. However, in most of the regions HVDC grid interconnection leads to a decrease the cost of electricity.

The total annualized cost of the system decreased from 387.6 b€ for the region-wide scenario to 320.6 b€ for the area-wide scenario. The total capex requirements dropped substantially from 3502 b€ for the region-wide open trade scenario to 2945 b€ for the area-wide open trade scenario. However, the total annualized cost and the total capex increased drastically for the integrated scenario by 498.2 b€ and 4668 b€, respectively, to cover the excess electricity needed for the desalination and industrial gas sectors. Additional costs of the HVDC transmission grid in the area-wide open trade scenario are compensated by a significant decline in installed capacities of electricity generation and storage capacities, which is enabled by lower efficiency losses, decrease in costs of energy transmission compared to energy storage and access to low cost electricity generation in other regions. It is important to mention that the HVDC transmission grid may not cover the electricity supply in rural areas in Mexico, since RE-based mini-grid applications based on PV systems, small wind power and micro-hydro plants might be a proper solution in addition to grid extension [61]-[65].

PV technologies followed by wind energy have the highest share of installed capacities for a 100% RE energy mix in the region-wide and the area-wide scenarios. The reason is that these technologies have well distributed FLH all over the sub-regions and are the least cost RE technologies in most of the cases. In comparison, the share of wind energy increased in the integrated scenario by 48% of total installed capacity, followed by PV by 45%. However, an increase in the share of wind could be inverted again after 2030, since the learning curve of PV continues [66], in particular in combination with batteries [67],[68]. On the other hand, the learning curve of wind is not so sharp, i.e. the share of PV is expected to grow year by year. Such an effect had been found for instance for the case of Saudi Arabia [70]. Besides, the installation of distributed small-scale and centralized PV plants is already profitable in numerous countries across the world and PV electricity generation cost will tend to be further reduced in the coming years [69].

A study carried out by IRENA [71] reveals that doubling the share of RE in the US by 2030 is achievable due to the best wind, solar, geothermal, hydro and biomass resources in the country. Apart from that, US has a vibrant culture of innovation, plentiful financing opportunities, and a highly skilled labors for making possible a transition from conventional energy system towards a sustainable energy system. Solar PV is considered as a main RE resources in the country and it is expected that 135 GW of solar PV would be installed by 2030. A rapid reduction in the price of solar PV technologies has also been discussed and a LCOE of 33-41 €/MWh for utility-scale and 63-74 €/MWh for rooftop installations is anticipate for the year 2030 in the IRENA study, which is only little higher than assumed in this research, but also in comparison to others [18]. In comparison to solar PV, onshore wind seems to have a better potential to become a leader of RE resources in the US with installed capacity of 314 GW by 2030, according to IRENA [71]. The LCOE for wind energy is assumed to become 37-44 €/MWh.

A high share of RE resources for the energy sectors have studied and discussed for different states and cities of the US, such as Washington State, Hawaii and San Francisco. Hawaii sets a target for achieving RE goal at 100% by 2045 [72]. In other words, Hawaii should be running only on electricity produced by RE resources in 30 years. Nevertheless, 70% of electricity in the State generated by oil in 2013, and coal accounted for 14% of total electricity production [73]. In addition to getting off of fossil fuels, policymakers want to save money which is spent for importing fossil fuels every year. San Francisco has a goal to achieve a 100% renewable electricity supply community wide by 2030 [74]. A study carried out by Jacobson et al. [75] analyzed the potential and consequences of Washington State's use of wind, water and solar (WWS) to produce electricity and electrolytic hydrogen for 100% of its all-purposes energy in different sectors by 2050, with 80-85% conversion by 2030. Even more, they expected to experience a reduction in power demand by around 40% due to electrification plus modest efficiency measures. Similar approach has studied for each of the 50 United States [5]. The results show that US all-purpose load would be met with 30.9% onshore wind, 19.1% offshore wind, 30.7% utility-scale PV, 7.2% rooftop PV, 7.3% CSP with storage, 3% hydroelectric, 1.25% geothermal, 0.37% wave power and 0.14% tidal power. As a result, a 100% renewable energy is a real policy option for North America and is not a wishful thinking anymore. Grid interconnection will also make this transition easier to achieve a fully sustainable energy system in the mid-term future.

The integrated scenario, focusing on the integration to cover current natural gas demand in the industrial sector (the gas demand for heat generation and residential use are not considered in this study) by flexible generation of SNG and renewable water demand by SWRO desalination. The availability of RE in North America is sufficient to cover additional electricity demand to produce 2134 TWh_{th} (218.4 bcm) of SNG and 83.6 billion m^3 of renewable water. Nonetheless, expected growth in electricity demand stimulates a rise in electricity cost due to inconvenient distribution and profiles of RE generation. Additional generation of electricity to cover 4245.6 TWh_{el} for gas synthesis and SWRO desalination promotes a supplementary installation of RE generation capacities of 1850.4 GW of PV and 1979.6 GW of wind energy. Furthermore, there is a noticeable increase in electrolyser units of about 328.7 GW (+289%) and a significant reduction in CCGT capacities of about 56.6 GW (-35.5%). This can be explained by flexibility provided by CCGT in the area-wide scenario in the supply side to be substituted by flexibility of electrolysers in the integrated scenario in the demand side.

In addition, the system generates excess heat as a by-product of different processes such as biogas and biomass CHP plants, waste-to-energy incinerators and gas turbines, as well as excess electrical energy which can be curtailed or converted to usable heat and stored in heat storage. The usable heat amount varies from 498 TWh_{th} (7% of total generated electricity) per year for the area-wide scenario up to 735 TWh_{th} (9.7% of total generated electricity) for the region-wide scenario. The amount of usable heat in the integrated scenario is 828 TWh_{th}, which is noticeably larger than for the area-wide scenario because of higher losses in gas turbines, methanation and electrolysis. Totally, the estimation of the integration benefit for the electricity, industrial gas and water sectors is around 56 b€ of the annual system cost (Figure 9). An additional reduction can be seen in the electricity generation by 774 TWh and the curtailed electricity by 208 TWh. Further, the cost of desalinated water seems to be quite affordable at 1.1 €/m^3 and the cost of electricity decreases by 21% to 42 €/MWh for the integrated scenario compared to the area-wide open trade scenario without sector integration.

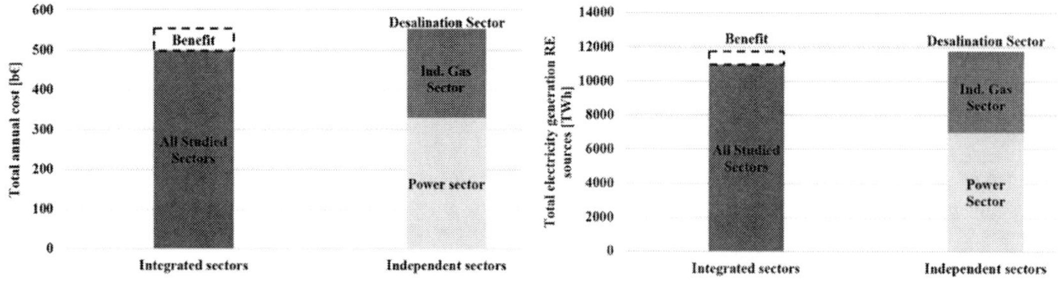

Figure 9: A comparison of the integrated sectors and independent sectors for both the total annual cost (left) and the total electricity generation from RE resources.

The findings for North America 100% RE-based energy system clearly reveal that the potential of the region for RE generation, in particular for solar and wind energy, is quite high. By applying this system to the region, not only can North America fulfill all the electricity demands in the region itself, but also could potentially export the rest of the production to other continents. The results of a fairly low LCOE in all the considered scenarios for the year 2030 added to the already existing RE policies and low carbon development plans can boost the development of a renewable power system in the region in the coming years. Several studies have been already discussed about various alternatives to reduce the production of fossil fuels and to achieve a low carbon based energy system [76],[77]. While, some studies such as IEA [12] have mostly addressed non-renewable options, such as nuclear energy, natural gas and coal carbon capture and storage (CCS). However, the LCOE of the alternatives are rather high compared to our findings, which are listed as follows [76]: 112 €/MWh for new nuclear (assumed for 2023 in the UK and Czech Republic), 112 €/MWh for gas CCS (assumed for 2019 in the UK) and 126 €/MWh for coal CCS (assumed for 2019 in the UK). Moreover, a report published by the European Commission [78] concludes that CCS technology is not likely to be commercially available before the year 2030. The findings for Europe are assumed to be also valid for North America in the mid-term. In addition, several studies for 100% RE by 2050, such as from WWF [79] and Greenpeace [80] state that nuclear fission produced dangerous waste that remains extremely toxic for many years and it cannot be stored safely anywhere in the world. Consequently, these studies have assumed a phase-out of nuclear energy due to sustainability reasons. Dittmar [81] also points out that the financial and human research and development resources spent for nuclear fusion cannot solve the energy problems in the world and even worse these resources are not available for research of pathways towards a low cost energy future.

The 100% renewable resource-based energy system options for North America demonstrated and described in this paper seem to be substantially cost-effective (about 63 - 67% lower cost) than the other alternatives, by comparison of the integrated scenario with nuclear energy and CCS technology, which have still further drawbacks. These include nuclear melt-down risk, nuclear terrorism risk, unsolved nuclear waste disposal, remaining CO_2 emissions of power plants with CCS technology, a diminishing conventional energy resource base and high health cost due to heavy metal emissions of coal fired power plants [20].

6 CONCLUSION

This study analyzes the technical and economic feasibility of a roadmap for North America to convert the energy infrastructure into a green and sustainable energy system powered by solar PV, wind energy, hydropower and other RE resources. Existing RE technologies can generate sufficient energy not only to cover all electricity demand in North America for the year 2030 on a cost level of 42-63 €/MWh$_{el}$, but also to potentially export the additional electricity to other countries. In addition, the electricity demand, gas demand and clean water demand can be covered by electricity generation based on RE sources as well using PtG and SWRO desalination technology, providing the region with 100% renewable synthetic natural gas and clean water supply. The advantage regarding the use of RE for electricity generation has been discussed through four scenarios; the region-wide, country-wide, area-wide and integrated scenarios.

Extreme dependency on fossil fuel resources to generate electricity, rather low prices of these resources and high share of nuclear power lead to a low share of RE in the region today, especially in the US and Mexico. However, the 'net zero' greenhouse gas emissions agreed at COP21 in Paris clearly guides the pathway towards sustainability. In order to cut the dependency of North America power sector on fossil fuels, different shares of variable RE technologies are taken into consideration for a 100% RE-based power system for the region. In the cost minimized design of the energy mix presented in this study, solar PV dominates the energy system in the electricity sector by 45-56% of total RE share, depending on the defined scenarios, in most sub-regions of North America, followed by wind energy by 29-48% of total RE share. For all the studied scenarios solar PV technology emerged as the main energy supply (in terms of GW of installed capacities) in most of the sub-regions; however, for the integrated scenario the role of PV decreases in sub-regions where wind sites are the least cost technology and high potential of wind resources are available. It is important to note that an increase in the share of wind could be inverted again after 2030, since the learning curve of PV continues, in particular in combination with batteries. However, the learning curve of wind is not so sharp, i.e. the share of PV expected to grow year by year.

Heat generated as a by-product of electricity, synthetic natural gas generation and curtailed electricity conversion can cover up to 828 TWh$_{th}$ of heat demand. The HVDC transmission power lines plays a substantial role within the renewable resource-based energy system since the established Super Grid enables a significant cost decrease, a cut-off of storage utilization, and a significant reduction of primary generation capacities. A remarkable decline can be seen in the case of A-CAES, where the need for storage dropped dramatically in the area-wide open trade scenario and the integrated scenario by around 99.9%. For the integrated scenario it was found that industrial SNG generation displaces SNG storage as seasonal storage for the electricity sector. Instead of gas turbine utilization in the case of an energy deficit, the system curtails the SNG generation in that system set-up as a major source of flexibility to the system.

Based on the presented results, current barriers to implement a 100% RE-based system are neither technical nor economic for the applied year. As such, they must be political and social. Such obstacles are likely due to the fact that most people are unaware of the potential of RE resources in their countries and how they will benefit from them. In addition, fossil fuels and nuclear energy are beneficial for many investors across the world and those investors resist to make this transition possible.

More research is needed for a better understanding of a fully integrated RE system in North America. However, this research work clearly indicates that a 100% renewable resources-based energy system is a real policy and low cost option.

Acknowledgements

The authors gratefully acknowledge the public financing of Tekes (Finnish Funding Agency for Innovation) for the 'Neo-Carbon Energy' project under the number 40101/14.

Supplementary Material

Supplementary data associated with this article can be found in the appendix of this paper.

7 REFERENCES

[1] United Nations, 2015. Probabilistic Population Projections based on the World Population Prospects: The 2015 Revision. Population Division, DESA, New York City, NY, http://esa.un.org/unpd/ppp/.

[2] [IEA] – International Energy Agency, 2015, Key World Energy Statistics, Paris.

[3] [UNFCCC] – United Nations Framework Convention on Climate Change, 2015. Adoption of the Paris Agreement – Proposal by the President, UNFCCC, Paris, December 12, http://unfccc.int/resource/docs/2015/cop21/eng/l09.pdf [accessed: 04.03.2016].

[4] Fthenakis V., Mason J.E. and Zweibel K., 2009. The technical, geographical, and economic feasibility for solar energy to supply the energy needs of the US. Energy Policy, 37(2), 387-399.

[5] Jacobson M.Z., Delucchi M.A., Bazouin G., Bauer Z.A., Heavey C.C., Fisher E., Morris S.B., Piekutowski D.J., Vencill T.A. and Yeskoo T.W., 2015. 100% clean and renewable wind, water, and sunlight (WWS) all-sector energy roadmaps for the 50 United States. Energy & Environmental Science, 8(7), 2093-2117.

[6] Jacobson M.Z., Delucchi M.A., Cameron M.A. and Frew B.A., 2015. Low-cost solution to the grid reliability problem with 100% penetration of intermittent wind, water, and solar for all purposes. Proceedings of the National Academy of Sciences, 112(49), 15060-15065.

[7] MacDonald A.E., Clack C.T., Alexander A., Dunbar A., Wilczak J. and Xie Y., 2016. Future cost-competitive electricity systems and their impact on US CO_2 emissions. Nature Climate Change, published online January 25, DOI: 10.1038/NCLIMATE2921.

[8] REN21, 2016. Renewables 2016 Global Status Report, Paris, www.ren21.net/wp-content/uploads/2016/06/GSR_2016_Full_Report_REN21.pdf.

[9] Werner Ch., Gerlach A., Breyer Ch., Masson G., 2016. Global Photovoltaics in 2015 – Analysis of Current Solar Energy Markets and Hidden Growth Regions, 32nd EU PVSEC, Munich, June 20-24.

[10] Makyhoun M., Edge R. and Esch N., 2015. Utility solar market snapshots; sustained growth in 2014, SEPA (solar electric power association), Washington, DC, US, www.solarelectricpower.org/media/322918/solar-market-snapshot-2014.pdf [accessed: 01.06.2016].

[11] Farfan J. and Breyer Ch., 2016. Structural changes of global power generation capacity towards sustainability and the risk of stranded investments, submitted.

[12] [IEA] – International Energy Agency, 2015. World Energy Outlook 2015. IEA Publishing, Paris.

[13] Breyer Ch., Bogdanov D., Gulagi A., Aghahosseini A., Barbosa L.S.N.S., Koskinen O., Barasa M., Caldera U., Afanasyeva S., Child M., Farfan J., Vainikka P., 2016. On the Role of Solar Photovoltaics in Global Energy Transition Scenarios, 32nd EU PVSEC, Munich, June 20-24.

[14] Kost C., Mayer J.N., Thomsen J., Hartmann N., Senkpiel C., Philipps S., Nold S., Lude S., Saad N. and Schlegl T., 2013. Levelized cost of electricity renewable energy technologies. Techno-Economic Assessment of Energy Technologies, Fraunhofer ISE, Germany.

[15] [IRENA] - International Renewable Energy Agency, 2016. The Power to Change: Solar and Wind Cost Reduction Potential to 2025, Abu Dhabi. www.irena.org/DocumentDownloads/Publications/IRENA_Power_to_Change_2016.pdf.

[16] PV-magazine, 2016. Mexico awards more than 1 GW of solar at US$40.50/MWh, Berlin, March 30, www.pv-magazine.com/news/details/beitrag/mexico-awards-more-than-1-gw-of-solar-at-us4050-mwh_100023944/#axzz4BIYC8nq7 [accessed: 16.06.2016].

[17] PV-magazine, 2016. Peru awards 185 MW of solar PV at US$48/MWh, Berlin, February 17, www.pv-magazine.com/news/details/beitrag/peru-awards-185-mw-of-solar-pv-at-us48-

mwh_100023273/#axzz4BIYC8nq7 [accessed: 16.06.2016].
[18] Lazard, 2016. Lazard's levelized cost of energy analysis – Version 9.0, New York, www.lazard.com/media/2390/lazards-levelized-cost-of-energy-analysis-90.pdf [accessed: 03.06.2016].
[19] Council of economic advisers, 2014. the all-of-the-above energy strategy as a path to sustainable economic growth, Executive office of the president of the United States, https://www.whitehouse.gov/sites/default/files/docs/aota_energy_strategy_as_a_path_to_sustainable_economic_growth.pdf [accessed: 01.06.2016].
[20] Epstein P.R., Buonocore J.J., Eckerle K., Hendryx M., Stout B.M., Heinberg R., Clapp R.W., May B., Reinhart N.L., Ahern M.M., Doshi S.K., Glustrom L., 2011. Full cost accounting for the life cycle of coal, Annals of the New York Academy of Sciences, 1219, 73-98.
[21] Energy and Mines Ministers' conference, 2013. Canada – a global leader in renewable energy, enhancing collaboration on renewable energy technologies, Yellowknife, Northwest Territories.
[22] IRENA [International Renewable Energy Agency], 2015. Renewable Energy Policy Brief: Mexico; IRENA, Abu Dhabi.
[23] Martinez A., Eurek K., Mai T. and Perry A., 2013. Integrated Canada-US Power Sector Modeling with the Regional Energy Deployment System (ReEDS), Contract, vol. 303, 50 pp.; NREL Report No. TP-6A20-56724, http://www.nrel.gov/docs/fy13osti/56724.pdf.
[24] Bogdanov D. and Breyer Ch., 2015. Eurasian Super Grid for 100% Renewable Energy power supply: Generation and storage technologies in the cost optimal mix, ISES Solar World Congress 2015, Daegu, Korea, November 8-12.
[25] Bogdanov D. and Breyer Ch., 2016. North-East Asian Super Grid for 100% Renewable Energy supply: Optimal mix of energy technologies for electricity, gas and heat supply options. Energy Conversion and Management, 112, 176-190.
[26] Gulagi A., Bogdanov D., Breyer Ch., 2016. Southeast Asia and the Pacific Rim Super Grid for 100% Renewable Energy Power Supply, submitted.
[27] Gulagi A., Aghahosseini, A., Bogdanov D., Breyer Ch., 2016. Comparison of the potential role of adiabatic compressed air energy storage (A-CAES) for a fully sustainable energy system in a region of significant and low seasonal variations, 10th International Renewable Energy Storage Conference, Düsseldorf, Germany.
[28] Barbosa L.S.N.S, Bogdanov D., Vainikka P., Breyer Ch., 2016. Hydro, wind and solar power as a base for a 100% Renewable Energy supply for South and Central America, submitted.
[29] Aghahosseini A., Bogdanov D., Ghorbani N., Breyer Ch., 2016, The role of a 100% renewable energy system for the future of Iran: Integrating solar PV, wind energy, hydropower and storage, 11th International Energy Conference, Tehran, May 30-31.
[30] Aghahosseini A., Bogdanov D., Breyer Ch., 2016, The MENA Super Grid towards 100% Renewable Energy Power Supply by 2030, 11th International Energy Conference, Tehran, May 30-31.
[31] Stackhouse P.W., Whitlock C.H., editors. 2008. Surface meteorology and solar energy (SSE) release 6.0, NASA SSE 6.0, earth science enterprise program. National Aeronautic and Space Administration (NASA), Langley; http://eosweb.larc.nasa.gov/sse/ [accessed 28.05.2015].
[32] Stackhouse P.W., Whitlock C.H., editors. 2009. Surface meteorology and solar energy (SSE) release 6.0 Methodology, NASA SSE 6.0, earth science enterprise program, National Aeronautic and Space Administration (NASA), Langley; http://eosweb.larc.nasa.gov/sse/documents/SSE6Methodology.pdf [accessed 28.05.2015].
[33] Stetter D., 2012. Enhancement of the REMix energy system model: global renewable energy potentials optimized power plant siting and scenario validation. Dissertation, Faculty of Energy-, Process and Bio-Engineering, University of Stuttgart.
[34] Luck M., Landis M., Gassert F., 2015. Aqueduct Water Stress Projections: Decadal projections of water supply and demand using CMIP5 GCMs, Washington DC, World Resources Institute.
[35] Caldera U., Bogdanov D., Breyer Ch., 2016. Local cost of seawater RO desalination based on solar PV and wind energy - A global estimate, Desalination, 385, 207-216.
[36] AAPG, 2015. Global Heat Flow Database. American Association of Petroleum Geologists, Tulsa, USA, www.datapages.com/gis-map-publishing-program/gis-open-files/globalframework/global-heat-flow-database.
[37] IASPEI, 2015. International Association of Seismology and Physics of the Earth's Interior, The International Heat Flow Commission, Colorado, USA. IHFC database http://www.heatflow.und.edu/index2.html.
[38] Chamorro C.R., García-Cuesta J.L., Mondéjar M.E., Pérez-Madrazo A., 2014a. An estimation of the enhanced geothermal systems potential for the Iberian Peninsula. Renewable Energy 66, 1-14.
[39] Chamorro C.R., García-Cuesta J.L., Mondéjar M.E., Pérez-Madrazo A., 2014b. Enhanced geothermal systems in Europe: An estimation and comparison of the technical and sustainable potentials. Energy 65, 250-263.
[40] Huenges E., 2012. Geothermal Energy Systems, Exploration, Development and Utilization, 2nd edition, Weinheim, Germany.
[41] German Biomass Research Centre, 2009. Regionale und globale räumliche Verteilung von Biomassepotenzialen. German Biomass Research Centre; Leipzig [in German].
[42] [IEA] - International Energy Agency, 2012. Technology roadmap – bioenergy for heat and power. IEA Publications; Paris.
[43] [IPCC] - Intergovernmental Panel on Climate Change, 2011. Special report on RE sources and CC mitigation. Intergovernmental panel on climate change; Geneva.
[44] [EIA] – U.S. Energy Information Administration, 2014. Utility Bundled Retail Sales- Commercial, Data from forms EIA-861- schedules 4A & 4D and EIA-861S, https://www.eia.gov/electricity/sales_revenue_price/pdf/table7.pdf [accessed: 20.03.2016].
[45] [EIA] – U.S. Energy Information Administration, 2014. Utility Bundled Retail Sales- Residential,

Data from forms EIA-861- schedules 4A & 4D and EIA-861S, https://www.eia.gov/electricity/sales_revenue_price/pdf/table6.pdf [accessed: 20.03.2016].
[46] Statistics Canada, 2013. Report on Energy Supply and Demand in Canada, catalogue no. 57-003-x, http://www.statcan.gc.ca/pub/57-003-x/57-003-x2015002-eng.pdf [accessed: 24.04.2016].
[47] SENER, 2014. Balance Nacional de Energia, Secretaria De Energia, Mexico, https://www.gob.mx/cms/uploads/attachment/file/44353/Balance_Nacional_de_Energ_a_2014.pdf [accessed: 02.02.2016].
[48] [EIA] – U.S. Energy Information Administration, 2015. Electric Power Annual 2013, Washington, DC.
[49] [EIA] – U.S. Energy Information Administration, 2014. Utility Bundled Retail Sales- Industrial, Data from forms EIA-861- schedules 4A & 4D and EIA-861S, Washington, DC, https://www.eia.gov/electricity/sales_revenue_price/pdf/table8.pdf [accessed: 20.03.2016].
[50] Northwest Territories Power Corporation, 2016. residential Electrical Rates for Northwest Territories, https://www.ntpc.com/customer-service/residential-service/what-is-my-power-rate [accessed: 22.03.2016].
[51] Government of the Northwest Territories, 2008. A review of electricity regulation, rates and subsidy programs in the Northwest Territories, http://www.pws.gov.nt.ca/pdf/Energy/Electrical%20Review%20Discussion%20Paper.pdf [accessed: 22.03.2016].
[52] CBC News, 2011. Electricity in Nunavut, http://www.cbc.ca/news/canada/electricity-in-nunavut-1.1042217 [accessed: 25.03.2016].
[53] The Yukon Electrical Company Limited, 2011. Industrial primary, http://www.yukonenergy.ca/media/site_documents/1024_39%20Industrial%20Primary.pdf [accessed: 23.03.2016].
[54] The Yukon Electrical Company Limited, 2011. Residential service, hydro, non-government, http://www.yukonenergy.ca/media/site_documents/1038_1160%20Residential%20Service%20Hydro,%20Non-Government.pdf [accessed: 23.03.2016].
[55] The Yukon Electrical Company Limited, 2011. Wholesale primary (YEC), http://www.yukonenergy.ca/media/site_documents/1046_42%20Wholesale%20Primary.pdf [accessed: 23.03.2016].
[56] Hydro Québec, 2015. Comparison of electricity prices in major North American cities. Montréal: Hydro Québec.
[57] [CFE] – Comision Federal De Electricidad, 2016. Tarifas para el suministro y venta de energía eléctrica (2015 - 2016) http://app.cfe.gob.mx/Aplicaciones/CCFE/Tarifas/Tarifas/tarifas_casa.asp [accessed: 25.03.2016].
[58] [CFE] - Comision Federal De Electricidad, 2016. Tarifas para el suministro y venta de energía eléctrica (2015 - 2016), http://app.cfe.gob.mx/Aplicaciones/CCFE/Tarifas/Tarifas/tarifas_negocio.asp [accessed: 26.03.2016].
[59] [CFE] - Comision Federal De Electricidad, 2016. Tarifas para el suministro y venta de energía eléctrica (2015 - 2016), http://app.cfe.gob.mx/Aplicaciones/CCFE/Tarifas/Tarifas/tarifas_industria.asp [accessed: 27.03.2016].
[60] Gerlach A., Werner Ch., Breyer Ch., 2014. Impact of Financing Cost on Global Grid-Parity Dynamics till 2030. 29th EU PVSEC, Amsterdam, September 22-26, DOI: 10.4229/29thEUPVSEC2014-7DO.15.4., www.researchgate.net/publication/266558306_Impact_of_Financing_Cost_on_Global_Grid-Parity_Dynamics_till_2030 [accessed: 15.01.2016].
[61] Breyer Ch, Werner C., Rolland S., Adelmann P., 2011. Off-Grid Photovoltaic Applications in Regions of Low Electrification: High Demand, Fast Financial Amortization and Large Market Potential. 26th EU PVSEC, Hamburg, September 5-9. Available online at: www.researchgate.net/publication/261172910_Off-Grid_Photovoltaic_Applications_in_Regions_of_Low_Electrification_High_Demand_Fast_Financial_Amortization_and_Large_Market_Potential [accessed: 22.02.2016].
[62] Breyer Ch., Gaudchau E., Gerlach A.-K., Hlusiak M, Cader C., Bertheau P., Wasgindt V., 2012. PV-based Mini-Grids for Electrification in Developing Countries - An overview on market potentials and business models. A study of the Reiner Lemoine Institut gGmbH on behalf of the cdw Stiftungsverbund gGmbH, December 11. www.researchgate.net/publication/261345306_PV-based_Mini Grids_for_Electrification_in_Developing_Countries_-_An_overview_on_market_potentials_and_business_models [accessed: 29.02.2016]
[63] Molyneaux L., Wagner L., Foster J., 2016. Rural electrification in India: Galilee Basin coal versus decentralised renewable energy micro grids, Renewable Energy 89, 422-436.
[64] Yadoo A. and Cruickshank H., 2012. The role for low carbon electrification technologies in poverty reduction and climate change strategies: A focus on renewable energy mini-grids with case studies in Nepal, Peru and Kenya, Energy Policy 42, 591-602.
[65] Sen, R. and Bhattacharyya, S.C., 2014. Off-grid electricity generation with renewable energy technologies in India: An application of HOMER, Renewable Energy 62, 388-398.
[66] Fraunhofer ISE, 2015. Current and Future Cost of Photovoltaics. Long-term Scenarios for Market Development, System Prices and LCOE of Utility-Scale PV Systems, study on behalf of Agora Energiewende, Freiburg and Berlin, www.agora-energiewende.de/fileadmin/Projekte/2014/Kosten-Photovoltaik-2050/AgoraEnergiewende_Current_and_Future_Cost_of_PV_Feb2015_web.pdf [accessed: 12.02.2016].
[67] Hoffmann W., 2014. Importance and evidence for cost effective electricity storage, 29th EU PVSEC, Amsterdam, September 22-26.
[68] Nykvist B. and Nilsson M., 2015. Rapidly falling costs of battery packs for electric vehicles, nature climate change, 5, 329–332.
[69] Breyer Ch., Gerlach A., 2013. Global Overview on Grid-Parity, Progress in Photovoltaics: Research and Applications 21, 121-136.
[70] Caldera U., Bogdanov D., Afanasyeva S., Breyer Ch., 2016. Integration of reverse osmosis seawater

desalination in the power sector, based on PV and wind energy, for the Kingdom of Saudi Arabia, 32nd EU PVSEC, Munich, June 20-24.
[71] [IRENA] - International Renewable Energy Agency, 2015. Renewable Energy Prospects: United States of America, REmap 2030 analysis. IRENA, Abu Dhabi. www.irena.org/remap.
[72] Clean Technica, 2015. 100% renewable energy goal for Hawaii: Governor signs bill, the US, http://cleantechnica.com/2015/06/11/100-renewable-energy-goal-hawaii-governor-signs-bill/ [accessed: 10.06.2016].
[73] DBEDT Hawaii State Energy Office, 2015. Hawaii Energy Facts & Figures, Hawaii, the US, www.energy.hawaii.gov.
[74] Go 100% Renewable Energy, n.d., San Francisco - 100% Renewable Power by 2030, the US, www.go100percent.org/cms/index.php?id=77&tx_ttnews[tt_news]=79&cHash=064a35afc66d04fe0785cee33ef04ba0 [accessed: 15.06.2016].
[75] Jacobson, M.Z., Delucchi, M.A., Bazouin, G., Dvorak, M.J., Arghandeh, R., Bauer, Z.A., Cotte, A., de Moor, G.M., Goldner, E.G., Heier, C. and Holmes, R.T., 2016. A 100% wind, water, sunlight (WWS) all-sector energy plan for Washington State. Renewable Energy, 86, 75-88.
[76] Agora Energiewende, 2014. Comparing the cost of low-carbon technologies: what is the cheapest option? report by Prognos AG on behalf of Agora Energiewende, p. 10-13. Available online at: www.prognos.com/fileadmin/pdf/publikationsdatenbank/140417_Prognos_Agora_Analysis_Decarbonisationtechnologies_EN.pdf [accessed: 20.02.2016].
[77] Tokimatsu K., Yasuoka R., Nishio M., 2016. Global zero emissions scenarios: The role of biomass energy with carbon capture and storage by forested land use. Applied Energy, in press.
[78] European Commission, 2014a. Integration of renewable energy in Europe. study prepared by KEMA Consulting, DNV GL – Energy, Imperial College and NERA Economic Consulting on behalf of DG Energy, Brussels, pp. 7-8 and 54.
[79] [WWF] – World Wide Fund for Nature, 2011. The energy report: 100% renewable energy by 2050, Switzerland, assets.panda.org/downloads/101223_energy_report_final_print_2.pdf [accessed: 12.12.2015].
[80] Teske S., et al. 2015. Energy [r] evolution - a sustainable world energy outlook 2015, Greenpeace International, GWEC and SolarPowerEurope, 5th edition, Amsterdam.
[81] Dittmar M., 2012. Nuclear energy: Status and future limitations, Energy, 37, 35-40.
[82] Pleßmann G., Erdmann M., Hlusiak M., Breyer Ch., 2014. Global energy storage demand for a 100% renewable electricity supply. Energy Procedia 46, 22-31.
[83] Komoto K., Ito M., Van der Vleuten P., Faiman D., Kurokawa K. (eds.), 2009. Energy from the desert – very large scale photovoltaic systems: socio-economic, financial, technical and environmental aspects. Earthscan, London.
[84] Urban W., Lohmann H., Girod K., 2009. Abschlussbericht für das BMBF-Verbundprojekt Biogaseinspeisung. Fraunhofer UMSICHT.
[85] European Commission, 2014a. ETRI 2014 - Energy technology reference indicator projections for 2010-2050. EC Joint Research Centre Institute for Energy and Transport, Petten, Netherlands.
[86] Vartiainen E., Masson G., Breyer Ch., 2015. PV LCOE in Europe 2015-2050. 31th EU PVSEC, Hamburg, September 14-18; 10.4229/31stEUPVSEC2015-7DO.15.1. www.researchgate.net/publication/281939918_PV_LCOE_in_Europe_2015-2050 [accessed: 17.06.2016].

Supplementary Material

Table 1. Financial assumptions for energy system components ([35],[66],[67],[82]-[86]).

Technology	Capex [€/kW]	Opex fix [€/(kW·a)]	Opex var [€/kWh]	Lifetime [a]
PV optimally tilted	550	8	0	35
PV single-axis tracking	620	9	0	35
PV rooftop	813	12	0	35
Wind onshore	1000	20	0	25
CSP (solar field)	528	11	0	25
Hydro run-of-river	2560	115.2	0.005	60
Hydro dam	1650	66	0.003	60
Geothermal energy	4860	87	0	30
Water electrolysis	380	13	0.0012	30
Methanation	234	5	0.0015	30
CO_2 scrubbing	356	14	0.0013	30
CCGT	775	19.4	0.001	30
OCGT	475	14.25	0.001	30
Steam turbine	600	12	0	30
Hot heat burner	100	2	0	30
Heating rod	20	0.4	0.001	30
Biomass CHP	2500	175	0.001	30
Biogas CHP	370	14.8	0.001	30
Waste incinerator	5240	235.8	0.007	20
Biogas digester	680	27.2	0	20
Biogas upgrade	250	20	0	20
	Capex [€/kWh]	Opex fix [€/(kWh·a)]	Opex var [€/kWh]	Lifetime [a]
Battery system	150	10	0.0002	10/20
Battery prosumer	150	10	0.0002	10
PHS	70	11	0.0002	50
A-CAES	31	0.4	0.0012	40
TES	24	2	0	20
Gas storage	0.05	0.001	0	50
	Capex [€/(kW$_{NTC}$*km)]	Opex fix [€/(kW$_{NTC}$·km·a)]	Opex var [€/kWh$_{NTC}$]	Lifetime [a]
HVDC line on ground	0.612	0.0075	0	50
HVDC line submarine	0.992	0.0010	0	50
	Capex [€/kW$_{NTC}$]	Opex fix [€/(kW$_{NTC}$·a]	Opex var [€/kWh$_{NTC}$]	Lifetime [a]
HVDC converter pair	180	1.8	0	50
	Capex [€/(m^3·a)]	Opex fix [€/(m^3 a)]	Opex var [€/m^3]	Lifetime [a]
Water desalination	2.23	0.09	0	30
	Capex [€/(m^3·h·km)]	Opex fix [€/(m^3·h·km·a)]	Opex var [€/m^3·h·km]	Lifetime [a]
Horizontal pumping and pipes	19.3	0.39	0	30
Vertical pumping and pipes	15.5	0.31	0	30

Table 2. Efficiencies and energy to power ratio of storage technologies.

Technology	Efficiency [%]	Energy/Power Ratio [h]	Self-Discharge [%/h]
Battery	90	6	0
PHS	85	8	0
A-CAES	70	100	0.001
TES	90	8	0.002
Gas storage	100	80*24	0

Table 3. Efficiency assumptions for energy system components for the year 2030 reference years.

	η_{el} [%]	η_{th} [%]
CSP (solar field)		51
Steam turbine	42	
Hot heat burner		95
Heating rod		99
Water electrolysis		84
Methanation		77
CO_2 scrubbing		78
CCGT	58	
OCGT	43	
Geothermal	24	
Biomass CHP	40	45
Biogas CHP	42	43
Waste incinerator	34	
Biogas upgrade		98

Table 4. Efficiency assumptions for HVDC transmission.

	Power losses
HVDC line	1.6 % / 1000 km
HVDC converter pair	1.4%

Table 5. Regional end-user grid electricity costs for the year 2030.

Region	Electricity costs [€/MWh]		
	Residential	Commercial	Industrial
Total area (weighted average)	169	174	154
CA-West	162	164	153
CA-East	151	179	172
US-New England & New York	205	194	164
US-Mid-Atlantics	186	178	145
US-Carolinas	186	175	151
US-Southern	182	179	154
US-TVA	166	169	127
US-Midwest	182	170	148
US-Central	175	167	145
US-Texas	176	162	137
US-Southwest	181	167	137
US-Northwest	158	144	126
US-California	210	205	187
US-Alaska	309	296	209
US-Hawaii	358	354	342
US-Gulf	160	161	131
MX-Noroeste (Northwest)	103	176	176
MX-Noreste (North)	104	175	175
MX-Centro-Occidente (Center)	105	186	186
MX-Centro & Sur-Sureste (South)	102	182	182

Table 6. Regional biomass potentials and geothermal energy potentials.

Region	Biomass potential [TWh_{th}/a]			Geothermal Potential [TWh_{th}/a]
	Solid waste	Solid biomass	Biogas sources	
Total area	97.8	1140.7	0.0	860.7
CA-West	6.3	155.7	0.0	176.1
CA-East	13.9	71.9	0.0	5.3
US-New England & New York	7.1	26.1	0.0	0.0
US-Mid-Atlantics	10.4	27.3	0.0	2.2
US-Carolinas	3.0	17.7	0.0	0.6
US-Southern	7.1	36.6	0.0	0.0
US-TVA	2.3	17.0	0.0	0.0
US-Midwest	10.6	127.3	0.0	124.9
US-Central	1.8	47.3	0.0	80.3
US-Texas	5.4	55.4	0.0	1.3
US-Southwest	7.2	104.4	0.0	21.1
US-Northwest	5.3	164.5	0.0	316.1
US-California	7.9	33.8	0.0	24.4
US-Alaska	0.2	135.4	0.0	15.2
US-Hawaii	0.3	2.3	0.0	0.0
US-Gulf	2.2	31.6	0.0	0.0
MX-Noroeste (Northwest)	0.6	16.2	0.0	9.9
MX-Noreste (North)	0.9	22.6	0.0	9.0
MX-Centro-Occidente (Center)	1.6	18.4	0.0	54.1
MX-Centro & Sur-Sureste (South)	3.7	29.2	0.0	20.0

Table 7. Regional biomass costs, calculated based on biomass sources mix in the region. Solid wastes cost are based on assumption of 50 €/ton gate fee paid to the MSW incinerator.

Region	Biomass costs [€/MWh_{th}]		
	Solid waste	Solid biomass	Biogas sources
Total area (weighted average)	-10.1	9.7	1.9
CA-West	-9.8	13.4	0.0
CA-East	-9.8	13.4	0.0
US-New England & New York	-10.2	10.5	0.6
US-Mid-Atlantics	-10.2	10.5	0.6
US-Carolinas	-10.2	10.5	0.6
US-Southern	-10.2	10.5	0.6
US-TVA	-10.2	10.5	0.6
US-Midwest	-10.2	10.5	0.6
US-Central	-10.2	10.5	0.6
US-Texas	-10.2	10.5	0.6
US-Southwest	-10.2	10.5	0.6
US-Northwest	-10.2	10.5	0.6
US-California	-10.2	10.5	0.6
US-Alaska	-10.2	10.5	0.6
US-Hawaii	-10.2	10.5	0.6
US-Gulf	-10.2	10.5	0.6
MX-Noroeste (Northwest)	-10.2	5.3	7.2
MX-Noreste (North)	-10.2	5.3	7.2
MX-Centro-Occidente (Center)	-10.2	5.4	7.2
MX-Centro & Sur-Sureste (South)	-10.2	5.5	7.2

Table 8. Average full load hours (FLH) and LCOE for single-axis tracking PV, optimally tilted PV, solar CSP and wind power plants in North America by 2030.

Region	Pop. [mio. Pop]	Electr. demand [TWh]	PV-single-axis FLH	PV optimally tilted FLH	CSP FLH	Wind FLH	PV single-axis LCOE [€/MWh]	PV optimally tilted LCOE [€/MWh]	CSP LCOE [€/MWh]	Wind LCOE [€/MWh]
Total area	494	6059	1977	1544	1861	2848	30	34	66	40
CA-West	11	189	1424	1113	1295	3115	40	45	90	34
CA-East	24	372	1555	1258	1410	3972	37	40	83	27
US-New England & New York	34	307	1597	1317	1440	2144	36	38	81	49
US-Mid-Atlantics	50	680	1616	1353	1417	2552	35	37	83	41
US-Carolinas	14	243	1754	1456	1575	2295	32	35	74	46
US-Southern	34	514	1903	1536	1715	2000	30	33	68	53
US-TVA	11	209	1732	1427	1543	2310	33	35	76	46
US-Midwest	51	746	1755	1408	1580	3944	32	36	74	27
US-Central	9	150	1983	1562	1850	4178	29	32	63	25
US-Texas	26	428	2168	1698	2039	3752	26	30	57	28
US-Southwest	35	543	2348	1780	2354	3264	24	28	50	32
US-Northwest	25	379	2160	1652	2069	3423	26	31	57	31
US-California	38	299	2347	1758	2386	2192	24	29	49	48
US-Alaska	1	7	1330	966	1229	3478	43	52	95	30
US-Hawaii	1	11	2468	1812	2337	2889	23	28	50	37
US-Gulf	11	211	1832	1496	1666	2013	31	34	70	53
MX-Noroeste (Northwest)	10	115	2475	1842	2548	1923	23	27	46	55
MX-Noreste (North)	17	320	2371	1853	2313	2808	24	27	51	38
MX-Centro-Occidente (Center)	28	163	2419	1837	2290	2698	24	27	51	39
MX-Centro & Sur-Sureste (South)	65	173	2298	1762	2173	2019	25	29	54	52

Table 9. Lower limits of installed capacities in North America sub-regions.

Region	Installed capacity [MW]				
	Solar PV	Wind	Hydro dams	Hydro RoR[3]	PHS[4]
Total area	20573.6	73779.6	125483.2	29165.2	18023.4
CA-West	0.0	2228.0	13687.7	5779.6	0.0
CA-East	1895.6	6723.5	35120.3	9300.9	174
US-New England & New York	580.7	2473.5	6291.4	166.2	1680.0
US-Mid-Atlantics	1183.0	2457.1	4001.5	805.8	5669.8
US-Carolinas	771.5	0.0	3940.8	179.0	2356.2
US-Southern	188.4	0.0	4768.4	233.0	1723.0
US-TVA	28.7	29.0	2893.4	332.0	0.0
US-Midwest	168.3	17760.3	3341.7	70.3	2312.0
US-Central	1.0	6809.0	708.1	29.0	260
US-Texas	125.0	13068.9	387.9	0.0	0.0
US-Southwest	1313.9	1192.7	2842.8	111.5	0.0
US-Northwest	9060.2	11948.4	24144.3	11363.9	944.5
US-California	5018.5	6456.4	8937.5	755.6	2903.9
US-Alaska	0.0	44.3	381.2	15.0	0.0
US-Hawaii	47.5	207.1	5.6	12.3	0.0
US-Gulf	0.9	0.1	1291.4	0.0	0.0
MX-Noroeste (Northwest)	46.8	10.6	931.6	9.6	0.0
MX-Noreste (North)	17.3	76.0	125.5	0.0	0.0
MX-Centro-Occidente (Center)	5.8	150.4	2940.7	1.4	0.0
MX-Centro & Sur-Sureste (South)	120.7	2144.3	8741.5	0.0	0.0

Table 10. Upper limits on installable capacities in the North America in units of GW_{th} for CSP and GW_{el} for all other technologies.

Region	area [1000 km²]	Limits [GW]					
		Solar CSP	Solar PV	Wind	Hydro RoR	Hydro dams	PHS
Total area	22035747	198321.7	99160.9	7404.0	43.7	188.2	36.0
CA-West	6827155	61444.4	30722.2	2293.9	8.7	20.5	0.0
CA-East	3152421	28371.8	14185.9	1059.2	14.0	52.7	0.3
US-New England & New York	327560	2948.0	1474.0	110.1	0.2	9.4	3.4
US-Mid-Atlantics	342403	3081.6	1540.8	115.0	1.2	6.0	11.3
US-Carolinas	222295	2000.7	1000.3	74.7	0.3	5.9	4.7
US-Southern	460040	4140.4	2070.2	154.6	0.3	7.2	3.4
US-TVA	213823	1924.4	962.2	71.8	0.5	4.3	0.0
US-Midwest	1599070	14391.6	7195.8	537.3	0.1	5.0	4.6
US-Central	594514	5350.6	2675.3	199.8	0.0	1.1	0.5
US-Texas	695673	6261.1	3130.5	233.7	0.0	0.6	0.0
US-Southwest	1311066	11799.6	5899.8	440.5	0.2	4.3	0.0
US-Northwest	2066206	18595.9	9297.9	694.2	17.0	36.2	1.9
US-California	423970	3815.7	1907.9	142.5	1.1	13.4	5.8
US-Alaska	1700131	15301.2	7650.6	571.2	0.0	0.6	0.0
US-Hawaii	28313	254.8	127.4	9.5	0.0	0.0	0.0
US-Gulf	397458	3577.1	1788.6	133.5	0.0	1.9	0.0
MX-Noroeste (Northwest)	317774	2860.0	1430.0	106.8	0.0	1.4	0.0
MX-Noreste (North)	444063	3996.6	1998.3	149.2	0.0	0.2	0.0
MX-Centro-Occidente (Center)	355014	3195.1	1597.6	119.3	0.0	4.4	0.0
MX-Centro & Sur-Sureste (South)	556798	5011.2	2505.6	187.1	0.0	13.1	0.0

[3] RoR = Run-of-River
[4] PHS = Pumped Hydroelectric Storage

Table 11. Annual industrial gas demand and water demand for the year 2030.

	Annual gas demand		Annual electricity demand for gas synthesis	Annual water desalination demand	Annual electricity demand for water desalination
	[Mtoe]	[TWh_{th}]	[TWh_{el}]	[Mil. m^3]	[TWh_{el}]
Total	183.5	2134.0	3783.7	83607.2	462.0
CA-West	7.1	82.9	149.3	230.7	1.4
CA-East	15.6	181.8	327.4	506.0	0.0
US-New England & New York	14.7	171.4	303.5	6266.4	0.0
US-Mid-Atlantics	21.6	251.1	444.8	9182.2	30.6
US-Carolinas	6.2	72.7	128.7	2657.6	0.0
US-Southern	14.7	171.0	303.0	6252.9	18.8
US-TVA	4.7	54.4	96.4	1989.5	0.0
US-Midwest	22.0	256.2	455.8	9368.6	43.2
US-Central	3.7	43.0	77.8	1570.8	8.6
US-Texas	11.3	130.9	233.7	4784.1	31.2
US-Southwest	15.0	174.2	310.4	6370.1	56.9
US-Northwest	11.0	127.6	225.9	4664.5	44.6
US-California	16.4	191.0	338.6	6983.9	29.5
US-Alaska	0.3	3.7	6.5	134.3	0.0
US-Hawaii	0.6	7.0	12.5	255.6	0.0
US-Gulf	4.5	52.9	94.0	1934.3	7.3
MX-Noroeste (Northwest)	1.1	13.4	22.6	1684.7	7.5
MX-Noreste (North)	2.0	22.8	40.3	2879.8	22.8
MX-Centro-Occidente (Center)	3.2	37.8	63.7	4761.5	45.2
MX-Centro & Sur-Sureste (South)	7.6	88.2	148.7	11129.8	114.5

Table 12. Total LCOE components in all sub-regions of North America.

Region-wide	LCOE primary [€/MWh]	LCOC [€/MWh]	LCOS [€/MWh]	LCOT [€/MWh]	LCOE total [€/MWh]	
Area weighted average	37.5	2.8	22.9	0.0	63.2	
CA-West	41.1	1.3	7.0	0.0	49.4	
CA-East	34.6	2.7	6.0	0.0	43.3	
US-New England & New York	45.4	2.9	22.2	0.0	70.5	
US-Mid-Atlantics	40.8	3.6	31.4	0.0	75.8	
US-Carolinas	41.8	4.1	29.0	0.0	74.9	
US-Southern	40.0	3.8	31.9	0.0	75.7	
US-TVA	41.2	2.7	29.4	0.0	73.3	
US-Midwest	33.6	2.5	20.9	0.0	57.0	
US-Central	34.2	1.5	21.6	0.0	57.3	
US-Texas	30.7	2.6	26.8	0.0	60.1	
US-Southwest	31.3	3.4	24.7	0.0	59.4	
US-Northwest	39.0	2.1	9.2	0.0	50.3	
US-California	41.4	2.9	20.2	0.0	64.5	
US-Alaska	41.2	1.6	8.8	0.0	51.6	
US-Hawaii	36.1	2.1	27.6	0.0	65.8	
US-Gulf	43.8	3.1	30.4	0.0	77.3	
MX-Noroeste (Northwest)	33.7	2.0	32.7	0.0	68.4	
MX-Noreste (North)	33.9	2.7	26.4	0.0	63.0	
MX-Centro-Occidente (Center)	39.4	2.8	18.6	0.0	60.8	
MX-Centro & Sur-Sureste (South)	40.7	0.9	13.3	0.0	54.9	
Country-wide	LCOE primary [€/MWh]	LCOC [€/MWh]	LCOS [€/MWh]	LCOT [€/MWh]	LCOE total [€/MWh]	export (-)/ import (+) [%]
Area weighted average	35.7	2.5	15.0	3.1	56.3	21.5
CA-West	41.0	1.3	7.0	0.0	49.3	0.2
CA-East	34.5	2.5	6.2	0.0	43.2	-0.1
US-New England & New York	46.3	1.9	12.7	2.8	63.7	17.7
US-Mid-Atlantics	49.5	1.7	12.7	8.9	72.8	48.6
US-Carolinas	43.9	2.6	14.6	2.8	63.9	18.1
US-Southern	38.9	1.9	20.3	2.9	64.0	16.4
US-TVA	44.0	5.8	16.8	26.3	92.9	72.3
US-Midwest	30.5	2.7	16.1	3.5	52.8	-35.9
US-Central	28.5	2.9	16.7	4.0	52.1	-37.4
US-Texas	30.4	2.9	17.5	2.2	53.0	-18.9
US-Southwest	32.5	2.9	17.3	1.2	53.9	4.4
US-Northwest	37.7	1.8	7.7	1.6	48.8	-13.6
US-California	34.3	3.1	14.6	3.0	55.0	15.6
US-Alaska	40.0	2.0	5.7	0.8	48.5	-0.9
US-Hawaii	36.4	2.0	24.8	0.6	63.8	1.6
US-Gulf	39.8	4.1	27.2	39.8	110.9	80.1
MX-Noroeste (Northwest)	30.5	1.4	32.3	1.6	65.8	9.2
MX-Noreste (North)	35.9	3.5	18.4	0.9	58.7	-3.3
MX-Centro-Occidente (Center)	39.9	1.9	13.6	1.0	56.4	0.6
MX-Centro & Sur-Sureste (South)	42.3	0.9	10.5	1.5	55.2	-0.5
Area-wide	LCOE primary [€/MWh]	LCOC [€/MWh]	LCOS [€/MWh]	LCOT [€/MWh]	LCOE total [€/MWh]	export (-)/ import (+) [%]
Area weighted average	34.6	2.1	12.4	3.9	53.0	27.4
CA-West	40.8	1.1	2.4	1.9	46.2	-8.3
CA-East	31.3	1.4	3.7	5.0	41.4	-62.4
US-New England & New York	40.1	2.5	19.5	16.7	78.8	61.8
US-Mid-Atlantics	43.6	1.1	18.1	19.4	82.2	66.1

US-Carolinas	42.5	1.7	15.3	5.4	64.9	36.4
US-Southern	40.9	2.2	17.2	2.8	63.1	18.2
US-TVA	44.6	5.2	14.1	21.0	84.9	68.8
US-Midwest	31.0	2.2	11.9	2.3	47.4	-22.1
US-Central	28.6	3.1	16.6	3.7	52.0	-34.2
US-Texas	31.0	2.5	18.4	1.6	53.5	-6.2
US-Southwest	32.3	2.8	18.3	1.1	54.5	6.1
US-Northwest	36.4	2.1	6.3	1.7	46.5	-14.1
US-California	34.5	3.0	13.3	3.1	53.9	17.3
US-Alaska	39.2	2.0	6.1	0.9	48.2	-1.7
US-Hawaii	36.4	1.9	24.9	0.5	63.7	1.4
US-Gulf	44.2	2.0	11.4	8.9	66.5	51.1
MX-Noroeste (Northwest)	30.8	2.3	26.1	1.9	61.1	8.8
MX-Noreste (North)	35.5	2.6	17.0	1.3	56.4	3.9
MX-Centro-Occidente (Center)	39.8	1.8	14.2	1.0	56.8	2.5
MX-Centro & Sur-Sureste (South)	40.5	1.0	11.7	1.3	54.5	-2.3

Integrated scenario	LCOE primary [€/MWh]	LCOC [€/MWh]	LCOS [€/MWh]	LCOT [€/MWh]	LCOE total [€/MWh]	export (-)/ import (+) [%]
Area wighted average	32.5	1.2	5.1	3.4	42.2	26.8
CA-West	38.5	0.2	1.4	0.6	40.7	4.5
CA-East	30.1	0.9	1.0	4.1	36.1	-62.2
US-New England & New York	39.9	1.0	19.2	38.2	98.3	81.5
US-Mid-Atlantics	42.0	0.3	9.2	14.1	65.6	65.3
US-Carolinas	40.5	1.1	6.5	3.6	51.7	30.2
US-Southern	35.6	1.1	9.3	2.6	48.6	22.8
US-TVA	37.3	3.1	17.4	43.1	100.9	83.9
US-Midwest	30.1	1.1	3.3	2.2	36.7	-29.5
US-Central	28.0	1.9	5.6	3.4	38.9	-46.3
US-Texas	30.2	1.4	4.5	1.6	37.7	-15.8
US-Southwest	31.8	1.9	6.4	0.7	40.8	0.5
US-Northwest	35.3	0.8	2.4	1.3	39.8	-14.9
US-California	29.0	1.0	4.8	2.5	37.3	21.1
US-Alaska	37.6	1.3	2.6	0.0	41.5	0.0
US-Hawaii	28.3	1.2	15.4	0.0	44.9	0.0
US-Gulf	35.2	1.1	11.4	16.7	64.4	70.4
MX-Noroeste (Northwest)	28.1	1.2	21.4	1.7	52.4	5.7
MX-Noreste (North)	34.9	1.7	13.0	1.1	50.7	0.3
MX-Centro-Occidente (Center)	37.6	2.0	7.5	1.1	48.2	-2.9
MX-Centro & Sur-Sureste (South)	35.1	0.8	6.5	1.0	43.4	7.7

Table 13. Prosumer electricity costs, installed capacities and electricity utilization in North America.

Prosumers parameters	Residential	Commercial	Industrial
Electricity price [€/kWh]	0.170	0.174	0.153
PV LCOE [€/kWh]	0.029	0.039	0.037
Self-consumption PV LCOE [€/kWh]	0.039	0.049	0.045
Self-consumption PV and Battery LCOE [€/kWh]	0.071	0.086	0.078
Self-consumption LCOE [€/kWh]	0.070	0.086	0.078
Benefit [€/kWh]	0.100	0.088	0.075
Installed capacities	Residential	Commercial	Industrial
PV [GW]	285.2	212.9	313.8
Battery storage [GWh]	480.2	389.8	541.0
Generation	Residential	Commercial	Industrial
PV [TWh]	425.8	319.0	492.1
Battery storage [TWh]	143.3	114.6	163.1
Excess [TWh]	92.4	51.9	70.7
Utilization	Residential	Commercial	Industrial
Self-consumption of generated PV electricity [%]	74.6	79.7	81.9
Self-coverage market segment [%]	17.0	16.8	15.1
Self-coverage operators [%]	85.0	84.0	75.3

Table 14. Overview on storage capacities, throughput and full cycles per year for the four scenarios in North America.

			Region-wide	Country-wide	Area-wide	Integrated
Storage capacities	Battery SC	[GWh_{el}]	1410.9	1410.9	1410.9	1410.9
	Battery system	[GWh_{el}]	1234.4	592.9	483.3	402.4
	PHS	[GWh_{el}]	34.9	34.9	34.9	24.5
	A-CAES	[GWh_{el}]	5821.6	185.9	1.3	4.8
	Gas	[GWh_{th}]	273396.6	229590.8	174299.2	221421.2
Throughput of storage	Battery SC	[TWh_{el}]	421.1	421.1	421.1	421.1
	Battery system	[TWh_{el}]	364.2	181.9	153.4	121.4
	PHS	[TWh_{el}]	7.9	7.6	8.4	5.2
	A-CAES	[TWh_{el}]	142.9	5.2	0.03	0.1
	Gas	[TWh_{th}]	555.3	494.2	378.4	40.6
Full cycles per year	Battery SC	[-]	298.4	298.4	298.4	298.4
	Battery system	[-]	295.0	306.8	317.3	301.7
	PHS	[-]	228.5	216.7	242.1	210.6
	A-CAES	[-]	24.6	28.0	25.7	25.8
	Gas	[-]	2.0	2.2	2.2	0.2

Table 15. Electricity transmission line parameters for the area-wide scenario for North America.

Region 1	Region 2	Length [km]	Capacity [GW]	Utilization [%]
CA-West	US-Northwest	899.8	4.4	66
CA-East	US-Midwest	742.5	25.6	62
US-New England & New York	CA-East	424.6	90.2	88
US-New England & New York	US-Mid-Atlantics	82.4	64.9	86
US-Mid-Atlantics	US-Carolinas	463.1	9.6	70
US-Mid-Atlantics	US-Midwest	660.0	4.8	60
US-Carolinas	US-Southern	337.7	2.5	58
US-Carolinas	US-TVA	333.3	8.4	71
US-Southern	US-TVA	485.1	16.2	70
US-TVA	US-Midwest	379.5	41.8	73
US-Midwest	CA-West	1486.1	7.0	71
US-Midwest	US-Gulf	838.2	5.7	66
US-Central	US-Midwest	705.1	10.4	63
US-Texas	US-Central	400.4	10.4	56
US-Southwest	US-Central	1036.2	10.0	69
US-Southwest	US-Northwest	1133.0	4.1	65
US-Northwest	US-Midwest	1789.7	2.6	71
US-California	US-Southwest	356.4	9.2	55
US-California	US-Northwest	1089.0	13.7	68
US-Alaska	CA-West	1530.1	0.2	63
US-Gulf	US-Southern	574.2	9.7	68
US-Gulf	US-Central	574.2	11.9	70
US-Gulf	US-Texas	316.8	12.8	60
MX-Noroeste (Northwest)	US-California	124.3	4.8	52
MX-Noroeste (Northwest)	MX-Noreste (North)	618.2	3.2	62
MX-Noreste (North)	US-Texas	658.9	8.5	65
MX-Noreste (North)	MX-Centro-Occidente (Center)	833.8	4.6	69
MX-Centro-Occidente (Center)	MX-Centro & Sur-Sureste (South)	294.8	5.6	58

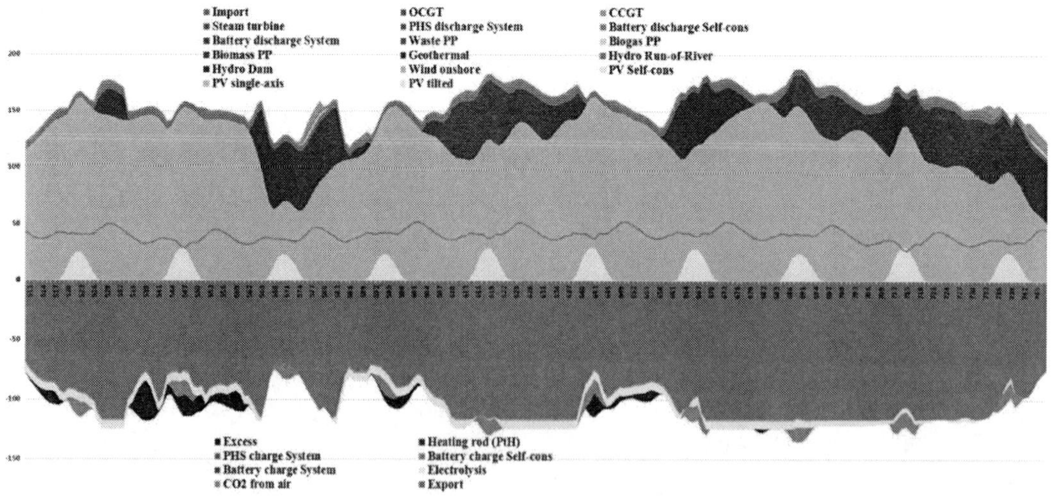

Figure 1. Hourly generation profile for a net exporter region, Eastern Canada.

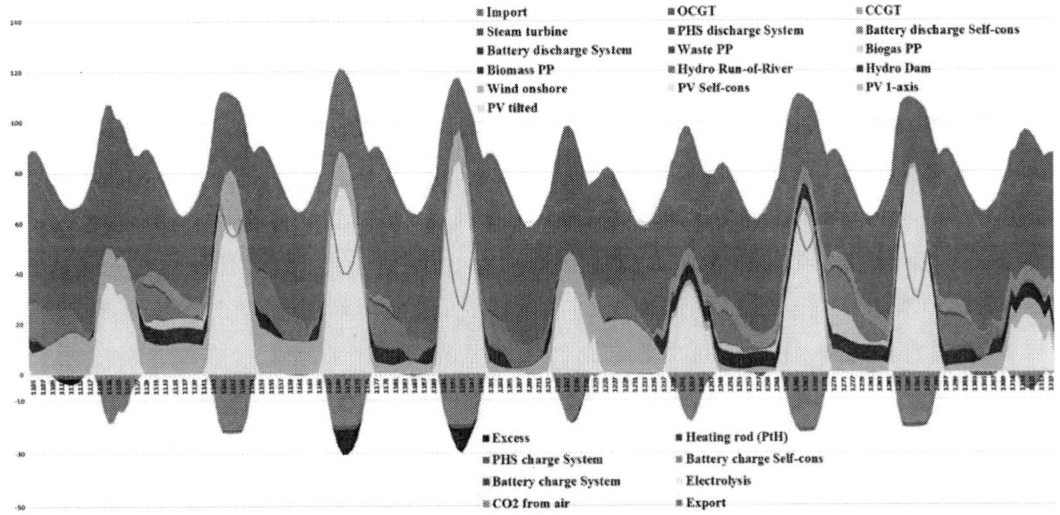

Figure 2. Hourly generation profile for a net importer region, US-Mid-Atlantics.

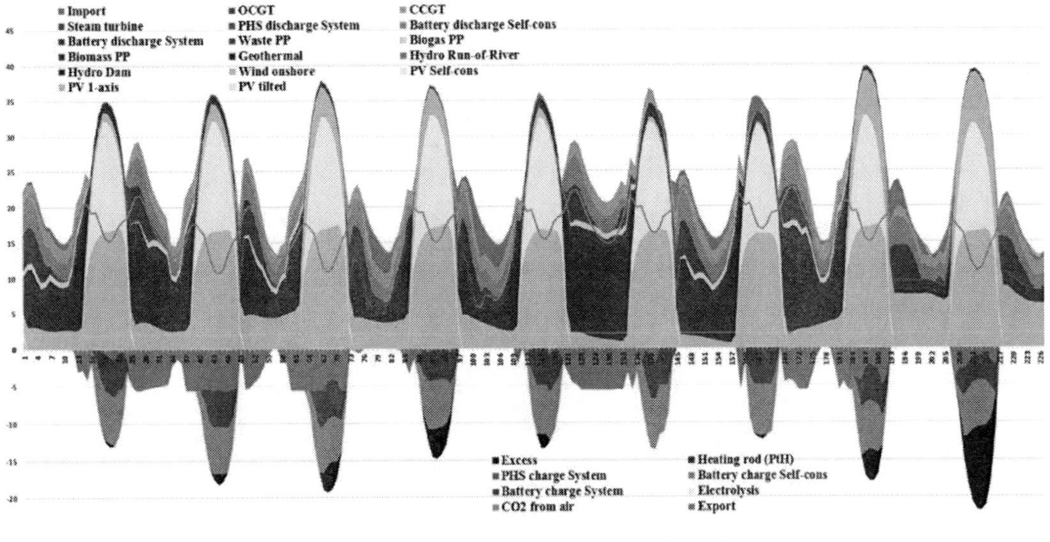

Figure 3. Hourly generation profile for Southern Mexico, example for a balancing region.

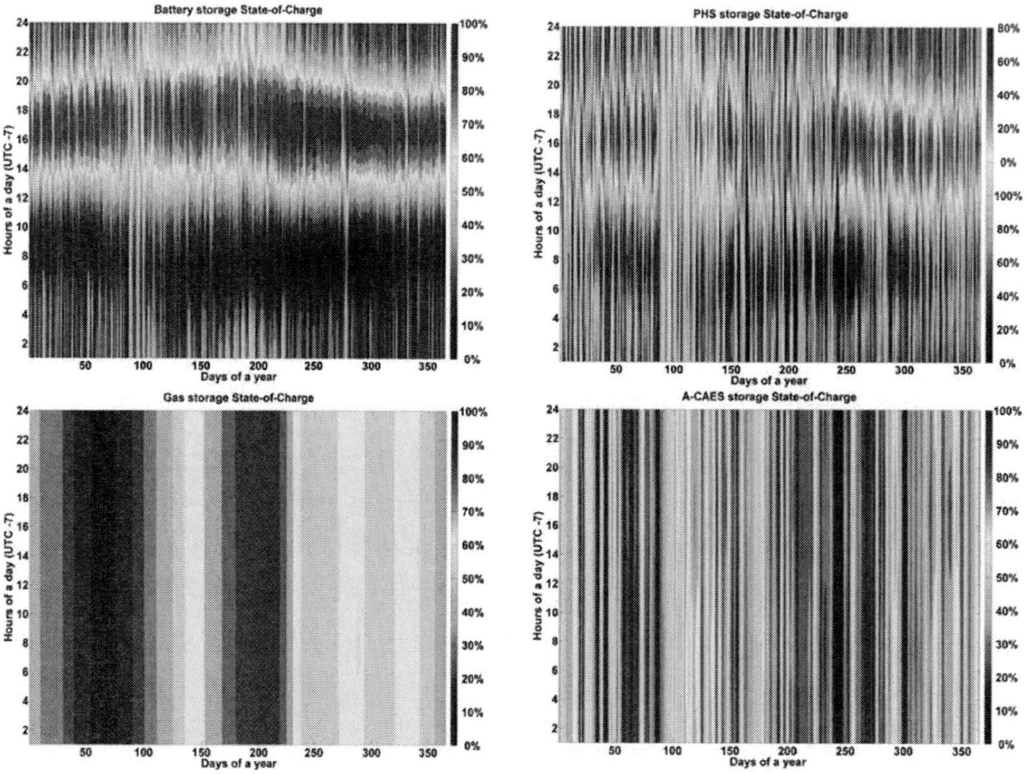

Figure 4. Aggregated yearly state-of-charge for storage in area-wide open trade scenario, battery (top left), PHS (top right), gas storage (bottom left) and A-CAES (bottom right).

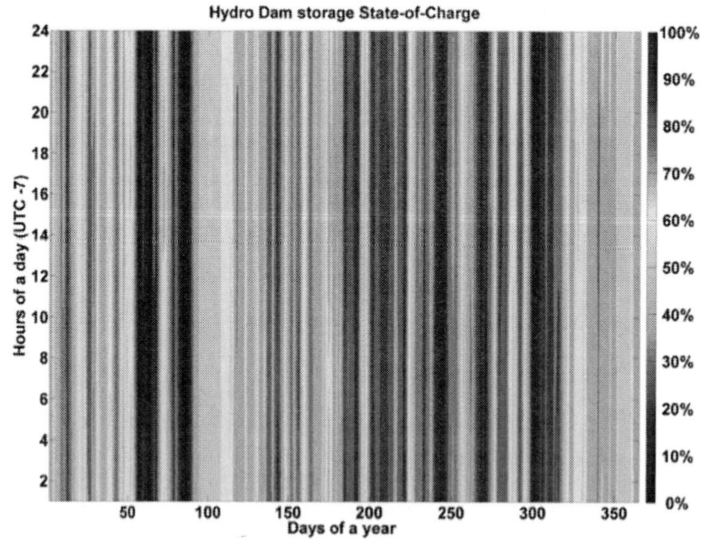

Figure 5. State-of-charge for hydro dams in the area-wide open trade scenario.

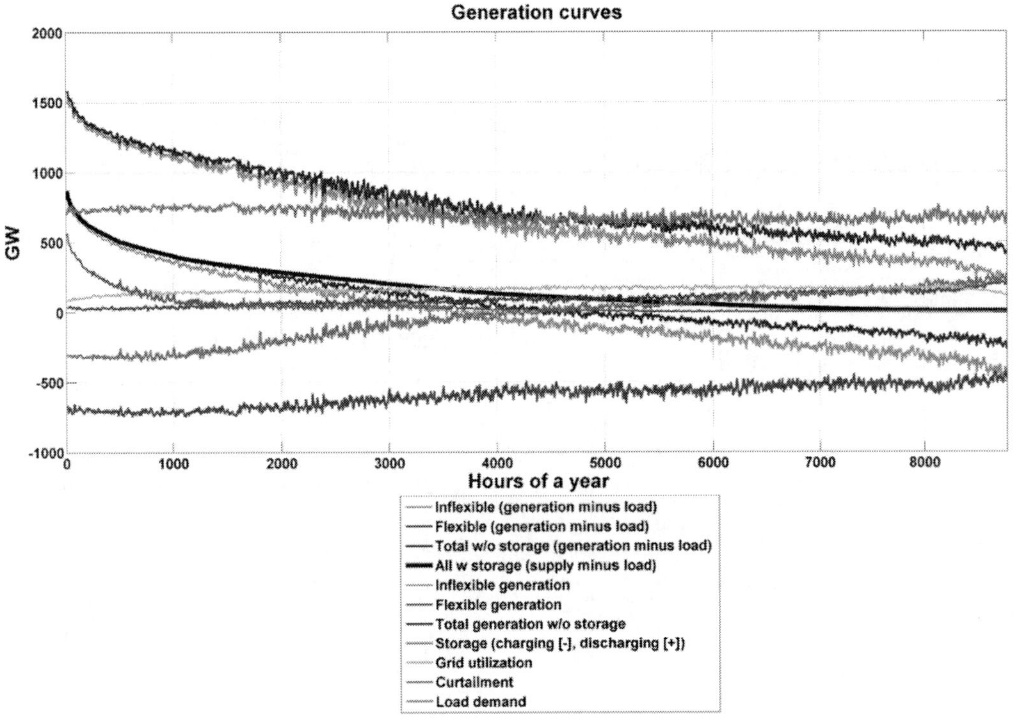

Figure 6. Electricity generation curves for a whole year for area-wide open trade scenario for North America.

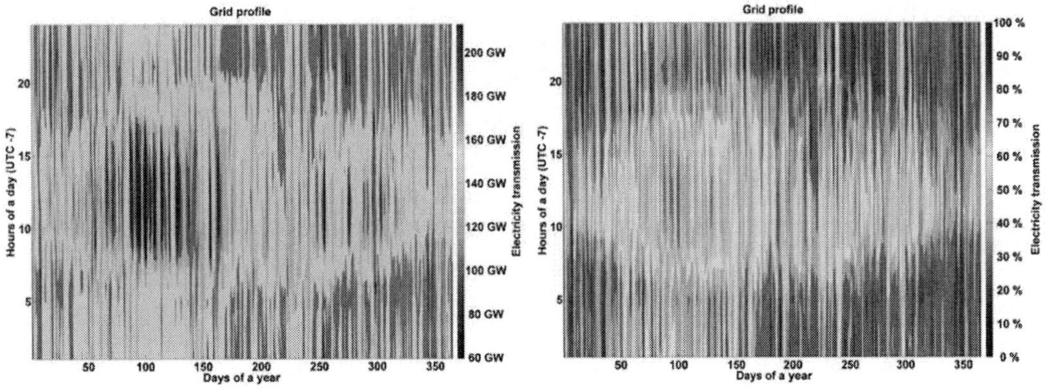

Figure 7. Profile for interregional electricity trade between regions with the absolute numbers (left) and with power line utilisation (right) for area-wide open trade scenario for North America.

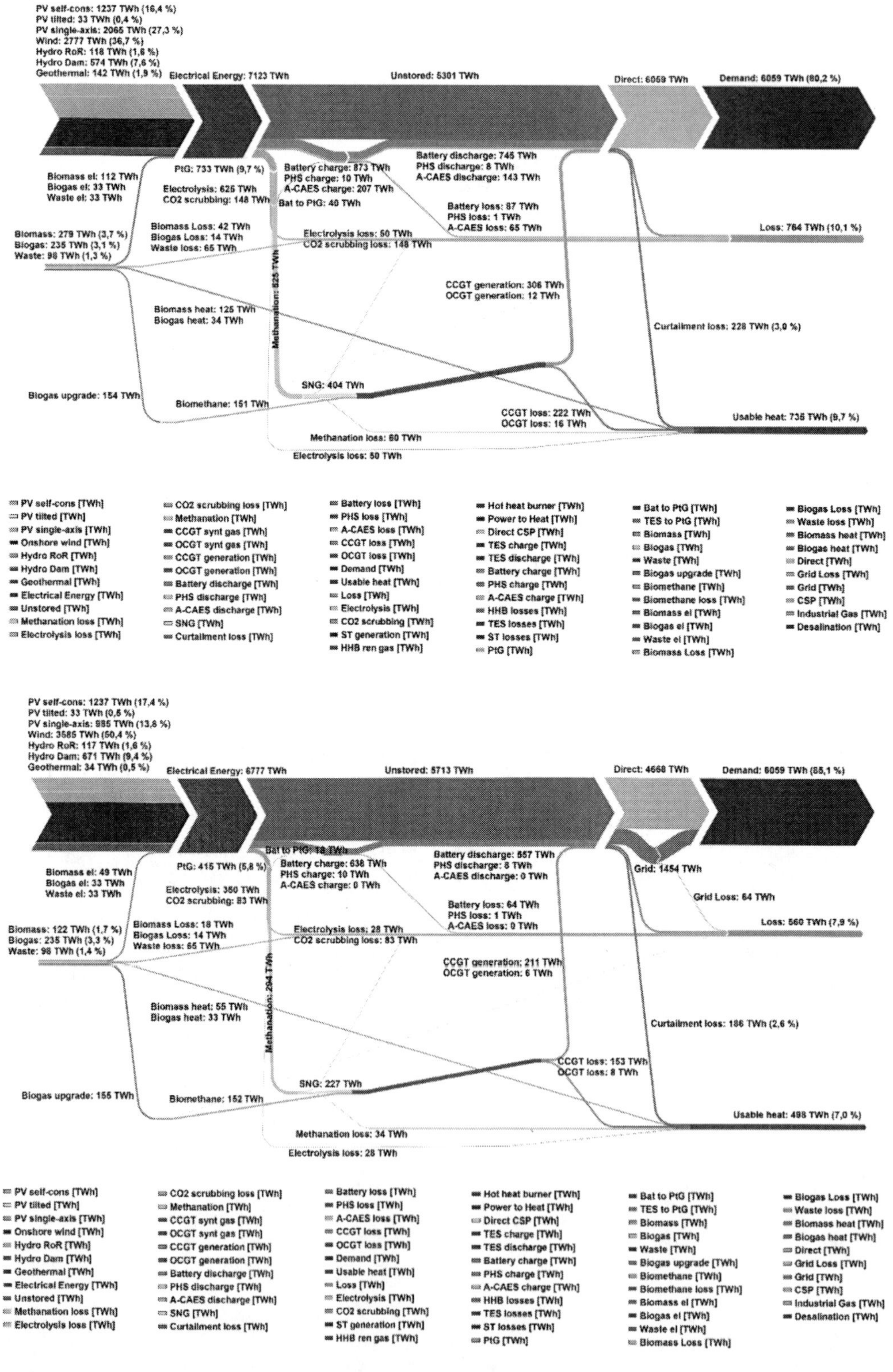

Figure 8. Energy flow of the system for the region-wide open trade scenario (top) and the area-wide open trade scenario (bottom) in 2030.

A FRAMEWORK MODEL FOR POST-SUBSIDY PV MARKET FORECAST

Adel El Gammal[1], Alexander Gerlach[2], Philippe Macé[1], Chris Werner [3], Gaëtan Masson[1]

[1] Becquerel Institute, 69-71 Rue D'Arlon, 1040 Brussels, Belgium
Phone +32 475 67 68 05, e-mail a.elgammal@becquerelinstitute.org
[2] Gerlach Consulting, 04109 Leipzig, Germany
Phone: +49 (0) 172 2483647, e-mail ag_office@yahoo.de
[3]Chris Werner Energy Consulting, 06846 Dessau, Germany
Phone: +49 (0) 176 64093993, e-mail chris.werner@cw-energyconsulting.de

ABSTRACT

The paper addresses the key question of assessing potential future PV development in every country. Especially after the phase out or in the absence of any financial policy support, subsidies or political barriers influencing PV development, a sustainable and thus highly predictable PV development is assumed. The method proposed in this paper is of major interest to anticipate the potential development of PV in emerging markets where, the absence of any significant financial support scheme has usually prevented the uptake of policy driven (and non-competitive based) PV market. Therefore, in most of these markets, PV has been installed solely due to its intrinsic competitiveness in comparison to other power generation technologies.
The method thus assumes that PV is transitioning towards a post-subsidy era and that the long-term PV development in any given country will be driven by its true competitiveness only. The paper presents an overview of PV development scenarios in 21 countries of which 4 highly contrasted examples are analysed in more details and will be a basis for evaluating a closer analysis of all other PV markets future developments.

Keywords

Emerging Markets, Logistic Growth Curve, PV Market Development, PV Penetration, Long-Term Scenario

1 MOTIVATION

PV markets have historically started to develop due to the influence of highly favourable policy incentives schemes intended to develop the renewable energy markets.
Representing a fast growing market estimated to exceed 60 GW this year [1], the PV market is currently scrutinized by a number of market research companies, with many reports being periodically published on the anticipated short-term PV market development [1].
Most of available market analyses however result from a similar methodological approach that typically consists of an anticipation of near future market behaviours based on the analysis of historical market developments and of the anticipated short term policy evolution, in order to derive the level of PV attractiveness and the likely short term development of the market.
Such method is by essence an incremental approach, i.e. it is based on the analysis of the current market situation and of short-term observable market drivers.
If such an approach might prove useful for short-term forecasts in already existing markets, it fundamentally fails to integrate the true long-term attractiveness of PV driven by its intrinsic competitiveness. Thus, the high long-term penetration potential of PV in the electricity system is often not considered. Moreover, this traditional approach is continuously applied to well established markets and not for countries without support policy and/or where no mature PV market exists.
PV is expected to become one of the least cost generation technologies in the future [2]. Therefore, a reliable long term projection of competitiveness driven PV market development is expected to be an essential decision-making tool for any stakeholder involved in defining long term energy strategy, regulatory framework or market development.
This paper describes the framework of a unique approach based on an analysis of the PV competitive long-term penetration potential in the electricity system of any given country. The development of cumulative installed PV capacities is calculated backwards until the year 2000, from which annual market installations are derived. This method proves to be particularly reliable and useful in countries where no significant PV market exists or where the instability and/or phase out of policy support makes any market development forecast difficult to ascertain.

2 METHODOLOGY

2.1 APPROACH

The method proposes, for any given country, to build a PV capacity development curve backwards from the long term PV penetration, by using a traditional logistic growth curve - that generally reflects the way markets tend to develop for an independent and competitive technology – by enforcing a number of constraints, as shown in Figure 1. To draw the curve, the calculated long-term penetration of PV is calculated for defining the long-term PV capacity, and historical data is used, when available for the short term.

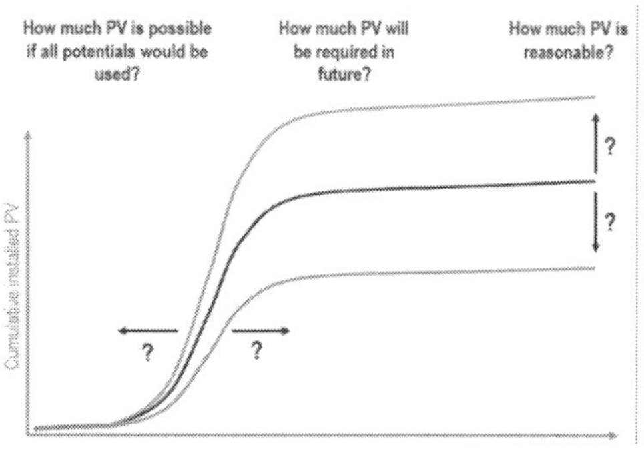

Figure 1: Schema of PV market development curves determined by external factors such as, available potential and long-term demand.

The 2050 long-term perspective in this paper assumes that under the only effect of its intrinsic competitiveness to other power generation, PV penetration in any given country will grow until it reaches a natural maximum, determined by several country dependent factors, and constrained by an absolute physical limitation resulting from grid constraints, the local attractiveness of other competitive energy resources and the need for generation continuity and flexibility.

To give a long-term perspective, 2050 has been chosen as it is widely used by official bodies and agencies as political and scientific milestone, when it comes to climate change (e.g. IPPC, ref [3]) and energy forecast (e.g. IEA, ref [4]) and further gives a well sized time frame of roughly 35 years after PV globally becoming one of the most attractive energy source.

2.2 HISTORICAL INSTALLATION DATA

Historical PV installation data for the years 2009 – 2014 for 190 countries provided by previous publications [5] has been utilized. Sources besides an international customs database, monitored by the 'Market Analysis and Research' section of the International Trade Centre, - an agency of UN's World Trade Organization - include IRENA, International Energy Agency Photovoltaic Power Systems Programme (IEA PVPS), journal du photovoltaique, the European Commission, Hanwha Q CELLS Market Intelligence as well as other public resources [6 – 13].

2.3 LONG-TERM PV POTENTIAL

To assess the long-term potential target of PV capacity for all countries, the methodology described in Figure 2 has been applied to determine first the level of power demand by 2050 within each country. The power demand forecast by country is derived from the IEA 2DS HIREN scenario [4]. The IEA 2DS HIREN scenario provides *gross electricity generation* forecasts in 2050. However, it only provides figures for a few countries, as well as for a number of cluster of nations. In order to derive country specific data, it has been assumed that a certain convergence of development levels and living standards would occur within each group of developed and developing countries, and that all countries within a specific cluster would reach a similar level of development by 2050, materialized by a comparable *gross electricity generation* per capita.

The *gross electricity generation* per country is then calculated on basis of the IEA 2DS HIREN data provided at cluster level and on the 2050 forecasted population of each country within the cluster, using the official population forecast from the World Bank [3].

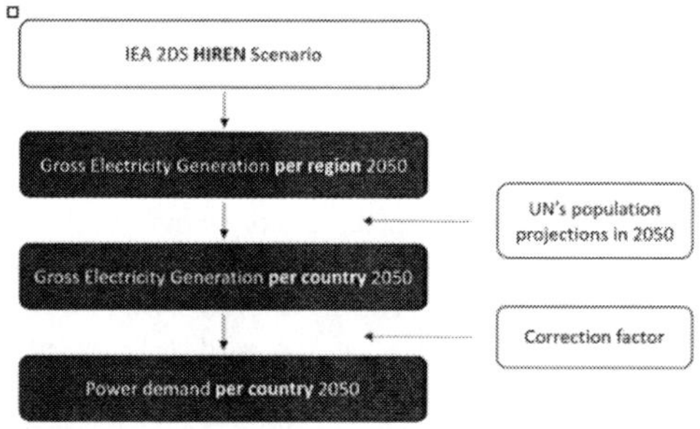

Figure 2: Methodology scheme to compute long-term power demand [3, 4].

Finally, the power demand is derived from the *gross electricity generation* using an efficiency factor representing the efficiency of power transport and distribution as a function of the development level of the country. The methodology provided in Figure 3 will assume two base scenarios for 2050, the 'HIGH scenario' and the 'LOW scenario', corresponding respectively to the assumed best and worst-case scenarios for PV penetration in 2050.

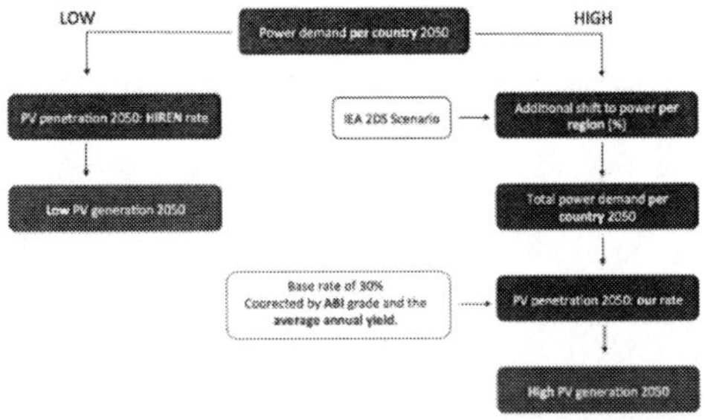

Figure 3: Development of LOW and HIGH 2050 scenarios based on differing assumptions regarding PV penetration in 2050.

To calculate both scenarios, PV capacity is directly derived from the calculated country specific power demand and the penetration rate provided by country and/or by cluster in the IEA 2DS HIREN scenario. In the original IEA scenario, PV should cover 16% of the global demand in 2050, according to varying penetration rates by region and/or by country. Due to the expected competitiveness of PV this scenario is assumed to be the 'LOW scenario' [4]. The 'HIGH scenario' is equally derived from the IEA 2DS HIREN scenario by adding two more aggressive assumptions.

First, it is assumed that the shift to power occurring by 2050 due to increasing shares of electric vehicles, electric heating and power-to-X applications will be significantly higher than the one expressed in the original IEA 2DS HIREN reference scenario. Electricity from Renewables is expected to outcompete other energy sources like oil and gas [2]. The 'HIGH scenario', thus assumes that countries will shift an additional 50%, 80% and 50% of their oil, coal and gas consumption to power respectively thereby significantly increasing the 2050 projected power demand compared to the 'LOW scenario'. These shifts to power were calculated considering average efficiency rates of 30% for oil, 40% for coal and 50% for gas technologies. Resulting increases in power demand vary significantly from country to by country.

Furthermore, the 'HIGH scenario' assumes a much higher penetration rate of PV of 30% on average according to factors representing:

- The general attractiveness of the country for PV investment (*Aggregate Business Index),*
- The general attractiveness of PV for the country.

For the first factor an *Aggregate Business Index (ABI)* is calculated as a weighted average of six indicators, extracted from the *Worldwide Governance Indicators (WGI)*, published by the World Bank [14]. These factors are listed in Figure 4.

World Governance Indicators
1. - Voice & Accountability
2. - Political Stability & No Violence
3. - Government Effectiveness
4. - Regulatory Quality
5. - Control of Corruption
6. - Rule of Law

Figure 4 : Overview of Worldwide Governance Indicators (WGI) used in the ABI calculation [14].

The level of ABI calculated for each country is used to adjust the base penetration rate by a correction factor representing the investment climate dependency on the governance quality and stability within a country, as shown in Figure 5.

ABI grade	Impact on PV penetration rate
A	+10%
B	=
C	-10%
D	-25%
E	-50%

Figure 5: Aggregate Business Index (ABI) impact on PV penetration rates based on Worldwide Governance Indicators (WGI), published by the World Bank [14].

Additionally, the PV attractiveness for a country has been considered dependent on the solar irradiation level. The adjustment factors used by level of PV yield are shown in Figure 6.

Yield (kWh/kWp)	Impact on PV penetration rate
800 - 1200	-20%
1200 - 1500	=
1500 - 1700	+20%
>1700	+40%

Figure 6: Impact of different yield impact on PV penetration rates in the year 2050.

Applying these correction factors to the base penetration of 30%, for each country, provides an estimated natural long-term penetration rate for PV in that country in the 'HIGH scenario'. The final PV penetrations rate in our 'HIGH scenario' range from 15% to 45%, the latter being considered as a physical limitation.

2.4 LOGISTIC GROWTH-FUNCTION

With the 2050 PV penetration goal calculated for each country and both scenarios the projected PV development curves have been plotted. A long-term growth can be described well by applying a logistic growth function in

Equation 1 [2].

$$f(t) = A + \frac{K - A}{(1 + Qe^{-B\cdot(t-M)})^{1/v}}$$

Equation 1: Logistic growth function in generalized form. Abbreviations: time (*t*), lower asymptote (*A*), upper asymptote (*K*), growth rate (*B*), parameter affecting near which asymptote maximum growth occurs (*v*), scaling parameter depending on f(0) (*Q*) and time of maximum growth (*M*). This growth function is used as a methodological tool to describe the long-term growth assumed for 'LOW and HIGH scenarios' in section 2.

The behaviour of the calculated logistic-growth curve was set to follow closely the historical PV development till 2014. A gap of ± 3% tolerance versus actual cumulated capacity was allowed in comparison to the 2014 cumulated installed capacity. The further progress of the curve towards 2050 was provided by the maximum PV capacity described in both the 'LOW and HIGH scenarios' for each country in combination with results on Residential Grid-Parity, Industrial Grid-Parity and Fuel-Parity from previous publications [15]. Without subsidy scheme, any type of parity, but Grid-Parity especially, provides the economic basis for the PV market.

This however does not automatically translate into spontaneous market growth. All market actors have to be aware of the opportunities PV represents and experience has to be developed on PV components, systems, installation and financing options. It is assumed that the lower a country ranks according to the Worldwide Governance Indicators [14], the higher are transaction costs in this market. A lower score corresponds to a reaction time of several years before PV is starting to become a major energy source in a country, and high installation rates are achieved. Thus as a matter of example, a reaction time of 4 years has been considered for developed countries like Sweden or Switzerland whereas countries such as the Democratic Republic of Congo or Nigeria are expected to need some 8 years before PV reaches the highest installation rates.

In several countries, sudden change of policy support has strongly impacted annual installation rates, such as in Germany or in the UK, and could even bring markets to a standstill [5], in countries such as Spain or Greece. Therefore, markets with an active policy support program have to be carefully extrapolated to account for the market development, which includes both policy driven installations, as well as installations resulting from emerging competitiveness of PV.

3 CASE STUDIES

For the purpose of this paper, 21 countries have been analyzed, which were chosen to be effectively representative of the majority of nations worldwide. The list has been formed by choosing a diverse mix of developed and emerging markets, island and continental countries, as well as nations with an established or without a renewable subsidy program. Table 1 shows the 21 countries and the respective cumulated maximum PV capacities for both the 'HIGH and LOW scenarios.'

In order to give an overview on a variety of countries and keeping in mind the ambitions of this paper, the case studies will focus on the HIGH scenario of the selected countries.

Country	Cumulated PV Capacity in 2050	
	Scenario - High [MWp]	Scenario - Low [MWp]
Bolivia	14.698	6.859
Congo DR	17.115	11.690
Fiji	313	112
Ghana	7.880	2.990
Ireland	14.118	3.810
Kazakhstan	18.493	3.818
South Korea	147.818	49.542
Malta	863	116
New Zealand	10.720	2.367
Dominican Rep.	10.584	4.199
Nicaragua	5.374	2.468
Nigeria	45.671	22.282
Pakistan	158.871	53.675
Poland	61.311	18.617
Portugal	16.711	3.383
Russia	201.038	12.205
Egypt	154.900	76.452
Peru	45.356	18.280
Spain	81.536	16.505
Sweden	71.008	0
Switzerland	14.578	4.664

Table 1: List of 21 countries, representative of the majority of nations worldwide and their respective forecasted 2050 maximum cumulated PV installations in the 'HIGH and LOW scenario.'

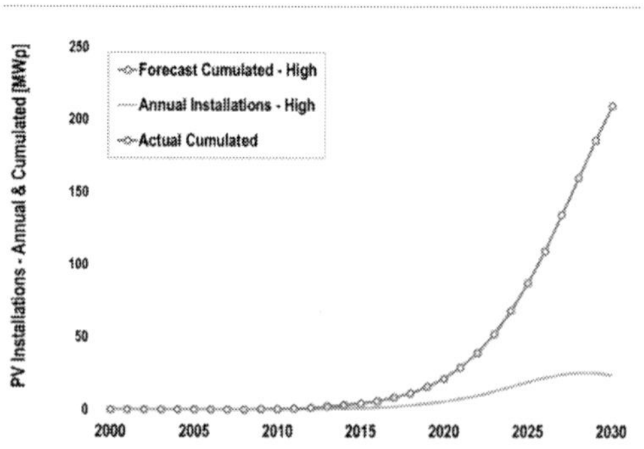

Figure 7: Results for the island nation of Fiji in the 'HIGH scenario', which will reach 313 MWp cumulated in 2050. In 2014, the capacity stood at 3.2 MWp. The orange curve shows the forecasted cumulative PV installation, while the yellow line states projections for annual PV added.

For the countries in Figure 7 – 10 the method described in section 2 provides excellent results and depicts the historical development with high accuracy. Depending on the individual characteristics of the country, PV market growth may start earlier or later. Figure 7 shows the projected PV development in the island country of Fiji. Typically, island nations are characterized by high energy prices due to imports of fossil fuel, enabling PV to be competitive earlier than in countries that claim relevant fossil fuel resources. By end of 2014 Fiji claimed 3.2 MWp PV installed. The projection expects some 313 MWp of PV power potential for Fiji to be reached by the year 2050. Following the methodology proposed in this paper that is purely based on PV competitiveness versus conventional power generation, PV is expected to take off in the coming years and reach some 21.5 MWp of cumulative PV installations by 2020 and 210 MWp by end of 2030. Highest annual installations are expected to take place roughly around 2025. For the end of 2015 this methodology projected 4.4 MWp installed cumulatively. Actual numbers reached by end of 2015 show a very close match with 4.6 MWp installed. For the end of 2016 it is expected that 6.0 MWp will be installed cumulatively.

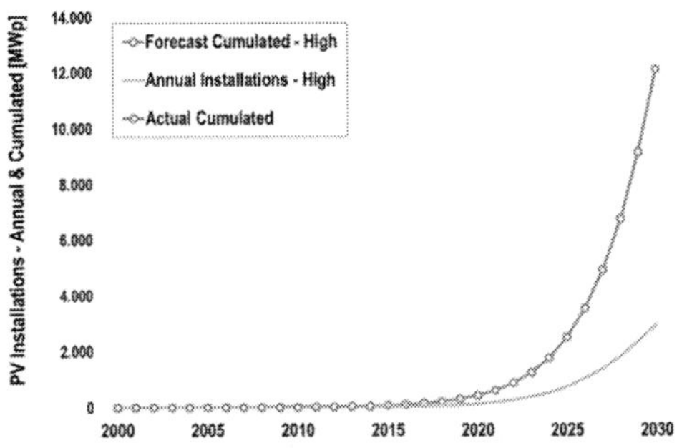

Figure 8: Results for the African nation of Nigeria with 175 million inhabitants in the 'HIGH scenario', which is expected to reach 45,700 MWp cumulative PV in 2050. In 2014, the installed capacity amounted to 45.6 MWp. The orange curve shows the forecasted cumulative PV installation, while the yellow line states projections for annual PV added.

Nigeria, shown in Figure 8, suffers as a developing country from high market transaction costs and therefore significant PV deployment may only start beyond 2020. There was 45.6 MWp installed by end of 2014, that has been expected to grow to 68.4 MWp by end of 2015, according to the present methodology. The real numbers for 2015 show that 68.1 MWp have been installed cumulatively by end of 2015, which matches perfectly the projected amounts [5]. The highest annual installations of 4,150 MWp are expected to be reached in 2033, with already high figures of several GWp of annual installations expected during the 2020s. This demand will be enhanced by drastic growth of power demand going along with rapid GDP growth and the attractiveness of PV compared to the opportunity cost of fossil fuel resources, if these are not exported to global markets. With 175 million inhabitants in 2015, the country could reach a cumulated PV capacity of 45,700 MWp by 2050.

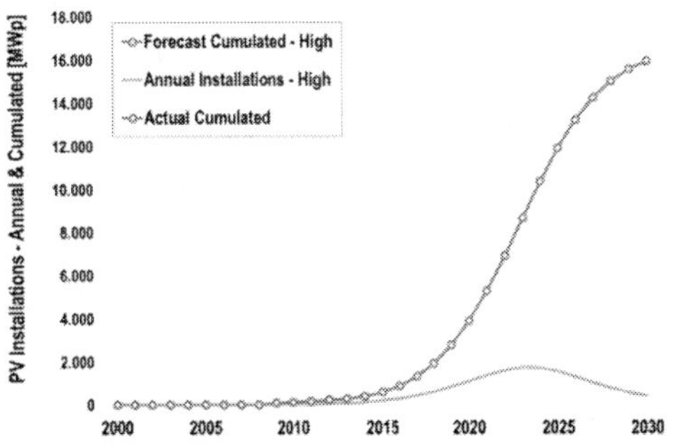

Figure 9: Results for the developed nation of Portugal with 10.5 million inhabitants in the 'HIGH scenario', which will reach 16,700 MWp cumulated in 2050. In 2014, the capacity stood at 402 MWp. The orange curve shows the forecasted cumulative PV installation, while the yellow line states projections for annual PV added.

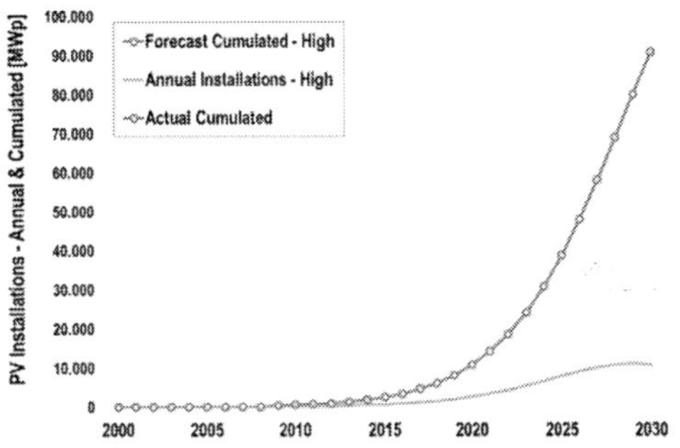

Figure 10: Results for the developed nation of South Korea ranking 11th regarding GDP in the 'HIGH scenario', which could reach 147,700 MWp, cumulated in 2050. In 2014, the capacity stood at 1,906 MWp. The orange curve shows the forecasted cumulative PV installation, while the yellow line states projections for annual PV added.

Within the category of developed countries, actual installations in Portugal and South Korea, shown in Figure 9 and 10, seem to follow the logistic growth curve quite closely. Furthermore, PV development in these countries has been very stable in recent years and thus very predictable, showing constant growth rates of about 30% and 40% respectively [5]. With high solar irradiations and relatively high conventional grid power prices solar power is today an attractive cost option ranging from the residential to the utility scale segments [15]. This leads to PV installations expected to take off in near future to reach 3,900 MWp of cumulative installed solar capacity by end of 2020, while 461 MWp have been installed till the end of 2015. It is estimated that Portugal would reach 16,700 MWp of PV installed by 2050 of which the majority will be installed during the early 2020s in a relatively steep growth rate. In comparison South Korea that shows a rather steep but flatter growth curve should reach about 10,800 MWp by 2020 with an overall 2050 target of 147,700 MWp installed cumulatively.

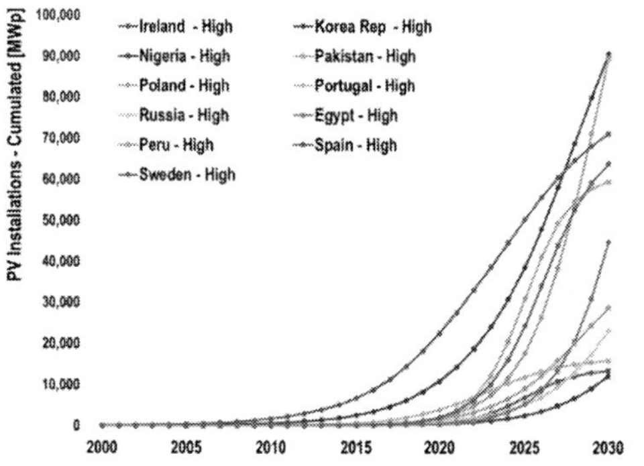

Figure 11: Overview of projected PV development for countries in this study with expected 2030 PV installations above 12,000 MWp. The graphs refer to cumulative installed PV capacity in each country per respective year.

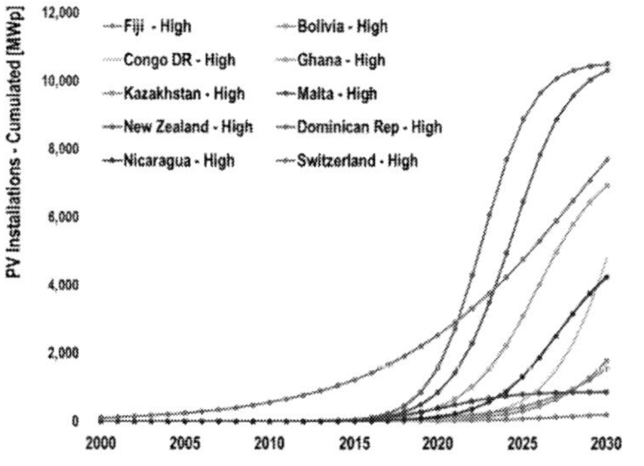

Figure 12: Overview of projected PV development for countries in this study with expected 2030 PV installations below 12,000 MWp. The graphs refer to cumulative installed PV capacity in each country per respective year.

Figure 11 and 12, show that Pakistan and South Korea are the markets, among the 21 countries of this study, that are expected to show highest PV cumulative installations by end of 2030 with an 89,000 MWp and 90,000 MWp respectively. Even though both countries are forecasted to reach comparable installation numbers by 2030, the way to achieve this projected point shows some different development. Whereas the South Korean market will, with a flatter development curve, show relevant annual installations already before 2020, the Pakistan market should be delayed in its development, even having already installed some 600 MWp in 2015 [5]. A main reason for this delay can be seen in the negative influence of business attractiveness factors that due to high level of corruption rank much lower than South Korean ones. Figure 12 shows that markets such as the Dominican Republic and New Zealand should develop very quickly their PV potentials during the early 2020s, and will already reach a slight flattening before 2030. This translates to only low figures of PV additions after 2030, with an installation focus of replacing and upgrading existing PV power plants. The same is valid for Malta, which, like the Dominican Republic and New Zealand, is and island country. This shows PV true and intrinsic competitiveness in comparison to high conventional energy costs, especially on remote islands that usually have to import high shares of their energy needs.

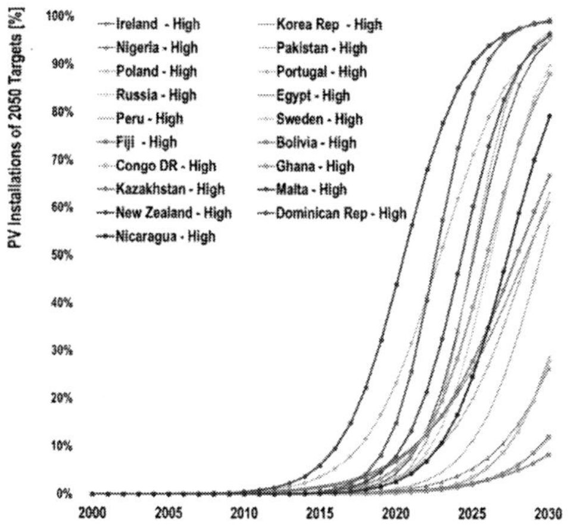

Figure 13: Projected PV installation of the 2050 PV targets identified per each year.

As already stated in Figure 11 and Figure 12, some countries show earlier or delayed takeoffs of their respective PV developments. As result of this competitiveness-based methodology, such delays or early starts give a close insight of the PV attractiveness versus the cost of conventional power generation systems. As an early starter, Malta has already reached 6%, and will reach 99% of its 2050 expected PV capacity already in 2030. This immediately results from PV power being less expensive than the conventional diesel power generation on the island – even though a new, however under-sized, grid connection to Sicily has been inaugurated recently. Further early starters are New Zealand, the Dominican Republic, Fiji, Portugal that all are expected to implement relevant parts of their expected 2050 PV targets before 2025. On the other end, countries that claim high amounts of fossil fuels like Kazakhstan, will show delays in their PV developments with reaching by 2030only 8.5% of their 2050 target equivalent to 1,500 MWp.

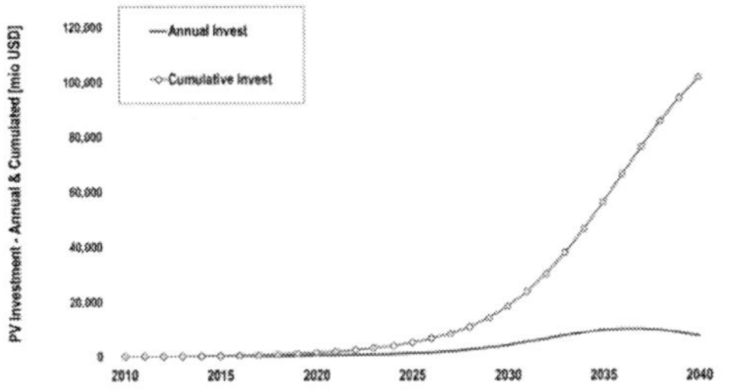

Figure 14: Projected PV investment in Russia until 2040 in million USD. The light blue graph refers to cumulative investments while the dark blue ones represents annual expected investments.

By combining the unique results of this methodology with data on historical data and future forecasts on PV system costs, an estimation of expected PV investments is possible. System costs are expected to decrease down to a global average of 1,200 USD/kWp by 2020, 714 USD/kWp by 2030 and 580 USD/kWp by 2040 [15]. In the example of Russia given in Figure 14, 101 billion USD are expected to be invested into PV until 2040 enabling the installation of 154,000 MWp of PV in that period. Highest annual investments are expected for 2036 and 2037 with 10 billion USD per each year.

4 FURTHER DEVELOPMENTS

The approach presented in this paper is based on a robust methodological framework and proves to be highly effective in a variety of contrasted situations.

However, it is believed that further refinements could be brought by implementing additional differentiation to the countries belonging to the same cluster, which in this paper, have been considered as reaching the same development level, and thus the same energy intensity per capita by 2050.

Secondly, if the method provides useful information on the electrification level and the contribution of PV, it currently does not allow to be prescriptive on the type of electrification/PV deployment that will take place. Additional considerations on the existing electrical system technology and governance, on the core economical activities of the country (agricultural, industrial, mining, etc.) and on the level of centralization of the country would further be required in order to describe the natural electrification path in more details and distinguish central vs. community vs. individual electrification.

Finally, considering the very promising results delivered by this paper, that also draws from previous publications, the authors intend to release later this year a report extending and detailing this analysis to all countries worldwide [2].

5 CONCLUSIONS

The paper shows both the applicability and the robustness of the method that constitutes a highly original approach to forecasting sustainable mid-term and long-term PV development, based on its sole economic competitiveness in any given country and further assuming all policy support will progressively phase out in the coming years. It provides as such a reliable instrument for policy makers in emerging countries to build advanced understanding of the most straightforward and economical path for electrification of their country.

Comparing expected PV market development calculated with the methodology proposed in this paper, before official installation data for 2015 were available, shows that for the majority of countries, actual 2015 cumulative installed PV capacity data match perfectly to the projected figures. As an example, this methodology projected some 53.6 MWp and 68.4 MWp installed by end of 2015 in Malta and Nigeria respectively. Latest installation data show a close match of 55.6 MWp and 68.1 MWp installed in these countries by end of 2015 in real [5]. This proves the great accuracy and suitability of this methodology for the prediction of mid-term and long-term PV development, especially for sustainable non-policy driven PV markets.

The paper will be used as the methodological framework for the development of an extensive study on emerging markets undertaken by the authors and planned to be released in Q4 2016.

REFERENCES

[1] [PV Market Alliance] – PVMA Global PV Market Report 2016 -2020, June 2016. www.pvmarketalliance.biz

[2] Gerlach A., Breyer Ch., Fischer M., Werner Ch., 2015. Forecast of Long-Term PV Installations – A discussion of Scenarios ranging from IEA to the Solar Economy, Proceedings 31th European Photovoltaic Solar Energy Conference, Hamburg

[3] [UN] – United Nations, Department of Economic & Social Affairs, Population Division. World Population Prospects, the 2015 Revision, New York,

[4] [IEA] – International Energy Agency, 2016. Energy Technology Perspectives 2016, 2DS & 2DS HIREN Scenarios

[5] Werner Ch., Gerlach A., Breyer Ch., Masson G., 2016. Global Photovoltaics in 2015 – Analysis of Current Solar Energy Markets and Hidden Growth Regions Proceedings 32nd European Photovoltaic Solar Energy Conference, Munich

[6] International Trade Centre Market Analysis Tool, 2015. Product Map, Market Analysis and Research Section, International Trade Centre, Geneva, www.trademap.org

[7] [PVMA] – PV Market Alliance, 2016. The PVMA Global PV Market Report, Brussels

[8] Tsagas I., 2016. Cyprus: Construction begins on five PV plants; storage to emerge, PV magazine, Berlin, 2016, April 22

[9] Lopez D., 2016. Panama reached 43 MW of installed solar PV in 2015, PV magazine, Berlin, 2016, March 18

[10] [IEA PVPS] - International Energy Agency Photovoltaic Power Systems Programme, 2016. Snapshot of Global PV Markets 2015, Paris

[11] Journal des Énergies Renouvelables, 2016, Photovoltaic Barometer, EurObserv'er, 2016 (April), Paris

[12] [REN21], 2016. Renewables 2016 – Global Status Report, REN21, Paris

[13] Hanwha Q CELLS Market Intelligence, 2011-2015. Market Research, Hanwha Q CELLS, Bitterfeld-Wolfen

[14] [IRENA] – International Renewable Energy Agency, 2016. Renewable Capacity Statistics 2016, Dubaix

[15] Gerlach A., Werner Ch., Breyer Ch., 2014. Impact of Financing Costs on Global Grid-Parity Dynamics till 2030, Proceedings 30th EU PVSEC, Amsterdam

UNSUBSIDISED PV MARKETS: HOW A NATIONAL PUBLIC-PRIVATE FINANCING PLATFORM COULD REDUCE THE IMPACT FROM LOW OIL PRICES

Monika BIERI[#*], Rupesh S. BAKER[+], Stephen TAY[*], Thomas REINDL[*]
[*]Solar Energy Research Institute of Singapore (SERIS), National University of Singapore, 7 Engineering Drive 1, Block E3A, #06-01, Singapore 117574
[+]Cleantech Energy Corporation Pte Ltd, Capital Square Three, 25 Church Street, #03-04, Singapore 049482
[#]E-mail: monika.bieri@nus.edu.sg Tel: +65 6601 3156, Fax: +65 6775 1943

ABSTRACT: For the case of Singapore, climate change is an important topic and carbon reduction is high up on the agenda. Besides energy efficiency initiatives, increasing solar photovoltaic (PV) adoption is another crucial driver for a sustainable future. As the Singapore government strives for a level-playing field among all energy players, solar PV is not directly subsidised. Having achieved grid parity in late 2012, the installed capacity has grown more than tenfold over the last five years. However, electricity prices are in a downward spiral, driven by an increasing overcapacity of conventional generation and gas-supply contracts that are pegged directly to oil prices. The economic viability for PV rooftop system investments, especially for liberalised energy intensive consumers, has deteriorated. This work looks into ways on how to use part of Singapore's large finance and real estate resources to frame national financing schemes in order to offer more favourable financing options. A national property-backed PV financing scheme followed by an on-bill financing scheme are seen as the best possible ways to bridge current difficulties of PV systems to remain competitive. Political backing is seen as the most important driver to facilitate such schemes.
Keywords: Rooftop, Economic Analysis, Funding and Incentives, Financing

1 INTRODUCTION

After having contributed to the global agreement in Paris to limit global warming in December 2015, Singapore's carbon reduction goals became more tangible. The low-lying city state's unconditional pledge from 2009 to reduce its absolute emissions from a business-as-usual figure of 77.2 MT in 2020 [1] compared to 2005 levels is an almost doubling of the original range (-7% to -11%) to now -16% [2]. In addition, Singapore announced to target a -36% emission intensity reduction from 2005 by 2030 with an absolute emissions peak in 2030 [3]. Globally, Singapore ranks favourably in terms of emission intensity (123rd out of 141 countries covered, with first place having the highest emission intensity) but less so in terms of emissions per capita (26th) [4]. While its emission intensity improved significantly over the years, due to a vibrant economy, emissions per capita remained at almost constant levels (Fig.1).

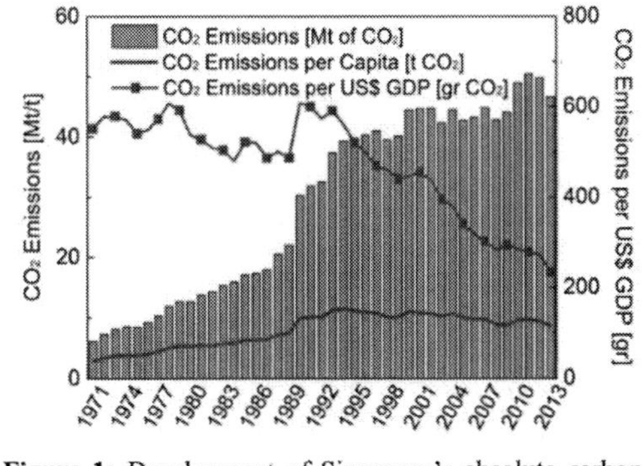

Figure 1: Development of Singapore's absolute carbon emissions, its carbon intensity and carbon per capita figures. Data source: International Energy Agency

The Singapore government has therefore started many years ago to promote energy efficiency in various areas [5]. In terms of conventional electricity generation, Singapore has successfully replaced its oil-fired power plants with gas [6], leading to a grid emission factor of 431 gCO_2/kWh in 2015 [7], one of the lowest compared to its neighbouring countries [8]. Therefore, in combination with energy efficiency improvements, increasing the share of renewable energies within its electricity generation portfolio is the other important pillar for Singapore to reduce carbon emissions [9]. Renewable energy options however are limited in Singapore as there are no significant wind, hydro and geothermal resources. In addition, high population density and land scarcity limits the sustainably-grown domestic biomass resource potential. Being situated in the tropical Sunbelt, Singapore however is blessed with an annual irradiance of ~1,630 kWh/m^2 [10]. Despite its limited space, an installed PV capacity of 3-4 GW_p could be possible by 2030 based on a conservative net usable area calculation [11]. This could reduce current mid-day system peak demand of ~7 GW [12] significantly and would result in ~4-6% annual carbon emission reductions [13]. The shift towards – locally available – renewable energies would also help to decrease the dependency on imported energy resources.

As of March 2016, 71 MW_p of solar PV capacity [14] has been deployed accounting for less than 1% of the total electricity demand. Thanks to lower PV installation cost, the levelised cost of electricity (LCOE) of solar energy continuously declined and reached ~18 SGD cents per kWh for a residential rooftop and ~13 SGD cents per kWh for an industrial PV rooftop system during 2Q16 [15]. While the reduced LCOE for solar PV has been highly promising, the economic viability of PV rooftop installations in Singapore's unsubsidised market has suffered significantly with the decline in oil prices and the increased overcapacity situation of conventional electricity generation assets. While grid parity was achieved in late 2012 for industrial-sized rooftop PV systems, it has been lost recently due to a downward spiral of electricity and oil prices (see Fig. 2, [16]). In an unsubsidised market, such as Singapore, solar PV needs to compete directly with the marginal cost of gas-fired power plants. Based on different future wholesale power price scenarios [16] industrial-sized rooftop PV reaches competitiveness against gas-fired power plants only in the years of 2019, 2023 and 2035 (see Fig. 3) when a continuous decline of rooftop PV LCOE is taken into account at 1.5% annually until 2030 and 0.5% thereafter.

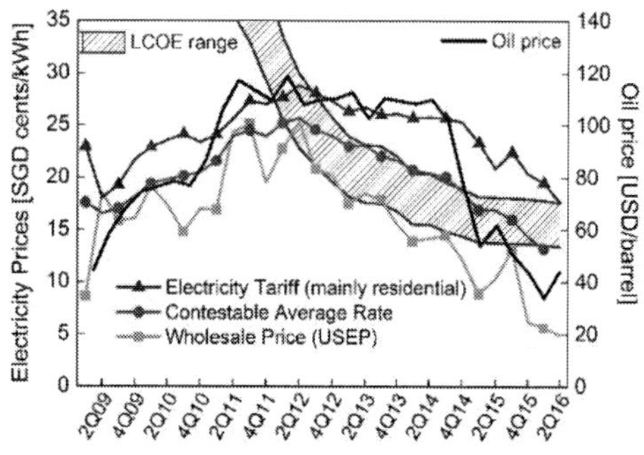

Figure 2: Historic LCOE range for residential and industrial-sized rooftop PV systems compared to Singapore's electricity price benchmarks.

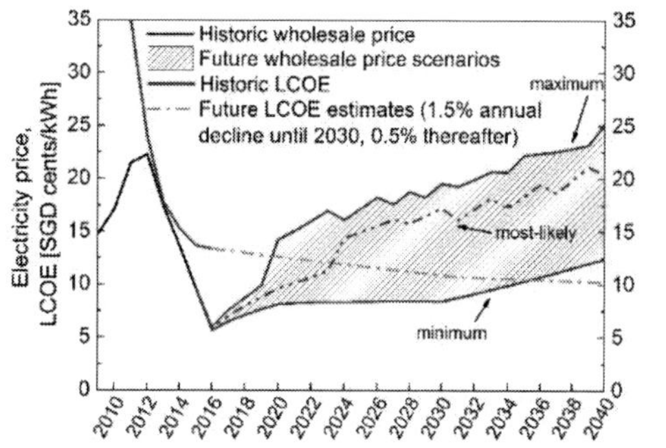

Figure 3: Future LCOE compared to future wholesale electricity price scenarios (2040 oil price at 75, 100 and 115, demand/supply balance reached in >2040, 2026 and 2021 for the minimum, most-likely and maximum scenario, respectively).

In line with the decreasing competitiveness, the economic viability has worsened for solar PV rooftop systems in Singapore. Back in September 2014, when oil prices were ~97 USD per barrel, the discounted payback period for a 1 MW_p industrial rooftop PV system ranged between 8-12 years and yielded an EIRR in the range of 14-20% using the minimum, most-likely and maximum future electricity average contestable price scenarios. Based on the 25-Jan-2016 oil price forward curve with spot oil prices averaging ~30 USD per barrel (lowest point recorded was at 26 USD per barrel as of 20-Jan-2016), the discounted payback period range went up to 15-20+ years with EIRRs ranging between 3% and 9%, which made an investment financially unattractive [17]. While the recent ~50% recovery in oil prices, reaching a monthly average of ~46 USD per barrel in May 2016 compared to the average in January [18], brought some relief, the discounted payback period now ranges at 12-19 years, with EIRRs between 7% and 14%, which are still not very attractrive levels [19]).

The long payback period, the most prominent financial term where local companies and the public is focused on, highlights the necessity for timely and appropriate actions from relevant stakeholders in the PV scene. Furthermore, the weakness in oil prices and continuing overcapacity makes it difficult for solar PV to compete outright with conventional gas-fired generation, and if no actions are taken, this will most likely delay the "solarisation of Singapore" considerably.

Singapore's energy policies are framed based on an unsubsidised level playing field for all types of generating sources. Therefore, direct supportive policies (e.g. feed-in-tariffs, mandatory green electricity certification, capital tax incentives etc.) which successfully promote renewable energies in other countries are excluded in this work as potential solutions. However, some "soft" support schemes were enacted in the past to facilitate the long-term energy vision of the country. For example, the (liquefied natural gas) LNG terminal, which began operation in May 2013, was only possible with a ten year price guarantee at prevailing long-run marginal cost (LRMC) offered to gas-fired power plants consuming LNG instead of piped gas [20]. In addition, solar energy is well supported indirectly via research, development and deployment (RD&D) funding, various test-bedding options (e.g. the "Living Laboratory Fund", LLF) and the SolarNova programme, announced in March 2014, which aims to install 350 MW_p on government buildings by 2020 [21]. Due to the fact, that more than 80% of a PV system's life cycle cost is related to the upfront investment [22] and in combination with Singapore's robust financing resources, this work analyses whether a national public-private financing platform could be a way to bridge the gap until electricity prices have recovered and make investments into rooftop PV systems economically more viable. While some pilot financing schemes have been tested, such as the "Building Retrofit Energy Efficiency Financing" [23] or the "Financing Programme for Energy Efficiency Projects" [24], none of the frameworks target yet a nation-wide platform using additional securities to make participation attractive for local banks. Furthermore, traditional well-functioning financing schemes, such as third-party financing solutions (e.g. solar leasing or power purchase agreements), crowd funding initiatives or asset-backed securities (ABS) transactions for single companies are excluded from this work, as the aim is to analyse novel renewable energy and energy efficiency financing concepts successfully implemented elsewhere (especially in the U.S.) and to investigate how the strengths of these concepts could be combined and tailored for the Singaporean energy market. The ultimate structure should create a win-win situation for all local stakeholders.

The benefits of having favourable financing are evaluated by detailed calculations based on real-world case studies. The access to favourable financing is paramount in order to motivate the adoption of renewable energies, knowing that inaction could jeopardise Singapore's solar PV installation target and its carbon reduction goals. This also holds true for other cities in the world that are keen on deploying solar but are facing challenging economic environments, which, apart from low oil prices, could also arise from high country risks or unfavourable exchange rates.

2 APPROACH

Based on a detailed discounted cash flow model, the benefits from favourable financing terms are first calculated and visualised. For this exercise the base case of a 1 MW_p industrial rooftop PV system is taken, as described under [15] where a 10-year loan at 4.5% p.a. interest rate was assumed. The calculations use three future contestable electricity price scenarios based on the 31-May-2016 oil price forward curve [16]. For this work, the following three cases of more favourable financing terms are analysed individually as well as combined: a) increasing financial leverage to 80% from 60% [25][1], b) financing at the government risk-free rate (reduction of the interest rate by 2%, which is the premium overlaid on risk-free rates) and c) lengthening of the debt term to 20 years (in line with the assumption of the minimum operational period of the PV system). As the U.S. is on the forefront with financing innovations, financial concepts such as the Property Assessed Clean Energy (PACE [26]) financing, the Warehouse for Energy Efficiency Loans (WHEEL [27]) and on-bill financing/repayment (OBF/OBR, discussed under [28]) are studied. This analysis allows assessment for each financial concept, potential opportunities and hurdles with respect to Singapore's specific policy and economic environment. The outcome is then used to build a possible national financing platform suitable for the Singaporean PV rooftop market and extendable to other global cities facing similar opportunities and challenges.

3 RESULTS AND DISCUSSION

3.1 Results

The change in the cumulative NPV profile from the base case under the most-likely future contestable price scenario [29] has first been investigated by changing the financing terms in isolation. The discounted payback period could be reduced to 12 and 9 years from 13 years in terms of reducing the interest rate and lengthening the debt term, respectively. Under the scenario of higher financial leverage, while the cumulative NPV in year 12 has improved from SGD -87k to SGD -22k, the discounted payback period remained at 13 years as higher amortization and interest payments in the first 10 years could not entirely offset the lower initial equity injection. It is observed that the different favourable fiancing terms altered the original annual cash flow profile by different magnitudes. As an example, while extending the debt term to 20 years yields the most benefit in terms of discounted payback period, it results in lower net present value (NPV) compared to the other two adjustments. As the debt payment is spread over the whole operational period, cash flows from later years benefit less from an expected recovery in power prices compared to the other two alternatives, outweighing even the worsening discounting effect in later years (Fig. 4). The range of discounted payback periods - based on the different future contestable price scenarios -are shown in Figure 5, and how it can be reduced by improving financing terms in isolation or in combination. At the extreme, a three years discounted payback period can be reached for all future oil price scenarios, by combining all three favourable financing terms. The break-even occurs in year 3 under all three future price sceanarios at the same time, as under these conditions, the first three year's of tax benefits from having 3 years accelerated depreciation helps all of the scenarios to offset the initial capital outlay in year 3 - and can even withstand inverter replacement impact in year 5.

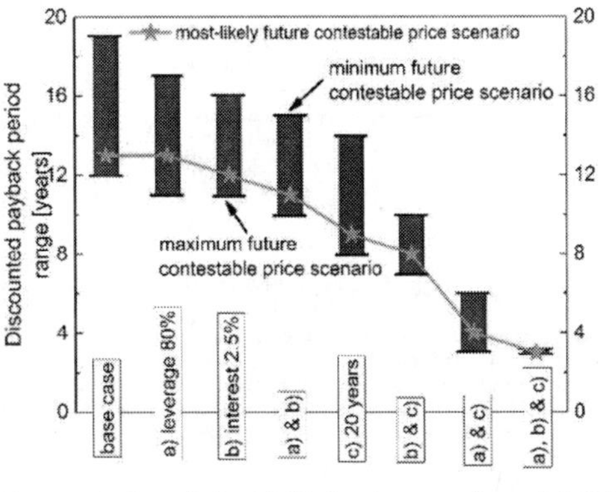

Figure 4: Cumulative NPV development for a 1 MW_p industrial rooftop PV system under the most-likely future contestable electricity price scenario, changing base case with either lengthening debt terms to 20 years (from 10 years), or reducing interest cost to 2.5% (from 4.5%) or increasing financial leverage to 80% (from 60%)

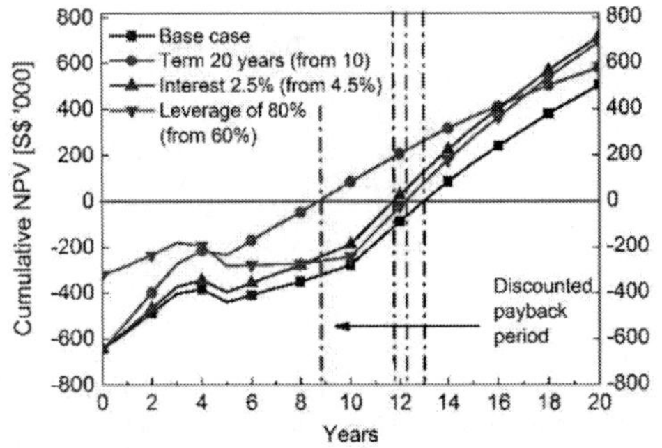

Figure 5: Range of discounted payback periods based on different future contestable electricity price scenarios and changing financing terms (base case: oil price of USD 49 per barrel)

It can be seen, that more favourable financing terms significantly improve the economic viability of PV rooftop systems in Singapore. However in order to motivate stakeholders to offer more favourable financing terms, additional credit enhancements need to be provided. This can be done either with additional financial securities (e.g. in terms of PACE financing, the loan is backed by a senior lien on the property value), more seniority of the payment (e.g. in regards to PACE non-payment triggers a foreclosure, in terms of OBR/OBF, see also Table I, non-payment would trigger a disconnection to the power grid) or additional guarantees (e.g. sponsors cover equity risk for the WHEEL programme, government steps-in in case of non-payment). The three financing schemes are summarised in Table I, including main opportunities and hurdles for Singapore.

Table I: Description of innovative financing schemes from the U.S. and potential opportunities and hurdles for Singapore

[1] *While the Home Energy Renovation Opportunity (HERO) program, one of the largest and most successful residential PACE program in the U.S. (see text), even provides 100% financing, for Singapore the authors thought 100% debt support is rather unrealistic and hence only an increase to 80% financial leverage has been taken into account to calculate the benefit from increasing the debt portion in the capital structure.*

Innovative financing schemes	Opportunities for Singapore	Hurdles for Singapore
PACE (property assessed clean energy): Property owner obtains loan which is backed by a specific property tax assessment, i.e. secured by a senior lien on the property which acts as strong debt collateral in the event property owner defaults on the tax payment. The repayments and interest payment on the loan is directly linked to the property tax payments. Failure to pay both, results in the same consequences, i.e. foreclosure, (it can be seen as a "on-tax-bill financing")	· The loan is backed by the property value which is high for the land-constraint city state · The PV system is attached to the property, hence barriers regarding change of owner or change of tenant is removed · The loan is paid back in monthly instalments with the property tax bill which makes it easy to collect · These type of loans can be bundled and create an own asset class to be securitised by local banks	· The tax authority needs to create a special tax assessment for solar PV systems · In addition, the government needs to change the tax law to empower the tax assessment for solar PV systems · Mortgage lenders need to agree on the seniority addition of the value of the PV system (i.e. the PACE loan is primary to a mortgage) · Possible increased risk for the property market as loan-to-value is increased with the inclusion of the PV system's loan
WHEEL (warehouse energy efficiency loan): Public-private financing platform providing favourable financing for homeowners or companies to perform energy efficiency investments and facilitating a secondary market by bundling the loans and creating a securitised product sold onwards to institutional investors. It represents a collaboration among national leaders in finance and energy, including banks, financing companies, municipalities, states, energy companies and non-profit organisations.	· The idea allows to build a national platform to include all important stakeholders which is important for a small place such as Singapore · It overcomes the scaling problem of too small investments · Securitisation into the secondary market are new business opportunities for local banks · Institutionalisation on national level minimises single-project or single counterparty risk for local banks	· Needs sponsors who take equity risk and facilitate first loans · Requires an empowered body to standardise loan process and is in charge of the administration · Requires clear quality standards, validation process and proper auditing
OBF/OBR (on-bill financing, on-bill repayment): This links the financing of an investment to an existing bill (e.g. utility, telecom etc.). While OBF means that private money is used for the investment, OBR uses public funds. The loan is backed by the consequences of non-payment of underlying bill (e.g. if power bill is not paid, cut off from the power grid might be the consequence)	· This could be linked to the grid fee payment, which is collected by Singapore Power, a 100% government-owned entity	· Singapore's generation and retail electricity industry is privatised and ~75% of the demand liberalised · The likelihood is low that Singapore government will enact a mandate to the private sector to offer favourable financing to PV

3.2 Discussions

In Singapore property values are high due to land scarcity, hence the PACE financing scheme, where the property value backs the loan to finance the rooftop PV system looks promising. While property prices have come down during the last two years, property indices for different property types remain ~90% higher when compared to 2005 [30]. The price for owning property in Singapore can be ~SGD 12k to 24k per square meter depending whether it is in the city centre or outside [31, 32]. Our base case 1 MW_p PV system uses ~7,500 m^2 rooftop space (depending on the efficiency of the panel and system design) and has an investment value of ~SGD 1.6 million which would result in ~SGD 210 per square meter terms. Therefore it seems that the value or additional investment of a rooftop PV system in Singapore is only a small fraction (~0.9-1.7%) relative to the total property value which would mean that increases in the loan-to-value ratio will be limited. In addition, property-backed finances links the PV system, which is built for the long-term, directly to the property and is immune against changing owners or changing tenant contracts. The property tax in Singapore is levied on a 10% of the annual value of industrial, commercial and rental properties [33], which would mean based on the value range above, a property tax payment ranges annually between ~SGD 1.2k to 2.4k per square meter. In terms of the 1 MW_p PV system base case, annual financing payment range between ~SGD 73k and SGD ~120k in case of a 20 or 10 year loan, respectively. This results in ~SGD 10 to 16 in square meter terms, again, a fraction of what is usually paid annually for property taxes.

However, a big stumbling block might be the fact that tax legislation needs to be changed and banks might not be receptive about being disadvantaged by the value of a PV system when tax payments ranking ahead of mortgage payments. This can be a controversial issue, as discussed well in [34] where in the U.S., a federal body suddenly decided on July 2010 that the government-sponsored mortgage buying entities are not allowed anymore to buy mortgages with senior liens to PACE programs. This essentially halted all residential PACE programs, and it only started again when the decision was successfully repealed. For Singapore, there might be some mitigation opportunities: firstly, with decreasing installation cost, as calculated above, a rooftop PV system's value appears to be only a small fraction of the property value and the annual property tax payment, secondly, the benefit for local banks to create a new asset class, which could be bundled and resold into the secondary market as a high-quality product (diversified and backed by multiple property estates) could act as an offset to its degradation of its mortgage repayment seniority ranking, and thirdly, while in the U.S., the PACE program varies between municipalities and states according to prevailing energy efficiency and renewable energy policies, in Singapore, a PACE-like financing scheme could be combined with the WHEEL's concept of creating a collaboration of all stakeholders together.

An alternative to the property-backed financing could be on-bill financing through Singapore Power (SP), who owns and operates the countries' electricity distribution and transmission system. Every electricity consumer, whether contestable or non-contestable, needs to pay the grid utilisation fee to SP. A favourable financing scheme could be attached to this monthly grid utilisation fee payment, with non-payment triggering a cut-off from the electricity grid (after some warnings). SP's credit worthiness is robust with AA/Aa2 credit ratings from Standard & Poor's and Moody's, respectively, both with stable outlooks [35]. Furthermore, SP is fully owned by the Singaporean Government through its investment corporation Temasek. The latest financial credit ratios of the SP PowerAssets Ltd entity (the subsidiary who owns the Singaporean electricity transmission and distribution grid [36]) show recently improved credit ratios (see Table II) and access to the capital market remains strong with high demand for their latest bond issuance [37]. SP's credit profile could easily embrace funding for new PV capacity additions. For example, a 1 GW_p addition at todays prices with a 80% debt funding would only be ~15% of SP's current operating assets and ~12% of its total asset base.

In all three concepts, there needs to be an oversight body, which facilitates the validation process of underlying loans and oversees qualitative and operative aspects of the PV systems. In any case, most important would be the political backing, which provides regulation and resources to facilitate either the property-backed-PV financing scheme, or the on-bill financing scheme.

Table II: Latest credit ratios of SP PowerAssets Ltd.

SGD million	Mar-15	Mar-14
Property, Plant and Equipment	8,327	7,975
Financial leverage[1)] (%)	43%	49%
Cash flow in relation to debt[2)] (%)	32%	19%
Interest coverage[3)] (times)	7.8x	4.0x

1) Net Debt / (Equity + Net Debt)
2) Funds from Operating / Net Debt
3) Funds from Operation / Interest

With respect to the former, a prerequisite would be the willingness of banks involved in the local mortgage market, to accept the added PV value as a lien senior to them. For the later option, SP would need to be willing to offer on-bill financing to its clients. Potential sponsors could be either quasi-government owned infrastructure companies or international corporations, already active in the Singaporean PV market with a strong, green corporate agenda. The facilitation and quality assurance could be achieved through independent third parties with strong technical background, such as SERIS. The offering of PV systems to end-customers via the national financing scheme should then be open to any player in the energy market (Fig. 6).

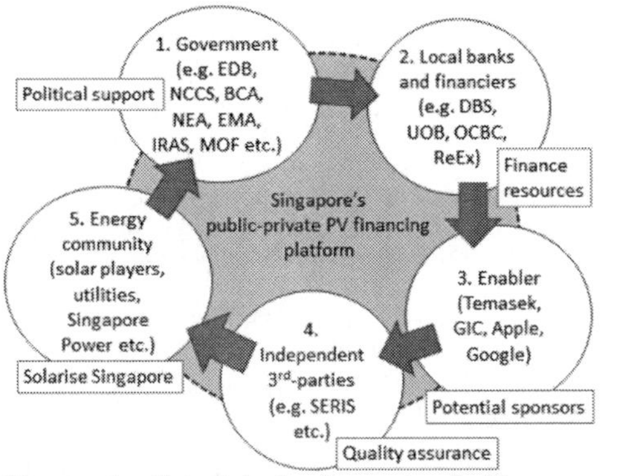

Figure 6: Potential involvement of Singaporean stakeholders in a national public-private PV financing platform[2]

4 CONCLUSION

More favourable financing terms increase the economic viability of PV rooftops in Singapore. With Singapore's considerable finance resources and real estate values, a national property-backed financing scheme for rooftop PV systems could act as a successful bridging policy tool until oil prices have recovered, the oversupply situation in electricity generation has eased, and solar PV rooftop investments yield higher economical values. In case rewriting the property tax law or negotiation for higher seniority with banks look too cumbersome, an on-bill financing scheme via SP could be a promising alternative. The results from this study could be extended to other cities with great solar potential, but facing similar financing challenges for increased adoption of solar PV.

REFERENCES

[1] National Climate Change Secretariat (NCCS) Government of Singapore. *Singapore's Emission Profile*. Available: https://www.nccs.gov.sg/climate-change-and-singapore/national-circumstances/singapores-emissions-profile, accessed 13.06.2016
[2] National Climate Change Secretariat (NCCS) Government of Singapore. *Goverment Seeks Views on Singapore's Climate Action Plans*. Available: https://www.nccs.gov.sg/news/government-seeks-views-singapore%E2%80%99s-climate-action-plans, accessed 13.06.2016
[3] National Climate Change Secretariat (NCCS) Government of Singapore, "Singapore's intended national determined contribution (INDC) and accompanying information," https://www.nccs.gov.sg/sites/nccs/files/Appendix%20I%20Singapore%27s%20INDC.pdf2015.
[4] International Energy Agency (IEA), "2015 Key World Energy Statistics," https://www.iea.org/publications/freepublications/publication/KeyWorld_Statistics_2015.pdf2015.
[5] Singapore Government. *Energy Efficient Singapore, An E2PO Initiative*. Available: http://www.e2singapore.gov.sg/, accessed 13.06.2016
[6] Solar Energy Research Institute of Singapore (SERIS) National Solar Repository (NSR). *Solar Economics Handbook*. Available: http://www.solar-repository.sg/singapore-electricity-market-generation, accessed 13.06.2016
[7] Energy Market Authority (EMA). *Statistics, Electricity Grid Emissions Factors and Upstream Fugitive Methane Emission Factor*. Available: https://www.ema.gov.sg/statistic.aspx?sta_sid=20140729MPY03nTHx2a1, accessed 13.06.2016
[8] Institute for Global Environment Strategies (IGES). (2016/03). *List of Grid Emission Factor*. Available: http://pub.iges.or.jp/modules/envirolib/view.php?docid=2136, accessed 13.06.2016
[9] National Climate Change Secretariat (NCCS) Government of Singapore. (2015). *Singapore's Submission to the United Nations Framework Convention on Climate Change (UNFCCC)*. Available: https://www.nccs.gov.sg/news/singapore%E2%80%99s-submission-united-nations-framework-convention-climate-change-unfccc, accessed 13.06.2016
[10] National Environment Agency (NEA), "Historic average computed based on NEA data: 1991-2000, 2010 to Aug 2014."
[11] Solar Energy Research Institute of Singapore (SERIS), "Solar Photovoltaic (PV) Roadmap for Singapore (A summary)," https://www.nccs.gov.sg/sites/nccs/files/Roadmap_Solar_20140729.pdfJuly 2014.
[12] Energy Market Authority (EMA). *Statistics, Monthly Peak System Demand*. Available: https://www.ema.gov.sg/statistic.aspx?sta_sid=201408047htU7faVzLaZ, accessed 13.06.2016
[13] Solar Energy Research Institute of Singapore (SERIS) National Solar Repository (NSR). *Solar*

[2] EDB = Singapore Economic Development Board, NCCS = National Climate Change Secretariat, BCA = Building & Construction Authority, NEA = National Environment Agency, EMA = Energy Market Authority, IRAS = Inland Revenue Authority of Singapore, MOF = Ministry of Finance, DBS = Development Bank of Singapore, UOB = United Overseas Bank, OCBC = Oversea-Chinese Banking Corporation, ReEx = ReEx Capital Asia, GIC = Government of Singapore Investment Corporation.

Economics Handbook, Signapore PV Market, Future annually avoided CO2 potential Available: http://www.solar-repository.sg/singapore-pv-market, accessed 13.06.2016
[14] Energy Market Authority (EMA). *Statistics, Installed capacity of grid connected solar PV systems*. Available: https://www.ema.gov.sg/statistic.aspx?sta_sid=20140730tzT7HYFaefda, accessed 13.06.2016
[15] Solar Energy Research Institute of Singapore (SERIS) National Solar Repository (NSR). *Solar Economics Handbook, PV adoption in Singapore*. Available: http://www.solar-repository.sg/pv-adoption-in-singapore, accessed 13.06.2016
[16] Solar Energy Research Institute of Singapore (SERIS) National Solar Repository (NSR). *Solar Economics Handbook, Future electricity price scenarios*. Available: http://www.solar-repository.sg/future-electricity-price-scenarios, accessed 13.06.2016
[17] Solar Energy Research Institute of Singapore (SERIS) National Solar Repository (NSR), "Solar Economics Handbook, January 2016 Edition," http://www.solar-repository.sg/up/SEHarchives/internal/13397/2016-02-01-Solar-Economics-Handbook-January-2016.pdf.
[18] U.S. Energy Information Administration (EIA). *European Brent Spot Price FOB*. Available: http://www.eia.gov/dnav/pet/hist/LeafHandler.ashx?n=PET&s=RBRTE&f=D, accessed 13.06.2016
[19] Solar Energy Research Institute of Singapore (SERIS) National Solar Repository (NSR). *Solar Economics Handbook, Grid parity and economic viability analysis*. Available: http://www.solar-repository.sg/grid-parity-and-economic-viability-analysis, accessed 13.06.2016
[20] Energy Market Authority (EMA), "EMA's Procedure for calculating the components of the vesting contracts, Version 2.1, page 1-1," ed, Dec-2013.
[21] Ministry of Trade and Industry Singapore (MTI). *SolarNova*. Available: http://www.mti.gov.sg/MTIInsights/SiteAssets/Pages/Budget-2014/SolarNova.pdf, accessed 13.06.2016
[22] Solar Energy Research Institute of Singapore (SERIS) National Solar Repository (NSR). *LCOE Calculator, using default value inputs to calculate the "total life cycle cost" pie chart*. Available: http://www.solar-repository.sg/lcoe-calculator/calculator.cfm accessed 23.06.2016
[23] Building and Construction Authority (BCA). (2016). *Pilot Building Retrofit Energy Efficiency Financing (BREEF) Scheme*. Available: https://www.bca.gov.sg/GreenMark/breef.html, accessed 14.06.2016
[24] Singapore Government. *Energy Efficient Singapore, An E2PO Initiative, Financing Programme for Energy Efficiency Projects*. Available: http://www.e2singapore.gov.sg/Incentives/Finance_Programme_for_Energy_Efficiency_Projects.aspx, accessed 14.06.2016
[25] Pace Market Links. (June 2016). *Largest PACE Bond Securitization Completed by Renovate America*. Available: http://pacemarketlinks.com/pace-content/industry-news/060916-largest-pace-bond-securitization-pml, accessed 13.06.2016
[26] National Renewable Energy Laboratory (NREL), "Property-Assessed Clean Energy (PACE) Financing of Renewables and Efficiency," http://www.nrel.gov/docs/fy10osti/47097.pdf2010.
[27] Mark Wolfe; Cisco DeVries, "Wheel Primer: A Sustainable Solution for Residential Energy Efficiency," https://www.naseo.org/Data/Sites/1/documents/committees/financing/documents/WHEEL_Primer.pdf2013.
[28] The State and Local Energy Efficiency Action Network (SEEAction), "Energy Efficiency Financing Program Implementation Primer," https://www4.eere.energy.gov/seeaction/system/files/documents/financing_primer_0.pdfJanuary 2014.
[29] Solar Energy Research Institute of Singapore (SERIS) National Solar Repository (NSR). *Solar Economics Handbook, Grid parity and economic viability analysis, slide 67*. Available: http://www.solar-repository.sg/grid-parity-and-economic-viability-analysis, accessed 20-06-2016
[30] Department of Statistics Singapore. *Building, Real Estate, Construction and Housing*. Available: http://www.singstat.gov.sg/statistics/visualising-data/charts#bldg, accessed 14.06.2016
[31] Numbeo. *Property Prices in Singapore*. Available: http://www.numbeo.com/property-investment/city_result.jsp?country=Singapore&city=Singapore, accessed 14.06.2016
[32] Global Property Guide. *Square Metre Prices Singapore compared to Continent*. Available: http://www.globalpropertyguide.com/Asia/singapore/square-meter-prices, accessed 14.06.2016
[33] Global Property Guide. (June 2015). *Rental income tax in Singapore is high*. Available: http://www.globalpropertyguide.com/Asia/Singapore/Taxes-and-Costs#property-tax, accessed 20.06.2016
[34] A. Justin Kirkpatrick and Lori S. Bennear, "Promoting clean energy investment: An empirical analysis of property assessed clean energy," *Journal of Environmental Economics and Management*, pp. 357-375, 2014.
[35] Singapore Power Group (SP). *Investor Relations, Credit Ratings*. Available: http://www.singaporepower.com.sg/irj/portal?NavigationTarget=navurl://b8e5b097207c246bee94f96a02680be3&windowId=undefined, accessed 14.06.2016
[36] Singapore Power Group (SP), "SP PowerAssets Ltd, Financial Statements for the year ending March 2015," www.singaporepower.com.sg.
[37] Jackie Horne FinanceAsia 20th, "Singapore delivers two benchmark bonds in one day," ed. http://www.financeasia.com/News/403944,singapore-delivers-two-benchmark-bonds-in-one-day.aspx, 18 November 2015.

ECONOMIC EVALUATION FOR STABLE ELECTRIC POWER SYSTEM WITH HIGH RATIO OF PHOTOVOLTAIC POWER SYSTEM - TOWARD MORE THAN 90% CO_2 EMISSIONS REDUCTION OF ELECTRIC POWER SYSTEM IN JAPAN

Toshihiro Inoue[1,2], Shigenobu Matsuda[1], Hiroshi Iwasaki[1], Ryuji Matsuhashi[1,2], Koichi Yamada[1,2]
[1] Center for low carbon society strategy, Japan science and technology agency
Sience Plaza 4F, 5-3, Yonbancho, Chiyoda-ku, Tokyo 102-8666, JAPAN
Phone:+81-3-6272-9271 FAX: +81-3-6272-9273
E-mail: inoue@jst-lcs.jp
[2] The University of Tokyo

ABSTRACT: This paper describes the future structure of Japanese electric power generation systems which can reduce life cycle CO_2 emissions by more than 90% based on 2010 and also provide economic electricity in 2050. It is important to clarify additional cost to establish a stable grid system with high ratio of variable renewable energy. Firstly, we calculate costs of present and future renewable energy systems using our quantitative methodology. To assess future renewable energy sources such as photovoltaics (PV), wind energy, geothermal, battery and so on, we designed manufacturing processes considering future technology progresses. For example, PV system cost will reach to 60 yen/W in 2030 according to this method. Secondly, we evaluate the cost of electric power systems using our multi-regional power generation model that takes into account the system stability. As a result of the cost minimized power generation model, it is possible to reduce CO_2 emissions to 50% based on 2010 without additional cost in 2030. To reduce CO_2 emissions by 90% in 2050, electricity cost will tremendously increase. In order to solve this problem, we should develop an economical and also low carbon PV system. Future economical and low carbon power systems are described using calculation results on new PV, battery, hydrogen, etc. which are key issues to be developed towards low cost and low carbon.

Keywords: Economic analysis, Life cycle CO_2, Energy options, Stable electric power system

1 INTRODUCTION

As Japanese government sets pathway to achieve GHG emissions reduction target 26% and 80% by 2030 and 2050 respectively in the plan for global warming countermeasures, long-term strategy to reduce dramatically CO_2 emission is now essential. However, technology improvement has been often discussed without economical and environmental evaluation.

The PV market size has tremendously increased over the last several years. In some region, PV systems are already competitive with existing power. However sudden installation increase of variable renewable energy sources causes higher electricity cost. A stable electricity system with high ratio of renewable energy is essential for the future.

The purpose of this study is to find electric power system with low cost and low life cycle CO_2 ($LCCO_2$) emissions using our optimized multi-regional power generation model from the present to 2050.

2 METHODOLOGY

2.1 Approach

The high cost of renewable energy prevents to achieve low carbon society. This study shows that technology improvement will promote high ratio of renewable energy especially PV system. Furthermore, we clarify the relations between electricity cost and CO_2 emissions under different composition of renewable energy supply. This methodology contributes to estimate quantitatively the economic effect of technology development and the issues to achieve CO_2 emissions reduction target.

Firstly, we have developed a calculation method for cost estimation of future renewable energy technologies. In this method we aggregated the costs of equipment, labor, raw materials, and utilities required to manufacture each type of renewable energy power system based on our original equipment database [1-3]. The cost of each renewable energy technology at present and in 2020 and 2030 was estimated by use of this method. The cost of electric power system in 2050 was calculated based on the technology level in 2030.

Secondly, we have developed dynamic optimized multi-regional power generation model which can provide future electricity costs and annual $LCCO_2$ emissions under the consideration of stable electricity supply. Figure 1 shows the structure of our model. The parameters for the model are electric energy supply, capacities of battery and hydrogen turbine under the given $LCCO_2$ emissions constraint and electricity demand.

This model is able to minimize the total power generation costs from 2010 to 2050. We adopted the above mentioned estimated cost of each renewable energy technology. This model use linear optimization programming with more than 480 thousand constraint equations and more than 840 thousand parameters. Japan is divided into 10 regions regarding main electric power companies (see Figure 2). In this model, we estimate power supply and demand data by using the hourly data of electric consumption of each region. Hourly output of PV and wind power generations are estimated using several tens of Japanese places' climate conditions data for one year. We classified daily demand curve into seven representative days. We suppose to use existing transmission line.

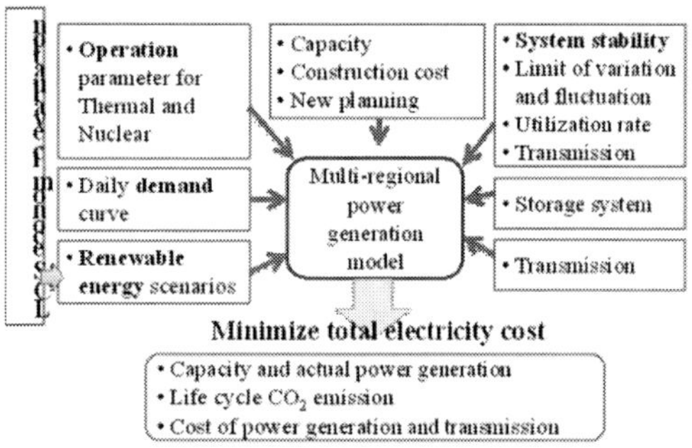

Figure 1: Structure of multi-regional power generation model

Figure 2: Region and Electricity companies in Japan

3 RESULTS AND DISCUSSION

3.1 Cost of renewable energy technology

Table I shows the calculation results of PV costs from 2015 to 2030. Technology development of high-efficiency tandem solar cell with a rational manufacturing process are realized for cost reduction of PV module. The cost of balance of systems (BOS) will decrease 70 yen/W to 20 yen/W by technology improvement and also market expansion. Total PV system cost will be as low as 60 yen/kWh in 2030.

Costs of other renewable energy technologies are also calculated as shown in table II[4].

Table I: Cost structure of future PV technology toward 2030

Technology level	present	2020	2030
Annual product [GW/y]	1	5	5
Lifetime [year]	20	20	30
Installed system cost [Yen/W]			
Module cost	60	50	40
BOS cost	70	50	20
Total PV system cost	130	100	60
Cost of electricity [Yen/Wh]	13	10	5

Table II: Estimated standard electricity cost of power generation in 2050[4]

	Standard electricity cost [Yen/kWh]	Standard operating condition [h/year]
Hydro	10.8	3500
Nuclear	8.8	6100
Coal	7.8	6100
LNG	11.8	6100
PV	5.7	1000
Wind	10.2	2000
Geothermal	8.0	6100
Biomass	10.9	6100
Battery system	6 yen/Wh	
Hydrogen fuel	3-10 yen/MJ	

3.2 Cost of electricity under given reduction rate of CO_2 emissions

Table III shows several scenarios under different rate of CO_2 emissions and also different electricity demands in 2050. In the scenario A, the PV and other RE technology developed till 2050 can reduce CO_2 emissions by 50% without increase of electricity cost based on the technology level in 2030.

In scenario B of the 80% CO_2 reduction, electricity cost of 11-12.8 yen/kWh will be acceptable level in Japan. However when the CO_2 reduction rate increases to 90% (scenario C), electricity cost becomes very high as 25.1 yen/kWh because of high usage of hydrogen electricity to stabilized as explained below. Next importance of saving electricity demand is shown in scenario D and scenario E. In these scenarios, electricity demand is reduced from 1000 TWh to 800 TWh. The electricity cost can be kept at a public acceptance level. Even in the case of 90% CO_2 reduction scenario D, the electricity cost becomes high as 27.1 yen/kWh in the 93% CO_2 reduction scenario E.

Figure 3 shows average daily electricity supply and demand curve of a summer weekday for the scenario C in 2050. This figure explains the feature of electricity production and consumption. As shown in figure 3, in a day-time excess electricity production over demand is used for both hydrogen production and storage by battery. In the night-time, electricity stored in battery is used for the balance of supply and demand. Electricity generated by hydrogen is used in a whole day to keep stable supply in a short time (Load frequency control;LFC and Governor-free control;GF) and also in an emergency.

We should consider fluctuation in the short term (from seconds to minutes), in a long term (from hours to days). For the short term fluctuation, LFC and GF can be controlled by battery and hydrogen electricity in the scenario C. The long term fluctuation can be controlled mainly by battery.

In Figure 4, scenario C shows large discrepancy between electricity production and consumption. There is a room to use LNG electricity in the case of 80% CO_2 reduction in scenario B. Because of the reduction rate of 90%, LNG electricity should be reduce to nearly zero. Then we should use hydrogen instead of LNG. However total efficiency from transportation to generation using hydrogen is low. Therefore we should generate excess electricity in scenario C(1) and causes high cost of electricity.

Table III: Cost and CO_2 emissions with different Electricity Supply in 2050 (Results of Cost Minimum Model Calculation)

Scenario	A	B	C	D	E
Scenario assumption:					
$LCCO_2$ emission reduction constraint					
	no	80%	90%	90%	min
Demand (TWh/y)	1000	1000	1000	800	800
Electricity production (TWh/y)					
Hydro	130	130	130	130	130
Coal	279	0	0	0	0
LNG	10	202	44	82	2
PV	409	483	746	413	626
Wind	12	57	212	31	193
Geothermal	218	228	255	223	205
Biomass	22	31	31	29	31
Total	1080	1131	1418	909	1145
Capacity of battery (GWh)					
	377	468	447	440	2076
Hydrogen use (TWh/y)					
	0	30	188	21	126
Cost of electricity (Yen/kWh)					
Low (H_2 3¥/MJ)	10.1	11.0	16.0	11.1	19.5
High (H_2 10¥/MJ)	10.1	12.8	25.1	12.4	27.1
Reduction rate of $LCCO_2$ (based on 2010)					
	50%	80%	90%	90%	93%

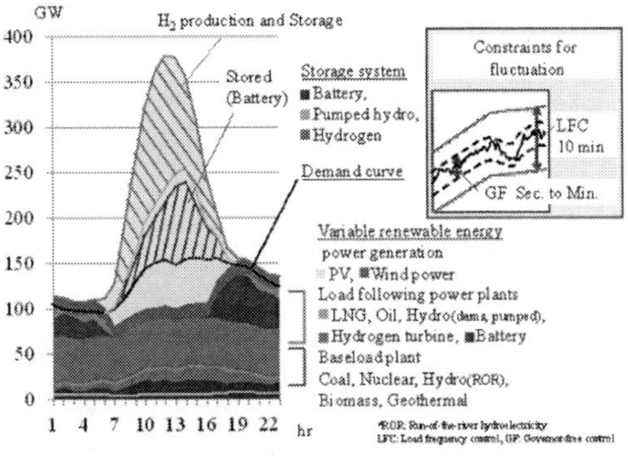

Figure 3: Average daily output of summer weekdays in 2050 in scenario C.

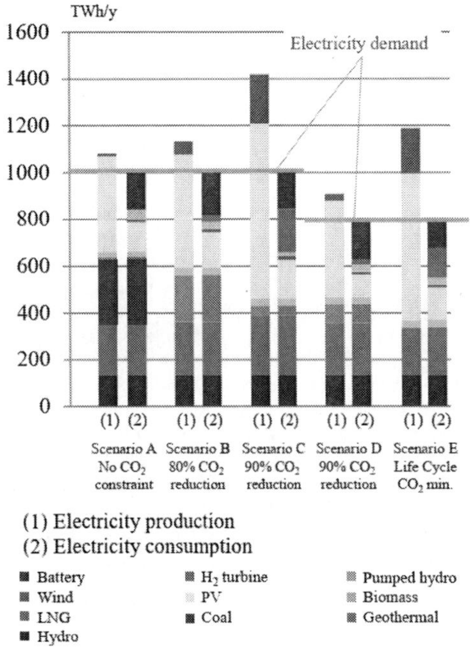

Figure 4: Share of electricity generation in 2050 considering energy storage and conversion

4 CONCLUSIONS

Our quantitative cost evaluation of PV systems clarified future cost to be economical. Future electric power systems with high renewable energy sources were designed and evaluated in terms of CO_2 emissions and economies using the cost minimized power generation model. For the model, our quantitative renewable energy and battery evaluation results were used. In 2050, economically competitive electric power systems with 90% reduction of CO_2 emissions were shown. Future economical and low carbon power systems are described using calculation results on new PV, battery, hydrogen, etc. which are key issues to be developed towards low cost and low carbon.

Reference

[1] K. Yamada, T. Inoue, K. Waki, "Future Prospects of Photovoltaic Systems for Mitigating Global Warming", World Engineers' Convention 2011 in Geneva, 2011

[2] "PV power systems: Quantitative technology scenarios, and science and technology roadmap based on elemental technology structure", Center for low carbon society strategy Japan science and technology agency, 2014 (in Japanese)

[3] T. Inoue, K. Waki, K.Yamada, "Cost estimates for PV manufacturing through 2030 –Potential for reducing costs using high-efficiency Si-based tandem solar cells", PVSEC, 2014

[4] In file data of Center for low carbon society strategy, Japan science and technology agency (to be published)

SOCIALISED SAVINGS FROM DECENTRALISED PHOTOVOLTAICS IN SOUTH PACIFIC MARKETS

Benedict O'Donnell
Heliocentric Solutions, 56 Vincent Square, SW1P 2NE, London, UK
+44 (0)207 828 28 20, benedict.odonnell@heocs.com

Hai Son Nguyen
Institut de Nanotechnologies de Lyon, Ecole Centrale de Lyon (UMR 5270)
36 av Guy de Collongue, 69134 Ecully, France

ABSTRACT: Socket parity has redefined the business case for rooftop photovoltaics, but it has yet to cause a perceivable increase in global photovoltaic sales. As subsidies for renewables falter across both established and emerging markets, the future of the photovoltaics sector grows increasingly reliant on sustainable niches independent of government support. Small rooftop photovoltaic systems may offer one such opportunity as building occupants in socket-paired markets now stand to save enough money on their utility bills to offset system costs under current market conditions. Such partial electricity autarky raises questions on how to mutualise grid fees and capacity costs fairly. Here, we present markets in which this need not be a challenge. In climates where electricity production peaks close to hours of high insolation, entire communities can benefit on a system level from installing rooftop photovoltaics and diminishing the amount of grid infrastructure needed to meet their peak load. Building on observations across an Australian electrical distribution network, we illustrate the opportunities for decentralised photovoltaics in solar synchronous electricity markets more broadly.

Keywords: Financing, Rooftop, Small Grid-connected PV Systems, Demand-Side, Distributed System Operators (DSO)

1 INTRODUCTION

As photovoltaic technology reaches socket parity in electricity markets across all continents, the practice of self-consuming electricity generated by photovoltaic systems installed on the consumers' rooftop is growing both more financially compelling and more common. The expansion of self-consumption challenges the business model of conventional energy markets. Reduced usage of centralised energy infrastructure shifts a disproportionate share of its cost to electricity consumers who cannot afford rooftop photovoltaic systems of their own. National governments commonly patch this market failure by introducing fees for self-consumption. The measure is generally unpopular, of questionable legality and against the spirit of longer-term climate policy. In this paper, we investigate markets in which the system benefits provided by self-consumed photovoltaics outweigh the grid fees that they redistribute so that all consumers can benefit from rooftop installations, not just their owners.

2 SOCKET PARITY

Socket parity describes an electricity market in which the levelised cost of energy generated by a photovoltaic system is lower than local electricity retail rates. While socket parity is a precondition for solar energy to be cheaper than electricity bought on the grid, it should not be confused with competitiveness. The main distinction is that the levelised costs of energy used to define socket parity assume that every kilowatt hour produced by a photovoltaic system is consumed by its owner. Mismatches between the time at which photovoltaic systems generate electricity and the time at which buildings consume energy contradict this assumption in most real-world installations.

2.1 Methodology

In previous work, we have compared building load profiles with local insolation measurements to calculate the overlap between the two curves and determine the fraction of solar energy generated by a rooftop photovoltaic system that could realistically be self-consumed [1].

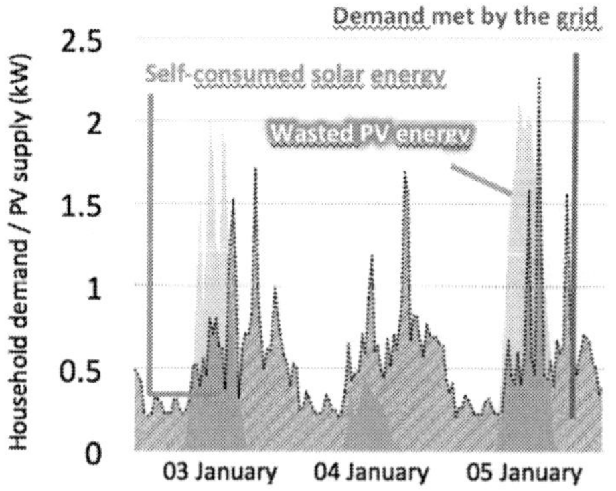

Figure 1: Calculation method for self-consumption by rooftop photovoltaic systems. Solid orange regions are self-consumed, blue regions are bought from the grid and light orange regions are wasted photovoltaic production.

For the purpose of our calculation, all electricity demand that coincided with generation from the photovoltaic system was deemed to be self-consumed. Electricity demand in times of insufficient insolation was purchased from the grid at standard utility retail rates. Any generation from the photovoltaic system that did not coincide with the building's electricity demand was discarded as a donation to the grid (or "wasted"). We refer to the fraction of electricity self-consumed over the total electricity consumed by a household over the year as its autarky rate, and the share of photovoltaic energy used over the photovoltaic system's total production as the self-consumption rate. Our initial financial analysis discarded storage systems on the basis of cost, meaning that only electricity generated at times of building demand was accounted for as self-consumed.

The average yearly spending on electricity of each household was calculated using the following relation:

$$Y.sp. = \frac{\sum_{year}^{20}\left[\frac{(Cons. - (PVcap \times yd)) \times r) + Sys.pr.}{(1 + WACC)^{y}}\right]}{20}$$

where *Cons* was the amount of electricity consumed by the household from the grid, *PVcap* the capacity of the photovoltaic system installed, *yd* the local yield of a photovoltaic system in kWh/kWp.m^2, *r* the local electricity retail rate, *Sys.pr.* the local price for photovoltaic systems in US$/kWp, *WACC* the weighted average cost of capital and *y* each year in the 20-year guaranteed lifetime of the photovoltaic system. Repeating the calculation for increasing photovoltaic system sizes revealed that, even without government incentives or storage solutions, German households could in principle reduce their averaged yearly spending on electricity by about 10% if they installed a rooftop photovoltaic system of 1 kWp capacity [1] (see the dotted black line in Figure 5 for an updated example).

2.3 Limitations

There is a systemic flaw built into this calculation in that it neglects capacity charges for the grid. In markets where utilities charge for energy consumption, rather than fixed service fees, households that purchase less electricity from the grid (for instance, because part of their demand is being met by a rooftop photovoltaic system) while requiring the same grid capacity (for periods when their photovoltaic system produces no energy) end up paying fewer grid fees for the same amount of infrastructure. Grid operators eventually recover their losses by increasing fees for all consumers, with the burden falling disproportionately on conventional consumers that do not use photovoltaics. In essence, from an overall system perspective, rooftop photovoltaics allow their owners to save money, while their neighbours lose out.

3 SOCIALISED SAVINGS

Rather than leave households to partially opt out of the grid, we present a socially equitable rollout plan for rooftop photovoltaics that makes financial sense at a system scale. Its essence is to only deregulate the deployment of rooftop photovoltaic systems in markets where they can progressively replace centralised energy assets at a profit. This would be the case of distribution networks where solar generation and grid consumption both peak at the same time. Under such conditions, decentralised photovoltaics could potentially help meet peak demand on the grid, substituting for generation, transmission and distribution infrastructure. If such markets exist, they would provide the ideal setting for decentralised photovoltaics to mature in while their price drops to levels at which they can compete on electricity markets more generally. To test the hypothesis, we have worked with the standard load profiles of Stadtwerk München in Germany (Figure 2.a), Enel Distribuzione (Figure 2.b) in Italy and Citipower in Australia (Figure 2.c), and compared them to local insolation patterns.

3.1 Solar synchronous markets

The drawback of installing rooftop photovoltaic systems at latitudes as high as those of most European electricity markets is that insolation levels differs greatly between summer and winter months. In contrast, electricity demand remains comparatively stable throughout the months, generally peaking at night time during the winter when photovoltaic systems tend to be inactive (Figure 2.a). As a consequence, tailoring photovoltaic system sizes to meet winter demand generates an excess of electricity in the summer, and tailoring photovoltaic systems to meet summer demand underexploits the solar resource throughout the rest of the year. Electricity markets such as Germany are out of phase with the potential of solar energy (Figure 2.a).

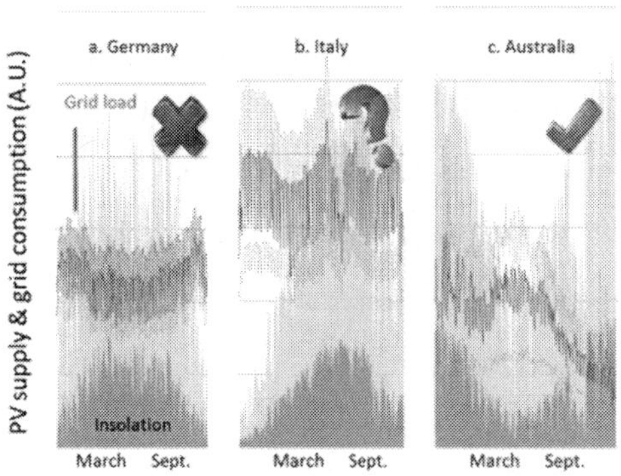

Figure 2: Daily average (yellow) and maximum (orange) insolation compared to daily average (solid blue) and maximum and minimum (dotted blue) demand for electricity distribution networks in Germany (a), Italy (b) and Australia (c).

In contrast, electricity demand in warmer climates, such as Australia, tends to peak in periods of extreme heat, rather than extreme cold (Figure 2.b). This may be related to the local consumption of air conditioning. In solar synchronous markets of this kind, the consumer base stand in principle better chances of benefitting from reducing the amount of infrastructure needed to meet their peak load by installing rooftop photovoltaics instead of centralised grid assets.

3.2 Beyond overlap in insolation and grid consumption

The financial benefits of solar synchronous markets are not limited to the overlap between load profiles and insolation curves. In the present study, the autarky and self-consumption rates of rooftop photovoltaic systems distributed evenly among the households connected to distribution networks in Germany, Italy and Australia remain similar (Figure 3).

What does affect autarky and self-consumption rates are the shape and constancy of the daily electricity demand. As such, electricity consumption patterns in commercial buildings and factories were observed to fit the generation curve of rooftop photovoltaic systems more tightly, but local differences in the peak hours of demand of residential consumers less so. The practical benefits of solar synchronous markets lie not in greater overlap between periods of overall insolation and grid consumption but in the overlap between insolation and the peak load on the grid.

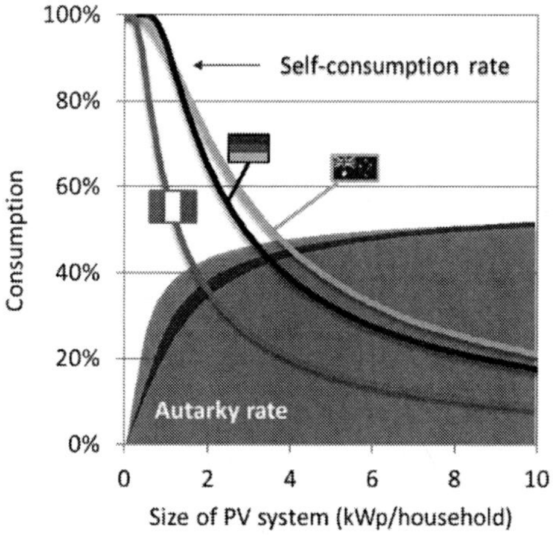

Figure 3: Autarky and self-consumption rates expected for photovoltaic systems installed on the rooftop of residential buildings in Germany, Italy and Australia.

3.3 Reducing peak load

The key factor of interests in the three markets investigated is the moment at which their demand peaks. In the case of Germany, demand on the grid peaks at night time, so solar can never help reduce the peak load on the grid. In Italy and Australia, the largest loads are distributed during the daytime, meaning that decentralised photovoltaics can in principle help supply them (Figure 4).

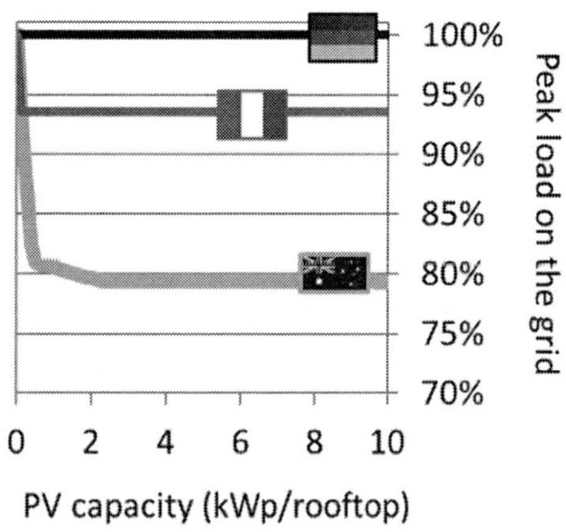

Figure 4: Peak load expected on electricity distribution networks in Germany, Italy and Australia given increased penetration of rooftop photovoltaics.

Assuming that both the load and insolation are distributed evenly among consumers, the peak yearly load carried on Enel Distribuzione's grid would fall to 94% of its present value with even a modest penetration of decentralised photovoltaics installed on the rooftop of all consumers. The peak yearly load carried on Citipower's grid would fall to 79% of its present value for a penetration of rooftop photovoltaics equivalent to a 2 kWp system installed on each consumer's rooftop. In all markets, night time loads eventually limit the impact of photovoltaics on shaving yearly peaks, the question is how many day time peaks decentralised photovoltaics can shave before that.

3.4 Systemic spending on electricity

Since the cost of grid infrastructure is determined by its capacity, reducing the peak load on the grid should also reduce the investment needed to build and maintain the grid. We assume a linear relation between the cost and capacity of the grid and that this cost is entirely collected through grid fees. We then can calculate the overall spending on electricity among the consumer base of Stadtwerk München, Enel Distribuzione and CitiPower so as to determine whether, at an overall system scale, their consumers would save or lose money by installing decentralised photovoltaics.

To incorporate the increase in grid rates that decentralised photovoltaic production will cause by reducing the usage of centralised infrastructure, we divide present grid fees by *1 – the autarky* of consumers with photovoltaic systems on their rooftop. This means that if consumers use the same grid less often, they have to pay more for it each time they use it. To incorporate the reduction in grid rates resulting from the reduced capacity of the grid, we multiply present grid rates by the ratio of the peak load on the grid using decentralised photovoltaics and the present peak load. This means that if consumers use a grid that is smaller than the one they currently have, it will cost them less money. The complete relation is given by:

$$\boldsymbol{Soc.r. = W.r. \ + T. + \left[\frac{G.r.}{1 - aut.}\right] \times \left[\frac{Pf}{Pi}\right]}$$

where *Soc.r.* are the new, socialised electricity rates, *W.r.* are wholesale electricity rates (assumed to be constant in this approximation), *T.* are taxes on electricity, *G.r.* are present grid rates, *aut.* is the energy autarky of consumers with photovoltaic systems on their rooftop, *Pf* is the new peak load on the grid when consumers benefit from distributed photovoltaic generation and *Pi* is the present (initial) peak load on the grid.

Using the same relation as in Section 2 and values reported in Table I for grid fees, weighted average costs of capital, photovoltaic yields, photovoltaic system prices and annual loads carried on these grids and households, we observe that the impact of decentralised photovoltaics on electricity spending averaged over all consumers varies markedly depending on the market (Figure 5).

In the German case, where self-consumption in individual buildings is expected to save 10% on electricity spending when grid capacity charges are ignored, we observe only marginal savings for small photovoltaic systems when spending is averaged over the entire consumer base. This is because decentralised photovoltaics do cannot reduce grid investments and system-level savings derive uniquely from the margins of rooftop photovoltaic system owners self-consuming their own electricity. The same is true of the Italian case, although the amount of money saved is higher as a result of lower grid fees reported by Enel Distribuzione [2]. Incidentally, we observed that when identical grid fees were used, the Italian case produced a systemic spending curve similar to that of Germany.

Table I: Financial parameters for investing in photovoltaics in Germany, Italy and Australia.

	SWM	Enel	CitiPower	Reference
Carries (TWh/y)	20	300	6.2	[2]–[4]
WACC (%)	9.3	8.5	8.5	[2], [4], [5]
Revenues (US$/kWh)	0.1	0.04	0.08	[2]–[4]
PV yield (MWh/y)	1.0	1.4	1.4	[6]
PV price (MWh/y)	1.8	1.9	1.5	[7]–[9]
Consumption (kWh/y)	3.5	2.7	5.8	[10]

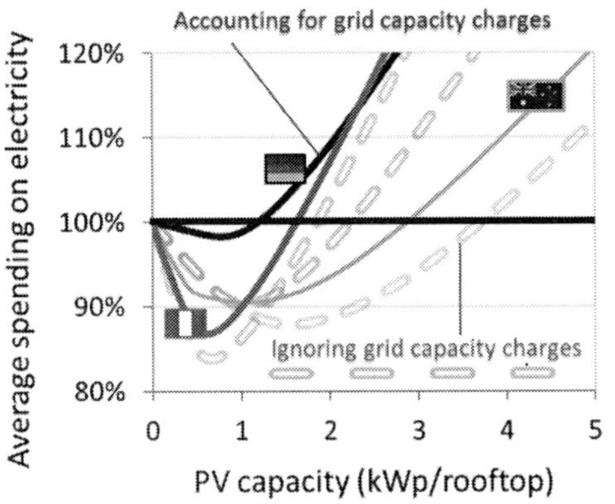

Figure 5: Spending on electricity expected for households connected to electricity distribution networks in Germany, Italy and Australia given increased penetration of rooftop photovoltaics. Individual household spending is given as dotted curves and averaged systemic spending as solid lines.

In the Australian case, substantial savings were observed on a systemic scale with photovoltaic systems as large as 2 kWp per rooftop. These savings would in principle benefit all consumers of CitiPower as they reduce the peak load for everyone, not just photovoltaic system owners. In all systems, above a given size of photovoltaic systems per building (between 0.5 – 2 kWp depending on the market), decentralised photovoltaics were observed to increase the overall spending of the consumer base on electricity.

4 DISCUSSION

This study has tested the hypothesis that, in solar synchronous markets, rooftop photovoltaics could in principle reduce the overall spending on electricity at a systemic level. The results show that the hypothesis is valid, leading to opportunities for consumers to reduce their bills, photovoltaic developers to exploit a subsidy-free market and grid operators to save on their infrastructure costs. The notion of socialised savings is handy when considering the overall cost of a system from the perspective of an urban planner, an informed and well organised consumer based, or a thrifty policy maker. Perhaps the greatest opportunity offered by decentralised photovoltaics is to grid operators in solar synchronous markets. Socialised savings could in principle translate into higher margins if distribution system operators were leasing the rooftop photovoltaic systems to their consumers. To materialise these opportunities into sound business cases, it is now important to refine some of the assumptions made in our calculations.

4.1 Polishing predictions

Comparing insolation curves with standard load profiles over an entire electricity distribution network leads to overly optimistic results in terms of their overlap. Abrupt fluctuations in the demand of individual buildings are averaged out when observing the consumption curves at a systemic level. Most importantly, these fluctuations may impose higher grid capacity needed by individual buildings to meet brief peaks in demand. Our findings would also benefit from verification under different climatic conditions, with load profiles of higher time resolution, covering a statistically representative sample of individual buildings over a statistically representative number of years.

Other weak points in our theory include the specifics of financial calculations in Sections 2 & 3. The relation between the cost of an electricity grid and its load is unlikely to be linear. Reducing grid capacity is unlikely to prove financially optimal. Instead socialised savings from decentralised photovoltaics may best be kept in mind when maintaining the grid or extending it. In most liberalised energy markets, grid operators are not allowed to own electricity generation assets. Our systemic savings calculation may need to be complexified if viewed from the perspective of other stakeholders owning the decentralised generation assets.

4.2 South Pacific markets

If the results in Section 3 hold upon closer scrutiny, they presage that, without subsidies, decentralised photovoltaics can save entire communities money when installed under the right market conditions. Optimal conditions to benefit from the socialised savings of decentralised photovoltaics would be a market where electricity rates and grid fees are high, demand peaks in summer near noon, consumption patterns vary little between summer and winter and capital is cheap. Many islands of the South Pacific fit this description in remarkable detail. Their proximity to the equator limits interseasonal variation in their consumption and insolation patterns. Their isolation increases grid and electricity rates. Their ongoing economic development is increasing electricity demand and grid upgrades. Their eligibility for foreign development aid (particularly from Australia and Japan) could also provide project-based capital at comparatively low interest rates. Electrical utilities in many South Pacific islands remain government controlled and vertically integrated, removing free market issues in terms of asset ownership. What information we have on their load profiles also suggests that local electricity demand peaks when it is sunny [11]. These advantages would suggest that South Pacific markets are well suited to pioneer a controlled roll-out of unsubsidised rooftop photovoltaics and to demonstrate how to benefit from unsubsidised solar without the drawbacks.

4.3 In store for the future

In our study, decentralised battery systems were ruled out on the basis of cost. At current market rates, their usage was systematically more expensive than the centralised grid. However, as this cost decreases, stationary storage and electric vehicle batteries offer exciting prospects to extend the reach of decentralised photovoltaics by storing electricity for periods of peak demand. This may transform the business case for rooftop photovoltaics in markets where electricity demand peaks at night. It is unlikely to resolve issues in electricity markets such as Germany where seasonal demand is out of phase with insolation profiles. No volume of batteries can store in summer the energy needed by the grid in winter. Still, the roll-out of unsubsidised rooftop photovoltaics offers a sustainable market within which all storage solutions can mature.

5 CONCLUSION

Depending on the location and the capacity of systems installed, rooftop photovoltaics can either increase or decrease the overall cost of a centralised grid to the consumer base paying for it. Wise planning under appropriate market conditions could reduce consumer bills or boost the margins of grid operators with today's photovoltaic technology, and unlock sales of unsubsidised photovoltaics on an unprecendented scale. South Pacific islands present a compelling sandbox within which to investigate these initial results. Further insight supported by cost-reductions in electricity storage technology could pave the way for a sustained roll-out of unsubsidised photovoltaics across the equator (in particular in India and South-East Asia) and, as longer term storage technologies mature, at progressively higher latitudes.

6 REFERENCES

[1] B. O'Donnell, "Tailoring Rooftop Photovoltaics for Self-Consumption," in *29th EU-PVSEC*, 2014, pp. 4062 – 4064.
[2] Enel, "Annual Report 2014," 2015.
[3] Stadtwerke München, "Annual report 2015," 2015.
[4] Austalian Energy Regulator, "State of the Energy Market 2013," 2013.
[5] EY, "Mapping power and utilities regulation in Europe," 2013.
[6] EU Joint Research Center, "PVGIS," 2016. [Online]. Available: http://re.jrc.ec.europa.eu/pvgis/index.htm.
[7] G. Altenhöfer-Pflaum, "National Survey Report of PV Power Applications in Germany 2014," 2015.
[8] S. (ENEA) Castello, F. (GSE) Tilli, and S. (RSE) Guastella, "National Survey Report of PV Power Applications in Italy 2014," 2015.
[9] W. Johnston, C. Taeni, and R. Egan, "National survey report of PV power applications In Austalia," pp. 1–24, 2013.
[10] Eurostat, "T_NRG Table," 2016. [Online]. Available: http://ec.europa.eu/eurostat/web/energy/data/main-tables.
[11] M. Isaka, L. Mofor, and H. Wade, "Pacific Lighthouses," 2013.

BIPV- POTENTIAL AND STATE OF THE ART IN AUSTRIA: HIGH FUTURE BIPV-DEMAND FOR 100% RENEWABLES - BUT LOW ACTUAL SHARE

Shokufeh Zamini[1], Astrid Schneider[1] - astrid.schneider@ait.ac.at, Marcus Rennhofer[1], Christoph Mayr[1], Karl A. Berger[1], Tim Selke[1], Barbara Beigelböck[1], Momir Tabacovic[2], Hubert Fechner[2]; Georg Reinberg[3]; Gernot Becker[4]
[1] AIT Austrian Institute of Technology, Giefinggasse 2, 1210 Vienna, Austria;
[2] FH Technikum Wien, Giefinggasse 6, 1210 Vienna, Austria;
[3] Architekturbüro Reinberg ZT GmbH, Lindengasse 39/8, 1070 Wien, Austria;
[4] ATB-Becker e.U, Dörferstraße 16, 6067 Absam, Austria

ABSTRACT: At the Austrian Institute of Technology a comprehensive, current overview of the BIPV market was created. It shows, that despite a very well developed industry the share of BIPV is very low with about 2% of the overall PV-installations. However about 85% of all PV-systems in Austria are installed on buildings. But the majority of all systems are roof top or roof parallel installations. The suitable area on roofs and facades, which could be used to install Photovoltaics is about 230 km^2 according to the Austrian PV-Roadmap [1], while the needed area to achieve a share of 27% on the energy demand is about 170 km^2. This means, that the biggest part of suitable and well solarized roofs and facades will be used for solar electricity production in the future. To achieve at the same time an ecologic and aesthetic building stock it will be obligatory, not to install highly Aluminium consuming, roof top structures with todays standard modules, which are – in the building context - often ugly, at the same time, but to change today's roof top share of PV-installations into beautiful, adapted, architecturally highly valuable and accepted Building Integrated Photovoltaic Systems.

Keywords: Building Integrated PV (BIPV), Module integration

1 SHAPE PV: THE SHAPE OF BIPV IN AUSTRIA TO COME

The project SHAPE PV - which is funded by the Austrian Ministry of Technology (bmvit) and led by the AIT-Austrian Institute of Technology in Vienna - looks into a research and development strategy for the BIPV-Sector in Austria. (project number: 848884)

The main working steps of the SHAPE-PV project are:

- Analysis of state of the art of BIPV in Austria
- International comparison of the BIPV sector
- Stake holder analysis: needs and hurdles regarding broad BIPV deployment
- Stakeholder workshop to figure out Research & development needs
- Proposal of a BIPV-project chain to facilitate and accelerate a broad BIPV deployment in Austria

The participation of the AIT and the FH Technikum Wien as a project partner of Shape PV in the IEA-PVPS Task 15 'Acceleration of BIPV' helps to optimize the work and puts it into an international worldwide framing regarding the actual effort and trend to push the BIPV-Sector including the standardization efforts and other means. At the same time the 'PV-Roadmap for Austria' [1] and the actual PV-Market survey for 2015 [2] were presented.

2 CURRENT STATUS OF BIPV IN AUSTRIA:

2.1 Installed Capacity of BIPV-Systems in Austria Market Share in 2010 – 2015

Based on the yearly market analysis [2], a comparison was done. Figure 1 shows that the installed BIPV (KW/year) at 2015 is in the same range as 2010. There was a slight increase in the years 2011-2014, which was in accordance to the trend of market world wide. The total installed PV showed an increase from 2010 to 2013 and then a decrease in agreement with global trend. The diagram shows that the PV market is stalling. The roof and facades integration systems installed in comparison with total installed is in the range of only a few percent.

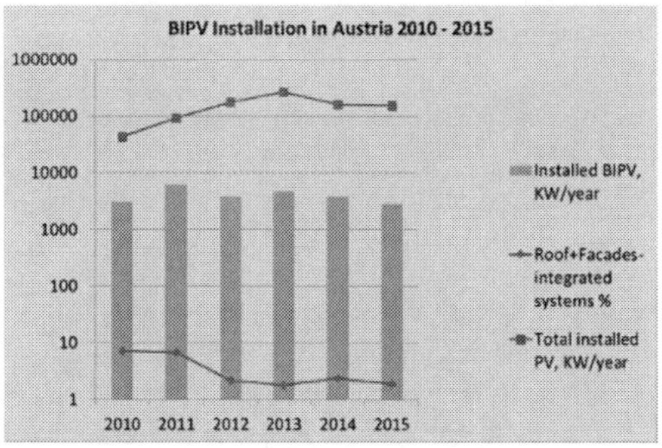

Figure 1: BIPV Installation in Austria, Graph: AIT-Austrian Institute of Technology, Data: Austrian Energy Market Data 2010-2015 [2].

2.2 Installed Capacity of BIPV-Systems in Austria 2015

Actual market survey data shows, that BIPV has a very low one digit share of the overall PV-market in Austria.

Figure 2 shows very clearly, that the building is the preferred place of PV-installation in Austria:

- 87.6 % of all PV-systems were installed on buildings
- Majority of systems with 85.7 % were installed as roof top systems
- But only 1.9 % are BIPV
- With 0.6 % the facades have a very small share
- Only 12.4 % are green field installations

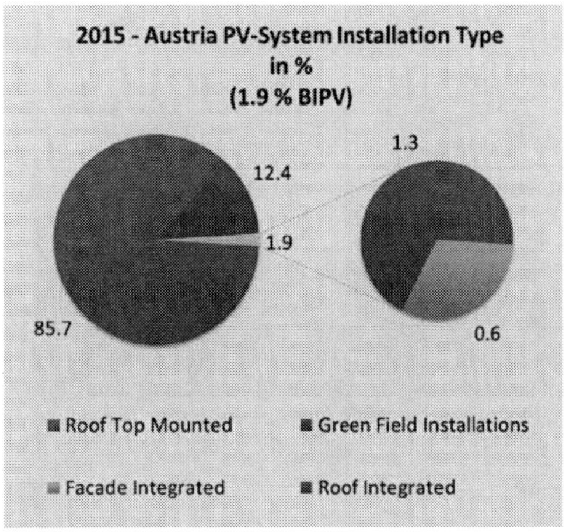

Figure 2: Share of BIPV in the PV-Systems, which were installed in the year 2015 in Austria (in % of installed capacity), Graph: AIT-Austrian Institute of Technology, Data: Austrian Energy Market Data 2010-2015 [2].

2.3 BIPV installations – international comparison

Based on the IEA, PVPS Task 1 Reports and Building integrated Photovoltaics a global strategic business report, a comparison for a few countries for the year 2014 was done. Figure 3 and 4 show that France and Switzerland are leaders in BIPV-installations per head of population. France is in absolute and relative numbers top, due to its strong BIPV-support.

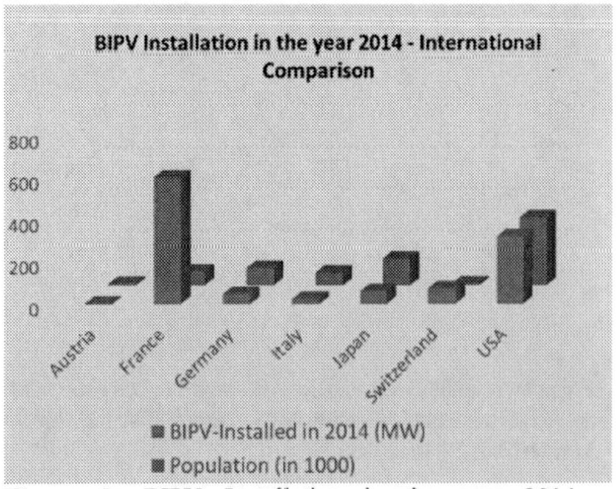

Figure 3: BIPV Installation in the year 2014 - International Comparison, Graph: AIT-Austrian Institute of Technology, Data: Data: IEA-PVPS National Surveys PV 2014 [3] and A Global Strategie Business Report [4].

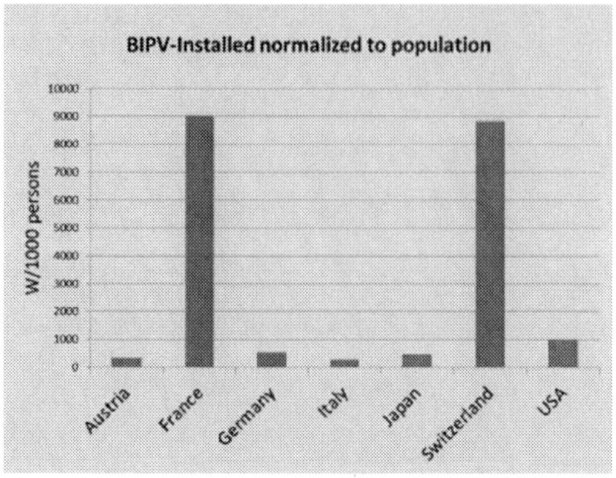

Figure 4: BIPV-Installed normalized to population Graph: AIT-Austrian Institute of Technology, Data: Data: IEA-PVPS National Surveys PV 2014 [3] and A Global Strategy Business Report [4].

3 BIPV Challenge, the MARRIAGE of an IPAD & a STONE

The great challenge in the field of Building integrated photovoltaics is to "Marry" consumer electronics similar product (Solar cell) with construction material (bricks, glass). This is schematically shown in figure 5.
The goal is to harmonize Electronic & Construction properties and purposes from global to local markets, rules and conditions. The differences in various levels are summarized in Table 1.

	Photovoltaics	Construction Material
Industry	Semi conductors /Coating /wafers	Stone, Steel, Concrete, Wood
Technology	Solar cells: Thin film, Crystalline, others	Stonewall, Roof structure
Standards	IEC	ISO

Table 1: Differences of Photovoltaics Technology and Construction material

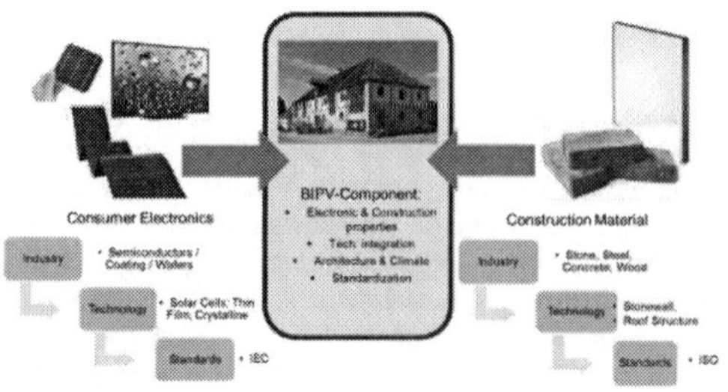

Figure 5: BIPV Challenge, the MARRIAGE of an IPAD & a STONE[6].

Moreover a BIPV Component should find a solution for Technical, architectural and climatic integration and be standardized.

4 ADVANCED PLAYERS IN INDUSTRY, ARCHITECTURE, RESEARCH & INSTALLATIONS = HIGH POTENTIAL

Innovative and advanced players in Industry, architecture and research lead to a high potential in BIPV market in Austria. In the following a few examples of projects, and players are named.

Figure 6: Skyscraper renovation with BIPV-façade: TU-Vienna, Project partners: TU Vienna, Prof. Dr. T. Bednar, ATB Becker, Schöberl & Pöll GmbH

Figure 7: Roof Garden + BIPV, Project: BOKU University – Ulrike Pitha, ATB Becker, Ertex, …
© Irene Zluwa

Figure 8: Skyscraper Design: BIPV-façade, Project: Chalabi Architects, Vienna

Figure 8: Office Building PV as Design Feature, Project: Reinberg Architects, Vienna

Figure 9: Efficient BIPV, AIT / POS Architects, Vienna

Figure 10: Ertex Solar – BIPV-Facade with LED, Fondation Pierre Arnault, Architect: Jean Pierre Emery

5 A FEW RESULTS FROM SHAPE-PV PROJECT

Shape-PV project focuses on three main goals:
Shape-PV: Concept is a study for a project chain dealing with the use of building-integrated photovoltaic (BIPV); in view of an overall, systematic analysis for implementing innovative BIPV in Austria the project (I) defines the requirements and future targets for BIPV firstly for the Austrian market and its stakeholders (with the perspective of the European market) (II) develops together with a group of key-players a necessary research and implementation roadmap and (III) drafts a concept for a first implementation phase. Existing know-how, current initiatives, obstacles and chances are analyzed systematically, the activities of potential stakeholders are submitted to a screening and all relevant partners are integrated. Based on these findings a chain of R&D follow-up projects will be defined which will have to be elaborated synergetically.

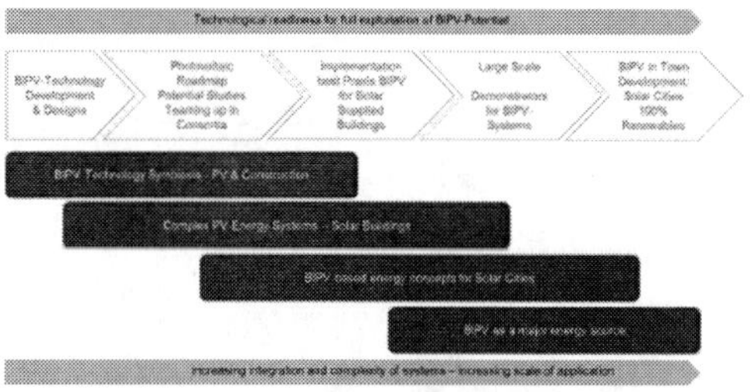

Figure 11: Shape-PV concept for large scale BIPV

Figure 11 shows the steps of the concept of the project. Increasing integration and complexity of systems relates to increasing the scale of application, i.e. from material and cells to modules, buildings, cities and BIPV as major energy source. The requirements for this target are developments and design, strategic research roadmap, and implementation of ideas, in building, and respectively in large scale and towns.

In the framework of the project, different groups of stakeholder are defined. Based on the questionnaire developed to identify the potentials and barriers from the viewpoint of stakeholder, the most relevant aspects were recognized.

45 PV projects, more than 10 BIPV feasibility studies, 31 BIPV research projects, 18 smart city projects, as well as more than 120 integrated/added PV projects on buildings are identified, analyzed, and summarized.
The results are categorized in 6 different groups:

- Innovative cells and components (laminating materials, cell contacts, lamination) for BIPV;
- Standard-BIPV Products;
- System technology, Mounting and Maintenance;
- Integration, Esthetics and Design (architectural aspects);
- Tools und Simulation, Know-how Transfer in the Market;
- BIPV-electrical energy concepts (i.e. Grid integration)

Based on above categories, and in cooperation with a group of stakeholders with more than 50 Persons, research topics for the future are identified. Research projects based on these topics, are going to build on the results of the projects already done and have synergy together, to accelerate BIPV Market in Austria.

6 References

[1] H. Fechner, C. Mayr, A. Schneider, M. Rennhofer, G. Peharz, Technologie-Roadmap für Photovoltaik in Österreich, Berichte aus Energie und Umweltforschung 15/2016, BMVIT 2016

[2] Peter Biermayr, Manuela Eberl, Monika Enigl, Hubert Fechner, Kurt Leonhartsberger (et al.), Innovative Energietechnologien in Österreich – Marktentwicklung, BMVIT 2016 (editions 2010 – 2015)

[3] International Energy Agency (IEA), Photovoltaics Power Systems Program (PVPS), National Survey Report of PV Power Applications in AUSTRIA, FRANCE, GERMANY, ITALY, JAPAN, SWITZERLAND, USA (2014)

[4] BUILDING INTEGRATED PHOTOVOLTAICS (BIPV), A Global Strategy Business Report, MCP-7004

[5] M. Rennhofer, S. Zamini, A. Schneider, C. Mayr, (et al.), "Shape PV – The Shape of BIPV to come", Otti BIPV-Symposium Bad Staffelstein 2016

[6] A. Schneider, "IEA PVPS Task 15 Acceleration of BIPV: Potential and Barriers", 6th SOPHIA Workshop PV-Module Reliability, Vienna, 2016

PROSPECTS IN SOLAR WATER DESALINATION: TOWARDS AFFORDABLE H_2O WITHOUT CO_2

Vasilis Fthenakis[1, 2], Adam A. Atia[1], Raed Bkayrat[3], Kim Choon Ng[4], Tawfiq Alghasham[5], Arslan Khalid[6], Divyam Nagpal[6], Sgouris Sgouridis[7]
[1]Center for Life Cycle Analysis, Columbia University,
926 S.W. Mudd Bldg. 500 West 120th Street, New York, NY 10027, vmf5@columbia.edu, aaa2269@columbia.edu,
[2]Renewable Energy Group, Brookhaven National Laboratory,
[3]First Solar International Middle East, Raed.Bkayrat@FIRSTSOLAR.COM,
[4] King Abdullah University of Science & Technology, Kim.NG@KAUST.EDU.SA,
[5] MEDAD Technologies, tawfiq.ghasham@medad-tech.com,
[6]IRENA, AKhalid@irena.org,
[7] Masdar Institute of Science & Technology, ssgouridis@masdar.ac.ae

ABSTRACT: This paper presents performance and cost analysis of small- and medium-scale seawater reverse osmosis (RO) desalination plants powered in full or in part by photovoltaic power systems in the Gulf Cooperative Council (GCC) region. Simulations of PV-RO desalination were conducted using HOMER Pro 3.6.1 and an Excel-based, hourly RO model. We describe a) small and medium size stand-alone systems operating 8, 10, 12 or 24 hours/day, which are powered by PV only, or PV and battery storage and diesel generators, and, b) combinations of PV and solar thermal and co-location of PV and conventional power plants. The stand-alone systems operating less than 24 hours included water storage. Our simulations show water production costs in the range of \$1.2-\$2.65 per m^3 of product water for small (i.e., $5m^3$/day) stand-alone PV-RO systems with PV LCOE values ranging from 12–18 ¢/kWh; water production costs are in the range of \$0.9-\$2.0/m^3 for medium size (e.g., 6500 m^3/day) systems with PV LCOE down to 6-8.8 US¢/kWh. Co-locating a 50 MW PV system with an existing power plant in Jordan resulted in LCOE being reduced from 8.8 ¢/kWh to 7.1 ¢/kWh. The potential for improved water production yield and cost savings in hybrid PV-solar thermal is also highlighted.
Keywords: Photovoltaics, PV, hybrid systems, reverse osmosis, desalination, MED-AD, SWRO, energy-water nexus

1 INTRODUCTION

Producing fresh water via water desalination is essential for arid, water-scarce regions, but expensive and energy-intensive compared to using conventional water sources. The cost of energy is a significant contributor to this high cost and the use of fossil fuels to power the desalination plants causes emissions of hazardous pollutants. Water desalination using thermal energy from the sun has a long history. Aristotle first described the concept of saline water evaporation for desalination. The earliest commercial solar thermal system on record is a solar still for a mining community in Las Salinas Chile, producing 22.7 m^3/d of fresh water from the evaporation of very high salinity (140,000 ppm) water; installed in 1872, it operated for about 40 years until the mine was exhausted [1]. Solar stills resemble greenhouses; they comprise a basin holding the saline water within a transparent enclosure. Solar energy enters the still through a sloping transparent glass or plastic roof and heats the saline water in the basin. The heated water evaporates and then condenses on the roof glass panels. The condensed droplets run down the panels and are collected for use as fresh water. Solar stills require a very large area (250-300 m^2/m^3 daily water production) and, therefore, are not viable for large-scale production. In contrast, a PV-RO system will require less than 8 m^2 per m^3/d of installed capacity. However, combinations of solar thermal and electric technologies have promise, and this is explored in our study.

The rapidly decreasing cost of photovoltaic (PV) electricity makes solar-powered desalination an affordable and ultimately sustainable option in arid countries with high insolation such as the Gulf Cooperative Council (GCC) nations, Northern Chile, the Southwest US, and other regions. The current worldwide water desalination capacity is approximately 90 million cubic meters per day [2], accounting for ~0.8% of the global fresh water demand. The desalination capacity in the GCC region is about 35 million m^3/day; seawater is the feed water in more than 85% of the GCC plants. Saudi Arabia (SA) has the biggest share, followed by United Arab Emirates (UAE), Kuwait, Qatar, Oman and Bahrain. The share of reverse osmosis (RO), which is a natural fit with PV, is increasing. As of 2014, RO accounted for approximately 60% of global desalination capacity (~74% of the capacity contracted over the last 10 years) but only 34% of the desalination capacity in the GCC countries. The region's stakeholders feel comfortable with the thermal technologies they have been employing for 30-35 years, and, only recently started adopting seawater RO (SWRO), but the trend in favor of RO is clear. Early on, RO was used only for brackish water as the high TDS content of seawater was excessively fouling the membranes. Subsequently, advances in higher-permeability membranes that reduced both energy consumption and the extent of membrane fouling, made RO technologies increasingly the preferred choice for desalination of both brackish water and seawater. Several projects combining RO and thermal desalination have been commissioned, helping RO in gaining more traction in the region [2]. This creates a pathway for decoupling electricity generation from fossil fuels in the co-generation multi-stage flash (MSF) and multi-effect desalination (MED) plants towards using solar electricity in desalination as RO's sole energy input. If such a trend continues, and the recent introduction of solar power plants in the region supports this, it will create a significant additional market for photovoltaics. Further prospects are created with PV electricity being combined with solar thermal energy in newly developed technologies, like adsorption desalination. In addition to grid integrated systems, solar desalination can be a solution for arid areas away from electricity grid, mobile sectors such as military and drilling services, refugee camps, and disaster emergency cases.

2 BACKGROUND AND MOTIVATION

In the current study we are examining cost reductions in small, stand-alone systems via smart combination of water and solar electricity storage. Also we identify potential reductions of the levelized cost of water (LCOW) production from seawater, via integration of PV into existing desalination plants, integration of PV and solar thermal technologies, and combinations of solar RO with conventional thermal desalination technologies. Such combinations could significantly decrease operating and maintenance (O&M), land and grid integration costs.

3 APPROACH AND METHODOLOGY

3.1 Stand-alone PVRO Plants

The objective of this analysis is to compare small-scale, stand-alone PV-RO systems that utilize water storage so that intermittent operation is possible. Small-scale scenarios were simulated for a daily water demand of 5 m3/day; combinations of PV and water storage were compared with PV, batteries, and water storage. To extend the analysis further, medium-scale scenarios were simulated subsequently to meet a daily water demand of 6,550 m3/day. For these scenarios, PV-battery power systems with water storage were compared with PV-battery-genset systems with necessary water storage. All simulations were conducted with HOMER, the Hybrid Optimization Model for Electric Renewables [3], in combination with an Excel-based hourly RO model developed by Atia with guidance from Fthenakis.

The project location used in all simulations was in Saudi Arabia at 20.67° N, 39.55° E coordinates. The annual horizontal global irradiation (GHI) at this site was estimated to be 2,128 kWh/m^2/yr. Table I gives the main technical and economic HOMER inputs for small- and medium-scale RO scenarios, which include costs assumed for the PV system (including BOS), diesel generators, and battery storage.

The main difference in our assumptions pertaining to small-scale and medium-scale scenarios, is the change in PV unit cost. The levelized cost of electricity (LCOE) for large-scale PV in the GCC has reached 5.8 US cents per kWh. Therefore, the PV system cost of \$922.50/kW$_{DC}$ in Table I reflects the cost that corresponds to that LCOE of 5.8 ¢/kWh; whereas a two times higher capital cost was assumed for very small systems. PV modules were modeled at 16% efficiency and configured for single-axis, E-W tracking, inclined at latitude-tilt. Furthermore, an 80% derating factor was assumed to account for inefficiencies in PV system performance due to soiling, ohmic losses, and other losses. The subsidized diesel fuel price is assumed to be 8 cents/liter while the global market price is assumed to be 70 cents/liter. Lithium-ion batteries with 1 kWh of energy storage and an 80% maximum depth of discharge were used in simulations. We assumed a battery system price of \$250/kWh which is less than present battery prices in the GCC but reflects a price forecasted for the near-future.

The average feedwater salinity assumed in each scenario is 40,000 ppm TDS, while the target product water quality was set to 213 ppm TDS. The RO plant model has a recovery of 35% using a single-pass system with an energy recovery device and booster pump (Figure 1).

Table I. Summary of HOMER input parameters.

Parameters	Parameter Values	
	Small-Scale RO	Medium-Scale RO
PV Capital Cost (\$/kW$_{DC}$)	2050	923
PV O&M Cost (\$/kW$_{DC}$/year)	36	36
PV Module DC Efficiency (%)	16	16
PV Lifetime (years)	25	25
Inverter Efficiency (%)	97.8	97.8
Diesel Gen Capital Cost (\$/kW$_{DC}$)	--	1009
Diesel Gen O&M Cost (\$/hr)	--	0.02
Subsidized Diesel Fuel Cost (\$/liter)	--	0.08
Unsubsidized Diesel Fuel Cost (\$/liter)	--	0.70
Diesel Gen Minimum Load Ratio (%)	--	25
Diesel Gen Lifetime (hours)	--	15000
Li-Ion Battery Specs (Ah/V/kWh)	166.7/6/1	166.7/6/1
Li-Ion Capital Cost (\$/battery; \$/kWh)	250	250
Li-Ion Replacement Cost (\$/battery)	250	250
Li-Ion O&M Cost (\$/battery/year)	3.5	3.5
Battery Lifetime (years)	15	15
Discount Rate (%)	5	5

In addition to the high-pressure pump and the booster pump, a seawater intake pump (not shown in Figure 1) is also included in all scenarios. Sea water, and therefore feed-water, temperature varies seasonally as shown in Figure 2. Four operational scenarios were considered for the RO plant's operational availability while holding the total daily water demand constant: 8, 10, 12 and 24 hours/day. This means that for a daily water demand of 5 m3/day, each one of these operational scenarios require the RO plant to produce enough water to meet that demand. Hence, the 8, 10, and 12 hour scenarios correspond to capacity increases by factors of 3, 2.4, and 2, respectively, relative to an RO plant that operates 24 hours per day. For further investigation a variable water demand that fluctuated monthly as a function of global horizontal irradiation was analyzed for comparison with the constant water demand scenario. Note that even in the constant water demand scenario the electricity load still fluctuates as seasonal variations of feedwater temperature were accounted for; an increase in feedwater temperature corresponds to an increase in membrane water permeability, enabling lower feed pressures and thus requiring a lower electric load.

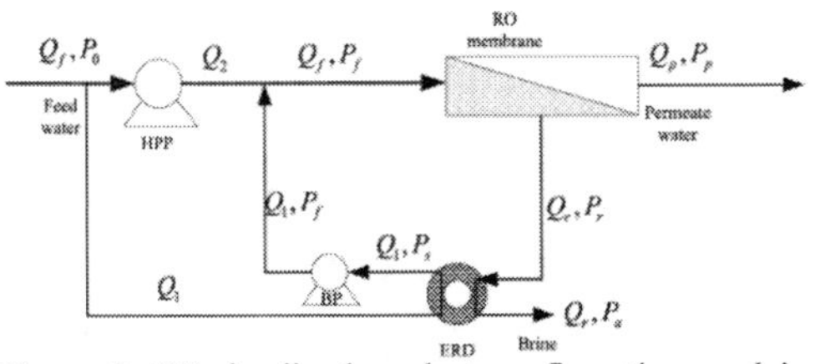

Figure 1. RO desalination plant configuration used in stand-alone PVRO scenarios **[4]**

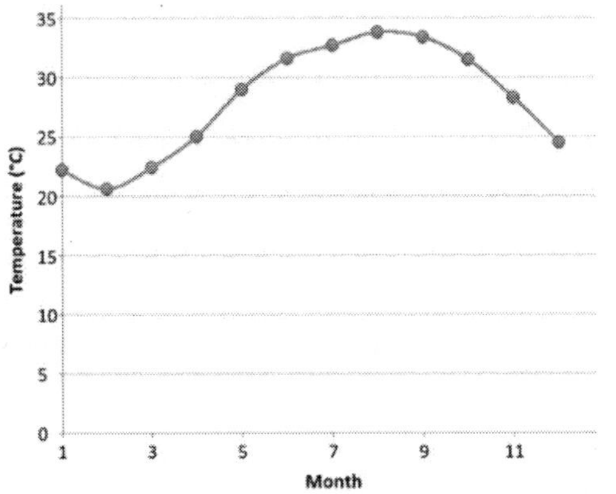

Figure 2. A monthly feedwater temperature profile was assumed for all simulations.

The high-pressure, seawater, and booster pump efficiencies were each set at 82%, 90%, and 85%. The pressure exchanger that was modeled for energy recovery was set to an efficiency of 95%. The main technical and economic parameters used in the RO model are summarized in Table II.

Table II. Summary of RO model input parameters.

Feedwater Salinity	40000 ppm
Feedwater Temperature	Seasonal Profile (see Figure)
Recovery Rate	35%
Daily Water Demand (Small-/Medium-Size Scenarios)	5/6550 m^3/day
RO Plant Lifetime	25 years
Base Unit Cost	1800 \$/m^3/day
Water Plant Owners' Cost Factor	5%
Water Plant Cost Contingency Factor	10%
Interest Rate	5%
Capital Recovery Factor	7%
Annual Water Plant Fixed Cost	2000 \$/yr
Water Storage Capital Cost	72.5 \$/m^3 [5]
Water Storage O&M Cost	2% of water storage capital cost [5]
Membrane Module Cost	990 \$/module
Membrane Module Area	37 m^2
Membrane Replacement Rate	33%
Membrane Water Permeability	4.25×10^{-4} L/m^2-bar-s
Membrane Salt Permeability	1.93×10^{-5} L/m^2-s
O&M Spare Parts Cost	0.02 \$/m^3
O&M Chemical Cost	0.0225 \$/m^3
Annual Labor Salary	36000 \$/yr-worker
Annual Management Salary	42000 \$/yr-manager

3.2 Hybrid Grid-connected Plants

Most of the recently constructed desalination plants use either thermally-driven MED or pressure-driven RO. Ongoing R&D work at KAUST shows the potential for greatly decreased the energy requirement of MED by employing nano-filtration and solar-energy driven adsorption cycles (AD) [6], [7]. This work shows the potential for reducing the water production cost to less than \$0.5/m^3. Also the hybrid technology can be retrofitted in existing thermally-driven plants. Moreover, cost reductions can be realized by adding PV in rights-of-way around existing power plants.

4 RESULTS AND DISCUSSION

4.1 Stand-alone PVRO Plants

For small-scale PVRO scenarios, a constant water demand of 5 m3/day was satisfied by either the 8, 10, 12, or 24 hour scenario, corresponding to RO capacities of 15, 12, 10, and 5 m^3/day, respectively; electric loads follow the same correspondence. With the exception of PV-alone scenarios, these operational scenarios mean that the RO plant will run for the given number of hours. However, for PV-alone cases, the electric load was allowed to extend for additional day hours to maximize PV power production for as many hours during the year as possible. Power systems comprising PV alone were compared with PV with batteries, and water storage was estimated in both configurations. Furthermore, for every simulation, initial water storage volume was set to full capacity; water storage was sized so that 100% of the water demand was satisfied throughout the first operational year. To add another dimension to the small-scale simulations, a seasonal water demand was also modeled as a function of the ratio of the monthly-average GHI over the peak GHI.

Table III shows the nominal capacities for all small-scale RO power systems. Figure 3 shows the LCOE values for these scenarios. For PV alone, the lowest LCOE is 19.4 cents/kWh, corresponding to the 8-hour, seasonal water demand scenario. The LCOW estimates and LCOW breakdown are shown in Figure 4 and Figure 5, respectively, for small-scale RO plants. The LCOW breakdown comprises RO capital cost, operational water production cost, and water storage cost, all in \$/m^3. Operational water production cost includes energy costs as well as costs for labor and maintenance. The minimum LCOW values correspond to constant water demand. For PV alone, the optimal scenario is the 8-hour RO scenario at \$1.75/m^3; for PV with batteries, the optimal scenario is the 24-hour RO scenario at \$1.14/m^3. From these results, it is evident that the balance between energy costs and the RO plant utilization rate (i.e., RO capital cost) is key to arriving at the optimal system. In this case, although PV-alone scenarios showed lower LCOEs than those of PV-battery scenarios, the RO plant utilization (fraction of annual load met) of a 24-hour PV-battery system leads to a more significant reduction in the LCOW.

Table III. Nominal capacities for small-scale scenarios

System	Scenario	Nominal Capacity	
		PV (kW)	Battery (kWh)
PV Only, Seasonal Load	8-hour	5	--
	10-hour	5	--
	12-hour	7	--
PV Only, Constant Load	8-hour	6.5	--
	10-hour	6	--
	12-hour	9.5	--
PV w/ Batteries, Seasonal Load	8-hour	3.25	12
	10-hour	3.5	8
	12-hour	2.75	12
	24-hour	3	14
PV w/ Batteries, Constant Load	8-hour	3.5	16
	10-hour	3.5	14
	12-hour	3.5	14
	24-hour	3	20

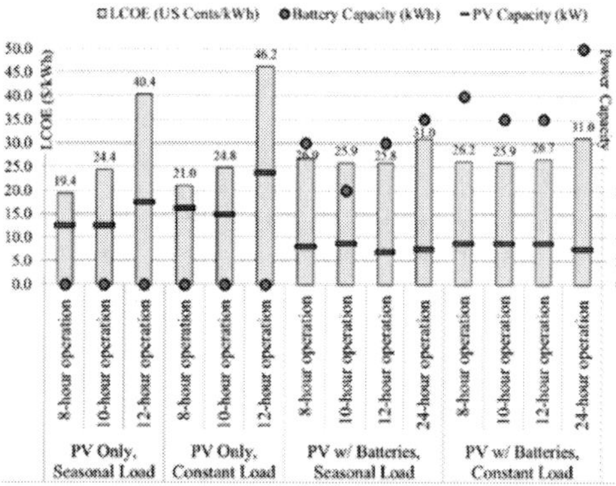

Figure 3. LCOE values and power system capacities for small RO plant scenarios

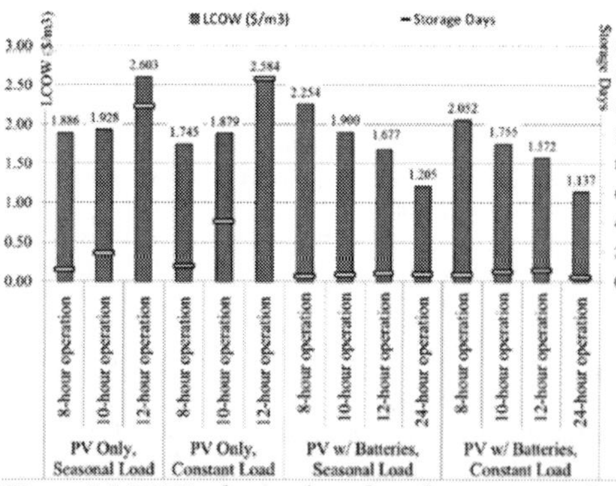

Figure 4. LCOW values and number of water storage days for small RO plant scenarios

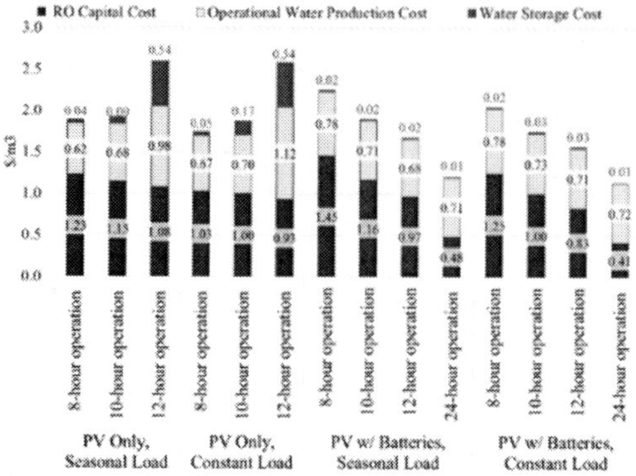

Figure 5. RO capital, operational, and water storage cost contributions to LCOW for small-scale scenarios

For medium-scale PVRO scenarios, a constant water demand of 6,550 m3/day is met by either the 8, 10, 12, or 24 hour scenario, corresponding to RO capacities of 19650, 15720, 13100, and 6550 m^3/day, respectively. Furthermore, power systems comprising PV with batteries were compared with systems powered by PV, batteries, and diesel generators (water storage included for all simulations). Table IV shows the nominal capacities for these power systems; the corresponding LCOE values are shown in Figure 6. It is shown that PV, batteries, and diesel generator systems result in lower LCOEs then simply PV with batteries, regardless of whether 8 or 70 ¢/L is assumed for diesel fuel cost. PV/battery systems require ~2 – 27 times the amount of nominal battery capacity than PV/battery/diesel systems (considering both fuel costs). Considering the relatively low battery capital costs that we assumed ($250/kWh) and a battery lifetime of 15 years, these results show batteries to be less cost-competitive than diesel generators. LCOW values and the LCOW breakdown are shown in Figure 7 and Figure 8, respectively, for medium-scale RO plants. These results indicate that energy costs cause the main disparity between water producion costs for PV/battery systems compared to PV/battery/diesel systems; furthermore, increasing the operational availability (e.g., from 8-hour to 24-hour) yields a decrease in LCOW due to the decrease in RO capital cost, which is an artifact of increased RO plant utilization.

Table IV. Nominal capacities for medium-scale scenarios

System	Scenario	Nominal Capacity		
		PV (MW)	Battery (MWh)	Genset (MW)
PV w/ Batteries	8-hour	5.6	18.8	--
	10-hour	6	15	--
	12-hour	5	18.4	--
	24-hour	5	23	--
PV, Batteries, & Diesel, 8-¢/L	8-hour	3.2	1.8	2.1
	10-hour	3	0.7	1.6
	12-hour	2.4	2	1.3
	24-hour	1.1	0.85	0.56
PV, Batteries, & Diesel, 70-¢/L	8-hour	3.5	2.3	2.1
	10-hour	4	0.7	1.6
	12-hour	3.1	3.7	1.3
	24-hour	2.5	10.2	0.56

Therefore, the minimum LCOW values correspond to each of the 24-hour scenarios: 0.88, 0.98, and 1.15 $/m3 for

PV/battery/diesel (8 ¢/L), PV/battery/diesel (70 ¢/L), and PV/battery, respectively.

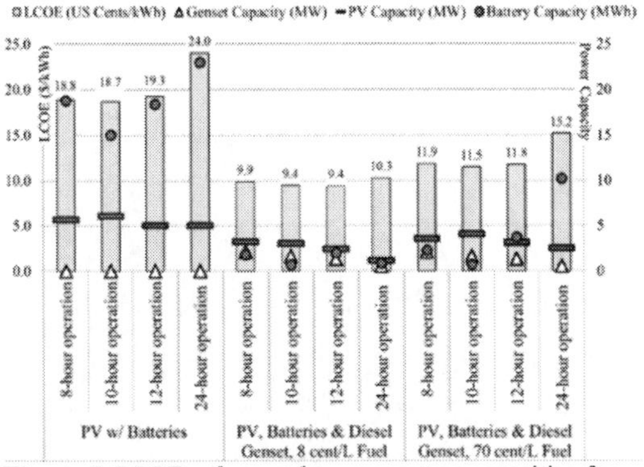

Figure 6. LCOE values and power system capacities for medium RO plant scenarios

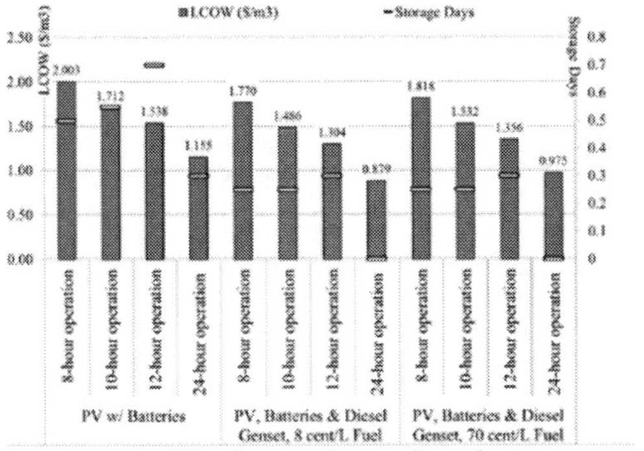

Figure 7. LCOW values an the number of water storage days for medium RO plant scenarios

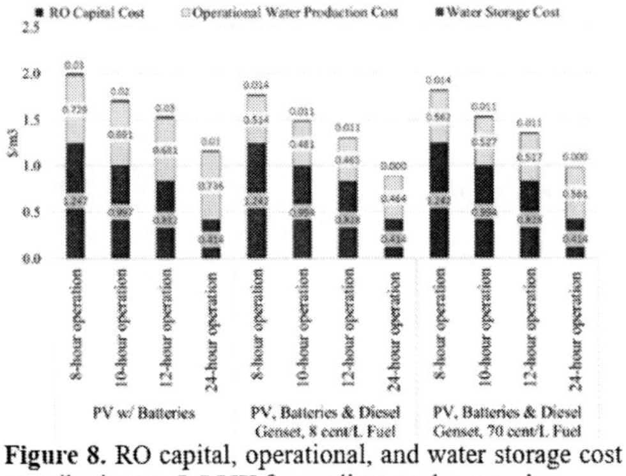

Figure 8. RO capital, operational, and water storage cost contributions to LCOW for medium-scale scenarios

4.1 Hybrid Grid-connected Solar Desalination Plants

The energy cost is the major constituent of the operating cost of water desalination and inexpensive solar energy makes desalination increasingly the most affordable option for providing fresh water in arid areas. The PV industry has accomplished phenomenal cost reductions through improving efficiency in the stages of manufacturing and deployment. The prices of photovoltaic modules have been reduced by a factor of three over the last five years, and together with increased system efficiencies, have brought the cost of solar electricity down to levels in parity with subsidized electricity rates in areas of high solar irradiation in the Middle East and other arid and sunny regions. On a non-subsidy basis, the cost of electricity from photovoltaics has become compatible to that of electricity in the GCC countries; this is exemplified by a contract to provide PV electricity for 5.84 US₵/kWh in Dubai. The residential electricity rates in Abu Dhabi are 6 US₵/kWh and they do not reflect full cost recovery. With current subsidized rates, adding 10% of PV electricity into an existing RO plant would influence the cost per cubic meter of water produced by 2.5%, while reducing the carbon emissions of the plant by 10%. Another possibility is the integration of PV and solar thermal utilizing recent developments in adsorption desalination technologies (Figure 9).

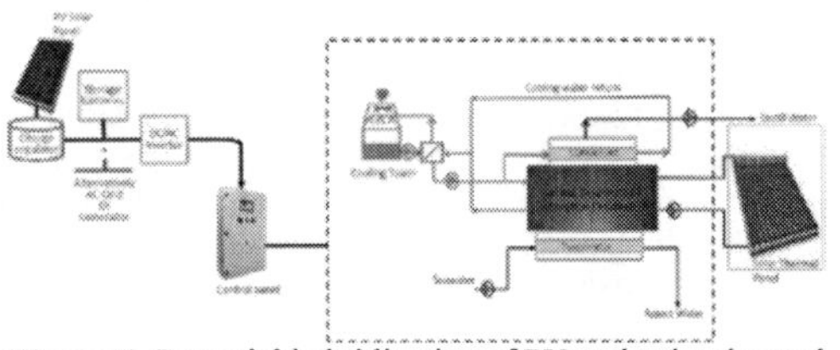

Figure 9. Potential hybridization of PV and solar thermal technologies

Also PV-RO can be combined in series with MED-AD [6], [7]. The MED-AD can take in RO reject brine because the evaporator can handle TDS up to 250,000 ppm, thereby increasing the water recovery to 85%. The brine is then transferred to a crystallizer to form salts. The electric-equivalent energy consumption of MED with improvements in exergy recovery can be reduced to less than 2.5 kWh/m^3[8].

The deterministic variability of solar energy is mitigated to a large degree by using tracking PV and the stochastic one is mitigated by power controls in modern solar farms that provide power factor control, voltage control, active power control, ramp-rate control and frequency droop response [9].

PV-hybrid systems could make a significant impact by allowing fuel-independence and reducing CO2 emissions while meeting water demand. However, there are institutional barriers related to solar desalination, and have mainly to do with introducing renewable energy in the region [10]. New policy initiatives, like Saudi Arabia's intent to replace fossil fuels with solar energy in most desalination plants by mid-century, the Masdar-led Initiative of RE deployment in the UAE, and the Global Clean Water Desalination Alliance [11], are the type of drivers needed for long-term sustainable growth in industry. Photovoltaic electricity is cost-competitive in the GCC based on direct costs (LCOE) and is likely cheaper than fossil fuel when full cost-accounting for resource (fuel and water) depletion and environmental impacts is considered [12], [13].

Conventional power generation can be integrated with solar power generation (solar thermal and photovoltaic) as part of the overall generation mix for centralized power plants. There are examples, such as Al Abdaliya Integrated Combined Cycle Solar (ICCS) power plant in Kuwait where 60 MW of solar thermal generation was added to a conventional 220 MW power plant. Solar thermal troughs have been used to generate steam which would consequently be injected in the low power steam cycle for combined cycle power plant and hence increase the power output capability and/or save fuel for the same capacity value. ICCS plants need thermal storage to allow for steam to be provided during off solar hours and even then require over sizing for days when there are low DNI values for the location or there are dust storms which severely impact their performance.

Solar PV power plants on the other hand represent an easier integration from a system stand point with conventional power plants; be it liquid or gas fired power plants. The case for hybrid PV power plants on the utility scale is gaining more and more momentum in recent years, driven by the fluctuation in the hydrocarbon pricing and the constant drop in solar PV LCOEs. Also, there is a greater realization that in large area electrical grids, the benefits of utility scale solar PV are maximized in terms of reducing electrical losses and reducing or eliminating the need for grid infrastructure upgrades by co-locating solar PV power plants next to or in the vicinity of conventional hydrocarbon based centralized power plants.

The value of integrating the simple solar output of PV power plants is also based on the fact that with little output profile adjustments, there is a direct complementarity between the output of thermal power plants and solar PV plants based on the reduction in ISO efficiency of the engines/turbines in these power plants during midday temperatures especially in areas like the GCC (Figure 10 The same complementarity applies to geothermal and PV as heat exchange is less efficient during the hot ambient hours when solar is at its best.

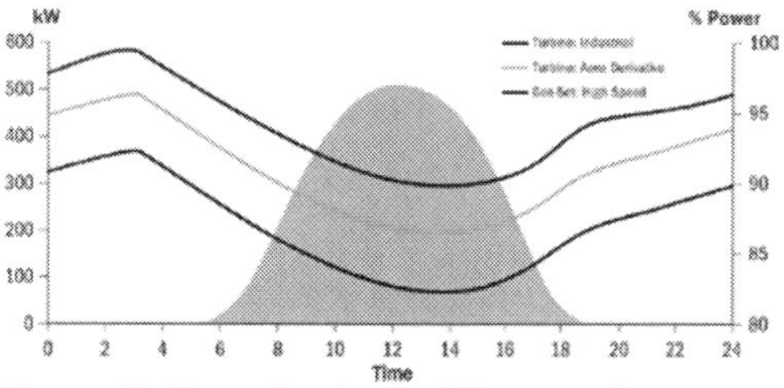

Figure 10. Thermal and solar PV power plant outputs illustrate a complementary relationship due to mid-day thermal power efficiency reductions .

There are three main synergies in co locating solar PV power plants with centralized power plants; those are reduction in O&M costs, electricity grid connection, and land use.

- Operation and Maintenance:
  - The O&M cost for hybrid PV power plants can be up to 50% lower than stand-alone solar PV power plants.
  - Solar PV plant scheduled and preventative maintenance can be coordinated with main conventional power plant.
  - There is minimal need to add any head count for supporting the O&M of co-located solar PV with other power plants.
- Electrical Grid Connection:
  - Overall capacity nameplate can stay the same for the conventional power plant as solar PV generation will act mainly as a fuel saver.
  - Typically 10% to 15% is available as additional capacity on MV/HV switch yard and transmission lines

Land Utilization:

- Solar PV plants can be built in areas that are deemed unutilized, such as HV transmission rights of way and zoning areas already designated to conventional plants. For example, the Al Manakher power plant in Jordan, shown in Figure 11, has a lot of unutilized land in the right-of-way between the HV transmission lines (red lines) coming out of the switch yard of the plant which can be used for PV plants.

Figure 11. Red lines show the HV transmission lines' right-of-way and land available between the transmission for solar PV power plants.

In short, solar PV power plants are simple in terms of integration requirements to conventional power plants and this integration results in a lower LCOE than that of PV power plants that aren't co-located with thermal power plants. The graph below shows the case for a proposed solar PV power plant in Al Manakher, Jordan, whereby a meaningful drop in the LCOE is achieved by co-locating the solar plant with a conventional power plant. The monetary benefits from such co-location, include a reduction of O&M costs, almost free land, use of an existing Special Purpose Vehicle (SPV) for asset management of the plant (e.g., managing O&M, accounting, revenue, reporting), and use of an existing substation.

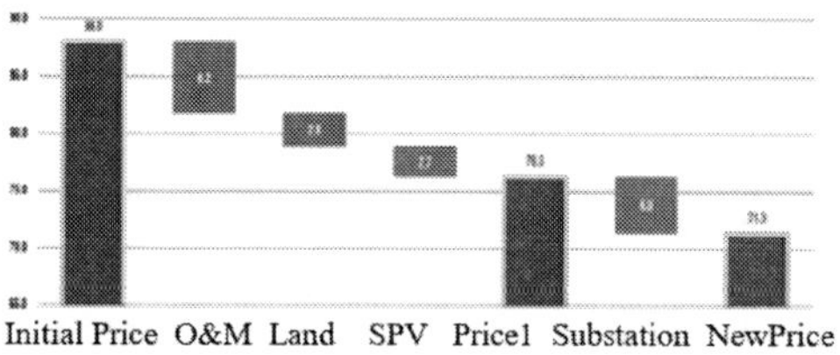

Figure 12. Price reduction due to co-locating PV and conventional power plant

The same impact on the hybrid LCOE can be derived from co locating solar PV power plants and seawater RO desalination plants where the synergies in O&M, grid connection (or feed in this case) and land utilization are achieved.

The above comparisons are based on direct cost only. A more complete cost accounting, including external fossil fuel subsidies, reveals that PV powered desalination is less expensive than that powered by diesel or natural gas.. A comprehensive study commissioned by the KAUST Industry Collaboration Program [14], showed that ground-mount PV and wind with capacity ratios of 28% are the least-cost renewable energy technologies Saudi Arabia. The LCOE from such plants is lower than the cost of electricity from oil- and gas-burning power plants if the opportunity costs (i.e., lost revenue reflecting fuel subsidies), are accounted for. The KICP 2014 study determined the LCOE of ground mounted PV to be in the range of \$0.07-\$0.15 /kWh and the LCOE of oil and gas power plants including opportunity costs to be \$0.09/kWh and \$0.18/kWh correspondingly.

4.3 Future Work

The work presented here for stand-alone PVRO systems assumes a constant load (with the exception of monthly variations) that can only be met when the power supply is equal to it or exceeds the load. In future work, modeling efforts will account for variable recovery rate, variable feed-pressure, and variable flow in order to optimize water production with a variable power supply. Also research is ongoing on synergistic combinations of renewable energy technologies for powering desalination plants without generating CO_2 emissions.

5 ACKNOWLEDGMENT & DISCLAIMER

The work of Adam Atia is supported by the National Science Foundation Graduate Research Fellowship Program under Grant No. DGE-11-44155. Vasilis Fthenakis and Raed Bkayrat serve at the board of the Global Clean Water Desalination Alliance and they acknowledge useful conversations with multiple members of the Alliance. Any opinions, findings, and conclusions or recommendations expressed in this material are those of the authors and do not necessarily reflect the views of the National Science Foundation or the Global Clean Water Desalination Alliance.

6 REFERENCES

[1] E. Delyannis, "Historic background of desalination and renewable energies," *Solar Energy*, vol. 75, no. 5, pp. 357–366, Nov. 2003.
[2] GWI, "Desalination Markets 2016," Global Water Intelligence, Oxford, UK, 2016.
[3] "HOMER - Hybrid Renewable and Distributed Generation System Design Software." [Online]. Available: http://www.homerenergy.com/. [Accessed: 30-Jun-2014].
[4] A. Jiang, J. Wang, L. T. Biegler, W. Cheng, C. Xing, and Z. Jiang, "Operational cost optimization of a full-scale SWRO system under multi-parameter variable conditions," *Desalination*, vol. 355, pp. 124–140, Jan. 2015.
[5] U. Caldera, D. Bogdanov, and C. Breyer, "Local cost of seawater RO desalination based on solar PV and wind energy: A global estimate," *Desalination*, vol. 385, pp. 207–216, May 2016.
[6] M. Wakil Shahzad and Kim Choon Ng*, "On the road to water sustainability in the Gulf," *Nature Middle East*, Apr. 2016.
[7] K. C. Ng, K. Thu, S. J. Oh, L. Ang, M. W. Shahzad, and A. B. Ismail, "Recent developments in thermally-driven seawater desalination: Energy efficiency improvement by hybridization of the MED and AD cycles," *Desalination*, vol. 356, pp. 255–270, Jan. 2015.
[8] M. W. Shahzad, K. Thu, K. C. Ng, and C. WonGee, "Recent development in thermally activated desalination methods: achieving an energy efficiency less than 2.5 kWhelec/m3," *Desalination and Water Treatment*, vol. 57, no. 16, pp. 7396–7405, Apr. 2016.
[9] M. Morjaria, D. Anichkov, V. Chadliev, and S. Soni, "A Grid-Friendly Plant: The Role of Utility-Scale Photovoltaic Plants in Grid Stability and Reliability," *IEEE Power and Energy Magazine*, vol. 12, no. 3, pp. 87–95, May 2014.
[10] S. Sgouridis, A. Abdullah, S. Griffiths, D. Saygin, N. Wagner, D. Gielen, H. Reinisch, and D. McQueen, "RE-mapping the UAE's energy transition: An economy-wide assessment of renewable energy options and their policy implications," *Renewable and Sustainable Energy Reviews*, vol. 55, pp. 1166–1180, Mar. 2016.
[11] Global Clean Water Desalination Alliance, "H2O Minus CO2 - Concept Paper," 2015. [Online]. Available: http://www.clca.columbia.edu/Global-Water.pdf.
[12] V. Fthenakis, A. A. Atia, M. Perez, A. Florenzano, M. Grageda, M. Lofat, S. Ushak, and R. Palma, "Prospects for photovoltaics in sunny and arid regions: A solar grand plan for Chile -Part I-investigation of PV and wind penetration," in *Photovoltaic Specialist Conference (PVSC), 2014 IEEE 40th*, 2014, pp. 1424–1429.
[13] A. A. Atia, V. Fthenakis, and R. Bkayrat, "Techno-economic evaluation of stand-alone, PV-powered, seawater desalination plants in Saudi Arabia," in *Proceedings of the 29th European Photovoltaic Solar Energy Conference*, Amsterdam, 2014, pp. 4075–4080.
[14] KICP, "KAUST Industry Collaboration Program (KICP) Strategic Study: Appraisal and Evaluation of Energy Utilization and Efficiency in Saudi Arabia: Supply and Demand Impacts, Business Opportunities and Technological and Economic Consideration," 2014.

THE IMPACTS OF THE INCREASING COSTS OF ELECTRICITY TARIFFS ON THE ECONOMIC FEASIBILITY OF DISTRIBUTED GENERATION WITH PV SYSTEMS IN BRAZIL

João P. C. Nascimento[2], Luis C. M. Blasques[1], André C. Nascimento[1], João T. Pinho[2]
[1] Grupo de Estudos Estratégicos em Energia Elétrica (G4E), Instituto Federal do Pará (IFPA), Brazil
Av. Almirante Barroso, 1155 – Marco, Zip Code 66.093-020, Belém, Pará, Brazil. Phone: +55-91-98111-4489
E-mail: blasques@ifpa.edu.br
[2] Grupo de Estudos e Desenvolvimento de Alternativas Energéticas (GEDAE), Universidade Federal do Pará (UFPA) PO Box 6010, Zip Code 66.025-772, Jurunas, Belém, Brazil

ABSTRACT: Despite the considerable increase of grid-connected photovoltaic systems' (GCPVS) applications in Brazil, since the publication, in 2012, of Normative Resolution number 482, that regulates the grid-connection of micro and mini distributed generation systems, economic issues are still the main obstacle to the faster dissemination of GCPVS installations in the country. In this context the present paper analyses the discounted payback period of GCPVS installed in the whole country, considering the variation of the electricity tariffs applied by the Brazilian utilities and the installation's investment costs. The study is applied to residential and commercial small-scale PV systems, connected to low voltage grids, and is accomplished with the aid of a software developed by the authors. The analysis considers the reality of the 27 Brazilian states, including Distrito Federal, the country's capital, selecting the main utility of each state. The results indicate a strong dependence of the system's economic feasibility on the electricity tariffs applied and on the investment costs, varying from an unfeasible situation to a payback period of less than 5 years, which is very positive considering the present scenario. The results also make possible a construction of a GCPVS economic feasibility Brazilian map and a three-dimensional graph, indicating the states and the conditions that present the best economic scenarios for the installation of GCPVS.
Keywords: Small Grid-connected PV Systems, Economic Analysis, Utilities, PV Market.

1 INTRODUCTION

Brazil is undergoing recent changes in the economic feasibility of distributed generation with grid-connected photovoltaic systems (GCPVS). These changes have two main factors: the publication of Normative Resolution number 482 in 2012 [1], that regulates the grid-connection of micro and mini distributed generation systems, and the crisis in the national electricity sector that caused a huge increase in the electricity tariffs applied by all Brazilian utilities in 2014 and 2015. Furthermore, a considerable variability in the investment costs, mainly associated to the PV modules' costs, contributes to an even more uncertain scenario.

Based on these assumptions, the present study focuses on two main aspects which are causing this irregular variation on the economic feasibility of GCPVS. First, the variation of the installation costs, either by the market expansion in Brazil, which tends to reduce costs, or by the recent increase of the exchange rates, which tends to increase final installation costs. Second, the significant increase in electricity tariffs applied by the Brazilian electricity utilities in the last two years, which is positive to accelerate the transition from a conventional generation-distribution relationship to a distributed generation scenario.

The analyses are made with the support of a software developed by the authors [2], which considers all the variables that affect the economic feasibility of GCPVS. The most important are: (i) the tariffs applied by the utilities and their possible annual increase rates; (ii) the investment costs of the PV system; (iii) the taxes that affect the final tariffs; and (iv) the load profile of the building where the system is to be installed.

The first two variables are selected to be the focus of this study, due to their stronger impact, especially at the present moment, when Brazil undergoes a big economic crisis. The electricity sector is in the spotlight of this crisis, since the country is strongly dependent on hydropower and in the last two years the Brazilian regions where some of the most important hydroelectric plants are located suffered an impactful drought. This motivated the Brazilian Government to stablish, in the beginning of 2015, a "tariff flag system" that imposes an increment to the normal tariffs applied by the utilities. Besides, the tariffs presented a cumulative increase of 42% from the beginning of 2014 to the middle of 2015 [3] (see Fig. 1).

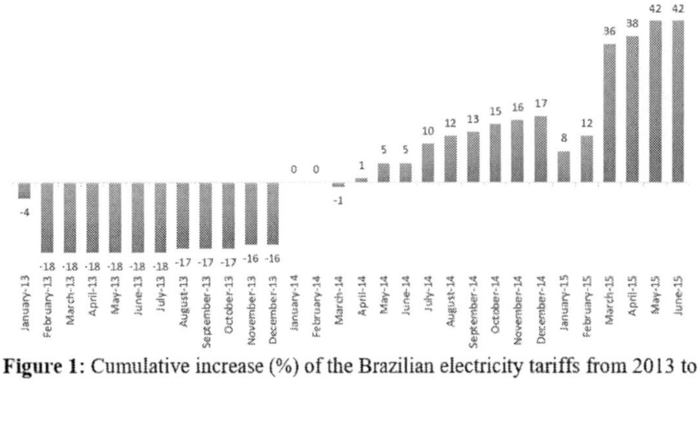

Figure 1: Cumulative increase (%) of the Brazilian electricity tariffs from 2013 to the middle of 2015 [3].

The economic crisis also affected the exchange rates. In the period from 2013 to 2015 the Brazilian currency, Real, experienced a strong devaluation. One American dollar was equivalent to two Reais in the beginning of 2013, and nowadays the rate is almost four Reais, an increase of nearly 100%. Since the PV market in Brazil is still strongly dependent on imported equipment, this also caused a negative impact on the investment costs of PV systems in Brazil.

Based on all these issues, the present work presents a comprehensive scenario of the economic feasibility of GCPVS in all Brazilian regions. The study takes the investment costs of the PV systems and the annual increase rates of the electricity tariffs as the base variables.

2 ECONOMIC FEASIBILITY ANALYSIS

This section presents the economic analysis based on a realistic scenario considering technical aspects (solar resource of each considered city) and the assumptions that cause the main fluctuations on the payback period.

The designed PV system considers the solar resource of each of the 27 Brazilian states (including Distrito Federal) and the real tariffs applied by the main utilities of each state. The load scenario is the same for all cities, a typical residence with annual consumption of 6,758 kWh, which indicates a different value of PV power installed (P_{inst}) for each city, according to its solar potential. Table 1 presents the results of the study, with the discounted payback period, in years (gray cells), for each considered scenario. The steps of the annual increase rates of the electricity tariffs (IT) are 3, 15 and 30%, and the initial investment costs of the PV system vary from US$ 1,250 to US$ 3,750, with an average value of US$ 2,500. The period of analysis is 25 years, the interest rate is 12% and the unfeasible (UN) fields of the table represent situations where the investment returns after this period.

Table I: Results of the economic feasibility study for all Brazilian states and their main utilities (gray cells represent the payback periods, in years).

Utility / State	IT %	Investment Costs US$/kWp			P_{inst} kWp
		1,250	2,500	3,750	
AES-SUL / Rio Grande do Sul	3	7.92	UN	UN	3.59
	15	6.33	12.5	22.08	
	30	5.5	9.33	13.58	
AmE / Amazonas	3	9.5	UN	UN	3.78
	15	7.25	14.92	UN	
	30	6.08	10.67	15.92	
BOA VISTA/ Roraima	3	13.33	UN	UN	3.82
	15	8.92	20.92	UN	
	30	7.25	13.17	21.5	
CEA / Amapá	3	UN	UN	UN	3.82
	15	13	UN	UN	
	30	9.58	19.83	UN	

Table I: (cont.)

Utility Company / State	IT %	Investment Costs US$/kWp			P_{inst} kWp
		1,250	2,500	3,750	
CEAL / Alagoas	3	8.33	UN	UN	3.38
	15	6.58	13.08	24	
	30	5.67	9.67	14.25	
CEB / Distrito Federal	3	7.75	UN	UN	3.32
	15	6.25	12.25	21.33	
	30	5.42	9.17	13.33	
CELESC / Santa Catarina	3	9.67	UN	UN	4.07
	15	7.33	15.17	UN	
	30	6.17	10.75	16.17	
CELG / Goiás	3	6.58	17.83	UN	3.25
	15	5.5	10.42	16.92	
	30	4.83	8.17	11.58	
CELPA / Pará	3	7	20.75	UN	3.69
	15	5.83	11.08	18.5	
	30	5.08	8.58	12.25	
CELPE / Pernambuco	3	8.33	UN	UN	3.27
	15	6.58	13.08	23.92	
	30	5.67	9.67	14.17	
CEMAR / Maranhão	3	7.25	22.67	UN	3.57
	15	6	11.5	19.42	
	30	5.17	8.83	12.67	
CEMIG / Minas Gerais	3	7.5	24.25	UN	3.94
	15	6.08	11.75	20.17	
	30	5.25	8.92	12.92	
EPISA / Piauí	3	7	20.83	UN	3.2
	15	5.83	11.17	18.58	
	30	5.08	8.58	12.25	
CERON / Rondônia	3	7.58	25	UN	3.25
	15	6.17	11.92	20.58	
	30	5.33	9.08	13.08	
COELBA / Bahia	3	9.17	UN	UN	3.5
	15	7	14.33	UN	
	30	5.92	10.33	15.42	
COELCE / Ceará	3	7.58	25	UN	3.35
	15	6.17	11.92	20.58	
	30	5.33	9.08	13.08	
COPEL / Paraná	3	9.75	UN	UN	4.04
	15	7.33	15.33	UN	
	30	6.17	10.83	16.33	
COSERN / Rio Grande do Norte	3	7.75	UN	UN	3.07
	15	6.25	12.25	21.33	
	30	5.42	9.17	13.33	

Table I: (cont.)

Utility Company / State	IT %	Investment Costs US$/kWp			P_{inst} kWp
		1,250	2,500	3,750	
ELETRO ACRE / Acre	3	7.42	24	UN	3.66
	15	6.08	11.75	20.08	
	30	5.25	8.92	12.83	
ELETRO PAULO / São Paulo	3	11.17	UN	UN	4.47
	15	8	17.33	UN	
	30	6.58	11.75	18.25	
CEM / Mato Grosso do Sul	3	6.92	20	UN	3.32
	15	5.75	10.92	18.17	
	30	5.08	8.5	12.08	
EMT / Mato Grosso	3	7.42	23.67	UN	3.54
	15	6	11.67	19.92	
	30	5.25	8.92	12.83	
EPB / Paraíba	3	7	20.75	UN	3.1
	15	5.83	11.08	18.5	
	30	5.08	8.58	12.25	
ESCELSA / Espírito Santo	3	7.75	UN	UN	3.52
	15	6.25	12.25	21.5	
	30	5.42	9.25	13.42	
ESE / Sergipe	3	8.33	UN	UN	3.44
	15	6.58	13.17	24.08	
	30	5.67	9.75	14.25	
ETO / Tocantins	3	7.92	UN	UN	3.57
	15	6.33	12.5	22.17	
	30	5.5	9.42	13.67	
LIGHT / Rio de Janeiro	3	6.67	18.33	UN	3.79
	15	5.58	10.58	17.17	
	30	4.92	8.25	11.67	

The next section presents a detailed discussion of the results of Table I. To support the analysis, a more comprehensive study is developed for one of the electricity utilities of Table I, CELPA, the electricity utility of the state of Pará. This detailed analysis considers a variation of the main parameters in shorter steps, in order to present results for a more extensive scenario, which are presented in the graph of Figure 2. The investment cost varies from US$ 1,250 to US$ 3,750 per kWp, the same values considered in Table I, but in this graph in steps of US$ 250. The three axes of the graph represent the payback period (result of the analysis), the interest rate (varying from 6 to 15% per year with steps of 3%), and the annual increase rates of the electricity tariffs (varying from 0 to 30% per year with steps of 3%).

3 RESULTS AND DISCUSSION

The results presented in Table 1 can motivate several discussions. First, one topic that is not the main subject of this work, is the variation of the PV installed power to supply the same load. The cities with lower solar potential present the systems with higher installed PV power, and the contrary is true for the cities with the highest solar potential. Moving on the analysis to the two main questions here discussed, the table cells with the lower investment costs, US$ 1,250 per kWp installed, are all feasible, except for the state of Amapa (CEA utility), which is coherent since this utility presents the lowest electricity tariff in the whole country. The lower payback period is verified for the state of Goiás that combines a good solar potential with a high electricity tariff. This is achieved for the scenario of investment costs of US$ 1,250 per kWp and 30% of annual increase rate of the electricity tariff (this scenario is called here "optimistic scenario"). However, for this same state, the here called "average scenario", with investment costs of US$ 2,500 and 15% of annual increase rate of the electricity tariff, indicates a payback period of 10.42 years, which is in the center of the national average nowadays. Finally, the "pessimistic scenario", with investment costs of US$ 3,750 and 3% of annual increase rate of the electricity tariff, indicates an unfeasible picture in all Brazilian sates. It is important to mention that this scenario is unlikely to occur in the country.

Figure 2 enables a more detailed analysis. It shows that for an investment cost equal to or lower than US$ 2,000 per kWp, the systems are all economically feasible, regardless of the other two parameters. It can also be noticed that for higher investment costs, the combination of high interest rates with low increase rates of the electricity tariffs, the systems' feasibility evolves exponentially to an unfeasible scenario. On the other hand, the combination of low interest rates with high increase rates of the electricity tariffs makes the systems economically feasible for any of the considered investment cost.

To summarize the results with the support of pictures, Figs. 3 and 4 show maps representing the two expected scenarios for Brazil for the next years, the average (Fig. 3) and the optimistic scenario (Fig. 4). The maps help conclude what is presently verified and what is expected to be found in Brazil in the short-term, respectively.

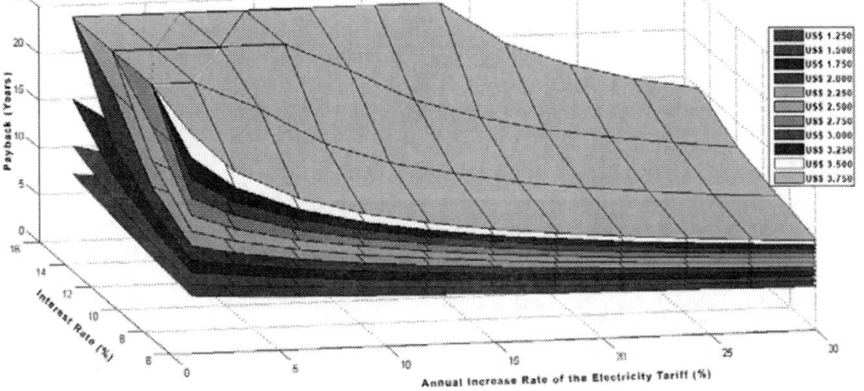

Figure 2: Graph of the payback varying with the investment costs, the interest rate and the annual increase rate of the electricity tariff.

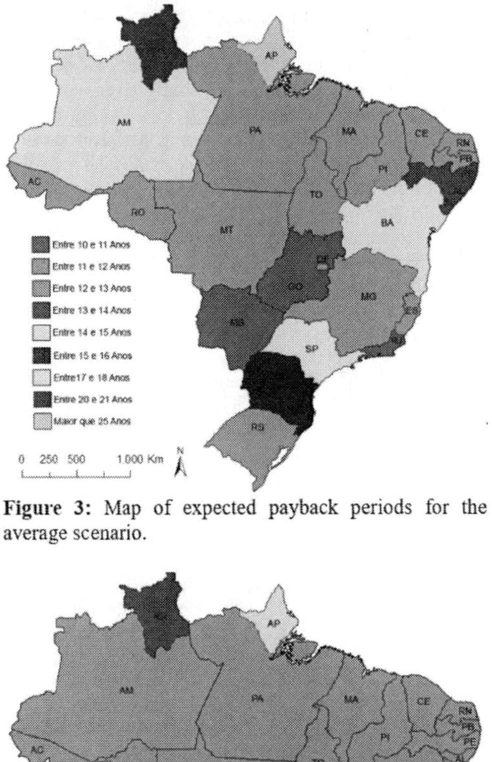

Figure 3: Map of expected payback periods for the average scenario.

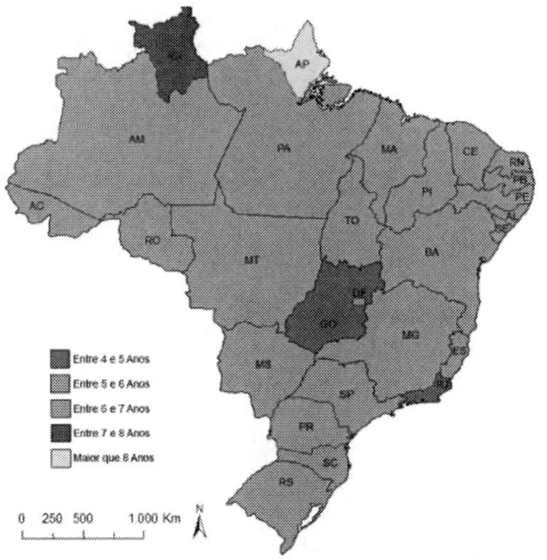

Figure 4: Map of expected payback periods for the optimistic scenario.

4 POSSIBLE POLICY MEASURES TO ACHIEVE THE OPTIMISTIC SCENARIO IN BRAZIL

It is obvious that the optimistic scenario needs strong policy measures to be a reality in Brazil in the upcoming years. Despite some advances carried out by the national electricity sector [4-6], some measures implemented in other countries [7-21] shall be considered and adopted in Brazil. A considerable progress was achieved in November 2015 with the update of Normative Resolution 482, through Normative Resolution number 687 [22]. Among the improvements brought by this latter Resolution, the most important are:

– The permission that two or more householders of the same condominium can install a single GCPVS and be all benefited from the energy credits;

– The PV energy credits, generated by the energy portion delivered to the grid and not used by the costumer in the same month had an expiration date extension from 36 to 60 months;

– The reduction of the maximum deadline the utilities have to approve the final connection of micro-PV systems (installed power equal or less than 75 kWp), from 82 to 34 days.

Another measure that has already been implemented in some of the Brazilian states and affects the economic feasibility of GCPVS is the exemption of the most significant tax (ICMS, Brazilian equivalent to VAT) in the amount of PV electricity generated and delivered to the grid. This issue was not considered by the Normative Resolution and required the signing of an agreement by the CONFAZ (Brazilian abbreviation for "National Council for Finance Policy") to allow such exemption. Unfortunately, only 14 of the 27 Brazilian states adhered to this agreement, namely Acre, Alagoas, Minas Gerais, Rio de Janeiro, Rio Grande do Sul, Goiás, Pernambuco, São Paulo, Bahia, Maranhão, Mato Grosso, Ceará, Tocantins, and Rio Grande do Norte.

Besides the mentioned measures, which are already implemented, a recent study published by the Greenpeace [4] indicates some possible public policies that could be applied to accelerate the economic feasibility of GCPVS in Brazil, which, if implemented, can surely accelerate the achievement of the here called optimistic scenario. Most of them are related to the increase of the costumers' purchase capability, since the considerable high investment costs of the PV systems are still the main obstacle to boost the GCPVS installations in Brazil.

The first measure is the possibility of workers to use the financial resources of the FGTS (Brazilian abbreviation of a workers′ unemployment fund) as investment costs to install their own GCPVS. The FGTS is a benefit that the formal contract employees receive in a special account from the employers, but can be only withdrawn in case of unemployment or retirement. The study indicates that it is more profitable to invest the resources from FGTS in a GCPVS than keeping it in the original account, which has very low profitability [4].

Another policy mentioned is the application of a discount in the IPTU (Brazilian Urban Building and Land Tax) for buildings with a GCPVS installation. This measure would cause a larger impact in big cities, where this tax is higher.

As a third measure, the study mentions the possibility of exemption of import taxes for all components of the GCPVS. By December 2015, only PV modules had this exemption and, despite this equipment being the most expensive part of the system, the inverters also have a significant share of the total cost, and nowadays the majority of the inverters are imported. The exemption of import taxes for all components would represent a decrease of 13% in the system's total costs [4].

Finally, another point highlighted by the authors of the present paper is the need of specific fund programs in Brazil to support the Brazilian citizens that want to install a GCPVS but do not have the total amount available. Even the public banks do not have loans and financing programs, specific for the PV sector. Some general initiatives are already being conducted by some banks, but there is a necessity to implement more impactful actions.

5 CONCLUSIONS

The results presented in this work are relevant to give an important view of the possible scenarios for the economic feasibility of GCPVS in Brazil. The impacts of the increasing electricity tariffs and the investment costs are significant, and even if the tariff costs stabilize, the reduction of the PV investment costs to about 1,250 US$/kWp, which is perfectly attainable in the short-term, can boost the PV installations all over the country. The authors considerer that a strong growth of the Brazilian PV market will attract the world leading PV module manufacturers to the country, helping to reduce the final costs for the consumers, which can make the "optimistic scenario" come true in the next few years.

Finally, even considering the average scenario for the present days, the perspectives are good. In this case, the payback periods vary from 10.4 years (best case) to a single case with an unfeasible scenario (Amapá State). Excluding the latter, the average payback period for the other 26 states is 12.7 years. Considering that the systems simulated are all small-scale systems, it is expected that an increase in the scale can guarantee a considerable reduction in this average payback period. The authors conclude that, considering the crisis scenario verified in the present moment in Brazil, GCPVS application in the country is feasible on average, which contradicts the results presented in other studies [6, 23].

6 REFERENCES

[1] ANEEL. Normative Resolution number 482. April 17th, 2012.

[2] J. P. C. Nascimento, L.C.M. Blasques. Software for the Analysis of the Impacts of Energy Efficiency, Electricity Tariff Systems and Taxes on the Economic Feasibility of Micro and Mini DG with GCPVS (in Portuguese: Software para Avaliação do Impacto da Eficiência Energética, do Enquadramento Tarifário e dos Impostos na Viabilidade Econômica da Micro GD Com SFCR). VI Brazilian Congress on Solar Energy, Belo Horizonte, April, 2016.

[3] DIEESE – Brazilian Inter-Trade Department of Statistic and Socioeconomic Studies. Technical note number 147 – Performance of the Brazilian Electricity Tariffs (in Portuguese: Nota Técnica 147 – Comportamento das Tarifas de Energia Elétrica no Brasil). August, 2015.

[4] GREENPEACE. Dawn - How encouraging photovoltaic solar energy can transform Brazil (in Portuguese: Como o incentivo à energia solar fotovoltaica pode mudar o Brasil). Available at <http://www.greenpeace.org/brasil/Global/brasil/documentos/2016/Relatorio_Alvorada_Greenpeace_Brasil.pdf> [accessed on 05.15.16].

[5] M. Mitscher, R. Rüther. Economic performance and policies for grid-connected residential solar photovoltaic systems in Brazil. Energy Policy 49 (2012) 688–694.

[6] C. Holdermann, J. Kissel, J. Beigel. Distributed photovoltaic generation in Brazil: An economic viability analysis of small-scale photovoltaic systems in the residential and commercial sectors. Energy Policy 67 (2014) 612–617.

[7] A. Orioli, A. D. Gangi. The recent change in the Italian policies for photovoltaics: Effects on the payback period and levelized cost of electricity of grid-connected photovoltaic systems installed in urban contexts. Energy 93 (2015) 1989-2005.

[8] J. Seel, G. L. Barbose, R. H. Wiser. An analysis of residential PV system price differences between the United States and Germany. Energy Policy 69 (2014) 216–226.

[9] M. Lee, T. Hong, C. Koo. An economic impact analysis of state solar incentives for improving financial performance of residential solar photovoltaic systems in the United States. Renewable and Sustainable Energy Reviews 58 (2016) 590–60.

[10] L. De Boeck, S. Van Asch, P. De Bruecker, A. Audenaert. Comparison of support policies for residential photovoltaic systems in the major EU markets through investment profitability. Renewable Energy 87 (2016) 42-53.

[11] K.H. Solangi, M.R. Islam, R. Saidura, N.A. Rahim, H. Fayaz. A review on global solar energy policy. Renewable and Sustainable Energy Reviews 15 (2011) 2149–2163.

[12] A. Batman, F. G. Bagriyanik, Z. E. Aygen, Ö. Gül, M. Bagriyanik. A feasibility study of grid-connected photovoltaic systems in Istanbul, Turkey. Renewable and Sustainable Energy Reviews 16 (2012) 5678–5686.

[13] S. Tudisca, A. M. D. Trapani, F. Sgroi, R. Testa, R. Squatrito. Economic analysis of PV systems on buildings in Sicilian farms. Renewable and Sustainable Energy Reviews 28 (2013) 691–701.

[14] F. Spertino, P. D. Leo, V.Cocina. Economic analysis of investment in the rooftop photovoltaic systems: A long-term research in the two main markets. Renewable and Sustainable Energy Reviews 28 (2013) 531–540.
[15] D.L. Talavera, G. Nofuentes, J. Aguilera. The internal rate of return of photovoltaic grid-connected systems: A comprehensive sensitivity analysis. Renewable Energy 35 (2010) 101–111.
[16] M. Ruppert, M. Hayn, V. Bertsch, W. Fichtner. Impact of residential electricity tariffs with variable energy prices on low voltage grids with photovoltaic generation. Electrical Power and Energy Systems 79 (2016) 161-171.
[17] S. Reichelstein, M. Yorston. The prospects for cost competitive solar PV power. Energy Policy 55 (2013) 117–127.
[18] P. P. Singh, S. Singh. Realistic generation cost of solar photovoltaic electricity. Renewable Energy 35 (2010) 563–569.
[19] K. Branker, M.J.M. Pathak, J.M. Pearce. A review of solar photovoltaic levelized cost of electricity. Renewable and Sustainable Energy Reviews 15 (2011) 4470– 4482.
[20] G. G. Pillai, G. A. Putrus, T. Georgitsioti, N. M. Pearsall. Near-term economic benefits from grid-connected residential PV (photovoltaic) systems. Energy 68 (2014) 832-843.
[21] A. Al-Salaymeh, Z. Al-Hamamre, F. Sharaf, M.R. Abdelkader aTechnical and economical assessment of the utilization of photovoltaic systems in residential buildings: The case of Jordan. Energy Conversion and Management 51 (2010) 1719–1726.
[22] ANEEL. Normative Resolution number 687. November 24th, 2015.
[23] S. Rodrigues, R. Torabikalaki, F. Faria, N. Cafôfo, X. Chen, A. R. Ivaki, H. M. Lima, F. M. Dias. Economic feasibility analysis of small scale PV systems in different countries. Solar Energy 131 (2016) 81-95.

POWER-SYSTEM-WIDE ANALYSIS OF THE BENEFITS OF RESERVE PROVISION FROM SOLAR PHOTOVOLTAICS IN SOUTH AFRICA

Tobias Bischof-Niemz[1], Joanne Calitz[2], Jarrad G. Wright[3]
[1, 2, 3] Council for Scientific and Industrial Research (CSIR), Energy Centre, P O Box 395, Pretoria 0001, South Africa
[1]Phone: +27 83 403 1108, E-Mail: TBischofNiemz@csir.co.za
[2]Phone: +27 12 842 7076, E-Mail: JRCalitz@csir.co.za
[3]Phone: +27 12 842 7269, E-Mail: JWright@csir.co.za

ABSTRACT: Reserve provision from standalone solar photovoltaic (PV) power plants has to date not been adopted in practice but has been conceptualised by a number of authors for variable renewable energy (RE) power plants like solar PV and wind. The authors have assessed the economic viability of this application in a developing country with low-levels of solar PV penetration thusfar (South Africa). The majority of power generation in South Africa is coal-fired, making up over 70% of the installed capacity. However, South Africa has abundant solar resources and has seen a dramatic decline in solar PV tariffs over the past few years. The value of reserve provision to the system from solar PV was determined by simulating the South African power system on an hourly basis in a least-cost unit commitment and dispatch model capable of co-optimising energy and reserves simultaneously. Preliminary results showed that purely reserve provision in the short term by new solar PV in South Africa is not yet economically viable as it requires a solar PV LCOE below 285 ZAR/MWh which is not likely to materialise in the short term.. However, results showed that a combination of reserve provision and energy generation by new solar PV capacity could be economically viable in the short term at solar PV LCOE below 430 ZAR/MWh. This research work is ongoing and will be expanded to include aspects of solar PV forecasting, investigations into the provision of other reserve types by solar PV simultaneously, reserve provision by other Renewable Energy (RE) technologies as well as further research into possible reserve costing and incentivisation.

Keywords: Large grid-connected PV systems, simulation, co-optimisation, cost reduction, developing countries, economic analysis, grid integration, reserves

1 INTRODUCTION

Reserve provision from standalone solar photovoltaic (PV) power plants has to date not been adopted in practice but has been conceptualised by a number of authors for variable renewable energy (RE) power plants like solar PV and wind [1]–[4]. The authors have assessed the economic viability of this application in a developing country with low-levels of solar PV penetration thusfar (South Africa). This is envisaged to create a new business case, prepare for better grid integration as well as increasing the viability of high solar PV penetration levels in future.

A brief background of the South African context, the requirement for reserves in a power system and reserve provision by solar PV is provided. This is followed by the approach taken to determine the economic viability of reserve provision by solar PV, the presentation of selected results, a brief discussion of the saliency of these results, suggested future research work and conclusions.

2 BACKGROUND

2.1 The South African context

The current South African power system is summarised in Table 1. It is predominantly coal based with small amounts of nuclear, hydro and pumped storage generation capacity. Only since 2014 has utility scale RE been deployed via the Renewable Energy Independant Power Producer Procurement Programme (REIPPPP) [5]. This has included solar PV, wind, Concentrated Solar Power (CSP), biomass, biogas and small hydro renewable energy (RE) technologies.

The REIPPPP has resulted in significant tariff reductions for solar PV in South Africa (Figure 1) to the point where planning assumptions for expected 2030 solar PV tariffs had already been reached in 2014 (reductions from 3500 ZAR/MWh to 840 ZAR/MWh in 3 years, annual average exchange rates in 2015 were EUR:ZAR: 14.2 and USD:ZAR: 12.8).

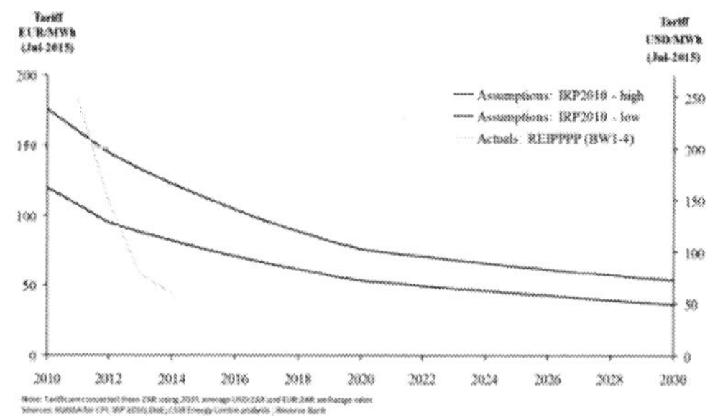

Figure 1: Actual solar PV tariffs in South Africa for bid-windows 1-4 of the REIPPPP

Electricity plans for South Africa (Figure 3) currently indicate that from the small base of 1 GW in 2015, solar PV is planned to expand to 8.4 GW of installed capacity by 2030 (~10% of total installed capacity) while providing ~2.8% of total energy [6]. However, indications in updated plans not promulgated (Figure 4) have been that solar PV could contribute 9.8 GW by 2030 and 25 GW by 2050 while contributing 4.9% and 9.6% of annual energy respectively [7].

Work in [8] showed that solar PV potential in just high density settlement areas of South Africa (rooftop solar PV) is ~73 GW and that there is ~220 GW of utility scale solar PV just in existing Environmental Impact Assessment Areas (EIAs) i.e. total of ~300 GW of easily accessible solar PV capacity.

The solar resource in South Africa is almost twice that of Germany (Figure 2). As shown in Table 1, Germany already have an installed solar PV capacity of ~40 GW [9]. South Africa currently only have ~1 GW of installed solar PV capacity.

Globally, solar PV installed capacity at the end of 2015 totalled 227 GW. This is dominated by China (45 GW), Germany (39 GW), Japan (35 GW), the United States (27 GW) and Italy (19 GW). In 2015, 50 GW of solar PV capacity was installed (mostly by China, Japan, the United States and United Kingdom) [10].

Considering the significant solar resource in South Africa, significant tariff reductions in the REIPPPP and solar PV adoption globally, the planned share for solar PV in South Africa's power system is not very ambitious and should likely be higher. Thus, the value of solar PV providing reserves (in addition to energy) is likely to be a beneficial research effort in demonstrating additional value to the power system.

Table 1: Summary of current South African and German power systems (2015)

Property	*Unit*	*RSA*	*DE*
Demand	[TWh]	~240	521
Peak demand	[GW]	34.1	83.1
Installed capacity	[GW]	54	193
Coal	[GW]	38	49
Gas/oil	[GW]	4	33
Nuclear	[GW]	2	11
Hydro (incl PS)	[GW]	5	6
Wind	[GW]	1	46
Solar PV	[GW]	1	39
CSP	[GW]	<1	-
Other	[GW]	<1	9

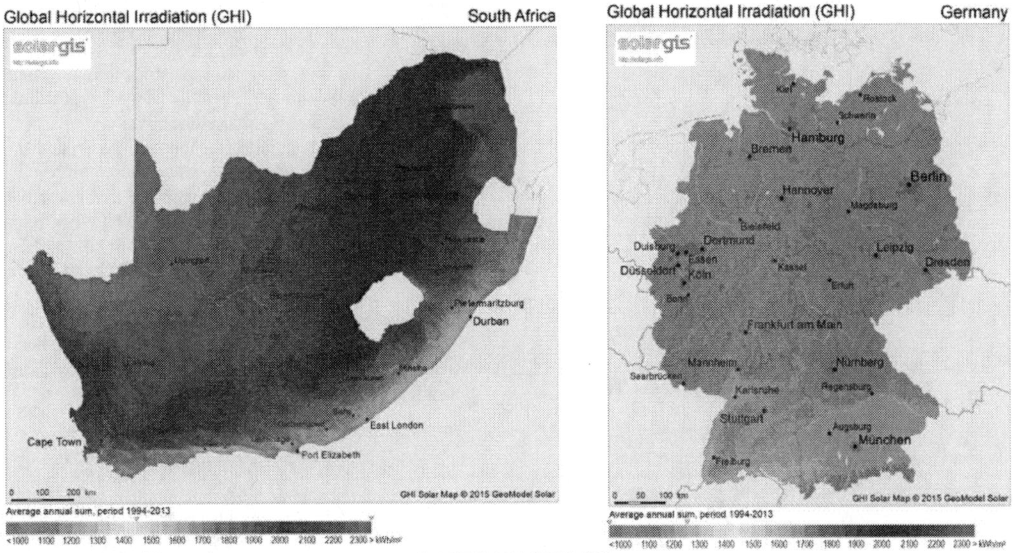

Figure 2: South Africa and Germany average annual GHI (1994-2013) [8]

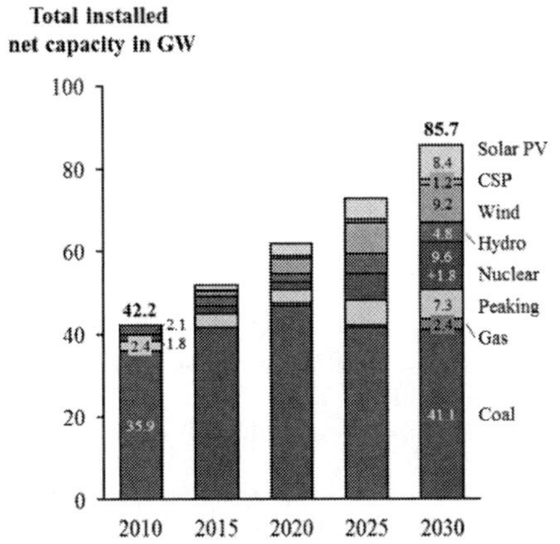

Figure 3: Power system expansion plans to 2030 in South Africa

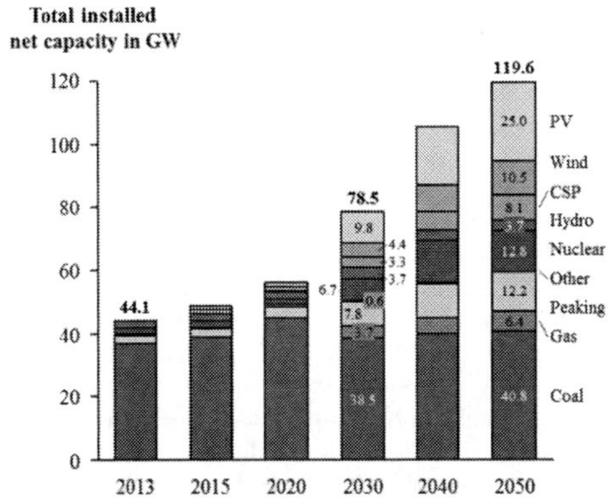

Figure 4: Updated power system expansion plans to 2050 in South Africa (not promulgated)

2.2 The requirement for reserves

The main task of any System Operator (SO) is to continuously balance electricity supply and demand in the power system. If supply and demand are not balanced at any particular time, the system frequency will deviate from the scheduled frequency (50 Hz or 60 Hz, country dependant). Significant frequency deviation could lead to damage of generators, transmission infrastructure, end-user equipment, result in under-frequency or over-frequency load shedding and possibly system collapse if insufficient action is taken. It is impossible for any SO to maintain system frequency at the exact scheduled frequency at all times as a result of uncertainties in demand forecasts and generation/ transmission availability. SOs also do not typically own generation infrastructure and as a result, it is generally accepted that the SO will maintain system frequency within a reasonable range by procuring a certain amount and type of reserves to ensure a balanced system. The nature of the three types of reserves most commonly applied and the system frequency are shown illustratively in Figure 5 (following a large disturbance i.e. a loss of supply). A similar graphic but with the inverse response from reserves (down) would represent a loss of system demand.

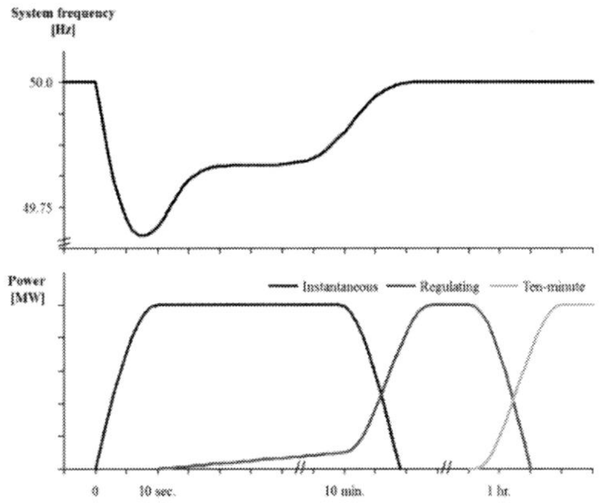

Figure 5: Illustration of system frequency and reserve response following a large system disturbance i.e. loss of supply (adjusted from [11])

Currently in South Africa, the five reserve types procured are summarised in Table 2 [12]. Operational reserves include Instantaneous (Up/Down), Regulation (Up/Down), Ten-minute and Emergency reserves (Supplemental reserves are not considered part of operating reserves). Emergency and Supplemental reserves are used very infrequently with the former only being used when the power system is in an abnormal state.

The provision of operating reserves in South Africa is mainly by coal-fired thermal power stations with smaller amounts being provided by pumped storage capacity and Open Cycle Gas Turbines (OCGTs). The drawbacks of the majority of reserves being provided by coal-fired generators are having to operate at part-load (reducing efficiency) as well as this relatively cheap generation fleet not being able to produce as much energy as possible.

Table 2: Reserve types and characteristics in South Africa

Type	*Response [min]*	*Full activation [min]*	*Sustain [min]*
Inst. Up/Down	-	10 sec.	10
Reg. Up/Down	10 sec.	10	60
Ten-minute	<10	10	120
Emergency	10-30	<30	120
Supplemental	<360	360	120

2.3 Reserve provision by solar PV

Solar PV plants can technically provide any reserve type as a result of their near instantaneous response time. Down reserves provision would be provided by the curtailment of solar PV from maximum available output while up reserves would be provided via constant curtailment and increase of output to maximum available output when activated by the SO. The primary technical constraints in this regard are:

i. Solar PV plants are operating at above zero levels i.e. it is day time.
ii. Size and geographic distribution of plants is large enough to allow for aggregation that sufficiently reduces uncertainty
iii. Contract duration of the reserve product is not too long (again, to reduce uncertainty)

Providing down reserves is feasible as it is merely the typical curtailment of solar PV production when required i.e. unless activated by the SO, there will be no effect on solar PV production. Providing up reserves would entail constant curtailment of production until activated by the SO and is thus more difficult to justify.

This study is focused only on the provision of Instantaneous Up reserves provided by solar PV plants in South Africa.

3. APPROACH AND ASSUMPTIONS

The value of reserve provision to the system from solar PV was determined by simulating the South African power system on an hourly basis in weekly steps in a least-cost unit commitment and dispatch model capable of co-optimising energy and reserves simultaneously. The simulation was only conducted to represent the short term (<5 years) with no new supply capacity assumed other than supply capacity already under construction.

Existing generator properties were modelled with publically available data and generally accepted technical assumptions were made where public data was not available. Hourly solar PV profiles for South Africa were obtained from [13] and scaled with installed capacity.

The system was modelled and total system production costs determined without any additional solar PV capacity. Following this, a pre-defined amount of solar PV capacity was added (which could provide either energy or reserves) and total system production costs were determined. This simulates the decision being considered; should a solar PV plant be built for reserve and/or energy provision in South Africa? The value added to the system with the provision of energy or reserves by the new solar PV capacity is then calculated as the total production cost difference between these two cases. The economic viability of solar PV providing reserves will be when the value provided to the system is greater than solar PV Levelised Cost of Electricity (LCOE).

System production costs include fuel costs, variable operation and maintenence (O&M) costs as well as unit start and and stop costs. All costs are in July 2015 South African rands (ZAR).

All operating reserve types were modelled as being mutually exclusive to align with [14], as shown in Table 2 and Table 3.

In three cases of new solar PV capacity being considered solar PV is assumed to be:

i. 500 MW installed capacity with no Instantaneous Up reserve provision (all energy)
ii. 500 MW of Instantaneous Up reserve provision at any time it is available
iii. 500 MW of Instantaneous Up reserve only between 10h00-15h00 daily.

The three demand scenarios considered to represent different demand growth scenarios in the short term are taken from [15] and shown in Table 4.

Table 3: South African reserve types and provision modelled

Type	*Risk, Summer peak [MW]*	*Risk, Winter peak [MW]*	*Risk, Summer off-peak [MW]*	*Risk, Winter off-peak [MW]*
Inst. Up/Down	500	500	800	800
Reg. Up/Down	500	600	500	600
Ten-minute	1200	1100	900	800

4. RESULTS AND DISCUSSION

For the demand scenarios assessed, the value of solar PV Instantaneous Up reserve provision is summarised in Table 4 and graphically in Figure 6. The simulation results showed that when the system is more constrained (higher demand scenario), there is more value provided by solar PV for both energy and reserve provision. In addition, solar PV provides more value to the system when generating than providing reserves for all demand scenarios. This ranged from 350-675 ZAR/MWh when solar PV provides no reserve (only generates) and 21-285 ZAR/MWh for 100% reserve provision. Thus, for 100% solar PV reserve provision to be economically viable, the LCOE would need to be less than 285 ZAR/MWh. For the case of a combination of solar PV energy production and reserve provision (35/65), the LCOE of solar PV would need to be at most 430 ZAR/MWh.

As can be seen, preliminary results showed that pure reserve provision in the short term by new solar PV capacity in South Africa is not yet economically viable as it requires a solar PV LCOE below 285 ZAR/MWh (not likely to materialise in the short term). However, results showed that a combination of reserve provision and energy generation by new solar PV capacity could be economically viable in the near future at solar PV LCOE below 430 ZAR/MWh.

Weekly patterns of the three reserve provision cases for solar PV is shown in Figure 7 to Figure 9. These three reserve provision cases provide 14%, 9% and 0% respectively of total Instantaneous Up reserves (of the required 6.54 TWh annually).

For illustration purposes, actual generation superimposed onto reserve provision (with available capacity) are shown for solar PV, thermal (coal-fired), pumped storage and OCGTs in Figure 10 to Figure 13 (for the case of solar PV reserve provision between 10h00-15h00 daily).

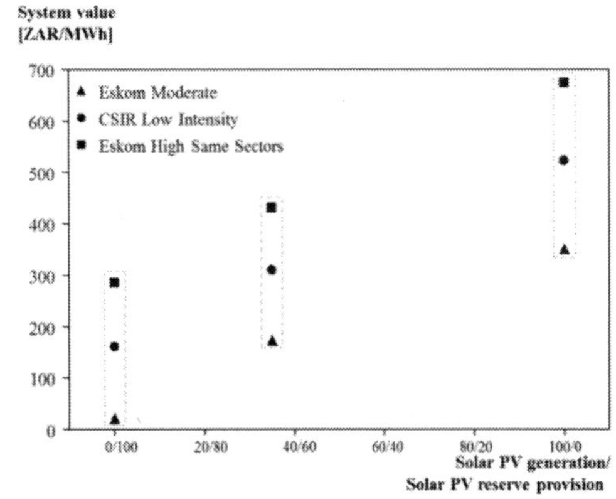

Figure 6: Range of system value provided for a range of generation to reserve provision ratios

5. FUTURE RESEARCH

The current research determined at what LCOE it is feasible for solar PV to provide reserves and/or energy. Future research will consider the case where solar PV capacity already exists and at what solar PV prices it is economically viable to provide varyinng levels of reserves.

Aspects of solar PV forecasting and the associated uncertainty which could impact the assumed amount of solar PV available for reserve provision in particular timeframes is not only critical for improved system operations but can be used to improve the business case for reserve provision by solar PV e.g. hour-ahead, day-ahead, week-ahead etc.

Investigations into the provision of other reserve types by solar PV simultaneously with Instantaneous Up reserves will be investigated. This will include Regulating and 10-Minute reserve types with up and down reserve considerations.

Similar to the research being undertaken on solar PV specifically, the provision of reserves by other RE technologies like wind and Concentrated Solar Power (CSP) are also under consideration. This will initially be considered independently (on a per technology basis) following which the integration of all RE technologies simultaneously will be pursued.

The incentive to provide reserves by any reserve provider will be driven inherently by policy and market structures. The associated incentives required are also under consideration with the specific context of South Africa in mind but generalisation to international jurisdictions with existing and planned high penetration of RE is being considered. As an example, reserve could be valued as the marginal cost to meet the next increment of reserve requirement in each interval at the optimal solution of the energy and reserves co-optimisation i.e. the shadow price of reserve provision in each interval.

6. CONCLUSIONS

The majority of power generation in South Africa is coal-fired, making up over 70% of the installed capacity. However, South Africa has abundant solar resources and has seen a dramatic decline in solar PV tariffs in recent years.

The value of reserve provision to the system from solar PV was determined by simulating the South African power system on an hourly basis in a least-cost unit commitment and dispatch model capable of co-optimising energy and reserves simultaneously.

Preliminary results showed that only reserve provision in the short term by new solar PV in South Africa is not yet economically viable as it requires a solar PV LCOE below 285 ZAR/MWh which is not likely to materialise in the short term.. However, results showed that a combination of reserve provision and energy generation by new solar PV capacity could be economically viable in the short term at solar PV LCOE below 430 ZAR/MWh.

Table 4: Summary of scenarios assessed for up to 500 MW solar PV reserve provision (Instantaneous Up)

Demand scenario	*Annual energy demand [TWh]*	*System peak demand [GW]*	*Annual reserve provision (total) [TWh]*	*Annual reserve provision (solar PV) [%]*	*Annual reserve provision (others) [%]*	*System value of solar PV [ZAR/MWh]*
Eskom Moderate	*256*	38.2	6.54	14	86	21
				9	91	174
				-	100	351
CSIR Low Intensity	*271*	40.4	6.54	14	86	161
				9	91	309
				-	100	522
Eskom High Same Sectors	*274*	41.0	6.54	14	86	285
				9	91	431
				-	100	674

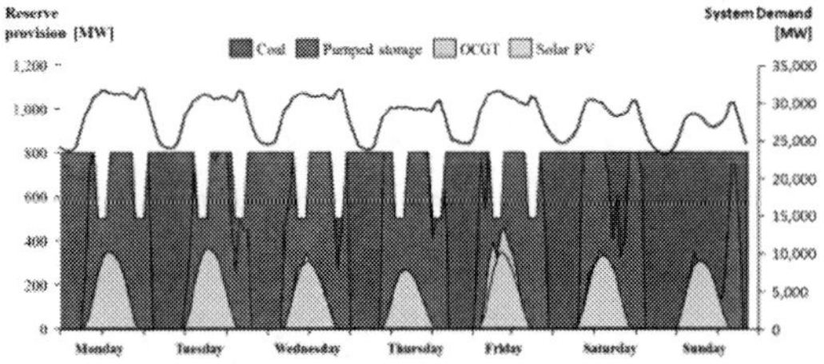

Figure 7: Sample week of reserve provision with solar PV as a reserve provider (Instantaneous Up)

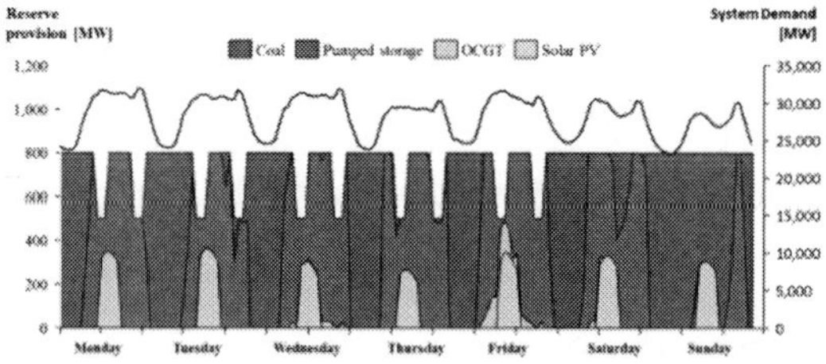

Figure 8: Sample week of reserve provision with solar PV as a reserve provider between 10h00-15h00 (Instantaneous Up)

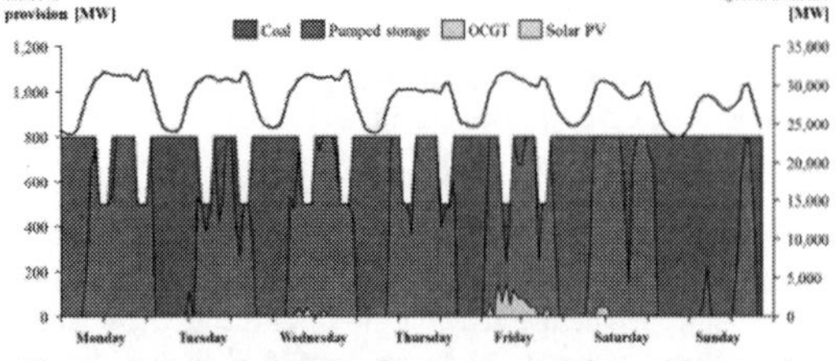

Figure 9: Sample week of reserve provision without solar PV as a reserve provider (Instantaneous Up)

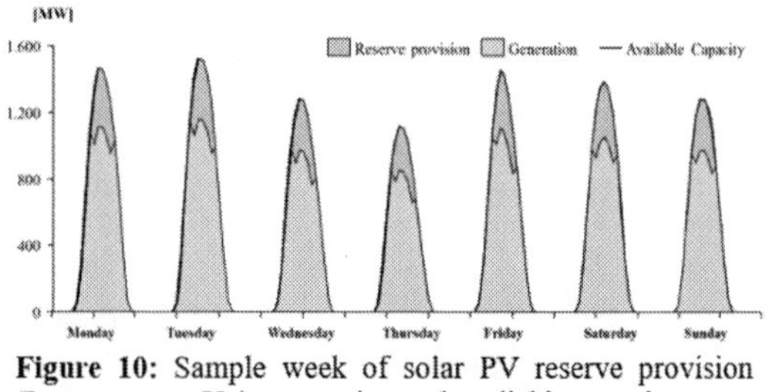

Figure 10: Sample week of solar PV reserve provision (Instantaneous Up), generation and available capacity

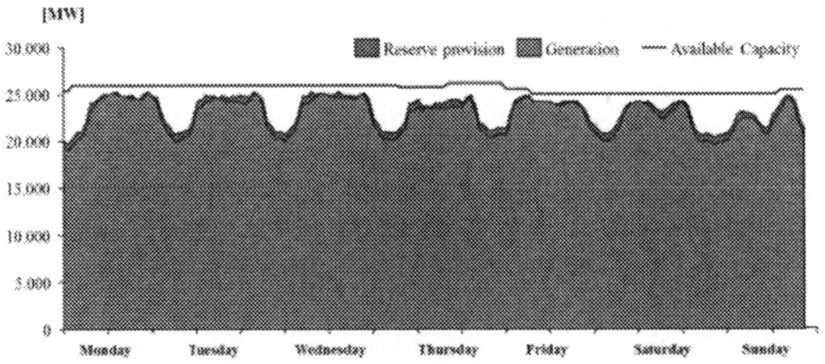

Figure 11: Sample week of thermal (coal fired) reserve provision (Instantaneous Up), generation and available capacity

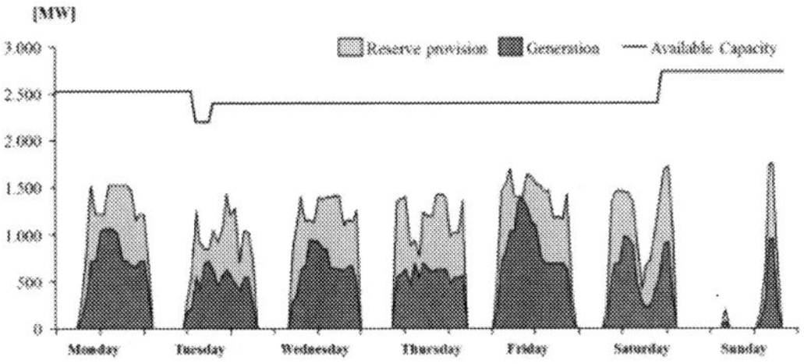

Figure 12: Sample week of pumped storage reserve provision (Instantaneous Up), generation and available capacity

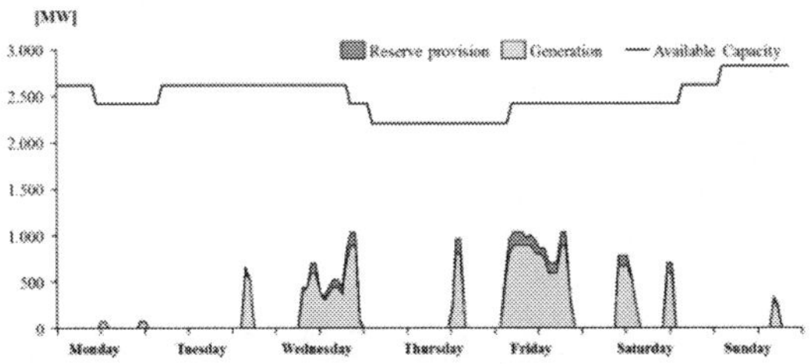

Figure 13: Sample week of OCGTs reserve provision (Instantaneous Up), generation and available capacity

REFERENCES

[1] L. Hirth and I. Ziegenhagen, "Balancing Power and Variable Renewables: A Glimpse at German Data," No. 13-154, 2013.

[2] L. Hirth and I. Ziegenhagen, "Balancing power and variable renewables: Three links," *Renew. Sustain. Energy Rev.*, vol. 50, pp. 1035–1051, 2015.

[3] B. Kirby, M. Milligan, and E. Ela, "Providing minute-to-minute regulation from wind plants," *Annu. Large-Scale Integr. Wind Power into Power Syst. Transm. Networks Offshore Wind Power Plant*, no. October, 2010.

[4] C. Fernandes, P. Frías, and J. Reneses, "Participation of intermittent renewable generators in balancing mechanisms: A closer look into the Spanish market design," *Renew. Energy*, vol. 89, pp. 305–316, 2016.

[5] Department of Energy, *State of Renewable Energy in South Africa*. 2015.

[6] South African Department of Energy, "Integrated Resource Plan for Electricity 2010-2030," 2011.

[7] South African Department of Energy, "Integrated Resource Plan for Electricity - Update Report," 2013.

[8] Council for Scientific and Industrial Research (CSIR), "Wind and Solar PV Resource Aggregation Study for South Africa," 2016.

[9] Fraunhofer ISE, "Net installed electricity generation capacity in Germany," *Energy Charts*, 2016. [Online]. Available: https://www.energy-charts.de/power_inst.htm. [Accessed: 09-Jun-2016].

[10] REN21, *Renewables 2016: Global Status Report*. 2016.

[11] J. Eto, J. Undrill, P. Mackin, R. Daschmans, B. Williams, B. Haney, R. Hunt, J. Ellis, H. Illian, C. Martinez, M. O'Malley, K. Coughlin, and K. Hamachi-LaCommare, "Use of Frequency Response Metrics to Assess the Planning and Operating Requirements for Reliable Integration of Variable Renewable Generation," no. December 2010, p. LBNL–4142E, 2010.

[12] RSA Grid Code Secretariat, "The South African Grid Code: System Operation Code (Version 9)," 2014.

[13] Council for Scientific and Industrial Research (CSIR), "Wind and Solar PV Resource Aggregation Study for South Africa," 2016. [Online]. Available: http://www.csir.co.za/Energy_Centre/wind_solar pv.html. [Accessed: 05-Apr-2016].

[14] Eskom System Operator, "Ancillary Services Technical Requirements for 2016/11-2020-21," 2014.

[15] Eskom Holdings SOC Limited, "Medium-Term System Adequacy Outlook (2016 to 2021)," 2016.

SUSTAINABLE DEVELOPMENT OF SOLAR PV SECTOR IN MENA- CASE STUDY JORDAN

Samer Zawaydeh 1,
1 Independent Engineer
Association of Energy Engineers, Assistant Director to International Membership Development Special Projects,
Masters' Student at the Environmental and Energy Engineering Master Program at German Jordanian University
PO Box 2681, Amman, Jordan
samer_zawaydeh@yahoo.com

ABSTRACT: The fluctuation on the oil prices in the international markets in 2008 coupled with the development of the PV industry led to the gradual integration of PV in the energy mix in many countries in MENA region. Stakeholders in the energy sector started developing the regulations necessary for the development of solar projects in their region. The renewable energy framework in Jordan will be presented with the learning lessons and the development that happened after implementing it for few years.

This paper will present the development of PV Policies and regulations in Jordan since 2008 and the status of installation of PV projects on utility scale and consumer levels. PV on grid and off grid is used at more than 20 applications and some of the most recent installations will be presented. The public-private partnership between the regulators, banking sector, contractors, engineers and suppliers was successful in developing a sustainable model for the continuous integration of more PV projects on grid.

Consumers are now able to meet 100% of the annual electricity needs using on grid PV projects both on site and off sites. The historical cost of generation of electricity in Jordan will be presented with the current electricity prices for different consumer sectors. The attractive payback period for different consumer sectors will be presented with learning lessons and recommendations based on the current prices of PV development.
The driving factors for the growth of the PV sector will be presented. This includes the prices of energy, the demand for more energy, and the energy intensity for a developing economy. The expected development of the electrical distribution and transmission networks development to accommodate the needs of consumers to integrate PV projects will also be presented.

With the continuous decline in the PV Levelized cost of energy, it is expected that more PV will be integrated into the electricity network. This will develop the energy services sector, contractors sector, and the industrial sector that will provide the necessary material that can be locally manufacturer.
Key findings about the development of the PV sector will show that it will contribute to the sustainable development of the energy sector and provide affordable clean energy, with reduced losses in transportation, distribution and transition. The local efforts will be part of the future energy routes and will contribute to the clean energy development in MENA.

Keywords: see PV, Projects, Regulation, Sustainability, Clean Energy

1 INTRODUCTION

The energy crisis in 2008 was the driver for change in the energy sector in Jordan. Decision makers wanted to change the energy mix and develop local energy sources like renewable energy to avoid the fluctuation of prices of energy in the international market.
The global increase in PV installation started around 2009 and it continuous to increase at double digits since then [1]
The energy status in Jordan is available in the annual report of the Ministry of Energy and Mineral Resources [2]. Jordan continues to import around 97% of its primary energy. Around 43% of the primary energy goes to transportation and remaining is used for electricity generation. The report shows that 88% of the primary energy is fossil fuel based.
Once the international oil price reached 147$/barrel [3] the country started paying five to six time more than the budgeted amount for energy. During the past six years, the average cost of energy was between 15-21% of the GDP.

2 REGULATOR FRAMEWORK

Policies in Jordan are developed by the government. The Ministry of Energy and Mineral Resources (MEMR) is responsible for the development of all Laws, Bylaws, Regulations and Instructions related to all types of energy. And the Ministry of Public Works and Housing MoPWH) is responsible for the development of the regulatory framework for buildings.
The Figure below shows a summary of the MEMR and MoPWH regulations.

2.1 MEMR

The regulatory framework started with the development of the Renewable Energy and Energy Efficiency Law (13) year 2012. This was later supported with the Bylaw (73) for the energy efficiency. The instructions for Energy Auditing were issued in 2015. Bylaw (10) is issued to provide customs and tax exemption. Bylaw (10) acts like a driver for investors and business people to work in the RE sector because there are not taxes. The Jordan Renewable Energy and Energy Efficiency Fund (JREEEF) bylaw (49) was developed for regulating projects like the renewable energy revolving loans to community based organizations. The Direct Proposal bylaw (50) was developed to support the development of RE projects by investors.

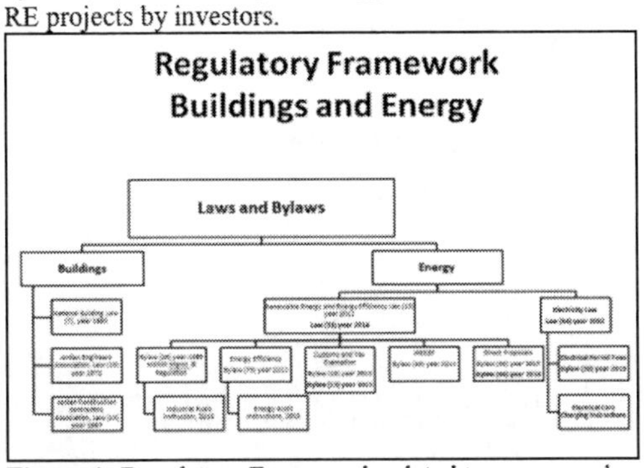

Figure 1: Regulatory Framework related to energy ordan

2.2 MoPWH
The Ministry has Laws for the regulating the Engineering Offices and Engineers in Jordan. In addition, a separate Law is issued for regulating the Contracting works.

A Solar Energy Code was issued in 2012. This will be further developed in the future to reflect the changes in technology and requirements of international and local standards.

2.3 Electricity Law
The law was issued in 2002 and it is used as reference to the connection of the new PV On Grid projects.

Jordan has over 4000MW of fossil fuel based generating stations as of the end of 2015 as shown in the figure below.
Currently there are four large size utility scale developments:

a. EOI#1 (12 developers) 200MW, 50MW commissioned and the remaining is under construction.
b. EOI#2 (4 developers) 50MW each, total 200MW, financial closing is expected end of 2016
c. Quaira, 100MW, awarded
d. Masdar 200MW, bidding stage

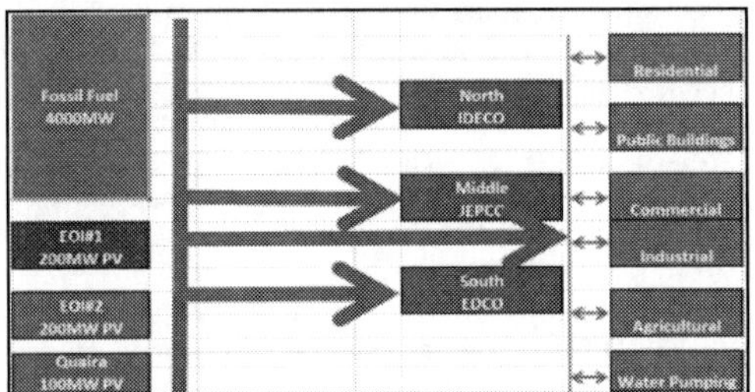

Figure 2: Electricity Flow Diagram

The transmission of electricity is done by National Electrical Power Company (NEPCO) which buys electricity from the generating stations and large customers, and sells it to the three distribution companies in Jordan; IDECO (North region), JEPCO (middle region) and EDCO (South and east region)

Large consumers can also connect to NEPCO directly. Consumers like; residential, public buildings, commercial, industrial, agricultural, and waster-pumping are now installing PV on grid projects, and providing electricity to the grid in addition to purchasing electricity from the grid.

Consumers are allowed to cover 100% of their annual electricity requirements. Once a consumer wants to install a PV project, they sum up their previous 12 months of electricity consumption reported in their electricity bills and divide that number by 1560KWh/year to determine the maximum system size (AC) that can be connected to grid. The system can be connected on site, if the consumer has enough space at their site. Or they can connect it using a wheeling scheme, where the consumer is consuming electricity at one location, and has their PV on grid system at another location, as shown in the figure below.

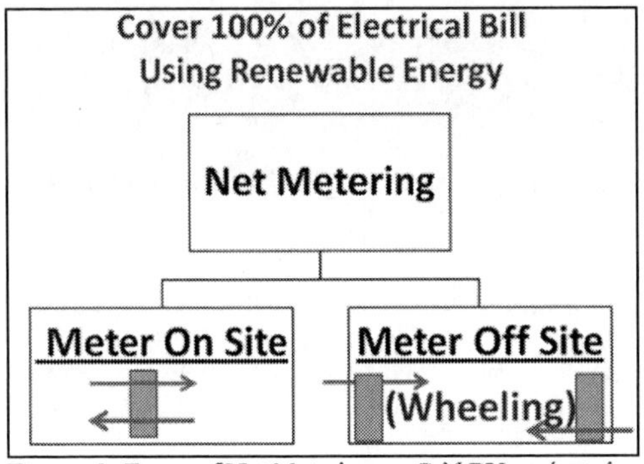

Figure 3: Types of Net Metering on Grid PV projects in Jordan

In addition, a recent development was made for the regulating of electrical charging stations for cars in Jordan.

The requirements of the electricity distribution companies for connecting PV Projects on grid are issued in guides. Two guides are available; one for connecting projects on site (net metering) and the other for connecting projects off site (wheeling)

3 UTILITY SCALE PROJECTS
Two types of utility scale projects are implemented in Jordan.

3.1 Expression of Interest (EOI)
A publicly announced call of Expression Of Interest (EOI) is issued by the Ministry of Energy and Mineral Resources (MEMR). The 1st EOI for 200MW Solar was issued in 2011 and sixty eight companies submitted their EOI for the development of RE projects on the 28th of July 2011. The EOI has a set price for purchasing of electricity from PV at 0.12 JD/KWh. That was the cost of generating electricity from conventional fossil fuel plant at that time. After study and elimination, twelve contractors were awarded projects for the development of PV projects between (10 – 50MW). Four projects (50MW) are already commission and it is expected that the remaining projects will be on grid by end 2016.
This is providing to be time consuming since the cycle took around five to six years. The government will be

purchasing electricity at a fixed rated of 0.12JD/KWh for 20 years according to the Power Purchase Agreements (PPA) signed with the developers.

3.2 Direct Proposals
Philadelphia Solar, the local manufacturer of PV Modules and Mounting Structures, applied for a license to build 10MW plants and they were successful in completing the plant and commissioning it in October 2015.
Since they are a local manufacturer with high local content, they are allowed an additional (15%) on the prices of electricity purchased from this project.

Figure 4: 1st Local Utility Scale Project

4 COMMERCIAL AND RESIDENTIAL PROJECTS
There are around 2000 projects on grid as of mid 2016. The three electricity distribution companies reported 135MW (1500 projects) installed on grid by end 2015. This includes the onsite generation and the offsite generation (wheeling)

The figure below shows an off-site and on-site generation project at the Royal Hashemite Court. This is a 5.6MW project that covers the annual electricity needs at three different locations by installing a PV system mounted over car park canopies at one location.

Figure 5: PV on Grid project at Royal Hashemite Court

5 ECONOMIC ADVANTAGE OF PV
The cost of generating electricity from fossil fuel resources in 2015 was 0.156 JD/KWh [4] (0.23USD/KWh), whereas the cost of generating electricity from PV is 0.049JD/KWh which was the average price of the four lowest financial bids for EOI#2.

The banking sector has developed awareness in renewable energy projects since 2010 [6]. The central bank for Jordan offers clean technology loans up to 10Million JD (14$ Million USD) at 2.3% interest rates, and the commercial banks can offer these loans at 5% to consumers.

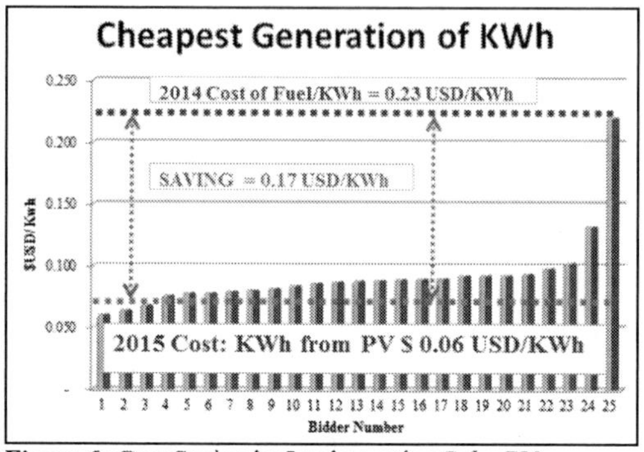

Figure 6: Cost Saving by Implementing Solar PV

The impact of installign 1KWp in Jordan is shown in the fiture below. Each 1KWp on average will produce 1560KWh/year. This figure is approved by regulators as the average figure to be used across Jordan by the distribution and the transmission companies. PV Modules have different efficiencies and degradation rates, and for the purpose of this study, the figure of 34000 KWh is used as the average production by 1KWp every 25 years.

PV Impact of $1KW_p$			
1	Produced Electricity	1560	KWh/year
2	Produced Electricity	34000	KWh/25years
3	Losses in Distribution and Transmission	17%	
4	Avoided Elec Losses in Distribution & Transmission	5780	KWh/25years
5	Losses in Generation	59%	
6	Avoided Electricity Losses in Generation	20060	KWh/25years
	Total Electricity Generation Avoided (2 + 4 + 6)	59840	KWh/25years
	Cost of Fuel for Generation 1KWh in 2013	0.1566	JD/KWh
	Cost of Fuel Avoid by installing 1KWp	9371	JD /25 years
2020 Impact of $3000MW_p$			
	Cost of Fuel Avoid	28112832000	JD /25 years
	Cost of Fuel Avoid	28113	Million JD /25 years

Figure 7: Impact of Installing 1KWp PV in Jordan

The overall average efficiency of the generating units in Jordan [4] is 41% (20,060 KWh/year/KWp can be avoided). This means that the overal losses if 59%. The losses in the distribution and transmission networks are 17% (5,780 KWh/year/KWp can be avoided).
When adding the actual generation of 1KWp to the avoided losses in generation and transmission of electricity, because the PV on site projects avoid these losses, the advantages of each 1KWp is the generation of 59840 KWh/KWp/25 years
When we multipy the cost of generation of electricity [4] (0.1566JD/Kwh), the economical advantages of each 1KWp is around 9371 JD/ 25 years or close to 14,000 USD/ KWp. This can be applied to the expected 3000MW of PV to be installed in the future.

The economical advantage without calculating the losses in the distribution, transmission and generation of electricity is shown in the figgure below. In addition the avoided CO2 emissions are also calculated based on the average 633grams CO_2/KWh, which is the reported 2012 emissioning from electricit in Jordan. The figures can be multilied by the average life span of 25 years and for a parametric size of Plants (1000MW), the expected savings are 6 Billion JD (8.4 Billion USD) and 25 Million Tons CO_2.

Economical and Environmental Impacts 1000 MW by 2020 For 25 Years

Installed Capacity	1,000,000	KW
Generation Capacity	1560	KWh/KWp
Generation Capacity	1,560,000,000	KWh/year
Cost of Generation	0.156	JD/KWh
Annual Saving (JD)	243,360,000.00	JD saving per year
Annual Saving (JD)	243.36	Million JD saving/year
Emissions Avoided	633	gram CO_2/KWh
Annual Emissions Avoided	987,480,000	Kg CO_2/year
Annual Emissions Avoided	987,480.00	ton CO_2/ year
25 YEARS Generation Capacity	39,000,000,000	KWh/25year
Cost of Generation	0.156	JD/KWh
25 YEARS Savings	6,084,000,000	JD saving per 25 years
25 YEARS Savings	6,084.00	Million JD saving/ 25year
Emissions Avoided	633	gram CO_2/KWh
25 YEARS Emissions Avoided	24,687,000,000	Kg CO_2/ 25 years
25 YEARS Emissions Avoided	24,687,000	ton CO_2/ 25 years

Figure 8: Cost Saving by Implementing Solar PV

6 STORAGE TECHNOLGY DEVELOPMENT

The commercial development of technology enabled the implementation of electricity storage that can receive electricity from renewable energy PV.

6.1 Electricity Generation Level

Because of the fluctuation of energy prices internationally, it was necessary to rely more on local resources. The IPP3 (270MW) combined cycle plant at Manakher in Jordan will install a 46MW PV power plant. In addition, 20MW of Lithium Ion battery storage will be implanted on grid.

6.2 Electrical Transportation

Electrical Cars are allowed in Jordan and encouraged as well because no customs and sales tax apply. The government has implement public projects (30 cars) and (100 cars) for ministries

The spreading of electric cars in the private sector started as well. It is important to keep in mind that the electricity needed to charge these vehicles should be generated from renewable energy sources; otherwise, it will not benefit the environment

7 DECARBONIZING THE ECONOMY

Carbon can be generated from different sources. Source 1 emissions are produced at site due to emission generating equipment like boilers. Source 2 emissions are produced off site for energy used on site like Electricity. Source 3 emissions are produced across the supply chain from material procurement to product delivery. This is shown in the figure below

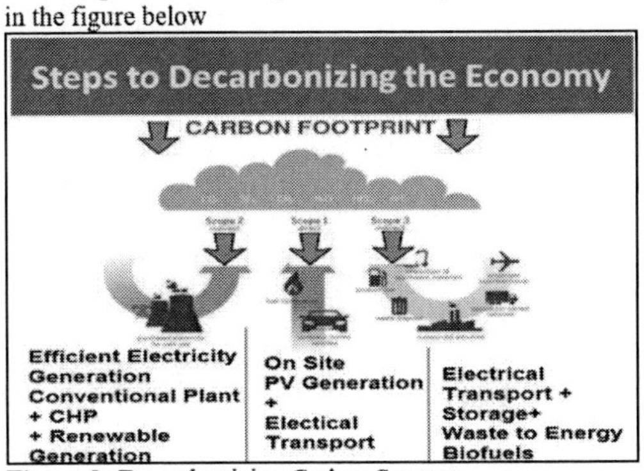

Figure 9: De-carbonizing Carbon Sources

DEVELOPMENT OF CLEAN TECHNOLOGY WORKFORCE

Solar PV and Wind are now the biggest sectors that are undergoing development. Sales, Design Installation and Maintenance jobs are opening up in the PV residential and commercial scale. Training of the current workforce and up-skilling their capacities with RE knowledge is ongoing. At the same time, there is a need for software development skills.

Regulators need knowledge of the latest technologies and how to connect it to grid. Grid impact studies are done by the distribution companies alone, and opportunities for development exist.

Storage of PV started, and the technology is developing, but training and education is not progressing at the same speed.

The figure below shows a list of the opportunities that exist in the market and the required workforce to support the development of the sector by qualified engineers and technicians at all levels across the supply chain.

Development of Clean Energy Workforce

Opportunities	Job Force Required
Renewable Energy PV + Wind	Sales, Design, Installation, Maintenance
Increase On site PV Generation	Marketing, Education, Training, Software
Increase % of Renewable on Grid	Regulators, Grid Studies, technicians,
RE Storage	Research, manufacturing, pilot projects
Connect to neighboring Countries	Contractors, Designers, Regulators
Wheeling	Engineers, Regulator development,
Grid Development	Construction, technicians, managers
Solar Water Heaters for Multi Family	Design, Manufacture, Install, Maintenance
Electric Transport	Sales, Marketing, Maintenance
Waste To Energy	Sales, Design, Installation, Maintenance
Thermal Insulation	Design, Manufacture, Install

Figure 10: Workforce Development Opportunities and Requirements

8 CONCLUSIONS

The Regulatory Framework was the basis for the start of development of the renewable energy sector in Jordan. And it continuous to be developed and expanded as per the market needs and future strategic plans for the integration of more renewable into the energy mix.

The development of the PV projects started in the sectors that were paying the highest electricity tariffs with the lowest return on investment, and it further developed into more sectors with the increasing economic challenges and the need to reduce the energy cost from the operational cost of services and products.

The PV sector flourished at the start of the development in 2012 with the start of hundreds of companies at all scales to sever the various sizes of PV projects.

The Expression of Interest Scheme took around six years from the date of the announcement and until the completion of construction. This development period can be reduced because of the learning curve that all the stakeholders underwent from the private, public sectors and regulators.

Setting a Feed In Tariff scheme for the purchase of renewable energy from investors and developers is not a sustainable act. Because the prices of equipment will continue to decrease and the efficiency will keep increase. Within the period of the 1st EOI in 2011 and the 2nd EOI in 2015, the price reduced from (0.12JD/KWh to 0.049 JD/KWh). Therefore, the future development of PV projects will be based on competitive bidding process at the time of the project.

The Solar PV Technology is continuously improving and this will make it more feasible for a larger spectrum of applications.

Solar PV Prices are decreasing and it is expected that after five years, by 2020, the prices will be lower by 50% and that will make it more feasible economically for consumers paying lower tariffs for electricity which will increase the number of potential customers and locations to install PV projects.

Grid Development is necessary and continuous training of the grid operators in the implementation of RE projects. In addition, the distribution grid is continuously upgraded with the necessary requirements when installation new Solar PV projects.

Solar PV market will continuous to grow, and this is a fast job creating activity in clean tech sector.

Energy efficiency remains a key factor in the development of the energy sector [5] and consumers must be wise in using their resources. Currently, some consumers are moving toward using more energy after the installation of PV systems, and this will require fossil fuel based electricity and future expansion of their PV system.

9 RECOMMENDATONS

Capacity Building in Grid Impact Studies is one of the most important activities that must be developed and integrated into the university level and professional training programs

Sharing of EU experience in the management and development of renewable energy PV projects will avoid huge losses and failures on equipment and plant in the future. This is needed.

RE remains a new field that is pioneered by few people. The decision makers at many economic sectors are not aware of the economic advantages of implementing RE projects and continuous awareness campaigns to different target groups is necessary.

Developing the quality standards for the RE sector is necessary for the growth of the market and implementation of the best practices to ensure better performance of the systems

Developing a solar career map for the different job that is needed by the PV sector is vital to the healthy and continuous growth of the market. Jobs are needed at the universities, training institutes, procurement, contractual, design and engineering offices, vocational training, manufacturing, construction, operation and maintenance.

5 REFERENCE

[1] Global Status Report 2016, REN21, http://www.ren21.net/status-of-renewables/global-status-report/

[2] Ministry of Energy and Mineral Resources (MEMR) Energy Facts and Figures report 2014

[3] Crude Brent Oil 10 year prices, http://www.nasdaq.com/markets/crude-oil-brent.aspx?timeframe=10y

[4] National Electrical Power Company (NEPCO) annual report, www.nepco.com.jo.

[5] "Energy Efficiency is the Challenge for Sustainable Development" 3rd International Conference of Renewable Energy, December 9, 2013, EngSamer A. Zawaydeh

[6] "Status of Green Finance in Jordan", Presented by Eng. SamerZawaydeh, Msc, PMP, CRM, REP, AEE Life Member at USAID Jordan Water Reuse and Environmental Conservation Project Green Finance Conference, at Royal Scientific Society December 10, 2013, Amman, Jordan

A PRATICAL ANALYSIS OF THE 58-15 MOROCCAN LAW ON RENEWABLE ENERGY

Rachid El Bachtiri
REEPER Group, PERE Lab, EST school; CED STI, FST; USMBA University; Fez
rachid.elbachtiri@usmba.ac.ma

ABSTRACT: This article analyses the convenience of the 58-15 law on renewable energy in Morocco for free production of electric power from photovoltaic stands. This law completes the 13-09 law on renewable energy. The maximum power permitted was and still fixed at 20 kW. The problem is that the 13-09 law required the connection of electrical generators from renewable to the medium, high or very high voltage on the public grid. We present the case of PV platform allowing this production with the highest voltage possible. Afterward, we will show some practical difficulties both technical and economic. The amendment made by the new version (58-15 law) states that the connection can be on the grid low voltage. But, this possibility is dependent on a regulatory text to expect. What mean that now (no text yet), there isn't any possibility to use such connection.
Keywords: renewable energy, legislation, photovoltaic, inverter.

1 INTRODUCTION

Many states have instituted several renewable energy goals and methods to achieve a balance of generation and consumption. Renewable energy power plants, energy efficiency, and smart grids are all part of their initiatives, but some states have developed more rigorous methods and goals to be achieved in long-term plans [1-2].

Morocco imports most its needs from different sources, including fuel, oil, and gas, and this dependency has a great impact on the energy bill. Furthermore, due to the economic and social growth that our country is undergoing nowadays, the demand for energy is increasing. On the other hand, Morocco has an important production potential constituted in particular by a large source of renewable energy estimated at 6,000 MW from wind, and 5 kWh/m^2/j from solar [3]. So the new national energy strategy is based on three main orientations: (1) availability of energy at the best cost for all the people in all the country, (2) the integration of Morocco in the Euro-Mediterranean area and the Maghreb, and (3) the development of renewable energy. Morocco has also organized the environment to attract companies and investors; a law was promulgated for renewable energy [4].

In this paper, we will recall the 13-09 law, and will present the changes introduced by the 58-15 law. Our study case concerns free electricity generation from renewable energy sources when the maximum cumulative power, by site or group of sites belonging to the same operator, is less than 20 kilowatts.

2 CASE STUDY

Our study is about a photovoltaic generation with 20 kW as peak power. We will compare the connection to the medium voltage (MV) and the connection to the low voltage (LV).

For the first scenario (MV), we will propose a multilevel inverter to convert DC voltage to AC voltage with reducton of harmonics [5-6]. In this sense, multi-level converters are increasingly preferred for medium- and high-power applications due to their ability to meet the increasing demand of power ratings and power quality associated with reduced harmonic distortion, lower electromagnetic interference, and higher efficiencies when compared with the conventional two-level topologies. The use of power electronic converters enables PV systems to provide more effective power production and to meet specific electrical grid code requirements. Obviously, a transformer is necessary, so we will present its characteristics. The general scheme is given in Figure. 1. The general scheme is given in Figurte. 1.

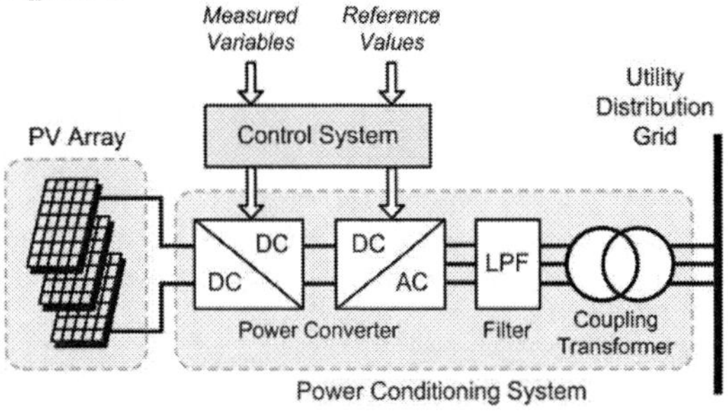

Figure 1: Global scheme of the system

For the second scenario (LV), we will present the simplification of the montage and the potential gain.

3 STRATEGY AND ACTIONS FOR USING RENEWABLE ENERGY IN MOROCCO

Morocco, which will host the next climate change conference COP22 in November next year, plans to derive more than half of its energy from renewable sources by 2030. Our country raised its renewable energy target from 42 per cent by 2020 to 52 per cent by 2030. In order to achieve this goal, renewable energy in all its forms is considered as a priority in the Moroccan energy strategy, which is based on a balanced mix of energy where clean fossil fuels are combined with renewable energies as well as energy efficiency. The main points of Morocco's energy policy are:

- The strengthening of energy supply security through the diversification of sources and resources, and the optimization of the energy balance and the mastery in the planning of capacities.
- The generalized access to energy through the availability of modern energy for the whole population and at competitive prices.
- Sustainable development through the promotion of renewable energies for the strengthening of the competitiveness of the productive sectors of the country, and environmental protection through the use of clean energy technologies, with a view to the limitation of the Greenhouse Gas emissions and the reduction of the strong pressure exercised on the forest cover.
- The strengthening of regional integration through the opening up to the Euro-Mediterranean energy markets and the harmonization of energy legislations and regulations.

Morocco has also organized the environment to attract companies and investors. The Government has promulgated some laws for renewable energy, and has created a dedicated agency to construct this solar plan, which is the Moroccan Agency for Solar Energy, and has considered the solar and wind projects as integrated projects. The Government wants to construct industry around solar and wind equipment, to attract industries and investors.

4 Law no. 13-09 on Renewable Energy

The Law No. 13-09 on renewable energy was promulgated by Dahir No. 1-10-16 of 26 Safar 1431 (February 11, 2010) published in Official Newsletter No. 5822 of the first Rabii II 1431 (March 18, 2010). It states: Development of renewable energy sources constitutes one of the priorities of national policy on energy, whose main aims are:

- strengthening the security of energy supply through diversification of sources and resources, optimization of energy balance and control of capacity planning;
- widespread access to energy, the availability of modern energy for all segments of the population and at competitive prices;
- sustainable development through the promotion of renewable energy, to strengthen the competitiveness of productive sectors of the country, preserving the environment through the use of clean energy technologies for limiting emissions of greenhouse gases and reduction of high pressure on forest cover;
- strengthening of regional integration through openness to Euro Mediterranean markets for energy and the harmonization of laws and regulations of energy.

To act in synergy with the national policy, this Act takes steps to develop and adapt the renewable energy sector with future technological developments and able to encourage private initiatives.

The new legislative framework for renewable energy sector, notably, set goals:

- promoting energy production from renewable sources, its marketing and export by the public or private entities;
- subjugation of facilities producing energy from renewable sources to the licensing or declaration;
- the right for an operator to generate electricity from renewable energy sources on behalf of a consumer or a group of consumers connected to the national grid medium voltage (MV), high voltage (HV) and extra high voltage (EHV), under an agreement whereby they undertake to remove and consume the electricity produced exclusively for their own use.

So, in order to achieve these objectives, the Act establishes a legal framework providing opportunities for construction and operation of electric power plants from renewable sources by natural or legal persons, public or private, specifying in particular the general principles they should follow, the legal regime including the marketing and export...

We will consider tow articles from chapter II (General principles):

Article 5: The power plants from renewable energy sources can only be connected to the MV, HV or EHV national grid. However, the connection of renewable energy power plants to the MV national grid is subject to terms and conditions prescribed by regulation.

Article 6: are established, operated and modified freely, power generation facilities as follow:

- Electrical plants from renewable energy sources when the maximum total capacity per site or group of sites belonging to the same operator, is less than 20 kilowatts;
- Thermal plants from renewable energy sources when the maximum cumulative power, by site or group of sites belonging to the same operator, is less than 8 megawatts.

5 LAW NO. 58-15 ON RENEWABLE ENERGY

Within the implementation framework of the roadmap for the development of photovoltaic energy, law No 58-15 amending and supplementing law No 13-09 on renewable energy was adopted by the first chamber of the Moroccan parliament dated on October 27th and by the second chamber on December 2015.

The 58-15 law modifies and complements the existing 13-09 law on renewable energy by four changes. The two changes we are interested in this study are:

1- Sale of the excess production.

Possibility of selling the excess electricity production from renewable sources as part of the facilities connected to the national network of high voltage (HV) and very high voltage (VHV) and of medium voltage and low voltage network in public grid. The operator shall not sell more than 20% as a surplus of the annual production.
The commercial terms and conditions of the surplus's redemption are set by regulation.

2- Opening of the low voltage renewable energy market.

The opening of access to low voltage distribution network would allow the development of industrial sector of small and medium facilities, including photovoltaic and creating jobs in the renewable energy sector. The access to the low voltage electric network will be subordinated to "the terms and conditions set by regulation".

6 SCENARIOS USING PHOTOVOLTAIC PLANT

In this section, we will determine a power PV plant with 20 kW as rated since article 6 of the law authorizes a freely production of $P_{max} = 20$ kW.

The first scenario considers the connection to the medium voltage [4], and the second scenario considers the connection to the low voltage.

6.1 PV Modules

The photovoltaic module considered is Siemens SM-50. Manufacturer data are:

Table I: Rates data

Characteristics	*Values*
Nominal voltage	12V
Maximum power	50W
Current at the maximum power point	3.05A
Voltage at the maximum power point	16.6V
Maximum voltage (open circuit)	21.4
Voltage temperature coefficient	-0.077 Volts/°C
Maximum system voltage	1000V (ISPRA) 600V (UL1703)

The total number of modules is: $N_t = 20\ 000/50 = 400$. By adopting a maximum voltage of 600 V, the number of modules to connect in series is $N_s = 600/12 = 50$. So the number of module chains to connect in parallel is $N_p = N_t/N_s = 8$.

Actually, PV plants should be designed to operate at their optimal operating point power for any temperature and solar radiation level, so a maximum power point tracker (MPPT) is used to maximize the PV output power. Therefore, the DC voltage will be: $U_{DC} = N_s \times V_{mpp} = 50 \times 16.6 = 830$ V. The corresponding current is $I_{DC} = N_p \times I_{mpp} = 8 \times 3.05 = 24.4$ A. Which gives the power $P_{DC} = U_{DC} \times I_{DC} = 20.252$ kW.

We remark that $P_{DC} > P_{max}$, but, we note that:

- The difference between the powers is low.
- The power P_{DC} assumes standard test conditions.
- There are losses in the inverter and the transformer.

When there isn't a load (no current delivered by the PV), the voltage output is, at 25 °C, $U_{DCoc} = N_s \times V_{oc} = 50 \times 21.4 = 1070$ V . Moreover, this voltage depends on temperature, and in extreme case when temperature is zero, it becomes: $U_{DCoc0} = U_{DCoc} + N_s \times \Delta T \times k_{TV} = 1070 + 50 \times 25 \times 0.077 = 1166.25$ V what is over the admissible voltage of 1000V. If we adopt a maximum voltage of $U_{DCmax} = 850$ V (see paragraph V bellow), we will have $N'_s = U_{DCmax} / (V_{co} + \Delta T \times k_{TV}) = 850 / (21.4 + 25 \times 0.077) = 36.44$ .Then, the number of parallel string becomes: $N'_p = N_t/N'_s = 400 / 36.44 = 10.97$. So we can opt for a solution with 15 strings consisting of 27 modules in series.

6.2 Power converter

PV arrays are known to be nonlinear, and there exists one operating point where the PV array generates maximum power [7]. In order to achieve maximum utilization efficiency of the PV array, the MPPT control technique, which extracts the maximum possible power from the PV array, is essential.

So we have chosen a 20kW Grid-tied PV Inverter from Chint Power Systems. CPS SC20KTL-O grid-tied PV inverter can be flexibly utilized in different types of commercial rooftops and utility applications. Its technical data are presented in the following table (Table II):

Table I: Rates data of the converter

Characteristics	*Values*
DC Input	
Max. DC Voltage	850Vdc
MPPT Voltage Range (Full Power)	430-800Vdc
Working Voltage Range	500-800Vdc
Max. DC Power	22kWp
Max. Input Current / Per String	42A / 14A
Max. Number of Strings	5
MPP Tracker	1
AC Output	
Output Power	20kW
Nominal Output Voltage	400Vac, 3 phase
Grid Voltage	320-418Vac
Nominal Grid Frequency	50/60Hz
Current THD	<2%
Power Factor	~1

The output voltage of PV inverter is never in MV range. Indeed, the maximum DC voltage provided by string PV modules can't go over 1 kV (North America standard states 600V).

6.3 MV connection scenario

In this case, it is necessary to consider the use of a LV/MV transformer. Due to undetermined behavior of distribution transformers in case of fluctuating power flow the rated power of the transformer was set to limit the PV-in feed [8]. Three phase MV voltage are widely standardized.

Beltransfo Company in Morocco offers some top of pole transformers with the adequate voltage (400V/22kV), but there apparent power is, at least, 50 kVA ! This company can produce our specific transformer (20 kVA) but it will cost more.

6.4 LV connection scenario

In this case, a transformer is not necessary which constitute a great economy and a high technical simplification. On other hand, there is a proximity to the grid comparing to the MV connection. Consequently cables are dramatically less short.

7 CONCLUSION

The purpose of this paper is to present a conception of a photovoltaic plant for producing freely electrical power to inject in the Moroccan public grid. It's clear that the act allowing this issue is very limited. Firstly, it was forbidden, in the 13-09 law, to use the low voltage grid for coupling the PV plant. Secondly, even though the 58-15 law completes and amends the first one, the regulatory text giving administrative procedure for coupling to LV is not promulgated yet. We have determined a technical solution for the realization of a 2O kW photovoltaic system. We have considered two scenarios: a medium voltage coupling and a low voltage coupling. It's clear that the first scenario (Connection to MV) will cost a lot, mainly if the MV grid is far from the installation.

Now, an economic study is needed to know the profitability of the investment in 20 kW photovoltaic productions in Morocco. This study must take into account, among other criteria, the cost of the installation and its maintenance, in addition to the kWh resale price. However, since the impact of renewable energy is obviously positive on the environment, the economy, and the social (employment issue) sector, we hope that the Moroccan Government will provide more encouragement in this domain.

8 REFERENCES

[1] A Izadian, N. Girrens, and P. Khayyer, "Renewable Energy Policies. A Brief Review of the Latest U.S. and E.U. Policies," *IEEE industrial electronics magazine*, September 2013.

[2] M. Liserre, T, Sauter, and J. Y. Hung, "Future energy systems: Integrating renewable energy sources into the smart power grid through industrial electronics," IEEE Ind. Electron. Mag., vol. 4, no. 1, pp. 18–37, 2010.

[3] A. Lorca et al, "Renewable Energies And Sustainable Development in the Mediterranean: Morocco and the Mediterranean Solar Plan (MSP)," *FEMISE Research*, http://www.femise.org/PDF/ci2010/FEM34-02.pdf

[4] R. El Bachtiri and E. Matagne, "A Technical Reading of the 13-09 Law on Renewable energy in Morocco," *IEEE International Renewable and Sustainable Energy Conference*, 07th-09th March 2013, Ouarzazate Morocco.

[5] A. Fri, R. El bachtiri and A. El Ghzizal "Comparative study of H-bridge multilevel inverters dedicated to PV," IEEE International Renewable and Sustainable Energy Conference, 07th-09th March 2013, Ouarzazate Morocco.

[6] A. Fri, R. El Bachtiri and A. El Ghzizal, "A Comparative Study of Various Multicarrier PWM Strategies Single Phase Cascaded Inverter Symmetric 5L For PV Systems" WOTIC'2013, 26th - 27th December 2013 ; Fez, Morocco.

[7] M. Salhi, and R. El Bachtiri, "A Maximum Power Point Tracking of Photovoltaic System Supplying a Resistor Load," International Journal of Research and Reviews in Electrical and Computer Engineering (IJRRECE), Vol. 1, No. 4, December 2011, pp. 1034–1039.

[8] G.Kerber, and R. Witzmann, "Loading Capacity of Standard Oil Transformers on Phtovoltaic Load Profiles," *World Renewable Energy Congress (WRECX)*, 2008 WREC, pp. 1198-1203.

MISSING GAPS IN THE CHALLENGE OF MASSIVE INTERVENTION OF GRID-CONNECTED PV SYSTEMS IN PERU

Rafael Espinoza[1], Carolina Luque[1], E.Muñoz-Cerón (⊠)[2] and J. de la Casa[3]

[1] Centro de Energías Renovables y Uso Racional de la Energía. Universidad Nacional de Ingeniería de Lima.
Avenida Tupac Amaru 210. Office B1-260. Pabellón Central 2° Piso. Rimac-Lima
cer.uni.edu.pe Phone: +51 511 3821058. E-mails: respinoza@uni.edu.pe ; fluquea@uni.edu.pe

(⊠)[2] *Corresponding author*. IDEA Research Group (Research and Development in Solar Energy)
Electrical Engineering Department. University of Jaén
Campus las Lagunillas s/n. Building A3, Office 234. 23071 Jaén, Spain
Phone.: +34 953 212 462. E-mail: emunoz@ujaen.es.

[3] IDEA Research Group (Research and Development in Solar Energy)
Electronics and Automation Engineering Department. University of Jaén.
E-mail: delacasa@ujaen.es

ABSTRACT. The electrical consumption of Peru has increased an average of 5.1% annually from 2002 to 2011 while the electrical power demand has shifted from 2700MW to 4800MW as a consequence of the increment in its GDP

In order to overcome this increment in the electricity demand and fostered by the fact that the generation of electricity by means of PV systems is becoming competitive with those traditional energy generation sources, in the recent years, Peru has shifted from being one of the leaders in the installation of Stand-Alone PV systems to become a potential one for large grid-connected PV plants. Nevertheless, in the pathway from small isolated PV "watts" to the last "MW-size" plants installed recently, there is a tremendous gap and technical challenge for the Peruvians, because these large PV systems are being installed by foreign companies using non-local technicians.

Therefore, the present work intends to propose the establishment of the technical background and conditions to encourage the development of the minimum requirements to promote a massive introduction of medium-size (100kW-1MW) distributed photovoltaic systems in Peru throughout the next years. In this direction, this proposal has been made under the technical and training support from the professional team part of the project "Emerging with the Sun" under the formal cooperation framework between the University of Jaen (Spain) and the Universidad Nacional de Ingenieria (Peru).

Keywords: Developing Countries, Dissemination, Education and Training, Global Issues, Policies and Strategies

1 CONTEXTUALIZATION

In early 1993 the Peruvian government encouraged private investment through the enactment of the law No. 25844 about "Electricity Concessions Law and Regulations" which generated the emergence of the electricity market associated with the National Interconnected Electrical System (SEIN) to foster system with a range of magnitude in the MW order. Thus, during the period 1995-2015 Peru monitors SEIN's behavior by observing performance indicators such as: Installed power, effective power, maximum demand, power Generation and Energy Consumption Per capita. These SEIN performance indicators can be observed in Figures 1, 2 and 3 for the period 1995-2015 [1][2].

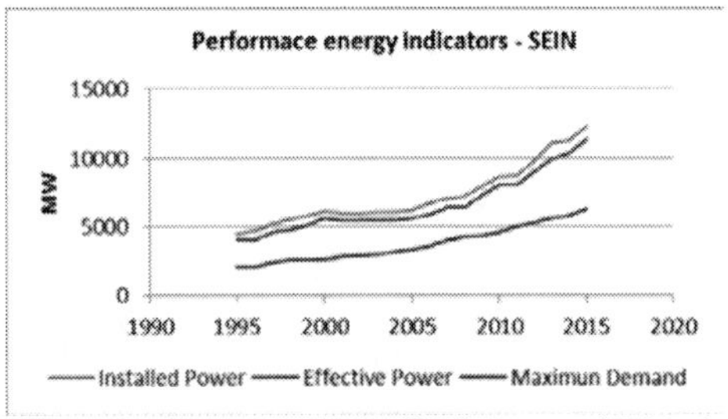

Figure 1: Intalled power (IP), effective power (EP) and maximun demand (MD) – SEIN [3]

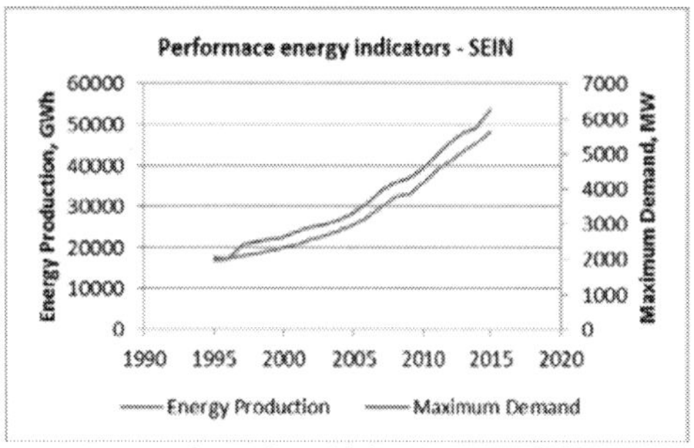

Figure 2: Maximun demand (MD) – SEIN [3]

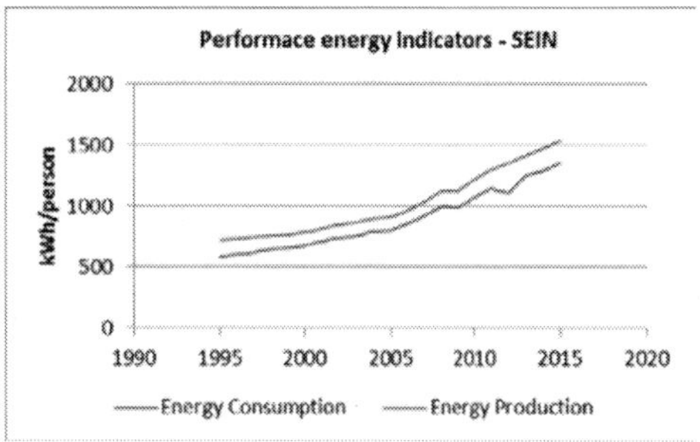

Figure 3: Energy consumption per person – SEIN [3]

Regarding the sectorial consumption profiles, in figure 4 the commercial-residential and industrial (excluding mining) profiles at the national geographic scenario can be observed.

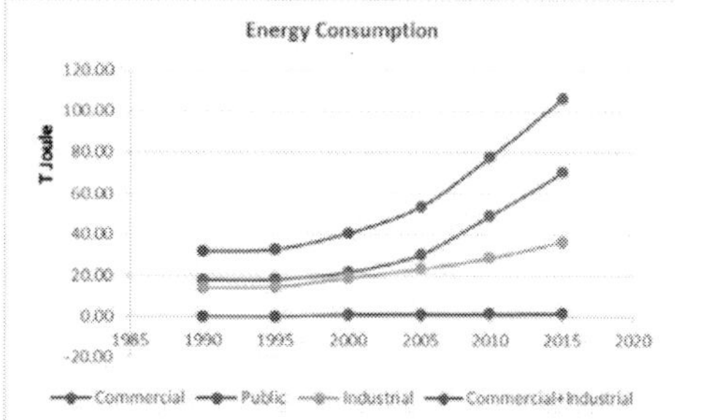

Figure 4. Historical change in EE consumption in the residential and commercial sectors, Public and Industrial. (Source, own adaptation from BNE-2013 [4])

Reinforcing the idea of the increment in the energy demand in Peru, Figure 5 to 7 show the distribution of the industrial, commercial and residential profiles in 10 of the 25 regions of Peru with greater sectorial consumption, where it is clearly noticeable the increment of power in all the sector and in almost all the regions.

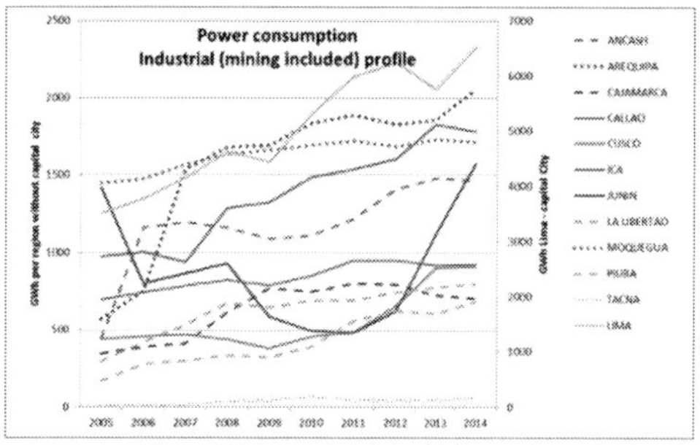

Figure 5: Power consumption industrial (not mining included) profile by regions [5] - [14]

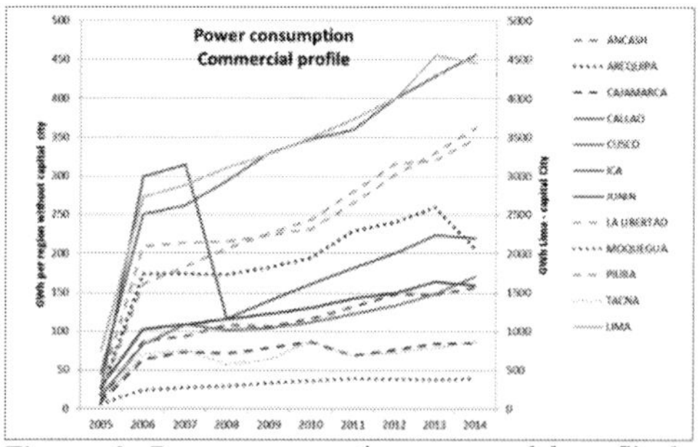

Figure 6: Power consumption commercial profile by regions [5] - [14].

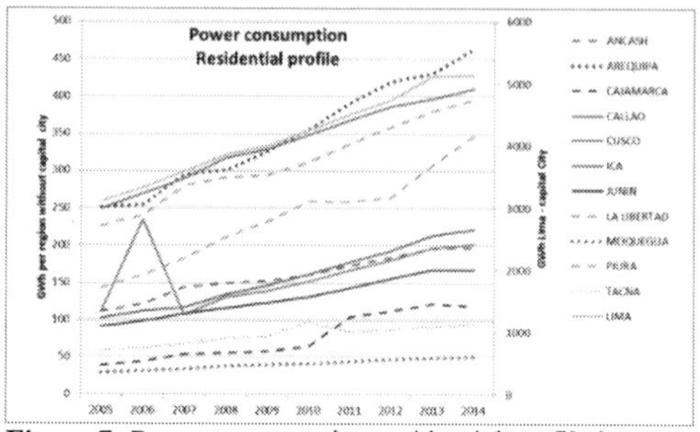

Figure 7: Power consumption residential profile by regions [5] - [14].

On the other hand, Peru has been working in the rural sector electrification since 1996 by means of residential photovoltaic systems also named as solar home system (SHS), whose individual power rated in the order of dozens of Watts, with an overall result of 2MWp installed to 25000 users with an average annual consumption of 8.4kWh/user-month.

Several rural electrification programs due to the participation of the private and public sector, have resulted that in 2014 there have been 17197 SHS installed, of which 3997 were promoted by the private sector (Acciona Microenergía). A detailed summary of the results are shown in Table 1.

Table I: Quantities of SHS operating in Peru from 1995 to 2014 [15].

Distribution Company	ELECTRO ORIENTE S.A.	ADINELSA	ELECTRO NORTE S.A.	ELECTRONOROESTE S.A.	ELECTRO SUR ESTE S.A.A.	ACCIONA MICROENERGÍA PERU	TOTAL SHS
Nº Users	746	6,238	1,549	2,102	2,585	3,977	17,197

Additionally, in 2015, the Peruvian government approved an energy project for the supply, installation and management of 150,000 SHS in rural areas, although due to formal differences between the government and the awarded company has made this project momentarily to stand by.

Regarding the installation of large power plants based on renewable sources in Peru, they have installed and connected power to SEIN of around 96MW which has been awarded by public auction procedure. During the period 2010- 2015, the Peruvian government has awarded seven projects of photovoltaic plants, 5 of these plants are in operation (96MW total installed capacity) and 2 additional plants (184.5 MW) are under implementation and interconnection with the SEIN.

The characteristic of these systems are collected in table II and III

Table II: Main characteristics of the PV plants connected to SEIN under the modality of energy auctions [16].

Item	Solar Plant	Awarded company	Anual Energy (MW.h)	Power (MW)	Price (Ctvs. US$/kW.h)	Status
1	Panamericana Solar (Ilo)	Consorcio Panamericana	50,676	20	21.50	Connected
2	Majes Solar 20T (Arequipa)	Grupo T Solar Global S.A.	37,630	20	22.25	Connected
3	Repartición Solar 20T (Arequipa)	Grupo T Solar Global S.A.	37,440	20	22.30	Connected
4	Tacna Solar	Consorcio Tacna	47,196	20	22.50	Connected
5	Moquegua Solar (Mariscal Nieto)	Moquegua FV S.A.C.	43,000	16	11.99	Connected

Table III: Characteristics of future generation plants with renewable energy sources corresponding to the fourth Energy auction [17].

Technology	Projects awarded	Total Power (MW)	Total Energy (GWh/año)	Price (USD/MWh)
Photovoltaic	2	184,5	523,4	48,09
Biomass	2	4,0	29,0	77,00
Wind power	3	162,0	738,6	37,79
Hidro	6	79,7	448,2	43,86

Besides the promotion of large grid-connected PV systems or the installation of SHS, there is none or scarce data regarding PV systems, either isolated or grid-connected, in the order of kW. One of the main reason is the lack of regulation for PV systems of such middle power magnitude.

In a different field, but necessary to include in the contextualization of the Peruvian situation, is with regards to educational programs. There are isolated efforts from public universities in vocational training in renewable energy since 1981. Recently in 2015 the government approved and published the National Environmental Education Plan, to continue an outline of "Promoting the use of clean energy technologies in mining activity to minimize environmental risks and impacts". Furthermore since 2014, two technical institutes education programs include renewable energy training lines.

2 MISSING GAPS

As it could be deducted from the previous section, Peru needs to overcome certain barriers, such as: Lack of promoting regulatory framework of PV technology connected to the grid, shortage of critical mass of professionals specialized in PV issues, shortage of local business supply in PV sector and Socio-Cultural condition to achieve conditions that allows a massive penetration of grid-connected photovoltaic systems to the grid.

Related to the absence of promoting regulatory framework about PV technology connected to the grid, there are currently five decrees laws (DL No. 25844-1993, DL No. 28546-2005, DL No. 28832-2006, DL1002-2008, DL1221-2015) that promote the expansion of renewable and unconventional energies but none of them considers mass installation of power generation based on photovoltaic technology.

The barrier of the lack of critical mass of professionals specialized in PV issues; in Peru there are only two technical institutes which have a course in renewable energy within a technical career (Tecsup-Lima-2014, SENATI-Lima-2014) and only two universities have a specialization in Solar Energy (UNSA-Arequipa-2014 and UNI-Lima-1981). Two other universities have a degress that addresses issues of renewable energies (UNS-Ancash-2005 and UNJ-Puno-2014) but they do not deeply cover the PV characteristics. Additionally, there is a Master of Renewable Energy (UNI-Lima-2012) and a PhD in Environment and Renewable Energy (UNSA-Arequipa-2013), but these programs are no sufficient enough for the mass potential of students graduating at high school level in Peru, which reach 2.5 million people per year.

The shortage of local business supply in PV sector is another missing gap detected. It is important to mention that companies which won the approval of implementation of PV projects for power plants (96MW) or rural electrification (2MW) are foreign companies with barely local intervention neither in the installation nor in the integration of PV systems. National engineering companies only accounts for an approximate total capacity of 18.65MW in the last 10 years. Other factors such as complicated mechanism to access to financing, as well as the lack of knowledge of the PV technology by financiers and end-user in regards to the energy and economic benefits of these systems, make this local business shortage one of the most resistant barrier.

Finally, the Socio-Cultural barrier has been noticed when collecting information for this paper. It was perceived in different socio-cultural sectors a remarkable conceptual distortion on the advantages and benefits of using PV systems. Besides, the lack of energy efficiency habits in the lifestyles of the population is also a barrier to the desired generalization of PV systems in Peru [18] [19].

3 DESCRIPTION OF THE PROJECT "EMERGING WITH THE SUN"

In order to start to overcome some of these barriers, professors from the University of Jaén (Spain) and the National Engineering University in Lima designed an international cooperation project whose main characteristics are summarized in the following lines.

3.1 Overall objective

To promote generation, dissemination and transfer of knowledge and technology to put forward solutions to critical problems of human, social and economic development in Peru.

3.2 Specific objective

Strengthen technological, teaching and research capacity in the field of generation of conventional electricity based on photovoltaic technology through scientific-material support and transfer of experience-knowledge by the University of jaén to the National Engineering University in Lima (uni) and its Renewable Energy Centre (CER-UNI).

3.3 Activities carried out under the project

1.-Installation of four 3kWp grid-connected PV (GcPV) systems for testing and training of local engineers. These GcPV systems are located in three diferent cities of Peru (Arequipa, Tacna and Lima) and in the case of Lima, two differents PV technologies are used, that is, monocrystalline and thin-film. Figures 8 to 12 show these PV systems

Figure 8: 3kW$_p$ GcPV made of monocrystalline PV modules and Steca inverter (Location: Lima)

Figure 9: $3kW_p$ GcPV made of monocrystalline PV modules and Steca inverter (Location: Arequipa)

Figure 10: $3kW_p$ GcPV made of monocrystalline PV modules and Steca inverter (Location: Tacna)

Figure 11: $3kW_p$ GcPV located in Lima based on thin-film PV modules and SMA inverter.

2.-Installation of facilities oriented towards research in the future National Renewable Energy Laboratory LANER-CER-UNI (see figures 12 and 13)

3.- Participation of PV researchers from UJA to the Masters, postgraduate and graduate courses in UNI.

4.- Implementation and active participation of PV researchers from UJA in technical seminars, training courses and seminars for the dissemination of renewable technology for students, local businesses, government agencies and others.

5.- Technical Scientific support to new Peruvian national universities to show their interest in joining the network and incorporate technology into their activities.

Figure 12: Monitoring system for PV modules quality performance evaluation.

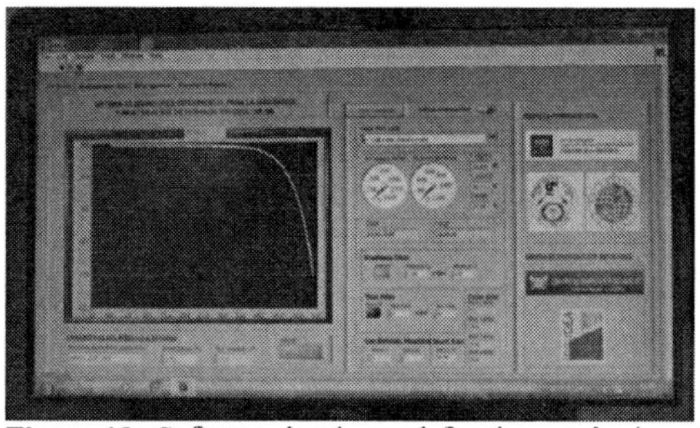

Figure 13: Software developmed for the monitoring of the systems.

3.4 Performance

A summary of the performance of the installed monocrystalline systems is shown in the following table

Parameter	Arequipa	Tacna	Lima
GcPV system efficiency	12-14%	11-14%	9-12%
Performance Ratio	75-89%	70-89%	75-98%
Reference Yield	3-7 kWh/kW	3.7-6.3 kWh/kW	0.95-5.8 kWh/kW
Final Yield	2.6-6 kWh/kW	2.6-5.6 kWh/kW	0.7-4.6 kWh/kW

3.5 Final benefits expected

- Reducing the time for "the appropriation of PV technology"
- Lead CER as benchmark in Peru in research, engineering and technology diffusion of conventional electricity generation from photovoltaic systems.
- With regard to the country: Creation in Peru of a critical mass of local engineers and technicians capable of leading the development of it.

4 CONCLUSIONS AND FUTURE ACTIONS

It may be concluded from this paper that the main barriers to the massive penetration of distributed generation with grid connected photovoltaic systems (DG-GcPV) in Peru are: Lack of promoting regulatory framework of PV technology connected to the grid, shortage of critical mass of professionals specialized in FV issues, shortage of local business supply in PV sector and Socio-Cultural condition.

According to the analysis undertaken in this paper, the following measures are suggested to contribute to overcome the aforementioned barriers:

- Strengthening the existing regulatory framework by means of introducing appropriate measures such as integration of RES and energy efficiency. It would also allow a progress in its dissemination, as there is still nearly a total lack of knowledge that the mere fact of using RES is a good practice for implementing energy efficiency. Likewise, the Latin American energy integration will boost the development of RES when Governments politically decide that those are the first choice for a clean, sustainable and secure regional energy development. Self-sufficiency and energy security are key components of a sustainable development.
- Generate business expectations among local companies in photovoltaic investments based on policies that promote a regulatory framework to support investments in DG-GcPV.
- Strengthen capacities and specialized technical skills in DG-GcPV for implementation and sustainability through technology transfer agreements between national and international institutions
- Promote wide dissemination of strategic information for actors of an eventual national program of development of renewable energies, actors such as public officials, political authorities, financial institutions, science and technology agencies and end users who would generate growth in demand for DG-GcPV.

5 ACKNOWLEDGEMENT

This work has been supported by the Andalusian International Cooperation Agency (Agencia Andaluza de Cooperación Internacional para el Desarrollo, AACID) in the framework of the project: "Emergiendo con el sol. Apoyo institucional al centro de energías renovables de la Universidad Nacional de Ingeniería en el campo de la generación de energía eléctrica empleando tecnología fotovoltaica".

6 REFERENCES

[1] Ministry of Energy and Mining, "Anuario Estadistico de Electricidad 1998," March 1999.

[2] Ministry of Energy and Mining, "Anuario estadistico de electricidad 1999," March 2000.

[3] Ministry of Energy and Mining, "Indicadores de sector electrico peruano 1995 - 2015," March 2016.

[4] Ministry of Energy and Mining, "Balance Electrico Nacional 2013," March 2014.

[5] Ministry of Energy and Mining, " Anuario Estadistico de Electricidad 2005," March 2006.

[6] Ministry of Energy and Mining, " Anuario Estadistico de Electricidad 2006," March 2007.

[7] Ministry of Energy and Mining, " Anuario Estadistico de Electricidad 2007," March 2008.

[8] Ministry of Energy and Mining, " Anuario Estadistico de Electricidad 2008," March 2009.

[9] Ministry of Energy and Mining, " Anuario Estadistico de Electricidad 2009," March 2010.

[10] Ministry of Energy and Mining, " Anuario Estadistico de Electricidad 2010," March 2011.

[11] Ministry of Energy and Mining, " Anuario Estadistico de Electricidad 2011," March 2012.

[12] Ministry of Energy and Mining, " Anuario Estadistico de Electricidad 2012," March 2013.

[13] Ministry of Energy and Mining, " Anuario Estadistico de Electricidad 2013," March 2014.

[14] Ministry of Energy and Mining, " Anuario Estadistico de Electricidad 2014," March 2015.

[15] Ministry of Energy and Mining, "Folleto de electrificacion rural con energias renovables", May, 2013.

[16] Peruvian Supervising Organism for the Investment in Energy and Mining - OSINERMIN, "Centrales de generación eléctrica con recursos energéticos renovables," *Boletin OSINERMIN*, May 2014.

[17] Peruvian Supervising Organism for the Investment in Energy and Mining - OSINERMIN, "Centrales de generación eléctrica con recursos energéticos renovables," *Boletin OSINERMIN*, May 2015.

[18] A. Yadoo and H. Cruickshank, "The role for low carbon electrification technologies in poverty reduction and climate changes strategies: A focus on renewable energy mini-grid with case studies in Nepal, Peru and Kenya," *Energy Policy 42,* pp. 591-602, 2012.

[19] L. Narvarte and E. Lorenzo, "Review of barriers to the dissemination of decentralized renewable energy systems," *Renewable and Sustainable Energy Reviews 58,* pp. 477-490, 2016.

Start-Up Factory Kumasi (Ghana) – Paderborn (Germany): Smart Tropical House

Tobias Klaus[1*], Eva Alexandra Schmitz[1], Björn Klusmann[1], Mark Piper[1], Ludwig Rongen[2], Rolf Gruber[2], Isaac Nkrumah[3], Kofi Ampong[3], Michael Donkor[3], Reuben Tamakloe[3], Stefan Krauter[1], and students from all participating universities

[1] University of Paderborn, Germany
[2] University of Applied Sciences Erfurt, Germany
[3] Kwame Nkrumah University of Science and Technology, Kumasi, Ghana

* corresponding author: tobias.klaus@uni-paderborn.de

Abstract

Students of Kwame Nkrumah University of Science and Technology (KNUST) in Kumasi, Ghana, and of Paderborn University are working on one renewable energy-related project each year. Those projects are aimed at improving on innovations or technical developments (renewable energy supply and/or energy efficiency) from young engineers and/or start-ups in Ghana. The shall enable the young entrepreneurs getting their products developed to maturity, and then introduced in the West African market. Within the projects, the technical expertise and knowledge will be improved through common extensive work at all participating universities - together with the inventor and the participating companies. Besides the technical development, market studies will be conducted, a business plan and a concept for market introduction will be developed.

The yearly sub-projects are treated similarly to industry-funded technological development projects as they take place at German universities.

Each project is structured in three phases:

- during summer: preparatory train-the-trainer workshops for the participating staff at KNUST, including the selection of projects and project coordination
- cooperative student seminars via video conferences
- semesterly meetings at KNUST, review of the projects, presentation to the project partners (inventor / initiator)

Keywords: Solar Home Sytems, Solar Architecture, System, Stand-alone PV Systems

1. Introduction

Start-Up Factory Kumasi-Paderborn is a project where students in an international cooperation collaborate in advising a young Ghanaian entrepreneur, who is active in the field of renewable energy and energy efficiency.
In the first year's project (2015/16), students work on the development of a Smart Tropical House, inspired by the passive house concept, where the electrical energy for cooling purposes (AC) and appliances shall provided via PV. As there is a strong architectural component, beside Paderorn University and Kwame Nkrumah University of Science and Technology, the University of Applied Sciences Erfurt is part of the project, with their students engaging in a small architecture competition. The building also includes in-house water purification, to account for the need of clean drinking water in ropical regions. The building itself is based on a freight container, allowing for a standardized and reproducible layout, fixed structure pricing, and air-tightness to minimize losses by air-conditioning.

Illustration 1: Design of the Smart Tropical House, Merdivan, Winter 2016.

The architecture (Ill. 1) was chosen due to its compactness, and on the basis of a small survey in Ghana. Similarly, the layout of he rooms hat to fit the needs and cultural practises of the people in Ghana (Ill. 2)

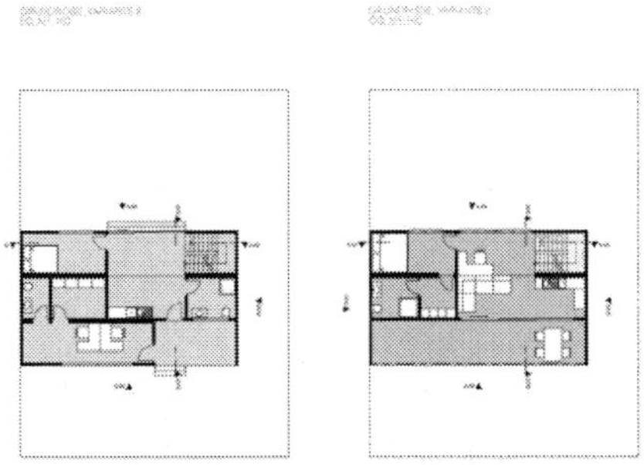

Illustration 2: Room Layout, Merdivan, Winter 2016.

2. CFD Simulation

Based on the chosen architectural design, a CFD simulation (Ill. 3) is conducted wich allows for a detailed thermal modelling of the interior of the house.

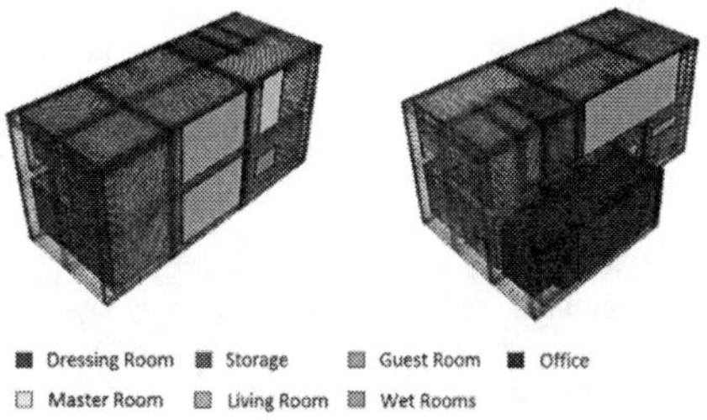

Illustration 3: CFD Modelling

For this modelling, a 3D steady-state CFD simulations using the commercial solver Star-CCM+. The thermal management and design of the passive house is accomplished by CFD, which provides detailed information on local temperatures and air flow in the house. *Ill. 3* shows the numerical domain. The container walls are transparent. The boundary conditions are chosen according to characteristic times during the day based on a representative daily routine for working days (i.e. 6 am, 1:30 pm, 6 pm, 10 pm). The representative daily routine for working days was developed with students and staff in Ghana. Settings for air condition vary by room and representative conform to this daily routine. Heat generation of electric devices, such as light bulbs and computers, is considered in the simulations according to VDI Directive 2007. The ambient temperature is set according to the hottest and the coldest day in Ghana. Maximum daily temperatures were obtained from the Climate Data Center of the DWD.

3. Water Filtering

For the water filtering, a series of different filtering techniques and devices were chosen: Water and More, Seccua Ultrafiltration, Grunbeck, carbon filters A and B, all being pressurized systems. A mechanical pump was used to drive water through them; in addition non-pressurized zeolite and ceramic filters were tested
A test was performed for the presence of fecal coliforms, E-Coli and the total coliform count for each of the samples.
All nine samples showed no fecal coliform and E-coli.
Six of the samples labelled ceramic, source water, carbon filter A, carbon filter B, Grunbeck and zeolite tested positive for the presence of total coliform.
Values of total coliform count for these samples were higher than the WHO recommended value of 0 cfu/ml for a safe drinking water and therefore requiring further treatment.

4. Outlook

Based on the results of the CFD analysis, the evaluation of different insulation materials, and the thermal properties of different paints, a solar system will be designed that allows the house to be energy-autonomous.

A SURVEY INFORMED MODELLING OF TEMPORAL DIFFUSION OF PV-BASED COMMUNAL GRIDS IN A RURAL DEVELOPING COMMUNITY

Nicholas Opiyo, Peter G. Taylor, and Rolf Crook
School of Chemical and Process Engineering
University of Leeds, Leeds, LS2 9JT, UK

ABSTRACT: An agent-based model (ABM) is developed in Netlogo as a tool for evaluating temporal diffusion of PV-based communal grids in a rural developing community. ABMs model individual entities within a complex system and the rules that govern the interactions of the entities within the system, to capture the overall effect of such interactions. Results show that given various choices, small PV mircogeneration systems would emerge as the prefered source of electrification for many households. Also, with increasing power demands, many household would prefer communal grids to the national utility grid. Furthermore, introduction of favourable government policies in the form of subsidies and mircocredit facilities, coupled with increasing social acceptance, would lead to increased PV installations and correspondingly to increased communal grid connections. Specifically, introduction of a subsidy of $0.15/kWh would result in an increase of 249% in communal grid connections. Increasing the sensing radius from 1 km to 2 km would result in an increase of 118% in communal grid connections.

Keywords: Communal Grids, PV Diffusion, Temporal, Agent-Based Model

1 INTRODUCTION

Development of electricity delivery infrastructures are path-dependent, meaning, each development decision and step affects subsequent steps, and the final outcome. Human actors are therefore the most important variables in any energy development plan as their decisions affect the way a system evolves. Proper policy-planning tools are therefore required to guide decision-makers on least cost rural electrification pathways. Many factors influence choices of technologies used in rural electrification, the main ones being availability of resources, cost, demand, social acceptance, and government policies. Different modelling tools and techniques have been applied in planning rural electrification paths in many countries. However, these often view this problem as a question of expansion of grid coverage through extensions of existing transmission and distribution networks and seldom address the unique and regionally-specific challenges presented by many developing nations. In most of sub-Saharan Africa for example, grid electricity is often unreliable, plagued with frequent blackouts, poor maintenance, and low quality of service. In these regions, expansions of the national grids often result in further strain on the systems and thus in further reduction in quality of services to those already grid-connected. Decentralized generation, transmission, and distribution of electricity based on locally available renewable energy resources offer potential solutions to low rural electrification rates in these communities.

Decentralized power generation systems are usually used as pre-grid electrification modulators, sensitizing and readying households for future grid connections, and are thus often abandoned once the latter arrives. This is more so with stand-alone systems which have limited power capacities. However, depending on local resources, capacities, designs and technologies used, communal grids could provide the final solutions to rural electrification in many developing nations and entrench green economies in the process. In fact, it is estimated that there will be almost 400TWh of installed green communal grid capacity by 2030, about 40% of new installed capacities towards universal electrification in developing nations [1]; communal grids offer many advantages over other options in that when compared to national grids, they are cheaper to put up, with shorter lead times, sized to match local demands, and are modifiable with increasing demands or changing technologies, while when compared to stand-alone systems, they offer access to higher power capacities and unlimited virtual storage, and could thus stimulate local microeconomic activities.

2 METHODOLOGY

In this work an agent-based model (ABM) is developed as a tool for evaluating temporal diffusion of PV-based communal grids in a rural sub-Saharan developing community to provide decision-makers with a user-friendly environment for PV-based rural electrification policy development, planning, and implementation. The model takes into account the complexities and limitations of solar electricity microgeneration technologies, decisions by human actors, geographical factors, and interaction between the three factors. ABMs seek to capture the overall macro-effects of different micro-decisions in a virtual world; they model individual entities within a complex system and the rules that govern the interactions of the entities within the system, to capture the overall effect of such interactions.

Survey data from a case study of Kendu Bay area of Kenya is used to inform the model. NetLogo was chosen for this work because it is user friendly, free, and easy to learn and modify, as is currently considered the best agent-based modelling software by researchers. The following agents are created in the model: a) a representation of the environment and the solar potential in it, b) the populations in it that require electricity, c) PV seeds that would use the environment to produce electricity, d) links to connected households to communal grids, and e) a central observer or stakeholder who determines the strategies and preferences for PV diffusions. Through these agents and the rules created for their interactions, the model sis used to simulate how step-decisions by human actors, based on prevailing local socio-economic factors and government policies, influence temporal diffusion of PV-based communal grids in a rural developing community.

Folowing the method of Opiyo et al. [2], a house without PV must first develop the idea to install PV given the following factors amongst others: a) cost, b) neighbourhood influence, c) government policies such as subsidies, and d) social pressure (advertisements). To calculate the idea to install PV (PVD), the above factors are combined as follows [9]

$$PVD = \sum_{i=1}^{n} W_i F_i \tag{1}$$

where n is the total numbers of factors, W_i is the weight associated with each factor F_i, and each F_i has a value between 0 and 1.

The idea to install PV returns true if

$$\frac{HPV_{SR}}{TH_{SR}} \times 100 > NT \tag{2}$$

where HPV_{SR} is the number of houses with PV within a given sensing radius (SR) or neighbourhood, TH_{SR} is the total number of houses within the same sensing radius, and NT is the neighbourhood threshold, minimum percentage of neighbours within the sensing radius who must have installed SHS for a household to develop the same.

If the idea to install PV returns true, cost comparison is done before a final decision is made. PV is installed if:

$$PVGC_{kWh} < AVC_{kWh} + S_{kWh} \tag{3}$$

where $PVGC_{kWh}$ is the PV generation cost per kWh, AVC_{kWh} is avoided cost per kWh, and S_{kWh} is subsidies per kWh.

Avoided cost per kWh, AVC_{kWh} is given by

$$AVC_{kWh} = NGC_{kWh} \tag{4}$$

where NGC_{kWh} is the prevailing national grid electricity cost per kWh.

PV generation cost per kWh ($PVGC_{kWh}$) is given by

$$PVGC_{kWh} = \frac{PVC_T}{(PVO \times PVL)} \tag{5}$$

where PVC_T is the total PV cost, PVO is the total PV output, and PVL is PV lifetime.
Total PV cost (PVC_T) is given by

$$PVC_T = (PVC_{m^2} - S_{m^2}) \times PV\varepsilon \times RS \times AF \tag{6}$$

where PVC_{m^2} is PV cost per m^2, S_{m^2} is subsidies per m^2, $PV\varepsilon$ is PV efficiency, RS is roof size available under PV Installation and ranges between 10m^2 and 50m^2, and AF is an accounting factor given by

$$AF = (1 + r)^{PVL} \tag{7}$$

where r is the interest rate and PVL is the PV lifetime.

If there are no functional subsidies policies, S_{m^2} and S_{kWh} are set to zero. Total PV output PVO is given by

$$PVO = I \times PQ \times PV\varepsilon \times RS \tag{8}$$

where $PV\varepsilon$ is PV efficiency, RS is roof size, PQ is the patch-quality (landscape quality) and depends on topography and prevailing climatic conditions, and I in W/m^2 is the insolation and is given by

$$I = 600 + \left[1000 \times pr \times \frac{(1 - \cos\theta)}{2}\right] \tag{9}$$

where θ is the orientation angle while pr is the particle reduction factor, a factor used to calculate reduction in aerosol particles which scatter solar radiation. It depends on location and prevailing climatic and weather conditions.

To be allowed to join a communal grid, a house must have installed PV of a given minimum capacity (power-threshold) and be within a given sensing radius of other homes with PV that meet the power-threshold. The idea to join a communal grid (CgD) returns true if

$$\frac{HCgPT_{CSR}}{TH_{CSR}} \times 100 > CgT \tag{10}$$

where $HCgPT_{CSR}$ is the number of houses with PV within a given communal grid sensing radius that meet the power-threshold ($CgPT$), TH_{CSR} is the total number of houses within the same sensing radius, and CgT is the communal grid neighbourhood threshold, minimum percentage of neighbours with PV that meet the $CgPT$ required within the sensing radius for a household to think about joining a communal grid. Communal grid sensing radius (CSR) is the radius within which a household can sense its neighbours. It depends on location and population density.

3 RESULTS AND DISCUSSION

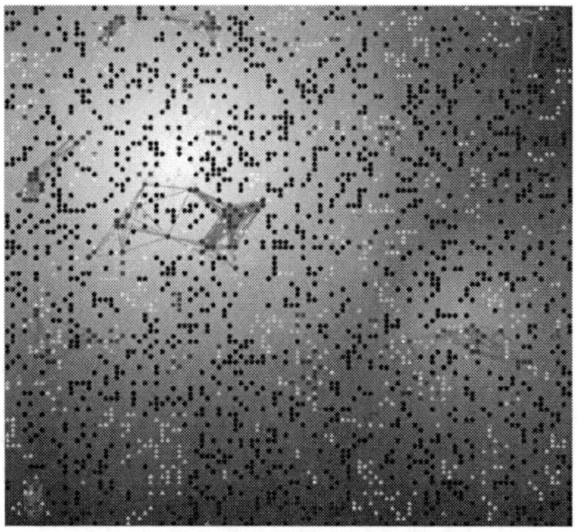

Figure 1: A View of the World after Simulations

Figure 2 shows a comparison of households with PV, households connected to communal grids, and households connected to the national utility grid after 25 years. Initially 347 households are allocated PV of various sized to mirror Kendu Bay area of Kenya. This represents 4.7% of all households in the area. Similarly, 229 households are connected to the national grid, representing 3.1% of all households. Communal grids are initially set at zero as none currently exists in Kendu Bay. After 5 years, the number of households connected to PV-based communal grids would have surpassed the number of households connected to the national-utility grid, indicating preferences for PV-based electrification systems. After 25 years, 4,325 households would have installed PV of various sizes, representing 44.1% of all households in the area, and 76.6% electrified households. Households connected to communal grids would have grown to 2,410, representing 24.6% of all households. On the other hand, households connected to the national grid would have only grown to 1,323, representing 13.5% of all households and 23.4% of all electrified households. It is noteworthy that after 25 years, 5,648 households would have been electrified, representing 57.6% of all households.

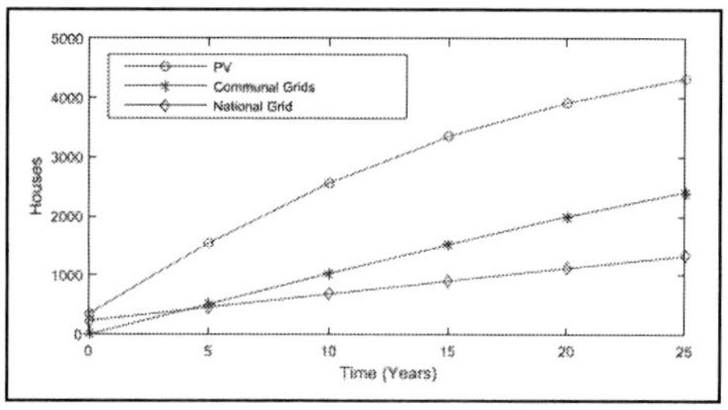

Figure 2: Different Electrification Topologies after 25 Years

3.1 Impacts of subsidies on PV diffusion

Studies show that positive government incentives and subsidies are largely responsible for rapid growths in PV installations in developed nations which have proper market and technical infrastructures necessary for implementations of such policies; Beise's research on diffusions of technologies across national boundaries found that positive government policies significantly stimulated diffusion of PV across many countries, underscoring the importance of government intervention in adaptations of new technologies [4]. A comparable study by Guidolin et al also found that positive government policies in forms of incentives and subsidies positively influenced PV diffusions across 11 countries [5]. Wustenhagen found that positive government policies, especially feed-in-tariffs (FiT), were largely responsible for the boom in PV installations in Germany [6]. Zhang et al found that in addition to positive regional governments' policies, costs also played significant roles in PV diffusion across Japan [7]. In developing nations on the other hand, limited resources restrict governments' abilities to subsidize new technologies as priorities are given to crucial projects. In this section, impacts of subsidies per kWh of PV electricity generated and subsidies per m^2 of PV Installed on the diffusions of PV microgeneration systems in Kendu Bay area are investigated.

Figure 3 shows a comparison of households with PV, households connected to communal grids, and households connected to the national utility grid after 25 years as a function of increasing subsidies/kWh of PV electricity generated. With no subsidies, 4,325 households would have installed PV after 25 years, representing 44.1% of all households. 2,410 households with PV would have connected to various communal grids, representing 24.6% of all households, while 1,323 households would have connected to the national utility grid, representing 13.5% of all households.

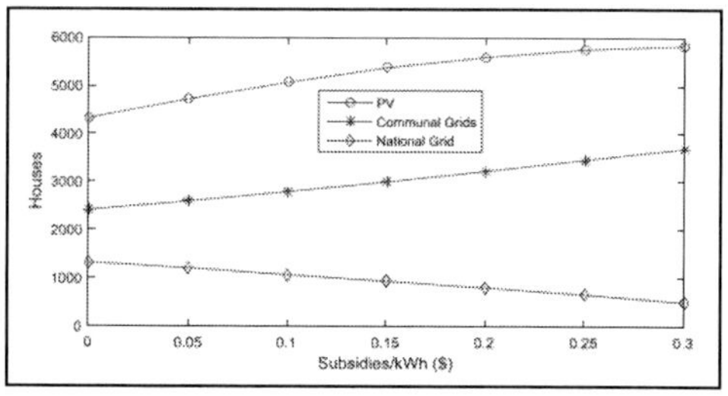

Figure 3: Impacts of Subsidies/kWh

With a subsidy of $0.30/kWh, 5,827 households would have installed PV after 25 years, representing 59.4% of all households, and an increase of 34.7% in PV installations. Similarly 3,685 households would have connected to communal grids, representing 37.6% of all households and an increase of 52.9% in communal grid connections. On the other hand, the number of households electriying through the national grid would have fallen to 499, representing 5.1% of all households.

Figure 4 shows a comparison of households with PV, households connected to communal grids, and households connected to the national utility grid after 25 years as a function of increasing subsidies/m^2 of PV installed. With no subsidies, 4,325 households would have installed PV after 25 years, representing 44.1% of all households. 2,410 households with PV would have connected to various communal grids, representing 24.6% of all households, while 1,323 households would have connected to the national utility grid, representing 13.5% of all households. With a subsidy of \$150/m^2, 6,012 households would have installed PV after 25 years, representing 61.4% of all households, and an increase of 39% in PV installations. Similarly 3,689 households would have connected to communal grids, representing 37.6% of all households and an increase of 53.1% in communal grid connections. On the other hand, the number of households electriying through the national grid would have fallen to 701, representing 7.2% of all households.

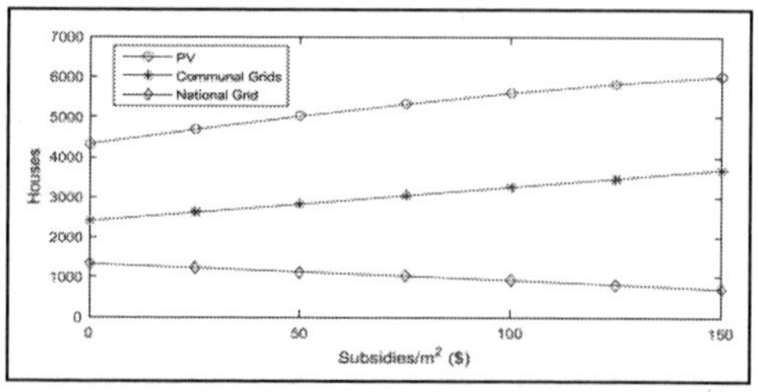

Figure 4: Impacts of Subsidies/m^2 after 25 Years

Table I below summarizes the above information.

Houses after 25 years	Subsidies/kWh ($)		
	0	0.15	0.30
Houses with PV	4325	5374	5827
Connected to Communal Grids	2410	2994	3685
Connected to National Grid	1323	937	499
Houses after 25 years	**Subsidies/m² ($)**		
	0	**75**	**150**
Houses with PV	4325	5321	6012
Connected to Communal Grids	2410	3046	3689
Connected to National Grid	1323	1027	701

Table I: Impacts of Subsidies on PV Diffusion

3.2 Impacts of neighborhood influence on PV diffusion

It is difficult to model the impacts of different non-quantitative social aspects on the adoption of a new technology. However, measurable parameter such as sensing-radius, the radius within which a household can 'sense' its neighbours, can be modelled and varied to explore the impacts of such parameters on the adaptation of a new technology.

Figure 5 shows a comparison of households with PV, households connected to communal grids, and households connected to the national utility grid after 25 years as a function of increasing sensing radius. At the default sensing radius of 500m, based on kendu Bay survey data, 4,325 households would have installed PV after 25 years, representing 44.1% of all households. 2,410 households with PV would have connected to various communal grids, representing 24.6% of all households, while 1,323 households would have connected to the national utility grid, representing 13.5% of all households. If the sensing radius is increased to 2 km, 5,204 households would have installed PV after 25 years, representing 53.1% of all households and an increase of 20.3% in PV installations. Similarly, households conected to communal grids would increase to 3,092 representing 31.5% of all households and an increase of 28.3% in communal grid connections. On the other hand, the number of households electriying through the national grid would have fallen to 831, representing 8.5% of all households

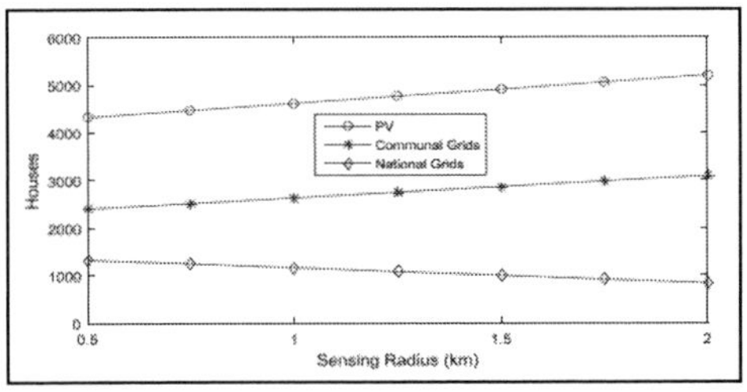

Figure 5: Impacts of Semsing Radius after 25 Years

3.3 Model Validation

Empirical PV installations data from 1985 to 2010 obtained from the Kenya's Ministry of Energy and Petroleum for two separate locations with similar geolocation data to Kendu Bay, i.e. Homa Bay and Ahero [8], are used to validate the model. The two towns were also chosen because in 1985, their populations of 31,063 (Homa Bay) and of 26,322 (Ahero) mirrored that of Kendu Bay (29,518) in 2014 [9]. Moreover, both areas are mainly occupied by people of the same ethnic and cultural backgrounds as Kendu Bay and would thus respond similarly to given social and cultural factors. The PV installations data are obtained by the government from retailers who also perform the tasks of installing and repairing the purchased systems. Figure 6 shows a plot of empirical and simulated PV installations data after 25 years. For Homabay and Ahero, 1985 is taken as year zero while 2010 is taken as year 25. For Kendu Bay, year zero is 2015 and year 25 is 2040. The plots show similar trends in PV installations, with Kendu Bay having mo PV installations due to falling PV costs. The plots validate the model.

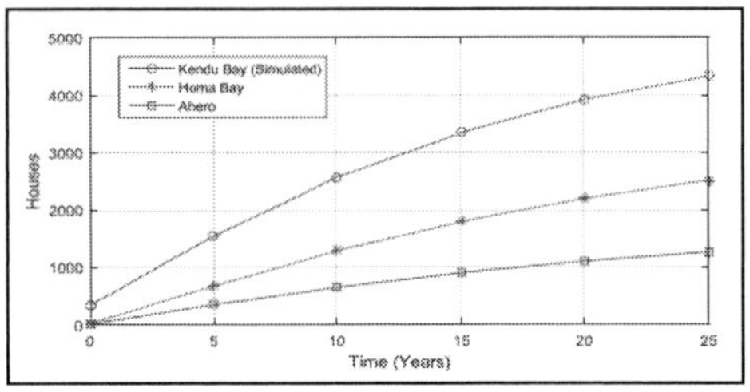

Figure 6: Empirical and Simulated PV Installations

4 CONCLUSION

In this work temporal diffusion of PV microgeneration systems in a typical rural Sub-Saharan developing community, Kendu Bay area of Kenya, are modelled and simulated using an agent-based model developed in Netlogo. Results show that after 25 years, more households will be electrified through PV systems of various sized than through the national grid. Specifically, results show that electrification rate in the area will steadily rise from current 7.8% to about 57.6%. Of these, 44.1% will be electrified through PV systems, with 24.6% being connected to communal grids. Another 13.5% will be electrified through the national grid. Introduction of subsidies would see further increases in households installing PV systems, leading to big rise in electrified households. Similarly, increasing neighbourhood influence, modelled as sensing radius, would lead to increased PV installations and subsequent communal grid connections.

4.1 Acknowledgement

This research is funded by Leeds International Research Scholarship

4.2 References

1. International Energy Agency (IEA), *World Energy Outlook*, 2012
2. Opiyo, N., Crook, R., Taylor, P., *A Survey Informed PV-Based Cost-Effective Electrification Options for Rural Sub-Saharan Africa*, Energy Policy **91**: 1-11, 2016
3. Zhao, J., Mazhari, E., Celik, N, Son, Y., *Hybrid Agent-Based Simulation for Policy Evaluation of Solar Power Generation Systems*, Simulat Pract Theory **19**: 2189-2205, 2011
4. Beise, M., *The International Adoption of Photovoltaic Energy Conversion: Is Japan a Lead*

Market? Discussion Paper Series **153**, Research Institute for Economics & Business Administration, Kobe University, 2004
5. Guidolin, M., Mortarino, C., *Cross-country Diffusion of Photovoltaic Systems: Modelling Choices and Forecasts for National Adoption Patterns*, Technological Forecasting and Social Change **77**: 279-296, 2010
6. Wustenhagen, R., Bilharz, M., *Green Energy Market Development in Germany: Effective Public Policy and Emerging Customer Demand*, Energy Policy **34**: 1681 – 1696, 2006
7. Zhang, Y., Song, J., Hamori, S., *Impact of Subsidy Policies on Diffusion of Photovoltaic Power Generation,* Energy Policy **39**: 1958 – 1964, 2011
8. Ministry of Energy and Petroleum, Republic of Kenya, *Solar Energy*, 2012
9. Central Bureau of Statistics, Office of the Vice President, and Ministry of Planning and National Development, *Economic Survey*, Nairobi, 2011

LARGE-SCALE (UTILITY-SCALE) PHOTOVOLTAIC POWER PLANTS 1980-2015, LITERATURE SURVEY

Denis Lenardic, independent expert, Jesenice, Slovenia
phone: +386 40 526 479, email: contact@pvresources.com

ABSTRACT: After the oil crisis in the 1970's, more attention was placed on renewable energy sources, so in the early 1980's the first large-scale (>100 kW) PV power plants, based on research and development projects in the USA and Europe, were commissioned. Over the last four decades, numerous scientific papers and other publications related to large scale PV projects were published, either as part of R&D scientific projects or as part of other technical/construction activities. The range of scientific/technical publications about large scale photovoltaic power plants spans from conference proceedings to magazine papers and also includes some other entries of broader interest, like books, GIS databases and wall maps for instance. The results of the literature survey, which includes more than 800 entries among all scientific papers in English and in German, related not only to large-scale photovoltaic power plants but also to other grid-connected PV projects/power plants, are presented in this paper.
Keywords: Large Grid-connected PV systems, R&D and Demonstration Programmes, Building Integrated PV (BIPV),

1 INTRODUCTION

After the successful implementation of PV systems in space programmes, rich experiences were used for the realisation of terrestrial PV systems. The most common applications in the early beginnings of terrestrial PV applications included small rural PV power plants, PV systems for telecommunication purposes, PV systems for water supply and similar applications [1, 2]. Based on national and international PV R&D programmes [3, 4, 5], first large scale PV power plants (>100 kW) were constructed at the end of the 1970's and commissioned in the early 1980's. During the last 35 years, the market for large-scale (utility-scale) PV has not only significantly grown but has also changed from an R&D driven market into an FiT (feed-in-tariffs) driven market – in recent years, it has been on the way to being a full grid-parity driven market. After the first successful projects, based on R&D projects in the USA and Europe, the second phase of market development driven by feed-in tariffs and initiated by the German 1,000 roofs solar programme followed in the 1990's. The next important milestone in market development was the German 100,000 roof programme, substituted later by the FiT law and used also as an example for similar programmes in Italy, Spain and in some other countries. The final period of large scale PV market development, which is still ongoing, can be considered as an LCoE/grid parity driven market. A detailed overview of most important market boost support schemes is given in the literature [6, 7]. Market development has also had a significant impact on the planning and construction of large scale PV power plants.

Over the last forty years, a lot of scientific papers and other publications related to large scale PV power plants were published. Only a few literature surveys related to large scale PV power plants have been published since the 1980's [8], so this paper should, together with supplemental documentation that is available on-line, give the reader a detailed and comprehensive overview regarding scientific and technical publications published over the last forty years. The main goal of the survey was the compilation of publication titles and other important references related to large-scale PV power plants that includes not only scientific papers but also other publications with scientific or significant technical relevance, thus targeting not only the scientific audience but also, at least to a minor extent, the broader technical community.

This literature survey includes publications and studies of more than 220 grid-connected PV projects worldwide, whereas the main motivation was to compile an overview or list of publications that can serve researches in field of large-scale grid connected PV systems as a time saving tool in case of searching for particular project related information.

2 CRITERIA FOR CONDUCTING THE SURVEY

One of the first large-scale PV power plants was the 100 kW Natural Bridges National Monument PV Power System, put into service in 1980 in Utah, USA [9]. In the USA, other large-scale PV power plants (>100 kW) were also commissioned soon after that. At the same time, a PV power plant on Pellworm Island in northern Germany, with a 300 kWp power capacity of the first stage/array [10], was planned. It was commissioned in 1983 and later expanded to 600 kWp, whereas wind generators were also added, so the Pellworm plant also became the largest hybrid PV power plant at that time. These power plants were much larger than power plants constructed and commissioned at the end of the 1970's, and from the author's point of view, they represent an important milestone – the year 1980 can thus be considered as the initial year of the large scale PV power plant period and is also used as the initial year for the presented literature survey. Project development in the 1980's in the USA, as most important market at that time, is presented in the literature [11]. Due to rapid market development, the meaning of the term *large-scale PV* changed over time. Unfortunately, for the terms *utility-scale* or *large-scale PV power plants* no common accepted definition is available. A short survey of some definitions used in different publications is available in the literature [12]. Based on the survey, the term *large-scale* used in this paper differs in different time periods – the following PV power capacity ranges can be considered as a rough guide related to the literature survey:

- 1980–1989, R&D driven period, 100 kW – 6 MW
- 1990–1999, toward FiT period, 200 kW – 6 MW

• 2000–2010, FiT driven period, 1 MW – 100 MW
• 2011–2015, toward grid parity, > 20 MW

In the literature survey, not only publications about large-scale ground-mounted, grid-connected PV power plants within the above specified power ranges are considered but also other PV projects of broader interest, like fundamental or preliminary studies, important project proposals, projects from the initial period, particularly important projects for some countries, some very large systems with batteries, large-scale BIPV systems, large-scale distributed BIPV projects on housing estates and similar projects. Some other entries, like databases, maps and similar entries, with high technical relevance are also included. However, inclusion in the survey is not necessarily an endorsement of high scientific relevance or quality.

3 DATA SOURCES USED IN THE SURVEY

3.1 Scientific event series

Over the course of time, various scientific events took place and numerous scientific papers and other publications regarding large-scale PV power plants were published. With the purpose of a detailed literature survey in the period 1975–2015, several series of proceedings were investigated:

• EU PVSEC – European Photovoltaic Solar Energy Conference Proceedings (1977–2015),
• IEEE PVSC – IEEE Photovoltaic Specialists Conference Proceedings (1975–2015),
• WCPEC – World Conference on Photovoltaic Energy Conversion Proceedings (1996–2014),
• ASES – American Solar Energy Society, Annual Meetings Proceedings (1980–2015),
• Symposium Photovoltaische Solarenergie, Bad Staffelstein (1986–2015),
• ISES Solar World Congress Proceedings (1977–2015).

At least partially, some other scientific event series are also considered in the survey:

• ISES EuroSun Proceedings,
• Australian and New Zealand Solar Energy Society, ANZSES, Annual Conference Proceedings,
• Canadian Solar Energy Society Conference Proceedings.

Conference proceedings were the main and most important source of information for conducting the survey. Over the past forty years, more than 200 scientific events in English and in German - where large-scale (utility-scale) PV power plants were at least also partially discussed - took place worldwide. The most important are the contributions of the EU PVSEC and IEEE PVSC conference proceedings, followed by events with high regional importance like ASES annual conference and the Symposium in Bad Staffelstein, Germany or the ANZSES annual conferences. For a detailed evaluation of conference papers and posters, please refer to Figure 3.

All of the above mentioned events are scientific events series in the English language – an exception is the Symposium in Bad Staffelstein, Germany, which is oriented towards the German PV industry, but it each year also includes numerous excellent papers with high scientific relevance and/or technical value, so papers from this event are for this reason also included in the survey.

3.2 Other scientific sources

Other scientific publications and databases researched and included in the survey are:

• IEEE Xplore Digital Library [13],
• Scientific workshops reports (AIAA, NASA...),
• Scientific journals (Progress in Photovoltaics, Solar Energy, Solar Cells...),
• Scientific books (Advances in Solar Energy, books based on IEA PVPS Task activities...),
• Scientific reports (IEA PVPS...),
• NASA's Energy, a Continuing Bibliography with Indexes [14].

3.3 Other data sources with high technical relevance

In the survey, other sources were also considered, under condition that they provide information with significant technical relevance. This group of sources includes:

• Magazine papers (Photovoltaics International, Solar Today, Photon...),
• Different reports (REN21...),
• Personal communication [15].

4 PUBLICATIONS INCLUDED IN THE SURVEY

4.1 Publications included in the survey

The survey includes publications published in English and in German. Publications or similar entries from the below listed groups are included in the survey:

• Conference proceedings,
• Conference papers and posters,
• Journal papers,
• Books,
• Workshops proceedings and reports,
• Other reports,
• Magazine papers,
• GIS databases,
• Other databases
• Maps in books and journals, wall maps.

4.2 Data included in the survey

The data included in the survey are listed below – some of these entries, like *Uniform Resource Locator* (URL) or *Digital Object Identifier* (DOI), are of course not available for each publication listed in the survey:

• Author(s), Editor(s),
• Year of publication,
• Title,
• Scientific event or journal,
• Volume, issue, pages,
• Publisher,
• URL,
• ISBN, ISSN,
• DOI.

The results of the survey are presented as entries listed in a searchable publicly available pdf file [16]. At the beginning of each section, relevant proceedings, studies and reports are listed, followed by conference and journal papers. For easier searching/navigation, entries are provided in groups spanning a five year period. For each paper, keywords and a project description, depending on the content, were provided by the author of this paper. Keywords include terms like: *project description, project evaluation, project proposal,* etc. An example of an entry as included in [16] is:

VERVAART, M., R. MERTEN, R. COOL and T. CADIN. 2010. Performance Analysis of the First Medium Size PV-Plant in Burkina Faso on the City Grid of Ouagadougou. In: 25th European Photovoltaic Solar Energy Conference Proceedings, Valencia, Spain, 6–10 September 2010, 3947–3951.

doi: 10.4229/25thEUPVSEC2010-4DO.11.6

keywords: project description, project evaluation

project: Ouagadougou, Burkina Faso

At the end of the survey, a list of the discussed PV projects, including the year of commissioning, is given [16]. The survey also includes a reference list and a list of some other, from a historical point of view, interesting sources. The number of PV projects, most of them subjects of R&D programmes in the period 1975–2015, described in different publications and included in the survey is presented on Figure 1.

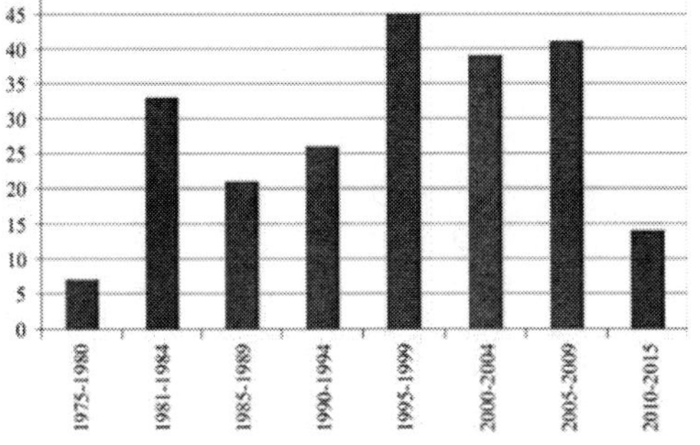

Figure 1: The number of commissioned PV projects, subjects of different R&D programmes, described in the publications considered in the survey in the period 1975–2015

5 RESULTS OF THE LITERATURE SURVEY

5.1 Scientific conference proceedings

Over the course of time, different scientific events took place and numerous scientific papers and other publications regarding large-scale PV power plants were published. With the purpose of a detailed literature survey, several series of proceedings in the period 1975–2015 were investigated; for details, see Figure 2.

In case of merged events like *World PV conference*, a single proceeding is usually available for all merged events; in such cases, it is listed under *other* entries.

5.2 Scientific papers and posters

More than 800 papers and conference posters related to PV power plant related projects were published over the last forty years. The events that contributed most of the high quality papers with significant scientific relevance are EU PVSEC, IEEE PVSC, the Symposium in Bad Staffelstein and the ASES Annual Conference event series. Other events like ISES Solar World Congress, EuroSun and the ANZSES Annual Conferences contributed few publications, but despite that, some high quality papers can be found even in the proceedings of these events. A summary of the published papers and posters is given in Figure 3. The short analysis shows the significance of particular scientific conference series to the contribution of papers and posters about the subject. For details, please see Figure 4. EU PVSEC, IEEE PVSC and, particularly in the period 1995-2010, also the Bad Staffelstein Symposium contributed the vast majority of scientifically relevant papers.

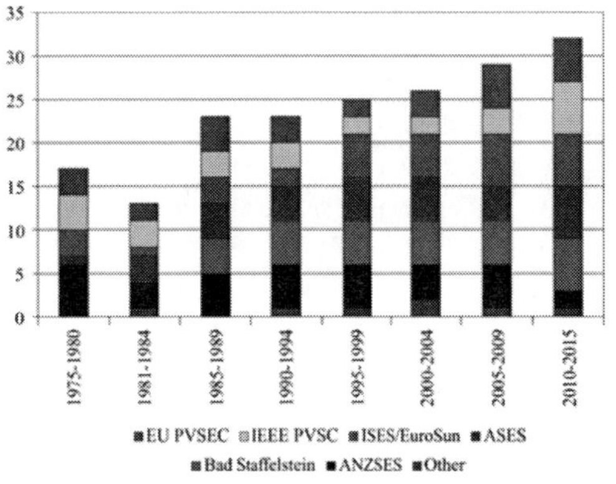

Figure 2: The number of scientific events/proceedings at least partially related to large-scale grid-connected PV projects in the period 1975–2015

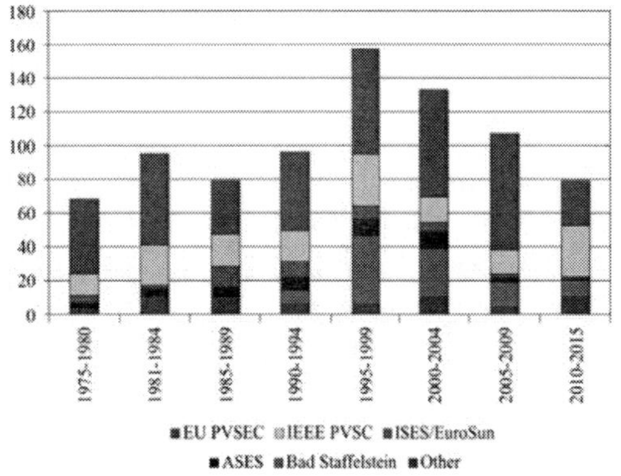

Figure 3: Published journal and conference papers and posters about PV power plants in the period 1975–2015. *Remark:* in case of joint events, each relevant group includes one half of the papers. EU PVSEC group also includes EC contractor meetings and EC driven PV project contributions in the 1980's and 1990's.

5.3 Reports and Workshops Proceedings

Workshops about early projects in the EU are a very important part of this group. Workshop proceedings include many interesting details like investment costs, construction and maintenance efforts, etc. that can help to understand the technical and economic development of PV power plants and markets over the last four decades from a historical point of view. Other groups of reports include REN21 reports about renewable energy development, targeting the general audience, which each year also include a short section about the largest PV power plants commissioned recently. Another group of reports are industrial and not scientific reports, but they include some technical details about large-scale PV power plants that are interesting from a technical point of view.

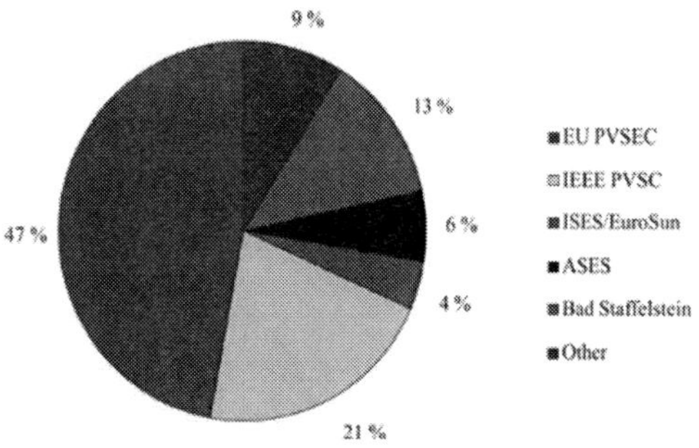

Figure 4: Contribution of scientific conference papers and posters of particular scientific event series in the period 1975–2015. *Remark:* EUPVSEC includes also EC contractor meetings and EC driven PV project meeting contributions in the 1980's and 1990's

5.4 Books

Books about large-scale PV power plants are the result of different R&D programmes, working tasks and similar activities. The books' range spans from historical activities to preliminary studies and applications of large scale ground mounted PV power plants as well as large-scale distributed PV power plants in urban areas. A very good overview about the early development in the USA is given in [11]. Books related to large scale PV power plants are mainly the result of the activities IEA PVPS Task 7, Task 8 and Task 10. The four volume series *Energy from the Desert* [17] is noteworthy as well as books about BIPV projects, including large-scale distributed BIPV systems [18, 19].

5.5 Databases

Two groups of databases with data related to large-scale PV projects are available. The first group includes databases with technical details or with performance related data of PV projects located worldwide. Examples of such databases are [20, 21]. GIS databases include large scale PV power plants and their locations, e.g. geographic coordinates. Other databases include technical data about large scale PV power plants – these databases were established on behalf of IEA PVPS Task 10 and Task 13 activities. An overview of available databases is given in Table 1.

5.6 Magazine Papers

Magazine papers are included in the survey only if they have significant technical relevance. Included are papers from Photovoltaics International, Photon, Solar Today and Refocus magazines. Papers from Photon and Refocus offer mainly a general overview, but papers from Photovoltaics International are more science oriented. Papers from Solar Today offer significant technical details and can be considered as relevant case studies. Additional technical project data sheets not included in the survey can also be found in SolarPro magazine.

5.7 Other entries

Other entries include maps with locations of large scale PV power plants published in books, journals and magazines as well as educational wall maps based on GIS databases, as given in Table I. In Table II, maps published in books and magazines as well as wall maps with locations of large-scale PV power plants are listed. Maps are particularly important because of their high educational value. If such maps are used in public lectures, they can also significantly boost the public perception regarding large-scale PV projects.

Table I: GIS and other (*)databases that include data related to large-scale PV power plants. For additional information and explanations, see [16]

Database owner	Contents
ITAS, KIT Karlsruhe	Locations of large- scale PV power plants worldwide, as in 2008, with a focus on Germany and Spain, as published on maps in [22].Database maintained by U. Dewald, PhD, ITAS, KIT
La Tene, Dublin	MW-ranged PV power plants in EU, UK and Ireland, 2016
Solargis, Bratislava [23]	Database with PV power plants in Slovakia, 2014
*ZSW, Stuttgart [24]	Ground mounted PV power plants in Germany, 2012
*PVdatabase [25]	Building integrated and urban photovoltaic solar energy projects and products
*IEA PVPS Task 13 [26]	Performance Database, web application, 2000–2005

Table II: Maps with locations of large scale PV power plants published in scientific books and journals as well as in magazines in period 1980–2015. Maps based on data for years given in table. For detailed information, see [16]

Author, publication	Map description
Treble et al. [27]	Proposed PV projects in EU, 1980
National Geographic [28]	Large-scale PV power plants worldwide (>100kW), 2004
ZSW, Stuttgart [24]	Ground-mounted PV power plants (>1MW) in Germany, 2014
U. Dewald, PhD, ITAS, KIT, Karlsruhe [22, 29]	PV power plants in EU countries and in the USA, 2008
Photon [30]	MW ranged PV Power plants in Spain, 2009
Journal Photovoltaique [31]	Large-scale PV power plants in France, 2012
La Tene Maps, Dublin [32]	Wall maps with locations of MW ranged PV Projects in EU, 2016, UK, 2015, and Ireland, 2016

5.8 Summary of the Results

The Results Summary is, for the sake of easier navigation, grouped into five year periods. For each period, relevant proceedings are listed followed by studies, reports and conference/journal papers. The results show a significant increase in publications in the feed in driven period. It also gives an overview about different scientific events – the main contributions are from EU PVSEC, followed by IEEE PVSC, Bad Staffelstein and the ASES Annual Conferences. The survey also includes large-scale distributed projects like PV on housing estates and similar projects. Other non-scientific entries include case studies related to large scale PV projects, with Solar Today as the main source of such information. In the appendix of the literature survey, a list of projects included in the survey is also given [16]. A summary of results of the Literature Survey is given in Table IIIa and IIIb.

Table IIIa: Literature survey, summary of results as included in [16] for period 1975–1994. A 1975–1980, B 1981–1984, C 1985–1989, D 1990–1994. For detailed information, see [16]

Period	A	B	C	D
Proceedings	17	14	23	23
Papers	68	95	79	96
Workshops	1	3		4
Studies	5	9	1	
Reports	1	7	2	3
Maps in books, journals	1			
Books				1

Table IIIb: Literature survey, summary of results as included in [16] for period 1995–2015. E 1995–1999, F 2000–2004, G 2005–2009, H 2010–2015. For detailed information, see [16]

Period	E	F	G	H
Proceedings	25	27	29	32
Papers	157	133	107	79
Reports	2	3	5	14
Maps in books, journals			4	2
Books		1	4	1
Magazine papers(*)	3	3	2	
Wall maps(**)				5
Other	1			

* papers from Photovoltaics International, Photon, Solar Today and Refocus included.
** available as paper maps or pdf files

6 POSSIBLE IMPROVEMENTS OF THE SURVEY RESULTS IN THE FUTURE

Despite thorough research work, some improvements of the survey results could be made in the future. Due to limited research time in Hannover, for a few papers (Symposium, Bad Staffelstein) only the lead author is listed in this survey. Additionally, some papers from the proceedings of the ANZSES annual conferences are not included, however proceedings are listed in the supplement file. An additional improvement would also be if the first names of the authors would be included in the future.

For easier navigation, keywords were defined by the author for each entry listed in the survey. Examples of keywords include: *project description, project overview, project evaluation, preliminary study, project proposal,* etc. Keywords were defined based on the paper's project relevance and could be used in the future in an electronic resource if entries would be transferred to an electronic database.

7 CONCLUSION

The presented literature survey includes more than 800 publications related to large-scale or utility-scale PV projects. The file with the results of the literature survey is a supplement to this paper and is publically available as a pdf file [16]. The main goal of the presented survey was the compilation of all available large-scale PV power plant related publications, serving, among other things, as a time saving resource for researchers searching for PV project related information like case studies or similar information. The majority of the entries have scientific relevance, but the survey also includes some entries with high educational value, useful in economic geography for instance, like GIS databases and maps with locations of large-scale PV projects. Further updates in the future and even an extension into searchable PV project oriented electronic databases are also possible. Updates could include some missing papers from national conferences like ANZSES for example. One serious obstacle in regard to the literature survey regarding large-scale PV systems was the lack of a common accepted definition for *large-scale* or *utility-scale* PV power plants. To overcome this limitation, a short proposal about the possible definition is under discussion with the main intention of starting a technical discussion about this subject on an international level.

The literature survey [16], a supplement to this paper, is a non-commercial pdf file available to the public, and it can be downloaded at the following link: http://www.pvresources.com/download/EUPVSEC2016_7DV.4.36.supplement.pdf.

8 ACKNOWLEDGEMENTS

Technische Informationsbibliothek (TIB), Hannover - for allowing me access to their library collection and for support during my stay in Hannover. Without TIB Hannover, this survey would have not been possible.

Central Technological Library at the University of Ljubljana (CTK), Interlibrary Loan Department - for providing an extensive amount of requested papers and other information during conducting the survey.

Mary Brunisholz, IEA-PVPS for providing the Task 6 workshop report, available only in the Task 6 archive.

Ulrich Dewald, PhD, Institut für Technikfolgen-abschätzung und Systemanalyse (ITAS), Karlsruher Institut für Technologie (KIT), for permission to use a map of Andalusia as an example of a literature survey entry on the conference poster.

National Geographic Creative, for permission to use a map as a survey example on the conference poster.

9 REFERENCES

[1] Palz, Wolfgang Ed. 2011. Power for the World. The Emergence of Electricity from the Sun. Singapore: Pan Stanford Publishing. ISBN 978-981-430-337-8.

[2] Loferski, Joseph, J. 1993. The first forty years: A brief history of the modern photovoltaic age. Progress in Photovoltaics Research and Applications, 1, 67–78. DOI 10.1002/pip.4670010109.

[3] Schueler, Donald, G. Ed. 1979. A bibliography of reports by the Sandia photovoltaic projects. Prepared by Sandia Laboratories for the U.S. Department of Energy. Albuquerque, New Mexico: Sandia Laboratories.

[4] JPL. 2013. Overall Array Design Requirements, Concepts, and Cost Optimization, Central-Station Ground Mounted Arrays and Support Structures. Jet Propulsion Laboratory, California Institute of Technology. http://www2.jpl.nasa.gov/adv_tech/photovol/Pub_arraytech.htm (26.3.2016).

[5] Palz, Wolfgang Ed. 1981. Photovoltaic Power Generation Proceedings of the Final Design Review Meeting on EC Photovoltaic Pilot Projects. Solar Energy R&D in the European Community Series C, Volume 1. Brussels, 30 November – 2 December 1981. Brussels: Commission of the European Communities, D. Reidel Publishing. ISBN 90-277-1386-3.

[6] Mendonça, Miguel, David Jacobs and Benjamin Sovacool. 2010. Powering the Green Economy. The feed-in tariff handbook. London: Earthscan. ISBN 978-1-84407-858-5.

[7] Mendonça, Miguel. 2009. Feed-In-Tariffs, Accelerating the Deployment of Renewable Energy. London: Earthscan. ISBN 978-1-84407-788-5.

[8] Sugimura, Russel, S. and Jane, M. Wood. 1987. Utility Application of Photovoltaic Power Generation: A Survey of Recent Literature. IEEE Transactions on Energy Conversion, EC-2, 4, 563–569. DOI 10.1109/TEC.1987.4765893.

[9] Lyon, E. F., L. L. Bucciarelli and E. A. Benoit. 1978. Design of the Natural Bridges National Monument 100 kW PV Power System. In: Conference Record of the 13th IEEE Photovoltaic Specialists Conference. Washington, D.C., June 5–8, 1978, 1268–1271.

[10]Cordes, V. and K. H. Korupp. 1980. 300 kW Photovoltaic Pilot Plant Pellworm/ Germany. In: Medium-Size Photovoltaic Power Plants. EEC/DOE Workshop Proceedings, Commissariat a l'Energie Solaire, Sophia-Antipolis, France, 23–24 October 1980, 55–58.

[11]DeMeo, Edgar, A. and Peter Steitz. 1990. The U.S. Electric Utility Industry's Activities in Solar and Wind Energy Survey and Perspective. In: Advances in Solar Energy, Böer, Karl, W. Ed., 6, Boulder, CO: American Solar Energy Society, New York, London: Plenum Press. ISBN 978-1-4613-9950-6. DOI 10.1007/978-1-4613-9948-3.

[12]Bollinger Mark and Joachim Seel. 2015. Utility-Scale Solar 2014, An Empirical Analysis of Project Cost, Performance, and Pricing Trends in the United States. Energy Analysis & Environmental Impacts Division, Lawrence Berkeley National Laboratory. https://emp.lbl.gov/sites/all/files/lbnl-1000917.pdf (14.2.2016).

[13]IEEE Xplore Digital Library, http://ieeexplore.ieee.org/Xplore/home.jsp (7.5.2016)

[14]NASA. 1982. Energy, a Continuing Bibliography with Indexes, 32, 33. NASA SP-7043(32), SP-7043(33).

[15]Brunisholz, Mary. 2016. Personal communication (March 2016).

[16]Lenardic, Denis. 2016. Large-Scale (Utility-Scale) PV Power Plants, Literature Survey. Supplemental Information to paper 7DV.4.36, presented on EUPVSEC 2016. http://www.pvresources.com/download/EUPVSEC2016_7DV.4.36.supplement.pdf (15.6.2016).

[17]Energy from the desert, Fact sheets and the summary of the research, http://www.iea-pvps.org/index.php?id=95&eID=dam_frontend_push&docID=2575 (7.5.2016).

[18]Snow, Mark and Prasad Deo Eds. 2005. Designing with Solar Power. London: earthscan. ISBN 1 844071 47 2.

[19]Gaiddon, Bruno, Henk, Kaan and Donna Munro Eds.2009. Photovoltaics in the Urban Environment, Lessons Learnt from Large-Scale Projects. London: earthscan. ISBN 978-1-84407-771-7.

[20]IEA PVPS Task 13. Performance Database web application, http://77.245.18.90/ (11.5.2016)

[21]PVdatabase. Building integrated and urban photovoltaic solar energy projects and products, http://www.pvdatabase.org/ (11.5.2016).

[22]Lenardic, Denis. 2008. Installations of PV Power Plants in 2008, Photovoltaics International, 3rd Edition, Semiconductor Media Publ., 148–153.

[23]Šuri, Marcel. 2016. Solargis. Personal communication (May 2015).

[24]ZSW. 2014. Solare Strahlungsenergie Wissenschaftlicher Bericht. Vorbereitung und Begleitung der Erstellung des Erfahrungsberichts 2014 gemäß § 65 EEG im Auftrag des Bundesministeriums für Wirtschaft und Energie. Vorhaben IIc. Stuttgart: ZSW, url: http://www.bmwi.de/BMWi/Redaktion/PDF/XYZ/zwischenbericht-vorhaben-2c,property=pdf,bereich=bmwi2012,sprache=de,rwb=true.pdf (7.5.2016)

[25]PVdatabase, http://www.pvdatabase.org (16.5.2016)

[26]IEA PVPS Task 13, Performance Database web application, http://77.245.18.90/ (16.5.2016)

[27]Treble, F. C., G. Grassi and W. Schnell. 1980. Photovoltaic Pilot Projects in the European Community. In: Third E.C. Photovoltaic Solar Energy Conference Proceedings, Cannes, France, 27–31 October 1980, 513–520.

[28]National Geographic. 2006. Collegiate Atlas of the World. ISBN 0-7922-3662-9.

[29]Lenardic, Denis, Sibylle Petrak and Ulrich Dewald. 2008. Large-Scale PV Power Plants, Annual Review 2008. ISBN 978-961-245-739-6.

[30]Rutschmann, Ines. 2009. Land der Megawattparks. Photon, 9, 40–45.

[31]Le Annic, N. and A. Guillier. 2012. L'Atlas des Grandes Centrales Photovoltaïques. Le Journal du photovoltaïque. Novembre 2012, 44–83.

[32]La Tene Maps. 2016, http://latenemaps.com (7.5.2016).

SOLAR-ERA.NET
ERA-NET ON SOLAR ELECTRICITY FOR THE IMPLEMENTATION OF THE SOLAR EUROPE INDUSTRY INITIATIVE

Stefan Nowak, NET Nowak Energy & Technology, Waldweg 8, 1717 St. Ursen, Switzerland,
T: +41 26 494 00 30, F: +41 26 494 00 34, stefan.nowak@netenergy.ch
Marcel Gutschner, NET Nowak Energy & Technology, marcel.gutschner@netenergy.ch
Stefan Oberholzer, Swiss Federal Office of Energy, stefan.oberholzer@bfe.admin.ch
Christoph Hünnekes, Forschungszentrum Jülich – PtJ, ch.huennekes@fz-juelich.de
Herman Bastek, Forschungszentrum Jülich – PtJ, h.bastek@fz-juelich.de
Martina Biedrawa, Forschungszentrum Jülich – PtJ, m.biedrawa@fz-juelich.de
Dorothea Winkler - Forschungszentrum Jülich – PtJ, d.winkler@fz-juelich.de
Melanie Schulte, Forschungszentrum Jülich – ETN, me.schulte@fz-juelich.de
Stefan Rabe, CEF-NRW, rabe@cef.nrw.de
Karin Wikman, TEKES Teknologian Ja Innovaatioiden Kehittaemiskeskus, Karin.Wikman@tekes.fi
Yvonnick Durand, Agence de l'Environnement et de la Maîtrise de l'Energie (ADEME), yvonnick.durand@ademe.fr
Manon Gerbaud, Agence de l'Environnement et de la Maîtrise de l'Energie (ADEME), manon.gerbaud@ademe.fr
José Herrero, Centro de Investigaciones Energéticas, Medioambientales y Tecnológicas (CIEMAT),
jose.herrero@ciemat.es
Juan Francisco Trigo Escalera, Centro de Investigaciones Energéticas, Medioambientales y Tecnológicas (CIEMAT),
juanfrancisco.trigo@ciemat.es
Severino Falcon Morales, MINECO, severino.falcon@mineco.es
Laurence Polain, Service Public de Wallonie (SPW), laurence.polain@spw.wallonie.be
Nicolas Delsaux, Service Public de Wallonie (SPW), nicolas.delsaux@spw.wallonie.be
Elsie De Clercq, Agentschap Innoveren & Ondernemen (VLAIO), elsie.declercq@vlaio.be
Anna Ostapczuk, Narodowe Centrum Badan i Rozwoju (NCBR), anna.ostapczuk@ncbr.gov.pl
Małgorzata Świderska, Narodowe Centrum Badan i Rozwoju (NCBR), malgorzata.swiderska@ncbr.gov.pl
Kaan Karaösz, Turkiye Bilimsel ve Teknolojik Arastirma Kurumu (TUBITAK), kaan.karaoz@tubitak.gov.tr
Ismail Dogan, Turkiye Bilimsel ve Teknolojik Arastirma Kurumu (TUBITAK), ismail.dogan@tubitak.gov.tr
Otto Bernsen, Ministerie van Economische Zaken, Landbouw en Innovatie / RVO, otto.bernsen@rvo.nl
Bouke Bussemaker, Ministerie van Economische Zaken, Landbouw en Innovatie, B.H.Bussemaker@minez.nl
Stathis Tselepis, Kentro Ananeosimon Pigon ke Exikonomisis Energeias (CRES), stselep@cres.gr
Christian Inglis, Innovate UK, Christian.Inglis@innovateuk.gov.uk
Graham Mobbs, Innovate UK, graham.mobbs@innovate.gov.uk
Leonidas Antoniou, Research Promotion Foundation (RPF), lanto@research.org.cy
Demetra Petsa, Research Promotion Foundation (RPF), dpetsa@research.org.cy
Adriana Agrimi, Regione Puglia, a.agrimi@regione.puglia.it
Dario Tornabene, Regione Sicilia, dario.tornabene@regione.sicilia.it
Fabio Montagnino, Regione Sicilia, fmontagnino@consorzioarca.it
Theodor Zillner, Bundesministerium für Verkehr, Innovation und Technologie (BMVIT), theodor.zillner@bmvit.gv.at
Ulrike Rohrmeister, Bundesministerium für Verkehr, Innovation und Technologie (BMVIT),
Ulrike.Rohrmeister@bmvit.gv.at
Elvira Lutter, Klima- und Energiefonds, elvira.lutter@klimafonds.gv.at
Gernot Wörther, Klima- und Energiefonds, gernot.woerther@klimafonds.gv.at
Pierre-Jean Rigole, Swedish Energy Agency, Pierre-Jean.Rigole@energimyndigheten.se
Tobias Walla, Swedish Energy Agency, tobias.walla@energimyndigheten.se

ABSTRACT: SOLAR-ERA.NET is a network that brings together more than 20 RTD and innovation programmes in the field of solar electricity technologies in the European Research Area. The network of national and regional funding organisations has been established in order to increase transnational cooperation between RTD and innovation programmes and to contribute to achieving the objectives of the Solar Europe Industry Initiative (SEII) through dedicated transnational activities (especially transnational calls). SOLAR-ERA.NET is an EU funded FP7 project running from 2012 to 2016. Through the support of the funding organisations, some 70 MEUR will have been mobilised for transnational RTD and innovation projects through four joint calls carried out. Intermediary results are presented for the transnational calls in this paper.

Keywords: R&D and Demonstration Programmes, Photovoltaic, Funding and Incentives, Strategy

1 INTRODUCTION AND CONTEXT

SOLAR-ERA.NET is a European network of national and regional funding organisations and RTD and innovation programmes in the field of solar electricity generation, i.e. photovoltaics (PV) and concentrating solar power (CSP) / solar thermal electricity (STE). SOLAR-ERA.NET shall contribute to reaching the objectives of the former Solar Europe Industry Initiative (SEII), recently replaced by the European Technology & Innovation Platforms (ETIP) for PV and STE, by carrying out the coordination and support actions for the implementation of the SEII between national and regional RTD and innovation programmes.

The SEII was a joint initiative of the industry sector, EC and member states and is embedded in the European Strategic Energy Technology Plan (SET-Plan) which aims to increase, coordinate and focus EU support on key low-carbon energy technologies in order to achieve Europe's 2020 energy objectives in the future.

2 NETWORK AND MISSION

2.1 Network

As the largest ERA-NET (network in the European Research Area) ever in the solar power field, SOLAR-ERA.NET involves more than 20 national and regional RTD and innovation programmes dealing with PV and CSP. This high level of involvement of most relevant stakeholders provides excellent outreach and allows for a solid coordination needed for an efficient and coherent approach in the highly diverse and versatile RTD landscape.

The network is composed of 19 organisations being programme owners and / or programme managers (see list of authors). Countries and regions participating in the SOLAR-ERA.NET consortium and / or in the trans-national calls are identified in Figure 1. SOLAR-ERA.NET is in principle open to other countries and programmes wishing to join in transnational calls.

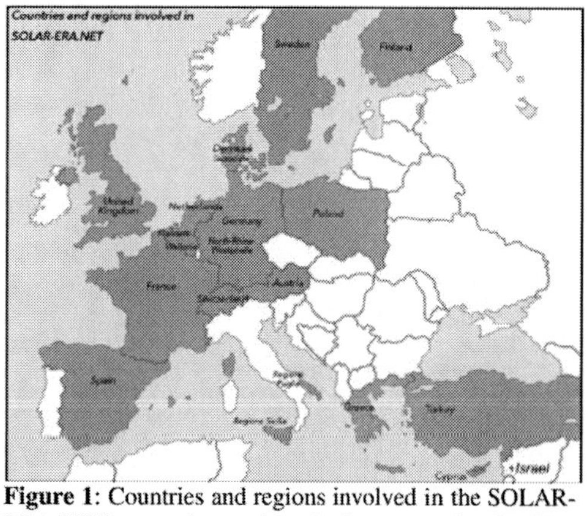

Figure 1: Countries and regions involved in the SOLAR-ERA.NET consortium and / or in the transnational calls

2.2 Missions and goals

SOLAR-ERA.NET has two fundamental missions. As a network supported by the EC within the ERA-NET scheme, the mission is to improve the coordination and cooperation between national and regional RTD programmes. In the context of the Solar Europe Industry Initiative (SEII), the mission of the network is to implement central parts of the SEII on a transnational level and thus contribute to achieving the goals defined in the SEII.

The more specific and essential goals and activities of SOLAR-ERA.NET are i) to launch joint calls for RTD proposals by national and regional RTD and innovation programmes and ii) to define and support the best joint activities, strategic information exchange and use of implementation tools.

By identifying and choosing SEII priority topics based on the Implementation Plans for PV respectively CSP for transnational calls, the SOLAR-ERA.NET network shall select and fund industrially relevant transnational RTD and innovation projects in the field of solar electricity technologies.

In quantitative, financial terms, SOLAR-ERA.NET shall result in a total funding volume by the participating national and regional programmes of approximately 50 MEUR for innovative projects.

3 ACTIVITIES, FACTS AND FIGURES

3.1 Set up of transnational calls

The network started in November 2012. Four sets of transnational joint calls have been launched so far in March 2013, January 2014, December 2014 and December 2015 with the participation of up to 17 countries covering up to 8 PV and 4 CSP topics.

Call topics in photovoltaics (PV) and concentrating solar power (CSP) are commonly defined by the SOLAR-ERA.NET consortium and adopted by the participating programmes and countries according to their national / regional priorities. The topics can be found in Table 1.

The transnational call is based on a 2-step/stage-procedure with i) a preproposal and ii) a full proposal stage. Preproposals are checked according to the national / regional regulations and then discussed in the SOLAR-ERA.NET consortium setting up lists of preproposals recommended respectively not recommended for stepping on to the full proposal phase.

Once the full proposals are evaluated by independent international experts as well as by national experts, a list of projects suggested for funding is established. The ultimate funding decisions are however taken by the national agencies / ministries.

The transnational SOLAR-ERA.NET joint calls are each allocated with a total public funding budget of approximately 12 million euros provided by the participating national and regional programmes and agencies. The joint call consortium is mainly composed of SOLAR-ERA.NET partners but can be extended by associate partners (like Denmark and Israel). Applicants may also come from other countries (like Ireland and Norway) but they do have to provide their own funding for their participation in the projects.

3.2 Participation in transnational calls

The four transnational SOLAR-ERA.NET calls found good interest in the solar power industry sector and research community. Some 180 preproposals were submitted involving more than 800 partners respectively contributions from 18 countries. The fourth joint call is currently in the second stage, i.e. the phase from preproposals to full proposals.

A slight majority (54%) of the organisations involved in the preproposals submitted belong to the categories of small and medium sized enterprises and large enterprises; research organizations have a share of 43%. Figure 2 shows the share of different types of organizations among the partners being involved in the preproposals submitted.

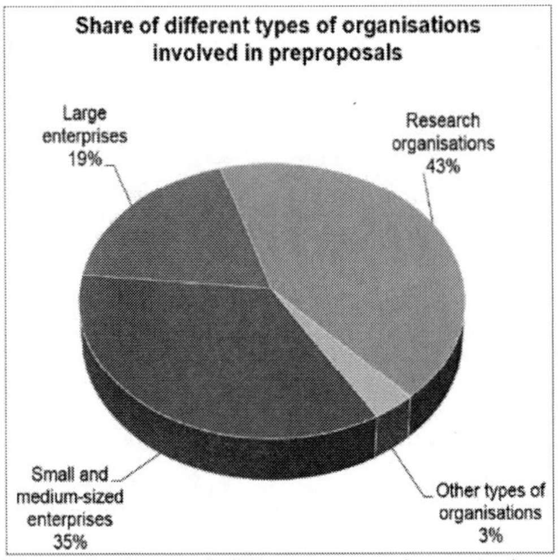

Figure 2: Share of different types of organisations involved in preproposals

66 full proposals were submitted in the first three joint calls. Figure 3 shows the number of coordinators and partners in the full proposals submitted by call as well as by country / region.

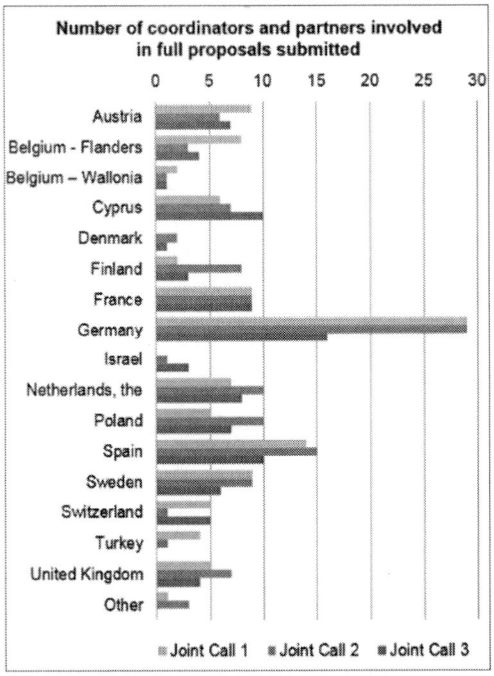

Figure 3: Number of coordinators and partners in full proposals submitted according to the countries / regions participating in the three transnational SOLAR-ERA.NET calls

Out of the 66 full proposals submitted, 57 were for PV topics and 9 for CSP topics. Table 1 shows the number of full proposals submitted per call topic. Some topics have been added in the second and third call.

Table 1: Three transnational SOLAR-ERA.NET calls - topics and numbers of full proposals submitted (number of full proposals per call and topic as well as total sum)

Transnational SOLAR-ERA.NET topics in calls 1, 2, 3 and all three together	1	2	3	Σ
PV1 Innovative processes for inorganic thin-film cells & modules	3	4	2	*9*
PV2 Dedicated modules for BIPV design and manufacturing	6	3	1	*10*
PV3 Grid integration and large-scale deployment of PV	3	2	4	*9*
PV4 High-efficiency PV modules based on next generation c-Si solar cells	3	5	4	*12*
PV5 Solar glass and encapsulation materials	2	3	1	*6*
PV6 Concentrator PV technology	-	3	3	*6*
PV7 Si feedstock, crystallization and wafering	-	2	1	*3*
PV8 Organic solar cells and other emerging concepts	-	-	2	*2*
CSP1 Cost reduction and efficiency increase in components	2	0	1	*3*
CSP2 Dispatchability through storage and hybridisation	2	1	1	*4*
CSP3 New fluids for CSP plants	0	1	1	*2*
CSP4 Innovative thermodynamic cycles	-	0	0	*0*
Total	*21*	*24*	*21*	*66*

4 PRELIMINARY AND EXPECTED RESULTS

4.1 Preliminary results

The transnational SOLAR-ERA.NET calls shows good interest and participation of the sector:

- 180 preproposals submitted with total project costs of 230 MEUR and funding requested of 160 MEUR
- 803 partners / contributions involved in preproposals
- 17 countries and regions participating in the joint call consortium
- 66 full proposals submitted (after three joint calls)
- 25 projects running (mainly out of first and second joint call)
- Funding committed so far for the 25 projects running: approximately 24 MEUR public funding for an approximate total project volume of 35 MEUR

Although some delays and insolvency issues occurred, SOLAR-ERA.NET is rather on track, dedicating about 10 to 12 MEUR of funding per set of transnational joint calls on innovative PV and CSP projects.

The 25 projects (22 projects with PV and 3 projects with STE) running are:

- SLAGSTOCK: Low-cost Sustainable Thermal Energy Storage Systems Made of Recycled Steel

Industry Waste
- LIMES: Light Innovative Materials for Enhanced Solar Efficiency
- INTESEM: Intelligent Solar Energy Management Pipeline from Cell to Grid
- NOVACOST: Non vacuum based strategies for cost efficient low weight chalcogenide photovoltaics
- NovaZOlar: All-non-vacuum processed ZnO-based buffer and window layers for CIGS solar cell technology
- InnoModu: Leadfree modules with low silver content and innovative busless cell grid
- AER II: Industrialization and System Integration of the Aesthetic Energy Roof Concept
- PV4FACADES: Photovoltaics for high-performance building-integrated electricity production using high-efficiency back-contact silicon modules
- BLACK: Black silicon and defect engineering for highly efficient solar cells and modules
- PV2GRID: A next generation grid side converter with advanced control and power quality capabilities
- THESEUS: Tandem High Efficiency Solar Cells Utilizing III-V Semiconductors on Silicon
- APPI: Atmospheric Pressure Processing for Industrial Solar Cells
- SNOOPI: Smart Network Control with Coordinated PV Infeed
- SPRINTCELL: Sulfide-based Ink for Printable Earth Abundant Solar Cell
- IPERMON : Innovative Performance Monitoring System for Improved Reliability and Optimized Levelized Cost of Electricity
- ACCESSCGIS: Atmospheric Cost Competitive Elemental Sulfo Selenization for CIGS
- PV me: organic PhotoVoltaic systems integrated in manufactured building elements
- U-light: Ultra lightweight PV modules and their applications in innovative PV systems achieving lowest levelized cost of electricity (LCOE)
- inGRID : High efficiency PV modules with cells interconnected by a novel conducting encapsulation layer
- MONOSCRIBE: Roll-to-Roll Monolithic Interconnection of Customizable Thin-film Solar Modules
- HighCast : High Performance Silicon Casting and Wafering
- EDITOR: Evaluation of the Dispatchability of a Parabolic Trough Collector System with Concrete Storage
- FunGlass : Multi-Functional Glass for PV Application
- SITEF: Silicone Fluid Test Facility
- PROOF: Competitive Industrialized Photovoltaic Roofing

4.2 Outlook

SOLAR-ERA.NET will likely result in some 40 RDI projects with up to 50 MEUR of public funding.

The successful projects demonstrate i) potential commercial impact / relevance to industrial and market needs / contribution to the Solar Europe Industry Initiative and added transnational value, ii) scientific and technological excellence and iii) high quality and efficiency of the implementation and the management. Further information on running projects from the transnational joint calls can be found on the solar-era.net website.

Having successfully submitted a proposal in the ERA-NET Cofund scheme, SOLAR-ERA.NET will launch another joint call by end of 2016. Allocated with a funding budget of 20 MEUR, the SOLAR-ERA.NET Cofund call will address 4 topics, i.e. innovative and low-cost PV manufacturing, advanced PV products and applications, PV system integration and CSP cost reduction and system integration, and support successful projects aiming at Technology Readiness Levels of 5 to 7

News on the future SOLAR-ERA.NET Cofund, will be channelled through the solar-era.net and funding agencies involved.

REFERENCES

Further information is available on the project website www.solar-era.net.

ACKNOWLEDGEMENTS
The support of the SOLAR-ERA.NET joint activities by the EC under grant agreement N° 321571 within the 7th Framework Programme is gratefully acknowledged.

SOLAR LIGHTING PRODUCTS AND SMALL DC APPLIANCES PUT TO THE MARKET TEST: A 3 YEAR HANDS-ON EXPERIENCE

Nicola Baggio, Francesco Minichiello
Futura Holding Limited, Via Riva del Pasubio 14, Cittadella (PD), Italy
Ph. +39 0495979802 Fax +39 049 5973490 n.baggio@futurasun.it

While in the developed countries the transition from AC to DC systems is a topic related to the introduction of electric vehicles and residential energy storage, in the low-electrified countries this choice is becoming more and more logical and cost-effective for the energy development.

Starting from small solar lighting systems, we have already met the target of powering small DC appliances such as TVs, fans, radios and computers with solar energy. Recently the order of magnitude has changed significantly: a few years ago the market of solar lanterns was about 1 Wp, this number grew to 10 Wp with Solar Home systems and now it is nearly 100 Wp with kits provided with additional appliances, in continuous current.

Public programs such as Lighting Global or Global Leap are playing an important role for standardization and technical verification of the products. However, today in most of the countries where these applications could bring greater environmental, social and economic benefits, there are many practical barriers, presented in this study, that impede the process of distribution of the products.

In this study, based on our direct experience in more than 40 markets in the last three years, we are going to present the most common problems that hinder various stakeholders: end users, manufacturers, importers and distributors.

In this field we analyze the role of the NGOs that can be extremely significant or counterproductive, depending on the approach that is being adopted. Simple donation mechanisms are likely to destroy the development of small local businesses that would need to be supported financially instead.

Finally, the study will present proposals to make the development of these markets healthier, while maintaining some understandable commercial tools, such as custom duties, which many countries, especially in Africa, cannot do without.

Keywords: Solar Lighting, Rural Electrification, Financing, PV Market

In the last few years photovoltaic systems have reached a lower electricity generation cost than fossil fuels. For medium/large grid-connected systems the turn-key price is below 1 €/Wp. The main technical issues are the unpredictability of the power output and the absence of production during night-time.
These issues shift the discussion to the availability of energy storage systems at competitive prices to guarantee a 24/7 access to energy.

In some way we came back to the origin of the photovoltaics when the systems were always developed together with batteries and energy management systems to allow off-grid energy supply.
The attention to the batteries for grid-connected systems is extremely high but in many countries, without specific subsidies, it is difficult to see a financial benefit from this application.

At the same time, the real market where PV, storage and specific DC appliances are sold, is the off-grid market.
Due to the fact that most of these products are sold in poor countries, where access to energy is limited or none, the average size is really small, even if the trend is to increase them.

The obstacles found in the distribution of these products are different from those in the common grid-connected markets.
In this study we would like to list the main issues we found in the last three years of experience in this field.

Customer's needs

For many years the diffusion of photovoltaic systems in poor countries was related to charity initiatives: in some cases the impact was positive for the local population but too often these actions were only "marketing-oriented" making the Westerns donors happy.
It is crucial to understand the needs and the tastes of the local customers at the base of the pyramid.
The access to energy is the primary need for community based services such as hospitals, water pumping systems, telecommunication antennas and schools. However, there are also individual needs related to lighting, recharging of mobile phone batteries, usage of TVs, fans, refrigerators, etc.
Until few years ago the cost of small-scale PV systems was 10 times higher than now. Therefore, the first application for individual use was a solar lantern with the peak power less than 1 Wp.
Also, today this product size represents the main market share. However, the cost reduction of solar cells, lithium batteries and led lights, as well as the contemporary quality improvement of these three technologies allowed to upgrade the system complexity in the last period, maintaining the price at acceptable level.
From the basic lighting needs, now a typical customer wants to have multiple lights and the possibility to charge phone batteries. Now a simple lantern is just a product for students and not for the entire families.

In the last two years, customers started to ask for photovoltaic applications for TVs and fans: these are not exactly "primary" needs but this reminds us of what our grandparents were looking for 60 years ago…the same!
These last requests are not only paving the way to the 100 Wp systems but also pushing the development of low consumption applications based on direct current (DC).
DC current is becoming a common denominator of the entire system: modules, batteries, led lights, mobile phones, tvs, fans, radios, etc.

There are also upper class customers looking for solutions able to satisfying their entire electricity needs (continuous blackouts, high electricity costs, etc.) with more complex off-grid systems. However, in this study we do not analyze them.

Market barriers

If the technical and economical aspects are not an issue any more, there are now other barriers that slow down the diffusion of the off-grid products.

Access to foreign currency

In many countries, especially in Africa, access to foreign currency is not easy at all. This means that, even with a real demand and a real purchasing potential, end-users and distributors/importers are not able to purchase the needed volumes of products. The scarcity of USD/EUR forces the distributors to purchase low-price products without guarantee, not because of the lower price for the end customer but because of the financial requirements.
The Lighting Africa program allows to solve this issue in Ethiopia with the help of a dedicated fund. Similar initiatives could be really useful in other countries too.

Custom duty

Only few countries exempt solar products from custom duty. The duty implicates an extra cost - up to 50% - for the end customer. It is difficult to understand the meaning of the duties for the products that have a demonstrated positive social, economic and sanitary impact.
In this sense, the current regulation in force in Kenya is a good example: custom barriers exist only for products without quality certification.

Lack of micro-finance

In most cases customers prefer to buy products with deferred payments. The total amount is relatively small, typically below 100 USD. Unfortunately, there are no homogeneous microfinance facilities. To bypass this issue, a lot of attention is paid to the pay-as-you-go (PAYG) system. However, this system, quite effective for medium-sized systems, is too expensive to be integrated into small applications.

Pre-shipping inspection certificates

Many countries require a pre-inspection certificate for the importation before the shipping from the country of origin. This could be reasonable for large volumes (to avoid illegal capital exports) but, in many cases, it can account for 10% of the total volume of small supplies. Below certain values this inspection could be removed or a specific financial support could be provided.

Export guarantee

Many medium-sized importers do not have the required financial liquidity to start their business. This issue could be solved by giving an export credit warranty to the manufacturer covering the risk of a missing payment. The credit facility could be managed by a European bank and it could offer a safe environment for the manufacturer to allow deferred payments. At the same time, by providing this warranty only to qualified manufacturers, it will be possible to exclude poor quality products.

Too many local subsidies

This is a serious issue because it creates additional confusion exactly where simplicity is essential.
There are too many funds for supporting of diffusion of small solar system products. However, in every country the mechanisms are different and sometimes very complex (i.e. with result-based subsidies). Both the manufacturers and the importers have to invest a lot of time and human resources to study and understand each mechanism. Sometimes these initiatives are related to the same "parent" grant from some external donor but applied in different way in different places.
Sometimes we have the doubt that more money is spent on "building" the mechanism itself and managing it than on the funds available for the stakeholders.

Controversial role of the NGOs and charities

Even if this issue is slowly being resolved, we have noted that too often the donations are used to give the products to the customers for free. This action is extremely meaningless because it discourages the creation of a local distribution and sales network and, at the same time, creates the expectation in the end customer that the products must be "free of charge".
It would be much better to allocate the funds collected from donors, to create rotative funds that could provide micro-finance to end customers or importers.

Conclusions

To create better conditions for the development of these markets, it is extremely important not only to promote the sale of small PV systems but also the diffusion of the confidence in a new distributed energetic model. To summarize our experience of the last three years, we can suggest the following actions:

1. Establishing a unique USD/EUR fund by foreign governments and international institutions in order to allow the importers to have access to credit lines
2. Removing custom duties for certified products
3. Simplifying of the requests for the pre-shipping certificates
4. Creating local funds with the donations of NGOs in order to promote micro-finance
5. Removing of all the local subsidies and using this money for harmonized micro-finance

These actions could lead to an enormous social benefit. Today we can say that PV energy is the most economical energy source in most places of the world. After decades of technical improving and cost reduction, it will be unintelligible to not remove the barriers, created not by some physical limits, but by human irrationality.

AN OVERVIEW OF PATENT APPLICATION DATA IN THE FIELD OF PHOTOVOLTAICS

Alberto Visentin[1], B. Erol Sağol[1], Mauro Boero[2], and Christian Königstein[2]
[1,2]European Patent Office, Directorate General 1
[1] Gitschinerstrasse 103, D-10969 Berlin
[2]Postbus 5818, NL-2280 HV Rijswijk, The Netherlands
e-mail: avisentin@epo.org, bsagol@epo.org, mboero@epo.org, ckoenigstein@epo.org

ABSTRACT: The present study is aimed at giving an update on patent activities in the photovoltaics (PV) field covering the past 45 years. In this study, the number of patent applications published during this time period, per year and for the whole period, in several of the most relevant solar cell technologies is analysed. Patent filing figures and the origin of patent applications are shown for the different types of solar cells including polymer and organic solar cells, and dye sensitized solar cells.

Keywords: PV patent activity

1 INTRODUCTION

The European Patent Office (EPO) is the executive body of the European Patent Organisation, an intergovernmental organisation that was created with the entry into force of the European Patent Convention (EPC) on 7 October 1977. In 2013 the EPO celebrated the 40th anniversary of the signature of the EPC, which forms its legal basis. The Organisation currently comprises 38 member states [1], two extension states – Bosnia-Herzegovina and Montenegro – and two validation states [6] – Morocco and Moldavia - which recognise the validity of European patent applications and patents upon request by the patentee.

Figure 1: Member states of the EPO (shown in red)

Patent applications at the EPO (which numbered over 270.000 in 2015) are assessed by the EPO's patent examiners *inter alia* on novelty and inventiveness. The examiner's search for prior art – to see if the invention is really novel or already known – requires a huge documentation in a large number of technical fields. The EPO's documentation comprises an internal database [5], containing 75.7 million published patents (corresponding to about 20 million patent family members) and 23.5 million technical literature papers. In 2015 alone, 3.58 million patent documents and about 700 000 literature documents were added. The documents are ordered using a new classification scheme called the Cooperative Patent Classification (CPC), which has been developed jointly by the EPO and the United States Patent and Trademark Office [2]. The classification scheme is built up in a hierarchical manner and based on the International Patent Classification (IPC) [3], allowing one to narrow down searches to a certain set of specifications and technical features describing the invention. The CPC classification scheme contains about 260.000 entries, allowing the precise classification of the documents – compared to about 73.400 entries in IPC. In the field of photovoltaics (PV), our database contains at present (May 2016) 115.000 documents, including 7200 non-patent literature documents. The 107.000 patent documents correspond to about 35.000 patent families.

This database also allows for statistical evaluations of the information supplied for each application, such as publication date and country of origin of the application. Due to the recently increasing contribution of organic solar cells to the field of PV and the growing number of companies producing such cells, an overview of the patent application history in the field is interesting.

2 METHODOLOGY

The presented statistics are extracted from our database using the CPC scheme to specify the fields of interest and cover the years 1970 - 2014. The numbers are based on a single count for every patent family, wherein the data are based on the priority date of the priority application, i.e. the date of first filing. Since patent applications are published 18 months after their first filing date, applications having a priority date after 1.1.2015 are not yet available and, therefore, cannot be considered. For example, an application having a priority date of 31/12/2014 will be published on 1/7/2016. Therefore, filing numbers are presented up until priority year 2014.

The geographical analyses are based on the priority patent, i.e. the country of first filing. The EPO member states [1] are taken into account as well as the US, Japan, Australia, Republic of Korea, Canada, Province of China Taiwan, China, Israel and New Zealand [4]. In addition, direct European ("EP") and international Patent Cooperation Treaty (PCT) first filings ("WO") are identified separately. When referring to patent applications in Europe, all patent applications filed with the EPO or with one of its member states are taken into account. However, when referring to EP Applications or EP Patents, only those applications are considered that have been filed with the EPO (as first or second filing).

3 RESULTS

The development of the total field of PV - including inorganic, organic and dye sensitized solar cells and modules - is shown in figure 2.

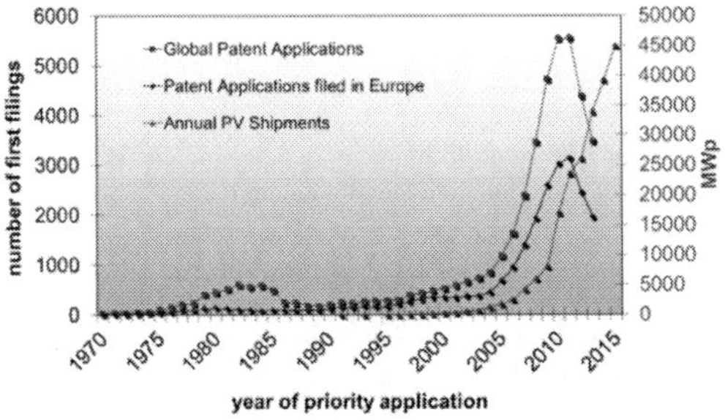

Figure 2: Annual Patent Applications (first filings) in PV

The graph shows a more or less general increase in the number of filings in the European states up to 2012. For the global number of filings - including patent applications that had been filed only outside Europe - a big temporary increase can be seen for the early 1980s, during the period of the second oil crisis. This increase is mainly attributed to US and Japanese patent applications that were not filed in Europe. Over the last two years a decrease in the number of filings has been observed – both globally and in Europe. It is partly due to a bias linked to a lack of data available, in particular for international applications where applicants have up to 30 months before entering a regional or national phase, plus the time it takes for us to receive the data and update the collection. The future will show whether this is trend is reversed in the future.

When comparing patent activity in the field of PV with activity in wind energy and nuclear energy (see Figure 3), a clear increase can be shown for the field of renewable energies (PV and wind), whereas nuclear energy (nuclear fission and nuclear fusion) did not show any substantial increase in activity.

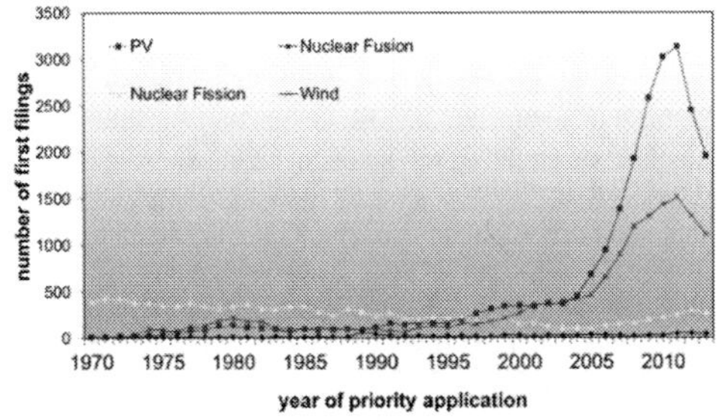

Figure 3: Patent Applications filed in Europe for PV, wind energy, nuclear fission and nuclear fusion

From 2012 onwards, we see again a downward trend in the number of filings for wind energy and nuclear fission.

Looking at patent activity in the field of PV (including all types of solar cells and PV modules), 60% of the patent applications filed in Europe, i.e. applications filed with the EPO or with one of its member states [1], originate from Germany, Japan and the US (Figure 4), with Germany still having the largest share of 25 %.

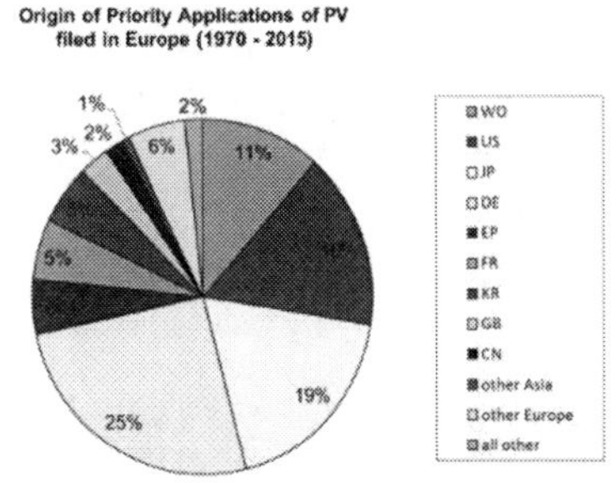

Figure 4: Origin of Patent Applications Filed in Europe - All Types of Solar Cells

A slightly different picture emerges when only those patent applications are taken into account, which are filed at the EPO.

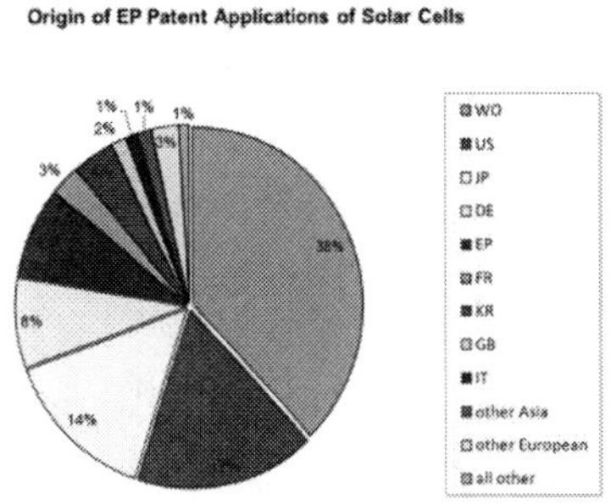

Figure 5: Origin of Patent Applications Filed at the EPO - All Types of Solar Cells

In that case, Germany has a much smaller share (8% of EP applications compared with 26% of all European applications, including national applications), while the shares of Japan (14% vs. 19%) and the US (17% vs. 16%) remain almost unchanged. This can be explained by the fact that most of the German patent applications are only filed nationally (i.e. in Germany), but not with the EPO. More than one third of the applications (38%) received by the EPO arrive via the international (PCT) route (WO). For those applications the country of origin is not accessible.

When looking at granted EPO patents (Figure 6), it can be seen that Germany has a larger share than in the patent applications (13% of the grants vs 8% of the applications). The US (14% of grants vs 17% of applications) has a similar share of applications and Japan has a larger share of grants than applications (17% of grants, 14% of applications).

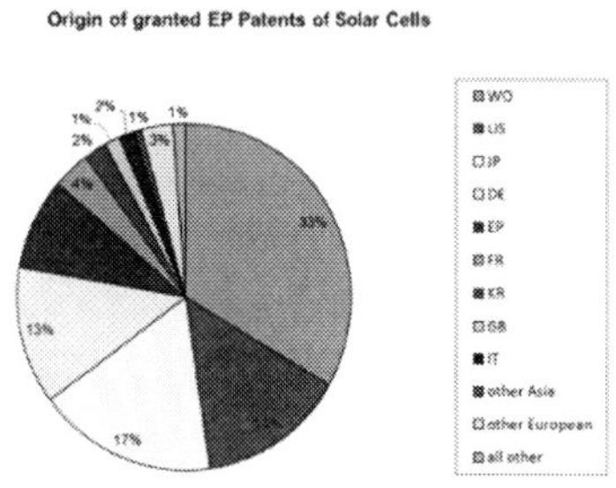

Figure 6: Origin of Granted EPO Patents - All Types of Solar Cells

Patent activity as presented in this study is usually based on patent applications. However, not every application becomes a valid (granted) patent in the end. At the EPO, on average, 50% of EP patent applications in the field of PV are granted, i.e. become a valid European patent. EP applications originating from Japan, Germany, and France are amongst those that have a higher success rate than the average (Figure 7).

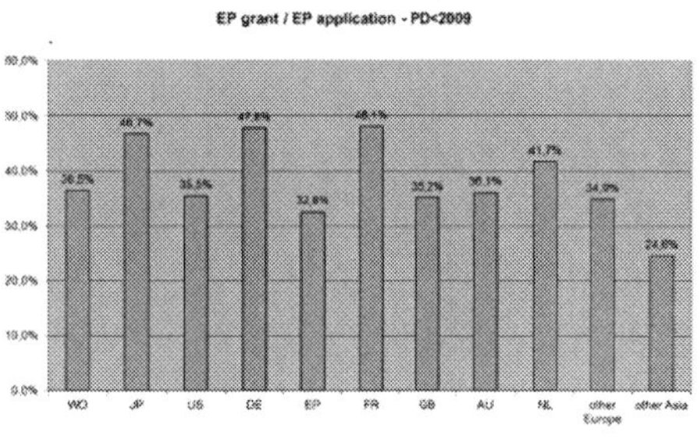

Figure 7: Granted EP Patents in % of EP Applications - All Types of Solar Cells

Up to now, filing figures have been considered for the whole period of 1970 – 2013. When analysing the filing figures for different decades (Figure 8), the evolution of the PV industry in different regions of the world can be seen. While in Germany and Japan a continuous increase in patent activity can be seen, France, the UK ("GB"), and the US showed a dip in the period 1984 – 1993.

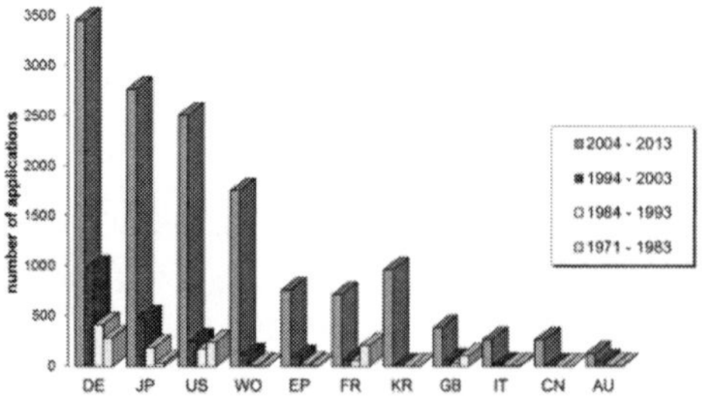

Figure 8: Patent Applications Filed in Europe for different decades

Asian countries other than Japan (i.e. China & Korea), showed significant filing figures only for the last decade (2004 – 2013). Both China and Korea had almost no patent activity in the field of PV before 2000. For the last decade, Korea has surpassed France, the UK and Italy, which had much higher filing figures than Korea in the decades before.

Apart from inorganic solar cells, organic solar cells and dye sensitized solar cells – also called second generation solar cells – are becoming more and more important. When looking at the development of filings in organic PV, it is interesting to compare the number of filings in dye sensitized solar cells with that of other organic solar cells, including layered and bulk hetero junction cells (Figure 9).

Both fields develop similarly from 1990 on, although the number of filings in dye sensitized cells was outgrown by filings in other organic cells for the first time in 2004. Since 2012 (year of priority application), a drop in filings could be observed for dye sensitized solar cells, whereas the filing figures of organic solar cells increased further, however, at a lower rate.

Looking at the filing figures in the field of dye sensitized solar cells, as presented in more detail in Figure 9, the first significant growth can be noted in the early 1990s. The increase in patent activity seems to be related to the work of Michael Grätzel et al. on TiO_2 cells using transition metal complexes as sensitizing dye (EP 0 333 641 A1).

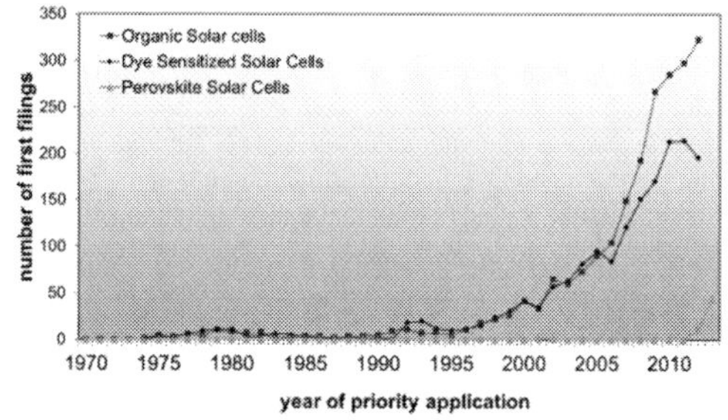

Figure 9: Patent Applications Filed in Europe - Organic and Dye Sensitized Solar Cells

In the field of organic solar cells, growing patent activity can be observed starting in the late 1990s, after the publication of the patent application by Sariciftci and Heeger (WO 94/05045) describing a bulk hetero junction solar cell based on a mixture of a polymer and a fullerene.

Recently a new type of solar cells emerged, which uses a metal organic perovskite (typically $CHNH_3PbI_{3-x}Br_x$) as active material. For these type of devices, steep increase of patent activity has been observed.

Whereas patent activity in the field of dye sensitized solar cells is dominated by Japan (Figure 10), patents in the field of organic solar cells originate from the US and Japan at an about equal share (Figure 11). As shown in Figure 11, by far the greatest part of filings in organic solar cells comes from the US (23%), followed by Japan (21%) and Germany (14%). Other major players are the UK (5%) and Korea (6%). Direct PCT filings ("WO") (13%) and direct EP filings (10%) cannot be attributed to a specific country. Both Japan and Korea are much stronger in the area of dye sensitized solar cells with 32% and 10%, respectively. In the area of dye sensitized solar cells, other big players are the US (11%) and Germany (6%) – again with major contributions from direct EP filings (8%) and PCT filings (18%).

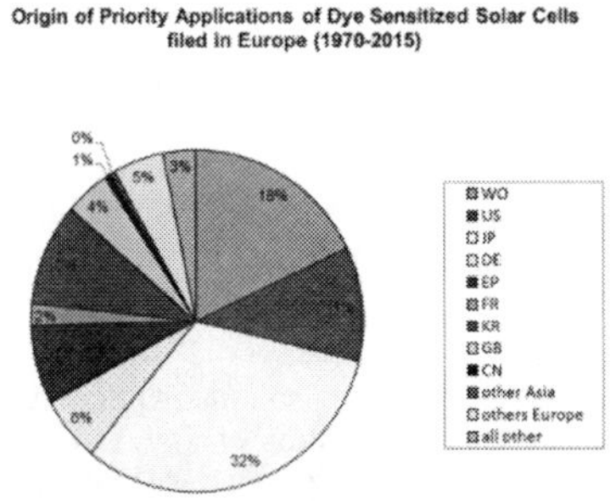

Figure 10: Origin of Patent Applications in Europe – Dye Sensitized Solar Cells

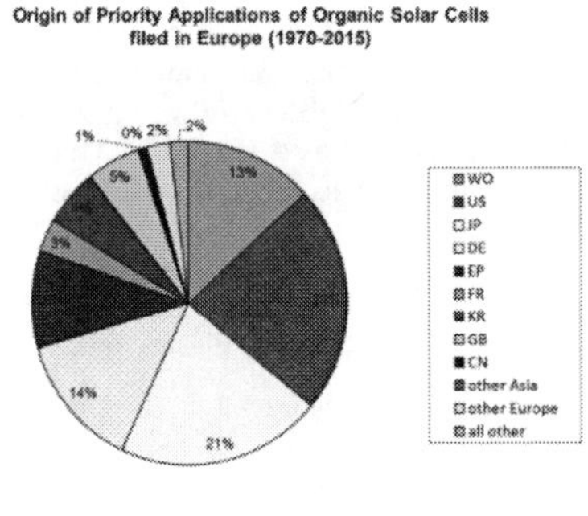

Figure 11: Origin of Patent Applications in Europe - Organic Solar Cells

A different picture is obtained for inorganic solar cells (see Figure 12), where Germany has the biggest share (28%) followed by JP (17%) and the US (16%).

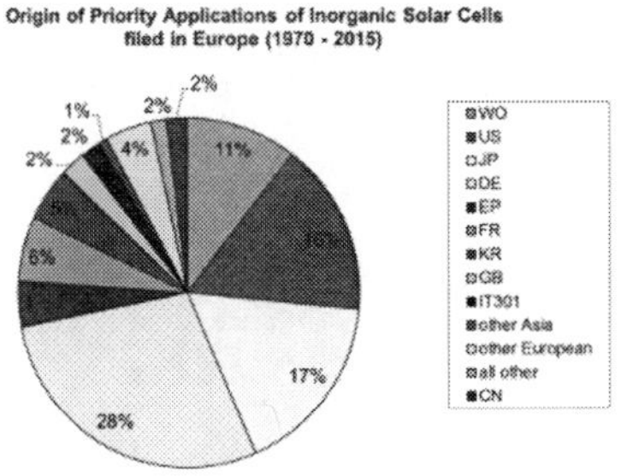

Figure 12: Origin of Patent Applications in Europe - Inorganic Solar Cells

6 SUMMARY

The tremendous success of PV is reflected in the increasing number of patent applications in Europe, which peaked in 2012. Since then, a decrease of applications can be observed. The origin of patent applications filed in Europe is nearly equally distributed between Europe, Japan, the US and other countries. Within Europe, most of the patent applicants come from Germany.

The number of applications relating to organic (polymer) solar cells and dye sensitized solar cells increased considerably over the past 15 years. Perovskite solar cells are expected to become the future boom field in PV patent applications with a doubling in application figures over the past two years.

References

[1] EPO member states: AL = Albania, AT = Austria, BE = Belgium, BG = Bulgaria, CH = Switzerland, CY = Cyprus, CZ = Czech republic, DE = Germany, DK = Denmark, EE = Estonia, ES = Spain, FI = Finland, FR = France, GB = United Kingdom, GR = Greece, HR = Croatia, HU = Hungary, IE = Ireland, IS = Iceland, IT = Italy, LI = Liechtenstein, LT = Lithuania, LU = Luxembourg, LV = Latvia, MC = Monaco, MK = Former Yugoslav Republic of Macedonia, MT = Malta, NL = Netherlands, NO = Norway, PL = Poland, PT = Portugal, RO = Romania, SE = Sweden, SI = Slovenia, SK = Slovakia, SM = San Marino, RS = Serbia, TR = Turkey; Extension States: BA = Bosnia and Herzegovina, ME = Montenegro

[2] Information on the Cooperative Patent Classification scheme is available at www.cpcinfo.org; A search using the classification scheme can be performed using Espacenet: http://worldwide.espacenet.com/classification?locale=en_EP

[3] Information on the International Patent Classification is available at http://web2.wipo.int/ipcpub/

[4] US = USA, JP = Japan, CN = China, AU = Australia, KR = Korea, TW = Taiwan, NZ = New Zealand, CA = Canada, IL = Israel, WO = PCT Patent Application.

[5] https://www.epo.org/searching-for-patents.html

[6] http://www.epo.org/about-us/organisation/validation-states.html

IS THE WORLD HEADING TOWARDS THE POST-PARIS AGREEMENT RENEWABLE (SOLAR) FUTURE QUICKLY ENOUGH?

Alexander Lagaaij
SucceSun, www.succesun.com
Breukelen, The Netherlands
+31(0)652715867, alexanderlagaaij@succesun.com

ABSTRACT: It is amazing to see how suddenly solar civilization started unfolding so rapidly after the signing of Paris Climate Agreement on 12 december 2015. Poised between the demands of quality and quantity, the photovoltaic industry in 2016 took on its global role. This paper is a thought-exercise in bringing the energy transition forward to facilitate the needed end to global CO2 emissions.
Keywords: photovoltaics, climate, energy transition, CO2 emissions reduction

1 INTRODUCTION

Looking back from 2023 it is amazing to see how suddenly the solar civilization started unfolding so rapidly after the signing of Paris Climate Agreement on 12 december 2015 by all the countries of the world. At that time world CO2 emissions were on the order of 40 billion tons per year.

2 THE GLOBAL SITUATION IN 2016

2.1 Solar Energy and CO2 emissions

Only with TeraWatts of solar energy does a real contribution to bringing down CO2 emissions occur (rule of thumb 1 TW = 1.5 - 3 Gigaton CO2/yr) [1].
In 2016 Europe which a decade earlier had started the real growth of solar energy was at a plateau, with about 100 GW installed, mitigating 300 million tons of CO2 per year (or less than 1% of global CO2 emissions).
In the rest of the world growth started taking off, with China leading, India moving ahead with speed, the United States on the go, and Africa and other continents picking up tempo.
The growth towards Terawatt went much quicker than anyone making the business cases and projections at that time imagined. [2]

2.2 'Looking back from 2023'

This, Fig.1, is what happened:

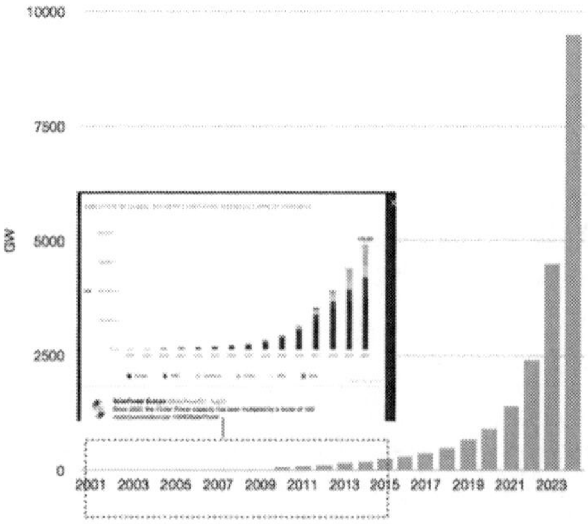

Figure 1: In the inset is the Solar Power Europe Global Market Outlook 2015-2019 graph on World installed capacity from 2000 until 2015 (179 GW). The orange bars of the grapph show the humongous growth that came since 2015, bringing the world to just below 5 TeraWatt in 2023. Expecting next year (2024) the doubling of this to finish the replacement of all fossil fuels by solar energy.

It is so clear now how PV plays a role in climate change mitigation.

Constructing the Solar Civilization new structures, towers, canopies, roofs and endless arrays were made from photovoltaics(pv). To integrate it into the world's cultures colorful, polyform pv was applied in landscapes, on farmland and on existing structures in cities.

Everywhere PV came to be seen, becoming a presence universally recognized and cherished for its climate change mitigating force.

2.3 CO2 emissions brought to zero

Here in Fig. 2 is the effect on the CO2 emissions :

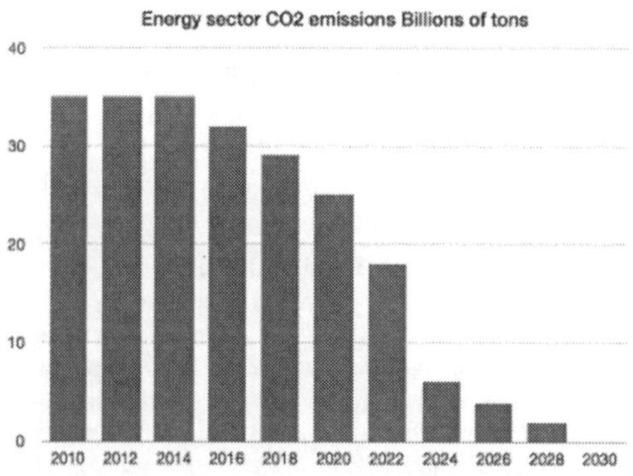

Figure 2: CO2 emissions with the 100-fold PV growth in 7/8 years, similar to Representative Concentration

Pathways (RCP) 2.6 (IPCC) [3], or '2DS' (2 Degree Scenario) IEA [4], but quicker. Extrapolation by the author.

3 SUPPLY AND DEMAND, ECONOMICS AND ECOLOGICS

Solar **supply** alliances with planetary scope started emerging from 2015 on: International Solar Alliance [5], Global Solar Council [6], and Solar United [7].

Solar **demand** coalitions were also growing: by citizens through public ownership of renewable energy (global citizens cooperation emerged!), by businesses through among others Corporate Sourcing of Renewables and RE100 [8] and the UNFCCC Lima-Paris Action Agenda taken forward to Marrakech and COP22 [9].

The issues of **quality** and **quantity** both needed to be addressed by the solar industry and 2016 saw some intense work to get these two aligned.

Multiple global intiatives in policy and finance followed the Paris Agreement, with more and more fossil fuel companies ready to change towards a renewable future bringing a new spirit of transition to the world.

Just in time to give the earth's atmosphere and oceans a chance to remain in the condition enabling life and human life on earth.

4 PRODUCTIVE CAPACITY

4.1 The PV industry and R&D community prepared for a large expansion/acceleration as mandated by The Paris Agreement of COP21.

Within a 10-, 100-worldwide expanded PV industry, as was needed to make any reasonable impact on CO2 reduction worldwide, there was ample room for PV actors (industry) on all continents.
The accelerated filling of this greater space by PV envisioned policies and actions necessary to enhance the growth of PV (industry and installations) everywhere.

Post Paris 12-12-2015 the financing became available to make PV industries grow. This further changed the economics of PV, the cost of which had been coming down with ca. 20% per year for decades. In 2016 the beginning of a level playing field started, with fossil fuel subsidies ending, CO2 pricing coming into effect and a recognition of the financial risks of climate change through the work of the Financial Stability Board Task Force on Climate-related Financial Disclosures [10].

4.2 Anticipation

Preparatory and forward-looking investigation of potential for acceleration of industrial production, distribution and implementation took off in the course of 2016.

Incorporation of research results: a unifying proposal for companies to unite in creating and setting up a factory to explore the latest technology together making advanced Silicon Solar cells, launched during EUPVSEC and InterSolar in Munich, 20-25 June 2016, became a showpiece of the solar industry global vision at COP22 in Marrakech. It was funded through climate related funds and divestments from fossil fuels. It was a crystallization point for demonstrating to the world where the solutions for decarbonization lie, in a cooperative show of what PV can bring in the great challlenge of climate change.

It showed how quickly the PV industry could accelerate and take on the responsible presence with global vision that it was destined to have.
A voice started coming from the global PV industry, rising above regional and continental differences in conditions, regulations, support and roll-out. It organized itself into a common presence proposing a way in mitigation of climate change.
Thus it was possible to accelerate the growth so that instead of in 15 years, it was in 5 years that PV became a major electricity source on this planet.
It was a great challenge succesfully defied to combine quality and the acceleration in quantity.

Major issues were: 1. Grid integration and storage; 2. Speeding up of the inclusion of new research through open cooperation worldwide (+ BIPV) 3. Need for production facilities and replacement of existing fossil fuel power sources/plants in the world. 4. Training of workers and citizens

The decision by Tesla to buy SolarCity announced on June 21st 2016, heralded in the worldwide take off of the integration of Solar energy, Battery Storage and Electric vehicles, that became the universal lever to end the dependence on fossil fuels.

4.3 Approach

The photovoltaic world was prepared for acceleration through

1. Alliance building with future orientation, International Solar Alliance, Global Solar Council (focus on quantity), Solar United (focus on quality)

2. Cooperation with the The LIMA-PARIS ACTION AGENDA to:
- Facilitate public and private sector uptake of identified policy options with high potential for mitigation and adaptation;
- Showcase existing significant partnerships and non-state actors;
- Stimulate new or enhanced action initiatives engaging state and non-state actors;
- Engage with key actors in assessing the impact of collaborative action to meet the contribution to the pathway towards a below 1.5°C goal;
- Communicate the benefit of pv enhancing climate action including economic growth and sustainable development.

3. Unification through cooperation with world-wide consumer and business demands for zero carbon energy.

4.4 Significance for life on this planet

It was the largest global experiment (with life or death of civilization outcome) to ward off climate change caused by the filling of the atmosphere with CO2 by accelerating implementation of solar energy [11],[12],[13],[14,],15], [16].

It worked through:
1. Aiming for a 100-fold increase in the shortest time, and preparing for acceleration;
2. Marrying advances of policy with technology and finance (in the words of UNFCCC's Christiana Figueres);
3. Focussing on getting solar energy in TeraWatts with an eye to CO2 mitigation 1 TW=3 Gton CO2/yr.

5. CONCLUSION

With photovoltaics put high on the World agenda in the course of 2016 a global effort in unfossilizing and solarifying the world energy system started taking place. It took all hands on deck and more to get this acceleration towards pv to come to the rescue of the planet.

6. REFERENCES

[1] SolarCity 2015 impact report http://www.solarcity.com/sites/default/files/SolarCity_Impact_Report_2015.pdf?062916= , Appendix A.

[2] RECGroup, CLIMATE CHANGE: CLOSING THE COP21 GAP BY GOING SOLAR http://www.recgroup.com/sites/default/files/documents/rec_cop21_study_final_web.pdf

[3] http://www.ipcc.ch/report/ar5/syr/

[4] http://www.iea.org/publications/freepublications/publication/TechnologyRoadmapSolarPhotovoltaicEnergy_2014edition.pdf

[5] http://intsolaralliance.org/

[6] http://www.globalsolarcouncil.org

[7] http://www.solar-united.org/

[8] http://www.cleanenergyministerial.org/Our-Work/Campaigns/Corporate-Sourcing-of-Renewables and http://there100.org/

[9] http://newsroom.unfccc.int/lpaa/ and http://climateaction.unfccc.int/

[10] https://www.fsb-tcfd.org/

[11] Jeremy Leggett, The Wining of the Carbon War, 2016, free download at http://www.jeremyleggett.net/

[12] Neville Williams, Sun Power, 2015.

[13] Tony Seba, Clean Disruption of Energy and Transportation, 2014.

[14] Mark Z. Jacobson et al., 100% Clean and Renewable Wind, Water, and Sunlight (WWS) All Sector Energy Roadmaps for 139 Countries of the World, december 2015:
http://web.stanford.edu/group/efmh/jacobson/Articles/I/CountriesWWS.pdf

[15] Alexander Lagaaij, Cop 21 and 100xSolar http://www.succesun.com/resources/100x-Solar-energy.pdf

[16] Alexander Lagaaij, Solar and the road to COP21 http://www.pv-tech.org/guest-blog/solar_and_the_road_to_cop-21

ECONOMIC FEASIBILITY ANALYSIS OF CHOICES FOR NOT CONNECTED CLIENTS: PHOTOVOLTAIC OFF-GRID SYSTEMS, COMPARED TO GRID BASED SYSTEMS.

M.A. Munoz-García[1], Huaxing Wang[1], G.P. Moreda[1], I. Collar[1]
[1] Dep. Agroforestry Engineering. Technical University of Madrid (UPM)
Avda. Complutense, s/n; 28040 Madrid, Spain
miguelangel.munoz@upm.es Phone: +34 913365458; Fax: +34 91336567

ABSTRACT: Off-grid photovoltaic systems are less constrained by national regulations than their grid-connected counterparts. Off-grid systems do not depend on feed-in tariff and the analysis of their cost-efficiency is more stable along time.

This work calculates the economic parameters for two different technologies of photovoltaic systems and compares the levelized cost of electricity (LCOE), with the price of the electricity, including the cost for the connection of the system related to the distance to the grid. The results show that the sustained decline in photovoltaic module's price contributes to subsequent decline in LCOE, whereas the increase in the price of grid electricity and materials related to the connection makes off-grid systems more cost-effective for remote areas.

Finally, to determine if the photovoltaic system could be an economic alternative for investments in remote regions, a comparison between the cost of grid extension and the LCOE of a high capacity photovoltaic system was undertaken.

Keywords: PV system, off-grid, LCOE, grid extension

1 INTRODUCTION AND OBJECTIVS

Despite grid parity has been reached in many countries, the regulation for grid-connected photovoltaic systems is completely different depending on the country and even on the region inside a country. This situation makes complicated to state a global conclusion about the economic feasibility of a particular investment. Given the decline in prices of the elements of photovoltaic systems, and after removal of national subsidies that were previously granted to electricity consumers like irrigation users, this work is considered useful for reducing costs in remote areas not connected to the grid as well as in the agriculture field. The analysis takes into account photovoltaic emerging technologies that will make solar panels more appropriate in some cases over others, depending on their performance. Thus, we seek a formula that determines the optimal solution for each case, based on objective and reliable data.

2 MATERIALS AND METHODS

This work includes not only the calculation of LCOE based on the prices of the system and its maintenance. It also compares different photovoltaic technologies, taking into account the different rates for the degradation and the area occupied by them. Even further, a comparison that includes distances to the grid and whether the connection is made at high or low voltage is considered in the cost comparison.

Levelized Cost of Electricity:

$$LCOE = \frac{\sum_{t=0}^{T}(I_t+O_t+M_t)/(1+r)^t}{\sum_{t=0}^{T} E*(1-d)^t/(1+r)^t}$$

I_t: the investment of year t;
O_t: operating cost of year t;
M_t: maintenance cost of year t;
E: energy generated;
d: degradation rate;
r: discount rate;
T: lifetime or analysis period.
Payback Period (PBP): PBP = C_0/C_t

Net Present Value (NPV):

$$NPV = \sum_{t=0}^{T}\frac{C_t}{(1+r)^t} - C_0$$

C_t: net cash flow during the period t;
C_0: total initial investment costs;
Internal Return Rate (IRR): IRR=r, when NPV=0;

According to the previous equations and using the data of two real installations: one in Madrid and another one in Colombia, the following data (Table 1 and Table 2) were obtained.

Annual global irradiation(kWh/m2)	1793
Peak Power(Wp)	4940(1610+1680+1650)
Inverter capacity	3*1200W
DC to AC ratio (includes clipping)	1.37
Area(m^2)	45.06
Annual output(kWh)	7442.13
Subsidies (€/kWh)	0.3
Annual income(€)	2232.64
Total cost (€)	23844
Cost per peak watt (€/Wp)	4.83
Payback period(years)	10.68
Internal return rate (IRR, %)	9.59

Table 1. Summary of PV system in Madrid, Spain, installed at the end of 2012.

We also analyzed which would be the LCOE of the connection of a consumer to the grid, when the grid is located far from the consumer, another analysis can be performed. In this case we compared the LCOE of two alternatives: off-grid photovoltaic system (comparing different nominal powers) and grid-connected consumer.

Annual global irradiation(kWh/m2)	1617
Peak Power(Wp)	3370(1620+1750)
Inverter capacity	2*1500W
DC to AC ratio (without clipping)	1.25
Area(m2)	39.05
Annual output(kWh)	4542.51
Industrial electricity price (€/kWh)	0.1
Annual income(€)	454
Total cost (€)	6000
Cost per peak watt(€/Wp)	1.82
Payback period(years)	13.21
Internal return rate (IRR, %)	6.46

Table 2. Summary of PV system in Popayán, Colombia, installed in June 2016.

Installed capacity	Cost (€/Wp)	Cost of installation (€/Wp)	Cost of inverter(€/Wn)	Discount rate (%)
1 kW	0.5	0.4	0.5	6
5 kW	0.4	0.4	0.35	6
10 kW	0.3	0.4	0.3	6
	Degradation rate(%/year)	O&M cost (%/total initial cost)	Lifetime (years)	LCOE (€/kWh)
1 kW	0.5	0.1	30	0.103
5 kW	0.5	0.1	30	0.086
10 kW	0.5	0.1	30	0.075

Table 3. Cost of different elements forming an a-Si based photovoltaic system, depending on the total installed capacity (taxes and transport excluded)

In the case of a grid-connected system, the cost of the extension (the line from the grid to the consumer) must be taken into account. Such cost depends on the voltage and on the method of transport: air or underground

	Cost of installation (€/m)
High voltage, underground	120
Medium voltage, air	20
Medium voltage, underground	80

Table 4. Cost of different types of connection to the grid.

Considering the previous values, the following results and conclusions were obtained.

According to Fig. 1, the distances could have a high effect on the real cost of the electricity.

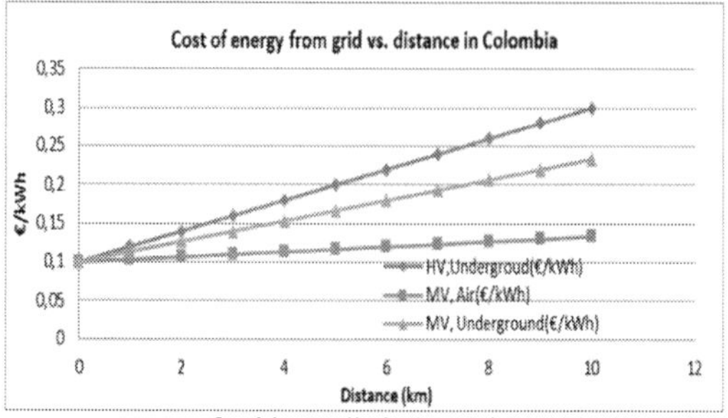

Figure 1. Cost of grid-supplied electricity, with the cost of the connection to grid included, as a function of the distance from the consumer to the grid.

3 RESULTS AND DISCUSSIONS

In the following graphs, the LCOE is analyzed for different mortgage loan interest rate, including the case of self-financing, i.e. no mortgage.

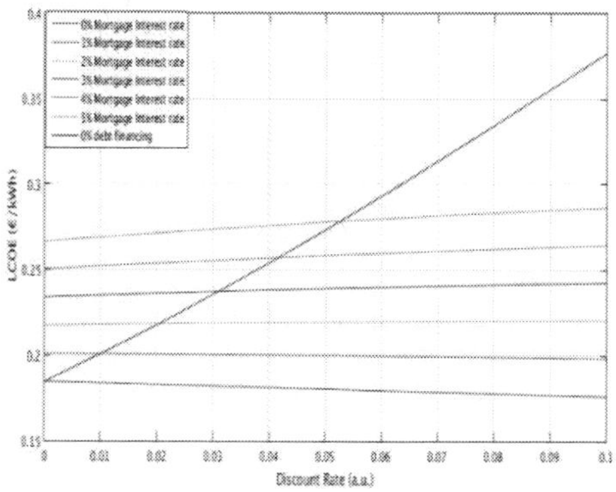

Figure 2. LCOE for 25 years lifetime of the PV system in Madrid varying mortgage interest rate and discount rate;, the inputs are: total cost of system=23844€, first year output = 7442.13kWh, degradation rate= 0.5%, operating and maintenance cost= 238.44 € per year, loan term= 25 years, the inverter's replacement in the fifteenth year=2384.4€.

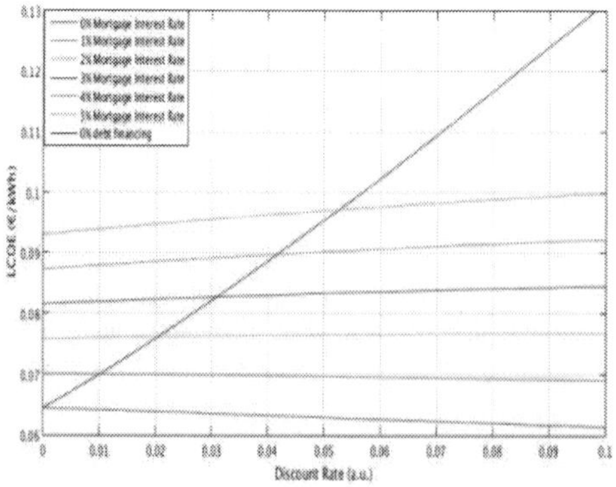

Figure 3. LCOE for 25 years lifetime of the PV system in Colombia varying mortgage interest rate and discount rate; the inputs are: total cost of system=6000€, first year output = 5364.5kWh, degradation rate= 0.5%, operating and maintenance cost= 60 € per year, loan term= 25

years, the inverter's replacement in the fifteenth year: 600€.

By comparing these two different PV systems, it can be observed that due to the decline in PV module's price, as well as the price of other essential components, the LCOE of the system in Colombia is rather lower than in Madrid.

However, from the point of view of a particular investor, the benefit of the investment could be more attractive. The latter is more related to the government incentive policy or the electricity price if there is no incentive. In Spain, subsides of PV system connected to the grid until 2013 would lead to a better PBP and IRR than in Colombia. Despite nowadays no feed-in tariffs are applied in either country the LCOE would be better than the price of the electricity from the grid.

ACKNOWLEDGMENTS

This project has been funded by Technical University of Madrid (UPM) and Santander Bank with their support to the solar plant in Popayán (Colombia) through the Cooperation Project COOP-XV-06. We want to acknowledge to TECNiCAFE (Technological park for the addition of value to coffee crop and to coffee culture) supported by SupraCafé, and also to Unicomfacauca and CIEMAT for their collaboration and their technical support.

REFERENCES

Šúri M., Huld T.A., Dunlop E.D. Ossenbrink H.A., 2007. Potential of solar electricity generation in the European Union member states and candidate countries. Solar Energy, 81, 1295–1305, http://re.jrc.ec.europa.eu/pvgis/

Muñoz-García, Miguel Angel and Moreda Cantero, Guillermo P. and Alonso Garcia, M. Carmen and Balenzategui, José Lorenzo (2013). *Electricity production using different photovoltaic technologies working at the same conditions , a real case.* "Journal of Beijing Institute of Technology", v. 22 (n. Suppl.); pp.. ISSN 1004-0579.

Muñoz-García, Miguel Angel and Chenlo, Faustino and Alonso Garcia, M. Carmen (2011). *Influence of initial power stabilization over crystalline-Si photovoltaic modules maximum power.* "Progress in Photovoltaics", v. 19 (n. 4); pp. 417-422. ISSN 1062-7995.

Muñoz-García, Miguel Angel and Moreda Cantero, Guillermo P. and Valle-García, Alvaro and Novillo, Jesus and Alonso Garcia, M. Carmen and Chenlo, Faustino (2012). *Grid-connected roof solar plants in educational institutions for testing and electricity generation: a case in Spain..* In: "27th European Photovoltaic Solar Energy Conference and Exhibition", Septiembre-2012, Frankfurt, Germany.

Valle-García, Alvaro and Muñoz-García, Miguel Angel and Moreda Cantero, Guillermo P. and Chenlo, Faustino (2012). *Success possibilities of grid parity in particular cases in the new spanish regulatory framework.* In: "27th European Photovoltaic Solar Energy Conference and Exhibition", Septiembre-2012, Frankfurt, Germany.

AUSTRALIAN PV SYSTEM MARKET ANALYSIS FOR THE POWERWALL

Sandy Rodrigues[a,b], Roham Torabikalaki[a], Fábio Faria[a], Helena G. Ramos[b], and F. Morgado-Dias[a]
[a] Instituto de Telecomunicações, Instituto Superior Técnico, Universidade de Lisboa, Portugal
[b] Madeira Interactive Technologies Institute, Caminho da Penteada, Funchal Portugal
[c] Universidade da Madeira, Caminho da Penteada, Funchal Portugal
sandy.rodrigues@tecnico.ulisboa.pt; morgado@uma.pt

ABSTRACT: Tesla chose Australia to be one of the first Powerwall Battery delivery countries in the world, because of its tariff structure, high solar radiation values, and high electricity costs. The Powerwall Battery solution was promoted as a low-cost attractive investment, therefore the purpose of this paper is to analyse four scenarios in five Australian states (ACT, SA, VIC, NSW, and QLD) to determine which scenario and respective state presents the highest investment attractiveness. Scenarios 1 and 2 consider an off-grid 5kW PV system with one Powerwall Battery and two Powerwall Batteries respectively. Scenario 3 considers a Powerwall Battery without a PV system connection, which is charged by the household grid and discharged back into the same grid. Scenario 4 studies an on-grid 3kW PV system without a Powerwall Battery practicing a 70% self-consumption regime. The results show that scenario 4 is the most attractive investment option with SA, VIC, and QLD making 3 times the investment within a 25-year investment period and in NSW the investment is doubled. Scenario 1 is currently viable in VIC and only viable after a 10% discount on the Powerwall Battery in QLD.
Keywords: Rooftop PV Systems, Economic Analysis, Powerwall Battery, Hybrid Inverter.

1 INTRODUCTION

Late 2015, the Tesla Company announced that Australia would be one of the first countries outside of the US to be supplied with the Powerwall Battery (PB) systems due to its desirable weather conditions, the high solar roof-top installation per capita and the high electricity prices that are practiced [1], [2]. The Tesla Energy Authorized Resellers in Australia include energy retail Origin Energy and solar installer company Natural Solar [3].

The Photovoltaic (PV) system and solar battery solutions may be considered as investments since they provide monetary returns through feed-in/grid injection tariffs and/or electricity bill savings making it possible to generate a payback on the investment. The feed-in tariffs were cut in Australia on January 1st 2016 and the price of the grid injected energy is currently lower than the electricity prices in almost all Australian states (except for the Northern Territory state, which has a grid injection tariff that is equal to the consumption price) [4], [5]. Now, the grid injection tariffs range from 0.04€/kWh to 0.08€/kWh which encourages the use of a 100% self-consumption regime because the electricity consumption prices are higher and more attractive than the grid injection prices.

The objective of this paper is to analyse the economic feasibility of four different self-consumption scenarios in five Australian states to determine which of the states is the most financially attractive, when it comes to investing in a PV system and/or a PB solution.

The number of Australian states used in this study was limited to those in which the private energy retailer Origin Energy operates the power generation stations. The five states are namely the Australian Capital Territory (ACT), South Australia (SA), Victoria (VIC), New South Wales (NSW), and Queensland (QLD) [6].

The Tesla Powerwall Battery system began its distribution and household usage in the beginning of 2016. This battery system is advertised to be revolutionary mainly due to its dimensions, plug and play features, storage capacity, and low price. However, this new product is surrounded by scepticism around its investment attractiveness, and investors want to know not only if this new battery system will generate an attractive profit on the investment, but also how long is the payback period.

This paper provides answers to these questions, as it indicates the Internal Rate of Return, Profitability Index, and Discounted Payback Period at the end of a 25-year investment period. Since different electricity prices are practiced in every state of Australia, one aim of this paper is to provide the reader with enough information about the different investments in the different states based on data that is up to date. This data includes all the economic parameters that can influence the PV system and solar battery system investments during a 25-year period such as the operations and maintenance (O&M) costs, inverter replacement cost rates (in years 10 and 20), electricity evolution rates, interest evolution rates, and others. This study also provides information about the types of households (depending on the number of people and total simultaneous power of appliances) that can be associated to an off-grid scenario.

Efficient and low-cost solar battery systems pave the way to efficient off-grid scenarios, which in turn contributes to reducing the CO_2 emissions and consequently reduce global warming.

The organization of the paper is the following: the methodology is in Section 2 and explains all methods used to evaluate the economic feasibility of the investments; Section 3 presents the results and discussion and finally Section 4 presents the conclusions.

2 METHODOLOGY

The Powerwall Battery (PB) is a lithium-ion battery pack used to store energy provided by a PV system or that comes from a household or business grid.

The PB can be used for load-shifting, PV storage, or backup power. The term load-shifting defines the energy management to move demand from peak to off-peak hours. Thus, the PB can be used for load-shifting by being charged by the household or business grid during the off-peak hours and discharged back into the same grid in the on-peak hours. The difference between the two tariffs (on and off-peak) provide electricity bill savings which pays back the investment made on the PB and is intended to eventually make a profit on the investment.

This paper studies the economic feasibility of four different PV systems and/or PB solution scenarios in five Australian states namely ACT, SA, VIC, NSW, and QLD. The four scenarios are as follows:

- Scenario 1 (S1) – Off-grid 5kW PV system w/ 1 PB
- Scenario 2 (S2) – Off-grid 5kW PV system w/ 2 PB
- Scenario 3 (S3) – PB, hybrid inverter, and AC/DC charger w/o a PV system
- Scenario 4 (S4) – On-grid 3kW PV system w/o a PB

2.1 Scenario Description

The PB scenarios consider a 7kWh Tesla PB in a residential setting.

The first two scenarios consist of a PB system linked to an off-grid 5kW PV system, in which scenario 1 uses one PB while scenario 2 uses two PBs. The reason behind choosing two off-grid scenarios with a different number of PB was to compare a system capable of supporting an energy consumption of a household with one or two people (one PB) and a system that would support an energy consumption of a household with three or four people (two PBs).

The Australian average daily energy consumption depends not only on the number of people in the household but also on the state since some states use more heating and cooling equipment than others due to climate conditions. Statistics about the average daily energy consumption of a household of up to four people in a household per state is available online [7]. The average daily consumption of a household of two people in any state of Australia is around 17kWh. It is known that a 5kW PV system has an average daily solar production of 17kWh [8] in Australia, and it is known that the PB is able to store a total of 6.4kWh of energy and output a maximum of 3.3kW continuous power [9]. Thus, an off-grid 5kW PV system with one PB can be associated to a household of two people given that the total power when all the appliances are working simultaneously is not more than 3.3kW. The PB limits this value, and if the simultaneous power is higher than 3.3kW, it is advised to use two PBs in the household.

Scenario 3 consists of only the PB, a hybrid inverter and an AC/DC 2kW charger, in which the PB is charged by the household grid and discharged back into the same grid without being connected to a PV system. The purpose of this solution is to buy electricity at a lower price (off-peak) and use it when the price is higher (on-peak), which saves money on the electric bill.

Scenario 4 considers an on-grid 3kW PV system practicing a 70% self-consumption regime without a PB connection. In this regime, 70% of the power from the PV system is injected into the household (self-consumption) grid while the rest of the solar power (30%) is injected into the distribution grid. This self-consumption regime reflects the consumption reality of a typical household during the day in Australia that has a 3kW PV system as shown in [10], which is based on industry averages from publically available data on Australian homes.

Scenarios 1 and 2 consist of a 5kW PV system, a PB system, and a hybrid inverter while Scenario 4 includes a 3kW PV system and a simple solar inverter (not hybrid inverter), which in turn makes this last system much cheaper. On-grid PV system prices are lower than off-grid PV system prices because of the solar subsidies associated with the on-grid PV system solutions and the low prices of simple solar inverters [11]. The PB solutions have high costs due to the high PB and hybrid inverter prices.

2.2 Economic Methods

The economic feasibility of the PV system and solar battery solution investments is determined by calculating the economic methods such as the Discounted Payback Period (DPBP), the Net Present Value (NPV), the Internal Rate of Return (IRR), and the Profitability Index (PI). The DPBP indicates the number of years needed to make a breakeven (return of money invested) on the investment in which after that period, a profit on the investment is made. The NPV shows if the earnings generated by the investment exceeds or not the anticipated costs and is used to calculate the PI. The PI indicates how much profit is made by the end of a 25-year period. The IRR evaluates the attractiveness of an investment and is used to compare the profitability between various investments.

In order to calculate the DPBP, the Simple Cash Flow (SCF) and Discounted Cash Flow (DCF) has to be first calculated. The subtraction between the cash inflow and the cash outflow calculates the SCF as shown in equation 1 [12].

$$SCF_y = \sum_{y=1}^{Y} (Ts \times Es)_y - \sum_{y=1}^{Y} (M)_y \qquad (1)$$

The investment time in years is presented by Y, the self-consumption tariff is Ts, the annual self-consumption of the PV system production is Es, and M represents the maintenance cost. The DCF calculates the money value over time, represents the SCF value in the future, and is updated with the interest rate r (equation 2).

$$DCF_y = \frac{SCF_y}{(1+r)^y} \qquad (2)$$

The DPBP is calculated by using the DCF values to calculate the number of years needed to make a breakeven on the investment and therefore considers the money value over time (equation 3).

$$DPBP = A + \frac{B}{C} \qquad (3)$$

Where,
A = Last period with a negative discounted cumulative cash flow;
B = Absolute value of the discounted cumulative cash flow at the end of the period A;
C = Discounted cash flow during the period after A [13].

The short-term investment has a payback period of less than 3 years, a medium term investment of 4 to 10 years and long-term investments of more than 10 years.

The Net Present Value (NPV) calculates the present value of all cash inflows with the present value of all cash outflows associated with an investment (equation 4).

$$NPV = \sum_{y=1}^{Y} \frac{C_y}{(1+r)^y} - C_0 \qquad (4)$$

Where Cy represents the yearly net cash flow, and C_0 represents the initial investment of the PV system.

The investment is profitable when the interest rate is lower than the IRR indicator [14]. The IRR is used for comparison across locations, as there is no need to consider the local interest rates [15]. The IRR formula is presented in equation 4 when NPV = 0, and r = IRR.

The Profitability Index (PI) indicates the investor if the investment should be accepted or rejected (equation 5). When the PI is equal to 1.00, there is a breakeven on the investment which means that by the end of a 25-year PV system investment period the investor receives the money invested. This type of investment is not profitable since the life-cycle of a PV system is 25 years. If the PI is equal to 2.00 or higher, the investment is considered attractive, since the investment is doubled by the end of a 25-year period, which is equal to having a 4% annual interest rate on the return of investment [16].

$$PI = \frac{\text{NPV}}{\text{Initial Investment}} + 1 \quad (5)$$

2.3 Economic Parameters

The economic methods mentioned in the previous section are influenced by the economic parameters shown below which are based on assumptions for the purpose of this work. Table I shows the economic parameters common to all states while Table II presents the economic parameters specific to each state.

The economic parameters common to all states include the maintenance and operations cost rate; the hybrid inverter replacement cost rates for scenarios 1, 2 and 4; the PB annual charge/discharge energy amount; the PB replacement cost in year 10 and in year 20; the hybrid inverter and PB efficiency loss; the PV degradation rate; the GST rate; the Australian currency exchange rate and the annual interest evolution rate.

Table I: Common Economic Parameters

Parameter Description	Value
Annual Maintenance and operations cost rate	1.0%
Scenario 1 - Hybrid Inverter replacement cost rate	12.0%
Scenario 2 - Hybrid Inverter replacement cost rate	10.0%
Scenario 4 - On-grid Inverter replacement cost rate	20.0%
PB Annual Charge/Discharge (kWh)	2336
Powerwall Replacement Cost year 10	4,110.12 €
Powerwall Replacement Cost year 20	2,935.80 €
Hybrid Inverter efficiency loss	7.0%
Powerwall efficiency loss	8.0%
PV Degradation	0.7%
GST included in all prices	10.0%
Currency exchange AUD to EUR (1/2/2016)	0.6524 €
Annual Interest Evolution Rate	6.25%

It is assumed that all inverters and PB are replaced in years 10 and 20. Since the PB prices are expected to decrease in the upcoming years, it is assumed in this work that the PB replacement has a 30% discount in year 10 and a 50% discount in year 20 while the inverter cost remains the same in both replacement years. All prices presented in Tables I and II have the GST (Australia Sales Tax Rate) included and are subtracted from all the Federal Subsidies related to solar energy which can be consulted in [11]. The currency exchange value from Australian Dollar to Euro is referenced from the 1st of February 2016. The interest evolution rate is an average value based on the past 25 years [18]. The cost rates mentioned in Table I are relative to the investment prices presented in Table II.

The economic parameters specific to each state (Table II) include the solar production of the 5kW and 3kW PV systems, the investment prices of each scenario, the mono and tri-hourly tariffs, the grid injection tariffs, the distribution charge savings made by the off-grid scenarios and the annual electricity evolution rate.

Table II, presents the solar production values for the 5kW and 3kW PV systems calculated by using RETSCREEN software version 4 [17]. The investment costs of each scenario were obtained by calculating an average of three quotes from three different companies in a given Australian state. All quotes were obtained in March 2016. Next to the investment costs are the investment costs per watt in euros (€/W).

The hourly electricity tariff prices are the standing offers currently practiced by the energy utility company Origin Energy [6]. The hourly tariffs used in this work are the mono-hourly and tri-hourly (also known as the Time of Use tariff - TOU) which is divided into on-peak, shoulder and off-peak tariffs. The on-peak tariffs are higher than the off-peak. The mono-hourly tariff is used in scenarios 1, 2 and 4 while the tri-hourly tariffs are used in scenario 3 where the PB is charged in the off-peak hours by the household grid and discharged back into the household grid in the on-peak hours. The savings made in scenario 3 result from the subtraction between the on-peak and off-peak tariff prices, which is presented by the term "Difference" in Table II for each state. Unfortunately SA does not practice a TOU tariff, therefore scenario 3 was not analysed in SA.

In the off-grid scenarios, the supply charges are considered as yearly savings. These supply charge expenses vary among the retailers and there is no regulated history data on them in order to make an average evolution rate, therefore the cost of these expenses are maintained throughout the 25-year period in economic feasibility calculations of this work.

Just like the interest evolution rate, the electricity evolution rate was calculated in the same way just that only over the past 18 years, which is limited by the information provided by the utility company Origin Energy [19].

Table II also presents the grid injection tariffs provided by the utility company Origin Energy for each of the states, and it is assumed that these are not associated to an evolution rate due to the lack of historical data.

Australia currently practices a net-metering scheme. The grid injection tariffs are associated to net metering grid connected systems where any excess energy is delivered back into the distribution grid through a bi-directional meter in which the import and export flows are measured and recorded in separate registers. This meter is interval based (30-minute blocks), and periodically reads the exported/imported electricity to and from the grid [20], [21].

Australia lowered the feed-in tariffs but maintained the very attractive solar initial investment subsidies associated to the PV systems [11].

Table II: Economic parameters and values associated to each state of Australia

		ACT	SA	VIC	NSW	QLD
Solar Production	**3kW**	4354 kWh/year	4164 kWh/year	3846 kWh/year	4040 kWh/year	4268 kWh/year
	5kW	7257 kWh/year	6940 kWh/year	6157 kWh/year	6704 kWh/year	7114 kWh/year
Investiment	**S1**	10,252.47 € 2.05 €/W	10,252.47 € 2.05 €/W	10,089.37 € 2.02 €/W	9,991.51 € 2.00 €/W	9,926.27 € 1.99 €/W
	S2	12,831.19 € 2.57 €/W	12,912.74 € 2.58 €/W	12,901.10 € 2.58 €/W	12,779.21 € 2.56 €/W	12,673.41 € 2.53 €/W
	S3	5,871.60 €	5,871.60 €	5.871.60 €	5.871.60 €	5.871.60 €
	S4	2,933.63 € 0.59 €/W	2,860.56 € 0.57 €/W	3,061.71 € 0.61 €/W	2,916.66 € 0.58 €/W	2,823.80 € 0.56 €/W
Electricity Hourly Tariff Price (€/kWh)	**Mono**	0.1127 €	0.2232 €	0.1804 €	0.1502 €	0.1596 €
	Tri On-Peak	0.1525 €	- €	0.3031 €	0.3302 €	0.2142 €
	Tri shoulder	0.1030 €	- €	0.1864 €	0.1322 €	0.1516 €
	Tri Off-Peak	0.0754 €	- €	0.1323 €	0.0746 €	0.1167 €
	Tri Difference	0.0771 €	- €	0.1707 €	0.2556 €	0.0975 €
Grid Injection tariffs		0.0391 €	0.0444 €	0.0326 €	0.0391 €	0.0391 €
Savings from Grid Distribution Costs		180 €	184 €	259 €	198 €	305 €
Annual Electricity Evolution rate		5.06%	2.43%	6.87%	5.37%	5.56%

3 RESULTS AND DISCUSSION

In this section there are four sets of results. The first refers to investment prices as they are at the moment (Table III), the second considers a 10% discount on only the PB solution itself and not to the whole PV system investment (Table VI). The third result set considers a 20% discount on the PB solution (Table V) and finally the fourth result set considers a 30% discount on the PB solution (Table VI). This work studies the effects of the discount on the PB up to 30% due to the fact that shortly after 2020 (year of the facility completion) the PB production costs are expected to be reduced by 30% because of the mass production provided by the Tesla Gigafactory [22]. These four sets of results provide information about the price that should be put in place for PB solutions to become an attractive investment.

Table III shows how the on-grid scenario (scenario 4) is attractive, particularly more in SA, VIC, and QLD which make 3 times (PI > 3.00) the investment, followed by NSW that more than doubles the investment by the end of a 25-year investment period. The attractive PI values (PI ≥ 2.00) are coloured in grey and the breakeven PI values (1.00 < PI < 2.00) are in bold letters. The non-attractive investments (red) have PI values that are lower than 1.00. The payback periods of these investments are attractive because they range from 5 to 8 years (medium term investment). This is due to the high solar energy subsidies that take place in Australia [11], associated to the PV system initial cost.

The next best investment in Table III is in Victoria with the off-grid 5kW PV system connected to one PB system (scenario 1), in which the investment is doubled by the end of a 25-year investment period. The payback period is long term in VIC and, SA has a quicker payback period even though it does not make at least double the investment by the 25th year. This is due to SA's low electricity evolution rate (2.43%) compared to the other states. Even though Victoria makes double the investment by the 25th year, the investment is viable but not considered a very good investment because of the 12-year payback period, which is longer than the 10-year warranty offered by the Tesla Company. This means that the investor will have to invest in the PB replacement without first making a breakeven on the total investment.

The IRR indicator value is always higher than the average Australian interest rate of 6.25% in all the depicted states within scenarios 1 and 4. The investment attractiveness results of scenario 1 are directly related to the electricity prices practiced in a given state because the investment costs and solar radiation values in the different states do not differ much from one another. All the other investments are long term (> 10 years) and/or have very little profit margins at the end of a 25-year investment period.

The 10% discount on the PB system, presented in Table IV, made very little improvements to the previous results, only adding Queensland to the attractive investment category. The investment is doubled by the 25th year in QLD and this payback period is considered a medium term investment, which takes place before the 10-year warranty of the PB. All the other scenarios with PB solutions remain as non-attractive investments.

The 20% discount on the PB system (Table V) also made very little improvements to the previous results except for the payback period in Victoria that lowered to 9 years instead of 12. The payback period takes place before the 10-year warranty period offered by Tesla for the PB making this an attractive investment. All the other PB scenarios in this table remain as non-attractive investments.

Considering the 30% discount on the PB system (Table VI) the changes to the previous results are also small, except for the fact that the payback period was lowered by one year in both Victoria and Queensland. All other PB scenarios remain as non-attractive investments in this table.

Table III: No Discount on the Powerwall Battery

	ACT			SA			VIC			NSW			QLD		
	IRR	**PI**	**DPBP**	**IRR**	**PI**	**DPBP**	**IRR**	**PI**	**DPBP**	**IRR**	**PI**	**DPBP**	**IRR**	**PI**	**DPBP**
S1	7%	**1,079**	23	13%	**1,719**	8	14%	2,126	12	11%	**1,527**	15	14%	**1,949**	12
S2	2%	0,591	26	7%	**1,094**	19	9%	1,389	18	5%	0,917	26	8%	**1,249**	18
S3	-12%	-0,094	26	-	-	-	5%	0,796	26	8%	**1,169**	22	-6%	0,080	26
S4	13%	**1,930**	11	23%	2,962	5	20%	3,205	7	17%	2,503	8	20%	3,002	7

Table IV: Powerwall Battery with 10% Discount

	ACT			SA			VIC			NSW			QLD		
	IRR	**PI**	**DPBP**	**IRR**	**PI**	**DPBP**	**IRR**	**PI**	**DPBP**	**IRR**	**PI**	**DPBP**	**IRR**	**PI**	**DPBP**
S1	7%	**1,117**	22	14%	**1,777**	8	14%	2,197	12	11%	**1,580**	15	14%	2,015	9
S2	3%	0,652	26	8%	**1,193**	17	10%	**1,510**	16	6%	**1,004**	25	9%	**1,361**	17
S3	-12%	-0,091	26	-	-	-	5%	0,886	26	9%	**1,296**	18	-5%	0,100	26
S4	13%	**1,930**	11	23%	2,962	5	20%	3,205	7	17%	2,503	8	20%	3,002	7

Table V: Powerwall Battery with 20% Discount

	ACT			SA			VIC			NSW			QLD		
	IRR	**PI**	**DPBP**	**IRR**	**PI**	**DPBP**	**IRR**	**PI**	**DPBP**	**IRR**	**PI**	**DPBP**	**IRR**	**PI**	**DPBP**
S1	8%	**1,158**	22	14%	**1,838**	8	15%	2,272	9	11%	**1,636**	14	15%	2,086	9
S2	3%	0,722	26	10%	**1,307**	16	11%	**1,651**	15	7%	**1,104**	23	10%	**1,492**	16
S3	-11%	-0,087	26	-	-	-	6%	0,996	26	10%	**1,451**	17	-4%	0,125	26
S4	13%	**1,930**	11	23%	2,962	5	20%	3,205	7	17%	2,503	8	20%	3,002	7

Table VI: Powerwall Battery with 30% Discount

	ACT			SA			VIC			NSW			QLD		
	IRR	**PI**	**DPBP**	**IRR**	**PI**	**DPBP**	**IRR**	**PI**	**DPBP**	**IRR**	**PI**	**DPBP**	**IRR**	**PI**	**DPBP**
S1	8%	**1,201**	19	15%	**1,902**	8	15%	2,351	8	12%	**1,696**	14	15%	2,161	8
S2	4%	0,806	26	11%	**1,442**	14	12%	**1,816**	14	8%	**1,222**	18	12%	**1,647**	15
S3	-10%	-0,083	26	-	-	-	7%	**1,133**	23	11%	**1,643**	15	-4%	0,155	26
S4	13%	**1,930**	11	23%	2,962	5	20%	3,205	7	17%	2,503	8	20%	3,002	7

4 CONCLUSION

We conclude that only scenario 1 (off-grid 5kW PV system connected to one PB system) PB system investment is viable at the moment and only in the state of Victoria. The state of Queensland is also viable in scenario 1 but only after applying a 10% discount on the PB solution. For the PB system investments to be viable in all the other states, the PB system investment costs have to be lowered much more than 30%.

Perhaps the prices that are now practiced are purposely high to limit the volume of manufacturing demand in this initial stage of the business. The high PB solution prices could also be influenced by the high labour costs of installation that are practiced in Australia.

Having stated earlier that the discounts were only applied to the PB system itself and not to the PV system as a whole, it is clear that the results show that all costs still need to lower in order for the PB investments to be more attractive. The investment costs include the PV system and installation costs as well as the operations and maintenance cost, the hybrid inverter costs, and the inverter replacement costs in years 10 and 20. The PB investments would benefit from the electricity cost rise, from the PB system cost reduction and from solar battery solution governmental subsidies.

Off-grid solutions do not seem plausible in the near future for any household. Therefore, the energy utilities do not have to fear the migration to solar power off-grid solutions for a long time unless the equipment costs lower drastically. This will make the energy utilities slow down their raises in the electricity prices, because the electric prices is one of the main economic parameters that help improve the PV/PB system investment attractiveness. Cheaper solar battery solutions promote highly attractive off-grid investments.

Scenario 4 (on-grid 3kW PV system with 70% self-consumption and 30% grid injection and without a PB solution) presents attractive investments in all states except for the Australian Capital Territory state ranging from double to triple the investment with a payback period between 5 and 8 years. This scenario represents a self-consumption regime of a typical household in Australia with 70% self-consumption and 30% grid injection. Scenarios 2 and 3 do not present viable results in any of the Australian states studied in this work.

Future work includes the study of the same Australian states comparing the PB to other solar batteries that are equivalent.

5 ACKNOWLEDGMENTS

This work was developed in the Instituto de Telecomunicações in Instituto Superior Técnico da Universidade de Lisboa, and supported by ARDITI - Agência Regional para o Desenvolvimento da Investigação Tecnologia e Inovação through the support provided by the FSE - Madeira 14-20. This support is gratefully acknowledged.

Also acknowledge the Portuguese Foundation for Science and Technology for their support through project PEst-OE/EEI/LA0009/2011.

6 REFERENCES

[1] S. Vorrath, "Tesla Targets Australia As 1st Market For Powerwall, Deliveries In 2015," Renew Economy, 2015. [Online]. Available: http://goo.gl/1s8YQN. [Accessed: 15-Mar-2016].

[2] J. Nicholas, "Why Tesla Is Smart To Launch Powerwall In Australia," Business Insider Australia, 2015. [Online]. Available: http://goo.gl/7F01fQ. [Accessed: 15-Mar-2016].

[3] S. Vorrath, "First Tesla Powerwall Installations In Homes Begin This Week In Australia," Renew Economy, 2016. [Online]. Available: http://goo.gl/gRsGC8. [Accessed: 15-Mar-2016].

[4] Energy Matters, "Feed-In Tariff: Solar And Wind Power In Australia - Energy Matters," Energy Matters, 2016. [Online]. Available: http://goo.gl/ypLfLe. [Accessed: 15-Mar-2016].

[5] Solar Choice, "When Do Solar Feed-In Tariffs End? NSW VIC QLD TAS SA ACT WA NT - Solar Choice," Solar Choice, 2016. [Online]. Available: http://goo.gl/1wl8xT. [Accessed: 15-Mar-2016].

[6] Origin Energy, "Energy Price Facts Sheet," Origin Energy, 2016. [Online]. Available: https://goo.gl/iXCYBI. [Accessed: 15-Mar-2016].

[7] Energy Made Easy, "Understand and compare your home electricity usage," Energy Made Easy, 2016. [Online]. Available: https://goo.gl/dyLIjs. [Accessed: 15-Mar-2016].

[8] Solar Choice Staff, "Average daily production for solar PV cells in Australia," Solar Choice, 2010. [Online]. Available: http://goo.gl/ERmVwU. [Accessed: 15-Mar-2016].

[9] Tesla Motors, "Powerwall - The Tesla Home Battery," Tesla Motors, 2016. [Online]. Available: https://goo.gl/4XE3Fd. [Accessed: 15-Mar-2016].

[10] M. Nigel, "Solar export calculator," Roof Juice Australia, 2015. [Online]. Available: http://roofjuice.com.au/solar-export-calculator/. [Accessed: 15-Mar-2016].

[11] Energy Matters, "Australian Solar Credits Subsidy," Energy Matters, 2016. [Online]. Available: http://goo.gl/Ov5Cwj. [Accessed: 15-Mar-2016].

[12] R. Bernal-Agustín, J. L., and Dufo-López, "Economical and environmental analysis of grid connected photovoltaic systems in Spain," Renewable Energy Journal, vol. 31, no. 8, pp. 1107–1128.

[13] Accountingexplained, "Discounted Payback Period Calculation | Formula | Examples," 2013. [Online]. Available: http://goo.gl/GL9kqp. [Accessed: 01-Mar-2016].

[14] A. Campoccia, L. Dusonchet, E. Telaretti, and G. Zizzo, "An analysis of feed'in tariffs for solar PV in six representative countries of the European Union," Solar Energy, vol. 107, pp. 530–542, 2014.

[15] T. Lang, E. Gloerfeld, and B. Girod, "Don't just follow the sun – A global assessment of economic performance for residential building photovoltaics," Renewable and Sustainable Energy Reviews, vol. 42, pp. 932–951, Feb. 2015.

[16] J. Obaidullah, "Profitability Index," Accounting Explained, 2013. [Online]. Available: http://goo.gl/Kx823F. [Accessed: 01-Feb-2016].

[17] Retscreen International, "Renewable energy project analysis software," 2014. [Online]. Available: http://www.retscreen.net/. [Accessed: 01-Jan-2015].

[18] IECONOMICS, "Interest Rate," 2016. [Online]. Available: http://www.tradingeconomics.com/. [Accessed: 20-Apr-2016].

[19] AEMO, "Average Price Tables," AEMO - The Australian Energy Market Operator. [Online]. Available: http://goo.gl/SAqWNu. [Accessed: 15-Mar-2016].

[20] M. Lasauce, "Your Gross or Net Solar Electricity Meter," Solar Choice, 2009. [Online]. Available: http://goo.gl/wF5xWW. [Accessed: 15-Mar-2016].

[21] Origin Energy, "Changing from Gross to Net Metering," Origin Energy, 2016. [Online]. Available: https://goo.gl/x8zhQs. [Accessed: 15-Mar-2016].

[22] Tesla Motors, "Tesla Gigafactory," Tesla Motors, 2016. [Online]. Available: https://www.teslamotors.com/gigafactory. [Accessed: 15-Mar-2016].

AUTHOR INDEX

AUTHOR INDEX

AUTHOR INDEX

AUTHOR INDEX

AUTHOR INDEX

AUTHOR INDEX

AUTHOR INDEX

AUTHOR INDEX

AUTHOR INDEX

AUTHOR INDEX

AUTHOR INDEX

AUTHOR INDEX

AUTHOR INDEX

AUTHOR INDEX

AUTHOR INDEX

AUTHOR INDEX

AUTHOR INDEX

AUTHOR INDEX

AUTHOR INDEX

AUTHOR INDEX

AUTHOR INDEX

AUTHOR INDEX

AUTHOR INDEX

AUTHOR INDEX

AUTHOR INDEX